P9-AOS-310

This text is printed on acid-free paper.

Disclaimer

The editor, authors, and the publisher have made every effort to provide accurate and complete information in this Handbook but the Handbook is not intended to serve as a replacement for professional advice. Any use of the information in this Handbook is at the reader's discretion. The editor, authors, and the publisher specifically disclaim any and all liability arising directly or indirectly from the use or application of any information contained in this Handbook. An appropriate professional should be consulted regarding your specific situation.

Library of Congress Cataloging-in-Publication Data

Handbook of human factors and ergonomics / [editor, Gavriel Salvendy].
—2nd ed.
p. cm.
Includes indexes.
ISBN 0-471-11690-4 (cloth : alk. paper)
1. Human engineering—Handbooks, manuals, etc. I. Salvendy, Gavriel
TA166.H275 1997
620.8′5—dc21 96-51093

Printed in the United States of America

10 9 8 7 6 5 4 3

HANDBOOK OF HUMAN FACTORS AND ERGONOMICS

Second Edition

Edited by
Gavriel Salvendy
Purdue University

A Wiley-Interscience Publication
JOHN WILEY & SONS, INC.
New York / Chichester / Weinheim / Brisbane / Toronto / Singapore

ABOUT THE EDITOR

Gavriel Salvendy, the NEC Professor of Industrial Engineering at Purdue University, is the author or co-author of over 300 research publications including over 150 journal papers and 20 books. His publications have appeared in six languages. Dr. Salvendy is the founding editor of the *International Journal on Human–Computer Interaction, International Journal of Cognitive Ergonomics,* and the *International Journal of Human Factors in Manufacturing.* He was the founding chair of the International Commission on Human Aspects in Computing, headquartered in Geneva, Switzerland. He was elected to the National Academy of Engineering "for fundamental contributions to and professional leadership in human, physical, and cognitive aspects of engineering systems." The recipient of the Mikhail Vasilievich Lomonosov Medal (founder medal) of the Russian Academy of Science, Dr. Salvendy also received an Honorary Doctorate from the Chinese Academy of Sciences "for great contributions to the development of science and technology and for the great influence upon the development of science and technology in China." He is the fourth person ever to receive this award in all fields of science and engineering. He is an honorary fellow and life member of the Ergonomics Society and fellow of Human Factors and Ergonomics Society, Institute of Industrial Engineers, and the American Psychological Association. Gavriel Salvendy has advised organizations in 23 countries on the human side of effective design, implementation, and management of advanced technologies in the workplace. He earned his Ph.D. in engineering production at the University of Birmingham, United Kingdom.

ADVISORY BOARD

CONTRIBUTORS

William B. Albery
Supervisory, Electronics Engineer
Combined Stress Branch
Biodynamics and Biocommunications Division
Crew Systems Directorate
USAF Armstrong Laboratory
Wright-Patterson Air Force Base, Ohio

Timothy R. Anderson
Researcher
Biodynamics and Biocommunications Division
Armstrong Laboratory
Wright-Patterson Air Force Base, Ohio

M. M. Ayoub
Horn Professor
Department of Industrial Engineering and Biomedical Engineering
Texas Tech University
Lubbock, Texas

Anitesh Barua
Doctoral Student
Center for Information Systems Management
Department of MSIS
The University of Texas at Austin
Austin, Texas

Wilhelm Bauer
Researcher
Fraunhofer Institute for Industrial Engineering (IAO), Stuttgart, and Institute for Human Factors and Technology Management (IAT)
University of Stuttgart
Stuttgart, Germany

Kevin B. Bennett
Professor
Department of Psychology
Wright State University
Dayton, Ohio

Carolyn K. Bensel
Principal Researcher
Behavioral Science Division
U.S. Army Natick Research, Development and Engineering Center
Natick, Massachusetts

Susan A. H. Benysh
Doctoral Student
School of Industrial Engineering
Purdue University
West Lafayette, Indiana

Charlie Billings
Associate Professor
Department of Industrial and Systems Engineering
Cognitive Systems Engineering Laboratory
The Ohio State University
Columbus, Ohio

Kenneth R. Boff
Division Chief
Human Engineering Division
Armstrong Laboratory, AL/CFH
Wright-Patterson Air Force Base, Ohio

Peter R. Boyce
Professor
Lighting Research Center
Rensselaer Polytechnic Institute
Troy, New York

Edward Boyle
Researcher
Logistics Research Division
Air Force Armstrong Laboratory
Wright-Patterson Air Force Base, Ohio

Martin Braun
Scientist
Fraunhofer Institute for Industrial Engineering (IAO), Stuttgart, and Institute for Human Factors and Technology Management (IAT)
University of Stuttgart
Stuttgart, Germany

John F. Brock
Vice President
InterScience America, Inc.
Leesburg, Virginia

Hans-Jörg Bullinger
Professor and Director
Fraunhofer Institute for Industrial Engineering (IAO), Stuttgart, and Institute for Human Factors and Technology Management (IAT)
University of Stuttgart
Stuttgart, Germany

Michael A. Campion
Professor
Krannert School of Management
Purdue University
West Lafayette, Indiana

Pascale Carayon
Associate Professor
Department of Industrial Engineering
University of Wisconsin—Madison
USA and Visiting Professor
Ecole des Mines de Nancy, FRANCE

John M. Carroll
Professor and Head
Department of Computer Science
Virginia Polytechnic Institute and State University
Blacksburg, Virginia

C. Melody Carswell
Professor
Department of Psychology
University of Kentucky
Lexington, Kentucky

Don B. Chaffin
The Johnson Professor of Industrial and Operations Engineering, and Director, Center for Ergonomics
The University of Michigan
Ann Arbor, Michigan

Ramnath Chellappa
Assistant Professor
Center of Information Systems Management
Department of MSIS
The University of Texas at Austin
Austin, Texas

Mark H. Chignell
Associate Professor
Department of Mechanical and Industrial Engineering
University of Toronto
Toronto, Ontario
Canada

William J. Cohen
Doctoral Student
Department of Industrial Engineering
University of Wisconsin—Madison
Madison, Wisconsin

Kevin Corker
Principal Scientist for Cognition
NASA—Ames Research Center
Moffett Field, California

Malcolm J. Crocker
Distinguished University Professor
Department of Mechanical Engineering
Auburn University
Auburn, Alabama

Sara J. Czaja
Professor
Department of Industrial Engineering
University of Miami
Coral Gables, Florida

Patrick G. Dempsey
Doctoral Student
Department of Industrial Engineering and Biomedical Engineering
Texas Tech University
Lubbock, Texas

Colin G. Drury
Professor
Department of Industrial Engineering
State University of New York—Buffalo
Amherst, New York

Ray Eberts
Associate Professor
School of Industrial Engineering
Purdue University
West Lafayette, Indiana

Robert G. Eggleston
Researcher
Fitts Human Engineering Division
Armstrong Laboratory
Wright-Patterson Air Force Base, Ohio

Carolanne Fisher
Design Scientist
U S West
Advanced Technologies
Denver, Colorado

John M. Flach
Associate Professor
Department of Psychology
Wright State University
Dayton, Ohio

Paul A. Green
Associate Research Scientist
Human Factors Division
Transportation Research Institute
University of Michigan
Ann Arbor, Michigan

Michael Griffin
Professor
Human Factors Research Unit
Institute of Sound and Vibration Research
University of Southampton
Southampton, England

Helen M. Haines
Doctoral Student
Department of Manufacturing Engineering and Operations Management
University of Nottingham
Nottingham, United Kingdom

John E. Harrigan
Professor of Architecture *emeritus*
New Castle, New Hampshire

Martin Helander
Professor
Department of Mechanical Engineering
Division of Industrial Ergonomics
Linköping Institute of Technology
Linköping, Sweden

Hal W. Hendrick
President
Error Analysis, Inc.
Englewood, Colorado

Ronald A. Hess
Professor
Department of Mechanical and Aeronautical Engineering
University of California—Davis
Davis, California

Manfred Hueser
Research Fellow
Fraunhofer Institute for Manufacturing Engineering and Automation (IPA)
Stuttgart, Germany

Raija Kalimo
Professor and Director
Department of Psychology
Finnish Institute of Occupational Health
Helsinki, Finland

Waldemar Karwowski
Professor and Director
Center for Industrial Ergonomics
Department of Industrial Engineering
University of Louisville
Louisville, Kentucky

Peter Kern
Professor
Fraunhofer Institute for Industrial Engineering, Stuttgart, and Institute for Human Factors and Technology Management (IAT)
University of Stuttgart
Stuttgart, Germany

Stephan A. Konz
Professor
Department of Industrial and Manufacturing Systems Engineering
Kansas State University
Manhattan, Kansas

Richard J. Koubek
Associate Professor
School of Industrial Engineering
Purdue University
West Lafayette, Indiana

Karl H. E. Kroemer
Professor and Director
Department of Industrial Systems and Engineering
Virginia Polytechnic Institute and State University
Blacksburg, Virginia

K. Ronald Laughery, Jr.
President
Micro Analysis and Design, Inc.
Boulder, Colorado

Kenneth R. Laughery, Sr.
Henry R. Luce Professor of Engineering Psychology
Psychology Department
Rice University
Houston, Texas

Mark R. Lehto
Associate Professor
School of Industrial Engineering
Purdue University
West Lafayette, Indiana

Kari Lindström
Professor and Director, Action Program on Healthy and Productive Workplace
Department of Psychology
Finnish Institute of Occupational Health
Helsinki, Finland

Yili Liu
Assistant Professor
Department of Industrial and Operations Engineering
University of Michigan
Ann Arbor, Michigan

Robert E. Llaneras
Associate Scientist
InterScience America
Leesburg, Virginia

Holger Luczak
Professor and Director
Institute of Industrial Engineering and Ergonomics
Aachen, Germany

Anthony E. Majoros
Principal Engineer
Advanced Transport Aircraft Systems
McDonnell Douglas Aerospace
Long Beach, California

William Marras
Professor
Department of Industrial Engineering
The Ohio State University
Columbus, Ohio

Jeffrey L. Maxey
Doctoral Student
Department of Industrial Engineering and Management Systems
University of Central Florida
Orlando, Florida

Grant R. McMillan
Principal Scientist
Fitts Human Engineering Division
Armstrong Laboratories
Wright-Patterson Air Force Base, Ohio

Gina Medsker
Assistant Professor
School of Business Administration
University of Miami
Coral Gables, Florida

Neville Moray
Professor
Université de Valenciennes
Laboratoire d'Automatique et Mecanique Industrielles et Humaines
Valenciennes, Cedex, France

Allen L. Nagy
Associate Professor
Department of Psychology
Wright State University
Dayton, Ohio

Jakob Nielsen
Distinguished Engineer
Strategic Technology
Sun Microsystems
Mountain View, California

Michael J. Paley
Doctoral Student
Department of Psychology
University of Connecticut
Storrs, Connecticut

Kyung (Ken) S. Park
Professor
Department of Industrial Engineering
Korea Advanced Institute of Science & Technology
Taejon, Korea

Annelise M. Pejtersen
Senior Research Scientist
Center for Cognitive Research
RISO
Roskilde, Denmark

Stephen M. Popkin
Doctoral Student
Department of Psychology
University of Connecticut
Storrs, Connecticut

Robert W. Proctor
Professor
Department of Psychological Sciences
Purdue University
West Lafayette, Indiana

Janet D. Proctor
Professor
Department of Psychological Sciences
Purdue University
West Lafayette, Indiana

Jens Rasmussen
HURECON
Smorum, Denmark

Ray Reaux
Assistant Professor
Department of Computer Science
Virginia Polytechnic Institute and State University
Blacksburg, Virginia

David Regan
Professor
Department of Psychology
York University
North York, Ontario, Canada

Suzanne H. Rodgers
Ergonomists/Human Factors Consultant
Rochester, New York

William B. Rouse
Enterprise Support Systems
Atlanta, Georgia

Andrew P. Sage
First American Bank Professor and Dean
School of Information Technology and Engineering
George Mason University
Fairfax, Virginia

Gavriel Salvendy
NEC Professor of Industrial Engineering
School of Industrial Engineering
Purdue University
West Lafayette, Indiana

Penelope Sanderson
Associate Professor
Swinburne Computer–Human Interaction Laboratory
Swinburne University of Technology
Hawthorn, Victoria, Australia 3122

William R. Santee
Researcher
Biophysics and Biomedical Modeling Division
U.S. Army Research Institute of Environmental Medicine
Natick, Massachusetts

Nadine B. Sarter
Associate Professor
Department of Industrial and Systems Engineering
Cognitive Systems Engineering Laboratory
The Ohio State University
Columbus, Ohio

Joseph Sharit
Associate Professor
Department of Industrial Engineering
State University of New York at Buffalo
Buffalo, New York

Thomas B. Sheridan
Ford Professor
Department of Mechanical Engineering
Massachusetts Institute of Technology
Cambridge, Massachusetts

Michael J. Smith
Professor and Chair
Department of Industrial Engineering
University of Wisconsin—Madison
Madison, Wisconsin

Kay M. Stanney
Assistant Professor
Department of Industrial Engineering and Management Systems
University of Central Florida
Orlando, Florida

Robert W. Swezey
President
InterScience America
Leesburg, Virginia

Eric Tang
Assistant Professor
Department of Industrial Engineering
Feng-Chia University
Taichung, Taiwan

Donald I. Tepas
Professor and Director
Department of Psychology
University of Connecticut
Storrs, Connecticut

Pamela S. Tsang
Professor
Department of Psychology
Wright State University
Dayton, Ohio

Gregg C. Vanderheiden
Professor and Director, Trace R&D Center
Department of Industrial Engineering
University of Wisconsin, Madison
Madison, Wisconsin

Patricia F. Waller
Professor and Director
Transportation Research Institute
University of Michigan
Ann Arbor, Michigan

Hans-Jurgen Warnecke
Professor and President
Fraunhofer Society (FhG)
Munich, Germany

John A. Waterworth
Assistant Professor
Department of Informatiks
Umeå University
Umeå, Sweden

Andrew B. Whinston
John B. Harbin Centennial Chair in Business
Center for Information Systems Management
Department of MSIS
Graduate School of Business
The University of Texas at Austin
Austin, Texas

Christopher D. Wickens
Professor and Director
Aviation Research Laboratory
University of Illinois
Savoy, Illinois

John R. Wilson
Professor of Industrial Ergonomics
Department of Manufacturing Engineering and Operations Management
University of Nottingham
Nottingham, United Kingdom

Glenn F. Wilson
Crew Systems Director
Fitts Human Engineering Division
Armstrong Laboratory
Wright-Patterson Air Force Base, Ohio

Michael S. Wogalter
Associate Professor
Psychology Department
North Carolina State University
Raleigh, North Carolina

David D. Woods
Professor
Department of Industrial and Systems Engineering
Cognitive Systems Engineering Laboratory
The Ohio State University
Columbus, Ohio

Barbara Woolford
Professor
Flight Crew Support Division
Johnson Space Center
National Aeronautics and Space Administration
Houston, Texas

Bernhard Zimolong
Professor
Institute of Psychology
Ruhr-Universität Bochum
Bochum, Germany

FOREWORD

With the rapid introduction of highly sophisticated computer, communication, and manufacturing systems, we are seeing dramatic changes in the ways people work and use technology. That is why every industry today should recognize the importance of human factors and ergonomics.

The practice of this science has done much to improve the interaction of people with their environment. Ergonomics is a valuable tool in the design of products that are safe, convenient, and user friendly. It is equally important in the creation of jobs and workplaces that increase worker safety and satisfaction. Ergonomics allows designers to comprehend the capabilities, limitations, and motivations of workers in order to improve efficiency and cut costs—especially those associated with human error and occupational injury or illness.

By applying human factors and ergonomics to our products, our office technology, and our manufacturing processes, we can enhance the satisfaction and enthusiasm of both consumers and producers.

Thus, the publication of this Second Edition of the *Handbook of Human Factors and Ergonomics* is very timely. It is a comprehensive guide that contains practical knowledge and technical background on virtually all aspects of physical, cognitive, and social ergonomics. As such, it can be a valuable source of information for any individual or organization committed to providing competitive, high-quality products and safe, productive work environments.

John F. Smith, Jr.
Chairman of the Board
Chief Executive Officer and President
General Motors Corporation

PREFACE

This Handbook is concerned with the role of humans in complex systems, the design of equipment and facilities for human use, and the development of environments for comfort and safety. The first edition of the Handbook was a major success and profoundly influenced the human factors profession. It was translated and published in Japanese and Russian, and won the Institute of Industrial Engineers Joint Publishers Book of the Year Award. It has received strong endorsement from top management; the late Elliot Estes, retired president of General Motors Corporation, who wrote in the Foreword to the first edition of the Handbook, indicated that "Regardless of what phase of the economy a person is involved in, this handbook is a very useful tool. Every area of human factors from environmental conditions and motivation to use of new communication systems . . . is well covered in the handbook by experts in every field."

In a literal sense, human factors and ergonomics is as old as the machine and the environment, for it was aimed at designing them for human use. However, it was not until World War II that human factors emerged as a separate discipline.

The field of human factors and ergonomics has developed and broadened considerably since its formal inception more than 50 years ago and has generated a body of knowledge in the following broad areas of specialization:

- The human factors function
- Human factors fundamentals
- Equipment, workplace, and environmental design
- Design for health and safety
- Performance modeling
- Human–computer interaction evaluation
- Selected applications of human factors

The foregoing list shows how broad the field has become. As such, this Handbook should be of value to all human factors and ergonomics specialists, engineers, industrial hygienists, safety engineers, and computer scientists.

Such a breadth of subject matter presents a serious challenge to represent successfully the entire field of human factors and ergonomics in a single handbook. I did not believe in 1993, when this all began, that any one person could properly select the subjects to be included in the Handbook without serious distortions to fit his or her own particular area of knowledge and bias. Accordingly, an advisory board composed of experts in the more important areas of human factors was invited to advise the editor in planning the contents of the Handbook. The advisory board members are listed on pages v and vi. I sincerely appreciate their excellent counsel and advice during the preparation of this Handbook. Nevertheless, any sampling deficiencies that remain are of course my own responsibilities.

The 4953 manuscript pages of the 60 chapters constituting the second edition of the Handbook were written by 112 people. In creating this Handbook, the authors gathered information from 5577 references and presented 264 tables and 599 figures to provide theoretically based and practically oriented material for use by both practitioners and researchers. In the second edition of the *Handbook of Human Factors and Ergonomics,* 37 chapters have been completely rewritten and 23 chapters are completely new—they cover subject areas which were not included in the first edition of the Handbook. It demonstrates how broadly the discipline of human factors and ergonomics evolved over the past decade.

The main purpose of this Handbook is to serve the needs of the human factors practitioner. Each chapter has a strong theory and science base and is heavily tilted toward application orientation. As such, a significant number of case studies, examples, figures, and tables are utilized to facilitate the usability of the presented material.

The many contributing authors came through magnificently. I thank them all most sincerely for agreeing so willingly to create this Handbook with me.

Each chapter submitted was peer reviewed. The following individuals kindly contributed to the review process:

William B. Albery
Kevin B. Bennett
Deborah Boehm-Davis
Kenneth R. Boff
Indranil Bose
Hans-Jorg Bullinger
John M. Carroll
Don B. Chaffin
Sarah J. Czaja
Marvin J. Dainoff
Colin G. Drury
John M. Flach
Jon D. Fricker
Qing Gong
Martin Helander
Hal W. Hendrick
Waldemar Karwowski
Stephan A. Konz
Mark R. Lehto
Nancy J. Lightner
Xianzhan Lin
Yili Liu
Aura Matias
Neville Moray
Jason D. Papastravrou
Kyung S. Park
Robert W. Proctor
Jens Rasmussen
Suzanne H. Rodgers
William B. Rouse
Stephan L. Sauter
Joseph Sharit
Tom B. Sheridan
Michael J. Smith
Kay M. Stanney
Robert W. Swezey
Christopher D. Wickens
John R. Wilson
Neil J. Zimmerman

I had the privilege of working with Robert L. Argentieri, our Wiley editor, who significantly facilitated my editorial work. I was fortunate to have during the preparation of this Handbook the most able contribution of my administrative assistants Kim Gilbert and Theresa Brown. Thanks also go to Ira Brodsky, our Wiley Production Editor; and Suzanne Ingrao, of Ingrao Associates for their work in producing the Handbook.

GAVRIEL SALVENDY

West Lafayette, Indiana
January 1997

CONTENTS

PART 1

THE HUMAN FACTORS FUNCTION

CHAPTER 1
THE HUMAN FACTORS PROFESSION

Martin G. Helander
Division of Industrial Ergonomics
Linköping Institute of Technology
58183 Linköping, Sweden

1.1 INTRODUCTION

Ergonomics is the scientific discipline concerned with the interaction between humans and artefacts and design of systems where people participate. It deals with design of systems that people use at work and in leisure, tools that are used and procedures and practices. The purpose of the design activities is to match systems, jobs, products and environments to the physical and mental abilities and limitations of people. A complementary way to make a system function is to train and educate the operator or the user of the system. Ideally, however, systems should be designed so that they are intuitive to use and do not require special training or education.

The word *ergonomics* comes from the Greek *ergo* (work) and *nomos* (rules, law). It was first used by Wojciech Jastrzebowski in a Polish newspaper in 1857 (Karwowski, 1991). One may argue that ergonomics is nothing new. Hand tools, for example, have been used since the beginning of mankind, and ergonomics was always a concern. Hand tools are an extension of the hand that concentrate and deliver power, and aid the human in tasks such as cutting, smashing, scraping and piercing. Various hand tools have been developed since the Stone Age, and the interest in ergonomic design can be traced back in history (Childe, 1944; Braidwood, 1951).

In the eighteenth century Ramazzini (1713) published a book, *The Diseases of Workers,* where he documented links between many occupational hazards and the type of work performed. He described, for example, the development of cumulative trauma disorder and believed that these events were caused by repetitive motions of the hand, by constrained body posture, and by excessive mental stress.

LaMettrie's controversial book *L'homme Machine* was published in 1748, at the beginning of the Industrial Revolution (Christensen, 1962). Two things can be learned from LaMettrie's writings. First, the comparison of human capabilities and machine capabilities was a sensitive issue already in the eighteenth century. Second, by considering how machines operate, one can also learn much about human behavior. Both issues remain debated in ergonomics in our days. The comparison of robots and humans has helped us understand how industrial tasks should be designed to better fit humans (Helander, 1995).

Rosenbrock (1983) pointed out that during the Industrial Revolution there were efforts to apply the concepts of a "human centred design" to tools such as the spinning-jenny and the spinning-mule. The concern was to allocate interesting tasks to the human operator, but let the machine do repetitive tasks.

At the beginning of the twentieth century, Frederick Taylor introduced the "scientific" study of work. Tasks were analyzed and opportunities were sought to enhance productivity by simplifying movement patterns of workers. Many of Taylor's studies analyzed the work of brick-layers. In the same tradition Frank and Lillian Gilbreth developed the time-and-motion study and the concept of dividing ordinary jobs into several small micro-elements called "therbligs" (Konz, 1990). Today there are sometimes objections against *Taylorism,* which has been seen as a tool for exploiting workers. Nonetheless, these methods are useful for measuring and predicting work activities. The time-and-motion study is a valuable tool if used for the right purpose.

The emphasis in *ergonomics* in the beginning of the twentieth century was largely on how to adapt people to their work. This resulted in much research to select, classify, and train suitable human operators. This approach was not very fruitful, and the current focus is on ergonomics design of environments and artefacts. The emphasis is on fitting the task to the person, not fitting the person to the task.

In Europe, ergonomics started with industrial applications in the 1950s, and used information from work physiology, biomechanics, and anthropometry for design of workstations and industrial processes. The focus was on the well-being of workers and manufacturing productivity. In the United States, *human factors engineering, human factors,* or *engineering psychology* developed from military problems. Human factors has it origin in experimental psychology and system engineering. The purpose is to enhance systems performance.

Ergonomics and human factors since the 1950s have proliferated to Asia, Africa, Latin America, and Australia (Luczak, 1997). In many industrially developing countries (IDCs), there are new problems with rapid industrialization, which has led to the situation where workers have difficulties adapting to new environments. The transition from a rural, agrarian life to urban industries comes at a cost. Workers are "paying a price" in terms of a tremendous increase of industrial injuries and in terms of stress. Many of these problems remain hidden, because there are rarely official statistics that can illuminate the true state of affairs. The industrialization may have come about too quickly, since many countries have difficulties in simultaneously changing other important aspects of their infrastructure (Vanwonterghem, 1994). There is hence a great need for ergonomics in IDCs, and one may expect a rapid growth of interest during the next few years when the problems are commonly recognized. Overall, there is a tremendous potential for growth of the ergonomics profession in IDCs.

1.2 A DEFINITION OF ERGONOMICS

Some 130 definitions of human factors and ergonomics are discussed in a report by Licht, Polzella, and Boff (1991). The following definition is inspired by Chapanis (1995):

- Ergonomics and human factors use knowledge of human abilities and limitations
- To the design of systems, organizations, jobs, machines, tools, and consumer products
- For safe, efficient, and comfortable human use.

Note that the main purpose of ergonomics is design. Ergonomics is thereby different from most of the bodies of knowledge that we use for supporting our discipline. Ergonomics is different from anthropology, cognitive science, psychology, sociology, and medical sciences, since their primary purpose is to understand and model human behavior, but not to utilize the knowledge for design.

Ergonomics may have more in common with engineering, which is also design oriented (Moray, 1994). This makes the practice of ergonomics/human factors an applied science. Ergonomics may be thought of as a technology, not merely a science. In fact, most of the ergonomics education takes place within engineering programs.

Ergonomics research can be both applied and basic. Good research must support itself on theories of human behavior and human functionality, and will also seek to develop new theories and discover lawfulness in human behavior. Theories developed in ergonomics can be generalized across different applications; the functionality of human perception, the implications of the limitations in short-term memory, and the speed of manual response remain. It would be an oversimplification to label ergonomics solely as an applied discipline. The beauty of our discipline is that we use basic research and we engage in basic research to seek support for design solutions. There is an exciting combination of basic and applied research in ergonomics.

Despite the historic differences between human factors and ergonomics in the type of knowledge used and in the goals for design, the two approaches are coming closer. In the Western world, physical workload is no longer so common. For example, in manufacturing, hard physical labor has been taken over by materials handling aids, mechanical processes, and automation. Legislation has also put limits to the amount of workload that employees can be exposed to. At the same time, the introduction of automation and computers in the workplace has transformed many factories, as well as offices, to systems of great complexity. The increased demands for human information processing warrants a human factors approach. Human factors and ergonomics have therefore come closer, and in the following section we use the terms without distinction.

1.2.1 Development of Ergonomics Since 1950

In exploring the field of ergonomics, there may be reasons to look at its evolution in Europe and the United States, both starting around 1950. In the United States human factors emerged as a discipline after World War II. There were many design problems encountered in the use of sophisticated war equipment, such as airplanes, radar and sonar stations, and tanks. Sometimes these caused human errors with grave consequences. For example, during the Korean War reputedly more U.S. pilots were killed during training than in war activities. This focused the interest on design of controls and displays in cockpits. How could information be better displayed in the cockpit, and how could controls be redesigned and integrated with the task so that they were easier to handle? Improvements were implemented, such as controls which combined several functions and made it easier to handle the airplane and auxiliary combat functions. As a result of these improvements and new pilot training programs, the number of fatalities in pilot training decreased to a fraction (5%) of what they had been. Ever since, much of the research in human factors in the United States has been sponsored by the Department of Defense. Consequently, the information in textbooks on human factors is heavily influenced by results from military research. This is, however, not a disadvantage. Most models, theories, and findings are applicable to the design of civilian systems.

Other U.S. federal agencies have sponsored research on civilian applications: the Federal Highway Administration (design of highways and road signs), NASA (physiological impact; human factors design of space stations), the National Highway Traffic Safety Administration (design of cars to improve crash worthiness, lighting system, and controls), the Department of the Interior (ergonomics in mining), the former National Bureau of Standards (safe design of consumer products), the National Institute of Occupational Health (ergonomic injuries at work, industrial safety, work stress), the Nuclear Regulatory Commission (design requirements for nuclear power plants), and the Federal Aviation Administration (aviation safety, air traffic control). In manufacturing, the Eastman-Kodak Company in Rochester, New York, was probably the first company to implement a substantial program around 1965.

In the former USSR, just as in the United States, the interest in ergonomics focused primarily on Department of Defense activities. There have been few applications on the industrial side, although the interest is quickly growing.

In Europe, ergonomics had a different history and is particularly well established in Belgium, France, Germany, Holland, Italy, Poland, the Scandinavian countries, Spain, and the United Kingdom. The interest in other countries is growing very quickly.

In many European countries, the driving factor for ergonomics has been its importance for worker safety, health, comfort, and convenience. Particularly in the Scandinavian countries and in Germany, labor unions have taken a strong interest, and they often dictate what type of production machinery to be purchased. Today, ergonomics in industry has the dual purpose of promoting both productivity and improved conditions of work. Several recent studies have pointed to significant improvements in productivity as a result of ergonomic measures taken in industry. With rather modest investments, benefit–cost ratios of 50:1 have been obtained (Helander, 1995).

1.3 THE SYSTEMS APPROACH TO HUMAN FACTORS AND ERGONOMICS

Usually, the study of ergonomics is conceptualized with a systems approach. Figure 1.1 gives an example of an environment-operator-machine system. This can be applied for design of work systems as well as machines/products—the two mainstream applications of ergonomics.

The operator is central to the system—he or she will first perceive the environment through visual and auditory perception, then consider the information, make a decision, which is then manifested as a response output. The perception is guided by the operator's attention. From all the millions of bits of information available, the operator is forced to choose the information most relevant to perform the task. Some attentional processes are subconscious (pre-attentive), some become automatic with training, and some are deliberate strategies.

For new or unusual tasks, decision making is a slow process. The operator will have to interpret the information, what alternatives for action there are, and to what extent those actions are relevant to achieve the goals of the task.

For routine tasks, decisions are more or less automatic and much quicker to accomplish than for unaccustomed tasks. Some researchers question whether decision making is an appropriate term. Other terms, such as "situated action" may be more appropriate to describe the automaticity in response that humans strive for.

The purpose of the operator's response is to manually control a machine (e.g., computer) or a tool (e.g., hammer) or an artefact (e.g., football). A second purpose may be to communicate visually or verbally with a fellow operator or a machine (voice control).

There are modulating factors, which affect the success of task performance. Operator competence and training, age, and motivation will affect all stages in information processing. An experienced operator will perceive the environment differently than a novice operator. He or she will focus on details of importance, filter out other information, and possibly also "chunk" the information into larger units, so that it is possible to make faster and more efficient decisions.

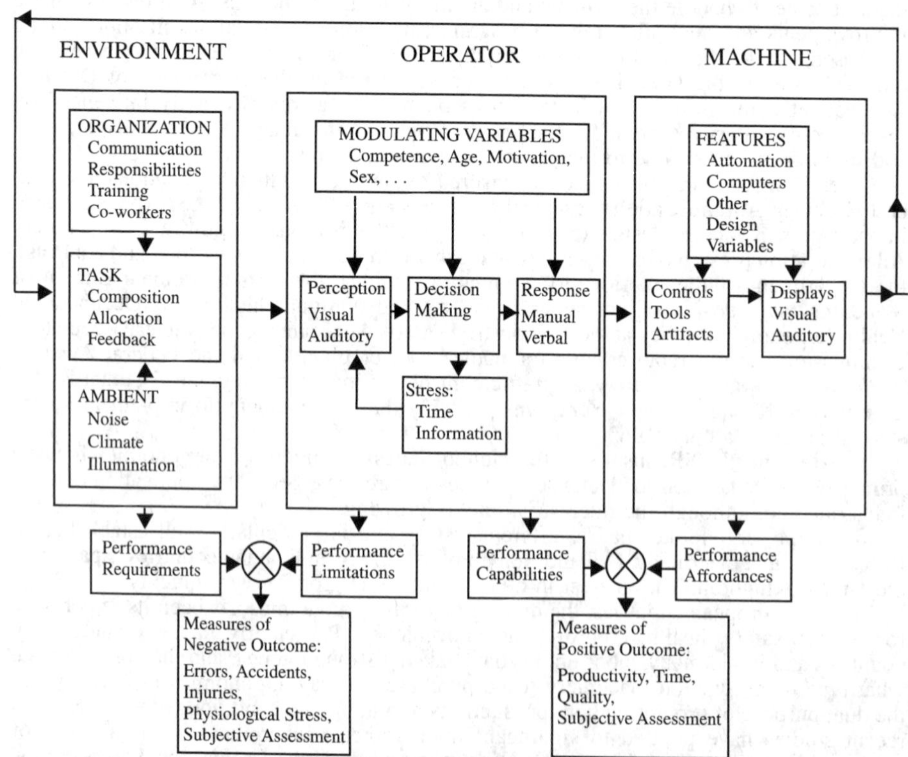

Figure 1.1 A systems approach to ergonomics.

Stress is an important variable that affects perception, decision making, and response selection. High stress levels are common when the time to perform the task is limited or when there is too much information to process. High stress levels lead to increased physiological arousal and can be monitored by using various physiological measures (e.g., heart rate, EEG, excretion of catecholamines). Under such conditions, the bandwidth of attention may narrow and operators develop "tunnel vision," thereby increasing the probability of operator error.

Figure 1.1 illustrates only the most important concepts and connections. In reality, human information processing is much more complex with many more feed-back loops and concepts (for an overview, see Wickens, 1992).

In this general and conceptual system, the subsystem "environment" is used to conceptualize the task as well as the context in which it is performed. It could be a steel worker monitoring an oven. Here organizational aspects are important, since they determine the task composition. Some tasks may be allocated to fellow workers and to supervisors, or even to computers. Such considerations will affect worker performance and how information is communicated between employees, customers, and computers. *Task allocation* is a central problem in human factors: How can one best allocate the work tasks among machines and workers so as to achieve both organizational goals and individual goals?

The operator receives various forms of feedback from his or her actions. There may be feedback from task performance, from coworkers, from management, and so forth. To enhance task performance, communication, and job satisfaction, such feedback must be efficient. This means that individuals must be given and must receive feedback on how well or how poorly they are doing, as well as feedback through communication.

The ambient environment describes the influence of environmental variables. For example, for a steel worker, noise stress and heat stress are usually high. This increases physiological arousal and physiological stress, thereby affecting task performance.

The importance of the social/organizational environment has been increasingly stressed during the last few years. This movement in ergonomics is referred to as "macroergonomics" (Hendrick, 1993). Ergonomics is undertaken in an organizational context, which deeply affects the appropriateness of alternative design measures. Company policies with respect to communication patterns, decentralization of responsibilities, training, and education have a great impact on ergonomics design. Perhaps because the military organization is fairly static, the importance of organizational context has not been emphasized in previous research. One exception is the sociotechnical research developed in the United Kingdom in the 1950s (e.g., the Tavistock group).

DeGreene (1973, p. 52) commented that "human factors has essentially ignored the motivational and morale problems and social stresses ensuing from a given systems design."

We should mention that organizational considerations are not necessary for leisure systems or consumer products.

The global purpose of ergonomics is to design an optimal system. A competent ergonomist would need to consider all three subsystems in Figure 1.1 as well as their interactions. It would not, for example, be appropriate to design displays and controls without consideration to the environment and the task.

The machine subsystem is broadly conceptualized in Figure 1.1. The term "machine" is in a sense misleading, since it symbolizes any artefact. The "machine" could be a computer, a VCR, or a tennis racket. Any of these artefacts is controlled by a human, and as a result there is a changing state which is "displayed"—it can be seen or heard. A pocket calculator will show the results of a calculation, the melting iron in a steel plant will change temperature and color, the toaster will pop the bread, and the sports fisher will catch a trout. All these scenarios can be designed to optimize systems performance.

It is important to note that the system in Figure 1.1 has feedback. Display information is fed back to the environment subsystem and becomes an important task variable. Ergonomics studies the effect of design on the performance of the human operator. To evaluate design, it is necessary to go around the loop. Ergonomics is in this respect different from other disciplines. In experimental psychology, for example, there is no requirement for study of dynamic systems.

1.3.1 System Goals

This system will be used to discuss two major goals of ergonomics: safety and productivity. In an industrial production system ergonomics is rarely a goal by itself. Safety and productivity are common goals. Ergonomics is rather a methodology that is used to achieve safety and productivity. As we mention above, the main purpose of ergonomics is the design of systems so that they are productive, safe, and also comfortable and enjoyable.

1.3.1.1 The Goal of Productivity

At the bottom of Figure 1.1 there are indicators which can be used to measure positive and negative outcomes of systems performance and dependent variables that can be used for these measures.

We may compare the performance capabilities of the human operator with the performance affordances of the machine (Norman, 1990). The purpose here is to design the machine so as to maximize task performance. Well-designed controls and displays "afford" better human performance. If the human performance capabilities are greater than the performance afforded by the machine, we may try to improve machine design.

The positive outcomes of systems performance can be quantified using *objective measures* such as "time to perform the task" and "quality." There are also *subjective variables,* that is, assessment by humans. Such assessments can be used to assess productivity as well as job satisfaction and operator comfort and convenience.

1.3.1.2 The Goal of Safety

The safety status of a system can be assessed by comparing performance requirements of the environment with performance limitations of the operator (Figure 1.1). If performance limitations of the operator are exceeded, there is an increased risk of human error with probabilities of an injury. Injuries and accidents are relatively rare in the workplace. Rather than waiting for accidents to happen, it may be necessary to predict safety problems by analyzing other dependent variables such as operator errors or obtain subjective assessments of safety risks or use physiological response variables to estimate if the operator is overloaded.

The goal is then to design a safer environment. This can be done most efficiently if the operator's reactions and actions can be interpreted by the designer. The operator's perception, decision making, and actions are therefore always central to ergonomics design.

From 1920 to 1960, safety research focused on accident proneness with the intent to "improve" the operator through careful selection and training. This research tradition produced little practical and theoretical value (Shaw and Sichel, 1975) and has been abandoned. The *accident proneness* concept is misleading. In fact automobile drivers who had accidents will often not incur future accidents. Likewise, factory workers who had accidents will often not have future accidents. The tendency to blame individuals for accidents is therefore counterproductive. What is needed is greater focus on environmental design and machine design to make environments and machines safer. For similar reasons safety training is only partially valid. Many accidents involve unexpected scenarios that are difficult to prepare for by training. For example, injuries due to manual materials handling are often due to a combination of unexpected events, and training of correct lifting techniques does not help (Kroemer et al., 1992).

In the machine/environment subsystems, there are several design elements which can be improved, for example:

1. The display (visual and auditory) of information
2. The allocation of tasks between workers and machines/computers
3. Optimization of the ambient environment, including illumination, noise, and climate
4. Organizational parameters such as competence of the workforce, allocation of responsibility, and the design of communication systems

Such design efforts are in ergonomics considered more efficient, since they have the potential of making training—and in this case safety training—unnecessary.

1.3.1.3 The Trade-Off between Safety and Productivity

As we have noted, there are two global goals of ergonomics: safety and productivity. There is also a trade-off relation between the two. This is referred to as "Speed-Accuracy Trade-Off" or SATO (Wickens, 1992). In manufacturing, this has the effect that operators may improve productivity, but at the cost of more errors (or decreased work quality).

Industrial managers who encourage employees to improve both productivity and quality most likely will not be successful, unless organizational parameters are improved.

1.4 DESIGN AS AN ACTIVITY

Many ergonomists are involved in design, and it is therefore of interest to analyze the steps that lead to a new design. A procedure for design activity is given in Figure 1.2. Systems goals for design are formulated based on requirements as they are formulated by users, markets, or organizations. User and Market requirements specify what types of products should be manufactured. User and Organizational requirements may specify how a job should be designed. There may also be Legal requirements, for example, that a robotics workplace should be safe or Company requirements that certain existing machines must be used for production. The latter are usually handled as constraints in design, rather than goals (Suh, 1990). Constraints are different from goals. Goals define the design space whereas constraints delimit the design space by invalidating certain designs.

Based on the systems goals, functional requirements are specified, and the task is then to design a new system, artefact, or job that will satisfy functional requirements.

There are two main activities in design work: *synthesis* and *analysis*. The synthetic stage is when designers use their knowledge and experience to come up with a design solution that will satisfy the functional requirements. This is a creative task. Studies of designers' decision making have produced some interesting results. Goel (1990) investigated engineers' design decisions for design of new products. He arrived at the surprising result that only 2% of the decisions were logical, in the sense that B follows from A. The remainder, 98%, were decisions based on associations and experiences. This means that designers will first hand try to apply what they know, and see if it works out. This trial-and-error-based procedure resembles case-based reasoning; that is, take a previous design solution or experience, modify it slightly so that it can fit the present scenario, then evaluate the results.

This rather erratic and irrational process in design work is also verified by Guindon's (1989) studies. She found that software programmers will jump back and forth between different tasks following their associations. She referred to this as "situated cognition"—do whatever is at hand and is practical to deal with—forget about logic.

This "irrational" process that is driven by associations and evaluations of trade-offs may be the basis for creativity. New combinations are generated and evaluated immediately by the designer and then either abandoned or retained. The associative and negotiative nature of creative cognition is probably why artificial intelligence (AI) has failed. Design supports based on AI cannot consider associations other than those based on logic, and they have a poor sense of evaluation (Gero, 1990).

To be successful in synthetic design, the ergonomist must have knowledge and experience that can produce good design solutions. Moray (1991) argued that ergonomists

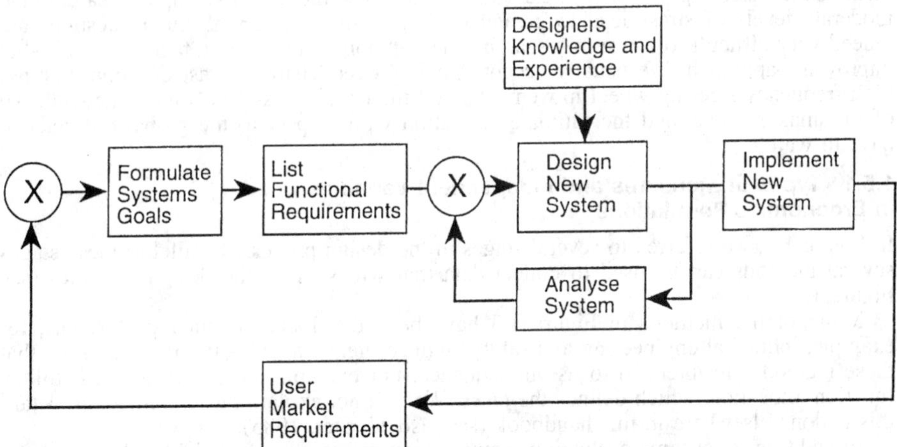

Figure 1.2 Procedure for design and redesign of a system.

must have an interdisciplinary background. For example, to come up with good design solutions in a manufacturing plant, the ergonomist must also be an engineer. In contrast, an occupational nurse, even though he or she is an expert at diagnosing occupational injuries and treating the patient, would not be able to propose a new machine in place of the machinery that caused the problem, nor evaluate the productivity consequences (Helander, 1995).

To suggest appropriate design solutions, a combination of ergonomics and domain knowledge is required. Landauer (1995) quoted studies on usability testing of human–computer interfaces. The average interface has 40 usability bugs (the range of variation is 17–140). One domain expert will on average identify 20% of the bugs during a test session with the software. One human factors expert will identify 40%, and one individual who is both a domain and human factors expert will identify 60% of the bugs. A combination of human factors and domain expertise is therefore desirable.

If the synthesis of design is difficult to structure, the analysis is not. Analytical methods can be proceduralized and computerized. There can be laboratory studies or studies of real-world performance. Typically the types of dependent variables referred to in Figure 1.1 are used: human performance data, physiological data, and subjective assessments.

Based on the outcome of the analysis an improved design may be proposed, the purpose of which is to improve operator performance and safety. The system is then implemented. Implementation gives an opportunity to collect real data in the real setting. Then it is possible to identify usability problems as well as market requirements.

The designer will sometimes find himself or herself in the situation where functional requirements have not been identified. This is often the situation in software design, where new software will be launched with the aspirations and hopes that it will sell. Rapid prototyping and beta-testing may be used to assess usability problems and predict market requirements, but realistic data will be available only when the product is put on the market. Then it will be feasible to formulate systems goals and functional requirements, and the feedback loop in Figure 1.2 will be closed.

1.5 THE MEASUREMENT PROBLEM

In Figure 1.1, we defined several global dependent variables such as operator errors and productivity which can be used to assess the state of the system. We have not defined independent variables. These are easier to research in laboratory experiments than in the real world (Chapanis, 1967). In a controlled experiment, the ergonomics researcher may investigate the effects of different levels of an independent variable, such as gender, level of noise, or type of feedback on some dependent variable. These are not the global variables mentioned above. At the subsystems level, dependent variables are usually those defined within the operator subsystem, such as legibility and visibility, speed and quality of decisions, and measures of manual response and physiological stress.

Unfortunately, there is a conceptual gap between these variables and global dependent variables. What is the effect of legibility of safety warnings on industrial safety? What is the effect of slow decision making on errors in control manipulation? Will quick movements of the steering wheel in a car (rather than slow movements) improve safety? Do moderate levels of stress lead to greater quality in manufacturing? Such questions are indeed very difficult to prove in research. The ergonomist must often take a logical—but unproven—approach. We usually accept that improved visibility, fast decision making, high-frequency steering wheel movements, and moderate stress levels are beneficial. All of this makes sense, and the ultimate validation with respect to the global criteria will have to wait.

1.5.1 Types of Measures and Independent Variables Used in Ergonomics Populations

In Figure 1.2, we referred to several stages in the design process. Parallel to these stages several methods can be used to collect data that will support the design activities (see Figure 1.3).

Many of the methods in Figure 1.3 have been developed in other professions, for example, industrial engineering and safety engineering. Chapanis (1995) pointed out that these methods are intended to use in sequence. For example, task analysis must follow function allocation, which defines the tasks. We will not explain those methods in detail; this is done elsewhere in this handbook (see also Meister, 1985).

In addition to systems evaluation methods, several dependent variables that are commonly used in ergonomics are listed in Table 1.1.

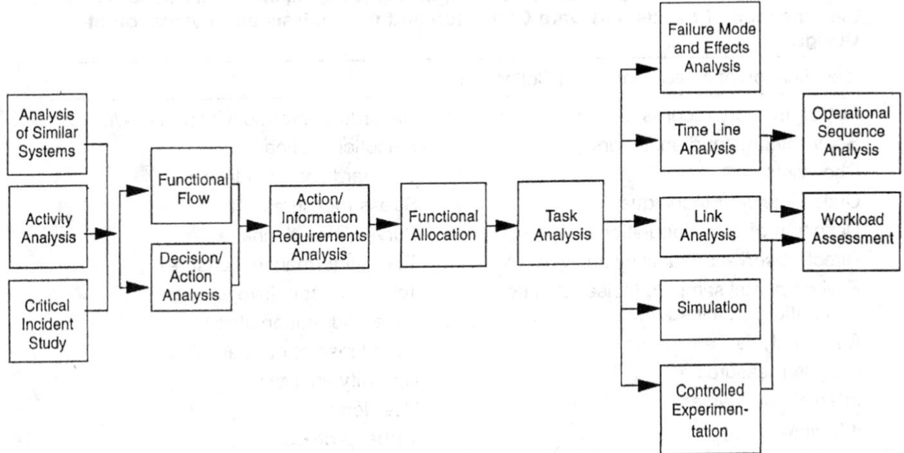

Figure 1.3 A sequence of methods that has been useful for ergonomics design (Chapanis, 1995).

1.6 THE GROWTH OF ERGONOMICS AROUND THE WORLD

Human factors and ergonomics started off in the Western world and has grown around the world. In many industrially developing countries, the emphasis has been on job design including biomechanics, heat stress, and work physiology. The economical conditions are different, and there is a greater tendency to use manual labor. As an example, Eriksson (1976) estimated 20 years ago that 200 workers at a road construction site in Bangladesh could move as much dirt manually as one construction vehicle, and the construction company costs were equivalent. Under such circumstances the national economy, as well as the workers' economy, will gain by using manual labor. Thus the interest and focus of ergonomics is different. However, with the introduction of computers in industrially developing countries, there has been a sudden shift in interest, and the problems of usability of complex systems is now universal. This shift is very logical, as the cost of computers relative to the cost labor is low.

1.6.1 The International Ergonomics Association

The International Ergonomics Association (IEA) is an international umbrella organization for ergonomics societies. IEA was conceived at a seminar organized in Leiden, Holland, by the European Productivity Agency in 1957. In 1961, the first triannual congress was held in Stockholm. The present membership of IEA is listed in Table 1.2.

During the period 1990 to 1995, the number of member societies in IEA doubled. In 1996, there were 35 federated member societies and two affiliated societies. A few societies cover several countries. Altogether about 17,000 ergonomists in 45 countries are affiliated with IEA through their professional societies.

One major goal of the IEA is to support ergonomics in countries where it has not developed and to inspire the formation of ergonomics societies. At present, IEA is well represented in all parts of the world except Latin America (one member), Africa (one member), and the Middle East (one member). However, several new societies are under formation in these parts of the world.

A survey among 25 IEA federated societies investigated the interests and activities of ergonomists around the world (Brown, Noy, and Robertson, 1995).

1.6.1.1 Membership Profile

Federated societies were asked to estimate the proportion of eligible ergonomists who belonged to their society. The average estimate was 61%. This would indicate that the current worldwide population of ergonomists is about 25,000. However, this is an underestimation since many professionals belong to other types of association.

Ergonomists come from a variety of professional fields. This mixed background is well demonstrated by the membership of societies, which typically consists of engineers, psy-

Table 1.1 Common Dependent Variables, Methodologies, and Tools Used for Identification of Needs and Data Collection and for Analysis and Evaluation of Design

Identification of Needs and Data Collection	
Accident/injury records	Records of productivity and quality
Body rhythms and shift work	Selection testing
Checklists	Standards and guidelines
Critical incident technique	Stress measures
Definition of user population	Survey/questionnaire
Direct observation/activity analysis	Task performance measures
Environmental samples (noise, climate, vibration, illumination)	Testing of cognitive abilities
Error analysis	Time and motion study
Fatigue measures	Time lapse photography
Intervention studies	Usability analysis
Interview	User log books
Operator performance (time and error)	Verbal protocol
Physical work activity (HR, EMG, oxygen uptake)	Video recording
Psychological Stress (EEG, GSR, HR variability, etc.)	Visibility/legibility
Readability	Vision/hearing testing
	Walkthrough
Analysis, Evaluation, and Design	
Anthropometric analysis	Mockups/rapid prototyping
Biomechanical analysis	Operational sequence analysis
Communication analysis	Organizational analysis
Control theory modeling	Predetermined time analysis
Cost/benefit analysis	Production systems analysis
Design reviews	Ratings
Failure mode analysis	Readability analysis
Fault tree analysis	Simulation
Function allocation	Systems analysis
General error modeling system (GEMS)	Task analysis
GOMS/cognitive walkthrough	The human error rate prediction (THERP)
Hazard analysis	Time-line analysis
Information analysis	Training needs analysis
Link analysis	Workload analysis

chologists, and individuals from the medical profession. Society members have received academic training in industrial engineering, ergonomics, safety, physiology, physiotherapy, and psychology. These educational programs accounted for about 70% of all members. Of the current membership 29% are academics, 27% practice in industry, 15% do research, 10% are private consultants, 8% work for the government, and 11% work in other occupations.

Despite the formal education of the membership in ergonomics or related fields, many societies claimed that the major problem facing the occupation was recognition by the government, industry, and the public. This was an issue of great concern both for professional societies and for individuals belonging to them. To improve this situation, societies seek new ways to enhance their profession. Certification and licensing of ergonomists are important issues. It is also important that ergonomists find ways of communicating between themselves.

To ensure continued growth of ergonomics, the federated societies suggested many specific needs which societies must consider. These are listed in Table 1.3.

Societies were then asked about the most important plans for new initiatives and actions. The results are given in Table 1.4.

Table 1.2 Founding Year of 35 IEA Member Societies and Membership in 1996

Federated Societies	Founded	Members
Association Española de Ergonomia	1966	110
Belgian Ergonomics Society	1986	142
Brazilian Ergonomics Association	1983	133
Chinese Ergonomics Society	1989	450
Croatian Ergonomics Society	1974	70
Czech Ergonomics Society	1990	160
Ergonomics Society	1949	1164
Ergonomics Society of Australia	1966	578
Ergonomics Society of the Federal Republic of Yugoslavia	1973	50
Ergonomics Society of Korea	1982	219
Ergonomics Society of South Africa	1984	56
Ergonomics Society of Taiwan	1993	210
Gesellschaft für Arbeitswissenschaft	1952	700
Hellenic Ergonomics Society	1991	33
Human Factors and Ergonomics Society	1957	5187
Human Factors Association of Canada	1968	623
Hungarian Ergonomics Society	1987	70
Indian Ergonomics Society	1987	135
Irish Ergonomics Society	1992	26
Israeli Ergonomics Society	1981	59
Japan Ergonomics Research Society	1964	2052
Nederlandse Vereniging voor Ergonomie	1962	565
New Zealand Ergonomics Society	1986	116
Nordic Ergonomic Society	1969	1370
Österreichische Arbeitsgemeinschaft für Ergonomie	1964	44
Polish Ergonomics Society	1977	369
Portuguese Association of Ergonomics	1992	70
Russian Ergonomics Association	1989	625
Slovak Ergonomics Association	1992	27
Società Italiana di Ergonomia	1968	120
Societé d'Ergonomie de Langue Française	1962	693
South East Asia Ergonomics Society	1983	70
Ukraine Ergonomics Society	1992	128
Affiliated Societies		
European Society of Dental Ergonomics	1990	31
Human Ergology Society	1970	230
	Total	16,685

1.6.2 Applications of Ergonomics

Table 1.5 compares the current applications of ergonomics to the early applications when a federated society was founded. Only the five most important areas are listed in the table. In the current applications anthropometry, work physiology, and psychology were replaced by safety, workload assessment, and human–computer interaction. This illustrates the current trend toward cognitive ergonomics.

Federated societies were also asked to indicate important emerging areas in ergonomic interest. The responses are given in Table 1.6.

A new image of ergonomics is emerging from this survey. There has been a shift in interest over the last 20 years. The diffusion of computer technology and complex machinery has had a tremendous impact. Cognitive ergonomics, usability studies, human reliability, and human–computer interaction are new top priories. Organizational design and the study of industrial change processes and continuous improvement are also im-

Table 1.3 Most Important Needs in Ergonomics in 25 Federated Societies (Brown, Noy, and Robertson, 1995.)

Specific Needs	Frequency
More training and education programs in ergonomics	5
Diffusing ergonomics into industry	4
Marketing of ergonomics	4
Financial resources	3
Increase membership of society	3
Greater involvement with European Societies and IEA	3
More members from different professions	3
Establishing network, interchange and support function	2
Recognize multidisciplinary approach to problem solving	2
Total	**25**

Table 1.4 Most Important New Initiatives in 25 Ergonomics Societies (Brown, Noy, and Robertson, 1995.)

Society's Plan	Frequency
International collaboration in research and information exchange	5
Active promotion of ergonomics research	4
Networking the society to improve communications	4
Marketing of the ergonomics profession	3
Actions to develop business	3
Certification of ergonomists	2
Regional meetings/workshops	2
Integration of Eastern European ergonomists	2
Total	**25**

Table 1.5 The Five Most Important Early and Current Applications of Ergonomics in 25 Ergonomics Societies (Brown, Noy, and Robertson, 1995.)

Importance	Early Applications	Current Applications
1	Anthropometry	Safety
2	Work physiology	Industrial engineering
3	Industrial engineering	Biomechanics
4	Biomechanics	Workload
5	Psychology	Human–computer interaction

portant. Biomechanics and work physiology are less dominating than they were in the past, except that there is a renewed interest in biomechanics since the frequency of musculoskeletal disorders has increased in industry.

It is interesting to note that this trend was valid not only for industrialized countries but also for industrially developing countries. It behoves our profession to take note of this new emphasis. Since the beginning of ergonomics around 1950, society and technology have developed tremendously. Brian Shackel (1991) characterized the development as follows:

1950s—military ergonomics
1960s—industrial ergonomics
1970s—consumer products ergonomics

Table 1.6 Important Emerging Areas in Ergonomics in 25 Ergonomics Societies Around the World

Topics	Frequency
Methodology to change work organization and design	7
Work-related musculoskeletal disorders	7
Usability testing for consumer electronic goods	6
Human–computer interface: Software	6
Organizational design and psychosocial work organization	5
Ergonomic design of physical work environment	4
Control room design of nuclear power plants	3
Training of ergonomists	3
Interface design with high technology	3
Human reliability research	3
Mental workload	3
Workforce cost calculation	3
Product liability	2
Road safety and car design	2
Transfer of technology to developing countries	2

1980s—human–computer interaction and software ergonomics
1990s—cognitive ergonomics and organization ergonomics

I predict that in the years 2000–2010 we will see an interest in eco-ergonomics. This will analyze how ergonomics methodology can be used to reduce pollution, decrease fuel consumption, reduce crime, reduce the size of very large cities, and so forth. Although these are not related to work activity, they can utilise ergonomics methodology for design and analysis of dynamic systems. This approach would study systems at the level of individuals to understand how individual behavior through the use of feedback can be structured and steered towards desirable goals.

Our profession is driven by design requirements from users, markets, industries, organizations, and governments. Ergonomics must be able to quickly respond to the changing needs of society. Training programs in ergonomics must be able to incorporate new areas of interest. Certification programs for ergonomists must be flexible enough to reconsider current needs, and teaching programs must incorporate new knowledge (Bullock, 1995).

REFERENCES

Braidwood, R. (1951). *Prehistoric Men.* Chicago, IL: Natural History Museum.

Brown, O. Jr., Noy, I., and Robertson, M. (1995). *Special Survey of IEA Federated Socities.* Santa Monica, CA: International Ergonomics Association, c/o Human Factors and Ergonomics Society.

Bullock, M. I. (1995). Harmonizing professional standards in ergonomics while recognizing diversity. *Ergonomics, 38,* 1558–1570.

Chapanis, A. (1967). The relevance of laboratory studies to practical situations. *Ergonomics, 10*(5), 557–577

Chapanis, A. (1995). Ergonomics in product development: A personal view. *Ergonomics, 38,* 1625–1638.

Childe, G. (1944). *The Story of Tools.* London: Cobbot.

Christensen, J. M. (1962). The evaluation of the systems approach in human factors engineering. *Human Factors, 4*(1).

DeGreene, K. B. (1973). *Sociotechnical Systems.* Englewood Cliffs, NJ: Prentice Hall.

Eriksson, R. (1976). Personal communication. International Labor Organization, Geneva, Switzerland.

Gero, J. (1990). University of Sydney. Personal communication. University at Buffalo, Buffalo, New York, April 1990.

Goel, V., and Pirolli, P. (1992). The structure of design problem spaces. *Cognitive Science, 16,* 395–429.

Guindon, R. (1990), Designing the design process: Exploiting opportunistic thoughts. *Human Computer Interaction, 5,* 305–344.

Helander, M. G. (1995). *A Guide to the Ergonomics of Manufacturing.* London: Taylor & Francis.
Hendrick, H. (1993). The IEA and International Ergonomics: Past, Present and Future. In: *Proceedings of the IEA/Russian conference—Ergonomics in Russia, the other independent states and around the world—past, present and future.* St. Petersburg, Russia: Russian Ergonomics Society.
Hendrick, H. W. (1995). Future directions in macroergonomics. *Ergonomics, 38,* 1617–1624.
Karwowski, W. (1991). Complexity, fuzziness, and ergonomic incompatibility issues in the control of dynamic work environments. *Ergonomics, 34,* 671–686
Konz, S. (1990). *Work Design: Industrial Ergonomics.* Worthington, OH: Publishing Horizons.
Kroemer, K., Kroemer, H., and Kroemer-Elbert, K. (1994). *Ergonomics: How to Design for Ease and Efficiency.* Englewood Cliffs, NJ: Prentice Hall.
Landauer, T. (1995). *The Trouble with Computers.* Cambridge, MA: MIT Press.
Licht, D. M., Polzella, D. J., and Boff, K. R. (1991). Human factors, ergonomics and human factors engineering. An analysis of definitions (Report No 89-01). Wright-Paterson AFB, OH: CSERIAC Program Office.
Luczak, H. (1997). Arbeitswissenschaft in internationalen Vergleich. In H. Luczak and E. Volpert, Eds., *Handbuch der Arbeitwissenschaft.* Stuttgart: Verlag Schaeffer-Poeschel.
Meister, D. (1985). *Behavioral Analysis and Measurement Methods.* New York: Wiley.
Moray, M. (1994). The future of ergonomics—The need for interdisciplinary integration. *Proceedings of IEA Congress, 1791–1793.* Santa Monica, CA: The Human Factors and Ergonomics Society.
Norman, P. (1990). *The Origin of Everyday Things.* New York: Doubleday.
Ramazzini, B. 1940 (1713). Wright, W. (Trans.). *The Diseases of Workers.* Chicago, IL: University of Chicago Press.
Rosenbrock, H. H. (1983). Seeking an Appropriate Technology. *Proceedings of IFAC Symposium on Systems Approach to Appropriate Technology Transfer, 127–134.* Vienna Austria: IFAC.
Shackel, B, (1991). Ergonomics from Past to Future: An Overview. In M. Kumashiro and E. D. Megaw, Eds, *Towards Human Work: Solutions to Problems in Occupational Health and Safety.* London: Taylor & Francis.
Shaw, L. and Sichel, H. (1975). *Accident Proneness.* New York: Pergamon Press.
Suh, M. (1990). *Principles of Design.* New York: Cambridge University Press.
Vanwonterghem, K. (1994). Personal communication. Brussels, Belgium.
Wickens, C. D. (1992). *Engineering Psychology and Human Performance.* New York: Harper Collins.

CHAPTER 2
SYSTEMS DESIGN AND EVALUATION

Sara J. Czaja
Department of Industrial Engineering
University of Miami
Coral Gables, FL 33124 USA

2.1 INTRODUCTION AND OVERVIEW

2.1.1 The Centrality of the Systems Concept to Human Factors

Human factors is generally defined as the study of human beings and their interaction with products, environments, and equipment in performing tasks and activities. Within this domain the central focus of study is the human–machine system. The objectives of human factors are to maximize human and system efficiency and human well-being and quality of life (Meister, 1991; Sanders and McCormick, 1993).

A human–machine system is some combination of one or more humans and one or more physical components to transform inputs into desired outputs. Further, this interaction takes place within an environment. The environment encompasses the specific task environment (e.g., lighting, temperature) and the social and organizational environment. The system may be simple, such as a human interacting with a tool, or it may be complex, such as a human monitoring an automated process. In all cases, human factors is concerned with the optimization of the interaction between the human and the physical component. Given that performance is studied in interactional terms, the *systems concept* is fundamental to human factors engineering.

The systems concept implies that components or elements of a system are only meaningful in terms of the whole system. Specifically each element of a system must be viewed in terms of its interaction with other elements of the system. This is in contrast to a reductionist approach, which focuses on a particular system component or element in isolation. The *reductionist* approach has traditionally been the "popular" approach to system design where the focus has been on the physical or technical components of the system with little regard to the behavioral component. In recent years, there have been a number of incidents, such as the accident at 3 Mile Island and the Exxon Valdez disaster, which have demonstrated the shortcomings of this approach. Butera (1984) reviewed 19 cases studies which evaluated the implementation of automation and found that in most cases the technology did not achieve its anticipated potential. Further, management, in most of these cases, did not consider human issues during the implementation process.

Applied to the field of human factors, the systems concept infers that performance must be evaluated in terms of the context of the human–machine system; equipment, environment, operating procedures, and goals. Human factors engineers generally agree that the overall efficiency of a system is determined by optimizing the performance of both the human and physical components. Traditionally the design engineer focuses solely on the technical component of the system, and the behavioral scientist focuses solely on the performance component. Human factors is unique in that it is concerned with both the behavioral and physical domains. The systems concept provides a unifying framework for the study of these domains (Meister, 1989).

The systems concept is both a theoretical construct and a methodological orientation; implicit in the belief in the systems concept is the adoption of the systems approach. The systems approach considers the interactions of all of the components of a system relative to system goals when evaluating a particular phenomena. A basic tenet of human factors is that the optimization of human and system efficiency requires the adoption of the systems approach where all major system components are given adequate consideration throughout the system design process. Designs which do not consider the human element will not achieve the maximum level of performance. Thus a central activity of human factors is the application of information regarding human performance to all phases of system development and design. For this reason a discussion of the role of human factors in system design and evaluation is central to a handbook on human factors engineering. This is especially true in today's era of computerization and automation where systems are becoming increasingly complex.

This chapter will discuss the role of human factors engineering in system design. The focus will be on the approaches and methodologies used by human factors engineers to integrate knowledge regarding human performance into the design process. The topics of system design is vast and encompasses many areas of specialization within human factors. Thus this chapter will introduce several concepts (e.g., task analysis, function allocation, measurement) which are covered in depth in other chapters of this handbook. Prior to discussing the design process, a brief history of the systems approach will be provided followed by a more detailed definition of the systems concept and the assumptions of the systems approach. The intent is to provide an overview of the system design process and to demonstrate the importance of human factors to systems design. Further, the chapter will introduce new approaches to system design which are being applied to complex, integrated systems.

2.1.2 A Brief History of the Systems Approach

The systems concept was initially a philosophy associated with thinkers such as Hegel who recognized that the whole is more than the sum of its parts. It was also a fundamental concept among the Gestalt psychologists who recognized the importance of "objectness" or wholeness to human perception (Meister, 1989). The systems approach, which evolved from systems thinking, was initially developed in the biological sciences and further refined by communication engineers in the 1940s. The adoption of this approach was bolstered during World War II when it was recognized that military systems were becoming too complex for humans to successfully operate. This discovery gave rise to the emergence of the field of human factors engineering and its emphasis on human–machine systems.

Several authors (Sheridan, 1985; Sanders and McCormick, 1993) have classified the history of human factors engineering and the study of human–machine systems into three time periods: (1) 1945–1960; (2) 1960–1970; (3) post-1970. The initial time period gave birth to the concept of human–machine systems. The focus of human factors engineers was primarily on aircraft (civilian and military) and weapon systems with some limited application in the automotive and communication industries. In this era, human factors was often equated with the study of "knobs and dials."

During the 1960s, systems theory became a dominant way of thinking within engineering and human factors engineers began to use modeling techniques, such as control theory, to predict human-system performance. A number of investigators were concerned with developing models of human performance and applying these models to system design. For example, signal detection theory was applied to human detection performance and the concept of the "ideal observer" evolved (Green and Swets, 1966). At the same time, the application of human factors expanded beyond the military and many companies began to establish human factors groups. The concept of the human–machine system also expanded as human factors engineers became involved with the design of consumer products and workplaces. In 1960, a human factors group was formed at Eastman Kodak. The group specialized in problems related to workplace and job design and also became involved in the design and evaluation of products (Eastman Kodak Company, 1983).

In the past several decades, rapid developments in computer technology and automation have generated renewed interest in the study of human–machine systems. A tremendous amount of research and development has been directed toward optimizing the human–machine interface (hardware and software). This is largely because the complexity of human–machine systems has increased as has the nature of the human–machine interaction. The primary emphasis in today's human–machine systems is on the exchange, storage, and processing of information where both the human and machine components engage in cognitive activities.

Advances in computers and automation have resulted in new challenges for human factors engineers and system designers, necessitating the development of new approaches for the system design process. In many work domains the deployment of computers and automation has changed the nature of the demands placed on the worker to a more conceptual cognitive level where the emphasis is on thinking, decision making, and problem solving as opposed to strength and dexterity (Rasmaussen, Pejtersen, and Goodstein, 1994). In order to design these types of work systems effectively, we need to apply knowledge regarding human information processing capabilities to the design process. The need for this type of knowledge has created a greater emphasis on issues related to human cognition within the field of human factors and has lead to the emergence of cognitive engineering (Woods, 1988). Cognitive engineering is an applied cognitive science which applies the knowledge of cognitive psychology and other related disciplines to the design of cognitive environments (e.g., decision support systems) (Woods and Roth, 1988).

Further need for knowledge of cognitive operations comes from the changing nature of the design process. Generally the design of complex systems requires the involvement of many individuals with different types and levels of knowledge who must consider and integrate technical information at each stage of the design process and make design decisions. Typically these decisions are made under time and resource constraints. Information must be supplied to designers in a way that supports the design process. Thus the issue of developing effective design support tools has emerged within the domain of system design. Questions arise regarding what types of information designers need and how this information should be formatted. The development of design tools must be based on an understanding of the design problem, designer characteristics, and human information processing abilities (Rouse and Boff, 1987).

A related issue, which has emerged in the design of human–computer (automation) systems, is the degree to which cognitive support should be provided to the human operator. Computers now offer the potential of assisting the human in the performance of cognitive activities, such as decision making, and a question arises as to what level machine power should be deployed to assist human performance so that the overall performance of the system is maximized. This question has added complexity, as in most complex systems the problem is not restricted to one operator but to two or more operators who cooperate and have access to different databases. The operator is part of a decision-making team who together with the computer system controls the process. As noted by several authors (e.g., Woods, 1988; Rasmussen and Goldstein, 1985), the answer to this question should not be machine driven but situation or problem driven where the technology is used to augment the capabilities of the human. Thus development of effective decision support requires an understanding of the problem domain and the characteristics of the human problem solver. As stated by Rasmussen and Goldstein (1985), building an effective symbiosis between humans and computer systems requires maximizing the information processing abilities and knowledge states of both entities so that they act together in a complementary fashion.

Tangential to the issue of decision support is the general issue of function allocation. This issue has taken on increased importance in complex technological systems, and efforts are being directed toward the development of strategies for optimizing function allocation decisions (Chapter 11). The central goal with respect to system design is achieving a balance between the human system and the technological system. Too often technology is viewed as a panacea and implemented without sufficient attention is to human and organizational issues. There are numerous examples where this technocentered approach has resulted in negative consequences such as declines in productivity and worker dissatisfaction. It has become apparent that this approach is inadequate and that human factors must be interjected at all phases of the system design process. It has also become apparent that we must broaden our view of system design and consider both micro- and macroergonomic issues. The introduction of technology not only changes the nature of an individual's task but also changes the relationship among system components and the underlying organizational structure. Thus we can no longer limit our efforts to the design of specific jobs and human–machine interfaces but also must consider the sociotechnical/organizational aspects of the system (Hendrick, 1991).

In sum, the nature of human–machine systems has changed drastically since the era of "knobs and dials," presenting new challenges and opportunities for human factors engineers. We are not only faced with designing and evaluating new types of systems and a wider variety of systems (e.g., health care systems, living environments) but also with many different types of operators. Many people with limited technical background and of varying ages are operating computer systems, which raises many new issues for system designers. For example, older workers may require different types of training or different work schedules to effectively interact with new technology or operators with a limited technical background may require a different type of interface than those who are more experienced. Emergence of these types of issues reinforces the need to include human factors in system design. The following section presents a general model of a system and serves as background to a discussion of the system design process.

2.2 DEFINITION OF A SYSTEM

2.2.1 General System Characteristics

A *system* is an aggregation of elements organized in some structure (usually hierarchical) to accomplish system goals and objectives. All systems have the following characteristics: interaction of elements, structure, purpose and goals, and inputs and outputs. Systems are usually composed of humans and machines and have a definable structure and organization and external boundaries which separate it from elements outside of the system. All of the elements within a system interact and function to achieve system goals. Further, each system component has an effect on the other components. It is through the system inputs and outputs that the elements of a system interact and communicate. Systems also exist within an environment (physical and social), and the characteristics of this environment have an impact on the structure and the overall effectiveness of the system (Meister, 1989, 1991). For example, in order to be responsive to today's highly competitive and unstable environment, manufacturing systems have to be flexible and dynamic. Traditional models of manufacturing systems which are based on long-term planning for production where production takes place in a sequence of separate processes in separate departments are no longer adequate. Instead production needs to be organized around simultaneous activities where there is decentralized decision making and quick and easy access to information (Drucker, 1998). This creates a need for a change in the organizational structure. Formal, hierarchical organizations do not effectively support distributed decision making and flexible production processes.

Figure 2.1 presents a general functional model of a system. According to this model, the system receives inputs from various sources, which are analyzed and a response is made in the form of outputs. These inputs may relate to the routine functioning of the system or to some problem state. For example, a production system receives an order for x number of units, this order is analyzed and evaluated, and the elements of the system interact to produce the desired number of units. A key feature of this model is the dynamic interaction between the system and its environment.

Generally all systems have the following components: (1) elements (personnel, equipment, procedures); (2) conversion processes (processes which result in changes in system states); (3) inputs or resources (personnel abilities, technical data); (4) outputs (e.g., number of units produced); (5) an environment (physical and social and organizational); (6) purpose and functions (the starting point in system development); (7) attributes (e.g., reliability); components and programs; (8) management, agents, and decision makers; and (9) structure. These components must be considered in the design and evaluation of every system. For example, the nature of the system inputs has a significant impact on the ability of the system to produce desired outputs. Inputs which are complex, ambiguous,

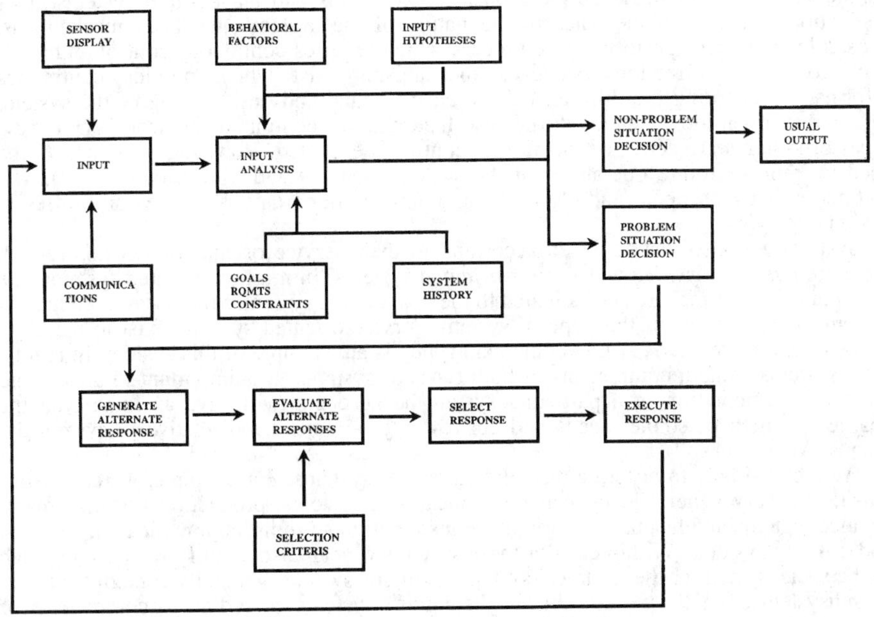

Figure 2.1 A general functional model of a system (adapted from Meister, 1991).

or unanticipated may lead to errors or time delays in information processing which in turn may lead to inaccurate or inappropriate responses. For example, if there is conflicting or confusing information on a patient's chart, a physician might have difficulty diagnosing the illness and prescribing the appropriate course of treatment.

Understanding the nature of system inputs is an important aspect of designing human–machine systems, especially for complex technological systems. This type of knowledge is needed to provide effective decision support for the operator. Different types of support are required for different types of decisions or problems. If the input is anticipated and familiar, people can solve problems using preplanned routines or "rule-based behavior" (Rasmussen, 1983). In this case, performance aiding might include training and practice on routines and the provision of retrieval aids, e.g., given a current set of symptoms an expert system might prompt the physician as to the typical medications used in treatment. If however, there is unanticipated variability in the input, or a novel or problem situation arises, the operator needs knowledge beyond the preplanned routine. In this case, performance aiding might take the form of supplying the operator with additional information or presenting alternative scenarios. In both cases a clear understanding of the problem-solving domain is required in order to design effective support (Woods, 1988).

Systems can also be characterized according to the nature of the system variables. Meister (1991) distinguishes between two types of system variables: *physical* variables and *behavioral* variables. Behavioral variables describe requirements for system operators and include factors such as task and function requirements, skill and training requirements, number and type of interdependencies, etc. Physical variables describe the physical and structural function of the system. These variables include the number of subsystems, the size and complexity of the system, the number and specificity of goals and missions, the requirements place on the system, the nature of feedback mechanisms, etc. The manner in which these variables are treated in system design determines the overall performance of the system (Meister, 1991). Decisions regarding function allocation or amount, type, and scheduling of feedback have an obvious impact on overall system efficiency.

2.2.2 System Classifications

There are various ways in which systems are classified. Systems can be distinguished according to degree of automation, functions and tasks, feedback mechanisms, system class, hierarchical levels, and combinations of system elements (Meister, 1991). A basic distinction is usually made between *open-loop* systems and *closed-loop* systems. This distinction is made on the basis of the nature of the system's feedback mechanisms. Closed-loop systems perform some process which requires continuous control and feedback for error correction. Feedback mechanisms exist which provide continuous information regarding the difference between the actual and desired state of the system. In contrast, open-loop systems do not use feedback for continuous control. When these types of systems are activated, no further control is executed. However, feedback can be used to improve future operations of the system (Sanders and McCormick, 1993). The distinction between open- and closed-loop systems is important as they require different design strategies.

Systems are also distinguished according to their service orientation. In this regard, there are *mission-oriented* and *service-oriented* systems. In mission-oriented systems, the needs of the personnel are subordinated to the goals of the system. Military and production systems are examples of this type of system. Service-oriented systems exist to meet the needs of clients or users. A governmental agency is an example of this system. In reality, most systems contain components of both types of systems. It is important to understand the service orientation of a particular system because this will have an impact on the degree to which personnel needs and desires may be considered relative to system demands (Meister, 1989).

We are also able to describe different classes of systems. For example, we can distinguish, at a very general level, between educational systems, production systems, maintenance systems and health care systems, transportation systems, communication systems, and military systems. Within each of these systems we can also identify subsystems, such as the social system or the technical system. Complex systems generally contain a number of subsystems. Finally we are able to distinguish systems according to components or elements. For example, we can distinguish among machine systems, human systems (biological systems), and human–machine systems. It is the latter type of system which is of interest to human factors engineers.

2.2.3 Human–Machine Systems

A human–machine system is some combination of humans and machines which interact to achieve the goals of a system. These systems are characterized by elements which interact, structure, goals, conversion processes, inputs, and outputs. Further, they exist in an environment and have internal and external boundaries. A simple model of a human–machine system is presented in Figure 2.2. As shown in Figure 2.2, the general systems model applies to human–machine systems; inputs are received and processed, and outputs are produced through the interaction of the system components. We can conceptualize Figure 2.2 as a model of a human–computer system where the human is

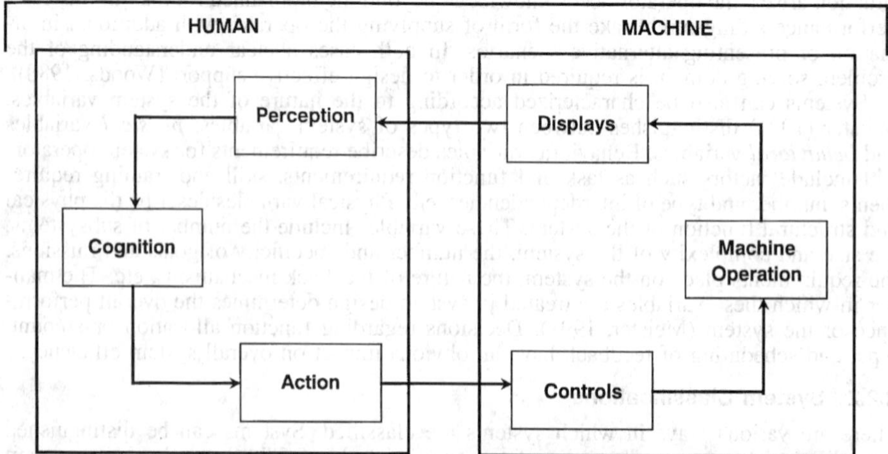

Figure 2.2 A simple model of human–machine system (adapted from Proctor and Van Zandt, 1994, Copyright © 1994, Allyn and Bacon. Reprinted by permission).

interacting with the computer to modify some data records. In this system, the human receives information, via the visual sense, from the computer display, processes the information and makes a decision regarding a response, and produces an output to the computer via a keyboard. The computer in turn processes this information and produces a response. These interactions take place within a particular environment, which as an impact on overall system performance. For example, if the lighting is inadequate, it may be difficult to read the information on the screen or if the system is part of a large open office environment it may be difficult for the operator to concentrate due to distractions. In both cases, the environment may contribute to an increased probability of operator error.

In recent years, the concept of the human–machine system has changed as machines have become more intelligent and capable of performing tasks formerly restricted to humans. Prior to the development of intelligent machines, the model of the human–machine interface was formed around a control relationship where the machine was under human control. In current human–machine systems (which involve some form of computer technology), the machine is intelligent and capable of extending the capabilities of the human. Computer/automation systems can perform routine, elementary tasks, complex computations, suggest ways to perform tasks, or engage in reasoning or decision making. In these instances, the human–machine interface can no longer be conceptualized in terms of a control relationship where the human controls the machine. A more accurate representation is a partnership where the human and the machine are engaged in a two-way cognitive interaction (Eggleston, 1987).

For example, in supervisory control, decision making is shared between the designer, the automatic control system, and the operator(s). The operator is part of a decision-making "team" where each member of the team has specified roles or tasks when making control decisions (Rasmussen and Goodstein, 1985). The nature of these roles depends on system design decisions. Figures 2.3*a* and 2.3*b* depict how the decision roles of the operator change according to level of automation in a supervisory control task for a nuclear reactor system. In 2.3*a*, the system is fully automated; the computer and the operator act in parallel, each with a diagnostic task. In 2.3*b*, the system is not fully automated and the operator and computer act in "series" with the diagnosis allocated to the operator.

Thus human–machine systems which involve "intelligent machines" can take many forms depending on the function allocation between the human and machine. Therefore, the specification of the human–machine relationship is an important design decision and the nature of this relationship must be considered in the system design process and treated as a design variable. The relationship must be such that the abilities of the components and the cooperation among the components is maximized. Further the human operator needs to develop a sense of trust for the machine component. This requires new approaches to system design as traditional models and approaches to system design do not consider the nature of the human–machine relationship especially at a cognitive level (Eggelston, 1987; DeGreen, 1991). It is no longer adequate to consider the human–machine relationship as one of control; instead we need to view it as a dynamic interaction. The impact of the changing nature of human–machine systems on the system design process and current approaches to system design will be discussed in a later section. However, before this topic is addressed, the concepts of system and human reliability will be introduced because these concepts are important to a discussion of system design and evaluation.

Obviously there are many different types of human–machine systems, and they vary greatly in size, structure, complexity, etc. While the emphasis in this chapter will be on work systems where computerization is an integral system component, we should not restrict our conceptualization of systems to large, complex technological systems in production or process environments. We also need to consider other types of systems such as a person using an appliance within a living environment, or a physician interacting with a heart monitor in an intensive care unit, or an older person driving an automobile within a highway environment. In all cases, the overall performance of the system will be improved with the application of human factors engineering to system design.

2.2.4 System Reliability

2.2.4.1 System Reliability

System reliability refers to the dependability of performance of system, subsystem, or system component in carrying out its intended function for a specified period of time.

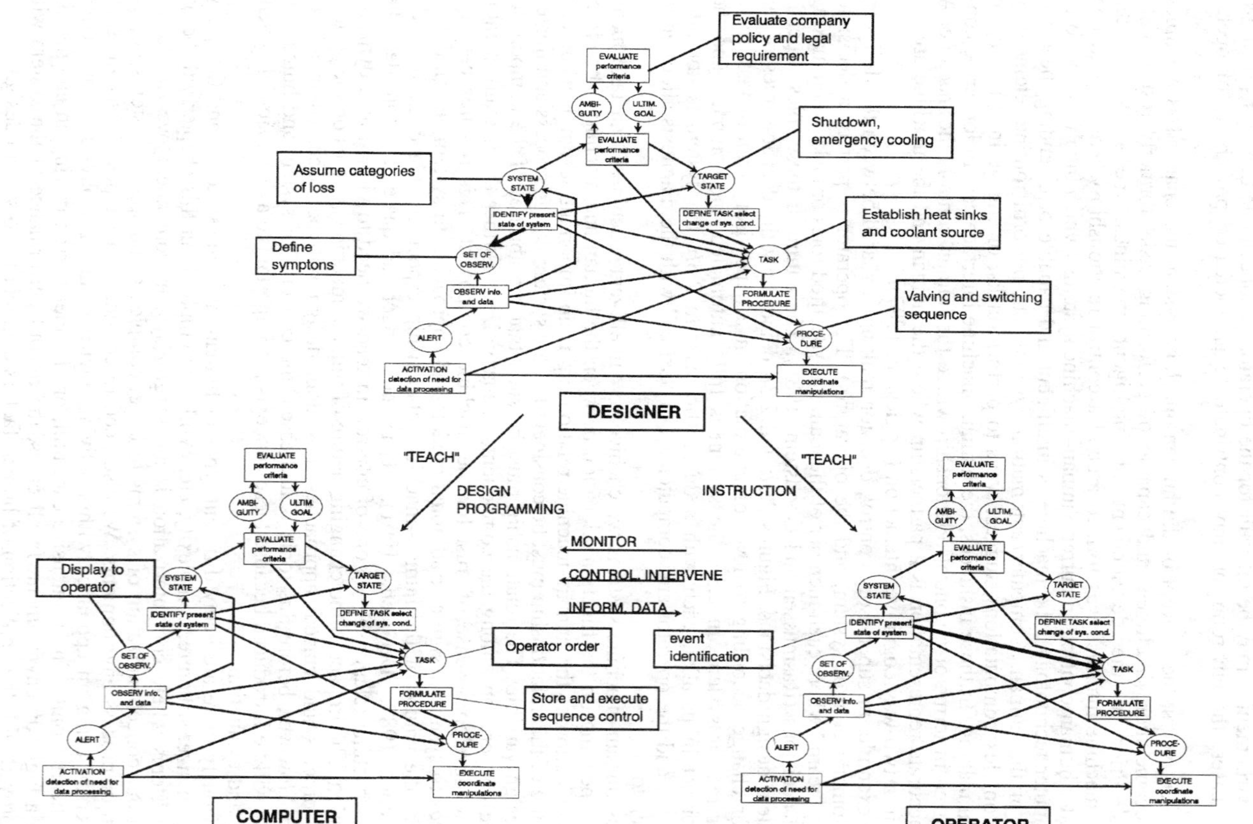

Figure 2.3a Decision roles in a fully automated nuclear system. The operator and process computer act "in parallel" in the control decision (adapted from Rasmussen, 1986).

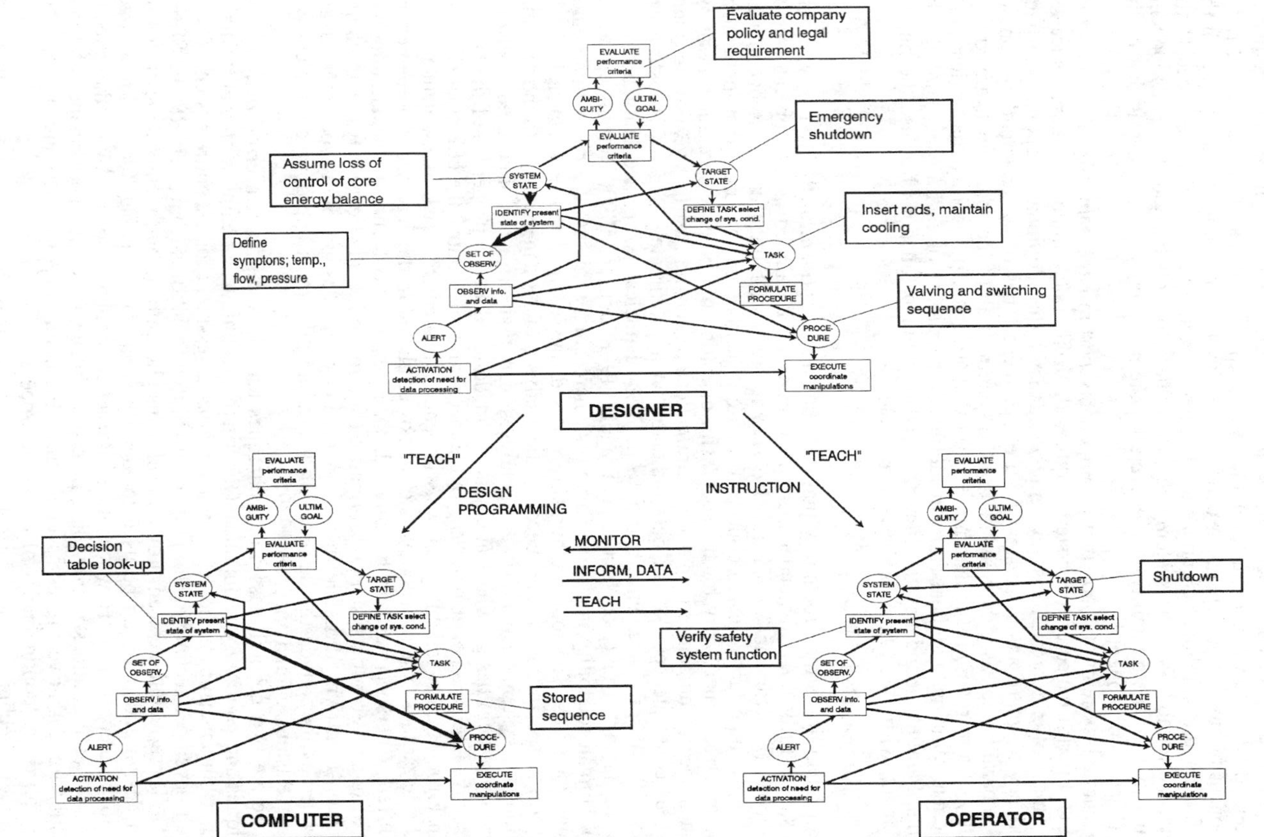

Figure 2.3b Decision roles in a partially automated nuclear system. The operator and the computer act "in series" in the control decision (adapted from Rasmussen, 1986).

Reliability is usually expressed as the probability of successful performance; therefore, the criteria for successful performance must be specified for the probability estimate to be meaningful (Proctor and Van Zandt, 1994). The overall reliability of a system depends on the reliability of the individual components and how they are combined within the system. Reliability of a component is the probability that it does not fail and is defined as r where $r = 1 - p$; p represents the probability of failure.

Generally components in a system are arranged in series, in parallel, or some combination of both. If the components are arranged in series, they must all operate adequately in order for the total performance of the system to be satisfactory. In this case, if the component failures are independent of each other, system reliability is the product of the reliability of the individual components. Further, as more components are added to the system, the reliability of the system decreases unless the reliability of these components is equal to 1.0. The reliability of the overall system can only be as great as that of the least reliable component.

In parallel systems, two or more components perform the same function such that successful performance of the system requires that only one component operate successfully. This is often referred to as "system redundancy"; the additional components provide redundancy to guard against system failure. For these types of systems, adding components in parallel increases the reliability of the system. If all of the components are equally reliable, system reliability is determined by calculating the probability that at least one component remains functional and considering the reliability of each of the parallel subsystems. Parallel redundancy is often provided for human functions because the human component, within a system, is the least reliable.

2.2.4.2 Human Reliability

Human reliability is the probability that each human component of the system will perform successfully for an extended period of time and is defined as 1—(operator error probability) (Proctor and Van Zandt, 1994). The study of human error has become an increasingly important research concern because it has become apparent that the control of human error is necessary for the successful operation of complex, integrated systems. The incidence of human error has risen dramatically over the past few years with many disastrous consequences. Recent data indicate that approximately 80% of industrial accidents, 50% of pilot accidents, and 50% to 70% of nuclear power accidents are attributable to human error (Dougherty and Fragola, 1988; Jensen, 1992; Rasmussen et. al, 1994). Human reliability is also becoming an increasingly important topic within health care environments. The topic of human error is discussed in detail in Chapter 11. It is briefly discussed in this chapter because the analysis of human error has important implications for system design. *Post-hoc* analyses of human error are often used in efforts to improve system design where attempts are made to adapt the properties of the system to lessen the likelihood of error or to provide mechanisms to control the effects of possible human error. Human reliability analysis is central to predictive risk assessment for potentially hazardous systems such as nuclear power plants. In the analysis of human error, two different perspectives can be adopted. One perspective focuses on examining the effect of human error on system performance (failure mode and effect analysis), the other focuses on identifying possible improvements in system design. Both types of analysis are needed as they guide the designer in determining the most appropriate error control strategy (Rasmussen and Vicente, 1989).

There are two general classes of techniques used to study human error, *quantitative* techniques and *qualitative* techniques. Quantitative techniques attempt to predict the likelihood of human error for the development of risk assessment for the entire system. There are well-developed techniques for human reliability prediction such as THERP and SLIM MAUD. These techniques can provide useful insights into human factors deficiencies in system design and thus can be used to identify areas where human factors knowledge needs to be incorporated. However, there are shortcomings associated with these techniques such as limitations in providing precise estimates of human performance abilities especially for cognitive processes (Wickens, 1992). Further, designers are not able to identify all of the contingencies of the work process.

Qualitative techniques emphasize the causal element of human error and attempt to develop and understanding of the causal events/factors contributing to human error. Clearly, when using these approaches, the circumstances under which human error is observed and the resultant causal explanation for error occurrence have important implications for system design. If the causal explanation stops at the level of the operator, remedial measures might encompass better training or supervision, e.g., a common solution for back injuries is to provide operators with training on "how to lift," overlooking opportunities for other, perhaps more effective, changes in the system such as modifications in management, work procedures, work planning, or resources.

Recent analysis of major accident events indicate that the root cause of many of these events can be traced to latent failures and organizational errors. In other words, human errors and their resultant consequences usually result from inadequacies in system design. An example is the crash at Dryden Airport in Ontario. The analysis of this accident revealed that the accident was linked to organizational failings such as poor training, lack of management commitment to safety, and inadequate maintenance and regulatory procedures (Reason, 1995). These findings indicate that when analyzing human error it is important to look at the entire system and the organizational context in which the error occurred.

Several researchers have developed taxonomies for classifying human errors into categories. These taxonomies are useful as they help identify the source of human error and strategies which might be effective in coping with error. Different taxonomies emphasize different aspects of human performance. For example, some taxonomies emphasize human actions, whereas others emphasize information processing aspects of behavior. Rasmussen and colleagues (Rasmussen, 1982; Rasmussen et al., 1994) developed a taxonomy of human errors from the analyses of human involvement in failures in complex processes. This schema is based on a decomposition of mental processes and states involved in erroneous behavior. Figure 2.4 presents a model for this taxonomy, which was developed for analyzing events from nuclear power plants. For the analysis, the events of the causal chain are followed backwards from the observed accidental event through mechanisms involved at each stage. As shown, the taxonomy is based on an analysis of the work system and considers the context in which the error occurred (e.g., workload, work procedures, shift requirements). This taxonomy has been applied to the analysis of work systems and has proven to be useful for understanding the nature of human involvement in accident events.

Reason (1990, 1995) has also developed a similar scheme for examining the etiology of human errors for the design and analysis of complex work systems. The model is based

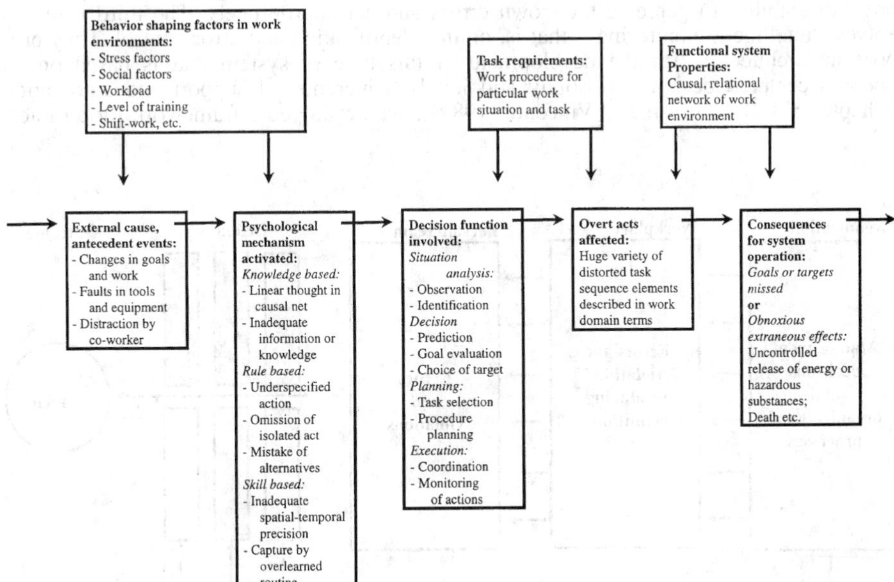

Figure 2.4 A classification scheme used for analyzing event reports from nuclear power plants. The scheme illustrates the decomposition of human activity to identify internal mental processes and states in the explanation of erroneous behavior (adapted from Rasmussen, Pejtersen, and Goodstein, 1994, Copyright © 1994, John Wiley and Sons. Reprinted by permission).

on a systems approach and describes a pathway for identifying the organizational causes of human error. As shown in Figure 2.5, the model presents two interrelated causal sequences for error events: (1) an active failure pathway where the failure originates in top management decisions and proceeds through error-producing conditions in the various workplaces to unsafe acts committed by workers at the immediate human–machine interface and (2) a latent failure pathway that runs directly from the organizational processes to deficiencies in the system's defenses. The model can be used to assess organizational safety health in order to develop proactive measures for remediating system difficulties and as an investigation technique for identifying the root causes of system breakdown.

The implications of error analysis for system design depend on the nature of the error as well as the nature of the system. Errors and accidents have multiple causes, and different types of errors require different remedial measures. For example, if an error involves deviations from normal procedures in a well-structured technical system, it is possible to derive a corrective action for a particular aspect of an interface or task element. This might involve redesign of equipment or of some work procedure to minimize the potential for the error occurrence. However, in complex dynamic work systems it is often difficult or undesirable to completely eliminate the incidence of human error. In these types of systems, there are many possible strategies for achieving system goals; thus it is not possible to specify precise procedures for performing tasks. Instead, operators must be creative and flexible and engage in exploratory behavior in order to respond to the changing demands of the system. Further, designers are not able to anticipate the entire set of possible events; thus it is difficult to build in mechanisms to cope with these events. This makes inevitable a certain amount of error.

Several researchers (Rouse and Morris, 1987, Rasmussen et al., 1994) advocate the design of "error tolerant systems," where the systems tolerates the occurrence of errors but avoids the consequences; there is a means to control the impact of error on system performance. Design of these interfaces requires an understanding of the work domain and the acceptable boundaries of behavior and modeling the cognitive activity of operators dealing with incidents in a dynamic environment. A simple example of this type of design would be a computer system which holds a record of a file so that it is not permanently lost if an operator mistakenly deletes the files. A more sophisticated example would be an intelligent monitoring system which is capable of varying levels of intervention.

Rouse and Morris (1987) describe an error tolerant system which provides three levels of support. Two levels involve feedback (current state and future state) and rely on the operator's ability to perceive their own errors and act appropriately. The third level involves "intelligent monitoring", that is, online identification and error control. They propose an architecture for the development of this type of system that is based on an operator-centered design philosophy and involves incremental support and automation (Chapter 57). Rasmussen and Vincente (1989) have developed a framework for an inter-

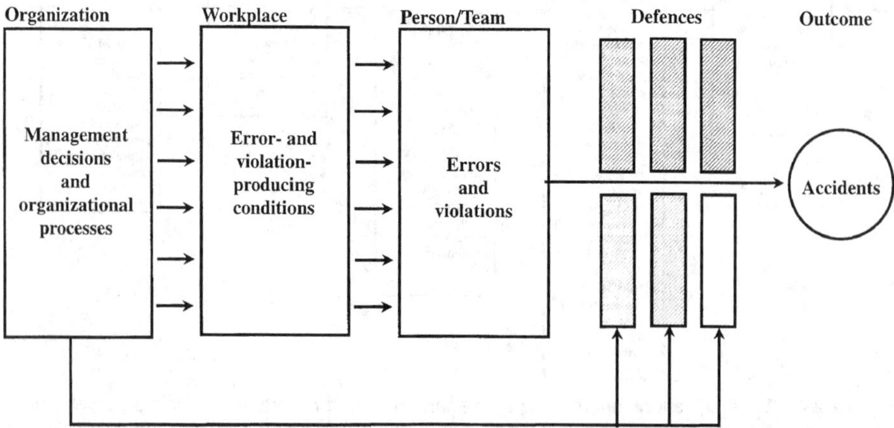

Figure 2.5 A systems model for examining the etiology of errors which identifies two causal pathways an active failure pathway and a latent failure pathway (adapted from Reason, 1995).

face which supports recovery from human errors. The framework is called "ecological interface design" and is based on an analysis of the work system. This approach will be described in more detail in a later section.

2.3 THE SYSTEM DESIGN PROCESS

2.3.1 Approaches to System Design

2.3.1.1 Overview

System design is usually depicted as a highly structured and formalized process characterized by stages in which various activities occur. These activities vary as a function of system requirements but they generally involve planning, designing, testing, and evaluating. A more detailed discussion of these activities will be discussed in a subsequent section. Generally the system design process is characterized as a "top-down" process where the process proceeds, in an interactive fashion, from broad molar functions to progressively more molecular tasks and subtasks. It is also a time-driven process and is constrained by cost, resources, and organizational and environmental requirements. The overall goal of system design is to develop an entity that is capable of transforming inputs into outputs to accomplish specified goals and objectives.

Meister (1991) distinguishes among three levels within system design: (1) the design process—how the system is designed, (2) the design philosophy—the conceptual framework of the design, and (3) the design architecture—the specification of the structure of the system and the human–machine interface. In recent years, within the realm of system design, a great deal of attention has been given to the design philosophy and the resultant design architecture as it has become apparent that new design approaches are required to design modern work systems. The design and analysis of modern dynamic work systems cannot be based on design models developed for work systems characterized by a stable environment and stable task procedures. Instead, the design approach is concerned with supplying resources to operators who operate in a dynamic workspace and change their work patterns according to changing environmental conditions. In other words, a structural perspective whereby we describe the behavior of the system in terms of cause-and-effect patterns and arrange system elements in cause-and-effect chains is no longer adequate. Instead, we need a framework for system design that represents all aspects of work systems in a coordinated and compatible fashion (Rasmussen et. al, 1994).

2.3.1.2 A Traditional Model of System Design

The traditional view of the system design process is that it is a linear sequence of activities where the output of each stage serves as input to the next stage. The stages generally proceed from the conceptual level to physical design through implementation and evaluation. Human factors inputs are generally considered in the design and evaluation stages (Eason, 1991). The general characteristics of this approach are that it represents a reductionist approach where various components are designed in isolation and made to fit together, it is dominated by technological considerations where humans are considered secondary components and the focus is on fitting the person to the system, and it is monodisciplinary in that different components of the system are developed on the basis of narrow functional perspectives (Kidd, 1991; Liker and Majchrzak, 1994). Generally this approach has dominated the design of overall work systems, such as manufacturing systems, as well as the design of the human–machine interface. For example, the emphasis in the design of human–computer systems has largely been on the individual level of the human–computer interaction without much attention to task and environmental factors which may affect performance. Hendrick (1991) maintains that the primary emphasis of human factors engineers has been on the micro-ergonomic aspects of design without sufficient attention to social and organizational issues.

The implementation of computers of automation into most work systems coupled with enhanced capabilities of technological systems has created a need for new approaches to system design. As discussed, there are many instances where technology has failed to achieve its potential resulting in failures in system performance with adverse and often disastrous consequences. These events have demonstrated that the traditional design approach is no longer adequate. In this vein Liker and Majchrzak (1994) identify nine features that need to be incorporated into a design process in order to effectively design the human–technology infrastructure. As shown in Table 2.1, features 1 to 4 are concerned with the content of the design requirements and mandate that the design process should

Table 2.1 Desired Features for System Design Processes

Integrated Human Infrastructure Design

The design process should provide a means for considering the fit between the formal organization, informal organization, individual characteristics, and linking mechanism to the environment.

Concurrent Human Infrastructure and Technical System Design

The design process should provide not only analytic tools for understanding the likely implications of any particular technical solution for people and the organization, but also the likely implications of different organizational arrangements for technology design.

Fit with Environment

The design process should consider the requirements and constraints imposed on the human infrastructure and technology by the larger environment of the system and its missions.

Specificity

The design process should provide precise information on implications so that engineers can alter specific technical designs and organizational planners can modify organizational scenarios as the design process proceeds.

Multiple-Scenario Generation

The design process should facilitate the development of multiple design scenarios that can then be compared.

Design Evaluation and Refinement

An interactive design procedure, which facilitates systematic comparisons among the scenarios, should then be used to gradually fine-tune the design based on "prototyping" and "testing" design concepts.

Life-Cycle User Involvement

The design process should provide a process for involving potential users of the new technology throughout the design and implementation process.

Facilitating Design Team Learning

The design process should increase the knowledge of the planners of the sociotechnical system about issues relevant to organizational-technology integration.

Facilitating Design Modification

The design process should enable redesign as circumstances, such as technology, change.

Source: (Liker and Majchrzak, 1994)

produce an integrated sociotechnical design in accordance with open system principles at a level of specificity required to implement the design. Features 5 to 9 are concerned with the actual design process. Overall, these features incorporate many of the aspects of concurrent engineering methodologies which have proven to be effective within manufacturing systems. They provide a useful framework for the development and analysis of design strategies. Clearly, the traditional method of design fails to incorporate most of these features.

Liker and Majchrzak (1991) review four design approaches including the sociotechnical systems approach, participatory ergonomics, human-centered human factors design, and computer models of integrated systems in terms of their potential effectiveness in designing human-technical systems. They conclude that these approaches hold promise as they have many of the characteristics outlined in Table 2.1. Further, they suggest that each approach has unique strengths and that perhaps a combination of these approaches would be most effective. A brief overview of these approaches and some other design approaches will be presented to provide some examples of alternative approaches to system design and demonstrate methodologies and concepts that can be applied to the design of current human–machine systems. This will be followed by a discussion of the specification application of human factors engineering to design activities.

2.3.1.3 Alternative Approaches to System Design

Sociotechnical Systems Approach

The sociotechnical systems approach, which evolved from work conducted at the Tavisock Institute, represents a complete design process for the analysis, design, and implementation of systems. The approach is based on open systems theory and emphasizes the fit between the social and technical systems and the environment. This approach includes methods for analyzing the environment, the social system, and the technical system. The overall design objective is the joint optimization of the social and technical systems (Pasmore, 1988). Some drawbacks associated with sociotechnical design are that the design principles are often vague and many times there is an overemphasis on the social system without sufficient emphasis on the design of the technical system.

In this regard, Lunstrom (1994) has devised a method for designing assembly systems that is based on a sociotechnical view. The method comprises 13 stages and involves an analysis of the background, problem solving when designing system alternatives, evaluation of alternatives, and selection of a suitable alternative so that a description of the assembly system is achieved, and production and implementation of the system. The method is flexible, systematic, and considers the design of the technical and social system. Each phase of the design is decomposed into detailed design steps in a tree structure and is modular in character so that designers can choose the appropriate level of detail.

The sociotechnical approach is rather popular and has been applied to many industrial settings generally resulting in positive outcomes. For example, the design of the Romeo Engine Plant of the Ford Motor Company was largely based on a sociotechnical approach. The plant proved to be one of the best engine plants within the Ford Motor Company. Berger (1994) presents data from three case studies aimed at testing the notion that balancing the technological, organizational, and human aspects is important to manufacturing development. The case studies involved medium-sized manufacturing companies in Sweden. The companies were concerned with reducing lead times and engaged in redesign efforts. The cases varied according to degree of balance achieved. None of the cases achieved a balance on all three aspects. Two of the cases emphasized the balance between the human and organizational aspects, and the other case emphasized the technical aspects largely ignoring the human and organizational aspects. All three of the cases showed considerable improvements in reducing lead times; therefore, few conclusions regarding the balancing aspects could be drawn in terms of an amount relationship. Berger concluded that a balanced approach needs to be integrated within the problem-solving sequence of the redesign effort and that a key feature in manufacturing development is the ability to rebalance according to the time course and the prevailing problem of the redesign process.

Macroergonomics represents an example of a sociotechnical design approach. The central focus of macroergonomics is on interfacing organizational design with the technology employed in the system to optimize human–system functioning. Macroergonomics considers the human–organization–environment–machine interface as opposed to microergonomics, which focuses on the human–machine interface. Macroergonomics is considered to be the driving force for microergonomics. Macroergonomic concepts have been applied successfully to manufacturing, service, and health care organizations as well as to the design of computer-based information systems (Hendrick, 1991).

Participatory Ergonomics

Participatory ergonomics is the application of ergonomic principles and concepts to the design process by individuals who are part of the work group and "users" of the system. These individuals are typically assisted by ergonomic experts who serve as trainers and resource centers. The overall goal of participatory design is to capitalize on the knowledge of users and incorporate their needs and concerns into the design process. Methods, such as focus groups, quality circles, and inventories, have been developed to maximize the value of user participation. Participatory ergonomics has been applied to the design of jobs and workplaces and to the design of products. For example, the quality circle approach was adopted by a refrigerator manufacturing company that needed a system-wide method for assessing the issues of aging workers. The assembly line for medium-sized refrigerators was chosen as an area for job redesign. The project redesign team involved the workers from the line as well as other staff members. The team was instructed with respect to the principles of ergonomics and design for older workers. The solution, proposed by the team, for improving the assembly line resulted in improved performance and also allowed the older workers to continue to perform the task (Imada, Nora, and Nagamachi, 1986). The design of current personal computer systems also typically involves user participation. Representative users participate in useability studies.

In general, participatory ergonomics does not represent a design process because it does not consider broader system design issues but rather focuses on individual components. However, the benefits of user participation should not be overlooked and should be a fundamental aspect of system design.

User-Centered Design

The user-centered design approach represents an approach where human factors are of central concern within the design process. It is based on an open-systems model and considers the human and technical subsystems within the context of the broader environment. User-centered approaches propose general specifications for system design, such as the system must maximize user involvement at the task level, the system should be designed to support cooperative work and allow users to maintain control over operations (see Liker and Majchrzak, 1994 for a more complete listing), and suggest a process for incorporating these specifications into the design. User-centered approaches have been used primarily in product production particularly in the area of human–computer interaction; however, the design principles and practices are also relevant to other domains.

Eason (1989) has developed an detailed process for user-centered design in which the system is developed in an evolutionary incremental fashion and the development of the social system compliments the development of the technical system. Eason maintains that the technical system should follow the design of jobs and that the design of the technical system must involve user participation and consider criteria for four factors: functionality, usability, user acceptance, and organizational acceptance. Once these criteria are identified, alternative design solutions are developed and evaluated. There are different philosophies with respect to the nature of user involvement. Eason emphasizes user involvement throughout the design process, whereas with other models the users are considered sources of data and the emphasis is on translating knowledge about users into practice. Advocates of the user participation approach argue that users should participate in the choice between alternatives because they have to live with the results. Advocates of the knowledge approach express concern about the ability of users to make informed judgments. Eason (1991) maintains that designers and users can form a partnership where both can play an effective role.

Computer-Supported Design

The design of complex technical systems involves the interpretation and integration of vast amounts of technical information. Further, design activities are typically constrained by time and resources and involve the contributions of many individuals with varying backgrounds and levels of technical expertise. In this regard, computer-based design support tools have emerged to aid the designer and support the design of effective systems. These systems are capable of offering a variety of supports including information retrieval, information management, and information transformation. The type of support warranted depends on the needs and the expertise of the designer (Rouse, 1987). A common example of this type of support is a CAD/CAM system.

Majchrzak and Gasser (1992) have developed a computer modeling system, ACTION, for manufacturing system designers. ACTION (which evolved from HITOP) is a knowledge-based design, decision support, and simulation software package that is to be used as a tool to aid in the design and planning of technology of discrete parts manufacturing systems. ACTION represents a design tool rather than a design process; however, the tool has some design methodology features. The design of this tool was based on an open-systems model, and the design methodology features support a concurrent organizational design approach and a supportive human infrastructure.

There are many issues surrounding the development and deployment of computer-based design support tools including the specification of the appropriate level of support, determination of optimal ways to characterize the design problem and the type of knowledge most useful to designers, and the identification of factors that influence the acceptance of these tools. A discussion of these issues is beyond the scope of this chapter. Refer to Rouse and Boff (1987) for an excellent review of this topic.

Cognitive Systems Approaches

Ecological interface design (EID) is a theoretical framework and method for designing interfaces for complex systems (Rasmussen and Vincente, 1989; Rasmussen et al., 1994). EID is based on a cognitive systems engineering approach and involves an analysis of the work domain and the cognitive characteristics and behavior tendencies of the individual. The analysis of the work domain is based on an abstraction hierarchy (means-end analysis) (Rasmussen, 1986) and relates to the specification of information content. The skills, rules, knowledge taxonomy (Rasmussen, 1983) is used to derive inferences for how information should be presented. The aims of EID are to support the entire range of activities that confront operators, including familiar, unfamiliar, and unanticipated events, without contributing to the difficulty of the task.

The EID approach offers designers a set of maps that can be used to guide the analysis of the design territory (the work domain and the operators). These maps can be tailored to the designers experience and expertise. Rasmussen and colleagues (Rasmussen et al., 1994) have successfully used this approach for the design of a library information retrieval system. An example of design maps developed for the library system are presented in Figure 2.6.

2.3.2 Incorporating Human Factors in System Design

One problem faced by human factors engineers in system design is convincing project managers, engineers, and designers of the value of incorporating human factors knowledge and expertise into the system design process. In many instances, human factors issues are ignored or human factors activities are restricted to the evaluation stage. This is referred to as the "too-little-too late" phenomenon (Lim, Long, and Silcock, 1992). Restricting human factors inputs to the evaluation stage limits the utility and effectiveness of human factors contributions. Either the contributions are ignored because it would be too costly or time consuming to alter the design of the system ("too late") or minor alterations are made to the design to pay lip service to human issues ("too little"). In either case there is limited realization of human factors contributions. In order for human factors to be effective, there needs to be continued involvement of human factors engineers throughout the design process.

There are a variety of reasons why human factors engineers are not considered as "equal partners" in a design team. One reason is that other team members (e.g., designers, engineers) have misconceptions about the potential contributions of human factors and the importance of human issues. They perceive, for example, that humans are flexible and can adapt to system requirements or that accommodating human issues will compromise the technical system. Another reason is that sometimes human factors inputs are of limited value to designers (Meister, 1989; Chapanis, 1995). The inputs are either so specific that they apply to a particular design situation and not to the design process in question or they are vague and overly general. For example, a design guideline which specifies that "older people need larger characters on computer screens" is of little value. How does one define "larger characters"? Obviously the type of input required depends on the nature of the design problem. Design of a kitchen to accommodate individuals in wheelchairs requires precise information such as counter height dimensions or required turning space. In contrast, guidelines for designing intelligent interfaces need to be expressed at the cognitive task level, independent of a particular technology (Woods and Roth, 1988). Thus one important task for human factors engineers is to ensure that design inputs are in a form that is useable and useful to designers. Williges and colleagues (Williges, Williges, and Han, 1992) demonstrate how integrated empirical models can be used as quantitative design guidelines. Their approach involved integrating data from four sequential experiments and developing a model for the design of a telephone-based information system.

To ensure that human factors will be systematically applied to system design, we need to market the potential contributions of human factors to engineers, project managers, and designers. One approach is to use case studies, relevant to the design problem, that illustrate the benefits of human factors. Case studies of this nature can be found in technical journals (e.g., Applied Ergonomics, Ergonomics in Design) and technical reports. Another approach is to perform a cost-benefit analysis. Estimating the costs and benefits associated with human factors is difficult because it is hard to isolate the contribution of human factors relative to other variables, baseline measures of performance are unavailable, or performance improvements are hard to quantify and link to system improvements. There are methods available to conduct this type of analysis. Cost-benefit analysis is discussed in detail in Chapter 49 of this handbook.

2.3.3 Applications of Human Factors to the System Design Process

2.3.3.1 Overview

System design can be conceptualized as a problem-solving process that involves the formulation of the problem, the generation of solutions to the problem, analysis of these alternatives, and selection of the most effective alternative (Rouse, 1985). There are var-

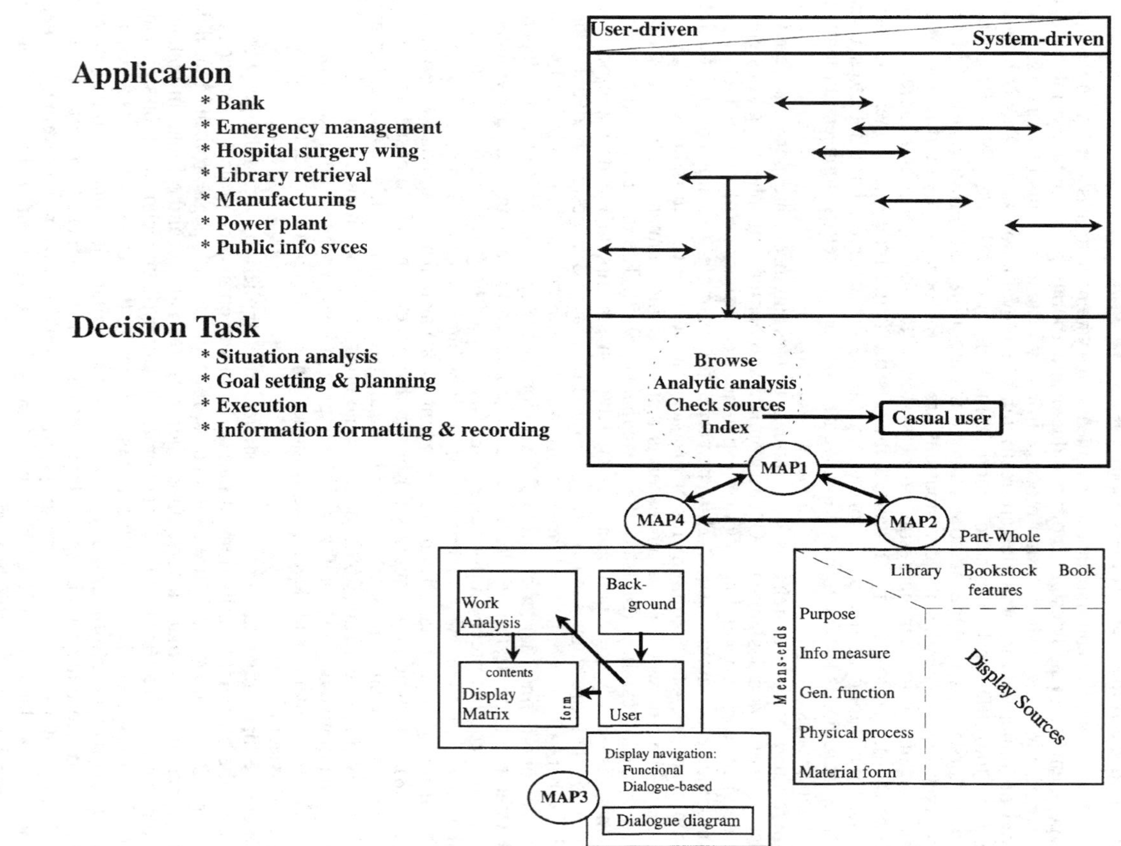

Figure 2.6 A set of design maps developed for a library information system. The maps characterize the territory through which the designer has to navigate (adapted from Rasmussen, Pejtersen, and Goodstein, 1994, Copyright © 1994, John Wiley and Sons. Reprinted by permission).

ious ways to classify the various stages in system design. Meister (1989), on the basis of a military framework, distinguishes among four phases: (1) system planning—the need for the system is identified and system objectives are defined; (2) preliminary design—alternative system concepts are identified, prototypes are developed and tested; (3) detail design—full-scale engineering development; (4) production and testing—the system is built and undergoes testing and evaluation. To maximize system effectiveness, human factors engineers need to be involved in all phases of the process. In addition to human factors engineers, a representative sample of operators (users) should also be included.

The basic role of human factors in system design is the application of behavioral principles, data, and methods to the design process. Within this role, human factors get involved in a number of activities. These activities include specifying inputs for job, equipment and interface design, human performance criteria, operator selection and training, and inputs regarding testing and evaluation. The nature of these activities are discussed, at a general level in the next section. A detailed discussion of most of these issues is provided in subsequent chapters. The intent of this discussion is to highlight the nature of human factors involvement in the design process.

2.3.3.2 System Planning

During system planning, the need for the system is established and the goals and objectives and performance specifications of the system are identified. Performance specifications define what the system must do to meet its objectives and the constraints under which the system will operate. These specifications determine the system's performance requirements. Human factors should be a part of the system planning process. The major role of human factors engineers during this phase is to ensure that human issues are considered in the specification of design requirements and the statement of system goals and objectives. This includes understanding personnel requirements, general performance requirements, the intended users of the system, user needs, and the relationship of system objectives relative to these needs.

2.3.3.3 System Design

System design encompasses both preliminary design and detailed design. During this phase of the process, alternative design concepts are identified and tested and a detailed model of the system is developed. To ensure adequate consideration of human issues, the involvement of human factors engineers is critical during this phase. The major human factors activities include: (1) function allocation, (2) task analysis, (3) job design, (4) interface design, (5) design of support materials, and (6) workplace design. The primary role of the human factors engineer is to ensure joint optimization of the human and technical systems.

Function Allocation

Function allocation is a critical step in work system design. This is especially true in today's work systems as machines are becoming more and more capable of performing tasks once restricted to humans. A number of studies have shown (e.g., Morris, Rouse, and Woods, 1985; Sharit, Chang, and Salvendy, 1987) that proper allocation of functions between humans and machines results in improvements in overall system performance.

Function allocation involves formulating a functional description of the system and subsequent allocation of functions among system components. A frequent approach to function allocation is to base allocation decisions on machine capabilities; automate wherever possible. While this approach may appear expedient, there are several drawbacks. In most systems not all tasks can be automated, and thus some tasks must be performed by humans. These tasks are typically "left-over" tasks. Allocating these tasks to humans generally leads to problems of underload, inattention, and job dissatisfaction. A related problem is that automated systems fail and humans have to take over. This can be problematic if the human is out of the loop or if their skills have become rusty due to disuse. In essence the machine-based allocation strategy is inadequate. As previously discussed, there are numerous examples of "technocentered" design.

It has become clear that a better approach is complementary where functions are allocated so that human operators are complemented by technical systems. This approach involves identifying how to couple humans and machines to maximize system performance. In this regard, there is much research aimed at developing methods to guide function allocation decisions. These methods include lists (e.g., Fitt's List), computer simulation packages, and general guidelines for function allocation (e.g., Price, 1985).

Recently the traditional static approach (humans are better at . . .) to function allocation has been challenged and dynamic allocation approaches have been developed. With dynamic allocation, responsibility for a task at any particular instance is allocated to the component most capable at that point in time. Hou, Lin, and Drury (1993) developed a framework to allocate functions between humans and computers for inspection tasks. Their framework represents a dynamic allocation framework and provides for a quantitative evaluation of the allocation strategy chosen. Morris et al. (1985) investigated the use of a dynamic adaptive allocation approach within an aerial search environment. They found that the adaptive approach resulted in an overall improvement in system performance.

Task Analysis

Task analysis is also a central activity in system design. Task analysis helps ensure that human performance requirements match operators' (users') needs and capabilities and that the system can be operated in a safe and efficient manner. The output of a task analysis is also essential to the design of the interface, workplaces, support materials, training programs, and test and evaluation procedures.

A task analysis is generally performed after function allocation decisions are made; however, sometimes the results of the task analysis alter function allocation decisions. A task analysis usually consists of two phases: a *task description* and a *task analysis*. The task description involves a detailed decomposition of functions into tasks which are further decomposed into subtasks or steps. The task analysis specifies the physical and cognitive demands associated with each of these subtasks.

There are a number of methods available for conducting task analysis. A description of these methods is provided in Chapter 12. Commonly used methods include flow process charts, critical task analysis, and hierarchical task analysis. Techniques for collecting task data include: documentation review, surveys and questionnaires, interviews, observation, and verbal protocols.

As the demands of tasks have changed and become more cognitive in nature, methods have been developed for performing *cognitive task analysis*. Cognitive task analysis attempts to describe the knowledge and cognitive processes involved in human performance in particular task domains. The results of a cognitive task analysis are important to the design of interfaces for intelligent machines. A common approach used to carry out a cognitive task analysis is a goal-means decomposition. This approach involves an analysis of the work domain to identify the cognitive demands inherent in a particular situation and building a model which relates these cognitive demands to situational demands (Roth, Woods, and Pople Jr., 1992). Another approach involves the use of cognitive simulation. Roth et al. (1992) have developed a cognitive simulation tool, Cognitive Simulation Environment, which has proven to be useful for helping to evaluate the cognitive activities associated with emergency events in nuclear power plants.

Job Design

The type of work a person performs is largely a function of job design. Jobs involve more than tasks and include work content, distribution of work, and work roles. Essentially a job represents an individual's perscribed role within an organization. Job design involves determining how tasks will be grouped together, how work will be coordinated among individuals, and how individuals will be rewarded for their performance (Davis and Wacker, 1987). To design jobs effectively, consideration must be given to workload requirements and to the psychosocial aspects of work (peoples' needs and expectations). This consideration is especially important in automated work systems where the skills and potential contributions of humans are often overlooked.

In terms of workload the primary concern is that work requirements are commensurate with human abilities and that individuals are not placed in situations of underload or overload as both situations can lead to performance decrements, job dissatisfaction, and stress. Both the physical and mental demands of the task need to be considered. There are well-established methods for evaluating the physical demands of tasks and for determination of work/rest schedules. The concept of mental workload is more esoteric. This issue has received a great deal of attention in the literature, and a variety of methods have been developed to evaluate the mental demands associated with a task (Chapter 13).

Consideration of operator characteristics is also an essential element of job design as the workforce is becoming more heterogeneous. For example, older workers may need different work/rest schedules than younger workers or may be unsuited to certain types of tasks. Individuals who are physically challenged may also require different job specifications.

In terms of psychosocial considerations, a number of studies have identified critical job dimensions. Generally these dimensions include task variety, task identity, feedback, autonomy, task significance, opportunity to use skills and challenge. As far as possible, these characteristics should be designed into jobs. Davis and Wacker (1987) have developed a quality of working life criteria checklist which lists job dimensions important to the satisfaction of individual needs. These dimensions relate to the physical environment, institutional rights and privileges, job content, internal social relations, external social relations, and career path.

A number of approaches to job design have been identified (Chapter 14). These include work simplification, job enrichment, job enlargement, job rotation, and team work design. The method chosen should depend on the actual design problem, work conditions, and individuals. However, it is generally accepted that the work simplification approach does not lead to optimal job design.

Interface Design

Interface design involves the specification of the nature of the human–machine interaction; that is, the means by which the human is connected to the machine. During this stage of design, the human factors specialist typically works closely with engineers and designers. The role of human factors is to provide the design team with information regarding the human performance implications of design alternatives. This generally involves three major activities: (1) gathering and interpreting human performance data; (2) conducting attribute evaluations of suggested designs; and (3) human performance testing (Sanders and McCormick, 1993). Human performance testing typically involves building mock-ups and prototypes and testing them with a sample of users. This type of testing can be expensive and time consuming. Recently the development of rapid prototyping tools has made it possible to speed up and compress this process. These tools have been used primarily in the testing of computer interfaces; however, they can be applied to a variety of situations.

Interface design encompasses the design of both the physical and cognitive components of the interface and includes the design and layout of controls and displays, information content, and information representation. Physical components include factors such as type of control or input device, size and shape of controls, control location, and visual and auditory specifications (e.g., character size, character contrast, labeling, signal rate, signal frequency).

Cognitive components refer to the information processing aspects of the interface (e.g., information content, information layout). As machines have become more intelligent, much of the focus of interface design has been on the cognitive aspects of the interface. Issues of concern include: determination of the optimal level of machine support, identification of the type of information users need, determination of how this information should be presented, and identification of methodologies to analyze work domains and cognitive activities. The central concern is developing interfaces which best support human task performance. In this regard, a number of approaches have evolved for interface design. Ecological interface design (Rasmussen and Vincente, 1989) is an example of recent design method.

There are a variety of sources of data on the characteristics of human performance that can serve as inputs to the design process. These include handbooks, textbooks, standards (e.g., ANSI), and technical journals. There are also a variety of models of human performance including cognitive models [e.g., GOMS (Card, Moran, and Newell, 1985], control theory models, and engineering models. These models can be useful in terms of predicting the effects of design parameters on human performance outcomes. As previously discussed, it is the responsibility of the human factors engineer to make sure that information regarding human performance is in a form that is useful to designers. It is also important, when using this data, to consider the nature of the task, the task environment, and the user population.

Design of Support Materials

This phase of the design process includes identifying and developing materials which facilitate the user's interaction with the system. These materials include job aids, instructional materials, and training devices and programs. All too often this phase of the design process is neglected or given little attention. A common example is the cumbersome manuals that accompany software packages or VCRs.

Support materials should not be used as a substitute for "good design" however; the design of effective support materials is an important part of the system design process. Users typically need training and support to successfully interact with new technologies and complex systems. Human factors principles need to be applied to the design of instructional materials, job aids, and training programs in order to maximize their effectiveness. Guidelines are available for the design of instructional materials and job aids. Bailey (1982) provides a thorough discussion of these issues. A great deal has also been written on design of training programs (Chapter 16).

Design of the Work Environment

The design of the work environment is an important aspect of work system design. Systems exist within a context, and the characteristics of this context impact on overall system performance.

The primary concern of workplace design is to ensure that work environment supports the operator and activity performance and allows the worker to perform tasks in an efficient, comfortable, and safe manner. Important issues include workplace and equipment layout, furnishings, reach dimensions, clearance dimensions, visual dimensions, and the design of the ambient environment. There are numerous sources of information related to workplace design and evaluation which can be used to guide this process. These issues are also covered in detail in other chapters of this handbook.

2.3.4 Test and Evaluation

Test and evaluation are critical aspects of system design and usually take place throughout the system design process. Test and evaluation provide a means for continuous improvement during system development. Human factors inputs are essential to the testing and evaluation of systems. The primary role of human factors is to assess the impact of system design features on human performance outputs. These outputs include objective outputs such as speed and accuracy of performance and workload, and subjective outputs such as comfort and user satisfaction. Human factors specialists are also interested in ascertaining the impact of human performance on overall system performance. Issues related to evaluation and the assessment of system effectiveness are covered in detail in Chapters 43 through 49 of this handbook.

Because the evaluation of systems and system components involves measurement of human performance in operational terms (relative to the system or subsystem in question) human factors engineers face a number of challenges when evaluating systems. Generally the standards of generalizability are higher for human factors research as the research results must be extended to real-world systems (Kantowitz, 1992). At the same time, it is often difficult to achieve an appropriate level of control. Unfortunately, in many instances the utility of test and evaluation results are limited because of deficiencies in test and evaluation procedures (Bitner, 1992).

In this regard, there are three key issues that need to be addressed when developing methods for evaluating system effectiveness: (1) subject representativeness, (2) variable representativeness, and (3) setting representativeness (Kantowitz, 1992). Subject representativeness refers to the extent to which subjects tested in the research study represent the population to which the research results apply. In most cases, the sample involved in system evaluation should represent the population of interest on relevant characteristics. Variable representativeness refers to the extent that the study variables are representative of the research question. It is important to select variables which capture the essential issues being assessed in the research study. Setting representativeness is the degree of congruence between the test situation in which the research is performed and the target situation in which the research must be applied. The important issue is the comparability of the psychological processes captured in these situations, not necessarily physical fidelity.

There are a variety of techniques available for conducting human factors research including experimental methods, observational methods, surveys and questionnaires, and audits. There is no one preferred method; each has its associated strengths and weaknesses. The method one chooses depends on the nature of the research question. It is generally desirable to use several methods in conjunction.

2.4 CONCLUSIONS

System design and development is an important area of application for human factors engineers. System performance will be improved by consideration of behavioral issues.

While much has been written on system design our knowledge of this topic is far from complete. The changing nature of systems coupled with the increased diversity of users presents new challenges for human factors specialists and affords many research opportunities. The goals of this chapter were to summarize some of the current issues in system design and to illustrate the role of human factors engineers within the system design process. Further, the chapter provides a framework for many of topics addressed in this handbook.

REFERENCES

Bailey, R. W. (1982). *Human Performance Engineering: A Guide for System Design.* Englewood Cliffs, New Jersey: Prentice Hall.

Berger, A. (1994). Balancing technological, organizational, and human aspects in manufacturing development. *The International Journal of Human Factors in Manufacturing, 4,* 261–280.

Bitner, A. C. (1992). Robust testing and evaluation of systems, *Human Factors, 34,* 477–484.

Butera, F. (1984). Designing work in automated systems: A review of case studies. In F. Butera and J. E. Thurman, Eds., *Automation and Work Design.* New York: Elsevier Science.

Card, S. K., Moran, T. P., and Newell, A. (1983). *The Psychology of Human-Computer Interaction.* Hillsdale, NJ: Lawrence Erlbaum.

Chapanis, A. (1995). Ergonomics in product development: A personal view. *Ergonomics, 38,* 1625–1638.

Davis, L. E., and Wacker, G. J. (1987). Job design, In G. Salvendy ed., *Handbook of Human Factors,* New York: John Wiley, 431–452.

DeGreen, K. B. (1991). Emergent complexity and person-machine systems. *International Journal of Man-Machine Studies, 35,* 219–234.

Dougherty, E. M., and Fragola, J. R. (1988). *Human Reliability Analysis.* New York: John Wiley.

Drucker, P. F. (1988). The coming of a new organization. *Harvard Business Review,* January–February, 45–53.

Eason, K. D. (1988). *Information technology and organizational change.* London: Taylor and Francis.

Eason, K. D. (1991). Ergonomic perspectives on advances in human-computer interaction. *Ergonomics, 34,* 721–741.

Eastman Kodak Company (1983) *Ergonomic Design for People at Work.* Belmont, CA: Lifetime Learning Publications.

Eggleston, R. G. (1987). The changing nature of the human-machine design problem: Implications for system design and development. In W. B. Rouse and K. R. Boff, Eds., *System Design, Behavioral Perspectives on Designers, Tools and Organizations.* New York: North-Holland, pp. 113–126.

Green, D. M., and Swets, J. A. (1966). *Signal Detection Theory and Psychophysics.* New York: John Wiley and Sons.

Hendrick, H. W. (1991) Ergonomics in organizational design and management. *Ergonomics, 34,* 743–757.

Hou, T., Lin, L., and Drury, C. G. (1993). An empirical study of hybrid inspection systems and allocation of inspection function. *The International Journal of Human Factors in Manufacturing, 3,* 351–367.

Imada, A. S., Nora, K., and Nagamachi, M. (1986). Participatory ergonomics: methods for improving individual and organizational effectiveness. In O. Brown Jr. and H. W. Hendrick, Eds., *Human Factors in Organizational Design and Management-II.* New York: North-Holland, 403–406.

Jensen, R. S. (1982). Pilot judgement: Training and evaluation. *Human Factors, 34,* 61–73.

Kantowitz, B. H. (1992). Selecting measures for human factors research, *Human Factors, 34,* 387–398.

Kidd, P. T. (1994). Skill-based automated manufacturing. In W. Karwowski and G. Salvendy, Eds., *Organization and Management of Advanced Manufacturing.* New York: John Wiley, pp. 165–195.

Liker, J. K., and Majchrzak, A. (1994). Designing the human infrastructure for technology. In W. Karwowski and G. Salvendy, Eds., *Organization and Management of Advanced Manufacturing.* New York: John Wiley, pp. 121–164.

Lim, K. Y., Long, J. B., and Silcock, N. (1992). Integrating human factors with the Jackson System Development method: An illustrated overview. *Ergonomics, 35,* 1135–1161.

Lundstrom, M. (1994). Designing assemble systems based on a socio-technical view. In G. E. Bradley and H. W. Hendrick, Eds., *Human Factors in Organizational Design and Management-IV.* New York: North-Holland, pp. 213–218.

Majchrzak, A., and Gasser, L. (1992). On using artificial intelligence to integrate the design of organizational and process changes in U.S. manufacturing. *AI and Society, 5,* 321–338.

Meister, D. (1989). *Conceptual Aspects of Human Factors.* Baltimore and London: John Hopkins University Press.

Meister, D. (1991). *Psychology of System Design.* Amsterdam-Oxford-New York-Tokyo: Elsevier.
Morris, N. M., Rouse, W. B., and Ward, S. L. (1985). Experimental evaluation of adaptive task allocation in an aerial search environment. In G. Mancici, G. Joahnnsen, and L. Martensson, Eds., *Proceedings of the 2nd IFAC/IFIP/IFORS/IEA Conference: Analysis, Design and Evaluation of Man-Machine Systems.* Varese, Italy, pp. 67–72.
Pasmore, W. A. (1988). *Designing Effective Organizations: The Sociotechnical Systems Perspective.* New York: John Wiley.
Price, H. (1985). The allocation of functions is systems. *Human Factors, 27,* 33–45.
Proctor, R. W. and Van Zandt, T. (1994). *Human Factors in Simple and Complex Systems.* Boston: Allyn and Bacon.
Rasmussen, J. (1982). Human errors. A taxonomy for describing human malfunctioning in industrial installations. *Journal of Occupational Accidents, 4,* 311–333.
Rasmussen, J. (1983). Skill, rules and knowledge; Signals, signs and symbols, and other distinctions in human performance models. *IEEE Transactions of Systems, Man, and Cybernectics,* SMC-13, 257–266.
Rasmussen, J. (1986). *Information Processing and Human-Machine Interaction: An Approach to Cognitive Engineering.* The Netherlands: North-Holland.
Rasmussen, J., and Goldstein, L. P. (1985). Decision support in supervisory control. In G. Mancici, G. Joahnnsen, and L. Martensson, Eds., *Proceedings of the 2nd IFAC/IFIP/IFORS/IEA Conference: Analysis, Design and Evaluation of Man-Machine Systems.* Varese, Italy, pp. 79–90.
Rasmussen, J., and Vincente, K. J. (1989). Coping with human errors through system design: Implications for ecological interface design. *International Journal of Man-Machine Studies, 31,* 517–534.
Rasmussen, J., Pejtersen, A. M., and Goodstein, L. P. (1994). *Cognitive Systems Engineering.* New York: John Wiley.
Reason, J. (1990). *Human Error.* New York: Cambridge University Press.
Reason, J. (1995). A systems approach to organizational error. *Ergonomics, 38,* 1708–1721.
Roth, E. M., Woods, D. D., and Pople, H. E. Jr. (1992). Cognitive simulation as a tool for cognitive task analysis. *Ergonomics, 35,* 1163–1198.
Rouse, W. B. (1985). On the value of information in system design: A framework for understanding and aiding designers. *Information Processing and Management, 22,* 217–228.
Rouse, W. B. (1987). Designers, decision making and decision support. In W. B. Rouse and K. R. Boff, Eds., *System Design: Behavioral Perspectives on Designers, Tools and Organizations.* New York: North-Holland, pp. 275–284.
Rouse, W. B., and Boff, K. R. [1987(a)]. Workshop themes and issues: The psychology of systems design. In W. B. Rouse and K. R. Boff, Eds., *System Design: Behavioral Perspectives on Designers, Tools and Organizations,* New York: North-Holland, pp. 7–18.
Rouse, W. B., and Boff, K. R. [1987(b)]. *System Design: Behavioral Perspectives on Designers, Tools and Organizations.* New York: North-Holland.
Rouse, W. B., and Morris, N. M. (1987). Conceptual design of error tolerant interface for complex engineering systems. *Automatica, 23,* 231–235.
Sharit, J., Chang, T. C., and Salvendy, G. (1987). Technical and human aspects of computer-aided manufacturing. In G. Salvendy, Ed., *Handbook of Human Factors.* New York: John Wiley, pp. 1694–1724.
Sanders, M. S., and McCormick, E. J. (1993). *Human Factors in Engineering and Design* (*7th Ed.*). New York: McGraw-Hill.
Sheridan, T. B. (1985). Forty-five years of Man-Machine systems: History and trends. In G. Mancici, G. Joahnnsen, and L. Martensson, Eds., *Proceedings of the 2nd IFAC/IFIP/IFORS/IAE Conference: Analysis, Design and Evaluation of Man-Machine Systems* Varese, Italy, pp. 1–10.
Wickens, C. D. (1992). *Engineering Psychology and Human Performance* (*2nd ed.*). New York: Harper-Collins.
Williges, R. C., Williges, B. H., and Han, S. (1992). Developing quantitative guidelines using integrated data from sequential experiments. *Human Factors, 34,* 399–408.
Woods, D. D. (1988). Commentary: Cognitive engineering in complex and dynamic worlds. In E. Hollnagel, G. Mancini, and D. D. Woods, Eds., *Complex Engineering in Complex Dynamic Worlds.* New York: Academic Press, pp. 115–129.
Woods, D. D., and Roth, E. M. (1988). Cognitive systems engineering, In M. E. Helander, Ed., *Handbook of Human-Computer Interaction.* New York: North-Holland, pp. 3–43.

PART 2

THE HUMAN FACTORS FUNDAMENTALS

CHAPTER 3

SENSATION AND PERCEPTION

Robert W. Proctor and Janet D. Proctor
Department of Psychological Sciences
Purdue University
West Lafayette, IN 47907 USA

3.1 INTRODUCTION

Human–machine interaction involves a continuous exchange of information between the operator(s) and the machine. The operator provides input to the machine, which acts on this input and displays information back to the operator regarding its status and the consequences of the input. The operator must process this information, decide what, if any, controlling actions are needed, and then provide new input to the machine. One important facet of this exchange of information between the machine and the operator is the displaying of the information from the machine as input to the operator. All such information must enter through the operator's senses and be organized and recognized accurately to ensure correct communication of the displayed information. Thus, an understanding of how people sense and perceive is essential for display design. An effective display is consistent with the characteristics and limitations of the human sensory and perceptual systems. These systems are also involved intimately in both the control of human interactions with the environment and of actions taken to control machines. However, because perceptual-motor skills is the topic of Chapter 7, we will focus primarily on the nature of sensation and perception in this chapter. Likewise, because other chapters focus on the applied topics of visual displays (Chapter 20), noise (Chapter 24), motion and vibration (Chapter 25), and illumination (Chapter 26), we will concentrate primarily on the nature of sensory and perceptual processes and the general implications for human factors and ergonomics.

Many classifications of sensory systems exist, but most commonly distinctions are made between five sensory modalities: vision, audition, olfaction, gustation, and somasthesis. The vestibular system, which provides the sense of balance, is also of importance. All sensory systems extract information about four characteristics of stimulation: (1) the sensory modality and submodalities (e.g., touch as opposed to pain); (2) the stimulus intensity; (3) the duration of the stimulation; and (4) its location (Martin, 1991). Each system has receptors that are sensitive to some aspect of the physical environment. These receptors are responsible for sensory transduction, or the conversion of physical stimulus energy into electrochemical energy in the nervous system. The sensory information for each sense is encoded in the activity of neurons and travels to the brain via specialized, structured pathways consisting of highly interconnected networks of neurons. For most modalities, two or more pathways operate in parallel to analyze and convey different kinds of information from the sensory signal. The pathways project to primary receiving areas in the cerebral cortex (see Figure 3.1), in most cases after passing through relay areas in the thalamus. From the primary receiving area, the pathways then project to many other areas within the brain.

Each neuron in the sensory pathways is composed of a cell body, dendritic trees at the input side, and an axon with branches at the output side. The neuron "fires" in an all-or-none manner, sending spike or action potentials down the axon away from the cell body. The rate at which the neuron fires varies as a function of the input that the neuron is receiving from other neurons (or directly from sensory receptors) at its dendrites. Most neurons exhibit a baseline firing rate in the absence of stimulation, usually on the order of 5 to 10 spikes/s, and information is signaled by deviations above or below this baseline rate. The speed of transmission of a spike along the fiber varies across different types of neurons, ranging from 20 to 100 m/sec. Immediately after an action potential occurs, the neuron is in a refractory state in which another action potential cannot be generated immediately. This sets an upper limit on the firing rate of about 1000 spikes/s.

Transmission between neurons occurs at small gaps, called synapses, between the axonal endings of one neuron and the dendrites of another. Communication at the synapse takes place by means of transmitter substances that have a facilitatory effect of increasing the firing rate of the neuron or an inhibitory effect of decreasing the firing rate. Because as many as several hundred neurons may have synapses with the dendrites of a specific neuron, whether the firing rate will increase or decrease is a function of the sum of the facilitatory and inhibitory inputs that the neuron is receiving. Which specific neurons

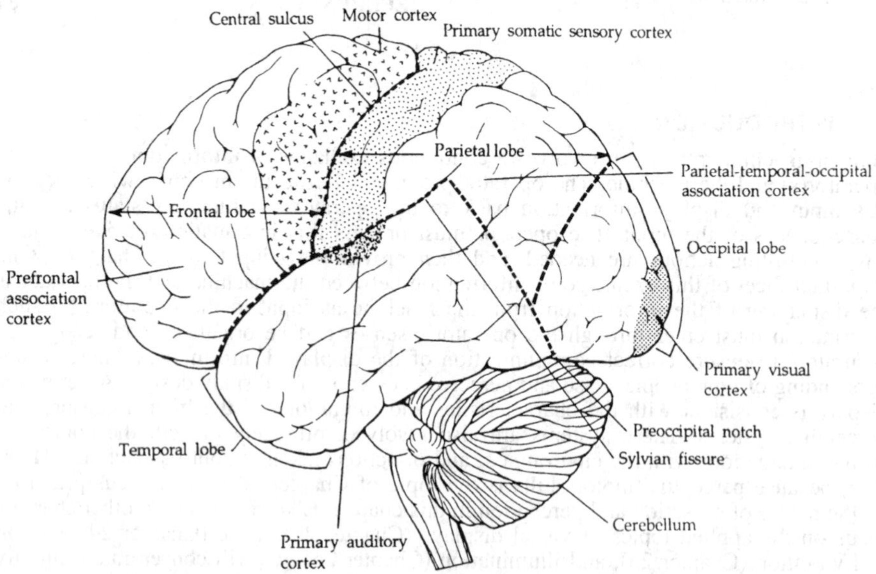

Figure 3.1 The primary sensory receiving areas (visual, auditory, and somatic) of the cerebral cortex and other important landmarks and areas. [Reprinted by permission of the publisher from E. R. Kandel and J. H. Schwartz, eds., *Principles of neural science* (Second Edition), p. 214. Copyright © 1985 by Elsevier Science Publishing.

provide excitatory and inhibitory inputs will determine the patterns of stimulation to which the neuron will be sensitive, i.e., to which the firing rate will either increase or decrease from baseline. The patterns may be either rather general (e.g., an increase in illumination) or quite specific (e.g., a pair of lines at a particular angle, moving in a particular direction). In general, increases in stimulus intensity result in increased firing rates for individual neurons and in a larger population of neurons that respond to the stimulus. Thus, intensity is coded by frequency and population codes.

The study of sensation and perception involves not only the anatomy and physiology of the sensory systems, but also behavioral measures of perception. Psychophysical data obtained from tasks in which observers are asked to detect, discriminate, rate, or recognize stimuli provide information about how the properties of the sensory systems relate to what is perceived. Behavioral measures also provide considerable information about the functions of higher-level brain processes for which current knowledge of the physiological bases is rudimentary. The sensory information must be interpreted by these higher-level processes, which include mental representations, decision making, and inference. Thus, perceptual experiments provide evidence about how the sensory input is organized into a coherent percept. In the next section, we review methods used to investigate sensory and perceptual processes.

3.2 METHODS FOR INVESTIGATING SENSATION AND PERCEPTION

Many methods have been, and can be, used to obtain data pertinent to understanding sensation and perception. The most basic distinction is between methods that involve anatomy and physiology as opposed to methods that involve behavioral responses. Because the former are not of much direct use in human factors and ergonomics, we will not cover them in as much detail as we do the latter.

3.2.1 Anatomical and Physiological Methods

A wide variety of specific techniques exist for analyzing and mapping out the pathways associated with sensation and perception. These include injecting tracer substances into the neurons, classifying neurons in terms of the size of their cell bodies and characteristics of their dendritic trees, and lesioning areas of the brain (see Wandell, 1995). Such techniques have provided a relatively detailed understanding of the sensory pathways.

One particular technique that has produced a wealth of information about the functional properties of specific neurons in the sensory pathways and their associated regions in the brain is *single-cell recording*. Such recording is typically performed on a monkey, cat, or other nonhuman species; an electrode is inserted that is sufficiently small to record only the activity of a single neuron. The responsivity of this neuron to various features of stimulation can be examined in order to gain some understanding of the neuron's role in the sensory system. By systematically examining the responsivities of neurons in a given region, it has been possible to determine much about the way that sensory input is coded. In our discussion of the sensory systems, we will have much opportunity to refer to the results of single-cell recordings.

Neuropsychological and psychophysiological investigations of humans have been used increasingly in recent years to evaluate issues pertaining to information processing. Neuropsychological studies typically examine patients who have some specific neurological disorder. Several striking phenomena have been observed that enhance our understanding of higher-level vision (Farah and Ratcliff, 1994). One example is hemispatial neglect, in which an individual with a lesion in one cerebral hemisphere—usually the right hemisphere—that has damaged the parietal visual pathway fails to detect or respond to stimuli in the contralesional field (Heilman and Valenstein, 1979). This is in contrast to people who have damage to the temporal visual pathway, and hence have difficulty recognizing stimuli. These and other results have been interpreted as suggesting that the parietal system determines where something is and the temporal system what that something is (Merigan and Maunsell, 1993).

Psychophysiological methods include recently developed neuroimaging techniques, like *functional magnetic resonance imaging* (fMRI) and *positron emission tomography* (PET), that have provided insight into the spatial and temporal organization of brain functions (Posner and Raichle, 1994). The most widely used psychophysiological method has involved the measurement of *event-related potentials* (ERPs) (Rugg and Coles, 1995). Electrodes are attached to a person's scalp to measure voltage variations in the electroencephalogram, and the ERP is those changes that are related to the brain's response to a stimulus. Those ERP components occurring within 100 ms after onset of a stimulus are sensory components that reflect transmission of sensory information to, and its arrival at, the sensory cortex. The latencies for these components differ across sensory modalities. Later components reflect other aspects of information processing. For example, a negative component called *mismatch negativity* is evident in the ERP about 200 ms after presentation of a stimulus event other than the one that is most likely. It is present regardless of whether the stimulus is in an attended stream of stimuli or an unattended stream, suggesting that it reflects an automatic detection of physical deviance. The latency of a positive component called the *P300* is thought to reflect stimulus evaluation time, that is, the time to update the perceiver's current model of the physical environment. Psychophysiological methods such as these provide tools that can be used to address many issues of concern in human factors. Among other things, these methods can be used to determine whether a particular experimental phenomenon has its locus in processes associated with sensation and perception or with those involving subsequent response selection and execution. Because of this diagnosticity, it has been suggested that ERPs and other psychophysiological measures may be used in the future to provide precise measurement of dynamic changes in mental workload (e.g., Backs, Ryan, and Wilson, 1994; Humphrey and Kramer, 1994).

3.2.2 Psychophysical Methods

The more direct concern in human factors and ergonomics is with behavioral measures, because our interest is primarily with what people can and cannot perceive and with evaluating specific perceptual issues in applied settings. Because many of the methods used for obtaining behavioral measures can be applied to evaluating aspects of displays, we will cover them in some detail. For even more thorough coverage, the reader is referred to textbooks on psychophysical methods by Gescheider (1985) and Baird and Noma (1978).

3.2.2.1 Psychophysical Measures of Sensitivity

Classical Threshold Methods

The goal of one class of psychophysical methods is to obtain some estimate of sensitivity to detecting either the presence of some stimulation or differences between stimuli. The classical methods were based on the concept of a threshold, with an absolute threshold representing the minimum amount of stimulation necessary for an observer to tell that a stimulus was presented on a trial and a difference threshold representing the minimal amount of difference in stimulation along some dimension required to tell that a comparison stimulus differs from a standard stimulus. Several techniques were developed by Fechner (1860) for finding absolute thresholds, with the methods of limits and constant stimuli being among the most widely used.

To find a threshold using the *method of limits,* equally spaced stimulus values along the dimension of interest (e.g., magnitude of stimulation) that bracket the threshold are selected (see Figure 3.2). In alternating series, the stimuli are presented in ascending or descending order, beginning each time from a different, randomly chosen starting value below or above the threshold. For the ascending order, the first response typically would be, "No, I do not detect the stimulus." The procedure is repeated, incrementing the stimulus value each time, until the observer's response changes to "Yes," and the average of that stimulus value and the last one to which a "No" response was given is taken as the threshold for that series. A descending series is conducted in the same manner, but from a stimulus above threshold, until the response changes from "Yes" to "No." The thresholds for the individual series are then averaged to produce the final threshold estimate. A particularly efficient variation of the method of limits is the *staircase* method (Cornsweet, 1962). For this method, rather than having distinct ascending and descending series started from randomly selected values below and above threshold, only a single continuous series is conducted in which the direction of the stimulus sequence—ascending or descending—is reversed when the observer's response changes. The threshold is then taken to be the average of the stimulus values at which these transitions occur. The staircase method has the virtue of bracketing the threshold closely, thus minimizing the amount of stimulus presentations that is needed to obtain a certain number of response transitions on which to base the threshold estimate.

The *method of constant stimuli* differs from the method of limits primarily in that the different stimulus values are presented randomly, with each stimulus value presented many different times. The basic data in this case are the percentage of "yes" responses for each stimulus value. These typically plot as an S-shaped psychophysical function (see Figure

Stimulus Intensity (arbitrary units)	A	D	A	D	A	D	A	D	A	D
15						Y				
14				Y		Y				Y
13				Y		Y				Y
12		Y		Y		Y				Y
11		Y		Y	Y	Y		Y		Y
10	Y	Y		Y	N	Y	Y	Y		Y
9	N	Y	Y	N	N	N	N	Y	Y	N
8	N	N	N		N		N	Y	N	
7	N		N		N		N	N	N	
6	N		N		N		N		N	
5	N				N		N		N	
4	N				N				N	
3	N								N	
2	N								N	
1	N									
Transition points =	9.5	8.5	8.5	9.5	10.5	9.5	9.5	7.5	8.5	9.5
Mean threshold value = 9.1										

Figure 3.2 Determination of a sensory threshold by the method of limits using alternating ascending (A) and descending (D) series.

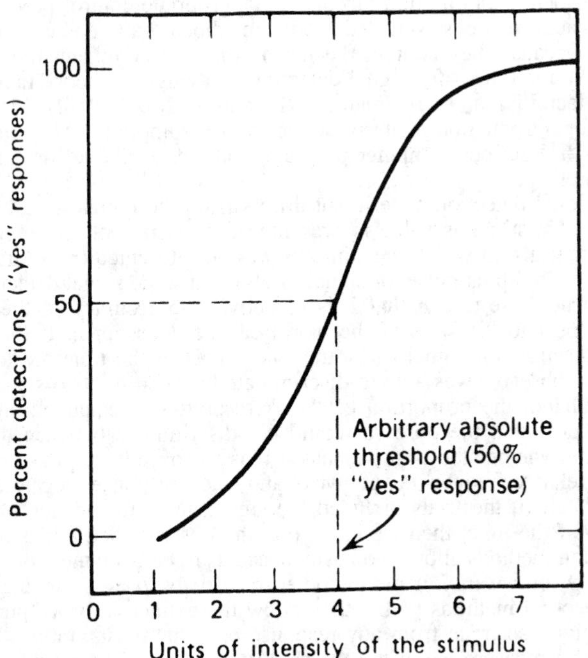

Figure 3.3 A typical S-shaped psychophysical function obtained with the method of constant stimuli. The absolute threshold is the stimulus intensity estimated to be detected 50% of the time. [Reprinted by permission of the publisher from H. R. Schiffman, *Sensation and perception: An integrated approach* (Fourth Edition), p. 23. Copyright 1996 © by John Wiley and Sons, Inc.]

3.3). The threshold is taken to be the estimated stimulus value for which the percentage of "yes" responses would have been 50%.

Both the methods of limits and constant stimuli can be extended to difference thresholds in a straightforward manner (see Gescheider, 1985). The most common extension is to use stimulus values for the comparison stimulus that range from being distinctly less than that of the standard stimulus to being distinctly greater. For the method of limits, ascending and descending series are conducted in which the observer responds "less," then "equal," and then "greater" as the magnitude of the comparison increases or vice versa as it decreases. The average stimulus value for which the responses shift from "less" to "equal" is the lower threshold and from "equal" to "greater" is the upper threshold. The difference between these two values is called the interval of uncertainty, and the difference threshold is found dividing the interval of uncertainty by 2. The midpoint of this interval is the point of subjective equality, and the difference between this point and the true value of the standard stimulus reflects constant error, or the influence of any factors that cause the observer to overestimate or underestimate systematically the value of the comparison in relation to that of the standard.

When the method of constant stimuli is used to obtain difference thresholds, the order in which the standard and comparison are presented is varied, and the observer judges which stimulus is greater than the other. The basic data then are the percentage of "greater" responses for each value of the comparison stimulus. The stimulus value corresponding to the 50th percentile is taken as the point of subjective equality. The difference between that stimulus value and the one corresponding to the 25th percentile is taken as the lower difference threshold, and the difference between the subjectively equal value and the stimulus value corresponding to the 75th percentile is the upper threshold; the two values are averaged to get a single estimate of the difference threshold.

Signal Detection Methods

Although many variants of the classical methods are still used, they are not as popular as they once were. The primary reason is that the threshold measures confound perceptual sensitivity, which they are intended to measure, with response criterion or bias (e.g., willingness to say "yes"), which they are not intended to measure. The threshold estimates also can be influenced by numerous other extraneous factors, although the impact of most of these factors can be minimized with appropriate control procedures. Alternatives to the classical methods, signal detection methods, have come to be preferred in many situations because they contain the means for separating sensitivity and response bias. Authoritative references for signal detection methods and theory include Green and Swets (1966), Macmillan and Creelman (1991), and McNicol (1972); Macmillan (1993) provides a briefer introduction to its principles and assumptions. Macmillan and Creelman's Appendix 6 describes computer programs that are available for calculating signal detection measures.

The typical signal detection experiment differs from the typical threshold experiment in that only a single stimulus value is presented for a series of trials and the observer must discriminate trials on which the stimulus was not presented (noise trials) from trials on which it was (signal-plus-noise, or signal, trials). Thus, the signal detection experiment is much like a true–false test in that it is objective; the accuracy of the observer's responses with respect to the state of the world can be determined. If the observer says "yes" most of the time on signal trials and "no" most of the time on noise trials, then we know that the observer was able to discriminate between the two states of the world. If, on the other hand, the proportion of "yes" responses is equal on signal and noise trials, then we know that the observer could not discriminate between them. Likewise, we can determine whether the observer has a bias to say one response or the other by considering the relative frequencies of "yes" and "no" responses regardless of the state of the world. If half of the trials included the signal and half did not, yet the observer said "yes" 70% of the time, then we know that the observer had a bias to say "yes."

Signal detection methods allow two basic measures to be computed, one corresponding to discriminability (or *sensitivity*) and the other to *response bias*. Thus, the key advantage of the signal detection methods is that they allow the extraction of a "pure" measure of perceptual sensitivity separate from any response bias that exists, rather than combining the two in a single measure, as in the threshold techniques. There are many alternative measures of sensitivity (Swets, 1986) and bias (Macmillan and Creelman, 1990), based on a variety of psychophysical models and assumptions. We will base our discussion around signal detection theory and the two most widely used measures of sensitivity and bias, d' and β.

Signal detection theory assumes that the sensory effect of a signal or noise presentation on any given trial can be characterized as a point along a continuum of evidence indicating that the signal was in fact presented. Across trials, the sensory effect will vary, such that for either type of trial the evidence will sometimes be higher (or lower) than at other times. For computation of d' and β, it is assumed that the resulting distribution of values is normal (i.e., bell-shaped and symmetric), or Gaussian, for both the signal and noise trials and that the variances for the two distributions are equal (see Figure 3.4). To the extent that the signal is discriminable from the noise, the distribution for the signal trials should be shifted to the right (i.e., higher on the continuum of evidence values) relative to that for the noise trials. The measure d' is therefore the distance between the means of the signal and noise distributions, in standard deviation units. That is,

$$d' = \frac{\mu_s - \mu_n}{\sigma}$$

where μ_s is the mean of the signal distribution, μ_n is the mean of the noise distribution, and σ is the standard deviation of both distributions. The assumption is that the observer will respond "yes" whenever the evidence value on any trial exceeds a criterion. The measure of β, which is expressed by the formula

$$\beta = \frac{f_s\,(C)}{f_n\,(C)}$$

where C is the criterion and f_s and f_n are the heights of the signal and noise distributions, respectively, is the likelihood ratio for the two distributions at the criterion. It indicates the placement of this criterion with respect to the distributions and thus reflects the relative bias to respond "yes" or "no."

Computation of d' and β is relatively straightforward. The placement of the distributions with respect to the criterion can be determined as follows. The *hit rate* is the

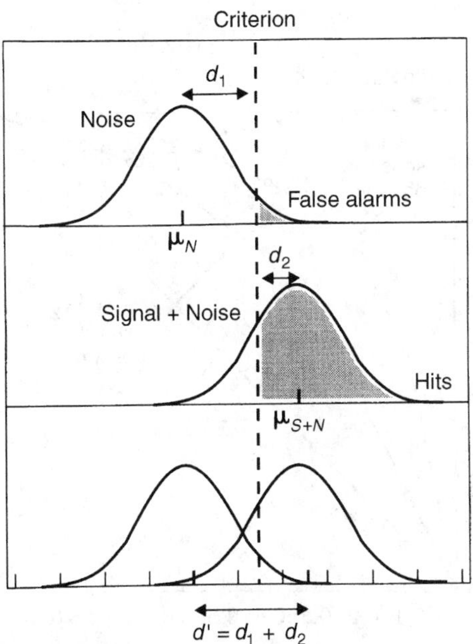

Figure 3.4 Equal variance, normal probability distributions for the noise and signal-plus-noise distributions on the sensory continuum, with a depiction of the proportion of false alarms, the proportion of hits, and the computation of d'. Bottom panel shows both distributions on a single continuum. (Reprinted by permission of the publisher from R. W. Proctor and T. Van Zandt, *Human factors in simple and complex systems*, p. 67. Copyright © 1994 by Allyn and Bacon.)

proportion of signal trials on which the observer correctly said "yes"; this can be depicted graphically by placing the criterion with respect to the signal distribution so that the proportion of the distribution exceeding it corresponds to the hit rate. The *false-alarm rate* is the proportion of noise trials on which the observer incorrectly said "yes." This corresponds to the proportion of the noise distribution that exceeds the criterion; when the noise distribution is placed to where the proportion exceeding the criterion is the false-alarm rate, then relative positions of the signal and noise distributions are depicted. Sensitivity, as measured by d', is the difference between the means of the signal and noise distributions, and this difference can be found by separately calculating the distance of the criterion from each of the respective means and then combining those two distances. Computationally, this involves converting the false-alarm rate and hit rate into standard normal z scores. If the criterion is located between the two means, d' is the sum of the two z scores. If the criterion is located outside of that range, the smaller of the two z scores must be subtracted from the larger to obtain d'. The likelihood ratio measure of bias, β, can be found from the hit and false-alarm rates by using a z table that specifies the height of the distribution for each z value. When β is 1.0, no bias exists to give one or the other response. A value of β greater than 1.0 indicates a bias to respond "no," whereas a bias less than 1.0 indicates a bias to respond "yes."

Although β has been used most often as the measure of bias to accompany d', several recent investigations have indicated that an alternative bias measure, C, is better (Corwin, 1994; Macmillan and Creelman, 1990; Snodgrass and Corwin, 1988). C is a measure of criterion location, rather than likelihood ratio. Specifically,

$$C = -0.5[z(H) + z(F)]$$

where H is the hit rate and F the false-alarm rate. C is superior to β on several grounds, including that it is less affected by the level of accuracy than is β and will yield a meaningful measure of bias when accuracy is near chance.

For a given d', the possible combinations of hit rates and false-alarm rates that the observer could produce through adopting different criteria can be depicted in a *receiver operating characteristic,* or ROC, curve (see Figure 3.5). The further an ROC is from the diagonal that extends from hit and false-alarm rates of 0 to 1.0, which represents chance

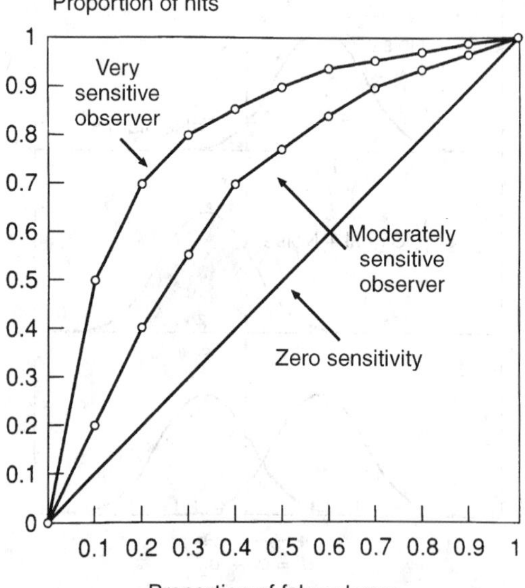

Figure 3.5 ROC curves showing the possible hit and false-alarm rates for different sensitivities. (Reprinted by permission of the publisher from R. W. Proctor and T. Van Zandt, *Human factors in simple and complex systems*, p. 69. Copyright © 1994 by Allyn and Bacon.)

performance (i.e., d' of 0), the greater the sensitivity. The procedure described above yields only a single point on the ROC, but in many cases it is advantageous to examine performance under several criteria settings so that the form of the complete ROC is evident (Swets, 1986). One advantage is that the estimate of sensitivity will be more reliable when it is based on several points along the ROC than when it is based on only one. Another is that the empirical ROC can be compared to the ROC implied by the psychophysical model that underlies a particular measure of sensitivity to determine whether serious deviations occur. For example, when enough points are obtained to estimate complete ROC curves, it is possible to evaluate the assumptions of equal variance, normal distributions on which the measures of d' and β are based. When plotted on z score coordinates, the ROC curve will be linear with a slope of 1.0 if both assumptions are supported; deviations from a slope of 1.0 mean that one distribution is more variable than the other, whereas systematic deviations from linearity indicate that assumption of normality is violated. If either of these deviations are present, alternative measures of sensitivity and bias that do not rely on the assumptions of normality and equal variance should be used.

In the cases in which a complete ROC curve is desired, several procedures exist for varying response criteria. The relative payoff structure may be varied across blocks of trials to make one or the other response more preferable; similarly, instructions may be varied regarding how the observer is to respond when uncertain. Another way to vary response criteria is to manipulate the relative probalilities of the signal and noise trials; the response criterion should be conservative when signal trials are rare and become increasingly more liberal as the signal trials become increasingly more likely. One of the most efficient techniques is to use rating scales (e.g., from 1, meaning very sure that the signal was not present, to 5, meaning very sure that it was present) rather than "yes"–"no" responses. The ratings are then treated as a series of criteria, ranging from high to low, and hit and false-alarm rates are calculated with respect to each.

Signal detection methods are powerful tools for investigating basic and applied problems pertaining not only to sensation and perception but also to many other areas in which an observer's response must be based on probabilistic information, such as distinguishing normal from abnormal X rays (Metz, 1989) or discriminating old from new words in a recognition memory test (Snodgrass and Corwin, 1988). Although most work on signal detection theory has involved discriminations along a single psychological continuum, it has been extended also to situations in which multidimensional stimuli are presumed to produce values on multiple psychological continua such as color and shape (e.g., Ashby and Townsend, 1986). Kadlec (1995) describes a computer program for performing such analyses, which have the benefit of allowing evaluation of whether the stimulus dimensions are processed in perceptually separable and independent manners and whether the decisions for each dimension are also separable. As these examples illustrate, signal detection methods can be extremely effective when used with discretion.

3.2.2.2 Psychophysical Scaling

Another concern in psychophysics is to construct scales for the relation between physical intensity and perceived magnitude. Scales can be constructed from three types of tasks that differ in the responses required of subjects (Gescheider, 1985, 1988). One way to build such scales is to do so from discriminative responses to stimuli that differ only slightly. Fechner (1860) established procedures for constructing psychophysical scales from difference thresholds. Later, Thurstone (1927) proposed a method for constructing a scale from paired comparison procedures in which each stimulus is compared to all others. Thurstonian scaling methods can be used even for complex stimuli for which physical values are not known. The second type of task involves having subjects divide the sensory continuum into two or more intervals that are subjectively equal. Such methods date back to the work of Plateau (1872).

Most recent work on scaling has followed the lead of Stevens (1975) in using direct methods that require some type of magnitude judgment (see Bolanowski and Gescheider, 1991, for an overview of this work). The technique of *magnitude estimation* is the most widely used. With this procedure, the observer is either presented a standard stimulus and told that its sensation is a particular numerical value (modulus) or is allowed to choose his or her own modulus. Stimuli of different magnitudes are then presented randomly, and the observer is to assign values to them proportional to their perceived magnitudes. These values then directly provide the scale relating physical magnitude to perceived magnitude. A technique called *magnitude production* can also be used, in which the observer is instructed to adjust the value of a stimulus to be a particular magnitude. Variations of magnitude estimation and production have been used to measure such things as emotional stress (Holmes and Rahe, 1967) and environmental odors and noise (Berglund, 1991).

Baird and Berglund (1989) have coined the term *environmental psychophysics* for the application of psychophysical methods such as magnitude estimation to applied problems of the type examined by Berglund (1991) that are associated with odorous air pollution and community noise. As Berglund puts it, "The method of ratio scaling developed by S. S. Stevens (1956) is a contribution to environmental science that ranks as good and important as most methods from physics or chemistry" (p. 141). When any measurement technique developed for laboratory research is applied to problems outside of the laboratory, special measurement issues may arise. In the case of environmental psychophysics, the environmental stimulus of concern typically is complex and multisensory, diffuse, and naturally varying, presented against an uncontrollable background. The most serious measurement problem is that often it is not possible to obtain repeated measurements from a given observer under different magnitude concentrations, necessitating that a scale be derived from judgments of different observers at different points in time.

Because differences exist in the way individuals assign magnitude numbers to stimuli, each individual's scale must be properly calibrated. Berglund and her colleagues have developed what they call the *master scaling procedure* to accomplish this purpose. The procedure has observers make magnitude judgments for several values of a referent stimulus, as well as for the environmental stimulus. Each observer's power function for the referent stimulus is transformed to a single master function (this is much like converting different normal distributions to the standard normal distribution for comparison). The appropriate transformation for each observer is then applied to her or his magnitude judgment for the environmental stimulus so that all such judgments are in terms of the master scale.

3.2.2.3 Other Techniques

Many other techniques have been used to investigate issues in sensation and perception. Most important are methods that use response times either instead of or in conjunction with response accuracy (see Luce, 1986; Welford, 1980). Reaction-time methods have a history of use approximately as long as that of the psychophysical methods, dating back to Donders (1868), but have been particularly widespread since about 1950. Simple reaction times require the observer to respond as quickly as possible with a single response (e.g., a keypress) whenever a stimulus event occurs. Alternative hypotheses of various factors that affect detection of the stimulus and the decision to respond, such as the locus of influence of visual masking and whether the detection of two signals presented simultaneously can be conceived of as an independent race, can be evaluated using simple reaction times.

Decision processes play an even larger role than in simple reaction tasks for go–no-go tasks, in which responses must be made to some stimuli but not to others, and for choice–reaction tasks, where there is more than one possible stimulus, more than one possible response, and the stimulus must be identified if the correct response is to be made. Methods such as the *additive factors logic* (Sternberg, 1969) can be used to isolate perceptual and decisional factors. This logic proposes that two variables whose effects are additive affect different processing stages but two variables whose effects are interactive affect the same processing stage. Variables that interact with marker variables whose effects can be assumed to be in perceptual processes but not with marker variables whose effects are on response selection or programming can be assigned a perceptual locus. Analyses based on the distributions of reaction times have gained in popularity in recent years; a computer program for performing such analyses is described by Heathcote (1996).

3.3 THE SENSORY SYSTEMS AND BASIC PERCEPTUAL PHENOMENA

The ways in which the sensory systems encode information have implications not only for the structure and function of the sensory pathways but, ultimately, also for the nature of human perception. They also place restrictions on the design of displays. Displays must be designed to satisfy known properties of sensory encoding (e.g., visual information that would be legible if presented in central vision will not be legible if the display is presented in the visual periphery), but they do not need to exceed the capabilities of sensory encoding. The sensory information that is encoded also must be represented in the nervous system. The nature of this representation also has profound implications for perception.

3.3.1 Vision

3.3.1.1 The Visual System

The sensory receptors in the eye are sensitive to energy within a limited range of the electromagnetic spectrum. One way of characterizing such energy is as continuous waves of different wavelengths. The visible spectrum ranges from wavelengths of approximately 370 nanometers (nm, billionths of a meter) to 730 nm. Any energy outside of this range, such as ultraviolet rays, will not be detected because they have no effect on the receptors. Light also can be characterized in terms of small units of energy called *photons*. Conceiving light in terms of wave length is important for some aspects of perception, such as color vision, whereas for others it is more useful to treat it in terms of photons. As with any system in which light energy is used to create a representation of the physical world, the light must be focused and a clear image created. In the case of the eye, the image is focused on the photoreceptors located on the retina, which lines the back wall of the eye.

The Focusing System

Light enters the eye (see Figure 3.6) through the cornea, which acts as a lens of fixed optical power and provides the majority of the focusing. The remaining focusing is performed by the lens, whose power varies automatically as a function of the distance from the observer of the object that is being fixated. Beyond a distance of approximately 3 m, the far point, the lens is relatively flat; for distances closer than the far point, muscles attached to the lens cause it to become progressively more spherical the closer the fixated object is to the observer, thus increasing its refractive power. The reason why this process, called *accommodation,* is needed is that without an increase in optical power for close objects, their images would be focused at a point beyond the retina and the retinal image would be out of focus. Accommodation is effective for distances as close as 20 cm (the near point), but the extent of accommodation, and the speed at which it occurs, decreases with increasing age, with the near point receding to approximately 100 cm by age 60. This decrease in accommodative capability, called presbyopia, can be corrected with reading glasses. Other imperfections of the lens system—myopia, where the focal point is in front of the receptors; hyperopia, where the focal point is behind the receptors; and

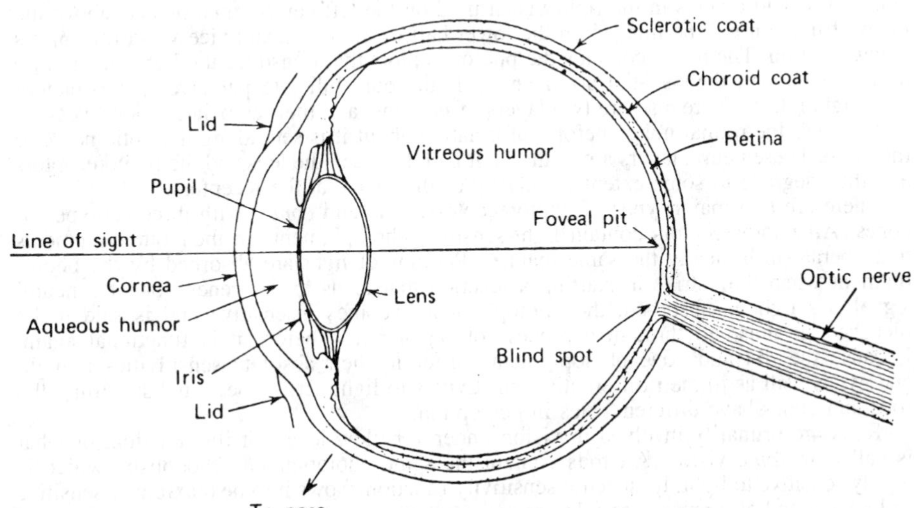

Figure 3.6 Structure of the human eye. Reprinted by permission of the publisher from H. R. Schiffman, *Sensation and perception: An integrated approach* (Fourth Edition), p. 52. Copyright 1996 © by John Wiley and Sons, Inc.)

astigmatism, where certain orientations are out of focus while others are not—also typically are treated with glasses.

Between the cornea and the lens, the light passes through the pupil, which can vary in size from 2 mm to 8 mm. The pupil size is large when the light level is low, to maximize the amount that gets into the eye, and small when the light level is high, to minimize the imperfections in imaging that arise when light passes through the extreme periphery of the lens system. One additional consequence of these changes in image quality as a function of pupil size is that the *depth of field,* or the distance in front of or behind a fixated object at which the images of other objects will be in focus also, will be greater when the pupil size is small than when it is large. In other words, under conditions of low illumination, accommodation must be more precise and work that requires high acuity, such as reading, can be fatiguing (Randle, 1988). The pupil size also varies from moment to moment as a function of arousal level and the amount of mental effort being devoted to a task (Kahneman, 1973), although the changes in size associated with these factors are less than those associated with the lighting level.

When the eyes fixate on an object at a distance of approximately 6 m or further, the lines of sight are parallel. As the object is moved progressively closer, the eyes turn inward and the lines of sight converge. Thus, the degree of *vergence* of the eyes varies systematically as a function of the distance of the object being fixated. The near point for vergence is approximately 5 cm, and if an object closer than that is fixated, the images at the two eyes will not be fused and a double image will be seen.

The natural resting states for accommodation and vergence, called *dark focus* and *dark vergence,* respectively, are intermediate to the near and far points (Leibowitz and Owens, 1975; Miller, 1990; Owens and Leibowitz, 1983). One view for which there is considerable support is that dark focus and vergence provide zero reference points about which accommodative and vergence effort varies (Ebenholtz, 1994). A practical implication of this is that less eye fatigue will occur if a person working at a visual display screen for long periods of time is positioned at a distance that corresponds approximately to the dark focus and vergence points. As with most other human characteristics of concern in human factors and ergonomics, considerable individual differences in dark focus and vergence exist. Individuals with far dark vergence postures tend to position themselves further away from the display screen than will those with closer postures (Heuer, et al., 1989), and they also show more visual fatigue when required to perform close visual work (Jaschinski-Kruza, 1991).

The Retina

If the focusing system is working properly, the image will be focused on the retina, which lines the back wall of the eye. Objects in the left visual field will be imaged on the right hemi-retina and objects in the right visual field on the left hemi-retina; objects above the point of fixation will be imaged on the lower half of the retina and vice versa for objects below fixation. The retina contains the photoreceptors that transduce the light energy into a nervous signal; their spatial arrangement limits our ability to perceive spatial pattern (see Figure 3.7). There also are two layers of neurons, and their associated blood vessels, that process the retinal image before information about it is sent along the optic nerve to the brain. These neural layers are in the light path between the lens and the photoreceptors and thus degrade to some extent the clarity of the image at the receptors.

There are two major types of photoreceptors, *rods* and *cones,* with three subtypes of cones. All photoreceptors contain light-sensitive photopigments in their outer segments that operate in basically the same manner. Photons of light are absorbed by the photopigment when they strike it, starting a reaction that leads to the generation of a neural signal. As light is absorbed, the photopigment becomes insensitive and is said to be bleached. It must go through a process of regeneration before it is functional again. Because the rod and cone photopigments differ in their absolute sensitivities to light energy, as well as in their differential sensitivities to light across the visual spectrum, the rods and cones have different roles in perception.

Rods are primarily involved in vision under very low levels of illumination, or what is called *scotopic* vision. All rods contain the same photopigment, rhodopsin, which is highly sensitive to light. Its spectral sensitivity function shows it to be maximally sensitive to light around 500 nm and to a lesser degree to other wavelengths. One consequence of there being only one rod photopigment is that we cannot perceive color under scotopic conditions. The reason for this is easy to understand. The rods will respond relatively more to stimulation of 500 nm than they will to 550-nm stimulation of equal intensity.

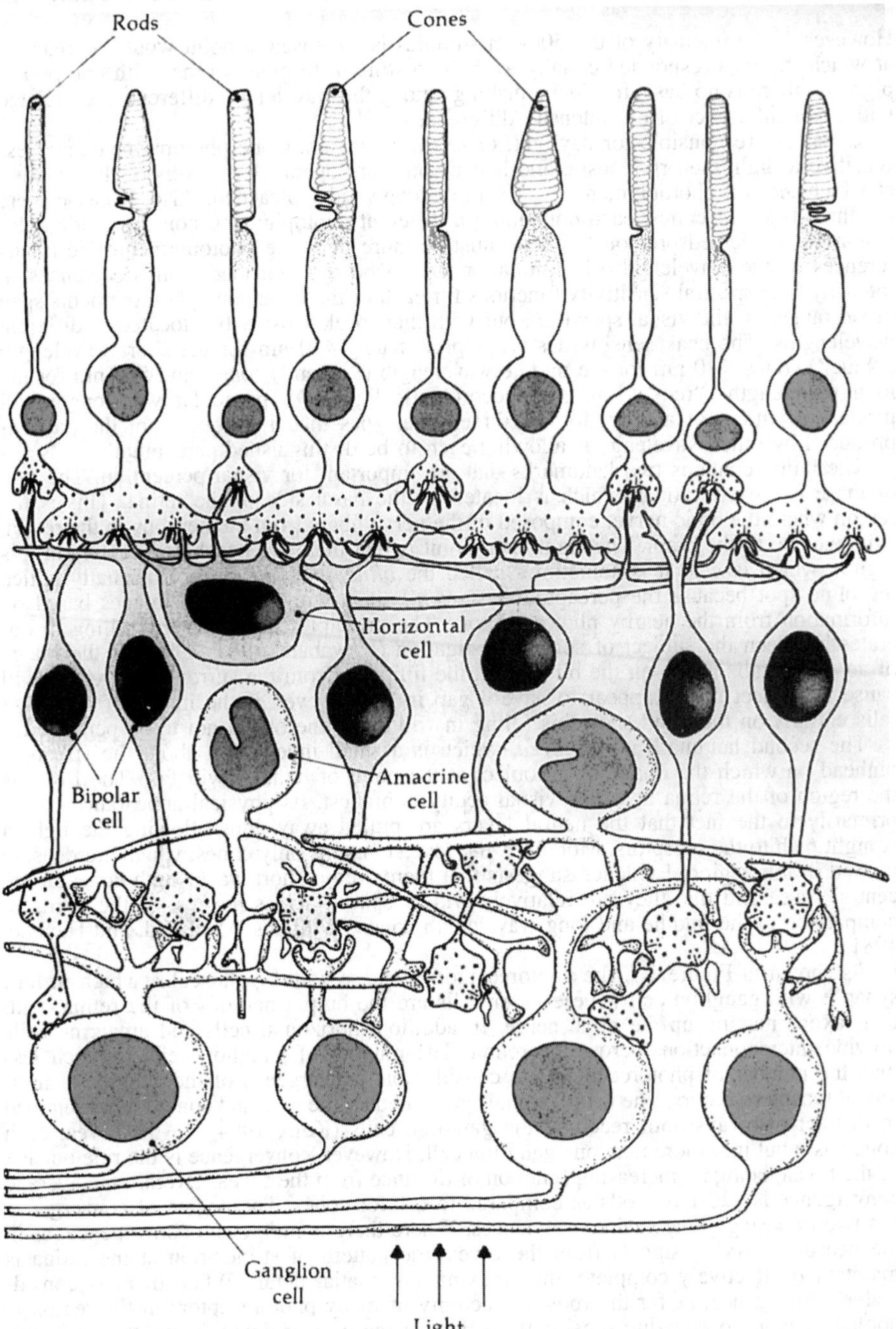

Figure 3.7 Schematic diagram of the neural structures and interconnections of the vertebrate retina. (Reprinted with permission from J. E. Dowling and B. B. Boycott, 1966, Organization of the primate retina: Electron microscopy, *Proceedings of the Royal Society* Series B, *166*, 80–111.)

However, if the intensity of the 500-nm stimulus is increased, a point would be reached at which the rods responded equally to the two stimuli. In other words, with one photopigment, there is no basis for distinguishing among the wavelength differences associated with color differences from intensity differences.

Cones are responsible for daylight, or *photopic,* vision. Cone photopigments are less sensitive to light than rhodopsin, and hence cones are operative at levels of illumination at which the rod photopigment has been effectively fully bleached. Also, because there are three types of cones, each containing a different photopigment, cones provide color vision. As explained previously, there must be more than one photopigment type if differences in the wavelength of stimulation are to be distinguished from differences in intensity. The spectral sensitivity functions for each of the three cone photopigments span broad ranges of the visual spectrum, but with their peak sensitivities located at different wavelengths. The peak sensitivities are approximately 440 nm for the short wavelength ("blue") cones, 540 nm for the middle wavelength ("green") cones, and 565 nm for the long wavelength ("red") cones. Monochromatic light of a particular wavelength will produce a pattern of activity for the three cone types that is unique from the patterns produced by other wavelengths, allowing each to be distinguished perceptually.

The retina contains two landmarks that are important for visual perception. The first of these is the optic disk, which is located on the nasal side of the retina. This is the region where the optic nerve, composed of the nerve fibers from the neurons in the retina, exits the eye to the brain. The significant point is that there are no photoreceptors in this region, which is why it is sometimes called the *blind spot.* We do not normally notice the blind spot because the perceptual system fills it in with fabricated images based on information from the nearby photoreceptors. The principles by which this filling in operates has been the subject of some investigation (Kawabata, 1984, 1990). If the image of an object falls partly on the blind spot, the filling in from the surrounding region will cause the object not to appear to have a gap in it. However, if the image of an object falls entirely on the blind spot, this filling in will cause the object not to be perceived.

The second landmark is the *fovea,* which is a small indentation about the size of a pinhead on which the image of an object at the point of fixation will fall. The fovea is the region of the retina in which visual acuity is highest. Its physical appearance is due primarily to the fact that the neural layers are pulled away, thus allowing the light a straight path to the receptors. Moreover, the fovea contains only cones, which are densely packed in this region. Evidence suggests that there are no short wavelength cones in the central fovea and that they are relatively widely spaced across the rest of the retina in comparison to the middle and long wavelength cones (Williams, MacLeod, and Hayhoe, 1981).

As shown in Figure 3.7, the photoreceptors synapse with bipolar cells, which in turn synapse with ganglion cells; these latter cells are the output neurons of the retina, with their axons making up the optic nerve. In addition, horizontal cells and amacrine cells provide interconnections across the retina. The number of ganglion cells is much less than the number of photoreceptors, so considerable convergence of the activity of individual receptors occurs. The neural signals generated by the rods and cones are maintained in distinct pathways until reaching the ganglion cells (Kolb, 1994). In the fovea, each cone has input into more than one ganglion cell. However, convergence is the rule outside of the fovea, being an increasing function of distance from the fovea. Overall, the average convergence is 120:1 for rods as compared to 6:1 for cones. The degree of convergence has two opposing perceptual consequences. Where there is little or no convergence, as in the neurons carrying signals from the fovea, the pattern of stimulation at the retina is maintained effectively complete, thus maximizing spatial detail. When there is considerable convergence, as for the rods, the activity of many photoreceptors in the region is pooled together, optimizing sensitivity to light at the cost of detail. Thus, the wiring of the photoreceptors is consistent with the fact that the rods operate when light energy is at a premium but the cones operate when it is not.

The ganglion cells show several interesting properties pertinent to perception. When single-cell recording techniques are used to measure their receptive fields (i.e., the regions on the retina to which they are sensitive to light energy), these fields are found to have a circular, center-surround relation for most cells. If light presented in a circular, center region causes an increase in the firing rate of the neuron, then light presented in a surrounding ring region will cause a decrease in the firing rate, and vice versa. What this means is that the ganglion cells are tuned to respond primarily to discontinuities in the light pattern within their receptive fields. If the light energy across the entire receptive field is increased, then there will be little if any effect on the firing rate. In short, the information extracted and signaled by these neurons is based principally on contrast, which is important for perceiving objects in the visual scene, and not on absolute intensity, which will vary as a function of the amount of illumination. Not surprisingly, the average receptive field size is larger for ganglion cells receiving their input from rods than for those receiving it solely from cones and increases with increasing distance from the fovea.

Although most ganglion cells have the center-surround receptive field organization, two pathways can be distinguished on the basis of other properties. The ganglion cells in the *parvocellular* pathway have small cell bodies and relatively dense dendritic fields. Many of these ganglion cells, called midget cells, receive their input from the fovea. They have relatively small receptive fields, show a sustained response as long as stimulation is present in the receptive field, and show a relatively slow speed of transmission. The ganglion cells in the *magnocellular* pathway have larger cell bodies and sparse dendritic trees. They have their receptive fields at locations across the retina, have relatively large receptive fields, show a transient response to stimulation that dissipates if the stimulus remains on, show a fast speed of transmission, and are sensitive to motion. Because of these unique characteristics, and the fact that these channels are kept separated later in the visual pathways, it has been thought that they contribute distinct information to perception. The parvocellular pathway is presumed to be responsible for pattern perception and the magnocellular pathway for high temporal frequency information, such as in motion perception and perception of flicker. The view that different aspects of the sensory stimulus are analyzed in specialized neural pathways has been popular in recent years.

The Visual Pathways

The optic nerve from each eye splits at what is called the *optic chiasma* (see Figure 3.8). The fibers conveying information from the nasal halves of the retinas cross over and go to the opposite sides of the brain, whereas the fibers conveying information from the temporal halves do not cross over. Functionally, the significance of this is that, for both eyes, input from the right visual field is sent to the left half of the brain and input from the left visual field to the right half. After the cortex, a relatively small subset of the fibers (approximately 10%) split off from the main tract and go to structures in the brain stem, the tectum, and then the pulvinar nucleus of the thalamus. This *tectopulvinar* pathway is involved in localization of objects and the control of eye movements.

Approximately 90% of the fibers continue on the primary *geniculostriate* pathway, where the first synapse is at the lateral geniculate nucleus (LGN). The distinction between the parvocellular and magnocellular pathways is maintained here. The LGN is composed of six layers, four parvocellular and two magnocellular, each of which receives input from only a single eye. Hence, at this level, the input from the two eyes has yet to be combined. Each layer is laid out in a retinotopic map that provides a spatial representation of the retina. In other respects, the receptive field structure of the LGN neurons is similar to that of the ganglion cells. The LGN also receives input from the cortex—in fact, relatively more synapses in the LGN originate from the cortex than from the retina (Sherman and Koch, 1990)—meaning that this is the first point in the pathway in which the brain can have some effect on the signals arriving from the sensory system.

From the LGN, the fibers go to the primary visual cortex, which is located in the posterior cortex. This region is also called the striate cortex, because of its stripes, area 17, and area V1. The visual cortex consists of six layers. The fibers from the LGN have their synapses in the fourth layer from the outside, with the parvocellular neurons sending their input to the bottom half and the magnocellular neurons to the top half. The neurons in layer 4, then send their output to other layers. In layer 4 the neurons have circular-surround receptive fields, but in other layers, they have more complex patterns of sensitivity. Also, whereas layer 4 neurons receive input from one or the other eye, in other layers most neurons respond to some extent to stimulation at either eye. A distinction can be made between simple cells and complex cells (e.g., Hubel and Wiesel, 1977). The responses of simple cells to shapes can be determined from their responses to small spots of light (e.g., if the receptive field for the neuron is mapped out using spots of light, the neuron will be most sensitive to a stimulus shape that corresponds with that receptive field), whereas those for complex cells cannot be. Simple cells have center-surround receptive fields, but they are more linear than circular; this means that they are orientation selective and will respond optimally to bars in an orientation that corresponds with that of the receptive field. Complex cells have similar linear receptive fields, and so are also orientation selective, but they are movement sensitive as well. These cells respond opti-

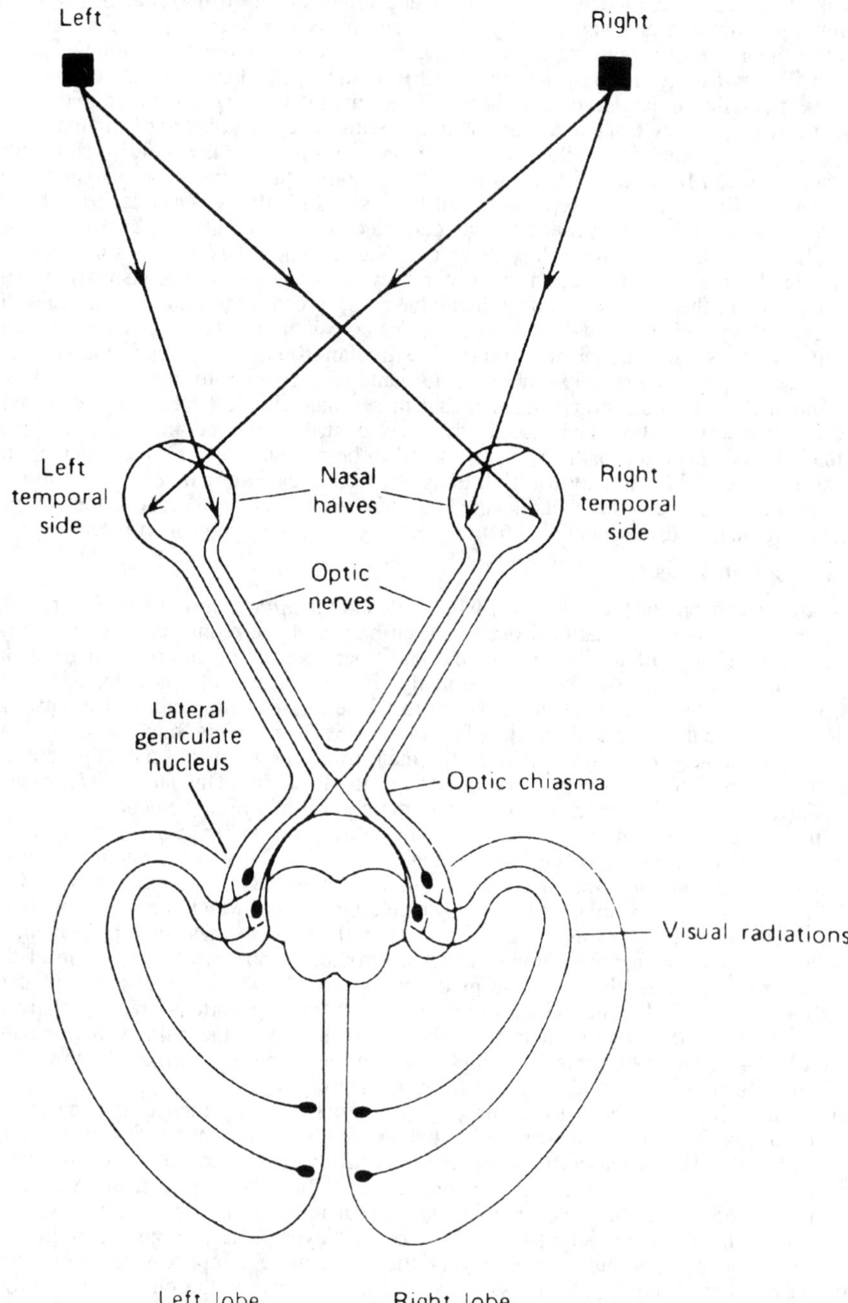

Figure 3.8 Schematic diagram of the human visual system showing the projection of the visual fields through the system. [Reprinted by permission of the publisher from H. R. Schiffman, *Sensation and perception: An integrated approach* (Fourth Edition), p. 71. Copyright 1996 © by John Wiley and Sons, Inc.]

mally not only when the bar is at the appropriate orientation, but also when it is moving. Some cells, which receive input from the magnocellular pathway, are also directionally sensitive; they respond optimally to movement in a particular direction. Certain cells, called hypercomplex cells, are sensitive to the length of the bar so that they will not respond if the bar is too long. Some neurons in the visual cortex also are sensitive to disparities in the images at each eye and to motion velocity. In short, the neurons of the visual cortex analyze the sensory input for basic features that are provide the information on which higher-level processes operate.

The cortex is composed of columns and hypercolumns arranged in a spatiotopic manner. Within a single column, all of the cells except for those in layer 4 will have the same preferred orientation. The next column will respond to stimulation at the same location on the retina, but will have a preferred orientation that is approximately 10° different than that of the first column. As we proceed through a group of about 20 columns, called a hypercolumn, the preferred orientation will rotate 180°. The next hypercolumn will show the same arrangement, but for stimulation at a location on the retina that is adjacent to that of the first. A relatively larger portion of the neural machinery in the visual cortex is devoted to the fovea, which is to be expected because it is the region for which detail is being represented.

After area V1, many other cortical areas have been found that are devoted to vision. Some of the neurons in the magnocellular pathway project directly to the medial temporal cortex, suggesting that it may play a role in motion perception. Other neurons in the magnocellular pathway converge with the parvocellular pathway in V1 (Nealey and Maunsell, 1994), and then separate into two pathways. The temporal pathway goes to the inferior portion of the temporal lobe and the parietal pathway to the posterior portion of the parietal lobe (Merigan and Maunsell, 1993). As mentioned earlier for the example of hemispatial neglect, it has been suggested that the temporal pathway is involved primarily in determining what the stimulus is and the parietal pathway is involved in determining where the stimulus is. Barber (1990) found that visual abilities pertaining to detection, identification, acquisition, and tracking of aircraft in air defense simulations clustered into subgroups consistent with the hypothesis that different aspects of visual perception are mediated by distinct subsystems in the brain.

3.3.1.2 Basic Visual Perception

Brightness

Brightness is that aspect of visual perception that corresponds most closely to the intensity of stimulation. To specify the effective intensity of a stimulus, we want to use photometric measures, which are calibrated to reflect human spectral sensitivity. A photometer can be used to measure either the *illuminance,* that is, amount of light falling on a surface, or *luminance,* that is, the amount of light generated by a surface. To measure illuminance, an illuminance probe is attached to the photometer and placed on the illuminated surface. The resulting measure of illuminance is in lumens per square meter (lm/m^2), or lux (lx). To measure luminance, a lens with a small aperture is attached to the photometer and focused onto the surface from a distance. The resulting measure of luminance is in candelas per square meter (cd/m^2).

Judgments of brightness are related to intensity by the power function

$$B = aI^{0.33}$$

where B is brightness, a is a constant, and I the physical intensity. However, brightness is not determined by intensity alone but also by several other factors. For example, at brief exposures on the order of 100 ms or less and for small stimuli, temporal and spatial summation occur. That is, a stimulus of the same physical intensity will look brighter if its exposure duration is increased or if its size is increased.

One of the most striking influences on brightness perception and sensitivity to light is the level of *dark adaptation*. When a person first enters a dark room, he or she is relatively insensitive to light energy. However, with time, dark adaptation occurs and sensitivity increases drastically. The time course of dark adaptation is approximately 30–45 min. Over the first few minutes, the absolute threshold for light decreases and then levels off. However, after approximately 8 min in the dark, it begins decreasing again, approaching maximum around the 30-min point. After 30 min in the dark, lights that were of too low intensity to be seen initially can be seen and stimuli that appeared dim now seem much brighter. Dark adaptation reflects primarily regeneration into a light-sensitive state of first the cone photopigments and then the rod photopigment. After becoming dark adapted, vision may be impaired momentarily when the person returns to photopic viewing conditions. Providing gradually changing light intensity in regions where light intensity would normally change abruptly, such as at the entrances and exits of tunnels, may help minimize such impairment (e.g., Oyama, 1987).

The brightness of a monochromatic stimulus of constant intensity will vary as a function of its wavelength because the photopigments are differentially sensitive to light of different wavelengths. The scotopic spectral sensitivity function is shifted toward the short wavelength end of the spectrum, relative to the photopic function. Consequently, if two stimuli, one short wavelength and one long, appear equally bright at photopic levels, the short wavelength stimulus will appear brighter at scotopic levels—a phenomenon called the Purkinje shift. Little light adaptation will occur when high levels of long wavelength light are present because the sensitivity of the rods to long wavelength light is low. Thus, it is customary to use red light sources to provide high illumination for situations in which a person needs to remain dark adapted.

It has become common practice to distinguish between brightness and lightness: Judgments of brightness are of the perceived intensity of a stimulus, whereas judgments of lightness are of perceived achromatic color along a black to white dimension. Both the brightness and lightness of an object are greatly influenced by the surrounding context. *Lightness contrast* is a phenomenon where the intensity of a surrounding area influences the lightness of a stimulus. The effects can be quite dramatic, with a stimulus of intermediate reflectance ranging in appearance from white to dark gray or black as the reflectance of the surround is increased from low to high. The more common phenomenon of lightness constancy occurs when the light intensity is increased across the entire visual field, as occurs when the level of illumination is increased. In this case, the absolute amount of light reflected to the eye by an object may be quite different, but the percept remains constant. Basically, lightness follows a constant ratio rule (Wallach, 1972): Lightness will remain the same if the ratio of light energy for a stimulus relative to its surround remains constant. *Lightness constancy* holds for a broad range of ratios and across a variety of situations, with brightness constancy obtained under a more restricted set of viewing conditions (Jacobsen and Gilchrist, 1988; Sewall and Wooten, 1991).

Although low-level mechanisms early in the sensory system likely contribute at least in part to constancy and contrast, more complex higher-level brain mechanisms do as well. Particularly compelling are demonstrations showing that the lightness and brightness of an object can vary greatly simply as a function of organizational and depth cues. Agostini and Proffitt (1993) demonstrated lightness contrast as a function of whether a target gray circle was organized perceptually with black or white circles, even though the inducing circles were not in close proximity to the target. Gilchrist (1977) used depth cues to cause a piece of white paper to be perceived incorrectly as in a back chamber that was highly illuminated or correctly as in a front chamber that was dimly illuminated. When perceived as in the front chamber, the paper was seen as white; however, when perceived as in the back chamber, the paper appeared to be almost black. Adelson (1993) showed that such effects are not restricted to lightness but also occur for brightness judgments. For example, when instructed to adjust the luminance of square a_1 in Figure 3.9 to equal that of square a_2, observers set the luminance of a_1 to be 70% higher than that of a_2. Thus, even relatively "sensory" judgments such as brightness are affected by higher-order organizational factors.

Visual Acuity and Sensitivity to Spatial Frequency

Visual acuity refers to the ability to perceive detail. Acuity is highest in the fovea, and it decreases with increasing eccentricities due in part to the progressively greater convergence of activity from the sensory receptors that occurs in the peripheral retina. Distinctions can be made between different types of acuity. *Identification acuity* is the most commonly measured, using a Snellen eye chart that contains rows of letters that become progressively smaller. The smallest row for which the observer can identify the letters is used as the indicator of acuity. 20/20 vision is regarded as normal. This means that the person being tested is able to identify at a distance of 20 ft letters of a size that a person with normal vision is expected to identify. A person with 20/100 vision can identify letters at 20 ft only as large as those that a person with normal vision could at 100 ft. *Vernier acuity* is a person's ability to discriminate between broken and unbroken lines, and *resolution acuity* is the ability to distinguish gratings from a stimulus that covers the

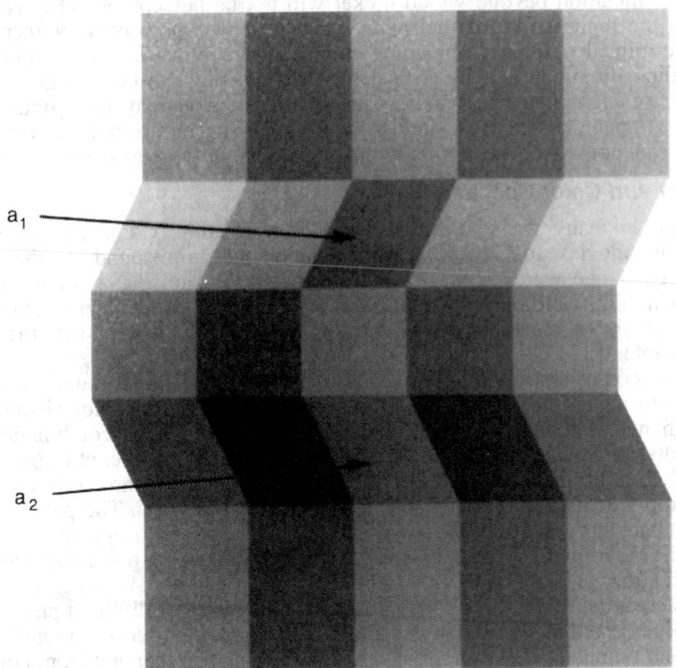

Figure 3.9 Effect of perceived shading on brightness judgments. The patches a_1 and a_2 are the same shade of gray, but a_1 appears much darker than a_2. (Reprinted with permission from E. H. Adelson, 1993, Perceptual organization and the judgment of brightness, *Science, 262*, 2042–2044. Copyright © 1993 by the Association for the Advancement of Science.)

same area but is of the same average intensity throughout. All of these measures are variants of *static acuity,* in that they are based on static displays. *Dynamic acuity* refers to the ability to resolve detail when there is relative motion between the stimulus and the observer. Dynamic acuity is usually poorer than static acuity (Morgan, Watt, and McKee, 1983), partly due to an inability to keep a moving image within the fovea (Murphy, 1978). A concern in measuring acuity is that the types are not perfectly correlated, and thus an acuity measure of one type may not be a good predictor of ability to perform a task whose acuity requirements are of a different type. For example, the elderly show typically little loss of identification acuity as measured by a standard test, but they seem to have impaired acuity in dynamic situations and at low levels of illumination (Kosnik, Sekuler, and Kline, 1990; Sturr, Kline, and Taub, 1990). Thus, an elderly person who does well on the standard static acuity test may be impaired when it comes to driving an automobile.

In recent years, spatial contrast sensitivity has been shown to provide an alternative way for characterizing acuity. A *spatial contrast sensitivity function* can be generated by obtaining threshold contrast values for discriminating sine wave gratings (for which the bars change gradually rather than sharply from light to dark) of different spatial frequencies from a homogeneous field. The contrast sensitivity function for a typical adult shows maximum sensitivity at a spatial frequency of about 3–5 cycles per degree of visual angle, with relatively sharp drop-offs at high and low spatial frequencies. Basically what that means is that we are not extremely sensitive to very fine or course gratings. Because high spatial frequencies pertain to ability to perceive detail and low to intermediate frequencies to the more global characteristics of visual stimuli, tests of acuity based on contrast sensitivity may be more analytic concerning aspects of performance that are necessary for performing specific tasks. For example, Evans and Ginsburg (1982) found contrast sensitivity at intermediate and low spatial frequencies to predict the detectability of stop signs at night.

Of particular concern in human factors and ergonomics is temporal acuity. Because many light sources and displays present flickering stimulation, we need to be aware of the rates of stimulation beyond which flicker will not be perceptible. The *critical flicker frequency* is the highest rate of flicker at which it can be perceived. Numerous factors influence the critical frequency, including stimulus size, retinal location, and the level of the surrounding illumination. The critical flicker frequency can be as high as 60 Hz for large stimuli of high intensity, but it typically is less. You may have noticed that the flicker of a computer display screen is perceptible when you are not looking directly at it, as a consequence of the greater temporal sensitivity in the peripheral retina.

Color Vision and Color Specification

Color is used in many ways to display visual information. Color can be used, as in televisions and movies, to provide a representation that corresponds to the colors that would be seen if one were physically present at the location that is depicted. Color is also used to highlight and emphasize, as well as to code different categories of displayed information. In these situations, such as these, we want to be sure that the colors are perceived as intended.

In a color mixing study, the observer is asked to adjust the amounts of component light sources to match the hue of a comparison stimulus. Human color vision is *trichromatic,* which means that any spectral hue can be matched by a combination of three primary colors, one each from the short, middle, and long wavelength positions of the spectrum. This trichromaticity is a direct consequence of having three types of cones that contain photopigments with distinct spectral sensitivity functions. The pattern of activity generated in the three cone systems will determine what hue is perceived. A specific pattern can be determined by a monochromatic light source of a particular wavelength or by a combination of light sources of different wavelengths. As long as the relative amounts of activation in the three cone systems are the same for different physical stimuli, they will be perceived as being of the same hue. This fact is used in the design of color televisions and computer monitors, for which all colors are generated from combinations of pixels of three different colors.

Another phenomenon of additive color mixing is that blue and yellow when mixed in approximately equal amounts yield an achromatic (e.g., white) hue, as do red and green. This stands in contrast to the fact that combinations involving one hue from each of the two complimentary pairs are seen as combinations of the two hues. For example, when blue and green are combined additively in similar amounts, the resulting stimulus appears blue-green. Each of the pairs that yield an achromatic additive mixture are called *complimentary colors.* That these hues have a special relation is evident in other situations as well. When a background is one of the hues from a pair of complimentary colors, it will tend to induce the opposing hue in a stimulus that would otherwise be perceived as a neutral gray or white. Likewise, if a background of one hue is viewed for awhile, and then the gaze is shifted to a background of a neutral color, an afterimage of the complimentary hue will be seen.

The complimentary color relations also appear to have a basis in the visual system, but in the neural pathways rather than in the sensory receptors. That is, considerable evidence indicates that the output from the cones is rewired into opponent processes at the level of the ganglion cells and beyond. If a neuron's firing rate increases when a blue stimulus is presented, it decreases when a yellow stimulus is presented. Similarly, if a neuron's firing rate increases to a red stimulus, it decreases to a green stimulus. The pairings in the opponent cells always involve blue with yellow and red with green. Thus, a wide range of color appearance phenomena can be explained by the view that the sensory receptors operate trichromatically but this information is subsequently recoded into an opponent format in the sensory pathways.

The basic color mixing phenomena are depicted in color appearance systems. A *color circle* can be formed by curving the visual spectrum, as done originally by Isaac Newton. The center of the circle represents white, and its rim represents the spectral colors. A line drawn from a particular location on the rim to the center depicts saturation, or the amount of hue that is present. Thus, for example, if one picks a monochromatic light source that appears red, points on the line represent progressively decreasing amounts of red as one moves along it to the center. The appearance for a mixture of two spectral colors can be approximated by drawing a chord that connects the two colors. If the two are mixed in equal amounts, then the point at the center corresponds to the mixture; if the percentages are unequal, the point is shifted accordingly toward the higher percentage spectral color. The hue for the mixture point can be determined by drawing a diagonal through it; its hue corresponds to that of the spectral hue at the rim, and its saturation corresponds to the proximity to the rim.

The color circle is too imprecise to be used to specify color stimuli, but a system much like it, the CIE (Commission Internationale de l'Eclairage) system, is the most widely used color specification system. The CIE provided a standardized set of color matching functions, $x(\lambda)$, $y(\lambda)$, and $z(\lambda)$, called the *XYZ* tristimulus coordinate system (see Figure 3.10). The tristimulus values for a monochromatic stimulus can be used to determine the proportions of three wavelengths (X, Y, and Z, corresponding to red, green, and blue, respectively) needed to match it. For example, the *x*, *y*, and *z* tristimulus values for a 500 nm stimulus are 0.0049, 0.3230, and 0.2720). The proportion of *X* primary can be determined by dividing the tristimulus value for *x* by the combined values for *x* plus *y* plus *z*. The proportion of *Y* primary can be determined in like manner, and the proportion of *Z* primary is simply 1 minus the *X* and *Y* proportions. The spectral stimulus of 500 nm thus has the following proportions: X = 0.008, Y = 0.539, and Z = 0.453.

The *CIE color space,* shown in Figure 3.10, is triangular rather than circular. Location in the space is specified according to the relative amounts of the three primary colors, *X*, *Y*, and *Z*. Only *X* and *Y* are used as the axes for the space because *X*, *Y*, and *Z* sum to 1.0. The spectral stimulus of 500 nm would be located on the rim of the triangle, in the

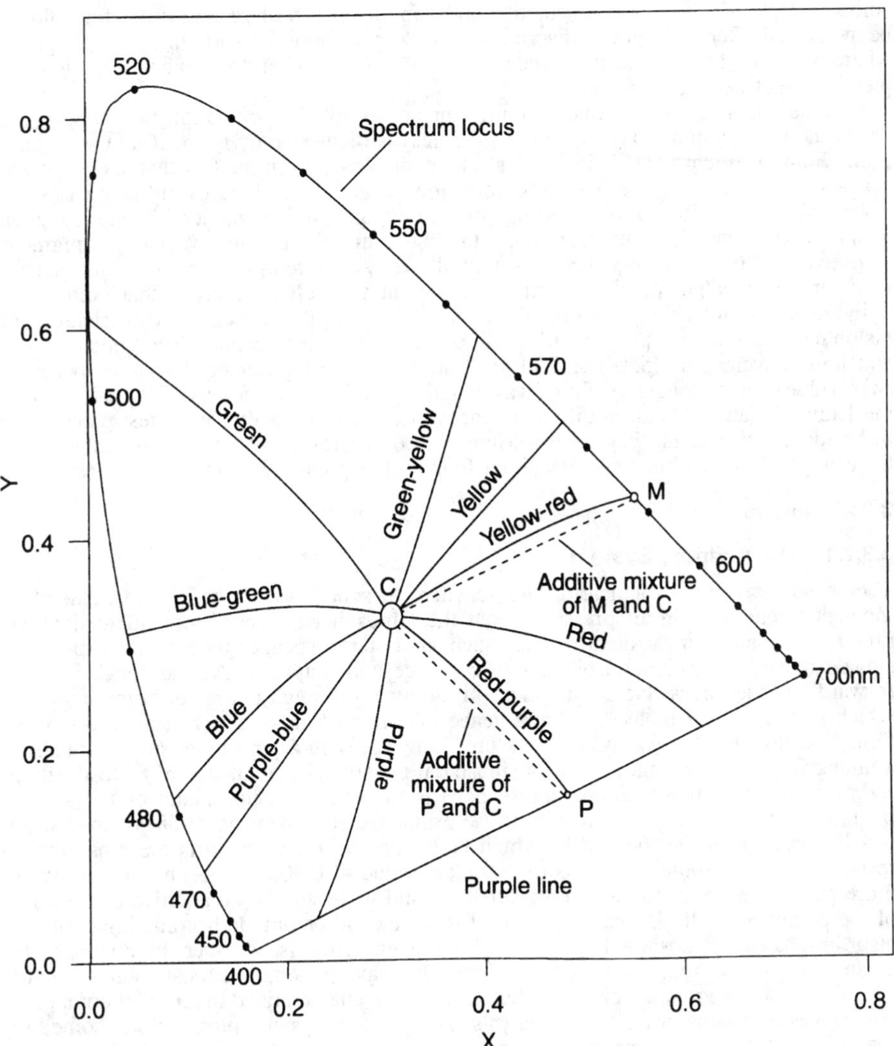

Figure 3.10 The CIE color space. The abscissa and ordinate indicate the proportions of X and Y primary colors, respectively.

upper left of the figure. Saturation decreases as proximity to the rim decreases, to an achromatic point labeled C in Figure 3.10. The location of mixtures of primaries in the CIE space can either be specified precisely by using the tristimulus values for each component spectral frequency or approximately by determining the coordinates at the location approximating its appearance.

Another widely used color specification system is the *Munsell Book of Colors*. This classification scheme is also a variant of the color circle, but adding in a third dimension that corresponds to lightness. In the Munsell notation, the word hue is used as normal, but the words value and chroma are used to refer to lightness and saturation, respectively. The book contains sheets of color samples organized according to their values on the three dimensions of hue, value, and chroma. Color can be specified by reporting the values of the sample that most closely match those of the stimulus of interest.

When using a colored stimulus, one important consideration is its location in the visual field (Hurvich, 1981). The distribution of cones varies across the retina, resulting in variations in color perception at different retinal locations. For example, because short wavelength cones are absent in the fovea and only sparsely distributed throughout the periphery, very small blue stimuli imaged in the fovea will be seen as achromatic and the blue component in mixtures will have little impact on the perceived hue. Cones of all three types decrease in density with increasing eccentricity, with the consequence that color perception becomes less sensitive and stimuli must be larger in order for color to be perceived. Red and green discrimination extends only 20°–30° into the periphery, whereas yellow and blue can be seen up to 40°–60° peripherally. Color vision is completely absent beyond that point.

Another consideration is that a significant portion of the population has *color blindness*. The most common type of color blindness is dichromatic vision. It is a sex-linked trait, with most dichromats being males. The name arises from the fact that such a person can match any spectral hue with a combination of only two primaries; in most cases this disorder can be attributed to a missing cone photopigment. The names tritanopia, deuteranopia, and protanopia refer to missing the short, middle, or long wavelength pigment, respectively. The latter two types (commonly known as red-green color blindness) are much more prevalent than the former. The point to keep in mind is that color blind individuals are not able to distinguish all of the colors that a person with trichromatic vision can. Specifically, people with red-green color blindness cannot differentiate middle and long wavelengths (520 nm–700 nm), and the resulting perception is composed of short (blue) versus longer (yellow) wavelength hues. Tests for color deficiencies include the Ishihara plates, which require differences in color to be perceived if test patterns are to be identified, and the Farnsworth-Munsell 100-hue test, in which colored caps are to be arranged in a continuous series about four anchor-point colors (Wandell, 1995).

3.3.2 Audition

3.3.2.1 The Auditory System

The sensory receptors for hearing are sensitive to sound waves, which are moment-to-moment fluctuations in air pressure about the atmospheric level. These fluctuations are produced by mechanical disturbances, such as a stereo speaker moving in response to signals that it is receiving from a music source and amplifier. As the speaker moves forward and then back, the disturbances in the air go through phases of compression, in which the density of molecules—and hence the air pressure—is increased, and rarefaction, in which the density and air pressure decrease. With a pure tone, such as made by a tuning fork, these changes follow a sinusoidal pattern. The frequency of the oscillations (i.e., the number of oscillations per second) is the primary determinant of the sound's pitch, and the amplitude or intensity is the primary determinant of loudness. Intensity is usually specified in *decibels* (dB), which is 20 log (p/p_0), where p is the pressure corresponding to the sound and p_0 is the standard value of 0.0002 dyne/cm^2. When two or more pure tones are combined, the resultant sound wave will be an additive combination of the components. In that case, not only frequency and amplitude become important but also the phase relationships between the components, that is, whether the phases of the cycles for each are matched or mismatched. The wave patterns for most sounds encountered in the world are quite complex, but they can be characterized in terms of component sine waves by means of a Fourier analysis. The auditory system must perform something like a Fourier analysis, since we are capable to a large extent of extracting the component frequencies that make up a complex sound signal, so that the pitches of the component tones are heard.

The Ear

A sound wave propagates outward from its source at the speed of sound (344 m/s), with the amplitude proportional to $1/(\text{distance})^2$. It is the cyclical air pressure changes at the ear as the sound wave propagates past the observer that starts the sensory process. The outer ear (see Figure 3.11), consisting of the pinna and the auditory canal, serves to funnel the sound into the middle ear; the pinna will amplify or attenuate some sounds as a function of the direction from which they come and their frequency, and the auditory canal amplifies sounds around 3 kHz because that is its resonant frequency. A flexible membrane, called the *eardrum* or tympanic membrane, separates the outer and middle ears. The pressure in the middle ear is maintained at the atmospheric level, by means of the Eustachian tube which opens into the throat, so any deviations from this pressure in the outer ear will result in a pressure differential that causes the eardrum to move. Consequently, the eardrum vibrates in a manner that mimics the sound wave that is affecting it. However, changes in altitude, such as those occurring during flight, can produce a pressure differential that impairs hearing until that differential is eliminated, which cannot occur readily if the Eustachian tube is blocked by infection or other causes.

Because the inner ear contains fluid, there is an impedance mismatch between it and the air that would greatly reduce the fluid movement if the eardrum acted directly on it. This impedance mismatch is overcome by a lever system of three bones (the ossicles) in the middle ear—the *malleus, incus,* and *stapes.* The malleus is attached to the eardrum and is connected to the stapes by the incus. The stapes has a footplate that is attached to a much smaller membrane, the oval window, that is at the boundary of the middle ear and the cochlea, the part of the inner ear that is important for hearing. Thus, when the eardrum moves in response to sound, the ossicles move, and the stapes produces movement of the oval window. Muscles attached to the ossicles tighten when sounds exceed 80 dB, thus protecting the inner ear to some extent from loud sounds by lessening their impact. However, because this acoustic reflex takes between 10 ms and 150 ms to occur, depending on the intensity of the sound, it does not provide protection from percussive sounds such as gunshots.

The *cochlea* is a fluid-filled, coiled structure (see Figure 3.12). It consists of three chambers, the vestibular and tympanic canals, and the cochlear duct that separates them at all but a small hole at the apex called the helicotrema. Part of the wall separating the cochlear duct from the tympanic canal is a flexible membrane called the *basilar membrane*. This membrane is narrower and stiffer nearer the oval window than it is nearer the helicotrema. The organ of Corti, the receptor organ that transduces the pressure changes to neural impulses sits on the basilar membrane in the cochlear duct. It contains

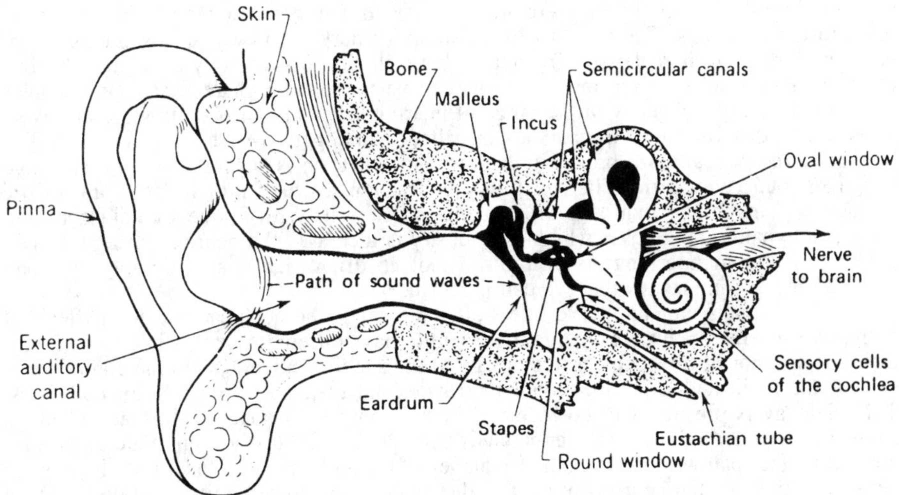

Figure 3.11 Structure of the human ear. [Reprinted by permission of the publisher from H. R. Schiffman, *Sensation and perception: An integrated approach* (Fourth Edition), p. 329. Copyright 1996 © by John Wiley and Sons, Inc.]

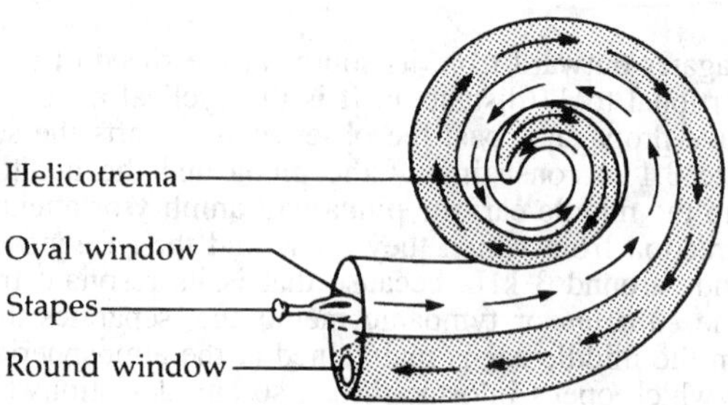

Figure 3.12 Schematic diagram of the cochlea. The direction of wave motion in the fluid is indicated by the arrows. [Reprinted by permission of the publisher from M. W. Matlin, *Sensation and perception* (Second Edition), p. 273. Copyright © 1988 by Allyn and Bacon.]

two groups of hair cells, whose cilia project into the fluid in the cochlear duct and either touch or approach the tectorial membrane, which is inflexible. When fluid motion occurs in the inner ear, the basilar membrane vibrates, causing the cilia of the hair cells to be bent. It is this bending of the hair cells that initiates a neural signal. One group of hair cells, the inner cells, consists of a single row of approximately 3500; the other group, the outer cells, is composed of approximately 12,000 hair cells arranged in three to five rows. Permanent hearing loss most often is due to hair cell damage that results from excessive exposure to loud sounds or to certain drugs.

Sound causes a wave to move from the base of the basilar membrane, at the end near the oval window, to its apex. Because the width and thickness of the basilar membrane vary along its length, the magnitude of the displacement produced by this traveling wave at different locations will vary. For low frequency sounds, the greatest movement is produced near the apex; as the frequency increases, the point of maximal displacement shifts toward the base. Thus, not only does the frequency with which the basilar membrane vibrates vary with the frequency of the auditory stimulus, but so does the location.

The Auditory Pathways

The auditory pathways after sensory transduction show many of the same properties as the visual pathways. The hair cells have synapses with the neurons that make up the auditory nerve. The ratio of hair cells to auditory nerve fibers is much greater for the inner hair cells than for the outer hair cells, suggesting that the inner hair cells provide the detailed information for hearing. In addition to afferent neurons that send information from the receptors to the brain, there are also efferent neurons that send signals from the brain to the hair cells. The neurons in the auditory nerve show frequency tuning. Each has a preferred or characteristic frequency, but will fire less strongly to a range of frequencies about the preferred one. Neurons can be found with characteristic frequencies for virtually every frequency in the range of hearing. The contour depicting sensitivity of a neuron to different tone frequencies is called a *tuning curve* (see Figure 3.13). The tuning curves typically are broad, indicating that a neuron is sensitive to a broad range of values, but asymmetric: The sensitivity to frequencies higher than the characteristic frequency is much less than that to frequencies below it. With frequency held constant, there is a dynamic range over which as intensity is increased the neuron's firing rate will increase. This dynamic range is on the order of 25 dB, which is considerably less than the full range of intensities that we can perceive.

The first synapse for the nerve fibers after the ear is the cochlear nucleus. After that point, two separate pathways emerge that seem to have different roles, as in vision. Fibers from the anterior of the cochlear nucleus go to the superior olive, half to the contralateral side of the brain and half to the ipsilateral side, and then on to the inferior colliculus. This pathway is presumed to be involved in the analysis of spatial information. Fibers from the posterior of the cochlear nucleus project directly to the contralateral inferior colliculus. This pathway analyzes the frequency of the auditory stimulus. From the inferior colliculus, most of the neurons project to the medial geniculate and then to the auditory cortex. Frequency tuning is evident for neurons in all of these regions, with some neurons responding to relatively complex features of stimulation. The auditory cortex has a ton-

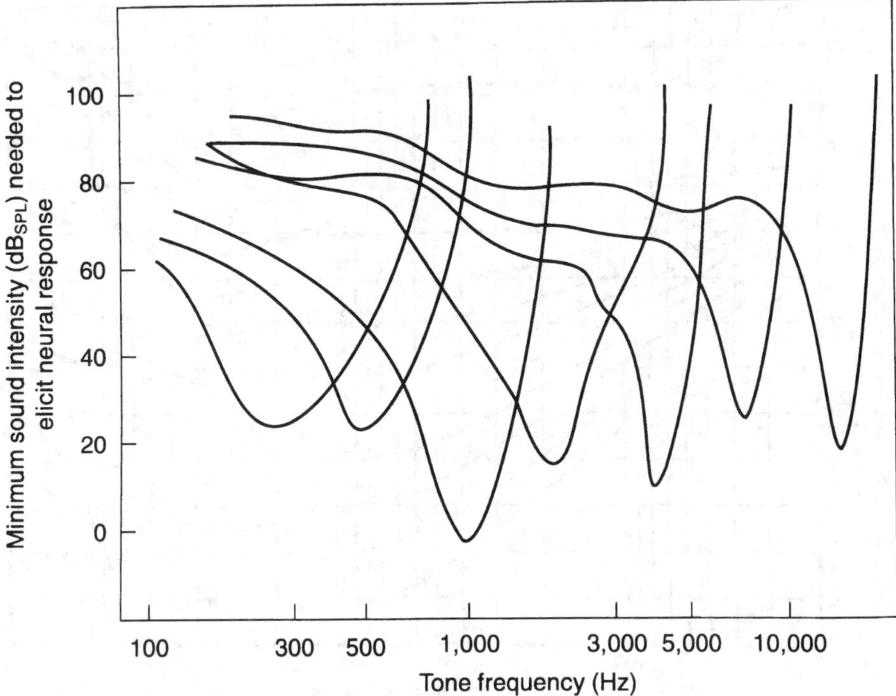

Figure 3.13 Frequency tuning curves for neurons in the auditory nerve. Each of the curves depicted has a different characteristic frequency. (Reprinted by permission of the publisher from R. W. Proctor and T. Van Zandt, *Human factors in simple and complex systems*, p. 102. Copyright © 1994 by Allyn and Bacon.)

otopic organization, in which cells responsive to similar frequencies are located in close proximity, and contains neurons tuned to extract complex information.

3.3.2.2 Basic Auditory Perception

Loudness and Detection of Sounds

Loudness for audition is the equivalent of brightness for vision. More intense auditory stimuli produce greater amplitude of movement in the eardrum, which produces higher amplitude movement of the stapes on the oval window, which leads to bigger waves in the fluid of the inner ear and, hence, higher amplitude movements of the basilar membrane. Thus, loudness is primarily a function of the physical intensity of the stimulus and its effects on the ear, although—as with brightness—it is affected by many other factors. The relation between judgments of loudness and intensity follows the power function

$$L = aI^{0.6}$$

where L is loudness, a is a constant, and I is physical intensity.

Just as brightness is affected by the spectral properties of light, loudness is affected by the spectral properties of sound. Figure 3.14 shows *equal loudness contours* for which a 1000 Hz tone was set at a particular intensity level and tones of other frequencies were adjusted to match its loudness. The contours illustrate that humans are relatively insensitive to low frequency tones below approximately 200 Hz and, to a lesser extent, to high frequency tones exceeding approximately 6000 Hz. The curves tend to flatten at high intensity levels, particularly in the low frequency end, indicating that the insensitivity to low frequency tones is a factor primarily at low intensity levels. This is why most audio amplifiers include a "loudness" switch for artificially enhancing the low frequency sounds

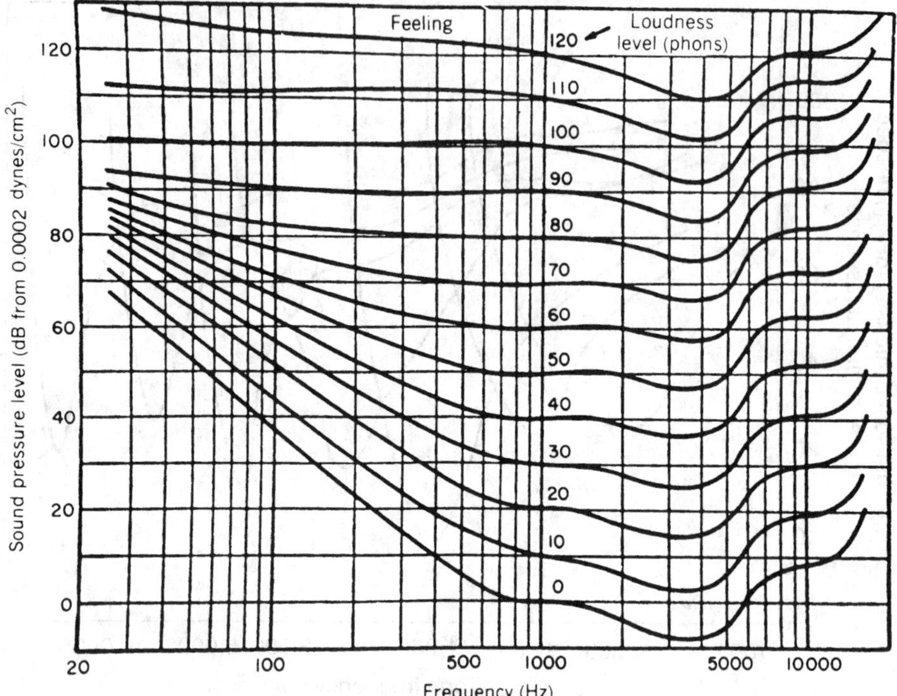

Figure 3.14 Equal loudness contours. Each contour represents the sound pressure level at which a tone of a given frequency sounds as loud as a 1000-Hz tone of a particular intensity. [Reprinted by permission of the publisher from H. R. Schiffman, *Sensation and perception: An integrated approach* (Fourth Edition), p. 353. Copyright 1990 © by John Wiley and Sons, Inc.]

when music is played at low intensities. The curves also show the maximal sensitivity to be in the range of 3000–4000 Hz, which is critical for speech perception. The two most widely cited sets of equal loudness contours are those of Fletcher and Munson (1933), obtained when listening through earphones, and of Robinson and Dadson (1956), obtained for free-field listening.

Temporal summation can occur over a period of approximately 200 ms, meaning that loudness is a function of the total energy presented for tones of this duration or less. The bandwidth, that is the range, of the frequencies in a complex tone is important for determining its loudness. With the intensity held constant, increases in bandwidth have no affect on loudness until a critical bandwidth is reached. Beyond the critical bandwidth, further increases in bandwidth result in increases in loudness.

Extraneous sounds in the environment can mask targeted sounds. This becomes important for situations such as work environments, in which audibility of specific auditory input must be evaluated with respect to the level of background noise. The degree of masking is dependent on the spectral composition of the target and noise stimuli. Masking only occurs from frequencies within the critical bandwidth. Of concern for human factors is that a masking noise will exert a much greater effect on sounds of higher frequency than on sounds of lower frequency. This asymmetry is presumed to arise primarily from the operation of the basilar membrane.

Pitch Perception

Pitch is the qualitative aspect of sound that is a function primarily of the frequency of a periodic auditory stimulus. The higher the frequency, the higher the pitch. The pitch of a note played on a musical instrument is determined by what is called its *fundamental frequency,* but the note also contains energy at frequencies that are multiples of the fundamental frequency. These are called *harmonics* or overtones. Observers can resolve perceptually the lower harmonics of a complex tone, but have more difficulty resolving the higher harmonics (Plomp, 1964). This is because the perceptual separation of the successive harmonics is progressively less as their frequency increases.

Pitch is also influenced by several factors in addition to frequency. A phenomenon of particular interest in human factors is that of the missing fundamental, which is that the fundamental frequency can be removed, yet the pitch of a sound remains unaltered. This suggests that pitch is based on the pattern of harmonics and not just the fundamental frequency. This phenomenon allows a person's voice to be recognizable over the telephone and music to be played over low fidelity systems without distorting the melody. The pitch of a tone also varies as a function of its loudness. Equal pitch contours can be constructed much like equal loudness contours by holding the stimulus frequency constant and varying its amplitude. Such contours show that as stimulus intensity increases, the pitch of a 3000-Hz tone remains relatively constant. However, tones whose frequencies are lower or higher than 3000 Hz show systematic decreases and increases in pitch, respectively, as intensity increases.

Two different theories were proposed in the nineteenth century to explain pitch perception. According to Ernest Rutherford's (1886) *frequency theory,* the critical factor is that the basilar membrane vibrates at the frequency of an auditory stimulus. This in turn gets transduced into neural signals at the same frequency such that the neurons in the auditory nerve respond at the frequency of the stimulus. Thus, according to this view, it is the frequency of firing that is the neural code for pitch. The primary deficiency of frequency theory is that the maximum firing rate of a neuron is restricted to about 1000 spikes/s. Thus, the firing rate of individual neurons cannot match the frequencies over much of the range of human hearing. Wever and Bray (1937) provided evidence that the range of the auditory spectrum over which frequency coding could occur can be increased by neurons that phase lock and then fire in volleys. The basic idea is that an individual neuron fires at the same phase in the cycle of the stimulus but not on every cycle. Because many neurons are responsive to the stimulus, some neurons will fire on every cycle. Thus, across the group of neurons, distinct volleys of firing will be seen that taken together match the frequency of the stimulus. Phase locking extends the range for which frequency coding can be effective up to 4000–5000 Hz. However, at frequencies beyond this range, phase locking breaks down.

According to Hermann von Helmholtz's (1877/1954) *place theory,* different places on the basilar membrane are affected by different frequencies of auditory stimulation. He based this proposal on his observation that the basilar membrane was tapered from narrow at the base of the cochlea to broad at its apex. This led him to suggest that it was composed of individual fibers, much like piano strings, that would resonate when the frequency of sound to which it was tuned occurred. The neurons that receive their input from a the location on the membrane affected by a particular frequency would fire in its presence, whereas the neurons receiving their input from other locations would not. The neural code for frequency thus would correspond to the particular neurons that were being stimulated. However, subsequent physiological evidence showed that the basilar membrane is not composed of individual fibers.

Von Békésy (1960) provided evidence that the basilar membrane operates in a manner consistent with both frequency and place theory. Basically, he demonstrated that waves travel down the basilar membrane from the base to the apex at a frequency corresponding to that of the tone. However, because the width and thickness of the basilar membrane vary along its length, the magnitude of the *traveling wave* is not constant over the entire membrane. The waves increase in magnitude up to a peak and then decrease abruptly. Most important, the location of the peak displacement varies as a function of frequency. Low frequencies have their maximal displacement at the apex; as frequency increases, the peak shifts systematically toward the oval window. Although most frequencies can be differentiated in terms of the place at which the peak of the traveling wave occurs, tones of less than 500 Hz to 1,000 Hz cannot be. Frequencies in this range produce a broad pattern of displacement, with the peak of the wave at the apex. Consequently, location coding does not seem to be possible for low frequency tones. Because of the evidence that frequency and location coding both operate but over somewhat different regions of the auditory spectrum, it is now widely accepted that frequencies less than 4000 Hz are coded in terms of frequency and those above 500 Hz in terms of place, meaning that at frequencies within this range, both mechanisms are involved.

3.3.3 The Vestibular System and the Sense of Balance

The vestibular system provides us with our sense of balance. It contributes to the perception of bodily motion and helps in maintaining an upright posture and the position of the eyes when head movements occur. The sense organs for the vestibular system are contained within a part of the inner ear called the *vestibule,* which is a hollow region of bone near the oval window (Kelly, 1991). The vestibular system includes the *otolith organs,* one called the utricle and the other the saccule, and three semicircular ducts (see Figure 3.11). The otolith organs provide information about the direction of gravity and linear acceleration. The sensory receptors are hair cells lining the organs whose cilia are embedded in a gelatin-like substance that contains otoliths, which are calcium carbonate crystals. Tilting or linear acceleration of the head in any direction causes a shearing action of the otoliths on the cilia in the utricle, and vertical linear acceleration has the same effect in the sacule. The semicircular canals are placed in three perpendicular planes. They also contain hair cells that are stimulated when relative motion between the fluid inside of them and the head is created, and thus respond primarily to angular acceleration or deceleration in specific directions.

The vestibular ganglion contains the cell bodies of the afferent fibers of the vestibular system. They project to the vestibular nuclear complex, a part of the medulla that is made up of four distinct nuclei, the lateral vestibular nucleus, the medial vestibular nucleus, the superior vestibular nucleus, and the inferior vestibular nucleus. Each of these nuclei serves a distinct role. The lateral nucleus seems to be involved in the control of posture, the medial and superior nuclei in vestibulo-ocular reflexes, and the inferior nucleus with integration of the vestibular input with inputs from the cerebellum.

Two functions of the vestibular system, one static and one dynamic, can be distinguished. The static function, performed primarily by the utricle and sacule, is to monitor the position of the head in space, which is important in the control of posture. The dynamic function, performed primarily by the semicircular ducts, is to track the rotation of the head in space. This tracking is necessary for reflexive control of what are called *vestibular eye movements.* If you maintain fixation on an object while rotating your head, the position of the eyes in the sockets will change gradually as the head moves. When your nose is pointing directly toward the object, the eyes will be centered in their sockets, but as you turn your head to the right the eyes will rotate to the left, and vice versa as the head is turned to the left. These smooth, vestibular eye movements are controlled rapidly and automatically by the brain stem in response to sensing of the head rotation by the vestibular system.

Exposure to motions that have angular and linear accelerations substantially different from those normally encountered, as occurs in aircraft, space vehicles, and ships, can produce erroneous perceptions of attitude and angular motion that result in *spatial disorientation* (Benson, 1990). Spatial disorientation accounts for approximately 35% of all general aviation fatalities, with most occurring at night when visual cues are either absent or degraded and vestibular cues must be relied on heavily. The vestibular sense also is key to producing *motion sickness,* as indicated by the fact that people who do not have a functional vestibular apparatus do not show motion sickness. The dizziness and nausia associated with motion sickness is generally assumed to arise from a mismatch between the motion cues provided by the vestibular system, and possibly vision, with the expectancies of the central nervous system. The vestibular sense also contributes to the related problem of simulator sickness that arises when the visual cues in a simulator or virtual reality environment do not correspond well with the motion cues that are affecting the vestibular system (Kennedy and Fowlkes, 1992).

3.3.4 The Somatic Sensory System

The somatic sensory system is composed of four distinct modalities (Martin and Jessell, 1991). *Touch* is the sensation elicited by mechanical stimulation of the skin; *proprioception* is the sensation elicited by mechanical displacements of the muscles and joints; *pain* is elicited by stimuli of sufficient intensity to damage tissue; and *thermal sensations* are elicited by cool and warm stimuli. The receptors for these senses are the terminals of the peripheral branch of the axons of ganglion cells located in the dorsal root of the spinal cord. The receptors for pain and temperature, called nociceptors and thermoreceptors, are bare (or free) nerve endings. Three types of nociceptors exist that respond to different types of stimulation. Mechanical nociceptors respond to strong mechanical stimulation; thermal nociceptors respond to extreme heat or cold; and polymodal nociceptors respond to several types of intense stimuli. Distinct thermoreceptors exist for cold and warm stimuli. Those for cold stimuli respond to temperatures between 1°C and 20°C below skin temperature, whereas those for warm stimuli to temperatures up to 13°C warmer than skin temperature.

The mechanoreceptors for touch have specialized endings that affect the dynamics of the receptor to stimulation. Some mechanoreceptor types are rapidly adapting and respond at the onset and offset of stimulation, whereas others are slow adapting and respond throughout the time that a touch stimulus is present. Hairy skin is innervated primarily by hair follicle receptors. Hairless (glabrous skin) receives innervation from two types, Meissner's corpuscles, which are fast adapting, and Merkel's disks, which are slow adapting. Pacinian corpuscles, which are fast adapting, and Ruffini's corpuscles, which are slow adapting, are located in the dermis, subcutaneous tissue that is below both the hairy and glabrous skin.

The nerve fibers for the skin senses have a center-surround organization of the type found for vision. The receptive fields for the Meissner corpuscles and Ruffini disks are smaller than those for the Pacinian and Ruffini corpuscles, suggesting that the former provide information about fine spatial differences and the latter about coarse spatial differences. The density of mechanoreceptors is greatest for those areas of the skin such as the fingers and lips for which two-point thresholds (i.e., the amount of difference needed to tell that two points rather than one are being stimulated) are low. Limb proprioception is mediated by three types of receptors, mechanoreceptors located in the joints, muscle spindle receptors in muscles that respond to stretch, and cutaneous mechanoreceptors. The ability to specify limb positions decreases when the contribution of any of these receptors is removed through experimental manipulation.

The afferent fibers enter the spinal cord at the dorsal roots and follow two major pathways, the dorsal-column medial-lemniscal pathway and the anterolateral (or spinothalamic) pathway. The lemniscal pathway conveys information about touch and proprioception. It receives input mainly from fibers with corpuscles and transmits this information quickly. It ascends along the dorsal part of the spinal column, on the ipsilateral side of the body. At the brainstem, most of its fibers cross over to the contralateral side of the brain and project to the medial lemniscus in the thalamus, and from there to the anterior parietal cortex. The fibers in the anterolateral pathway ascend along the contralateral side of the spinal column and project to the reticular formation, midbrain, or thalamus, and then to the anterior parietal cortex and other cortical regions. This system is primarily responsible for conveying pain and temperature information.

The somatic sensory cortex is organized in a spatiotopic manner, much as is the visual cortex. That is, it is laid out in the form of a homunculus representing the opposite side of the body, with areas of the body for which sensitivity is greater, such as the fingers and lips, having relatively larger areas devoted to them. There are four different, independent spatial maps of this type in the somatic sensory cortex, with each map receiving its inputs primarily from the receptors for one of the four somatic modalities. The modalities are arranged into columns, with any one column receiving input from the same modality. When a specific point on the skin is stimulated, the population of neurons that receive innervation from that location will be activated. Each neuron has a concentric excitatory-inhibitory, center-surround receptive field, the size of which varies as a function of the location on the skin. The receptive fields are smaller for regions of the body in which sensitivity to touch is highest. Some of the cells in the somatic cortex respond to complex features of stimulation, such as movement of an object across the skin.

Vibrotaction has proven to be an effective way for transmitting complex information through the tactile sense (Verrillo and Gescheider, 1992). When mechanical vibrations are applied to a region of skin, such as the tips of the fingers, the frequency and location of the stimulation can be varied. For frequencies of less than 40 Hz, the size of the contactor area does not influence the absolute threshold for detecting vibration. For higher frequencies, the threshold decreases with increasing size of the contactor, indicating spatial summation of the energy within the stimulated region. Except for very small contactor areas, sensitivity reaches a maximum for vibrations of 200 Hz–300 Hz. A similar pattern of less sensitivity for low frequency vibrations than for high frequency vibrations is evident in equal sensation magnitude contours (Verrillo, Fraioli, and Smith, 1969), much like the equal loudness contours for audition. With multicontactor devices, which can present complex spatial patterns of stimulation, masking stimuli presented in close temporal proximity to the target stimulus can degrade identification (e.g., Craig, 1982), as in vision and audition. However, with practice, pattern recognition capabilities with these types of devices can become quite good. As a result, they can be used successfully as reading aids for the blind and to a lesser extent as hearing aids for the hearing impaired (Summers, 1992).

A distinction is commonly made between *active* and *passive* touch (Gibson, 1966). Passive touch refers to situations in which the individual does not move her or his hand, and the touch stimulus is applied passively, as in vibrotaction. Active touch refers to situations in which the individual intentionally moves the hand to manipulate and explore an object. According to Gibson, active touch is the most common mode of acquiring tactile information in the real world and involves a unique perceptual system, which he called haptics. Pattern recognition with active touch typically is superior to that with passive touch (Appelle, 1991). However, the success of passive vibrotactile displays for the blind indicates that much information can also be conveyed passively.

3.3.5 Gustation and Olfaction

The taste of a good meal and the smell of perfume can be quite pleasurable. On the other hand, the taste of rancid potato chips or the smell of manure or of a paper mill can be quite noxious. In fact, odor and taste are quite closely related, in that the taste of a substance is highly dependent on the odor it produces. This is evidenced by the changes in taste that occur when a cold reduces olfactory sensitivity. In human factors, both sensory modalities can be used to convey warnings. For example, ethylmercaptan is added to natural gas to warn of gas leaks because humans are quite sensitive to its odor. Also, as mentioned in the section on psychophysical scaling, there is concern with environmental odors and their influence on people's moods and performance.

The sensory receptors for taste are groups of cells called *taste buds.* They line the walls of bumps on the tongue that are called *papillae,* as well as being located in the throat, the roof of the mouth, and inside the cheeks. Each taste bud is composed of several receptor cells in close arrangement. The receptor mechanism is located in projections from the top end of each cell that lie near an opening called a *taste pore.* Sensory transduction occurs when a taste solution comes in contact with the projections. The fibers from the taste receptors project to several nuclei in the brain and then to the insular cortex, located between the temporal and parietal lobes, and the limbic system.

In 1916, Henning proposed a taste tetrahedron in which all tastes were classified in terms of four primary tastes: sweet, sour, salty, and bitter. This categorization scheme has been accepted since then, although not without opposition. Schiffman and Erickson (1993) summarize research indicating that there are many sensations that fall outside of the range of these four tastes. They suggest that there is a broad range of transduction mechanisms, including ion channels and transepithelial currents, in addition to the taste-bud receptors.

For smell, molecules in the air that are inhaled affect receptor cells located in the *olfactory epitheleum,* a region of the nasal cavity. An *olfactory rod* extends from each receptor and goes to the surface of the epitheleum. Near the end of the olfactory rod is a knob from which olfactory cilia project. These cilia are thought to be the receptor elements. Different receptor types apparently have different receptor proteins that bind the odorant molecules to the receptor. The axons from the smell receptors project to the olfactory bulb, located in the front of the brain, via the olfactory nerve. From there, the fibers project to a cluster of neural structures called the olfactory brain.

Olfaction shows several functional attributes (Engen, 1991). For one, a novel odor will almost always cause apprehension and anxiety. As a consequence, odors are useful as warnings. However, odors are not very effective at waking someone from sleep, which is illustrated amply by the need for smoke detectors that emit a loud auditory signal, even though the smoke itself has a distinctive odor. There also seems to be a bias to falsely detect the presence of odors and to overestimate the strength when the odor is present. Such a bias ensures that a miss is unlikely to occur when an odor signal is really present. The sense of smell shows considerable plasticity, with associations of odors to events readily learned and habituation occurring to odors of little consequence.

3.4 HIGHER LEVEL PROPERTIES OF PERCEPTION

3.4.1 Perceptual Organization

The stimulus at the retina consists of patches of light energy that affect the photoreceptors. Yet, we do not perceive patches of light. Rather, we perceive a structured world of meaningful objects. The organizational processes that affect perception go unnoticed in everyday life, until we encounter a situation in which we initially misperceive the situation in some way. When we realize this and our perception now is more veridical, we become aware that the organizational processes can be misled.

Perceptual organization is particularly important for the design of any visual display. If a symbol on a street sign is incorrectly organized, it may well go unrecognized. Similarly, if a warning signal is grouped perceptually with other displays, then its message may be lost. The investigation of perceptual organization was initiated by a group of German psychologists called *Gestalt* psychologists earlier in this century whose mantra was, "The whole is more than the sum of the parts." The demonstrations they provided to illustrate this point were sufficiently compelling that it is now accepted by all perceptual psychologists.

The overriding principal of perceptual organization, according to the Gestalt psychologists, is the *law of prägnänz.* The basic idea of this law is that the organizational processes will produce the simplest possible organization. The first step in perceiving a figure requires that it be separated from the background. Any display that is viewed will be seen as a figure or figures against a background. The importance of *figure-ground* organization is illustrated clearly in figures with ambiguous figure-ground organizations, as in the well-known Ruben's vase (see Figure 3.15). Such figures can be organized with either the light region or the dark region seen as figure. When a region is seen as figure, the contour appears to be part of it. Also, the region seems to be in front of the background

Figure 3.15 Ruben's vase, for which two distinct figure-ground organizations are possible. [Reprinted by permission of the publisher from H. R. Schiffman, *Sensation and perception: An integrated approach* (Third Edition), p. 288. Copyright 1996 © by John Wiley and Sons, Inc.]

and takes on a distinct form. When the organization changes so that the region is now seen as the ground, its perceived relation with the other region reverses.

Clearly, when designing displays, one wants to construct them in such a way that the figure-ground organization of the observer will correspond with what is intended. Fortunately, research has indicated factors that influence figure-ground organization. Symmetric patterns tend to be seen as figure over asymmetric ones; a region that is surrounded completely by another tends to be seen as figure and the surrounding region as background; convex contours tend to be seen as figure in preference to concave contours; the smaller of two regions tends to be seen as figure and the larger as ground; and a region oriented vertically or horizontally will tend to be seen as figure relative to one that is not so oriented.

In addition to figure-ground segregation being crucial to perception, the way that the figure is organized is important as well (see Figure 3.16). The most widely recognized *grouping principles* are proximity—display elements that are located close together will tend to be grouped together; similarity—display elements that are similar in appearance, for example, shape and color, will tend to be grouped together; continuity—figures will tend to be organized along continuous contours; closure—gaps in forms will tend to be filled in perceptually; and common fate—elements with a common motion will tend to be grouped together. Differences in orientation of stimuli seem to provide a particularly distinctive basis for grouping. As illustrated in Figure 3.17, when stimuli differ in orientation, ones of like orientation are grouped and perceived separately from ones of a different orientation. This relation lies behind the customary recommendation that displays for check reading be designed so that the pointers on the dials all have the same orientation when working properly. When something is not right, the pointer on the dial will be at an orientation different from that of the rest of the pointers, and it will "jump out" at the operator.

Two additional grouping principles (see Figure 3.18) have been described by Rock and Palmer (1990). The principle of connectedness is that lines drawn between some elements but not others will cause the connected elements to be grouped perceptually.

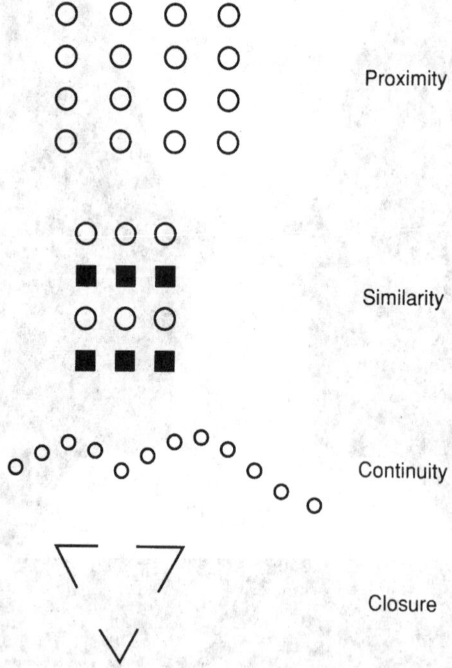

Figure 3.16 The Gestalt organizational principles. (Reprinted by permission of the publisher from R. W. Proctor and T. Van Zandt, *Human factors in simple and complex systems*, p. 134. Copyright © 1994 by Allyn and Bacon.

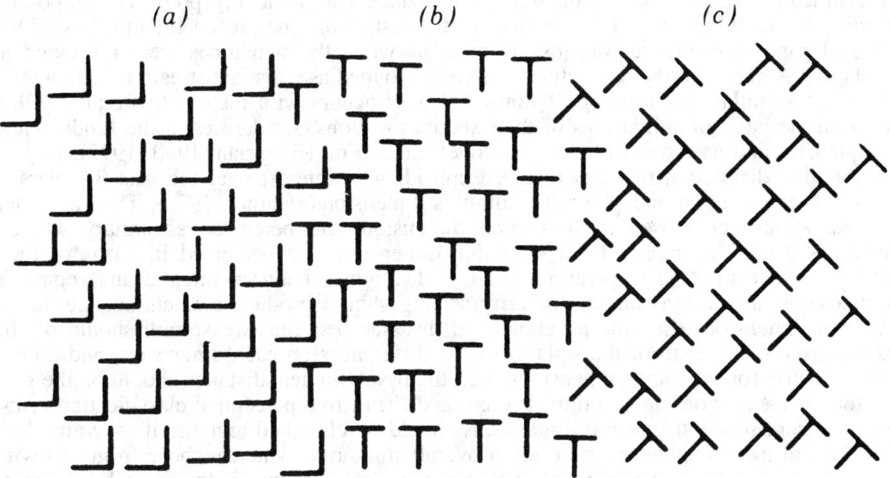

Figure 3.17 Tilted Ts group more distinctly from upright Ts than do backward L characters.

The principle of common region is that a contour drawn around display elements will cause those elements to be grouped together. Palmer (1992) has demonstrated several important properties of grouping by common region. When multiple, conflicting regions are present, the smaller enclosing region seems to dominate the organization; for nested, consistent regions, the organization appears to be hierarchical. Grouping by common region breaks down when the elements and background region are at different perceived depths, as does grouping by proximity (Rock and Brosgole, 1964), suggesting that such grouping occurs relatively late in processing, after at least some depth perception has occurred.

Although most work on perceptual organization has been conducted with visual stimuli, there are numerous demonstrations that the principles apply as well to auditory stimuli (Julesz and Hirsh, 1972). Grouping by similarity is illustrated in a study by Bregman and Rudnicky (1975) in which listeners had to indicate which of two tones of different frequency occurred first in a sequence. When the two tones were presented in isolation, performance was good. However, when preceded and followed by a single occurrence of

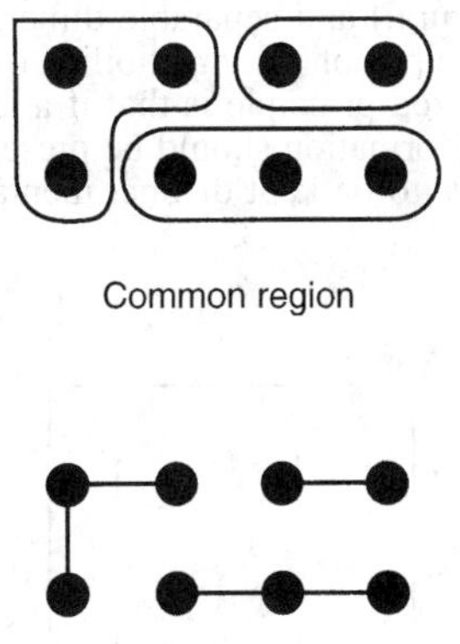

Figure 3.18 Grouping by connectedness and by common region. (Reprinted by permission of the publisher from R. W. Proctor and T. Van Zandt, *Human factors in simple and complex systems*, p. 137. Copyright © 1994 by Allyn and Bacon.)

a distractor tone of lower frequency, performance was relatively poor. The important finding is that when several occurrences of the distractor tone preceded and followed the critical pair, performance was just as good as when the two tones were presented in isolation. Apparently, the distractor tones were grouped as a distinct stream based on their frequency similarity. Grouping of tones not only occurs with respect to frequency, but also on the basis of similarities of their spatial positions, similarities in the fundamental frequencies and harmonics of complex tones, and so on (Bregman, 1990, 1993).

Another distinction that has received considerable interest over the past 30 years is that between *integral* and *separable* stimulus dimensions (Garner, 1974). The basic idea is that stimuli composed from integral dimensions are perceived as unitary wholes, whereas stimuli composed from separable dimensions are perceived in terms of their distinct dimensions. The operations used to distinguish between integral and separable dimensions are that (a) direct similarity scaling should produce a Euclidian metric for integral dimensions (i.e., the psychological distance between two stimuli should be the square root of the sum of the squares of the differences on each dimension) and a city-block metric for separable dimensions (i.e., the psychological distance should be the sum of the differences on the two dimensions); and (b) in free perceptual classification tasks, stimuli from sets with integral dimensions should be classified together if they are close in terms of the Euclidian metric (i.e., in overall similarity), whereas those from sets with separable dimensions should be classified in the same category if they match on one of the dimensions (i.e., the classifications should be in terms of dimensional structure; Garner, 1974). Perhaps most important for human factors, speed of classification with respect to one dimension is unaffected by its relation to the other dimension if the dimensions are separable but shows strong dependencies if they are integral. For integral dimensions, classifications are slowed when the value of the irrelevant dimension is uncorrelated with the value of the relevant dimension but speeded when the two dimensions are correlated.

Based on these criteria, dimensions such as hue, saturation, and lightness, in any combination, or pitch and loudness are classifed as integral, and size and lightness or size and angle as separable (e.g., Grau and Kemler Nelson, 1988; Shepard, 1991). A third classification, called configural dimensions (Pomerantz, 1981), has been proposed for dimensions that maintain their separate codes but have a new relational feature that emerges from their specific configuration. For example, as illustrated in Figure 3.19, a diagonally oriented line can be combined with the context of two other lines to yield an emergent triangle. Configural dimensions behave much like integral dimensions in speeded classification tasks, although the individual dimensions are still relatively accessible. Recently, there has been controversy over whether the sets of dimensions described above as integral (i.e., hues, saturation, and brightness; loudness and pitch) are truly integral in the sense of being processed as unitary wholes, or whether the dimensions are still psychologically distinct (see Melara, Marks, and Potts, 1993a, 1993b; Kemler Nelson, 1993). If the latter turns out to be the case, there may not be a fundamental difference between the integral and configural dimensions.

Wickens and his colleagues have tried to extend the distinction between interactive dimensions (integral and configural) and separable dimensions to display design by proposing what they call the principle of "compatibility of proximity" (e.g., Barnett and Wickens, 1988). This hypothesized principle is that if a task requires that information be integrated mentally, then that information should be presented in an integral or integrated display; if the information needs to be kept distinct mentally, then it should be presented

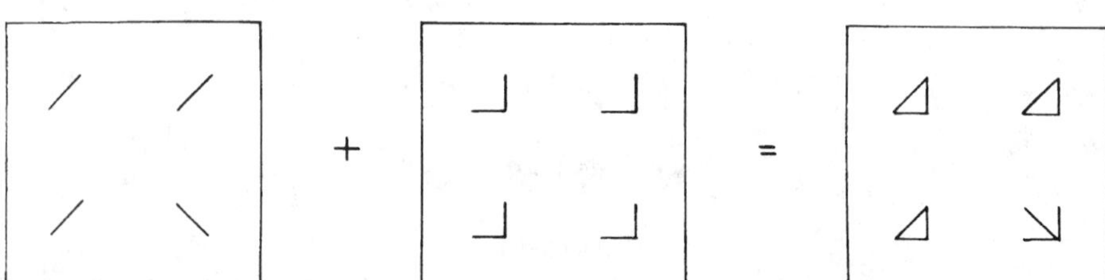

Figure 3.19 Configural dimensions. The top context helps in discriminating the line whose slope is different from the rest.

in separable dimensions. Results of studies evaluating this principle have been mixed, with one of the principal violations being that displays intended to be separable often enable better integrated task performance than integral displays (Sanderson, Haskell, and Flach, 1992). This may often be because the separated displays have unintended configural properties. This underscores that a major limitation in applying the perceptual organizational principles to display design is that the factors that determine which organization will predominate in a complex arrangement are not well articulated.

3.4.2 Spatial Orientation

We live in a three-dimensional world and, hence, must be able to perceive locations in space relatively accurately if we are to survive. Many sources of information come into play in the perception of distance and spatial relations, and the consensus view is that the perceptual system constructs the three-dimensional representation using this information as cues.

3.4.2.1 Visual Depth Perception

Vision is a strongly spatial sense and provides us with the most accurate information regarding spatial location. In fact, when visual cues regarding location conflict with those from the other senses, the visual sense typically wins out—a phenomenon called *visual dominance*. There are several areas of human factors in which we need to be concerned about visual depth cues. For example, accurate depth cues are crucial for situations in which navigation in the environment is required; misleading depth cues at a landing strip at an airfield may cause a pilot to land short of the runway. For another, a helmet-mounted display, viewed through a monocle, will eliminate binocular cues and possibly provide information that conflicts with that seen by the other eye. As a final example, it may be desired that a simulator depict three-dimensional relations relatively accurately on a two-dimensional display screen.

One distinction that can be made is between *oculomotor cues* and *visual cues*. The oculomotor cues are accommodation and vergence angle, both of which we discussed earlier in the chapter. At relatively close distances, vergence and accommodation will vary systematically as a function of the distance of the fixated object from the observer. Therefore, either the signal sent from the brain to control accommodation and vergence angle or feedback from the muscles could provide cues to depth. Evidence suggests that distance judgments based on vergence angle alone can be quite accurate at ranges of up to 6 to 9 m, but judgments based only on accommodation are inaccurate (e.g., Morrison and Whiteside, 1984; Sekuler and Blake, 1994). Bourdy, Cottin, and Monot (1991) suggest that errors in distance appreciation for automobile drivers at night may in part be linked to inaccurate vergence settings. They found that individuals who underconverge in darkness overestimate distance, whereas individuals who overconverge underestimate distance.

Visual cues can be partitioned into binocular and monocular cues. The binocular cue is *retinal disparity,* which arises from the fact that the two eyes view an object from different locations. An object that is fixated falls on corresponding points of the retinas. This object can be regarded as being located on an imaginary curved plane, called the *horopter,* and any other object that is located on this plane will also fall on corresponding points. For objects that are not on the horopter, the images will fall on disparate locations of the retinas. The direction of disparity, uncrossed or crossed (i.e., whether the image from the right eye is located to the right or left of the image from the left eye), is a function of whether the object is in back of or in front of the horopter, respectively, and the magnitude of disparity is a function of how far the object is from the horopter. Thus, retinal disparity provides information with regard to the locations of objects in space with respect to the surface that is being fixated.

Retinal disparity is a strong cue to depth, as witnessed by the effectiveness of 3-D movies and stereoscopic static pictures, which are created by presenting slightly different images to the two eyes to create disparity cues. Anyone who has seen the 3-D Muppet Movie at MGM Studios in Florida or the popular autostereogram posters realizes how compelling these effects can be. In addition to enhancing the perception of depth relations in displays of naturalistic scenes, stereoptic displays may be of value in assisting scientists and others in evaluating multidimensional data sets. Wickens, Merwin, and Lin (1994) found that a three-dimensional data set could be processed faster and more accurately to answer questions that required integration of the information if the display was stereoptic than if it was not.

The fundamental problem for theories of stereopsis is that of matching. Disparity can be computed only after corresponding features at the two eyes have been identified. When viewing the natural world, each eye receives the information necessary to perceive contours and identify objects, and stereopsis could occur after monocular form recognition. However, one of the more striking findings of the past 30 years is that there do not have to be contours present in the images seen by the individual eyes in order to perceive objects in 3-D. This phenomenon was discovered by Julesz (1971), who used random dot stereograms in which a region of dot densities is shifted slightly in one image relative to the other. Although a form cannot be seen if only one of the two images is viewed, when each of the two images are presented to the respective eyes, a three-dimensional form emerges. Random dot stereograms have been popularized recently through the Magic Eye series of figures that utilize the autostereogram variation of this technique in which the disparity information is incorporated in a single, two-dimensional display. That stereopsis can occur with random dot stereograms suggests that the matching of the two images can be based on dot densities.

Not all features in the two images of a stereogram can be matched; unpairable features arise at half occlusions (where the object in the image to one eye covers a background region that is visible in the image to the other eye). These half occlusions have tended to be treated as noise by most models of stereopsis, with various processing strategies proposed for how this noise is eliminated from the process of computing disparity. The assumption has been that these half occlusions are interpreted only after the computation of disparity is completed. However, Anderson and Nakayama (1994) recently presented evidence that the half occlusions are detected early in processing and contribute to the construction of the three-dimensional representation.

There are many static, or pictorial, monocular cues to depth. These cues are such that individuals with only one eye or who lack the capability of detecting disparity differences are still able to interact with the world with relatively little loss in accuracy. The monocular cues include retinal size—larger images appear to be closer—and familiar size—for example, a small image of a car provides a cue that the car is far away. The cue of interposition is that an object that appears to block part of the image of another object that is located in front of it. Although interposition provides information that one object is nearer than another, it does not provide information about how far apart they are. Another cue comes from shading. Because light sources typically project from above, as with the sun, the location of a shadow provides a cue to depth relations. A darker shading at the bottom of a region implies that the region is elevated, whereas one at the top of a region provides a cue that it is depressed. Aerial perspective refers to blue coloration that appears for objects that are far away, such as is seen when viewing a mountain at a distance. Finally, linear perspective, is that parallel lines receding into the distance, such as train tracks, converge to a point in the image.

Gibson (1950) emphasized the importance of texture gradient, which is a combination of linear perspective and relative size, in depth perception. If one looks at a textured surface, such as a brick walkway, the parts of the surface (i.e., the bricks) become smaller and more densely packed in the image as they recede into the distance. The rate of this change is a function of the orientation of the surface in depth with respect to the line of sight. This texture change specifies distance on the surface, and an image of a constant size will be perceived to come from a larger object that is farther away if it occludes a larger part of the texture. Certain color gradients, such as a gradual change from red to gray, provide effective cues to depth as well (Truscianko, Montagnon, and le Clerc, 1991).

There are plenty of cues to depth for a stationary observer. However, they become even richer once the observer is allowed to move. When you maintain fixation on an object and change locations, as when looking out a train window, objects in the background will move in the same direction in the image as you are moving, whereas objects in the foreground will move in opposite directions. This cue is called *motion parallax.* When you move straight ahead, the optical flow pattern conveys information about how fast your position is changing with respect to objects in the environment. There are also numerous ways in which displays with motion can generate depth perception (Braunstein, 1976).

Of particular concern for human factors is how the various depth cues are integrated. Bruno and Cutting (1988) varied the presence or absence of four cues, relative size, height in the projection plane, interposition, and motion parallax. They found that the four cues combined additively in one direct and two indirect scaling tasks. That is, each cue supported depth perception, and the more cues present, the more depth was revealed. Bruno and Cutting interpreted these results as suggesting that a separate "minimodule" processes each source of depth information. In agreement with this view, when depth cues conflict, depth perception is degraded (Rogers and Collett, 1989). Thus, depth perception will be impaired when depth cues are minimal or when they conflict.

Because the size of the retinal image of an object varies as a function of the distance of the object from the observer, perception of size is intimately related to perception of distance. When accurate depth cues are present, good *size constancy* is shown. That is, the perceived size of the object does not vary as a function of the changes in retinal image size that accompany changes in depth. Similarly, rotation of an object in depth does not alter perception of the shape of the object in most cases, a phenomenon called *shape constancy.* The size-distance and shape-slant invariance hypotheses state that size and shape constancy result from being able to scale the retinal image size in relation to changes in depth. One implication of this view is that size and shape constancy will break down and illusions appear when depth cues are erroneous. There are numerous illusions of size, as for example, the Ponzo illusion (see Figure 3.20), in which one of two stimuli of equal physical size appears larger than another, that are due at least in part to misleading depth cues. Misperceptions of size and distance can also arise when depth cues are minimal, as when flying at night.

3.4.2.2 Sound Localization

The cues for sound localization involve disparities at the two ears, much as disparities of the images at the two eyes are cues to depth. Two different sources of information, interaural intensity and time differences, have been identified. Both of these cues vary systematically with respect to the position of the sound relative to the listener. At the

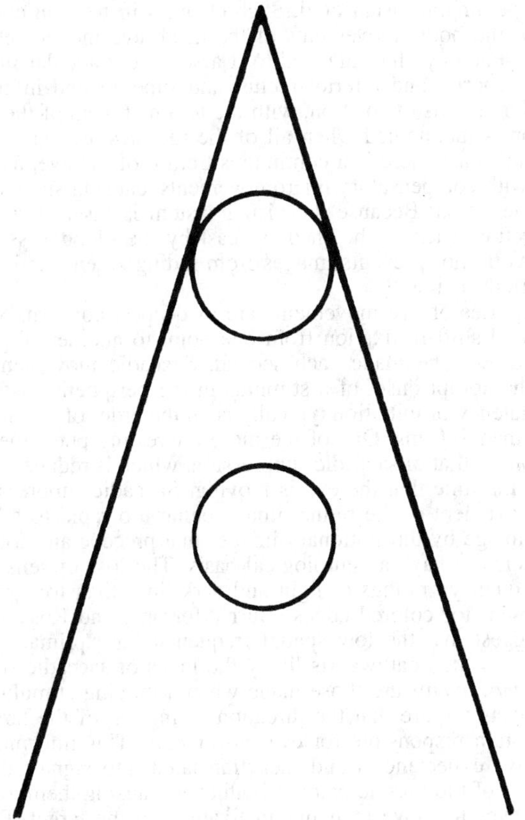

Figure 3.20 The Ponzo illusion. The top circle appears larger than the lower circle due to the linear perspective cue.

front and back of the listener, the intensity of the sound and the time at which it reaches the ears will be equal. As the position of the sound along the azimuth, that is, relative to the listener's head, is moved progressively toward one side or the other, the sound will become increasingly louder at the ear closest to it relative to the ear on the opposite side, and it also will reach the ipsilateral ear first. The interaural intensity differences are due primarily to a sound shadow created by the head. Because the head produces no shadow for frequencies less than 1000 Hz, the intensity cue is most effective for relatively high frequency tones. In contrast, interaural time differences are most effective for low frequency sounds. Localization accuracy is poorest for tones between 2000 and 4000 Hz, and this is thought to be because neither the intensity nor time cue is very effective in this intermediate frequency range (Stevens and Newman, 1934).

Both the interaural intensity and time difference cues are ambiguous because the same values can be produced by stimuli in more than one location. As an example, both differences are zero for locations directly in front of or behind the listener. Because both cues are ambiguous at the front and back, front–back confusions of the location of brief sounds will often occur. Confusions are relatively rare in the natural world because head movements and reflections of sound make the cues less ambiguous than they are in the typical localization experiment (e.g., Guski, 1990; Makous and Middlebrooks, 1990). As with vision, misleading cues can cause erroneous localization of sounds. Caelli and Porter (1980) illustrated this point by having listeners in a car judge the direction from which a siren occurred. Localization accuracy was particularly poor when all but one window were rolled up, which would alter the normal relation between direction and the cues.

3.4.3 Eye Movements and Motion Perception

Because details can be perceived well only at the fovea, the location on which the fovea is fixated must be able to be changed regularly and rapidly if we are to maintain an accurate perceptual representation of the environment and to see the details of new stimuli that appear in the peripheral visual field. Such changes in fixation can be brought about by displacement of the body, movements of the head, eye movements, or some combination of the three. Each eye has attached to it a set of extraocular muscle pairs: medial and lateral rectus, superior and inferior rectus, and superior and inferior obliques. Each pair controls a different axis of rotation, with the two members of the pairs acting antagonistically. Fixation is maintained when all of the muscles are active to similar extents. However, even in this case there is a continuous tremor of the eye, as well as slow drifts that are corrected with compensatory micromovements, causing small changes in position of the image on the retina. Because the visual system is insensitive to images that are stabilized on the retina, such as the shadows cast by the blood vessels that support the retinal neurons, this tremor prevents images from fading when fixation is maintained on an object for a period of time.

Two broad categories of eye movements are of deepest concern. *Saccadic eye movements* involve a rapid shift in fixation from one point to another. Typically, four or five saccadic movements will be made each second. Saccadic movements can be initiated automatically by the abrupt onset of a stimulus in the peripheral visual field or by conscious intent. The latency of initiation typically is on the order of 200 ms, and the duration of movement less than 100 ms. One of the more interesting phenomena associated with these eye movements is that of saccadic suppression, which is reduced sensitivity to visual stimulation during the time that the eye is moving. Saccadic suppression does not seem to be due to the movement of the retinal image being too rapid to allow perception nor to masking of the image by the stationary images that precede and follow the eye movement. Rather, it seems to have a neurological basis. The loss of sensitivity is much less for high spatial frequency gratings of light and dark lines than for low spatial frequency gratings, and is absent for colored edges (Burr, Morrone, and Ross, 1994). Because lesioning studies suggest that the low spatial frequencies are primarily conveyed by the magnocellular pathway, this pathway is likely the locus of saccadic suppression.

Smooth pursuit movements are those made when a moving stimulus is tracked by the eyes. Such movements require that the direction of motion of the target be decoded by the system in the brain responsible for eye movements. This information must be integrated with cognitive expectancies and then translated into signals that are sent to the appropriate members of the muscle pairs of both eyes, causing them to relax and contract in unison and the eyes to move to maintain fixation on the target. Pursuit is relatively accurate for relatively slow moving targets, with increasingly greater error occurring as movement speed increases.

Eye movement records provide precise information about where a person is looking at any time. Such records have been used to obtain evidence about strategies for determining where successive saccades are directed when scanning a visual scene and about the extraction of information from the display (see Abernethy, 1988, for a review). Because direction of gaze can be recorded on-line by appropriate eye-tracking systems, eye-gaze computer interface controls have considerable potential applications for persons with physical disabilities and for high workload tasks (e.g., Goldberg and Schryver, 1995). It is tempting to equate direction of fixation with direction of attention, and in many cases that may be appropriate. However, there is considerable evidence that attention can be directed to different locations in space while fixation is held constant (e.g., Sanders and Houtmans, 1985), indicating that direction of fixation and direction of attention are not always one and the same.

Movements of our eyes, head, and body produce changes in position of images on the retina, as does motion of an object in the environment. How we distinguish between motion of objects in the world and our own motion has been an issue of concern for many years in perception. We have already seen that many neurons in the visual cortex are sensitive to motion across the retina. However, detecting changes in position on the retina is not sufficient for motion perception, since those changes could be brought about by our own motion, motion of an object, or some combination of the two. Typically, it has assumed that the position of the eyes is monitored by the brain, and any changes that can be attributed to eye movements are taken into account. According to inflow theory, first suggested by Sherrington (1906), it is the feedback from the muscles controlling the eyes that is monitored. According to outflow theory, first proposed by Helmholtz (1909/1962), it is the command to the eyes to move that is monitored. Evidence, such as that the scene appears to move when an observer who has been paralyzed tries to move her or his eyes (which do not actually move; Matin, et al., 1982; Stevens, et al., 1976), has tended to support the outflow theory.

Sensitivity to motion is affected by many factors. For one, motion can be detected at a slower speed if a comparison, stationary object is also visible. When a reference object is present, changes of as little as 0.03° sec can be perceived (Palmer, 1986). However, this gain in sensitivity for detecting relative motion is at the potential cost of attributing the motion to the wrong object. For example, it is common for movement of a large region that surrounds a smaller object to be attributed to the object, a phenomenon that is called *induced motion* (Mack, 1986). The possibility for misattribution of motion is a concern for any situation in which one object is moving relative to another.

Induced motion is one example of a phenomenon in which motion of an object is perceived in the absence of motion of its image on the retina. The phenomenon of apparent, or stroboscopic, motion is probably the most important of these. This phenomenon of continuous perceived motion occurs when discrete changes in position of stimulation on the retina take place at appropriate temporal and spatial separations. It appears to be attributable to two processes, a short-range process and a long-range process (Petersik, 1989). The short-range process is presumed to reflect relatively low level directionally sensitive neurons that respond to small spatial changes that occur with short interstimulus intervals. The long-range process is presumed to reflect higher-level processes and to respond to stimuli at relatively large retinal separations presented at interstimulus intervals as long as 500 ms. Apparent motion is not only responsible for the motion produced in flashing signs, but also for motion pictures and television, in which a series of discrete images is presented.

3.4.4 Pattern Recognition

The organizational principles and depth cues determine form perception, that is, what shapes and objects will be perceived. However, for the information in a display to be conveyed accurately, the objects must be recognized. If there are words, they must be read correctly; if there is a pictograph, the pictograph must be interpreted accurately. In other words, good use of the organizational principles and depth cues by a designer does not ensure that the intended message will be conveyed to the observer.

Concern with the way in which stimuli are recognized and identified is the domain of pattern recognition. Most, but not all, research on pattern recognition has been conducted with verbal stimuli. The initial step in pattern recognition is typically presumed to be feature analysis. If visual, alphanumeric characters are presented, they are assumed to be analyzed in terms of features such as a vertical line segment, a horizontal line segment, and so on. Such an assumption is generally consistent with the evidence that neurons in the primary visual cortex respond to specific features of stimulation. Moreover, confusion matrices obtained when letters are misidentified indicate that an incorrect identification is most likely to involve a letter with considerable feature overlap with the one that was actually displayed (e.g., Townsend, 1971).

Letters are composed of features, but they in turn are components of the letter patterns that form syllables and words (see Figure 3.21). The role played by letter-level information in visual word recognition has been the subject of considerable debate. Numerous findings have suggested that in at least some cases, letter-level information is not available prior to word recognition. For example, Healy and colleagues have found that when people perform a letter detection task while reading a prose passage, the target letter is missed more often when it occurs in a very high frequency word such as *the* than when it appears in lower frequency words (e.g., Healy, 1994; Proctor and Healy, 1995). Their results have shown that this "missing-letter" effect is not just due to skipping over the words while reading. To account for this phenomenon and other data, they have proposed a set of unitization hypotheses, according to which very high frequency words are identified prior to their component letters, with letter processing terminated once the word is identified (see Figure 3.21). Numerous studies have provided evidence for the need to distinguish the five levels of reading units shown in Figure 3.21 and for other claims of the unitization hypotheses, such as that the particular units identified by an individual depends on the familiarity of those units.

Although there is agreement that letter-level information is often difficult to access within a word context, there is not agreement that this relative inaccessibility is due to word recognition occurring without identification at the letter level. For example, Johnson and Pugh (1994) note that the evidence suggesting that the component letter level is

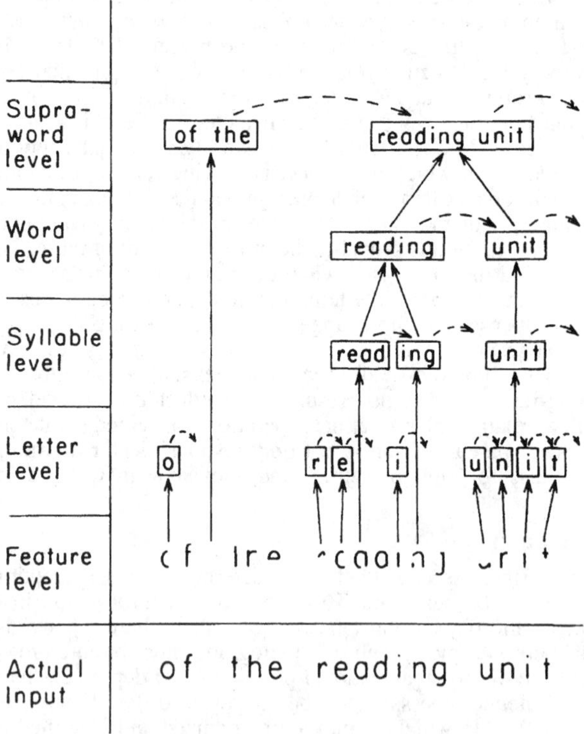

Figure 3.21 Levels of representation in reading a short passage of text. The operation of the unitization hypotheses is illustrated by the bypassing of levels that occurs for "of the." (From A. F. Healy, 1994, Letter detection: A window to unitization and other cognitive processes in reading text, *Psychonomic Bulletin & Review, 1*, 333–344. Copyright © 1994 by the Psychonomic Society, Inc.)

bypassed in word recognition may be an artifact of using tasks that require decisions and responses to be made with regard to the letters that would not be a normal part of reading. They propose a cohort model of visual word recognition in which visual access to lexical entries occurs in a bottom-up sequence that always involves the letter level. It is assumed that there is considerable variation in the rates at which individual letters are encoded, meaning that information about the component letters is made available gradually. The initially encoded feature information activates a cohort of candidate letters for each letter position, and these cohorts get reduced as additional features become available until the letter in a particular position is identified (see Figure 3.22). The initially identified letter information then activates a cohort of candidate word encodings consistent with this information, and identification of subsequent letters gradually eliminates members of the cohort until the word is identified. Cohort models similar to Johnson and Pugh's have been particularly popular as accounts for speech perception, due to the fact that there is considerable performance data indicating that an initial set of candidates is made available and then reduced (e.g., Marslen-Wilson, 1987), although there is more controversy in speech perception regarding the nature of the component segments and the process that produces them.

The primary emphasis in the models just described is on bottom-up processing from the sensory input to recognition of the pattern, but it is clear that pattern recognition is also influenced by top-down, nonvisual information of several types (Massaro and Cohen, 1994). These include orthographic constraints on the spelling patterns, regularities in the mapping between spelling and spoken sounds, syntactic constraints regarding which parts of speech are permissible, semantic constraints based on coherent meaning, and pragmatic constraints derived from the assumption that the writer is trying to communicate effectively. Interactive activation models, in which lower-level sources of information are modified by higher levels, have been popular (e.g., McClelland and Rumelhart, 1981).

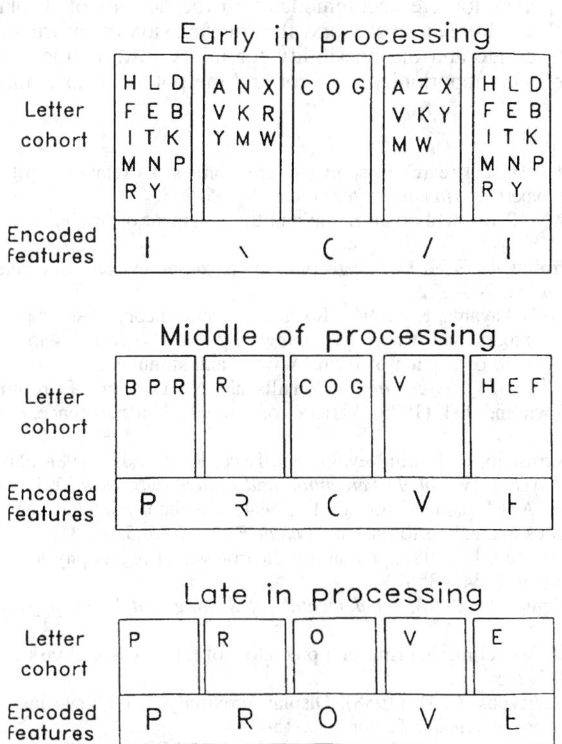

Figure 3.22 Letter cohorts activated by encoded features from early to late in processing. (Reprinted with permission from N. F. Johnson and K. R. Pugh, 1994, A cohort model of visual word recognition, *Cognitive Psychology, 26*, 240–346. Copyright © 1994 by Academic Press.)

However, Massaro and his colleagues (e.g., Massaro and Cohen, 1994) have been quite successful in accounting for a range of reading phenomena with a model, which they call the fuzzy logical model of perception, in which the multiple sources of information are assumed to be processed independently—rather than interactively—and then integrated.

Reading can be viewed as a prototypical pattern recognition task. The implications of the analysis of reading are that multiple sources of information, both bottom-up and top-down, are exploited. For accurate pattern recognition, the possible alternatives need to be physically distinct and consistent with expectancies created by the context.

3.5 SUMMARY

In this chapter, we have reviewed much of what is known about sensation and perception. Any such review must necessarily exclude certain topics and be limited in the treatment given to the topics that are covered. Excellent introductory texts that provide more thorough coverage include Coren, Ward, and Enns (1994), Goldstein (1996), Schiffman (1996), and Sekuler and Blake (1994). More advanced treatments of most areas are included in Volume 1 of *Stevens' Handbook of Experimental Psychology* (Atkinson, Herrnstein, Lindzey, and Luce, 1988) and Volume 1 of the *Handbook of Perception and Performance* (Boff, Kaufman, and Thomas, 1986). The *Engineering Data Compendium: Human Perception and Performance* (Boff and Lincoln, 1988) is an excellent resource for information pertinent to many human engineering concerns. Also, throughout the text, we have provided references to texts and review articles devoted to specific topics. These and related sources should be consulted to get an in-depth understanding of the relevant issues pertaining to any particular application involving perception.

Virtually all concerns in human factors and ergonomics involve perceptual issues to at least some extent. Whether dealing with instructions for a consumer product, control rooms for chemical processing or nuclear power plants, interfaces for computer software, guidance of vehicles, office design, and so on, information of some type must be conveyed to the user or operator. To the extent that the characteristics of the sensory systems and the principles of perception are accommodated in the designs of displays and the environments in which the human must work, the transmission of information to the human will be fast and accurate and the possibility for injury low. To the extent that they are not accommodated, the opportunity for error and the potential for damage are increased.

REFERENCES

Abernethy, B., (1988). Visual search in sport and ergonomics: Its relationship to selective attention and performer expertise. *Human Performance, 4*, 205–235.

Adelson, E. H. (1993). Perceptual organization and the judgment of brightness. *Science, 262*, 2042-2044.

Agostini, T., and Proffitt, D. R. (1993). Perceptual organization evokes simultaneous lightness contrast. *Perception, 22*, 263–272.

Anderson, B. L., and Nakayama, K. (1994). Toward a general theory of stereopsis: Binocular matching, occluding contours, and fusion. *Psychological Review, 101*, 414–445.

Appelle, S. (1991). Haptic perception of form: Activity and stimulus attributes. In M. A. Heller and W. Schiff, Eds., *The Psychology of Touch*. Hillsdale, NJ: Lawrence Erlbaum Associates.

Ashby, F. G., and Townsend, J. T. (1986). Varieties of perceptual independence. *Psychological Review, 102*, 154–179.

Atkinson, R. C., Herrnstein, R. J., Lindzey, G., and Luce, R. D., Eds., (1988). *Stevens' Handbook of Experimental Psychology (Vol. 1: Perception and Motivation)*. New York: John Wiley & Sons.

Backs, R. W., Ryan, A. M., and Wilson, G. F. (1994). Psychophysiological measures of workload during continuous manual performance. *Human Factors, 36*, 514–531.

Baird, J. C., and Berglund, B. (1989). Thesis for environmental psychophysics. *Journal of Environmental Psychology, 9*, 345–356.

Baird, J. C., and Noma, E. (1978). *Fundamentals of Scaling and Psychophysics*. New York: John Wiley.

Barber, A. V. (1990). Visual mechanisms and predictors of far field visual task performance. *Human Factors, 32*, 217–233.

Barnett, B. J., and Wickens, C. D. (1988). Display proximity in multicue information integration: The benefits of boxes. *Human Factors, 30*, 15–24.

Békésy, G. von (1960). *Experiments in Hearing*. New York: McGraw-Hill.

Benson, A. J. (1990). Sensory functions and limitations of the vestibular system. In R. Warren and A. H. Wertheim, Eds., *Perception and Control of Self-Motion*. Hillsdale, NJ: Lawrence Erlbaum Associates.

Berglund, M. B. (1991). Quality assurance in environmental psychophysics. In S. J. Bolanowski and G. A. Gescheider, Eds., *Ratio Scaling of Psychological Magnitude.* New York: John Wiley.

Boff, K. R., and Lincoln (1988). *Engineering Data Compendium: Human Perception and Performance.* Wright-Patterson AFB, Dayton, OH: Harvey G. Armstrong Medical Research Laboratory.

Boff, K. R., Kaufman, L., and Thomas, J. P., Eds. (1986). *Handbook of Perception and Performance (Vol. 1: Sensory Processes and Perception).* New York: John Wiley.

Bolanowski, S. J., and Gescheider, G. A., Eds. (1991). *Ratio Scaling of Psychological Magnitude* Hillsdale, NJ: Lawrence Erlbaum Associates.

Bourdy, C., Cottin, F., and Monot. A (1991). Errors in distance appreciation and binocular night vision. *Opthalmic and Physiological Optics, 11,* 340–349.

Braunstein, M. L. (1976). *Depth Perception through Motion.* New York: Academic Press.

Bregman, A. S. (1990). *Auditory Scene Analysis: The Perceptual Organization of Sound.* Cambridge, MA: MIT Press.

Bregman, A. S. (1993). Auditory scene analysis: Hearing in complex environments. In S. McAdams and E. Bigand, Eds., *Thinking in Sound: The Cognitive Psychology of Human Audition.* New York: Oxford University Press.

Bregman, A. S., and Rudnicky, A. I. (1975). Auditory segregation: Stream or streams? *Journal of Experimental Psychology: Human Perception and Performance, 1,* 263–267.

Bruno, N., and Cutting, J. E. (1988). Minimodularity and the perception of layout. *Journal of Experimental Psychology: General, 117,* 161–170.

Burr, D. C., Morrone, M. C., and Ross, J. (1994) Selective suppression of the magnocellular visual pathway during saccadic eye movement, *Nature, 371,* 511–513.

Caelli, T., and Porter, D. (1980). On difficulties in localizing ambulance sirens. *Human Factors, 22,* 719–724.

Coren, S., Ward, L. M., and Enns, J. T. (1994). *Sensation and Perception* (Fourth Edition). Fort Worth, TX: Harcourt Brace.

Cornsweet, T. N. (1962). The staircase method in psychophysics. *American Journal of Psychology, 75,* 485–491.

Corwin, J. (1994). On measuring discrimination and response bias: Unequal numbers of targets and distractors and two classes of distractors. *Neuropsychology, 8,* 110–117.

Craig, J. C. (1982) Vibrotactile masking: A comparison of energy and pattern maskers. *Perception & Psychophysics, 31,* 523–529.

Donders, F. C., (1868/1969). On the speed of mental processes (W. G. Koster, Trans.), *Acta Psychologica, 30,* 412–431.

Ebenholtz, S. M. (1994). Accommodative hysteresis as a function of target-dark focus separation. *Vision Research, 32,* 925–929.

Engen, T. (1991). *Odor Sensation and Memory.* New York: Praeger.

Evans, D. W., and Ginsburg, A. P. (1982). Predicting age-related differences in discriminating road signs using contrast sensitivity. *Journal of the Optical Society of America, 72,* 1785–1786.

Farah, M. J., and Ratcliff, G., Eds. (1994). *The Neuropsychology of High-Level Vision: Collected Tutorial Essays.* Hillsdale, NJ: Lawrence Erlbaum Associates.

Fechner, G. T. (1860/1966). *Elements of Psychophysics* (Vol. 1), H. E. Adler, Trans.), New York: Holt, Rinehart, and Winston.

Fletcher, H., and Munson, W. A. (1933). Loudness, its definition, measurement, and calculation. *Journal of the Acoustical Society of America, 5,* 82–108.

Garner, W. (1974). *The Processing of Information and Structure.* Hillsdale, NJ: Lawrence Erlbaum Associates.

Gescheider, G. A. (1985). *Psychophysics: Method, Theory, and Application* (Second Edition). Hillsdale, NJ: Lawrence Erlbaum Associates.

Gescheider, G. A. (1988). Psychophysical scaling. In M. R. Rosenzweig and L. W. Porter, Eds., *Annual Review of Psychology* (Vol. 39), Palo Alto, CA: Annual Reviews.

Gibson, J. J. (1950). *The Perception of the Visual World.* Boston: Houghton Mifflin.

Gibson, J. J. (1966) *The Senses Considered as Perceptual Systems.* Boston: Houghton Mifflin.

Gilchrist, A. L. (1977). Perceived lightness depends on perceived spatial arrangement. *Science, 195,* 185–187.

Goldberg, J. H., and Schryver, J. C. (1995). Eye-gaze-contingent control of the computer interface: Methodology and example for zoom detection. *Behavior Research Methods, Instruments, & Computers, 27,* 338–350.

Goldstein, E. B. (1996). *Sensation and Perception* (Fourth Edition). Belmont CA: Wadsworth.

Grau, J. W., and Kemler Nelson, D. G. (1988). The distinction between integral and separable dimensions: Evidence for the integrality of pitch and loudness. *Journal of Experimental Psychology: General, 117,* 347–370.

Green, D. M., and Swets, J. A. (1966). *Signal Detection Theory and Psychophysics.* New York: John Wiley & Sons. Reprinted, 1974, by Krieger, Huntington, NY.

Guski, R. (1990). Auditory localization: Effects of reflecting surfaces, *Perception, 19,* 819–830.

Healy, A. F. (1994). Letter detection: A window to unitization and other cognitive processes in reading text. *Psychonomic Bulletin & Review, 1*, 333–344.

Heathcote, A. (1966). RTSYS: A DOS application for the analysis of reaction time data. *Behavior Research Methods, Instruments, & Computers, 28*, 427–445.

Heilman, K. M., and Valenstein, E. (1979). Mechanisms underlying hemispatial neglect. *Annals of Neurology, 5*, 166–170.

Helmholtz, H. von (1954). On the sensation of tone as a psychological basis for the theory of music (2nd ed.) A. J. Ellis (trans. and ed.). New York: Dover. (Original work published 1877)

Helmholtz, H. von, (1962). *Treatise on Physiological Optics* (J. C. P. Southall, Ed. and Trans.), New York: Dover. (Original work published 1909)

Henning, H. (1916). Die qualitätsreibe des geschmacks. *Zeitschrift für Psychologie, 74*, 203–219.

Heuer, H., Hollendiek, G., Kroger, H., and Romer, T. (1989). The resting position of the eyes and the influence of observation distance and visual fatigue on VDT work. *Zeitschrift fur Experimentelle und Angewandte Psychologie, 36*, 538–566.

Holmes, T. H., and Rahe, R. H. (1967). The social readjustment rating scale. *Journal of Psychosomatic Research, 11*, 213–218.

Hubel, D. H., and Wiesel, T. N. (1977). Functional architecture of macaque monkey visual cortex. *Proceedings of the Royal Society of London, 198*, Series B, 1–59.

Humphrey, D. G., and Kramer, A. F. (1994). Toward a psychophysiological assessment of dynamic changes in mental workload. *Human Factors, 36*, 3–26.

Hurvich, L. M. (1981). *Color Vision*. Sunderland, MA: Sinauer Associates.

Jacobsen, A., and Gilchrist, A. L. (1988). The ratio principle holds over a million-to-one range of illumination. *Perception & Psychophysics, 43*, 1–6.

Jaschinski-Kruza, W. (1991). Eyestrain in VDU users: Viewing distance and the resting position of ocular muscles. *Human Factors, 33*, 69–83.

Johnson, N. F., and Pugh, K. R. (1994). A cohort model of visual word recognition. *Cognitive Psychology, 26*, 240–346.

Julesz, B. (1971). *Foundations of Cyclopean Perception*. Chicago: University of Chicago Press.

Julesz, B., and Hirsh, I. J. (1972). Visual and auditory perception: An essay of comparison. In E. E. David and P. B. Denes, Eds., *Human Communication: A Unified View*. New York: McGraw-Hill.

Kadlec, H., (1995): Multidimensional signal detection analyses (MSDA) for testing separability and independence: A Pascal program. *Behavior Research Methods, Instruments, & Computers, 27*, 442–458.

Kahneman, D. (1973). *Attention and Effort*. Englewood Cliffs, NJ: Prentice-Hall.

Kawabata, N. (1984). Perception at the blind spot and similarity grouping. *Perception & Psychophysics, 36*, 151–158.

Kawabata, N. (1990). Structural information processing in peripheral vision. *Perception, 19*, 631–636.

Kelly, J. P. (1991). The sense of balance. In E. R. Kandel, J. H. Schwartz, and T. M. Jessell, Eds., *Principles of Neural Science*. Amsterdam: Elsevier.

Kemler Nelson, D. G. (1993). Processing integral dimensions: The whole view. *Journal of Experimental Psychology: Human Perception and Performance, 19*, 1105–1113.

Kennedy, R. S., and Fowlkes, J. E. (1992). Simulator sickness is polygenic and polysymptomatic. *International Journal of Aviation Psychology, 2*, 23–38.

Kolb, H. (1994). The architecture of functional neural circuits in the vertebrate retina. *Inv. Opthalmo. Vis. Sci., 35*, 2385–2403.

Kosnik, W. D., Sekuler, R., and Kline, D. W. (1990). Self-reported visual problems of older drivers. *Human Factors, 5*, 597–608.

Leibowitz, H. W., and Owens, D. A. (1975). Anomalous myopias and the intermediate dark focus of accommodation. *Science, 189*, 646–648.

Luce, R. D. (1986). *Response Times: Their Role in Inferring Elementary Mental Organization*. New York: Oxford University Press.

Mack, A. (1986). Perceptual aspects of motion in the frontal plane. In K. R. Boff, L. Kauffman, and J. P. Thomas, Eds., *Handbook of Perception and Performance, Vol I: Sensory Processes and Perception*. New York: John Wiley.

Macmillan, N. A. (1993). Signal detection theory as data analysis method and psychological decision model. In G. Keren and C. Lewis, Eds., *A Handbook for Data Analysis in the Behavioral Sciences: Methodological Issues*. Hillsdale, NJ: Lawrence Erlbaum Associates.

Macmillan, N. A., and Creelman, C. D. (1990). Response bias: Characteristics of detection theory, threshold theory, and "nonparametric" indexes. *Psychological Bulletin, 107*, 401–413.

Macmillan, N. A., and Creelman, C. D. (1991). *Detection Theory: A User's Guide*. New York: Cambridge University Press.

Makous, J. C., and Middlebrooks, J. C. (1990). Two-dimensional sound localization by human listeners. *Journal of the Acoustical Society of America, 87*, 2188–2200.

Marslen-Wilson, W. D. (1987). Functional parallelism in spoken word recognition. *Cognition, 25*, 71–102.

Martin, J. H. (1991). Coding and processing of sensory information. In E. R. Kandel, J. H. Schwartz, and T. M. Jessell, Eds., *Principles of Neural Science*. Amsterdam: Elsevier.

Martin, J. H., and Jessell, T. M., (1991). Modality coding in the somatic sensory system. In E. R. Kandel, J. H. Schwartz, and T. M. Jessell, Eds., *Principles of Neural Science*, Amsterdam: Elsevier.

Massaro, D. W., and Cohen, M. M. (1994). Visual, orthographic, phonological, and lexical influences in reading. *Journal of Experimental Psychology: Human Perception and Performance, 20*, 1107–1128.

Matin, L., Picoult, E., Stevens, J., Edwards, M., and McArthur, R. (1982). Oculoparalytic illusion: Visual-field dependent spatial mislocations by humans partially paralyzed with curare. *Science, 216*, 198–201.

McClelland, J. L., and Rumelhart, D. E. (1981). An interactive activation model of context effects in letter perception: Part I. An account of basic findings. *Psychological Review, 88*, 375–407.

McNicol, D. (1972). *A Primer of Signal Detection Theory*. London: Allen & Unwin.

Melara, R. D., Marks, L. E., and Potts, B. C. (1993a). Early-holistic processing or dimensional similarity? *Journal of Experimental Psychology: Human Perception and Performance, 19*, 1114–1120.

Melara, R. D., Marks, L. E., and Potts, B. C. (1993b). Primacy of dimensions in color perception. *Journal of Experimental Psychology: Human Perception and Performance, 19*, 1114–1120.

Merigan, W. H., and Maunsell, J. H. R. (1993). How parallel are the primate visual pathways? *Annual Review of Neuroscience, 16*, 369–402.

Metz, C. E. (1989). Some practical issues of experimental design and data analysis in radiological ROC studies. *Investigative Radiology, 24*, 234–245.

Miller, R. J. (1990). Pitfalls in the conception, manipulation, and measurement of accommodation. *Human Factors, 32*, 27–44.

Morgan, M. J., Watt, R. J., and McKee, S. P. (1983). Exposure duration affects the sensitivity of vernier acuity to target motion. *Vision Research, 23*, 541–546.

Morrison, J. D., and Whiteside, T. C. D. (1984). Binocular cues in the perception of distance of a point source of light. *Perception, 13*, 555–566.

Murphy, B. J. (1978). Pattern thresholds for moving and stationary gratings during smooth eye movement. *Vision Research, 18*, 521–530.

Nealey, T. A., and Maunsell, J. H. R. (1994). Magnocellular and parvocellular contributions to the responses of neurons in macaque striate cortex. *Journal of Neuroscience, 14*, 2069–2079.

Owens, D. A., and Leibowitz, H. W. (1983). Perceptual and motor consequences of tonic vergence. In C. M. Shor, and K. J. Cuiffreda, Eds., *Vergence Eye Movements: Basic and Clinical Aspects*. Boston: Butterworth.

Oyama, T. (1987). Perception studies and their applications to environmental design. *International Journal of Psychology, 22*, 447–451.

Palmer, J. (1986). Mechanisms of displacement discrimination with and without perceived movement. *Journal of Experimental Psychology: Human Perception and Performance, 12*, 411–421.

Palmer, S. E. (1992). Common region: A new principle of perceptual grouping. *Cognitive Psychology, 24*, 436–447.

Petersik, J. T. (1989). The two-process distinction in apparent motion. *Psychological Bulletin, 106*, 107–127.

Plateau, J. A. F. (1872). Sur la mesure des sensations physiques, et sur la loi que lie l'intensité des sensations à l'intensité de la cause excitante. *Bulletins de l'Academie Royale des Sciences, des Lettres, et des Beaux-Arts de Belgique, 33*, 376–388.

Plomp, R. (1964). The ear as frequency analyzer. *Journal of the Acoustical Society of America, 36*, 1628–1636.

Pomerantz, J. R. (1981). Perceptual organization in information processing. In M. Kubovy and J. R. Pomerantz, Eds., *Perceptual Organization*. Hillsdale, NJ: Lawrence Erlbaum Associates.

Posner, M. I., and Raichle, M. E. (1994). *Images of Mind*. New York: Scientific American Library.

Proctor, J. D., and Healy, A. F. (1995). Acquisition and retention of skilled letter detection. In A. F. Healy and L. E. Bourne, Jr., Eds., *Learning and Memory of Knowledge and Skills: Durability and Specificity*. Thousand Oaks, CA: Sage.

Randle, R. (1988). Visual accommodation: Mediated control and performance. In D. J. Oborne, Ed., *International Reviews of Ergonomics* (Vol. 2). London: Taylor & Francis.

Robinson, D. W., and Dadson, M. A. (1956). A re-determination of the equal-loudness relations for pure tones. *British Journal of Applied Physics, 7*, 166–181.

Rock, I., and Brosgole, L. (1964). Grouping based on phenomenal proximity. *Journal of Experimental Psychology, 67*, 531–538.

Rock, I., and Palmer, S. (1990). The legacy of Gestalt psychology. *Scientific American, 263* (6), 84–90.

Rogers, B. J., and Collett, T. S. (1989). The appearance of surfaces specified by motion parallax and binocular disparity. *Quarterly Journal of Experimental Psychology, 41A*, 697–717.

Rugg, M. D., and Coles, M. G. H., Eds. (1995). *Elecrophysiology of Mind: Event-Related Brain Potentials and Cognition*. New York: Oxford University Press.

Rutherford, W. (1886). A new theory of hearing. *Journal of Anatomy and Physiology, 21,* 166–168.

Sanders, A. F., and Houtmans, M. J. M. (1985). Perceptual processing modes in the functional visual field. *Acta Psychologica, 58*, 251–261.

Sanderson, P. M., Haskell, I., and Flach, J. M. (1992). The complex role of perceptual organization in visual display design theory. *Ergonomics, 35*, 1199–1219.

Schiffman, H. R. (1996). *Sensation and Perception: An Integrated Approach* (Fourth Edition). New York: John Wiley.

Schiffman, S. S., and Erickson, R. P. (1993). Psychophysics: Insights into transduction mechanisms and neural coding. In S. A. Simon and S. D. Roper, Eds., *Mechanisms of Taste Transduction.* Boca Raton, FL: CRC Press.

Sekuler, R., and Blake, R. (1994). *Perception* (Third Edition) New York: McGraw-Hill.

Sewall, L., and Wooten, B. R. (1991). Stimulus determinants of achromatic constancy. *Journal of the Optical Society of America A, 8*, 1794–1809.

Shepard, R. N. (1991). Integrality versus separability of stimulus dimensions. In G. R. Lockhead and J. R. Pomerantz, Eds., *The Perception of Structure*. Washington, DC: American Psychological Association.

Sherman, S. M., and Koch, C. (1990). Thalamus. In G. M. Shepherd, Ed., *The Synaptic Organization of the Brain* (Third Edition). New York: Oxford University Press.

Sherrington, C. S. (1906). *Integrative Action of the Nervous System.* New Haven, CT: Yale University Press.

Snodgrass, J. G., and Corwin, J. (1988). Pragmatics of measuring recognition memory: Applications to dementia and amnesia. *Journal of Experimental Psychology: General, 117*, 34–50.

Sternberg, S. (1969). The discovery of processing stages: Extensions of Donders' method. In W. G. Koster, Ed., *Attention and Performance II*, Amsterdam: North-Holland.

Stevens, J. K., Emerson, R. C. Gerstein, G. L., Kallos, T., Neufeld, G. R., Nichols, C. W., and Rosenquist, A. C. (1976). Paralysis of the awake human: Visual perceptions. *Vision Research, 16,* 93–98.

Stevens, S. S. (1956). The direct estimation of sensory magnitudes—loudness. *American Journal of Psychology, 69*, 1–25.

Stevens, S. S. (1975). *Psychophysics*. New York: John Wiley.

Stevens, S. S. and Newman, E. B. (1934). The localization of pure tone. *Proceedings of the National Academy of Science, 20*, 593–596.

Sturr, F., Kline, G. E., and Taub, H. A. (1990). Performance of young and older drivers on a static acuity test under photopic and mesopic luminance conditions. *Human Factors, 32*, 1–8.

Summers, I. R., Ed., (1992). *Tactile aids for the hearing impaired.* London: Whurr Publishers.

Swets, J. A. (1986). Indices of discrimination or diagnostic accuracy: Their ROCs and implied models. *Psychological Bulletin, 99*, 100–117.

Thurstone, L. L. (1927). A law of comparative judgment. *Psychological Review, 34*, 273–286.

Townsend, J. T. (1971). Theoretical analysis of an alphabetic confusion matrix. *Perception & Psychophysics, 9*, 40–50.

Truscianko, T., Montagnon, R., and le Clerc, J. (1991). The role of colour as a monocular depth cue. *Vision Research, 31*, 1923–1930.

Verrillo, R. T., and Gescheider, G. A. (1992). Perception via the sense of touch. In I. R. Summers, Ed., *Tactile Aids for the Hearing Impaired*. London: Whurr Publishers.

Verrillo, R. T., Fraioli, A. J., and Smith, R. L., (1969). Sensation magnitude of vibrotactile stimuli. *Perception & Psychophysics, 6*, 366–372.

Wallach, H. (1972). The perception of neutral colors. In T. Held and W. Richards, Eds., *Perception: Mechanisms and models: Readings from Scientific American.* San Francisco: W. H. Freeman.

Wandell, B. A. (1995). *Foundations of Vision*. Sunderland, MA: Sinauer.

Welford, A. T., Ed. (1980). *Reaction Times*. London: Academic Press.

Wever, E. G., and Bray, C. W. (1937). The perception of low tones and the resonance-volley theory. *Journal of Psychology, 3*, 101–114.

Wickens, C. D., Merwin, D. F., and Lin, E. (1994). Implications of graphics enhancements for the visualization of scientific data: Dimensional integrality, stereopsis, motion, and mesh. *Human Factors, 36*, 44–61.

Williams, D., MacLeod, D. I. A., and Hayhoe, M. (1981). Punctate sensitivity of the blue sensitive mechanism. *Vision Research, 21*, 1357–1375.

CHAPTER 4

INFORMATION PROCESSING

Christopher D. Wickens
Aviation Research Laboratory
University of Illinois
Savoy, IL 61874 USA

C. Melody Carswell
Department of Psychology
University of Kentucky
Lexington, KY 40506 USA

Information processing lies at the heart of human performance. In a plethora of situations in which humans interact with systems, the operator must perceive information, transform that information into different forms, and take actions on the basis of the perceived and transformed information. These characteristics apply regardless of whether "information processing" is defined in terms of the classic "open loop" information-processing model that derives from much of psychological research (Figure 4.1(*a*)) or the closed-loop model of Figure 4.1(*b*), which has its roots both within control engineering (e.g., Baron et al., 1970; McRuer, 1980; Pew and Baron, 1978), and more recent conceptualizing in ecological psychology (Flach et al., 1995; Hancock et al., 1995). In either case, **transformations** must be made on the information as it flows through the human operator. These transformations take time, and may be the source of error. Understanding their nature, their time demands, and the kinds of errors that result from their operation is critical to predicting and modeling human–system interaction.

This chapter describes characteristics of the different important *stages* of information processing, from perception of the environment, to acting upon that environment. We will try to do so in a way that is neither too specific to any particular system nor so generic that the relevance of the information processing model to system design is not evident. We begin by contrasting three ways in which information processing has been treated in applied psychology, and then we will describe processes and transformations related to **attention, perception, memory** and **cognition, action selection,** and **multiple task performance**.

4.1 THREE APPROACHES TO INFORMATION PROCESSING

The classic "information-processing" approach to describing human performance, owes much to the seminal work of Sternberg (1969), Broadbent (1958, 1972), Neisser (1967), Posner (1978), and others in the decades of the 50s, 60s, and 70s, who applied the metaphor of the digital computer to human behavior. In particular, as characterized by the representation in Figure 4.2, information was conceived as passing through a finite number of discrete stages. These stages were identifiable, not only by experimental manipulations, but also by converging evidence from brain physiology. Thus, for example, it makes sense to distinguish a perceptual stage, from one involving the selection and execution of action, because of the morphological distinctions between sensory and motor cortex.

There is also a human factors rationale for the stage distinction made by information-processing psychology. This is because different task or environmental factors appear to differentially influence processing at the different stages, a distinction that has certain design implications. For example, the **aging** process appears to affect the selection and execution of actions, more than the speed of perceptual encoding (Strayer et al., 1987). Different stressors may differentially affect different stages of processing (Broadbent, 1972; Hockey, 1984, 1986), and different sources of workload may have different influences on the different stages (Wickens, 1992). Decision-making biases can be

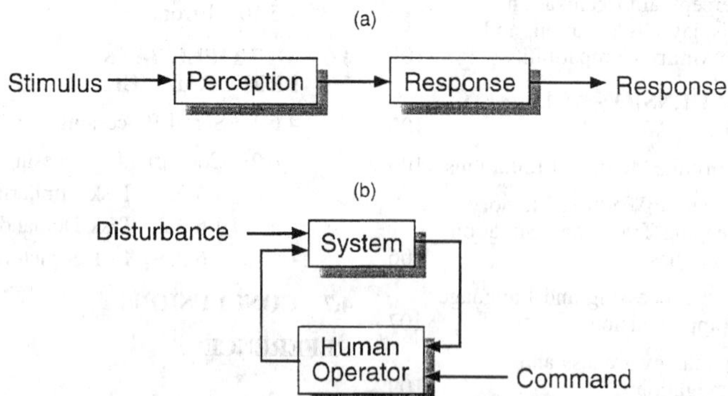

Figure 4.1 Two representations of information processing: (*a*) a traditional open-loop representation from cognitive psychology; (*b*) closed-loop system, following the tradition of engineering feedback models.

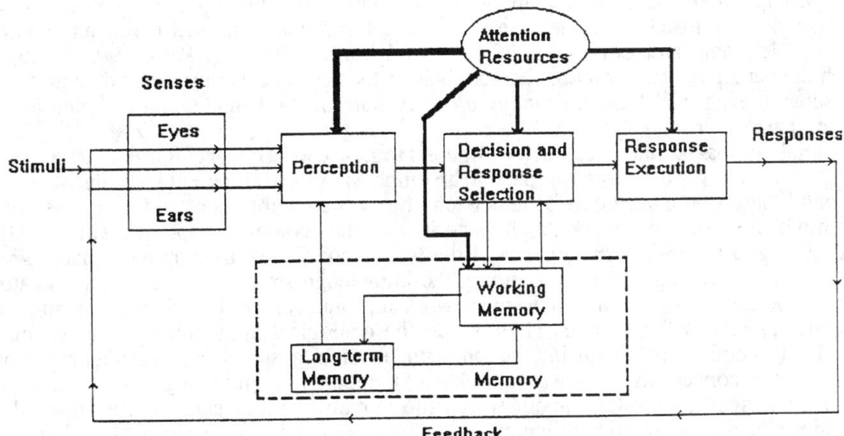

Figure 4.2 A model of human information processing (adapted from Wickens, 1992).

characterized by whether they influence perception, diagnosis, or action selection (Wallsten, 1980; Wickens, 1992), and the different stages may also be responsible for the commission of qualitatively different kinds of errors (Reason, 1990; see Section 4.5.10).

In contrast to the stage approach, the **ecological** approach to describing human performance provides much greater emphasis on the integrated flow of information through the human, rather than on the distinct, analyzable stage sequence (Flach et al., 1995; Gibson, 1979; Hancock et al., 1995; Warren and Wertheim, 1990). The ecological approach also emphasizes the human's integrated interaction with the *environment,* in a way that the stage approach does not because the latter characterizes information processing in a "context-free" manner. Accordingly the ecological approach focuses very heavily on modeling the perceptual characteristics of the environment to which the user is "tuned" and responds, in order to meet the goals of a particular task. Action and perception are inexorably linked, since, to act is to change what is perceived, and to perceive is to change the basis of action, in a manner consistent with the closed-loop feedback representation shown in Figure 4.1(*b*).

As a consequence of these properties, the ecological approach is most directly relevant to describing human behavior in interaction with the natural environment (e.g., walking or driving through natural spaces, or directly manipulating objects). However, as a direct outgrowth, this approach is also quite relevant to the design of controls and displays that mimic characteristics of the natural environment—the concept of **direct manipulation interfaces** (Hutchins et al., 1985). As a further outgrowth, the ecological approach is relevant to the design of interfaces that mimic characteristics of how users **think** about a physical process, even if the process itself is not visible in a way that can be directly represented. In this regard, the ecological approach has been used as a basis for designing effective displays of energy conversion processes such as those found in a nuclear reactor (Bennett, Toms, and Woods, 1993; Moray et al., 1994; Vicente and Rasmussen, 1992; Vicente, Christofferson and Pereklita, 1995).

Because of its emphasis on interaction with the natural (and thereby familiar) environment, the ecological approach is closely related to other approaches to performance modeling that emphasize people working with domains and systems with which they are highly familiar. This feature characterizes for example, the study of **naturalistic decision making** (Klein and Crandall, 1995; see Chapter 38), which is often set up in contrast to the study of **classical decision making** (Wallsten, 1980; Yates, 1993), an approach more often structured within the framework of the information-processing model.

Both the stage-based approach and the ecological approach have a great deal to offer to human factors, and the position we take in this chapter is that aspects of each can and should be selected, as they are more appropriate for analysis of the operator in a particular system. For example, the ecological approach is highly appropriate for modeling vehicle control, but less so for describing processes in reading, understanding complex instructions under stress, or dealing with highly symbolic logical systems (e.g., the logic of digital computers or decision tree analysis; see Chapter 38). Finally, both approaches can be harmoniously fused, as is the case with the optimal control model of manual control (Baron, Kleinman, and Levison, 1970; Pew and Baron, 1978; Wickens, 1992, 1986) in which the perceptual-motor loop, characteristic of the ecological approach is dichotomized into separate stages of human *estimation* of the state of the vehicle, and *computation* of the optimal response.

A final approach, that of **cognitive engineering,** or **cognitive ergonomics** (Rasmussen et al., 1995), adopts a hybrid approach. The emphasis of cognitive engineering is, on the one hand, based on a very careful understanding of the environment and task constraints within which an operator works, a characteristic of the ecological approach. On the other hand, as suggested by the prominence of the word "cognitive," the approach places great emphasis on modeling and understanding the knowledge structures that expert operators have of the domains in which they must work and indeed, the knowledge structures of computer agents in the system. Thus, while the ecological approach tends to be more specifically applied to human interaction with physical systems (and particularly those that obey the constraints of Newtonian physics), cognitive engineering is relevant to the design of almost any system, about which the human operator can acquire knowledge, including the very symbolic computer systems whose logic has no physical analogy.

Whether human performance is approached from an information-processing, an ecological, or a cognitive engineering point of view, we assert here that in almost any task, a certain number of mental processes, involved in selecting, interpreting, retaining, or responding to information, may be implemented; and it is understanding the vulnerabilities of these processes, and capitalizing, where possible, on their strengths, that can provide an important key to effective human factors of system design.

In this chapter, we adopt as a framework the information-processing model depicted in Figure 4.2 (Wickens, 1992). Here information is sensed, and that information received by our sensory system is *perceived*—that is, provided with some meaningful interpretation based on memory of past experience (Section 4.3). That which is perceived may be directly responded to, through a process of action selection (decision of what act to take) and execution (Section 4.5). Alternatively, it may be stored temporarily in working memory, a system which may also be involved in thinking about, or transforming information that was not sensed and perceived, but was internally generated (e.g., mental images, rules, Section 4.4). Working memory is of limited capacity, but is closely related to our large capacity long-term memory, a system which stores vast amounts of facts about the world, knowledge, and procedures but is not always fully available for retrieval.

As noted in the figure, and highlighted in the ecological approach, actions generally produce feedback which is then sensed to complete the closed-loop cycle. In addition, human attention plays two critical roles in the information processing sequence. As a selective agent, it chooses and constrains this information that will be perceived (Section 4.2). As a task management agent, it constrains what tasks (or mental operations) are performed (Section 4.6).

4.2 SELECTING INFORMATION

Since Broadbent's (1958) classic book, it has been both conventional and important to model human information processing as, in part, a filtering process. This filtering is assumed to be carried out by the mechanisms of *human attention* (Damos, 1991; Kahenman, 1973; Parasuraman, Davies, and Beatty, 1984). Attention, in turn, may be conceptualized as having three **modes;** selective attention chooses what to process in the environment; focused attention characterizes the efforts to sustain processing of those elements while avoiding distraction from others, and *divided attention* characterizes the ability to process more than one attribute or element of the environment at a given time.

We discuss below the human factors implications of the selective and focused attention modes, and will discuss those of divided attention in more detail in Sections 4.3.6 and 4.6.

4.2.1 Selective Attention

The aspects of the environment that are selected for further processing are driven by both **top-down** and **bottom-up** processes (Yantis, 1993). Top-down processes refer to those that are generated from past experience by cognitive mechanisms in the brain. Bottom-up processes are characteristics of the environmental stimuli themselves. Human selective attention is said to be driven heavily by the top-down process of a **mental model.** That is, the typical skilled performer in any environment, whether a pilot, a vehicle driver, a radiologist, or a map reader, has gained experience knowing what parts of the environment to look at when, in order to gain the greatest information from the environment. In dynamic environments, such as that represented by the aircraft instrument panel, skilled pilots tend to look most at those sources that change most frequently (contain the most information in terms of bits/sec, see Section 4.5.1; Harris and Christhilf, 1980; Senders, 1964; 1980). However, they also tend to sample information sources with a frequency proportional to the **cost of not sampling.** That is, sampling is guided by the question: "If I fail to look at a source, and something critical occurs there, how severe will the penalty be if I miss it?" (Carbonnell, Ward, and Senders, 1968; Moray, 1986; Sheridan, 1972). While it is easiest to monitor the direction of selective attention by observing the position of the eyeball, there is evidence also that we selectively **listen** to auditory channels, on the basis of their perceived information content and importance (Moray, 1975).

Bottom-up influences on selective attention are of two types. First, **salient** sources, like loud, bright, or flashing (intermittent) events, will fairly automatically call attention to themselves (Yantis, 1993). That is, they guarantee selective attention (whether this is desired by the task—a loud warning alarm—or not—a distracting voice). What source is selected, however, is also dictated, in part, by the *cost* of accessing that source, the **information access cost** (Wickens, 1994; Wickens and Carswell, 1995). Thus, it is more costly in terms of time and effort to move the head than to move the eyes; just as it is more effortful to walk across the room for a book, rather than simply reach across the desk. In the long run, these added costs of accessing more "distant" information sources can be considerable, and hence, they influence the tendency of operators to select one source rather than another for processing. As an example, it is assumed that selective attention to flight instruments will be easier if they are positioned head up (in *head-up displays*) so that visual scanning is not required to access their content, than if they are presented head down (Weintraub and Ensing, 1992; Wickens and Long, 1995).

The combined considerations of information access cost (IAC) and the mental model, have been incorporated in principles of the layout of multiple display elements, to be discussed further in Section 4.3.6, and in Chapter 21 Design of Displays. If the operator's selective attention must often travel between two (or more) sources on a display, because they are both frequently used, or are related in the operator's mental model of the task, the best display layout will be one that minimizes the travel distance (IAC) between those sources. Adherence to this constraint characterizes the manner in which the layout of the aircraft instrument panel has evolved (O'Hare and Roscoe, 1990). Errors of selective attention have also had disastrous consequences, such as when aircraft altitude is not appropriately monitored leading to controlled flight into terrain (Wiener, 1977), or when critical instruments on the control panel of the Three Mile Island nuclear power plant were not sampled at the appropriate time (Adams, 1989).

4.2.2 Focused Attention

While selective attention dictates where attention should travel, the goal of focused attention is to maintain processing of the desired source, and avoid the distracting influence of potentially competing sources. The primary source of breakdowns in focused attention are certain physical properties of the visual environment (clutter) or the auditory environment (noise), that will nearly guarantee some processing of those environments, whether such processing is desired or not. Thus any visual information source within about 1° of visual angle of a desired attentional focus will disrupt processing of the latter to some extent (Broadbent, 1982). Any sound within a certain range of frequency and intensity of an attended sound will have a similar disruptive effect on auditory focused attention (see Chapter 3). However, even beyond these minimum limits of space and frequency, information sources can be disruptive of focused attention if they are salient.

4.2.3 Discrimination

A key to design that can address issues of both selective and focused attention is concern for the **discrimination** (or its inverse, the confusion) between information sources. Making sources discriminable, by space, color, intensity, frequency, or other physical differences, has two benefits. First, it will allow the display viewer to **parse** the world into its meaningful components on the basis of these physical features, thereby allowing selective attention to operate more efficiently (Treisman, 1986, 1988). For example, the air traffic controller who views on her display all of the aircraft within a given altitude range depicted in the same color can easily select all of those aircraft for attention, to ascertain which ones might be on conflicting flight paths. Parsing via a discrimination will be effective as long as all elements that are rendered physically similar (and therefore are parsed together) share some characteristic that is relevant for the user's *task* (as in the preceding example, all aircraft at the same altitude represent potential conflicts).

Second, when elements are made more discriminable by some physical feature, it is considerably easier for the operator to **focus** attention on one and ignore distraction from another, even if the two are close together in space (or are similar in other characteristics). Here again, in our air traffic control example, it will be easier for the controller to focus attention on the convergent pattern of two commonly colored aircraft, if other aircraft are colored differently, than if all are depicted in the same hue.

Once information is selected by attention, it then forms the basis for further processing, to interpret its meaning, and then formulate plans of actions. We consider now this next stage of processing, the interpretation through perception, and the implications of perception to display design.

4.3 PERCEPTION AND DATA INTERPRETATION

4.3.1 Detection as Decision Making

The first piece of advice offered in most display design checklists is that the designer should be absolutely certain that the display code will be detectable in the environment for which it is intended (e.g., Sanders and McCormick, 1993; Travis, 1991). Assuring the detectability of the relevant information might seem to be simply a matter of knowing enough about the limits of the operator's sensory systems to choose appropriate levels of physical stimulation, for example, appropriate wavelengths of light, frequencies of sound, or concentrations of an odorant. Chapter 3 reviews human sensitivity to the presence and variation of different physical dimensions, and these data must be considered limiting factors in the design of displays. Yet, the detectability of any critical signal is also a function of the operator's goals, knowledge, and expectations. In short, detection must also be viewed as active interpretation.

The interpretive nature of signal detection becomes most apparent when we consider that operators may make two types of detection errors. As shown in Figure 4.3, operators may occasionally miss a signal when it is present and, importantly, they may also respond as if a signal is present when it is not (i.e., a "false alarm"). Signal Detection Theory (SDT) provides one model of the processing that can lead to false alarms as well as to misses (Green and Swets, 1988; Tanner and Swets, 1954). SDT conceptualizes the detection task as one in which the operator is trying to decide whether any momentary amount of sensory stimulation actually reflects the presence of the signal or is simply irrelevant "noise." This noise may consist of stimulation that is external to the operator, as when a momentary reflection on a windshield is mistaken for an oncoming vehicle, or the noise may be internal to the operator, as when we mistake a "floater" (debris in the vitreous humor of the eye) for a rapidly moving aircraft. Because noise of both sorts is always present, and is always fluctuating in intensity, some detection errors are inevitable.

In order to deal with the uncertainty that is inherent in detection, SDT proposes that operators choose some level of sensory excitation that will serve as a response criterion.

		State of the world	
		Signal	Noise
Response	Yes	Hit	False alarm
	No	Miss	Correct rejection

Figure 4.3 Joint Contingent Events Used in Signal Detection Theory Analysis.

If the momentary level of stimulation that they experience is above this criterion, then they will respond as if a signal is present, with all the consequences this action might entail. An operator who sets a low or "risky" criterion will respond as if a signal is present more frequently than one who sets a higher, more "conservative" criterion. Adopting a low criterion will ensure that the operator rarely misses a signal; however, the reduction in misses is at the expense of increased false alarms. Setting a higher criterion, in contrast, results in fewer false alarms at the expense of increased misses. SDT provides a way of measuring the criterion set by individual operators, which, in turn, allows the isolation of factors that influence criterion choice.

To the extent that the total segregation of signal from noise is impossible in a system, the designer must focus on manipulating those factors that ensure that operators set the optimal response criterion for their given detection task. SDT formally demonstrates that overall detection errors, misses and false alarms combined, are minimized if the response criterion is shifted downward (i.e., is made more risky) with increases in signal likelihood. Subjects performing laboratory detection tasks tend to adjust their response criteria to the direction prescribed by SDT; however, they do *not* tend to adjust it far enough (Green and Swets, 1988).

The effects of signal probability observed in the lab have also been observed in operational settings. Lusted (1976) found that physicians' criteria for detecting particular medical conditions were influenced by the base rate of the abnormality. Likewise, sheet metal inspectors adjusted their criteria for fault detection based on estimated defect rates (Drury and Addison, 1973). Yet, plant inspectors do not adjust their criteria enough when defect rates fall below 5% (Harris and Chaney, 1969). If an operator's failure to adjust the response criterion results from inadequate knowledge about actual signal probabilities, then the presentation of this information may encourage a more optimal criterion choice. To facilitate the failure detection performance of process control operators, for example, Moray (1981) has suggested that warning indicators be scaled by the probability of their normal value. Another way to influence the operator's criterion, at least when the criterion appears to be too conservative, is to inflate signal probability by occasionally introducing additional, artificial signals into the event stream (Baker, 1961; Wilkinson, 1964), for example, by adding faulty products into the product stream viewed by the quality control inspector (making sure the added faults are tagged for later removal in case they are missed!).

A second factor that should influence the location of the response criterion, according to SDT, is the relative costs associated with misses and false alarms, and the relative benefits of correct responses. As an extreme example, if there were dire consequences associated with a miss and absolutely no costs for false alarms, then the operator should adopt the lowest criterion possible, and simply respond as if the signal is there at *every* opportunity. Usually, however, circumstances are not so simple. A missed air space conflict by the air traffic controller or a missed tumor by the radiologist may have enormous costs, possibly in terms of human lives. Alternatively, the interventions that result from false alarms also have costs, for example, air traffic delays or unnecessary surgery. The operator must adjust his or her response criterion downward to the degree that misses are more costly than false alarms.

Green and Swets (1988) have found that response criteria are more sensitive to the changes in costs and benefits of the various signal detection outcomes than they are to changes in signal probabilities. Ideally, knowledge of both signal probabilities and the costs and benefits of different action outcomes should be reinforced through explicit training and appropriate incentives in order for the operator to establish the response criterion that maximizes overall detection performance.

It should be noted that the principles of optimizing signal detection performance can be applied regardless of whether the signal is very faint, such as the slightest smell of smoke, or is physically quite intense, such as the sound of a blaring fire alarm. Even though the detectability of an alarm's sound may be quite high, with virtually no misses or false alarms, the sound may be viewed eventually as either signal (when a fire is really present) or noise (when there is no fire). The task of the office worker who detects the alarm then becomes one of combining the odds of an actual fire with knowledge about the outcomes of heeding or ignoring the alarm to determine how much evidence (e.g., how long the alarm must continue) before evacuating the building. If the alarm is overly sensitive to many harmless situations, and is activated often, the worker's mental estimate of the signal (true fire) will drift downward as his or her response criterion shifts upward (Getty et al., 1995; Sorkin, 1988). The subjective costs of heeding the alarm when no fire occurs (an "alarm false alarm"), in terms of lost productivity or additional overtime, may also increase. This situation may ultimately lead workers to disregard the alarm altogether, a serious safety risk. Improving the sensitivity of the alarm itself is a reasonable design response. However, the redesign must, according to SDT, be accompanied by a modification of (1) workers' perception of signal probability (i.e., the increased probability that a fire is present when the alarm sounds), and (2) incentives for responding to the alarm that offset the perceived costs of leaving one's work. Another approach is to design alarms that themselves indicate intermediate states of uncertainty regarding the presence of the alarm condition, a solution which appears to increase the sensitivity of the overall human-machine system (Sorkin, 1990).

4.3.2 Expectancy, Context, Redundancy, and Identification

We have seen that the operator's response to the question "Is the target out there?" is influenced by his or her beliefs about the target's likelihood. Similarly, responses to the more general question "*What* is out there?" are also a function of the operator's expectations. The effect of such expectancies on speeded word recognition was demonstrated by Tulving, Mandler, and Baumal (1964). Their subjects were presented with a truncated sentence, and were then presented with a target word that completed the sentence. The authors found that the time necessary to identify the final target word was reduced as more words from the sentence were revealed in the previous display. Presumably, this manipulation increased subjects' expectations for some words and reduced them for others. Although these results were obtained with stimuli presented sequentially, similar results have been obtained with stimuli presented simultaneously, for example, when the identification of a letter embedded in a word is faster than the identification of the same letter in isolation (Reicher, 1969; Wheeler, 1970). Likewise, Palmer (1975) found that caricature facial features, such as ears and eyes, required less physical detail for recognition when they were embedded in a face rather than when presented alone. It also seems that object identification is enhanced by the complex context provided by naturalistic scenes, for example, photographs of city streets or offices (Biederman et al., 1981).

It might seem counterintuitive that recognition of targets improves with the number of nontarget display elements, with extra letters, features, words, and objects. However, the additional items in the stimulus array may increase the odds that the operator will recognize some portion of the display, which in turn provides cues for the processing of additional items. That is, the redundancies inherent in natural language, familiar objects, and common scenes allow operators to develop expectancies about additional display elements (see Massaro, 1979; McClelland and Rumelhart, 1981, for detailed models of context effects). These expectancies can, in turn, be used to offset degraded stimulus conditions such as poor print reproductions, faulty lighting, brief stimulus exposures, presentation to peripheral vision, or even the momentary diversion of attention. While designers must try to enhance the distinctiveness of target stimuli from alternatives, they must also consider the use of redundant contextual information to further aid discrimination whenever recognition errors lead to severe consequences. Thus a vocal 'A,' 'B,' and 'C' are made more discriminable by making each the initial sound of a two-syllable word: 'Alpha,' 'Bravo,' and 'Charlie.'

One additional effect of operator expectancies on recognition performance is related to the orientation of the target object relative to the operator. Some target perspectives are more commonplace than others, for example viewing a chair from above and slightly off to the side rather than directly from the front. Palmer, Rosch, and Chase (1981) found that the time required to identify objects increased with the distance of the target from its most likely orientation. Such data have implications for the choice and design of icons, such as those routinely found in computer menus, highway signs, maps, and building directories. For maximum discriminability, icons should represent objects as shown in typical perspective.

4.3.3 Judgments of 2-D Position and Extent

Both detection and identification tasks, as described above, are categorical judgments. In the case of detection, the operator must choose from two alternatives—"signal" or "noise." For identification, the operator must choose one classification from what may sometimes be a very large number of alternatives, for example in identifying which numeral has appeared on the screen or what type of vehicle is on the horizon. In addition to such "what" questions, operators must often answer questions of "how much" and "where." These judgments are critical for manual control and locomotion (see Section 4.5), as well as for the interpretation of maps, graphs, and dynamic analog indicators. In the next two subsections, we focus mainly on spatial judgments of static formats before turning to their dynamic counterparts.

It has been known for some time that the spatial judgments required for the reading of even the most familiar graphical formats are prone to systematic distortions. For example, Graham (1937) found that people tended to overestimate the values represented by bar heights in a bar graph. This finding was particularly true with shorter bars and those furthest from the *y*-axis. Working with line graphs, Poulton (1985) found that point-reading errors seemed to reflect a perceptual "flattening" of the line as its distance from the *y*-axis increased. Thus, in the typical line graph with the *y*-axis located on the left side, points along an increasing function are underestimated, and this underestimation increases the further the point is from the *y*-axis. Systematic distortions have also been obtained for pie charts where the percentage represented by slices subtending 0°–90° (and 180°–270°) are overestimated, while other angles are underestimated (Spence and Krizel, 1994).

These distortions in graphs may be special cases of geometric illusions such as those reviewed by Coren and Girgus (1978) and Gregory (1980). Poulton (1985), for example, has ascribed the perceptual flattening of lines in graphs to the Poggendorf illusion (see Figure 4.4). Many such illusions may be the result of a perceptual system that processes, and enhances, differences in stimuli rather than their absolute values. Thus, a slight difference in the slope of two line segments may be magnified perceptually, making acute angles appear larger than they really are (Blakemore, 1980). Small angles in a pie chart may be distorted in this manner, as might the position of a line with respect to a nearby axis in a line graph. Whatever the cause of these point-reading distortions, design modifications can reduce their severity. For example, Poulton (1985) found that adding a redundant *y*-axis to the right side of his graphs effectively reduced point-reading errors. It should be noted, however, that the presence of illusions is not always harmful. In fact, some designers have used illusions of size to reduce traffic accidents. Shinar, Rockwell, and Malecki (1980) painted a pattern similar to that used to induce the Wundt illusion (see Figure 4.4) on a roadway leading to a dangerous, obscured curve. After the roadway was painted, drivers tended to reduce their speed before encountering the curve, presumably because the pattern made the road seem more narrow.

The most systematic work on the size and nature of errors in the perception of graphical displays, rather than focusing on how accurately we can read precise data values, has looked at how precisely we can make comparisons among data values. Cleveland and McGill (1984, 1985, 1986; Cleveland, 1985) developed a list of the physical dimensions that are commonly used to code data values in graphs and maps. These dimensions were ordered, as shown in Figure 4.5, in terms of the accuracy with which they could be used by people to make relative magnitude judgments (e.g., "What percentage is Point A of Point B?"). Dimensions at the top of the figure were used more accurately than those at the bottom. Therefore, Cleveland and McGill advise designers to use position on common scales rather than, for example, volumes or positions on misaligned scales, to represent data whenever possible. A meta-analysis of the comparative graphics literature (Carswell, 1992) revealed that the ordering shown in Figure 4.5 fared well when predicting how well graph users could make simple comparisons or read specific data points. However, this ordering fared less well when used to predict performance in tasks that required the user to identify overall patterns (e.g., "is there an increasing trend?"). In addition, each

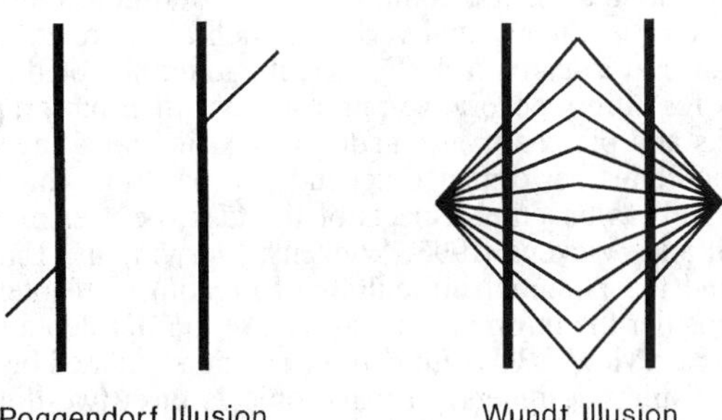

Figure 4.4 Two perceptual illusions influencing the perception of graphs.

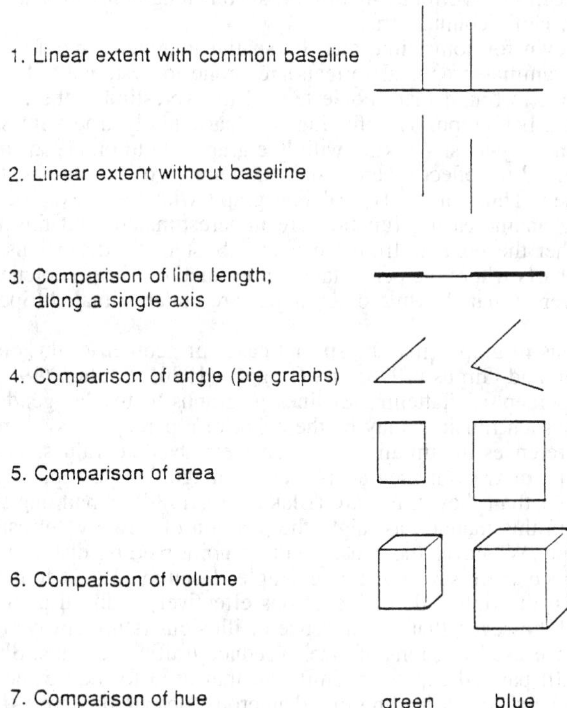

Figure 4.5 Different graphical dimensions for making comparative judgments (from Wickens, 1992; used by Cleveland and McGill, 1985).

step down the ordering did not correspond to an equal decrement in performance. The meta-analysis revealed that the performance differences associated with position, length, and angle judgments were small. However, large differences were encountered when volume and area judgments were compared with any of the higher-ranking dimensions. Area and volume were used much less accurately than position, length, or angle.

4.3.4 Judgments of Distance and Size in 3-D Space

Given that research on graphical perception has found that we are generally poor judges of differences in volumes (e.g., Carswell, 1992; Cleveland and McGill, 1985), it might be somewhat surprising that so many displays use volumetric properties, or spatial relations in three dimensions, to represent information of various types. As with any other analog display, perspective displays can be used to directly represent spatial variation in three dimensions, as when charting the position of planes in air traffic displays or the shape and location of mountains in a contour map. In addition, the three spatial dimensions can be used to represent nonspatial variables such as the relationship of mood, time of day, and performance in a search task. Finally, it should be noted that perspective may be added to displays for purely decorative purposes, as when bar graphs are transformed into a series of blocks and pie charts are made to look like hovering discs. The extent to which perspective facilitates performance depends, in part, on which of these functions the graph serves, as well as on other aspects of the task the user must perform with the display (e.g., Haskell and Wickens, 1993; Wickens, Merwin, and Lin, 1994).

First, to understand the potential difficulties of creating and interpreting perspective displays, we must consider the information that we use to infer depth (see Wickens, Todd, and Seidler, 1989, for a review). *Pictorial depth cues* are so named because they are often used by artists (and display designers) to make objects on a two-dimensional canvas or VDU appear three-dimensional. Figure 4.6 illustrates and describes eight such cues.

In general, the impression of depth increases with the number of pictorial depth cues that are available, as long as these cues are consistent with a single interpretation of the

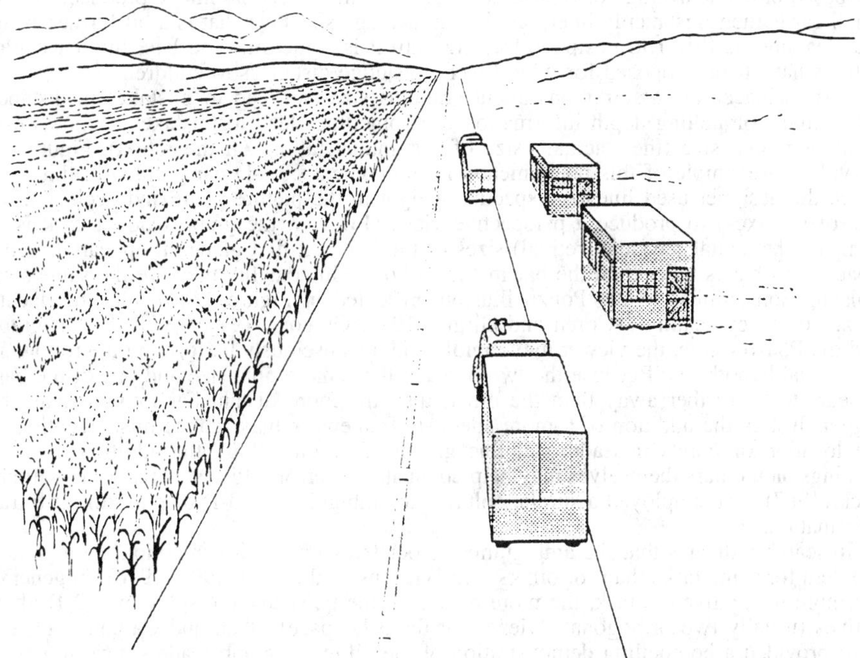

Figure 4.6 Perceptual cues for depth perception (from Wickens, 1992).

spatial relation of objects in the display (Berbaum, Tharp, and Mroczek, 1983). Bruno and Cutting (1988), for example, found that relative size, height in the picture plane, occlusion, and motion parallax had an additive effect on perceived depth. If all these cues are present and consistent, it is relatively easy to determine which of two parked cars is closer: It is the vehicle that creates a larger retinal image, is lower in the picture plane, partially occludes the other, and appears to move further across the visual field with movements of the observer. This additive effect of consistent pictorial cues points to the wealth of information used by the visual system to infer location, and once again highlights the importance of redundancy.

In addition to the pictorial depth cues, there are several *physiological depth cues* that result from the intrinsic characteristics of the observer's visual system. Accommodation involves feedback from the muscles that control the shape of each eye's lens, and convergence involves feedback about the coordinated movement of the two eyes. A third physiological cue, retinal disparity, is of particular importance to display designers and, like convergence, it depends on the placement and interaction of our two eyes. Because the eyes are offset, they receive slightly different views of the world. Comparisons of the differences between the two views provides yet another clue to the distance of various objects. Binocular disparity can be simulated by the display designer to create the impression of depth. It is the cue long used in Victorian parlor stereoscopes (i.e., "stereopticons"), where pictures taken from slightly different horizontal locations are projected to the two eyes separately, creating the retinal disparity that would result if the person were actually looking at the real 3D scene. Modern advances in display technology have produced dynamic displays that utilize retinal disparity to create the illusion of depth (Yeh and Silverstein, 1992). Of particular interest to designers, it appears that retinal disparity combines additively with the pictorial cues to provide yet a stronger impression of three-dimensionality (Rogers and Collett, 1989; Van der Meer, 1979).

To the extent that they provide evidence that is consistent with a single hypothesis about the relative location of stimuli, additional depth cues are desirable. A poverty of depth cues can lead to false hypotheses that may prove costly, and sometimes tragic. A classic example of depth misperception involves the increased incidence of small automobiles in rear-end collisions (Eberts and MacMillan, 1985). In this case, the cue of familiar size gives misleading information to the driver, who infers the distance to the car based on some average or typical car size. The smaller retinal image produced by the small car is then mistakenly interpreted as an average-sized car that is a bit further away, thus leading the driver to postpone braking until it is sometimes too late. Similar explanations have been proposed for why drivers misjudge distances to children.

Just as inaccurate assumptions about stimulus size can lead to mistakes in distance estimation, compelling depth information can sometimes lead to mistakes in the assessment of retinal size (the "actual" size of a stimulus falling on the retina). Figure 4.7 provides an example of this phenomenon in the context of 3-D bar graphs. In this situation, the designer used linear perspective, height in the picture plane and occlusion (of one of the axes) to produce a perspective view. However, the graph reader's task is to compare the actual physical (retinal) sizes of the bars. At first glance it appears that the "back" left bar is taller than the bar in "front" of it. In actuality they are the same size. This illusion, similar to the Ponzo illusion presented in Figure 4.7(*b*), may be due to "size constancy scaling" (Coren and Girgus, 1978, Gregory, 1977). With the bar graph and the Ponzo figure, the viewer may rapidly and unconsciously be forming the following perceptual hypothesis: Because the two bars are the same retinal size and one of the bars appears to be further away than the other, then the more distant bar must actually be bigger. Just as the addition of a more adequate framework may reduce misperception of the location of points in traditional line graphs (Poulton, 1986), the addition of scale makings on the bars themselves may help attenuate this error. Ellis, McGreevy, and Hitchcock (1987) have employed a similar solution to ambiguities in 3-D displays of air traffic information.

Research indicates that the ambiguities associated with depth perception are a greater problem for some tasks than for others (see Wickens et al., 1989, 1994, 1996). In general, the more integrative the task, the more beneficial the perspective display over 2-D alternatives (usually two orthogonal "views" of the 3-D space). Shah and Carpenter (1995) have provided a compelling demonstration of the difficulty graph readers have in developing integrated mental representations of the relationship among three variables when this information is presented in traditional 2-D data graphics. Yet, 2-D views of the data tend to be used more accurately for tasks requiring more precise readings of specific values. As noted above in the 3-D bar graph shown in Figure 4.7(*a*), accurate comparisons of the size of bars in different depth planes is difficult. Still, it is immediately apparent that the overall pattern of the data is one of a two-way interaction between variables A and B (a global, integrative conclusion). Given the task dependency of the effectiveness of 3-D displays, one alternative is to provide 3-D formats along with 2-D "slices" of the space. Another alternative, particularly relevant when the use of depth is naturally compatible with the 3-D nature of the information represented, as in air traffic control, is to use the perspective formats with the addition of the artificial frameworks described above as an aid when specific size or height judgments are necessary.

Overall, the most productive use of perspective displays appears to be when global integrations need to be made about more than two data dimensions, when an integrated representation of these variables needs to be maintained, and when the to-be-represented

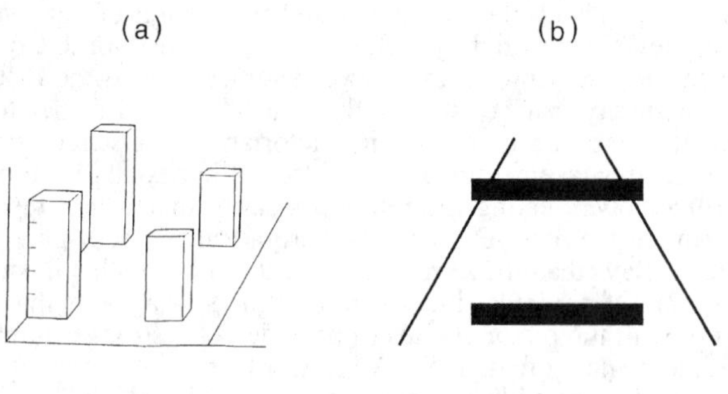

Figure 4.7 The role of size constancy in depth perception in creating illusions: (*a*) the distorted over estimation of the size of the more distant bar graphs; (*b*) the Ponzo illusion, illustrating the greater perceived length of the more distant bar.

dimensions are themselves spatial in nature. In situations when perspective is to be used, the display will be more compelling and more accurately interpreted with the increasing number of redundant depth cues. Realizing, however, that the addition of many of these cues may be too expensive or otherwise impractical to implement, research has indicated that some of the cues are more critical than others. When two depth cues provide conflicting information about the relative distance of two objects, one of the cues typically proves dominant (Dosher, Sperling, and Wurst, 1986). In general, the most dominant cues have proven to be interposition, binocular disparity, and motion parallax (Wickens et al., 1989).

4.3.5 Dynamic Displays, Mental Models, and Analog Compatibility

While many of the graphs and maps discussed above are static displays, many other analog displays are frequently or continuously updated. These dynamic displays are sometimes used in data graphics to represent an additional variable in both 2-D and 3-D formats (Cleveland and McGill, 1988). However, they are a mainstay in aviation, process control, and manufacturing environments. Because these indicators often require rapid responses under stressful conditions, a foremost concern is their compatibility with the operator's mental model of the displayed variables.

There are three ways in which a dynamic analog display can be compatible or incompatible with the operator's mental representation of the underlying system parameters being presented. The first of these is applicable to both static and dynamic displays because it deals with the fundamental issue of which physical dimension to use to represent to-be-displayed values. We have already seen that the precision with which the dimension can be used to make comparative judgments is one factor to consider in making this selection. In addition, we must consider whether there are any physical dimensions that are already associated with the analog concept to be displayed. A direct example, noted in the discussion of 3-D displays, is that it may be particularly advantageous to use variations in 3-D display space to represent an entity that actually varies in real 3-D position, such as an aircraft. When the spatial metaphor is direct, as in this case, the display fulfills what Roscoe (1968) calls the "principle of pictorial realism." In cases where the variable to be represented is nonspatial, there still may be associations between the concept and certain physical codes. For example, temperature is generally associated with variation in vertical position, with "high" temperatures being warmer than "low" temperatures. Color is another possible choice to represent temperature, because colors are often described as varying from "warm" to "cool."

A second compatibility concern is an extension of pictorial realism and is again relevant to both static and dynamic displays. The designer must maintain the expected mapping of data values to levels of the code dimensions. Thus, choosing a vertical pointer to represent altitude is not sufficient to ensure compatibility. The designer must also ensure that higher positions on the pointer are associated with higher altitudes as well. A violation of this principle can lead to performance decrements as indicated by Antes and Chang (1990) in their study of map reading. Darker areas on a map are usually associated with higher values of a mapped variable such as population density or disease rate. A design that used less shading to represent more of a variable resulted in degraded memory for information presented in the map and longer times spent studying the map legend.

Finally, in dynamic displays, the designer must be concerned with whether the direction of movement represented in the display is compatible with the direction of movement expected by the operator. At first it would seem that ensuring compatibility of the code and its mapping would ensure movement compatibility; however, the static and dynamic aspects of compatibility sometimes can be in opposition. The classic example is that of how an aircraft's bank and pitch are represented in the traditional "moving horizon" indicator (Roscoe, 1981, 1968). This display is compatible with the principle of pictorial realism in that a tilted horizon is what the pilot actually sees when he or she looks out the windscreen while banking. Yet, this same display violates motion compatibility. A leftward bank, in the pilot's mental model of the aircraft, should result in a leftward (or counterclockwise) rotation. However, the moving horizon moves in the opposite (clockwise) direction to represent a leftward bank. To resolve this conflict of compatibility principles, the frequency-separated display (Fogel, 1959; Johnson and Roscoe, 1972) was proposed. This display distinguishes dynamic phases of flight, when the pilot is actively changing the bank angle of the aircraft, from relatively static phases, when the pilot is maintaining a particular bank angle while making a turn. The frequency-separated display maintains motion compatibility during the more dynamic phase, while static pictorial realism becomes the dominant compatibility concern at other times.

4.3.6 Perceptual Organization, Display Organization, and Proximity Compatibility

To this point, we have been concerned mainly with aspects of perception relevant to the design of individual indicators. However, indicators are rarely presented in isolation. For graphs, many data values are usually shown in a single framework, and multiple frameworks may be presented on a single page or screen. For dynamic displays, any one indicator is usually displayed within a context of other indicators. The Lockheed L-1011 aircraft, for example, has more than 800 such display elements. As noted in Section 4.2, the sheer number of indicators can lead to clutter, increased information access cost, and, in extreme situations, information overload. To understand how the operator deals with this wealth of information, we must consider how we routinely combine disparate sensory elements to form the higher-order entities we call "objects" or "groups." This is the problem of perceptual organization. The designer's task may be viewed as organizing displays so that the natural laws of perceptual organization support rather than hinder the user's acquisition of information. In Section 4.2, we noted how "parsing" the display space by salient perceptual features could facilitate the processing of multi-element displays.

The *Proximity Compatibility Principle* deals specifically with the issue of display organization (Barnett and Wickens, 1988; Carswell and Wickens, 1987; Wickens and Andre, 1990; see Wickens and Carswell, 1995, for review). In general, the principle holds that those indicators or displayed data values that are conceptually related or that need to be used in combination should belong to the same perceptual group. In short, related information should be perceptually proximate. Kosslyn (1994) describes the difficulty experienced by graph readers when they must isolate an element from one perceptual group in order to compare it with an element from another. In addition to avoiding such "parsing costs," Wickens and Carswell (1995) have argued that there are a variety of other information-processing advantages that arise when displays are designed to encourage task-compatible perceptual organization. These advantages include (1) the perception of emergent features and (2) object-based parallel processing.

Emergent features are relational properties of a group of display elements that are not properties of any of the elements in isolation (Garner, 1978; Pomerantz, 1981; Pomerantz and Pristach, 1989). These emergent features are often rapidly detected by the user and may sometimes be used as direct response cues for the task at hand (Bennett and Flach, 1992; Buttigieg and Sanderson, 1991; Sanderson, Flach, Buttigieg, and Casey, 1989). For example, a series of vertical, moving-pointer displays that are placed side by side may produce the emergent feature of pointer alignment. If pointer alignment represents a critical system state, for example if alignment indicates that everything is operating normally, then the emergent feature itself becomes a higher-order display code. That is, detecting the emergent feature provides a shortcut to the reading of each individual pointer and the effortful checking of each value against every other. Note that if the indicators were not exactly the same in scale design, had different baselines, or were located at distant parts of the display panel with many intervening displays, then alignment would not be available to use as a cue.

Figure 4.8 illustrates a number of ways that separate indicators can be arranged so as to increase the probability of producing emergent features. These design heuristics are based on the Gestalt laws of perceptual organization which state that proximity, similarity, and good continuation of potentially separate perceptual entities will cause those entities to group (see Pomerantz and Kubovy, 1986, for review of these and other gestalt principles). First, increasing spatial proximity among indicators will increase the salience of emergent features (Pomerantz and Schwaitzberg, 1975). Connecting the frameworks of separate indicators, for example adding static line segments between separate meters, also has proven beneficial for the production of emergent features that aid information integration (Dashevsky, 1964).

Perhaps the most important design modification that can be made to increase the number and salience of emergent features, however, is to increase the **similarity** of the indicators. One way to increase the similarity is to make sure that each dimension that is being varied belongs to objects with similar features, for example when bar height is used to represent data values and the bars are all the same color, width, and orientation (see Garner, 1978, for evidence that feature similarity enhances the production of emer-

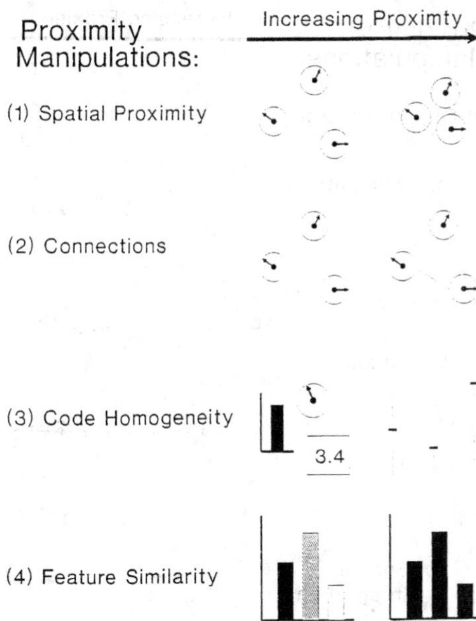

Figure 4.8 Illustrates various ways of creating close display proximity between pairs or triads of indicators. Close proximity configurations are on the right (from Wickens and Carswell, 1995).

gent features). Finally, to ensure that there is the possibility of emergent features at all, display codes must be homogenous. That is, multiple meters or multiple bars or multiple color indicators must be used to represent separate data values; the codes cannot be mixed. Carswell and Wickens (1996) have found drastic reductions in performance of a simple integration task (speeded "which is greater?" comparisons) when such mixed formats are used.

An extreme manipulation of perceptual grouping is the creation of object displays. Object displays include any arrangement of elements that make the different data values appear to be part of a single perceptual object. As Figure 4.9 shows, this can be accomplished in a number of ways. One of the most common manipulations is to add lines or contours directly between the data-varying dimensions (as opposed to connections added between the frameworks, discussed above). Line graphs are an example of this kind of object display, and as such they have been found to facilitate information integration when compared to isolated bars or points (e.g., Carswell and Wickens, 1995; Hollands and Spence, 1992; Schutz, 1961a,b). Multiple indicators can also be arranged so that the addition of line segments creates a closed object. The "polygon," "star," or "polar plot" display that has received attention in both process control (e.g., Woods, Wise, and Hanes, 1981) and aviation (e.g, Beringer and Chrisman, 1991) is an example of this type of object display. Such object displays may contain a wide variety of emergent features, such as global shape, symmetry, and area, that may prove to be useful response cues if directly mapped onto task-relevant variables by the display designer.

A very different type of object display, illustrated in the bottom of Figure 4.9, can be created by spatially integrating mixed codes. So, for example, the vertical position and orientation of a single boundary line can be used to represent the pitch and roll of an aircraft, and the color, vertical position, and horizontal position of a single point can be used to represent information about an applicant's performance on three different aptitude tests on a multivariate scatterplot. Such object displays do not take advantage of emergent features, but instead may encourage (or even force) operators to process information about multiple variables in parallel (Kahneman and Treisman, 1984; Treisman, 1993; Treisman, Kahneman, and Burkell, 1983). Thus one cannot attend to the location of the data point without also registering its color. Object displays with such mixed codes may prove particularly useful when operators must identify a particular conjunction of data values

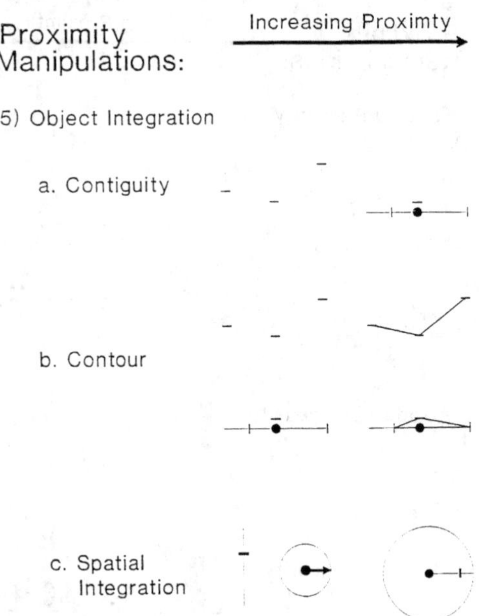

Figure 4.9 Three ways of configuring dimensions to produce object-based proximity. Close proximity displays are on the right (from Wickens and Carswell, 1995).

rather than perform mathematical integrations or comparisons on the data values (Carswell and Wickens, 1996).

As the above descriptions of display arrangements suggest, the application of the proximity compatibility principle to display design requires a detailed understanding of the task or tasks for which the display(s) will be used. Wickens and Carswell (1995) provide a more detailed description of task characteristics important in the selection of display arrangements. In general, however, if data values must be integrated, or are related functionally or conceptually, then they should be grouped in one of the ways described above. If two or more values are unrelated or must be used independently, then they should be assigned to different perceptual groups.

However, in actuality, most tasks require some type of information integration at some stage of processing. The problem facing the designer is to determine what information is integrated and how it is to be combined. For example, by proposing specific mental operations that are combined to perform common map- and graph-reading tasks, several recent models may provide guidance for designers (e.g., Cleveland, 1990; Gillan and Lewis, 1994; Herrmann and Pickle, in press; Hollands, 1992; Lohse, 1991; Pinker, 1990; Simkin and Hastie, 1987). For example, reading a single data point from a line graph involves independent processing of the target value in relation to all other values. At the same time, however, it involves information integration (comparison) with an axis. Thus data points that are far from the axis will be read less accurately than those near the axis (Poulton, 1985). Data points must also, in most graph-reading tasks, be integrated with the labels that identify them. The importance of this integration argues strongly against the common practice of placing identifying information in separate "keys" or "legends" (Milroy and Poulton, 1978). Instead, labels should be spatially integrated, if possible, with the data points themselves.

An emerging theme within the general approach of cognitive engineering (Section 4.1), and one that is closely related to concepts of proximity compatibility, object displays, and emergent features, is the theme of *Ecological Interface Design* (Bennett, Toms, and Woods, 1993; Kirlik, Miller, and Jagacinski, 1993; Vicente and Rasmussen, 1992). The focus of the EID approach is directly upon understanding the constraints of dynamic systems components that are directly related to an operators' monitoring, decision, and control tasks. Based on this cognitive task analysis, effort is then expended to designing displays in which system variables are depicted in a way that the critical constraints can be directly perceived in a manner that is natural and intuitive to the system expert. Figure 4.10 provides one such example of a display designed by Moray et al. (1994) to support the perception of constraints in a nuclear power process control plant (see Chapter 58).

4.4 COMPREHENSION AND COGNITION

In our discussion of perception and display design, we have treated many of the operator's perceptual tasks as decision-making, problem-solving, or reasoning tasks. Detection involves decisions about criterion setting. Identification involves estimations of stimulus probabilities. Size and distance judgments in 3-D space involve the formulation of perceptual hypotheses. For the most part, however, these processes occur rapidly and automatically and, as a result, we are generally not aware of them. In this sense, "perceptual reasoning" is a far cry from the effortful, deliberate, and often time-consuming process that we are very aware of when trying to troubleshoot an ailing computer, find our way through an unfamiliar airport, understand a legal document, or choose among several product designs. In the following we will first describe the limits of our working memory before discussing its relevance to the higher order cognitive tasks of maintaining situation awareness, comprehending text, spatially navigating, planning and problem solving. As we will see, the parameters of working memory constrain, sometimes severely, the strategies we can deploy to understand and make choices in a dynamic environment.

4.4.1 Working Memory Limitations

Working memory refers to the limited number of ideas, sounds, or images that we can mentally maintain and manipulate at any time. The concept has its roots in William James's (1890) "primary memory" and Atkinson and Shiffrin's (1971) "short-term store." All three concepts share the distinction between information that is available in the conscious "here-and- now" (working, short-term, or primary memory) and information that we are not consciously aware of until it is called upon from a more permanent storage system (long-term or secondary memory).

Unlike items in long-term memory, items in working memory are rapidly lost if no effort is made to maintain them (Brown, 1959; Peterson and Peterson, 1959). For example, decay rates of less than 20 sec have been obtained for verbally delivered navigation information (Loftus, Dark, and Williams, 1979) as well as for visuospatial radar information (Moray, 1986). However, even when tasks require minimal delays, working memory is still severely limited in terms of its capacity. George Miller (1956) suggested that this capacity, the "memory span," is limited to about five to nine independent items. The

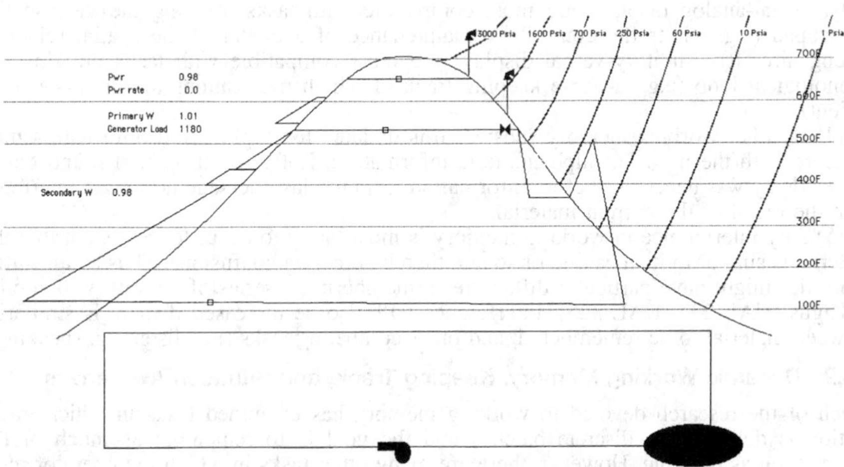

Figure 4.10 The Rankine Cycle display for monitoring the health of a nuclear power generating plant. The jagged line indicates the trajectory of the plant parameters (steam pressure and temperature) as they follow the constraints of the thermodynamic laws (from Moray et al., 1994). This display is proposed but not yet implemented for operational evaluation.

qualifier "independent" is critical, however, because physically separate items that are stored together as a unit in long-term memory may be rehearsed and maintained in working memory as a single entity—a "chunk." Thus, a long-distance telephone number that consists of 11 numbers (e.g., 1—904-638-1803) might seem to be beyond the limits of working memory. However, if the caller is familiar with the area code and the prefix, then each of these numeric groups counts as one chunk instead of three, and the entire number is reduced to a more manageable seven chunks.

Taking into account these limitations in duration and capacity, as well the nature of errors made in tasks with high working memory demands, Baddeley (1990; 1986) has proposed a three-part model of working memory structure. First, there are two "slave" subsystems, the "phonological loop" and "visuospatial scratchpad" (Logie, 1995), that maintain verbal and spatial-pictorial information, respectively. These subsystems are used by a "central executive" that transforms information within each store, transfers information from one store to the other, and integrates information in these stores with information in long-term memory. This conceptualization highlights a number of potential reasons for failures in working memory:

(1) The capacity of either slave may be exceeded, resulting in a loss of information that may be necessary to perform an ongoing task. The design implication is to avoid, whenever possible, codes that infringe on the limits of these systems. However, when longer codes are necessary, there are several ways to reduce memory loss, most involving designs that encourage chunking. For example, parsing material into 3- to 4-item units may increase chunking and subsequent recall (Wickelgren, 1984). Thus, 3546773 is more difficult to recall than 354-6773. In addition, information for different tasks may be split between the two slave systems so that neither the visuospatial scratchpad nor the phonological loop is overburdened. More will be said about such interventions when we turn to the discussion of multitask performance (see Section 4.6).

(2) Information from either store may be lost if there are delays longer than a few seconds between receiving the information and using it. Thus, systems should not be designed so that the user must perform several operations before being able to perform a "memory dump." For example, voice mail systems should always allow users to select a menu option as soon as it is presented rather than forcing them to wait until all the options have been read to make their choice. Methods of responding should be simplified as well, so that the user does not have to retain their choice for long periods of time while trying to figure out how to execute it (see Section 4.5.5 on s-r compatibility).

(3) Information may need to be transferred from one subsystem into the other before further transformations or integrations can be made, thus reducing the resources available for the main processing goal. Wickens, Sandry, and Vidulich (1983) and Wickens, Vidulich, and Sandry-Garza (1984) have provided evidence that the display format should be matched to the working memory subsystem that is used to perform the task. Specifically, visual-analog displays are most compatible with tasks utilizing the visuospatial scratchpad (e.g., air traffic controllers' maintenance of a model of the spatial relations among aircraft). Auditory-verbal displays are most compatible with tasks utilizing the phonological loop (e.g., a nurse keeping track of which medications to administer to a patient).

(4) If either working memory subsystem is updated too rapidly, old information may interfere with the new. For alphanumeric information, Loftus et al. (1979) found that a 10-sec delay was necessary before information from the last message no longer interfered with the recall of the current material.

(5) Any interference in working memory is most likely to occur to the extent that the material is similar in its meaning or sound, thereby creating confusions. Thus an air traffic controller might have particular difficulties remembering a series of aircraft with similar callsigns (UAL 235, UAL 325). Interference will also be increased if there is similarity between material to be remembered, and other competing tasks (i.e., listening, speaking).

4.4.2 Dynamic Working Memory, Keeping Track, and Situation Awareness

Much of the research devoted to working memory has examined tasks in which information is delivered in discrete batches and the goal is to remember as much of the information as possible. However, there are many other tasks in which the operator must deal with continuous information updates with little expectation of perfect retention. Moray (1980) studied several running memory tasks that simulated the demands of a more continuous input stream, and he found the typical memory span to be less than five chunks. In some cases it was difficult for subjects to keep track of items more than two places back in the queue. Yntema (1963) demonstrated that the way information is organized has a direct impact on supervisors' abilities to keep track of values of multiple attributes of several objects (e.g., status and descriptions of several aircraft). Supervisors had greater success keeping track of a few objects that varied on many different (and discriminable) attributes than keeping track of variation in a few attributes for many objects. In the former case, there are fewer opportunities for confusion than in the latter case, and confusion is a minor source of disruption in working memory. Memory in such tasks is particularly helped when each item is associated with a unique and constant location in space (e.g., a spatially consistent window) (Hess and Detweiler, 1995).

This earlier research on running memory anticipates current interest in *situation awareness* (Human Factors, 1995). Endsley (1995) defines situation awareness as the "perception of the elements of the environment within a volume of time and space, the comprehension of their meaning and the projection of their status in the near future." Thus the pilot, chef, and pedestrian must all keep track of a multitude of changing dynamic stimuli and events in their environment, they must determine the relevance of those stimuli to their current task and overall goals, and they must project the status of the most relevant stimuli in the near future. Of the factors affecting situation awareness, the limitations of working memory play a large role (Adams, Tenney, and Pew, 1995; Wickens, 1996). While working memory is critical to situation awareness, it is the integration of information in working memory with information about earlier phases of the task, along with the operator's general knowledge of system functioning (his or her mental model) that determines comprehension of the current situation, and prediction of the future. This interplay between general domain knowledge stored in long-term memory, situation-specific memory, and current working memory is also emphasized by Adams, Tenney, and Pew (1995), who compare situation awareness to the processing involved in language comprehension, particularly the processing of stories.

4.4.3 Text Processing and Language Comprehension

By considering the constraints of working memory, we can begin to understand the strategies that we use to comprehend both spoken and written language. These strategies may, in turn, help us determine what makes some conversations or text passages more difficult than others to understand. Of course factors influencing the detectability and discriminability of the individual speech sounds (phonemes) and written symbols (letters) will limit the extent to which language can be meaningfully processed. However, it should be recalled that easily comprehensible phrases or sentences can also influence the detectability of the individual words. See Section 3.2 for a discussion of the effect of context on identification.

In order to provide estimates of the difficulty of texts, "readability" metrics have been developed that assume that a more difficult text has, on average, longer words and sentences. Indeed, longer words generally may be less frequently used, and hence less familiar, and longer sentences may place greater demands on working memory. However, language can be more or less comprehensible for a variety of other reasons. Kintsch and Vipond (1979), for example, used traditional readability indices to compare the comprehensibility of the political speeches of candidates in the 1952 presidential campaign. Eisenhower's speeches were generally reputed to be simpler than those of Stevenson, yet the readability indices indicated that Stevenson's used shorter words and sentences. This contradiction between public opinion and the formal metrics corresponds to our experience that some sentences with a few short words can still be very confusing. We now discuss some of the additional factors that can directly affect comprehensibility.

Kintsch and colleagues (e.g., Kintsch and Keenan, 1973; Kintsch and Van Dijk, 1978) have proposed a model of text comprehension that identifies several complexity-adding features. First, they argue that the complexity of a sentence is actually determined by the number of underlying ideas, or *propositions*, it contains rather than by the number of words. Although a few specific words may be carried forward in working memory for brief periods, it is the underlying propositions that are used to relate information in different phrases and sentences. Kintsch and Van Dijk estimate that only four such propositions can be held in working memory at one time. Thus, the reader must be selective in the choice of propositions to retain. According to the model, readers tend to favor the most recent propositions and those they believe to be most central to the overall text message.

Problems arise in comprehension when newly encountered propositions cannot be easily related to the propositions active in working memory. Such problems often occur when readers attempt to integrate information across sentence boundaries. Consider, for example, the following sentences:

1. When the battery is weak, a light will appear.
2. You will see it at the top of the display panel.

Readers must make the *bridging inference* that the second sentence is telling them where to look for the light rather than where to find the battery. This inference, in turn, depends on their general knowledge of displays—specifically, the fact that lights rather than batteries tend to appear on display panels. A second type of integration failure occurs when a concept introduced earlier in the text is not used again until some sentences, paragraphs, or pages later. For instance, if the battery mentioned above was first introduced a paragraph or two before its current use, with no reference in the intervening text, then the reader would have to pause to search long-term memory (or scan the text itself) to determine precisely *what* battery was being discussed. Having to perform such a *reinstatement search* is yet another negative consequence of delaying the use of information in working memory.

One general goal in striving for comprehensibility is to avoid the need to make bridging inferences or perform reinstatement searches. However, it is clearly impossible to remove the need to make some inferences, and it is probably undesirable given that such elaborations may make the information more memorable. One goal of the text designer is simply to assist the reader in making the appropriate inferences. One important way that this can be done is by providing adequate *context* prior to the presentation of target information. Because inferences draw on the reader's knowledge of particular topics, it is useful to allow the reader to access the relevant knowledge structures in long-term memory at the outset. Bransford and Johnson (1972) provide a powerful demonstration of the importance of providing context in the form of pictures or descriptive titles presented just prior to textual material. In one case, a series of instructions on how to wash clothes was presented (Bower, Clark, Lesgold, and Winzenz, 1969) with and without the prior context of a title "washing clothes." Most of the material was well recalled when the title was present, but when the title was removed, the loss in the ability to understand and recall the instructions was dramatic.

Other factors that increase the processing demands of verbal material include the use of negations and lack of congruence between word orders and logical orders. With regard to negations, research indicates that it takes longer to verify a sentence such as "the circle is not above the star" compared to "the star is above the circle" (Carpenter and Just, 1975; Clark and Chase, 1972). Results further suggest that the delay is due to something other than the time necessary to process an additional word (i.e., "not"). Instead, it appears that listeners or readers first form a representation of the objects in the sentence based on the order of presentation (e.g., circle-before-star in the sentence "the circle is not above the star." However, to make their mental representation congruent with the meaning of the negation, they must perform a transformation of orders (i.e., to end up with a circle that is *not* before/above the star). Similar logic is used to explain why subjects have trouble processing statements in which the logical order represented by the sentence is inconsistent with the physical ordering of the words (DeSoto, London, and Handel, 1965). Returning to our battery instructions once again, the underlaying causal sequence assumed by most people would be that a weak battery would trigger a warning light. To be consistent with this causal order, it would be better to state that "If the battery is weak, then the light will come on" rather than "If the light comes on, then the battery is weak."

Finally, the physical parsing of sentences on a page, sign, or computer screen can also influence the comprehensibility of verbal messages. Just and Carpenter (1987) have suggested that although meaning is continually extracted as we arrive at each word in a sentence, there is a pause for the overall integration at the end of the constituent phrases. Consistent with this idea, Graf and Torrey (1966) found enhanced comprehension for sentences that were broken into several different lines of text, when the end of each line corresponded to the end of a phrase. Thus, instructions or warnings that must appear on several different lines (or as a few words on several successive screens) should be divided by phrases rather than, for example, on the basis of number of letters. "Watch your step . . . when exiting . . . the bus" will be more quickly understood than "Watch your . . . step when . . . exiting the bus."

4.4.4 Spatial Awareness and Navigation

Language comprehension sometimes taxes working memory, particularly the phonological rehearsal loop and central executive. However, as we saw when discussing problems with negation, people may use text to generate representations of spatial relations. This facet of text and language comprehension has been particularly prominent in recent discussions of the "situation models" that we develop when reading or listening to a story. We now turn to a task that relies more heavily, for many people, on the capacity limits of the "visuospatial scratchpad" (Logie, 1995)—navigating through our worlds, both real (finding our way through a maze of looping suburban streets and cul-de-sacs; Whitaker and CuQlock-Knopp, 1995) and virtual (searching a complex computer-displayed multidimensional data base).

4.4.4.1 Geographical Knowledge

Thorndyke (1980) has studied the knowledge that people use when finding their way through an environment. Of particular interest is Thorndyke's claim that increased familiarity with an area causes changes in more than the amount of detail contained in our mental representation of that area stored in long-term memory. In addition, the actual type of mental representation (analog versus verbal/symbolic), as well as its frame of reference, may evolve in a predictable way. After an initial encounter with a city, neighborhood, or building, we may develop *landmark knowledge*. The person with landmark knowledge, if told that his or her destination is beside the "telephone tower," will visually scan the environment until spotting something that appears to be the tower and will then strike off in its direction. Thus, the newcomer had the knowledge necessary to recognize the landmark, but had no knowledge about its location. For the person with landmark knowledge alone, wayfinding would be impossible if the landmarks were obscured. This problem has become commonplace as once-salient landmarks have become obscured by new and often taller structures. Guidance signs to landmarks have become a familiar antidote to the problem, but these in turn add to the visual clutter and confusion that may greet a first-time visitor to a new area. The problem for urban planners, then, is to ensure that landmarks (both natural and designed) remain easily visible and distinctive in order to serve their navigational function for years to come.

With more experience traveling about an area, we typically develop an ordered series of steps that will get us from one location to another. These sets of directions, called *route knowledge*, tend to be verbal in nature, stated as a series left/right turns (e.g., Go left on Woodland until you get to the fire station. Then take a left . . .). Navigation along these routes may be rapid and very automatic; however, limited knowledge of the higher order relations among different routes and landmarks still limits navigational decision making, for example making it difficult to figure out "short cuts." With stili more extensive wayfinding experience, or with specific map study, *survey knowledge* may be acquired. Survey knowledge is an integrated representation of the various routes and landmarks that preserves their spatial relations. This analog representation is usually referred to as a "cognitive map."

The type of representation—route versus survey—that best supports performance in various way-finding tasks, like so many other aspects of mental (and display) representation, depends on the nature of the task or problem. Thorndyke and Hayes-Roth (1982) compared route training (actual practice navigating between specific points in a large building) to survey training (study of the building plan). Route training appeared to facilitate subjects' estimates of route distance and orientation, while survey training appeared to facilitate judgments of absolute (Euclidean) distance and object localization.

4.4.4.2 Navigational Aids

While we can often navigate through environments on the basis of our acquired knowledge stored in long-term memory, whether route, survey, or even landmark, there are many other circumstances in which we require displayed *navigational aids* which are perceived. These aids may take on a wide variety of forms, ranging in the degree to which guidance to a target is supported; from tightly guided flight directors in aircraft, and turn signs on highways, to route lists, to electronic maps that highlight one's current position, to simple paper maps. Furthermore, electronic maps can vary in the extent to which they rotate in the direction of travel, and both electronic and paper maps can vary in terms of whether they present the world in planar or perspective view (see Section 4.3.4).

To understand which forms of maps support the best spatial information processing to accomplish navigation, it is important to consider briefly the stages involved in this process. The navigator must engage in some form of *visual search* of both the navigational aid (to locate the final destination and intermediate goals), and of the environment or a displayed representation thereof (to locate features that establish the current location and orientation). The navigator then must establish the extent to which the former and the latter are congruent. That is, establish the extent to which "where I am" (located and oriented) agrees with the intermediate goal of "where I want to be." Establishing this congruence may require any number of different *cognitive transformations* that add both time and effort to the navigational task (Aretz, 1991; Aretz and Wickens, 1992; Huey and Wickens, 1993; Wickens and Prevett, 1995; Wickens et al., 1996).

An example of two of these transformations is represented in Figure 4.11, which represents the information processing of a pilot, flying south through an environment depicted on a north-up contour map. To establish navigational congruence, the pilot must *mentally rotate* the map to a track up orientation, and then *envision* the contour representation of the 3-D terrain, to determine its congruence with the forward view. Both of these information transformations are effortful, time consuming, and provide sources for error. In particular, those sources involved with mental rotation of maps have been well documented (Aretz, 1991; Eley, 1988; Levine, 1982; Peruch and Savoyant, 1991; Warren, Rossano, and Wear, 1990; Wickens, Liang, Prevett, and Olmos, 1996).

Different transformations may be required when other navigational aids are provided. For example, verbal descriptions of landmarks will also require some transformations to evaluate against their visible 3-D spatial counterparts. Transformations may also be required to "zoom in" to a large-scale map (Kosslyn, 1980, 1987), in order to establish its congruence with a close-in view of a small part of the environment. Modeled in terms of processing operations such as search and transformations, one can then determine the form of navigational aids that would be of benefit for certain tasks. For example, elec-

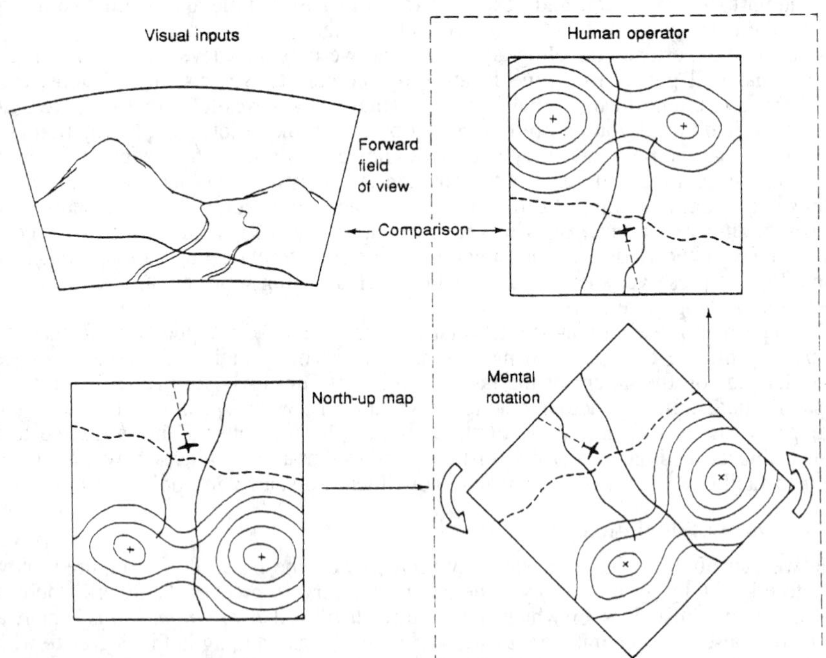

Figure 4.11 Illustrates the mental rotation required to compare the image seen in an ego-referenced forward field of view (top) with a world-referenced north up map (below) when the aircraft is heading south. The map image is mentally rotated (right) to bring it into lateral congruence with the forward field of view. It is then envisioned in 3-D to compare with the forward field of view.

tronic maps are beneficial if they highlight the navigator's current location, thus obviating visual search of the map. Highlighting landmarks on the map, which are salient in the visual world, will correspondingly reduce search.

Rotating maps in a track-up orientation will facilitate navigation by eliminating mental rotation (Aretz, 1991; Wickens et al., 1996). Presenting guidance information in a "3-D" format, like one would see looking ahead into the environment itself will also reduce the magnitude of any sort of transformations and considerably improve navigational performance (Haskell and Wickens 1993; Wickens and Prevett, 1995). The benefits of a 3-D view will be enhanced if the viewpoint of the display corresponds to the same "zoom in" viewpoint as that occupied by the navigator, looking forward, rather than a viewpoint that is behind and from outside [Wickens and Prevett, 1995; see Figure 4.12; compare (*a*) and (*b*) with (*c*)].

Expressing navigational guidance in terms of command **route lists** (e.g., "turn left at *X*; go three blocks until *Y*") will also eliminate the need for many spatial cognitive transformations that may be imposed when spatial maps are used, because the language of command is thereby directly expressed in the language of action. Such congruence can account for the benefits of route lists over spatial maps in certain ground navigation tasks (Streeter et al., 1985; Wetherell, 1979). A second advantage to such route lists is that they can be presented verbally, and mentally represented in working memory in a phonetic or verbal code, thus reducing competition for the visual spatial processing resources involved in many aspects of environmental scanning and vehicle navigation (Section 4.6).

Extremely direct levels of navigational guidance that eliminate most or all levels of mental transformations (e.g., a flight director, a 3-D forward looking display shown in Figure 4.12, or a verbal route list), will provide for effective navigation while on route. However, such displays may do a disservice to the navigator who suddenly finds himself lost, disoriented, or required to make a spontaneous departure from the planned route. Very often those features that make a navigational display best for guidance, will harm its effectiveness to support the spatial situation awareness which is necessary for a successful recovery from a state of geographical disorientation (Wickens, 1996; Wickens and Prevett, 1995).

4.4.5 Planning and Problem Solving

Our previous discussion has focused on cognitive activities that were heavily and directly driven by information in the environment (e.g., text, maps, or material to be retained in working memory). In contrast, the information-processing tasks of planning and problem solving are much less directly tied to perceptual processing, and are more critically de-

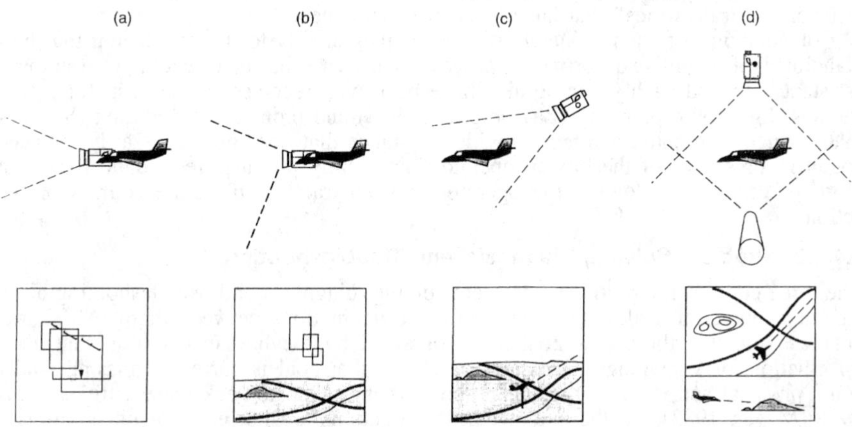

Figure 4.12 Various display viewpoints in an aircraft display that require varying degrees of transformations to compare with a pilot's direct view forward from the cockpit. The four panels below schematically illustrate what would be seen by the pilot with the viewpoint above. The transformation in (*a*) is minimal, in (*b*) and (*c*) is modest, and in (*d*) is large. Views (*a*) and (*b*), however, reduce global situation awareness because less of the space is portrayed.

pendent on the interplay between information available in (and retrieved from) long-term memory, and information-processing transformations carried out in working memory.

4.4.5.1 Planning

The key to successful operation in many endeavors (Miller, Galanter, and Pribram, 1960) is to develop a good plan of action. When such a plan is formulated, steps toward the goal can be taken smoothly without extensive pauses between subgoals. Furthermore, developing contingency plans will allow selection of alternative courses of actions should primary plans fail. As an example, pilots are habitually reminded to have contingency flight plans available, should the planned route to a destination become unavailable because of bad weather.

Planning can typically depend on either of two types of cognitive operations (or a blend of the two). The planner may depend on **scripts** (Schank and Abelson, 1977) of typical sequences of operations that they have stored in long-term memory, on the basis of past experience. In essence, one's plan is either identical to or involves some minor variation on the sequence of operations that one has carried out many times previously. Alternatively, planning may involve a greater degree of guess work, and some level of **mental simulation** of the intended future activities (Klein and Crandall, 1995; see Chapter 38). For example, in planning how to attack a particular problem, one might play a series of "what if" games, imagining the consequences of action, based again on some degree of past experience. Hence an air traffic controller, in planning how to manage a potential future conflict situation, might mentally simulate the future trajectories of the affected aircraft under different proposed maneuvers, to see if the intended commands would resolve the conflict, and would stay clear of other aircraft.

Consideration of human performance issues and some amount of experimental data reveals three characteristics of planning activities. First, they place fairly heavy demands on working memory, particularly as plans become less script-based, and more simulation-based. Hence, planning is a task that is vulnerable to competing demands from other tasks. Under high workload conditions, planning is often the first task to be dropped, and operators become less **proactive** and more **reactive** (Hart and Wickens, 1990). The absence of planning is often a source of poor decision making (Orasanu, 1993). Second, perhaps because of the high working memory demands of planning, in many complex settings, people's **planning horizon** tends to be fairly short, working no more than one or two subgoals into the future (Tulga and Sheridan, 1980). To some extent, however, this characteristic may be considered as a reasonably adaptive one in an uncertain world, because many of the contingency plans for a long time horizon in the future would never need to be carried out, and hence, probably not worth the workload cost of their formulation. Finally, given the dependency of script-based planning on long-term memory, many aspects of planning may be biased by the **availability heuristic,** discussed in more detail in Chapter 38 (Tversky and Kahneman, 1974). That is, one's plans may be biased in favor of "trajectories" that have been tried with success in the past.

Consideration of such vulnerabilities leads inescapably to the conclusion that human planning is a cognitive information-processing activity that can benefit from automated assistance. Indeed such planning aids have been well received in the past, for activities such as flight route planning (Layton et al., 1994) and industrial scheduling (Sanderson, 1989). Such automated planners provide assistance that does not necessarily replace the cognitive processes of the human operator, but merely provides redundant assistance to those processes, in allowing the operator to keep track of plausible courses of future action.

4.4.5.2 Problem Solving, Diagnosis, and Troubleshooting

The three cognitive activities of problem solving, diagnosis and troubleshooting all have similar connotations, although there are some distinctions between them. All share the characteristic that there is a goal to be obtained by the human operator, that actions, information, or knowledge necessary to achieve that goal is currently missing, and that some physical action or mental operation must be taken to seek these entities (Levine, 1988; Mayer, 1983). To the extent that these actions are not easy, or not entirely self-evident, the processes are more demanding.

Like planning, the actual cognitive processes underlying the diagnostic troubleshooting activities, can involve some mixture of two extreme approaches. On the one hand, sometimes situations can be diagnosed (or solutions to a problem reached) by a direct match between the features of the problem observed and patterns previously experienced and stored in long-term memory. Such a "pattern matching" technique, analogous to the role of scripts in planning, can be carried out rapidly, with little cognitive activity, and is often highly accurate (Rasmussen, 1981). This is a pattern of behavior often seen in the study of naturalistic decision making (Klein and Crandall, 1995; see Chapter 38). It is a pattern characterizing what Rasmussen et al. (1995) describe as "skill-based behavior."

At the other extreme, when solving complex and novel problems which one has never experienced before, a series of diagnostic tests often must be performed, their outcomes considered, and based on these outcomes, new tests or actions taken, until the existing state of the world is identified (diagnosis) or the problem is solved. Such an iterative procedure is typical in medical diagnosis (Shalin and Bertram, in press). The updating of belief in the state of the world, on the basis of the test outcomes, may or may not approach prescriptions offered by guidelines for "optimal information integration," such as Bayes theorom (Yates, 1992; see Chapter 38). It is a pattern of behavior characterizing what Rasmussen et al. (1995) refer to as "knowledge-based behavior."

In between these two extremes are hybrid approaches that depend to varying degrees on information already stored in long-term memory on the basis of experience. For example, the sequence of administering tests (and the procedures for doing so) may be well learned in long-term memory, even if the outcome of such tests is unpredictable, and must be retained or aggregated in working memory. This pattern may be described as "rule-based behavior" (Rasmussen et al., 1995). Furthermore, the sequence and procedures may be supported by (and therefore directly perceived from) external **checklists,** relieving cognitive demands still further. The tests themselves might be physical tests, such as the blood tests carried out by medical personnel, or they may involve the same "mental simulation" of "what if" scenarios that were described in the context of planning (Klein et al., 1993).

As with issues of planning, so also with diagnosis and problem solving there are three characteristics of human cognition that impact that efficiency and accuracy of such processes. First, as these processes become more involved with mental simulation, and less with more automatic pattern matching, their cognitive resource demands grow, and their vulnerability to interference from other competing tasks increases in a corresponding fashion (see also Chapter 13, Workload). Second, as we noted, past experience, reflected in the contents of long-term memory, can often provide a benefit for rapid and accurate diagnosis or problem solutions. But at the same time, such experience can occasionally be hazardous by trapping the troubleshooter to consider only the most **available** hypotheses; that is, those that have been recently or frequently experienced, and hence well represented in long-term memory (Tversky and Kahneman, 1974). In problem solving, this dependence on familiar solutions in long-term memory has sometimes been described as **functional fixedness** (Adamson, 1952; Levine, 1988).

Third, the diagnostic/troubleshooting process is often thwarted by phenomena alternatively referred to by similar terms as the **confirmation bias, cognitive tunneling,** or **functional fixedness** (Levine, 1988; Wickens, 1992). These terms describe a state in which the troubleshooter tentatively formulates one hypothesis of the true state of affairs (or the best way to solve a problem), and then excessively continues on that track even when it is no longer warranted. This may be done by seeking only evidence to confirm that the chosen hypothesis is correct (the confirmation bias), or by simply ignoring other competing and plausible hypotheses (cognitive tunneling).

Collectively then, the joint cognitive processes of planning and problem solving (or troubleshooting), depending as they do on the interplay between working memory and long-term memory, reflect both the strengths and the weaknesses of human information processing. The output of each process is typically a **decision,** to undertake a particular course of action, to follow the plan, to choose a treatment based upon the diagnosis, or to formulate a solution to the problem. The cognitive processes involved in such decision making are discussed extensively in Chapter 38, as are some of the important biases and heuristics in diagnosis discussed more briefly above. In the present chapter, we turn now to characteristics of those actions that are typically selected rapidly, and without great uncertainty of their outcome.

4.5 ACTION SELECTION

4.5.1 Information and Uncertainty

In the previous sections, we have discussed different stages at which the human processes *information* about the environment. When we turn to the stage of action selection and execution, a key concern addresses the *speed* with which information is processed, from perception to action. How fast, for example, can we expect the pilot to react to an emergency signal, or how rapidly can we expect the postal worker to sort letters. Borrowing from terminology in communications, we describe information-processing speed in terms of the *bandwidth*, or the amount of information processed per unit time. In this regard, a unit of information is defined as a *bit*. One bit can be thought of as specifying between one of two possible alternatives; 2 bits as one of four alternatives; 3 bits as one of eight, or in general, # bits (conveyed by an event) = Log_2N, where N = the number of possible environmental events that could occur in the relevant task confronting the operator. In the pages below, we will see how information influences the bandwidth of human processing after we describe a taxonomy of human actions.

The speed with which people perform a particular action depends jointly on the *uncertainty* associated with the outcome of that action and on the *skill* of the operator in the task at hand. Rasmussen (1986) and Rasmussen et al. (1995) have defined a behavior-level continuum that characterizes three levels of action selection and execution in a manner parallel to its description of three levels of problem solving (Section 4.4.5.2). *Knowledge-based behavior* characterizes the action selection of the unskilled operator, or the skilled operator working in a highly complex environment facing a good deal of uncertainty. In the first case, we might consider the vehicle driver trying to figure out how to navigate through an unfamiliar city; in the second case, we consider the nuclear reactor operator trying to diagnose an apparent system failure. This is very much the sort of behavior discussed in our previous section on problem solving.

Rule-based behavior typically characterizes actions that are selected more rapidly, based on certain well-known rules. These rules map environmental characteristics (and task goals) to actions, and their outcomes are fairly predictable: "If the xxx conditions exist, then do yyy."

The response of rule-based behavior is fairly rapid, but is still "thought through" and may be carried out within a time-frame of a few seconds. Working memory is required. Finally, *skilled-based behavior* is very rapid and nearly *automatic* in the sense that little working memory is required and the action may be initiated within less than a second of the triggering event. Skill-based behavior, for example, characterizes the response of the operator monitoring a robot to an emergency light; the movement of the fingers to a key to type a letter; or the response of the pilot to an emergency ground proximity warning which says "pull up, pull up."

Human factors designers are very interested in the system variables that affect the speed and accuracy of behavior of all three classes. Typically those variables affecting knowledge-based behavior are discussed within the realm of problem solving and decision making (see Section 4.4 and Chapter 38), while we discuss below, the variables that influence rule- and skill-based behavior (see Wickens, 1992, for a more detailed discussion).

4.5.2 Complexity of Choice

Response times for either rule- or skill-based behavior become longer if there are more possible choices that could be made, and therefore, more information transmitted per choice (Hick, 1952; Hyman, 1953). The rule-based decision to go left or right at a Y fork in the road is simpler (and made more rapidly) than at an intersection where there are three or four alternative paths. Menu selections take longer on a page where there are more menu options, and each stroke on a typewriter (26 letter options) takes longer to initiate than each depression of a Morse code key (two options). As a guideline, designers should not give users more choices of action than are essential, particularly if time criticality is at issue.

However, this does not necessarily mean that simpler choices (fewer bits/choice) are necessarily better. Indeed, generally an operator can transmit more total information per unit time with a few complex (information-rich) choices than several simple (information-poor) choices. This *decision complexity advantage* (Wickens, 1992) can be illustrated by two examples: First, an option provided by a single computer menu with eight alternatives (one complex decision) can be selected faster than an option provided by three consecutive selections from three, two-item menus (three simple decisions; see Figure 4.13). Second, voice input, in which each possible word is a choice from a potentially large vocabulary (high complexity) can transmit more information per unit time than typing, with each letter indicating one of only 26 letters (less complex); and typing in turn can transmit more information per unit time than can Morse code. The general conclusion drawn from

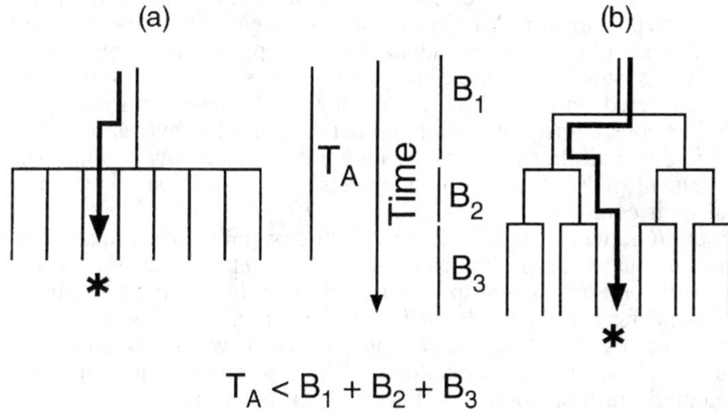

Figure 4.13 The decision complexity advantage. (*a*) illustrates the time required for a single high complexity choice. (*b*) illustrates the (longer) total time required for three "simple" (low complexity) choices. The total amount of information transmitted is the same in both cases.

these examples and from other studies (Wickens, 1992) points to the possible advantage of incorporating keys, or output options that can select from a larger number of possible options—like special service "macro" keys, keys that represent common words, or "chording" keyboard devices (Gopher and Raij, 1988) that allow a single action selection (a chord depression using several fingers simultaneously) to select from one of a very large number of possible options.

4.5.3 Probability and Expectancy

People respond more slowly (and are more likely to respond erroneously) to signals and events that they do not expect. Generally, such events are unexpected (or surprising) because they occur with a low probability in a particular context. Low probability events, like events with a greater number of alternatives, are also said to **convey more information,** and the greater information content requires more time for processing (Hyman, 1953; Fitts and Posner, 1967). For example, system failures usually occur rarely and as such, are often responded to slowly or inappropriately. A similar status may characterize the driver's response to the unexpected appearances of a pedestrian on a freeway, a traffic light that changes sooner than expected, or the pilot's response to an aircraft that suddenly appears on the runway which had previously been cleared for landing. More serious than the slower response to the unexpected event is the potential failure to detect that event altogether (see Section 4.3.1). It is for this reason that designers should ensure that annunciators of rare events are made salient and obtrusive or redundant (to the extent that the rare event is also one that is important for the operator's task) (see Section 4.2).

4.5.4 Practice

Practice has two benefits to action selection. First, practice can move knowledge-based behavior into the domain of rule-based behavior (see Chapter 5) and can move rule-based actions into the domain of skill-based ones. The novice pilot may need to think about what action to take when the stall warning sounds, whereas the expert should respond automatically and instinctively. In this sense, practice increases both speed and accuracy. Second, practice will provide the operator with a sense of *expectancy* that is more closely calibrated with the actual probabilities and frequencies of events in the real world. Hence, frequent events will be responded to more rapidly by the expert; but ironically, expertise may lead to *less* efficient processing of the rare event. This may not be the case for the novice, for which the rare event is not perceived as unexpected.

4.5.5 Spatial Compatibility

The compatibility between a display and its associated control has two components. One relates to the **location** of the control relative to the display; the second to how the display reflects (or commands) control **movement.** In its most general form, the principle of location compatibility dictates that the location of a control should correspond to the location of a display. There are several ways of describing this correspondence. Most directly this correspondence is satisfied by the principle of **collocation,** which dictates that each display should be located adjacent to its appropriate control. But this is not always possible in systems when the displays themselves may be closely grouped together (e.g., closely clustered on a CRT panel) or may not be easily reached by the operator because of other constraints (e.g., common visibility needed by a large group of operators). Then the compatibility principle of **congruence** takes over, which states that the spatial arrangement of a set of two or more displays should be congruent with the arrangement of their controls.

The distinction between "left" and "right" in designing for compatibility can be expressed either in relative terms (indicator A is to the left of indicator B) or in absolute terms, relative to some prominent axis. This axis may be the body midline (i.e., distinguishing left hand from right hand), or it may be a prominent visual axis of symmetry in the system, like that bisecting the cockpit on a twin seat airplane design. When left-right congruence is violated, the operator may have a tendency to activate the incorrect control, particularly in times of stress (Fitts and Posner, 1967).

Sometimes an array of controls are to be associated with an array of displays (e.g., four engine indicators). Here, congruence can be maintained (or violated) in several ways. Compatibility will be best maintained if the control and display arrays are parallel. It will be reduced if they are orthogonal (Figure 4.14; i.e., a vertical display array with a horizontal left-right or fore-aft control array). But even where there is orthogonality, compatibility can be improved by adhering to two guidelines: (1) the left end of a horizontal array should map to the near end of a fore-aft array; (Figure 4.14(*b*)) (2) the particular display (control) at the end of one array should map to the control (display) at the end of the other array to which it is closest (Andre and Wickens, 1990). It should be noted in closing, however, that the association of the top (or bottom) of a vertical array with the right (or high) level of a horizontal array is not a strong one. Therefore, ordered compatibility effects with orthogonal arrays will not be strong if one of those arrays is vertical. Hence, some augmenting cue should be used to make sure that the association

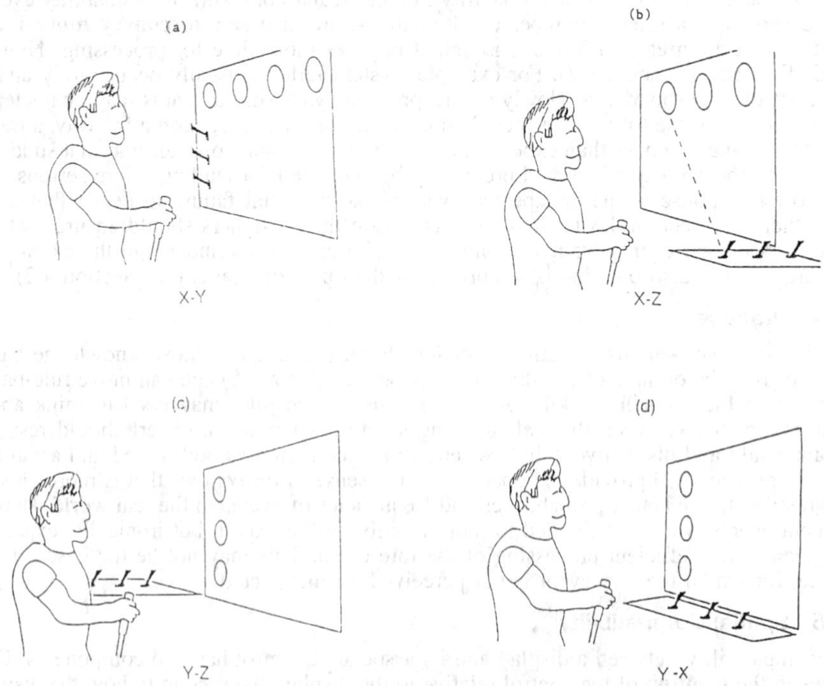

Figure 4.14 Different possible orthogonal display-control configurations.

between the appropriate ends of the two arrays is clearly articulated (e.g., a common color code on both, or a painted line between them; Osborne and Ellingstad, 1987).

The movement aspect of SR compatibility may be defined as **intention-response-stimulus compatibility** or **IRS-compatibility.** This means that the operator formulates an intention to do something: increase, activate, set, turn something on, adjust a variable, etc. Given that intention, the operator makes a response, an adjustment. Given that response, some stimulus is (or should be) displayed as feedback from what has been done (Norman, 1988). There is a set of rules for this kind of mapping between an intention to respond, a response, and the display signal. The rules are based on the idea that people generally have a conception of how a quantity is ordered in space. As we noted in Section 4.3.5, when we think about something increasing, we think about a movement of a display that is either upwards, to the right, forward, or clockwise. Both control and display movement should then be congruent in form and direction with this ordering. These guidelines are shown in Figure 4.15 (Wickens, 1984). Whenever one is dealing, for example, with a rotary control, operators have certain expectations about how the movement of that control will be associated with the corresponding movement of a display. These expectancies may be defined as *stereotypes,* and there are three important stereotypes.

The first is the **clockwise increase stereotype:** a clockwise rotation of a control or display signals an increasing quantity (Figure 4.15(*c*),*d*). The **proximity of movement stereotype** says that with any rotary control, the arc of the rotating control element that is closest to the moving element in a display is assumed to move in the same direction as that display element. In panel (*c*) of Figure 4.15, rotating the control clockwise is assumed to move the needle to the right, while rotating it counterclockwise is assumed to move the needle to the left. It is as if the human's "mental model" is one that assumes that there is a mechanical linkage between the rotating object and the moving element, even though that mechanical linkage may not really be there.

The important point is that it is very easy to come up with designs of control display relations that conform to one principle and violate another. Panel (*e*) shows a moving vertical scale display with a rotating indicator. If the operator wants to increase the quantity, he or she rotates the dial clockwise. That will move the needle on the vertical scale up, thus violating the proximity of movement stereotype. The conflict may be resolved

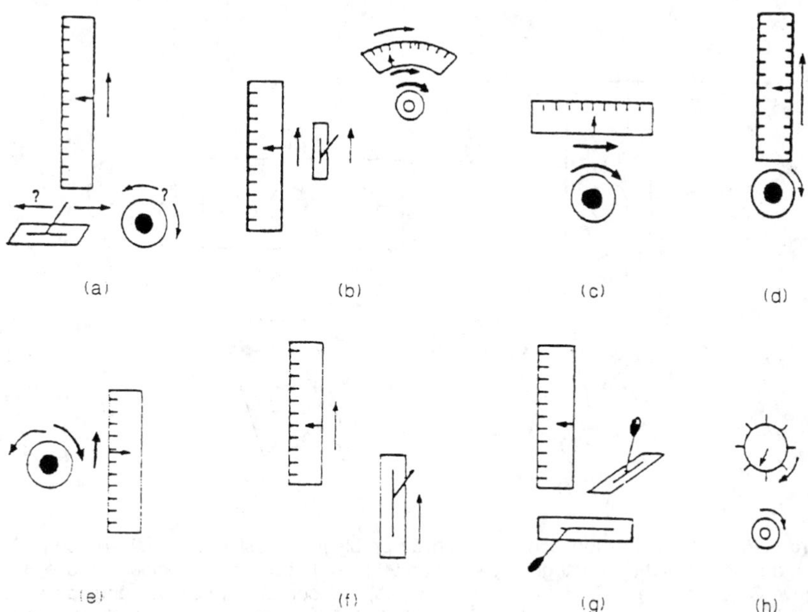

Figure 4.15 Examples of population stereotypes in control-display relations. These are discussed in the text (from Wickens, 1984).

by putting the rotary control on the right side rather than the left side of a display. We have now created a display-control relationship that conforms to both the proximity of movement stereotype as well as the clockwise to increase stereotype.

The third component of movement compatibility relates to **congruence.** Just as with location compatibility, so movement compatibility is also preserved when controls and displays move in a congruent fashion: linear controls parallel to linear displays [(*f*), but not (*g*)], and rotary controls congruent with rotary displays [(*b*) and (*h*)]. Note, however, that (*h*) violates proximity of movement]. When displays and controls move in orthogonal directions, as in (*g*), the movement relation between them is ambiguous. Such ambiguity, however, often can be reduced by placing a modest "cant" on either the control or display surface, so that some component of the movement axes are parallel. This approach may be employed equally with regard to ambiguity of movement compatibility, or ambiguity of location compatibility as shown in Figure 4.16 (Wickens, 1992).

4.5.6 Modality

Skilled responses in most human–machine systems are typically executed either by the hands or the voice. With increasingly sophisticated automated voice recognition systems, the latter option is becoming progressively more feasible. While the particulars of voice control are addressed in more detail in Chapter 22, at least three characteristics of voice control are relevant in the context of information processing.

(1) Voice options allow more possible responses to be given, in a shorter period of time, without imposing added time-consuming manual components (i.e., keys) (although this requires more sophisticated software in the voice recognition algorithms). Providing more options, enabling more complex decisions to be selected is, as we have seen in Section 4.5.2, a positive benefit.

(2) Voice options represent more *compatible* ways of transmitting symbolic or verbal information, than can be achieved by spatially guided manual options (Wickens, Vidulich, and Sandry-Garza, 1984). But at the same time, voice responses make relatively poor candidates for transmitting continuous analog-spatial information, particularly in dynamic situations (e.g., tracking; Wickens, Zenyuh, Culp, and Marshak, 1985).

(3) Voice options are valuable in environments when the eyes, and in particular the hands, are otherwise engaged; but conversely voice options can be problematic in environments in which a large amount of other verbal activity is required, either by the user or by other people in the nearby workspace. The former causes competition for processing

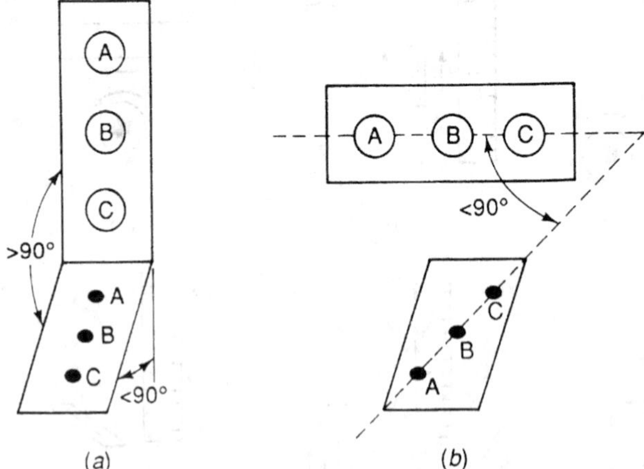

Figure 4.16 Solutions of location compatibility problems by using cant. (*a*) The control panel slopes downward slightly (an angle greater than 90°), so that control A is clearly above B, and B is above C, just as they are in the display array. (*b*) The controls are slightly angled from left to right across the panel, creating a left-right ordering that is congruent with the display array. The approach is also effective for creating movement compatibility if the ordering A, B, C in the figure is replaced by movement of a control (display) from A → C.

resources within the operator, while the latter creates the possibility of confusion (see Section 4.6 and Chapter 13).

4.5.7 Response Discriminability

Whenever a set of manual responses are specified, any increases in the similarity between them (decreases in discriminability) will increase the likelihood of confusion. Thus, movement of the stick to either one of two forward positions is a response choice that has greater opportunity for confusion than movement in either a forward or backward position. Correspondingly, two buttons that look alike are more confusional (and hence, error prone) than are two that are differently colored or shape coded. While making controls physically distinct from each other may sometimes destroy a sense of aesthetics in design, such distinctions will generally lead to improved human reliability (Norman, 1988). Incidentally, increased similarity between voice control options will produce the same increase in error likelihood, although here the mediating agent is the voice recognition agent (whether human or computer) rather than the human responder.

4.5.8 Feedback

The quality of feedback provided by control manipulation (or action expression) is often critical to the speed of information transmission (Norman, 1988). Indeed sometimes problems of poor response discriminability (Section 4.5.7) can be addressed and at least partially remedied by providing clear, salient, and immediate *feedback* as to which (of several confusable) response alternatives have been chosen. This feedback may be in the form of a visible light, or an auditory or tactile "click" as the control reaches its appropriate destination.

It turns out, however, that salient feedback is not always necessary or even desirable. In particular, expert users rely far less on feedback than novices (e.g., the skilled typist, when transcribing, rarely looks at the keyboard or the screen). Thus, if the feedback is salient (and hence intrusive), as is often the case with auditory feedback, it may be distracting to the expert. This will be particularly true whenever the feedback is *delayed,* a quality that is especially disruptive for continuous performance like data transcription or voice translation (Smith, 1962).

4.5.9 Continuous Control

Our discussion in the previous section focused on the selection of discrete actions, such as a keypress or lever movement. Equally important are the *continuous* movements of some indicator to reach targets in space. These movements may refer, for example, to the movement of the hand to a point on a touch screen, the movement of a cursor to a window or word on a computer screen, or the movement of a pointer to a set point on a meter. Generically then, we can speak of these skills involving movement of a *cursor* to a *target.*

To an even greater extent than the discrete movements discussed in the previous section, performance of these continuous movement skills is critically dependent on *visual feedback*, depicting the difference between the current cursor location and the desired target. Performance on control tasks in which a cursor is moved a distance, D, into a target of width, W, is well described by *Fitts' Law* (Fitts, 1966; Jagacinski, 1989), an equation of the form:

$$\mathrm{MT} = a + b \operatorname{Log}_2 \frac{2D}{W}$$

This very robust law can accurately predict the movement of all sorts of devices, from microscopic pointers (Langolf et al., 1976) to cursor movements by mice (Card, 1981) and to foot movement around a set of pedals (Drury, 1975). The basis of Fitts' Law lies in the processing of visual feedback, such that the movement toward a target is maintained at a rate that is inversely proportional to the momentary distance of the "cursor" away from the target.

Just as Fitts' Law nicely describes the continuous movement toward a static target, it can also characterize the movement of a cursor toward a *continuously moving target*, a process typically described as *tracking* (McRuer, 1980; Poulton, 1974; Wickens, 1986). When operators engage in continuous tracking, however, whether keeping a car in the center of the highway, flying an airplane down the glidepath, or moving one's viewpoint through a "virtual environment" via some control device, interest is more focused on minimizing the error away from the target than on the time required to reach the target. Also, concern is less with the amplitude of the movement than it is with variables such as the *frequency* with which corrections must be made (the signal bandwidth), the *complexity and lag* of the dynamics of any system mediating between hand movement and cursor movement, and the manner in which feedback is displayed. These issues extend well beyond the scope of the current chapter and are covered in more depth in Chapter 38.

4.5.10 Errors

The previous discussion has focused primarily on the **time** required to process and respond to various items of information. Yet in many systems, the occurrence of errors are more critical than the occurrence of delays in processing. That is, the loss, rather than the delay of information transmission is the factor of greatest concern. Although errors are treated extensively in Chapter 6, we wish here to highlight the manner in which different classes of errors can be categorized according to the framework of information, processing, as depicted in Figure 4.2 (Norman, 1981; Reason, 1990).

First, *mistakes* represent errors of the earlier stages of information processing, in which incorrect action is carried out as a result of a failure to **understand** the nature of a situation. This may result from a breakdown in perception, in working memory, or from insufficient knowledge to interpret the available cues. Second, while a situation may be diagnosed and understood correctly, errors may result from a failure to appropriately apply the correct *rules* for selection of a response (Reason, 1990). Third, errors may result from *slips* of action, when the correct response is intended, but an incorrect action is actually released (Norman, 1981). Slips of this sort are typically the result of poor human factors design, such as incompatible control-display relationships, coupled with an operator who is well skilled, and performing a task in a highly automated mode, thereby not carefully monitoring his or her own action selections.

Errors can be directly attributed also to the breakdown of *memory*. As noted in Section 4.4, short-term memory breakdowns may lead to forgetting or confusing material, whereas errors of prospective memory may lead operators to forget to perform some action that was previously intended.

Usually the conditions that are associated with slower processing are those that also produce more errors, and hence design remediation based on measures of processing speed will be productive in improving overall system performance. However, in certain circumstances, a strategic adjustment in how an operator performs a task will lead to an inverse relationship between speed and accuracy; the so-called **speed accuracy tradeoff.** In this case, a "set" to respond rapidly will lead to more, rather than fewer errors (Wickelgren, 1977). An example here would be the effects of time stress in emergencies, which may lead to hasty and error prone actions in the processing of information.

4.6 MULTIPLE TASK PERFORMANCE

Many task environments require operators to process information from more than one source and to perform more than one task at a time (Damos, 1991). Such environments are as diverse as that confronting the secretary, who talks on the phone while typing; the vehicle driver, who listens to the radio while steering the car and searching for a road sign; or the maintenance technician, who performs and observes diagnostic tests while keeping active hypotheses rehearsed in working memory.

In such multiple task environments, we may distinguish between three qualitatively different *modes* of multiple task behavior: *perfect parallel* processing, in which two (or more) tasks are performed concurrently as well as either is performed alone; *degraded concurrent* processing, in which both tasks are performed concurrently, but one or the other suffers, relative to its single task level; and *strict serial* processing, during which the operator performs only one task at a time. Each of these different modes are observed under different circumstances and have somewhat different implications for design.

4.6.1 Serial Processing

Performing tasks in a serial mode generally presents no human factors concerns unless performance of one task or the other is undesirably delayed because of those sequential constraints. Such a delay might characterize the behavior of a pilot who fails to check the aircraft altimeter sufficiently often because he is engaged in other visual tasks, leading to dangerous "altitude busts" (Raby and Wickens, 1994). Typically then the interests of human factors personnel in sequential task performance is in modeling the *choice* process whereby the operator chooses to perform one task (and, by necessity, neglect another) at any given moment in time. This choice process is often modeled by queuing theory (Kleinman and Pattipati, 1991), or variants thereof (Carbonnell, Ward, and Senders, 1968; Sheridan, 1972; Moray, 1986), that specify when a task *should* be sampled (performed) as a function of that task's *importance* (cost of not performing it) and the *frequency* with which it should be carried out. When evaluated against these optimal benchmarks, human performance appears to be reasonably optimal, subject to the constraints of working memory. That is, humans may sometimes forget the precise value of a task status at the time that it was last sampled, and hence sample it more frequently than necessary, compared to an "optional" performer with perfect memory.

However, "reasonably" optimal is not the same thing as "perfectly" optimal, and others have focused interest on the occasional breakdowns in optimality that do occur. Thus, a different approach to human multiple task performance is to focus on the accidents and incidents that have apparently resulted from failures of effective task management (Adams, Tenney, and Pew, 1991; Funk, 1991; Hart and Wickens, 1990; Huey and Wickens, 1993; Orasanu, 1993; Raby and Wickens, 1994; Laudeman and Palmer, 1995); that is, what causes people to neglect a task.

Here the answers, based on empirical research, are not entirely clear, although three prominent factors do appear to emerge. First, visible or auditory reminders to do a task increase the likelihood that that task will be done, compared to circumstances in which task initiation must be based on prospective memory alone (Norman, 1988). The vulnerability of such memory highlights the value of checklists as visual reminders for the pilot (or other operators) to carry out certain actions at certain times (Degani and Wiener, 1993). Second, heavy involvement (high workload) with one task may lead an operator to neglect a second one, and perhaps fail to return to an activity at a time when that return should be critical. This deficiency may be addressed through task or workload management training programs (Wiener, Kanki, and Helmreich, 1993). Third, there are prominent individual differences between operators in both the kind and effectiveness of task management strategies that they deploy (Schutte and Trujillo, 1996; Orasanu, 1993; Raby and Wickens, 1994).

4.6.2 Concurrent Processing

In contrast to sequential processing, the understanding of concurrent processing, whether in degraded mode or perfect parallel mode depends on somewhat different mechanisms. These mechanisms are more closely related to the structure of the information-processing sequences within the tasks themselves than to the operator's knowledge of task importance and priority [although there are interactions between these two influences (Gopher, 1992)]. Here human factors interest is in the task features that can enable any sort of concurrent processing to emerge from serial processing, and that can furthermore enable that concurrent processing to be perfect rather than degraded. Three characteristics appear to influence this degree of success: task similarity, task demand, and task structure (Wickens, 1992, 1991), although how these influences are exerted is somewhat complex.

4.6.2.1 Task Similarity

A high degree of *similarity* between two tasks may induce confusion, just as similarity between perceptual signals will cause confusion, or high similarity between items held in working memory will increase the degree of interference between them, or high similarity between two response devices will increase the possibility of confusion of actions (Section 4.5.7).

On the other hand, making certain aspects of two tasks more similar *may* allow the tasks to be better integrated, fostering *more* effective concurrent processing. This may involve combining two axes of motion on a tracking task into the X and Y deviations of a single cursor (Chernikoff and LeMay, 1963; Fracker and Wickens, 1989), or using similar rules to map stimuli (events) to responses (actions) (Duncan, 1979; Fracker and Wickens, 1989).

This latter aspect of similarity—the similarity of rules—also facilitates the ability of the operator to *switch* attention between two tasks in sequential fashion. That is, it is easier to keep doing alternate versions of the same task, than it is to switch between different tasks, as if there is some "overhead" penalty for switching rules (Rogers and Monsell, 1995).

4.6.2.2 Task Demand

Easier tasks are more likely to be performed concurrently (and perfectly) than are more difficult or demanding tasks. We argue that easier tasks are generally more automated and consume less "mental effort" or resources than more difficult ones (see Chapter 13), although precisely what these resources are remains less clearly defined.

4.6.2.3 Task structure

Certain structural differences between two time-shared tasks increase the efficiency of their concurrent processing, as if the two tasks demand entirely (or partially) separated "resources," within the human processing system. Thus it is easier to distribute the two tasks across multiple resources, than to perform them within a single resource (Wickens, 1991, 1992).

These resources appear to be defined by perceptual modality (auditory versus visual), processing code (verbal versus spatial), and processing stage (perceptual cognitive operations versus response operations). It is because of the separate perceptual resources that designers have chosen to offload the heavy visual processing load of pilots and vehicle drivers with some information presented on auditory channels. It is because of the separate spatial and verbal code resources that spatially guided manual responses may be more effective than vocal responses when operators must also rehearse verbal material. But vocal responses will be more effective than spatially guided manual ones when the operator must concurrently perform another spatial task like tracking. As another example, we saw in Section 4.4.4 that the spatial cognitive processes involved in navigation and vehicle control are better time-shared with the verbal input of a memorized verbal route list than with the spatial input of a memorized map. Finally, it is because of the separate stage-defined resources that we are often able to effectively time share responding operations (e.g., talking) with perceptual ones (e.g., scanning).

4.7 CONCLUSIONS

In conclusion, the systems with which people must interact vary vastly in their complexity, from the simple graph or tool, to things like nuclear reactors and the pattern of air traffic flow. As a consequence, they vary drastically in the type and degree of demands imposed upon the varying information-processing components we have discussed in this chapter. In some cases, systems will impose demands on components that are quite vulnerable; working memory, predictive capabilities, and divided attention, for example. At other times, they may impose upon human capabilities that are a source of great strength, particularly if these sources rely upon the vast store of information that we retain in long-term memory; information that assists us in pattern recognition, top-down processing, chunking, and developing plans and scripts on the basis of past experience are examples. The acquisition of material to be stored in long-term memory, through training and learning, is the subject of the next chapter.

In addition to facilitating the performance of experts in many ways, long-term memory has a second implication for the practice of human factors. This is that predictions of human performance in many systems can only be partially based on an understanding of the generic information-processing components described in this chapter. An equal, and sometimes greater partner in this prediction is extensive **domain knowledge** regarding the particular system with which the human is interacting. As several of the chapters in this handbook address, the best prediction of human performance must be based upon the intricate interaction between the information-processing components discussed here, the domain knowledge employed by the human operator, and the physical environment within that domain, including the tools with which the operator works. The reader will find all of these issues covered from multiple perspectives in the following chapters of this handbook.

REFERENCES

Adams, J. A. (1989). *Human factors engineering*. New York: Macmillan.

Adams, M. J., Tenney, Y. J., and Pew, R. W. (1991). *Strategic workload and the cognitive management of advanced multi-task systems* (SOAR CSERIAC 91-6). Wright-Patterson AFB, OH: Crew System Ergonomics Information Analysis Center.

Adams, M .J., Tenney, Y. J., and Pew, R. W. (1995). Situation awareness and the cognitive management of complex systems. *Human Factors*, *37*(1), 85–104.

Adamson, R. E. (1952). Functional fixedness as related to problem solving: A repetition of three experiments. *Journal of Experimental Psychology*, *44*, 288–291.

Andre, A. D., and Wickens, C. D. (1990). *Display control compatibility in the cockpit: Guidelines for display layout analysis.* University of Illinois Institute of Aviation Technical Report (ARL-90-12/NASA-A3I-90-1). Savoy, IL: Aviation Res. Lab.

Antes, J. R., and Chang, K. (1990). An empirical analysis of the design principles for quantitative and qualitative area symbols. *Cartography and Geographic Information Systems, 17,* 271–277.

Aretz, A.J. (1991). The design of electronic map displays. *Human Factors, 33*(1), 85–101.

Aretz, A. J., and Wickens, C. D. (1992). The mental rotation of map displays. *Human Performance, 5,* 303–328.

Atkinson, R. C., and Shiffrin, R. M. (1971). The control of short-term memory. *Scientific American, 225,* 82–90.

Baddeley, A. D. (1986). *Working Memory.* Oxford: Oxford University Press.

Baddeley, A. D. (1990). *Human memory: Theory and practice.* Hove, UK: Lawrence Erlbaum.

Baker, C. H. (1961). Maintaining the level of vigilance by means of knowledge of results about a secondary vigilance task. *Ergonomics, 4,* 311–316.

Barnett, B. J, and Wickens, C. D. (1988). Display proximity in multicue information integration: The benefit of boxes. *Human Factors, 30,* 30, 15–24.

Baron, S., Kleinman, D., and Levison, W. (1970). An optimal control model of human response. *Automatica, 5,* 337–369.

Bennett, K. B., and Flach, J. M. (1992). Graphical displays: Implications for divided attention, focused attention, and problem solving. *Human Factors, 34,* 513–533.

Bennett, K. B., Toms, M. L., and Woods, D. D. (1993). Emergent features and graphical elements: Designing more effective configural displays. *Human Factors, 35*(1), 71–98.

Berbaum, K., Tharp, D., and Mroczek, K. (1983). Depth perception of surfaces in pictures: Looking for conventions of depictions in Pandora's box. *Perception, 12,* 5–20.

Beringer, D. B., and Chrisman, S. E. (1991). Peripheral polar-graphic displays for signal/failure detection. *International Journal of Aviation Psychology, 1,* 133–148.

Biederman, I., Mezzanotte, R. J., Rabinowitz, J. C., Francolin, C. M., and Plude, D. (1981). Detecting the unexpected in photo interpretation. *Human Factors, 23,* 153–163.

Blakemore, C. (1980). The baffled brain. In R. L. Gregory and E. H. Gombrich (Eds.), *Illusion in Nature and Art.* New York: Charles Scribner's Sons.

Bower, G. H., Clark, M. C., Lesgold, A. M., and Winzenz, D. (1969). Hierarchical retrieval schemes in the recall of categorical word lists. *Journal of Verbal Learning and Verbal Behavior, 8,* 323–343.

Bransford, J. D., and Johnson, M. K. (1972). Contextual prerequisites for understanding: Some investigations of comprehension and recall. *Journal of Verbal Learning and Verbal Behavior, 11,* 717–726.

Broadbent, D. (1958). *Perception and Communications.* New York: Permagon Press.

Broadbent, D. (1972). *Decision and Stress.* New York: Academic Press.

Broadbent, D. E. (1982). Task combination and selective intake of information. *Acta Psychologica, 50,* 253–290.

Brown, J. (1959). Some tests of the decay theory of immediatememory. *Quarterly Journal of Experimental Psychology, 10,* 12–21.

Bruno, N., and Cutting, J. E. (1988). Minimodularity and the perception of layout. *Journal of Experimental Psychology: General, 117,* 161–170.

Buttigieg, M. A., and Sanderson, P. M. (1991). Emergent features in visual display design for two types of failure detection tasks. *Human Factors, 33,* 631–651.

Carbonnell, J. R., Ward, J. L., and Senders, J. W. (1968). A queuing model of visual sampling: Experimental validation. *IEEE Trans. on Man-Machine Systems, MMS-9,* 82–87.

Card, S. K. (1981). The model human processor: A model for making engineering calculations of human performance. In R. Sugarman (Ed.), *Proceedings of the 25th Annual Meeting of the Human Factors Society.* Santa Monica, CA: Human Factors Society.

Carpenter, P. A., and Just, M. A. (1975). Sentence comprehension: A psycholinguistic processing model of verification. *Psychological Review, 82*(1), 45–73.

Carswell, C. M. (1992). Reading graphs: Interactions of processing requirements and stimulus structure. In B. Burns (Ed.), *Percepts, concepts, and categories* (pp. 605–647). Amsterdam: Elsevier Science Publishers.

Carswell, C. M., and Wickens, C. D. (1987). Information integration and the object display: An interaction of task demands and display superiority. *Ergonomics, 30,* 511–527.

Carswell, C. M., and Wickens, C. D. (1995). Mixing and matching lower-level codes for object displays: Evidence for two sources of proximity compatibility. *Human Factors, 38.*

Chernikoff, R., and LeMay, M. (1963). Effect of various display-control configurations on tracking with identical and different coordinate dynamics. *Journal of Experimental Psychology, 6,* 95–99.

Clark, H. H., and Chase, W. G. (1972). On the process of comparing sentences against pictures. *Cognitive Psychology, 3,* 472–517.

Cleveland, W. S. (1985). *The Elements of Graphing Data.* Monterey, CA: Wadsworth.

Cleveland, W. S. (1990). A model for graphical perception. *Proceedings of the Statistical Graphics Section*. American Statistical Association (pp. 30–35).

Cleveland, W. S., and McGill, R. (1984). Graphical perception: Theory, experimentation, and application to the development of graphic methods. *Journal of the American Statistical Association, 70*, 531–554.

Cleveland, W. S., and McGill, R. (1985). Graphical perception and graphical methods for analyzing scientific data. *Science, 229*, 828–833.

Cleveland, W. S., and McGill, R. (1986). An experiment in graphical perception. *International Journal of Man-Machine Studies, 25*, 491–500.

Cleveland, W. S., and McGill, R. (1988). *Dynamic graphics for statistics*. Monterey, CA: Wadsworth.

Coren, S., and Girgus, J. S. (1978). *Seeing is deceiving: The psychology of visual illusions*. Hillsdale, NJ: Erlbaum.

Damos, D. (Ed.) (1991). *Multiple Task Performance*. London: Taylor and Francis.

Dashevsky, S. G. (1964). Check-reading accuracy as a function of pointer alignment, patterning, and viewing angle. *Journal of Applied Psychology, 48*,344–347.

Degani, A., and Wiener, E. L. (1993). Cockpit checklists: Concepts, design, and use. *Human Factors 35*(4), 345–360.

DeSoto, C. B., London, M., and Handel, S. (1965). Social reasoning and spatial paralogic. *Journal of Personal and Social Psychology, 2*, 513–521.

Duncan, J. (1979). Divided attention: The whole is more than the sum of its parts. *Journal of Experimental Psychology: Human Perception and Performance, 5*, 216–228.

Drury, C. (1975). Application to Fitts' law to foot pedal design. *Human Factors, 17*, 368–373.

Drury, C. G., and Addison, S. L. (1973). An industrial study of the effects of feedback and fault density on inspection performance. *Ergonomics, 16*, 159–169.

Eberts, R. E., and MacMillan, A. G. (1985). Misperception of small cars. In R. E. Eberts and C. G. Eberts, Eds., *Trends in ergonomics/human factors II* (pp. 33–39). North Holland, Netherlands: Elsevier Science Publishers, B. V.

Eley, M. G. (1988). Determining the shapes of land surfaces from topographical maps. *Ergonomics, 31*, 355–376.

Ellis, S. R., McGreevy, M. W., and Hitchcock, R. J. (1987). Perspective traffic display format and airline pilot traffic avoidance. *Human Factors, 29*, 371–382.

Endsley, M. R. (1995). Measurement of situation awareness in dynamic systems. *Human Factors, 37*(1), 65–84.

Flach, J. M., Hancock, P. A., Caird, J., and Vicente, K.J., Eds., (1995). *Global Perspectives on the Ecology of Human-Machine Systems*. Hillsdale, NJ: Lawrence Erlbaum.

Fitts, P. M. (1966). Cognitive aspects of information processing III: Set for speed versus accuracy. *Journal of Experimental Psychology, 71*, 849–857.

Fitts, P. M., and Peterson, J. R. (1964). Information capacity of discrete motor responses. *Journal of Experimental Psychology, 67*, 103–112.

Fitts, P. M., and Posner, M. A. (1967). *Human Performance*. Pacific Palisades, CA: Brooks/Cole.

Fogel, L. J. (1959). A new concept: The kinalog display system. *Human Factors, 1*, 30–37.

Fracker, M. L. and Wickens, C. D. (1989). Resources, confusions, and compatibility in dual axis tracking: Displays, controls, and dynamics. *Journal of Experimental Psychology: Human Perception and Performance, 15*, 80–96.

Funk, K. (1991). Cockpit task management: Preliminary definitions, normative theory, error taxonomy, and design recommendations. *The International Journal of Aviation Psychology, 1*(4), 271–286.

Garner, W. R. (1978). Selective attention to attributes and to stimuli. *Journal of Experimental Psychology: General, 107*, 287–308.

Getty, D. J., Swets, J. A., Pickett, R. M., and Gonthier, D. (1995). System operator response to warnings of danger: A laboratory investigation of the effects of the predictive value of a warning on human response time. *Journal of Experiment Psychology: Applied, 1*(1), 19–33.

Gibson, J. J. (1979). *The ecological approach to visual perception*. Boston: Houghton Mifflin.

Gillan, D. J., and Lewis, R. (1994). A componential model of human interaction with graphs: I. Linear regression modeling. *Human Factors, 36*(4) 419–440.

Gopher, D. (1992). The skill of attention control: Acquisition and execution of attention strategies. In D. Meyer and S. Kornblum (Eds.), *Attention and performance XIV: Synergies in experimental psychology, artificial intelligence, and cognitive neuroscience—A silver jubilee*. Cambridge, A: MIT Press.

Gopher, D., and Raij, D. (1988). Typing with a two hand chord keyboard—will the QWERTY become obsolete? *IEEE Trans. on System, Man, and Cybernetics, 18*, 601–609.

Graham, J. L. (1937). Illusory trends in the observation of bar graphs. *Journal of Experimental Psychology, 20*, 597–608.

Green, D. M., and Swets, J. A. (1988). *Signal Detection Theory and Psychophysics*. New York: Wiley.

Gregory, R. L. (1977). *Eye and Brain.* London: Weidenfeld and Nicolson.

Gregory, R. L. (1980). The confounded eye. In R. E. Gregory and E. H. Gombrich (Eds.), *Illusion in nature and art.* New York: Charles Scribner's Sons.

Hancock, P., Flach, J., Caird, J., and Vicente, K. (Eds.) (1995). *Local Applications of the Ecological Approach to Human-Machine Systems.* Hillsdale, NJ: Lawrence Erlbaum Associates.

Harris, D. H., and Chaney, F. D. (1969). *Human Factors in Quality Assurance.* New York: Wiley.

Harris, R. L., and Christhilf, D. M. (1980). What do pilots see in displays? *Proceedings of the 24th Annual Meeting of the Human Factors Society.* Santa Monica, CA: Human Factors Society.

Hart, S. G., and Wickens, C. D. (1990). Workload assessment and prediction. In H. R. Booher (Ed.,), *MANPRINT: An Approach to Systems Integration* (pp. 257–296). New York: Van Nostrand Reinhold.

Haskell, I. D., and Wickens, C. D. (1993). Two- and three-dimensional displays for aviation: A theoretical and empirical comparison. *International Journal of Aviation Psychology, 3*(2), 87–109.

Herrmann, D., and Pickle, L. W. (in press). A cognitive subtask model of statistical map reading. *Visual Cognition.*

Hess, S. M., and Detweiler, M. C. (1995). The effects of response alternatives on keeping-track performance. *Proceedings of the 39th Annual Meeting of the Human Factors and Ergonomics Society* (pp. 1390–1394). Santa Monica, CA: Human Factors and Ergonomics Society.

Hick, W. E. (1952). On the rate of gain of information. *Quarterly Journal of Experimental Psychology, 4,* 11–26.

Hockey, R. (1984). Varieties of attentional state: The effects of environment. In R. Parasuraman, R. Davies, and J. Beatty (Eds.), *Varieties of Attention* (pp. 449–479). New York: Academic Press.

Hockey, G. R. J. (1986). Changes in operator efficiency as a function of environmental stress, fatigue, and circadian rhythms. In K. R. Boff, L. Kaufman, and J. P. Thomas (Eds.), *Handbook of Perception and Human Performance, Vol. II* (pp. 44–1 to 44–49). New York: Wiley.

Hollands, J. (1992). Ergonomic data display. *Human Factors Bulletin, 35*(7), 4–5.

Hollands, J. G., and Spence, I. (1992). Judgments of change and proportion in graphical perception. *Human Factors, 34*(3), 313–334.

Huey, B. M., and Wickens, C. D., Eds., (1993). *Workload Transition: Implications for Individual and Team Performance.* Washington, DC: National Academy Press.

Human Factors Journal (1995). Whole issue on Situation Awareness, *37*(June).

Hutchins, E. L., Hollan, J. D., and Norman, D. A. (1985). Direct manipulation interfaces. *Human-Computer Interaction, 1*(4), 311–338.

Hyman, R. (1953). Stimulus information as a determinant of reaction time. *Journal of Experimental Psychology, 45,* 423–432.

Jagacinski, R. J. (1989). Target acquisition: Performance measures, process models, and design implications. In G. R. McMillan, D. Beevis, E. Salas, M. H. Strub, R. Sutton, and L. Van Breda, Eds., *Applications of Human Performance Models to System Design* (pp. 135–150). New York: Plenum Press.

James, W. (1890). *The Principles of Psychology, Vol. 1.* New York: Holt.

Johnson, S. L., and Roscoe, S. N. (1972). What moves, the airplane or the world? *Human Factors, 14,* 107–129.

Just, M. A., and Carpenter, P. A. (1987). *The Psychology of Reading and Language Comprehension.* Boston: Allyn and Bacon.

Kahneman, D., and Treisman, D. (1984). Changing views of attention and automaticity. In R. Parasuraman, R. Davies, and J. Beatty (Eds.), *Varieties of Attention* (pp. 29–61). New York: Academic Press.

Kahenman, D. (1973). *Attention and Effort.* Englewood Cliffs, NJ: Prentice Hall.

Kintsch, W., and Keenan, J. (1973). Reading rate and retention as a function of the number of propositions in the base structure of sentences. *Cognitive Psychology, 5,* 257–274.

Kintsch, W., and Van Dijk, T. A. (1978). Toward a model of text comprehension and reproduction. *Psychological Review, 85,* 363–394.

Kintsch, W., and Vipond, P. (1979). Reading comprehension and readability in educational practice and psychological theory. In L. G. Nilsson (Ed.), *Perspectives on Memory Research.* Hillsdale, NJ: Erlbaum.

Kirlik, A., Miller, A., and, Jagacinski, R. J. (1993). Supervisory control in a dynamic and uncertain environment: A process model of skilled human-environment interaction. *IEEE Trans. on Systems, Man, and Cybernetics, 23*(4), 929–952.

Klein, G., and Crandall, B. W. (1995). The role of simulation in problem solving and decision making. In P. Hancock, J. Flach, J. Caird, and K. Vicente, Eds., *Local Applications of the Ecological Approach to Human-Machine Systems.* Hillsdale, NJ: Lawrence Erlbaum Associates.

Klein, G. A., Orasanu, J., Calderwood, R., and Zsambok, E. (Eds.)(1993). *Decision Making in Action: Models and Methods.* Norwood, NJ: Ablex Pub. Corp.

Kleinman, D. L. and Pattipati, K. R. (1991) A review of the engineering models of information processing and decision making in multi-task supervisory control. In D. Damos (ed.) *Multiple Task Performance*. London: Taylor & Francis, 35–68.

Kosslyn, S. M. (1980). *Image and Mind*. Cambridge, MA: Harvard University Press.

Kosslyn, S. M. (1987). Seeing and imagining in the cerebral hemispheres: A computational approach. *Psychological Review, 94*, 148–175.

Kosslyn, S. M. (1994). *The Elements of Graph Design*. New York: Freeman and Co.

Langolf, C. D., Chaffin, D. B., and Foulke, S. A. (1976). An investigation of Fitts' law using a wide range of movement amplitudes. *Journal of Motor Behavior, 8*, 113–128.

Laudeman, I. V., and Palmer, E. A. (1995). Quantitative measurement of observed workload in the analysis of aircrew performance. *The International Journal of Aviation Psychology, 5*(2), 187–198.

Layton, C., Smith, P. J., and McCoy, C. E. (1994). Design of a cooperative problem-solving system for en-route flight planning: An empirical evaluation. *Human Factors, 36*(4), 94–119.

Levine, M. (1982). You-are-here maps: Psychological considerations. *Environment and Behavior, 14*, 221–237.

Levine, M. (1988). *Effective Problem Solving*. Englewood Cliffs, NJ: Prentice Hall.

Levison, W. H., Elkind, J. I., and Ward, J. L. (1971). *Studies of multivariable manual control systems: A model for task interference*. NASA Contractors Report NASA CR-1746. Washington, DC: NASA.

Loftus, G., Dark, V., and Williams, D. (1979). Short-term memory factors in ground controller/pilot communication. *Human Factors, 21*, 169–181.

Logie, R. H. (1995). *Visuo-Spatial Working Memory*. East Sussex, UK: Lawrence Erlbaum.

Lohse, J. (1991). A cognitive model for the perception and understanding of graphs. In S. P. Robertson, G. M. Olson, and J. S. Olson (Eds.), *Human Factors in Computing Systems - Reaching Through Technology, CHI '91 Conference Proceedings* (pp. 137–144). New York: The Association for Computing Machinery, Inc.

Lusted, L. B. (1976). Clinical decision making. In D. Dombaland J. Grevy (Eds.), *Decision Making and Medical Care*. Amsterdam: North Holland.

Massaro, D. W. (1979). Letter information and orthographic context in word perception. *Journal of Experimental Psychology: Human Perception and Performance, 5*, 595–609.

Mayer, R. E. (1983). *Thinking, problem solving, cognition*. San Francisco: W.H. Freeman.

McClelland, J. L., and Rumelhart, D. E. (1981). An interactive activation model of context effects in letter perception: Part I. An account of basic findings. *Psychological Review, 88*, 375–407.

McRuer, D. (1980). Human dynamics in man-machine systems. *Automatica, 16*, 237–253.

Miller, G. A. (1956). The magical number seven plus or minus two: Some limits on our capacity for processing information. *Psychological Review, 63*, 81–97.

Miller, G. A., Galanter, E., and Pribram, K. H. (1960). *Plans and the Structure of Behavior*. New York: Holt, Rinehart, and Winston.

Milroy, R., and Poulton, E. C. (1978). Labeling graphs for increasing reading speed. *Ergonomics, 21*, 55–61.

Moray, N. (1975). A data base for theories of selective listening. In P. M. A. Rabbitt and S. Dornic Eds., *Attention and Performance V* (pp. 119–135). London: Academic Press, Inc.

Moray, N. (1981). The role of attention in the detection of errors and the diagnosis of errors in man-machine systems. In J. Rasmussen and W. Rouse (Eds.), *Human Detection and Diagnosis of System Failures*. New York: Plenum Press.

Moray, N. (1986). Monitoring behavior and supervisory control. In K. R. Boff, L. Kaufman, and J. P. Thomas, Eds., *Handbook of Perception and Performance, Vol. 2* (pp. 40–1 to 40–51). New York: Wiley and Sons.

Moray, N. (1969) *Selective Processes in Vision and Hearing*. London: Hutchinson.

Moray, N., Lee, J., Vicente, K., Jones, B. G., and Rasmussen, J. (1994). A direct perception interface for nuclear power plants. *Proceedings of the 38th Annual Meeting of the Human Factors and Ergonomics Society* (pp. 481–485). Santa Monica, CA: Human Factors and Ergonomics Society.

Neisser, U. (1967). *Cognitive Psychology*. Englewood Cliffs, NJ: Prentice Hall.

Norman, D. A. (1981). Categorization of action slips. *Psychological Review, 88*, 1–15.

Norman, D. (1988). *The Psychology of Everyday Things*. New York: Basic Books.

O'Hare, D., and Roscoe, S. N. (1990). *Flightdeck Performance: The Human Factor*. Ames, IA: Iowa State University Press.

Orasanu, J. M. (1993). Decision-making in the cockpit. In E. L. Wiener, B. G. Kanki, and R. L. Helmreich, Eds., *Cockpit Resource Management* (pp. 137–173). San Diego, CA: Academic Press.

Osborne, D. W., and Ellingstad, V. S. (1987). Using sensor lines to show control-display linkages on a four burner stove. *Proceedings of the 31st Annual Meeting of the Human Factors Society* (pp. 581–584). Santa Monica, CA: Human Factors Society.

Palmer, S. E. (1975). The effects of contextual scenes on the identification of objects. *Memory and Cognition, 3*, 519–526.

Palmer, S. E., Rosch, E., and Chase, P. (1981). Canonical perspective and the perception of objects. In J. Long and A. Baddeley (Eds.), *Attention and Performance IX*. Hillsdale, NJ: Erlbaum.

Parasuraman, R., Davies, R., and Beatty, J., Eds. (1984). *Varieties of Attention*. New York: Academic Press.

Peruch, P., and Savoyant, A. (1991). Conflicting spatial frames of reference in a locating task. In R. H. Logie and M. Denis (Eds.), *Mental Images in Human Cognition* (pp. 47–55). Amsterdam: Elsevier Science Pub., B.V.

Peterson, L. R., and Peterson, M. J. (1959). Short-term retention of individual verbal items. *Journal of Experimental Psychology*, 58, 193–198.

Pew, R. W., and Baron, S. (1978). The components of an information processing theory of skilled performance based on an optimal control perspective. In G. E. Stelmach (Ed.), *Information processing in motor control and learning*. New York: Academic press.

Pinker, S. (1990). A theory of graph comprehension. In R. Freedle (Ed.), *Artificial Intelligence and the Future of Testing* (pp. 73–126). Hillsdale, NJ: Erlbaum.

Pomerantz, J. R. (1981). Perceptual organization in information processing. In M. Kubovy and J. R. Pomerantz (Eds.), *Perceptual Organization* (pp. 141–180). Hillsdale, NJ: Erlbaum.

Pomerantz, J. R., and Kubovy, M. (1986). Theoretical approaches to perceptual organization. In K. R. Boff, L. Kaufman, and J. P. Thomas (Eds.), *Handbook of Perception and Human Performance Vol. II* (pp. 36–1 to 36–46). New York: Wiley and Sons.

Pomerantz, J. R., and Pristach, E. A. (1989). Emergent features, attention, and perceptual glue in visual form perception. *Journal of Experimental Psychology: Human Perception and Performance, 15*, 635–649.

Pomerantz, J. R., and Schwaitzberg, S. D. (1975). Grouping by proximity: Selective attention measures. *Perception and Psychophysics, 18*, 355–361.

Posner, M. I. (1978). *Chronometric explorations of the mind*. Hillsdale, NJ: Lawrence Erlbaum.

Poulton, E. C. (1974). *Tracking Skills and Manual Control*. NewYork: Academic press.

Poulton, E. C. (1985). Geometric illusions in reading graphs. *Perception and Psychophysics, 37*, 543–548.

Raby, M., and Wickens, C. D. (1994). Strategic workload management and decision biases in aviation. *International Journal of Aviation Psychology, 4*(3), 211–240.

Rasmussen, J. (1981). Models of mental strategies in process control. In J. Rasmussen and W. Rouse (Eds.), *Human Detection and Diagnosis of System Failures*. New York: Plenum Press.

Rasmussen, J. (1986). *Information Processing and Human-Machine Interaction: An Approach to Cognitive Engineering*. New York: North Holland.

Rasmussen, J., Pejtersen, A-M, and Goodstein, L. (1995). *Cognitive Engineering: Concepts and Applications*. New York: Wiley.

Reason, J. (1990). *Human Error*. New York: Cambridge University Press.

Reicher, G. M. (1969). Perceptual recognition as a function of meaning fulness of stimuli materials. *Journal of Experimental Psychology, 81*, 275–280.

Rogers, B. J., and Collett, T. S. (1989). The appearance of surfaces specified by motion parallax and binocular disparity. *Quarterly Journal of Experimental Psychology: Human Experimental Psychology, 41*, 697–717.

Rogers, D., and Monsell, S. (1995). Costs of a predictable switch between simple cognitive tasks. *Journal of Experimental Psychology: General, 124*, 207–231.

Roscoe, S. N. (1968). Airborne Displays for Flight and Navigation. *Human Factors, 18*, 321–332.

Roscoe, S. N. (1981). *Aviation psychology*. Iowa City: University of Iowa Press.

Sanders, M. S., and McCormick, E. J. (1993). *Human Factors in Engineering and Design* (7th ed.). New York: McGraw Hill.

Sanderson, P. M. (1989). The human planning and scheduling role in advanced manufacturing systems: An emerging human factors domain. *Human Factors, 31*, 635–666.

Sanderson, P. M. Flach, J. M., Buttigieg, M. A., and Casey, E. J. (1989). Object displays do not always support better integrated task performance. *Human Factors, 31*, 183–198.

Schank, R. C., and Abelson, R. (1977). *Scripts, Plans, Goals, and Understanding*. Hillsdale, NJ: Lawrence Erlbaum.

Schutte, P. C., and Trujillo, A. C. (1996). Flight crew task management in nonnormal situations. *Proceedings of the 40th Annual Meeting of the Human Factors and Ergonomics Society*. Santa Monica, CA: Human Factors & Ergonomics Society.

Schutz, H. G. (1961a). An evaluation of formats for graphic trend displays—Experiment II. *Human Factors, 3*(2), 99–107.

Schutz, H. G. (1961b). An evaluation of methods for presentation of graphic multiple trends—Experiment III. *Human Factors, 3*(2), 108–119.

Senders, J. W. (1964). The human operator as a monitor and controller of multidegree of freedom systems. *IEEE Trans. on Human Factors in Electronics, HFE-5*, 2-6.

Senders, J. W. (1980). *Visual scanning processes.* Unpublished doctoral thesis, University of Tilburg, Netherlands.
Shah, P., and Carpenter, P. A. (1995). Conceptual limitations in comprehending line graphs. *Journal of Experimental Psychology: General, 124*, 46–61.
Shalin, V. L., and Bertram, D. A. (in press). Functions of expertise in a medical intensive care unit. *Journal of Experimental and Theoretical Artificial Intelligence: Special Issue on Expertise.*
Sheridan, T. (1972). On how often the supervisor should sample. *IEEE Trans. on Systems, Science, and Cybernetics, SSC-6*, 140–145.
Shinar, D., Rockwell, T. H., and Malecki, J. A. (1980). The effects of changes in driver perception on road curve negotiation. *Ergonomics, 23*, 263–275.
Simkin, D., and Hastie, R. (1987). An information processing analysis of graph perception. *Journal of the American Statistical Association, 82*, 454–465.
Smith, K. U. (1962). *Delayed Sensory Feedback and Balance.* Philadelphia: Saunders.
Sorkin, R. D. (1988). Why are people turning off our alarms? *Journal of the Acoustical Society of America, 84*, 1107–1108.
Spence, I., and Krizel, P. (1994). Children's perceptions of proportions in graphs. *Child Development, 65*, 1189–1209.
Streeter, L. A., Vitello, D., and Wonsiewicz, S. A. (1985). How to tell people where to go: Comparing navigational aids. *International Journal on Man-Machine Studies, 22*, 549–562.
Sternberg, S. (1969). The discovery of processing stages. Extension of Donders' method. *Acta Psychologica, 30*, 276–315.
Strayer, D. L., Wickens, C. D., and Braune, R. (1987). Adult age differences in the speed and capacity of information processing: II. An electrophysiological approach. *Psychology and Aging, 2*, 99–110.
Tanner, W. P., and Swets, J. A. (1954). A decision-making theory of visual detection. *Psychological Review, 61*, 401–409.
Thorndyke, P. W. (1980, December). *Performance Models for Spatial and Locational Cognition* (Technical Report R-2676-ONR). Washington, DC: Rand.
Thorndyke, P. W., and Hayes-Roth, B. (1982). Differences in spatial knowledge acquired from maps and navigation. *Cognitive Psychology, 14*, 560–589.
Travis, D. (1991). *Effective Color Displays: Theory and Practice.* New York: Academic Press.
Treisman, A. (1986). Properties, parts, and objects. In K. R. Boff, L. Kaufman, and J. P. Thomas (Eds.), *Handbook of Perception and Human Performance, Vol. II* (pp. 31–1 to 35–70. New York: John Wiley.
Treisman, A. (1988). Features and objects: The fourteenth Bartlett memorial lecture. *The Quarterly Journal of Experimental Psychology, 40A*, 201–237.
Treisman, A. (1993). The perception of features and objects. In A. Baddeley and N. Weiskrantz (Eds.), *Attention, selection, awareness, and control: A tribute to Donald Broadbent.* Oxford: Clarendon Press.
Treisman, A., Kahneman, D., and Burkell, J. (1983). Perceptual objects and the cost of filtering. *Perception and Psychophysics, 33*, 527–532.
Tulving, E., Mandler, G., and Baumal, R. (1964). Interaction of two sources of information in tachistoscopic word recognition. *Canadian Journal of Psychology, 18*, 62–71.
Tversky, A., and Kahneman, D. (1974). Judgment under uncertainty: Heuristics and biases. *Science, 185*, 1124–1131.
Tulga, M. K., and Sheridan, T. B. (1980). Dynamic decisions and workload in multitask supervisory control. *IEEE Trans. on Systems, Man, and Cybernetics, SMC-10*, 217–232.
Van der Meer, H. C. (1979). Interaction of the effects of binocular disparity and perspective cues on judgments of depth and height. *Perception and Psychophysics, 26*, 481–488.
Vincente, K., Christofferson, K. & Pereklita, A. (1995) Supporting operator problem solving through ecological interface design. *IEEE Transactions in Systems Man & Cybernetics* SML-25, 529–545.
Vicente, K., and Rasmussen, J. (1992). Ecological interface design: Theoretical foundations. *IEEE Trans. on Systems, Man, and Cybernetics, 22*, 589–606.
Wallsten, T. S. (Ed.) (1980). *Cognitive Processes in Choice and Decision Behavior.* Hillsdale, NJ: Lawrence Erlbaum Associates.
Warren, D. H., Rossano, M. J., and Wear, T. D. (1990). Perception of map-environment correspondence: The roles of features and alignment. *Ecological Psychology, 2*, 131–150.
Warren, R., and Wertheim, A. H. (Eds.), (1990). *Perception and Control of Self-Motion.* Hillsdale, NJ: Lawrence Erlbaum.
Weintraub, D. J., and Ensing, M. J. (1992). Human factors issues in head-up display design: The book of HUD. *SOAR CSERIAC State of the Art Report 92-2.* Dayton, OH: Crew System Ergonomics Information Analysis Center, Wright Patterson-AFB.
Wetherell, A. (1979). Short-term memory for verbal and graphic route information. *Proceedings of the 23rd Annual Meeting of the Human Factors Society.* Santa Monica, CA: Human Factors Society.

Whitaker, L. A., and CuQlock-Knopp, V. G. (1995). Human exploration and perception in off-road navigation. In P.A. Hancock, J. M. Flach, J. Caird, and K. J. Vicente (Eds.), *Local Applications of the Ecological Approach to Human-Machine Systems*. Hillsdale, NJ: Lawrence Erlbaum Associates.

Wheeler, D. (1970). Processes in word recognition. *Cognitive Psychology, 1*, 59–85.

Wickelgren, W. A. (1977). Speed accuracy tradeoff and information processing dynamics. *Acta Psychologica, 41*, 67–85.

Wickens, C. D. (1986). The effects of control dynamics on performance. In K. Boff, L. Kaufman, and J. Thomas (Eds.), *Handbook of Perception and Performance, Vol. II* (pp. 39–1 to 39–60). New York: Wiley and Sons.

Wickens, C. D. (1991). Processing resources and attention. In D. Damos (Ed.), *Multiple Task Performance*. London: Taylor & Francis.

Wickens, C. D. (1992). *Engineering Psychology and Human Performance* (2nd Ed.). New York: HarperCollins.

Wickens, C. D. (1994). *Context, computation, and complexity: Applications to aviation display system design*. University of Illinois Institute of Aviation Final Technical Report (ARL-94-7/NASA-A3I-94-2). Savoy, IL: Aviation Res. Lab.

Wickens, C. D. (1996). *Situation awareness: Impact of automation and display technology* (Keynote address). NATO AGARD Aerospace Medical Panel Symposium on Situation Awareness: Limitations and Enhancement in the Aviation Environment, AGARD CP-575. Neuilly-Sur-Seine, France.

Wickens, C. D., and Andre, A. D. (1990). Proximity compatibility and information display: Effects of color, space, and objectness on information integration. *Human Factors, 32*, 32, 61–77.

Wickens, C. D., and Carswell, C. M. (1995). The proximity compatibility principle: Its psychological foundation and its relevance to display design. *Human Factors, 37*(3).

Wickens, C. D., Liang, C-C, Prevett, T., and Olmos, O. (in press). Electronic maps for terminal area navigation: Effects of frame of reference and dimensionality. *International Journal of Aviation Psychology,6*,241–271.

Wickens, C. D., and Long, J. (1995). Object- vs. space-based models of visual attention: Implications for the design of head-up displays. *Journal of Experimental Psychology: Applied, 1*(3), 179–194.

Wickens, C. D., Merwin, D. H., and Lin, E. (1994). Implications of graphics enhancements for the visualization of scientific data: Dimensional integrality, stereopsis, motion, and mesh. *Human Factors, 36*(1), 44–61.

Wickens, C. D., and Prevett, T. (1995). Exploring the dimensions of egocentricity in aircraft navigation displays: Influences on local guidance and global situation awareness. *Journal of Experimental Psychology: Applied, 1*(2), 110–135.

Wickens, C. D., Todd, S., and Seidler, K. S. (1989). *Three-dimensional displays: Perception, implementation, and applications*. CSERIAC—State of the Art Report, CSERIAC SOAR #89-001, Wright-Patterson AFB: Crew System Ergonomics Information Analysis Center.

Wickens, C. D., Vidulich, M., and Sandry-Garza, D. (1984). Principles of S-C-R compatibility with spatial and verbal tasks: The role of display-control location and voice-interactive display-control interfacing. *Human Factors, 26*, 533–543.

Wickens, C. D., Zenyuh, J., Culp, V., and Marshak, W. (1985). Voice and manual control in dual task situations. *Proceedings of the 29th Annual Meeting of the Human Factors Society*. Santa Monica, CA: Human Factors Society.

Wiener, E. L. (1977). Controlled flight into terrain accidents: System-induced errors. *Human Factors, 19*, 171–181.

Wiener, E. L., Kanki, B. G., and Helmreich, R. L. (Eds.) (1993). *Cockpit Resource Management*. San Diego, CA: Academic Press.

Wilkinson, R. T. (1964). Artificial "signals" as an aid to an inspection task. *Ergonomics, 7*, 63–72.

Woods, D., Wise, J., and Hanes, L. (1981). An evaluation of nuclear power plant safety parameter display systems. In *Proceedings of the 25th Annual Meeting of the Human Factors Society*. Santa Monica, CA: Human Factors Society.

Yantis, S. (1993). Stimulus driven attentional capture. *Current Directions in Psychological Science, 2*, 156–161.

Yates, J. F. (1990). *Judgment and Decision Making*. Englewood Cliffs, NJ: Prentice Hall.

Yeh, Y. Y., and Silverstein, L. D. (1992). Spatial judgments with monoscopic and stereoscopic presentation of perspective displays. *Human Factors, 34*, 583–600.

CHAPTER 5
LEARNING

Richard J. Koubek
Susan A. H. Benysh
School of Industrial Engineering
Purdue University
West Lafayette, IN 47907-1287 USA

Eric Tang
Department of Industrial Engineering
Feng-Chia University
Taichung, Taiwan

5.1 INTRODUCTION

The goal of this chapter is to explain what and how changes to knowledge occur with expertise. To this end, the chapter first distinguishes between the various types of knowledge that people use in the workplace and in daily life. The second section of this chapter explores the changes that occur in each of these knowledge types of as people gain expertise. Having laid the groundwork of knowledge types and the changes which occur, the third chapter section discusses a variety of approaches to *how* these changes take place. Finally, a comprehensive framework is presented which integrates the variety of approaches to skill acquisition which can be used to describe learning for each of the types of knowledge. Chapter 16, "Instructional Training Models," describes how these learning mechanisms are implemented in training technologies. There are many excellent references that discuss learning in great detail, two of which are Proctor and Dutta's (1995) *Skill Acquisition and Human Performance* and Healy and Bourne's (1995) *Learning and Memory of Knowledge and Skills*.

5.2 TYPES OF KNOWLEDGE AND SKILLS

Traditionally, knowledge has been classified into either declarative or procedural type knowledge (Kyllonen and Alluisi, 1987). Consequentially, the Information Processing (IP) models that have been presented have addressed only these types of knowledge (Anderson, 1983; Kyllonen and Alluisi, 1987). Recently, it has been hypothesized, and supported by research, that a particular type of knowledge, *conceptual* knowledge, exists (Byrnes and Wasik, 1991; Hiebert, 1987; Koubek, 1991; Resnick, 1987; Ward et al., 1990). Most likely, conceptual knowledge resides in the procedural and declarative knowledge bases. However, for the purposes of discussion, it is presented independently. Therefore, following is a discussion of the three major types of knowledge: *declarative*, *conceptual*, and *procedural*.

5.2.1 Declarative Knowledge

Declarative knowledge is considered to be stable factual knowledge regarding the world. It is knowledge about facts, concepts, and principles. These are the bits of information that memory and knowledge is built on. These bits are data that define *what* something is and examples of declarative knowledge are many. For instance, one bit of knowledge is "the sky," another can be the color "blue." These declarative bits of knowledge, by themselves, cannot define a relationship such as "the sky is blue." The following section will discuss development of relationships. The following table adapted from Kyllonen and Alluisi (1987) summarizes the characteristics of declarative knowledge by examining the characteristics of declarative memory.

Characteristics of declarative knowledge include being reportable, each fact being distinguishable, and can be given independently of other facts. Associative priming is another characteristic of declarative knowledge in which one can prepare a user to anticipate a given word by priming. An example of this would be a user that is driving a car and receiving directions from a passenger. In this situation, the user is primed for driving instructions. When the passenger says "right," the user will be primed to go in that direction versus interpreting it as the passenger affirming a statement.

5.2.2 Conceptual Knowledge

Conceptual knowledge, also known as metaknowledge or relational knowledge, consists of core concepts for a specific domain and the interrelations between those concepts. This

Table 5.1 Characteristics of Declarative Memory

Primary function	Stores meaning of inputs
Capacity	Unlimited
Contents	Semantic codes (primarily) Spatial codes Acoustic codes Motor codes Temporal codes
Information units	Facts Concepts Principles
Learn/forget process	Learning by being told (advice-taking) Encoding (encoding specificity principle) Storage (very slow decay) Retrieval

Source: Modified from Kyllonen and Alluisi (1987).

relational representation consists of two or more declarative knowledge bits that are mentally linked through some sort of relation. This is the knowledge in which one knows *that*. An example of conceptual knowledge is "I know that the sky is blue." Keating and Crane (1990) (as well as Ornstein and Naus, 1985) characterize "concepts" in the form of relational representations. Some of the main constructs that characterize conceptual knowledge include semantic nets, hierarchies, and mental models. Subtypes of concepts include taxonomic categories, causal models, and spatial representations.

5.2.3 Procedural Knowledge

Procedural knowledge is knowledge of "how to do something" (Byrnes, 1992). It is typically modeled as production rules which represent the steps that an individual performs to complete a task. Generally, these rules are condition/action pairs, in which the framework is **IF** *condition*, **THEN** *action*. Production rules are discussed in detail in Section 5.4. The process of developing procedural knowledge can be considered to be the linking of production rules from a large number of small rules to a small number of large rules. This development also replaces the variables within the production rules with constants. The linking can only be performed when there is a goal present.

With respect to knowledge acquisition, declarative knowledge comes from encoding information from the environment, a direct representation of facts from the surroundings into the memory. Procedural knowledge acquisition is encoded by practicing the information that exists in declarative knowledge, strengthening certain links, such that they have a goal-oriented meaning associated with them. Finally, individuals are better at remembering and retaining procedural knowledge than declarative knowledge.

Novices usually begin with only declarative knowledge and, over time, learn by linking declarative facts together in order to successfully complete a task or attain a goal. These resulting procedures are not only the detailed actions that need to be performed to attain a goal, but also the order in which the actions should be performed. Once the links are established, the user may have procedularized knowledge. Once knowledge is procedularized, it is thought to be relatively stable in long-term memory. Some of the constructs that characterize procedures are skills, strategies, and rules. Furthermore, subtypes of procedures may include memory strategies, mathematical algorithms, and grammatical rules. The following table from Kyllonen and Alluisi (1987) summarizes the characteristics of procedural knowledge by examining the characteristics of procedural memory.

There have been many cases in which procedural skill and conceptual understanding have been found to be independent. Ward et al. (1990) discusses those cases where individuals show a conceptual understanding of some topic but lack procedural skill. Resnick (1987) discusses individual cases where individuals exhibit procedural skill but lack a conceptual understanding. These distinctions between conceptual and procedural knowledge have been applied to many aspects of cognition, including memory (Anderson, 1983; Schacter, 1989), language (Anderson, 1983), propositional reasoning (Byrnes, 1988; Keating, 1988), causality (Inhelder and Piaget, 1980), and mathematics (Byrnes and Wasik, 1991; Hiebert, 1987). Also, psychophysiological studies suggest that these two types of knowledge are localized in different areas of the brain, with damage to one region producing deficits in only one system (Squire, 1987).

5.3 KNOWLEDGE TYPES AND EXPERTISE LEVELS

The transition from novice to expert has many steps that must be completed by every learner. Some may not be able to complete all the steps due to lack of abilities or the mental workload required by the task (see Section 5.4.2, *Factors Affecting Knowledge Acquisition*). Also, some learners, depending on their abilities, may be able to rapidly

Table 5.2 Characteristics of Procedural Memory

Primary function	Permanent store of how-to knowledge
Capacity	Unlimited
Contents	Same as declarative memory
Learn/Forget process Learning by doing (practice) General operators (problem solving)	Generalization Discrimination Strengthening

Source: Modified from Kyllonen and Alluisi (1987).

proceed through the steps, obtaining expertise in a more efficient manner. The general changes that occur in the different types of knowledge are discussed below.

5.3.1 Assessing Expertise Level

The task of categorizing a person by their level of expertise can be accomplished in a number of ways. Two common methods are categorizing by experience and by amount of knowledge. Categorization by task experience requires some basic assumptions:

1. The individual learns about the task before beginning the task.
2. The individual learns during task completion, is always learning, and is motivated to become more effective and efficient.
3. All mistakes will eventually be eradicated over time with practice.
4. Eventually the individual correctly completes the task each time.

These assumptions can usually be made without fault, although the task being performed must be taken into consideration. For instance, the simplicity of the task will lend itself to the process of learning; a simpler task will be easier to learn than a more complex or difficult one. Also, the time elapsed between task completion will have an effect. Consider a task that is only performed once a year, regardless of its complexity, some re-learning must occur for the individual to perform it a subsequent year. Further, the individual may not become an expert in the task, even if they do it once a year for 50 years. On the other hand, for a simple task it may be perfectly reasonable to state that an expert level of performance will be achieved within 50 trials performed consecutively.

A second method for classifying expertise level is by the individual's amount of knowledge. Knowledge can be quantified by a wide variety of knowledge elicitation techniques, such as verbal reports, clustering methods, or scaling methods (Benysh, Koubek, and Calvez, 1993) or through an experimental simulation. For a majority of these techniques, the individual's knowledge is measured and evaluated against the mean and standard deviation of the expected population. If these statistical characteristics are unknown, they can be determined by testing a large sample population.

It is very difficult to determine the appropriate knowledge ratings or cut-off values for the novice and the expert categories due to the variety of potential tasks and specific knowledge required for a task. Some researchers have used the 50% mark to determine user's classifications; those with performance in the lower 50% of the population distribution are considered novices, and those users with performance scores that rate in the higher 50% are considered experts (Biswas and Sherrell, 1993; Diehl and Mikulecky, 1981; Stramler, 1993). Some researchers have adopted the technique classifying the lower 25% of the population as novices while the upper 25% are considered experts. The users that fall between these regions are considered to be in transition and may be referred to as competent users (Eberts, 1994).

5.3.2 Changes in Knowledge with Respect to Expertise Level

5.3.2.1 Novices versus Experts

According to Glaser and Chi (1988), the main difference between experts and novices is that experts possess a readily accessible and organized body of facts and procedures that can be applied to problems concerning their specified domain. Generally, novices require more information directly in their Working Memory (WM), or short-term memory, during task completion and take longer to answer a problem (De Groot, 1965; Chase and Simon, 1973). According to Chase and Simon (1973), the main differences between experts and novices are that experts can identify many patterns, experts will scan first to find these patterns, and experts use more pointers and more efficient chunks in WM. Chase and Simon also state that it is the organization of the stored information that permits experts to determine the proper problem-solving techniques almost instantaneously. Additionally, they have a better set of problem-solving techniques from which to select (De Groot, 1965; Miller, 1956). Therefore, it seems that both contents *and* structure of knowledge are what makes the difference in the outcome between novices and experts.

5.3.2.2 Experts in More Detail

Generally, experts are grouped into one category; however, recent research is starting to indicate a split between two types of expertise. These two classifications of expertise are procedural experts and conceptualized experts.

As previously indicated, an aspect of procedural knowledge is procedularization, the substitution of constants or functions for the variables within the production rules. As the learner becomes more procedularized, a set of actions becomes fixed. Then, the learner composes the rules, linking them, and eventually task completion becomes automatic. An expert in procedural knowledge is one for whom the learning curve has flattened out. The power law of practice, which is widely accepted, indicates that performance will always continue to improve with practice, although at a decreasing rate. More information on the power law and learning curves can be obtained by referring to Chapter 13 in this Handbook. Additionally, when the individual cannot decrease errors or time in performing a task, they are believed to have highly procedularized knowledge.

Conceptual knowledge, on the other hand, is often referred to as meta-knowledge. This is the broad or general knowledge that one possesses about a specific domain, commonly referred to as the elements contained within the knowledge structure (Koubek, 1991). It has been found (Adelson, 1984) that users with high conceptual knowledge have more abstract representations while those with low conceptual knowledge have more concrete knowledge representations. This can be seen in earlier work by Adelson (1981), where users were separated into two groups, low and high conceptual knowledge. The users were required to organize randomized computer code by any manner. The low conceptual knowledge users organized by key control categories, while the high conceptual knowledge users organized the code into working programs. Also, it has been found that those with high conceptual knowledge have a knowledge structure that is stable, well organized, and simpler than for those with low conceptual knowledge (Schvaneveldt et al., 1985). These studies showed that there is a difference between low and high conceptual knowledge. These results have been similarly produced across a wide variety of domains, including chess (Chase and Simon, 1973; De Groot, 1965), word processing (Koubek and Mountjoy, 1991), problem solving, (Hardiman et al., 1989; Chi et al., 1981), and medical diagnosis (Hobus et al., 1987).

Given that an expert can possess procedural or conceptual knowledge, Wilson and Corlett (1990) go one step further to state that all experts can be placed into one of three categories. *Academics* are the first category. These experts regard their knowledge domain as being logically organized and use theory to solve problems (high conceptual knowledge). The second categorization of experts defined by Wilson and Corlett is the *samurai*. This type of expert uses pure performance to solve problems (high procedural knowledge). The final classification of experts is practitioners who engage in activity every day that uses both theory and performance to solve problems (high conceptual and procedural knowledge).

5.4 HOW INDIVIDUALS ACQUIRE KNOWLEDGE

With an understanding as to the various knowledge types as well as the different levels of expertise, this section will discuss modeling the knowledge acquisition process as well as factors that aid or impede acquisition. The most commonly used form of modeling for memory and knowledge purposes is production systems (Lehto and Miller, 1986). Therefore, general production systems will be presented, followed by Anderson's ACT, a very specific and thorough modeling technique that is often cited in literature. Finally, various factors affecting the acquisition of knowledge are discussed.

5.4.1 Modeling Changes in Knowledge by Production Systems

Essentially, a production system is a scheme for specifying the function of an information processing system (Anderson, 1983; Lehto and Miller, 1986; Newell, 1973). A production system is usually exemplified as a computer program developed with artificial intelligence programming techniques with the emphasis on representing procedural knowledge (Lehto and Miller, 1986).

Production systems include a set of productions and data structures. Each production (also referred to as a production rule) consists of a condition and a resulting action; **IF** *condition*, **THEN** *action*. The condition must be true before the action can be performed. An example of a production rule would be "IF light is green, THEN go." The condition is "light is green" and the action is "go." This production is very simple, but productions are frequently more complicated, with compound condition expression, a series of actions, as well as containing nested task goals. For instance, a slightly more complicated production could be "IF light is green AND the road is clear AND the car is still running, THEN continue on toward destination."

Data structures are expressions that encode information relevant to the task. It is this information which is used to determine whether a production's condition is true or false. Typically, this information comes from the task environment or from the individual's long-term memory. Continuing the above driving example, the data structure would contain all inputs surrounding the driver (i.e., traffic light color, driving conditions, radio station, locations of other drivers) in addition to items from long-term memory (i.e. status of vehicle, traffic laws, destination).

A very simplified example of the functioning of a production system is shown in Figure 5.1. Once an initial data structure is established, the production rules are scanned in search for one with a condition expression which evaluates to true. However, since not all the information in the data structure is relevant to any specific production, selective attention is employed. Selective attention is a focusing or narrowing of attention, acting like a filter, to allow only important stimuli access into the working memory and thus be used in the evaluation of the condition expression. Once a rule is found (or activated), its respective action is performed. The action may be a physical or mental act or operation or it may modify the data structure's contents (resulting in other production's conditions becoming valid or invalid). More complex production systems can involve nested rule activation. In general, the production system, as a set of production rules, is constantly changing due to changing data structures and satisfying or negating productions.

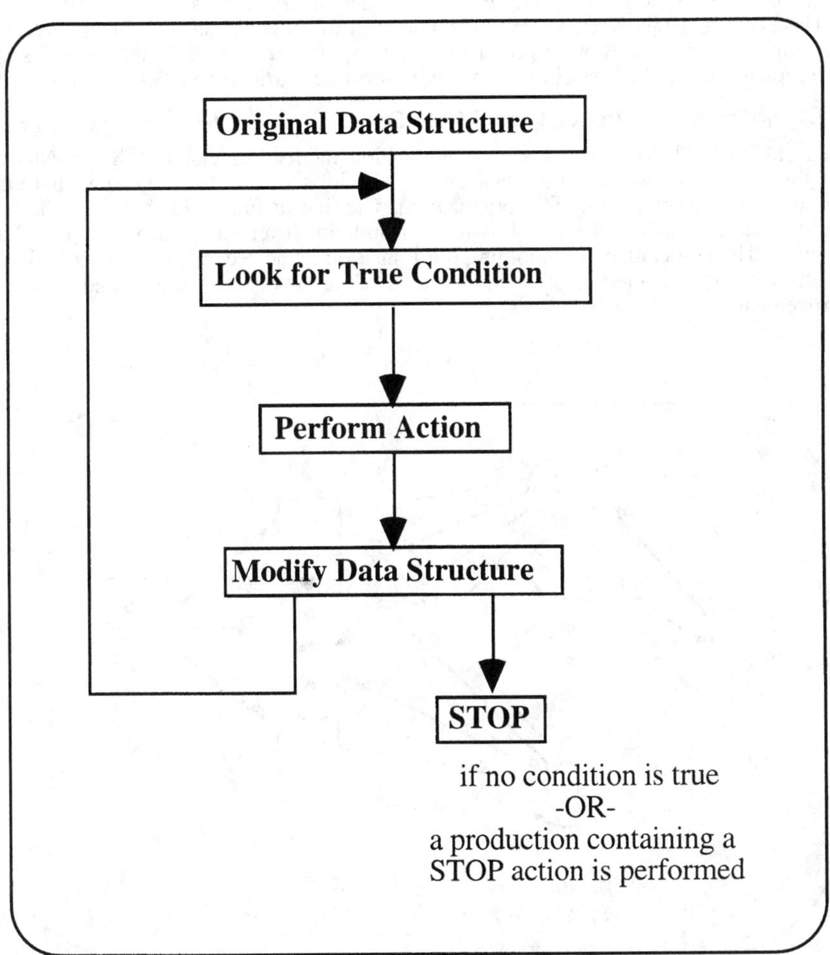

Figure 5.1 A simplified model of information processing via a production system.

Within a person's knowledge structure (discussed further in Section 5.4.2.2), there exist many production rules linking one bit of information to another. Also, there can be a range of rule transfer from one task to another. Additionally, between these bits of information, there can exist one or more links, each path or link based on a different goal. This leads to the Principle of Adaptation (Newell, 1973):

> Other things equal, the subject will adopt that production system that more closely obtains his goals. (p. 494)

The individual is assumed to perform a task once they are instructed, choosing the production system that best fits their determined goal. Newell also postulated that due to the unknown variables in one's environment the individual does not come to the task with a predetermined method of how to perform the task. Thus, the organization of task completion is based on the demands of the task and the environment in which it is performed (Anderson, 1983; Duncker, 1945; Hayes-Roth and Hayes-Roth, 1979; Newell, 1973; Norman, 1968). One selects inputs based on goals, as well as selecting the goals based on the inputs. This principle and its assumptions are demonstrated in Figure 5.2 (Duncan, 1990). Furthermore, it is important to realize that there are internal (memory) and external (environmental) constraints on the user that will limit or affect the production system that is selected (Newell, 1973).

There have been many different types of production system models proposed, including PSG (Production System version G) by Newell and Simon (1972) (Newell, 1972), Anderson's (1982) ACT, Soar (Newell, 1990); and others (Duncker, 1945; Hayes-Roth and Hayes-Roth, 1979, Norman, 1968). Though all of these models can be credited with following the basic definition of production systems, Anderson's ACT and Newell's Soar are complete examples. Therefore, discussion on these particular models will ensue.

5.4.1.1 Adaptive Control of Thought (ACT)

Based on Fitts's (1964) three-step skill acquisition model, Anderson (1982) created the ACT theory which describes a three-stage model with a series of mechanisms that effect data during the development of automatization. The first of Fitts's three steps is the basic cognitive stage, which is the step in which an individual rehearses information for skill execution. The associative or smoothing out transition is the second stage. The final stage of Fitts's model, defined as the autonomous stage, involves gradual and continued improvement.

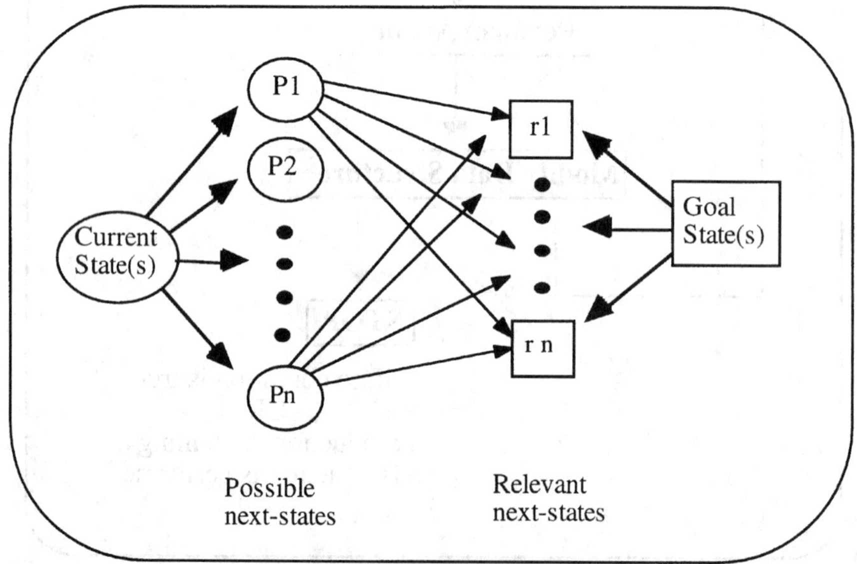

Figure 5.2 Activation of candidate goals. The current state activates possible next-states, while the goal state activates relevant next-states. [Modified from Duncan (1990).]

These three stages, further refined, became the basis for Anderson's (1982) ACT theory, which describes the progression of a learner from low to high proceduralization of knowledge, by execution of generalization, discrimination, and strengthening mechanisms on the production rules. In this manner, the production system representing procedural knowledge is continually honed to be more efficient. This set of mechanisms is reported to account for each stage in the cognitive skill acquisition process. The following figure will be useful in discussing the stages associated with Anderson's model.

Before the acquisition of new knowledge, a user is already equipped with some general-purpose production rules (Anderson, 1982). These rules are transferred from other knowledge areas that may or may not be related to the relevant task. These rules have variable condition/action pairs, which will be modified as the individual progresses through the various stages. This transition from variable to constant production rules is called proceduralization. This term used by Anderson, although related, is not to be confused with procedural knowledge, which are linked rules used to complete a goal.

The first stage in the ACT theory is the conversion of declarative input. A learner in this stage will rehearse, in working memory, the information needed to complete the task. This stage is represented as a propositional network, imposing structure on declarative facts, allowing for completion of a task. Individuals at this stage are considered novices in proceduralized knowledge. The advantage to this stage is that the pieces of knowledge are easy to modify. The main disadvantage, however, is that substantial interpretation of the knowledge is required.

Referring to Figure 5.3, the transition stage that occurs between declarative and procedural knowledge is knowledge compilation, when an individual's declarative knowledge begins to become structured by rules. In this stage, the user's method for processing information will be modified to procedural rule-based knowledge, if both the individual's ability and practice allow. Proceduralization of knowledge eliminates the need for retrieving information from the long-term memory by creating production rules with constants. In addition to proceduralization, compilation, which is the second effect at this stage, is the linking and combining of sequential rules. This can only occur when the rules have a common goal. A benefit of this stage is that compilation and proceduralization decrease the time required to complete the task.

The final stage proposed by the ACT theory is the result of both the proceduralization and compilation processes. This stage uses facts from the declarative database to access and perform the productions required. This stage also involves tuning the learner's production system. The representations of knowledge in this stage are the productions, discussed earlier, containing a specific and constant condition/action pair linked together. The primary advantage of reaching this stage is that the tasks are more effectively performed, thus taking less time and producing fewer errors. A fundamental disadvantage is

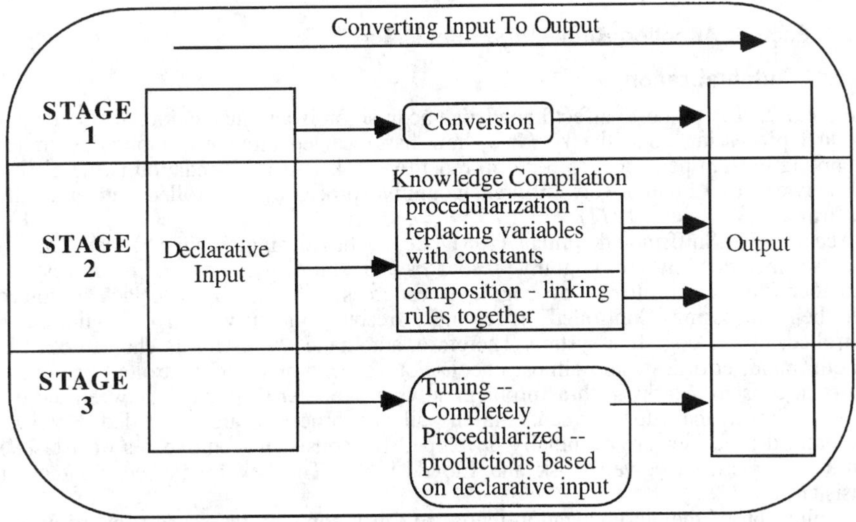

Figure 5.3 Anderson's (1982) ACT Theory. [Modified]

that a procedure, once the condition/action pairs are made constant, is difficult to alter for specific or unusual situations.

In summary, ACT's procedural knowledge is a higher-level or skill-based knowledge in which declarative facts are already compiled and proceduralized. Procedural knowledge can only be manifested in an individual's performance (Anderson, 1993). Therefore, proceduralization and compilation explain why experts have fewer errors (due to the immediate link with relevant information) and a lower completion time (no retrieval into the long-term memory). A super expert is defined as an individual whose knowledge has been so completely compiled that task completion is automatic. This high level of expertise lowers the load on working memory and increases the number of simultaneous tasks a user can perform. Other advantages upon reaching this stage are that there is no interpretation phase required, it is a more direct way to perform a task, and the information is handled in a fast, natural, and easy manner, thus further decreasing task completion time. However, the problems with performing a task in this stage are that the production rules cannot be modified and unique or infrequent tasks do not become as highly proceduralized.

5.4.1.2 Soar

Soar is organized entirely as a production system. The long-term memory, for both program and data, consists of parallel-acting condition-action rules, much like ACT. Soar is a symbolic computational system (Newell, 1990; 1992). The purpose of Soar is to manipulate and evaluate symbol structures according to a main or master symbol structure. Soar organizes all tasks into problem spaces. In this problem space, operators are selectively applied to current states to attain desired states. The task is accomplished when all desired states are attained. Problem solving proceeds in a sequence of decision cycles that select problem spaces, states, and operators. Each decision cycle accumulates knowledge into a long-term recognition memory. This memory continually matches against elements in the working memory, updating the current state and retrieving preferences that encode knowledge about the next step. Access of recognition memory is involuntary, parallel, and rapid.

If Soar does not know how to proceed in a problem space, an impasse will occur. Soar responds to an impasse by creating a subgoal in which a new problem space can be used to acquire the needed knowledge. If lack of knowledge prevents progress in a new space, another subgoal is created. Creation of subgoals continues if lack of knowledge persists, thus creating a goal hierarchy. All the incoming perception and outgoing motor commands flow through the highest level of problem space. Once an impasse is resolved through problem solving, the chunking mechanism adds new productions to recognition memory by encoding the results of the problem solving. This is an attempt to avoid the same impasse in the future. The purpose of chunking is to construct new productions (chunks of simple productions) that capture new knowledge developed by Soar (in the working memory) to resolve difficulties.

5.4.2 Factors Affecting Knowledge Acquisition

5.4.2.1 Automatization

One factor that is found to affect knowledge acquisition is automatization, also known as the dual processing code theory. This view has received attention among researchers attempting to train personnel to perform repetitive tasks. At its foundation is the distinction between two qualitatively different cognitive processes: controlled and automatic (Shiffrin and Schneider, 1977).

According to Shiffrin and Dumais (1981), key distinguishing features between the two processes include the use of cognitive resources and control. Automatized processes have been found to require little or no cognitive resources and are subject to lack of control over their processing. Controlled processing is resource intensive and is monitored, or controlled, by the cognitive system. Therefore, automatic processing is characterized by smooth, rapid, effortless, and almost unconscious execution while controlled processing occurs in conscious tasks such as problem solving. Another distinction between the two processes lies in their development. Initially all new processes are controlled. However, automatized processes are eventually developed for consistent components of a task, if significant practice is given (Fisk and Lloyd, 1987). The key to this development is consistency.

Application of the automatization theory to fairly complex tasks has shown implications for training and cognitive skill acquisition. Myers and Fisk (1987) simulated the acquisition of skill in a telecommunications job which requires identification of patterns of information and execution of the appropriate response. Their findings indicated that performance on both consistent and variable task components increased with practice; however, the consistent task yielded significantly greater performance gains. Also, they found that as practice accrued on the consistent task, between-subjects variance decreased. This important finding suggests that automatized processes are ability insensitive, while controlled process are more affected by individual variance in that ability.

Fleishman and Hempel (1954) have shown similar results for repetitive tasks; while abilities are important for acquiring the skill, as practice accrues, performance becomes less dependent on the initial ability set possessed by personnel. Therefore, to acquire knowledge and perform problem-solving tasks, Fleishman's ability approach takes on increasing importance, while it is less important in tasks for which significant practice on repetitive task has accrued.

Because automatized tasks require minimal amounts of cognitive resources, the resources initially invested in performance of these tasks are freed for other purposes, such as learning at higher levels. Of course, automation is not the only way to free resources, since resources may also be freed by learning new principles and by doing the task more efficiently. For example, as individuals acquire skill, they become more selective with regard to the use of available information needed in decision-making tasks (Salvendy and Seymour, 1973). Cognitive resources can also be freed through the acquisition and application of higher-order heuristics and learning strategies. These strategies, often termed *metacognition*, provide personnel with the plans and procedures used to acquire new knowledge.

5.4.2.2 Learning Hierarchy

The manner in which information is presented has been found to affect learning and the type of knowledge that is attained from the information. Examining and defining this type of presentation is called the learning hierarchy. Furthermore, each learner has a presentation manner which is best suited for them. These topics will be discussed in detail in the following sections.

Key Principles

Gagné (1965) was one of the first people to differentiate intellectual skills acquisition processes into different categories of learning. He considered six types of learning, arranged in a hierarchy from lower levels of mental processes to higher levels, with the assumption that each higher level of learning depends on the mastery of the one below it. In other words, the learner cannot achieve a specific higher order of learning without first acquiring all lower levels below the targeted level. This assumption implies that rather than an eclectic process, the intellectual skills acquisition is a sequential process. The six-level learning hierarchy proposed in his later work (Gagné, 1985) is shown in Figure 5.4.

Associations and chains: At the bottom of Gagné's learning hierarchy are the basic forms of learning: association and chaining. An association is a stimulus (S) with a response (R). In this stage, a learner acquires a precise response to a discriminated stimulus. Chaining is the connection of a set of S → R associations in sequence. One cannot expect a chain to be learned in an optimal way unless the learner is already able to carry out the S → R pairs that constitute the links. More advanced forms of learning, such as intellectual skills, are built on these two basic forms of learning.

Discriminations: Identifying the difference between variations in some particular property of an object is called discrimination, or, when more than one property is involved, multiple discrimination. Discrimination learning is often concerned with the distinctive features of stimulus objects in which the learner will begin to distinguish the differences among similar stimuli and responses so that the S → R pairs learned in the previous stages can be treated differently.

Concepts: The next stage of learning is called concept learning. Individuals learn to make a common response to collections of stimuli that may be very different in terms of appearance or some physical properties. In order to achieve concept learning, learners need to put these stimuli into a class by some criterion and respond to an instance or event of the class as a member of that class. According to Gagné, there are two kinds of concepts. One is the concrete concept, which is normally observable by one of the five senses. The other is the defined concept, which is described by abstract terms or definitions.

Rules: Concept learning, discussed above, leads to rule learning, which is the most typical form of an intellectual skill. In simple terms, a rule is a chain of two or more

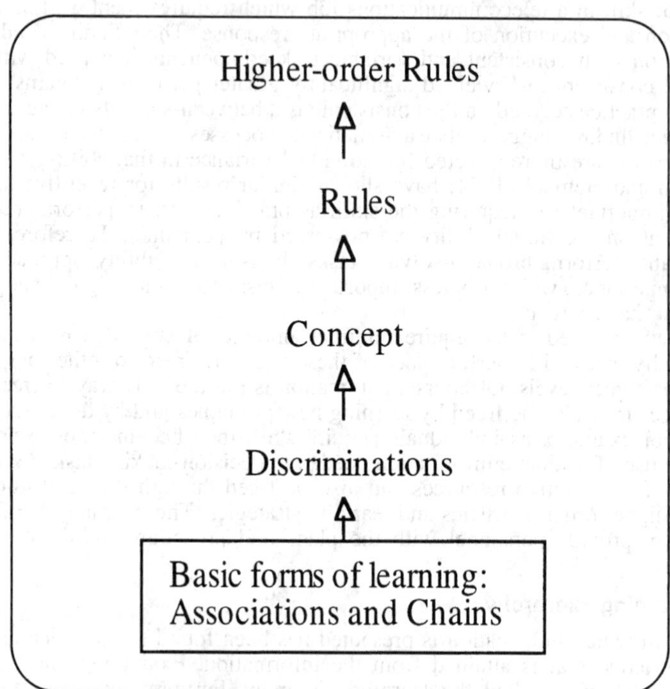

Figure 5.4 Gagné's basic forms of learning and intellectual skill learning hierarchy. [Adapted from Gagné (1985).]

concepts, which can be represented by the form of "IF A, THEN B" where A and B are concepts. Note that rules and associations differ in that the objects in rules are concepts, whereas the objects in associations are concrete stimuli. Since most of the intellectual skill learning is based on rule learning, it is the focus of recent learning theories such as ACT (Anderson, 1982) (Section 5.4.1).

Higher-order rules: Through learning, rules may be combined into more complex rules, called higher-order rules. Anderson called this *composition* in stage 2. Higher-order rules often result from the learner's deliberation during a problem-solving situation. In attempting to solve a novel problem, a learner may put several rules from different domains together to form a higher-order rule. Therefore, it is also a process that yields new learning.

According to Gagné (1985, 1987), one can train for either high or low conceptual knowledge. The training for the high conceptual knowledge is by a top-down process, in which the abstract concepts are first presented, and then proceed down the hierarchy of information to the lower levels containing detailed knowledge. In contrast, training for low conceptual knowledge would present information through a bottom-up training process, with details presented and then followed by abstract concepts.

Mechanisms for Facilitating Learning

Throughout the learning hierarchy, a distinction is made between features which are *external* and *internal* to the individual. External conditions are the methods and stimuli used to instruct the individual on a particular task. Internal conditions for learning an intellectual skill consist of the previously learned skills that are components of the new skill and the processes that will be used to recall them and put them together in a new form. In order to reach a specified stage on the learning hierarchy, all knowledge acquired previous to the targeted stage is a prerequisite.

Gagné's learning theory provides a comprehensive view of hierarchical learning in a sequential manner as well as internal and external learning conditions. Although the internal and external conditions are known to be required for moving up the learning hierarchy, the hierarchy itself does not explicitly contain mechanisms for moving up the learning levels. Nonetheless, Gagné's learning hierarchy is based on information-processing theories.

More than two decades ago, Atkinson and Shiffrin (1968) developed an information-processing theory of memory which changed the view of interaction between cognitive processes and learning, though this was, by no means, the first model to represent cognitive processes in terms of information flow. In this view, information from the outside world is said to first enter a short-term sensory store. From here, the information is transferred into working memory. Information can flow to the response generator or sometimes be transferred to long-term memory where the information is permanently stored. While Atkinson and Shiffrin's model has been refined over the years, the basic ideas in their model are still relevant to modern learning theories. The basic concepts of Atkinson and Shiffrin's model include: (1) knowledge is represented in some form of long-term memory thus reducing the burden in working memory, and (2) information flow can represent cognitive processes (Hunt, 1989). These two concepts can be seen in many modern learning theories, including Gagné's learning theory. Therefore, the mechanisms in the information-processing theory adopted in Gagné's learning theory such as encoding, storage, and retrieval can be considered as the learning mechanisms in Gagné's learning theory. For details for information-processing theories, please refer to Chapter 4 in this handbook.

The information-processing theory provides general learning mechanisms for Gagné's learning hierarchy. These general learning mechanisms can be applied throughout all levels on the hierarchy. However, the rapid development of advanced technology requires humans to learn new skills at a continually increasing pace, and puts the emphasis on learning processes at higher levels on the learning hierarchy. Recent research regarding learning processes has focused particularly on higher learning levels, especially in a problem-solving domain (Anderson, 1989, 1993; Anderson et al., 1990; Card et al., 1983; Laird et al., 1987; Larkin, 1981). These levels of learning can be mapped to Gagné's rule and higher-order rule levels on the learning hierarchy. These models hypothesize on the differentiation of procedural and declarative knowledge and their interaction in human memory and cognition (Willingham et al., 1989). These issues are detailed in the next section of this chapter.

Other than procedural and declarative knowledge, Gagné suggests another aspect of higher learning levels: *cognitive strategies*, which are the skills that are used by learners to regulate their own internal process to combine previously learned rules to solve a novel problem. This can also be called *metacognitive skills* and *conceptual knowledge*. From the above definition, it is clear that cognitive strategies are a learning mechanism commonly used in Gagné's problem-solving level on the learning hierarchy. Problem solving is not simply a matter of applying previously learned rules, it is also the process that yields new cognitive strategy. During problem-solving, the learner tries to select and combine previously learned rules to achieve a goal. Once the goal is achieved, the learner not only solves the problem, he or she may also have learned a new rule or a new way of regulating the rules or thinking. This new way of regulating rules is a new cognitive strategy. Therefore, although problem solving involves only old rules, the learner may change permanently after solving the problem. Cognitive strategies are largely independent of learning contents or domains. Thus, once a cognitive strategy is learned, it may be transferred to other kinds of learning in a different knowledge domain.

5.4.2.3 Knowledge Structures

Knowledge structure theories propose that the manner in which humans structure their knowledge about the domain is a significant determinant of performance (Enkawa and Salvendy, 1989; Gibson and Salvendy, 1990; Ye and Salvendy, 1991). Numerous studies have shown differences between the knowledge structures of novices and experts and have suggested this as a potential explanation of performance differences (Adelson, 1981; Barfield, 1986; Egan and Schwartz, 1979; Hardiman, Dufresne, and Mestre, 1989; Hobus, Schmidt, Boshuizen, and Patel, 1987; Murphy and Wright, 1984; Schoenfeld and Herrmann, 1982).

In a recent set of papers, Koubek and Salvendy (1989, 1991) have proposed three levels of knowledge structures corresponding to the level of operator skill: surface feature, task specific, and abstract/hierarchical. The *surface feature* structures, found in novices, are composed of the explicit, physical, salient domain features. Since no abstracted concepts are used, such a structure does not provide the operator the capability to reason about their domain in any but the most basic manner.

As humans become skilled, they develop a more conceptual, yet *task specific* framework which allows for more complex cognitive activity, such as decision making, inferences, and extrapolation. However, when personnel with this structure face a problem-solving task, they evoke only those parts of the structure that appear directly relevant to a narrow subset of the task at hand. Their structure, which contains conceptual information, is not yet organized in a principled manner which allows the operator to see the more broad implications and impending features from the domain for their particular problem. As such, the solution path is narrow, with a depth-first flavor to the search strategy. This task specific knowledge structure coincides with high procedural knowledge. The majority of individuals do not progress beyond this level. Performance at a level which has been termed super-expert is dependent on the operator not only possessing a conceptual understanding of the domain, but also that these concepts be organized in a hierarchical manner under increasingly abstracted concepts, or principles. This highest knowledge structure is termed abstract/hierarchical. This coincides with high conceptual knowledge.

This theoretical approach suggests a focus on training personnel to develop a high conceptual understanding of the domain and coincides with the Gagné's learning approach. Neither of these approaches however, directly addresses the acquisition of high-performance cognitive and physical skills as modeled by the automatization approach. Also, within the knowledge structures, domain specific knowledge which is required for the effective performance of a task needs to be acquired. This domain specific knowledge is unaccounted for in the knowledge structure approach.

5.4.2.4 Abilities

Abilities are relatively enduring attributes of an individual's performance. A specific task is said to require certain abilities in order to perform to an established criterion (Fleishman and Quaintance, 1984). Many lists of human abilities have been suggested in the past (Fleishman, 1975; French et al., 1963; Harman, 1975). The objective is to identify and define the fewest independent ability categories which can describe performance for various tasks in a meaningful way. According to this objective and the empirical results, ability can be distinguished from skill, with ability referring to "a more general capacity of the individual related to performance in a variety of human tasks," and skill defined as "the level of proficiency on a specific task or group of tasks" (Fleishman and Quaintance, 1984, pp. 162–163). Probably the most significant work in this area has been done by Fleishman and Quaintance, who have isolated a validated taxonomy of 52 abilities by using a series of interlocking experimental factor-analytic studies. These 52 abilities cover cognitive, perceptual, psychomotor, and physical areas of performance.

Research (Salvendy, 1969) has shown that the relationship of an ability to task performance systematically changes during practice trials. This indicates that some abilities are more important during early learning while others are more important during performing. After analyzing numerous experiments with a great variety of tasks, Fleishman and Quaintance concluded that

> (1) the particular combinations of abilities contributing to performance on a task may change as practice on this task continues; (2) these changes are progressive and systematic and eventually become stabilized; (3) in perceptual-motor tasks, for example, the contribution of non-motor abilities (e.g., verbal or spatial) may play a role early in learning, but their contribution relative to motor abilities decreases systematically with practice; (4) there is also an increase in a factor specific to the task itself, not common to the more general abilities. (Fleishman and Quaintance, 1984, p. 337)

Based on numerous studies, it appears that ability requirements for initial stages of learning and final stages of performing are different. This information can be used to predict the final performance and help to understand the emphasis that should be placed on various stages of learning. As time-based competition and rapid development of new technology is emerging, learning new materials and the short time period for this change-over may be mentally stressful and demanding (Seppälä et al., 1992). Thus, it is profitable to predict the learning and final performing capability of workers separately so that a short life-cycle job can be assigned to one who has better learning abilities, whereas a long life-cycle job can be assigned to one who has better final performance abilities. The taxonomy of abilities approach provides an important factor which accounts for individual differences in learning. It also serves as a critical internal condition in Gagné's learning hierarchy.

5.4.2.5 Mental Workload

Mental workload is intrinsically complex and multifaceted. Many techniques have been developed to predict workload, assess the workload imposed by systems, or assess the workload experienced by the humans (Hancock and Meshkati, 1988). Although there is not a consensus regarding the definition of mental workload, most people accept that mental workload is defined by the fact that mental resources have a limited capacity. For details about mental workload, please refer to Chapter 13 of this Handbook. In the context of learning processes, it deals with the limited human capacity for the process of stimulus selection. Since learning involves processing the presented information, when a learner focuses on one information source, they may ignore information from other sources if the limit is reached. This situation is called selective attention. Alternatively, if the first information source is not very demanding, the learner may simultaneously process another information source, which is known as divided attention. Many theories, based on the phenomenon of attention, explain how the limited capacity of human attention affects human performance.

Kahneman (1973) proposed that there is a single undifferentiated, sharable pool of resources available to all tasks and mental activities, which is known as single-resource theory. According to Kahneman's model, attention is a limited capacity resource which can be applied to different tasks in accordance with some allocation policy. The limitation of single-resource theory is that it cannot account for several aspects of the data from dual-task studies (Wickens, 1980, 1984). For example, when using a dual-task experimental paradigm, sometimes the performance on the secondary tasks does not reflect the required attention. In other words, the trade-off of performance between primary and secondary tasks does not show up. An alternative view, multiple-resource theory with three dichotomous dimensions, is proposed by Wickens (1984), which argues that people have several pools of resources with different capacities. Therefore, tasks that require different pools of resources are less likely to interfere with each other. Further, since the capacity of each pool of resources depends on each individual, the performance is a function of task demands and operator capacities.

Certain amounts of cognitive resources are required to learn and perform a task to an established standard. In terms of learning, these cognitive resources are the internal conditions for the learner to properly finish learning. Following either the single resource pool theory or the multiresource pool theory, since the total mental capacity is fixed, as more cognitive resources are invested into learning, the corresponding mental workload increases. When the mental workload is too high and is beyond the critical region, the learning process suffers or may even fail.

5.5 INTEGRATION OF LEARNING APPROACHES

Each of the above-cited approaches provides different aspects of the learning and skill acquisition processes. However, due to different emphases, these theories generally only cover a part of the whole learning process. In order to understand the multifaceted nature of learning processes, it is beneficial to integrate these theories as a hybrid model. Koubek et al. (1994) present a hybrid learning model which consists of the following model components: Gagné's (1985) learning hierarchy, Schneider and Shiffrin's (1977) automaticity, Koubek and Salvendy's (1991) three levels of knowledge structures, Anderson's (1982) knowledge compilation, Wickens's (1992) multiple-resource theory, and Fleishman and Quaintance's (1984) taxonomy of abilities.

Among these model components, Gagné's learning theory provides the view of hierarchical learning. Wickens's multiple-resource theory and Fleishman and Quaintance's taxonomy of abilities fit into the internal learning conditions of Gagné's learning theory and can be considered sources and constraints during skill acquisition. While skill acquisition requires cognitive resources, Schneider and Shiffrin's automaticity, or dual code theory, can be integrated as a dynamic view of returning cognitive resources after automatization of a task. When learning by moving up Gagné's hierarchy, the development of a higher level of knowledge structure facilitates the efficient use of domain knowledge. Koubek and Salvendy's three levels of knowledge structures therefore can be mapped to the learning hierarchy to expedite learning processes. Finally, Anderson's knowledge compilation can be incorporated into the learning hierarchy to elaborate the details of knowledge transformation at the higher levels of learning. A conceptual hybrid skill acquisition model (outlined in Figure 5.5) combines the previous models. Also taken into consideration are the rationale for the interactions among various theories of skill acquisition (these are discussed in Section 5.5.1).

According to Koubek et al. (1994), the learning process requires dynamic interactions between ability requirements, knowledge structures, resources requirements, and other model components. These interactions between model components are the most important information provided by the integrated model. The main interactive behaviors, which also are the rationale behind the conceptual framework, are briefly discussed below.

5.5.1 Learning Hierarchy and Ability Taxonomy

Three pieces of information are needed for developing the interactions between learning hierarchy and taxonomy of abilities.

1. Two phases can be determined by Gagné's learning hierarchy. One is the processes of attaining new skills and moving up the hierarchy. This is represented by thick arrow lines in Figure 5.5. The other is the process of performing skills on the obtained learning levels. This is represented by the text on the learning hierarchy in Figure 5.5. The first can be called the learning phase and the second the performing phase.
2. From Gagné's learning hierarchy, moving up the learning hierarchy requires that the internal conditions are available. These internal conditions include all previous stages of learning as well as the abilities required by the next level.
3. Research (Salvendy, 1969) has shown that abilities important for initial stages of learning are often different from those required for skilled task performance.

Combining the above three pieces of information, it can be seen that although some overlap may exist, the abilities required during the learning phase to move up to the next targeted level may not be the same as the abilities required to perform the task either at the original level or next targeted level. The abilities required for learning are known as *learning* abilities and abilities required for performing are termed *performance* abilities. From Figure 5.5, it can be shown that in each level of learning, the performance abilities are a subset within learning abilities. This means that during learning of the next targeted level, both the performance abilities for the targeted level and the learning abilities associated with that level are required. It is assumed that the boundaries between learning abilities and performance abilities are blurred due to the fact that the learning process itself is continuous. As for the performance abilities, given that all previous stages of learning are prerequisite to attaining the next level, it can be further assumed that the performance abilities required are cumulative when moving up the learning hierarchy. However, the required ability levels may vary at different levels of learning.

The interaction between learning hierarchy and taxonomy of abilities suggests that a general set of performing and learning abilities may exist. However, most current job and task analysis methodologies such as PAQ (McCormick and Jeanneret, 1988) are not very sensitive to cognitive task components (Koubek et al., 1994).

5.5.2 Learning Hierarchy, Automaticity, and Cognitive Resources

Learning hierarchy, automatization, and cognitive resources are all interrelated in a skill acquisition process. In order to perform a task or learn a new skill, a certain amount of resources is required. As learning accrues, more cognitive resources are required for moving up along the learning hierarchy. On the other hand, repeatedly performing a consistent task gradually releases cognitive resources through automatization. Thus, the individual who is learning a new skill may spend most of their resources at a certain level in the beginning. Due to lack of resources, they are not able to move up to the next level at this time. After the learner becomes more skilled at this level, a portion of their cognitive resources will be released through automatization of the consistent components. The learner now may have enough cognitive resources for learning at the next higher level. This process will continue until either ability limits or cognitive resource limits are met, or the task reaches its optimal learning level. In this process, the cognitive resources of learners are not wasted, but rather are continuously re-invested to acquire new states

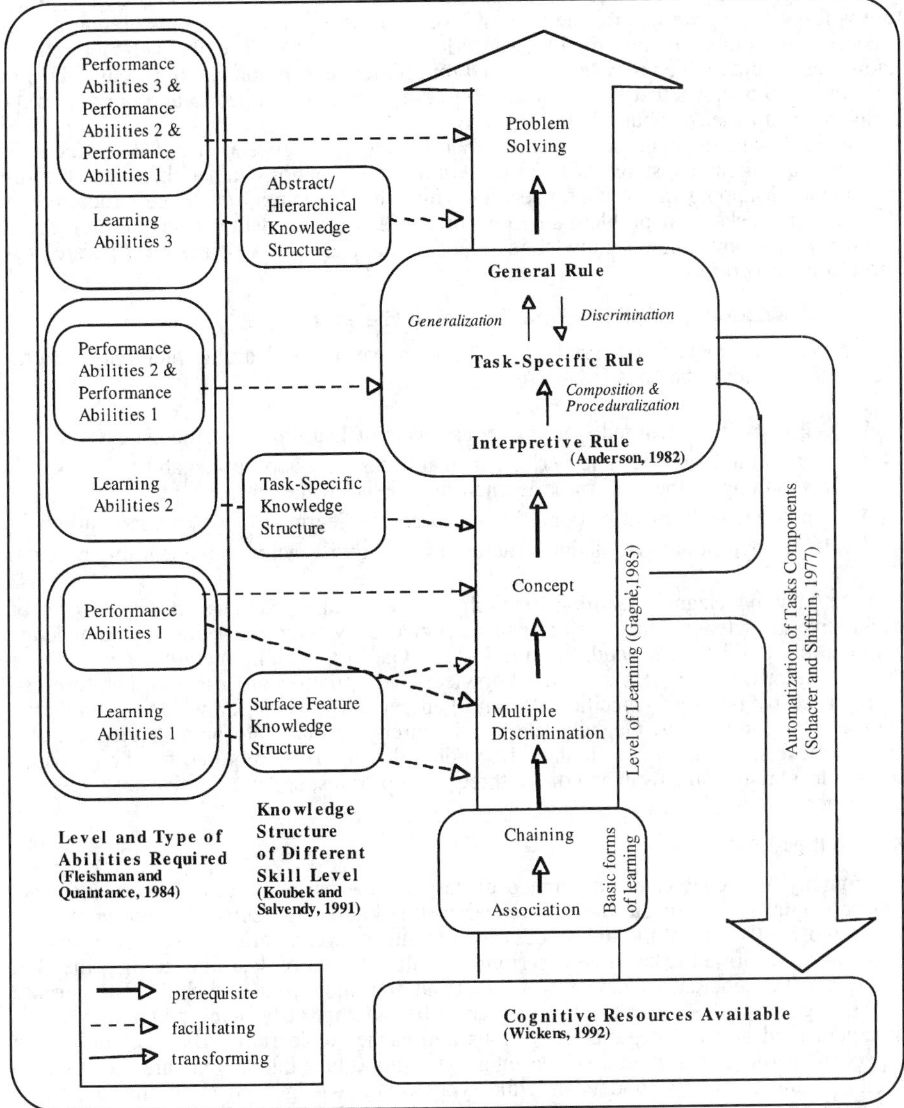

Figure 5.5 Hybrid conceptual framework for skill acquisition processes.

of learning. From Figure 5.5, cognitive resources can be viewed as flowing through the learning hierarchy and cycling back to the original resources pool through automatization for further learning.

5.5.3 Knowledge Structure, Learning Hierarchy and Ability Taxonomy

Koubek and Salvendy's (1991) three levels of knowledge structures can be mapped to the learning hierarchy to improve learning processes. As learning progresses up the hierarchy and the tasks become increasingly cognition intensive, more deep knowledge and reasoning activities will be involved in rule-based related tasks. The original surface feature knowledge structure becomes insufficient to perform these tasks. Learners need to develop a more abstract, higher level of knowledge structure to cope with the increasingly complex domain knowledge. Thus, the task-specific knowledge structure will be developed gradually. If the learning process continues, the task-specific knowledge structure will not be appropriate for the highest level of learning since problem-solving related tasks require both deep and broad views of domain knowledge. The abstract/hierarchical knowledge structure then may be developed for efficient use of domain knowledge. Therefore, Figure 5.5 shows that the knowledge structures can be mapped to appropriate levels of learning in order to obtain better performance.

One issue which should be clarified is that the mapping between knowledge structures and learning hierarchy suggests a set of optimal combinations rather than absolute requirements. Mapping means that a learner with a task-specific knowledge structure can still perform tasks in a problem-solving domain. However, in this case, the degree of efficiency of using domain knowledge is not as high as one with abstract/hierarchical knowledge structure.

5.5.4 Knowledge Compilation and Learning Hierarchy

According to Koubek et al. (1994), in order to move up the learning hierarchy, several requirements should be satisfied.

1. All knowledge learned from previous levels of learning is prerequisite.
2. The training abilities and performance abilities with appropriate ability levels corresponding to the skill to be learned must exist in the learner.
3. An adequate amount of cognitive resources for learning new skills is required.
4. If an appropriate knowledge structure exists, it will facilitate the learning process.

By reviewing Gagné's learning hierarchy, the defined concept is actually a form of declarative knowledge and therefore can be replaced by interpretive rules in knowledge compilation. On the other hand, the rule level in Gagné's learning hierarchy accounts for the procedural knowledge part in the knowledge compilation process and therefore can be replaced by both task-specific rules and general rules. Thus, knowledge compilation can be incorporated into Gagné's learning hierarchy to elaborate both defined concept and rule levels. In Figure 5.5, it should be noted that the rules required for the problem-solving level may come from any of the three types of rules, i.e., interpretive, task-specific, and general rules.

5.6 SUMMARY

In summary, this chapter has attempted to describe the various types of knowledge that can be acquired, to indicate the differences in this knowledge between novices and experts, and finally, to outline the process for moving between various levels of expertise. The chapter is not an exhaustive description of all research on learning. Rather, the goal is to present a coherent framework for understanding knowledge and skills. The purpose of this organizing framework is to make an initial attempt at bringing parsimony to this divergent field and to integrate the various approaches to learning. By examining each aspect of learning in isolation of the others, the important interactions are lost. A successful training program understands the types of knowledge and skills that are to be acquired, and their interactions, to develop technologies for efficiently executing the learning mechanisms discussed in this chapter.

In conclusion, learning facts and skills is a multifaceted issue that impacts nearly all aspects of work life. A proper understanding of these principles is key to developing and maintaining a qualified workforce in today's constantly changing environment.

REFERENCES

Adelson, B. (1984). When novices surpass experts: The difficulty of a task may increase with expertise. *Journal of Experimental Psychology, 10*(3), 483–495.

Adelson, B. (1981). Problem-solving and the development of abstract categories in programming languages. *Memory & Cognition, 9*, 422–433.

Anderson, J. R. (1993). *Rules of the Mind.* NJ: Erlbaum.

Anderson, J. R. (1989). A theory of the origins of human knowledge. *Artificial Intelligence, 40*, 313–351.

Anderson, J. R. (1983). *The Architecture of Cognition.* Cambridge: Harvard University Press.

Anderson, J. R. (1982). Acquisition of cognitive skill. *Psychological Review, 89*(4), 369–406.

Anderson, J. R., Boyle, C. F., Corbett, A. T., and Lewis, M. W. (1990). Cognitive modeling and intelligent tutoring. *Artificial Intelligence, 42*, 7–49.

Atkinson, R. C., and Shiffrin, R. M. (1968). Human memory: A proposed system and its control processes. In K. W. Spence, Ed., *The Psychology of Learning and Motivation*. New York: Academic Press.

Barfield, W. (1986). Expert-novice differences for software: Implications for problem solving and knowledge acquisition. *Behaviour and Information Technology, 5,* 15–29.

Benysh, D. V., Koubek, R. J., and Calvez, V. (1993). A comparative review of knowledge structure measurement techniques for interface design. *International Journal of Human-Computer Interaction, 5*(3), 211–237.

Biswas, A., and Sherrell, D. L. (1993). The influence of product knowledge and brand name on internal price standards and confidence. *Psychology and Marketing, 10*(1), 31–46.

Byrnes, J. P. (1992). The conceptual basis of procedural learning. *Cognitive Development, 7,* 235–257.

Byrnes, J. P. (1988). Formal operations: A systematic reformulation. *Developmental Review, 8,* 1–22.

Byrnes, J. P., and Wasik, B. A. (1991). The role of conceptual knowledge in mathematical procedural learning. *Developmental Psychology, 27,* 777–786.

Card, S. K., Moran, T. P., and Newell, A. (1983). *The Psychology of Human-Computer Interaction.* NJ: Erlbaum.

Chase, W. G., and Simon, H. A. (1973). The mind's eye in chess. In W. G. Chase, Ed., *Visual Information Processing*. New York: Academic Press.

Chi, M. T. H., Feltovich, P. J., and Glaser, R. (1981). Categorization and representation of physics problems by experts and novices. *Cognitive Science, 5,* 121–152.

De Groot, A. (1965). *Thought and Choice in Chess*. Mouton: The Hague.

Diehl, W., and Mikulecky, L. (1981). Making written information fit workers' purposes. *IEEE Transactions on Professional Communications, 24*(1), 5–9.

Duncan, J. (1990). Goal weighting and the choice of behaviour in a complex world. *Ergonomics, 33,* 1265–1279.

Duncker, K. (1945). On problem solving. *Psychological Monographs, 58* (Whole No. 270), 1–113.

Eberts, R. E. (1994). *User Interface Design.* Englewood Cliffs, NJ: Prentice-Hall.

Egan, D. E., and Schwartz, B. J. (1979). Chunking in recall of symbolic drawings. *Memory & Cognition, 7,* 149–158.

Enkawa, T., and Salvendy, G. (1989). Underlying dimensions of human problem solving and learning: implications for personnel selection, training, task design and expert systems. *International Journal of Man-Machine Studies, 30,* 235–254.

Fisk, A. D., and Lloyd, S. J. (1987). The role of stimulus-to-rule consistency in learning rapid application of spatial rules. School of Psychology. Atlanta: Georgia Institute of Technology.

Fitts, P. M. (1964). Perceptual-motor skill learning. In A. W. Melton, Ed., *Categories of Human Learning*. New York: Academic Press.

Fleishman, E. A. (1975). Toward a taxonomy of human performance. *American Psychologist, 30,* 1127–1149.

Fleishman, E. A., and Hempel Jr., W. E. (1954). Changes in factor structure of a complex psychomotor task as a function of practice. *Psychometrika, 19*(3), 239–252.

Fleishman, E. A., and Quaintance, M. K. (1984). *Taxonomies of Human Performance*. Orlando, FL: Academic Press.

French, J. W., Ekstrom, R. B., and Price, L. A. (1963). *Manual for Kit of Reference Tests for Cognitive Factors.* NJ: Educational Testing Service.

Gagné, R. B. (1985). *The Conditions of Learning and Theory of Learning*. New York: CBS College Publishing.

Gagné, R. B. (1987). *Instructional Technology: Foundations*. Hillsdale, NJ: Erlbaum.

Gagné, R. M. (1985). *The Conditions of Learning and Theory of Instruction* (4th Ed.). New York: Holt, Rinehart and Winston.

Gagné, R. M. (1965). *The Conditions of Learning*. New York: Holt, Rinehart and Winston.

Gibson, D., and Salvendy, G. (1990). Knowledge representation in human problem solving: Implications for expert system design. *Behaviour and Information Technology, 9*(3), 191–200.

Glaser, R., and Chi, M. T. H. (1988). In the nature of expertise. In M. T. H. Chi, R. Glaser, and M. J. Farr, Eds., Hillsdale, NJ: Lawrence Erlbaum.

Hancock, P. A., and Meshkati, N. (1988). *Human Mental Workload.* New York: Elsevier Science Publishing Company.

Hardiman, P. T., Dufrensne, R., and Mestre, J. P. (1989). The relation between problem categorization and problem solving among experts and novices. *Memory & Cognition, 17,* 627–638.

Harman, H. H. (1975). *Final Report of Research on Assessing Human Abilities (PR-75-20).* Princeton, NJ: Educational Testing Service.

Hayes-Roth, B., and Hayes-Roth, F. (1979). A cognitive model of planning. *Cognitive Science, 3,* 275–310.

Healy, A. F., and Bourne, L. E. Jr., Eds. (1995). *Learning and Memory of Knowledge and Skills.* Thousand Oaks, California: Sage.

Hiebert, J. (1987). *Conceptual and Procedural Knowledge: The Case of Mathematics.* Hillsdale, NJ: Erlbaum.

Hobus, P. P. M., Schmidt, H. G., Boshuizen, H. P. A., and Patel, V. L. (1987). Contextual factors in the activation of first diagnostic hypotheses: Expert-novice differences. *Medical Education*, 21, 471–476.

Hunt, D. (1989). Cognition and Learning. In *Understanding Literacy and Cognition, Theory, Research, and Application.*, C. K. Leong and B. S. Randhawa, Eds., New York: Plenum Press.

Inhelder, B., and Piaget, J. (1980). Procedures and Structures. In D. R. Oldson, Ed., *The Social Foundations of Language and Thought.* New York: W. W. Norton.

Kahneman, D. (1973). *Attention and Effort.* NJ: Prentice Hall.

Keating, D. P. (1988). Byrnes' reformulation of Piaget's formal operations: Is what's left what's right? *Developmental Review, 8*, 376–384.

Keating, D. P., and Crane, L. L. (1990). Domain-general and domain-specific processes in proportional reasoning: A commentary on the Merrill-Palmer Quarterly special issue on cognitive development. *Merrill-Palmer Quarterly, 36*, 411–424.

Koubek, R. J. (1987). *Toward a Model of Knowledge Structure and a Comparative Analysis of Knowledge Structure Measurement Technique.* Unpublished doctoral dissertation, Purdue University, School of Industrial Engineering, W. Lafayette, Indiana.

Koubek, R. J., Clarkston, T. P., and Calvez V. (1994). The training of knowledge structures for manufacturing tasks: An empirical study. *Ergonomics, 37*(4), 765–780.

Koubek, R. J., and Mountjoy, D. N. (1991). The impact of knowledge representation on cognitive-oriented task performance. *International Journal of Human-Computer Interaction, 3*, 31–48.

Koubek, R. J., and Salvendy, G. (1989). The implementation and evaluation of a theory for high level cognitive skill acquisition through expert systems modelling techniques. *Ergonomics, 32*(11), 1419–1429.

Koubek, R. J., and Salvendy, G. (1991). Cognitive performance of super-experts on computer program modification tasks. *Ergonomics, 34*(8), 1095–1112.

Kyllonen, P. C., and Alluisi, E. A. (1987). Learning and Forgetting Facts and Skills. In G. Salvendy, Ed., *Handbook of Human Factors.* New York: John Wiley & Sons.

Laird, J. E., Rosenbloom, P. S., and Newell, A. (1987). SOAR: An architecture for general intelligence. *Artificial Intelligence, 33*, 1–64.

Larkin, J. (1981). Enriching formal knowledge: A model for learning to solve textbook physics problems. In J. R. Anderson, Ed., *Cognitive Skills and Their Acquisition.* Hillsdale, NJ: Erlbaum.

Lehto, M. R., and Miller, J. M. (1986). *Warnings, Volume 1, Fundamentals, Design, and Evaluation Methodologies.* Ann Arbor, MI: Fuller Technical Publications.

McCormick, E. J., and Jeanneret, P. R. (1988). Position Analysis Questionnaire (PAQ). In G. Salvendy, Ed., *Handbook of Industrial Engineering.* New York: John Wiley and Sons.

Miller, G. A. (1956). The magical number seven, plus or minus two: Some limits on our capacity for processing information. *Psychological Review, 63*, 81–97.

Murphy, G. L., and Wright, J. C. (1984). Changes in conceptual structure with expertise: Differences between real-world experts and novices. *Journal of Experimental Psychology: Learning, Memory and Cognition, 10*, 144–155.

Myers, G. L., and Fisk, A. D. (1987). Training consistent task components: Application of automatic and controlled processing theory to industrial task training. *Human Factors, 29*(3), 255–268.

Newell, A. (1972). A theoretical exploration of mechanisms for coding the stimulus. In A. W. Melton and E. Martin, Eds., *Coding Processes in Human Memory.* Washington, DC: Winston.

Newell, A. (1973). Production Systems: Models of Control Structures. In W. G. Chase, Ed., *Visual Information Processing.* New York: Academic Press.

Newell, A. (1990). *Unified Theories of Cognition.* Cambridge: Harvard University Press.

Newell, A. (1992). Unified theories of cognition and the role of Soar. In J. A. Michon and A. Akyurek, Eds., *Soar: A Cognitive Architecture in Perspective.* Boston: Kluwer Academic.

Newell, A., and Simon, H. A. (1972). *Human Problem Solving.* NJ: Prentice-Hall.

Norman, D. A. (1968). Toward a theory of memory and attention. *Psychological Review, 75*, 522–536.

Ornstein, P. A., and Naus, M. J. (1985). Effects of the knowledge base on children's memory strategies. In H. Reese, Ed., *Advances in Child Development and Behavior* (Vol. 19, pp. 113–148). Orlando: Academic Press.

Proctor, R. W., and Dutta, A. (1995). *Skill Acquisition and Human Performance.* Thousand Oaks, CA: Sage.

Resnick, L. B. (1987). Constructing knowledge at school. In L. S. Liben, Ed., *Development and Learning: Conflict or Congruence.* Cambridge: Harvard University Press.

Salvendy, G. (1969). Learning fundamental skills—A promise for the future. *AIIE Transactions, 1*(4), 300–305.

Salvendy, G., and Seymour, W. D. (1973). *Prediction and development of industrial work performance.* New York: John Wiley and Sons.

Seppälä, P., Tuominen, E., and Koskinen, P. (1992). Impact of flexible production philosophy and advanced manufacturing technology on organization and jobs. *The International Journal of Human Factors in Manufacturing*, *2*(2), 177–192.

Schacter, D. L. (1989). Memory. In M. I. Posner, Ed., *Foundations of Cognitive Science*. MA: MIT Press.

Schoenfeld, A. H., and Herrmann, D. J. (1982). Problem perception and knowledge structure in expert and novice mathematical problem solvers. *Journal of Experimental Psychology: Learning, Memory and Cognition, 8*, 484–494.

Schvaneveldt, R. W., Durso, F. T., Goldsmith, T. E., Breen, T. J., Cooke, N. M., Tucker, R. G., and De Maio, J. C. (1985). Measuring the structure of expertise. *International Journal of Man-Machine Studies, 23*, 699–728.

Shiffrin, R. M., and Dumais, S. T. (1981). The development of automatism. In J. R. Anderson, Ed., *Cognitive Skills and their Acquisition.* Hillsdale, NJ: Erlbaum.

Shiffrin, R. M., and Schneider, W. (1977). Controlled and automatic human information processing II. Perceptual learning, automatic attending, and a general theory. *Psychological Review, 84*, 127–190.

Squire, L. R. (1987). *Memory and Brain.* Oxford: Oxford University Press.

Stramler, J. (1993). *The Dictionary for Human Factors/Ergonomics.* Boca Raton, FL: CRC Press.

Ward, S. L., Byrnes, J. P., and Overton, W. F. (1990). Organization of knowledge and conditional reasoning. *Journal of Educational Psychology, 82*, 832–837.

Wickens, C. D. (1980). The Structure of Attentional Resources. In R. Nickerson, Ed., *Attention and Performance VIII.* Hillsdale, NJ: Erlbaum.

Wickens, C. D. (1984). Processing resources in attention. In R. Parasuraman and R. Davies, Eds., *Varieties of Attention.* New York: Academic Press.

Wickens, C. D. (1992). *Engineering Psychology and Human Performance* (2nd Ed.). New York: Harper Collins.

Wilson, J. R., and Corlett, E. N. (1990). *Evaluation of Human Work.* New York: Taylor & Francis.

Willingham, D. B., Nissen, M. J., and Bullemer, P. (1989). On the development of procedural knowledge. *Journal of Experimental Psychology: Learning, Memory, and Cognition, 15*(6), 1047–1060.

Ye, N., and Salvendy, G. (1991). Cognitive engineering based knowledge representation in neural networks. *Behaviour and Information Technology, 10*(5), 403–418.

CHAPTER 6
HUMAN ERROR

Kyung S. Park
Department of Industrial Engineering
Korea Advanced Institute of Science & Technology
Taejon, 305-701 Korea

6.1 INTRODUCTION

The current trend toward automation and centralization leads to large facilitates in industrial installation harboring the potential for large-scale accidents with enormous economic losses, damage to equipment, and danger to the environment. The same trend is altering the nature of human involvement in the human–machine systems, shifting the human contribution more toward the operation and maintenance of machines. A considerable amount of time is spent by the operator in monitoring the system.

Due to its very nature, automation has given humans a host of new problems with serious consequences. For example, with regard to monitoring, problems of maintaining vigilance can exist. The amount of cognitive information processing required of the operator has also vastly increased with the growing scale and complexity of the systems to be monitored, especially when quick human intervention is needed to handle unusual disturbances that render the system unstable leading to hazardous conditions.

The subject of human error identification, particularly misdiagnosis during abnormal events in systems, has become an increasing concern in high-risk industries. There also has been a slight shift of emphasis from quantifying errors, to understanding human errors at a deeper, psychological level, trying to identify their causes, which could be used to identify what errors should appear in the future risk analysis. The current trend in technological development increases the demand for systematic consideration of the effects of human errors.

Recently, it has also been recognized that management and organizational influences can contribute significantly to risk and human error level. Much effort has been put into protecting industries from such influences. However, quantification of the impact of such influences on human reliability is not advocated since it is as yet relatively unproven (Kirwan, 1994).

6.1.1 The Ubiquity of Human Errors

Every human–machine system contains certain functions that must be performed by people. Even the so-called fully automated systems need human interventions in monitoring and maintaining. If the variability in human performance is recognized as inevitable, then it is easy to understand that when humans are involved, errors will be made, regardless of the level of training, experience, or skill.

As the human–machine systems are required to become more reliable, human influence becomes more and more important. The effort that is sometimes spent in designing ultrareliable equipment is often negated by human error (Hagen and Mays, 1981). It is essential to understand the error characteristics of all the components of a system which includes the human beings involved.

Human errors are said to occur when the performance is outside predefined tolerance limits. Typically, the errors are manifested as a failure to perform a required action; or its performance in an incorrect manner, out of sequence, or at an incorrect time.

6.1.1.1 Human Initiated Failures

Slightly different from the trivial errors of omission or commission are the human initiated failures, which typically prevent the system from accomplishing its mission, if the nature of the failure is serious enough. Human errors that do not result in system failure are often reversible, whereas errors causing human initiated failures cannot be reversed, because failed machines usually cannot restore themselves.

The high incidence of human error in the operation of human–machine systems is well documented by many investigators. For examples, Willis (1962) states that "40% of the problems uncovered in missile testing derived from the human element. Sixty-three and six-tenths percent of the (shipboard) collisions, flooding, and grounding could be blamed upon human error. Reports produced by the United States Air Force indicate that human error was responsible for 234 of 313 aircraft accidents during 1961." Shapero et al. (1960), Cooper (1961), and Meister (1962) have all indicated that a substantially high percentage of all system failures (anywhere from 20% to 80% or more depending on various circumstances) results from human error.

The remainder of the failures are presumably from normal machine wearout, design deficiencies not directly related to machine operators, or other causes which can be ascribed only remotely to personnel (Meister, 1971). However, the distinction between the human initiated failures and other failures cannot be drawn clearly, because the ultimate responsibility for reliable design and safe preventive maintenance practices can be traced only to the human component in the overall human–machine system.

6.1.1.2 Human Errors in Production

During fabrication and assembly of a product, production workers can make errors that could later cause failures and problems for users. Failure to torque various nuts and bolts correctly by under- or over-torquing them may lead to loosening or cracking. Failures to keep electrical connectors clean, and free of wire strands or solder splashes can cause short circuits when the system is energized. The following list is a sample of production errors (and ratios) reported by Rook (1962), based on 23,000 production defects examined.

- Component wired backwards (0.001)
- Transposition of wires (0.0006)
- Wrong component (or value) used (0.0002)
- Solder joint omitted (0.00005)
- Component omitted (0.00003)

Considering the thousands of individual components assembled into a product, the cumulative effect of these small error ratios can result in serious scrap/reject frequencies.

6.1.1.3 Inspection Errors and Product Quality

Similar factors which are considered responsible for production error also contribute to inspection error. The inspection work, by its very nature, requires an intense degree of perceptual and cognitive processing on the part of the inspector. McCornack (1961) reports an average inspection effectiveness of about 85%. However, Harris and Chaney (1969) report a range from 20% to 80%, and point out that they seldom find over 50% to 60% of the defects being detected at any point in time by a single inspector.

Poor inspection may increase production costs by causing machine failures, process interruptions, or by material wasted. However, the worst part of it is that a small production defect in a complex modern human–machine system which had escaped the inspector's surveillance can result in the loss of numerous lives and millions of dollars.

To improve the accuracy of the inspection performance, the cognitive discriminability of the defects must be increased.

6.1.2 Importance of Human Error in System Safety

Human inattention and negligence in machine operations look like innocent mistakes. But as the human–machine systems become more complex, the chances of these innocent mistakes and the seriousness of their consequences are ever growing. Valve misoperation, for example, is very commonplace, but has caused many industrial accidents, including the radioactive leakage accident at Three Mile Island on March 28, 1979.

However, practically all accidents can be prevented using the methods and practices well within the abilities of every normal human being, that is, if he will but 'pay his attention.' This continuing wastage of our human and material resources from accidents should receive greater attention, particularly since most of it is preventable. People should not be entrapped to irreversible consequences of trivial errors which would be condoned as typical traits of normal human behavior in their everyday life—"to err is human."

Although Three Mile Island was beginning to be written off as an isolated instance, and the need for human reliability analysis seemed to be weakening, the subsequent accidents such as Bhopal (1984), Challenger (1986), and Chernobyl (1986) rekindled the interest in human reliability issues.

6.1.2.1 Industrial Risk Management

Although notable technological developments have taken place in systems design, accidents, sometimes disastrous and large scale, are still occurring. The so-called high-risk industries are still not particularly safe from human error. This suggests the need for tools of properly assessing the risks attributable to human error and for methods of reducing the impact of the error on the system. Human reliability analysis achieves these objectives by identifying *what* errors can occur, and deciding *how likely* the errors are to occur. If appropriate, human reliability is enhanced by *reducing* this error likelihood.

The systematic consideration of human error in system designs can improve safety and productivity. Human reliability analysis can also enhance the profitability and availability of systems via human error reduction.

Human reliability analysis techniques for human-error identification, prediction, and reduction are best applied in their most challenging context, i.e., within the detailed qualitative and quantitative risk-assessment of large-scale systems. The risk assessments, which are carried out for all large hazardous systems, consider hardware and software failures, and environmental events, as well as humans errors.

The risk assessment involves the evaluation of conceivable risks to worker and public safety, and the damage to the plant and the environment. The risk assessment identifies ways in which hazards can occur during system operation and maintenance, which can lead to accidents. The risk assessment then calculates the probabilities and frequencies of such events, and determines their consequences. The summarized risks estimated are compared against industry/regulatory criteria. If the summarized risk breaches any of the criteria, the system's design must be improved so as to reduce the risk factor to an acceptable level.

There are a number of qualitative tools for carrying out safety-management audits (see Kirwan, 1994). They use performance indicators, which correlate with general safety performance. Such indicators may range from safety policies to the spare-parts inventory. Usually, such audit tools organize their indicators into groups and have a means of aggregating or scoring their ratings to derive a total-safety index rating.

6.2 THE NATURE OF HUMAN ERROR

6.2.1 Human Behavior in a Human–Machine System

In human–machine systems, human activities are required for monitoring, adjusting, maintenance, and other normal operations as well as for coping with unusual disturbances that place a system at risk. In general, each human, individually or as a member of a team, must perceive information (from displays, environmental conditions, or procedural instructions) about the state of the system or subsystem for which he is responsible. Then he must process that information to determine what action or inaction he should take, and then take it.

In this chain of events, man essentially serves the three basic functions with the support of the fourth function: human memory. They are depicted schematically in Figure 6.1.

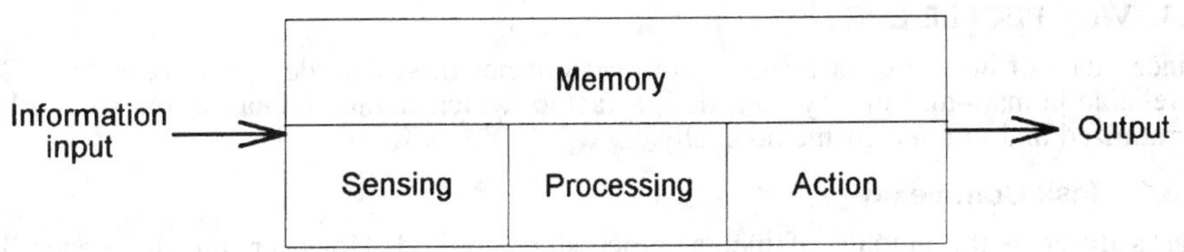

Figure 6.1 A simplified model of the human component in a human–machine system. (Adapted from Park, 1987, with the permission of Elsevier.)

Human error occurs when any element in this chain of events is broken due to [Payne and Altman (1962)]:

1. Input errors—errors of sensory or perceptual input,
2. Mediation errors—errors of mediation or information processing, or
3. Output errors—errors in making physical responses.

Inexpensive yet powerful microcomputers are frequently used for the selection and preprocessing of the data to represent the conditions of the system confronting the user. This introduces excellent potential for matching tasks to human abilities and preferences, but also accompanies the risk of the user losing control when unforeseen situations arise (Rasmussen, 1987). In particular, when information is selected and preprocessed, it is important to consider the different modes of human failure and the kind of information needed in order to recover from slips and mistakes.

Much of human error results from inadequacies in system designs which create favorable conditions for error occurrence (Meister and Rabideau, 1965). Space does not allow the detailed review of all human factors theories. However, understanding the basic human capabilities is essential in creating a reliable environment that allows for inherent human psychomotor limitations.

6.2.2 What Is Human Error?

Human error can be formally defined as "a failure on the part of the human to perform a prescribed act (or the performance of a prohibited act) within specified limits of accuracy, sequence, or time, which could result in damage to equipment and property or disruption of scheduled operations" (Hagen and Mays, 1981). It is an out-of-tolerance action, or deviation from the norm, where the limits of acceptable performance are defined by the system.

The major cause of human error is inherent human variability. A human being is variable by nature; no one does anything the same way twice. Sheer variability results in random fluctuations of performance which are sometimes great enough to produce error, which can only be controlled by acquiring skill through training.

6.2.3 Characteristics of Human Error

Although, there are certain similarities between humans (with multiple organs and functions) and machines (with multiple components and functions) in terms of their proneness to failure, which lead to the parallelism of the methods of analysis in each, the human failure process has its own peculiarities, too.

Probably, the most important difference is that the human errors randomly recur, whereas hardware failure condition is irreversible by itself. Human errors that do not result in system failure are often reversible. Hardware reliability is typically concerned with the first failure.

A second difference is that a human continually improves his or her performance from learning unlike his or her machine counterpart. Learning and adaption during performance will be significant features of many situations.

The human performance and stress follows a nonlinear relationship: When the stress is moderate, the performance level is highest. Also, the human performance may not be independent of the past performance record (autocorrelation), especially when the human has any preset low performance goal. Therefore, the parameters of the human variables should be obtained under conditions close to operational reality, considering the actual physical, emotional, intellectual, and attitudinal characteristics of the person to operate the machine (Peters and Hussman, 1959).

6.3 WHY PEOPLE ERR

Since much of human error results from inadequacies in system design, in order to build a reliable human–machine system, design factors which induce human errors should be scrutinized and eliminated methodically.

6.3.1 Task Complexity

Tasks differ in the amount of mental processing required. However, humans generally have similar performance limitations, and process information similarly. These universal capacity limitations cause people to make more errors in more complex tasks.

Capacity limitations in short-term memory and recall problems in long-term memory strongly affect human performance reliability. Complex task sequences in a specific order overstrain human memory. Written procedures and detailed checklists can be used to unburden the operators of memorizing all the task elements and their correct sequential order.

6.3.2 Error-Likely Situations

Error-likely situations are identified as work situations where the *human engineering* (*HE*) is so poor that errors are likely to occur (Swain and Guttmann, 1983). These situations overtax operators in a manner that is not compatible with their capabilities, limitations, experience, and expectations. For instance, any design that violates a strong population stereotype could be considered error likely.

This work situation approach is rooted in the HE design philosophy that the system should be fitted to the man, not vice versa. The work situation approach emphasizes the identification of error-inducing conditions and their remediation. This approach assumes that errors are more likely to occur for reasons other than operator's faults. Thus, accident proneness applies to work situations, not people.

Situational task and equipment characteristics that predispose operators to increased errors include the following (Meister, 1971):

- Inadequate work space and layout,
- Poor environmental conditions,
- Inadequate HE design,
- Inadequate training and job aids, procedures,
- Poor supervision.

6.3.3 Behavioral Characteristics

The individual variables that might be associated with high error rates in various types of tasks cover virtually the entire range of human characteristics. They are human attributes such as skills and attitudes that the worker brings to the job. Some examples of behavioral factors include: age, sex differences, intelligence, perceptual abilities, physical condition, strength/endurance, task knowledge, training/experience, skill level, motivation/attitude, emotional state, stress level, and social factors.

Stress and inexperience are such influential behavioral factors that their combination can increase an operator's error probability by a factor of as much as 10 (Table 6.1).

6.4 HUMAN ERROR TAXONOMY

Human error can be approached from two viewpoints: *prospective* and *retrospective.* The prospective approach assumes that human errors are random phenomena to be predicted quantitatively in the context of system reliability evaluation.

In contrast, the retrospective approach assumes that human errors can be traced to causes and contributing factors for qualitative diagnosis (Rouse and Rouse, 1983). Once isolated, these may be eliminated or ameliorated. The classification of human errors is an important tool for the purpose of qualitative diagnosis. Classification systems are useful

Table 6.1 Effects of Stress and Experience on Human Error Probability in Performing Routine Tasks

	Increase in Error Probability	
Stress	Skilled	Novice
Very low	×2	×2
Optimum	×1	×1
Moderately high	×2	×4
Extremely high	×5	×10

Source: Adapted from Miller and Swain, 1986, with the permission of John Wiley.

in identifying what incident happened and then preventing its recurrence, as well as in identifying human errors which may have an impact upon a system's goals.

6.4.1 Behavior-Oriented Classification

The study of human cognitive processes and the related error mechanisms has gained increasing interest over the last decade. Much interest focuses on human reliability in discretionary tasks in less structured work systems for which the traditional, mechanistic, engineering approach developed for process systems may be inadequate. The behavior-oriented schemes classify human behavior independent of a specific task or application area.

Payne and Altman's (1962) three behavior components were subsequently incorporated into a two-way classification scheme by Rook (1962), the other dimension of the system describing the intent in performing the action that resulted in the error, as in Table 6.2.

A classification scheme by Swain (1963) relates human output to system requirements without regard to internal processes. An error of omission occurs when an operator omits a step in a task, or the entire task. An error of commission occurs when an operator does the task, but does it incorrectly. This is a broad category encompassing selection errors, sequence errors, time errors, and qualitative errors.

6.4.1.1 The Step-Ladder Model

The *step-ladder* model of dynamic decision making proposed by Rasmussen and his colleagues (Rasmussen, 1976; Rasmussen and Jensen, 1974) is an information-processing model that has been used as the basis for error classification. The step-ladder model proposes that there is a normal and expected sequence of information-processing stages that people need to follow when performing a problem-solving or decision-making task, but that there are many situations where people do not perform according to the ideal case. Borrowing a concept first used by Gagné (1962), Rasmussen noticed that people "shunt" certain mental operations where varying amounts of demanding processing can be avoided depending on the operator's familiarity with the task. In highly familiar circumstances, operators were not believed to perform each stage of processing and, on the basis of an analysis of performance-related errors in a nuclear power plant domain, Rasmussen and Jensen (1974) identified several types of behavioral shortcuts that were commonplace in maintenance operations. To exemplify these, Rasmussen developed an eight-stage model of information processing, as shown in Fig. 6.2.

The central theme of the step-ladder model is the idea that shunting between cognitive stages (shown by the dotted paths in Fig. 6.2) is an efficient form of information-processing behavior because it reduces the amount of cognitive effort in the performance of a task. However, Rasmussen also pointed out that the strategy can increase a person's vulnerability to making errors because it is overly reliant on the appropriateness of past experience.

6.4.1.2 The Skill-, Rule-, and Knowledge-Based Model

Rasmussen adopts a human information-processing point of view. His taxonomic approach involves discriminating among three levels of human behavior: skill-based, rule-based, and knowledge-based (Rasmussen et al., 1981; Rasmussen et al., 1992). Further, he distinguishes among causes, mechanisms, and modes of human error. His overall goals include developing a comprehensive classification scheme for reporting events involving human error.

Table 6.2 Rook's System of Error Classification

	Behavior Component		
Intent in Performing Act	Input(I)	Mediation(M)	Output(O)
A Intentional	AI	AM	AO
B Unintentional	BI	BM	BO
C Omission	CI	CM	CO

Source: Adapted from Rook, 1962.

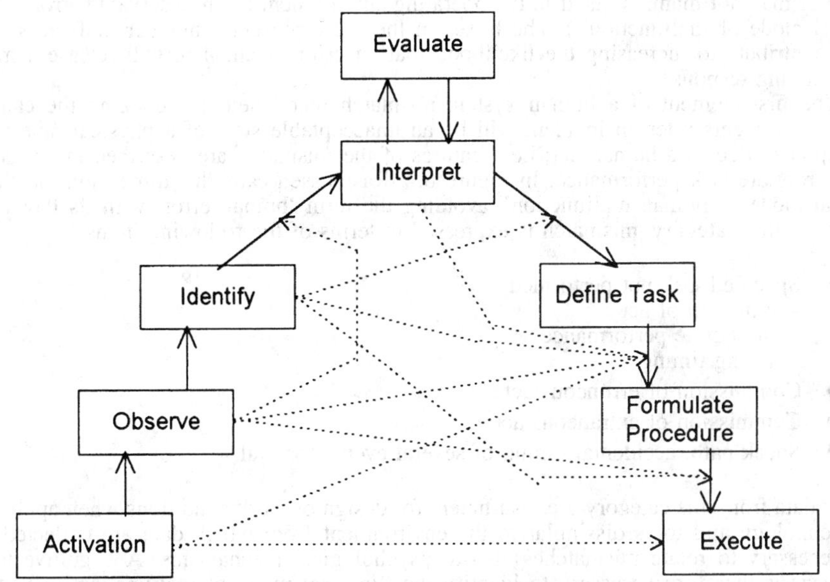

Figure 6.2 Rasmussen's step-ladder model of decision making. (Adapted from Rasmussen, 1976, with the permission of John Wiley.)

Skill-based behavior involves a performance controlled by stored patterns of behavior, and the operator reacts to stimuli with little conscious effort or consideration. Rule-based behavior involves a performance in familiar settings, using stored or readily available rules. Knowledge-based behavior involves event-specific behavior that is based on a knowledge of the system. This level of behavior usually requires 'higher level' cognitive processes such as problem solving, goal selection, and planning. The model, which is essentially qualitative in nature, outlines the hierarchical 'step-ladder' processes associated with actions and decision making (Rasmussen, 1976). Any plan developed using knowledge-based behavior will require rule-based procedures (e.g., coming from memory) and skill-based operation (to execute the procedure).

Figure 6.3 shows the overall framework for the model. The figure shows a sevenfold classification of factors relevant to human error in summary form. The boxes in the bottom indicate that the error process typically begins with the occurrence of an event in the environment ("causes of human malfunction") which causes the release of a psychological "failure mechanism." This in turn invokes a malfunction in human behavior which

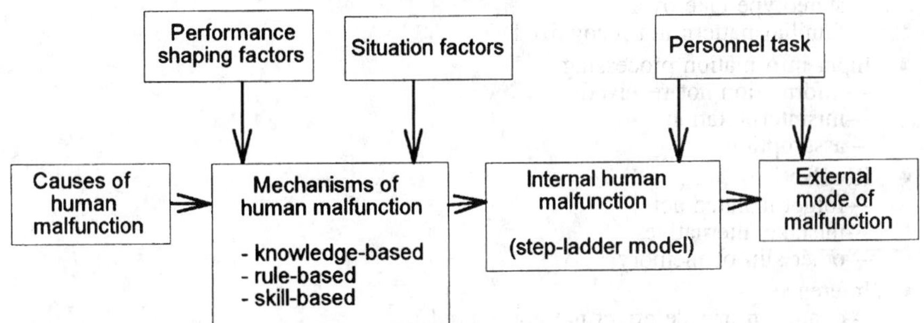

Figure 6.3 Framework for skills-rules-knowledge model. (Adapted from Rasmussen et al., 1981; Rasmussen et al., 1992, with the permission of John Wiley.)

may or may not manifest itself in the operating environment as an observable error ("external mode of malfunction"). The boxes in the top represent a number of factors that can contribute to increasing the likelihood that an error mechanism will release a malfunctioning response.

The first element of a human–system mismatch met when backtracking the causal course of events after an incident will be an unacceptable state of a physical object or component due to a human act; i.e., features of the mismatch are described in terms of inappropriate task performance. In Figure 6.2, Rasmussen calls this dimension the "external mode of human malfunction" avoiding the term 'human error' with its flavor of guilt. In this category, mismatch is expressed in terms of the following items.

- Specified task not performed
  —omission of act
  —inaccurate performance
  —wrong timing
- Commission of erroneous act
- Commission of extraneous act
- Sneak-path, accidental timing of several events for faults

If data from this category are insufficient for design of work conditions when applying a technology and tools dissimilar to the environment from which data are collected, it is necessary to relate mismatches also to psychological mechanisms. A cognitive task analysis is therefore necessary to identify the "internal mode of malfunction," i.e., the element of the internal cognitive decision process which was affected, either by being improperly performed or bypassed by a habitual shortcut. This analysis is possible to the level of detail represented in the decision sequence of the step-ladder model in Figure 6.2.

Rasmussen et al. (1992) note that a malfunction in each decision function can be caused by several different psychological mechanisms related to the three levels of cognitive control (skill-based, rule-based, and knowledge-based) and give the following examples. At the level of highly skilled performance, typical error mechanisms can result in inadequacies in the control of movements such as a lack of spatial or temporal precision or of topographic coordination or a lack of precision in exerting physical force. At the rule-based level, typical error categories are related to memory characteristics, e.g., forgetting to perform isolated acts. An example is forgetting to switch back to normal operation after a test. Another category is mistaking alternatives such as left and right, up and down, + and −, etc. Another type of error results from *functional fixation* caused by the normally effective cue–action correlations when they are used during abnormal situations. An example is associating abnormal instrument readings to an inaccurate meter calibration. This category of "mechanisms of human malfunction" includes the following items.

- Discrimination,
  —stereotype fixation
  —familiar shortcut
  —stereotype take-over
  —familiar pattern not recognized
- Input information processing
  —information not received
  —misinterpretation
  —assumption
- Recall
  —forget isolated act
  —mistake alternatives
  —other slip of memory
- Inference,
  —condition or side effect not considered
- Physical coordination
  —motor variability
  —spatial misorientation

The "causes of human malfunction" include external events (distraction, etc.), excessive task demand (force, time, knowledge, etc.), operator incapacitation (sickness, etc.), and intrinsic human variability. This category should identify the possible external causes of the inappropriate human actions.

The identification of the task performed is important to characterize the circumstances during which the malfunction occurred. Included in this category of "personnel task" are equipment design, procedure design, fabrication, installation, inspection, operation, test and calibration, maintenance/repair, logistics, administration, and management. "Situation factors" include task characteristics, physical environment, and work time characteristics. "Performance shaping factors" are subjective goals and intentions, mental load/resources, and other affective factors.

The model also identifies *what* error occurred ("external mode of malfunction"), *who* committed the error ("personnel task"), *how* it occurred ("internal human malfunction"), and *why* it occurred ("performance shaping factors," etc.). Although it was designed as a retrospective technique, it may be applied predictively, in combination with a task analysis/description.

This model, a useful framework for classifying errors, paved the way for later techniques. It investigates the causes of human error (i.e., the internal mode of malfunction), not just the overt or externally observable manifestation of error. However, the approach provides little guidance regarding assignment of operator errors to specific categories.

6.4.1.3 Generic Error Modeling Systems (GEMS)

Reason (1987) uses a cognitive model similar in form to the step-ladder model (Rasmussen, 1976) as the technical basis for the Generic Error Modeling System (GEMS) to develop a context-free model of human error. The GEMS model concerns cognitive error modes. Though partly model-based, it is largely taxonomic in nature. The emphasis on cognitive factors in contrast to environmental- or context-related factors is expected to permit the error classification embodied within GEMS to be applied to the analysis of error in a variety of industrial situations (e.g., nuclear power, process-control, and aviation).

GEMS makes the following general assumptions concerning the architecture of the cognitive system (Reason, 1987):

1. Cognitive control processes operate at various levels and exert their influence simultaneously over widely differing time spans.
2. The higher levels of this control system can function over both long time spans and a wide range of circumstances. These higher levels are primarily concerned with setting goals, with selecting the means to achieve them, and with monitoring progress toward these objectives. The products of this high-level planning and monitoring activity are available to consciousness.
3. Higher level agencies can govern the order in which lower level processors are brought into play, but this control function is only intermittently exercised.
4. A substantial part of cognitive activity is governed by lower level processors (schemata, scripts, frames, heuristics, and rules) capable of independent function. Typically, these low-level processors operate over brief time spans in response to very specific data sets.
5. The successful repetition of any human activity results in the gradual devolution of control from the higher to the lower levels.
6. The control of human action arises from the interaction between two control modes: the *attentional* and the *schematic* modes. The former is closely identified with consciousness and working memory. The schematic mode can process familiar information rapidly, in parallel, and without conscious involvement or effort. It has no known limits on its capacity, but is ineffective in the face of novel or unforeseen circumstances.
7. Central to GEMS is Rasmussen's (1976, 1981) distinction between the skill-based, rule-based, and knowledge-based levels of performance. At the skill-based levels, the informational content is the form of signals, and performance is governed by stored patterns of preprogrammed instructions (schemata) represented as analogue structures in a time–space domain. At the rule-based level, performance is guided by signs relating to stored rules or productions (of the IF-THEN form). The knowl-

edge-based level comes into play in novel situations for which actions must be planned on-line, through the manipulation of symbols. The skill- and rule-based levels most involve the schematic control mode, while the attentional mode predominates at the skill-based level.

8. The predictable varieties of human error have their origins in useful, functional, and adaptive processes. Systematic error forms are inextricably bound up with things at which the human cognitive system excels.

GEMS extends an important feature of the step-ladder model which relates to the assumption that there are three levels of cognitive control that can be distinguished on the basis of a person's familiarity with a task or situation. It attempts to present an integrated picture of the error mechanisms operating at all three levels of performance: skill-based as well as rule-based and knowledge-based. It assists the analyst in understanding the process by which an operator moves from the skill-based, automatic level of task implementation to the rule-based level, and to a higher, knowledge-based diagnosis.

The switching rules governing the shifts of control between the skill-based, rule-based, and knowledge-based levels are summarized from Reason (1987) as follows. The skill-based level of GEMS relates to the execution of highly routinized activities in familiar surroundings. The rule-based level is engaged when an attentional check ("OK?") upon progress detects a problem, i.e., a situation that cannot be handled by the routines set in train by the current plan. A primary feature of GEMS is that rule-based efforts at problem solving will always be tried first. If the deviation is minor, and appropriate corrective rules are readily found, this phase may be terminated by a rapid return to the skill-based level. With more difficult problems, however, the cycle may be repeated several times.

According to the simple logic of GEMS, the switch from the rule-based to the knowledge-based level occurs when the problem solver realizes that none of his or her repertoire of rule-based solutions is adequate to solve the problem. With knowledge-based processing, the focus of problem solving shifts away from immediate state considerations to some mental model of the system as a whole. Progress at this level tends to involve the search for suitable analogies, or diagnostic theories, to fit the current situation. The discovery of such an analogy usually brings with it a set of remedial possibilities. This cycling between the knowledge-based and rule-based levels can be repeated several times as various theories are explored.

Activity at the knowledge-based level can be terminated by the discovery of an adequate solution to the problem. This solution is likely to constitute a new plan of action involving a fresh set of skill-based routines. There are powerful cognitive forces at work to encourage the problem solver to accept inadequate or incomplete solutions as being satisfactory at this point. The consequences of this error may or may not be detected rapidly by an "OK?" check. Once such error has been detected; however, the system will again switch into the rule-based mode. In this way, the focus of control will shift continuously between all three performance levels.

GEMS classifies errors into two categories: *slips* and *lapses* at the one level, and *mistakes* at the other level. It postulates that slip-type errors occur most frequently at the skill-based level, whereas mistake-type errors usually occur at the higher, rule- and knowledge-based levels. In this context, a slip or lapse is considered unintentional, whereas a mistake (or rule violation) is generally thought of as an actual error of judgment or intention. GEMS is basically a classification system, although it can be computerized as a model.

Errors at the three levels of performance are further distinguished by the nature of the psychological and situational factors which combine to shape characteristic error forms. However, these distinctions are quite subtle since many of these error-shaping factors operate at more than two levels. The error types and major error-shaping factors at each level are summarized in Table 6.3.

The GEMS system is useful in that it defines a set of cognitive-error modes, including biases in judgment. The technique gives guidance on the types of error-shaping factors which are likely to apply to the above two categories of error, such as mind-set, overconfidence, and an incomplete mental model. However, the guidance on how to choose these underlying errors is quite limited and relies on the assessor's own judgment. It largely depends on the analyst's insight and resourcefulness to attribute particular error-shaping factors to any individual task step and to prescribe countermeasures.

Table 6.3 The Generic Error Modeling System (GEMS)

Performance Level	Error-Shaping Factors
Skill based	1. Recency and frequency of previous use
	2. Environmental control signals
	3. Shared schema properties
	4. Concurrent plans
Rule based	1. Mind-set ("It's always worked before")
	2. Availability ("First come best preferred")
	3. Matching bias ("like relates to like")
	4. Oversimplification (e.g., "halo effect")
	5. Overconfidence ("I'm sure I'm right")
Knowledge based	1. Selectivity (bounded rationality)
	2. Working-memory overload (bounded rationality)
	3. Out of sight, out of mind (bounded rationality)
	4. Thematic 'vagabonding' and 'encysting'
	5. Memory prompting/reasoning by analogy
	6. Matching bias revisited
	7. Incomplete/incorrect mental model

Source: Adapted from Reason, 1987, with the permission of John Wiley.

6.4.2 Task- or System-Oriented Classification

Task- and system-oriented schemes apply to particular areas, either at the specific task level or system level (covering various tasks).

Several classification schemes reflect a particular task orientation. For example, the classification scheme of Nawrocki et al. (1973) reflects the specific task of keyboard operation and communication with an information system.

System-oriented classification schemes use relatively broad categories covering a series of tasks within a particular system. Most of the schemes in this group deal with the aircraft area.

6.4.3 Two-Level Classification Scheme

Rouse and Rouse (1983) propose a human error classification scheme involving two levels: *general* categories and *specific* categories. In Table 6.4, general categories discriminate among the six behavioral processes within which human error occurs (behavior-oriented); specific categories define the particular characteristics of erroneous decisions or actions (task-oriented).

6.5 ANALYTIC HUMAN RELIABILITY

The prospective approach to human error is interested in its quantification. This approach has evolved, in particular, within nuclear power and, consequently, has been directed toward 'operator errors.' The approach is particularly relevant for repetitive tasks in work situations when the task sequence is paced by a stable, technical system as is typically the case for certain tasks in process industries.

Human error probabilities or human reliabilities are useful not only in estimating human–machine system reliability, but also in other HE activities such as allocating functions between man and machine (based on reliabilities with and without the human), quantifying the error likelihood and consequences of human engineered equipment, and estimating the success of personnel training programs.

6.5.1 Measurement of Human Error

Human performance concerns both time-discrete and time-continuous tasks. Data on reliability are expressed in terms of failures per event for the first and failures per unit time for the second.

6.5.1.1 Human Error Probability in Discrete Tasks

A task is discrete if its content is well predefined with a definite beginning and ending (not unfolding in time).

Table 6.4 Two-Level Classification Scheme

General Category	Specific Category
1. Observation of system state	**a.** excessive
	b. misinterpreted
	c. incorrect
	d. incomplete
	e. inappropriate
	f. lack
2. Choice of hypothesis	**a.** inconsistent with observations
	b. consistent, but unlikely
	c. consistent, but costly
	d. functionally irrelevant
3. Testing of hypotheses	**a.** incomplete
	b. false acceptance of wrong hypothesis
	c. false rejection of correct hypothesis
	d. lack
4. Choice of goal	**a.** incomplete
	b. incorrect
	c. unnecessary
	d. lack
5. Choice of procedure	**a.** incomplete
	b. incorrect
	c. unnecessary
	d. lack
6. Execution of procedure	**a.** step omitted
	b. step repeated
	c. step added
	d. steps out of sequence
	e. inappropriate timing
	f. incorrect discrete position
	g. incorrect continuous range
	h. incomplete
	i. unrelated inappropriate action

Source: Adapted from Rouse and Rouse, 1983. Copyright © 1983 IEEE, with permission.

The basic unit of human reliability in discrete tasks is the *human error probability* (HEP), which is the probability of an error occurring during a specified task. The time allotted is either implicit or unspecified. (Sometimes, the term human error *rate* is used in this situation, but the term is a misnomer in that rate is often used in the sense of frequency per unit of time.)

HEP is estimated from the ratio of errors committed to the total number of opportunities for that error:

$$\text{HEP} \cong \hat{p} = \frac{\text{number of human errors}}{\text{total number of opportunities for the error}} \tag{1}$$

The successful performance probability of a task (or task reliability) can be generally expressed as 1 − HEP. This is essentially equivalent to the measure of achieved equipment reliability. Thus, when we speak of the reliability of performance of an elemental human task, we are speaking of the probability of successful performance per demand.

Table 6.5 presents general HEP estimates derived from various existing data sources and modified by the independent judgments of two human reliability analysts. Some of the estimates were based directly on data collected on tasks. In other cases, the tasks

Table 6.5 General Operator HEP Estimates[a]

Estimated HEPs	Activity
.0001	Selection of a key-operated switch rather than a nonkey switch (this value does not include the error of decision where the operator misinterprets the situation and believes the key switch is the correct choice).
.001	Selection of a switch (or pair of switches) dissimilar in shape or location to the desired switch (or pair of switches), assuming no decision error; for example, the operator actuates the large-handled switch rather than the small switch.
.003	General human error of commission, e.g., misreading the label and therefore selecting the wrong switch.
.01	General human error of omission when there is no display in the control room of the status of the item omitted, e.g., failure to return the manually operated test valve to the proper configuration after maintenance.
.003	Errors of omission, where the items being omitted are embedded in a procedure rather than at the end as above.
.03	Simple arithmetic errors with self-checking but without repeating the calculation by redoing it on another piece of paper.
$1/x$	Given that an operator is reaching for an incorrect switch (or pair of switches), he selects a particular similar-appearing switch (or pair of switches), where x is the number of incorrect switches (or pairs of switches) adjacent to the desired switch (or pair of switches); the $1/x$ applies up to five or six items; beyond that point the HEP would be lower because the operator would take more time to search; with up to five or six items, the operator does not expect to be wrong and therefore is more likely to be less deliberate in searching.
.1	Given that an operator is reaching for the wrong motor-operated valve (MOV) switch (or pair of switches), he fails to note from the indicator lamps that the MOV is already in the desired state and merely changes the status of the MOV without recognizing that he had selected the wrong switch (or pair of switches).
~1.0	Same as above, except that the state of the incorrect switch (or pair of switches) is not the desired state.
~1.0	If an operator fails to operate correctly one of two closely coupled valves or switches in a procedural step, he also fails to correctly operate the other valve.
.1	The monitor or inspector fails to recognize the initial error by the operator; with continuing feedback of the error on the annunciator panel, this high HEP would not apply.
.1	Personnel on different work shifts fail to check the condition of the hardware unless required by a checklist or a written directive.
.5	Monitor fails to detect undesired position of valves, etc., during general walk-around inspections, assuming that no checklist is used.
.2 ~ .3	General HEP given very high stress levels where dangerous activities are occurring rapidly.
$2^{(n-1)}y$	Given severe time stress, as in trying to compensate for an error made in an emergency situation, the initial HEP, y, for an activity doubles for each attempt, n, after a previous incorrect attempt, until the limiting condition of an HEP of 1.0 is reached or until time runs out; this limiting condition corresponds to an individual becoming completely disorganized or ineffective.
~1.0	The operator fails to act correctly in the first 60 sec after the onset of an extremely high stress condition, e.g., a large loss-of-coolant accident (LOCA).
.9	The operator fails to act correctly after the first 5 min after the onset of an extremely high stress condition.
.1	The operator fails to act correctly after the first 30 min in an extreme stress condition.
.01	The operator fails to act correctly after the first several hours in a high stress condition.
y	Seven days after a large LOCA, there is a complete recovery to the normal HEP, y, for any task.

[a] Unless otherwise indicated, estimates of HEPs assume no undue time pressures or stresses related to accidents.

Source: Adapted from NUREG-75/014, WASH-1400, 1975, with the permission of USNRC.

were broken down into smaller bits of behavior that could be combined more readily with existing data or with the experience of the analysts. Then, the estimates of HEPs for the individual behavioral units were combined into estimates of HEPs for larger units of behavior.

The nominal HEPs are to be modified by assigning higher value to unfavorable situations to reflect such factors as psychological stress, quality of HE of controls and displays, quality of training and practice, presence and quality of written instructions and method of use, dependence of human actions, type of display feedback, and personnel redundancy. (More on this in Section 6.6.4, THERP). A brief review of the methodologies is given in Hagen (1976).

An aid to applying such data to similar operational situations at hand is the well-known phenomenon of behavior constancy (Swain, 1964). Human behavior is constant in many tasks on different machines, i.e., there is behavior similarity despite equipment dissimilarity.

6.5.1.2 Human Error Rate in Continuous Tasks

Tasks such as vigilance (scope monitoring), stabilizing, tracking (automobile operating), etc. are known as the time-continuous tasks, in which the task content unfolds continuously in time. The reliability modeling of such tasks are analogous to the classical time continuous reliability modeling.

Since the human errors are of randomly recurring type (with increasing error counts), general error processes are very difficult to model mathematically. However, when the error process is independent of the past history (with independent increments), the process becomes Poisson type.

By analogy to the arrival rate (or intensity function), define the human error rate $\lambda(t)$ as

$$\begin{aligned} \lambda(t)dt &= \text{Prob}\{\text{at least an error in } [t, t + dt)\} \\ &= E[\text{number of errors in } [t, t + dt)] \end{aligned} \tag{2}$$

in a narrow interval dt near t.

If the $\lambda(t)$ is a constant, λ, it is called a renewal rate, and the error process is homogeneous. The constant human error rate is estimated by dividing the errors committed by the total task duration:

$$\lambda \cong \hat{\lambda} = \frac{\text{number of errors}}{\text{total task duration}} \tag{3}$$

6.5.2 Human Reliability

Reliability is the antithesis of error likelihood. *Human reliability* is then defined as the probability that a person's performance will be error free for a specified duration. The basic unit of human reliability is the HEP or error rate defined in the previous section. By combining these measures in assorted ways, the analyst can calculate a total reliability figure for human performance in some specific task or job.

6.5.2.1 Human Reliability in Repetitive Discrete Tasks

In a series of repetitive trials of a given task, a human can fail to perform a prescribed act (or perform a prohibited act), thus causing a human error. In this context, it is sometimes of interest to know his or her reliability of completing a prescribed sequence of successive trials.

Assume that the error probabilities are stationary and independent of the past performance under some favorable conditions. Given the HEP per trial, p (estimated by $\hat{p}$), the human interval) reliability that a prescribed sequence of successive trials from n_1 through n_2 are completed without any error is

$$R(n_1, n_2) = (1 - p)^{n_2 - n_1 + 1}. \tag{4}$$

When the HEPs are time-varying as specified by p_i, possibly due to fatigue or learning, the process is nonstationary. Park (1985) discusses a mathematical model describing human reliability with learning as a nonstationary Bernoulli process.

If the HEPs are dependent on past performance, the error process is non-Markovian. No general theory on the 'non-Markov chain' exists yet. The total reliability figure may still be obtained by multiplying the terms involving the conditional HEPs associated with the trials, but it may be difficult to obtain the data for dependent HEPs as well as to manipulate them.

6.5.2.2 Human Reliability in Time-Continuous Tasks

In the context of a time-continuous task, it is sometimes of interest to know the reliability of performing the given task successfully during a specified interval. Here, the human error condition is treated as an event without any duration.

Assume that the error rates are stationary and independent of the past performance (Poisson error process) under some favorable conditions. Given the error rate, λ (estimated by $\hat{\lambda}$), the human (interval) reliability that a given task of specified duration (t_1, t_2) is performed successfully without any error is

$$R(t_1, t_2) = \exp[-\lambda(t_2 - t_1)] \tag{5}$$

When the error rate is time-varying as specified by $\lambda(t)$, possibly due to vigilance effect, fatigue, or learning, the process is nonstationary. Park (1985) generalizes Equation (5) as

$$R(t_1, t_2) = \exp[-\int_{t_1}^{t_2} \lambda(t)\, dt] \tag{6}$$

and discusses a mathematical model describing human reliability with learning as a nonhomogeneous Poisson process.

The error process can be completely general and dependent on the past performance, with nonexponential interarrival distributions. Except in a few special cases, such processes are not generally tractable.

6.5.2.3 Personnel Redundancy

Most studies of human reliability have been limited to the behavior of individuals operating independently. However, where personnel backup is anticipated for some of the tasks, it is necessary to account for the redundancy effect of additional surveillance on the increase in task reliability that would accrue. For example, to minimize the possibility of human error in carrying out any procedure involving a nuclear device, the Department of Defense has developed the *two-man concept* (Hammer, 1976).

If the human performance reliability of one operator is R_1, the two-man team reliability, R_2, is

$$R_2 = 1 - (1 - R_1)^2 \tag{7}$$

In estimating human reliabilities, however, a slight modification of the hardware redundancy assumption may be necessary. Unlike the devoted backup machine, a second individual may not always be available or attentive to back up the first individual. Therefore, when two men are working together to perform a task, their team reliability may be somewhat less than the ideal value derivable from Equation (7). In this case, a weighted average of R_1 and R_2 may be used to consider part-time personnel redundancy (Pontecorvo, 1965).

6.5.3 Synthesis of Machine Reliability and Human Reliability

Human performance concerns both time-discrete and time-continuous tasks. Human reliabilities are expressed in terms of demand reliability for the first and of time reliability for the second. By combining these reliabilities appropriately with machine reliability, a total human–machine system reliability figure for the human performance in some specific task or job can be calculated, as will be illustrated by a simple hypothetical example.

Example

Consider a radar operator who continuously monitors the plan-position indicator (PPI) display for aircraft-warning. From time to time, he fails to detect a blip due to inattention.

The machine fails at the rate of 0.01/h without repair. Because of the vigilance effect, the operator error rate is estimated to be $\lambda(t) = 0.1t/\text{h}$. He is also required to manipulate a bank of switches once every hour with an estimated HEP of 0.03 per demand. What is the human–machine system reliability between $t_1 = 1$ and $t_2 = 2$ h if the human and machine performances as well as the two human tasks are independent?

For the machine to function successfully during the mission duration, it must operate at $t = 1$ and operate continuously from $t = 1$ to 2 h (therefore, it must not fail during [0, 2]). However, the operator's errors during [0, 1] are of no concern in this problem. Thus, the system reliability is, from Equation (6),

$$\begin{aligned} R_s(1, 2) &= R_m(0, 2)\, R_h(1, 2) \\ &= e^{-0.01(2)}(1 - 0.03)^1 \exp[-\int_1^2 0.1t\,dt] \\ &= e^{-0.02}(0.97)e^{-0.15} = 0.82. \end{aligned}$$

6.6 PREDICTIVE TECHNIQUES FOR HUMAN ERROR PROBABILITIES

The best way of obtaining an HEP in a task is to directly observe the errors committed in the task. One major problem with this approach is that it is only applicable to those installations and tasks for which empirical data are available. Thus, the general approach is to divide human behavior in a system into small behavioral units, find data for these subdivisions and then recombine them to estimate the error probabilities for the task.

Numerous techniques have been proposed for predicting the performance of the human component of the human–machine system. Most of the techniques take one of two general approaches: analysis of historical data or computer-simulation of behavioral processes.

Some of the techniques are briefly reviewed below, but space does not allow the description of all human reliability predictive techniques that have been developed. Meister (1973) compares 22 methods available for quantitatively predicting operator and technician performance. For more recent descriptions and reviews of the methodologies, refer to Park (1987) or Sayers (1988).

6.6.1 Critical Incidence Technique

Generally, less attention is paid to those designs and conditions that could be hazardous but to which no actual accidents were attributed. Consequently, many of these near accidents or critical incidents go unreported. However, such information and data on critical incidents can provide clues that reflect design defects or behavioral idiosyncrasies which may be valuable in developing preventive measures.

As applied to accident research, the critical incident technique (CIT) enables the human-reliability engineer to determine previously experienced difficulties by interviewing personnel to collect information on involvements in accidents or near misses, operational errors, and unsafe conditions and practices. These descriptions can be used beneficially to investigate human–machine relationships in existing systems, to ascertain the critical job requirements, and to use the information acquired to improve them.

6.6.2 Human Error Databases

The most serious problem in the human reliability field is the lack of data. A prerequisite for a database is a taxonomic structure for organizing the data. Unfortunately, due to the diversity in approaches to human reliability analysis and application areas, universal consensus on an agreeable taxonomy is next to impossible. Another difficulty is the wide variety of the data sources. There are four basic sources of human error data: the field, simulator activities, laboratory experiments, and expert judgment. Field data, collected in the industrial setting, are the most applicable to the tasks studied in the human reliability analysis, but at the same time the most difficult to collect. Expert judgment is the least valid of the data sources. Subject-matter experts judge the relative or absolute likelihood of error for several tasks, and these responses are 'calibrated' to produce usable HEPs by mathematical scaling techniques.

The early attempt to create various databases virtually failed, and with a few exceptions, it has, until recently, remained a relatively unfruitful endeavor for various reasons (see Kirwan et al., 1990). Only a handful of human-error databases have been developed, and the ones that are available are not sufficiently comprehensive to serve many purposes.

Topmiller et al. (1984) performed a detailed review of existing human reliability databanks. In general, there are two sources of such information, namely empirical observations and human estimates.

6.6.2.1 Empirical Task Data

The Data Store developed in 1962 by the American Institute for Research (Payne and Altman, 1962) is organized around common controls and displays (e.g., knobs, joysticks, meters) particularly for tasks in the operation of electronic equipment. The data include performance time and reliability under the variations of the illustrated dimensions.

The second source is referred to as SHERB [Sandia Human Error Rate Bank (Rigby, 1967)], which is a compilation of HEP data based on the THERP (described below). This body of data consists of HEPs for many industrial tasks. The sample list of production errors in Section 6.1.1.2 gives a few examples.

6.6.2.2 Judgmental Task Data

Where empirical data on tasks are not available, it may be useful to resort to human estimates about certain task parameters for use in a system development process. To illustrate such a process, in the estimation of the HEPs of tasks, Pontecorvo (1965) asked personnel familiar with the requirements of the gross task to rate 60 tasks on a 10-point scale ranging from those that would be performed with the least error to those with the greatest error. Using the Data Store reliability estimates that were applicable to 29 of the 60 tasks rated, extrapolated to field conditions, Pontecorvo developed a regression line in logarithmic form. From the regression equation, it was possible to derive quantitative estimates of the reliability of all 60 tasks.

6.6.3 Task Criticality Analysis

Although human performance reliability is important for total system reliability, errors in performance of all tasks may not be necessarily equal in their effects on system performance. Thus, another task parameter might be criticality.

In one phase of their elaboration of human performance reliability, Pickrel and McDonald (1964) elicited ratings of error effect severity, setting ranges of rating values for the following classification: safe, marginal, critical, and catastrophic. Using such data as a springboard, they derived the error criticality ratings reflecting both frequency and severity. They recommend that efforts for failure reduction be concentrated on those errors with high criticality ratings. An overview of their error analysis procedures is excerpted in Park (1987).

6.6.4 Technique for Human Error Rate Prediction (THERP)

The THERP is a predictive technique for HEP in human reliability analysis (Rook, 1962; Swain, 1963; Swain and Guttmann, 1983). Here, the term "rate" is a misnomer for "probability."

6.6.4.1 Human-Reliability-Analysis Event Trees

THERP models events as a sequence of binary decision branches. At each node, the task is done either correctly or incorrectly. Except for those in the first branching, the probabilities assigned to all tree limbs are conditional probabilities. A conditional probability will be assigned to the first branching if it too is a carry-over from some other tree, or represents a task based on some previous event likelihood.

Let script letters denote tasks and events. Capital letters are used to denote failures. Lowercase letters are used to indicate success (or desirable states). Figure 6.4 illustrates a very simple example of an event tree representing the performance of two tasks, $\mathcal{A}$ and $\mathcal{B}$. Note that because task $\mathcal{A}$ is always performed first, the probabilities associated with task $\mathcal{B}$ are all conditional on the outcome of $\mathcal{A}$. The interdependence of tasks $\mathcal{A}$ and $\mathcal{B}$ are represented by the symbols $b|a$, $B|a$, $b|A$, and $B|A$.

Once the event tree is constructed and the estimates of the conditional probabilities of success or failure are assigned to each limb, the probability of each path through the tree can be calculated.

6.6.4.2 Data Sources for Human Error Probabilities

An interim HEP databank in Swain and Guttmann (1983) can serve as a major source of these estimates. The HEPs in Table 6.5 also have been incorporated into the predictive

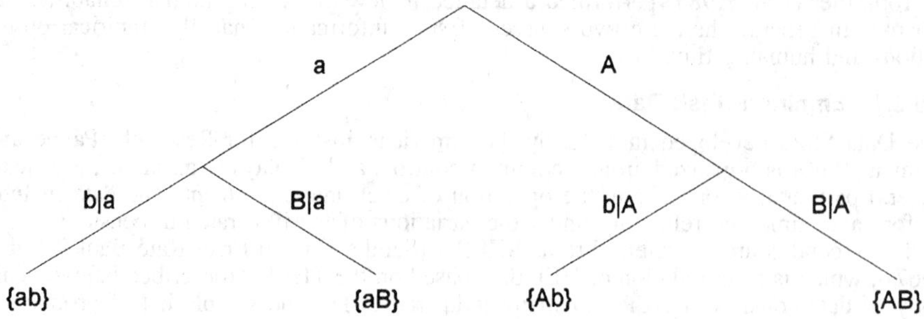

Figure 6.4 Human-reliability-analysis event tree. (See text for index.)

HEP estimates that are applicable to many other industrial settings. However, the tabled HEPs are *nominal* HEPs, i.e., estimates of HEPs for tasks or activities prior to consideration of plant-specific performance shaping factors.

Once the influence of plant-specific and task-specific behavioral factors such as stress, skill level, administrative control, etc. are taken into account, a *basic* HEP is formed that represents the probability of human error for that task in isolation. See Park and Jung (1996) for the probabilistic treatment of performance shaping factors.

A *conditional* HEP is a modification of the basic HEP to account for influences of other tasks or events that may include the preceding task elements, tasks, and the number of people involved in performing the task.

6.6.4.3 Dependence

One of the complications in modeling tasks as sequences of behaviors is the dependence among the task elements. Dependency can occur between people working on a task and also within an individual as several related tasks are performed. If a mechanic overtorques a bolt, chances are high that he will also overtorque the other bolts remaining in the sequence. In errors of omission, the entire sequence of tasks so highly related may be represented by one HEP for omission (completely dependent). Failure to take dependence into account can underestimate task error probabilities.

THERP accounts for the dependence by dividing the continuum of complete independence to complete positive dependence into five discrete levels of dependence according to the degree of dependence ($\%_{dep}$): zero (0%), low (5%), medium (15%), high (50%), and complete (100%). Positive dependence assumes that failure on the first task increases the probability of failure on the second task. The same relationship holds for probability of success.

For each level, a formula for obtaining the weighted average of 1.0 and its original value is provided to calculate the conditional probability of failure (or success) on task $\mathfrak{N}$ given the failure (or success) on the previous task $\mathfrak{N} - 1$:

$$\begin{aligned} \text{Prob}\{N|N-1\} &= (\%_{dep})1.0 + (1 - \%_{dep})\,\text{Prob}\{N\} \\ \text{Prob}\{n|n-1\} &= (\%_{dep})1.0 + (1 - \%_{dep})\,\text{Prob}\{n\} \end{aligned} \quad (8)$$

The formulas alter the conditional probability from its original value (at zero dependence) to 1.0 at complete dependence.

6.6.4.4 Outputs of THERP and Sensitivity Analysis

The outputs from the human reliability analysis using THERP will consist of estimates of task success and failure probabilities that can be used in design trade-off studies or in probabilistic risk assessments.

Because of the uncertainty associated with most of the data on human performance probabilities, it is often useful to perform a sensitivity analysis with different values of the estimated HEPs, dependence levels, stress levels, or other indices of human performance to determine the effects of such variation on system outcomes. Swain and Guttmann (1983) have used two types of sensitivity analysis: pessimistic HEPs (e.g., 0.1 for all tasks) are used in the worst-case analysis; optimistic HEPs (e.g., 0.0001 for all tasks) are used in the best-case analysis.

6.6.5 Operator-Action Tree Methodology

Another aspect of human reliability analysis concerns the diagnosis and decision errors (e.g., of nuclear power plant operators after the initiation of an accident).

In one line of development, Hall et al. (1982) focused on critical task sequences and analyzed the operators' roles in a critical course of events by means of an *operator-action tree* (*OAT*). An OAT represents the branching of paths toward success or failure depending on the operators' choice at various stages in decision making. Performance estimates in the form of HEPs are then incorporated into an event tree or fault-tree analysis.

The OAT approach recognizes three activities for the human response to an environmental event: (1) perception, (2) diagnosis, and (3) response. The basic OAT shown in Figure 6.5 is based on the potential for error in each of the three activities. However, the OAT approach assumes that the time available for diagnosis dominates the operator failure probability. A log-log time-reliability correlation model showing the relationship between probability of failure and time available was developed to represent the diagnosis errors.

6.6.6 Fault-Tree Analysis of Intermittent Events

Fault-tree analysis is a method of system reliability/safety analysis that has rapidly gained favor with analysts of complex systems. A fault tree shows a logical description of the cumulative effect of faults within the system. The effects of a set of basic faults are propagated through logical AND or OR gates until they reach a specified total system failure.

In elementary applications of the fault-tree methodologies, the basic fault conditions (such as component failures) are assumed to remain unchanged, once occurred. However, when the method is to be applied to industrial safety problems with recurring human errors, not only the occurrence rates, but also the effect of hazard durations must be taken into consideration. Park (1986) presents a bottom-up methodology for sequential reduction of fault-tree end-branches with intermittent events regulated by AND, OR, XOR, or NOT gates. The methodology can be used in conjunction with ordinary events with permanent state change.

6.6.7 Success Likelihood Index Method Using Multi-Attribute Utility Decomposition (SLIM-MAUD)

SLIM-MAUD (Embrey et al., 1984) is a computerized technique deriving from decision analysis. The SLIM is essentially a method of defining preferences among a set of items (i.e., human-error tasks). MAUD is a sophisticated approach to ensure that the 'expert panel' carrying out the SLIM exercise defines the preferences without biases. As a single combined technique, SLIM-MAUD defines the preferences representing the relative likelihoods of different errors as a function of various *performance-shaping factors* (PSFs). This represents a psychologically dominated approach, as opposed to engineering- and reliability-based approaches such as THERP.

The expert panel would typically comprise, for example, two operators with a sufficient operational experience, one human-factors analyst, and a reliability analyst who is familiar

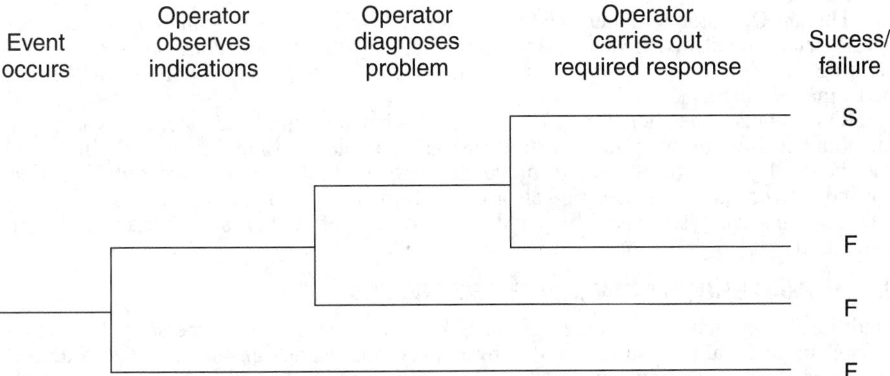

Figure 6.5 Basic operator action tree. (Adapted from Hall et al., 1982, with the permission of USNRC.)

with the system. The panel is initially asked to identify a set of important PSFs such as the level of training, the time available for the task, or the quality of the human–system interface design. PSFs are defined as any factors relating to the individual(s), environment, or task which affect performances positively or negatively. The panel is then asked to determine to what extent each PSF is optimal or suboptimal for that task in the situation being assessed. The rating is made on a scale of 1 (worst) to 9 (optimal). Similarly, relative importance weightings of the PSFs are obtained and normalized so as to add up to unity.

The SLIM proposes that the degree of preference can be represented as a function of the sum of the weightings multiplied by their ratings for each task error. The SLIM creates a relative scale representing the likelihoods of different errors, called the *success-likelihood index* (*SLI*). This index is transformed into HEPs using a logarithmic relationship of the form

$$\mathrm{Log(HEP)} = a \cdot \mathrm{(SLI)} + b \tag{9}$$

based both on experimental data and on a theoretical argument for the paired-comparisons technique (Hunns, 1982). If two tasks with known HEPs are included in the set of errors which are being quantified, then the unknown parameters of the equation can be solved by simultaneous equations. Then, the other HEPs can be determined.

This is the rationale underlying the SLIM. In practice, this method is somewhat more complex. For the ease of use, the technique is computerized. In the computerized version, known as SLIM-MAUD, the simple summary of weightings and ratings is refined in several ways, according to the more detailed mathematical requirements of multi-attribute utility theory.

6.6.8 Digital Simulation for Human Reliability Prediction

The digital simulation technique differs from the analytic methods in that it models operator performance within a behavioral structure reflecting psychological factors such as stress.

6.6.8.1 Monte Carlo Simulation

Siegel and Wolf (1961) have pioneered this approach in HE. The behavior and the states that govern human performance are inherently variable, so a probabilistic model is needed to describe them. The purpose of the technique is to indicate where the system may over- or under-load its personnel and to determine whether the average operator can complete all required tasks on time.

The technique has been expanded into a comprehensive human–machine reliability model for allocation, early prediction, and evaluation of achieved human reliability under U.S. Navy sponsorship (LaSala et al., 1978). The Navy approach represents a twofold expansion to human–machine reliability rather than just human reliability, thus implying a continuity between human and hardware reliability concepts. The technique also has been suggested for maintenance activities (MAPPS) in power plants (Siegel et al., 1984).

6.6.8.2 Deterministic Simulation

The Human Operator Simulator (HOS) (Wherry, 1969) simulates human–machine operations deterministically. Instead of random sampling, the model relies on equations describing relationships between parameters (e.g., memory and habit strength) and performance outputs.

HOS outputs a detailed time-history log of what the simulated operator was doing, including such items as the action start time and completion time, the anatomy involved, and the display/control function or procedure involved in the action. Although HOS does not indicate errors or the effect of error on performance, its output can be used in timeline and link analyses and compared with system effectiveness criteria for system evaluation.

6.7 HUMAN ERROR PREVENTIVE TECHNIQUES

Traditional approaches to reducing production error relied heavily on personnel selection, placement, and training, supplemented by motivational campaigns such as "zero defect" movements. These approaches, however, leave much to be desired as cost-effective techniques for human error prevention.

6.7.1 Improve the Work Situation

The first step toward reducing human error is to identify its causes correctly. All too often, the operators are blamed for making errors, producing defects, and initiating accidents, when, in fact, the poorly designed work situation itself is error inducing.

Typically, a human engineer, systems safety engineer, or similarly trained specialist examines the situation to identify error-likely conditions. When HE deficiencies are identified, the specialist can assess the impact on errors and recommend design changes. The changes may involve modifications of equipment design, work methods, operating procedures, job aids, performance feedback, layout, environmental conditions, etc.

An alternative approach to identifying situation-caused errors involves worker participation. One version is called the *error-cause removal* (*ECR*) program and consists of six basic elements (Swain, 1973). Perhaps, the ultimate in worker participation programs to reduce production errors and defectives may be found in Japanese quality control (QC) circles originated in 1963. The QC circle is a small group of work leaders and production workers who help solve quality problems.

6.7.2 Change the Personnel

When the work situation is satisfactory and the tasks are reasonable, but the operator still makes frequent errors, the poor performance may be due to individual factors such as inadequate skills, deficient vision, poor attitude, etc. Tasks involving highly skilled performance or decision making with considerable responsibility (and risk) usually require certification (e.g., driver's or pilot's license), which should be renewed regularly.

Operators with poor motivation or emotional problems can commit numerous unintentional errors and compromise safety. Physical and mental aptitudes are often crucial to good matching between the human and task. Sometimes, job rotation may help operators find a good match.

6.7.3 Reduce the System Impact

In most cases, human errors can be reduced to a low level using previously described techniques. However, if human error remains above the tolerable level, then the system vulnerability to human error impact must be reduced. In other words, the system must be designed to be tolerant of human errors.

Redundancy is the key to such a forgiving system. Machines can monitor human performance, and vice versa. Personnel redundancy can be used in critical operations. 'Hot' and 'cold' spares may stand by to backup the primary system. Such a system experiences graceful degradation instead of violent failure. Systems so designed can have so many checks, redundancies, and safety rules, that several serious human errors must be made in a particular sequence to cause a severe accident.

REFERENCES

Cooper, J. (1961). Human-initiated failures and malfunction reporting. *IRE Trans. Human Factors in Electronics, HFE-2,* 104–109.

Embrey, D., Hymphreys, P., Rosa, E., Kirwan, B., and Rea, K. (1984). SLIM-MAUD: An Approach to Assessing Human Error Probabilities Using Structured Expert Judgement, NUREG/CR-3518 (BNL-NUREG-51716). Department of Nuclear Energy, Brookhaven National Laboratory, Upton, New York 1173. For the Office of Nuclear Regulatory Research, U.S. Nuclear Regulatory Commission, Washington, DC 20555.

Gagné, R. (Ed.) (1962). *Psychological Principles in System Development.* New York: Holt, Rinehart & Winston.

Hagen, E. (Ed.) (1976). Human reliability analysis, control and instrumentation. *Nuclear Safety, 17*(3), 315–326.

Hagen, E., and Mays, G. (1981). Human factors engineering in the U.S. nuclear arena, *Nuclear Safety, 22*(3), 337–346.

Hall, R., Fraglola, J., and Wreathall, J. (1982). Post Event Human Decision Errors: Operator Action Tree/Time Reliability Correlation. Brookhaven National Laboratory, U.S. Nuclear Regulatory Commission, NUREG/CR-3010, Washington DC.

Hammer, W. (1976). *Occupational safety management and engineering.* Englewood Cliffs, NJ: Prentice-Hall.

Harris, D., and Chaney, F. (1969). *Human factors in quality assurrance.* New York: John Wiley.

Hunns, D. (1982). The method of paired comparisons. In A. Green, Ed., *High Risk Safety Technology.* Chichester: John Wiley.

Kirwan, B. (1994). *A guide to practical human reliability assessment.* London: Taylor & Francis.

Kirwan, B., Martin, B., Rycraft, H., and Smith, A. (1990). Human error data collection and data generation. *Int. J. Quality & Reliability Management, 7.4,* 34–66.

LaSala, K., Siegel, A., and Sontz, C. (1978). Man-machine reliability—a practical engineering tool. *IEEE Proc. Annual Reliability & Maintainability Symposium,* 389–394.

McCornack, R. (1961). Inspector Accuracy: A Study of the Literature (Report SCTM-53-61(14)), Sandia Corporation, Albuquerque, NM.

Meister, D. (1962). The problem of human-initiated failures. Proceedings, 8th National Symposium Reliability and Quality Control, 234–239.

Meister, D. (1971). *Human Factors: Theory and Practice.* New York: John Wiley.

Meister, D. (1973). A critical review of human performance reliability predictive methods. *IEEE Trans. Reliability, R-22*(3), 116–123.

Meister, D., and Rabideau, G. (1965). *Human Factors Evaluation in System Development.* New York: John Wiley.

Miller, D., and Swain, A. (1986). Human error and human reliability. In G. Salvendy, Ed., *Handbook of Human Factors/Ergonomics.* New York: John Wiley.

Nawrocki, L., Strub, M., and Cecil, R. (1973). Error categorization and analysis in man-computer communication systems. *IEEE Trans. Reliability, R-22*(3), 135–140.

NUREG-75/014, WASH-1400, 1975, Reactor Safety Study—An Assessment of Accident Risks in U.S. Commercial Nuclear Power Plants, U.S. Nuclear Regulatory Commission, Washington DC, Appendix III—Failure Data.

Park, K. S. (1985). Human reliability with probabilistic learning in discrete and continuous tasks: Conceptualization and modeling. *Microelectronics & Reliability, 25*(1), 157–166.

Park, K. S. (1986). Fault tree analysis of intermittent events. *Int. J. Systems Science, 17*(8), 1133–1138.

Park, K. S. (1987). Human reliability: Analysis, prediction, and prevention of human errors. *Advances in Human Factors/Ergonomics, 7,* New York: Elsevier.

Park, K. S., and Jung, K. (1996). Considering performance shaping factors in situation-specific human error probabilities. *Int. J. Industrial Ergonomics, 18*(4), 325–331.

Payne, D., and Altman, J. (1962). An Index of Electronic Equipment Operability: Report of Development, Report AIR-C-43-1/62, American Institutes for Research, Pittsburg.

Peters, G., and Hussman, T. (1959). Human factors in system reliability. *Human Factors, 1*(2), 38–42.

Pickrel, E., and McDonald, T. (1964). Quantification of human performance in large, complex systems. *Human Factors, 6*(6), 647–662.

Pontecorvo, A. (1965). A method of predicting human reliability. *Proceedings, 4th Annual Reliability & Maintainability Symposium,* Spartan Books, Washington, DC, 337–342.

Rasmussen, J. (1976). Outlines of a hybrid model of the process operator. In Sheridan and Johannsen, Eds., *Monitoring Behavior and Supervisory Control.* New York: Plenum Press.

Rasmussen, J. (1987). The definition of human error and a taxonomy of technical system design. In J. Rasmussen, K. Duncan, and J. Leplat, Eds. *New Technology and Human Error.* Chichester: John Wiley.

Rasmussen, J., and Jensen, A. (1974). Mental procedures in real life tasks: A case study of electronic troubleshooting. *Ergonomics, 17,* 293–307.

Rasmussen, J., Pedersen, O. M., Mancini, G., Griffon, M., and Gagnolet, P. (1981). Classification system for reporting events involving human malfunctions. Risø Nat. Lab., Roskilde, Denmark, RISØ-M-2240, March.

Rasmussen, J., Pejtersen, A., and Goodstein, L. (1992). Cognitive Engineering—Concepts and Applications, Part 1: Concepts. New York: John Wiley.

Reason, J. T. (1987). Generic error modeling system: A cognitive framework for locating common human error forms. In J. Rasmussen, K. Duncan, and J. Leplat, Eds., *New Technology and Human Error.* Chichester: John Wiley.

Rigby, L. (1967). *The Sandia Human Error Rate Bank (SHERB).* Paper presented at Symposium on Man-Machine Effectiveness Analysis: Techniques and Requirements, Human Factors Society, Los Angeles Chapter, June 15.

Rook, L. (1962). Reduction of Human Error in Industrial Production (Report SCTM-93-62(14)), Sandia Corporation, Albuquerque, NM.

Rouse, W., and Rouse, S. (1983). Analysis and classification of human error. *IEEE Trans. Systems, Man, and Cybernetics, SMC-13*(4), 539–549.

Sayers, B., Ed. (1988). *Human Factors and Decision Making: Their Influence on Safety and Reliability.* London: Elsevier.

Shapero, A., Cooper, J., Rappaport, M., Schaeffer, K., and Bates, C. (1960). Human Engineering Testing and Malfunction Data Collection in Weapon System Test Programs (Wright Air Development Division Technical Report 60-36, February).

Siegel, A., and Wolf, J. (1961). A technique for evaluating man-machine system designs. *Human Factors, 3*(1), 18–28.

Siegel, A., Barter, W., Wolf, J., and Knee, H. (1984). Maintenance Personnel Performance Simulation (MAPPS) Model: Description of Model Content, Structure and Sensitivity Testing, NUREG/CR-3626(2), ORNL/TM-9041/V2, Washington, DC.

Swain, A. (1963). A Method for Performing a Human Factors Reliability Analysis (Monograph SCR-685). Sandia National Laboratories, Albuquerque, NM.

Swain, A. (1964). Some problems in the measurement of human performance in man-machine systems. *Human Factors, 6*(6), 687–700.

Swain, A. (1973). An error-cause removal program for industry. *Human Factors, 15*(3), 207–221.

Swain, A., and Guttmann, H. (1983). *Handbook of Human Reliability Analysis with Emphasis on Nuclear Power Plant Applications,* U.S. Nuclear Regulatory Commission Technical Report NUREG/CR-1278, Washington, DC.

Topmiller, D., Eckel, J., and Kozinsky, E. (1984). Human Reliability Data Bank for Nuclear Power Plant Operations, NUREG/CR-2744, U.S. Nuclear Regulatory Commission, Washington DC.

Wherry, R. (1969). The development of sophisticated models of man-machine system performance. In G. Levy, Ed., Symposium, Applied Models of Man-Machine Systems Performance, Rep. NR69H-591, North American Aviation/Columbus (AD 697 939). Available from NTIS, Springfield, VA. 22151.

Willis, H. (1962). The Human Error Problem. Paper presented at American Psychological Association Meeting, Report M-62-76, Martin-Denver Company, Denver, CO.

CHAPTER 7

PERCEPTUAL MOTOR SKILLS AND HUMAN MOTION ANALYSIS

David Regan
Department of Psychology
York University
North York, Ontario, M3J1P3 Canada

7.1 INTRODUCTION

7.1.1 The Generalizability of Empirical Studies on Perceptual Motor Skills

Driving a car, flying a helicopter, and hitting a tennis ball are all examples of skilled visually guided motor actions that are limited by the ability of an individual's visual pathway to process the visual information that is relevant to the particular task, as well as by that individuals's motor capabilities.

If the findings of an experimental study of performance of a particular task carried out in a particular visual environment are to be generalized to a wide variety of tasks and visual environments, it is necessary to have a general quantitative model of human visual function. At the present time, such a model is not even on the horizon. But this does not necessarily mean that every one of the indefinitely large number of different visual environments requires a separate study. In the absence of a general quantitative model, a conceptual framework, even though of less than completely general applicability, might provide a useful guide to extending the applicability of data collected in one particular visual environment to a wider range of visual environments.

An attempt to develop such a conceptual framework can be understood as follows. Consider the process of learning a skill of eye–limb coordination such as hitting a tennis ball. To successfully hit the ball the player must, several hundred milliseconds ahead of time, judge exactly when the ball will arrive, and also its direction of motion in three-dimensional space. It has been shown mathematically that one ratio between retinal image variables correlates with the time until the ball will arrive and that other ratios correlate with the ball's direction of motion (see Sections 7.5 and 7.7). Suppose that an important part of the skill of hitting the ball stems from learned computations carried out on the outputs of neural mechanisms sensitive to these ratios. If any given ratio-sensitive mechanism filters out all visual information other than the particular ratio to which it responds, it will be "blind" to everything other than its own particular ratio, and if the several ratio-sensitive mechanisms operate with negligible cross-talk then, once learned, the skill could readily transfer to a wide variety of visual environments, as indeed is the case. This line of argument leads to a conceptual approach, which, even though it is far from being a general quantitative model, might prove useful in generalizing over some limited part of the visual psychophysical data on eye–limb coordination.

The basic idea can be summarized as follows [a detailed review is available in Regan (1982)].

1. The human visual system's responses to a limited number of visual dimensions are independent of each other (e.g., color, luminance, spatial frequency, orientation, changing size, motion in depth, time to contact).
2. Each of these selective sensitivities shows considerable intersubject variability.
3. Intersubject variability is far from perfectly correlated across the set of selective sensitivities.

The implications for visually guided, goal-directed motor action of this so-called channeling or modular approach are as follows. First, some tasks involve very few of the limited number of selective sensitivities, especially when the visual environment is impoverished (e.g., landing an aircraft in fog). Consequently, for these tasks laboratory tests of the relevant sensitivities may well predict intersubject variations in performance. Second, since different tasks involve different selective sensitivities, the level of performance in one laboratory test will not predict performance in all tasks (i.e., "specific tests for specific tasks"). Third, in some individuals the various selective sensitivities may depart considerably from the requirement of perfect independence. This would cause skilled eye–limb coordination in a complex or cluttered visual environment to be inferior to eye–limb coordination in a simple visual environment.

7.2 SEEING AN OBJECT

7.2.1 Five Forms of Contrast That Can Render an Object Visible and Recognizable

While discussing subtleties, it is easy to overlook the obvious. For example, it is obvious that to judge and respond to an object is motion, it is first necessary to see the object. This point is not academic: In a significant proportion of aircraft accidents, the major cause has been shown to be the pilot's failure to see an external object or terrain feature before it is too late ("controlled flight into terrain"). Enquiries into highway accidents also not infrequently describe an apparent failure to see a pedestrian or another vehicle and go on to implicate possible causes of low visibility such as inadequate street lighting, fog, snow, or glare (caused, for example, by low sun). Therefore, before going on to discuss visual responses to an object's motion, it will be appropriate to review briefly the factors that allow us to see the object.

If an object's retinal image does not differ from the retinal image of its surroundings, the object is said to be perfectly camouflaged and is quite invisible. Turning this around, there are five kinds of difference between the retinal images of an object and its surroundings that—alone or in combination—can render an object visible. These are differences in luminance, color, motion, texture, and depth (reviewed in Bergen, 1991 and Regan, 1991a).

When considering the ability to see (i.e., detect) an object that is rendered visible by luminance contrast, it is important to recognize that the neural mechanisms that process high-contrast and low-contrast retinal images are not entirely the same, and that standard tests of vision (e.g., the Snellen test) assess only the ability to see and recognize high-contrast objects. Even a normally sighted individual's sensitivity to low-contrast objects cannot be predicted completely on the basis of Snellen acuity, and in some patients the distinction between visual sensitivities to low-contrast and high-contrast objects is very large.[1] Several tests are available for assessing visual sensitivity to low-contrast objects including the Pelli-Robson contrast sensitivity chart (Pelli, Robson, and Wilkins, 1988) and the Regan low-contrast acuity charts (Regan, 1988).

Color differences (chromatic contrast) can increase an object's visibility and can even make it "pop out" from a visually cluttered scene (e.g., a red apple whose background is the green leaves of a tree). However, only in rare cases is chromatic contrast entirely responsible for an object's visibility. An individual's ability to use chromatic contrast to see an object can be tested straightforwardly by means of equiluminant patterns such as the Ishihara plates.

Helmholtz (1909/1962) seems to have been the first to note that relative motion (which he termed *motion parallax*) is sufficient to render an object visible. This form of object perception may be important in low-level helicopter flight where an object that is virtually invisible when hovering (e.g., a protruding grassy terrain feature with a grassy surround) springs into view when the aircraft is moving. The best values of those spatial discriminations for motion-defined form that have been investigated—vernier acuity, aspect ratio (two-dimensional shape) discrimination, and orientation discrimination—are hardly, if at all, inferior to the comparable abilities for luminance-defined form (Regan, 1991a). However, even in control subjects, the ability to recognize motion-defined shapes (letters) does not correlate perfectly with the ability to recognize similar luminance-defined shapes (Simpson and Regan, 1995). This dissociation can be very marked in patients with damage to white matter connections (i.e., myelinated axons of neurons) within the brain, some of whom may be effectively blind to motion-defined form while retaining visual acuity of 20/20 or better (Giaschi, Regan, Kothe, Hong, and Sharpe, 1992; Regan, 1991a). The dissociation is even evident in the "clinically unaffected" eye of individuals with a history of unilateral amblyopia, a not uncommon childhood problem (Giaschi, Regan, Kraft, and Hong, 1992). These findings in neuro-ophthalmological patients have been taken to indicate that different neural mechanisms process motion-defined form and luminance-defined form. A motion-defined letter test offers a simple means of assessing an individual's ability to recognize motion-defined form.[2]

Objects can be rendered visible and recognizable by texture contrast alone (reviewed in Bergen, 1991). This form of object perception may be important in low-level aviation, and effective rendering of ground texture presents a problem in the design of flight simulators. A form of texture contrast that has attracted considerable attention from laboratory researchers is orientation contrast, where an object is distinguished from its surroundings

[1]For example, some patients with multiple sclerosis who have 20/20 Snellen acuity are virtually blind to low-contrast objects. On the highway, a typical low-contrast object would be an approaching truck in fog (Regan, 1991b).

[2]Software to generate motion-defined letters is available with instruction sheets gratis from D. Regan. Send a blank 5.25 or 3.5 disc in a cardboard disc envelope. The test will run only on an IBM PC type 386 or 486 (or clone) with an ATI VGA "Wonder XL" graphics card.

by the different orientations of short lines inside and outside the object's boundaries. The best value of orientation discrimination, positional acuity and shape discrimination for texture-defined form is little, if at all, inferior to the comparable (i.e., with matched spatial sampling) value for luminance-defined form (Regan, Hajdur, and Hong, 1996; Regan, 1995a). However, even in control subjects, the ability to recognize texture-defined letters does not correlate perfectly with the ability to recognize luminance-defined letters (Simpson and Regan, 1995). This dissociation can be very marked in patients with damage to white matter connections within the brain, a finding that has been taken to indicate that different neural mechanisms process texture-defined form and luminance-defined form (Regan and Simpson, 1995). A texture-defined letter test offers a simple means of assessing an individual's ability to recognize motion-defined forms.[3]

Objects can be rendered visible and recognizable by binocular disparity alone (so-called cyclopean form), though this form of perception operates only at ranges of up to a few meters. Nevertheless, the effectiveness of binocular stereoscopic vision in breaking the otherwise-perfect camouflage of prey or predator has been suggested as the major evolutionary advantage of developing a pair of forward-looking eyes. Just as for color-defined form, the spatial discriminations that have been investigated for cyclopean form are hardly, if at all, inferior to comparable abilities for luminance-defined form. This applies to vernier acuity (Morgan, 1986), orientation discrimination (Mustillo et al., 1988; Regan, 1991a), and aspect ratio (shape) discrimination (Regan, 1991a). When a cyclopean target moves within the frontoparallel plane, discrimination thresholds for its direction and speed of motion are little inferior to comparable values for monocularly visible targets (Phinney, Boyd, Winget, and Patterson, 1995), and the same holds for a cyclopean target moving in depth (Portfors-Yeomans and Regan, 1997b).

7.2.2 The Reduction of Object Visibility by Glare

It is well known that particles in the atmosphere scatter light and that this scatter causes an object's retinal image to lose contrast as the object's distance increases, so that beyond some critical distance the object cannot be seen. Duntley (1948) provides a quantitative treatment of this effect.

Light scattered within the eye (i.e., veiling glare) can also render an object invisible even at close range. Object visibility problems associated with veiling glare caused by either low sun or oncoming headlamps are familiar to highway drivers. In aviation also, veiling glare produced by the sun is a potent cause of failure to see an object. Formal discussion of the optical factors associated with glare are available (Ijspeert, deWaard, van den Berg, and deJong, 1990; Van den Berg, 1991; Westheimer and Liang, 1995). Veiling glare is most serious when a bright light source is almost in line with the object one is attempting to see, a fact that, from the earliest days over the trenches of the Western Front in 1914, has been exploited by pilots engaged in air-to-air combat (Lewis, 1936; Yeats, 1934). Light scattered within the eye falls diffusely on the retinal image and dilutes the contrast of the retinal image, so that all objects become less visible and some objects may be rendered invisible (Nadler, Miller, and Nadler, 1990). Multiple point-like opacities within the lens of the eye considerably increase susceptibility to glare sources as does damage to the cornea and even a light-colored iris [an appreciable amount of light leaks through the iris (Van den Berg, 1991)]. A fact that is important for aviation and highway driving is that there is marked intersubject variability in susceptibility to glare, even within 19- to 25-year-old individuals with normal visual acuity (Regan, 1991c).

We have recently developed a simple technique that is sufficiently sensitive to quantify individual differences in susceptibility to glare even in young, normally sighted individuals. This technique is to measure visual acuity using low-contrast letter charts, first in minimal-glare conditions and then in the presence of a glare source. The ratio of the two visual acuities is a measure of the individual's susceptibility to glare (Bailey and Bullimore, 1992; Elliott and Bullimore, 1993; Regan, 1991c).

[3]Software to generate texture-defined letters is available with instruction sheets gratis from D. Regan. Send a blank 5.25 or 3.5 disc in a cardboard disc envelope. The test will run only on an IBM PC type 386 or 486 (or clone) with an ATI VGA "Wonder XL" graphics card.

7.3 OBJECT MOTION WITHIN A FRONTOPARALLEL PLANE[4]

7.3.1 Detection of Motion That Is Contained Within a Frontoparallel Plane

Early research on the detection of real (as distinct from apparent) motion parallel to the frontal plane was restricted to coherently moving targets such as a bright spot or a bright line (reviewed in Graham, 1965; Kaufman, 1974; Nakayama, 1985). It was soon found that the absolute threshold was roughly 1 to 2 arc min/sec when the observer fixated the moving target, provided that there were one or more clearly visible stationary reference marks close to the moving target. When all stationary reference marks were removed from the field of view, detection threshold rose to roughly 10 to 20 arc min/sec. Leibowitz (1955a,b) suggested that a candidate explanation for this effect of stationary reference marks might be that, at very slow speeds, motion is inferred from a change in relative position—much as we infer that a clock's hour hand was moved because it is now in a different place from where it once was—rather than by detecting a rate of change of position, i.e., motion per se. Other candidate explanations for the difference between detection thresholds for relative and absolute motion are the existence of relative motion detectors (Section 7.4.1) and ocular tracking of an isolated moving object.

With the aim of removing relative position cues to motion (c.f. the hour-hand case just mentioned) so as to uncover true motion detection thresholds, Nakayama and Tyler (1981) stimulated the eye with a field of random dots undergoing a differential shearing motion. A comparison of motion and position thresholds showed both to be very low, each requiring a differential displacement of only 5 arc sec. In comparison to position thresholds, motion threshold was 10 times lower at the lowest spatial frequencies, but was higher at high spatial frequencies.

A more recent approach to uncovering true motion detection thresholds is to use a signal-to-noise measure of detection threshold. This line of thought can be understood by first considering the distinction between a coherent and an incoherent motion stimulus. An example of a coherent motion stimulus is a patch of dots, all moving at the same speed in the same direction. The patch of dots moves as a whole, and the motion of the patch is completely specified by the motion of any arbitrary individual dot. Completely incoherent motion means that each dot moves in a random direction. Van Doorn and Koenderink (1982a,b) degraded the coherence of motion until global motion could no longer be detected. Their procedure was to vary the relative contrasts of superimposed dot patterns undergoing coherent and incoherent motion. More recently, several authors have used a different approach in which dots within the pattern have limited lifetimes (Newsome and Pare, 1988; Williams and Sekuler, 1984); in this approach, a variable proportion of the dots in the pattern move coherently while the rest are randomly repositioned on each frame. For example, Williams and Sekuler (1984) developed a random dot cinematogram in which all dots drew their successive displacements from a rectangular distribution characterized by some particular directional range. On any frame of the display, the direction in which any single dot moved was independent of its own history and independent of the movements of other dots. When the distribution of directions covered a broad range of directions (for example, 270°–360°) observers saw only the local, pseudo-random motions of individual dots. When the distribution of directions covered a narrower range (for example, 90°–180°) observers saw, in addition to local random motions, a global, coherent flow in the general direction of the mean of the distribution. Williams and Sekuler (1984) measured the probability with which global motion was seen as a function of the range of the directional distribution. They found that as the distribution changed, so did the percept, from random, incoherent motions to global flow, or vice versa.

Evidence that motion analyzers within the human visual pathways are selective to the direction of motion was reported by Addams (1834) who found that when, after staring fixedly at a waterfall, he gazed at stationary rocks, the rocks appeared to drift upwards (the classic motion aftereffect or waterfall illusion). The directional bandwidth of these motion analyzers was measured by Levinson and Sekuler (1980) who adapted to an isotropic pattern of random dots moving uniformly in one direction. Adaptation elevated

---

[4]The frontal plane is a plane through the first nodal points of the two eyes and perpendicular to a plane containing the two nodal points and the point of fixation. A frontoparallel plane is any plane parallel to the frontal plane (Cline, Hofstetter, and Griffin, 1989, p. 535).

the luminance detection threshold for subsequently viewed moving dots, and the threshold elevation was maximal for test dots moving in the adapting direction. The tuning curve for directional sensitivity was broad, with some threshold elevation being found even when test and adapting directions differed by 45° (Figure 7.1).

The directionally tuned motion analyzers postulated by Levinson and Sekuler (1980) seem to be selective for binocular disparity as well as for direction. There is evidence that an object's leftward motion excites different neural motion-sensitive analyzers depending on whether the object is closer or further than the plane of binocular convergence, and that this is also the case for rightward motion (Regan et al., 1995).

Several authors have investigated whether motion thresholds or motion perception depend on the direction of motion [e.g., away from the fovea versus toward the fovea. Findings are complex and there is considerable disagreement (reviewed in Raymond, 1994)].

So far, we have discussed the processing of motion in the central visual field. It has long been known that the detection threshold for real movement is higher in peripheral than in central vision (Johnson and Leibowitz, 1974; Leibowitz et al., 1972; McCoglin, 1960; Tyler and Torres, 1972). For example, Regan and Beverley (1983) measured thresholds for detecting 2 Hz sinusoidal oscillatory motion in 13 subjects. A plot of threshold (linear axis) versus log eccentricity was approximately linear, rising from roughly 1 arc min peak-to-peak at an eccentricity of 4° to 10 arc min peak-to-peak at an eccentricity of 24°. Thresholds were similar for square targets of 0.5°, 1.0°, and 2.0° side length.

Reichardt's (1961) mathematical model of a motion detector was a major factor in transforming the status of motion research from a rather uncoordinated collection of observations into a research area with widespread agreement as to priorities. The basic idea, illustrated in Figure 7.2A, is that the visual pathway compares the signals that arise from different points on the retina after delaying one of the signals. Suppose two photoreceptors (P) at distance Δx apart are connected to a comparator (e.g., a multiplier), but that the signal from the left photoreceptor is subjected to a constant delay τ sec. The

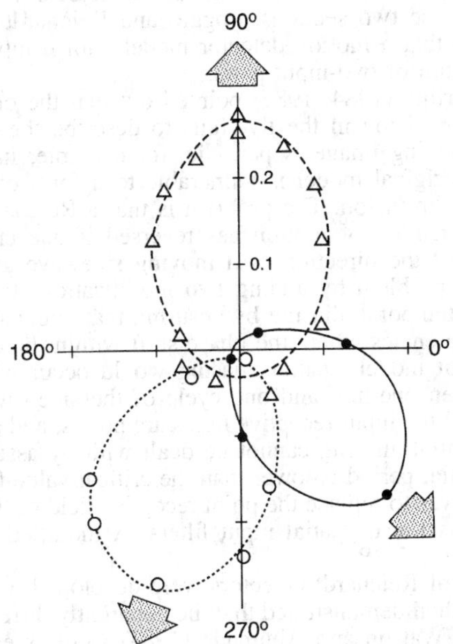

Figure 7.1 Log threshold elevation for moving dot patterns as a function of the direction of drift. Azimuthal angle in these polar coordinate graphs corresponds to the direction of stimulus motion, whereas radial distance indicates log threshold elevation. The adapting directions used in three separate experiments are shown by the arrows. [Modified from Levinson, E., and Sekuler, R. (1980). A two-dimensional analysis of direction-specific adaptation. *Vision Research, 20,* 103–108 and first printed in Regan, D. (1982). Visual information channeling in normal and disordered vision. *Psychological Review, 89,* 407–444. Reprinted with permission.]

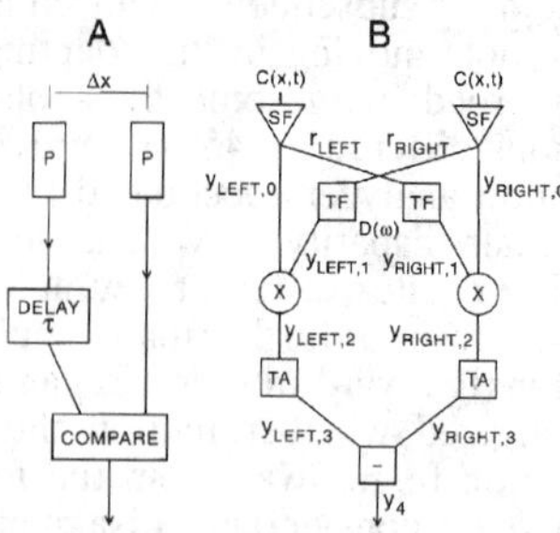

Figure 7.2 (A) The "delay and compare" principle of a Reichardt motion detector. The rightwards-motion subunit only is shown. (B) The elaborated Reichardt motion detector. The input is a luminance pattern with contrast $C(x, t)$. It is sampled by linear spatial filters (receptive fields, SFs) with spatial responses r_{left} and r_{right} centered at locations x_{left} and x_{right}, $y_{i,H}$ (H = left, right) represents the signal at various stages (i) for the left and right subunits. TF indicates a linear, time-invariant filter with Fourier transform $D(\omega)$, X indicates a multiplication unit, TA indicates a temporal integration operation and—indicates a unit that subtracts its left input from its right input. [From Van Santen, J. P. H., and Sperling, G. (1985). Elaborated Reichardt detectors. *Journal of the Optical Society of America, A2,* 300–321. Reprinted with permission.]

comparator will give a strong output when the signals arriving from the two photoreceptors are identical at every instant. This situation is achieved when a spatial pattern is moving over the retina from left to right at the particular speed that causes it to reach the first photoreceptor exactly τ sec before it reaches the second photoreceptor. This speed is equal to $(\Delta x / \tau)$. Patterns moving faster or slower will generate a lower output from the comparator, and patterns moving in the opposite direction will produce no output at all. A complete Reichardt detector comprises a section for rightward motion (Figure 7.2A) plus a mirror-image section for leftward motion. The detector's output is the difference between the outputs of the two sections. Poggio and Reichardt (1973), developing the basic idea, have shown that a motion detector model with n inputs and a single output can be reduced to the sum of two-input pairs.

Van Santen and Sperling (1984, 1985) pointed out that the original Reichardt model (designed to account for vision in the fly) fails to describe the facts of human motion perception when the moving image is periodic, for example, in the case of a moving sinewave grating. The original model is vulnerable to a form of aliasing that is hardly evident, if at all, in human vision. The problem is that a Reichardt detector can erroneously signal that the direction of motion has reversed if one changes the temporal or spatial frequency but not the direction of a moving sinewave grating. Van Santen and Sperling overcame this problem by adding two modifications to the original Reichardt detector. They avoided temporal aliasing by ensuring that attenuation reduced the output to zero at temporal frequencies where the phase shift within the detector exceeded 180°. In the original Reichardt model, spatial aliasing would occur when the distance Δx in Figure 7.2A was between one-half and one cycle of the sinewave grating. Since in the original Reichardt model the input receptive fields are points, and a point has a flat spatial frequency spectrum, spatial aliasing cannot be dealt with by assuming zero sensitivities to sinewaves whose spatial period is lower than the critical value for aliasing. Van Santen and Sperling's solution was to replace the point receptive field of the inputs with extended receptive fields acting as linear spatial input filters. A modified Reichardt filter of this type is illustrated in Figure 7.2B.

Various elaborations of Reichardt detectors were developed by Van Santen and Sperling (1984, 1985), who then demonstrated that the apparently different models of Adelson and Bergen (1985) and Watson and Ahumada (1983) were essentially special cases of elaborated Reichardt detectors as Fourier-analytic motion detectors (Chubb and Sperling, 1988).

There is, however, a class of stimuli that produces a clear percept of motion, yet is not detected by a Reichardt detector. This class of stimuli has been called *non-Fourier motion,* and the hypothetical detector of the various kinds of non-Fourier motion are referred to as non-Fourier motion detectors (Chubb and Sperling, 1988). Examples of non-Fourier motion stimuli include draft-balanced and microbalanced stimuli (Chubb and

Sperling, 1988), the motion of a boundary defined by shearing motion, the motion of a boundary defined by texture contrast, and the motion of cyclopean form.

7.3.2 Discrimination of Speed

The *just-noticeable difference* (*JND*) in speed is a broad U-shaped function of reference speed at all retinal eccentricities. In foveal vision, the just-noticeable difference is approximately constant from roughly 3°/sec to 60°/sec (Figure 7.3). As viewing eccentricity is progressively increased, the just-noticeable difference grows progressively larger at low reference speeds, but remains roughly constant at high reference speeds. The lowest value of the just-noticeable difference in threshold is approximately 5%, and this increases only slightly with eccentricity (by roughly 50% from 0° to 50° eccentricity) (McKee, 1981; McKee and Nakayama, 1984; Orban et al., 1985).

7.3.3 Discrimination of the Direction of Motion

To avoid confounding the orientation and direction of motion of a target, Pasternak and Merigan (1984) used a random dot pattern rather than a grating. They reported that the just-noticeable difference in direction varied with reference speed for all three observers. At low speeds (0.4°/sec) the just-noticeable difference was large (15°–25°), but fell as speed was increased until it reached a limitingly low value of approximately 4.6° at a reference speed of 1°/sec, and thereafter remained constant up to the highest speed tested (5°/sec). DeBruyn and Orban (1988) also used a dot pattern but reported that the just-noticeable difference in direction was as low as 1.8° over the speed range 16–64°/sec. These reported JNDs differ quite considerably. Whether the disagreement is due to differences in methodology (apparent motion versus continuous motion) or to intersubject differences is not clear.

Ball and Sekuler (1979, 1987) found that subjects were able to discriminate smaller differences between two directions of dot motion when the mean direction of motion was horizontal or vertical than when the mean direction was oblique. The just-noticeable difference in direction fell considerably with practice. This improvement was greatest along the training direction, but some improvement was also seen for directions up to 45° away from the training direction. The effect of training persisted up to at least 10 weeks after training ceased. Ball and Sekuler (1987) expressed this just-noticeable difference in terms of the d' measure of Green and Swets (1966). Before training, a difference in direction of 3.0° gave a d' of 0.51 for horizontal or vertical motion and 1.55 for oblique motion. After training, the corresponding values of d' were 1.60 and 3.10.

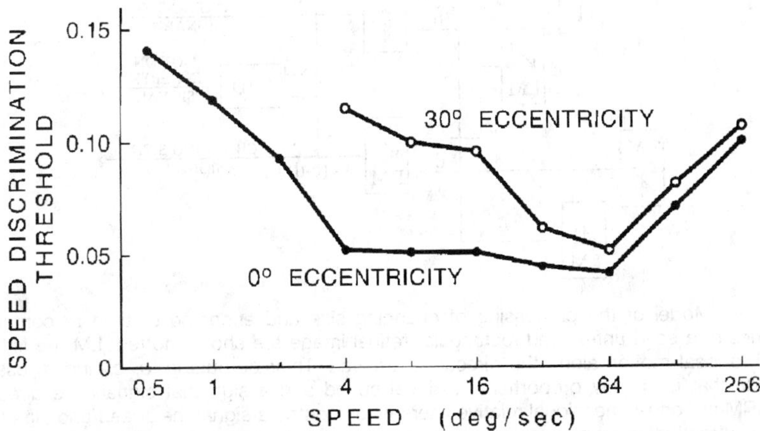

Figure 7.3 The just-noticeable difference in speed expressed as a Weber fraction is plotted as ordinate versus reference speed in degrees per second. Solid symbols are for foveal vision [i.e., 0° eccentricity) and open symbols are for peripheral vision (30° eccentricity). The target was a high-contrast slit of fixed (0.1°)] width. [Modified from Orban, G. A., Calenbergh, F. van, DeBruyn, B., and Maes, H., 1985, Velocity discrimination in central and peripheral visual field. *Journal of the Optical Society of America, A2,* 1836–1847. Reprinted with permission.]

Phinney et al. (1995) measured the just-noticeable difference in the direction of motion for a cyclopean target. At a reference speed of 6°/sec and presentation duration of 1.0 sec, the JND for the cyclopean target was approximately 6° compared with a JND of 1° for a monocularly visible target. On the other hand, this difference between JNDs for the cyclopean and monocularly visible targets might have been caused by the greater difficulty of seeing the cyclopean target. Phinney et al. noted that the JNDs for both cyclopean and monocularly visible targets depended on presentation duration, shorter presentations favoring the monocularly visible target.

7.4 DETECTION OF AN OBJECT'S MOTION IN DEPTH

7.4.1 The Detection of Motion in Depth Using Monocular Retinal Image Information Alone

As a rigid sphere moves in a straight line directly toward an observer's eye, the sphere's retinal image expands isotropically (i.e., without changing shape), thus providing a monocular retinal image correlate of motion in depth. Conversely, when a retinal image expands without changing shape, an impression of motion in depth can be created (Wheatstone, 1852).

The way in which the visual system processes the expansion of an object's retinal image is subtle. In following the line of argument reviewed next, it will be helpful to bear in mind the distinction between the perception of changing-size and the perception of motion in depth.

In a series of papers, Regan and Beverley investigated differences between the way in which the human visual system processes isotropic and nonisotropic expansion of an object's retinal image (reviewed in Regan et al., 1995). First they investigated whether visual responses to changing-size can be explained in terms of local detectors of unidirectional motion. Their experimental evidence showed that this is not the case.

As already discussed in Section 7.3.1, there is a substantial literature on detectors of local unidirectional motion (LM filters in Figure 7.4) that are often modeled as "Reichardt Detectors" or as "Modified Reichardt Detectors." Both the original and modified Reichardt detectors, however, respond to absolute rather than relative motion.

Evidence for a neural mechanism sensitive to unidirectional *relative* motion was obtained as follows. There were two adapting stimuli: a bright solid rectangle whose op-

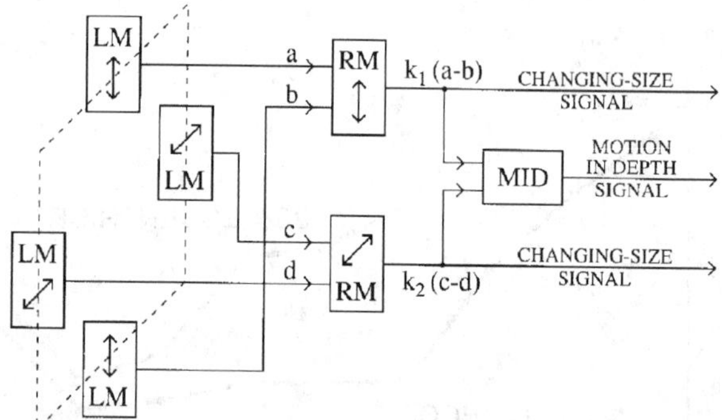

Figure 7.4 Model of the processing of changing size and encoding of time to contact. The boundaries of a solid untextured rectangular retinal image are shown dotted. LM are filters that respond to local motion along the directions arrowed. Their outputs (*a, b, c,* and *d*) assume a magnitude that is linearly proportional to local speed and a sign that signals the direction of motion. RM are one-dimensional relative filters whose outputs signal the speed and sign (expansion vs. contraction) of relative motion along some given retinal azimuth. MID is a two-dimensional relative motion filter that is strongly excited by expansion of the retinal image provided that expansion is isotropic and homogeneous, i.e., when $k_1(a - b) = k_2(c - d)$. We assume here that the amplitude of the output of filter MID is equal to any one of its inputs from RM filter. If so, output amplitude is inversely proportional to time to contact. [Modified from Figure 1 in Beverley and Regan (1979) and Figure 2 in Regan (1986b). From Regan and Hamstra (1993). Dissociation of discrimination thresholds for time to contact and rate of angular expansion. *Vision Research, 33,* 447–462. Reprinted with permission.]

posite vertical edges oscillated in antiphase (so that the rectangle's location and height remained constant while its width oscillated), and the same rectangle but with opposite vertical edges oscillating in phase (so that its width and height remained constant while its location oscillated from side to side). After adapting to antiphase oscillations, visual sensitivity to antiphase oscillations was considerably reduced, while sensitivity to inphase oscillations was little affected. A second finding was that adapting to inphase oscillations had little effect on sensitivity to either antiphase or inphase test oscillations.

The crucial point here is that the oscillations of either of the rectangle's edges were identical for the two adapting stimuli: The two adapting stimuli differed only in the relation between movements of the two vertical edges. Clearly, the results of this experiment could not be explained entirely in terms of adaptation of local responses to unidirectional motion (LM filters in Figure 7.4). The conclusion was that the differential loss of sensitivity was caused by adapting to a relationship between the velocities of the rectangle's two vertical edges, i.e., adapting to the *relative velocity* of the two vertical edges computed in Figure 7.4 by RM filters. Further evidence for a neural mechanism sensitive to relative unidirectional motion has subsequently been published (Freeman and Harris, 1992).

Now we turn to experiments on visual responses to isotropic changes of retinal image size. First, I describe a two-dimensional variant of the study just described. Figure 7.5A,B shows the two stimuli. After adapting to isotropic oscillations in the size of a bright solid square (Figure 7.6B), detection threshold for size oscillations was increased fivefold (continuous line, Figure 7.6C),[5] but detection threshold for oscillations of the square's location were unaffected (dotted line, Figure 7.6C). In addition, both thresholds were little affected by adapting to oscillations of the square's location (continuous and dotted lines, Figure 7.6D). Again, the crucial point was that the oscillations of the square's edges (all four in this case) were identical for the two adapting stimuli in Figure 7.6A,B: The two adapting stimuli differed only in the relationship between movements of opposite edges. Just as for the one-dimensional case reviewed above, these findings for the two-dimensional case could not be explained in terms of adaptation of local detectors of unidirectional motion. The most parsimonious explanation was that the reduction of sensitivity to isotropic oscillations of size was due to adaptation of a pair of relative motion filters (RM in Figure 7.4), one driven by the square's vertical edges and one by the square's horizontal edges.

The finding that sensitivity to *relative* motion can be reduced by adaptation that has comparatively little effect on sensitivity to absolute motion is evidence that these relative motion filters exist as physical neural entities. The outputs of these RM filters signal changing-size and [if $k_1(a - b)$ differs from $k_2(c - d)$ in Figure 7.4], changing-shape. The processing carried out by the RM filters is remarkably linear, as we see next.

The particular velocity relationship to which these relative motion filters were sensitive and the linearity of the changing-size processing was indicated in a second study in which there was only two test stimuli—the isotropic size oscillation depicted in Figure 7.5B and the inphase oscillation depicted in Figure 7.5A. Eleven different adapting stimuli were used on different days. The antiphase oscillation depicted in Figure 7.5B simulated oscillatory motion along the Z-axis in Figure 7.6A, and the inphase oscillation depicted in Figure 7.5A simulated oscillatory motion along the x-axis in Figure 7.6A. The 11 adapting oscillations were different combinations of these two test oscillations. The particular combination depicted in Figure 7.6A corresponds to oscillations along a direction inclined at angle β to the line of sight. The 11 adapting trajectories illustrated in Figure 7.6B all had the same antiphase component of oscillation (i.e., the same simulated speed of motion along the line of sight), namely, a 2 Hz antiphase triangular wave of amplitude 6 arc min peak-to-peak. The adapting trajectories were created by adding to the constant antiphase component a 2 Hz inphase triangular wave. For the two most oblique trajectories, this inphase oscillation had an amplitude of 48 arc min peak-to-peak.

Results are shown in Figure 7.7. The first finding was unsurprising. The continuous line in Figure 7.7 shows that adaptation had no effect on detection threshold for the inphase test stimulus when the adapting stimulus contained no inphase component (0.0 on the abscissa), but produced a progressively greater threshold elevation as more and more inphase component was added to the adapting stimulus. The crucial finding, shown by the dashed line in Figure 7.7, was that adaptation had the same effect (to within less

[5]Thresholds for detecting isotropic oscillations of retinal image size range from 0.25 to 1.0 arc min over a range of oscillation frequencies between 0.3 and 3.0 Hz.

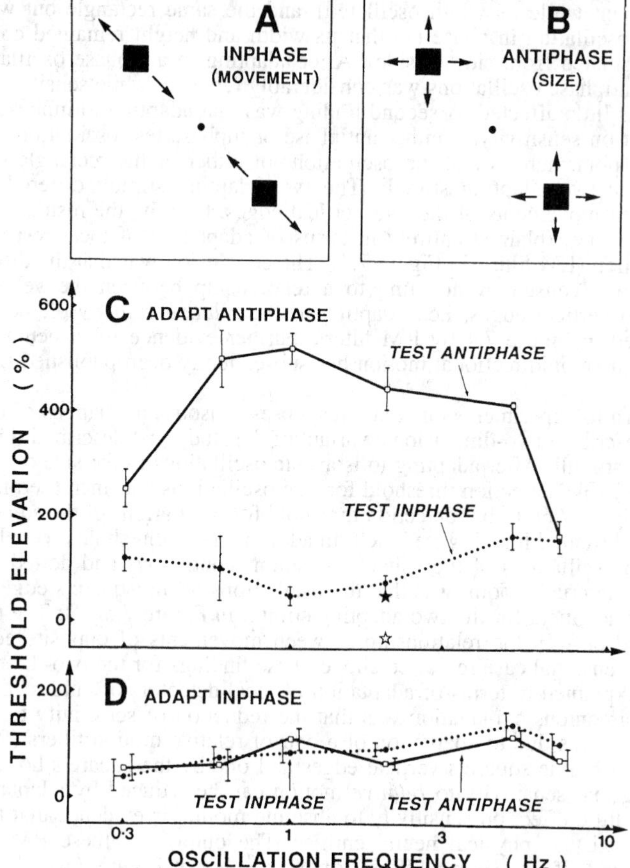

Figure 7.5 (A) and (B) the stimulus was two identical bright solid squares on either side of a fixation spot. Opposite edges oscillated either inphase (A) or in antiphase (B). (C) elevations of detection threshold produced by adapting to 2 Hz antiphase oscillations (ordinate) are plotted versus the oscillation frequency of the test stimulus (abscissa). Results for antiphase and inphase text stimuli are shown by the continuous and dotted line, respectively. (D) as in C but the adapting stimulus was 2-Hz inphase oscillation. [From Regan, D., and Beverley, K. I. (1978). Looming detectors in the human visual pathway. *Vision Research, 18,* 415–421. Reprinted with permission.]

than 5%) on detection threshold for the antiphase test stimulus whatever the amplitude of the inphase component of the adapting stimulus—even when the adapting stimulus contained 8 times more inphase component than antiphase component (48 vs. 6 arc min). Another way of stating this finding is that the eye can abstract the line-of-sight component of an object's motion independently of the object's trajectory. This finding implies that the changing-size mechanism responds linearly to the algebraic difference between the velocities of the square's opposite edges to within an accuracy of 5%.[6,7]

[6]Strictly accurate sensitivity to the algebraic difference in velocities is observed (to within 5%) only when the object's retinal image is subject to positional jitter of roughly the same amount as is produced by natural eye movements during free-head viewing. The accuracy is *less* when the subject maintains precise fixation with the head on a bite bar. This finding illustrates that noise does not necessarily degrade the performance of the human visual system.

[7]It has been stated incorrectly that the response of our proposed changing-size detectors is produced by a nonlinear combination of the responses of motion elements (Sekuler, 1992; Simpson, 1988). Subtraction of the velocities of opposite edges as proposed by Regan and Beverley obeys the principle of superimposition and is a linear process. However, as discussed below, an expanding retinal image

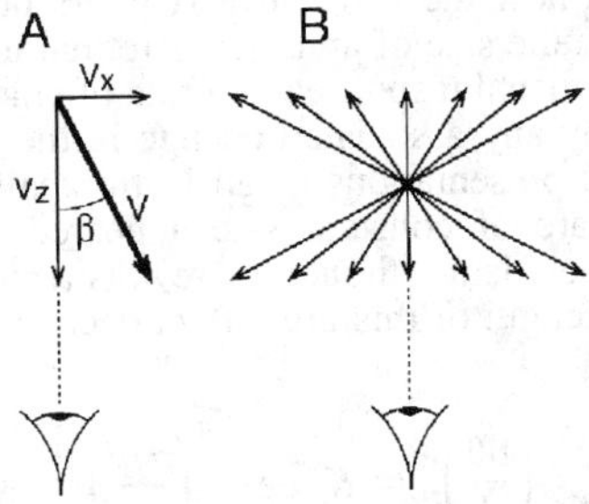

Figure 7.6 (A) Motion along an arbitrary direction of motion in depth is equivalent to the vector sum of a velocity component (V_x) along the x axis and a velocity component (V_z) along the z axis. In monocular vision, the V_x component of an untextured object's motion can be simulated by moving the opposite edges of the object's retinal image in the same direction at the same speed, and the V_z component can be simulated by moving the opposite edges of the object's retinal image in opposite directions at the same speed. (B) The trajectories shown have different components of motion parallel to the x axis, but identical components parallel to the z axis. [Modified from Regan, D., and Beverley, K. I. (1980). Visual responses to changing size and to sideways motion for different directions of motion in depth: linearization of visual responses. *Journal of the Optical Society of America, 11,* 1289–1296. Reprinted with permission.]

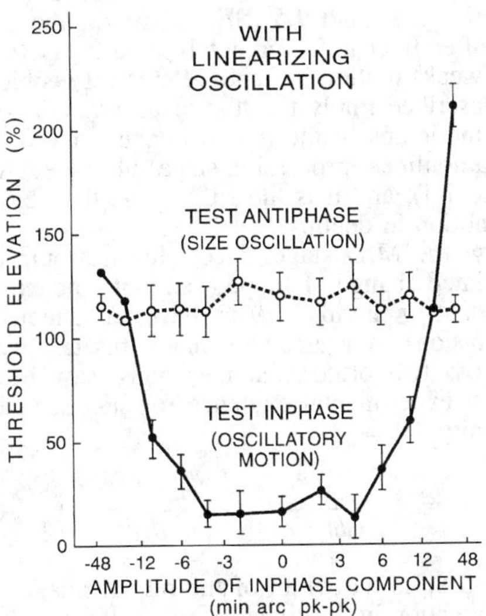

Figure 7.7 Threshold elevations produced by separately adapting to 11 different oscillations in depth, all of which had the same antiphase component of oscillation (isotropic oscillation of size), but different inphase components of oscillation as illustrated in Figure 6B. The dotted line plots elevations of threshold for detecting antiphase test stimulus oscillations. The continuous line plots elevations of threshold for detecting inphase test stimulus oscillations. Vertical bars show ±1 standard error. [Modified from Regan, D., and Beverley, K. I. (1980). Visual responses to changing size and to sideways motion for different directions of motion in depth: linearization of visual responses. *Journal of the Optical Society of America, 11,* 1289–1296. Reprinted with permission.]

---

does not produce a strong *motion-in-depth signal* unless expansion of the object's retinal image is isotropic (i.e., unless different meridia in the object's retinal image signal the same time to collision), and this processing is nonlinear.

So far, we have discussed how the LM and RM filters in Figure 7.4 can account for visual responses to changes in the size of an object's retinal image. We have not yet taken into account the findings that visual responses to a rate of change of the size of the retinal image are strongly affected by any associated change in the shape of the retinal image.

After adapting to repeated presentations of an isotropically expanding square, a subsequently presented test square of constant size appeared to be of constant size—no changing-size aftereffect was evident—though there was a strong illusion that the square was moving in depth. The strength of this aftereffect decayed exponentially according to the equation

$$\left(\frac{d\theta}{dt}\right)_{N} = K_{MD} \exp\left(\frac{-t}{\tau_{MD}}\right) \tag{1}$$

where $(d\theta/dt)_N$ was the rate of change of size required to null the instantaneous aftereffect t sec after the adaptation ceased. The decay time constant (τ_{MD}) was 24 (SE = 3), 30 (SE = 6), 34 (SE = 4), and 54 (SE = 3) sec for four observers.

After adapting to repeated presentations of a square whose expansion was not isotropic (the height was constant, and the width shrank), subjects reported that a test square of constant size appeared to be expanding in width. While this changing-size aftereffect remained visible, no motion-in-depth aftereffect was perceived. The strength of the changing-size aftereffect decayed exponentially according to the equation

$$\left(\frac{d\theta}{dt}\right)_{N} = K_{CS} \exp\left(\frac{-t}{\tau_{CS}}\right) \tag{2}$$

where $(d\theta,dt)_N$ was the rate of change of angular width required to null the instantaneous aftereffect t sec after adaptation ceased. The decay time constant (τ_{CS}) was 7.5 (SE = 0.7), 8.1 (SE = 0.8), 6.4 (SE = 0.4), and 9.5 (SE = 0.6) for the same four subjects. Only after the changing-size aftereffect had completely died away (at about 25 sec after adaptation ceased) was a (weak) motion-in-depth aftereffect visible.

These findings just described imply that the visual system compares the rates of expansion along different meridians of the retinal image.[8] It was proposed that this comparison is carried out at a nonlinear processing stage subsequent to the LM and RM filters (the MID stage in Figure 7.4), and it is the MID stage that, by generating an output z, causes a perception of motion in depth.

Next, we discuss how the MID stage selects for isotropic expansion. The test for expansion with no associated change of shape (i.e., isotropic expansion) is that the value of $\theta/(d\theta/dt)$ is independent of meridian (θ is the instantaneous size and $d\theta/dt$ the instantaneous rate of expansion). A rough version of this test would be to compare the values of $\theta/(d\theta/dt)$ across two orthogonal meridians, say the horizontal and vertical meridians. There is direct experimental evidence for such a neural process. The neural process tests for the identity

$$\frac{\theta_V}{(d\theta_V/dt)} = \frac{\theta_H}{(d\theta_H/dt)} \tag{3}$$

where θ_V is the height and θ_H is the width of the retinal image. In the format of Figure 7.4, we can satisfy this equation by writing

$$k_1\,(a - b) = k_2\,(c - d) \tag{4}$$

and

$$k_1 = \frac{K}{\theta_1} \tag{5}$$

and

[8]By meridian we mean the angular orientation of a great circle of the eye containing the line of sight. For brevity we will refer to the coplanar great circle of the left and right eyes as the horizontal meridian, and the perpendicular meridian as the vertical meridian (Cline et al., 1989, p. 432).

$$k_2 = \frac{K}{\theta_2} \tag{6}$$

where K is a constant.

As mentioned already, there is evidence that, provided that Equation (3) is satisfied, the MID stage produces a motion in depth signal z. If we assume that the amplitude of this signal is equal to the amplitude of any one of the several equal inputs, i.e.,

$$z = K\,\theta_V/(d\theta_V/dt) = K\,\theta_H/(d\theta_H/dt), \quad \text{etc.} \tag{7}$$

then

$$z = \frac{K}{T} \tag{8}$$

where T is the time to collision and K is a constant. In words, "the magnitude of the motion-in-depth output of the MID filter is inversely proportional to time to collision." This hypothesis would make sense in terms of visually guided motor action. For example, in highway driving, the larger the output of the local MID filter, the more urgent the demand for evasive action. Again, when walking, driving, or flying through a space containing several objects, locomotion could be guided by the strategy of continuously adjusting the direction of motion so as to prevent the output of any local MID filter from rising above a safe level.

The hypothesis expressed in Equation (8) leads to the prediction that altering the perceived speed of motion in depth produced by a given looming stimulus should change the estimated time to collision. This prediction is fulfilled when the effectiveness of a looming stimulus is changed by mismatching texture dynamics (Vincent and Regan, 1996; Section 7.7.1). A second prediction—that adaptation to looming should change the estimated time to collision—has not yet been tested.

We suggested that intersubject differences in visual sensitivity to looming might be used to predict intersubject differences of performance in certain flying tasks where visual sensitivity to expansion of the retinal image is important. This idea was tested in two studies, one using a flight simulator and the other using telemetry-tracked real aircraft. Pilots first carried out a laboratory test that measured the ability to track changes in the size of a square in the situation that the square's location on the display monitor was randomly jittered. The purpose of this jitter was to check the degree to which visual sensitivity to changing size was independent of translational motion of the retinal image. The rationale of the jitter was that, if there was appreciable crosstalk between visual responses to changing size and to translational motion, then the turbulence and vibration encountered in a real high-performance aircraft would jitter the retinal image and thereby degrade the accuracy of judgments based on changing size. Laboratory measurements of visual sensitivity to image expansion were found to correlate with flying performance in low-level flight, close maneuvering, and formation flight (Kruk, Regan, Beverley and Longridge, 1981, 1983; Kruk and Regan, 1983).

The changing-size square used in the field research just described was untextured bright-on-dark. In a separate laboratory study, we investigated the effects of texturing the square (Beverley and Regan, 1983). When the rate of change of magnification of the texture exactly matched the ratio of expansion of the square (as in a real-world situation) the drive to the human motion-in-depth system was stronger than when the square was untextured, but only slightly stronger. But when the rate of expansion of texture was less than the rate of expansion of the square, the motion-in-depth system was stimulated much more weakly. This laboratory finding suggests that judgments of time to collision with a simulated approaching object might be inaccurate unless the rate of expansion of the object's texture elements matched the rate of expansion of the object's size. A laboratory test of this prediction is reviewed in Section 7.7.1.

7.4.2 The Detection of Motion in Depth Using Binocular Retinal Image Information Alone

In the contexts of aviation and highway driving, it is often assumed that binocular retinal image information is of negligible importance because the viewing distances involved are usually considerably larger than the few meters over which the visual perception of relative depth is supported by binocular disparity. The fallacy in this argument is that it does not take into account the distinction between the binocular visual processing of static depth and the binocular visual processing of motion in depth. The distinction is twofold. First, while the relative disparity of an object depends on the square of the viewing distance, its rate of change of disparity involves an additional factor—its velocity along a line passing between the eyes [Equation (11)]. In Equation (11), speed is pitted against viewing distance: A reduction in sensitivity produced by an increase of viewing distance can to some extent be countered by increasing the object's speed. Second, an individual's visual sensitivity to changing disparity cannot be directly predicted from that same individual's visual sensitivity to static disparity.[9]

In comparing the importance of binocular and monocular cues, it should also be remembered that the monocular correlate of an object's motion in depth (i.e., $d\theta/dt$) grows relatively weaker in comparison with the binocular cue ($d\delta/dt$) as the linear size of the object is decreased. This can be understood as follows. In Figure 7.8A,B, an object of width $2s$ is at distance D at time $t = 0$ and distance $(D - \Delta D)$ at time $t = \Delta t$). The object is moving at instantaneous speed V_z along a straight line that passes through point C, midway between the eyes. For convenience, the resulting changes in angular subtense (i.e., $\Delta\theta = \theta_2 - \theta_1$) and binocular disparity ($\Delta\delta$) are depicted separately in Figure 7.8. It was shown previously that the relation

$$\frac{(d\theta/dt)}{(d\delta/dt)} \approx 2s/I \tag{9}$$

can be derived by combining the following two equations

$$\frac{d\theta}{dt} \approx \frac{2s}{D^2}\left(\frac{dD}{dt}\right) \tag{10}$$

and

$$\frac{d\delta}{dt} \approx \frac{I}{D^2}\left(\frac{dD}{dt}\right) \tag{11}$$

where I is the observer's interpupillary separation and $D \gg I$ and $D \gg s$ (Regan and

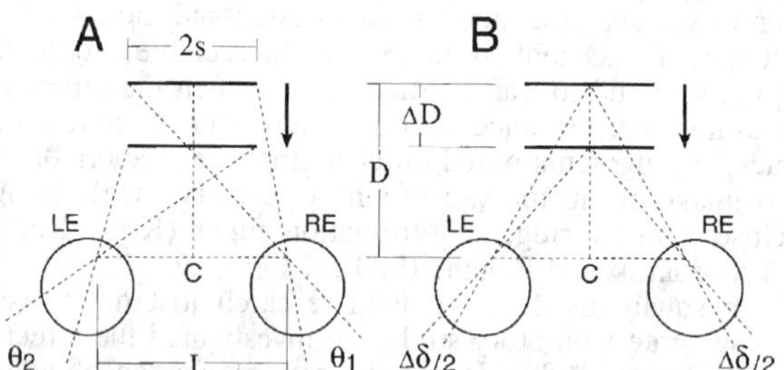

Figure 7.8 LE, RE: Left eye, right eye. Point C is midway between the eyes. An object of width $2s$ is moving along a straight line towards point C midway between the eyes. At time t the object is at a distance D from point C. A short time later at time $(t + \Delta t)$ the object is at a distance $(D - \Delta D)$ from point C. The change of disparity during interval Δt is $\Delta\delta$ and the change in angular subtense is $(\theta_2 - \theta_1)$. For convenience the change in angular subtense is depicted in A and the change in disparity in B. [Modified from Regan, D., and Beverley, K. I. (1979). Binocular and monocular stimuli for motion in depth: Changing-disparity and changing-size feed the same motion-in-depth stage. *Vision Research, 19,* 1331–1342. Reprinted with permission.]

[9]An extreme illustration of this point is that some individuals with normal stereoacuity for static disparity are "blind" to changing disparity (Hong and Regan, 1989; Richards and Regan, 1973). A proposed explanation is that static disparity and changing-disparity are processed by different cortical neurons (Cynader and Regan, 1978; Spileers, Orban, Gulyas, and Maes, 1990).

Beverley, 1979a). Counterintuitively, the object's distance D does not enter into Equation (9), though its absolute linear width ($2s$) does.

The weakness of the monocular correlate of motion in depth grows even more pronounced than indicated by Equation (9) when the approaching sphere's angular subtense is very small. This is because, for a sphere of very small angular subtense, retinal image size is little affected by the sphere's distance; in this situation, Equation (9), an equation based on geometrical optics, is rendered invalid by a fact of physical optics (Ditchburn, 1976). Thus, it is not necessary to go as far as the limiting case of a point object for the $d\theta/dt$ monocular correlate of motion in depth to vanish leaving only the $d\delta/dt$ binocular cue.

Whether or not a given individual can in practice actually use the binocular retinal image information about an object's motion in depth depends on the object's velocity as well as the viewing distance, and also on that individual's sensitivity to rate of change of binocular disparity.

The temporal tuning of the stereomotion system for suprathreshold disparity oscillations was measured as follows. A stationary stimulus bar was placed 1° to the right of the fixation point, and a reference bar was placed 1° to the left. The disparity of the stimulus bar was oscillated sinusoidally about zero. First, the reference bar was placed 20 min arc in front of the fixation plane (curve marked 20 in Figure 7.9). The oscillation frequency of the stimulus bar was set at 1.0 Hz, and the amplitude of oscillation adjusted until the nearest depth of the stimulus bar matched the depth of the reference bar. The experiment was repeated then for a series of oscillation frequencies (filled symbols). Then the entire experiment was repeated with the reference bar set at 10, 5, and 2.5 min arc disparity (Regan et al., 1995). Figure 7.9 shows that, for suprathreshold oscillations, the stereomotion system has a lowpass characteristic with a half-sensitivity point of approximately 2 Hz.

The suprathreshold data shown in Figure 7.9 are unequivocal in the sense that observers were confident that they were matching depth, and the depth percept disappeared when one eye was closed. To obtain the stereomotion system's attenuation characteristic for threshold oscillations is less straightforward because, at near-threshold amplitudes of disparity oscillation, the percepts seen in monocular and binocular view are qualitatively similar. To circumvent this potential problem, threshold data were obtained by stimulating the left eye with an F_1 Hz oscillation and the right eye with an F_2 Hz oscillation. This "beat" technique assigns different "signatures" to monocular and binocular motion information. In particular, oscillations of frequency ($F_1 - F_2$) Hz could only exist after convergence of signals from left and right eyes (Regan et al., 1995). The threshold characteristic obtained in this way resembled the suprathreshold characteristic shown in Figure 7.9. In this "beat" experiment, it was quite evident that disparity oscillations of 3–4 Hz could not be seen in the situation that monocular oscillations were quite evident (e.g., with $F_1 = 14$ Hz and $F_2 = 18$ Hz). This is because the effect of frequency on the

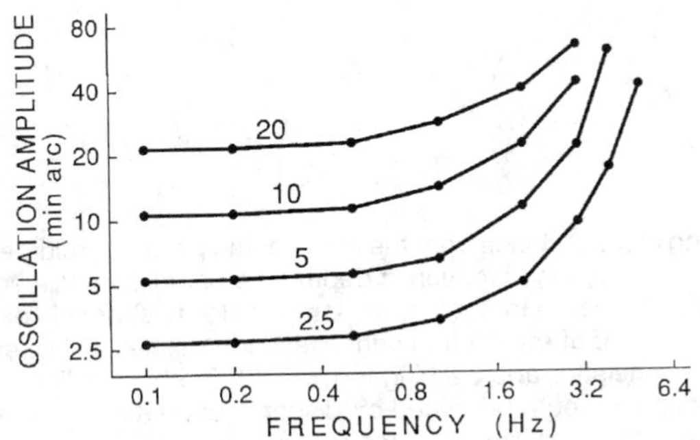

Figure 7.9 Temporal characteristic for suprathreshold sinusoidal oscillations of disparity. Ordinates plot disparity oscillation amplitudes that produce a constant perceived amplitude of depth oscillation. The test bar oscillated about the fixation plane. A stationary matching bar was located 20, 20, 5, and 2.5 min arc disparity in front of the fixation plane (see numbers on the curves). (Regan, D., and Beverley, K. I. (1973). Some dynamic features of depth perception. *Vision Research, 13,* 2369–2379. Reprinted with permission.]

perception of disparity oscillations is very different from its effect on the perception of oscillations in the frontoparallel plane. The latter characteristic is bandpass with a peak sensitivity at 1–4 Hz and extends to 20–25 Hz (Tyler and Torres, 1972; Regan et al., 1995).

There is psychophysical evidence that the binocular stimulus for motion in depth perception is a *rate of change of relative disparity* rather than a rate of change of absolute disparity. In particular, motion-in-depth sensation collapses when no reference marks are visible in the visual field (for a spatially extended target), even when oscillations of absolute disparity are large. This failure of motion-in-depth perception is not due to ocular tracking. Interestingly, an unreferenced disparity change drives vergence eye movements quite effectively, even though it fails to generate any percept of motion in depth (Erkelens and Collewijn, 1985a,b; Tyler, 1975).

A secondary question is whether motion-in-depth perception is possible in the absence of monocularly visible motion. In everyday life, the motion of a real object produces monocularly visible motion as well as disparity change, so a special kind of stimulus is needed to address this question. Dynamic random noise, introduced by Julesz, is such a stimulus. He noted that the perception of motion in depth can be produced by changing the disparity of a target that cannot be seen monocularly (Julesz, 1971, p. 184). This conclusion has been confirmed by several other authors (for example, Cumming and Parker, 1994; Norcia and Tyler, 1984; Regan et al., 1995). Motion thresholds for stereo motion generated by monocularly invisible and monocularly visible targets were compared by Gray and Regan (1995a,b), who reported that detection thresholds for 0.75 Hz oscillations in depth for the cyclopean target were 2.65 (SE = 0.17), 2.52 (SE = 0.11), and 3.24 (SE = 0.29) min arc disparity peak-to-peak for three subjects. Each measurement was repeated 18 times. Corresponding data for the noncyclopean target were 2.61 (SE = 0.12), 2.74 (SE = 0.12), and 2.98 (SE = 0.31) min arc for the same three subjects, again with 18 repeats for each measurement. For none of the three subjects did the threshold for the cyclopean target differ significantly from the threshold for the monocularly visible target. Thus, these data provided no support for the hypothesis that monocularly visible motion provides any contribution to the detection of motion in depth. Cumming and Parker (1994) have reached similar conclusions using a different method.

The broken line in Figure 7.10B indicates that detection sensitivity for disparity oscillations is *low* for the direction passing midway between the eyes, and increases for more oblique directions up to at least 10°. It has been proposed that disparity changes produced by an approaching object are processed through four channels, each tuned to a different range of directions of motion in depth (Figure 7.10A). The evidence for this proposal is as follows: (1) Detection sensitivity for oscillatory motion in depth is depressed by adapting to oscillations of motion in depth, and (2) the magnitude of the elevation depends on the directions in three-dimensional space of both adapting and test

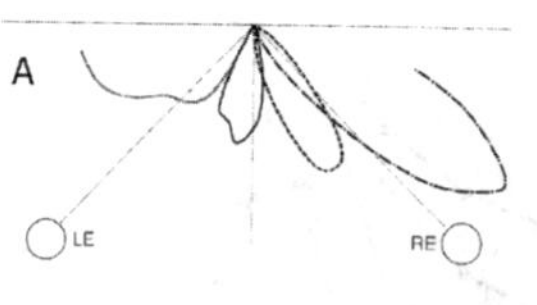

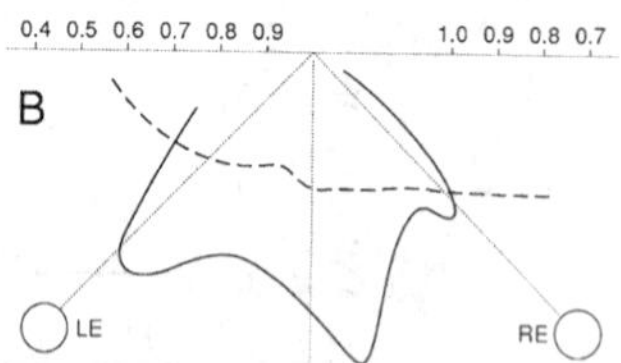

Figure 7.10 Increasing radial distance from the origin of the polar coordinates denotes increasing sensitivity. Azimuth is the direction of motion in depth. The particular scale chosen (azimuth angle proportion to the ratio of the left and right retinal image velocities) greatly exaggerates the difference between the line of sight of the left and right eyes (dotted lines). (a) Directional tuning curves for the stereomotion channels for approaching motion. (b) The broken line plots detection sensitivity for oscillatory motion in depth produced by disparity oscillations. The continuous line plots discrimination sensitivity for the direction of motion in depth. The right-hand numbers give the radial calibration of detection threshold in min arc. The left-hand numbers give the radial calibration of directional discrimination threshold in min arc. [Data in (a) are replotted from Beverley and Regan (1975); data in (b) are replotted from Beverley and Regan (1975). From Regan, D., 1992, Visual judgements and misjudgements in cricket, and the art of flight. *Perception, 21,* 91–115. Reprinted with permission.]

oscillations (Regan et al., 1995). Estimated sensitivity curves for the four channels are shown in Figure 7.10A. Two channels process directions passing between the eyes, while the remaining very large range of directions is left to the other two channels.

Ocular convergence was held constant in this experiment. It remains to be shown whether the results hold in the situation that vergence tracking is allowed.

7.4.3 The Incidence of Stereomotion Blindness Among Normally Sighted Individuals

A substantial portion of normally sighted individuals have areas of the binocular visual field that are selectively insensitive, or even blind, to motion in depth. These stereomotion field defects were first reported for oscillating disparity (Richards and Regan, 1973), but more recently have been found for unidirectional motion in depth also (Hong and Regan, 1989).

Of the six subjects whose binocular stereomotion fields for oscillating disparity were published up to 1988, five showed clear visual field defects (stereomotion scotomata) that varied in size from about 1° across to more than an entire quadrant (Regan, Erkelens, and Collewijn, 1986; Richards and Regan, 1973). In a later study, the database was extended to a further 21 subjects, of whom 8 out of 21 had field defects, and only 6 had full symmetrical fields extending to at least 10° eccentricity for both crossed and uncrossed disparities (Hong and Regan, 1989).

7.5 DISCRIMINATING THE DIRECTION OF AN OBJECT'S MOTION IN DEPTH

7.5.1 Discriminating the Direction of Motion in Depth Using Monocular Cues Alone

Considering how widely it is known that some one-eyed gamesplayers and aviators have reached the highest levels of achievement (Bose, 1990; Mohler and Johnson, 1971), it is surprising that formal psychophysical studies of monocular discrimination of the direction of motion in depth are so sparse.

To avoid complications caused by the object's rotation, we first consider the special case of a rigid spherical object. As illustrated in Figure 7.6A,B, different directions of an object's motion in depth can be simulated monocularly by moving the opposite edges of the object's retinal image with a combination of inphase and antiphase motion (the inphase component simulates the V_x component of velocity and the antiphase component simulates the V_z component of velocity). Referring to Figure 7.11A,B, two monocularly available correlates of the object's direction of motion in depth can be written as follows:

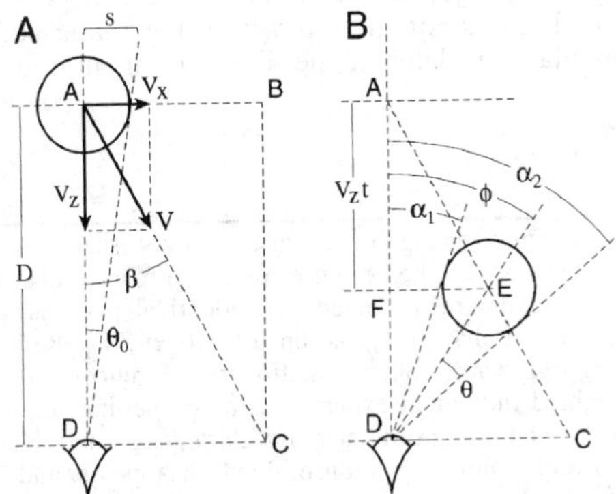

Figure 7.11 (A) A sphere of radius s whose center is at A moves at constant speed V along a straight line toward C. D is the first nodal point of the eye, and angle ADC is 90°. Distance DC is equal to ns where n is a scaling factor. The situation is shown at time $t = 0$. (B) At time t the sphere's center has reached point E. [From Regan, D., and Kaushal, S. (1994). Monocular judgment of the direction of motion in depth. *Vision Research, 34*, 163–177. Reprinted with permission.]

$$ns \approx \frac{(d\phi/dt)}{(d\theta/dt)} \tag{12}$$

and

$$ns \approx \frac{[1 + (d\alpha_1/dt)/(d\alpha_2/dt)]}{[1 - (d\alpha_1/dt)/(d\alpha_2/dt)]} \tag{13}$$

where $d\phi/dt$ is the translational velocity of the retinal image, $d\theta/dt$ is its rate of expansion and $d\alpha_1/dt$ and $d\alpha_2/dt$ are the velocities of opposite edges of the retinal image (Bootsma, 1991; Regan et al., 1995).

In Equations (12) and (13), we express the direction of motion in depth in terms of the distance *DC* in Figure 7.11, and we express the distance *DC* as a multiple (n) of the sphere's radius (s). Although Equations (12) and (13) are mathematically equivalent, it does not necessarily follow that they are equivalent physiologically. We further note that, although the value of the object's absolute size ($2s$) is required to estimate its direction of motion on the basis of monocular information alone, Equation (9) indicates that the value of s can, in principle, be obtained from binocular retinal image information.

We measured monocular discrimination threshold for the direction of motion in depth by generating a bright square whose expansion and translational motions exactly mimicked the retinal image of a rigid object moving along a straight line at constant speed (Regan et al., 1995). When measuring discrimination threshold for objects passing to one side of the eye, the reference stimulus and all the test stimuli passed to the same side of the eye, thus removing the direction of translational motion as a cue to the direction of movement in depth. We removed the speed of translational motion as a reliable cue to the direction of motion in depth by randomizing the magnitudes of $d\phi/dt$ and $d\theta/dt$ in a yoked manner so as to keep the ratio $(d\phi/dt)/(d\theta/dt)$ constant. Subjects were able to discriminate the simulated direction of motion in depth even though the only cue to the task was the ratio $(d\phi/dt)/(d\theta/dt)$. Threshold was remarkably acute. The best value of discrimination threshold Δn ranged from 0.2 to 0.4. In our situation (square side length of 0.7 cm starting at 143 cm), this is equivalent to an angular discrimination threshold $\Delta\beta$ for the direction for motion in depth (β in Figure 7.11A) of 0.03° to 0.06°. Threshold was approximately the same for the vertical, horizontal, and oblique meridian out to at least $n = 4$.

7.5.2 Discriminating the Direction of Motion in Depth Using Binocular Cues Alone

In Figure 7.12, we consider a point object at A moving along the direction shown by the thick arrow. A stationary point object at (arbitrary) location B provides a reference.[10] The disparity of A relative to B is δ. For motion within the plane containing left and right eyes, there are two binocular correlates of the direction of motion in depth. They can be written as follows:

[10]As noted in Section 7.4.2, a rate of change of disparity is a necessary, but not a sufficient stimulus for the perception of motion in depth: One or more reference marks must be visible. For example, suppose that the left and right eyes are presented with identical patterns, and the two patterns are oscillated from side to side in antiphase (e.g., when the left eye's pattern is moving leftward, the right eye's pattern is moving rightward). When a stationary reference mark or background is visible to both eyes, a normally sighted individual experiences a compelling impression that the pattern is moving in depth relative to the stationary reference mark or background. However, if the background or landmark is removed, all impression of motion in depth disappears, and the pattern appears to be stationary, even though the retinal images on the two eyes are moving considerably. Observers report that, although a large amount of motion is visible when one eye is closed, no motion is seen when both eyes are open. This dramatic effect is not due to the eyes tracking the moving images: The rate of change of absolute retinal disparity can be large without producing any impression of motion. Furthermore, vergence eye movements have no effect on visual sensitivity to motion in depth (Regan et al., 1995).

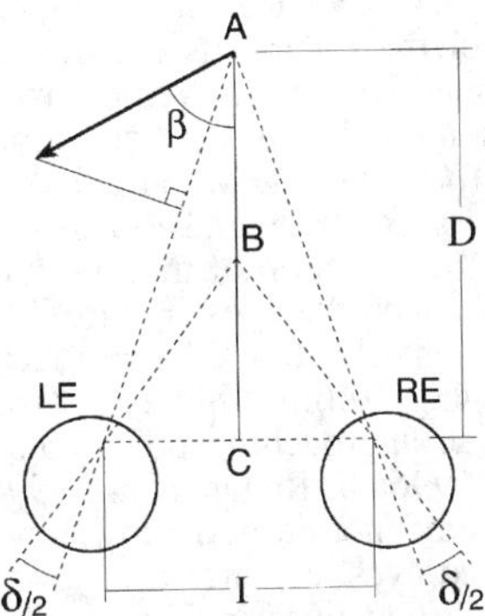

Figure 7.12 Geometry of binocularly viewed motion in depth. LE, RE: left eye, right eye. The heavy arrowed line represents the velocity of an object at point A moving in a straight line at constant speed V. Point C is midway between the eyes. Angle δ is the disparity of point A relative to stationary reference mark B. [From Regan, D., Hamstra, S. J., Kaushal, S., Vincent, A., Gray, R., and Beverley, K. I. (1995). Visual processing of an object's motion in three dimensions for a stationary or a moving observer. *Perception, 24,* 87–103. Reprinted with permission.]

$$\beta \approx \tan^{-1}\left[\frac{I\{(d\phi/dt)_R + (d\phi/dt)_L\}}{2D(d\delta/d)}\right] \approx \tan^{-1}\left[\frac{I(d\phi/dt)}{D(d\delta/dt)}\right] \tag{14}$$

and

$$\beta \approx \tan^{-1}\left(\frac{I\{[(d\phi/dt)_R/(d\phi/dt)_L] + 1\}}{2D\{[(d\phi/dt)_R/(d\phi/dt)_L] - 1\}}\right) \tag{15}$$

where $(d\phi/dt)_R$ and $(d\phi/dt)_L$ are, respectively, the translational velocities of the left and right retinal images, $d\phi/dt$ is the translational velocity of the binocularly fused retinal image, $d\delta/dt$ is the rate of change of disparity, I is the interpupillary separation, and D is the viewing distance (Regan et al., 1995).[11]

In experimental studies of binocular judgments of the direction of motion in depth it is important to ensure that observers use binocular cues rather than monocular cues. This can be checked by using a multifactor design so that, after the experiment is completed, a statistical analysis can be used to confirm that the observer based all responses on the designated cue and effectively ignored all other cues (Portfors-Yeomans and Regan, 1997a,b). This approach was also used to show that observers can perfectly unconfound and discriminate simultaneous variations in the direction and the speed of motion in depth.

Although Equations (14) and (15) are equivalent mathematically, they are not necessarily equivalent physiologically. At the physiological level, it is in principle possible that, for motion within the horizontal meridian, the direction of motion in depth is encoded binocularly entirely in terms of $(d\phi/dt)/(d\delta/dt)$, or entirely in terms of $(d\phi/dt)_R/(d\phi/dt)_L$, or in terms of some combination of these two binocular correlates of the direction of motion in depth.

For motion in depth contained within the vertical meridian, the value of $(d\phi/dt)_R/(d\phi/dt)_L$ provides no correlate to the direction of motion in depth. Therefore, binocular discrimination of the direction of motion in depth in the vertical meridian must be based entirely on discrimination of the ratio $(d\phi/dt)/(d\delta/dt)$. For some subjects, there is no difference between discrimination of the direction of motion in depth within the horizontal

[11]In Equation (14), the ratio of velocities $(d\phi/dt)/(d\delta/dt)$ is confounded with the ratio of displacements $(\phi_1 - \phi_2)/(\delta_1 - \delta_2)$, where ϕ_1 and ϕ_2 are the values of ϕ at the start and end of the movement, and δ_1 and δ_2 are the values of δ at the start and end of the movement. Similarly, in Equation (15) the ratio of velocities $(d\phi/dt)_R/(d\delta/dt)_L$ is confounded with the ratio of displacements during the movement $[(\phi_1 - \phi_2)_R/(\phi_1 - \phi_2)_L]$.

and vertical meridians (Yeomans and Regan, 1995; Portfors-Yeomans and Regan, 1997a,b). This suggests that, for these subjects the direction of motion in depth within the horizontal meridian is encoded largely or even entirely in terms of $(d\phi/dt)/(d\delta/dt)$. Other evidence suggests that some subjects who are comparatively insensitive to $d\delta/dt$ can learn to use the $(d\phi/dt)_R/(d\phi/dt)_L$ cue when it is available (i.e., for motion within the horizontal meridian) (Portfors-Yeomans and Regan, 1997a).

Binocular discrimination of the direction of motion in depth is most acute for a direction passing close to midway between the eyes (Figure 7.10B, continuous line): The just-noticeable difference in direction is 0.15° to 0.22° for a viewing distance of roughly 1.5 m. Discrimination threshold is roughly 10 times higher for trajectories inclined at 10° to this line, i.e., trajectories that take an approaching point object wide of the observer's head (Portfors-Yeomans and Regan, 1996a,b; Regan et al., 1995). As mentioned already, binocular sensitivity to the presence of motion in depth follows an inverse pattern, being low for direction passing between the eyes and increasing for directions passing wide of the eyes (compare the continuous dashed lines in Figure 7.10B). This inverse effect can be understood in terms of the idea, illustrated in Figure 7.10A, that there are several binocular neural mechanisms, each of which prefers a different direction of motion in depth. If binocular direction-in-depth discrimination is determined by the relative activity of these mechanisms, then directional discrimination can be modeled in opponent process or line element terms as suggested by Beverley and Regan (1975). To a first approximation, this "relative activity" type of hypothesis predicts that directional discrimination would be best along the approximate direction where the sensitivity curves of adjacent channels show the greatest difference in slope. This direction is approximately midway between the eyes.

An experimental demonstration that the direction of motion in depth can be encoded entirely in terms of $(d\phi/dt)/(d\delta/dt)$ for motion within the meridian that contains the eyes was obtained by viewing a dynamic random noise pattern of the kind devised by Julesz (1971). The pattern contained a cyclopean rectangle i.e., a rectangle that was clearly evident in binocular view, but was perfectly camouflaged when viewed by the left or right eye alone. Thus, no monocular motion signals were available. Nevertheless, large changes in the perceived direction of motion in depth of the rectangle could be produced by altering the ratio $(d\phi/dt)/(d\delta/dt)$ (Regan et al., 1995). A subsequent study explored just-noticeable differences in the perceived direction of the motion in depth of a cyclopean rectangle (Portfors-Yeomans and Regan, 1997b). By using a multifactor design and subjecting the data to multiple regression analysis, it was confirmed that subjects based their directional discriminations entirely on the variable $(d\phi/dt)/(d\delta/dt)$, and totally ignored simultaneous variations in $(d\phi/dt)$, $(d\delta/dt)$, disparity excursion and presentation duration. The just-noticeable difference in the direction of motion in depth was significantly higher for cyclopean than for monocularly visible targets, but the difference was not great (1.1 to 2.5 times higher) and might, at least in part, have been explained by the lower visibility of the cyclopean target. By altering the viewing distance from 0.6 m to 6.0 m, it was shown that discriminations were based entirely on retinal image variables. In particular, the ocular convergence angle [and hence viewing distance D in Equations (14) and (15) had no effect on the observer's discriminations].

7.6 DISCRIMINATING THE SPEED OF AN OBJECT'S MOTION IN DEPTH

7.6.1 Discriminating the Speed of Motion in Depth Using Monocular Cues Alone

Equation (10) can be rewritten as

$$V_z \approx \frac{D^2}{2s}\left(\frac{d\theta}{dt}\right) \tag{16}$$

where D is the distance and s the radius of the sphere in Figure 7.11A and $2(d\theta/dt)$ the angular rate of expansion of the sphere's retinal image (Regan and Beverley, 1979a). In practice, however, V_z cannot be estimated more accurately than can absolute distance D, and it seems to be the case that, when D is greater than a few m, its absolute value can be estimated only crudely (Collewijn and Erkelens, 1990).

On the other hand, and already as discussed in Section 7.4.1, it has been proposed that the perceived speed of the illusory motion in depth produced by isotropically expanding an object's retinal image is determined by time to collision (T) in Equation (18) rather than by V_z.

7.6.2 Discriminating the Speed of Motion in Depth Using Binocular Cues Alone

When a rigid sphere approaches a point midway between the eyes, its two retinal images move in opposite directions at the same speed, thus producing a rate of change of binocular disparity (Figure 7.8B). Evidence that the human visual pathway contains a neural mechanism sensitive to the rate of change of disparity includes the effect of adapting to changing disparity in normally sighted observers, and the finding that, in some individuals, areas of the visual field that have normal discrimination for static disparity and to translational motion are blind to changing-disparity (Hong and Regan, 1989; Regan, Erkelens, and Collewijn, 1986; Regan et al., 1995; Richards and Regan, 1973).

Equation (11) can be rewritten as

$$V_z \approx \frac{D^2}{I}\left(\frac{d\delta}{dt}\right) \tag{17}$$

For distant objects, the absolute speed (V_z) cannot be estimated accurately because, beyond a few m, the value of D can be estimated only crudely (Collewijn and Erkelens, 1990). On the other hand, a difference between two values of V_z can be discriminated with considerable precision; discrimination threshold can be as low as 7%—comparable with speed discrimination threshold for frontal plane motion—and trial-to-trial variations in the speed of motion in depth can be almost perfectly dissociated from simultaneous trial-to-trial variations in the direction of motion in depth (Portfors-Yeomans and Regan, 1995a,b; Yeomans and Regan, 1995).

The studies just cited were restricted to monocularly visible targets. A comparison of just-noticeable differences in the speed of motion in depth for cyclopean and monocularly visible targets showed them to be the same (0.10–0.13 Weber fraction). Both also were independent of the direction of motion over a range of directions between a line passing midway between the eyes to a direction 12° from that line in the horizontal or vertical meridians (Portfors-Yeomans and Regan, 1995b).

7.7 JUDGING THE TIME TO COLLISION WITH AN APPROACHING OBJECT

7.7.1 Accuracy and Precision of Time to Collision Judgments Using Monocular Cues Alone

As an object on a collision course approaches a pilot, a driver, or a baseball player the retinal image of the object expands continuously (Figure 7.8A). The rate of expansion does not, however, signal the time to collision: A small object's retinal image would expand at a lower rate of degrees per second than a large object's retinal image, even though they were at the same distance from the eye and moving with the same time to collision. A retinal image correlate of time to collision is, however, available. The distinguished astronomer Fred Hoyle (1957) pointed out that if an object is approaching along the line of sight, then

$$T \approx \theta/(d\theta/dt) \tag{18}$$

where T is the time to collision, and θ is the angle subtended at the eye by the object.[12]

Following Lee (1976), several authors have suggested that humans use the geometrical fact expressed in Equation (18) in everyday life, for example, in sport, highway driving, and aviation. The participants in some studies had access to monocular information only (DeLucia, 1991; Kruk et al., 1981; Lee and Lishman, 1977; McLeod and Ross, 1985; Regan and Vincent, 1995; Schiff and Detwiler, 1979; Sekuler, 1992). In other studies, however, although the discussion of results focused entirely on the monocular information expressed in Equation (18), valid binocular information was available, and it is difficult to be sure that observers ignored this binocular information (Bootsma and van Wieringen, 1990; Kruk and Regan, 1983; Lee, Lishman, and Thomson, 1982; Lee, Young, Reddish,

[12]Providing that the object is moving at constant speed, is rigid, and (except for in some special cases, e.g., a sphere) is nonrotating, and the value of θ is small (less than about 10°).

Lough, and Clayton, 1983; Savelsbergh, Whiting, and Bootsma, 1991; Schiff and Detwiler, 1979; Warren, Young, and Lee, 1986).

There are four issues here which I will discuss in the rest of this section. First, to what extent are the precision and accuracy of time to collision estimates affected by the object's angular size and rate of expansion? Second, how *precisely* can time to collision be estimated, and in particular what is the just-noticeable difference in time to collision? Third, with what *absolute accuracy* can time to collision be estimated? Fourth, what visual situations lead to failure in the accuracy and precision of estimating time to collision?

If humans are able to use Equation (18) to judge time to collision with an approaching object when they have no prior knowledge of the object's size, they must be able to distinguish between trial-to-trial variations in the ratio $\theta/(d\theta/dt)$ and simultaneous trial-to-trial variations in the values of θ and/or $d\theta/dt$. We tested this assumption in the following way. We simulated an approaching object by presenting a bright solid square on a monitor and arranging that the rate of expansion of the square was isotropic, and exactly mimicked the retinal image produced by a rigid object approaching at constant speed along the line of sight. Observers were presented with 64 different combinations of time to collision, rate of expansion, and starting size in random sequence, and instructed to look directly at the square. The 64 test stimuli were organized in an 8×8 square array so that the ratio $\theta/(d\theta/dt)$ varied along the x axis of the array, but rate of expansion $(d\theta/dt)$ was constant along the x axis. Rate of expansion varied along the y axis of the array, but the ratio $\theta/(d\theta/dt)$ was constant along the y axis. Starting size varied in the same way along both x and y axes. Test stimuli were presented in random order.

The observer's task was to signal whether the time to collision was sooner or later than the mean of the 64 test stimuli. Observer's responses were stored in an 8×8 response array. The response array was analyzed in two ways. First, in Figure 7.13A all responses were collapsed onto the x axis, thus creating a plot of "sooner" responses versus the task-relevant variable—time to collision $[\theta/(d\theta/dt)]$—with time to collision expressed relative to the mean rate of expansion of the set of 64 stimuli (1.0 on the ordinate signifies the mean time to collision). Second, in Figure 7.13B all responses were collapsed onto the y axis, thus creating a plot of "sooner" responses versus the task-irrelevant variable—rate of expansion $(d\theta/dt)$—with rate of expansion expressed relative

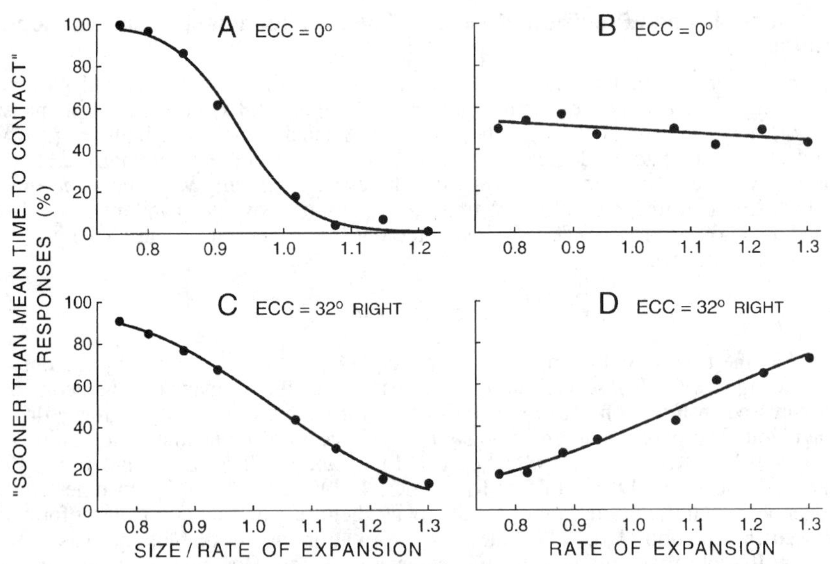

Figure 7.13 Psychometric functions for simulated approaching target viewed foveally (0° eccentricity, A, B) or at 32° eccentricity in the peripheral field (C, D). In A and C, subject responses were analyzed with respect to time to collision (i.e., size/rate of expansion), and in B and D re-analyzed with respect to rate of expansion. See text for details. [Modified from Regan, D., and Vincent, A. (1995). Visual processing of looming and time to contact throughout the visual field. *Visual Research, 35,* 1845–1857. Reprinted with permission.]

to the mean rate of expansion of the set of 64 stimuli (1.0 on the ordinate signifies the mean rate of expansion). If the observer had based responses on trial-to-trial variations in time to collision and ignored variations in both rate of expansion and starting size, then Figure 7.13A would be a steep psychometric function and Figure 7.13B would be flat. If the observer had based responses on trial-to-trial variations in rate of expansion and ignored variations in both time to collision and starting size, then Figure 7.13B would be a steep psychometric function and Figure 7.13A would be flat. If the observer had based responses entirely on starting size, then Figure 7.13A,B would be identical. The results shown in Figure 7.13A,B indicate that, when presented with 64 different combinations of time to collision, rate of expansion, and starting size, the observer could totally ignore trial-to-trial variations in rate of expansion and starting size. We concluded that the human visual pathway contains a neural mechanism sensitive to the ratio $\theta/(d\theta/dt)$ that is insensitive to θ and $d\theta/dt$ (Regan and Vincent, 1995).

For foveal vision, the just-noticeable difference in time to collision, estimated from the psychometric function shown in Figure 7.13A and expressed as a Weber fraction was 6.7 (SE = 0.4)%, 5.5 (SE = 0.4)%, and 9.3 (SE = 0.6)% for the three observers.[13] Note that this measure gives the *precision* of estimating time to collision (but gives no indication of *accuracy*).

Figure 7.13C,D illustrates that the independence between the processing of time to collision and rate of expansion falls off as viewing eccentricity is increased. Figure 7.13C,D was recorded in the same way as Figure 7.13A,B, but the stimulus was presented at an eccentricity of 32° rather than 0°. The two psychometric functions are clearly much less different than in Figure 7.13A,B. The major reason for this considerable failure of independence is that, in peripheral vision, trial-to-trial variations in rate of expansion produce illusory variations in time to collision, while that is not the case when the subject gazes directly at the approaching object (Regan and Vincent, 1995). Stepwise multiple regression analysis showed that, at an eccentricity of 32°, the variable $\theta/(d\theta/dt)$ accounted for only 45% of the total variance, while the two variables $\theta/(d\theta/dt)$ and $d\theta/dt$ together accounted for 85% of the total variance. In contrast, the variable $\theta/(d\theta/dt)$ alone accounted for 89% of the total variance in foveal vision.

The implication of the Figure 7.13 data for flying or driving performance is that judgments of time to collision with an approaching object will be more reliable when a pilot or driver looks directly at the approaching object. This may create a problem when a pilot or driver is faced with the problem of avoiding collision with two or more objects approaching simultaneously along very different directions, because it is not possible to look directly at two objects at the same time. A flight simulator or car simulator study of this potential hazard is lacking.

Now we turn to the accuracy with which observers can estimate time to collision. In a recent laboratory study, we simulated an approaching spherical object that was visible only during the initial 1.0 sec of its trajectory (Gray and Regan, 1996; 1997). Some fixed time after the simulated object had disappeared the subject heard a brief tone. The subject's task was to press one of two buttons depending on whether collision with the simulated object would have occurred before or after the tone. Depending on the subject's response, a computer altered the simulated object's time to collision [i.e., $\theta/(d\theta/dt)$] so as to converge on the 50% "before" probability of response. This procedure followed tracking laws described by Levitt (1971). Measurements were carried out for times to collision ranging from 1.7 to 3.25 milliseconds. The mean difference between the estimated and the actual time to collision was 2.5% to 12% (underestimation).

In Section 7.4.1, I noted that judgments of time to collision with a simulated object might be inaccurate unless the rate of expansion of the object's texture elements matched the rate of expansion of the object's size. In a recent laboratory study, we measured estimates of time to collision with a simulated approaching textured object (Vincent and Regan, 1997). Time to collision was nominally 2000 millisecond. The simulated object was visible only during the early part of its trajectory. The just-noticeable difference in time to collision was 8.0% for one subject and 8.9% for a second subject. Estimated time to collision was shorter than 2000 millisecond when texture expanded faster than size,

[13]This kind of discrimination threshold is distinct from an observer's ability to judge which two approaching objects will arrive first. Todd (1981) found that observers' responses were over 90% correct for a 150 msec difference in time to collision and did not reach chance until the difference was 10 msec.

and longer when texture expanded more slowly than size. For example, when texture expanded at 0.9 times the rate of expansion of size, the estimated time to collision was 2178 millisecond. To place this within the context of training collision avoidance skills in a flight simulator, consider the case of low-level helicopter flight at 62 m/sec (120 knots). A 178 millisecond error in time corresponds to an 11 m error in distance. If the laboratory findings described above extrapolate to complex visual scenes, they imply that it is important in flight simulator displays to match the dynamics of texture to the dynamics of the textured object. It may be that, even though texture adds realism to *static* scenes, in some dynamic scenes the overall effect of texture might degrade rather than enhance transfer of training unless the texture dynamics are precisely correct. An in-the-field investigation of this possibility is lacking.

In Section 7.4.1, I also noted that prolonged inspection of a looming stimulus produces adaptation that can reduce the apparent speed of motion in depth caused by a given value of $\theta/(d\theta/dt)$. This raises the possibility that adaptation might cause errors in the accuracy of time to collision estimates. No information on this point is currently available.

7.7.2 Accuracy and Precision of Time to Collision Judgments Using Binocular Cues Alone

As discussed in Section 7.4.2, the impression that an object is moving directly toward the observer's head can be produced by changing the object's binocular disparity, and many studies on this phenomenon (stereomotion) have been published. In Figure 7.8B, if the speed (V_z) of the approaching object is constant, it will collide with the head in T sec, where $T = D/V_z$. Hence, rearranging Equation (11)

$$T \approx \frac{I}{D(d\delta/dt)} \tag{19}$$

(Regan, 1995c). However, although an object that appears to be moving directly toward the head will presumably collide with the head at some future instant, there is a paucity of reports on human ability to judge time to collision on the basis of binocular retinal information alone. Among the possible reasons for this omission is that viewing distance enters into Equation (19), and the weight of evidence is that we are poor at judging the absolute distances of objects further than a meter or so away (Collewijn and Erkelens, 1990). As well, apparent speed of stereomotion depends, not only on the value of $d\delta/dt$, but also on the presence of reference marks in the visual field, so that time to collision judgments might be different in different visual environments.

Most previous studies on visual judgments of time to collision either eliminated binocular disparity information altogether (DeLucia, 1991; McLeod and Ross, 1985; Regan and Vincent, 1995; Schiff and Detwiler, 1979; Sekuler, 1992; Todd, 1981) or, when disparity information has been available, it was confounded with monocular information (Bootsma and van Wieringen, 1990; Lee, Lishman, and Thomson, 1982; Lee et al., 1983; Warren, Young and Lee, 1986; Savelsbergh et al., 1991) or disparity has been confounded with ocular vergence (Heuer, 1993a,b). On the other hand, circumstantial or suggestive evidence that binocular retinal image information might aid judgment of time to collision has been scattered through the literature over a long period. For example, one hint was provided by Bannister and Blackburn (1931), who ranked 258 Cambridge undergraduates into "poor" and "good" categories according to their ability at ball games and found that the group who were ranked "good" had a larger interpupillary distance than the group ranked as "poor". More recently, using high-speed photography it was found that, when catching a ball with one hand, the temporal organization of finger flexions occurred only after the ball was closer than 6 ft from the hand. These finger flexions are necessary if the ball is to be retained in the catcher's grip. They were disrupted when the lights were switched off 275 msec before the ball arrived (Alderson, Sully, and Sully, 1974; Whiting, Alderson, and Sanderson, 1973). Binocular vision seems to be important also at distances relevant to highway driving. Cavallo and Laurent compared the accuracy of time to collision judgments using binocular vs. monocular vision on a circuit under actual driving conditions. Accuracy was greater for binocular judgments providing that viewing distance was less than about 75 m, but errors were still considerable (time to collision was consistently underestimated by at least 30%). However, at the considerably greater distances associated with landing a jet aircraft, occluding one eye during the landing approach had no detrimental effect on landing performance (Grosslight et al., 1978; Lewis and Kriers, 1969; Lewis et al., 1973; Plaffman, 1948).

In a laboratory study, observers were required to judge, using binocular retinal information only, whether a simulated approaching object would arrive before or after a brief auditory tone (Gray and Regan, 1996, 1997). At a close viewing distance (1.6 m), two observers each made a total of 48 estimates of time to collision. Times to collision ranged from 1.7 to 3.25 msec. The mean error was 2% to 12% (overestimation). This finding suggests that, at close viewing distances, accurate estimates of time to collision can be made on the basis of binocular retinal image information only.

7.7.3 Accuracy and Precision of Time to Collision Judgments Using a Combination of Monocular and Binocular Cues

Except for very small objects (for which, as noted in Section 7.4.2, the only available information about time to contact is binocular information), both monocular and binocular information will be available in everyday visual situations.

In a laboratory study, estimates of time to collision were more accurate when both binocular and monocular retinal image information was available than when only monocular information was available (Gray and Regan, 1996, 1997).

One possible reason for errors in estimating time to contact in the situation that disparity changes while size remains constant is that the two cues are providing conflicting information about the simulated object's motion in depth. This conflict may also explain why some subjects report an illusory change of size in this situation (Regan and Beverley, 1979a). A conflict of information when size changes at constant disparity might explain our observation that the speed of illusory motion in depth is less when the expanding target is viewed binocularly than when the same target is viewed monocularly. A similar argument might lead one to expect errors in judging time to collision when an observer with normal binocular vision views the expanding target monocularly. Whether errors would be less for an observer who lost the use of an eye in early life remains to be shown.

7.7.4 Interaction Between Perceived Self-Motion and Object-Motion in Relation to Collision Avoidance

A compelling impression of self-motion along a straight path can be reduced by causing an extended retinal image to expand radially outward from a focus, an effect called linearvection. Linearvection and circularvection (the visually induced illusion of self-rotation) are reviewed by Howard (1982). Early research supported this idea that peripheral stimulation is necessary to produce vection, but later work showed that this is incorrect. Vection can be produced by either central or by peripheral stimulation, provided that the central stimulus is perceived as a more distant surface seen through a surround. Whether presented centrally or peripherally, the effectiveness of a vection stimulus increases with the area of the stimulus (Anderson and Braunstein, 1985; Delorme and Martin, 1986; Howard and Heckman, 1989; Post, 1988; Shaver, Telford, and Frost, 1991; Telford and Frost, 1991).

The laboratory research on discriminating the speed and judging the time to collision with an approaching object that has been cited so far has been restricted to the case of a stationary observer. However, in everyday life there are many situations in which these judgments must be made when the observer as well as the external object are moving (e.g., when braking a car to avoid a rear-end collision). It has been reported that thresholds for detecting object motion are elevated by the perception of self-motion, whether the self-motion is real or illusory, and that the estimate of safe following distances for cars in convoy should allow for this by adding an "additional 300 msec to the hitherto acceptable range of reaction times which varies between 0.6 and 1.0 sec" (Probst, Krafczyk, Brandt, and Wist, 1984).

As to judgments of time to collision there is evidence that time to collision is underestimated by 200–300 msec when there is a vection-induced perception of self-motion in the cases of highway driving and helicopter flight (Cavallo and Laurent, 1988; Kruk and Regan, 1995; Schiff and Detwiler, 1979).

7.8 JUDGING THE DIRECTION OF SELF-MOTION ON THE BASIS OF THE RETINAL FLOW PATTERN

This section reviews the substantial body of published research on visually guided locomotion including the special cases of running on foot, highway driving, and aviation. Although Sections 7.4 to 7.7 focused on how a stationary observer can judge the direction of motion of an isolated object that is approaching the observer through a three-dimensional visual environment, the monocular and binocular cues to the direction of motion and time to collision expressed by Equations (12) to (15) and (18) and (19) are also available when an observer moves through a stationary three-dimensional visual environment. However, discussions of visually guided locomotion have commonly been framed in terms of optic or retinal flow patterns. Without doubt, one reason for this bias is the lasting influence of Gibson (1950, 1966, 1979). Gibson pointed out that an observer's motion through the three-dimensional visual environment produces a continuous deformation of the retinal image of the environment, and that this deformation is quite different from uniform isotropic expansion (as produced, for example, by a zoom lens). He went on to suggest that, from the continuous deformation, a human observer can obtain information not only about the three-dimensional structure of the environment, but also about his or her direction of self-motion through the environment. Gibson's explicit and clear enunciation of this idea was the inspiration for a substantial body of research during the last 25 years.

Before we review this topic, it will be helpful to distinguish several classes of hypothesis—"Local Process" hypotheses assert that some local feature in the retinal image is used to estimate the moving observer's destination, while "Global Process" hypotheses assert that judgments of destination are based on analysis of a large fraction of the visual field. "Quick Look" hypotheses assert that a brief sampling of the retinal image is sufficient to estimate the moving observer's destination, while "Visual Search" hypotheses assert that judgments are based on a comparison of multiple samples of the retinal image, each with a different direction of gaze.

It is important to distinguish between *optic flow* and *retinal flow.* Optic flow, a term coined by Gibson, can be regarded as equivalent to the pattern of flow in the image formed in a pinhole camera that is moving through the three-dimensional environment while maintaining its orientation at a constant angle to its direction of motion. The flow pattern within an actual retinal image (i.e., retinal flow) can be quite different because the eye may well not maintain a constant orientation with respect to its direction of motion. In that case, the flow pattern on the retina is the vector sum of the flow pattern caused by motion of the eye (optic flow component) and a bodily motion of the entire retinal image caused by rotation of the eye in its orbit. Gibson and others have assumed a priori that the visual system can separate these two components (Gibson, 1950, 1954), but experimental investigations have shown that this is not generally the case [see Cutting, Springer, Braren, and Johnson (1992) for a discussion of this point]. The distinction between optic flow and retinal flow can be illustrated by a simple example. Figure 7.14 is a well-known sketch from Gibson (1950) showing that a pilot's direction of self-motion is[14] given by the focus of expansion in the optic flow pattern. Gibson proposed that a pilot can use this fact to guide locomotion. However, Figure 7.14 depicts the *retinal* flow only if the pilot's direction of gaze is held at a constant angle to the direction of self-motion, and to do this it is (except in special cases) necessary to know the direction of self-motion—and that is what the pilot is trying to find in the first place. What can happen when a pilot fixates in an arbitrary point in the environment is illustrated in Figure 7.15. That figure makes the obvious point that the point fixated is stationary on the retina. Also, and less obviously, the pilot's destination may no longer be a focus of expansion, and the arbitrary point fixated may become a focus of expansion.

When the visual environment extends in depth and one's vehicle is traveling at a constant angle with respect to the direction it is pointing (as in everyday highway driving), the problem of judging one's direction of locomotion reduces to a trivial application of motion parallax. If the driver uses the vehicle as a gunsight, lining up one mark on the vehicle with a distant point while the head is held still, the distant point is the destination when there is no retinal flow across the line of sight. According to Langewiesche (1944), this stratagem was in common use in the early days of aviation. The stratagem can be regarded as a kind of vernier task and can be performed with high accuracy. This kind of cue was available to observers in a number of laboratory studies on visually guided locomotion (Johnston et al., 1973; Llewellyn, 1971). Laboratory simulations indicate that an accuracy of about 0.5° can be achieved in this special visual situation (Priest and Cutting, 1985; Regan and Beverley, 1985; Warren and Hannon, 1988, 1990; Warren,

[14]In laboratory research, the word *heading* is often used to mean the direction of locomotion. However, in aviation and sailing, *heading* means the direction in which the observer's vehicle is pointing, and in aviation, sailing, and sometimes even in highway driving, a vehicle travels in a direction quite different from the direction in which it is pointing.

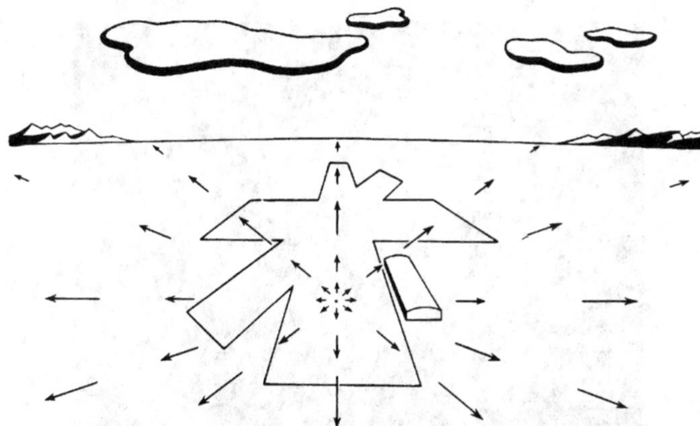

Figure 7.14 The flow pattern on the retinae of a pilot during a landing in the situation that there is a constant angle between the pilot's direction of gaze and direction of motion. [From Gibson, J. J. (1950). *The Perception of the Visual World.* Houghton Mifflin: Boston. Reprinted with permission.]

Morris, and Kalish, 1988; and Cutting 1986) showed that visual guidance is precise when using motion parallax even when moving along a curved path.

Now we turn from the special case where the "gunsight" stratagem is valid to the more general case where it is not because no part of the observer's vehicle is within the observer's field of view or, even if the vehicle parts are visible, they provide no useful information about the task (e.g., in a car that is spinning on ice).

At this point the following three remarks about performance in the everyday world are appropriate. First, I have been unable to find any field investigation of the accuracy with which (in a single brief look) an observer can judge his or her direction of motion in a variety of visual scene geometries while being conveyed in a vehicle whose orientation is dissociated from the direction of motion. There is a paucity of field studies designed to identify real-world environments in which directional judgments *fail:* It is not even known whether an observer is capable of judging the direction of self-motion when the vehicle's orientation is continuously changing. Nevertheless, many authors have assumed that human observers can indeed judge the direction of self-motion on the basis of a single brief look, and have carried out laboratory and theoretical studies to identify the visual cues on which judgments might be based. Second, the flow patterns used in many laboratory studies have comprised motions of dots of constant size, i.e., the simulated objects did not obey the laws of projective geometry that allow the monocular cues expressed in Equations (12) and (13) to be used. Third, in most laboratory studies the flow patterns contained no binocular information (see Section 7.5.2 for a discussion of binocular cues to direction).

Longuet-Higgins and Prazdny (1980) demonstrated theoretically that information about the direction of self-motion is provided by motion parallax between two sharp-edged objects at different depths in the same visual direction. Because eye rotation affects such overlapping images in the same way, any relative motion between them must be caused by the observer's self-motion. Their difference vector passes through the direction of locomotion, so that several overlapping pairs of elements specify the direction of locomotion. Rieger and Lawton (1985) extended the concept of edge parallax to cover the case of relative motion between nonoverlapping elements within a restricted region of space (Rieger and Toet, 1985).

The theoretical work reviewed so far demonstrates that sufficient visual information for judging the direction of locomotion is, in principle, available when the external environment is structured in depth. There is experimental evidence that, at least in the laboratory situation, observers can use this differential motion to effectively remove the component of retinal flow caused by pursuit eye movements (Warren and Hannon, 1990). This leads us to ask whether, in principle, sufficient visual information is available when the external environment has a constant depth and, if so, whether human observers can use this information. In one experiment, eye rotation was simulated during the approach

Figure 7.15 Expanding flow patterns similar to those in the retinal image for an observer moving through the outside world. The multiple exposure in the middle picture was taken with a camera moving toward the head while pointing directly at the head. The center of expansion coincided with the direction of the camera's motion. The lowest multiple exposure was taken with a camera moving toward the head while pointing to one side (arrow). The center of the expanding flow pattern in this picture did not coincide with the direction of the camera's motion, but rather with the direction of the camera's "gaze." [From Regan, D., and Beverley, K. I. (1982). How do we avoid confounding the direction we are looking with the direction we are moving? *Science, 215,* 194–196. Reprinted with permission.]

to a picket fence (simulated by a sinewave grating). When the picket fence was flat (i.e., at constant distance), the observers' judgment of their future point of impact on the fence was subject to large error. When the simulated picket fence was convex in depth, however, subjects could judge the future point of impact with the fence with high accuracy (considerably better than 1.0°), because the point of impact coincided with the point of maximum div **V**. (In the case of the flat fence, div **V** did not vary greatly over the entire fence) (Regan, 1995). Taken together with a previous study that showed that the immediate neighborhood of the center of expansion of a radially expanding flow pattern (where div **V** rose to a maximum) produced a sharp maximum in excitation of the changing-size detectors described in Section 7.4.1, this finding indicates that human observers can identify the location of the maximum div **V** within a flow pattern. However, it does not follow that the identification of a maximum div **V** is a candidate strategy for identifying the direction of locomotion, because, depending on the three-dimensional scene geometry, the direction of locomotion may or may not coincide with a maximum in div **V**. For example, when moving in a straight line inclined to the surface of a plane, there is a local maximum of div **V** at a point halfway between the point of impact and the closest point on the plane (Regan and Beverley, 1982; footnote 6). Also, the retinal flow pattern may contain more than one local maximum of div **V**.

If objects in the three-dimensional visual scene are scattered symmetrically with respect to the direction of locomotion, the retinal flow pattern is symmetrical when an observer gazes along the direction of locomotion and becomes progressively more asymmetric as the angle between the direction of gaze and the direction of locomotion is increased (Richards, 1975). In such a visual environment, the observer can discover his or her direction of locomotion by adjusting the direction of gaze until the retinal flow pattern is symmetrical (Richards, 1975). Template matching offers a possible means of testing the symmetry of wide-field pattern of retinal flow (Regan, 1985; Warren and Saunders, 1995). On the grounds that neurons in the parietal lobe of monkey cortex have large receptive fields and respond best to motion directed either radially away from the fovea or radially toward the fovea (Motter, Steinmetz, Duffy, and Mountcastle, 1987; Steinmetz, Motter, Duffy, and Mountcastle, 1987), it has been suggested that pooled neural activity in the human analog of this cortical region is a candidate template (Regan, 1991a). Template models of this kind would not provide accurate information about the direction of locomotion when the observer was moving through an asymmetric visual environment, e.g., when driving with a high wall on one side of the car and a wide open area on the other side of the car. On the other hand, there is little, if any, quantitative data collected in the field on how the accuracy of judging the direction of self-motion is affected when the symmetry of the visual environment is progressively degraded.

So far, we have discussed the motion of an observer through a rigid environment, yet there are often independently moving objects in the space through which humans attempt to navigate. Warren and Saunders (1995) simulated this kind of visual environment and found that judgments of the direction of self-motion were affected when a moving opaque object was within 6° of the observer's destination. They accounted for their results in terms of a center-weighted template model and suggested that their findings might indicate a strategy of taking into account, not only the direction of self-motion, but also objects immediately ahead: "... it may be more important to perceive one's heading relative to the vehicle in front..... than the roadway itself."

Cutting et al. (1992) have pointed out that, in the everyday world, the visual cues used to guide locomotion may be different for low retinal flow speeds (e.g., when walking) and for high retinal flow speeds (e.g., when driving at high motorway speeds or flying at very low altitude). At walking speeds, Cutting et al. (1992) favor a model in which eye movements are used to search for the direction of gaze for which motion parallax between near and far objects is zero (akin to the "gunsight" strategy already discussed). At the other extreme, in high-speed, low-level flight a pilot sees the out-of-cockpit scene as blur lines radiating from the aircraft's destination (Whiteside and Samuel, 1970). Cutting et al. (1992) cited a report by Calvert (1954), who filmed the face of a driver of a vehicle starting from rest. He noted that, as speed increased, the driver scanned less and less until at high speed the driver stared fixedly in the direction he wished to go, i.e., at the center of the radial pattern of blur lines, presumably because the eye rotates rapidly and unpleasantly if the driver fixates on any point other than the destination. In this special situation, with direction of gaze at a constant angle to the direction of locomotion, the focus of expansion coincides with the destination, as illustrated in Figure 7.14. Intersubject differences in the ability to discriminate the rate of expansion of a radially expanding flow pattern has been reported to correlate with intersubject differences in the performance of pilots flying jet aircraft at low level (Kruk and Regan, 1983).

7.9 VISUALLY GUIDED, GOAL-DIRECTED MOTOR ACTION

Although no definitive answer to the question can be expected, it may not be without merit to ask what vision is for. If one assumes that the purpose of the visual system is to provide visual guidance for goal-directed motor action, then it makes sense to regard the visual and motor pathways as a single entity and to anticipate that some features of the neurophysiological and neuroanatomical organization of the visual pathway might be best understood in terms of the organization of the motor pathway and, as well, that some functional aspects of visual processing might be best understood in terms of particular goal-directed motor actions.[15]

7.9.1 Reaction Time, Cueing, and Stimulus-Response Compatibility

So far, we have discussed both the process of detecting (i.e., seeing) an object and the further stages of processing its spatial characteristics and motion. These processes can be distinguished from the evaluation of the visual information, and the decision making that underlies the consequent motor action (or withholding of motor action).

An individual's motor reaction time is fastest when the individual knows in advance what information will be presented and what response should be made (simple reaction time). Simple reaction time for a visual stimulus is approximately 180 msec under optimal conditions; simple reaction times for auditory and tactile stimuli are approximately 40 msec shorter, again under optimal conditions (relevant data are summarized in Teichner and Krebs, 1972). In the simple reaction-time situation, the decision-making component is essentially absent. Motor reaction time is longer when a decision is required (choice reaction time).

Here we should briefly step aside to outline how communication theory has been applied to analysis of goal-oriented motor action. Within the context of *information theory* (Gabor, 1946) and *communication theory* (Shannon and Weaver, 1949), the term *information* has a more restricted and precise definition than in everyday usage. Subject to this severe restriction of meaning, information can be quantified in a way that *does not depend on the physical units of the stimulus.*

Information is something we gain when an event tells us something we did not already know. The amount of information gained is measured by the amount of uncertainty that is lost (and is therefore related to the concept of entropy in physics) and quantified in "bits" (i.e., binary digits). This can be understood as follows. Suppose that one wishes to find a book in a library, and knows already that the book is located on one of two shelves (i.e., a binary choice). If the librarian points out which of the two alternative shelves it is on, the two alternatives (2^1) are reduced to one, and one binary digit of information has been transmitted. If there had been four shelves to choose from, the librarian would have transmitted two bits of information (2^2) in pointing out the correct shelf. In general, N bits of information are transmitted when 2^N equally likely alternatives are reduced to one. A slightly more complex formula is required when all alternatives are not equally likely (Allusi, 1970; Attneave, 1954).

Communication theory was developed in the context of radio and telephone communication (Shanon and Weaver, 1949). It provides a quantitative description of the rate of flow of information along a passive communication channel. In Shannon and Weaver's conceptual formalism, a message is selected from an "information source" depicted in Figure 7.16, and encoded for transmission over a "passive communication channel" for subsequent decoding into a form appropriate for its final "destination." Communication theory has been used as a conceptual tool in many areas other than the radio and telephone communication area within which it was developed, including sensory psychophysics, cognitive psychology, and neuroscience. Its success in these other areas has been less than complete. One reason for these (partial) failures is that communication theory exists within the conceptual formalism illustrated in Figure 7.16, so that, for example, the "destination" concept is not valid in some attempted applications of the theory; in other

---

[15]A second surmise about the purpose of the visual pathway is to aid the planning of future goal-directed motor action, and again one might envisage that this purpose would manifest itself in the organization of visual encoding and memory.

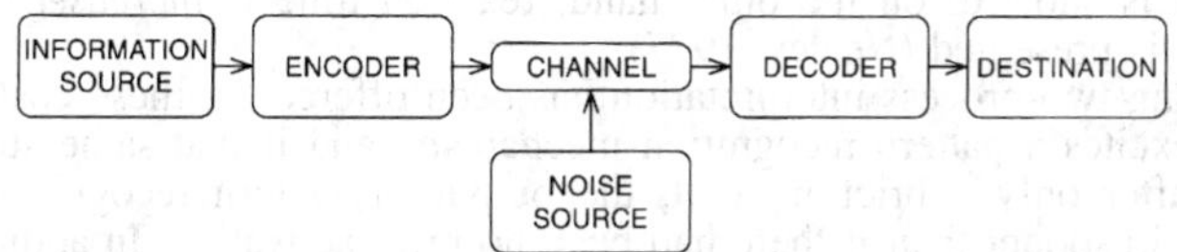

Figure 7.16 Schematic diagram of the flow of information through a passive communication channel. [Modified from C. E. Shannon, and Weaver, W. (1949). *The Mathematical Theory of Communication.* Urbana, IL: University of Illinois Press. Reprinted with permission.]

attempted applications, the effects of a two-way flow of information are not well described in terms of the one-way flow of Figure 7.16.

As already mentioned, motor reaction time increases when the subject is required to make a decision, as, for example, when the number of possible stimuli is greater than one and/or the number of possible responses are greater than one (choice reaction time). According to the Hick-Hyman empirical equation, motor reaction time (RT) is a logarithmic function of the number of alternative stimuli and responses. In particular,

$$\mathrm{RT} = a + b \log_2 N \tag{20}$$

where a and b are constants and N is the number of choices. Although the logarithmic relation expressed in Equation (20) is based on empirical data, base 2 logarithms are used rather than base 10 or natural logarithms with the intent of creating a link to theory, and in particular with the binary digit (bit) measure of information, and also with the communication theory illustrated in Figure 7.16.

In the case that N stimuli have equal probabilities of occurrence (p), Equation (20) can be rewritten as

$$\mathrm{RT} = a + b \log_2 (1/p) \tag{21}$$

where $p = (1/N)$.

This version of the Hick-Hyman equation brings us to the question whether the relative probability of a stimulus determines reaction time rather than the total number of stimuli being the only important factor. A relevant finding is that, when the N stimuli have different probabilities of occurrence (and provided that these possibilities remain constant), reaction time to the more probable stimuli are shorter than reaction times to the less probable stimuli (Fitts, Peterson, and Wolfe, 1963). To calculate the average reaction time to a set of N stimuli which have difference probabilities of occurrence, the reaction times to individual stimuli must be weighted by their respective probabilities

$$\mathrm{RT} = a + b \textstyle\sum_{i=1} p_i \log_2 (1/p_i) \tag{22}$$

(Hyman, 1953). A link with communication theory is that the second term is equal to the average information in a set of events (Shannon and Weaver, 1949).

In its most general form, the Hick-Hyman equation can be written

$$\mathrm{RT} = a + b\,\mathrm{H_T} \tag{23}$$

where $\mathrm{H_T}$ is the information transmitted.

Providing a cue that indicates which of the N target stimuli is next can shorten reaction time, but may not be without cost. At first sight it may seem surprising that, even if the cue has zero reliability (zero predictive value) there is a modest reduction in the reaction time to a subsequently presented target stimulus, provided that the target and cue are identical spatially or semantically (e.g., a letter A cued by a letter A). This effect is present less than 100 msec after the cue. There is no cost if some others target stimulus is presented, presumably because the cue does not summon attention (Posner and Snyder, 1975). If the cue has no spatial or semantic resemblance to the subsequent target stimulus (e.g., cueing a letter A with a green spot of light), there is no early benefit. If the target stimulus is presented more than 300 msec after the cue, reaction time to a correctly cued target stimulus is shorted; on the other hand, reaction time is *increased* if a "wrong" target stimulus is presented (Neeley, 1977).

The following two-process interpretation has been offered for these cost/benefit data. A visual cue excites a pattern recognition mechanism, and if that same stimulus is presented again after only a brief interval, the output of pattern recognition mechanism reaches threshold sooner than if there had been no prior activation. In addition, and after an interval of about 300 msec, the cue summons an attentional process that speeds the motor response to the expected target stimulus, but slows the motor response to a "wrong" target stimulus (Keele, 1986; Neeley, 1977; Posner and Snyder, 1975).

The Hick-Hyman equation has considerable generality. For example, it can be used to analyze the speed-accuracy tradeoff (Hick, 1952). On the other hand, it fails to deal with the effects of stimulus-response compatibility on the speed with which choices are made (see below), or the effects of practice (Gopher and Donchin, 1986).

The importance of stimulus-response compatibility has been described in many experimental contexts. For example, in the Fitts and Seeger (1953) study, there were three modes of motor response (R1-R3 in the leftmost vertical column in Figure 7.17) and three spatial arrangements of stimulus lights (only one S is shown in Figure 7.17). Eight lights were arranged round the circumference of a circle. The stimulus was the illumination of one of the eight lights, and the subject was instructed to move a stylus from the center in the direction of the illuminated light. Response mode R1 allowed the stylus to be moved radially using one hand. Response mode R2 also allowed the use of one hand, but the stylus could be moved only horizontally or vertically, so that oblique movements had to be achieved by a combination of horizontal and vertical movements. Response mode R3 allowed only horizontal and vertical movements, but the horizontal movements were achieved using a stylus held in one hand while the vertical movements were achieved using a second stylus held in the other hand.

The numerical data set out in Figure 7.17 show how reaction times and errors (in brackets) depended on the relation between the stimuli and motor responses. This phenomenon is called *stimulus-response* (*S-R*) compatibility.

Although the existence of this phenomenon may seem intuitively obvious, it is often ignored in the practical design of appliances and machines with a consequent increase in the accident rate experienced by the users of these artifacts. A classic example is the

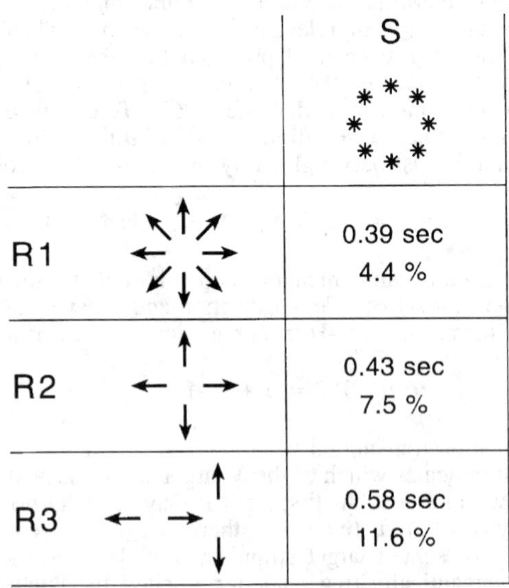

Figure 7.17 Choice reaction time and errors as a function of stimulus-response compatibility. Subjects were required to respond in each of three response modes ($R_1 - R_3$) to the stimulus displays. See text for details. [Modified from Fitts, P. M., and Seeger, C. M. (1953). S-R compatibility: spatial characteristics of stimulus and response codes. *Journal of Experimental Psychology, 46,* 199–210. Reprinted with permission.]

importance of a close spatial correspondence between the burners and control knobs of a kitchen range.

7.9.2 Motor Actions That Are Modified by Visual Feedback During the Execution of the Motor Action

In this section, we discuss motor actions in which the interval between the instant that relevant visual information is first available and the completion of the motor act is sufficiently long that visually guided modification of the motor action is possible after its initiation.

That the time to complete a limb movement of this kind depends both on the distance moved and the accuracy required was known from early work on time and motion analysis in the industrial setting, and extensive tabular data were available (Niebel, 1962), but the relationship was not expressed in the form of an equation until 1954. Fitts and his colleagues (Fitts, 1954; Fitts and Peterson, 1964) found that, when the required precision of the movement was fixed, movement time increased with the logarithm of distance moved; and, for a fixed distance of movement, movement time increased with the logarithm of required *precision* (precision was defined as the reciprocal of target width). These findings were encapsulated in Equation (24), known as Fitts' law.

$$MT = K + C \log_2 (2D/W) \tag{24}$$

where D is the distance of movement and W is the target width. The factor 2 is inserted to avoid negative logarithms when D is (a little) less than W. Constant K is obtained empirically. It is a fixed delay that depends on the limb used for the motor action. For example, when the hand is used, a typical value of K is 0.18 sec (Knight and Salvendy, 1992). Fitts (1954) termed this quantity $\log_2 (2D/W)$ the index of movement difficulty (*ID*).

Although Equation (24) is empirical, it has been discussed in terms of the Communication Theory conceptual format illustrated in Figure 7.16, and in those terms the reciprocal of C can be regarded as the information capacity of the response control "channel" (hence, the choice of base 2 rather than base 10 or natural logarithms).

The value of K is higher for a series of to-and-fro movements than for a single discrete movement (probably because of dwell time at the target before movement reverses), and in addition the value of I/C falls by about 25% (Keele, 1986).

In an investigation of Equation (24), Longulf, Chaffin, and Foulke (1976) investigated short-distance repetitive movements performed under a low-power stereo microscope, a situation common in microscopic assembly work. Wrist movements transported a small pin over 1.3 cm and fingers moved 0.25 cm. For comparison, arm movements of 5 cm to 30 cm were carried out under normal vision. The required precision (target size) was varied in all three cases. Equation (24) described the data, but the value of C varied, being 105 msec/bit for the arm movement, 43 msec/bit for the wrist movement, and 26 msec/bit for the finger movement. The values of K also varied in the same sequence.

The laboratory research just discussed can be summarized by stating that the total time required to execute a motor response to a visual stimulus that is well above detection threshold can be taken as the sum of the following three intervals: (1) perceptual delay; (2) decision delay; (3) movement duration. Hence, from Equations (23) and (24),

$$\text{Total time} = [a] + [b\, H_T] + [K + C \log_2 (2D/W)] \tag{25}$$

An application of Equation (25) is described in Knight and Salvendy (1992) from which the following is taken.

> *Consider a task in which a set of choices is presented on a computer-controlled VDT equipped with a touch-sensitive screen. Such screens sense the position at which an operator's finger touched the display. The operator can indicate to the computer a selection by touching the screen with an appropriate "target" area associated with each displayed choice. The operator must reach toward the display screen and carefully control the position of his or her finger as it contacts the display screen. Critical elements that influence performance capability in this task include the size of the target area associated with each choice, as well as the distance of the VDT face from the operator. Assume the target is a square with sides* **W** *cm long. The operator must position his or her finger in two dimensions.*

Thus, both vertical and horizontal positioning information must be transmitted through the operator's body and into the computer via the touch screen. We may assume that the vertical and horizontal positioning tasks are independent for an engineering level of analysis. This means the ID contribution to performance time is doubled. Furthermore, assume there are **N** *choices displayed on the VDT. Thus, the operator's task is to decide which of* **N** *choices should be selected, then move his or her finger to the selected choice, position the finger properly, and press the appropriate target area of the screen. Then the hand is returned.*

This description makes clear the relevance of traditional task movements elements to modern human–computer interaction tasks. The critical parameters of this task model include:

The number of choices, N.
The width of the target, W (in each dimension).
The amplitude, D, of the movement needed to reach the VDT screen.

Table 7.1 shows how performance is related to the model's parameters. We have assumed perceptual delays (identifying the target area on the display) are about 0.25 sec. b is assumed to be 220 msec/bit, and the operator always presses the intended choice's target. K is assumed to be 0.177 sec, and C is 0.1 sec/bit.

Although the preceding discussion has focused on discrete movements, the concepts apply to continuous tasks (such as tracking) as well. Delays inherent in the human information processing system impose an upper limit on the frequency of continuous information to which the operator can respond. The operator can deal effectively only with continuous signal components below 1 Hz. In responding to continuous, time-varying input signals, the operator can be modeled as an intermittent servomechanism making a series of discrete corrective movements about twice per second." Further empirical examples are discussed in Holding (1981), Salvendy (1987), Salvendy and Seymour (1973), and Welford (1968, 1976).

Table 7.1 Idealized Choice-Entry Time as a Function of Task Parameters N, D, and W

N	Decision H_T (bits)	D (in.)	W (in.) (Width of Target)	Index of Movement Difficulty $\log_2 (2D/W)$	Total Time (sec)
4	2	12	0.25	6.5	2.19
4	2	12	0.5	5.5	1.99
4	2	12	1.0	4.5	1.79
4	2	18	0.25	7.2	2.31
4	2	18	0.5	6.2	2.11
4	2	18	1.0	5.2	1.91
4	2	24	0.25	7.6	2.39
4	2	24	0.5	6.6	2.19
4	2	24	1.0	5.6	1.99
8	3	12	0.25	6.6	2.41
8	3	12	0.5	5.6	2.21
8	3	12	1.0	4.6	2.01
8	3	18	0.25	7.2	2.53
8	3	18	.5	6.2	2.33
8	3	18	1.0	5.2	2.13
8	3	24	.25	7.6	2.61
8	3	24	0.5	6.6	2.41
8	3	24	1.0	5.6	2.21

Fitts' law describes total movement time in cases where the duration of movement is long enough to allow motor action to be affected by visual feedback. How long it takes for visual information to affect motor action is a controversial question, and it may well be that the answer depends on the motor task (reviewed in Carlton, 1992).

A popular method of estimating the time required to make use of visual feedback is to switch off the lights some time after initiation of a motor action. This approach leads to estimates of 200–290 msec (Beggs and Howarth, 1970; Keele and Posner, 1968). Carlton (1981) revised this estimate to 135 msec, pointing out that it is not until some time after the initiation of a motor action that the moving body part is sufficiently close enough to the target that visual feedback can be used.

Several authors (Bootsma, 1991; Lee et al., 1983; Regan, Beverley, and Cynader, 1979) have suggested that the reaction time for modifying a highly practiced movement that is already in progress might be considerably shorter than the laboratory estimates of 180–300 msec for the reaction time to an unpredicted event (Keele and Posner, 1968; Teichner and Krebs, 1972), but findings on this point are inconsistent (Carlton, 1991). So far as the higher reaches of human motor performance are concerned, it has been estimated that international-level cricket batsmen can continue to monitor an unpredictable change in the ball's direction of motion to a time considerably less than the 230 msec remaining after a 90 mph delivery had bounced (Regan, et al., 1979; Regan, 1972). However, in an experimental study of this visually guided motor action, McLeod (1987) measured the time required for highly skilled batsmen to adjust their bat swing in response to an unpredictable deviation in the direction of the ball after bouncing (this deviation was caused by a small peg placed under the mat). He found the minimum time for a detectable correction to be not less than 210 msec after the deviation had occurred, considerably longer than the 135 msec visual feedback time estimated by Carlton (1981). However, McLeod noted that his estimate might have been artifactually lengthened by the inertia of the bat.

7.9.3 Intermittent versus Continuous Visual Feedback

This takes us to the question of intermittent versus continuous control of motor action (reviewed by Elliott, 1992). There is evidence that moving an arm to a target occurs in two or more phases: During the first phase, the arm is moved without visual feedback until it is near the target; during later phases, visual feedback is used to guide accurate placement onto the target (Jeannerod and Prablanc, 1983; Newell, Carlton, and Carlton, 1982; Meyer, Smith, Kornblum, Abrams, and Wright, 1990).

By using high-speed photography to film experts catching tennis balls with one hand, Alderson et al. (1974) obtained evidence that the first phases of the catch placed the correctly oriented hand in line with the approaching ball. The last motor phase of the catch, a ballistic closing of the fingers took less than 70 msec and demanded timing accuracy to within 16–30 msec. If the lights were extinguished less than 275 msec before the ball hit the hand, the final stage of flexing the fingers was mistimed.

A two-phase organization also seems to be used in the long jump—a locomotor task. Lee, Lishman, and Thomson (1982) found that the early part of the run (roughly 15 strides) was similar from trial-to-trial, but the last five strides showed considerable variability as though it were only during the final phase of the run that visual feedback was used to compensate for positional errors.

Next, we consider tasks of visually guided, goal-directed motor action in which, even if the feedback of visual information is not truly continuous, the information being fed back is updated so often that the feedback is effectively continuous. Catching a flyball in the game of baseball (or a high-lofted hit in the game of cricket) is a much-discussed task of this kind. The catcher may have to run up to 30 m. How does the fielder run so as to be within reach of the ball when a high lofted hit plunges to the ground? Before rejecting it, we should first consider the hypothesis that, during the first instants after the ball leaves the bat, the fielder judges where and at what future instant the ball will hit the ground and runs accordingly without altering this judgment. On the face of it, this hypothesis has several unsatisfactory features. In the spirit of the computational approach to vision (Marr, 1982) we can ask whether the information required to make such a judgment is necessarily available to a fielder during the first instants of the ball's flight. This seems unlikely. It would be difficult to explain how the fielder could instantly allow for the effect of aerodynamic drag on the ball's flight. Since drag is a simple function of speed only over a limited range of speeds, and the speed at which the function changes (the "drag crisis") depends on the exact condition of the ball's surface (Achenbach, 1974;

Appendix 2; Davies, 1949; Frohlich, 1984, Regan, 1992), its effect on the point at which the ball returns to the ground would depend not only on the balls initial speed and surface condition, but also on the initial angle of elevation in a complex manner. Also, if there is any appreciable difference between air movements at ground level and at the maximum altitude reached by the ball—as may well be the case in an enclosed stadium—all the necessary information is not initially available. Nevertheless, expert fielders do catch high lofted balls even in blustery conditions.

A second hypothesis can be traced from Chapman (1968) through Todd (1981) and Michaels and Oudejans (1992) to Babler and Dannemiller (1993) and Dienes and McLeod (1993). Chapman suggested that fielders run along a path that cancels the acceleration of the ball's retinal image. Two limitations of Chapman's analysis are that it assumed a parabolic trajectory and it required the fielder to run at constant speed. However, even if the ball's trajectory deviates somewhat from the parabolic curve that it would follow in a vacuum, the general strategy can still work because it is essentially an error-nulling feedback strategy (reminiscent of a velodyne servomechanism designed to maintain a constant motor speed), in which the changes in the fielder's running are driven by the departure of optical acceleration from zero. The most recent version of this hypothesis has been expressed as follows: "when fielders run backwards or forwards to catch a ball they run at a speed which keeps d^2 (tan α)/dt^2 zero, where α is the angle of elevation of gaze from catcher to ball.... whatever the effect of aerodynamic drag on the ball's trajectory" (McLeod and Dienes, 1993). In a field test of the hypothesis, the movements of a skillful fielder were recorded as he ran to catch cricket balls fired into the air at 20 to 25 m/sec at an elevation of 45° (McLeod and Dienes, 1993). The balls came directly toward the fielder. Experimental findings supported the hypothesis. The fielder waited for roughly 0.5 sec after the ball was fired, quickly reached a velocity where d^2 (tan α)/dt^2 was equal to zero and from then on modified his running speed up to the point of catching so as to keep d^2 (tan α)/dt^2 close to zero.

According to McBeath, Shaffer, and Kaiser (1995), the d^2(tan α)/dt^2 hypothesis is flawed in several respects. First, there is a body of evidence that humans have both poor sensitivity to acceleration and poor discrimination of acceleration (Calderone and Kaiser, 1989; Gottsdanker, Frick, and Lockard, 1961; Runeson, 1974; Schmerler, 1976). Second, McBeath et al. noted that the ball's lateral motion had been treated as a nuisance variable that was irrelevant except possibly to make the catcher's task more difficult, and that the experimental study reported by McLeod and Dienes (1993) had been restricted to balls fired directly at the fielder. McBeath et al. (1995) claimed that fielders find such balls more difficult to catch than the balls that do not travel directly toward them because it prevents the use of a preferred strategy that requires some lateral motion. This strategy is to run along a curved path with a U-shaped speed function so as to create an apparently linear ball trajectory. This strategy derives its error feedback signal from judgments of the curvature of the ball's path (for which discrimination threshold is very low), rather than by judgments of acceleration. Field measurements in which fielders caught fly balls fired in different directions supported this hypothesis. Note that this hypothesis resembles the "optical acceleration cancellation" strategy described above in that it is also an error-cancellation feedback strategy, allows some correction for a nonparabolic trajectory and moderate changes in trajectory caused by the Magus effect or blustery conditions.

7.9.4 Ballistic Motor Actions

By definition, a ballistic motor action is one for which no modification of the catcher's motor action is possible after its initiation because the interval between the instant at which relevant information is first available and the completion of the action is so short that there is insufficient time to change the velocity of a moving body part (or body part plus implement), or even that there is insufficient time to allow the updated visual information to be processed by the brain.

Even when visual feedback is not involved, there is a trade-off between accuracy and speed; faster movements are less accurate. The trade-off, however, is not as described by Fitts' law; its basis is variations of the force required to start and stop the motor action; to move a body part over a fixed distance within a given duration requires larger accelerating and decelerating forces for a movement of shorter duration than for a movement of longer duration. In a series of laboratory studies, Schmidt and his colleagues instructed subjects to apply short duration force pulses to a lever that did not move. The force was transduced into a deflection of a light spot, and the subjects were instructed to apply a force pulse that would move the light spot to a designated location. The force pulses were too short to allow the motor action to be modified by visual feedback (spot position) during the performance of the motor action (Schmidt, Zelaznik, and Frank, 1978; Schmidt, Zelaznik, Hawkins, Frank, and Quinn, 1979; Sherwood and Schmidt, 1980). Figure 7.18A,B shows how the standard deviation of the force applied varied with the force required for small forces ranging from 2.0 to 10 N (panel A) and for larger forces ranging from 20 to 120 N (panel B). Over this range of forces, the data were described by Equation (26), often referred to as Schmidt's law,

$$F_{SD} = k + c\ (D/T) \tag{26}$$

where F_{SD} is the standard deviation of the applied force, k and c are empirical constants, D is the distance of movement, and T is the duration of movement. Equation (26) is valid for movements of 140–200-msec duration, and for distances up to 30 cm. Schmidt's law

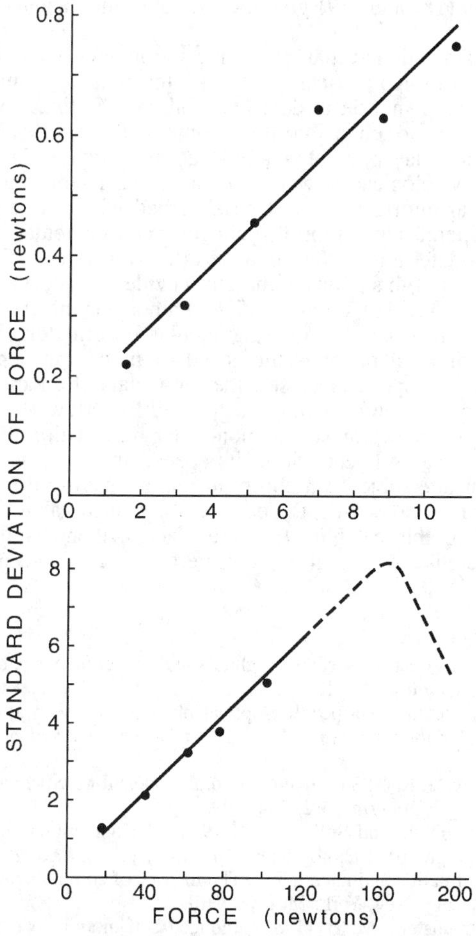

Figure 7.18 Subjects hit a fixed lever, and the location of a light spot indicated the force applied. The duration of the force was so short that the visual feedback could not be used during the motor action. The standard deviation of the force actually applied over many repetitions is plotted as ordinate. The force required by the experimenter is plotted as abscissa. (A) small forces ranging from 2 to 10 N. (B) larger forces ranging from 20 to 120 N. Equation (25) (continuous line) is fitted to the data points. The dashed line in B depicts how Schmidt's law breaks down for high forces. [Modified from Schmidt, R. A., Zalaznik, H. N., Hawkins, B., Frank, J. S., and Quinn, J. T. (1979). Motor output variability: A theory for the accuracy of rapid motor acts. *Psychological Review, 86,* 415–451. Reprinted with permission.]

breaks down for movements lasting longer than about 200 msec; there is considerably less variability than predicted by Equation (26), presumably because visual feedback can be used to aid accuracy (Zelaznik, Shapiro, and McColsky, 1981). Schmidt's law also breaks down when the required force exceeds 60% to 70% of the maximum force that subjects can exert; as the force is increased beyond the 120 N limit in Figure 7.18B there comes a point where the ordinate peaks, beyond which the standard deviation *decreases* as force is increased (dashed line in Figure 7.18B).

The so-called reflex catch in the game of cricket may well be a true ballistic motor action. This kind of catch is exemplified by the sharp catch made by a slipfielder who stands behind and close to the batsman. This rapidly moving ball passes well outside the reach of the fielder who completes the catch (using his bare hands) with his entire body off the ground in a horizontal position. Both monocular and binocular information is available, but the only visual information available to determine the catcher's motor action is the information gathered during the first instants of the ball's flight. Equations (12) and (13) give the available monocular retinal image correlates of the direction of motion, and Equations (14) and (15) give the available binocular retinal image correlates. As to time to collision, Equations (18) and (19) give the available monocular and binocular retinal image information.

It has already been mentioned that, providing the angular subtense of the ball is not too small, it follows from geometrical optics that the relative amplitudes of monocular and binocular correlates of motion in depth depends on the linear width of the ball, but the viewing distance does not enter into this geometrical consideration [Equation (9)]. It has also been mentioned that the relative *effectiveness* of monocular and binocular information depends on physiological as well as geometrical factors. In the present context, since all relevant visual information must be absorbed within a very short interval, any differences in the temporal integration of monocular and binocular information might be an additional factor in determining their relative effectiveness (see also Deary and Mitchell, 1989 for a discussion of this point). Laboratory evidence indicates that, as presentation duration increases from 0.25 sec through 0.5 sec, binocular information about motion in depth becomes more effective relative to monocular information (Regan and Beverley, 1979). Another physiological factor is the speed of motion in depth. Here, laboratory evidence indicates that, as speed increases, the binocular information becomes more effective relative to monocular information (Regan and Beverley, 1979). Thus, for a reflex catch, the importance of binocular versus monocular information may be determined by a tradeoff between the ball's linear width, its speed, and the time of flight. Lastly, an important point is that intersubject variability in relative sensitivity to the monocular and binocular correlates of motion in depth seems to be considerable (Regan and Beverley, 1979). The demands of this catching task are made evident by the fact that, even at international level, there is a wide gulf in motor performance between the great slipfielders and everyone else.

REFERENCES

Achenbach, E. (1974). The effects of surface roughness and tunnel blockage on the flow past spheres. *Journal of Fluid Mechanics, 65,* 113–125.

Addams, R. (1834). An account of a peculiar optical phenomenon seen after looking at a moving body. *London and Edinburgh Philosophical Magazine and Journal of Science,* 3rd Series, *5,* 373–374.

Adelson, E. H., and Bergen, J. R. (1985). Spatiotemporal models for the perception of motion. *Journal of the Optical Society of America,* A2, 284–299.

Alderson, G. J. K., Sully, D. J., and Sully, H. G. (1974). An operational analysis of a one-handed catching task using high-speed photography. *Journal of Motor Behaviour, 6,* 217–226.

Allusi, E. A. (1970). Information and uncertainty: The metrics of communication. In K. B. DeGreene, Ed., *Systems Psychology.* New York: McGraw Hill.

Andersen, G. J., and Braunstein, M. L. (1985). Induced self-motion in central vision. *Journal of Experimental Psychology: Human Perception and Performance, 11,* 122–132.

Attneave, F. (1954). Some informational aspects of visual perception. *Psychological Reviews, 61,* 183–193.

Bailey, I., and Bullimore, M. A. (1992). Measuring the effects of glare. *Optometry and Vision Science, 69,* 593–594.

Ball, K., and Sekuler, R. (1979). Masking of motion by broadband and direction filtered noise. *Perception and Psychophysics, 26,* 206–241.

Ball, K., and Sekuler, R. (1987). Direction-specific improvement in motion discrimination. *Vision Research, 27,* 953–965.

Bannister, H., and Blackburn, J. M. (1931). An eye factor affecting proficiency at ball games. *British Journal of Physiology, 21,* 382–384.

Beggs, W. D. A., and Howarth, C. I. (1970). Movement control in a repetitive motor task. *Nature, 225,* 752–753.

Bergen, J. R. (1991). Theories of visual texture perception. In D. Regan, Ed., *Spatial Vision.* London: Macmillan, pp. 114–134.

Beverley, K. I., and Regan, D. (1975). The relation between discrimination and sensitivity in the perception of motion in depth. *Journal of Physiology, 249,* 387–398.

Beverley, K. I., and Regan, D. (1983). Texture changes versus size changes as stimuli for motion in depth. *Vision Research, 23,* 1387–1400.

Bootsma, R. J. (1991). Predictive information and the control of action: What you see is what you get. *International Journal of Sports Psychology, 22,* 271–278.

Bootsma, R. J., and van Wieringen, P. C. W. (1990). Timing an attacking forehand drive in table tennis. *Journal of Experimental Psychology: Human Perception and Performance, 16,* 21–29.

Bose, M., 1990, *A History of Indian Cricket.* London: A. Deutsch Ltd.

Calderone, J. B., and Kaiser, M. K. (1989). Visual acceleration detection: Effect of sign and motion orientation. *Perception and Psychophysics, 45,* 391–394.

Calvert, E. S. (1954). Visual judgements in motion. *Journal of the Institute of Navigation, London, 27,* 233–251.

Carlton, L. G. (1981). Processing visual feedback information for movement control. *Journal of Experimental Psychology: Human Perception and Performance, 7,* 1019–1030.

Carlton, L. G. (1992). Visual processing time and the control of movement. In L. Proteau and D. Elliott, Eds., *Vision and Motor Control.* New York: Elsevier, pp. 3–31.

Cavallo, V., and Laurent, M. (1988). Visual information and skill level in time-to-collision estimation. *Perception, 17,* 623–632.

Chapman, S. (1968). Catching a baseball. *American Journal of Physics, 36,* 868–870.

Chubb, C., and Sperling, G. (1988). Drift-balanced random stimuli: A general basis for studying non-Fourier motion perception. *Journal of the Optical Society of America A5,* 1986–2007.

Cline, D., Hofstetter, H. W., Griffin, J. R. (1989). *Dictionary of visual science.* Radnor, Pennsylvania: Chilton, 432, 535.

Collewijn, H., Erkelens, C. J. (1990). Binocular eye movements and the perception of depth. In Kowler, Ed., *Eye Movements and their Role in Visual and Cognitive Processes,* Amsterdam: Elsevier, 213–261.

Cumming, B. G., and Parker, A. J. (1994). Binocular mechanisms for detecting motion in depth. *Vision Research, 34,* 483–495.

Cutting, J. E. (1986). *Perception with an eye for motion.* Cambridge, MA: M.I.T. Press.

Cutting, J. E., Braren, P. A., Johnson, S. H., and Springer, K. (1992). Wayfaring on foot from information in retinal, not optical flow. *Journal of Experimental Psycholgy: General, 121,* 41–72.

Cynader, M., and Regan, D. (1978). Neurones in cat parastriate cortex sensitive to the direction of motion in three-dimensional space. *Journal of Physiology, 274,* 549–569.

Davies, J. M. (1949). The aerodynamics of golf balls. *Journal of Applied Physics, 20,* 821–828.

Deary, I. J., and Mitchell, H. (1989). Inspection time and high-speed ball games. *Perception, 18,* 789–792.

De Bruyn, B., and Orban, G. A. (1988). Human velocity and direction discrimination measured with random dot patterns. *Vision Research, 28,* 1323–1335.

Delorme, A., and Martin, C. (1986). Roles of retinal periphery and depth periphery in linear vection and visual control of standing in humans. *Canadian Journal of Psychology, 40,* 176–187.

DeLucia, P. R. (1991). Pictorial and motion-based information for depth perception. *Journal of Experimental Psychology: Human Perception and Performance, 17,* 738–748.

Dienes, Z., and McLeod, P. (1993). How to catch a cricket ball. *Perception, 22,* 1427–1439.

Ditchburn, R. W. (1976). *Light,* 3rd Ed. London: Academic Press, pp. 184–186.

Duntley, S. Q. (1948). Visibility of distant objects. *Journal of the Optical Society of America, 38,* 232–237.

Elliott, D. B. (1992). Intermittent versus continuous control of manual aiming movements. In L. Proteau and D. Elliott Eds., *Vision and Motor Control,* New York: Elsevier, pp. 33–48.

Elliott, D. B., and Bullimore, M. A. (1993). Assessing the reliability discriminative ability and validity of disability glare tests. *Investigative Ophthalmology and Visual Science, 34,* 108–119.

Erkelens, C. J., and Collewijn, H. (1985a). Motion perception during dichoptic viewing of moving random-dot stereograms. *Vision Research, 25,* 583–588.

Erkelens, C. J., and Collewijn, H. (1985b). Eye movements and stereopsis during dichoptic viewing of moving random-dot stereograms. *Vision Research, 25,* 1689–1700.

Fitts, P. M. (1954). The informational capacity of the human motor system in controlling the amplitude of movements. *Journal of Experimental Psychology, 47,* 381–391.

Fitts, P. M., and Seeger, C. M. (1953). S-R compatibility: Spatial characteristics of stimulus and response codes. *Journal of Experimental Psychology, 46,* 199–210.

Fitts, P. M., Peterson, J. R., and Wolfe, G. (1963). Cognitive aspects of information processing. II. Adjustments to stimulus redundancy. *Journal of Experimental Psychology, 65,* 423–432.

Fitts, P. M., Peterson, J. R., and Wolfe, G. (1964). Information capacity of discrete motor responses. *Journal of Experimental Psychology, 67,* 103–112.

Freeman, T. C. A., and Harris, M. G. (1992). Human sensitivity to expanding and rotating motion: Effects of complementary masking and directional structure. *Vision Research, 32,* 81–87.

Frohlich, G. (1984). Aerodynamic drag crisis and its possible effect on the flight of baseballs. *American Journal of Physics, 52,* 325–334.

Gabor, D. (1946). Theory of communication. *Journal of the IEEE, 93,* 429–456.

Giaschi, D., Regan, D., Kraft, S., Hong, X. H. (1992). Defective processing of motion in the fellow eye of unilateral amblyopes. *Investigative Ophthalmology and Visual Science, 33,* 2483–2489.

Giaschi, D., Regan, D., Kothe, A. C., Hong, X. H., and Sharpe, J. A. (1992). Motion-defined letter detection and recognition in patients with multiple sclerosis. *Annals of Neurology, 31,* 621–628.

Gibson, J. J. (1950). *The Perception of the Visual World.* Boston: Houghton Mifflin.

Gibson, J. J. (1954). The visual perception of objective motion and subjective movement. *Psychology Review, 61,* 304–314.

Gibson, J. J. (1966). *The senses considered as perceptual systems.* Boston: Houghton Mifflin.

Gibson, J. J. (1979). *An ecological approach to visual perception.* Boston: Houghton Mifflin.

Gopher, D., and Donchin, E. (1986). Workload—An examination of the concepts. In K. R. Boff, L. Kaufman and J. P. Thomas, Eds., *Handbook of Perception and Human Performance, Vol. 2,* New York: Wiley, (Chapter 41, pp. 1–49).

Gottsdanker, R. M., Frick, J. W., and Lockard, R. B. (1961). Identifying the acceleration of visual targets. *British Journal of Psychology, 52,* 31–42.

Graham, N. (1989). *Visual pattern analyzers.* New York: Oxford University Press.

Gray, R. D., and Regan, D. (1995a). Cyclopean motion perception produced by oscillations of size, disparity and location. *Investigative Ophthalmology and Visual Science, 36* (4), S369.

Gray, R. D., and Regan, D. (1995b). Cyclopean motion perception produced by oscillations of size, disparity and location. *Vision Research, 35,* 655–666.

Gray, R. D., and Regan, D. (1997). Estimating time to collision using binocular retinal information alone, monocular retinal information alone, and a combination of the two. *Journal of Experimental Psychology: Human Perception and Performance,* submitted.

Green, D. M., and Swets, J. A. (1966). *Signal Detection Theory in Psychophysics.* New York: John Wiley.

Grosslight, J. H., Fletcher, H. J., Masterton, R. B., and Hagen, R. (1978). Monocular vision and landing performance in general aviation pilots: Cyclops revisited. *Human Factors, 20,* 127–133.

Helmholtz, H von (1962). Handbook of physiological optics. New York: Dover, pp. 295–296. First edition of the German original was published in 1866. English translation by JPC Southall for the Optical Society of America (1924) from the 3rd German edition of Handbuch der physiologischen optik. Hamburg: Voss (1909).

Heuer, H. (1993a). Directional discrimination of motion in depth based on changing target vergence. *Vision Research, 33,* 2153–2156.

Heuer, H. (1993b). Estimate of time to contact based on changing size and changing target vergence. *Perception, 22,* 549–563.

Hick, W. E. (1952). On the rate of gain of information. *Quarterly Journal of Experimental Psychology, 4,* 11–26.

Holding, D. H. (1981). *Human skill.* New York: John Wiley.

Hong, X. H., and Regan, D. (1989). Visual field defects for unidirectional and oscillatory motion in depth. *Vision Research, 29,* 809–819.

Howard, I. P. (1982). *Human Visual Orientation.* New York: John Wiley.

Howard, I. P., and Heckmann, T. (1989). Circular vection as a function of the relative sizes, distances, and positions of two competing visual displays. *Perception, 18,* 657–665.

Hoyle, F (1957). *The black cloud.* London: Penguin Books, pp. 26–27.

Hyman, R. (1953). Stimulus information as a determinant of reaction time. *Journal of Experimental Psychology, 45,* 188–196.

Ijspeert, K. J., deWaard, P. W. T., van den Berg, T. J. T. P., and deJong, P. T. V. M. (1990). The intraocular straylight functions in 129 healthy volunteers: dependence on angle, age and igmentation. *Vision Research, 30,* 699–707.

Jeannerod, M., and Prablanc, C. (1983). The visual control of reaching movements. In J. E. Desmedt, Ed., *Motor Control Mechanics in Man,* Paris: Massou, pp. 13–29.

Johnson, C. A., and Leibowitz, H. W. (1974). Practice, refractive error and feedback as factors influencing peripheral motion thresholds. *Perception and Psychophysics, 15,* 1276–1280.

Johnson, I. R., White, G. R., and Cumming, R. W. (1973). The role of optic expansion patterns in locomotor control. *American Journal of Psychology, 86,* 311–324.

Julesz, B. (1971). *Foundations of Cyclopean Perception.* Chicago: Chicago University Press.

Kaufman, L. (1974). *Sight and mind.* New York: Oxford University Press, 374–392.

Keele, S. W. (1986). Motor control. In, K. R. Boff, L. Kaufman, and J. P. Thomas, Eds., *Handbook of Perception and human Performance, Vol. 2,* New York: Wiley, Chapter 30, 1–60.

Keele, S. W., and Posner, M. I. (1968). Processing of visual feedback in rapid movements. *Journal of Experimental Psychology, 77,* 155–158.

Knight, J. T., and Salvendy, G. (1992). Psychomotor work capabilities. In G. Salvendy Ed., *Handbook of Industrial Engineering, 2nd ed.* New York: John Wiley.

Kruk, R., and Regan, D. (1995). Collision avoidance: A helicopter simulator study. *Aviation, Space and Environmental Medicine,* 67, 111–114.

Kruk, R., and Regan, D. (1983). Visual test results compared with flying performance in telemetry-tracked aircraft. *Aviation, Space and Environmental Medicine, 54,* 906–911.

Kruk, R., Regan, D., Beverley, K. I., and Longridge, T. (1981). Correlations between visual test results and flying performance on the Advanced Simulator for Pilot Training (ASPT). *Aviation, Space and Environmental Medicine, 52,* 455–460.

Kruk, R., Regan, D., Beverley, K. I., and Longridge, T. (1983). Flying performance on the Advanced Simulator for Pilot Training and laboratory tests of vision. *Human Factors, 25,* 457–466.

Langewiesche, W. (1944). *Stick and Rudder.* New York: McGraw Hill.

Langolf, G. D. (1976). An investigation of Fitts's law using a wide range of movement amplitudes. *Journal of Motor Behaviour, 8,* 113–128.

Lee, D. N., and Lishman, J. R. (1977). Visual control of locomotion. *Scandinavian Journal of Psychology, 18,* 224–230.

Lee, D. N., Lishman, J. R., and Thomson, J. A. (1982). Visual regulation of gait in long jumping. *Journal of Experimental Psychology: Human Perception and Performance, 8,* 448–459.

Lee, D. N., Young, D. S., Reddish, D. E., Lough, S., and Clayton, T. M. H. (1983). Visual timing in hitting an accelerating ball. *Quarterly Journal of Experimental Psychology, 35A,* 333–346.

Leibowitz, H. W. (1955a). The relation between the rate threshold for the perception of movement and luminance for various durations of exposure. *Journal of Experimental Psychology, 49,* 209–214.

Leibowitz, H. W. (1955b). Effect of reference lines on the discrimination of movement. *Journal of the Optical Society of America, 45,* 829–830.

Levinson, E., and Sekuler, R. (1980). A two-dimensional analysis of direction-specific adaptation. *Vision Research, 20,* 103–107.

Levitt, H. (1971). Transformed up-down methods in psychoacoustics. *Journal of the Acoustical Society of America, 49,* 467–477.

Lewis, C. (1936). *Sagittarius Rising,* London: Davis, Report 1977.

Lewis, C. E. Jr. and Kriers, G. E. (1969). Flight research program: XIV Landing performance in jet aircraft after the loss of binocular vision. *Aerospace Medicine, 44,* 957–963.

Longuet-Higgins, H. C., and Prazdny, K. (1980). The interpretation of a moving retinal image. *Proceedings of the Royal Society of London, Series B,* 208, 385–397.

Marr, D. (1982). *Vision: A computational Investigation into the human representation and processing of visual information.* San Francisco: Freeman.

McBeath, M. K., Shaffer, D. M., and Kaiser, M. K. (1995). How basebal outfielders determine where to run to catch fly balls. *Science, 268,* 569–573.

McCoglin, F. H. (1960). Movement thresholds in peripheral vision. *Journal of the Optical Society of America, 50,* 775–779.

McKee, S. P. (1981). A local mechanism for differential velocity detection. *Vision Research, 21,* 491–500.

McKee, S. P., and Nakayama, K. (1984). The detection of motion in the peripheral visual field. *Vision Research, 24,* 25–32.

McLeod, P. (1987). Reaction time and high-speed ball games. *Perception, 16,* 49–59.

McLeod, P., and Dienes, Z. (1993). Running to catch a ball. *Nature, 362,* 23.

Meyer, D. E., and Smith, J. E. K., Kornblum, S., Abrams, R. A., and Wright, C. E. (1990). Speed-accuracy tradeoffs in aimed movements: Towards a theory of rapid voluntary action. In M. Jeannerod, Ed., *Attention and Performance 13.* Hillsdale, N.J.: Erlbaum, 173–226.

Michaels, C. F., and Oudejans, R. R. F. (1992). The optics and actions of catching fly balls: zeroing out optical acceleration. *Ecological Psychology, 4,* 199–222.

Mohler, S. R., and Johnson, B. H. (1971). *Wiley Post, His* Winnie Mae, *and the World's First Pressure Suit.* Washington, D.C.: Smithsonian Institute Press, 15.

Morgan, M. (1986). Positional acuity without monocular cues. *Perception, 15,* 157–162.

Motter, B. C., Steinmetz, M. A., Duffy, C. J., and Mountcastle, V. B. (1987). Functional properties of parietal visual neurons: mechanisms of directionality along a single axis. *Journal of Neuroscience, 7,* 154–176.

Mustillo, P., Francis, E., Oross, S., Fox, R., and Orban, G. (1988). Anisotropies in global stereoscopic discrimination. *Vision Research, 28,* 1315–1321.

Nadler, M. P., Miller, D., and Nadler, D. J., Eds. (1990). *Glare and contrast sensitivity for clinicians.* New York: Springer.

Nakayama, K. (1985). Biological image motion processing: A review. *Vision Research, 25,* 625–660.

Nakayama, K., and Tyler, C. W. (1981). Psychophysical isolation of movement senstivity by removal of familiar position cues. *Vision Research, 21,* 427–433.

Neeley, J. H. (1977). Semantic priming and retrieval from lexical memory: roles of inhibitionless spreading activation and limited capacity attention. *Journal of Experimental Psychology: General, 106,* 226–254.

Newell, K. M., Carlton, L. G., and Carlton, M. J. (1982). The relationship of impulse to response timing error. *Journal of Motor Behaviour, 14,* 24–45.

Newsome, W. T., and Pare, E. B. (1988). A selective impairment of motion processing following lesions of the middle temporal visual are (MT). *Journal of Neuroscience, 8,* 2201–2211.

Niebel, B. (1962). *Motion and time study, 3rd ed.* Homewood, IL: Richard D. Irwin.

Norcia, A. M., and Tyler, C. W. (1984). Temporal frequency limits for stereoscopic apparent motion processes. *Vision Research, 24,* 395–401.

Orban, G. A., Calenbergh, F. van, DeBruyn, B., and Maes, H. (1985). Velocity discrimination in central and peripheral visual field. *Journal of the Optical Society of America, A2,* 1836–1847.

Pasternak, T., and Merigan, W. (1984). Effects of stimulus speed on direction discrimination. *Vision Research, 24,* 1349–1355.

Pelli, D. G., Robson, J. G., and Wilkins, A. J. (1988). The design of a new letter chart for measuring contrast sensitivity. *Clinical Vision Sciences, 2,* 187–199.

Peper, L., Bootsma, R. J., Mestre, D. R., and Bakker, F. C. (1994). Catching balls: how to get the hand in the right place at the right time. *Journal of Experimental Psychology: Human Perception and Performance, 20,* 591–612.

Pfaffman, C. (1948). Aircraft landings without binocular cues: a study based upon observations made in flight. *American Journal of Psychology, 61,* 323–335.

Phinney, R., Boyd, C., Winget, A., Patterson, R. (1995). Direction-thresholds for stereoscopic (non-Fourier) motion perception. *Investigative Ophthalmology and Visual Science, 36,* S666.

Portfors-Yeomans, C. V., and Regan, D. (1997a). Discrimination of the direction and speed of motion in depth from binocular information alone. *Journal of Experimental Psychology: Human Perception and Performance,* in press.

Portfors-Yeomans, C. V., and Regan, D. (1997b). Discriminating the direction of motion in depth of cyclopean form. *Vision Research,* in press.

Portfors-Yeomans, C. V., and Regan, D. (1997c). Just-noticeable differences in the speed of motion in depth and of motion within a frontoparallel plane for cyclopean form: Effects of the direction of motion and the sign of disparity. *Journal of Experimental Psychology: Human Perception and Performance,* in press.

Posner, M. I., and Snyder, C. R. R. (1975). Facilitation and inhibition in the processing of signals. In P. M. A. Rabbitt and S. Dornic, Eds., *Attention and Performance, Vol. 5,* New York: Academic.

Post, R. B. (1988). Circular vection is independent of stimulus eccentricity. *Perception, 17,* 737–744.

Probst, T., Krafczyk, S., Brandt, S., and Wist, E. R. (1984). Interaction between perceived self-motion and object-motion impairs vehicle guidance. *Science, 225,* 536–538.

Raymond, J. E. (1994). Directional anisotropy of motion sensitivity across the visual field. *Vision Research, 34,* 1029–1037.

Regan, D. (1985). Visual flow and direction of locomotion. *Science, 227,* 1063–1065.

Regan, D. (1988). Low contrast letter charts and sinewave grating tests in ophthalmological and neurological disorders. *Clinical Vision Science, 2,* 235–250.

Regan, D. (1991a). Spatial vision for objects defined by colour contrast, binocular disparity and motion parallax. In D. Regan, Ed., *Spatial Vision.* London: Macmillan, 135–178.

Regan, D. (1991b). Spatiotemporal abnormalities of vision in patients with multiple sclerosis. In D. Regan, Ed., *Spatial Vision.* London: Macmillan, 239–249.

Regan, D. (1991c). Prentice Medal Lecture. Specific tests and specific blindnesses: keys, locks and parallel processing. *Optometry and Vision Science, 68,* 489–512.

Regan, D. (1992). Visual judgments and misjudgments in cricket, and the art of flight. *Perception, 21,* 91–115.

Regan, D. (1995a). Orientation discrimination for bars defined by orientation texture. *Perception, 24,* 1131–1138.

Regan, D., (1995c). Spatial orientation in aviation: Visual contributions. *Journal of Vestibular Research, 5,* 455–471.

Regan, D., and Beverley, K. I. (1983). Visual fields for frontal plane motion and for changing size. *Vision Research, 23,* 673–676.

Regan, D., and Simpson, T. L. (1995). Multiple sclerosis can cause visual processing deficits specific to texture-defined form. *Neurology, 45,* 809–815.

Regan, D., and Vincent, A. (1995). Visual processing of looming and time to contact throughout the visual field. *Vision Research, 35,* 1845–1857.

Regan, D., Beverley, K. I., and Cynader, M. (1979). The visual perception of motion in depth. *Scientific American, 241,* 136–151.

Regan, D., Erkelens, C. J., and Collewijn, H. (1986). Visual field defects for vergence eye movements and for stereomotion perception. *Investigative Ophthalmology and Visual Science, 27,* 806–819.

Regan, D., Hajdur, L. V., and Hong, X. H. (1996). Two-dimensional aspect rato discrimination and one-dimensional width and height discriminations for shape defined by orientation texture. *Vision Research,* in press.

Regan, D., Hamstra, S. J., Kaushal, S., Vincent, A., Gray, R., and Beverley, K. I. (1995). Visual processing of an object's motion in three dimensions for a stationary or a moving observer. *Perception, 24,* 87–103.

Reichardt, W. (1961). Autocorrelation, a principle for the evaluation of sensory information by the central nervous system. In W. A. Rosenblith, Ed., *Sensory Communication.* Cambridge: MIT Press, 303–317.

Richards, W. (1971). Motion detection in man and other animals. *Brain Behaviour & Evolution, 4,* 162–181.

Richards, W., and Regan, D. (1973). Brightness contrast and evoked potentials. *Journal of the Optical Society of America, 63,* 606–11.

Rieger, J. H., and Lawton, D. T. (1985). Processing differential imagemotion. *Journal of the Optical Society of America, A2,* 354–360.

Rieger, J. H., and Toet, L. (1985). Human visual navigation in the presence of 3D rotations. *Biological Cybernetics, 52,* 377–381.

Runeson, S. (1974). Constant velocity—not perceived as such. *Psychological Research, 37,* 3–23.

Salvendy, G. (1987). *Handbook of human factors.* New York: John Wiley.

Salvendy, G., and Seymour, W. D. (1973). *Prediction and development of industrial work performance.* New York: John Wiley.

Savelsbergh, G. J. P., Whiting, H. T. A., and Bootsma, R. J. (1991). Grasping Tau. *Journal of Experimental Psychology: Human Perception and Performance, 17,* 315–322.

Sherwood, D. E., and Schmidt, R. A. (1980). The relationship between force and force variability in minimal and near-maximal static and dynamic contractions. *Journal of Motor Behaviour, 12,* 75–89.

Schiff, W., and Detwiler, M. L. (1979). Information judged in impending collision. *Perception, 8,* 647–658.

Schmerler, J. (1976). The visual perception of accelerated motion. *Perception, 5,* 167–185.

Schmidt, R. A., Zalaznik, H. N., and Frank, J. S. (1978). Sources of inaccuracy in rapid motor acts. In G. M. Stelmach, Ed., *Information Processing in Motor Control and Learning.* New York: Academic Press.

Schmidt, R. A., Zalaznik, H. N., Hawkins, B., Frank, J. S., and Quinn, J. T. (1979). Motor output variability: A theory for the accuracy of rapid motor acts. *Psychological Reviews, 86,* 415–451.

Sekuler, A. (1992). Simple-pooling of unidirectional motion predicts speed discrimnination for looming stimuli. *Vision Research, 32,* 2277–2288.

Shannon, C. E., and Weaver, W. (1949). *The mathematical theory of communication.* Urbana: University of Illinois Press.

Shaver, S. W., Telford, L., and Frost, B. J. (1991). The role of stereoscopic depth cues in centrally-mediated linear vection. *Investigative Ophthalmology and Visual Science, 32,* 696.

Simpson, T., and Regan, D. (1995). Test-retest variability and correlations between tests of texture processing, motion processing, visual acuity and contrast sensitivity,. *Optometry and Visual Science, 72,* 11–16.

Simpson, W. A. (1988). Depth discrimination from optic flow. *Perception, 17,* 497–512.

Spileers, W., Orban, G. A., Gulyas, B., and Maes, H. (1990). Selectivity of cat area 18 neurons for direction and speed. *Journal of Neurophysiology, 63,* 936–954.

Steinmetz, M. A., Motter, B. C., Duffy, C. J., and Mountcastle, V. B. (1987). Functional properties of visual parietal neurons: radial organization of directionality within the visual field. *Journal of Neuroscience, 7,* 177–191.

Teichner, W. H., and Krebs, M. J. (1972). Laws of simple visual reaction time. *Psychological Review, 79,* 344–358.

Telford, L., and Frost, B. J. (1991). The role of kinetic depth cues in centrally-mediated linear vection. *Investigative Ophthalmology and visual Science, 32,* 830.

Todd, J. T. (1981). Visual information about moving objects. *Journal of Experimental Psychology: Human Perception and Performance, 7,* 795–810.

Tyler, C. W. (1975). Characteristics of stereomovement suppression. *Perception and Psychophysics, 17,* 225–230.

Tyler, C. W., and Torres, J. (1972). Frequency response characteristics for sinusoidal movement in the fovea and periphery. *Perception and Psychophysics, 12,* 232–236.

van den Berg, T. J. T. P. (1991). On the relation between glare and straylight. *Documenta Ophthalmologica, 78,* 177–182.

van Doorn, A. J., and Koenderink, K. J. (1982a). Temporal properties of the visual detectability of moving white spatial noise. *Experimental Brain Research, 45,* 179–188.

van Doorn, A. J., and Koenderink, K. J. (1982b). Spatial properties of visual detectability of moving white spatial noise. *Experimental Brain Research, 45,* 189–195.

Van Santen, J. P. H., and Sperling, G. (1984). A temporal covariance model of motion perception. *Journal of the Optical Society of America, A1,* 451–473.

Van Santen, J. P. H., and Sperling, G. (1985). Elaborated Reichardt detectors. *Journal of the Optical Society of America, A2,* 300–321.

Vincent, A., and Regan, D. (1997). Judging the time to collision with a simulated textured object: effect of mismatching rates of expansion of size and of texture elements. *Perception and Psychophysics,* in press.

Warren, W. H., and Saunders, J. A., (1995). Perceiving heading in the presence of moving objects. *Perception, 24,* 315–331.

Warren, W. H., and Hannon, D. J. (1988). Direction of self-motion is perceived from optical flow. *Nature, 336,* 162–163.

Warren, W. H., and Hannon, D. J. (1990). Eye movements and optical flow. *Journal of the Optical Society of America, A7,* 160–169.

Warren, W. H., Young, D. S., and Lee, D. N. (1986). Visual control of step length during running over irregular terrain. *Journal of Experimental Psychology: Human Perception and Performance, 12,* 259–266.

Warren, W. H., Morris, M. W., and Kalish, M. (1988). Perception of translational heading from optical flow. *Journal of Experimental Psychology: Human Perception and Performance, 14,* 644–660.

Watson, A. B., and Ahumada, A. J. (1985). Model of human visual-motion sensing. *Journal of the Optical Society of America, A2,* 322–341.

Welford, A. T. (1968). *Fundamentals of skill.* London: Methuen.

Welford, A. T. (1976). *Skilled Performance: Perceptual and Motor Skills.* Glenview, IL: Scott, Foresman.

Westheimer, G., and Liang, J. (1995). Influence of ocular light scatter on the eye's optical performance. *Journal of the Optical Society of America, A12,* 1417–1424.

Wheatstone, C. (1852). Contributions to the physiology of vision. II. *Philosophical Transcripts of the Royal Society, 142,* 1–18.

Whiteside, T. C., and Samuel, C. D. (1970). Blur zone. *Nature, 225,* 94–95.

Williams, D. W., and Sekuler, R. (1984). Coherent global motion percepts from stochastic local motions. *Vision Research, 24,* 55–62.

Yeats, V. M. (1934). *Winged victory.* London: J. Cape, Report 1971.

Yeomans, C. V., and Regan, D. (1995). Direction discrimination and speed discrimination of motion in depth using binocular cues only. *Investigative Ophthalmology and Visual Science, 36,* S813.

Zelaznik, H. N., Shapiro, D. C., and McColsky, D. (1981). Effects of a secondary task on the accuracy of single aiming movements. *Journal of Experimental Psychology: Human Perception and Performance, 7,* 1007–1018.

CHAPTER 8
ENGINEERING ANTHROPOMETRY

Karl H. E. Kroemer
Department of Industrial and Systems Engineering
Virginia Polytechnic Institute and State University
Blacksburg, VA 24061-0118 USA

8.1 INTRODUCTION: THE HUMAN IS A SYSTEM ELEMENT

Data describing the size of the human body (and "biomechanical" information on strength, power, mobility, endurance, or fatigue) are available to the engineer just as is information on the technical components of the operator-task-equipment-environment (OTEE) system (see Figure 8.1). Engineers must be familiar with anthromechanic data and with their applications, so that they can design work systems, equipment, tools, and jobs for proper fit to the human, to achieve safe and efficient operation.

For the planned design of manned systems, or for the evaluation of existing ones, the following logic sequence applies.

1. *Task analysis.* Allocate duties to either person or machine, considering the operational environment. For this:
   - **(a)** review operator dimensions and capabilities essential for the design task, and
   - **(b)** review machine specifications needed.
2. *Operator–machine interface design.* Using computer models, templates, and drawings, carefully design the person–equipment interface.
3. *Test of the designed system.* Using prototypes or models, evaluate the ease and efficiency of task performance. Redesign, if necessary.

Figure 8.1 indicates this sequence. There is every reason to treat information on the human in the same systematic and conscious manner as information on machine characteristics; in fact, the human is the most important part in a OTEE system and should therefore be considered first.

8.2 DEVELOPMENT AND SCOPE OF ANTHROPOMETRY

At the end of the thirteenth century, Marco Polo described the different body builds of the human tribes that he saw on his world travels. Hence, physical anthropology as a recording and comparing science is often traced to him. Blumenbach's *On the Natural*

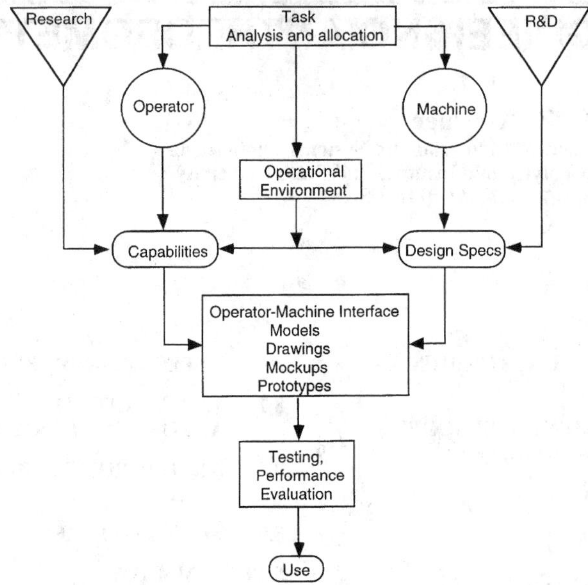

Figure 8.1 Ergonomic approaches to design of OTEE (operator-task-equipment-environment) systems.

Differences in Mankind contained complete anthropometric data available around 1800. Quetelet applied statistics to anthropometric and somatographic surveys in the middle 1800s. At about the same time, a new offspring (now called "biomechanics") indicated that engineers were becoming aware of data on the human body and its mechanical properties for design purposes.

The rapidly increasing interest in body size information made standardization of measuring methods desirable. For this, the body is placed into certain repeatable "erect" standing and sitting postures, and bony landmarks on the body are identified from which to take measurements. Conventions of anthropologists in Monaco in 1906 and six years later in Geneva established standardization of measurement methods, described in Martin's *Lehrbuch der Anthropologie* (first edition in 1914) which became the authoritative textbook and handbook for many decades. New engineering applications, newly developed measuring techniques, and advanced statistical procedures gave reason to update standardization in the 1960s (by Hertzberg) and 1980s (by ISO Technical Committee 159). In the United States, the armed forces and NASA have been at the forefront of applying anthropometric data, often with computerized design models, an approach followed by industry.

Based on anthropometry, biomechanics (see Chapter 9) explains the characteristics of a biological system, especially of the human body, in mechanical terms. The origins reach back to the middle 1600s: Alfonso Giovani Borelli's model of the human body depicts the long bones as straight links connected at mechanical articulations representing body joints; muscles are the "engines" spanning one or two body mechanical articulations and moving the system components. This link-joint-muscle model still underlies current approaches. High-performance aircraft, space travel, automobile crash experiments, orthopedics, and sports sciences have benefited much from the mechanical consideration of human body. Modern anthropometric data often describe link lengths (Kroemer, Kroemer, and Kroemer-Elbert, 1994).

Anthropometric data are needed for general design standards and specific requirements, used both in the design of new systems and in the evaluation of existing ones so that products, machines, tools, and activities "fit" the human operator and user, with the purpose of achieving "ease and efficiency."

8.3 APPLICATIONS: HUMAN–EQUIPMENT INTERFACES

The primary application areas for anthromechanic data are in design of

- Work task
- Work space
- Equipment, tools, controls and other interfaces
- Protective clothing.

Design "contours" are established by the outline of the human body. These contours determine, for example, the sizes of openings *large* enough so that the human body can pass through them (doors, hatches, etc.) or through which body segments must fit for maintenance purposes, or of openings in machines so *small* that human body parts (e.g., fingers) cannot pass so that they may not be injured.

Human–equipment interfaces are related to points ("landmarks") on the human body. For example, the eye establishes a design specification for the "line" or "field" of sight, for example, in aircraft or computer workstation design; the "eye ellipse" (all possible locations of the driver's eyes) is used in automobile design. Another major interface is established by "butt and back" for design of seats in vehicles, or offices. The foot establishes the heel rest or "package origin" in automobile design, determining the location of pedals. Finally, the hand is a very important design parameter in terms of the outer reach, or preferred manipulation envelope.

Examples of anthromechanical applications are the "cockpits" in airplanes, automobiles, earthmoving equipment, tanks, submarines, and surface ships. Other typical applications are hand-operated controls, helmets, face masks, and survival clothing for fire fighting or mine rescues. The design of space crafts in the United States and in the Soviet Union depends much on human data.

8.4 TERMINOLOGY, STANDARDIZATION, AND MODELS

Figure 8.2 shows the measuring planes and often used terms in describing the human body. Figure 8.3 shows the Ear-Eye line (passing through the ear hole and the junction of the eye lids) as reference for head tilt posture.

The following terms and analogies are used in anthromechanics:

Height: A straight-line, point-to-point vertical measurement.

Breadth: A straight-line, point-to-point horizontal measurement running across the body or segment.

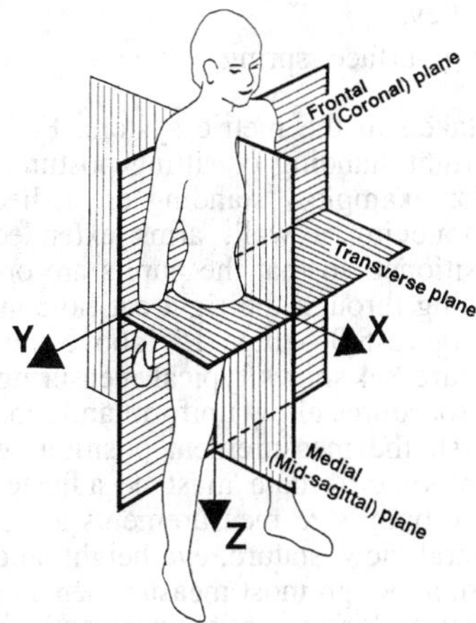

Figure 8.2 Reference planes used in anthropometry. If the person stands upright (erect), the three planes are usually set to meet in the center of mass of the body.

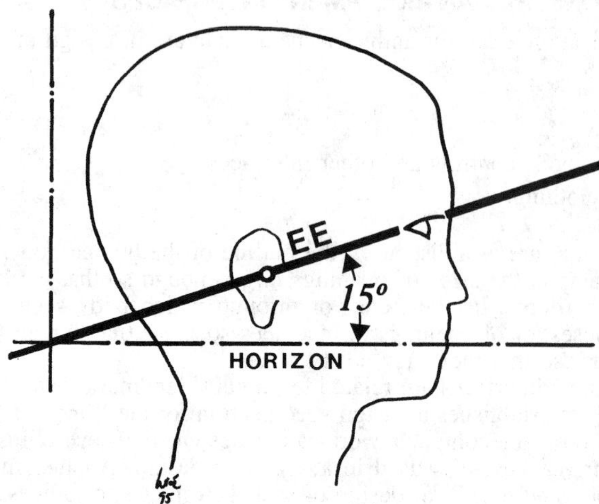

Figure 8.3 Ear–Eye line. If the head is held upright (erect), the EE Line is tilted 15° above the horizon.

Depth: A straight-line, point-to-point horizontal measurement running fore-aft.

Distance: A straight-line, point-to-point measurement between landmarks.

Circumference: A closed measurement following a body contour, hence usually not circular.

Curvature: A point-to-point measurement following the contour, usually not circular.

Bones: Structural members, central axes, lever arms.

Contour: Surface of geometric bodies.

Flesh: Volume, mass.

Joint: Bearing surface and articulation.

Joint lining, synovia: Lubricant.

Muscle: Motor, damper, or lock.

Nerve: Control and feedback circuit.

Organ: Generator, consumer.

Tendon: Cable transmitting pull force.

Tendon sheath: Sleeve, pulley.

Tissue: Elastic load-bearing surface, spring.

Traditionally, measures are taken in the metric system. For standardized measurement, the body is placed in an upright standing or sitting posture, with the extremities at 0°, 90°, or 180° to the trunk, for example, "standing erect, heels together; butt, shoulder blades, and back of head touching a wall; arms extended straight forward, fingers straight. . ." The head is positioned so that the pupils are on the same horizontal level, and the "Ear-Eye line" (running through the right ear hole and the outer junction of the lids of the right eye)—see Figure 8.3—tilted 15° above the horizon (eyes about 2 cm higher than the ear hole). Figure 8.4 shows typical measuring postures.

Although these standard procedures allow uniform and repeatable measurements, there remain several problems. First, the measurement positions employed are not the ones found at work. Hence, anthropometric data must be adjusted to reflect actual working postures. Second, many of the body size measurements are not spatially related to each other; for example, in the lateral view, stature, eye height, and shoulder height are not in the same frontal planes. Third, although most measurements are taken to bony landmarks, some are taken along soft tissues; this may cause problems, for example, in the fitting of face masks. Bony landmarks establish more reliable measuring points than soft tissue contours, and their relations to the skeletal links and articulations of the human body can be determined.

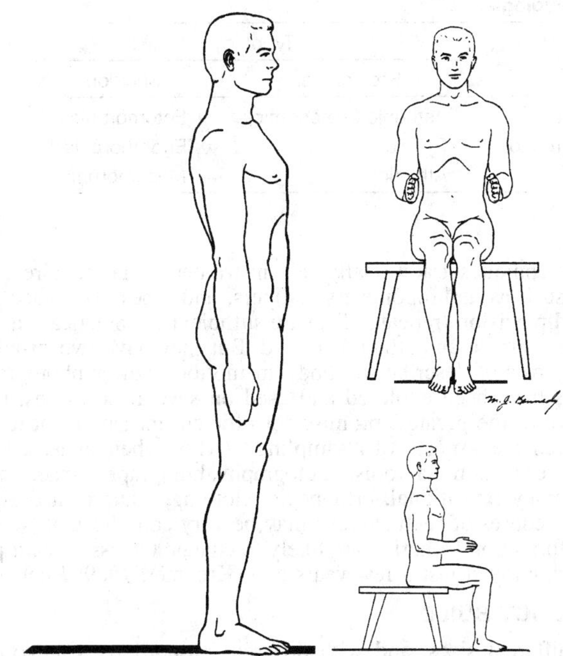

Figure 8.4 Typical postures used in anthropometric data gathering.

Anthropometric (as well as biomechanic) information is dependent on body postures, and motions. Static data, for example, on size, reach, or mass properties (such as location of the center of mass, or moment of inertia) or on isometric muscle strength depend decidedly on body posture. Dynamic data, such as on motion capabilities, energy exertion, or speed and accuracy of movements in the work space, are functions of the specific kinematic and kinetic conditions and are often hardly related to data obtained under static conditions. No general procedures or specific "recipes" are available for the conversion of static into dynamic data, although some data guidance is available from textbooks (Chaffin and Andersson, 1991; Kroemer, Kroemer and Kroemer-Elbert, 1994; NASA, 1979, 1989; Oezkaya and Nordin, 1991).

Computerized models of the human body, especially if incorporated into design software programs, can be of great help, or greatly deceiving. Some programs rely on outdated or insufficient anthropometric information; some programs, advertised as incorporating dynamic information, in reality rely on limited static information. The fact that a CAD program includes a "human model" does not guarantee that it is an appropriate or correct model. On the other hand, computerization does allow incorporation of rather complex and comprehensive information even for dynamic conditions; if well done, such a model of the human can be much more comprehensive, and much more easily used, than hard copy data. It is the responsibility of the designer to ascertain that the computer model used is complete and correct.

8.5 ANTHROPOMETRIC SURVEYS AND MEASUREMENT TECHNIQUES

Classical anthropometric surveys were not performed for engineering application purposes; only during the last few decades were surveys specifically aimed at providing design data, particularly for the military. Therefore, body size information on civilian populations is scarce, whereas soldiers are well described.

Large-scale anthropometric surveys are very expensive, time-consuming, and difficult to perform. However, on the basis of existing anthropometric knowledge and experience one can conduct highly directed smaller investigations in which only key dimensions are measured; then, other dimensions are derived statistically from those key dimensions. For this, regression models and other matching procedures have proved to be effective.

Table 8.1 Body Typologies

Descriptors	Typology	
	Kretschmer	Sheldon
Lean, slender, fragile	Asthenic (leptosomic)	Ectomorphic
Stocky, stout, soft, round	Pyknic	Endomorphic
Muscular	Athletic	Mesomorphic

Measurement techniques traditionally use instruments that require physical contact with the subject, such as anthropometers, calipers, and tapes. Their accuracy (reflecting the true value) and precision (repeatability) are sufficient for practical purposes, and their application is rather simple and straightforward. But they have two disadvantages: They are slow, and they cannot describe the body in motion. Newer photographic and holographic techniques have been explored and used on several occasions. Currently, laser-based techniques offer the greatest promise for efficient measurement techniques. There is interplay between the aspects of "sampling" (who, when, where, how many), the measuring procedures (traditional tools, photography, holography, space-spotting), and the purposes of the survey (nutritional, orthopedic, clothing, equipment design). Hence, organization and procedures of such surveys may be very complicated. New approaches to anthropometric information gathering are likely to change the assessment procedures very much from those employed just a few years ago (Kroemer, 1989; Roebuck, 1995).

8.6 TYPES OF BODY BUILD

People come in different sizes, and with different proportions. Categorization of body builds into different types is called *somatotyping.*[1] An early scheme was developed by Hippocrates (about 400 BC) who thought that body types, and particularly temperaments, were determined by body fluids. The psychiatrist Ernst Kretschmer developed in the 1920s a three-type system intended to describe character traits. His basic body types are somewhat similar to the ones used in the 1940s by W. H. Sheldon, whose ratings were meant to be purely descriptive of the body proportions. Sheldon typology was originally based on intuitive assessment, not on actual measurements, which were included later by his disciples. Table 8.1 describes the body types and the terms used to describe them. Unfortunately, these somatotypes have not proved to be reliable predictors of performance or capabilities, and hence are of little use for engineers.

8.7 VARIABILITY OF ANTHROPOMETRIC DATA

There are many causes for and symptoms of variability of anthropometric data. In addition to measuring or data treatment errors, they can be divided into three groups: (1) *interindividual* variations, (2) *intraindividual* variations, and (3) *secular* variations.

DNA combinations are a major cause of interindividual size variability: About 2.4×10^9 possible chromosome combinations exist. The individual genetic endorsement comprises the genotype (cellular differentiation) and the phenotype, which determines the biologically measurable characteristics. The environment may influence body size by altitude, temperature, sunlight, and topography including soil type. Nutrition has indirect and direct effects: Overeating increases body sizes in the direction of obesity, lack of nutrition leads to slenderness and to a reduction in height. The effects of aging are obvious. During the growing years, stature, weight, and accompanying dimensions increase. During early adulthood, in the twenties, many dimensions become reasonably stable. With increasing age, certain dimensions begin to change again, heights are reduced, and circumferences and weight increase. Table 8.2 approximates changes in key body dimensions with age.

Because the U.S. population is a composite of many different ethnic origins, it is difficult and in many cases pointless to explore body size differences associated with ethnic origin. In general terms, differences between U.S. white and black ethnic groups are relatively small, but there may be distinctly different measures between these two groups and people of oriental origin (Gordon, Churchill, Clauser et al., 1989). Some of

[1]From the greek *soma,* body.

Table 8.2 Approximate Changes in Stature with Age

	Change (cm)	
Age (years)	Females	Males
1 to 5[a]	+36	+36
5 to 10	+27	+25
10 to 15	+22	+30
15 to 20	+7	+11
20 to 35[b]	0	0
35 to 40	−1	0
40 to 50	−1	−1
50 to 60	−1	−1
60 to 70	−1	−1
70 to 80	−1	−1
80 to 90	−1	−1

[a] Average stature at age 1: females 72 cm, males 74 cm.

[b] Average maximal stature: females 163 cm, males 176 cm.

Source: Kroemer, Kroemer, and Kroemer-Elbert, 1994

these differences, however, are of no consequence for the design of equipment for the total U.S. population. United States population statistics include *all* ethnic groups—see Table 8.3. For other national and ethnic group data, see, e.g., Kroemer, Kroemer, and Kroemer-Elbert (1994); Pheasant (1996); and Roebuck (1995).

There are also some differences in body sizes among different professions. However, the often postulated and occasionally demonstrated differences between white and blue collar groups are not very clear and of little or no consequence for designs that aim to fit the whole population. The same holds true for left–right asymmetry, referring to differently developed body segments depending on one's preference to perform activities with the left or right hand or foot. However, about 10% of the U.S. population are left-handed, meaning that the left hand is preferred for such activities as hand-writing or hand-tool use. This may warrant special designs.

There are, of course, differences in body dimensions between males and females, as shown in Table 8.3 and Figure 8.5. However, in many measures there is also significant overlap, for example, in lower leg length and in buttock circumference. Thus, although there are gender disparities that may be of importance for special design purposes, other dimensions are sufficiently similar to allow "unisex" designs. A typical example is the adjustment range of arm rests on chairs, as reflected by the anthropometric dimension "elbow height above seat" in Table 8.3.

There are also transient diurnal body size changes. For example, a person's body weight may vary by up to 1 kg per day owing to changes in body water content. Stature may be reduced at the end of the day by up to 5 cm, mostly because of changes in postures and in thickness of spinal disks. Leaning erect against a wall during measurement as opposed to free standing may increase stature by up to 2 cm. Measuring stature in the prone position can increase stature up to 3 cm as compared to standing. Circumferences also change with different conditions: For example, hip and buttock circumference are quite different while sitting as compared to standing, chest circumference changes with breathing, biceps circumference reflects muscle flexion. Garments can change body space: For example, "foundations" worn by females influence certain circumferences; short sleeve clothing as compared to outdoor winter clothes establishes quite different contours and work spaces.

"Secular" changes in adult body sizes have been observed in the United States since the Civil War. There are striking developments since the 1940s or 1950s: Stature has increased by about 1 cm per decade, weight by as much as 2 kg per decade. One can speculate whether these body size changes will continue at the same speed, but the ob-

Table 8.3 Body Dimensions of U.S. Civilian Adults, Female/Male, in cm

	Percentiles			Std. Deviation
	fem. 5th male	fem. 50th male	fem. 95th male	fem. S. male
HEIGHTS				
STANDING				
(*f* = above floor, *s* = above seat)				
Stature ("height")^f	152.78 /164.69	162.94/175.58	173.73 /186.65	6.36 / 6.68
Eye height^f	141.52 /152.82	151.61/163.39	162.13 /174.29	6.25 / 6.57
Shoulder (acromial) height^f	124.09 /134.16	133.36/144.25	143.20 /154.56	5.79 / 6.20
Elbow height^f	92.63 / 99.52	99.79/107.25	107.40 /115.28	4.48 / 4.81
Wrist height^f	72.79 / 77.79	79.03/ 84.65	85.51 / 91.52	3.86 / 4.15
Crotch height^f	70.02 / 76.44	77.14/ 83.72	84.58 / 91.64	4.41 / 4.62
SITTING				
Height (sitting)^S	79.53 / 85.45	85.20/ 91.39	91.02 / 97.19	3.49 / 3.56
Eye height^S	68.46 / 73.50	73.87/ 79.20	79.43 / 84.80	3.32 / 3.42
Shoulder (acromial) Ht^S	50.91 / 54.85	55.55/ 59.78	60.36 / 64.63	2.86 / 2.96
Elbow height^S	17.57 / 18.41	22.05/ 23.06	26.44 / 27.37	2.68 / 2.72
Thigh height^S	14.04 / 14.86	15.89/ 16.82	18.02 / 18.99	1.21 / 1.26
Knee height^S	47.40 / 51.44	51.54/ 55.88	56.02 / 60.57	2.63 / 2.79
Popliteal height^f	35.13 / 39.46	38.94/ 43.41	42.94 / 47.63	2.37 / 2.49
DEPTHS				
Forward (thumbtip) reach	67.67 / 73.92	73.46/ 80.08	79.67 / 86.70	3.64 / 3.92
Buttock-knee distance (sitting)	54.21 / 56.90	58.89/ 61.64	63.98 / 66.74	2.96 / 2.99
Button-popliteal distance (sitting)	44.00 / 45.81	48.17/ 50.04	52.77 / 54.55	2.66 / 2.66
Elbow-fingertip distance	40.62 / 44.79	44.29/ 48.40	48.25 / 52.42	2.34 / 2.33
Chest depth	20.86 / 20.96	23.94/ 24.32	27.78 / 28.04	2.11 / 2.15

BREADTHS				
Forearm-forearm breadth	41.47 / 47.74	46.85/ 54.61	52.84 / 62.06	3.47 / 4.36
Hip breadth (sitting)	34.25 / 32.87	38.45/ 36.68	43.22 / 41.16	2.72 / 2.52
HEAD DIMENSIONS				
Head circumference	52.25 / 54.27	54.62/ 56.77	57.05 / 59.35	1.46 / 1.54
Head breadth	13.66 / 14.31	14.44/ 15.17	15.27 / 16.08	0.49 / 0.54
Interpupillary breadth	5.66 / 5.88	6.23/ 6.47	6.85 / 7.10	0.36 / 0.37
FOOT DIMENSIONS				
Foot length	22.44 / 24.88	24.44/ 26.97	26.46 / 29.20	1.22 / 1.31
Foot breadth	8.16 / 9.23	8.97/ 10.06	9.78 / 10.95	0.49 / 0.53
Lateral malleolus height[f]	5.23 / 5.84	6.06/ 6.71	6.97 / 7.64	0.53 / 0.55
HAND DIMENSIONS				
Circumference, metacarpale	17.25 / 19.85	18.62/ 21.38	20.03 / 23.03	0.85 / 0.97
Hand length	16.50 / 17.87	18.05/ 19.38	19.69 / 21.06	0.97 / 0.98
Hand breadth, metacarpale	7.34 / 8.36	7.94/ 9.04	8.56 / 9.76	0.38 / 0.42
Thumb breadth	1.86 / 2.19	2.07/ 2.41	2.29 / 2.65	0.13 / 0.14
Interphalangeal				
WEIGHT (in kg)	39.2*/ 57.7*	62.01/ 78.49	84.8*/ 99.3*	13.8*/12.6*

*Estimated (from Kroemer, 1981)

From Kroemer, Kroemer, and Kroemer (1994), Ergonomics. Englewood Cliffs, NJ: Prentice Hall. Adapted from U.S. Army data reported by Gordon, Churchill, Clauser, Bradtmiller, McConville, Tebbetts, and Walker (1989).

Note: In this table, the entries in the fiftieth percentile column are actually "mean" (average) values. The fifth and ninety-fifth percentile values are from measured data, not calculated (except for weight). Thus, the values given may be slightly different from those obtained by subtracting 1.65 S from the mean (fiftieth percentile), or by adding 1.65 S to it.

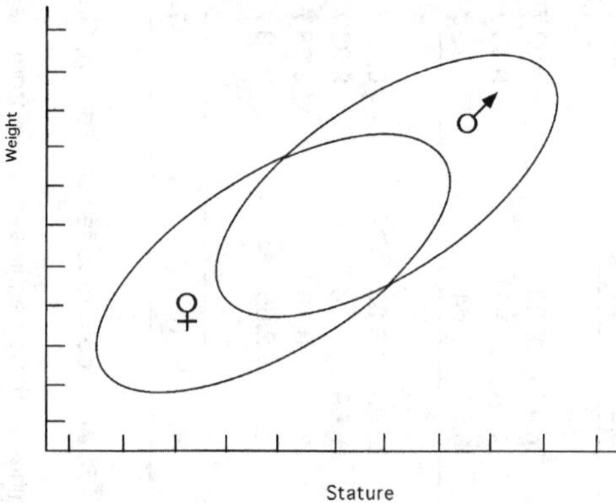

Figure 8.5 Typical stature and weight distributions for females and males. Approximately 95% of each group is contained within the ellipse.

servation period with reliable body measures is too short to be certain; however, there is reason to believe that the size changes depend on interactions between genetic and environmental factors such as nutrition, health, care, and sanitation. This poses an interesting design task: Clothing and other equipment that must fit the body tightly need to be adjusted to accommodate short-term secular changes, such as circumferences and weight; however, changes in longitudinal dimensions, such as in stature, are too slow to be of importance for most industrial applications; very few pieces of equipment, or tools, have to be fitted so exactly that the rather minute changes in anthropometry would require design standard changes within relatively the short use periods, such as 5 or 10 years, during which the products will be in use. Altogether, inter-individual variability changes are far wider than most intra-individual or secular variations.

8.8 ANTHROPOMETRIC DESIGN DATA

A most recent and complete data set on body size of the U.S. Army soldiers was published by Gordon, Churchill, Clauser et al. in 1989. With no comprehensive and current information on the American civilian population available, the military data set is the best possible estimate of present body data on U.S. civilians. Fortunately, there is reason to believe that American Army soldiers are quite representative of the U.S. civilian population, better than Navy or Air Force personnel. However, soldiers usually do not represent the extremes of body dimensions, such as very tall or very short, or very heavy. Thus, as discussed elsewhere (Kroemer, Kroemer, and Kroemer-Elbert, 1994; Roebuck, 1995), the U.S. Army data contained in Table 8.3 can be used to estimate body dimensions of the U.S. civilian population, with some caution to be applied to the extreme ends of the distributions, especially with respect to height–weight combinations.

As Table 8.3 indicates, the average stature (50th percentile) for females is nearly 163 cm (5′4″), for males nearly 176 cm (about 5′9″). This is a bit less than has been reported in other publications, which apparently relied on data taken from the U.S. Air Force or Navy subpopulations. More information on body dimensions of populations throughout the world has been reported by Kroemer, Kroemer, and Kroemer-Elbert (1994) and by Pheasant (1986).

Most anthropometric (and many biomechanical) data follow reasonably well a normal (Gaussian) distribution. Hence, one can equate the fiftieth percentile with the average (mean) value. Normal distributions can be described using parametric statistics: average $\bar{x}$, standard deviation S, and sample size N. The most important formulas to calculate data cut-offs or ranges are given in Tables 8.4 and 8.5. Of course, they apply only if the variable x is normally distributed.

A percentile is the value of a variable (e.g., eye height) below which is a known percentage of values (say, 5%) and above which is the rest (in this case, 95%). If one designs, for example, to fit persons between the fifth and ninety-fifth percentiles, one knows that this design will fit the central 90%, but it is too large for 5% and too small for another 5% of the intended user population. The use of percentiles is a major step

Table 8.4 Primary Statistics to Describe Normal Distributions

Mean (average)	$\bar{x} = \frac{\Sigma x}{N}$ coincides with median (50th percentile) and mode
Range	$x \max - x \min$
Standard deviation	$S = (\text{variance})^{1/2} = \left[\frac{\Sigma(x - \bar{x})^2}{N}\right]^{1/2}$ or $\left[\frac{\Sigma(x - \bar{x})^2}{N - 1}\right]^{1/2}$
Coefficient of correlation	$R = \frac{S_{xy}}{(S_x S_y)^{1/2}}$
Adding mean values:	$\bar{Z} = \bar{X} + \bar{Y}$ $S_Z = S_x^2 + S_y^2 + 2\mathbf{r}\, \mathbf{S}_x \mathbf{S}_y$
Subtracting mean values:	$\bar{Z} = \bar{X} - \bar{Y}$ $S_Z = S_x^2 + \mathbf{S}_y^2 - 2\mathbf{r}\, \mathbf{S}_x \mathbf{S}_y$

away from the simple but false assumption that one could design for the mythical average person. No such phantom has ever existed. It has been shown repeatedly that a person average in one dimension (say, stature) is not very likely to be average in other dimensions (for example, weight, leg length, arm circumference). A person "average" in many or all dimensions simply does not exist; using this ghost as a design principle is inexcusable. Of course, one cannot make the assumptions either that there would be persons who have all fifth percentile dimensions, or eightieth, for example. A similar faulty assumption underlies the idea that one could use constant proportions such as alleging that leg length must be at a given percentage of stature, or that there is a constant ratio between weight and stature. Such proportioning assumes again that one can simply use average dimensions for both numerator and denominator of that ratio—this obviously can lead to a multiplication of errors.

A suitable design procedure is to select carefully the most critical dimension and to determine the appropriate percentile accommodation cutoff points at the low and high ends of that distribution. Though one often designs for a symmetrical distribution about the mean (such as designing for the fifth through ninety-fifth percentiles), there may be reason to select a nonsymmetrical user population, such as ranging from the twenty-fifth to the sixty-seventh percentile. Table 8.5 contains the multiplication factors and proce-

Table 8.5 Calculation of Percentiles Using Multiples of the Standard Deviation

Factor	Percentile p Associated With		
k	Below mean $x_b = \bar{x} - \text{kS}$	Above mean $x_a = \bar{x} + \text{kS}$	Central percent Included in the range x_b to x_a
2.576	0.5	99.5	99
2.326	1	99	98
2.06	2	98	96
1.96	2.5	97.5	95
1.88	3	97	94
1.65	5	95	90
1.28	10	90	80
1.04	15	85	70
1.00	16.5	83.5	67
0.84	20	80	60
0.67	25	75	50
0.32	37.5	62.5	25
0	50	50	0

Examples: For calculating the fifth percentile value, use $k = 1.65$, i.e., $X_{5_p} = \bar{x} - 1.65\ \text{S}$. The ninetieth percentile is at $X_{90_p} = \bar{x} + 1.28\ \text{S}$

dures to calculate percentile points. Table 8.6 helps to determine percentiles for composite data sets.

An example for the use of anthropometric data for the design of equipment is presented in Table 8.7. Here, the heights of the seat, of a support surface, and of a foot-rest were calculated for computer workstations, assuming that the first percentile female through ninety-ninth percentile male user had to be accommodated. The starting point for the calculations was "popliteal height, sitting" (from Table 8.3) to determine the necessary height of the seat. Thigh clearance height was added[2] to seat height for determining the minimal height of the opening under the support surface height. The entries in Table 8.7 show that different adjustment ranges and design dimensions result from the three design strategies used.

Of course, one must usually make assumptions and must do with data that may not be quite appropriate or reliable but are the "best available" or one's "best estimate." For example: In the calculation of adjust heights for computer workstations, the assumption was made that a person adjusting the seat to the highest position in order to accommodate long legs also may have very thick thighs (which is not necessarily true; see above) thus needing maximal clearance under the table; and that the computer workstations were to fit very small females as well as very big males (which may be an excessively wide design range). However, if such assumptions are rationally made and clearly stated, one can discuss the appropriateness in a reasoned manner and hence either continue to use these assumptions, or readjust them. The point is that exact anthromechanic data are available and should be used in the same exacting way as is done usually in engineering design.

8.9 BODY POSTURES

Standardized upright and stretched body postures help to measure body size; however, they are not convenient, comfortable, or voluntarily assumed at work. What, then, is a healthy and appropriate work posture? The evolution from four-legged to upright walking

Table 8.6 Percentile Values of Composite Populations

Given: Two samples, a and b; $a\% + b\% = 100\%$ of the composite population

To determine at what percentile of the composite population is a specific value of x, one proceeds stepwise. (For this, one needs to know mean $\bar{x}$ and standard deviation S of the distribution of variable x for both samples a and b.)

Step 1: Determine factors k associateed with x in the samples a and b. For sample a:

$$x_a = \bar{x}_a - k_a S_a \qquad \text{if } x_a < \bar{x}_a$$

$$x_a = \bar{x}_a + k_a S_a \qquad \text{if } x_a > \bar{x}_a$$

$$k_a = \frac{|x_a| - |\bar{x}_a|}{S_a}$$

Similarly, for sample b:

$$k_b = \frac{|x_b| - |\bar{x}_b|}{S_b}$$

Step 2: Obtain factor k associated with x in the combined population:

$$k = ak_a + bk_b$$

Step 3: Determine percentile p associated with k; use Table 8.5. If percentiles for each x are known in each group, one may simply add the proportioned percentiles:

$$p = ap_a + bp_p$$

[2]Measured (as opposed to calculated) fifth or ninety-fifth values were added without formally addressing covariation. Table 8.4 contains relevant formulas.

Table 8.7 Effects of Three Design Strategies on Heights of Equipment (in Centimeters) for Computer Workstations (Based on 1st Percentile Female and 99th Percentile Male Anthropometry)

Equipment Height	Strategies: Adjust Heights of Seat and of Support Surface	Strategies: Adjust Seat Height but Keep Support Surface Fixed	Strategies: Keep Height of Seat Fixed but Adjust Support Surface
Support surface[a]			
Maximum	73.0	73.0 (fixed)	73.0
Minimum	50.5		66.3
Adjustment range	22.5		6.7
Seat[b]			
Maximum	51.2	57.9	51.2 (fixed)
Minimum	35.4	51.9	
Adjustment range	15.8	6.7	
Footrest			
Maximum	Not needed	22.4	15.8
Minimum		0.0	0.0
Adjustment range		22.4	15.8

Source: Adapted from K. H. E. Kroemer, Design parameters for video display terminal stations. *Journal of Safety Research,* 14: 131–136 (1983).

[a] Assuming 2 cm table thickness.

[b] Including 2 cm for heels.

has been blamed for current spinal problems. Orthopedists and parents used to exhort others to "sit straight," but relaxed leaning against a backrest is usually more agreeable. Although many medical, physiological, orthopedic, and anatomical questions still must be answered, biomechanical considerations and practical experiences support the following observations:

- A person should be allowed to change body postures often and freely. Forcing an operator into a given posture that must be maintained over considerable time always becomes uncomfortable, if not unbearable, rather quickly.
- Walking and standing allow exertion of strong hand forces and permits large body motions, but the total weight of the body including the legs must be transmitted through the feet, whereas in sitting only the weight of the upper body must be transmitted to the seat while the feet rest on the floor.
- Sitting is preferable to standing, provided the work to be performed does not require much hand force or extensive body movements. Large forward foot forces can be exerted while sitting with a solid backrest.
- When sitting, a comfortable chair must be provided. Though the concept of comfort is difficult, certain attributes are obvious: to allow change of sitting postures, transmit force through accommodating but not spongy surfaces, provide a large and adjustable back support with suitable upholstery, and permit easy adjustments in seat height and angles.

Under normal conditions, for healthy persons, a body position is "suitable" if it fulfills two requirements: The first is that the weight of body segments must be transmitted, in terms of a kinematic chain, in the least strainful way. For a seated person this means, for example, that headrest, full-size backrest, armrests, and the seatpan should be receiving the weight of head, arms, trunk, and thighs, while the weight of the lower legs and feet is transmitted to the floor or a suitable footrest. Examples are the driver workplace and the computer workstation. The second requirement applies if forces must be exerted toward outside objects, usually with hands and/or feet; they shall be counteracted at the shortest possible distance by the reaction forces provided on support surfaces. The typical example for this is the vehicle driver whose foot forces are primarily taken up by the lower portion of the backrest.

For material handling, particularly lifting and lowering of loads, the most adequate posture is to bend the knees and to lean only slightly forward, and to straighten legs (and back) while lifting the load in front of, and close to, the trunk (Ayoub and Mital, 1989; NIOSH, 1981; Waters, Putz-Anderson, Garg, and Fine, 1993). This reduces the biomechanical strain of body components to a minimum. Dangerous postures in manual material handling are particularly those in which the back is severely bent, and in which sideward twisting of the spinal column occurs. To avoid overexertion injuries, it is most effective to provide ergonomically designed work tasks and equipment that make stainful lifting, lowering, pushing, pulling, and carrying unnecessary (Snook and Ciriello, 1991). Training people how to "lift correctly" is much less effective than ergonomically correct job design.

8.10 SUMMARY

Ergonomic information about the human body, both in size and biomechanical capabilities, is just as important for the design of tools, equipment, workstations, and tasks, as is information about the technical components. Such human factors information is available, and its proper consideration ensures "usability" in terms of safety, performance, and ease of use.

REFERENCES

Ayoub, M. M., Mital, A. (1989). *Manual Materials Handling.* London, UK: Taylor & Francis.

Chaffin, D. B., and Andersson, G. B. J. (1991). *Occupational Biomechanics (2nd ed.)* New York: Wiley.

Gordon, C. C., Churchill, T., Clauser, C. E., Bradtmiller, B., McConville, J. T., Tebbetts, I., and Walker, R. A. (1989). *1988 Anthropometric Survey of U.S. Army Personnel: Summary Statistics Interim Report* (Technical Report NATICK/TR-89-027). Natick, MA; U.S. Army Natick Research, Development and Engineering Center.

Greiner, T. M. (1991). *Hand Anthropometry of U.S. Army Personnel* (Technical Report TR-92/011). Natick, MA: U.S. Army Natick Research, Development and Engineering Center.

Kroemer, K. H. E. (1983). An isoinertial technique to assess individual lifting capability. *Human Factors, 25,* 493–506.

Kroemer, K. H. E. (1989). Engineering Anthropometry. *Ergonomics, 32,* 767–784.

Kroemer, K. H. E., Kroemer, H. J., and Kroemer-Elbert, K. E. (1994). *Ergonomics. How to Design for Ease and Efficiency.* Englewood Cliffs, NJ: Prentice Hall.

Kroemer, K. H. E. Marras, W. S., McGlothlin, J. D., McIntyre, D. R., and Nordin, M. (1990). Assessing Human Dynamic Strength. *International Journal of Industrial Ergonomics, 6,* 199–210.

NASA (1978). *Anthropometric Source Book,* 3 vols. (NASA Reference Publication 1024). Houston, TX: NASA.

NASA (1989). *Man-Systems Integration Standards* (Revision A). (NASA-STD 3000). Houston, TX: L. B. J. Space Center.

NIOSH, Ed. (1981). *Work Practices Guide for Manual Lifting* (Technical Report 81-122). Cincinnati, OH: NIOSH.

Oezkaya, N., and Nordin, M. (1991). *Fundamentals of Biomechanics.* New York: Van Nostrand Reinhold.

Pheasant, S. (1996). *Bodyspace.* London, UK: Taylor and Francis.

Putz-Anderson, V., (1988). *Cumulative Trauma Disorders: A Manual for Musculoskeletal Diseases of the Upper Limbs.* London, UK: Taylor & Francis.

Roebuck, J. A. (1995). *Anthropometric Methods: Designing to Fit the Human Body.* Santa Monica, CA: Human Factors and Ergonomics Society.

Snook, S. H., Ciriello, V. M. (1991). The Design of Manual Handling Tasks: Revised Tables of Maximum Acceptable Weights and Forces. *Ergonomics, 34,* 1197–1213.

Waters, T. R., Putz-Anderson, V., Garg, A., and Fine, L. J. (1993). Revised NIOSH Equation for the Design and Evaluation of Manual Lifting Tasks. *Ergonomics, 36,* 749–776.

Weimer, J. (1994). *Handbook of Ergonomics and Human Factors Tables.* Englewood Cliffs, NJ: Prentice Hall.

CHAPTER 9

BIOMECHANICS OF THE HUMAN BODY

William S. Marras
Biodynamics Laboratory
The Ohio State University
Columbus, OH 43210 USA

9.1 SCOPE OF BIOMECHANICS

9.1.1 Definitions

Biomechanics is an interdisciplinary field that utilizes information from both the biological sciences and engineering mechanics to assess the function of the body. The function of interest in the human body in an ergonomics context is typically the mechanical loading or activity of the musculoskeletal system. The objective of biomechanical analyses is to enable one to quantitatively assess the loading and behavior of a body structure and compare this loading or behavior to the tolerance or capacity of the body. The characteristic that distinguishes biomechanics from other types of ergonomic analyses is that the comparison is quantitative in nature. By comparing the joint or tissue loading or activity to the tolerance or capacity of the structure, this quantification permits one to assess human function in terms of levels of risk related to a task. Thus, the quantification of human function permits one to design work tasks so that they minimize the risk of musculoskeletal injury or illness to the body.

The portion of biomechanics dealing with ergonomics issues is typically called industrial or occupational biomechanics. Chaffin and Andersson (1991) have defined occupational biomechanics as "the study of the physical interaction of workers with their tools, machines, and materials so as to enhance the worker's performance while minimizing the risk of musculoskeletal disorders." This chapter will concern itself with human biomechanics in this framework.

9.1.2 Significance of Biomechanics in the Control of Musculoskeletal Disorders

Musculoskeletal disorders are responsible for the greatest amount of activity limitation in the United States. The American Academy of Orthopedic Surgeons (1992) reports that during 1988 the number of musculoskeletal impairments nationwide reached nearly 30 million. They report musculoskeletal disorders as being responsible for the greatest number of physician office visits in the United States. In other words musculoskeletal disorders are frequently the cause of employee absences from the workplace. The distribution of musculoskeletal disorders has also been studied (Praemer et al., 1992). Back or spine impairments are responsible for over half (51.7%) of musculoskeletal disorders. The next most often affected portion of the body was the lower extremity or hip (37.3%) followed by the upper extremity or shoulder (11.0%).

The cost of these impairments is also substantial. For example, back injuries represent 22% of workers' compensation cases yet are responsible for 31% of compensation payments (National Safety Council, 1991). It has been estimated that the total cost to society of low back disorders alone ranges from 25 to 95 billion dollars per year (Cats-Baril and Frymoyer, 1991). Thus, prevention of work related musculoskeletal disorders can have a major impact on society and the industrial and manufacturing environment.

Some musculoskeletal disorders may not involve mechanical factors at all. For example, musculoskeletal disorders may be due to arthritis or other naturally occurring diseases. Work may serve as a catalyst to simply exacerbate the symptoms. There are also studies to suggest that musculoskeletal disorders may be a result of psychological or psychosocial factors (Bigos et al., 1986). However, many of the work-related musculoskeletal disorders are a result of mechanical damage to the musculoskeletal system. For example, specific types of mechanical loading constitute by far the greatest known risk factors for disc prolapse (Kelsey et al., 1984; Mundt et al., 1993) and for occupationally related low back disorder (Marras et al., 1993, 1995). It has been concluded that "spine mechanical failure (studies) must surely hold the key to understanding the underlying cause of much back pain" (Adams and Dolan, 1995) Thus, even though some nonmechanical factors may somewhat influence musculoskeletal injury, it is believed that biomechanical analyses could explain a major component of musculoskeletal injury and provide a valuable tool for the assessment and control of work-related disorders.

9.1.3 Assumptions

A major assumption of biomechanics is that the body behaves according to the laws of Newtonian mechanics. That is not to say that we consider workers as machines. But it does suggest that we must accept that mechanically, workers' musculoskeletal systems must behave as a mechanical system so that forces and loads imposed upon the body can be quantified. By definition, "mechanics is the study of forces and their effects on masses" (Kroemer, 1987).

Within Newtonian mechanics bodies behave according to three principles or "laws." We assume that in biomechanics the body is also subject to these principles. Newton's *first law* states that "a body persists in a state of rest or uniform motion in a straight line unless it is acted upon by another force." The *second law* states that "force is a product of mass and acceleration." Newton's *third law* stated that "to every action there is an equal and opposite reaction."

It should be noted that Newton's laws assume that motion is present. The study of mechanics that includes motion is called *dynamics*. If the resultant force in Newton's laws is zero then this special case involves no motion of the structure and this becomes a *static* analysis.

9.1.4 When is Biomechanics Appropriate?

It is important to understand the conditions under which biomechanics is useful for workplace assessments. Biomechanical assessments are most useful when the magnitude of the loads imposed upon the body are suspected of exceeding the tolerance of a structure. Our knowledge of instantaneous biomechanical loading is well developed. Thus, conditions where nonfatiguing exertions are present can be best analyzed by biomechanical means. However, when repetition rates become high or either local or whole body fatigue becomes an issue biomechanical analyses can still be useful in defining the loads imposed upon the body. However, the tolerance of the bodies tissues to such loads may not be well known. As our knowledge about tissue tolerance becomes more complete, biomechanical analyses of repetitive and fatiguing exertions will become more accurate.

9.2 BASIC BIOMECHANICAL PRINCIPLES

9.2.1 Acute versus Cumulative Trauma

Two types of trauma can affect the human body. *Acute* trauma refers to an application of force that exceeds the tolerance of the structure during an infrequent application of force. Acute trauma is typically associated with large exertions of force. For example, an acute trauma can occur when a worker is asked to lift an extremely heavy object as when moving a piano. *Cumulative* trauma, on the other hand, refers to the repeated application of force to a structure that tends to wear down a structure, thus lowering the structure tolerance to the point where the tolerance is exceeded through a reduction of the tolerance limit. Therefore, cumulative trauma represents more of a "wear and tear" on the structure. This type of trauma is becoming far more common in the workplace because more repetitive jobs are becoming common in industry, and it is the mechanism of concern for many ergonomics evaluations.

The sequence of events that occurs as a result of cumulative trauma is shown in Figure 9.1. This figure shows that cumulative trauma is initiated by manual exertions that are either frequent of prolonged. This repetitive or prolonged application of force affects either the tendons or muscles of the body. If the tendons are affected, the following sequence occurs. Tendons are subject to mechanical irritation. During repetitive work, tendons are repeatedly exposed to high levels of tension and groups of tendons can rub against each other. This mechanical irritation can cause the tendons to inflame and swell. This swelling will stimulate the activities of the *nociceptors* (pain sensors) surrounding the structure and will signal the central processing mechanism (brain) that there is a problem via the perception of pain. In response to this pain the body will attempt to control the problem via two mechanisms. First, the muscles surrounding the irritated area will increase their level of coactivation in an attempt to minimize the motion of the tendon since this motion will stimulate the nociceptors and result in further pain. Second, in an attempt to reduce the friction occurring within the tendon the body will increase its production of synovial fluid within the tendon sheath. However, given the limited volume of the tendon and the tendon sheath this increased production of synovial fluid often exacerbates the problem by further increasing the volume of the tendon and in turn further stimulating the surrounding nociceptors. This process results in chronic joint pain and a series of musculoskeletal reactions such as reduced strength, reduced tendon motion, and reduced mobility. Collectively, these reactions result in a functional disability.

A similar process occurs if the muscles are affected by cumulative trauma. Muscles can become problematic when they become fatigued. Fatigue can lower the tolerance to stress and can result in muscle *microtrauma*. This microtrauma typically means the muscle is partially torn and the tear will cause capillaries to rupture and result in swelling, edema, or inflammation at the site of the tear. This in turn causes pain through the stimulation of nociceptors. The body reacts by cocontracting the surrounding musculature and thereby

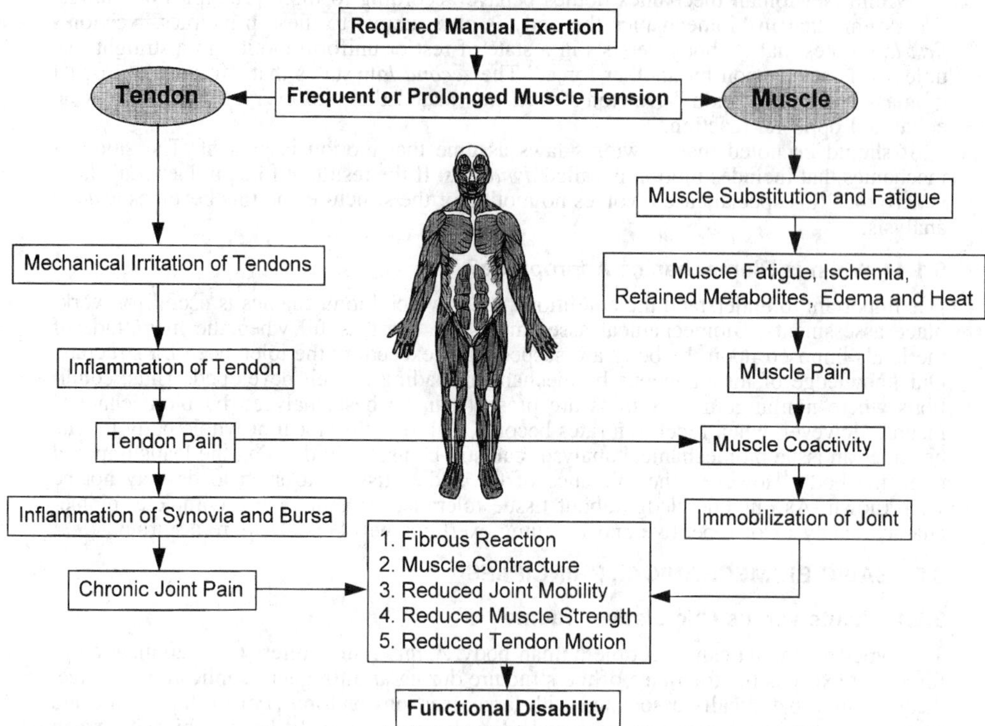

Figure 9.1 Sequence of events in cumulative trauma disorders. (Adapted from Occupational Biomechanics, Chaffin and Anderson, copyright © 1991, John Wiley and Sons, Inc. Reprinted by permission of John Wiley and Sons, Inc.)

minimizing the motion of the joint. This results in the same series of musculoskeletal reactions that result from tendon irritation (i.e., reduced strength, reduced tendon motion, and reduced mobility). The ultimate result of this process is once again a functional disability.

Even though the cumulative trauma process is somewhat similar between tendons and muscles there is a large difference in the time required to heal from the damage to a tendon compared to a muscle. Both tendons and muscles repair themselves via blood flow to the damaged structure. Blood provides nutrients for repair as well as dissipates waste materials. However, the blood flow to a tendon is just a fraction of that supplied to a muscle. Thus, given an equivalent strain to a muscle and a tendon the muscle will heal rather rapidly (at most about 10 days if not re-injured), whereas the tendon could take months to accomplish the same level of repair. For this reason, ergonomists must be particularly vigilant in the assessment of work places that could pose a danger to the tendons of the body.

9.2.2 Moments

Biomechanical loadings are not determined simply by the amount of weight supported by the body. The position of the weight relative to the axis of rotation of the body joint of interest defines the imposed load on the body. This position relative to the axis of rotation is called a *moment*. A moment is defined as the product of force and distance. Thus, a 50 Newton mass (Nm) held at a horizontal distance of 75 cm (0.75 m) from the shoulder joint imposes a moment of 37.5 Nm (50 N * 0.75 m) on the shoulder joint. Whereas, the same weight held at a horizontal distance of 25 cm from the shoulder joint imposes a moment or load of only 12.5 Nm (50/n * 0.25 m) on the shoulder. Thus, the load on a joint is a function of where the load is held relative to the joint and the mass of the weight held. Load is not simply a function of weight.

9.2.3 External versus Internal Loading

There are two types of forces or loadings that can affect the body while it is performing work tasks. *External* loads refer to those forces that are imposed on the body as a result of gravity acting upon an external object being manipulated by the worker. In Figure 9.2*a*, the tool held in the worker's hand is subject to the forces of gravity and imposes a 44.5 N (10 lb) external load at a distance from the joint of 30.5 cm (12 in.) on the elbow joint. However, this external load is counteracted by an *internal* load that is supplied by the muscles of the body. This figure shows that internal load (muscle) acts at a distance relative to the elbow joint that is much closer to the fulcrum than the external load (tool). Hence, the internal load or force is at a biomechanical disadvantage and must be much larger (534 N or 120 lb) than the external load (44.5 N or 10 lb) in order to keep the musculoskeletal system in equilibrium. It is typically the internal loading that contributes most to cumulative trauma of the musculoskeletal system during work. The sum of the external load and the internal load defines the total loading experienced at the joint, and it is a common occurrence that the internal loads far exceeds the external load. Therefore, when evaluating a workstation the ergonomist must not only consider the externally applied load but must be particularly sensitive to the magnitude of the internal forces that can load the musculoskeletal system.

9.2.4 Lever Systems in the Body

Three types of lever systems are represented in the human body. All parts of the musculoskeletal system can be represented by one of these classes of levers and aid in the quantification of loads imposed upon the body during work. *First-class* levers are those that have a fulcrum in the middle of the system, an imposed load on one end of the system and the restorative or internal load imposed on the opposite end of the system. Figure 9.3 shows an example of such a first-class lever system in the human body represented by the head and neck. In this example, the spine serves as the fulcrum. The center of gravity of the head is forward or anterior to the spine and represents the applied load. The restorative or internal load is supplied by the neck muscles (trapezius).

A *second-class* lever system can be found in the lower extremity. As shown in Figure 9.4, a second-class lever system is one where the fulcrum is on one end of the lever, the

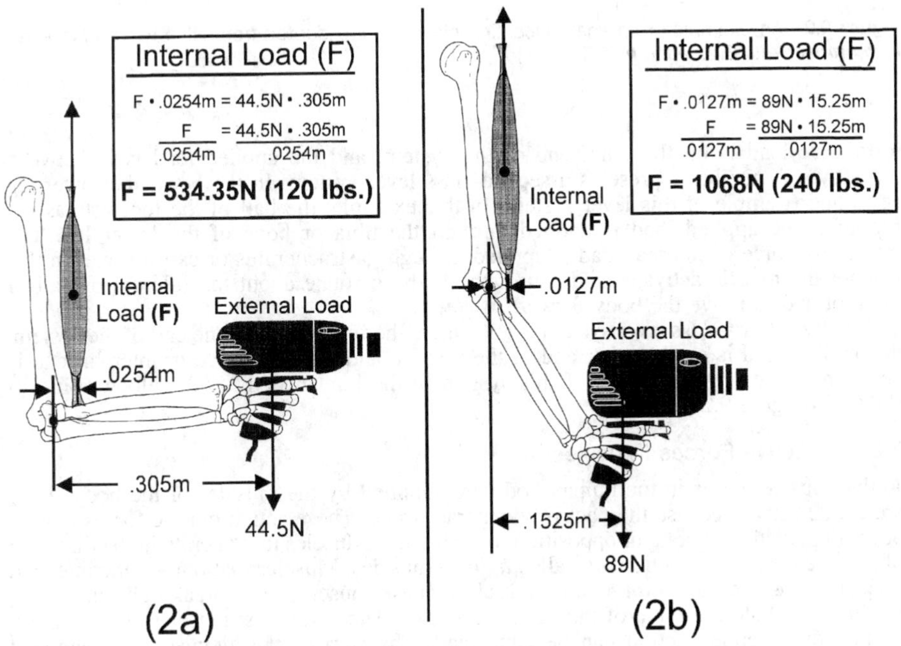

Figure 9.2 An example of an anatomical third-class lever.

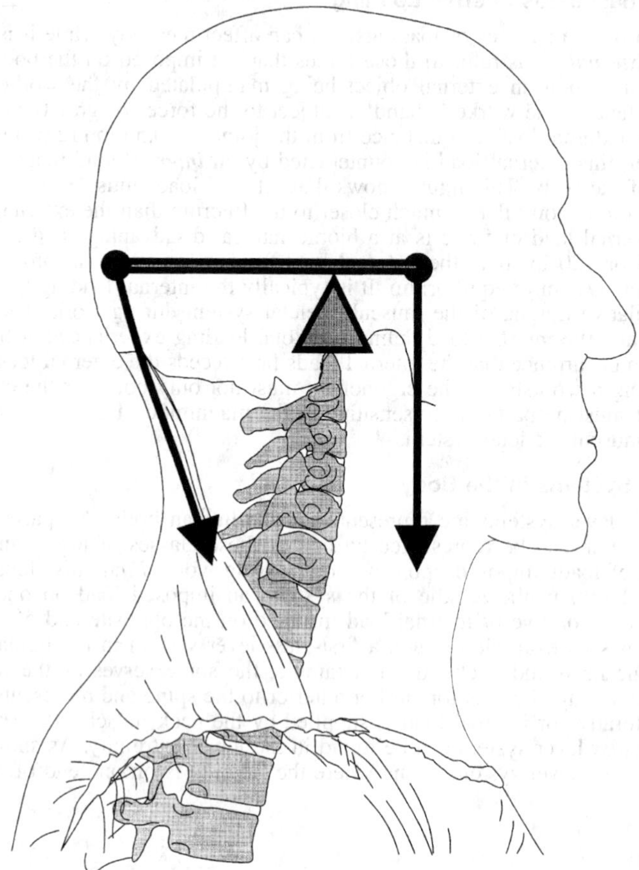

Figure 9.3 An example of an anatomical first-class lever. (Adapted from NIOSH, *The Industrial Environment—Its Evaluation and Control.*)

restorative load is on the other end of the system, and the applied load is in-between these two. A pry bar represents a second-class lever system. In the body, the lower leg is a good example of this lever system. In this example, the ball of the foot acts as the fulcrum. The applied load is applied through the tibia or bone of the lower leg. The restorative force or internal load is applied through gastrocnemius or calf muscle. In this manner the muscle activates and causes the body to rotate about the fulcrum or ball of the foot and to move the body forward.

Finally, a third-class lever system is one where the fulcrum is on one end of the system, the applied load is on the other end of the system, and the restorative or internal load is between the two. An example of this system in the human body is the elbow joint, as shown in Figure 9.2.

9.2.4.1 Active Forces (Muscles)

Active muscle forces in the human body are supplied by the muscles of the body. They are called active because they actively generate force. These active muscle forces always occur in pairs that operate in opposition to each other. Muscles that operate in the direction of the intended line of action are called *agonist* muscles. Muscles that oppose the intended line of action are called *antagonist* muscles. These muscles are "breaking" muscles in that they mediate the force of the agonist muscles. The antagonist muscles are necessary so that the intended action can be controlled. The sum of the agonist and antagonist muscle forces define the active loads experienced by the joint. Thus, any lever system

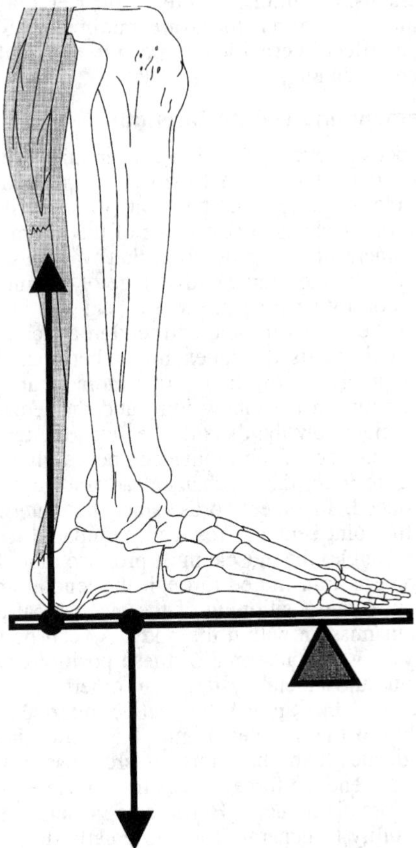

Figure 9.4 An example of an anatomical second-class lever. (Adapted from NIOSH, *The Industrial Environment—Its Evaluation and Control.*)

will have a multiple muscle system and in order to accurately assess the loads experienced by the system all muscle activities must be considered.

9.2.4.2 Passive Forces

Forces not related to the activity of the muscles can also load the musculoskeletal system and will contribute to the acute as well as cumulative trauma experienced by the system during work. Passive loads or forces are loads that are generated by the soft tissues but are not a direct result of muscle action. The most common type of passive load is generated by the ligaments of the body. Ligaments are viscoelastic and can store energy when stretched. Thus, when a portion of the body is loaded beyond the point where muscles are effective the load is supported by the ligaments. An example of such a load would be a stoop-type lift. If one was to lift an object off the floor without bending the knees and fully flexing the trunk, little back muscle activity would be necessary to support the object. However, the load would be supported by the ligaments of the spine. These ligaments can impose loadings on the joints just as the muscles. However, the loads are a result of stored energy in the ligaments. Thus, the activity of passive loading on the musculoskeletal system must be considered if one is to accurately assess loads on the musculoskeletal system under all working conditions.

9.2.5 Work Postures and Internal Loading

The previous discussion has discussed how important it is to understand the relationship between the external loads imposed upon the body and the internal loads generated by the force-generating mechanisms within the body. Proper ergonomic design involves designing workplaces so that the internal loads are minimized. Several properties of the work environment that can affect internal loading can be manipulated in order to achieve this minimization of internal loading.

9.2.5.1 Lever Arrangement and Length-Strength

The arrangement of the body's leverage system can greatly affect the magnitude of the internal load required to support the external load. The arrangement of the lever system can influence the magnitude of the external moment (force · distance) imposed upon the body as well as affect the biomechanical advantage of the internal forces. Consider again the biomechanical arrangement of the elbow joint that was presented in Figure 9.2. One can gain an appreciation for the mechanical advantage of the internal force generated by the biceps muscle and tendon by keeping one's arm bent at a 90° angle (as shown in the figure 9.2*a*) and palpating the biceps muscle and tendon attachment to the lower arm. If one palpates the tendon and inserts the index finger between the joint center and the tendon, one can gain an appreciation for the internal moment arm distance. Now with the index finger still inserted between the elbow joint and the tendon, if the arm is slowly straightened one can appreciate how the distance between the tendon and the joint center of rotation is significantly reduced. If the moment imposed about the elbow joint is held constant, as shown in Figure 9.2*b*, the mechanical advantage of the internal force generator is significantly reduced. In other words, since the internal moment or distance between the tendon and the joint center is reduced (compared to the situation where the elbow is positioned at a 90° angle) the muscle must produce more force in order to support the external load. This force is transmitted through the tendon and can increase the risk of cumulative trauma. Hence, the positioning of the mechanical lever system can greatly affect the internal load transmission within the body. As can be seen, the same task can be performed in a variety of ways but some of these positions are much more costly in terms of loading of the musculoskeletal system than others.

Another important relationship among the active internal force generators is the length–strength relationship of the muscles. Figure 9.5 shows the nature of this relationship. This relationship indicates that when muscles are close to their resting length they have the greatest capacity to generate force. But as they deviate from this resting position, their capacity to generate force is reduced. Hence as a muscle is stretched or as a muscle becomes very short, the ability to generate force is greatly diminished. This relationship is a function of the arrangement of the muscle fibers and can greatly influence work design. The practical significance of this relationship is that a given tension on a muscle can either tax the muscle greatly or be a minimum burden on the muscle. What might be considered a moderate force for a muscle at the resting length can become the maximum force a muscle can produce when it is in a stretched or contracted position, thus increasing the risk of muscle strain. When this relationship is considered in conjunction with the mechanical load place on the muscle and tendon via the arrangement of the lever

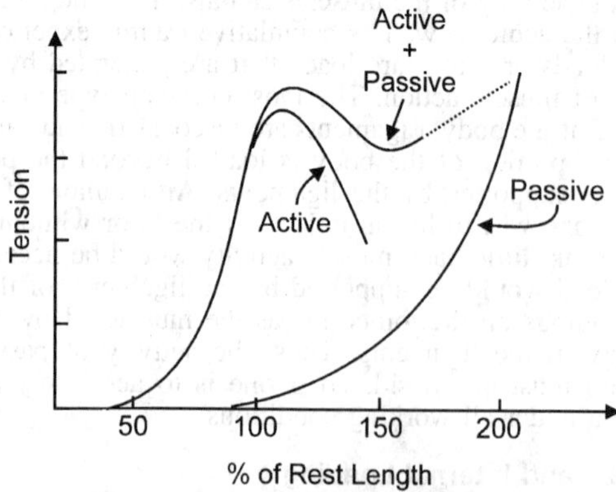

Figure 9.5 Length–tension relationship for a human muscle. (Adapted from Basmajian and DeLuca, *Muscles Alive*, Fifth edition copyright © 1985, Williams and Wilkins.)

system the position of the joint arrangement becomes a major factor in the design of the work environment. It is also common for the length–strength relationship to interact synergistically with the lever system. Figure 9.6 shows the effect of elbow position on the force generation capability of the elbow. This figure shows that position can have a dramatic effect on force generation. As discussed, this position can also have a great effect on internal loading of the joint.

9.2.5.2 The Force-Velocity Relationship

Motion can have a profound influence on the ability to generate force and load the biomechanical system. Motion can be either a benefit to the biomechanical system if momentum is properly used or it can increase the load on the system if the worker is not utilizing momentum. In general, we know that the effects of motion on a muscle is to reduce the force generation capacity of the muscle. Hence, the faster the muscle is moving the greater the reduction in force capability of the muscle. This effect is considered in many dynamic ergonomic biomechanical models.

9.2.5.3 The Strength-Endurance Relationship

From a biomechanical standpoint it is important to realize that strength is also associated with a temporal component. A worker could generate a great deal of strength during a one-time brief exertion. However, if the worker is expected to exert a maximal level of strength either for a prolonged period of time or repeatedly, the amount of force that the worker can generate is dramatically reduced. Figure 9.7 shows this relationship for the static endurance case. The dashed line in this figure indicates that if a person is asked to generate maximum muscle force, maximum force output is only generated for a very brief period of time. Then, as time increases, strength output decreases exponentially and levels off at about 20% of maximum after about 7 min. Similar trends occur under repeated dynamic conditions. In biomechanical terms, this indicates that if it is determined that a task requires a large portion of a worker's strength, one must consider how long that portion of the strength is required in order to ensure that the work does not strain the musculoskeletal system.

9.2.5.4 Percent of Strength Required

The percentage of strength required during a task also dictates the endurance capacity of the worker. This general relationship is shown via the family of curves, shown in Figure 9.8. This relationship indicates that the greater the percentage of strength required by a task, the less the amount of time a worker can be expected to maintain that level of strength. Thus, workers should be required to exert as small a percentage of their maximum strength as possible.

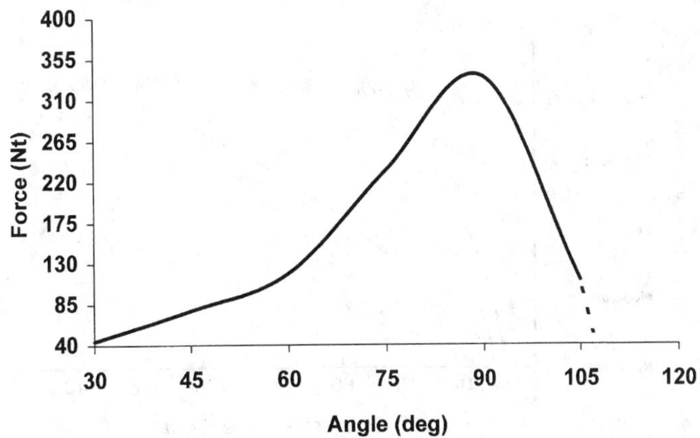

Figure 9.6 Length–tension diagram produced by flexion of the forearm in pronation. (Adapted from *Occupational Biomechanics*, Chaffin and Andersson, copyright © 1991, John Wiley and Sons, Inc. Reprinted by permission of John Wiley and Sons, Inc.)

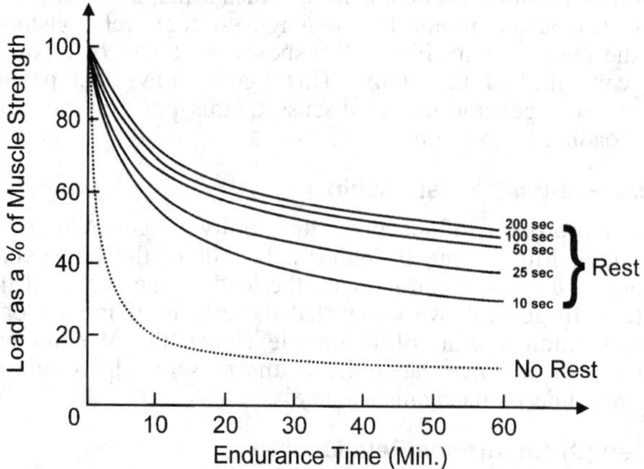

Figure 9.7 Forearm flexor muscle endurance times in consecutive static contractions. (Adapted from *Occupational Biomechanics*, Chaffin and Andersson, copyright © 1991, John Wiley and Sons, Inc. Reprinted by permission of John Wiley and Sons, Inc.)

9.2.5.5 Rest Time

Rest time has a profound effect on the ability to exert force. Figure 9.9 outlines the energy restoration process for a muscular contraction. This figure indicates that *adenosine triphosphate* (*ATP*) (the major energy source within cells) is required to produce a significant muscular contraction. ATP changes form to *adenosine diphosphate* (*ADP*) once a muscular contraction has occurred. This ADP must be converted to ATP in order to enable another muscular contraction. This conversion can occur with the addition of oxygen to the system. If oxygen is not present, then the system goes into oxygen debt and there is insufficient ATP for a muscular contraction. Thus, this flow chart indicates that oxygen is a key ingredient to maintain a high level of muscular exertion. Oxygen is delivered to the target muscles via the blood flow. However, under static exertions the blood flow is reduced and there is a subsequent reduction in the blood flow to the muscle. This restriction of blood flow and oxygen was responsible for the rapid decrease in force generation as a function of time in Figure 9.7. The solid lines shown in Figure 9.7 represent varying degrees of rest associated with a fatiguing exertion. As more and more rest time is permitted, an increase in force is realized since more oxygen is delivered to the muscle and

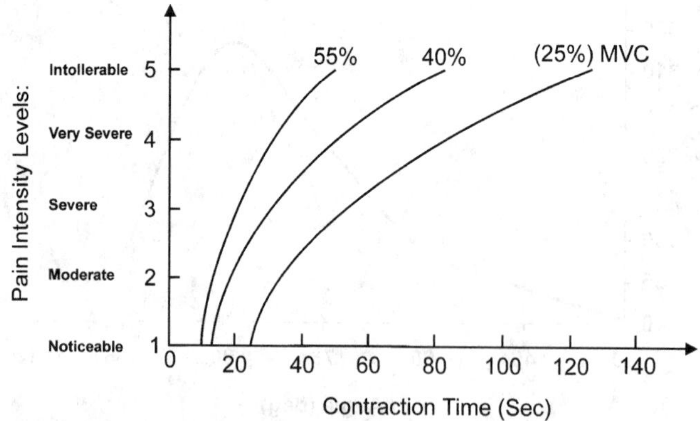

Figure 9.8 Mean time of appearance of various pain intensities for three static load levels. (Reprinted from *Occupational Biomechanics*, Chaffin and Andersson, copyright © 1991, John Wiley and Sons, Inc. Reprinted by permission of John Wiley and Sons, Inc.)

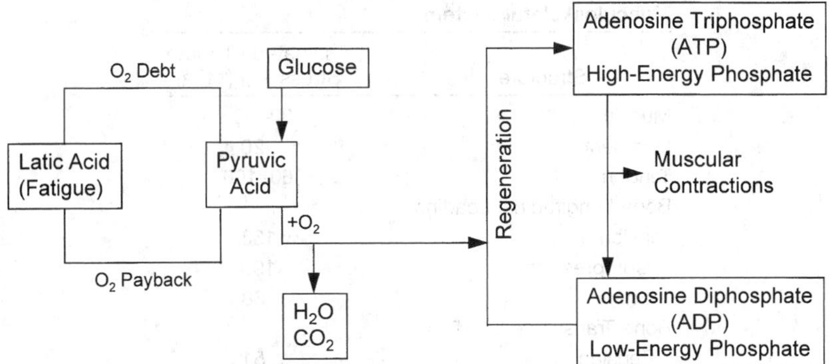

Figure 9.9 The body's energy system during work. (Adapted from Grandjean, *Fitting the Task to the Man,* Taylor & Francis, publisher.)

more ADP can be converted to ATP. However, this relationship also shows that any more than about 50 sec of rest, under these conditions, does not result in a significant increase in force generation capacity of the muscle. Hence, in order to optimize the strength capacity of the worker and minimize the possibility of muscle strain more frequent brief rest periods would be more beneficial than lengthy, infrequent rest periods.

9.2.6 Tissue Tolerance

Up to this point this chapter has been concerned with defining the factors that influence the loads applied to the structures of the body. A biomechanical analysis usually consists of comparing these applied loads against the ability of these structures to withstand or tolerate these loads. This section will briefly review the knowledge base associated with these tolerances.

9.2.6.1 Muscle, Ligament, Tendon, and Bone Capacity

The precise tolerance of human tendon under working conditions is difficult to estimate. Such measures depend on many factors such as strain rate, natural variability, and many unknown factors. In addition, it is not possible to measure these tolerances under human in vivo conditions. Therefore, most of the published estimates of tissue tolerance have come from various animal and/or theoretical sources.

Muscle appears to be the structure that has the lowest tolerance in the human musculoskeletal system. The ultimate strength of muscle has been estimated at 32 MPa (Hoy et al., 1990). It is generally accepted that the muscle will rupture prior to the tendon in a healthy tendon (Nordin and Frankel, 1989). Hence there is a safety margin between the muscle failure point and that of the tendon of about two (Nordin and Frankel, 1989) to three times (Hoy et al., 1990). Tendon stress has been estimated at between 60 and 100 MPa (Hoy et al., 1990; Nordin and Frankel, 1989).

Tolerances for the ligaments and bones of the musculoskeletal system have also been estimated. Ligamentous ultimate stress is approximately 20 MPa. The ultimate stress of bone is dependent on the direction of loading. It can range from as low as 51 MPa in transverse tension to over 190 MPa in longitudinal compression. Table 9.1 summarizes the ultimate strength of bone loaded in different loading conditions as well as the tolerance of muscle, ligaments, and tendon.

9.2.6.2 Disc/Endplate and Vertebrae Tolerance

Repeated microfracture of the vertebral end plane is thought to impair the nutrient flow to the disc fibers and thereby lead to fiber degeneration. Thus, it is thought that if one can determine the level at which the endplate experiences a microfracture, one can then minimize the effects of cumulative trauma and disc degeneration within the spine. Several studies of disc endplate tolerance have been performed. Figure 9.10 shows the levels of end plate tolerance that have been used to establish safe lifting situations at the worksite (NIOSH, 1991). This figure shows that for those under 40 years of age end plate damage begins to occur at about 350 Kg (3432 N) of compressive load on the spine. When the

Table 9.1 Tissue Tolerance of the Musculoskeletal System

Structure	Estimated Ultimate Stress (σ_u) (MPa)
Muscle	32–60
Ligament	20
Tendon	60–100
Bone Longitudinal Loading	
Tension	133
Compression	193
Shear	68
Bone Transverse Loading	
Tension	51
Compression	133

Source: Adapted from Ozkaya and Nordin, 1991.

compressive load reaches 650 Kg (6375 N), approximately 50% of those exposed to the load will experience vertebral end plate microfracture. If the compressive load on the spine is further increased to 950 Kg (9317 N) of loading, almost all of those exposed to the loading will experience a vertebral end plate microfracture. It should also be noted that the tolerance distribution shifts to lower levels with increasing age.

Even though this distribution of risk has been widely used as the tolerance limits of the spine it should be pointed out that others have identified different limits of vertebral end plate tolerance. Jager, Luttmann, and Laurig (1991) have reviewed 13 studies of spine compressive strength and identified significantly different risk limits. Their summary of these spine tolerance limits is shown in Table 9.2. These researchers have also been able to describe the vertebral compressive strength based upon an analysis of 262 values collected from 120 persons. They have related the compressive strength of the lumbar spine according to the following regression equation:

$$\text{Compressive strength (kN)} = (7.26 + 1.88\,G) - 0.494 + 0.468\,G) \cdot A + (0.042 + 0.106\,G) \cdot C - 0.145 \cdot L - 0.749 \cdot S$$

where:

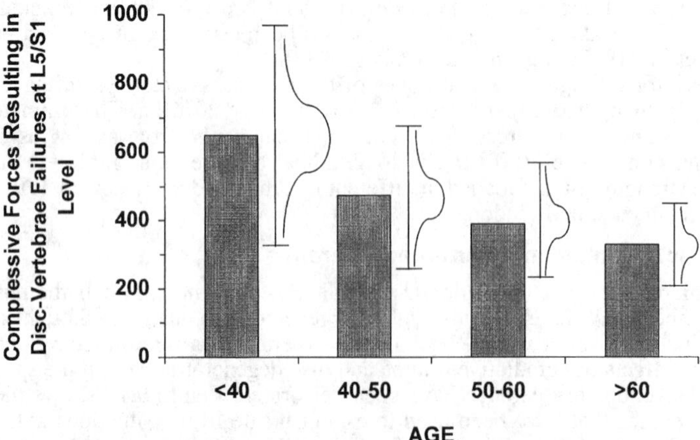

Figure 9.10 Mean and range of disc compression failures by age. (Adapted from NIOSH, *Work Practices Guide for Manual Lifting.*)

Table 9.2 Lumbar Spine Compressive Strength

		Strength in kN	
Population	n	Mean	s.d.
Females	132	3.97	1.50
Males	174	5.81	2.58
Total	507	4.96	2.20

Source: Jager et al., 1991.

A = age in decade
G = gender coded as 0 for female or 1 for male
C = cross-sectional area of the vertebrae in cm^2
L = the lumbar level unit where 0 is the L5/S1 disc, 1 represents the L5 vertebrae, etc. through 10, which represents the T10/L1 disc
S = the structure of interest where 0 is a disc and 1 is a vertebrae

This description indicates that the decrease in strength with the lumbar level is about 0.15 kN per adjacent vertebrae and that the strength of the vertebrae is about 0.8 kN lower than the strength of the discs (Jager et al., 1991). Using this equation, these researchers were able to describe 62% of the variance among the samples.

It has also been established that the tolerance limits of the spine varies as a function of frequency of loading (Brinckmann and associates, 1988). Figure 9.11 shows how the spine tolerance varies as a function of spine load level and frequency of loading.

9.3 BIOMECHANICS IN THE WORKPLACE

Now that the basic concepts and principles of biomechanics relevant to ergonomics situations have been established, this section will apply these principles to various regions of the body that are typically affected by work.

9.3.1 Shoulder

Shoulder region pain is second only to low back injury and neck pain in clinical frequency, and the reporting of such problems is on the rise. The shoulder is one of the more complex

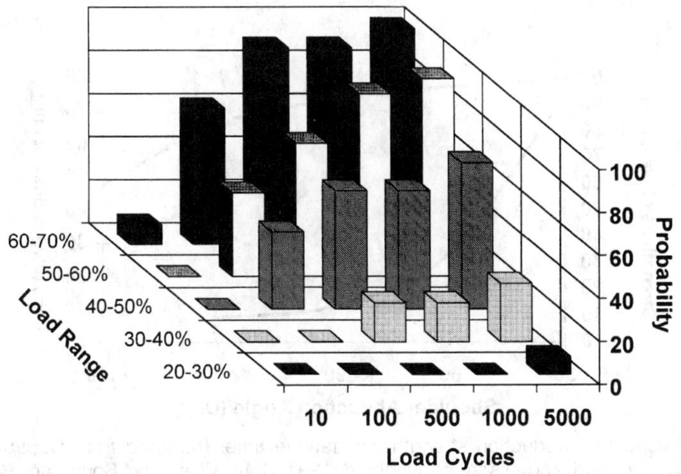

Figure 9.11 Probability of a motion segment to be fractured in dependence on the load range and the number of load cycles. (Adapted from Brinkman et al., 1988.)

structures of the body with numerous muscles and ligaments crossing the shoulder joint complex. Because of this complexity, surgical repair, once the shoulder is injured, can be problematic. It is often the case that repairing a tissue in the shoulder region necessitates damaging surrounding tissue. Thus, the best course of action is to ergonomically design workstations so that risk of initial injury is minimized.

Optimal workplace design is typically defined in terms of preferred posture during work. Shoulder abduction, defined as the elevation of the shoulder in the lateral direction, is often of concern when work is performed overhead. Figure 9.12 summarizes the strength and fatigue characteristics of the shoulder while in varying degrees of abduction. This figure indicates that shoulder strength remains relatively high from 30° through 90° of shoulder abduction. However, fatigue increases rapidly as the shoulder is abducted above 30°. Thus, even though strength is sufficient in some abducted postures, fatigue becomes the limiting factor. Hence, the only position of the shoulder that is acceptable from both a strength and fatigue standpoint is a shoulder abduction of at most 30°.

Shoulder flexion has been studied mainly as a function of fatigue. Chaffin (1973) has shown that even slight shoulder flexion can influence time to fatigue of the shoulder musculature. Figures 9.13 and 9.14 show the effects of vertical height of the work and horizontal distance of the work from the body during shoulder flexion while seated upon fatigability of the shoulder musculature. In the vertical flexion/extension case (Figure 9.13), fatigue occurs more rapidly as the worker's arm becomes more elevated. This trend is probably due to the fact that the muscles are farther from the neutral position as the shoulder becomes more elevated, thus affecting the length–strength relationship (Figure 9.5) of the shoulder muscles. Figure 9.14 shows that as horizontal distance is increased, the time to reach significant fatigue is decreased. This trend is due to the fact that as a load is held further from the body, more of a moment (force · distance) must be supported by the shoulder. Thus, the shoulder muscles are simply doing more work when the load is held further from the body and they fatigue quicker. Studies have also shown that providing an elbow support significantly increases the endurance time in these postures.

9.3.2 Neck

Neck problems are also typically associated with sustained postures of the neck. In general, the more upright the head the less muscle strength is required and the less the fatigue that will be experienced in the neck region. This effect is shown in Figure 9.15. This trend shows that once the head is tilted forward by 30° or more from vertical the time to fatigue increases rapidly in the neck musculature. Biomechanically, as the head is flexed forward, the center of mass of the head moves forward relative to the base of support of the head (spine). Thus, as the head is moved forward, more of a moment is being imposed about the spine which necessitates increased activation of the neck musculature. Therefore, when the head is not flexed forward, but is relatively upright, the neck can be positioned in such a way that minimal muscle activity is required of the neck muscles.

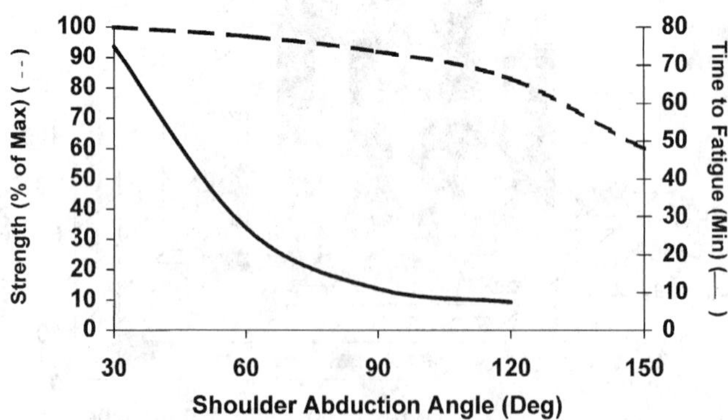

Figure 9.12 Shoulder abduction strength and fatigue time. (Adapted from *Occupational Biomechanics*, Chaffin and Andersson, copyright © 1991, John Wiley and Sons, Inc. Reprinted by permission of John Wiley and Sons, Inc.)

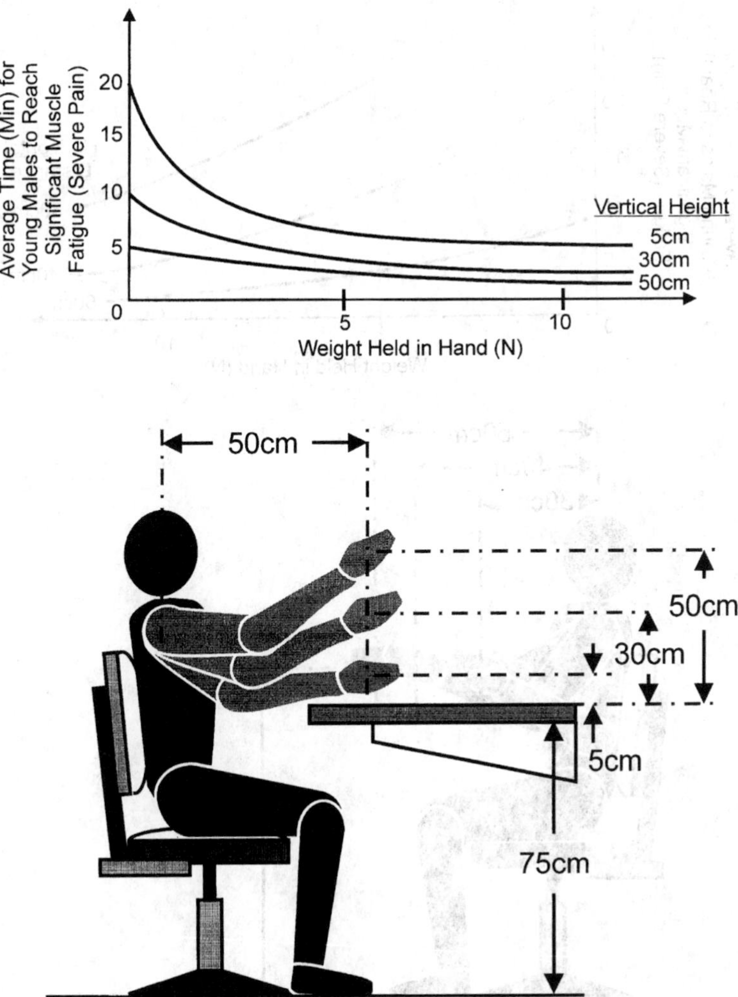

Figure 9.13 Expected time to reach significant shoulder muscle fatigue for varied arm flexion postures. (Adapted from *Occupational Biomechanics*, Chaffin and Andersson, copyright © 1991, John Wiley and Sons, Inc. Reprinted by permission of John Wiley and Sons, Inc.)

9.3.2.1 Neck–Shoulder Trade-Offs and Work Height

Biomechanical considerations in workplace design often necessitate the consideration of trade-offs between joints of the body. These trade-offs are necessary because it is often the case that what is advantageous for one part of the body is disadvantageous for another part of the body. Thus, many biomechanical considerations in the ergonomic design of the workplace requires one to think through the various trade-offs and rationales for various design options.

One of the most common trade-off situations in ergonomic design is a trade-off between accommodating the shoulders and accommodating the neck. This situation is often resolved by considering the type of work required of the worker. Figure 9.16 shows the recommended height of the work as a function of the type of work to be performed. In precision work visual acuity is of prime importance in order for the worker to be able to accomplish the work task. Thus, if the work is too low the head must be flexed to accommodate the visual requirements of the job and this could result in neck discomfort. Therefore, the work is typically raised to a relatively high level (95–110 cm above the

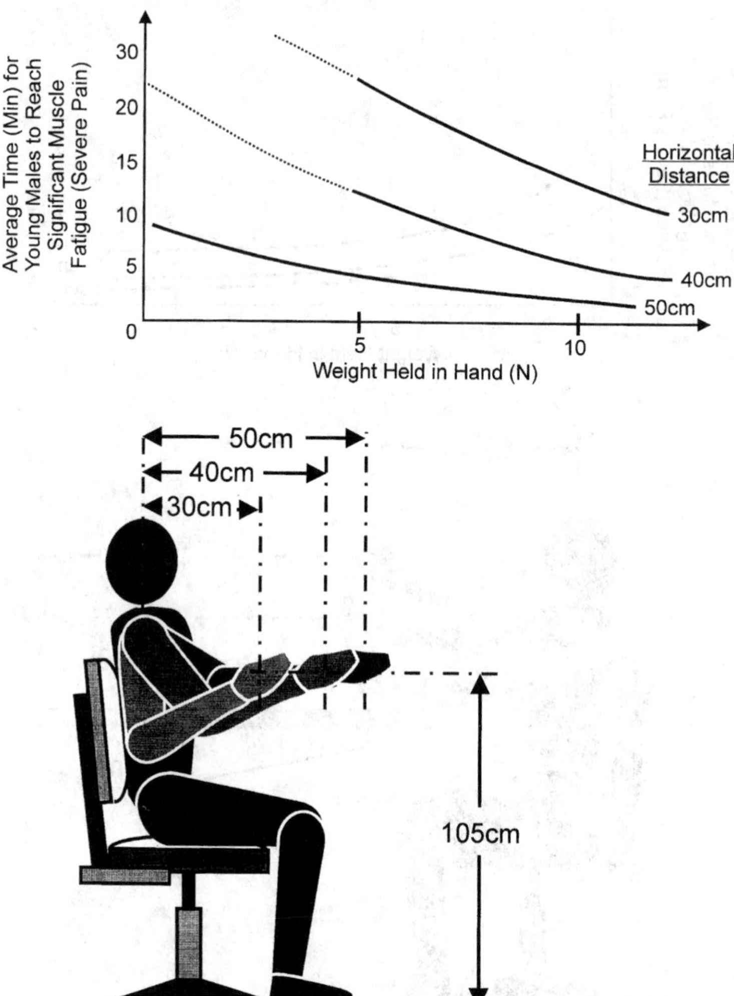

Figure 9.14 Expected time to reach significant shoulder muscle fatigue for different forward arm reach postures. (Adapted from *Occupational Biomechanics*, Chaffin and Andersson, copyright © 1991, John Wiley and Sons, Inc. Reprinted by permission of John Wiley and Sons, Inc.)

floor). This position solves the neck problem but creates a problem for the shoulders since they must be abducted. Thus, a trade-off must be considered. The shoulders are sacrificed in order to accommodate the neck since the visual requirements of the job are great. The logic in this decision also dictates that the shoulder problems can be minimized by providing wrist or elbow supports at the workplace.

The other extreme of this range of working heights involves heavy work. The greatest demand of the worker in heavy work is for a high degree of arm strength. Visual requirements in this type of work are typically minimal. Thus, in this situation neck posture is typically sacrificed in favor of shoulder and arm posture. Heavy work is performed at a height of 70–90 cm above floor level. In this position the elbow angles are close to 90°, which maximizes strength (Figure 9.6), and the shoulders are not abducted, so fatigue is minimized. The neck is not in an optimal position but the logic dictates that the visual demands of a heavy task would not be substantial, and thus the neck should not be flexed for prolonged periods of time. Light work is a mix of slight visual demand with small amounts of strength. In this case, work is a compromise between shoulder position and

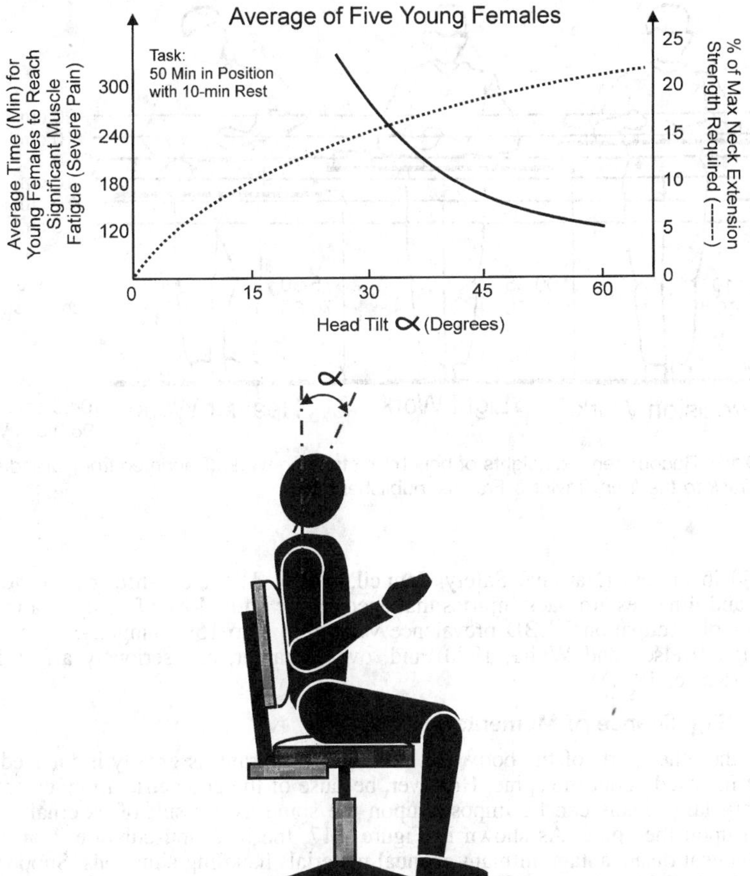

Figure 9.15 Neck extensor fatigue and muscle strength required vs. head tilt angle. (Adapted from *Occupational Biomechanics*, Chaffin and Andersson, copyright © 1991, John Wiley and Sons, Inc. Reprinted by permission of John Wiley and Sons, Inc.)

visual accommodation. Thus, the height of the work is between the precision work height level and the heavy work height level. Work is performed at a level of between 85 and 95 cm off the floor under light work conditions.

9.3.3 The Back

Back pain has been described as one of the most common and significant musculoskeletal problems in the United States that leads to substantial amounts of morbidity, disability, and economic loss (Hollbrook et al., 1984; Praemer et al., 1992). Back disorders were responsible for the loss of half a billion lost workdays in 1988 with 22 million cases reported that year (Guo, 1993). Among people under 45 years of age, LBD is the leading cause of activity limitation and can also affect up to 47% of workers with physically demanding jobs (Andersson, 1991). The prevalence of LBD has increased by 2700% since 1980 (Pope, 1993). The costs associated with LBD are enormous. Early estimates of lost wages alone amount to 4 billion dollars annually (Frymoyer et al, 1983). Webster and Snook (1989) estimated that in 1986 the average direct costs of LBD was $6,800 per case. More recent estimates of total costs to society range from 25 to 95 billion dollars per year (Cats-Baril and Frymoyer, 1991).

It has been clear for some time that the risk of LBD is associated with industrial work (Andersson, 1981). Thirty percent of occupation injuries in the United States are caused by overexertion, lifting, throwing, holding, carrying, pushing, and/or pulling objects that

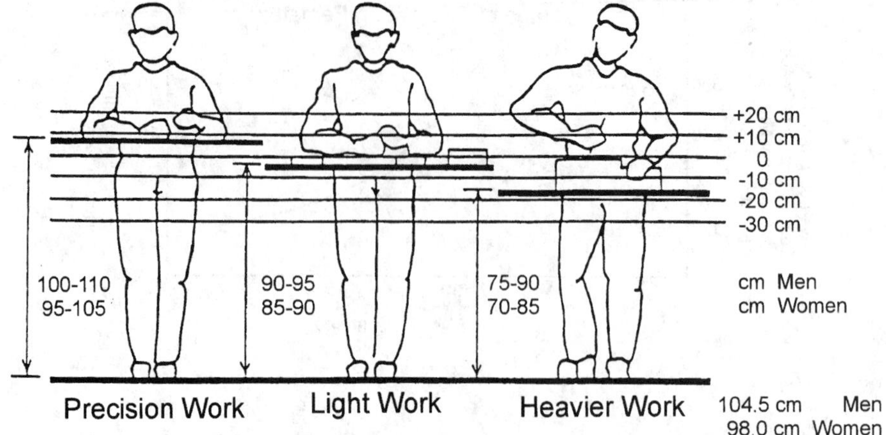

Figure 9.16 Recommended heights of bench for standing work. (Reprinted from Grandjean, *Fitting the Task to the Man,* Taylor & Francis, publisher.)

weigh 50 lb or less (National Safety Council, 1989). About one-fifth of all workplace injuries and illnesses are back injuries that account for up to 40% of compensation costs. Estimates of occupational LBD prevalence vary from 1 to 15% annually, depending on occupation (Kelsey and White, 1980) and, over a career, can seriously affect 56% of workers (Rowe, 1981).

9.3.3.1 Significance of Moments

As with the other parts of the body, the loading of the trunk is greatly influenced by the moment imposed about the spine. However, because of the geometric arrangement of the trunk very large loads can be imposed upon the spine as a result of external moments imposed upon the spine. As shown in Figure 9.17, the back musculature is at a severe biomechanical disadvantage in many manual materials handling situations. Supporting an external load of 222 N (about 50 lb) at a distance of 1 m from the spine imposes a load of 222 Nm of external moment about the spine. However, since the spine supporting musculature are in relatively close proximity relative to the external load, the trunk musculature must exert extremely large forces (4440 N or 998 lb) to simply hold the external

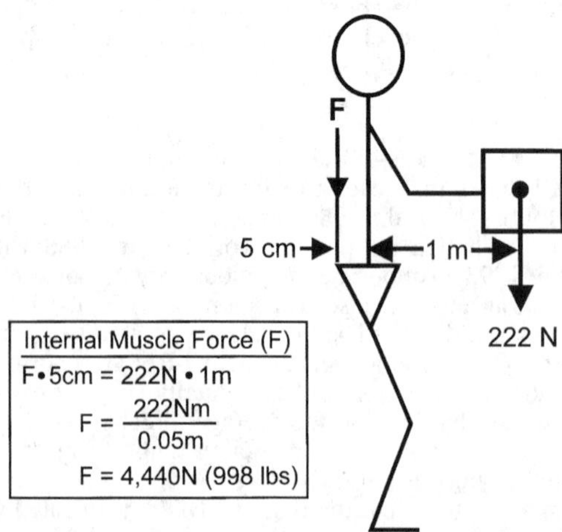

Figure 9.17 Calculation of internal muscle force for a given load.

load in equilibrium. These internal loads can be far greater if dynamic motion of the body is considered. Thus, the most important concept to consider in workplace design from a back protection standpoint is to keep the moment arm at a minimum. This also has implications for lifting styles. Since the externally applied moment has such a significant influence on internal loading, the lifting style is of far less concern compared to the magnitude of the applied moment. Hence, the correct lifting style is whatever style permits the worker to bring the center of mass of the load as close to the spine as possible.

9.3.3.2 Seated Workplaces

Loads on the lumbar spine are always greater when one is seated compared to a standing posture. This is because the posterior elements of the spine form an active load path when one is standing. However, when seated these elements are disengaged and more of the load passes through the intervertebral disc. Thus, when work is performed while seated, there is a greater risk of damaging the disc. Hence, it is important to consider the design features of a chair that may influence disc loading. Figure 9.18 shows the results of pressure measurements made in the intervertebral disc of workers as the back angle of the chair and the amount of lumbar support were varied. This figure indicates that both design features have a significant effect on disc pressure. Disc pressure decreases when the backrest angle is increased. However, increasing this angle in the workplace is often not practical. The figure also shows that increasing lumbar support can also significantly reduce disc pressure. This is probably due to the fact that as lumbar curvature (lordosis) is reestablished the posterior elements play more of a role in providing an alternative load path, as is the case when standing in the upright position.

9.3.3.3 MMH and the Control of Risk

It is well established that the type of work involved in an occupation is closely associated with the risk of suffering a *lower back disorder* (*LBD*). In particular, *manual materials handling* (*MMH*) activities, specifically lifting, dominate occupationally related LBD risk.

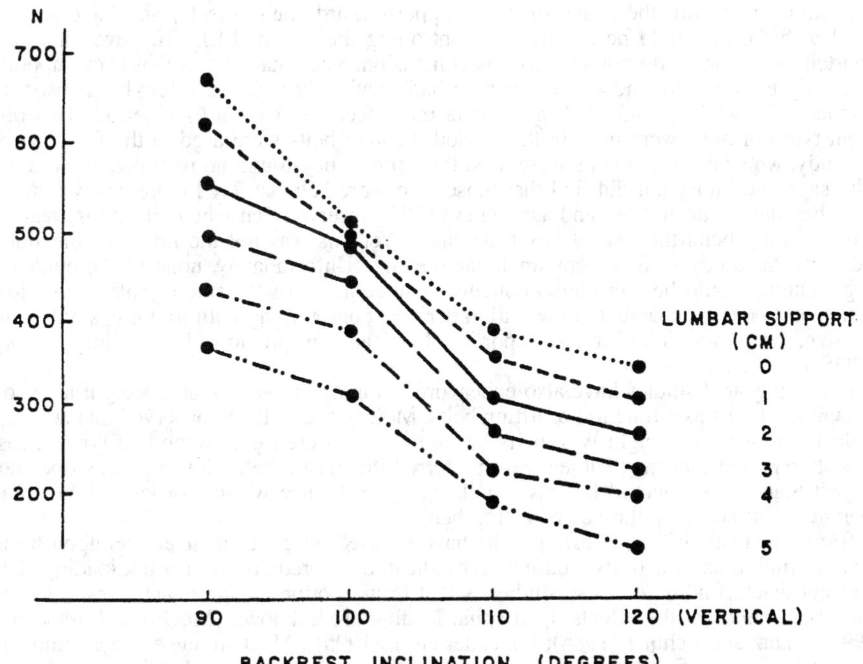

Figure 9.18 Disc pressures measured with different backrest inclinations and different size lumbar supports. (Adapted from *Occupational Biomechanics*, Chaffin and Andersson, copyright © 1991, John Wiley and Sons, Inc. Reprinted by permission of John Wiley and Sons, Inc.)

It is estimated that lifting and MMH account for 50 to 75% of all back injuries (Bigos et al., 1986; Snook, 1989; Spengler et al., 1986). It has been assumed that back pain is discogenic and has a mechanical origin (Nachemson, 1975). Retrospective studies have found increased degeneration in the spines of cadaver specimens who had previously been exposed to physically heavy work (Videman et al., 1990). Hence, this suggests that occupationally related LBDs are often associated with spine loading.

Lift Style

Much of the literature suggests that lift style does not play a significant role in the risk of low back disorder compared to other workplace controllable biomechanical factors such as the imposed moment about the spine. As mentioned earlier, it is believed that the correct working posture is whatever posture permits the worker to bring the center of gravity of the load as close a possible to the spine. Biomechanical analyses (Park and Chaffin, 1974) have demonstrated that there is no one lift style that is correct for all body types. Individual anthropometry plays a significant role defining the best lift style for an individual. For this reason the National Institute of Occupational Safety and Health (NIOSH, 1981) has concluded that lift style should not be a consideration when assessing the risk of occupationally related low back disorder. It should also be noted that biomechanically the internal moment of the trunk has a greater mechanical advantage when lumbar lordosis is preserved in the lifting style (Anderson and Chaffin, 1986; McGill, 1986).

Lift Belts

Lifting belts have been used with increasing frequency over the past decade. However, there exists a great deal of controversy as to whether lifting belts are beneficial or a liability during manual materials handling. A review of the literature surrounding the usage of lifting belt usage provides no clear answer as to the usefulness of the belts. Reviews by both McGill (1993) and NIOSH (1994) conclude that there are few definitive, well-executed studies upon which to base an opinion regarding these devices.

Epidemiological studies have generally been limited in their scope and have often resulted in findings that were confounded by other factors such as training, the type of belt used, or the "Hawthorne Effect." Walsh and Schwartz (1990) found a reduction in LBD injury rate with the usage of back supports (hard shell corsets) and have recommended that they would be effective at controlling the risk of LBD. However, the data reported in this paper do not support this conclusion. The data suggest that back supports were only effective for those workers who had previously suffered a low back disorder. Mitchell et al. (1994) evaluated injury data retrospectively over a 6-yr period. Two different types of belts were used in this period. Leather belts were used in the first 2 yr of the study, whereas, velcro belts were used thereafter. They found no relationship between belt usage and injury but did find that those who wore belts suffered more costly injuries once they did occur. Riddell and associates (1992) observed that when one stops wearing (fabric) lifting belts the risk of LBD increases. Yet this was not the intent of the study and, thus, the study suffers from small sample size. Unfortunately, none of the epidemiologic studies could be considered conclusive since many of the studies suffer from low participation rates, inadequate observation periods, confounding with training, small sample size, low back injury rates, reporting bias, and/or previous back injury history (NIOSH, 1994).

Psychophysical studies have also endeavored to assess the acceptable weight a person was willing to lift as a function of lifting belts. McCoy et al. (1988) observed that subjects could lift 19% more weight with belts but found no difference between belt types. Most subjects reported that they subjectively preferred the elastic belt. However, this does not suggest that workers would be less at risk for a LBD since we do not know how spine tolerance is affected by the use of lifting belts.

Biomechanical studies of lifting belts have focused on either their effects upon trunk motion, trunk muscle activity, and indirect indicators or predictions of trunk loading. The most consistent finding of these studies is that trunk motion is significantly restricted by belt usage, although the effects upon spine loading are unknown [McGill and associates (1994); Lantz and Schultz (1986); Lavender et al. (1995)]. Most studies were also unable to identify any significant reduction in muscle activity, and thus spine loading, when using belts.

There are physiological reasons to be concerned with the use of lifting belts. One study has shown that lifting belts can significantly increase blood pressure. This could become problematic for workers who have a compromised cardiovascular system.

The literature indicates that there is a large amount of conflicting evidence as to the benefits or liabilities associated with the use of back belts. Most reviews of the scientific literature have concluded that if there is a benefit to back belts it is probably for those who have previously experienced a low back disorder. It has also been suggested that belts should only be used for a limited period of time. Until more definitive studies are performed it is prudent to use caution when considering the use of back belts during lifting and to have an occupational physician screen the worker for potential cardiovascular problems.

1981 and 1991 NIOSH Lifting Guides

In an attempt to help industry to determine whether a manual materials handling task is safe or risky, NIOSH has developed lifting guides. The first lifting guide was developed in 1981 (NIOSH, 1981) and applies to lifting situations where the lifts are performed in the sagittal plane and where the motions are slow and smooth. Two limits are defined by this guide. The first limit is called the *action limit* (*AL*) and represents the point at which risk of low back disorder begins along a risk continuum. The weight of an object to be lifted on a job is compared to the AL. If the weight of the object is below that of the AL, the job is considered safe. The general form of the AL is defined according to Equation (1).

$$AL = k\,(HF)\,(VF)\,(DF)\,(FF) \tag{1}$$

where:

AL = the action limit in kilograms (kg) or pounds.
k = load constant (40 kg or 90 lb) which is the greatest weight a subject could lift if all lifting conditions are optimal.
HF = horizontal factor defined as the horizontal distance from a point bisecting the ankles to the center of gravity of the load at the lift origin. Defined algebraically as $15/H$ (metric) or $6/H$ (US units).
VF = vertical factor or height of the load at lift origin. Defined algebraically as $(0.004)|V\text{-}75|$(metric) or $1 - (0.01)|V\text{-}30|$(US units).
DF = distance factor or the vertical travel distance of the load. Defined algebraically as $0.7 + 7.5/D$ (metric) or $0.7 + 3/D$ (US units).
FF = frequency factor or lifting rate defined algebraically as $1 - F/F\text{max}$. F = average frequency of lift, F_{max} is shown in Table 9.3.

This equation assumes that if the lifting conditions are ideal a worker could safely lift the load constant (40 kg or 90 lb). However, if the lifting conditions are not ideal the allowable weight is discounted by the four factors HF, VF, DF, and FF. These four factors are also shown graphically in Figures 9.19 through 9.22. These figures show the HF or the moment has the most dramatic effect on acceptable lifting conditions. VF and DF relate to the length strength relationship of the trunk musculature. FF attempts to consider the cumulative effects of repetitive lifting.

The second limit defined by the guide is the *maximum permissible limit* (MPL). The MPL represents a level of considerable risk, defined in part as a significant chance of vertebral endplate microfracture (Figure 9.10). Equation (2) shows that this limit is simply a function of the AL.

Table 9.3 F_{max} Table

Period	Standing $V > 75$ (3)	Stooped $V \leqslant 75$ (3)
1 hr	18	15
8 hr	15	12

Source: Reprinted from NIOSH, *Work Practices Guide for Manual Lifting.*

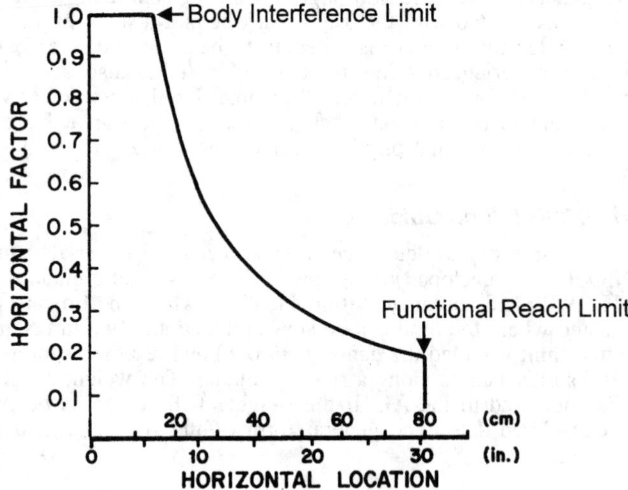

Figure 9.19 Horizontal factor varies between the body interference limit and the limit of functional reach. (Adapted from NIOSH, *Work Practices Guide for Manual Lifting*.)

$$MPL = 3\ (AL) \tag{2}$$

The weight that the worker is asked to lift is compared to the AL and MPL. If the weight falls below the AL, then the work is considered safe and no adjustments are necessary. If the weight falls above the MPL, then the work is considered significantly risky and engineering changes (adjusting HF, VF, DF) are required to reduce the AL and MPL.

If the weight falls between the AL and MPL, then either engineering changes or administrative changes (selecting workers who are less likely to be injured or rotating workers) is suggested.

In 1991, NIOSH introduced the revised lifting equation to address those lifting jobs that violate the sagittally symmetric lifting assumption (Waters et al., 1993). In this re-

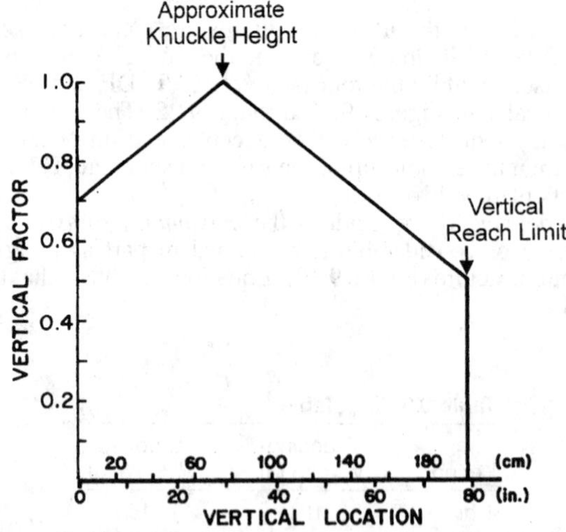

Figure 9.20 Vertical factor varies both ways from knuckle height. (Adapted from NIOSH, *Work Practices Guide for Manual Lifting*.)

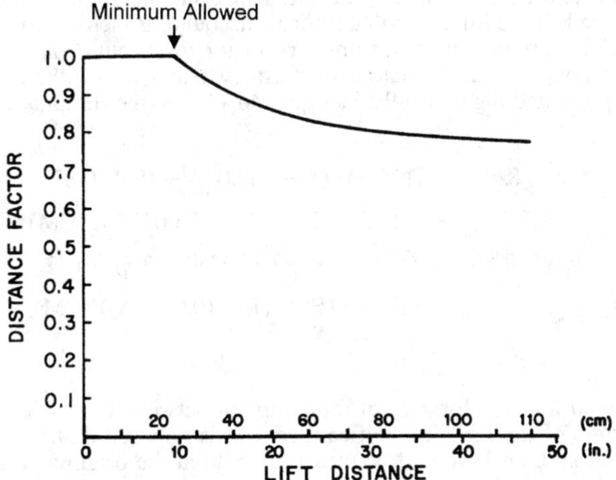

Figure 9.21 Distance factor varies between a minimum vertical distance moved of 25 cm to a maximum distance of 200 cm. (Adapted from NIOSH, *Work Practices Guide for Manual Lifting.*)

vision, the concept of AL and MPL was replaced with a concept of a *lifting index* (LI). The LI is defined according to Equation (3).

$$\mathrm{LI} = \frac{\mathrm{L}}{\mathrm{RWL}} \tag{3}$$

where:

L = load weight or the weight of the object to be lifted.
RWL = recommended Weight Limit for the particular lifting situation.
LI = lifting index used to estimate relative magnitude of physical stress for a particular job.

The 1991 revised equation suggests that if the LI is greater than 1.0, an increased risk exists for suffering a lifting related low back disorder. The RWL is somewhat similar to the AL equation [Equation 1] in the 1981 guide in that it contains factors that discount the allowable load as a function of horizontal distance, vertical location of the load,

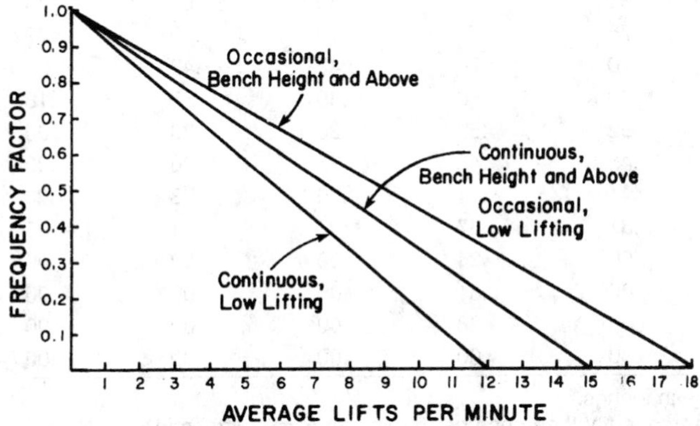

Figure 9.22 Frequency factor varies with lifts/minute and the F_{max} curve. (Reprinted from NIOSH, *Work Practices Guide for Manual Lifting.*)

vertical travel distance, and frequency of lift. However, the form of these discounting factors has changed. In addition, two additional discounting factors have been included. These additional factors include an asymmetry factor to account for asymmetric lifting conditions and a coupling factor that accounts for whether or not the load has handles. The RWL is represented algebraically in Equation (4) for metric units and in Equation (5) for US units.

$$\text{RWL(Kg)} = 23\ (25/H)\ (1 - (0.003|V-75|))\ (0.82 + 4.5/D))\ (FM)\ (1 - (0.0032A))\ (CM) \tag{4}$$

$$\text{RWL(lb)} = 51\ (10/H)\ (1 - (0.0075|V-30|))\ (0.82 + 1.8/D))\ (FM)\ (1 - (0.0032A))\ (CM) \tag{5}$$

where:

H = horizontal location forward of the midpoint between the ankles at the origin of the lift. If significant control is required at the destination, then H should be measured both at the origin and destination of the lift.
V = vertical location at the origin of the lift.
D = vertical travel distance between origin and destination of the lift.
FM = frequency multiplier shown in Table 9.4.
A = angle between the midpoint of the ankles and the midpoint between the hands at the origin of the lift.
CM = coupling multiplier ranked as good, fair, or poor and described in Table 9.5.

Note that the load constant has been significantly reduced compared to the 1991 equation and that adjustments for load moment, muscle length-strength relationships, and

Table 9.4 Frequency Multiplier Table (FM)

Frequency Lifts/min (*F*)‡	Work Duration ≤1 hr		>1 but ≤2 hr		>2 but ≤8 hr	
	V < 30†	V ≥ 30	V < 30	V ≥ 30	V < 30	V ≥ 30
≤0.2	1.00	1.00	.95	.95	.85	.85
0.5	.97	.97	.92	.92	.81	.81
1	.94	.94	.88	.88	.75	.75
2	.91	.91	.84	.84	.65	.65
3	.88	.88	.79	.79	.65	.55
4	.84	.84	.72	.72	.45	.45
5	.80	.80	.60	.60	.35	.35
6	.75	.75	.50	.50	.27	.27
7	.70	.70	.42	.42	.22	.22
8	.60	.60	.35	.35	.18	.18
9	.52	.52	.30	.30	.00	.15
10	.45	.45	.26	.26	.00	.13
11	.41	.41	.00	.23	.00	.00
12	.37	.37	.00	.21	.00	.00
13	.00	.34	.00	.00	.00	.00
14	.00	.31	.00	.00	.00	.00
15	.00	.28	.00	.00	.00	.00
>15	.00	.00	.00	.00	.00	.00

†Values of *V* are in inches.
‡For lifting less frequently than once per 5 min, set *F* = .2 lifts/min.
Source: Reprinted from NIOSH, *Applications Manual for the Revised NIOSH Lifting Equation.*

Table 9.5 Coupling Multiplier

	Coupling Multiplier	
Coupling Type	V > 30 in. (75 cm)	V ≥ 30 in. (75 cm)
Good	1.00	1.00
Fair	0.95	1.00
Poor	0.90	0.90

Source: *Applied Manual for Revised NIOSH Equation,* NIOSH, 1994.

cumulative loading are still integral parts of the equation. Studies have shown that, for the most part, the 1991 revision of the equation yields a more conservative (protective) prediction of work related low back disorder risk.

2D/3D Static Models

Other biomechanically based models have been developed to help assess manual materials handling in the workplace. One of the first static models was developed by Chaffin at the University of Michigan (1969). Both two-dimensional (2D) as well as three-dimensional (3D) static models (Chaffin and Muzaffer, 1991) have been developed and computerized to help assess the risk of injury during manual materials handling activities. Both models compare the moments imposed upon the various joints of the body due to the object lifted in a particular static posture. These models then compare these imposed moments about each joint with the strength capacity of each joint as defined by a database of over 3000 workers. In this manner the proportion of the population capable of performing a particular static exertion can be predicted. These models also assume that a single equivalent muscle (internal force) supports the external moment about each joint. Hence, by considering the contribution of the externally applied load and the internally generated force spine compression acting on the lumbar discs can be predicted. The predicted compression can then be compared to the tolerance limits of the vertebral end plate. The 2D version of this computer model assumes (as does the 1981 NIOSH Lifting Guide) that all lifts occur in the sagittal plane. It also assumes that no significant motion occurs during the exertion since it is a static model. Figure 9.23 shows the output screen for this computer model where the lifting posture, lifting distances, strength predictions, and spine compression are shown. The 3D version of the computer model works in a similar manner; however, the sagittally symmetric lifting assumption is relaxed.

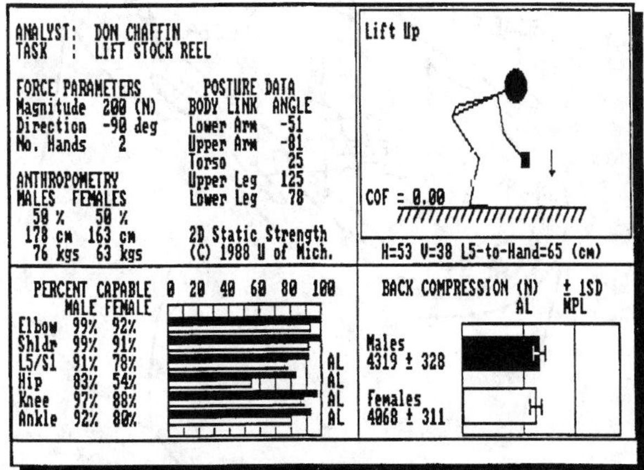

Figure 9.23 The 2D static strength prediction model. (Adapted from *Occupational Biomechanics*, Chaffin and Andersson, copyright © 1991, John Wiley and Sons, Inc. Reprinted by permission of John Wiley and Sons, Inc.)

Multiple Muscle System Models—Optimization and Neural Networks

In an effort to estimate more accurately the loads on the lumbar spine, especially under complex static postures, multiple muscle system models of the trunk have been developed. These models attempt to assess more realistically spine loading through a more physiologically correct representation of the trunk muscles orientations. Figure 9.24 shows a typical representation of the trunk internal forces (muscles) and their positions relative to the spine. In this system, the influence of each trunk muscle is considered in terms of how it contributes to the internally generated moments that support the spine and how the muscle force magnitudes load the spine in terms of compression, shear, and torsional forces on the vertebral body. The three forces and three moments imposed by the external load upon the spine can be used to predict the activities of the trunk muscles in this case. However, since there are only six governing equations (three moment and three force equations) and ten unknown muscle forces, the system becomes statically indeterminate, and, hence, the muscle forces and spine loading cannot be uniquely defined. In order to overcome this problem of indeterminacy, many attempts have been made to predict the muscle activities through alternative means. One such method has been through the use of optimization techniques such as linear optimization (Schultz et al., 1982; Bean, Chaffin, and Schultz, 1988). Unfortunately, due to the limitations of linear optimization these models are unable to predict coactivation of the antagonist muscles which could contribute to the loading of the spine (Marras, 1988). More recent attempts have also tried to use neural networks to predict muscle coactivation (Nussbaum, 1995). These attempts do have promise for predicting the muscle coactivation under selected conditions; however, these systems also have not been able to demonstrate that they could adapt to changing (dynamic) loading conditions. Another promising means to predict muscle activity in a multiple muscle system has been through the use of stochastic predictions of the muscle activity (Mirka and Marras, 1993).

EMG-Assisted Multiple Muscle System Models

Most of the previously mentioned models have had extreme difficulty in attempting to assess the effects of dynamic trunk motion upon loading of the spine. People recruit their muscles in various manners when moving dynamically. For example, when moving slowly

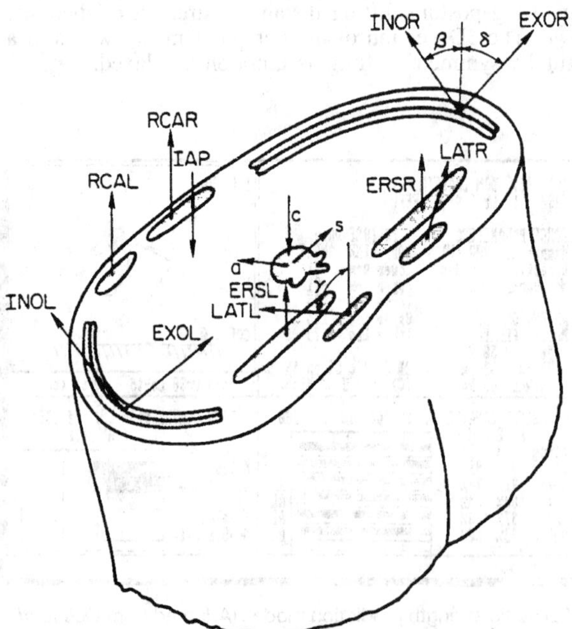

Figure 9.24 Cross-sectional view of the human trunk at the lumbrosacral junction. (Adapted from Schultz and Andersson, 1981.)

the agonist muscle may dominate the muscles activities during a lift. However, when moving cautiously, asymmetrically, or rapidly there may be a great deal of antagonistic coactivation. Unfortunately, in industrial work these latter dynamic conditions are typically the rule rather than the exception during lifting. For these reasons it has been virtually impossible to predict the instantaneous coactivation and resultant loading on the spine during dynamic trunk exertions. In order to circumvent these difficulties, biologically assisted models have been developed. The most common of these models have used *electromyography* (*EMG*) (measuring the electrical activity of a skeletal muscle by means of an electrode inserted into the muscle or placed on the skin) to develop EMG-driven multiple muscle system models. These models use information about the electrical activity of the muscles to determine individual muscle force and the subsequent spine loading. These models have been developed and tested under bending and twisting dynamic motion conditions and have been validated (Granata and Marras, 1993; Marras and Granata, 1995; Marras and Reilly, 1988; McGill and Norman, 1985; McGill and Norman, 1986; Marras and Sommerich, 1991a, b; Reilly and Marras, 1989). Figure 9.25 shows how such models can assess the effects of lifting dynamics on spine loading. These models can predict the loads on the lumbar spine under many complex dynamic lifting conditions. However, the limitation of such models is that it is often difficult to instrument workers with EMG under work conditions.

Dynamic LMM Risk Model

An alternative means to assess the risk of low back disorders associated with dynamic lifting at the worksite has been developed recently (Marras et al., 1993, 1995). Figure 9.26 shows a trunk goniometer (lumbar motion monitor or LMM) that has been used to document the trunk motion patterns of workers at the workplace. The LMM has been used to document the trunk motion patterns of over 400 workers in jobs that have been associated historically with varying degrees of risk. Multiple logistic regression models were developed that best discriminate between a high risk and a low risk of low back disorder based upon characteristics of the job and worker trunk motion patterns. Figure 9.27 shows the risk model that can be used for workplace design and assessment purposes. Five factors are present in this model and have been scaled relative to their association with risk of low back disorder. These factors include: (1) frequency of lifting, (2) load moment (load weight multiplied by the distance of the load from the spine), (3) average twisting velocity (measured by the LMM), (4) maximum sagittal flexion angle through the job cycle (measured by the LMM), and (5) maximum lateral velocity (measured by the LMM). The figure shows that by considering the combination of these five factors, the probability of high-risk group membership can be predicted. This model has been shown to have a high degree of predictability (odds ratio = 10.7) compared to previous attempts to assess work-related low back disorder risk. This model can be used to assess current risk associated with the design of a material handling tasks or it can be used to assess the expected risk associated with modifications or redesign of a job. In these cases,

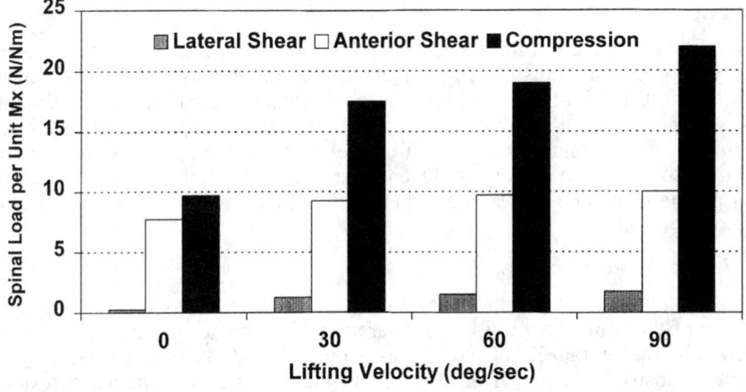

Figure 9.25 Spine loading “cost” as a function of dynamic trunk loading.

Figure 9.26 The lumbar motion monitor (LMM) used to assess trunk motion on the job.

it would be necessary to "mock up" the workplace and test a worker performing five or six repetitions of the job. The advantage of this assessment is that the evaluation provides information about risk that would take years to derive from historical accounts of incidence rates.

9.3.4 Wrists

The wrist has been an area of increased interest to ergonomists in recent years. According to the Bureau of Labor Statistics, repetitive trauma has increased from 18% of occupational illnesses in 1981 to 63% of occupational illnesses in 1993. Even though these numbers appear alarming, one must bear in mind that occupational illnesses represent only 6% of all occupational injuries and illnesses and include illnesses unrelated to musculoskeletal disorders such as noise-induced hearing loss. Nonetheless, there are industries (i.e. meat packing, poultry processing) where cumulative trauma to the wrist has reached epidemic proportions.

9.3.4.1 Anatomy and Loading

In order to understand how cumulative trauma occurs to the wrist, one must appreciate the anatomy of the upper extremity. There are very few power-producing muscles in the hand itself. The thenar muscle of the thumb is one of the only power-producing muscles in the hand. Most of the power-producing muscles of the hand actually reside in the forearm. Force is transmitted to the fingers through a network of long tendons (tendons attach muscles to bone). These tendons pass from the muscles in the forearm through the

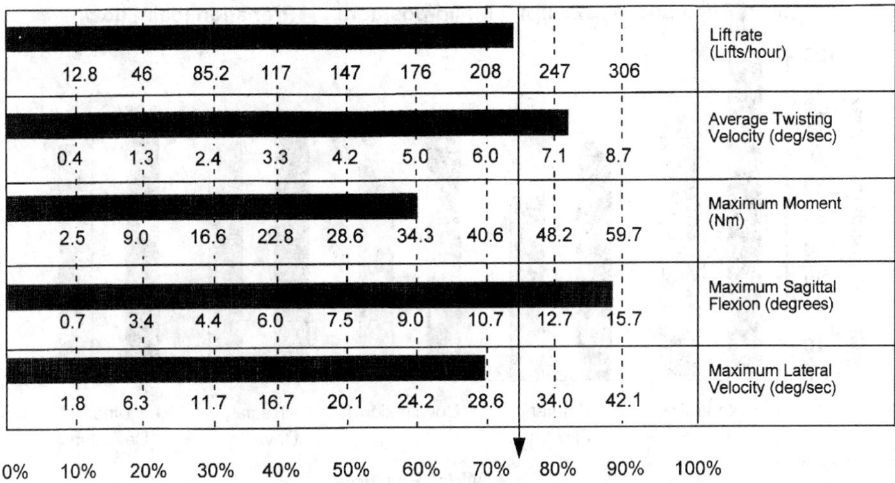

Figure 9.27 The LMM risk model.

wrist (with many of them passing through the carpal canal), through the hand, and to the fingers. These tendons are secured at various points along this path with ligaments that keep the tendons in close proximity to the bones. This system allows a hand that is very small and compact yet can generate significant amounts of force. The price the system pays for this design is friction. Since the forearm muscles must transmit force over a long distance, a great deal of tendon travel must occur. This can result in increased tendon friction under repetitive motion conditions, and the sequence of events outlined in Figure 9.1 can occur. Thus, the key to understanding tasks that lead to cumulative trauma disorders is to understand those workplace factors that adversely affect the internal force generating (muscles) and transmitting (tendons) structures.

9.3.4.2 Biomechanical Risk Factors for the Wrist

Several risk factors for cumulate trauma have been documented for the wrist. First, deviations of the wrist, or deviated wrist postures, are known to reduce the volume of the carpal tunnel, thus increasing the level of friction experienced by the tendons. Thus, deviated wrist postures can increase the risk of a cumulative trauma disorder. This is true not only because the volume in the carpal canal is reduced under these conditions, but the strength of the hand is dramatically reduced by deviations in the wrist posture. Figure 9.28 shows that any deviation from neutral significantly decreases the grip strength of the hand. This trend is due to the fact that the length–strength relationship (Figure 9.5) of the forearm muscles is affected once the wrist is bent. Thus, deviated wrist postures not only increase tendon travel and friction, but also increase the percentage of muscle activity and relative percentage of muscle force necessary for a given grip strength requirement. Therefore, the muscles are working at a level that is higher than necessary and the risk of muscle strain increases.

Second, frequency or repetition has been identified as a risk factor for *cumulative trauma disorders* (*CTD*s) (work related musculo-skeletal injuries resulting from repeated microtrauma) (Silverstein et al., 1986, 1987). Industrial studies have indicated that increased frequency of repeating a wrist motion increases the risk of developing a CTD. Any repeated motion with a cycle time of less than 30 sec is considered a candidate for CTD risk.

Third, the force applied by the hands and fingers is considered a risk factor. The greater the force required by the work the greater the risk of CTD. Greater hand forces require greater tension in the tendons and subsequently greater tendon friction and tendon travel. Related to force is wrist acceleration. Studies have shown that repetitive jobs that require greater wrist acceleration are associated with greater CTD incident rates (Marras and Schoenmarklin, 1993; Schoenmarklin et al., 1994). Since force is a product of mass and acceleration, those jobs that increase the angular acceleration of the wrist joint result in

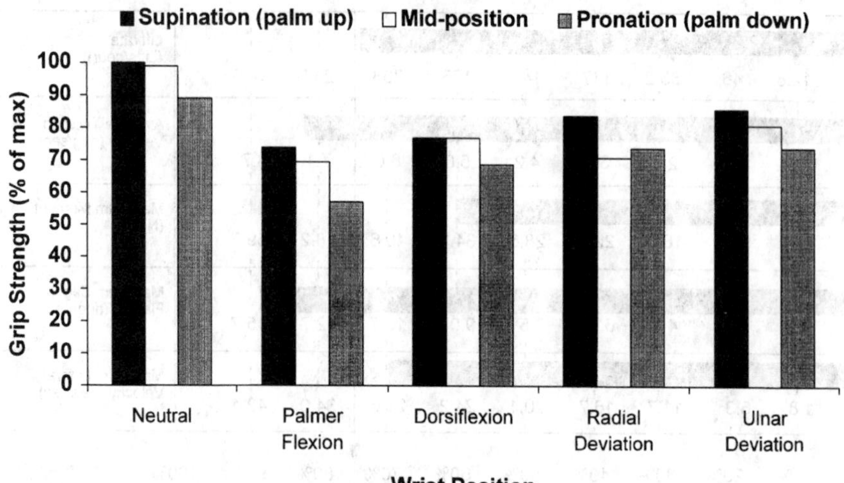

Figure 9.28 Grip strength as a function of wrist and forearm position. (Adapted from Sanders and McCormick, *Human Factors in Engineering and Design* copyright © 1993 McGraw Hill. Reproduced with permission of the McGraw-Hill Companies.)

greater tension and force transmitted through the tendons of the wrist. Thus, wrist acceleration is another way of imposing force on the wrist structure.

Fourth, the anatomy of the hand is such that the median nerve of the hand becomes very superficial at the palm. Thus, direct stimulation of the palm through pounding or striking an object with the palm, as is done often in assembly work, can directly stimulate the median nerve and cause symptoms of CTDs even though the work may not be repetitive in nature.

9.3.4.3 Grip Design

Another factor that can dramatically affect the activity of the internal force transmission system (tendon travel and tension) is the design of the grip associated with a tool. In particular the grip opening and shape have a major influence on the available grip strength. Figure 9.29 shows grip strength capacity as a function of the separation distance of the grip opening. This figure indicates that there is a very narrow range of grip openings that can maximize grip strength. In addition, if the grip opening deviates from this ideal range by as little as an inch (a couple of centimeters), then grip strength is dramatically reduced. This trend is also due to the length–strength relationship of the forearm muscles.

The narrow range of the ideal grip opening along with the dramatic drop in strength when deviated from this ideal range also suggest that the pulley system of the wrist magnifies any biomechanical effects. Figure 9.29 also suggests that the ideal grip opening is also a function of hand anthropometry. Thus, the worker's hand size as well as preferred hand can influence risk. Thus, proper design of the handles is crucial in ergonomic workplace design.

The shape of the grip can also affect the strength of the wrist. Figure 9.30 shows how variations in the design of screwdriver handles can affect the maximum force that can be exerted. Here again the biomechanical origin of this trend is probably related to the length–strength relationship of the forearm muscles. The handle designs that result in less strength probably permit the wrist to twist or permit the grip to slip, thus resulting in a deviation from the ideal forearm muscle length–strength position.

9.3.4.4 Gloves

The use of gloves can also significantly influence the generation of grip strength as well as the cumulative trauma experienced by the wrist structures. When gloves are worn while performing hand-intensive work, three effects are common. First, the grip strength generated by the worker is typically reduced. This reduction is typically of the magnitude of about 10% to 20%. When one uses gloves, the coefficient of friction between the hand and the tool is usually decreased, which in turn permits some amount of slippage of the

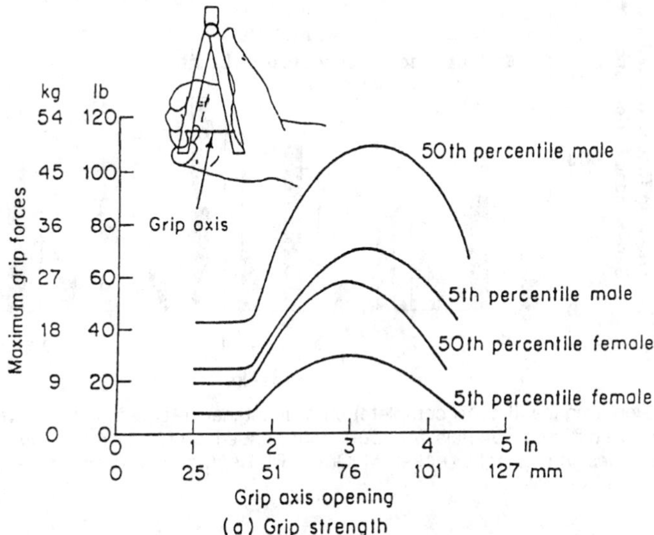

Figure 9.29 Grip strength as a function of grip opening and hand anthropometry. (Adapted from Sanders and McCormick, *Human Factors in Engineering and Design* copyright © 1993 McGraw Hill. Reproduced with permission of the McGraw-Hill Companies.)

hand on the tool surface. This slippage can result in a deviation from the ideal muscle length and thus a reduction in strength. The degree of slippage and thus the degree of strength loss depends on how well the gloves fit the hand. The worse the fit, the more the strength loss.

Second, even though the externally applied force (grip strength) is typically reduced with glove usage, the internal forces are typically greater when using a glove compared to not using a glove. Studies have shown that for a given grip strength the muscle activity is significantly greater when using gloves compared to a bare-handed condition (Sudhakar et al., 1988).

Third, performance is generally affected negatively when workers wear gloves. Figure 9.31 shows the increase in time required to perform work tasks while wearing gloves of four different materials compared to performing the task bare-handed. This figure indicates that the components of the task can require up to 70% more time when wearing gloves.

These three effects have shown that there are biomechanical costs associated with the use of gloves. Less strength is available to the worker, more internal force is generated, and worker productivity is affected. However, this does not suggest that gloves should

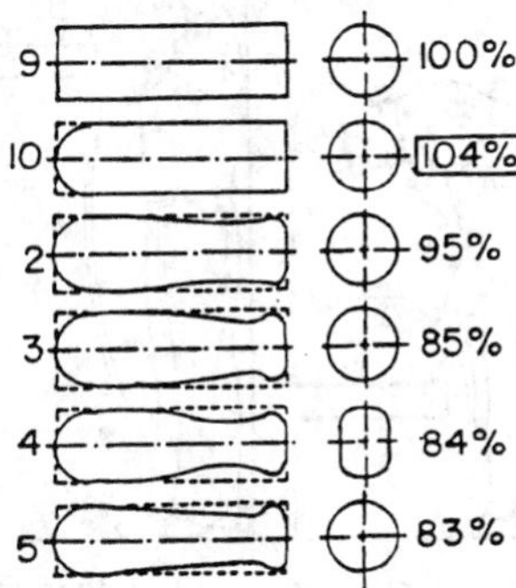

Figure 9.30 Maximum force which could be exerted on a screwdriver as a function of handle shape. (Reprinted from Konz, *Work Design*.)

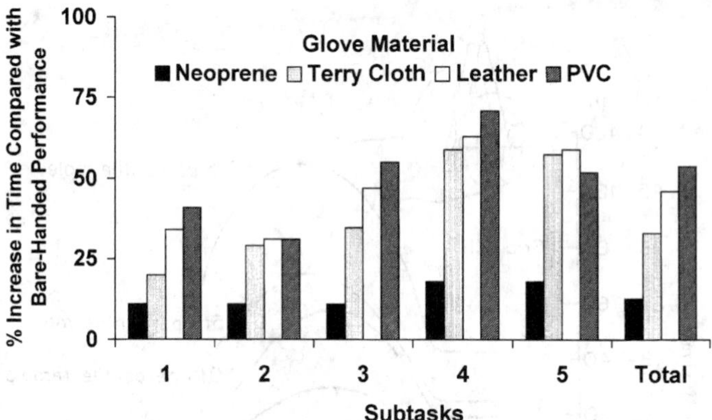

Figure 9.31 Performance (time to complete) on a maintenance-type task while wearing gloves constructed of five different materials. (Adapted from Sanders and McCormick, *Human Factors in Engineering and Design* copyright © 1993 McGraw Hill. Reproduced with permission of the McGraw-Hill Companies.)

not be worn. Obviously, when hand protection is needed, gloves should be considered. However, protection should only be considered for the parts of the hand that require it. For example, if the palm of the hand requires protection, fingerless gloves should be considered. If the fingers need protection but there is no risk to the palm of the hand, then grip tape wrapped around the fingers should be considered. In addition, different styles and sizes of gloves will fit workers differently. Thus, gloves produced by different manufacturers and of different sizes should be available to the worker to minimize the effects mentioned above.

9.3.4.5 Design Guidelines

This discussion of wrist biomechanics should make it clear that there are many factors that can affect the biomechanics of the wrist and the risk of CTDs. This implies that proper ergonomic design of a work task cannot be accomplished by simply providing the worker with a so-called ergonomically designed tool. Because ergonomics is associated with matching the workplace design to the worker's capabilities it is not possible to design an ergonomic tool without considering the workplace design simultaneously. As shown in Figure 9.32, what might be an ergonomic tool for one work situation may be improper

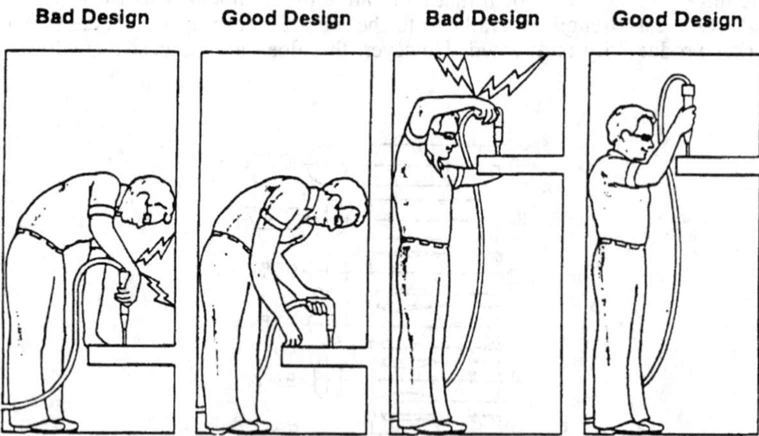

Figure 9.32 The correct tool for one position may be incorrect for a different position of the body. (Reprinted from UAW, *Strains and Sprains*.)

for use while a worker is assuming another work posture. Therefore, there are no ergonomic tools—there are just ergonomic situations, and what may be an ergonomically correct tool in one situation may not be ergonomically correct in another work situation. Thus, workplace design should be performed with care and trade-offs between different parts of the body must be considered and reasoned out logically by taking into consideration the various biomechanical trade-offs. Given these considerations the following components of the workplace should be considered when designing a workplace that minimizes risk of CTD to the wrist:

- Maintain a neutral wrist posture.
- Minimize tissue compression.
- Avoid actions that repeatedly impose force on the internal structures.
- Minimize the required wrist acceleration.
- Consider the impact of glove use, hand size, and left-handed workers.
- Consider safety issues and potential misuse of tools.

9.4 SUMMARY

This chapter has shown that it is important to consider the biomechanical implications of workplace design. Biomechanical design is important when a particular job is suspected of imposing large or repetitive forces on a particular structure of the body. It is particularly important to recognize that the internal structures of the body, such as muscles, are the primary loaders of the joint and tendon structures. Thus, in order to evaluate the risk of injury from a particular task, one must consider the contribution of both the external loads and internal loads upon the structure. A large compliment of models and assessment methods have been developed that systematically consider the internal loading imposed on the worker due to workplace layout and task requirements. Proper use of these models and methods involves recognizing the limitations and assumptions of each technique so that they are not applied inappropriately. When properly used these assessments can help assess the risk of work-related injury and illness.

REFERENCES

Adams, M. A., and Dolan, P. (1995). Recent Advances in Lumbar Spinal Mechanics and their Clinical Significance, *Clinical Biomechanics*, *10(1)*, 3–19.

Anderson, C. K., and Chaffin, D. B. (1986). A biomechanical evaluation of five lifting techniques. *Applied Ergonomics*, *17(1)*: 2–8

Andersson G. B. (1981). Epidemiologic aspects of low back pain in industry. *Spine*, *6*, 53–60

Andersson, G. B. (1991). The epidemiology of spinal disorders. In: J. W. Frymoyer, T. B. Ducker, N. M. Hadler, J. P. Kostuik, J. N. Weinstein, and T. S. Whitecloud, Eds., *The Adult Spine.* New York: Raven Press, pp. 107–146.

Basmajian, J. V., and De Luca C. J. (1985). *Muscles Alive: Their Funcitons Revealed by Electromyography* (5th ed.). Baltimore, MD: Williams and Wilkins.

Bean J. C., Chaffin D. B., and Schultz, A. B. (1988). Biomechanical model calculation of muscle forces: A double linear programming method. *J. Biomechanics*, *21(1)*, 59–66.

Bigos S. J., Spengler D. M., Martin N. A., Zeh J., Fisher L., Nachemson A., and Wang, M. H. (1986). Back injuries in industry: A retrospective study. II. Injury factors. *Spine*, *11(3)*, 246–251

Brinckmann, P., Biggemann, M., and Hilweg, D. (1988). Fatigue fracture of human lumbar vertebrae. *Clinical Biomechanics*, *3*: Supplement 1, S1–S23.

Cats-Baril, W., and Frymoyer, J. W. (1991). The economics of spinal disorders. In: Frymoyer J. W., Ducker T. B., Hadler N. M., Kostuik J. P., Weinstein J. N., and Whitecloud T. S., Eds., *The Adult Spine.* New York: Raven Press, pp. 85–105.

Chaffin D. B. (1969). A computerized biomechanical model: development of and use in studying gross body actions. *Journal of Biomechanics*, *2*, 429–441.

Chaffin, D. B., and Andersson, G. B. (1991). *Occupational Biomechanics*, New York: John Wiley.

Chaffin D. B., and Baker, W. H. (1970). A biomechanical model for analysis of symmetric sagittal plane lifting. *AIIE Transactions*, *II(1)*, 16– 27.

Chaffin, D. B. (1973). Localized Muscle Fatigue Definition and Measurement, *Journal of Occupational Medicine*, *15(4)*, 346–354.

Chaffin D. B., and Muzaffer E. (1991). Three-dimensional biomechanical static strength prediction model sensitivity to postural and anthropometric inaccuracies. *IIE Transactions*, *23(3)*, 215–227.

Frymoyer, J. W., Pope, M. H., Clements, J. H., Wilder, D. G., MacPherson, B., and Ashikaga, T. (1983). Risk factors in low back pain: An epidemiologic survey. *J. Bone Joint Surg.*, *65A*, 213–216.

Granata, K. P., and Marras, W. S. (1993). An EMG-assisted model of loads on the lumbar spine during asymmetric trunk extensions, *J. Biomechanics*, *26*(*12*), 1429–1438

Grandjean, E. (1982). *Fitting the Task to the Man: An Ergonomic Approach*, London: Taylor and Francis, Ltd.

Guo, H. R. (1993). Back pain and U.S. workers (NIOSH report). Paper presented at American Occupational Health Conference, April 29.

Hollbrook, T. L., Grazier, K., Kelsey, J. L., and Stauffer, R. N. (1984). The frequency of Occurrence, Impact and Cost of Selected Musculoskeletal Conditions in the United States. Chicago, IL: American Academy of Orthopaedic Surgeons, pp. 24–45.

Hoy M. G. Zajac F. E. and Gordon, M. E. (1990). A musculoskeletal model of the human lower extremity: The effect of muscle, tendon, and moment arm on the moment-angle relationship of the musculotendon actuators at the hip, knee, and ankle. *Journal of Biomechanics*, *23*(2), 157–169

Jager, M., Luttmann, A., and Laurig, W. (1991). Lumbar load during one-handed bricklaying, *International Journal of Industrial Ergonomics*, *8*(*3*), 261–277

Kelsey, J. L., and White, A. A. III. (1980). Epidemiology and impact on low back pain. *Spine*, *5*(*2*), 133–142

Kelsey, K. L., Githens P. B., White, A. A. III, Holford T. R., Walter S. D., O'Conner, T., Ostfeld, A. M., Weil, U., Southwick, W. O., and Calogero, J. A. (1984). An epidemiologic study of lifting and twisting on the job and risk for acute prolapsed lumbar intervertebral disc. *J Ortho Res*, *2*(*1*): 61–66

Konz, S. A. (1983). *Work Design: Industrial Ergonomics* (*2nd ed.*). Columbus, OH: Grid Publishing.

Kroemer, K. H. E. (1987). Biomechanics of the human body. In G. Salvendy, Ed., *Handbook of Human Factors*. New York: John Wiley.

Lantz, S. A., and Schultz, A. B. (1986). Lumbar spine orthosis wearing. I. restrictions of gross body motion. *Spine*, *11*(*8*): 834–837.

Lavender, S. A., Thomas, J. S., Chang, D., and Andersson, G. B. (1995). The effects of lifting belts, foot movement and lifting asymmetry on trunk motions. *Human Factors*, *37*(*4*):844–853.

Marras, W. S. (1988). Predictions of forces acting upon the lumbar spine under isometric and isokinetic conditions: A model–experimental comparison. *Int. J. Ind. Ergonomics 3*, 19–27.

Marras, W. S., and Granata, K. P. (1994). A biomechanical assessment and model of axial twisting in the thoraco-lumbar spine. *Spine, 20*(*13*): 1440–1451

Marras, W. S., Lavender, S. A., Leurgans, S. E., Rajulu, S. L., Allread, W. G., Fathallah, F. A., and Ferguson, S. A. (1993). The role of dynamic three-dimensional trunk motion in occupationally-related low back disorders: The effects of workplace factors, trunk position and trunk motion characteristics on risk of injury. *Spine*, *18*(*5*), 617–628

Marras, W. S., Lavender, S. A., Leurgans, S. E., Rajulu, S. L., Allread, W. G., Fathallah, F. A., and Ferguson, S. A. (1995). Biomechanical risk factors and trunk motion. *Ergonomics*, *38*(*2*): 377–410.

Marras, W. S., and Reilly, C. H. (1988). Networks of internal trunk-loading activities under controlled trunk-motion conditions. *Spine*, *13*(*6*), 661–667

Marras, W. S., and Schoenmarklin, R. W. (1993). Wrist motion in industry. *Ergonomics*, *36*(*4*), 341–351.

Marras, W. S., and Sommerich, C. M. (1991a). A three-dimensional motion model of loads on the lumbar spine: I model structure. *Human Factors*, *33*(2), 123–137

Marras, W. S., and Sommerich, C. M. (1991b). A three-dimensional motion model of loads on the lumbar spine: II model validation. *Human Factors*, *33*(*2*), 139–149

McCoy, M. A., Congleton, W. L., Johnston, W. L., and Jiang, B. C. (1988). The role of lifting belts in manual lifting. *International Journal of Industrial Ergonomics*. *2*, 259–256.

McGill, S. M., and Norman, R. W. (1985). Dynamically and statically determined low back moments during lifting. *J. Biomechanics*, *8*(*12*), 877–885

McGill, S. M., and Norman, R. (1986). Partitioning the L4-L5 dynamic moment into disc, ligamentous, and muscular components during lifting. *Spine*, *11*, 666–678

McGill, S. M. (1993). Abdominal belts in industry: A position paper on their assets, liabilities and use. *American Industrial Hygiene Association Journal*, *54*(*12*): 752–754.

McGill, S., Seguin, J., and Bennett, G. (1994). Passive stiffness of the torso in flexion, extension, lateral bending, and axial rotation: Effects of belt wearing and breath holding. *Spine*, *19*(*6*): 696–704.

Mitchell, L. V., Lawler, F. H., Bowen, D., Mote, W., Asundi, P., and Purswell, J. (1994). Effectiveness and cost-effectiveness of employer-issued back belts in areas of high risk for low back injury. *Journal of Medicine, 36*(*1*): 90–94.

Mirka, G. A., and Marras, W. S. (1993). A stochastic model of trunk muscle coactivation during trunk bending. *Spine*, *18*(*11*), 1396–1409.

Mundt, D. J., Kelsey, J. L., Golden, A. L., et al. (1993). An epidemiologic study of non-occupational lifting as a risk factor for herniated lumbar intervertebral disc. *Spine*, *18*(*5*), 595–602.

National Institute for Occupational Safety and Health. (1981). Work practices guide for manual lifting. Department of Health and Human Services (DHHS), Cincinnati, OH, D. W. Badger and D. J. Habes, Eds., Publication No. 81-122.

National Institute for Occupational Safety and Health. (1994). Workplace Use of Back Belts. Department of Health and Human Services (DHHS), Cincinnati, OH. Publication No. 94-122.

Nachemson, A. (1975). Towards a better understanding of low-back pain: a review of the mechanics of the lumbar disc. *Rheumatology and Rehabilitation, 14*, 129–143.

National Safety Council. (1989). *Accident Facts, 1989*, Chicago, IL.

National Safety Council. (1991). *Accident Facts*, Chicago, IL.

National Instiitute for Occupational Safety and Health. (1973). U.S. Department of Health and Human Services, Centers for Disease Control. *The Industrial Environment-Its Evaluation and Control.* U.S. Government Printing Office, Washington, D.C.

Nordin, M., and Frankel, V.M. (1989). *Basic Biomechanics of the Musculoskeletal System* (*2nd ed.*). Philadelphia: Lea and Febiger, p. 67.

Nussbaum, M. A., Chaffin, D. B., and Martin, B. J. (1995). A back-propogation neural network model of lumbar muscle recruitment during moderate static exertions. *Journal of Biomechanics, 28*(*9*), 1015–1024.

Ozkaya, N., and Nordin, M. (1991). *Fundamentals of Biomechanics, Equilibrium, Motion and Deformation.* New York: Van Nostrand Reinhold.

Park, K. S., and Chaffin, D. B. (1974). A Biomechanical evaluation of two methods of manual load lifting, *AIIE Transactions*, *6*(*2*), 105–113.

Pope, M. H. (1993). Muybridge lecture, International Society of Biomechanics XIVth Congress, Paris, France, July 5.

Praemer, A., Furner, S., and Rice, D. P. (1992). *Musculoskeletal Conditions in the United States*, Park Ridge, IL: American Academy of Orthopaedic Surgeons.

Reddell, C. R., Congleton, J. J., Huchingson, R. D., and Mongomery, J. F. (1992). An evaluation of a weightlifting belt and back injury prevention training class for airline baggage handlers. *Applied Ergonomics*, *23*(*5*): 319–329.

Reilly, C., and Marras, W. (1989). Simulift: A simulation model of the human trunk motion. *Spine, 14*(*1*), 5–11.

Rowe, M. L. (1981). Low back disability in industry: An updated position. *Journal of Occupational Medicine*, *13*(*10*), 476–478.

Sanders, M. S., and McCormick, E. J. (1993). *Human Factors in Engineering and Design.* McGraw-Hill, New York.

Schoenmarklin, R. W., Marras, W. S., and Leurgans, S. E. (1994). Industrial wrist motions and risk of cumulative trauma disorders in industry. *Ergonomics*, *37*(*9*), 1449–1459.

Schultz, A. B., and Andersson, G. B. J. (1981). Analysis of loads on the lumbar spine. *Spine*, *6*, 76–82.

Schultz, A. B., Andersson, G. B. J., Haderspeck, K., Ortgren, R., Nordin, R., and Bjork, R. (1982a). Analysis and measurement of the lumbar trunk loads in tasks involving bends and twists. *J. Biomechanics*, *15*, 669–675

Silverstein, B. A., Fine, L. J., and Armstrong, T. J. (1986). Hand wrist cumulative trauma disorders in industry. *Journal of Industrial Medicine*, *43*, 779–784.

Silverstein, B. A., Fine, L. J., and Armstrong, T. J. (1987). Occupational factors and carpal tunnel syndrome. *American Journal of Industrial Medicine*, *11*, 343–358.

Snook, S. H. (1989). The control of low back disability: The role of management. *Manual Materials Handling: Understanding and Preventing Back Trauma.* Akron, OH: American Industrial Hygiene Association.

Spengler, D. M., Bigos S. J., Martin, B. A., et al. (1986). Back injuries in industry: A retrospective study, I. Overview and costs analysis. *Spine*, *11*, 241–245.

Sudhakar, L. R., Schoenmarklin, R. W., Lavender, S. A., and Marras, W. S. (1988). The effects of gloves on grip strength and muscle activity. *Proceedings of the Human Factors Society 32nd Annual Meeting.* October 24–28, Anaheim, CA.

UAW International Union. (1982). *Strains and sprains: A workers's guide to job design.* Detroit, MI: UAW.

Videman, T., Nurminen, M., and Troup, T. D. G. (1990). Lumbar spinal pathology in cadaveric material in relation to history of back pain, occupation, and physical loading. *Spine*, *15*(*8*), 728–740

Walsh, N. E., and Schwartz, R. K. (1990). The influence of prohylactic orthoses on abdominal strength and low back injury in the workplace. *American Journal of Physical Medicine and Rehabilitation*, *69*(*5*), 245–250.

Waters, T. R., Putz-Anderson, V., Garg, A., and Fine, L. J. (1993). Revised NIOSH equation for the design and evaluation of manual lifting tasks. *Ergonomics*, *36*(*7*), 749–776

Webster, B. S. and Snook, S. H. (1989). *The Cost of Compensable Low Back Pain.* Hopkinton, MA: Liberty Mutual internal report.

CHAPTER 10

WORK PHYSIOLOGY—FATIGUE AND RECOVERY

Suzanne H. Rodgers
Ergonomists/Human Factors Consultant
Rochester, NY 14622 USA

10.1 INTRODUCTION

The physiological basis of ergonomics/human factors has had less focus in the United States than in other countries because of the strong emphasis, until recently, on performance rather than on the prevention of illness and injury on jobs. The dimension that physiology adds to the biomechanical assessment of task or job suitability is *time*. Instead of looking at strength alone, we add the time factor to look at strength over time, or *endurance*. The time factor includes not only the duration of the effort itself, but also the amount of time spent on that task during the shift. Local muscle fatigue can be predicted by the pattern of effort and relaxation of the active muscles during a task. Whole body fatigue is described by the distribution of those tasks, in time, over the shift. It is best described by energy utilization, or oxygen consumption, relative to aerobic work capacities and by the accumulation of lactic acid in the arterial blood.

If one watches people working, one can see that their pattern and style of work is influenced by the demands of the task/job and by their work capacities. This measure is the percent of maximum strength, or maximum voluntary contraction (% MVC), that the active muscles use. The higher the strength required, the less time the task can be maintained because of accumulating fatigue. Even light strength requirements can become fatiguing, if they are sustained for long times, as can be attested to by office workers and inspectors.

Fatigue is important because it is fatigue that people *sense* when they sustain effort for minutes and hours. If the effort is excessive, many people cannot perform the jobs, and those who can are more fit than most of the workforce. If the effort is not excessive but the repetition rate is high or the effort holding time is sustained beyond a few seconds, then fatigue will be the cue that the job or task has to be improved. On self-paced jobs, the worker who is experiencing discomfort from fatigue will find ways to reduce the stress by either changing to a less fatigued muscle group (e.g., alternating hands on a tool), using larger muscle groups (e.g., using arm rotation instead of wrist rotation), or increasing secondary work (e.g., checking with the team leader about a quality issue). On externally paced jobs, the worker will often speed up the task in order to reduce the accumulated fatigue and to get additional recovery time during the task cycle. This can affect quality performance and may make it harder to enforce the use of slower assist devices. In addition to monitoring local muscle fatigue during tasks, most people learn to monitor total workload for the shift and to relate acceptable values to their 24-hr energy demands. Consequently, any task that takes high effort levels will be balanced with a lighter, recovery task such that the effort level at the end of the shift is within the acceptable energy cost range (see 10.3.2). Because people monitor their local muscle and whole body fatigue less well when their interest and emotions are highly engaged, one cannot separate physical fatigue from mental work. A good observer can document "physiological behavior" while watching people in production, office, and other service jobs. It provides strong clues for identifying which tasks are contributing to potential injuries and illnesses and where the ergonomics interventions will be most effective.

A causative relationship between fatigue and injury/illness is not scientifically proven to date, although most people agree that a fatigued muscle is more subject to injury than a fresh one. Whether repetitive motions disorders are *caused* by fatiguing tasks is not certain either (Kilbom, 1988), although circulatory factors probably contribute to them. The material discussed in this chapter is based on a recent review of published papers, but also on more than 25 year's experience studying jobs in industry and office settings. Physiological and psychophysical data were collected in many of those studies, and techniques for analyzing jobs for potential fatigue are based on that experience and on physiological information.

A healthy dialogue on the mechanisms of muscular and whole body fatigue has developed in the European Community as it strives to develop common health and safety standards for acceptable workload (Dul et al., 1991, 1994; Mathiassen and Winkel, 1991, 1992; Winkel and Mathiassen, 1994). Some of this information is included in Sections 10.2 and 10.3 despite the lack of resolution of all of the issues to date. There are opportunities to develop better techniques to assess the physiological demands of work; examples of research needs are included to stimulate the researchers to fill in these blanks in our understanding.

10.2 LOCAL MUSCLE FATIGUE AND RECOVERY

The physiological acceptability of a task for a majority of the workforce is determined by several factors. A primary one is whether the necessary capacity is available to do the task for the time it must be done. That capacity includes the necessary strength, the ability to sustain that strength for the required seconds of effort, and the ability to recover quickly from any fatigue that may accumulate before the next effort is needed. If fatigue cannot be fully repaid before the next effort, the ability to recover quickly at the end of the task is needed. Much of the attention to ergonomics in industrial work has been related to strength and to the biomechanical aspects of effort generation (ANSI, 1993). That information (see Chapter 9) can be used in determining how much is too much effort, but a biomechanically acceptable task may not necessarily be a safe one when the effort pattern is considered. One has to look at the time of the effort and the time available to recover from the effort to know if a task is acceptable for a few minutes, a few hours, or for the majority of an 8-, 10-, or 12-hr shift. The higher the fatigue rate, the less likely people will be able to sustain the work for hours instead of minutes.

10.2.1 Aerobic and Anaerobic Muscle Work

The ability of a muscle to contract and relax depends on many factors. At the cellular level, the cells need to have ATP (adenosine triphosphate), CrP (creatinine phosphate), calcium ions (Ca^{++}), oxidative, and nonoxidative enzymes working at their optimal hydrogen ion concentration (pH = 7.4), glycogen or glucose, and other substrates and transfer agents to allow actin and myosin to both bind and break apart (Vollestad and Sejersted, 1988). At the cell membrane, there have to be an appropriate ion balance and transmittters for generation of the electrical signal (excitation-contraction coupling) (Pagala et al., 1984). And in the central nervous system, there have to be discharges to the motorneurms and recruitment of fibers for sustaining the effort level in the active muscles (Petrofsky et al., 1981). The cellular level needs are satisfied by keeping blood flowing close enough to the muscle cells to permit delivery of substrates and oxygen and removal of waste products and metabolites, such as carbon dioxide and lactate. The oxidative metabolism of muscle cells helps keep the muscle cell membrane integrity so it can respond to electrochemical excitation at the myoneural junction (Westerblad et al., 1990). The central control of muscle tension is dependent on feedback from the muscle and joint sensors. Anything that changes the environment or sensitivity of these receptors can also affect muscle tension and can contribute to a person's "sense" of fatigue or overload. Each muscle has the ability to sustain very short, heavy efforts with its resting supply of ATP and CrP in the cell, but repeating or sustaining that effort becomes a problem if the blood supply to the cell is restricted or stopped. Once the blood supply drops below the levels needed to support the complete oxidation of glucose to carbon dioxide and water (the Kreb's cycle part of glucose metabolism), the glycolytic breakdown of sugar results in an accumulation of lactic acid in the cell. As the acid accumulates and reduces the pH of the muscle cell, the metabolic enzymes are not in their optimal functional range, potassium (K^+) and calcuim (Ca^{++}) balance is affected, and there is a fairly rapid degradation in the ability of the muscle cells to sustain contraction (Bergstrom and Hultman, 1991; Westerblad et al., 1990).

Short periods of anaerobic muscle work are not problematic, but they should be avoided if one is doing the task for an hour or more continuously. Anaerobic work is particularly to be avoided if the task is done for a majority of the shift (Bjorksten and Jonsson, 1977). If a fatiguing effort is sustained for hours, the worker can experience symptoms that range from a dull ache or burning at the end of the shift to muscle cramping and swelling that is painful and prevents comfortable use of the muscles for several hours, until they have recovered. Muscle effort can be sustained in aerobic work, although long periods of light effort for postural support (e.g., head down to read) can contribute to a soreness discomfort at the end of the shift that may become chronic (Feely et al., 1995).

10.2.2 Circulatory Effects of Increased Muscle Tension

The distribution of blood flow in an active muscle is controlled by the autonomic nervous system and by local metabolites. As the demand for blood flow increases to allow sustained or frequent contraction of the active muscle fibers, those arterioles open to distribute the flow to their capillaries. There the oxygen can diffuse out to the muscle cell mitochondria and keep ATP available for contraction. They also supply glucose to cells that have reduced their resting glycogen (starch) levels, and they assure that calcium, phosphate, and other ions are kept in balance to allow the muscle fibers to both contract and relax (Miller et al., 1987; Westerblad et al., 1990).

The heavier the muscle effort level, the more its blood flow can be compromised. If the muscle fibers are contracting strongly and relaxing alternately (dynamic work), some arterial blood flow may be re-established during the relaxation period (Lind et al., 1981; Sjogaard et al., 1988). If muscle tension stays ≥10% MVC (Bystrom and Fransson-Hall, 1994), some local redistribution of blood flow to the most active muscle fibers may occur because of local build-up of metabolites, but the blood flow will remain compromised . If the involved muscle mass is large, an increase in systemic blood pressure and heart rate may overcome this increased peripheral resistance in the working muscles (Lind and McNichol, 1967; Petrofsky and Lind, 1975). This is initiated by a reflex from the arteriolar "gates" and is probably affected by rising lactate levels in the blood as well. The increased blood pressure counteracts some of the muscle pressure effect to increase blood flow to the muscle cells, and the increased heart rate will assure that more oxygen is delivered per minute. At muscle tensions below 10% MVC, blood flow is not affected (Sjogaard et al., 1988).

Even with this adaptive response to re-establish the compromised blood flow to the statically loaded muscle, there will be some residual discomfort because the lactate increases the osmotic pressure of the cells and draws fluid into the interstitial tissue, causing soreness from swelling (Sjogaard, 1986). If static postures are maintained for several hours a day and for 5 days a week, resolution of the soreness may not be complete and a chronic condition can develop.

10.2.3 Work and Recovery Times at Different Work Intensities

The description of an industrial worker as an "athlete" is both useful and misleading. Athletes are fit and flexible and have good techniques to minimize strain during their work periods. These are all good traits for industrial workers. Wellness programs and training in lifting techniques and stretching exercises are being used by most progressive companies as ways to reduce workers' risk for muscle and joint strains and sprains and cardiovascular overload.

Most athletes, however, work at high percents of their maximum strengths or endurance capabilities when they are active, and those peak loads are interrupted regularly by short breaks that permit some recovery time. They also seldom sustain activity for more than 4 hr a day and do not necessarily work 5 days in a row before having a day or two off. Industrial and service workers have to sustain activity for 8 to 12 hr a day (of which 1 to 1.5 hr may be break time). If a task is heavy, they usually do not have a built-in recovery period afterwards, or even within it, and they have to pace themselves to avoid premature fatigue. As will be seen later in Sections 10.2.5 and 10.3.2, the longer one works, the lower the average percent of maximum strength or percent of aerobic capacity one can use.

10.2.3.1 Global Job Design Guidelines to Reduce Fatiguing Muscle Effort

Heavy efforts in localized muscles that are sustained for 6 to 40 sec will require very long recovery times (Monod and Scherrer, 1957; Rohmert, 1973a). Light efforts, especially static, postural efforts, develop significant recovery time needs if they exceed 8% of maximum voluntary muscle strength and are sustained more than 20 sec continuously (Bjorksten and Jonsson, 1977; Hagberg, 1981). With this in mind, the following goals should be set for the design of muscular efforts:

1. If a task involves heavy stress on a muscle group, make it as short as possible (or reduce the load). This assumes that the effort is not excessive, just heavy.
2. If a task involves static loading of a muscle group, find ways to reduce the effort time by changing posture, providing supports, or speeding up the process (e.g., a tool change).
3. Avoid high-frequency, high-effort tasks.

4. Avoid moderate or high efforts that are sustained for 10 to 15 sec before relaxation unless they are done less than once in 5 min.

10.2.3.2 Recovery Time Allowances and Maximum Effort Times

Other interactions of effort level, effort duration, and frequency may be assessed using these variables and estimating the amount of fatigue accumulating from static work/recovery curves. Monod and Scherrer (1957) and Rohmert (1960a,b) published isometric fatigue curves for several muscle groups and gave formulae for predicting recovery times from different effort levels (expressed as a % of maximum voluntary contraction forces). Figure 10.1 shows the basic fatigue curve from these studies. Figure 10.2 shows continuous static, continuous static/dynamic, and intermittent static fatigue curves for the same muscle group (Hagberg, 1981). Rohmert (1973a,b) and Laurig (1973) have published formulae for isometric and dynamic work recovery time allowances, which are as follows:

(a) For static muscular work:

$$RA = 18(t/T)e^{+1.4} \cdot [(f/F) - 0.15]e^{+0.5} \cdot 100\%$$

where RA = rest allowance, as a % of working time, in seconds
t = time of static effort in seconds
T = total time of work cycle in seconds
f = force exerted in the static effort, in Newtons
F = Maximum static force from that muscle group in that posture, in Newtons
e is an exponent term

(b) For heavy, dynamic muscular work:

$$RA = 1.9 \cdot (t)e^{+0.145} \cdot \left[\left(\frac{\text{Neff}}{\text{Nendurance limit}}\right) - 1\right]e^{+1.4} \cdot 100\%$$

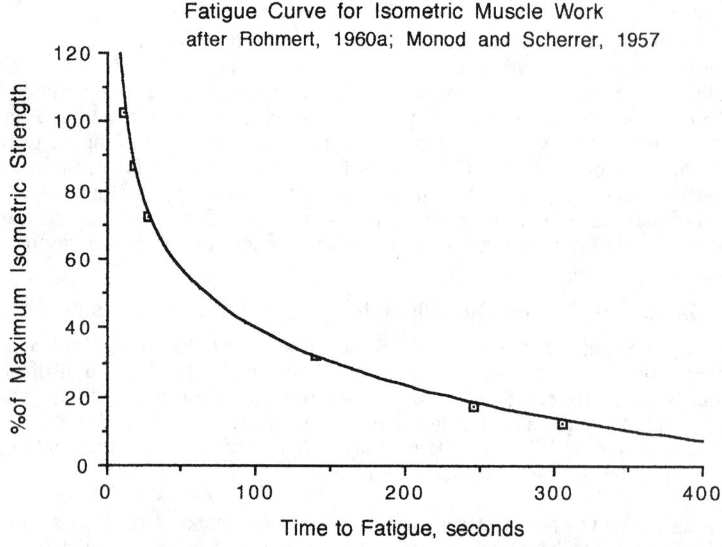

Figure 10.1 *Fatigue Curve for Isometric Work.* The time to fatigue is shown for different muscle effort intensities, expressed in % MVC, or % of maximum voluntary contraction strength. Fatigue is defined as the point when the effort intensity can no longer be maintained. (Source: Developed from data in Monod and Scherrer, 1957 and Rohmert, 1960a with modifications based on more recent research—see text).

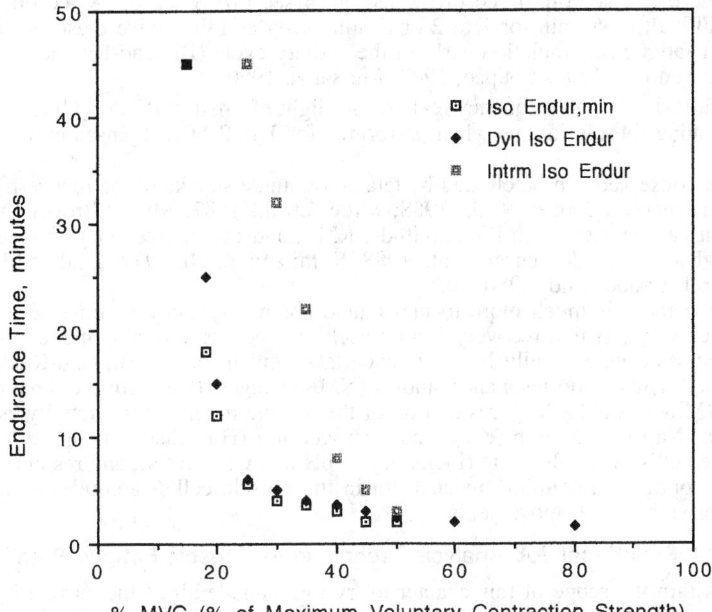

Figure 10.2 *Endurance Time versus Effort Intensity (% MVC) for Isometric and Dynamic Exercise.* The elbow flexors were studied on 9 people using isometric and dynamic exercise, the latter with the muscle statically loaded during the movement. The intermittent isometric exercise studies were done on 4 subjects' elbow flexors. Endurance time improved only with the intermittent pattern of work. (Source: Developed from data in Hagberg, 1981).

where the terms are similar to those in (a) on page 40, except that work is expressed as the endurance requirement compared to maximum endurance time for the active muscles, in seconds.

(c) For active light muscular work:

$$T = \left(\frac{\text{Fschw}}{\text{Fmax}}\right)\% \cdot 0.9 + [\ f(l/\text{min})e^{+0.5}] \cdot 0.77$$

where T = the maximum working period in minutes

F_{schw} = the measured force, in Newtons

F_{max} = the maximum force generated by those muscles in the same posture, in Newtons

f = the muscle activation frequency, in activations per minute

Since the publication of these curves and formulae, futher research has suggested that these rest allowances are underestimated for sustained lighter efforts and for short but highly repetitive efforts. Using electromyography (EMG), electrical stimulation, and nuclear magnetic resonance (NMR) techniques to assess the fatigue state of muscles before and after different work and rest cycles, the following has been found:

(a) Static efforts at 15% MVC cannot be sustained indefinitely. Static efforts less <8% MVC appear to need little recovery time, although some researchers place the nonfatiguing isometric effort level at 2% MVC (Jorgensen et al., 1988; Westgaard, 1988). For dynamic repetitive handgrip work, a 17% MVC effort appears to be highest level (Bystrom and Fransson-Hall, 1994).

(b) Intermittent, short, and moderately heavy static efforts (40–50% MVC) that are repeated with a 50% or 100% recovery cycle (4 sec on, 2 sec off; 4 sec on, 4 sec off; 6 sec on, 4 sec off; 8 sec on, 4 sec off; 8 sec on, 8 sec off; etc.) are difficult to sustain for 1 to 2 hr continuously and the active muscles show signs of fatigue accumulation early in the activity cycle (Bigland-Ritchie et al., 1986; Davenport, 1990; Loupee, 1989; Messaros, 1990).

(c) Very short, high frequency (>15/min), light efforts (<30% MVC) exhibit a continuing muscle fatigue when performed for 1 to 2 hr (Jorgensen et al., 1988).

The fatigue observed can be elicited by tetanic or single shocks to the recovering muscle group or its nerve (Cooper et al., 1988; Miller et al., 1987; Milner-Brown and Miller, 1986) or it can be seen in EMG amplitude, RMS, and center frequency changes (Cook, 1990; Hagberg, 1981; Jorgensen et al., 1988; Komi and Tesch, 1979; Lind and Petrofsky, 1979; Petrofsky and Lind, 1980a,b).

Although there is much more to understand about this recovery time need, most of the research suggests that recovery from muscle fatigue has several components. Muscle tension recovers quite rapidly but endurance takes much longer (Bigland-Ritchie et al., 1986). Nuclear magnetic resonance studies (NMR) suggest that there are three phases of recovery (Miller et al., 1987): restoration of the muscle membranes's activity as affected by sodium (Na^+), potassium (K^+), and hydrogen ion (H^+) distributions; restoration of the muscle cell's metabolic state (including its pH and CrP levels); and restoration of the excitation-contraction coupling mechanism in the muscle cell's sarcoplasmic reticulum, perhaps through Ca^{++} movements.

10.2.3.3 A Functional Job Analysis Technique for Muscle Fatigue Prediction

It is not within the scope of this chapter to try to reconcile all of the current knowledge on local muscle fatigue mechanisms, but rather to try to identify ways of predicting fatigue in jobs with some degree of accuracy. Table 10.1 shows one attempt to do this (Rodgers, 1987, 1988, 1992). It was originally based on the Rohmert curves for static efforts (Figure 10.3). Because of the newer research findings, quantification of the fatigue is confined to categories; actions to improve the jobs come from identifying the most significant factors

Table 10.1 Functional Job Analysis Form. Ratings Per Body Part of Effort Intensity, Duration, and Frequency Are Used to Predict Muscle Fatigue on Occupational Tasks

		Functional Job Fatigue Analysis				
Body Part		Effort Level	Continuous Effort Time	Efforts/Minute	Priority	
Neck		_____	_____	_____	_____	Effort Categories
Shoulders	R	_____	_____	_____	_____	(assume one 3s effort in 5 min)
	L	_____	_____	_____	_____	1 = light
Back		_____	_____	_____	_____	2 = moderate 3 = heavy
Arms/elbows	R	_____	_____	_____	_____	Continuous Effort
	L	_____	_____	_____	_____	Time Categories
Wrists/hands/fingers	R	_____	_____	_____	_____	1 = <6 s
	L	_____	_____	_____	_____	2 = 6 to 20 s 3 = >20 and <30 s
Legs/knees	R	_____	_____	_____	_____	Efforts/Minute
	L	_____	_____	_____	_____	Categories 1 = <1/min
Ankles/feet/toes	R	_____	_____	_____	_____	2 = 1 to 5/min
	L	_____	_____	_____	_____	3 = >5–15/min

Priority for Change (time to recover after 5 min of continuous work on the task)

*Moderate =	123, 132, 213, 222, 231, 232, 312	30 to 90 s fatigue accumulation
**High =	223, 313, 321, 322	90 s to 3 min.
**Very High =	323, 331, 332	>3 min.

Source: Modification from Rodgers (1987, 1988, 1992).

*Usually acceptable for 1 hr continuously; rotate with lighter task beyond 2 hr, or modify to lower the priority.

**Make job improvement to lower the priority; not good job rotation candidates.

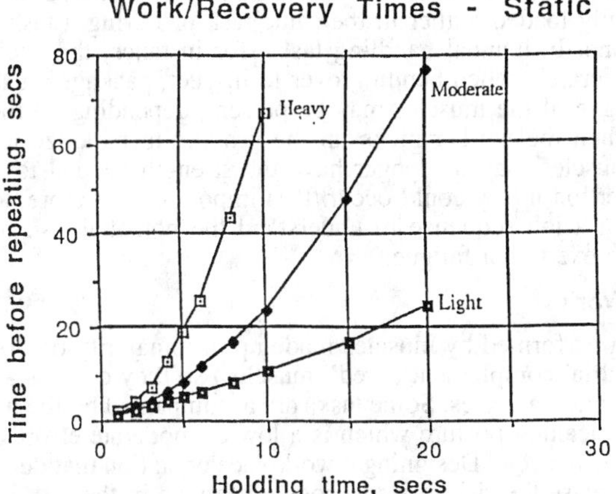

Figure 10.3 *Work and Recovery Times for Static Work.* The interaction is shown between effort intensity (light = 30%, moderate = 50%, and heavy = 80% MVC), duration, and frequency (indirectly indicated as time before repeating, which includes work and recovery time). Time before repeating may be the cycle time on a repetitive task but is determined by muscle activity, not time/motion analysis (Source: Drawn from data in Rohmert, 1973a).

in the risk for fatigue, e.g., effort levels, effort time, or effort frequency. Moderate and heavy effort levels are rated after identifying risk factors for each body part that relate to the postures used and to the capacities of the active muscles. A psychophysical rating of muscle effort levels independent of effort duration and frequency can be used to rate this parameter (see Section 10.2.5.4). Effort durations and frequencies are based on muscle group activities rather than on a task cycle; they are most easily assessed while viewing a videotape of the job after the operations have been studied on the job site. A case study using this analysis is given in Section 10.3.3.2 .

The priority for changing a task is based on how much fatigue accumulates in 5 min of work. If a longer time is needed for the task, the total time of continuous work will determine how serious a problem there is on the job. For example, if a task involves a high force exerted for more than 6 sec continuously at a rate of three times/min, one can estimate that about 3 min of fatigue could have accumulated after 5 min of work. If the task is done for 20 min straight, the fatigue accumulation recovery time need is at least 12 min, and it is unlikely the task can be sustained safely by most workers. In the past, the stronger person, for whom the force was a moderate effort, would be put on this job. The ergonomic goal of designing for most workers changes the emphasis. If one can modify the effort level to moderate, fatigue accumulation should be less than a minute in 5 min of work; at the end of 20 min, the recovery time need would be 4 min. If the next task is not fatiguing, the debt should be able to be repaid without stopping work. Used with caution as a way to estimate potentially fatiguing tasks, this analysis method has helped improve communications between production workers and engineers during the setting of job standards. By including recovery time needs in the work design process, one can better determine the workload and put a priority on reducing the more fatiguing tasks.

10.2.4 Application of Local Muscle Fatigue Data to Job Risk Assessment

Although research is still needed to establish the exact relationship between fatigue and musculoskeletal illness and injury, it is probable that a fatigued muscle that can no longer sustain the effort called for by a task will be at risk for strain/sprain injuries or for cumulative trauma disorders. Because the degree of fatigue can be expected to relate to the degree of risk in physically demanding jobs, using a fatigue-rating system to prioritize the need to address ergonomic problems on a job seems warranted.

Most jobs are a combination of static and dynamic tasks. The static tasks are often postural, and the dynamic tasks can be repetitive. A common finding on jobs where musculoskeletal problems have been seen is that the muscles involved in a dynamic task have also been statically loaded, either in the same task or during a task performed just before the dynamic one. In manual handling tasks, for instance, the back muscles may have been loaded moderately when bending over to inspect, package, or label a product. Moderate to high fatigue of the muscles may be present, depending on the time to complete that activity. When the worker picks up the product to palletize it or put it on a conveyor, the back muscles may no longer have the strength needed to support the lift safely, and an overexertion injury could occur. It is important, therefore, to look at all of the tasks performed in a job sequence to understand the potential risk for an injury or illness associated with exertional fatigue.

10.2.4.1 Postural Work

Postural work is often performed by muscles made up of a majority of slow twitch fibers with a high mitochondrial complement ("red" muscles) so they can sustain effort longer than fast twitch, or "white" muscles. Some tasks are accompanied by the need to maintain a bent forward or side-leaning posture which is a low or moderate effort but is sustained throughout the task or task cycle. Designing a workplace or station that delivers everything to the worker may increase the risk of static postural fatigue in the task if that means he or she cannnot get up to change postures intermittently.

Since the nonfatiguing static effort level is 8% MVC, it is probable that some fatigue will accumulate in most postural muscles during work. How much accumulates and how long it takes to recover after a particular task is completed will determine long-term and short-term comfort for the worker. As can be seen in the relationship between effort level and time in Figure 10.1, the longer the posture has to be sustained, the lower the effort level has to be to avoid significant fatigue.

To reduce the effects of sustained static postural work, one can support the muscle or body part, or design tasks to include activities that provide recovery for the most active muscle groups. For example, if a worker is using a magnifying glass or microscope for prolonged periods, the job should included a task of procurring the slide trays or products and disposing of them to the next operation. This ability to get away from the magnifying device and to get up from the chair can provide a welcome interruption for the back, neck, hand, and arm muscles and will stimulate circulation to them until the next batch is ready to inspect. The best recovery activity for steadily loaded muscles is light, dynamic work, although this does not assure complete recovery (Hansson et al., 1992).

10.2.4.2 Repetitive Tasks

For repetitive tasks where muscles are not activated more than 15 times a minute, the fatigue model reflects subjective responses on discomfort surveys. Once the repetition rate exceeds 30 activations/min, light effort may be associated with repetitive motions disorders over time (ANSI, 1993). This is probably a "wear and tear" phenomenon and is most prevelant when the task has to be sustained for the majority of the shift rather than for a few hours. Physiologically, one can predict that a muscle group that is being triggered more than once every 4 sec is probably not fully relaxed during the recovery period between contractions (Jonsson, 1988). Most movements take at least 2 sec and transfer of materials by gripping or applying force is more likely to take 3 to 4 sec (Eastman Kodak, 1986). As repetition frequency increases, there is a tendency to put more effort into the task and to accelerate the load faster, both of which add to the risk of injury (see Chapter 9). The trade-off is that the reduced holding time should reduce the recovery time needs for the muscle.

An important component of fatigue in upper extremity disorders is static loading of the shoulders (Mathiassen et al., 1993) that can affect blood flow to the hands and to the median and ulnar nerves. The shoulder tension can be associated with too high a work height, an extended reach to use a mouse when it is positioned ahead of the keyboard, or rotation of the shoulder to the side when the mouse or a tablet is used on a side table or extension. Fatigue of the shoulder muscles in these tasks may also change the trunk posture and result in the worker leaning to one side and putting pressure on the elbow, another risk factor for repetitive motions disorders of the upper extremities.

At repetition frequencies greater than 15/min, the Rohmert curves describing static effort and recovery times are no longer predictive of the potential fatigue development. The more appropriate dynamic work curves may underestimate the fatigue, too. Current research is aimed at establishing the actual relationship between work and recovery at activation frequencies between 10 and 30/min. The mechanism of the fatigue is still not clear, so prediction of actual recovery time needs for different work patterns at higher effort frequencies remains imprecise (Mathiassen and Winkel, 1992).

10.2.4.3 Manual Handling Tasks

Manual handling tasks include lifting and lowering items and pushing and pulling equipment or parts, usually done as part of a transfer of the parts or product between operations. In high-volume production, manual handling tasks may be done under time pressure, e.g., loading or unloading production machines or conveyors. In these jobs, the handler has to meet the needs of the production process and may have little control over his or her work pattern. If the items handled are moderately heavy (or greater), and if the frequency is high, the potential for fatigue is increased and may contribute to musculoskeletal discomfort or injury. One way the worker has to reduce the discomfort is to minimize the holding time. That may result in a tendency to throw the product into place instead of taking the extra second or two to set it on a pallet or conveyor more precisely. If the product is light to light-moderate in weight and handled frequently, most workers will minimize fatigue by picking up more than one item at a time, thereby reducing the effort frequency. This is not a viable option if the items are bulky, so their strategy with the bulkier items is to move them very rapidly, as described earlier.

For pushing and pulling tasks, fatigue is related to the time of exerting the effort, most often to the distance traveled. If a loaded cart has to be moved 50 m (164 ft) to the next operation, the handler would be wise to postpone a subsequent heavy lifting task a few minutes until the active muscles have had an opportunity to recover. Using the recovery allowance formula curves in Section 10.2.3, one can predict how long a recovery period is needed to, at least, re-establish effective blood flow to the muscles, and wash out the accumulated metabolites. The risk of injury is of concern when a cart or truck has to be maneuvered into a fairly tight location after it has been pushed from another area for more than 90 sec (See Chapter 34 for guidelines on the design of materials handling tasks).

10.2.5 Methods to Study Local Muscle Fatigue

Several methods are used to quantify fatigue of the musculoskeletal system. Early studies included constant holding of a maximum or submaximum effort and watching the decay of strength over time (Figure 10.1). Later studies created severe muscle fatigue and then used electrical stimulation of the muscle group or its nerve (from implanted electrodes or surface electrodes) to determine how quickly the muscle was able to recover. Chemistries of the venous blood coming from the active muscles have been done to estimate the degree of fatigue present, and systemic arterial blood lactates were measured when the active muscle mass was large. More recently, the emphasis has been on looking for RMS or center frequency spectrum shifts in muscle electromyography (EMG) patterns over time as the active muscles are fatiguing. The research methods for studying fatigue in active muscles are not always practical, so many ergonomists have relied on self-reporting of fatigue or discomfort by the workers, often using a 10-pt psychophysical scale (Borg, 1982). The positive and negative aspects of each of these methods are summarized below.

10.2.5.1 Fatigue and Recovery Curves for Static and Dynamic Work

The decay in strength with continuous effort at a given percent of maximum strength describes the "fatigue curve" for static efforts shown in Figure 10.1 (Kahn and Monod, 1989). This is strongly affected by the subjects' motivational level since fatigue was defined as the time when a given level of effort could no longer be achieved. Fatigue studies are not frequently done because it is an hour or more before the same subject could be tested again and get repeatable results. Taking a muscle to exhaustion results in a change in its chemistry such that it will not recover quickly. The nature of that fatigue is still being investigated and appears to relate to ionic balances in the cell and to oxidative metabolism needs (Bergstrom and Hultman, 1991; Westgaard, 1988).

Using animal and human models, researchers have used direct tetanizing electrical stimulation of a voluntarily fatigued muscle or its nerve to assess its fatigue state during the recovery period (Bergstrom and Hultman, 1991; Bigland-Ritchie et al., 1986; Duchateau and Hainaut, 1984; Hainaut and Duchateau, 1989; Pagala et al., 1984). In humans, electrical stimulation results did not always agree with studies done where the fatigue was created voluntatily and recovery was measured by how long it took to be able to do the same task for the same length of time. Whatever the mechanism, this method does identify residual fatigue in muscles that can respond to short maximal exertions (Vollestad and Sejerstad, 1988). Older men were found to recover somewhat slower and to relax a muscle more slowly after a short stimulation (twitch) compared to young men when the muscle was fatigued with an intermittent, tetanizing stimulus for 10 min (Klein et al., 1988).

Dynamic work can be studied by having people repeat a work pattern until they are no longer able to sustain it (Bystrom and Franssom-Hall, 1994; Edwards, 1986; Lind et al., 1982). Because of the intermittent restoration of circulation during muscle relaxation, the endurance time is longer than that seen in continuous static work (Hagberg, 1981), as can be seen in the curves in Figure 10.2 (Section 10.2.3.2). Questions remain about the etiology of the fatigue that accumulates in dynamic work, and there may be different mechanisms depending on the repetition rate. EMG studies suggest that a shift to lower frequencies may indicate some fatigue at the muscle membrane rather than a problem with ion transfers inside the muscle cells (Komi and Tesch, 1979; Zwarts et al., 1987). Because fatigue occurs in dynamic work, the rate and duration of that work will determine how much recovery time is needed to re-establish the resting state of the muscle. There are still too few studies that relate the fatigue rates to tasks that are done for a majority of the shift (Baidya and Stevenson, 1989; Christensen, 1986; Tesch et al., 1985; Mathiassen et al., 1993; Petrofsky and Lind 1978b; Sundelin and Hagberg, 1989), including 12-hr shifts, so interpreting the risk of injury or illness on such tasks is more difficult.

To assess the time needed to recover from a fatiguing effort, one can remeasure the strength and endurance at regular intervals after a known fatiguing task (Lind, 1959; Milner et al., 1986). The problem with this approach is that the length of time a strength exertion is held is critical, because one can often do short efforts but not longer ones before full recovery has take place. For example, a forearm flexor muscle group is taken to fatigue with a static effort at 50% of maximum isometric strength for more than 1 min. The muscle strength was tested with 2- to 3-sec maximum efforts during the next several minutes. Because the oxygen in the muscle cell and arm blood supply can provide enough ATP for short contractions (up to 6 sec), it could appear that the muscle is "recovered" within 15 to 20 sec. However, if the same fatiguing effort was used, the muscle strength would drop off after much less than 60 sec of continuous effort, which would demonstrate that recovery had not been complete in the first 15 to 20 sec of relaxation. Because repeating the fatiguing work during the recovery period makes it impossible to track the rest of the recovery response from the first fatiguing condition, this research takes a long time. Repeatablility is harder to achieve since each recovery point tested could have started with a different state of muscle fatigue.

10.2.5.2 Psychophysical Measures of Local Muscle Fatigue

Psychophysical methods to assess the amount of discomfort a person has on a task have the advantage of requiring little equipment and being relatively easy to administer to large numbers of volunteers. The original methods were administered very carefully (Borg and Lindblad, 1976; Ciriello et al., 1990; Stevens, 1975), but less care is being exhibited as the techniques gain wider use, so study results must be carefully critiqued and not relied on too heavily for guideline setting. Nonetheless, it is clear that people can indicate how much discomfort they have on a good word scale, usually associated with numbers from 0 to 10, as long as the instructions are good and the top and bottom of the scale are defined. For example, in the *Large Muscle Group Activity Scale* published by Borg in 1982, the top value of 10 was defined as "the worst pain you have ever experienced," which made it more likely that it would not be used to describe most occupational discomforts. For fatiguing tasks, such as maximal aerobic capacity testing on a bicycle ergometer, some 9s or 10s could be used if the top of the scale is defined as "so uncomfortable you want to throw up" (Williams, 1990).

Because most people sense not just muscle effort but also effort in time (effort duration and frequency), Borg's 10-pt scale can give an estimate of the discomfort that relates to the fatigued state of the muscle. Where this type of scale will be less accurate is in assessing when the muscles have fully recovered. Once the muscle tension is decreased and the blood flow is re-established, the person may feel the discomfort is gone yet may not have the endurance needed to repeat the task. With experience, the worker learns how to pace the work to reduce fatigue accumulation, but the full effect of the work may not be evident until after the task is performed (Jorgensen et al., 1988). Work with a high mental investment or situations where other motivators disguise the fatigue can contribute to a person's lack of realization of accumulating fatigue on some tasks. Poor communications or feedback and other psychosocial stressors may increase the sense of fatigue or discomfort on tasks, probably by lowering the sensitivity or threshold of the stress response. Thus, the rating of fatigue during recovery activities will be less accurate than ratings of fatigue during musculoskeletally demanding work.

10.2.5.3 Electromyographic Studies of Local Muscle Fatigue

The use of electromyography (EMG) to study muscle fatigue has found new emphasis with compter technologies that permit electronic, rather than mechanical counter, data collection and processing (Basmajian and DeLuca, 1985; Cook, 1990; Dolan et al., 1995; Miller et el., 1987; Petrofsky et al., 1982; DeLuca, 1984; Seidel et al., 1987; Soderberg, 1989; van Dieen et al., 1993; Zwarts et al, 1987). It is known that the center frequency spectrum of the muscle action potentials (usually monitored with skin electrodes) shifts to the low frequency side when an active muscle is fatiguing during static or dynamic work . Portable EMG systems are now available (Cook, 1990) that permit the worker to do the job without interference while data is stored in a computer chip, to be processed later using RMS and amplitude wave analysis programs.

There is some concern about the sensitivity of this technique for detecting fatigue when several muscle groups are involved in a task. Because the most quickly fatiguing muscle may lie below others, it might not be reflected very strongly in the surface EMG signal (Basmajian and DeLuca, 1985), giving the impression that fatigue is not present when the worker can clearly sense discomfort that he or she relates to "fatigue." For static muscle loading tasks where the skin stays over the same muscles during the work, the EMG technique is usually more accurate than when the joint is extended, flexed, or rotated and the skin electrodes keep moving over different parts of the muscle (Soderberg, 1989). Needle electrodes reduce this artifact but are able to pick up only a small part of the active muscle fibers, and are probably not acceptable for on-site studies of work; only noninvasive techniques are routinely acceptable.

More recent studies of muscle fatigue and recovery show that the EMG amplitude increase and center frequency shift to the low frequency side of the power spectrum occur quite early in isometric and dynamic work patterns where fatigue accumulates and often are reduced at the end of 1 hr of continuous work (Davenport, 1990; Loupee, 1989; Messaros, 1990). This may represent recruitment of more slow twitch ("red") fibers with time, synchronization of the central discharge to the motorneurons as fatigue accumulates, or an increase in muscle temperature with sustained work (Komi and Tesch, 1979; Lind and Petrofsky , 1979; Petrofsky and Lind, 1980a,b; and Petrofsky et al., 1982). The psychophysical rating of fatigue continues to increase even when these EMG responses are declining, suggesting that internal cell metabolism is a limiting factor and is not fully reflected in the EMG changes. The amplitude and center frequency of the EMG power spectra recover within 60 sec from isometric or dynamic contraction at 33% of MVC of the brachioradiales (Petrofsky and Lind, 1980a), while the muscle is far from recovered in that time period (Hagberg, 1981; Kuorinka, 1988). For postural loads at 5 to 20% MVC as one might see in office settings for the neck and shoulders, there is a low fatigue level that may not be seen in EMG changes at all, yet where electrical stimulation of the muscles shows some loss of strength or endurance. These loads are often associated with upper extremity and trunk injuries and illnesses, so Westgaard (1988) cautions that a biomechanical analysis of the stress may be more accurate than looking for EMG power spectra shifts. Seidel et al. (1987) suggest that for low-level efforts (20 to 40% of MVC), the autoregressive (AR) time series models of the fifteenth order and spectral densities may better predict the fatigue.

A primary advantage of EMG for studying jobs and predicting effort levels and fatigue is that it demonstrates the pattern of discharge of the muscle fibers (Jonsson, 1988). From the EMG pattern one can pick up continued activity during a relaxation period for muscles that are activated very repetitively, for instance. This may explain why dynamic tasks at high frequencies (>15 activations/min) may be largely "static" in the sense that the muscles do not relax completely between contractions.

EMG used very carefully is a valuable tool for assessing job demands and identifying alternative designs of workplaces, equipment, tools, or jobs. Comparative EMG studies of the demands of tasks done with two different tools or with positioners versus no postural support can give quantification of the value of making an ergonomic change in the workplace. Improper controls during EMG measurement, however, can result in data that is uninterpretable or misleading in assessing the physical job demands.

10.2.6 Fatigue Research Needs

Despite a strong interest in fatigue at the beginning of the century, when the Harvard Fatigue Laboratory in the United States and the Industrial Fatigue Council in Great Britain were doing studies focused on reducing physical fatigue, there is much more to learn about all types of fatigue in the workplace. With automation and increased production rates, workers are experiencing mental and perceptual stress as well as being put into jobs whose content has been altered. What used to be recovery tasks in some assembly operations may now be automated (because it is easy to automate the simpler tasks). Consequently, the assemblers are doing more complex operations most of the shift, and fatigue may result.

There are several areas where research would help clarify what patterns of work (and job design) will be less fatiguing and will be, therefore, sustainable for extended hours or 12-hr shift schedules. We need clarification of the interaction between work and recovery on highly repetitive tasks (>15/min). We also need to better understand how much light effort work improves recovery time after a static, fatiguing effort.

As repetition rate increases for a given muscle group, two things tend to be seen in the EMG trace. First, the time of effort decreases but the amplitude of the signal increases, suggesting an increased velocity of contraction and less sustained tension at the peaks. Second, the signal does not fall to zero between contractions, and there is a residual tension that usually increases significantly when one exceeds 25 to 30 activations per minute (Jonsson, 1988). This pattern will vary in different muscle types and movements, but the overall response is to see a reduction in movement time, higher peak load, and a continued but lighter effort in the relaxation period between contractions. Recent studies have begun to explore this part of the repetition range where the earlier research (Rohmert, 1973) tends to underestimate the fatigue accumulated during a task. One would like to know the upper limits of repetition for different effort levels and durations of effort (holding time in seconds) so nonfatiguing task designs can be predicted by job planners, so they know when they should provide additional assistance or should look toward automation to reduce the risk for repetitive motions disorders or excessive fatigue.

Many companies and service organizations suggest implementing "exercise breaks" to reduce fatigue and the risk of injury/illness on jobs where postural stress is significant or where the same muscles are used in repetitious activities (high-volume production). The exercises are either stretches or light, dynamic movements that are intended to stimulate circulation to the fatiguing muscles and connective tissues and "realign" the musculokskeletal system. Although the theory behind the exercises makes sense, it is not widely researched. Some studies to quantify the effectiveness of different types of work pauses, active or passive, would be helpful. Some research suggests that short rest pauses in sustained isometric tasks may actually increase the risk of injury or discomfort by causing swelling in the muscle cells (Sjogaard et al., 1988).

For example, should one do 2 min of stretching and light, dynamic work after 50 to 90 min of data entry, or would 30 sec be as effective? Or does one need 5 min to recover from sustained neck and shoulder loading? Since not all workplaces are ergonomically designed, what is the appropriate additional recovery time need for sustained typing or assembly in an ergonomically deficient workplace? What are the effects of interruptions (such as phone calls or trips to the copier) on fatigue time or discomfort for the more heavily loaded postural muscles? Subjective and objective measures of the impact of dynamic activities or interruptions during posturally demanding tasks should help the ergonomist assign a productivity and comfort benefit to ergonomic redesign of the job setting.

10.3 TOTAL WORKLOAD AND FATIGUE/RECOVERY NEEDS

If there is a little *local muscle* fatigue on a job, one can find people who can do it safely as long as they can control their own work pace. People monitor fatigue locally and energy cost, or whole body fatigue, globally. The worker senses his or her fatigue as the task continues and adjusts the pace, work and recovery tasks, and effort levels (by changing muscle groups) to keep the total cost down. Acceptable muscle efforts and patterns of work for 1 or 2 hr may not be acceptable for 8, 10, or 12 hr a day or for 5, 6, or 7 days a week (Ciriello et al., 1990). In many instances, it is the fatiguing activity that contributes the most to overload in a job's total effort level. By fixing that task, people can work more efficiently and should be at less risk for an overexertion injury.

For jobs where work pace is externally determined (by conveyor rates, production pressure, incentive pay systems, etc.), whole body fatigue is more likely to be seen if the tasks are not ergonomically designed. A common example is a job where a production machine is loaded and unloaded manually, and the handler cannot control the machine's rate. If the amount of in-line inventory at each end of the machine is held small (as management programs such as "Just -In-Time Manufacturing" and "Demand Flow Technology" require, SME, 1995), the handler is forced into a lifting fixed pace to get the trays or parts in and out of the machine and cannot alter that to meet his or her needs for work and recovery time. The usual result is more effort, if only from walking between the ingoing and outgoing ends of the machine, and a greater potential for whole body fatigue over the shift.

Total workload acceptability decreases as hours of work increase either with increased overtime or extended hours shift schedules (e.g.,12-hr shifts). While the latter may have more days off between shifts, the fatigue at the end of three 12-hr shifts in a row can be limiting and may make the older worker susceptible to an overexertion injury if a fourth day in a row is worked during a production push. Heat affects acceptable total workload since the blood flow needed for body temperature regulation is not available to carry oxygen and substrates to the working muscles.

10.3.1. Aerobic Capacities for Work

For many years, acceptable workloads were based on whole body aerobic capacity, usually measured by sequentially increasing the workload on a treadmill or bicycle ergometer (Astrand, 1960; Robinson, 1938). Upper body aerobic capacity averaged 70% of whole body capacity and was usually assessed using an arm cranking task (Astrand and Rodahl, 1986). More recently, additional work capacities have been identified in lifting tasks (Petrofsky and Lind, 1978; Sharp et al., 1988), with capacity decreasing as the frequency of lifting increases. An optimal workload defines the interaction of lifting frequency and load weight when the lifting capacity is neither limited by movement time nor by strength nor by fatigue accumulation (Petrofsky and Lind, 1978b). For extended work periods, acceptable workloads are 33% of whole body or upper body maximal aerobic capacities for an 8-hr shift and 28% for a 12-hr shift (Eastman Kodak, 1986; NIOSH, 1981). For frequent lifting tasks, the acceptable workload is closer to 27% for an 8-hr shift (Legg and Pateman, 1980). The relationship between percent of aerobic capacity and total time of work is shown in Figure 10.4 .

It is generally agreed that one must know which muscles are active when assessing workload acceptability and know their aerobic capacities. If the job involves low lifting, walking, climbing, and other large muscle group tasks, whole body aerobic capacity should be used for the maximum value. If only the upper body musculature is active in a task or job, then upper body aerobic capacity should be the denominator to assess the percent of maximal capacity used. If the task is primarily lifting and extends from floor to shoulder height, the workload can be expressed as a percent of whole body aerobic capacity, but the acceptable 8-hr workload will be a lower percent of maximum (27%).

10.3.1.1 Whole Body Capacities

To assess the aerobic demands of a job in the field, one has to measure the oxygen consumption and heart rates of a person doing the job. The person studied on the job has to be "calibrated" on a standard exercise test, such as graded exercise on a treadmill or on a bicycle ergometer. Protocols for capacity testing range from a four- or five-stage test (Eastman Kodak, 1986) to a one-stage test based on the original Harvard Fatigue Lab's Step Test (Brouha, 1949; Master, 1934) or a one-level stress test on the bicycle ergometer (Astrand, 1960) to predict cardiovascular fitness and aerobic capacity. One-stage tests may not be very accurate for predicting maximal aerobic capacity, nor are estimates from settings on the treadmill (speed and grade) or bicycle ergometer (tension and cycling rate).

The range of whole body and upper body aerobic capacities measured on a treadmill submaximal exercise test in an industrial workforce is given in Figure 10.5 (Eastman Kodak, 1986). The units are ml O_2 consumed per kilogram of body weight per minute. Resting oxygen usage is about 3.5 ml $O_2 \cdot$ kg $BW^{-1} \cdot min^{-1}$, which is about 0.25 l O_2/min or 1 kcal/min. Healthy, young Army recruits have aerobic capacities in the range of 45 to 55 ml $O_2 \cdot$ kg $BW^{-1} \cdot min^{-1}$ (Sharp et al., 1988). The industrial workforce, particularly in the 1990s, has become older; the average age has increased from 39 to 43 yr in many large companies and many more women are doing entry level jobs than was the

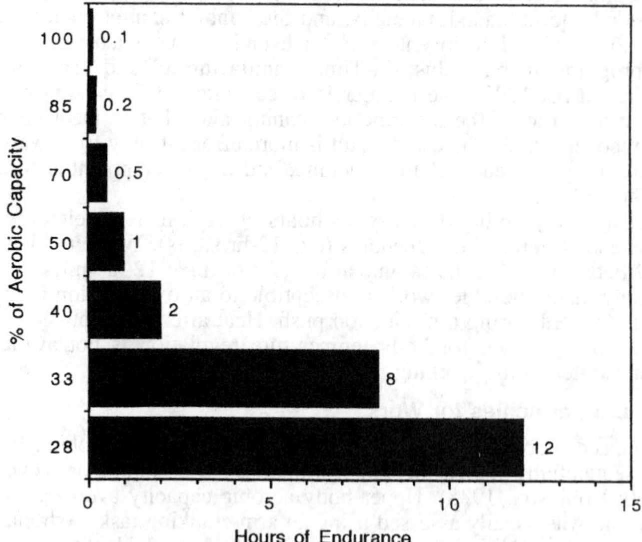

Figure 10.4 *Percent of Aerobic Capacity versus Hours of Work.* The relationship between aerobic workload (as a percentage of maximal aerobic capacity) and hours of work is shown, giving guidelines for the longest continuous work periods for each level of effort. These are used to assess the suitability of different workloads and peak activities and apply across most muscle groups. (Source: Developed from information in Astrand and Rodahl, 1970; Rodgers, 1978; and Eastman Kodak, 1986).

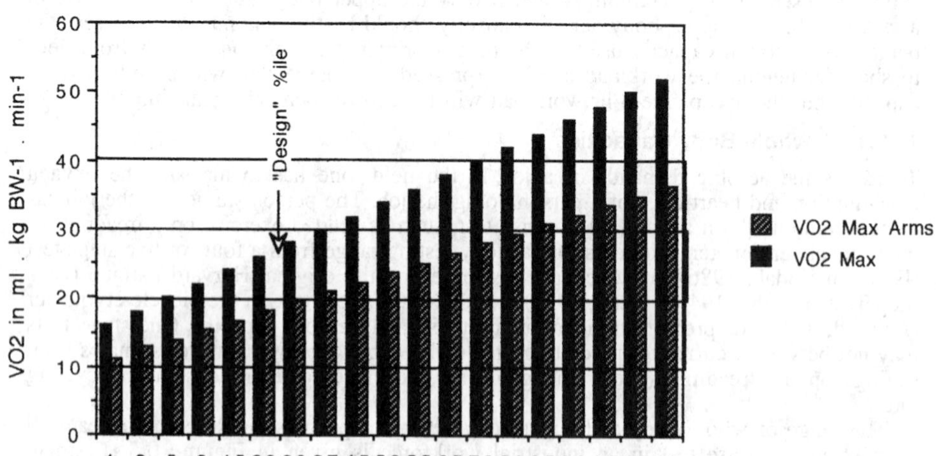

Figure 10.5 *Whole Body and Upper Body Aerobic Capacities by Percentiles.* The aerobic capacities are based on oxygen consumption and heart rate measurements during a 4- to 5-stage progressive treadmill test of industrial workers. Upper body aerobic capacities are shown as 70% of whole body values. (Source: From data in Eastman Kodak, 1986).

case in the 1960s and 1970s. What used to be an acceptable workload for a healthy 20-yr-old male in earlier years is not appropriate for men 43 yr old and for most women because their maximal aerobic capacities are lower.

10.3.1.2 Upper Body Capacities

Upper body work is more strenuous (per liter of oxygen used) than whole body work because there is less muscle mass available to produce the forces (Astrand et al, 1965). On the average, maximal upper body aerobic capacity is 70% of whole body aerobic capacity (Astrand and Rodahl, 1986), although the range may be from 63 to 78% depending on which of the upper body muscles are most heavily loaded and on the fitness of the people studied (Eastman Kodak, 1986). Acceptable workload for upper body work is determined by using the same percent of maximal aerobic capacity versus time curve (see Figure 10.4 in Section 10.3.1), but the maximum is now 70% of the whole body aerobic capacity. Figure 10.5 shows upper body aerobic capacities for an industrial workforce.

10.3.2 Relationship of Acceptable Workload to Hours of Work

Most of the guidelines for acceptable workload assume an 8-hr-day, 5-day-week work cycle. Although that is still the most common schedule, new work schedules that include overtime or extended hours are being used routinely in some companies or professions. In addition, the downsizing trend to cut costs has resulted in increased numbers of workers accepting voluntary overtime schedules on which they are working many of their "recovery" days, even on the 12-hr schedule.

Figures 10.6 and 10.7 give whole body and upper body workloads for 1-, 8-, and 12-hr tasks by potential workforce aerobic capacity percentiles. This information can be used to predict what percent of a male/female workforce would find a job acceptable. For example, if a person has a whole body aerobic capacity of 40 ml $O_2 \cdot kg\ BW^{-1} \cdot min^{-1}$ and is doing a task that involves only arm work, his or her true capacity for that work is 40 x 0.7, or 28 ml $O_2 \cdot kg\ BW^{-1} \cdot min^{-1}$. If the task being done requires an energy output of 14 ml $O_2 \cdot kg\ BW^{-1} \cdot min^{-1}$, it takes 14/28, or 50% of upper body capacity. If the task is done with the whole body, it would require 14/40, or 35% of whole body aerobic

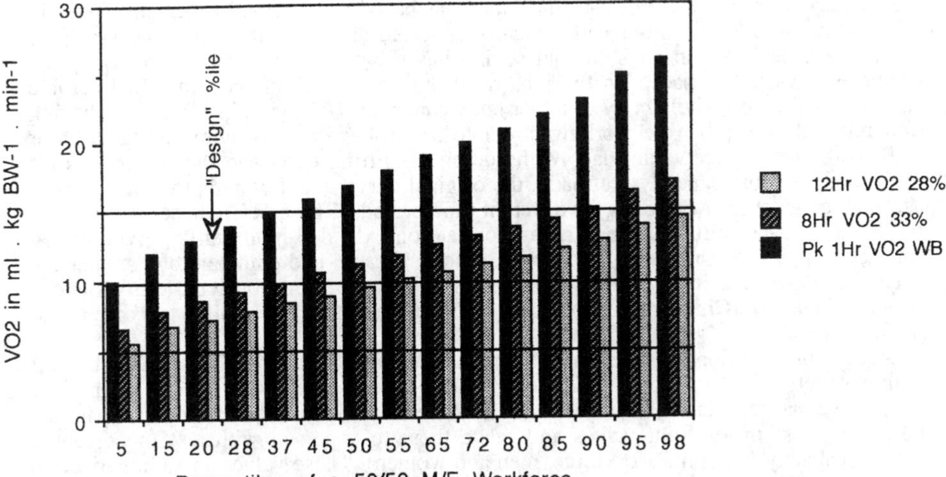

Figure 10.6 *Workload Limits by Hours for Whole Body Aerobic Capacity Percentiles.* The upper limits for acceptable aerobic workloads in whole body work are shown for peak loads (1 hour), sustained loads (8 hours), and extended hours loads (12 hours). The design percentile represents the highest average workload for a person with a 25th percentile aerobic capacity in a mixed 50/50 male/female population. (Source: Developed from data in Eastman Kodak, 1986).

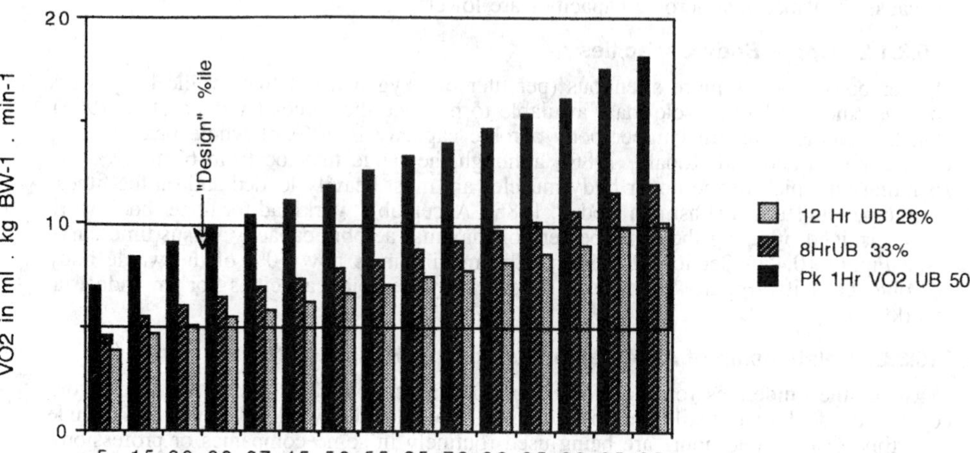

Figure 10.7 *Workload Limits by Hours for Upper Body Aerobic Capacity Percentiles.* This is the same as Figure 10.6 but for upper body work.

capacity. In the first case, the acceptable continuous duration would be 1 hr before going to a lighter task (from Figure 10.4). In the second case, the task could be sustained for more than 4 hr but less than 8 hr a shift. Using a design level where most people can do the work without risking overload (averaging 9 ml $O_2 \cdot$ kg $BW^{-1} \cdot min^{-1}$ for 8 hr of whole body work), one can reduce the probability that production fluctuations and shift schedule changes will contribute to health and safety problems on these jobs.

10.3.2.1 Acceptable Sustained Workloads

The original curve on which Figure 10-4 is based showed a percent of aerobic capacity used versus hours of work for 8-hr days and was based on observed acceptable loads in a laboratory setting and on-site in a factory (Astrand and Rodahl, 1970). Measurement of the "heavier" jobs in an industrial setting has shown that chosen workloads by the people who were self-paced on the jobs ran from 25 to 35% of their maximal aerobic capacities, based on whole body or upper body capacities (Rodgers, 1978). People having difficulty with the job were working closer to 35 to 45% of maximum during the 8-hr shift. If the lower acceptable loads with continuous lifting tasks and the corrections for upper body aerobic capacity are made, the original curve is quite predictive of observed acceptable industrial workloads for different time periods. The predicted acceptable loads for 10- and 12-hr shifts appear to be appropriate on extended hour shifts providing that the recovery days are not used as overtime. More research and epidemiology studies are needed to determine if any longer-term health effects are found on long-term extended hours schedules (Chapter 32 includes more information on shift schedules and performance).

Acceptable workload as a percent of maximal aerobic capacity is 33% for an 8-hr shift with a variety of effort types, 27% for continuous lifting over 8 hr, and 28% for 12-hr shifts. Average workloads, in ml $O_2 \cdot$ kg $BW^{-1} \cdot min^{-1}$, for industrial or occupational tasks use these upper limit values and relate them to a person with a 25th percentile aerobic capacity (potential workforce, men and women). These values are summarized in Table 10.2 (Rodgers, Yates, and Garg, 1990; Waters et al., 1993). It is important to remember that these values are averages and include break times as well as work times. They also are workloads that will be acceptable for 75% of a 50/50 mixed male/female potential workforce. Peak loads, greater than the average load, are acceptable if they do not exceed the guidelines in Figure 10-4, and if they are balanced by lighter work or rest to keep the 8-hr or 12-hr average within the recommended limits.

Table 10.2 Upper-Limit Aerobic Demands for Lifting Tasks. The Recommended Highest Workloads for Aerobic Work Design of Repetitive Lifting Tasks Based on Hours of Work. The Design Value is for a 25th Percentile Person's Maximal Aerobic Work Capacity and Is Based on a 50/50 Mixed Male/Female Industrial Workforce

		Duration of Lifting		
		1 hr	2 hr	8 hr
Low lifting	ml O2 · kg BW^{-1} · min^{-1}	13.5	10.8	9.0
	Kcal/min	4.7	3.7	3.1
Lifting above knuckle height	ml O2 · kg Bw^{-1} · min^{-1}	9.5	7.5	6.3
	Kcal/min	3.3	2.7	2.2

Source: Based on material in Rodgers, et al. (1991) and Waters, et al. (1994).

Table 10.3 Heart Rate Elevations in Several Occupational Tasks. Occupational Tasks Are Categorized by Workload Intensity Based on Measured Heart Rate Elevations above Resting

Heart Rate Elevations in Several Occupational Tasks				
Effort Level	Light	Moderate	Heavy	Very Heavy
HR elevation				
Whole body	10–20	21–45	46–75	>75
Upper body	10–15	16–30	31–50	>50
Tasks	Typing, Data Entry (U)	Packing, Small (U)	Industrial Cleaning	Landscaping
	Record Keeping (U)	Punch Press Operation	Carpentry	Loading Coal
	Small Parts Assembly (U)	Metal Working (U)	Plastering	Handling Cases, >25 lbm >4/min.
	Drill Press Operation (U)	Painting	Sweeping/Mopping Floors	Cement Mixing
	Sitting, Reading	Driving a Truck/Car	Gardening	Stone Masonry
	Monitoring	Sewing (U)	Packing, Large (U)	Smelting Work
	Standing	Ironing (U)	Sheet Metal Work	Agricultural Work
	Drafting	Washing Windows	Laundry Operations (U)	
	Inspecting	Bench Work	Truck and Auto Repair	
		Sorting Scrap	Road Paving	
		Walking	Metal Casting	
		Crane Operation (U)		
		Cafeteria Work (U)		
		Machine Tending		

U = Upper body work.
Source: Reprinted with permission from Rodgers (1985).

To determine the aerobic workload on tasks or jobs, one should measure oxygen utilization with a portable ventilation/oxygen detection system. Tables of the energy costs of jobs are available for estimating workload (Table 10.3), but these do not replace direct measurement since patterns of work vary greatly within a job title (Astrand and Rodahl, 1986; AIHA Technical Committee on Ergonomics, 1971; Eastman Kodak, 1986; Lehmann, 1962; Rohmert, 1987). The relative components of static and dynamic work influence both the oxygen consumption and heart rate responses of the worker (Sanchez et al., 1979). Therefore, ergonomic improvements to reduce local muscle fatigue will also improve the job by reducing its cardiovascular stress and increasing the level of workload acceptability.

In many instances, it is difficult to measure oxygen consumption on a job and to relate it to a person's maximal aerobic capacity. An estimate of the percent of maximal aerobic capacity used on the job can be made by taking heart rate measurements at rest and on the job. The heart rate (HR) elevation above the seated, resting rate can be divided by the available range of heart rate elevation (Predicted maximum HR minus resting HR) to get a percent maximal HR Range. Predicted maximum HR is loosely predicted using the formula, 220 − age in years.

$$100 \cdot (HR_{work} - HR_{rest})/[(220 - \text{age in years}) - HR_{rest}] = \%\text{ of Heart Rate Range}$$

This estimate cannot be used if there are psychological or environmental factors on the job that can elevate the heart rate independent of the physical workload (Eastman Kodak, 1986; Rodgers, 1988). A case study is given in Section 10.3.3.1 to illustrate how to apply this information in assessing the suitability of different workloads on a lifting task.

10.3.2.2 Influence of Pattern of Work on Fatigue and Recovery Time Needs

Even if a job has several heavy, potentially fatiguing tasks, a worker often has the latitude to vary the way the job is done in order to modify the fatigue and finish the shift without feeling exhausted. By understanding how work and recovery time needs intersect, a job planner can design jobs with enough variety in the tasks to let the workers prevent excessive fatigue through alternating them. The needed recovery times begin to exceed the work times at 65% to 70% of maximum aerobic capacity, all of which are under an hour (Kamon, 1975; Scheen et al., 1981). To minimize fatigue on a heavy task, one should break it up into shorter segments and do a light task in between to speed recovery. If all of the heavy tasks are done together ("to get them over with"), the muscles could be fatigued to the point where the person may be at risk for an overexertion injury for a large part of the shift. Anaerobically fatiguing workloads are inefficient. They can be identified physiologically by an upward sloping heart rate pattern as the task progresses, never achieving a steady value for the effort level performed (Brouha, 1973). Aerobic work heart rates are quite steady during the heavier work, only creeping upward after hours of work.

Because there is a natural work break after 2 hr of sustained work in most production jobs, there is a tendency to rotate jobs based on the schedule, not on the physiological needs. Job rotations on a week on/week off, every other day, or half-day schedule may provide the "relief" activity too late to prevent significant fatigue development. Flexibility in allowing the production people to find their own best rotation schedule is the ideal situation, giving them the quality, quantity, and safety standards to meet but not forcing them into a single way of achieving them. Local muscle fatigue will also contribute to a person's choice of the acceptable continuous work duration before rotation to a lighter task.

10.3.2.3 Twelve-Hour Shifts and Overtime—Influence on Acceptable Workload

The research on 12-hr effort levels is sparse, but the popularity of these shift schedules from a psychosocial standpoint makes them likely to remain in industry. Older shift workers have commented that the fourth day in a row on a 12-hr shift schedule can be very difficult to sustain, so schedules that have no more than 3 days in a row are usually recommended (cf, the "Firefighters' Schedule" with 3 days (or nights) on, 3 days off (McCreary, 1992). See Chapter 32 for more discussion of shift schedules and performance.

For physically demanding jobs, the shift pattern can influence muscle fitness and training, particularly around vacation time. If a worker takes off 2 weeks with a total of two "weekends" and seven shifts, there may be enough muscle detraining during the vacation to increase the risk for an overexertion injury or significant fatigue upon return to work. Another factor in determining fatigue and acceptable workload relates to the numbers of days when overtime is worked. If overtime hours are routine, the true work demands should be assessed using the 10- or 12-hr percent of aerobic capacity values (see Figure 10.4). The effect of 6 or 7 days of work instead of 5 is probably more related to recovery time needs for healing or restoring musculoskeletal physical and chemical stresses than it is for tolerating the percent of aerobic capacity used. Particularly in repetitive tasks, one can accumulate small pulls and tears of the active muscles and tendons and accumulate chemical and metabolic imbalances that may not be fully repaired or returned to the resting state between shifts or over the weekend (Moore, 1992). If the muscles and joints are inflamed, scar tissue could be deposited, and that may restrict the motion of tendons in their sheathes (in the carpal tunnel, for example) or create local irritation or swelling (Rowe, 1987). The relationship between these medical findings and the hours of work on certain jobs has been noted anecdotally, but research is needed to determine if there is additional risk that can be predicted scientifically.

10.3.2.4 Self-Regulation of Workload to Reduce Fatigue

One reason why one might not see overexertion injuries or fatigue in extended-hours schedules or overtime work is that the worker will pace the effort to reach his or her acceptable work demands level. If this cannot be done, people without the needed capacities will move to other, less stressful jobs. Few studies have been done where productivity gains have been analyzed when a shift schedule is changed from 8 to 12 hr. In many observed instances, there have been improvements in efficiency associated with the reduction of one shift changeover, but the actual working time has been less than the extra 4-hr segment should have provided. Secondary work, such as checking product or part quality, getting product, talking with fellow workers, and setting up the workplace often increases with increased hours, as does personal time for getting a drink of water, using the facilities, taking a stretch break, and getting a snack (Eastman Kodak, 1986).

If the worker is captive to a production machine or cannot get away from the workplace, taking control of the job by rejecting good parts, recycling them, or by unintentionally creating a need to get up to clean something up or to check on a problem has also been observed. Even in automated systems, workers can find ways to control their workload and stress levels in order to keep the task and job acceptable. From extensive observations of people at work, it has been apparent that people will seldom let themselves be overtaxed by a job if they have any way of controlling it, unless a stronger drive is present (such as increased compensation).

10.3.2.5 Effects of Heat and Cold on Acceptable Workload

Part of physical work is the generation of body heat because few tasks are done with more than 15% efficiency (Dutta and Taboun, 1989; Larson, 1974). Eighty-five percent of the energy expended appears as body heat, and this must be dissipated so the core temperature does not rise significantly. The body defends core temperature in the cold through shivering and piloerection of the body hairs, shunting of blood flow from the skin and extremities, and the behavioral use of increased insulation in the clothing. In hot environments, blood circulation to the skin and extremities is increased to dissipate heat, and sweating keeps the skin temperature low so heat can be transferred out readily and cooling occurs through evaporation (Brouha, 1973; Eastman Kodak, 1983).

Physical effort in hot and humid environments results in the skin receiving much of the blood flow that working muscles need in order to deliver oxygen for ATP synthesis and substrates for intermediary metabolism. Therefore, work capacity and maximal aerobic capacities are less in hot environments than they are in temperate or cold environments. In a 38°C (100°F), moist environment, approximately 40% of the cardiac output (in liters per minute leaving the left side of the heart) goes to the skin for temperature regulation compared to 10% in temperate conditions (Brouha, 1973). Between 24° and 40°C (75 to 105°F) dry bulb ambient temperatures, each additional 1°C (1.8°F) elevation in ambient temperature is equivalent to the physiological stress of a 1% increase in the percent of maximal aerobic capacity used (Kamon, 1975). Work capacity is reduced as the resting heart rate moves higher, lowering the difference between the maximum predicted heart rate and the resting value (Eastman Kodak, 1973).

In a colder environment, the facilitation of conductive, convective, and radiative heat loss from the body may actually improve the work capacity in moderately heavy to heavy tasks because the usual heat build-up will be reduced and more of the cardiac output will be available for the working muscles (Eastman Kodak, 1983). The pattern of work will be important, however, since heavy work in the cold can result in sweating and loss in the insulation value of clothing. If the heavy work is followed by rest or light work in the same cold environment, too much heat might be lost and shivering could ensue. This would reduce work capacity somewhat by using energy to create the shivering response. See Chapter 27 for more discussion of environmental heat and cold effects on performance.

10.3.3. Application of Workload and Capacity Data to Job Risk and Demands Evaluations

Two examples are given below to show how local muscle work/recovery guidelines, workload, and capacity data can be used to assess the risk for overload on tasks or jobs. A few assumptions are made in determing the risk for overload on a job:

1. The aerobic capacities used to establish guidelines for the demands of tasks/jobs are for a person with a 25th percentile aerobic capacity. This is 27 m O_2/kg BW^{-1}/min^{-1} whole body work and 19 for upper body. (Eastman Kodak, 1986). This should be acceptable for about 75% of the women and about 90% of the men. By pacing themselves, people with less capacity can still do the job safely.
2. The risk for overload is determined by identifying what percent of a potential workforce (50/50 male/female) would find the job acceptable. This is assessed for the overall average demands of the job or task and for its peak demands. If less than 75% of the population would find the job acceptable, the job should be made lighter through varying tasks or reducing the load on the heavier tasks.
3. For jobs where there is not a significant environmental or psychological load (heat or high pace pressure), an estimate of percent of maximum aerobic capacity can be made from the percent of range of the heart rate (HR) measured:

$$\frac{\text{HR Work} - \text{HR Rest}}{\text{HR Predicted Max} - \text{HR Rest}} \cong \frac{\text{VO}_2\text{ Work} - \text{VO}_2\text{ Rest}}{\text{VO}_2\text{ Max} - \text{VO}_2\text{ Rest}}$$

   Heart rate is measured in beats per minute and oxygen usage is given in ml $O_2 \cdot$ kg $BW^{-1} \cdot min^{-1}$. This is an estimate of the percent of maximal aerobic capacity used and can be quite inaccurate if other factors contribute to heart rate elevation or reduction, such as a heat load, a high-stress situation, or strong cognitive requirements on the job. It is most accurate for moderately heavy or heavy work where the heart rate more directly relates to the delivery of oxygen to the muscles.
4. The examples used are from jobs where oxygen consumption was measured directly. If this measurement is difficult to make (e.g., firefighting), one can also use metabolic demands estimation methods (Garg et al., 1978) or tables of job demands where a comparable effort can be found (AIHA Technical Committee on Ergonomics, 1971; Eastman Kodak, 1986; Lehmann, 1962) to predict a job's requirements. Table 10.3 classifies jobs and tasks in effort categories associated with expected elevations in heart rate (Rodgers, 1985). However, these are necessarily less accurate than direct measurement.

10.3.3.1 Frequent Lifting—Whole Body Work

The job entails unloading cases of product (weight range = 6 to 9 kg, or 13–20 lb) at the end of a cartoning machine. The average number of cases shipped per 8-hr shift is 3200 and, when the line runs well, they flow at a rate of 8/min. The cases are stacked five tiers high on floor pallets, each of which contains 40 cases when shipped. Final strapping and preparation of the pallet load for shipping is done in another operation, and that same person is responsible for moving out the full pallets at the end of the machine.

When this job was studied, the measured oxygen consumption over 10 min of steady work was 15 ml $O_2 \cdot$ kg $BW^{-1} \cdot min^{-1}$. Another sample that was taken when there was a temporary shut-down of the machine (about one-fifth of the sample time) was 12 ml $O_2 \cdot$ kg $BW^{-1} \cdot min^{-1}$. The engineering and production data showed that the cartoning machine line was "up" about 6.75 to 7 hr per shift. An evaluation of all job tasks showed that the handler at the end of the line performed light to moderate effort tasks (paper work, walking, handling supplies) outside of the handling activity. So, the average job demands were taken as 12 ml $O_2 \cdot$ kg $BW^{-1} \cdot min^{-1}$ and the peak demands were taken as 15. Continuous work time for the average load was 7^+ hr; for the peak demands, the longest single *continuous* handling period measured was 1 hr. Table 10.4 shows this information and the results of the analysis.

Because the pallets were on the floor, whole body aerobic capacity is involved in the handling task as it was for most of the residual tasks. The "design" aerobic capacity for whole body work is 27 ml $O_2 \cdot$ kg $BW^{-1} \cdot min^{-1}$ (Figure 10.5). Therefore, one can divide the average workload (12 ml $O_2 \cdot$ kg $BW^{-1} \cdot min^{-1}$) by the aerobic capacity to find that the job requires 44% of capacity. This compares to an acceptable value of 27% for a continuous lifting task. By taking the desired percentage and the "design" aerobic capacity, we find that the "design" load would be 0.27 x 27, or 7.3 ml $O_2 \cdot$ kg $BW^{-1} \cdot min^{-1}$ for the 8-hr shift lifting task. One could also look at the measured load and determine what percent of the population would find it acceptable (Figure 10.6), or 12 /aerobic capacity 27%. The aerobic capacity needed to perform the job within recommended safe limits would be 44 ml $O_2 \cdot$ kg $BW^{-1} \cdot min^{-1}$. This is the aerobic capacity of the 85th percentile of a 50/50 mixed male/female industrial population (Figure 10.5). In other words, 85% of the potential workforce may be at risk on the job unless they can regulate their work and recovery activities to reduce the load.

To evaluate the 1 hr peak load, the curve in Figure 10.4 indicates that 50% of maximum aerobic capacity is sustainable for 1 hr (although the actual value for a continuous lifting task may somewhat lower). If the peak load is 15 ml $O_2 \cdot$ kg $BW^{-1} \cdot min^{-1}$, then the handler would have to have a whole body aerobic capacity of 30, which is higher than the "design" capacity for these jobs (Figure 10.5). If the cartoning machine were to be more efficient and the continuous handling time increased to 2 hr, only 40% of capacity

Table 10.4 Analysis of the Suitability of a Lifting Job Workload. The Measured Workload on a Lifting Job Is Related to Workforce Aerobic Capacities For Several Work Durations

Description of Job Demands	Job/Task Demands ml $O_2 \cdot$ kg $BW^{-1} \cdot min^{-1}$	Needed Aerobic Capacities ml $O_2 \cdot$ kg $BW^{-1} \cdot min^{-1}$	% Aerobic Capacity
15 min steady lifting—WB	15	18.8 (<25th %ile WB)	80
20 min lifting and shutdown (1/3)—WB	12	16 (<25th %ile WB)	75
7+ hr lifting/plus light work—avg load	12	44* (85th %ile WB)	27
1 hr lifting continuously—peak load	15	30* (35th %ile WB)	50
7+ hr "design" workload	7.3	27 (25th %ile WB)	27
2 hr of lifting continuously	15	37.5* (65th %ile WB)	40
7+ hr of lifting above knees	8.8	32.6* (47th %ile WB)	27
7+ hr of lifting and rotating with the fork truck job every 2 hr	7.5	27.8* (28th %ile WB)	27
7+ hr of lifting above waist level—UB and light WB activities	8.4	31.1* (85th %ile UB)	27
3+ hr of UB lifting and 3+ hr of WB lifting and 1+ hr of other lighter WB tasks	8	29.6* (30th %ile WB) (80th %ile UB) (55th %ile UB/WB)	27
5 min steady lifting UB or WB—peak load	15 WB 9 UB	17.6 (<25 %ile WB) 10.6 (<25 %ile UB)	85 85
11+ hr of UB and WB lifting and light WB tasks	8	36.4* (55th %ile WB) (98th %ile UB) (80th %ile UB/WB)	22

*Exceed "design" aerobic capacities for whole body (WB) or upper body (UB) work.

would be acceptable. The needed aerobic capacity for the peak 1-hr load would be 38 ml $O_2 \cdot$ kg $BW^{-1} \cdot min^{-1}$, the capacity of a 65th percentile person, or where only 35% of the potential workforce would find it acceptable (Figure 10.5).

Because one is over the design limit for the average and peak loads, one should look at ergonomic strategies to reduce the individual load or to reduce the job demands. Rotation of handlers each hour would keep the peak loads from becoming a problem, providing that the job to which they rotate is not as demanding as this one. In this instance, the job with fork truck driving, pallet-strapping and light handling was available for rotation where the total work load was under 7 ml $O_2 \cdot$ kg $BW^{-1} \cdot min^{-1}$ and the peak loads were very short (less than 5 min). To reduce the job demands on the primary handling task without reducing the line's productivity, the approach was to reduce the energy cost of handling by reducing the amount of low lifting. This was done by raising the pallets on fixed platforms that were 51 cm (20 in.) above the floor. The range of handling was now above the knees and up to shoulder height. This was not difficult for most people to perform because the cases were not very heavy. The energy cost of low lifting is about 40% greater than the cost of lifting around waist level because the body weight has to be raised and lowered each time a case is transferred (Eastman Kodak, 1986; NIOSH, 1981). By raising the pallets 51 cm (20 in.) higher, there was a 30% reduction in the energy cost, reducing the value to less than 9 ml $O_2 \cdot$ kg $BW^{-1} \cdot min^{-1}$. This made the job suitable for most workers; by rotating it with the other job, one could accommodate the needs of people with less than 33 ml $O_2 \cdot$ kg $BW^{-1} \cdot min^{-1}$ aerobic capacity (Figure 10.5).

One potential concern with raising the pallet was that about one-half of the handling now would be done in the upper body zone where aerobic capacity is 70% of whole body capacity. To address this concern, a calculation was done to estimate the demands with this combination of low and high lifting. Table 10.4 shows this breakdown. The longest continuous time of low lifting or of higher lifting was now 5 min, assuming that the handler builds two pallets at a time. Handling still proceeded for 1 hr before an interruption, but the energy cost varied.

For a 5-min peak load, one can use more than 85% of aerobic capacity, providing the demands average out to 45–50% after a 1 hr and to 27% after 8 hr. If the low lifting still averaged 15 ml $O_2 \cdot$ kg $BW^{-1} \cdot min^{-1}$ and the higher lifts were 40% less at 9 ml $O_2 \cdot$ kg $BW^{-1} \cdot min^{-1}$, the peak loads are within the 85% maximal aerobic capacity guidelines even when upper body capacity (19 ml $O_2 \cdot$ kg $BW^{-1} \cdot min^{-1}$) is used as the baseline (Figures 10.5 and 10.7). For the overall load, one can average the upper and lower body costs and express them as a percent of the average design capacities. The nonlifting 30% lighter work tasks used in the first calculation are also included to get the average aerobic demands. This would give an upper body peak demand of 15 + 9 = 24/2 = 12 ml $O_2 \cdot$ kg $BW^{-1} \cdot min^{-1}$, and a total job average demand of 8 ml $O_2 \cdot$ kg $BW^{-1} \cdot min^{-1}$. From Figure 10.7, it can be seen that an upper body peak demand of 12 is acceptable to about 50% of the workforce, and an 8-hr demand of 8 would be at the "design" level if it were all whole body work or at the 60th percentile aerobic capacity level if it were all upper body work. Since the high and low lifting is evenly divided over the shift, the averaged maximal aerobic capacities would be 27 + 19 46/2, or 23 ml O2 · kg BW-1 · min -1. With the upper body work included, this exceeds the design guidelines for an 8-hr day of 27% of maximal capacity seen in Figure 10.4 (8/23 = 35%). Rotating every other hour with the lighter job should keep it nearer to the desired range, however.

A question arose about asking people to work overtime on the improved job: Would it be too strenuous? From the curve in Figure 10.4 it can be seen that the acceptable load drops from about 33% to 28% between 8-hr and 12-hr shifts for whole body work; so one can assume that the drop is proportional for lifting tasks, giving a acceptable load target of 22% of aerobic capacity. If the average workload is still 8 ml $O_2 \cdot$ kg $BW^{-1} \cdot min^{-1}$, and both upper and whole body aerobic capacities are used, then people who can perform this job safely for 12 hr will be those who have whole and upper body aerobic capacities that average 8/.22, or 36 ml $O_2 \cdot$ kg $BW^{-1} \cdot min^{-1}$. Since the upper and whole body capacities are used equally during the handling task, the needed capacities would be 42 ml $O_2 \cdot$ kg $BW^{-1} \cdot min^{-1}$ for whole body work and 29 ml $O_2 \cdot$ kg $BW^{-1} \cdot min^{-1}$ for upper body work. These are the aerobic capacities of a 75th to 80th percentile person (male and female data combined, Figure 10.5); so 80% of the potential workforce would not find the extra hours acceptable as designed. They could do the job if they were permitted to reduce the total handling time either by job rotation or by doing a different, lighter job on the 4 hr of overtime or extended-hours shift schedule.

10.3.3.2 Sustained Upper Body Work—Paced Assembly

The job entailed touch-up spray painting of panels as they moved through a paint line on a conveyor. Breathing protection was provided to protect the worker, and they did the spraying from outside the enclosed line. This meant they stood next to a well-ventilated opening in the enclosure and reached forward 51 cm (20 in.) to access the far corners of the panels and to orient the spray gun properly for even application of the paint. The interval between panels that needed touch-up ran from one every three panels to one every nine panels, depending on the panel size and its final location in the product. The conveyor line rate was 10 panels per minute, giving a maximum of 10 sec per panel to do the touch-up painting because they could work a little ahead of and a little behind the line speed. On the average they took 5 to 6 sec per touch-up. The panels that took longer also came more frequently; they signaled a production problem or a failure in the automated spray booth prior to this job. The job was performed throughout the 8-hr shift.

Both local muscle fatigue and whole body fatigue were experienced by these workers. Static loading of their legs, back, arm, shoulder, hand, and fingers were common complaints and they were addressed first. Using the rating system shown in Table 10.1, the body part fatigue was assessed (Table10.5). Static postural loading was seen in the neck, back, right shoulder, legs, and feet and related to the need to reach to get the spray gun in the best position for touching up the panel. In addition, the right shoulder was abducted (rotated out) in order to reach the near side of the panel and adducted to catch the back side of the panel once it had passed the operator, raising that effort level to heavy on the painting arm. Wrist, hand and finger stress related to the grip and triggering forces used to control the spray gun and to the time it took to complete a panel. The longer, more frequent, touch-up requirements were measured because they represented a predictable situation in the job (occurring about 20% of the time in a typical shift). When the less frequent problems were evaluated, the main differences were in continuous effort duration and in frequency, usually dropping the Priority for Change rating down one category.

Since slowing down the paint conveyor on this job was not a popular option, several ways were used to reduce the stress on the operator and to lessen the perceived discomfort from fatigue. The dominant hand finger stress was greatest and was addressed by using a less strong spring in the spray gun which could be activated by more than one finger, bringing the effort level down to moderate. The shoulder and arm stress were reduced by providing a removable elbow support at the enclosure opening so the spraying arm could be rested between touch-ups. This reduced continuous holding time in all but the most difficult touch-up operations, reducing the priority for change to moderate. The wrist and hand stress was primarily related to the sustained holding with rotation and awkward wrist angles. By providing a slight jog in the conveyor line as it ran past the touch-up station, the panel could be rotated for better access to the corners, relieving the wrist and hands. In addition, the enclosure's opening was modified slightly to make it easier to reach the far parts of the panels and to simultaneously provide about 1–2 more seconds to do the touch-up in a good posture.

Table 10.5 Fatigue Ratings for a Spray Painting Job. The Priority For Changing the Job Is Evaluated by Rating Body Part Fatigue on the Task through Ratings of Effort Intensity, Duration, and Frequency. The High (H) and Very High (VH) Fatigues Indicate the Need for Ergonomic Improvements to the Job. The Moderate (M) Fatigue Suggests that the Total Time of Doing the Task Should Be Assessed and Either Ergonomic Improvements or Job Rotation Every Hour or Two to a Less Fatiguing Task Should be Instituted

	Effort Level	Duration of Effort	Effort Frequency	Priority for Change
Neck	2	3	2	M
Shoulder (R)	2	3	2	M
	3	2	2	H
Back	2	3	2	M
Arms (R)	2	3	2	M
Hands and wrist (R)	2	2	3	H
Fingers (R)	3	1–2	3	H–VH
Legs and feet	2	3	2	M

Once the high and very high local fatigue problems had been addressed, there still had to be some attention to the moderately fatiguing muscles. The operators were going home at the end of the shift with a residual discomfort in the upper body, particularly. A job that is done 8 hr per shift and produces moderate fatigue in several muscle groups is a good one to rotate with a job that has lower stress on the upper extremities. The constant loading of the muscles, even at a moderate effort level, increases oxygen utilization and therefore, the workload for the upper body. The unergonomic job, as first observed, took an average of 8.5 ml $O_2 \cdot$ kg $BW^{-1} \cdot min^{-1}$. The "design" upper body aerobic capacity is 19 (Figure 10.5), and the 8-hr demands would need an aerobic capacity of $0.33x = 8.5$ or $x = 26.5$ ml O_2. The 8.5 ml $O_2 \cdot$ kg $BW^{-1} \cdot min^{-1}$ represents about 45% of upper body aerobic capacity at the design level (Figure 10.5). About 65% of the mixed 50/50 male/female workforce would not be able to meet these job demands for 8 hr a day. based on these calculations and using the curves in Figure 10.7.

To improve the over-all job (after the initial improvements), the first step was to set a policy that when the workload for touch-up spraying increased, as judged by the operator, the department would either add a second person to the line (having provided another opening) or the operators would rotate every hour to a less demanding upper extremity task. In parallel with that approach, a review of the automated paint booth was made to reduce the probability that large amounts of touching-up would have to be done. Better guidelines on when to reject panels from the previous operation and to send them back through the line again also reduced the need for continuous and very frequent touch-up work. By reducing the amount of touching up and the long continuous periods of postural load, the job became acceptable for most people.

10.3.3.3 Workload, Capacities, and Rehabilitation to Work

The "design" aerobic capacities and acceptable workloads are based on an older female population but may be greater than those of a person who has a chronic illness or is recovering from an acute illness or injury. The *percentage of capacity* that is acceptable for the whole or upper body does not change in setting guidelines for job demands, but aerobic capacities can be lower than 27 (or 19) ml $O_2 \cdot$ kg $BW^{-1} \cdot min^{-1}$. Work capacity is usually measured as "tolerance" for performing a lifting or otherwise moderately heavy task at a submaximal level. In rehabilitation programs, the aerobic challenge often used is a step test, bearing some resemblance to the first level in the Harvard Step Test protocol (Master, 1934). How well the worker tolerates 5 min of this high frequency, moderate load stepping task is sometimes measured by pulse or heart rate changes and sometimes by observation of work difficulty or the time before stopping.

As was mentioned earlier in Section 10.2.1.1, a single-stage stress test will not give an accurate picture of a person's aerobic capacity, although it will identify people with very low or very high capacities. To get a more accurate picture of aerobic capacity, one first has to get a good resting value of oxygen consumption and heart rate; this is usually taken before the exercise after 5 min of sitting, but it should also be obtained about 10 min after the exercise, also sitting (Eastman Kodak, 1986). Anxiety may affect the heart rate values before the test so the recovery heart rate is important to measure. For the exercise test, one needs to gather data from at least two exercise levels where 40% of maximal aerobic capacity is exceeded in order to define the slope of the oxygen consumption versus heart rate relationship. Submaximal tests are the norm in rehabilitation and industrial physiology because the goal is not to overexert the worker/patient on the capacity test. With a good resting value and two moderately heavy to heavy workloads one can predict maximum aerobic capacities using the predicted maximal heart rate values from stress test studies (Astrand and Rodahl, 1986). The actual capacities of the worker could be under- or overpredicted by 15% even with this more careful approach to capacity assessment.

The importance of getting an accurate assessment of a worker's aerobic fitness depends on how the data will be used. If a conclusion will be made that the person cannot return to work because of inadequate work capacity, a very accurate measure of work capacity should be made. Reasonable accommodations on the job, similar to ones described earlier in Section 10.3.3, can make the job more suitable for everyone. These may make the difference in how soon the person can return to work if he/she is also participating in a fitness program as part of the return-to-work program.

The advantage of choosing a "design" work aerobic capacity that accommodates a majority of the potential workforce (Figure 10.5) is that it is easier to build up an ill or injured worker to that level than to try to make them as fit as the most fit people in the workforce. Work restrictions that limit the continuous time a person can do any moderately heavy to heavy task to 1 hr, for example, and then rotate them to a lighter task, are often superior to half-day restrictions, where a recovering person spends four hours straight on a job and then goes home. Such work pattern considerations in the design of jobs are best "designed" by the affected worker together with the team leader or supervisor of the department, and they will speed an injured or ill worker's ability to return to a full and productive work schedule. Above all, any ways of reducing the fatiguing tasks will make the work situation better for all workers and make it less likely that they will have to take time off from work when they are not feeling fit.

10.3.4 Workload Research Needs

Several research needs were identified earlier in this discussion (Section 10.3). There are several things we do not know about extended hours of work and their effects on long-term wellness. There is not much data on how the acceptable upper and whole body work demands interact with time of day. Although one can measure aerobic capacities at different times of day and see similar cardiovascular and metabolic responses (Eastman Kodak, 1986), workers have a psychophysically measurable perception that the effort is greater in the hours after midnight and before 6 A.M. Two areas for research that might come from this observation are to study the interaction of physical and perceptual/mental/emotional effort and to determine the effect of age on acceptable workload independent of differences in aerobic capacity levels. Heart rate is used to assess acceptable load in many physically demanding tasks, but blood pressure may be as important and is not often monitored. For tasks with moderately heavy static muscle loading for more than 20 sec continuously, blood pressure should be monitored (Lind and McNichol, 1967). The total cardiovascular stress in postural work needs to be defined. Finally, a better understanding is needed of the interaction of local muscle fatigue and total workload and how they affect the acceptable shift workloads for all workers.

REFERENCES

AIHA Technical Committee on Ergonomics (1971). Ergonomic guide to assessment of metabolic and cardiac costs of physical work, *American Industrial Hygiene Association Journal, 32,* 560–564.

ANSI (1993). Z365 Standard Committee meeting discussion, Control of Cumulative Trauma Disorders (National Safety Council).

Astrand, I. (1960). Aerobic work capacity in man and women with special reference to age, *Acta Physiologica Scandinavica, 49 (Supplement 169)*, 1–92.

Astrand, P-O., Ekblom, B., Messin, R., Saltin, B., and J. Stenberg (1965). Intra-arterial blood pressure during exercise with different muscle groups, *Journal of Applied Physiology, 20,* 253–256.

Astrand, P-O., and Rodahl, K. (1986). *Textbook of Work Physiology* (3rd ed.), New York: McGraw-Hill; also 1970 (1st ed).

Baidya, K. N., and Stevenson, M. G. (1989). Local muscle fatigue in repetitive work. *Ergonomics, 31*(2), 227–229.

Basmajian, J. V., and DeLuca, C. J. (1985). *Muscles Alive: Their Functions Revealed by Electromyography* (5th ed.), Baltimore, M.D.: Williams and Wilkins.

Bergstrom, M., and Hultman, E. (1991). Relaxation and force during fatigue and recovery of the human quadriceps muscle: Relations to metabolite changes. *Pflugers Archive—European Journal of Physiology, 418*(*1–2*), 153–160.

Bigland-Ritchie, B., Furbush, F., and Woods, J. J. (1986). Fatigue of intermittent submaximal voluntary contractions: central and peripheral factors. *Journal of Applied Physiology, 61*(2), 421–429.

Bjorksten, M. J., and Jonsson, B. (1977). Endurance limit of force in long-term intermittent static contractions. *Scandinavian Journal of Work, Environment, and Health, 3,* 23–27.

Borg, G. (1982). Psychophysical basis of perceived exertion. *Medicine and Science in Sports and Exercise, 14,* 377–381.

Borg, G., and Lindblad, I. (1976). The Determination of Subjective Intensities in Verbal Descriptions of Symptoms, Report # 75. Stockholm: Institute of Applied Psychology.

Brouha, L. (1943). The step test: A simple method for measuring physical fitness for muscular work in young men. *Research Quarterly, 14,* 31–36.

Brouha, L. (1967). *Physiology in Industry* (2nd ed). Oxford, Great Britain: Pergammon Press.

Bystrom, S., and Fransson-Hall, C. (1994). Acceptability of intermittent handgrip contractions based on physiological response. *Human Factors, 36*(*1*), 158–171.

Christensen, H. (1986). Muscle activity and fatigue in the shoulder muscles of assembly-plant employees. *Scandinavian Journal of Work and Environmental Health, 12*(6), 582–587.

Ciriello, V. M., and Snook, S. H. (1983). A study of size, distance, height, and frequency effects on manual handling tasks. *Human Factors, 25(5)*, 473–483.

Ciriello, V. M., Snook, S. H., Blick, A. C., and Wilkinson, P. L. (1990). The effects of task duration on psychophysically-determined maximum acceptable weights and forces. *Ergonomics 33*(2), 187–200.

Cook, T. M. (1990). EMG amplitude and "fatigue" during repetitive muscle use. (Cited in Davenport, 1990).

Cooper, R. G., Edwards, R. H., Gibson, H., and Stokes, M. J. (1988). Human muscle fatigue: Frequency dependence of excitation and force generation. *Journal of Physiology (London) 397,* 585–589.

Davenport, M. E. (1990). The Effects of Varying Work-Rest Cycles on the Performance of a Repetitive Hand-Gripping Task. Master's Thesis in the Physical Therapy Department at the Graduate College of the University of Iowa (T. M. Cook, Ph.D., Advisor).

DeLuca, C. J. (1984). Myoelectrical manifestations of localized muscular fatigue in humans [Review]. *Critical Reviews Biomedical Engineering, 11*(*4*), 251–279.

Dolan, P., Mannion, A. F., and Adams, M. A. (1995). Fatigue of the erector spinae muscles. A quanititative assessment using "frequency banding" of the surface electromyography signal. *Spine, 20*(*2*), 149–159.

Duchateau, J., and Hainaut, K. (1985). Electrical and mechanical failures during sustained and intermittent contractions in humans. *Journal of Applied Physiology, 58*(*3*), 942–947.

Dul, J., Douwes, M., and Smitt, P. (1991). A work/rest model for static postures. In Y. Queinnec and F. Daniellou, Eds., *Designing for Everyone: Proceedings of the 11th IEA Congress,* London: Taylor and Francis, London. pp. 93–95.

Dul, J., Douwes, M., and Smitt, P. (1994). Ergonomic guidelines for the prevention of discomfort of static postures based on endurance data [Comment], *Ergonomics, 37*(*5*), 807–815.

Dutta, S. P., and Taboun S. (1989). Developing norms for manual carrying tasks using mechanical efficiency as the optimization criterion. *Ergonomics, 32*(*8*), 919–943.

Eastman Kodak Company Human Factors Group (1983). *Ergonomic Design for People at Work, Volume 1.* New York: Van Nostrand Reinhold.

Eastman Kodak Company Ergonomics Group (1986). *Ergonomic Design for People at Work, Volume 2.* New York: Van Nostrand Reinhold.

Edwards, R. H. (1986). Interaction of chemical with electromechanical factors in human skeletal muscle fatigue. *Acta Physiologica Scandinavica Supplement, 556,* 149–155.

Feely, C. A., Seaton, M. K., Arfken, C. L., Edwards, D. F., and Young, V. L. (1995). Effects of work and rest on upper extremity signs and symptoms of workers performing repetitive tasks. *Journal of Occupational Rehabilitation, 5*(*3*), 145–156.

Garg, A., Chaffin, D. B., and Herrin, G. (1978). Prediction of metabolic rates for manual materials handling jobs. *American Industrial Hygiene Association Journal, 39*(*8*), 661–674.

Hagberg, M. (1981). Muscular endurance and surface electromyography in isometric dynamic exercise. *Journal of Applied Physiology 51*(*1*), 1–7.

Hainaut, K., and Duchateau, J. (1989). Muscle fatigue, effects of training and disuse [Review]. *Muscle Nerve, 12*(*8*), 660–666.

Hansson, G-A., Stromberg, U., Larsson, B., Ohlsson, K., Balogh, I., and Moritz, U. (1992). Electromyographic fatigue in neck/shoulder muscles and endurance in women with repetitive work. *Ergonomics, 35*(*11*), 1341–1352.

Jonsson, B. (1988). The static load component in muscle work. *European Journal of Applied Physiology, 57*(*63*), 305–310.

Jorgensen, K., Fallentin, N., Krogh-Lund, C., and Jensen, B. (1988). Electromyography and fatigue during prolonged, low-level static contractions. *European Journal of Applied Physiology, 57*(*3*), 316–321.

Kahn, J. F., and Monod, H. (1989). Fatigue induced by static work. [Review], *Ergonomics, 32*(*7*), 839–846.

Kamon, E. (1975). The ergonomics of heat and cold. *Texas Reports on Biology and Medicine, 33*(*1*), 145–182.

Kilbom, A. (1988). Isometric strength and occupational muscle disorders. *European Journal of Applied Physiology, 57*(*3*), 322–326.

Klein, C., Cunningham, D. A., Paterson, D. H., and Taylor, A. W. (1988). Fatigue and recovery contractile properties of young and elderly men. *European Journal of Applied Physiology, 57*(*6*), 684–690.

Komi, P. V. and Tesch, P. (1979). EMG frequency spectrum, muscle structure, and fatigue during dynamic contractions in man. *European Journal of Applied Physiology, 42,* 41–50.

Kuorinka, I. (1988). Restitution of EMG spectrum after muscle fatigue. *European Journal of Applied Physiology, 57*(*3*), 311–315.

Larson, L. A., Ed. (1974). *Fitness, Health, and Work Capacity, International Standards for Assessment.* New York: Macmillan.

Laurig, W. (1973). Suitability of physiological indicators of strain for assessment of active light work. *Applied Ergonomics* (cited in Rohmert, 1973b).

Legg, S. J. and Pateman, C. (1984). A physiological study of the repetitive lifting capabilities of healthy young males. *Ergonomics, 27,* 259–272.

Lehmann, G. (1962). *Praktische Arbeitsphysiologie* (Second Edition). Stuttgart: Georg Thieme Verlag.

Lind, A. R. (1959). Muscle fatigue and recovery from fatigue induced by sustained contractions. *Journal of Physiology* (*London*), *147,* 162–171.

Lind, A. R., Dahms, T. E., Williams, C. A., and Petrofsky, J. S. (1981). The blood flow through the "resting" arm during hand-grip contractions. *Circ. Res., 48* (*6, Pt 2*), 1104–1109.

Lind, A. R. and McNichol, G. W. (1967). Circulatory responses to sustained grip contractions performed during exercise, both rhythmic and static. *Journal of Physiology* (*London*), *192,* 595–607.

Lind, A. R., and Petrofsky, J. S. (1979). Amplitude of the surface electromyogram during fatiguing isometric contractions. *Muscle Nerve, 2*(*4*), 257–264.

Lind, A. R., Rochelle, R. R., Rinehart, J. S., Petrofsky, J. S., and Burse, R. L. (1982). Isometric fatigue induced by different levels of rhythmic exercise. *European Journal of Applied Physiology, 49*(*2*), 243–254.

Loupee, S. J. (1989). Surface EMG Amplitude and Fatigue During Continuous and Intermittent Submaximal Tasks, Master's Thesis in the Physical Therapy Department at the Graduate College of the University of Iowa (T. M. Cook, Ph.D., Advisor).

Master, M. A. (1934). Two step test of myocardial function. *American Heart Journal, 10,* 495.

Mathiassen, S. E., and Winkel, J. (1991). Quantifying variation in physical load using exposure-vs-time data [Review]. *Ergonomics, 34*(*12*), 1455–1468.

Mathiassen, S. E., and Winkel, J. (1992). Can occupational guidelines for work-rest schedules be based on endurance time data? *Ergonomics, 35*(*3*), 253–259.

Mathiassen, S. E., Winkel, J., Sahlin, K., and Melin, E. (1993). Biochemical indicators of hazardous shoulder-neck loads in light industry. *Journal of Ocupational Medicine, 35*(*4*), 404–407.

McCreary, C. (1992). Report on 12-hour shift schedule studies, unpublished.

Messaros, A. J. (1990). *EMG Amplitude During Fatiguing Submaximal Contractions With Varied Rest Times.* Master's Thesis in the Physical Therapy Department at the Graduate College of the University of Iowa (T. M. Cook, Ph.D., Advisor).

Miller, R. G., Giannini, D., Milner-Brown, H. S., Layzer, R. B., Koretsky, A. P., Hooper, D., and Weiner, M. W. (1987). Effects of fatiguing exercise on high-energy phosphates, force and EMG: Evidence for three phases of recovery. *Muscle Nerve, 10*(*9*), 810–821.

Milner, N. P., Corlett, N., and O'Brien, C. (1986). A model to predict recovery from maximal and submaximal isometric exercise. In Corlett, N., Manenica, I., and Wilson, J., Ed., *The Ergonomics of Working Postures.* London: Taylor and Francis, pp. 126–135.

Milner-Brown, H. S., and Miller, R. F. (1986). Muscle membrane excitation and impulse propagation velocity are reduced during muscle fatigue. *Muscle Nerve, 9*(*4*), 367–374.

Monod, H., and Scherrer, J. (1957). Capacite de travail statique d'un groupe musculaire synergique chez l'homme. *Comptes Rendus Societe de Biologie Paris, 151,* 1358–1362.

Moore, J. S. (1992). Function, structure, and responses of components of the muscle-tendon unit. In J. S. Moore and A. Garg, Eds., *Ergonomics, Occupational Medicine: State of the Art Reviews, 7*(*4*), 713–740.

NIOSH (National Institutes for Occupational Safety and Health) (1981). *Work Practices Guide for Manual Lifting,* DHHS/NIOSH Publication #81-122. Washington, DC: Government Printing Office.

Pagala, M. K., Namba, T., and Grob, D. (1984). Failure of neuromuscular transmission and contractility during muscle fatigue. *Muscle Nerve, 7*(*6*), 454–464.

Petrofsky, J. S., Glaser, R. M., and Phillips, C. A. (1982). Evaluation of the amplitude and frequency components of the surface EMG as an index of muscle fatigue. *Ergonomics, 25,* 213–223.

Petrofsky, J. S., and Lind, A. R. (1975). Ageing, isometric strength and endurance, and cardiovascular responses to static effort. *Journal of Applied Physiology, 38,* 91–95.

Petrofsky, J. S., and Lind, A. R. (1978a). Comparison of metabolic and ventilatory responses of men to various lifting tasks and bicycle ergometry. *Journal of Applied Physiology* (*Respiratory, Environmental, and Exercise Physiology*) *45*(*1*), 60–63.

Petrofsky, J. S., and Lind, A. R. (1978b). Metabolic, cardiovascular, and respiratory factors in the development of fatigue in lifting tasks. *Journal of Applied Physiology, 45*(*1*), 64–68.

Petrofsky, J. S., and Lind, A. R. (1980a). Frequency analysis of the surface electromyogram during sustained isometric contractions. *European Journal of Applied Physiology, 43*(2), 173–182.

Petrofsky, J. S., and Lind, A. R. (1980b). The influence of temperature on the amplitude and frequency components of the EMG during brief and sustained isometric contractions. *European Journal of Applied Physiology, 44,* 189–200.

Petrofsky, J. S., Phillips, C. A., Sawka, M. N., Hanpeter, D., Lind, A. R., and Stafford, D. (1981). Muscle fiber recruitment and blood presssure response to isometric exercise. *Journal of Applied Physiology, 50*(*1*), 32–37.

Robinson, S. (1938). Environmental studies of physical fitness in relation to age, *Arbeitsphysiologie, 10,* 251–323.

Rodgers, S. H. (1985). *Working with Backache.* Fairport, NY: Perinton Press.

Rodgers, S. H. (1987). Recovery time needs for repetitive work. In S. H. Rodgers, Ed., *Seminars in Occupational Medicine,* Vol. 2, # 1, March 1987. New York: Thieme Medical Publishers, pp. 19–24.

Rodgers, S. H. (1988). Job evaluation in worker fitness determination. Chapter 5 in *Worker Fitness and Risk Evaluations.* March 1988 edition of *Occupational Medicine: State of the Art Reviews,* J. Himmelstein, M.D. and G. Pransky, M.D., Philadelphia: Hanley & Belfus Publishers.

Rodgers, S. H. (1992). A functional job analysis technique. Chapter in *Ergonomics.* October/December 1992 edition of *Occupational Medicine: State of the Art Reviews,* pp. 679–711. J. S. Moore, M.D. and A. Garg, Ph.D. Eds., Philadelphia: Hanley & Belfus Publishers.

Rodgers, S. H. (1978). Metabolic indices in materials handling tasks. In *Safety in Manual Materials Handling.* C. G. Drury, Ed., DHEW/NIOSH Publication #78-185. Cincinnati: DHEW/NIOSH.

Rodgers, S. H., Yates, J. W., and Garg, A. (1991). The physiological basis of the manual lifting guidelines. In *Scientific Documentation for the Revised 1991 NIOSH Lifting Equation.* NIOSH Document #PB91-226274, May 1991. Washington, DC: NTIS Department of Commerce.

Rohmert, W. (1960a). Ermittlung vor Erhohlenspausen fur statische Arbeit des Menschen. *International Zeitschrift Einschlissen Angewandte Arbeitshysiologie, 18,* 123–164.

Rohmert, W. (1960b). Zür Theorie der Erhohlenspausen bei dynamischen Arbeit. *International Zeitschrift Einschlissen Angewandte Arbeitshysiologie, 18,* 191–212.

Rohmert, W. (1973a). Problems in determining rest allowances. Part 1: Use of modern methods to evaluate stress and strain in static muscular work. *Applied Ergonomics, 4,* 91–95.

Rohmert, W. (1973b). Problems in determining rest allowances. Part 2: Determining rest allowances in different human tasks. *Applied Ergonomics, 4,* 158–162.

Rohmert, W. (1987). Physiological and psychological workload measurement and analysis, Chapter 3.5. In G. Salvendy, Ed., *Handbook of Human Factors.* New York: John Wiley.

Rowe, M. L. (1987). The diagnosis of tendon and tendon sheath injuries. *Seminars in Occupational Medicine, 2,* 1–6.

Sanchez, J., Monod, H., and Chabaud, F. (1979). Effects of dynamic, static, and combined work on heart rate and oxygen consumption. *Ergonomics, 22(8),* 935–943.

Scheen, A., Juchmes, J., and Cession-Fossion, A. (1981). Critical analysis of the "anaerobic threshold" during exercise at constant workloads. *European Journal of Applied Physiology, 46,* 367–377.

Seidel, H., Beyer, H., and Brauer, D. (1987). Electromyographic evaluation of back muscle fatigue with repeated sustained contractions of different strengths. *European Journal of Applied Physiology, 56(5),* 592–602.

Sharp, M. A., Harman, E., Vogel, J. A., Knapik, J. J., and S. J. Legg (1988). Maximal aerobic capacity for repetitive lifting: Comparison with three standard exercise testing modes. *European Journal of Applied Physiology, 57,* 753–760.

Sjogaard, G. (1986). Intramuscular changes during long-term contraction. In Corlett, N., Manenica, I., and Wilson, J., Eds., *The Ergonomics of Working Postures.* London: Taylor and Francis, pp. 126–135.

Sjogaard, G., Savard, G., and Juel, C. (1988). Muscle blood flow during isometric activity and its relation to muscle fatigue. *European Journal of Applied Physiology, 57(3),* 327–335.

SME (Society of Manufacturing Engineers) (1995). *Manufacturing Insights Case Study Videos* (SME, Dearborn, MI, 1-800-773-4SME).

Soderberg, G. L., Ed., (1989). *Manual of Surface Electromyography for Use in the Occupational Setting,* Morgantown, WV: NIOSH.

Stevens, S. S. (1975). *Psychophysics—Introduction to Its Perceptual, Neural, and Social Prospects.* New York: John Wiley.

Sundelin, G., and Hagberg, M. (1989). The effects of different pause types on neck and shoulder EMG activity during VDU work. *Ergonomics, 32(5),* 527–537.

Tesch, P. A., Wright, J. E., Vogel, J. A., Daniels, W. L., Sharp, D. S., and Sjodin, B. (1985). The influence of muscle metabolic characteristics on physical performance. *European Journal of Applied Physiology, 54(3),* 237–243.

van Dieen, J. H., Toussaint, H. M., Thissen, C., and van de Ven, A. (1993). Spectral analysis of erector spinae EMG during intermittent isometric fatiguing exercise. *Ergonomics, 36(4),* 407–414.

Vollestad, N. K. and Sejersted, O. M. (1988). Biochemical correlates of fatigue. A brief review [Review]. *European Journal of Applied Physiology, 57(3),* 336–347.

Waters, T. R., Putz-Anderson, V., Garg, A., and Fine, L. J. (1993). Revised NIOSH equation for the design and evaluation of manual lifting tasks. *Ergonomics, 36(7),* 749–776.

Westerblad, H., Lee, J. A., Lamb, A. G., Bolsover, S. R., and Allen, D. G. (1990). Spatial gradients of intracellular calcium in skeletal muscle during fatigue. *Pflugers Archives—European Journal of Physiology, 415(6),* 734–740.

Westgaard, R. H. (1988). Measurement and evaluation of postural load in occupational work situations. *European Journal of Applied Physiology, 57(3)*, 291–304.

Williams, I. M. (Lindblad) (1995). Personal communication.

Winkel, J., and Mathiassen, S. E. (1994). Assessment of physical work load in epidemiologic studies: concepts, issues and operational considerations [Review]. *Ergonomics, 37(6)*, 979–988.

Zwarts, M. J., Van Weerden, T. W., and Haenen, H. T. (1987). Relationship between average muscle fibre conduction velocity and EMG power spectra during isometric contraction, recovery and applied ischemia. *European Journal of Applied Physiology, 56*(2), 212–216.

PART 3

JOB DESIGN

CHAPTER 11

ALLOCATION OF FUNCTIONS

Joseph Sharit
Department of Industrial Engineering
State University of New York at Buffalo
Buffalo, NY 14260-0250 USA

11.1 INTRODUCTION

11.1.1 Problem Definition and Scope

Allocation of functions generally refers to a systems design problem that concerns assigning system functions to human and machine agents. This problem statement, however, is an oversimplification of what in fact has become an increasingly complex issue, especially in view of developments in computer hardware and software that are blurring conventional distinctions between humans and intelligent machines. Successful resolution of function allocation problems now often requires having not only an understanding of fundamental issues concerning the relative limitations and capabilities of human and machine resources, but also of a number of subtle considerations that reflect the increased complexity in human–machine interaction brought about by enhancements in technology.

Furthermore, many of the complex contexts within which human–machine collaborations occur are not likely to be anticipated by designers, arguing for an increased emphasis on ensuring that mechanisms are in place for evaluating the implications of allocation of function decisions and for suggesting changes in the light of system performance. These are challenging issues in their own right whose resolution will likely require the application of improved methods for generating scenarios as well as methods such as cognitive task analysis for evaluating the deeper implications of allocation decisions.

The general approach taken toward allocation of functions in this chapter is to view this issue within a continuum that is bounded on one extreme by a purely *static* design perspective and that extends to a perspective characterized by *dynamic* function allocation; this continuum is, in turn, embedded within an iterative systems design process that includes sociotechnical perspectives and other considerations. A static design policy implies a fixed allocation strategy as a basis for assigning all functions, whereas in dynamic function, allocation assignments are dictated by ongoing events. In anticipation of continued developments in technology, special emphasis will be placed on demonstrating the effects of increased technology on *movement* within this continuum. Moreover, we will examine in detail a number of studies that expose various subtle considerations that can potentially impact the suitability and viability of dynamic function allocation arrangements.

Examples are provided to clarify a number of the ideas and concepts associated with the various approaches to function allocation. Some of these examples are revisited in order to track the course of movement through the static–dynamic continuum, as well as to illustrate how considerations related to the systems design process impact the function allocation design stage. Although allocation of functions represents a design problem, implying that its solution should, in practice, be top-down, our discussion will proceed in the opposite direction—the more detailed methods associated with static and dynamic function allocation will be initially addressed, followed by the presentation of increasingly broader perspectives within which these methods are potentially embedded.

Prior to formally initiating discussion on the various approaches to function allocation, several other issues and topics related to allocation of functions are addressed. First, we will attempt to clarify the distinction that is often made between the terms *task* and *function*, especially in view of the widespread use of the terms *task allocation* and *function allocation.* Next, we turn our attention to the concepts of system definition and description. Our intention is not to present an exhaustive account or formal discussion of these concepts, but rather to emphasize the flexibility the analyst has in conceptualizing a system and how this conceptualization can influence issues related to allocation of functions. Finally, we summarize a number of analytic techniques for function analysis, noting on occasion the relationship between these techniques and the function allocation problem.

11.1.2 Functions and Tasks

The terms function and task are often used interchangeably by human factors specialists. If the contexts in which these terms are used are made known, in most cases this practice does not lead to confusion. However, it is probably worthwhile to clarify some distinctions in the use of these terms.

During the systems design process (see Sections 11.1.3, 11.3.2, 11.6, and 11.8.3), the purposes or objectives of the system as well as system requirements and constraints are generally designated in the mission statement. These objectives translate initially into broadly defined functions (e.g., transportation, monitoring) that gradually become more well defined through the iterative elaboration of the design process (e.g., movement of parts between conveyors and automated guidance vehicles, monitoring of part quality). Meister (1991) contrasts *functions,* which describe relatively "molar behaviors" (such as analyzing and detecting), from *tasks,* which describe more detailed behaviors needed to carry out the functions, but admits that explicit rules are not available for determining whether behaviors comprise a function or task.

Function definitions are particularly useful in the functional design stage of systems design, where an entirely new system can be fully defined strictly in terms of functional statements. Each statement describes the function's objective, point of occurrence in the system, and its inputs, outputs, and constraints; at this stage, however, functions should not be defined in terms of human or machine performance (Price, 1985). By keeping the conception of a function relatively abstract, designers can minimize the intrusion of preconceived notions concerning systems design. The realization of these functions can then be deferred to a more appropriate stage of systems design—the stage in which allocation of function decisions are made, at which point biases related to human and machine performance are expected.

Just as we often think in terms of task categories and subtasks, functions can likewise be aggregated and decomposed resulting in functional categories and subfunctions. Moreover, as is the case for tasks, there are numerous ways to characterize functions. For example, they can be simple or complex, instantaneous or prolonged. In aviation, deploying the landing gear would be considered an instantaneous simple function, whereas arriving at a particular destination on time (which requires performing the function of following a scheduled flight plan) would represent a complex prolonged function.

In general, the iterative process of design (Price, 1985) forces many functions to be defined at increasingly greater levels of detail, often obscuring distinctions between functions and tasks. Consequently, one often encounters references to task allocation as opposed to function allocation; in these cases, the function is sufficiently well defined to pose as a task or subtask. Applying the iterative design process to the function allocation stage of design typically results in task descriptions, where a given task may be composed of a number of detailed functions derived from different functional categories (see Section 11.6.3). A corollary of this perspective is that many task analytic methods (see Chapter 12) are more appropriate for analysis of existent systems or systems in the later stages of design, where function specification has evolved to the point at which functions can be treated as tasks or task components. Therefore, at some level of system analysis, tasks and functions can be viewed as *isomorphic* in the sense that task descriptions can be transformed into functional requirements.

In this chapter, we generally refer to "function" when addressing function allocation problems. When we do use the term "task," it will be in a context that is intended to be consistent with the foregoing discussion.

11.1.3 System Definition and Description

Considering that concise definitions of systems generally do not exist, it is often more meaningful to define a system in terms of its various aspects. These may include: its purpose (the mission statement), evolutionary forces driven by preexistent infrastructures and sociocultural factors, the process by which the system was designed and engineered, the functions performed by the system including organizational and management functions, the agents capable of performing these functions, the knowledge and skills required of these agents (and in the case of humans, their preferences), the interrelationships between agents, and the manner in which the system influences and is influenced by environmental factors (which to some extent represents a measure of the system's adaptability). Given their links to specific phases of the systems design and development process (see Chapter 2), cause and effect relationships between these system attributes are expected.

For existing systems or tentative designs, the value in undertaking this exercise in system definition will often lead to derived properties that better organize and shape our understanding of the system. For example, we may now find ourselves thinking in terms of system complexity as perhaps reflected in the physical configuration of the system, the number of goals and agents, temporal constraints related to when information becomes available and how rapidly one must respond to events, and the types of couplings between agents. Taking a somewhat different stance, it may prove more useful to characterize the system in terms of degree of automation and workload (Kantowitz and Sorkin, 1987; see Chapter 57), where the emphasis might be on whether sufficient attentional resources are available for the human to coordinate his or her own activities as well as activities involving interaction with automation, human attitudes toward automation and their implications for human and system performance, and the consequences of error. Another useful characterization might be based on the concept of information flow, where the emphasis now would be on the information necessary for supporting system objectives, the form and content of this information, the agents who must act on this information and the time constraints associated with these actions, and the communication channels through which this information flows.

System representations based on partitioning techniques offer yet another means for accessing important system properties. By decomposing a system into logical subsystems and exploring their interdependencies, new insights into system functions may emerge. For example, it may be convenient to subdivide a city's entire transit rail system into quadrants based on the locations at which passengers are served. Couplings between subsystems may depend on how many transfer points exist which, in turn, could affect scheduling strategies in dealing with blockages and delays. In the case of Buffalo's metrorail transit system, which services only a single linear route approximately 10 kilometers (km) in length, the decomposition of the system into two subsystems—a tunnel portion and a surface (downtown) portion—has had important implications for the allocation of functions. For example, although the technology existed for automating train operation along this route, the presence of the surface subsystem, which implied the need to negotiate pedestrians and vehicles, prohibited this choice. Other factors such as the perceived presence of employees were also likely to influence the decision to employ human operators to the extent that these factors were viewed as consistent with the system mission statement—in this case, of providing high-quality, safe, reliable, convenient, and cost-effective public transportation service.

A more formal partitioning technique is the abstraction hierarchy (AH) proposed by Rasmussen (1986) in which the entire system is described at each of several levels of abstraction. Each level taken by itself can be decomposed to describe relationships among system entities and processes at that level of abstraction. System processes at one level provide explanations of the processes at the level above (*how* they operate) and reasons for the processes at the level below (*why* they operate), resulting in a chain of causal relationships that can account for many features of human problem solving and decision making (Figure 11.1 and Table 11.1). The AH, though often difficult to construct, provides a basis for design of human–machine interfaces that considers the entire problem space within which the human acts. Designs based on the AH representation would therefore include *all* the information available—not merely those items which the designer had decided in advance would be useful, and not only at one level of abstraction (Vicente

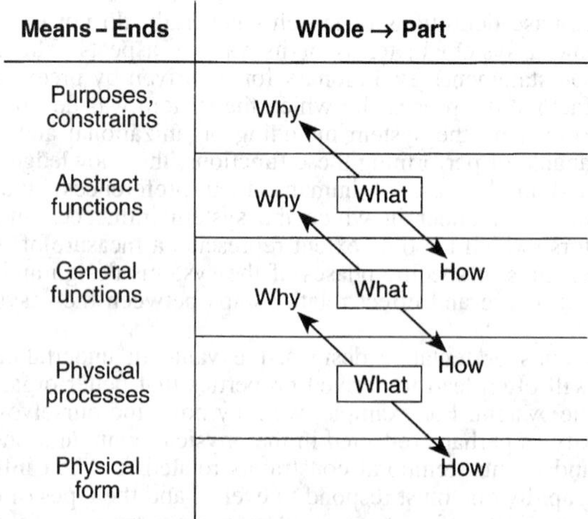

Figure 11.1 WHY, WHAT, and HOW in the means-ends space. (From Rasmussen, Pejtersen, and Goodstein, 1994, reprinted by permission of John Wiley.)

Table 11.1 The Classes within the Means-Ends Hierarchy

Means-Ends Relations	Properties Represented
Purposes and Constraints	Properties necessary and sufficient to establish relations between the performance of the system and the reasons for its design, that is, the purposes and constraints of its coupling to the environment. *Categories are in terms referring to properties of environment.*
Abstract Functions and Priority measures	Properties necessary and sufficient to establish priorities according to the intention behind design and operation: Topology of flow and accumulation of mass, energy, information, people, monetary value. *Categories in abstract terms, referring neither to system nor environment.*
General Functions	Properties necessary and sufficient to identify the 'functions' which are to be coordinated irrespective of their underlying physical processes. *Categories according to recurrent familiar input-output relationships.*
Physical Processes and Activities	Properties necessary and sufficient for control of physical work activities and use of equipment: to adjust operation to match specifications or limits; to predict response to control actions; to maintain and repair equipment. *Categories according to underlying physical processes and equipment*
Physical Form and Configuration	Properties necessary and sufficient for classification, identification, and recognition of particular material objects and their configuration: for navigation in the system. *Categories in terms of objects, their appearance and location.*

From Rasmussen, Pejtersen, and Goodstein, 1994, reprinted by permission of John Wiley.)

and Rasmussen, 1992). In principle, this affords the human with better opportunities for handling abnormal and fault conditions, which ultimately can influence function allocation decisions.

These various system representations can be viewed as devices or frameworks for organizing the description of the system. In so doing, they provide both a means for identifying and clarifying system functions as well as alternative contexts for addressing issues fundamental to the allocation of these functions. Shifting between these alternative system descriptions as one iterates through the systems design process (Price, 1985) provides opportunities for more creative solutions to the function allocation problem. Along the way, more formal systems analysis and modeling techniques generally associated with function analysis (see Section 11.2) will likely need to be applied that compliment these system descriptions. At some point in the design cycle, methods associated with task analysis (see Chapter 12) and cognitive task and work analysis (Gordon, 1995; Rasmussen, Pejtersen, and Goodstein, 1994; Vicente, 1995) may need to be employed in order to support specific design goals and test hypothesized function allocation arrangements.

The coupling between system representation and cognitive task analysis is illustrated by Roth and Mumaw (1995) who describe the use of a function-based cognitive task analysis methodology for identifying design requirements in developing a new human–machine interface for a "first-of-a-kind" process control plant. Initially, the range and complexity of the system's tasks are mapped out through a function-based goal–means decomposition of the system that results in a goal–means representation of system functions. This representation reflects the relationships between system goals and processes, and enables identification of issues concerning information and display requirements for supporting operator functions. Cognitive task analysis is then performed in order to derive design solutions for displays, procedures, and training by determining the specific information requirements for supporting operator decision and control activities. This is accomplished by overlaying questions associated with an operator decision-making model proposed by Rasmussen (1986) onto the elements or "nodes" comprising the goal–means representation.

11.2 TECHNIQUES FOR FUNCTION ANALYSIS

Although a system's functional requirements can, in principle, be derived from the system mission statement and, if sufficient information is available, from the broader-based definition of the system that might include some scheme for system representation (see Section 11.1.3), elaboration of these functions is generally required for system realization. Toward this end, a host of analytic techniques has been developed over the years that models the structural and dynamic properties of systems. These relatively formal methods of systems analysis have generally become synonymous with "functional analysis" to the extent that they help identify and resolve issues in systems design and redesign, including the issue of allocation of functions.

Following the line of discussion presented earlier concerning distinctions between functions and tasks, it should not be surprising to find that for many problems the boundary between function analysis and task analysis is not always very clear. In general, techniques associated with function analysis are more likely to be useful for detecting and evaluating problems with broader implications for system performance such as those involving workload distribution requirements and production delays. Consequently, when applied to the design of a new system, the application of these methods can accelerate the iterative design process. The output of these methods can also serve as a basis for the application of various task analytic techniques including cognitive task analysis which, by virtue of their finer grain of analysis, can offer further insights into the function allocation design problem.

The majority of these techniques essentially represent methods of system decomposition. Laughery and Laughery (1987) refer to three categories of function analysis techniques: flow analysis, time-line analysis, and network analysis. Methods of flow analysis include flow process charts, operational sequence diagrams, and other charting techniques such as FAST (Functional Analysis Systems Technique). These methods attempt to represent the sequence of system events, whether involving humans, machines, or material resources (including information) that are required in order to meet system objectives. By qualitatively *tracing* these events or activities over time, the order in which functions that represent or support these events or activities can be determined and analyzed.

Using different symbols for representing functions and events such as a machine operation, transportation of materials, or production delays, flow process charts can be used to chart the sequence of operations associated with a machine (e.g., an automated guidance vehicle as it travels between various manufacturing subsystems and automated storage facilities), a material (e.g., a raw workpart from entry into a job shop through its assembly into a finished product), or an operator's task (where the method can serve as a task analysis). From the standpoint of function allocation, flow process charts can help determine whether functions are necessary, in which case allocation decisions to human or machine entities may need to be rethought or improved, perhaps through hybrid human–machine arrangements. When applied to complex systems, these methods first chart the system at its most general level, and through hierarchical decomposition progressively chart the lower-order system functions.

Methods of flow analysis also exist that emphasize the information requirements for decision making by modeling the information-decision-action sequences (Laughery and Laughery, 1987). In a health care facility, where a series of critical decisions concerning a patient may need to be made at different points within the facility based on information that is not always complete or unambiguous, this type of analysis could suggest alterations in the allocation of information acquisition, documentation, and communication functions that target the various health care personnel and computer-based information systems.

Another well-known function analysis technique in this category derives from an approach referred to as *value engineering* (VE) (DeMarle and Shillito, 1992) that was originally developed to study ways in which cost-reduction changes could be made without compromising product quality or performance. Fundamental to the VE methodology is an information phase in which function analysis techniques are used to document each of the functions associated with a product (or its components) selected for analysis, constraints dictating product design (e.g., materials or procedures) are challenged, and the importance and cost of each of the functions are quantified through value measurement techniques. The function analysis technique typically employed by VE practitioners is FAST (Functional Analysis Systems Technique), a hierarchically organized structural technique that visually displays the interrelationships between all functions that must be performed to achieve the basic function. The steps involved in developing a FAST diagram are summarized in Table 11.2. When applied to complex "products" such as intelligent vehicle highway systems, the derivation of cost and importance values could impact the function allocation configuration, for example, by identifying specific components of a navigational system that might prove counterproductive for certain classes of product users.

Although temporal information is not specifically included in any of these methods of flow analysis, there is nothing in their construction that would prohibit this additional feature. Techniques such as Gantt charts (Laughery and Laughery, 1987) that graphically illustrate the time courses of functions and tasks could conceivably be incorporated into flow-charting methods in order to address potential workload problems that could have

Table 11.2 Development of a FAST Diagram

Rules for Function Description

1. Determine the user's needs for a product or service. What are the qualities, traits, or characteristics that define what the product must be able to do? Why is the product needed?
2. Use only one verb and one noun to describe a function. The verb should answer the question "What does it do?"; the noun should answer "What does it do it to or with?" Where possible, nouns should be measurable and verbs should be demonstrable or action oriented.
3. Avoid passive or indirect verbs such as *provides, supplies, gives, furnishes, is, prepares*. Such verbs contain very little information.
4. Avoid goal-like words or phrases such as *improve, maximize, minimize, optimize, prevent, least, most, 100%*.
5. List a large number of two-word combinations and then select the best pair. Teams can be used to derive a group definition of function. Examples are:
   a. Light bulb: "Emit light"
   b. Coffee cup: "Hold liquid"
   c. Screwdriver: "Transfer torque," or, if a painter uses it to open cans, it would be "transfer linear force." Function depends on the user's intended use.

Steps in Developing a FAST diagram

1. All the functions performed by the product and its elements are defined by using the two-word function description. Each function is written on a separate small card to facilitate construction of the diagram. The cards are placed on a surface where they are visible and easily accessible and can be moved about. It is convenient to use a table covered with paper to which the cards can eventually be fastened with tape.
2. The card that best describes the basic function is selected from among the many cards. Included in the initial listing of functions will be cards that describe secondary functions. Although many secondary functions may be present, some may have questionable value and may add unnecessary cost. They are prime candidates for improvement and innovation.
3. A branching tree structure should be created from the basic function. This is best done by personal analogy by assuming that you are the item under analysis and asking the question "How do I (verb) (noun)?" A more depersonalized branching question would be, "How does (the product) do this?" In any case answers to the question are placed to the immediate right of the basic function. The "how" question will result in branching and is repeated until branching has stopped and the function order is in logical sequence. The individual cards are a convenient aid to arranging and laying out the logical order of the how questions.
4. The logic structure is verified in the reverse direction by asking the question "Why do I (verb) (noun)?" for each function in the logic sequence. The how-why questions are used to test the logic of the entire diagram. Answers to the how and why logic sequence questions must make sense in both directions; that is, answers to how questions must logically flow from left to right, and answers to why questions must read logically from right to left.
5. A "critical function path" may result from the logic sequence of the basic and secondary functions. It is composed only of those functions that must be performed to accomplish the basic function. Any function not on this path is a prime target for redesign, elimination, and cost reduction.
6. The FAST diagrams are usually bounded on both ends by "scope lines," which delineate the limits of responsibility of the study. For example, if one is value analyzing an overhead projector, the FAST diagram would be expanded up to the point where current is conducted to the device. "Generate electricity" is outside the scope of the study.

Source: DeMarle and Shillito (1992), reprinted by permission of John Wiley.

important implications for function allocation. Returning to the health care facility example, the use of this technique may reveal that, for certain procedures performed in the trauma unit, the trauma unit nurse may not have sufficient time to manually document patient progress, suggesting that some of this information may need to be documented by other means, perhaps through downloading of information being collected and analyzed by automatic patient monitoring systems. These techniques may also reveal that information arriving from laboratory tests will often arrive too late to be of use in an operating room, forcing a reconsideration of the protocol by which this information is transferred—e.g., by suggesting the use of digitizing and fax technologies. An added benefit of timeline techniques arises when they are applied to functions that are temporal by definition such as scheduling. For example, Pinedo (1992) describes a computer-generated Gantt chart that assists the human scheduler by enabling machine-generated schedules to be edited. The impact of these changes on system performance can then be followed, and extensive "what-if" analysis can be performed, resulting in a complimentary human–computer scheduling system.

Finally, a variety of techniques exist for analyzing the relationships among system entities, events, and processes (Laughery and Laughery, 1987). Particularly relevant to the problem of allocation of functions, whether for hypothesizing initial function allocation design solutions or for testing and evaluating these hypotheses, are a number of computer simulation modeling techniques (Price, Maisano, and Van Cott, 1982) such as SAINT (System Analysis of Integrated Network of Tasks). By incorporating artificial intelligence techniques, these computer simulation models have become much more sophisticated, encompassing knowledge-based modules capable of representing human expertise and cognitive tendencies that enable the propensity for human error to be explored in a variety of situations (Woods and Roth, 1988).

In complex systems such as nuclear power plants, where the consequences of either human error or inadequacies in the automatic control systems can be catastrophic, these computer simulation models can just as well be used as a basis for selecting between alternative allocation of function design strategies. This perspective was adopted by Schryver and Knee (1987) who developed a computer simulation model that consisted of a dynamic integration of a SAINT task network model, a continuous nuclear power plant process model, and a knowledge-based module that provided predictions of cognitive workload, human error, and performance. Through sensitivity analysis performed on the relevant parameters, the capability exists for this technique to provide iterative fine-tuning of the function allocation design process that can serve to rule out candidate hybrid function allocation arrangements involving the human and automatic control systems that compromise safety criteria to an unacceptable degree. Although these types of models have often been viewed as suspect due to alleged deficiencies in the databases and knowledge-based modules on which their logic and inferential mechanisms are based, their ability to capture the dynamic interrelationships between system events and activities, which, to a large extent, define a system's complexity, should continue to make them an important class of techniques for systems design and function allocation. For further details on techniques of function analysis the reader is referred to the reviews by Laughery and Laughery (1987) and Kadota and Sakamoto (1992).

11.3 STATIC ALLOCATION OF FUNCTIONS

11.3.1 Historical Perspective

Historical perspectives on the topic of allocation of functions generally begin by alluding to lists intended for summarizing the relative advantages of humans and machines with respect to a variety of functions. The list provided by Fitts (1951) shown in Table 11.3 has served as a template for the generation of more elaborate lists (e.g., Bekey, 1970; Nof, Knight, and Salvendy, 1980) that have since come to be known as "Fitts lists." The use of such lists or of any other method that lends itself to the a priori assignment of functions to humans and machines comprises the static allocation of functions perspective. In applying this design strategy, functions are generally considered in isolation, although Fitts cautioned against using the criteria embodied in his list as the sole basis for allocating functions. If the advantage in performing a particular system function rested with either the human or machine, then that function was allocated accordingly. However, if the sum total of such allocation decisions resulted in imbalances in work assignments so that humans were potentially overloaded or shortages in machinery were readily apparent, functions would be shifted across these two resource classes until a viable solution was reached.

Table 11.3 The Original Fitts List

Humans appear to surpass present-day machines with respect to the following:

1. Ability to detect small amounts of visual or acoustic energy.
2. Ability to perceive patterns of light or sound.
3. Ability to improvise and use flexible procedures.
4. Ablility to store very large amounts of information for long periods and to recall relevant facts at the appropriate time.
5. Ability to reason inductively.
6. Ability to exercise judgment.

Present-day machines appear to surpass humans with respect to the following:

1. Ability to respond quickly to control signals, and to apply great force smoothly and precisely.
2. Ability to perform repetitive, routine tasks.
3. Ability to store information briefly and then to erase it completely.
4. Ability to reason deductively, including computational ability.
5. Ability to handle highly complex operations, i.e., to do many different things at once.

Source: Fitts (1951), reprinted by permission of the Ohio State University Research Foundation.

The contention that such lists had minimal impact on design practice due to their overly general and nonquantitative nature (Price, 1985) is true to the extent that design engineers prefer quantitative solutions, which is why approaches to human reliability assessment such as THERP (Swain and Guttmann, 1983), with its quantitative treatment of human error, was readily accepted in the nuclear industry. This approach to function allocation, however, very much embodied an engineering design perspective that emphasizes decomposition followed by aggregation. Consequently, its solutions were highly articulatable to, and consistent with the thinking of design engineers.

11.3.2 Example: Flightdeck Design

Design of the flightdeck represents an area in which allocation of functions plays an important role. In addition to illustrating the role of allocation of functions within the systems design process, it demonstrates the changes in functions that occur with increased automation as well as the implications of these changes.

The design of the earlier single-function, single-indicator mechanical analog instruments (i.e., static displays) as well as the electronic flight instrumentation systems (EFIS) introduced in the mid-1970s can be decomposed into information analysis and display format phases (Banks, Hunter, and Noviski, 1985). The purpose of the first phase is to clearly define the information needs of the pilot in support of his or her tasks, which in turn requires: (1) defining the overall system goals (e.g., meeting flight schedules); (2) scenario identification, functional analysis, and allocation of functions; (3) task analysis; and (4) identification of informational requirements for each individual function. Based on the system goals, operational mission scenarios (such as "minimal ATC interaction") representing the planned use of the system are constructed, often through the use of timeline presentations that plot the events and tasks that define a mission scenario against time in as much detail as is needed for further analysis. These scenarios form the basis for identifying the function requirements (e.g., the need to deploy the landing gear or to follow a scheduled flight plan). Once a list of functions have been derived, these functions must be allocated to the human or machine. The tasks encompassing these functions that must be sequentially performed by the pilot are then identified through task analysis (see Chapter 12) and analyzed in order to determine the data that must be provided to ensure that all functions and tasks are carried out successfully. Ultimately, this information is used in the subsequent display formatting phase.

The progression from static displays to EFIS can be represented by movement within the static–dynamic function allocation continuum from a predominantly static design perspective toward dynamic allocation of functions. These changes arose and continue to be manifest as a result of advances in automation and computer technology; they parallel a shift in emphasis from *comparability* to *complimentarity* between humans and machines (Jordan, 1961). We now need to reevaluate system goals in the light of new functions that emerge due to considerations of complimentarity, which, in turn, requires the generation of new mission scenarios capable of capturing the relevant contexts.

Functions previously allocated to humans based on a fixed policy could now be allocated to the machine. Classic examples include flight path guidance through activation of auto-controllers and preprogrammed navigational systems, access to computational support, and the capability for configuring cockpit displays as a function of environmental conditions through selecting and deselecting features that enable control of what and how much information should appear on a particular CRT (Weiner, 1985). When the decisions to allocate these functions to the machine are made on-line, function allocation can be considered dynamic. However, these allocation decisions are made at the discretion of the human. As we shall see in Section 11.5.7, with further movement along the static–dynamic continuum, we will confront the more idealized version of dynamic function allocation whereby the ongoing responsibility for allocation decisions can be made by either agent.

Automation can also introduce new functions that are allocated to the machine according to static design policies. Although these functions are typically activated in response to certain conditions, implying the presence of a dynamic component in the allocation process, when considered within the context of the systems design process the activation and allocation of these functions are fixed. Examples include stick pushers that automatically assume control from the human to prevent a stall (Kantowitz and Casper, 1985) as well as an array of warning and alerting systems (Weiner, 1985). These functions brought about by automation in turn generate new functions. In the analysis of scenarios, we would now have to consider the impact on the human of having control overridden by automation and on monitoring the warning and alerting systems, including the human's ability to evaluate the validity of the warning. Likewise, in the case of human-driven automation, the ability to configure displays results in new memory functions potentially imposed on the pilot that address the need to know what is *not* being displayed and how to obtain this information if needed.

It is interesting to note that in comparing the design process for EFIS with that associated with static displays, the same basic four steps outlined above comprising the informational analysis phase still occur, although their content changes. For example, the conditions under which the pilot can or should dynamically allocate certain functions to auto-controllers must be determined from a scenario analysis that will necessarily be much more complex. The primary change that does occur in the design process is in the display design phase. In the design of static displays, the entire "data world" is presented to the pilot. In contrast, the flexibility afforded to crew members by EFIS results in information being integrated, primarily as a function of information needed for basic guidance and control and optional information that is generally based on the particular phase (e.g., descent) of flight.

11.3.3 Example: Flexible Manufacturing Systems

Automation of manufacturing processes involving operations such as the machining of metal parts proceeded more gradually as compared to automation in other areas. The term *islands of automation* arose during the early stages of automation in manufacturing as a result of the ability to establish an infrastructure capable of supporting automated activities, but not for the capability of effectively computer-integrating this automation. Flexible manufacturing systems (FMSs) represented just such an achievement. Consisting of predominantly computer numerically controlled processing stations interconnected by automated material handling and storage systems and controlled by an integrated supervisory computer control system, FMSs were capable of exploiting the use of classification schemes that take advantage of part similarities to process a variety of different part types simultaneously (Greenwood, 1988).

Their implementation also raised an important human factors design issue concerning defining the human role in these systems. One appealing characterization of this role that is in line with human–machine interaction in other highly automated systems has been referred to as *supervisory control* (Sheridan, 1993). This paradigm emphasizes the need for an individual to exercise and manage monitoring, interpretive, diagnosis, decision-making, and corrective manual intervention functions. Although in principle adopting a human supervisory control job description for these systems seems logical, in practice FMSs vary enormously and slight changes in degree and type of automation and in product type and demand often translates into significant differences in the need for human involvement.

One approach toward assessing the extent of this involvement is through an allocation of functions analysis. Hwang, Barfield, Chang, and Salvendy (1984) did just this, adopting an essentially static function allocation perspective in comparing the relative abilities and limitations of humans and automation in carrying out the various functions associated with operation and control of an FMS. Referring to their analysis (Table 11.4), notice that the information need not be compiled for the sole purpose of specifying strict comparability criteria. For example, in "shop scheduling" the clear implication is that the human and computer need to work together to improve upon "good" starting solutions derived from computer heuristics.

In addition to the obvious benefits such as identifying functions that a particular component should or should not be exclusively assigned to, what are the advantages of performing this type of analysis? For one, it provides an assessment of the state-of-the-art in automation and computer control. The inclusion of technological forecasts in these lists (Table 11.4) could then be used to anticipate potential changes in allocation of functions, enabling systems to evolve rather than be "shocked" by new technology. It can also help bridge the gap between static and dynamic function allocation by identifying conditions under which system performance would benefit from a complimentary human–machine arrangement, and in the process help define a logical supervisory control role for the human.

This type of analysis, however, needs to be approached cautiously. The combinatorial complexity of discrete-event systems such as FMSs often precludes quantitative predictions of process outputs from system inputs and can greatly complicate the analysis and assessment of work contexts. Consequently, this factor can hamper not only the process of function allocation, but also the ability to extend findings regarding either comparability or complimentarity of humans and automation from one system to another.

11.4 DYNAMIC ALLOCATION OF FUNCTIONS

11.4.1 Problem Formulation

Strictly defined, dynamic allocation of functions refers to an allocation policy that can be altered at any given point in time during system performance. While computer systems for personal use and other systems may now feature schemes (e.g., the capability for inferring user problems and responding accordingly) that are consistent with those that underlie dynamic allocation of functions in more complex environments (Norcio and Stanley, 1989), we will be interested primarily in application areas in which allocation of function decisions can have significantly more far reaching implications for both systems design and performance.

The notion of allocating decision-making responsibility between human and computer across the multiple tasks that might comprise the activities of a pilot or supervisory controller of an industrial process control system was formalized by Rouse (1976). Recognizing the possibility for many of the tasks to be performed by either the human or computer (i.e., automation), Rouse suggested that function allocation be based on which agent was free to attend to that task, and furthermore, that this determination should be done dynamically, as dictated by the situation. Not only was this scheme supposed to result in more efficient task performance (analogous to having customers standing in lines designated for service by multiple line servers), but it presumably would allow the human to be better able to assume a supervisory role by not being perpetually consumed about the status of tasks that were supposed to be performed by the machine.

Rouse's general mathematical formulation of this problem requires: (1) identifying the N tasks to be performed by the human and computer; (2) characterizing each task in terms of the "states" in which the task is in; (3) characterizing the observations by the human or computer in terms of the best estimates of these states (which takes into account partial, noisy, or missed observations); and (4) based on these observations, the agent must trade off the benefits of attending to some task i by performing some set of actions required of this task against the cost of ignoring the $N-1$ other tasks; the agent must consider that over the course of time it takes to attend to task i events could occur in the $N - 1$ other tasks that are of greater importance than those that caused attention to be diverted to task i.

Evaluating system performance measures such as the average time a task had to wait to become completed and the percentage of allocation decisions (i.e., decisions to service a task) performed by the computer requires a host of assumptions. These include the provision of joint probability distributions associated with: (1) the occurrence of task events; (2) the time between events in each of the tasks; and (3) the times required to service the tasks. In addition, a priority scheme needs to be established governing the

Table 11.4 Capabilities of Human and Computer in the Operation and Control of Flexible Manufacturing Systems

Flexible Manufacturing System	Computer	Human
Physical Operation		
System loading and unloading	Can direct robot to handle up to 900 kg. Gives part type, part number and quantity Shows diagram on-line of how parts are to be placed Difficult to determine workpiece orientation	Maximum arm load: less than 30 kg; varies drastically with type of movement, direction of load Power: 2 h.p. ≅ 10s 0.5 h.p. ≅ 120s 0.2 h.p. ≅ *continuous 5 kc/* min subject to fatigue; may differ between static and dynamic conditions Checks part information, notifies computer that part has been loaded
Workpiece setup on machine	Very difficult to set up if the workpiece is not prefixtured on a pallet Machine vision is required Uses locking device when the workpiece is fixtured on a pallet	Slow, but good pattern recognition skills for complex parts
Tool setup (off line)	High precision on complex geometry is not available	Human reliably performs this task Positioning accuracy: relatively (with feedback)-good: 0–1 mm; absolute (without feedback)-very poor: 3–8 cm Fitts' law governs movement
Tool replacement	Automatically changes tool magazine: 5 min required Single tool changed in 2 sec Computer knows when a tool magazine is due and instructs a robot crane to perform this job	Inserts tool manually Checks tools when signal is received from computer
Shop scheduling	Computer heuristics can provide 'good' schedule Combinatorial problem, no guarantee on optimal solution Computer simulation evaluates scheduling Plots the number of pallets required during the scheduling period	Human intuition helps to determine algorithm Human and computer help determine optimal schedule Human inputs workload data and uses computer simulation to see the effects of scheduling

Table 11.4 ***(Continued)***

Flexible Manufacturing System	Computer	Human
Equipment Control		
Machine and tool control	CNC controls axle and spindle motors, tool changer, pallet changer, etc. DNC computer downloads part program to CNC Faster and more accurate than human Preprograms cutter motion	Human interacts with computer to modify online program Human supervises machine tool control, primarily monitoring machine status
Robot control	Robot controller controls the motion robot Sensing devices, e.g., vision camera, texture sensor, strain gauge, etc., provide feedback data System control computer downloads robot control program and operation instructions Very limited learning or environment adaptation	Human trains robot via computer which requires human intelligence transfer to robot Human good at pattern recognition Amenable to learning and flexible adaptation to the shop environment
Inspection device control	Computer can control automatic measurement and gauging devices Presents inspection results used for quality control analysis Limited range which can be optimized under relevant needs Can handle position tolerance below ± 0.03 mm	Human sets criteria for inspection device Good absolute judgment Smallest detectable threshold 10^{-6} ml
Flow Control		
Workpiece input rate control (0.5 to 10) pieces/hour	Ability to process information is very high, limited only by channel rate Computer uses production schedule and capacity information to determine input rate Physical workpiece input is accomplished by workpiece loading function	Limited ability to process information (10–20 bits/s) Human monitors and identifies unusual flow rates Human determines the rate limit, maximum and minimum

Table 11.4 (*Continued*)

Flexible Manufacturing System	Computer	Human
Traffic control	Computer can optimize the transportation of workpieces, fixtures and tools in the system The correctness of traffic control is dependent upon sensory feedback. When feedback information is not correct, computer may make mistake AVG. powered conveyor, tow line, overhead trolley, etc., are used to transport items in the system	Human has good comparative judgment to correct traffic control problems Human 'programs' program controller to transport items in the system Human good at detecting incorrect traffic control Can identify unusual flow rates, mean response latency 113–528 ms
System Monitoring		
Tool life	Computer records the time a tool has been used. When the cumulative time exceeds the rated life, computer informs the control system to change tool. Sensing devices monitor the tool wear and catastrophic tool breakage; not easy to sense the minute changes in tool wear.	Human helps supervise tool life by entering tool transactions into database Has control over tool life, i.e., can alter program if catastrophe occurs
Quality control	Computer analyses product quality change, detects machine wear and other condition changes. Reports to human for decision-making Computer compares the finished part geometry with the desired geometry and calculates the adjustments needed to correct	Human makes decisions for prodcut quality control Can adopt and identify new plans for quality control
System status	Checks and updates the status of the entire system at least once every minute Computer can perform diagnostic tests rapidly and accurately Reports workpiece jam, or redirects workpiece if possible	Human monitors overall system status with flexible decision-making abilities Human queries computer when system is down Computational capability slow—5 bits/s

Table 11.4 (*Continued*)

Flexible Manufacturing System	Computer	Human
Other Information-Handling Functions		
Machine control program storage	Computer can store vast amount of part programs and robot programs Program can be downloaded at a speed of 10 bytes/s to several million bytes/s Reliability can be increased by using redundant storage device	Human reviews computer output of program storage Human decides if new programs or variables need to be added to program Human increases reliability of program storage as back-up decision-maker
Tool and workpiece inventory records	Maintains a large database Updated data can be automated or manually handled	Human updates database for tool and inventory, adds tool to inventory
Job priority determination	Computer can apply a predetermined rule to set priority	Human writes software which determines job priority
Emergency handling	Computer can shut off the system based on a preprogrammed sequence	Reaction time 0.25 to 0.33s Good error detection/correction at cost of redundancy
Maintenance	Can keep a maintenance schedule Updates status database at each time increment	Human determines maintenance standards Human performs periodic maintenance by visual spot checks
Repair	Can diagnose source of problem in order of priority The most serious problem causes an immediate shutdown; less serious ones alerts operator	Human deals with repair problems manually System is dependent on human involvement for repair. Human initiates a hold on system and corrects the problem

Source: Hwang, Barfield, Chang, and Salvendy (1984), reprinted by permission of Taylor & Francis Ltd.

order in which tasks are performed which will depend on costs associated with delaying performance of any particular task. Finally, assumptions concerning the performance characteristics of both the human and computer need to be made that address speed–ability mismatches between these two agents.

In solving this problem using computer simulation, the handling of these assumptions is a relatively trivial matter. Through selective manipulation of these parameters, "computer experiments" can be performed in order to determine the implications of dynamically allocating decision-making responsibility. Whether the results of such simulation experiments can be used to gain insight into this problem is, however, questionable. By virtue of being "context free," there are no real implications for incorrect actions in performing these tasks, or for that matter of the human or computer having meaningful knowledge of the tasks. Moreover, this type of study can easily assume that both the human and computer cannot attempt to perform a task that the other is already performing, implying that each is aware of what the other is doing.

In essence, this study established a platform of issues that would help shape future directions in research on dynamic task allocation. One such study is presented in the next example and adheres faithfully to Rouse's (1976) general problem formulation. This research represents an attempt not only to map these ideas into an experimental context, but also to address the issue of constructing models of human performance to enable the machine to infer the human's state of knowledge and thereby minimize interference between agents during task performance.

11.4.2 Example: A Process Control Scenario

In this process control scenario (Greenstein and Revesman, 1986; Revesman and Greenstein, 1986), the human operator's task consisted of monitoring nine displays that provided ongoing information concerning nine independent machines. The objective was to detect machine failures and indicate whether a repair action should be taken; other actions could not be taken during the time it took for a machine to become repaired. Machine failures were signified by increased variance in the displayed output, which continued to increase as long as the machine was in the failed state. The output contained noise, implying that some degree of learning was necessary in order to distinguish patterns reflecting failed states from those representing normal functioning. To facilitate learning of the task, immediate feedback was provided concerning the correctness of a repair action. In addition to the machine output data, information was available on the time between failures, repair time, and the cost per second of allowing a machine to remain in the failed state (Figure 11.2). The overall objective was to minimize the cost of failed machines, which involved trading off the cost of making a repair to a machine—in which case potential costly failures of other machines are necessarily put on hold—with the cost associated with allowing that machine to remain in a failed state.

If the computer had available to it a model that can predict the likelihood with which the human will detect machine failure events, and that can subsequently use these likelihoods to predict what actions the human will take, then the computer could use this information to "work around" the human. Implementing this type of dynamic function allocation arrangement is precisely what these researchers had in mind. The model they developed consisted of two stages. An *empirical* failure detection stage applied linear discriminant analysis techniques to features of the displayed output to derive the probabilities with which the human would identify each of the displays as representative of a failed machine. Incorporating knowledge of these event detection probabilities as well as knowledge of the mathematical processes governing machine failures, a stochastic dynamic programming technique was then employed in a *normative* action selection stage to determine the optimal control action (which minimized the expected cost) at a particular point in time (Figure 11.3).

Having found the model to be a reasonably accurate predictor of human performance (Greenstein and Revesman, 1986), the next question was whether this model-based communication offers a suitable alternative to explicit communication between human and computer (Revesman and Greenstein, 1986). Models, after all, are not perfect. In the case of model-based dynamic function allocation, the implication is that redundant actions, whereby both agents attempt the same action, can never be totally avoided. This study explored concomitantly the effects of having available: (1) computer-to-human commu-

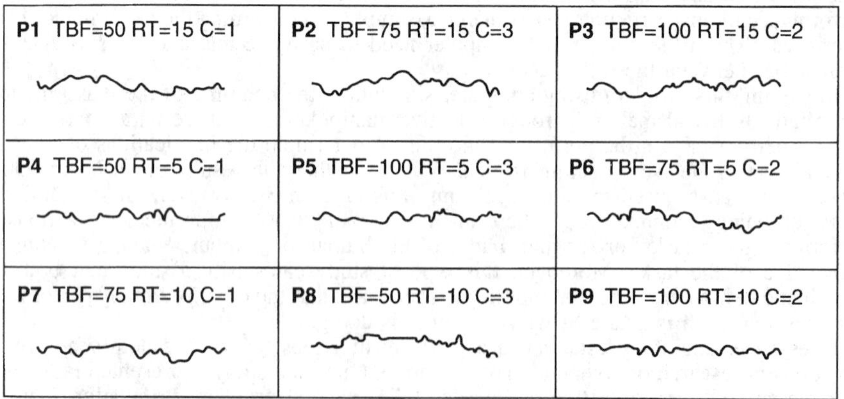

Figure 11.2 Sample screen displays in the experimental process control scenario. (From Greenstein and Revesman, 1986, © IEEE.)

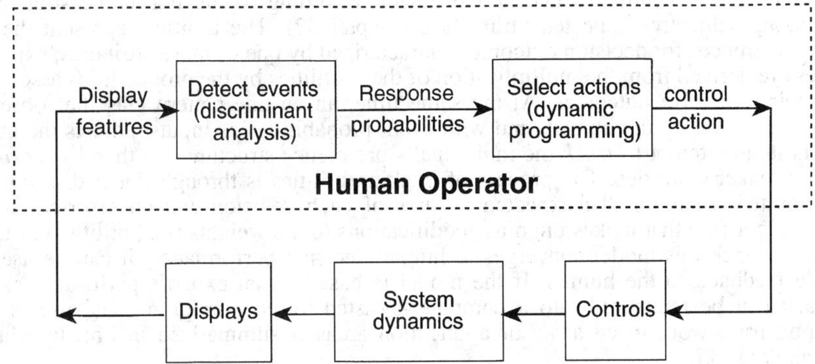

Figure 11.3 Block diagram of the stages comprising the model-based system used in the experimental process control scenario. (From Greenstein and Revesman, 1986, © IEEE.)

nication (the computer's intention to take a repair action on a machine was established through the appearance of an asterisk in the corner of the associated display); and (2) model-based communication to allow the computer to predict the human's actions (as compared to use of only stage 2 of the model which did not have available to it the predictions of human repair actions from stage 1). Results indicated that even when computer-to-human communication was available, the availability of model-based communication significantly reduced the number of redundant actions taken by the human (note that ideally, with perfect reception of computer-to-human communication, there should be no redundant actions on the part of the human). Perhaps most disconcerting was that use of the model had no significant impact on overall (i.e., combined human and computer) system performance. Moreover, although the model reduced the number of redundant actions by the overall system to a larger extent when computer-to-human communication was not available than when it was available, the drop from 17 to 7.31 when explicit communication was available over the 15-min experimental trial, while statistically significant, could prove disastrous in real-world scenarios. Even the implication that model-based communication offers a viable alternative to situations where explicit dual human–computer communication could, due to attentional demands, degrade human performance needs to be more carefully considered within the detailed task contexts characteristic of real-world environments.

This model-based perspective to dynamic function allocation is nevertheless impressive in its objective of capturing a meaningful representation of human behavior. From the standpoint of allocation of functions, we need to consider whether it is worthwhile pursuing applications for which this approach would prove not only experimentally feasible but practical as well. Otherwise, its true meaningfulness is likely to lie elsewhere. In particular, as Revesman and Greenstein (1986) suggest, these models may find useful applications in computer aiding and even training. While perhaps not bringing us back full circle to issues of comparability, this perspective to model-based communication would occupy a more realistic ground between strictly static and dynamic function allocation perspectives.

11.4.3 Adaptive User Models and Dynamic Function Allocation

Instead of an emphasis on establishing decision-making responsibility, the model-based perspective can be used to provide a basis for computer-aiding decision-making functions, with the human clearly responsible for system performance and consequently in the position to dictate whether model-based advice should be taken. A potentially useful side effect of this arrangement, especially if the model-based system is *adaptive*, is that the model may very well be sufficiently competent to assume decision making and control function responsibilities in the event the human cannot do so. One such class of models, referred to as *adaptive user models*, is described by Freedy, Madni, and Samet (1985).

In each of these applications, the basic idea is to capture the human's *preference structure* by observing the information available to the human as well as the decisions made by the human on the basis of that information. A multiattribute utility model is used to predict the human's decision behavior according to the normative strategy of maximizing subjective expected utility (see Chapter 37). The utilities represent the human's preferences for decision outcomes characterized by one or more attributes; expected utilities are derived from the multiplication of these utilities by the probabilities associated with their respective outcomes. At the same time, an on-line pattern classifier observes the choices made by the human, and with event probabilities as inputs, adjusts the utility weights in an attempt to *track* the individual's preference structure and thereby *learn* the decision maker's utilities. The process of tracking utilities is through linear discriminant functions that represent the expected utilities of each decision and an error-correction training algorithm that makes ongoing modifications to the weights (i.e., utilities) of these functions. Once this model converges to human decision performance, it can be used to provide feedback to the human. If the model is based on an expert's performance, this feedback can be integrated into a computer-assisted training program. The use of this modeling framework in an array of application areas is summarized in Freedy, Madni, and Samet (1985).

11.5 SUBTLE CONSIDERATIONS IN DYNAMIC FUNCTION ALLOCATION

Determining how and to what extent model-based automation should be used raises a new set of issues in the allocation of functions. Without a deeper understanding of the model's structure and dynamics, the environmental contexts within which model-based automation is supposed to operate (including the implications for incorrect or redundant actions), and human behavioral tendencies (especially behaviors stemming from the presence of model-based automation), it is difficult to establish the usefulness of this automation—whether as a parallel player in multitask situations, as an adaptive aid capable of assuming human functions under conditions where such action is deemed reasonable, or as aid or controller acting completely under the discretion of the human. As complex as each of these issues can be, it is the interplay between these issues that is critical to understanding the dynamic function allocation problem. In the following sections, several of these issues will be discussed.

11.5.1 Model Structure and Limits in Model-Based Systems

What was the basis for the success of the adaptive user models employed by Freedy, Madni, and Samet (1985)? While knowledge of the precise details associated with the application areas is usually necessary for answering this question, the search for more general explanations can go a long way toward establishing the validity and bounds of such model-based systems. In this case, some of the answers may be found through examination of the structural characteristics of the model within the general context of the types of problems it was applied to.

First, the fundamental model is linear and, as with many linear models, inherently robust. Furthermore, for decision-making situations in which alternatives are clearly delineated and described by linear representations, use of a multiattribute utility representation of the human's preference structure provides a normative basis for decision making—a prescription for how decisions should be made—that is not susceptible to the biases that often lead human decision makers to "bad" choices. The particular characteristics of the decision-making situation are therefore critical.

In the applications documented by Freedy, Madni, and Samet (1985), the situations represent the dynamic counterpart to the static problems for which the expected utility ranking of decision alternatives provides a rational basis for decision making (French, 1986). Clearly, these models would not be able to accommodate decision making in *naturalistic* settings (see Chapter 37) where the characteristics of the problem are too ill-structured to allow the relationships between input information and decisions to be represented parametrically.

11.5.2 Fundamental Limits in Modeling

In alerting the behavioral science community to the limitations of laboratory experiments for solving real-world problems, Chapanis (1967) noted that these experiments represent models that offer at best approximations to real-life situations. With the development of model-based systems—whether in the form of rules, algorithms, networks, or equations—that have sufficient intelligence to either aid the human or perform functions previously allocated only to humans, we are faced with the same prospect: that these models are fundamentally limited.

The taxonomy of Rouse and Hammer (1991) identifies ten types of model limitations categorized into two classes. The first class of limitations considers internal limitations. In these situations, the model may be incomplete, inaccurate, incompatible, or incorrect. The second class of limitations concerns model inputs and outputs. The latter form the basis for inputs to a decision module that governs the functional behavior of the system (e.g., enabling the landing gear of an aircraft), while inputs could comprise sensor readings and outputs from other models in the overall model-based system. Inputs to the model may take on unanticipated values or have missing or extra elements. Likewise, outputs may require unanticipated values (i.e., require more information for input into the decision module than the model's output can provide) or have missing or extra elements. Where these limitations reflect underspecification, it is unlikely that real-world situations are being modeled; where these limitations reflect overspecification of the inputs and outputs between the model-based system and the real world, situations that are irrelevant likely are being modeled. While overspecification may appear innocuous and perhaps at worst may slow system performance down, it is symptomatic of more serious design problems in the model.

Assuming (as Rouse and Hammer do) that these types of modeling limitations are unavoidable, then priority should be given to methods for detecting unacceptable consequences of modeling limitations, diagnosing their causes, and compensating for their problems that can address both *internal* and *external validity* issues. The approach taken by Rouse and Hammer on this matter reflects their interest in modeling limitations in knowledge-based systems that attempt to model the competency or intelligence of human experts in particular domains. This approach is based on developing a method for examining the overall structure of the knowledge bases underlying the intelligent model. The method used is called "plan-goal graphs," which we will have more to say about when we revisit flightdeck design.

11.5.3 Environmental Context and Model Limitations

In the example on flexible manufacturing systems (FMSs) in Section 11.3.3, we noted how these types of systems could be characterized by combinatorial complexity and emphasized how this system feature needed to be accounted for in static allocation of function designs. This system property can also affect design of model-based systems used for both computer aiding as well as for arrangements that allow computer takeover of decision-making functions. Sharit and Elhence (1990) experimentally investigated the tool replacement decision function in a simulated FMS—whether a tool should be replaced prior to processing a part at a machine—with the intention of determining the implications of various cooperative human–computer arrangements on system performance. Although this is not a function one would normally assign to humans in FMSs (at least not on a continuous basis), the intention was to examine human performance under various decision support arrangements in order to assess human capabilities relative to model limitations in processing information in such combinatorially complex environments. By adapting Bayes' policy (see Chapter 37) with respect to minimizing tool-related costs, the computer could automatically assume this function at each "decision point." However, these decisions would constitute at best a *locally optimal* decision policy whereby each decision would be implemented independently of the others.

In addition to minimizing tool-replacement costs, the human was required to maximize part throughput—the extent to which parts are processed by the system. Achieving increased throughput as compared to the model-based decision maker was expected; after all, Bayes' policy was directed only at tool-replacement costs. However, the ability for a human–computer arrangement to also improve upon Bayes' policy would indicate the possibility for the human to use information concerning the *context* in which tool-replacement decisions are made with respect to both of these performance objectives. (Whether contextual information was being used could be verified to some extent by rerunning the simulation using the human's decisions as inputs and examining the state of the system at each decision point.) Furthermore, due to the general difficulty for model-based controllers to capture and utilize contextual information in combinatorially complex systems, this information could provide a basis for developing a more desirable model-based approach to tool replacement (Sharit and Elhence, 1989).

The question was whether a particular type of human–computer arrangement would be advantageous in promoting this type of global or contextual processing in the human. While automatic control was governed by Bayes' policy, the human performed the task under three arrangements: (1) total responsibility for decisions with no decision support from the computer; (2) at each decision point the computer provided three recommendations concerning whether to replace the tool based on three decision policies—Bayes, minimax, and optimism—as well as the expected loss in the case of Bayes and the maximum and minimum losses in the cases of minimax and optimism, respectively; and (3) computer control with human override, whereby the human had the option at each decision point to override the computer's (i.e., Bayes') decision policy. Somewhat surprisingly, the arrangement in which decision making rested exclusively with the human proved most effective. The hypothesis offered for this result was that the uncertainty concerning the local economic consequences of decisions made this arrangement more conducive to attending to global or contextual considerations, perhaps by encouraging more "active participation in the control loop" (Kessel and Wickens, 1982) that enabled relevant cues related to system state to be more readily recognized and integrated into the decision process. The decision support that was provided could have proven counterproductive by virtue of *biasing* the human toward more local considerations; in combinatorially complex environments, this generally would not be desirable. In supervisory control, where multiple functions need to be considered, an understanding of what constitutes an appropriate human–computer cooperative arrangement for each function could serve as a starting point for implementing an overall dynamic function allocation scheme.

11.5.4 Example: An Aerial Search Scenario

In a study by Morris and Rouse (1988), the issues of model-based limitations as well as the human's attitude toward a model-based assistant had important implications for design of dynamic function allocation arrangements. In their study, subjects were required to perform two tasks: a target identification task and a tracking task. The display consisted of an intercoastal waterway with varying proportions of water, and the objective was to distinguish and identify target boats from other boats that were present in a spotting window on the display. The computer was also capable of performing the target identification task, but its performance depended on the amount of water in the display which varied over time. Performance was expected to favor the human when the proportion of water to terrain in the spotting window was low, enabling the human to take advantage of perceptual abilities. The human was also required to simultaneously perform a tracking task, thereby simulating the basic functions comprising aerial search.

In one experiment, spotting task difficulty (as determined by terrain composition), tracking difficulty, and availability of the computer to perform the spotting task function at the discretion of the human were manipulated. Of particular interest was whether the human was making appropriate allocation decisions by deferring control of the spotting task to the computer aid when the human required this assistance. To determine when such assistance was indeed appropriate, multiple regression models were developed predicting the human's spotting performance. These models were then used to *infer* when the aid should have been used. Most of the discrepancies observed between predicted and actual use of the computer aid resulted from an unwillingness by the human to allocate the spotting task to the computer when in fact this was the predicted course of action.

In a subsequent experiment, an automatic dynamic function allocation feature was available; the multiple regression models in the previous experiment served as the basis for dynamic function allocation decisions. In this experiment, tracking difficulty was held constant. However, in addition to spotting task difficulty, three function allocation arrangements were investigated: (1) the unavailability of a computer aid; (2) the ability for the human to allocate the spotting task to the computer; and (3) the automatic takeover of the spotting task by the computer. The latter case occurred on the basis of comparing the human's predicted performance to the computer's predicted performance, with the computer informing the human prior to assuming control of the spotting task. Our interest is in determining whether recognizable patterns of function allocation were apparent, and, based on the subjective reports of the subjects, whether we can gain insight into the more general problem of designing cooperative human–machine systems. Not surprisingly, activation of the automatic aid was found to be relatively consistent over bays, where the proportion of water in the spotting window was high. When the human was "in charge" of allocating the spotting task to the computer, function allocation was less consistent. For example, the computer was occasionally assigned the task over channels where the terrain characteristics favored the human's perceptual skills, and disregarded as an assistant over bays where the computer's probabilistic-based approach to identifying targets was more reliable.

Perhaps most interestingly, the availability of the computer enhanced human unaided (i.e., manual) performance on the spotting task. Furthermore, this benefit occurred when the human was in charge of function allocation—no such benefit was observed when function allocation of the spotting task was automatic (i.e., model based). Recall that in terms of consistency there were some advantages to having an automatic function allocation policy. The fact that dynamic function allocation decisions as made by either the human or the computer each offered unique benefits obviously does not help simplify the design problem.

Despite the obvious benefits to be gained by enhancing the performance of the computer (through better models) and the human (through better selection and training), we can be assured that human perceptions toward its model-based assistant or partner will continue to lend uncertainty to the dynamic function allocation design problem. As Morris and Rouse suggest, if task accuracy is important to the human, he or she may not want to defer control to the computer unless the aid is perceived as superior (to himself or herself), and making this assessment is likely to be complicated by many individual and task-related factors. The cooperative arrangement could be more fine-tuned by having the computer suggest its intentions and wait for approval prior to its acting, or simply by allowing it to act but enable overriding control by the human. Ultimately, the balance of authority in dynamic function allocation is likely to be determined by: (1) the nature of the task, including the consequences of incorrect decisions or actions by the human and machine; and (2) the attitudes of designers and users, which can be expected to be influenced by the perceived (and to some extent, assessed) limitations of both human and model-based performance.

11.5.5 Knowledge and Confidence Factors

Decisions by the human to allocate functions to the machine, or even to override automatic takeover of functions, may also depend on the degree to which the human understands the model underlying machine performance. Knowledge of the model comprises more than just an understanding of its dynamics. It also includes an ongoing assessment of its capabilities based on feedback the user has been able to attribute to the model's performance. Assuming that extremely high degrees of reliability have not yet been established for the model-based system, it would appear reasonable to suggest that humans would tend to be more suspicious of underlying models whose composition is perceived as vague and highly esoteric as compared to models that appear more accessible. At the very least, an understanding of the model's dynamics allows feedback on machine performance to be more accurately evaluated, potentially providing the human with a better understanding of the conditions warranting model-based execution of functions.

A possible implication of this line of reasoning is that dynamic function allocation is likely to be exercised with greater *confidence* if machine performance is governed by rule-based expert systems; that is, by structures familiar to humans and that can also provide explanatory systems documenting how conclusions are arrived at. A common application of such assistants is in troubleshooting (Bond, 1987), where the machine generally functions in an advisory capacity designed to support humans but can also function in an unaided capacity. However, as Roth and Woods (1989) caution, if development of these "models of expertise" and their associated advisory components are not consistent with cognitive systems engineering principles (Woods, 1988), a variety of problems leading to "brittle" human–machine performance are likely to be encountered whereby the range of conditions under which the model-based system can function appropriately is fairly narrow. For example, if the classes of problems that fall outside the system's expertise are not clearly indicated by the designers of the expert system—a common symptom of a "piecemeal" approach to knowledge acquisition—workers are likely to lose confidence in the machine due to observing frequent failures.

Confidence in machines should be enhanced if the underlying models are designed to support an active human role. This can be accomplished, for example, by allowing the human to monitor the model's behavior, redirect its path in pursuing a line of diagnosis, or assess whether it was encountering situations beyond its boundaries of competence. Disclosure of this kind is facilitated in machines governed by rule-based expert systems, and serves to expand the function set, with the human now capable of editing the machine's knowledge base or even redirecting its diagnostic path (Roth and Woods, 1989).

However, as expert systems become more complex and consequently more *opaque* to the user, especially in application domains that impose severe time constraints that do not afford the luxury of relatively involved human–machine dialogue, we should again expect the issue of confidence in the model-based machine to influence the prospect for achieving successful dynamic function allocation. This is one of a number of issues in dynamic function allocation that we may need to confront when we revisit flightdeck design. Prior to doing so, we will examine an issue related to confidence—that of trust.

11.5.6 Trust and Dynamic Function Allocation

In complex systems, human supervisory controllers can often assume manual control from, and assign control to, automatic controllers. For example, in aviation, pilots can assume manual control from auto-controllers as well as allocate to them flight control functions. Lee and Moray (1992; see Chapter 58) were interested in the relationship between the human's decisions to allocate functions between automatic and manual control and the *trust* the human has in the automatic controller. From the standpoint of human–machine system interaction, the construct of trust was presumed to encompass dimensions that reflect the human's: (1) ability to *predict* the performance of the automatic systems, for example, based on trial-and-error performance; (2) *understanding* of the basis for how the automatic systems are governed, which corresponds to system dependability; and (3) *faith* in the system, which may reflect the human's more deeply held beliefs concerning the system's capabilities.

In their experiment, subjects were required to control a simulated orange juice pasteurization plant through either manual control or by assigning automatic control to each of three subsystems. Each subject operated the plant for 60 trials across three days. On the sixth trial of the second day and for all 20 trials on the third day, one of the subsystems failed to respond correctly. Subjects were divided into four groups based on the magnitude of this subsystem fault. Following each trial, their feelings concerning their trust in the system were evaluated through a series of computer-displayed questions related to predictability, dependability, and faith in the system (i.e., the hypothesized dimensions of trust) as well as to their overall trust in the system. Results indicated that while the subjects were generally able to quickly adapt to the occurrence of the chronic fault (day 3), as demonstrated by their performance, their trust in the system recovered more slowly. Also, while differences in the magnitude of the fault did not affect human performance, they did have a significant effect on trust, indicating that trust did not simply depend on the predictability dimension but that dependability was also a factor.

It should be noted that in this study, system faults also had the effect of disrupting the human's manual control of the system. One implication from this study was that the tendency to engage automatic control in response to such faults may be due more to the human's loss of confidence in manual control of the system than a result of increased trust in the system. In these situations, a delicate balance appears to exist between trust and self-confidence that can influence allocation decisions concerning the choice between manual or automatic control. Whether faults affect manual as well as automatic control and whether they are transient or chronic are just some of the factors we may now have to consider, especially in process control scenarios, if we want to gain further insight into the human's choice of control. Ultimately, the balance between trust in automatic control and self-confidence in assuming manual control is likely to be manifest in: (1) the human's variability in using manual and automatic control (which reflects the tendency to stay with a particular allocation strategy); and (2) the balance between manual and automatic control demonstrated under steady-state conditions (which reflects the degree of reliance on either manual or automatic control).

Overall, the more subtle considerations that have been discussed in the previous sections make one wonder whether we will ever be able to suggest, let alone adopt hard and fast rules concerning when humans should assume manual control or assign functions to automatic control. Given our general understanding of human capabilities and limitations and continued development of cognitive task analytic techniques that can help analyze and predict human behaviors within the contexts in which systems operate, increases in sophistication and reliability of model-based systems may ultimately lead to such suggestions. In any case, sociocultural and economic factors will likely continue to play a role in dynamic function allocation design guidelines, especially when the consequences of inappropriate responses are severe.

11.5.7 Example: Revisiting Flightdeck Design

The introduction of computer technology into the flightdeck provided very fundamental computer aiding in the form of integrated displays that impacted human monitoring and control functions. The human, however, was still very much in charge. Automation in the cockpit has continued to evolve by replacing more human functions, primarily those dealing with monitoring and controlling, with machine functions. Most of these developments first occur in military aircraft, with the benefits generally extending to commercial aviation at some later time. The Defense Advanced Research Project Agency's (DARPA) Pilot's Associate (PA) program represented an attempt to take cockpit automation to a completely new level by incorporating artificial intelligence techniques. These techniques can significantly increase the degree of autonomy in machine functions and thereby the potential for adaptive aiding, causing a dramatic shift in the function allocation balance equation in ways that may not be entirely clear.

One of the subcontractors to this program, Search Technology Inc., advocated an approach to automation in which the model-based system was tailored to the individual pilot's style and needs (Rouse, Geddes, and Hammer, 1990). Although the product of this research was intended for the Advanced Tactical Fighter being designed by Lockheed Corporation, the groundwork for development of this model-based system was actually based on a more generalized interface architecture for aiding humans in complex systems described by Rouse, Geddes, and Curry (1988). The overall architecture relies on a real-time dynamic knowledge base from which the states of the world (e.g., weather), the system (e.g., information on current and upcoming operational phases), and the human (e.g., the human's activities, information-processing resources given these activities, and intended goals and activities) can be assessed and predicted. This is indeed a tall order, and requires a multitude of closely linked software-based modules. The purpose of the aiding component of this interface was to provide aiding that adapts to the human's current needs and abilities as demands change or exceed human abilities. As was the case in some of the previous adaptive aiding approaches we discussed that underlie dynamic function allocation, this capability requires a model of the user. However, now we want such a model to be able to infer human intentions as reflected in the human's *goals*.

The roles of the intent interpretation module and other modules within the pilot–vehicle interface (pvi) developed for the PA program are illustrated in Figure 11.4. The ability to infer the pilot's intentions from such raw inputs as switch activations, engine status readings, and stick movements was based on a modeling structure we referred to earlier (Section 11.5.2) called *plan-goal* graphs. Basic actions by pilots in controlling the aircraft would be considered low level plans whose goals can be inferred by moving up the graph. These goals could then comprise the plans to achieve yet higher-order objectives. Through rule-based artificial intelligence techniques, the nodes making up this graph are linked in order to ultimately determine the pilot's intentions. This information can directly form the basis of displayed information, or be input to other modules as a basis for establishing adaptive aiding (Figure 11.4). Note that in our previous example on flightdeck design, computer-based displays were *normalized*, requiring that all pilots adapt to the same display. By incorporating a sophisticated model-based system, the function of display generation is now equipped with sufficient intelligence to enable displays to be *customized* to each pilot.

Knowledge of the pilot's intentions opens up many opportunities for the pvi. For example, it can detect control input errors through the error module and, following their evaluation, inform the pilot of these potential errors if the adaptive aiding module determines that it is in the best interest to so at that time. The pilot, however, is still in charge, and correcting these errors may be allowable if the pilot has authorized such actions by the pvi. Clearly, with such a design concept the adaptive aiding module can greatly influence dynamic function allocation. For example, if the pvi determines that the pilot does not intend to perform a task, then the task could get allocated to an automatic system. Based on assessments from the resource and performance modules, it could reallocate functions, perhaps in the form of "partitioning" (Andes, 1990)—for instance, attitude control might be allocated to the pilot while altitude control is allocated to the computer.

Model-based systems with such impressive potential capabilities are, however, far from immune to many of the considerations at issue in dynamic function allocation. For example, the consequences of model limitations would probably be accentuated if the technologies emerging from the PA program are adapted to commercial aviation, as is ultimately expected. High levels of automation in the current airbuses have been implicated in a variety of crashes that have been attributed to pilots misinterpreting what flightdeck computers are doing. In trying to override these computers, "wrestling matches" are, in effect, started that attempt to compensate for pilot actions (*The Economist*, 1995). Unlike the Boeing and McDonald Douglas Corporations, the design philos-

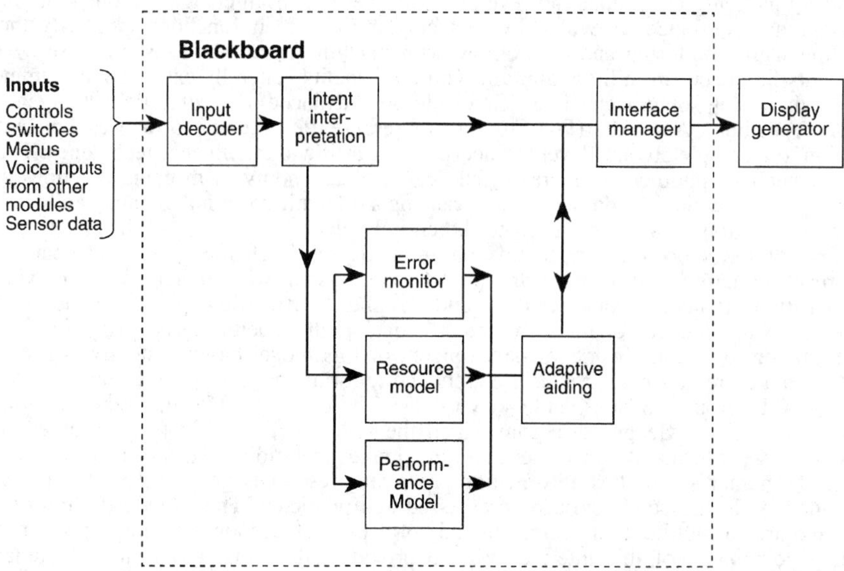

Figure 11.4 The pilot–vehicle interface (pvi) consists of expert-system software modules that interpret a pilot's actions, such as flipping a switch, and relays them via an input decoder to a plan–goal graph (not shown) in the intent interpretation module. The plan–goal graph is used to infer the pilot's intent, information which is passed to the interface manager or to three other modules: the error monitor, which checks for flying mistakes; the resource model, which estimates the pilot's mental workload; and the performance model, which predicts how long a pilot will take to complete a task. Adaptive aiding determines whether the pilot needs automated assistance from inputs from the error monitor, the resource model, or the performance model. The interface manager decides what display information the pilot needs before going to the display generator outside the pvi; a blackboard lets modules share information. (From Rouse, Geddes, and Hammer, 1990, © IEEE.)

ophy of Airbus Industrie is to shift the balance of function allocation toward its computers, allowing them to maintain the plane's speed or altitude at values they determine to be appropriate regardless of the pilot's intentions, creating the opportunity for unpredictable human–computer interactive sequences. In a similar manner, due to modeling limitations adaptive aiding based on intent inferencing could cause confusion for the pilot if under certain conditions the pilot's intentions do not match those inferred by the model. We also need to be concerned with how such mismatches could affect pilot trust and confidence and ultimately system performance, and the extent to which training can overcome these problems by making the complexity underlying this technology more transparent to the human.

11.6 ALLOCATION OF FUNCTIONS AS PART OF THE SYSTEMS DESIGN PROCESS

11.6.1 A Systems Design Methodology for Allocating Functions

Many human factors and ergonomics specialists have more or less acknowledged that systematic methods do not exist for allocating functions to humans and machines, especially in highly complex systems. Implicit to the general perspective of a static–dynamic *continuum* that has been taken in this chapter toward this issue is the assumption that function allocation is not generally a "zero-sum" problem whereby the choice between allocating functions to either human or machine is necessarily clear (Price, 1985). This idea is graphically illustrated in Figure 11.5, where the "goodness" in response to some performance demand is scaled from unsatisfactory (U) to excellent for both the human (h) and automation (a). Demands that fall into the U_{ah} region indicate the need for system redesign; those falling into the U_h or U_a regions are biased toward static allocation design perspectives favoring the machine or human, respectively; and demands

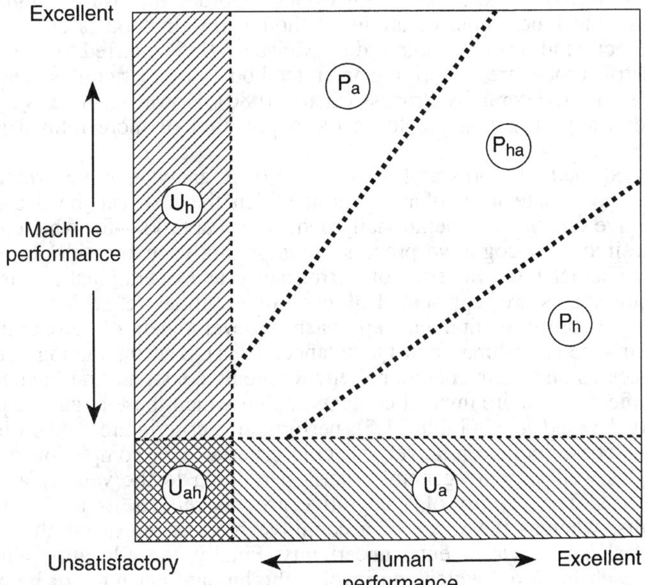

Figure 11.5 Decision matrix for allocation of functions. (From Price, Maisano, and Van Cott, 1982, reprinted by permission of the first author.)

in the P_a, P_h, and P_{ha} (where both human and machine can perform the function reasonably well) regions will offer the most design options, including the potential for dynamic function allocation. Whether such options ultimately are chosen will depend on considerations of the type we have previously encountered as well as on factors such as cost and time to implement.

Price's approach to allocation of functions is to treat this problem within a systems design perspective that is linked to formal procedures for system development. A strong emphasis is placed on assembling the appropriate mutidisciplinarian teams, performing a rigorous analysis of functions, hypothesizing function allocation design solutions, and iterating the design cycle in attempting to converge toward an acceptable solution. A number of rules are offered for supporting the design solution hypothesis stage. One such rule, "mandatory allocation," is meant to alert the designer to the need for recognizing exceptional circumstances in making allocation decisions. Aside from legal or labor stipulations in mandating control of certain functions, there may be other considerations. For example, limitations in the reliability of the engineered system and consequences of error may make the human responsible for implementing certain shutdown operations.

A second rule of particular interest refers to the need to allocate functions for "affective or cognitive support." Providing affective support acknowledges emotional requirements of humans in relation to their work, such as the need for perceiving their work as valuable and that they are exerting a reasonable level of control over work activities. Cognitive support encompasses many factors related to task and job design (see Chapter 14). In this discussion, it refers to design considerations that promote the development and updating of an adequate mental model (Gentner and Stevens, 1983), and that ensure an appropriate level of active involvement in the task so that human capabilities are maximized while at the same time reducing as much as possible the potentially negative consequences of human limitations.

11.6.2 Quantifying the Initial Cognitive Function Allocation Hypothesis

Although the use of Figure 11.5 can be helpful in formulating an initial allocation hypothesis (Price, 1985), in some cases analysts may desire a more quantitative assessment. Papantonopoulos and Salvendy (1995) have proposed a formal method for allocating cognitive functions that is potentially useful for evaluating allocation decisions that are likely to be subjected to iterative design cycles in order to ensure compatibility with higher-level sociotechnical considerations. Although this method is consistent with the approach to function allocation suggested by Meister (1985) referred to in Section 11.8.3, the psychometric procedure underlying their method is more rigorous, being based on Saaty's Task Analytic Hierarchy Process (Saaty, 1980). However, there is nothing inherent in Meister's approach that precludes the application of more refined psychometric methods.

The proposed method works as follows. First, task analysis is performed in order to determine an appropriate level of analysis at which the task can be decomposed into relevant cognitive functions—operationally defined as subtasks—for which allocation decisions are desired. The cognitive processes comprising each function are subsequently identified and characterized in terms of performance and operational requirements. Performance requirements are represented by cognitive abilities (e.g., deductive reasoning) that comprise the ability requirements approach to development of taxonomic systems of human performance (Fleishman and Quaintance, 1984). The interactions between these cognitive processes and their operational environments constitute the operational requirements and reflect the environmental contexts within which these cognitive processes are performed. At the next *level* (Table 11.5), performance criteria are defined for each cognitive process. These include relatively specific criteria (e.g., completion time and cost-effectiveness) as well as more general criteria pertaining to the validity and scope of a cognitive process. The *validity* of a cognitive process reflects its tendency to produce accurate results, whereas the *scope* of a cognitive process represents the capability for coping with task and environmental uncertainty. Finally, the allocation alternatives are identified for each function, which can involve the human, machine, or both.

The process of modeling the structure of a decision problem (Table 11.5) in and of itself is likely to result in a number of benefits, and probably accounts for experimental findings (Papantonopolous and Salvendy, 1995) indicating that subjects exposed to this process were more "balanced" (i.e., less biased) in assessing design trade-offs associated with making allocation decisions than subjects who were not provided with this modeling structure. In addition to these potential benefits, this model also provides the means for quantifying allocation decisions through application of the Analytic Hierarchy Process (see Chapter 37). Pairwise ratings are elicited through a ratio scale that, through the derivation of importance weights, provide the basis for prioritization of the cognitive processes (level 2) for performance of a specified function (level 1). Ratings are also elicited reflecting the extent to which each allocation alternative (level 4) can best meet the performance criteria (level 3) for each cognitive process. Multiplying the ratings (represented as normalized weights) for each decision alternative on each cognitive process by the importance weights of these processes provides a quantitative basis for assessing the relative advantages of each allocation alternative.

11.6.3 Example: Revisiting Flexible Manufacturing Systems

To illustrate some of the ideas in Price's (1985) approach to allocation of functions and, in particular, the principles of providing cognitive support and mandatory allocation, we revisit flexible manufacturing systems (FMSs). What we have to say, however, applies to

Table 11.5 A Multilevel Model for Assessing the Allocation of Cognitive Processes

Level 1	Identification of functions	Determine an appropriate level of analysis for task decomposition Perform task analysis
Level 2	Identification of cognitive processes required for each function	Identify performance requirements Identify operational requirements
Level 3	Specification of performance criteria	Specific criteria (e.g., cost, completion time) General criteria (validity, scope)
Level 4	Specification of function allocation decision alternatives	Human Machine Various H/M combinations

Source: Papantonopoulos and Salvendy (1995).

any type of cellular manufacturing arrangement (of which FMSs represent perhaps the most sophisticated type) that applies group technology concepts to support flexible manufacturing of one or more part families (Groover, 1987). This example will also provide the opportunity to examine the process of functional design (see Section 11.1.2).

Suppose the goals of a company are to determine the number of workers needed to operate such systems as well as to allocate functions to each of these workers. Consistent with the systems approach to design, there would be a strong initial emphasis on functional design and analysis. In order to identify possible functions, the overall system could be characterized in terms of relevant subsystems. In FMSs or FMCs (flexible manufacturing cells), examples of subsystems include: the scheduling subsystem, parts setup (i.e., set up and loading of parts onto machines), process operations (on parts and materials), the material handling subsystem (within and between cells or systems), the in-process inspection subsystem, tool and fixture inspection, tracking, and replacement susbsystems, part tracking and part quality tracking subsystems, and the part programming subsystem.

For each of these subsystems, general categories of functions now can be established and the primary functions within each category identified. Applying this process across all subsystems could conceivably lead to well over 500 individual functions that require allocation decisions. For example, functional categories within the scheduling subsystem include monitoring the adequacy of an assigned schedule; planning of daily, weekly, and monthly schedules; and dynamic rescheduling. For the material handling subsystems, functional categories include monitoring for adequacy, on-line adjustments, active (repair during breakdowns) and preventive (scheduled) maintenance, and functions associated with improving the performance of these subsystems. Individual functions associated with the dynamic rescheduling category for the scheduling subsystem might consist of information acquisition on potential problem sources, generation of alternative schedules, evaluation of results of new schedules, and integration of information on the effectiveness of new schedules into databases for possible use in future situations.

In theory, each of these individual functions can be performed by either human or machine. If the system has not been implemented, then a static function allocation design perspective (e.g., using a Fitts list approach) can often serve as a starting point. For operational systems constrained by a pre-existent technological infrastructure or by product demands that require at least a moderate degree of human participation, a number of the functions may be assigned to humans by default, by virtue of the inability (for whatever reason) to automate these functions. Although such a situation is far from ideal, the number and diversity of functions allocated to humans may still be sufficient to warrant the creation of meaningful jobs.

The process of function allocation as embedded within a systems design perspective can be refined further through iteration of the functional design and analysis stage in order to consider implications for human workload, task priority, and motivating potential as reflected in the degree of control the human has over each task. Essentially, these considerations serve to bridge the gap between function analysis and job design, and require that potential human functions be assessed in terms of factors such as: the frequency with which each task or function must be performed; the routineness of the task; attentional demands; temporal constraints; the degree to which skill-, rule-, or knowledge-based task performance is required; the degree to which activities are linked to individuals in other systems; the importance of the task to the organization's goals; and the relative degree of human and computer control with respect to human decision-making activities. Rules can then be developed that evaluate each of these factors for the various categories of human functions (e.g., information acquisition, tracking, monitoring) across all relevant subsystems, providing an overall assessment of job demands and capacity requirements. A number of these ideas reside in the HITOP-A system (Majchrzak and Glasser, 1992), a manufacturing systems design tool that, in addition to task allocation and job design, can also address such issues as training and organizational design.

At this stage, mandatory allocation policies based on the decision-making activities the organization desires the human to exercise control over can be determined. The organization may stipulate that the human assume primary responsibility for improvement and active maintenance functions while the computer, by virtue of its expert knowledge, be given greater authority in performing preventive maintenance. In further iterating the function allocation design cycle, the focus may shift to defining *clusters* of functions with the purpose of directing allocation decisions toward satisfying principles such as providing affective and cognitive support. For example, it may not make sense to dissociate the part tracking, part and material handling system monitoring, planning, schedule generation, and dynamic rescheduling functions if knowledge from any one of these functions is needed by the human for the development and maintenance of a mental model capable of supporting the other functions. If there is some compelling reason to allocate one or more of these functions to the machine (e.g., if a dynamic rescheduling algorithm with proven reliability exists), then care must be taken to ensure that appropriate human–machine interfaces have been designed with respect to these functions. Otherwise, the human's job can easily become *fragmented*, potentially resulting in decrements not only in performance but also in perceived job value.

The process of clustering functions may, in turn, force us to rethink our function analysis. For example, it may lead to the identification of new functions involving collaboration between maintenance personnel and persons tentatively assigned the functions above in order that the latter individuals can better plan schedules by incorporating models of machine unavailability. This collaboration may involve direct communication links or may be achieved indirectly, through new functions of computer-based record keeping required of maintenance personnel that can be easily accessed, interpreted, and integrated by workers performing scheduling and monitoring functions.

A closer examination of temporal demands, task priorities, and task complexity in yet further iteration of the function allocation design cycle may indicate the need for enhanced computerized decision support (see Chapter 42) and perhaps the possibility for implementing dynamic function allocation. This implies the possible need for trading off human autonomy and control over decision making with the potential consequences of exceeding human information-processing capacities.

Finally, if in fact the initial function analysis revealed the capability for automating many of the requisite functions and the possible need for a single person serving in a supervisory control capacity, then the predominant function allocation design effort can be concentrated on dynamic function allocation issues such as adaptive aiding. Overall, this exercise hopefully demonstrated that the resolution of function allocation problems in terms of movement within the boundaries of a static–dynamic continuum is very much consistent with a systems design perspective that embeds the process of function allocation within iterative design cycles.

11.6.4 Example: Visual Inspection Systems

Economic pressures have forced many industries to abandon sampling inspection plans in favor of 100% inspection in order to achieve quality standards that are generally beyond the detection capabilities of human inspectors. These demands, trends toward product miniaturization, as well as other factors have encouraged design of automated inspection systems. However, depending on the product, these systems may fare even worse than human inspectors (Drury and Sinclair, 1983). Complicating this problem is the fact that many inspection tasks represent performance demands that neither human nor machine are particularly well suited for; from a systems design standpoint (Figure 11.5), this predicament implies the need to rethink the design of inspection systems. Surprisingly, the idea of jointly utilizing human and machine capabilities in performing inspection tasks has not been considered in the past.

The notion of a *hybrid* inspection system was pursued by Hou, Lin, and Drury (1993). Design of such a system first required that the primary functions of an inspection task be identified. A task analysis of the inspection task indicated that search and decision making represented the two most important and difficult functions in inspection. With three possible inspection systems—human, machine (i.e., computer), and hybrid—and two distinct and sequentially performed functions (search and decision making), nine possible design alternatives exist. Taking into account the relative advantages of computers in performing search functions and of humans in performing decision-making functions, four systems were ultimately selected for investigation. In pure human inspection, the computer serves only to present the images to the human inspector. In the first of the two hybrid systems, the computer performs the search task; flaws that it detects are shown to the human who decides on their status. In the other hybrid scheme, the computer performs both the search and decision-making functions; however, if the computer is uncertain about its decision, the human intervenes in the decision-making process. Finally, a fully automated system was also tested.

The products inspected consisted of computer-generated 3-D height maps of electronic circuit boards. Flaws were denoted by missing components, wrong-sized components, and misaligned components; only a single flaw of any of these types could occur in a given board. Three factors were experimentally investigated: product complexity based on the number of components on the board, contrast level between circuit board components and background, and visual noise in the displayed image. Each of these factors was investigated at three levels, giving rise to 27 inspection scenarios. Three subjects were assigned to each of the three inspection system designs that called for human involvement and were required to inspect 20 circuit boards in each of the 27 conditions. The performance criteria consisted of speed (seconds per board inspected) and accuracy (hits and false alarms).

Given the different strengths and limitations of current approaches to computer-based inspection, two different computer algorithms were developed, providing for two different types of fully automated systems (giving rise to a total of five different inspection systems). In addition, each was better suited and therefore implemented in one of the two hybrid systems: A rule-based template matching system was used for computer search in the case where the human was solely responsible for decision making and a neural net model-based system was used when decision-making responsibility was shared by the human and computer. In the latter case, when the model could not classify patterns in terms of acceptance or rejection, they were passed on to the human for further examination.

With inspection accuracy as the criterion, the computer-search human decision hybrid system proved to be the best function allocation policy across most conditions. The template-matching automated system, however, performed the inspection task the fastest. In order to consider both criteria, a cost-based evaluation function that expresses the inspection cost per board as a weighted function of the speed and accuracy performance criteria, was used to generate iso-cost contours that enabled system function allocation policies that *dominated* others to be identified. These results indicated that the computer-search human-decision hybrid system was faster and more accurate than either the purely human or the joint human–computer decision-making inspection system.

The possibility also existed for predicting the most appropriate inspection system for *new* (i.e., untested) inspection situations by modeling the relationships between performance and the various inspection conditions (i.e., factor levels) in an approach analogous to that adopted by Morris and Rouse (1988). A neural net model was used for predicting inspection accuracy and time and, through the use of the cost-based evaluation function, for predicting inspection system allocation. When compared to actual allocations observed, this model-based approach to dynamic inspection system allocation did not achieve results that would prove useful in practice, although it was far superior to a scheme in which components were allocated to inspection systems on a random basis. Even if model-based limitations are ultimately overcome, there are many other factors that would need to be evaluated before considering abandoning the static allocation result in favor of a dynamic allocation policy. However, for this application area, achieving a useful *hybrid* static system allocation policy is in and of itself a significant improvement over the previous more conventional static allocation policies in which inspection was performed either manually or through automation. The notion of a model-based dynamic *system allocation* policy, whereby the system can be comprised of either the human or machine agent as well as hybrid arrangements, offers yet another perspective to dynamic function allocation, and one that evolves naturally from the iterative function allocation design cycle in systems design.

11.7 GUIDELINES FOR ALLOCATION OF FUNCTIONS IN MANUFACTURING ENVIRONMENTS

Despite strong arguments that can be offered in support of the unique role of humans in manufacturing (especially when the information requirements for decision making represent high levels of detail) as well as for the need for automating various operations, in the more broad-based "hybrid" manufacturing systems a large number of activities will likely continue to exist that could be performed by either humans or machines. Recognizing that the allocation of functions in these environments is a multilevel process, where initial allocation decisions must be subsequently considered within the context of job design (as discussed in Section 11.6.3) and workgroup objectives, Mital, Motorwala, Kulkarni, Sinclair, and Siemieniuch (1994a, 1994b) have developed a set of systematic guidelines intended for making initial function allocation decisions.

The centerpiece of this approach consists of an array of interlinked flow charts. Initialization of the function allocation process occurs within a flow chart that requires answers to a set of generic questions directed at each of six classes of activities that can be performed by either humans or machines—material handling, material processing, inspection/monitoring/supervision, assembly, packaging, and shipping. The questions address the following: (1) the requirements of complex decision making and experience; (2) whether humans are physically capable of performing the activity; (3) safety considerations to determine the potential for injury to humans; and (4) economic considerations.

Further elaboration of the function allocation process results from links to more specialized flow charts that provide details concerning safety analysis, economic analysis, and the relative capabilities and limitations of humans and automated equipment for performing each of the six activities. For example, the safety analysis identifies the major risks resulting from various hazards in manufacturing operations and the methods required for reducing these hazards. If any of these risks exceed acceptable limits and means are not available for reducing these risks to acceptable levels, then the activity is allocated to automated equipment.

The activity-specific flow chart for the materials processing activity class is shown in Figure 11.6. The circle containing the symbol 'A2' refers to the second of the six classes of activities listed in a master flow chart; this flow chart is referred to upon both entry into and exit from each of the activity-specific flow charts. The circle containing the symbol '1' refers to a list provided by Mital et al. (1994a) of day-to-day activities that occur on the shopfloor from which the six activity-specific classes were derived, and implies the need to automate one or more activities. Note that safety and economic considerations, which are analyzed in separate flow charts, are cross-referenced in this (and other) activity-specific flow charts in the form of questions that address these concerns. Although the authors recommend that these flow charts be treated as checklists, they caution that the information required and the criteria used for making the decisions referred to in the flow charts cannot always be specified because these are often situation specific and require a level of detail that would undermine the benefits general guidelines typically provide.

11.8 SOCIOTECHNICAL PERSPECTIVES TO FUNCTION ALLOCATION

11.8.1 Micro- and Macroergonomic Design

If we choose to characterize the discipline of human factors and ergonomics as a human–systems interface technology, distinctions between microergonomics and macroergonomics can be made on the basis of the various ergonomic interface technologies they encompass. Microergonomics is generally assumed to incorporate human–machine interface technology, human–environment interface technology, and user-interface technology (or software ergonomics) while macroergonomics comprises the organization–machine interface (Hendrick, 1991). In theory, macroergonomics represents a sociotechnical systems approach to work systems design that, through its top-down perspective, essentially embodies the interface technologies that comprise microergonomics. Design of an organization's structure requires consideration of the critical aspects of these microergonomic subsystems. Taking this idea a step further, it may be reasonable to assume that the sociotechnical systems approach to organizational design *prescribes* microergonomics design (Hendrick, 1991). This idea is certainly gaining momentum in the area of human reliability analysis and error prevention (Center for Chemical Process Safety, 1994; Reason, 1990; Sharit, 1993) where there is a growing emphasis on how organizational structure and management policies ultimately translate into situational contexts conducive to human error and violations. It is also typified in continuous quality and safety assurance programs that have established effective data-collection mechanisms and feedback channels for worker communication.

Sociotechnical system perspectives have not been entirely ignored to this point. The view of allocation of functions as embedded within the systems design process, and many of the considerations that emerged in our revisit of flexible manufacturing were consistent with this perspective. A methodological framework for function allocation that is, however, more faithful to this perspective has been developed (Clegg, Ravden, Corbett, and Johnson, 1989) as part of the ESPRIT program (European Strategic Program for Research and Development in Information and Technology). Their work also involved computer-integrated manufacturing systems, and their methodology encompassed many of the features of Price's (1985) approach (see Section 11.6.1). The key difference, in their opinion, is that their method considers broader organizational issues within the context of system requirements specification and the development of corresponding scenarios that serve to demonstrate how the system could be operated. Prior to presenting this approach, it may be instructive at this point to consider some of the shortcomings of other "methods" for

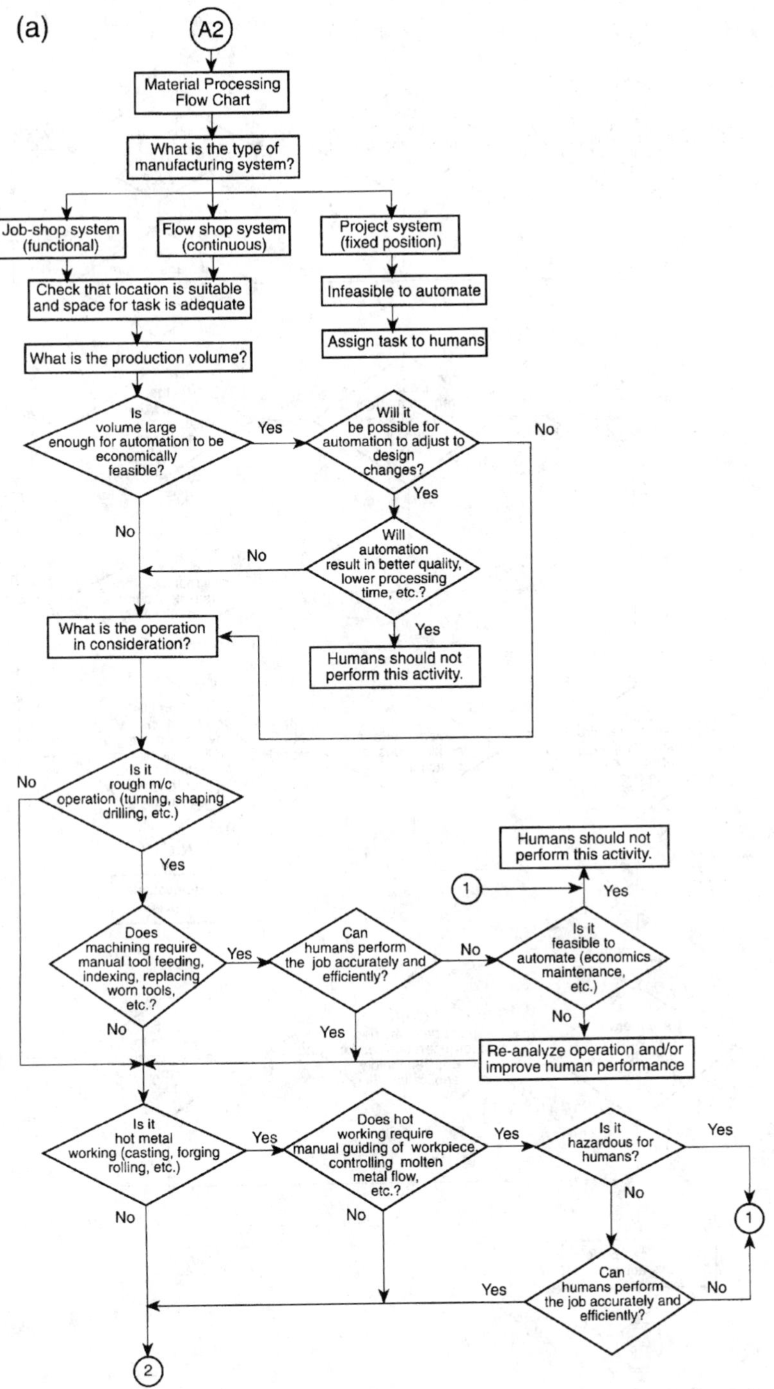

Figure 11.6 The flow chart for the materials processing activity class. (From Mital, Motorwala, Kulkarni, Sinclair, and Siemieniuch, 1994a, reprinted by permission of Elsevier Science.)

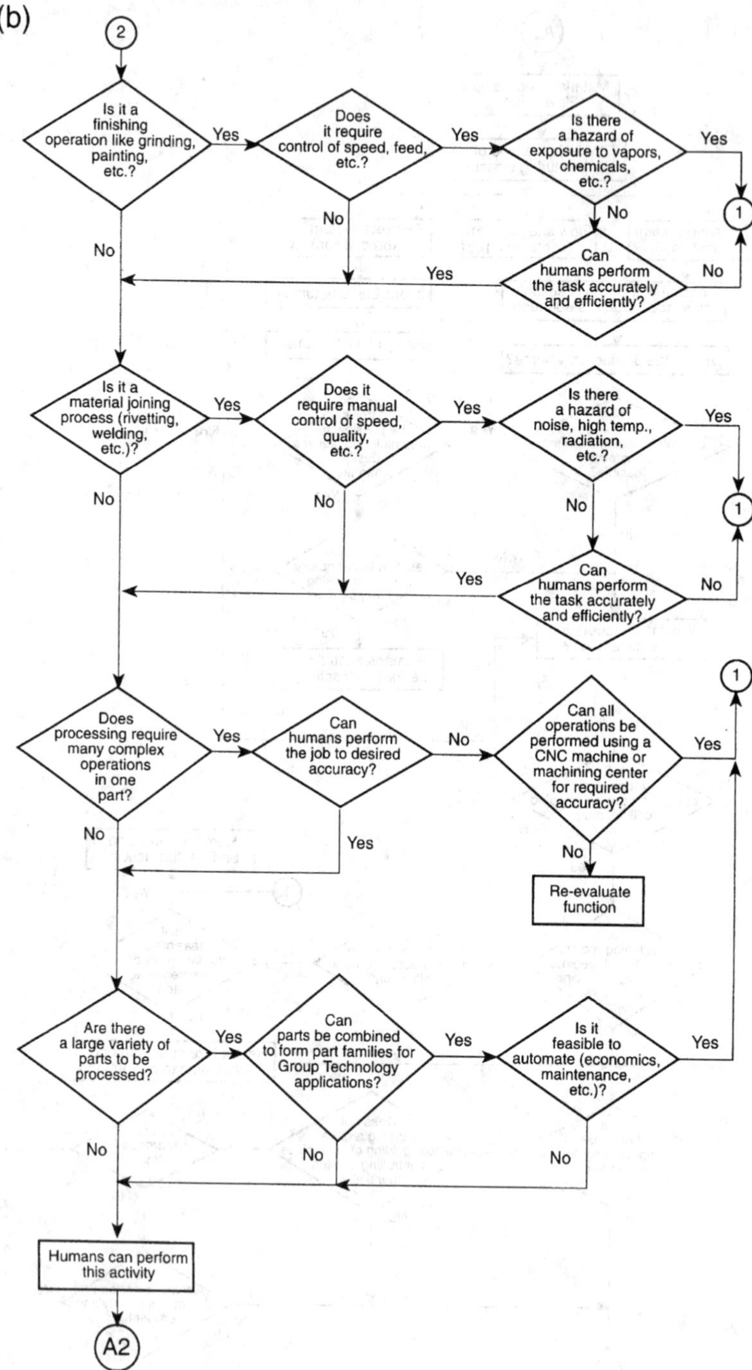

Figure 11.6 (Continued).

allocating functions, as viewed by Clegg et al. Although some of these are, more or less, implied throughout this chapter, a synthesis of these shortcomings is appropriate at this juncture given that they potentially reflect an absence in links between microergonomic issues and macroergonomic design.

11.8.2 Some Criticisms of Function Allocation Methods

When the allocation of functions to machine appears both technically feasible and affordable, designers tend to favor employing technology over humans, perhaps due the belief that humans are too fallible to be entrusted with the task of maintaining safe and reliable operations in complex systems. Unfortunately, designers can easily underestimate system complexity, especially in the early stages of design, and fail to anticipate numerous situational contexts that would not only benefit from, but require human involvement. Furthermore, if critical human skills are not taken into account at the function allocation stage of systems design, improvements in the technical system that could benefit from these skills as well as from organizational requirements are also left unspecified early in design, and therefore become more difficult to achieve. This "technology-driven" approach can also adversely shape the human's job by virtue of constraining the set of functions available to the human, and thereby creating the potential for demand-capacity mismatches (Kantowitz and Sorkin, 1987).

Static allocation of function approaches that often rely on lists as a basis for allocation decisions (see Section 11.3.1) are criticized by Clegg et al. on a number of grounds. First, as any sociotechnical perspective would be apt to note, they tend to ignore the broader organizational and cultural issues, as well as financial, economic, and psychological considerations, which may suggest changes in function allocation to the human that were not implied by an analysis based strictly on comparability. Moreover, the performance data on which the lists are based can often be called into question, especially if the data is lacking in generalizability—whether from the laboratory to the work environment, or from one job setting to the next. This reinforces what we have been advocating to this point, namely that an analysis of human capabilities and limitations for the purpose of supporting function allocation decisions should, especially in complex systems, be performed within the context of a detailed cognitive task analysis of work scenarios. Criticisms raised by Price (1985) pertaining to the assumption that allocation decisions as determined by the static function allocation perspective represent a zero-sum problem, and that by neglecting to embed allocation decisions within the iterative process of systems design important implications for function allocation are likely to be lost, have already been discussed in Section 11.6.1.

Dynamic function allocation schemes are criticized by Clegg et al. for emphasizing primarily performance considerations while ignoring other potentially relevant factors. More generally, they argue that these approaches do not provide systematic guidance concerning when allocations should be made or how they should be implemented in practice. Finally, Clegg et al. feel Price's approach, while systematic, still overly emphasizes performance criteria, and ultimately results in allocations to either humans or machines—i.e., based on comparability.

Some of these criticisms are, however, not entirely fair. The issue of when dynamic function allocations should be made is essentially subsumed within issues motivating the consideration of this function allocation perspective. Also, how dynamic function allocations can be achieved in practice will depend largely on the extent to which many of the considerations discussed in Section 11.5 (e.g., model-based limitations) can be resolved and the particular application area. With respect to Price's approach, comparability between human and machine is not explicitly advocated, and sufficient flexibility for incorporating dynamic function allocation solutions is available through reexamination of allocation decisions during the iterative course of design.

11.8.3 A Broader Perspective to Allocation of Functions

The methodology for allocating functions proposed by Clegg, Ravden, Corbett, and Johnson (1989) alluded to in Section 11.8.1 was guided by a general set of requirements (Table 11.6) that were believed would promote a method that would prove more usable in practice to the extent that these requirements could be satisfied. The overall method is summarized in Figure 11.7 in terms of seven phases. Phase 1 is essentially the mission statement in systems design. Although phase 2 is also fundamental to conventional systems design, in this method it is supplemented by: (1) a set of human factors requirements directed at considerations such as allocation of functions but that also included such

Table 11.6 Requirements of Function Allocation Methods

(a) ***Systematic:*** Any method should have some procedures or steps through which the designer can work in a specified order. This does not prescribe rigidity, but does allow for a methodical approach. This is highly valued by designers.

(b) ***Criterion-based:*** Choices and decisions should be guided by specified criteria, which can be stated and defined. As with criteria generally, their specification provides an opportunity for debate and for refinement. This is especially important in multidisciplinary design teams.

(c) ***Multidimensional:*** This refers to the need for including considerations which are not solely related to short run performance efficiency. Financial, psychological and wider organizational aspects may be important, and they should be included. This notion also incorporates the idea that some of these considerations may conflict with one another, and therefore the method needs to build in recognition of this.

(d) ***Capable of handling large- and small-scale functions:*** The method needs to help with decisions about major system functions such as transportation in a large system, and also with those relating to more specific aspects, such as how a particular part moves from one location to another.

(e) ***Iterative:*** Recognizing that designers work iteratively, it is important that the method can handle iterations, and indeed allow for changes in decisions as design progresses. Furthermore, given the way designers work, the method should allow for quick iterations.

(f) ***Linked to earlier and later design decisions:*** It should be clear how the method fits into early choices about the requirements specification of the system and how people want the system to work, and also how it relates to later detailed choices, for example about designing jobs for human operators. This is the next phase in the design process whereby the tasks and roles for humans are allocated to individuals and groups. This linking into the design process more generally is potentially very useful for designers.

(g) ***Face valid in an organizational context:*** The method should seem valid to its potential users (i.e., the designers) when faced with the practical difficulties of allocating functions in real settings.

(h) ***Promote participation:*** A method allocating functions should allow and encourage non-experts to participate in these decisions. This makes the issue less of a black art within the control of the cognoscenti. In practice we believe this will be fostered by adoption of the previous requirements.

Source: Clegg, Ravden, Corbett, and Johnson (1989), reprinted by permission of Taylor & Francis Ltd.

factors such as organizational structure, job design, and health and safety; and (2) a set of alternative scenarios that were developed to illustrate to system designers how the system could be operated, and which reflected to varying extents the degree to which human factors requirements were being met. Cost–benefit analyses were also performed on the scenarios adopted as well as on those rejected in order to clarify to designers the relative strengths of the proposed solutions.

It is worth noting that the idea of exploring design alternatives could conceivably be applied as well directly to the function allocation problem, whereby the alternative scenarios represent different function allocation arrangements. Assuming one has established a set of system criteria (e.g., cost, safety, maintainability), the relative weights of these criteria as well as of the design alternatives could be evaluated through psychometric methods to ultimately provide a quantitative function allocation design solution. This approach to allocation of function was in fact suggested by Meister (1985).

The functional specification in phase 3, which precedes function allocation, also requires consideration of alternative system designs, though initially these would not need to be as detailed as the scenarios proposed in phase 2. System design details could be filled in following selection of a preferred alternative, which, in turn, is influenced by human factors requirements and scenario selection from phase 2. Next, the functions identified are checked against constraints enabling initial mandatory allocations to be made. Aside from the more commonly encountered constraints based on legal considerations (e.g., a governing agency dictating that system shutdown should be performed manually), mandatory allocations could arise by default; for example, the objectives in phase 1 may stipulate incorporating technology (such as a robot) that results in allocation decisions for functions associated in various ways with that technology (e.g., positioning a part on a pallet in order for the robot to recognize it).

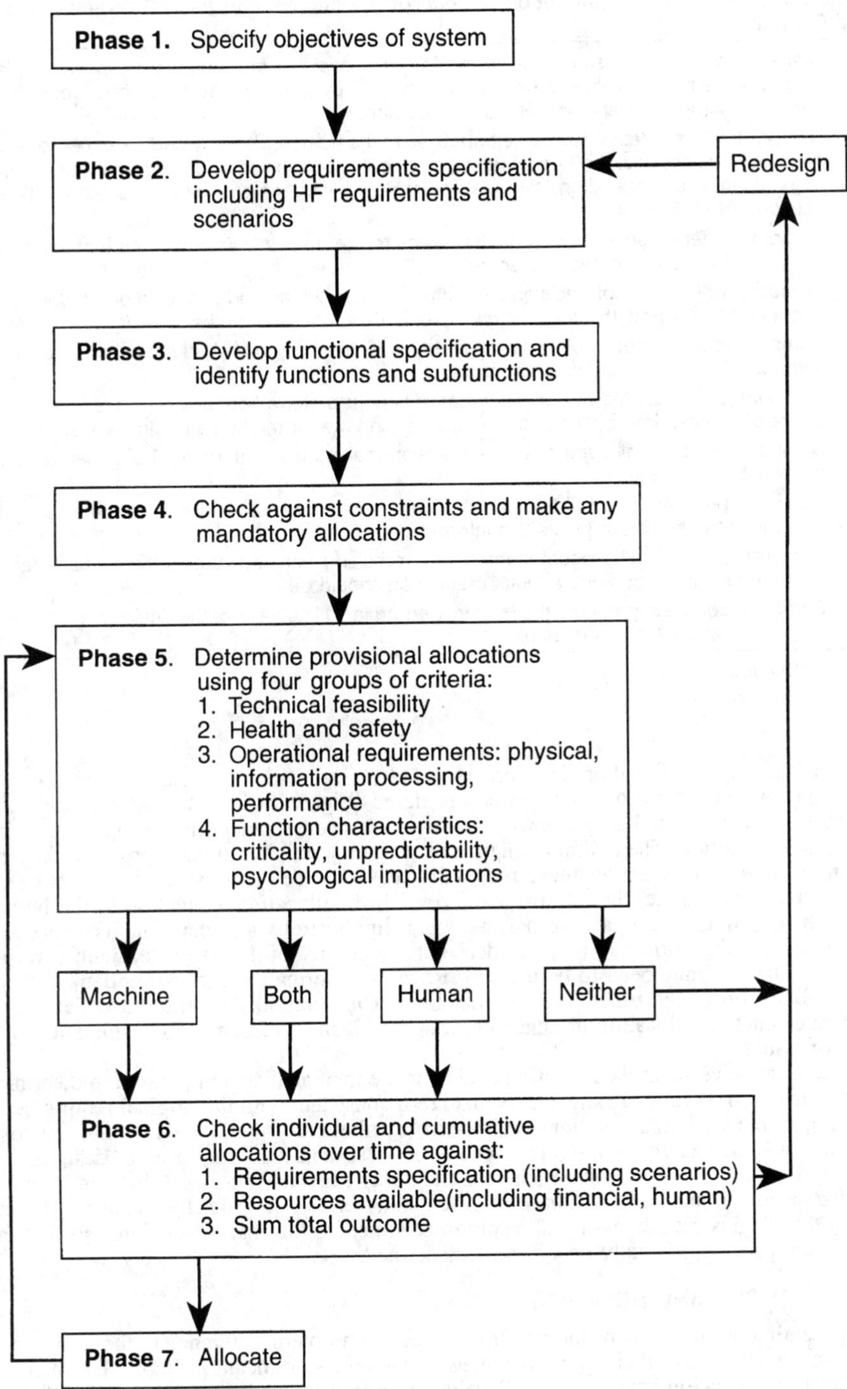

Figure 11.7 A broad-based methodology for allocation of functions. (From Clegg, Ravden, Corbett, and Johnson, 1989, reprinted by permission of Taylor and Francis Ltd.)

Table 11.7 Levels in the Distribution of Decision-Making Responsibility Between Human and Computer

1. Human is solely responsible for all aspects of the decision-making problem including planning, structuring, modeling, generation of decision options, and selection of a decision alternative; computer merely executes the decision.
2. Computer offers a set of decision alternatives to the human who may choose either to ignore them or to select one; the human is still involved in all aspects of carrying out the decision (e.g., planning any operations that may be required) with the exception of deterministic execution of the decision.
3. Computer offers a set of decision alternatives to the human from which the human selects an option for the computer to execute.
4. Computer offers a set of decision alternatives to the human and suggests one of them; the human either accepts this suggestion or decides on one of the other alternatives offered.
5. Computer offers a set of decision alternatives to the human and suggests one that it will carry out, but only if the human approves.
6. Computer generates the decision alternatives and a preferred option and makes the human aware of its intention to carry out this decision in time for the human to intervene.
7. Computer performs all aspects of the decision-making task, informing the human after the fact what it has done.
8. Computer performs all aspects of the decision-making task, but informs the human after the fact only if the human requests this information.
9. Computer performs all aspects of the decision-making task and informs the human after the fact, but only if the computer thinks the human should be told.
10. Computer completely ignores the human, performing all aspects of the decision making task only if and when it thinks it should.

Source: Sheridan (1980).

The majority of allocation decisions are likely to be made in phase 5 where for each function four categories of criteria are considered (Figure 11.7). Fitts list approaches to function allocation are likely to prove useful only when considering the operational requirements criterion. The assimilation of information pertaining to these groups of criteria in order to arrive at tentative allocations will necessarily be based on the judgment of the designers. For example, the advantages derived from allocating a function to the human through its provision of positive psychological implications and the benefits that could accrue due to the human's ability to deal with unpredictability associated with carrying out that function may need to be traded off against information-processing demands that favor allocation of the function to the machine. The solution may entail a collaborative dynamic function allocation arrangement that could, in turn, demand consideration of a host of issues.

These tentative function allocation decisions are then analyzed in phase 6 to determine if they result in working arrangements that are inconsistent with the original requirements specification. Other considerations such as the distribution of workload and financial costs are also assessed in this phase. The results of these analyses may prompt designers to rethink the design process, in which case they may need to consider a different requirements specification. The final phase represents an iterative loop back to phase 5 (Figure 11.7) that begins by addressing allocation decisions for the major functions, so that the allocation process gradually becomes more defined.

11.9 CONCLUDING REMARKS

In presenting an overview of the function allocation problem, reference to the term *methods* was generally avoided. By no means was this omission meant to imply that methods of function allocation have not been developed; indeed they have, and a number of them have been discussed in this chapter. The reason for favoring expressions such as "approaches" and "perspectives" was to emphasize the more general sociotechnical and systems design frameworks within which this problem should be initiated and governed, as well as the more subtle considerations that ultimately dictate where along the static–dynamic continuum solutions to function allocation problems are likely to lie.

The particular emphasis given to issues in dynamic function allocation is primarily in anticipation of continued developments in computer hardware and software and other technologies that potentially translate into rather exotic model-based human–machine arrangements. Though the selection of an appropriate arrangement from the types of options listed in Table 11.7 may currently be far from straightforward (Levis, Moray, and Hu, 1994), through further identification and understanding of the more subtle considerations that underlie dynamic function allocation and through further development and application of techniques such as cognitive task analysis and human error prediction, our understanding of this problem should gradually improve.

However, this understanding will likely be undermined if a concurrent effort is not undertaken toward understanding the impact on function allocation of organizational and sociocultural factors. Finally, the identification and implementation of a successful function allocation design strategy will still very much depend on factors such as the generation of creative system descriptions, the selection and use of appropriate function analysis techniques, and incorporating procedures consistent with the basic principles associated with the process of systems design.

REFERENCES

Andes, R. C. Jr. (1990). Adaptive aiding automation for system control: Challenges to realization. *Proceedings of the Topical Meeting on Advances in Human Factors Research on Man/Computer Interactions: Nuclear and Beyond,* 304–310.

Banks, W. W., Hunter, S. L., and Noviski, O. J. (1985). *Human Factors Engineering: Display Development Guidelines.* Lawrence Livermore Laboratory, UCID-20560.

Bekey, G. A. (1970). The human operator in control systems. In K. B. DeGreene, Ed., *Systems Psychology,* New York: McGraw Hill.

Bond, N. A. (1987). Maintainability. In G. Salvendy, Ed., *Handbook of Human Factors.* New York: John Wiley.

Center for Chemical Process Safety (1994). *Guidelines for Preventing Human Error in Process Safety.* New York: American Institute of Chemical Engineers.

Chapanis, A. (1967). The relevance of laboratory studies to practical situations. *Ergonomics, 10,* 557–577.

Clegg, C., Ravden, S., Corbett, M., and Johnson, S. (1989). Allocating functions in computer integrated manufacturing: A review and a new method. *Behaviour and Information Technology, 8,* 175–190.

DeMarle, D. J., and Shillito, M. L. (1992). Value engineering. In G. Salvendy, Ed., *Handbook of Industrial Engineering* (Second Edition). New York: John Wiley.

Drury, C. G., and Sinclair, M. A. (1983). Human and machine performance in an inspection task. *Human Factors, 25,* 391–399.

The Economist (1995). So who's the pilot here? April 8, 71.

Fitts, P. M., Ed. (1951). *Human Engineering for an Effective Air-Navigation and Traffic-Control System.* Columbus Ohio: Ohio State University Research Foundation.

Freedy, A., Madni, A., and Samet, M. (1985). Adaptive user models: Methodology and applications in man-computer systems. In W. B. Rouse, Ed., *Advances in Man-Machine Systems Research* (Volume 2). Greewich and London: JAI Press.

French, S. (1986). *Decision Theory: An Introduction to the Mathematics of Rationality.* New York: Halsted Press.

Gentner, D., and Stevens, A. L. (1983). *Mental Models.* Hillsdale, NJ: Lawrence Erlbaum Associates.

Gordon, S. E. (1995). Cognitive task analysis using complimentary elicitation methods. *Proceedings of the Human Factors and Ergonomics Society 39th Annual Meeting* (San Diego, California), 525–529.

Greenstein, J. S., and Revesman, M. E. (1986). Development and validation of a mathematical model of human decision making for human-computer communication. *IEEE Transactions on Systems, Man, and Cybernetics, 16,* 148–154.

Greenwood, N. R. (1988). *Implementing Flexible Manufacturing Systems.* New York: John Wiley.

Groover, M. P. (1987). *Automation, Production Systems, and Computer Integrated Manufacturing.* Englewood Cliffs, NJ: Prentice-Hall.

Hendrick, H. W. Ergonomics in organizational design and management. *Ergonomics, 34,* 743–756.

Hou, T-H, Lin, L., and Drury, C. G. (1993). An empirical study of hybrid inspection systems and allocation of inspection functions. *International Journal of Human Factors in Manufacturing, 3,* 351–363.

Hwang, S-L., Barfield, W., Chang, T-C., and Salvendy, G. (1984). Integration of humans and computers in the operation and control of flexible manufacturing systems. *International Journal of Production Research,* 22, 841–856.

Jordan, N. (1963). Allocation of functions between man and machines in automated systems. *Journal of Applied Psychology*, *47*, 161–165.
Kadota, T., and Sakamoto, S. (1992). Methods analysis and design. In G. Salvendy, Ed., *Handbook of Industrial Engineering* (Second Edition). New York: John Wiley.
Kantowitz, B. H., and Casper, P. A. (1988). Human workload in aviation. In E. L. Weiner and D. C. Nagel, Eds., *Human Factors in Aviation*. San Diego: Academic Press.
Kantowitz, B. H., and Sorkin, R. D. (1987). Allocation of functions. In G. Salvendy, Ed., *Handbook of Human Factors*. New York: John Wiley.
Kessel, C. J., and Wickens, C. D. (1982). The transfer of failure-detection skills between monitoring and controlling dynamic systems. *Human Factors*, *24*, 19–60.
Laughery, K. R. Sr., and Laughery, K. R. Jr. (1987). Analytic techniques for function analysis. In G. Salvendy, Ed., *Handbook of Human Factors*. New York: John Wiley.
Lee, J., and Moray, N. (1992). Trust, control strategies and allocation of function in human-machine systems. *Ergonomics*, *35*, 1243–1270.
Levis, A. H., Moray, N., and Hu, B. (1994). Task decompositoin and allocation problems and discrete event systems, *Automatica*, *30*, 203–216.
Majchrzak, A., and Gasser, L. (1992). HITOP-A: A tool to facilitate interdisciplinary manufacturing systems design. *International Journal of Human Factors in manufacturing*, *2*, 255–276.
Meister, D. (1985). *Behavioral Analysis and Measurement Methods*. New York: John Wiley.
Meister, D. (1991). *Psychology of System Design*. Amsterdam: Elsevier.
Mital, A., Motorwala, A., Kulkarni, M., Sinclair, M., and Siemieniuch, C. (1994a). Allocation of functions to humans and machines in a manufacturing environment: Part I—Guidelines for the practitioner. *International Journal of Industrial Ergonomics, 14*, 3–29.
Mital, A., Motorwala, A., Kulkarni, M., Sinclair, M., and Siemieniuch, C. (1994b). Allocation of functions to humans and machines in a manufacturing environment: Part II—The scientific basis (knowledge base) for the guide. *International Journal of Industrial Ergonomics, 14*, 33–49.
Morris, N. M., and Rouse, W. B. (1988). Studies of dynamic task allocation in an aerial search environment. *IEEE Transactions on Systems, Man, and Cybernetics*, *18*, 376–389.
Nof, S. Y., Knight, J. L., and Salvendy, G. (1980). Effective utilization of industrial robots—a job and skill analysis approach. *AIIE Transactions*, *12*, 216–225.
Norcio, A. F., and Stanley, J. (1989). Adaptive human-computer interfaces: A literature survey and perspective. *IEEE Transactions on Systems, Man, and Cybernetics*, *19*, 399–408.
Papantonopoulos, S. A., and Salvendy, G. (1995). A decision analytic model for cognitive task allocation. *Unpublished Report.* West Lafayette, Indiana: Purdue University.
Pinedo, M. (1992). Scheduling. In G. Salvendy, Ed., *Handbook of Industrial Engineering* (Second Edition). New York: John Wiley.
Price, H. E. (1985). The allocation of functions in systems, *Human Factors*, *27*, 33–45.
Price, H. E., Maisano, R. E., and VanCott, H. P. (1982). The allocation of functions in man-machine systems: A perspective and literature review. *NUREG-CR-2623*, Oak Ridge, Tennessee: Oak Ridge National Laboratory.
Rasmussen, J. (1986). *Information Processing and Human-Machine Interaction: An Approach to Cognitive Engineering*. New York: North-Holland.
Rasmussen, J., Pejtersen, A. M., and Goodstein, L. P. (1994). *Cognitive Systems Engineering*. New York: John Wiley.
Reason, J. (1990). *Human Error*. New York: Cambridge University Press.
Revesman, M. E., and Greenstein, J. S. (1986). Application of a mathematical model of human decision making for human-computer communication. *IEEE Transactions on Systems, Man, and Cybernetics*, *16*, 142–147.
Roth, E. M., and Woods, D. D. (1989). Cognitive task analysis: An approach to knowledge acquistion for intelligent system design. In G. Guida and C. Tasso, Eds., *Topics in Expert System Design*. Amsterdam: Elsevier Science Publishers.
Roth, E. M., and Mumaw, R. J. (1995). Using cognitive task analysis to define human interface requirements for first-of-a-kind systems. *Proceedings of the Human Factors and Ergonomics Society 39th Annual Meeting*. California: San Diego, 520–524.
Rouse, W. B. (1976). Adaptive allocation of decision making responsibility between supervisor and computer. In T. B. Sherian and G. Johannsen, Eds., *Monitoring Behavior and Supervisory Control*. London: Plenum Press.
Rouse, W. B., and Hammer, J. M. (1991). Assessing the impact of modeling limits on intelligent systems. *IEEE Transactions on Systems, Man, and Cybernetics*, *21*, 1549–1559.
Rouse, W. B., Geddes, N. D., and Curry, R. E. (1988). An architecture for intelligent interfaces: Outline of an approach to supporting operators of complex systems. *Human-Computer Interaction*, *3*, 87–122.
Rouse, W. B., Geddes, N. D., Hammer, J. M. (1990). Computer-aided fighter pilots. *IEEE Spectrum*. March, 38–41.
Saaty, T. L. (1980). *The Analytic Hierarchy Process*. New York: McGraw Hill.

Schryver, J. C., and Knee, H. E. (1987). Integrated operator-plant process modeling and decision support for allocation of function. *Proceedings of the Human Factors Society 31st Annual Meeting*. New York City, 815–819.

Sharit, J. (1993). Human reliability modeling. In K. B. Misra, Ed., *New Trends in System Reliability Evaluation*. Amsterdam: Elsevier Science Publishers.

Sharit, J., and Elhence, S. (1989). Computerization of tool-replacement decision making in flexible manufacturing systems: A human-systems perspective. *International Journal of Production Research*, *27*, 2027–2039.

Sharit, J., and Elhence, S. (1990). Allocation of tool-replacement decision-making responsibility in flexible manufacturing systems. *International Journal of Industrial Ergonomics*, *5*, 29–46.

Sheridan, T. B. (1980). Theory of man-machine interaction as related to computerized automation. In E. J. Kompass and T. J. Williams, Eds., *Man-Machine Interfaces for Industrial Control*. Barington, IL: Technical Publishing Company.

Sheridan, T. B. (1993). *Telerobotics, Automation and Human Supervisory Control*. Cambridge, Mass.: The MIT Press.

Swain, A. D., and Guttmann, H. E. (1983). *Handbook of Human Reliability Analysis with Emphasis on Nuclear Power Plant Applications*, NUREG/CR-1278 (Washington, D.C.).

Vicente, K. J. (1995). Task analysis, cognitive task analysis, cognitive work analysis: what's the difference? *Proceedings of the Human Factors and Ergonomics Society 39th Annual Meeting* (San Diego, California), 534–537.

Vicente, K. J., and Rasmussen, J. (1992). Ecological interface design: Theoretical foundations. *IEEE Transactions of Systems, Man, and Cybernetics*, *22*, 589–607.

Weiner, E. L., and Nagel, D. C., Eds. (1988). *Human Factors in Aviation* San Diego: Academic Press.

Woods, D. D., and Roth, E. M. (1988). Cognitive systems engineering. In M. Helander, Ed., *Handbook of Human-Computer Interaction*. Amsterdam: North Holland.

Woods, D. D., Roth, E. M., and Pople, H. (1988). Modeling human intention formation for human reliability assessment. *Reliability Engineering and System Safety*, *22*, 169–200.

CHAPTER 12
TASK ANALYSIS

Holger Luczak
Institute of Industrial Engineering and Ergonomics
D-52 062 Aachen, Germany

12.1 IDEA AND IMPORTANCE OF TASK ANALYSIS

12.1.1 Etymology

The fundamental idea of task analysis lies in a science-based and purpose-oriented method or procedure to determine, what kind of elements the respective task is composed of, how these elements are arranged and structured in a logical, or/and timely order, how the existence of a task can be explained or justified, what the driving force to generate it was, and how the task or its elements can be aggregated to another entity, composition, or compound. Hence the idea of analysis is just the same as that in natural sciences: The classical analysis of a material, for example, gives evidence, on what environmental conditions it exists as gas, liquid, or solid, if it is an acid or base, what mixture of basic materials it is, what chemical elements are inside, what kind of structure the molecules have, a.s.o., how it can be combined with other materials and elements, and how its properties can be changed.

Speaking more generally in scientific terms, the notion *task analysis* is composed by two words. The senseful interpretation of the word's meaning in an etymological approach may lead to a deeper understanding of the term.

Analysis means in its greek origin the decomposition, resolution, or separation of anything which is compound. In science this means a procedure to investigate and to discover material or ideal facts. This procedure is characterized by a practical or intellectual decomposition of a whole into its parts, a compound into its elements. The objective of analysis is to differentiate essential features and relations from irrelevant and insignificant properties, to separate necessary elements from accidental, common from individual. By this procedure it is possible to penetrate from a global view to the discovery of the essence of an object, its causes and its constituents.

This way is only practicable if analyzed properties and relations are not considered separately but seen in their context. That means analysis has to be combined with synthesis; both procedures make up a dialectical unit (according to textbooks of philosophy and metascience, for example Klaus and Buhr, 1975).

The word *task* cannot be described in a *Handbook of Human Factors* in some simple definitions beforehand. But it may be allowed to follow the ideas of one of the ancestors in our scientific discipline as an initial solution for the problem. Gilbreth (1919) describes the word task as follows:

1. A tax, an assessment, an impost
2. Labor imposed, especially a definite quantity or amount of labor; work to be done; one's stint; that which duty or necessity imposes; duty or duties collectively
3. A lesson to be learned; a portion of study imposed by a teacher
4. Work undertaken, an undertaking
5. Burdensome employment, toil

From a discussion of these variants, Gilbreth (1919) specifies:

The task, under Scientific Management differs from the task under Traditional Management in that

1. The tools and surrounding conditions with which the work shall be done are standardized.
2. The method in which the work shall be done is prescribed.
3. The time that the work shall take is scientifically determined.
4. An allowance is made for rest from fatigue.
5. The quality of the output is prescribed.

To this set of conditions and prescriptions he adds that the task idea applies to any member of the organization (any type of individual work), and to the work of the organization as a whole, in which individual tasks are elements of organization tasks.

Proceeding from the "ancestor" to a recent survey of task analysis and the descriptive decomposition categories used for the term, it is apparent, that the core idea of Gilbreth (1919) and Kirwan and Ainsworth (1992) is almost the same, but the variety and diversity of irradiations into different fields of human factors/ergonomics with different concepts has drastically increased (Table 12.1).

From these etymological considerations the conclusion can be drawn that "task analysis" is a very important concept in human factors and ergonomics, because it has a lot

Table 12.1 Taxonomy of Descriptive Decomposition Categories Which Have Been Used in Various Studies, As Compiled by Kirwan and Ainsworth (1992)

Description of task	Task difficulty
Description	Task criticality
Type of activity/behavior	Amount of attention required
Task/action verb	**Performance on the task**
Function/purpose	Performance
Sequence of activity	Time taken/starting time
Requirements for undertaking task	Required speed
Initiating cue/event	Required accuracy
Information	Criterion of response adequacy
Skills/training required	**Other activities**
Personnel requirements/manning	Subtasks
Hardware features	Communications
Location	Coordination requirements
Controls used	Concurrent tasks
Displays used	**Outputs from the task**
Critical values	Output
Job aids required	Feedback
Nature of the task	**Consequences/Problems**
Actions required	Likely/typical errors
Decisions required	Errors made/problems
Responses required	Error consequences
Complexity/task complexity	Adverse conditions/hazards

to do with the essentials of human work: its goals and intentions, its prescriptions and standardizations, its measurements and methods, its tools and conditions, its qualitative and quantitative output.

12.1.2 Utility

As technical systems become more sophisticated and pressure to reduce manpower in them increases, there is a severe risk that unique human skills and abilities may not be used as effectively as they could, thus degrading the potential performance of a system. Therefore, task analysis as one of the main analysis techniques for human–machine systems design (Beevis et al., 1992) plays an important role in different project development phases, as could be demonstrated in empirical studies of 33 military acquisition or development projects (Table 12.2).

Another index for the importance of task analysis in human factors and engineering may be derived by a frequency analysis of usage of different human engineering analysis techniques in the five project development phases, as performed by Beevis et al. (1992) as well.

Figure 12.1 clearly shows the importance of different methods/techniques, that are known or can be subsumed under the heading of "task analysis." Though an analysis of relative frequencies largely depends on the explanation of how task analysis differs from the other types of analysis mentioned, and what kind of method can be assigned to what domain, the figures are quite convincing for the unprejudiced spectator.

The importance of the field can be emphasized by a growing amount of survey literature: Proceeding from the status as defined in the previous edition of this handbook (Drury et al., 1987), surveys appeared by Beevis et al. (1992) who used a system life-cycle approach for their report, by Kirwan and Ainsworth (1992), who emphasized empirical techniques and practical examples, by Diaper (1989), who stressed the application field for human-computer-interaction, by Landau and Rohmert (1987) and Oesterreich and Volpert (1987), who used a paradigm-driven approach, by van Ouwerkerk, Meijman, and Mulder (1994), who assigned it to a disciplinary perspective, and by de Keyser (1991) and Leplat (1993), who emphasized a national contribution and historical perspective. Surely other national surveys exist, but are not within the scope of the language capabilities of the author and access via literature data banks.

The present approach to a survey on task analysis differs to a certain degree from the prementioned reports: Tasks can be seen as the intersection of two main perspectives in "work sciences"—to use this comprehensive notion of the pluridisciplinary field of our scientific efforts.

In a bottom-up view tasks have to be executed by persons with individual demands, traits, and capacities in an individual or cooperative setting (Figure 12.2). This is the perspective of mostly person-related disciplines. In a top-down view the tasks are result of a breakdown of company goals to a level, where organizational units, groups of personnel, and individual persons with their jobs contribute to these goals. This is the perspective of mostly organization-related disciplines. In the present survey both perspectives will be combined systematically with a logical segmentation of disciplinary perspectives,

Table 12.2 Mean Rate of Use of Different Categories of Analysis Technique in Five Project Phases, in 33 Projects (total for each category divided by 33), As Reported by Beevis et al. (1992)

Category of Analysis Technique Used	Project Development Phase: Analysis of Existing Systems	Planning New Systems	Preliminary Design	Design	Test and Evaluation	Mean Overall Usage
Mission analysis	0.51	0.54	0.54	0.48	0.39	0.49
Function analysis	0.70	0.97	0.97	0.90	0.48	0.80
Function allocation	0.52	0.90	0.82	0.76	0.42	0.68
Task analysis	1.96	1.70	1.97	1.94	1.90	1.89
Performance prediction	0.90	1.00	1.18	0.88	1.12	1.02
Mean across all stages	0.92	1.02	1.10	0.99	0.86	0.98

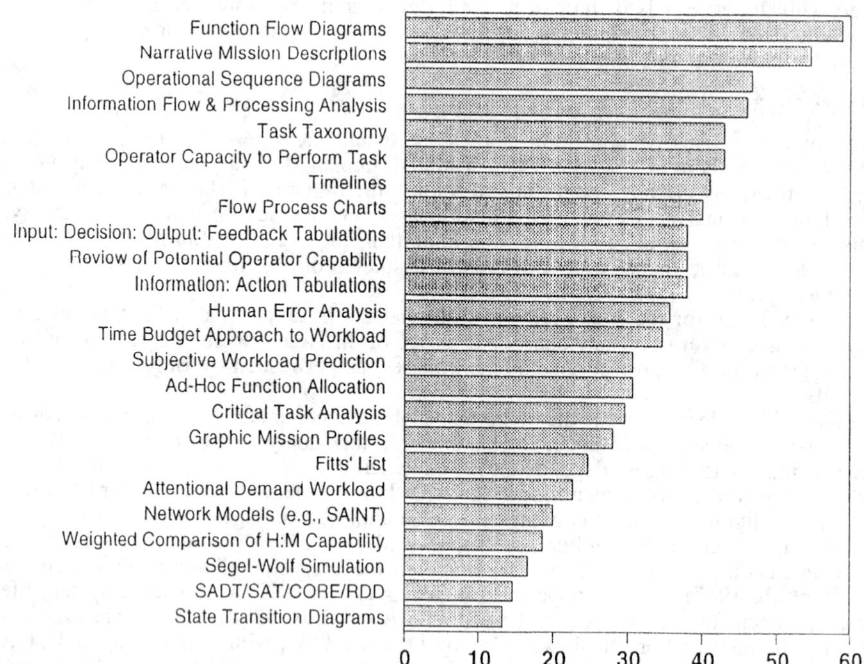

Figure 12.1 Overall number of applications of 24 human engineering analysis techniques in five phases in 33 projects, as reported by Beevis et al. (1992).

their understanding of the notion task, their contribution to analytical paradigms and procedures, as well as to evaluation and design principles.

12.1.3 Segmentation

It belongs to the "best practise" customs in scientific procedures to decompose a complex topic into segments which can be analyzed afterwards in an isolated manner. There is no doubt that a profound analysis of a subject can be performed only at the cost of reducing the limits of extents of the tackled problem. However, only a profound analysis shows exactly the size, components, and relations of a problem-field with all its ramifications and interrelationships. The use of a microscope, for instance, reveals the microstructure of a physical object and is therefore indispensable for its analysis. But it reveals only one facet in the two-dimensional plane with very limited extent: the other areas and dimensions remain in the darkness. Considering these aspects of decomposition and precise analysis it seems to be necessary to identify the *essential* components of the object studied. This requires us to regard the respective object being embedded in its functional context in order to find out the appropriate separation and classification of components. With respect to task analysis this is not a trivial task at all, because the structural and procedural context of the notion "task" has to be taken into account.

An ordering model by Luczak and Volpert (1989) is oriented toward different levels for description of structures and processes, which result when a working person is followed with different analytical approaches in time on the basis of hours, days, or weeks (Figure 12.3). Considering procedural aspects of work processes, the following levels can be distinguished:

V1. Activity of sensomotory automatisms of a person, i.e., elementary operations in sequencing and control of movements

V2. Goal-oriented, consciously controlled action of a person

V3. Motive-related activity of persons, whose concrete results are produced by the sequential and logical arrangement of action

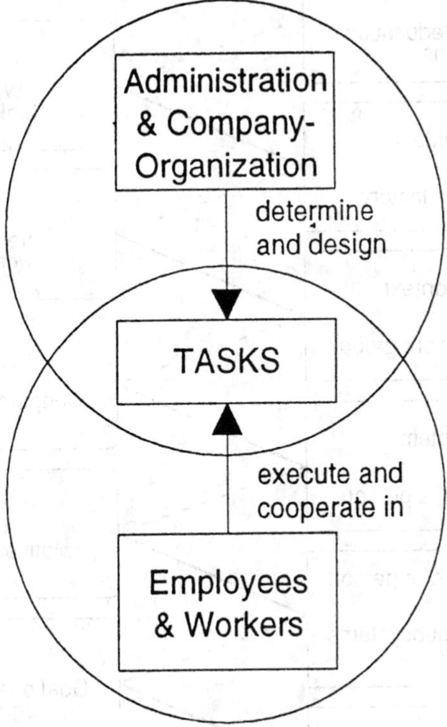

Figure 12.2 Two main perspectives on tasks in the domain of work sciences.

V4. Cooperative work in groups, where the working person has to tune his or her activity to the activities of other persons

V5. Organization within the company (employers, employees) and between companies to define the roles and orientations, to which the working person has to contribute with tasks implicitly or explicitly

V6. Work-oriented political actions, which shall maintain or modify the frame for the actors within the company, that may have severe consequences for any working person

In a structure-oriented form, the levels can be distinguished top-down in the following manner:

S7. Political and societal organization of work

S6. Forms of industrial relations and organization

S5. Forms of cooperation in groups and human relations

S4. Forms and types of work and personal activities (individual work)

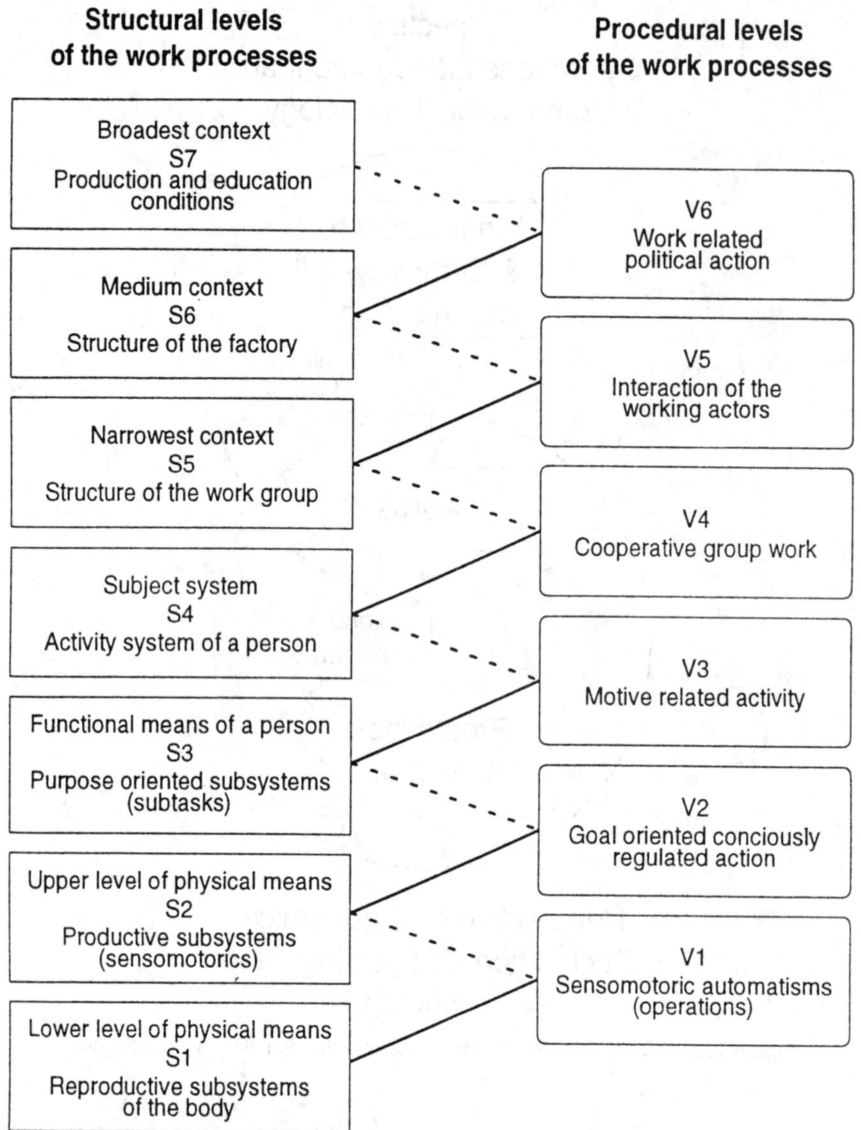

Figure 12.3 Structural and procedural levels of the work process (Luczak, Volpert et al., 1989).

S3. Subtasks and workplaces
S2. Operations with tools and working means
S1. Vegetative systems and environmental factors

It can be seen easily that those levels have to do a lot with the orientation of work-related scientific disciplines. Thus, level S1 is "conditio sine qua non" of occupational physiology and health, but this discipline can widen its approach to levels S2 and S3 as well. Level S7 describes the specialty of political economy and macroeconomics in the sense of labor policy, but excursions to level S6 (industrial relations and organization) and even to level S5 (human relations and group work) may be possible. Levels S2 and S3 may be the focus of industrial engineering (motion study) and ergonomical design, whereras levels S5 and S6 form the core of personnel management and occupational sociology. An integrative approach to work can be assigned from level S3 to level S5.

Following this segmentation of "work sciences," the understanding of the notion task, the respective analytical and purposeful procedures and techniques can be easily subdivided into meaningful contexts:

1. The implications of tasks with autonomous bodily functions and with the work-situational environment can be summarized in terms of "physiological costs."
2. The operations and movements with tools and at equipment at an elementary sensomotory level can be combined in a context "time consumption."
3. The work processes at workplaces or its elements can be described in a context "single- and multitasks in human–machine interaction."
4. The personal work and types of work perspective can be outlined in a context "combination of tasks to a job."
5. The forms of cooperation in a working group as task structure can be reported in a context "task structure and group organization."
6. The inner-company and intercompany organizational determination of task structures is emphasized in a separate analytical context as well.
7. The use of analytical knowlege from task analysis for societal purposes (education, labor politics, etc.) deserves its own context.

12.2 TASK ANALYSIS CONSIDERING PHYSIOLOGICAL COSTS

In this perspective of task analysis, the task elements are subdivided and ordered according to the physiological costs they provoke in different organismic systems. Thus, the principle of redefinition for the notion "task" is the impact on organismic systems and physiological functions. The principle of decomposition into subtasks is the type and amount of costs in terms of strain or possibility of disease. The approach may be top-down by biomechanical or informational modeling as well as "opportunistic" in terms of a bottom-up procedure to identify and order detailed symptoms (Luczak 1991).

12.2.1 Physical Perspective

One of the approaches with a very long scientific tradition is the physiological study of energy expenditure with different task elements and their composition to an adequate sequence of operations, for instance, in lifting, carrying, moving, and lowering work objects, or even the assignment of ranges of metabolic rates to jobs or a bundle of professional activities.

The French chemist Lavoisier discovered in 1789 that energy expenditure increases while humans work. He proposed comparing, by oxygen consumption, the tasks and activities of any philosopher, author, or composer with any heavy muscular worker. Historically, therefore, this seems to be one of the first proposals for a scientific task analysis on the basis of a reproducible method. Though Lavoisier's idea was erroneous with respect to mental tasks, the history of occupational physiology demonstrates that due to energy-expenditure measurement hundreds of different tasks and more than 100 professions can be distinguished quantitatively. Today we dispose of summarizing tables in different languages (Katsuki, 1960; Passmore and Durnin, 1967; Spitzer and Hettinger, 1964), which review the results of several hundred references about individual experimental work of different research groups or schools of occupational physiology all over the world.

Although the number of tasks and professions with an available amount of data is high, the generation of new jobs by task composition avoids to built complete overviews. Therefore estimating tables are used to conclude from task elements in terms of body positions and movements (A) combined with a differentiation of body segments working (B) to energy expenditure (Table 12.3). By this form of task analysis interesting questions for evaluation and design can be solved, for example (Rohmert, and Luczak, 1979):

1. How much of a worker's food consumption is due to his or her work? In other words, the question of nutrition during times of scarce food production can be tackled.

Table 12.3 Analyzing and Estimating Energy Expenditure Per Minute

A Body Position or Movement	kcal/min (net)	*B* Type of Work		kcal/min (net)
Sitting	0.3	Hand work	Light	0.3–0.6
			Medium	0.6–0.9
Kneeling	0.5		Heavy	0.9–1.2
Crouching	0.5	One arm works	Light	0.7–1.2
			Medium	1.2–1.7
Standing	0.6		Heavy	1.7–2.2
Stooping	0.8	Both arms work	Light	1.5–2.0
			Medium	2.0–2.5
Walking	1.7–3.5		Heavy	1.5–3.0
Climbing (without load, inclination 10°)	0.75 per metre height	Whole-body work	Light	2.5–4.0
			Medium	4.0–6.0
			Heavy	6.0–8.5
			Very heavy	8.5–11.5

Estimated energy expenditure = $A + B$

2. For what type of task or tool does the energy requirement become a minimum, and the degree of efficiency a maximum? Thus, best practices of task execution can be derived.
3. What is the tolerable amount of daily energy expenditure? Thus, limits of tolerability and trade-off functions between physical work load and working time can be derived.

All these questions of task analysis can be answered and solved within the range of "physiological costs" that do not cause exhaustion, disorders, or even diseases. But many task analysis procedures tend to identify task elements and task factors related to a nuisance or disease. A task-analytical method of movement and posture analysis on the basis of biomechanical considerations and practical knowledge was developed in Scandinavia, the so-called OWAS-method. In the perspective of limiting of physiological costs and prevention of diseases, the method describes the following (Stoffert 1985):

- Types of working postures and "frozen" movements
- Their relative frequency in a task structure
- The segments in the work process, to which the tasks/types can be assigned
- The necessity of design interventions
- The distribution of posture/movements to bodily segments
- The weights handled or forces exerted

Respective data are collected by the principle of multimoment studies. The special approach of the OWAS method is the evaluation and combination of data in matrices of postures and movements (Figure 12.4), the analysis of relative frequencies, and the derivation of necessities and measures for design:

1. Posture is normal: No measures necessary
2. Posture is stressing: Measures to be taken soon
3. Posture is stressful: Measures to be taken as soon as possible
4. Posture is heavily straining: Measures to be taken immediately

The primary goal of work protection to prevent occupational diseases by task analysis of working postures and movements can be accomplished by the method. Thus, it is one good example for an analytical procedure to reduce physiological costs.

Though the OWAS method is available as a computer program for practical use, another good example can be reported here from the point of view of computer-based task analysis: the expert system ERGON-EXPERT, as developed by Laurig and Rombach (1989) for lifting tasks and material handling tasks. The specialty of the method is not only the expert-system approach, but the hierarchical and sequential representation of

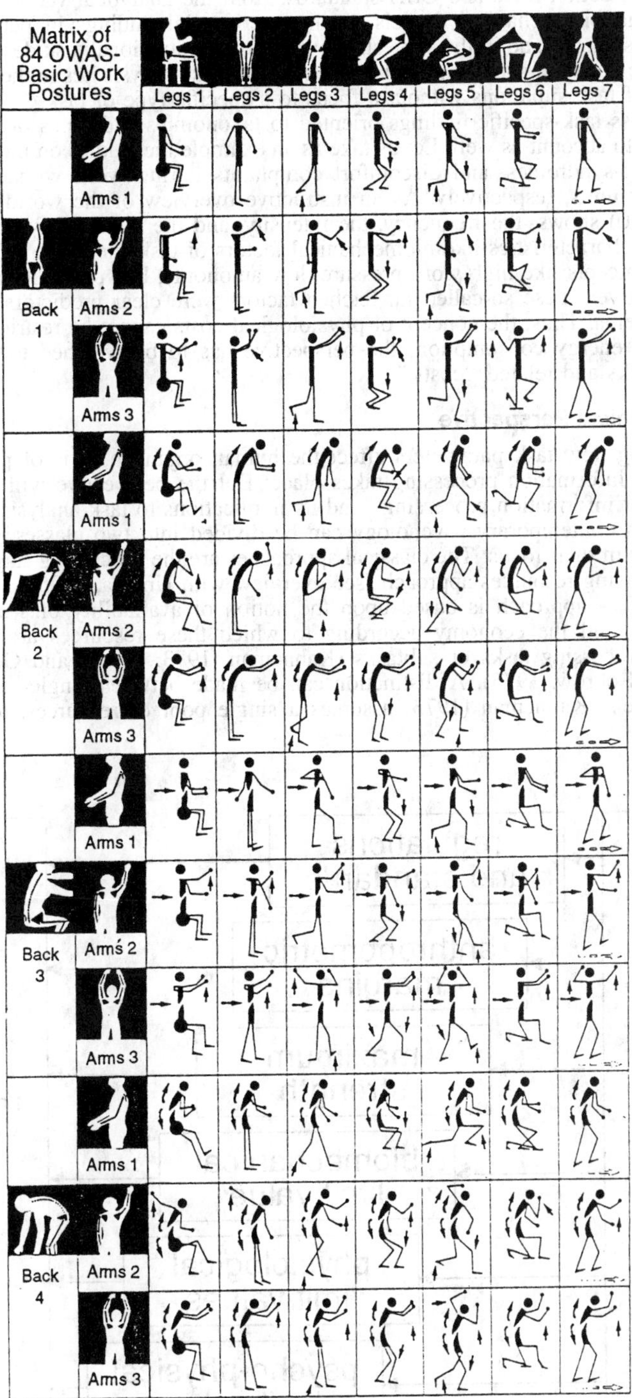

Figure 12.4 Principle of the OWAS method.

knowledge in the knowledge base. Thus, proceeding from an identification of task features and attributes of the working person the analysis is detailed and compared with knowledge from labor protection rules and OSH standards, from the analytical variables in anthropometrical and biomechanical models, from experimental findings in performance, in organismic responses, and in the subjective evaluation of situational task description (Figure 12.5). The utility of such computer-based tools of task analysis lies in the applicability to a variety of task situations without a focus to a specific type of task.

Nevertheless task-specific findings oriented to taxonomies and types of injuries have to be taken into account as well. Let us take as an example the studies on musculoskeletal disorders, RSI-syndromes, and discomfort complaints during work with computers or visual display units, respectively. As an instructive overview of the world literature by Bammer (1990) shows, the frequency, the intensity, and the kinds of symptoms clearly relate to task characteristics like biomechanical factors of task execution as well as such far-fetched concepts like high work pressure, low autonomy, low peer cohesion, and low task variety: Even these so-called far-fetched factors were clear predictors of musculoskeletal problems. Thus, the concept of physiological costs cannot be restricted to physical tasks and energy consumption; the perspective has to be widened to information-processing tasks and related "costs."

12.2.2 Psychical Perspective

When studying how task parameters affect the human organism, it is of particular importance how information processing takes place: But the perspective will be restricted to the "cost of information processing" and its implications to task analysis. Conceptual frameworks in contemporary psychology can be divided into two classes, which differ fundamentally in their focus: The classical approaches are the processing stages approach and the processing resources approach (see overview with Unema, 1995).

The resources approach is based upon the notion of availability of scarce processing resources, and the economy according to which these resources are allocated to information-processing tasks or subtasks (Kahneman, 1973; Navon and Gopher, 1979; Norman and Bobrow, 1975). A distinction can be made between single- and multiple-resource models. Kahneman (1973) assumes a single pool of resources, to be termed

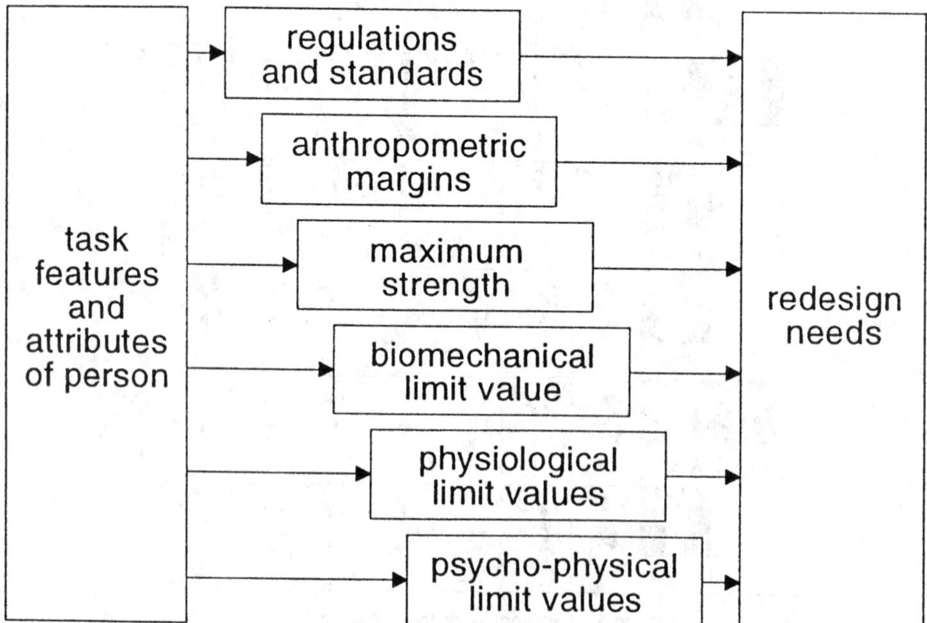

Figure 12.5 Structure of a knowledge base in an expert system to derive from task features evaluations of physiological costs, as developed by Laurig and Rombach (1989).

processing capacity, *attention*, or *effort*, which is controlled by feedback. When an ongoing task requires an increased amount of capacity in order to process the present information, there is an increase of arousal, and with this, an increase in attention. Thus, arousal and effort are usually not determined prior to action: They vary continously, depending on the load which is imposed by personal activity at any instant of time. The total amount of supplied resources varies with physiological arousal and is controlled by the demands. The amount of effort invested in a task is not subject to voluntary control; the level of physiological arousal is indicated by a variety of variables (Kahneman, 1973; Luczak, 1987), which react to a certain degree task-specific (e.g., Beatty, 1982). The processing resources for any system are limited. When several processes in task execution compete for the same resources there will be a decrement in performance: "Resources are such things as processing effort, various forms of memory capacity, and communication channels. Resources are always limited. If several processes request a portion of the same available resource, this resource must be allocated among them" (Norman and Bobrow, 1975).

Since the observation, that manipulation of some task parameters affects the joint performance of two simultaneously performed tasks, while other manipulations affect only the performance on one task, the notion of *multiple resources* has emerged (e.g., Navon and Gopher, 1979; Norman and Bobrow, 1975; Wickens, 1980). It was found that an interference occurs under the condition of resource competition, when performing one task while simultaneously performing a second, e.g., when both tasks require manual responses to aurally presented stimuli. No (or less) response competition would be expected when one task manually responds to visual stimuli, and the other orally responds to acoustical stimuli. On the basis of such observations, Wickens (1984) proposed a model of multiple resources. He distinguishes between the following:

1. *Stages of processing*: The resources needed for perceptual and central processing seem to be functionally distinguishable from those that are involved with response selection and execution. This becomes apparent when response complexity does not interfere with a mainly perceptually oriented secondary task.
2. *Sensory modalities*: When capacity has to be shared between two tasks involving two different sensory modalities (in particular, auditory vs. visual), a better performance is achieved when modalities are different (Treisman and Davies, 1973).
3. *Processing codes*: Spatial and verbal codes are differentially represented in working memory. In particular, the processing of spatial codes seems to take place in the right cerebral hemisphere, whereas the left hemisphere seems to be the main resource for processing verbal information (see also Posner and Petersen, 1990; Tucker and Williamson, 1984).

Multiple-resource models bear the intrinsic appeal of applicability in real work situations. A major point of critique of multiple resource models is, that they may explain any state of affairs and are very difficult to refute. There is no principle way of deciding whether one kind of resource should be added or removed from the model. Furthermore, it is very difficult to assess the individual contribution of each of two or more simultaneously presented tasks to the resulting change in resource depletion by physiological parameters. The combination of any two tasks may as well result in cumulative as in compensatory or indifferent superpositions of workload (Hertting-Thomasius, 1992).

Structural models have their roots in the work of Donders (1868) and Sternberg (1968) and are based upon the idea that performance variability is due to differences in the number and complexity of signal transformations in order to produce a response. Since the logics of subtraction (Donders) and additive factors (Sternberg) have been discussed extensively elsewhere (e.g., Luczak, 1992), an extensive discussion in the present context is therefore considered redundant (see Section 12.3.1.1).

The idea that energetic (processing) resources can be allocated to specific processing structures bridges the gap between structural and resource concepts of information processing. The differences between these conceptual frameworks are mainly the result of a different interest in the outcome of such studies. Whereas linear stage models are primarily conceived in order to study the architecture of human information-processing devices, resource models have the implicit (though often distant) goal to assess human performance limits and breakdown. In other words, both conceptual frameworks address different kinds of problems. Despite these elementary differences, it is not a priori impossible to reconcile both frameworks. Although it is quite hazardous to attain perfect integration, a promising attempt has been made by Sanders (1983) and Gopher and Sanders (1984). The cognitive energetic stage model proposed by Sanders (1983) encompasses the notions of resources as structure-specific energizing units for computational processes, which are considered to be organized in a serial way (Figure 12.6). Starting point of the model is the assumption that duration of a processing stage is affected by both the state of the subject and the computational demands of the task at hand.

The basic model relies on four computational stages in the traditional choice reaction process that appear sufficiently established by studies using the additive factors paradigm. These stages are as follows:

- Stimulus preprocessing (which is affected by, e.g., signal intensity)
- Feature extraction (affected by, e.g., signal quality)
- Response choice (affected by, e.g., stimulus response compatibility)
- Motor adjustment (affected by, e.g., time uncertainty)

The model is based upon the notion that the efficiency of the computational mechanisms is influenced by the amount of "energy" provided. The exact nature of this energy resource is not explicated, however—the model only describes the mechanisms providing it. These mechanisms are termed after the attention-controlling mechanisms, as described by Pribram and McGuinness (1975, 1976). Pribram and McGuinness distinguish three basic processes in the control of attention: one that regulates the arousal from input, a second that regulates the preparation for action (activation), and a third that operates to coordinate arousal and activation. This third mechanism demands "effort," and is hence called so.

Sleep loss (Sanders and Reitsma, 1982a,b) and knowledge of results (Wilkinson and Colquhoun, 1968) have been demonstrated to modulate the willingness to invest effort in the task. The evaluation mechanism is assumed to be a continuous process of supervision of performance. It reflects the level of performance on a cognitive level. The use of criteria enables the evaluation mechanism to estimate the adequacy of present performance. In the case of a discrepancy between performance level and performance criteria, the mechanism exerts a modulatory influence on the energetical mechanisms. Whether this mechanism acts on the basis of economy (save energetical cost) or on the basis of performance maximizing may be the result of a trade-off between speed and accuracy, and may be influenced both by motivational (incentives) and emotional factors (frustration).

The model has no prediction value with respect to task execution, but it is highly explanatory in terms of interaction of processes involved in task execution and their distribution considering physiological costs. At that, it gives a first insight into the steering of cognitive components in task execution.

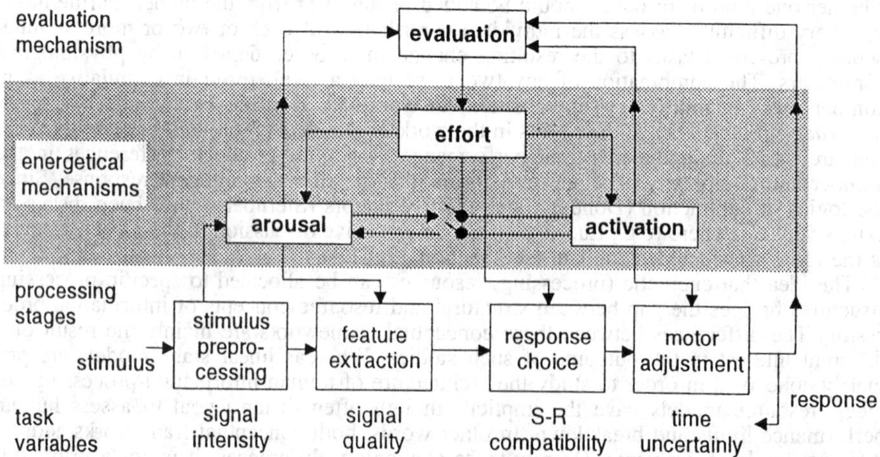

Figure 12.6 Cognitive energetic linear stage model, as proposed by Gopher and Sanders (1984).

12.3 TASK ANALYSIS CONSIDERING TIME CONSUMPTION

This kind of task analysis is centered around planning of tasks with respect to task elements, which use a definite amount of time for their execution or allow a description of variables affecting time consumption. Thus, the principle of redefinition of the notion "task" is—on a microstructural level—human performance modeling in the sense of work amount per time unit. The principle of decomposition into subtasks and operations is the possibility of finding descriptive variables, which can be coupled to time consumption for the respective task elements in quantitative terms. The approach may be model based, like in the Donders and Sternberg paradigm for information processing or like in the acceleration, velocity and movement studies for mechanical processes. It may be empiristic like in the systems of predetermined times, or it may be heuristic like in the human processor model.

12.3.1 Mental and Informational Processes

12.3.1.1 The Roots: Donders and Sternberg

One of the first experimental settings to systematically separate mental processes—subdivision of a task into subtasks—stems from the Dutch scientist Franz Donders (1868). He was fascinated by the Helmholtz discovery that neural and mental operations cost time, and do not—as presumed until that time—perform instantly. Donders asked whether the speed of thinking was measurable. To answer this question, he invented a method to separate task elements in reaction time. This method is based on the assumption (which is no longer shared today by most scientists) that mental processes are performed serially in steps and that throughput/lead times of separate steps are additive. His method contains three reaction time tasks (Table 12.4):

Task A is a simple reaction time task with one stimulus and one response possibility.

Task B is a multiple-choice task with at least two stimuli and two assigned responses.

Task C contains two (or more) stimuli but only one response, which means, for example, that a response has to be performed whenever a left light flashes, but nothing has be done on the right light signal. Donders presumed that task B contains three subtasks:

1. Simple reaction
2. Stimulus categorization
3. Response choice

Thus, task A is a component of B: Multiple-choice reaction times are longer than simple reaction times, because they incorporate two additional subtasks, and task C contains all subtasks except for the response choice organization.

Comparing of the time consumption of the different tasks, an easy calculation of processing times is possible:

- Simple reaction = A
- Stimulus categorization = C − A
- Response choice = B − C

This method of subtraction by Donders is an historical example for the decomposition of informational subtasks with respect to time consumption. However, it is based on the

Table 12.4 The Donders Paradigm

Task	Number of Stimuli	Number of Responses	Subtasks/Processes Involved/Times
A	1	1	Simple reaction time
B	Several	Several	Simple reaction time Stimulus categorization Response choice
C	Several	1	Simple reaction time Stimulus categorization

(weak) assumption that different subtasks can be characterized and combined in this type of serial process model and only in this model. Thus, the model does not allow a transfer of results to other informational processes, but demonstrates a creative idea in task analysis.

At the 100th "birthday" of the reaction time studies by Donders, Sternberg (1969) presented a memorandum with some of his investigations on tasks and reaction times, which reanimated the interest in this type of studies. Sternberg postulated three basic assumptions about mental processes as they are implicitly formulated in reaction time studies:

1. There are sequential functional steps (subtasks) between stimulus and response, whose lead times sum up to the reaction time of a task.
2. The steps (subtasks) are independent, that is, the time consumption of a step is independent of the time consumption of the preceding step.
3. If the statistical distribution of lead times in different steps is known, the statistical distribution of summed up reaction times can be derived.

In contrary to Donders paradigm, in the Sternberg paradigm the assumption must be fulfilled that in a specific task all steps of information processing (subtasks) are involved. On the basis of the variation of two or more factors in a respective experimental design, the effect of these variables on time consumption in specific steps of information processing can be diagnosed by measurements of reaction time. Because Sternberg presumes that the influence of the factors is additive, his method may be called *method of additive factors*. If several factors influence the same step (subtask), it is probable (but not necessary) that an interaction occurs, which means that the influence is no longer additive, but perhaps cumulative.

The typical Sternberg task is characterized as follows:

- The stimulus material consists of numbers or characters.
- All numbers or characters together form the stimulus set.
- An arbitrary subset is chosen, the "positive set" (mostly one to four different items). The subject has to remember this subset.
- All other items form the "negative set."
- The subject is confronted with one or more test stimuli and has to decide whether the test set contains an element of the positive set.
- The subject has to give one out of two responses afterwards: a positive answer whenever an element of the positive set is identified, a negative answer if not.
- The reaction time between stimulus presentation and pressing a respective response key is measured.

The task is so simple that it can be performed perfectly (no speed accuracy trade-off). Figure 12.7 shows the processing steps of a typical Sternberg task (item identification) and schematically the results of measurements of reaction times (RT) depending on the number of items in the positive set.

The slope of the curves is determined by the time increment used for a simple comparison in the step "serial comparison." The respective time consumption of stimulus encoding, of binary choice, and of response organization is the intersection of the curves with the RT-axis. Thus, by a factorial design of experiments in the Sternberg paradigm, the time consumption depending on, for instance, readability of stimuli, type and extent of positive sets, modes of response organization, and relative frequency of response can be analyzed in quantitative terms.

However, this paradigm loses its analytical value as soon as the assumption of a serial step-wise information processing must be given up, because parallel or cascade processing contradicts the measurement of reaction time as the main variable, that gives an unmasked picture of information handling difficulty.

Nevertheless, practical methods of task analysis rely on the same paradigm, for instance, the Work Factor Mento System.

12.3.1.2 WF-Mento

The Work-Factor Mento System is the only *system of predetermined times* (*SPT*) with worldwide recognition that seems to allow the analysis of time consumption of repetitive

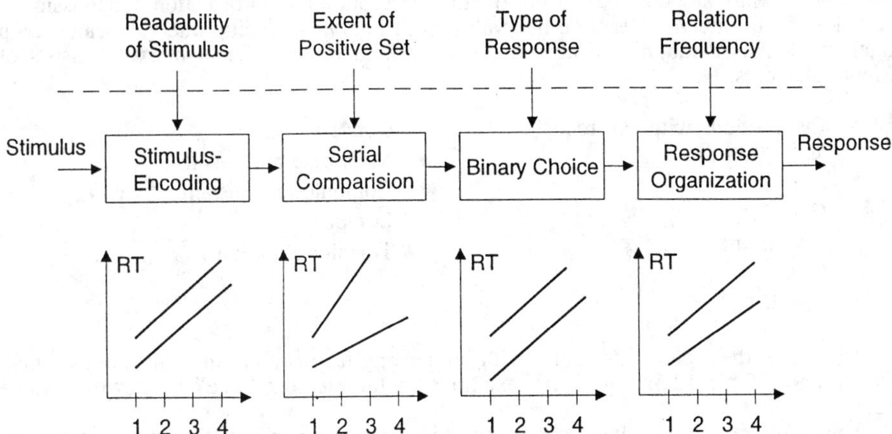

Figure 12.7 Processing steps in a Sternberg task with experimental factors (above) and typical results (below).

informatory processes. The development and differentiation of the WF-standard element "mental processes," as described firstly by Quick et al. (1965), shall separate mental procedures of information handling in the stadium of planning or correction of tasks into so-called Mento-elements, to which predetermined times according to an analysis of influencing factors can be assigned. Like in other SPT, the time consumption of a complete task is calculated by adding time elements of the subtasks/operations. WF-Mento does not consider cognitive processes or creative work in informational terms; its application is limited to tasks with transmission of information perception to an action, for instance, in inspection of parts or quality assurance of products (Luczak and Samli, 1986; Samli, 1987).

The approach of WF-Mento is based on a phase or step model of human information processing: analyzing, for instance, the process of task execution, when a lamp flashes, on which the operator has to press a specific key, means to identify the kind and importance of the respective signal, means to decide which key or key combination has to be pressed, means to initiate the action by respective nervous impulses (see Figure 12.8).

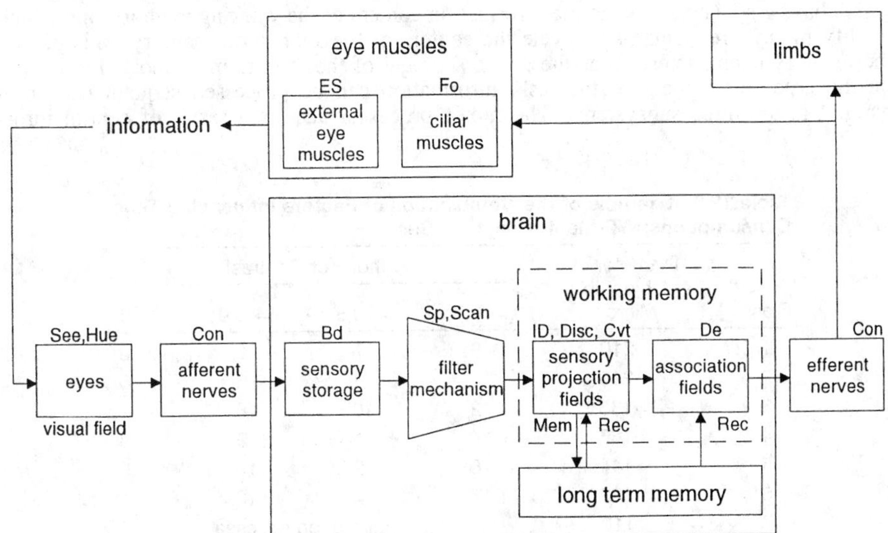

Figure 12.8 Model of information processing used for WF-Mento.

In WF-Mento, any complex of information perception and information transmission is called a Mento-interval. Reaction intervals, check intervals, reading intervals, calculation intervals can be distinguished according to the kind of task. Each interval consists of Mento-elements, like

- Eye motions with eye focus and eye shift
- See
- Conduct
- Discriminate
- Span
- Identify
- Convert
- Memorize and recall
- Compute and sustain
- Decide
- Transfer attention

To each of these elements, tables with influencing factors and time units are assigned, for instance, Table 12.5: "See." An example for the element "See" is shown in Table 12.5.

Thus, inspection tasks and tasks of checking/surveying information items of limited amount and complexity especially are the classic application domains of WF-Mento. For this type of tasks, computer-supported planning tools were developed recently on the basis of WF-Mento (Fechner, 1994). Like other SPT, the tasks should be repetitive and characterized by a certain time pressure. Nevertheless, the method did not reach the popularity of SPT for motion time analysis, especially in manual assembly.

12.3.1.3 Human Processor Model

Even in newer models of information processing—for instance, the "human processor model"—which are the basis for task analysis in human–computer interaction—for instance the GOMS—Goals-Operators-Methods-Selection Rules model, time consumption τ and its prediction is essential (Card, Moran, and Newell, 1983). The authors combine this aspect with storage capacities μ of the short-term and long-term memory, with time constants for the correct storage reproduction τ and κ, the coding of information in the respective storages and memories (Figure 12.9).

The values in the brackets are derived from a literature review as max./min. indications, the values in front of the brackets are denominated as typical for the respective type of process. The model contains three subsystems (sensory, cognitive, motory), which have their storages and processors, which interact in the time domain with certain task-specific time consumption. The information is held via the sensory processor for a short time in the visual/acoustic storage, is encoded and transfered to the short-time memory /working memory. The cognitive processor processes the information in the working memory and, for this purpose, accesses the information content in the long-term memory, which have associative links to the information content of the working memory, and which modify in any "recognize-act" cycle the content of the short-term memory. This part of the memory is characterized as the active segment of the long-term memory. Depending on the type and content of the task, information can be processed sequentially or in parallel in the three subsystems. The motor processor acts as a result of central infor-

Table 12.5 Example of the Combination of Factors Influencing Time Consumptions; WF-Mento-Element: See

Class	Diameter of Object in Micron	Index of Contrast			
		⇒ 0,7	⇒ 0,8	⇒ 0,9	> 0,9
1	>210	8	7	6	6
2	⇒210	10	8	6	6
3	⇒188	12	10	8	7
4	⇒166	14	11	9	8
5	⇒144	16	12	10	9
6	⇒122	17	14	12	11
7	<110	Amplification necessary			

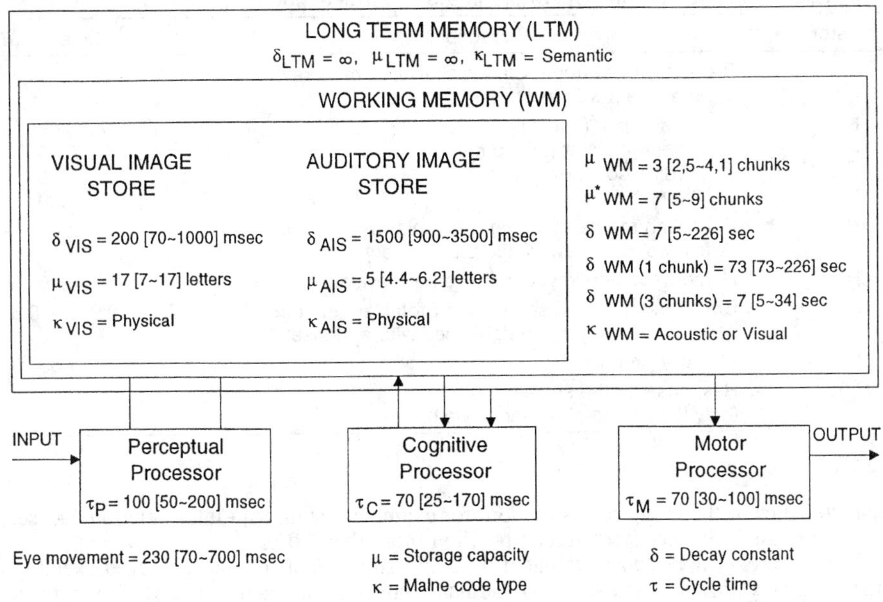

Figure 12.9 The model human processor, as developed by Card, Moran, and Newell (1983).

mation processing. Thus, the model does not focus on time consumption alone, because it "knows" other variables determining task execution. Nevertheless, it is a consequent development of models of information processing and task execution defining subprocessors, subtasks, subvariables, etc. to give a prognosis on the execution time of the whole task from a respective task analysis: It stands in the tradition of Donders, Sternberg, Quick, and others with its intentions and its level of analysis.

12.3.1.4 Keystroke-Level Model

Closely related to the GOMS model and the model human processor, which were designed by the same authors (Card, Moran, and Newell, 1980, 1983), the keystroke-level model describes the time it takes an experienced user to perform a task with a given method on an interactive computer system. Task execution time is a function of four physical-motor operators (K, P, H, D), one mental operator (M), and one system response operator (R), which are outlined in Table 12.6.

An encoding method is predetermined for specifying the series of operators in a task prior to applying the equation

$$T_{execute} = T_K + T_P + T_H + T_D + T_M + T_R$$

Given a large task, like editing a comprehensive document, the user will decompose it into a series of small, cognitively manageable, quasi-independent tasks, so-called unit tasks. The complete task and the interactive system influence the structure of these unit tasks: Nevertheless they can be separated respecting the memory limits of human cognition.

A unit task has two parts: (1) acquisition of the task and (2) execution of the task acquired. During acquisition the user builds a mental representation of the task, during execution he calls on the system facilities to accomplish the task. The total time of these two parts is

$$T_{unit\ task} = T_{acquire} + T_{execute}$$

Afterwards the times for a complete task are calculated by the sum of the times of its constituent unit tasks. Especially the acquisition time for a unit task depends on the

Table 12.6 Description of the Operators in the Keystroke Model

Operator	Description	Time (sec)
	Press key or button (includes shift or control keys). Time varies with skill:	
K	Best typist (135 wpm)	0.08
	Average typist (55 wpm)	0.20
	Typing complex codes	0.75
	Worst typist	1.20
P	Point with mouse to target on display (follows Fitts' Law, range 0.8–1.5 sec)	1.0
H	Home-hands-on keyboard (or other device)	0.40
$D(n_d, l_d)$	Draw *n* straight-line segments of total length l cm (assumes drawing straight lines with a mouse)	$0.9n_d + 0.6l_d$
M	Mentally prepare	1.35
$R(t)$	Response by the system (only if it causes the user to wait)	t

characteristics of the larger task situation, for example, manuscript interpretation 2–3 sec, routine design 5–30 sec, creative composition more than 30 sec.

The keystroke model was evaluated by comparing calculated and observed execution times in many different situations of human–computer interaction (HCI), for instance, text editing, graphical design, etc. The method was primarily used to calculate system benchmarks and to predict the effects of different ways of task execution. Its use is restricted to error-free behavior of experienced users in routine-oriented HCI without creative task elements.

12.3.2 Motory Processes

Gilbreth (1911, 1917) seems to be the creator of motion study in work sciences: In 1885, he performed systematical motion studies at a mason work by decomposing a complex movement cycle into motion elements. This technology was improved by his so-called micromotion study, by which the "Therbligs," 17 basic motion elements, resulted from which all industrial work processes should be composed. Another essential Gilbreth idea was that motion and time are two integrated components of a working process. This insight was taken over by the founders of the system of predetermined times: They assigned process times to the motion elements by analyzing the width and difficulty of motions.

12.3.2.1 Systems of Predetermined Times

On the basis of this knowledge in the time span from 1930 to 1950, the SPT were developed in the United States (Maynard et al., 1948; Quick et al., 1965). Today, these SPT are applied in industry where relatively small products are produced in large series, for instance, in electrical and mechanical production, especially in manual assembly.

The SPT, today mostly represented by the Work Factor and Methods Time Measurement approach, consist of rules for task analysis and time values, with which working times for limb operation, especially manual assembly and machine tool operation, can be determined beforehand, that is, in the planning and design phase of a workplace. The analysis of a movement cycle consists of the following steps:

1. *Analysis of movement pattern*: decomposition of the movement cycle into motion elements. All possible movements are mapped to eight types of basic motions: Reach, Grasp, Turn, Move/Bring, Position, Release, Disengage, and Mental processes.
2. *Time analysis*: determination of time-influencing variables/factors for any motion element. The kind and importance of recognized factors differs essentially between the SPT. All systems recognize length of motion in Reach and Move, the extent of the object in Grasp, and the tolerance in Position as important factors.
3. Determination of the time value from the tables of motion times on the basis of movement and time analysis with respect to the factors identified.

Table 12.7 shows as an example the motion times for Reach depending on different movement cases (A to E), of distance moved, and combinations.

To demonstrate this task analysis procedure a simple symmetrical assembly of both hands is shown in Figure 12.10.

The upper part of the figure contains all necessary measurements at the workplace and the work object. In the lower part, separate lines are used for analysis of any motion element. Work Factor divides the element Position into subelements, whereas MTM is much simpler in this respect.

The importance of the SPT as a method of task analysis does not focus on time consumption and planning of working time alone. It is the possibility of workplace design, the planning of work method, and the estimation of effects of tools and work objects design that made the SPT so attractive for engineers and ergonomists. At that, the possibility of generating macros for tasks in specific companies with certain production technologies and for definite branches of industries with their types of orders and order processing makes the impact and utility of SPT as task analysis tools. In spite of the fact that they seem to be "overaged" and outdated, from the point of view of usefulness they seem to be "forever young."

12.3.2.2 Movement Studies in Scientific Management/Industrial Engineering

Systematic investigations of work movements at industrial workplaces and in the laboratory have been performed for decades—almost 100 years—under the heading "scientific management." Though in ancient times in Egyptian, Greek and Roman culture "reports" about the planning and logistics of building construction imply to a certain degree task analytical components (Luczak, 1991a), the breakthrough to a systematical approach of task analysis was performed by Taylor (Copley, 1923).

The algorithm of analysis comprises several steps (Barnes, 1963; Mundel, 1947, 1950):

1. Divide the task of a worker into simple elementary movements.
2. Identify superfluous movements and eliminate them.

Table 12.7 Example of a MTM-Table of Motion Times

Methods-Time Data								
Table I-Reach								
			Leveled Time TMUs					
Case	Description	Distance Moved (in.)	A STD.	A Hand in Mot.	A with CD or B	B Hand in Mot.	C or D	E
A	Reach to object in fixed location or to object in other hand or on which other hand rests	1			2.1		3.6	
		2			4.3		5.9	
		3			5.9		7.3	
		4	6.1	4.9	7.1	4.3	8.4	6.8
B	Reach to single object in location which may vary slightly from cycle to cycle	5	6.5	5.3	7.8	5.0	9.4	7.4
		6	7.0	5.7	8.6	5.7	10.1	8.0
		7	7.4	6.1	9.3	6.5	10.8	8.7
		8	7.9	6.5	10.1	7.2	11.5	9.3
C	Reach to object in group	9	8.3	6.9	10.8	7.9	12.2	9.9
		10	8.7	7.3	11.5	8.6	12.9	10.5
		12	9.6	8.1	12.9	10.1	14.2	11.8
		14	10.5	8.9	14.4	11.5	15.6	13.0
D	Reach to very small object or where accurate grasp is required	16	11.4	9.7	15.8	12.9	17.0	14.2
		18	12.3	10.5	17.2	14.4	18.4	15.5
		20	13.1	11.3	18.6		19.8	16.7
		22	14.0	12.1	20.1		21.2	18.0
E	Reach to indefinite location to get hand in position for body balance or next motion or out of way	24	14.9	12.9	21.5		22.5	19.2
		26	15.8	13.7	22.9		23.9	20.4
		28	16.7	14.5	24.4		26.3	21.7
		30	17.5	15.3	25.8		26.7	22.9

Source: Maynard et al., 1948

Working procedure: Simultaneous-symmetrical assembly of bolts

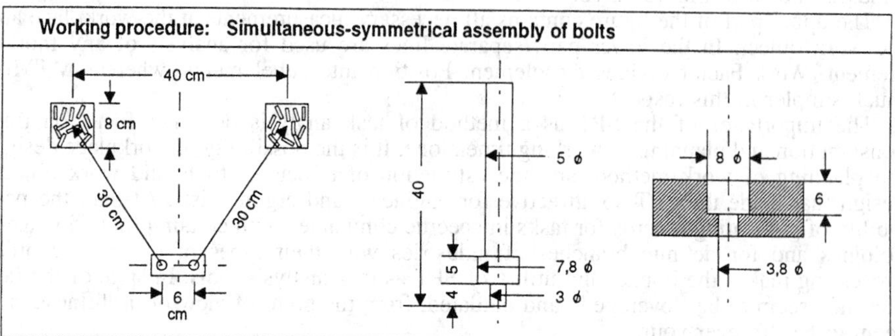

WF-Analysis

Left hand					Right hand				
No.	Description	Analysis	ZE singl.	ZE total	ZE total	ZE singl.	Analysis	Description	No.
1	Hl Container	A 30 D	65	65					1
2	Gr Bolts	cyl 5 X 40 si	52	117					2
3	Vrn Bolts	V3F2, 5 X 50%	24	141					3
4	simu	50% of 24	12	153					4
5	Tp Bolts	A 30 SD	85	238					5
6	Mt	X 8 Dv = 0.375, La = 6cm	69	307					6
7	Ts	V 1 A2, 5S	28						7
8	La	30% of 28	8						8
9	simu	50% of 34	17						9
10	Es	A 2.5	18						10
11	Mt	X 8 Dv = 0.975, La = 6cm	103	410					11
12	Ts	V 1 A2, 5S	26						12
13	La	30% of 26	8						13
14	Simu	50% of 34	17						14
15	Ar	A 2.5 S	26						15
16	Es	A 2.5 P	26						16
17	Lz . Bolts	1/2 F 2.5	8	418	418	418	like left hand		17

MTM-Analysis

No.	Left hand	Analysis	TMU	Analysis	Right hand	No.
1	towards container	R 30 C	14.1	R 28 B		1
2	Bolts	G 4 B	9.1			2
3			2.0	R 2 C		3
4			9.1	G 4 B		4
5	in device	M 30 C	15.1	M 30 C		5
6		G 2		G 2		6
7		P 1 S D	11.2			7
8			11.2	P 1 S D		8
9		RL 1	2.0	RL 1		9
		total	73.8	= 443 ZE		

1 ZE = 0.0001 min 1 TMU = 0.0006 min 6 ZE = 1 TMU

Figure 12.10 Example of a WF and MTM analysis of a manual assembly task.

3. Study, step by step, how the experienced worker performs subtasks and movements, and choose by time measurement the fastest and best method.
4. Describe and codify any elementary movement with its time consumption.
5. Investigate and note the percentage of time to be allowed for delays, break-down etc., for respecting the learning curve, for rest pauses and rest-intervals to avoid fatigue.

From this analytical material design, consequences can be drawn:

6. Sum up different combinations of elementary times, which are used in task execution in the same sequence frequently, which means generation of macros for tasks.
7. With the help of such macros, the time consumption for almost any task can be determined easily.
8. The analysis opens the eyes for design deficits in task execution, such as tools, machinery, operational characteristics, and thus allows a standardization of working conditions, tools, and machines.

These principles of the algorithm are still the basis of modern standardized work study systems, for example, the REFA system in Germany (REFA, 1978), and of new developments in time-management systems of different complex tasks, for instance, on a strategic level, meeting market demands by an "economy of speed," on a dispositive level, the approach of "simultaneous engineering" and "just in time" connection of organizational units in a company, and on an operative level, the planning and steering of productive processes by "production planning and control systems" (Theiß, 1996).

12.3.3 Empirical Techniques

More complex forms of task analysis on a somewhat higher level of aggregation in relation to time consumption refer to the flow of information or the sequence of activities, based on industrial engineering ideas, like the formalisms of GANTT-charts, CPM (Critical Path Method), or PERT (Program Evaluation and Review Technique). Examples of this type are time-line analysis, flow-process charts, operational-sequence diagrams, link analyses, etc.

Time-line analysis, according to the American National Standards Institute is "an analytical technique for the derivation of human performance requirements which attends to both the functional and the temporal loading for any given combination of tasks" (Kirwan and Ainsworth, 1992). The charts show clearly any tasks/activities, conducted serially or in parallel, as well as those where the timing of operator's response is dictated by external events (Meister, 1985). An example of a multi-operator time-line analysis is given in Figure 12.11. The technique can be applied to one or more operators with a rather restricted amount of different tasks and activities for not losing the overview, and for different precision levels on the time axis in terms of seconds up to hours.

Flow-process charts use, in addition to time-line analysis, a taxonomy of types of tasks, as executed regularly for instance in industrial goods production (Table 12.8), where

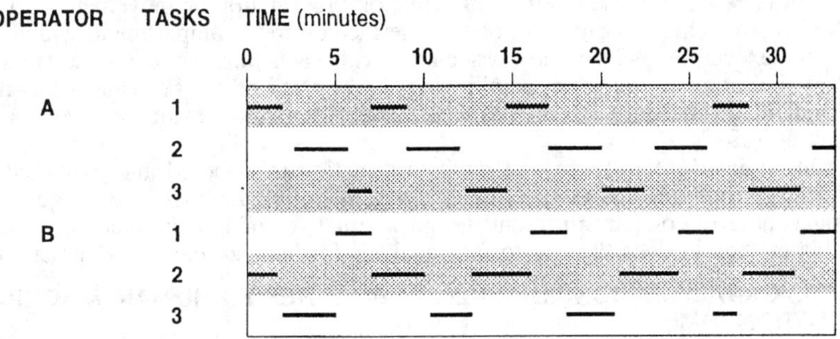

Figure 12.11 Example of a multi-operator time-line analysis.

Table 12.8 Flow Process Chart Task Types and Symbols

Symbol	Description
○	**Operation**—an operation occurs when an object, person, or information is intentionally changed.
⇨	**Transportation**—transport occurs when an object, person, or information moves or is moved from one location to another.
□	**Inspection**—an inspection occurs when an object, person, or information is examined or tested for identification, quality, or quantity.
D	**Delay**—a delay occurs when an object, person, or information waits for the next planned action.
▽	**Storage**—storage occurs when an object, person, or information is kept under control and authorization is required for removal.
○ in □	**Combined operation**—inspection is performed with an operation.
⇨ in ○	**Combined operation**—an operation is performance while a product is in motion.

they were originally developed for motion study (Gilbreth and Gilbreth, 1921). With other taxonomies the method can be adopted easily to different application contexts.

Operational-sequence diagrams are extended forms of flow-process charts, which provide a graphic representation of the flow of information, decision, and activities in a system, using a set of basic graphic symbols and an associated grammar (see Beevis et al., 1992). Especially the flow of information is emphasized by the technique, with symbols for the transmission, receipt, processing, and use of previously stored information.

In *link analysis*, the operational sequences are superimposed with relative weights of "costs" or "resource consumption" to optimize a design solution. Thus, not only "time consumption" is mentioned, but "strength," "distance," and other resources may be scaled too, and even coupled in a multidimensional scaling for the treatment of link data (Siegel, Wolf, and Pilitis, 1982; Laughery and Laughery, 1987).

A *link* is any operational sequence-oriented connection between two elements of an interface or crew compartment, either information presentation or movement of a person or its limbs. An element is any item of the interface or crew compartment, which is used during task execution. "Costs" are associated with each link based on factors such as frequency, importance, time, etc. Minimizing the "costs" of a task sequence is the aim of the analytical procedure. More details and some illustrative examples are reported in Beevis et al. (1992).

As can be easily seen, more and more information is gathered and exploited in the spectrum from time-line analysis to link analysis. With link analysis, a technique is available that is able to compare different design alternatives of human–machine interfaces. Thus, this method bridges the gap to task analysis for human–machine interaction.

12.4 TASK ANALYSIS CONSIDERING APPROACHES TO HUMAN–MACHINE INTERACTION (HMI)

The human operator in complex systems was modeled in various task settings to predict performance depending on design variants. These models of task execution are mainly called *human performance models* (Pew and Baron, 1983) which can be defined as a formal, often quantitative description of the behavior of one or more people in interaction with equipment. Mostly such models are task-specific in the sense that they are oriented to processing stages or processing levels like perception/sampling/diagnosis/monitoring, like regulating/steering/controlling, like decision making/planning/maintaining, and the like (Stein, 1992).

The redefinition of the notion "task" in this context is oriented toward the following criteria:

1. Making highly detailed explanations (predictions) on human operator performance elements in highly constrained task environments in systems
2. Referring to mathematical, logical or grammar-oriented structures of behavior analyses
3. Respecting explicitly or implicitly the behavior of the technical system or its components in informatory terms
4. Emphasizing the information flow, its elements and algorithmic components as primary perspective of task execution
5. Being design oriented in the sense of identifying performance limits and standards, stability reserves and stability criteria, support functions, and support characteristics for human tasks execution

Some theoretical approaches of this type focus on "single" tasks, others are "multiple"-task oriented, and even an approach with the promise to attain "complete" tasks can be reported.

12.4.1 Single-Task Models of HMI

Single-task models or models of individual tasks mostly rely on a theoretical description of one task category in HMI. The most advanced and most frequently applied models in this context are information theory and control theory, which provide the theoretical and mathematical description language for task representation.

12.4.1.1 Monitoring and Decision Making

According to Moray (1986), a human operator is *monitoring* a system when he scans an array of displayed information without taking any action to change the system state to update his knowledge for an appropriate decision preparation. Decision making implies a selection of an appropriate alternative from a set of possibilities, based on sources of information about the system state. A model-oriented classification of related operator tasks implies (Stein, 1992):

- Pure monitoring
- Independent decision (binary or multivalued) based on sequential or nonsequential observations
- Dependent decisions (dynamic) based on heterogenous types of observations
- Heterogenous types of observation and decisions, often embedded in sequences of other task components.

A model by Senders (1983), based upon information theoretic concepts, describes the way in which an operator divides his attention among a number of instruments, while he monitors them. Fundamentally, it assumes that a human operator's fixation frequency for a particular instrument depends on its information generation rate, and, assuming fixed channel capacity with the observer, the attentional demand for any in a set of instruments can be calculated in terms of time percentages and durations.

In newer types of the model, different sampling strategies and signal characteristics are recognized. A connection of this task model to link analysis is apparent and the application context of structuring complex displays in HMI is the same, respecting especially task-inherent monitoring characteristics.

Close to Senders's visual sampling approach, Carbonell (1966) introduces queuing theory concepts, considering operators' actions as well, and thus moves to more realistic task settings. In the same direction, the extension of Smallwood (1967) can be seen, who respects instrument amplitude thresholds.

The more the focus shifts from pure monitoring to decision components of a task, the more information theory concepts may apply to task analysis (see surveys in Luczak, 1975; Boff et al., 1986; Boff and Lincoln, 1988; Stein, 1992). Whenever the number of input and output alternatives in a task, their frequency, or bandwidth is changed, the information content is affected. According to Hicks's law, information-processing latency is a linear function of information content, derived from the number of stimulus-response pairs or better of transinformation. Error frequency and mental workload are affected in the same direction.

When dependent decisions on heterogenous types of observations come into play, the theory of Markov processes or the theory of automates may be the basis of task analyses (see Luczak, 1975) in human–machine interaction. The electrical networks analogy may be used too, as described by Drury et al. (1987), especially in the form of signal flow graphs or other instruments from the theory of graphs, algorithms, and networks to which we will come back later in this article.

12.4.1.2 Manual Control

Because of the detailed description in other parts of this handbook, the comments can be kept very short. Human performance modeling arose several decades ago, when signal processing and control-theoretic concepts were applied to the task component that is called *manual control* (Tustin, 1947; McRuer, 1980; see also the recent survey with Stein, 1992).

The quasilinear model approach is used to predict closed-loop system response as a function of the controlled system dynamics and the input stimuli (Figure 12.12). Results are presented in the frequency domain, and estimates of phase and gain margins can be deduced. Performance predictions for compensatory tracking tasks using a single display and controller are highly accurate. Thus, applications in vehicle control systems (aircraft, ships, cars) to derive stability reserves from a five-parameter quasilinear human operator model and the knowledge of the transfer function of the technical system can be mentioned (Figure 12.12). Workload and associated problems (fatigue, stresses) can also be predicted, based on experimental findings (see, e.g., Luczak, 1975, 1979). The control theory paradigm has been so useful in quantifying control-related human behavior in many tasks, that it has become a common technique mostly for engineering practitioners.

A further development in this context is the optimal control model (Kleinman, Baron, and Levison, 1971) which uses modern control and estimation theory, assuming that a well-trained and motivated human controller behaves optimally to achieve a specified performance criterion, subject to certain internal constraints on the human's information-processing capabilities and subject to the operator's understanding of task objectives. Computer tools for the analysis are available (Allen et al., 1989).

Extensions in terms of theory and applications in specific task situations include supervisory control (Moray, 1986), monitoring, and even decision-making tasks (Stein, 1989, 1992).

12.4.2 Multiple-Task Models of HMI

The growing interest in multiple-task performance and related models depends on many factors. Due to increasing automation and the use of advanced information technology, the human's role in vehicle and process control is shifting from a direct and continously active involvement toward a supervisory control structure, where human–machine interaction is exercised through the mediation of a computer. Supervisory control indicates a hierarchy or coordinated set of human activities that includes initiating, monitoring, detecting events, recognizing, diagnosing, adjusting, and optimizing processes in systems that are otherwise automatically controlled. In many of these situations, the human operator has to accomplish several tasks at the same time and allocate his or her attention and effort appropriately among them. Industrial process monitoring, multiprocess scheduling, aircraft piloting, and air traffic control are among the more obvious examples (Stein, 1992).

12.4.2.1 Supervisory Control

One of the best documented and accessible models is the human supervisory control model.

According to Sheridan (1992), the simplest model of supervisory control might be that of nested control loops where one or more inner loops are automatic and the outer one is manual. In aerospace vehicles the innermost of four nested loops is typically called

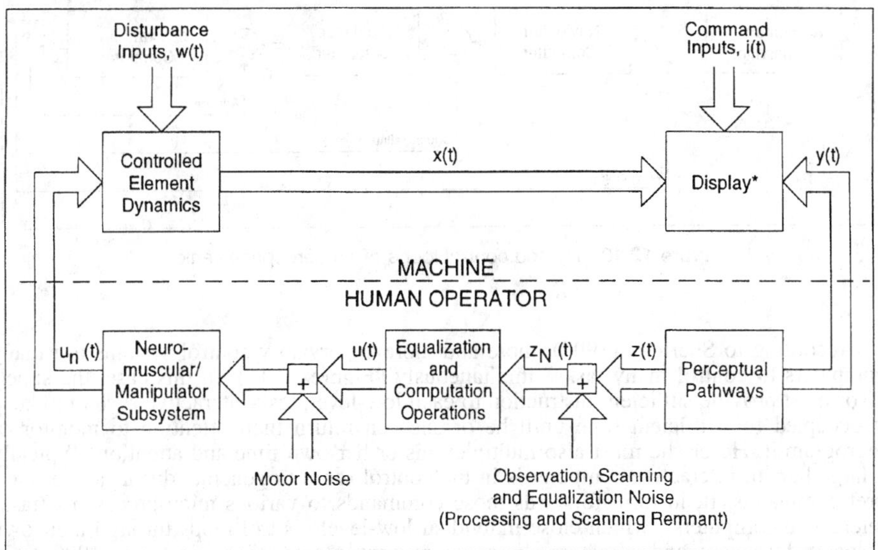

1. The Five Parameter Quasilinear Model:

HUMAN OPERATOR		GAIN		TIME DELAY		SERIES EQUALIZATION		NEUROMUSCULAR ACTUATION SYSTEM
$Y_p(j\omega)$	=	K_p	·	$e^{-j\omega\tau}$	·	$\frac{1 + j\omega T_L}{1 + j\omega T_I}$	·	$\frac{1}{1 + j\omega T_N}$

$Y_p(j\omega)$: human operator describing function

T_L / T_I : lead/lag time constants

T_N : neuromuscular time constant

2. The Crossover Model or McRuer's Law:

HUMAN OPERATOR		CONTROLLED ELEMENT		CROSSOVER TERM
$Y_p(j\omega)$	·	$Y_C(j\omega)$	=	$\frac{\omega_c}{j\omega} e^{-j\omega\tau_e}$

$Y_p(j\omega)$: human operator describing function

$Y_C(j\omega)$: controlled element transfer function

ω_c : crossover frequency

τ_e : effective time delay

Figure 12.12 Task model in control theory terms.

control, the next *guidance*, and the next *navigation*, each having a set point determined by the next outer loop (Figure 12.13). Conventional manual control models can be extended to such multiloop systems. The outer loop in an aerospace vehicle includes the human operator, who, given the mission goals, programs in the destination. When driving a car the functions of navigation, guidance, and control are all performed by the person, and can be seen to correspond roughly to knowledge-based, skill-based, and rule-based behavior (Rasmussen, 1983).

The qualitative functional model of supervisory control shows the various cause–effect loops or relationships among elements of the system, and emphasizing its symmetry as viewed from top and bottom (human, task) of the hierarchy (Figure 12.14).

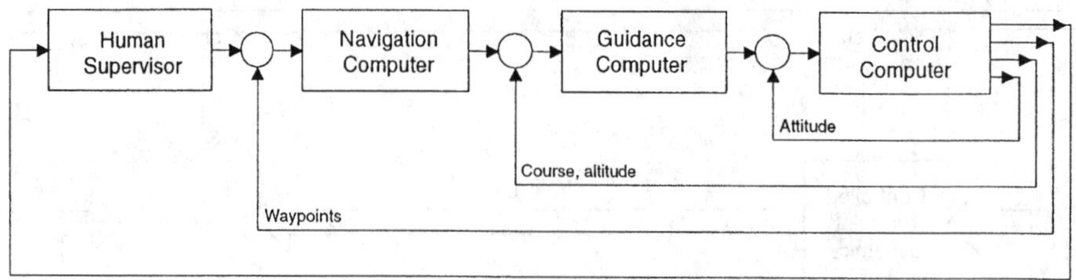

Figure 12.13 Nested control loops of an aerospace vehicle.

According to Sheridan (1992), more and more supervisory control systems use many computers to control many tasks simultaneously (Figure 12.15). In this case, the supervisor must provide sufficient instruction to each low-level, task-interactive system to keep it occupied for sufficient time until he or she can return their attention to monitor or reprogram it. He or she must also multiplex his or her own time and attention. Typically a large human-interactive computer is in the control room to generate displays and interpret commands. It, in turn, forwards those commands to various microprocessors (task-interactive computers), which close individual low-level control loops through their own associated sensors and effectors. In some process plants, there are over 1000 task-interactive computers (TICs), some serving simply as conventional feedback control stations and some performing higher-level decision functions.

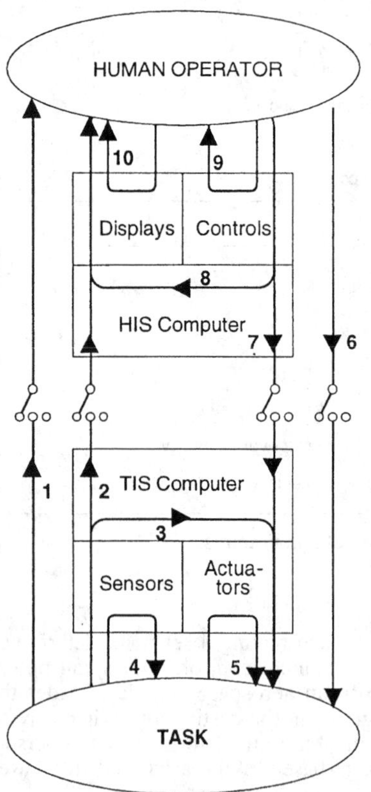

1. Task is observed directly by human operator's own senses.
2. Task is observed indirectly through artificial sensors, computers and displays. This TIS feedback interacts with that from within HIS and is filtered or modified.
3. Task is controlled within TIS automatic mode.
4. Task is affected by the process of being sensed.
5. Task affects actuators and in turn is affected.
6. Human operator directly affects task by manipulation.
7. Human operator affects task indirectly through a controls interface, HIS/TIS computers, and actuators. This control interacts with that from within TIS and is filtered or modified.
8. Human operator gets feedback from within HIS, in editing a program, running a planning model, etc.
9. Human operator orients him- or herself relative to control or adjusts control parameters.
10. Human operator orients him- or herself relative to display or adjusts display parameters.

Figure 12.14 Multiloop model of supervisory control (HIS: human-interactive subsystem; TIS: task-interactive subsystem).

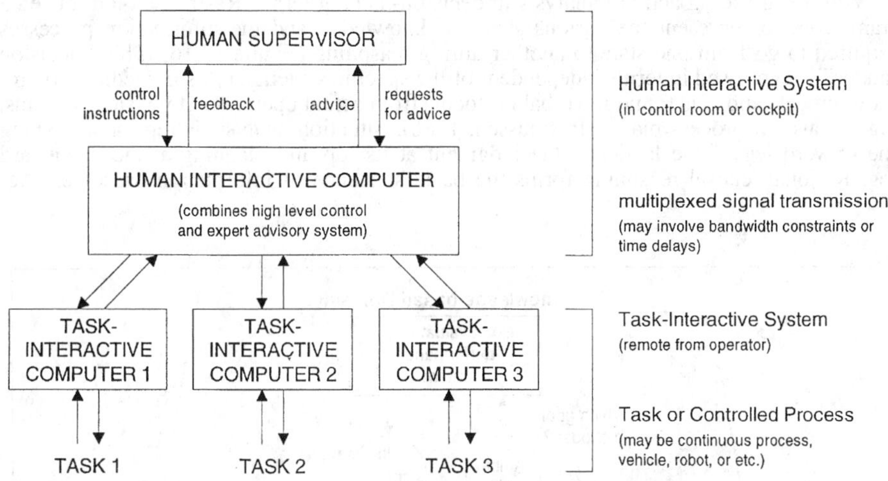

Figure 12.15 Supervision of multiple computers and tasks.

These models are of practical use in two ways. They can be used for system design to estimate the human's ability to coordinate the set of tasks which are proposed for the job. Although the accuracy of specific predictions produced by these models will be highly dependent on the degree to which a task analysis produces reasonable estimates of the arrival and service distributions, etc. The models are also useful for sensitivity analyses (i.e., comparing two or more task sets or comparing two or more parameterizations of a single-task set), in which case the value of the results is less dependent on the accuracy of the parameter estimates.

A variety of multitask models, in addition to the example presented here, are reported by Stein (1992). Most of them are either very specialized, for example, vehicle remote control, or they refer to crew cooperation, a topic of work organization, discussed later.

12.4.2.2 Complex Decision Situations in Cognitive Models

The operators' tasks in highly automated systems contain more and more planning and decision components. The analysis and registration of these components include especially *mental activities* by time-line analysis, process flow charts, operations-sequence diagrams, action-information tables, and similar techniques (Beevis, 1992; Diaper, 1989; Kirwan and Ainsworth, 1991). These methods, however, analyze task-oriented behavior only at the level of skill-based and rule-based action patterns, as Rasmussen, Pejtersen, and Goodstein (1994) would object.

Classical decision models describe the decision process as an analytical comparison of alternatives of action. Options are evaluated on the basis of a combination and weighting of situational characteristics. This seems to be an unrealistic approach, because decisions in realistic situations mostly must be made under time pressure, high risk, and a dynamically changing situation with alternating goals (Klein et al., 1993a,b).

In the following, some models and techniques of cognitive task analysis will be described in more detail which represent the most advanced status of task analysis in this domain; others can be only scratched because of the variety of this field.

Rasmussen: Decision Ladder

According to Rasmussen, a cognitive task analysis is embedded in a general activity analysis, which relates the requirements posed by the work environment to the cognitive resources and subjective performance criteria of the staff members (Rasmussen et al., 1994). Three stages of activity analysis are distinguished: (1) An analysis in terms of the work domain serves to define prototypical work situations and functions. (2) Next, these activities are identified in decision-making terms, that is, control functions that the actors carry out. (3) Finally, the activity is described in cognitive terms by identifying the mental strategies that can be used by the actors in exercising control.

With regard to an activity analysis in decision-making terms, Rasmussen introduced a framework to represent the various states of knowledge and the information processes required to go from one state to another during reasoning (Figure 12.16). This "decision ladder" is expressed in terms independent of the specific system and its immediate control requirements and was found in verbal protocols from actual operation of physical systems, e.g., industrial process plants (Rasmussen, 1986). Situation analysis is represented along the upward leg of the ladder, value judgment at its top and planning at the downward leg. Rational, causal reasoning forms the basic sequence of information processes. Ac-

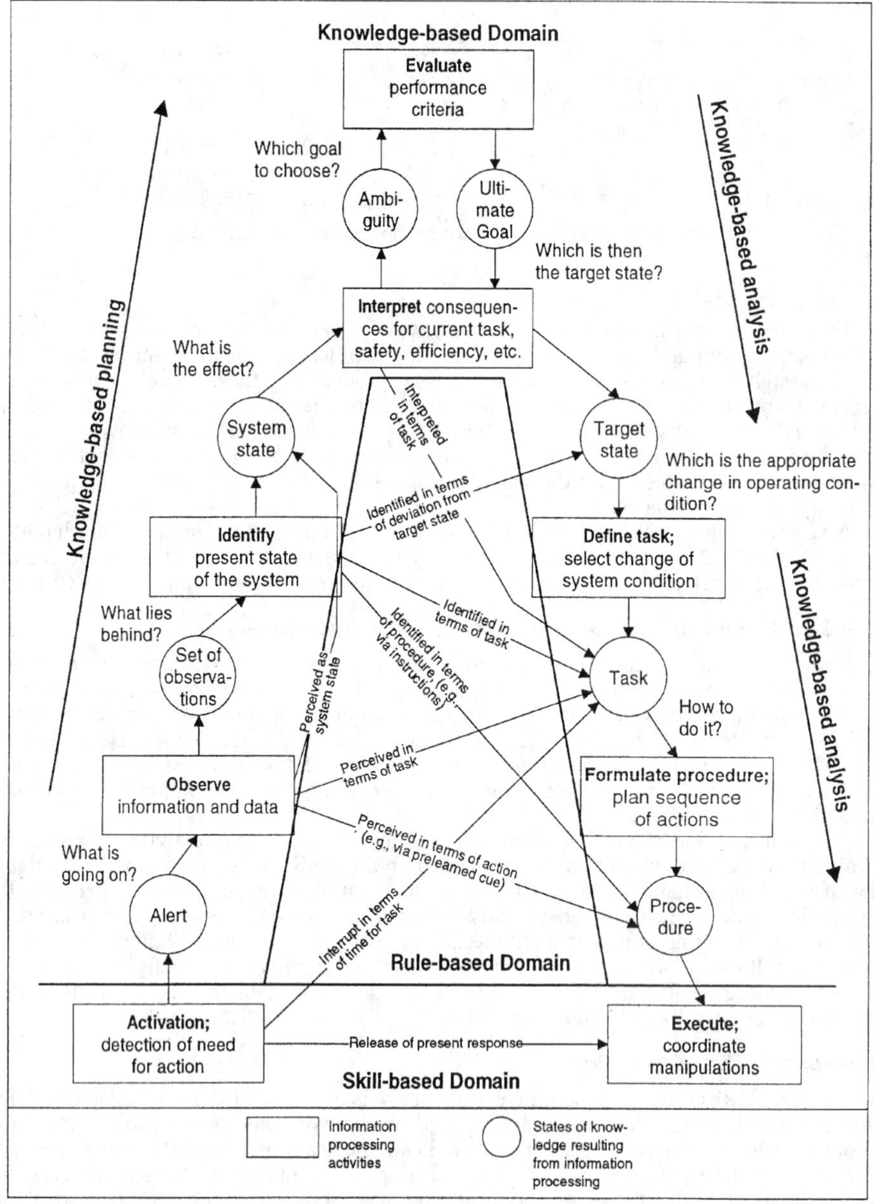

Figure 12.16 Schematic map of Rasmussen's (1986) decision ladder: Rational, causal reasoning connects the states of knowledge. Stereotyped processes can bypass intermediate stages using heuristic shortcuts.

cording to Rasmussen, this case refers to a knowledge- or model-based mode of cognitive control which is used, e.g., by novices or experts during unfamiliar situations. The decision process depends on knowledge about both causal and intentional properties of the work system. The problem space for reasoning might be represented by a Means–Ends Abstraction Hierarchy (see Rasmussen, 1986). A significant characteristic of formal decision making is separation in time. On the other hand, natural or heuristic decision making is better characterized by a continuous control of activities where separate decision events are difficult to differentiate. Heuristic decision making is applied by skilled actors in a familiar work situation and relies on rule-based mode of cognitive control. Thus, "shortcuts" based on procedural know-how occur in the basic decision sequence. Familiar states of the environment are associated with actions that have been effective in previous encounters (Rasmussen, 1993). At the lowest level of human performance, skill-based control is characterized by the ability to subconsciously generate the movement patterns required for interaction with a familiar environment by means of an internal, dynamic world model. This level is typical of the master, or expert, that has been fascinating philosophers and artists through the ages (Rasmussen, 1993).

Methodically for an analysis in decision-making terms, Rasmussen and his coworkers recommend interview techniques, where the analyst participates in the work process, preferably with a tape recorder. Thus, the activities are structured and categorized into a combination of the decision function elements defined by the decision ladder. In this way it is possible to identify the information requirements of the different typical, recurrent activity elements, and to find all the sections of the activity records belonging within the same category of mental activity in order to have material for detailed analysis of the mental strategies. Mental strategies itself are further analyzed by an individual behavioral trace through time on task. Thus, the various information-processing strategies that can be used for decision functions are identified. See Rasmussen references (Hollnagel et al., 1981; Sanderson and Fisher, 1994) for a detailed description of this kind of analysis.

The decision ladder was applied in a variety of contexts. The use of this model as a sketch pad to record the actual decision paths during different diagnostic sessions in a hospital is outlined in Rasmussen et al. (1994). In the context of new technologies and human error, the decision ladder was applied for analysis of the phases which can be supported in a driving accident (Leplat and Rasmussen, 1987). Lind (1987) considered the decision model as a framework for representation of control knowledge which could be used in the design of knowledge-based systems. Ultimately, he maps the classical control cycle of AI production systems into the model in order to indicate how to extend the cycle for future interpreting and evaluation capabilities.

Card, Moran, and Newell: Goals, Operators, Methods, and Selection Rules (GOMS)

The GOMS model was developed by Card et al. (1983) based upon previous work in the domain of human problem solving by Newell and Simon (1972). GOMS has been developed as a cognitive engineering model. Therefore its primary purpose is to be used in design of human–computer interfaces, rather than a foundation of scientific theories. GOMS models are applicable to routine cognitive skills comparable with Rasmussen's rule-based mode of cognitive control, where the focus is on procedural knowledge. With reference to Rasmussen's "decision ladder," GOMS covers all cognitive processes in principle, but has a strong emphasis on the planning aspects represented at the downward leg of the ladder. The model is used to outline the cognitive performance of a person by decomposing a problem into hierarchical subgoals and goal stacks. The basic GOMS model is best for making qualitative predictions about differences between tasks where users make no or few errors. By associating times or time distributions with each operator, such a model is also able to make total time or statistical predictions. Depending on the grain of analysis, several variations of GOMS models can be explored to make quantitative predictions, e.g., from Unit Task to Keystroke-Level models. An application example on the Unit Task level is given in Figure 12.17.

The four basic elements of GOMS are a set of goals, a set of operators, a set of methods for achieving the goals, and a set of selection rules for choosing among competing methods for goals. A goal is a symbolic structure defining a state of affairs to be achieved and determining a set of possible methods to accomplish it. The dynamic function of a goal is to provide a memory point to which the system can return upon failure or error and from which information can be obtained about what is desired, what methods are available, and what has been already tried (Card et al., 1983). Operators are elementary

UNIT TASK LEVEL

Model UT:

GOAL: EDIT MANUSCRIPT	
EDIT UNIT TASK	• repeat until no more unit tasks

FUNCTIONAL LEVEL

Model F1:

GOAL: EDIT MANUSCRIPT	
• GOAL: EDIT UNIT TASK	• repeat until no more unit tasks
• • GOAL: ACQUIRE UNIT TASK	• • if task not remembered
• • • GET NEXT PAGE	• • • if at end of manuscript page
• • • GET NEXT TASK	
• • GOAL: EXECUTE UNIT TASK	• • if an edit task was found
• • • LOCATE LINE	• • • if task not on current line
• • • MODIFY TEXT	
• • • VERIFY EDIT	

Model F2:

GOAL: EDIT MANUSCRIPT	
• GOAL: EDIT UNIT TASK	• repeat until no more unit tasks
• • GOAL: ACQUIRE UNIT TASK	• • if task not remembered
• • • GET NEXT PAGE	• • • if at end of manuscript page
• • • GET NEXT TASK	
• • GOAL: EXECUTE UNIT TASK	• • if an edit task was found
• • • GOAL: LOCATE LINE	• • • if task not on current line
• • • • (select USE QS METHOD)	
• • • • USE LF METHOD)	
• • • GOAL: MODIFY TEXT	
• • • • (select USE S COMMAND	
• • • • USE M COMMAND)	
• • • • VERIFY EDIT	

Figure 12.17 Application example of GOMS for the unit task level.

perceptual, motor, or cognitive acts whose execution is necessary to change any aspects of the user's mental strategy or to affect the task environment. Methods describe procedures for accomplishing goals in terms of operators or other goals. Operators that are often used together as an element are grouped into methods. Ultimately, to model user behavior the selection of methods must be integrated. Therefore selection rules are specified through the basic flow structures of GOMS like: IF(condition is true in the task situation), THEN use method M.

Card et al. (1983) methodically evaluated the GOMS model with an observation technique of user behavior in close laboratory analogue of the task commonly performed. In general, the user's behavior was recorded with video cameras at proper positions; keystrokes and system's responses were time-stamped and recorded on a computer file. These experimental data were used to derive the operators and methods chosen for each task and the reasons for choosing them.

GOMS has received much more empirical testing than any other analytical model of human–computer interaction tasks (Gugerty, 1993). Models have been developed for a wide variety of contexts, including the use of simple text editors (Card et al., 1983; Douglas and Moran, 1983; Singley and Anderson, 1988), spreadsheets (Olsen and Nilsen, 1988) or hypertext applications (Carmel et al., 1992). GOMS served as a basis for further development of cognitive task-analysis models like CCT or NGOMSL (Kieras and Polson, 1985; Kieras, 1988). An application example of GOMS is shown in Figure 12.17 for the unit task level.

Klein: Recognition-Primed Decisions (RPD)

In contrast to traditional analytical models of decision making, Klein introduced a Recognition-Primed Decision (RPD) model that shows how people can use experience to avoid the limitations of analytical strategies (Klein et al., 1986; Klein, 1989; Klein, 1993a). According to Klein (1993a), it fuses two processes—situation assessment and mental simulation—and asserts that people use situation assessment to generate a plausible course of action and use mental simulation to evaluate that course of action. Again, with reference to Rasmussen's decision ladder, the RPD model covers all cognitive processes in principle but has an emphasize on action and focuses on serial evaluation of options. The model distinguishes three cases (Figure 12.18): In the first case—*Simple Match*—the situation is recognized and the obvious reaction implemented. In the second case—*Developing a Course of Action*—the decision maker performs some conscious evaluation of the reaction in terms of a mental simulation. The third case—*Complex RPD Strategy*—is one in which the evaluation reveals flaws requiring modification, or the option is judged inadequate and rejected in favor of the next most typical reaction. Because of the importance of such evaluation, the decision is primed by the way the situation is recognized and not completely determined by that reaction.

RPD focuses on situation assessment rather than judgment of feasable options on the one hand and relies on satisficing rather than optimizing—finding the first option that works, not neccessarily the best—on the other. Thus, a recognitional strategy enables the decision maker to be continually prepared to initiate action by committing to the option being evaluated. However, formal strategies require the decision maker to wait until the analyses are completed before finding out which option was rated the highest (Klein, 1993a). The RPD modeling approach is appropriate when the decision maker is experienced, when time pressure is great, and when conditions are less stable. Analytical or formal decision making is preferred when the available data is abstract and alphanumeric rather than perceptual, when the problems are very combinatorial, when there is a dispute between different constituencies, and when there is a strong requirement to justify the course of action chosen.

Methodically, the RPD Model was developed using both qualitative and quantitative methods of investigation in real task situations, military exercises, computer simulations, and classroom planning exercises. This includes semistructured interviews, on-site observation, and protocol analysis (Klein, 1993a).

Studies of natural decision making in a variety of domains, e.g., fireground command (Klein et al., 1986), U.S. Army Armored Disvision personnel (Klein, 1989), battle planning (Thordsen et al., 1990), neonatal intensive care nursery (Crandall and Calderwood, 1989), and chess tournament play (Calderwood et al., 1988) served as a basis for the development of the RPD model (Klein, 1993a). In a survey of five empirical studies the measured proportion of decision points handled using RPD strategies versus concurrent deliberation of options varied from 80% for urban fireground commanders to 42% for tank platoon leaders (Klein, 1989). Noble (1993) proposed a model to support the development of situation assessment aids that relates to RPD directly. Finally, Klein (1993b)

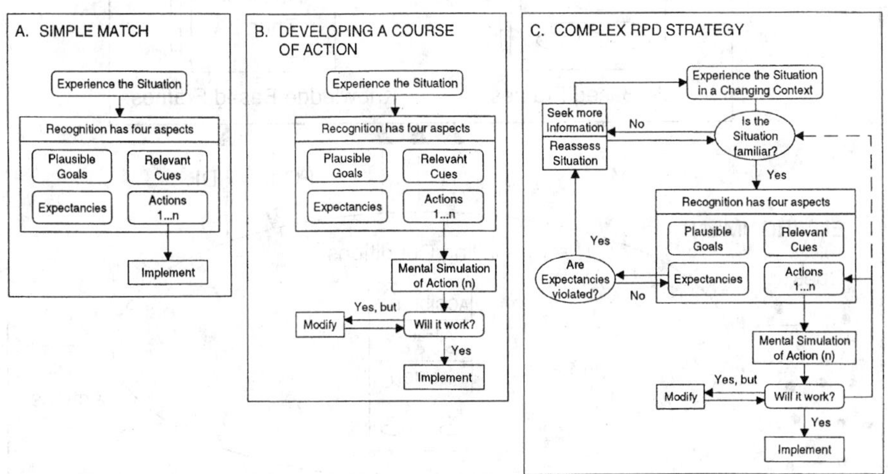

Figure 12.18 Illustration of decision stages of the Recognition Primed Decision model as indicated by Klein (1993a).

attempts to synthesize five cognitive decision models including Rasmussen's framework into a superprocess model of naturalistic decision making.

Cacciabue: Cosimo—Cognitive Simulation Model

According to Cacciabue and Kjaer-Hansen (1993), Cosimo focuses on decision making in accident situations and emphasizes in its current state of development, the selection and implementation of emergency procedures taken by an operator. The emergency procedures are activated and selected based on heuristic knowledge and preset frames for interpretation of cues of data from the environment and for actions to cope with the emergency situations (Figure 12.19). The operator, modeled by Cosimo, is considered to be highly skilled in controlling the complex physical system. Usually, in emergency situations, the operators are bounded by strict time restrictions and are requested to respond immediately. As a consequence, they are expected to bypass the stages of determining the exact system state and to elaborate investigation of future system states.

In Cosimo, the activation phase may involve activities generated as a consequence of alarms received from the interface to the physical system or from monitoring data displayed to the operator. The cues of data formed by the environment are decoded in relation to the physical components from which they were obtained and attributed with subjective and semantic interpretations. In an emergency situation, the operator may select a preset frame for action, or schemata, based on a partial match between the subjectively interpreted cues and the preconditions specified in the frame. To make a partial verification of the selected frame, the operator may, however, seek further information to support the immediate selection of a frame for action. If the frame can get the required support of belief, the specification for action is activated. During the simulations of accident situations, Cosimo behaves according to the above scenario and consequently reflects the lower levels of the "ladder" model.

Specifically, it accounts for the three information-processing activities of: *activation*, *observation*, and *execution*.

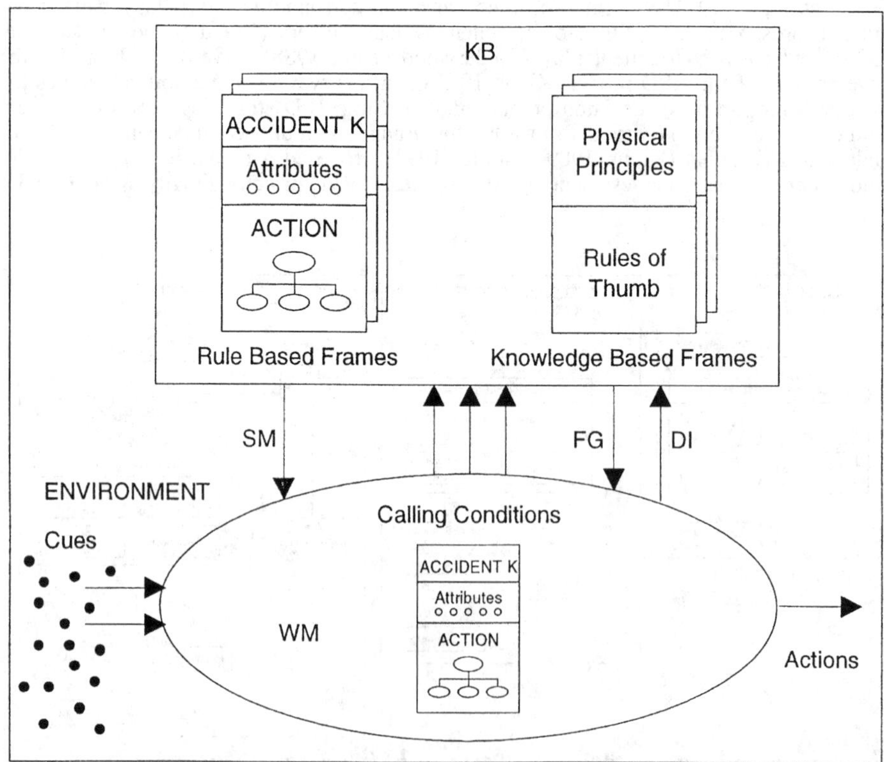

Figure 12.19 COSIMO theoretical representation (Cacciabue and Kjaer-Hansen, 1993).

Hollnagel: Contextual Control Model

Hollnagel (1993) argues—in the same context of dynamic process control—that the modeling of cognition should attempt to reproduce key phenomena of a task in practical terms rather than replicate the mechanism of a specific theory. According to his literature review, most cognitive models refer to procedural prototypes, and he outlines the alternative to focus on contextual control: A model is proposed which explicitly describes the way in which actions are chosen to meet the requirements of a situation. Control is described in terms of different control modes which match different performance characteristics, for instance, scrambled/opportunities/tactical/strategic controls and the transition between modes of control according to task complexity in the situation encountered.

Grammar Techniques

Grammar techniques concentrate on the operationalization of users' knowledge. They are based on grammar notations which allow you to generate the "legal" sentences within a given language. With respect to interactive appplications of user and system, these sentences correspond with action sequences. The language corresponds to the command language of an interface or to a representation and interpretation of this command language.

The system that is designed by the task analyst is a set of rules in a given grammar which transforms the commands into action sequences to be executed.

If a certain task (for example, "delete a word in a text") is given, the respective rules will develop the task execution into a sequence of actions. Mappings in between commands or tasks and the respective actions are unambiguous.

Two techniques of this class are the BNF-grammar by Reisner (1981) and the TAG-grammar by Payne and Greene (1986). The TAG-grammar (Task-Action-Grammar) translates global tasks to specific and simple tasks which are presumed to be recognized by a potential user as such in the represented form. Thus, the technique has a certain relation to a cognitive representation, because semantic aspects of user understanding of the tasks and commmands are analyzed.

Semiotic Models

In the early years of software-development segments of human–machine interaction were investigated empirically to generate appropriate support functions. Today, a holistic view of the system based on a sound analysis of tasks to be performed is a common strategy. Nevertheless, a model frame is necessary for this approach. Based on works of Morris (1946) on signal theory, who distinguished between a semantic, syntactic, and lexical level, a distinction in the analysis of user interfaces is commonly used in a conceptual framework with four levels: the input-output level, the dialogue level, the function level, and the task level. The application of this idea in the design of software systems is far advanced: For instance, the VDI-Richtlinie 5005 (guideline of design for bureau-communication systems, authorized by the German association of engineers) distinguishes as design goals "competency-improvement," "flexibility of action," "task-orientation," and "adaptability," and recommends a detailed analysis of these criteria on the four levels of abstraction top-down:

- Task
- Function
- Operation
- Input-output

to break down the global task in an application situation into subtasks and respective software support. Applications include, for example, the design of CAD systems to support creative-informational work of designers.

Figure 12.20 shows the elaboration of a designer's subtask, e.g., the ensurance of a centrification. First of all, on the pragmatic level the user has to define the goals as to design a product part in terms of designing for manufacturing. After the formulation to ensure the centrification, on the semantic or functional level, a chamfer might be chosen as one possible solution. Also, position and dimensions of the chamfer are defined on that level. In a design department, normally the technical drawing represents the defined syntax for the technical communication. Therefore the syntax of the interaction process consists of the generation of lines, the selection of geometrical elements as borders, and

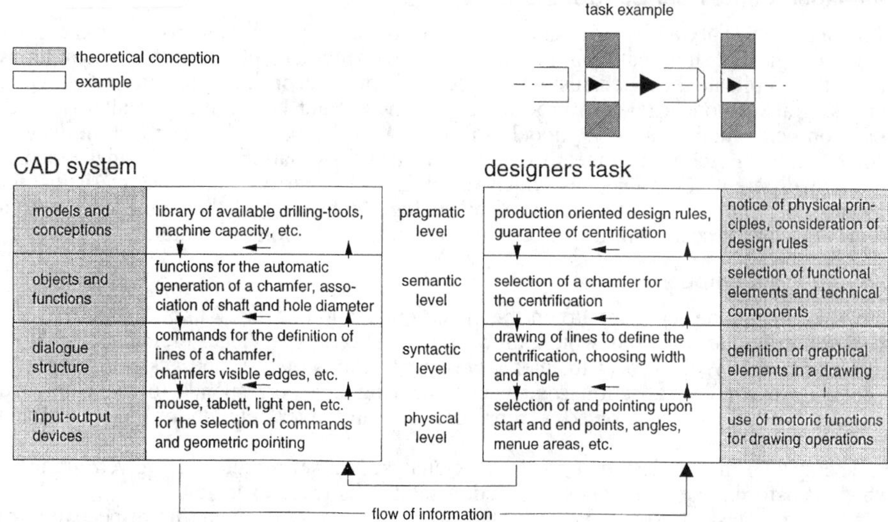

Figure 12.20 Semiotic four-level model of communication, as applied to a design task (Springer et al., 1991).

the definition of start and end points and angles of the chamfer lines. On the physical level, movements must be carried out for a selection of commands or the identification of elements. These movements are initiating commands for generation or manipulation of the chamfer lines to represent the border lines of the centrification; that's the syntactical level of the CAD system. The functionality of the CAD software generates the lines by electronic data and, if possible, connects, for example, the diameter of the shaft and the drill-hole semantically. Finally, for example, on the pragmatic level of the CAD system, the system could check for the existence of tools for the production of such a chamfer, or the production costs.

Shown by this example, it is possible to describe all tasks, activities, and operations on the user's side and of the CAD system with the help of such a hierarchically structured systematic (Springer et al., 1991).

Empirical Framework

To get a better understanding of mental representations of the operator, it is not definitely necessary to dispose of a model of task execution in terms of human problem solving (Rasmussen, 1986) or in terms of situational control specifications (Hollnagel, 1991). A purely empirical method of task description with a high relevance for the cognitive components was presented by Sebillotte (1991), oriented to operators' objectives. The method is based on semidirected interviews of operators with an informal part and a directed part. In the informal part, the operators talk about their work without interruption to find out where the spontaneous level of their representations are situated. In the directed parts of the interview the "why and how" is used making it possible to distinguish between the operator-intended goals and their way of attaining these goals. The extraction of task-related information, the determination of different tasks and their decomposition to sub-tasks, the use of a formalism of description for task representation, and the reference to psychological/cognitive models are part of the method. It "works" in air traffic control and document preparation for send out, as demonstrated by the author.

Some Others

Many other analytical procedures for this type of task analysis in HMI are available:

- *Concept mapping* is an interactive interview method for the formulation of a user-oriented problem clarification. A graphical "concept map" consists of *nodes* and *links*. The nodes are described as circles and contain concepts which represent objects, actions, and events. The links are designed as lines and describe by words the relation between concepts (Glowin and Nowak, 1984; Thordson, 1991).

- *Critical decision method* is a retrospective interview technique by which—on the basis of selected events—the goals, the options of action, the indices, and context elements are investigated which are connected to a concrete decision taking (Klein et al., 1989; Klinger and Gomes, 1993).
- *Cognitive walkthroughs* allow—by a detailed procedure and with help from scenario generation—to simulate mentally the problem-solving process of a person (Nielsen and Mack, 1994).
- *Hierarchical task analysis* comprises the analysis of complex tasks via a description of less complex subtasks. Tasks are registered in the form of a hierarchy of actions and presuppositions which are necessary to reach goals (Diaper, 1989).
- *Task analysis for knowledge description* uses a grammar of knowledge representation to describe the amount and context of knowledge to execute a task. The derived information can be mapped to a hierarchy of task description (Payne and Greene, 1986).

As mentioned, all these methods of task analysis were developed more or less to get design criteria for user interfaces of knowledge-based systems. Thus, human–computer interaction was the main driving force to generate more advanced methods of cognitive task analysis (Redding, 1990; Cooke, 1994; Ziegler, 1988, 1994; Merkelbach and Schraagen, 1994).

12.4.3 Relations Between Task and Error Analysis

Cognitive task analysis and error analysis are strongly related because "slips are the window to the mind," as Norman (1979) states. In spite of the fact that empirical techniques of error and reliability analysis from the past refer to human variability of behavior (Swain and Guttmann, 1983), a cognitive view predominates (Reason, 1990; Rouse, 1990; Wehner, 1996). This becomes apparent especially in an essay on "error analysis, instrument and object of task analysis" by J. Leplat (1989), which is followed further in its argumentation.

The notion of error implies the notion of a standard and of deviations from it. Standards are set by the goals/objectives of the task and the modes of its execution. From this definition Leplat (1989) derives five representations of the task:

T1: The prescribed task = equivalent to the order set by a superior or by an organization, the task instructions, the rationale of a best-way of execution;

T2: The implied prescribed task = intricate personal representation of the person of his or her task, made clear by the question "What do you think you are expected to do?"

T3: The task the operator sets himself = personal redefinition of goals, of considered conditions, of instructions, of preferred subtasks in logical and timely order.

T4: The task actually carried out = activity schedule of the operator, visible or analyzable work elements to be executed.

T5: The representation of the performed task as identified by the question "What have you done?"

In relation to these five representations of the task, the deviations can be derived and discussed:

1. Deviations between a prescribed task and a task which is assumed to have been prescribed: The primary source of these deviations is the poor definition or understanding of the prescribed task. These deviations Δ T1 T2 often produce "mistake" type errors.
2. Deviations between the task assumed to have been prescribed and the task the operator prescribes herself: The deviations Δ T2 T3 indicate that the subject is not doing what is expected of her. This type of error may be named "misinformation" or "mismatch."
3. Deviations between the prescribed task and the task the subject prescribes herself: These deviations Δ T1 T3 show that the subject has not intended to carry out what

is prescribed; especially imposed precautions and restrictions/requirements are not respected which may cause errors of a *misfit* type.

4. Deviations between the task the subject prescribes himself and the task he actually carries out: These deviations Δ T3 T4 often give rise to errors known as *slips* with a correct aim but an inadequate plan of action.
5. Deviations between the task actually carried out and the subjects' representation of this task may occur when automized systems or high inertia procedure control dynamics do not react as intended according to a current action sequence. Thus, this Δ T4 T5 error is mostly due to system intransparancy. FMEA—failure mode and effect analysis—may be an approach to this error type.

It is obvious by this conceptual analysis of the notion task and the notion error, how strongly the concepts interfere. This may apply to error recovery as well: an essential condition of recovery is that the error is noticeable and noticed, i.e., that the task includes the possibility of providing feedbacks which may be seen as the Δ Tx Ty as well. Based on these fundamental schemata of deviation in terms of task-reinterpretation differences, Leplat and coworkers (1990) refer to empirical and computer-based methods and representation techniques for error analysis.

A practical method of error analysis is *THERP* (Technique for Human Error Rate Prediction), which predicts the impact of human operator errors on system operation. THERP produces two kinds of performance measures (Swain and Guttmann, 1983):

- *Task reliability*: An estimate that a task will be completed successfully
- *Recovery factors*: Estimates of the probability of detecting and correcting incorrect task performance in time to avoid undesirable consequences on system performance.

Thus, THERP permits the estimation of the probability that an operation will result in an error, and the probability that an error will result in system failure. The technique uses conventional reliability engineering approaches with modifications to acknowledge the greater variability, unpredictability, and interdependence of human performance: The basis of THERP is the preparation of a human reliability analysis event tree, to which probabilities of success or failure are attached. Those probabilities are modified by performance-shaping factors, which reflect the effect of stress/skill/environmental conditions on operator reliability. THERP was widely applied in the task analysis and operation of huge technical systems, especially nuclear power plants. In less rigorous and detailed approaches than THERP the errors are identified on the basis of an operator task analysis in empirical terms, and reference to an error classification scheme, like Hammer (1972) with his list of 34 causes of primary errors; Norman (1983) with his design rules based on task-oriented error analysis; and Rouse (1990) with a provision of a classification scheme for 31 types of error in operating systems. Reason (1987) proposes a Generic Error Modelling System (GEMS) based on Rasmussen's (1983) categories of skill-, rule- and knowledge-based behavior.

12.4.4 "Complete" Tasks in Action Regulation Theory and Related Concepts

An approach for the analysis of tasks, especially as basis for the design of complete tasks (overcoming work partition) and meaningful jobs, was developed in central Europe on the basis of the theory of action regulation (Dunckel, 1996; Hacker, 1986, 1993; Oesterreich and Volpert, 1987; Volpert, 1987). Analogous concepts are, for instance, hierarchical task analysis (Sebillotte, 1988, 1989) and some more specialized approaches for the analysis of human–computer interaction, for example, job structure analysis (Downs, 1989) or task-action grammers (Payne and Greene, 1986). Following a description by Hacker (1993), one of the main protagonists of action regulation theory, the essential common denominators of these approaches are the intentionality of human activity, the hierarchies of goals organizing activity, and the reconceptualization of schema theory in terms of task analysis (e.g., Lindsay and Norman, 1977; Norman, 1982). The action that is the measure to be taken to cope with a given situation becomes the central mode of a network of relationships or connectivities, e.g., with the objects or instruments of an action. This framework and concepts task analysis emphasizes:

- The key role of goals as anticipations, intentions, and frames of reference for the feedback of results

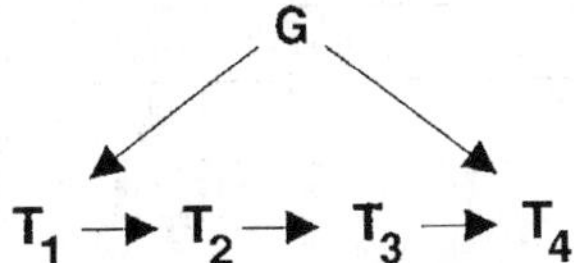

Figure 12.21 Relation between goals and transformations.

- The regulation of activities by hierarchies of goals, which are implemented sequentially
- The hierarchical multilevel structure of the feedback units of actions
- The regulation function of mental representations which simultaneously will consist of propositional, iconic, and motor (kinesthetic) parts

An illustrative visualization of these ideas is presented by Volpert (1982, 1987), as seen in Figure 12.21.

In an analogy to a technical control loop, the self-set or preset goal G is reached by transformations $T_1 - T_4$ in a gradual approach. G is the idea of the status of work or work environment after the transformation, and the T_x are the sequential operations in time from start to the intended completion of the task. The T_x are the visible surface of task execution, whereas the G forms the invisible structure of intentions which are determined by anticipating the work result to be attained by the T_x-operations. In principle, this model of task analysis is equivalent to the idea of Miller, Galenter, and Pribram (1960) that human task execution is organized in TOTE (Test-Operate-Test-Exit) units, which—on different levels—take care that a global control loop forms the goals for a more detailed goal setting unit, which in itself is responsible for the steering of operations. Thus, hierarchical coupling of control loops is the basic idea of TOTE units *and* action regulation theory. Another fundamental principle is the anticipation of the intended work result as the primary goal. In terms of control theory, we would speak of a feed-forward principle which is combined with feedback principles on a more detailed level (see Figures 12.22 and 12.23).

Two other important elements of the concept of action regulation theory are the "plan," namely, a hierarchically organized sequence of instructions for the steering of actions (see Figure 12.23), and the "image," which means the accumulated set of knowledge of a person about herself and the environment. In newer publications about action-regulation theory, the plan and the image are combined in the "operative image-system" (operatives Abbildsystem = OAS). OAS contain all relevant information for the execution of tasks in terms of work objects, work means, and the necessary action sequence with respect to task goals. Inadequate OAS may be the cause of bad performance, trial and error procedures, and mistakes.

As shown, the structure of action regulation is hierarchically organized. The execution of real operations, however, can only take place successively (Figure 12.23). Proceeding from this hierarchical structure different levels of regulation were identified or defined.

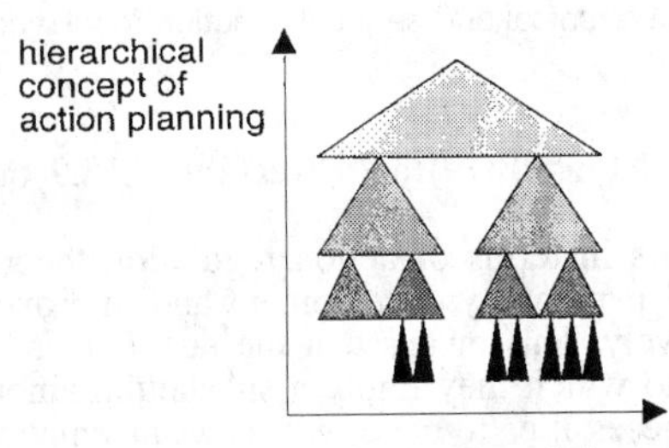

Figure 12.22 Hierarchical and sequential action regulation.

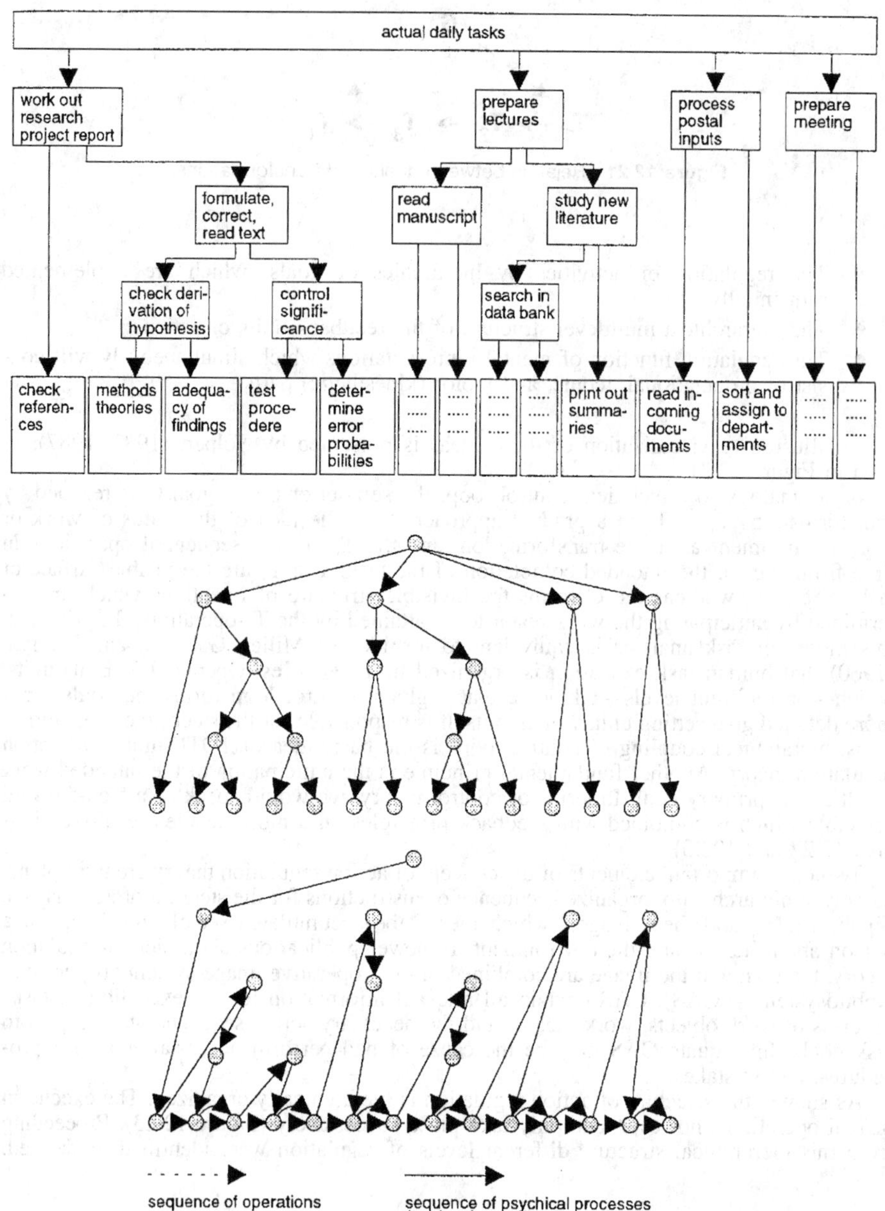

Figure 12.23 Example for a hierarchical and sequential action regulation (professors tasks).

In between three (Figure 12.24 on p. 379) and five (Table 12.9 on p. 379) levels are distinguished.

The purpose of task analysis in terms of action regulation theory is to evaluate the possibilities and restrictions of personality development and improvement by a potential or given task and job respectively. This means that the task demands are analyzed from the perspective of the degree to which they imply a substantial amount of time for self-set and creative activities, of necessities to cope with new information contents and possibilities to extend ones own knowledge, and in general to develop qualification in a continous improvement process not only in performance-related terms but in social-

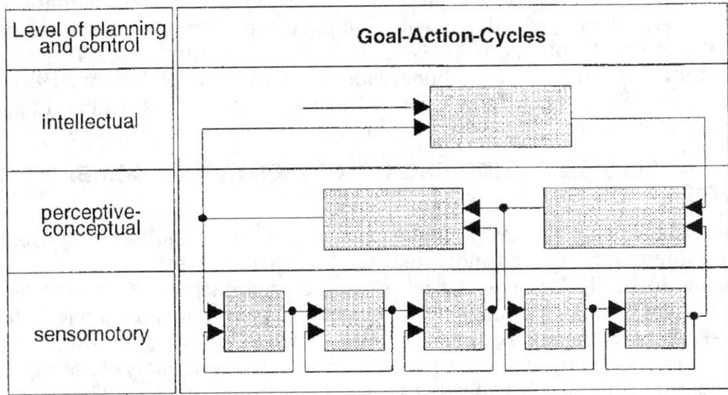

Figure 12.24 Coordination of goal-action cycles.

interactive terms as well. In other words, action regulation theory is a basis for the creation and establishment of interesting and meaningful jobs.

Thus, practice-oriented methods of analysis, evaluation, and design indications were developed from the theoretical concept (summarizing overview with Bonitz, 1995). Especially the scope of action is classified by the VERA method for tasks in industrial production (Volpert et al., 1983, 1989), the regulation frictions and hindrances by the RHIA method (Leitner, Volpert et al., 1987), and the special references to creative and administrative work in the "bureau" is made by the KABA method (Dunckel, Volpert et al., 1993). Especially the personal aspect is represented by the TBS (= Tätigkeitsbewertungssystem), a person-oriented and subjective method of rating of personality devel-

Table 12.9 Five-Level Concept of Regulation Requirements

	Short definitions of 5 levels and 10 steps of regulation requirements
Level 5	*Establishing new working processes*
Step 5	New interacting working processes, their coordination and material conditions to be planned.
Step 5R:	As for Step 5: the new working proc sses are complements to processes already in operation, to which as few changes as possible should be made.
Level 4	*Coordinating several working processes*
Step 4:	Several sub-goal plans (in the sense of Step 3) for interacting parts of the working process are to be coordinated with one another.
Step 4R:	Although only sub-goal planning is required, conditions for other sub-goal plans (not to be formulated by the worker himself) must be considered here.
Level 3	*Sub-goal planning*
Step 3:	Only a roughly determined sequence of sub-activities can be planned in advance. Each of them requires the worker to make plans of his own (in the sense of Step 2).
Step 3R:	A sequence of sub-activities is determined in advance. Each sub-activity requires the worker to make plans of his own.
Level 2	*Action planning*
Step 2:	The sequence of work steps must be planned in advance; the planning, however, only extends to the result of the work.
Step 2R:	The sequence of work steps is pre-ordained. However, it varies repeatedly to such an extent that it has to be mentally processed in advance.
Level 1	*Sensory-motor regulation*
Step 1:	No conscious planning is required for the projection of the sequence of work movements to be regulated, although a different tool occasionally has to be used.
Step 1R:	As for Step 1, but only the same tools are required in each case.

opment potentials (Hacker, Iwanowa, and Richter, 1983). The design intention is emphasized in the ATAA (= Analyse von Tätigkeitsstrukturen und zur prospektiven Arbeitsgestaltung bei Automation/Analysis of job/task structures and prospective work-design in automation projects) (Wächtler, Modrow-Thiel, and Roßmann, 1989). In conclusion, this type of task analysis and derived methods had a considerable impact on the discussion of "humanization of work" in Europe.

12.5 TASK ANALYSIS CONSIDERING COMPONENTS AND COMBINATIONS OF A ONE-PERSON JOB

At this level of task analysis, several approaches superimpose and interfere: Coming back to the work system approach an individual working person—the human—interacts with its work means to fulfill the goals set by a task description for the transformation of an input to an output. This idea is the core of a task analysis on this level and is underlying most methods and techniques. Nevertheless the nomenclature is rather diverse: It implies beneath the term task analysis the terms position analysis, job analysis, occupation analysis, and others. The relation between these different terminologies may be seen as follows (Figure 12.25).

Following the biographical qualification path of an individual the process of building up competencies by different education, training, and experience phases may be investigated by a career analysis, until the demands of an actual position can be met. Thus, competencies are those psychical and physical capacities, which a person must possess to cope with certain tasks. On the other hand, demands are those working conditions, imposed on the working person, which influence the working behavior, for instance, task-oriented information to be perceived and processed by the person and so stimulates competencies.

A position consists of a group of *tasks*, which are performed/executed by a certain person. Thus, within an organization/company the number of employees is equal to the number of *positions*.

The term *job* can be understood as a synonym to position or as a generic term: A job consists of a group of similar positions within an organization, the same company, the same organizational unit. In a job, one or more persons can be employed.

An *occupation* is a group of similar jobs in different companies and organizations.

Thus, tasks, positions, jobs, and occupations stand in structural relation with each other in the sense of agglomeration and division: A position and job analysis method and

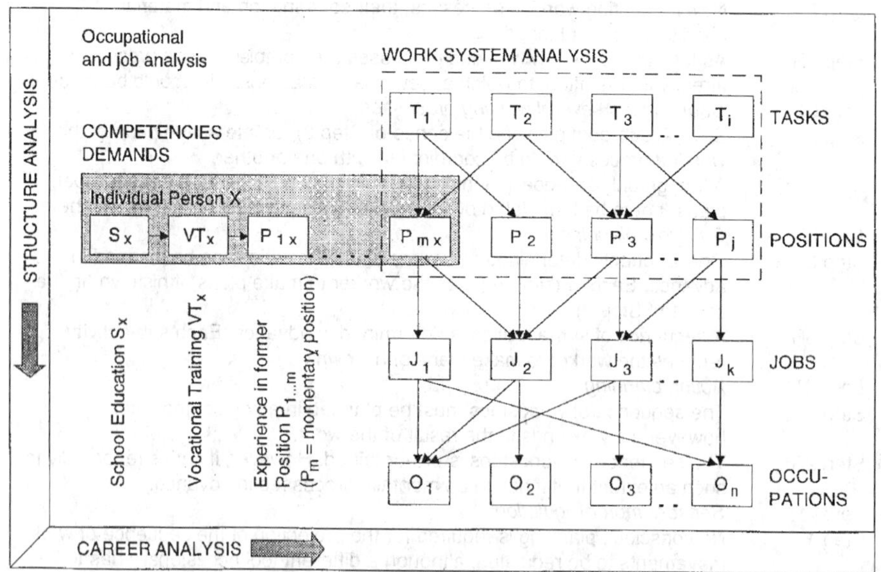

Figure 12.25 Relation between task analysis and other concepts (modified from Frieling, 1974).

technique may imply mainly aspects of task analysis on different levels of aggregation (micro-, meso-, macroformulation and definition of tasks).

12.5.1 PAQ—Position Analysis Questionnaire

The father of this approach of analysis can be seen in McCormick, who, together with his coworkers, developed the Position Analysis Questionnaire (PAQ) (McCormick, Jeanneret, and Mecham, 1969).

Following the description by Drury et al. (1987) in a former issue of this handbook, in its final form (Form B), the PAQ has the structure shown in Table 12.10. The PAQ consists of 187 job elements (items) that characterize or imply various types of basic human behaviors that are involved in jobs in general. The elements are organized into six major divisions. These divisions and their subdivisions are given in Table 12.10 along with the number of job elements in each. Some examples of job elements are as follows:

- Use of written materials
- Use of touch
- Coding/decoding
- Use of precision tools
- Use of keyboard devices

Table 12.10 Structure of Position Analysis Questionnaire

Area of Coverage	Number of Questions		
1. Information Input	Total 35		
1.1 Sources of job information		20	
1.1.1 Visual sources of job information			14
1.1.2 Nonvisual sources of job information			6
1.2 Discrimination and perceptual activities		15	
1.2.1 Discrimination activities			8
1.2.2 Estimation activities			7
2. Mediation Processes	Total 14		
2.1 Decision making and reasoning		2	
2.2 Information processing activities		6	
2.3 Use of stored information		6	
3. Work Output	Total 50		
3.1 Use of physical devices		29	
3.1.1 Hand tools			6
3.1.2 Other hand devices			5
3.1.3 Stationary devices			1
3.1.4 Control devices			9
3.1.5 Mobile and transportation equipment			8
3.2 Integrative manual activities		8	
3.3 General body activities		7	
3.4 Manipulation/coordination abilities		6	
4. Interpersonal Activities	Total 36		
4.1 Communications		10	
4.2 Miscellaneous interpersonal relationships		3	
4.3 Amount of personal contact		1	
4.4 Types of personal contact		14	
4.5 Supervision and coordination		8	
4.5.1 Supervision given			7
4.5.2 Supervision received			1
5. Work Situation and Job Context	Total 18		
5.1 Physical working conditions		12	
5.2 Psychological and sociological aspects		6	
6. Miscellaneous aspects	Total 36		
6.1 Work schedule, method of pay, and apparel		21	
6.2 Job demands		12	
6.3 Responsibility		3	

- Specified workplace
- Number of personnel supervised

In the analysis of jobs with the PAQ, the analyst makes a response for each job element, using a specific scale. There are six types of scales, as follows: extent of use, amount of time, importance to the job, possibility of occurrence (used only with hazards), applicability, and special scales. All are five-point scales except the applicability scale (which requires merely a presence or absence decision). Examples of three scales are given below.

Amount of Time	Extent of Use	Importance
Does not apply	**Does not apply**	**Does not apply**
1. Infrequent/rarely	1. Nominal/very infrequent	1. Very minor
2. Under 1/3 of the time	2. Occasional	2. Low
3. Between 1/3 and 2/3	3. Moderate	3. Average of the time
4. Over 2/3 of the time	4. Considerable	4. High
5. Almost continually	5. Very substantial	5. Extreme

Examples of these scales would be question 92 (Section 3.3) asking for the amount of time spent walking, question 17 (Section 1.1.2) asking for the extent of use of tactual information sources, and question 175 (Section 6.2) asking for the importance of keeping a specified work pace. Special scales are used only rarely, for example, a 6-pt scale for number of personnel supervised in question 134 (Section 4.5.1). In general, a great effort has been made to simplify the collection of data, resulting in a more workable instrument. The inter-rater reliability in one study was found to be reasonably high, with a mean coefficient of 0.79.

A factor analysis of data for 2200 jobs resulted in the identification of 45 factors (called job dimensions) that represent the "structure" of jobs in the labor force. Certain examples of these job dimensions are as follows:

- Interpreting what is sensed
- Watching devices/materials for information
- Processing information
- Performing handling-related manual activities
- Exchanging job-related information
- Being in a stressful/unpleasant environment
- Being alert to changing conditions

Some of these factors are clearly identified to be "task descriptive" and can serve various purposes in connection with related personnel-management functions (such as those involved in what is sometimes referred to as the *personnel subsystem*). Its primary uses in organizations have been in such areas as personnel selection (it provides the basis for deriving estimates of aptitudes, thus eliminating the need for conventional test validation procedures), personnel development and training, the development of career ladders, performance appraisal (the job dimension scores provide the basis for such appraisal), and job evaluation (it serves as the basis for establishing rates of pay without the need for conventional job evaluation procedures). Over 400 organizations have used the PAQ for such personnel-related purposes or for research, involving the analysis of over 100,000 positions representing about 3000 different jobs. A few organizations have used the PAQ in combination with other task analysis techniques in connection with their personnel-management programs.

12.5.2 AET—Arbeitswissenschaftliches Erhebungsverfahren zur Tätigkeitsanalyse

A technique derived rather directly from the PAQ approach and other works (Landau, Luczak, and Rohmert, 1975; Landau, 1978) is the AET method, an acronym for "Arbeitswissenschaftliches Erhebungsverfahren zur Tätigkeitsanalyse" or "ergonomic job analysis technique" (Rohmert and Landau, 1983).

This job analysis starts from a model of work activity and assesses relevant aspects of the following:

- The work object
- The work resources
- The working environment
- The work tasks
- The work requirements

with regard to stress and strain considerations. Therefore, the basis of understanding job analysis is composed of the theoretical model of the man-at-work system and the concept of the simultaneous distinction and interdependence of stress and strain (Rohmert, 1987).

The AET is structured in the three parts according to the theoretical concept described above: tasks, conditions of carrying out these tasks, and the resulting demands upon the workperson (see Table 12.11).

In part A (work system analysis) types and properties of work objects, the equipment to be used and the physical, social, and organizational working environment are documented on nominal and numerical scales. The behavior requirements approach is covered in part B (task analysis). On 31 numerically scaled items, tasks relating to material work objects, to abstract work objects, and to human-related tasks are rated. Performing a task (as rated in part B) under the conditions documented in part A leads to job demands, which are evaluated in part C (job demand analyses) in three sections: perception, decision, and response/activity.

Every AET item consists of an (underlined) question outlining the state of affairs to be grasped—under certain circumstances with examples as classification aids—and of the indication of a code for classifying this feature.

Item rating can be done only by means of the corresponding code. The code of significance (S) and the exclusive code (E) are used for indicating the level of stress, the code "amount of time" for the duration (D), and the code of frequency (F) for characterizing the temporal distribution and position of stress sections. Alternative and nominal scales have special codes for documenting the working system characteristics (cf. McCormick et al., 1969). The items of the job demand analysis, which could possibly be difficult to answer for a practician who is not sufficiently trained in ergonomics, contain additional classification aids in the form of "task or activity scales." The activity scale, based upon previously investigated data, contains a series of grades of mainly illustrative tasks and activities.

The analysis of a job is done in the form of an observation/interview, which means that the necessary analytical data are collected first by interviewing the incumbent and his supervisor. Priority must be given to the analyst's observation. The interview, however, is used to ascertain those job characteristics that could not be determined by observation.

The concept of work analysis as a universal *wideband* method allows the solution of quite a variety of practical and scientific problems. Illustrative examples are given in Landau and Rohmert (1981).

12.5.3 TAI-Tätigkeitsanalyse inventar

One of the latest and most comprehensive developments in this context is the TAI (Tätigkeitsanalyseinventar-Inventory of job analysis), a method of task-related work analysis with a broad background from all aspects of occupational psychology (Facaoaru and Frieling, 1991; Frieling et al., 1990). The TAI serves for an analysis and documentation of energetical, sensomotory, and informational demands and stressors on several theoretical considerations:

1. An approach of information theory with a phase model of human problem solving and information transmission: Perception, processing, output. The method is mainly restricted to the first and last phase, because cognitive processes are not directly observable, whereas information-perception and -output can be coded by an experienced rater.
2. A multilevel concept, according to which any segment of human action can be analyzed on a concrete, a functional, and a process-oriented level. The *concrete* level relates to the work objects, the work equipment and tools/machinery, the sources of information, the work documents and the work results (concrete, written, calculated). On the *functional* level tasks as well as generation and transmission of energy and information are considered. Objectives of analysis on the *process* level are the intensity, the difficulty, the knowledge demands, and aggra-

Table 12.11 Contents of the AET

Part A—Work System Analysis

1. Work objects
   - 1.1 Material work objects (physical condition, special properties of the material, quality of surfaces, manipulation delicacy, form, size, weight, dangerousness)
   - 1.2 Energy as work object
   - 1.3 Information as work object
   - 1.4 Man, animals, plants as work objects
2. Equipment
   - 2.1 Working equipment
     - 2.1.1 Equipment, tools, machinery to change the properties of work objects
     - 2.1.2 Means of transport
     - 2.1.3 Controls
   - 2.2 Other equipment
     - 2.2.1 Displays, measuring instruments
     - 2.2.2 Technical aids to support human sense organs
     - 2.2.3 Work chair, table, room
3. Work environment
   - 3.1 Physical environment
     - 3.1.1 Environmental influences
     - 3.1.2 Dangerousness of work and risk of occupational diseases
   - 3.2 Organizational and social environment
     - 3.2.1 Temporal organization of work
     - 3.2.2 Position in the organization of work sequence
     - 3.2.3 Hierarchical position in the organization
     - 3.2.4 Position in the communication system
   - 3.3 Principles and methods of remuneration
     - 3.3.1 Principles of remuneration
     - 3.3.2 Methods of remuneration

Part B—Task Analysis

1. Tasks relating to material work objects
2. Tasks relating to abstract work objects
3. Man-related tasks
4. Number and repetitiveness of tasks

Part C—Job Demand Analysis

1. Demands on perception
   - 1.1 Mode of perception
     - 1.1.1 Visual
     - 1.1.2 Auditory
     - 1.1.3 Tactile
     - 1.1.4 Olfactory
     - 1.1.5 Proprioceptive
   - 1.2 Absolute/relative evaluation of perceived information
   - 1.3 Accuracy of perception
2. Demands for decision
   - 2.1 Complexity of decision
   - 2.2 Pressure of time
   - 2.3 Required knowledge
3. Demands for response/activity
   - 3.1 Body postures
   - 3.2 Static work
   - 3.3 Heavy muscular work
   - 3.4 Light muscular work, active light work
   - 3.5 Strenuousness and frequency of movements

vating conditions of task execution for bodily and information components of processing.

3. Approaches of cognitive ergonomics, different concepts for task taxonomies, for behavioral analysis, and for functional analysis.
4. The stress-strain concept with a differentiation of stresses according to type/kind, amount/height, and time/duration.

The complete TAI comprises more than 2000 ratings. It delivers, for instance, multidimensional profiles of task-demands, which allow to compare time relations, frequency, complexity, amount of used documents, etc. (see Figure 12.26).

The TAI was developed as a project in the German "Humanization of Work" initiative of the federal government. Because of the number of items and its strong reference to various scientific bases, it is seen mainly as a tool in project-driven applied research.

12.5.4 MJDQ—Multimethod Job Design Questionnaire

The MJDQ is the result of an eclectic approach to work design that extracted ≈700 design rules from the literature, categorized them, and condensed them to ordered design principles (Campion 1985, 1988; Campion and Thayer, 1985). Combined by the following approaches of

- Motivation theory in organizational psychology
- Mechanistic considerations of classical engineering sciences
- Biological thoughts of occupational physiology
- Sensomotory findings of experimental and engineering psychology

Task-related information is gathered via a five-step rating scale as well; especially the possibility of execution of tasks can be evaluated, weaknesses can be diagnosed, and design guidelines can be derived. Thus, eclecticism from the major disciplines of work sciences led to a comprehensive tool.

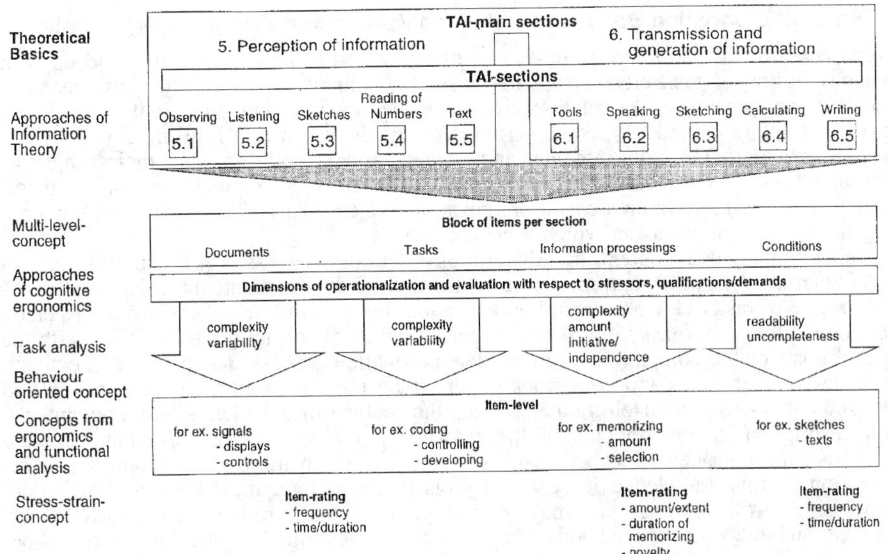

Figure 12.26 Structure of the TAI: Relation between theoretical concepts and respective implementation.

12.6 TASK ANALYSIS IN WORK ORGANIZATION

12.6.1 Redefinition

In research about working groups mainly three classes of variables can be distinguished: presuppositions in the form of situational conditions, consequences, and characteristics of the group. One of the main presuppositions or independent variables for group work is the type of task, the autonomy of the group-work in task execution, and the organizational context—for instance, task-related links between group members. The main dependent variables are performance and group stability. The main intervening variables that define the working group as an entity (Leplat, 1994; Nerdinger and Rosenstiel, 1996) are:

- Several persons with cohesion
- Direct interaction and cooperation
- Own role differentiation and standards
- Performing complete tasks and subtasks within the order processing of a larger organization

Thus, a redefinition of the "task" in this context implies some principle components:

1. Tasks are complete in the sense of being composed of preparatory, performing, and control elements, and in that they offer a variety of demands.
2. Tasks are characterized by participative leadership in the sense of equal distribution of dispositive elements and autonomy.
3. Tasks allow qualification, personality development, and continuous improvement (Kaizen) and thus have a dynamic character.
4. Tasks imply cognitive demands, require the handling of complexity and problem solving in information processing, and allow choices according to individual preferences.

With regard to these considerations it is more appropriate to speak about "task structure analysis" instead of "task analysis," when discussing organizational impacts on tasks and their analytical documentation and evaluation.

12.6.2 Differentiation Between Work Organization and Company Organization

Approaches of task structure analysis cannot be limited to perspectives that end up with the description of task elements in the individual job of a person. Analysis means to identify the cause-to-effect relationships in a phenomenon and thus finding the background structure behind the visible surface. It is obvious that task fulfillment is bound to a great extent to the environmental and organizational conditions and their subjective interpretation by a worker. Thus, task structures in an organizational context are determined by at least two main areas of organization, the company organization and the work organization of the group in which a person is active.

In a companywide context, it is the aim of order and business process analysis as well as of enterprise structure/corporation structure analysis to investigate the objectively available scope of tasks of a person in his or her job, the possibilities of the working person to redefine his or her tasks in this scope, and to adjust his or her goals of action with the global goals of the company in the complete production process. To analyse this context, a review of the orders and conditions under which the orders are planned, steered, and worked out is necessary. Helpful methods are the analysis of documents (order documents, job description in the job plan of the enterprise, product lists, and product structure analysis), the interview of experts and working persons in the company, and the representation of this knowledge in software tools (CIM architectures, CAx tools of order processing). At that, what a task may be or may not be, is determined by the sociocultural context and the societal demands (Figure 12.27). Nevertheless, the knowledge about "analysis" in this area is large and tends to explode with the introduction of corporate information systems. Thus, only these types of analysis can be reported here which have the objective to include "human factors" aspects in the analytical procedure. Nevertheless, the review will be selective and exemplary.

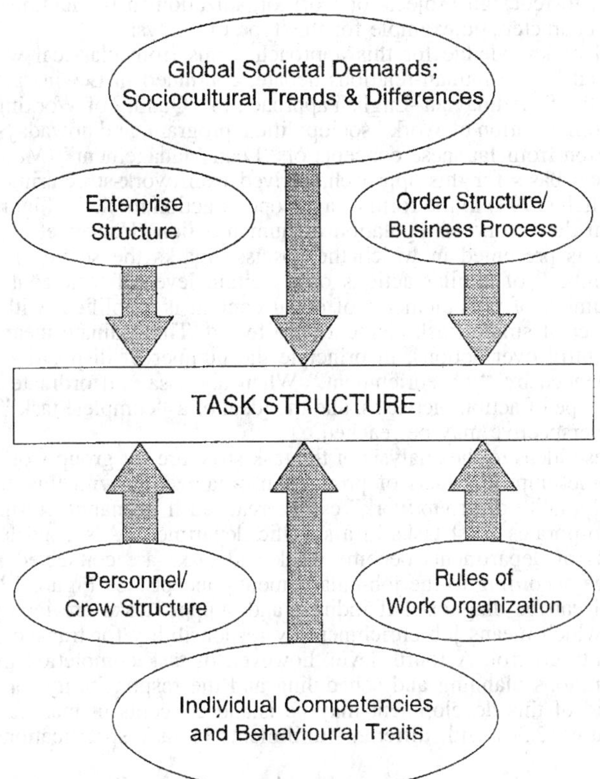

Figure 12.27 Views on task structure by work organization and company organization.

The company-oriented view of organzation does not contain the whole truth. The processing of orders is assigned mostly to working groups. This means that human relations may determine how a group of personnel or crew with a certain qualification and motivation structure redefines what has to be done by whom, respecting competencies and individual traits as well as implicit or explicit rules of work organization. On this level of task structure analysis in the work organization, two methodological approaches interfere: A *condition-oriented* and a *person-oriented* approach.

In the condition-oriented approach, the working conditions are analyzed as determinants of the task structure. The individual specialties of working persons, the person-depending differences in the redefinition and interpretation of the task are neglected, and the level and amount of cooperation and communication, the execution characteristics, and the objective works are emphasized. In the person-oriented approach, the individual differences in the interpretation of tasks, their importance for the person, and the mechanisms between task structure and motivational/behavioral structure are investigated. Both approaches were widely used in the European programs for the Humanization of Work, which were focused on work organization.

It seems obvious that task structure analysis in the context of work organization serves the goal to improve design of tasks. Thus, the intention to modify a status is implicitly the basis of analytical procedures. Basically a status 1 before design and a status 2 after design have to be compared, or the transient procedure from S1 to S2 has to be mapped in a dynamic model. In the first case, the methods may be called *comparative-static*, in the second case, *dynamic*.

12.6.3 Condition-Oriented Comparative Static Methods

12.6.3.1 Empirical Project-Oriented Task Structure Analysis for Redesign

Referring to the elements and principles outlined in the introduction to this chapter on task structure analysis in work organization, the application of task analytical approaches is mostly bound to redesign projects of work organization in production systems. They may serve as a characteristic example for this type of analysis.

A good deal of knowledge for this approach stems from classical work structuring concepts, initiated by the human-relations debate, continued in Lewins group dynamics, rediscovered in the Scandinavian–English approach to "Quality of Working Life," and in the German "Humanization of Work" sociopolitical program, and nowadays serving again as a re-importation from Japanese concepts of "Lean-Management" (Moldaschl, 1996).

The theoretical basis for this approach, derived from work-structuring concepts, was developed by Ulich (1994) in the form of a "scope of action model" (Figure 12.28) based upon ideas from the theory of regulation of human action (Hacker et al., 1993). A coordinate system is presented in which the abscissa marks the scope of activity which means the "number" of similar actions on a certain level of task content. With "job rotation" the number of task elements of equal content is amplified, with "job enlargement" the number of similar task elements is affected. The ordinate means the scope of decision and control over actions, in principle the number of dispositive elements in a task to be influenced by "job enrichment." When abscissa and ordinate are influenced positively, the scope of action increases, and the ideal of a "complete task" in hierarchical and sequential perspective may be reached.

Applying these ideas to the analysis of the task structure for group work in production means to combine empirical facts of production organization with this concept: A first level of higher "quality of group work" can be reached if in manufacturing or assembly the rotation incorporates most tasks in a specific department. A second level is reached when limitations of departments become weak and tasks are connected along the production sequence according to the job-enlargement concept (see Figure 12.29).

A third level can be diagnosed if indirect and support tasks are incorporated in the task spectrum, which means job enrichment by responsibility for transportation, maintenance, and quality control. A fourth level, however, of task completion implies the disposition of operations planning and scheduling and the responsibility for deliveries just in time. The end of this development may be island concepts in manufacturing departments (see Figure 12.30) with different fulfillments of task specifications and requirements.

In general many task structure analysis techniques exist. Considering the "Humanization of Work" program in Germany and related projects only, Frieling et al. (1982) documented 87 different approaches. So the report here may serve as a somewhat representative example for task structure analysis in production systems that can be expanded to other areas of work organization (Luczak, 1995).

12.6.3.2 Graphical Analysis and Representation Techniques

If tasks are combined to group work structures, it seems to be necessary to map the tasks in a respective model, to adopt them to design specifications, and to evaluate them. For

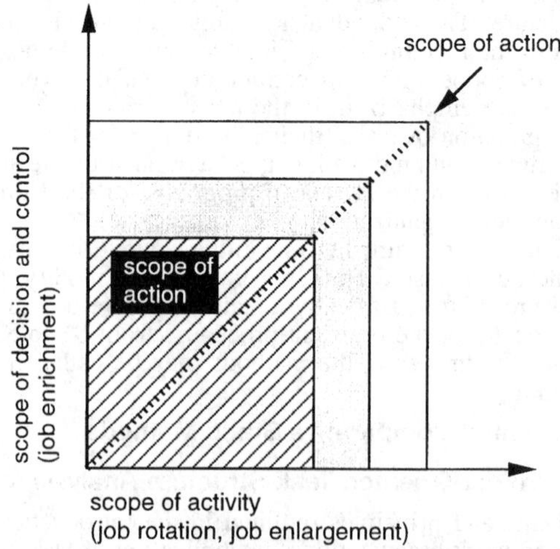

Figure 12.28 Scope of action model (Ulich, 1994).

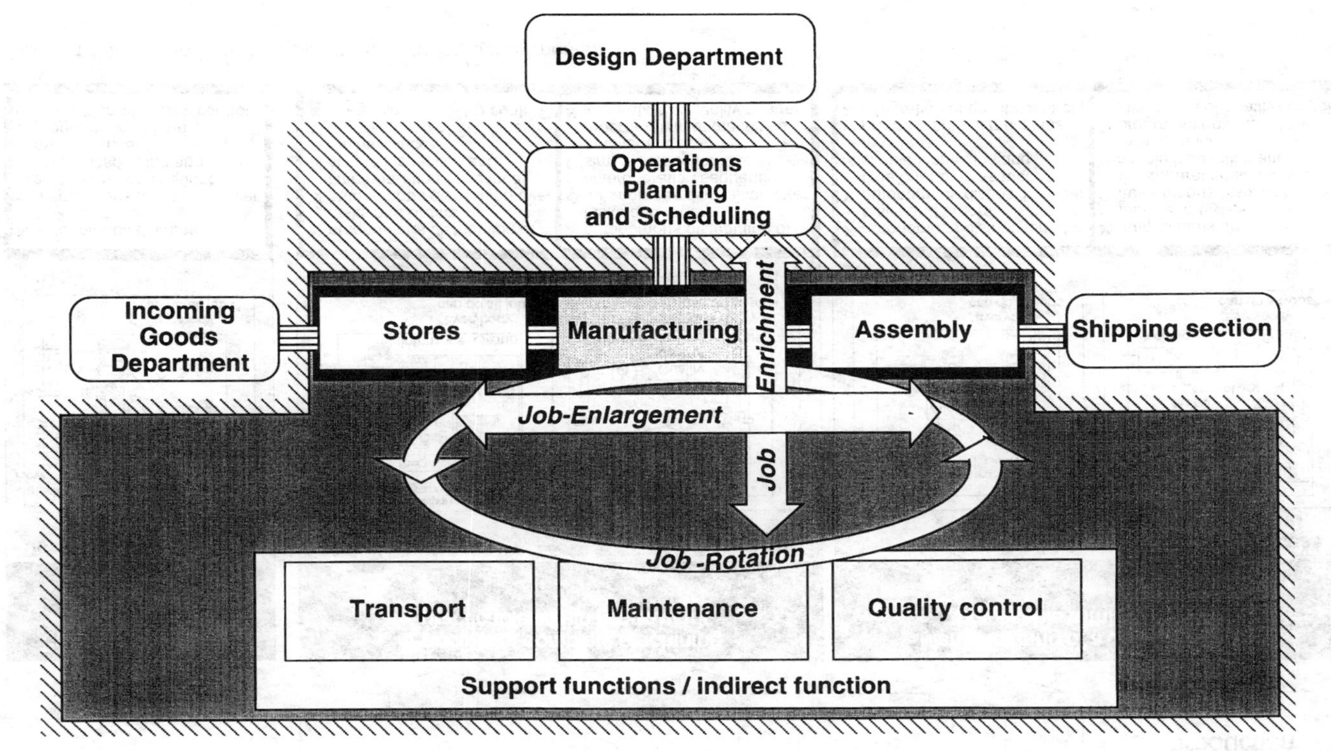

Figure 12.29 Task-structuring concepts to design autonomous group work.

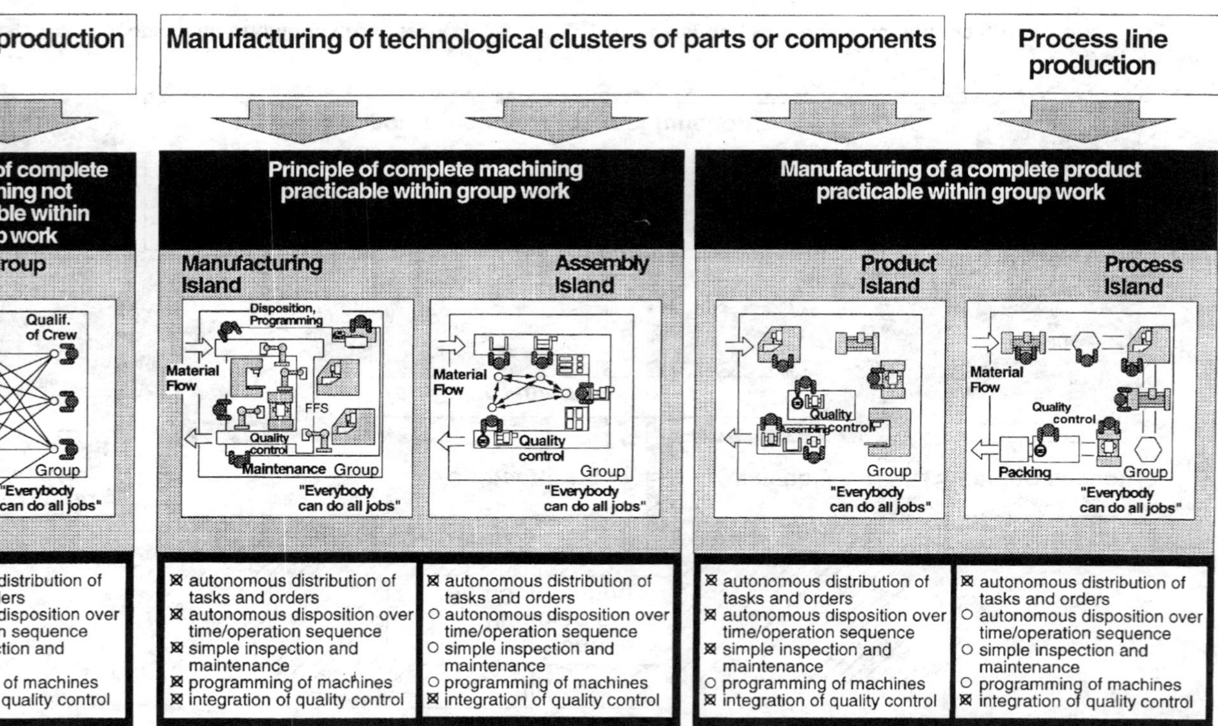

Figure 12.30 Island concepts in manufacturing departments.

this purpose of task-structuring, graphical description languages can be used, which describe sequential subtasks/actions as well as informational relations. Most languages of this type stem from software development to ease user-designer communication when developing software products to support task execution of a user. Mostly corporation/company-oriented as well as group-oriented orders and respective tasks/functions are described, assigned to elements (task-executors: human, technical equipment), and related via different forms of connections (material, information) on adequate levels of abstraction. Those methods are primarily oriented to the development of information systems; however, they are powerful tools of task structure analysis as well. Various tools exist. One of them will be described in detail, more in an additional survey.

A computer-supported method of task structure analysis is MSG (Modellgestütztes System zur Gestaltung von Gruppenarbeit) (Schumann, 1995; Luczak, 1995). MSG is characterized by two components: a modeling unit with which a real work structure can be mapped graphically onto a task structure model and an evaluation unit with which the task structure model is investigated by indices and items, derived from action regulation theory (Hacker, 1986). At the basis of the modeling unit are RFA-networks (Role, Function, Action), a subclass of Petri-Nets that deliver a powerful tool for the description and graphical representation of complex human–machine systems (Oberquelle, 1987). RFA-networks are mainly characterized by a breakdown of a complex task situation into roles, functions (task bundles), and actions (subtasks) as well as the possibility to combine these elements via material and information flows and communicative and cooperative relations.

The most elementary components of RFA-networks are actions and positions. Actions symbolize human activities or machine operations and are networked horizontally via material and information flows. Via these relations physical objects, which are characterized by attributes, are transported. The objects or their attributes, respectively, are modified by the actions. The different locations of objects in the organization sphere (for instance, storage, status of completeness, and composition) are represented by the positions.

Actions can be grouped vertically in tasks which are assigned to a person. A set of tasks defines the role of this person. The communication and cooperation between several persons is represented by relations between the tasks. In MSG the original typology of RFA-networks was complemented by an object-oriented description of actions and relations. As actions are elements of tasks, features/characteristics were assigned to the actions which are necessary to describe the demands of the complete task/job in group organization (Figure 12.31). These characteristics are the step of regulation requirements (see Table 12.9), the phase of regulation, feedback, and autonomy. The step of regulation requirements characterizes the cognitive demands in a six-step ordinal scale, following Volpert et al. (1989). The phase of regulation can be described on a four-alternative nominal scale by identifying if an action is preparatory, executive, concurrent, or assessive afterwards. Feedback is described by length of the control loop for the evaluation of work results in a three-step ordinal scale: The highest step is used with a direct feedback from the working process, the lowest by an external feedback from a subsequent production system. With the degree of autonomy the dependency from imposing and imperative interventions by colleagues, supervisors, and restrictive working conditions is scaled in a five-step ordinal scale between "no degrees of freedom" until "self-set goals and chosen ways to problem solving."

A detailed description of communicative and cooperative relations between the group members is possible on a four-step scale. A reference catalogue of actions, derived from empirical investigations in machining and assembly is used to build up the RFA network.

The evaluation unit calculates for the complete task the "range of cognitive demands," "range of completeness of subtasks," "range of autonomy," "range of feedback," "possibility of communication," "possiblity of cooperation," "range of responsibility," and the "range of possibility for qualification." This results in profiles of different criteria being assigned to task elements and the whole task. As MSG is a CA-group task-tool the variations in task content can be immediately performed by adding and deleting task elements from the reference catalogue. The effects of these design interventions are represented immediately in the visible network structure and in the profile calculation for the evaluative criteria.

Beneath MSG, which is based on RFA-networks, more tools are disposable, as outlined and evaluated in Table 12.12. GRAI, Petri-Nets, and SADT are described in more detail in the next chapter in relation to company-wide organizational task structure analysis, because that is their main field of application. Nevertheless, they may be used for task structure analysis in the group work context, too.

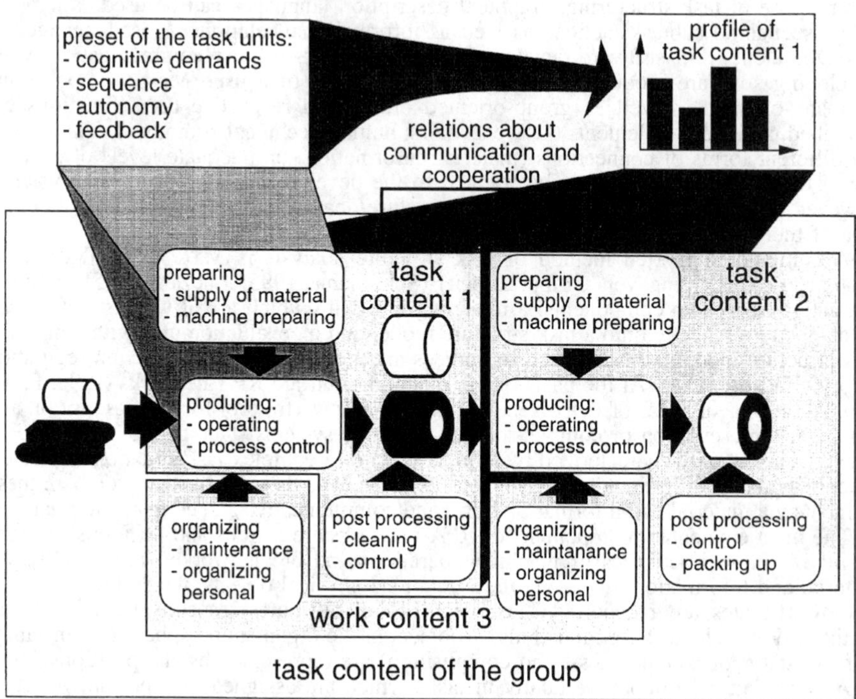

Figure 12.31 Principle elements of MSG, a software tool for the design and evaluation of task content.

BENORSY (Essig et al., 1981) is a tool for user-oriented system revision (= *Ben*utzer-*or*ientierte *Sy*stemrevision). The objective of the concept is an improvement of the work situation of a user plus ensuring performance standards of the computer-using corporation. The graphical description distinguishes two types of knots, human or machine actions (instances), and storages (channels). An activity is initiated by the presence of objects and rules as input variables. The activity delivers output objects, which can be input objects for subsequent activities. By breaking down an activity into partial activities any desirable level of task description can be reached.

NIAM (Nijssen Information Analysis Method) is especially useful for data modeling. Predefined symbols and a differentiated description of related pairs allow modeling of complex systems in their internal structure (Nijssen and Halpin, 1989). Relation pairs can be described with cardinal numbers according to their importance and can be connected logically.

However, these representation techniques do not dispose of an evaluation unit, like MSG, in which the task content is scaled in different human-oriented dimensions.

12.6.4 Person-Oriented, Comparative Static Methods

12.6.4.1 JDS—Job Diagnostic Survey

The JDS, a person-oriented method of job analysis, is, however, centered around the task of a person in its organizational context. It diagnoses the subjective perception of a work situation by a working person with the aim to identify characteristics of the work situation and the job with respect to the motivational incentives and thus to evaluate changes in work organization. The theoretical basis of the JDS is the "Job Characteristics Model" of motivation by Hackman and Oldham (1975, 1976, 1980). According to the model, intrinsic motivation is determined by three conditions: (1) The person should have "knowledge of results" of his or her own work to be able to compare them with self-set or preset standards. (2) The person must feel responsible for the results of work. (3) The person must perceive the work as important and valuable. It is presumed that mainly five

Table 12.12 Comparison of Functional and User-Oriented Requirements for Software Tools to Represent Task Structures in Work Organization

Functional Requirements	BENORSY	GRAI Method	NIAM	Petri Networks	RFA Networks	SADT
Relevance						
Modeling of production systems	possible	possible	possible with restrictions	possible	possible	possible with restrictions
Modeling of the task structures	possible with restrictions	possible with restrictions	possible with restrictions	possible with restrictions	possible	possible
Reproduction of diversified relations	none	none	possible	none	possible	none
Reproduction of object flows	possible	possible	possible	possible	possible	possible
Reproduction of passive and active elements	possible	possible	possible	possible	possible	not possible
Reliability						
Restrictions of application	none	none	information processor only	none	none	none
Transferability of model representations	possible	possible	possible with restrictions	possible	possible	possible with restrictions
User-Oriented Requirements	**BENORSY**	**GRAI Method**	**NIAM**	**Petri Networks**	**RFA Network**	**SADT**
Manageability						
Complexity of the method	low	high	high	medium	high	high
Complexity of the representation	high	high	high	high	medium	high
Diversified symbolism	possible with restrictions	none	none	none	possible	none
Textual description of the elements	possible	possible	possible	possible	possible	possible
Mode of the procedure	top-down	top-down	top-down	top-down	bottom-up top-down	top-down
Adaptability						
Needed modification of the modeling	medium	high	high	medium	low	high
Expandability of the symbolism	not possible	not possible	not possible	not possible	possible	not possible
Robustness						
Practicability of plausibility controls	possible	no indications	possible	possible	possible	possible
Efforts for modifications and corrections	medium	medium	high	medium	medium	high

attributes (core dimensions) determine the perception of knowledge of results, responsibility, and importance:

1. Skill variety
2. Task identity
3. Task significance
4. Autonomy
5. Feedback from the job itself and from agents

The JDS is available as a questionnaire with seven segments and 83 items which are scaled in seven-step Likert scales with verbal description. Any variable is scaled by several items in two different formats to reduce method and raters variance. The results are analyzed by distributing the items with a definite key to the core dimensions and moderating variables. Beneath the singular variables a "motivation potential score" is calculated:

$$\text{MPS} = \left[\frac{\text{skill variety} + \text{task identity} + \text{task significance}}{3}\right] * \text{autonomy} * \text{feedback}$$

which gives an index of the effect of task arrangement and task execution in a given social and organizational work context.

12.6.4.2 SAA/STA—Subjektive Arbeitsanalyse/Subjektive Tätigkeitsanalyse

Another instrument with a person-oriented evaluation of the work situation and considerable implications on task analysis is the "subjective work analysis" (SAA = Subjektive Arbeitsanalyse), as developed by Ulich (1994) and coworkers (Udris and Alioth, 1980). It is based on two aspects: The first aspect implies "alienation" with the categories "heteronomy versus self-regulation," "senselessness versus transparency," "dequalification versus action-oriented competency," and "social isolation versus social engagement." The second aspect contains the strain/task stress of the working person with the categories "qualitatively insufficient demands" and "quantitatively and qualitatively excessive demands." To operationalize these aspects six main indices (scope of action, transparency, responsibility, qualification, social structure, work stress) with two to three subindices and three to five items per subindex are formulated verbally in five-step scales in the form of a questionnaire. From the mean values and the distribution characteristics of items and indices, an analytical evaluation of a task in perspective of a working person can be derived absolutely as well as relatively in terms of a comparison of values before and after a redesign.

From the same Swiss school of occupational psychology, another person-oriented task-evaluation procedure was developed, the "subjective job analysis" (STA = Subjektive Tätigkeitsanalyse) (Ulich, 1981; Frei et al., 1993). Employees classify works and tasks in their work system according to these six criteria: "freedom of decisions," "changes in demands," "possibilities of qualification," "mutual personal support," "sensefulness of the tasks/works for the company," "possibilities of personal development" on a 0- to 10-pt scale. Thus, potentials for the amelioration of tasks can be identified in a group work approach.

12.6.4.3 IPA—Interaction-Process-Analysis

Additional information can be gathered by expert-oriented methods of observation from a position out of the group. One of the main task-related methods is the *I*nteraction-*P*rocess-*A*nalysis (Baks, 1950; Baks and Cohen, 1982), which scales the frequency of specific behavioral attitudes in group processes in 12 categories. The group processes are subdivided into "task orientation" and "socio-emotional orientation." With this method, reliable information about cooperation and task-execution-characteristics in working groups can be gathered.

12.6.5 Dynamic Simulation of Tasks in Work Organization

Another set of models allows the simulation of tasks, task-allocation, and respective performance in the time domain. With the description of data, which characterize the actual status of a complex human–machines-system by an analysis of its mission, its order-processing, and its environmental and situational conditions, a data set is available that allows you to calculate and investigate the reaction of the system under various conditions of task arrangement. An appropriate method is system simulation over time-spans. For this purpose the result of empirical task analyses has to be coded in a way that, via a dynamical description component, a continuous simulation in a dynamic model and respective simulation language becomes possible. With the help of such simulations development processes of the system and its future status can be investigated. Some models and simulation approaches for task simulations in work organization will be discussed, but only in a selective an exemplary manner to demonstrate the possibilities of the approach for task analysis.

12.6.5.1 Personnel-Oriented Simulations and Crew Modelings

These modelings emphasize the cooperation between working persons in a mission scenario or a production scenario (Luczak 1995).

To evaluate the crew size and the task assignment to different crew members for a "ship of the future" a simulation model was developed. The objective of the investigation was to determine on what operation conditions (mission scenario with defined task sets) and what technological (automation of workplaces with different degrees of task-allocation to the technical system) and organizational (assignment of tasks to hierarchies and personnel qualifications) measures, a container ship, sailing in a regular liner service (repetitiveness of tasks) can be handled by a reduced crew (Schwier and Luczak 1986). To do this, a DEMOS (Discrete Event Modelling On Simula) system of simulation was used. The basis for the simulation runs was an empirical inventory of about 600 different tasks and activities to which time data were assigned concerning the regular of stochastic frequency of occurence and probability density function of length. The simulation model thus comprises all crew members with their qualifications, all tasks and activities with their regularity, their length/duration, their priority according to the operational conditions, and their relation to the different technical and functional systems, and their organizational context in terms of cooperation.

Bringing together tasks and personnel allows the formulation of a dynamic manning model, in which the size and structure of the crew can be varied in relation to work organization.

Technological determinants that reduce the amount and frequency of tasks can be simulated too, such as improved nautical equipment and the status of automation in machine operating. Thus, stability considerations under different scenarios of shiphandling for the task execution in due time, even in emergencies, result from this simulation.

A congruent approach was followed for the organization of group work in design departments with the introduction of CAD systems (Luczak et al., 1990; Steidel, 1994).

In production and assembly organization, this person-focused and task-analytical simulation approach was applied as well (Zülch, 1992; Zülch and Grobel, 1990), especially in the planning of manufacturing islands by simulation of work organization (Zülch and Ernst, 1990).

Most simulations of this type are based upon special software developments of the authors, because a "subjective" access to task simulation via the behavior of workers is not foreseen in most simulation program packages.

Nevertheless, some software tools exist which seem to be especially applicable in the context of task analysis for simulation purposes (see overview in Table 12.13 for the German production engineering and production organization sphere).

12.6.5.2 Generic Tools of Task Structure Simulation

Beneath the specific simulation tools for a segment of tasks in an industrial context, generic tools of task structure simulation exist which are designed to map a variety of task situations. Some of the best-known and well-tested simulation tools in this class are SAINT and SLAM (Kraiss, 1981).

SLAM is a block-oriented simulation language which allows the generation of event-related and process-related simulation models (Chubb et al., 1987). The blocks can be parametrized in a way that they can be assigned to tasks. In the process-oriented models a network of knots (blocks) and branches is designed which represents the task context. Within the blocks the properties of a task are specified by the parameters as further detailed in FORTRAN subprograms. This network model is transformed in an input model for the SLAM processor, where it is translated into the SLAM programming language. Within the SLAM processor the application model can then be simulated. In a discrete,

Table 12.13 Task Simulation Tools in Production Organization

Reference / Acronym of the Respective Tool	Focus of Application	Characteristics	Variables of Simulation	Person Task-Oriented Properties
Heinz/Lange 1992 / SIGRID	manufacturing and assembly	flow simulation of defined orders/tasks in production systems	work means, work flow, steering strategy, quality traits, disturbances	degrees of performance
Seliger, Feige, Wang 1993 / MOSYS	assembly only	flow simulation of defined orders in a workplace layout	persons, work means, work flow, steering strategy, cooperation, disturbances	extent of tasks, number of persons, working time, degrees of performance
Zülch 1993 / FEMOS (SIMULAST)	manufacturing assembly, indirect tasks	flow simulation of defined orders in organizational units with defined sequence	persons, work means, work flow, steering strategies, disturbances	extent of tasks, use of work means, number of persons, time on task
Bullinger et al. 1990 / PERSIMO	manufacturing assembly	flow simulation of defined orders on the basis of a functional layout	persons, work means, work flow, disturbances	extent of tasks, number of persons, time on task

event-oriented view singular states of the system can be investigated depending on the previous events in a step-by-step approach. For any possible event, a process logic must be defined that serves for the step-by-step simulation of sequences of events. SLAM was applied, for instance frequently to task simulation in production engineering and in logistics (Noche et al., 1993).

SAINT (System Analysis for Integrated Networks of Tasks) is a simulation method which was especially developed for the modeling, simulation, and analysis of complex human–machines systems (Seifert and Döring, 1981). The basis of the method is a detailed empirical task analysis which describes the properties of the tasks, their logical and sequential couplings. With this set of task-related information a network is built up, in which the knots represent processes or tasks, and branches symbolize the relationships among processes. In addition to that, the task executors (personnel, technical means), the descriptive variables for the status of the system in the form of a set of time-dependent equations, and constraints for the network are to be described. This network model is then transformed into the simulation language. The simulation program then simulates all system processes, as defined by the network, and delivers an evaluation of system performance under various conditions of a mission. With respect to task analysis the types of tasks to be activated, the frequency and duration of task execution, the length of queues and the lead times for tasks, the possibilities of task allocation to different crew members and the necessary crew size are to be investigated. Mostly in military applications the power of SAINT could be demonstrated, concerning the analysis of the understanding of system reactions in standardized and extreme conditions (see examples in McMillan et al., 1989).

SAINT and SLAM are strongly related to each other: Commercial developments of SAINT resulted in the SLAM simulation language (Pritsker, 1986). The philosophy of the techniques relies on project-scheduling approaches like CPM and PERT.

12.7 TASK ANALYSIS DERIVED FROM COMPANY ORGANIZATION AND INTERCOMPANY RELATIONS

12.7.1 Redefinition

The most important approach to company organization is the analysis and definition of tasks to be fulfilled in the respective system. The separation and ordering of tasks as well as their combination with goal-oriented structures is the core of organizational design of a company and its results, the integrated structure of the company and the process flows within this structure. Mostly economical and social sciences contributed in the past to this form of task analysis (Kosiol, 1962). Thus, the redefinition of the notion task means the following:

- The orientation of any activity toward global goals of the company
- The "objective" identification of complexes of activities and their logical arrangement
- The result of a top-down labor or work partition, especially under the perspective of decision competencies
- An element in the statical structure of the company
- A process segment in the business and production processes of the company

Especially in the European/German management science approach, a differentiation between statics and dynamics of task synthesis, task assignment, task combination is obvious. Nevertheless, these approaches rely on the same analytical principles. As criteria of analysis stable task characteristics were derived:

1. *Execution* (Verrichtung) which means a description of purposeful activities
2. *Object* at which the execution is applied
3. *Phase* to order the multifarious types of execution according to—for instance—planning, decision, realization, and control
4. *Purpose* aims at a differentiation between direct and indirect tasks; in administrative and productive components; in steering, supporting, and modifying tasks; etc
5. *Rank* means the information and decision content of the task and the respective arrangement in hierarchies

12.7.2 Company Structure and Task Structure

Isolated or combined, these aspects of task analysis determine the structure of the company.

One-dimensional structures result if a decision task is broken down into subtasks according to *one* criterion: *Activity* (execution)-oriented structures (functional organization) result if the company is broken down into segments like Buying/Procuring, Making/Production, or Sales/Delivery. Product (object)-oriented structures result if the company is ordered around main product lines. Market-oriented structures result if competencies/tasks are ordered with respect to regions, customer groups, or branches.

Multidimensional structures try to combine task perspectives in decision processes from different viewpoints: Thus, staff-oriented structures, matrix-structures, and outsourcing components may result. Especially cross-sectional structures, like controlling or logistics, try to overcome the weaknesses of one dimensional structures by introducing a totally different, but comprehensive viewpoint into task execution on the company level (Frese, 1996).

Thus, the complexity of interrelations between tasks is reduced by clustering and ordering them around principles which lead to an institutional structure of the company. On the other hand, "operations management" and "business process redesign" (Hammer and Champy, 1993) are necessary to sequentially order the activities in task fulfillment steps, in process elements, and in process chains. In recent years, many concepts of organizational design were based on the core idea of process management. Processes in this sense are sequences which combine activities, superimposing functions to generate a product or a service delivery. They contain a flow/transformation of material, information, operations, or decisions with a definite start and end, a structural composition in a way, that a value for an internal or external customer is generated. As can be easily concluded, process management in the mentioned sense is task structure analysis under the perspective of customer demands (see Figure 12.32).

Beneath these modern concepts of process management, the classics of operations management rely on task analysis as well (Gaitanides, 1996):

1. *Task allocation as performance adjustment*: task elements must be combined in a way of personal and time-oriented synthesis, so that the performance output of a person or job is synchronized with the output of others. A performance adjustment

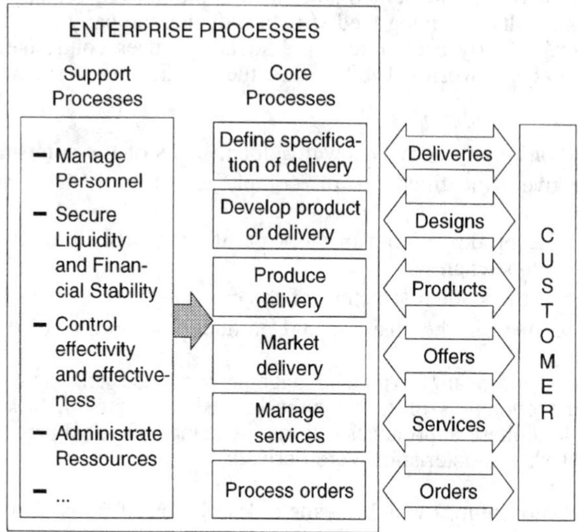

Figure 12.32 "Business Processes" as complex tasks in a process management approach (Gaitanidis, 1996).

is reached when the average performance or speed of workers is equal. Especially in time-phased continuous production lines, task-assignment is a powerful strategy to reach performance adjustment. In algorithms for the minimization of operation times or the number of jobs per unit, combinatory calculation of task elements play a dominant role.

2. *Task allocation as grouping problem*: grouping problems refer to the following:
   - arrangement of technical work means, for instance, of machines according to line or shop floor production
   - products/work objects in the sense of a determination of the batch size and thus the relation between set-up tasks and repetitive machining tasks
   - working subjects and their arrangement in shifts, working groups, task forces, repair crews, etc., as outlined in the previous chapter.
3. *Task allocation as sequence problem in order processing*: the distribution of orders in a machine park and their sequential arrangement, known as job shop scheduling problem to minimize lead times, and the definition of priority rules for operation sequences is a well-known problem of operations research. It is task structure analysis in the sense, that mathematical algorithms for the combination of task elements are available to optimize a goal function or to meet target planning.
4. *Task allocation as transport problem*: a necessary consequence of task partition is the local separation of workplaces which have to be connected by transportation tasks. Minimizing transportation times, -ways, or -costs is the aim of specific algorithms which prescribe batch sizes and routes for transportation.

It is obvious that the basis of most company-oriented reorganization processes is task analysis, in the sense of empirical subdivision and partition of tasks *and* in the sense of analytical combination of task elements to task design of higher order as well. Even "modern" concepts of organizational design like "lean production" or "quality management" cannot be understood and applied without task analysis.

12.7.3 Concepts of Modeling Information Systems in Companywide Business and Production Processes

Today the success of enterprises in competitive markets is determined to a great extent by the use of information systems. Thus, the analysis of enterprises with respect to market-oriented flexibility, product innovation, and service delivery is focused on the business and production processes on the one hand *and* the related information systems with flexibly adaptable application software, databases, computer networks, hardware, etc. on the other hand. For the purpose of their analyses in, for instance, CIM architectures a lot of models, modeling languages, and representation techniques have been developed that may include task analysis or end up with a successive breakdown process of the overall company goals at the level of tasks and activities. In this case, different types of analyses available are multifarious, manifold, and numerous. Some of them are used in macroergonomics and organizational aspects of human factors as well as in communication and software engineering and business process reengineering or operations management.

To give a short survey on this approach of task analysis, four different modeling languages are shortly described and evaluated with respect to their power of modeling and analyzing tasks in the sense that they fulfil task criteria. Naturally they were selected beforehand with respect to task analysis, a fact that can be demonstrated afterwards in an evaluative overview.

The methods/techniques mentioned are GRAI, SADT, IDEF, and Petri-Nets.

12.7.3.1 GRAI Method

The GRAI method is a technique for the analysis and design of production management systems which comprises the decision system, the information system, and the physical system of an enterprise (Doumeingts, 1984a,b; Bünz, 1987). In a hierarchical structure the decision system steers actively the production process, whereas the information system only connects the decision system and the physical system and serves as a communicative device within the decision frame (Figure 12.33). The decision system is hierarchically ordered in cells which contain functions of similar planning horizon and cycle time. Their cells are interrelated by a decision frame and information relations which are mapped in a GRAI grid. The decision frame is documented by "thick arrows," supportive information or orders are "thin arrows" (Figure 12.34). The GRAI grids represent the global relations in the decision system, and the GRAI networks represent the sequences in the form of a decision table. A GRAI grid is a two-dimensional matrix which is ordered according to functional and temporal aspects. The flows within the decision cells can be mapped in detail in GRAI networks (Figure 12.35). With the GRAI method, properties

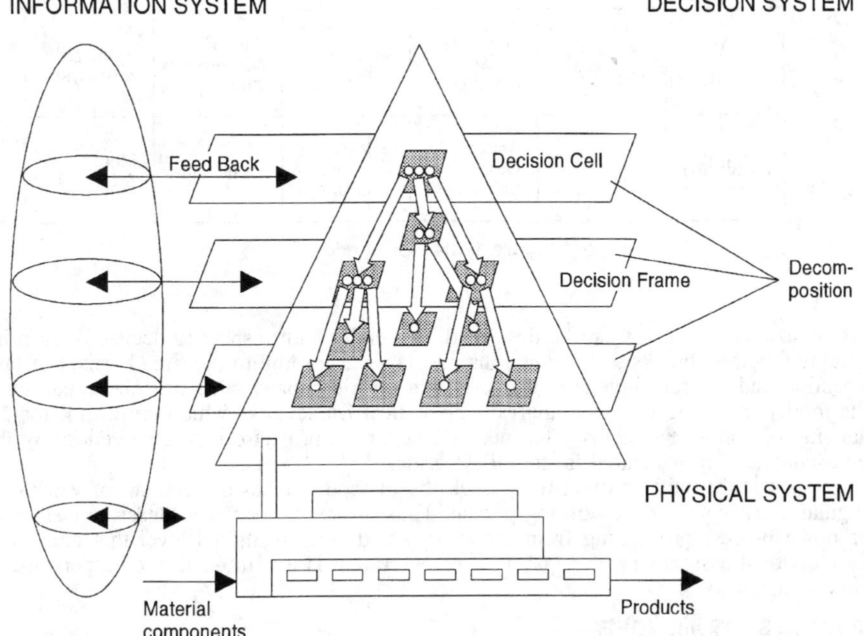

Figure 12.33 Decomposition of a production system in GRAI.

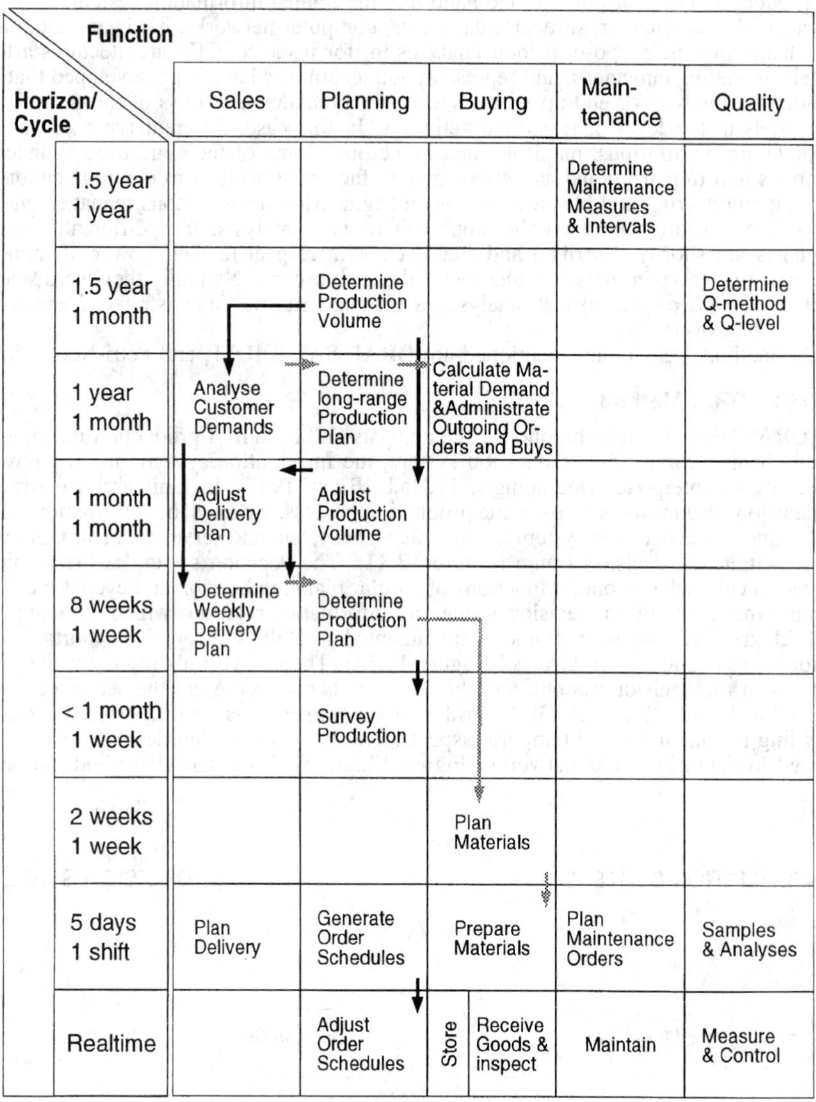

Function
Horizon/ Cycle
Sales
Planning
Buying
Main- tenance
Quality
1.5 year
1 year
Determine Maintenance Measures & Intervals
1.5 year
1 month
Determine Production Volume
Determine Q-method & Q-level
1 year
1 month
Analyse Customer Demands
Determine long-range Production Plan
Calculate Material Demand &Administrate Outgoing Orders and Buys
1 month
1 month
Adjust Delivery Plan
Adjust Production Volume
8 weeks
1 week
Determine Weekly Delivery Plan
Determine Production Plan
< 1 month
1 week
Survey Production
2 weeks
1 week
Plan Materials
5 days
1 shift
Plan Delivery
Generate Order Schedules
Prepare Materials
Plan Maintenance Orders
Samples & Analyses
Realtime
Adjust Order Schedules
Store
Receive Goods & inspect
Maintain
Measure & Control

Figure 12.34 GRAI grids.

of organizational systems can be described, especially with respect to decisions. In principle, the method breaks down functions into tasks according to the time horizon of task execution and the repetitiveness of cycles, but on a company-wide organizational level. The models resulting from the analysis are statical frameworks which are useful for the planning of software systems. The method supports a project-oriented procedure by the time-sequences implemented in the GRAI grids.

At least at the shift and real-time level, the method reaches the content of work of a singular work system or a working person. Thus, it has the power to describe tasks in a top-down process proceeding from a company-wide organizational level that ends with the individual assignment of a task to a resource that is a combination of a person and work means.

12.7.3.2 SADT and IDEF

A widely used graphical tool for software development is SADT—Structured Analysis and Design Technique—which simplifies communication among participants of software

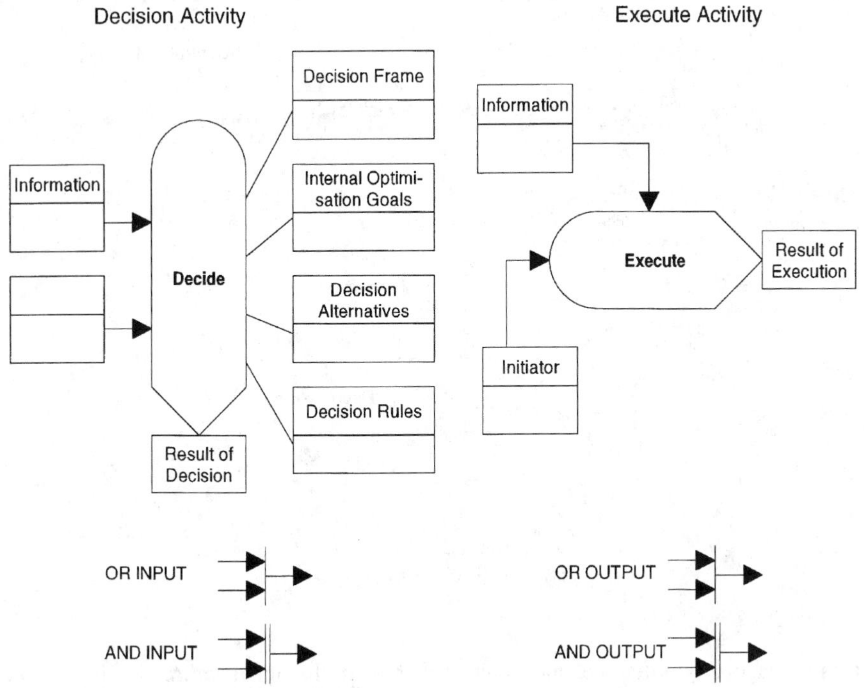

Figure 12.35 GRAI networks.

projects (Marca and McGrowan, 1987). Its field of application is spread from real-time communication problems to commercial and organizational plannings. SADT builds up diagrams with boxes and links. Boxes represent components of the system which can be subdivided into more detailed components on another level. Altogether the levels form a hierarchical graph. Each box is coupled to its environment mainly by interface links (Figure 12.36), except the "mechanism" relationships. Those represent the support function for a specific activity, for instance, a program or an expert, a necessary tool or device (Ross, 1985; Smith and Wang, 1988). By this level-oriented breakdown process the complete enterprise with its material, information, and process flows can be mapped to details on specific-level tasks and hierarchies, workplaces, and their personal activities. Thus, SADT supports two model concepts:

- Modeling of activities that can be taken over by men, machines, computers, or algorithms
- Modeling of data, objects, and devices

In principle, all aspects that are necessary conditions of task analysis are comprised in the concept, except the time domain with possibly dynamical alterations. SADT and IDEF 0 are equivalent.

At the end of the seventies, the IDEF-modeling technique was developed on the basis of SADT (Harrington, 1985; Ross, 1985; Smith, 1988) which comprises three aspects:

- A functional model (IDEF 0) which primarily describes activities
- A data model (IDEF 1) which uses data and objects
- A dynamic model (IDEF 2)

IDEF is based on a concept for definition of system structures and uses a decomposition method for structural analysis and a graphical tool, respectively. Thus, the principle of the method is similar to task analysis, because it develops production systems into a hierarchy of activities for a better understanding of the whole system and its interfaces by a top-down approach. The representation of this structure is done by graphical dia-

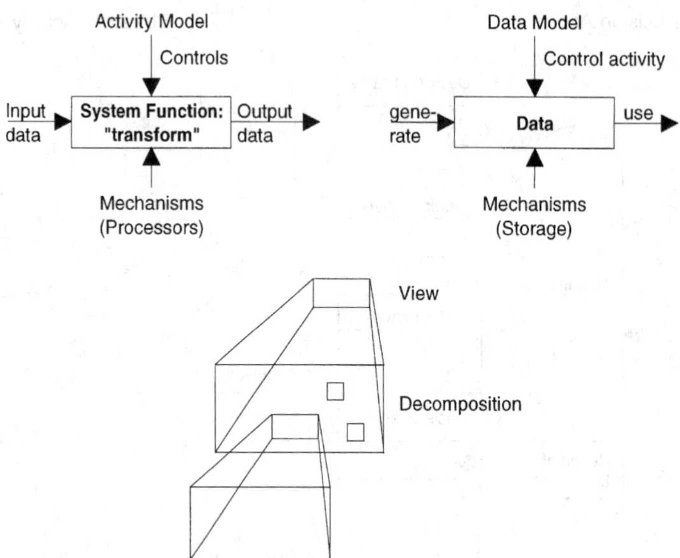

Figure 12.36 SADT diagrams.

grams in diagonally ordered elements and relations. In the functional model, for instance, the elements/boxes signify functions, the horizontal flows determine the input-output relations, bottom-up arrows represent necessary tools/mechanisms, and the top-down arrows represent control aspects (cf. Figure 12.36).

In the data model, for instance, the boxes represent the data and the horizontal arrows functions. A specialty of IDEF is that in IDEF 2 even dynamic task execution can be outlined, in addition to the possibilities of SADT.

12.7.3.3 Petri-Nets

Whereas GRAI is specifically oriented to production systems and SADT was originally conceived for administrative processes, Petri-Nets are general methods of system description (Petri, 1979; Reisig, 1985). A main characteristic of Petri-Nets is the description of concurrent processes in a causal order and the mapping of dynamic processes by a static system structure. With the help of tokens, the causal structure of a process is described in a network of branches and knots. Conditions describe the status of a system or a process; events, called actions or transitions, too, effect the transfer from a status to another status/condition. According to the type of tokens, different types of networks can be developed, like condition-event networks, position-transition networks, predicate-transition networks, etc. (see the example in Figure 12.37). Especially for the analysis of production systems (Möhrle, 1990) and of CIM systems (Lescak, 1988) Petri-Nets were widely used in the past. As the example in Figure 12.37 demonstrates, the level of description may be oriented to the workplace of an individual working person with the logical and sequential conditions of task execution, with supervisory and intervention processes of machine operation that are assigned to a working person as individual task.

12.7.4 Summary

Coming back to an overview of different methods/modeling languages it may be helpful to see them in the perspective of a combination or distillation of criteria, derived from task definitions. Following Campanion and Medsker (1992), "A task is a set of actions performed by a worker that transforms inputs into outputs through the use of tools, equipment, or work aids." That means that a task is described by a serial execution (action, function, process), the object or its status before and after execution, means (tools, mechanisms, aids), and the process owner (human, computer, autonomous system). In addition to these basic presuppositions (necessary conditions) additional attributes can be defined for the modeling which may be relevant for different goals/purposes. Hackman (1969,

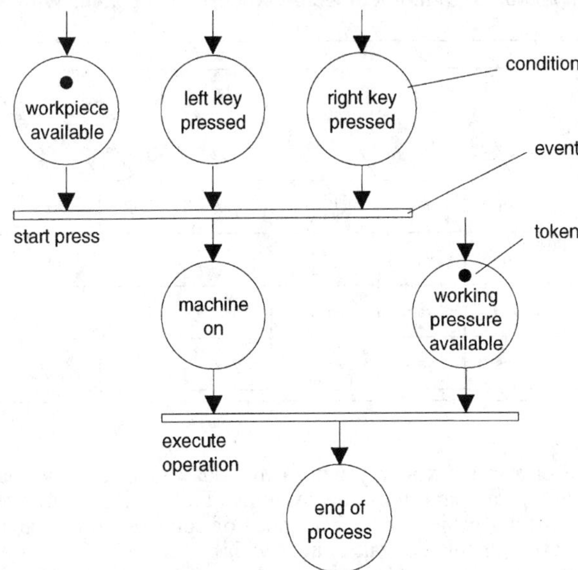

Figure 12.37 Example for a Petri-Net: Segment from a condition-event network.

1970), for instance, says that a task may be self-set or self-imposed (decision aspect), and that it consists of a stimulus complex and a set of rules which specify what has to be done according to the stimuli. These instructions indicate the necessary operations to reach goals (communication aspect). Miller (1967) states the opinion that each task activitiy consists of the following:

1. An indicator on which the activity-relevant indication appears
2. The indication or cue which calls for a response
3. The control object to be activated
4. The activation or manipulation to be made
5. The indication of response adequacy, or feedback

This means that in addition to the mentioned aspects instructions/goals of a task, the information flow and the executors (person, group) as well as the transmittors must be mapped: This relates to the structures of communication and cooperation in an organizational context, even with respect to the time conditions.

With this set of characteristics, the different modeling languages can be evaluated in an overview (see Table 12.14). As can be easily seen, the majority of essential criteria is met by the four languages, described earlier.

Nevertheless, there are many more possibilities of descriptive modeling languages to be chosen for specific purposes which cannot be presented here in more detail. Software engineering of production, business, and military processes seems to be a powerful source of development of task-analytic techniques that must be acknowledged by human factors/ergonomics as well (Oberquelle, 1987; Mertins et al., 1994).

12.7.5 Outlook to CSCW and Teleworking

The prementioned tools must be seen primarily in the context of efforts to realize "lean production." Together with the idea of a "lean organization," business processes and administrative chains come into the focus of reorganization under the heading "lean office." Here new systems of information technology with a certain impact on task analysis and task taxonomies must be at least mentioned: Computer Supported Cooperative Work with Workgroup-Computing and Workflow-Automation (Götzer, 1994). Models of the coordination process (Malone 1990) and psychological considerations about the nature of cooperative work (Leplat 1994), especially, lead to an understanding of what kind of

Table 12.14 Comparison of Different Selected Modeling Languages With Respect to Task Criteria

Modeling Languages	Serial Execution	Objects	Means/Tools	Time Conditions	Decisions	Organizational Units	Communication	Cooperation
GRAI	yes	yes	yes	comparative static	yes	yes	yes	no
IDEF	yes	yes	yes	dynamic and static	not explicity	no	yes	no
SADT	yes	yes	yes	static	no	no	yes	no
Petri-Nets	yes	yes	yes	static model, dynamic process	yes	in principle	hardly possible	no

tasks in the sense of a task taxonomy with a limited amount of task characteristics can be supported by groupware and workflow systems (Table 12.15). However, development initiatives of new, mostly project-related methods of cooperative task analysis are driven by this software development (Luczak et al., 1995a).

Another aspect of CSCW is *teleworking*, which means within this context the task-assignment and task-oriented cooperation among locally distributed working persons. Identifying task characteristics for this drastically increasing domain of ergonomics—in terms of importance and frequency—means to separate those tasks with a high development potential from those with hindrances (Table 12.16). Up to the level of semistructured tasks, telecooperation has a good prognosis: A similar result is reported by Gray et al. (1993), who defines the following task characteristics:

- Qualitatively and quantitatively defined or agreed task contents and time plans
- High degree of autonomy with partners
- Task requires longer periods of calm concentration
- Communication demands can be met by CSCW/Teleworking Systems
- Rare/infrequent exchange of physical resources (work objects, means)
- Physical face-to-face communication unnecessary

Table 12.15 Task Taxonomy for Workgroup Computing and Workflow Automation

	Workgroup-Computing	Workflow-Automation
Information Technology	Groupware systems	Workflow Systems
Focus of deployment	Support cooperative group tasks	Manage complex administrative processes with high priorities
Type of tasks	Decisions, project and teamwork	Time and event-driven administration of processes
Execution of tasks/operations	Unstructured or weakly structured subtasks in ad-hoc execution	Well-structured and predetermined operations
Characteristics	Coordination of tasks by a document management system and tools of administrative communication + infrastructure	Operations management by linking of different systems to an automation of a process-oriented application system
Object of view	Coworker	Task or operation
Orientation to	Team, groups, workplans	Company/corporation
Initiation of activities organization steering and control	} User	} System
Computer architecture	Primarily decentralized	Primarily centralized

Table 12.16 Task Characteristics to Identify Tasks for Teleworking

	Unstructured Tasks	Semistructured Tasks	Structured Tasks
Complexity	high	medium	low
Planning possibilities	low	medium	high
Information demands	vague	problem-dependent	predetermined
Communication demands	very high	changing in time and context	low
Cooperation partners	changing, undetermined	changing, but determined	permanent, equal
Way of solution	open	planned but open for variants	predetermined

From this specification of task characteristics, global tasks/jobs can be deduced and design options can be evaluated for telecommunication systems (Luczak et al., 1995b).

12.8 TASK ANALYSIS FOR SOCIETAL DEMANDS

The objective conditions of working life which were developed in an evolutionary process by society, represent action/activity demands for the individual working person which are—at the same time—prescriptions and offers. By demanding a certain action/activity, society makes it possible as well. Thus, in the societal context, tasks are generated for the individual. Only those demands of society may be classified as tasks which are derived from a labor partition between different segments and groups of society, its functional entities as global production units or members of a class of working people with a specific class-related spectrum of activities/actions.

Therefore task-related information is grouped around societal categories and necessities. The grouping structure known as "task structure analyses" highly/considerably differs from the level of aggregation found on the company or work-group level. The main fields of application of task-analytical knowledge lie in the following task-clustering aspects:

1. Determination of task similarities are relevant for society in the following perspectives: In most constitutions or subsequent legal regulations the principle of equal pay for equal work is written down. This is not only a problem of employers' representations and unions, but ranks highly in social politics to avoid discrimination of certain groups in society (women, races, etc.). Thus, task-analytical information is a presupposition of legally acceptable payment systems. The AET, for instance, was originally developed on the basis of a research contract with the federal minister of work and social affairs in Germany to objectively identify womens discrimination by pay-roll formulation (Rohmert, Rutenfranz, and Luczak, 1976). Facts about similarities of tasks are used in the context of career guidance as well. The process of guidance by career advisors relies on grouped task structure information with respect to professions and possible changes of a person between them.
2. The determination of curricula and of vocational training contents is based on "task inventory approaches" (AT&T, 1980) mostly for the large organizations in the public and pivate sector, especially the military services (Christal and Weißmüller, 1976), but for the task-oriented development of the school system for vocational trainings (Troll, 1987) in relation to technological innovation and even for university curricula as well. In the 70s and 80s, for example, German universities of technology mostly introduced new curricula of production and industrial engineering on the basis of empirical task analysis and task clusterings (e.g., Luczak and Schwier, 1983). Systematic task-analytical approaches are urgently needed to follow-up the specialization of jobs with technological development and the differentiation of the spectrum of professions. In the United States, the Occupational Analysis Inventory, based on the theoretical and methodological concept of the PAQ, was developed (Cunningham et al., 1971) for this purpose, whereas the

aspect of interest for a profession from a person is covered by the Job Activity Preference Questionnaire (Mecham et al., 1972).

3. The development of valid selection and placement tests which can guarantee the same employment chances for any person are another public interest of society in task analysis. In the United States, for example, Uniform Guidelines for Employment Selection Procedures (USEEOC, USCSC, USDOL, and USDOJ, 1978) were developed which name in detailed form the importance and specifications for task-analytical techniques which can validate the tests.
4. The classification of professions and their change with technological and economical development is a central interest of any national economy to get a survey over the labor market, for example, in the form of a "Dictionary of Occupational Titles," or for segments of the industrial relations and branch-oriented workforce potential analysis. These applications mostly rely on task analysis techniques from the more detailed levels—especially those from the PAQ type—or formulate properties of professional work in a verbal description of tasks (see, e.g., Karg and Staehle, 1982). Classifications of professions on the basis of task analysis are used in other various contexts, too: The health status in different profession groups can be documented more easily and more detailed, when specific task information is available for occupational safety and health authorities in the state. Thus, interventions and regulations can be formulated task-specific. This implies the identification of accidents according to frequency and severity as well and its relation to activities and tasks executed in these professional contexts. Impacts on accident prevention authorities, on the design of the social security system, and on the national insurance systems for the risks of working life are obvious.
5. International comparisons, as initiated by OECD (1977) and ILO (1979), try to use task information as social indicators worldwide. Thus, the kind and character of tasks, their organization, and their grouping to jobs and professions are indices of the status of development of a national economy in international comparative studies. OECD and ILO try to harmonize the content of empirical studies and research programs nationwide in the form of perspective and recommended lists to get the possibility for international comparison of the "quality of working life" or similar aggregations. Task-related information is at least one-fourth of the items of the mentioned lists and thus a constituent part of global evaluation of societal demands and societal status.

12.9 CONCLUSION AND GENERALIZED REMARKS ON "METHODS" AND "TECHNIQUES" OF TASK ANALYSIS

The numerous instruments reported in the previous survey indicate that the term task analysis has to be interpreted as a method, with which data are collected, ordered in a systematical manner, and evaluated by relevant criteria in relation to the intention of the analysis. That means that different characteristics of task analysis are based upon different intentions and purposes. The differentiation with respect to the notions "method" and "technique" of task analysis are useful for a conclusive overview. Following, for instance, Frieling (1974), a *method* of task analysis requires that

- A theoretical model for this analysis be available that allows an interpretation of the data collected within a broader framework with possibly generalized practical consequences
- Elements can be differentiated into central/peripheral and/or common/specific categories
- Cost to effectiveness considerations play an important role in the collection, processing, and evaluation of data
- Standardizations of the procedure with respect to objectivity, validity, reliability, and utility are taken into account
- Nominal scaling in terms of verbal description be discarded in favor of quantitative scales; at least an ordinal level of measurement should be strived for.

A *technique* of task analysis, however, is present whenever information is gathered and described with the help of an instrument or procedure in a certain systematical manner. In practice, different techniques of data collection and processing can be used within a method of task analysis. Thus, Kirwan and Ainsworth (1992) differentiate among

- Collection techniques
- Representation techniques
- Simulation techniques

Following this differentiation will make clear that within a technique a model basis and explicit evaluation procedure for possible conclusions about the impacts on humans is not explicitly foreseen.

The method also indicates the "person" and the "purpose" to motivate evaluation and design standards.

12.9.1 Collection Techniques

In principle, all information gathered about tasks depends on an idea about the person who is executing the task, whether this information is situation related (objective) or person related (subjective). Clearly spoken, also data about the situation or condition of the task imply necessarily a picture of the person previewed or used for the possible execution, thus making a differentiation into subjective or objective task analysis techniques somewhat questionable. Nevertheless, some data categories can only be obtained from the person directly, whereas others are data about the person.

The most common and simple technique of data collection is to *ask* the person about the task, to perform a questionnaire, or to let the person comment on the task execution in form of verbal protocols.

Additionally, the person can be *observed* during task execution by the analyst, by a camera, by time-line-oriented automatic event registration, or any access that is based on perception (not only visual).

A third approach is the collection of *documents* about the task, which are available in the company, such as working orders, planning and steering data for the work process, assignments and charts of the organizational structure, cooperation and communication orders, descriptions of tools and machines, including computers, and all other information material, which is stored in paper or in a data repository of a computer, related to task execution, its goals and purposes as well as its economic inputs and outputs.

Finally, direct person-related *measurements* must be mentioned, such as *physiological* variables, which are affected by the task elements, *behavioral* traits of the working person, its activity schedule and personal expertise or mastership in execution, and *psychophysical* reactions during or after task execution. Practically, these measurements imply a lot of person-related variation on top of the task-related variation in focus. Nevertheless, the part of variance or variation in the measurements characteristic for the task and its execution is mostly considerable.

12.9.2 Representation Techniques

The consequence and result of data collection may easily end in a "data cemetery" of "numbers" and "words" of description. To bring a structure into these data sets is the purpose of representation techniques.

Normally the data about the prescription or about the execution of the task or its elements are ordered sequentially or arranged along the time-line in a reproducible and systematical manner. This implies *first* the use of description decomposition categories, which represent the essential features of the task/task elements, *second* the application of ordering principles in a context-specific manner, mostly time based and logic based, and *third* an algorithmical or even mathematical formulation of the interrelationships as the most precise and most advanced representation tool. Normally the second and third step are a presupposition for the generation of design variants by simulation techniques.

Task decomposition is a structured way of expanding the information from a task description into a series of more detailed statements about particular issues which are of interest to an analyst (Kirwan and Ainsworth, 1992). This can be done model based or paradigm based: Thus, decomposition may not be a "simple" technique of representation, but really a scientific procedure with a generalizable idea about the "nature of work," or with a theory about the function and malfunction of task components, or with a model of interaction between the elements. These models may be simple additive models in balancing energy consumption by calories materials or time segments, for example, or sophisticated models of cognitive levels of interaction with a large-scale technical system.

Nevertheless, task decomposition can be done and is done frequently in a very practical manner, for which Kirwan and Ainsworth (1992) developed an empirical taxonomy of descriptive categories that have been used in various studies (Table 12.17).

In principle, three main problems are connected to task representation by decomposition: That is, *first*, the choice of categories that are relevant for the description; *second*, the principle of decomposition, for instance, hierarchical, sequential, simultaneous, location-operational, etc. breakdown principles; and, *third*, the depth or level of aggregation on which the breakdown begins and where it ends. All of these problems have to be considered before deciding on an ordering representation technique, which means in what kind of network the decomposed parts can be introduced and mapped.

For example, one of the earliest representation techniques referenced in the literature is the *flow diagram*, which is found in task analysis in the form of a descriptive technique for the following:

a. Sequential operations and movements in combination with time and motion analysis, for instance, for manual assembly
b. Logical combinations of the perception and operation of displays and controls in link analysis to demonstrate and evaluate the connections via frequency and importance weightings at a workplace in human–machine interaction
c. Allocation of functions to human and machine as programs for the execution of processes in a work system context with personnel qualification and crew size consequences
d. Interaction between the workforce in a work organization and company organization via process analysis and mapping in GANTT charts, networks like PERT/GERT, or Petri-Nets and other computer-oriented tools like workflow and workgroup management systems in packages of CSCW

Thus the flow diagram approach stretches over a wide spectrum of task representation possibilities.

In principle, the idea of the analyst about the utility of the respective flow diagram technique and its analytical background determines the level of aggregation or decomposition of the technique used. Basically flow diagrams may be action oriented, event oriented, signal- and information oriented, function- and relation oriented, and/or condition oriented when condition-depending sequences are mapped. The scope of variety in one representation technique must be considered by the analyst.

With respect to mathematics as a representation technique, for example, tasks with a continuous information input and output flow can be represented by a transfer function from control theory. In the mathematical representation of this concept human information-processing tasks are described by the methods of differential equations, frequency analysis, and matrix-calculus. The structural analogy for task representation is a linear electrical network.

In a simple case, a scalar input variable $x(t)$ is connected with a scalar output $y(t)$ by a differential equation:

$$y(t) + T_{\mathrm{n}}\frac{dy(t)}{dt} = Kx(t - \tau)$$

This equation means that the input variable $x(t)$ initiates a human action after a reaction time τ. This action becomes effective via an amplification factor K, which has an impact on the output variable $y(t)$ itself and its derivation $\frac{dy}{dt}$ with an additional time constant T_{n}.

The structure of such an equation and the coeffficents are *task specific*, which means, that the human task of connecting input and output is represented by a mathematical formula. Multiple inputs and outputs are represented in a matrix calculus.

Directly related are control theory, information theory, the theory of graphs, algorithms, automates, etc. (Luczak, 1975). New hopes are set on fuzzy set theory, for example, for a vague representation of tasks.

12.9.3 Simulation Techniques

If a task representation via a network or via a mathematical formulation exists, it is quite easy to do the next step into the direction of simulation, which normally means that a computer model is used to "play" with the task description for purposes of demonstration and training, of experimenting numerically and logically with inputs and outputs of the computer model, and of doing R & D in the direction of composing new task designs in terms of different task configurations, team structures and complements, operating procedures for the identification of deadlocks and bottlenecks, etc. Many illustrative examples are mentioned in the previous text.

Coming back to the notion "method of task analysis" and the requirements formulated by Frieling (1974), two additional conditions for a method must be fulfilled if a theoretical model shall lead to a fruitful contribution to human factors evaluation and design: A "picture of the person" or an "image of the human" and an "idea of the purpose" and the "imagination of the application context." These requirements are respected by using the level-concept of segmentation of "work sciences," as explained before. But on each level the philosophy of task analysis is found to respect the limitations of human performance, to overcome the weaknesses of humans by design, to support the humans by the identification of their strengths, and a widening of their possibilities in bottleneck situations.

REFERENCES

Allen, R. W., McRuer, D. T., and Thompson, P. M. (1989). Dynamic system analysis programmes with classical and optimal control applications of human performance models. In G. R. McMillan, D. Beevis, E. Sales, M. H. Strub, R. Sutton, and L. van Breda, Eds., *Application of Human Performance Models to System Design.* New York: Plenum.

AT&T, Ed. (1980). *The work performance survey system: A procedure guide.* Basking Ridge, NJ: AT&T.

Baks, R. F. (1950). *Interaction Process Analysis.* Reading, Mass: Addison-Wesley.

Baks, R. F., and Cohen, S. P. (1982). *Symlog—Ein System für die mehrstufige Beobachtung von Gruppen.* Stuttgart: Klett-Cotta.

Bammer, G. (1990). Review of current knowledge—musculoskeletal problems. In L. Berlinguet and D. Berthelette, Eds., *Work with display units 89.* Amsterdam: North Holland.

Barnes, R. M. (1963). Motion and time study—design and measurement of work. New York: John Wiley.

Baron, J. (1984). A control theoretic approach to modelling human supervisory control of dynamic systems. In W. B. Rouse, Ed., *Advances in man-machine-systems research.* Greenwich, CT: JAI Press.

Beatty, J. (1982). Task evoked pupillary responses, processing load, and the structure of processing resources. *Psychol. Bulletin, 91*(2), 276–292.

Beevis, D., Bost, R., Döring, B., Nordö, E., Papin, J. P., Schuffel, H., and Streets, D. (1992). Analysis Techniques for Man-Machine-System Design. *Technical Report* AC/243 (Panel 8) TR/7, Volumes 1 and 2, NATO Defense Research Group, Panel 8, RSG 14, Brussels.

Boff, K. R., Kaufman, L., and Thomas, J. P. Eds., (1986). *Handbook of Perception and Human Performance.* New York: John Wiley.

Boff, K. R., and Lincoln, J. E. (1988). *Engineering Data Compendium—Human Perception and Performance.* Harry G. Armstrong Aerospace Medical Research Laboratory, Wright-Patterson Air Force Base, OH.

Bonitz, D. (1995). *Evaluation von Arbeitssystemen.* Frankfurt: Peter Lang.

Bünz, D. (1987/88). Die GRAI-Methode zur Analyse und zum Entwurf von Produktionsmanagement-Systemen, *CIM-Management* 4,2, 43–37, and 5, 2, 56–59.

Bullinger, H. J., Rally, P., and Schweizer, W. (1990). Simulation flexibler Arbeits- und Nutzungszeitmodelle, *VDI-Zeitschrift 132*, *4*, 55–59.

Cacciabue, P. C., and Kjaer-Hansen, J. (1993). Cognitive Modelling and Human-Machine Interaction in Dynamic Environments. *Le Travail Humain, 56*(*1*), 1–26.

Calderwood, R., Klein, G. A., and Grandall, B. W. (1988). Time pressure, skill, and move quality in chess. *American Journal of Psychology, 101*, 481–493.

Campion, M. A. (1985). The multimethod job design questionaire (MJDQ). *Psychological Documents, 15* (*1*), 12.

Campion, M. A. (1988). Interdisciplinary Approaches to Job Design: A Constructive Replication with Extensions. *Journal of Applied Psychology, 73*, 467–481.

Campion, M. A., and Medsker, G. J. Job Design. (1992). In Salvendy, G. E., Ed., *Handbook of Industrial Engineering,* New York: John Wiley.

Campion, M. A., and Thayer, P. W. (1985). Development and Field Evaluation of an Interdisciplinary Measure of Job Design. *Journal of Applied Psychology, 70*, 29–43.

Carbonell, J. R. (1966). A queuing model of many-instrument visual sampling, *IEEE Transactions on Human Factors in Electronics, 4*, 157–174.

Card, S. K., Moran, T. P., and Newell, A. (1980). The keystroke level model for user performance time with interactive systems. *Communications of the ACM, 23*, 396–410.

Card, S. K., Moran, T. P., and Newell, A. (1983). *The Psychology of Human-Computer Interaction.* Hillsdale: Lawrence Erlbaum Assoc.

Carmel, E., Crawford, S., and Chen, H. (1992). Browsing in Hypertext: A Cognitive Study. *IEEE Transactions on Systems, Man and Cybernetics, 5*, 865–884.

Christal, R. E., and Weißmüller, J. J. (1976). New Comprehensive Data Analysis Program (CODAP) for Analyzing Task Factor Information. *Lackland AF HRL International Professional Papers,* No. TR–76–3.

Chubb, G. P., Laughery, K. R., and Pritzker, A. A. D. (1987). Simulating Manned Systems. In G. Salvendy, Ed., *Handbook of Human Factors.* New York: John Wiley.

Cooke, N. J. (1994). Varieties of knowledge elicitation techniques. *International Journal of Human-Computer-Studies, 41*, 801–849.

Copley, F. B. (1923). *Frederick W. Taylor.* New York: Harper & Brothers.

Crandall, B., and Calderwood, R. (1989). *Clinical Assessment Skills of Experienced Neonatal Intensive Care Nurses.* Yello Springs, OH: Klein Associates Inc.

Cunningham, J. W., Tuttle, T. C., Floyd, J. R., and Bates, J. A. (1971). The development of the Occupational Analysis Inventory: An "ergometric" approach to an educational problem. In Copy, Ed., *Center Research Monograph No. 6*, Raleigh, NC: Center for Occupational Education.

Diaper, D. (1989). *Task Analysis for Human-Computer Interaction.* New York: Wiley.

Donders, F. C. (1868). On the speed of mental processes. *Acta Psychologica, 30*, 412–431.

Douglas, S. A., and Moran, T. P. (1983). Learning text editor semantics by analogy. In Janda, A., Ed., *Human Factors in Computing Systems.* New York: ACM Press, 207–211.

Doumeingts, G. (1984a). *Methode GRAI: méthode de conception des systemes en productivité (Thèse).* Bordeaux: Université Bordeaux I.

Doumeingts, G. (1984b). Methodology to Design Computer Integrated Manufacturing and Control by Manufacturing Units. In M. Rembold, Ed., *Methods and Tools for Computer Integrated Manufacturing.* Berlin: Springer.

Downs, C. G. (1988). Representing the structure of jobs in job analysis. *International Journal of Man-Machine-Studies, 28*, 363–390.

Drury, C. G., Paramore, B., VanCott, H. P., Grey, S. M., and Corlett, E. N. (1987). Task Analysis. In G. Salvendy, Ed., *Handbook of Human Factors.* New York: John Wiley.

Dunckel, H., Volpert, W., Zölch, M., Kreutner, U., Pleiss, C., and Hennes, K. (1993). *Kontrastive Aufgabenanalyse im Büro.* Stuttgart: Teubner Verlag.

Dunckel, H. (1996). Aufgabenstruktur. In H. Luczak, W. Volpert, T. Müller, Eds., *Handbuch der Arbeitswissenschaft.* Stuttgart: Poeschel Verlag.

Essig, H., Heibeg, H.-W., Kühn, M., and Wolf, A. (1981). *BENORSY—Ein formalisiertes Verfahren zur benutzerorientierten Systemrevision* (Forschungsbericht DV–81–001, Hamburg).

Facaoaru, C., and Frieling, E. (1991). Zur Problematik der Erfassung informatorischer und sensumotorischer Anforderungen und Belastungen—dargestellt an der Entwicklung des Tätigkeitsanalyseinventars (TAI), *Zeitschrift für Arbeitswissenschaft, 45(3)*, 146–155.

Fechner, W. (1994). Ein objektorientiertes Konzept zur Planung visueller Prüf- und Kontrolltätigkeiten, *VDI-Fortschritt-Berichte*, Reihe 20/Nr. 115 (VDI-Verlag, Düsseldorf).

Frei, F., Hugentobler, M., Alioth, A., Duell, W., and Ruch, L. (1993). *Die kompetente Organisation. Qualifizierende Arbeitsgestaltung—die europäische Alternative.* Stuttgart: Schäffer Poeschel.

Frese, M. (1996). Aufbauorganisation. In H. Luczak, W. Volpert u.M.v. T. Müller, Eds., *Handbuch der Arbeitswissenschaft,* Stuttgart: Poeschel.

Frieling, E. (1974). *Psychologische Probleme der Arbeitsanalyse - Dargestellt an Untersuchungen zum Position Analysis Questionnaire (PAQ)—.* (Doctoral Diss., München.)

Frieling, E. et al. (1982). *Bestandsaufnahme arbeitsanalytischer Methoden in Forschungsvorhaben aus dem Bereich "Arbeitsorganisation."* München: Universität.

Frieling, E., Facaoaru, C., Benedix, J., Pfeus, H., and Sonntag, K. (1990). *Tätigkeitsanalyse inventar.* Kassel: Copied Manuscript.

Gaitanides, M. (1996). Ablauforganisation. In H. Luczak, W. Volpert u.M.v. T. Müller, Eds., *Handbuch der Arbeitswissenschaft.* Stuttgart: Poeschel.

Gilbreth, F. B. (1911). *Motion Study.* New York: Van Nordstrand.

Gilbreth, F. B. (1917). *Applied Motion Study.* New York: Macmillan.

Gilbreth, L. M. (1919). *The Psychology of Management.* New York: Macmillan.

Gilbreth, F. B., and Gilbreth, L. M. (1921). Process Charts. *Trans. Americ. Soc. Mech. Eng., 43/1818*, 1029–1050.

Götzer, K. G. (1994). *Lean Office: Effektive Büroarbeit durch neue Technologien.* München: Computerwoche Verlag.

Gopher, D., and Sanders, A. F. (1984). S-OH-R: Oh Stages! Oh Resources! In W. Prinz and A. Sanders, Eds., *Cognition and Motor Processes.* Berlin: Springer.

Glowin, R., and Novak, J. D. (1984). *Learning how to learn.* New York: Cambrigde University Press.

Gray, M., Hodson, N., and Gordon, G. (1993). *Teleworking explained.* Chichester: John Wiley & Sons.

Gugerty, L. (1993). The use of analytical models in human-computer-interface design, *International Journal of Man-Machine Studies, 38,* 625–660.

Hacker, W. (1986). *Arbeitspsychologie* (Deutscher Verlag der Wissenschaften, Berlin).

Hacker, W. (1993). Occupational psychology between basic and applied orientation—some methodological issues. *Le Travail Humain, 56, 2–3,* 157–169.

Hacker, W., Iwanowa, A., and Richter, P. (1993). *Tätigkeitsbewertungssystem.* Berlin: Psychologisches Zentrum, Sektion Psychologie der Humboldt-Universität.

Hackmann, J. R. (1969). Towards understanding the role of tasks in behavioral research. *Acta Psychologica, 31,* 97–128.

Hackman, J. R. (1970). Tasks and task performance in research on stress. In J. E. McGrath, Ed., *Social and psychological factors in stress.* New York: Holt, Rinehart, Winston.

Hackman, J. R., and Oldham, G. R. (1975). Development of the Job Diagnostic Survey. *Journal of Applied Psychology, 60,* 159–170.

Hackman, J. R., and Oldham, G. R. (1976). Motivation through the design of work: Test of a theory. *Organizational Behaviour and Human Performance, 16,* 57–71.

Hackman, J. R., and Oldham, G. R. (1980). *Work Redesign.* Reading, MA: Addison Wesley.

Hammer, M., and Champy, J. (1993). *Reengineering the Corporation.* New York: Harper Collins.

Hammer, W. (1972). *Handbook of system and product safety.* Englewood Cliffs, NJ: Prentice-Hall.

Harrington, J. (1984). *Understanding the manufacturing process.* New York: Marcel Dekker.

Heinz, K., and Lange, W. D. (1992). Simulation gruppenorientierter Fertigungsstrukturen. *CIM-Management, 81*(2), 33–39.

Hertting-Thomasius, R. (1992). *Untersuchungen zur Belastungssuperposition einer informatorischen und einer sensumotorischen Belastung am Beispiel einer gekoppelten Montage- und Kontrolltätigkeit.* Dissertation (Technical University, Berlin).

Hollnagel, E. (1993). Models of Cognition: Procedural Prototypes and Contextual Control. *Le Travail Humain, 56(1),* 27–51.

Hollnagel, E., Pedersen, O. M., and Rasmussen, J. (1981). *Notes on Human Performance Analysis.* Roskilde, Denmark: Riso National Laboratory, Riso-M-2285.

ILO (International Labour Organization) (1979). *New Forms of Work Organization.* Geneva: ILO.

Kahnemann, D. (1973). *Attention and Effort.* Englewood Cliffs: Prentice Hall.

Karg, P., and Staehle, H. W. (1982). *Analyse der Arbeitssituation—Verfahren und Instrumente.* Freiburg: Haufe.

Katsuki, S. (1960). *Relative metabolic rate of industrial work in Japan.* Metabolic Rate, Food and Nutrition Committee in Japan.

deKeyser, V. (1991). Work analysis in French language ergonomics: origins and current research trends. *Ergonomics, 34(6),* 653–669.

Kieras, D. E. (1988). Towards a practical GOMS model methodology for user interface design. In M. Helander, Ed., *Handbook of Human Computer Interaction.* Amsterdam: Elsevier.

Kieras, D. E., and Polson, P. G. (1985). An approach to the formal analysis of user complexity. *International Journal of Man-Machine Studies,* 22, 223–274.

Kirwan, B., and Ainsworth, L. K. (1992). *A guide to task analysis.* London: Taylor & Francis.

Klaus, G., and Buhr, M. (1975). *Philosophisches Wörterbuch; Stichwort: Analyse.* Berlin: deb—das europäische Buch.

Klein, G. A. (1989). Recognition-primed decisions. In W. B. Rouse, Ed., *Advances in man-machine systems research.* Greenwich, CT: JAI Press.

Klein, G. A. (1993a). A Recognition-primed Decision (RPD) Model of Rapid Decision Making. In G. A. Klein, J. Orasanu, R. Calderwood, and C. E. Zsambok, Eds., *Decision Making in Action: Models and Methods.* Norwood, NJ: Ablex Publishing.

Klein, G. A. (1993b). Twenty Questions—Suggestion for Research in Naturalistic Decision Making. In G. A. Klein, J. Orasanu, R. Calderwood, and C. E. Zsambok, Eds., *Decision Making in Action: Models and Methods.* Norwood, NJ: Ablex Publishing.

Klein, G. A., Calderwood, R., and Clinto-Cirocco, A. (1986). Rapid decision making on the fire ground. *Proceedings of the Human Factors Society 30th Annual Meeting, 1,* pp. 576–580.

Klein, G. A., Calderwood, R., and Macgregor, D. (1989). Critical Decision Method for Elicitating Knowledge. *IEEE Transaction on Systems, Man, and Cybernetics, 19,* 3.

Kleinman, D. L., Baron, S., and Levison, W. H. (1971). A control theoretic approach to manned-vehicle systems analysis. *IEEE Trans. Autom. Control, 16,* 824–832.

Klinger, D. W., and Gomes, M. E. (1993). Cognitive Systems Engineering Application for Interface Design. Designing for Diversity. *Proceedings of the Human Factors and Ergonomics Society 37th Annual Meeting in Seattle, Washington, October 11–15, 1993,* by The Human Factors and Ergonomics Society, Ed., Santa Monica, CA, Volume 1, pp. 16–20.

Kosiol, E. (1962). *Organisation der Unternehmung*. Wiesbaden: Gabler.

Kraiss, K. F. (1981). Analytical evaluation of manned systems with task network models. In J. Moraal and K.F. Kraiss, Eds., *Manned systems design*. New York: Plenum Press.

Landau, K. (1978). *Das Arbeitswissenschaftliche Erhebungsverfahren zur Tätigkeitsanalyse—AET.* Doctoral Dissertation, Darmstadt.

Landau, K., Luczak, H., and Rohmert, W. (1975). Arbeitswissenschaftlicher Erhebungsbogen zur Tätigkeitsanalyse. In W. Rohmert and J. Rutenfranz, Eds., *Arbeitswissenschaftliche Beurteilung der Belastung und Beanspruchung an unterschiedlichen industriellen Arbeitsplätzen*. Bonn: BMAS.

Landau, K., and Rohmert, W. (1981). *Fallbeispiele zur Arbeitsanalyse—Ergebnisse zum AET-Einsatz.* Bern: Huber.

Landau, K., and Rohmert, W. (1987). Aufgabenbezogene Analyse von Arbeitstätigkeiten. In U. Kleinbeck and J. Rutenfranz, Eds., *Arbeitspsychologie*. Göttingen: Hogrefe.

Laughery, K. R. Sen. and Laughery, K. R. Jr. (1987). Analytic techniques for function analysis. In G. Salvendy, Ed., *Handbook of Human Factors*. New York: John Wiley.

Laurig, W., and Rombach, V. (1989). Expert systems in ergonomics: requirements and an approach. *Ergonomics, 32(7)*, 795–811.

Leitner, K., Volpert, W., Grainer, B., Weber, W., and Hennes, K. (1987). *Analyse psychischer Belastung in der Arbeit. Das RHIA-Verfahren.* Köln: Verlag TÜV Rheinland.

Leplat, J., and Rasmussen, Y. (1987). Analysis of human errors in industrial incedents and accidents for improvement of work safety. In J. Rasmussen, K. Duncan, and J. Leplat, Eds., *New Technology and Human Error*. Chichester, UK: John Wiley.

Leplat, J. (1989). Error analysis, instrument and object of task analysis. *Ergonomics, 32(7)*, 813–822.

Leplat, J. (1993). L'analyse psychologique des travail: quelques jalons historiques. *Le Travail Humain, 56*, 2–3, 115–131.

Leplat, J. (1994). Collective activity in work: Some lines of research. *Le Travail Humain, 57(3)*, 209–226.

Leplat, J., Ferssac, G. de, Cellier, J. M., Nekoit, M., and Oudiz, A. (1990). *Les Facteurs Humains de la Fiabilité dans les Systèmes Complexes*. Marseille: Editions Octares Entreprises.

Lescak, M., and Eggert, H. (1988). *Petri-Netz-Methoden und Werkzeuge: Hilfsmittel zur Entwurfsspezifikation und -validation von Rechensystemen.* Berlin: Springer.

Lind, M. (1987). Decision Models and the Design of Knowledge-based Systems. In J. Rasmussen, B. Brehmer, and J. Leplat, Eds., *Distribution Decision Making. Cognitive Models for Cooperative Work.* New York: John Wiley.

Lindsay, P. H., and Norman, D. A. (1977). *Human Information Processing*. New York: Academic Press.

Luczak, H. (1975). *Untersuchungen informatorischer Belastung und Beanspruchung des Menschen.* Fortschritts-Bericht VDI-Z *10(2)*. Düsseldorf: VDI-Verlag.

Luczak, H. (1979). *Arbeitswissenschaftliche Untersuchungen von maximaler Arbeitsdauer und Erholungszeiten bei informatorisch-mentaler Arbeit nach dem Kanal- und Regler-Mensch-Modell sowie superponierter Belastungen am Beispiel Hitzearbeit.* Fortschritts-Bericht VDI-Z *10(6)*. Düsseldorf: VDI-Verlag.

Luczak, H. (1987). Psychophysiologische Methoden zur Erfassung psychophysischer Beanspruchungszustände. In U. Kleinbeck and J. Rutenfranz, Eds., *Arbeitspsychologie, Band D/III/1, Enzyklopädie der Psychologie*. Göttingen: Hogrefe Verlag.

Luczak, H. (1991). Die Bewegungsanalyse in der Arbeitswissenschaft, In U. Boenick and M. Näder, Eds., *Gangbildanalyse*. Duderstadt: Mecke Druck.

Luczak, H. (1991). Work under extreme conditions. *Ergonomics, 34(6)*, 687–720.

Luczak, H. (1992). *Arbeitswissenschaft*. Berlin: Springer Verlag.

Luczak, H. (1995). Macroergonomic anticipatory evaluation of work organization in production systems. *Ergonomics, 38(8)*, 1571–1599.

Luczak, H., and Schwier, W. (1983). Anforderungen an Produktionsingenieure. *Zeitschrift für Arbeitswissenschaft, 37(1)*, 54–62.

Luczak, H., and Samli, S. (1986). Zur Validität von WF-Mento. In R. Hackstein, F.-J. Heeg, and F. von Below, Eds., *Arbeitsorganisation und neue Technologien*. Berlin: Springer.

Luczak, H., Volpert, W., Raeithel, A., Schwier, W. u.M.v. Müller, Th., and Rötting, M. (1989). *Arbeitswissenschaft, Kerndefinition—Gegenstandskatalog, Forschungsgebiete*. Eschborn: RKW.

Luczak, H., Reuschenbach, T., and Steidel, F. (1990). Simulation von Konstruktionsabteilungen im Hinblick auf CAD-Organisationsmodelle. In E. Zahn, Ed., *Organisationsstrategie und Produktion*. München: gfmt-Verlag.

Luczak, H., Herbst, D., Schlick, C., Stahl, J., and Springer, J. (1995a). Kooperative Konstruktion und Entwicklung. In R. Reichwald and H. Wildemann, Eds., *Kreative Unternehmen—Spitzenleistungen durch Produkt- und Prozeßinnovationen*. Stuttgart: Schäffer-Poeschel.

Luczak, H., Springer, J., Herbst, D., and Schlick, C. (1995b). Telecooperation for locally distributed working persons. In *Proceedings of the IEA world conference 1995/3rd Latin American Congress/7th Brazilian Ergonomics Congress*. ABERGO, Rio de Janeiro.

Malone, Th., and Crowston, K. (1990). What is Coordination Theory and How Can it Help Design Cooperative Work Systems? *Proceedings of the Conference on Computer Supported Cooperative Work*, October 7–10, Los Angeles, CA and New York, pp. 357–370.

Marca, D. A., and McGrowan, C. C. (1987). *SADT—Structured Analysis and Design Technique*. New York: McGraw-Hill.

Maynard, H. B., Stegemerten, G. J., and Schwab, J. L. (1948). *Methods-Time Measurement*. London: McGraw-Hill.

McCormick, E. J., Jeanneret, P. R., and Mecham, R. C. (1969). *The development and background of the Position Analysis Questionaire (Report #5)*. Lafayette: Occupational Research Center, Purdue University.

McMillan, G. R., Beevis, D., Salas, E., Strub, M. H., Sutton, R., and van Breda, L. (1989). Applications of human performance models to systems design. *Defense Research Series*, Volume 2. New York: Plenum Press.

McRuer, D. T. (1980). Human dynamics in man-machine-systems. *Automatica*, *16*(*3*), 237–253.

Mecham, R. C., Harris, A. F., McCormick, E. J., and Jeanneret, P. R. (1972). *Job Activity Preference Questionnaire* West Lafayette, IN: Copy.

Meister, D. (1985). *Behavioural Analysis and Measurement Methods*. New York: John Wiley.

Merkelbach, E. J. H. M., and Schraagen, J. M. C. (1994). A framework for the analysis of cognitive tasks. *Rapport TNO-TM* B-13, Soesterberg.

Mertins, K., Süssenguth, W., and Jochem, R. (1994). *Modellierungsmethoden für rechnerintegrierte Produktionsprozesse*. München: Hanser.

Miller, R. B. (1967). Task taxonomy: Science or technology. *Ergonomics*, *10*, 167–176.

Miller, G. A., Galenter, E., and Pribram, K. H. (1973). *Pläne und Strukturen des Verhaltens*. Stuttgart: E. Klett Verlag.

Möhrle, M. (1990). *Petrinetze in der Produktionstechnik: Integration von Planung, Simulation und Steuerung von Produktionsanlagen*. Doctoral Dissertation, Bochum.

Moldaschl, M. (1996). Ursachen und Folgen der Einführung von Gruppenarbeit. In H. Luczak, W. Volpert u.M.v. T. Müller, Eds., *Handbuch der Arbeitswissenschaft*. Stuttgart: Poeschel.

Moray, N. (1986). Monitoring behaviour and supervisory control. In K. R. Boff, L. Kaufman, and J. P. Thomas, Eds., *Handbook of Perception and Human Performance*. New York: John Wiley.

Morris, C. W. (1946). *Signs, Language and Behavior*. New York: Prentice-Hall.

Mundel, M. E. (1947). *Systematic Motion and Time Study*. New York: Prentice-Hall.

Mundel, M. E. (1950). *Motion and Time Study—Principles and Practice*. New York: Prentice-Hall.

Navon, D., and Gopher, D. (1979). On the economy of the human processing system. *Psychol. Review*, *86*(*3*), 214–255.

Nerdinger, F. W., and Rosenstiel, L.v. (1996). Grundbegriffe und Methoden der Forschung über Kooperation in Arbeitsgruppen. In H. Luczak, W. Volpert u.M.v. T. Müller, Eds., *Handbuch der Arbeitswissenschaft*. Stuttgart: Poeschel.

Newell, A. L., and Simon, H. (1972). *Human Problem Solving*. Englewood Cliffs, NJ: Prentice Hall.

Nielsen, J., and Mack, R. L. (1994). *Usability Inspection Methods*. New York: John Wiley.

Nijssen, G. M., and Halpin, T. A. (1989). *Conceptual Schema and Relational Database Design*. New York: Prentice Hall.

Noble, D. (1993). A Model to Support Development of Situation Assessment Aids. In G. A. Klein, J. Orasanu, R. Calderwood, and C. E. Zsambok, Eds., *Decision Making in Action: Models and Methods*. Norwood, NJ: Ablex Publishing Corp.

Noche, B., Bernhard, W., Krauth, J., Meyer, R., and Wenzel, S. (1993). Simulationsinstrumente im Überblick. In A. Kuhn, A. Reinhardt, and H. P. Wiendahl, Eds., *Handbuch der Simulationsanwendungen in Produktion und Logistik*. Braunschweig: Vieweg.

Norman, D.A. (1979). *Slips of the mind and an outline for a theory of action*. CHIP-Center for Human Information Processing, San Diego.

Norman, D. A. (1982). *Learning and Memory*. San Francisco: Freeman.

Norman, D. A. (1983). Design rules based on analysis of human error. *Communications of the ACM*, *26*(*4*), 254–258.

Norman, D. A., and Bobrow, D. G. (1975). On data limited and resource limited processes. *Cognitive Psychology*, *7*, 44–64.

Oberquelle, H. (1987). *Sprachkonzepte für benutzergerechte Systeme*. Berlin: Springer.

OECD (Organization for Economical Cooperation and Development) (1977). *Basic Disaggregations of Main Social Indicators*. Paris: OECD.

Oesterreich, R., and Volpert, W. (1987). Handlungstheoretisch orientierte Arbeitsanalyse. In U. Kleinbeck and J. Rutenfranz, Eds., *Arbeitspsychologie (Enzyklopädie der Psychologie, Themenbereich D, Serie III, Bd. 1)*. Göttingen: Hofgrefe.

Olsen, J. R., and Nilsen, E. (1988). Analysis of the cognition involved in spreadsheet software interaction. *Human-Computer Interaction, 3*, 309–350.

Ouwerkerk, van, R. J., Meijman, T. F., and Mulder, G. (1994). *Arbeidspsychologische Taakanalyse*. Utrecht: Lemma.

Passmore, R., and Durnin, J. V. G. A. (1967). *Energy, Work and Leisure*. London: Heinemann.

Payne, S. J., and Greene, T. R. G. (1986). Task-Action Grammars: A model of the mental representation of task languages. *Human-Computer-Interaction, 2*, 93–133.

Petri, C. A. (1979). Über einige Anwendungen der Netztheorie. *Informatik-Fachberichte*, Berlin: Springer.

Pew, R. W., and Baron, S. (1983). Perspectives on human performance modelling. *Automatica, 19(6)*, 663–676.

Posner, M. I., and Petersen, S. E. (1990). The attention system of the human brain. *Annual Review of Neuroscience, 13*, 25–42.

Pribram, K. H., and McGuinness, D. (1975). Arousal, activation and effort in the control of attention. *Psychol. Review, 82(2)*, 116–149.

Pribram, K. H., and McGuinness, D. (1976). Arousal, Aktivierung und Anstrengung: gesonderte neurale Systeme. *Zeitschrift für Psychologie, 184(3)*, 382–403.

Pritsker, A. A. B. (1986). *Introduction to simulation and SLAM II*. West Lafayette, IN: Systems Publishing Corporation.

Quick, J. H., Duncan, J. H., and Malcolm, J. A. (1965). *Das Work-Factor-Buch*. München: C. Hanser.

Rasmussen, J. (1976). Outlines of a hybrid model of the process operator. In T. B. Sheridan and G. Johannsen, Eds., *Monitoring Behaviour and Supervising Control*. New York: Plenum Press.

Rasmussen, J. (1983). Skills, rules, knowledge; signals, signs and symbols; and other distinctions in human performance models. *IEEE Transactions on Systems, Man and Cybernetics, SMC 13*, 3.

Rasmussen, J. (1986). *Information Processing and Human-Machine Interaction. An Approach to Cognitive Engineering*. Amsterdam: North Holland.

Rasmussen, J. (1993). Deciding and Doing: Decision Making in Natural Contexts. In G. A. Klein, J. Orasanu, R. Calderwood, and C. E. Zsambok, Eds., *Decision Making in Action: Models and Methods*. Norwood, NJ: Ablex Publishing Corp.

Rasmussen, J., Pejtersen, A., and Goodstein, L. P. (1994). *Cognitive Systems Engineering*. New York: John Wiley.

Reason, J. (1987). Generic error-modelling system (GEMS): A cognitive framework for locating common human error forms. In J. Rasmussen, K. Duncan, and J. Leplat, Eds., *New technology and human error*. Chichester: John Wiley.

Reason, J. (1990). *Human Error*. New York: Cambridge Univ.

Redding, R. W. (1990). Taking cognitive task analysis into the field: Bridging the gap from research to application. *Proceedings of the Human Factors Society Annual Meeting*. Volume II, 1304–1308.

REFA, Ed. (1983). *Methodenlehre des Arbeitsstudiums, Datenermittlung*. München: C. Hanser.

Reisig, W. (1985). *Petri-Nets: An Introduction*. Berlin: Springer.

Reisner, Ph. (1981). Formal grammar and human factors design of an interactive graphics system. *IEEE Transactions on Software Engineering, 7*, 229–240.

Rohmert, W. (1987). Physiological and psychological work load measurement and analysis. In G. Salvendy, Ed., *Handbook of Human Factors*. New York: John Wiley.

Rohmert, W., Rutenfranz, J., and Luczak, H. (1975). *Arbeitswissenschaftliche Beurteilung der Belastung und Beanspruchung an unterschiedlichen industriellen Arbeitsplätzen (Forschungsbericht)*. Bonn: Bundesministerium für Arbeit und Sozialordnung.

Rohmert, W., and Luczak, H. (1979). Stress, Work and Productivity. In V. Hamilton, V., and D. Warburton, Eds., *Human Stress and Cognition*. Chichester: John Wiley.

Rohmert, W., and Landau, K. (1983). *A new technique for job analysis*. London: Taylor & Francis.

Ross, D. T. (1985). Application and Extensions of SADT. *IEEE Computer Magazine, 18(4)*, 25–34.

Rouse, W. B. (1990). Designing for human error: Concepts for error tolerant systems. In H. R. Booker, Ed., *MAINPRINT: An approach to system integration*. New York: Van Nordstrand Reinhold.

Samli, S. (1987). *Arbeitswissenschaftliche Untersuchungen zum Work-Factor-Mento Grundverfahren*. Doctoral Dissertation D 83, Berlin.

Sanders, A. F. (1983). Towards a model of stress and human performance. *Acta Psychologica, 74*, 124–167.

Sanders, A. F., and Reitsma, W. D. (1982a). Sleep loss and covert orienting of attention. *Acta Psychologica, 52*, 137–152.

Sanders, A. F., and Reitsma, W. D. (1982b). The effect of sleep loss on processing information in the functional visual field. *Acta Psychologica, 51*, 149–162.

Sanderson, P., and Fisher, C. (1994). Exploratory Sequential Analysis: Foundations. *Human Computer Interaction, 9,* 251–317.

Schlaich, K. (1967). *Vergleich von beobachteten und vorbestimmten Elementarzeiten manueller Willkürbewegungen bei Montagearbeiten.* Berlin: Beuth.

Schumann, R. (1995). *Entwicklung eines modellgestützten Systems zur Gestaltung teilautonomer Arbeitsgruppen in der Fertigung von Automobilzulieferbetrieben.* Doctoral Dissertation, Aachen.

Schwier, W., and Luczak, H. (1986). Simulation studies on workload and size of the operation crew of a container ship. In H. P. Willumeit, Ed., *Human Decision Making and Manual Control.* Amsterdam: North Holland.

Sebillotte, S. (1988). Hierarchical planning as a method for task analysis: The example of office task analysis. *Behaviour and Information Technology, 7(3),* 275–293.

Sebillotte, S. (1989). *Action schemata in professional activities. Rapport de Recherche 1059.* Rocquencourt: INRIA.

Sebillotte, S. (1991). Décrire les tâches selon les objectives des opérateurs—de l'interview à la formalisation. *Le Travail Humain, 54(3),* 193–223.

Seifert, D. J., and Döring, B. (1981). SAINT—ein Verfahren zur Modellierung, Simulation und Analyse von Mensch-Maschine-Systemen. *Angewandte Systemanalyse, 2(3),* 127–135.

Seliger, G., Feige, M., and Wang, Y. (1993). Simulationsunterstützte Planung von Gruppenarbeit in der Montage. *Zeitschrift für Wirtschaftliche Fertigung, 88,* 14–16.

Senders, J. W. (1983). *Visual Scanning Processes.* Tilburg, NL: Tilburg University Press.

Sheridan, T. (1992). *Telerobotics, automation and human supervisory control.* Cambridge, MA: MIT-Press.

Siegel, A. I., Wolf, J. J., and Pilitis, J. (1982). A new method for the scientific layout of workspaces. *Applied Ergonomics, 13(2),* 87–90.

Singley, M. K., and Anderson, J. R. (1988). A keystroke analysis of learning and transfer in text editing. *Human-Computer Interaction, 3,* 223–274.

Smallwood, R. D. (1967). Internal models and the human instrument monitor. *IEEE Transactions on Human Factors in Electronics, 8(3),* 181–187.

Smith, G. W., and Wang, M. (1988). Modelling CIM-Systems. *Butterworth, 1(1),* 13–17, and *(3),* 169–178.

Spitzer, H., and Hettinger, Th. (1964). *Tafeln für den Kalorienumsatz bei körperlicher Arbeit.* Berlin: Beuth.

Springer, J., Langner, T., Luczak, H., and Beitz, W. (1991). Experimental comparison of CAD-systems by stressor variables. *International Journal of Human Computer Interaction, 3(4),* 375–405.

Steidel, F. (1994). *Modellierung arbeitsteilig ausgeführter, rechnerunterstützter Konstruktionsarbeit—Möglichkeiten und Grenzen personenzentrierter Simulation.* Doctoral Dissertation, Berlin.

Stein, W. (1989). Models of human monitoring and decision making in vehicle and process control. In G. R. McMillan, D. Beevis, E. Sales, M. H. Strub, R. Sutton, and L. v. Breda, Eds., *Application of Human Performance Models to System Design.* New York: Plenum.

Stein, W. (1992). *Models of the Human Operator in the Control of Complex Systems.* Research Institute for Human Engineering, Wachtberg-Werthoven/Germany.

Sternberg, S. (1968). The discovery of processing stages: extension of Donders' method. *Acta Psychologica, 30,* 276–315.

Stoffert, G. (1985). Analyse und Einstufung von Körperhaltungen bei der Arbeit nach der OWAS-Methode. *Zeitschrift für Arbeitswissenschaft, 39(1),* 31–38.

Swain, A. D., and Guttmann, H. E. (1983). *Handbook of Human Reliability Analysis with Emphasis on Nuclear Power Plant Applications.* Sandia National Laboratories, NUBERG/CR-1278, US Nuclear Regulatory Commission, Washington, DC.

Theiß, E. (1996). Zeitmanagement. In H. Luczak, W. Volpert u.M.v. T. Müller, Eds., *Handbuch der Arbeitswissenschaft.* Stuttgart: Poeschel.

Thordsen, M. L. (1991). A Comparison of Two Tools for Cognitive Task Analysis: Concept Mapping and the Critical Decision Method. *Proceedings of the Human Factors Society 35th Annual Meeting* by the Human Factors and Ergonomics Society, Ed., Santa Monica, CA.

Thordsen, M. L., Galushka, J., Klein, G. A., Young, S., and Brezovic, C. P. (1990). *A Knowledge Elicitation Study of Military Planning.* Technical Report No. 876 (U.S. Army Research Institute for the Behavioural and Social Sciences, Alexandria, VA).

Treisman, A. M., and Davies, A. (1973). Divided Attention to ear and eye. In S. Kornblom, Ed., *Attention & Performance IV.* New York: Plenum Press.

Troll, L. (1987). Berufs- und Tätigkeitsanalyse mit standardisierten Rastern. In K. H. Sonntag, Ed., *Arbeitsanalyse und Technikentwicklung.* Köln: Bachem.

Tucker, D. M., and Williamson, A. (1984). Asymmetric neural control systems in human self-regulation. *Psychological Review, 91(2),* 185–215.

Tustin, A. (1947). The nature of the operators response in manual control and its implications for controller design. *J. Inst. Electr. Engrs., 94,* Part II A, 190–202.

Udris, I., and Alioth, A. (1980). Fragebogen zur "Subjektiven Arbeitsanalyse." In E. Martin, U. Ackermann, I. Udris, and K. Oegerli, Eds., *Monotonie in der Industrie. Schriften zur Arbeitspsychologie*. Bern: Huber.

Ulich, E. (1981). Subjektive Tätigkeitsanalyse als Voraussetzung autonomie-orientierter Arbeitsgestaltung. In F. Frei, and E. Ulich, Eds., *Beiträge zur psychologischen Arbeitsanalyse. Schriften zur Arbeitspsychologie*. Bern: Huber.

Ulich, E. (1994). *Arbeitspsychologie* (vdf, Zürich).

Unema, Pieter J. A. (1995). *Eye Movements and Mental Effort*. Aachen: Shaker.

Université de Bordeaux I (1992). *The GRAI-Method*. Bordeaux: Laboratoire d'Automatique et de Productivité.

U.S. Equal Employment Opportunity Commission (USEEOC), U.S. Civil Service Commission (USCSC), U.S. Department of Labour (USDOL), U.S. Department of Justice (USDOJ) (1978). Uniform guidelines on employee selection procedures. *Federal Register*, August 25, 1978, *43*(*166*), 38290–38315

Volpert, W. (1982). The Model of Hierarchical-Sequential Organization of Action. In W. Hacker, W. Volpert, and M. v. Cranach, Eds., *Cognitive and Motivational Aspects of Action*. Amsterdam: North Holland.

Volpert, W. (1987). Psychische Regulation von Arbeitstätigkeiten. In U. Kleinbeck and J. Rutenfranz, Eds., *Arbeitspsychologie (Enzyklopädie der Psychologie, Themenbereich D, Serie III, Bd. 1)*. Göttingen: Hofgrefe.

Volpert, W., Oesterreich, R., Gablenz-Kolakovic, S., Krogoll, T., and Resch, M. (1983). *Verfahren zur Ermittlung von Regulationserfordernissen in der Arbeitstätigkeit (VERA)—Analyse von Planungs- und Denkprozessen in der industriellen Produktion*. Köln: Verlag TÜV Rheinland.

Volpert, W., Kötter, W., Gohde, H.-E., and Weber, W. G. (1989). Psychological evaluation and design of work tasks: Two examples. *Ergonomics*, *32*(*7*), 881–890.

Wächtler, H., Modrow-Thiel, B., and Roßmann, G. (1989). *Persönlichkeitsförderliche Arbeitsgestaltung. Die Entwicklung des arbeitsanalytischen Verfahrens ATAA*. München: Hampp.

Wehner, Th. (1996). Fehler und Fehlhandlungen. In H. Luczak, W. Volpert u.M.v. T. Müller, Eds., *Handbuch der Arbeitswissenschaft*. Stuttgart: Poeschel.

Wickens, C. D. (1980). The structure of attentional resources. In R. Nickerson, Ed., *Attention & Performance IX*. Hillsdale, NJ: Erlbaum.

Wickens, C. D. (1984). Processing resources in attention. In R. Parasuraman and D. R. Davies, Eds., *Varieties of Attention*. New York: Academic Press.

Wilkinson, R. T., and Colquhoun, W. P. (1968). Interaction of alcohol with incentive and with sleep deprivation. *J. Exp. Psychol.*, *76*(*4*), 623–629.

Ziegler, J. (1988). Aufgabenanalyse und Funktionsentwurf. In H. Balzert, Ed., *Einführung in die Softwareergonomie*. Berlin and New York: de Gruyter.

Ziegler, J. (1994). Aufgabenorientierte Systemgestaltung. In E. Eberleh, Ed., *Einführung in die Software-Ergonomie*. Berlin and New York: de Gruyter.

Zülch, G. (1992). Einbeziehung der Arbeitsorganisation in die Simulation von Produktionsstrukturen. In VDI-Gesellschaft für Produktionstechnik, Ed., *Rechnergestützte Fabrikplanung*. Düsseldorf: VDI.

Zülch, G. (1993). Simulation aufbauorganisatorischer Veränderungen in Produktionsunternehmen. *Information Management*, *8*, 27–29.

Zülch, G., and Ernst, W. (1990). Personenbezogene Simulation zum Planen von Fertigungsnestern. *Die Arbeitsvorbereitung*, *27*, 220–224.

Zülch, G., and Grobel, T. (1990). Analyse von Produktionssystemen mit Hilfe der Simulation. *VDI-Z*, *132*(*10*), 176–182.

CHAPTER 13
MENTAL WORKLOAD

Pamela Tsang
Department of Psychology
Wright State University
Dayton, OH 45435 USA

Glenn F. Wilson
***Fitts* Human Engineering Division**
Armstrong Laboratory
Wright-Patterson AFB, OH 45433-7022 USA

13.1 MENTAL WORKLOAD FUNDAMENTALS

Operator mental workload is an important consideration when designing new systems or upgrading existing ones. Excessively high levels of mental workload can lead to errors and system failure, whereas underload can lead to complacency and eventual errors (e.g., Braby, Harris, and Muir, 1993). Modern complex systems can place very high demands on the human operator. Automation is often introduced to alleviate the heavy demand on the operator or to reduce the number of operators required to control a system. However, too much automation can keep the operator out of the loop of the operating system and reduce the operator's capability to recover from unusual events. It is now recognized that automation often redistributes, rather than reduces, the workload within the system (e.g., Lee and Moray, 1992; Parasuraman and Mouloua, 1996; Wiener, 1988). The impact on any newly introduced automated elements on system performance and must be carefully studied.

There exists a large body of literature on the topic of mental workload. The purpose of this chapter is not to add another review to those already existing, but rather to provide a guide to this literature and provide a strategy for mental workload assessment. We will provide an overview of the area and discuss methods that can be used to develop the requisite knowledge and strategies that should be useful in assessing mental workload.

While there is not one agreed upon definition of mental workload, a succinct definition of mental workload is offered by O'Donnell and Eggemeier (1986): "The term workload refers to that portion of the operator's limited capacity actually required to perform a particular task" (p. 42-2). The theoretical assumption underlying this definition is that the human operator has limited processing capacity or processing resources (e.g., Kahneman, 1973). Greater task difficulty increases the requirement for mental processing resources (Norman and Bobrow, 1975). If the processing demands of a task or tasks exceed available capacity, performance decrements result. Having some indications of the expended capacity or workload would therefore be useful for ensuring adequate performance.

Gopher and Donchin (1986) added that,

> *The term workload is used to describe aspects of the interaction between an operator and an assigned task. Tasks are specified in terms of their structural properties; a set of stimuli and responses are specified with a set of rules that map responses to stimuli. There are, in addition, expectations regarding the quality of the performance, which derive from knowledge of the relation between the structure of the task and the nature of human capacities and skills. . . . workload is invoked to account for those aspects of the interaction between a person and a task that cause task demands to exceed the person's capacity to deliver. . . . mental workload is clearly an attribute of the information processing and control systems that mediate between stimuli, rules, and responses. Mental workload is an attribute of the person-task loop, and the effects of workload on human performance can therefore be examined only in relation to a model of human information processing.* (p. 41-3)

13.1.1 Historical Perspective

The origins of contemporary interest in mental workload can be traced to two workshops held in the late 1970s. In 1977, the NATO Special Panel on Human Factors sponsored a workshop on mental workload. Participants of the multidisciplinary workshop included engineers, psychologists, and human factors specialists. In 1979, Moray edited a book based on this meeting titled *Mental Workload, Theory and Measurement.* The volume was divided into four main sections: experimental psychology and mental workload, control engineering and workload measurement, physiological psychology and mental workload, and applied psychology and mental workload. The goals for the workshop were to: (1) come to a consensus on a definition of workload, (2) integrate the existing knowledge about workload and formulate an application-oriented theory or procedure of workload assessment, and (3) find the relationships among workload measurement techniques (Johannsen, 1979).

The previous year, a Symposium on Mental Workload was held at the XXI International Congress of Psychology in Paris. Papers from this symposium were published in 1978 as a special issue of *Ergonomics* in English (Leplat and Welford, 1978) and in *Le Travail Humain* in French. These meetings served as catalysts for much of the subsequent research.

Despite the lack of a universally accepted definition, a great deal of progress has been made in the ensuing years. The concept of workload has been related to a number of theoretical psychological frameworks (e.g., Gopher and Donchin, 1986; Welford, 1978; Wickens, 1992). A multitude of behavioral, physiological, subjective, and analytic measures have been developed by psychologists and engineers to measure workload. The lack of a consensus on a workload definition has not deterred practitioners from assessing and quantifying mental workload. Despite the fuzziness of the concept and imperfection of the assessment techniques, practitioners have found measuring workload to be useful. A perusal of the workload literature attests to the pervasive use of workload assessment (see suggested sources below).

13.1.2 Current Perspective

A decade after the NATO workshop, Moray (1988) commented on the state of affairs in the field of mental workload. He concluded that among the different approaches to assessing mental workload, subjective measures had made the most progress. Moray perceived a lack of progress in theory development but did not find the vigor of the field to have diminished. In a 1989 chapter on Engineering Psychology in the *Annual Review of Psychology*, Gopher and Kimchi included mental workload as one of three topics they addressed in depth. However, in the next *Annual Review of Psychology* chapter on Engineering Psychology, Howell (1993) concluded that activities in mental workload had declined in the 1990s. Several factors can account for this apparent trend.

In terms of quantity, Howell (1993) is correct that there are fewer publications in the recent years than during the "heyday" of mental workload (around 1979–1989). On the other hand, workload assessment is by no means a bygone practice. A quick check of the current publications (see the Annotated Selected Bibliography section that follows) suggests that there is a healthy level of activity in the field. The number of review chapters on the topic also attests to the large body of existing literature.

One reason for the decreased number of publications is that the field is maturing. Tools have been developed over the years that are now being applied regularly. This does not mean that all of the questions have been successfully answered and that further development is not needed. Secondary task techniques and most physiological measures remain difficult to use unless one has access to sophisticated expertise and the necessary equipment. Although Gopher and Donchin (1986), Wickens (1992), and others have provided useful theoretical frameworks for the concept of workload, further theoretical workload development is bounded by the existing knowledge of human information processing. It may be that significant theoretical development in mental workload is unlikely until significant advances are made in the psychological theories. In the Final Report of the Experimental Psychology Group of the 1977 NATO workshop, Johannsen, Moray, Pew, Rasmussen, Sanders, and Wickens (1979) posited that a complete theory of human operator workload will be either: (a) the end product of a total theory of human performance or (b) simply a description of how people feel when doing a task. The field of workload has matured in the latter sense more than in the former sense, partly because there is not yet a complete theory of human performance.

Another reason for the decline in the number of workload publications may be tied closely to recent technological developments, especially those in display capability and automation. Gopher and Kimchi (1989) have communicated most lucidly the intimate relationship between engineering psychology and technology: "The link between engineering psychology and technology not only affects the content and priorities of research work in this field, but also strongly influences its life cycle and pace" (p. 432). For example, the needs of World War II resulted in a flourish of research in sustained attention and theoretical development of attention in general (e.g., Broadbent, 1958). In the 1970s, the growing complexity of systems was placing increasing cognitive demands on the operator to the point where human capabilities became the limiting factor of system performance. At the same time, the advent of increasingly automated systems had fundamentally changed the role of the operator from manual control to supervisory monitoring. These changes led to the need for the concept of mental workload to benchmark automated systems that require decreasing physical but increasing cognitive effort. In the 1990s, development in automation continues to escalate. There are many reasons for automation (see, for example, Wiener and Curry, 1980), one of which is to expand system capability. The underlying assumption is that, with the aid of automation, the system can now perform functions that operators without automation would not be able to perform. Another assumption is that the automated system can now perform the same functions with fewer operators. In either case, the operator's supervisory and managerial responsibility multiplies. This leads to concerns for operators being able to maintain an overall appreciation of the whole situation. To take advantage of, or even just to cope with, highly automated complex systems, Adams, Tenney, and Pew (1991) suggest that an expanded view of workload as a strategic task management problem would be necessary. In very much the same way that the necessity for the concept of workload was demanded by practical needs, the concept of situation awareness is an outgrowth of a necessity to understand how human operators manage dynamic and complex systems. Therefore much recent research has been devoted to the discussion of the concept of situation awareness. This change of emphasis may have potentially detracted effort from workload activities.

13.1.2.1 Mental Workload and Situation Awareness

In a recent article, Pew (1994) stated that situation awareness has replaced workload as the buzzword of the 1990s. But can the concept of situation awareness replace that of mental workload? Endsley (1993) does not see much relationship between the two concepts. On the other hand, Hendy (1995) and Wickens (1995) believe that the concepts of situation awareness and mental workload are clearly distinct and yet related to each other. First, both are affected by visual display support and automation aids. Second, the attainment or maintenance of a high level of situation awareness could be demanding. The need for a high level of situation awareness could compete with the actual task demands for the same limited resources (attentional resources in Wickens's framework and time in Hendy's framework). Thus, although effort devoted to the task may enhance situation awareness, this would also add to the workload already incurred by the task. Excessive workload in turn could adversely affect situation awareness.

Space limitation does not allow more in-depth examination of the relationship between workload and situation awareness. But there seems to be a consensus in the literature that one concept does not replace, nor encompass, the other, even though the two concepts are affected by many of the same human variables (such as limited processing capacity and the severe limit of working memory) and system variables (such as task demands and technological support). Wickens (1995) also points out that, just as several types of measures are needed to reveal a complete picture of the workload condition, several types of measures may be needed to reveal the level of situation awareness attained by the operator. While new techniques for assessing situation awareness have been proposed (e.g., Fracker, 1989), attempts have also been made to adopt workload assessment techniques to measure situation awareness (e.g., Taylor, 1990; Vidulilch, Stratton, Crabtree, and Wilson, 1994). There is little question that valuable lessons can be learned from the study of mental workload. For example, many of the criteria that have been developed for evaluating workload metrics apply just as well for evaluating situation awareness metrics. Similarly, methodologies that have been developed to attain validity and reliability of the workload measures could apply to situation awareness measures.

In brief, technological changes have brought about a practical need to know about the situation awareness of the operator. There seems to be a consensus that the concepts of situation awareness and mental workload are not interchangeable. The utility of assessing situation awareness therefore does not diminish that of mental workload. In fact, parallel studies of both could help sharpen their respective definitions and stimulate new understanding.

13.1.2.2 Assessing Mental Workload in Complex Environments

Technology, economics, and better understanding of human performance characteristics have acted together to produce systems that are highly complex and use a great deal of automation. These modern systems can place new levels of demand upon human operators. Operators may be required to manage complex displays, process large amounts of information, and make decisions about the appropriate action. A reaction to the escalating complexity in automated systems is a recent publication, *State of the Art Report: Strategic Workload and the Cognitive Management of Advanced Multi-Task Systems*. Here, Adams et al. (1991) provide one of the most comprehensive and stimulating treatments of workload management in modern complex human–machine systems. Adams et al. state that, "The study of multi-task management raises a broad spectrum of psychological issues that go beyond those associated with divided attention, time-sharing, or dual task performance. This includes concerns about situation awareness, interruptions, prioritization, and scheduling, maintenance of the queue, and the development of expertise" (p. 12).

To address all of these cognitive processes, Adams et al. (1991) propose a theoretical framework that is based on several principal psychological theories. Central to these psychological theories is the tenet that human behavior is goal directed. Because of the dynamic nature of complex systems, Adams et al. suggest that "the overall mission be envisaged as a multilevel goal hierarchy, with competing tasks lying on different branches of the overall structure" (pp. 35–36). Further, the mission activities are to be analyzed in terms of their goals, rather than by the conventional time-line analysis which is unlikely to be able to capture the nature of the dynamic demands encroached upon the operator. Because modern complex systems typically involve multitasks, a special concern is the operator's "strategic management" of the different tasks. Adams et al. point out that a key component to strategic workload management is the ability to prioritize tasks and to anticipate the future, both of which depend on the operator's situation awareness.

13.1.2.3 Theoretical Frameworks of Mental Workload

Moray, Dessouky, Kijowski, and Adapathya (1991) proposed that scheduling theory should be the normative model for strategic behavior of task selection and switching between them. Moray et al. used a modified version of Tulga and Sheridan's (1980) task in which several tasks were presented, each requiring different processing time and each having different due dates. Only one task could be engaged at any one time. The results showed that subjects in general did not follow the optimal rule of scheduling theory and knowledge of the optimal rule neither improved performance nor reduced subjective workload. Moray et al. concluded that the findings of nonoptimal performance with relatively simple laboratory tasks suggested that people in general will have great difficulty in optimally scheduling more complex real tasks, when they have little time to plan their strategies.

More encouraging findings are reported in a different paradigm where subjects were asked to modulate their time-shared performance (or simultaneous performance of more than one task) according to certain task priorities. This was observed with laboratory tasks (e.g., Gopher, 1993; Gopher and Navon, 1980; Gopher and North, 1977; Tsang and Wickens, 1988; Tsang, Velazquez, and Vidulich, 1996; Wickens and Gopher, 1977) as well as in simulation studies (e.g., Raby, Wickens, and Marsh, 1990; Segal and Wickens, 1991), with college students, aircraft pilots, and noncollege adults, across different age groups (e.g., Tsang, Shaner, and Schnopp-Wyatt, 1995; Wickens, Braune, and Stokes, 1987). The laboratory task studies have been criticized as mere dual-task studies that do not address problems at a level of complexity that is realistic to operational systems. However, simulated landing approaches were performed with additional discrete tasks under different workload conditions in one simulation study (Raby et al., 1990) and a helicopter simulation was performed with secondary and tertiary tasks in another (Segal and Wickens, 1991). Regardless of whether they were laboratory or simulation studies, they show that subjects have good, though not perfect, voluntary control over how they allocate their effort or attentional resources to the different tasks according to their relative priorities.

What accounts for the discrepancy between the highly nonoptimal behavior observed in the task switching study and the mildly nonoptimal behavior observed in the time-sharing studies? The degree of nonoptimality observed in Moray et al.'s study was inferred from the disparity between actual human behavior and that prescribed by a normative model. Wickens (1989) pointed out that a fundamental assumption underlying the kind of normative models used in Moray et al. is inconsistent with a rather large body of empirical findings. This is the assumption that the human operator can only be engaged in one task at any one time. To deal with varying multitask demands, the human operator can prioritize and select which task to perform first and decide when to switch to a different task, but the operator is primarily a single-channel operator (see also Baron, Zacharias, Muralidharan, and Landraft, 1980; Chu and Rouse, 1979; Corker, Davis, Papazian, and Pew, 1986; Wherry, 1976). Wickens (1989) went on to cite a host of findings and theoretical positions that are contradictory to such an assumption.

First, examples of successful simultaneous performance of more than one task abound. Second, a simple change in the structure of the one of the task components without altering its difficulty could significantly change the degree of interference between the time-shared tasks. For example, a simple change in the response modality from a manual response to a vocal response can substantially reduce task interference (e.g., Brooks, 1968; McLeod, 1977; Tsang and Wickens, 1988). Wickens (1984) referred to this phenomenon as the structural alteration effect. Third, a definite difficulty increase in one of the task components does not always produce greater interference between the time-shared tasks (e.g., Isreal, Chesney, Wickens, and Donchin, 1980; Wickens, Sandry, and Vidulich, 1983). Wickens (1984) referred to this phenomenon as the difficulty insensitivity effect. To explain these effects, Wickens proposed a multiple resource model (1980, 1987). The formulation of this model was based on an expansive systematic review of dual task studies.

Very broadly, Wickens' model says that there are specific attentional resources for different types of cognitive processing. According to this model, attentional resources are defined along three dimensions: (a) in terms of stages of processing, perceptual/central processing requires resources different from those used for response processing; (b) in terms of processing codes, spatial processing requires resources different from those used for verbal processing; and (c) in terms of input/output modalities, visual and auditory processing requires different processing resources, while manual and speech responses also require different processing resources. When performing multiple tasks, the higher the similarity in the resource demands among the task components, the more severe the competition for similar resources, the higher the level of workload would result.

There are two particularly important features of the multiple resource model that distinguishes it from the single-channel scheduling theory. First, the multiple-resource model allows for parallel processing of more than one task at one time. Second, the multiple-resource model allows graded allocation of processing resources to the time-shared tasks according to task demands. With extremely difficult tasks (when resource demands exceed availability), one would also need to select which one to perform or else face performance degradation. But selection is not a requirement, as stipulated in the scheduling theory. The meaning of prioritization is therefore quite different for the two models. According to the scheduling theory, the high-priority task will be performed first and only when the task priority has changed will switching to another task occur. According to the multiple-resource model, more resources will be allocated to the high-priority task. Whether any additional tasks can be accommodated depends on the relative priority of the time-shared tasks and whether there are any spare resources not needed by the high-priority task. Further, the amount of spare resources available depends on the similarity of the resource demands of the component time-shared tasks. This set of assumptions of the multiple-resource model is receiving a growing body of empirical support.

Currently, limitations of the multiple-resource model are also being studied. Progress toward discovering principles to account for patterns of dual task interference that cannot be explained by the multiple-resource model is being made (e.g., Fracker and Wickens, 1989; Netick and Klapp, 1994; Navon and Miller, 1987). The multiple-resource model has progressed from a purely academic psychological model in the late 1970s to becoming a commonly applied performance model in the 1990s. The acceptance of this model, as well as the understanding of its limitations has provided practical guidelines for selecting the appropriate mental workload metrics. This model has influenced subjective (e.g., Eisen and Hendy 1987; Tsang and Velazquez, 1996), performance (O'Donnell and Eggemeier, 1986), physiological (e.g., Kramer, Sirevaag, and Braune, 1987; Sirevaag, Kramer, Wickens, Reisweber, Strayer and Grenell, 1993; Wickens, Kramer, Vanasse and Donchin, 1983), as well as analytic (e.g., Alridch, Szabo, and Bierbaum, 1989; North and Riley, 1989) assessment techniques. In this respect, we are less disappointed than Moray (1988) with the progress of the theoretical development of mental workload and its measurements. Still, further theoretical development is needed to integrate the large body of existing findings and to explain the intricate relationship among the different workload metrics. We are also in need of more reliable and valid workload metrics that have both evaluative and predictive value.

13.2 WORKLOAD METRICS

Measures of mental workload have traditionally been divided into four groups based upon the nature of the data collected: subjective ratings, operator performance, psychophysiological measures, and analytic methods. Each of these is thought to provide valuable information about the mental workload experienced by an operator. Modern systems are quite complex and no one measure of mental workload can be expected to index all of the relevant aspects that bear upon the mental workload of the operator. In most situations, multiple measures will produce more accurate and more complete information about the mental workload (Nygren, 1991; Wickens, 1995; Wierwille and Eggemeier, 1993). As outlined below, each group of measures has its strengths and weaknesses and a thoughtful combination of measures can lead to a more complete picture of the operator's mental workload. There is a large body of literature covering the use of the various workload measures in laboratory and field situations. The reader is referred to recent reviews that are listed under the Annotated Selected Bibliography section.

13.2.1 Properties of Workload Measures

There are several properties that should be considered when selecting workload measures. These properties are: sensitivity, diagnosticity, intrusiveness, validity, reliability, ease of use, and operator acceptance. Several other sources contain in-depth discussions of these properties and should be consulted (Eggemeier, Wilson, Kramer, and Damos, 1991; Kramer, 1991; Lysaght et al., 1989; O'Donnell and Eggemeier, 1986). Since the various workload measures have different properties, one should have a good understanding of the properties of each measure so that the most appropriate choice(s) can be made.

13.2.1.1 Sensitivity

Sensitivity refers to how well a measure detects changes in the mental workload experienced by an operator. Often the degree of sensitivity of a measure depends on the level of workload experienced by the operator. For example, performance measures are thought to be insensitive at very low levels of workload where there is sufficient spare capacity to meet the task demands even if they increase. Performance measures are more sensitive at higher levels of workload where the limit of the operator capacity is being reached and performance deficits are expected to occur with further increase of task demand. With this information, one might apply subjective and/or psychophysiological measures in situations where the workload demands are expected to be in the low range and add performance measures if the workload is expected to be high.

13.2.1.2 Diagnosticity

Diagnosticity refers to how precisely a measure can reveal the nature of the workload. Wickens's (1984, 1991) multiple resource model can be used to estimate which aspect of the human information processing system is being overloaded (O'Donnell and Eggemeier, 1986). For example, if the performance of a secondary task that is thought to utilize response-processing resources deteriorates, then the output aspects of the primary task should be examined for potential workload problems. On the other hand, if the workload measure suggests excessive visual workload, then the displays of the system need to be examined. This sort of analysis can be very helpful in identifying "choke" points in the system. Experimental design can be used to help diagnose the causes of workload changes. By varying only one aspect of the task at a time, it may be possible to infer, from changes in the workload, whether a particular aspect is responsible for the nonoptimal level of workload. Diagnosticity is therefore especially useful when one needs to ascertain which components in a system are causing performance decrements such that the appropriate solution can be devised.

13.2.1.3 Intrusiveness

Intrusiveness is an important consideration in operational environments as well as in simulator and laboratory settings. If taking the measure interferes with the performance of the task, then a contaminated workload measure is obtained. Besides interfering with the primary task(s), the measurement may actually add to the workload since in some cases it may be considered an additional task. For example, subjective reports that are required during the performance of a complex task can use mental resources that should be allotted to the primary task and therefore can add to the mental workload experienced by the operator. This is, of course, undesirable and must be avoided. Systems that are equipped to record operator and system performance do not interfere with primary task performance. For example, modern aircraft are equipped with data busses that permit the recording of a large number of variables during flight. These data can then be analyzed after the flight and provide extensive and detailed information about the operator's inputs and system performance during a test flight. Unfortunately, most systems that are evaluated do not have these capabilities.

13.2.1.4 Validity

Validity has to do with whether the workload measure is measuring mental workload. Although construct validity is important in general, it is also the most difficult to verify. This is because mental workload itself is difficult to define operationally and there is ongoing debate as to whether it is a theoretical construct. However, two other validity benchmarks should be considered whenever possible: concurrent validity and predictive validity. Different measures (or assessment techniques) should yield similar results under similar workload conditions. Concurrent validity is when the different workload measures correlate in a systematic fashion. For example, subjective ratings tend to increase as performance deteriorates. Although certain dissociations among workload measures are known, these dissociations are not haphazard. The different workload measures should yield a consistent picture of the actual task demands. Predictive validity is an especially important consideration for the analytic methods. For most of the analytic methods, subject matter experts' (SME) workload estimates are used to predict system workload. Although the SMEs are asked to provide the estimate in a structured way (usually based on an explicit model of mental workload and detailed task/mission analysis), they are providing an analytical as opposed to an experiential estimate. Their estimates are therefore in particular need of validation with the actual workload evaluation. The validation is typically evaluated by obtaining the correlation between the SME's estimates and the workload measures obtained from the actual operators (e.g., Aldrich et al., 1989; Christ, Hill, Byers, Iavecchia, Zaklad, and Bittner, 1993). An excellent discussion on the importance and the process of validation of human factors work in general is offered by Hopkin (1993).

13.2.1.5 Reliability

Reliability has to do with whether the workload measures are stable and consistent over a period of time. Similar workload levels should be obtained under similar workload conditions. That is, they should be replicable. Measures that fluctuate widely independently of the task demands and workload conditions have no predictive value and are therefore not useful for evaluative, diagnostic, or design purposes. One common test of the reliability of a workload measure is to calculate the test–retest correlation. Under similar testing conditions, the metrics obtained at Time A should correlate well with those obtained at Time B (see, for example, Tsang and Vidulich, 1994; Vidulich and Tsang, 1987).

13.2.1.6 Ease of Use

Ease of use refers to how easy it is to collect and analyze the data associated with a measure. All measures require time and effort to collect, but the demands on the experimenter and operator vary. Subjective measures are easy to administer and are relatively simple to analyze. The procedures for implementation are generally not complex, but care should be taken in the preparation of the instructions to the subjects. Subjects tend to take the workload evaluation much more seriously if they understand the objectives of the evaluation and its importance. Primary task analysis may be more difficult since most systems do not provide performance measures and the system under evaluation must be specially modified in order to record performance data. Analyzing and interpreting the information from data bus recorders requires special training. Secondary task measures require skill to properly analyze the mental resources required by the primary task so that the appropriate secondary task can be selected. Physiological measures require specialized equipment and expertise in data collection, analysis, and interpretation. However, recent advances in physiological recording equipment and analysis software have greatly enhanced the capability of collecting psychophysiological data.

13.2.1.7 Operator Acceptance

Operator acceptance is crucial in the successful use of mental workload measures. If the operators do not feel comfortable with the selected measures, they may refuse to cooperate with data collection. It is always useful to fully explain the nature of the measures and answer any questions that the operators may have about the procedures and later use of the data. Individual anonymity must be assured in order to gain and maintain cooperation.

A coding scheme should be used that does not permit the linking of data to any individual operator. This is necessary for several reasons. Operators may be concerned that problems with the system may be blamed on them and not on the system. In many situations operators are very competitive and knowledge of performance or workload scores can be detrimental to the overall evaluation. Subjective measures are usually well accepted because of their face validity and the operators appreciate the opportunity to provide their evaluation of the system.

13.2.2 Types of Workload Measures

There are four common types of workload measures: subjective, performance, psychophysiological, and analytic measures. Subjective measures are designed to elicit the perceptions of the operator about the mental workload of the system. Performance measures are viewed as part of the system performance of interest. Psychophysiological measures record changes in the operator's body that are related to the demands of the task being performed. Performance and physiological measures sometimes can be recorded continuously during test and evaluation sessions. Analytic methods are modeling efforts used for both predictive and evaluative purposes.

13.2.2.1 Subjective Measures

The subjective method requires operators to rate the level of mental effort that they feel is required to accomplish a task. This method reflects the direct opinion of the operator about the mental effort required in the context of the task environment and the skill and experience level of the operator. This method has been extensively used to assess workload in a wide range of application areas (Eggemeier and Wilson, 1991). The widespread use of the subjective method is, in part, due to its ease of use, face validity, and operator acceptance.

Rating scales are frequently used to collect subjective data. Typically, operators are asked to select a term or phrase that best describes their feelings or they are asked to provide a number that represents the level of mental effort. In addition to structured rating scales, questionnaires for a specific evaluation have also been used. These questionnaires can be constructed to directly address issues relevant to the system being evaluated. Open-ended questions and interviews with operators can also provide useful information about a system.

Tsang and Vidulich (1994) discuss the different approaches to subjective workload assessment. Table 13.1 categorizes the more common subjective rating scales along these different approaches. First, a particular rating scale either asks for a single unidimensional rating concerning the overall workload level or ratings on multiple dimensions for each task condition. Whereas *unidimensional* ratings are easier to obtain, only *multidimensional* ratings can provide diagnostic information pinpointing the nature of the workload. Second, the ratings are either obtained *immediately* after performing the task or *retrospectively* after having experienced all of the task conditions. Third, the ratings are either *absolute*

Table 13.1 Subjective Assessment Approaches

Subjective Rating Instruments	Approaches		
	Unidimensional vs. Multidimensional	Absolute vs. Relative	Immediate vs. Retrospective
Bedford	U	A	I
MCH	U	A	I or R
Psychophysical	U	RS	R
SWORD	U	RR	R
NASA-TLX	M	A	I
SWAT	M	A	I
Workload Profile	M	A	R

Notes. MCH = Modified Cooper-Harper; U = undimensional, M = multidimensional; A = absolute rating, RS = relative rating to a single standard, RR = redundant relative rating; I = immediate, R = retrospective.

or *relative*. For relative ratings, operators are asked either to compare the task condition of interest to a *single* standard or to multiple task conditions (the *redundant* method).

Vidulich and Tsang (1994) have performed a series of studies that examine the different approaches to subjective assessment. Although specific instruments were compared in these studies, the goal was not to determine which instrument was superior. Rather, the objective was to determine which assessment approach can elicit the most and accurate workload information. Tsang and Vidulich (1994) found that performing relative judgments retrospectively (SWORD—Subjective Workload Dominance, Vidulich, Ward, and Schueren, 1991) was superior to rating each task condition immediately after performing it (psychophysical scaling, Gopher and Braune, 1984). Further, redundant relative comparisons produced more sensitive ratings than did relative comparisons to a single reference task. Tsang and Velazquez (1993) found that, compared to an immediate absolute instrument [Bedford (Roscoe, 1987)], a relative, retrospective instrument (psychophysical scaling) was more sensitive to task demand manipulation and had higher concurrent validity with performance. Tsang and Velazquez (1996) also found that a subjective multidimensional technique (Workload Profile) provided diagnostic workload information that could be subjected to quantitative analysis. Vidulich and Tsang (1987) found a redundant relative instrument (SWORD) to be more sensitive and more reliable than an absolute instrument (TLX, Hart and Staveland, 1988). Collectively, these studies suggested a relative-retrospective approach advantage. It is important, however, to note that only partial results are in and more systematic comparisons will be needed to obtain a complete picture.

For more detailed information for the specific subjective workload metrics, the readers are referred to the Annotated Selected Bibliography section. Here, we will provide references that have examined the various subjective instruments in a comparative fashion. Eggemeier and Wilson (1991) provide a good description of several of the common instruments. It is also a rich source of references for comparative studies of the different subjective instruments. More recent comparative studies have been performed by Christ et al. (1993); Hendy, Hamilton, and Landry (1993); Hill, Iavecchia, Byers, Bittner, Zaklad, and Christ (1992); Nygren (1991); and Wierwille and Eggemeier (1993).

Although space limitation does not permit a detailed description of the different subjective instruments, recent developments concerning two of the most popular techniques will be presented. The NASA Task Loading Index (TLX, Hart and Staveland, 1988) and the Subjective Workload Assessment Technique (SWAT, Reid and Nygren, 1988) permit the operator to rate the task on the basis of several dimensions. In both cases, the operator provides an absolute rating immediately after task performance. The subscales for the TLX are mental demand, physical demand, temporal demand, performance, effort, and frustration level. The SWAT subscales are time load, mental effort load, and psychological stress load. Both instruments also take into account individual differences by asking each operator to indicate which subscale affects their workload rating the most. Prior to evaluating the task conditions, a pair-wise comparison of the TLX subscales are performed by the operators. A weighting scheme is then derived for each operator and used to generate a weighted average rating (across subscales) for the task condition. For the SWAT, evaluation of each of the three dimensions is rated on a three-point scale. Three possible levels of workload for each of the three dimensions of workload produce 27 possible workload levels. Operators indicate their preference by rank ordering these 27 workload levels. A conjoint measurement analysis is used to assign a unique workload score for each of these 27 levels. The ratings that the operator provides for each task condition on the three subscales are then translated to the workload scores derived from the conjoint analysis. While these are the standard procedures for collecting the TLX and SWAT ratings, there is recent debate concerning the merits of the TLX weighted average and the SWAT derived workload score. Nygren (1991) examined the psychometric properties of both the TLX and SWAT ratings and found no psychometric basis for calculating the weighted average for the TLX rating. Empirically, Christ et al. (1993) and Hendy et. al. (1993) did not find the weighted average to be superior to simple averages of the raw ratings. Moroney, Biers, and Eggemeier (1995) also question the usefulness of the derived SWAT score as opposed to simple averages of the raw ratings from the three subscales.

One distinction the subjective measures enjoy is their potential to be used as projective measures. Whereas it is difficult to collect performance or physiological data from a hypothetical scenario or from a system yet to be realized, operators (usually subject matter experts) could be given a detailed description of a task condition or the functionality of a new system and then asked for a subjective evaluation of workload. Standard subjective instruments have been used in a projective fashion [e.g., Pro-SWAT (Detro, 1985; Kuperman, 1985; Masline and Biers, 1987; Reid, Shingledecker, Hockenberger, and Quinn, 1984), Pro-TLX (Christ et al., 1993), and SWORD (Vidulich et al., 1991)]. Instruments that are designed specifically for projective use are mostly analytic methods and are considered below. Since projective methods are not yet widely used, only limited data are available. However, these data suggest that projective ratings obtained from subject matter experts tend to correlate well with subjective ratings obtained from operators after they have experienced the task conditions.

Subjective measures as a class of workload measures have several strengths. Subjective measures have high face validity and widespread operator acceptance since it provides the operators with a direct method of giving their opinion about the system under evaluation. The main reason that subjective measures are the most commonly used measure of workload is ease of use. Subjective measures are also said to have high transferability to new system or new task conditions because its unit of measurement is not task dependent. They are most useful for comparison purposes.

Subjective measures have been shown to be sensitive to a variety of task demand manipulations (e.g., Wierwille and Eggemeier, 1993). Vidulich (1988) examined the results from a series of single-task and dual-task studies. These results support Ericsson and Simon's (1980) theory that subjective verbal reports are most sensitive to the information available in the working memory (see also Nisbett and Wilson, 1977), while less sensitive to differences in demands associated with response execution processing. Vidulich (1988) further found that although subjective ratings were sensitive to the overall level of workload of simultaneous multiple tasks, they failed to reflect the demand of the individual components of multiple tasks. In terms of diagnosticity, multidimensional subjective metrics have been shown to provide diagnostic information (e.g., Tsang and Velazquez, 1996; Wierwille and Eggemeier, 1993). In addition, subjective measures have been shown to be reliable and have significant concurrent validity with performance measures (e.g., Casalie and Wierwille, 1983; Gopher and Braune, 1984; Tsang and Velazquez, 1993; Tsang and Vidulich, 1994; Wierwille and Connor, 1983; Wierwille, Rahimi, and Casali, 1985). Subjective ratings are nonintrusive if subjects are not asked for their ratings during task completion. However, collecting subjective ratings can be intrusive if subjects are asked to rate a segment of a complex scenario before the entire task is completed. Further, it has been pointed out that the knowledge of having to evaluate the workload could possibly affect the workload itself and may alter the way that the task is performed. One potential solution is to delay the collection of the subjective reports until after the completion of the task so that there is no interference with the primary task.

As with all workload measures, there are drawbacks associated with subjective measures. These measures can potentially be susceptible to memory problems if the ratings are made after the performance of the task(s). Research on the effects of memory loss caused by delays in reporting has shown that a 15–30 min delay does not significantly interfere with the ratings, while 48 hr does alter the ratings (Corwin et al, 1989; Eggemeier and Wilson, 1991; Moroney, et al., 1995). In Vidulich and Tsang's evaluations, the delay in reporting was generally longer than 15 min, but less than 3 to 4 hr. They have repeatedly demonstrated the superiority of the retrospective approach. Use of videotape playback to serve as a memory aid while the subjective ratings are collected has been successfully used also (Corwin et al., 1989; Hunn, 1992).

Subjective estimates of mental workload may also be susceptible to operator bias about the system or component being evaluated. They could be influenced by the operator's past experience and degree of familiarity with the task or system being evaluated. Operator ego can also be a factor at the higher levels of workload where task performance is difficult but hard for the operator to admit that they had trouble operating the system. It is therefore advisable to always carefully explain to the operators the importance and value of their honest response. Last, an important consideration with subjective measures is their potential to dissociate with other types of workload measures. This topic will be addressed below.

13.2.2.2 Performance Measures

Performance measures of mental workload use operator behavior to determine workload. Deteriorated and/or erratic performance may indicate that workload is at or is approaching unacceptable levels. This assumption is grounded in the idea that humans have limited processing capacity or resources (e.g., Kahneman, 1973) and task performance requires processing capacity (Norman and Bobrow, 1978). Since human operators have finite capacity to deal with the demands of a task, as task demands increase, performance will deteriorate and at some point they would not be capable of maintaining adequate performance. With the primary task method, the actual performance of the operator and system is monitored and changes are noted as the demands of the task vary. For example, flight path deviation may increase in conditions of strong cross winds and may further increase if emergency conditions require close monitoring of engine state indicators.

Another performance-based workload metric is the secondary task measure. This is used in systems where primary task performance is difficult to obtain or is not available. This is often the case in many real-world systems such as automobiles, ships, and airplanes that do not have performance recording capability. Second, with highly automated systems in which the primary role of the operator is that of monitoring and supervising, little observable performance would be available for analysis. Third, primary task measures may not be sensitive to very low workload levels because operators could increase their effort in order to maintain a stable level of performance (e.g., O'Donnell and Eggemeier, 1986). Adding a secondary task will increase the overall task demand to a level where performance measures may be more sensitive.

With the secondary task method, the operator is required to perform a second task concurrently with the primary task of interest. It is explained to the operators that the primary task is more important and the primary task performance must be performed to the best of their ability whether or not it is performed with the secondary task. Operators are to use only their spare capacity (not needed by the primary task) to perform the secondary task. Since the primary and secondary tasks would compete for limited processing resources, changes in the primary task demand should result in changes in the secondary task performance as more or less resources become available for the secondary task (Eggemeier and Wilson, 1991; Gopher and Donchin, 1986). Laboratory tasks are often used as secondary tasks such as memory and mental arithmetic tasks. Alternatively, the embedded secondary task method may be used (Shingledecker, 1987; Vidulich and Bortolussi, 1988). With this method a naturally occurring part of the overall task is used as the secondary task. Since these secondary tasks are part of the overall task, they do not appear artificial to the operator and should not interfere with the primary task performance in a contrived way. For example, during an aircraft flight with varying levels of workload, response to aircraft communications could be used as a secondary measure. Changing levels of primary task demand would be expected to affect the response latency to radio messages.

An important consideration in the selection of a secondary task is the type of task demand of both the primary and secondary tasks. Secondary task performance will only be a sensitive workload measure of the primary task demand if the two tasks compete for the same processing resources. According to the multiple-resource model described above, the greater the dissimilarity of the resource demands of the time-shared tasks, the lower the degree of the interference there would be between the two tasks. Although low degree of interference usually translates to higher level of performance (which of course is desirable), this is not compatible with the goal of workload assessment. A fundamental assumption of the secondary task method is that the secondary task will compete with the primary task for limited processing resources. It is the degree of interference that is used for inferring the level of workload. Therefore, care must be taken to ensure that the selected secondary task demands resources similar to the primary task.

There are two approaches to selecting an appropriate secondary task. If the primary task demand is known, one can select from among the more commonly used secondary tasks with well-established demand characteristics. Excellent sources of references are O'Donnell and Eggemeier (1986); Ogden, Levine, and Eisner (1970); and Wickens, Hyman, Dellinger, Taylor, and Meador (1986). If the task demand of the primary task has not been established or when no established secondary task can be implemented, then additional research will be needed. This research is most likely to involve using multiple secondary tasks and carefully examining the pattern of interference between the primary and secondary tasks. Readers are advised to first consult references on dual-task methodology and multiple-resource assumptions before launching such a research effort. Damos (1991), Gopher and Donchin (1986), Guttentag (1989); and Wickens (1992) serve as good starting references.

One advantage associated with the primary task measures is their face validity. It is the most direct, "objective" measure. Among the different workload metrics, we have highest confidence in the primary task measure especially when it is part of the actual system performance. Except for extremely low workload conditions, primary task performance in general is sensitive to a variety of task demand manipulations. Performance measures are reliable to the extent that performance has stabilized at the time of workload assessment. For reasons discussed above, primary task performance is not always available. When available, the primary task is specific to the system evaluated. This poses a limitation on the generality of the workload results. Further, because different primary tasks may have different units of measurement, this could pose a scaling problem when comparisons or an aggregate of different primary task workload measures are needed.

The addition of a secondary task can circumvent the problem of insensitivity to extremely low workload level. A decided advantage of the secondary task is its diagnosticity value. Because tasks of dissimilar resource demands interfere little with each other, the pattern of interference observed with secondary tasks of known demand characteristics could pinpoint the type of processing resources demanded by the primary task. Knowledge of the type of demand of the task of interest is valuable information for optimizing system performance. For the same reason that secondary task measures can be diagnostic, they are selectively sensitive. That is, they are only sensitive workload measures if they tap similar resources as the primary task.

One drawback of the secondary task method is that the addition of an extraneous task to the operational environment may not only add to the workload, but may fundamentally change the processing of the primary task. That is, it could be intrusive. The resulting workload metric would then be nothing more than an experimental artifact. The embedded secondary task technique was proposed to circumvent this difficulty. In some situations, such as piloting jet fighter aircraft, task shedding is an accepted and taught strategy that is used when primary task workload becomes excessive. Tasks that can be shedded can perhaps serve as naturally lower priority embedded secondary tasks in a less intense workload evaluation situation. However, a naturally lower priority operational task may not always be available.

Another drawback is that using the secondary task method requires considerable background knowledge and experience to properly conduct a secondary task evaluation and to interpret the results. For example, care must be taken to control for the operator's attention allocation strategy, so as to ensure that the operator is treating the primary task as a high-priority task. The use of secondary tasks may also entail additional software and hardware development.

13.2.2.3 Psychophysiological Measures

Psychophysiological assessment methods measure changes in the operator physiology that are associated with cognitive task demands. As task demands change and the operator adjusts the level of mental activity associated with task performance, there are associated changes in the operator's physiological systems. This seems obvious when brain activity is considered since the brain is responsible for gathering information about the outside world, evaluating this information and initiating responses to make adjustments to the system being controlled. Changes associated with different workload demands have been reported in the cardiac, ocular, respiratory, and brain systems (see Kramer, 1991; O'Donnell and Eggemeier, 1986; and Wilson and Eggemeier, 1991 for reviews).

Heart rate has the longest history of use in assessing operator state with the first reported use during flight in 1917 (reported in Roscoe, 1992). Numerous reports in the aviation literature have noted increases in heart rate associated with increased mental workload. These include landing at airports with different levels of difficulty (Nicholson, Hill, Borland, and Ferres, 1970), with different gradients of the landing approach (Roscoe, 1975), and with different segments of surface attack missions (Comens, Reed, and Mette, 1987; Wilson, 1993). Heart rate can be used as a debriefing tool to locate mission segments having high or low rates for further evaluations (Rokicki, 1987). Differences among the workload of crew members during flights have also been reported (Roscoe, 1984; Wilson, 1993). Heart rate has also been used to study automobile driving in normal individuals, cardiac patients, and race car drivers (Taggart, Gibbons, and Somerville, 1969), and driving under different levels of difficulty (Helander, 1975). Race boat drivers (Johnson, 1980) and telemarketing workers (France and Ditto, 1989) have also been studied using heart rate. Even though mental activity and physical activity both produce changes in cardiac activity, useful data are provided even if the effects of these two variables cannot be easily parceled (e.g., Backs, Ryan, and Wilson, 1994).

The variability of the heart rhythm has been suggested as a measure of mental effort (Mulder, 1988; Mulder, 1980; Vicente, Thornton, and Moray, 1987). Laboratory studies have reported decreased variability of the heart rhythm under conditions of increased mental load (for reviews, see Grossman, 1992; Kramer, 1991; Mulder, 1992; Porges and Byrne, 1992). In applied settings, heart rate variability has been reported to decrease during segments of flight having higher mental workload such as take-off and landing and as a result of automobile driver fatigue and traffic density (Egelund, 1982; Itoh, Hayashi, Tsukui, and Saito, 1989). However, other studies have not reported systematic relationships between heart rate variability and mental workload (Wilson, 1992; Veltman and Gaillard, 1996). Further, Jorna (1992) suggests that heart rate variability decreases from no task to task conditions, but further increases in task demands do not produce related further decreases in heart rate variability. At this time the utility of heart rate variability in real-world situations has not been determined (Kramer, 1991; Wilson and Eggemeier, 1991).

Endogenous eye blink rate has been shown to decrease with increased visual demand such as driving an automobile in city as opposed to country environments (Lecret and Pottier, 1971; Pfaff, Fruhstorfer, and Peter, 1976), landing an aircraft, and target detection and weapons delivery in air-to-ground aircraft missions (Wilson, 1993; Wilson and Fullenkamp, 1990; Wilson, Fullenkamp, and Davis, 1994). Since visual input is disabled during eye closure, reduced blinks help to maintain continuous visual input when high levels of visual attention are required. During a low altitude parachute extraction maneuver, which requires a C-130 aircraft to fly 10 ft above ground and deliver several tons of cargo, an inhibition of blinking was found. This inhibition was preceded by a very slow, rhythmic pattern of blinking which is not seen in other situations (Wilson, 1992). These examples show that eye blinks are sensitive to the visual demands of a situation. However, under conditions of high rates of scanning of instruments, higher rates of blinking may be found since blinks often punctuate the end of an episode of information intake (Fogarty and Stern, 1989). There is a tendency to blink as the eye moves from instrument to instrument, for example. Interpretation of eye blink data requires that the context from which they were recorded be considered. If the task requires gathering information from one source, such as the case in landing an aircraft, then blink rates would be expected to decrease. However, if the task requires scanning in order to obtain the information, such as from an instrument panel with multiple instruments, then increased blink rates would be expected, especially if the required information is difficult to find.

Two types of brain activity have been used to study the effects of mental load: ongoing EEG and evoked potentials. When ongoing EEG activity is recorded, spectral analysis is typically performed and changes in the energy levels of the EEG bands are analyzed. Sterman, Schummer, Dushenko, and Smith (1988) and Sterman and Mann (1995) reported decreased alpha band power with increased task difficulty during actual and simulated aircraft flights. Brookings, Wilson, and Swain (1995) found increases in theta band activity due to increased workload produced by two manipulations of air traffic controller workload in a simulated air traffic control task. In a study designed to manipulate situational awareness, Vidulich et al. (1994) reported changes in theta band activity as the difficulty of the task increased. This was found over the frontal and central scalp sites.

Evoked potentials are the smaller electrical brain signals that are associated with processing information from discrete stimuli or the cognitive activity associated with these stimuli. Since these small signals are embedded in the background EEG, several stimuli are sequentially presented and ensemble averaging is obtained across stimuli in order to extract the small evoked potential waveform. The evoked potential waveform consists of several peaks and troughs that typically occur between 100 and 500 ms following stimulus presentation, although earlier and later components may also be present. The so-called P-300 component is found in the evoked responses to task-relevant stimuli. The P300 component is one of the largest components and has been reported to change amplitude with mental workload demands in several laboratory and in simulated flight studies (Fowler, 1994; Kramer, Sirevaag, and Braune, 1987; Sirevaag et al., 1993; Wickens et al., 1983). Evoked potentials have been collected in one flight study (Wilson, Fullenkamp, and Davis, 1994). In another paradigm, probe-evoked potentials are elicited from unattended, irrelevant stimuli and have been used to assess the effects of mental workload demands. These stimuli are not relevant to the task the operator is performing, but the hypothesis is that if resource demands are high then the brain will process the irrelevant information in these probes differently and the evoked responses will be different from those collected during low demand periods (Kramer, Trejo, and Humphrey, 1995; Wilson and McCloskey, 1988). This is a promising paradigm that deserves further research.

EEG is particularly sensitive to numerous artifacts that are present in real-world environments such as eye blinks, head and body movements, muscle activity, and speech.

Also, several stimuli must be presented in order to increase the signal-to-noise ratio sufficiently for the evoked potentials to be discernible. These two factors make the recording of acceptable EEG and evoked potentials difficult in the real world.

For the most part, psychophysiological measures do not intrude into the performance of the primary task. For example, the collection of heart rate data does not require any additional responses from the operator. Once the recording electrodes are in place, the heart rate data are collected without interference to the operator. This is also true for eyeblink, respiration, and brain wave data. Evoked responses may require the injection of external stimuli into the situation and have the potential of interfering with the primary task performance.

Because psychophysiological measures are continuously available they can be used to monitor moment to moment changes in an operator's response to task demands. Psychophysiological measures are acquired by attaching small electrodes to the operator, the detected potentials are amplified and recorded so that these signals then can be processed. Small, operator-worn, multichannel physiological recorders are available that permit the operator to go about their normal jobs without interference. These units are lightweight and battery powered so that the operator is unencumbered.

13.2.2.4 Analytic Methods

Analytic methods rely on modeling of the workload situation. Mathematical, engineering, and psychological models are all represented in some of the analytic workload methods. While only a few examples can be provided here, readers are referred to Bi and Salvendy (1994); Eisen and Hendy (1987); Hamilton, Bierbaum, and Fulford (1991); McMillan, Beevis, Salas, Strub, Sutton, and Van Breda (1989); Moray et al. (1991); Sarno and Wickens (1995); and Wickens, Larish, and Contorer (1989) for an orientation to the analytic methods. One distinct advantage of analytic methods is that each parameter and assumption of the model must be made explicit. This serves several important functions: (a) it fosters careful consideration of the relevant input and output variables to be included in the model, (b) it provides specific predictions that could be subjected to empirical testing and comparisons, (c) it facilitates communication of findings. As mentioned earlier, all analytic methods are designed to be used in both a predictive and evaluative fashion. The development and use of analytic methods are partly driven by the recognition of the inordinately high cost of designing and building a new system only to find that workload would not be at an acceptable level. Workload prediction is very much a primary goal of these analytic methods. Because many of the analytic methods provide a workload metric based principally on subject matter expert inputs in a predictive fashion, they are in particular need of validation.

Most of the recent analytic methods recognize the dynamic nature of the complex multitask environment in which workload measurement is most needed. For example, although time-line analysis (which by itself assumes serial, single-channel processing) is a major component in TAWL [Task Analysis/Workload (Hamilton et al., 1991)], TLAP [Time-Line Analysis and Prediction (Parks and Boucek, 1988)], and W/INDEX [Workload Index (North and Riley, 1989)], all of these methods allow for multiple concurrent processes and assume multiple-processing channels or multiple-processing resources. These analytic methods have all been applied in modern complex systems. For example, PROCRU [Procedure-Oriented Crew Model (Baron and Corker, 1988)] was designed primarily for investigating the procedural and system design changes in commercial aircraft operations in the approach-to-landing phase of flight; TLAP was applied in the Boeing 757 and 767 airplane programs, TAWL has been applied to the Army LHX helicopter, the AH-64, the UH60A, and the CH-47D aircraft; W/INDEX has been used to evaluate the Army Apache and LHX helicopter designs, the Advanced Tactical Fighter (ATF) design, and the National Aerospace Plane crewstation design.

Because analytic methods are not as commonly used and most have been developed only recently, few investigations have been performed to assess them and to compare them with other workload metrics. An exception is a recent article by Sarno and Wickens (1995). Sarno and Wickens first described three analytic methods—TLAP, VCAP, and W/INDEX, and delineated the differences in their parameters, their assumptions, and their predictions. Then, an experiment designed to examine time-sharing efficiency of performing multiple complex tasks was reported. The three analytic methods and the time-line analysis method were evaluated as to how well each model fit the performance data obtained from the experiment. Several interesting conclusions of this comparison were drawn. First, all three analytic methods accounted for over 50% of the performance variance but the time-line method accounted for only a small percentage of the performance variance that was not significantly different from zero. Sarno and Wickens pointed out that one difference between the time-line analysis method and the other three methods is that the time-line analysis method was the only one that did not assume multiple resources or multiple channels of processing. Among the three analytic methods, VCAP accounted for the least performance variance. Sarno and Wickens suspected that this could be attributed to the red-line assumption (there was a cap to the workload level allowable) made in the VCAP method and not the others. When the red-line assumption was removed from the VCAP method, the variance accounted for increased substantially, while the variance decreased when the red-line assumption was added to W/INDEX. Sarno and Wickens also proposed that the reason W/INDEX accounted for the most variance was its consideration of resource conflict. In W/INDEX, performing concurrent tasks that demanded similar processing resources would impose a greater "penalty" on the workload than time-sharing tasks demanding dissimilar resources. When the resource conflict assumption was added to VCAP and TLAP, variance accounted for increased. These conclusions are notable not so much for determining which analytic method was superior. They are notable because they help identify which parameters and assumptions are most appropriate for workload prediction and evaluation purposes. This information should be most useful for developing new, or refining existing, analytic methods.

In brief, because many analytic methods are recently developed, they have not been tested as much as some of the other workload metrics. The workload models upon which the workload predictions are based are in particular need of validation. Most analytic methods are not as easy to apply as the subjective techniques. On the other hand, most of them do not require sophisticated equipment. Although elaborate software is involved in these analytic techniques, generally they can be run on common computer platforms.

13.2.3 Subjective versus Objective Measures

Additional considerations for selecting the appropriate measures are provided by Kantowitz (1992) and Muckler and Seven (1992). Muckler and Seven (1992) hold that, "the distinction between 'objective' and 'subjective' measurement is neither meaningful nor useful in human performance studies" (p. 441). They contend that all measurements contain a subjective element as long as the human is part of the assessment. Not only is there subjectivity in the data obtained from the human subject, the human experimenter also imparts his or her subjectivity in the data collection, analysis, and interpretation. Thus, performance measures are not all objective nor are subjective ratings entirely subjective. Muckler and Seven further propose that the selecting of a measure (or a set of measures) be guided by the information needs. After an initial set of candidate measures has been identified, the different measures can be compared by considering their relative strengths (such as their reliability and diagnosticity). It is recognized that no perfect measures exist. Tradeoffs among the different measures must therefore be evaluated carefully. Kantowitz (1992) echoes Muckler and Seven's recommendation that measure selection be based on how well the chosen measures would meet the need of the assessment. Kantowitz further advocates that the measures be theory-based. Kantowitz made an analogy between theory and the blueprint of a building. Trying to interpret data without the guidance of a theory is like assembling bricks randomly when constructing a building. To elaborate, Kantowitz points out that the understanding of both the substantive theory of human information processing and the psychometric theory of the measurements are helpful. The former dictates what one should measure, and the latter suggests ways of measuring them.

13.2.4 Benefits and Perils of Multiple Workload Measures

Most real-world tasks are by nature complex and/or involve multiple tasks. This presents problems to the investigator interested in measuring the mental workload associated with these tasks. For example, measures that provide global mental workload information may fail to provide information about the individual task components. On the other hand, measures that are sensitive to only a specific aspect or dimension of the mental workload may not provide accurate estimates of the overall workload, especially if the workload component to which they are sensitive is behaving differently from the overall workload. For example, it is possible that the overall demands of a task are quite high but the visual demands of the task are low. If the eye blink measures are not sensitive to the nature of the demands of a particular task, the eye blink measures would not show high workload even though the actual overall workload level is high.

The performance of complex or multiple tasks has not received the same amount of research emphasis that has been devoted to the single or dual task situation. One reason for this is that complex tasks do not provide the level of experimental control that is desired in research settings. For example, it is difficult to know at any point in time just which aspect of the task a truck driver or an airline pilot may be attending to. Without this knowledge it is difficult to use the traditional methods of analysis to determine the effects of experimental manipulations. However, most real-world situations involve multiple-task performance and must be evaluated. Since these evaluations must be completed, methods and theories will have to be developed to address the multiple-task environments.

Multiple-workload measures are generally recommended because workload is a multifaceted concept. As outlined elsewhere in this chapter, the available measures differ in their sensitivity and diagnosticity to the different aspects of mental workload. Some measures provide information about the global workload levels, whereas others are quite specific in the nature of the information that they can provide. Carefully selected multiple measures should provide a well-rounded evaluation of the workload of a given task. However, as discussed below, multiple measures can dissociate and provide seemingly contradictory or different information about the mental workload of a task or system. The nature of these dissociations can be very helpful in evaluating the nature of the workload of the system if the evaluator is knowledgeable about the measures that were used.

13.2.4.1 Dissociations Among Workload Measures

When different workload measures suggest different trends for the same workload situation, the workload measures are said to dissociate. Given that mental workload is a multidimensional concept, and the various workload measures may be differentially sensitive to the different workload dimensions, some dissociations among workload measures are to be expected. Measures having qualities of general sensitivity (such as certain unidimensional subjective estimates) respond to a wide range of task manipulations but may not provide diagnostic information about the individual contributors to the workload. Measures having selective sensitivity (such as most primary task measures) respond only to specific manipulations. In addition, different measures may be differentially sensitive at different ranges of workload levels. If workload measures are found to dissociate, then the nature of the dissociation needs to be examined because this information is invaluable for interpreting the characteristics of the workload in the task under evaluation. The potential for dissociations also means that the selection of workload measures must be carefully considered for each situation since measures sensitive to the dimensions relevant to the given situation must be used if the evaluation of workload is to be meaningful. Without this preliminary analysis, the measures that are used may not be sensitive to any of the relevant dimensions or sensitive to only a subset of them, leading to erroneous conclusions.

Researchers have shown that performance and subjective measures tend to dissociate under the following conditions (Eggemeier and Wilson, 1991; Vidulich, 1988; Vidulich and Wickens, 1986; Wickens, 1992; Yen and Wickens, 1988): (1) Performance and subjective measures tend to dissociate under low workload conditions (e.g., Eggemeier, Crabtree, Zingg, Reid, and Shingledecker, 1982). As mentioned previously, primary task measures are not sensitive to low levels of workload, whereas subjective measures would be more sensitive. Performance would therefore not change with variation of low demand, whereas subjective measures would. (2) The two measures tend to dissociate when subjects are performing data-limited tasks. If subjects are already expending their maximum resources, increasing task demand would further deteriorate performance but would not affect the subjective ratings because they are probably already at a ceiling level. (3) Greater effort would generally result in higher subjective ratings, however, greater effort could also improve performance (Vidulich and Wickens, 1986). Increased subjective ratings associated with improved performance is considered a dissociation because normally, increased subjective ratings and degraded performance together are considered to indicate increased workload. (4) Subjective ratings are particularly sensitive to the number of tasks that the subjects have to time-share. For example, subjective ratings for performing a pair of easy tasks (that results in good performance) would be higher than those for performing a difficult single task (that results in poor performance) (Yeh and Wickens, 1988). (5) Performance measures are sensitive to the severity of the resource competition (or the similarity of the type of processing resources demanded) between the time-shared tasks, but subjective measures are less so (Yeh and Wickens, 1988). (6) Given that subjects only have access to information available in their consciousness (Ericson and Simon, 1980), subjective ratings are more sensitive to central processing demand (such as working memory demand) than to demands that are not represented well consciously, such as response execution processing demand. Subjective and performance measures therefore would tend to dissociate when the main task demands lie in response execution processing (Vidulich, 1988). An excellent reference is McCoy, Derrick, and Wickens (1983) who provided realistic examples of how performance and subjective ratings may dissociate in system evaluations and discuss how these dissociations can be interpreted in meaningful ways.

Another potential source of dissociation is discussed in the next section. This is when a workload measure is affected by factors other than mental demands. For example, heart rate can be influenced by physical activity that is not necessarily related to mental task demands; subjective ratings may be influenced by prior experiences (Moroney et al., 1995).

13.2.4.2 Factors Unrelated to Task Demand that Affect Workload Measures

Workload measures provide information about the interaction between the task and the operator in a given situation. The context of the situation must be considered carefully. Issues of concern include training, operator state, and operator biases. Training can be very important when testing a new system. Typically, the operators evaluating a new system have a great deal of experience with older systems and none with the new system or system component being tested. It is desirable to provide as much training as possible with the new system so that the operator is familiar with the operation of the new system and the evaluation will not be contaminated with learning and familiarity issues. If the operator is just learning the system, they may not fully appreciate all of the nuances of the new system and therefore not be able to take full advantage of its capabilities. Further, the mental workload will be higher as the operator is learning the new system and learning effects may contaminate the workload data. Since it is not always possible to provide the operators with all of the training that is necessary, one must provide appropriate controls in the experimental design and be aware of these influences when interpreting the data.

Operator state includes variables such as fatigue and motivation. Different operator states can produce different estimates of mental workload and performance for the same system state. A fatigued operator may perceive the task as more difficult or may not be motivated to use the same criteria as when they are rested. Personal matters can also affect the operator's performance and how they perceive the system being evaluated. Illness can also interfere with mental workload since it affects the state of the operator and how they interact with the environment and how they perceive the outcome of this interaction.

Operators may have biases that they bring with them to a testing situation or that develop during the testing. For example, an operator may dislike an existing system and any new system is automatically favored over the older system. This can manifest in better performance and glowing subjective ratings and reports about the new system. In other situations, an operator may dislike a new system based on certain attributes that are irrelevant to mental workload or system performance and provide negative subjective evaluations. These biases are usually unconscious on the part of the operators. One should be aware of these potential biases and try to find out if they are present in a testing situation.

Some measures may have several determinants. Heart rate, for example, can become higher due to increased mental load and also due to increased physical activity. Often, these two causes occur together since increased mental activity may lead to increased physical activity. However, it may be possible to separate these effects (e.g., Backs et al., 1994). In any case, the increased heart rate can be taken as an indication of increased mental workload whether or not it is accompanied by increased physical activity. One, however, needs to be careful in interpreting the magnitude of the workload increase if it is accompanied by increased physical activity.

13.2.5 Practical Use of Workload Measures

In system test and evaluation a common demand is for a workload "redline." Managers of acquisition programs would like to have an absolute or "redline" criteria with which they could assess the acceptability of new or modified systems similar to the criteria used by structural engineers which can specify the G limits of an aircraft wing. If the limit is exceeded the wing of the aircraft will break. Because of the multifaceted nature of humans, it is not possible to provide a similar measure or set of equations for mental workload. At present, we do not have sufficient understanding of all of the variables that contribute to human performance. At least two courses of action have been suggested to answer the redline problem. One is to use past experience to derive a level of subjective workload that is used to indicate a definite problem area (Reid and Colle, 1988; Rueb, Vidulich, and Hassoun, 1994). Based upon past research the Subjective Workload Assessment Technique (SWAT) was used to establish the level at which performance had been found to begin to degrade. This "redline" was determined to be a SWAT score of 40 ± 10. Exceeding this score does not predict immediate operator failure, but does define a point at which the system must be placed under close scrutiny because performance degradation is highly probable. This work provides a good start on defining a workload "redline" but more research is required to provide a universal scheme for defining the redline.

A second approach to the redline problem is to use data collected from existing and acceptable systems to provide a standard or criterion against which new or upgraded systems can be compared (Ruggiero and Fadden, 1987). For example, subjective performance and/or psychophysiological data are first collected with an existing system. Then tasks that are common to both the existing and new systems are performed with the new system and similar measures are recorded. Data from the old and new systems are then compared and indications of increased workload are used to focus attention on areas of concern in the new system. While this is a relative measure, it can be used to indicate whether the new system is as good as, superior to, or inferior to the existing system. This strategy could also be used when evaluating upgraded systems.

13.3 GUIDELINES FOR SELECTING A WORKLOAD ASSESSMENT STRATEGY

In the past two decades, a multitude of laboratory, simulation, and field studies on mental workload have been performed. Although the usefulness of workload assessment has been demonstrated, it is also clear that workload measurements do not provide all the answers concerning the design, safety, and effectiveness of a human–machine system. Along with the empirical studies, much effort has also been devoted to understanding the underlying constructs of mental workload so as to improve the existing assessment techniques or to develop new ones. One outcome of these activities is that one can now choose from an array of workload techniques. Another outcome of these activities is that we have learned more about the boundaries of the workload construct as well as the unique efficacies and inadequacies of the different approaches to measuring mental workload. Though the work is far from being complete, there is now a body of knowledge that can be put to practical use. From this, we have developed the following guidelines for selecting an appropriate strategy for workload assessment and evaluation.

The philosophy underlying these guidelines is that conducting workload assessment, especially that of a fairly complex system, does require some knowledge and training in the area. At the minimum, the following knowledge is necessary: Knowledge about the boundaries of the construct of mental workload, knowledge of the various mental workload assessment approaches that are available, knowledge of the strengths and drawbacks of the different approaches and techniques, knowledge about the accurate procedure for the technique(s) chosen, and an understanding of the limits of generality and applicability of the results. Through the existing body of literature and cumulative experience in the process of measuring mental workload, one can methodically acquire the knowledge that is needed and conduct a profitable assessment of workload. The present chapter itself will not instill all the necessary knowledge. Rather, the present chapter aims at serving as a source of information for where and how the necessary knowledge and tools can be acquired. Accepting the fact that no workload technique is perfect and different approaches and techniques may yield different information, the following guidelines suggest a methodical approach for selecting the appropriate assessment technique. Table 13.2 outlines the major steps involved in selecting the appropriate workload assessment strategy.

13.3.1 Delineate the Objective of the Mental Workload Assessment

The objective of the workload assessment should be one of the main determinants in selecting the appropriate workload metric(s). There are several broad categories of objectives for assessing workload. The first is prediction. To be able to predict the level of mental workload of planned system is particularly pertinent during the system design phase (see, for example, Aldrich et al., 1989; Christ et al., 1993; Kuperman, 1985). The objective here is to predict whether the workload of a new system might be at a satis-

Table 13.2 Major Steps in Selecting the Appropriate Workload Assessment Technique

Delineate the objective(s) of mental workload assessment
• Prediction, Evaluation, Diagnosis
Perform a task/mission/system analysis
Evaluate the resources available
• Time and costs constraints, Equipment, Expertise
Select the type(s) of workload measure to be used
• Performance, Physiological, Subjective, Analytic
Select the specific workload assessment technique(s)
Familiarize oneself with the chosen technique(s)
Formulate the design for evaluation
Reexamine the appropriateness of the selected workload measure(s)

factory level. Other instances when workload prediction might be helpful is when the system configurations may be modified (such as due to the incorporation of new automated components), when the operational procedure may be changed, function allocation may be altered, or when the personnel may be changed. There are techniques designed specifically for predicting workload [e.g., network modeling (Laughery, 1989); VCAP (McCracken and Aldrich, 1984); W/INDEX (North and Riley, 1989; Wickens et al., 1989)] and several standard techniques have been applied in the projective fashion (e.g., Christ et al., 1993; Kuperman, 1985; Madni and Lyman, 1983).

The second objective for assessing mental workload is evaluation. The most common example is comparing workload of alternative, either existing or prototype, systems (e.g., Anderson and Toivanen, 1970; Vidulich and Bortolussi, 1988). Clearly, the system with the more satisfactory level of workload that delivers the same degree of effectiveness would be the more desirable system. For example, the decision of which prototype to produce could be based partly on the results of workload assessment. A particularly notable example is the evaluation of the necessary crew size of wide-bodied jet aircraft (Wiener, 1985). Another reason for workload evaluation is to determine the demands across different phases or conditions of the task/mission (e.g., Johannsen and Rouse, 1983; Kreifeldt, 1980)

The third reason for assessing workload is diagnosis. One potential use of workload measurements is for diagnosing the potential cause of a nonoptimal system. Intuition and contemporary theoretical frameworks of mental workload all predict a degradation of performance with excessive workload. Less obvious but empirically supported is the finding that an extremely low level of workload could also bring about nonoptimal performance (e.g., Warm and Parasuraman, 1987). At a more micro level, the appropriate workload assessment can potentially isolate the troublesome spot in the system. In addition to learning about the level of the overall workload, it might be important to ascertain more precisely what exactly is overly demanding. A unidimensional subjective rating does not provide such information, but several multidimensional ratings (e.g., Reid and Nygren, 1988; Tsang and Velazquez, 1996) and appropriately selected secondary task measures would (e.g., Brown, 1965; Dougherty, Emery, and Curtin, 1964; Isreal, Wickens, Chesney, and Donchin, 1980; Wickens et al., 1986).

13.3.2 Perform a Task/Mission/System Analysis

With the advent of automation, growing complexity of human–machine systems, and the ensuing changes in the role of the human operators, simple time-line analysis will not suffice. Because the role of the human operator becomes predominantly that of monitoring and supervising, the degree to which the operator is performing some observable functions is not an adequate index of cognitive workload. This is not to say that time-line analysis cannot be useful, but its usefulness is limited especially in the context of dynamic, complex systems (see Adams et al., 1991; Gopher and Donchin, 1986).

Methodical analysis of what tasks are required to achieve the missions given the system constraints is always necessary (e.g., Aldrich et al., 1989; Christ et al., 1993). Now in vogue, but with their application effectiveness in complex system not yet clear, are the techniques of cognitive task analysis (e.g., Glaser, Lesgold, Lajoie, Eastman, and Greenberg, 1985; Klein, 1990) and knowledge elicitation (e.g., Geiwitz, Klatzky, and McCloskey, 1988; Kitto and Boose, 1989; Olson and Biolsi, 1991).

As discussed above, one distinguishing role of the operators in a dynamic complex environment is their multitask management role. Adams et al. (1991) emphasize that human behavior is goal directed and advocate that missions be analyzed in terms of goals. An individual task can be considered a subgoal in support of broader goals. Adams et al. contrast this approach to time-line analysis as follows: Analyzing mission activities in terms of goals

> *. . . distinguishes between those events and activities that happen to occur closely in time in the particular sample being evaluated, and those which must, for reasons of logical or functional interdependence, occur closely in time in any realistic scenario. This distinction between necessary and circumstantial coincidence is invaluable when consideration is turned to the generality or generalization of the scenario.* (p. 37)

13.3.3 Assess the Constraints and Resource Availability

Other major determinants for the choice of a workload metric include the existing constraints and resource availability. Time and cost constraints are usually the primary concerns. In general, among the different types of workload measures, subjective ratings, especially unidimensional ratings [e.g., the Bedford scale developed by Roscoe and Ellis (1990)], can be obtained most quickly with the least cost. However, in an established facility with resident expertise, performance and physiological measures can also be obtained without inordinate hardship. Otherwise, physiological measures require the most sophisticated equipment and software. Certain performance measures (such as real-time manual control measures) also require specialized equipment and software. Laboratory assessment of real-time performance, as well as simulator performance, requires tailored programming which has little commercial support. Performance and physiological measures could also require additional computer power to handle large databases.

Besides the time/costs constraints and the equipment/software availability, an important consideration is the availability of expertise in workload assessment. All assessment techniques require some training in experimental design, data collection, and data analysis, training in data interpretation, and the translating of results to practical use. The latter requires thorough familiarity not only with the assessment technique itself, but the system to which the data apply. Although all forms of data analysis and data interpretation require some formal training, that required for collecting and analyzing physiological data is especially substantial.

13.3.4 Select the Type(s) of Workload Measures

The relative strengths and weaknesses of the different types of workload measures are presented above. These strengths and weaknesses need to be weighed with the specific objective of the workload assessment, the task/mission/system analysis, the constraints, and the resources available. Several attempts have been made to use subjective and performance measures to predict workload (e.g., Aldrich et al., 1989; Wickens et al., 1989; Wickens and Yeh, 1986); the authors are not aware of any attempt with physiological measures. On the other hand, many physiological measures are particularly strong in their diagnostic value, pinpointing the nature of the task demand (e.g., the P300 evoked potential is sensitive only to perceptual/central processing demands). Performance measures, selected properly, could also provide diagnostic information. For subjective measures, only multidimensional ratings afford diagnostic information (e.g., Reid and Nygren, 1988; Tsang and Velazquez, 1996).

After the task/mission/system analysis, if it is expected that the workload will be extremely high or extremely low, performance measures are unlikely to yield useful information (O'Donnell and Eggemeier, 1986). If the workload is already expected to be high, and the operators are expected to carry out a variety of tasks in a dynamic environment, the addition of a secondary task is likely to be more intrusive than informative. Although, some researchers have found the use of an embedded secondary task to be useful (e.g., Kantowitz and Casper, 1988; Vidulich and Bortolussi, 1988). Though most physiological measures are considered to be nonintrusive, accurate pupil dilation measures, for example, could not be obtained in a vibrating environment with less than optimal lighting control such as the cockpit.

A special workload condition is one of workload transition when workload is suddenly increased after long periods of underload (Huey and Wickens, 1993). A promising approach to capture the effects of transitions is the adoption of continuous measures. Continuous physiological measures such as heart rate may be particularly suitable for capturing transitory changes in workload. Continuous performance measures have also been proposed (Levison, 1979). In a flight simulator, flight control data can serve as a continuous as well as an instantaneous measure of workload. Retrospective subjective ratings could also be used. Antin and Wierwille (1984) developed a battery of tasks aimed at capturing instantaneous mental workload and proposed that carefully chosen primary task performance measures and on-line subjective opinion measures can be promising.

13.3.5 Select the Specific Workload Assessment Technique(s)

After having selected the type of measure, further selection of a specific technique is necessary. Again, the selection here has to be based on the objective of the workload assessment, the task/mission/system involved, the constraints, and the resources available. For example, if it is decided that secondary task performance measures are to be used for diagnostic purposes, then several (at least two) secondary tasks should be carefully selected to be paired with the primary task. The secondary tasks should be sensitive to different resources. For example, performance of the Sternberg memory task has been shown to be sensitive to the central processing demands (e.g., Wickens et al., 1986) and performance of a continuous tracking task has been shown to be sensitive to primarily response processing demands (e.g., Wickens, Gill, Kramer, Ross, and Donchin, 1981). If the two different secondary tasks produce significantly different effects on the primary task performance, the predominant demand on the primary task would be the same as that of the secondary task that caused the most interference.

One also has a host of subjective techniques available should the decision be to collect subjective ratings. Tsang and Vidulich (1994) and Vidulich and Tsang (1986) examined the different approaches to assessing subjective workload and proposed several dimensions that can be used to categorize the different subjective instruments. Unidimensional techniques (e.g., the psychophysical scale proposed by Gopher and Braune, 1984) of course would have no diagnostic value in terms of pinpointing the nature of the demand. However, they provide the most globally sensitive measures. Relative techniques that require subjects to rate the workload of one task condition relative to another (e.g., SWORD) would be particularly suitable for comparison purposes.

Psychophysiological measures have been shown to possess both general and specific sensitivity to mental workload. Heart rate provides a global index of task demands. Blink rate, in some situations, has been shown to be a sensitive indicator of visual demand. Respiration rate also changes with the mental demands of a task but is also susceptible to artifacts during speech. Evoked potentials, the P300 component, is specifically sensitive to perceptual/central processing demands, whereas spectral analysis of brain waves have proven useful for assessing the level of mental engagement in cognitive tasks.

13.3.6 Familiarize Oneself with the Chosen Technique(s)

As much as possible, consult investigations that have applied the same technique chosen prior to the administration of the technique. For example, seemingly subtle differences in the instructions to subjects could elicit drastically different results. Some of the more theoretical and conceptual treatment of workload issues relevant to the technique(s) chosen should also be studied. This up-front effort is well spent, considering that interpretation of the results is seldom straightforward (see also Kantowitz, 1992).

An example of the subtlety of the methodological complexity of the secondary task technique can be provided. Ideally, the demands of the different secondary tasks chosen for the workload evaluation should be approximately equal in objective difficulty. However, consider the task of equating the objective difficulty of a tracking task and a memory task. While these are not easy problems to solve, there are sound experimental psychological methodologies that could be very helpful. A second example can be provided. Preferably, the nature of the demands of the secondary tasks have already been validated. Because the memory task and the tracking task are widely used in the psychological as well as applied literature, their demand characteristics are well established. Other tasks may not enjoy the same distinction. Again, there are experimental methodologies that could help validate the nature of the demand of the tasks. However, this would involve additional time and effort that have to be budgeted into the workload evaluation project. Since it would not be possible to go into detailed discussion for the different techniques, an annotated bibliography is provided below to orient readers to the major sources of information.

13.3.7 Formulate the Design for Evaluation

Thorough consideration should be paid to the experimental design and procedures. Tradeoffs between within-subject designs and between-subjects design, the appropriate subject sample (e.g., novice vs. experts, college students vs. operational personnel), presentation order of the different evaluation conditions, potential learning/training effects, clarity of the instructions provided to the subjects, and the proper incentives for the subjects should all be considered carefully. Standard experimental controls should be implemented when appropriate. A well-thought-out design and procedure takes longer to formulate, but not necessarily longer to implement. Again, the effort spent up front is well worth the more generalizable and more interpretable results that will be obtained.

13.3.8 Reexamine the Appropriateness of the Selected Workload Measures

After the initial selection one might wish to reevaluate the appropriateness of the choice of technique after one has some in-depth familiarity with the technique. For example, one may find that the only feasible appropriate secondary task has not been validated in terms of its nature of demand, and yet time or other resource limitations do not allow the necessary research to establish its demand nature. An alternative workload technique should then be considered.

Of course, one could have multiple objectives in assessing workload and there is no reason to confine oneself to obtaining one measure or one type of measures. In fact, Eggemeier and Wilson (1991) and Wierwille and Eggemeier (1993) advocate using multiple measures as much as feasible. The strongest argument for such an approach is that we have now learned that the different measures could provide different information. Even their dissociation can be informative and revealing in some ways.

13.4 ANNOTATED SELECTED BIBLIOGRAPHY

Presented in this section is annotated bibliographic information that should be useful for gaining a basic familiarity with the existing literature. While this is not an all-inclusive list, it should provide a starting point. Four categories are presented, the first provides fundamental information about the field and should provide a good basis for working in this area. The second deals with guidelines for applying the different types of workload measures. This information is useful when it is time to actually design the evaluation and start data collection. The third category provides information about existing bibliographies on the topic of mental workload. These are sources that have compiled lists of references to the workload literature. The final category provides details on where to gain access to existing workload measure software, manuals, and related information.

13.4.1 Workload Fundamentals

Three chapters on the topic of mental workload can serve as an excellent starting point for those unfamiliar with workload assessment. The authority on the subject is still the two chapters which appeared in the 1986 *Handbook of Perception and Human Performance: Volume II. Cognitive Processes and Performance*. The chapter by Gopher and Donchin (1986) provides a theoretical framework, a stimulating discussion on whether mental workload is an intervening variable or a hypothetical construct, a definition of workload that contains elements shared by many other definitions that have been proposed, and an overview of the general types of workload measures that are available. The chapter by O'Donnell and Eggemeier (1986) offers a comprehensive review of the different types of workload measures and a detailed illustration of specific techniques within each type of measure. The chapter begins with a discussion on the properties for selecting workload assessment techniques. Assumptions underlying each type of workload measure, detailed procedures, and known strengths and weaknesses of the more common techniques are presented. A less detailed, but an excellent integration of theory and application is the chapter on "Attention, Time-Sharing, and Workload" in Wickens (1992).

Two recent publications reflect the more current concerns of assessing mental workload. Adams et al. (1991) present a comprehensive treatment of workload issues faced by operators in a dynamic multitask environment. A thoughtful, theoretical framework drawn from several contemporary theoretical psychological approaches is presented. Promising interventions that would be helpful to the dynamic multitask operators are proposed. Both Adams et al. (1991) and Huey and Wickens (1993) pay special attention to the potential problems brought about by technology and, in particular, automation. Huey and Wickens in particular address the problem of workload transition such as a sudden increase in workload due to an automation malfunction after long periods of underload.

Reviews of the psychophysiological workload literature can be found in Kramer (1991) and Wilson and Eggemeier (1991) in a book on multiple-task analysis (Damos, 1991). These chapters describe the various physiological measures and contain examples of both laboratory and applied research as well as discussions about the merits of the various measures. Topics such as sensitivity and diagnosticity are discussed as is the notion of multiple measures. Readers interested in psychophysiological measures are referred to several special issues of journals that have been devoted to the application of psychophysiological methods to human factors problems in general and to workload assessment specifically. The titles and citations of the special issues follow: Cognitive psychophysiology, *Human Factors*, 1987, Vol. 29 (2); Cardiorespiratory measures and their role in studies of performance, *Biological Psychology*, 1992, Vol. 34 (2, 3); Psychophysiological measures in transport operations, *Ergonomics*, 1993, Vol. 36 (9); EEG in basic and applied settings, *Biological Psychology*, 1995, Vol. 41 (1 & 2); Psychophysiology of workload, *Biological Psychology*, 1996 Vol. 42.

13.4.2 Application Guidelines

A detailed step-by-step guide for several common workload assessment techniques for each type of workload measure can be found in the *Engineering Data Compendium: Human Perception and Performance* (1988). This Compendium is also a good source of references for empirical studies that have used the various specific techniques. Lysaght et al. (1989) provide an extensive review of the existing workload techniques and guidelines for selecting the appropriate techniques for specific applications. Wierwille and Eggemeier (1993) consolidated the cumulative findings from a vast amount of work on mental workload and recommended general guidelines for selecting an assessment strategy. The article serves also as a rich source of references for both the development and application of an array of workload assessment techniques.

13.4.3 Bibliographies of Mental Workload Publications

Wierwille and Williges's (1980) bibliography contains over 600 citations based on searches performed in 1979. Schmidt and Nicewonger's (1988) bibliography presents over 200 citations, which cover the period from 1980 to 1986. Moray's (1988) article, "Mental Workload Since 1979," offered a review of the progress in the field of mental workload and commented on its state of affairs. The review provided a bibliography of over 400 references. Another bibliographic listing of mental workload research is provided by Hancock, Mihaly, Rahimi, and Meshkati (1988). Although papers in Proceedings of scientific meetings is a popular forum for publishing workload research and findings, Hancock et al. reported that more citations can be found in journal articles than any other forum. The three main Proceedings were the *Proceedings of the Annual Meeting of the Human Factors and Ergonomics Society*, the *Proceedings of the Conference on Manual Control*, and the *Proceedings of the Symposium on Aviation Psychology*. The six leading journals are *Human Factors; Ergonomics; Aviation, Space, and Environmental Medicine; IEEE Systems, Man, and Cybernetics; Journal of Experimental Psychology*; and *Acta Psychologia*. Two new relevant journals have emerged since Hancock et al.'s review: *The International Journal of Aviation Psychology*, and the *Journal of Experimental Psychology: Applied*.

The following are more specialized bibliographies. Clements (1977) provided an early bibliography on workload research using the primary task performance technique. Williges and Wierwille (1979) reviewed the behavioral measures of air crew mental workload. Gawron, Schflett, and Miller (1989) examined workload measures that are applicable to in-flight workload. As Moray (1988) pointed out, work on mental workload issues has been centered predominantly on aerospace. Moray attributed this to the "spectacular increase in automation" in both military and civilian aircraft (p. 124).

A recent NATO AGARD Advisory Report, number 324, *Psychophysiological Assessment Methods* by Caldwell et al. (1994) contains a theoretical overview, reviews of the literature using the various measures, and a discussion of the areas where these measures have been applied. In addition to this review, there is an extensive bibliography and a number of appendices which describe each measure and discuss the technical issues of data collection and analysis for each one. A register of over 700 psychophysiological researchers in the United States and Europe is also included which lists their areas of interest and addresses. The Advisory Report can be obtained from NASA CASI, in Linthicum Heights, MD.

13.4.4 Sources for Software, Manuals, and Additional Information

Table 13.3 provides pertinent references to some of the more common assessment techniques. The *Engineering Data Compendium: Human Perception and Performance* mentioned above and its CD-ROM version (*Computer Aided System Human Engineering: Performance Visualization System, CASHE:PVS*) can be obtained from:

CSERIAC Program Office
ATTN: Products and Services
AL/CFH/CSERIAC
2255 H Street, Bldg 248
Wright-Patterson AFB, OH 45433-7022
email: CSERIAC@falcon.al.wpafb.af.mil
WWW: http://www.dtic.dla.mil/iac/cseriac/cseriac.html

CSERIAC distributes the NASA-TLX (computer and paper-and-pencil versions) and SWAT (computer version) for a nominal cost. A user guide for the SWAT is also available (Reid, Potter, and Bressler, 1987). Users can also download the software and manuals from the World Wide Web at no cost. A user guide for the Task Analysis/Workload

Table 13.3 Reference for Some of the More Common Workload Assessment Techniques

Types of Measures/Techniques	References
Subjective	
Bedford	Roscoe (1987), Roscoe and Ellis (1990)
Modified Harper-Cooper	Wierwilli and Casali (1983)
NASA-TLX	Hart and Staveland (1988)
Psychophysical Scaling	Gopher and Braune (1984)
SWAT	Reid and Nygren (1988)
SWORD	Vidulich et al. (1991)
Workload Profile	Tsang and Velazquez (1996)
Performance	
AGARD STRESS Battery	AGARD (1989)
Choice Reaction Time	Kalsbeek and Sykes (1967), Krol (1971)
Criterion Task Set	Shingledecker (1984)
Multi-Attribute Task Battery	Comstock and Arnegard (1992)
Time Estimation/Interval Production	Wierwille and Connor (1983)
Mental Arithmetic	Brown and Poulton (1961), Harms (1986)
Sternberg Memory	Wickens et al. (1986)
Tracking Task	Levison (1979), Wickens (1986)
Psychophysiological EEG/ERPS	*Biological Psychology* (1995), *40* (1 & 2)
Eye Blinks	Stern and Dunham (1990)
Heart Rate Variability	*Biological Psychology* (1992), *34* (2 & 3)
Respiration	*Biological Psychology* (1992), *34* (2 & 3)
Analytic	
PROCRU	Baron and Corker (1988)
Queuing Theory	Moray et al. (1991)
Time-Line Analysis	Kirwin and Ainsworth (1992)
TAWL	Hamilton, et al. (1991)
TLAP	Parks and Boucek (1989)
VCAP	McCracken and Aldrich (1984)
W/INDEX	North and Riley (1989)

(TAWL) (Hamilton et al., 1991) and its Operator Simulation System (TOSS) used for performing all the database management and model execution functions are published by Anacapa Sciences, Inc., P.O. Box, 489, Fort Rucker, AL 36362-5000.

The paucity of references in Table 13.3 for the performance techniques is not an accurate reflection of the level of research and applications that have been conducted. Rather, primary task measures are system specific and under the multiple-resource framework, as are the secondary task measures. Therefore, there are few standardized primary and secondary tasks. The Criterion Task Set listed in Table 13.3 is a set of different tasks methodically chosen to represent different types of human information processing. Ogden, Levine, and Eisner (1979) provide a good coverage of the secondary task technique. See also the Compendium compiled by Boff and Lincoln (1988), Section 7.719.

13.5 ACKNOWLEDGMENTS

A portion of the chapter was prepared while Pamela Tsang was spending her sabbatical with the Operator Performance and Safety Analysis Division at the Volpe National Transportation Systems Center in Cambridge, MA. The authors thank the following individuals for their comments on earlier drafts: William Albery, Rickard Backs, Julie Cohen, Bruce Humm, Eric Nadler, Gavriel Salvendy, Mike Vidulich, and an anonymous reviewer.

REFERENCES

Adams, M. J., Tenney, Y. J., and Pew, R. W. (1991). *State of the art report: Strategic workload and the cognitive management of advanced multi-task systems.* BBN Report 7650. (Crew Systems Ergonomics Information Analysis Center, City, Wright-Patterson Air Force Base, OH.)

AGARD (1989). *Human performance assessment methods.* AGARDograph-AG-308. Neuilly Sur Seine, France: Advisory Group for Aerospace Research and Development.

Aldrich, T. B., Szabo, S. M., and Bierbaum, C. R. (1989). The development and application of models to predict operator workload during system design. In G. R. McMillan, D. Beevis, E. Salas, Strub, M. H., Sutton, R., and Van Breda, L., Eds., *Applications of Human Performance Models to System Design.* New York: Plenum.

Anderson, P. A., and Toivanen, M. L. (1970, March). *Effects of Varying Levels of Autopilot Assistance and Workload on Pilot Performance in the Helicopter Formation Flight Mode.* Technical Report JANAIR 680610. U.S. Office of Naval Research, Washington, DC.

Antin, J. F., and Wierwille, W. W. (1984). Instantaneous measures of mental workload: An initial investigation. *Proceedings of the Human Factors Society 28th Annual Meeting.* Santa Monica: Human Factors Society, pp. 6–10.

Backs, R. W., Ryan, A. M., and Wilson, G. F. (1994). Psychophysiological measures of workload during continuous manual performance, *Human Factors*, *36*, 514–531.

Baron, S., and Corker, K. (1988). Engineering-based approaches to human performance modeling. In G. R. McMillan, D. Beevis, E. Salas, M. H. Strub, R. Sutton, and L. Van Breda, Eds., *Applications of Human Performance Models to System Design.* New York: Plenum, pp. 203–218.

Baron, S., Zacharias, G., Muralidharan, R., and Landraft, R. (1980). PROCRU: A model for analyzing flight crew procedures in approach to landing. *Proceedings of Eighth IFAC Work Congress* (Vol. XV), pp. 71–76 (also, NASA Report No. CR152397).

Bi, S., and Salvendy, G. (1994). Analytical modeling and experimental study of human workload in scheduling of advanced manufacturing systems. *The International Journal of Human Factors in Manufacturing*, *4*, 205–234.

Boff, K. R., and Lincoln, K. R. (1988). *Engineering Data Compendium: Human Perception and Performance.* Armstrong Aerospace Medical Research Laboratory, Wright-Patterson Air Force Base, OH.

Braby, C. D., Harris, D., and Muir, H. C. (1993). A psychophysiological approach to the assessment of work underload, *Ergonomics*, *36*, 1035–1042.

Broadbent, D.E. (1958). *Perception and Communication.* London: Pergamon, London.

Brooks, L. (1968). Spatial and verbal components of the act of recall. *Canadian Journal of Psychology*, *22*, 349–368.

Brookings, J., Wilson, G. F., and Swain, C. (1996). Psychophysiological responses to changes in workload during simulated air traffic control. *Biological Psychology*, *42*, 361–378.

Brown, I. D. (1965). A comparison of two subsidiary tasks used to measure fatigue in car drivers. *Ergonomics*, 8, 467–473.

Brown, I. D., and Poulton, E. C. (1981). Measuring the spare mental capacity of car drivers by a subsidiary task. *Ergonomics*, *4*, 35–40.

Caldwell, J. A., Wilson, G. F., Centiguc, M., Gaillard, A. W. K., Gundel, A., Lagarde, D., Makeig, S., Myhre, G., and Wright, N. A. (1994). *Psychophysiological Assessment Methods.* (AGARD Advisory Report AGARD-AR-324). Paris, France: AGARD.

Casalie, J. G., and Wierwille, W. W. (1983). A comparison of rating scale, secondary-task, physiological, and primary-task workload estimation techniques in a simulated flight task emphasizing communications load. *Human Factors, 25*, 623–642.

Christ, R. E., Hill, S. G., Ayers, J. C., Iavecchia, H. M., Zaklad, A. L., and Bittner, A. C. (1993). *Application and Validation of Workload Assessment Techniques.* Technical Report 974. U.S. Army Research Institute for the Behavioral and Social Sciences, Alexandria, VA.

Chu, Y. Y., and Rouse, W. B. (1979). Adaptive allocation of decision making responsibility between human and computer in multitask situations. *IEEE Transactions on System, Man, and Cybernetics, SMC-9*, 769–778.

Clements, W. F. (1977). *Annotated Bibliography of Procedures Which Assess Primary Task Performance in Some Manner as the Basic Element of a Workload Measurement Procedure.* Technical Report 1104-2. Mountain View, CA: Systems Technology, Inc.

Comens, P., Reed, D., and Mette, M. (1987). Physiologic responses of pilots flying high-performance aircraft. *Aviation, Space, and Environmental Medicine, 58*, 205–210.

Comstock, J. R., and Arnegard, R. J. (1992). *The Multi-Attribute Task Battery for Human Operator Workload and Strategic Behavior Research.* NASA Tech. Memorandum 104174. NASA Langley Research Center, Hampton, VA.

Corker, K., Davis, L., Papazian, B., and Pew, R. (1986). *Development of an Advanced Task Analysis Methodology and Demonstration for Army-NASA Aircrew/Aircraft Integration.* Report No. 6124. Cambridge, MA: Bolt, Beranek & Newman.

Corwin, W. H., Sandry-Garza, D., Biferno, M. H., Boucek, G. P., Jr., Logan, A. L., Jonsson, J. E., and Metalis, S. A. (1989). *Assessment of Crew Workload Measurement Methods, Techniques and Procedures: Volume I—Process, Methods, and Results.* WRDC-TR-89-7006. Wright Research and Development Center, Air Force systems Command, Wright-Patterson Air Force Base, OH.

Damos, D. L. (1991). Dual-task methodology: Some common problems. In D. L. Damos, Ed., *Multiple Task Performance.* Washington, DC: Taylor & Francis, pp. 101–120.

Detro, S. D. (1985). Subjective assessment of pilot workload in the advanced fighter cockpit. *Proceedings of the Third Symposium on Aviation Psychology.* Columbus, OH: Ohio State University, Department of Aviation.

Dougherty, D. J., Emery, J. H., and Curtin, J. G. (1964). *Comparison of Perceptual Workload in Flying Standard Instrumentation and the Contact Analog Vertical Display.* JANAIR-D228-421-019. Ft. Worth, TX: Bell Helicopter (DTIC 610617).

Egelund, N. (1982). Spectral analysis of heart rate variability as an indicator of driver fatigue. *Ergonomics, 25*, 663–672.

Eggemeier, F. T., Crabtree, M. S., Zingg, J. J., Reid, G. B., and Shingledecker, C. A. (1982). Subjective workload assessment in a memory update task. *Proceedings of the 26th Human Factors Society Annual Meeting.* Santa Monica, CA: Human Factors Society, pp. 643–647.

Eggemeier, F. T., and Wilson, G. F. (1991). Subjective and performance-based assessment of workload in multi-task environments. In D. L. Damos, Ed., *Multiple Task Performance.* London: Taylor & Francis, pp. 217–278.

Eggemeier, F. T., Wilson, G. F., Kramer, A. F., and Damos, D. L. (1991). Workload assessment in multi-task environments. In D. Damos, Ed., *Multiple Task Performance*, London: Taylor & Francis, pp. 207–216.

Eisen, P. S., and Hendy, K. C. (1987). *A Preliminary Examination of Mental Workload, its Measurement and Prediction.* DCIEM No. 87-RR- 57. Downsview, Ont., Canada: Defence and Civil Institute of Environmental Medicine.

Endsley, M. A. (1993). Situation awareness and workload: Flip sides of the same coin. *Proceedings of the Seventh International Symposium on Aviation Psychology.* Columbus, OH: Ohio State University, Department of Aviation, pp. 906–911.

Ericsson, K. A., and Simon, H. A. (1980). Verbal reports as data. *Psychological Review, 87*, 215–251.

Fogarty, C., and Stern, J. A. (1989). Eye movements and blinks: Their relationship to higher cognitive processes. *International Journal of Psychophysiology, 8*, 35–42.

Fowler, B. (1994). P300 as a measure of workload during a simulated aircraft landing task. *Human Factors, 36*, 670–683.

Fracker, M. L. (1989). Attentional allocation in situation of awareness in map displays. *Proceedings of the Human Factors Society 33rd Annual Meeting.* Santa Monica, CA: Human Factors Society, pp. 1396–1399.

Fracker, M. L., and Wickens, C. D. (1989). Resources, confusions, and compatibility in dual-axis tracking: Display, controls, and dynamics. *Journal of Experimental Psychology: Human Perception and Performance, 15*, 80–96.

France, C., and Ditto, B. (1989). Cardiovascular responses to occupational stress and caffeine in telemarketing employees. *Psychosomatic Medicine, 51*, 145–151.

Gawron, V. J., Schflett, S. G., and Miller, J. C. (1989). Measures of in-flight workload. In R.S. Jensen, Ed., *Aviation Psychology.* Brookfield, MA: Gower, pp. 240–287.

Geiwitz, J., Klatzky, R. L., McCloskey, B. P. (1988). *Knowledge Acquisition for Expert Systems: Conceptual and Empirical Comparisons.* Santa Barbara, CA: Anacapa Sciences.

Glaser, R., Lesgold, A., Lajoie, S., Eastman, R., Greenberg, L. (1985). *Cognitive Task Analysis to Enhance Technical Skills Training and Assessment.* Contract F41689-83-C-0029. Air Force Human Resources Lab, Brooks AFB, TX.

Gopher, D. (1993). The skill of attention control: Acquisition and execution of attention strategies. In D. Meyer and S. Kornblum, Eds., *Attention and performance XIV.* Hillsdale, NJ: Lawrence Erlbaum, pp. 299–322.

Gopher, D., and Braune, R. (1984). On the psychophysics of workload: Why bother with subjective measures? *Human Factors, 26,* 519–532.

Gopher, D., and Donchin, E. (1986). Workload—An examination of the concept. In *Handbook of Perception and Human Performance: Volume II. Cognitive Processes and Performance.* K. R. Boff, L. Kaufman, and J. Thomas, Eds., New York: John Wiley, Chap. 41.

Gopher, D., and Kimchi, R. (1989). Engineering psychology. *Annual Review of Psychology, 40,* 431–455.

Gopher, D., and Navon, D. (1980). How is performance limited: Testing the notion of central capacity. *Acta Psychologica, 46,* 161–180.

Gopher, D., and North, R. (1977). Manipulating the conditions of training in time-sharing performance. *Human Factors, 19,* 583–593.

Grossman, P. (1992). Respiratory and cardiac rhythms as windows to central and autonomic biobehavioral regulation: Selection of window frames, keeping the panes clean and viewing the neural topography. *Biological Psychology, 34,* 131–162.

Guttentag, R. E. (1989). Age differences in dual-task performance: Procedures, assumptions, and results. *Developmental Review, 9,* 146–170.

Hamilton, D. B., Bierbaum, C. R., and Fulford, L. A. (1991). *Task Analysis/Workload (TAWL) User's Guide: Version 4.0.* Fort Rucker, AL: Anacapa Sciences.

Hancock, P. A., Mihaly, T., Rahimi, M., and Meshkati, N. (1988). A bibliographic listing of mental workload research. In P. A. Hancock and N. Meshkati, Eds., *Human Mental Workload.* New York: North-Holland, pp. 329–382.

Harms, L. (1986). Drivers' attentional response to environmental variations: A dual-task real traffic study. In A. G. Gale, M. H. Freeman, C. M. Haslegrave, P. Smith, and S. P. Taylor, Eds., *Vision in Vehicles.* Amsterdam: North-Holland, pp. 131–138.

Hart, S. G., and Staveland, L. E. (1988). Development of NASA-TLX (Task Load Index): Results of experimental and theoretical research. In P. A. Hancock and N. Meshkati, Eds., *Human Mental Workload.* Amsterdam: North Holland, pp. 139–183.

Helander, M. (1975). Physiological reactions of drivers as indicators of road traffic demand. In *Transportation Research Board's Driver Performance Studies.* Washington DC: Transportation Research Board. TRB/TRR-530, 1–17.

Hendy, K. C. (1995). Situation awareness and workload: Birds of a feather? In *Proceedings (CP-575) of the AGARD AMP Symposium on Situation Awareness: Limitations and Enhancements in the Aviation Environment.* Neuilly Sur Seine, France: Advisory Group for Aerospace Research & Development, pp. 21-1–21-7.

Hendy, K. C., Hamilton, K. M., and Landry, L. N. (1993). Measuring subjective workload: When is one scale better than many? *Human Factors, 35,* 579–601.

Hill, S. G., Iavecchia, H. P., Byers, J. C., Bittner, A. C., Jr., Zaklad, A. L., and Christ, R. E. (1992). Comparison of four subjective workload rating scales. *Human Factors, 34,* 429–440.

Hopkin, V. D. (1993). Verification and validation: Concepts, issues, and applications. In J. A. Wise, V. D. Hopkin, and P. Stager, Eds., *Verification and Validation of Complex Systems: Human Factors Issues.* New York: Springer-Verlag, pp. 9–34.

Howell, W.C. (1993). Engineering psychology in a changing world. *Annual Review of Psychology, 44,* 231–263.

Huey, B. M., and Wickens, C. D., Eds. (1993). *Workload Transition.* Washington, DC: National Academy Press.

Hunn, B.P. (1992). Enhancing the sensitivity of subjective ratings using video cued recall and graphed ratings. *Proceedings of the Fourteenth Biennial Applied Behavioral Sciences Symposium.* USAF Academy, CO, pp. 126–130.

Isreal, J. B., Chesney, G. L., Wickens, C. D., and Donchin, E. (1980). P300 and tracking difficulty: Evidence for multiple resources in dual task performance. *Psychobiology, 17,* 57–70.

Isreal, J., Wickens, C. D., Chesney, G., and Donchin, E. (1980). The event-related brain potential as a selective index of display monitoring load. *Human Factors,* 22, 211–224.

Itoh, Y., Hayashi, Y., Tsukui, I., and Saito, S. (1989). Heart rate variability and subjective mental workload in flight task validity of mental workload measurement using H.R.V. method. In M. J. Smith and G. Salvendy, Eds., *Work with Computers: Organizational, Management, Stress and Health Aspects.* Amsterdam: Elsevier, pp. 209–216.

Johannsen, G. (1979). Workload and workload measurement. In N. Moray, Eds., *Mental Workload, Theory and Measurement.* New York: Plenum, pp. 3–12.

Johannssen, G., Moray, N., Pew, R., Rasmussen, J., Sanders, A., and Wickens, C. (1979). Final report of experimental psychology group. In N. Moray Ed., *Mental Workload: Its Theory and Measurement.* New York: Plenum, pp. 101–114.

Johannsen, G., and Rouse, W. B. (1983). Studies of planning behavior of aircraft pilots in normal, abnormal, and emergency situation. *IEEE Transactions on Systems, Man, and Cybernetics, SMC-13*, 267–278.

Johnson, C. (1980). Heart rates in boat racers. *The Physician and Sportsmedicine, 8(6)*, 86–93.

Jorna, P. G. A. M. (1992). Spectral analysis of heart rate and psychological state: A review of its validity as a workload index. *Biological Psychology, 34*, 237–258.

Kahneman D. (1973). *Attention and Effort.* Englewood Cliffs, NJ: Prentice Hall.

Kalsbeek, J. W. H., and Sykes, R. N. (1967). Objective measurement of mental load. In A. F. Sanders, Eds., *Attention and Performance.* Amsterdam: North-Holland.

Kantowitz, B. H. (1992). Selecting measures for human factors research. *Human Factors, 34*, 387–398.

Kantowitz, B. H., and Casper, P. A. (1988). Human workload in aviation. In E. L. Wiener and D. C. Nagel, Eds., *Human Factors in Aviation.* New York: Academic Press, pp. 171–188.

Kirwin, B., and Ainsworth, L. K. (1992). *A Guide to Task Analysis.* London: Taylor & Francis.

Kitto, C. M., and Boose, J. H. (1989). Selecting knowledge acquisition tools and strategies based on application characteristics. *International Journal of Man-Machine Studies, 31*, 149–160.

Klein, G. A. (1990). Knowledge engineering: Beyond expert systems. *Inf. Design Technology, 16*, 27–41.

Kramer, A. (1991). Physiological metrics of mental workload: A review of recent progress. In D. Damos, Ed., *Multiple Task Performance.* London: Taylor & Francis, pp. 279–328.

Kramer, A. F., Sirevaag, E., and Braune, R. (1987). A psychophysiological assessment of operator workload during simulated flight missions. *Human Factors, 29*, 145–160.

Kramer, A. F., Trejo, D., and Humphrey, D. (1995). Assessment of mental workload with task-irrelevant auditory probes. *Biological Psychology*, 83–101.

Kreifeldt, J. G. (1980). Cockpit displayed traffic information and distributed management in air traffic control. *Human Factors, 22*, 671–691.

Krol, J. P. (1971). Variations in ATC-workload as a function of variations in cockpit workload. *Ergonomics, 14*, 585–590.

Kuperman, G. G. (1985). *Projective Application of the Subjective Workload Assessment Technique to Advanced Helicopter Crew System Designs.* AFAMRL-TR-85-014. Air Force Aerospace Medical Research Laboratory, Wright-Patterson Air Force Base, OH.

Laughery, K. R. Jr. (1989). Task network modeling as a basis for analyzing operator workload. *Proceedings of the Human Factors Society 33rd Annual Meeting.* Santa Monica, CA: Human Factors Society, pp. 110–114.

Lecret, F., and Pottier, M. (1971). La vigilance facteur de securite dans la conduite automobile. *Le Travial Humain, 34*, 51–68.

Lee, J., and Moray, N. (1992). Trust, control strategies and allocation of function in human-machine systems. *Ergonomics, 35*, 1243–1270.

Leplat, J., and Welford, A. T. (1978). Special issue: Symposium on mental workload. *Ergonomics, 21*, 141–233.

Levison, W. H. (1979). A model for mental workload in tasks requiring continuous information processing. In N. Moray, Ed., *Mental Workload, Theory and Measurement.* New York: Plenum, pp. 189–218.

Lysaght, R. J., Hill, S. G., Dick, A. O., Plamondon, B. D., Linton, P. M., Wierwille, W. W., Zaklad, A. L., Bittner, A. C. Jr., and Wherry, R. J. Jr. (1989). *Operator Workload: Comprehensive Review and Evaluation of Workload Methodologies.* ARI Technical Report 851. U.S. Army Research Institute for the Behavioral and Social Sciences, Alexandria, VA. AD A212 879.

Madni, A.M., and Lyman, J. (1983). Model-based estimation and prediction of task-imported mental workload. *Proceedings of the Human Factors Society 27th Annual Meeting.* Santa Monica, CA: Human Factors Society, pp. 314–318.

Masline, P. J., and Biers, D. W. (1987). An examination of projective versus post-task subjective workload ratings for three psychometric scaling techniques. *Proceedings of the Human Factors Society 31st Annual Meeting.* Santa Monica: Human Factors Society, pp. 77–80.

McCoy, T. M., Derrick, W. L., and Wickens, C. D. (1983). Workload assessment metrics—What happens when they dissociate? *Proceedings of the Second Aerospace Behavioral Engineering Technology Conference,* P-132. Warrendale, PA: Society of Automotive Engineers, pp. 37–42.

McCracken, J. H., and Aldrich, T. B. (1984). *Analyses of Selected LHX Mission Functions: Implications for Operator Workload and System Automation Goals.* Technical Note ASI 479-024-84(B). Fort Rucker, AL: Anacapa Sciences.

McLeod, P. (1977). A dual-task response modality effect: Support for multiprocessor models of attention. *Quarterly Journal of Experimental Psychology, 29*, 651–667.

McMillan, G. R., Beevis, D., Salas, E., Strub, M. H., Sutton, R., and Van Breda, L., Eds. (1989). *Applications of Human Performance Models to System Design*. New York: Plenum.

Moray, N., Ed. (1979). *Mental Workload, Theory and Measurement*. New York: Plenum.

Moray, N. (1988). Mental workload since 1979. *International Review of Ergonomics, 2*, 123–150.

Moray, N., Dessouky, M. I., Kijowski, B. A., and Adapathya, R. (1991). Strategic behavior, workload, and performance in task-scheduling. *Human Factors, 33*, 607–629.

Moroney, W. F., Biers, D. W., and Eggemeier, F. T. (1995). Some Measurement and Methodological Considerations in the Application of Subjective Workload Measurement Techniques. *The International Journal of Aviation Psychology, 5*, 87–106.

Muckler, F. A., and Seven, S. A. (1992). Selecting performance measures: "objective" versus "subjective" measurement. *Human Factors, 34*, 441–456.

Mulder, G. (1980). *The Heart of Mental Effort*. Ph.D. Thesis. University of Groningen, Groningen.

Mulder, L. J. M. (1988). *Assessment of Cardiovascular Reactivity by Means of Spectral Analysis*. Ph.D. Thesis. Rijksuniversiteit Groningen, Groningen.

Mulder, L. J. M. (1992). Measurement and analysis methods of heart rate and respiration for use in applied environments. *Biological Psychology, 34*, 205–236.

Navon, D., and Miller, J. (1987). The role of outcome conflict in dual-task interference. *Journal of Experimental Psychology: Human Perception and Performance, 13*, 435–448.

Netick, A., and Klapp, S.T. (1994). Hesitations in manual tracking: A single-channel limit in response programming. *Journal of Experimental Psychology: Human Perception and Performance, 20*, 766–782.

Nicholson, A. N., Hill, L. E., Borland, R. G., and Ferres, H. M. (1970). Activity of the nervous system during the let-down, approach and landing: A study of short duration high workload. *Aerospace Medicine, 41*, 436–446.

Nisbett, R. E., and Wilson, T. D. (1977). Telling more than we can know: Valid reports on mental processes. *Psychological Review, 84*, 231–259.

Norman, D. A., and Bobrow, D. G. (1975). On data-limited and resource-limited processes. *Cognitive Psychology, 7*, 44–64.

North, R. A., and Riley, V. A. (1989). A predictive model of operator workload. In G. R. McMillan, D. Beevis, E. Salas, M. H. Strub, R. Sutton, and L. Van Breda, Eds., *Applications of Human Performance Models to System Design*. New York: Plenum, pp. 81–90.

Nygren, T.E. (1991). Psychometric properties of subjective workload measurement techniques: Implications for their use in the assessment of perceived mental workload. *Human Factors, 33*, 17–31.

O'Donnell, R. D., and Eggemeier, F. T. (1986). Workload assessment methodology. In K. R. Boff, L. Kaufman, and J. Thomas, Eds., *Handbook of Perception and Human Performance: Volume II. Cognitive Processes and Performance*. New York: John Wiley, Chap. 42.

Ogden, G. D., Levine, J. M., and Eisner, E. J. (1970). Measurement of workload by secondary tasks. *Human Factors, 21*, 529–548.

Olson, J. R., Biolsi, K. J. (1991). Techniques for representing expert knowledge. In K. A. Ericsson and J. Smith, Eds., *Toward a General Theory of Expertise*. Cambridge: Cambridge University Press, pp. 240–85.

Parks, D., and Boucek, G. (1989). Workload prediction, diagnosis, and continuing challenges. In G. R. McMillan, D. Beevis, E. Salas, M. H. Strub, R. Sutton, and L. Van Breda, L., Eds., *Applications of Human Performance Models to System Design*. New York: Plenum, pp. 47–64.

Parasuraman, R. and Mouloua, M., Eds. (1996). *Automation and Human Performance*. Hillsdale, NJ: Lawrence Erlbaum.

Pew, R. W. (1994). Situation awareness: The buzzword of the '90s. *Gateway, V(1)*, 1–4.

Pfaff, U., Fruhstorfer, H., and Peter, H. H. (1976). Changes in eye-blink duration and frequency during car driving. *Pfugers Archives, 363*, R 21.

Porges, S. W., and Byrne, E. A. (1992). Research methods for measurement of heart rate and respiration. *Biological Psychology, 34*, 93–130.

Raby, M., Wickens, C. D., and Marsh, R. (1990). *Investigation of Factors Comprising a Model of Pilot Decision Making: Part 1. Cognitive Biases in Workload Management Strategy.* Tech. Rep. ARL-90-7/SCEEE-90-1. Savoy, IL: University of Illinois, Aviation Research Laboratory.

Reid, G. B., and Colle, H. A. (1988). Critical SWAT values for predicting operator overload. *Proceedings of the 32nd Human Factors Society Annual Meeting*. Santa Monica, CA: Human Factors Society, pp. 1414–1418.

Reid, G. B., and Nygren, T. E. (1988). The Subjective Workload Assessment Technique: A scaling procedure for measuring mental workload. In P. A. Hancock and N. Meshkati, Eds., *Human Mental Workload*. Amsterdam: North Holland, pp. 185–218.

Reid, G. B., Potter, S. S., and Bressler, J. R. (1987). *User's Guide for the Subjective Workload Assessment Technique (SWAT)*. Technical Report AAMRL-TR-87. Armstrong Aerospace Medical Research Laboratory, Wright-Patterson Air Force Base, OH.

Reid, G. B., Shingledecker, C. A., Hockenberger, R. L., and Quinn, T. J. (1984). A projective application of the subjective workload assessment technique. *Proceedings of the 1984 IEEE National Aerospace and Electronics Conference*, pp. 824–826.

Rokicki, S. M. (1987). Heart rate averages as workload/fatigue indications during OT&E. *Proceedings of the Human Factors Society 31st Annual Meeting.* Santa Monica, CA: Human Factors Society, pp. 784–785.

Roscoe, A. H. (1975). Heart rate monitoring of pilots during steep-gradient approaches. *Aviation, Space, and Environmental Medicine, 11*, 1410–1413.

Roscoe, A. H. (1984). Assessing Pilot Workload in Flight. *AGARD Conference Proceedings No.373-Flight Test Techniques.* Neuilly Sur Seine, France: Advisory Group for Aerospace Research & Development, pp. 12-1–12-13.

Roscoe, A. H. (1987). In-flight assessment of workload using pilot ratings and heart rate. In *The Practical Assessment of Pilot Workload.* AGARD-AG-282. Neuilly Sur Seine, France: Advisory Group for Aerospace Research & Development, pp. 78–82.

Roscoe, A. H. (1992). Assessing pilot workload. Why measure heart rate, HRV and respiration? *Biological Psychology, 34*, 259–288.

Roscoe, A. H., and Ellis, G. A. (1990, March). *A Subjective Rating Scale for Assessing Pilot Workload in Flight: A Decade of Practical Use.* Technical Report TR 90019. Farnborough, UK: Royal Aerospace Establishment.

Rueb, J. D., Vidulich, M. A., and Hassoun, J. A. (1994). Use of workload redlines: A KC-135 crew-reduction application. *The International Journal of Aviation Psychology, 4*, 47–64.

Ruggiero, F., and Fadden, D. M. (1987). Pilot subjective evaluation of workload during a flight test certification programme. In *The Practical Assessment of Pilot Workload.* AGARDograph No. 282. Neuilly Sur Seine, France: Advisory Group for Aerospace Research & Development, pp. 32–36.

Sarno, K. J., and Wickens, C. D. (1995). Role of multiple resources in predicting time-sharing efficiency: Evaluation of three workload models in a multiple-task setting. *The International Journal of Aviation Psychology, 5*, 107–130.

Schmidt, J. K., and Nicewonger, H. M. (1988). *An Annotated Bibliography on Operator Mental Workload Assessment.* Technical Note 7-88. U.S. Army Human Engineering Laboratory, Aberdeen Proving Ground, MD.

Segal, L. D., and Wickens, C. D. (1991). *TASKILLAN II: A Study of Pilot Strategies for Workload Management.* Technical Report ARL-91-1/NASA-91-1. Savoy, IL: University of Illinois, Aviation Research Laboratory.

Shingledecker, C. A. (1984). *A Task Battery for Applied Human Performance Assessment Research.* Technical Report AFAMRL-TR-84-071. Air Force Aerospace Medical Research Laboratory, Wright-Patterson Air Force Base, OH.

Shingledecker, C.A. (1987). In-flight workload assessment using embedded secondary radio communications tasks. In *The Practical Assessment of Pilot Workload.* AGARDograph No. 282. Neuilly Sur Seine, France: Advisory Group for Aerospace Research & Development, pp. 11–14.

Sirevaag, E. J., Kramer, A. F., Wickens, C. D., Reisweber, M., Strayer, D. L., and Grenell (1993). Assessment of pilot performance and mental workload in rotary wing aircraft. *Ergonomics, 36*, 1121–1140.

Sterman, M. B., and Mann, C. A. (1995). Concepts and applications of EEG analysis in aviation performance evaluation. *Biological Psychology, 40*, 115–130.

Sterman, M. B., Schummer, G. J., Dushenko, T. W., and Smith, J. C. (1988). Electroencephalographic correlates of pilot performance: Simulation and in-flight studies. *AGARD Conference Proceedings No. 432, Electric and Magnetic Activity of the Central Nervous System: Research and Clinical Applications in Aerospace Medicine.* Neuilly Sur Seine, France: Advisory Group for Aerospace Research & Development, pp. 31-1–31-16.

Stern, J. A., and Dunham, D. N. (1990). The ocular system. In J. T. Cacioppo and L. G. Tassinary, Eds., *Principles of Psychophysiology.* Cambridge: Cambridge University Press, pp. 513–553.

Taggart, P., Gibbons, D., and Somerville, W. (1969). Some effects of motor-car driving on the normal and abnormal heart. *British Medical Journal, 4*, 130–134.

Taylor, R. M. (1990, April). Situation awareness rating technique (SART): The development of a tool for aircrew systems design. In *Situation Awareness in Aerospace Operations.* AGARD-CP-478. Neuilly Sur Seine, France: Advisory Group for Aerospace Research & Development, AD-A223939, pp. 3-1–3-17.

Tsang, P. S., Shaner, T. L., and Schnopp-Wyatt, E. N. (1995). *Age, Attention, Expertise, and Time-Sharing Performance.* Technical Report EPL-95-1. Dayton: Wright State University, Engineering Psychology Laboratory.

Tsang, P. S., and Velazquez, V. L. (1996). Diagnosticity and multidimensional subjective workload ratings, *Ergonomics, 39*, 358–381.

Tsang, P. S., and Velazquez, V. L. (1993). Subjective workload profile. *Proceedings of the 7th International Symposium on Aviation Psychology.* Columbus, OH: Ohio State University Association of Aviation Psychologists, pp. 859–864.

Tsang, P. S., Velazquez, V. L., and Vidulich, M. A. (1996). The viability of resource theories in explaining time-sharing performance. *Acta Psychologica, 91*, 175–206.

Tsang, P. S., and Vidulich, M. A. (1994). The roles of immediacy and redundancy in relative subjective workload assessment. *Human Factors, 36*, 503–513.

Tsang, P. S., and Wickens, C. D. (1988). The structural constraints and strategic control of resource allocation. *Human Performance, 1*, 45–72.

Tulga, M. K., and Sheridan, T. B. (1980). Dynamic decisions and workload in multitask supervisory control. *IEEE Transactions and Systems, Man, and Cybernetics*, SMC-10, 217–232.

Veltman, J. A., and Gaillard, A. W. K. (1996). Physiological indices of workload in a simulated flight task. *Biological Psychology, 42*, 323–342.

Vicente, K. J., Thornton, D. C., and Moray, N. (1987). Spectral analysis of sinus arrhythmia: A measure of mental effort. *Human Factors, 29*, 171–182.

Vidulich, M. A. (1988). The cognitive psychology of subjective mental workload. In P. A. Hancock and N. Meshkati, Eds., *Human Mental Workload.* New York: North-Holland, pp. 219–229.

Vidulich, M. A., and Bortolussi, M. R. (1988). Control configuration study. *Proceedings of the American Helicopter Society National Specialist's Meeting: Automation Applications for Rotocraft.* Atlanta Southeast Region AHS, pp. 1–10.

Vidulich, M. A., Stratton, M., Crabtree, M., and Wilson, G. (1994). Performance-based and physiological measures of situational awareness. *Aviation, Space, and Environmental Medicine, 65*(5, Supplement), A7-A12.

Vidulich, M. A., and Tsang, P. S. (1986). Techniques of subjective workload assessment: A comparison of SWAT and the NASA-Bipolar methods. *Ergonomics, 29*, 1385–1398.

Vidulich, M. A., and Tsang, P. S. (1987). Absolute magnitude estimation and relative judgment approaches to subjective workload assessment. *Proceedings of the Human Factors Society 31st Annual Meeting.* Santa Monica, CA: Human Factors Society, pp. 1057–1061.

Vidulich, M. A., Ward, G. F., and Schueren, J. (1991). Using subjective workload dominance (SWORD) technique for projective workload assessment. *Human Factors, 33*, 677–692.

Vidulich, M. A., and Wickens, C. D. (1986). Causes of dissociation between subjective workload measures and performance: Caveats for the use of subjective assessments. *Applied Ergonomics, 17*, 291–296.

Warm, J. S., and Parasuraman, R., Eds. (1987). Vigilance: Basic and applied. *Human Factors, 29*, 623–740.

Welford, A. T. (1978). Mental workload as a function of demand, capacity, strategy and skill. *Ergonomics, 21*, 151–167.

Wherry, R. J. (1976). The human-operator simulator—HOS. In T. B. Sheridan and G. Johansen, Eds., *Monitoring and Supervisory Control.* New York: Plenum.

Wickens, C. D. (1980). The structure of attentional resources. In R. S. Nickerson, Ed., *Attention and Performance VIII.* Hillsdale, NJ: Lawrence Erlbaum Associates, pp. 239–257.

Wickens, C. D. (1984). Processing resources in attention. In R. Parasuraman and D. R. Davies, Eds., *Varieties of Attention.* San Diego, CA: Academic, pp. 63–102.

Wickens, C. D. (1986). The effects of control dynamics on performance. In K. R. Boff, L. Kaufman, and J. Thomas, Eds., *Handbook of Perception and Human Performance: Volume II. Cognitive Processes and Performance.* New York: John Wiley, Chap. 39.

Wickens, C. D. (1987). Attention. In P. A. Hancock and N. Meshkati, Eds., *Human Mental Workload.* New York: North-Holland, pp. 29–80.

Wickens, C. D. (1989). Models of multitask situations. In G. R. McMillan, D. Beevis, E. Salas, M. H. Strub, R. Sutton, and L. Van Breda, Eds., *Applications of Human Performance Models to System Design.* New York: Plenum, pp. 259–273.

Wickens, C. D. (1991). Processing resources and attention. In D. Damos, Ed., *Multiple Task Performance.* London: Taylor and Francis, pp. 1–34.

Wickens, C. D. (1992). *Engineering Psychology and Human Performance*, Second Edition. New York: Harper Collins.

Wickens, C. D. (1995). Situation awareness: Impact of automation and display technology. In AGARD conference Proceedings 575: *Situation Awareness: Limitations and Enhancements in the Aviation Environment.* Neuilly Sur Seine, France: Advisory Group for Aerospace Research & Development, pp. K2-1–K2-13.

Wickens, C. D., Braune, R., and Stokes, A. (1987). Age differences in the speed and capacity of information processing: 1. A dual-task approach. *Psychology and Aging, 2*, 70–78.

Wickens, C. D., Gill, R., Kramer, A., Ross, W., and Donchin, E. (1981, October). The processing demands of higher order of manual control. *Proceedings of the 17th Annual Conference on Manual Control* (Jet Propulsion Lab, La Canada), CA81-95.

Wickens, C. D., and Gopher, D. (1977). Control theory measures of tracking as indices of attention allocation strategies. *Human Factors, 30*, 599–616.

Wickens, C. D., Hyman, F., Dellinger, J., Taylor, H., and Meador, M. (1986). The Sternberg memory search task as an index of pilot workload. *Ergonomics, 29*, 1371–1383.

Wickens, C. D., Kramer, A. F., Vanasse, L., and Donchin, E. (1983). The performance of concurrent tasks: A psychophysiological analysis of the reciprocity of information processing resources. *Science*, *221*, 1080–1082.

Wickens, C. D., Larish, I., and Contorer, A. (1989). Predictive performance models and multiple task performance. *Proceedings of the Human Factors Society 33rd Annual Meeting*. Santa Monica, CA: Human Factors Society, pp. 96–100.

Wickens, C. D., Sandry, D., and Vidulich, M. A. (1983). Compatibility and resource competition between modalities of input, central processing, and output. *Human Factors*, *25*, 227–248.

Wickens, C. D., and Yeh, Y. Y. (1986). A multiple resource model of workload prediction and assessment. *Proceedings of the IEEE Conference*, pp. 1044–1048.

Wiener, E. L. (1985). Beyond the sterile cockpit. *Human Factors*, *27*, 75–90.

Wiener, E. L. (1988). Cockpit automation. In E. L. Wiener and D. C. Nagel, Eds., *Human Factors in Aviation*. San Diego: Academic, pp. 433–462.

Wiener, E. L., and Curry, R. E. (1980). Flight-deck automation: Promises and problems. *Ergonomics*, *23*, 995–1011.

Wierwille, W. W., and Casali, J. G. (1983). A validated rating scale for global mental workload measurement applications. *Proceedings of the Human Factors Society 27th Annual Meeting*. Santa Monica, CA: Human Factors Society, pp. 129–133.

Wierwille, W. W., and Connor, S. A. (1983). Evaluation of twenty workload assessment measures using a psychomotor task in a moving-base aircraft simulator. *Human Factors*, *25*, 1–16.

Wierwille, W. W., and Eggemeier, F. T. (1993). Recommendations for mental workload measurement in a test and evaluation environment. *Human Factors*, *35*, 263–282.

Wierwille, W. W., and Williges, B. H. (1980). *An Annotated Bibliography on Operator Mental Workload Assessment*. NATC-TR-27R-80. Patuxent River, MD: Naval Air Test Center.

Wierwille, W. W., Rahimi, M., and Casali, J. G. (1985). Evaluation of 16 measures of mental workload using a simulated flight task emphasizing mediational activity. *Human Factors*, *25*, 1–16.

Williges, R., and Wierwille, W. W. (1979). Behavioural measures of aircrew mental workload. *Human Factors*, *21*, 549–574.

Wilson, G. F. (1992). *Progress in the Psychophysiological Assessment of Workload* AL-TR-1992-0007. Armstrong Laboratory, Wright-Patterson AFB, OH.

Wilson, G. F. (1993). Air-to-ground training missions: a psychophysiological workload analysis. *Ergonomics*, *36*, 1071–1087.

Wilson, G. F., and Eggemeier, F. T. (1991). Physiological measures of workload in multi-task environments. In D. Damos, Ed., *Multiple-Task Performance*. London: Taylor and Francis, pp. 329–360.

Wilson, G. F., and Fullenkamp, P. (1990). A comparison of pilot and workload during training missions using psychophysiological data. *Proceedings of the Western European Association of Aviation Psychologists, Volume II: Stress and Error in Aviation*, pp. 27–34. Aldershot Hants, UK: Avery Technical.

Wilson, G. F., Fullenkamp, P., and Davis, I. (1994). Evoked potential, cardiac, blink and respiration measures of workload in air-to-ground missions. *Aviation, Space, and Environmental Medicine*, *65*, 100–105.

Wilson, G. F., and McCloskey, K. (1988). Using probe evoked potentials to determine information processing demands. *Proceedings of the Human Factors Society 32nd Annual Meeting*. Santa Monica, CA: Human Factors Society, pp. 1400–1403.

Yeh, Y. Y., and Wickens, C. D. (1988). Dissociation of performance and subjective measures of workload. *Human Factors*, *30*, 111–120.

CHAPTER 14
JOB AND TEAM DESIGN

Gina J. Medsker
School of Business Administration
University of Miami
Coral Gables, FL 33124 USA

Michael A. Campion
Krannert School of Management
Purdue University
West Lafayette, IN 47907-1287 USA

14.1 INTRODUCTION

14.1.1 Job Design

Job design is an aspect of managing organizations that is so commonplace it often goes unnoticed. Most people realize the importance of job design when an organization or new plant is starting up, and some recognize the importance of job design when organizations are restructuring or changing processes. But fewer people realize that job design may be affected as organizations change markets or strategies, managers use their discretion in the assignment of tasks on a daily basis, people in the jobs or their managers change, the work force or labor markets change, or there are performance, safety, or satisfaction problems. Fewer yet realize that job design change can be used as an intervention to enhance organizational goals (Campion and Medsker, 1992).

It is clear that many different aspects of an organization influence job design, especially an organization's structure, technology, processes, and environment. These influences are beyond the scope of this chapter, but they are dealt with in other references (e.g., Davis, 1982; Davis and Wacker, 1982; see also Chapter 18). These influences impose constraints on how jobs are designed and will play a major role in any practical application. However, it is the assumption of this chapter that considerable discretion exists in the design of jobs in most situations, and the job (defined as a set of tasks performed by a worker) is a convenient unit of analysis in both developing new organizations or changing existing ones (Campion and Medsker, 1992).

The importance of job design lies in its strong influence on a broad range of important efficiency and human resource outcomes. Job design has predictable consequences for outcomes including the following (Campion and Medsker, 1992):

- productivity
- quality
- job satisfaction
- training times
- intrinsic work motivation
- staffing
- error rates
- accident rates
- mental fatigue
- physical fatigue
- stress
- mental ability requirements
- physical ability requirements
- job involvement
- absenteeism
- medical incidents
- turnover
- compensation rates

According to Louis Davis, one of the most prolific writers on job design in the engineering literature over the last 35 years, many of the personnel and productivity problems in industry may be the direct result of the design of jobs (Davis, 1957; Davis, Canter and Hoffman, 1955; Davis and Taylor, 1979; Davis and Valfer, 1965; Davis and Wacker, 1982, 1987). Unfortunately, people mistakenly view the design of jobs as technologically determined and inalterable. However, job designs are actually social inventions. They reflect the values of the era in which they were constructed. These values include the economic goal of minimizing immediate costs (Davis et al., 1955; Taylor, 1979) and theories of human motivation (Steers and Mowday, 1977; Warr and Wall, 1975). These values, and the designs they influence, are not immutable givens, but are subject to modification (Campion and Medsker, 1992; Campion and Thayer, 1985).

The question then becomes: What is the best way to design a job? In fact, there is no single best way. There are several major approaches to job design, each derived from a different discipline and reflecting different theoretical orientations and values. This chapter describes these approaches, their costs and benefits, and tools and procedures for developing and assessing jobs in all types of organizations. It highlights trade-offs which must be made when choosing among different approaches to job design. This chapter also compares the design of jobs for individuals working independently to the design of work for teams, which is an alternative to designing jobs at the level of individual workers. This chapter presents the advantages and disadvantages of designing work around individuals compared to designing work for teams and provides advice on implementing and evaluating the different work design approaches.

14.1.2 Team Design

The major approaches to job design typically focus on designing jobs for individual workers. However, the approach to work design at the level of the group or team, rather than at the level of individual workers, is gaining substantially in popularity, and many U.S. organizations are experimenting with teams (Guzzo and Shea, 1992; Hoerr, 1989; Majchrzak, 1988). New manufacturing systems (e.g., flexible, cellular) and advancements in our understanding of team processes not only allow designers to consider the use of work teams, but often seem to encourage the use of team approaches (Gallagher and Knight, 1986; Majchrzak, 1988).

In designing jobs for teams, one assigns a task or set of tasks to a team of workers, rather than to an individual, and considers the team to be the primary unit of performance. Objectives and rewards focus on team, not individual, behavior. Depending on the nature of its tasks, a team's workers may be performing the same tasks simultaneously or they may break tasks into subtasks to be performed by individuals within the team. Subtasks can be assigned on the basis of expertise or interest, or team members might rotate from one subtask to another to provide variety and increase breadth of skills and flexibility in the work force (Campion and Medsker, 1992; Campion, Cheraskin, and Stevens, 1994).

Some tasks are of a size, complexity, or, otherwise seem to naturally fit into a team job design, whereas others may seem to be appropriate only at the individual job level. In many cases, though, there may be a considerable degree of choice regarding whether one organizes work around teams or individuals. In such situations, the designer should consider advantages and disadvantages of the use of the job and team design approaches with respect to an organization's goals, policies, technologies, and constraints (Campion, Medsker, and Higgs, 1993).

14.2 JOB DESIGN APPROACHES

This chapter adopts an interdisciplinary perspective on job design. Interdisciplinary research on job design has shown that different approaches to job design exist. Each is oriented toward a particular subset of outcomes, each has disadvantages as well as advantages, and trade-offs among approaches are required in most job design situations (Campion, 1988, 1989; Campion and Berger, 1990; Campion and McClelland, 1991, 1993; Campion and Thayer, 1985). The four major approaches to job design are reviewed below. Table 14.1 summarizes the job design approaches and Table 14.2 provides specific recommendations. The team design approach is reviewed in Section 14.3.

14.2.1 Mechanistic Job Design Approach

14.2.1.1 Historical Development

The historical roots of job design can be traced back to the idea of the division of labor, which was very important to early thinking on the economies of manufacturing (Babbage, 1835; Smith, 1776). Division of labor led to job designs characterized by specialization and simplification. Jobs designed in this fashion had many advantages, including reduced learning time, saved time from not having to change tasks or tools, increased proficiency from repeating tasks, and development of specialized tools and equipment.

A very influential person for this perspective was Frederick Taylor (Taylor, 1911; Hammond, 1971). He explicated the principles of scientific management which encouraged the study of jobs to determine the "one best way" to perform each task. Movements of skilled workmen were studied using a stopwatch and simple analysis. The best and quickest methods and tools were selected, and all workers were trained to perform the job the same way. Standard performance levels were set, and incentive pay was tied to the standards. Gilbreth also contributed to this design approach (Gilbreth, 1911). With time and motion study, he tried to eliminate wasted movements by the appropriate design of equipment and placement of tools and materials.

Surveys of industrial job designers indicate that this "mechanistic" approach to job design has been the prevailing practice throughout this century (Davis et al., 1955; Taylor, 1979). These characteristics are also the primary focus of many modern day writers on job design (e.g., Mundel, 1985; Niebel, 1988). The discipline base for this approach is early or "classic" industrial engineering.

14.2.1.2 Design Recommendations

Table 14.2 provides a brief list of statements which describe the essential recommendations of the mechanistic approach. In essence, jobs should be studied to determine the most efficient work methods and techniques. The total work in an area (e.g., department) should be broken down into highly specialized jobs assigned to different employees. The tasks should be simplified so skill requirements are minimized. There should also be repetition in order to gain improvement from practice. Idle time should be minimized. Finally, activities should be automated or assisted by automation to the extent possible and economically feasible.

14.2.1.3 Advantages and Disadvantages

The goal of this approach is to maximize efficiency, both in terms of productivity and utilization of human resources. Table 14.1 summarizes some human resource advantages and disadvantages that have been observed in research. Jobs designed according to the mechanistic approach are easier and less expensive to staff. Training times are reduced. Compensation requirements may be less because skill and responsibility are reduced. And because mental demands are less, errors may be less common. Disadvantages include the

Table 14.1 Advantages and Disadvantages of Different Job Design Approaches

APPROACH/Discipline Base (example references)	Illustrative Recommendations	Illustrative Advantages	Illustrative Disadvantages
MECHANISTIC/Classic Industrial Engineering (Gilbreth, 1911; Niebel, 1988; Taylor, 1911)	Increase in specialization simplification repetition automation Decrease in spare time	Decrease in training staffing difficulty making errors mental overload and fatigue mental skills and abilities compensation	Increase in absenteeism boredom Decrease in satisfaction motivation
MOTIVATIONAL/Organizational Psychology (Hackman and Oldham, 1980; Herzberg, 1966)	Increase in variety autonomy significance skill usage participation feedback recognition growth achievement	Increase in satisfaction motivation involvement performance customer service catching errors Decrease in absenteeism turnover	Increase in training staffing difficulty making errors mental overload and fatigue stress mental skills and abilities compensation
PERCEPTUAL-MOTOR/Experimental Psychology, Human Factors (Salvendy, 1987; Sanders and McCormick, 1987)	Increase in lighting quality display and control quality user-friendly equipment Decrease in information processing requirements	Decrease in making errors accidents mental overload and fatigue stress training staffing difficulty compensation mental skills and abilities	Increase in boredom Decrease in satisfaction
BIOLOGICAL/Physiology, Biomechanics, Ergonomics (Astrand and Rodahl, 1977; Grandjean, 1980; Tichauer, 1978)	Increase in seating comfort postural comfort Decrease in strength requirements endurance requirements environmental stressors	Decrease in physical abilities physical fatigue aches and pains medical incidents	Increase in financial cost inactivity

Note: Advantages and disadvantages based on findings in previous interdisciplinary research (Campion, 1988, 1989; Campion and Berger, 1990; Campion and McClelland, 1991, 1993; Campion and Thayer, 1985). Table adopted from Campion and Medsker (1992).

Table 14.2 Multimethod Job Design Questionnaire

(Specific Recommendations from Each Job Design Approach)
Instructions: Indicate the extent to which each statement is descriptive of the job using the scale below. Circle answers to the right of each statement.

Please Use the Following Scale:

(5) Strongly agree
(4) Agree
(3) Neither agree nor disagree
(2) Disagree
(1) Strongly disagree
( ) Leave blank if do not know or not applicable

Mechanistic Approach					
1. Job specialization: The job is highly specialized in terms of purpose, tasks, or activities.	1	2	3	4	5
2. Specialization of tools and procedures: The tools, procedures, materials, etc., used on this job are highly specialized in terms of purpose.	1	2	3	4	5
3. Task simplification: The tasks are simple and uncomplicated.	1	2	3	4	5
4. Single activities: The job requires you to do only one task or activity at a time.	1	2	3	4	5
5. Skill simplification: The job requires relatively little skill and training time.	1	2	3	4	5
6. Repetition: The job requires performing the same activity(s) repeatedly.	1	2	3	4	5
7. Spare time: There is very little spare time between activities on this job.	1	2	3	4	5
8. Automation: Many of the activities of this job are automated or assisted by automation.	1	2	3	4	5
Motivational Approach					
9. Autonomy: The job allows freedom, independence, or discretion in work scheduling, sequence, methods, procedures, quality control, or other decision making.	1	2	3	4	5
10. Intrinsic job feedback: The work activities themselves provide direct and clear information as to the effectiveness (e.g., quality and quantity) of job performance.	1	2	3	4	5
11. Extrinsic job feedback: Other people in the organization, such as managers and co-workers, provide information as to the effectiveness (e.g., quality and quantity) of job performance.	1	2	3	4	5
12. Social interaction: The job provides for positive social interaction such as team work or co-worker assistance.	1	2	3	4	5
13. Task/goal clarity: The job duties, requirements, and goals are clear and specific.	1	2	3	4	5
14. Task variety: The job has a variety of duties, tasks, and activities.	1	2	3	4	5

Table 14.2 (*Continued*)

15. Task identity: The job requires completion of a whole and identifiable piece of work. It gives you a chance to do an entire piece of work from beginning to end.	1	2	3	4	5
16. Ability/skill level requirements: The job requires a high level of knowledge, skills, and abilities.	1	2	3	4	5
17. Ability/skill variety: The job requires a variety of knowledge, skills, and abilities.	1	2	3	4	5
18. Task significance: The job is significant and important compared with other jobs in the organization.	1	2	3	4	5
19. Growth/learning: The job allows opportunities for learning and growth in competence and proficiency.	1	2	3	4	5
20. Promotion: There are opportunities for advancement to higher level jobs.	1	2	3	4	5
21. Achievement: The job provides for feelings of achievement and task accomplishment.	1	2	3	4	5
22. Participation: The job allows participation in work-related decision making.	1	2	3	4	5
23. Communication: The job has access to relevant communication channels and information flows.	1	2	3	4	5
24. Pay adequacy: The pay on this job is adequate compared with the job requirements and with the pay in similar jobs.	1	2	3	4	5
25. Recognition: The job provides acknowledgement and recognition from others.	1	2	3	4	5
26. Job security: People on this job have high job security.	1	2	3	4	5
Perceptual/Motor Approach					
27. Lighting: The lighting in the work place is adequate and free from glare.	1	2	3	4	5
28. Displays: The displays, gauges, meters, and computerized equipment on this job are easy to read and understand.	1	2	3	4	5
29. Programs: The programs in the computerized equipment on this job are easy to learn and use.	1	2	3	4	5
30. Other equipment: The other equipment (all types) used on this job is easy to learn and use.	1	2	3	4	5
31. Printed job materials: The printed materials used on this job are easy to read and interpret.	1	2	3	4	5
32. Work place layout: The work place is laid out such that you can see and hear well to perform the job.	1	2	3	4	5
33. Information input requirements: The amount of information you must attend to in order to perform this job is fairly minimal.	1	2	3	4	5

Table 14.2 **(*Continued*)**

34. Information output requirements: The amount of information you must output on this job, in terms of both action and communication, is fairly minimal.	1	2	3	4	5
35. Information processing requirements: The amount of information you must process, in terms of thinking and problem solving, is fairly minimal.	1	2	3	4	5
36. Memory requirements: The amount of information you must remember on this job is fairly minimal.	1	2	3	4	5
37. Stress: There is relatively little stress on this job.	1	2	3	4	5
Biological Approach					
38. Strength: The job requires fairly little muscular strength.	1	2	3	4	5
39. Lifting: The job requires fairly little lifting, and/or the lifting is of very light weights.	1	2	3	4	5
40. Endurance: The job requires fairly little muscular endurance.	1	2	3	4	5
41. Seating: The seating arrangements on the job are adequate (e.g., ample opportunities to sit, comfortable chairs, good postural support, etc.).	1	2	3	4	5
42. Size differences: The work place allows for all size differences between people in terms of clearance, reach, eye height, leg room, etc.	1	2	3	4	5
43. Wrist movement: The job allows the wrists to remain straight without excessive movement.	1	2	3	4	5
44. Noise: The work place is free from excessive noise.	1	2	3	4	5
45. Climate: The climate at the work place is comfortable in terms of temperature and humidity, and it is free of excessive dust and fumes.	1	2	3	4	5
46. Work breaks: There is adequate time for work breaks given the demands of the job.	1	2	3	4	5
47. Shift work: The job does not require shift work or excessive overtime.	1	2	3	4	5
For jobs with little physical activity due to single work station add:					
48. Exercise opportunities: During the day, there are enough opportunities to get up from the work station and walk around.	1	2	3	4	5
49. Constraint: While at the workstation, the worker is not constrained to a single position.	1	2	3	4	5
50. Furniture: At the workstation, the worker can adjust or arrange the furniture to be comfortable (e.g., adequate legroom, foot rests if needed, proper keyboard or work surface height, etc.).	1	2	3	4	5

Source: Table adopted from Campion (1988). See supporting reference and related research (e.g., Campion and McClelland, 1991, 1993; Campion and Thayer, 1985) for reliability and validity information. Scores for each approach are calculated by averaging applicable items.

Table 14.3 Team Design Measure

Instructions: This questionnaire consists of statements about your team, and how your team functions as a group. Please indicate the extent to which each statement describes your team by circling a number to the right of each statement. Please Use the Following Scale:

(5) Strongly agree
(4) Agree
(3) Neither agree nor disagree
(2) Disagree
(1) Strongly disagree
( ) Leave blank if do not know or not applicable

Self-Management					
1. The members of my team are responsible for determining the methods, procedures, and schedules with which the work gets done.	1	2	3	4	5
2. My team rather than my manager decides who does what tasks within the team.	1	2	3	4	5
3. Most work-related decisions are made by the members of my team rather than by my manager.	1	2	3	4	5
Participation					
4. As a member of a team, I have a real say in how the team carries out its work.	1	2	3	4	5
5. Most members of my team get a chance to participate in decision making.	1	2	3	4	5
6. My team is designed to let everyone participate in decision making.	1	2	3	4	5
Task Variety					
7. Most members of my team get a chance to learn the different tasks the team performs.	1	2	3	4	5
8. Most everyone on my team gets a chance to do the more interesting tasks.	1	2	3	4	5
9. Task assignments often change from day to day to meet the workload needs of the team.	1	2	3	4	5
Task Significance (Importance)					
10. The work performed by my team is important to the customers in my area.	1	2	3	4	5
11. My team makes an important contribution to serving the company's customers.	1	2	3	4	5
12. My team helps me feel that my work is important to the company.	1	2	3	4	5
Task Identity (Mission)					
13. The team concept allows all the work on a given product to be completed by the same set of people	1	2	3	4	5
14. My team is responsible for all aspects of a product for its area.	1	2	3	4	5
15. My team is responsible for its own unique area or segment of the business.	1	2	3	4	5

Table 14.3 (*Continued*)

Task Interdependence (Interdependence)					
16. I cannot accomplish my tasks without information or materials from other members of my team.	1	2	3	4	5
17. Other members of my team depend on me for information or materials needed to perform their tasks.	1	2	3	4	5
18. Within my team, jobs performed by team members are related to one another.	1	2	3	4	5
Goal Interdependence (Goals)					
19. My work goals come directly from the goals of my team.	1	2	3	4	5
20. My work activities on any given day are determined by my team's goals for that day.	1	2	3	4	5
21. I do very few activities on my job that are not related to the goals of my team	1	2	3	4	5
Interdependent Feedback and Rewards (Feedback and Rewards)					
22. Feedback about how well I am doing my job comes primarily from information about how well the entire team is doing.	1	2	3	4	5
23. My performance evaluation is strongly influenced by how well my team performs.	1	2	3	4	5
24. Many rewards from my job (pay, promotion, etc.) are determined in large part by my contributions as a team member.	1	2	3	4	5
Heterogeneity (Membership)					
25. The members of my team vary widely in their areas of expertise.	1	2	3	4	5
26. The members of my team have a variety of different backgrounds and experiences.	1	2	3	4	5
27. The members of my team have skills and abilities that complement each other.	1	2	3	4	5
Flexibility (Member Flexibility)					
28. Most members of my team know each other's jobs.	1	2	3	4	5
29. It is easy for the members of my team to fill in for one another.	1	2	3	4	5
30. My team is very flexible in terms of membership.	1	2	3	4	5
Relative Size (Size)					
31. The number of people in my team is too small for the work to be accomplished. (Reverse scored)	1	2	3	4	5
Preference for Team Work (Team Work Preferences)					
32. If given the choice, I would prefer to work as part of a team rather than work alone.	1	2	3	4	5
33. I find that working as a member of a team increases my ability to perform effectively.	1	2	3	4	5
34. I generally prefer to work as part of a team.	1	2	3	4	5

Table 14.3 **(*Continued*)**

Training					
35. The company provides adequate technical training for my team.	1	2	3	4	5
36. The company provides adequate quality and customer service training for my team.	1	2	3	4	5
37. The company provides adequate team skills training for my team (communication, organization, interpersonal, etc.).	1	2	3	4	5
Managerial Support					
38. Higher management in the company supports the concept of teams.	1	2	3	4	5
39. My manager supports the concept of teams.	1	2	3	4	5
Communication/Cooperation between Work Groups					
40. I frequently talk to other people in the company besides the people on my team.	1	2	3	4	5
41. There is little competition between my team and other teams in the company.	1	2	3	4	5
42. Teams in the company cooperate to get the work done.	1	2	3	4	5
Potency (Spirit)					
43. Members of my team have great confidence that the team can perform effectively.	1	2	3	4	5
44. My team can take on nearly any task and complete it.	1	2	3	4	5
45. My team has a lot of team spirit.	1	2	3	4	5
Social Support					
46. Being in my team gives me the opportunity to work in a team and provide support to other team members.	1	2	3	4	5
47. My team increases my opportunities for positive social interaction.	1	2	3	4	5
48. Members of my team help each other out at work when needed.	1	2	3	4	5
Workload Sharing (Sharing the Work)					
49. Everyone on my team does their fair share of the work.	1	2	3	4	5
50. No one in my team depends on other team members to do the work for them.	1	2	3	4	5
51. Nearly all the members of my team contribute equally to the work.	1	2	3	4	5
Communication/Cooperation within the Work Group					
52. Members of my team are very willing to share information with other team members about our work.	1	2	3	4	5
53. Teams enhance the communications among people working on the same product.	1	2	3	4	5
54. Members of my team cooperate to get the work done.	1	2	3	4	5

Source: Table adopted from Campion, Medsker, and Higgs (1993). See reference and related research (Campion, Papper, and Medsker, 1995) for reliability and validity information. Scores for each preference/tolerance are calculated by averaging applicable items.

fact that extreme use of the mechanistic approach may result in jobs so simple and routine that employees experience low job satisfaction and motivation. Overly mechanistic, repetitive work can lead to health problems such as repetitive motion disorders.

14.2.2 Motivational Job Design Approach

14.2.2.1 Historical Development

Encouraged by the human relations movement of the 1930s (Hoppock, 1935; Mayo, 1933) people began to point out the negative effects on worker attitudes and health of the overuse of mechanistic design (Argyris, 1964; Blauner, 1964). Overly specialized, simplified jobs were found to lead to dissatisfaction (Caplan, Cobb, French, Van Harrison, and Pinneau, 1975) and adverse physiological consequences for workers (Johansson, Aronsson, and Lindstrom, 1978; Weber, Fussler, O'Hanlon, Gierer, and Grandjean, 1980). Jobs on assembly lines and other machine paced work were especially troublesome in this regard (Salvendy and Smith, 1981; Walker and Guest, 1952). These trends led to an increasing awareness of employees' psychological needs.

The first efforts to enhance the meaningfulness of jobs involved the opposite of specialization. It was recommended that tasks be added to jobs, either at the same level of responsibility (i.e., job enlargement) or at a higher level (i.e., job enrichment) (Ford, 1969; Herzberg, 1966). This trend expanded into a pursuit of identifying and validating characteristics of jobs that make them motivating and satisfying (Griffin, 1982; Hackman & Oldham, 1980; Turner & Lawrence, 1965) This approach considers the psychological theories of work motivation (e.g., Steers and Mowday, 1977; Vroom, 1964) thus this "motivational" approach draws primarily from organizational psychology as a discipline base.

A related trend following later in time but somewhat comparable in content is the sociotechnical approach (Emory and Trist, 1960; Pasmore, 1988; Rousseau, 1977). It focuses not only on the work, but also on the technology itself and the relationship of the environment to work and organizational design. Interest is less on the job, and more on roles and systems. Keys to this approach are work system and job designs which fit their external environment and the joint optimization of both social and technical systems in the organization's internal environment. Though this approach differs somewhat in that consideration is also given to the technical system and external environment, it is similar in that it draws on the same psychological job characteristics which affect satisfaction and motivation. It suggests that as organizations' environments are becoming increasingly turbulent and complex, organizational and job design should involve greater flexibility, employee involvement, employee training, and decentralization of decision making and control, and a reduction in hierarchical structures and the formalization of procedures and relationships (Pasmore, 1988).

Surveys of industrial job designers have consistently indicated that the mechanistic approach represents the dominant theme of job design (Davis et al., 1955; Taylor, 1979). Other approaches to job design, such as the motivational approach, have not been given as much explicit consideration. This is not surprising because the surveys only included job designers trained in engineering-related disciplines, such as industrial engineering and systems analysis. It is not necessarily certain that other specialists or line managers would adopt the same philosophies, especially in recent times. Nevertheless, there is evidence that even fairly naive job designers (i.e., college students in management classes) also adopt the mechanistic approach in job design simulations. That is, their strategies for grouping tasks were primarily the similarity of such factors as activities, skills, equipment, procedures, or location. Even though the mechanistic approach may be the most natural and intuitive, this research has also revealed that people can be trained to apply all four approaches to job design (Campion and Stevens, 1991).

14.2.2.2 Design Recommendations

Table 14.2 provides a list of statements which describe recommendations for the motivational approach. It suggests a job should allow a worker autonomy to make decisions about how and when tasks are to be done. A worker should feel his or her work is important to the overall mission of the organization or department. This is often done by allowing a worker to perform a larger unit of work, or to perform an entire piece of work from beginning to end. Feedback on job performance should be given to workers from the task itself, as well as from the supervisor and others. Workers should be able to use a variety of skills and to personally grow on the job. This approach also considers the social, or people-interaction, aspects of the job: jobs should have opportunities for participation, communication, and recognition. Finally, other human resource systems should contribute to the motivating atmosphere, such as adequate pay, promotion, and job security systems.

14.2.2.3 Advantages and Disadvantages

The goal of this approach is to enhance psychological meaningfulness of jobs, thus influencing a variety of attitudinal and behavioral outcomes. Table 14.1 summarizes some of the advantages and disadvantages found in research. Jobs designed according to the motivational approach have more satisfied, motivated, and involved employees who tend to have higher performance and lower absenteeism. Customer service may be improved, because employees take more pride in work and can catch their own errors by performing a larger part of the work. In terms of disadvantages, jobs too high on the motivational approach require more training, have greater skill and ability requirements for staffing, and may require higher compensation. Overly motivating jobs may also be so stimulating that workers become predisposed to mental overload, fatigue, errors, and occupational stress.

14.2.3 Perceptual/Motor Job Design Approach

14.2.3.1 Historical Development

This approach draws on a scientific discipline which goes by many names, including human factors, human factors engineering, human engineering, man–machine systems engineering, and engineering psychology. It developed from a number of other disciplines, primarily experimental psychology, but also industrial engineering (Meister, 1971). Within experimental psychology, job design recommendations draw heavily from knowledge of human skilled performance (Welford, 1976) and the analysis of humans as information processors (see Chapters 3–6). The main concern of this approach is on efficient and safe utilization of humans in human–machine systems, with emphasis on selection, design, and arrangement of system components to take account of both human abilities and limitations (Pearson, 1971). It is more concerned with equipment than psychology, and more concerned with human abilities than engineering.

This approach received public attention with the Three Mile Island incident where it was concluded that the control room operator job in the nuclear power plant may have placed too many demands on the operator in an emergency situation thus predisposing errors of judgment (Campion and Thayer, 1987). Government regulations issued since then require nuclear plants to consider "human factors" in their design (U.S. Nuclear Regulatory Commission, 1981). The primary emphasis of this approach is on perceptual and motor abilities of people. (See Chapters 20–22 for more information on equipment design.)

14.2.3.2 Design Recommendations

Table 14.2 provides a list of statements describing important recommendations of the perceptual/motor approach. They refer to either equipment and environment or to information-processing requirements. Their thrust is to consider mental abilities and limitations of humans, such that the attention and concentration requirements of the job do not exceed the abilities of the least capable potential worker. Focus is on the limits of the least capable worker because this approach is concerned with the effectiveness of the total system, which is no better than its "weakest link." Jobs should be designed to limit the amount of information workers must pay attention to and remember. Lighting levels should be appropriate, displays and controls should be logical and clear, work places should be well laid out and safe, and equipment should be easy to use. (See Chapters 56–61 for more information on human factors applications.)

14.2.3.3 Advantages and Disadvantages

The goals of this approach are to enhance reliability, safety, and positive user reactions. Table 14.1 summarizes advantages and disadvantages found in research. Jobs designed according to the perceptual/motor approach have lower errors and accidents. Like the mechanistic approach, it reduces mental ability requirements of the job, thus employees may be less stressed and mentally fatigued. It may also create some efficiencies, such as reduced training time and staffing requirements. On the other hand, costs from the excessive use of the perceptual/motor approach can include low satisfaction, low motivation, and boredom due to inadequate mental stimulation. This problem is exacerbated by the fact that designs based on the least capable worker essentially lower a job's mental requirements.

14.2.4 Biological Job Design Approach

14.2.4.1 Historical Development

This approach and the perceptual/motor approach share a joint concern for proper person–machine fit. The major difference is that this approach is more oriented toward biological considerations and stems from such disciplines as work physiology (see Chapter 10), biomechanics (i.e., study of body movements, see Chapter 9) and anthropometry (i.e., study of body sizes, see Chapters 8 and 23). Although many specialists probably practice both approaches together as is reflected in many texts in the area (Konz, 1983) a split does exist between Americans who are more psychologically oriented and use the title "human factors engineer," and Europeans who are more physiologically oriented and use the title "ergonomist" (Chapanis, 1970). Like the perceptual-motor approach, the biological approach is concerned with the design of equipment and work places, as well as the design of tasks (Grandjean, 1980).

14.2.4.2 Design Recommendations

Table 14.2 lists important recommendations from the biological approach. This approach tries to design jobs to reduce physical demands to avoid exceeding people's physical capabilities and limitations. Jobs should not require excessive strength and lifting, and again, abilities of the least physically able potential worker set the maximum level. Chairs should be designed for good postural support. Excessive wrist movement should be reduced by redesigning tasks and equipment. Noise, temperature, and atmosphere should be controlled within reasonable limits. Proper work/rest schedules should be provided so employees can recuperate from the physical demands.

14.2.4.3 Advantages and Disadvantages

The goals of this approach are to maintain employees' comfort and physical well-being. Table 14.1 summarizes some advantages and disadvantages observed in research. Jobs designed according to this approach require less physical effort, result in less fatigue, and create fewer injuries and aches and pains than jobs low on this approach. Occupational illnesses, such as lower back pain and carpal tunnel syndrome, are fewer on jobs designed with this approach. There may be lower absenteeism and higher job satisfaction on jobs which are not physically arduous. However, a direct cost of this approach may be the expense of changes in equipment or job environments needed to implement the recommendations. At the extreme, costs may include jobs with so few physical demands that workers become drowsy or lethargic, thus reducing performance. Clearly, extremes of physical activity and inactivity should be avoided, and an optimal level of physical activity should be developed.

14.3 THE TEAM DESIGN APPROACH

14.3.1 Historical Development

An alternative to designing work around individual jobs is to design work for teams of workers. Teams can vary a great deal in how they are designed and can conceivably incorporate elements from any of the job design approaches discussed. However, the focus here is on the self-managing, autonomous type of team design approach, which is gaining considerable popularity in organizations and substantial research attention today (Guzzo and Shea, 1992; Hoerr, 1989; Swezey and Salas, 1992; Sundstrom, DeMeuse, and Futrell, 1990). Autonomous work teams derive their conceptual basis from motivational job design and from sociotechnical systems theory, which in turn reflect social and organizational psychology and organizational behavior (Cummings, 1978; Davis, 1971; Davis and Valfer, 1965). The Hawthorne studies (Homans, 1950) and European experiments with autonomous work groups (Kelly, 1982; Pasmore, Francis and Haldeman, 1982) called attention to the benefits of applying work teams in other than sports and military settings. Although enthusiasm for the use of teams had waned in the 1960s and 1970s due to research discovering some disadvantages of teams (Buys, 1978; Zander, 1979), the 1980s brought a resurgence of interest in the use of work teams and it has become an extremely popular work design in organizations today (Hoerr, 1989; Sundstrom et al., 1990). This renewed interest may be due to the cost advantages of having fewer supervisors with self-managed teams or the apparent logic of the benefits of teamwork.

14.3.2 Design Recommendations

Teams can vary in the degree of authority and autonomy they have. For example, manager-led teams have responsibility only for the execution of their work. Management designs the work, designs the teams, and provides an organizational context for the teams. However, in autonomous work teams, or self-managing teams, team members design and monitor their own work and performance. They may also design their own team structure (e.g., delineating interrelationships among members) and composition (e.g., selecting members). In such self-designing teams, management is only responsible for the teams' organizational context (Hackman, 1987). Although team design could incorporate elements of either mechanistic or motivational approaches to design, narrow and simplistic mechanistically designed jobs would be less consistent with other suggested aspects of the team approach to design than motivationally designed jobs. Mechanistically designed jobs would not allow an organization to gain as much of the advantages from placing workers in teams.

Figure 14.1 and Table 14.3 provide important recommendations from the self-managing team design approach. Many of the advantages of work teams depend on how teams are designed and supported by their organization. According to the theory behind self-managing team design, decision making and responsibility should be pushed down to the team members (Hackman, 1987). If management is willing to follow this philosophy, teams can provide several additional advantages. By pushing decision making down to the team and requiring consensus, the organization will find greater acceptance, understanding, and ownership of decisions (Porter, Lawler, and Hackman, 1987). The perceived autonomy resulting from making work decisions should be both satisfying and motivating. Thus, this approach tries to design teams so they have a high degree of self-management and all team members participate in decision making.

The team design approach also suggests that the set of tasks assigned to a team should provide a whole and meaningful piece of work (i.e., have task identity as in the motivational approach to job design). This allows team members to see how their work con-

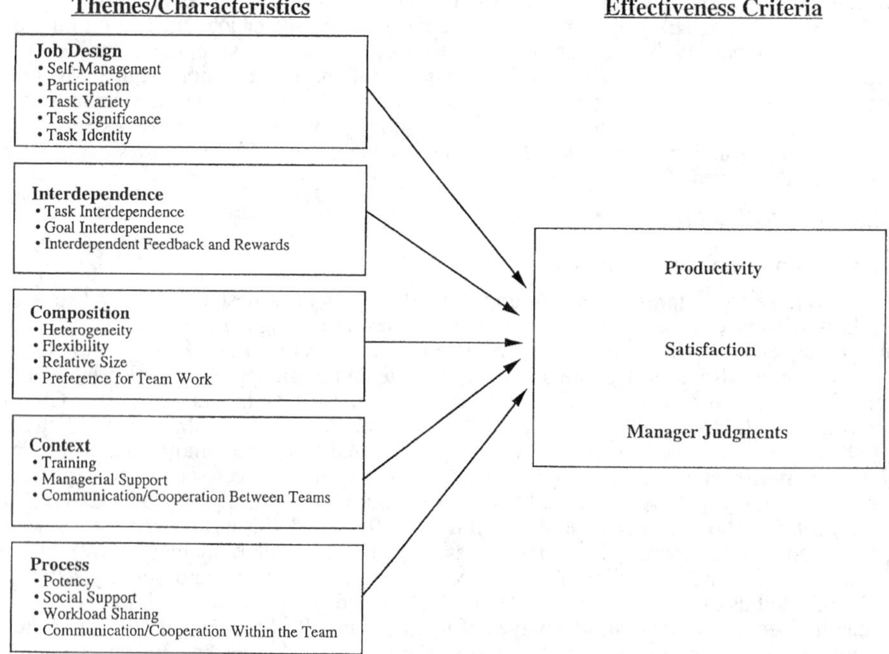

Figure 14.1 Characteristics related to team effectiveness.

tributes to a whole product or process, which might not be possible with individuals working alone. This can give workers a better idea of the significance of their work and create greater identification with the finished product or service. If team workers rotate among a variety of subtasks and cross-train on different operations, workers should also perceive greater variety in the work (Campion, Cheraskin, and Stevens, 1994).

Interdependent tasks, goals, feedback, and rewards should be provided to create feelings of team interdependence among members and focus on the team as the unit of performance, rather than on the individual. It is suggested that team members should be heterogeneous in terms of areas of expertise and background so their varied knowledge, skills, and abilities (KSAs) complement one another. Teams also need adequate training, managerial support, and organizational resources to carry out their tasks. Managers should encourage positive group processes including open communication and cooperation within and between work groups, supportiveness and sharing of the workload among team members, and development of positive team spirit and confidence in the team's ability to perform effectively.

14.3.3 Advantages and Disadvantages

Table 14.4 summarizes advantages and disadvantages of team design relative to individual job design. To begin with, teams designed so members have a heterogeneity of KSAs can help team members learn by working with others who have different KSAs. Cross-training on different tasks can occur, and the work force can become more flexible (Goodman, Ravlin, and Argote, 1986). Teams with heterogeneous KSAs also allow for synergistic combinations of ideas and abilities not possible with individuals working alone, and such teams have generally shown higher performance, especially when task requirements are diverse (Goodman et al., 1986; Shaw, 1983).

Social support can be especially important when teams face difficult decisions and deal with difficult psychological aspects of tasks, such as in military squads, medical teams, or police units (Campion and Medsker, 1992). In addition, the simple presence of others can be psychologically arousing. Research has shown that such arousal can have a positive effect on performance when the task is well learned (Zajonc, 1965) and when

Table 14.4 Advantages and Disadvantages of Work Teams

Advantages	Disadvantages
Team members learn from one another	Lack of compatibility of some individuals with team work
Possibility of greater work force flexibility with cross-training	Additional need to select workers to fit team as well as job
Opportunity for synergistic combinations of ideas and abilities	Possibility some members will experience less motivating jobs
New approaches to tasks may be discovered	Possible incompatibility with cultural, organizational, or labor management norms
Social facilitation and arousal	
Social support for difficult tasks and situations	Increased competition and conflict between teams
Increased communication and information exchange between team members	More time consuming due to socializing coordination losses, and need for consensus
Greater cooperation among team members	Inhibition of creativity and decision-making processes; possibility of groupthink
Beneficial for interdependent work flows	Less powerful evaluation and rewards; social loafing or free-riding may occur
Greater acceptance and understanding of decisions when team makes decisions	Less flexibility in cases of replacement, turnover, or transfer
Greater autonomy, variety, identity, significance, variety, and feedback possible for workers	
Commitment to the team may stimulate performance and attendance	

Source: Table adopted from Campion and Medsker (1992)

other team members are perceived as evaluating the performer (Harkins, 1987; Porter et al., 1987). With routine jobs, this arousal effect may counteract boredom and performance decrements (Cartwright, 1968).

Another advantage of teams is that they can increase information exchanged between members through proximity and shared tasks (McGrath, 1984). Increased cooperation and communication within teams can be particularly useful when workers' jobs are highly interrelated, such as when workers whose tasks come later in the process must depend on the performance of workers whose tasks come earlier or when workers exchange work back and forth among themselves (Mintzberg, 1979; Thompson, 1967).

In addition, if teams are rewarded for team effort, rather than individual effort, members will have an incentive to cooperate with one another (Leventhal, 1976). The desire to maintain power by controlling information may be reduced. More experienced workers may be more willing to train the less experienced when they are not in competition with them. Team design and rewards can also be helpful in situations where it is difficult to measure individual performance or where workers mistrust supervisors' assessments of performance (Milkovich and Newman, 1993).

Finally, teams can be beneficial if team members develop a feeling of commitment and loyalty to their team (Cartwright, 1968). For workers who do not develop high commitment to their organization or management and who do not become highly involved in their job, work teams can provide a source of commitment. That is, members may feel responsible to attend work, cooperate with others, and perform well because of commitment to their work team, even though they are not strongly committed to the organization or the work itself.

Thus, designing work around teams can provide several advantages to organizations and their workers. Unfortunately, there are also disadvantages to using work teams and situations in which individual-level design is preferable to team design. For example, some individuals may dislike team work and may not have necessary interpersonal skills or desire to work in a team. When selecting team members, one has the additional requirement of selecting workers to fit the team, as well as the job. (Section 14.4.3 provides more information on the selection of team members; see also Chapter 16 for general information on personnel selection.)

Individuals can experience less autonomy and less personal identification when working on a team. Designing work around teams does not guarantee workers greater variety, significance, and identity. If members within the team do not rotate among tasks or if some members are assigned exclusively to less desirable tasks, not all members will benefit from team design. Members can still have fractionated, demotivating jobs.

Team work can also be incompatible with cultural norms. The United States has a very individualistic culture (Hofstede, 1980). Applying team methods that have been successful in collectivistic societies like Japan may be problematic in the United States. In addition, organizational norms and labor–management relations may be incompatible with team design, making its use more difficult.

Some advantages of team design can create disadvantages, too. First, though team rewards can increase communication and cooperation and reduce competition within a team, they may cause greater competition and reduced communication between teams. If members identify too strongly with a team, they may not realize when behaviors which benefit the team detract from organizational goals and create conflicts detrimental to productivity. Increased communication within teams may not always be task-relevant either. Teams may spend work time socializing. Team decision making can take longer than individual decision making, and the need for coordination within teams can be time consuming.

Decision making and creativity can also be inhibited by team processes. When teams become highly cohesive they may become so alike in their views that they develop "groupthink" (Janis, 1972). When groupthink occurs, teams tend to underestimate their competition, fail to adequately critique fellow team members' suggestions, not appraise alternatives adequately, and fail to work out contingency plans. In addition, team pressures distort judgments. Decisions may be based more on persuasiveness of dominant individuals or the power of majorities, rather than on the quality of decisions. Research has found a tendency for team judgments to be more extreme than the average of individual members' predecision judgments (Janis, 1972; McGrath, 1984). Although evidence shows highly cohesive teams are more satisfied with their teams, cohesiveness is not necessarily related to high productivity. Whether cohesiveness is related to performance depends on a team's norms and goals. If a team's norm is to be productive, cohesiveness will enhance

Table 14.5 When to Design Jobs Around Work Teams

1. Are workers' tasks highly interdependent, or could they be made to be so? Would this interdependence enhance efficiency or quality?
2. Do the tasks require a variety of knowledge, skills, abilities such that combining individuals with different backgrounds would make a difference in performance?
3. Is cross-training desired? Would breadth of skills and work force flexibility be essential to the organization?
4. Could increased arousal, motivation, and effort to perform make a difference in effectiveness?
5. Can social support help workers deal with job stresses?
6. Could increased communication and information exchange improve performance rather than interfere?
7. Could increased cooperation aid performance?
8. Are individual evaluation and rewards difficult or impossible to make or are they mistrusted by workers?
9. Could common measures of performance be developed and used?
10. It is technically possible to group tasks in a meaningful, efficienct way?
11. Would individuals be willing to work in teams?
12. Does the labor force have the interpersonal skills needed to work in teams?
13. Would team members have the capacity and willingness to be trained in interpersonal and technical skills required for team work?
14. Would team work be compatible with cultural norms, organizational policies, and leadership styles?
15. Would labor–management relations be favorable to team job design?
16. Would the amount of time taken to reach decisions, consensus, and coordination not be detrimental to performance?
17. Can turnover be kept to a minimum?
18. Can teams be defined as a meaningful unit of the organization with identifiable inputs, outputs, and buffer areas which give them a separate identity from other teams?
19. Would members share common resources, facilities, or equipment?
20. Would top management support team job design?

Source: Table adopted from Campion and Medsker (1992). Affirmative answers support the use of team job design.

productivity; however, if the norm is not one of commitment to productivity, cohesiveness can have a negative influence (Zajonc, 1965).

The use of teams and team-level rewards can also decrease the motivating power of evaluation and reward systems. If team members are not evaluated for individual performance, do not believe their output can be distinguished from the team's, or do not perceive a link between their personal performance and outcomes, social loafing (Harkins, 1987) can occur. In such situations, teams do not perform up to the potential expected from combining individual efforts.

Finally, teams may be less flexible in some respects because they are more difficult to move or transfer as a unit than individuals (Sundstrom et al., 1990). Turnover, replacements, and employee transfers may disrupt teams. And members may not readily accept new members.

Thus, whether work teams are advantageous depends to a great extent on the composition, structure, reward systems, environment, and task of the team. Table 14.5 presents questions which can help determine whether work should be designed around teams rather than individuals. The more questions answered in the affirmative, the more likely teams are to be beneficial. If one chooses to design work around teams, suggestions for designing effective teams are presented in Section 14.4.3.

14.4 IMPLEMENTATION ADVICE FOR JOB AND TEAM DESIGN

14.4.1 General Implementation Advice

14.4.1.1 Procedures to Follow

There are several general philosophies that are helpful when designing or redesigning jobs or teams:

1. As noted previously, designs are not inalterable or dictated by technology. There is some discretion in the design of all work situations, and considerable discretion in most.
2. There is no single best design, there are simply better and worse designs depending on one's design perspective.
3. Design is iterative and evolutionary and should continue to change and improve over time.
4. Participation of workers affected generally improves the quality of the resulting design and acceptance of suggested changes.
5. The process of the project, or how it is conducted is important in terms of involvement of all interested parties, consideration of alternative motivations, and awareness of territorial boundaries.

Procedures for the Initial Design of Jobs or Teams

In consideration of process aspects of design, Davis and Wacker (1982) suggest four steps:

1. ***Form a steering committee.*** This committee usually consists of a team of high-level executives who have a direct stake in the new jobs or teams. The purposes of the committee are to: (a) bring into focus the project's objective, (b) provide resources and support for the project, (c) help gain cooperation of all parties affected, and (d) oversee and guide the project.
2. ***Form a design task force.*** The task force may include engineers, managers, job or team design experts, architects, specialists, and others with relevant knowledge or responsibility relevant. The task force is to gather data, generate and evaluate design alternatives, and help implement recommended designs.
3. ***Develop a philosophy statement.*** The first goal of the task force is to develop a philosophy statement to guide decisions involved in the project. The philosophy statement is developed with input from the steering committee and may include the project's purposes, organization's strategic goals, assumptions about workers and the nature of work, and process considerations.
4. ***Proceed in an evolutionary manner.*** Jobs should not be overspecified. With considerable input from eventual job holders or team members, the work design will continue to change and improve over time.

According to Davis and Wacker (1982), the process of redesigning existing jobs is much the same as designing original jobs with two additions. First, existing job incumbents must be involved. Second, more attention needs to be given to implementation issues. Those involved in the implementation must feel ownership of and commitment to the change and believe the redesign represents their own interests.

Potential Steps to Follow

Along with the steps discussed above, a redesign project should also include the following five steps:

1. ***Measuring the design of the existing job or teams.*** The questionnaire methodology and other analysis tools described in Section 14.5 may be used to measure current jobs or teams.
2. ***Diagnosing potential design problems***. Based on data collected in step 1, the current design is analyzed for potential problems. The task force and employee involvement are important. Focused team meetings are a useful vehicle for identifying and evaluating problems.
3. ***Determining job or team design changes.*** Changes will be guided by project goals, problems identified in step 2, and one or more of the approaches to work design. Often several potential changes are generated and evaluated. Evaluation of alternative changes may involve consideration of advantages and disadvantages identified in previous research (see Table 14.1) and opinions of engineers, managers, and employees.
4. ***Making design changes.*** Implementation plans should be developed in detail along with back-up plans in case there are a difficulties with the new design.

Communication and training are keys to implementation. Changes might also be pilot tested before widespread implementation.

5. ***Conducting a follow-up evaluation.*** Evaluating the new design after implementation is probably the most neglected part of the process in most applications. The evaluation might include the collection of design measurements on the redesigned jobs/teams using the same instruments as in step 1. Evaluation may also be conducted on outcomes, such as employee satisfaction, error rates, and training time (Table 14.1). Scientifically valid evaluations require experimental research strategies with control groups. Such studies may not always be possible in organizations, but often quasi-experimental and other field research designs are possible (Cook and Campbell, 1979). Finally, the need for adjustments are identified through the follow-up evaluation. (For examples of evaluations, see Section 14.5.8 and Campion and McClelland, 1991, 1993)

14.4.1.2 Individual Differences Among Workers

It is a common observation that not all employees respond the same to the same job. Some people on a job have high satisfaction, while others on the same job have low satisfaction. Clearly, there are individual differences in how people respond to work. Considerable research has looked at individual differences in reaction to the motivational design approach. It has been found that some people respond more positively than others to highly motivational work. These differences are generally viewed as differences in needs for personal growth and development (Hackman and Oldham, 1980).

Using the broader notion of preferences/tolerances for types of work, the consideration of individual differences has been expanded to all four approaches to job design (Campion, 1988; Campion and McClelland, 1991) and to the team design approach (Campion, Medsker, and Higgs, 1993; Campion, Papper, and Medsker, 1995). Table 14.6 provides scales that can be used to determine job incumbents' preferences/tolerances. These scales can be administered in the same manner as the questionnaire measures of job and team design discussed in Section 14.5.

Although consideration of individual differences is encouraged, there are often limits to which such differences can be accommodated. Jobs or teams may have to be designed for people who are not yet known or who differ in their preferences. Fortunately, though evidence indicates individual differences moderate reactions to the motivational approach (Fried and Ferris, 1987), the differences are of degree but not direction. That is, some people respond more positively than others to motivational work, but few respond negatively. It is likely that this also applies to the other design approaches.

14.4.1.3 Some Basic Choices

Hackman and Oldham (1980) have provided five strategic choices that relate to implementing job redesign. The note that little research exists indicating the exact consequences of each choice, and correct choices may differ by organization. The basic choices are:

1. ***Individual versus team designs for work.*** An initial decision is to either enrich individual jobs or create teams. This also includes consideration of whether any redesign should be undertaken and its likelihood of success.
2. ***Theory based versus intuitive changes.*** This choice was basically defined as the motivational (theory) approach versus no particular (atheoretical) approach. In the present chapter, this choice may be better framed as choosing among the four approaches to job design. However, as argued earlier, consideration of only one approach may lead to some costs or additional benefits being ignored.
3. ***Tailored versus broadside installation.*** This choice is between tailoring changes to individuals or making the changes for all in a given job.
4. ***Participative versus top-down change processes.*** The most common orientation is that participative is best. However, costs of participation include the time involved and incumbents' possible lack of a broad knowledge of the business.
5. ***Consultation versus collaboration with stakeholders.*** The effects of job design changes often extend far beyond the individual incumbent and department. For example, a job's output may be an input to a job elsewhere in the organization. The presence of a union also requires additional collaboration. Depending on considerations, participation of stakeholders may range from no involvement, through consultation, to full collaboration.

Table 14.6 Preferances / Tolerances for the Design Approaches

Instructions: Indicate the extent to which each statement is descriptive of your preferences and tolerances for types of work on the scale below. Circle answers to the right of each statement.

Please Use the Following Scale:

(5) Strongly agree
(4) Agree
(3) Neither agree nor disagree
(2) Disagree
(1) Strongly disagree
( ) Leave blank if do not know or not applicable

Preferences / Tolerance for Mechanistic Design					
1. I have a high tolerance for routine work.	1	2	3	4	5
2. I prefer to work on one task at a time.	1	2	3	4	5
3. I have a high tolerance for repetitive work.	1	2	3	4	5
4. I prefer work that is easy to learn.	1	2	3	4	5
Preferences / Tolerances for Motivational Design					
5. I prefer highly challenging work that taxes my skills and abilities.	1	2	3	4	5
6. I have a high tolerance for mentally demanding work.	1	2	3	4	5
7. I prefer work that gives a great amount of feedback as to how I am doing.	1	2	3	4	5
8. I prefer work that regularly requires the learning of new skills.	1	2	3	4	5
9. I prefer work that requires me to develop my own methods, procedures, goals, and schedules.	1	2	3	4	5
10. I prefer work that has a great amount of variety in duties and responsibilities.	1	2	3	4	5
Preferences / Tolerances for Perceptual / Motor Design					
11. I prefer work that is very fast paced and stimulating.	1	2	3	4	5
12. I have a high tolerance for stressful work.	1	2	3	4	5
13. I have a high tolerance for complicated work.	1	2	3	4	5
14. I have a high tolerance for work where there are frequently too many things to do at one time.	1	2	3	4	5
Preferences / Tolerances for Biological Design					
15. I have a high tolerance for physically demanding work.	1	2	3	4	5
16. I have a fairly high tolerance for hot, noisy, or dirty work.	1	2	3	4	5
17. I prefer work that gives me some physical exercise.	1	2	3	4	5
18. I prefer work that gives me some opportunities to use my muscles.	1	2	3	4	5
Preferences / Tolerances for Team Work					
19. If given the choice, I would prefer to work as part of a team rather than work alone.	1	2	3	4	5
20. I find that working as a member of a team increases my ability to perform effectively.	1	2	3	4	5
21. I generally prefer to work as part of a team.	1	2	3	4	5

Source: Table adopted from Campion (1988) and Campion, Medsker, and Higgs, (1993).

Note: See reference for reliability and validity information. Scores for each preference/tolerance are calculated by averaging applicable items. Interpretations differ slightly across the scales. For the Mechanistic and Motivational designs, higher scores suggest more favorable reactions from incumbents to well designed jobs. For the Perceptual/Motor and Biological approaches, higher scores suggest less unfavorable reactions from incumbents to poorly designed jobs.

14.4.1.4 Overcoming Resistance to Change in Redesign Projects

Resistance to change can be a problem in any project involving major changes. Failure rates of new technology implementations demonstrate a need to give more attention to the human aspects of change projects. It has been estimated that between 50% and 75% of newly implemented manufacturing technologies in the United States have failed, with a disregard for human and organizational issues considered to be a bigger cause for the failures than technical problems (Majchrzak, 1988; Turnage, 1990). The number one obstacle to implementation was considered to be human resistance to change (Hyer, 1984).

Based on the work of Majchrzak (1988), Gallagher and Knight (1986), and Turnage (1990), guidelines for reducing resistance to change include the following:

1. ***Involve workers in planning the change.*** Workers should be informed of changes in advance and involved in the process of diagnosing current problems and developing solutions. Resistance is decreased if participants feel the project is their own and not imposed from outside and if the project is adopted by consensus.
2. ***Top management should strongly support the change.*** If workers feel management is not strongly committed, they are less likely to take the project seriously.
3. ***Create change consistent with worker needs and existing values.*** Resistance is less if change is seen to reduce present burdens, offer interesting experience, not threaten worker autonomy or security or be inconsistent with other goals and values in the organization. Workers need to see the advantages to them of the change. Resistance is less if proponents of change can empathize with opponents (recognize valid objections and relieve unnecessary fears).
4. ***Create an environment of open, supportive communication.*** Resistance will be lessened if participants experience support and have trust in each other. Resistance can be reduced if misunderstandings and conflicts are expected as natural to the innovation process. Provision should be made for clarification.
5. ***Allow for flexibility.*** Resistance is reduced if the project is kept open to revision and reconsideration with experience.

14.4.2 Implementation Advice for Job Design and Redesign

14.4.2.1 Methods for Combining Tasks

In many cases, designing jobs is largely a function of combining tasks. Some guidance can be gained by extrapolating from specific design recommendations in Table 14.2. For example, variety in the motivational approach can be increased by simply combining different tasks in the same job. Conversely, specialization from the mechanistic approach can be increased by only including very similar tasks in the same job. It is also possible when designing jobs to first generate alternative task combinations, then evaluate them using the design approaches in Table 14.2.

A small amount of research within the motivational approach has focused explicitly on predicting relationships between combinations of tasks and the design of resulting jobs (Wong, 1989; Wong and Campion, 1990). This research suggests that a job's motivational quality is a function of three task level variables, as illustrated in Figure 14.2.

1. ***Task design.*** The higher the motivational quality of individual tasks, the higher the motivational quality of a job. Table 14.2 can be used to evaluate individual tasks, then motivational scores for individual tasks can be summed together. Summing is recommended rather than averaging because both the motivational quality of the tasks and the number of tasks are important in determining a job's motivational quality (Globerson and Crossman, 1976).
2. ***Task interdependence.*** Interdependence among tasks has been shown to be positively related to motivational value up to some moderate point; beyond that point increasing interdependence has been shown to lead to lower motivational value. Thus, for motivational jobs, the total amount of interdependence among tasks should be kept at a moderate level. Both complete independence and excessively high interdependence should be avoided. Table 14.7 contains the dimension of task interdependence and provides a questionnaire to measure it. Table 14.7 can be used to judge the interdependence of each pair of tasks that are being evaluated for inclusion in a job.

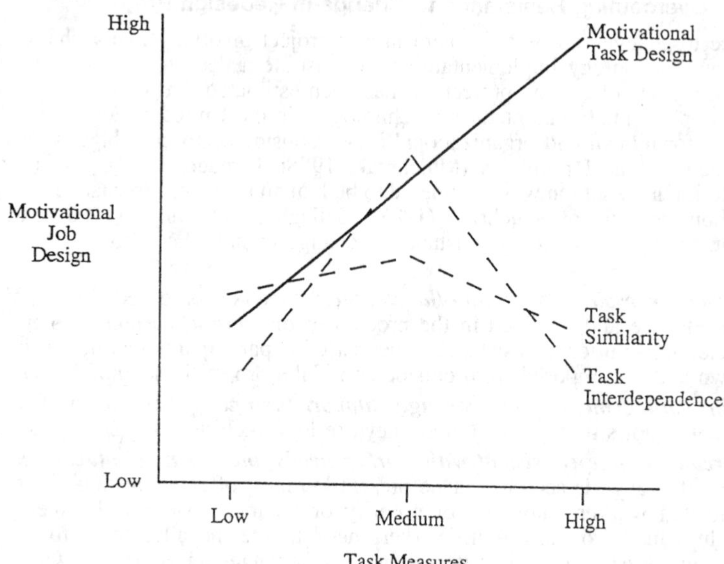

Figure 14.2 Effects of task design, interdependence, and similarity on motivational job design.

3. ***Task similarity.*** Similarity among tasks may be the oldest rule of job design, but beyond a moderate level, it tends to decrease a job's motivational value. Thus, to design motivational jobs, high levels of similarity should be avoided. Similarity at the task pair level can be judged in much the same manner as interdependence by using dimensions in Table 14.7 (see the note to Table 14.7).

14.4.2.2 Trade-Offs Among Job Design Approaches

Although one should strive to construct jobs that are well designed on all the approaches, it is clear design approaches conflict. As Table 14.1 illustrates, benefits of some approaches are costs of others. No one approach satisfies all outcomes. The greatest potential conflicts are between the motivational and the mechanistic and perceptual/motor approaches. They produce nearly opposite outcomes. The mechanistic and perceptual/motor approaches recommend jobs that are simple, safe, and reliable, with minimal mental demands on workers. The motivational approach encourages more complicated and stimulating jobs, with greater mental demands. The team approach is consistent with the motivational approach, and therefore also may conflict with the mechanistic and perceptual/motor approaches.

Because of these conflicts, trade-offs may be necessary. Major trade-offs will be in the mental demands created by the alternative design strategies. Making jobs more mentally demanding increases the likelihood of achieving workers' goals of satisfaction and motivation, but decreases the chances of reaching the organization's goals of reduced training, staffing costs, and errors. Which trade-offs will be made depends on outcomes one prefers to maximize. Generally, a compromise may be optimal.

Trade-offs may not always be needed, however. Jobs can often be improved on one approach while still maintaining their quality on other approaches. For example, in a recent redesign study, the motivational approach was applied to clerical jobs to improve employee satisfaction and customer service (Campion and McClelland, 1991). Expected benefits occurred along with some expected costs (e.g., increased training and compensation requirements), but not all potential costs occurred (e.g., quality and efficiency did not decrease).

One strategy for minimizing trade-offs is to avoid design decisions which influence the mental demands of jobs. An example of this is to enhance motivational design by focusing on social aspects (e.g., communication, participation, recognition, feedback, etc.). These design features can be raised without incurring costs of increased mental demands. Moreover, many of these features are under the direct control of managers.

Table 14.7 Dimensions of Task Interdependence

Instructions: Indicate the extent to which each statement is descriptive of the pair of tasks using the scale below. Circle answers to the right of each statement. Scores are calculated by averaging applicable items.

Please Use the Following Scale:

(5) Strongly agree
(4) Agree
(3) Neither agree nor disagree
(2) Disagree
(1) Strongly disagree
( ) Leave blank if do not know or not applicable

Inputs of the Tasks					
1. Materials/supplies: One task obtains, stores, or prepares the materials or supplies necessary to perform the other task.	1	2	3	4	5
2. Information: One task obtains or generates information for the other task.	1	2	3	4	5
3. Product/service: One task stores, implements, or handles the products or services produced by the other task.	1	2	3	4	5
Processes of the Tasks					
4. Input–output relationship: The products (or outputs) of one task are the supplies (or inputs) necessary to perform the other task.	1	2	3	4	5
5. Method and procedure: One task plans the procedures or work methods for the other task.	1	2	3	4	5
6. Scheduling: One task schedules the activities of the other task.	1	2	3	4	5
7. Supervision: One task reviews or checks the quality of products or services produced by the other task.	1	2	3	4	5
8. Sequencing: One task needs to be performed before the other task.	1	2	3	4	5
9. Time sharing: Some of the work activities of the two tasks must be performed at the same time.	1	2	3	4	5
10. Support service: The purpose of one task is to support or otherwise help the other task get performed.	1	2	3	4	5
11. Tools/equipment: One task produces or maintains the tools or equipment used by the other task.	1	2	3	4	5
Outputs of the Tasks					
12. Goal: One task can only be accomplished when the other task is properly performed.	1	2	3	4	5
13. Performance: How well one task is performed has a great impact on how well the other task can be performed.	1	2	3	4	5
14. Quality: The quality of the product or service produced by one task depends on how well the other task is performed.	1	2	3	4	5

Source: Table adopted from Wong and Campion (1991). See reference and Wong (1989) for reliability and validity information.

Note: The task similarity measure contains 10 comparable items (excluding 4, 6, 8, 9, and 14, and including an item on customer/client). Scores for each dimension are calculated by averaging applicable items.

The independence of the biological approach provides another opportunity to improve design without incurring trade-offs with other approaches. One can reduce physical demands without affecting mental demands of a job. Of course, the cost of equipment may need to be considered.

Finally, adverse effects of trade-offs can often be reduced by avoiding designs that are extremely high or low on any approach. Or, alternatively, one might require minimum acceptable levels on each approach. Knowing all approaches and their corresponding outcomes will help one make more informed decisions and avoid unanticipated consequences.

14.4.2.3 Other Implementation Advice for Job Design

Davis and Wacker (1982, 1987) have provided a list of criteria for grouping tasks, part of which is reproduced below. The list represents a collection of criteria from both the motivational (e.g., 1, 5, 9) and mechanistic (e.g., 2, 8) approaches. Many of the recommendations could also be applied to designing work for teams.

1. Each task group is a meaningful unit of the organization.
2. Task groups are separated by stable buffer areas.
3. Each task group has definite, identifiable inputs and outputs.
4. Each task group has associated with it definite criteria for performance evaluation.
5. Timely feedback about output states and feedforward about input states are available.
6. Each task group has resources to measure and control variances that occur within its area of responsibility.
7. Tasks are grouped around mutual cause–effect relationships.
8. Tasks are grouped around common skills, knowledge, or data.
9. Task groups incorporate opportunities for skill acquisition relevant to career advancement.

Based on experience redesigning jobs in AT&T, Ford (1969) advocated "work-itself workshops." These are basically workshops of managers and employees trained in motivational job design who then attempt to come up with ways to improve jobs. Ford provides the following advice for these workshops:

1. Start with a meeting with senior management.
2. Work within a single department at first.
3. Gain commitment.
4. Pick a job to focus on.
5. Conduct workshop meetings.
6. Facilitate creative thinking.
7. Deal with visitors to the job site.
8. Search for a natural module of work.
9. Deal with resistance due to expense.
10. Individualize feedback.

Griffin's (1982) advice is geared toward the manager considering a job redesign intervention in his or her area. He notes the manager may also rely on consultants, task forces, or informal discussion groups. Griffin suggests nine steps:

1. Recognition of a need for change.
2. Selection of job redesign as a potential intervention.
3. Diagnosis of the work system and content on the following factors:
   a. Existing jobs.
   b. Existing work force.
   c. Technology.
   d. Organization design.

   e. Leader behaviors.
   f. Team and social processes.
4. Cost/benefit analysis of proposed changes.
5. Go/no-go decision.
6. Establishment of a strategy for redesign.
7. Implementation of the job changes.
8. Implementation of any needed supplemental changes.
9. Evaluation of the redesigned jobs.

14.4.3 Implementation Advice for Team Design

14.4.3.1 Deciding on Team Composition

Research encourages heterogeneous teams in terms of skills, personality, and attitudes because it increases the range of competencies in teams (Gladstein, 1984) and is related to effectiveness (Campion, Papper, and Medsker, 1995). However, homogeneity is preferred if team morale is the main criterion, and heterogeneous attributes must be complementary if they are to contribute to effectiveness. Heterogeneity for its own sake is unlikely to enhance effectiveness (Campion, Medsker, and Higgs, 1993). Another composition characteristic of effective teams is whether members have flexible job assignments (Campion, Medsker, and Higgs, 1993; Sundstrom et al., 1990). If members can perform different jobs, effectiveness is enhanced because they can fill in as needed.

A third important aspect of composition is team size. Evidence suggests the importance of optimally matching team size to team tasks to achieve high performance and satisfaction (Campion, 1993). Teams need to be large enough to accomplish work assigned to them, but may be dysfunctional when too large due to heightened coordination needs (O'Reilly and Roberts, 1977; Steiner, 1972) or increased social loafing (McGrath, 1984; Wicker, Kirmeyer, Hanson, and Alexander, 1976). Thus, groups should be staffed to the smallest number needed to do the work (Goodman et al., 1986; Hackman, 1987; Sundstrom et al., 1990).

14.4.3.2 Selecting Team Members

With team design, interpersonal demands appear to be much greater than with traditional individual-based job design (Lawler, 1986). A team-based setting highlights the importance of employees being capable of interacting in an effective manner with peers, because the amount of interpersonal interactions required is higher in teams (Stevens and Campion, 1994a,b). Team effectiveness can depend heavily on members' "interpersonal competence," or their ability to successfully maintain healthy working relationships and react to others with respect for their viewpoints (Perkins and Abramis, 1990). There is a greater need for team members to be capable of effective interpersonal communication, collaborative problem solving, and conflict management (Stevens and Campion, 1994a,b).

The process of employment selection for team members places greater stress on adequately evaluating interpersonal competence than is normally required in the selection of workers for individual jobs. To create a selection instrument for evaluating potential team members' ability to work successfully in teams, Stevens and Campion (1994a,b) reviewed literature in areas of sociotechnical systems theory (e.g., Cummings, 1978; Wall, Kemp, Jackson, and Clegg, 1986), organizational behavior (e.g., Hackman, 1987; Shea and Guzzo, 1987; Sundstrom et al., 1990), industrial engineering (e.g., Davis and Wacker, 1987; Majchrzak, 1988), and social psychology (e.g., McGrath, 1984; Steiner, 1972) to identify relevant knowledge, skills, and abilities (KSAs). Table 14.8 shows the 14 KSAs identified as important for teamwork.

These KSAs have been used to develop a 35-item, multiple-choice employment test, which was validated in two studies to determine how highly related it was to team members' job performance. The job performance of team members in two different companies was rated by both supervisors and co-workers. Correlations between the test and job performance ratings were significantly high, with some correlations exceeding .50. The test was also able to add to the ability to predict job performance beyond that provided by a large battery of traditional employment aptitude tests. Thus, these findings provide support for the value of the teamwork KSAs and a selection test based on them (Stevens and Campion, 1994a). Table 14.9 shows some example items from the test.

Table 14.8 Knowledge, Skill, and Ability (KSA) Requirements for Teamwork

I. Interpersonal KSAs
   A. Conflict Resolution KSAs
      1. The KSA to recognize and encourage desirable, but discourage undesirable, team conflict.
      2. The KSA to recognize the type and source of conflict confronting the team and to implement an appropriate conflict resolution strategy.
      3. The KSA to employ an integrative (win–win) negotiation strategy rather than the traditional distributive (win–lose) strategy.
   B. Collaborative Problem Solving KSAs
      4. The KSA to identify situations requiring participative group problem solving and to utilize the proper degree and type of participation.
      5. The KSA to recognize the obstacles to collaborative group problem solving and implement appropriate corrective actions.
   C. Communication KSAs
      6. The KSA to understand communication networks, and to utilize decentralized networks to enhance communication where possible.
      7. The KSA to communicate openly and supportively, that is, to send messages which are (a) behavior- or event-oriented, (b) congruent, (c) validating, (d) conjunctive, and (e) owned.
      8. The KSA to listen nonevaluatively and to appropriately use active listening techniques.
      9. The KSA to maximize consonance between nonverbal and verbal messages, and to recognize and interpret the nonverbal messages of others.
      10. The KSA to engage in ritual greetings and small talk, and a recognition of their importance.

II. Self-management KSAs
   D. Goal Setting and Performance Management KSAs
      11. The KSA to help establish specific, challenging, and accepted team goals.
      12. The KSA to monitor, evaluate, and provide feedback on both overall team performance and individual team member performance.
   E. Planning and Task Coordination KSAs
      13. The KSA to coordinate and synchronize activities, information, and task interdependencies between team members.
      14. The KSA to help establish task and role expectations of individual team members, and to ensure proper balancing of workload in the team.

Aside from written tests, there may be other ways teamwork KSAs could be measured for purposes of selection. For example, interviews may be especially suited to measuring interpersonal attributes (e.g., Arvey and Campion, 1982). There is evidence that a structured interview specifically designed to measure social (i.e., nontechnical) KSAs can have validity with job performance and predict incrementally beyond traditional employment tests (M. Campion, J. Campion, and Hudson, 1993).

Assessment center techniques might also lend themselves to measuring teamwork KSAs. Group exercises have been used to measure leadership and other social skills with good success (Gaugler, Rosenthal, Thornton, and Benston, 1987). It is likely that existing team exercises, such as group problem-solving tasks, could also be modified to score teamwork KSAs.

Selection techniques using biodata may be another way to measure teamwork KSAs. Many items in biodata instruments reflect previous life experiences of a social nature, and recruiters interpret biodata information on applications and resumes as reflecting attributes such as interpersonal skills (Brown and Campion, 1993). A biodata measure developed to focus on teamwork KSAs might include items on teamwork in previous jobs, team experiences in school (e.g., college clubs, class projects), and recreational activities of a team nature (e.g., sports teams and social groups).

14.4.3.3 Designing the Teams' Jobs

This aspect of team design involves team characteristics derived from the motivational job design approach. The main distinction is in level of application rather than content

Table 14.9 Example Items from the Teamwork KSA Test

1. Suppose you find yourself in an argument with several co-workers about who should do a very disagreeable, but routine task. Which of the following would likely be the most effective way to resolve this situation?
   A. Have your supervisor decide, because this would avoid any personal bias.
   *B. Arrange for a rotating schedule so everyone shares the chore.
   C. Let the workers who show up earliest choose on a first-come, first-served basis.
   D. Randomly assign a person to do the task and don't change it.
2. Your team wants to improve the quality and flow of the conversations among its members. Your team should:
   *A. use comments that build upon and connect to what others have said.
   B. set up a specific order for everyone to speak and then follow it.
   C. let team members with more to say determine the direction and topic of conversation.
   D. do all of the above.
3. Suppose you are presented with the following types of goals. You are asked to pick one for your team to work on. Which would you choose?
   A. An easy goal to ensure the team reaches it, thus creating a feeling of success.
   B. A goal of average difficulty so the team will be somewhat challenged, but successful without too much effort.
   *C. A difficult and challenging goal that will stretch the team to perform at a high level, but attainable so that effort will not be seen as futile.
   D. A very difficult, or even impossible goal so that even if the team falls short, it will at least have a very high target to aim for.

* Correct answers.

(Campion and Medsker, 1992; Shea and Guzzo, 1987; Wall et al., 1986). All the job characteristics of the motivational approach to job design can be applied to team design.

One such characteristic is self-management, which is the team level analogy to autonomy at the individual job level. It is central to many definitions of effective work teams (e.g., Cummings, 1978, 1981; Hackman, 1987). A related characteristic is participation. Regardless of management involvement in decision making, teams can still be distinguished in terms of the degree to which all members are allowed to participate in decisions (McGrath, 1984, Porter et al., 1987). Self-management and participation are presumed to enhance effectiveness by increasing members' sense of responsibility and ownership of the work. These characteristics may also enhance decision quality by increasing relevant information and by putting decisions as near as possible to the point of operational problems and uncertainties.

Other important characteristics are task variety, task significance, and task identity. Variety motivates by allowing members to use different skills (Hackman, 1987) and by allowing both interesting and dull tasks to be shared among members (Davis and Wacker, 1987). Task significance refers to the perceived significance of the consequences of the team's work, either for others inside the organization or its customers. Task identity (Hackman, 1987), or task differentiation (Cummings, 1978), refers to the degree to which the team completes a whole and meaningful piece of work. These suggested characteristics of team design have been found to be positively related to team productivity, team member satisfaction, and managers' and employees' judgments of their teams' performance (Campion, Medsker, and Higgs, 1993; Campion, Papper, and Medsker, 1995).

14.4.3.4 Developing Interdependent Relations

Interdependence is often the reason teams are formed (Mintzberg, 1979) and is a defining characteristic of teams (Salas, Dickinson, Converse, and Tannenbaum, 1992; Wall et al., 1986). Interdependence has been found to be related to team members' satisfaction and team productivity and effectiveness (Campion, Medsker, and Higgs, 1993, Campion, Papper, and Medsker, 1995).

One form of interdependence is task interdependence. Team members interact and depend on one another to accomplish their work. Interdependence varies across teams, depending on whether the work flow in a team is pooled, sequential, or reciprocal (Thompson, 1967). Interdependence among tasks in the same job (Wong and Campion, 1991) or between jobs (Kiggundu, 1983) has been related to increased motivation. It can also increase team effectiveness because it enhances the sense of responsibility for others' work (Kiggundu, 1983) or because it enhances the reward value of a team's accomplishments (Shea and Guzzo, 1987).

Another form of interdependence is goal interdependence. Goal setting is a well-documented, individual-level performance improvement technique (Locke and Latham, 1990). A clearly defined mission or purpose is considered to be critical to team effectiveness (Campion, Medsker, and Higgs, 1993; Campion, Papper, and Medsker, 1995; Davis and Wacker, 1987; Hackman, 1987; Sundstrom et al., 1990). Its importance has also been shown in empirical studies on teams (e.g., Buller and Bell, 1986; Woodman and Sherwood, 1980). Not only should goals exist for teams, but individual members' goals must be linked to team goals to be maximally effective.

Finally, interdependent feedback and rewards have also been found to be important for team effectiveness and team member satisfaction (Campion, Medsker, and Higgs, 1993; Campion, Papper, and Medsker, 1995). Individual feedback and rewards should be linked to a team's performance in order to motivate team-oriented behavior. This characteristic is recognized in many theoretical treatments (e.g., Hackman, 1987; Leventhal, 1976; Steiner, 1972; Sundstrom et al., 1990) and research studies (e.g., Pasmore et al., 1982; Wall et al., 1986).

14.4.3.5 Creating the Organizational Context

Organizational context and resources are considered in all recent models of work team effectiveness (e.g., Guzzo and Shea, 1992; Hackman, 1987). One important aspect of context and resources for teams is adequate training. Training is an extensively researched determinant of team performance (for reviews see Dyer, 1984; and Salas et al., 1992), and training is included in most interventions (e.g., Pasmore et al., 1982; Wall et al., 1986). Training is related to team members' satisfaction, and managers' and employees' judgments of their teams' effectiveness (Campion, Medsker, and Higgs, 1993; Campion, Papper, and Medsker, 1995).

Training content often includes team philosophy, group decision making, and interpersonal skills, as well as technical knowledge. Many team-building interventions focus on aspects of team functioning that are related to the teamwork KSAs shown in Table 14.8. A recent review of this literature divided such interventions into four approaches (Tannenbaum, Beard, and Salas, 1992)—goal setting, interpersonal, role, and problem solving—which are similar to the teamwork KSA categories. Thus, these interventions could be viewed as training programs on teamwork KSAs. Reviews indicate that the evidence for the effectiveness of this training appears positive despite the methodological limitations that plague this research (Buller and Bell, 1986; Tannenbaum et al., 1992; Woodman and Sherwood, 1980). It appears that workers can be trained in teamwork KSAs. (See Chapter 16 for more information on team training.)

Regarding how such training should be conducted, there is substantial guidance on training teams in the human factors and military literatures (Dyer, 1984; Salas et al., 1992; Swezey and Salas, 1992). Because these topics are thoroughly addressed in the cited sources, they will not be reviewed here.

Managers of teams also need to be trained in teamwork KSAs, regardless of whether the teams are manager-led or self-managed. The KSAs are needed for interacting with employee teams and for participating on management teams. It has been noted that managers of teams, especially autonomous work teams, need to develop their employees (Cummings, 1978; Hackman and Oldham, 1980; Manz and Sims, 1987). Thus, training must not only ensure that managers possess teamwork KSAs, but that they know how to train employees on these KSAs.

Managerial support is another contextual characteristic. Management controls resources (e.g., material and information) required to make team functioning possible (Shea and Guzzo, 1987), and an organization's culture and top management must support the use of teams (Sundstrom et al., 1990). Teaching facilitative leadership to managers is often a feature of team interventions (Pasmore et al., 1982). Finally, communication and cooperation between teams is a contextual characteristic because it is often the responsibility of managers. Supervising team boundaries (Cummings, 1978) and externally integrating teams with the rest of the organization (Sundstrom et al., 1990) enhance effectiveness. Research indicates that managerial support and communication and cooperation between work teams are related to team productivity and effectiveness and to team members' satisfaction with their work (Campion, Medsker, and Higgs, 1993; Campion, Papper, and Medsker, 1995).

14.4.3.6 Developing Effective Team Process

Process describes those things that go on in the group that influence effectiveness. One process characteristic is potency, or the belief of a team that it can be effective (Guzzo and Shea, 1992; Shea and Guzzo, 1987). It is similar to the lay-term "team spirit." Hackman (1987) argues that groups with high potency are more committed and willing to work hard for the group, and evidence indicates that potency is highly related to team members' satisfaction with work, team productivity, and members' and managers' judgments of their teams' effectiveness (Campion, Medsker, and Higgs, 1993; Campion, Papper, and Medsker, 1995).

Another process characteristic found to be related to team satisfaction, productivity, and effectiveness is social support (Campion, Medsker, and Higgs, 1993; Campion, Papper, and Medsker, 1995). Effectiveness can be enhanced when members help each other and have positive social interactions. Like social facilitation (Harkins, 1987; Zajonc, 1965), social support can be arousing and may enhance effectiveness by sustaining effort on mundane tasks.

Another process characteristic related to be satisfaction, productivity, and effectiveness is workload sharing (Campion, Medsker, and Higgs, 1993; Campion, Papper, and Medsker, 1995). Workload sharing enhances effectiveness by preventing social-loafing or free-riding (Harkins, 1987). To enhance sharing, group members should believe their individual performance can be distinguished from the group's, and that there is a link between their performance and outcomes.

Finally, communication and cooperation within the work group are also important to team effectiveness, productivity, and satisfaction (Campion, Medsker, and Higgs, 1993; Campion, Papper, and Medsker, 1995). Management should help teams foster open communication, supportiveness, and discussions of strategy. Informal, rather than formal, communication channels and mechanisms of control should be promoted to ease coordination (Majchrzak, 1988; Bass and Klubeck, 1952). Managers should encourage self-evaluation, self-observation, self-reinforcement, self-management, and self-goal setting by teams. Self-criticism for purposes of recrimination should be discouraged (Manz and Sims, 1987).

14.5 MEASUREMENT AND EVALUATION OF JOB AND TEAM DESIGN

The purpose of an evaluation study for either a job or team design is to provide an objective evaluation of success and to create a tracking and feedback system to make adjustments during the course of the design project. An evaluation study can provide objective data to make informed decisions, help tailor the process to the organization, and give those affected by the design or redesign an opportunity to provide input. An evaluation study should include measures which describe the characteristics of the jobs or teams so that it can be determined whether or not jobs or teams ended up having the characteristics which they were intended to have. An evaluation study should also include measures of effectiveness outcomes an organization hoped to achieve with a design project. Measures of effectiveness could include such *subjective* outcomes as employee job satisfaction or employee, manager, or customer perceptions of effectiveness. Measures of effectiveness should also include *objective* outcomes such as cost, productivity, rework/scrap, turnover, accident rates, or absenteeism. Additional information on measurement and evaluation of such outcomes can be found in Part 6 of this handbook.

14.5.1 Using Questionnaires to Measure Job and Team Design

One way to measure job or team design is by using questionnaires or checklists. This method of measuring job or team design is highlighted because it has been used widely in research on job design, especially on the motivational approach. More importantly, questionnaires are a very inexpensive, easy, and flexible way to measure work design characteristics. Moreover, they gather information from job experts, such as incumbents, supervisors, and engineers and other analysts.

Several questionnaires exist for measuring the motivational approach to job design (Hackman and Oldham, 1980; Sims, Szilagyi, and Keller, 1976), but only one questionnaire, the *Multimethod Job Design Questionnaire* measures characteristics for all four approaches to job design. This questionnaire (presented in Table 14.2) evaluates the quality of a job's characteristics based on each of the four approaches. The *Team Design Measure* presented in Table 14.3) evaluates the quality of work design based on the team approach.

Questionnaires can be administered in a variety of ways. Employees can complete them individually at their convenience at their work station or some other designated area, or they can complete them in a group setting. Group administration allows greater standardization of instructions and provides the opportunity to answer questions and clarify ambiguities. Managers and engineers can also complete the questionnaires either individually or in a group session. Engineers and analysts usually find that observation of the work site, examination of the equipment and procedures, and discussions with any incumbents or managers are important methods of gaining information on the work before completing the questionnaires.

Scoring for each job design approach or for each team characteristic on the questionnaires is usually accomplished by simply averaging the applicable items. Then scores from different incumbents, managers, or engineers describing the same job or team are combined by averaging. Multiple items and multiple respondents are used to improve the reliability and accuracy of the results. The implicit assumption is that slight differences among respondents are to be expected because of legitimate differences in viewpoint. However, absolute differences in scores should be examined on an item-by-item basis, and large discrepancies (e.g., more than one point) should be discussed to clarify possible differences in interpretation. It may be useful to discuss each item until a consensus rating is reached.

The higher the score on a particular job design scale or work team characteristic scale, the better the quality of the design in terms of that approach or characteristic. Likewise, the higher the score on a particular item, the better the design is on that dimension. How high a score is needed or necessary cannot be stated in isolation. Some jobs or teams are naturally higher or lower on the various approaches, and there may be limits to the potential of some jobs. The scores have most value in comparing different jobs, teams, or design approaches, rather than evaluating the absolute level of the quality of a job or team design. However, a simple rule of thumb is that if the score for an approach is smaller than three, the job or team is poorly designed on that approach and it should be reconsidered. Even if the average score on an approach is greater than three, examine any individual dimension scores that are at two or one.

Uses of Questionnaires in Different Contexts

1. ***Designing New Jobs or Teams.*** When jobs or teams do not yet exist, the questionnaire is used to evaluate proposed job or team descriptions, work stations, equipment, and so on. In this role, it often serves as a simple design checklist. Additional administrations of the questionnaire in later months or years can be used to assess the longer-term effects of the job or team design.

2. ***Redesigning Existing Jobs or Teams or Switching from Job to Team Design.*** When jobs or teams already exist, there is a much greater wealth of information. Questionnaires can be completed by incumbents, managers, and engineers. Questionnaires can be used to measure design both before and after changes are made to compare the redesign with the previous design approach. A premeasure before the redesign can be used as a baseline measurement against which to compare a postmeasure conducted right after the redesign implementation. A follow-up measure can be used in later months or years to assess the long-term difference between the previous design approach and the new approach.

If other sites or plants with the same types of jobs or teams are not immediately included in the redesign but are maintained with the older design approach, they can be used as a comparison or "control group" to enable analysts to draw even stronger conclusions about the effectiveness of the redesign. Such a control group allows one to control for the possibilities that changes in effectiveness were not due to the redesign but were in fact due to some other causes such as increases in workers' knowledge and skills with the passage of time, changes in workers' economic environment (i.e., job security, wages, etc.), or workers trying to give socially desirable responses to questionnaire items.

3. ***Diagnosing Problem Job or Team Designs.*** When problems occur, regardless of the apparent source of the problem, the job or team design questionnaires can be used as a diagnostic device to determine if any problems exist with the design of the jobs or teams.

14.5.2 Choosing Sources of Data

1. ***Incumbents.*** Incumbents are probably the best source of information for existing jobs or teams. Having input can enhance the likelihood that changes will be accepted, and involvement in such decisions can enhance feelings of participation thus increasing motivational job design in itself (see item 22 of the motivational scale in Table 14.2). One should include a large number of incumbents for each job or team because there can be slight differences in perceptions of the same job or team due to individual differences (discussed in Section 14.4.1). Evidence suggests that one should include at least five incumbents for each job or team, but more are preferable (Campion, 1988; Campion and McClelland, 1991; Campion, Medsker and Higgs, 1993; Campion, Papper and Medsker, 1995).

2. ***Managers or Supervisors.*** First-level managers or supervisors may be the next most knowledgeable persons about an existing work design. They may also provide information on jobs or teams under development. Some differences in perceptions of the same job or team will exist among managers, so multiple managers should be used.

3. ***Engineers or Analysts.*** Engineers may be the only source of information if the jobs or teams are not yet developed. But also for existing jobs or teams, an outside perspective of an engineer, analyst, or consultant may provide a more objective viewpoint. Again, there can be differences among engineers, so several should evaluate each job or team.

It is desirable to get multiple inputs and perspectives from different sources in order to get the most reliable and accurate picture of the results of the job or team design.

14.5.3 Long-Term Effects and Potential Biases

It is important to recognize that some effects of job or team design may not be immediate, others may not be long lasting, and still others may not be obvious. Initially, when jobs or teams are designed, or right after they are redesigned, there may be a short-term period of positive attitudes (often called a "Honeymoon Effect"). As the legendary Hawthorne studies indicated, changes in jobs or increased attention paid to workers tends to create novel stimulation and positive attitudes (Mayo 1933). Such transitory elevations in affect should not be mistaken for long-term improvements in satisfaction, as they may wear off over time. In fact, with time, employees may realize their work is now more complex and should be paid higher compensation (Campion and Berger, 1990).

Costs which are likely to lag in time also include stress and fatigue, which may take a while to build up if mental demands have been increased excessively. Boredom may take a while to set in if mental demands have been overly decreased. In terms of lagged benefits, productivity and quality are likely to improve with practice and learning on the new job or team. And some benefits, like reduced turnover, simply take time to estimate accurately.

Benefits which may potentially dissipate with time include satisfaction, especially if the elevated satisfaction is a function of novelty rather than basic changes to the motivating value of the work. Short-term increases in productivity due to heightened effort rather than better design may not last. Costs which may dissipate include training requirements and staffing difficulties. Once jobs are staffed and everyone is trained, these costs disappear until turnover occurs. So these costs will not go away completely, but they may be less after initial start-up. Dissipating heightened satisfaction but long-term increases in productivity were observed in a recent motivational job redesign study (Griffin, 1989). These are only examples to illustrate how dissipating and lagged effects might occur. A more detailed example of long-term effects is given in Section 14.5.8.

A potential bias which may confuse the proper evaluation of benefits and costs is spillover. Laboratory research has shown that the job satisfaction of employees can bias perceptions of the motivational value of their jobs (O'Reilly, Parlette, and Bloom, 1980). Likewise, the level of morale in the organization can have a spillover effect onto employees' perceptions of job or team design. If morale is particularly high, it may have an elevating effect on how employees or analysts view the jobs or teams; conversely, low morale may have a depressing effect on views. The term *morale* refers to the general level of job satisfaction across employees, and it may be a function of many factors including management, working conditions, wages, and so on. Another factor which has an especially strong effect on employee reactions to work design changes is *employment security*. Obviously, employee enthusiasm for work design changes will be negative if they view them as potentially decreasing their job security. Every effort should be made to eliminate these fears. The best method of addressing these effects is to be attentive to their potential existence and to conduct longitudinal evaluations of job and team design.

In addition to questionnaires, there are many other analytical tools which are useful for work design. The disciplines which contributed the different approaches to work design have also contributed different techniques for analyzing tasks, jobs, and processes for design and redesign purposes. These techniques include job analysis methods created by specialists in industrial psychology, variance analysis methods created by specialists in sociotechnical design, time and motion analysis methods created by specialists in industrial engineering, and linkage analysis methods created by specialists in human factors. This section briefly describes a few of these techniques to illustrate the range of options. The reader is referred to the citations for detail on how to use the techniques.

14.5.4 Job Analysis

Job analysis can be broadly defined as a number of systematic techniques for collecting and making judgments about job information. Information derived from job analysis can be used to aid in recruitment and selection decisions, determine training and development needs, develop performance appraisal systems, and evaluate jobs for compensation, as well as to analyze tasks and jobs for job design. Job analysis may also focus on tasks, worker characteristics, worker functions, work fields, working conditions, tools and methods, products and services, and so on. Job analysis data can come from job incumbents, supervisors, and analysts who specialize in the analysis of jobs. Data may also be provided by higher management levels or subordinates in some cases.

Considerable literature has been published on the topic of job analysis (Ash, Levine & Sistrunk, 1983; Gael, 1983; Harvey, 1991; U.S. Department of Labor, 1972). Some of the more typical methods of analysis are briefly described below:

1. Conferences and Interviews. Conferences or interviews with job experts, such as incumbents and supervisors, are often the first step. During such meetings, information collected typically includes job duties and tasks, and knowledge, skill, ability (KSA), and other worker characteristics.
2. Questionnaires. Questionnaires are used to collect information efficiently from a large number of people. Questionnaires require considerable prior knowledge of the job to form the basis of the items (e.g., primary tasks). Often this information is first collected through conferences and interviews, and then the questionnaire is constructed and used to collect judgments about the job (e.g., importance and time spent on each task). Some standardized questionnaires have been developed which can be applied to all jobs to collect basic information on tasks and requirements. An example of a standardized questionnaire is the Position Analysis Questionnaire (McCormick, Jeanneret & Mecham, 1972).
3. Inventories. Inventories are much like questionnaires, except they are simpler in format. They are usually simple checklists where the job expert checks whether a task is performed or an attribute is required.
4. Critical Incidents. This form of job analysis focuses only on aspects of worker behavior which are especially effective or ineffective.
5. Work Observation and Activity Sampling. Quite often job analysis includes the actual observation of work performed. More sophisticated technologies involve statistical sampling of work activities.
6. Diaries. Sometimes it is useful or necessary to collect data by having the employee keep a diary of activities on his or her job.
7. Functional Job Analysis. Task statements can be written in a standardized fashion. Functional job analysis suggests how to write task statements (e.g., start with a verb, be as simple and discrete as possible, etc.). It also involves rating jobs on the degree of data, people, and things requirements. This form of job analysis was developed by the U. S. Department of Labor and has been used to describe over 12,000 jobs as documented in the Dictionary of Occupational Titles (Fine & Wiley, 1971; U.S. Department of Labor, 1977).

Very limited research has been done to evaluate the practicality and quality of various job analysis methods for different purposes. But analysts seem to agree that combinations of methods are preferable to single methods (Levine, Ash, Hall & Sistrunk, 1983).

Current approaches to job analysis do not give much attention to analyzing teams. For example, the Dictionary of Occupational Titles (U.S. Department of Labor, 1972) considers "people" requirements of jobs, but does not address specific teamwork KSAs. Likewise, recent reviews of the literature mention some components of teamwork such as communication and coordination (e.g., Harvey, 1991), but give little attention to other teamwork KSAs. Thus, job analysis systems may need to be revised. Teamwork KSAs are more likely to emerge with conventional approaches to job analysis because of their unstructured nature (e.g., interviews), but structured approaches (e.g., questionnaires) will have to be modified to query about teamwork KSAs.

14.5.5 Variance Analysis

Variance analysis is a tool of sociotechnical design used to identify areas of technological uncertainty in a production process. Variance analysis aids the organization in designing jobs so jobholders can control variability in their work. A variance is defined as an unwanted discrepancy between a desired state and an actual state and is a deviation that falls outside a specified range of tolerance. The variance concept is applied to the technical system and involves five steps (Davis and Wacker, 1982):

1. List variances that could impede the production or service process.
2. Identify causal relationships among variables. Job designers can use information about dependencies and points of interrelatedness in order to cluster tasks and link jobs.
3. Identify and focus on key variances whose control is most critical to successful outcomes.
4. Construct a table of key variance control which contains brief descriptions of variances.
5. Construct a table of skills, knowledge, information, and authority needed so workers can control key variances.

Chapters 12 and 13 in this handbook provide more information about task and workload analysis.

14.5.6 Time and Motion Analysis

Industrial engineers have created many techniques for use in the study of job design which help job designers visualize operations in order to improve efficiencies. A considerable literature exists on the topic (e.g., Mundel, 1985; Niebel, 1988). Some of the methods are briefly described below.

Process charts graphically represent separate steps or events that occur during performance of a task or series of actions. Charts usually begin with inputs of raw materials and follow the inputs through transportation, storage, inspection, production, and finishing. Charts use symbols for different types of operations. Examples of different types of process charts include Operation Process Charts, which show a chronological sequence of operations, inspections, time allowances, and materials used in a process from arrival of raw material to packaging of the finished product. Another type of process chart is a Worker and Machine Process Chart which combines operations of both the worker and equipment and shows idle time and active time for both. These charts are used to analyze only one work station at a time.

Flow diagrams differ from process charts because they utilize drawings of an area or building in which an activity takes place. Flow diagrams help designers visualize the physical layout of the work. Lines are drawn to show the path of travel. Process chart symbols and notations can be included to describe the process.

Possibility guides are tools for systematically listing all possible changes suggested for a particular activity or output. They assist in examining consequences of suggestions to aid in selecting the most feasible changes. Suggestions are recorded and are coded as to what classes of change they affect: job, equipment, process, product design, or raw materials.

Network diagrams are better for use in describing complex relationships than the above techniques. They are useful for situations where: (a) dependencies are tangled and do not progress uniformly, (b) the output has many components, (c) many of the components are service-type outputs, (c) the relationships among the steps of the process with respect to time are of vital importance, or (d) the process is too complex or large in scope for the usual process chart analysis. In network diagrams, a circle or square represents a "status" which is a partial or complete service or substantive output. Heavy lines are "critical paths" which determine the minimum time in which a project can be expected to be completed.

14.5.7 Linkage Analysis

Linkage analysis is a technique used by human factors specialists to represent relationships between components in a work system (Sanders and McCormick, 1987). Components can be either people or things and the relationships between them are called "links." Links fall into three classes as listed below with examples:

1. Communication links.
   a. Visual (person to person or equipment to person).
   b. Auditory, voice (person to person, person to equipment, or equipment to person).
   c. Auditory, nonvoice (equipment to person).
   d. Touch (person to equipment).
2. Control links.
   a. Control (person to equipment).
3. Movement links (movements from one location to another).
   a. Eye movements.
   b. Manual movements, foot movements, or both.
   c. Body movements.

Information collected about links generally includes how often components are linked, in what sequence links occur, and the importance of links. Once obtained, linkage data can be summarized in link tables, adjacency layout diagrams, and spatial operational sequences (SOS) diagrams. Designers of physical work arrangements use these tools to represent relationships between components so that they can better understand how to place these components in advantageous locations in order to minimize lengths between frequent or important links. With complex systems involving many components, quantitative analysis techniques, such as linear programming, can be used.

14.5.8 Example of an Evaluation of a Job Design

Studies conducted by Campion and McClelland (1991, 1993) are described as an illustration of an evaluation of a job redesign project. They illustrate the value of considering an interdisciplinary perspective. The setting was a large financial services company. The units under study processed the paperwork in support of other units which sold the company's products. Jobs had been designed in a mechanistic manner such that individual employees prepared, sorted, coded, and computer input the paper flow.

The organization viewed the jobs as too mechanistically designed. Guided by the motivational approach, the project intended to enlarge jobs by combining existing jobs in order to attain three objectives: (1) enhance motivation and satisfaction of employees, (2) increase incumbent feelings of ownership of the work, thus increasing customer service, and (3) maintain productivity in spite of potential lost efficiencies from the motivational approach. The consequences of all approaches to job design were considered. It was anticipated that the project would increase motivational consequences, decrease mechanistic and perceptual/motor consequences, and have no affect on biological consequences (Table 14.1).

The evaluation consisted of collecting detailed data on job design and a broad spectrum of potential benefits and costs of enlarged jobs. The research strategy involved comparing several varieties of enlarged jobs with each other and with unenlarged jobs. Questionnaire data were collected and focused team meetings were conducted with incumbents, managers, and analysts. The study was repeated at five different geographic sites.

Results indicated enlarged jobs had the benefits of more employee satisfaction, less boredom, better quality, and better customer service; but they also had the costs of slightly higher training, skill, and compensation requirements. Another finding was that all potential costs of enlarging jobs were not observed, suggesting that redesign can lead to benefits without incurring every cost in a one-to-one fashion.

In a two-year follow-up evaluation study, it was found that the costs and benefits of job enlargement changed substantially over time, depending on the type of enlargement. Task enlargement, which was the focus of the original study, had mostly long-term costs (e.g., lower satisfaction, efficiency, and customer service, and more mental overload and errors). Conversely, knowledge enlargement, which emerged as a form of job design since the original study, had mostly benefits (e.g., higher satisfaction and customer service, and lower overload and errors).

There are several important implications of the latter study. First, it illustrates that the long-term effects of job design changes can be different than the short-term effects. Second, it shows the classic distinction between enlargement and enrichment (Herzberg, 1966) in that simply adding more tasks did not improve the job, but adding more knowledge opportunities did. Third, it illustrates how the job design process is iterative. In this setting, the more favorable knowledge enlargement was discovered only after gaining experience with task enlargement. Fourth, as in the previous study, it shows that it is possible in some situations to gain benefits of job design without incurring all the potential costs, thus minimizing the trade-offs between the motivational and mechanistic approaches to job design.

14.5.9 Example of an Evaluation of a Team Design

Studies conducted by the authors and their colleagues are described here as an illustration of an evaluation of a team design project (Campion, Medsker, and Higgs, 1993; Campion, Papper, and Medsker, 1995). They illustrate the use of mutliple sources of data and multiple types of team effectiveness outcomes. The setting was the same financial services company as in the example job design evaluation above. Questionnaires based on Table 14.3 were administered to 391 clerical employees in 80 teams and 70 team managers in the first study (Campion, Medsker, and Higgs, 1993) and to 357 professional workers in 60 teams (e.g., systems analysts, claims specialists, underwriters) and 93 managers in the second study (Campion, Papper, and Medsker, 1995) to measure teams' design characteristics. Thus, two sources of data were used, team members and team managers, to measure the team design characteristics.

In both studies, effectiveness outcomes included the organization's employee satisfaction survey, which had been administered at a different time than the team design characteristics questionnaire, and managers' judgments of teams' effectiveness, measured at the same time as the team design characteristics. In the first study, several months of records of team productivity were also used to measure effectiveness. Additional effectiveness measures in the second study were employees' judgments of their team's effectiveness, measured at the same time as the team design characteristics, managers judgments of teams' effectiveness, measured a second time three months after the team design characteristics, and the average of team members' most recent performance ratings.

Results indicated that all of the team design characteristics had positive relationships with at least some of the outcomes. Relationships were strongest for process characteristics, followed by job design, context, interdependence, and composition characteristics (see Figure 14.1). Results also indicated that when teams were well designed according to the team design approach, they were higher on both employee satisfaction and team effectiveness ratings than less well designed teams.

Results were stronger when the team design characteristics data were from team members, rather than from the team managers. This illustrates the importance of collecting data from different sources to gain different perspectives on the results of a team design project. Collecting data from only a single source may lead one to draw different conclusions about a design project than if one obtains a broader picture of the team design results from multiple sources.

Results were also stronger when outcome measures came from employees (employee satisfaction, team member judgments of their teams), managers rating their own teams, or productivity records, than when they came from other managers or from performance appraisal ratings. This illustrates the use of different types of outcome measures to avoid drawing conclusions from overly limited data. This example also illustrates the use of separate data collection methods and times for collecting team design characteristics data versus team outcomes data. A single data collection method and time in which team design characteristics and outcomes are collected from the same source (e.g., team members only) on the same day can create an illusion of higher relationships between design characteristics and outcomes than really exist. Although it is more costly to use multiple sources, methods, and administration times, the ability to draw conclusions from the results is far stronger if one does.

REFERENCES

Argyris, C. (1964). *Integrating the individual and the organization.* New York: John Wiley.

Arvey, R. D., and Campion, J. E. (1982). The employment interview: A summary and review of recent research. *Personnel Psychology, 35,* 281–322.

Ash, R. A., Levine, E. L., and Sistrunk, F. (1983). The Role of Jobs and Job-Based Methods in Personnel and Human Resources Management. In K. M. Rowland and G. R. Ferris, *Research in Personnel and Human Resources Management* (Vol. 1). Eds., Greenwich, CT: JAI Press.

Astrand, P. O., and Rodahl, K. (1977). *Textbook of work physiology: Physiological bases of exercise* (Second Edition). New York: McGraw-Hill.

Babbage, C. (1835). On the Economy of Machinery and Manufacturers. Reprinted in *Design of Jobs* (Second Edition). L. E. Davis and J. C. Taylor, Eds., Santa Monica, CA: Goodyear.

Bass, B. M., and Klubeck, S. (1952). Effects of seating arrangements and leaderless team discussions. *Journal of Abnormal and Social Psychology, 47,* 724–727.

Blauner, R. (1964). *Alienation and freedom.* Chicago: University of Chicago Press.

Brown, B. K., and Campion, M. A. (1994). Biodata phenomenology: Recruiters' perceptions and use of biographical information in personnel selection. *Journal of Applied Psychology, 79,* 897–908.

Buller, P. F., and Bell, C. H. (1986). Effects of team building and goal setting on productivity: A field experiment. *Academy of Management Journal, 29,* 305–328.

Buys, C. J. (1978). Humans would do better without groups. *Personality and Social Psychology Bulletin, 4,* 123–125.

Campion, M. A. (1988). Interdisciplinary approaches to job design: A constructive replication with extensions. *Journal of Applied Psychology, 73,* 467–481.

Campion, M. A. (1989). Ability requirement implications of job design: An interdisciplinary perspective. *Personnel Psychology, 42,* 1–24.

Campion, M. A., and Berger, C. J. (1990). Conceptual integration and empirical test of job design and compensation relationships. *Personnel Psychology, 43,* 525–554.

Campion, M. A., Campion, J. E., and Hudson, J. P. (1994). Structured interviewing: A note on incremental validity and alternative question types. *Journal of Applied Psychology, 79,* 998–1002.

Campion, M. A., Cheraskin, L., and Stevens, M. J. (1994). Job rotation and career development: Career-related antecedents and outcomes of job rotation. *Academy of Management Journal, 37,* 1518–1542.

Campion, M. A., and McClelland, C. L. (1991). Interdisciplinary examination of the costs and benefits of enlarged jobs: A job design quasi-experiment. *Journal of Applied Psychology, 76,* 186–198.

Campion, M. A., and McClelland, C. L. (1993). Follow-up and extension of the interdisciplinary costs and benefits of enlarged jobs. *Journal of Applied Psychology, 78,* 339–351.

Campion, M. A., and Medsker, G. J. (1992). Job Design, In *Handbook of Industrial Engineering,* G. Salvendy, Ed., New York: Wiley.

Campion, M. A., Medsker, G. J., and Higgs, A. C. (1993). Relations between work group characteristics and effectiveness: Implications for designing effective work groups. *Personnel Psychology, 46,* 823–850.

Campion, M. A., Papper, E. M., and Medsker, G. J. (1995). Relations between work team characteristics and effectiveness: A replication and extension. *Personnel Psychology, 49,* 429–452.

Campion, M. A., and Stevens, M. J. (1991). Neglected questions in job design: How people design jobs, influence of training, and task-job predictability. *Journal of Business and Psychology, 6,* 169–191.

Campion, M. A., and Thayer, P. W. (1985). Development and field evaluation of an interdisciplinary measure of job design. *Journal of Applied Psychology, 70,* 29–43.

Campion, M. A., and Thayer, P. W. (1987). Job design: Approaches, outcomes, and trade-offs. *Organizational Dynamics, 15(3),* 66–79.

Caplan, R. D., Cobb, S., French, J. R. P., Van Harrison, R., and Pinneau, S. R., (1975). *Job demands and worker health: Main effects and occupational differences.* HEW Publication No. (NIOSH) 75–160. U.S. Government Printing Office, Washington, DC.

Cartwright, D. (1968). The nature of team cohesiveness. In D. Cartwright and A. Zander, Eds., *Team Dynamics: Research and Theory* (Third Edition). New York: Harper & Row.

Chapanis, A. (1970). Relevance of physiological and psychological criteria to man-machine systems: The present state of the art. *Ergonomics, 13,* 337–346.

Cook, T. D., and Campbell, D. T. (1979). *Quasi-experimentation: Design & analysis issues for field settings.* Chicago: Rand-McNally.

Cummings, T. G. (1978). Self-regulating work teams: A sociotechnical synthesis, *Academy of Management Review, 3,* 625–634.

Cummings, T. G. (1981). Designing Effective Work Groups. In P. C. Nystrom and W. H. Starbuck, Eds., *Handbook of Organization Design,* Vol. 2, New York: Oxford University Press.

Davis, L. E. (1957). Toward a theory of job design. *Journal of Industrial Engineering, 8,* 305–309.

Davis, L. E. (1971). The coming crisis for production management: Technology and organization. *International Journal of Production Research, 9,* 65–82.

Davis, L. E. (1982). Organization Design. In G. Salvendy, Eds., *Handbook of Industrial Engineering.* New York: Wiley.

Davis, L. E., Canter, R. R., and Hoffman, J. (1955). Current job design criteria. *Journal of Industrial Engineering, 6*(2), 5–8, 21–23.

Davis, L. E., and Taylor, J. C. (1979). *Design of Jobs* (Second Edition) Santa Monica, CA: Goodyear.

Davis, L. E., and Valfer, E. S. (1965). Intervening responses to changes in supervisor job design. *Occupational Psychology, 39,* 171–189.

Davis, L. E., and Wacker, G. L. (1982). Job design. In G. Salvendy, Eds., *Handbook of Industrial Engineering.* New York: Wiley.

Davis, L. E., and Wacker, G. L. (1987). Job design. In G. Salvendy, Eds., *Handbook of Human Factors.* New York: Wiley.

Dyer, J. (1984). Team research and team training: A state-of-the-art review. In F. A. Muckler, Ed., *Human Factors Review,* Santa Monica, CA: Human Factors Society.

Emory, F. E., and Trist, E. L. (1969). Sociotechnical systems. In C. W. Churchman and M. Verhulst, Eds., *Management Sciences, Models, and Techniques* (Vol. 2). London: Pergamon Press.

Fine, S. A., and Wiley, W. W. (1971). *An introduction to functional job analysis.* W. E. Upjohn Institute for Employment Research, Kalamazoo, MI.

Ford, R. N. (1969). *Motivation through the work itself.* New York: American Management Association.

Fried, Y., and Ferris, G. R. (1987). The validity of the job characteristics model: A review and meta-analysis. *Personnel Psychology, 40,* 287–322.

Gael, S. (1983). *Job analysis: A guide to assessing work activities.* San Francisco, CA: Jossey-Bass.

Gallagher, C. C., and Knight, W. A. (1986). *Team technology production methods in manufacture.* Chichester: Ellis Horwood Limited.

Gaugler, B. B., Rosenthal, D. B., Thornton, G. C., and Benston, C. (1987). Metaanalysis of assessment center validity (Monograph). *Journal of Applied Psychology, 72,* 493–511.

Gilbreth, F. B. (1911). *Motion study: A method for increasing the efficiency of the workman.* New York: Van Nostrand.

Gladstein, D. L. (1984). Groups in context: A model of task group effectiveness. *Administrative Science Quarterly, 29,* 499–517.

Globerson, S., and Crossman, E. R. (1976). Nonrepetitive time: An objective index of job variety. *Organizational Behavior and Human Performance, 17,* 231–240.

Goodman, P. S., Ravlin, E. C., and Argote, L. (1986). Current thinking about teams: Setting the stage for new ideas. In P. S. Goodman and Associates, Eds., *Designing Effective Work Teams.* San Francisco: Jossey-Bass.

Grandjean, E. (1980). *Fitting the tasks to the man: An ergonomic approach.* London: Taylor & Francis.

Griffin, R. W. (1982). *Task design: An integrative approach.* Glenview, IL: Scott-Foresman.

Griffin, R. W. (1989). Work redesign effects on employee attitudes and behavior: A long-term experiment. *Academy of Management Best Papers Proceedings* (Washington, DC), 214–219.

Guzzo, R. A., and Shea, G. P. (1992). Group performance and intergroup relations in organizations. In M. D. Dunnette and L. M. Hough, eds., *Handbook of Industrial and Organizational Psychology* (Vol. 3), Palo Alto, CA: Consulting Psychologists Press.

Hackman, J. R. (1987). The design of work teams. In *Handbook of Organizational Behavior,* J. Lorsch, Eds., New Jersey: Prentice-Hall.

Hackman, J. R., and Oldham, G. R. (1980). *Work Redesign.* Reading, MA: Addison-Wesley.

Hammond, R. W. (1971). The History and Development of Industrial Engineering. In H. B. Maynard, eds., *Industrial Engineering Handbook* (Third Edition), New York: McGraw-Hill.

Harkins, S. G. (1987). Social loafing and social facilitation. *Journal of Experimental Social Psychology, 23,* 1–18.

Harvey, R. J. (1991). Job analysis. In M. D. Dunnette and L. M. Hough, eds., *Handbook of Industrial and Organizational Psychology,* Vol. 2, Second Edition, Palo Alto, CA: Consulting Psychologists Press.

Herzberg, F. (1966). *Work and the nature of man.* Clevcland, OH: World.

Hoerr, J. (1989). The payoff from teamwork. *Business Week,* July 10, 56–62.

Hofstede, G. (1980). *Culture's consequences.* Beverly Hills, CA: Sage.

Homans, G. C. (1950). *The human group.* New York: Harcourt, Brace, and World.

Hoppock, R. (1935). *Job satisfaction.* New York: Harper and Row.

Hyer, N. L. (1984). Management's guide to team technology. In N. L. Hyer, Ed., *Team Technology at Work,* Dearborn, MI: Society of Manufacturing Engineers.

Janis, I. L. (1972). *Victims of groupthink.* Boston: Houghton-Mifflin.

Johannson, G., Aronsson, G., and Lindstrom, B. O. (1978). Social psychological and neuroendocrine stress reactions in highly mechanized work. *Ergonomics, 21,* 583–599.

Kelly, J. (1982). *Scientific management, job redesign, and work performance.* London: Academic Press.

Kiggundu, M. N. (1983). Task interdependence and job design: Test of a theory. *Organizational Behavior and Human Performance, 31,* 145–172.

Konz, S. (1983). *Work design: Industrial ergonomics* (Second Edition). Columbus, OH: Grid.

Lawler, E. E. (1986). *High-involvement management: Participative strategies for improving organizational performance.* San Francisco: Jossey-Bass.

Leventhal, G. S. (1976). The Distribution of Rewards and Resources in Teams and Organizations. In L. Berkowitz and E. Walster, Eds., *Advances in Experimental Social Psychology* (Vol. 9). New York: Academic Press.

Levine, E. L., Ash, R. A., Hall, H., and Sistrunk, F. (1983). Evaluation of job analysis methods by experienced job analysts. *Academy of Management Journal, 26,* 339–348.

Locke, E. A., and Latham, G. P. (1990). *A theory of goal setting and task performance.* Englewood Cliffs, NJ: Prentice Hall.

Majchrzak, A. (1988). *The human side of factory automation.* San Francisco: Jossey-Bass.

Manz, C. C., and Sims, H. P. (1987). Leading workers to lead themselves: The external leadership of self-managing work teams. *Administrative Science Quarterly, 32,* 106–129.

Mayo, E. (1933). *The Human Problems of an Industrial Civilization.* New York: Macmillan.

McCormick, E. J., Jeanneret, P. R., and Mecham, R. C. (1972). A study of job characteristics and job dimensions as based on the Position Analysis Questionnaire (PAQ), *Journal of Applied Psychology, 56,* 347–368.

McGrath, J. E. (1984). *Teams: Interaction and performance.* Englewood Cliffs, NJ: Prentice-Hall.

Meister, D. (1971). *Human factors: Theory and practice.* New York: John Wiley.

Milkovich, G. T., and Newman, J. M. (1993). *Compensation* (Fourth Edition). Homewood, IL: Business Publications.

Mintzberg, H. (1979). *The structuring of organizations: A synthesis of the research.* Englewood Cliffs, NJ: Prentice-Hall.

Mundel, M. E. (1985). *Motion and time study: Improving productivity* (Sixth Edition). Englewood Cliffs, NJ: Prentice-Hall.

Niebel, B. W. (1988). *Motion and Time Study* (Eighth Edition). Homewood, IL: Irwin.

O'Reilly, C., Parlette, G., and Bloom, J. (1980). Perceptual measures of task characteristics: The biasing effects of differing frames of reference and job attitudes. *Academy of Management Journal, 23,* 118–131.

O'Reilly, C. A., and Roberts, K. H. (1977). Task group structure, communication, and effectiveness. *Journal of Applied Psychology, 62,* 674–681.

Pasmore, W. A. (1988). *Designing effective organizations: The sociotechnical systems perspective.* New York: John Wiley.

Pasmore, W., Francis, C., and Haldeman, J. (1982). Sociotechnical systems: A North American reflection on empirical studies of the seventies. *Human Relations, 35,* 1179–1204.

Pearson, R. G. (1971). Human Factors Engineering. In H. B. Maynard, Ed., *Industrial Engineering Handbook* (Third Edition) New York: McGraw-Hill.

Perkins, A. L., and Abramis, D. J. (1990). Midwest Federal Correctional Institution. In J. R. Hackman, Ed., *Groups That Work (and Those That Don't).* San Francisco: Jossey-Bass.

Porter, L. W., Lawler, E. E., and Hackman, J. R. (1987). Ways teams influence individual work effectiveness. In R. M. Steers and L. W. Porter, Eds., *Motivation and Work Behavior* (Fourth Edition), New York: McGraw-Hill.

Rousseau, D. M. (1977). Technological differences in job characteristics, employee satisfaction, and motivation: A synthesis of job design research and sociotechnical systems theory. *Organizational Behavior and Human Performance, 19,* 18–42.

Salas, E., Dickinson, T. L. Converse, S. A., and Tannenbaum, S. I. (1992). Toward an understanding of team performance and training. In R. W. Swezey and E. Salas, Eds., *Teams: Their Training and Performance,* Norwood, NJ: Ablex.

Salvendy, G., Ed., (1987). *Handbook of human factors.* New York: John Wiley.

Salvendy, G., and Smith, M. J., Eds. (1981). *Machine pacing and occupational stress.* London: Taylor & Francis.

Sanders, M. S., and McCormick, E. J. (1987). *Human factors in engineering and design* (Sixth Edition). New York: McGraw-Hill.

Shaw, M. E. (1983). Team composition. In Vol. 1. H. H. Blumberg, A. P. Hare, V. Kent, and M. Davies Eds., *Small Teams and Social Interaction.* New York: John Wiley.

Shea, G. P., and Guzzo, R. A. (1987). Teams as human resources. In K. M. Rowland and G. R. Ferris, Eds., *Research in Personnel and Human Resources* Vol. 5. Greenwich, CT: JAI Press.

Sims, H. P., Szilagyi, A. D., and Keller, R. T. (1976). The measurement of job characteristics. *Academy of Management Journal, 19,* 195–212.

Smith, A. (1776). *An inquiry into the nature and causes of the wealth of nations.* Reprinted by R. H. Campbell and A. S. Skinner, Eds. Indianapolis: Liberty Classics.

Steers, R. M., and Mowday, R. T. (1977). The motivational properties of tasks. *Academy of Management Review, 2,* 645–658.

Steiner, I. D. (1972). *Group process and productivity.* New York: Academic Press.

Stevens, M. J., and Campion, M. A. (1994a). *Staffing teams: Development and validation of the teamwork-KSA test.* Paper presented at the annual meeting of the Society of Industrial and Organizational Psychology (Nashville, TN).

Stevens, M. J., and Campion, M. A. (1994b). The knowledge, skill, and ability requirements for teamwork: Implications for human resource management. *Journal of Management, 20,* 503–530.

Sundstrom, E., DeMeuse, K. P., and Futrell, D. (1990). Work teams: Applications and effectiveness. *American Psychologist, 45,* 120–133.

Swezey, R. W., and Salas, E. (1992). *Teams: Their training and performance.* Norwood, NJ: Ablex.

Tannenbaum, S. I., Beard, R. L., and Salas, E. (1992). Team building and its influence on team effectiveness: An examination of conceptual and empirical developments. In K. Kelley, Ed., *Issues, Theory, and Research in Industrial and Organizational Psychology.* Amsterdam: Elsevier.

Taylor, J. C. (1979). Job design criteria twenty years later. In L. E. Davis and J. C. Taylor, eds., *Design of Jobs* (Second Edition) New York: Wiley.

Taylor, F. W. (1911). *The principles of scientific management.* New York: Norton.

Thompson, J. D. (1967). Organizations in action. New York: McGraw-Hill.

Tichauer, E. R. (1978). *The biomechanical basis of ergonomics: Anatomy applied to the design of work situations.* New York: Wiley.

Turnage, J. J. (1990). The challenge of new workplace technology for psychology, *American Psychologist, 45,* 171–178.

Turner, A. N., and Lawrence, P. R. (1965). *Industrial jobs and the worker: An investigation of response to task attributes.* (Harvard Graduate School of Business Administration, Boston.

U.S. Department Of Labor (1972). *Handbook for analyzing jobs* (U.S. Government Printing Office, Washington, DC).

U.S. Department Of Labor (1977). *Dictionary of occupational titles* (Fourth Edition) (U.S. Government Printing Office, Washington, DC).

U.S. Nuclear Regulatory Commission (1981). *Guidelines for control room design reviews* (NUREG-0700) (Nuclear Regulatory Commission, Washington, DC).

Vroom, V. H. (1964). *Work and motivation.* New York: Wiley.

Walker, C. R., and Guest, R. H. (1952). *The man on the assembly line.* Cambridge, MA: Harvard University Press.

Wall, T. B., Kemp, N. J., Jackson, P. R., and Clegg, C. W. (1986). Outcomes of autonomous workgroups: A long-term field experiment. *Academy of Management Journal, 29,* 280–304.

Warr, P., and Wall, T. (1975). *Work and well-being.* Harmondsworth, MD: Penguin.

Weber, A., Fussler, C., O'Hanlon, J. F., Gierer, R., and Grandjean, E. (1980). Psychophysiological effects of repetitive tasks. *Ergonomics, 23,* 1033–1046.

Welford, A. T. (1976). *Skilled performance: Perceptual and motor skills.* Glenview, IL: Scott-Foresman.

Wicker, A., Kirmeyer, S. L., Hanson, L., and Alexander, D. (1976). Effects of manning levels on subjective experiences, performance, and verbal interaction in groups. *Organizational Behavior and Human Performance, 17,* 251–274.

Wong, C. S. (1989). *Task Interdependence: The Link Between Task Design and Job Design.* Doctoral Dissertation, Purdue University, West Lafayette, IN.

Wong, C. S., and Campion, M. A. (1991). Development and test of a task level model of job design. *Journal of Applied Psychology, 76,* 825–837.

Woodman, R. W., and Sherwood, J. J. (1980). The role of team development in organizational effectiveness. A critical review. *Psychological Bulletin, 88,* 166–186.

Zajonc, R. B. (1965). Social facilitation. *Science, 149,* 269–274.

Zander, A. (1979). The study of group behavior over four decades. *Journal of Applied Behavioral Science, 15,* 272–282.

CHAPTER 15
PARTICIPATORY ERGONOMICS

John R. Wilson
Helen M. Haines
Institute for Occupational Ergonomics
Department of Manufacturing Engineering and Operations Management
University of Nottingham
Nottingham, NG7 2RD, United Kingdom

15.1 INTRODUCTION

15.1.1 Participation and Ergonomics

This chapter is concerned with participatory ergonomics. This, as will be explained below, can be seen as the use of participative techniques within ergonomics enquiry and intervention *and* as the implementation of ergonomics within a participative framework or organization.

Participatory ergonomics might be regarded as a relatively recent phenomenon; others may view it to have existed since the inception of ergonomics as a discipline, but without formal title or definition. Whichever, there is considerable and growing interest in the connections between participation and ergonomics and this should not surprise us.

For ergonomists, participation—as a philosophy and a process—has always been of interest, partly due to the strong ergonomics emphasis on application. People-centered approaches are endemic, for instance, user trials in consumer product testing (e.g., McClelland, 1995), user-centered design in human–computer interaction (Chapter 46) and human–centered technology programs in manufacturing industry (Chapters 18 and 56). Ergonomists utilized user trials in their work long before computer systems developers 'discovered' the benefits of user representatives in rapid prototyping, storyboarding, and laboratory or field experiments. Most ergonomists also take a broad view of work redesign and of improving the work environment, arguing that the physical aspects of work neither can nor should be divorced from a consideration of psychosocial and organizational factors. Even in a work redesign program with the physical environment established as of greatest concern, to ignore other aspects of work may diminish acceptance and success of any changes made due to the central role of workforce attitudes (Wilson and Grey, 1990).

What perhaps distinguishes ergonomics from other disciplines is the paramount need to input data and knowledge into practical applications; thus, anything which can assist this—and the participation of potential users must, on the face of it, do so—will be attractive to ergonomists. Furthermore, the functionally and temporally isolated nature of their activities for many ergonomics practitioners (McNeese et al., 1995) can promote the perceived need for participative ergonomics practice. Finally, the personal philosophy of many ergonomists and their antipathy to the Taylorist approach in work design means that a participative approach within their own professional activities is very attractive to them.

15.1.2 Growth in Participatory Ergonomics

One timely measure of a discipline's growth of interest in a topic or subdiscipline is the numbers of relevant presentations at international conferences. Scientific journal publications may be subject to long delays, contain an overrepresentation of academic contributions, be limited by problems of commercial confidentiality or shortage of time, and—in some cases—suffer from editorial bias toward 'respectable' laboratory research. Conference presentations, on the other hand, may reflect a wider base of interest—geographically and professionally, they can be more timely and less restricted by confidentiality or time resource problems, and can cover all areas of interest and types of presentation from controlled data collection to speculative discussion. By this measure participatory ergonomics has certainly grown; from a panel discussion and a scattering of papers at IEA '85, each succeeding Congress of the International Ergonomics Association in 1988, 1991, and 1994 has seen an increased number of special sessions of papers presented and discussion panels. The authoritative Ergonomics Abstracts from the Ergonomics Information and Advisory Centre shows a marked increase also, say from the early 1980s to the early 1990s.

If participatory ergonomics has grown—in terms of the number of reported initiatives and hopefully in influence—in the past 10 years, then we need to ask why. In part this is to do with a widespread general revival of interest in participation which itself may be attributed to a number of circumstances. Despite reverses and obvious exceptions, the social and political climate worldwide can be seen to have become more participative over the past 40 or 50 years; this still has a long way to go of course and improvements are very slow and difficult, but the general public and the workforce in many countries will not accept management practices which were widespread half a century ago. Unemployment in industrially developed countries—whether caused structurally by governments or part of the economic cycle—will slow the process of democratization and participation but will not irredeemably reverse it. Disappointment with technical investments and recognition that motivation and performance are complex issues has spawned human-centered manufacturing initiatives with emphasis upon an educated, involved, and responsible workforce. Greater emphasis on attributes of quality, flexibility and customer service, rather than merely quantity of output, supports greater workforce participation. It is also possible that today's technologies, with less emphasis on rigid, complex, and large-scale machinery and systems (ship yards, steelworks) or Fordist assembly lines and more on information systems and product-based manufacture (or cells), are more fertile ground for participation. Certainly modern organizations, which are IT-based, flatter, and more decentralized, are likely to be more supportive of participation than traditional hierarchical, authoritarian, and physical activity-based factories (see also Gaudier, 1988).

The above reasons are all germane to the growth in participatory ergonomics as well as to participation in general. To them we must add also the reasons already explained of why participation and ergonomics are such compatible bedfellows. More pragmatically, we can also note developments in ergonomics and health and safety legislation. For instance, the European Framework Directive (89/391/EEC) on "the introduction of measures to encourage improvements in the safety and health of workers at work" contained specific provisions about consultation and participation of workers (Article 11). Much of the remainder of the Directive can be shown also to have a strong relationship to, and implications for, participative processes. The so-called Daughter Directives—for instance, on manual handling or display screen equipment—also make provision for a joint approach to health and safety by employers and workers. Although all the specific provisions have not yet always been translated into national legislation, the changing legislative scene has already had considerable impact for ergonomics in Europe.

Finally, the possible systemic advantages of participation, particularly in the light of the types of problems and challenges facing ergonomists today, almost dictate the need for participation in general and for participatory ergonomics in particular. The issue is taken up again later in this chapter.

In summary, the reasons for growth in participatory ergonomics generally, and for its use in specific implementations or investigations, might be neatly, if cynically, described as **Need, Greed,** or **Vision.**

15.1.3 Definition of Participatory Ergonomics

For reasons alluded to above and expanded below, a neat definition of participatory ergonomics is elusive. An analogy might be drawn with views of *just-in-time (JIT)* manufacturing (Oliver, 1990). Like JIT, participation and participatory ergonomics will be seen by different people at different times as a philosophy (even a "religion"), an approach or strategy, a program, a set of tools and techniques, or a requirement to be met by those tools and techniques. Perhaps because the notion of participation, and thus participatory ergonomics, seems so self-evident, few formal definitions exist. The dictionary tells us that to participate is to have a share or to take part in something *(Chambers 20th Century Dictionary)*; thus participatory ergonomics will, by extension, consist of 'stakeholders' taking part in an ergonomics initiative or sharing ergonomics knowledge and methods. Stakeholders include anyone affected by the process or consequent changes and is a broader group than 'users' or 'workers'.

Already here we can see a strong, if sometimes subtle, distinction between:

- Participatory ergonomics in which stakeholders contribute to an ergonomics initiative (e.g., redesign or investigation), where the focus is on use of participation to enhance the specific ergonomics initiative, and
- Participatory ergonomics where ergonomics and other techniques are used to strengthen a participative approach to work.

This distinction can be seen in the description of participatory ergonomics by Noro (1991) as both a new technology for disseminating ergonomics information and also as a procedure whereby ergonomists work together with non-ergonomists on a company-wide basis. For this to happen, Noro identifies the need for 'fusion', assuming cooperation and rational resolution of trade-offs. In the same book, his co-editor Imada (1991) sees participatory ergonomics as one perspective in systems (or macro) ergonomics, which requires end-users as the beneficiaries of ergonomics to be involved in developing and implementing technology. Similarly, Lewis et al (1988) state that "The rationale behind participatory ergonomics is to involve the end-user in the change process so that he/she becomes an advocate and an active change agent rather than a passive recipient of the process."

More recently, Nagamachi (1995) defines participatory ergonomics as "the workers' active involvement in complementary ergonomic knowledge and procedures in their workplace . . . supported by their supervisors and managers in order to improve their working conditions and product quality" (p. 371). Sen (1988) sees ergonomics participative change as embracing models of "create, catalyse and care," "design, develop and demonstrate," and "implement, involve, and improve." The first ensures the optimum climate for change, the second provides a method to ensure feasibility and perception of quality for the new system, and the third optimizes the chances of successful change and subsequent enhancements.

We have proposed a definition of participation in the context of ergonomics management programs at work, as "**The involvement of people in planning and controlling a significant amount of their own work activities, with sufficient knowledge and power**

to influence both processes and outcomes in order to achieve desirable goals" (Wilson, 1995a). Within this definition is encompassed all the situations identified above—ergonomics as a part of a participative program, participative techniques used to enhance an ergonomics initiative, and participation to assist with change implementation—and also the macro- and microlevels of participatory ergonomics as defined below. Influence on both processes and outcomes is needed due to the intimate connection between the process and content of change, further emphasized by the need for involvement in both planning and controlling work activities.

The key part of the definition is the requirement that potential participants have "sufficient knowledge and power." This is not to suggest that participants will have unlimited freedom; there will almost always be limits on what they can do and on the suggestions they would like to implement. However, limits on what is achievable should always be made clear at the outset, in terms of what is a "significant amount" of the activities, what the "desirable goals" are, and thus what "knowledge and power" are needed for participants to be sufficiently influential. It is the perceived absence of any real knowledge and power, and thus lack of belief that they can influence outcomes, that can lead to nonparticipation or half-hearted involvement and cynicism on the part of stakeholders (see below).

15.1.4 Participative Work Organization

We cannot discuss participatory ergonomics without mentioning the concurrent growth of participation in overall organization management. Many traditions in work organization theory and practice have elements of participation as a central notion, almost all in reaction to the autocratic and nonparticipative ethos of scientific management or "Taylorism." Although job enrichment as proposed by Herzberg (1966) perhaps was not participative (for instance, Herzberg and others recommended implementation without consultation), adaptations since have been much more participative in nature. Sociotechnical systems theory and its incorporation into group-based work designs has also stressed participative development and implementation (e.g., Chapter 14). In his influential work, Robert Karasek has emphasized the participatory work process as combining increases in social interaction and support with greater opportunities for decision making and control over one's work activities, in order to enhance worker health and productivity (e.g., Karasek and Theorell, 1990). In the same tradition, but perhaps with the narrower focus of systems design, Mumford (1991) proposes participation as a vital facilitating process, with contribution to company performance and employee welfare.

Participation under various guises has had something of a renaissance in human resources management and operations management circles. High-performing (or so-called world class) companies are said to have employee empowerment as one of the enablers to be successful, and this has often been in tandem with teamwork as another enabler. People Involvement programs have grown out of quality circles and action groups as a part of Total Quality Management. Participative management is more and more accepted as *the* effective approach to change and to organization of work, and as a part of human centered systems generally. Over a decade ago in a collection of case histories of change in UK companies, the Work Research Unit (1982) found that 'participation' (although defined differently in different cases) was a characteristic feature of successful change. The Ingersoll Engineers (1987) Report into manufacturing technology implementation also found the biggest single barrier to technology change to be a failure to involve people closely and to generate feelings of ownership.

It is difficult at times to separate the impact of particular programs of participatory ergonomics from that of general participative-based management. The former are often discussed within the context of programmes of total quality, quality of work life, continuous improvement or *Kaizen,* production groups, quality circles, and the like (e.g., Liker et al., 1991; Nagamachi, 1991, 1995; Noro, 1991; Zink, 1991). There is a need for caution of course. Many commentators, workers, and trade unions (as well as ergonomists) are skeptical about some aspects of the 'world class company'. It is difficult sometimes to distinguish exactly what is meant by Total Quality Management *in practice,* and workforce involvement sometimes only seems to be to the extent that this suits management. Thinking of the definition of participatory ergonomics proposed earlier, participants may not be given any real knowledge or power or may not be permitted really to influence outcomes.

However, although there is a need for caution, ergonomists should embrace the opportunities this widespread interest in participation can open up. Ergonomics can take advantage of current concerns with involvement and empowerment, with support and interaction, and can become an integral part of total quality programs and of the strategic mix required by a high-performing company.

15.2 DIMENSIONS OF PARTICIPATORY ERGONOMICS

15.2.1 Models and Structures for Participatory Ergonomics

Does participatory ergonomics have theoretical models or application frameworks, to parallel those of motivation and job design for instance? Are they needed? As a result of an IEA Congress round table session in 1991, Vink et al. (1992) concluded that there is no single, unifying model or theoretical framework for participatory ergonomics. However, Liker et al. (1989) describe six models of participation each of which is a combination of two dimensions. The first dimension is taken from the work of Vroom and Yetton (1973), which outlines three modes of participation: First, where managers make decisions having consulted with individuals; second, where managers do the same having had group consultations, and third, where managers and staff negotiate joint decisions. The other dimension comes in the form of Coch and French's distinction between "direct" and "representative" participation (Coch and French, 1948). In the former case, each worker contributes to the consultative process, whereas in the latter case an individual is selected as the spokesperson. Liker et al.'s six-model framework ranges, therefore, from a situation where managers consult individual worker representatives yet still make the decisions themselves (model I) to one where all employees both participate in consultations and have a role in the decision-making process (model VI).

In practice, ergonomics programs reported in the literature have a variety of different structures. For example, in a comparative analysis of participatory ergonomics programs in U.S. and Japanese manufacturing plants, Liker et al. (1989) note that while the U.S. cases consisted of multifunction groups of representative workers, lead either by engineers or by area managers (in one instance a steering committee was also set up), the Japanese cases were more hierarchical. There, steering committees made the decisions while staff members (e.g., industrial engineers, production supervisors) were responsible for some of the analysis work, equipment design, training, and so on. Quality circles provided data and improvement suggestions.

In a paper on marketing participatory ergonomics, Brown (1990) advocates a "contingency" model of participation arguing that no single approach will be effective in all situations (see also Gjessing et al., 1994). Brown describes the role of the ergonomist, therefore, as one of influencing management to utilize the most appropriate participatory techniques, and marketing as the "critical first step" in the adoption and implementation of participatory ergonomics. It is only when management has been sold the idea that participative approaches will improve the way in which their organization functions that they will begin to utilize it.

There seems to be a number of dimensions across which any participatory ergonomics initiative might be defined. We have previously distinguished three dimensions: discrete or continuous use; remotely or directly coupled; and actual or representative "users" (Wilson, 1991a). Taking this and Liker et al.'s (1989) models, we believe now that there are more dimensions than this, and understanding them will both enable insight into the different forms of participatory ergonomics and also inform appropriate development and use of participative techniques (Wilson and Haines, 1996). These dimensions and their "scales" are shown in Figure 15.1, and are discussed in some detail below.

15.2.2 Level

The first dimension of participatory ergonomics is the level along a spectrum that we term *macroparticipation* to *microparticipation.* At the macro end, there is the notion of participative working and the participative organization; worker cooperatives are perhaps at the extreme. At the microlevel, we might see one-off uses of a participative technique, in redesign of a problematic workstation for example. In many cases, we will be working simultaneously with participatory ergonomics at different points along this spectrum, and indeed often with microlevel activities within a macrolevel strategy. It is an interesting debate as to whether a macrolevel participative process should be implemented before utilizing techniques at a microlevel, or whether use of the latter will actually help put participative strategies in place.

15.2.3 Focus

Related in part to the level of participation is its focus. If the ergonomics is at the macrolevel, then participation will be applied across an **organization** or at least a **work**

Dimension	"Scale"
Level	Macro ______________________________ Micro
Focus	Organisation ... Worksystem ... Jobs ... Workstation ... Product
Purpose	Work organisation ______ Design ______ Implementation
Timeline	Continuous ______________________________ Discrete
Involvement	Full direct _______ Partial direct ________ Representative
Coupling	Direct ______________________________ Remote
Requirement	Voluntary ______________________________ Necessary

Figure 15.1 Dimensions of participatory ergonomics. (*Source:* Wilson and Haines, 1996.)

system (see also Chapter 18); if we are at the microlevel, then some participative techniques may be applied within the design or redesign of a single **workstation** (see Chapter 23), or **product,** with **job** or **team** design (Chapter 14) in the middle of the scale.

15.2.4 Purpose

One of the great difficulties in any rational discussion of participation within a company is that the underlying purpose is often misunderstood. Is participatory ergonomics to be used to **implement** a particular change or to be *the* method of **work organization** whether under conditions of change or not? Use of participation in **design** might be seen as a particular case of implementation or as another distinct purpose. Of course, when change implementation has been participative, then greatest benefit will be derived if the mechanisms put in place continue as a part of the change. This emphasises again the fact that the process of change is more important than the content; a flexible and robust process can support adaptation of any change content. The "principle of compatibility" (e.g., Cherns, 1987) from sociotechnical systems theory also proposes benefits when the procedures for change become part of the change itself.

15.2.5 Timeline

Implementation of any change, or the set-up of a participative process (see below) should take place at a pace that is neither so fast that participants are left behind or feel excluded nor so slow that it falls into disrepute. A major influence is whether the process has a **continuous** or **discrete** time line: Is participation to be used as an everyday part of an organization's activities or is it applied from time to time as a one-off exercise? Although relatively weak types of participation, job attitude surveys might be seen as discrete and quality circles as continuous initiatives.

15.2.6 Involvement

A basic distinction in participatory ergonomics is whether participation is **direct** or **representative.** The latter occurs primarily in research laboratories, where new products or workstations will be evaluated with subjects selected from the general population so as to reflect the likely users. Representation in occupational settings might come from trade unions. Direct participation itself might be full or partial. **Full direct** participation is when all stakeholders directly affected become participants; a work team in a manufacturing cell deciding on how the group will allocate responsibilities and handle the interface with the organization is an example. It is more likely that resource restrictions and large numbers of potential contributors will dictate representation by a subgroup of those affected, or even by a small number of 'champions'; this is **partial direct** participation, and an example is when a mixed group of operators, supervisors, engineers, health and safety staff, and ergonomists are formed into an action team to tackle upper-limb disorders (cumulative trauma disorders).

15.2.7 Coupling

Participative methods can be **directly** or **remotely coupled.** The former involves relatively direct application, with very little filtering, of participants' views, recommendations, and process outcomes. Work groups redesigning their own workplace are an example. Remote coupling involves some translation of participants' views; suggestion schemes or newsletters, although weak types of participation, are examples of this and company-wide questionnaires are slightly stronger examples.

15.2.8 Requirements

A final classification of participatory ergonomics which might permit differentiation between appropriate approaches and techniques is whether the participation is **voluntary, imposed,** or **necessary.** Voluntary participation is the most usual form, in the sense that participation works best where the workforce volunteers its contributions and is involved in setting up the process. Imposed participation is seen, for instance, in companies with compulsory quality circles or production groups schemes, where involvement in troubleshooting and continuous improvement is an obligatory part of job specification and roles; it is arguable whether such imposed participation is truly participative. Finally, participation from necessity is found where resources allow no other way of implementing change or of working, typically in industrially developing nations (Kogi, 1991; Kogi et al., 1988).

15.2.9 Examples of Participation with Different Dimensions

In order to illustrate the different dimensions of participatory ergonomics, Table 15.1 includes summaries of particular well-known generic examples with their specifications along the 'scales' of each dimension (Wilson and Haines, 1996).

15.3 THE VALUE OF PARTICIPATORY ERGONOMICS

Many of the advantages of a participatory ergonomics approach are implicit in the intimate connection between ergonomics and participation which was described above. Indeed, reported views of ergonomists often seem to assume axiomatically that participation is always the best approach. This is not really sufficient, however, and we must be aware of both its strong and weak points.

It should be emphasized strongly that to provide unequivocal support for the benefits of participation is not easy. Bernoux (1994) reports a survey of managers who attributed 10% of increased productivity to participative measures, but none could actually substantiate this figure. Gjessing et al. (1994) examined evidence for the effectiveness of work involvement in efforts to reduce work-related injury and disease, and found that "Reports documenting the importance of these approaches in cause-effects terms, as well as defining factors of major consequence to successful outcomes are not numerous. Indeed, field studies in this area do not allow for easy isolation of these variables and their manipulation or comparisons with adequate control or nontreatment conditions." This is true of cost-benefit analysis in ergonomics and in work organization generally (e.g., Chapter 49, Corlett, 1988; Simpson and Mason, 1995). Nonetheless, efforts must be made to demonstrate general benefits of participatory ergonomics, to evaluate specific applications and hence also to provide material for the general case. Such evaluations should not be only in

Table 15.1 Generic examples of participatory ergonomics described by different dimensions

Dimension	User Testing for Product or Hci	Self-directed Work Teams	Safety Representatives	Participative Management Programmes
Level	Micro	Micro + Macro	Micro (in Macro?)	Macro
Focus	Product	Jobs/Work systems	General	General
Purpose	Design	Work Organisation	Implementation	Implementation (Work Organisation)
Continuity	Discrete	Continuous	Discrete	Continuous
Involvement	Population representation	Full direct	Partial direct ('champions')	Full/partial direct
Coupling	Remote	Direct	Direct or remote	Direct & remote
Requirement	Voluntary	Voluntary (Imposed?)	Imposed	Voluntary

(*Source:* Wilson and Haines, 1996.)

financial terms; accurate economic estimates for all factors are very difficult to produce and anyway some positive outcomes might not have direct (or even observable indirect) economic consequences.

Most proponents of participatory ergonomics stress two types of direct potential benefit, one linked more to 'participation in implementation' and one to 'participation in design' (see the purpose dimension of participatory ergonomics above). First, use of participative techniques in change analysis, development, and implementation may generate greater feelings of solution ownership among those affected and involved, and thus may breed a greater commitment to the changes being implemented (e.g., Imada and Robertson, 1987; Wilson, 1995b). Second, at the simplest level of analysis the involvement of current job holders or system users in redesign should improve the information and idea generation in the design process. Participants are, after all, those who should know most about the good and bad points of the current situation or design. This should thus result in a better (effective, healthy, safe, satisfying) development as well as a more implementable one (e.g., Noro, 1991).

Related to these first two advantages, participation may often be a learning experience for designers/planners and users/workers alike. For the former it can improve the current and subsequent designs or implementations; the very departure from conventional thinking has been said to improve design effectiveness (Sanoff, 1985). For the latter, involvement in a development or implementation process can mean faster and deeper learning of a new system and hence decreased training costs and improved performance (e.g., see Wilson and Grey Taylor, 1995). Widening the scope of potential benefits, participatory ergonomics can—if successful—sow the seeds for its own extension. By spreading interest, understanding and expertise (within acceptable limits) in ergonomics among those involved, and by attracting interest and desire for involvement among work colleagues, solutions can be generalized and ergonomics transferred elsewhere throughout an organization (Daniellou and Garrigou, 1992; Daniellou et al, 1990).

With the contemporary perspective of human centered manufacturing within advanced manufacturing systems and organizations, a further gain for companies is the fuller use of a scarce resource—the skills and knowledge of their employees. Often unacknowledged, or tacit (Leplat, 1990), these competencies underpin the need for a responsible and involved workforce; knowledge and ability, opportunity and motivation are all necessary if people are to be more involved in their work and their company. Participation generally, and participatory ergonomics in particular, will not only assist and support involvement, they are vital prerequisites. Mambrey et al. (1987) found that people who participate directly show more self-confidence, competence, and independence, and attach more importance to self-determination than do their colleagues. Such attributes may, of course, be seen as the causes not the effects, as the individual characteristics necessary to motivate someone to participate rather than as the outcomes of participation, but Mambrey and his colleagues do report them as gains.

Finally, in the context of effective change and of job satisfaction, the relationship between participative processes and groups or teams is interesting. Basic workplace ergonomics and health and safety interventions can be carried out very efficiently and effectively through work teams, participating together to identify problems and implement improvements. Moreover, involvement in such initiatives can help give the work teams a directed purpose and strengthen the team basis and functioning. Group working is a highly appropriate mechanism for enrichment of job designs and to reduce job strain via social interaction and support (Karasek and Theorell, 1990; Wall et al., 1990). Furthermore, as already indicated, the sociotechnical principle of compatibility dictates that it is desirable for the groups which develop ideas for new workplaces, work tasks and work structures to subsequently form into work teams themselves—the change process enhancing and becoming the content.

15.4 DISADVANTAGES, LIMITATIONS AND PROBLEMS IN PARTICIPATORY ERGONOMICS

Participatory ergonomics seems so appropriate to most ergonomists, and often so attractive to many employers and producers, that we must strike a note of caution. Participation of whatever type and at whatever level is rarely easy to promote and support. At times it may be an inappropriate option. Its misapplication may even be harmful—either at the time or into the future—in terms of stakeholders' attitudes or system performance for instance. As Garrigou et al. (1995) explain, ". . . although participation has been very popular over the last 15 years, it seems that a certain number of participatory experiences ended in failure or disappointment at the level of company managements or staff representatives . . . results expected in matters of production were not achieved . . . [or] . . . working conditions were not really improved . . . [or] . . . operators and their representatives complained of having been 'cheated' and having obtained no benefit from the participation . . . " [p. 313]. Therefore, all ergonomists (and other applied human scientists) must understand the potential disadvantages of a participative approach, the limitations of the techniques and structures, and the problems and difficulties that will be faced. It is not always easy to obtain this understanding because there is not widespread publication or communication of failed experiments in participation (or in many endeavors!). Even where experiments have been successful, there is little motivation to isolate and publicize those aspects that might have been improved.

Participatory ergonomics may fail, or its implementation be problematic or limited, for several broad reasons. The manner of its application might be **cynical** or **half-hearted.** Possibly connected to this there might be **individual** or **group resistance,** or **individuals may be unwilling or unable to participate.** Finally, there are **disadvantages** or **problems intrinsic to the nature and structure of participative efforts.** These broad groups of reasons will be discussed in turn.

Earlier we saw that participation can vary along such dimensions as level, focus, and degree of involvement. As a consequence, initiatives labeled "participatory ergonomics" might be genuine on both counts (i.e., the remit is ergonomics and stakeholders are participating), but at the other end of the spectrum they might be mere exercises in information provision or might even be manipulative programs with a covert agenda. Several writers distinguish types of participation which appear across such a spectrum.

Reuter (1987) defines nonmutually exclusive participation procedures as: decentralization of planning (groups of potential users plan *aspects* of the system); limited participation (selected users are asked about specific problems for part of the system); advocacy planning (one person acts on behalf of the group); leader representation (one of a large unorganized group represents the interests of all, through 'popular' support not institutional power); ombudsman; hearings (e.g. planning enquiries); information (given to 'passive' users); advisory councils; a forum; seminars held with all interested groups; and user initiatives. Further, Reuter quotes Arnstein's (1969) scale, summarized as:

- User control
- Delegation; some decision-making authority, following bargaining
- Partnership; decisions by users and 'owners'.
- Placating; allowing a (powerless) few to participate
- Consultation
- Information (one-way)
- User therapy; 'curing' user pathological ideas
- User manipulation; public relations exercises

Cynical use of participative methods may yield some limited dividends in the short term but will not provide any of the potential greater long term benefits "As long as participation is only concerned with expropriating the knowledge of those concerned, . . . with acceptance (the ability and readiness to accept optimum utilization), and with efficiency . . . the interest is solely that of the management and of system developers employed by it" (Fuchs-Kittowski and Wenzlaff, 1987, p. 5). Such use may partially explain participation's ambiguous tradition within industrial relations as identified by Forrester (1986). He sees both the criteria and context of participation as being viewed with suspicion by trades unions, citing the case of quality circles, often viewed as being used by management to bypass procedural channels and to weaken trade unionism.

It might not only be that resistance is forthcoming from trade unions and the workforce. Management can also see participation as a threat to their right to manage (Mumford, 1991) rather than an aid, and examples of cynical application may be as much to do with operational obstruction from such a source as with any willful intent of the originators.

Resistance to participation links with a second broad group of problems and drawbacks, those to do with the readiness or willingness of the people involved. It is undeniable that at least some in every population group or workforce will not want to participate. This might be genuinely because certain individuals lack, or feel that they lack, sufficient motivation, energy, knowledge, and ability to contribute; technical training may be insufficient (Garrigou et al., 1995), workers might show too much deference to management or technical specialists (Mumford, 1991), or they might feel the required contribution is outside their sphere of knowledge or influence (cognitive limits according to Bernoux, 1994).

However, Neumann (1989) prefers to look for explanations of why people don't participate from the structure and environment of participation itself rather than from the personality or attitudes of people. Thus she identifies structural, relational, and societal explanations for nonparticipation. Structural explanations are that the organization is not suited to participation—real decisions are taken elsewhere, there is no support or reinforcement or encouragement from job designs, training, or reward systems (see the strictures on cynical use above). Relational explanations include poor or discouraging management, strictly hierarchical structures and conflicts with life outside work. Societal explanations are that participation at work might challenge deeply held beliefs, personal (conflict avoidance) strategies, and run contrary to expectations of adversarial relationships.

Finally, even if we discount difficulties resulting from inappropriate, unwilling, or cynical implementation of participation, problems and limitations may still be found which are the outcomes of the particular participative approach and techniques employed. Participation is not an easy option, and hard choices often have to be made. Even where participative processes are implemented with care, commitment and clear goals, certain general disadvantages or particular problems have been identified. For instance, an argument may be raised against the group-based nature of participation. Group decision making is often said to take only the safe options, or to degenerate to the lowest common denominator; this, we feel, merely emphasizes the importance of appropriate structures and methods for group decision making.

Planning and developing new systems, workplaces or organizations participatively *may* be slower, more complex, require greater effort, and thus be more costly. (Of course the first three of these could be blessings in disguise if they lead to a more reflective, universal, and broad-based approach in development.) Some cases have found: that the process causes disturbance or ripple effects in the organization that are dysfunctional; that the participative process does not increase room for maneuver (and due to the efforts required may actually reduce it); that the motivation, knowledge, and energy of potential participants may not be sufficient to carry the process; that the outcomes are not always greater job enrichment or higher acceptance of new technology; and that some situations dictate that user influence is low or that at least not everyone involved can practically and directly participate (e.g., Diani and Bagnara, 1984; Eason, 1989; Griffith, 1985; Mambrey et al., 1987; Reuter, 1987; Soderberg, 1985).

An important area of potential advantages of participatory ergonomics identified above is the idea that the very process of participation has an effect far beyond its original focus and boundaries; the systemic or ripple effect is one which has impact upon other parts of the organization, either through the content or the process of participation strategies. However, these systemic effects may give rise to problems also. We may find that increasing the autonomy of assembly work groups detracts from the work of the purchasing or maintenance departments; enlarged job content for word processing staff may reduce the requirement for personal assistants, clerks, or even information specialists; and improved work environment resulting from participation may make workers in other offices or departments envious or dissatisfied. Related to this last problem, the very application of the process of participation may have harmful systemic effects, in that other groups may wish to be similarly involved and this may not always be feasible or desirable. Whether organizational ripple effects turn out to be beneficial or calamitous may depend in large measure on the care with which companies deal with them and with the original participative process. A recent case has highlighted a further age-old problem of work systems change, participative or not; at what point in the implementation should the 'change agent', or 'facilitator', leave the scene? If they leave too late, then true participation may be stifled; if they leave too early, then a vacuum may be created in which nobody feels any ownership for the process or change (Wilson, 1995b). There is, finally, a more general important area of difficulty. In common with many work design and ergonomics initiatives, it is often difficult to 'prove' any results. It may be hard to show that participation has brought about a better system or system change, or that more autocratic methods would have led to far more problems (see also Shipley, 1990). Such lack of evidence, compounded by the paucity of good evaluated case studies, may contribute to a lack of face validity in the eyes of management.

In summary, while there is much to be said in support of participatory ergonomics for the well-being of both individuals and producers or employers, we cannot just uncritically accept participation as appropriate in every case. However regrettable, there are organizations which are not ready for participative practices. Child (1984) identified circumstances where participation was seen simply as a waste of time and others where it was used actually to obstruct change. Therefore, careful consideration must be given to the time, place, circumstances, and participants, as well as to the structure and techniques (we return to this later in the chapter). Nonetheless, a healthy respect for the limits to participation and awareness of all the problems and pitfalls should not prevent us from employing it wherever possible.

15.5 PARTICIPATIVE TOOLS AND TECHNIQUES

A broad discussion of participative techniques is made difficult due to the diversity of ways in which participation may be applied (the level and focus defined earlier). While Imada and Nagamachi (1995) see the role of tools and techniques as useful in "channeling participants' unique knowledge and skills to solve problems and make improvements," it is obvious that participation applied as a way to organize work activities at a macrolevel will demand very different sorts of techniques from participative change or analysis of a workstation at a microlevel. Furthermore, it could be argued that any technique that is valuable within ergonomics generally might be useful within participation. Nevertheless, we might be able to distinguish between those techniques that have been borrowed, adapted, or developed with participatory ergonomics specifically in mind, and those which may be applied within a participative exercise as one would apply such techniques in many circumstances (and therefore can be of any type of analysis or solution generation technique). Other chapters in this book are full of examples of the latter techniques; the reader is also referred to Meister (1985) and Wilson and Corlett (1995). Table 15.2 shows a number of techniques reported as useful for, or actually used within, participatory ergonomics initiatives. Just a selection of these are discussed below.

If we look first at participation as a macroergonomic strategy, then we may be looking for tools and techniques which help us to facilitate and support the participative process, to 'sell' participation to stakeholders and/or to maximize training, teamworking, and interpersonal skills.

An obvious issue in the selection and use of appropriate tools and techniques concerns participants' expertise. Depending on the structure chosen for the participatory initiative, 'experts' may play a number of different roles. They may restrict their intervention to largely supporting or guiding the participatory process, or they may be involved with a variety of other workers as members of a multifunction group. At the more 'hands-on' end of the scale they may undertake much of the analysis themselves, maybe through the use of a research team. An example of the latter is given by Kuorinka and Patry (1995) where, in a study prompted by an increase in the incidence of WRULDs in a poultry processing plant, analysis of the problem was undertaken by a project group. According to the authors, the participatory group's role was primarily the validation and implementation of the project group's proposals. Clearly the techniques used in this instance may vary from those we would teach to 'nonexpert' participants in order for them to analyse workplaces, generate new ideas, and evaluate changes for themselves. Nagamachi (1995) has described a number of tools developed specifically for this purpose. These include a device designed to identify the factors necessary for job redesign (labeled a JDLC chart—Job Redesign for Life Cycle), another for assessing working postures and a two-dimensional mannequin to assist in the selection of an appropriate work height.

A number of authors have highlighted the need to provide training as part of participatory ergonomics initiatives. Gjessing et al. (1994) describe how members of ergonomics committees or problem solving groups should be given training in teamwork and interaction skills to allow them to perform effectively as a group. They also state that management may need instruction in how to relate to workers who are now making decisions.

Training in the use of the ergonomics techniques themselves is also important. If techniques are complicated the training required may well be time-consuming and companies must be made aware of this, yet it can still be the case that the costs of training may be more than offset by savings resulting. For example, Kuorinka and Patry (1995) describe how a mixed group of workers, engineers, and occupational health representatives investigated a particularly hazardous task within a steel-making plant (replacing the fire-resistant lining of processing vessels). A great deal of time and effort was spent not only teaching the group motion-time-method (MTM) analysis so that they could look at

Table 15.2 Methods and techniques useful or used in participatory ergonomics

Method or Technique	Primary Purpose	Reference (for description or use of technique or method)
Task forces	Participation organisation	Liker et al, 1991; Wilson, 1994
Change management process	Participation organisation	Buchanan and Boddy, 1992
Team formation, building	Participation organisation	Kuorinka and Patry, 1995; Caccamise, 1995; West, 1994
Team training	Preparation and support	Gjessing et al, 1994
"Train-the-trainers"	Preparation and support	Corlett, 1991; Silverstein et al, 1991
Stakeholder analysis	Preparation and support	Burgoyne, 1994; West, 1994
Task analysis, functional task decomposition	Problem analysis	Kirwan and Ainsworth, 1992; McNeese et al, 1995
Work study techniques (e.g. MTMM)	Problem analysis	Kuorinka and Patry, 1995; Noro, 1991
"Statistical analysis measurements" (SAM)	Problem analysis	Liker et al, 1991
JDLC chart	Problem analysis	Nagamachi, 1991
Posture analysis tool	Problem analysis	Nagamachi, 1991
Pareto analysis	Problem analysis	Imada, 1991
Cause-and-effect diagram	Problem analysis	Imada, 1991
"Five ergonomic viewpoints"	Problem analysis	Noro, 1991
Link analysis	Problem analysis	Imada, 1991; Kirwan and Ainsworth, 1992
Activity analysis	Problem analysis and situation prediction	Garrigou et al, 1995
"Quality circles"	Problem analysis and idea generation	Nagamachi, 1991
Round robin questionnaire	Creativity stimulation and idea generation	O'Brien, 1981; Wilson, 1991a
Word map	Creativity stimulation and idea generation	O'Brien, 1981; Wilson, 1991a
Silent drawing/assessment	Creativity stimulation and idea generation	O'Brien, 1981; Wilson, 1991a
Brainstorming techniques	Idea generation and concept development	West, 1994
Focus groups	Idea generation and concept development	Caplan, 1990
Shared Experience Events	Idea generation	O'Brien, 1981
Delphi technique	Idea generation and concept evaluation	Linstone and Turoff, 1975
Interviews, questionnaires	Problem analysis, idea generation	Oppenheim, 1992
Checklists	Problem analysis and concept evaluation	Rawling, 1991; Meister, 1984; Sinclair, 1995
AKADAM	Problem analysis and concept evaluation	McNeese et al, 1995
Role playing techniques and simulation games	Idea generation and concept evaluation	Ruohomaki, 1995
Design Decision Group	Idea generation and concept evaluation	Wilson, 1991a
Problem Solving Group	Idea generation and concept evaluation	Wilson, 1995b
Story boarding	Concept development	McNeese et al, 1995
Concept mapping	Concept development	McNeese et al, 1995
Layour modelling and mock-ups	Concept evaluation	Wilson, 1991
'Metaplan'	Process recording	Frei et al, 1993

existing work methods and plan new ones, but also on team building—investments which the management considered justified by the short run-in time of the redesigned process. As a final point, it is interesting to note that "train-the-trainer" programs are becoming increasingly used in ergonomics program management as well as for health and safety in general (see Corlett, 1991; Silverstein et al., 1991).

By and large, however, it may be beneficial to avoid using overly complex or technical analytical tools, for as Liker et al. (1989) point out, the ability of organizations to remain self-sufficient goes down as the sophistication of ergonomics analysis tools rises. Certainly if one of the aims of participatory ergonomics is to free institutions from a reliance on the outside expert the use of highly complex tools may prove self-defeating. This may certainly restrict the use of computer-based modeling and simulation tools for the time being, the sophistication here lying not so much in their use as in the interpretation of results.

In the work of Imada (1991) we find a number of tools described in some detail, including well-known problem analysis tools such as pareto analysis, cause-and-effect diagram (or 'fishbone chart'), and link analysis. He also discusses the use of world maps, round-robin questionnaires, layout modeling and mock-ups—techniques which also form part of Design Decision Groups (Wilson, 1991b). These are derived from the work of O'Brien (1981), who adapted theories and techniques from market research and the literature on creativity and innovation. O'Brien encouraged participants to use a range of "thinking tools" aimed at getting them to visualize their ideas. All of these tools or techniques form part of what he called *Shared Experience Events (SEEs)*. Activities included splitting groups into discussants and listeners (where the latter subgroup provide feedback only when the first group has finished), the use of dynamic agendas, and the use of drawing materials to facilitate concept development and the communication of ideas.

Design Decision Groups have been used in a number of applications including the design of retail checkouts and library issue desks. The entire DDG process is summarized in Figure 15.2 (see Wilson, 1991b for a full explanation of each stage). Lehtela and Kukkonen (1991) have subsequently used a modified version of the DDG in designing a telephone exchange where, in particular, the benefits of "reference visits" by some of the participants to other exchanges using different systems was highlighted. They suggest that people tend to base designs largely on their own experience and that an awareness of solutions adopted in other situations serves to broaden their outlook. DDGs have been utilized also as part of a much wider systematic approach to engineering and design throughout an organization Systematic Ergonomics in Engineering Design or SEED (Aikin, Rollings, and Wilson, 1994; Sullivan and Wilson, 1994).

A frequent criticism of DDGs is that they are used only with white collar or technical staff. However, in one study an adapted DDG (Problem Solving Group Procedure) was used with a group of crane drivers working at an incinerator plant. In this instance, problems with the design of the workplace were addressed by the drivers themselves and they also sourced and costed their solutions within a budget set by management. The process is illustrated in Table 15.3 where it should be noted that the final stage is that of continual improvement (Wilson, 1995b).

Other techniques we might add to this nonexhaustive discussion of participatory techniques are brainstorming and group discussions. Caplan (1990) describes how focus groups (an approach commonly used in market research) can be used to address ergonomics problems. Typically, a small number of selected participants discuss a set of specific issues, facilitated by a moderator. Caplan describes how this approach can be particularly useful in coming up with solutions for workplace or product design problems.

Clearly, group discussions form a useful participatory technique. However it is important to find some method of adequately recording the salient points or decisions reached during the discussion process. This may involve nominating 'notetakers' or using video or audiotape recording equipment. In their example of sociotechnical systems design using a participatory approach, Frei et al. (1993) describe how the metaplan technique was used to organize and document discussions. Large pin boards are covered with paper on which 4×8 cards of different colors are pinned. Any ideas and problems are listed and regrouped on the board according to a theme or category. After each meeting the cards are photographed in their last position and photocopies distributed to the group members.

Finally we should consider the use of role playing techniques and simulation games. For instance, Pankakoski (1995) describes one such approach whereby a real situation is simulated and participants act as themselves rather than assuming the role of another.

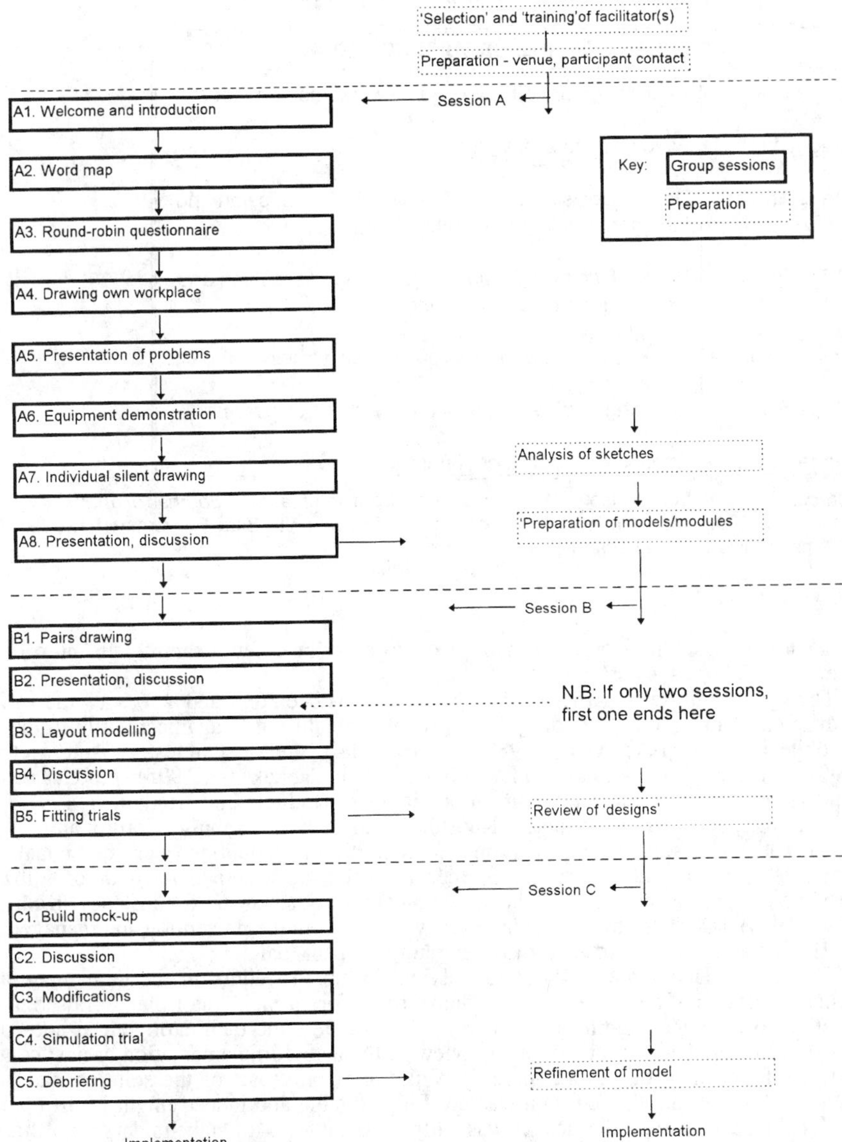

Figure 15.2 Design Decision Group Process (*Source:* Wilson 1991a, reprinted with permission from Taylor and Francis, London.)

Drawing upon the principles of the "Work Flow Game" (Ruohomaki, 1995)—developed to be "a concrete and effective method for visualizing abstract and often complicated administrative processes"—Pankakoski argues that by bringing together representatives from various levels of an organization, both work processes and work flow can be analyzed, developed, and tested before implementing any actual changes. Furthermore, she notes that the planning of the game is also done participatively, with a representative group of people working in conjunction with a consultant. During the game itself, all the participants sit in a circle simulating the work processes, while other observers note down problems and possible solutions. At the end there is a debriefing session where the entire game is analyzed.

15.6 CASE STUDIES AND RELEVANT LESSONS

Much of the literature on participatory ergonomics is composed of case studies, and we clearly can't cover all that has been published. In this section our aim is to refer the

Table 15.3 Stages of the Problem Solving Group Procedure

PROBLEM SOLVING GROUP PROCEDURE	
Stage One:	Familiarisation with tasks, jobs, workplace and team
Stage Two:	Field visits to similar sites
Stage Three:	Design Decision Group (A) - Drawing, discussion and idea generation (in neutral setting)
Stage Four:	Design Decision Group (B) - Building, discussion and idea testing (in neutral setting)
Stage Five:	Workplace and work environment simulation at the site
Stage Six:	Sourcing and costing of solutions by participants
Stage Seven:	Continual improvement after the change

(*Source:* Adapted from Wilson, 1995b: reprinted from *International Journal of Industrial Ergonomics, 15,* 1995, with kind permission of Elsevier Science-NL, Sara Burgerhartstraat 25, 1055 KR Amsterdam, The Netherlands.)

reader to some of this work and then go on to describe a few examples in more detail where particular issues are raised.

The successful use of participatory ergonomics has been reported across a wide variety of different applications, including design for manufacturing (Nagamachi, 1992), the automobile industry (Liker et al., 1989; Moore, 1994), the steel industry (Algera et al., 1990), office work (Kukkonen and Koskinen, 1993; Pankakoski, 1995), product design (Andersson, 1989), the telecommunications industry (Lehtela and Kukkonen, 1991), design of emergency service vehicles (Kuorinka et al., 1994); banking (Hornby and Clegg, 1992); and the design of control rooms for complex installations (van der Schaaf and Kragt, 1992; Pikaar et al., 1990). Furthermore, the constructive potential of utilizing participatory approaches has also been recognized in other areas; for instance, Kuorinka (1993, 1994) describes how participation may be useful in the prevention and management of VDU-related musculoskeletal disorders and low back pain.

The use of DDGs and a Problem Solving Group Procedure in the case mentioned earlier—redesign of a crane operators control room for an incinerator plant—raised many questions of practical significance (Wilson, 1995b). Here the facilitators left as soon as a first solution had been specified; upon review of the actual implementation of new control pods we found that, due to a supplier's bankruptcy, only part of the solution had been installed. We felt that the redesign was not fully suitable and indeed might itself increase risk of injury. However, the design was fully acceptable to all stakeholders who did not wish us to change anything. They were also so pleased by the process of change that a spontaneous continuous improvement of control room ergonomics, involving operators, engineers, and supervisors, had already initiated several new adaptations. This case also involved an agreed budget set at the outset and costing of all suggested alternatives by the participants within the process itself, which greatly helped acceptance of solution and process by all concerned.

Support for a participative approach to ergonomic problem solving at a microlevel is provided by Moore (1994) in a case study from the automotive industry. Following an evaluation by an external ergonomist and a subsequent unsuccessful intervention by company engineers, a multifunction group was set up to address injuries associated with flywheel truing. The flywheel truing committee followed a step-by-step methodology which included defining a number of goals. Brainstorming techniques were used, and committee members voted on which solutions should be evaluated. Postintervention evaluation showed the success of the chosen solution and the problem-solving approach. The author concluded that this participatory approach to ergonomics problem solving was more effective and more efficient than the previous sole reliance on outside ergonomists or engineering staff.

In another case study, this time from the steel industry, Algera et al. (1990) describe some of the benefits of taking a participatory approach to systems design. They state that by using as much expertise as is available about steel production (particularly the workforce's tacit knowledge), better decisions can be reached. In particular, early user involvement means that there is more time available for addressing issues such as training or staffing so that a "*planned* organizational change process" can be set in motion. However, they also noted a number of associated problems including the fact that, given the demands made by their continuing production tasks, some participants found it difficult to devote sufficient time to the design process. Furthermore, the authors identified a communication gap between different subsections of the participating group. However, it was felt that the use of simulations and modeling techniques helped to overcome this obstacle.

An 'ideal' participatory ergonomics approach to reduce mental and physical workload is described by Vink et al. (1995). Based on a number of studies, a stepwise approach was developed to reduce workload in office work which involves a high degree of worker participation at each of five phases of the process: securing commitment and formulating goals, the analysis of work, selection of solutions, implementation, and the evaluation of the changes. The authors argue that this was a highly effective approach resulting in positive changes to the working environment and a commitment to a continuing program of improvement. However, they do acknowledge that the process was time-consuming—a year being spent getting to the end of the implementation stage.

Well-known studies conducted by a number of French ergonomists set out to provide workers with confidence and knowledge to enable them to analyse their own work-related problems and to develop appropriate solutions, for example, in chemical plant control room design (Boel et al., 1985), posture and health analyses (Montreuil and Laville, 1985), and shiftwork systems (Teiger and Laville, 1987). Groups of workers, ergonomists, and technical staff were formed with a task force to oversee the project and disseminate information. Through the analysis work undertaken by study groups (administration of questionnaires and direct observation), other members of the workforce became interested in ergonomics issues and in suggesting improvements. The success of these project lies in the widespread acceptance of ergonomics and in the role workers can play in developing solutions (see also Daniellou et al., 1990).

At the level of ergonomics programs in their efforts to reduce injury levels in the meatpacking industry the Occupational Safety and Health Administration (OSHA, 1991) has produced Ergonomics Program Management Guidelines for Meatpacking Plants, which recommends a participatory approach. In order to show how such an approach may be used, NIOSH produced a report detailing three intervention projects in the meatpacking industry (Gjessing et al., 1994). In addition to describing the way in which the participation was structured and implemented, the report summarizes what can be learned from these cases. This includes the need for strong in-house direction and support and expertise (which emphasizes the need for training in ergonomics and teamwork). Multifunction teams should include workers from the areas under investigation, plus supervisors, engineers, and maintenance personnel. It may be useful to provide second-level groups to incorporate management (which can facilitate approval for decisions). However, all the workforce should be kept up to date on the progress of the work. Team members confidence can be boosted through attacking more straightforward problems first, and it is important that they are provided with access to the information necessary for problem solving. Finally, realistic goals for the program should be established and methods identified for evaluating results (see also Wilson, 1994).

15.7 REQUIREMENTS FOR PARTICIPATORY ERGONOMICS

Earlier in this chapter, we outlined the claimed or reported advantages of participatory ergonomics and also the possible limitations, difficulties, and disadvantages. It is by examining both these positive and negative aspects, by considering reported cases—hopefully the failures as well as comparative successes—and by review of related literature in change management and systems implementation, that we can identify requirements for participatory ergonomics.

Before examining these requirements, two cautionary notes are in order. First, as explained throughout this chapter, **participatory ergonomics can operate at many levels,** from single design or implementation exercises up to the basis of a shopfloor culture; different requirements must therefore be of different relevance in various types of initiative.

Second, we must **beware of easy prescriptions.** What might seem a set of eminently sensible and achievable guidelines to the outside consultant or advisor may prove impossibly impracticable, demanding or time consuming for an individual manager and her or his colleagues to follow. This is particularly so when we remember that real people in real organizations will invariably be practising participatory initiatives while still fulfilling their normal roles and functions in the organization.

Despite these qualifications, however, it is possible to offer guidance to those who seek to implement participation approaches and techniques in general, and those of participatory ergonomics in particular. In outline, the process of participatory ergonomics involves the stages and requirements summarized in Table 15.4 (see also Wilson, 1995a).

15.7.1 Establishing a Climate for Participation

When the organisation has economic difficulties or when a climate is hostile it is difficult or even impossible to implement any change including change to a participative way of working. While participation might be clutched at by a management desperate after the failure of other routes, and may even seem to offer a lifeline, generally participatory ergonomics should not be introduced in circumstances of conflict, unrest, or great uncer-

Table 15.4 Stages and Requirements for Participatory Ergonomics

Stage	Requirements
1. Climate for participation	Not during times of uncertainty, hostility, crisis. Better in open, nonhierarchical, noncentralized organizations. Group work and enriched jobs in place already offers support Devolve Ergonomics: Foundation/Spread/ Embed
2. Support and resources	Support throughout the organisation, and T-Us Key decision-maker commitment Expect "rational" and "irrational" resistance Allocate sufficient finance and personnel No unreasonable time constraints Initial, flexible, goals, and criteria
3. Facilitator	Not necessarily project "owner" Sensitive yet robust and decisive Unbiased yet knowledgeable Flexible and consistent Respected Know when to leave the process
4. Set-up	Timing and starting point are critical: process has to have initiation and evolution Interdependence of motivation, ability and confidence of participants Participation to be voluntary where possible Potential benefits should be promoted during set-up
5. Processes and methods	Participants can be 'self-selecting'; otherwise, use volunteers, elections, or selection Size of group: 6-10 (or 12) Need facilitator, chairperson, resources assistant Methods to be well costed, cost-effective and flexible Process to be iterative, relaxed, and allow compromises Methods to promote creativity, learning, confidence, and decision making Include solution costing methods Need to have direction without being overtly directed
6. Outcomes	Testable solutions or plans Continual improvements Spread of expertise in the organisation Process to become internal and ordinary, not external and extraordinary

(*Source:* Adapted from Wilson, 1995b and Wilson and Haines, 1996.)

tainty. Furthermore, participation should not be implemented when the survival of the organization is at stake; the extra pressures this might impose as well as the need to concentrate upon normal operational activities would detract from the participative activities.

There is little, if any, empirical evidence of which organizational attributes most suit participatory ergonomics. It does seem sensible though to suppose that open and less hierarchical structures, a history of good labor relations, a tradition of consultation, well-established communication channels, and job designs which emphasize control, responsibility, and teamwork will be more fertile ground for a participative system.

Given these attributes and attendant requirements, we can see how an organizational commitment to a full participative program can itself help provide the suitable climate for specific participative initiatives. This fits the philosophy and approach of Ergonomics Devolution (Wilson, 1994). First is the provision of a **foundation,** involving adequate support and resources (see next section) and a structure of review committees and task forces. The work of ergonomics within this foundation should be prospective as well as retrospective. The second element, to **spread** ergonomics, necessitates awareness raising, development of tools and techniques suitable for the organization, and a structured training program to support widespread and skilled use of the techniques. The third element is to **embed** the approach and practices within the organization. In this sense the participatory ergonomics is organic, in that workers' participation in work analysis and design itself is supportive of the participative process and helps further spread as well as embed the program.

15.7.2 Providing Support and Resources

Participatory ergonomics is like any other strategy—it needs solid foundations and good management. These foundations should provide initial *and* continuing support; the resources required will be human, financial, and temporal.

Any prescriptions for implementation of change stress the need for support from the top from the outset, usually taken to mean that the process and direction of change must be knowingly approved at top management or Board level and that this approval must be translated into active support when and where necessary. The support must also be sufficient to set up a program with sufficient resources to see it through, with the inclusion of key decision makers and others with the skills required, and given an adequate financial budget. Perhaps most important, the participatory ergonomics program should not be unduly time-constrained.

These requirements illustrate very well the point made earlier about outsiders giving impracticable or overdemanding prescriptions. It is all very well setting relaxed time scales at the outset of a project, but circumstances will often dictate that various stages in a participatory ergonomics program might have to be brought forward or put back according to operational and market constraints.

Not only should support be secured "from the top"; all parts of the organization that will be affected should at least know what will be involved; at best they should be enthusiastic and at worst neutral about the contributions necessary. Some resistance will always be expected—resistance to change is almost a fact of life in social systems—but such resistance is not always irrational; open discussion of the basis and causes of any resistance as early as possible can help considerably. As far as the stakeholders are concerned, support must be sought among both the job holders and also their trade unions; many participative changes will have implications for terms and conditions of employment and may be subject to collective bargaining.

Again, the question of obtaining support throughout an organization, and of the timing of this, is in danger of being seen somewhat glibly as an obvious requirement. Just when to inform the people affected by an impending technology change or participative program can be a very difficult decision. If done too late, rumors will already have spread and opposition hardened. If done too early, and if, as is likely, delays occur, then the process is likely to fall into disrepute as people wait for something to happen.

Finally, quality support can be given only if realistic flexible goals are set at the outset, constraints are clear, and these are communicated to all relevant people.

15.7.3 Selection of Facilitator

One great difficulty for promoting participation and for the transfer of participative solutions or participatory ergonomics from one situation to another, is the one-off nature of many cases. Nowhere is this more so than within one of the most critical factors for success—the nature of the facilitator (or change agent—Buchanan and Boddy, 1992). Sometimes the facilitator will be the "owner" of the project or the person who wants the program. Often, though, it will be another person from inside or outside the organization.

Criteria for selection of the facilitator make apparent what a difficult role this is to fill. They should be sensitive to different people's needs yet robust and decisive where necessary; they should be seen as unbiased (difficult for an insider) yet knowledgeable (difficult for an outsider); not seen as inconsistent yet able to change roles (e.g., information generator, arbitrator, archivist, process driver) when necessary; respected and appreciated for their contribution, yet aware of when to leave the process if necessary. This last dilemma was neatly illustrated in the crane control room case referred to earlier (Wilson, 1995b).

15.7.4 Setting Up Participatory Ergonomics

Although we might distinguish between the process of setting up a participatory ergonomics initiative and the structures through which it works, in practice these should be compatible. At best, the implementation or set-up process becomes the basis of the operational structure; a mixed-function task force to implement shopfloor workplace ergonomics might become the first investigatory and intervention team, for instance.

Setting up the participative ergonomics program is, as pointed out above, difficult to time. Also, great care has to be taken about how much is actually set up by the project "owner", facilitator, or group chairperson and how much of the structure evolves once initial mechanisms have been put in place.

Participants should expect and want to be involved, although this may have to be promoted by an introductory training and awareness program which emphasizes the voluntary nature of the exercise and stresses the potential benefits to participants, to their colleagues, and to the organization. At the same time, all potential participants should be made aware, in an open fashion, of the constraints on what they can do. Any unavoidable limits to the influence of participants or to their access to information must be made clear, and budgets agreed at the outset.

In our definition of participatory ergonomics at the start of this chapter, we highlighted the need for participants to be motivated and knowledgeable; the third main requirement is for them to be confident. These three requirements are interdependent. As participants gain in knowledge and ability, so they will slowly gain confidence both in their own contributions and also in the participative process itself, seeing for themselves that they are actually influencing events and outcomes. In turn this will motivate them and others to be more involved both in the current and also in subsequent participative initiatives. This close association between knowledge/ability, confidence, and motivation should not surprise us; it underpins much modern theory of motivation at work (e.g., Expectancy Theory—Vroom, 1964).

Where the change agent should start—by motivating people to make initial contributions or by providing training to collaborate in problem-solving teams—will depend on individual cases and on previous history. If participatory ergonomics has already been used successfully and openly in the organization, then some confidence in the process will be available, with consequent motivation to contribute. If, however, previous experience in the company has been that participation has been a cynical exercise to obtain acceptance, or a PR stunt, or has been applied ineptly or inappropriately, then there will be little chance of potential participants having the motivation or confidence to take part in any new schemes without considerable efforts in consultation and training.

Illustration of these needs for motivation and confidence comes from two well-known case studies of 30 or 40 years ago. The first is the work of Coch and French (1948) where a U.S. pyjama factory faced a number of problems of employee resistance, poor absenteeism, and labor turnover records, and an unskilled female workforce with few career aspirations. Relatively minor changes were introduced with the help of outside consultants, and three groups of employees were formed: A 'control group' was given no say in changes, a 'representation' group had some employee representatives who could discuss the changes, and a 'total participation' group all discussed the changes before implementation and contributed to the structure of the changes. The first group showed hostility and production restrictions of 10% to 20%; the representation group—after an initial period of resistance—showed cooperation, increased morale, and production rose by 10% to 15%; and the total participation group showed the same improvements but did so immediately. Two months later the remaining members of the control group (a third had left the company) became a participation group themselves, with the same positive responses. The authors concluded that there is no natural resistance to change, only strategic resistance when stakeholders are not in control of the consequences.

In a subsequent Norwegian experiment involving one of the earlier authors, the experience was reported to be a complete failure (French et al., 1960). The two crucial reasons identified for this were that (1) those involved had not first been convinced of the need for change; and (2) confidence is crucial, and people won't get involved in change if they have no guarantees that the new situation will not work against their best interests.

15.7.5 Participative Processes and Methods

Just as the role of facilitator is difficult to manage, with several seemingly contradictory requirements, so also with the actual participatory ergonomics processes. Whatever the level and focus, participation will by definition involve team-based activities; this might vary from a product (design) team or a workplace ergonomics group through to a participative structure with steering committee and project teams. In some cases, for instance where the workforce of one product cell or a control room analyse and redesign their own work environment, the team will be self-defining and self-selecting (although some may wish not to be involved). In other cases, the participative group will have to be formed, for instance where a new technical system is to be selected and introduced. Here the choice is to call for volunteers and perhaps screen them or to hold "elections" or to select. Often a mixture of these strategies might be employed, but all hold dangers of alienation or of disruptive influences in the group, and great care must be taken.

As for team size, this will again depend in part on the focus and level of participation. For group work generally a size of 6 to 8 is often the recommended norm; given that a greater mix of skills might be needed in a project team or that subgroups with specific tasks might have to be formed, then the preferred maximum number might be 10 or even 12. The facilitator will of course be a member, but may or may not be the chairperson at meetings. Often (e.g., in DDGs) a resources assistant is needed as well.

Different processes may be implemented depending on the remit of the group. Different methods and techniques will be employed for idea generation, concept formation, system development, selection and evaluation, and implementation. Methods will also vary according to the size of the project and whether the purpose is design, implementation, or work organization. These methods should be well costed, cost-effective within the process resources, and as flexible as possible. The whole process, and sessions within it, should be iterative, relaxed, and allowance made for compromise—between participants or between original and subsequent ideas or outcomes. Methods work best which themselves promote the creativity, learning, and confidence that participatory ergonomics requires. We have also found that use of methods which allow potential solutions to be costed, within the budgets agreed at the outset, can greatly aid acceptance of outcomes by stakeholders, management and workers. Perhaps most critical and most difficult, the process must be nondirective but must have direction. In other words, the facilitator must not be seen to be pushing her or his own agenda too hard or to be constraining participants too much. At the same time, the process must not become an excuse for inaction or a "talking shop"; discussing last night's football or soap opera can make up a small relaxing part of the group's time but the group must be kept to the point of the exercise, in a subtle, tolerant, and open way. This is not an easy task, emphasizing again the key role, and attributes required, of the facilitator.

15.7.6 Outcomes

The outcomes of participatory ergonomics obviously should first of all include solutions or plans appropriate to the remit of the exercise, such as prototype product, workstation layout, phased technology implementation, or teamwork structure. Moreover, these solutions should be testable; it must be possible to evaluate the outcomes against the criteria for the project, preferably within the participative process. This evaluation should account for the acceptance of the process and outcomes among stakeholders, as well as the quality of the solution per se. Further assessment might be made of the process in terms of its efficiency and timeliness.

Mention was made earlier of the systemic consequences (good and bad) of participatory ergonomics. In the context of outcomes, we would wish to see continual improvements after the focal exercise. Participants should be interested and able enough at the least to continue to collaborate on related issues; this is particularly the case when the focus is workplace ergonomics and health and safety interventions. At best this will spread ergonomics expertise in general across the organization, with relatively unprompted use of some of the less complex methods and techniques. In time, participatory ergonomics should become a part of the everyday culture of the company.

15.8 CONCLUSION

Ergonomists around the world have made great efforts in recent years to enable the widespread and successful implementation of ergonomics at work. These applied ergonomists are taking ergonomics out of the pages of textbooks and making it more available for practical use by company employees themselves. There is a strong argument for a participative approach to ergonomics, both participation at an individual level whereby workers are involved in workplace and work assessment and in providing solutions to any health and safety problems, and participation at an organizational level whereby engineering, production, medical, and other staff develop their own programs for ergonomics awareness and action. There are limits to such participation; some problems and initiatives will require professional (outside) ergonomics expertise, but much ergonomics can and should be carried out internally.

We have already described the challenge for Ergonomics Devolution as being to:

- motivate people throughout a company, the management, workforce and technical specialists, to become their own "ergonomists" within a total ergonomics program
- give these people the tools and techniques and related training to make ergonomics interventions, ensuring that training comes at the right time for all relevant groups
- ensure they have enough understanding to know when and how to use these tools and techniques, **and when they cannot and must call in specialist assistance.** (Wilson, 1995)

Successful participatory ergonomics requires awareness of its limitations and deficiencies as well as support for its strengths. Participatory ergonomics will not succeed in all circumstances, and we should not pretend that it will. There is current debate about the practical significance of seemingly positive outcomes of participation in general (Wagner, 1994). Sometimes the outcomes will be limited or only satisfactory; systemic problems may be unearthed; 'proof' of its efficacy can be difficult to find and report. Clearly there is a need at times for specialist assistance and advice, and perhaps even rejection of the participative process outcomes; how do we do this while remaining true to the fundamental principles of what is participation in a real sense?

For the theorists as well as the practitioners, participatory ergonomics holds fundamental challenges as well. Where are the theory and models to draw from? What evidence is there for the validity, reliability, and generalizability of methods and techniques? These and other questions must be addressed, but within an on-going worldwide movement of participatory ergonomics. One is tempted to paraphrase (and misquote) - "Occupational ergonomics is too important to be left only to the ergonomists."

ACKNOWLEDGMENTS

Many colleagues, too numerous to mention individually, have worked with us on participatory ergonomics initiatives; we have learned from all of them. We also thank two anonymous reviewers for their helpful comments. The Health and Safety Executive have funded part of the work underlying some of this chapter under grant no. 3496/R49.004; their support and assistance is gratefully acknowledged, but all views and opinions are the authors' own. Finally, we are immensely grateful to those in a variety of organizations who have contributed to our explorations of participatory ergonomics down the years.

REFERENCES

Aikin, C., Rollings, M., and Wilson, J. R. (1994). Providing a foundation for ergonomics: Systematic Ergonomics in Engineering Design (SEED). In: Proceedings of the 12th Congress of the International Ergonomics Association. Toronto, August, 1994, Vol. 5, 276–278.

Algera, J. A., Reitsma, W. D., Scholtens, S., Vrins, A. A. C., Wijnen, C. J. D. (1990). Ingredients of ergonomic intervention: How to get ergonomics applied, *Ergonomics, 33,* 5, 557–578.

Andersson, E. R. (1989). A systems approach to product design and development - an ergonomic perspective. *International Journal of Industrial Ergonomics, 6,* 1–8.

Arnstein, S. R. (1969). A ladder of citizen participation. *Journal of American Institute of Planners,* November. (Note: This is referenced on the basis of a citation from another source, Reuter (as mentioned in text). Therefore no full reference available.

Bernoux, P. (1994). Participation: A review of the literature. *European Participation Monitor,* Part 9, pp. 6–11.

Boel, M., Daniellou, F., Desmores, E. and Teiger, C. (1985). Real work analysis and workers' involvement. In *Ergonomics International 85, Proceedings of the 9th Congress of the International Ergonomics Association,* Bournemouth, pp. 235–237.

Brown, O. Jr. (1990). Marketing participatory ergonomics: current trends and methods to enhance organizational effectiveness. *Ergonomics, 33,* 5, 601–604.

Buchanan, D., and Boddy, D. (1992). *The Expertise of the Change Agent.* London: Prentice Hall.

Burgoyne, J.G. (1994). Stakeholder analysis. In *Qualitative Methods.* In C. Cassell and G. Symon, Eds., *Organizational Research: A Practical Guide,* New York: Sage.

Caccamise, D. J. (1995). Implementation of a team approach to nuclear criticality safety: The use of participatory methods in macroergonomics. *International Journal of Industrial Ergonomics, 15,* 397–409.

Caplan, S. (1990). Using focus group methodology for ergonomic design, *Ergonomics, 33,* 527–533.

Cherns, A. (1987). Principles of sociotechnical design revisited, *Human Relations, 40,* 153–162.

Childs, J. (1984). *Organisation: A guide to problems and practice.* London: Harper & Row.

Coch, L., and French, J. (1948). Overcoming resistance to change. *Human Relations, 2,* pp. 512–532.

Corlett, E. N. (1988). Cost-benefit Analysis of Ergonomic and Work Design Changes. In D. J. Oborne, Ed., *International Reviews of Ergonomics. 2,* 85–103. London: Taylor and Francis.

Corlett, E. N. (1991). Ergonomics Fieldwork: An Action Programme and Some Methods. In M. Kumashiro and E. D. Megaw, Eds., *Toward Human Work: Solutions to Problems in Occupational Health and Safety,* London: Taylor and Francis.

Daniellou, F., Kerguelen, A., Garrigou, A., and Laville, A. (1990). Taking future activity into account at the design stage: participative design in the printing industry. In C. M. Haslegrave, J. R. Wilson, and E. N. Corlett, Eds., *Work Design in Practice,* London: Taylor and Francis.

Daniellou, F., and Garrigou, A. (1992). Human factors in design: Sociotechnics or ergonomics? In M. Helander and M. Nagamachi, Eds., *Design for Manufacturability and Process Planning,* London: Taylor and Francis. pp. 55–63.

Diani, M., and Bagnara, S. (1984). Unexpected consequences of participative methods in the development of information systems: The case of office automation. In Grandjean, E., Ed., *Ergonomics and Health in Modern Offices,* pp. 227–232. Grandjean, E., Eds., London: Taylor and Francis.

Eason, K. (1989). New systems implementation. In J. R. Wilson, and E. N. Corlett, Eds., *Evaluation of Human Work: A Practical Ergonomics Methodology.* London: Taylor and Francis.

Forrester, K. (1986). Involving workers: Participatory ergonomics and the trade unions. In D. J. Oborne, Ed., *Contemporary Ergonomics.* London: Taylor and Francis.

Frei, F., Hugentobler, M., Schurman, S., Duell, W., and Alioth, A. (1993). *Work Design for the Competent Organization,* (Chapter 7), pp. 103–116. Greenwood Press, Westport, CT.

French, J., Israël, I., and As, D. (1960). An Experiment on Participation in a Norwegian Factory. *Human Relations, 14,* 1–19.

Fuchs-Kittowski, K., and Wenzlaff, B. (1987). Integrative participation, a challenge to the development of informatics. In P. Doherty et al., Eds., *System Design for Human Development and Productivity: Participation and Beyond.* North-Holland, Amsterdam, pp. 3–18.

Garrigou, A., Daniellou, F., Carballeda, G., and Ruaud, S. (1995). Activity analysis in participatory design and analysis of participatory design activity. *International Journal of Industrial Ergonomics, Special Issue: 'Participatory Ergonomics, 15, 5,* 311–329.

Gaudier, M. (1988). Workers participation within the new industrial order: A review of literature. *Labour and Society, 13,* 313–332.

Gjessing, C. C., Schoenborn, T. F., and Cohen, A. (1994). Participatory Ergonomics Interventions in Meatpacking Plants. *DHHS (NIOSH)* Publication No. 94-124.

Griffith, R. W. (1985). Moderation of the effects of job enrichment by participation: A longitudinal field experiment. *Organizational Behaviour and Human Decision Processes, 35,* 73–93.

Herzberg, F. (1966). *Work and the Nature of Man.* New York: Thomas Crowell.

Hornby, P., and Clegg, C. (1992). User participation in context: a case study in a UK bank. *Behaviour & Information Technology, 11, 5,* 293–307.

Imada, A. S., and Robertson, M. M. (1987). Cultural perspectives in participatory ergonomics. *Proceedings of the Human Factors Society 31st Annual Meeting.* 1018–1022.

Imada, A. S. (1991). The rationale and tools of participatory ergonomics. In K. Noro and A. S., Imada, Eds., *Participatory Ergonomics.* London: Taylor and Francis, pp. 30–51.

Imada, A. S., and Nagamachi, M. (1995). Introduction to a special issue on participatory ergonomics. *International Journal of Industrial Ergonomics, 15.*

Ingersoll Engineers (1987). *Technology in Manufacturing.* Warwickshire: Rugby.

Karasek, R., and Theorell, T. (1990). *Healthy Work: Stress, Productivity and the Reconstruction of Working Life.* Basic Books.

Kirwan, B., and Ainsworth, L. K. (1992). A Guide to Task Analysis. London: Taylor and Francis.

Kogi, K., Phoon, W-O., and Thurman, J. E. (1988). *Low-Cost Ways of Improving Working Conditions: 100 Examples from Asia.* International Labour Office, Geneva.

Kogi, J. (1991). Participatory training for low-cost improvements in small enterprises in developing countries. In K. Noro and A. S., Imada, Eds., *Participatory Ergonomics,* London: Taylor and Francis, pp. 73–81.

Kukkonen, R., and Koskinen, P. (1993). *User participation in workplace design,* In H. Luczak, A. Cakir, and G. Cakir, Eds., Work with Display Units 92. New York: Elsevier.

Kuorinka, I. (1993). Prevention of VDU-related musculoskeletal problems: a shift from a specialist's to a participatory approach. In H. Luczak, A. Cakir, and G. Cakir, Eds., *Work with Display Units 92.* New York: Elsevier.

Kuorinka, I., Cote, M-M, Baril, R., Geoffrion, R., Giguere, D., Dalzall, M-A., and Larue, C. (1994). Participation in workplace design with reference to low back pain: a case for improvement of the police patrol car. *Ergonomics, 37,* 1131–1136.

Kuorinka, I., and Patry, L. (1995). Participation as a means of promoting occupational health. International Journal of Industrial Ergonomics, 15, 365–370.

Lehtela, J., and Kukkonen, R. (1991). Participation in the purchase of a telephone exchange-a case study. Designing for Everyone. *Proceedings of the Eleventh Congress of the International Ergonomics Association,* Y. Queinnec and F. Daniellou, Eds. London: Taylor and Francis.

Leplat, J. (1990). Skills and tacit skills: A psychological perspective. *Applied Psychology: An International Review, 39,* 143–154.

Lewis, H. B., Imada, A. S., Robertson, M. M. (1988). Xerox leadership through quality: Merging human factors and safety through employee participation. *Proceedings of Human factors Society 32nd Annual Meeting.* Anaheim, California, pp. 756–759.

Liker, J. K., Nagamachi, M., and Lifshitz, Y. R. (1989). A comparative analysis of participatory ergonomics programs in U.S. and Japan manufacturing plants. *International Journal of Industrial Ergonomics, 3,* 185–189.

Liker, J. K., Joseph, B. S., and Ulin, S. S. (1991). Participatory ergonomics in two US automotive plants. In Noro K. and A. S., Imada, Eds., *Participatory Ergonomics,* London: Taylor and Francis, pp. 97–139.

Linstone, H. A., and Turoff, M. (1975). *The Delphi Method: Techniques and Applications.* Reading, MA: Addison-Wesley.

McClelland, I. (1995). Product assessment and user trials. In J. R., Wilson, and E. N. Corlett, Eds., *Evaluation of Human Work,* London: Taylor and Francis.

McNeese, M.D., Zaff, B.S., Citera, M., Brown, C. E., and Whitaker, R. (1995). AKADAM: Eliciting user knowledge to support participatory ergonomics. *International Journal of Industrial Ergonomics, Special Issue: Participatory Ergonomics, 15, 5,* 345–365.

Mambrey, P., Opperman, R. & Tepper, A. (1987). Experiences in participative systems design. In *System Design for Human Development and Productivity: Participation and Beyond,* Doherty, P. et al., Eds., Amsterdam: North-Holland, pp. 345–358.

Meister, D. (1985). *Behavioural Analysis and Measurement Methods.* New York: John Wiley.

Montreuil, S. and Laville, A. (1986). Cooperation between ergonomists and workers in the study of posture in order to modify work conditions. In E. N. Corlett, J. R. Wilson, and I. Manenica (Eds.), *The Ergonomics of Working Postures,* London: Taylor and Francis, pp. 293–304.

Moore, J. S. (1994). Flywheel truing-a case study of an ergonomic intervention. *American Industrial Hygeine Association Journal, 55(3),* 236–244.

Mumford, E. (1991), Participation in systems design-what can it offer? In B. Schackel and S. J. Richardson, Eds., *Human Factors for Informatics Usability,* Cambridge University Press, 267–290.

Nagamachi, M. (1991). Application of participatory ergonomics through quality-circle activities. In Noro K. and A. S. Imada, Eds., *Participatory Ergonomics.* London: Taylor and Francis, pp. 139–165.

Nagamachi, M., and Yamada, Y. (1992). Design for manufacturing through participatory ergonomics. In M. Helander and M. Nagamachi, Eds., *Design for Manufacturing,* London: Taylor and Francis.

Nagamachi, M. (1995). Requisites and practices of participatory ergonomics. *International Journal of Industrial Ergonomics, Special Issue: 'Participatory Ergonomics, 15, 5,* 371–379.

Neumann, J. (1989). Why people don't participate when given the chance. *Industrial Participation, 601* (Spring), 6–8.

Noro, K. (1991). Concepts, methods and people. In K. Noro, and A. S. Imada, Eds., *Participatory Ergonomics,* London: Taylor and Francis, pp. 3–30.

O'Brien, D. D. (1981). Designing systems for new users. *Design Studies, 2,* 139–150.

Occupational Safety and Health Administration (1991). Ergonomics Program. Management Guidelines for Meatpacking Plants (Reprint). OSHA Report 3123 US Department of Labor.

Oliver, N. (1990). The JIT jigsaw. *Manufacturing Engineer.* March 29–30.

Oppenheim, A. N. (1992). *Questionnaire Design, Interviewing and Attitude Measurement* (Second Edition). London: Pinter Publishers.

Pankakoski, M. (1995). Experiences on participation-the simulation game for participative design. Paper presented at course on participative approaches to workplace design. NIVA Nordisk, Mustio Manor, Finland.

Pikaar, R. N., Thomassen, P. A. J., Degeling, P., and van Andel, H. (1990). Ergonomics in control room design, *Ergonomics, 33, (3),* 589–600.

Rawling, R. G. (1991). Participative approaches to the design of physical office environments. In K. Noro and A. Imada, Eds., *Participatory Ergonomics,* London: Taylor and Francis, pp. 53–72.

Reuter, W. (1987). Procedures for participation in planning, developing and operating information systems. In P. Doherty et al., Eds., *System Design for Human Development and Productivity: Participation and Beyond.* Amsterdam: North-Holland, pp. 271–276.

Ruohomaki, V. (1995). A simulation game for the development of administrative work processes. In Saunders, D., Ed., *The Simulation and Gaming Yearbook.* London: Kogan Page.

Sanoff, H. (1985). The application of participatory methods in design and evaluation. *Design Studies, 6(4),* 178–180.

Sen, T. K. (1988). Participative group techniques. In G. Salvendy, Ed., *Handbook of Human Factors,* Chichester: John Wiley, pp. 453–469.

Shipley, P. (1990). Participation ideology and methodology in ergonomics practice. In J. R. Wilson and E. N. Corlett, Eds., *Evaluation of Human Work: A Practical Ergonomics Methodology,* London: Taylor and Francis.

Silverstein, B. A., Richards, S. E., Alcser, K., and Schurman, S. (1991). Evaluation of in-plant ergonomics training. *International Journal of Industrial Ergonomics, 8,* 179–193.

Simpson, G., and Mason, S. (1995). Economic analysis of ergonomics. In Wilson, J. R. and Corlett, E. N., Eds., *Evaluation of Human Work: A Practical Ergonomics Methodology,* Second Edition. Taylor and Francis, London.

Sinclair, M. A. (1995). Subjective assessment. In J. R. Wilson and E. N. Corlett Eds., *Evaluation of Human Work* (Second Edition). London: Taylor and Francis, pp. 69–100.

Soderberg, I. (1985). Office work and office automation influence over the work environment and work organization. Report 1985: 27, National Board of Occupational Safety and Health, Solna, Sweden.

Sullivan, A., and Wilson, J. R. (1994). Ergonomics Design and Review Manual (restricted). Commonwealth Industrial Gases Ltd., Chatswood, NSW.

Tieger, C., and Laville, A. (1987). How ergonomic methods can be used by non-ergonomists. *Paper presented at 2nd International Occupational Ergonomics Symposium, Applied Methods in Ergonomics,* Zadar, Yugoslavia.

U.S. Department of Labor, Occupational Safety and Health Administration (1990). Ergonomics Program Management Guidelines for Meatpacking Plants (DOL/OSHA Pub No 3123).

van der Schaaf, T. W., and Kragt, H. (1992). Redesigning a control room from an ergonomics point of view: a case study of user participation in a chemical plant. In H. Kragt, Eds., *Enhancing Industrial Performance.* London: Taylor and Francis, pp. 165–178.

Vink, P., Lourijsen, E., Wortel, E., Dul, J. (1992). Experiences in participatory ergonomics: results of a roundtable session during the 11th IEA Congress, Paris, July 1991. *Ergonomics, 35 (2),* 123–127.

Vink, P., Peeters, M., Grundemann, R. W. M., Smulders, P. G. W., Kompier, M. A. J., and Dul, J. (1995). A participatory approach to reduce mental and physical workload. *International Journal Industrial Ergonomics, 15,* 389–396.

Vroom, V. H. (1964). *Work and Motivation.* Chichester: John Wiley.

Vroom, V. H., and Yetton, P. W. (1973). *Leadership and Decision Making.* University of Pittsburgh Press, Philadelphia.

Wagner III, J. A., (1994). Participatory effects of performance and satisfaction: a reconsideration of research evidence. *Academy of Management Review, 19,* 312–330.

Wall, T. D., Corbett, J. M., Clegg, C. W., Jackson, P. R., and Martin, R. (1990). Advanced manufacturing technology and work design: Towards a theoretical framework. *Journal of Organizational Behaviour, 11,* 201–219.

West, M. (1994). *Effective Teamwork.* The British Psychological Society, Leicester, UK.

Wilson, J. R., and Grey, S. M. (1990). But, what are the issues in work redesign? In: C. M. Haslegrave, J. R. Wilson, and E. N. Corlett, Eds., *Work Design in Practice,* London: Taylor and Francis.

Wilson, J. R. (1991a). Design Decision Group - A participative process for developing workplaces. In K. Noro and A. Imada Eds., *Participatory Ergonomics.* London: Taylor and Francis.

Wilson, J. R. (1991b). A framework and a foundation for ergonomics? *Journal of Occupational Psychology, 64,* 67–80.

Wilson, J. R. (1994). Devolving ergonomics: The key to ergonomics management programmes. *Ergonomics, 37,* 579–594.

Wilson, J. R., and Corlett, E. N., Eds. (1995). *Evaluation of Human Work* (Second Edition). London: Taylor and Francis.

Wilson, J. R., and Grey Taylor, S. M. (1995). Simultaneous engineering for self directed work teams implementation: A case study in the electronics industry. *International Journal of Industrial Ergonomics, 16,* Nov.–Dec.

Wilson, J. R., and Haines, H. M. (1995). Towards a model of participatory ergonomics. Institute for Occupational Ergonomics Report No. IOE/95/120. University of Nottingham.

Wilson, J. R. (1995a). Ergonomics and participation. In J. R. Wilson & E. N. Corlett, Eds., *Evaluation of Human Work: A Practical Ergonomics Methodology,* Second Edition. Taylor and Francis, London: pp. 1071–1096.

Wilson, J. R. (1995b). Solution ownership in participative work design: The case of a crane control room. *International Journal of Industrial Ergonomics, 15,* 329–344.

Work Research Unit, 1982, Meeting the Challenge of Change: Guidelines. London: HMSO.

Zink, K. J. (1991). Participatory ergonomics-some developments and examples from West Germany. In K. Noro and A. S. Imada, Eds., *Participatory Ergonomics.* London: Taylor and Francis, pp. 165–180.

CHAPTER 16

MODELS IN TRAINING AND INSTRUCTION[1]

Robert W. Swezey
Robert E. Llaneras
InterScience America, Inc.
Leesburg, VA 22075 USA

[1]The authors would like to thank Lisa L. Swezey for her capable help during the preparation of this chapter.

16.1 THE UTILITY OF MODELS IN TRAINING DESIGN, DEVELOPMENT AND EVALUATION

The term "model" has a variety of meanings. Basically, a model is a representation of a phenomenon—an event, process, or physical entity. A model attempts to incorporate at least the essential, defining features or variables of the phenomenon it represents, and to show relationships among those features or variables. The different connotations of the term arise from the variety of forms a model may take. For example, a model may be a conceptual representation consisting of definitions, constructs, and propositions about relationships (a theoretical model). It may be a mathematical formulation of variables and their relationships (an analytic or predictive model). It may be a computer-based representation of a dynamic process (a simulation model); or it may be a physical replica, a "mockup" or facsimile of some kind. Some models, such as complex training simulators, are hybrids involving two or more of these forms (DeBor and Swezey, 1989). Models of various kinds are important tools for use in considering training-oriented developments. In general, theoretical models may be used to help guide research. Other types of models can provide a way of performing activities that minimize the penalty for error.

A considerable amount of effort in the training world has been devoted to techniques which are variously called *simulation, modeling*, and *gaming*. A good discussion of the nature of these terms is in a book chapter by Chapanis and Van Cott (1972), where it is suggested that the term "gaming" is fairly widely understood to apply to the independent representation of a contest in which there are antagonists or opponents. The term, therefore, applies in the discussion of war games, business games, computer games, and the like as applied to competition-oriented training activity.

The terms simulation and modeling, however, appear to be less specific. In the training context, these terms may apply to physical items of equipment which in some way represent other pieces of equipment and which are used for training purposes. In a scientific context, the terms may apply to sets of mathematical or statistical equations which are used to describe processes or systems. To human or animal researchers, modeling may refer to the mimicking of the behavior of a person or organism. Another area which has been suggested as an aspect of the term modeling is experimentation. Experiments, it may be argued, are special types of models. They, like models, are finite representations of larger parts of the real world. These terms are, therefore, ambiguous. They apply to a variety of divergent concepts and often mean entirely different things to different people.

What these terms have in common is that all are, at some level of detail or another, analogies. They are all designations of objects, events or systems, which correspond in some particular way to the thing being depicted. Thus maps, Markov chains, cues, aircraft trainers, war games, experiments, and regression equations are all models, in a physical or in a symbolic sense, or both. A map of the United States, for example, is a physical model in that its shape is the same shape as the actual landmass. It is symbolic in that terrain features may be depicted by lines, colors, and words.

In other contexts (Swezey, 1977–78) it has been suggested that training-oriented models should have several unique characteristics.

First they must reliably *represent* a real situation. Here representation is meant to imply that the model portrays certain important situational characteristics in a way which allows for understanding of the portrayed characteristics. Thus, for example, a regression equation represents characteristics of a situation quantitatively, and in a manipulable format.

Second, models must allow for retention of *control* over represented characteristics. This central aspect may in some cases define the difference between a model and an operational environment. In the operational environment, a situation may often be essentially uncontrolled, and the requirement for planned variation is the differentiating characteristic of the model.

Third, models may be designed to deliberately *omit* certain situational characteristics. Reasons for the deliberate omissions may include several factors: (1) certain situational aspects may be considered unimportant. For example, the requirement to close a vehicle's door may be omitted in a model of driving behavior because it is not considered important for training purposes. (2) Situational aspects may be omitted in training-oriented models because they are considered to be too dangerous, prohibitively costly to represent, or because they are otherwise not feasible. Such is the case in weapons system models where actual firing never occurs. (3) Characteristics may be omitted in order to eliminate unpredictability. This in fact is often the rationale used to argue for controlling aspects of models. Elimination of unpredictable variance by controlling situational variables can be an important benefit of models over operational contexts.

The major advantage of modeling, however, is that, to the extent that one has a real-world representation at some acceptable level of fidelity, he can then manipulate the representation and measure it instead of the real world. A major problem involves determining the level of fidelity. The problem is a difficult one, since it is not necessarily true, for example, that a higher fidelity of modeling yields greater transfer-of-training (see below). Another problem is that modeling in some cases introduces unique characteristics which are peculiar to the model or simulation, and do not reflect the real world. Such is the case with driving simulators which induce motion sickness to a greater extent than do actual vehicles. What we have here in some cases, is a modeling equivalent of an experimental Hawthorne effect.

It is apparent therefore, that the validity of a model cannot be assumed, no matter how elegant the representation, unless it is demonstrated under controlled, systematic conditions.

As has been discussed elsewhere, a fundamental problem in the area of training-oriented models, is to optimize an equation which includes at least three components: fidelity, transfer-of-training, and cost (Biel, 1966; Miller, cited by Biel). Basically, the idea is to create a situation which accurately locates the fidelity level that is appropriate for creating transfer from training to job performance, at a point where additional training transfer increments are not justified in terms of added costs. This is the point of diminishing returns where, if fidelity is increased, additional increases in transfer-of-training are not proportional to increases in cost. It is apparent that a variety of variables external to the situation may affect this equation. These include: quality and type of instruction, the nature of the skills and knowledges involved, the severity of the standards required for successful task performance, the criticality of the task, and trainee characteristics. Thus it appears that a multivariate problem exists, having several general types of independent variables: those external to the model (such as the ones mentioned) as well as modeling-oriented inputs such as fidelity, design variables, and simulator complexity. Dependent measures include acquisition, retention, transfer, and cost factors.

One major problem faced by those involved with training development is resolving this situation optimally under whatever training context is present. This is no small trick, and we in the field continue to investigate new methods and approaches for attacking this problem in our increasingly complex world. Readers interested in training and instruction may also wish to consult Chapters 5 (Learning Facts and Skills), 12 (Task Analysis), 41 (Simulation), and 55 (Multimedia) in this volume.

16.2 MODELS OF ACQUISITION

Significant development and application of models of the acquisition process have occurred over the last three decades. Learning curves have been applied to establish and control wage incentive plans (Broadston, 1968), compare the difficulty of different jobs (Thompson, 1963), provide a metric for personnel selection and prediction of job success (Downs, 1970; Reilly and Chao, 1982), and designate training criteria (Johnson, 1980). There does not, however, appear to be any one universal function for describing skill acquisition; the best fit to a given set of data is likely to vary as a function of the type of task, as well as its levels of difficulty and complexity, among other factors. Nevertheless, as noted by Lane (1987), researchers continue to study the shape and characteristics of acquisition curves since these data have practical utility for managing training, including estimating how long it will take to attain a given asymptotic level, determining when sufficient training or practice has been provided, and tracking individual performance. It is important, for example, to be able to determine at what place on an acquisition curve an individual's score is located in order to ascertain whether training in a given segment is essentially complete, and to avoid needless additional training.

16.2.1 Learning Curves and Related Issues

The literature on learning and skill acquisition is broad. Although efforts to derive universal laws which explain learning have existed for decades, no unified theoretical concept exists to explain skill acquisition. Nevertheless, numerous attempts to quantify the literature in this area have been conducted. Thurstone (1919), for example, derived and tested many learning functions, and Hull (1952) proposed a large number of highly complex mathematical laws of learning which included explicit predictions about the shape of acquisition functions.

In an attempt to summarize this area, Newell and Rosenbloom (1981) performed extensive analyses of the forms of acquisition curves from dozens of studies. Prior to this analysis, learning was generally posited to be an exponential function of practice in which performance improved by a constant fraction with each additional trial. Hull's (1943) mathematical equation characterizing habit strength, for example, was expressed in exponential terms. As illustrated in Figure 16.1, three additional classes of functions for describing performance changes during learning also exist. These are termed the *power* function, the *hyperbolic* function, and the *logistic* function (Lane, 1987). For power functions, changes in performance between trials decrease systematically as the number of trials increase. Power law functions are further distinguished by the property of negative acceleration; each unit of practice produces a smaller and smaller improvement in performance. In contrast, the exponential function generally decreases or increases much more rapidly producing a much steeper curve than the power function. Hyperbolic functions, considered to be a special case of a generalized power function, produce a steadily increasing (or decreasing) curve with a slope that changes across successive trials. Logistic functions are characterized by S-shaped curves which rise slowly during the early stages of learning, accelerate rapidly during the middle stages, then level off to become progressively flatter near the final stages. Acquisition functions in each of the four families

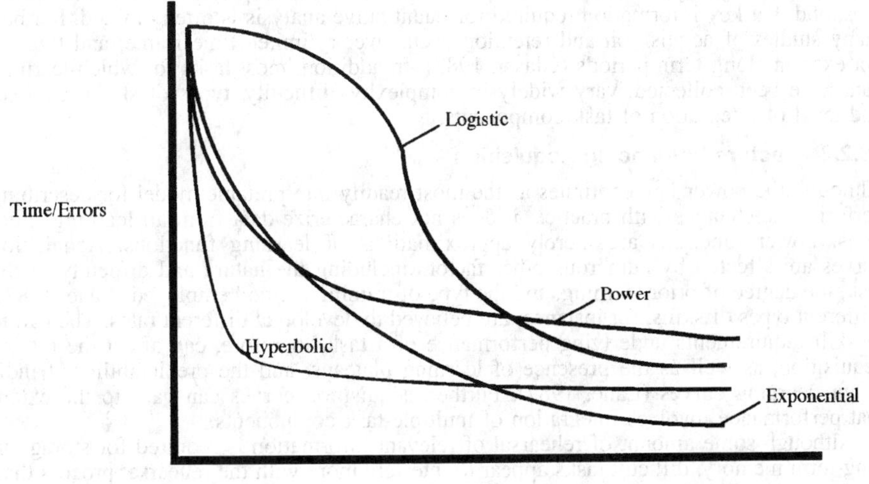

Figure 16.1 Comparison of theoretical learning curve functions.

typically assume one of three general shapes; positively accelerated, negatively accelerated, and a hybrid possessing both positively and negatively accelerated components (Lane, 1987). A variety of functions have been used to fit acquisition curves. Reaction time data, for example, typically fit with power and exponential functions (Newell and Rosenbloom, 1981); data derived via performance ratings, however, typically fit logistic and exponential functions (Spears, 1985). All such functions tend to show fast initial acquisition rates with a gradual approach to an asymptote.

Most learning functions appear to be described by power functions (Newell and Rosenbloom, 1981). This implies that learning never stops entirely, but as practice continues, smaller and smaller benefits are realized. Thus, performance improves rapidly early in practice, but soon approaches a state of diminishing returns where each additional increment in performance requires longer and longer practice intervals. Many types of dependent measures show this power relationship, including time and error rate performance. This empirical law has been known for some time (Snoddy, 1926) and is widely recognized (Fitts and Posner, 1967; Welford, 1968). Although the power law is generally viewed as associated with acquisition of perceptual and motor skills, it also appears to hold for learning of other kinds, including memory tasks and problem solving (Anderson, 1980). Neves and Anderson (1981) found that performance as measured by total time, total number of steps, and time per step in solving geometry type proofs all followed power functions, decreasing rapidly during early practice and continuing to improve after extensive practice. Predictions from the power law have also been shown to fit data from a number of other contexts, including operating a cigar rolling machine (Crossman, 1959), perceptual tracking (Stevens and Savin, 1962), playing card games (Newell and Rosenbloom, 1981), and detecting letter targets (Neisser, Novick, and Lazar, 1963). Anderson (1982) has argued that the reason for this generalizability is that all changes brought on as a function of practice depend upon simple associative learning.

Newell and Rosenbloom (1981), concluded that: (a) although both exponential and power functions are negatively accelerated, the rate of improvement described by the power function (a relatively slow and gradual decrease) accurately characterizes learning in most contexts examined; (b) models of the acquisition process must consider asymptote and experience parameters; and (c) the power law generalizes to many types of cognitive as well as perceptual and motor data. In a similar review, Lane (1987) examined findings from the literatures on skill acquisition, learning, retention, and transfer in order to identify the extent to which existing research on acquisition and learning provided evidence of learning curves which could support quantitative predictions. As part of his effort, Lane also identified variables which systematically affect the rate of acquisition. Lane concluded that despite the existence of regularities in the shape of acquisition curves, existing data do not appear to support performance prediction, since parameters known to influence skill acquisition vary widely. Lane noted that the literature is primarily descriptive in nature. The majority of tasks are simple or otherwise not representative of real-world jobs, and that key information required for quantitative analysis is often omitted. Further, many studies of acquisition and retention occur over a limited time frame, and thus do not examine long-term periods (Glaser, 1982). In addition, most tasks for which learning data have been collected, vary widely in complexity, difficulty, types of skills required, and level of integration of task components.

16.2.2 Factors Influencing Acquisition

Although the power law continues as the most readily interpretable model for describing performance changes with practice, it does not characterize data from all learning situations. Power functions are merely approximations of learning functions. Acquisition curves are affected by numerous other factors including the nature and difficulty of the task, the degree of prior learning, and the type of training method employed (Lane, 1987). Different types of skills, for instance, are believed to develop at different rates. The nature of skill requirements underlying performance of a task, therefore, can affect the rate of acquisition, as well as the presence of learning plateaus, and the predictability of individual learning curves (Lane, 1987). Further, acquisition curves can vary to the extent that performance involves integration of multiple-task components.

Although some amount of rehearsal of relevant information is required for storage in long-term memory, difficult tasks appear to interfere more with the rehearsal process than do easy tasks (Waugh and Norman, 1965). They also require greater amounts of short-term memory capacity and appear to produce positively accelerated curves during skill acquisition (Lane, 1987). Gay (1973), for instance, found that as problems increased in difficulty, students required more time to process information (the time elapsing between presentation of an example and advancement to the next problem was positively related to rule difficulty). The difficulty of learning material is often described either in terms of time or in terms of the number of trials required for mastery. Easy tasks often lead to ceiling effects, while excessively difficult tasks often demonstrate an insensitivity to practice (Mazur and Hastie, 1978; Bahrick, Fitts, and Briggs, 1957). Several factors contribute to the difficulty of learning an unfamiliar item including meaningfulness, concreteness, familiarity, and associations with other information in the memory store (Carter and McCarthy, 1988). Concrete nouns, for example, have an average recall rate of 58%, versus 36% for abstract nouns (Rodgers, 1969). Additional considerations include: (1) the intrinsic difficulty of the item, (2) interactions among previously learned items and new items to be learned, (3) interactions among groups of items to be learned at the same time, and (4) the interaction between groups of items to be learned in sequence (Higa, 1966). In general, individuals find material difficult to learn if it has no relation, association, or similarity to previously acquired material. Thus, difficult tasks appear to require greater amounts of short-term memory capacity than do easy tasks (Gay, 1973; Waugh and Norman, 1965). Increasing task difficulty by manipulating task requirements (e.g., requiring recall rather than recognition, using fixed-pace rather than self-paced practice, and/or presenting learning material in isolation rather than in context) appears to decrease acquisition and increase forgetting (Crothers and Suppes, 1967). It has also been found that task difficulty drives learning rate, which in turn determines time required to achieve proficiency (Krueger, 1947).

The point in time or level of performance chosen to terminate a training period is another important factor in determining the form of resulting acquisition curves (Jones, 1985). Employing arbitrary lengths for an acquisition period may serve to reveal only part of the actual acquisition curve, thus the shape of the curve during that portion may provide a misleading impression (Spears, 1985). Johnson (1980) developed a procedure which can be used to determine the point at which training can be terminated. This formula involves estimating the incremental benefits of additional training from the slope of the learning curve at a given point in time. The benefit is then compared to the incremental cost of additional training, expressed as either quantitative values or qualitative estimates derived from scaling. Training is terminated when the slope of the function is equal to or less than the benefit/cost ratio. Another means of determining the cut-off point for training involves the concept of mastery level training, proposed by Bloom (1974). This technique establishes the degree of mastery required at a known fixed level, and manipulates variables such as time and conditions so that all trainees attain the preestablished level. This view, often termed "Criterion Referenced Instruction," is designed to specify the required performance at a fixed criterion point, rather than at some artificial point which may capitalize on both random and cyclical changes in performance (Hayes and Pereboom, 1959). Criteria for mastery should, however, be defined on the basis of something beyond training performance (i.e., actual task or job performance), using empirical data to determine what level of mastery leads to corresponding levels of post-training performance.

Evidence suggests that mastery of a knowledge domain requires different training events than those used to develop mastery of procedural skills (Gagné, Briggs, and Wager, 1989). Much of the published literature on skill and knowledge acquisition, therefore, involves comparisons of performance under two or more alternate training methods in hope of specifying optimal training conditions for various skills and knowledge. As a result, various principles (e.g., sequencing, scheduling, pacing, course length) for improving both training and acquisition performance have been identified; each having different effects on acquisition parameters. Studies with intervals of no-practice, for example, tend to show higher intercepts, slopes, and asymptotic values than those with continuous practice, over the same number of trials (Mazur and Hastie, 1978; Stevens and Savin, 1962). Of the many factors considered in motor learning, methods of feedback and practice schedules have received the most attention. Perhaps one of the most significant findings concerns knowledge of results (response produced feedback supplemented by information about the outcome of the response) which, with the exception of practice, is one of the most critical variables which influences skill acquisition positively (Schendel, Shields, and Katz, 1978). Kinetic feedback, for example, has been shown to improve motor skill acquisition and performance beyond levels obtained when only knowledge of results are presented (Newell, 1976). In general, data indicate that control of movements continues to rely on intrinsic feedback (inherent in producing a response) even after extensive practice (Kohl and Shea, 1992).

The distribution of practice can also have a significant influence of skill acquisition. Massed practice generally tends to decrease the level of performance acquired in a fixed number of task repetitions (Digman, 1959; Reynolds, 1952). A recent meta-analysis, conducted by Lee and Genovese (1988), indicated that massed practice results in poorer learning than distributed practice, although the effect is much weaker for retention tests than during acquisition. Some evidence suggest that different principles of distribution of practice are required for discrete versus continuous tasks (Lee and Genovese, 1988). Shea and Morgan (1979) also investigated different practice schedules, consisting of random versus blocked schedules, in the context of acquiring a multicomponent motor skill. Learners practiced three tasks, one at a time in distinct blocks of trials, or randomly intermixed so that a given task was never practiced on successive trials. These researchers found a clear advantage for the blocked practice group over the random practice group; the former was significantly faster. However, performance assessed after a ten-day retention interval favored the random practice group. Shea and Morgan concluded that conditions of practice that promote good acquisition performance may not necessarily produce effective learning (refer to Schmidt and Bjork, 1992, for a comprehensive discussion of this topic).

A major difficulty in deriving general laws of acquisition (McKeachie, 1974; Powers, 1976) concerns the issue of individual differences. Different ability levels can result in dramatically different acquisition curves which vary in the rate, intercept, and asymptote under a single method of instruction; yet these curves often become similar when individualized methods of instruction are used (Cronbach and Snow, 1976; Frederickson, 1969). Further, higher ability students, regardless of method of instruction, typically have higher initial performance levels, and thus tend to improve at slower rates. In addition, aptitude typically tends to interact with method of instruction so that some individuals learn better with one method or treatment than with another (Cronbach, 1957; Frederickson, 1969). Instability during the early stages of complex skill development have been attributed to four sources: (1) large within-subject variation, (2) differences in initial performance levels, (3) differences in rates of acquisition of skills, and (4) changes in ability requirements across various stages of learning (Schneider, 1985).

16.2.3 Learning Plateaus

Plateaus, or periods of no significant improvement in learning, have received attention in the literature since they offer a way to interpret the process of acquisition. Bryan and Harter (1899) interpreted plateaus as representing transition periods where lower level skills become automated and subsequently integrated into higher order skills. The idea that skill acquisition involves transitions between fundamentally different ways of processing tasks is evident in many contemporary models of acquisition. Some researchers have suggested that three distinct performance components are required to describe the process of acquisition given the existence of plateaus; one each for early, middle, and late periods of practice in which regular improvement occurs over long time periods (Taylor and Smith, 1956). Although much evidence supports the presence of plateaus in a variety of task situations (Spears, 1983; Kao, 1937), some researchers (Keller, 1958) have suggested that artificial plateaus may result as a function of data collection methods. Glover (1966) has suggested that plateaus can be reduced by greater attention to instruction during the learning period, by using more individualized approaches to instruction, and by demonstrating correct methods rather than relying on discovery or trial and error approaches. Plateaus do not necessarily represent stages of learning.

16.2.4 Distinguishing Between Learning and Acquisition Performance

Many researchers have assumed that the conditions of acquisition that speed a learner's rate of improvement, or result in more effective performance during practice, will lead to the most effective learning. According to Schmidt and Bjork (1992), however, this view ignores the critical distinction between relatively permanent effects of a manipulation and effects which may be transient or temporary. A given training manipulation may have either or both of these effects. Variables such as distribution of practice, task difficulty, and feedback during training can cause increases or decreases in initial acquisition, with no effect (or even an opposite effect) on retention or transfer. Augmented feedback, for example, has been shown to influence acquisition rather than learning, and performance increments often tend to vanish when augmentation is removed (Cormier, 1984; Kinkade, 1963). Similarly, introduction of short periods of rest between trials (characteristics of distributed training) tends to increase performance across trials and thus decrease the rate (slope) of the learning curve as a function of number of trials; yet spacing of repetitions generally yields better long-term retention than does massed practice (Glenberg and Lehmann, 1980; Underwood, 1961). According to these researchers, in order to qualify as a valid learning effect, experimental manipulations must either demonstrate permanence across time or the ability to survive the removal of the manipulation in question. They argue, therefore, that the actual effectiveness of learning is measured by the level of retention.

Much literature has indicated that it is necessary to separate acquisition from learning in order to evaluate the effects of the many variables presumed to influence skill acquisition (Cormier, 1984; Salmoni, Schmidt, and Walter, 1984; Schmidt and Bjork, 1992). Schmidt and Bjork (1992) and others have described the process by which characteristics of a task create a learning environment in which performance is depressed but learning is apparently enhanced. The implications of this phenomenon, known as "contextual interference" (Battig, 1966; Shea and Morgan, 1979) for training, are that variations of task conditions, manipulation of task difficulty, or other means of inducing extra effort by the trainee are likely to be beneficial for retention and transfer. Recent research by Schneider, Healy, Ericsson, and Bourne (1995), in a series of experiments investigated effects of contextual interference within the context of cognitive rule-based learning, a skill underlying many applied tasks found in the aviation, military, and computer fields. Results mirrored those found in the verbal learning and motor skill acquisition literature, indicating that a random practice schedule leads to the best retention performance after initial acquisition, but it hinders initial learning. Thus, training conditions designed to achieve a particular training objective (long-term retention, transfer, and/or resistance to altered contexts) are not necessarily those that maximize performance during acquisition. This concept also has significant implications for predicting trainee performance. Although a typical method employed to predict trainee performance involves monitoring acquisition data during training, using immediate past performance data as predictors to estimate future retention performance, this strategy may not appropriately index performance on the job or outside of the training environment. Further, initial performance of complex skills tends to be unstable and often is a poor indicator of final performance. Correlations between initial and final performance levels for a grammatical reasoning task, for example, reached only .31 (Kennedy, Jones, and Harbeson, 1980), a moderate level at best. This strategy therefore, may lead to inconsistent and/or erroneous training prescriptions. Contextual interference in which manipulations of task difficulty produce decreased training performance but enhanced retention and transfer, reinforces the need to distinguish between training and post-training performance.

16.2.5 Stages of Learning

Skill acquisition is believed by many to proceed in accordance with a number of stages or phases of improvement. Although the number of stages and their labels differ among researchers, the existence of such stages is supported by a huge amount of research and theoretical development during the past 30 years. Traditional research in learning and memory posits a three-stage model for characterizing associative type learning involving the process of differentiating between various stimuli or classes of stimuli to which responses are required (stimulus discrimination), learning the responses (response learning), and connecting the stimulus with a response (association). Empirical data suggest that this process is most efficient when materials are actively processed in a meaningful manner, rather than merely rehearsed via repetition (Craik and Lockhart, 1972; Cofer, Bruce, and Reicher, 1966).

Anderson (1982) has also proposed a three-stage model of skill acquisition, distinguishing among cognitive, associative, and autonomous stages. The cognitive stage corresponds to early practice in which the learner exerts effort to comprehend the nature of the task and how it should be performed. In this stage, the learner often works from instructions or an example of how a task is to be performed (i.e., modeling or demonstration). Performance may be characterized by instability and slow growth, or by extremely rapid growth, depending on task difficulty and degrees of prior experience of the learner. By the end of this stage, learners may have a basic understanding of task requirements, rules, and strategies for successful performance; however, these may not be fully elaborated. During the associative stage, declarative knowledge associated with a domain (e.g., facts, information, background knowledge, and general instruction about a skill acquired during the previous stage) is converted into procedural knowledge, which takes the form of what are called "production rules" (condition-action pairs). This process is known as, "knowledge compilation." It has been estimated that hundreds, or even thousands, of such production rules underlie complex skill development (Anderson, 1990). Novice and expert performance is believed to be distinguished by the number and quality of production rules; experts are believed to possess many more elaborate production rules than novices (Larkin, 1981). Thus, Anderson has distinguished between declarative knowledge, which is cast in a propositional network of facts, and procedural knowledge, which is represented by production rules. For example, in the context of statistics, students must learn symbols and statistical terms (declarative knowledge), as well as demonstrate the ability to calculate and compute various values (procedural knowledge). Anderson's system focuses primarily on the nature of activities in the declarative and procedural stages of knowledge acquisition. Performance under the associative stage, while still primarily under voluntary control, is fluid and error free, and is characterized by rapid progress. Thus, the learning curve is very steep during most of the stage, and appropriate stimulus-response relationships are established. Here also, strategies are refined, preliminary motor programs are developed, and integration of task components and whole task practice has begun. In Fitts and Posner's (1967) terms, this phase is characterized by strengthening associations between stimuli and responses. Performance continues to improve through the process of tuning, which involves refinement of procedures through mechanisms of generation, discrimination, and strengthening (Anderson, 1982; Rumelhart and Norman, 1978). By the end of this stage, performance begins to level off as an initial asymptote occurs.

The autonomous stage occupies the longest practice period and is characterized by gradual improvements over long sequences of task repetitions. Practice during this stage serves to shift control from overtly voluntary processes to low effort (or "automatic") control of performance. Thus, underlying the development of automaticity is a shift in operational modes from controlled to automatic processing (Schneider and Shiffrin, 1977). This enables tasks to be accomplished without conscious monitoring, and without interference from secondary tasks (Ackerman, 1987). According to Fitts (1990), this automaticity may take months or years to develop. Automaticity is theorized to result in improved performance, principally via increased efficiency in the use of attentional or processing resources (Schneider, 1984). Although Fitts's (1964) stages were originally developed to described in the context of motor skill acquisition, they apply equally well to a wide range of task types, including those with cognitive components (Lane, 1987). According to this model, skill undergoes dramatic qualitative changes during the acquisition process in which declarative knowledge plays a significant role initially, but is replaced by procedural knowledge in the form of production rules, and is subsequently automated requiring significantly less cognitive effort. It should be noted that these three phases overlap and that the progression from one to the other is a continuous rather than a discrete process. This production system approach to acquisition has been successfully used to model the acquisition and transfer of procedural information acquired from written instructions in textual formats (Kieras, 1985), as well as modeling high-performance skill acquisition which is characterized by improvement over long periods of training, initial instability, and false asymptotes (Schneider, 1985).

Similarly, Rumelhart and Norman (1978) recognized three kinds of learning processes: (1) the acquisition of facts in declarative memory (accretion), (2) the initial acquisition of procedures in procedural memory (restructuring), and (3) the modification of existing procedures to enhance reliability and efficiency (tuning). Kyllonen and Alluisi (1987) have reviewed these concepts in relations to learning and forgetting facts and skills. Briefly, new rules are added to established production systems through the process of accretion, rules are fine-tuned during the process of tuning, and subsequently reorganized into more compact units during the restructuring process. This viewpoint is consistent with schema theory, whereby learning consists of processing new information into existing schemata, adding knowledge to the propositional networks, and modifying the interrelationships among key variables to reflect inferences from the new information. Schemata, therefore, provide the organizational mechanisms by which new information is related to previous learning, serving as the framework for understanding new material. The mechanisms posited by schema theory to account for automatic behavior, though cast in terms of restructuring and tuning of schemata, predict a course of automaticity development that is almost empirically indistinguishable from that based on the capacity and attentional mechanisms.

Rasmussen (1979) has distinguished among three categories, or modes, of skilled behavior: *skill-based, rule-based,* and *knowledge-based.* Skill-based tasks are composed of simple stimulus-response behaviors which are well learned by extensive rehearsal, and highly automated. Rule-based behavior is guided by conscious control and involves the application of appropriate procedures based on unambiguous decision rules. This process involves the ability to recognize specific well-defined situations which call for one rule rather than the other. Knowledge-based skills are used in situations in which familiar cues are absent and clear and definite rules do not always exist. Successful performance involves the discrimination and generalization of rule-based learning. Unlike Anderson's model, Rasmussen proposed that performers can move among these modes of performance as dictated by task demands. This general framework is useful in conceptualizing links between task content and the type of training required for proficiency in complex tasks.

Gagné (1965) specified eight types or categories of learning, each with its own rules. These categories are: signal learning, stimulus-response learning, chaining, verbal associations, multiple discrimination, concept learning, principle learning, and problem solving. Gagné arranged each in a hierarchy from simple to complex on the assumption that each level of higher order learning depends on the mastery of the one below it. He further theorized that learning is different for each type. Thus, although the outcomes of rule-learning and problem solving are the same, the processes by which the learning takes place is uniquely different. Gagné further extended this model to include even more types of learning and approaches (Gagné, 1974).

All of the above models incorporate a hierarchy of simple processes and rules which must be individually mastered through practice and eventually integrated into a smoothly executed set of task performance procedures. According to Schneider and Detweiler (1988), models of skill acquisition must specify how attentional and memory resources are utilized to accomplish tasks, as well as how these requirements change with practice. Most models, however, do not explicitly deal with the use of cognitive resources, nor do they detail the manner in which practice affects such resources.

16.2.6 Acquisition Guidelines

As discussed, the typical curve relating performance to practice is a negatively accelerated function in which there is a decreasing amount of improvement per trial with increasing trials of practice (Fitts and Posner, 1967; Glover, 1966). However, parameters for estimating group acquisition curves (rate, initial experience level, asymptote) are highly task and content specific. Variability and time course of progress for skill acquisition tend to be a function of the materials and skills being taught, the methods used in a particular training setting, and the entrance skill level of the trainees. While curve shape can often be anticipated, the time course over which acquisition runs, and thus the curve parameters, is generally not predictable from prior knowledge of task characteristics.

The literature strongly supports the presence of a series of stages within the skill acquisition process; three-stage models provide a satisfactory general representation of the time course of acquisition. In general, skills and knowledge are accumulated gradually across successive segments and phases of training and practice. Finally, performance and learning are not the same thing; concepts such as "contextual interference" illustrate the distinctions between performance and learning.

The following list of guidelines provide guidance for instructional designers and developers in terms of facilitating the knowledge and skill acquisition process.

- Practice should continue for some time after mastery to increase retention (Jones, 1985; Schendel et al., 1978).
- Practice beyond mastery should be limited in duration to the minimum number of trials required to attain the component integration associated with automaticity (Cormier, 1984).
- Incorporation of some means of inducing extra effort by the learner (variation of task conditions during initial learning, manipulation of task difficulty) are likely to benefit retention and transfer (Cormier, 1984; Shapiro and Schmidt, 1982; Shea and Morgan, 1979).
- Present information to promote consistent processing, including the use of analogies, and adaptive training (Schneider, 1985).
- Design the task to allow many trials of critical skills in a short period of time (Schneider, 1985).

- Avoid overloading temporary memory; use of job aids and augmentation cues can alleviate working memory load (Schneider, 1985).
- Vary aspects of the task that vary in the operational setting (Schneider, 1985).
- Maintain active participation and high levels of motivation throughout training (Schneider, 1985).
- Skills for which practice is terminated prior to the initiation of automaticity are not retained as well as those for which practice is continued into overlearning (Jones, 1985).
- Use the "drill and practice" method of instruction to bring learners to a level of automaticity on lower-level subskills, so that learners can more readily perform the task (Bransen, Raynor, Coxx, and Hannum, 1978).
- Procedural instruction should be accompanied by conceptual information only if the goal of training is to generalize the skills to equipment and/or other situations which were not specifically addressed during training (Swezey, Perez, and Allen, 1991).
- Provide performance measurement relative to the learning process (Oosterveld, 1980).
- Arrange the order of presentation of instruction to require prerequisite knowledge and skills before a new set is undertaken (Caro, 1973).
- Early in training, present external cues that will bring out the desired responses in the student's actions (Branson, Raynor, Coxx, Furman, King, and Hannum, 1975).
- Augmenting cues should not distract from the relevant cues in the task (Eberts and Brock, 1984).
- Provide warm-up exercises and sample items (Hamel and Clark, 1986).
- Explain the criteria for evaluation of performance (Keller, 1987).
- Developing consistent processing can be accomplished in a variety of ways, including the use of analogies, the provision of specialized representations of the problem, and adaptive training (Schneider, 1985).
- Break difficult tasks into small, meaningful parts which are easily learned (Hamel and Clark, 1986).
- The learner should be provided with a simple overview of the domain of instruction, which epitomizes the conceptual model employed, before engaging in specific task details (Robinson and Knirk, 1984).
- The time required for a typical session (or lesson) should be within the attention span of the target audience (Hamel and Clark, 1986).
- Where possible, supply students with diagrams, pictures, charts, graphs, rhymes, key words, and other association devices which the student can use to relate what he already knows to what he is trying to learn (Branson et al., 1975).
- Providing learners with frequent feedback on their performance will allow them to modify their responses to improve the learning process. Knowledge of what to do and how to do it should be linked with what happened and how it happened (Smith and Beringer, 1987).
- Feedback on good performance is as important as feedback on poor performance, because both indicate the performances' appropriateness and effectiveness (Davis, Gaddy, and Turney, 1985).
- Vary the schedule of reinforcement in terms of both interval and quantity (Keller, 1987).
- Early stages of training and practice should incorporate Knowledge of Results. KOR should be withdrawn slowly and only after skill/knowledge has been acquired (Williams, 1974).
- Practice should be distributed so as not to fatigue the learner (Eberts and Brock, 1984).
- Facilitate acquisition of training material by relating it to on-the-job duties, responsibilities, advancement, or survival (Branson et al., 1975).
- Teaching consistent procedural rules associated with a task should facilitate the learning of cognitive skills in various settings, including the classroom (Fisk and Lloyd, 1988).

16.3 MODELS OF RETENTION

Instructional designers must consider not only how to achieve more rapid, high-quality training, but also how well the skills taught during training will endure after acquisition. Further, what has been learned must be able to be successfully transferred or applied to a wide variety of tasks and job specific settings.

16.3.1 Retention Functions

Retention of learned material is often characterized as a monotonically decreasing function of the retention interval, falling sharply during the time immediately following acquisition, and declining more slowly as additional time passes (Ebbinghaus, 1913; McGeoch and Irion, 1952; Wixted and Ebbesen, 1991). There is general consistency in the loss of material over time. Subjects who have acquired a set of paired items, for example, consistently forget about 20% after a single day, and approximately 50% after one week (Underwood, 1966). Bahrick (1984), among others, demonstrated that although large parts of acquired knowledge may be lost rapidly, significant portions can also endure for extended intervals, even if not intentionally rehearsed.

Evidence suggests that the rate at which skills and knowledge decay in memory varies as a function of the degree of original learning; decay is slower if material has previously been mastered than if lower level acquisition criteria were imposed (Loftus, 1985). The slope and shape of retention functions also depend on the specific type of material being tested, as well as upon the methods used to assess retention. As meaningfulness of material to the student increases, for example, rate of forgetting appears to slow down. Further, recognition performance may be dramatically better than recall, or vice versa, depending simply on how subjects are instructed (Tulving and Thomson, 1973; Watkins and Tulving, 1975). Also, various attributes of the learning environment such as conditions of reinforcement, characteristics of the training apparatus, and habituation of responses appear to be forgotten at different rates (Parsons, Fogan, and Spear, 1973). Retention functions reported in the literature vary widely in terms of the time span over which they occur; measured in terms of seconds, minutes, hours, days, weeks, months, and years. Many analytic studies have examined retention within the context of verbal learning. Extensive reviews and analysis of information in this area may be found in Murdock (1974), Crowder (1976), and Kintsch (1970).

16.3.2 Relationship Between Acquisition and Retention

Although learning and retention are independent concepts, the transition from acquisition to retention may be characterized as continuous rather than discrete. The boundary between acquisition and retention (whether a practice trial is considered to occur during, or after acquisition) therefore, often depends on arbitrary definitions. In general, distinctions between acquisition and retention are defined either in terms of operational specifications, or in terms of functional performance differences. Some studies operationally define retention, for example, as any duration which is longer than the intertrial interval employed at the time of training. There are variables which produce marked differences in the rate of learning which appear to have no residual influence over long retention intervals. For example, learning is enhanced when the information to be learned is high in "meaningfulness"; but once the material is learned, meaningfulness has little influence on retention. Also, the distribution of practice (spacing effect) has little influence on learning responses, but can markedly facilitate retention (Wickelgren, 1972). Thus, some variables affect learning and retention in decidedly different ways.

16.3.3 Measuring Retention

Retention, like learning and motivation, cannot be observed directly; it must be inferred from performance following instruction or practice. To date, no uniform measurement system for indexing retention has been adopted by the applied community. General indices of learning and retention used in the research literature over the past 100 yr include a huge variety of methods for measuring recall, relearning, and recognition. Luh (1922), for instance, used five different measures to index retention: recognition, reconstruction, written reproduction, recall by anticipation, and relearning. Luh found that while retention curves for all measures were negatively accelerated, they had different rates of decline, demonstrating that different ways of assessing retention can yield different results.

Retention appears to be lowest for recall measures, intermediate with relearning, and highest with recognition measures (Postman and Rau, 1957). Although such results suggest that recognition measures invariably result in higher retention scores than do recall measures, other evidence suggests that differences among response measures are a function of the conditions under which the measures are obtained (Bahrick, 1964). The degree of superiority of recognition over recall, for example, is a direct function of the similarity of the alternatives to the correct items (McNulty, 1965). Savings scores used as measures of relearning also provide different results across different tasks, and existing evidence about relearning suggests that it is difficult to predict the rate of relearning as a function of original learning (Schendel et al., 1978). Some consistent data in the area of relearning motor skills suggest that the time to retrain is consistently less than 50% of the original learning time, and that retraining time increases with longer retention intervals, more difficult tasks, and procedural rather than continuous tasks (Farr, 1987).

With regard to the time interval for measurement, some researchers have gauged the effects of practice on learning immediately after practice, whereas others have used various periods of delay. The latter approach has been generally viewed as superior since it provides a mechanism for the temporary effects associated with an experimental manipulation (increased motivation to perform, or fatigue) to dissipate (Schmidt and Bjork, 1992). Such measurement issues have resulted in the delineation of several critical methodological factors that should be employed when evaluating long-term retention studies, including the assessment of performance after a delay interval rather than merely relying on acquisition data, and the incorporation of longer retention intervals (weeks, months, and even years). Consistent with this perspective is work performed by Bahrick and associates (Bahrick 1984; Bahrick, Bahrick, and Wittlinger, 1975; Bahrick and Hall, 1991). These investigators have pioneered a unique approach for assessing real-world skill/knowledge retention which allows retention to be evaluated over significantly longer intervals than do traditional evaluation research efforts. The method uses a cross-sectional statistical strategy which draws samples of individuals who have acquired the same skills/knowledge at different times in the past, while controlling for the degree of original learning, and amount of rehearsal. This approach has been used to study retention of such information as buildings and street names, spatial locations, Spanish knowledge acquired as long as 50 years previously, and retention of college level course content (Bahrick, 1979; Conway, Cohen, and Stanhope, 1991). Results from long-term retention research has lead Bahrick, among others, to conclude that although a significant amount of what influences long-term retention is determined by acquisition processes, acquisition performance alone is not an accurate predictor of long-term retention. Bahrick's data also emphasizes that the quality and extent of training at the time of initial learning can significantly impact long-term retention.

16.3.4 Factors Influencing Retention

The list of variables known to reliably influence rate of forgetting of learned knowledge and skills is relatively short. In a recent review, Farr (1987) surveyed the literature on long-term retention and identified a list of variables known to influence long-term retention from laboratory to applied contexts, including: degree of original learning, characteristics of the learning task, and the instructional strategies used during initial acquisition.

Degree of Original Learning

According to Farr (1987) the single largest determinant of the magnitude of retention appears to be the degree of original learning. In general, the greater the degree of learning, the slower will be the rate of forgetting (Underwood and Keppel, 1963). This relationship is so strong that it has prompted some researchers to argue that any variable that leads to high initial levels of learning will facilitate skill retention (Hurlock and Montague, 1982; Prophet, 1976). Systematic relationships between degree of original learning or practice, and retention have occurred over a wide variety of contextual domains, including motor tasks (Gardlin and Sitterley, 1972; Schendel et al., 1978), procedural tasks (Hagman and Rose, 1983), and verbal learning tasks (Bahrick, 1984). Correlations between level of performance attained on the last acquisition trial and the recall tasks, for instance, have ranged from .74 to .81 (Underwood, 1964). Such relationships suggest that differences among subjects in terms of original learning attained are far more powerful determinants of retention than are differences in rate of forgetting. The degree of original learning has also been shown to be far more important for long-term retention than is the time at which additional trials are given (Schendel and Hagman, 1982). Thus, one way to diminish decay is to strengthen the degree of original learning by increasing the amount of practice or extending the number of practice trials beyond mastery (overlearning). Others include requiring learners to overcome intratask or contextual interference (Battig, 1972; Shea and Morgan, 1979), and providing opportunities for deeper levels of elaborative processing involved in acquiring material for later retrieval (Bower and Karlin, 1974; Craik and Lockhart, 1972).

Characteristics of the Learning Task

The retention, organization, and complexity of the material to be learned also appears to have a powerful influence on retention. The efficiency of information acquisition, retrieval, and retention appears to depend to a large degree on how well the learning material has been organized (Cofer et al., 1966). Conceptually organized material appears to show considerably less memory loss and to be more generalizable than material which is not so organized (Swezey et al., 1991). Blocking stimulus items according to membership in common categories has also been shown to enhance memory (Moely and Shapiro, 1971). Tasks which are amenable to organization are learned at a faster rate than less structured tasks, and are better retained (Schendel et al., 1978). The argument has been made, therefore, that conditions at encoding that allow items to be cognitively interrelated enhance both learning and memory (Tversky, 1973). Continuous motor tasks appear to be more easily retained than many other types of tasks (Annett, 1979; Naylor and Briggs, 1961; Prophet, 1976). This is believed to result from enhanced task integration and organization. Although task complexity has been shown to be a primary predictor of the retention of procedural tasks, when properly organized, even procedural tasks, which generally show rapid declines in retention, may also exhibit high retention levels (Swezey et al., 1991). In a study by Gardlin and Sitterley (1972), subjects who were trained for relatively short time periods showed improved retention if they performed a procedural task with high levels of organization.

It also appears that certain "difficult" training conditions may foster various kinds of processing activities that are useful for effective retention. Shea and Morgan (1979), for example, investigated effects of random, as compared with blocked, practice sequences on acquisition and retention of motor skills. Retention was measured after a 10-min delay, and again after a 10-day delay and under either the same or different contextual conditions as acquisition. Results indicated that the blocked practice group performed considerably faster than the random group during the acquisition trials. In contrast to the acquisition findings, however, retention results showed that performance following the random acquisition condition was superior to performance following the blocked acquisition condition.

Instructional Strategies

Research has identified numerous variables that fall under the banner of strategies for skill and knowledge acquisition and retention, including spaced reviews, massed/distributed practice, part/whole learning, and feedback. An important variable with respect to forgetting is spaced practice. The *spacing effect* (the dependency of retention on the spacing of successive study sessions) suggests that material be studied at widely spaced intervals if retention is required. Similarly, research comparing distributed practice (involving multiple exposures of material over time) versus massed practice (requiring concentrated exposure in a single session) has occurred in a wide variety of contexts, including the acquisition of skills associated with aircraft carrier landings (Wightman and Sistrunk, 1987), word processing skills (Bouzid and Crawshaw, 1987), perceptual skills associated with military aircraft recognition (Jarrard and Wogalter, 1992), and learning and retention of second-language vocabulary (Bloom and Shuell, 1981). Most research in this area, however, has emphasized acquisition of verbal knowledge and motor skills. In general, research on the issue of distribution of practice has emphasized effects on acquisition, and found that distribution of practice is superior to massed practice in most learning situations, long rest periods are superior to short rest periods, and short practice sessions between rests yield better performance than do long practice sessions (Rea and Modigliani, 1988). However, no consistent relationships between these variables and long-term retention have emerged. Some evidence suggests that the spacing of study does appear to affect retention (Bahrick, 1984; Hagman and Rose, 1983). For long intervals, retention performance may be improved when learning is based upon slowly decaying retention functions (characteristic of spaced practice). At short intervals, however, retention performance appears to improve when learning has not had long to decay (characteristic of massed practice). Thus, material that is studied twice close together is forgotten more rapidly than if the two study sessions are separated further in time.

Techniques which help learners to build mental models which they can use to generate retrieval cues, recognize externally provided cues, and/or generate or reconstruct information have also generally facilitated retention (Kieras and Bovair, 1984). Gagné (1978) identified three general instructional strategies for enhancing long-term retention, including reminding learners of currently possessed knowledge which is related to the to-be-learned material, ensuring that the training makes repeated use of the information presented, and providing for and encouraging elaboration of the material during acquisition as well as the retention interval. Hurlock and Montague (1982) found that complex procedural skills benefit from mnemonic devices and contextual cues, as well as strategies for organizing and categorizing complex information. Another practice strategy, linked with the development of automated behavior, deals with consistent versus random (or varied) practice. Results of consistent versus varied practice generally show that high degrees of consistent practice leads to automation and better retention performance (Logan, 1988).

Other factors which have been shown to induce forgetting include proactive or retroactive interference by competing material that has been previously or subsequently acquired, and the events encountered by individuals between learning and the retention test. Information acquired during this interval may impair retention, while simple rehearsal or re-exposure may facilitate retention. Additional factors influencing skill and knowledge retention include: the length of the retention interval, the methods used to assess retention, and individual difference variables among trainees. The absolute amount of skill/knowledge decay, for example, tends to increase with time, while the rate of forgetting declines over time (Schendel et al., 1978). Researchers have also postulated a critical period for skills loss during a nonutilization period at between 6 mon and 1 yr after training (O'Hara, 1990). Psychomotor flight skills, for example, are retained many months longer than procedural flight skills (Prophet, 1976), and decay of flight skills begins to accelerate after a six-mon nonutilization period (Ruffner, Wick, and Bickley, 1984). According to Farr (1987), this area lacks a comprehensive theoretical framework and is cluttered with ambiguous and often inconsistent results.

16.3.5 Theoretical Underpinnings of Forgetting and Retention

The learning and memory literature identifies three fundamental causes of forgetting: (1) decay, which asserts that memories simply weaken as a function of time, (2) interference, which hypothesizes that competition from other material blocks the retrieval process, and (3) retrieval-cue mismatch which claims that access to cues that would serve to retrieve information are unavailable at the time of retrieval.

Decay Theory

Decay theory proposes that forgetting is produced by neural processes that progress at a steady rate independent of what material has been learned. Decay theory assumes that the role of rehearsal or practice is to strengthen memory traces (in terms of the degree to which cues can activate memory records) and thus make them more resistant to decay. Conrad (1967), for example, has considered decay as a loss of discriminative characteristics, and recall as a process involving discrimination of available traces. Research, however, has demonstrated that decay cannot be the only cause of forgetting since different amounts of forgetting have been shown to occur over the same delay period.

Interference Theory

According to interference theory, forgetting is a direct result of the negative influence of other learning; without other learning, no forgetting would occur. Postman (1961, 1971) and Underwood (1964), among others, have identified two basic types of interference. The first, *retroactive* interference, results when new associations interfere with old ones. The second, where old associations interfere with the formation of new associations, is referred to as *proactive* interference. Both kinds of interference increase with the amount of material to be learned (Bower and Hilgard, 1981; Gagné, 1977). Interference is often the cause of increased difficulty in learning, as additional information to be learned is added. More information is generally more difficult to learn than less information, and requires more time to learn as well (Calfee and Atkinson, 1965).

Procedural Reinstatement

Evidence suggests that how well material is remembered depends in part on how well cues associated with the material can be regenerated (Tulving and Psotka, 1971). Healy, Fendrich, Crutcher, Wittman, Gesi, Ericsson, and Bourne (1992) proposed a theoretical framework, termed *procedural reinstatement*, which integrates much of the work on long-term retention. According to this model, retention will occur to the extent that the specific procedures acquired during learning can be reinstated at the time of recall. This conceptualization is consistent with both Kolers and Roediger's (1984) hypothesis that the durability of memory depends on the extent to which learning procedures are reinstated at testing, as well as the encoding specificity principle (Tulving and Thomson, 1973), which asserts that no cue can be effective unless the to-be-remembered item is specifically encoded in memory with respect to that cue at the time of learning. The procedural reinstatement hypothesis is also is consistent with the phenomena of state-dependent learning, which suggests that retrieval depends upon faithful reproduction of the context of that learning, and that changing the context may result in failures to transfer the learning.

16.3.6 Context-Dependent Memory

Virtually all major theories of memory have considered context (the background or setting in which learning takes place) to be a major determinant of retention. If the context in which the material must be retrieved differs from that in which the information was acquired, it is likely that accuracy of retention will be impaired. A decrement in retention often occurs when the context of testing differs from that of training. This is referred to as the *contextual change effect.* It has been demonstrated that recall is improved when testing is conducted in the same environment in which leaning originally took place (Dallett and Wilcox, 1968). Tulving (1975) also noted that memory performance is best when the cues present at test match those that were available with the material at the time of acquisition. Godden and Baddeley (1975) demonstrated contextual effects in an experiment with deep-sea divers who learned instructional material while on the surface and tested them either under water or on the surface. They found retention performance was better on the surface than underwater, suggesting that memory was dependent on the match between context at encoding and the context at retrieval. Thus, similar contexts produce better retention than dissimilar ones (Metcalfe, 1985).

16.3.7 Task and Skill Classification

Task taxonomies have many implications for training, including prediction of retention across range of taxonomized tasks. Classifying tasks according to underlying skill requirements provides a means of consolidating and comparing research on skill degradation and retention. Numerous approaches and analytic methodologies designed to organize tasks, and expose underlying skills and behaviors, exist in the research literature, including those developed by Bloom (1956) and by Berliner, Angell, and Shearer (1964). Driskell, Willis, and Cooper (1992), for example, dichotomized skills into physical and cognitive components, and have proposed that longer retention intervals more adversely impact cognitive skills than physical task components. Similarly, Naylor and Briggs (1961), distinguished between discrete and continuous motor responses, and suggested that continuous responses should demonstrate superior retention to discrete responses. One result that has emerged from this line of research is that skills which are based upon cognitive-knowledge processes as opposed to those based upon perceptual-motor processes, are particularly susceptible to decay (Childs and Spears, 1986; Mengelkoch, Adams, and Gainer, 1971; Stammers, 1981). Adams and Hufford (1962), for example, found that after a 10-mon retention interval, pilot flight control skills were equivalent to their skills prior to the retention interval. In contrast, procedural tasks are forgotten over retention intervals of days, weeks, or months (Gardlin and Sitterley, 1972; McDonald, 1984). Certain cognitive, perceptual, and motor skills, such as letter detection (Healy, Fendrich, and Proctor, 1990), data entry tasks (Fendrich, Healy, and Bourne, 1991), and mental multiplication (Fendrich, Healy and Bourne, 1993), have demonstrated remarkable long-term retention. Visual search, an important component of many applied tasks such as air traffic control and radar operations, has consistently shown superior retention when compared with other types of tasks. Fisk and Hodge (1992), for example, examined retention of skilled search techniques over intervals of one yr and found minimal declines, although retention performance was a function of the manner in which training was organized. Students who learned these skills in a consistently mapped condition (where individuals make the same responses to stimuli across training trials) retained proportionately larger amounts of their original learning (approximately 85%) than those in a varied mapping condition (where the response requirements across practice trials change). Healy et al. (1990) reported that subject's detection skills in a visual search task demonstrated no significant forgetting, even after a one-mon retention interval. Similarly, data collected by Cooke, Durso, and Schvaneveldt (1994) suggest that performance associated with visual search declines minimally, even after a nine-yr period of nonuse. In contrast, skills involving substantial cognitive/procedural components have been shown to undergo greater and more rapid decay over time than control-oriented skills (Childs and Spears, 1986). Childs, Spears, and Prophet (1983), for instance, found that cognitive/procedural errors occurred routinely over a period of nonutilization of flight skills. Most of the literature in the skill retention arena has concentrated on relatively simple tasks (Adams, 1987). The extent to which decision making and knowledge-based skills decay is poorly understood (O'Hara, 1990).

Researchers have also distinguished between two types of classification systems; one for organizing task requirements, and another which structures the basic learning processes presumed to underlie task performance (Glaser and Resnick, 1972). Task analytic techniques (Stammers, Carey, and Astley, 1990), which involve decomposing a task into subtasks or elements, are often used as a basis for structuring the analysis for either or both types of systems. Fleishman and Quaintance (1984) have provided a framework for organizing and integrating various task descriptive approaches in both applied and research communities. These authors discriminated among four basic schemes for classifying human tasks: (1) a behavior description approach, which is based upon observations and descriptions of what operators actually do while performing a task; (2) a behavioral requirements approach, which emphasizes behaviors that should be demonstrated or are assumed to be required in order to correctly perform a task; (3) an abilities requirement approach, which describes tasks in terms of the abilities required by the performer; and (4) a task characteristics approach, which treats the task as a set of conditions that elicits performance. Taxonomies within each of these general classes can be used to identify important relationships associated with learning rates, performance levels, and individual differences. The utility of task classification systems, however, is largely an empirical issue. According to Marmie and Healy (1995), for example, classifications of tasks based upon their organization (which specify the number and type of task components comprising a task as well as how the components are combined) may have more value in predicting differential retention than continuous/discrete type distinctions.

Long-term retention of skills and knowledge over periods of nonuse is particularly relevant to military preparedness. Numerous techniques for organizing and predicting retention of tasks and skills have been formulated and demonstrated, based upon such taxonomic advances. One such technique, called the User's Decision Aid (UDA) developed by Rose, Czarnolewski, Gragg, Austin, Ford, Doyle, and Hagman (1984), relies on a systematic framework derived from task characteristics related to the internal structure, complexity, and cohesiveness of the task. Task properties inventoried by this method are designed to tap the organizational complexity characteristics of tasks, and primarily relate to factors associated with task difficulty (e.g., number of task steps, task step sequencing, amount of inherent feedback). The UDA instrument yields a retention rating score and has proven to be useful for reliably predicting retention associated with procedural type tasks.

16.3.8 Retention Guidelines

The literature supports numerous prescriptive guidelines for enhancing retention. Included among these are the following:

- To maintain any given knowledge or skill, refresher training sessions should be provided at intervals as long as the desired maintenance period (Bahrick, 1979).
- If multiple tasks are to be trained, it is preferable to use random rather than blocked scheduling of tasks during training (Landauer and Bjork, 1978; Schmidt and Bjork, 1992; Shea and Morgan, 1979).

- Performance feedback should be provided on an intermittent schedule that is gradually reduced across training, as opposed to after every trial (Schmidt and Bjork, 1992).
- Use of periodic retrieval practice benefits retention after a long delay interval (Bahrick, Bahrick, Bahrick, and Bahrick, 1993).
- Encourage active generation of learning material, rather than passive presentation (Healy and Sinclair, 1994).
- Retention is enhanced by prior familiarity; therefore, relate new learning to previous experience (King, 1992).
- Procedures used during training should be designed to match the procedures found in the operational setting (Healy et al., 1992).
- Measurement of the occurrence of recall responses requires that the trainee provide direct, overt evidence of recall (e.g., written or oral) of the fact, principle, or procedure (Willis and Peterson, 1961).
- Provide trainees with practice retrieving the appropriate response procedure from memory during training; do not rely exclusively on job aids to reduce memory reliance (Schneider, 1991).
- Training should promote the development of automaticity if speed of responding is an important aspect of performance (Rickard, 1994).
- The use of a training device that requires the trainee to provide cuing and feedback from memory is effective in increasing the retention of procedure-following skills over long periods of time (Johnson, 1981).
- Organize materials on an increasing level of difficulty; that is, structure the learning material to provide an attainable challenge (Keller, 1987).
- Although overlearning facilitates retention, it does so at a decreasing rate (Naylor and Briggs, 1961).
- Minimize overtraining, train the student to the level of proficiency required on the job (Caro, 1973).
- Present skills or concepts, which a student will be required to demonstrate at a later time, in a form permitting the student to recycle through the material until a performance criterion has been reached (Matlick, Swezey, and Epstein, 1980).
- The student should be required to attempt a new task immediately following a demonstration, without delay, so that the correct procedure is the most recent stimulus in the student's memory (Stonge and Becker, no date).
- Stimulate recall of prior learning. Help trainees tie what they are going to learn to what they already know (Sheppard, 1987).
- Previously mastered material should be subsequently tested for retention (Kessler, Macpherson, and Mirabella, 1987).
- Test frequently even in situations where the learner is attempting to acquire a skill primarily by observing (Kyllonen and Alluisi, 1987).
- Learning new tasks slowly will increase the student's ability to learn a large amount of material. Provide ample time for practice problems, exercises, review, and questions (Kyllonen and Alluisi, 1987).
- Practice on a particular task makes that task less susceptible to interference from a second task (Matlick et al., 1980).
- Procedural tasks are forgotten more quickly. Provide opportunity for frequent practice of procedural tasks (Johnson, 1981; Oosterveld, 1980).
- With very complex tasks, instruction in principles yields better results than laying down a detailed drill, while with simpler tasks, the drill is at least equally effective (Eberts and Brock, 1984).
- In training for recall of lengthy or difficult procedures, provision should be made for minimizing feedback on adequacy of performance following each "single-step" or "basic unit" response, especially during early stages of learning (Willis and Peterson, 1961).
- Encourage mental practice of skills (Smith and Beringer, 1987).

- Measures of skill retention after some period of inactivity appear to be better determinants of training program success than performance during or immediately following training (Schendel and Hagman, 1991).

16.4 MODELS OF TRANSFER

The topic of transfer-of-training is integrally related to other generic training issues such as learning, memory, retention, cognitive processing, and conditioning; these fields in combination make up a large subset of the subject matter of applied psychology. In general, the term *transfer-of-training* applies to the way in which previous learning effects new learning or performance. The central question is, how previous learning "transfers" to a new situation. The effect of previous learning may function either to improve or to retard new learning. The first of these is generally referred to as *positive transfer*, the second is known as *negative transfer* (if new learning is unaffected by prior learning, *zero transfer* is said to have occurred). Many training programs are based upon the assumption that what is learned during training will transfer to new situations and settings, most notably the operational environment. Although U.S. industries are estimated to spend upwards of $100 billion annually on training and development, only a fraction of these expenditures (not more than 10%) are thought to result in performance transfer to the actual job situation (Georgenson, 1982). Researchers, therefore, have sought to determine fundamental conditions or variables which influence transfer-of-training, and to develop comprehensive theories and models that integrate and unify knowledge about these variables.

16.4.1 Transfer Research

Two major historical viewpoints on classical theories of transfer exist. These are known as "identical elements" and "transfer-through-principles." The "identical elements" theory (first proposed by Thorndike and Woodworth, 1901) suggests that transfer occurs in situations where identical elements exist in both original and transfer situations. Thus, in a new situation, a learner presumably takes advantage of what the new situation offers that is in common with the learner's earlier experiences. Alternatively, the "transfer-through-principles" perspective suggests that a learner need not necessarily be aware of similar elements in a situation in order for transfer to occur. This position suggests that previously used "principles" may be applied to occasion transfer. A simple example involves the principles of aerodynamics learned from kite flying by the Wright brothers, and the application of these principles to airplane construction. Such was the position espoused by C. H. Judd (1908), who suggested that what makes transfer possible is not the objective identities between two learning tasks, but the appropriate generalization in the new situation of "principles" learned in the old. Hendriksen and Schroeder (1941) demonstrated this transfer-of-principles philosophy in a series of studies related to the refraction of light. Two groups were given practice shooting at an underwater target until each was able to hit the target consistently. The depth of the target was then changed, and one group was taught the principles of refraction of light through water while the second was not. In a subsequent session of target shooting, the trained group performed significantly better than did the untrained group. Thus, it was suggested that it may be possible to design effective training environments without a great deal of concern about similarity to the transfer situation, as long as relevant underlying "principles" are utilized. Actually, according to Ellis (1965), Thorndike and Woodworth did not intend for the identical elements view to be considered specific to stimulus and response components. Their elements, in fact, consisted primarily of items such as general principles and attitudes, as opposed to specific components.

An early stimulus-response theory of transfer was proposed by Hilgard (1962). Hilgard cited an experiment by Bruce (1933), in paired associate learning, as the basis for a stimulus-response analysis. Bruce arrived at four conclusions: (1) if new stimuli are similar to the original ones, and the responses remain identical, high positive transfer will result; (2) if the new stimuli are dissimilar from the original ones, but the responses remain constant, slight positive transfer will result; (3) if stimuli are identical, and responses are similar (but not identical), negligible transfer will result; and (4) if stimuli are identical, but responses are dissimilar, negative transfer will occur. Thus, according to Hilgard, if a subject attempts new learning in a situation where the new stimuli, or responses, or both, have some resemblance to the original stimuli or responses, then the presumption would be that a gradient of generalization would apply which would be similar in concept to those studied in simple conditioning experiments. Two empirical generalizations, termed the *Bruce-Wylie Laws*, are grounded in this basic theory (Bruce, 1933; Wylie, 1919). These laws state that: (1) the amount of transfer depends on the degree of similarity between situations, and (2) the direction of transfer depends on the similarity of the two responses. Many empirical findings (discussed later) can be interpreted in terms of this general framework. The results of such experiments have led to a variety of further efforts and models intended to specify in greater detail the conditions that facilitate positive transfer of training.

The major point here, is that the problem of transfer of training is indeed complex. It is, for example, difficult to determine the extent to which the complex stimulus and response elements that exist in most operational situations are similar or dissimilar, until adequate description and measurement techniques are established. Further, most complex environments do not result simply in either positive or in negative transfer. Most environments are sufficiently complex that their components interact in ways that produce negative and/or positive transfer for each component (as well as each two-way, three-way, and *n*-way interaction). Goldstein (1974) has cited as an example the situation encountered in shifting from a mechanical to an electric typewriter. In such a case, some positive effects exist (due to the knowledge of the QWERTY keyboard) and some negative effects exist (due to the differing sensitivity of keys). Howell and Goldstein (1971) have suggested, for instance, that it would do little good to have positive transfer on the overall performance of operating an aircraft when negative transfer existed on one critical element, such as altimeter reading.

16.4.2 Theoretical Transfer Models

Beyond the early transfer theories, a variety of basic theoretical approaches have emerged which describe many of the complex relationships among training effectiveness, fidelity, type of training, and similar variables to transfer. The Osgood (1949) model of transfer is perhaps the best known model which has been used to address variances in amount of transfer with gradients of similarity between operational and training settings. Osgood described this relationship using a three-dimensional surface, relating stimulus similarity on one axis, response similarity on the second, and amount and direction of transfer on the third dimension. According to this view, transfer is a function of both stimulus and response similarity. With identical stimuli, the effect of variation in required responses passes from maximum transfer at identical responses, through zero to negative transfer as antagonistic responses are reached. With identical responses, transfer drops to zero as stimulus similarity decreases. On the other hand, with antagonistic responses, transfer rises to zero from negative as stimulus similarity decreases. Thus, according to this model, a training device is most effective when it duplicates the operational equipment, of little (if any) use when it is very different from the equipment, and of detrimental value when moderately similar (and, hence moderately dissimilar) to the equipment.

A number of experiments have attempted to replicate Osgood's transfer surface; however, these studies have not yielded consistent results (Bugelski and Cadwallader, 1956; Dallett, 1962). Bugelski and Cadwallader (1956), for example, in testing this model obtained very similar results for positive transfer predictions, with contradictory results regarding negative transfer suggesting the model to be deficient in predicting negative transfer. Although this model appears to characterize transfer performance with simple tasks, such as learning verbal paired associates, the transfer surface does not adequately predict transfer performance with complex tasks (Hammerton, 1977; Holding, 1976). Numerous studies of real-world learning contradict the Osgood theory, demonstrating that transfer can actually be increased by reducing fidelity of simulation and thus similarity between the training and transfer contexts (Grunwald, 1968; Trollip and Ortony, 1977).

A model developed by Miller (1954) attempted to describe relationships among simulation fidelity and training value in terms of cost. Miller hypothesized that as the degree of fidelity in a simulation increases, the costs of the associated training also increases. At low levels of fidelity, very little transfer value can be gained with incremental increases in fidelity. However, at greater levels of fidelity, larger transfer gains can be made from small increments in fidelity. Thus, Miller hypothesized a point of diminishing returns, where gains in transfer value are outweighed by higher costs. According to this view, changes in the requirements of training should be accompanied by corresponding changes in the degree of fidelity in simulation if adequate transfer is to be provided. Although Miller did not specify the appropriate degree of simulation for various tasks, subsequent work (Alessi, 1988) suggests that the type of task as well as the trainee's level of learning are two parameters which interact with Miller's hypothesized relationships. Therefore, to optimize the relationship between fidelity, transfer, and cost, one must first identify the amount of fidelity of simulation required to obtain large amount of transfer and the point where additional increments of transfer are not worth the added costs.

Another model, developed by Kinkade and Wheaton (1972), distinguishes among three components of simulation fidelity: equipment fidelity, environmental fidelity, and psychological fidelity. Equipment fidelity refers to the degree to which a training device duplicates the "appearance and feel" of the operational equipment. This characteristic of simulators has also been termed *physical fidelity*. Environmental, or functional, fidelity refers to the degree to which a training device duplicates the sensory stimulation received from the task situation. Psychological fidelity [a phenomenon which Parsons (1980), has termed *verisimilitude*] addresses the degree to which the trainee perceives the training device as being a duplicate of the operational equipment (equipment fidelity) and of the task situation (environmental fidelity). Kinkade and Wheaton postulated that the optimal relationship among levels of equipment, environmental, and psychological fidelity varies as a function of the stage of learning. Thus, different degrees of fidelity may be appropriate at different stages of training. These authors distinguished among three principal stages of learning:

(1) ***Procedures training***. At this early stage of learning, the trainee not only does not benefit from high degrees of equipment and environment fidelity, but can be confused if provided with very realistic presentation of environmental conditions.
(2) ***Familiarization training***. During this stage of initial skill acquisition, the demand for higher levels of fidelity increases. The need for higher levels of environmental fidelity increases at a more rapid rate than for equipment fidelity.
(3) ***Skill training***. During the final stages of learning, fidelity requirements continue to increase. Here too, the demand for higher environment fidelity surpass that for higher equipment fidelity.

According to this view, a single level of fidelity will yield differential amounts of transfer depending on the stage of training. Further, to some extent, effects of fidelity on transfer depend on whether one refers to "physical" fidelity or to "functional" fidelity. Although, functional (environmental) fidelity should be considered as being more important than physical (equipment) fidelity, different levels of functional (environmental) fidelity are appropriate at different stages of learning.

Historically, predictions of transfer effects in applied settings have been based on the results of basic research in learning within the framework of stimulus-response theories (e.g., Hull, 1921; Osgood, 1949; Thorndike, 1932). Contemporary research on information processing and memory processes involved in encoding and retrieving information during both initial task acquisition and retention offers new ways for explaining and predicting transfer effects. Literature in this area has identified four major information-processing factors which contribute to the prediction of transfer of training: (1) relationships between retrieval cues and encoded information; (2) study-phase retrieval of information, which permits the integration and abstraction of both sets of information; (3) organizational strategies and schemata, which enhance stimulus processing; and (4) the automation of performance with consistent stimulus training (Cormier, 1984). According to this framework, positive transfer is promoted to the extent that cuing relationships between the training environment and transfer environment are distinctive, variations during initial training result in the formation of higher order concepts which can be applied in the transfer setting, the transfer task can be logically related to the organizational plan in use during training, and consistent relationships exist between the training and transfer contexts.

Many cognitive theories of transfer (Anderson, 1976; Bransford and McCarrel, 1974; Mayer, 1975) are not in direct opposition to S-R theories (like the identical elements theory), but view transfer in more complex ways. Cormier (1984), for example, has contended that the principle of identical elements can be reconceptualized by using encoding and retrieval processes. Similarly, Kolers and Roediger (1984) proposed that memory representations cannot be divorced from the procedures that were used to acquire them, and that transfer outside of the training context depends critically on the extent to which the learned procedures are reinstated at the time of recall. This characterization of memory is consistent with Tulving and Thompson's (1973) encoding specificity principle, which postulates that memory performance will be best when the retrieval operations required in the operational transfer setting match or overlap the encoding operations employed during learning. The implications for transfer are clear: If acquired knowledge or skills are to be transferred, the procedures used during initial acquisition should be reinstated at the time of recall. From the cognitive view, it is not physical fidelity per se that contributes to high positive transfer, rather it is the presence of retrieval information which has a high cuing capacity. Low fidelity devices should be effective in producing transfer as long as they provide the essential cuing relationships between the stimulus attributes of the task environment and the appropriate responses. Thus, it is not necessarily the device, therefore, that should be simulated, but the operations and tasks related to it.

16.4.3 Fidelity as a Component of Transfer

There is a clear lack of consensus in the literature on how to define simulation fidelity (Hays and Singer, 1988). Uses of the term have been numerous and have included such descriptions as: equipment fidelity, environmental fidelity, psychological fidelity, functional fidelity, and physical fidelity (Hays, 1980). In general, definitions of fidelity suffer from ambiguity (investigators use either different terms to describe the same type of fidelity or the same terms to describe different types of fidelity) and lack of user orientation (little guidance is provided to the user on how fidelity may be determined or measured). As various people have pointed out, a dictionary definition of fidelity merely means "duplication." Unfortunately, degrees of similarity or difference are not easily specified in most transfer-of-training environments, and definitions of fidelity in the literature have often been confusing and contradictory among themselves as well as lacking in the precision necessary to offer the user practical guidance.

The relationship between degree of fidelity and amount of transfer is complex. Fidelity and transfer relationships have been shown to vary as a function of many factors, including instructor ability, instructional techniques, types of simulators, student time on trainers, and measurement techniques (Hays and Singer, 1988). Nevertheless, training designers often favor physical fidelity in a training device, rather than the system's functional characteristics, assuming that increasing levels of physical fidelity are associated with higher levels of transfer. The presumption that similarity facilitates transfer can be traced to the "identical elements" theory of transfer first proposed by Thorndike and Woodworth (1901). In fact, numerous studies show that similarity does not need to be especially high in order to generate positive transfer-of-training (see Hays, 1980; Provenmire and Roscoe, 1971; Valverde, 1973). Hays's (1980) review of fidelity research in military training found no evidence of learning deficits due to lowered fidelity, and others have shown an advantage for lower fidelity training devices, suggesting that the conditions of simulation which maximize transfer are not necessarily those of maximum fidelity. One possible reason for this may be that widespread use of such well-known techniques as corrective feedback and practice to mastery in simulators and training devices, which may act to *increase* transfer, may also act to *decrease* similarity. A second possibility is that introduction of highly similar (but unimportant) components into a simulation diverts a student's attention away from the main topics being trained, thereby causing reduced transfer, as is the case when manipulating the fidelity of motion systems in certain types of flight simulators (Lintern, 1987; Swezey, 1989).

A recent review by Alessi (1988) devotes considerable attention to debunking the myth that high fidelity necessarily facilitates transfer. Having reviewed the literature in the area, Alessi concludes that transfer-of-training and fidelity are *not linearly related* but instead may follow an *inverted*-U-shaped function similar to that suggested by the Yerkes-Dodson Law (see Swezey, 1978b; Welford, 1968), which relates performance to stress. According to this view, increases in similarity first cause an increase and then a corresponding decrease in performance, presumably as a function of increasing cognitive load requirements associated with decreased stimulus similarity. Alessi, like Miller (1974), proposed that fidelity effects depend largely on the instructional level of the learner, and posited this inverted U-shape function to describe fidelity relationships during early learning stages; as students progress they will benefit from increasing fidelity. An alternative view, based on the early Skaggs-Robinson hypothesis of transfer (Robinson, 1927) suggests that the relationship between fidelity and transfer performance might follow a *normal*-U-shaped function whereby transfer decreases with decreasing similarity to a certain point, after which transfer becomes negative and then neutralizes as similarity continues to decrease (Swezey, 1989). From this perspective, high similarity (fidelity) between the training and operational equipment is predicted to result in high positive transfer, moderate similarity should result in negative transfer, and low similarity would lead to zero transfer.

As indicated in this section, transfer-of-training is a complex area. Fidelity of simulation (however defined), though a factor in facilitating transfer, is certainly not the sole determinant. General research in the area of simulation fidelity and transfer has indicated high physical fidelity is not always necessary for transfer to occur (Crawford and Crawford, 1978; Johnson, 1981), in fact, high physical fidelity is sometimes a detriment to learning due to information overload (Boreham, 1985; Roscoe, 1971; Valverde, 1973).

16.4.4 Measurement of Transfer-of-Training

The previous discussion concerned modeling of the transfer process. In this section, we briefly describe several measurement and evaluation approaches for assessing proficiency levels resulting from transfer-of-training.

Several types of experimental designs and computational formulas used to measure amount and direction of transfer exist in the literature (Ellis, 1965; Gagné, Forster, and Crowley, 1948; Murdock, 1957). Most studies express transfer as a percentage. The four transfer formulas described below, and listed in Table 16.1, involve making comparisons between an experimental and control group (whose performance data serve as a standard) against performance on a transfer task.

In this context, Micheli (1972) defined a "time-savings" performance measure, as the percentage of transfer determined by improvement in performance of the real system. In this model, percent transfer is based upon improvement in performance on operational tasks or on savings in time to reach a specified performance level on an operational task. Micheli's formula, therefore, compares the difference between the experimental and control groups with the control group itself. A second formula, proposed by Gagné et al.

Table 16.1 Examples of Transfer Formulas

Author/Developer	Formula
Micheli (1972)	$\text{Percent Transfer} = \frac{Z_c - Z_e}{Z_c} \times 100$ where: Z_c = performance or time required on the operational (or transfer) task by the control group Z_e = performance or time required on the operational (or transfer) task by the experimental group
Gagné et al. (1948)	$\text{Percent Transfer} = \frac{Z_c - Z_e}{\text{Max} - Z_c} \times 100$ where: Z_c = performance or time required on the operational (or transfer) task by the control group Z_e = performance or time required on the operational (or transfer) task by the experimental group Max = maximum possible score
Murdock (1957)	$\text{Percent Transfer} = \frac{Z_e - Z_c}{Z_e + Z_c} \times 100$ where: Z_c = performance or time required on the operational (or transfer) task by the control group Z_e = performance or time required on the operational (or transfer) task by the experimental group
Roscoe (1971)	$\text{Transfer Effectiveness Ratio} = \frac{Y_c - Y_e}{X_e}$ where: Y_c = time required by a control group to reach some criterion of proficiency in the operational (transfer task) Y_e = time required by an experimental group to reach some criterion of proficiency in the operational (or transfer task) X_e = the training device hours received by the experimental group

(1948) compares the difference between the experimental and control group with the *maximum amount* of improvement possible on the transfer task. Murdock (1957) evaluated a number of formulas and concluded that a balanced measure which is systematic about zero (ranging from a value of 100% positive transfer to 100% negative transfer) offers a purely empirical measure of transfer. In order to generate such as measure, Murdock (1957) used performance of both experimental and control groups in the denominator of the formula to derive a percentage figure. Finally, Provenmire and Roscoe (1971) and Roscoe (1971) departed from conventional measures to develop a *Transfer Effectiveness Ratio* (*TER*). The TER is the savings in transfer as a relation of the difference between transfer performance of control and experimental groups, and a similar measure taken on the time to learn for an experimental group. By employing a time factor, TER attempts to measure effectiveness by expressing savings as a function of amount of time in training. TER expresses savings in time or errors to reach criterion performance levels on the operational system as a function of the time spent in training on the simulator. The basic difference between estimates of percent transfer (e.g., Micheli, 1972) and the TER, is that the former ignores amount of pretraining required.

Since the numerator in each of the four types of transfer formulas (listed in Table 16.1) represents the mean difference between the experimental and the control groups' scores, these differences must be reliable. Unreliability can significantly affect the interpretation of any transfer efficiency formulas. Further, difference scores do not reflect where trainees began or where they finished. An alternative, proposed by Gagné et al. (1948), is to use the raw score values to express transfer in conventional analyses.

Measures frequently used in the application of transfer formulas include: (1) the number of trials required to reach a given level of mastery, (2) the amount of time required to reach a given level of mastery, (3) the level of mastery reached after a given amount of time or number of trials, and (4) the number of errors made in reaching a given criterion of mastery. Results of different transfer studies can easily depend on ways in which transfer is measured; therefore, it is important to identify how transfer is measured and calculated when comparing the magnitude and direction obtained in different studies (for additional information on the measurement of transfer, refer to Andreas, 1960; Gagné, Forster, and Crowley, 1948; Murdock, 1957; Osgood, 1953; Woodworth and Schlosberg, 1954).

16.4.5 Applying Transfer-of-Training Models to Training Device Assessment

Operational equipment and training devices are designed for distinctly different purposes. In many cases, training devices can be specifically designed to provide such instructional benefits as immediate feedback, reinforcement for correct responses, hierarchical sequencing of learning objectives, measurement of achievement, and other positive features in complex learning environments. Inclusion of such features often represents improved learning situations, not commonly found when using operational equipment for training. One major purpose of training devices, therefore, is to impart skills or knowledge to trainees in a fashion such that they can be directly transferred to appropriate operational situations. A number of techniques have been developed for estimating effectiveness of training devices as well as for the related problem of assessing transfer-of-training from training devices and simulators to operational equipment (Evans and Swezey, 1980; Gagné, 1962). Two distinct classes of modeling approaches have emerged: *empirical* and *analytic* models. The former approach employs the conduct of actual laboratory and field studies to assess the transfer effects of training devices, while the latter relies on parameter estimates to provide information on the extent to which a training device is potentially (as opposed to actually) effective in producing positive transfer to operational equipment.

A great deal of research on transfer, particularly in the military, has relied on actual experiments utilizing a classical two group transfer-of-training paradigm to assess transfer. Typically, such studies compare an experimental group that learns a task and then transfers to a second task, with a control group that receives an equivalent amount of training on the second task. If the experimental group performs better than the control group on the second task, positive transfer is presumed to have occurred.

Although models which purport to predict potential transfer without actually conducting transfer experiments have existed for years (Osgood's transfer surface for instance), they generally have not been implemented in practical ways in applied environments. Typically, predictions generated by analytic models are based on information about a training device and related system characteristics. Numerous reviews of these methods exist in the literature (Harris and Ford, 1983; Knerr, Nadler, and Dowell, 1983). One view, by Miller (1974), attempted to identify functional training requirements, gross specifications of kinds of devices, and ways in which tasks can be grouped for training on a single device. These inputs were designed to form a matrix where decisions regarding training strategies corresponding to stages of training and trainer types could be considered. Another approach, that of Demaree, Norman, and Matheny (1965) utilized a method for analyzing operational situations in order to reach recommendations for functional requirements and for specific training device equipment. The major components are termed, *Training Equipments Requirements Data.* In this approach, the Training Equipments Requirements Data are compared to a variety of training effectiveness characteristics for each task to be trained. Example characteristics include: level of equipment fidelity required to instruct a task, fidelity of trainee responses, fidelity of required feedback, etc.

Caro (1970) presented a model known as Equipment-Device Task Commonality Analysis (TCA), which emphasized the description of task elements and hardware in order to predict potential transfer-of-training. This model was developed concurrently with a second empirical methodology, Altman's (1970) transfer-of-training prediction model, which suggests that three kinds of transfer exist. When summed, these measures derive a measure of "Net Transfer," an item of obvious interest in device evaluation. A third model was proposed by Wheaton and Mirabella (1972), who calculated a series of quantitative task indices for use in multiple regression equations in an attempt to predict learning acquisition and transfer as a function of time and errors. This approach successfully demonstrated that it is possible to relate variations in quantitative task indices to trainee performance. Holman (1979) discussed a measure called the *Cumulative Transfer Effectiveness Ratio* (*CTER*), often used in the classical two-group transfer-of-training design used to assess the training benefits of periodic training in a simulator. The paradigm of the CTER is essentially a task by trials to criterion, or trials to time to learn.

One of the major efforts involving analytic approaches to predicting training device effectiveness resulted in the TRAINVICE family of models (Hirshfeld and Kochevar, 1979; Narva, 1979; Swezey and Evans, 1980; Wheaton, Fingerman, Rose, and Leonard, 1976). The earliest TRAINVICE model (Wheaton et al., 1976) was developed to combine judgmental data concerning several major variables into a figure of merit, tau (T), which purports to assess training device to operational equipment transfer. The Wheaton et al. model essentially consists of three parameters: (1) the transfer potential of the training device, (2) the learning deficit for which the device must compensate, and (3) an analysis of the adequacy of the training techniques employed by the device. These parameters combine in a multiplicative format to derive a figure of merit which attempts to quantify perceived training transfer. Underlying these models is the basic assumption that if the training device closely approximates the actual equipment (in terms of both its physical appearance and functioning), then its training effectiveness will be high. While various preliminary validation studies have been undertaken for the TRAINVICE approach (Faust, Swezey, and Unger, 1984; Tufano and Evans, 1982), relatively little field work has been done to determine their predictive properties and practical utility. Preliminary laboratory studies indicated that the early TRAINVICE models were relatively weak predictors of transfer-of-training potential, and that they were too time consuming, cumbersome, and technically complex to be of value for most practical applications (Faust et al., 1984). Although slightly better results were obtained when the TRAINVICE model developed by Wheaton et al. (1976), was applied to the evaluation of two antitank gunnery training devices (Swezey, 1983), this model also proved to be difficult to implement, lacked adequate provisions for dealing with negative transfer, and presumed that physical and functional similarity between the device and the operational equipment are equally important. A rigorous transfer-of-training study remains to be conducted.

Interest in speeding the results and reducing the administrative errors associated with such models lead to the development of a related series of analytic techniques developed by the Army, known as the *Device Effectiveness Forecasting Technique* (Rose, Wheaton, and Yates, 1985). DEFT consists of a series of interactive menu-driven computer programs which permit evaluation of alternative training devices at three levels of detail depending on the availability of predictor variables and the degree of diagnostic analysis required. DEFT converts information about various facets of a training system into forecasts of device effectiveness based on expert ratings of trainee and task characteristics, functional and physical similarity between the proposed device and the operational equipment, and instructional characteristics of the device. Although not empirically validated, DEFT represents a systematic procedure for assessing training device effectiveness which is sensitive to variations in the quality and quantity of input information as well as the practical aspects of implementation such as efficiency and ease-of-use.

Still another approach to the prediction of transfer from simulators to operational equipment is termed, *Comparison-Based Prediction.* Comparison-based prediction has been described by Klein (1982) as an application of reasoning by analogy to the evaluation of training devices. This process involves the identification of a comparison case and the application of what is known about that case to the target domain. Inferences regarding the effectiveness of a device are generated based upon the effectiveness of other similar devices, while adjusting for important differences. Klein and his coworkers have described comparison-based prediction as the middle ground between subjective estimates by experts and the application of formal models. It is designed to make use of the capabilities of experts rather than forcing them to make judgments for which they are ill suited. Since such methods do not typically rely on extensive data from the proposed device, they can often be applied early in a device's development cycle.

Adams (1979) has postulated a novel way to evaluate the training worth of simulators (in particular, flight simulators) based upon known inadequacies of conventional evaluation methods. Specifically, Adams reacted to the transfer-of-training and rating methods of simulator evaluation. The transfer-of-training view assumes that competence in the operational environment is required as evidence of a simulator's training value. The rating method bases this evaluation on (a) an evaluation of hardware and software specifications by engineers, and (b) an evaluation of the simulator's similarity to the operational equipment by experienced operators or users. According to Adams, both methods are flawed.

The transfer-of-training method of simulator evaluation is criticized on two fronts. First, the feasibility of conducting a transfer-of-training experiment for simulators of advanced aircraft or other expensive equipment is often prohibitively high. Second, when they are conducted, there are grave problems for both experimental and control groups. The control subjects must be raised, by some means, to a minimum level of proficiency in the equipment in question so that they can operate it well enough to generalize meaningful performance measures. They do not, therefore, provide the no-practice group baseline necessary for computation of percent transfer for the new simulator on which the experimental group will be trained. The experimental group, on the other hand, must perform well on the first trial after simulator training or they may not be allowed to use the equipment (e.g., fly an airplane) again. Hence the experiment cannot be performed unless high positive transfer is an implied guarantee. The rating method is flawed for somewhat different reasons, including the assumption that the amount of transfer-of-training is positively related to the rated similarity between the simulator and aircraft, and the ability of the rater to discern whether the source of poor simulation lies with human skill deficiencies or deficiencies of the simulator.

Based on these findings, Adams (1979) postulated a novel approach to training device evaluation which, he believed, may preclude the aforementioned shortcomings. He argued that a simulator, or any other system, need not necessarily be tested if it is based on reliable scientific laws and the success of other systems based on the same laws. Thus, in this view degree of adherence to known scientific principles becomes the evaluation criterion.

In general, analytic models attempt to assess the extent to which the similarity between a device and its parent equipment is sufficient to permit adequate transfer. The development of predictive models to evaluate training device/system effectiveness may eventually eliminate, or minimize, costly empirical evaluations. Presently, however, additional research and development is needed to overcome problems in at least the following four areas: (1) the theoretical constructs of the models; (2) their mathematical formulations; (3) measurement issues involving the validity, reliability, and precision of the models; and (4) their convenience of application and acceptability to the user. User acceptability of such models is a major area of concern. In the case of the TRAINVICE series of models, preliminary attempts were made at revision in order to enhance user acceptability (although considerable work remains to be accomplished in this area). It should also be noted that although transfer-of-training is an important component of device effectiveness, it is not the only one; device effectiveness is definitely a multidimensional issue.

16.4.6 Current Issues and Problems in Transfer

Current issues and problems in transfer span four general areas: (1) measurement of transfer, (2) variables influencing transfer, (3) models of transfer, (4) application of our knowledge of transfer to applied settings. Research on transfer needs to be conducted with more relevant criterion measures for both generalization and skill maintenance. Hays and Singer (1988) provide a comprehensive review of this domain. A large proportion of the empirical research on transfer has concentrated on improving the design of training programs through the incorporation of learning principles, including identical elements, teaching of general principles, stimulus variability, and various conditions of practice. A critique of the transfer literature, conducted by Baldwin and Ford (1988), also indicated that research in the area needs to take a more interactive perspective that attempts to develop and test frameworks to incorporate more complex interactions among training inputs. Many studies examining training design factors, for example, have relied primarily on data collected using college students working on simple memory or motor tasks with immediate learning or retention as the criterion of interest. Generalizability, in many cases, is limited to short-term simple motor and memory tasks.

16.4.7 Guidelines for Enhancing Transfer

Transfer tends to increase, nonlinearly, with increasing training time and with increasing degrees of similarity between training and real tasks. However, this is not a simple relationship; many factors, such as the experience and ability level of the learner (Campione, Brown, Ferrara, Jones, and Steinberg, 1985) as well as the type of training procedures employed during initial learning (Hagman, 1980) have been shown to affect the magnitude and direction of transfer. The following guidelines provide some specific information on factors known to have a direct influence on transfer:

- Maximize the similarity between the teaching and the testing situation (Ellis, 1965).
- Provide adequate experience with the original task, since most research shows that adequate practice is essential for positive transfer (Ellis, 1965)
- Provide for a variety of stimulus situations so that a student may generalize his knowledge (Ellis, 1965).
- Label or identify important or critical features of the task (Ellis, 1965).
- Ensure that general principles are understood by the learner (Ellis, 1965).
- Use a rote or algorithmic approach if near transfer is the goal of instruction (Brooks and Dansereau, 1983).
- Training which incorporates visual demonstrations provides positive transfer to real-world situations (Oosterveld, 1980).
- Examples provided during instruction should progressively become more similar to real task performance as training progresses (Branson et al., 1975).
- Design instruction to enhance generalization of what is learned by including varied applications (Smith and Beringer, 1987; Margolis and Kroes, 1975).
- Design instruction to provide opportunities to apply learned rules in a variety of new situations in which the learner has not previously been trained to apply the rules (Branson et al., 1975).
- Gradually decrease the amount of cues, prompts, and guides such that no guides or prompts (that would not be found on the job) remain at the end of training (Branson et al., 1975; Willis and Peterson, 1961; Eberts and Brock, 1984).
- To promote generalizability of skills, trainees should practice in a wide variety of situations (Branson et al., 1975; Eberts and Brock, 1984; Kyllonen and Alluisi, 1987).
- State explicitly how the instruction relates to future activities of the learner (Keller, 1987).
- Allow a student to use a newly acquired skill in a realistic setting as soon as possible (Keller, 1987).
- Provide instruction in context (Kyllonen and Alluisi, 1987).
- Cue stimuli in training should be as nearly identical as possible to cue stimuli which will be encountered in subsequent transfer to job conditions (Willis and Peterson, 1961).
- Determining the level of fidelity for device features should take into account, at least generally, the following: task type, task difficulty, specific skills required to perform the task, trainee sophistication, stage of training, and supportive instructional features (Kessler et al., 1987).

- Although high variability of examples generally aids transfer, it may hinder initial learning, especially if the number of examples is small (Bassok and Holyoak, 1987).
- Training device designers should consider at least the following when selecting an appropriate level of fidelity: the specific tasks and task situations which should be simulated (the extent of simulation required), the degree of physical correspondence to the operational environment required to achieve effective training, and the extent to which deliberate departures from reality should be introduced to enhance the instructional value of the simulator or training device (Smode, 1972).
- Functional fidelity appears to be a strong influence in transferring knowledge from the learning environment to the operational setting (Fink and Shriver, 1978; Mallory and Elliott, 1978).
- Departures from physical fidelity, even along dimensions that seem central to the task, do not always lead to a decrement in transfer (Matheny, Williams, and Dougherty, 1953).
- Near transfer (to very similar situations) may be enhanced through high fidelity, while far transfer to dissimilar environments appears to be enhanced by variation which would tend to decrease fidelity (Cronbach and Snow, 1976).
- Transfer effectiveness may be more a function of how a training device is used than a function of the simulator's fidelity (Micheli, 1972).

16.5 MODELS OF INSTRUCTION

We begin our discussion of models of instruction with reference to Goldstein's comprehensive work on training in organizations (1993). In that document, Goldstein refers to what is generally known as the "Systems Approach" to training. Among many other things, this view includes four components: (1) feedback is used to continually modify the instructional process, thus training programs are never completely finished products, but are always adaptive to information concerning the extent to which the programs meet stated objectives; (2) recognition exists that complicated interactions occur among components of training such as media, individual characteristics of the learner, and instructional strategies. Thus, the systems view is concerned with the complexity of these interactions in a total systems context; (3) the systems approach to training provides a framework for reference to planning; and (4) this view provides acknowledgment that training programs are merely one component of a larger organizational system involving personnel issues, organization issues, and corporate policies among many other variables. Thus, in developing needs analyses for initiating training program development, it is, according to Goldstein, necessary to consider and analyze not only the tasks that comprise the training, but also characteristics of the trainees involvement called "Person Analysis," and characteristics of the organizations involved, called "Organization Analysis." A thorough description of the need analysis process for training development can be found in Goldstein (1993).

The Systems Approach to instruction is essentially derived from systematic procedures developed in the areas of behavioral psychology. Two generic learning theory principles underlie technological development in this area. First, the requirement for a precise description of the specific behaviors necessary to perform a task, and second, use of feedback or reinforcement for action. The Systems Approach to instruction has provided the generic frame of reference for developing procedural techniques in instructional design (Landa, 1974). Gagné and Briggs (1974), for instance, have extended this view suggesting several typical activities used in the instructional system development process. These include:

1. Identification and analysis of specific needs.
2. Definition of goals and objectives.
3. Identification of alternative ways to meet needs.
4. Design of system components.
5. Analysis of: (a) resources required, (b) resources available, and (c) constraints.
6. Action to remove or modify constraints.
7. Development of instructional materials.
8. Design of student assessment procedures.

9. Field testing and evaluation.
10. Adjustment, revisions, and further evaluation.
11. Summative evaluation.
12. Operational installation.

This and previous work by Gagné (1965) served as a partial basis for the military's instructional systems development (ISD) movement (Branson et al., 1975) and the subsequent development and refinement of procedure for criterion referenced instruction and measurement technologies (Glaser, 1963; Mager, 1972; Swezey, 1981).

16.5.1 The Instructional Systems Development (ISD) Model

For discussion of the Systems Approach to training development we review the Branson et al. (1975) ISD model. This model is possibly the most widely used and comprehensively developed model in this area. It has been in existence for over 20 yr and has revolutionized the design of instruction in many military and civilian contexts. The Branson et al. ISD model is a culmination of various military models developed independently. The evolutionary premise behind the model is that training be developed to address specific behavioral objectives (identified by task analysis), that criterion tests be developed to address the behavioral objectives, and finally that training be developed essentially to teach the student to pass the tests, and thus to achieve the requisite criteria.

The ISD method has five basic phases: analysis, design, development, implementation, and control.

Phase One—Analysis—Has Five Steps

1. Task analysis. Develop a list of detailed tasks, the conditions under which the tasks are to be performed, the skills and/or knowledges required for their performance, and the standards which indicate when successful performance has been achieved. This often involves visiting actual job locations and observing each job task in detail.
2. Select pertinent tasks/functions. Note, by observation of an actual operation, the extent to which each identified task is actually performed, the frequency with which it is performed, and the percentage of a worker's time devoted to it. An assessment is then made about each task concerning its importance to the overall job.
3. Construct job performance measures from the viewpoint that job performance requirements are the basis for making decisions about instructional methods.
4. Analyze any training programs which already exist to determine their usefulness to the program being developed.
5. Select the instructional setting/location for each selected task for instruction.

Phase Two—Design—Has Four Steps

1. Describe student entry behavior. Classify at the individual task level the categories into which each entry behavior falls (i.e., cognitive, procedural, decision-making, problem-solving). A check is made as to whether each task activity is already within the performance repertoire of the trainees (something the potential student already knows) or is a specialized behavior (which the student will have to learn). Consider the use of pretests for possible advanced placement or exemption from the course.
2. Develop objectives which translate the job performance measures from Phase One into terminal learning objectives. Consider the behavior to be exhibited, the conditions under which the learning will be demonstrated, and what standards of performance should be considered acceptable. This step translates job tasks from Phase One into criterion (goal) objectives and enabling objectives (ways to meet the goals) for which instruction is designed.
3. Develop criterion-referenced tests (CRTs). CRTs are tests which measure what a person needs to know, or do, in order to perform his or her job. Ensure at least one test item measures each instructional objective. Determine whether the testing should be done by pencil and paper, by student demonstration (practical exercise), or by other methods. These criterion-referenced tests are an important element of the method. They perform three essential functions.

- They help to construct the training by defining detailed performance goals.
- They aid in diagnosis of whether the student has absorbed the training because the tests are, in fact, operational definitions of the performance objectives.
- They provide a validation test for the training program.

4. Determine sequence and structure of the training. Develop instructional strategies and the sequencing of the steps of instruction for the course.

Phase Three—Development—Has Five Steps

1. Specify specific learning events and activities.
2. Specify the instructional management plan. This includes the responsibility of the instructors, and the delivery systems. Determine how the instruction will be delivered and how the student will be guided through it: by group mode, self-paced instructional mode, or combinations of the two. Determine how the curriculum will be presented, (for example, lectures, training aids, simulators, job performance aids, television, demonstration, and audiovisual methods).
3. Review/select existing instructional materials for possible inclusion in the curriculum.
4. Develop new instructional materials.
5. Validate the instruction by having it reviewed by experts or job incumbents, or by trying it out on typical students.

Phase Four—Implementation—Has Two Steps

1. Implement the instructional management plan with emphasis on ensuring that all materials, equipment, and facilities are available and ready for use. Before beginning instruction, schedule students and train instructors.
2. Conduct the instruction under the prescribed management plan in the designated setting under actual conditions with the final course materials.

Phase Five—Control—Has Three Steps

This phase determines whether a graduate's job performance meets the established job performance standards. It is composed of three steps.

1. Conduct an internal evaluation to monitor course effectiveness, to assess student progress, to assess the quality of the course, and to compare results with the original learning objectives.
2. Conduct an external evaluation to determine how graduates perform on the job (i.e., after graduation).
3. Revise the course as appropriate.

Table 16.2 shows the output for each Phase of the ISD model.

16.5.2 Cognitive Viewpoints

Tannenbaum and Yukl (1992) have provided a discussion of the state-of-the art with respect to developments in the field of cognitive psychology as they impact training. In this context, the work of Anderson (1985, 1987) and associates, concerning the development of stages of skill acquisition is well known. These include declarative knowledge (that is, knowledge about facts and things), knowledge compilation (knowledge about how to integrate facts), and procedural knowledge (knowledge about how to do things). This work has resulted in the development of a variety of additional cognitive constructs designed to address the way instruction may proceed with respect to current cognitive theory in knowledge acquisition. One important contribution to this literature is the view that so-called tacit knowledge (that is, knowledge about how, when, and why to do things) facilitates transfer from learning to job performance situations.

According to Tannenbaum and Yukl, three areas of cognitive psychology have significantly influenced the design of training involving processes required to perform complex knowledge-based tasks; these are: (1) automatic processing, (2) mental models and schemata, and (3) metacognition. Automatic processing occurs to the extent that a trainee can perform complex tasks without direct resort to thought processes involving the tasks. This presumably occurs because the skills associated with these tasks have previously been

Table 16.2 ISD Model Outcomes

I	1	A list of tasks performed in a particular job.
	2	A list of tasks selected for training.
	3	A job performance measure for each task selected for instruction.
	4	An analysis of the job analysis, task selection, and performance measure construction for any existing instruction to determine if these courses are usable in whole or in part.
	5	Selection of the instructional setting for task selected for instruction.
II	1	A learning objective for and a learning analysis of each task selected for instruction.
	2	Test items to measure each learning objective.
	3	A test of entry behaviors to see if the original assumptions were correct.
	4	The sequencing of all dependent tasks.
III	1	The classification of learning objectives by learning category and the identification of appropriate learning guidelines.
	2	The media selections for instructional development and the instructional management plan for conducting the instruction.
	3	The analysis of packages of any existing instruction that meets the given learning objectives.
	4	The development of instruction for all learning objectives where existing materials are not available.
	5	Field-tested and revised instructional materials.
IV	1	Documents containing information on time, space, student and instructional resources, and staff trained to conduct the instruction.
	2	A completed cycle of instruction with information needed to improve it for the succeeding cycle.
V	1	Data on instructional effectiveness.
	2	Data on job performance in the field.
	3	Instructional system revised on basis of empirical data.

trained to the point of overlearning. Repeated practice of such skills, and overlearning the performance requirements associated with them, makes it possible to transfer such skills more easily to applied situations, because conscious cognitive processing is eliminated. This may occur, for instance, in such tasks as checking rear view mirrors before pulling into a passing lane while driving. The terms "mental models" and "schemata" (referred to elsewhere in this chapter) are viewed as being essentially synonymous, and refer to the extent to which learning may be more easily transferred (or acquired) in situations where a trainee has previously been able to develop an accurate mental representation of necessary prerequisite material required for task performance. According to Tannenbaum and Yukl, retention can be facilitated often by use of encoding methods such as mnemonics, cues that relate information to the learner's existing knowledge, the use of relevant category or taxonomic systems, coding guidelines, analogies, diagrams, and prescriptive job aids, among other things. Use of such techniques can presumably foster the development of appropriate mental models for use in applied situations.

The term *metacognition* refers to the mental processes involved in acquiring knowledge, and the parallel recognition by the learner that the knowledge acquisition itself is actually proceeding. Via this technique, learners can presumably recognize and monitor their own progress, and are therefore more readily able to evaluate what they do and/or do not know at a given point in the learning cycle. Research has suggested that provision of feedback about performance per se may be of dubious value if learners are not in fact operating according to a correct mental model about how a task is performed. According to Tannenbaum and Yukl, metacognitive skills, along with other types of cognitive learning strategies (such as "if-then rules" and "backward learning") are ways of distinguishing proficient from nonproficient learners, or masters from novices in a content domain. It is apparent that considerable further work is needed to examine the utility of cognitive concepts in applied training contexts.

16.5.3 Technology-Based Systems

The use of technology-based training systems can ease many training problems. One benefit of such systems has been their significant life cycle cost savings over human-based systems (Orlansky and String, 1977, 1981). In a recent review, Swezey, Perez, and Allen (1988) suggested three objectives for research in instructional delivery system technology. First, one should seek knowledge about the instructional effectiveness of various systems. Second, one should seek to increase one's understanding of how such systems affect learning. Third, one should try to enhance the practice of training and education via improved techniques.

Although numerous studies have focused on one or more of these objectives for nearly a century, the total yield in terms of understanding instructional approaches, guiding utilization, or improving training has been disappointing. One possible cause is that much of the research in technology-based training has been concerned with demonstrating the superiority of one delivery system over another, as opposed to determining what aspects of delivery systems are variously effective in teaching different types of tasks. In addition, researchers have not typically been concerned with identification of appropriate instructional strategies to enhance the effectiveness of delivery systems (Swezey et al., 1991).

Given that many studies in this area have concerned gross comparisons among delivery systems, it is not surprising that few direct recommendations can be extracted from the literature with respect to the selection of appropriate delivery systems for teaching the performance of specific tasks or skills, although different delivery systems may have unique characteristics and features that could be enhanced by the use of appropriate instructional strategies or by appropriate manipulations of instructional variables. To ignore these variables and strategies is to negate the unique characteristics of specific delivery systems.

Cost implication issues aside, data deriving from the field of instructional technology that clearly demonstrate the superiority of technology-based media or that can be used directly to design training programs, media, and devices is sparse and confusing (Montague, Wulfeck, and Ellis, 1983; Swezey et al., 1988).

This may again partly result from the fact that many studies addressing instructional delivery are applied research efforts that result in direct media (or method) comparisons. For example, Pieper, Swezey, and Valverde (1970) found that an individually paced, learner-controlled training program improved performance of both high- and medium-ability trainees over conventional lecture methods for an Air Force electronic maintenance task. Similarly, Holmgren, Hilligoss, Swezey, and Eakins (1979) found significant advantages for self-paced over group-paced instruction in an Army situation involving five electromechanical tasks. On the other hand, Swezey (1978a) found that self-paced instruction in a Navy mechanical maintenance task provided no performance advantage over traditional lecture methods; and Unger, Swezey, Hays, and Mirabella (1984a, 1984b, 1984c) showed similar results for two Army Maintenance Training and Evaluation Simulation System (AMTESS) devices across both electronic and mechanical maintenance tasks. One possible reason for this confusion has been suggested by the work of Mirabella and Wheaton (1974). They found that performance differences across training methods interacted with task complexity; performance on simple tasks showed no differences across three training methods. Performance differences did, however, occur on more complex tasks.

It seems reasonable to assume that if existing experiments in any area employ noncomparable tasks, noncomparable conditions, and/or confounded variables, they could not be expected to yield consistent results. Further, difficulties remain with the development of reliable and valid measures of task complexity. This may explain why researchers have not used it as an independent variable in more studies (see Streufert and Swezey, 1986, for a general discussion of this issue).

16.5.4 Networked Technologies and Equipment Simulation

As technology develops, progress is being made in linking various types of instructional devices. One such innovation is the linking of videodisc players with microcomputers to create interactive videodisc instruction.

Fletcher (1990) has recently reviewed much of the literature in this area. In general, he has found that videodisc instruction was often more effective than conventional instruction (e.g., lecture, videotaped demonstration, text, programmed text, on-the-job training) with respect to both knowledge acquisition and job performance. The more the videodisc instruction included interactive features, such as tutorials, the greater its effectiveness. Fletcher also noted that the available studies provide little insight into the relative contributions of various features of interactive technology to learning. He concluded that more research is needed to identify design alternatives that contribute to various learning objectives.

Equipment simulators have been used extensively for training aircraft pilots. Jacobs, Prince, Hays, and Salas (1990), and Andrews, Waag, and Bell (1992) have extensively reviewed this research. In general, it appears that simulator training combined with training on the actual aircraft is more effective than training on the aircraft by itself. According to Tannenbaum and Yukl (1992) the realism of simulators is being enhanced greatly by continuing developments in videodisc technology, speech simulation, and speech recognition; and advancements in networking technology have opened up new possibilities for large-scale simulator networking. Although simulators have been used in the past for training small teams in the military, simulator networking allows larger groups of military trainees to practice their skills in large scale interactive simulations (Alluisi, 1991; Thorpe, 1987; Weaver, Bowers, Salas, and Cannon-Bowers, 1995). In these circumstances, the opponent is not merely a computer but other teams, and thus trainees are often highly motivated by the competition. In such situations, as in the actual environment, the simulations are designated to operate directly, as well as to react to unpredictable developments. Learning presumably occurs at many levels, from operators of different simulators to managers who coordinate combined and multifaceted operations. A thorough history of the development of one such complex, networked system, known as SIMNET, has been provided by Alluisi (1991). Comprehensive, system-based evaluations of these technologies, however, remain to be performed.

16.5.5 Instructional Guidelines

In the area of instructional development, Campbell (1988) has offered five generic guidelines as follows:

1. The instructional events that comprise the training method should be consistent with the cognitive, physical, or psychomotor processes that lead to mastery.
2. The learner should be induced to produce the capability actively (e.g., practice behaviors, recall information from memory, apply principles in doing a task).
3. All available sources of relevant feedback should be used, and feedback should be accurate, credible, timely, and constructive.
4. The instructional processes should enhance trainee self-efficacy and trainee expectations that the training will be successful and will lead to valued outcomes.
5. Training methods should be adapted to differences in trainee aptitudes and prior knowledge.

Further specific guidelines (adapted from Branson et al., 1975, and from Swezey and Evans, 1980) follow:

- Positive rewards of the student's correct applications of the rules learned is required in the early stages of training.
- Reduce forgetting by providing periodic practice or refresher training for infrequently used material.
- At the beginning of the training, the instructor or the material/media should clearly inform the trainee of the learning objectives; that is, what the trainee is expected to be able to do by the completion of training.
- During practice, practical applications and practice tests, provide the student with immediate knowledge of results about his or her correct and incorrect answers.
- Emphasize distinctive features which can be remembered in the form of mental "pictures" instead of abstract words. When possible, supply students with diagrams, pictures, charts, graphs, rhymes, acronyms, keywords, self-instructions, common associations, and other association devices like these to which the student can relate the material he is trying to learn.
- Provide the student with practice in recognizing examples from the full range of patterns produced by a given object. Make the examples more similar as training

progresses. At the end of training, the similarities in the examples should be the similarities that exist in the real world.

- Different trainees will have different rates and styles of learning the material. Provide flexibility in the time allowed.
- Require the student to overlearn the original material; that is, the student should continue to practice the required tasks after the point that simple mastery of the task has been met.
- Reduce forgetting by providing periodic opportunity to recall and apply infrequently used material.
- Clearly relate the learning objectives and learning activities to operational tasks, which the trainee must perform in future real-world assignments.
- Break down the overall learning task into manageable steps or units when any of the following conditions exist: (a) lower ability students, (b) complex material, and (c) overall task composed of many small parts.
- Change the order of presenting material during practice so that each training item will be learned equally well.
- Allow for self-paced practice and provide the student with knowledge of the results of his or her identifications.
- Provide examples of correct performance of the task where appropriate.
- Relate the learning objectives and learning activities to operational tasks which the trainee must perform in future real-world assignments.
- Do not allow a student to leave one phase or level of the learning task until he or she has achieved the required level of mastery.
- For the most effective learning of decision making, the student must already have learned technical knowledge which will allow him or her to identify what the problem really is, make a list of the most reasonable solutions, and then determine which of the solutions would be best.
- The student will learn best if he or she is not afraid of making incorrect decisions in the training situations; this is particularly true in the early stages of training and in very complex decision-making processes. Materials and instructors should, therefore, attempt to decrease student fears to a low level.
- Give the students examples of these two types of actions which are to be avoided when making decisions: (a) response biases, that is, the tendency to make a "favorite" decision or use a "favorite" solution regardless of the real nature of the problem, and (b) perceptual sets, that is, the tendency to generalize problems or view several types of problems as if they were all the same when, in fact, they are quite different.
- Use high interest, attention-getting features of the learning materials throughout the training. Keep student attention by using learning activities which require active student participation.
- If the real-world job presents the trainee with very similar features of the job situation which requires the trainee to remember different knowledge for each feature, make sure that the trainee learns the difference between these features before he or she is taught which body of knowledge to associate with each feature.
- If it is necessary for the student to learn similar bodies of subject matter, then directly compare the bodies when they are first presented so that the student can tell them apart (or) separate their presentation by as much time as possible to avoid confusion between them.
- Prevent forgetting by showing the meaningfulness of the material to the learner's job environment and duties. Emphasize the organization and structure of the material.
- Provide for learner practice on parts (specific components) of the task for: (a) simple task-practice in entirety, and (b) complex task-practice in parts and then in entirety.
- Reward performances which are closer to the goal than the preceding performances. In this manner, the student's performance will become successfully closer to the desired performance (shaping).

- Allow the trainee to practice part-skills and provide feedback to train him or her to perfect the movement.
- Integrate the part-skills into a smooth sequence.
- Distractions and interference should be similar to what will be found on the job.
- Learning material to be recalled and used should be as it would be found on the job.
- Learning material should be as complex as that to be used on the job.
- Use no guides or prompts that would not be found on the job.
- Toward the end of training, increase stress and miscellaneous interruptions, distractions, and "noise" to the level that will appear on the job.
- Require enough practice trials of the learner to produce the correct performance; he should especially practice parts he is having difficulty with until he can demonstrate the correct procedure.
- Cross-train the learner so that he or she may perform other voice communication tasks and be able to act as a replacement for other members of his team.

16.6 MODELS INVOLVING PRESENTATION AND STRATEGY SELECTION

Two factors known to effect quality and effectiveness of training are the devices used to deliver the instruction (media) and the methods or strategies used to convey the instructional content (Robinson and Knirk, 1984; Romiszowski, 1974). One of the main tasks in the development of training, therefore, is the selection and implementation of appropriate instructional methods and media (Goldstein, 1993). Often, instructional strategy and media selection decisions are made by persons who themselves may be variously knowledgeable about such decisions. Efforts to increase the reliability and quality of instructional decisions have led to the development of numerous media selection models (Reiser and Gagné, 1982); however, relatively few models or guidelines exist to help in identifying appropriate training strategies.

16.6.1 Media Models

With growing numbers of media alternatives available to instructional designers, decisions on this topic have become increasingly important. Consequently, various procedures have been developed to aid instructional designers with such decisions (Higgins and Reiser, 1985; Levie, 1975). Implicit in most media selection models are the assumptions that: (1) some media will be more effective than others in certain situations, and (2) instructional designers are aware of the relative effectiveness of various media forms across instructional settings. Yet, few practical, valid, and dependable guidelines exist for making such choices, at least on the basis of instructional effectiveness (Levie, 1975; Higgins and Reiser, 1985).

Although media effectiveness is not a new issue, it continues to be of considerable interest due to the heightened promise associated with new instructional technologies. According to Salomon and Clark (1977), media research has often consisted of simple intermedia comparison studies which compare the relative effectiveness of one or two specific media in presenting similar instructional content. Such studies have stemmed from the commonly held belief that a so-called optimal medium existed, through which instruction across a variety of domains could be presented. Thus, efforts were concentrated toward discovering the "best" instructional delivery medium. Repeatedly, however, researchers comparing one form of mediated instruction with another, have failed to find differences in instructional effectiveness in terms of concept acquisition, retention, and transfer.

16.6.1.1 Media Research

One vital function of technology assessments and media research is to provide an evolving bases for estimating the utility of particular technology approaches for learning. Researchers have meta-analyzed virtually every significant media and technology trend of the last two decades, including effects of programmed instruction (Kulik, Kulick, and Cohen, 1980), computer-based instruction (Kulik and Kulick, 1987), and visual-based instruction (Cohen, Ebeling, and Kulik, 1981). The majority of studies individually show no significant differences and when pooled yield, at best, only slight advantages for innovative technologies. Kulik reports that the size of these statistically significant gains is in the order of 1.5 percentage points on a final exam. Cohen et al. (1981), for example, conducted a meta-analysis of 65 studies in which student achievement using traditional classroom-based instruction was compared with performance across a variety of video-based instructional media, including films, multimedia, and educational TV was compared. They found that only one in four studies (approximately 26%) reported significant differences favoring visual-based instruction. The overall effect size reported by Cohen et al. (1981) is relatively small compared to studies of computer-based instruction, which are two to three times as large, or studies of interactive video, which are three to four times as large.

Computer-based instruction and interactive video both represent areas where some data reporting advantages have emerged; specifically, reductions in the time required to reach threshold performance levels (Fletcher, 1990; Kulik and Kulick, 1987). Results of a meta-analysis of 28 studies, conducted by Fletcher (1990), suggests that interactive video-based instruction increases achievement by an average of .50 standard deviations over conventional instruction (lecture, text, on-the-job training, videotape). This is comparable to results reported by McNeil and Nelson (1991), who performed a similar meta-analysis of 63 studies on interactive video and found an average effect size of .53. In general, higher levels of interactivity were associated with higher levels of achievement. Further interactive video appeared to be equally effective for knowledge-based outcomes (learning facts and concepts) and performance outcomes. Similarly, meta-analytic studies on effects of computer-based instruction generally indicate that students learn more when they receive computer-based instruction versus traditional lecture. However, studies vary on the size of the gains to be expected, suggesting that some types of computer-based instruction work better than others. When used as vehicles for tutoring (in which the computer manages training by presenting material, evaluating responses, adapting material to past performance, etc.), for example, the average effect size is .38 (Baker and O'Neil, 1994). Kulik has argued that researchers need to advance beyond general conclusions about CBI and make statements regarding its effectiveness in specific applications and types of CBI.

One obstacle confronting intermedia comparison studies concerns matters of methodological design. According to Salomon and Clark (1977), strict comparisons among media have, in many cases, been constrained by inadequate control of contaminating variables. For example, increased student motivation, or more carefully designed instruction favoring one particular medium over another, have been common contaminants, effectively reducing the internal validity of many of these studies. For example, the average preparation time for 1 hr of conventional university instruction is about 1.7 hr, about 5.3 hr with closed-circuit television-based courses, and 9.6 hr with open circuit courses (Wells, 1976). Perhaps more importantly, media comparison research has generally lacked operational definitions which accurately distinguish among generic media types. For example, a motion picture may have various forms, including a filmed lecture or an animated cartoon, each of which could itself be considered as a separate and distinct medium. Even in well-controlled studies, consistent differences which definitively favor one medium over another, are rare (Salomon and Clark, 1977). Thus, efforts toward discovering a "best" delivery medium have been recognized as limited and, for the most part, abandoned. Subsequent efforts have therefore, tended to devote resources toward understanding various media attributes.

Results of media research led Clark (1983) to conclude that no media, in their own right, have any influence on learning effectiveness, but are mere vehicles for more or less well designed instruction. According to Clark (1983), "the best current evidence is that media are mere vehicles that deliver instruction but do not influence student achievement any more than the truck that delivers our groceries causes changes in our nutrition" (p. 445). Similarly, Schramm (1977) commented that "learning seems to be affected more by what is delivered than by the delivery system." Although students may prefer more sophisticated and elegant media forms (e.g., color versus black and white television), learning is largely unaffected by these features. It is not the delivery medium that determines the instructional effectiveness of training; all media can deliver either excellent or ineffective instruction. Research shows that it is the instructional methods that are embedded within the medium that facilitate or hinder learning. Thus, delivery media are considered to be equally effective provided that are capable of carrying the instructional methods required to achieve the training objective (Clark, 1988).

16.6.1.2 Media Attributes

While research addressing intermedia comparisons has generally resulted in equivocal conclusions, it is nevertheless possible that special attributes of a specific medium may make it more adaptable to one instructional function or another. Therefore, recent work has tended to de-emphasize general media comparison studies, in favor of examining specific attributes of various media which may potentiate instructional effectiveness. Media attributes are characteristics which define the kinds of information that may be useful within a particular medium, such as color or sound. Thus, the role of a particular medium in presenting instructional material may be viewed as a function of the various stimulus dimensions it can present to the trainee. If a particular medium lacks the capability to present one or more necessary attributes, then presentations via that medium would not be expected to result in adequate learning.

Media attribute research represents a shift of focus from the instructional device per se to attributes involved with media in general. Within this perspective, an important issue is to determine which stimulus attributes are useful in conveying instructional information across various tasks, domains, and situations. Considering media in terms of attributes offers some clear advantages, both to the researcher and to the instructional designer. Romiszowski (1988) has distinguished between essential media characteristics (ones which control the clarity of the message) and optional media characteristics (which improve the quality of the presentation). The media attribute approach overcomes many methodological issues associated with direct media comparisons, because it requires that researchers operationalize and focus upon critical aspects of media presentations. It also permits generalization beyond the content domain in which the experimental attributes were embedded. Typically, media attribute studies have addressed such dimensions as: the amount of realistic pictorial detail, chroma (i.e., color versus black and white), auditory effects, and dimensionality, among many others.

16.6.1.3 Media Realism and Fidelity

Interest in the representation of stimulus characteristics, such as motion, has evolved from the belief that instructional presentations are more effective when objects, situations, and events are realistically represented. That is, it has been presumed that "high fidelity" instructional materials are generally more effective than are lower fidelity abstract materials. *Fidelity* refers to the similarity between a representation and the actual object, event, or process (Alessi, 1988). Dale's (1969) cone of experience, which classified instructional media according to how concrete or abstract the experiences are for the learner, is an early expression of the idea that implied that the degree of fidelity is positively related to learning. However, other studies have demonstrated that high fidelity materials are not necessarily better than lower fidelity materials for all instructional purposes (Swezey, 1978b). The relationship between fidelity and instructional effectiveness has also been shown to depend on other variables, such as level of experience of the learner (Alessi, 1988). Similarly, Dale (1969) argued that it is the prior experience of the student that determines the appropriate level of abstraction of media. Novice learners, for instance, may become confused with high fidelity representations and require a more focused learning environment which can be provided by lower fidelity representations. Experienced learners, on the other hand, may tend to learn more effectively from high fidelity representations. In fact, learning in some situations may be enhanced when presentations are reduced in fidelity rather than increased, such that processing resources are directed only to the essential features relevant to the learning task. Irrelevant or excessive details do not always facilitate learning (Gorman, 1973; Borg and Schuller, 1979). Models or pictures of the heart, for instance, which possess excessive detail have been shown to be less effective than appropriately detailed line drawings in some situations (Dwyer, 1978). Thus, whether visual detail, such as motion, enhances or degrades learning may, among other possibilities, depend on the learner's level of expertise, amount of knowledge, aptitude for learning, and other variables.

Research in which motion has been experimentally manipulated as a visual presentation variable is limited. Nonetheless, some studies have compared learning from motion-based media, such as films and videotape, with learning from equivalent static versions, such as slides and filmstrips (Allen and Weintraub, 1972; Blake, 1977; Dwyer, 1969; Jeon, 1976; Laner, 1955; and Spangenberg, 1973). Spangenberg (1973), for example, studied the effect of hands-on practice of a procedural skill involving the disassembly of a machine gun. Motion was found to affect both speed and accuracy of the task. Studies of motion and procedural learning reveal instances where motion appears to benefit learning compared to use of static or textual methods. This research suggests that visually presented motion appears to be an important factor in instruction when:

1. It is a defining characteristic of a concept to be taught.
2. The activity involves natural human movements.
3. The activity requires simultaneous motion in different directions.
4. The activity is unfamiliar to the learner.
5. The activity is not readily described accurately with words.
6. The material contains figure–ground relationships that are critical and potentially "complex."
7. The learner population consists of low aptitude, uneducated, or retarded individuals; or where other cultural/language barriers exist.

Two additional factors which have been posited to account for motion's capacity to facilitate learning are: (1) its abilities to draw attention to crucial aspects in a display, and (2) the extent to which it facilitates the internalization of learner actions (Spangenberg, 1973).

16.6.1.4 Animation

Animation combines graphic illustration with simulated motion to represent objects, processes, or concepts. Common uses for animation include: directing the student's attention, presenting information, enhancing practice, and increasing a lesson's appeal (Rieber, 1990). Examples of the use of animation include the movements of parts of a machine (Laesecke, 1990) and the flow of fuel through a diesel engine (Swezey, Llaneras, and Salas, 1992). The instructional effectiveness of animation is unclear. Wetzel, Radtke, and Stern (1994) examined 16 studies incorporating the use of animation and found that less than 25% reported positive effects attributable to animation. Park and Gittelman (1992) compared performance on electronic troubleshooting problems between individuals trained with an animated and a static graphic lesson on basic electronic circuit theory. Although both forms of instruction provided equivalent content, the animated version yielded significantly fewer errors than the static version in completing a practice and follow-up test. Rieber and Kini (1991) noted that the conditions under which animation is expected to be more effective than static graphics are not well understood; however, the principle advantage appears to lie in animation's ability to represent change or movement over time. Rieber (1990) recommended that animation should be incorporated into instruction when its attributes are congruent to the learning task, and when learners are novices or unfamiliar with the content area.

16.6.2 Media Selection Models

Media selection models vary with regard to the physical form they take (flowchart, matrix, or worksheet), the ways in which they classify media, and the media factors they consider (Reiser and Gagné, 1982). When a flowchart is used, the media selection process consists of a progressive narrowing of media choices, while media decisions provided by a matrix are deferred until all selection criteria are displayed. Media selection worksheets typically present a tabular array of media characteristics against criteria. Selection criteria generally focus on media attributes, such as the ability to present sound or motion, characteristics of the intended learners, the instructional setting, and the learning task. Regardless of format, most models suggest that media should be selected for individual instructional events or objectives. Figure 16.2 illustrates various factors which influence the media selection process, including the instructional content, type of learning task, characteristics of the trainees, practical design considerations, as well as instructional method or strategy.

Although the most common form of media selection model is the matrix, the level of complexity and ease of use of these models varies widely. Early examples of matrix models include those of Gagné (1965), Allen (1974), Briggs (1970), McConnell (1974), and Tiffin and Combes (1980). Allen's matrix, for example, is based upon a review of hundreds of research articles, and examines the effectiveness of a specific medium for the type of learning measured. Media are classified according to high, medium, or low achievement in that type of learning. Examples of flowcharts include Romiszowski (1988), Anderson (1976), and Reiser and Gagné (1982). Anderson used seven complex flowcharts to capture the media selection process. Worksheet models guide the selection process via a checklist (Durham, Gearhart, and Austin, 1974; Locatis and Atkinson, 1984). Durham et al., for example, developed a seven-step procedure which involves identifying and classifying behaviors into a skill taxonomy, selecting media which support the type of learning, and matching the media to the instructional methods employed.

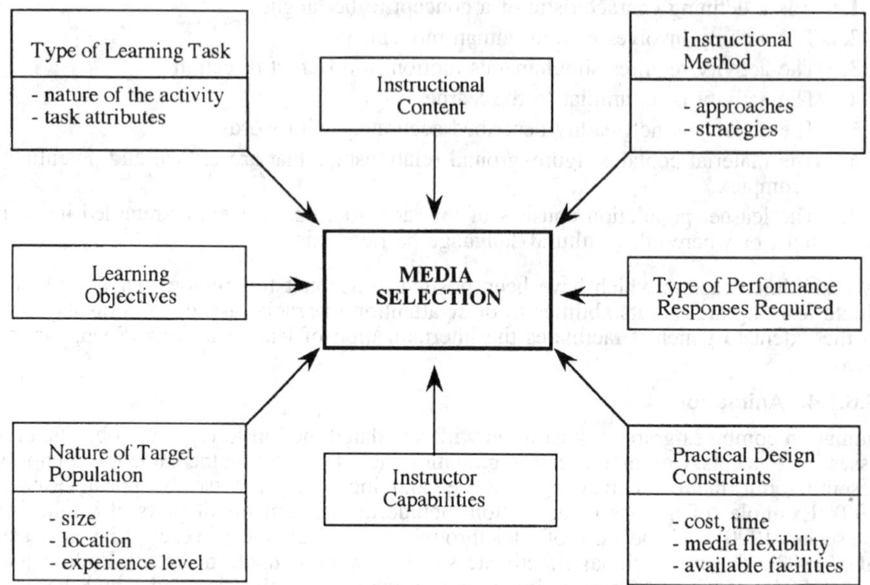

Figure 16.2 Factors influencing media selection.

Typically, media categories are defined on the basis of stimulus characteristics of the media and include audio, print, still visual, motion visual, and real objects. Figure 16.3 provides one example of such a taxonomy. Some define these categories along other dimensions such as the types of responses they will accept (Tosti and Ball, 1969), their feedback characteristics (Reiser and Gagné, 1982), and their ability to accommodate other events of instruction (Gropper, 1976). Most media selection models also include some means of classifying learning outcomes such as intellectual skills, verbal information, motor skills, attitudes, and cognitive strategies (Gagné, 1977). Yet other models base media decisions upon the events of instruction such as informing the learner of the objective, presenting the stimulus content, providing guidance, eliciting performance, providing feedback, assessing performance, and enhancing retention and transfer.

Although a number of media selection models have been developed, information regarding the usefulness of these models is limited. There is thus limited empirical evidence concerning the relative merits of media selection models.

16.6.3 Instructional Strategy Models

Although there is little disagreement that similar instruction can in many cases be delivered by a variety of media, it may sometimes be difficult, inefficient, or impossible to implement specific instructional methods across media alternatives (Petkovich and Tennyson, 1985). Media enable and constrain instructional methods, and the methods draw on and potentiate the capabilities of the media. Further, characteristics of media are only potential capabilities that may or may not be used in different instances of instruction (Baker and O'Neil, 1994).

Teaching methods are important variables in the effectiveness of instructional delivery systems. Instructional methods define how the process of instruction is expected to occur, that is, what information is presented, in what level of detail, how the information will be organized, how information is used by the learner, and how they receive guidance and feedback. The choice of a particular instructional method will often limit the choice of presentation media. Media selection decisions, therefore, should be guided by: (1) capacity of the media to accommodate the instructional method, (2) compatibility of the media with the user environment, and (3) trade-offs that must be made between media effectiveness and costs.

Conventional instructional strategies such as providing advance organizers (Allen, 1973), identifying common errors (Hoban and Van Ormer, 1950), and emphasizing critical

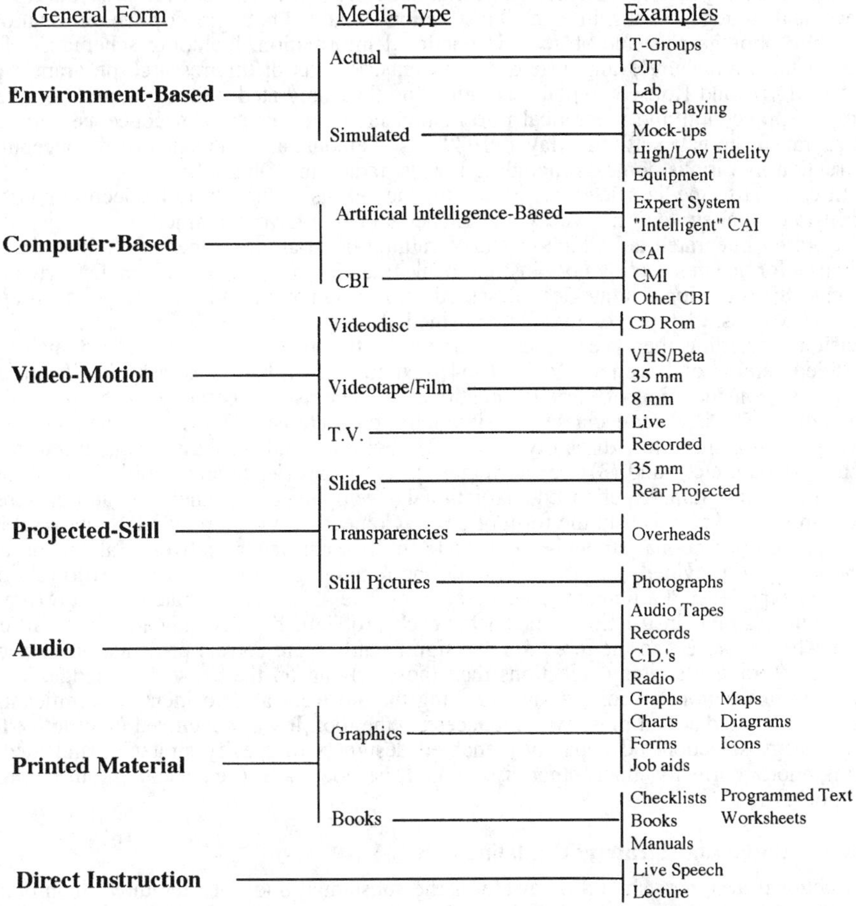

Figure 16.3 Sample instructional media classification taxonomy.

elements in a demonstration (Travers, 1967) work well in many media forms including video-based media. Practice questions embedded within instructional sequences also appear to be a potent variable supported by interactive video- and computer-based instruction. Benefits of embedded questions apparently occur whether responses to the questions are overt or covert, or whether the material is preceded or followed by the questions (Teather and Marchant, 1974). Compared to research on media, relatively little research exists which examines application of instructional methods.

Morris and Rouse (1985) have suggested that strategies used to teach troubleshooting skills should combine both theoretical and procedural knowledge. Swezey et al. (1991) addressed these issues using a variety of instructional training strategies. Three training strategies, termed *procedural*, *conceptual*, and *integrated*, were defined. The procedural instruction condition incorporated strictly prescriptive information, which completely specified the detailed step-by-step procedures required to troubleshoot two component systems of a diesel engine. The instruction contained information pertaining to the start-up, diagnosis, and inspection of the engine, and included both photographs of the equipment and simultaneous verbal narrative. The conceptual instruction condition provided training on the generic structure and function of the relevant diesel engine components—as opposed to system specific procedural information. Thus, subjects who received instruction in this condition were required to develop their own troubleshooting heuristics based upon their understanding of the system. Diagrams, schematics, and flow charts, rather than equipment specific photographs, were used in this condition. The "integrated" training strategy included information extracted from both the procedural and conceptual strategies. Thus, integrated instruction contained both prescriptive information, including photographs, and structural/functional information, including schematics, for each of the engine's relevant system components. Results of this research program supported Morris and Rouse's hypothesis indicating that integrated training strategies which combine procedural and theoretical information facilitate retention of concepts, and improve transfer to new settings. Mayer (1989) also demonstrated the benefits of conceptual instruction in various contexts including brakes, radar, and Ohm's Law.

In contrast to media selection, sparse guidance exists in the form of selection models for instructional strategies. Swezey and Llaneras (1992) have examined the validity and utility of an integrated aid which provides training program developers with prescriptive guidance for use in selecting *both* instructional strategies and media. The aid distinguishes among 18 instructional strategies, classified into presentation and study/practice strategies, as well as 14 types of instructional media. Outputs of the model are based upon decisions regarding the types of tasks for which training is to be developed (using a modified version of Berliner et al.'s (1964) taxonomy as a basis for task classification), as well as consideration of various situational characteristics operating in the particular job setting. Three general classes of situational characteristics are considered: (1) task attributes (motion, color, difficulty, etc.); (2) environmental attributes (e.g., hazardous, high workload, etc.); and (3) user attributes (level of experience and familiarity with the task). In this program, experienced instructional development personnel were administered problem-solving exercises in the form of training objectives which required them to select appropriate instructional strategies and media from a list of alternatives. Half of the individuals were provided with the aid, while the remaining personnel were left to rely on personal experience. Participants specified appropriate presentation strategies, study/practice strategies, and instructional media for each problem. Results indicated that participants who had access to the job aid made significantly more correct presentation strategy and study/practice strategy selections than those relying on their own knowledge. With regard to instructional media, designers using the aid were able to identify significantly more correct media than those without access to the tool. It was concluded that the media and strategy selection aid apparently enabled designers to specify strategies and media forms more correctly than otherwise would be the case (i.e., lacking any aiding technology).

16.6.4 Media and Strategy Guidelines

Instructional designers should be aware of the substantial body of literature which demonstrates that adequate portrayal of the content domain is without doubt the primary issue of importance in instructional delivery. It is important for instructional designers to be aware of the myriad factors and characteristics which impact training effectiveness, and not let technological issues or predetermined opinions per se drive training decisions. The following guidelines are intended to provide a framework for assessing the utility of instructional strategies and media for specific applications.

- With very complex tasks, instruction in principles yields better results than laying down a detailed drill, while with simpler tasks, the drill is at least equally effective (Eberts and Brock, 1984).
- Simulation as a method of instruction can be effective for teaching many tasks and skills such as perceptual motor skills, conceptual tasks, and team functions (Spangenberg, 1973).
- Viability of a device as an instructional delivery system should not be based solely upon transfer-of-training hypotheses but also upon such considerations as: cost, maintenance requirements, instructional flexibility, and ease of use (Kessler et al., 1987).
- Simulation devices need not replicate the actual equipment if improved training or maintenance can be provided by lower fidelity approximations (Criswell, Swezey, Allen, and Hays, 1984).
- Computer-based media must have the capabilities for review and branching, an instruction should include skill diagnosis and remediation (Hamel and Clark, 1986).

- Student performance records should be produced and stored by all automated training devices. Information collected should include: student ID, training attempted and passed, time elapsed, error scores, and correct steps for each training segment (Criswell et al., 1984).
- Because different trainees have different rates and styles of learning, it is better to use instructional techniques which can accommodate the student (Branson et al., 1975).
- Visual presentation may be used to facilitate reception and evaluation of information by trainees in complex tasks (Shriver, Shriver, and Bunderson, 1984).
- Sound provides helpful cues and should be included in instructional presentations; however, in most cases, these cues can be satisfied with relatively low fidelity but extremely timely executions (Oosterveld, 1980).
- Teaching machines or programmed learning techniques may be used for maximal control of both stimulus and reinforcement variables (Willis and Peterson, 1961).
- In computer-based training, the utility of such features as record and playback, sequence control, adaptive training, procedure monitoring, and malfunction insertion must be determined on an individual basis for each training requirement (Cream, 1978).
- Instructional flexibility of a training device, an attribute that should be sought in device design, can be assessed by the extent to which the device supports: (a) training alterations at low cost and in little time, (b) different student progress rates, and (c) changes in the order of training (Kessler et al., 1987).
- In computer-based training, students should have the opportunity to control the rate of presentation of frames (Hamel and Clark, 1986).
- Part-task training is effective when used with large and difficult tasks (Holding, 1987).
- In training for recall of lengthy or difficult procedures, provision should be made for minimizing feedback on adequacy of performance following each "single-step" or "basic unit" response, especially during the early stages of learning (Willis and Peterson, 1961).
- Training features of a device should accommodate various readiness levels of students so that the students can engage in a device-trained task at a level their skills accommodate (Kessler et al., 1987).
- Providing progress charts which display actual learning times to predicted learning times, when shown on student request, motivate students to complete the instructional course in less time (Gagné and Dick, 1983).
- Where recurrent or refresher training is being provided to experienced trainees, an element of competition tends to motivate trainees to better and faster learning (Oosterveld, 1980).
- Instructional programs should use such techniques as graphics, sound effects, challenge, and curiosity to increase and maintain student interest (Eberts and Brock, 1984; Kyllonen and Alluisi, 1987).
- Vary the format of instruction (information presentation, practice, testing), as well as the medium of instruction (platform delivery, film, video, print) according to the attention span of the audience (Keller, 1987).
- Device training features should limit the number of stimuli presented at any one time to avoid requiring inordinate attention from students (Unger et al., 1984a).
- Device training features should reflect principles of good instructional practices. Courseware should not submit to hardware limitations when the net result will be diminished instructional effectiveness (Kessler et al., 1987).
- Color appears to benefit learning in situations where color plays a significant role in aiding discriminations (Chute, 1980).
- Less experienced and low ability students show the greatest benefit from illustrations (Levie and Lentz, 1982; Peeck 1987; Winn, 1989).
- High fidelity device features are desirable for new and less experienced students. Advanced students often profit from lower fidelity approaches such as reading schematics (Criswell, Unger, Swezey, and Hays, 1983).

- Most automotive and electronic maintenance tasks can be taught with two-dimensional media. Three-dimensional high fidelity is often not required (Kessler et al., 1987).
- Pictures can have a positive effect on prose learning when they provide a meaningful supplement to text information (Alesandrini, 1984; Dwyer, 1978; Pressley and Miller, 1987).
- Determining the level of fidelity for device features should take into account, at least generally: (a) task type, (b) task difficulty, (c) specific skills required to perform the task, (d) trainee sophistication, (e) stage of training, (f) device role in the overall training framework, (g) user acceptance of the device, and (h) supportive instructional features (Kessler et al., 1987).
- Learning from text will be aided by relevant illustrations, particularly for poor readers. Illustrations may function best in aiding long-term retention. (Levie and Lentz, 1982).

16.7 MODELS OF TEAMWORK

What constitutes a "team"? The definition of this term is important in order that one can adequately differentiate a team from other groups of individuals (Salas, Bowers, and Cannon-Bowers, 1995). Boguslaw and Porter (1962) have defined the word "team" as a "relationship," where people generate and use work procedures to make possible interactions with machines, procedures, and other people in pursuing system objectives. Smillie, Shelnutt, and Bercos (1977) have defined a team as consisting of small groups, usually comprising 2 to 11 persons, who normally perform tasks in an interactive and interdependent manner. Dieterly (1978) defined a team as a "distinguishable" set of individuals who function together to accomplish a specific objective, and Thorndyke and Weiner (1980) used the word to describe a group of individuals working cooperatively to achieve a common objective (definitions adapted from Knerr, Nadler, and Berger, 1980). A more recent definition, provided by Dyer (1984), has defined a team as people linked together by a common goal, dedicated to reaching a specific role assignment, while working interdependently; this definition was expanded by Orasanu and Salas (1993) to include: working together at small tasks within the scope of a larger task, while contributing personal expertise and knowledge, often under intense workloads accompanied by high stress factors.

Using such definitions, teams can be found virtually everywhere in today's society. Although purposes and capacities vary, teams operate in many situations and places in our everyday lives.

Because teams are so prevalent, it is important to understand how they function (Swezey and Salas, 1992). Teamwork can take on various orientations, including behavioral and cognitive perspectives. Behaviorally oriented views of the teamwork process deal with aspects involved in coordinated events among members which lead to specific functions (i.e., tasks, missions, goals, or actions)—what a team actually does in order to accomplish its required output.

Tuckman (1965) has suggested that groups develop while progressing through four stages, termed *Forming*, *Storming*, *Norming*, and *Performing*. Morgan, Glickman, Woodard, Blaiwes, and Salas (1986) incorporated these four stages into their Team Evolution and Maturation (TEAM) model. This model, shown in Figure 16.4, expands upon Tuckman's work by indicating that teams progress through phases of growth depending on how long the team has been functioning together as a group, the individual abilities involved, what makes up the task, and the team's ability to adjust to the environment. Morgan et al.'s model touches upon aspects of team development from both individual and team perspectives. Not all individuals and teams necessarily pass through all phases of the Morgan et al. model. Morgan et al. (1986) further defined differences between "taskwork" and "teamwork" skills within the context of their model. Taskwork skills were viewed as being targeted toward an *individual's* ability to perform what is expected within the team context. Teamwork skills involve a *group's* ability to achieve cohesiveness and perform duties together via communication, awareness of the environment, and ability to adapt to different situations.

Cognitive views of the teamwork domain deal with a team's shared processing characteristics, such as joint knowledge about a task, joint ability to perform aspects of the work, motivation levels, personalities, and thought processes (Salas, Cannon-Bowers, and Blickensderfer, 1995). The extent to which "mental models" are shared among team

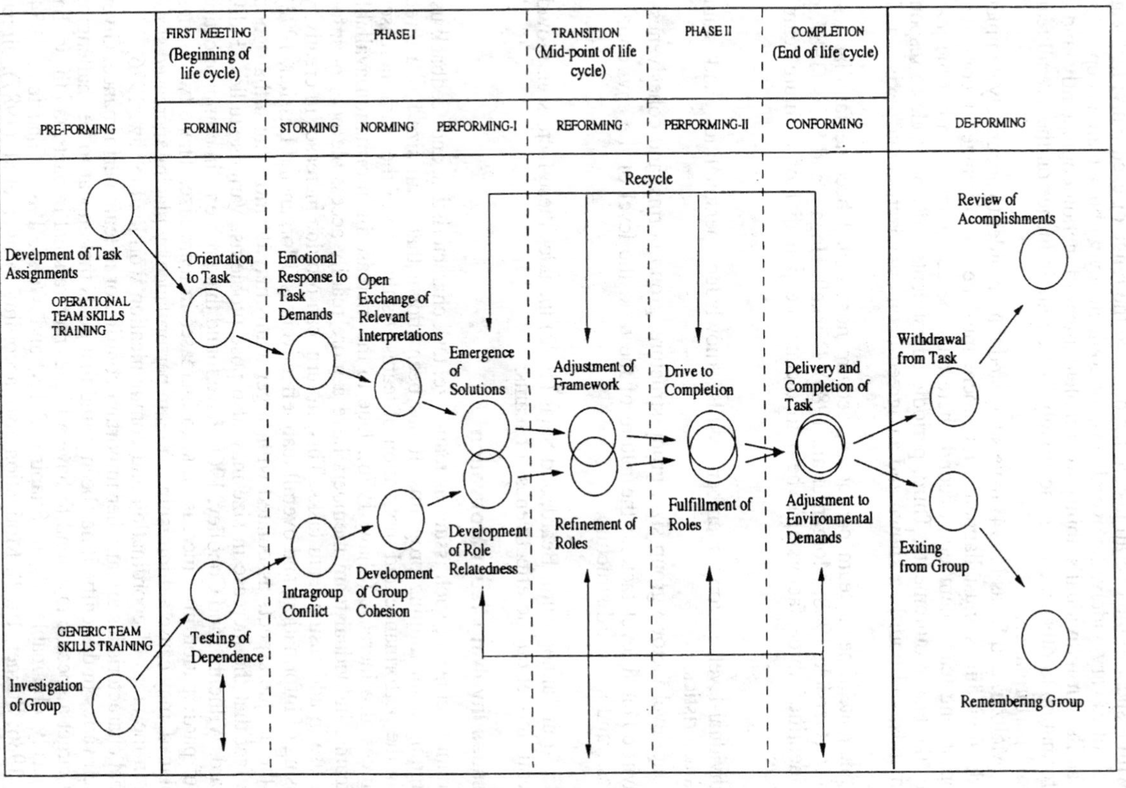

Figure 16.4 Team Evolution and Maturation model reproduced from Morgan et al. (1986). *Journal of Washington Academy of Sciences* by permission of the publisher.

members may help in the understanding and training of teamwork skills (Cannon-Bowers and Salas, 1990). Various teamwork researchers have suggested that in order for a member of a team to coordinate effectively with others, one must have shared knowledge of team requirements (Adelman, Zirk, Lehner, Moffett, and Hall, 1986). When each member of a team has definite knowledge of the shared requirements to be carried out in order to perform a task, they are said to share a joint mental model of the needs to be accomplished by the team (Cannon-Bowers, Salas, and Converse, 1990; Klein, 1989; Rouse, Cannon-Bowers, and Salas, 1993). This in turn, can help to coordinate a team effectively. Task-work in this context, addresses an individual's ability to understand his or her specific set of duties with respect to the overall task at hand. Since the range of each individual team member's abilities vary widely, the overall level of competency within a team is dependent on shared skills. An individual's ability to complete a team function competently, and to convey information to other members accurately, is a strong influence on the overall team structure and performance.

Swezey, Meltzer, and Salas (1994) have suggested that in order to accurately complete a team task, four things must happen: First, there must be an exchange of competent information among team members; Second, there must be coordination within the team structure of the task requirements; Third, periodic adjustments must be made to respond to task demands; and Fourth, a known and an agreed-upon organization must exist within the group.

Given adequate time, a team can adapt, acquire norms, and become familiar with specific roles (Gersick, 1988; Morgan et al., 1986). Driskell, Hogan, and Salas (1987, p. 95) have identified three factors that help to dictate the potential of a future team's performance:

1. Individual-level factors: members skills, knowledge, personalities, and status characteristics.
2. Group-level factors: group size, group structure, group norms, and cohesiveness.
3. Environmental-level factors: the nature of the task, the level of the environment stress, and a reward structure.

Figure 16.5 from Tannenbaum, Beard, and Salas (1992) includes these factors, in providing an overall model on the effectiveness of teams.

16.7.1 Issues Involving Team Motivation

It is important to consider such issues as team size, task characteristics, and feedback use when designing team training programs. It has been shown that team size has a direct influence in the performance and motivation of teams. As the size of the team increases, team resources also increase (Shaw, 1976). Due to the increase in information available to larger teams, individual team members have a more readily accessible pool of information for use in addressing team tasks. This, in turn, can provide increases in creativity, the processing of information, and overall team effectiveness (Morgan and Lassiter, 1992). However, the unique aspects associated with larger teams may do more harm than good, due to the fact that the increase in size may also pose problems. Various studies (Indik, 1965; Gerard, Wilhelmy, and Conolley, 1968) have found that problems in communication and level of participation may increase due to increases in team size. These two issues, when combined, may cause a diminishing of team performance by placing increased stress on a team in the areas of coordination and communication workload (Shaw, 1976).

Task goal characteristics are another important contributor to team performance. Given a task which varies in difficulty, it has been found that goals that appear to be challenging and which target a specific task tend to have a higher motivational impact than those that are more easily obtainable (Ilgen, Shapiro, Salas, and Weiss, 1987; Gladstein, 1984; Goodman, 1986; Steiner, 1972). Motivation also, according to Ilgen et al. (1987), directs the amount of time an individual is willing to put forth in accomplishing a team task.

The use of feedback is another factor which influences team motivation and performance. Feedback helps to motivate an individual concerning recently performed tasks (Salas, Dickinson, Converse, and Tannenbaum, 1992). Researchers concur that the feedback should be given within a short period of time after the relevant performance (Dyer, 1984; Nieva, Fleishman, and Rieck, 1978). The sequence and presentation of the feedback may also play significant roles in motivation. Salas et al. (1992) have noted that during the early stages of training, feedback should concentrate on *one* aspect of performance; how-

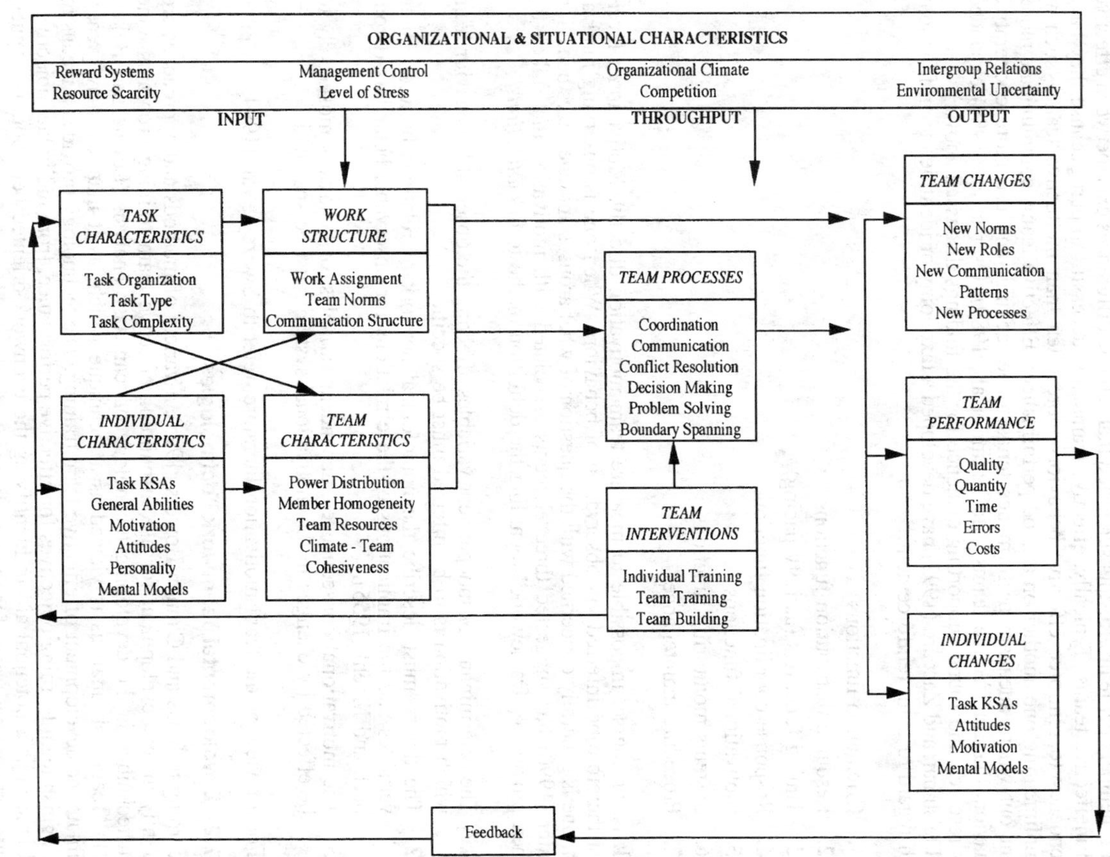

Figure 16.5 Team Performance and Training model reproduced from Tannenbaum et al. (1992). *Journal of Washington Academy of Sciences* by permission of the publisher.

ever, later in time it should concentrate on *several* training components. It has been found that sequencing helps team training in that teams adjust to incorporate their feedback into the task(s) at hand (Briggs and Johnston, 1967).

McCallum and associates (McCallum, Oser, Morgan, and Salas, 1989; Oser, McCallum, Salas, and Morgan, 1989) have conducted studies to identify behavioral components associated with team evolution and maturation over elongated timespans. These studies were suggested by the view that successful teams possess specific types of various behavioral characteristics (i.e., interaction, communication, and coordination behaviors), whereas unsuccessful teams display different types of these characteristics. In the McCallum et al. (1989) study, critical team behaviors were analyzed for 13 teams having 8–12 members per team. Results suggested that reliable differences exist between effective and ineffective teams. Typically, effective teams are successful in part because they have a tendency to reinforce effective behaviors more often than ineffective teams (such as complimenting one another on a good performance). Effective teams communicate more, point out where adequate performance tends to be lacking, and use more cooperative behaviors than ineffective teams (McCallum et al., 1989). Oser et al. (1989) expanded this research and found supporting evidence for the findings discussed above.

Fleishman and Zaccaro (1992) have developed a taxonomy of identified team functions with regard to team reinforcement:

1. Orientation functions
2. Resource distribution functions
3. Timing functions (activity pacing)
4. Response coordination functions
5. Motivational functions
6. Systems monitoring functions
7. Procedure maintenance

It is, of course, impossible to measure team motivation directly. Motivational levels must therefore be inferred via observation of behaviors. When one is observing motivation, one is generally concerned with the question of what arouses and energizes behavior. Kanfer (1990) has suggested three elements associated with motivation that, taken together, provide a "framework" regarding the definition of the term *motivation*:

1. The determinants or independent variables affecting the observable behavior from which motivation is to be inferred must be specific.
2. The theory must describe the nomological network of relations between latent variables and the implications of these relations for observable behaviors (Cronbach and Meehl, 1955).
3. The motivational consequences, dependent variables, or behaviors most likely to be effected by changes in the motivational system must be specified.

For a review of the team motivation literature, refer to Swezey et al. (1994).

16.7.2 Developmental Teamwork Technologies

According to Salas and Cannon-Bowers (1995), a variety of methods and approaches are currently under development for use in building effective team training programs. One such technique is the developing technology of team task analysis (Levine and Baker, 1991). Team task analysis is viewed as a technique which, as it matures, may greatly facilitate the development of effective team training. The technology provides a means to distinguish team learning objectives for effective performance, from individual objectives, and is seen as a potential aid in identifying the teamwork-oriented behaviors (i.e., cues, events, actions) necessary for the development of effective team training programs.

A second area is team performance measurement. To be effective, team performance measurement, must assess the effectiveness of various teamwork components (such as team member interactions) as well as intra-team cognitive and knowledge activities (e.g., decision-making and communication) in the context of assessing overall performance of team tasks. Hall, Dwyer, Cannon-Bowers, Salas, and Volpe (1994) have suggested the need for the integration of team performance outcomes into any teamwork measurement system. As team feedback is provided, it is also necessary to estimate the extent to which an individual team member is capable of performing his or her specific tasks within the team. Thus, any competently developed team performance measurement system needs to be capable of addressing both team capabilities and individual capabilities separately and in combination.

Teamwork simulation exercises are a third developing technology cited by Salas and Cannon-Bowers (1995). The intent of such simulations is to provide trainees with direct, behavioral cues designed to trigger competent teamwork behaviors. Essential components of such simulations include detailed scenarios or exercises where specific teamwork learning objectives are operationalized and incorporated into training.

Salas and Cannon-Bowers (1995) have also identified three generic training *methods* for use with teams. These are termed *information-based, demonstration-based* and *practice-based* methods.

The information-based method involves the presentation of facts and knowledge via the use of such standard delivery techniques as lectures, discussions, overheads, etc... Using such group-based delivery methods, one can deliver relevant information simultaneously to large numbers of individuals. The methods are easy to use and costs are usually low. Information-based methods may be employed in such areas as helping teammates understand what is expected of them, what to look for in specific situations, knowledge exchange areas, and many others.

Demonstration-based methods are *performed*, rather than *presented*, as in information-based methods. That is, they offer students an opportunity to observe behaviors of experienced team members, and thus the behaviors expected of them. Such methods help to provide shared mental models among team members, as well as examples of how one is expected to handle oneself within complex, dynamic, and multifaceted situations (Salas and Cannon-Bowers, 1995).

Practice-based methods are implemented via participatory activities such as role-playing, modeling, and simulation techniques, among others. These methods provide opportunities to practice specific learning objectives and to receive feedback information. With such methods, trainees can "build" upon previous practice attempts until achieving the desired level(s) of success.

A final category of developmental teamwork technologies according to Salas and Cannon-Bowers (1995) involves teamwork training implementation strategies. These authors have discussed two such strategies: *cross-training* and *team coordination training*. In their view, cross-training, the idea that all team members be trained to understand and perform each other's tasks, is an important strategy for use in integrating inexperienced members into experienced groups, since many teams are not permanent. The second strategy, team coordination training, addresses the issue that each team member has specific duties and that those duties, when performed together, are designed to provide a coordinated output. Team coordination training involves the use of specific task-oriented strategies to implement coordination activities and has been successfully applied in aviation and medial fields.

16.7.3 Team Training Guidelines

It is to the benefit of training developers to obtain and draw upon the many years of research information available in order to help develop team training programs. Information of this sort can be found in compiled lists of prescriptive guidelines and checklists based upon such guidelines (Swezey et al., 1992). Such guidelines help to target specific ways to design training programs. Though many different aspects of team training exist, the following guidelines (extracted from Swezey and Salas, 1992) may be helpful to training developers.

- Every team member should be able to describe how his or her own special (or subsidiary) team(s) will satisfy its task(s) in the execution of the team mission (Lewis, Hritz, and Roth, 1983).
- Team goals should be clearly and objectively stated (Dyer, 1984; Hall and Rizzo, 1975).
- Requiring articulation of concrete goals by all team members periodically during training will serve to reinforce their importance (Swezey and Salas, 1992).
- Teams should be trained under conditions that approximate the operational environment (Dyer, 1984; Gaddy, 1987).
- To the extent possible, training environments should include important constraints which actually operate in the operational setting (McIntyre, Morgan, Salas, and Glickman, 1988).

- The optimum size of a team depends on its task. The optimum size for most problem-solving discussions seems to be reached with five or six members, but for many production tasks, the optimum size may be two (Bass, 1982).
- Teams should be trained on the following specific details of team tasks: how individual member roles interrelate, how the performance of the team depends on the specific performances of the individuals, and what makes the team task different from the sum total of individual tasks (McIntyre et al., 1988).
- Every team member should be able to recognize and smoothly execute expected transitions among team activities (Lewis et al., 1983).
- The nature of tasks influences the appropriate number of team members in any activity. Have no more than six members in a group when performing problem-solving or decision-making tasks (Swezey and Salas, 1992).
- Divide teams into groups of observers and participants for various time periods during training. Observers can then watch and take notes on interaction patterns of team members as the latter perform a task. Then observers should inform the team of their interaction patterns as observed, and specify improved interaction patterns (Swezey and Salas, 1992).
- Team members having responsibilities for different task areas may wish to interview each other to promote group interaction and better understanding of the tasks, responsibilities, and difficulties of fellow teammates (Swezey and Salas, 1992).
- Every team member should be able to recognize and list all members of the team by position (Lewis et al., 1983).
- Accurate expectations of the contributions of other team members to overall team performance should be developed, both for the overall team and for special or subsidiary teams (Cream, 1978; Lewis et al., 1983).
- Employ positive reinforcement techniques throughout training and promote their proliferation beyond training to the work environment. Develop a system of rewards for those who exhibit supportive behavior towards teammates (Swezey and Salas, 1992).
- Establish homogenous groupings of team members for some teamwork activities and heterogenous groupings for other activities during training. The combination gives members a chance to work with individuals of similar and differing capabilities and backgrounds (Swezey and Salas, 1992).
- Every team member should recognize the authority of the team leader, both for the entire team and for all relevant subsidiary or special teams (Lewis et al., 1983).
- Team leaders should verbalize their plans for achieving the team goal (Helmreich, 1982).
- All people in leadership positions should recognize that team skills are at least as important as task skills. This can be demonstrated via specifically articulated media such as videotapes of teams performing in actual environments (McIntyre et al., 1988).
- Leadership training in the context of team training should be skill based. That is, it should reflect those critical skills and behaviors appropriate for the team's task requirements (Swezey and Salas, 1992).
- Team training should include discussions about what teams should expect from the leader and what the leader should expect from team members. This should include establishing criteria by which team members can determine whether or not the team leader is effective and can take control of a situation when necessary (Swezey and Salas, 1992).
- Team training should specify the measures the leader should take in familiarizing himself or herself with the tasks and responsibilities of all team members. Basically, the leader should be competent in all team member functions to the extent practicable (Swezey and Salas, 1992).
- Tasks that involve requirements for frequent communication should be given priority in team training (Gaddy, 1987).
- It is important that individual contributions be acknowledged as essential parts of total team performance (NATO, 1980).

- Task sequences, methods of communication, and potential barriers to communication should be identified and included when planning for team coordination training (Cream, 1978).
- Feedback of performance information among team members is especially important in task conditions that are changing rapidly and where direct equipment or instrument feedback is unavailable (Davis et al., 1985).
- Team training should include techniques for training individuals to analyze their own errors, to sense when the team or individual team members are overloaded, and to adjust their behavior when overloads occur (Dyer, 1984).
- In operational situations where it is not possible for one team member to compensate for the deficiencies of another, the use of direct feedback to individual team members is desirable (Dyer, 1984; Hall and Rizzo, 1975).
- When conducting the task/skill analysis, identify the unique communication requirements among team members for each specific team task (Swezey and Salas, 1992).
- Team skill training should focus on "emergent" tasks since these tasks involve unpredictable events that cannot be fully specified in advance (Gaddy, 1987).
- Every team member should be able to recognize unexpected events and to describe actions which he or she would expect to take when an unexpected event interferes with, or changes, the team's purpose, structure, or dependency situation (Lewis et al., 1983).
- Design tasks that have several specific and sensible avenues for solution. Require that team members demonstrate the ability to perform the task through several different approaches (Swezey and Salas, 1992).
- Construct activities which are designed to "go wrong" at specific points. Team members must confront and resolve the anomaly by using rational and available means (Swezey and Salas, 1992).
- Cross-train students so that they may perform other activities and thus act as replacements for other team members (Branson et al., 1975; Cream, 1978; Dyer, 1984; Lewis et al., 1983).
- Develop assessment instruments to accurately evaluate trainees' mastery of skills, both before and after training. The use of criterion-based job performance tests, and/or rating scales, which assess mastery of skill or knowledge components of tasks are especially encouraged (Swezey and Salas, 1992).
- In a later phase of team-training activity, have members rotate to different phases, in which they are "practical" novices in the subject domain. Teammates from one phase may be used to explain the process of that phase of work to the (now) novice, and vice versa, before it will be possible to execute the task (Swezey and Salas, 1992).
- Team performance measurement should consider the level of team development, the quality of team performance, and the effectiveness of person-to-person interaction within the team (NATO, 1980).
- Quantify dynamic forces such as turnover, change of assignment, and change in status which affect the performance of the team (Swezey and Salas, 1992).
- Team training should include systematic procedures for providing feedback information to the trainee(s) while they are learning team skills (Hall and Rizzo, 1975).
- Teams should be trained as entire units (Davis et al., 1985; Dyer, 1984).
- Team training should be sequenced according to increasing task complexity, as well as by the degree of teamwork required (Davis et al., 1985; Dyer, 1984).
- Where appropriate and useful, divide team members according to tasks and train members individually in their specific task domains (Swezey and Salas, 1992).
- The use of modeled vignettes in specific situations helps provide examples of good team skills for team members to model (Davis et al., 1985).
- Good team leaders tend to identify common factors that inhibit team member motivation (McIntyre et al., 1988).
- Like-minded individual team members can often work together more easily, but are likely to be less creative. Members with diverse attitudes will generate more conflict, but will also more often hammer out more creative solutions (Bass, 1982).

- Team members should be encouraged to show verbal and physical (i.e., verbal compliments and "pats on the back") signs of support for other members and the team as a whole (Morgan et al., 1986; Oser et al., 1989).
- Members of successful teams tend to praise the accomplishments of fellow teammates (Morgan et al., 1986).
- Effective team members make positive statements to motivate the team (Oser et al., 1989).

REFERENCES

Ackerman, P. L. (1987). Individual differences in skill learning: An integration of psychometric and information processing perspectives. *Psychological Bulletin, 10*, 3–27.

Adams, J. A. (1979). On the evaluation of training devices. *Human Factors, 21(6)*, 711–720.

Adams, J. A. (1987). Historical review and appraisal of research on the learning, retention, and transfer of human motor skills. *Psychological Bulletin, 101*, 41–74.

Adams, J. A., and Hufford, L. E. (1962). Contributions of a part-task trainer to the learning and relearning of a time-shared flight maneuver. *Human Factors, 4*, 159–170.

Adelman, L., Zirk, D. A., Lehner, P. E., Moffett, R. J., and Hall, R. (1986). Distributed tactical decision making: Conceptual framework and empirical results. *IEEE Transactions on Systems, Man, and Cybernetics, SMC-16*, 794–805.

Alesandrini, K. L. (1984). Pictures and adult learning. *Instructional Science, 13*, 63–77.

Alessi, S. M. (1988). Fidelity in the design of instructional simulations. *Journal of Computer-Based Instruction, 15*(2), 40–47.

Allen, W. H. (1973). Research in educational media. In J. W.Brown, Ed., *Educational Media Yearbook* NY: R.R. Bowker.

Allen, W. J. (1967, January 1974). Media stimulus and types of learning. *Audiovisual Instruction.* Reprinted (1974), In *Selecting Media for Learning: Readings from Audiovisual Instruction*, Association for Educational Communications and Technology, Washington, DC.

Allen, W., and Weintraub, R. (1972). *The motion variable in film presentation.* O.E. Final Report, Project No. 81123, University of Southern California, Dept. of Cinema, Los Angeles.

Alluisi, E. A. (1991). The development of technology for collective training: SIMNET, a case history. *Human Factors, 33(3)*, 343–362.

Altman, J. W. (1970). Behavior and accidents. *Journal of Safety Research, 2*, 109–122.

Anderson, J. R. (1976). *Language, Memory, and Thought.* Hillsdale, NJ: Erlbaum Associates.

Anderson, J. R. (1980). *Cognitive Psychology and Its Implications.* San Francisco, CA: Freeman.

Anderson, J. R. (1982). Acquisition of cognitive skill. *Psychological Review, 89*, 369–406.

Anderson, J. R. (1985). *Cognitive psychology and its implications* (Second Edition). NY: Freeman.

Anderson, J. R. (1987). Skill acquisition: Compilation of work-method problem solutions. *Psychological Review, 94*, 192–210.

Anderson, J. R. (1990). *Cognitive psychology and its implications* (Third Edition). New York: W. H. Freeman.

Anderson, R. H. (1976). *Selecting and developing media for instruction.* New York: Van Nostrand Reinhold.

Andreas, B. G. (1960). *Experimental Psychology.* New York: Wiley.

Andrews, D. H., Waag, W. L., and Bell, H. H. (1992). Training technologies applied to Team Training: Military examples. In R. W. Swezey and E. Salas, Eds., *Teams: Their Training and Performance.* Norwood, NJ: ABLEX.

Annett, J. (1979). Memory for skill. In M. M. Gruneberg and P. E. Morris, Eds., *Applied Problems in Memory.* London: Academic Press.

Bahrick, H. P. (1964). Retention curves: Facts or artifacts. *Psychological Bulletin, 61*, 188–194.

Bahrick, H. P. (1979). Maintenance of knowledge: Questions about memory we forgot to ask. *Journal of Experimental Psychology: General, 108*, 296–308.

Bahrick, H. P. (1984). Semantic memory content in permastore: Fifty years of memory for Spanish learned in school. *Journal of Experimental Psychology: General, 113*, 1–29.

Bahrick, H. P., Bahrick, L. E., Bahrick, A. S., and Bahrick, P. E. (1993). Maintenance of foreign language vocabulary and the spacing effect. *Psychological Science, 4*, 316–321.

Bahrick, H. P., Bahrick, P. O., and Wittlinger, R. P. (1975). Fifty years of memory for names and faces: A cross-sectional approach. *Journal of Experimental Psychology: General, 104*, 54–75.

Bahrick, H. P., Fitts, P. M., and Briggs, G. E. (1957). Learning curves—facts or artifacts? *Psychology Bulletin, 54*, 156, 268.

Bahrick, H. P., and Hall, L. K. (1991). Lifetime maintenance of high school mathematics content. *Journal of Experimental Psychology: General, 104*, 54–75.

Baker, E. L., and O'Neil, H. F. (1994). *Technology Assessment in Education and Training.* Hillsdale, NJ: Lawrence Erlbaum Associates.

Baldwin, T. T., and Ford, J. K. (1988). Transfer-of-training: A review and directions for future research. *Personal Psychology, 41*, 63–105.

Bass, B. (1982). Individual capability, team performance, and team productivity. In M. Dunnette and E. E. Fleishman, Eds., *Human performance and productivity: Human capability assessment.* Hillsdale, NJ: Lawrence Erlbaum Associates.

Bassok, M., and Holyoak, K. J. (1987). *Schema-based interdomain transfer between isomorphic algebra and physics problems.* Manuscript in preparation. University of Pittsburgh, Learning Research and Development Center, Pittsburgh, PA.

Battig, W. F. (1966). Facilitation and interference. In E. A. Bilodeau, Ed., *Acquisition of Skill.* New York: Academic Press.

Battig, W. F. (1972). Intratask interference as a source of facilitation in transfer and retention. In R. F. Thompson and J. F. Voss, Eds., *Topics in learning and performance.* New York: Academic Press.

Berliner, D. C., Angell, D., and Shearer, J. W. (August 1964). *Behaviors, measures, andinstruments for performance evaluation in simulated environments.* Paper presented at a symposium and workshop on the qualification of human performance, Albuquerque, NM.

Biel, W. C. (1966). Training programs and devices. In R. M. Gagne, Ed., *Psychological Principles in System Development.* New York: Holt, Rinehart and Winston, 343–386.

Blake, T. (1977). Motion in instructional media: Some subject-display mode interactions. *Perceptual and Motor Skills, 44*, 975–985.

Bloom, B. S., Ed. (1956). *A taxonomy of educational objectives. Handbook I: Cognitive Domain.* New York: McKay.

Bloom, B. S. (1974). Time and learning, *American Psychologist, 29*, 682–688.

Bloom, K. C., and Shuell, T. J. (1981). Effects of massed and distributed practice on the learning and retention of second-language vocabulary. *Journal of Educational Research, 74(4)*, 245–248.

Boguslaw, R., and Porter, E. H. (1962). Team functions and training. In R. M. Gagné, Ed., *Psychological Principles in System Development.* New York: Holt, Rinehart and Winston.

Boreham, N. C. (1985). Transfer-of-training in the generation of diagnostic hypothesis: The effect of lowering fidelity of simulation, *British Journal of Educational Psychology, 55*, 213–223.

Borg, W. R., and Schuller, C. F. (1979). Detail and background in audiovisual lessons and their effect on learners. *Educational Communication and Technology Journal, 27*, 31–38.

Bouzid, N., and Crawshaw, C. M. (1987). Masses versus distributed word processor training. *Applied Ergonomics, 18*, 220–222.

Bower, G. H., and Hilgard, E. R. (1981). *Theories of Learning* (Fifth Edition). Englewood Cliffs, NJ: Prentice-Hall.

Bower, G. H., and Karlin, M. B. (1974). Depth of processing of faces and recognition memory. *Journal of Experimental Psychology, 103*, 751–757.

Bransford, J. D., and McCarrel, N. S. (1974). A sketch of cognitive approach to comprehension: Some thoughts about understanding what it means to comprehend. In W. B. Weimer and D. S. Palermo, Eds., *Cognition and the symbolic processes.* Hillsdale, NJ: Erlbaum.

Branson, R. K., Rayner, G. T., Coxx, J. L., Furman, J. P., King, F. J., and Hannum, W. J. (1975). *Interservice Procedures for Instructional Systems Development: Executive Summary and Model.* ADA019486, U.S. Army Combat Arms Training Board, Fort Benning, GA.

Briggs, L. J. (1970). Handbook of procedures for the design of Instruction. American Institute for Research, *Monograph No. 4*, Pittsburgh.

Briggs, G. E., and Johnston, W. A. (1967). *Team training.* NAVTRADEVCEN-1327-1, AD-608309, Naval Training Device Center, Port Washington, NY.

Broadston, J. A. (1968). Learning curve wage incentives, *N.A.A. Management Accounting, 49*, 15–23.

Brooks, L. W., and Dansereau, D. F. (1983). Effects of structural schema training and text organization on expository prose processing. *Journal of Educational Psychology, 75*, 511–520.

Bruce, R. W. (1933). Conditions of transfer-of-training. *Journal of Experimental Psychology, 16*, 343–361.

Bryan, L. B., and Harter, N. (1899). Studies on the telegraphic language: The acquisition of a hierarchy of habits. *Psychological Review, 6*, 345–375.

Bugelski, B. R., and Cadwallader, T. C. (1956). A reappraisal of the transfer and retroaction surface. *Journal of Experimental Psychology, 52(6)*, 360–366.

Calfee, R. C., and Atkinson, R. C. (1965). Paired-associate models and the effects of list length. *Journal of Mathematical Psychology, 2*, 254–265.

Campbell, J. P. (1988). Training design for performance improvement. In J. P. Campbell, R. J. Campbell, and Associates, Eds., *Productivity in Organizations.* San Francisco: Jossey-Bass, 177–216.

Campione, J. C., Brown, A. L., Ferrara, D. A., Jones, R. S., and Steinberg, E. (1985). Breakdowns in flexible use of information: Intelligence related differences on transfer following equivalent learning performance. *Intelligence, 9*, 297–315.

Cannon-Bowers, J. A., and Salas, E. (1990). *Cognitive psychology and team training: Shared mental models in complex systems.* In K. Kraiger, Chair, Cognitive Representations of Work. Symposium conducted at the annual meeting of the Society for Industrial/Organizational Psychology. Miami, FL.

Cannon-Bowers, J. A., Salas, E., and Converse, S. A. (1990). Cognitive psychology and team training: Training shared mental models and complex systems. *Human Factors Society Bulletin, 33*, 1–4.

Caro, P. W. (1970). *Equipment-device task commonality analyses and transfer-of-training.* HumRRO Technical Report 70-7, AD709534. Human Resources Research Organization, Alexandria, VA.

Caro, P. W. (1973). Aircraft simulators and pilot training. *Human Factors, 15(6)*, 502–509.

Carter, R., and McCarthy, M. (1988). *Vocabulary and Language Teaching.* London and New York: Longman.

Chapanis, A., and Van Cott, H. P. (1972). Human Engineering Tests and Evaluations. In H. P. Van Cott and R. G. Kinkade, Eds., *Human Engineering Guide to Equipment Design.* U.S. Government Printing Office, Washington, DC, pp. 701–728.

Childs, J. M., and Spears, W. D. (1986). Flight-skill decay and recurrent training. *Perceptual and Motor Skills, 62*, 235–242.

Childs, J. M., Spears, W. D., and Prophet, W. W. (1983). *Private pilot flight skill retention 8, 16,and 24 months following certification.* Report No. DOT/FAA/CT-83/34. Federal Aviation Administration, Washington, DC.

Chute, A. G. (1980). Effect of color and monochrome versions of a film on incidental and task-relevant learning. *Educational Communication and Technology Journal, 28*, 10–18.

Clark, R. E. (1983). Reconsidering research on learning from media. *Review of EducationalResearch, 53*, 445–459.

Clark, R. E. (1988). Who's the fairest of them all? instructional medium, that is. *CBTDirections, 32.*

Cofer, C. N., Bruce, D. R., and Reicher, G. M. (1966). Clustering in free recall as a function of certain methodological variations. *Journal of Experimental Psychology, 71*, 858–866.

Cohen, P. A., Ebeling, B. J., and Kulik, J. A. (1981). A meta-analysis of outcome studies of visual-based instruction. *Educational Communication and Technology Journal, 29*, 26–36.

Conrad, R. (1967). Interference or decay over short retention intervals? *Journal of Verbal Learning and Verbal Behavior, 6*, 49–54.

Conway, M. A., Cohen, G., and Stanhope, N. (1991). On the very long-term retention of knowledge acquired through formal education: Twelve years of cognitive psychology. *Journal of Experimental Psychology: General, 120*, 395–409.

Cooke, N. J., Durso, F. T., and Schvaneveldt, R. W. (1994). Retention of skilled searchafter nine years. *Human Factors, 36(4)*, 597–605.

Cormier, S. M. (1984). *Transfer-of-training: An interpretive review.* Technical Report 608,Army Research Institute, Alexandria, VA.

Craik, F. I. M., and Lockhart, R. S. (1972). Levels of processing: A framework for memory research. *Journal of Verbal Learning and Verbal Behavior, 11*, 671–684.

Crawford, A. M., and Crawford, K. S. (1978). Simulation of operational equipment with a computer-based instructional system: A low cost training technology. *Human Factors, 20(2)*, 215–224.

Cream, B. W. (1978). A strategy for the development of training devices, *Human Factors, 20(2)*, 145–148.

Criswell, E. L., Swezey, R. W., Allen, J. A., and Hays, R. T. (1984). *Human factors analysis of two prototype army maintenance training and evaluation simulation system (AMTESS) devices.* Technical Report 652, U.S. Army Research Institute for the Behavioral and Social Sciences, Alexandria, VA.

Criswell, E. L., Unger, K. W., Swezey, R. W., and Hays, R. T. (1983). *Army maintenance training and evaluation simulation system (AMTESS) device development and features.* Technical Report 589. U.S. Army Research Institute for the Behavioral and Social Sciences, Alexandria, VA.

Cronbach, L. J. (1957). The disciplines of scientific psychology. *American Psychologist, 12*, 671–684.

Cronbach, L. J., and Meehl, P. E. (1955). Construct validity in psychological tests, *Psychological-Bulletin, 52*, 281–302.

Cronbach, L. J., and Snow, R. E. (1976). *Aptitude and Instructional Methods* (Irvington, NY).

Crossman, E. R. F. W. (1959). A theory of the acquisition of speed skill. *Ergonomics, 2,* 153-166.

Crothers, E., and Suppes, P. (1967). *Experiments in Second-Language Learning.* New York: Academic Press.

Crowder, R. G. (1976). *Principles of learning and memory.* Hillsdale, NJ: Erlbaum.

Dale, E. A. (1969). *Audiovisual Methods in Teaching* (Third Edition). New York: Holt, Rinehart, and Winston.

Dallett, K. M. (1962). The transfer surface re-examined, *Journal of Verbal Learning and Verbal Behavior, 1*, 91–94.

Dallett, K., and Wilcox, S. G. (1968). Contextual stimuli and proactive inhibition. *Journal of Experimental Psychology, 78*, 475–480.

Davis, L. T., Gaddy, C. D., and Turney, J. R. (1985). *An Approach to Team Skills Training of Nuclear Power Plant Control Room Crews.* Columbia, MD: General Physics Corporation.

DeBor, J., and Swezey, R. (1989). *Man-machine interface issues in nuclear power plants.* NUREG/CR-5348, SAIC-89/1114.

Demaree, R. G., Norman, D. A., and Matheny, W. G. (1965). *An experimental program for relating transfer-of-training to pilot performance and degree of simulation.* NAVTRADEVCEN-7619-1, Contract N61339-1388, AD471806. Fort Worth, TX: Life Sciences, Inc.

Dieterly, D. L. (October 1978). Team performance: A model for research. In *Proceedings of the Human Factors Society 22nd Annual Meeting*, Human Factors Society. Santa Monica, CA, pp. 486–492.

Digman, J. M. (1959). Growth of a motor skill as a function of distribution of practice. *Journal of Experimental Psychology, 57*, 311–316.

Downs, S. (1970). Predicting training potential. *Personnel Management, 2*, 26–28.

Driskell, J. E., Hogan, R., and Salas, E. (1987). Personality and group performance. *Review of Personality and Social Psychology, 9*, 91–112.

Driskell, J. E., Willis, R. P., and Copper, C. E. (1992). The effect of overlearning on retention, *Journal of Applied Psychology, 77*, 615–622.

Durham, N. H., Gearhart, R. G., and Austin, J. H. (1974). *Selecting instructional media and instructional systems.* Charles County Community College, La Plata, MD.

Dwyer, F. M. (1978). *Strategies for improving visual learning.* State College, PA: Learning Services.

Dwyer, M. (1969). The instructional effect of motion in varied visual illustrations. *Journal of Psychology, 73*, 167–172.

Dyer, J. L. (1984). Review on team training and performance: A state-of-the-art review. In F. A. Muckler, Ed., *Human factors review.* Santa Monica, CA: The Human Factors Society, Inc.

Ebbinghaus, H. (1913). *Memory: A Contribution to Experimental Psychology.* H. A. Ruger and C. E. Bussenius, Trans., Columbia University, NY. (Original work published 1885).

Eberts, R., and Brock, J. F. (1984). Computer applications to instruction. In F. Muckler, Ed., *Human Factors Review.* Santa Monica, CA: The Human Factors Society, Inc.

Ellis, H. C. (1965). *The transfer of learning.* New York: Macmillian.

Evans, R., and Swezey, R. (1980). *Development of a user's guidebook for TRAINVICE II.* Final Report for Contract No. MDA-903-79-C-0428. U.S. Army Research Institute, Alexandria, VA.

Farr, M. J. (1987). *The Long-Term Retention of Knowledge and Skills: A Cognitive and Instructionalperspective.* New York: Springer-Verlag.

Faust, D. G., Swezey, R. W., and Unger, K. W. (1984). *Field application of TRAINVICE: A study of four models designed to predict training device transfer-of-training potential.* Final Report for Contract No. MDA-903-79-C-0177. U.S. Army Research Institute, Alexandria, VA.

Fendrich, D. W., Healy, A. F., and Bourne, L. E., Jr. (1991). Long-term repetition effects for motoric and perceptual procedures. *Journal of Experimental Psychology: Learning, Memory, and Cognition, 17*, 137–151.

Fendrich, D. W., Healy, A. F., and Bourne, L. E., Jr. (1993). Mental arithmetic: Training and retention of multiplication skill. In C. Izawa, Ed., *Cognitive Psychology Applied.* Hillsdale, NJ: Erlbaum.

Fink, C. D., and Shriver, E. L. (1978). *Simulations for maintenance training: Some issues, problems and areas for future research.* Res. Pub. AFHL-TR-78-27. Air Force Human Resources Laboratory, Lowry Air Force Base, CO. (As cited by Hunt and Rouse, 1981.)

Fisk, A. D., and Hodge, K. A. (1992). Retention of trained performance in consistent mapping search after extended delay. *Human Factors, 34*(2), 147–164.

Fisk, A. D., and Lloyd, S. J. (1988). The role of stimulus-to-rule consistency in learning rapid application of spatial rules. *Human Factors, 30*, 35–49.

Fitts, P. M. (1964). Perceptual-motor skill learning. In A. W. Melton, Ed., *Categories of Human Learning.* New York: Academic Press.

Fitts, P. M. (1990). Factors in complex skill training. In M. Venturino, Ed., *Selected reading in Human Factors.* Santa Monica, CA: Human Factors Society. (Original work by Fitts published 1962).

Fitts, P. M., and Posner, M. I. (1967). *Human Performance.* Monterey, CA: Brooks/Cole.

Fleishman, E. A., and Quaintance, M. K. (1984). *Taxonomies of human performance: The description of human tasks.* Orlando, FL: Academic Press.

Fleishman, E. A., and Zaccaro, S.J. (1992). Toward a taxonomy of team performance functions. In R. W. Swezey and E. Salas, Eds., *Teams: Their training and performance.* Norwood, NJ: ABLEX.

Fletcher, J. D. (July 1990). *Effectiveness and cost of interactive videodisc instruction in defense training and education.* IDA Paper P-2372. Institute for Defense Analyses, Alexandria, VA.

Frederickson, C. H. (1969). Abilities, transfer and information retrieval in verbal learning. *Multivariate Behavioral Research Monographs*, No. 69-2.

Gaddy, C. (1987). *A Practitioner's Perspective on Team Skill Training in Industry.* Paper presented at the meeting of the American Psychological Association, NY.

Gagné, E. D. (1978). Long-term retention of information following learning from prose. *Review of Educational Research, 48*, 629–665.

Gagné, R. M. (1962). Simulators. In R. Glaser, Ed., *Training Research and Education*. University of Pittsburgh Press, Pittsburgh, 223–246.

Gagné, R. M. (1965). *The Conditions of Learning* (First Edition). New York: Holt, Rinehart, and Winston.

Gagné, R. M. (1974). *Essentials of Learning for Instruction*. New York: Holt, Rinehart, and Winston.

Gagné, R. M. (1977). *The Conditions of Learning* (Third Edition). New York: Holt, Rinehart, and Winston.

Gagné, R. M., and Briggs, L. J. (1974). *Principles of Instructional Design*. New York: Holt, Rinehart and Winston.

Gagné, R., Briggs, L. J., and Wager, W. W. (1989). *Principles of Instructional Design*. New York: Holt, Rinehart, and Winston.

Gagné, R. M., and Dick, W. (1983). Instructional psychology. *Annual Review of Psychology, 34*, 261–295.

Gagné, R. M., Forster, H., and Crowley, M. E. (1948). The measurement of transfer-of-training. *Psychological Bulletin, 45*, 97–130.

Gardlin, G. R., and Sitterley, T. E. (1972). *Degradation of Learned Skills: A Review and Annotated Bibliography*, D180-15081-1, NASA-CR-128611. Boeing Company, Seattle, WA.

Gay, L. R. (1973). Temporal position of reviews and its effect on the retention of mathematical rules. *Journal of Educational Psychology, 64*(2), 171–182.

Georgenson, D. L. (1982). The problem of transfer calls for partnership. *Training and Development Journal, 36*(10), 75–78.

Gerard, H. B., Wilhelmy, R. A., and Conolley, E. S. (1968). Conformity and group size. *Journal of Personality and Social Psychology, 8*, 79–82.

Gersick, C. J. G. (1988). Time and transition in work teams: Towards a new model of group development. *Academy of Management Review, 31*, 9–41.

Gladstein, D. L. (1984). Groups in context: A model of task group effectiveness. *Administrative Science Quarterly, 29*, 499–517.

Glaser, R. (1982). Instructional psychology: Past, present and future. *American Psychologist, 37(3)*, 292–305.

Glaser, R. B. (1963). Instructional technology and the measurement of learning outcomes: Some questions. *American Psychologist, 18*, 519–521.

Glaser, R., and Resnick, L. B. (1972). Instructional psychology. *Annual Review of Psychology, 23*, 181–276.

Glenberg, A., and Lehmann, T. (1980). Spacing repetitions over 1 week. *Memory and Cognition, 8*, 528–538.

Glover, J. H. (1966). Manufacturing progress functions: II. Selection of trainees and control of their progress. *International Journal of Production Research, 5(1)*, 43–59.

Godden, D. R. and Baddeley, A. D. (1975). Context-dependent memory in two natural environments: On land and under water. *British Journal of Psychology, 66*, 325–331.

Goldstein, I. L. (1974). *Training: Program development and evaluation*. Monterey, CA: Brooks/Cole.

Goldstein, I. L. (1993). *Training in Organizations* (Third Edition). Pacific Grove, CA: Brooks/Cole.

Goodman, P. S., Ed. (1986). *Designing Effective Work Groups* San Francisco: Jossey-Bass.

Gorman, D. A. (1973). Effects of varying pictorial detail and presentation strategy on concept formation. *AV Communication Review, 21*, 337–350.

Gropper, G. L. (1976). A behavioral perspective on media selection. *AV Communication Review, 24*, 157–186.

Grunwald, W. (1968). *An investigation of the effectiveness of training devices with varying degrees of fidelity*. Ph.D. Thesis, University of Oklahoma.

Hagman, J. D. (1980). *Effects of training task repetition and transfer of maintenance skill*. Norman, OK. RR 1271, Army Research Center, Alexandria, VA.

Hagman, J. D., and Rose, A. M. (1983). Retention of military tasks: A review. *Human Factors, 25*, 199–213.

Hall, E. R., and Rizzo, W. A. (1975). *An assessment of U.S. navy tactical team training*. TAEG Rep. No. 18. Orlando, FL: Training Analysis and Evaluation Group.

Hall, J. K., Dwyer, D. J., Cannon-Bowers, J. A., Salas, E., and Volpe, C. E. (1994). Toward assessing team tactical decision making under stress: The development of a methodology for structuring team training scenarios. *Proceedings of the 15th Annual Interservice/Industry Training Systems and Education Conference*. National Security Industrial Association (Washington, DC). 87–98.

Hamel, C. J., and Clark, S. L. (1986). *CAI evaluation checklist: Human factors guidelines for the design of computer-aided instruction*. NAVTRASYSCEN TR86-002, Orlando, FL.

Hammerton, M. (1977). Transfer and simulation. In *Institute of measurement and control Human Operations and Simulation,* Institute of Measurement and Control, London.

Harris, J. H., and Ford, P. (1983). *Application of transfer forecast methods to armor training devices*. FR-TRD(VA)-83-3. HumRRO, Alexandria, VA.

Hayes, K. J., and Pereboom, A. C. (1959). Artifacts in criterion-reference learning curves. *Psychological Review, 66(1)*, 23–26.

Hays, R. T. (1980). *Simulator fidelity: A concept paper*. Technical Report 490. Army Research Institute, Alexandria, VA.

Hays, R. T., and Singer, M. J. (1989). *Simulation fidelity in training system design*. New York: Springer-Verlag.

Healy, A. F., Fendrich, D. W., Crutcher, R. J., Wittman, W. T., Gesi, A. T., Ericsson, K. A., and Bourne, L. E., Jr. (1992). The long-term retention of skills. In A. F. Healy, S. M. Kosslyn, and R. M. Shiffrin, Eds., *From learning processes to cognitive processes: Essays in honor of William K. Estes*, Volume 2, Hillsdale, NJ: Erlbaum. 87–118.

Healy, A. F., Fendrich, D. W., and Proctor, J. D. (1990). Acquisition and retention of a letter-detection skill. *Journal of Experimental Psychology: Learning, Memory, and Cognition, 16*, 270–281.

Healy, A., and Sinclair, G. (1994). *Long-term retention of trained skills*. Technical Report 94-001. Naval Air Warfare Center Training Systems Division.

Helmreich, R. L. (August, 1982). *Pilot selection and training*. Paper presented at the meeting of the American Psychological Association, Washington, DC.

Hendriksen, G., and Schroeder, W. H. (1941). Transfer-of-training in learning to hit a submerged target. *Journal of Educational Psychology, 32*, 205–213. Cited in Hilgard, E. R., 1962.

Higa, M. (1966). The psycholinguistic concept of difficulty and the teaching of foreign language vocabulary. *Language Learning, 15*, 167–179.

Higgins, N., and Reiser, R. (1985). Selecting media for instruction: An exploratory study. *Journal of Instructional Design, 8*(2), 6–10.

Hilgard, E. R. (1962). *Introduction to Psychology* (Third Edition). New York: Harcourt, Brace and World.

Hirshfeld, S., and Kochevar, J. (1979). *Training device requirements documents guide*. PM TRADE, Naval Training Equipment Center. Orlando, FL.

Hoban, C. F., and Van Ormer, E. B. (1950). *Instructional film research, 1918–1950*. Tech. Rep. No. SDC 269-7-19. U.S. Naval Special Devices Center, Port Washington, NY. ERIC Document Reproduction Service No. ED 647 255.

Holding, D. H. (1976). An approximate transfer surface. *Journal of Motor Behavior, 8*, 1–9.

Holding, D. H. (1987). Concepts of training. In G. Salvendy, Ed., *Handbook of Human Factors*. New York: John Wiley and Sons.

Holman, G. J. (1979). *Training effectiveness of the CH-47 flight simulator*. Research Report No. 1209. U.S. Army Research Institute, Alexandria, VA.

Holmgren, J., Hilligoss, R., Swezey, R. W., and Eakins, R. (1979). *Training effectiveness and retention of training extension course (TEC) instruction in the Combat Arms*. Research Report 1208. U.S. Army Research Institute, Alexandria, VA.

Howell, W. C., and Goldstein, I. L. (1971). *Engineering psychology: Current perspectives in research*. New York: Appleton-Century-Crofts.

Hull, C. L. (1921). Quantitative aspects of the evaluation of concepts: An experimental study. *Psychology Monographs, 28*, No. 123.

Hull, C. L. (1943). *Principles of Behavior*. New York: Appleton Century Crofts.

Hull, C. L. (1952). *Essentials of Behavior*. New Haven, CT: Yale University Press.

Hurlock, R. E., and Montague, W. E. (1982). *Skill Retention and Its Primary Implications for Navy Tasks: An Analytical Review*. NPRDC SR 82-21. Navy Personnel Research and Development Center, San Diego, CA.

Ilgen, D. R., Shapiro, J., Salas, E., and Weiss, H. (1987). *Functions of group goals: Possible generalizations from individuals to groups*. Tech. Rep. 87-022. Naval Training Systems Center, Orlando, FL.

Indik, B. P. (1965). Organizational size and member participation: Some empirical test of alternatives. *Human Relations, 18*, 339–350.

Jacobs, J. W., Prince, C., Hays, R. T., and Salas, E. (1990). *A meta-analysis of the flight simulator training research*. NAVTRASYSCEN TR-89-006. Naval Training Systems Center, Orlando, FL.

Jarrard, S. W., and Wogalter, M. S. (1992). Recognition of non-studies visual depictions of aircraft: Improvement by distributed presentation. *Proceedings of the Human Factors Society 36th Annual Meeting*, pp. 1316–1320.

Jeon, U. (1976). *Effectiveness of image and motion variables in motor skill learning*. Dissertation Abstracts International, The Florida State University.

Johnson, S. L. (1980). A cost analysis method of establishing trainer criteria. *Ergonomics, 23*, 1137–1145.

Johnson, S. L. (1981). Effect of training device on retention and transfer of a procedural task. *Human Factors, 23(3)*, 257–272.

Jones, M. B. (1985). *Nonimposed overpractice and skill retention*. Final Report, Contract MDA-903-83-K-0246 for Army Research Institute, Alexandria, VA.

Judd, C. H. (1908). The relation of special training and general intelligence. *Educational Review, 36*, 28–42.

Kanfer, R. (1990). Motivation theory and industrial and organizational psychology. In M. D. Dunnette and L. M. Hough, Eds., *Handbook of industrial and organizational psychology* (Volume 1). Palto Alto, CA; Consulting Psychologists Press.

Kao, D. L. (1937). Plateaus and the curve of learning in motor skill. *Psychological Monographs, 49(3)*, Whole No. 219.

Keller, F. S. (1958). The phantom plateau. *Journal of Experimental Analysis of Behavior, 1*, 1–13.

Keller, J. M. (1987). Development and use of the ARCS model of instructional design. *Journal of Instruction Development, 3*(10), 2–10.

Kennedy, R. S., Jones, M. B., and Harbeson, M. M. (1980). Assessing productivity and well-being in Navy work-places. In *Proceedings of the 13th Annual Meeting of the Human Factors Association of Canada*. Human Factors Association of Canada, Ottawa, Canada, pp. 8–13.

Kessler, J. J., Macpherson, D., and Mirabella, A. (1987). *Army maintenance and evaluation simulation system (AMTESS) lesson learned*, Research Report, U.S. Army Research Institute for the Behavioral and Social Sciences, Alexandria, VA.

Kieras, D. E. (1985). *The role of prior knowledge in operating equipment from written instructions*, Report No. 19, FR-85/ONR-19, University of Michigan, Ann Arbor, MI.

Kieras, D. E., and Bovair, S. (1984). The role of a mental model in learning to operate a device, *Cognitive Science, 8*, 255–273.

King, C. L. (1992). *Familiarity effects on the retention of spatial, temporal, and item information in course schedules.* Unpublished doctoral dissertation, Colorado State University, Ft. Collins, CO.

Kinkade, R. G. (1963). *A differential influence of augmented feedback on learning and on performance.* AMRL-TDR-63-12. Aerospace Medical Research Laboratory, Dayton, OH.

Kinkade, R. G., and Wheaton, G. R. (1972). Training Device Design. In H. P. Van Cott and R. G. Kinkade, Eds., *Human engineering guide to equipment design*. U.S. Printing Office, Washington, DC.

Kintsch, W. (1970). *Learning, Memory, and Conceptual Processes*. New York: John Wiley.

Klein, G. A. (1982). The use of comparison cases. *IEEE 1982 Proceedings of the International Conference on Cybernetics and Society*, 88–91.

Klein, G. A. (1989). Recognition-primed decisions. In W. B. Rouse, Ed., *Advances in Man-Machine Systems Research* (Volume 5). Greenwich, CT: JAI Press, 47–92.

Knerr, C., Nadler, L., and Berger, D. (1980). *Toward a Naval taxonomy*. Contract No. N00014-80-C-0781. Litton Mellonics, Arlington, VA.

Knerr, M. A., Nadler, L., and Dowell, S. K. (1983). Comparison of training transfer and effectiveness models. *Proceedings of the Human Factors Society 27th Annual Meeting*. San Diego, CA: Human Factors Society.

Kohl, R. M., and Shea, C. H. (1992). Pew (1966) revisited: Acquisition of hierarchical control as a function of observational practice. *Journal of Motor Behavior*, 24, 247–260.

Kolers, P. A., and Roediger, H. L. (1984). Procedures of mind. *Journal of Educational Psychology, 53*, 250–253.

Krueger, W. C. F., 1947, Influence of difficulty of perceptual motor task upon acceleration of curves of learning. *Journal of Educational Psychology, 38*, 51–53.

Kulik, J. A., and Kulik, C. C. (1987). Review of recent research literature on computer-based instruction. *Contemporary Educational Psychology, 12*, 222–230.

Kulik, J. A., Kulik, C. L., and Cohen, P. A. (1980). Effectiveness of computer-based college teaching: A meta-analysis of comparative studies. *Review of Educational Research, 50*, 525–544.

Kyllonen, P. C., Alluisi, E. A. (1987). Learning and forgetting facts and skills. In G. Salvendy, Ed., *Handbook of Human Factors*. New York: John Wiley.

Laesecke, A. (1990). Computer animation as a teaching aid: The sterling cycle. *Cryogenics, 30*, 367–370.

Landa, L. N. (1974). *Algorithmization in learning and instruction*. Englewood Cliffs, NJ: Educational Technology Publications.

Landauer, T. K., and Bjork, R. A. (1978). Optimum rehearsal patterns and name learning. In M. M. Gruneberg, P. E. Morris, and R. N. Sykes, Eds., *Practical aspects of memory*. London: Academic Press, 625–632.

Lane, N. E. (1987). *Skill Acquisition Rates and Patterns: Issues and Training Implications*. New York: Springer-Verlag.

Laner, S. (1955). Some factors influencing the effectiveness of an instructional film. *British Journal of Psychology, 46*, 280–294.

Larkin, J. (1981). Enriching formal knowledge: A model for learning to solve textbook physics problems. In J. R. Anderson, Ed., *Cognitive Skills and Their Acquisition*. Hillsdale, NJ: Erlbaum.

Lee, T. D., and Genovese, E. D. (1988). Distribution of practice on motor skill acquisition: Learning and performance effects reconsidered. *Research Quarterly for Exercise and Sport, 59*, 277–287.

Levie, W. H. (1975). How to understand instructional media. *Viewpoints, Vol. 51(5)*, 25–42.

Levie, W. H., and Lentz, R. (1982). Effects of text illustrations: A review of research. *Educational Communication and Technology Journal, 30*, 195–232.

Levine, E. L., and Baker, C. V. (1991). Team task analysis: A procedural guide and test of the methodology. In E. Salas, Chair, *Methods and tools for understanding teamwork: Research with practical implications.* Symposium presented at the 6th Annual Conference of the Society for Industrial and Organizational Psychology, St. Louis, MO.

Lewis, C. M., Hritz, R. J., and Roth, J. T. (1983). *Understanding and improving teamwork: Concepts for understanding teams and teamwork.* Report 1, ARI Contract No. MDA903-81-C-0198. Applied Science Associates, Valencia, PA.

Lintern, G. (1987). Flight simulation motion systems revisited. *Human Factors Society Bulletin, 30(12)*, 1–3.

Locatis, C. N., and Atkinson, F. D. (1984). *Media and technology for education and training.* Bell and Howell Publication. Columbus, OH: Charles E. Merrill.

Loftus, G. R. (1985). Evaluating forgetting curves. *Journal of Experimental Psychology: Learning, Memory, and Cognition, 11*, 397–406.

Logan, G. D. (1988). Toward an instance theory of automatization. *Psychological Review, 95*, 492–527.

Luh, C. W. (1922). The conditions of retention. *Psychological Monographs*, Whole No. 142, Vol. XXXI, No. 3.

Mager, R. F. (1972). *Goal analysis.* Belmont, CA: Lear Siegler/Fearon.

Mallory, W. J., and Elliott, T. K. (1978). *Measuring troubleshooting skills using hardware-free simulation.* Res. Pub. AFHL-TR-78-47. Air Force Human Resources Laboratory, Lowry Air Force Base, Denver, CO. As cited in Hunt and Rouse, 1981.

Margolis, B., and Kroes, W. (1975). *The human side of accident prevention.* Springfield, IL: Charles C. Thomas.

Marmie, W. R., and Healy, A. F. (1995). The long-term retention of a complex skill: Part-whole training of tank gunner simulation exercises. In A. F. Healy and L. E. Bourne, Jr, Eds., *Learning and memory of knowledge and skills.* Newbury Park, CA: Sage.

Matheny, W. G., Williams, A. C., Jr., and Dougherty, D. J. (September, 1953). *The effect of varying control forces in the P-1 trainer upon transfer to the T-6 aircraft.* Report No. AFPTRC-TR-53-31. Air Force Personnel and Training Research Center, Lowry Air Force Base, Denver, CO.

Matlick, R. K., Swezey, R. W., and Epstein, K. I. (1980). *Alternative models for individualized armor training, Part I, interim report: Review and analysis of the literature.* U.S. Army Research Institute for the Behavioral Sciences, Alexandria, VA.

Mayer, R. E. (1975). Information processing variables in learning to solve problems. *Review of Educational Research, 45*, 525–541.

Mayer, R. E. (1989). Models for understanding, *Review of Educational Research, 59*, 43–64.

Mazur, J. E., and Hastie, R. (1978). Learning as accumulation A reexamination of the learning curve. *Psychological Bulletin, 85*, 1256–1274.

McCallum, G. A., Oser, R., Morgan, B. B., Jr., and Salas, E. (August, 1989). *An investigation of the behavioral components of teamwork.* Paper presented at the American Psychological Association Convention, New Orleans, LA.

McConnell, J. T. (1974). If the media fits, use it! In *Selecting Media for Learning: Reading from audiovisual instruction*, AECT, Washington, DC.

McDonald, L. B. (1984). *Passive acoustic analysis training problems: Final report* (Volume 1). NAVTRAEQUIPCEN 83-M-0844-2. McDonald and Associates, Inc., Orlando, FL.

McGeoch, J. A., and Irion, A. L. (1952). *The psychology of human learning* (Second Edition). New York: Longmans, Green and Co.

McIntyre, R. M., Morgan, B. B., Jr., Salas, E., and Glickman, A. A. (1988). Teamwork from team training: New evidence for the development of teamwork skills during operational training. *Proceedings of the 10th Interservice/Industry Training Systems Conference*, 21–27.

McKeachie, W. J. (1974). Instructional psychology. *Annual Review of Psychology, 25*, 161–193.

McNeil, B. J., and Nelson, K. R. (1991). Meta-analysis of interactive video instruction: A 10 year review of achievement effects. *Journal of Computer Based Instruction, 18*, 1–6.

McNulty, J. A. (1965). An analysis of recall and recognition process in verbal learning. *Journal of Verbal Learning and Verbal Behavior, 4*, 430–436.

Mengelkoch, R. F., Adams, J. A., and Gainer, C. A. (1971). The forgetting of instrument flying skills, *Human Factors, 13(5)*, 397–405.

Metcalfe, J. (1985). Levels of processing, encoding specificity, and CHARM. *Psychological Review, 92*, 1–38.

Micheli, G. S. (1972). Analysis of the transfer-of-training, substitution and fidelity of simulation of training equipment. TAEG Report No. 2. Naval Training Equipment Center, Orlando, FL.

Miller, G. G. (1974). *Some considerations in the design and utilization of simulators for technical training.* Technical Report AFHRL-TR-74-65. Air Force Human Resources Laboratory.

Miller, R. B. (1954). *Psychological considerations for the design of training equipment.* Pittsburgh, PA: American Institutes for Research.

Miller, R. B. *Psychological considerations in the design of training equipment.* Wright-Patterson Air Force Base, Ohio, Wright Air Development Center, WADC Technical Report 54-563. Cited by Biel, 1966.

Mirabella, A., and Wheaton, G. R. (1974). *Effects of task index variations on transfer-of-training criteria.* Tech. Report 72-C-0126-1. Naval Training Equipment Center, Orlando, FL.

Moely, B. E., and Shapiro, S. I. (1971). Free recall and clustering at four age levels: Effects of learning to learn and presentation method. *Developmental Psychology, 4*, 490.

Montague, W. D., Wulfeck, W. H., and Ellis, J. A. (1983). Quality CBI depends on quality instructional design and quality implementation. *Journal of Computer-Based Instruction, 10*, 90–93.

Morgan, B. B., Glickman, A. S., Woodard, E. A., Blaiwes, A. S., and Salas, E. (1986). *Measurement of team behaviors in a Navy environment.* Technical Report No. NTSC TR-86014. Naval Training Systems Center, Orlando, FL.

Morgan, B. B., Jr., and Lassiter, D. L. (1992). Team composition and staffing. In R. W. Swezey and E. Salas, Eds., *Teams: Their training and performance.* Norwood, NJ: ABLEX.

Morris, N. M., and Rouse, W. B. (1985). Review and evaluation of empirical research in troubleshooting. *Human Factors, 27(5)*, 503–530.

Murdock, B. B. (1957). Transfer designs and formulas. *Psychological Bulletin, 54*, 313–326.

Murdock, B. B., Jr. (1974). *Human memory: Theory and data.* Hillsdale, NJ: Erlbaum.

Narva, M. A. (1979). *Formative utilization of a model for the prediction of the effectiveness of training devices.* Research Memorandum 79-6. U.S. Army Research Institute for the Behavioral and Social Sciences, Alexandria, VA.

NATO (1980). Oosterveld, W. J., Chairman, *Working Group 10*, AGRAD-AR159 report.

Naylor, J. C., and Briggs, G. E. (1961). *Long-term retention of learned skills: A review of the literature.* ASD TR 61-390. Ohio State University, Laboratory of Aviation Psychology, Columbus, OH.

Neisser, U., Novick, R., and Lazar, R. (1963). Searching for ten targets simultaneously. *Perceptual and Motor Skills, 17*, 955–961.

Neves, D. M., and Anderson, J. R. (1981). Knowledge compilation: Mechanisms for the automatization of cognitive skills. In J. R. Anderson, Ed., *Cognitive Skills and Their Acquisition.* Hillsdale, NJ: Erlbaum.

Newell, A., and Rosenbloom, P. S. (1981). Mechanisms of skill acquisition and the law of practice. In J. R. Anderson, Ed., *Cognitive skills and their acquisition.* Hillsdale, NJ: Erlbaum.

Newell, K. M. (1976). Knowledge of results and motor learning. *Exercise and Sport Science Reviews, 4*, 195–228.

Nieva, V. F., Fleishman, E. A., and Rieck, A. (1978). *Team dimensions: Their identity, their measurement, and their relationships.* Contract No. DAHC19-78-C-0001. Response Analysis Corporation, Washington, DC.

O'Hara, J. M. (1990). The retention of skills acquired through simulator-based training. *Ergonomics, 33(9)*, 1143–1153.

Oosterveld, W. J., Chairman (1980). *Working group 10*, AGARD-AR-159 Report, NATO.

Orasanu, J., and Salas, E. (1993). Team decision making in complex environments. In G. Klein, J. Orasanu, R. Calderwood and C. Zsambok, Eds., *Decision making in action: Models and methods.* Norwood, NJ: ABLEX, 327–345.

Orlansky, J., and String, J. (1977). *Cost Effectiveness of Computer-Based Instruction in Military Training.* IDA Paper P-1375, Institute for Defense Analysis, Arlington, VA.

Orlansky, J., and String, J. (1981). *Cost Effectiveness of Maintenance Simulators for Military Training.* IDA Paper P-1568, Institute for Defense Analysis.

Oser, R., McCallum, G. A., Salas, E., and Morgan, B. B., Jr. (1989). *Toward a definition of teamwork: An analysis of critical team behaviors.* Tech. Rep. No. 89-004. Naval Training Systems Center, Orlando, FL.

Osgood, C. E. (1949). The similarity paradox in human learning: A resolution. *Psychological Review, 56*, 132–143.

Osgood, C. E. (1953). *Method and Theory in Experimental Psychology.* New York: Oxford Univ. Press.

Park, O., and Gittelman, S. S. (1992). Selective use of animation and feedback in computer-based instruction. *Educational Technology Research and Development, 40*, 27–38.

Parsons, H. M. (1980). *Aspects of a research program for improving training and performance of Navy teams.* Human Resources Research Organization.

Parsons, P. J., Fogan, T., and Spear, N. E. (1973). Short-term retention of habituation in the rat: A developmental study from infancy to old age. *Journal of Comparative Psychological Psychology, 84*, 545–553.

Peeck, J. (1987). The role of illustrations in processing and remembering illustrated text. In D. M. Willows and H. A. Houghton, Eds., *The psychology of illustration: Vol. 1. Basic research.* New York: Springer-Verlag, 115–151.

Petkovich, M. D., and Tennyson, R. D. (1985). A few more thoughts on Clark's "learning from media". *Educational Communication and Technology Journal, 33*, 146.

Pieper, W. S., Swezey, R. W., and Valverde, H. (1970). *Learner-centered instruction (LCI): Volume VII. Evaluation of the LCI approach.* Tech. Report AFHRL-TR-70-1, Air Force Human Resources Laboratory, Wright-Patterson AFB, OH.

Postman, L. (1961). The present status of interference theory. In O. N. Cofer, Ed., *Verbal Learning and Verbal Behavior.* New York: McGraw-Hill, 152–179.

Postman, L. (1971). Transfer, interference, and forgetting. In J. W. King and L. A. Riggs, Eds., *Woodworth and Schlosberg's Experimental Psychology*, (Third Edition). New York: Holt, Rinehart and Winston, 293–312.

Postman, L., and Rau (1957). Retention as a function of the method of measurement, *University of California Publications in Psychology, 8*, 217–270.

Powers, D. E. (1976). *Instructional strategies and individual differences: A selective review and summary of literature.* Report No. 4 on DARPA Contract MDA-903-74-C-0290, Educational Testing Service, Princeton, NJ.

Pressley, M., and Miller, G. E. (1987). Children's listening comprehension and oral prose memory. In D. M. Willows and H. A. Houghton, Eds., *The psychology of illustration: Vol. 1. Basic research.* New York: Springer-Verlag, 87–114.

Prophet, W. W. (1976). *Long-term retention of flying skills: A review of the literature.* HumRRO Final Report 76-35, ADA036977. Human Resources Research Organization, Alexandria, VA.

Provenmire, H. K., and Roscoe, S. N. (1971). An evaluation of ground-based flight trainers in routine primary flight training, *Human Factors, 13*(2), 109–116.

Rasmussen, J. (1979). *On the structure of knowledge: A morphology of mental models in man-machine systems context.* Report M-2192, Riso National Laboratory, Denmark.

Rea, C. P., and Modigliani, V. (1988). Educational implications of the spacing effect. In M. M. Gruenberg, P. E. Morris, and R. N. Sykes, Eds., *Practical aspects of memory: Current research and issues, Vol. I*, pp. 402–406, Chichester, UK: John Wiley.

Reilly, R. R., and Chao, G. R. (1982). Validity and fairness of some alternative employee selection procedures. *Personnel Psychology, 35*(*1*), 1–62.

Reiser, R. A., and Gagné, R. M. (1982). Characteristics of media selection models. *Review of Educational Research, 52*, 499–512.

Reynolds, B. (1952). Correlation between two psychomotor tasks as a function of distribution of practice on the first, *Journal of Experimental Psychology, 43*, 341–348.

Rickard, T. C. (1994). *Bending the Power Law: A Quantitative Model of the Transition From Algorithm to Retrieval.* Doctoral dissertation in preparation, University of Colorado, Boulder, CO.

Rieber, L.P. (1990). Animation in computer-based instruction. *Educational Technology Research and Development, 38*, 77–86.

Rieber, L. P., and Kini, A. S. (1991). Theoretical foundations of instructional applications of computer-generated animated visuals. *Journal of Computer-Based Instruction, 18*, 83–88.

Robinson, E. R. N., and Knirk, F. G. (1984). Interfacing learning strategies and instructional strategies in computer training programs. In F. A. Muckler, Ed., *Review of Human Factors.* Santa Monica, CA: Human Factors Society, 209–238.

Robinson, E. S. (1927). The "similarity" factor in retroaction. *American Journal of Psychology, 39*, 297–312 (cited in E. R. Hilgard and G. H. Bower, 1966, *Theories of learning*, (Third Edition), Appleton-Century-Crofts, NY).

Rodgers, T. S. (1969). On measuring vocabulary difficulty an analysis of item variables in learning Russian-English vocabulary pairs. *International Review of Applied Linguistics (IRAL), 7*, 327–343.

Romiszowski, A. J. (1974). *The Selection and Use of Instructional Media.* London: Kogan Page.

Romiszowski, A. J. (1988). *The Selection and Use of Instructional Media* (Second Edition). New York: Kogan Page.

Roscoe, S. N. (1971). Incremental transfer effectiveness. *Human Factors, 13*, 561–567.

Rose, A. M., Czarnolewski, M. Y., Gragg, F. E., Austin, S. H., Ford, P., Doyle, J., and Hagman, J. D. (1984). *Acquisition and Retention of Soldering Skills.* Contract MDA-903-81-C-AA01. AIR Final Report FR88600. American Institutes for Research, Washington, DC.

Rose, A. M., Wheaton, G. R., and Yates, L. G. (1985). *Forecasting device effectiveness: Volume I. Issues*, ARI Technical Report No. 680.

Rouse, W. B., Cannon-Bowers, J. A., and Salas, E. (1993). The role of mental models in team performance in complex systems. *IEEE Transactions on Systems, Man, and Cybernetics, 22*(*6*), 1296–1308.

Ruffner, J., Wick, W., and Bickley, W. (1984). Retention of helicopter flight skills: Is there a "critical period" for proficiency loss? *Proceedings of the Human Factors Society 28th Annual Meeting*, Santa Monica, CA: Human Factors Society.

Rumelhart, D. E., and Norman, D. A. (1978). Accretion, tuning and restructuring: Three modes of learning. In J. W. Cotton and R. L. Klatzky, Eds., *Semantic Factors in Cognition.* Hillsdale, NJ: Erlbaum.

Salas, E., Bowers, C. A., and Cannon-Bowers, J. A. (1995). Military team research: 10 years of progress. *Military Psychology, 7(2)*, 55–75.

Salas, E., and Cannon-Bowers, J. A. (1995). Methods, tools, and strategies for team training. In M. A. Quinones and A. Dutta, Eds., *Training for 21st Century Technology: Applications of Psychological Research*. Washington, DC: APA Press.

Salas, E., Cannon-Bowers, J. A., and Blickensderfer, E. L. (1995). Team performance and training research: Emerging principles. *Journal of the Washington Academy of Sciences, 83(2)*, 81–106.

Salas, E., Dickinson, T. L., Converse, S. A., and Tannenbaum, S. I. (1992). Toward an understanding of team performance and training. In R. W. Swezey and E. Salas, Eds., *Teams: Their Training and Performance*. Norwood, NJ: ABLEX.

Salmoni, A. W., Schmidt, R. A., and Walter, C. B. (1984). Knowledge of results and motor learning: A review and critical appraisal. *Psychological Bulletin, 95*, 355–386.

Salomon, G., Clark, R. (1977). Reexamining the methodology of research on media and technology in education. *Review of Educational Research, 47(1)*, 99–120.

Schendel, J., and Hagman, J. (1982). On sustaining procedural skills over a prolonged retention interval, *Journal of Applied Psychology, 67*, 605–610.

Schendel, J., and Hagman, J. (1991). Long-term retention of motor skills. In J. Morrison, Ed., *Training for Performance: Principles of Applied Human Learning*. New York: John Wiley.

Schendel, J., Shields, J., and Katz, M. (1978). *Retention of motor skills: Review*. Technical Paper 313, U.S. Army Research Institute for the Behavioral and Social Sciences, Alexandria, VA.

Schmidt, R. A., and Bjork, R. A. (1992). New conceptualizations of practice: Common principles in three paradigms suggest new concepts for training. *Psychological Science, 3(4)*, 207–217.

Schneider, V. I. (1991). *The effects of contextual interference on the acquisition and retention of logic rules*. Unpublished dissertation, University of Colorado, Boulder, CO.

Schneider, V. I., Healy, A. F., Ericsson, K. A., and Bourne, L. E., Jr. (1995). The Effects of Contextual Interference on the Acquisition and Retention of Logical Rules. In A. F. Healy and L. E. Bourne, Jr., Eds., *Learning and Memory of Knowledge and Skills*. Newbury Park, CA: Sage.

Schneider, W. (1984). *Toward a model of attention and the development of automatic processing*. HARL-ONR-8402. Human Attention Research Laboratory, University of Illinois, Champaign, IL.

Schneider, W. (1985). Training high performance skills: Fallacies and guidelines. *Human Factors, 27*, 285–300.

Schneider, W., and Detweiler, M. (1988). The role of practice in dual-task performance: Toward workload modeling in a connectionist/control architecture. *Human Factors, 30*, 539–566.

Schneider, W., and Shiffrin, R. M. (1977). Controlled and automatic human information processing: I. Detection, search and attention. *Psychological Review, 84*, 1–66.

Schramm, W. (1977). *Big Media, Little Media*. Beverly Hills, CA: Sage.

Shapiro, D. C., and Schmidt, R. A. (1982). The schema theory: Recent evidence and developmental implications. In J. A. S. Kelso and J. E. Clark, Eds., *The Development of Movement Control and Coordination*. New York: John Wiley, 113–150.

Shaw, M. E. (1976). *Group Dynamics: The Psychology of Small Group Behavior*. New York: McGraw-Hill.

Shea, J. F., and Morgan, R. L. (1979). Contextual interference effects on the acquisition, retention and transfer of a motor skill. *Journal of Experimental Psychology: Human Learning and Memory, 5*, 179–187.

Sheppard, S. B. (1987). Documentation for software systems. In G. Salvendy, Ed., *Handbook of Human Factors*. New York: John Wiley.

Shriver, E. L., Shriver, S. B., Bunderson, V. (1984). *The Cognitive Functional Context Training System*. Alexandria, VA: Kinton.

Smillie, R. J., Shelnutt, J. B., and Bercos, J. (1977). *Task report: Human factors research*. Fort Benning, GA: Litton Mellonics.

Smith, M. J., Beringer, D. B. (1987). Human factors in occupational injury evaluation and control. In G. Salvendy, Ed., *Handbook of Human Factors*. New York: John Wiley.

Smode, A. F. (1972). *Training device design: Human factors requirements in the technical approach*. Final Report, NAVTRAEQUIPCEN 71-C-0013-1, Dunlop and Associates.

Snoddy, G. S. (1926). Learning and stability. *Applied Psychology, 10*, 1–36.

Spangenberg, R. (1973). The motion variable in procedural learning. *AV Communication Review, 21(4)*, 419–436.

Spears, W. D. (1983). *Processes of skill performance: A foundation for the design and use of training equipment*. NAVTRAEQUIPCEN 78-C-0013-4. Naval Training Equipment Center, Orlando, FL.

Spears, W. D. (1985). Measurement of learning and transfer through curve fitting. *Human Factors, 27*, 251–266.

Stammers, R. (1981). Skill retention and control room tasks. *Proceedings of the Human Factors Society 25th Annual Meeting*, Santa Monica, CA: The Human Factors Society, 168–172.

Stammers, R. B., Carey, M. S., and Astley, J. A. (1990). Task analysis. In J. R. Wilson and E. N. Corlett, Eds., *Evaluation of Human Work.* Taylor and Francis, 134–160.

Steiner, I. D. (1972). *Group process and productivity.* New York: Academic Press.

Stevens, J. C., and Savin, H. B. (1962). On the form of learning curves. *Journal of Experimental Analysis of Behavior, 5,* 5–18.

Stonge, J. R., and Becker, J. H., [n.d.], *The modular Integrated Training System: A Total Integrated Maintenance Training Concept.* Grumman Display/Trainer Products.

Streufert, S., and Swezey, R. W. (1986). *Complexity, managers and organizations.* New York: Academic Press.

Swezey, R. W. (1977–78). Future Directions in Simulation and Training. *J. Educational Technology Systems, 6(4),* 285–292.

Swezey, R. W. (1978a). *Comparative Evaluation of an Advanced Naval Engineering Maintenance Training Program.* Technical Report NAVTRAEQUIPCEN 77-C-0150. Naval Training Equipment Center, Orlando, FL.

Swezey, R. W. (1978b). Retention of printed materials and the Yerkes-Dodson law. *Human Factors Society Bulletin, 21,* 8–10.

Swezey, R. W. (1981). *Individual Performance Assessment an Approach to Criterion-referenced Test Development.* Reston, VA: Reston Publishing.

Swezey, R. W. (1983). Application of a transfer-of-training model to training device assessment. *Journal of Educational Technology Systems, 11(3).*

Swezey, R. W. (1989). Generalization, fidelity and transfer-of-training. *Human Factors Society Bulletin, 32(6),* 4–5.

Swezey, R. W., and Evans, R. A. (May 1980). *Guidebook for users of TRAINVICE II,* SAI-80-065, Science Applications, Inc, McLean, VA. Also published as a U.S. Army Research Institute Technical Report under Contract No. MDA-903-79-C-0428, 1983.

Swezey, R. W., and Llaneras, R. E. (1992). Validation of an aid for selection of instructional media and strategies. *Perceptual and Motor Skills, 74,* 35.

Swezey, R. W., Llaneras, R. E., and Salas, E. (1992). Ensuring teamwork: A checklist for use in designing team training programs. *Performance and Instruction, 31*(2), 33–37.

Swezey, R. W., Meltzer, A. L., and Salas, E. (1994). Issues involved in motivating teams. In H. F. O'Neil Jr. and M. Drillings, Eds., *Motivation: Theory and Research.* Hillsdale, NJ: Lawrence Erlbaum. pp. 141–170.

Swezey, R. W., Perez, R. S., and Allen, J. A. (1988). Effects of instructional delivery system and training parameter manipulations on electromechanical maintenance performance. *Human Factors, 30(6),* 751–762.

Swezey, R. W., Perez, R. S., and Allen, J. A. (1991). Effects of instructional strategy and motion presentation conditions on the acquisition and transfer of electromechanical troubleshooting skill. *Human Factors, 33(3),* 309–323.

Swezey, R. W., and Salas, E. (1992). Guidelines for use in team training development. In R. W. Swezey and E. Salas, Eds., *Teams, their training and performance.* Norwood, NJ: ABLEX.

Tannenbaum, S. I., Beard, R. L., and Salas, E. (1992). Team building and its influence on team developments. In K. Kelley, Ed., *Issues, theory, and research in psychology.* Amsterdam: Elsevier, 117–153.

Tannenbaum, S. I., and Yukl, G. (1992). Training and development in work organizations. *Annual Review of Psychology, 43,* 399–441.

Taylor, J. G., and Smith, P. C. (1956). An investigation of the shape of learning curves for industrial motor tasks. *Journal of Applied Psychology, 40,* 142–149.

Thompson, D. A. (July-August 1963). The development of a quantified difficulty measure for precision work. *Journal of Methods-Time Measurement,* 8–18.

Thorndike, E. L. (1932). *The Fundamentals of Learning.* New York: Teachers College.

Thorndike, E. L., and Woodworth, R. S. (1901). The influence of improvement in one mental function upon the efficiency of other functions. *Psychological Review, 8,* 247–261.

Thorndyke, P. W., and Weiner, M. G. (1980). *Improving Training and Performance of Navy Teams: A Design for a Research Program.* Contract No. R-2607-ONR. Santa Monica, CA: Rand Corporation.

Thorpe, J. A. (1987). The new technology of large scale simulation networking: Implications for mastering the art of nonfighting. *Proceedings of the 15th Annual Interservice/Industry Training Systems and Education Conference,* Alexandria, VA.

Teather, D. C. B., and Marchant, H. (1974). Learning from film with particular reference to the effects of cueing, questioning, and knowledge of results. *Programmed Learning and Educational Technology, 11,* 317–327.

Thurstone, L. L. (1919). The learning curve equation. *Psychological Monographs, 26,* No. 114.

Tiffin, J. W., and Combes, P. (1980). *Processo de Selecao de Meios.* Portuguese language version of paper prepared for the Organizations of American States, Multinational Project in Educational Technology, Ministry of Education and Culture (FUNTEVE), Rio de Janeiro, Brazil.

Tosti, D. T., and Ball, J. R. (1969). A behavioral approach to instructional design and media selection. *AV Communication Review, 17*, 5–25.

Travers, R. M. (1967). *Research and theory related to audiovisual information transmission.* U.S. Office of Education contract No. OES-16-006. Western Michigan University, Kalamazoo (ERIC Document Reproduction Service No. ED 081 245).

Trollip, S., and Ortony, A. (1977). Real-time simulation in computer-assisted instruction. *Instruct, Sci., 6*, 135–149.

Tuckman, B. W. (1965). Developmental sequences in small groups. *Psychological Bulletin, 63*, 384–399.

Tufano, D. R., and Evans, R. A. (1982). *The prediction of training device effectiveness: A review of army models* (Technical Report), U.S. Army Research Institute for the Behavioral and Social Sciences, Alexandria, VA.

Tulving, E. (1975). Ecphoric processing in recall and recognition. In J. Brown, Ed., *Recall and recognition.* London: Wiley.

Tulving, E., and Psotka, J. (1971). Retroactive inhibition in free-recall: Inaccessibility of information available in the memory store. *Journal of Experimental Psychology.*

Tulving, E., and Thomson, D. M. (1973). Encoding specificity and retrieval processes in episodic memory. *Psychological Review, 3*, 112–129.

Tversky, B. (1973). Encoding processes in recognition and recall. *Cognitive Psychology, 5*, 275–287.

Underwood, B. J. (1961). Ten years of massed practice on distributed practice. *Psychological Review, 68(4)*, 229–247.

Underwood, B. J. (1964). Degree of learning and the measurement of forgetting. *Journal of Verbal Learning and Verbal Behavior, 3*, 112–129.

Underwood, B. J. (1966). *Experimental Psychology* (Second Edition). (Appleton-Century-Crofts, NY).

Underwood, B. J., and Keppel, G. (1963). Retention as a function of degree of learning and letter-sequencing interference. *Psychological Monographs, 77*, Whole No. 567.

Unger, K., Swezey, R. W., Hays, R., and Mirabella, A. (1984a). *Army maintenance training and evaluation simulation system (AMTESS) device evaluation: Volume I. Overview of the study effort.* Technical Report 642. U.S. Army Research Institute, Alexandria, VA.

Unger, K., Swezey, R. W., Hays, R., and Mirabella, A. (1984b). *Army maintenance training and evaluation simulation system (AMTESS) device evaluation: Volume II. Transfer-of-training assessment of two prototype devices.* Technical Report 643. U.S. Army Research Institute, Alexandria, VA.

Unger, K., Swezey, R. W., Hays, R., and Mirabella, A. (1984c). *Army maintenance training and evaluation simulation system (AMTESS) device evaluation: Volume III. Qualitative evaluation of two prototype devices.* Technical Report 644. U.S. Army Research Institute, Alexandria, VA.

Valverde, H. H. (1973). A review of flight simulator transfer-of-training studies. *Human Factors, 15*, 510–523.

Watkins, M. J., and Tulving, E. (1975). Episodic memory: When recognition fails. *Journal of Experimental Psychology, General, 1*, 5–29.

Waugh, N. C. and Norman, D. A. (1965). Primary memory. *Psychological Review, 72*, 89–104.

Welford, A. T. (1968). *Fundamentals of skill.* London: Methuen and Co.

Wells, S. (1976). Evaluation criteria and the effectiveness of instructional technology in higher education. *Higher Education, 5*, 253–275.

Wetzel, C. D., Radtke, P. H., and Stern, H. W. (1994). *Instructional effectiveness of video media.* Hillsdale, NJ: Lawrence Erlbaum Associates.

Weaver, J. L., Bowers, C. A., Salas, E., and Cannon-Bowers, J. A. (1995). Networked simulations: New paradigms for team performance research. *Behavior Research Methods, Instruments, and Computers, 27(1)*, 12–24.

Wheaton, G. R., Fingerman, P. W., Rose, A. M., and Leonard, R. (1976). *Evaluation of the effectiveness of training devices: Elaboration and application of the predictive model.* Research Memorandum 76-26, U.S. Army Research Institute for the Behavioral and Social Sciences, Alexandria, VA.

Wheaton, G. R., and Mirabella, A. (1972). *Effects of task index variations on training effectiveness criteria.* Final Report, NAVTRAEQUIPCEN 71-C-0059-1. U.S. Naval Training Equipment Center, Orlando, FL.

Wickelgren, W. A. (1972). Trace resistance and the decay of long-term memory. *Journal of Mathematical Psychology, 9*, 418–455.

Wightman, D. C., and Sistrunk, F. (1987). Part-task training strategies in simulated carrier landing final-approach training. *Human Factors, 29*, 245–254.

Williams, I. D. (1974). Practice and augmentation in learning. *Human Factors, 16(5)*, 503–507.

Willis, M. P., and Peterson, R. O. (1961). *Deriving training device implications from learning theory principles, Vol I: Guidelines for training device design, development and use.* U.S. Naval Training Device Center.

Winn, W. (1989). The design and use of instructional graphics. In H. Mandl and J. R. Levin, Eds., *Knowledge Acquisition From Text and Pictures*. Amsterdam: North-Holland, 125–144).

Wixted, J. T. and Ebbesen, E. B. (1991). On the form of forgetting. *Psychological Science, 2,* 409–415.

Woodworth, R. S., and Schlosberg, H. (1954). *Experimental Psychology*. New York: Holt, Rinehart, and Winston.

Wylie, H. H. (1919). An experimental study of transfer of response in the white rat. *Behavioral Monographs, 3*(No. 16).

CHAPTER 17
COMPUTER-BASED INSTRUCTION

John F. Brock
InterScience America, Inc.
Leesburg, VA 22075 USA

17.1 INTRODUCTION

In 1987, Eberts and Brock reviewed Computer-Based Instruction (CBI) for that year's *Handbook of Human Factors* (Salvendy, 1987). That article focused on human learning, instructional methods, and the appropriate use of computers to teach. Much of the research reported at that time had compared various computer-based instructional programs to more traditional methods (live persons, advanced media, on-the-job training). Since that review, the number of computers in use has increased exponentially. The power, speed, and versatility of computers continue to increase so fast that anything written about them for this article will be out of date by publication. The term *computer-based instruction* has come to cover not only instruction delivered by an obvious central processing unit (CPU), but also by such accessories as videodisc, CD-ROM, and various levels of simulation. Even recent reviews (e.g., Szabo and Montgomerie, 1992) have failed to account for the impact of the telecommunications revolution currently underway.

Computers provide ways to exploit human learning capabilities. The learning capabilities themselves have not changed. Swezey and Llaneras (1997) describe various instructional and learning models. CBI provides a lever that can be applied to those models to improve human performance.

CBI is not intrinsically good. If instructional programs are not well designed, if student needs are not met, if incorrect or incomplete content is presented, and if student performance is not measured, then all that the computer does is provide an efficient means for bad instruction to be distributed. CBI can be more interesting than conventional instruction; it can be more engaging, more entertaining, more individualized, and more exciting. Nevertheless, if the result of the instruction is not measurably improved human performance, it does not make any difference.

17.1.1 Scope of the Review

Most of the current uses of computer-based instruction are still based on the programmed instructional models developed right before, during, and after World War II (e.g., Pressey, 1950; Skinner, 1954). Early applications of these models allowed students to complete their learning experiences at their own paces (e.g., Keller, 1968); flexible time was the primary source of any individualization. This chapter will *not* review these earlier approaches. Modern computers are capable of much more sophisticated individualization techniques. It is the discussion of these techniques that define the scope of the review.

This review will describe the various forms of computer technologies that are being applied to instruction and presume to predict what lies ahead. Computer technology was once an expensive option to meet specific human performance training needs. It has become a ubiquitous method to teach everything from how to use a computer application to how to drive a car or fly an airplane. As for instructional options, the question has changed from *whether* to use a computer to train to *how*.

Not surprisingly, many problems associated with human–computer interface are also part of the CBI discussion. One great advantage of a book—if one can read—is that it is easy to use. The reader simply opens the book and begins to read. Whereas computers can make improvements over books in some useful ways (e.g., indexing, seeking out key words or text), if getting access to material is unwieldily or complicated, the program risks being rejected without serious trial. HCI design of CBI has special properties that will be discussed below.

17.2 EFFECTIVENESS OF COMPUTER-BASED INSTRUCTION

The most fundamental question about CBI is: does it work? Recent research, which has applied meta-analytic techniques to answer that question, suggests that it does (Fletcher, 1995; Kulik, 1994; Kulik and Kulik, 1991). Meta-analysis is a technique first proposed by Glass (1976). This approach begins by collecting studies which are relevant to whatever issue is being studied. It then applies a quantitative measure (effect size) to tabulate the outcomes of all the collected studies, including those with results that were not statistically significant. Finally, statistical processes synthesize the various quantitative measures and produce findings from the analysis.

Fletcher (1995) correctly notes that the measure of effect size is essentially a measure of standard deviations and may not be meaningful to persons not familiar with statistical analysis. Therefore, in a 1995 paper, he accompanies effect sizes with percentiles based on the notion that an effect size of, for instance, 0.50 is roughly equivalent to raising the performance of 50th percentile students to that of 69th percentile students. These three reports (Fletcher, 1995; Kulik, 1994; Kulik and Kulik, 1991) are this chapter's principal sources for the answer to the basic question of effectiveness.

Kulik and Kulik (1991) performed a meta-analysis of the findings of 254 controlled evaluation studies. The studies covered learners who ranged from kindergarten pupils to adults. The meta-analysis revealed a statistically significant increase of examination scores by students who received their instruction from a computer program. However, these effects varied greatly. Effects were larger (1) in published studies than in unpublished studies, (2) where different teachers dealt with the experimental and control groups, and (3) in studies of short duration.

In a later report, Kulik (1994) reported on 12 meta-analytic studies of CBI and reported the following:

1. Students normally learn more in classes in which they receive CBI. CBI effects were lowest in elementary and high school science courses (Willett, Yamashita,

and Anderson, 1983) and highest in special education classes (Schmidt, Weinstein, Niemiec, and Walberg, 1985).

2. Students learn in less time with CBI. The average reduction in instructional time was 34% in the studies of college instruction and 24% in the studies of adult education programs (Kulik and Kulik, 1991).
3. Students like classes in which they receive CBI.
4. Students develop more positive attitudes toward computers when they receive help through CBI.
5. However, CBI effects on the attitude of students toward subject matter were near zero.

In a report to a North Atlantic Treaty Organization research study group, Fletcher (1995) reports similar findings for a meta-analysis of CBI applications in military and industrial applications (e.g., Noja, 1987, 1991; Wisher, 1987; Yasutake, 1987). Reduced time to train continues to be a major CBI effect (Fletcher, 1990; Johnston and Fletcher, 1995; Orlansky and String, 1977).

It seems safe to conclude that computer-based instruction can be an effective medium for improving human performance. However, it is also clear that much of the effects of the CBI may be only coincidental to the actual delivery medium. Content, instructional design attributes, and student attitudes toward technology all affect the human performance outcomes of a CBI intervention. As Eberts and Brock (1984,1987) warn in earlier reviews, there is no evidence that computers automatically improve the learning process.

17.3 INSTRUCTIONAL COMPONENTS OF CBI

17.3.1 Drill and Practice

Perhaps the simplest approach to CBI is drill and practice. Students are exposed to some instructional material, they are then questioned and the computer tells them if they are right or wrong. If wrong, they either receive the same information again or have it supplemented in some way. One of the earliest examples of this kind of program was developed at Stanford University for the teaching of mathematics to children in elementary school (Suppes and Morningstar, 1969). The system had the following components: (1) a pretest for evaluating the student's entering ability, (2) an assignment component that routed the student to the appropriate instructional material based upon the pretest results, (3) a record-keeping component for maintaining up to date records of each student's performance, (4) a component that determined what instruction the student should receive next based on his or her performance in the module just completed, and (5) a post-test to measure the student's progress.

17.3.2 Tutorial process

Another early technique in CBI can be best described as the tutorial method. The tutorial method, as its name suggests, attempts to recreate a one-on-one tutorial between an all knowing teacher (in this case, the computer) and the learner. In research reported by Bloom (1984), it is suggested that the achievement differences between students taught in a 30-student classroom and those taught one-on-one by an individual instructor may be as great as two standard deviations. If a computer can do the same function as the human instructor, the results may be as significant. As Fletcher (1995) points out, such an approach is a classic economic solution; it substitutes the labor of human instructors with the capital of CBI.

If a computer is to provide the tutorial skills of a live teacher, it must possess some fundamental characteristics that are often associated with intelligence: subject matter expertise, knowledge of the student's state of knowledge about that subject matter, a growing knowledge of the students unique learning characteristics, and the ability to instruct efficiently. A comprehensive review of Intelligent Tutoring Systems can be found in Psotka, Massey, and Mutter (1988).

The fundamental attributes of an intelligent tutoring system (ITS) must be that it can make inferences about student knowledge and can interact intelligently with students (Mandl and Lesgold, 1988). Much of the research in the use of these systems was done in the 1970s and 1980s and relied on either very powerful minicomputers or very special stand alone machines (typically LISP processors in the United States). Modern personal computer software, almost as a matter of course, incorporates many design features of ITS into tutorial programs to teach specific application use.

17.3.2.1 Expert Knowledge

The first component of a complete ITS is expert knowledge. This is the subject matter that is to be learned. The subject matter can be represented in many ways. In early applications, knowledge was stored in a kind of black box. In a program called SOPHIE I (Brown, Burton, and Bell, 1975; Burton and Brown, 1979), the student could receive an answer to any question he or she posed about electrical measurement values for any point in a complex circuit. However, the student could not learn *why* the values were as reported. Later work attempted to store the subject matter in a "glass box," (Goldstein and Papert, 1977). Essentially, this was an attempt to represent knowledge in a way that most closely matched human capability.

17.3.2.2 Learner Modeling

ITS also has a learner modeling component. The system must be able dynamically to represent the emerging knowledge and skill of the learner. It must infer student knowledge from the student's interactions with the system. Carr and Goldstein (1977) developed an *overlay model* that represented the learner's knowledge as a subset of the expert's knowledge. The program then moved toward removing the gap between the two knowledge representations. An alternative approach is to include *deviations* from expertise in the model of the learner (Brown and Burton, 1978). Both Anderson (1983) and White and Frederiksen (1986) describe approaches to deviation modeling for ITS.

17.3.2.3 Tutorial Planning

The tutorial planning component of an ITS designs and regulates the instructional interactions with the learner. It is obviously closely linked to the learner model. This is the component that selects the appropriate pedagogical intervention. In early intelligent training systems, programs tended to be either didactic (the system told the learner what to do) or discovery oriented. In modern systems, both alternatives tend to coexist: Learners have either an overt choice or the system will go to one or the other method based on the learner's performance.

17.3.2.4 Communications

The fourth component is communications. In a later section of this chapter, various student/computer interactions will be discussed in some detail. The superior ITS will communicate in the language of the learner as opposed to some obscure code or instructions. It is common to have multimodal communications between learners and intelligent training systems: alphanumeric, graphic, still and motion video, and audio. ITS is often attached to commercial software applications (word processing, spreadsheets, database management, communications). These tutorial programs, which come bundled with the software and are usually available with a mouse click or two, provide step-by-step instructions, graphic demonstrations, and audio supplements.

17.3.3 Simulation

One of the earliest applications of computers to training was in the use of simulation. Powerful computers were (and still are) needed to create realistic replications of airplane characteristics in flight, which was the earliest application of simulation for training. The Department of Defense and, later, the entire aerospace industry invested millions of dollars in attempts to train pilots safely and less expensively than in actual airplanes. For the most part, these attempts have been successful (Orlansky, Dahlman, Hammon, Metzko, Taylor, and Youngblut, 1994; Orlansky and String, 1977).

Computer-based simulation has also been used for training maintenance tasks (Orlansky and String, 1981), driving tasks (Brock, 1990; Triggs and Drummond, 1991), and many military and aerospace systems activities (Orlansky, et al., 1994). Until very recently, simulation for training was seen as an attempt to replicate real-world conditions and provide for learning and practice opportunities without real-world consequences (e.g., falling out of the sky if the student makes a flying error).

However, the personal computer has led to simulation games that are only just now being seen as possessing more serious attributes. *SimCity, SimLife, SimAnt,* or *SimHealth* require the player to build a community, an ecosystem, or a health system. Amazingly complex, these games require the player to construct and test hypotheses, recover from failure, predict consequences, and make real-time decisions. Much more complex simulations can be found on the Internet, although rarely with well-defined instructional objectives. Games, after all, are intended to be fun.

Computer-based simulations were originally limited to linear activities and rules (e.g., the laws of aerodynamics, electron flow in a circuit card). Modern games allow their users to simulate changing the laws of physics, operating in zero gravity, or assuming the persona of another person or even an animal or plant (Turkle, 1996). Clearly, users of these games are learning *something*.

Woolf (1996) reports on combining simulations with animation and intelligent tutoring are combined to produce realistic problems for the learners. She reports on the Cardiac Tutor, developed at the University of Massachusetts, which provides video windows which look and act like a patient, and also provides realistic vital sign readings (EKG, blood pressure). The student can also request specific procedures and elaboration of the provided information.

17.3.4 Games

The use of games for instructional purposes is not new to CBI (e.g., Malone, 1980). Games that structure content and problems so that they are fun have been a part of education and training since the first competitive spelling bee. However, the entertainment value of computers cannot be overstated.

Games can lack the physical fidelity of simulation. However, there is evidence that games can produce processes in the students that will lead to improved performance in the real world. Gopher, Weil, and Bareket (1994) describe a spaceship game originally developed by Mane and Dorchin (1989) which improved actual flight performance. Brock (1996) and Blank (1995) report on the design of a driving game that is intended to teach young drivers to recognize and counter risk. In all cases, the game like qualities include rapid feedback on performance, accumulation of points, and positive game consequences from doing well.

17.4 DESIGN OF COMPUTER-BASED INSTRUCTION

Just putting instruction into a computer is insufficient to produce improved or even any learning. Just as the intelligent training systems discussed above have certain essential components, so does all CBI. First, of course, is the content. Swezey and Llaneras (1997) describe several models which one could follow to ensure correct content. Second is the way the program interacts with the student: what does the student see and hear and how does he or she respond to or initiate activity with the program? The third component is the actual delivery medium: the display surface (computer monitor or television screen) and the pipeline (telephone line, cable t.v., software from a store). The fourth component is the explicit instructional techniques embedded in the training program. The remainder of this chapter will address these latter issues by describing instructional transactions, interface design, and specific technological applications.

17.4.1 Instructional Transactions

Grabowski (1995) proposes that much CBI follows an instructional pattern that can become, in her words, "deadly." There are no guarantees that just because some instruction is delivered by computer it will automatically be interesting, much less engaging. Merrill (1994) has proposed that the developers of computer-based instruction should build their designs around transactions between the system and the student. Grabowski (1995) has listed three conditions required in a CBI transaction: mental activity by the learner, some level of response from the learner, and a computer response that is unique to the learner's activities in the first part of the transaction.

Merrill (1987, 1995) proposes that the primary function of instruction is to promote and guide the active metal processing of the student. This is a level of computer activity that goes beyond the typical expository presentation of material. Merrill (1987,1995) also proposes that some kind of task performance by the learner is critical to learning. If the real tasks cannot be presented, than the instruction needs to represent it in some way. He is essentially suggesting that CBI designers should think about student/system transactions rather than a series of instructional frames. He stresses that an instructional transaction is a dynamic interaction between a program and a student that includes an *exchange* of information.

Mental activity is not a term that lends itself to operational definition. Anderson and Meyer (1988) have suggested that simply attracting attention would be sufficient to constitute a communication interaction. Grabowski (1995) argues that such techniques as asking a learner to think about an example, note similarities or differences, summarize, create an analogy, or rearrange information would lead to a higher level of activity and, therefore, more learning.

The response that the system demands from the learner must be congruent with the activity and must also measure how much a student is learning. Designers must carefully decide what responses from the student will lead the system to infer that the learning of the correct material is occurring or has occurred. If the subject matter is arithmetic, for instance, this could become a matter of deciding how many problems of a particular kind the student will have to correctly answer.

If the subject is too complex to be measured by that kind of direct sampling, then the inference system must be more carefully designed. As Brock (1994) has pointed out in the context of mediated driver training, inferring a level of skill performance from measuring knowledge is very difficult. Grabowski (1995) correctly points out that a system can ask a learner to engage in all kinds of higher level problem solving with only a simple key press as a student response. This is a very simple overt response that allows the computer program to infer covert activity. This is not to suggest that this approach is incorrect, only that it needs to be explicitly built into the CBI design process.

The computer must respond to complete the transaction. It must recognize the keystroke or mouse arrow location as providing information about the state of learning of the student. It must make the inferences required in the design and then provide to the student the correct information, instructions, graphics, and scoring based upon that inference. The response must, at a minimum, inform the student as to the correctness of the student's actions. However, an interaction rich in information is to be desired (e.g., why the student is incorrect, new material to guide the student to a correct action, positive comments if the student actions were correct, affective context guiding the student to the next instructional event). It is the design of the actual interface between the student and the computer that determines that richness.

17.4.2 Interface Design

Screen and interface design should be concurrent with the overall instructional development process of a computer-based instructional program (Jones, 1993; Jones and Okey, 1995). There are several research studies that provide the characteristics of a good interface for CBI. These characteristics will be described in the following paragraphs. For a more detailed and specific discussion of this area, the reader is referred to Jones (1989).

17.4.2.1 Concepts of User Interface Design

Browsing

Students should be able to reach material in the program through a variety of controls. They could select a specific topic, a specifically identified word or phrase within a document (hypertext), or randomly search through the program. However, browsing should not be indiscriminate or uncontrolled (Jones, 1989; Laurel, Oren, and Don, 1992). A key to an efficient browsing program for CBI is that students need to be able to explore the program for new information, know when they have found it, know where they found it, and be able to find it again (Jones and Okey, 1995).

Changes in State

Students often approach CBI with real concerns about making mistakes or doing harm to the program or even to the computer. It is important that CBI be designed so that visual cues give the students information that a particular event or action is occurring (Nicol, 1990). Animation, movement of text or pictures, or specific icons can all provide information to the student that the program is changing in state. Screen wipes to the left and right can be used to show the student that he or she is moving to a new screen. Zooms, dissolves, and fades can signal a new topic. Such visual dynamics both give the learner information and reassure him or her that the activity on the screen is predictable.

Closure

Jones (1989) describes closure as having two aspects. The first is the organization of instructional information into manageable segments; students should not be overwhelmed by the amount of information in a program. The second aspect is to ensure that students know that a particular segment has been selected or completed. Closure helps students organize information and provides a specific marker of progress in what may seem to be an infinite process (Jones and Okey, 1995).

Interactive Tools for Interactive Tasks

Reingold (1990) reported that the controls a student has for a particular CBI program should be consistent with the particular instructional content and approach. If the program is about driving a car, then the control mechanisms on the screen should look like a dashboard in an automobile.

Interface Consistency

Laurel, Oren, and Don (1992) have found that whatever the interface to a CBI program is, it should be consistent across the entire program. This means that no matter where in the program the student may be, he or she can stop, return, move around, or reach supplementary materials using the same keystroke or mouse commands. The emphasis in any CBI program should be learning the content of the program, not learning how to use the CBI.

Media Integration

Media integration deals with allowing students to search for and retrieve instructional content across various media (Laurel, Oren, and Don, 1992). An arithmetic program might have elements of graphics, still or moving video, sound, and more traditional text and numbers. The student should be able to move among these various media with little effort.

Media Biases

Laurel, Oren, and Don (1992) found evidence that some media are more credible than others. Text is seen as most credible; video re-creations of events are seen as less credible. Students also expect to learn different material from various media options; video or audio should contain original instruction, not just a rehash of text material.

Metaphors

Metaphors can significantly reduce the time needed for the student to learn how to use the computer-based instructional program. Designers can use known functions or processes to show the students how the program works. Jones (1989) gives as examples such things as books, bookshelves, space exploration, or buildings with different rooms to help the students organize and access the program's content. Designers should ensure that selected metaphors are, in fact, meaningful to the target student population.

Progressive Disclosure

Progressive disclosure is a general design guideline for human–computer interface (HCI). In the context of CBI, it refers specifically to keeping information within the CBI environment presented to the student in small, manageable segments (Jones, 1989). This requires that the system can track the student's progress and therefore correctly control the instructional content and pace.

17.4.3 Learner-Centered Design

Central to learner-centered design of instructional programs is the idea that people learn best when they are engrossed in a topic and motivated to seek out new knowledge and skills because they need them to solve a problem (Norman and Spohrer, 1996). This shift in the design of CBI is away from linear, designer-controlled chunks of instruction. It emphasizes, instead, the unique needs of each learner, and to a very large extent it allows the learner to drive the instructional process. As Norman and Spohrer (1996) point out, it is the latest computer technologies which provide the instructional designer the power and flexibility to design such a system.

In their overview article on learner-centered design for *Communications of the ACM*, Norman and Spoher (1996) found three dimensions against which to evaluate learner-centered design programs: (1) engagement, (2) effectiveness, and (3) viability. They then find that most studies of learner-centered design reveal that engagement of the learners is a major strength of the programs. Data on effectiveness were limited to opinions of students and teachers. They also conclude that the long-term benefits of learner-centered programs (viability), because of the limited applications of the technique so far, are not known.

17.4.3.1 Construction Tool Kits

Computer-based instruction has always been most effective at teaching the use of computers. First, it is a domain that is typically well understood by the designer of the instruction. Second, one can accurately simulate computer activities on the computer. One form of learner-centered design has led to programs in which students actually design some kind of computer program and, in the process, learn both programming and the content of the to-be-designed instruction. For instance, Kafai (1996) describes a project in which 9 and 10 year olds designed instructional games to teach algebra. The students learned how to program the computers, but also demonstrated mastery of simple algebra.

In another example of this approach, Eden, Eisenberg, Fischer, and Repenning (1996) report on a series of studies with graduate and undergraduates in the computer sciences program at the University of Colorado. The students must develop programs for specific applications, thereby learning both the programming and the application. The authors report that the engagement of the students in choosing tasks and goals for the projects and the freedom the students have over the entire process leads to enthusiasm (they use the word, "passion") which they infer produces better learning.

17.4.3.2 Scaffolding Techniques

Scaffolding is an educational term that refers to providing support to learners while they engage in activities that are normally beyond their abilities (Jackson, Stratford, Krajcik, and Soloway, 1996). Jackson, et al. (1996) report success at teaching college level material to high school students using this technique.

Kolodner, Hmelo, Narayanan, Carlson, Rappin, Hubscher, Turns, and Newsletter (1996) report similar success combining computer-based scaffolding with collaboration among the members of problem-solving teams. The approach begins with the assumption that collaboration among students can lead to solving more interesting and complex problems than can an individual working alone. The computer's roles in this learning activity is to both structure the problem and keep the records of the problem-solving activity. Students record facts, ideas, or hypotheses, what needs to be learned, and action plans. This particular program, called Multiple Case-Based Approach to Generative Environments for Learning (McBagel), also provides tools to support the problem solving as needed (e.g., case histories, spreadsheets, relevant Web pages).

The scaffolding metaphor is apt. As the students become more proficient, the computer provides less and less support. Although this is cited as an example of learner-centered design, the design is based upon the *performance* of the students, rather than on whim or particular likes or dislikes. Rosson and Carroll (1996) use a similar approach to the teaching of object-oriented programming. The scaffolding examples begin with sample problems that can be solved by the beginning student. Problem complexity is gradually revealed in steps that leverage and reinforce not only the specific problem solution, but the intrinsic structure of the problem-solving process.

Jackson, Stratford, Krajcik, and Soloway (1996) describe a scaffolding approach that has allowed high school students to learn college level science concepts by using a tool called Model-It. Essentially, the tools allow the students to begin with concepts and relationships they know (e.g., the relationship between water pollution and river plant life) and to construct models of more complexity (in the example, an ecosystem). Again, the complexity of the problem increases and the computer support decreases based upon the performance of the student. The authors report on more than 1200 student-hours of Model-It usage data. Although quantitative performance data are not reported, they do show that the students create reasonable models of complex systems within short amounts of time. The authors also believe that the students learn not just about the systems they model, but what it means to model, as well.

17.5 SPECIFIC TECHNOLOGICAL APPLICATIONS

In 1983, R. E. Clark said, "The best current evidence is that media are mere vehicles that deliver instruction but do not influence student achievement any more that the truck that delivers our groceries causes changes in our nutrition." (p. 445). The evidence reported above suggests that Clark may not be correct when the media under investigation exploits the possibilities of the computer. Nonetheless, just as improved highways have enabled more people to enjoy the nutrition of the groceries carried in Clark's trucks, computers are now delivering instruction world wide. This section of the chapter will discuss CBI delivery methods studied in some detail. Although the specific technologies may someday be replaced, what has been learned about their capabilities and limitations should provide general guidance to the developer of future CBI systems.

17.5.1 Interactive Multimedia

17.5.1.1 Definitions

Multimedia has become a term of such common usage that it should be defined as narrowly as possible for this chapter. Any combination of media is multimedia: a motion picture with sound, a film strip with an audio cassette, or a computer with a sound card. Within the computer industry, it has come to mean a system with a CD-ROM capability. Heinich, Molenda, and Russell (1993) provide the following definitions:

> *Multimedia—sequential or simultaneous use of a variety of media formats in a given presentation or self study program.*
> *Multimedia system—A combination of audio and visual media integrated into a structured, systematic presentation.*
> *Computer Multimedia system—A computer hardware and software system for the composition and display of presentations that incorporate text, audio, and still and motion images.*

17.5.1.2 Interactive Videodisc

Most of the research of multimedia learning systems has focused on Interactive Videodisc (IVD), since it is the oldest of the technologies capable of providing complete interactive, multimedia instruction. As these words are written, the computer press is predicting major changes to the traditional videodisc format before the end of 1997. What will change, however, are elements like size, storage capability, speed, and price. From the end-user's point of view, any changes will be imperceptible: He or she will be viewing standardized and individualized instruction composed of random access to high quality still and motion video, high fidelity audio, overlaying graphics, and text or hypertext.

The University of Nebraska Videodisc Design and Production Group has developed a scheme for describing videodisc capabilities by levels. The U.S. Department of Defense, a major supplier of funds for IVD research and development, adopted the same scheme (Fletcher, 1990). The same levels could be applied to any technology capable of providing IVD standards of instruction. The levels are:

Level 0 A videodisc system intended for linear play without interruption. This kind of system is typically found for entertainment purposes or for high resolution video presentations. A linear videotape would be an example of Level 0 multimedia.

Level I A videodisc system with still/freeze frame, picture stop, frame and chapter search, dual channel audio, but no programmable memory. All functions are intended to be initiated by manual inputs from the student's keypad.

Level II A videodisc system with on-board, programmable memory. The videodisc player's memory is programmed by digital dumps from audio channel two of the videodisc or by manual entry from the videodisc player's keypad. Inputs are made from the keypad or from a device that emulated the keypad.

Level III A videodisc system in which the videodisc player is interfaced to an external computer. The videodisc player acts as a computer peripheral with its functions under the computer's control.

Level IV A videodisc system in which the videodisc is interfaced to an external computer. The videodisc functions both as an optical storage device for digital information and as the source of analog picture and sound. The video frames on the videodisc store digital data intended to be read and processed by the computer.

From the student's point of view, these are distinctions without a difference. Whereas these levels may be important to the software designers, the student is aware only of how well the program provides an opportunity to learn. If a major advantage of CBI is its ability to distribute instruction, interactive videodisc has failed. It is not accessible to most users of computers or even to users of computer-based instruction.

However, in its applications, IVD has been shown as an effective method for teaching. The future of multimedia lies in combining the technological capabilities of IVD with a technology that can be widely and inexpensively distributed. CD-ROM, a popular entertainment and information storage device that runs on personal computers, may be that technology. So may some future component of the Internet. It may be the next generation videodisc, rumored to play on a normal home computer. Or it may be some technology being developed in a garage which will change the face of computing as much as the personal computer did twenty years ago. Whatever the eventual technology, the lessons learned from the IVD studies will provide the model for the *instructional* components.

Fletcher and Rockway (1986) identified five new instructional applications for the Department of Defense IVD. Four of these five still apply, but can be expanded to apply to any multimedia instructional strategy.

Surrogate Travel

In the simplest terms, the videodisc provides a view of a particular place (city, terrain, environment) and the student can navigate through it. The student controls the sequence of views through a series of choices (turn right, go up, stop, veer to the left). The typical control for this kind of application is a joystick. The key difference between a surrogate travel IVD and something like Flight Simulator ©Microsoft , is that the latter is produced through sophisticated graphics rather than through actual video images. One example for use could include preparing a military team for a specific operation by having them rehearse walking or driving through a videodisc model of an actual area. Another use would have experienced drivers make decisions in unique and controlled environments. In each case, the instruction would not be in the psychomotor or perceptual skills of driving or walking, but rather the decisions one must make under those circumstances. King and Reeves (1985) report some success with this technique in training military operations in urban terrain; similar programs for experienced truck drivers (Blank, 1995; Brock, 1996) and novice automobile drivers (Brock, 1996; Lonero, Clinton, Brock, Wilde, Laurie, and Black, 1995) either have been developed or are currently under development.

Interactive Movies

This technique allows the student to function as the director. He or she decides when it is time for close up, when it is time for slow motion or reverse, and whether instruction should be abstract (e.g., line drawings) or concrete (the actual system, thing, or event). The essential element of this approach is how much control the student has over the way the material is presented.

Microtravel

Such a program will allow a student to visit places a person cannot normally go. One early Army demonstration allowed the student to take a tour of a Jeep engine while it was running (Fletcher, 1990). Microtravel could allow a student to interactively travel the human arterial system or the piping system of a nuclear power plant. So far, instances of this kind of multimedia usage have tended to be linear and noninteractive; they have also tended to be more entertaining (or the unfortunate phrase, edutaining) than pedagogical.

Low-Cost Portable Simulation

One major advantage of combining high video quality with the power of advanced computers is that specific objects and events can be realistically simulated. For instance, an object as small as a voltmeter or as large as an airplane can be represented photographically. Students can access either and, in fact, could use the simulated voltmeter to make primitive electronic checks on the represented airplane—all in the two-dimensional space of the multimedia presentation. Additionally, if a student makes a mistake, the system allows for pedagogical interventions to shape performance. As computers become more powerful, less costly, and even move pervasive, the ability to merge sound instruction with visually realistic representations of the world provides huge opportunities for CBI applications

Effectiveness

Interactive Videodisc has been a popular instructional medium for study. It is an early application of multimedia (sound, video, graphics, motion, interactivity) and has been shown to be effective as measured against more traditional instruction. It has also been shown to be effective in many different domains ranging from electronic maintenance (e.g., Spencer, 1983; Williams and Harold, 1985), CPR instruction (Lyness, 1987), biology (Bunderson, Baillio, Olsen, Lipson, and Fisher, 1984), and physics (Stevens, 1984) to interpersonal skills development (Schroeder, Dyer, Czerny, Youngling, and Gillotti, 1986), and foreign language instruction (Crotty, 1984; Verano, 1987).

In his comprehensive review, Fletcher (1990) concluded that IVD improved student achievement of about 0.50 standard deviations over more conventional instruction. He found that it was a more effective instructional technique for teaching both knowledge and skills. Surprisingly, his study also concluded that IVD was less costly than more traditional methods, albeit in a military setting. Student acceptance was also high.

One enigma about the videodisc (including interactive videodiscs) is that its technological superiority has never been reflected in the marketplace. The videotape cassette recorder has become the home entertainment standard, although the videodisc provides superior picture quality. Now it appears that IVD is about to be surpassed by the Compact Disc Read Only Memory (CD-ROM).

17.5.1.3 CD-ROM

The CD-ROM stems from the audio compact disc; its physical dimensions and characteristics are the same. Unlike the audio disc, the CD-ROM can contain audio, computer, and video data. A single disc can contain more than 650 megabytes of user data. This capacity is equivalent to 200,000 pages of *printed* text. Video storage, even using sophisticated compression techniques, can consume large amounts of disc real estate. The speed at which data can be retrieved from a CD-ROM is increasing, as is the number of personal computers equipped with CD-ROM drives.

Until very recently, the CD-ROM provided a superior storage medium for text and still photographs. Motion video was grainy and jerky and the interactivity found in IVD was not possible. That is changing. Personal computers are currently being sold which provide the hardware and software necessary to achieve the same technological benefits one finds with IVD.

There is also a CD-I (compact disc, interactive), which is a multimedia link to a conventional television set. It has failed to develop in the marketplace but also provides many capabilities of the IVD.

The minimum requirements for a CD-ROM to deliver the capabilities of a fully interactive multimedia system are defined by the computer for playback, rather than the disc itself. The Multimedia PC Marketing Council has set the following minimum standards:

Intel Pentium 75MHz CPU
8 MB RAM
Graphics support for 65,536 colors at 640 X 480 pixel resolution
MPEG1 digital video playback with scaling
Audio support for 16 bit MSADPCM digital sound
4x CD-ROM
530 MB Hard Drive
Mouse
Microsoft Windows © 3.11

It remains to be seen whether the CD-ROM, or some future version of it, becomes the standard for multimedia instruction. One major disadvantage of CD-ROM is that programming it to have the flexibility and video quality of IVD is arduous and exceedingly difficult (Blank and McCord, 1996).

17.5.2 Distance Learning

The U.S. Department of Education defines distance learning as follows:

> *The application of telecommunications and electronic devices which enable students and learners to receive instruction that originates from some distant location. Typ-*

ically, the learner is given the capacity to interact with the instructor or program directly and given the opportunity to meet with the instructor or a periodic basis.

17.5.2.1 Traditional Distance Learning

Distance learning has considerable history in American education. Correspondence courses using traditional mail have existed almost since the beginning of reliable mail service. Television and radio have been used for a variety of distance learning schemes. In a sense, even sending out films to classrooms represented distance learning—the instructor, demonstrator, or even cartoon character took over, although briefly, for the classroom teacher. There were rare instances, before satellites were highly developed, where courses were beamed to classrooms from a high flying airplane (NSTA, 1990). Having lecturers in one place and remote students viewing the lecture over closed-circuit television is now commonplace for universities. In more advanced settings, the students can interact with the lecturer using two-way communications.

17.5.2.2 Criteria

The National Science Teachers Association has established a set of criteria for the teaching of science by distance learning. The criteria that it sets out should serve well beyond the science curriculum.

Interaction

It is critical that the instructor and the learners can continuously adjust the learning situation. This is true whether the instructor is a remote person or a computer. Development of critical thinking skills requires constant feedback in both directions. Questions have to be asked and answered; problems need to be solved. Meaningful interaction among the learners is also key.

Flexibility

Instructional design must be flexible enough to allow for student and instructor to adjust any particular course of instruction. Each learner and learning situation is unique; flexibility must be maintained to adapt to individual and local differences.

Manipulative Experiences

Distance learning must allow for student activity beyond listening and taking notes. Hands-on activities must be part of the curriculum.

Competency

The teacher (human or computer) must be competent in the subject matter. Competence must go beyond the ability to deliver information. It includes conceptual knowledge of the subject, problem-solving skills, and appropriate pedagogical background. Learners must have ready access to this level of competence.

Variety of Appropriate Resources

The teacher or program can adapt to the variety of learning conditions in any local situation. Instructions must have relevance to the individual lives of the students. When distance learning is used, the source of instruction should give learners supplementary resource materials specifically designed in print or other appropriate format to support distance learning content.

Appropriate Technology

Technology should be selected for how well it will serve the needs of distance learning. The use of technology for its own sake is never justified. No single technology will likely meet all the needs of a curriculum.

Evaluation

As in any instructional program, evaluation must be an ongoing process. Program effectiveness and student achievement must be measured. Both formative and summative kinds of evaluation should be used to guide the improvement of instruction.

17.5.2.3 Distance Learning Trends

Distance learning was originally coined to refer to a television or radio connection between remote students and an instructor in a studio or classroom equipped with cameras and microphones. It is expanding worldwide to include computer links between a person or program and students. Students download or chat over Internet lines. One provider of Internet training claims 50,000 students taking computer and general interest courses.

It is too early to tell if the criteria set out above can be met by a completely automated distance CBI system. In the meantime, the convenience of learning in one's home or office when the time is available holds great promise the distance learning model.

17.5.2.4 Just-in-Time Instruction

A major advantage of distance learning is that instructional material can be delivered when it is needed by the learner. The notion comes from the just-in-time delivery methods pioneered by the Japanese and now standard in much of the world. Just as bumpers are delivered to the car assembly plant just-in-time to be installed on the cars coming off the assembly line, instructional materials are delivered to the learner when he or she needs them.

Perelman (1994), in a general interest article, provides several examples of this approach in industry. One computer company would bring all of its salespersons to a central location for training on new products. In 1990, they switched to downloading the instruction so that each salesperson could reach the material when he or she needed it. The company estimated that it saved nearly $5,000,000 in the first year just in travel and housing costs. The salespeople were happier because they did not have to leave their families or their customers. Learning improved because the learners could select the relevant instruction when it was needed.

17.6 THE FUTURE OF CBI

The Internet, this great and chaotic linkage of computers world wide, will not go away. How orderly it will become is still in doubt. To refer to the first point in this chapter, one direction CBI is going is toward the ability to deliver *bad* instruction more places and more quickly. One of the great appeals of the Internet and the World Wide Web is its chaos. No one is in charge.

The president of Oracle, a major computer software company, predicts a $500.00 computer in the near future which will provide full access to the Internet, presumably including training programs (ASAP, 1996). In the interview where the prediction was made, he also notes that the United States spends ten times as much money on education as we do on movies.

One major software and hardware manufacturer is developing a "school system," which will include physical locations with advanced CBI and simulator systems, but will also include a library of computer-based instructional programs available over the Internet. Users will lease the training over the net before and after they attend the physical school. Potential customers include both large and small employers (who can avoid major capital expenditures), individuals interested in a new career or advancing in their current one, and other schools and training institutions who would lease selected CBI on an as-needed basis.

The early question about CBI was *whether* to select it as an instructional medium. The research of the past decade has focused on *how* to most effective use CBI. The questions for the next decade will be *when* to use it and *where.* It will be the learner making those decisions.

REFERENCES

Anderson, J. R. (1983). *The Architecture of Cognition*. Cambridge, MA: Harvard University Press.

Anderson, J. A., and Meyer, T. P. (1988). *Mediated Communication: A Social Action Perspective.* Newbury Park, NJ. Sage Publications.

ASAP (1996). Larry Ellison samurai interview, *Forbes ASAP,* April, 54–55.

Blank, D. (1995). Personal Communication, February.

Blank, D., and McCord, R. (1996). Personal Communication, April.

Bloom, G. S. (1984). The 2 sigma problem: The search for methods of group instruction as effective as one-to-one instruction. In H. J. O'Neil, Ed., *Computer-Based Instruction: A State of the Art Assessment.* Toronto: Academic Press.

Brock, J. F. (1994). Instructional technology: A review. The AAA Foundation for Traffic Safety, Washington, D.C.

Brock, J. F. (1996). Simulation in training commercial drivers. Paper Presented at the Second Annual Meeting of the Driver Training and Development Alliance, April 18 and 19.

Brock, J. F. (1990). *Use of simulator technologies for training/test commercial drivers. Phase I, Task 1: Analyze current simulator application.* Task Memorandum Report, Federal Highway Administration, Washington, D.C.

Brown, J. S., Burton, R. R., and Bell, A. G. (1975). SOPHIE: A step toward creating a reactive leaning environment. *International Journal of Man-Machine Studies, 7*, 675–696.

Brown, J. S., and Burton, R. R. (1978). Diagnostic models for procedural bugs in basic mathematical skills. *Cognitive Skills, 2*, 155–192.

Bunderson, C. V., Baillio, B., Olsen, J. B., Lipson, J.I., and Fisher, K. M. (1984). Instructional effectiveness of an intelligent videodisc in biology. *Machine-Mediated Learning, 1*, 175–215.

Burton, R. R., and Brown, J. S. (1979). An investigation of computer coaching for informal learning activities. *International Journal of Man-Machine Studies, 11*, 5.24.

Carr, B., and Goldstein, I. (1977). *Overlays: A theory of modeling for computer-aided instruction.* Massachusetts Institute of Technology, Artificial Intelligence Laboratory, Cambridge, MA.

Clark, R. E. (1983). Reconsidering research on learning from media, *Review of Educational Research. 53*, 445–459.

Crotty, J. M. (1984). *Instruction via an intelligent videodisc system versus classroom instruction for beginning college French student: A comparative experiment. AFIT/CI/NR 84-51D.* Air Force Institute of Technology, Wright-Patterson, AFB, OH.

Eberts, R. E., and Brock, J. F. (1984). Computer applications to instruction: Issues and principles for design, 1984. *First Annual Review of Human Factors, 1984.* F. A. Muckler, Ed. Santa Monica, CA: Human Factors Society, pp. 239–284.

Eberts, R. E., and Brock, J. F. (1987). Computer assisted and computer managed instruction. In G. Salvendy, Ed., *Handbook of Human Factors.* New York: John Wiley.

Eden, H., Eisenberg, M., Fischer, G., and Repenning, A. (1996). Making learning a part of life. *Communications of the ACM, 39(4)*, (April 40–42).

Fletcher, J. D. (1990). *Effectiveness and cost of interactive videodisc instruction in defense training and education.* IDA PAPER P-2372. Institute for Defense Analysis, Alexandria, VA., July.

Fletcher, J. D. (1995). *Does this stuff work? Some findings from applications of technology to education and training.* Institute for Defense Analysis, Alexandria, VA.

Fletcher, J. D., and Rockway, M. R. (1986). Computer-based training in the military. In J. A. Ellis, Ed., *Military Contributions to Instructional Technology.* New York: Praeger Publishers.

Glass, G. V. (1976). Primary, secondary, and meta-analysis of research. *Educational Researcher, 5*, 3–8.

Goldstein, I. P., and Papert, S. (1977). Artificial intelligence, language, and the study of knowledge. *Cognitive Science, 1*, 1–21.

Gopher D, Weil, M., and Bareket, T. (1994). Transfer of skill from a computer game trainer to flight. *Human Factors, 36 (3)*, 387–405.

Grabowski, B. (1995). Instructional transactions in CBI—A discussion and application of Merrill's definition, Internet Instructional Technology Forum, (http:⁄29.7.160.78/Documents/T.Transactions/Transactions.html) June 13.

Guzdial, M., Klodner, J., Hmelo, C., Narayanan, H., Carlson, D., Rappin, N., Hubscher, R., Turns, J., and Newsletter, W. (1996). Computer support for learning through complex problem solving. *Communications of the ACM, 39 (4)*, April 43–45.

Jackson, S. L., Stratford, S. J., Krajcik, J., and Soloway, E. (1996). A learner-centered tool for students building models. *Communications of the ACM, 39 (4)*, April 48–49.

Johnston, B. B., and Fletcher, J. D. (1995). *Effectiveness of Computer Based Instruction in Miliary Training.* Institute for Defense Analysis, Alexandria, VA.

Jones, M.G., and Okey, J. R. (1995). Interface design for computer-based learning environments, Instructional Design Forum (http:⁄129.7.160.78/Documents/IVDGUIDE/IVDGUIDE.html) February 21.

Jones, M. G. (1993). *Guidelines for screen design and user-interface design in computer-based learning environments.* Dissertation Abstracts International, *54 (9)*, 308a–309a.

Jones, M. K. (1989). *Human-computer interaction: A design guide.* Englewood Cliffs, NJ: Educational Technology Publications.

Kafai, Y. B. (1996). Software by kids for kids. *Communications of the ACM, 39 (4)*, April 38–39.

Keller, F. S. (1968). Good-by teacher . . . *Journal of Applied Behavioral Analysis, 1*, 79–89.

Kulik, J. A. (1994). Meta-analytic studies of findings on computer-based instruction. In *Technology Assessment in Education and Training*, E. L. Baker and H. F. O'Neil Jr., Eds., Hillsdale, NJ: Lawrence Erlbaum Associates, Publishers.

Kulik, C-L. C., and Kulik, J. A. (1991). Effectiveness of computer-based instruction: An updated analysis. *Computers in Human Behavior, 7*, 75–94.

Laurel, B., Oren, T., and Don, A. (1992). Issues in multimedia design: Media integration and interface agents. In M. M. Blattner and R. B. Dannenberg, Eds., *Multimedia Interface Design.* New York: ACM Press, pp. 53–64.

Lonero, L., Clinton, K., Brock, J., Wilde, G., Laurie, I., and Black, D. (1995). *Novice Driver Education Model Curriculum Outline*. AAA Foundation for Traffic Safety, Washington, D.C., March.

Lyness, A. L. (1987). Performance and norms of time for adult learners instructed in CPR by an interactive videodisc system. Paper presented at the Fifth Annual Conference on Research in Nursing Education, San Francisco, CA. ERIC No. ED 281 986.

Mandl, H., and Lesgold, A. (1988). *Learning Issues For Intelligent Tutoring Systems*. New York: Springer-Verlag.

Mane, A., and Donchin, E. (1989). The Space Fortress game. *Acta Psychologica, 71*, 17–22.

Merrill, M. D. (1987). Prescriptions for an authoring system. *Journal of Computer-Based Instruction, 14 (1)*, 1–8.

Merrill, M. D. (1994). *Instructional design theory*. Englewood Cliffs, NJ: Educational Technology Publications.

Nicol, A. (1990). Interfaces for learning: What do good teachers know that machines don't? In *The Art of Human-Computer Interface Design*. Maidenhead Birkshire: Pergammon Infotech Limited, pp. 113–123.

Noja, G. P. (1987). New frontiers for computer-aided training. In R. J. Seidel and P. D. Weddle, Eds., *Computer-Based Instruction in Military Environments*. New York: Plenum Press.

Noja, G. P. (1991). DVI and system integration: A further step in ICAP/IMS technology. In R. J. Seidel and P. R. Chatelier, Eds., *Advanced Technologies Applied to Training Design*. New York: Plenum Press.

Norman, D. A., and Spohrer, J.C. (1996). Learner-centered education. *Communications of the ACM, 39 (4)*, 24–27.

NSTA (1990). *National Science Teacher's Association criteria for effective distance learning*, Arlington, VA.

Orlansky, J. and String, J. (1977). *Cost effectiveness of computer-based instruction in military training*, (IDA PAPER P-1375) (Institute for Defense Analysis, Arlington, VA

Orlansky, J., Dahlman, C. J., Hammon, C. P., Metzko, J., Taylor, H. L., and Youngblut, C. (1994). *The value of simulation for training*. IDA PAPER P-2982. Institute for Defense Analysis, Alexandria, VA.

Perelman, L. J. (1994). Kanban to Kanbrain, *Forbes ASAP*, June, 85–95.

Pressey, S. L. (1950). Development and appraisal of devices providing immediate automatic scoring of objective tests and concomitant self-instruction. *Journal of Psychology, 29*, 417–447.

Psotka, J., Massey, L. D., and Mutter, S. A. (1988). *Intelligent tutoring systems*. Hillsdale, NJ: Lawrence Erlbaum Associates Publishers.

Reingold, H. (1990). An interview with Don Norman. In *The Art of Human-Computer Interface Design*. Maidenhead Birkshire: Pergammon Infotech Limited, pp. 113–123.

Rosson, M. B., and Carroll, J. M. (1996). Scaffolded examples for learning object-oriented design. *Communications of the ACM, 39 (4)*, 46–47.

Salvendy, G. (1987). *Handbook of Human Factors*. New York: John Wiley.

Schmidt, M., Weinstein, T., Niemiec, R., and Walberg, H. J. (1985). Computer-assisted instruction with exceptional children: A meta-analysis of research findings. Paper Presented at the Annual Meeting of the American Educational Research Association, Chicago.

Schroeder, J. E., Dyer, F. N., Czerny, P., Youngling, E. W., and Gillotti, D. P. (1986). *Videodisc interpersonal skills training and assessment (VISTA): Overview and findings, Volume I*. ARI Technical Report 703. U.S. Army Research Institute Field Unit, Ft. Benning, GA.

Skinner, B. F. (1954). Science of learning and the art of teaching. *Harvard Educational Review, 24*, 86–97.

Spencer, J. (1983). *IVIS Benchmark Study*. Educational Services Division, Digital Equipment Company, Bedford, MA.

Stevens, S. M. (1984). *Surrogate Laboratory Experiments: Interactive Computer/Videodisc Lessons and Their Effect on Students' Understanding of Science,* Doctoral dissertation, University of Nebraska, Lincoln, NE. University Microfilms No. 8428209.

Suppes, P., and Morningstar, M. (1969). Computer-assisted instruction: Two computer-assisted instruction programs are evaluated. *Science, 166*, 343–350.

Swezey, R. W., and Llaneras, R. E. (1997). Models in training and instruction, In G. Salvendy, Ed., *Handbook of Human Factors and Ergonomics*. New York: John Wiley.

Szabo, M., and Montgomerie, T. C. (1992). Two decades of research on computer-managed instruction. *Journal of Research on Computing in Education, 25 (4)*, Fall, 113–133.

Turkle, S. (1996). Sex, lies, and avatars. *Wired, 4.04*, April, 106–110, 158–165.

Verano, M. (1987). *Achievement and Retention of Spanish Presented Via Videodisc in Linear, Segmented, and Interactive Modes*. Doctoral dissertation, University of Texas, Austin, TX. DTIC No. AD-A185 893.

White, B. Y., and Frederiksen, J. R. (1986). Intelligent tutoring systems based upon qualitative model evolutions. *Proceedings of the Fifth National Conference on Artificial Intelligence*, 313–319.

Willett, J. B., Yamashita, J. J., and Anderson, R. D. (1983). A meta-analysis of instructional systems applied in science teaching. *Journal of Research in Science Teaching,* 20, 405–417.

Williams, K. E., and Harold, L. J. (1985). *AN/BQQ-5 interactive video disc training effectiveness evaluation final report*, Naval Training Equipment Center, Orlando, FL.

Wisher, R. A. (1987). The development and test of a hand-held computerized training aid, In R. J. Seidel and P. D. Weddle, Eds., *Computer-Based Instruction in Military Environments.* New York: Plenum Press.

Woolf, B. P. (1996). Intelligent multimedia tutoring systems. *Communications of the ACM, 39* (*4*), 30–31.

Yasutake, J. Y. (1987). Implementation of computer-based training: A system evaluation and lessons learned. In R. Seidel and P. Weddle, Eds., *Computer-Based Instruction in Military Environments.* New York: Plenum Press.

CHAPTER 18

ORGANIZATIONAL DESIGN AND MACROERGONOMICS

Hal W. Hendrick
Error Analysis, Inc.
Englewood, CO 80111 USA

18.1 INTRODUCTION

Traditionally, human factors or ergonomics has tended to focus on the design of specific jobs, work groups, and related human–machine interfaces. Typically, in system design, the operations to be required of the system to accomplish its purposes are identified. These operations, in turn, are analyzed to identify the specific functions that constitute them. Ideally, if not before, the human factors specialist comes into the design process at this point; and based on his or her professional knowledge of human performance capabilities, limitations, and other characteristics, assists in allocating these functions to humans or machines. From this point in the process, the human factors specialist's knowledge of human–machine interface technology is applied to task analysis, designing specific jobs, integrating jobs into work groups, and then designing specific human–machine interfaces, including controls, displays, work-space arrangements, and work environments. More recently, this task also has come to include ergonomic design of software.

Although applied within a systems analysis framework, most of the above described activities actually are at the individual, team, or at best, subsystem level. In short, they represent what herein shall be referred to as human factors applications at the *microergonomic* level. The focus of this chapter is on the application of human factors in system design at the *macroergonomic* or overall organizational level. Conceptually, it is entirely possible to do an outstanding job of microergonomically designing a system's components, modules, and subsystems, yet fail to reach relevant system effectiveness goals or criteria because of inattention to the *macroergonomic* design of the system (Hendrick, 1984a).

18.1.1 The Tavistock Studies

The classic example of this problem was the introduction of the *longwall* method of mining in a British deep-seam coal mine. The traditional mining system was largely manual in nature. It utilized teams of small, fairly autonomous groups of miners. Control over work was exercised by the group itself. Each miner performed a variety of tasks; and most jobs were interchangeable among workers. The workers derived considerable satisfaction out of being able to complete the entire "task," and could readily satisfy social needs on the job. Sociotechnically, the psychosocial characteristics, the characteristics of the external culture, the task requirements, and the system's organizational design were congruent. The more automated, technologically advanced longwall method replaced this costly manual method of mining. No longer restricted to working a short face of coal, miners now could extract coal from a long wall. However, this new and more *technologically* efficient system resulted an organizational design that was not congruent with the psychosocial and cultural characteristics of the work force. Rather than small groups, shifts of 10 to 20 men were required. Workers were restricted to narrowly defined tasks. Opportunities for social interaction were limited. Job rotation was not possible. There now was a high degree of interrelationship among the tasks of the three shifts; and problems from one shift carried over to the next, thus holding up labor stages in the extraction process. This complex and highly rigid work system design was very sensitive to both productivity and social disruptions. Instead of improved productivity, low production, absenteeism, and intergroup rivalry became common (DeGreene, 1973). Later, in studies of other coal mines by the Tavistock Institute of Human Relations in London (Trist, Higgin, Murray, and Pollock, 1963), this conventional longwall method was compared with a *composite* longwall method in which the system's organizational design utilized a combination of the new technology and features of the old psychosocial work structure of the manual system. In comparison with the conventional longwall method, the composite longwall work system reduced the interdependence of the shifts, increased the variety of skills utilized by each worker, and permitted self-selection by workers of their team members. Production was significantly higher than for either the conventional longwall or the old manual system. Absenteeism and other measures of poor morale dropped dramatically.

Based on the Tavistock Institute studies, Emery and Trist (1960) concluded that *different organizational designs can utilize the same technology*. The key is to select the organizational design, or subset of designs, that is most effective in terms of the (a) people who will constitute the human portion of the system, and (b) relevant external environments; and then employ the available technology in a manner that achieves congruence.

Although any given technology can utilize different organizational designs, technology, *once employed in the design of a system*, does constrain the subset of possible designs. With the introduction of computers, microelectronics, and related automation into managerial, administrative, production, logistical, marketing, and other facets of our modern complex systems, the system's organizational designs become more constrained then in more traditional labor-intensive work systems. Because of this progressively increasing automation, it increasingly has become important to first determine the optimal *macroergonomic* design of the work system before fully proceeding with the microergonomic design of human–machine modules, subsystems, and interfaces. In short, at least *conceptually,* "a top-down ergonomic approach is essential to insure that the dog wags the tail, and not vise-versa" (Hendrick, 1984a).

18.1.2 Some Basic Concepts

In this section, the major dimensions of organizational design and their use in the macroergonomic design of systems are reviewed. In order to provide a common framework, a few basic concepts are clarified first.

18.1.2.1 The Meaning of Organization

An organization may be defined as "*the planed coordination of two or more people who, functioning on a relatively continuous basis and through division of labor and a hierarchy of authority, seek to achieve a common goal or set of goals*" (Robbins, 1983, p. 5). Breaking this definition down, we can note the following.

1. Planned coordination of collective activities implies *management*. Activities do not just emerge but are premeditated. Implicit in planned coordination in complex, high-technology organizations is a need for information and decision support systems to facilitate management of the system. The design of these support systems must be compatible with the organization's design.

2. Because organizations are made up of more than one person, individual activities must be designed and functionally allocated so as to be complementary, balanced, harmonized, and integrated to ensure an effective, efficient, and relatively economical functioning work *system.*
3. Organizations accomplish their activities and functions through a division of labor and a hierarchy of authority. Thus, organizations have *structure*. How this structure is designed is likely to be critical to an organization's functioning.
4. The collective activities and functions of an organization are oriented toward achieving a common goal or set of goals. From a system design standpoint, this implies that criteria for assessing the effectiveness of an organization's design exist and should be explicitly identified, weighted, and utilized in evaluating feasible alternative designs for the overall organization.

18.1.2.2 Organizational Structure

We noted in subparagraph 3 above that the concept of organization, with its division of labor and hierarchy of authority, implies *structure.* An organization's structure may be conceptualized as having three major components: *Complexity, formalization,* and *centralization* (Robbins, 1983).

1. *Complexity* refers to the degree to which organizational activities are differentiated, and the extent to which integrating mechanisms are utilized to coordinate and facilitate the functioning of the differentiated components.
2. *Formalization* is the degree to which an organization relies on rules and procedures to direct the behavior of the people within a given work system.
3. *Centralization* refers to the extent to which the locus of decision-making authority is either centralized or dispersed within the work system hierarchy.

The nature of each of these three components, including operational measures and design guidelines for each, is discussed below.

18.1.2.3 Organizational Design

Organizational design refers to the design of an organization's work system structure and related processes to achieve the organization's goals. From a macroergonomics standpoint, as part of the system design process, organizational design involves (1) identifying the system's purpose or goals; (2) making explicit the relevant measures of organizational effectiveness, weighting them, and subsequently utilizing these OE measures as criteria for evaluating feasible alternative structures; (3) systematically developing the design of the three major components of organizational structure; (4) systematically considering the system's *technology, personnel,* and *relevant external environment* variables as moderators of organizational structure; and (5) deciding on the general *type* of organizational structure for the system. These five "steps" of the organizational design process are considered in turn and constitute a major portion of this chapter. Although presented as though this is a sequential linear process, in reality, these five steps usually are carried out in an overlapping, iterative, and nonlinear manner.

18.2 ORGANIZATIONAL GOALS AND EFFECTIVENESS CRITERIA

18.2.1 Goals Classification Schemes

Goals may be defined as the *objectives* of the organization. They are the desired states of affairs that organizations are designed to achieve. Goals may be defined by *criteria, focus,* and *time frame* (Szilagyi and Wallace, 1990).

18.2.1.1 Classification by Criteria

The following six criteria are among the most frequently used (Raia, 1974; Szilagyi and Wallace, 1990).

1. **Productivity**. Productivity goals usually are expressed as levels of output per unit or per worker across the organization. Units produced per employee per day, costs per unit of production, or income generated per employee are commonly used operational measures of productivity.

2. **Market.** Market goals can be operationally defined in a variety of ways. Examples might be to increase the market share for a given product by 10% (market share goal), or to sell a specific number of units next year (output-oriented goal).
3. **Resources.** Organizations sometimes establish goals concerning changes in their resource base. Examples might include reducing the company's long-term debt by 20 million dollars within 5 yr (financial base goal), increase the production capacity by 30% (physical resource goal), or decrease turnover by 5% next year (human resources goal).
4. **Profitability**. Profitability usually is expressed as a ratio, such as net income, earnings per share, or return on investment.
5. **Innovation**. Because of rapid technology change, increasingly important to many organizations is the development of new products to maintain their competitive position. An innovation goal might be to develop a new, more efficient manufacturing process within 3 yr, or to develop a new computer having a specified increased data processing capacity by a given date.
6. **Social responsibility.** In part, because of culturally based psychosocial changes in the work force, social responsibility goals are becoming increasingly important to an organization's effectiveness. These goals might center around such factors as improving the quality of work life or reducing pollution.

During the past decade, *health* and *safety criteria* have emerged as important goals for many organizations. An example of these criteria might be to reduce lost time from accidents and injuries by 50% within the next 5 yr; or reduce the incidence of upper extremity work-related musculoskeletal disorders by 70% within 3 yr.

Achieving the goals within each of these classes of criteria may be facilitated or inhibited by the organization's design.

18.2.1.2 Classification by Focus

Szilagyi and Wallace (1990) note that classifying goals by focus entails describing the *nature of the action* that will be taken. They identify three frequently used categories.

1. **Maintenance goals.** Maintenance goals usually are stated as the specific level of activity or action that is to be sustained over time. An example for an airline would be to have at least 80% of its aircraft in service at one time.
2. **Improvement goals.** Any goal that includes an action verb is likely to be an improvement goal, as it indicates a specific change that is wanted. Examples might be "increasing" market share, "decreasing" accidents, or "improving" return on investment.
3. **Development goals.** Development goals are similar to improvement goals, but refer to some form of growth, expansion, learning, or advancement. Examples might be increasing the number of new products introduced, increasing the educational level of managers, or increasing plant capacity.

18.2.1.3 Classification by Time Frame

As we shall see below, when one considers environmental influences on complexity, classification of goals by time frame can be very useful in the organizational design process. Using this approach, goals usually are classified as long, intermediate, or short term.

1. **Short-term goals.** Short-term goals usually concern those that cover a period of 12 months or less. Frequently, production goals take this form.
2. **Intermediate-term goals**. Intermediate goals usually span a period of from 1 to 3 yr. Often, sales organizations have goals with an intermediate time orientation.
3. **Long-term goals**. Long-term goals typically cover a period of more than 3 yr. Research and development goals most often fall into this category.

18.2.2 Hierarchical Nature of Goals and Organizational Design

When goals are translated into objectives, they become *ends.* In analyzing these goals, system designers must evaluate alternatives as to *how* they will be achieved, or the *means.*

The means selected at one hierarchical level become the goals or ends for the next lower level. The hierarchical flow of means–ends for a system has strong design implications for the structural differentiation of the organization. To a significant extent, the division of labor within the organization will be a direct outcome of the means–ends analysis (Szilagyi and Wallace, 1990). For example, if the overall goal for a new system is to improve transportation in a large metropolitan area, a number of alternatives could be considered, such as new highways or a subway, streetcar, bus, light rail, or monorail system. The choice of system type, in turn, serves as the ends at the next hierarchical level, for it strongly affects selection of the types of organizational units that will be involved in operating and maintaining the system (i.e., from a work system design standpoint, selection of the units would be based on answering the question, "What types of functions are required to best accomplish the type of system selected?"). The approaches (means) selected by the system designers for meeting the operational and maintenance goals within the units, in turn, serve as the goals for the next, or subunit, level of organization. The approaches (means) chosen for meeting these ends at the subunit level, in turn, affect selection of the functions to be designed into the system and, hence, the grouping of activities into subunits. The system design choices for enabling these subunits to meet their goals, in turn, affect their division of labor.

18.2.3 Organizational Effectiveness Criteria

If, as part of the system design team, human factors specialists are to participate in the macroergonomic design of organizations, they must have a means of evaluating the relative *effectiveness* of various structural arrangements that appear feasible. During the 1960s and early 1970s, numerous studies were carried out to identify criterion measures of *organizational effectiveness* (*OE*). A review of these studies by Campbell (1977) identified 30 different criteria of OE. These are presented in Table 18.1.

18.2.4 OE Criteria Implications for Organizational Design

One thing that is reflected by the 30 criteria in Table 18.1 is that no single measure of OE is sufficient. Different organizational functions are likely to require evaluation based on different sets of characteristics. The task of the human factors specialist, working with other members of the organizational design team, is to establish which combination of OE criteria are relevant for evaluating each aspect of a proposed organizational design, and to weigh them in terms of their importance to system functioning. Those OE criteria that are selected then must be operationally defined in terms that are meaningful for the particular system (e.g., aircraft manufacturer, public utility, oil refinery) and system sub-functions (e.g., sales, marketing, production).

Second, the selection of specific OE criteria and their relative weighting often are value judgments. The stated goals for a system (themselves often a reflection of value judgments) can help considerably in the selection of OE criteria because some criteria are direct reflections of these goals. Others, however, are less tied to a specific, explicitly stated goal. For example, OE criterion 15 from Table 18.1, *flexibility/adaptation,* may be very important to the effectiveness of a particular organization or subunit in responding to its external environment, but it may not be explicitly stated. Rather it may be implicit across several goals as possibly important to their attainment.

A third factor to note is that a number of criteria may represent *competing values*. For example, *flexibility/adaptation* versus *stability,* and *participation* and *shared influence* versus *control* represent several of the more obvious potentially conflicting pairs. Striking the right balance between these competing values in the design of a particular system's organizational structure may be critical to its success (see Kuorinka and Forcier, 1995, or Smith and Sainfort, 1989 for a more detailed discussion of balance theory and its application in work system design). Factors that can aid in making these value judgments are presented in the next two "steps" in the organization design process.

18.3 DESIGNING THE DIMENSIONS OF ORGANIZATIONAL STRUCTURE

As was previously noted in defining "organizational structure," the structure of a human–machine system can be conceptualized as having three core dimensions. These are *complexity, formalization,* and *centralization* (Bedeian and Zammuto, 1991; Robbins, 1983; Stevenson, 1993).

Table 18.1 Criteria and Measures of Organizational Effectiveness

1. *Overall effectiveness.* The general evaluation that takes into account as many criteria facets as possible. It is measured usually by combining archival performance records or by obtaining overall ratings or judgments from persons thought to be knowledgable about the organization.
2. *Productivity.* Usually defined as the quantity or volume of the major product or service that the organization provides. It can be measured at three levels: individual, group, and total organization via archival records or ratings or both.
3. *Efficiency.* A ratio that reflects a comparison of some aspect of unit performance to the costs incurred for that performance.
4. *Profit.* The amount of revenue from sales left after all costs and obligations are met. Percentage return on investment or percentage return on total sales are sometimes used as alternative definitions.
5. *Quality.* The quality of the primary service or product provided by the organization that may take many operational forms, which are determined largely by the kind of product or service provided by the organization.
6. *Accidents.* The frequency of on-the-job accidents resulting in lost time.
7. *Growth.* Represented by an increase in such variables as total work force, plant capacity, assets, sales, profits, market share, and number of innovations. It implies a comparison of an organization's present state with its own past state.
8. *Absenteeism.* The usual definition stipulates unexcused absences, but even within this constraint there are a number of alternative definitions (e.g., total time absence versus frequency of occurrence).
9. *Turnover.* Some measure of the relative number of voluntary terminations, which is almost always assessed via archival records.
10. *Job satisfaction.* Has been conceptualized in many ways but the modal view might define it as the individual's satisfaction with the amount of various job outcomes that he or she is receiving.
11. *Motivation.* In general, the strength of the predisposition of an individual to engage in goal-directed action or activity on the job. It is not a feeling of relative satisfaction with various job outcomes but is more akin to a readiness or willingness to work at accomplishing the job's goals. As an organizational index, it must be summed across people.
12. *Morale.* The model definition seems to view morale as a group phenomenon involving extra effort, goal communality, commitment, and feelings of belonging. Groups have some degree of morale, whereas individuals have some degree of motivation (and satisfaction).
13. *Control.* The degree, and distribution, of management control that exists within an organization for influencing and directing the behavior of organization members.
14. *Conflict/cohesion.* Defined at the cohesion end by an organization in which the members like one another, work well together, communicate fully and openly, and coordinate their work efforts. At the other end lies the organization with verbal and physical clashes, poor coordination, and ineffective communication.
15. *Flexibility/adaptation.* Refers to the ability of an organization to change its standard operating procedures in response to environmental changes.
16. *Planning and goal setting.* The degree to which an organization systematically plans its future steps and engages in explicit goal-setting behavior.
17. *Goal consensus.* Distinct from actual commitment to the organization's goals, consensus refers to the degree to which all individuals perceive the same goals for the organization.
18. *Internalization of organizational goals.* Refers to the acceptance of the organization's goals. It includes the belief that the organization's goals are right and proper.
19. *Role and norm congruence.* The degree to which the members of an organization agree on such things as desirable supervisory attitudes, performance expectations, morale, role requirements, and so on.
20. *Managerial interpersonal skills.* The level of skill with which managers deal with supervisors, subordinates, and peers in terms of giving support, facilitating constructive interaction, and generating enthusiasm for meeting goals and achieving excellent performance.
21. *Managerial task skills..* The overall level of skills with which the organization's managers, commanding officers, or group leaders perform work-centered tasks and tasks centered on work to be done and not the skills employed when interacting with other organizational members.

Table 18.1 Continued

22. *Information management and communication.* Completeness, efficiency, and accuracy in analysis and distribution of information critical to organizational effectiveness.
23. *Readiness.* An overall judgment concerning the probability that the organization could successfully perform some specified task if asked to do so.
24. *Utilization of environment.* The extent to which the organization interacts successfully with its environment and acquires scarce and valued resources necessary to its effective operation.
25. *Evaluations by external entities.* Evaluations of the organization, or unit, by the individuals and organizations in its environment with which it interacts. Loyalty to, confidence in, and support given the organization by such groups as suppliers, customers, stockholders, enforcement agencies, and the general public would fall under this label.
26. *Stability.* The maintenance of structure, function, and resources through time and, more particularly, through periods of stress.
27. *Value of human resources.* A composite criterion that refers to the total value or total worth of the individual members, in an accounting or balance sheet sense, to the organization.
28. *Participation and shared influence.* The degree to which individuals in the organization participate in making the decisions that affect them directly.
29. *Training and development emphasis.* The amount of effort that the organization devotes to developing its human resources.
30. *Achievement emphasis.* An analog to the individual need for achievement referring to the degree to which the organization appears to place a high value on achieving major new goals.

Source: Adapted with permission from John P. Campbell, "On the Nature of Organizational Effectiveness," in P. S. Goodman, J. M. Pennings, and Associates, Eds., *New Perspectives on Organizational Effectiveness.* San Francisco: Jossey-Bass, 1977, pp. 36–41.

18.3.1 Complexity

Complexity refers to the degree of *differentiation* and *integration* that exists within an organization. Differentiation refers to the extent to which the structure is segmented into parts; integration refers to the number of devices or mechanisms that exist to integrate the segmented parts for purposes of communication, coordination, and control.

18.3.1.1 Differentiation

Organizational structures embody three major types of differentiation. These are *horizontal differentiation, vertical differentiation,* and *spatial dispersion.* Increasing any one of these three increases an organization's complexity.

Horizontal Differentiation

Horizontal differentiation refers to the degree of job specialization and departmentalization that is designed into the organization's structure. Job specialization leads to greater complexity because it requires more sophisticated and expensive methods of control. Yet specialization is common to virtually all high-technology work systems because of the inherent efficiencies in the division of labor. Adam Smith (1970/1876) demonstrated this point over 200 yr ago. He noted that 10 men, each doing particular tasks (job specialization), could produce about 48,000 straight pins per day. However, if each man worked separately and independently, performing all of the production tasks, those 10 workers combined would be lucky to make 200.

Division of labor creates groups of specialists. The way these specialists are grouped is known as *departmentalization.* Departments can be designed into a system on the basis of (1) function, (2) simple numbers, (3) product or services, (4) client, (5) geography, and (6) process. Most large corporations will use all six (Robbins, 1983).

Vertical Differentiation

The measure of vertical differentiation is the number of levels separating the chief executive position from the jobs directly involved with the system's output. In general, as the size of an organization increases, the need for vertical differentiation also increases (Mileti, Gillespie, and Haas, 1977). In one study, size alone was found to account for 50% to 59% of the variance (Montanari, 1976). A key factor underlying this size-vertical differentiation relationship is *span of control.* Any one supervisor is limited in the number of subordinates that he or she can direct effectively (Robbins, 1983). Thus, as the number of positions increases, the number of first-level supervisory jobs that must be designed into the work system also increases. This increase, in turn, increases the number of the second-, third-, and higher level managerial positions required. For example, if the span of control is eight, and the organization has 512 worker positions, the number of first-level supervisory jobs is 64, of second-level managerial positions is 8, and third level, 1 (e.g., the chief executive officer or plant manager). If the organization has 4096 employees, the number of supervisory levels increases to four (i.e., 512 first-level supervisors, 64 second-level managers, 8 third-level, and 1 fourth-level).

Although span of control limitations underlie the size-vertical differentiation relationship, these limitations can be quite varied, depending on a number of factors. For example, as span of control can be increased, the number of hierarchical levels required can be reduced for a given number of employees, making the structure of the organization flatter and broader in shape. A major factor affecting the span of control that is desirable is the degree of *professionalization* of employee positions. In general, as the degree of professionalism (education and skill requirements) designed into employee jobs increases, the better the employees will be able to function autonomously with only a minimum of supervision. Thus, the span of control for a given supervisor can be increased. Other factors, such as the degree of formalization, type of technology, psychosocial variables, and environmental factors also can affect vertical differentiation. These are discussed individually later.

Spatial Dispersion

Spatial dispersion refers to the performance of a system's activities in multiple locations. In a sense, it is an extended dimension to horizontal and vertical differentiation in that it is possible to separate both power and task centers geographically. Spatial dispersion measures include (1) the number of geographic locations within the organization, (2) the average distance of the separated places from the organization's headquarters, and (3) the number of employees in these separated locations in relation to the number in the headquarters. (Hall, Haas, and Johnson, 1967). In general, complexity increases as the number of geographically separated units, the average distance of these units from the headquarters, and the proportion of employees in these separated units increases.

The Horizontal, Vertical, Spatial Differentiation Relation

When one looks at many very large or very small organizations, it would seem that there is a high intercorrelation among the three types of differentiation. Large high-technology corporations, such as ITT, General Motors, and Exxon, are characterized by a high level of all three kinds of differentiation. On the other hand, the corner grocery store, local service station, and neighborhood dry cleaner have little of any kind of differentiation. However, between these two extremes there is little systematic relationship among the types of differentiation. Some fairly large organizations, such as an army battalion, are characterized by a high degree of vertical differentiation, but relatively little horizontal differentiation or geographical dispersion; universities typically have little vertical differentiation or geographical dispersion, but a high degree of horizontal differentiation (Hage and Aiken, 1967); some small retail chains have little horizontal or vertical differentiation, but geographically dispersed units. In short, the optimal level of each type of differentiation for an organization has to be individually assessed in terms of the variables that affect it. Some of these factors we already have noted in describing the measurement and nature of each type of differentiation. Other key variables are discussed in Section 18.5.

18.3.1.2 Integration

As the differentiation of an organization increases, the need for integrating devices also increases. The need for these mechanisms increases because with greater differentiation of the organization's activities, the difficulty of communication, coordination, and control also increases. Some of the more common integrating mechanisms that can be designed into the work system are formal rules and procedures, liaison positions, committees, teams, system integration offices, and computerized information and decision support systems. Vertical differentiation, in itself, is a form of integration mechanism because a manager at one level serves to coordinate and control the activities of several lower-level managerial or worker positions.

Once the differentiation aspects of the organization's design have been determined, the human factors specialists must pay particular attention to the resultant integration needs of the work system. In part, the nature of the integration devices that will be utilized will be determined by the degree of *formalization* and *centralization* utilized in the organization's design. This is because formalization and centralization are, themselves, integrating mechanisms. The extent to which computer-based information and decision support systems are included in the system's technology also will be a factor; for these too are integrating mechanisms. They thus form part of the organization's work system design, and care must be taken to ensure that they are macroergonomically designed to be an integrated, compatible part of the system's organizational structure. Too many integrating mechanisms tend to stifle effective organizational functioning; too few will result in poor coordination and control.

The type of technology to be utilized in the system, personnel subsystem factors, and external environment variables also help to determine the optimal number and types of integrating mechanisms that should be designed into the work system structure. These factors will be considered in Section 18.5.

18.3.2 Formalization

For ergonomic design purposes, formalization can be defined as the degree to which jobs within the organization are standardized. In highly formalized designs, jobs allow for little employee discretion over what is to be done, when it is to be done, and how it is to be accomplished (Robbins, 1983). There are explicit job descriptions, extensive rules, and clearly defined procedures covering work processes. Often, the design of the system's hardware and human–machine interfaces, in themselves, restrict employee freedom to exercise discretion. Where formalization is low, jobs are designed to allow employees considerable freedom to exercise discretion. Employee behavior thus is relatively unprogrammed, and usually allows for considerably greater use of one's mental capacities. Consequently, low formalization usually necessitates greater education or training and experience as part of the individual job requirements.

In general, the simpler and/or more repetitive the jobs to be designed into the system, the greater the value of formalization for effective system integration and functioning. The greater the professionalism (education, training, or experience requirements) designed into the jobs, the less the need for, or utility of high formalization. In fact, it is likely to inhibit both employee motivation and effective functioning. Related to this is the kind of work to be performed. For example, production jobs with stable repetitive characteristics lend themselves to relatively high formalization; whereas research and development or sales, which may require considerable flexibility, innovation, or responsiveness to change, are likely to be stifled by a high degree of formalization. The degree of predictability and stability or uncertainty and change in the relevant external environments of the organization and its constituent units also affect the degree of formalization that should be ergonomically designed into the work system. In general, the greater the degree of environmental instability and uncertainty, the greater is the need for low formalization and a high level of professionalism. More will be said about this in Section 18.5.

In summary, from a macroergonomics perspective, in considering the classic trade-offs between selection, training, and the hardware skill requirements to be designed into the system, the human factors specialist must consider (1) the relative stability of the external environments with which the organization and its constituent units interact, and (2) the related degree of formalization and professionalism that is optimal for effective system functioning. With the widespread introduction of computer-based information and decision-support systems, these two interrelated considerations are as important to the design of system software as they are for hardware.

18.3.3 Centralization

Centralization refers to the degree to which formal decision making is concentrated in an individual, unit, or level (usually high in the organization), thus permitting employees (usually low in the organization) minimal input into the decisions affecting their jobs (Robbins, 1983). It should be noted that centralization essentially is concerned with *decision discretion.* Where decisions are delegated downward, but policies or other formalized mechanisms exist in the system to constrain the discretion of lower-level positions, there is, in reality, increased centralization. Conversely, the transference of information in systems requires filtering. Decisions often are made at intermediate hierarchical levels as to what information gets passed upward to higher management. The greater the extent to which the system is designed to permit information to be reduced, summarized, selectively omitted, or embellished, the less is the extent to which the actual decision is concentrated and controlled by a centralized decision maker. In short, the less is the actual degree of centralization.

The filtering of information to the decision maker, discussed above, illustrates that from a systems standpoint the actual making of the decision, in itself, does not determine the true degree of centralization. Rather, it is the degree of control that one holds over the decision-making *process* that is the true measure of centralization. With this in mind, it is important to note that the actual decision only indicates the *intended* action. As the decision is passed down through the hierarchical levels to those responsible for its implementation, it too may undergo filtering. Where conditions, which perhaps are unknown to the decision maker, require safeguards to prevent implementation of an inappropriate decision, *formal* filtering of this kind actually can be a desirable ergonomic design feature. This would be particularly true in situations where the decision implementation might pose a health or safety hazard, or otherwise be highly detrimental to the organization. However, from an ergonomic design standpoint, it also is important to *prevent* downward filtering when it is likely to dilute the desired level of true centralization.

In general, centralization is desirable (1) when a comprehensive perspective is required, such as in strategic decision making; (2) when it provides significant economies; (3) for financial, legal, and other decisions where they clearly can be done more efficiently when centralized; and (4) when operating in highly stable and predictable relevant external environments. Decentralized decision making is desirable (1) when the design of a given manager's job will result in taxing or exceeding human information processing and decision-making capacity; (2) in order to enable the organization to respond rapidly to changing or unpredictable conditions at the point where change is occurring; (3) to provide more detailed "grass roots" input to into "decisions;" (4) to provide employees with greater motivation and intrinsic job satisfaction; and (5) to reduce stress and related health effects emanating from a lack of personal control by allowing them to participate in decisions that affect their jobs; (6) to gain greater employee commitment to, and support for, decisions by involving employees in the process; (7) to more fully utilize the mental capacities and detailed job knowledge of the employees; and (8) to provide greater training opportunity for lower level managers. More will be said about technological, personnel, and environmental factors as determinants of centralization in Section 18.5.

18.3.4 Relationship of Complexity, Formalization, and Centralization

18.3.4.1 Centralization and Complexity

In general, research indicates an inverse relationship between centralization and complexity (Child, 1972; Hage and Aiken, 1967; Robbins, 1983). For example, as the number of occupational specialties and related training requirements designed into a work system increase, the expertise required to make sound decisions also increases, thus forcing decentralization for effective system functioning. Conversely, the simpler and more repetitive the jobs, the greater is the utility of centralized decision making (consistent with not carrying it to the point of causing adverse employee motivational and stress effects).

18.3.4.2 Centralization and Formalization

No clear, simple relationship has been found between centralization and formalization. Other factors tend to moderate these relationships. for example, with a high degree of job professionalization, both low formalization and low centralization tend to be needed (Hage and Aiken, 1967). However, the *type* of decision moderates this relationship. Professionals expect decentralization of decisions that affect their work directly, together with a low level of formalization to enable them to respond to unique or changing situations; but *not* (1) decisions concerning personnel issues (e.g., salary and performance appraisal procedures) where the predictability that comes with standardization is desired; and (2) strategic decision making when it is enhanced by the more comprehensive perspective of centralization and which has little direct impact on their daily activities (Robbins, 1983).

18.3.4.3 Formalization and Complexity

The relationship between formalization and complexity tends to be a function of (1) the *direction* of differentiation; and (2) the degree of professionalization (Robbins, 1983). High horizontal differentiation, if achieved by increasing the number and kinds of routine repetitive tasks, results in the need for a high degree of formalization (Pugh, Hickson,

Hinings, and Turner, 1968). If it is achieved by increasing the number and kinds of highly skilled, complex positions (professionalization), then low formalization should be optimal, along with decentralized decision making (Hage, 1965). High vertical differentiation usually involves designing in an increased number of managerial and technical specialists (professionalization) and nonroutine tasks. Thus, for optimal functioning, low formalization should be incorporated into the organization's design for these positions (Hage, 1965; Robbins, 1983).

18.4 A MACROERGONOMIC APPROACH TO ORGANIZATIONAL DESIGN

18.4.1 A Technology Perspective of Human Factors or Ergonomics

Most frequently, we define human factors or ergonomics descriptively. For example, the International Ergonomics Association defines *ergonomics* as "integrating knowledge derived from the human sciences to match jobs, systems, products and environments to the physical and mental abilities and limitations of people" (IEA brochure, 1995). Similar descriptive definitions are widely used by ergonomists and ergonomics societies throughout the world. While highly useful *descriptively*, these definitions tell us little *operationally* about ergonomics as a *unique* profession. As with other disciplines, one way to more clearly define ergonomics is in terms of its unique *technology*. From this perspective ergonomics may be defined as *human–system interface technology* (Hendrick, 1991). As a *science*, ergonomics is concerned with developing knowledge about human performance capabilities, limitations and other characteristics as they relate to the design of the interfaces between people and other system components. As a *practice*, ergonomics concerns the application of human–system interface technology to the design or modification of systems to enhance safety, comfort, effectiveness, and quality of life. At present, this unique technology has at least four identifiable major components: Human–machine interface technology or hardware ergonomics, environment–machine interface technology or environmental ergonomics, user–system interface technology or software ergonomics, and organization–machine interface technology or macroergonomics. A brief description of these four technical areas follows.

18.4.1.1 Hardware Ergonomics or Human–Machine Interface Technology

Originally known as man–machine interface technology, this technology represents the focus of ergonomics during the first three decades of our profession. It primarily concerns the study of human physical and perceptual characteristics; and the application of that knowledge to the design of controls, displays and workspace arrangements.

18.4.1.2 Environmental Ergonomics or Human–Environment Interface Technology

This second ergonomics technology is concerned with human capabilities and limitations with respect to the demands imposed by various environmental modalities (e.g., light, heat, noise, vibration, etc.). It is applied to the design of human environments to minimize environmental stress on human performance, including comfort, health and safety, and to enhance productivity.

18.4.1.3 Software Ergonomics or User–System Interface Technology

This third technology is a relatively new development in our profession, having come into being with the development of the silicon chip in the 60s and the modern computer revolution which followed. Because this relatively new technology primarily is concerned with how people conceptualize and process information, it often is referred to as "cognitive ergonomics." The major application of this technology is to the design or modification of system software, including information and decision support systems, to enhance its usability.

18.4.1.4 Macroergonomics or Organization–Machine Interface Technology

Macroergonomics is the newest part of the human factors profession. The central focus of the first three technologies has been the individual operator and operator teams or subsystems. Thus, the primary application of these technologies has been at the microergonomic level. In contrast, because it deals with the overall structure of the work system as it interfaces with the system's technology, the organization–machine aspect of the human–system interface tends to be *macro* in its focus; hence, it is referred to as "macroergonomics." Conceptually, it is a top-down, sociotechnical systems approach to organizational and work systems design, and the design of related human–machine, human–environment, and user–system interfaces (Hendrick, 1986a,b).

Although conceptually top-down, the actual design process is likely to be nonlinear, iterative, and involve activity at all organizational levels. For example, one also must come to understand the nature and interactions of individual subsystems as part of the building and optimizing of the larger macro-organizational system (Passmore,1988).

18.4.2 Pitfalls of Conventional Approaches to Organizational Design

Based on assessments of over 200 organizational units, Hendrick (1995) has identified three highly interrelated work system design practices which frequently underlie dysfunctional work system development and modification efforts. These are *technology-centered design,* a *left-over approach to function and task allocation,* and *a failure to integrate the organization's sociotechnical system characteristics into its work system design.*

18.4.2.1 Technology Centered Design

When a new technology is developed, designers typically focus on incorporating it within some form of hardware or software to achieve some desired transformation or outcome. If consideration is given to the persons who must operate, maintain, or be serviced by the hardware and software, it typically takes the form of determining what skills, knowledge, and training will be necessary to utilize the technology. Often, even this kind of consideration is not systematic or well thought through ergonomically. When ergonomic factors *are* considered, it usually takes the form of designing human–system interfaces for the *already designed* hardware and software to minimize human error and improve physical comfort. Rarely are the intrinsic motivational aspects of the jobs, psychosocial characteristics of the work force, or other related organizational and work system design factors considered. Yet these are the very factors that are critical to improving organizational effectiveness (Hendrick, 1994b).

A technology-centered approach also frequently leads to treating persons as impersonal components of the system—both in designing or modifying organizations and in implementing the results. The outcome is likely to be both poor design analyses and decisions and ineffective implementation. As the organizational change literature repeatedly has shown, failure to actively involve employees throughout the change planning and implementation process invariably leads to a lack of commitment and, often, to overt or passive-aggressive resistance.

18.4.2.2 The "Left Over" Approach to Function and Task Allocation

In system design, when a purely technology-centered approach is taken, the focus is on assigning to the "machine" any functions or tasks which its technology can enable it to accomplish. Then, whatever is "left over" is allocated to the persons who must operate, maintain, or be serviced by the system. This, by far, has been the most common means of function and task allocation in exploiting new technology. Because this approach fails to adequately consider the characteristics of the work force and related external environmental factors in allocating functions and tasks, the result usually is a suboptimally designed work system.

A fundamental empirically derived principle of sociotechnical system design is that optimal effectiveness of the system requires *joint optimization* of the personnel and technological subsystems; and that this is achievable only through *joint design* of the two. When we attempt to optimize just the technical subsystem, we force the personnel subsystem to have to accommodate (DeGreene, 1973). In essence, this becomes a "fitting square pegs into round holes" situation.

Put in ergonomic terms, joint optimization requires a *human-centered* approach. Bailey (1989) refers to this method of function and task allocation as a *humanized task* approach, and states that "this concept essentially means that the ultimate concern is to design a job that *justifies* using a person, rather than a job that merely can be done by a human. With this approach, functions are allocated and the resulting tasks are designed to make full use of human skills and to compensate for human limitations. The nature of the work itself should lend itself to internal motivational influences. The leftover functions are allocated to the computers" (p. 190).

18.4.2.3 Failure to Integrate an Organization's Sociotechnical Characteristics into its Work System Design

The primary structural and process characteristics of sociotechnical systems first were empirically identified in the classic longwall coal mining studies by the Tavistock Institute in the UK over four decades ago (DeGreene, 1973), described briefly at the beginning of this chapter. From this literature, four major sociotechnical system elements may be identified: The (a) personnel subsystem, (b) technological subsystem, (c) organizational structure, and (d) external environment (Hendrick, 1986, 1987, 1991). Of particular note, these four elements interact with one another. A change to any one element affects the other three; and, if not properly planned for, often in unanticipated and dysfunctional ways. Further, as with biological and other kinds of systems, the whole is more than the simple sum of its parts.

Since these classic sociotechnical system studies were conducted, the critical dimensions of the *technology, personnel,* and *external environment* have been identified in terms of their relation to specific characteristics of organizational and work system structure, and empirical models of these relationships have been developed. By applying these models, it is possible to determine the optimal amount of such things as the following: Number of hierarchical levels, degree of departmentalization and specialization, centralization or decentralization of decision making (i.e., where what types of decisions should be made), and the degree of formalization of work processes. These models can be applied to diagnosing any complex work system and more optimally designing its organizational and work system structure and processes. Some of the most commonly used empirically developed models and their use in determining specific organizational design characteristics are described in Section 18.5.

Unfortunately, as first was documented in the Tavistock Institute studies, cited earlier, a technology centered approach to organizational and work system design does *not* adequately consider the relevant sociotechnical system variables—not only in terms of productivity, but also in terms of their effects on employee self-worth, stress, satisfaction, and related health and safety. As a result, work systems thus designed are most often *suboptimal.*

18.4.3 Criteria for an Effective Organizational Design Approach

Based on the above, what is needed is a design approach to work systems that accomplishes the following: First, it should be human centered; second, it should use a humanized task approach to function and task allocation; and third, it should adequately consider the relevant sociotechnical system variables in terms of their implications for organizational and work system design, and the related design of jobs, specific work processes, and human–system interfaces. One strategy that can satisfy all three of these criteria is a *macroergonomic* approach.

18.4.4 A Macroergonomic Approach

As was noted in 18.4.2 above, macroergonomics is a top-down sociotechnical systems approach to the design of organizations, work systems, jobs, and related human–machine, user–system, and human–environment interfaces (Hendrick, 1986; 1991). It is top-down in that it begins with an analysis of the relevant sociotechnical system variables in terms of their implications for the design of the overall structure of the work system and related processes. In short, it involves a systematic analysis of the key characteristics of the organization's technology, personnel subsystem, and relevant external environments which permeate the organization and upon which it is dependent for its survival and success. Once the key characteristics of the overall work system have been determined, they, in turn, prescribe many of the characteristics that need to be microergonomically designed into the individual jobs, specific work processes, and related human–machine and user–system interfaces. The result is a *fully harmonized* work system at both the macro- and microergonomic level.

Macroergonomics is human centered in that decisions regarding the structure of the organization and work system require consideration of the worker's professional and psychosocial characteristics, and of the relevant characteristics of the external environment to which these humans must effectively respond (in addition to consideration of the key characteristics of the technology to be employed). By the same token, consideration of these characteristics also make it a humanized task approach to function and task allocation.

As discussed in Section 18.7, although *conceptually* it is a top-down approach, a primary methodology of macroergonomics is employee involvement via the methods of *participatory ergonomics*. This approach further enhances the human centeredness of the design process and takes advantage of the detailed knowledge of employees at each organizational level. The application of participatory ergonomics also tends to enhance employee acceptance and commitment to the changes.

In summary, macroergonomics provides us with an organizational design strategy for developing or improving work systems that overcomes the dysfunctional shortcomings of historically used design practices. As a top-down sociotechnical systems approach to organizational design, it meets the criteria of being human centered, utilizing a humanized task approach to function and task allocation, and systematically considering the key sociotechnical system variables that have been found related to effective work system design.

18.5 SOCIOTECHNICAL SYSTEM COMPONENTS AS MODERATORS OF ORGANIZATIONAL DESIGN

Based on the Tavistock Institute Studies, mentioned earlier, Emery and Trist (1960) coined the term *sociotechnical systems* to convey more adequately the nature of complex human–machine–environment systems. The sociotechnical system concept views organizations as open systems engaged in transforming inputs into desired outcomes (DeGreene, 1973). *Open* systems means that sociotechnical organizations have permeable boundaries exposed to the environments in which they exist. These environments thus permeate or enter the organization along with the inputs to be transformed. The primary ways in which environmental changes enter the organization are through the people who work in it, through its marketing or sales function, and through its materials or other input functions (Davis, 1982).

As transformation agencies, organizations are in constant interaction with their environment. They receive inputs from their environment and transform these into desired outputs, and export outputs to their environment. Organizations bring two critical factors to bear on the transformation process: Technology in the form of a *technological subsystem*, and people in the form of what, in human factors parlance, is known as the *personnel subsystem*. The design of the technological subsystem defines the *tasks* to be performed. The design of the personnel subsystem prescribes the *ways* in which tasks are performed. Each interacts with the other at every human–machine interface. The technological and personnel subsystems thus are *mutually interdependent*. Both subsystems operate under *joint causation*, meaning that they both are affected by causal events in the environment. The technological subsystem, once designed, is relatively stable and fixed. It thus falls to the personnel subsystem to adapt further to environmental change. Joint causation leads to the related sociotechnical system concept of *joint optimization*. Joint optimization means that because the technological and personnel subsystems respond jointly to causal events, optimizing one subsystem and fitting the second to it results in suboptimization of the joint *system*. Thus, maximizing overall system effectiveness requires jointly optimizing *both* subsystems. Accordingly, the need for joint optimization requires the *joint design* of the technical and personnel subsystems in order to develop the best possible fit between the two, given the objectives and requirements of each, and of the overall system (Davis, 1982). Inherent in this joint design is developing an optimal structure for the overall system.

As is inferred above, the design of an organization's structure and related processes involves consideration of three major sociotechnical system components that interact and affect optimal organizational design. These are the (1) *technological subsystem*, (2) *personnel subsystem*, and (3) *relevant external environments* that permeate the organization. Each of these components has been studied in relation to its effects on the three organizational design dimensions, described in Section 18.3, and empirical models have emerged. These models can be used as ergonomic tools in analyzing organizations and developing or modifying their designs.

18.5.1 Technological Subsystem Characteristics

Technology, as a determinant of organizational design, has been operationally classified in four distinctly different ways. These are (1) by the mode of production, or *production technology*; (2) by the action individuals perform upon an object to change it, or *knowledge-based technology*; (3) by the strategy selected for reducing uncertainty, which is determined by the technology; or *technological uncertainty;* and (4) by the degrees of automation, work-flow rigidity, and quantitative specificity of evaluation of work activities, or *work-flow integration*. From each of these classification schemes, themselves empirically derived, a major generalizable model of the technology-organizational design relationship has been empirically derived.

18.5.1.1 Woodward: Production Technology

The classic series of studies of technology as a determinant of organizational structure was conducted by Joan Woodward and her associates (1965). She and her colleagues studied 100 manufacturing firms in South Essex, England, having at least 100 employees. The organizations varied greatly in terms of size, type of industry, managerial levels (2 to 12); span of control (2 to 12 at the top, 20 to 90 at the first-line supervisory level); and ratio of line employees to staff personnel (less than 1:1 to more than 10:1). Through interviews, observations, and review of company records, the following, among other factors, were noted for each firm: (1) the organization's mission and significant historical events; (2) the manufacturing processes and methods utilized; and (3) the organization's success, including changes in market share, relative growth or stagnation within its industrial field, and fluctuation of its stock prices.

As part of her analysis Woodward identified three modes of technology: (1) *unit,* (2) *mass,* and (3) *process* production. These modes were seen as representing categories on a scale of increasing *technological complexity.* At the least complex end were the unit and small batch producers. These firms produce custom-made products. Next were the large batch or mass production firms, such as those that produce automobiles and other more or less standardized products using predictable, repetitive production steps. The organizations highest in production complexity were long run, heavily automated process production firms, such as oil and chemical refineries. Three important organizational structure variables were found to increase as technological complexity increased. First, as *technological complexity increased, the degree of vertical differentiation also increased.* For each class of technology, the successful firms tended to have the median number of hierarchical levels for that category. This optimum for unit producers was three, for mass it was four, and for process, six. The less successful firms in each category had a noticeably greater or lesser number of levels. Second, *as technological complexity increased, the optimal ratio of administrative support staff to industrial line personnel increased.* Third, *as technological complexity increased, the span of control of the top-line managers increased.* Woodward's findings for the successful firms in each technology mode are summarized in Table 18.2 for the three organizational design dimensions, and are summarized below.

Unit production firms had low complexity. First line supervisors had relatively narrow spans of control and there was little line and staff differentiation; jobs were widely rather than narrowly defined; and formalization and centralization both were low.

Mass production units had high complexity. First line supervisors had relatively broad spans of control. There was clear line and staff differentiation, and jobs were narrowly defined. Formalization and centralization both were high.

Process production units had high vertical differentiation with wide spans of control, including the lower supervisory levels. There was little line and staff differentiation. Formalization and Centralization both were low.

Table 18.2 Summary of Woodward's Findings on the Design Features of Effective Organizations

Organizational Structure	Mode of Production		
	Unit	Mass	Process
Complexity			
Vertical differentiation	Low	Moderate	High
Horizontal differentiation	Low	High	Moderate
Formalization	Low	High	Low
Centralization	Low	High	Low

Several follow-up studies have lent support to Woodward's findings (Harvey, 1968; Zwerman, 1970). It should be noted that Woodward implies causation when, in fact, her methodology really only establishes correlation. In applying Woodward's findings, another caution is in order: Her data were collected from within a single culture and at a particular point in time. In a different culture, or at some other time, the sociocultural and other environmental factors might be very different, and thus result in somewhat different interactions with production mode in terms of their influence on organizational design.

18.5.1.2 Perrow: Knowledge-Based Technology

A major shortcoming of Woodward's model is that it applies only to manufacturing firms, which constitute less than half of all organizations. Perrow (1967) has empirically developed a more generalizable model of the technology-organizational design relationship using a *knowledge-based* rather than a *production* classification scheme. He began by defining technology as the action one performs upon an object in order to change the object. This action requires some form of technical knowledge. Using this approach, he identified two underlying dimensions of knowledge-based technology. The first of these concerns the number of exceptions encountered in one's work, or *task variability.* The second has to do with the type of search procedures one has available for responding to task exceptions, or *task analyzability.* These procedures can range from "well defined" to "ill defined." At the "well-defined" end of the continuum, solving problems involving task exceptions can be accomplished using logical and analytical reasoning. At the "ill-defined" end there are no readily available formal search procedures, and one must rely on experience, judgment, and intuition. The combination of these two dimensions, when dichotomized, yields a two-by-two matrix having four cells. As shown in Table 18.3, each cell represents a different knowledge-based technology.

1. *Routine* technologies have few exceptions and well-defined problems. Mass production units fall within this category. Routine technologies are best accommodated through standardized coordination and control procedures and, accordingly, are associated with high formalization and centralization.
2. *Nonroutine* technologies have many exceptions and difficult to analyze problems. Combat aerospace operations would be an example. These technologies require flexibility, and thus should be decentralized and have low formalization.
3. *Engineering* technologies have many exceptions, but can be handled using well-defined rational-logical processes. They therefore lend themselves to moderate centralization, but require the flexibility that is achievable through low formalization.

Table 18.3 Perrow's Technology Classification

<table>
<tr><td colspan="2"></td><td colspan="2">Task Variability</td></tr>
<tr><td colspan="2"></td><td>Routine with Few Exceptions</td><td>High Variety with Many Exceptions</td></tr>
<tr><td rowspan="2">Problem Analyzability</td><td>Well Defined and Analyzable</td><td>Routine</td><td>Engineering</td></tr>
<tr><td>Ill Defined and Unanalyzable</td><td>Craft</td><td>Nonroutine</td></tr>
</table>

4. *Craft* technologies typically involve relatively routine tasks, but problems rely heavily on the experience, judgment, and intuition for decision. Thus, problem solving must be done by those with the particular expertise. This requires decentralization and low formalization.

Perrow's model has been supported by considerable empirical research, both in the private and public sectors (e.g., Hage and Aiken, 1969; Magnusen, 1970; Van deVen and Delbecq, 1974).

18.5.1.3 Thompson: Technological Uncertainty

Thompson (1967) has demonstrated that the type of technology determines a *strategy* for reducing uncertainty, and that the specific structural arrangements facilitate uncertainty reduction. He identified three types of technologies, based on the tasks that an organization performs: (1) *long linked*, (2) *mediating,* and (3) *intensive.*

1. A *long-linked* technology accomplishes its tasks by a sequential interdependence of its units, as might be illustrated by an automobile assembly line. Because it involves a fixed sequence of repetitive steps, the major uncertainties in this type of organization are at the input and output sides. Thus, management responds to uncertainty by controlling inputs and outputs. This is accomplished through planing and scheduling, and suggests a moderately complex and formalized structure.
2. A *mediating* technology is one that links clients on both the input and output sides, thus performing a mediating or interchange function. In short, mediators—such as banks, utility companies, and the post office—link units that otherwise are independent. A mediating technology is characterized by a pooled or parallel interdependence of the different units. These otherwise independent units are bound together by rules, regulations, and standard operating procedures. They thus function best with low complexity and high formalization.
3. An *intensive* technology is one that represents a customized response to a diverse set of contingencies. It involves a variety of techniques that are drawn upon to transform an object from one state to another. The particular techniques employed are, at least in part, based upon feedback from the object itself. The classic example is a hospital, where the object being transformed is the patient. The techniques that can be employed are varied, and are selected largely on the patient's condition, and responses to previously used techniques. The major uncertainty is the object itself. The flexibility of response, such as many alternatives, is a must for effective system functioning. An intensive technology thus operates best with a structure characterized by high complexity and low formalization.

Thompson's model, unfortunately, has not been fully tested empirically. The lack of data thus makes it impossible to draw a definitive conclusion regarding the model's validity (Robbins, 1983). The one study of consequence analyzed 297 subunits for 17 business and industrial firms (Mahoney and Frost, 1974). This study provided partial support for Thompson's model by demonstrating that long-linked and mediating technologies were associated closely with formalization and advanced planning, whereas intensive technologies were characterized by mutual adjustments to other units.

18.5.1.4 Aston Studies: Work-Flow Integration

After studying a wide range of both manufacturing and service organizations, a team of researchers at the University of Aston in the UK concluded that technology can be defined in terms of three basic characteristics: *automation of equipment* or the extent to which work activities are performed by machines, *work-flow rigidity* or the extent to which the sequence of work activities is inflexible, and *specificity of evaluation* or the degree to which work activities can be assessed by specific, quantitative means. Because these three characteristics were found to be highly related, they were combined into a single scale labeled *work-flow integration* (Hickson, Pugh, and Pheysey, 1969). Work-flow integration was found related to organizational structure, but only weakly, and primarily within smaller organizations. In general, as work-flow integration increases, specialization, formalization, and decentralization of operational authority also increase for optimal functioning.

Perhaps the two most important outcomes of the Aston Studies were as follows. First, organizational *size* operates as a moderator of the effects of work-flow integration, with work-flow integration (technology) effects being stronger in relatively smaller organizations. Second, while technology does affect organizational structure, it is only one of several influences, and appears to have significantly less impact than the other sociotechnical system elements (i.e., the key characteristics of the personnel subsystem and relevant external environments, discussed later in this section). Of particular importance, the Aston studies further demonstrated that the so-called *technological imperative*—the view that technology has a compelling influence on structure, and thus should determine work system design—clearly overstates the case (Baron and Greenberg, 1990); a finding first documented in the Tavistock Institute studies of the longwall method of coal mining, discussed at the beginning of this chapter.

18.5.1.5 Other Technological Considerations

In recent years, we have seen the introduction of major advances in computer and communications technology in two forms that have major implications for organizational design. These are advanced information technologies (AIT) and computer integrated manufacturing (CIM). AIT has tended to result in facilitating the efficiency of decentralizing operational or *tactical* decision making, while enhancing the efficiency of centralized *strategic* decision making (Bedeian and Zammuto, 1991). Computer-based AIT links employees electronically, and thus enables them to more readily participate in the operational decision making process. In so doing, AIT enhances the efficiency of decentralization and greater professionalism; it also enables lower level managers and employees to have more *indirect* influence on strategic decision making, since they often structure the databases which form the input for the highly centralized strategic decisions.

CIM necessarily results in a very high level of integration of work-flow processes and, thus, a high level of interdependence among differentiated units. One result of this interdependence is an increased need for effective integrating mechanisms across functional units. CIM also tends to increase the need for market-based unit grouping (e.g., product task team grouping), especially during the product design phase (Bedeian and Zammuto, 1991; Drucker, 1988). This type of grouping, and when it should be used, is discussed further in Section 18.6.

18.5.2 Personnel Subsystem Characteristics

Critical to an organization's design are at least three major characteristics of the personnel subsystem. These are (1) the *degree of professionalism*, (2) *demographic characteristics*, and (3) *psychosocial aspects* of the work force.

18.5.2.1 Degree of Professionalism

Robbins (1983) notes that formalization can take place on the job or off. When it is done on the job, formalization is *external* to the employee. The rules, procedures, and human–system interfaces are designed to limit employee *discretion*. This often characterizes unskilled and semi-skilled positions, and is what is meant by the term "formalization." Professionalism, on the other hand, creates *internal* formalization of behavior through a socialization process that is an integral part of training or education. That is, persons learn the values, norms and expected behavior patterns of the job *before* entering the organization. Thus, from a macroergonomic standpoint, there is a trade-off between formalization of the work system and professionalization of the jobs in the organizational design process. Where jobs are designed to require persons with professional training or education, the work system needs to be designed and integrated to allow for low formalization and, thus, considerable employee discression. In fact, it often is the need to have employees that can deal with unique or unanticipated situations that creates the need for more professionalized jobs.

18.5.2.2 Demographic Factors

Many demographic characteristics of the work force that will form the organization's personnel subsystem could potentially interact with the organization's design. Those that are most striking within the United States, and often in other industrialized countries, are (1) the recent large increase in the number of women in the work force; (2) the "greying" of the work force; (3) demographic shifts in the psychosocial characteristics and, in some areas of the United States; (4) the broadening of the cultural diversity of the work force.

Women

Although women are entering the work force in progressively increasing numbers, there is as yet no clear indication as to how these demographic changes will or should affect organizational design.

"Greying" of the Work Force

Presently, within the United States and elsewhere, as the post-World War II baby boom bulge moves through their working careers, the average age of the work force is increasing at the rate of about 6 mon/yr. This trend began in the late 1970s, and will continue through the 1990s, resulting in an older, more experienced, more mature and better trained work force. As a result of this shift, the work force has been becoming more professionalized; and organizational designs need to accommodate by becoming less formalized and decentralizing more of the decision making if employees are to feel fully utilized and remain motivated towards their work. Designers of future systems and modifiers of old ones will need to consider these factors carefully, particularly in designing those functions and jobs where high formalization and centralization traditionally have characterized the organization's structure.

Value System Shifts

Yankelovich (1979), based on extensive longitudinal studies of work force attitudes and values, notes that those workers born after World War II have very different views and feelings about work than their predecessors, and noted that these conceptions and values would affect profoundly work systems in America in the next two decades. Yankelovich refers to this group of workers as the "new breed." This new breed has three principal values that distinguish these workers from those of the mainstream of older workers: (1) the increasing importance of leisure, (2) the symbolic significance of the paid job, and (3) the insistence that jobs become less depersonalized and more meaningful. According to Yankelovich, when asked what aspects of work are more important, the "new breed" person stresses "being recognized as an individual," and "the opportunity to be with pleasant people with whom I like to work." From an organizational design standpoint, these values translate into a need for more decentralized and less formalized organizational structures, and attendant greater professionalism designed into individual jobs and human–system interfaces then has characterized traditional bureaucracies. Because decentralization and lack of formalization permit greater participation in the decision-making process by employees, these design characteristics both allow for greater individual recognition and respect for an employee's worth and enhance meaningful social relationships on the job.

18.5.2.3 Psychosocial Factors

Harvey, Hunt, and Schroder (1961) have identified a higher-order structural personality dimension, concreteness-abstractness of thinking, or *cognitive complexity* as underlying different conceptual systems for perceiving reality. In general, the degree to which a given culture or subculture (1) encourages by its child-rearing and educational practices *active* exposure to new experiences or diversity; and (2) provides—through affluence, education, communications media, and transportation systems—*opportunities* for exposure to diversity, the more cognitively complex the persons of that culture will become. Active exposure to diversity increases one's opportunity to develop new conceptual categories, and shades of grey within categories in which to store experiential data, or *differentiation*. With an active exposure to diversity one also learns new rules and combinations of rules for *integrating* information and deriving more insightful conceptions of problems and solutions. Thus, persons who experience an active exposure to considerable diversity develop a high degree of conceptual differentiation and integration, or abstract functioning. Conversely, relatively closed-minded approaches to new experience and/or a lack of exposure to considerable diversity leads to the development of rather limited differentiation and integration in one's conceptualizing of reality, or concrete functioning.

Relatively concrete functioning consistently has been found to be characterized by a high need for structure and order and for stability and consistency, a low tolerance for ambiguity, closedness of beliefs, absolutism, authoritarianism, paternalism, and ethnocentrism. Concrete persons tend to see their views, values, norms, and institutional structures as relatively static and unchanging. Conversely, abstract adult functioning is characterized by a low need for structure and order and for stability and consistency, a high tolerance for ambiguity, openness of beliefs, relativistic thinking, empathy, and a strong people orientation. Abstract persons also have a dynamic conception of their world and *expect* their views, values, norms, and institutional structures to change. (Harvey, 1963; Harvey et al., 1961).

Hendrick (1979, 1981, 1990), not surprisingly, has found evidence to suggest that relatively concrete work groups and managers function best under moderately high centralization, vertical differentiation and formalization, or mechanistic organizational designs. In contrast, although they can perform well in mechanistic organizations, cognitively complex or abstract work groups and managers prefer more organic organizational designs, characterized by relatively low levels of vertical differentiation, formalization, and centralization.

Because of (1) the shift since World War II to more permissive child-rearing practices that tend to encourage relativistic thinking and an openness to new experiences, (2) a higher level of general education, (3) greater affluence, and (4) the development of far superior communications and transportation systems since World War II, the majority of the work force born after World War II are functioning at a more cognitively complex level than the mainstream of their older colleagues. As the World War II work force moves out of our organizations, and more of the "new breed" move in, the trend toward a progressively more cognitively complex work force will continue (Hendrick, 1981). The result likely will be a progressively increasing demand for greater worker participation in decision making, more professionalized work, and less formalization—a trend that already is very apparent.

18.5.2.4 Cultural Diversity

In a number of urban areas in the United States, such as the Los Angeles Basin, immigration has resulted in the development of work forces that are far more culturally diverse than those that existed two decades ago. In recent years, it has become apparent that unless organizations accommodate to this diversity, it will adversely affect employee motivation and commitment (Jackson, 1992; Thomas, 1991). While much of the accommodation required has to do with changing organizational culture, decentralizing some aspects of decision making to allow greater employee control within their work groups appears to be one structural change that can facilitate this accommodation process; included here is the use of participatory ergonomics in designing or modifying work systems.

18.5.2.5 Personnel Subsystem Implications for Organizational Design

Many of the available data on personnel subsystem determinants of organizational design are in the form of either attitude survey results or projections from psychosocial and demographic studies. In spite of their somewhat tenuous nature, there is a convergence of these data dealing with different subsystem dimensions that lends credence to the conclusion: They all indicate a need for organizations of the future to be as vertically undifferentiated, decentralized, and lacking in formalization as their technology and environments will permit.

18.5.3 External Environment Characteristics

Critical not only to the functioning of organizations, but to their very survival, is their ability to adapt to their environment. In terms of open systems theory, organizations require monitoring and feedback mechanisms to follow and sense changes in their relevant task environments, and a capacity to make responsive adjustments. *Relevant task environments* refers to that part of the organization's external environment that is comprised of the firm's critical constituencies that can positively or negatively influence the organization's effectiveness. Negandhi (1977), based on field studies of 92 industrial firms in five different countries, has identified five external environments that significantly impact on organizational functioning.

1. **Socioeconomic.** Particularly the degree of stability, nature of the competition, and availability of materials and workers.
2. **Educational.** The availability of facilities and programs, and the educational level and aspirations of workers.

3. **Political.** The degree of stability, and the governmental attitudes toward (a) business (friendliness versus hostility), (b) control of prices, and (c) "pampering" of industrial workers.
4. **Cultural.** Social status and caste system, values and attitudes toward work, management, etc., and the nature of trade unions and union–management relationships.

The specific task environments that are relevant are different for each organization in type, qualitative nature, and importance. The particular weighted combination of relevant task environments for a given organization can be thought of as its *specific task environment.* A major determinant of an organization's specific task environment is its *domain,* or the range of products or services offered and market share (Robbins, 1983). Domain is important because it determines the points at which the organization depends on its specific task environment (Thompson, 1967).

A second major determinant of an organization's environment is its *stakeholders.* The stakeholders in the firm's environment include stockholders, lenders, members of the organization, customers, users, governmental agencies, and the local community. Each of these has interests in the organization, and a potential for action that could significantly affect the organization's future. The design of the organization must be capable of responding to the objectives and related actions of its stakeholders. For example, as a transformation agency, the organization's technological subsystem has to respond to technical, economic, market, and other aspects of the environmental domain; has to meet governmental regulations for pollution, safety, etc.; has to meet the needs of its stockholders; and has to satisfy the needs of its workers for interesting work, careers, and quality social relationships on the job, among others (Davis, 1982).

Of primary importance to organizational design is the fact that all relevant task environments vary along two highly critical dimensions: *Change* and *complexity* (Duncan, 1972). Degree of change refers to the extent to which a given task environment is dynamic or remains stable over time. The degree of complexity refers to whether the components of an organization's specific task environment are few or many in number. For example, does the firm interact with many or few customers, suppliers, competitors, governmental agencies, etc.? These two environmental dimensions in combination determine the *environmental uncertainty* of an organization. Figure 18.1 illustrates this relationship and gives

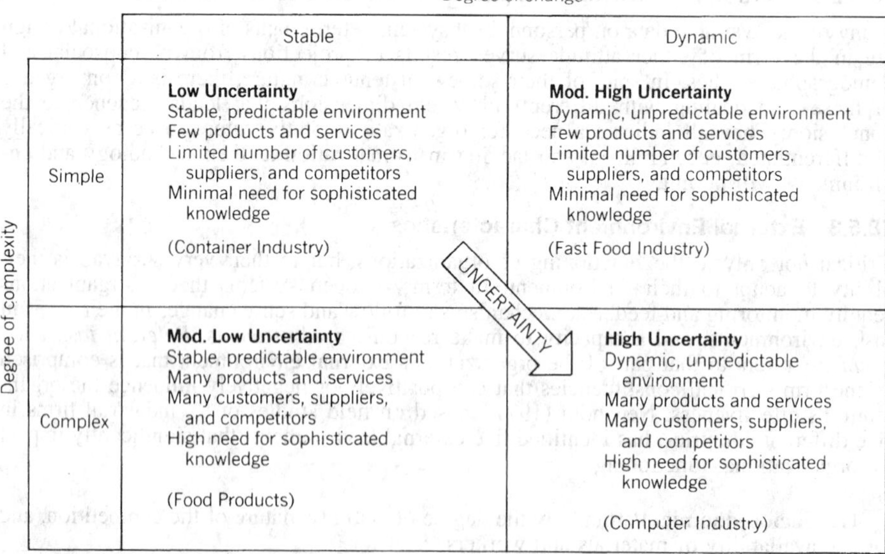

Figure 18.1 Environmental dimensions. Adapted from Robert Duncan, "What Is the Right Organization Structure? Decision Tree Analysis Provides the Answer," *Organizational Dynamics,* Winter 1979, p. 63.

the environmental characteristics and a representative industry for four different levels of uncertainty.

Two major generalizable models have been empirically derived for assessing environmental dimensions as determinants of organizational structure. One model focuses on environmental uncertainty. The other treats environmental uncertainty as one of several key environmental dimensions affecting structure, albeit the most important.

18.5.3.1 Burns and Stalker: Environmental Uncertainty

From their studies of 20 English and Scottish industrial firms, Burns and Stalker (1961) found that the type of organizational structure that worked effectively in a relatively stable and simple organizational environment was very different from that required for a more dynamic and complex environment. For stable environments *mechanistic* structures, characterized by high complexity, formalization, and centralization were the most efficient. These structures typically had routine tasks, programmed behaviors, and were slow in their capacity to respond to change. For unstable, more complex environments, *organic* structures, characterized by flexibility and adaptability, were the more successful. In organic structures, emphasis is placed on lateral rather than vertical communication, influence based on expertise and knowledge rather than authority and position, information exchange rather than directives from above, loosely defined responsibilities, and conflict resolution by interaction rather than by superiors. Thus, an organizational structure with low vertical differentiation, decentralization, and low formality is optimal (Burns and Stalker, 1961). Similar findings were implicit in Emery and Trist's (1965) analysis of the effects of environmental instability on sociotechnical systems. A comparison of the organizational design features of mechanistic versus organic systems is shown in Table 18.4.

18.5.3.2 Lawrence and Lorsh: Subunit Environment and Design Complexity

In a complex organizational environment, firms typically develop specialized units to deal with specific parts of the environment, or what was referred to earlier as the different relevant task environments. Lawrence and Lorsh (1969) conducted field studies to deter-

Table 18.4 Design Features of Effective Mechanistic and Organic Organizations

Mechanistic	Design Feature	Organic
	Complexity	
High: small spans of control	Vertical differentiation	Low: large spans of control
High: highly specialized tasks	Horizontal differentiation	Low: low task specialization
High: many required because of high differentiation and low autonomy	Integrating mechanisms	Low: few required because of low differentiation and high autonomy
	Centralization	
High: decision-making reserved for management		Low: decision-making relegated to lowest level at which competency and skill exists
	Formalization	
High: low autonomy and high differentiation require formal rules and procedures		Low: high autonomy and low differentiation require few formal rules and procedures

mine which type of organizational design was best for coping with different economic and market environments. They studied companies in various industries including plastics, food, and containers, which varied considerably in their degree of environmental uncertainty. Based on their study of these organizations, Lawrence and Lorsh identified five major variables that can be assessed regarding subunit environments to determine the optimal degree of horizontal differentiation. These are (1) *uncertainty of information* (low, moderate, high); (2) *time span of feedback* (short, medium, or long); (3) pattern of *goal orientation* (focus of tasks); (4) pattern of *time orientation* (short, medium or long); and (5) pattern of *interpersonal relationships* (task or social). In general, the more dissimilar the functions on one or more of these dimensions, the stronger the likelihood that they should be differentiated into separate subunits for effective functioning (Lawrence and Lorsch, 1969).

Lawrence and Lorsch also found that the greater the differentiation, the greater the need for integrating mechanisms. They noted that differentiation tends to encourage different viewpoints and thus greater conflict. Integrating mechanisms are needed to resolve these conflicts to the organization's benefit. Lawrence and Lorsch further noted that the more interdependent the tasks of the major subunits, the more information processing is required for effective integration.

Their research also found the level of environmental uncertainty to be of foremost importance in selecting the structure appropriate for effective functioning. Subunits with more stable environments (e.g., production) tended to have high formalization, whereas those operating in less predictable environments (e.g., research and development) had low formalization.

In summary, based on Lawrence and Lorsch's research, whenever an organization's design does not fit its mission, external environment, or resources, its functioning is likely to suffer. A mismatch between the organization's task and degree of differentiation results in a loss of relevant information. Differentiation, shifts in mission or resources, and environmental change each can create integration problems unless adequate integration mechanisms are ergonomically designed into the organization's structure and related processes. High levels of interdependence between subunits requires particularly careful ergonomic attention to information-processing mechanisms to ensure effective integration.

18.5.4 Relation of Macro- to Microergonomic Design of Work Systems

Through a macroergonomic approach to determining the optimal design of an organization's structure, many of the characteristics of the jobs to be designed into the system, and of the related human–machine and user–system interfaces, already have been prescribed. Some examples are as follows (Hendrick, 1991).

1. Horizontal differentiation decisions prescribe how narrowly or broadly jobs must be designed and, often, how they should be departmentalized.
2. Decisions concerning the level of formalization and centralization will dictate (a) the degree of routinization and employee discretion to be ergonomically designed into the jobs and attendant human–machine and user–system interfaces; (b) the level of professionalism to be designed into each job; and (c) many of the design requirements for the information, communications, and decision support systems, including what kinds of information are required by whom, and networking requirements.
3. Vertical differentiation decisions, coupled with those concerning horizontal differentiation, spatial dispersion, centralization, and formalization will prescribe many of the design characteristics of the managerial positions, including span of control, decision authority and nature of decisions to be made, information and decision support requirements, and qualitative and quantitative educational and experience requirements.

In summary, effective macroergonomic design drives much of the microergonomic design of the work system, and thus ensures *optimal ergonomic compatibility* of the system components with the system's overall structure. In sociotechnical system terms, this approach enables joint optimization of the technical and personnel subsystems from top to bottom throughout the organization. The result is greater assurance of optimal *system* functioning and effectiveness, including productivity, safety, comfort, intrinsic employee motivation, and quality of work life.

In contrast, a purely microergonomic approach has a high probability of creating systems in which the personnel subsystem is forced to adapt to the system's technology and structure in a "pounding square pegs into round holes" fashion. Beginning with the previously noted classic longwall coal mining studies by the Tavistock Institute (Trist and Bamforth, 1951), the organizational literature is full of examples which consistently have shown that this lack of compatibility not only directly adversely affects system productivity and efficiency, but employee motivation, commitment, and intrinsic job satisfaction as well (e.g., see Argyris, 1971 for a summary of the findings of a number of prominent U.S. researchers).

A widely accepted view among system theorists and researchers is that organizations are *synergistic*: That the whole is more, or less, than the simple sum of its parts. Because of this synergism the following tend to occur in our complex organizations (Hendrick, 1995).

18.5.4.1 When Systems Have Incompatible Organizational Designs

When organizational structures are grossly incompatible with their sociotechnical system characteristics, and/or jobs and human–system interfaces are incompatible with the organization's structure, the whole is *less than* the sum of its parts. Under these conditions, we can expect the following to be poor: (1) productivity, especially *quality* of production; (2) accident rates and adherence to safety standards and procedures; and (3) motivation and related aspects of quality of work life (e.g., stress). Further, these detriments may be *greater* than a simple sum of their parts would indicate.

18.5.4.2 When Systems Have Effective Macro- and Microergonomic Designs

When organizations have been effectively designed from a macroergonomics perspective, and that effort is carried through to the microergonomic design of jobs and human–machine, human–environment, and user–system interfaces, then production quality, safety, and quality of work life will be much *greater* than the simple sum of the parts would indicate.

18.5.4.3 Implications for the Potential of Organizations

Assuming the above two propositions are indeed true, then macroergonomic approaches have the potential to improve productivity, safety, health, and the quality of work life *exponentially,* rather than linearly. For example, quality measures, accident rates, scrap rates, and employee job satisfaction indices should not show the typical 10% to 20% improvement often seen as the result of typical *successful* microergonomic and organizational development efforts. Instead, improvements of 60% to 90% more typically should occur (and, in some cases, much greater) and be sustainable. There is some recent empirical support for this expectation. In several industrial studies in Japan and the United States by Nagamachi and Imada (1992) using macroergonomic approaches, reductions in both industrial injuries and vehicle accidents by 76% to over 90% were achieved. Using a macroergonomic approach to organizational change and total quality management, Rooney et al. (1993) recently reported a greater than 70% reduction in lost time accidents and injuries at the L. L. Bean Corporation.

18.5.5 Consideration of Job Design Characteristics in Macroergonomic Design

In macroergonomically designing the overall organization and work system, it is important to continually be aware of what the impact of these macro-design decisions is likely to be on the design of the individual jobs within the organization. Over two decades ago, Hackman and Oldham (1975) empirically identified five specific job characteristics which, especially for growth oriented employees, appear to be critical to intrinsic job motivation, employee self-worth, stress reduction, and satisfaction. These are task *variety* or having different (meaningful) things to do in one's work, *identity* or sense of job wholeness, *significance* or perceived job meaningfulness, *autonomy* or control over one's work, and *feedback* or knowledge of results. Today, I/O psychologists commonly identify the absence of these characteristics as often resulting in dehumanized jobs that reduce psychological meaningfulness, a sense of responsibility, and knowledge of results; and that frequently appear to lead to high stress, demotivated employees, job dissatisfaction, absenteeism, and reduced productivity (Organ and Bateman, 1991).

Since their identification, the importance of these job characteristics has been found repeatedly across many types of organizations and work situations. For example, Bammer

(1990) conducted a meta-analysis of field studies of work-related neck and upper limb disorders among computer operators reported internationally during the 1980s. Results of her analysis showed no consistent relationship of nonwork factors to employee musculoskeletal disorders. The data on biomechanical factors led her to conclude that (microergonomic) efforts to effect biomechanical improvements are important and should be encouraged; but, by themselves, these improvements are insufficient to reduce work related musculoskeletal disorders. The factors that did consistently relate to musculoskeletal disorders across studies were work organization variables. Bammer concluded that "improvements in work organization to reduce pressure, and to increase task variety, control, and the ability for employees to work together must be the main focus of prevention and intervention." She further notes that "ironically, such improvements in work organisation generally also lead to increased productivity" (Bammer, 1993, p. 35).

In essence, what Bammer identified as the key correlates of work-related musculoskeletal disorders (WMSDs) were Hackman and Oldham's job characteristics plus the opportunity to satisfy social needs on the job. Bammer's conclusions are further supported by a more recent major U.S. study by NIOSH in collaboration with U.S. West Communications and the Communications Workers of America (Hales et al., 1992). Prior to this study, in an attempt to reduce WMSDs, U.S. West had made a large investment in microergonomically improving VDT operator workstations and lighting. Over a year after these improvements were completed, there was no significant reduction in the incidence rate of WMSD symptoms—a finding that adds credence to the importance of macroergonomically considering the design of the work system, and not just the microergonomic design of workstations, to effect real organizational improvement.

In summary, the literature suggests that, in designing the overall organization and work system, one should be careful to ensure that the design will enable building Hackman and Oldham's five job characteristics into the specific jobs. Bammer's meta-analysis also suggests that attention should be given to ensuring that persons can satisfy social needs on the job.

There are many other facets to job design; and these may be found in Chapter 16.

18.6 CHOOSING THE RIGHT STRUCTURAL FORM

Macroergonomics involves more than considering how sociotechnical system variables should shape the basic dimensions of the work system. Ultimately, these factors must be integrated into an overall structural form. To this end, a variety of *types* of overall structural forms are available to system designers. Just as the design of the individual dimensions of organizational structure can enhance or inhibit organizational functioning, and just as each of these design decisions has particular ergonomic design implications, the same also is true for the type of overall structural form chosen for the work system. This section considers four general types of organizational structure most commonly found, and discusses the advantages and disadvantages of each. Finally guidelines for determining when each type is, and is not likely to be appropriate are provided.

The four general types of overall organizational structure are the (1) classical ***machine bureaucracy***, (2) ***professional bureaucracy***, (3) ***matrix organization***, and (4) ***free-form design*** (Robbins, 1983). It should be noted that large, complex organizations may, and often do have relatively autonomous units with different overall forms. In general, the smaller the organization, the greater is the likelihood that it will have a single overall type of work system structure.

18.6.1 Classical or Machine Bureaucracy

This form of work system has its roots in two streams of thought: ***scientific management*** and the ***ideal bureaucracy***.

18.6.1.1 Scientific Management

The end of the nineteenth century was characterized by the accumulation of resources and a rapidly developing technology in American and European industry. During this period, labor became highly specialized, and industrial engineers were called upon to help design organizations and optimize efficiency. One of these engineers, Frederick W. Taylor (1911), stood out among his colleagues and has had a major impact on the shaping of classical organizational theory (and organizations throughout the world) through his concept of *scientific management.* The essence of Taylor's concepts of work system organization are implicit in his four basic principles of managing (Szilagyi and Wallace, 1990, p. 662).

First. Develop a science for each element of man's work that replaces the old rule-of-thumb method.

Second. Scientifically select and train, teach, and develop the workman. In the past he chose his own work and trained himself as best he could.

Third. Hardily cooperate with the men in order to ensure all of the work is being done in accordance with the principles of the science that has been developed.

Fourth. Provide equal division of work and responsibility between the management and the workmen. The management takes over all work for which they are more qualified than the workmen. In the past, almost all the work and the greater part of the responsibility were thrown upon the men.

As one can see, Taylor advocated scientific analysis, rather than pure common sense and intuition, as the basis for job and work system design.

18.6.1.2 Ideal Bureaucracy

Although heavily influenced by Taylor's concepts, the classical bureaucratic design was conceptualized by Max Weber at the beginning of the twentieth century. Weber recommended that organizations adhere to the following work system design principles (1946, p. 214).

1. All tasks necessary to accomplish organizational goals must be divided into highly specialized jobs. A worker must master his trade, and this expertise can be more readily achieved by concentrating on a limited number of tasks.
2. Each task must be performed according to a "consistent system of abstract rules." This practice allows the manager to eliminate uncertainty due to individual differences in task performance.
3. Offices or roles must be organized into hierarchical structure in which the scope of authority of superordinates over subordinates is defined. This system offers the subordinates the possibility of appealing a decision to a higher level of authority.
4. Superiors must assume an impersonal attitude in dealing with each other and subordinates. This psychological and social distance enables the superior to make decisions without being influenced by prejudices and preferences.
5. Employment in a bureaucracy must be based on qualifications, and promotion is to be decided on the basis of merit. Because of this careful and firm system of employment and promotion, it is assumed that employment will involve a lifelong career and loyalty from employees.

Weber assumed that strict adherence to these organizing principles was the "one best way" to achieve organizational goals. By implementing a structure that emphasized efficiency, stability, and control, Weber believed organizations could obtain optimal efficiency (Szilagyi and Wallace, 1990).

Collectively, the theoretical principles of Taylor and Weber resulted in what today is known as the *machine bureaucracy* type of organizational structure. Its basic structural characteristics are as follows (Robbins, 1983).

1. **Division of labor.** Each person's job is narrowly defined, and consists of relatively routine and well-defined tasks.
2. **A well-defined hierarchy.** A relatively tall, clearly defined, formal hierarchical structure of position and offices in which each lower office is under supervision and control of a higher one. Tasks and functions tend to be grouped by function. Line and staff functions are clearly distinguished and are kept separate.
3. **High formalization.** A dependence on formal rules and procedures to ensure uniformity and to regulate employee behavior.
4. **High centralization.** Decision making is reserved for management. There is relatively little employee decision discretion.

5. **Career tracks for employees.** Members are expected to pursue a career within the organization; and career tracks form part of the organizational design for all but the most unskilled positions.

18.6.1.3 Advantages

The primary advantages of the machine bureaucracy are efficiency, stability, and control over the organizations functioning. Narrowly defined jobs with a clear set of routinized tasks minimize the likelihood of error, better enable individuals to know their own function and the roles of others, require comparatively few prerequisite skills, and minimize training time and costs. Formalization ensures stability and a smooth, integrated pattern of functioning. Centralization ensures control and further enhances stability.

18.6.1.4 Disadvantages

There are at least two major disadvantages to the machine bureaucracy form of work system. First, this form tends to result in jobs that are lacking in intrinsic motivation, and that fail to utilize adequately the mental and psychological capacities of the workers. Second, machine bureaucracies tend to be inefficient in responding to environmental change and nonroutine situations.

18.6.1.5 When to Use

When (1) the education and skill levels of the available labor pool are relatively low, (2) system operations tend to be repetitive or otherwise can largely be routinized, and (3) the relevant external environments are comparatively simple, stable, and/or predictable (i.e., are not highly uncertain).

To the extent that the above stated conditions do not exist, one of the other three forms of overall organizational structure is likely to be more effective.

18.6.2 Professional Bureaucracy

Professionalism was defined earlier as the degree of training and education required by the design of specific jobs and related human–system interfaces. The professional bureaucracy is one that is designed so as to rely on a relatively high degree of professionalism in the jobs that comprise the work system. Its major differences from the machine bureaucracy design are that jobs are more broadly defined, somewhat less routinized, and allow for greater employee decision discretion (Robbins, 1983). Thus, there is less need for formalization, and *tactical* decision making is decentralized. Like machine bureaucracies, positions are grouped functionally, are hierarchical, and *strategic* decision making often remains centralized.

18.6.2.1 Advantages

There are at least three major advantages to the professional, as compared with the machine bureaucracy. First, professional bureaucracies can more effectively cope with complex environments and nonroutine tasks. Second, jobs tend to be more intrinsically motivating and to better utilize the mental and psychological capabilities of employees. Third, less managerial control and tactical decision making are required, thus freeing management to give greater attention to long-range planning and strategic decision making.

18.6.2.2 Disadvantages

Professional bureaucracies are not as efficient as machine bureaucracies for coping with relatively simple, stable environments. They require a more highly skilled work force, and associated training time and expense. Control is less tight, and both line and staff functions are likely to be less clear. The management skills required also tend to be more sophisticated (i.e., a greater reliance on tolerance for ambiguity, and on persuasive and facilitation skills rather than on a simple and direct authoritarian style.

18.6.2.3 When to Use

A professional bureaucracy is preferred when the external environment is fairly complex, somewhat unstable, and there is an available applicant pool of professionalized workers. This form of design is less optimal than machine bureaucracies for highly repetitive operations with a simple, stable environment; or if the available management pool is highly authoritarian and concrete, rather than cognitively complex in its functioning.

18.6.3 The Concept of Adhocracy

Although more severe for machine bureaucracies, a major disadvantage of both forms of bureaucracy is that they tend to be inefficient in responding to highly complex or dynamic relevant external environments. Primarily because of this shortcoming, two more recent forms of organization have evolved. These are the *matrix* and *free-form* designs. Collectively, these two newer forms are known as *adhocracy* designs. An adhocracy can be described as a "rapidly changing adaptive, temporary system organized around problems to be solved by groups of relative strangers with diverse professional skills" (Bennis, 1969, p. 45). In terms of their structural dimensions, adhocracies are characterized by moderate to low complexity, low formalization, and decentralization (Robbins, 1983). Because adhocracies are staffed primarily by professionals, horizontal differentiation tends to be high. In contrast, vertical differentiation tends to be low. This low vertical differentiation reflects low formalization, decentralized decision making, and the low need for supervision because of the high level of professional staffing. Instead of formal rules and procedures, flexibility and rapidity of response are emphasized. It is for this reason that a high level of professionalism and the absence of layers of administration are necessary.

18.6.3.1 Advantages

Adhocracies had their origin in the task forces of World War II. The military created ad hoc teams to accomplish specific missions, and then disbanded them when the mission was completed. The roles played by team members were flexible and often interchangeable. Subunits could be added or deleted as required. As a result of these features, these teams had the ability to react quickly, efficiently, and creatively to a dynamic environment, and thus be highly effective. These same characteristics and resultant advantages characterize modern adhocracies (Robbins, 1983). When the ability to be adaptive and innovative, and to respond rapidly to changing situations and objectives, is paramount; and when these responses require collaboration of persons possessing different specialties, the adhocracy forms of organization are more effective than the bureaucratic forms (Robbins, 1983).

18.6.3.2 Disadvantages

All adhocracies have at least three major disadvantages. These are (1) *conflict*, (2) *social and psychological stress*, and (3) *inherently inefficient structures* (Robbins, 1983).

1. **Conflict.** Because there are no clear boss–subordinate relationships the lines of authority and responsibility are ambiguous. Thus conflict is an integral part of adhocracy.
2. **Psychological and Sociological Stress.** Because the structure of teams or units is temporary, work role interfaces also are not stable. The establishment and dismantling of human relationships is a slower psychosocial process and is stressed any time there is significant organizational change. In particular, concrete functioning employees are likely to be strained by these stresses, and find it difficult to cope.
3. **Inherent Inefficiency.** By comparison with bureaucratic forms of organization, adhocracies lack both the precision and the expediency that comes with routinization of function and structural stability. It is only where these inefficiencies are more than offset by the gains in efficiency in terms of responsiveness and/or innovation that an adhocratic design should be considered.

18.6.4 The Matrix Design

Of the two basic forms of adhocracy, the matrix design has been the more widely used. Basically, the matrix design combines departmentalization by function with departmentalization by product line or project. Functional departments, typical of bureaucracies, exist and tend to be lasting. Unlike bureaucracies, however, members of the functional departments are farmed out to project or product teams as new projects or product lines develop, and the combined technical expertise of the individual departments is required. As the need for a given department's professional input to the interdisciplinary team is no longer required, or the level of effort reduces, individuals return to their "home" department or transfer to another team. The product or project's manager supervises the team's interdisciplinary effort, but each team member also has a functional department supervisor. The matrix design thus breaks a fundamental design concept of bureaucracy, *unity of command.* A typical matrix design is depicted in Figure 18.2.

18.6.4.1 Advantages

The primary advantage of matrix designs is that they afford the best of two worlds: The stability and professional support of depth of functional departmentalization; and the interdisciplinary response capability of ad hoc teams, such as characterize free-form designs.

18.6.4.2 Disadvantages

Over and above those characteristics of all adhocracy designs that cause problems, the major disadvantage of matrix organizations is that employees must serve two bosses. One is their functional department head, who tends to be relatively long term oriented and somewhat remote from the team member's immediate tasks; the other is the project team director who tends to be comparatively short term oriented but immediate to the employees' present tasks. Serving two masters with overlapping supervisory responsibility and different goals, responsibilities, and orientations frequently creates conflict and can disrupt organizational functioning. A second major problem for the employee is that when assigned too long to a project or projects, the employee may have difficulty keeping technically current and may lose contact with his or her functional department. Both of these consequences can adversely affect the employee's career.

18.6.4.3 When to Use

The matrix organization is particularly well suited for responding to dynamic and complex relevant external environments where both interdisciplinary responsiveness and providing for functional depth in individual disciplines are considered essential.

18.6.5 Free-Form Designs

The newest of our four general types of organizational structure is the *free-form.* In its pure form, the free-form organization resembles an amoeba in that it continually changes its shape in order to survive (Szilagyi and Wallace, 1990). The focus of free form designs is responsiveness to change in highly dynamic, complex, and competitive environments.

In free-form designs, functional departmentalization is replaced by a *profit center* arrangement. Profit centers are results oriented and are managed by teams. Collectively, they constitute the organization. Accordingly, free-form designs are characterized by very low formalization and hierarchical differentiation; and decision making is highly decentralized. A very heavy reliance is placed on professionalism as expressed through participation, internalized formalization, and autonomy. As with matrix organizations, project teams are created, changed, and disbanded as required to meet organizational goals and problems. Unlike matrix organizations, there is no underlying functional departmentalization or "home" structure. To function effectively, managers and employees alike need to possess a great deal of personal flexibility, tolerance for ambiguity, and ability to handle change in free-form organizations. They thus need to be at least somewhat cognitively complex or abstract in their conceptual functioning.

18.6.5.1 Advantages

The primary advantage and basic purpose of free-form designs is the ability to respond to highly competitive, complex, and dynamic relevant external environments with speed and innovation.

18.6.5.2 Disadvantages

The free-form organization has essentially the same disadvantages as matrix adhocracies, only to a greater extent. Conflict, sociopsychological stress, and inherent administrative inefficiency are an integral part of a continuously changing and amorphous organizational structure. It thus requires a highly professionalized work force to succeed.

18.6.5.3 When to Use

Free-form designs should be considered whenever (1) the organization's success or survival critically depends on speed of response and innovation, and (2) the available work force and management pool are highly professionalized. These features tend to characterize small-to-medium-sized high-technology organizations operating in highly dynamic,

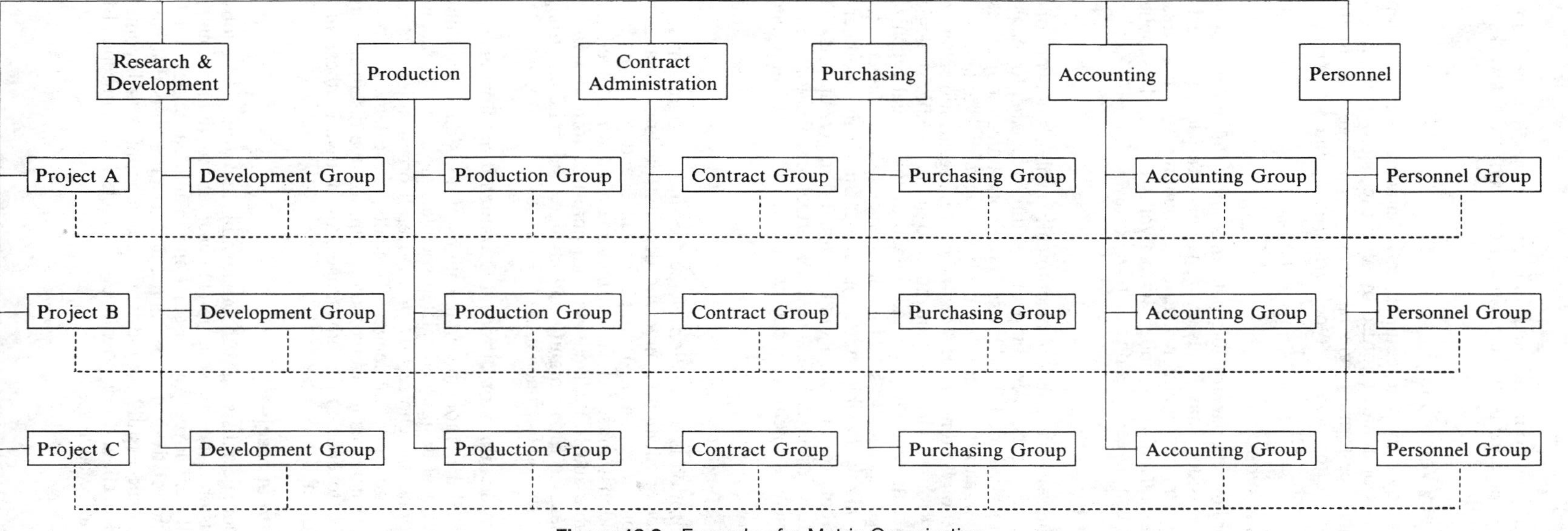

Figure 18.2 Example of a Matrix Organization

complex, and competitive environments; and semi-autonomous "outlaw" subunits of large bureaucratic or matrix organizations (e.g., the Lockheed "skunkworks"; the Apple MacIntosh design group).

18.6.6 New Adhocracy Variations

In response to both the dynamic and complex environments faced by many of today's organizations, and the opportunities afforded by recent technological advances, new variations of the matrix and free-form designs have begun to emerge. Among these, in order of structural flexibility, are the *modular* type, *virtual* type, and *barrier free* type (Dess, Rasheed, McLaughlin, and Priem, 1995).

18.6.6.1 The Modular Type

The modular type of organizational structure outsources nonvital functions while retaining full strategic control. Outsiders may be used to manufacture parts, handle logistics, and perform housekeeping services, maintenance, or accounting activities. Essentially, the "organization" is a central hub surrounded by networks of outside suppliers and specialists. Modular parts can be easily added or taken away.

18.6.6.2 The Virtual Type

The virtual type of organization consists of a continually evolving network of independent companies—suppliers, customers, even competitors—linked together to share skills, costs, and access to one another's markets, and pursue common strategic objectives. Participating firms in a virtual organization may be involved in multiple alliances In its purest form, a virtual organization need not have a central office, organization chart, or hierarchy. Participating firms may form a virtual organization to attain specific strategic objectives, and when they are met, disband. A major advantage of the virtual organization is that each subunit firm brings a particular set of competencies to the alliance, thus creating a more competitive, yet highly flexible (virtual), entity. Unlike the modular firm which maintains full strategic control, participants in the virtual firm give up part of their control and accept interdependent destinies. In essence, the virtual organization uses a collectivist strategy that better enables them to cope with environmental uncertainty and be competitive.

18.6.6.3 Barrier Free Type

The barrier free organization is a variation of the free-form design. Emphasis is placed on a project or product orientation, democratic decision making, organizational rather than individual development, rewarding team performance, and very loose or fluid boundaries between organizational units. Yet, it still may maintain some functionally oriented units that are more stable and formalized. As with pure free form designs, a high level of professionalism, tolerance for ambiguity and change, and building strong trust relationships are critical to successful functioning.

18.6.7 Ergonomic Implications of Different Organizational Forms

In Section 18.5, we noted the ergonomic implications of decisions regarding the specific dimensions of organizational design and related sociotechnical system factors. Over and above these considerations, the choice of *type* of overall organizational structure brings its own unique set of ergonomic challenges. Typically, these have to do with how to design human–machine interfaces to overcome or minimize the disadvantages of a given organizational form and to enhance its advantages. Illustrative of these human factors challenges are the following.

18.6.7.1 Machine Bureaucracies

How can human factors specialists help in broadening the structure of jobs and in increasing employee decision discretion so as to (1) make jobs more intrinsically interesting and to utilize more fully employee psychosocial and mental capacities; yet (2) maintain the stability, control, and relatively low personnel and training costs that have made this form of organization so successful? As the work force continues to become better educated, older and experienced, and less cognitively concrete in its conceptual functioning, these issues will have to be addressed.

18.6.7.2 Professional Bureaucracies

How can human factors specialists help in designing human–system interfaces so as to take full advantage of the greater skills and internal formalization of a professionalized work force? In particular, how can human factors specialists help professional bureaucracies to become more responsive to dynamic environments and nonroutinized problems?

18.6.7.3 Matrix Organizations

How can human factors specialists help facilitate the dual objectives of interdisciplinary project team membership and functional departmental membership? In particular, what kinds of inputs to human–system interface design can human factors specialists make to minimize the problems of dual membership and attendant disruptions of organizational functioning?

18.6.7.4 Free-Form Designs

How can human factors specialists contribute to human–system interface design so as to effectively accommodate these interfaces to the flexibility and fluidity of task assignment which these organizational designs require. Can human factors specialists also contribute to making these organizational structures more inherently efficient in the light of their design fluidity?

By using a macroergonomic approach, such as described herein, and carrying through on macroergonomic solutions to the microergonomic level of design, human factors specialists should be capable of making important contributions to meeting the above described challenges. In doing so, the field of human factors should contribute directly to organizational theory and to advancing the state of the art of organizational design. A number of methods have been adapted or developed for macroergonomics to enable us to meet these challenges.

18.7 COMMON MACROERGONOMIC METHODS

As a formally recognized subdiscipline of human factors, macroergonomics is less than two decades old, yet a number of methodologies have been successfully adapted for its use from both microergonomics and organizational psychology. In addition, unique technologies also have emerged. For example, from a review of the Proceedings for the first three international symposia on Human Factors in Organizational Design and Management (Hendrick and Brown, 1984; Brown and Hendrick, 1986, Noro and Brown, 1990) the following unique technologies were identified (Hendrick, 1991): participatory ergonomics (eight papers); user systems analysis (five papers); systems analysis modeling (four papers); ergonomic work analysis (two papers); work systems design (two papers); usability test methodology; function analysis modeling; fuzzy concepts as a macroergonomic tool; modified garbage can model for evaluating organizational design alternatives; task allocation charting; systematic organizational design methodology (SORD) for designing organizational units; use of CAD to simulate an organization; and organizational requirements definition tools (ORDIT) for assisting in specifying organizational requirements for information technology systems. In addition, many of the studies reported in these proceedings adapted and used traditional organizational survey methods, such as field observation, organizational questionnaire surveys, and interviews.

These traditional organizational survey methods, along with the most commonly used macroergonomic method, *participatory ergonomics*, and a hybrid method widely used in France and Brazil that actually predates participatory ergonomics as a formal methodology, *ergonomic work analysis,* will be described briefly in this section. More extensive and detailed information on organizational survey methods can be found in introductory texts on survey design, such as *Organizational Surveys* by R. Dunham and F. Smith (Scott, Foresman, Glenview IL, 1979).

18.7.1 The Ergonomist as an Organizational Change Agent

One of the major methodological impacts of macroergonomics has been to change the ergonomist's role from one of primarily being a (microergonomics) design team specialist to that of macroergonomic management consultant and organizational change agent (Kourinka and Forcier, 1995; Hendrick, 1991). Organizational design changes frequently require the assistance of persons with an outside perspective and special expertise to facilitate the change process. Traditional types of organizational change agents have included such professionals as *organizational development specialists*, who focus primarily on organizational processes; and analysis from the top types, such as *operations research specialists,* who focus on structural analysis and change. More recently, the need for *human–system interface* change agents, who can combine expertise in the behavioral sciences and industrial and systems engineering with specialized training in the *organizational* aspects of human–system interface design, has emerged. These *macroergonomics* specialists thus find themselves operating as consultants to management regarding ways to assess current organizational designs and, as required, making recommendations for improving them. This need becomes even more evident when an organization is introducing new technology; which, as we know from the sociotechnical systems literature, disrupts and changes the various human–system interfaces. A human–system interface expert thus is needed to assist management in determining what macroergonomic changes are needed, and in planning for and implementing these work system design changes.

18.7.2 Field Study Method

One of the oldest and most widely used methods for studying organizational or work system designs is the *field study.* The field study is variously referred to as systematic or naturalistic observation and as real-life research. All of these terms, when taken together, provide a good description of this method. It involves going out into the organization (field) to systematically observe events as they occur in their natural environment. The goal of this approach is to attempt to identify structural and process characteristics of the organization's design that either enhance or inhibit effective organizational functioning; and to glean insight into how to modify the design to either sustain effective functioning or, more often, to correct the design to improve functioning. In some cases, data gained from interviews, questionnaire surveys, various performance measures, and employee or customer complaints may provide a particular focus for the field observations.

The major advantage of the field study is *realism.* One is not relying purely on opinions of employees, customers, or managers, but is observing first-hand what actually happens. However, there are several disadvantages to this approach. First, the presence of the observer, in itself, constitutes the interjection of a new variable which may distort the true realism of the situation. Accordingly, special care is needed to ensure that the observer's presence has a negligible effect on normal operations. Second, it can be very difficult to determine which of many possible variables, of which the human–system interfaces are but one subset, or variable interactions actually are underlying causes of the observed activities. Finally, one must wait for things to happen which have significant implications for the organization's design; and those things may not happen very often, or for some time.

A variation of the field study which helps to overcome these last two disadvantages is the *field experiment.* The field experiment is similar to the field study, except that one deliberately manipulates one or more of the organizational design variables and sees what effect it has on organizational functioning. This may be done in a particular part of the organization, and, if it proves effective, it can then be implemented on a larger scale. Because of its greater efficiency, the field experiment has become increasingly popular—particularly where other data sources have suggested changes to the organization's design.

18.7.3 Organizational Questionnaire Surveys

Perhaps the most widely used method for evaluating organizational functioning and identifying deficiencies in the design of an organization's structure and processes is the written survey questionnaire. Some form of periodic organizational survey, based on the use of some form of organizational assessment questionnaire, is used by most large corporations and many smaller ones. Many organizations have in-house staff competence for such work and numerous consulting, and research firms offer professional services to clients (Blackstrom and Hursh-Cesar, 1981). A number of generic survey questionnaires have been developed and validated for assessing such things as organizational climate, supervisory or managerial behavior and practices, and job design characteristics. Companies often use these, but then identify specific survey topics and develop their own questions to assess those issues. The sources of these additional topics can be quite varied and often include data from interviews, field study observations, various types of company and personnel performance records, or simply managerial intuition or interest.

A very popular variation of the organizational questionnaire survey method is the *survey feedback* method. With this method, data gained from the survey is summarized and subgrouped statistically and naratively by organizational level, department, project, and/or by other meaningful subgroupings. These data, along with comparative data for the company as a whole, similar results from previous surveys, etc., are then "fed back" to the individual organizational units. Most commonly, the unit manager, who has received special training on how to use survey feedback data, then participates with the unit's employees in reviewing the data and interpreting what it really means. Where the data indicates deficiencies, the unit's employees work participatively with the manager to identify underlying causes and proposed solutions. Problems with the design of the organization's structure or processes can be included. At times, one or more organizational consultants may observe or participate in the process. The feedback data from these surveys, and the interpretation of these data by the individual managers and organizational units, can be excellent input for a macroergonomic analysis of the work system.

Constructing an organizational survey questionnaire that can yield useful data is a very tricky process, and should not be attempted without the help of a trained expert, such as an I/O psychologist. One further caution, it is essential to pilot test the survey, including instructions, etc., with at least a small representative sample of the employees to be surveyed prior to finalizing the questionnaire. Invariably, unforeseen problems related to the clarity of instructions and interpretation of the wording of some of the questions are identified during the pilot test.

18.7.3.1 Advantages

The main advantages of the survey method are as follows (Seashore, 1987):

1. **Anonymity and confidentiality.** Questionnaire surveys can be conducted with assured anonymity or confidentially for individuals, thus allowing (although not guaranteeing) candor in responding about sensitive issues without fear of punishment. As a result, important information that is not part of existing management information systems or other data sources may be obtained.
2. **Quantitative analysis.** With a sufficiently large respondent sample, quantification and statistical analysis of the data becomes possible. The quality of the data can be assessed, subgroups can be compared, variations as well as consensus uniformities can be displayed, and correlations among groups and with other variables can be determined.
3. **Replication.** By using standardized formats and questions, the survey can be readministered over time and the results compared to previous surveys to detect changes. This can be particularly valuable in assessing the impact of work system design changes.
4. **Cost and time effectiveness.** A questionnaire survey can elicit, at relatively low cost and within a short period of time, information from a large group of employees covering a wide range of topics.
5. **Representativeness.** Respondents readily can be selected in a way that ensures representativeness, or absence of serious sample bias.
6. **Linking of results with other data.** In some situations (with consent of the respondents and tight confidentiality provisions) the survey data can be linked to "open" records, such as absences, accidents, productivity, etc., as well as to organizational design changes to allow the assessment of associations between what people say and what actually happens.

18.7.3.2 Limitations and Risks

1. **Ambiguity of purpose.** The initiators of a survey, usually an upper management or staff group, may find it impossible to reach sufficient consensus about the purposes of the survey, the priorities of content, or the procedures for post-survey review and interpretation of the results (Sirota, 1974). Management (and unions) may already be so committed to the policies and work system design factors at issue that the results of the survey are unlikely to have any real effect. Such lack of response can ultimately undermine the whole process.
2. **Distrust.** Some initial level of trust in management and union leadership is needed for employees to volunteer freely, accept assurances of anonymity or confidentiality, respond seriously to the questions, or have confidence that sensible and considered interpretations and actions will follow.

3. **Unacceptable topics.** In any organization, certain topics of high interest may be disallowed on grounds that the employees have little information, context, or experience as a basis for forming opinions, or on grounds that the issue is too controversial. For example, management may choose not to solicit employee opinions on practices which, for legal reasons, cannot be altered; unions may not want to have available open information on employees views of issues currently under negotiation.
4. **Organizational disturbance.** The effective conduct of a survey requires consultation and information exchange with employees, scheduling of "nonwork" activities such as feedback sessions, and possible absence from the workplace for taking the questionnaire. Also, the survey is unavoidably a public event that may call attention to quiescent issues or induce unrealistic expectations about subsequent actions. Any of these potentially can disrupt normal operations.

18.7.3.3 The Utility of Surveys

Surveys can serve a variety of purposes in the organization analysis, design, and change implementation process. Some of the more common are as follows (Seashore, 1987).

1. **Prediction.** The survey results may better enable managers to be aware of employee intentions and/or to predict employee behavior or actions in response to either no change, or proposed changes to the organization.
2. **Explanation.** Surveys can be quite useful in helping determine the underlying causes of negative employee attitudes or behavior, or other dysfunctional aspects of the operation; or of positive employee attitudes and performance.
3. **Monitoring change.** Whenever organizational changes are implemented, it is essential to be able to monitor their effectiveness. It is equally important to quickly identify and correct unintentional and dysfunctional side effects of the change. With major changes to an organization's design, unintentional effects invariably happen, and if not identified and corrected, may well undermine the whole change effort. The questionnaire survey offers a relatively inexpensive and timely means for carrying out a major facet of this monitoring process.
4. **Evaluating programs.** Often related to "3" above, a major use of surveys can be to evaluate the impact or effectiveness of change programs, and to provide a timely basis for fine-tuning a change program during the course of the implementation process.
5. **Deciding.** At times, management may be faced with more than one design change alternative for correcting some deficiency; or with several incompatible lines of action. The questionnaire survey can provide an excellent participative means for quickly and economically deciding such issues. An additional side benefit is that, by participating directly in the decision process, employees are more likely to actively support the decision, even if they personally preferred an alternative solution.

18.7.4 Interview Surveys

A frequently used organizational survey method, and one which often is used in conjunction with survey questionnaires or field observations, is the interview. In organizational assessment, the most frequently used interview method is the *stratified semistructured interview.* The interview is *stratified* in that a stratified representative sample of the employee and managerial groups of interest are interviewed. This is done to ensure that perceptions and opinions will be solicited from the full range of employee functions and levels within the organization, and thus avoid too limited or too biased a set of responses. Often, because they are limited in number and have proportionately greater influence than individual lower-level employees on organizational functioning, a considerably greater proportion of the managers than workers will be included in the stratified sample. The interviews are semistructured in that the interviewer usually has a good idea of at least a number of the topics on which he or she wants input, so has some prepared questions; but follow-up questions will depend on the nature of the responses the interviewee gives to the prepared questions; and the interviewee may bring up important topics which were not anticipated by the interviewer.

18.7.4.1 Advantages

The major advantages of the interview is that it (1) enables the interviewee to observe both verbal and nonverbal responses to questions, and sometimes the nonverbal responses provide more relevant data than the verbal; (2) allows the interviewee to follow up on initial responses and explore unanticipated issues or unexpected responses in greater depth than is possible with a questionnaire; and (3) assuming the interviewer is effective in establishing rapport with the interviewee, the interviewee frequently will become more ego involved in the process than in responding to a comparatively impersonal written questionnaire, and thus put more thought and effort into his or her responses.

18.7.4.2 Disadvantages

The major disadvantages of the interview are time and cost. First, there is the cost of either training or hiring a skilled interviewer(s), which can be considerable if more than a very few persons are to be interviewed. Second, whereas a questionnaire survey typically can be administered to a large group within one-half to 2 hr, each interview typically will take an hour or more. Thus, conducting a series of interviews with a representative stratified sample can take considerable time to complete. A third disadvantage is that anonymity for the interviewee is lost—at least with the interviewer. Unless the interviewer is successful in fostering a high level of trust and rapport, the interviewee is likely to play it "safe" and only give what he or she believes to be the socially acceptable or "party line" answer.

18.7.4.3 When to Use Interviews

1. **When a high level of rapport and trust is possible.** Interviews should only be used as a survey data gathering technique when conditions are likely to enable the interviewees to have trust in the interviewer and the overall process. For this reason, outside consultants, skilled in conducting organizational assessment interviews often are used. Interviewing employees on their "turf" (i.e., their place of work, break area, etc.) rather than in the manager's office, etc. also can greatly facilitate creating trust and rapport, as opposed to creating an environment where the employee becomes conditioned to give the "safe" answer. Finally, it is important *not* to try and interview employees when it will seriously disrupt their work, or otherwise be "punishing" (e.g., after work hours when there is no financial renumeration for the additional "work" time).
2. **When the size of the organizational unit being surveyed is small.** When the organizational unit of interest is small, it may be practical to forgo use of a survey questionnaire and rely solely on the interview for gathering employee and manager opinions and perceptions.
3. **As an initial step in gathering information for constructing a questionnaire survey.** In starting the process of constructing a written questionnaire survey, management, or responsible staff typically will have some ideas of what to include, but often the range of their perceptions is limited. Accordingly, interviews often are used to survey a limited representative sample of employees and managers, and the resultant data serves as input for constructing at least part of the questionnaire. A related strategy is to do a fairly extensive stratified semistructured interview survey, and then construct and administer a written questionnaire survey to determine if the perceptions and opinions identified from interviews are shared by the total employee population.
4. **As a follow-up step for interpreting the results of a written questionnaire survey, or for developing implementation strategies based on the survey results.** Sometimes, interpreting the real meaning of survey results may be difficult. Using the survey results as the basis for follow-up interviews, a representative sample of the employees can be utilized to help interpret the results. Alternately, follow-up interviews can be used to solicit approaches or means for implementing the outcomes suggested by the written survey data.

18.7.4.4 Use of Focus Groups

As noted above, a major disadvantage of the interview survey is that it is costly in terms of both time and money. One way of modifying the interview survey to reduce these disadvantages is to interview homogeneous groups, rather than individuals. Because it retains many of the advantages of the individual interview approach while reducing the major disadvantages, the focus group interview approach has become an increasingly popular organizational survey method.

Depending on circumstances, having persons who work in the same department or type of jobs interviewed together can be a significant advantage or disadvantage. When the issues under discussion require responses which individuals would not want their colleagues to hear, it is a distinct disadvantage, as honest responses may be stifled by the others' presence. On the other hand, hearing a colleague's responses may stimulate the thinking of other participants in the focus group, thus yielding additional useful data that might not have surfaced in individual interviews. In fact, when the participants do not feel threatened, this often happens during focus group interviews.

18.7.5 Participatory Ergonomics

Of the various *ergonomics* methodologies, the one that has been most widely used in macroergonomic analysis and organizational design interventions is *participatory ergonomics.* Because workers at all levels typically have the best knowledge about how organizational structure and processes *really* work, and how these factors affect their work lives, they often are the best source of raw data. Similarly, because they are the ones that are directly affected by design changes, they are often in the best position to determine what those changes should be, using the macroergonomics specialist as a facilitator and knowledge resource. One of the beneficial side effects of participatory ergonomics is that it often facilitates important beneficial changes in the organizational culture that are related to the design changes. It is this cultural change that often appears to enable the design change to be successful—and to yield enduring results in improved organizational functioning.

Because the methodology of participatory ergonomics is covered in detail in Chapter 15 of this handbook, it will not be described here. However, a unique hybrid variation of participatory ergonomics that also incorporates some characteristics of the field study approach, and which preceded participatory ergonomics as a formal methodology by several decades, is worth mentioning. That method is known as *ergonomic work analysis* (EWA).

18.7.6 Ergonomic Work Analysis

EWA was pioneered more than 40 years ago in France by Pacaud (1949) and Ombredane and Faverge (1955), and subsequently refined by Wisner and his colleagues at the Conservatoire National des Arts et Metiers in Paris. The EWA methodology of activity analysis involves careful systematic observation and exhaustive checks of the worker's behavior in critical situations, and then confronts the worker with his or her own behavior in order to obtain pertinent explanations, often evoking the cognitive unconscious (i.e., it helps enable the worker to consciously make explicit aspects of the situation of which he or she was unaware). EWA is directed toward a particular objective: knowing and transforming obstacles of all types that hinder and prevent satisfactory activities (Wisner, 1995). As such, EWA can be a particularly useful method for identifying and correcting dysfunctional aspects of organizational design, some of which may be affecting the employee's activities without the employee's conscious awareness. For a description in English of the current methodology of EWA, see De Keyser (1991).

18.7.7 Planning for Organizational Design Change Implementation

Of all the characteristics that have distinguished successful from unsuccessful change efforts in organizations, the most striking has been the extent to which the change effort has been carefully planned. Some reasons for the critical nature of planning that often have been noted empirically are as follows (Kuorinka and Forcier, 1995).

18.7.7.1 Avoiding Unintended Ripple Effects

As noted earlier, a change to any sociotechnical system element affects the others, and often in unintended and dysfunctional ways. Careful planning is essential to anticipating these possible interaction effects; and for developing strategies for enhancing those that are desirable and, especially, blunting those that are dysfunctional.

18.7.7.2 Overcoming Resistance to Change

For a variety of reasons, change efforts invariably meet with employee resistance. Some of the more common reasons are fear of economic loss, potential disruption of social relationships, inconvenience, and especially, fear of uncertainty. In addition, if the changes threaten to disrupt a group's norms or sense of importance, their is likely to be group as well as individual resistance.

Although there are no formulas for overcoming resistance to change, there are some useful guidelines (Szilagyi and Wallace, 1990).

1. Any change program needs to integrate the needs and goals of the organization with those of employees. This will enable employees to see the *personal* benefit of the change, and thus reduce their resistance.
2. As Dalton (1969) has empirically noted, a prestigious person should introduce the change. This lends credibility to the need for change and enhances confidence that it can be carried out successfully.
3. Employees should be actively involved in the change process. This provides employees with some sense of control and gets them ego involved in the process, thereby reducing their fears and gaining their active support. Lack of active, planned employee involvement is a common cause of failed change efforts.
4. Provide frequent feedback to employees on progress in the change process, and on what can be expected in the future. This process of frequent communication and knowledge sharing begins with providing the necessary information for employees to be able to understand the need for change, providing knowledge about the changes required, providing understanding of the implementation process, and being up front and direct about the consequences of the change.
5. Provide incentives for compliance. This might be financial, such as salary or fringe benefit, incentives, desired education and training opportunities; or could involve noneconomic incentives, such as greater flexibility of work hours or greater job autonomy. Regardless of other incentives, frequent positive reinforcement from managers should always be in integral part of the change strategy.
6. One of the simplest and most effective ways for managers to overcome resistance to change is to provide positive empathic support. Such activities as simply being a good listener, providing tutoring, training, and other personal growth opportunities, making an extra effort to maintain close personal contact, and giving time off after a particularly heavy work period all can go a long way toward reducing resistance.

18.7.7.3 Harmonizing Change Among Organizational Elements

When an organization has either no plan, or a poorly conceived one for its change efforts, there invariably will be confusion among the various organizational elements as to specific goals, roles, responsibilities, and strategies for effecting the change. Careful planning can clarify goals, roles responsibilities, and strategies, thereby ensuring a reasonably integrated, harmonized change effort.

18.7.7.4 Determining What Should be Changed

Failure to plan adequately can result in focusing on the wrong *what* of change. All too often, with poor planning there is a tendency to focus on that which is obvious, easily quantifiable, or easy to change, and to overlook or avoid that which really needs changing.

18.7.7.5 Determining How to Change

Determining how to go about implementing change requires careful analysis and consideration of how power is to be utilized, how personal or impersonal the relationship approach will be, and the tempo or speed and depth of the process to be followed (Szilagyi and Wallace, 1990).

18.7.7.6 Determining the Success of the Change Effort

One of the most frequently overlooked areas in poorly planned change efforts is how to evaluate the effectiveness of the change, including (1) what progress is being made, (2) what aspects of the process need modification as the effort progresses, and (3) upon completion of stages of the change process, and at the end, what were the "lessons learned" that may be useful in future change efforts.

18.7.7.7 Other Considerations in Planning for Change

Szilagyi and Wallace (1990) have noted several other factors to be taken into account when planning for change. These include (1) altering social ties, (2) enhancing employee self-esteem, (3) gaining employee internalization, and (4) ensuring that training required by the change process is provided, and that the new learning will transfer to the changed organizational setting.

REFERENCES

Argyris, C. (1971). *Management and Organizational Development.* New York: McGraw-Hill.

Bailey, R. W. (1989). *Human Performance Engineering* (Second Edition). Englewood Cliffs, NJ: Prentice-Hall.

Baron, R. A., and Greenberg, J. (1990). *Behavior in Organizations: Understanding and Managing the Human Side of Work* (Third Edition). Boston: Allyn and Bacon.

Bedeian, A. G., and Zammuto, R. F. (1991). *Organizations: Theory and Design.* Chicago: Dryden Press.

Bammer, G. (1990). Review of current knowledge—musculoskeletal problems. In L. Berlinguet, and D. Berthelette, Eds., *Work With Display Units 89,* Amsterdam: Elsevier, pp. 113–120.

Bammer, G. (1993). Work-related neck and upper limb disorders—social, organisational, biomechanical and medical aspects. In L. A. Gontijo, and J. de Souza, Eds., *Segundo Congresso Latino Americano e Sexto Seminario Brasileiro de Ergonomia.* Ministerio do Trabalho Fundacentro/SC, Florianopolis, Brazil, pp. 23–38.

Bennis, W. G. (1969). Post bureaucratic leadership. *Transaction,* July–August, 45.

Blackstrom, C. H., and Hursh-Cesar, G. (1981). *Survey Research.* New York: John Wiley.

Brown, O. Jr., and Hendrick, H. W., Eds. (1986). *Human Factors in Organizational Design and Management-II.* Amsterdam: North Holland.

Burns, T., and Stalker, G. M. (1961). *The Management of Innovation.* London: Tavistock.

Campbell, J. P. (1977). On the nature of organizational effectiveness. In P. S. Goodman, J. M. Pennings, et al., Eds., *New Perspectives on Organizational Effectiveness.* San Francisco: Jossey-Bass.

Child, J. (1972). Organization structure and strategies for control. A replication of the Aston study. *Administrative Science Quarterly,* June, 163–177.

Dalton, G. D. (1959). *Influence and Organizational Change.* Paper read at a conference on organizational behavior models, Kent, OH: Kent State University.

Davis, L. E. (1982). Organizational Design. In G. Salvendy, Ed., *Handbook of Industrial Engineering,* New York: John Wiley.

DeGreene, K. (1973). *Sociotechnical Systems.* Englewood Cliffs, NJ: Prentice-Hall.

De Keyser, V. (1991). Work analysis in French Language ergonomics: origin and current research trends, *Ergonomics, 34,* 653–669.

Dess, G. G., Rasheed, A. M., McLaughlin, K. J., and Priem, R. L. (1995). The new corporate architecture. *The Academy of Management Executive, 9,* 7–20.

Drucker, P. (1988). The coming of the new organization. *Harvard Business Review, 88,* 47.

Duncan, R. B. (1972). Characteristics of organizational environments and perceived environmental uncertainty, *Administrative Science Quarterly, 17,* 313–327.

Emery, F. E., and Trist, E. L. (1965). The causal texture of organizational environments. *Human Relations, 18,* 21–32.

Emery, F. E., and Trist, E. L. (1960). Sociotechnical systems. In C. W. Churchman, and M. Verhulst, Eds., *Management Science* (Volume 2). Oxford: Pergamon.

Hackman, J. R., and Oldham, G. (1975). Development of the job diagnostic survey. *Journal of Applied Psychology,* 159–170.

Hage, J. (1965). An axiomatic theory of organizations. *Administrative Science Quarterly*, June, 303.

Hage, J., and Aiken, M. (1969). Routine technology, social structure, and organizational goals. *Administrative Science Quarterly,* September, 72–91.

Hales, T., Sauter, S., Petersen, M., Putz-Anderson, V., Fine, L., Ochs, T., Schleifer, L., and Bernard, B. (1992). *US West Communications (USWC): Phoenix, Arizona; Minneapolis, Minnesota; Denver, Colorado: Health Hazard Evaluation Report.* Cincinnati, OH: NIOSH, Centers for Disease Control. (HETA Report, 89-299-2230).

Hall, R. H., Haas, J. E. and Johnson, N. J. (1967). Organizational size, complexity and formalization. *Administrative Science Quarterly, 12,* 72–91.

Harvey, E. (1968). Technology and the structure of organizations. *American Sociological Review,* April, 247–259.

Harvey, O. J. (1963). System structure, flexibility and creativity. In O. J. Harvey, Ed., *Experience, Structure and Adaptability.* New York: Springer.

Harvey, O. J., Hunt, D. E., and Schroder, H. M. (1961). *Conceptual Systems and Personality Organization.* New York: Wiley.

Hendrick, H. W. (1979). Differences in group problem solving behavior and effectiveness as a function of abstractness. *Journal of Applied Psychology, 64,* 518–525.

Hendrick, H. W. (1981). Abstractness, conceptual systems, and the functioning of complex organizations. In G. England, A. Negandhi, and B. Wilpert, Eds., *The Functioning of Complex Organizations.* Cambridge, MA: Oelgeschalger, Gunn and Hain, pp. 25–50.

Hendrick, H. W. (1984a). Wagging the tail with the dog: Organizational design considerations in ergonomics. In *Proceedings of the Human Factors Society 28th Annual Meeting*, Santa Monica, CA: Human Factors Society, 899–903.

Hendrick, H. W. (1984b). Cognitive complexity, conceptual systems, and organizational design and management: Review and ergonomic implications. In H. W. Hendrick, and O. Brown, Jr., Eds., *Human Factors in Organizational Design and Management.* Amsterdam: North Holland, pp. 15–26.

Hendrick, H. W. (1986). Macroergonomics: a conceptual model for integrating human factors with organizational design. In O. Brown, Jr., and H. W. Hendrick, Eds., *Human Factors in Organizational Design and Management-II.* Amsterdam: North Holland, pp. 467–478.

Hendrick, H. W. (1990). Perceptual accuracy of self and others and leadership status as functions of cognitive complexity. In K. E. Clark, and M. B. Clark, Eds., *Measures of Leadership.* West Orange, NJ: Leadership Library of America, pp. 511–520.

Hendrick, H. W. (1991). Human factors in organizational design and management. *Ergonomics, 34,* 743–756.

Hendrick, H. W. (1995). Future directions in macroergonomics, *Ergonomics, 38,* 1617–1624.

Hendrick, H. W., and Brown, O. Jr., Eds. (1984). *Human Factors in Organizational Design and Management.* Amsterdam: North Holland.

Hickson, D., Pugh, D., and Pheysey, D. (1969). Operations technology and organizational structure: An empirical reappraisal. *Administrative Science Quarterly, 26,* 349–377.

Jackson, S. E. (1992). *Diversity in the Work Place.* New York and London: Guilford Press.

Kerr, C., and Roscow, J. M., Eds. (1979). *Work in America: The Next Decade.* New York: Van Nostrand.

Keidel, R. W. (1994). Rethinking organizational design. *The Academy of Management Executive, 8 (4),* 12–30.

Kuorinka, I., and Forcier, L., Eds. (1995). *Work Related Musculoskeletal Disorders (WMSDs): A Reference for Prevention.* London: Taylor & Francis.

Lawrence, P. R., and Lorsch, J. W. (1969). *Organization and Environment.* Homewood, IL: Irwin.

Lawrence, P. R., Kolodny, H., and Davis, S. (XXXX) The human side of the matrix. *Organizational Dynamics,* 43–61.

Magnusen, K. (1970). *Technology and Organizational Differentiation: A Field Study of Manufacturing Corporations.* Unpublished doctoral dissertation. University of Wisconsin, Madison, WI.

Mahoney, T. A., and Frost, P. J. (1974). The pole of technology in models of organizational effectiveness. *Organizational Behavior and Human Performance,* 122–138.

Mileti, D. S., Gillespie, D. S., and Haas, J. E. (1977). Size and structure in complex organizations. *Social Forces, 56,* 208–217.

Montanari, J. R. (1976). *An Expanded Theory of Structural Determinism: An Empirical Investigation of the Impact of Managerial Discression on Organizational Structure.* Unpublished doctoral dissertation. University of Colorado, Boulder, CO.

Nagamachi, M., and Imada, A. S. (1992). A Macroergonomic approach for improving safety and work design. *In Proceedings of the Human Factors Society 26th Annual Meeting.* Santa Monica, CA: Human Factors Society, 859–861.

Negandhi, A. R. (1977). A model for analysing organization in cross cultural settings: A conceptual scheme and some research findings. In A. R. Negandhi, G. W. England, and B. Wilpert, Eds., *Modern Organization Theory.* Kent State, OH: University Press.

Noro, K., and Brown O. Jr., Eds. (1990). *Human Factors in Organizational Design and Management-III* Amsterdam: North Holland.

Noro, K., and Imada, A. (1991). *Participatory Ergonomics.* London: Taylor & Francis.

Omberdane, A. and Faverge, J. M. (1955). *L'analyse du travail.* Paris: Presses Universitaires de France.

Organ, D. W., and Bateman, T. S. (1991). *Organizational Behavior.* Homewood, IL: Irwin.

Pacaud, S. (1949). Reserches sur le travail de telephonistes. Etude psychologique d'un metier. *Le Travail Humain, 12,* 46–65.

Passmore, W. A. (1988). *Design Effective Organizations.* New York: John Wiley.

Perrow, C. (1967). A framework for the comparative analysis of organizations. *American Sociological Review, 32,* 194–208.

Pugh, O. S., Hickson, D. J., Hinings, C. R., and Turner, C. (1968, June). Dimensions of Organizational Structure, *Administrative Science Quarterly,* 75.

Raia, A. (1974). *Managing by Objectives.* Glenview, IL: Scott, Foresman.

Robbins, S. R. (1983). *Organization Theory: The Structure and Design of Organizations.* Englewood Cliffs, NJ: Prentice-Hall.

Rooney, E. F., Morency, R. R., and Herrick, D. R. (1993). Macroergonomics and total quality management at L. L. Bean: A case study. In R. Neilsen, and K Jorgensen, Eds., *Advances in Industrial Ergonomics and Safety.* London: Taylor & Francis, pp. 493–498.

Seashore, S. E. (1987). Surveys in organizations. In G. Salvendy, Ed., *Handbook of Human Factors,* New York: John Wiley, pp. 313–328.

Sherman, S. (1992). Are strategic alliances working? *Fortune,* September 21, 77–78.

Sirota, D. (1974). Why managers don't use attitude survey results. In S. W. Gellerman, Ed., *Behavioral Science in Management.* Baltimore: Penguin Books.

Smith, A. (1970). *The Wealth of Nations.* London: Penguin, London. Originally published in 1876.

Smith, M. J., and Sainfort, P. C. (1989). A balance theory of job design for stress reduction. *International Journal of Industrial Ergonomics, 4,* 67–79.

Stevenson, W. B. (1993). Organizational design. In R. T. Golembiewski, Ed., *Handbook of Organizational Behavior.* New York: Marcel Dekker, pp. 141–168.

Szilagyi, A. D., Jr., and Wallace, M. J., Jr. (1990). *Organizational Behavior and Performance* (Fifth Edition). Glenview, IL: Scott Foresman.

Taylor, F. W. (1911). *Principles of Scientific Management.* New York: Harper.

Thomas, R. R., Jr. (1991). *Beyond Race and Gender.* New York: AMACOM.

Thompson, J. D. (1967). *Organizations in Action.* New York: McGraw-Hill.

Trist, E. L., and Bamforth, K. W. (1951). Some social and psychological consequences of the longwall method of coal-getting. *Human Relations, 4,* 3–38.

Trist, E. L., Higgin, G. W., Murray, H., and Pollock, A. B. (1963). *Organizational Choice.* London: Tavistock.

Van de Van, A. H., and Delbecq, A. L. (1979). A task contingent model of work-unit structure. *Administrative Science Quarterly*, June, 183–197.

Weber, M. (1946). *Essays on Sociology*. (trans. H. H. Grath and C. W. Mills), New York: Oxford.

Wisner, A. (1995). The Etienne Grandjean Memorial Lecture. Situated cognition and action: implications for ergonomic work analysis and Anthropotechnology. *Ergonomics, 38,* 1542–1557.

Woodward, J. (1965). *Industrial Organization: Theory and Practice*. London: Oxford University Press.

Yenkelovich, D. (1979). *Work Values and the New Breed.* New York: Van Norstrand Reinhold.

Zwerman, W. L. (1970). *New Perspectives on Organization Theory.* Westport, CT: Greenwood.

CHAPTER 19

SOCIALLY CENTERED DESIGN

Kay M. Stanney
Jeffery Maxey
Department of Industrial Engineering and Management Systems
University of Central Florida
Orlando, FL 32816-2450 USA

Gavriel Salvendy
School of Industrial Engineering
Purdue University
West Lafayette, IN 47907-1287 USA

19.1 OVERVIEW

The purpose of this chapter is to illustrate the evolution in the design of tasks and jobs, and its sequential interdependence on past design eras. Tayloristic *system-centered* design, which considers humans as resources and aims at productive systems design, is a prerequisite consideration to *human-centered* design, which considers the abilities and limitations of humans as well as the likes and dislikes. While human-centered design provides opportunities for psychosocial growth of the individual, system-centered design focuses on system performance and productivity. This paper introduces socially centered design, which considers social, organizational, and anthropological factors in the design of work. In order for a socially centered design to be effective, it must integrate system- and human- centered design principles into the socially centered design process. First the evolution of design from system to human to socially centered design is reviewed. Then the impact of socially centered design on the objectives and components of system design are considered. Finally, contextual components and models of socially centered design are introduced, with consideration given to the practical use of these models in advanced technologies.

19.2 INTRODUCTION

Like other dynamic enterprises, the approaches applied to the design of systems that involve human components have evolved to reflect the salient concerns of successive eras. Initially, design focused on technical systems innovations and the division of tasks between humans and the machine, i.e., the system-centered approach. More recently, the focus shifted to issues of usability and learnability, i.e., fitting the interface to the person to maximize both system and user performance. This is the user-centered approach. Now designers are confronting the problem of designing for groups of users and organizations. This has led them to socially centered design, which is concerned with "border issues," i.e., design variables that reflect socially constructed and maintained world views which both drive and constrain how people can and will react to and interact with a system or its elements.

Socially centered design fills the void between system-centered design (i.e., macroergonomic design of the overall organization and work system structure, as well as process interfaces with the system's environment, people, and technology) and user-centered design (i.e., microergonomic design of specific jobs and related human–machine, human–environment, and user system or software interfaces). It is thus a medialergonomic design, which focuses on group and intergroup structures and processes. Specifically, socially centered design is interested in users as their performance is influenced by their knowledge and understanding of the social context in which they perform their work.

19.3 EVOLUTION OF DESIGN

19.3.1 Traditional 'Tayloristic' System-Centered Design

System-centered design grew from a desire to enhance the standard of living through increased productivity. Through higher productivity, all aspects of life could presumably be improved: work, housing, nourishment, clothing, health, education, transportation, leisure, and government. The issue was how to effectively increase productivity, with the choices being to either work more hours or work more efficiently (Konz, 1983). Working efficiently was preferred because it was more socially acceptable (an early focus on social concerns) and had the potential for greater overall gains. System designers thus set out to design efficient technologies that could improve the output per unit of input of land, materials, machines, and labor. A primary premise of this traditional system-centered design approach is that all natural organizations (i.e., systems that exhibit predictable, organized relationships, are hierarchical in structure, and evolutionary in nature) follow organizational rules and are subject to a common set of analysis methods (Leslie, 1986). These systematic methods, known as the *Systems Development Lifecycle*, are followed to design more effective systems. While there are many variants of this traditional waterfall style lifecycle (e.g., Kendall and Kendall, 1988; Konz, 1983; Leslie, 1986; Parkin, 1980; Semprevivo, 1982), all include:

1. Problem/opportunity identification and definition
2. Generation of systems alternatives

3. Analysis and selection of the "optimal" system alternative
4. Systems design and implementation
5. Post implementation and maintenance

Throughout the design lifecycle, system analysts examine the complex network of interactions within their organizations to develop new and improved technical systems that meet increased productivity goals (Davis and Wacker, 1982; Semprevivo, 1982). Emphasis is given to the formal specification of roles for both humans and machines (see Chapter 11, *Allocation of Functions*). While these role assignments recognize the high-level creative and adaptive capabilities of workers, sources of human variability are primarily regarded as deterrents to meeting task objectives. System designers thus attempt to "design out" human variability through such mechanisms as training programs and job assignments (i.e., fitting the person to the task).

Under the traditional system-centered design paradigm, social aspects are considered in so far as they influence the functioning of an organization in meeting its productivity goals. Emphasis is placed on the design of social support systems that permit workers to effectively and efficiently perform their intended tasks, including: training procedures and programs; skill maintenance programs; aids for recruitment (i.e., selecting the "right" employee); wage incentive programs; safety; and facilities for rest, eating, and hygiene. The idea is to keep the workers (i.e., one of the fundamental "cogs" in the wheel of productivity) well "oiled" so that their output can be maximized.

A serious cost associated with the system-centered approach is the ramifications resulting from human errors caused by mismatches between the way systems are designed and the information-processing capabilities of humans. No place has this mismatch been more clearly delineated than in Steven Casey's *Set Phasers on Stun and Other True Tales of Design, Technology, and Human Error* (1993). This book presents examples of the human causalities that can result from incompatibilities between system designs and the manner in which humans perceive, think, and act on information. The costs associated with the traditional system-centered design approach have led to a rethinking of design philosophy. While system performance and productivity are important, human safety and well-being are essential. The need for a new design approach was recognized, one that maintained the importance of productivity while carefully factoring in user functions and capabilities. This new user-centered approach needed to consider not only the functionality of the system, but also the abilities, limitations, and preferences of humans in order to design effective user-system interfaces (Rouse, 1991).

19.3.2 Organizational-Centered Design

Organizational-centered design is one attempt to overcome the shortcomings of the traditional Tayloristic system-centered approach. This, more recent, macroergonomic approach to system-centered design gives more systematic and careful consideration to a system's key personnel and external environmental characteristics in designing the overall organizational and work system structure and processes (see Chapter 18 for a detailed discussion on *Organizational Design and Macroergonomics*; other innovations in system-centered design can be found in Chapter 2). The shortcoming of the organizational-centered approach is that, by itself, it does not adequately address group and intergroup interactions or socially centered design factors.

19.3.3 User-Centered Design

The fundamental difference between traditional Tayloristic system-centered and user-centered design is the emphasis given to users. Traditional system-centered design treats users as just another resource to be assigned and optimized to meet operational goals. User-centered design, on the other hand, considers users' roles and responsibilities as the key design objectives to be met and supported by advancing technologies. These objectives include enhancing human abilities, overcoming human limitations, and fostering user acceptance. The purpose of system design thus becomes one of enabling users to be in control of advancing technologies.

Through a development of user theories and principles, the user-centered design approach has attempted to meet these objectives. For example, in order to systematically study and describe the cognitive structure and performance of humans as they interact with complex systems, models such as the *Model Human Processor* and *GOMS* (Goals,

Operators, Methods, and Selection Rules) were developed (Card, Moran, and Newell, 1983). At a functional level, one approach has been to model the stages of user activity completed when operators interact with a system to achieve a goal (Norman and Draper, 1986). State-transition networks (Kieras and Polson, 1985) and grammars (Payne and Green, 1986; Reisner, 1981) have also been developed to support user modeling at the functional level. The general purposes of these user models are to provide a means to (1) predict the human performance consequences of system design alternatives, (2) evaluate the suitability of design alternatives to support and enhance human abilities and limitations, and (3) generate design principles and guidelines that enhance human performance and overcome human limitations.

This new focus on the user has also led to modifications in the *Systems Development Lifecycle* (Eberts, 1994). The additional stages required by user-centered design include: predefintion market analyses of users' needs and perceptions; task analysis (see Chapter 12, *Task Analysis*); user testing and evaluation; and postimplementation user surveys. User acceptance and satisfaction are thus fostered by implementing a cooperative design process (see Chapter 15, *Participatory Ergonomics*). Here users become a fundamental participant throughout the design process (Preece, Rogers, Sharp, Benyon, Holland, and Carey, 1994) and usability engineering practices are applied to ensure that users can access and use the functionality of a system in meeting their task objectives (Nielsen, 1993).

The main issues with the user-centered design approach are that it has heavily focused on "idiot-proofing" a system in order to avoid human error (Brown, 1986) and has largely ignored the social dynamics of the work setting (Buxton, 1994; Preece et al., 1994). Specifically, in optimizing the individualistic characteristics of the users' work environment, users have been isolated and bound to their workstations or computers (Buxton, 1994). When users experience trouble with their systems, this isolation hinders their willingness (perhaps due to a sense of ineptitude) and ability to garner assistance from others (Brown, 1986). The impact is that the structural organization of the workplace dictated by advances in system technologies (i.e., one worker–one computer), has likely impeded collaboration and sharing of system knowledge among users.

Thus, while optimizing the user–system interface by focusing on individualistic definitions of tasks and user's roles and responsibilities, the potentially myopic user-centered approach neglects the symbiotic nature of organizational activities. The importance of social factors was recognized during both the system-centered (e.g., Davis, 1982; Konz, 1983) and user-centered (e.g., Brown, 1986; Eggleston, 1987) design eras; however, the multiplicity of ways that social factors can influence the system design process have not been well understood. "In particular, it (was) unclear what it means to take the social into account and then incorporate it into the design process" (Jirotka and Goguen, 1994). Presently, system design practices are continuing to evolve to provide a set of design techniques that are based on an understanding of the social organizations in which systems exist (see Figure 19.1).

19.3.4 Socially Centered Design

The focus of socially centered design is to extract system design requirements from the moment-to-moment interactional work practices of individuals within an organization (Jirotka et al., 1994; Whiteside, Bennett, and Holtzblatt, 1988). Ethnographic studies (or naturalistic evaluations) are seen as a means of identifying the skills of people within an organization, with task analysis being used to reveal how activities are performed in cooperation with and for the express benefit of others. This information can then be used to design systems that support existing work practices.

One of the underlying assumptions of socially centered design is that by taking into consideration organizational and social factors, such as informal work practices and shared artifacts, the system design process will be enhanced because work generally transpires in a social setting (Grudin, 1990). Another fundamental assumption is that social interaction is orderly and can therefore be understood (Goguen, 1994). Goguen suggests that system designers can achieve this understanding by developing a sense of the intrinsic categories and methods (i.e., informal processes) members of an organization assign themselves during their interactions. Further, socially centered design recognizes that the roles and responsibilities of workers are situated (i.e., emergent rather than individualistic, local to a particular time and place, contingent on the current situation, embodied in a physical and social context, open to revision, and vague), being defined in relation to the social interaction in which they occur. The phases of the *Systems Development Lifecycle* similarly become situated and lose their formalized waterfall composition. The problem def-

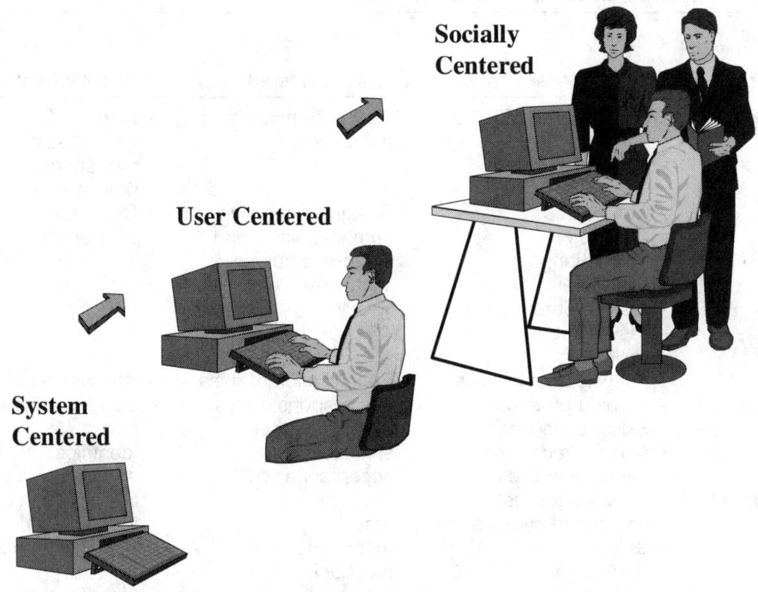

Figure 19.1 Evolution of the scope of system design.

inition *emerges* from the discourse between system designers and users in their natural settings. System alternatives are *open* and subject to change. The definition of the "optimal" system is *local* and *contingent*, being dependent on the context in which it is defined. Under socially centered design, the "design of systems starts with an understanding of (users') work and ends with technological artifacts that support that work and its goals" (Wixon and Comstock, 1994).

The evolution in system design has thus progressed through several stages, starting with a focus on system productivity, to a broader focus on the user-system interaction, to the comprehensive socially centered approach that considers the user-environment interaction (see Figure 19.1). Each stage has brought its own set of additional concerns, while maintaining the importance of factors considered fundamental by earlier design eras (see Table 19.1). Thus, while socially centered design considers additional contextual variables, it also is concerned with factors identified in the system-centered approach such as specification of proper hardware and software. Socially centered design should not be seen as a substitution for these earlier approaches, but rather a supplement to the overall system design process.

19.4 OBJECTIVE OF SYSTEM DESIGN

The overall objective of system design is to develop usable systems. Yet, the nature of the system features and attributes that render a system "usable" depend on the context in which the system is used (Bevan and MacLeod, 1994). The ISO (1993) has defined usable systems as those that are effective, efficient, and well received by specified users accomplishing intended goals in a *particular environment*. Under such a definition, system usability is not a static characteristic which can be instantaneously defined; it is a dynamic quality that is dependent on the context (i.e., environment) in which the work is performed (Brooke, 1994). Failing to understand individual and group logic and their importance in day-to-day performance may make the acceptance and adoption of new system level solutions very unlikely or difficult at best (e.g., Grudin, 1990). This explains why traditional empirical evaluations of system usability may indicate a new design is successful, while the product may fail in the marketplace (Wixon and Comstock, 1994). The traditional interpretations of usability may therefore require modification to incorporate context.

Attempts to incorporate context into the measurement of usability have generally focused on incorporating a range of context within an evaluation's experimental conditions

Table 19.1 Factors Considered in System Design

Additional Factors Considered	System Centered	User Centered	Socially Centered
Equipment	- Specification of hardware & software - Maintenance	- Design to minimize human error	- Design of technological artifacts that support work practices
Task	- Procedures - Methods - Instructions - Tools - Area layouts - Task inputs - Task output	- Design to enhance human abilities and overcome human limitations	- Design of cooperative activities
Users	- Specification of roles - Training procedures - Skill maintenance - Aids for recruitment - Wage incentives	- Specification of roles and responsibilities that foster user satisfaction and acceptance	- Specification of situated roles and responsibilities - Identification of direct and indirect users
Environment	- Facilities for rest, eating and hygiene - Safety - Work hours - Job function	- Performance prediction, monitoring and feedback - Workplace conditions and design	- Specification of informal work practices and shared artifacts - Identification of: • intrinsic categories and methods • group work practices • organizational structure

(Bevan et al., 1994). The socially centered system design process begins with a contextual inquiry (Wixon et al., 1994) to determine the user, product, task, and environmental parameters potentially influencing system design. This investigation will generally involve some form of direct observation such as field studies or ethnographic observation. Designers are expected to become immersed in one or more communities of practice to observe, record, and analyze what users actually do in applying artifacts to achieve goals. In applying techniques like interviews, focus groups, detailed questionnaires, or measuring specific aspects of work performance, the immersed designers seek to adopt the users' frame of reference or view of reality to understand the how and why of artifact use. Additionally, such data is organized and interpreted in the larger context of the technology and science in which artifacts exist.

Once the contextual inquiry is complete, system- and user-centered design approaches are used to optimize the user-system interface design across the range of relevant contexts. Usability is evaluated interactively and iteratively, as users perform tasks in the specified contexts. The measures used to quantify usability criteria (i.e., effectiveness, efficiency, and satisfaction) remain, however, defined at the individual user level. As socially centered design evolves, traditional measures of usability should become more context sensitive and index performance in terms of context variables known to constrain or drive usability.

19.4.1 Effectiveness

System effectiveness is traditionally evaluated in terms of the quality of performance, i.e., completeness and accuracy (Eberts, 1994; Nielsen, 1993). For socially centered design, a more comprehensive definition of effectiveness can be achieved by defining system performance quality as an outcome of both individual and group interaction within a given context (Bevan et al., 1994). Effectiveness is thus a byproduct of organizational factors such as work practices, artifacts, individual differences, aesthetic environmental factors, and cultural factors.

19.4.2 Efficiency

Under socially centered design, efficiency (e.g., performance time) should be measured in a given context. Thus, a system's efficiency would be defined relative to other systems, users, tasks, and/or environments. Efficiency measures would likely not generalize outside of the particular context in which they had been defined and measured.

19.4.3 Satisfaction

Satisfaction is traditionally defined in terms of how pleasant a system is to users. Socially centered design takes a broader perspective on satisfaction, defining it as the ability of a system to satisfy the needs of direct and indirect users in performing their intended tasks in a specified environment. Thus, in measuring this criterion, data is required for both types of users. The implication is that unless a system is satisfying to both direct and indirect users, the system may not be well used or well regarded even if it is effective and efficient. Satisfaction may thus be considered the most important usability criteria (Raskin, 1994).

19.5 COMPONENTS OF SYSTEM DESIGN

Socially centered design reflects the particular context in which the system under design either exists or is expected to operate. This context derives from the equipment, users, tasks, and environment (see Figure 19.2) that make up an organization (Bevan, et. al., 1994).

19.5.1 Equipment

The equipment consists of the basic technology (i.e., product identification and description, main application areas, and major functions) and the system specifications (i.e., hardware, software, materials, and artifacts) to be designed.

19.5.2 Users

The personal details, skills, and knowledge, and personal attributes of users influence how they will interact with a system. Design should take into consideration such factors as the primary and secondary (or indirect) users intended to operate the system, their history of system and task experience, system knowledge, organizational experience, training, qualifications, linguistic ability, general knowledge, age, gender, physical capabilities and limitations, attitude and motivation.

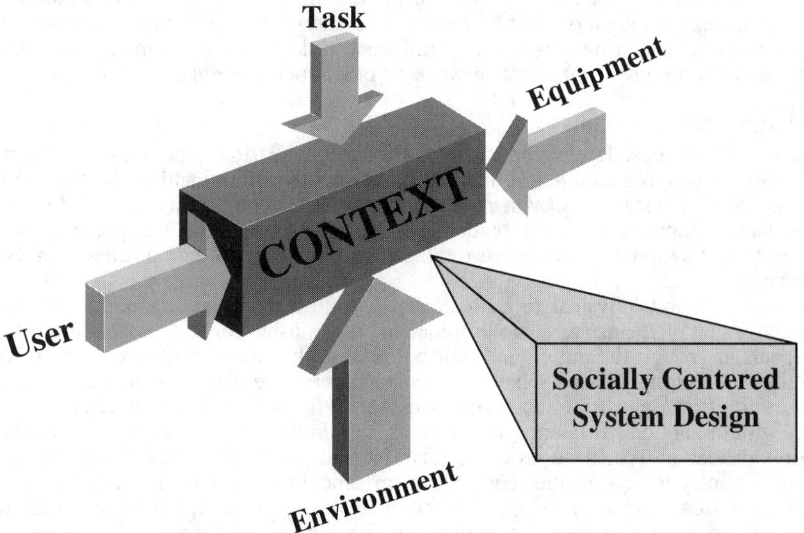

Figure 19.2 Components of socially centered design.

19.5.3 Task

Several factors can influence the system task requirements, including: the task breakdown or flow; goals and objectives; frequency; duration; flexibility; physical and mental demands; dependencies; output; and error related risks. Tasks differ in terms of dimensions such as complexity (e.g., number and interrelatedness of the steps or actions to be performed), difficulty (e.g., a task is experienced by performers as taking more effort or concentration to complete than other tasks), criticality (e.g., how important is a task to system success), demands on immediate and long-term memory, the distribution and coordination of actions among performers within and between systems, or its focus (e.g., planning, creation, communication, decision making, or psychomotor).

19.5.4 Environment

Environmental factors also influence socially centered system design. Influential organizational factors include structural factors (i.e., work hours, group working, job function, work practices, assistance, interruptions, management and communications structure, remuneration), attitudes, and cultural factors (i.e., policy on computers, organizational aims, and industrial relations). Other factors include:

- The technical environment within the organization, which is defined by the system configuration, hardware, software, and materials.
- The job design, which is characterized by its flexibility, performance monitoring and feedback, pacing, autonomy, and discretion.
- The workplace conditions, which are defined by the atmospheric, auditory, thermal, visual, and environmental conditions.
- The workplace design, which is made up of space and furniture, user positioning, and workplace location.
- The safety of the workplace, which is determined by the health hazards, protective clothing, and equipment provided by the organization.

19.6 CONTEXTUAL COMPONENTS OF SOCIALLY CENTERED DESIGN

There is general agreement (e.g., Hollingshead and McGrath, 1995; Kraemer and Pinsonneault, 1990; McLeod, 1992) that the unique variables considered by socially centered design comprise those input or contextual variables that are expected or that have been shown to influence group processes. These variables include the context itself, both as a global or comprehensive factor and the separate component variables that collectively comprise the context: the artifact and its related technology, the cooperative task activities associated with artifact use, the structure of the organizations or groups in which artifact use takes place, situational or cultural factors, and interpersonal characteristics (see Figure 19.3). These factors, in turn, potentially influence task level- or group-related outcomes associated with the use of artifacts in work or other social settings.

19.6.1 Context

Context is ubiquitous. It envelopes the individual at all times, both as a distinct and separate entity and as a member of a group, comprises the stimuli and events that surround and exist in the environment (Mowday and Sutton, 1993), and influences all that is perceived, learned, and recalled (Zimbardo, 1985). Because context is ever present, its effect on thought and action may not be recognized or, if so, may not be considered or viewed important.

Context has both physical (e.g., location in time and space and associated sensory cues) and social elements (e.g., the presence of, relationship to, and expectations of other individuals or groups, the individual's current role, or the community of practice to which the individual belongs). Both separately and together these elements may act to provide opportunities for or constrain individual behavior. Why is this? The answer appears to lie in part with the notion of schema, which is a general purpose representation of a concept, situation, or idea in terms of a set of characteristic attributes. Specific objects (e.g., a desk chair or a library table), people (e.g., Abraham Lincoln or Richard Feynman), situations (e.g., Custer's Last Stand or the Cold War), or relationships (e.g., marriage, motherhood, or student and mentor) are examples of particular schema that are coded in terms of the values they have for each schema attribute (Anderson, 1990). Schema are an enforced form of cognitive economy (Bourne and Ekstrand, 1985). They allow an individual to

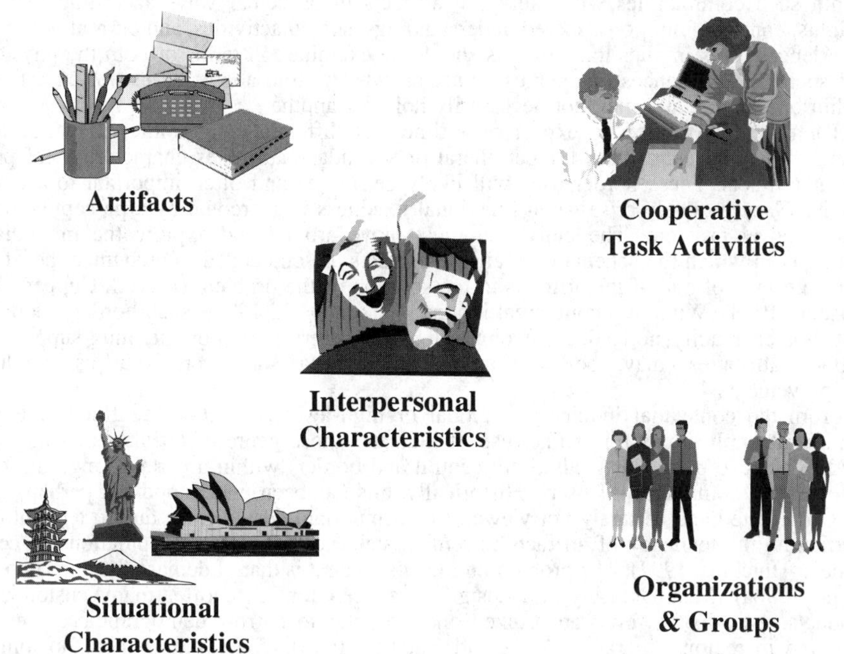

Figure 19.3 Contextual variables.

treat many different but similar objects or events as if they were equivalent. Further, their tendency to be organized in hierarchies assists in accessing related information when it is required.

19.6.2 The Artifact and Its Technology

Artifacts are the designed and created objects (e.g., hammers, televisions, computers) or systems (the U.S. Constitution, the air traffic control system, Internet) that humans apply and use to influence the world around them. Artifacts may be simple in form (e.g., a sewing needle) or very complex in design (e.g., the Space Transportation System). Typically, an artifact is a solution to some requirement or problem. Its application allows its users to perform tasks that could not be performed previously or perform an existing task more quickly, more efficiently, or more effectively.

Artifacts and their associate technology reflect the manner in which task performance at the user and group levels is supported through an artifact's functionality or technical characteristics. Over time and across many instances of a given context, users expect artifacts that belong to a particular type or genre to possess specific sets of attributes (like a schema) and that particular instances of a genre will reflect specific attribute values to include their physical form and operating context. In fact this is what designers often do in creating their designs. They take advantage of a genre and try to evoke it, while users try to recognize what genre has been invoked so they know what to do (or can be done) next. Thus, genre implies both context and use if it is recognized.

The effect of genre is often seen when users are provided artifacts out of their normal context and are asked to use them to solve novel problems (e.g., Duncker, 1945; Maier, 1931). Users in such novel contexts are generally able to recognize only the "fixed" use of an artifact according to its conventional form and context and are unable to re-represent it quickly to fit the novel context. In these situations users are said to be functionally fixed. Thus, they come to count on an artifact's continuity of use (Brown and Duguid, 1994a; Osgood, 1953) in addition to the temporal continuity of its attributes.

As Brown and Duguid (1994b) and Brown, Collins, and Duguid (1989), and Lave (1993) have all observed, many of the learning and performance situations that involve the use of artifacts occur or are situated within particular communities of practice. Users within such communities will likely use artifacts in particular ways according to past insights, common intuitions, shared understandings, shared activities, and current or recent experiences. This implies that artifacts should be examined in terms of both the physical and social circumstances of their use. Further, what is true about the use of an artifact within one community may not necessarily hold for another.

There is good reason to expect over time that different communities may develop "border" or noncentral uses for peripheral or secondary attributes characteristic of particular artifacts. These border uses will likely be unique and often important to a community. They probably arise when functional fixedness is overcome by a perceptive user with a novel problem. The individual must move around and explore the immediate hierarchy in which the schema of available artifacts are embedded to recognize the fit of some attribute of one of the artifacts to the solution of the problem (Flavell, Copper, and Loiselle, 1958). Without a contextual inquiry (Wixon et al., 1994), such border solutions may not be readily noticed or recognized by designers as important, thus supporting artifact attributes may be removed through what are perceived as product "improvements."

From the contextual inquiry, the global or high-level context is identified and has significant implications for artifact use. It indicates those genre of artifacts that may be used to achieve particular goals (both central and border) within a certain range of conditions for a community of users. Historically, this has been understood but perhaps not well articulated and certainly not viewed as an important design issue due to technology inertia and the tendency of artifacts to remain stable over time and populations of consumers (Hutchins, 1994). The problem in today's context is that of demassification (Brown et al., 1994a). Manufacturing and design have been forced to offer many customized products instead of a few standardized ones to cater to narrow demographic segments scattered in regional markets (Ayers and Butcher, 1993). The result is there no longer may be a guarantee of artifact continuity to include border uses. In turn, shared practices may not be easily maintained, especially if customization is extended downwards to the individual user so that only a very few common core or border attributes exist to support member coordination of the community's internal behaviors (Gaver, 1994). Such circumstances highlight the need for embracing socially centered design practices.

19.6.3 Cooperative Task Activities

Task characteristics or dimensions define or reflect the job or objective that the artifact is used in whole or part to accomplish. In organizations, large tasks tend to be devolved to groups of people working together in complex ways to achieve overall task objectives (Dobson, Blyth, Chudge, and Strens, 1994). In the domain of technology support of such group work, three task factors have been found to play an important role in group performance (Kraemer and Pinsonneault, 1990): degree of complexity, the nature of the task (e.g., financial versus personnel oriented), and level of uncertainty associated with either the information used to perform the task or the task's outcome. For example, groups have been found to be more efficient with difficult or complex tasks such as designing a computer program where creativity is required compared to single individuals (Hare, Blumberg, Davies, and Kent, 1994).

To date, though, much of what is known about tasks comes from studies of individuals performing as individuals. While taxonomies of the tasks that groups perform have been developed from the work of social psychology, e.g., McGrath (1984) and Steiner (1972), existing task typologies do not appear to handle group activities such as learning to use an artifact to achieve a common goal, coordinating individual activities to achieve group objectives, or getting to know or understand the capabilities, limitations, and personal characteristics of group members (Olson, Card, Landauer, Olson, Malone, and Leggett, 1993).

Hollingshead and McGrath (1995) suggest that the influence of the task factor on group performance, when this is mediated in some manner by technology, manifests itself in terms of the degree to which effective group performance depends on both the transmission of information about the task and transmission of information about the values, interests, and personal commitments of group members. This would also include any local idiosyncratic methodologies and how-to-do-it data. They note that this view reflects Daft and Lengel's (1986) notion of the degree of information richness required for successful task accomplishment.

19.6.4 The Organization and Group

In order to implement plans and achieve common objectives, group members must create and maintain meaningful relationships among themselves. The structure of these relationships can take on many different aspects. From the literature of organizational behavior and group psychology, Kraemer and Pinsonneault (1990) note that five structure variables (i.e., patterned relationships that exist among members of a group) have been found to be very important in affecting group processes (see Table 19.2). These include: norms (Asch, 1951; Sherif, 1935; Hare, 1976; Homans, 1958; Jacobs and Campbell, 1961; MacNeil and Sherif, 1976; Zimbardo, 1985); power relationships (Hare, 1976; Maier, 1950; Napier, 1969; Sample and Wilson, 1965); status relationships (Hare, 1976; Hare, et. al, 1994); cohesiveness (Hare, 1976); Hare, et. al., 1994); and group density (Egerbladh, 1976; Laughlin, Kerr, Davis, Halff, and Marciniak, 1975; Meister, 1976; Paulus, Annis, Seta, Schkade, and Mathews, 1976; Yetton and Bottger, 1983; Zander, 1979). Thus, "organizational requirements are those which come out of a system being placed in a social context rather than those deriving from the functions to be performed or the tasks to be assisted" (Dobson et al., 1994). These requirements have led to a recent organizational change to "nonterritorial work environments," where the traditional office layout has been replaced with an open environment in which the walls isolating workers have "come tumbling down" (Macht, 1995).

19.6.5 Situational Characteristics

Situational characteristics reflect the various social networks and relationships that exist among group members independent of their membership in the group as well as the group's location on the learning curve, e.g., just formed as a group with minimal skills versus formed for a long time with well-developed usage skills. Like context as a global factor, all groups through their members reflect the "weak ties" of the local social world (e.g., family, educational, business, professional, work, religious, political, ethnic, or national affiliations) to which everyone belongs (Granovetter, 1973).

Further, group members are embedded in a primary culture that can be expected to have a significant influence on the individual's ability and viewpoints. For example, Irvine and Berry (1988) conclude that while all populations have the same perceptual and cognitive processes to include the same potential for development, ecological and cultural factors will dictate what will be learned at what age. The net effect is that different cultural environments can be expected to lead to different patterns of ability. This is known as the law of Cultural Differentiation.

Table 19.2 The Five Structure Variables Affecting Group Processes

Structure Variable	Characteristics
Norms	Informal, covert regulators of interactions; expectations about what should be done in a situation, to include how members ought to behave; emerge in a group by diffusion or crystallization (member expectations converge over time)
Power relationships	Distance from seats of power; the basis (legitimate or not) for exercising power, or the ability to provide rewards, be coercive, or provide expert assistance; group members pay more attention to high power than low power individuals
Status relationships	Reflects those characteristics (age, gender, or ethnicity) around which evaluations and beliefs are organized; affect group members' ability to influence and participate in group activities
Cohesiveness	The attractiveness of a group to its members; willingness of members to belong and stay together as a group; cohesive groups tend to work harder regardless of outside supervision, take fewer risks, gain new information more efficiently, and have better recall of interrupted tasks
Group density	Groups' size within a given working space to include the relative interpersonal distances among group members; the optimum group size for many group discussion tasks is five members (in smaller groups individuals are too prominent and in larger groups there are fewer opportunities to speak and more control is required)

Cultural situational differences are especially evident in social and work environments. There is overwhelming evidence of differences in basic psychological processes between collectivistic and individualistic contexts (Kagitcibasi and Berry, 1989). For example, in responding to social scenarios, Chinese subjects emphasized communal feelings, social usefulness, and acceptance of authority compared to Australian subjects who emphasized competitiveness, self-confidence, and freedom (Forgas and Bond, 1985). This is in turn reflected in the behavior of the members of these cultures.

19.6.6 Interpersonal Characteristics

Interpersonal characteristics reflect the attitudes, behaviors, and motivations of individual group members (Gifford and Gallagher, 1985; Hare, 1976; Hare, Blumberg, Davies, and Kent, 1993). Influential types include (see Table 19.3): dominance-submission (e.g., assertiveness, dominance, aggressiveness, extroversion, and esteem); friendly-unfriendly (e.g., sociability, affiliation, or loneliness); authoritarianism; field-independence/field-dependence (Carli and Guerra, 1974; Greene, 1976; Greene, 1979; Oltman, Goodenough, Witkin, Freedman, and Friedman, 1975; Witkin and Goodenough, 1977); and gender (Anderson and Blanchard, 1982; Baird, 1976; Hare, 1976; Maccoby & Jacklin, 1974; Meeker and Weitzel-O'Neill, 1977; Shaw, 1981; Wood, 1987; Wood, Polek, and Aiken, 1985). These characteristics, however, are not necessarily independent of or orthogonal to each other. As discussed by Digman (1990) and Goldberg (1993), there is clear evidence that a five-factor model of personality (i.e., the enduring collection of traits that characterize an individual) appears to account for the underlying relationships among the characteristics listed in Table 19.3. These factors or dimensions are surgency (or extraversion), agreeableness (or pleasantness), conscientiousness (or dependability), emotional stability (vs. neuroticism), and intellect (or openness to experience). Not only has this model been found to be robust across time, cultures, and data sources, but there is reason to believe that these dimensions represent the important elements of the social terrain that humans have been selected to attend to and act upon (Buss, 1991). For example, Buss (1989) has suggested that the dimensions of surgency and agreeableness operate together through a collection of traits to influence how well or poorly individuals within a group will be able to identify position within a social hierarchy and then be able to form reciprocal alliances to obtain advantages or avoid punishment.

Thus, it is not surprising that groups whose members have compatible personalities tend to be more cohesive, more efficient, and more productive than groups with incompatible members. This appears to occur because incompatibilities lead to the disruption or inhibition of interactions among the group members. Further, compatible groups that perform under participatory, as compared to supervisory, leadership have been found to

Table 19.3 Interpersonal Characteristics

Interpersonal Types	Characteristics
Dominance/submission	Assertive, dominant, aggressive, outgoing, high-esteem individuals typically participate more in group and interpersonal situations
Friendly/unfriendly	Sociability, affiliation, or loneliness; the tendency to participate in group conversation is significantly related to a desire to affiliate but interacts with level of friendship, level of dependence, and seating arrangement
Authoritarianism	Authoritarian-oriented individuals tend to dominate, direct, and behave punitively when in positions of power or authority but tend to be submissive and conformist when acting in a subordinate role
Field-independence(FI)/ Field-dependence(FD)	FI function autonomously, generally are impersonal, and keep their distance both physically and emotionally; FD attend more to social cues, tend to have an interpersonal orientation, show strong interest in others, and prefer close interaction distances
Gender	Likely to make a difference in group productivity only when group tasks suit the interests, experience, or ability of one gender more than the other; males are typically task oriented and engage in more instrumental behavior; females tend to be expressive and engage in more socioemotional behavior

have higher productivity levels (Hewett, O'Brien, & Hornik, 1974). On the other hand, when moderate levels of tension exist within a group as a consequence of incompatibilities, productivity may be higher for incompatible versus compatible groups (Hill, 1975).

19.7 MODELS OF SOCIALLY CENTERED DESIGN

Models of socially centered design reflect the methodologies applied to acquire data about users and their artifacts so as to reflect both the global social context and its constituent elements. In general, the focus of these methodologies should be on the group processes (e.g., task, communication, decision, and required/imposed usage strategies and interpersonal relations) involved in artifact use. This will often involve addressing issues such as which artifact attributes are employed for what purposes (both core and border uses); the scope and nature (level and types of interactions) of working relationships among users to include the manner in which particular artifacts support these relationships; how work, communication, and the information flow is mediated (e.g., through face-to-face or technology-based contact); the extent to which particular artifact attributes facilitate, slow, or impede worker contact and how important this is to worker and group performance; and the influence of the representations (e.g., schema) inherent in an artifact on the individual and group work effort, including the ability of workers to coordinate in time and space (Hollingshead and McGrath, 1995). Several different models have been used in an attempt to elicit such contextual information.

19.7.1 Group Studies

Traditional sociological group studies have applied the scientific method, where a predictive theory is formulated and objectively tested in a controlled environment (Goguen, 1994). If objectivity is juxtaposed with situatedness, however, the hypotheses formulated and tested are no longer independent of the testing context. Sociological studies can overcome this issue by formulating and testing hypotheses in a variety of contextual settings. This, however, becomes a timely and resource intensive approach. Relevant summaries of the sociological group study approach are discussed by Goguen and Linde (1993).

19.7.2 Ethnographic Studies

Ethnographic studies are seen as an alternative to the empiricism of the sociology approach. Such evaluations not only relieve the constraints imposed by the scientific hypothesis-testing paradigm, they also acknowledge the significance of examining the influence of the work environment on the effectiveness of system design (Preece et al., 1994). The ethnographic approach sees system design activities as part of a socially organized interactive environment, rather than a conglomeration of evaluations of individualist, isolated units of task and user behaviors (Jirotka et al., 1994). With this approach, work practices and organizational interactions are identified through anthropological studies of the work setting. Ethnomethodology examines how competent members of an organization coordinate their behaviors, specifically in delimiting the categories and methods used to render their activities intelligible to one another (Goguen, 1994).

An ethnographic researcher becomes immersed in the work environment being studied, intensely observing its inhabitants' cultural activities, belief systems, rituals, institutions, and artifacts (Monk, Nardie, Gilbert, Manteir, and McCarthy, 1993). Based on this immersive examination, the researcher extracts the key behavioral and contextual characteristics of the work environment which influence performance. More specifically, the researcher and the intended users of the system develop a shared understanding of the nature of the work to be performed, the users' needs in performing that work, and the contextual components which support the completion of this work. This information is then fed into the design process, in which design solutions are generated that are grounded in users' experience (Whiteside et al., 1988). The effectiveness of a design concept thus emerges from users' interpretation of and response to system capabilities while interacting with the system in their daily work. It is important to note that this interpretation is deeply tied to the context in which the work is performed. Thus, design solutions based on these ethnographic results may not generalize to other situations. Relevant summaries of the ethnographic approach are discussed by Greenbaum and Kyng (1991), Sharrock and Anderson (1991), and Sharrock and Button (1991).

19.7.3 Cooperative Work Studies

Cooperative work studies involve observational investigations of groups working cooperatively in the social and cultural settings in which they occur. Generally, two or more individuals are studied, who knowingly collaborate on a common goal, require communication to support goal achievement, and who must in some manner coordinate their work activities (Olson et al., 1993; Meister, 1976). Cooperative work studies may or may not involve face-to-face contact, communication may involve only the transmission or receipt of data and not verbal mediated exchanges. Studies of cooperative work typically focus on structural variables reflecting the group (e.g., its size, organization, interactions, composition, and internal processes) and external situational factors that reflect the local operating environment. Relevant summaries of the small group and team performance literature are provided by Druckman and Bjork (1994), Hare (1976), Hare et al. (1994), and Meister (1976). Also see Chapter 53, *Social Computing: Computer Supported Cooperative Work and Groupware.*

19.7.4 Anthropomorphic Studies

Anthropomorphic studies generally examine human-to-human communication patterns and then try to emulate these relationships in the human–system interface design. The focus is to develop schema (i.e., communication modes, grammars, cognitive strategies) of the manner in which people perform tasks. The anthropomorphic approach is discussed by Eberts (1994).

19.8 SOCIALLY CENTERED DESIGN IN ADVANCED TECHNOLOGIES

Many emerging technologies (e.g., multimedia, computer-supported cooperative work, interactive television, virtual reality) involve some form of human–computer interaction (HCI). In the field of HCI, considerable emphasis has been given to the study of workers completing experimentally derived tasks in experimental settings. While the limitations of this approach are widely identified, there have been few HCI studies in real-world settings (Luff, Heath, and Greatbatch, 1994). Currently, as technologies have evolved to be more interactive in nature, there has been a move away from experimental approaches to more interpretive investigations (Preece et al., 1994). HCI designs are now being derived from contextual inquires, with more traditional usability techniques being reserved for evaluation and refinement of the ensuing designs. Further, as advancing technologies emerge, HCI designers must struggle with the issue of how to generate new innovative technological artifacts that meet users' requirements. The techniques of socially centered design may prove to be instrumental in this development process.

19.8.1 Multimedia

"Multimedia is (system) design that makes better and broader use of the human's capabilities to receive and transmit information" (Buxton, 1994) as they achieve specified task goals in a particular environmental context. While multimedia currently focuses on the technological integration of several input and output media, a socially centered approach would redirect focus from the media, which is transitory in nature, to the context in which this media is applied. More effective use of multimedia could be derived through contextual inquires that uncover the work practices, artifacts, individual differences, environmental factors, and cultural factors that this technology supports.

19.8.2 Computer-Supported Cooperative Work

Computer-supported cooperative work (CSCW) emerged from the recognition that many technological systems in use in cooperative environments were failing to meet the needs of their users (Randall, Hughes, and Shapiro, 1994). With the advent of relatively inexpensive and interconnected computing systems, the subgenre of CSCW developed to meet these cooperative needs.

Studies in CSCW tend to focus on the effect of computers or other electronic technology on group and individual task performance. The most recent summary of this literature (Hollingshead and McGrath, 1995) concludes that the impact of technology on group or team performance derives jointly from the technology's functionality, group and member attributes, the tasks being performed, and the interactions between these variables. Further, the conclusion is reached that technology's impact is contingent in part on the detailed history of the group under study, its specific tasks, and the circumstances of task performance. This argues strongly for data collection and analysis approaches that capture such detail in developing design solutions tailored to a specific social and operating context. In order to create robust design solutions that transcend contexts, an examination of a wide cross section of groups and their social and operating context also becomes essential.

19.8.3 Interactive Communication

Socially centered design may prove to be particularly influential in the design of emerging interactive networks, particularly for interactive television (I-TV). While in the past, television was designed to broadcast one program to millions of homes simultaneously, I-TV designs will consist of "nanocasting" millions of different programs that are called upon at times chosen by each individual viewer (Kelly, 1995). If programs, including entertainment, invitational advertisement, and education, do not interest viewers and fit into their lifestyles, they will not be selected for viewing. Designers of I-TV, who are freed from the need to create media that appeals to the masses, will thus need to tap into the lives of their viewers to identify their unique skills and interests if they are to capture targeted audiences. The techniques of ethnographic studies will be well suited to such investigations.

19.8.4 Virtual Reality

The emerging technology of virtual reality (VR) is inherently more social than past non-interactive technologies. First, users are immersed in a particular context or virtual world, with its own set of interactive artifacts and situational characteristics. Second, networked virtual environments allow *social* proximity despite geographic or temporal separation (Buxton, 1994). Users can interact with one another, collaboratively solving problems or sharing common experiences in a virtual context. With this prominence on contextual factors in virtual environments, the assumptions that system designers make concerning the types of users who will adopt the system, the types of tasks to be performed, and the environmental characteristics in which these activities will be done become critical to effective virtual world interaction. If these assumptions fail to represent expectations, users may be unable to effectively interact with artifacts and collaborators in the virtual environment. The usability of VR system design may thus heavily rely on the techniques of socially centered design.

19.9 CONCEPTUAL MODEL OF SOCIALLY CENTERED DESIGN

Socially centered design is concerned with the design of socio-technical systems (i.e., computer systems and related technological artifacts that interact with individual users and organizational groups). This design approach involves the functional and user specifications of past design eras (i.e., system- and user-centered, respectively), while considering additional organizational factors which derive from the day-to-day interactions of workers in their natural work settings. Socially centered design recognizes that these factors are embedded, and must therefore be elicited, from the organizational structure and its policies, through contextual inquiries.

Socially centered design begins with the understanding that the quality and effectiveness of task performance achievable through design reflects the immediate and long-term social and operating context in which any given design solution may be applied. Figure 19.2 illustrates that this context is a composite of the particular levels (or range of levels) of task, equipment, user, and environmental variables in effect at the time of design and projected to be in effect over the anticipated lifecycle of the design solution. The context in effect at the time of design will constrain what resources (items of technology and equipment, relevant social processes, or knowledge and understanding of users and their behavior) can be applied and what kinds of solutions can be offered. It will also constrain whether and how a design solution will be used and assimilated into a community of practice.

Figure 19.4 presents a flow diagram of the socially centered design process. The process begins with the traditional functional and user specifications; however, these requirements are considered emergent, deriving from interactions between system designers and organizational users, open to change, and local to a particular context. These early requirements serve solely as an impetus for an initial system design, which is further specified through contextual inquiries. These inquiries identify the task requirements, equipment/system specifications, environmental factors, and both direct and indirect user characteristics that must be supported by the system design. All of these variables con-

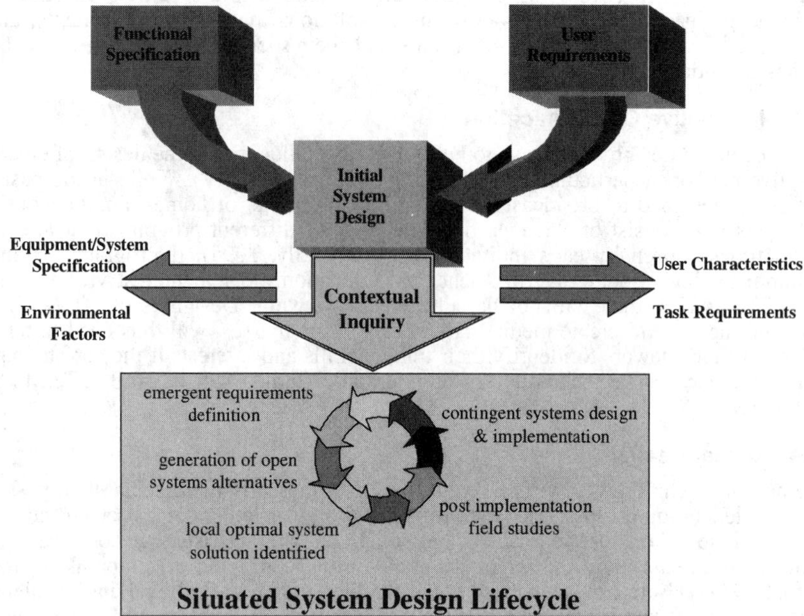

Figure 19.4 Flow diagram of the socially centered design process.

tribute to the context in which the design solution is created and evolves. From an ongoing contextual inquiry, information is fed directly into the situated system design lifecycle. The requirements definition emerges from the social interaction between system designers and users in their natural settings. Equipment, user, task, and environmental information from the contextual inquiry are used to formulate systems design alternatives which are open and subject to change. A systems design solution is selected which is a local optimal, particular to the situation studied during the contextual inquiry. During development and implementation the effectiveness of the design solution is contingent on the interpretive definitions resulting from the contextual inquiry and thus may require changes as those interpretations and requirements evolve. Postimplementation field studies are used to evaluate the contextual fit of the design solution into the organizational setting.

Figure 19.4 also indicates that the socially centered design process is *eclectic*, that is, it focuses on different kinds of data at various points within the cycle of design and redesign. At different times, the designer or the design team is interested in different questions. As such, different data collection and analysis methodologies will be needed to provide the answers required to support the solution selection process. In applying these methodologies, however, the overriding objective is to ensure that the effect of context variables is continually assessed or accounted for.

In making the transition from more traditional design approaches to the socially centered approach there are several principles to keep in mind. These are presented in Table 19.4.

19.10 CONCLUDING REMARKS

The identification of contextual factors that influence system design has begun to specify the range of ways that tasks can be embedded within interactions in a given work setting. In work environments, activities are performed in collaboration with and often for the express purpose of others. Traditional approaches to system design ignore the importance of these situational factors. The socially centered design approach suggests a reconsideration of design practices, with an emphasis given to contextual inquires that identify the importance of context variables, including: technological artifacts; cooperative task activities; the impact of both organizational and cultural factors; situational characteristics; and the importance of interpersonal skills to effective system design. This approach also suggests a respecification of the traditional usability measures, including effectiveness, efficiency, and satisfaction, which takes into consideration the implications of context.

Table 19.4 Socially Centered Design Guidelines

Identify:

- The details of the local context—its structure and physical arrangement and the social parameters that drive interactions, especially work, planning, and decision making.
- Which artifacts are shared within and between contexts and their importance to users.
- The informal work practices as well as formal work practices.
- The degree of interaction among users (frequent, occasional, almost never).
- The importance of contact when it is made, even if infrequent (very, modest, none at all).
- How orderly social interaction is (i.e., how structured) within and between contexts.
- The role people assign to themselves and others and how these roles are supported by the system (if at all).
- Which tasks require cooperation and which can be completed independently.

Do:

- Recognize that user requirements are contextually driven; their intelligibility resides in understanding their relationship to numerous work and organizational practices.
- Recognize that optimality depends on the current context.
- Use group performance measures as well as individual measures to assess usability (i.e., effectiveness, efficiency, and satisfaction).
- Recognize that usability is driven primarily from satisfaction which is not objective; it comes from a particular frame of reference—thus it is essential to understand the context from which this point of view is derived.
- For situated variables, develop pointers to their contexts so they are not "lost."
- Recognize that the situated system design lifecycle is iterative and has an unspecified ending point; it may continue indefinitely.

Don't:

- Stay in the lab, get into the field as soon as possible.
- Adhere dogmatically to the waterfall system design lifecycle.
- Forget that all system activities are embedded in a context of both strong and weak social ties.
- That roles and responsibilities range from being vague to very well defined.
- That roles and responsibilities are situated and may vary with changes in individuals, the local context, prevailing policies and social climate.
- Forget that macroergonomics or overall organization and work system design variables also impact socially centered design and may need to be addressed.

REFERENCES

Anderson, J. (1990). *Cognitive psychology and its implications*. New York: W. H. Freeman.

Anderson, L., and Blanchard, P. (1982). Sex differences in task and social-emotional behavior. *Basic and Applied Social Psychology*, *3*, 109–139.

Asch, S. (1951). Effects of group pressure upon the modification and distortion of judgments. In H. Guetz, Ed., *Groups, Leadership, and Men*. Pittsburgh, PA: Carnegie Press.

Ayers, R., and Butcher, D. (1993). The flexible factory revisited. *American Scientist*, *81*, 448–459.

Baird, J. (1976). Sex differences in group communication: A review of relevant research. *Quarterly Journal of Speech*, *62*, 179–192.

Bevan, N., and MacLeod, M. (1994). Usability measurement in context. *Behaviour & Information Technology*, *13*(1), 132–145.

Bourne, L., and Ekstrand, B. (1985). *Psychology: Its Principles and Meanings* (Fifth Edition). New York: Holt, Rinehart, & Winston.

Brooke, J. (1994). Designing flexible and adaptable interfaces. In L. MacDonald and J. Vince, Eds., *Interacting with Virtual Environments*. Chichester: John Wiley.

Brown, J. S. (1986). From Cognitive to Social Ergonomics. In D. A. Norman and S. W. Draper, Eds., *User Centered System Design: New Perspectives On Human-Computer Interaction*. Hillsdale, NJ: Lawrence Erlbaum Associates.

Brown, J., Collins, A. and Duguid, P. (1989). Situated cognition and the culture of learning. *Educational Researcher*, *18*, 32–42.

Brown, J., and Duguid, P. (1994a). Borderline issues: Social and material aspects of design. *Human-Computer Interaction*, *9*, 3–36.

Brown, J. and Duguid, P. (1994b). Patrolling the border: A reply. *Human-Computer Interaction*, *9*, 137–149.

Buss, D. M. (1989). *A strategic theory of trait usage: Personality and the adaptive landscape.* Paper presented at Invited Workshop on Personality Language, University of Groningen, Groningen, Netherlands.

Buss, D. M. (1991). Evolutionary personality psychology. In M. R. Rosenzweig and L. W. Porter, Eds., *Annual Review of Psychology* (Volume 42, pp. 459–491). Palo Alto, AC: Annual Reviews.

Buxton, W. A. S. (1994). Human Skills in Interface Design. In L. MacDonald and J. Vince, Eds., *Interacting with Virtual Environments.* Chichester: John Wiley.

Card, S. K., Moran, S. K., and Newell, A. (1983). *The Psychology of Human-Computer Interaction.* Hillsdale, NJ: Lawrence Erlbaum Associates.

Carli, R., and Guerra, G. (1974). Cognitive style and interpersonal perception. *Archivio di Psicologia, Neurologia e Psichiatria, 35*(1), 7–25.

Casey, S. (1993). *Set Phasers On Stun and Other True Tales of Design, Technology, and Human Error.* Santa Barbara: Aegean.

Daft, R., and Lengel, R. (1986). Information richness: A new approach to managerial behavior and organizational design. *Research in Organizational Behavior, 6*, 191–233.

Davis, L. E., and Wacker, G. J. (1982). Job Design. In G. Salvendy, Ed., *Handbook of Industrial Engineering.* New York: John Wiley.

Digman, J. M. (1990). Personality structure: Emergence of the five-factor model. In M. R. Rosenzweig and L. W. Porter, Eds., *Annual Review of Psychology* (Volume 41, pp. 417–440). Palo Alto, CA: Annual Reviews.

Dobson, J. E., Blyth, A. J. C., Chudge, J., and Strens, R. (1994). The ORDIT approach to organizational requirements. In M. Jirotka and J. Goguen, Eds., *Requirements Engineering: Social and Technical Issues* (pp. 87–106). London: Academic Press.

Duncker, K. (1945). On problem solving. L. S. Lees, Trans. *Psychological Monographs, 58*, 270.

Eberts, R. E. (1994). *User Interface Design.* Englewood Cliffs, NJ: Prentice Hall.

Egerbladh, T. (1976). The function of group size and ability level on solving a multidimensional complementary task. *Journal of Personality and Social Psychology, 34*, 805–808.

Eggleston, R.G. (1987). The Changing Nature of the Human-Machine Design Problem: Implications fir System Design and Development. In W. B. Rouse and K. R. Boff, Eds., *System Design: Behavioral Perspectives on Designers, Tools, and Organizations.* New York: North-Holland.

Flavell, J., Cooper, A., and Loiselle, R. (1958). Effect of the number of preutilization functions on functional fixedness in problem solving. *Psychological Reports, 4*, 343–350.

Forgas, J., and Bond, M. (1985). Cultural influences on the perception of interaction episodes. *Personality and Social Psychology Bulletin, 11*, 75–88.

Gaver, W. (1994). Commentary on borderline issues: 8. Grounding social behavior. *Human-Computer Interaction, 9*, 70–74.

Gifford, R., and Gallagher, T. (1985). Sociability: Personality, social context, and physical setting. *Journal of Personality and Social Psychology, 48*, 1015–1023.

Goguen, J. A. (1994). Requirements Engineering as the Reconciliation of Social and Technical Issues. In M. Jirotka and J. Goguen, Eds., *Requirements Engineering: Social and Technical Issues* (pp. 165–199). London: Academic Press.

Goguen, J. A., and Linde, C. (1993). Techniques for requirements elicitation. In S. Fickas and A. Finkelstein, Eds., *IEEE First International Symposium on Requirements Engineering 93* (pp. 152–164). Los Alamitos: IEEE Press.

Goldberg, L. R. (1993). The structure of phenotypic personality traits. *American Psychologist, 48*, 26–34.

Granovetter, M. (1973). The strength of weak ties. *American Journal of Sociology, 78*, 1360–1380.

Greenbaum, J., and Kyng, M., Eds. (1991). *Design at Work: Cooperative Design of Computer Systems.* Hillsdale, NJ: Lawrence Erlbaum.

Greene, L. (1979). Psychological differentiation and social structure. *Journal of Social Psychology, 109*, 79–85.

Greene, L. (1976). Body image boundaries and small group seating arrangements. *Journal of Consulting and Clinical Psychology, 22*, 244–249.

Grudin, J. (1990). The computer reaches out: The historical continuity of interface design. *Proceedings of CHI '90: Empowering People.* J. C. Chew and J. Whiteside, Eds. (ACM, New York), pp. 261–268.

Hare, P. (1976). *Handbook of Small Group Research* (Second Edition). New York: The Free Press.

Hare, P., Blumberg, H., Davies, M., and Kent, V. (1994). *Small Group Research: A Handbook.* Norwood, NJ: Ablex.

Hewett, T., O'Brien, G., and Hornik, J. (1974). The effects of work organization, leadership style, and member compatibility upon the productivity of small groups working on a manipulative task. *Organizational Behavior and Human Performance, 11*, 283–301.

Hill, R. (1975). Interpersonal compatibility and workgroup performance. *Journal of Applied Behavioral Science, 11*, 210–219.

Hollingshead, A., and McGrath, J. (1995). Computer-assisted groups: A critical review of the empirical research. In R. A. Guzzo and E. Salas, Eds., *Team Effectiveness and Decision Making in Organizations* (pp. 46–78). San Francisco: Jossey-Bass.

Homans, G. (1958). Social behavior as exchange. *American Journal of Sociology, 63*, 597–606.

Hutchins, E. (1994). Commentary on borderline issues: 10. In search of a unit of analysis for technology use. *Human-Computer Interaction, 9*, 78–81.

Irvine, S., and Berry, J., Eds. (1988). *Human Abilities in Cultural Context.* New York: Cambridge University Press.

ISO (1993). ISO CD 9241-11.2 *Ergonomic Requirements for Office Work with Visual Display Terminals (VDTs)- Guidance on Usability Specification and Measures.* Geneva, Switzerland: International Organization of Standardization.

Jacobs, R., and Campbell, D. (1961). The perpetuation of an arbitrary tradition through several generations of a laboratory microculture. *Journal of Abnormal and Social Psychology, 62*, 649–658.

Jirotka, M., and Goguen, J., Eds. (1994). *Requirements Engineering: Social and Technical Issues.* London: Academic Press.

Kagitcibasi, C., and Berry, J. (1989). Cross-cultural psychology: Current research and trends. *Annual Review of Psychology, 40*, 493–531.

Kelly, R. V. (1995). The ten myths of interactive television. *Virtual Reality Special Report, 2*(2), 29–36.

Kendall, K. E., and Kendall, J. E. (1988). *Systems Analysis and Design.* Englewood Cliffs, NJ: Prentice Hall.

Kieras, D. E., and Polson, P. (1985). An approach to formal analysis of user complexity. *International Journal of Man Machine Studies, 22*, 365–394.

Konz, S. (1983). *Work Design: Industrial Ergonomics* (Second Edition). New York: John Wiley.

Kraemer, K., and Pinsonneault, A. (1990). Technology and groups: Assessment of the empirical research. In J. Galegher and R. Kraut, Eds., *Intellectual Teamwork: Social and Technological Foundations of Cooperative Work*, Hillsdale, NJ: Erlbaum.

Laughlin, P., Kerr, N., Davis, J., Halff, H., and Marciniak, K. (1975). Group size, member ability, and social decision schemes on an intellective task. *Journal of Personality and Social Psychology, 31*, 522–535.

Lave, J. (1993). Situating learning in communities of practice. In L. Resnick, J. Levine, and S. Teasley, Eds., *Socially Shared Cognition.* Washington, DC: American Psychological Association.

Leslie, R. E. (1986). *Systems Analysis and Design: Methods and Invention.* Englewood Cliffs, NJ: Prentice Hall.

Luff, P., Heath, C., and Greatbatch, D. (1994). Work, interaction and technology: The naturalistic analysis of human conduct and requirements analysis. In M. Jirotka and J. Goguen, Eds., *Requirements Engineering: Social and Technical Issues* (pp. 259–288). London: Academic Press.

Maccoby, E., and Jacklin, C. (1974). *The Psychology of Sex Differences.* Stanford, CA: Stanford University Press.

Macht, J. (1995). When the walls come tumbling down. *Inc. Technology, 17*(9), 70–72.

MacNeil, M., and Sherif, M. (1976). Norm change over subject generations as a function of arbitrariness of prescribed norms. *Journal of Personality and Social Psychology, 34*, 762–773.

Maier, N. (1950). The quality of group decisions as influenced by the discussion leader. *Human Relations, 3*, 155–174.

Maier, N. (1931). Reasoning in humans: II. The solution of a problem and its appearance in consciousness. *Journal of Comparative Psychology, 12*, 181–194.

McGrath, J. (1984). *Groups: Interaction and Performance.* Englewood Cliffs, NJ: Prentice-Hall.

McLeod, P. (1992). An assessment of the experimental literature on electronic support of group work: Results of a meta-analysis. *Human-Computer Interaction, 7*, 257–280.

Meeker, B., and Weitzel-O'Neill, P. (1977). Sex roles and interpersonal behavior in task-oriented groups. *American Sociological Review, 42*, 91–105.

Meister, D. (1976). *Behavioral Foundations of System Development.* New York: John Wiley.

Monk, A., Nardie, B., Gilbert, N., Manteir, M., and McCarthy, J. (1993). Mixing oil and water? Ethnography versus experimental psychology in the study of computer-mediated communication. *Proceedings of INTERCHI '93: Bridges Between Worlds* by S. Ashlunds, K. Mullet, A. Henderson, E. Hollnagel, and T. White, Eds. Reading, MA: Addison-Wesley, pp. 3–6.

Mowday, R., and Sutton, R. (1993). Organizational behavior: Linking individuals and groups to organizational contexts. *Annual Review of Psychology, 44*, 195–229.

Napier, H. (1969). Group learning: Note on undivided vs. divided task information. *Psychological Reports, 24*(3), 847–848.

Nielsen, J. (1993). *Usability Engineering.* Boston: Academic Press.

Norman, D. A., and Draper, S. W., Eds. (1986). *User Centered System Design: New Perspectives On Human-Computer Interaction.* Hillsdale, NJ: Lawrence Erlbaum Associates.

Olson, J., Card, S., Landauer, T., Olson, G., Malone, T., and Leggett, J. (1993). Computer-supported co-operative work: Research issues for the 90s. *Behaviour & Information Technology, 12*(2), 115–129.

Oltman, P., Goodenough, D., Witkin, H., Freedman, N., and Friedman, F. (1975). Psychological differentiation as a factor in conflict resolution. *Journal of Personality and Social Psychology, 32*, 730–736.

Osgood, C. (1953). *Method and Theory in Experimental Psychology.* New York: Oxford University Press.

Parkin, A. (1980). *Systems Analysis.* Cambridge, MA: Winthrop.

Paulus, P., Annis, A., Seta, J., Schkade, J., and Matthews, R. (1976). Density does affect task performance. *Journal of Personality and Social Psychology, 34*, 248–253.

Payne, S. J., and Green, T. R. G. (1986). Task-action grammars: A model of the mental representation of task languages. *Human-Computer Interaction, 2*, 93–133.

Preece, J., Rogers, Y., Sharp, H., Benyon, D., Holland, S., and Carey, T. (1994). *Human-Computer Interaction.* Wokingham, England: Addison Wesley.

Randall, D., Hughes, J., and Shapiro, D. (1994). Steps toward a partnership: Ethnography and system design. In M. Jirotka and J. Goguen, Eds., *Requirements Engineering: Social and Technical Issues* (pp. 241–258). London: Academic Press.

Raskin, J. (1994). Intuitive equals familiar. *Communications of the ACM, 37*(9), 17–18.

Reisner, P. (1981). Formal grammar and human factors design of interactive graphics system. *IEEE Transactions on Software Engineering*, SE-7, 229–240.

Rouse, W. B. (1991). *Design for Success: A Human-Centered Approach to Designing Successful Products and Systems.* New York: John Wiley.

Sample, J., and Wilson, T. (1965). Leader behavior, group productivity, and rating of least preferred co-worker. *Journal of Personality and Social Psychology, 1*(3), 266–290.

Semprevivo, P. C. (1982). *Systems Analysis: Definition, Process, and Design* (Second Edition). Chicago: Science Research Associates.

Sharrock, W., and Anderson, B. (1991). Epistemology: Professional scepticism. In G. Button, Ed., *Ethnomethodology and the Human Sciences* (pp. 51–76). Cambridge University Press.

Sharrock, W., and Button, G. (1991). The social actor: Social action in real time. In G. Button, Ed., *Ethnomethodology and the Human Sciences* (pp. 137–175). Cambridge University Press.

Shaw, M. (1981). *Group Dynamics: The Psychology of Small Group Behavior* (Third Edition). New York: McGraw-Hill.

Sherif, M. (1935). A study of some social factors in perception. *Archives of Psychology, 27*(187).

Steiner, I. (1972). *Group Process and Productivity.* New York: Academic Press.

Whiteside, J., Bennett, J., and Holtzblatt, K. (1988). Usability Engineering: Our Experience and Evolution. In M. Helander, Ed., *Handbook of Human-Computer Interaction.* Amsterdam: North-Holland.

Witkin, H., and Goodenough, D. (1977). Field dependence and interpersonal behavior. *Psychological Bulletin, 84*, 661–689.

Wixon, D. R., and Comstock, E. M. (1994). Evolution of Usability at Digital Equipment Corporation. In M. E. Wiklund, Ed., *Usability in Practice.* Boston: Academic Press.

Wood, W. (1987). Meta-analytic review of sex differences in group performance. *Psychological Bulletin, 102*, 53–71.

Wood, W., Polek, D., and Aiken, C. (1985). Sex differences in group task performance. *Journal of Personality and Social Psychology, 48*, 63–71.

Yetton, P., and Bottger, P. (1983). The relationship among group size, member ability, social decision schemes, and performance. *Organizational Behavior and Human Performance, 29*, 145–159.

Zander, A. (1979). The psychology of group processes. *Annual Review of Psychology, 30*, 417–451.

Zimbardo, P. (1985). *Psychology and Life.* Glenview, IL: Scott, Foresman.

PART 4

EQUIPMENT, WORKPLACE, AND ENVIRONMENTAL DESIGN

CHAPTER 20
VISUAL DISPLAYS

Kevin B. Bennett
Allen L. Nagy
John M. Flach
Department of Psychology
Wright State University
Dayton, OH 45435 USA

20.1 INTRODUCTION

Advances in computer science and artificial intelligence currently provide new forms of computational power with the potential to support human problem solving. One use of this computational power is to provide an expert system or an automatic assistant which provides "advice" to the human operator at the appropriate times. For example, there has been some progress in the use of production systems and neural networks as the drivers for decision support. An alternative, complementary use is to integrate information graphically (or more generally "perceptibly"). Here computational power is used to create and manipulate representations of the target world, rather than to create autonomous machine problem solvers. Perhaps the most general term that has been applied to this endeavor is *representation aiding* (Woods, 1991; Woods and Roth, 1988; Zachary, 1986). Representation aiding offers a unique opportunity to improve overall performance of human–machine systems. The technologies needed to produce computer graphics are mature, and when designed properly representation aids maintain the flexibility of the human in the loop and improve the capability of the overall system to respond to unforeseen circumstances. The challenge in providing effective representation aids centers around how best to use these technological capabilities to support human decision making and problem solving. Although this chapter focuses on concepts and principles of design to meet this challenge, we see representation aiding and machine problem solvers as complementary tools in the designer's tool chest and we expect that for very complex systems both approaches will be necessary.

In contrast to most other treatments of display design, we did not provide a "cook book" of detailed guidelines and recommendations (primarily because they tend to be conflicting and difficult to apply). Instead, we chose to describe a set of general heuristics for display design. Because these heuristics are necessarily abstract, we have made the discussion more concrete by illustrating them within the context of a simple domain. We describe how the heuristics apply to that domain and annotate our written descriptions with concrete graphic examples. Our goal is to transfer functional knowledge of display design to practitioners.

We begin our discussion with a description of basic physiological, perceptual, and technological considerations in display design. These considerations are the foundation for display design, and represent the baseline conditions that must be met for a display to be effective. We next consider four alternative approaches to display design. Each approach emphasizes a different conceptual aspect of the display design puzzle, and each approach has both strengths and weaknesses. A fifth approach is outlined; this approach draws from the strengths of the previous approaches and incorporates new considerations that are particularly relevant to the design of displays for complex, dynamic domains. We end the chapter by considering the limitations of our discussion and examples and additional challenges for display design.

20.2 PHYSIOLOGICAL, PERCEPTUAL, AND TECHNOLOGICAL CONSIDERATIONS

This section considers fundamental aspects of the visual system and visual perception that are relevant for display design. Information on the surface of a display is most often represented by a difference in perceived brightness or a difference in perceived color between the information carrying stimuli and the background of the display field. This section is concerned primarily with the detection and perceived appearance of these differences. Although this chapter is focused primarily on emissive displays, it is useful to begin by discussing some of the differences between reflective and emissive displays and the implications of these differences for visual perception. Emissive displays, such as the CRT, generate the light that is used to produce text, symbols, or pictures that carry information. Reflective displays such as road signs, pages in a textbook, and the speedometer in an automobile do not produce any light, but reflect some portion of the light that falls on them. Though emissive displays are much more versatile and flexible in some respects, it is probably safe to say that the use of reflective displays to present information was, and still is, far more common. With regard to the visual system and visual perception, there are some fundamental differences between reflective and emissive displays. We will begin by examining properties of achromatic, or colorless, displays that illustrate these differences and later in this section take up chromatic displays.

20.2.1 Reflective Displays

The surface of a reflective display reflects some portion of the light energy that falls on it in many different directions. The percentage of light reflected, known as the "reflectance" of the surface, and the dependence of this percentage on the wavelength of the light, known as the "spectral reflectance function" of the surface, are determined by the physical properties of the surface (Nassau, 1983). We will begin by discussing surfaces with flat spectral reflectance functions that reflect approximately the same percentage of light for all wavelengths. Images are placed on the surface by changing the properties of the surface in local regions. For example, suppose a printer for a personal computer deposits black ink on a grey page so as to form text. The grey page reflects a percentage, perhaps 50%, of the light energy at each wavelength falling on it. The ink deposited on the page appears very dark because it reflects only a small percentage, for example, 5%, of the light energy falling on it. Suppose an observer views this page tacked to the wall painted uniformly white so that the surface of the wall has a reflectance of 90%.

The reflectance of surfaces varies with the angle of incidence of the illumination and the angle at which the reflectance is measured. Reflectances of surfaces can be described with two components, a specular component and a diffuse component (Hunter and Herold, 1987; Shafer, 1985). The specular component is "mirror-like" in that a large proportion of the light is reflected off at an angle equal to the angle of incidence. The diffuse component is characterized by light reflected off in all directions. Shiny surfaces, like mirrors, have a large specular component and a small diffuse component while matte surfaces, like a velvet cloth, have a large diffuse component and a small specular component. For simplicity we will ignore these complexities here. The graph shown in Figure 20.1(*a*) illustrates idealized spectral reflectance curves for the page, the ink, and the wall. Real spectral reflectance curves would only approximate flat curves. Surfaces with flat curves are neutral in the sense that they do not change the spectral quality of the light that falls on them.

In order to characterize the light reflected back from the surface, we need to know something about the light falling on the surface. A typical spectrum for sunlight is shown in Figure 20.1(*b*), where the relative energy is plotted as a function of wavelength. This

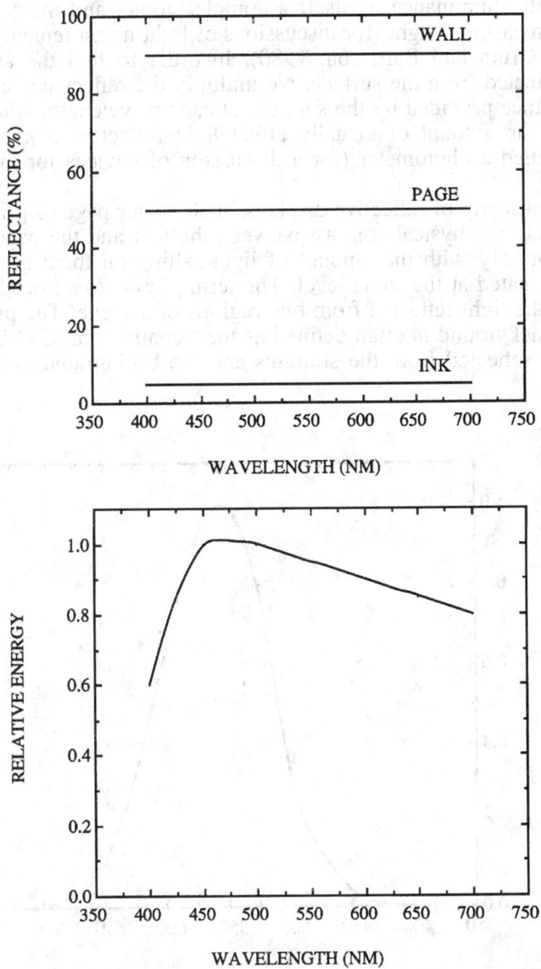

Figure 20.1 Figure 20.1(*a*) shows idealized spectral reflectance curves for the ink, the page, and the wall in the example described in the text. Figure 1(b) shows the relative energy at each wavelength in sunlight.

spectrum is referred to as typical, because the spectrum for sunlight varies with time of day, time of year, latitude, and atmospheric conditions. Not all of the energy in sunlight is effective in generating a visual response. Some wavelengths of light are more likely to be absorbed by the receptors in the eye, the rods and cones, than others. A function describing the relative effectiveness of different wavelengths for photopic or cone vision (see Figure 20.2) was standardized by the CIE in 1924 (see Wyszecki and Stiles, 1982). This function, which is known as the photopic luminosity function, has served as a standard in science and industry ever since.

A similar function for scotopic or rod vision was standardized in 1951 (see Wyszecki and Stiles, 1982). Since most displays are viewed under photopic conditions, we will concentrate on cone vision here. In order to get a measure of the visual effectiveness of the light energy from the sun, we multiply the energies at each wavelength in Figure 20.1(*b*) by the value of the photopic luminosity function at that wavelength. The sum or integral of these weighted energies, multiplied by a constant to convert the energy units to a convenient unit of visual effectiveness, is known as the luminance of the source. A commonly used unit for luminance today is the Candela per square meter (Cdm^2).

For our purposes, the more important measure is the amount of light that actually falls on the wall, the page, and the ink. This quantity is known as *illuminance*, the amount of visually effective light that actually falls on a surface in space. We will assume that the wall is evenly illuminated so that this measure is the same across the wall, the text, and the page. A common unit of illuminance is the Lux. The measurement of luminance, and the related quantity illuminance, is itself a complex topic, and many different types of units are used in measuring light (for discussions of light measurement see Wyszecki and Stiles, 1982 and Grum and Bartleson, 1980). In order to find the amount of visually effective light reflected from the surface, we multiply the reflectance at each wavelength times the illuminance provided by the sunlight at each wavelength. Alternately, we could measure directly the amount of visually effect light reflected in a particular direction using a device called a photometer (for a discussion of devices for measuring light see Post, 1992).

An important property of reflective displays, such as our page of printed text mounted on the wall, is that the physical contrast between the text and the page, or the page and the wall, does not vary with the amount of light falling on them as long as all of the surfaces are illuminated at the same level. The term *physical contrast* is used to refer to the difference in the light reflected from two regions of a scene. The physical contrast of a stimulus on a background is often defined as the "contrast ratio," $\Delta L/L$, the difference between the light reflected from the stimulus and the background divided by the back-

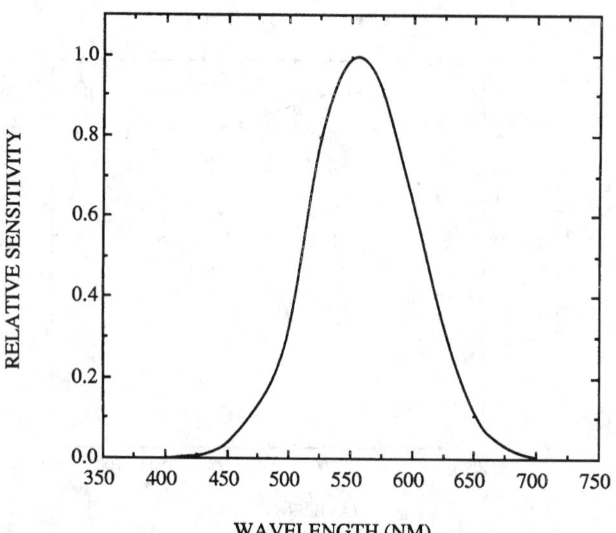

Figure 20.2 The CIE 1924 photopic luminosity curve.

ground level. In our example the physical contrast between the text and the page could be specified as the difference in the amounts of light reflected by the ink and by the page divided by the amount of light reflected by the page. Note that as the amount of light falling on the wall is changed, the physical contrast ratios calculated for the text and the page, the text and the wall, and the page and the wall will remain constant (see Figure 20.3). The reader can demonstrate this by setting up the contrast ratios and demonstrating that the light level, which appears in both the numerator and the denominator of the contrast ratio, will cancel out and the contrast ratios are determined by the reflectances alone.

The human visual system appears to have evolved to take advantage of the reflective properties of surfaces. One of the earliest relationships established in the study of visual perception is that the intensity difference between a stimulus and a background necessary

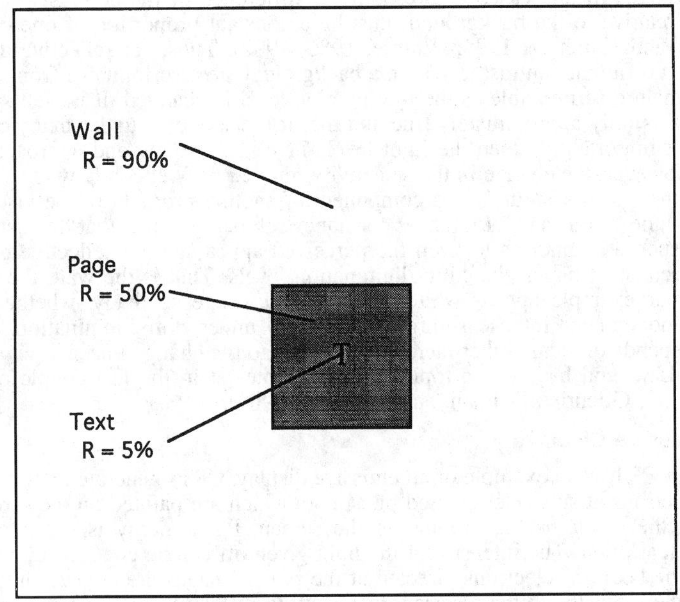

Contrast Ratios:

$$\text{Text / Page} = \frac{.5 \text{ X I} - .05 \text{ X I}}{.5 \text{ X I}} = \frac{.45}{.50} = .90$$

$$\text{Page / Wall} = \frac{.9 \text{ X I} - .5 \text{ X I}}{.9 \text{ X I}} = \frac{.40}{.90} = .44$$

$$\text{Text / Wall} = \frac{.9 \text{ X I} - .05 \text{ X I}}{.9 \text{ X I}} = \frac{.85}{.90} = .94$$

Figure 20.3 Calculation of contrast ratios for the page, the text, and the wall. The values of "r" indicate the reflectances of the three surfaces in the figure. The symbol "I" in the equations represents the illumination level which is identical for all three surfaces in the figure and therefore cancels out of the equations.

for detection of the stimulus is a constant proportion of the intensity of the background field. This rule, known as Weber's Law, is often written in equation form as $\Delta I = k * I$. Here ΔI refers to the difference between intensity of the stimulus and the intensity of the background, k is the proportionality constant or the Weber Fraction, and I is the intensity of the background field. Weber's Law indicates that the visual system becomes less sensitive to differences between the stimulus and the background as the intensity of the background field increases. That is, in order to keep the stimulus detectable, the difference between the stimulus and the background must be increased as the background is increased. Notice, however, that if we rearrange Weber's Law by dividing both sides of the equation by I, we get $(\Delta I/I) = k$.

At threshold, the difference between the intensities of the stimulus and the background (ΔI) divided by the background intensity (I) is constant. This is exactly the situation for the reflective displays described above. It means that if the text on a page is detectable at any light level, then it will remain detectable as the light level is changed. A somewhat different form of Weber's Law also applies to the discrimination of two stimuli presented on a background. In this case, at threshold the difference in the contrasts between the two stimuli relative to the background must be a constant proportion of one of the contrasts (Nagy and Kamholz, 1995; Whittle, 1986, 1992). Thus, for reflective displays, if two stimuli at different contrast levels on a background are discriminable from each other they will remain discriminable as the illumination level is changed. It is well known that Weber's Law is only approximately true and that it breaks down under many conditions, perhaps most importantly when the light levels involved are low and approach absolute threshold. However, the change in the sensitivity implied by Weber's Law is an important property of the visual system. It is a component of another property of the visual system known as lightness constancy. Lightness constancy refers to the fact that the visual system operates in such a manner as to keep the perceived appearance of reflective objects approximately constant under changing illumination levels. That is the wall, the page, and the text in our example appear white, grey, and black, respectively, whether they are viewed outdoors under intense sunlight or indoors under dim illumination. Lightness constancy depends on many other factors in addition to the change in sensitivity indicated by Weber's Law, and has been a topic of intense interest in the last couple of decades (Adelson, 1993; Gilchrist, Delman, and Jacobson, 1983).

20.2.2 Emissive Displays

We will use a CRT as an example of an emissive display. CRTs generate light by shooting beams of electrons at substances called phosphors which are painted on the screen of the CRT. When the electrons hit a point on the screen, light energy is given off by the phosphor at that point. The intensity of the light given off can be changed by varying the strength of the beam of electrons directed at the point. Images are created on the screen by varying the intensity of the electron beam hitting different points on the screen. The physical contrast between different regions of the screen can be defined in the same manner as for reflective displays.

Suppose that we mount the CRT on the white wall and use it to generate a page of dark text on a grey page. Suppose also that we adjust the CRT so that the page gives off 50 units of light and the text gives off 5 units of light. The physical contrast ratio between the text and the page would be 0.90 as it was for the reflective display (see Figure 20.3). Suppose that the white wall is illuminated initially so that 90 units of light are reflected from it. Also suppose, for the moment, that the surface of the CRT reflects none of this light. In this case the contrast ratios between the three surfaces would be the same as in our first example with the reflective page of text, and we might expect that the CRT display would look very similar to the reflective display.

Note what happens as the illumination falling on the wall is increased, however. The intensity of the light reflected from it increases, but the intensities of the lights from the text and the page on the CRT do not change. The contrast ratio between the text and the page on the CRT remains constant, but the contrast ratios between the page and the wall, and the text and the wall increase. Thus, we might expect that the appearances of the text and the page to change considerably as the light level falling on the wall is changed. If we regard the text and the page as individual incremental stimuli against the large background provided by the wall, then Weber's Law suggests that their discriminability will decrease as the light reflected from the wall increases. The decrease in discriminability occurs because the difference in contrast ratios decreases with increasing light level. In this case the decrease in the sensitivity of the visual system with increasing background light level reduces the ability to detect the difference between the text and the page which remains constant.

Any light which is reflected from the glass face of the CRT will reduce the discriminability of the text on the page even further, because it will be reflected from both the region containing the dark text and the region containing the page. The reflected light actually reduces the physical contrast between the text and the page and makes them even less discriminable. Thus, emissive displays behave quite differently than reflective displays in natural environments. These differences do not present much of a problem when emissive displays are placed in a constant environment such as an office illuminated by a fixed light source. However, when emissive displays are placed in natural environments in which the illumination level may vary by a factor of a million or more, the problems caused by the varying contrast ratios are evident. For example, this problem occurs when emissive displays are used in aircraft. The detectability and the appearance of elements within the display may vary dramatically. In order to keep the appearance of the text and the page constant, the light levels given off by the CRT must be adjusted in accord with the change in the illumination of the wall.

20.2.3 Factors Affecting Perceived Contrast

Besides the physical contrast, there are many other factors such as adaptive state, location in the visual field, eye movements, and the interpretation of the perceived illuminant which affect the perceived contrast of a stimulus against a background. One of the most important of these factors is stimulus size. In the last few decades this problem has been investigated very successfully with an approach based on Fourier analysis (for extensive reviews see DeValois and DeValois, 1990; Ginsburg, 1986; Graham, 1989; and Olzack and Thomas, 1986). Fourier analysis suggests that any pattern of light and dark on the retina can be described as a sum of sinusoidal components of different frequency and amplitude. The application of this idea to visual perception involves measuring an observer's sensitivity to a number of sinusoidal patterns of different spatial frequency (see Figure 20.4). These repetitive spatial patterns of light and dark are known as gratings.

Spatial frequency is essentially a measure of the size of the bars in the pattern. The spatial frequency of the pattern is defined as the number of cycles that occur in 1° of visual angle. As spatial frequency increases there are more cycles per degree of visual angle and the bars become smaller. Visual angle is used as the unit of size because it gives a measure of the size of the image on the retina (e.g., a book 12 in. long makes a larger image on the retina when it is held up close to the eye than when it is held far away). In order to get a measure of the size of an image on the retina the distance between an object and the observer's eye must be considered. Thus, the visual angle subtended by an object is defined as twice the arcTan of the height/2 divided by the distance (see Figure 20.5).

Sensitivity is measured by finding the physical contrast level at which a given pattern of light and dark is just detectable. In order to give a measure of sensitivity, the reciprocal of the threshold is calculated by dividing one by the threshold contrast. The measure of physical contrast typically used in this approach is slightly different than the contrast ratio described above, and is called the "Michelson Contrast." It is defined as $L\text{max} - L\text{min}$ divided by $L\text{max} + L\text{min}$, where Lmax is defined as the maximum luminance level in the pattern and Lmin is defined as the minimum luminance in the pattern. The curve described by plotting contrast sensitivity against the spatial frequency of the grating pattern is called the "contrast sensitivity function."

A typical contrast sensitivity function for photopic or cone vision obtained from a human observer is shown in the Figure 20.6. The curve shows that when spatial frequency is low (i.e., the bars are large), the sensitivity to contrast is low. As spatial frequency is increased the sensitivity increases up to spatial frequencies of about 5 to 10 cycles per degree. With further increases in spatial frequency (i.e., smaller and smaller bars), sensitivity falls off rapidly until at a spatial frequency of approximately 50 cycles per degree a grating of 100% contrast (the highest physical contrast obtainable) is not visible. Spatial patterns of even greater frequency also are not visible. Thus, very fine patterns are visible only if the spatial frequency is below 50 cycles per degree and they are very high contrast.

Over the last few decades many physical factors such as overall light level, number of cycles present in the pattern, and the location of the pattern in the visual field have been shown to affect the contrast sensitivity function. The shape of the curve as well as the overall sensitivity can vary considerably. The shape and height of the curve are affected by several components within the visual system that play a role in determining the

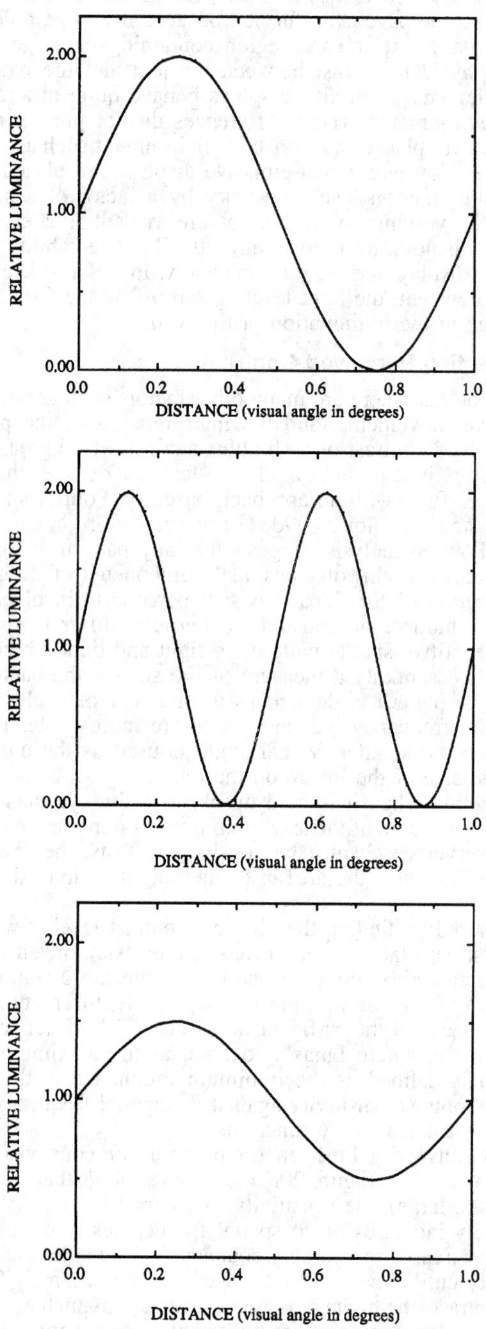

Figure 20.4 Plots showing the variation in luminance for sinusoidal patterns. Figure 20.4(*a*) illustrates a spatial frequency of 1 cycle per degree at a contrast of 100%. Figure 20.4(*b*) illustrates a spatial frequency of 2 cycles per degree at a contrast of 100%. Figure 20.4(*c*) illustrates a spatial frequency of 1 cycle per degree at a contrast of 50%.

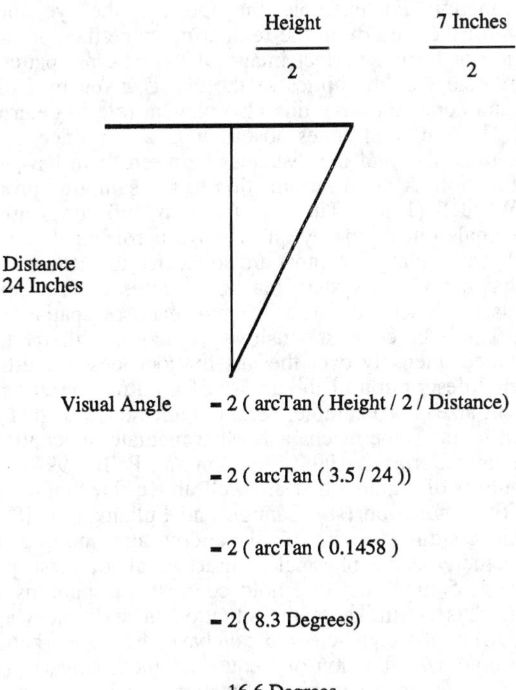

Height
2
7 Inches
2
Distance
24 Inches
Visual Angle = 2 (arcTan (Height / 2 / Distance)
= 2 (arcTan (3.5 / 24))
= 2 (arcTan (0.1458)
= 2 (8.3 Degrees)
= 16.6 Degrees

Figure 20.5 Calculation of visual angle, as described in the text.

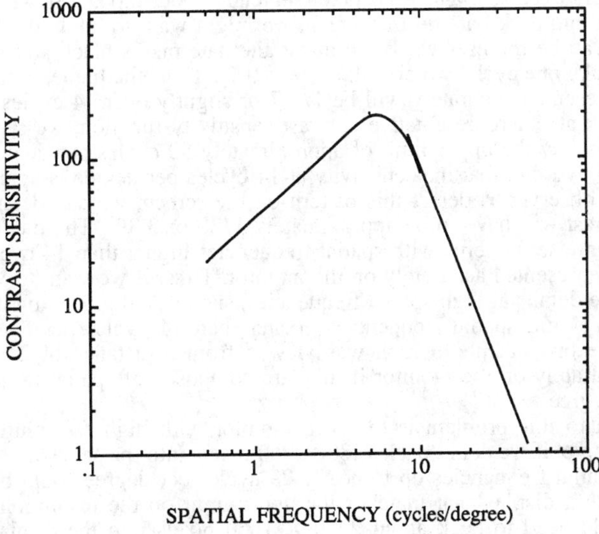

Figure 20.6 A typical plot of a contrast sensitivity function for a human observer. [Based on data given in DeValois and DeValois (1990)].

contrast sensitivity function. For example, the optics of the eye, the lens and cornea, which form an image of the pattern on the retina, influence the contrast sensitivity function, because they do not form a perfect image of the external pattern on the retina. A good introductory treatment of the optics of the eye is given by Millodot (1982). The distribution of rods and cones on the retina also plays a role in determining the contrast sensitivity function. The rods and cones absorb light and initiate neural signals in the visual system. Thus, their size and the distances between them have some affect on the contrast sensitivity function. A good introduction to the sampling properties of rods and cones is given by Wandell (1995). The way the rods and cones are connected to the neurons that carry signals out of the eye also plays a role in determining the contrast sensitivity function, because many receptors are connected to each neuron. Psychophysical evidence suggests that the visual system may be organized into approximately five to seven neural channels, each sensitive to a different band of spatial frequencies (Olzack and Thomas, 1986). Thus, the contrast sensitivity function is the result of many factors which have been studied intensely over the last few decades. Nevertheless, it is a very useful and fundamental description of the ability of a human observer to detect contrast in patterns of different size. For example, recent studies suggest that the recognition of text may be mediated by the same mechanisms that mediate the contrast sensitivity function (Alexander, Xie, and Derlacki, 1994; Solomon and Pelli, 1994).

The perceived contrast of patterns that are well above threshold is not simply related to the contrast sensitivity function (see Cannon and Fullenkamp, 1991). That is, if we measure the threshold contrast for sinusoidal patterns at a number of different spatial frequencies and then increase the physical contrast of all of these patterns so that the contrast for each one is 5 times the threshold contrast, the patterns will not appear to have equal contrasts. This is similar to the situation in audition where equal loudness curves for tones of different frequencies do not have the same shape as the audibility curve, a plot of threshold as a function of frequency, and change shape as the loudness level is raised. Thus, the contrast sensitivity function can be used to predict whether a pattern of a given spatial frequency is visible, but it cannot be used to predict accurately the perceived contrast of patterns that are well above threshold. For example, if a display designer wants to equate the perceived contrast of patterns of different size that are well above threshold, the contrast sensitivity function cannot be used to do this accurately.

The notions of visual angle, spatial frequency, and contrast sensitivity that were briefly introduced above are very useful in thinking about both reflective and emissive displays. Here we will concentrate on emissive displays. Consider a standard CRT display that is 9.5 in. wide and 7 in. high. Assume that this CRT has 640 columns of pixels each containing 480 rows (standard 640 × 480 resolution). If the observer views this display from a distance of 2 ft, then the screen subtends about 22.4° horizontally and 16.6° vertically (see Figure 20.5), and each pixel subtends about 0.035°. If we want to make patterns of light and dark bars on the screen, we might want to know the highest spatial frequency that can be represented. If we make alternate pixels black and white we need two pixels to make one cycle, which will subtend .07°. Thus, the highest spatial frequency that can be represented accurately will be 1/.07 or slightly over 14 cycles per degree.

Looking back at our representative contrast sensitivity function, we see that this frequency is well below the upper limit of approximately 50 cycles per degree. Looking at the vertical axis, we find that the sensitivity at 14 cycles per degree is approximately 30. In order for an observer to detect this pattern on the screen, we can determine that the Michelson contrast will have to be approximately 1/30 or 3.3%. These calculations also tell us something else. Patterns with spatial frequencies higher than 14 cycles per degree just cannot be represented accurately on the monitor. Thus, if we want to view an image with a lot of fine details at high spatial frequencies, such as a digitized photograph which subtends 9.5 by 7 in., spatial frequencies greater than 14 cycles per degree that were visible when the original photograph was viewed from a distance of 2 ft, will not be represented accurately on the monitor if they are composed of spatial frequencies above 14 cycles per degree.

One solution to this problem is to use a monitor with higher resolution or smaller pixels. For example, if we could pack 1280 × 960 pixels into the same 9.5 × 7 in. screen, patterns with spatial frequencies up to nearly 29 cycles per degree could be represented. In order to make a display that matches the upper limit on the resolution of the visual system we would need to pack about 2240 × 1660 pixels into the display. A 9.5 × 7 CRT with this resolution would permit the presentation of patterns with spatial frequencies up to 50 cycles per degree at a viewing distance of 2 ft. This would be very difficult to accomplish with present technology, making the display and the computer hardware that drives it very expensive.

It is also possible to portray patterns with spatial frequencies greater than 14 cycles per degree on the original CRT by moving the observer farther away so that each pixel subtends a smaller visual angle. The drawback to this approach is that the entire display field now subtends a smaller portion of the field of view. For example, if we move the observer back to a distance of about 4 ft, patterns with spatial frequencies up to nearly 29 cycles per degree could be portrayed on the screen. This example helps to illustrate a fundamental trade-off in emissive displays, the trade-off between field-of-view and resolution. With a fixed number of pixels, this trade-off is always present in an emissive display. If the pixels are spread over a larger viewing area the resolution will be poor. If they are packed into a smaller viewing area the resolution will improve, but the field of view will decrease.

The resolution of an emissive display may be limited either by the display itself, or by the hardware that drives it, that is the video card in a computer or the signals generated on a television cable. The detail in an image, or the spatial frequencies that can be portrayed, and the field of view that is visible, will be limited by this resolution and the size of the screen.

20.2.4 Color

Though black and white pictures carry much of the information in the real world, they do not carry information about color. Color in images is certainly important for aesthetic reasons, but in addition to the aesthetic qualities it brings to an image color serves two important basic functions (Boynton, 1990). First, chromatic contrast between two regions in image can add to the luminance contrast between these regions to make the difference between the regions much more noticeable, especially when the luminance contrast is small. Second, since color is perceived to be a property of an object (though in fact it also depends on illumination as we will see), it is useful in identifying objects, searching for them, or grouping them. Boynton (1990) regards the second function of color, which he describes as related to categorical perception, as the more important one.

It is probably because of these categorical properties, that color is often used as a coding device and as a means of segregating information in visual displays (see Widdel and Post, 1992).

Several excellent treatments of the basics of human color vision and the science of specifying colors for applications are available (Boynton, 1992; Pokorny and Smith, 1986; Post, 1992; Wyszecki and Stiles, 1982). So a very brief review will be given here. Normal human color vision depends on the presence three types of cone receptors in the retina. These cones differ in the type of light absorbing pigment contained in them. One of these pigments absorbs best, meaning the greatest percentage of the light falling on it, in the short-wavelength region of the spectrum; hence the cone containing it is referred to as the "S" cone. The second pigment absorbs best in the middle of the spectrum, and the cone containing it is referred to as the "M" cone. The third pigment absorbs best at slightly longer wavelengths than the "M" pigment and the cone containing it is referred to as the "L" cone.

The differences in the signals generated in these cones by a given light provide some information about the spectral content of the light. For example, a light source which gives off more energy in the long-wavelength portion of the spectrum than in the middle or short wavelength regions would tend to stimulate the L cones more than the other two cone types. On the other hand, a light source that gives off more energy in the short-wavelength region would tend to stimulate the S cones more than the other two types. The differences in the stimulation of the cone types serve as a means for discriminating between the lights, and result in the perception of color.

Since there are only three types of cones, normal human color vision is said to be three dimensional or trichromatic. Furthermore, since there are only three signals from different types of cones in the visual system, it follows that only three numbers are needed to specify the perceptual quality of a color. Much effort has gone into developing systems of specifying colors with three numbers such that they represent the perceptual qualities of the stimulus in useful ways. The fact that only three numbers are needed to specify the chromatic quality of a stimulus also means that there are many physically different stimuli that stimulate the three cones in the same way and thus appear to be the same color. Stimuli which are physically different but appear to be the same are called "metamers."

Consider the reflective display example given above. Suppose that we print the text on our gray page using red ink rather than black ink. The ink appears red because it tends to absorb short and middle wavelength light that falls on it while reflecting long wavelength light. A spectral reflectance curve showing the percentage of light reflected as a function of wavelength for red ink might look like the curve shown in Figure 20.7. To get the light reflected back from the ink we multiply the reflectance at each wavelength times the energy at each wavelength. In order to calculate the luminance of this light, we would weight the reflected energy at each wavelength by the photopic luminosity function and integrate or sum over the entire curve as we did for achromatic stimuli above. However, the text appears to differ from the grey page and the white wall in color as well as in lightness. In order to characterize this difference, we would like some means of measuring the colors of the text, the page, and the wall. The most widely used system for doing this is based on the CIE 1931 chromaticity diagram. This diagram is based on color matches of normal human observers. A good introduction to the color matching experiment and the development of chromaticity diagrams can be found in Boynton (1992). In the color matching task, observers were asked to adjust the intensities of three primary lights that were mixed together in a single stimulus field so as to match the colors of a wide variety of other lights presented in another stimulus field. The CIE chromaticity diagram uses three numbers (related to the intensities of the primaries needed to make a match in the color matching experiment) to represent the color or, more specifically, the "chromaticity" of a stimulus. These numbers are called the *chromaticity coordinates* of the color and are referred to as x, y, and z. The color matching data were normalized so that the values of these three chromaticity coordinates add up to 1 for any real color. As a result only two of the chromaticity coordinates need to be given to specify a color, because the third can always be obtained by subtracting the sum of the other two from 1. Therefore, all colors can be represented in a two-dimensional diagram like the CIE 1931 diagram shown in Figure 20.8, where only x and y are plotted. Many measuring instruments have been developed and are commercially available for measuring the CIE coordinates of a color (see Post, 1992 for some discussion of these).

The chromaticity coordinates specify the chromatic properties of a color but do not specify its appearance, because the appearance of the color can change with many viewing conditions that do not change its chromaticity coordinates. For example, the size of the stimulus, in terms of visual angle, can affect the color appearance even though the chromaticity coordinates of the ink used to make it do not change (Poirson and Wandell, 1993). This is a severe limitation on the meaning and usefulness of the CIE chromaticity diagram. One would like to have a system in which the appearance of the color is spec-

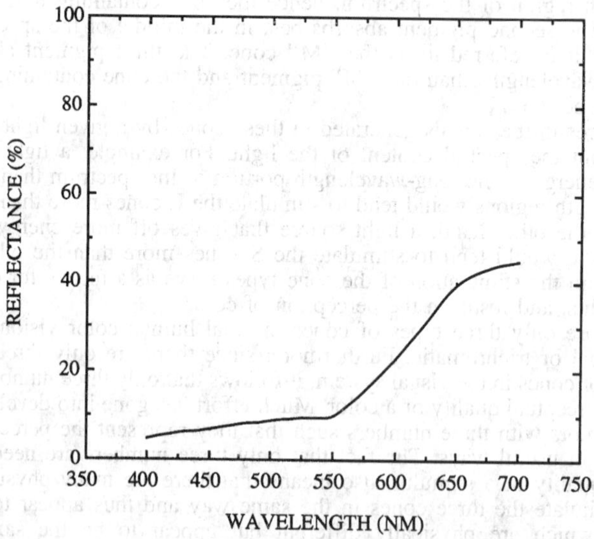

Figure 20.7 A spectral reflectance curve for red ink.

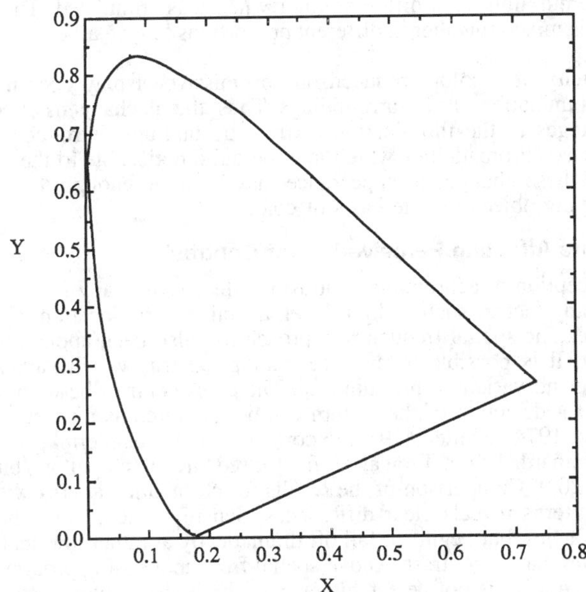

Figure 20.8 The CIE 1931 chromaticity diagram. [Plotted from data given in Wyszecki and Stiles (1982)].

ified, but this is a very difficult problem that has not yet been solved. Nevertheless, the specification of colors in the chromaticity diagram is still very useful, because any two stimuli with the same chromaticity coordinates will appear to be identical in color when viewed under the same conditions. What the chromaticity coordinates specify is how to make a color that will appear the same as a given sample under the same viewing conditions.

The chromaticity coordinates of a reflective display change with the chromaticity of the light used to illuminate it. The change occurs because the amount of light reflected back from an object at each wavelength depends in part on the amount of light falling on it. Therefore, when the chromaticities of objects, or dyes, or paints are specified they are usually given with reference to a standard light source (for a discussion of standardized light sources see Wyszecki and Stiles, 1982). One might expect that the change in the chromaticity coordinates accompanying a change in the light source would change the color appearance of a reflective display. Such changes in light source are actually quite common. As noted above, the spectral quality of daylight changes with time of day, atmospheric conditions, season, and location on earth. A large variety of artificial light sources are commercially available, and these can differ considerably in the spectral quality of the light given off. However, these changes do not generally result in large changes in the appearances of objects, because mechanisms within the visual system act to maintain a constant color appearance despite these changes in illumination. Color constancy has generally been shown to be less than perfect (Arend and Reeves, 1986; Brainard and Wandell, 1992). However, it appears to work well enough to prevent confusing changes in the appearance of reflective objects. The visual mechanisms mediating color constancy have been of intense interest over the past few decades (D'Zmura and Lennie, 1986; Maloney and Wandell, 1986). Selective adaptation within the three-cone mechanisms is thought to be one of the major mechanisms mediating color constancy (Worthy and Brill, 1986) much as the change in sensitivity described by Weber's Law plays a role in lightness constancy.

While mechanisms of color constancy work to maintain a constant appearance in reflective displays they actually work against the maintenance of a constant appearance in emissive displays, much as mechanisms of lightness constancy worked against the constant appearance of black and white emissive displays. Color CRTs take advantage of the fact that human color vision is trichromatic by using only three different phosphors.

Each phosphor emits light of a different color when it is stimulated. The light from the three phosphors is mixed together in different proportions to give all other colors including white.

The chromaticity of a color produced on an emissive display does not change with changes in the illumination of the surroundings. Thus, the mechanisms of color constancy, activated by changes in the illumination of the surroundings, introduce changes in the appearance of these chromaticities which may be quite noticeable to the observer. Under some conditions these changes in appearance may be large enough to cause some confusion in identifying objects on the basis of color.

20.2.4.1 Factors Affecting Perceived Color Contrast

Much as the perception of achromatic contrast is affected by many factors, color contrast is affected by many factors such as light level, adaptive state, location in the visual field and stimulus size. The spatial frequency approach has also been applied to the detection of color contrast. It is possible to produce grating patterns which vary sinusoidally in color with little or no variation in luminance. The color contrast between the bars of the grating required for detection of the pattern can be measured as a function of the spatial frequency (Kelly, 1974; Mullen, 1985; Noorlander and Koenderink, 1983; Sekiguchi, Williams, and Brainard, 1993). Typical results for red/green and yellow/blue gratings are shown in Figure 20.9. Comparison of the results for chromatic patterns with those shown for luminance patterns reveals clear differences. Sensitivity to color contrast is high at low spatial frequencies, but begins to fall off dramatically at rather low spatial frequencies as compared to luminance contrast. Above spatial frequencies of approximately 12 cycles per degree color contrast is not detectable even at the highest color contrasts producible. Thus, chromatic contrast information is limited to fairly low spatial frequencies, or large patterns, as compared to luminance contrast information. Within this range of spatial frequencies the color appearance of the bars of a pattern that is well above threshold is also affected by spatial frequency (Poirson and Wandell, 1993). As the spatial frequency of the pattern is increased the apparent color contrast between the bars is reduced. Thus, the detectability of color contrast and the color appearance of stimuli is dramatically affected by stimulus size.

20.3 FOUR ALTERNATIVE APPROACHES TO DISPLAY DESIGN

The previous sections have discussed physiological, perceptual, and technological considerations in designing visual displays. This has been the traditional focus for human factors research: to design displays that are legible. For example, the knowledge that a

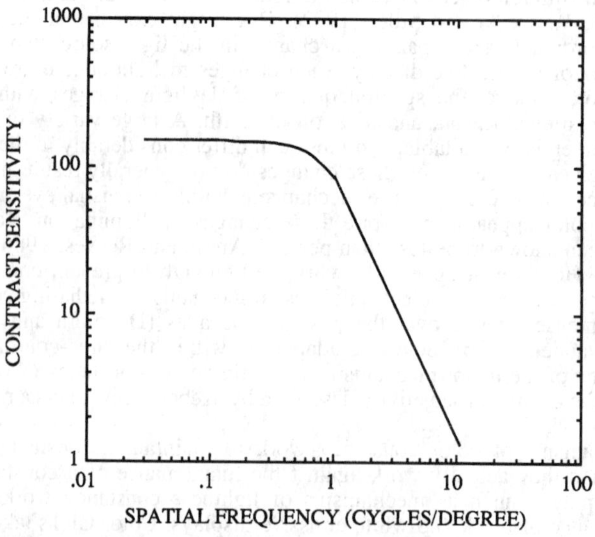

Figure 20.9 A typical plot of contrast sensitivity for isoluminant chromatic gratings. [Based on data given in Mullen (1985)].

user will be seated a particular distance from a particular type of display under a particular set of ambient lighting conditions can be used to determine the appropriate size and luminance contrast that will be necessary for the characters to be seen. Thus, the previous considerations provide us with an understanding of the baseline conditions of display design that must be met (are necessary) for an individual to use a display.

Although these considerations are necessary for the design of effective displays, they are not sufficient. Compliance with these considerations will make the *data* required to complete domain tasks available, but may not provide the *information* necessary to support an observer in decision making and action. Woods (1991) makes an important distinction between design for "data availability" and design for "information extraction." Designs that consider only data availability often impose unnecessary burdens on the user: to collect relevant data, to maintain these data in memory, and to integrate these data mentally to arrive at a decision. These mental activities require extensive knowledge, tax limited cognitive resources (attention, short-term memory) and therefore increase the probability of poor decision making and errors.

Our discussion of design for information extraction will begin with a consideration of four broadly defined approaches to display design. Each approach is complementary in the sense that it approaches the display design problem from different conceptual perspectives (i.e., graphical arts, psychophysical, attention-based, and problem solving/ decision making).

20.3.1 Aesthetic Approach

Tufte (1983, 1990) reviews the design of displays from an aesthetic, graphic arts perspective. Tufte (1983) describes principles of design for "data graphics" or "statistical graphics" which are expressly designed to present quantitative data. One principle is the "data-ink ratio": a measurement of the relative salience of data versus nondata elements in a graph. It is computed by determining the amount of ink that is used to convey the data and dividing this number by the total amount of ink that is used in the graphic. A higher data-ink ratio (a maximum of 1.0) represents the more effective presentation of information. A second measure of graphical efficiency is *data density.* Data density is computed by determining the number of data points represented in the graphic and dividing this number by the total area of the graphic. The higher the data density the more effective the graphic. Other principles include eliminating graphical elements that interact (e.g., moire vibration), eliminating irrelevant graphical structures (e.g., containers and decorations), and aesthetics (e.g., effective labels, proportion, and scale).

The two versions of a statistical graphic that are shown in Figure 20.10(*a*) and 20.10(*b*) illustrate several of Tufte's principles. The version in Figure 20.10(*a*) is poorly designed, while the version in Figure 20.10(*b*) is more effectively designed. In Figure 20.10(*b*) the irrelevant data container (the box) that surrounds the graph in Figure 20.10(*a*) has been eliminated. In addition, several other nondata graphical structures have been removed (grid lines). In fact, these grid lines are made conspicuous by their absence in Figure 20.10(*b*). Together, these manipulations produce both a higher data-ink ratio and a higher data density for the version in Figure 20.10(*b*). In Figure 20.10(*a*) the "striped" patterns on the bar graphs produce an unsettling moire vibration and have been replaced in Figure 20.10(*b*) with gray-scale patterns. In addition, the bar graphs in Figure 20.10(*b*) have been visually segregated by spatial separation. Finally, the three-dimensional perspective in Figure 20.10(*a*) complicates visual comparisons and has been removed in Figure 20.10(*b*).

Tufte (1990) broadens the scope of these principles and techniques by considering nonquantitative displays as well. Topics that are discussed include micro/macro designs (the integration of global and local visual information), layering and separation (the visual stratification of different categories of information), small multiples (repetitive graphs that show the relationship between variables across time, or across a series of variables), color (appropriate and inappropriate use of), and narratives of space and time (graphics that preserve or illustrate spatial relations or relationships over time). The following quotations summarize many of the key principles.

- It is not how much information there is, but rather, how effectively it is arranged (p. 50).
- Clutter and confusion are failures of design, not attributes of information (p. 51).
- Detail cumulates into larger coherent structures... Simplicity of reading derives from the context of detailed and complex information, properly arranged. A most unconventional design strategy is revealed: to clarify, add detail (p. 37).

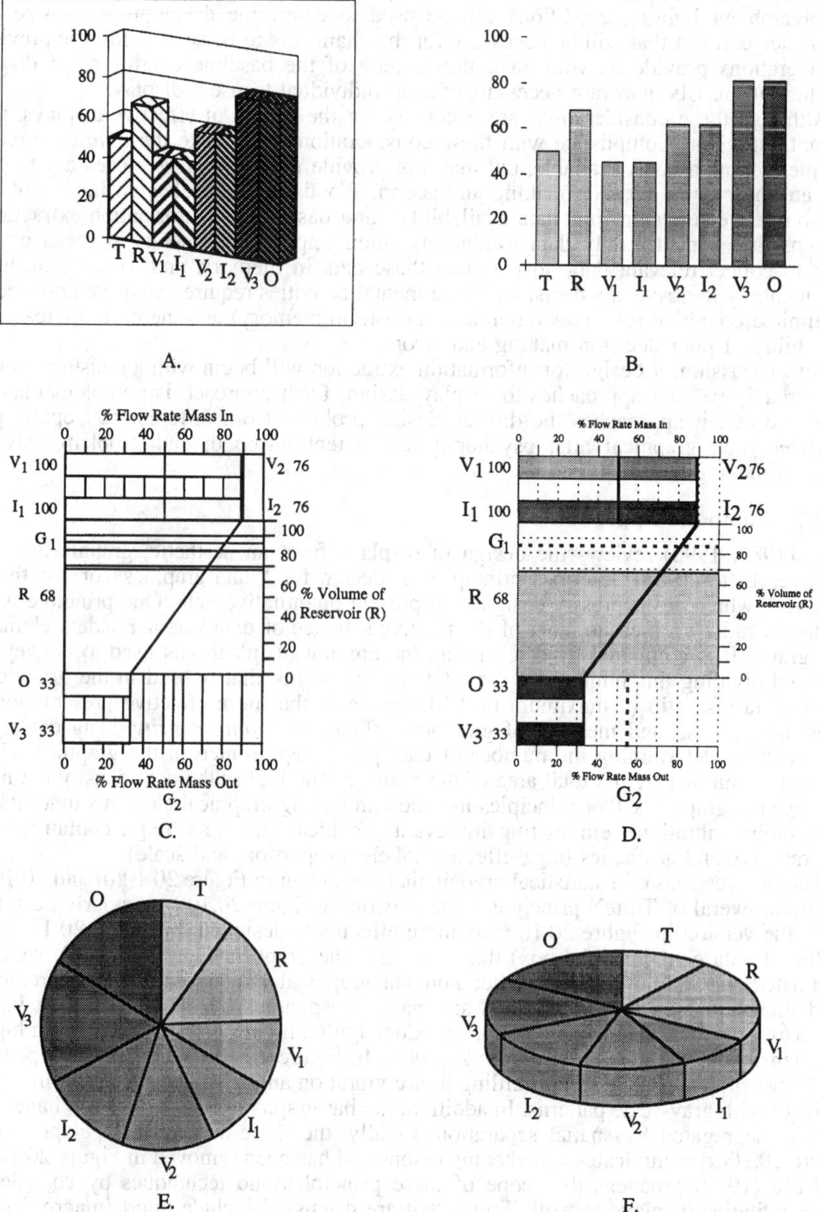

Figure 20.10 Six alternative mappings. Figure 20.10(*a*) and 20.10(*b*) represent alternative versions of a separable (bar graph) graphical format that is less effective (*a*) and more effective (*b*) mappings. Similarly, Figure 20.10(*c*) and 20.10(*d*) represent alternative versions of a configural display format that is less (*c*) and more effective (*d*), primarily due to layering and separation. Figure 20.10(*e*) and 20.10(*f*) represents the least effective mappings.

- Micro/macro designs enforce both local and global comparisons and, at the same time, avoid the disruption of context switching. All told, exactly what is needed for reasoning about information (p. 50).
- Among the most powerful devices for reducing noise and enriching the content of displays is the technique of layering and separation, visually stratifying various

aspects of the data. . . What matters—inevitably, unrelentingly — is the proper relationship among information layers. These visual relationships must be in relevant proportion and in harmony to the substance of the ideas, evidence, and data displayed (pp. 53–54).

This final principal, layering and separation, is graphically illustrated in Figure 20.10(*c*) and 20.10(*d*). These two versions of the same display vary widely in terms of the visual stratification of the information that they contain. In Figure 20.10(*c*) all of the graphical elements are at the same level of visual prominence; in Figure 20.10(*d*) there are at least three levels of visual prominence. The lowest layer of visual prominence is associated with the nondata elements of the display. The various display grids have thinner, dotted lines, and their labels have also been reduced in size and made thinner. The medium layer of perceptual salience is associated with the individual variables. The graphical forms that represent each variable have been gray-scale coded, which contributes to separating these data from the nondata elements. Similarly, the lines representing the system goals (G1 and G2) have been made bolder and dashed. In addition, the labels and digital values that correspond to the individual variables are larger and bolder than their nondata counterparts. Finally, the highest level of visual prominence has been reserved for those graphical elements which represent higher-level system properties (e.g., the bold lines that connect the bar graphs). The visual stratification could have been further enhanced through the use of color. The techniques of layering and separation will facilitate an observer's ability to locate and extract information.

To summarize, Tufte (1983, 1990) addresses the problem of presenting three-dimensional, multivariate data on flat, two-dimensional surfaces (primarily focusing on static, printed material) very admirably. He attacks the problem from a largely aesthetic perspective and provides numerous examples of both good and bad display design that clearly illustrate the associated design principles. Although there are aspects of dynamic display design for complex domains that are not considered, the principles can be generalized.

20.3.2 Psychophysical Approach

Cleveland and his colleagues have also developed principles for the design of statistical graphics. However, in contrast to the aesthetic conceptual perspective of Tufte, Cleveland has used a psychophysical approach. As an introduction consider the following quotation (Cleveland, 1985, p. 229):

When a graph is constructed, quantitative and categorical information is encoded by symbols, geometry, and color. Graphical perception is the visual decoding of this encoded information. Graphical perception is the vital link, the raison d'etre, of the graph. No matter how intelligent the choice of information, no matter how ingenious the encoding of the information, and no matter how technologically impressive the production, a graph is a failure if the visual decoding fails. To have a scientific basis for graphing data, graphical perception must be understood. Informed decisions about how to encode data must be based on knowledge of the visual decoding process.

In their efforts to understand graphical perception Cleveland and his colleagues have considered how psychophysical laws (e.g., Weber's law, Stevens' law) are relevant to the design of graphic displays. For example, psychophysical studies using magnitude estimation have found that judgments of length are less biased than judgments of area or volume. Therefore, visual decoding should be more effective if data has been encoded into a format that requires length discriminations, as opposed to area or volume discriminations. Cleveland and his colleagues have tested this, and similar intuitions, empirically. Their experimental approach was to take the same quantitative information, to provide alternative encodings of this quantitative information (graphs which required different "elementary graphical-perception tasks"), and to test observers' ability to extract the information.

The results of these experiments provided a rank-ordering of performance on basic graphical perception tasks: position along a common scale, position along identical, nonaligned scales, length, angle/slope, area, volume, and color hue/color saturation/density (ordered from best to worst performance, Cleveland, 1985, p. 254). Guidelines for display design were developed based on these rankings. Specifically, graphical encodings should be chosen that require the highest ranking graphical perceptual task of the observer during the visual decoding process. For example, consider the three graphs illustrated in Figures 20.10(*b*), 20.10(*e*), and 20.10(*f*). For decoding information contained in the Figure 20.10(*b*) an observer is required to judge position along a common scale (in this case, the vertical extent of the various bar graphs). For Figure 20.10(*e*) the observer is required to judge angles and/or area. Finally, to decode the information in Figure 20.10(*f*), the observer is required to judge volume (note that because of the three dimensional representation angles and area are no longer valid cues). According to the rankings, Cleveland and his colleagues would therefore predict that performance would be best with the bar chart, intermediate with the pie chart, and worst with the three-dimensional pie chart.

20.3.3 Attention-Based Approach

A third perspective on display design is to consider the problem in terms of visual attention and object perception. From this conceptual perspective, designers have a number of interface resources at their disposal for encoding information in graphical displays (e.g., chromatic contrast, luminance contrast, the integration of individual variables into geometrical objects, and animation). A great deal of basic research has attempted to identify the factors that control the distribution of attention to visual stimuli. The results have important theoretical and practical implications for display design.

Understanding these implications requires a brief consideration of the continuum of attention demands that operators might face in complex, dynamic domains. At one end of the attention continuum are tasks that require selective responses to specific elements in the display ("focused" tasks). This might refer to a response contingent on the height of a single bar in a bar graph or on the position of a pointer on a radial display. At the opposite end of this continuum are tasks that require the distribution of attention across many features that must be considered together in order to choose an appropriate response ("integration" tasks). For example, the response might be contingent on the relative position of numerous bars within a bar graph. Thus, tasks can be characterized in terms of the relative demands for selective attention to respond to specific features with specific actions and distributed or divided attention in which multiple display elements must be considered together in order to choose the appropriate actions.

Attention-based approaches to display design have examined how the design of visual representations can help to meet the cognitive load posed by this continuum of attention demands. Garner (Garner, 1970, 1974; Garner and Felfoldy, 1970) and Pomerantz (Pomerantz, 1986; Pomerantz and Pristach, 1989; Pomerantz, Sager, and Stoever, 1977) have used the speeded classification task to examine the dimensional structure of stimuli. Carswell and Wickens (1990) have generalized these results by investigating perceptual dimensions that are representative of those found in visual displays. Three qualitatively different relationships between stimulus dimensions have been proposed: "separable," "integral," and "configural" (Pomerantz, 1986).

Separable dimensions. A separable relationship is defined by a lack of interaction among stimulus dimensions. Each dimension retains its unique perceptual identity within the context of the other dimension. Observers can selectively attend to an individual dimension and ignore variations in the irrelevant dimension. On the other hand, no new properties emerge as a result of the interaction among dimensions. Thus, performance suffers when both dimensions must be considered to make a discrimination. This pattern of results suggests that separable dimensions are processed independently. An example of separable dimensions are color and shape: the perception of color does not influence the perception of shape, and vice versa.

Integral dimensions. An integral relationship is defined by a strong interaction among dimensions such that the unique perceptual identities of individual dimensions are lost. Integral stimulus dimensions are processed in a highly interdependent fashion: a change in one dimension necessarily produces changes in the second dimension. In their discussion of two integral stimulus dimensions, Garner and Felfoldy (1970, p. 237) state that "in order for one dimension to exist, a level on the other must be specified." As a result of this highly interdependent processing, a redundancy gain occurs. However, focusing attention on the individual stimulus dimensions becomes very difficult, and performance suffers when attention to one (selective attention) or both (divided attention) dimensions are required. An example of an integral stimulus is perceived color: It is a function of both hue and brightness.

Configural dimensions. A configural relationship refers to an intermediate level of interaction between perceptual dimensions. Each dimension maintains its unique perceptual identity, but new properties are also created as a consequence of the interaction between them. These properties have been referred to as "emergent features." Using parentheses as our graphic elements will allow us to demonstrate several examples of emergent features. Depending on the orientation, a pair of parentheses can have the emergent features of vertical symmetry, () and)(, or parallelism,)) and ((. Pomerantz and Pristach (1989, p. 636) state that "Emergent features may be global (i.e., not localized to any particular position within the figure), such as symmetry or closure, or they may be local, such as vertices that result from intersections of line segments." There are two significant aspects of performance with configural dimensions. First, relative to integral and separable stimulus dimensions there is a smaller divided attention cost, suggesting that performance can be enhanced when both dimensions must be considered to make a discrimination. The second noteworthy aspect of this pattern of results is that there is an *apparent* failure of selective attention (see Bennett and Flach, 1992, for a further discussion of why this failure may be apparent, and not inherent).

20.3.3.1 Proximity Compatibility Principle

Wickens and his colleagues (e.g., Wickens and Carswell, 1995) have applied the results of the visual attention research to the problem of display design. Their principle of "proximity compatibility" emphasizes the relationship between task demands and the graphical form of a display. *Perceptual proximity* (display proximity) refers to the perceptual similarity between information sources in a display. Perceptual proximity can be defined along several dimensions including: (1) spatial proximity (e.g., physical distance—near or far); (2) chromatic proximity (e.g., the same or different colors); (3) physical dimensions (e.g., information is encoded using the same or different physical dimensions—length versus volume); (4) perceptual code (e.g., digital versus analog); and (5) geometric form (e.g., object versus separate displays). For example, when individual variables are mapped into a closed geometric form the display is high in display proximity; when each variable has its own unique representation (e.g., a bar graph) the display is low in proximity.

Processing proximity (mental proximity) refers to the continuum of attentional demands, that is, to the extent to which information from the various sources in a display must be (or need not be) considered together to accomplish a particular task. There are three major categories of processing proximity: integrative processing, nonintegrative processing, and independent processing. Information from different sources must be explicitly combined in integrative processing, and this represents a high level of processing proximity. Integrative processing includes both computational processing (involving numerical operations) and boolean processing (involving logical operations). Nonintegrative processing represents an intermediate level of processing proximity and involves "some other features of similarity instead of (or in addition to) their need for combination" (Wickens and Carswell, p. 476). Examples include: (1) metric similarity (similarity of units), (2) statistical similarity (extent of covariation), (3) functional similarity (semantic relatedness), (4) processing similarity (similarity of computational procedures), and (5) temporal similarity (temporal proximity). Finally, *independent processing* refers to the situation where different information sources need not be considered together (in fact, one information source is independent from another).

Briefly stated, the principle of proximity compatibility maintains that the *display proximity* should match the *task proximity*. Performance on integrated tasks (high mental proximity) is predicted to be facilitated by displays that have high perceptual proximity (e.g., object display). Similarly, performance on focused tasks (low mental proximity) is predicted to be facilitated by displays that have low perceptual proximity (e.g., bargraph displays).

20.3.3.2 Implications

Researchers continue to investigate the potential trade-offs between display type [object—Figure 20.10(*d*) versus separate—Figure 20.10(*b*)] and task type (integrated versus focused). Initially, a straightforward trade-off was predicted: object displays would produce superior performance for integration tasks, while separable displays would produce superior performance for focused tasks. In general, laboratory research comparing performance differences between object and separate displays when integration tasks must be completed has revealed significant advantages for object displays (Bennett and Flach, 1992). However, there is a general consensus that these performance advantages are not attributable to objectness, per se (Bennett and Flach, 1992; Bennett, Toms, and Woods,

1993; Buttigieg and Sanderson, 1991; Sanderson, Flach, Buttigieg, and Casey, 1989; Wickens and Carswell, 1995). Instead, the quality of performance at integration tasks is dependent on the quality of the mapping between the emergent features produced by a display and the inherent data relationships that exist in the domain (this point will be discussed at length in subsequent sections).

There is much less consensus on the second major prediction regarding the potential costs for configural displays (relative to separable displays) when individual variables must be considered. We believe that a single display may support performance at both integration and focused attention tasks (Bennett and Flach, 1992). The attention and object perception literature (in particular, the principle of configurality) leaves open the possibility that a single geometric display may be designed to support performance for both distributed and focused attention tasks. One way to consider objects is as a set of hierarchical features (including elemental features, configural features, and global features) that vary in their relative salience. For example, Treisman (1986) observed that "if an object is complex, the perceptual description we form may be hierarchically structured, with global entities defined by subordinate elements and subordinate elements related to each other by the global description" (p. 35.54). Observers may focus attention at various levels in the hierarchy at their discretion, and in particular, there may be no inherent cost associated with focusing attention on elemental features. From a practical standpoint, any potential costs associated with low-level data can be eliminated outright by annotating the graphical representation with digital information.

20.3.4 Problem-Solving and Decision-Making Approach

The fourth perspective on display design that will be discussed is problem solving and decision making. Recently, there has been an increased appreciation for the creativity and insight that experts bring to human–machine systems. Under normal operating conditions, an individual is perhaps best characterized as a decision maker. Depending on the perceived outcomes associated with different courses of action, the amount of evidence that a decision maker requires to choose a particular option will vary. In models of decision making, this is called a *decision criterion*. Under abnormal or unanticipated operating conditions, an individual is most appropriately characterized as a creative problem solver. The cause of the abnormality must be diagnosed, and steps must be taken to correct the abnormality (i.e., an appropriate course of action must be determined). This involves monitoring and controling system resources, selecting between alternatives, revising diagnoses and goals, determining the validity of data, overriding automatic processes, and coordinating the activities of other individuals. Thus, the literature on reasoning, problem solving, and decision making has important insights for display design.

There is a vast literature on problem solving, ranging from the seminal work of the Gestaltists (e.g., Wertheimer, 1959), the paradigmatic contributions of Newell and Simon (1972), to contemporary approaches. For the Gestalt psychologists, perception and cognition (more specifically, problem solving) were intimately intertwined. The key to successful problem solving was viewed as the formation of an appropriate gestalt, or representation, that revealed the "structural truths" of a problem. For example, Wertheimer (1959, p. 235) states that "Thinking consists in envisaging, realizing structural features and structural requirements..." The importance of a representation is still a key consideration today; it is probably not an overstatement to conclude that the primary lesson to be learned from the problem solving literature is that the representation of a problem has a profound influence on the ease or difficulty of its solution.

Historically, decision research has focused on developing models that describe the generation of multiple alternatives (potentially all alternatives), the evaluation (ranking) of these alternatives, and the selection of the most appropriate alternative. By and large, perception was ignored. In contrast, recent developments in decision research, stimulated by research on naturalistic decision making (e.g., Klein, Orasanu, and Zsambok, 1993) has begun to give more consideration to the generation of alternatives in the context of dynamic demands for action. Experts are viewed as generating and evaluating a few "good" alternatives. The emphasis is on recognition (e.g., how is this problem similar, or dissimilar, to problems that I have encountered before?). As a result, perception plays a dominant role. This change in emphasis has increased awareness of perceptual processes and dynamic action constraints in decision making.

These trends have, either directly or indirectly, led researchers in interface design to focus on the representation problem. Perhaps the first explicit realization of the power of graphic displays to facilitate understanding was the STEAMER project (Hollan, Hutchins,

McCandless, Rosenstein, and Weitzman, 1987; Hollan, Hutchins, and Weitzman, 1984), an interactive inspectable, training system. STEAMER provided alternative conceptual perspectives—"conceptual fidelity" of a propulsion engineering system through the use of analogical representations. In addition, the current approach to the design of human–computer interfaces (direct manipulation—Hutchins, Hollan, and Norman, 1986; Shneiderman, 1986, 1993) can be viewed as an outgrowth of this general approach. More recently scientific visualization (the role of diagrams and representation in discovery and invention) is being vigorously investigated (Brodie, Carpenter, Earnshaw, Gallop, Hubbold, Mumford, Osland, and Quarendon, 1992; Earnshaw and Wiseman, 1992). Thus, the challenge for display design from this perspective is to provide appropriate representations that support humans in their problem solving endeavors.

20.4 THE REPRESENTATION AIDING APPROACH TO DISPLAY DESIGN

It should be noted that in the aesthetic, psychophysical, and attention-based approaches, there is little consideration given to a domain or problem behind the display. It was not necessary for us to describe the "problem" behind the displays shown in Figure 20.10. However, the correspondence between the visual structure in a representation and the constraints in a problem is fundamental to the problem solving and decision making approaches. Recently, a number of research groups have recognized that effective interfaces depend on both the mapping from human to display (the coherence problem) and the mapping from display to a work domain or problem space (the correspondence problem). Terms used to articulate this recognition include direct perception (Moray, Lee, Vicente, Jones, and Rasmussen, 1994), ecological interface design (Rasmussen and Vicente, 1989), representational design (Woods, 1991), or semantic mapping (Bennett and Flach, 1992). Woods and Roth (1988) have illustrated the problem of interface design in terms of a "cognitive triad" that we illustrate in Figure 20.11. The three components of the cognitive system triad are: (1) the cognitive demands produced by the domain of interest, (2) the resources of the cognitive agent(s) that meet those demands, and (3) the representation of the domain through which the agent experiences and interacts with the domain (the interface). Thus, the focus of our approach is not on information-processing characteristics, graphical forms, events, trajectories, tasks, or procedures per se. Instead, the focus is on the interactive and mutually constraining relationships between the individual, the interface, and the domain (labeled coherence and correspondence in Figure 20.11). The overall level of human–machine system performance is determined by the quality of these relationships.

20.4.1 The Correspondence Problem: The Semantics Of Work

Correspondence refers to the issue of content—what information should be present in the interface in order to meet the cognitive demands of the work domain? Correspondence is defined neither by the domain itself nor the interface itself: It is a property that arises from the interaction of the two. Thus, in Figure 20.11 correspondence is represented by the labeled arrows that connect the domain and the interface. One convenient way to conceptualize correspondence is as the quality of the mapping between the interface and the work space, where these mappings can vary in terms of the degree of specificity (consistency, invariance, or correspondence). As we will demonstrate, within this mapping there can be a one-to-one correspondence, a many-to-one, a one-to-many, or a many-to-many mapping between the information that exists in the interface and the structure within the work space.

20.4.1.1 Rasmussen's Abstraction Hierarchy

Addressing the issue of correspondence requires a deep understanding and explicit description of the "semantics" of a work domain. Rasmussen's abstraction hierarchy (1986) is a theoretical framework for describing domain semantics in terms of a nested hierarchy of functional constraints (including goals, physical laws, regulations, organizational/structural constraints, equipment constraints, and temporal/spatial constraints). One way to think about the abstraction hierarchy is that it provides structured categories of information (i.e., the alternative conceptual perspectives) that an individual must consider in the course of accomplishing system goals. Consider the following passage from Rasmussen (1986, p. 21):

> *During emergency and major disturbances, an important control decision is to set up priorities by selecting the level of abstraction at which the task should be*

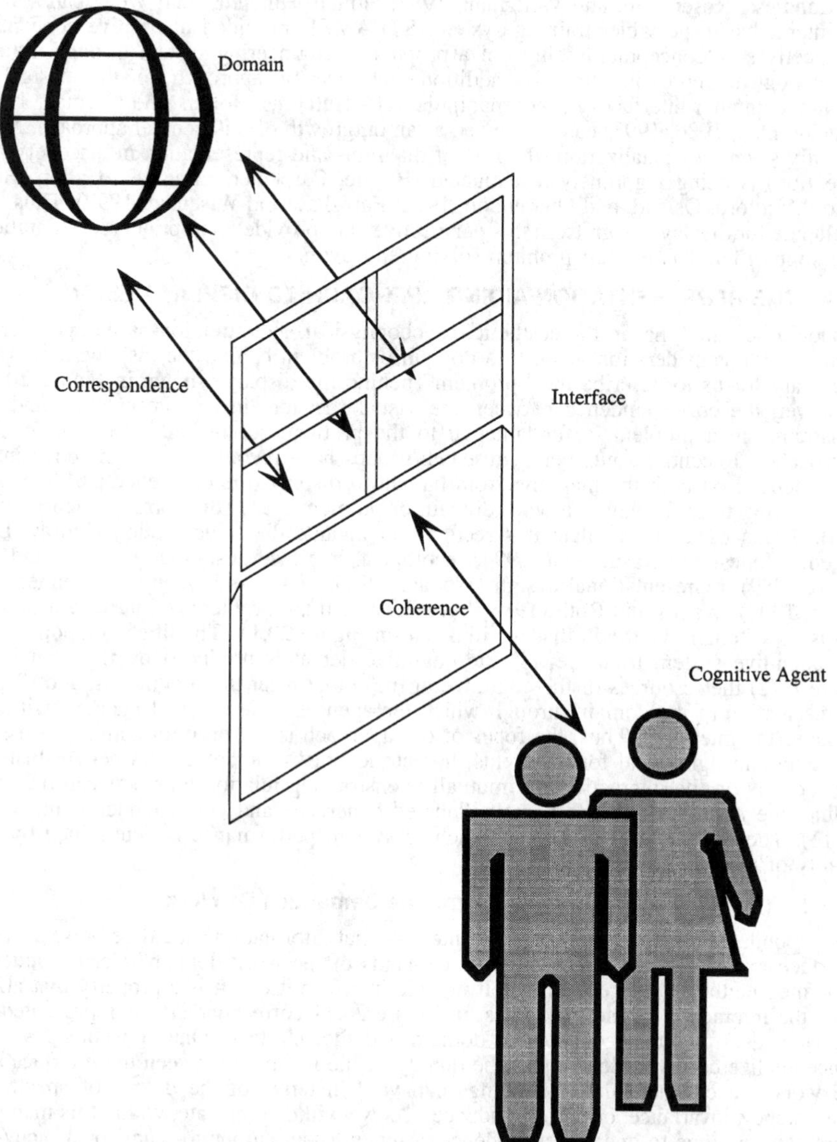

Figure 20.11 The "cognitive triad": A cognitive systems engineering perspective. Any domain produces cognitive demands that must be met by the cognitive agents interacting with (or controlling) the domain. The cognitive agent possesses cognitive resources that must be used to meet these demands. The interface is the medium (or representation) through which the cognitive agent views and controls the domain. The effectiveness of an interface is determined by both correspondence and coherence.

initially considered. In general, the highest priority will be related to the highest level of abstraction. First, judge overall consequences of the disturbances for the system function and safety in order to see whether the mode of operation should be switched to a safer state (e.g., standby or emergency shutdown). Next, consider whether the situation can be counteracted by reconfiguration to use alternative functions and resources. This is a judgment at a lower level of function and equipment. Finally, the root cause of the disturbance is sought to determine how it can be corrected. This involves a search at the level of physical functioning of parts and components. Generally, this search for the physical disturbance is of lowest priority (in aviation, keep flying—don't look for the lost light bulb!).

Thus, in complex domains, situation awareness requires the operator to understand the process at different levels of abstraction. Further, the operator must be able to understand constraints at one level of abstraction in terms of constraints at other levels. The correspondence question asks whether the hierarchy of constraints that define a work domain are reflected in the interface.

20.4.2 The Coherence Problem: The Syntax of Form

Coherence refers to the mapping from the representation to the human perceiver. Here the focus is on the visual properties of the representation. What distinctions within the representation are discriminable to the human operator? How do the graphical elements fit together or coalesce within the representation? Is each element distinct or separable? Are the elements absorbed within an integral whole, thus losing their individual distinctness? Or do the elements combine to produce configural or global properties? Are some elements or properties of the representation more or less salient than other elements or properties?

In general, coherence addresses the question of how the various elements within a representation compete for attentional and cognitive resources. Just as work domains can be characterized in terms of a nested hierarchy of constraints, so too can complex visual representations be perceived as a hierarchy of nested structures, with local elements combining to produce more global patterns or symmetries.

20.4.3 The Mapping Problem

In human–machine systems, a display is a representation of an underlying domain, and the user's tasks are defined by that domain, rather than by the visual characteristics of the display itself. Thus, whether a display will be effective or not is be determined by both correspondence and coherence. More specifically, the effectiveness of the display is determined by the quality of the mapping between the constraints that exist in the domain and the constraints that exist in the display. The display constraints are defined by the spatiotemporal structure (the visual appearance of the display over time) that results from the particular representation chosen.

Three rather fuzzy distinctions might be useful when thinking about the types of representations that might be used to accomplish the mapping of domain constraints to constraints within the interface—analogical, configural, and metaphorical. Analogical representations might be considered when the constraints of the work domain are fundamentally spatial. For example, STEAMER used an analogical representation of the spatial layout of the feedwater system to show the connections among component processes. Also, the standard flight display for representing pitch and roll (attitude) is an analog to the spatial relations between the aircraft and the horizon. In general, where the domain constraints themselves are naturally spatial, designers should consider whether the interface might provide a direct analog of these constraints.

Configural representations use geometric relations to represent constraints that are not spatial. A simple example is using an axis in a graph to represent time. In configural representations the geometrical display constraints will generally take the form of symmetries—equality (e.g., length, angle, area), parallel lines, colinearity, or reflection. In addition, Gestalt properties of closure and good form are useful. These display constraints will produce the emergent features that were discussed in Section 20.3.3. The core problem in implementing effective configural displays is to provide visual representations that are perceived as accurate reflections of the abstract domain constraints: Are the critical domain constraints appropriately reflected in the geometrical constraints in the display? Are breaks in the domain constraints (e.g., abnormal or emergency conditions) reflected by breaks in the geometrical constraints (e.g., emergent features such as nonequality, nonparallelism, nonclosure, bad form)? Only when this occurs will the cognitive agent be able to obtain meaning about the underlying domain in an effective fashion. One source of ideas for configural displays is the graphical representations that engineers use to make design decisions. For example, Beltracchi (1987, 1989; see also Moray, Lee, Vicente, Jones, and Rasmussen 1994; Rasmussen, Pejtersen, and Goodstein, 1994) has designed a configural display for controlling the process of steam generation in nuclear power plants based on the Temperature/Entropy graphic used to evaluate thermodynamic engines (Rankine Cycle Display).

Metaphorical representations use spatial or symbolic relations from other, more familiar, work domains as metaphors; the goal is to enhance the transfer of skills from one domain to another. Perhaps the most obvious example is the "desktop" metaphor that is used in personal computer systems. Another example is the BookHouse metaphor, developed by Goodstein and Pejtersen (Goodstein and Pejtersen, 1989; Pejtersen, 1992) to facilitate library information retrieval. Rasmussen, Pejtersen, and Goodstein (1994, pp. 289–291) describe the metaphor and its justification:

> *The use of the BookHouse metaphor serves to give an invariant structure to the knowledge base ... Since no overall goals or priorities can be embedded in the system, but depend on the particular user, a global structure of the knowledge base reflects subsets relevant to the categories of users having different needs and represented by different rooms in the house ... This gives a structure for the navigation that is easily learned and remembered by the user ... The user "walks" through rooms with different arrangements of books and people ... It gives a familiar context for the identification of tools to use for the operational actions to be taken. It exploits the flexible display capabilities of computers to relate both information in and about the data base, as well as the various means for communicating with the data base to a location in a virtual space ... This approach supports the user's memory of where in the BookHouse the various options and information items are located. It facilitates the navigation of the user so that items can be remembered in given physical locations that one can then retraverse in order to retrieve a given item and/or freely browse in order to gain an overview.*

Whether analogical, configural, metaphorical, or combined representations are used, the key to successful design is the quality of the mapping. The visual salience of the information in the display must reflect the relative importance of that information in terms of the work domain. For analogical displays the spatial analogs must scale appropriately to the real task constraints. For configural displays the geometric symmetries must correspond to higher order constraints on the process. For metaphorical displays, the intuitions and skills elicited by the representational domain must map appropriately to the target domain.

20.5 AN EXAMPLE-BASED TUTORIAL OF THE REPRESENTATION AIDING APPROACH

The concepts and principles of display design that have been introduced thus far include correspondence, coherence, process constraints, display constraints, and the mappings between process and display constraints. These concepts and principles are necessarily abstract, and in order for them to be useful for display design they must be presented in a clear and unambiguous fashion. In this section, we provide a tutorial that illustrates these concepts and principles through a series of concrete examples. We begin with an analysis of a simple system from the domain of process control; the goal is to provide a description of the associated process constraints. We then consider various types of displays that could be devised for the system. The goal is to consider the alternative mappings between process constraints and geometric (display) constraints that are provided by each representation, and in particular, the implications for correspondence and coherence. The representations are chosen to illustrate the continuum of visual forms from separable, through configural, to integral geometries. We then examine one representation in greater detail and discuss the implications of this mapping for normal and abnormal operating conditions. We end the section with a set of practical guidelines for display design.

20.5.1 A Simple Domain From Process Control

The process is a generic one that might be found in process control, and it is represented graphically in the lower portion of Figure 20.12. There is a reservoir (or tank, represented by the large rectangle in the middle of the figure) that is filled with a fluid (for example, coolant). The volume, or level, of the reservoir (R) is represented by the filled portion of the rectangle. Fluid can enter the reservoir through the two pipes and valves located above the reservoir; fluid can leave the reservoir through the pipe and valve located below the reservoir. We will categorize the information in this simple process using a simple dis-

Sample Process

Low Level Data (Process variables)	High Level Properties (Process constraints)	
T – Time	K_1 – $I_1 - V_1$	Relation between commanded flow (V) and actual flow (I or O)
V_1 – Setting for Valve 1	K_2 – $I_2 - V_2$	
V_2 – Setting for Valve 2	K_3 – $O - V_3$	
V_3 – Setting for Valve 3		
I_1 – Flow rate through Valve 1		
I_2 – Flow rate through Valve 2		
O – Flow rate through Valve 3		
R – Volume of reservoir	K_4 – $\Delta R - ((I_1 + I_2) - O)$	Relation between reservoir volume (R), mass in $(I_1 + I_2)$, and mass out (O).
G_1 – Volume goal	K_5 – $R - G_1$	Relation between actual states (R, O) and goal states (G_1, G_2)
G_2 – Output goal (demand)	K_6 – $O - G_2$	

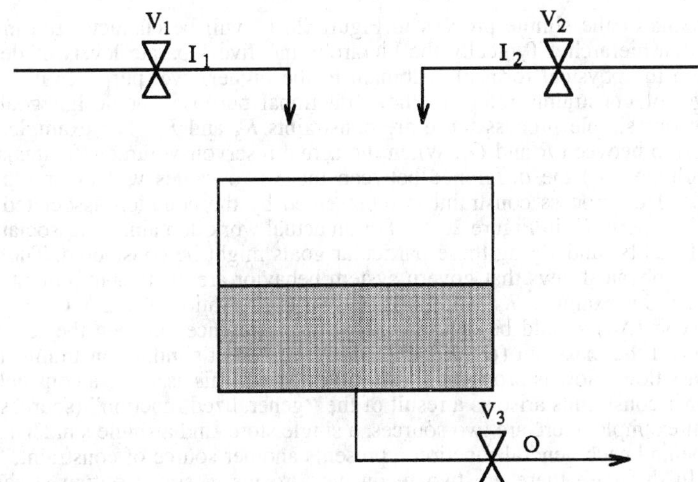

Figure 20.12 A simple domain from process control that has a reservoir for storing mass, two input streams that increase the volume of mass in the reservoir, and a single output stream that decreases the volume. The low-level data (the measured domain variables), the high-level properties (constraints that arise from the interaction of these variables and the physical design) and the domain goals (requirements that must be met for the system to be functioning properly) are listed.

tinction in which the term "low-level" data refers to local constraints or elemental state variables that might be measured by a specific sensor. The term "higher level properties" will be used to refer to more global constraints that reflect relations or interactions among multiple variables.

Low-level data (process variables). There are two goals associated with this simple process. First, there is a goal (G_1) associated with *R*, the level of the reservoir. The reservoir should be maintained at a relatively high level to ensure that sufficient resources are available to meet long-term increases in demanded output flow rate (O). The second goal (G_2) refers to the specific rate of output flow that must be maintained in order to meet an external demand. These goals are achieved and maintained by adjusting three valves (V_1, V_2, and V_3) that regulate flow through the system (I_1, I_2, and O). Thus, this simple process is associated with a number of process variables that can be measured directly: these low-level data are listed in the upper, left-hand portion of Figure 20.12 (V_1, V_2, V_3, I_1, I_2, O, G_1, G_2, and R).

High-level properties (process constraints). In addition, there are relationships between these process variables that must be considered when controlling the process (see the upper, right-hand portion of Figure 20.12). The most important high-level properties are goal-related: Does the actual reservoir volume level (R) match the goal of the system (G_1)—K_5? Does the actual system output flow rate (O) match the flow rate that is required (G_2)—K_6? Even for this simple process some of the constraints or (high-level properties) are fairly complex. For example, an important property of the system is mass balance. The mass balance is determined by comparing the mass leaving the reservoir (O, the output flow rate) to mass entering the reservoir (the combined input flow rates of I_1 and I_2). This relationship determines the direction and the rate of change for the volume inside the reservoir (ΔR). For example, if mass in and mass out are equal then mass is balanced, ΔR will equal 0.00, and R will remain constant.

Controlling even this simple process will depend on a consideration of both high-level properties and low-level data. As the previous example indicates, decisions about process goals (e.g., maintaining a sufficient level of reservoir volume) generally require consideration of relationships between variables (is there a net inflow, a net outflow, or is mass balanced?), as well as the values of the individual variables themselves (what is the current reservoir volume?).

20.5.1.1 An Abstraction Hierarchy Analysis

The constraints of the simple process in Figure 20.12 will be characterized in terms of the abstraction hierarchy. Typically the hierarchy has five separate levels of description, ranging from the physical form of a domain to the higher level purposes it serves. The highest level of constraints refers to the "functional purpose" or design goals for the system. For our simple process these are constraints K_5 and K_6. For example, consider the relationship between R and G_1. When the actual reservoir volume (R) equals the goal reservoir volume (G_1) the difference between these two values will assume a constant value (0.00). This process constraint is represented by the equation associated with the higher level property K_5 in Figure 20.12. For an actual work domain, the associated values (costs and benefits) underlying these particular goals might be considered. The "abstract functions" or physical laws that govern system behavior are another important source of constraints. In our example, K_4 reflects the law of conservation of mass. Change of mass in the reservoir (ΔR) should be determined by the difference between the residual mass in ($I_1 + I_2$) and the mass out (O). K_1, K_2, and K_3 represent similar constraints associated with the mass flow. Flow is proportional to valve setting (this assumes a constant pressure head). Further constraints arise as a result of the "generalized function" (sources, storage, sink). In our example, there are two sources, a single store, and a single sink. The physical processes behind each general function represents another source of constraint, "physical function." In this case there are two feedwater streams, a single output stream, and a reservoir for storage. Similarly, the moment-to-moment values of each variable (T, V_1, V_2, V_3, I_1, I_2, O, and R) should be considered at the level of physical function. Finally, the level of "physical form" provides information concerning the physical configuration of the system, including information related to causal connections, length of pipes, position of valves on pipes, and size of the reservoir. All of these constraints will be satisfied if the process is being controlled in a proper fashion.

To summarize, an abstraction hierarchy analysis provides information about the hierarchically nested constraints that constitute the semantics of a domain, and therefore defines the information that must be present in the interface for an individual to perform successfully. The product of this analysis (interrelated categories of information) provides a structured framework for display development, as we will demonstrate shortly. It should be emphasized that this analysis and description is independent of the interface, and therefore differs from traditional task analysis. Although space limitations do not permit a complete discussion, we view abstraction hierarchy analysis and task analysis (traditional or "cognitive") as complementary processes that are necessary for the development of effective displays.

20.5.2 Coherence and Correspondence: Alternative Mappings

In this section, we provide six examples that illustrate alternative mappings between domain semantics and representations (displays) for our simple process (see Figure 20.13). The discussion is organized in terms of the distinction between integral, configural, and separable dimensions that was outlined in Section 20.3.3. One goal is to illustrate what these terms, originally coined in the attention literature, mean in the context of display design for complex systems. A second goal is to focus on the quality of the mapping that each display provides, especially with respect to the ability of each display to convey information at the various levels of abstraction (see Section 20.4.1.1 and the previous section). To illustrate the quality of the mapping explicitly, we have provided a summary listing (at the right of each display in Figure 20.13) that sorts the associated process constraints into two categories ("P" and "D"). Process constraints that are represented directly in the display (that is, which can be "seen") have been placed in the P category (*P*erceived). Process constraints that are not represented directly, and must be computed or inferred, are placed in the D category (*D*erived). Process constraints that are related to physical structure are represented by the theta symbol (ϕ) process constraints related to the functional structure are represented by the symbol ($\int$).

Separable displays. Figure 20.13(*a*) represents a separable display which contains a single display for each individual process variable present. Each display is represented in the figure by a circle, but no special significance should be attached to the symbology: The circles could represent digital displays, bar graphs, etc. For example, four instantiations of this display are shown in Figures 20.10(*a*), (*b*), (*c*), and (*f*). In Figure 20.10(*a*) and (*b*), the display constraints are the relative heights of the bars in response to changes in the underlying variables.

In terms of the abstraction hierarchy the class of displays represented by Figure 20.13(*a*) provides information only at the level of physical function: Individual variables are represented directly. Thus, there is not likely to be a selective attention cost for low-level data. However, there is likely to be a divided attention cost, because the observer must derive the high-level properties. To do so, the observer must have an internalized model of the functional purpose, the abstract functions, the general functional organization, and the physical process. For example, to determine the direction (and cause) of ΔR would require detailed internal knowledge about the process, since no information about physical relationships (ϕ) or functional properties ($\int$) is present in the display.

Simply adding information about high-level properties does not change the separable nature of the display. In Figure 20.13(*b*), a second separable display has been illustrated. In this display the high-level properties (constraints) have been calculated and are displayed directly, including information related to functional purpose (K_5 and K_6) and abstract function (K_1, K_2, K_3, and K_4). This does offload some of the calculational requirements (e.g., ΔR). However, there is still a divided attention cost. Even though the high-level properties have been calculated and incorporated into the display, the relationships among and between levels of information in the abstraction hierarchy are still not apparent. The underlying cause of a particular system state still must be derived from the separate information that is displayed. Thus, while some low-level integration is accomplished in the display, the burden for understanding the causal structure still rests in the observer's stored knowledge.

Configural displays. The first configural display, illustrated in Figure 20.13(*c*), provides a direct representation of much of the low-level data that is present in the display in Figure 20.13(*a*). However, it also provides additional information that is critical to completing domain tasks: information about the physical structure of the system (ϕ). This "mimic" display format was first introduced in STEAMER (Hollan, Hutchins, and Weitzman, 1984), and issues in the animation of these formats have been investigated more recently (Bennett, 1993; Bennett and Madigan, 1994; Bennett and Nagy, 1996).

The mimic display is an excellent format for representing the generalized functions in the process. It has many of the properties of a functional flow diagram or flowchart. The elements can represent physical processes (e.g., feedwater streams) and, by appropriately scaling the diagram, relations at the level of physical form can be represented (e.g., relative positions of valves). Also, the moment-to-moment values of the process variables can easily be integrated within this representation. This display not only includes information with respect to generalized function, physical function, and physical form, but the organization provides a visible model illustrating the relations across these levels of abstraction. This visual model allows the observer to "see" some of the logical constraints

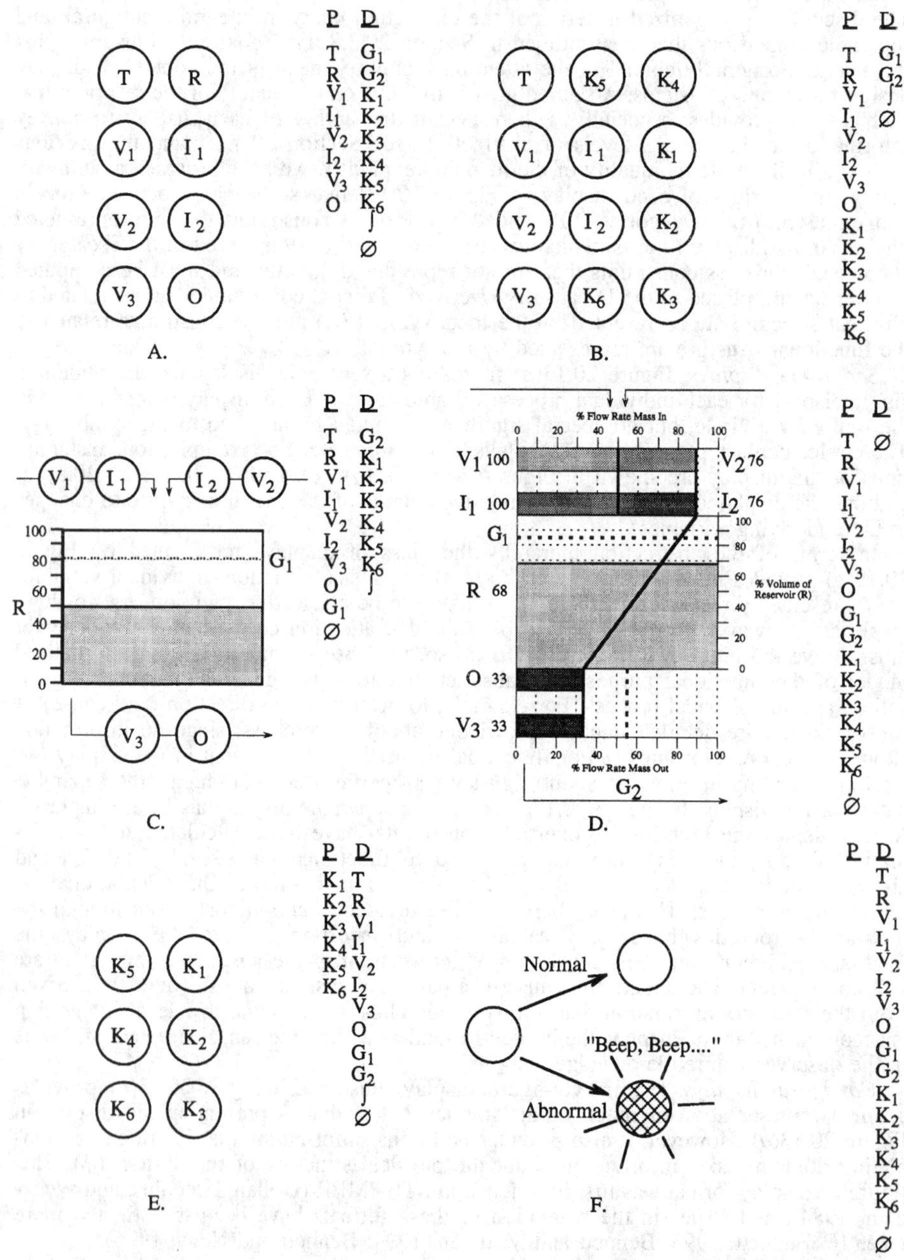

Figure 20.13 Six alternative mappings for the domain constraints described in Figure 20.12. The circles represent generic separable displays which could be bar graphs, pie charts, or digital displays. The data and properties outlined in Figure 20.12 have been placed in two categories for each mapping: "P" for data that can be perceived directly from the display and "D" for data that must be derived from the display by the observer. Figure 20.13(a) and 20.13(b) represent separable mappings, Figure 20.13(*c*) and 20.13(*d*) represent configural mappings, and Figure 13(*e*) and 13(*f*) represent integral mappings. These mappings illustrate how the terms separable, configural, and integral have a different meaning when applied to display design (as opposed to their meaning in the attention literature).

that link the low-level data. Thus, the current value of I_2 can be seen in the context of its physical function (feedwater stream 2) and its generalized function (source of mass) and in fact, its relation to the functional purpose in terms of G_1 is also readily apparent from the representation.

Just as in the displays listed in Figures 20.13(*a*) and 20.13(*b*), there is not likely to be a cost in selective attention with respect to the lower level data. However, although information about physical structure illustrates the causal factors that determine higher level system constraints, the burden of computing these constraints (e.g., determining mass balance) rests with the observer. Thus, what is missing in the mimic display is information about abstract function (information about the physical laws that govern normal operation).

The second configural display, illustrated in Figure 20.13(*d*), is slightly more complex (the logic is similar to Vicente, 1991) and will be described in detail before discussing the quality of the mapping that it provides. The valve settings V_1 and V_2 are represented as back-to-back horizontal bar graphs that increase or decrease in horizontal extent with changes in settings. The measured flow rates (I_1 and I_2) have the same configuration of graphical elements and are located below the valve settings in the display. The horizontal bar graphs depicting valve settings and flow rates for a particular pipe (e.g., V_1 and I_1) are connected with a vertical, bold line [in Figure 13(*d*) both of the lines are perpendicular because the settings and flow rates are equal in both input streams]. The volume of the reservoir (R) is represented by a horizontal bold line and as the filled portion of the rectangle inside the reservoir. The value of R can be read from the scale on the right side of the display and the associated digital value on the left; in Figure 13(*d*) the value of R is 68. The associated reservoir volume goal (G_1) is represented by the bold horizontal dashed line (approximately 85). The flow rate of the mass leaving the reservoir is represented by the horizontal bar graph labeled "O" at the bottom of the display; the corresponding valve setting is represented by the bar graph labeled "V_3." These two bar graphs are also connected by a bold vertical line. The mass output goal (G_2) is represented by the bold vertical dashed line (approximately 55). The relationship between mass in ($I_1 + I_2$) and mass out (O) is highlighted by the bold angled line which connects the corresponding bar graphs.

Unlike the displays that have been discussed previously, this configural display integrates information from all levels of the abstraction hierarchy in a single representation, making extensive use of the geometrical constraints of equality, parallel lines, and colinearity. The general functions are related through a funnel metaphor with input (source) at the top, storage in the center, and output (sink) on the bottom. The abstract functions are related using equality and the resulting colinearity across the bar graphs. For example, the constraints on mass flow (K_1, K_2, K_3) are represented in terms of equality of the horizontal extent of the bars labeled V_1/I_1, V_2/I_2, and V_3/O. In addition, the constraints relating rate of volume change and mass balance (K_4) are represented by the horizontal extent of $I_1 + I_2$ relative to the horizontal extent of O, and these relationships are highlighted by the bold line connecting these bars. Thus, the mass balance is represented by the symmetry between the input bar graphs and the output bar graphs; the slant of the line connecting them should be proportional to rate of change of mass in the reservoir. Constraints at the level of functional purpose are illustrated by the difference between the goal and the relevant variable. For example, the constraint on mass inventory (K_5) is shown using relative position between the hatched area representing volume within the reservoir and the bold, dashed, and horizontal line representing the goal level G_1.

This configural display, while not a direct physical analog, preserves important physical relations from the process (e.g., volume and filling). In addition, it provides a direct visual representation of the process constraints and connects these constraints in a way to make the "functional" logic of the process visible within the geometric form. As a result, performance for both selective (focused) and divided (integration) tasks is likely to be facilitated substantially.

Integral displays. Figure 20.13(*e*) shows an "integral" mapping in which each of the process constraints are shown directly, providing information at the higher levels of abstraction. However, the low-level data must be derived. In addition, there is absolutely no information about the functional processes behind the display and therefore the display does not aid the observer in relating the higher level constraints to the physical variables. Because there would normally be a many-to-one mapping from physical variables to the higher order constraints it would be impossible for the observer to recover information at lower levels of abstraction from this display.

Figure 20.13(*f*) shows the logical extreme of this continuum. In this display, the process variables and constraints are integrated into a single "bit" of information that indicates whether or not the process is working properly (all constraints are at their designed value). It should be obvious that while these displays may have no divided attention costs, they do have selective attention costs and they also provide little support for problem solving when the system fails.

Summary

This section has focused on issues related to the quality of mapping between process constraints and display constraints. Even the simple domain that we chose for illustrative purposes has a nested structure of domain constraints: There are multiple constraints that are organized hierarchically both within and between levels of abstraction. The six alternative displays achieved various degrees of success in mapping these constraints. The principle of correspondence is illustrated by the fact that these formats differ in terms of the amount of information about the underlying domain that is present. The display in Figure 20.13(*f*) has the lowest degree of correspondence, while the displays in Figure 20.13(*b*) and 20.13(*d*) have the highest degree of correspondence. These two displays are roughly equivalent in correspondence, with the exception of the two goals that are present in Figure 20.13(*d*) but absent in Figure 20.13(*b*). Although these two displays are roughly equivalent in correspondence, it should be clear from the prior discussion that they are definitely not equivalent in terms of coherence. Figure 20.13(*d*) allows an individual to perceive information concerning the physical structure, functional structure, and hierarchically nested constraints in the domain directly, a capability that is not supported by the format in Figure 20.13(*b*). The coherence of Figure 20.13(*d*) will be explored in greater detail in the following section. This section has also illustrated the duality of meaning for the terms integral, configural, and separable. In attention these terms refer to the relationship between perceptual dimensions, as described in Section 20.3.3; in display design, these terms more appropriately refer to the nature of the mapping between the domain and the representation.

20.5.3 Representation Aiding: Normal and Abnormal Operating Conditions

In the previous section, we outlined differences in correspondence and coherence that resulted from six alternative mappings for our simple domain. In this section, we explore issues related to coherence in greater detail, focusing on Figure 20.13(*d*) and the implications of the mapping for performance under both normal and abnormal, or emergency operating conditions. To begin, we discuss the facilitating role that graphical constraints representing information in the abstraction hierarchy (in particular, abstract function—the physical laws that govern normal operation) can play under normal conditions. Properly designed configural displays will provide a powerful representation for control: Breaks in the domain constraints will generally be seen as breaks in display constraints (e.g., nonsymmetries) and will suggest appropriate control inputs. This information is, perhaps, even more important for detecting faults (e.g., a leak). The possibility that these types of displays can change the fundamental nature of the behavior that is required on the part of the operator will also be entertained. Finally, the implications for the reduction of errors (more likely to occur under abnormal or emergency conditions) will be discussed.

The mapping between domain constraints and geometrical constraints that is provided in the configural display shown in Figure 20.13(*d*) provides a powerful representation for control under normal operating conditions. In Figure 20.14(*a*) the display is shown with values for system variables indicating that all constraints are satisfied. The figure indicates that the flow rate is larger for the first mass input valve (I_1, V_1) than for the second (I_2, V_2) but that the two flowrates added together match the flowrate of the mass output valve (O, V_3). In addition, the two system goals (G_1 and G_2) are being fulfilled.

In contrast, Figures 20.14(*b*)–(*d*) illustrate failures to achieve system goals. In these displays, not only is the violation of the goal easily seen, but each system variable is seen in the context of the control requirements. Thus, in Figure 20.14(*b*) it is apparent that the K_5 constraint is not being met (the actual level of the reservoir is higher than the goal). It is also apparent that the K_4 constraint is broken. The orientation of the line connecting mass in ($I_1 + I_2$) and the mass out (O) utilizes the funnel metaphor to indicate that a positive net inflow for mass exists (mass in is greater than mass out). In essence, the deviation in orientation of this line from perpendicular is an emergent feature corresponding to the size of the difference. Under these circumstances control input is required immediately: An adjustment at valve 1 and/or valve 2 will be needed to avoid overflow from the reservoir. The observer can see these valves in the context of the two system goals; the representation makes it clear that these are the appropriate control inputs to make. For example, although adjusting valve 3 from 54 to a value greater than 70 would also cause the reservoir volume to drop, it is an inappropriate control input because goal 2 would then be violated.

In Figure 20.14(*c*) the situation is exactly the same, with one exception: There is a negative net inflow for mass, as indicated by the reversed orientation of the connecting line. Under these circumstances the operator can see that no immediate control input is required. Because mass in is less than mass out, the reservoir volume is falling, and this is exactly what is required to meet the G_1 reservoir volume goal. Of course, a control input will be required at some point in the future (mass will need to be balanced when the reservoir level approaches the goal). Similarly, in Figure 20.14(*d*) the observer can see that the K_5 and K_6 constraints are broken, and that an adjustment to valve 3 (a decrease in output) is needed to meet the output requirements (G_2) and the volume goal (G_1).

Thus, in complex, dynamic domains it is the pattern of relationships between variables, as reflected in the geometric constraints, that determines the significance of the data that is presented. It is this pattern that ultimately provides the basis for action, even when the

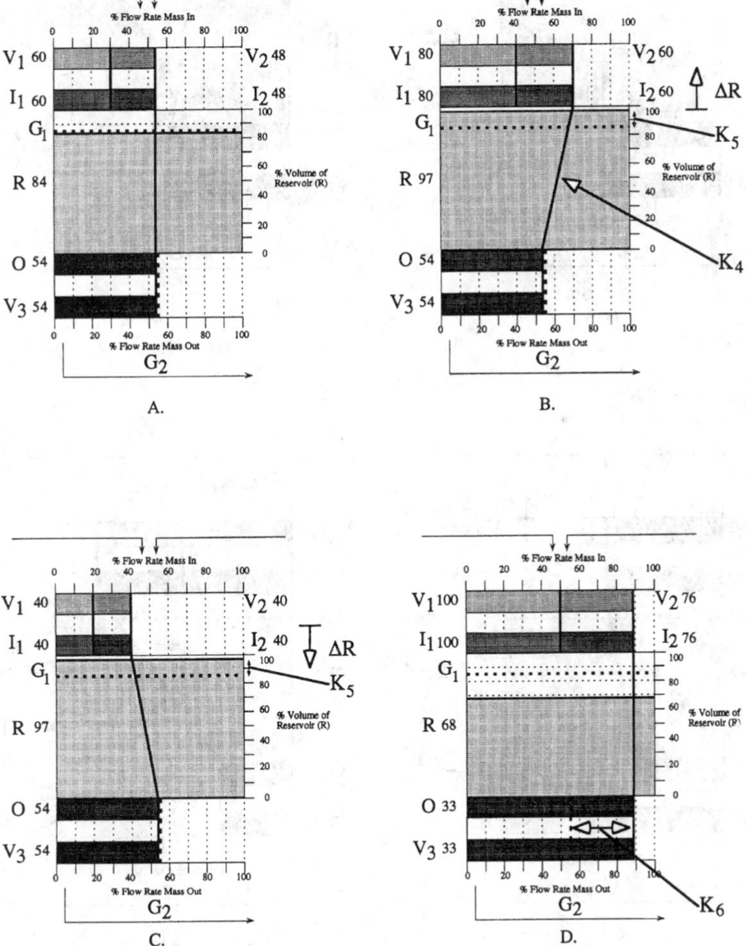

Figure 20.14 Illustration of the mapping between the domain constraints (data, properties, goals) and the geometric constraints (visual properties of the display, including emergent features such as symmetry and parallelism) under relatively normal operating conditions.

action hinges upon the value of an individual variable. When properly designed, configural displays will directly reflect these critical data relationships and suggest the appropriate control input.

A similar logic applies for operational support under abnormal or emergency conditions. As in the previous figure, Figure 20.15(*a*) represents a configuration with all system constraints being met. In Figure 20.15(*b*) the first constraint (K_1) is broken. There are two aspects of the display geometry indicating that the flowrate (I_1) does not match the commanded flow, or valve setting (V_1). First, the horizontal extent of the two bar graphs in the top left portion of the display are not equal and this relationship is emphasized by the bold line connecting the two graphs (similar to the connecting line for mass balance). There are a number of potential causes for this discrepancy, which include (1) a leak in

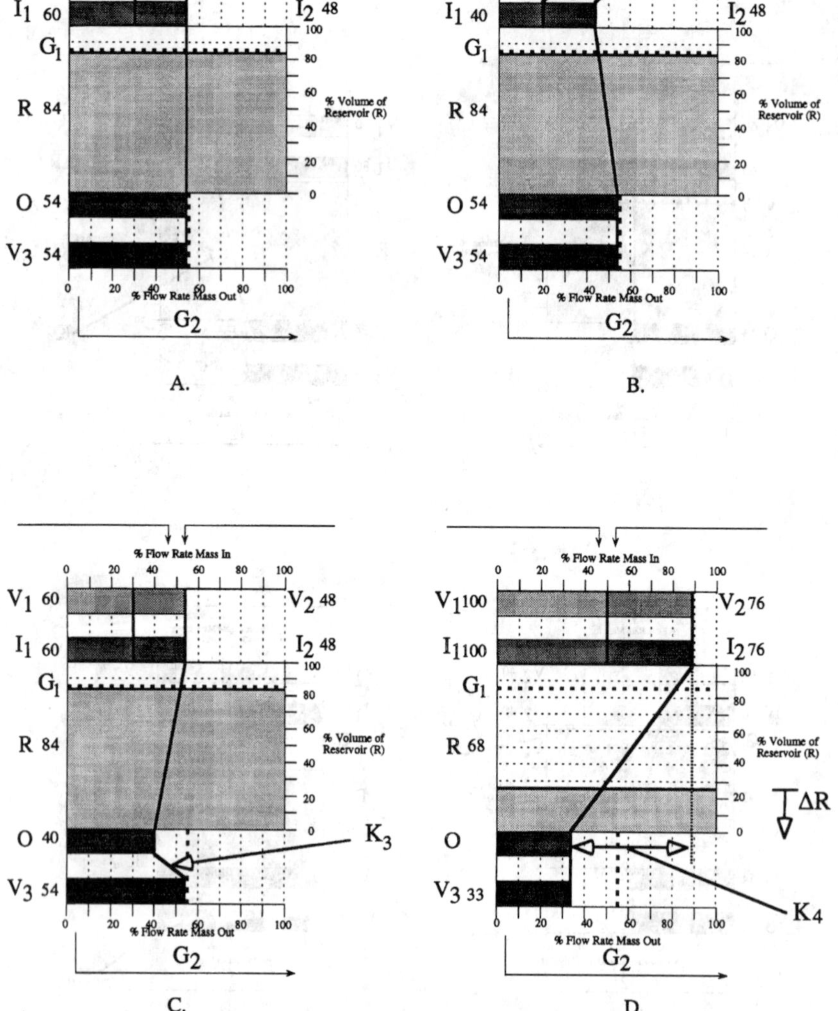

Figure 20.15 Illustration of the mapping between the domain constraints (data, properties, goals) and the geometric constraints (visual properties of the display, including emergent features such as symmetry and parallelism) under abnormal or emergency operating conditions.

the valve, (2) a leak in the pipe prior to the point at which the flow rate is measured, or (3) an obstruction in the pipe. In contrast, the fact that the line connecting V_2 and I_2 is not perpendicular (but is parallel to the first connector line) does not indicate that the K_2 constraint is broken. Instead, this is an indication that the commanded and actual mass flows in the second mass input stream are equal (and therefore that the discrepancy is isolated in the first mass input stream). A similar mapping between geometrical constraints and domain constraints represents a fault in the K_3 constraint, as illustrated in Figure 20.15(*c*).

Figure 20.15(*d*) illustrates changes in the visual display (breaks in the geometrical constraints) that are associated with a a fault in the system (a break in the mass balance constraint, K_4). In this example there is a positive net inflow of mass, which is normally associated with an increase in the volume of the reservoir (again, suggested by the funnel metaphor). However, in this case the mass inventory is falling, as we have indicated in the diagram by the downward-pointing arrow located near the ΔR symbol (this is difficult to represent in a static diagram, but would be clearly seen on a dynamic display). Again, there are several potential explanations for this fault. The most likely explanation is that there is a leak in the reservoir itself; however, there could be a leak in the pipe between the reservoir and the point at which the flow measurement is taken. It should be noted that while the nature of the fault can be seen (e.g., leak or blockage in feedwater line) this representation would not be very helpful in physically locating the leak within the plant (e.g., locating valve 1).

These examples illustrate that properly designed displays can change the fundamental type of behavior that is required of an operator under both normal and abnormal operating conditions. With separable displays (e.g., the separable configurations illustrated in Figure 20.13) the operators are required to engage in knowledge-based behaviors: They must rely upon internal models of system structure and function (and therefore use limited capacity resources—working memory) to detect, diagnose, and correct faults. As a result, the potential for errors is increased dramatically. In contrast, properly designed configural displays present externalized models of system structure and function through geometric constraints. This allows operators to utilize skill-based behaviors (e.g., visual perception and pattern recognition) that do not require limited capacity resources. As a result, the potential for errors will be dramatically decreased. As Rasmussen and Vicente (1989) have noted, changing the required behavior from knowledge-based behavior to rule-based or skill-based behavior is a goal for display design.

Properly designed configural displays will also reduce the possibility of "under-specified action errors" (Rasmussen and Vicente, 1989). In complex, dynamic domains individuals can form incorrect hypotheses about the nature of the existing problem if they do not consider the relevant subsets of data (Woods, 1988). Observers may focus on these incorrect hypotheses and ignore disconfirming evidence, showing a kind of "tunnel vision" (Moray, 1981). Observers may also exhibit "cognitive hysteresis," and fail to revise hypotheses as the nature of the problem changes over time (Lewis and Norman, 1986). Configural displays that directly reflect the semantics of a domain can reduce the probability of these types of errors by forcing an observer to consider relevant subsets of data.

20.5.4 Practical Guidelines

In conclusion, we believe that the application of the this approach to display design will improve overall human–machine performance through the development of configural displays that support normal control, as well as fault detection, diagnosis, and repair. The potential for errors will be dramatically decreased, because the critical information for control is represented directly in the interface. This in turn dramatically reduces the requirement for knowledge-based reasoning on the basis of internalized models. To summarize, we offer three general heuristics for graphic display design.

1. Each relevant process variable should be represented by a distinct element within the display. If precise information about this variable is desirable, then a reference scale or supplemental digital information should be provided.
2. The display elements should be organized so that the emergent properties (symmetries, closure, parallelism) that arise from their interaction correspond to higher order constraints within the process. Thus, when process constraints are broken (i.e., a fault occurs), the corresponding geometric constraints are also broken (the display symmetry is broken).

3. The symmetries within the display should be nested (from global to local) in a way that reflects the hierarchical structure of the process. High-order process constraints (e.g., at the level of functional purpose or abstract function) should be reflected in global display symmetries; lower order process constraints (e.g., functional organization) should be reflected in local display symmetries.

20.6 CHALLENGES OF COMPLEX SYSTEMS

The simple process described above is convenient for a tutorial introduction to some of the important decisions that must be made when designing a graphical representation. However, this example greatly underestimates the complexity seen in many advanced, human–technological systems (nuclear power, air traffic control, advanced tactical aviation, command and control centers for managing military and space operations, minimally invasive and remote surgery, etc.). These systems typically have multiple modes of operation (each with different constraints and boundary conditions) and require multiple windows into the process. In these systems, the goal remains the same, to make the real constraints of the work process (at all levels of abstraction) visible to the human operator. The designer must still address the problems of correspondence (so that all relevant process constraints are represented in the interface) and coherence (so that the representation is comprehendible to the human operator). For these complex systems, however, it will not be possible to achieve both correspondence and coherence with a single graphic display. Thus, the added problem of navigating through multiple views (i.e., windows, screens, pages) must be addressed.

A principal threat to these complex systems is "mode error" (Woods, 1984). A mode error occurs when the operator loses track of dynamic changes in the operating constraints governing a process. The operator responds to one set of constraints (i.e., mode), when a different set of constraints is, in fact, governing the process. The design challenge is to coordinate the multiple windows necessary for a complete representation with the changing operational modes. In simple terms, how can the interface be designed to ensure that the appropriate window is always coupled with the appropriate mode; to ensure that the important information is salient at the appropriate times.

Two classes of solutions might be considered for dealing with the navigation problems that typically lead to mode errors—computational and graphical. Computational solutions or adaptive interfaces include an inference engine that automatically manages the representation. This computational engine automatically adjusts the representation based on inferences about the state of the system and the state of the operator. Projects such as the "pilot's associate" are examples of attempts to design automatic systems to aid operators to navigate through the representations associated with a complex work domain. However, the focus of this chapter is on graphical solutions. For this reason, we will use the remaining space to briefly consider some graphical approaches to this problem.

Woods (1984) introduced the term *visual momentum* to refer to the cognitive costs associated with switching from one reference frame to another. If visual momentum is high, then the cost of switching views is low. In this case, the new display is consistent with expectations created by the prior display. If visual momentum is low, then there is a high cost of switching. That is, the new display is not consistent with expectations and the cognitive system must effectively recalibrate before information can be extracted from the new display. To ensure high visual momentum, the design of each graphical display must be considered relative to the other displays that operators may be using. Are the graphical conventions (coordinates, scales, directions, motions, colors, S-R mappings, etc.) used in one display consistent with those in another?

A graphical device that Woods (1984) has suggested to increase visual momentum is the use of landmarks. Landmarks are graphical elements that provide an orientation point that relates one display to another. Just as a tall building or mountain that is visible from many different parts of the landscape might help a person to orient to the geography, graphical landmarks can be designed with the objective of aiding the operator to orient within the functional landscape of the work domain. For example, Aretz (1991) used a shaded wedge within an electronic map display as a landmark to specify the region within the map that corresponded to the head-up forward view of the pilot.

Another graphical device to help operators navigate across multiple display pages is a map or overview display. This display can be implemented as a separate window or as an embedded landmark in all windows. This overview might use a flow diagram or hierarchical tree structure to show functional links among the multiple display pages.

The BookHouse interface designed by Goodstein and Pejtersen (1989) uses a spatial metaphor in which rooms in a "house" are set-up for different categories of users. This spatial metaphor allows the operator to apply natural abilities for navigating in three-dimensional spaces to the task of navigating in the more abstract space of a library database. In the BookHouse, the three-dimensional space is implemented in a two-dimensional display. Virtual reality systems now offer the possibility for effective three-dimensional (3-D) representations. With these systems, designers have the opportunity to maximize the transfer of natural human ability to orient and navigate in 3-D environments to more abstract environments; and to combine natural 3-D representations with imagery obtained by advanced sensor systems. For example, virtual displays for minimally invasive surgery are being designed that integrate the 3-D image of the patient's anatomy with information obtained by MRI scans and other advanced imaging technologies. Thus, virtual 3-D spatial metaphors might provide another technique for integrating complex information from distributed sensors into a coherent representation.

The central theme of this chapter is that problem solving can be critically influenced by the nature of visual representations. Building effective representations requires designers to go beyond the simple psychophysical questions of data availability to the more complex questions of information availability, where information refers to the specification of domain constraints and boundary conditions. This specification depends both on the mapping from display to human (i.e., coherence) and on the mapping from display to domain (i.e., correspondence).

20.7 ACKNOWLEDGMENTS

The authors would like to thank Brian Tsou for discussions and comments on earlier drafts, and the helpful comments provided by reviewers. Funding in support of this work was provided to Kevin Bennett by the Ohio Board of Regents (Wright State University Research Challenge Grant # 662613). John Flach was partially supported by a grant from the Air Force Office of Scientific Research during the preparation of this manuscript. Opinions expressed are those of the authors and do not represent an official position of any of the supporting agencies.

REFERENCES

Adelson, E. H. (1993). Perceptual organization and the judgement of brightness. *Science, 262*, 2042–2044.

Alexander, K. R., Xie, W., Derlacki, D. J. (1994). Spatial frequency characteristics of letter identification. *J. Opt. Soc. Am. A, 11*, 2375–2382.

Arend, L. E., and Reeves, A. (1986). Simultaneous color constancy. *J. Opt. Soc. Am. A, 3*, 1743–1751.

Aretz, A. J. (1991). The design of electronic map displays. *Human Factors, 33(1)*, 85–101.

Beltracchi, L. (1987). A direct manipulation interface for heat engines based upon the Rankine cycle. *IEEE Transactions on Systems, Man, and Cybernetics, 17(3)*, 478–487.

Beltracchi, L. (1989). Energy, mass, model-based displays, and memory recall. *IEEE Transactions on Nuclear Science, 36(3)*, 1367–1382.

Bennett, K. B. (1993). Encoding apparent motion in animated mimic displays. *Human Factors, 35(4)*, 673–691

Bennett, K. B., and Flach, J. M. (1992). Graphical displays: Implications for divided attention, focused attention, and problem solving. *Human Factors, 34(5)*, 513–533.

Bennett, K. B., and Madigan, E. (1994). Contours and borders in animated mimic displays. *International Journal of Human-Computer Interaction, 6(1)*, 47–64.

Bennett, K. B., and Nagy, A. (1996). Spatial and temporal considerations in animated mimic displays. *Displays, 17(1)*, 1–14.

Bennett, K. B., Toms, M. L., and Woods, D. D. (1993). Emergent features and configural elements: Designing more effective configural displays. *Human Factors, 35(1)*, 71–97.

Boynton, R. M. (1990). Human color perception. In K. N. Leibovic, Ed., *Science of Vision* (pp. 211–253. New York: Springer-Verlag.

Boynton, R. M. (1992). *Human Color Vision*. Washington, D.C.: Optical Society of America.

Brainard, D. H., and Wandell, B. A. (1992). Asymmetric color-matching: How color appearance depends on the illuminant. *J. Opt. Soc. Am. A, 9*, 1433–1448.

Brodie, K. W., Carpenter, L. A., Earnshaw, R. A., Gallop J. R., Hubbold, R. J., Mumford, A. M., Osland, C. D., and Quarendon, P., Eds. (1992). Scientific Visualization: Techniques and Applications. Berlin: Springer-Verlag.

Buttigieg, M. A., and Sanderson, P. M. (1991). Emergent features in visual display design for two types of failure detection tasks. *Human Factors, 33(6)*, 631–651.

Cannon, M. W., and Fullenkamp, S. C. (1991). A transducer model for contrast perception. *Vision Research, 31*, 983–998.

Carswell, C. M., and Wickens, C. D. (1990). The perceptual interaction of graphical attributes: Configurality, stimulus homogeneity, and object integration. *Perception and Psychophysics, 47*, 157–168.

Cleveland, W. S. (1985). *The Elements of Graphing Data*. Belmont, CA: Wadsworth.

DeValois, R. L., and DeValois, K. K. (1990). *Spatial Vision*. New York: Oxford University Press.

D'Zmura, M., and Lennie, P. (1986). Mechanisms of color constancy. *J. Opt. Soc. Am. A, 3*, 1662–1672.

Earnshaw, R. A., and Wiseman, N., Eds. (1992). An Introductory Guide to Scientific Visualization. Berlin: Springer-Verlag.

Garner, W. R. (1970). The stimulus in information processing. *American Psychologist, 25*, 350–358.

Garner, W. R. (1974). *The Processing of Information and Structure*. Hillsdale, NJ: Lawrence Erlbaum Associates.

Garner, W. R., and Felfoldy, G. L. (1970). Integrality of stimulus dimensions in various types of information processing. *Cognitive Psychology, 1*, 225–241.

Gilchrist, A., S. Delman, and Jacobson, A. (1983). The classification and identification of edges as critical to the perception of reflectance and illumination. *Percept. and Psychophys., 33*, 425–436.

Ginsburg, A. (1986). Spatial filtering and visual form perception. In K. R. Boff, L. Kaufmann, and J. P. Thomas, Eds., *Handbook of Human Perception and Performance* (pp. 34/1–34/41). New York: John Wiley.

Goodstein, L.P., and Pejtersen, A.M. (1989). *The BOOK HOUSE: System—Functionality and Evaluation* (Riso-M-2793). Roskilde, Denmark: Riso National Laboratory.

Graham, N. V. S. (1989). *Visual Pattern Analyzers*. New York: Oxford University Press,

Grum, F., and Bartleson, C. J. (1980). *Optical Radiation Measurements, Volume 2: Colorimetry*. New York: Academic Press.

Hollan, J. D., Hutchins, E. L., and Weitzman, L. (1984). Steamer: An interactive inspectable simulation-based training system. *The AI Magazine, Summer*, 15–27.

Hollan, J. D., Hutchins, E. L., McCandless, T. P., Rosenstein, M., and Weitzman, L. (1987). Graphical interfaces for simulation. In W. B. Rouse, Ed., *Advances in Man-Machine Systems Research* (Volume 3, pp. 129–163). Greenwich, CT: JAI Press.

Hunter, R.S., and Herold, R. W. (1987). *The Measurement of Appearance*. New York: John Wiley.

Hutchins, E. L., Hollan, J. D., and Norman, D. A. (1986). Direct manipulation interfaces. In D. A. Norman, and S. W. Draper, Eds., *User Centered System Design* (pp. 87–124). Hillsdale, NJ: Lawrence Earlbaum Associates.

Kelly, D. H. (1974). Spatio-temporal frequency characteristics of color-vision mechanisms. *J. Opt. Soc. Am., 64*, 983–990.

Klein, G. A., Orasanu, J., and Zsambok, C. E., Eds. (1993). *Decision Making in Action: Models and Methods*. Norwood, NJ: Ablex Publishing Corp.

Lewis, C., and Norman, D. A. (1986). Designing for error. In D. A. Norman, and S. W. Draper, Eds., *User Centered System Design*. Hillsdale, NJ: Lawrence Earlbaum Associates.

Maloney, L. T., and Wandell, B. A. (1986). Color constancy: A method for recovering surface reflectance. *J. Opt. Soc. Am. A, 3*, 29–33.

Millodot, M. (1982). Image formation in the eye. In H. Barlow and J. D. Mollen, Eds., *The Senses* (pp. 46–61). New York: Cambridge University Press.

Moray, N. (1981). The role of attention in the detection of errors and the diagnosis of failures in man-machine systems. In J. Rasmussen and W. B. Rouse, Eds., *Human Detection and Diagnosis of System Failures*. New York: Plenum Press.

Moray, N., Lee, J., Vicente, K. J., Jones, B. G., and Rasmussen, J. (1994). A direct perception interface for nuclear power plants. *Proceedings of the Human Factors and Ergonomics Society 38th Annual Meeting* (pp. 481–485). Santa Monica, CA: Human Factors and Ergonomics Society.

Mullen, K. (1985). The contrast sensitivity of human color vision to red-green and blue-yellow chromatic gratings. *J. Physiol., 359*, 381–400.

Nagy, A. L., and Kamholz, D. (1995). Luminance discrimination, color contrast, and multiple mechanisms. *Vision Research, 35*, 2147–2155.

Nassau, K. (1983). *The Physics and Chemistry of Color*. New York: John Wiley.

Newell, A., and Simon, H. A. (1972). *Human Problem Solving*. Englewood Cliffs, NJ: Prentice-Hall.

Noorlander, C., and Koenderink, J. J. (1983). Spatial and temporal discrimination ellipsoids in colour space. *J. Opt. Soc. Am., 73*, 1533–1543.

Olzack, L., and Thomas, J. (1986). Seeing spatial patterns. In K. R. Boff, L. Kaufmann, and J. P. Thomas, Eds., *Handbook of Human Perception and Performance* (pp. 7/1–7/56). New York: John Wiley.

Pejtersen, A. M. (1992). The Book House. An icon based database system for fiction retrieval in public libraries. In B. Cronin, Ed., *The Marketing of Library and Information Services 2*. (pp. 572–591). London, UK: ASLIB.

Poirson, A. B., and Wandel, B. A. (1993). The appearance of colored patterns. *J. Opt. Soc. Am. A, 12*, 2458–2471.

Pokorny J., and Smith, V. (1986). Colorimetry and color discrimination. In K. R. Boff, L. Kaufmann, and J. P. Thomas, Eds., *Handbook of Human Perception and Performance* (pp. 8/1–8/51). New York: John Wiley.

Pomerantz, J. R. (1986). Visual form perception: An overview. In H. C. Nusbaum and E. C. Schwab, Eds., *Pattern Recognition by Humans and Machines* (Volume 2, Visual Perception, pp. 1–30). Orlando, Florida: Academic Press.

Pomerantz, J. R., and Pristach, E. A. (1989). Emergent features, attention, and perceptual glue in visual form perception. *Journal of Experimental Psychology: Human Perception and Performance, 15(4)*, 635–649.

Pomerantz, J. R., Sager, L. C., and Stoever, R. J. (1977). Perception of wholes and of their component parts: Some configural superiority effects. *Journal of Experimental Psychology: Human Perception and Performance, 3*, 422–435.

Post, D. (1992). Colorimetric measurement, calibration, and characterization of self-luminous displays. In H. Widdel and D. L. Post, Eds., *Color in Electronic Displays* (pp. 299–312). New York: Plenum Press.

Rasmussen, J. (1986). *Information Processing and Human-Machine Interaction: An Approach to Cognitive Engineering.* New York: Elsevier Publishing Co., Inc.

Rasmussen, J., Pejtersen, A. M., and Goodstein, L. P. (1994). *Cognitive Systems Engineering*. New York: John Wiley.

Rasmussen, J., and Vicente, K. (1989). Coping with human errors through system design: Implications for ecological interface design. *International Journal of Man-Machine Studies, 31*, 517–534.

Sanderson, P. M., Flach, J. M., Buttigieg, M. A., and Casey, E. J. (1989). Object displays do not always support better integrated task performance. *Human Factors, 31(2)*, 183–198.

Sekiguchi, N., Williams, D. R., and Brainard, D. H. (1993). Aberration free measurements of the visibility of isoluminant gratings. *J. Opt. Soc. Am. A, 10*, 2105–2117.

Shafer, S. A. (1985). Using color to separate reflection components. *Color Res. and Appl., 10*, 210–218.

Shneiderman, B., Ed. (1993). *Sparks of Innovation in Human Computer Interaction.* Norwood, NJ: Ablex.

Shneiderman, B. (1986). *Designing the User Interface*. Reading, MA: Addison-Wesley.

Solomon, J. A., and Pelli, D. G. (1994) The visual filter mediating letter identification. *Nature* (London), 369, 395–397.

Treisman, A. M. (1986). Properties, parts, and objects. In K. Boff, L. Kaufmann, and J. Thomas (Eds.), *Handbook of Perception and Human Performance* (pp. 35/1–35/70). New York: John Wiley.

Tufte, E. R. (1990). *Envisioning Information.* Chesire, Connecticut: Graphics Press.

Tufte, E. R. (1983). *The Visual Display of Quantitative Information.* Chesire, Connecticut: Graphics Press.

Vicente, K. J. (1991). *Supporting Knowledge Based Behavior Through Ecological Interface Design.* (Tech. Report EPRL-91-1). Urbana Champaign, IL: Engineering Psychology Research Laboratory and Aviation Research Laboratory. University of Illinois.

Wandel, B. A. (1995). *Foundations of Vision.* Sunderland, Mass.:Sinauer.

Wertheimer, M. (1959). *Productive Thinking*. New York: Harper & Row.

Whittle, P. (1986). Increments and decrements: Luminance discrimination. *Vision Research, 26*, 1677–1692.

Whittle, P. (1992). Brightness, discriminability, and the "Crispening Effect." *Vision Research, 32*, 1493–1508.

Wickens, C. D., and Carswell, C. M. (1995). The proximity compatibility principle: Its psychological foundation and its relevance to display design. *Human Factors, 37(3)*, 473–494.

Widdell, H., and Post, D. L. (1992). *Color In Electronic Displays.* New York: Plenum Press.

Woods, D. D. (1984). Visual momentum: A concept to improve the cognitive coupling of person and computer. *International Journal of Man-Machine Studies, 21*, 229–244.

Woods, D. D. (1988). Coping with complexity: The psychology of human behavior in complex systems. In L. P. Goodstein, H. B. Andersen, and S. E. Olsen, Eds., *Mental Models, Tasks and Errors: A Collection of Essays to Celebrate Jens Rasmussen's 60th Birthday.* New York: Taylor Francis.

Woods, D. D. (1991). The cognitive engineering of problem representations. In G. R. S. Weir and J. L. Alty, Eds., *Human-Computer Interaction and Complex Systems* (pp. 169–188). London, UK: Academic Press.

Woods, D. D., and Roth, E. M. (1988). Cognitive systems engineering. In M. Helander, Ed., *Handbook of Human-Computer Interaction* (pp. 1–41). Amsterdam: Elsevier Science Publishers B. V. (North-Holland).

Worthy, J. A., and Brill, M. H. (1986). Heuristic Analysis of Color Constancy. *J. Opt. Soc. Am. A*, *3*, 1708–1712.

Wyszecki, G., and Stiles, W. S. (1982). *Color Science*. (Second Edition) New York: John Wiley.

Zachary, W. (1986). A cognitively based functional taxonomy of decision support techniques. *Human-Computer Interaction*, *2*, 25–63.

CHAPTER 21

CONTROLS

Hans-Jörg Bullinger
Peter Kern
Martin Braun
Fraunhofer Institute for Industrial Engineering (IAO), Stuttgart, and Institute for Human Factors and Technology Management (IAT)
University of Stuttgart
D-70569 Stuttgart 80 Germany

21.1 INTRODUCTION

Controls constitute interface elements in the human–machine system through which a human transfers mechanical energy or information to the technical system for performing automatic control functions. The human receives the haptic and proprioceptive information required to perform a task from the control. Design, arrangement, and task of the controls have a considerable influence on the strain to which humans are subjected as well as on the effectiveness and safety of the system. The influencing factors of the "operating effectivness" are represented in Figure 21.1. Owing to the interaction between human, control, and the technical system, it is obvious that controls cannot be regarded as machine elements, as it is carried out in many standards and regulations.

The design, selection, and arrangement of the controls must be made with special consideration of the criteria of ergonomics. The design dimensions of the control such as shape, size, material, and surface as well as the control task must be compatible with the anatomical, anthropometric, and physiological marginal conditions of man. Anthropometrical parameters must be taken into account with regards to dimensioning the control (size, shape) and for positioning of the control (grasp area of the hand–arm system HAS and/or step area of the foot–leg system FLS). Anatomical marginal conditions, such as the motion range of the joints, are of importance to the maximum actuating range, as are physiological data to the transmission of forces and torques. Furthermore, the human parameters that vary from one individual to the next must also be accounted for in the ergonomic design and arrangement of the controls. It is the intention of this chapter to give the designer of technical systems (work equipment, computer systems, and vehicles, with which energy and/or information must be transferred from human to the technical system via controls) guidance in the selection and arrangement of controls. The ergonomically suitable control can be selected only after the designer has defined the control task. General requirements for the definition of the design dimensions—shape, size, material, and surface—that are important to the manufacturer of controls, are discussed in Section 21.3.

21.2 CLASSIFICATION OF CONTROLS AND DEVICES

As guidance for the designer with regards to the selection of controls, a classification and evaluation of the wide of controls and devices can only be made in connection with the

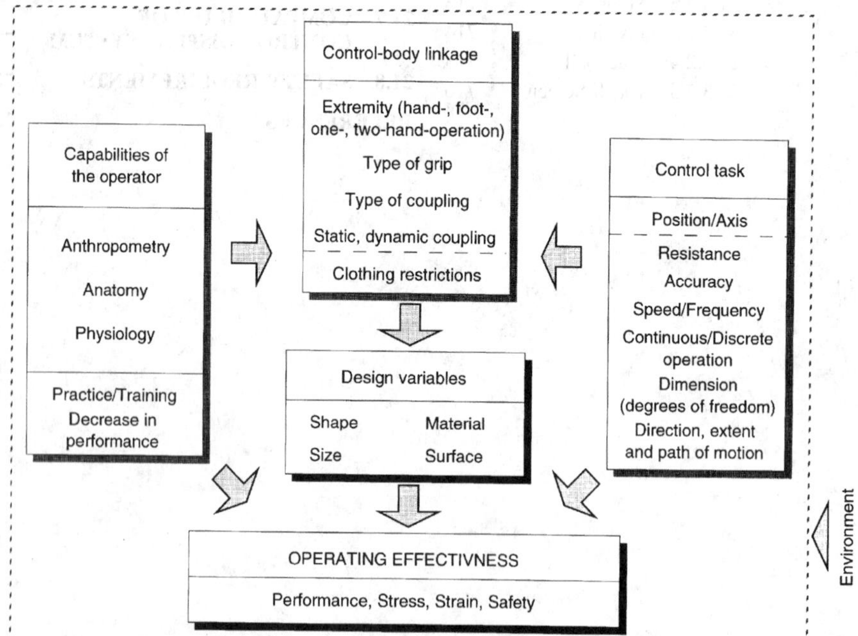

Figure 21.1 Factors influencing operating effectiveness

control task and the coupling conditions that exist between the contact element of the control and the extremity involved.

21.2.1 Control Task

Resistance, accuracy, and speed are regarded as the most important parameters of the control task. These performance parameters are particularly influenced by the type of control and by the dimensioning and positioning of the control. The maximum isometric muscle forces of humans in the HAS and FLS motion range (experimentally determined by Rohmert (1966), Caldwell (1959), and others) constitute the basis for the transmission of forces to controls. The most important influential variables regarding the maximum isometric actuating moments, besides human muscular capacity, are technical parameters such as the force application point, direction of force, and the type of coupling. Actuation of a control should take place within the maximum force output range. In hand- and foot-operated controls, the actuating forces or moments can be increased if the direction of force transmission is chosen so that the body can find a support (backrest) or if it is possible to translate the body's weight into actuating forces. The maximum isometric actuating moments for spoked handwheels (Mainzer, 1982) are shown in Figure 21.2 by way of example. As can be seen from Figure 21.2, the body's weight can be applied if the handwheel is arranged at low positions, resulting in hight actuating moments. However, it is possible to translate the body's weight effectively only if the handwheels have a larger diameter; this is not so for small diameters, which can be seen from the rather flat curve in Figure 21.2.

The actuating moment must not be, however, the exclusive criterion for the arrangement of controls, because, as shown in the above example, the high actuating moment is produced only in an unfavourable posture, thus requiring additional static work. In the case of a frequent actuation of the control, the actuation resistance, on account of the continuous performance limit of man, should not exceed 15% of the maximum force. If the actuation times are very short (≤5 sec), the resistance can account for approximately 50% of the maximum force. When designing safety equipment, it must be ensured, that regarding the force to be developed, even the weakest worker (fifth percentile) will still be able to actuate the control. When electric, pneumatic, and hydraulic servo-amplifiers are used, provision will have to be made for a resistance, to obtain a tactile and proprioceptive feedback (see Section 21.5) and to dampen disturbing influences caused by the operator and by the environment and imparted to the control.

The actuating accuracy depends on the following factors: type of control, type of coupling, control dynamic, forcing function (type of track), motion range, environmental

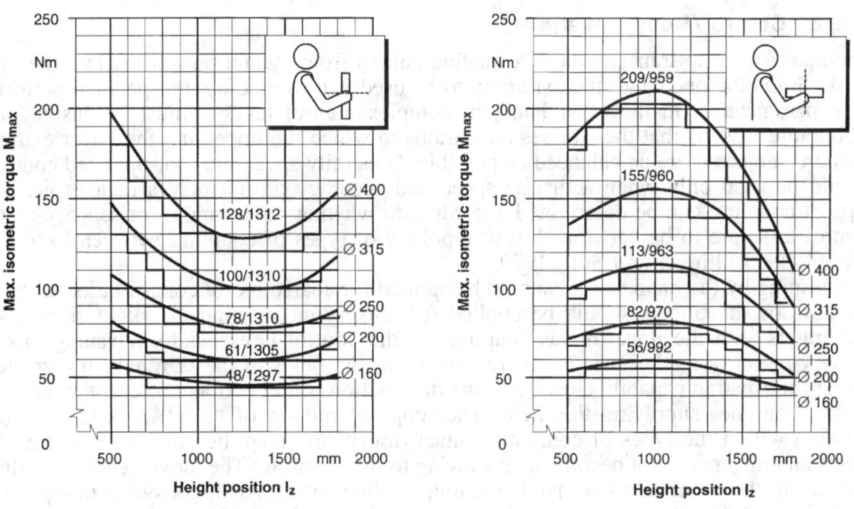

Figure 21.2 Isometric torques of handwheels ($T = f${control diameter, height, direction}). (*Source:* Mainzer, 1982.)

influences, and so on. The actuating speed depends primarily on the resistance and the type of control used. The highest speeds can be obtained over a large motion area by using a hand or a finger crank, because no regrasping of the control will be necessary due to the static coupling. The dimensions, that is, the crank radius in the above example, and the arrangement of the control element are further parameters. The motion speed of the HAS, which is dependent on the direction of motion on a horizontal plane and is a function of the motion amplitude, approximately corresponds to an ellipse, owing to physiological and biomechanical marginal conditions (Schmidtke and Stier, 1960). These findings have shown that motions with the right arm can be made more quickly from bottom left to top right than motions from bottom right to top left, which can be attributed to the biomechanical properties of the elbow joint. Considering the motion amplitude, this speed-to-accuracy relationship can be best described by the empirical relation of Fitts (1964). (Fitts's Law: $MT = a + b^2 \log(2A/W)$, where MT is movement time, A is amplitude, W is target size, and a and b are constants.) This relation is also applicable to the motions of controls.

21.2.2 Geometrics of Control Movements

The degrees of freedom of a control planned in the design are defined by the dimensionality. On the one hand, a multidimensional control task can be performed by several one-dimensional controls or, on the other hand, by one multidimensional control. Thus, for example, the longitudinal and transverse two-dimensional dynamics of a manually shifted motor vehicle is controlled by four one-dimensional controls and one two-dimensional gear-shift lever. However, it would be possible also to perform this task with one two-dimensional control lever, but this would entail greater technical complexity. In multidimensional control elements, problems are posed by disturbance in the various directions.

According to the *type of movement*, a distinction is made between translationally actuated and rotationally actuated controls. Depending on the scope of movement, rotational (handwheel) and quasirotational control movements (lever) may occur. The performance parameters are essentially determined by the *direction of movement*, which results from the orientation of the control axis and the position of the control to the worker (front-, side-, height-position).

Information input by controls can be either *continuous* or *discontinuous* (discrete), that is, in steps or stages. In the case of digital input, it must be differentiated whether a control can assume two positions, using mechanical stops, or several positions. To set the intermediate positions, provision will have to be made for mechanical ratchets or for an appropriate displacement resistance of characteristics, so that the position can be safely identified. For rotational one-hand controls, the control positions should be located between 15° and 45°.

21.2.3 Control/Body Linkage

The man is able to translate the information gained from mental processing into the FLS or HAS. On the one hand, the extremity to be used is governed by the specified performance parameters: On the other hand, in complex control tasks it must be ensured for ergonomic reasons that the stresses and strains to which the upper and the lower extremities are subjected are as balanced as possible. Generally speaking, foot-operated controls should be used only when actuating-speed and accuracy do not play a major role. The type of coupling can be determined according to whether coarse-motor or sensory-motor control tasks are to be performed. (For typology of types of coupling between hand and control, see Bullinger and Solf, 1979.)

Clasping by the hand or foot should be applied for coarse-motor control tasks, whereas gripping by the fingers should be applied for sensory-motor control tasks. Gripping by the fingers with the fingertips as coupling on the control is especially advantageous in sensory-motor tasks, because it is here that the modal fields of the skin have the greatest tactile information capability, owing to the distribution of the peripheral receptors and the central neurones supplying this field. The scope of motion of the HAS is steadily decreasing within the types of coupling-contact (touch) to grasp, because the joints of the hands and fingers cannot become active owing to the coupling. The movements are further reduced in the event of a two-hand coupling on the control. Analogous to hand-operated control are different types of coupling between foot and control. Actuation is possible with the forefoot, with the heel, or with the whole foot. Here, the type of coupling is also governed by the control task to be handled.

The transmission of energy to the control element may be *positive* or *frictional.* In frictional coupling, the power is transmitted through the friction forces that exist between the control and the hand or foot, which the necessary normal forces produced by the coupling forces. In frictional coupling, the coefficient of friction that exists between control element and hand is important (see Section 21.3); this coefficient is determined by the material and the surface structure of the control element. A positive coupling permits greater actuating force to be applied, but at the same time it also confines the possible movements of the HAS.

Frictional coupling is favorable for dynamic coupling on the control (regrasping). If the scope of movement of the HAS is not sufficient for the required actuating distance, *static coupling* changes to *dynamic coupling*. In dynamic coupling, regrasping is necessary on the control, so that the required movement can be performed. If the possibilities of movement are exhausted in static coupling, a larger scope of movement can be achieved only by movements of the trunk, resulting in additional stress and strain for the worker. The range of static movement of a rotationally actuated control by gripping it with five fingers is determined with approximately 3 rads, by the scope of pronation and supination movements of the proximal and distal ratio-ulnar joint, with the axis of rotation arranged in the sagittal-horizontal axis. For dynamic coupling, special requirements must be made with respect to the design parameters shape, size, and surface (Kern, Muntzinger, and Solf, 1984).

On the basis of the variables discussed above, the most important hand- and foot-operated controls are represented in Figure 21.3. The controls have been arranged into three main groups, based on whether a rotational, quasirotational (swiveling), or translational movement of the control is involved. Besides the characteristic dimensions, the figure also includes information regarding the permissible actuating forces. The evaluation of the controls with respect to the criteria, actuating movement (continuous, discrete), accuracy, speed, and so on is made by means of qualitative statements.

The large number of ergonomic findings available in various forms dealing with the properties of hand and foot controls often make it difficult for the designer to find the optimum control for specific requirements. Figure 21.3 is based on the most important ergonomic findings relevant to controls. The selection is aggravated by the interactions that exist among the different variables.

21.3 DESIGN PARAMETERS OF CONTROLS AND COMPUTER INPUT DEVICES

21.3.1 Control Elements

The general design principles and the marginal conditions to be taken into account are discussed now for some controls, by way of example. The following information is particularly important to the manufacturer of controls. As shown in Figure 21.1, the design variables—shape, size, material, and surface—are important factors influencing the operating effectiveness, with the characteristics performance (force, speed, accuracy), stress and strain of the worker, and safety criteria. For defining the design variables, the control task, the coupling conditions, and the capabilities and traits of humans are important. Important influencing factors for the characteristics of the longitudinal and cross-sectional form of the control are mainly the anatomy and the type of coupling, and the anthropometry of fingers and hand for the dimensions of the control.

The positioning of the control relative to the worker (side, top, front, orientation) is regarded as a further influencing variable with respect to the shape of the control. Because controls are manufactured as standardized elements without accounting for the position to be determined by the designer, it is necessary that the shape of the control should satisfy the ergonomic criteria for all prevailing positions. For these reasons, anthropomorphic shapes of controls with recessed grips for the phalanges are not suitable as standard controls. Anthropomorphic shapes of controls are also unfavorable for dynamic coupling on the control, because access to the control is only possible at quite specific points, thus impairing quick regrasping on the control. Further design hints regarding the shape and size are now given for one-hand-operated controls, for both coarse, and sensory-motor control tasks—and for one-hand- and two-hand-operated disk handwheels. Details on material and surface are also given. These details are of general validity and can also be applied to other controls illustrated in Figure 21.3.

For a one-hand operation, an approximate ellipsoid of revolution has proved to be the optimum shape for the transmission of large torques (Muntzinger, 1986). The shape of the control illustrated in Figure 21.4 permits the fingers and palm to couple on the rear

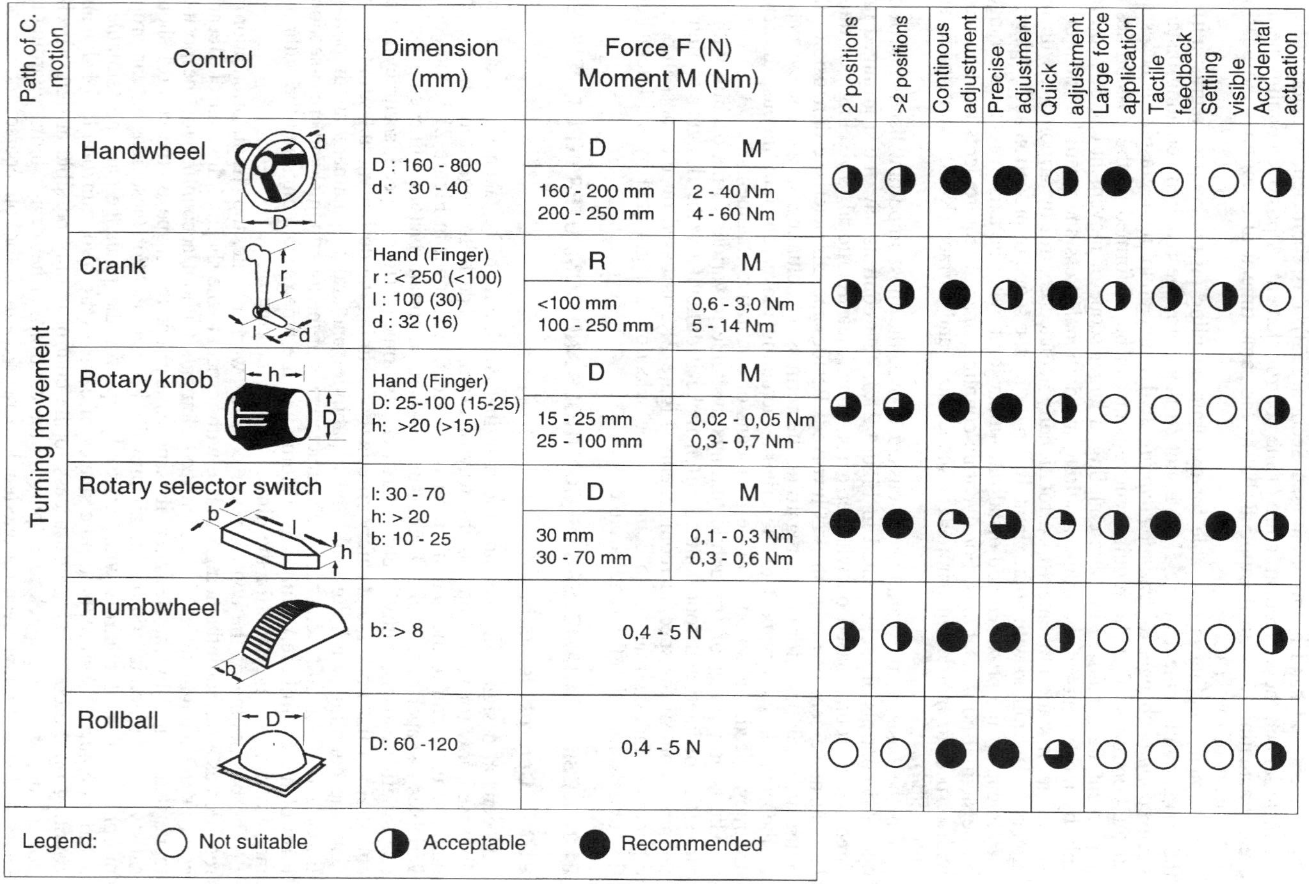

Path of C. motion	Control	Dimension (mm)	Force F (N) Moment M (Nm)		2 positions	>2 positions	Continous adjustment	Precise adjustment	Quick adjustment	Large force application	Tactile feedback	Setting visible	Accidental actuation
Turning movement	Handwheel	D : 160 - 800 d : 30 - 40	D 160 - 200 mm 200 - 250 mm	M 2 - 40 Nm 4 - 60 Nm	◐	◐	●	●	◐	●	○	○	◐
	Crank	Hand (Finger) r : < 250 (<100) l : 100 (30) d : 32 (16)	R <100 mm 100 - 250 mm	M 0,6 - 3,0 Nm 5 - 14 Nm	◐	◐	●	◐	●	◐	◐	◐	○
	Rotary knob	Hand (Finger) D: 25-100 (15-25) h: >20 (>15)	D 15 - 25 mm 25 - 100 mm	M 0,02 - 0,05 Nm 0,3 - 0,7 Nm	◕	◕	●	●	◐	○	○	○	◐
	Rotary selector switch	l: 30 - 70 h: > 20 b: 10 - 25	D 30 mm 30 - 70 mm	M 0,1 - 0,3 Nm 0,3 - 0,6 Nm	●	●	◔	◕	◔	◐	●	●	◐
	Thumbwheel	b: > 8	0,4 - 5 N		◐	◐	●	●	◐	○	○	○	◐
	Rollball	D: 60 -120	0,4 - 5 N		○	○	●	●	◕	○	○	○	◐

Legend: ○ Not suitable ◐ Acceptable ● Recommended

Figure 21.3 Hand- and foot-operated control devices and their operational characteristics and control functions.

Path of C. motion	Control	Dimension (mm)	Force F (N) Moment M (Nm)	2 positions	>2 positions	Continous adjustment	Precise adjustment	Quick adjustment	Large force application	Tactile feedback	Setting visible	Accidental actuation
Turning movement	Lever 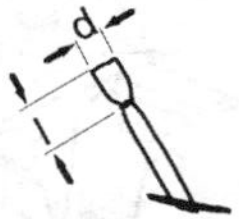	d : 30 - 40 l : 100 - 120	10 - 200 N	●	●	●	◕	◑	●	◑	◑	○
	Joystick 	s : 30 - 150 d : 10 - 20	5 - 50 N	●	●	●	●	◑	◔	◑	◑	○
	Toggle switch 	b : >10 l : >15	2 - 10 N	●	◑	○	○	●	○	●	●	○
	Rocker switch 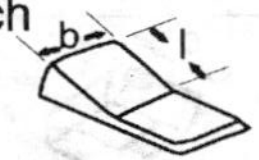	b : >10 l : >15	2 - 8 N	●	○	○	○	●	○	●	●	◕
	Rotary disk 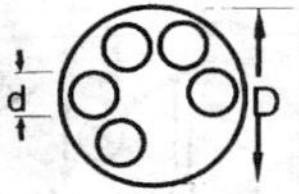	d : 12 - 15 D : 50 - 80	1 - 7 N	●	◔	◔	○	◔	○	○	○	◕
	Pedal 	b : 50 - 100 l : 200 - 300 l : 50 - 100 (Forefoot)	Sitting : 16 - 100 N Standing : 80 - 250 N	◕	◔	◕	◑	●	●	◑	○	○

Legend: ○ Not suitable ◑ Acceptable ● Recommended

Figure 21.3 (Continued)

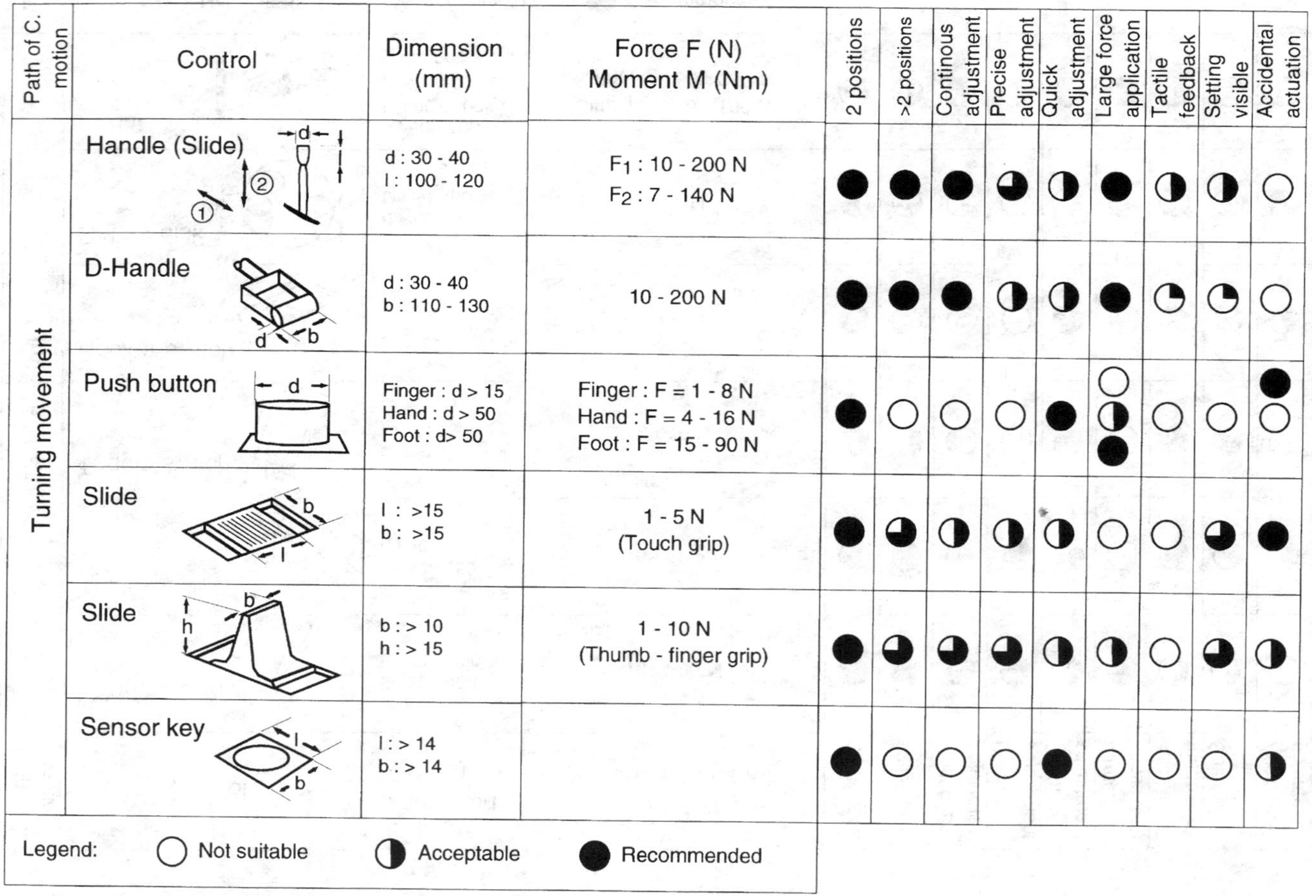

Path of C. motion	Control	Dimension (mm)	Force F (N) Moment M (Nm)	2 positions	>2 positions	Continous adjustment	Precise adjustment	Quick adjustment	Large force application	Tactile feedback	Setting visible	Accidental actuation
Turning movement	Handle (Slide)	d : 30 - 40 l : 100 - 120	F_1 : 10 - 200 N F_2 : 7 - 140 N	●	●	●	◕	◑	●	◑	◑	○
	D-Handle	d : 30 - 40 b : 110 - 130	10 - 200 N	●	●	●	◑	◑	●	◔	◔	○
	Push button	Finger : d > 15 Hand : d > 50 Foot : d> 50	Finger : F = 1 - 8 N Hand : F = 4 - 16 N Foot : F = 15 - 90 N	●	○	○	○	●	○ ◑ ●	○	○	● ○
	Slide	l : >15 b : >15	1 - 5 N (Touch grip)	●	◕	◑	◑	◑	○	○	◕	●
	Slide	b : > 10 h : > 15	1 - 10 N (Thumb - finger grip)	●	◕	◕	◕	◑	◑	○	◕	◑
	Sensor key	l : > 14 b : > 14		●	○	○	○	●	○	○	○	◑

Legend: ○ Not suitable ◑ Acceptable ● Recommended

Figure 21.3 (Continued)

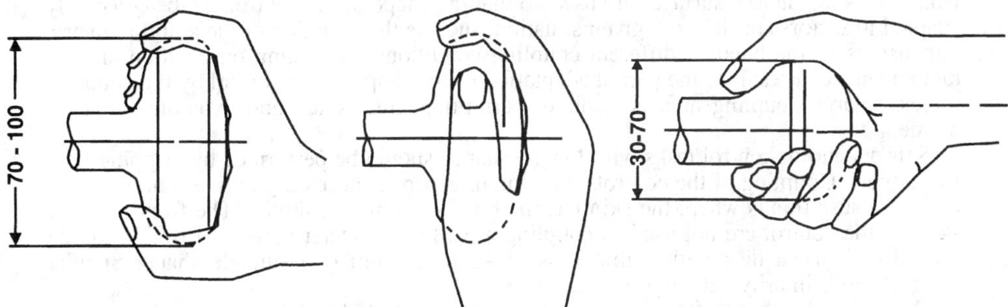

Figure 21.4 Shapes for one-hand-operated controls.

and front sides of the control and thus great actuating moments can be produced. This shape of the control element applies to a diameter range of between 70 and 100 mm. Control diameters above 100 mm do not permit a further increase in the actuating moment, because the end phalanges can no longer couple on the rear side of the control element. As shown in Figure 21.4, this shape is very favorable for the arrangement of the control element in both the sagittal-horizontal axis and in the horizontal-frontal axis. For the diameter range of 30 to 70 mm, the shape of the control changes, because the gripping by five fingers (with the fingers equally spaced) changes to a four-finger grip (thumb extended) (Figure 21.4, right).

Extra-fine profiles have proved to be advantageous with respect to the transmission of power in unfavorable surrounding conditions such as when the hand or the control is very dirty. Figure 21.5 shows the maximum actuating moments of one-hand-operated controls in relation to the shape of the control under favorable and unfavorable surrounding conditions (oil-contaminated). As the figure shows, the greatest actuating moment can be transmitted with the positively actuated dihedral part C_2 and the profiled control C_5, when the controls and/or the hands are dirty. Concerning the dynamic coupling on the control, the unfavorable regrasping in C_2 is regarded as a disadvantage. When controls are used in which heavy dirt can be expected, it must be ensured that the shape of the control is so defined that the deposit of dirt and fluids will not be possible in the coupling zone. Under favorable surrounding conditions, the control C_6 coated with a pressure-anthropomorphic material will be advantageous (coat thickness 6 mm, diameter 110 mm).

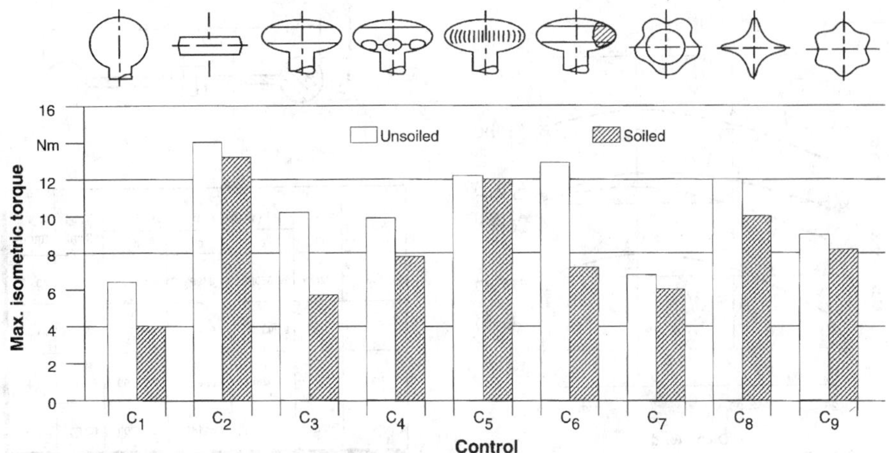

Figure 21.5 Torques exerted at various control alternatives (D = 80 mm) under soiled and unsoiled conditions. (From Kern, Muntzinger, and Solf, 1984.)

Highly elastic material surfaces are used so that the shape of the control can be constantly adapted to factors specific to a given situation, such as the variation of the anthropometric dimensions of the hand or different coupling conditions, depending on position and arrangement. As a result of the partial adaptation of the shape of the control by the coupling forces, a large coupling area is ensured, permitting great static and dynamic actuating moments.

Sensory-motor control tasks with low resistance should be performed by gripping with five fingers. Coupling of the controls with the fingertips is necessary in fine-motor control tasks, because this is where the skin has the highest tactile sensitivity. The front and rear sides of the control are not used as coupling area. In a diameter range of from 50 to 125 mm, a cylindrical disk with a thickness of 20 to 25 mm is a suitable shape. Smaller controls are primarily actuated by the fingers.

Designing the shape of one-hand- and two-hand-operated disk handwheels is regarded as problematic. For safety aspects, disk handwheels must have a closed stay. The relationship between maximum isometric actuating moment and the rim diameter has been approximated on the basis of experimental tests by a regression function of the second degree for different control arrangements (A_x sagittal-horizontal axis; A_y horizontal-frontal axis; I_{zK}, I_{zE}, I_{zS} positioning at knee, elbow, and parting level) (Kern et al., 1984). The regression function permits a rim diameter of approximately 45 mm to be derived ($dM/dd = 0 \rightarrow d_{opt}$). The results are shown in Figure 21.6. For reasons of economy and weight, the calculated rim diameter cannot be realized. Considerable savings of material and weight can be achieved by taking suitable design measures in which the stay of the control can be used as coupling area (control shape; see Figure 21.7). In the design, it will also have to be taken into account that the shape must satisfy criteria of ergonomics and safety for all relevant positions and arrangements. As shown in Figure 21.7, the shape of the control satisfies ergonomic criteria for both two-hand operation in the sagittal-horizontal axis and for one-hand operation in the horizontal-frontal axis; that is, a large coupling area and thus minimum strain on the hand is guaranteed in all control positions.

In the case of frictional coupling of the extremity with the control, the power is transmitted by friction forces. The magnitude of the friction force F_F can be determined by the law of friction ($F_F = \mu \cdot F_N$, where μ is the coefficient of friction and F_N the normal force). In a frictional coupling, the normal force will be developed by the coupling force of the hand or fingers. The coefficient of friction is determined by anatomical-physiological parameters such as the size of the coupling area, surface structure, and the skin's degree of moisture on the one hand, and by the material's properties such as surface roughness and profiles on the other hand. In frictional coupling—contrary to positive power transmission to the control—the material must also be selected under the aspect of the frictional behavior that exists between hand/fingers and the control. The reason why special importance is attached to the selection of the materials is that unsuitable

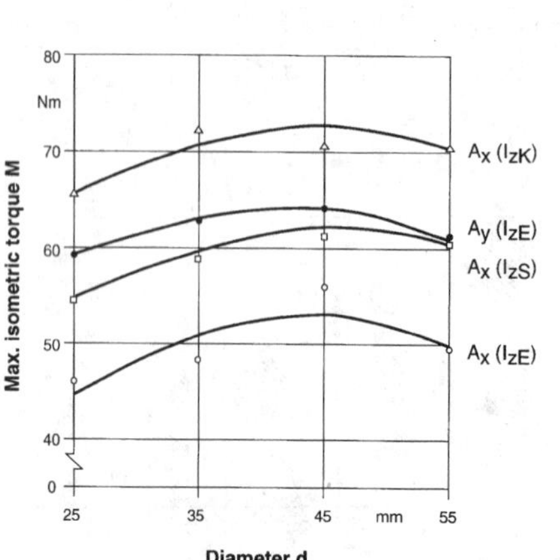

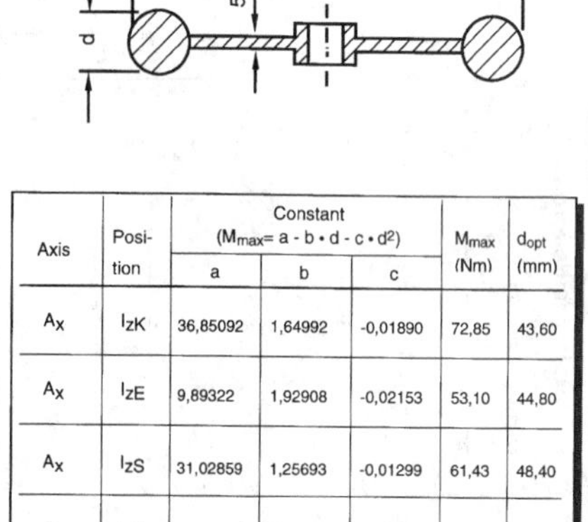

Axis	Position	Constant (M_{max} = a - b • d - c • d²) a	b	c	M_{max} (Nm)	d_{opt} (mm)
A_x	I_{zK}	36,85092	1,64992	-0,01890	72,85	43,60
A_x	I_{zE}	9,89322	1,92908	-0,02153	53,10	44,80
A_x	I_{zS}	31,02859	1,25693	-0,01299	61,43	48,40
A_y	I_{zE}	34,66092	1,41146	-0,01709	63,80	41,30

Figure 21.6 Torques exerted at handwheels (disk type) with different rim diameters. (From Kern, Muntzinger, and Solf, 1984.)

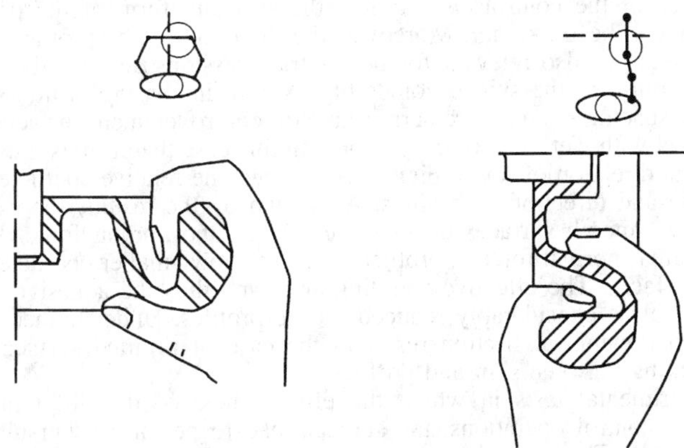

Figure 21.7 Shape for handwheels (disk type) and coupling conditions for different handwheel arrangements. (From Kern, Muntzinger, and Solf, 1984.)

materials and surface structures of controls very quickly lead to heavy strain and destruction of the upper skin layers. Figure 21.8 represents the coefficients of friction, standardized to the material of plexiglass, of 29 common materials for hand contact (Bullinger, Kern, and Solf, 1979). To determine the coefficient of friction of the materials, the samples were moved over the stretched hand at a defined normal force of 40 N. Figure 21.8 shows at the top the coefficients of static and sliding friction (μ_1, μ_2). The deformation of the skin until the setting in of sliding friction is shown at the bottom. For smooth materials, a high correlation could be seen between the surface roughness and the coefficient of friction. When the control elements are actuated with the fingers, the friction force, among other matters, is dependent on the direction. If the friction force is transmitted along the fingers, there will be smaller coefficients of friction than in force transmission at right angles to the fingers. In translationally actuated controls, the force will be transmitted longitudinally and transversely, and predominantly at right angles to the fingers in rotationally actuated controls.

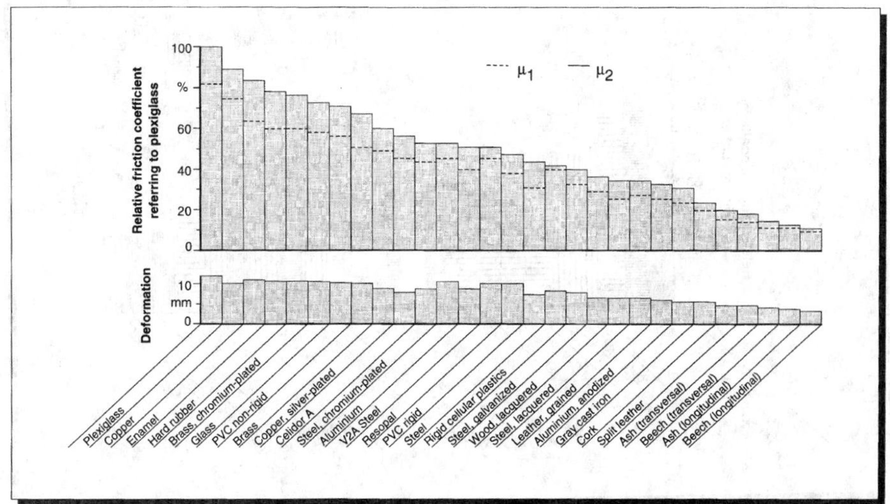

Figure 21.8 Friction coefficient between hand and different materials. (From Bullinger, Kern, and Solf, 1979.)

If the surface of the controls is profiled, the size and form of the profiles and the profile spacing will be important. Moreover, the direction of the profiles relative to the hand or the fingers is also relevant for power transmission. Because the strain on, and the danger of injury to, the skin becomes higher with increasing profile size, only fine profiles (profile spacing <3 mm) are permissible for control element surfaces if the control must be operated with wet, oily, or dirty hands. In this case the profiles must be vertically orientated to the direction of force. Figure 21.9 shows the relative coefficients of friction versus the form and direction of profiles. As shown by the results, the effectiveness of the profiles over smooth surfaces depends heavily on the normal force. Under smaller loads per unit area (normal forces), profiled surfaces show smaller coefficients of friction than smooth surfaces. The effective coupling area, and thus the adhesive force between hand and material, is considerably reduced by the profiles. Under higher loads per unit area, greater coefficients of friction result in the case of profiled surfaces, owing to a certain interlocking between skin and profiles.

During experimental tests in which the effectiveness of the different profiles was checked on real control operations, the subjects had to perform a pursuit tracking test (Kern et al., 1984). The control element (diameter = 70 mm, length = 80 mm) was alternately operated by two hands over a movement range of five revolutions (see Figure 21.10). The tracking error was used as an evaluation variable for the control quality, and the error rate as a measurement for the continuity of the control movement. Figure 21.11 shows the tracking error relative to the shape of the control, its profile, and the resistance. A significant influence could be evidenced regarding the profile and shape of the control. The largest tracking error was noted concerning the hexahedral part C_2 over the total resistance range. The nonprofiled cylinder C_1 is favorable only if the resistance is very low. Minor errors arise in the profiled controls at higher resistances (0.4 to 0.9 Nm). As can be seen in Figure 21.11, there is a characteristic pattern for the profiled controls over the resistance range. The smallest tracking error can be noticed at the medium resistance range. At high resistances, the abrupt movements due to static friction can be seen as a cause for the low control efficiency and, at low resistances, the missing proprioceptive information and the transfer of inadvertent movements (also compare Seibt, 1971).

If relative movements occur between control and hand during operation, the *thermal conductivity* of the control material will also have an influence on the operating effectiveness. The thermal conductivity of relevant control materials may vary between 0.15

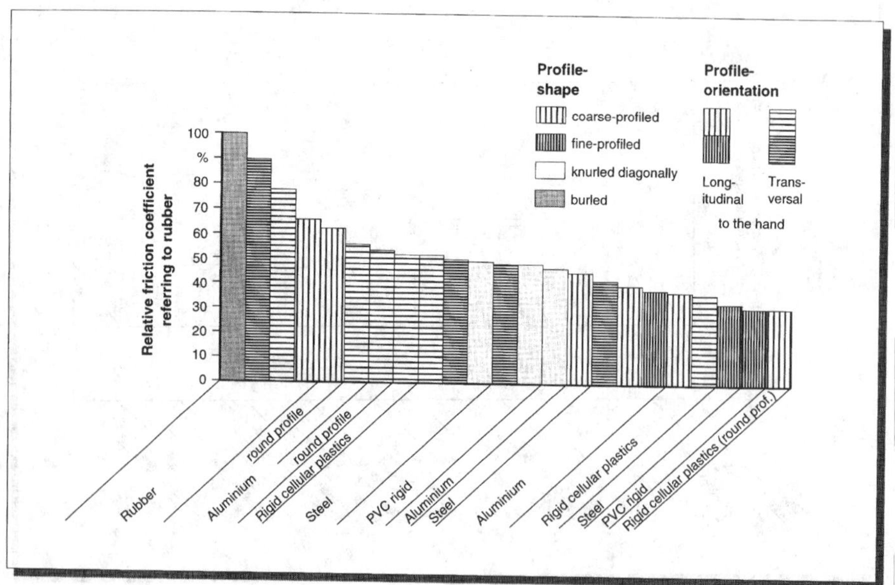

Figure 21.9 Friction coefficient between hand and different surface profiles. (From Bullinger, Kern, and Solf, 1979.)

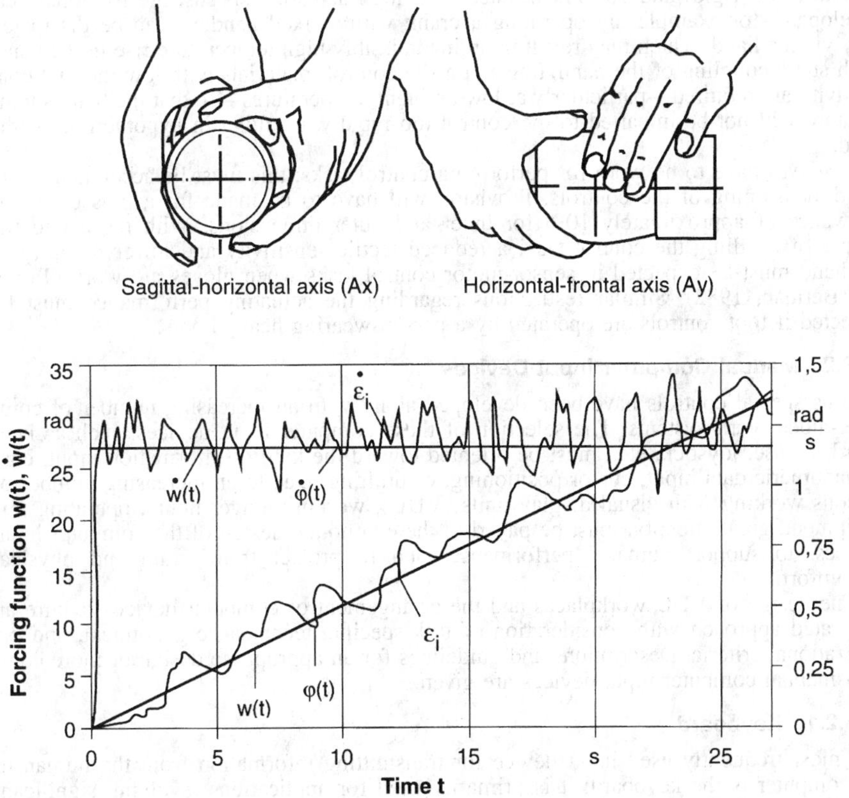

Figure 21.10 Two-hand-operated control and the forcing/response function of the pursuit tracking task. (From Kern, Muntzinger, and Solf, 1984.)

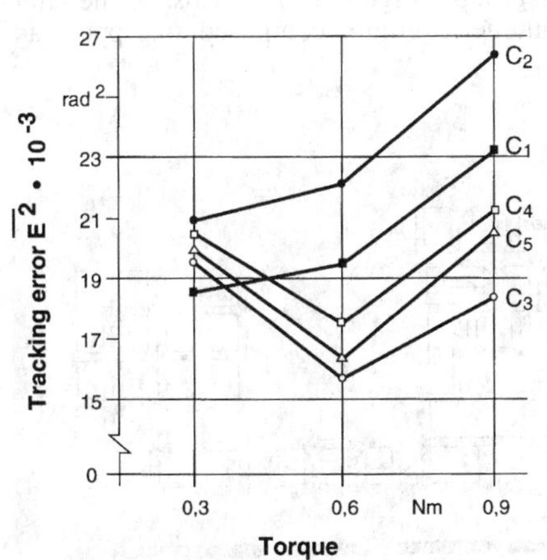

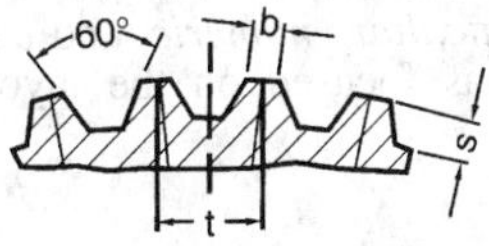

Control	Parameters of the surface t (mm)	b (mm)	s (mm)
C_1	Cylinder	(R_Z = 1,27 µm)	
C_2	Hexathedral control	(R_Z = 1,27 µm)	
C_3	1	0,4	0,6
C_4	3	0,6	1,2
C_5	4	0,8	2,2

Figure 21.11 Effects of control surface and resistance on tracking error. (From Kern, Muntzinger, and Solf, 1984.)

W/Km (PVC rigid) and 70 W/Km (steel). At high actuating speeds, the frictional heat developed—for example, by operating a crank with a fixed handle—can be dissipated only via the hand, which may result in an inadmissibly high temperature rise in the hand. With static coupling of the hand/fingers on the control, materials with low thermal conductivity are required—particularly at low ambient temperatures—so that the hand's temperature will not be imparted to the control too rapidly, resulting in hypothermia of the hand.

If *gloves* have to be worn for performing control tasks, this must be accounted for in the dimensioning of the controls. Provision will have to be made for a plus or minus allowance of approximately 10% for inner and outer dimensions. With respect to the degree of fulfilling the control task, a reduced tactile sensitivity and lower mobility of the hand must be expected in sensor-motor control tasks when gloves are worn (Taylor and Berman, 1982). Similar restrictions regarding the actuating performance must be expected if foot controls are operated by a person wearing heavy boots.

21.3.2 Manual Computer Input Devices

Various special controls have been developed along with an increasing number of computer-aided work systems. The selection of these computer input devices, such as keyboard, mouse, joystick, etc., must be oriented toward the kind of information input, e.g., alphanumeric data input, cursor positioning, or drafting. Due to an increasing number of persons working with visual display units (VDU), working longer hours, operating this equipment, great attention must be payed to the ergonomic design of the computer input devices, to support humans' performance, and to protect from strain and physical discomfort.

The design of VDU workplaces and the arrangement of computer devices require an integrated approach with consideration of task-specific, technical, ergonomical, and organizational criteria. Descriptions and guidelines for an appropriate design of more common manual computer input devices are given.

21.3.2.1 Keyboard

The most frequently used input device for transmitting information from the human to the computer is the keyboard; it is primarily used for applications involving significant amounts of alphanumeric data. The role of the present keyboard has been to a great extent taken from the outdated mechanical typewriter. The historical evolution of keyboards hardly took into consideration any ergonomical requirements. The construction of current electronic keyboards gives special consideration regarding the shape and layout of the keyboard on the account of secured ergonomical findings.

Keyboard Layout

The layout of the keys are considered according to their importance, function, frequency of use, and sequence of use. The standard keyboard usually includes *alphanumeric*, *function*, *auxiliary numeric*, and *cursor control key* groups (Figure 21.12). Most of the attention has focused on the layout of the alphanumeric group, composed of upper- and

Figure 21.12 Keyboard (QWERTZ layout) for computer input, including alphanumeric, function, numeric, and cursor control keys.

lowercase alphanumeric characters, i.e., numerals, punctuation marks, and special symbols. The function keys provide rudimentary functions such as mode changes (shift key) and communication (enter key). Frequently used functions such as these are typically included in the periphery of the alphanumeric keyset. Other specialized functions, e.g., for the editing of text may be offered by additional keypads on the same keyboard. To get some idea of the fundamental guidelines given for the keyboard layout, Rohmert and Haider (1981), Läubli, Fleischer, and Krueger (1989), and ISO 9241-4 (1994) are recommended.

With regards to the layout design of the alphanumeric keypad, there are a number of existing formats. The *QWERTY layout* for instance (named after the topmost upper-left alpha keys) is widely used in English-speaking countries and specifies the location of the alphabetic characters, as well as most of the numerals and symbols. Variations to this layout include the *QWERTZ* (Figure 21.12) in Germany and the *AWERTY* in France.

However, several disadvantages have been recognized with the QWERTY layout, and thus a number of alternative layouts have been formatted. One of the largest criticisms regarding the QWERTY layout is an overloaded use of the left-hand side of the keyboard and hence an overload on several of the fingers (Greenstein and Arnaut, 1987). Added to this is the excessive row hopping from one row to another which often occurs when having to type out frequently required sequences.

One such attempt to improve the QWERTY layout is the *Dvorak layout*, arranged according to the basis of sequential frequencies which occur in the English language. Criteria of keyboard design take into consideration the higher amount of work from the right hand to the left hand, the arrangement of most frequently used keys in one row, and therefore the minimalization of finger motions from row to row, respectively, awkward reaches.

An *alphabetical layout* would enhance the speed and accuracy for the nonexperienced user. The alphabetical layout provides a familiar, easily understandable structure that aids the user in the quest for the desired keys. Studies have shown that there are no practical performance advantages from one alphabetical layout relative to the QWERTY layout (Norman and Fisher, 1982).

An auxiliary *numeric keypad* provides an efficient adjunct to the alphanumeric keyset in applications requiring significant amounts of numeric data entry. The generally accepted keypad for numeric data entry consists of 10 keys arranged in a three-by-three matrix. There is a differentiation between the two keypad arrangements for the calculator layout and the telephone layout, which are shown in Figure 21.13. The practiced user will be able to use this format with the same degree of efficiency as with the telephone or the calculator layout.

A *cursor control keyset* provides a key-oriented means to control the current cursor position. With cursor control keys it is possible to direct the cursor on the display screen to left, right, up and down, and occasionally, to a home position.

In special keyset arrangements the location of functions keys are determined by general considerations such as of importance, frequency of use and sequence of use. As well as the guidelines for function key layout tend to be highly application-specific. Additional considerations regarding the layout design of some function keys may be the result of inadvertant operations.

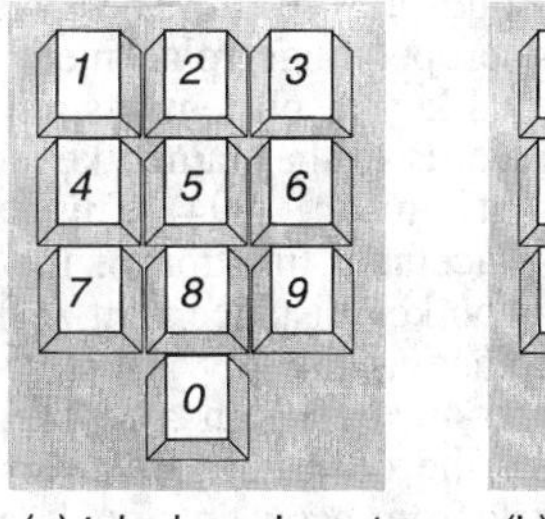

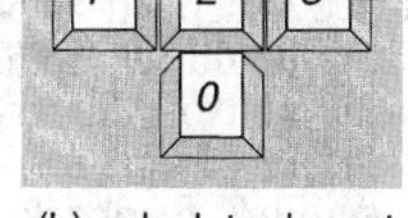

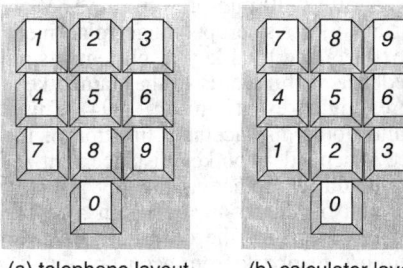

(a) telephone layout (b) calculator layout

Figure 21.13 Numeric keypad layouts: telephone layout (left) and calculator layout (right).

Fixed-Function and Variable-Function Keyboards

In many applications, an initial consideration regarding the selection of a keyboard concerns whether the keys shall have fixed (e.g., handheld calculators) or variable functions (e.g., terminal keyboard, shift keys).

The simple applicability and minimal software support of the *fixed-function keyboards* mean that all available functions can be determined easily by just scanning the keys. However, this is opposed by the unmethodical key layout and large-scale implementation of functions. On the whole the selection of fixed key functions hold down well, when functions must be executed quickly, when one set is frequently employed to carry out the task and when the correct selection of the functions is critical to the operation of the system.

Key labels of *variable-function keyboards* can be varied in one of three ways: Shift keys may be used, permitting the user to shift a key's function among several fixed alternatives; labeled overlays appropriate to a given mode of operation may be positioned above the keys; or, the functions of the keys can be placed under software control where the user is informed of the key-function relationships. Variable-function keyboards hold the advantage in contrast to fixed-function keyboards of equivalent power as fewer keys are needed. Therefore, less visual search and movement of the hand–arm system is required. Software-controlled key-function relationships can easiliy be modified with changes to software. Disadvantageous to variable-function keyboards is the execution of some functions with shift keys, where more than one key must be pressed. With overlays or software-controlled labeling, the user must select and attach the appropriate overlay to the respective software being used. In general, the selection of variable-function keyboards is appropriate, when subsets of functions are frequently used, when the pacing of entries is not forced, and when a relatively sophisticated prompting and feedback are available. Additionally, software-controlled key functions seem particularly appropriate to applications that experience continual modification.

Keyboard Dimensions, Mechanics, and Feedback

Most of the standard keyboards currently available are characterized by a straight-lined arrangement of the key rows (compare Figure 21.12). Consequently, a determination of the keyboard case inevitably occurs.

Regarding the required flexibility and individual adjustment corresponding to the working posture and work task, the keyboard must be separate from the visual display unit. The keyboard should be designed so that it is freely positionable relative to the VDU and that it does not slip on the desk surface. The width of the keyboard should be as narrow as possible, thereby minimizing the twist angle between keyboard and the script holder. Rohmert and Haider (1981) recommend a keyboard width of 370–540 mm. The recommended height of keyboard constructions regarding the middle row should be no higher than 30 mm, but must not exceed a maximum height of 50 mm. For the angle of inclination, a scope of 5–15° is recommended. In order to reduce the static work of the body concerning posture at a keyboard height of more than 30 mm, the keyboard should be provided with hand-seating. The seating's width should correspond to the one of the keyboard, its depth should not exceed the values of 50–100 mm and—corresponding to the keyboard—it should be adjustable in height and inclination.

The size and spacing of keys on general purpose alphanumeric keyboards are largely based on design conventions, rather than on empirical data. Key diameters of 12–15 mm with center-to-center spacings of 18–20 mm are typical. The keytop is typically square with a slightly concave surface to assist proper finger placement. Key activation forces generally vary from 0.25 to 1.5 N and total key displacements are between 1 and 5 mm.

The major source of feedback with regards to the trained keyboard user is the tactile feedback by moving and pressing a key (Thomson, 1994). This tactual feedback can be recognized as a corresponding force-displacement function of the keys or their pressure point. Because the key force is dependent on key displacement, it has been recommended that a rapid build-up force should take place as the key is pressed, with a reduction in the required force in the region of activation, of which a second increase in the force follows. Visual feedback is important for the correction of errors and during keyboard training. Supplementary auditory feedback from acoustic signals can also enhance feedback, and consequently the performance.

Alternative Keyboard Design

The introduction of the electronic keypad has made it possible to do away with the traditional formal characteristic principles of the mechanical typewriter. This had lead to multiple possibilities which have been carried out for the adaptation of hardware for the anthropometrical and mental conditions of the user (Cakir, 1995) (also compare Barry, 1993).

In Figure 21.14, the exemplary keyboard portrayed is that of an alphanumeric variety, which is split into two keypads left and right, set at angles. The key areas are layed out so that they are facing both laterally and frontally at approximately 58 The keyboard has an integrated numeric block at the right-hand alpha field. A separate numeric block can be used if required. In contrast to the usual keyboard design, the slightly angled key areas lead toward a natural abduction-free handholding (Ilg, 1987). By avoiding the forced position of both hands, one has counteracted the increased risks of muscle strain and posture disorders.

21.3.2.2 Mouse

With the introduction of graphical user systems the role of the cursor system and the cursor control devices such as the mouse, joystick, trackball, etc. has considerably increased. In graphical user systems, a function usually is activated by pointing the cursor on to the required icon (graphical interacted object), and, if necessary, confirm or mark by clicking the function button.

A mouse is a handheld cursor control device, which is normally attached to the computer by a wire. Mice are small containers with rollers or sensors on the underside, which can be fit under the palm or fingertips. By moving a mouse across a surface (e.g., mouse pad), the cursor's position will change accordingly on the screen. On the upperside, a mouse may offer up to three press buttons, allowing the user to carry out functions such as changing menus and confirming inputs. Mice are typically used only as peripheral devices and not as the sole input device. Their features are best suited for pointing and selecting tasks with regards to menu-operated applications, and dragging graphical objects around the screen, less so for drawing tasks. Mice are not suited for single character data entry.

Functioning

With the utilization of mice, one can decide whether to use the mechanical or the optical mouse. The principle functioning of the *mechanical mouse* is based on the conversion of a movement in an electrical signal. Therefore, principally, two rollers are mounted on the

Figure 21.14 Ergonomic keyboard design (Mini-Ergo).

bottom of the device, oriented at right angles to each other. Movement of the mouse, and consequently, rotation of the rollers, leads to voltage output from potentiometers, which is used to calculate the cursor's position coordinates on the screen. Usually, the rollers are driven by a smart ball, mounted in the bottom of the mouse. Mechanical mice work on any surface, but it is recommended to use a mouse pad to raise the friction coefficient and to reduce soiling. Mice will only work in the relative mode.

With regards to the *optical mouse,* optical sensors are inbuilt, which emit pulses as the mouse is moved across a special grid. The position of the mouse is calculated by the position on the lines, which are counted as the mouse crosses the grid. The control-display gain of the mouse can be varied by changing the gridline spacing. Because optical mice need no moving parts, they can not pick up debris from the surface. An optical mouse can only be used, however, in conjunction with the utilization of a grid. This could mean that the resolution may be lower than that with the use of a mechanical mouse (Greenstein and Arnaut, 1987).

Parameters

Cursor control device design should permit easy use of the preferred hand for both right- and left-handed users. Therefore, cursor control devices such as mice are usually mounted on a separate module from the keyboard so that it can be placed on either side of the keyboard. For better handling, the shape and surface of some types of mice are ergonomically designed to fit the human's hand.

As previously stated, mice usually have one or more buttons that may be pressed for any number of various functions. There are two modes in which the buttons on the mouse can be utilized: multiple depressions on one button and a depression of multiple buttons. For a task in which one item is repeatedly selected, performance is faster for clicks on a single button rather than on multiple buttons (Price and Cordova, 1983). However, for a task involving several actions, performance using a number of buttons is better than repeated clicking on the same button.

The integration of cursor movement with a selection of buttons has an advantage due to its flexibility when simple cursor controls are concerned. In turn, this makes the mouse quite compatible with visual interfaces and graphical manipulation (Shneiderman, 1987). It is normal for the operator to locate and move the mouse around, keeping his or her eyes on the screen. The control-display gain for many mice can be modified. Additionally, mice are inexpensive in comparison to other devices.

Disadvantages of the mouse include the fact, that a mouse requires some space in addition to that allotted to a keyboard. Thus, it is not compatible with some portable computers. To save desk space, the mouse can be picked up and repositioned. As already mentioned, a mouse can only be operated in relative mode, a feature that may limit its usefulness for drawing tasks (Mims, 1984). Other features that limit the use of a mouse for graphic applications include an inability to trace drawings or to handprint characters.

21.3.2.3 Joystick

A joystick is a cursor control device consisting of a control lever which is routed vertically into a fixed base. For better handling, the lever can be moulded into a handgrip that makes the joystick more comfortable to use. Joysticks generally offer one or two buttons which may be pressed to release various functions.

Joysticks are especially used in concordance with computer applications for moving the position on the screen and changing icons or interacted objects portrayed on the screen. They are most suited to tasks regarding tracking or to pointing tasks that do not require a great deal of precision (Mims, 1984). Joysticks can be used in regards to the placing and moving of symbols, and in that of menu selection if rate control is used. A joystick in absolute mode may also be used for line drawing, if high accuracy and speed are not required.

Functioning

The three basic operating mechanisms a joystick can work are displacement, force operation, or digital joystick. With a *displacement joystick* (or isotonic joystick), the user directs the lever in any one direction and the displayed cursor moves proportionally. Movements of the lever are detected by potentiometers and translated to a voltage output. By taking the user's hand off the joystick, it will either remain where it is or a set of springs will return it to the center. Some joysticks provide the possibility to switch between these two modes. Additionally, the friction force and force-displacement relationship in many displacement joysticks may be adjusted. Displacement joysticks typically operate in absolute mode; that is, the cursor position corresponds to the joystick position. (For control dynamics, see also Section 21.5.)

A *force-operated* (or isometric) joystick is a rigid lever that does not noticeably move in any direction. Rather, strain gauges measure pressure applied to the joystick. Thus, cursor position is dependent on the force applied to the joystick, not to the position of the device. As with the displacement joystick, the force-operated joystick will respond to pressure in any direction. From the pressure's direction, respectively, its amount, both direction of cursor movement and cursor speed are calculated. When the pressure decreases to zero, the cursor stops moving.

Switch-activated or *digital joysticks* work similiar to displacement joysticks in that the lever itself can be moved; generally, however, only movement in eight directions will be detected. Movement of the joystick generates voltage output by closing one or more switches that are connected to the base of the stick. When the joystick is released, springs return the lever to center and the switches are no longer activated. Switch-activated joysticks are typically operated such that the cursor moves with a constant velocity in the same direction as the stick displacement, but does not return to center when the joystick is released (Ohlson, 1978).

Parameters

As with other computer input devices, the gain of the joystick may be changed. Joystick gain is typically greater than the gain of other devices, due to the difference between the display's size and the restricted joystick displacement range. It is difficult to use a joystick in absolute mode because, owing to its gain, movements of the hands are magnified on the display (Foley, van Dam, Feiner, and Hughes, 1990).

Due to the suitability of isometric joysticks for rate and higher-order control systems, and isotonic joysticks for position control systems, an isotonic or a spring-centered control with a low spring rate should be used for cursor control, which constitutes a discrete positioning task.

Through several technical modes, a three-dimensional capability of the joystick can be achieved. With a table-fixed joystick, a rotatable knob can be placed on the top of joystick. Alternatively it can be done by allowing the stick to be twisted in clockwise or counterclockwise direction.

The advantages of the joystick include the fact that it requires only a small fixed amount of desk space; in a small version, it also can be fit into a keyboard. The joystick may be used for extended periods of time with little fatigue, if a handrest is provided. Because of its widely spread utilizations, there are many models of joysticks available.

Low accuracy and low resolution are the main disadvantages of the joystick. Joysticks can not be used to input single characters and to trace or digitize drawings.

21.3.2.4 Trackball

The trackball (or roll ball) comprises of a moulded container within which a fixed ball is housed. The ball is relatively large and can be rotated by the user in any given direction using the fingertips. As a cursor control device, the trackball holds a similar operation to the mouse.

The trackball comes into its own when dealing with cursor movements over long screen distances especially as its control sensitivity can be varied; this means that it is suited for rapid cursor positioning with high accuracy (Greenstein and Arnaut, 1987).

Functioning

As with the mouse, there are two different types of trackballs: mechanical and optical trackballs. With *mechanical trackballs,* the rotation of the ball leads to the movement of two shaft encoders. This in turn causes voltage output to be generated from internal potentiometers. The output signals from the potentiometers correspond to rotational changes of the ball carried out by the user, and the cursor's coordinate position is moved accordingly.

When rotating an *optical trackball,* optical encoders generate signals or pulses. These signals are used to determine increments in rotation in each direction on the cartesian axes. Both cursor distance and cursor direction are calculated.

Parameters

There are several features of the trackball that may be adjusted; so frictional forces present during rotation may be varied (Scott, 1982). Additionally, the cursor movement can become a nonlinear function of the ball's rotational velocity; this allows the use of a low control-display gain at low rotational velocities coupled with progressively higher gains as rotational velocity increases. The gain function of the trackball may then be adjusted for optimal use, both in gross movement and fine positioning; that is, rapid movements of the ball result in large changes in cursor position per rotation, while slower movements result in smaller changes in cursor position (Greenstein and Arnaut, 1987). Thus, the trackball may be fairly flexible, permitting accurate positioning and rapid movements.

The trackball has several advantages: It is comfortable to use for an extended period of time as it allows the user to rest his or her forearm. In addition, the trackball provides direct tactile feedback from the balls rotations. With its above-mentioned features, a trackball allows for very rapid cursor movement at high resolution. Since a trackball requires only a small fixed space, it can be integrated into or next to the keyboard to minimize the interference of cursor positioning on keyboard-intensive tasks.

It is disadvantageous with trackballs, that they are typically more expensive than other cursor control devices. A trackball cannot be used for tracing or input of hand-drawn characters and drawings.

21.3.2.5 Touch Screen

Touch screens (or touch-sensitive devices) are display devices that permit the user to input a selected position on the display by touching the desired location or item with a finger or pointer. Touch tablets are useful for cursor movement applications without any digitizing.

Touch screens can provide effective software-driven analogues to electromechanical control devices such as push buttons. Users benefit if the devices are programmed to simulate familiar physical controls. Touch screens are effective for approximate positioning tasks, for example menu selection; they are best suited when working with already displayed data. They are useful in reducing workload in situations where the possible types of inputs are limited and well defined (Plaisant and Sears, 1993). Touch screens are not suited for tasks requiring precise positioning, such as drawing and graphical input.

Functioning

There are two basic principles by which touch screen devices work. One method uses an overlay that responds to pressure. With the other method the device is activated when the finger or the pointer interrupts a signal. Representative kinds of touch technologies are briefly described:

Resistance techniques use two transparent membranes embedded with a grid of electrodes. The pressure of a touch causes the electrodes in this location to make contact with each other, completing a circuit representing that location. Disadvantages concern the unreliability of touches, which are sometimes undetected. Additionally, membranes and electrodes may obscure parts of the display. As touch screens use an overlay, parallax may become a problem.

Optical techniques detect the location at which the operator's finger or a pointer interrupts a grid of (infrared) light over the surface of the screen. Thus, no overlays are used with this device and as a result, no display obstruction or decrease in image quality is experienced. Because activation of the device occurs when the light beams are interrupted and not when the screen is actually touched, inadvertent activation may become a problem.

Capacitive techniques detect the capacitance introduced by the operator's finger or pointer at the touched location. Here a conductive film is used, which is deposited on the back of a glass overlay. Although the display is not obscured by anything in this method, the overlay may reduce the amount of light that is transmitted through the screen.

Acoustic techniques apply ultrasonic waves, detecting the touch location by the timing of echoes reflected off the finger or pointer. The ultrasonic waves are generated by transducers on a glass plate, placed over the display screen. Acoustic touch screen devices allow higher resolution than optical devices. However, the device may be inadvertently activated by dirt or scratches on the glass.

For further information on touch screen technologies, see Schulze and Snyder (1983).

Parameters

Touch devices allow greater flexibility than electromechanical controls by providing the capability to reconfigure the number, shape, size, location, and labels of the touch-sensitive fields under software control. Options can be displayed only when needed, rather than having to be present continuous.

The resolution of a touch screen is theoretically whittled down to a pixel, but the size of the user's finger or a pointer and the parallax potential of the screen limit the effective resolution to 6 mm at best. To reduce errors by activating the wrong button, large touch-sensitive areas or soft buttons with a minimal size of 20 mm high and width are recommended.

When the user touches a soft button, auditory or visual feedback, like beep signals or displaying the activated touch with inverse colors, permits the user to verify the activation of the intended button.

One of the most obvious advantages of touch screen devices is that the input device and the output device are the same. Thus, there are direct eye–hand coordination and direct relationship between the user's input and the displayed output. Possible inputs are limited by what is displayed on the screen; thus, no memorization of commands is required, and input errors are minimized.

Disadvantages of the touch screen device concern the possible limited resolution owing to the size of the operator's finger in relation to the touch-sensitive fields. Therefore, touch screen devices are inappropriate for selection of small items. Because only one input device element may be used at a time, data entry will also be slower than with a keyboard. If arm support during touch device usage is not provided, the role of the touch device is restricted to intermittent, short-duration applications. Continuous, intensive touch device usage may cause undue muscle strain, fatigue, impaired performance, and errors. A further disadvantage may be that the finger or arm can obscure the screen. With all touch devices, problems such as dirt and smudges on the screen, or the overlay of the fingertip touching the display may arise.

21.3.2.6 Graphic Tablet

Graphic tablets consist of a flat panel, placed on the desk in front of the VDU. The surface of the tablet represents the display. Input selection can be done by using a stylus, a puck, or a finger. Graphic tablets may be used for drawing and tracing purposes; they are also appropriate in situations in which the user is required to select or point to an item from a menu or an array. A graphic tablet is virtually the only input device that may be used for drafting or hardcopy data entry (Ohlson, 1978); thus, it is well suited for CAD applications. Because of its inherent graphic nature and the fact that only one input device element may be used at one time, a touch tablet is inappropriate for discrete data entry.

Functioning

There are two major categories existing regarding graphic tablets: The *digitizing tablets* (or digitizers) work through the use of a special stylus or puck, which is attached to the tablet by a wire, producing signals indicating coordinate values for cursor positioning. With *touch tablets* (or touch-sensitive tablets) a special stylus is not required; the tablet responds to a touch by any stylus. The main function principles of graphic tablets are described:

Matrix-encoded tablets work by the use of electrical or magnetical fields, produced by conductors or wires in the tablet. As the special stylus or puck is passed over the tablet surface, it detects signals, which are used to determine cursor coordinates. This method provides a high cursor control resolution.

With a *voltage-gradient tablet*, a conductive sheet forms the surface of tablet. A potential is applied to the point of the stylus. With the decrease in potential on the plate, measured at the stylus' location, cursor's position coordinates are calculated, using the distance from the borders of the tablets as a reference.

Electroacoustic digitizing tablets use electrical pulses generated on the tablet. These pulses are detected by the stylus. The time delay between pulse generation and reception is used to calculate the cursor position.

Another acoustic technique, not requiring any special stylus, is the *acoustic touch tablet*, where ultrasonic waves are produced on a glass surface. When these waves are interrupted by a stylus, they are reflected back and detected. The delay between the wave generation and reception is used to determine cursor position coordinates.

Multilayer tablets use two or three conducting sheets, where an electrical potential is generated, when pressing these sheets together using a passive stylus or finger. The generated signal becomes the base for calculating the cursor coordinate values.

For further informationon on graphic tablets, see Foley et al. (1990) and Mims (1984).

Parameters

In some ways, graphic tablets are similar to touch screen devices; however, they are far more flexible in their use. The size of a tablet can vary from a small keyboard component to an entire digitizing table. A maximum size of 420 mm width and 300 mm depth is recommended. With graphic tablets, some parameters, such as the control-display gain and the method of cursor movement, may be modified in wide ranges (Greenstein and Arnaut, 1987).

For avoiding incorrect entries, it is recommended to include some sort of feedback mechanism, like an audible click from a button on the stylus or tablet. Permitting that an operator presses a confirmation button before data entry is finalized, inadvertent inputs may decrease.

Graphic tablets are advantageous for the drafting or hardcopy of data entry; therefore, the display and the tablet may be positioned separately according to users' preference. The movement required and the control-display relationship are natural to many users, in order to achieve high efficiency with data input.

As a disadvantage, touch tablets may not necessarily provide a high positioning accuracy. On the whole, digitizers have a higher resolution than touch tablets, in part because of the small tip on the stylus. Large graphic tablets take up space on the work table. For tablets requiring a stylus, there may be a problem with loss or breakage of the stylus.

21.3.2.7 Light Pen

A light pen is a stylus, whose point contains an optical system as well as a light-sensor photocell. The light pen generates information when it is pointed at the display screen. It is connected by a wire to the terminal which provides a signal, permitting the screen location to be detected where the pen is pressed.

Light pens are typically used for cursor positioning or menu selection by pointing the stylus at the desired option. Apart from this, lines can be drawn and icons can be marked out, moved, and replaced. Therefore, light pens suit themselves especially for the operation of computer aided drawing and construction.

Functioning

The position of the light pen on the display screen can be determined as follows: At high speed (unperceivable to the human eye), every pixel starting from the upper-left-hand side of the screen is switched sequentially from dark to bright. The position of the light pen corresponds to the position of the lit pixel, which in turn releases an electronic pulse to the photocell. Light pens are equipped with either a shutter or finger-operated mechanical switch, which, when pressed, allows light to reach the photocell. In this manner, inadvertent activation is avoided.

Because the light pen is activated by the increasing brightness of the cathode-ray-tube (CRT) phosphor, it may typically only be used with CRT displays.

Parameters

Basic modification features of the light pen concern the pen's field of view and the type of activation switch (Foley et al., 1990). The light pen may be activated either in the pointing mode or in the tracking mode. In the pointing mode, a figure or character may be selected by pointing to a spot on the display and then engaging the light pen. In tracking mode, the light pen is used to position a cursor or cross-hair present on the display screen. Therefore, the operator aims the light pen at the cross-hair and then moves the pen at steady rate. As long as the image remains in the light pen's field of view, a line will be traced. If cross-hair is lost, tracking will be interrupted (Greenstein and Arnaut, 1987).

Because the display screen used in combination with a light pen is the same for data input and output, advantageously a direct information relationship is provided. The light pen does not require additional desk space and allows natural pointing and drawing movements for data input.

There are, however, some disadvantages with regards to the light pen; one being that the light pen must be held up against the screen, which will become tiresome over long periods of usage; also, the use of a light pen may obscure parts of the display from the user. The light pen lacks in resolution capabilities and may be activated inadvertently. Light pens are highly dependent on the hardware and software within which they can be used.

21.3.2.8 Special Devices

In addition to the manual computer input devices discussed earlier, there are many alternative devices and techniques available. In this case, it often deals with special types of the described input devices, which would be too much to cover here.

For use in *virtual environments* (VE), several computer input devices have been created. These input devices, such as spaceball, flying joystick, etc., are often further developments of conventional devices, in which extended forms of interaction within three-dimensional virtual environments have been taken into consideration. Other developments, such as the data glove, utilize user gestures as a basis for data input. For further informations on VE-interaction techniques and devices, see Chapters 22 and 52.

21.3.2.9 Overview

An overview of the described manual computer input devices is given in Table 21.1. This table summarizes some main features, like uses and recommendations.

21.4 ARRANGEMENT OF CONTROLS

The arrangement of controls is governed by human capabilities and traits such as anatomy, anthropometry, and physiology, on the one hand, and by the characteristics of the technical system to be manipulated, on the other hand. The most important criteria to be taken into account concerning the arrangement of controls are the following:

Movement range of the HAS or FLS

Movement-physiological marginal conditions (requirements made on actuating accuracy, speed, force, torque)

Coupling conditions

Frequency and importance of information input

Possibilities of visual feedback

Sequence of the process to be controlled (sequence of activities)

Spatial compatibility regarding the technical system or the displays

Safeguards against inadvertent operation

Operation while sitting or standing

As concerning the first two aspects, it must be considered whether the controls will have to be actuated by women or men, or by both, and what percentile range will be relevant to the collective (grasping reach capability of the HAS and FLS for the different percentiles of men and women). For the transmission of greater forces and torques, for example, by means of handwheels, provision will have to be made for operation while standing, and for informative control tasks, provision will have to be made for operation while sitting. This largely eliminates the need for static posture energy. Depending on the type of control used, its preferred vertical position will be between the elbow level and the shoulder level. Rotational controls actuated by both hands should be positioned in the median plane; one-hand actuated controls should be positioned in the sagittal plane (shoulder joint). Controls that must be actuated very accurately should be arranged in the grasp-

Table 21.1 Manual Computer Input Devices (Based on Brown, 1988)

Device	Uses	Recommended For	Not Recommended For
Alphanumeric Keyboard	Entering of text and numbers; Select	General purpose entry device	Selection by typing slower than by pointing
Numeric Keyboard	Entering of numbers	Fast entry of massed numbers; Calculations	Infrequent numbers; Mixed text and numbers
Cursor Control Keys	Discrete cursor movement	Tasks requiring short cursor movements	Extensive or fine cursor movements
Mouse	Point; Drag; Move cursor	Tasks requiring little keyboard use	Frequent mouse to keyboard changes
Joystick	Move cursor; Track; Select	Task with intensive cursor positioning	Frequent changes to and from keyboard
Trackball	Track: Select; Move cursor	Integrating graphics with keyboard entries	Frequent changes to and from keyboard
Touch Screen	Select	Infrequent use; Coarse pointing	Continuous use; Precise pointing
Graphic Tablet	Draw; Trace; Move cursor	Drafting or hardcopy data entry	Precise positioning (touch tablets)
Light Pen	Move cursor; Select; Draw	Infrequent use; Tasks with little keyboard use	Continuous use; Frequent changes to and from keyboard

ing reach of the forearms. An armrest will increase the accuracy of control movements and avoids user's fatigue and undue muscle strain.

In the design of control panels and control rooms containing a large number of controls and displays, further criteria will have to be taken into account to minimize unnecessary mental processes and to avoid time-consuming successive movements. When arranging several controls, the interaction that exists between these controls (functions) will have to be considered in addition. A distinction can be made between two principles. In a grouping according to the sequence of the process to be controlled, the controls will have to be arranged in the order of their operation. The arrangement in this *sequential grouping* should be made for an operating sequence from left to right or from top to bottom. In a *functional grouping*, the controls with the same functions will have to be clustered, with the spatial compatibility having to be considered additionally. Further improvements regarding the operating effectiveness can be expected if the operating frequency is used as an arrangement criterion. Thus, frequently used controls should be arranged in the central movement or actuating range and infrequently operated controls in the peripheral range. Vital controls, such as emergency stop switches, must also be arranged in the central area.

To permit a perceptive and cognitive organization of the visual field, controls should be combined to make up a matrix-like control field (Neumann and Timpe, 1976). The grouping effect can be achieved by suitable spacings and/or by color marking and boundary lines. If two controls have to be operated simultaneously, the controls will have to be grouped in such a way that operation with different extremities (right hand–left hand, foot–hand) is possible. The spatial correlation between controls and displays should also be ensured. The direct spatial allocation of displays and controls is optimal. The control is located directly below or adjacent to the associated display. If this allocation cannot be implemented for reasons of accessibility of the controls, it is recommended that a separate display and control field be used. The spatial organization of the controls within the grasping reach of the operator must be identical to the spatial organization of the displays within the visual field. For the design of control panels, vehicle cabins, aircraft cockpits, and so on, see specialized literature, such as Schmidtke (1991).

To permit error-free operation of the controls without inadvertently actuating any neighboring controls, certain distances between the controls must be observed, depending on the type of control used. In Table 21.2, the minimum and, at the same time, optimum distances between two neighboring controls are listed for the most important controls (Grandjean, 1991). If the operator wears gloves, provision will have to be made for a corresponding allowance. If the rear side has to be used as coupling area in one-hand- and two-hand-operated handwheels, a freespace between 20 and 35 mm will have to be provided between the rear side of the control and the technical system, depending on the control element involved.

Integrated controls can be of advantage for different reasons, such as for reducing the number of controls, of simultaneous actuation of different functions. For simultaneous actuation of two functions (continuous and discrete), a lever will be suitable, for example, that is operated by grasp, and that has an integrated rocker switch or push button to be operated with the thumb. If different control sensitivities are required for coarse and fine adjustment, knobs of different diameters can be arranged on the same axis of rotation. However, provision should not be made for more than three concentric knobs. For the dimensioning of concentric knobs, see McCormick (1976). To satisfactorily perform the

Table 21.2 Space between Adjacent Controls (*Source:* Grandjean, 1991)

Control	Extremity	Distance between Controls (mm) Minimum	Optimum
Push button	Finger	20	50
Toggle switch	Finger	25	50
Lever	Hand	50	100
Lever	Both hands	74	125
Handwheel	Finger	20	50
Knob and rotary selector switch	Hand	25	50
Pedal	One foot	50	100

control tasks "quick turning" over a wide actuating range and "precise actuation," a handwheel will be suitable that is provided with a crank handle, which may be collapsible.

It is also possible to combine controls with scales. For control tasks with several discrete positions, the control often assumes the shape of a hand (rotary selector switch). The positions are coded on the panel with digits, with a reasonable contrast being required between panel, hand (control), and digits. If the number of control positions amounts to twelve or less, the coding should correspond to that of a watch. This permits the operator to perceive the control position quickly, especially if there are several controls. Another variant is the use of a disk with a scale fixed to the rear side of the control and a marking on the panel. The dimensions must be so defined that the view of the scale or display cannot be obstructed by the fingers during operation.

21.5 CONTROL DYNAMICS

There are many technical systems in which a direct linkage of control and system is not possible, because the actuating forces required for controlling the system do not lie within human capabilities. Therefore, the system is often controlled indirectly by means of suitable servo-amplifier devices. The decoupling of the control and the system permits the displacement-resistance characteristic to be optimized and it thus permits an improved utilization of *proprioceptive* and *haptic* information. Proprioceptive perception is made through the sensors of the extrafusal muscles, the tendons, and the joints. By an existing actuation resistance, the displacement feedback is supplemented by an additional force feedback. Resistance is also necessary to reduce control errors resulting from tremoring of the hands and mechanical vibrations. The amount of resistance can be derived from the measurable status variables of the system or from simulated data of the system model. Owing to the immense technical complexity of such active controls, the generation of a control "feel" can often be realized only by displacement, rate, and acceleration-proportional actuating forces.

A nonlinear transfer characteristic is also possible if, for example, static and sliding friction or backlash exists in the control. Generally speaking, the transfer characteristics of the system will have to be taken into account for achieving an optimum adaptation of the control element dynamics.

21.5.1 Linear Mechanical Transfer Characteristics

A high degree of *inertia* of the controls causes a great time constant and impairs quick directional changes and accurate control movements. Owing to the existing inertia of the arm, Poulton (1974) does not think it necessary to apply additional inertial forces in the operation of joysticks. In connection with other dynamic resistances, particularly with a spring-centered control, the inertia also impairs the manual tracking performance. In rotationally actuated controls, however, inertia may improve the continuity of the control process at the expense of a reduced adjusting speed.

Rate-proportional resistance (viscous damping) generally increases the control performance by smoothing the actuating movement and, like all dynamic resistances, reduces human disturbances (tremor, movements of the hand) and environmental interference (mechanical vibrations, gravitational forces). Viscous damping has proved to be most effective in connection with elastic force, and most unfavorable in combination with inertia. A favorable effect of viscous damping is paticularly achieved in connection with controlled systems of zero order (Rühmann, 1978). In a two-dimensional step tracking (position control system), Kraiss (1970) has been able to give evidence that the damping of a joystick, in contrast to elastic forces, has an insignificant influence on the fine movement structure. In controlled systems of higher order, a control damped by viscous means will result in very poor control performance.

Elastic resistance (spring loading) supports the positional proprioception, supplies information regarding the zero point, and thus is of special advantage to the optimum utilization of the secondary control loop of the human motor system. The zero point information is of great importance to rate control systems and acceleration control systems. For reasons of safety, spring load has further advantages, because the control automatically returns to its zero point as soon as it is released. To achieve a favorable control action, the spring rate should not be dimensioned too low. Very small restoring forces result in overriding, whereas very great restoring forces lead to underriding. By contrast, a certain amount of initial stress in the control in its zero position causes a reduction in the control performance (Bahrick, Fitts, and Schneider, 1955). A combination of elastic resistance and heavy inertia will also have an adverse effect on the control action. To reduce overriding, 9 to 20 N is specified as the minimum value for restoring resistance in joysticks and hand levers, at maximum displacement. 130 N is recommended as the upper value that should not be exceeded, so that excessive muscular tension of the operator and underriding effects will be avoided. In pedals, the actuation time increases with rising restoring force. When the design of the restoring force remains within reasonable limits and the accuracy of maintaining a specific pedal position is the evaluation criterion instead of actual time, then there are no influences on the performance.

Isotonic (free of restoring forces) and *isometric* (free of displacement) controls are regarded as special cases. The controlled variable $y(t)$ is proportional to the displacement on the one hand, and to the force on the other hand. The advantages of the isometric control are the shorter deceleration times of the neuromuscular system, because no movements have to be made, the low degree of cross-talk between the operating directions, and the reduced fault susceptibility to mechanical vibrations. Nevertheless the advantage becomes evident only in controlled systems of higher order in the high-frequency range. Preference should be given to isotonic and spring-loaded elements in position control systems and low-frequency systems. The absence of the zero-point information in an isotonic control is the main influencing factor for the loss of performance in the control of systems of higher order.

21.5.2 Nonlinear Mechanical Transfer Characteristics

Nonlinear transfer characteristics are due to *static* and *sliding friction*, *mechanical backlash*, and *nonlinear spring characteristics*. Friction is composed of a static part and a sliding part (Coulomb's friction). There is no interrelation between Coulomb's friction and the kinematic parameters of the control; that is, the proprioceptors do not receive any information. Static friction has an especially adverse effect on the control performance at the reversal points of the actuating movements, owing to the large static part. Abrupt movements are induced on account of the high resistance that is temporarily effective. If the scope of movements is very small, precise actuation is thus not possible. This effect is particularly applicable in the case of low moments of inertia and great control gain. Adequate static friction has the advantage that the hand or foot can rest on the control without actuating it. The results obtained in various experimental tests with respect to the influence of Coulomb's friction on the tracking performance or adjusting time are not uniform. For knobs, it is true that the coarse adjusting time considerably increases with growing friction, whereas the fine adjusting time does not depend on the existing friction. Consequently, friction in particular has a negative effect when speed is of greater importance in a control task and accuracy is of less importance.

Backlash at any rate impairs the control accuracy. The control error is approximately linear to the existing backlash and is even magnified by additional system friction. If backlash exists, the negative effect is intensified with the increasing magnification factor of the controlled system. Among others, Gibbs (1962) has provided evidence of this in adjusting tests made with a joystick. This negative influence is retained in varying the control-display ratio (C/D ratio), the order of the technical system, and the actuating organ (thumb, hand, forearm). In all cases, the highest control performance is obtained for the backlash-free joystick. The above findings are applicable to all types of controls; for knobs, the influence of backlash is smaller, however. The higher the requirements on control accuracy get, the worse the adverse effects of backlash are. Besides nonlinearities regarding the restoring force, such as backlash and friction, nonlinearities may occur regarding the controlled variable (dead zone or dead space). A dead zone may be advantageous in systems of higher order (Rühmann, 1993). The dead zone of a control may extend over the total range of movement (on/off controls).

For *nonlinear degressive spring characteristics*, more favorable values are obtained only for the fine adjustment time. For a system of zero order and first order, the coarse adjustment time remains invariant between linear and degressive spring characteristic (Rothbauer, 1978). A degressive spring characteristic supplies better zero-point information to the operator, and this is important in rate and acceleration control systems. In discrete controls such as keyboards and push buttons, the resistance must increase and finally drop sharply, to indicate that actuation has been performed. This proprioceptive feedback is frequently supplemented by an acoustic feedback.

21.5.3 Control Sensitivity

On account of their adaption capability, humans can largely adapt their actions to the control sensitivity (gain). The functional relation between tracking error and gain represents an U-shaped pattern with a wide area of nearly constant control quality. An increase in the tracking error occurs only with a relatively high or low gain. The selection of gain depends primarily on whether emphasis is laid on criteria of speed or criteria of accuracy. The system dynamics must be known for the adaption of the control sensitivity. To be able to compensate large deviations, a higher gain will be necessary in systems of higher order. The amount of gain is determined by the stability of the total system. In nonlinear gains—that is, a small displacement of the control will receive little gain and a high displacement of the control will receive large gain—improvements become evident regarding the control performance and the operating activity. Results show that use of such nonlinear control gain can improve tracking performance by 10% and reduce the operators input frequency by 25% (Krüger, 1982).

21.5.4 Active Controls

Active controls supply information regarding the system status via the control to the operator. The increase in the restoring moment in an automobile's steering, depending on the road speed, for example, constitutes an active control. With the feedback of suitable status variables and the control element dynamics deduced from this, a higher degree of performance and reduced stress and strain can be achieved in comparison with passive controls. Active controls are only advantageous in high-frequency, complex systems (Bolte, 1991).

The functioning of active controls is based on the fact that the operator transfers a force to the control, whereas the displacement of the control is controlled by the technical system. An active control can also be realized by force zero point shift (Röger, 1978). The displacement or force zero point shift of the control is affected by an actuator and is proportional to the tracking error. This principle induces the operator to take control action in a quite specific direction. In addition to the transfer of quantitative information, an active control can also be used to transfer qualitative information, for example, mechanical vibrations as warning information.

21.6 CONTROL CODING

The objective of coding controls is to guarantee their quick and safe identification. The coding is based on the capability of visual and tactile discrimination of controls with different control functions. Coding can be realized by *color, shape, surface structure, dimension, position, text,* and *symbols*. For reasons of esthesiophysiology, color coding is not suitable as a primary type of coding, because the perception of colors depends heavily on the lighting conditions at a workplace and because disturbances of the color sense of humans are relatively frequent. By using a coding combination in which, for example, symbol coding and color coding are superimposed, the disadvantages of one type of coding can mostly be compensated. Colors have an inherent element of meaning that must be taken into account in the coding (e.g., red—danger; yellow—caution, attention). Shape and surface coding of the controls permits easy visual and tactile identification (for shape and surface coding, see Damon, Stoudt, and McFarland, 1966; and Woodson, Tillman and Tillman, 1992). Tactile information is retained, even without visual contact and under favorable lighting conditions.

If great transfer forces are involved, the shape must satisfy the anatomical and anthropometrical marginal conditions, despite the coding, so as not to cause overstrain of the hand and fingers. For these reasons, very complex shapes—for example, to give the control a symbolic meaning—can be used only with very low resistances. In size coding, the controls used for the same control tasks are given identical dimensions. Here, a maximum of three sizes should be used, with a minimum difference of 20% from one size to the next. When gloves have to be worn, the tactile perception is impaired in shape and surface coding. In identical technical systems, an equivalent position coding is required to avoid transfer effects when the operator changes from one system to another. Coding by text and symbols is based on the functional identification of controls and is effective only under good lighting and visual checking conditions. Brief signs and common abbreviations, which are easily comprehensible without any learning, are suitable for this kind of coding. Only such symbols are permissible that, if partly covered, do not allow another meaning to be implied. Alphanumeric characters have a high degree of unambiguity, but they are linked with the language involved and call for a comparative amount of space. The unique allocation of the coding to the control must be ensured; it is recommended that the markings appear above the control. Text and symbol coding results in a wide variety of coding possibilities. For the optimum size and line thickness of alphanumeric characters relative to the visual distance, see ISO 9241-12 (1994). The minimum symbol size for the normal-vision operator is a function of contrast, surrounding brightness, distance of observation, the geometric structure of the symbols, and occasionally matrix display resolution.

21.7 COMPATIBILITY OF CONTROL-DISPLAY-SYSTEM

To ensure high effectiveness and safety of the system and a reduction of the response time and learning phase, a high degree of *compatibility* is required between control, technical system, and display. Compatibility exists if the direction of movement of the control coincides with the direction of movement of the technical system or the observable system variables (stimulus-response compatibility). Thus, on clockwise turning of the steering wheel of an automobile, a change of direction to the right is expected. The objective of compatible system design is to take into account the generally existing or acquired stereotypes of humans. If movement and perception stereotypes are taken into account—which under certain circumstances may vary in the population, as, for example, in the case of left-handers—decoding steps and thus mental processing work will not be necessary for the operator. For the direction of movement of the control and the function or response of the technical system, recommendations are given in Table 21.3. The operation of a valve is an exception to these movement-effect stereotypes. To open the valve, it must be turned anticlockwise; to close it, it must be turned clockwise. The specifications contained in Table 21.3 may differ in national standards and regulations.

In complex technical systems, the system action can often be manipulated only through displays. Between control and display and between display and system action, design principles will also have to be observed, so that the required compatibility will be guaranteed. The movement of the controls and that of the associated display element should be identical in their directions. For example, if a control is turned to the right, the hand on a dial must move clockwise; on a horizontal scale, it must move to the right, and on a vertical scale, it must move upward. A rising value should bring about a displacement of the hand in the clockwise direction on a dial, a displacement to the right on a horizontal scale, and a displacement upward on a vertical scale. In the case of a combination of a rotational control and a vertical scale, it is also important whether the rotational control is arranged to the right or the left of the scale. If operated in clockwise direction and with the control arranged on the right side, the hand or pointer will move upward and, if arranged on the left side it will move downward. Displays with fixed pointer and movable scale, in connection with controls, inevitably lead to incompatibilities. Therefore, such combinations should not be used. Further design principles must be observed if the display and controls are arranged on different levels. In this connection, see, for example, Schmidtke and Rühmann (1993). Besides the stimulus-response compatibility, the position compatibility is also important. Position compatibility exists if a meaningful spatial allocation has been established between controls, displays, and the associated sensors and effectors (see Section 21.4).

Control-display ratio (C/D ratio) as a further important design variable is the ratio between the deflection of the display (pointer) and the displacement of the control. If pointer and control element are performing a translational movement, the following will apply to the C/D ratio: $R = C/D$, where C = displacement of control part (mm), and D = deflection of display marker (mm). With an optimum determination of the C/D ratio,

Table 21.3 Recommended Control Movements

Function	Control Action
On	Up, right, forward, pull (switch knobs)
Off	Down, left, rearward, push (switch knobs)
Right	Clockwise, right
Left	Counterclockwise, left
Up	Up, rearward
Down	Down, rearward
Retract	Rearward, pull, counterclockwise, up
Extend	Foreward, push, clockwise, down
Increase	Right, up, forward
Decrease	Left, down, rearward

it will be possible to reduce the adjusting time drastically. Generally speaking, a very large C/D ratio will be required for exacting requirements of accuracy, and a small C/D ratio for quick actuating activities over a large area. In fine adjustment, the actuating time will exponentially rise with decreasing C/D ratio. The large display deflection in less control movement will cause multiple overshooting in the target area, resulting in an increase in the adjusting time. Because adjusting tasks mostly constitute coarse and fine adjustment, a reasonable compromise will have to be found. A C/D ratio 0.1 to 0.4 is stated for rotationally actuated controls, and of 2.5 to 4 for levers (Chapanis and Kinkade, 1972). Although the diameter of a control does not have any influence, display size and time delays of the technical system (system dynamics) and tolerance band of the reference value are important influencing factors that have to be taken into account in determining the C/D ratio.

21.8 SAFETY REQUIREMENTS

Spurious actuation of a control may be caused by the operator, by unauthorized persons, or by surrounding influences such as mechanical vibrations or the falling down of objects. Inadvertent actuation of a control by the operator may be caused by slipping off a control and actuating a neighboring control, because of inadequate distance between one control and the next, getting one's clothing caught, and/or supporting oneself by holding onto a control. Inadvertent actuation of a control may also result from wrong operation by the operator, for example, due to unfavorable coding. The design principles discussed above should therefore also be seen in the light of safety aspects.

The marginal conditions of safety must be taken into consideration as early as in the definition of the design parameters (Section 21.3). Shape and surface of the control must be designed so that slipping off a control is prevented in order to avoid injury to the worker and spurious actuation. A push button may, for example, be provided with a concave top or with extra-fine profiles. If controls have to be mounted on rotating shafts, provision will have to be made for clutches, so that the transmission of power can be interrupted. For actuation, it will first be necessary to apply an axial force. If this can not be realized, controls will have to be used, in which getting clothing caught is not possible; that is, disk handwheels with a closed stay must be used instead of cranks and spoked handwheels. In the case of important controls, in which an inadvertent actuation would endanger persons and the system, provision will have to be made for additional safeguards. Such undesirable actuation can be prevented or minimized by the following design measures, which are based on various cause-and-effect principles such as the interruption or complication of the flux of force.

- Covering of the control
- Recessing or shielding of the control
- Controls must not be arranged within the supporting and main movement area of arms and legs
- Direction of control movement is not identical to the direction of movement of the body or its extremities
- Provision must be made for large tripping forces
- Making provision for frictionally actuated instead of positively actuated controls (e.g., finger slide instead of push button)
- Detachable controls (controls must not be interchangeable)
- Two-dimensional control (one degree of freedom for unlocking), lockable control, and multifunctional control (one control for unlocking or for simultaneous operation, two-hand control, dead man's control)

Moreover, *keys* are to be regarded as special forms of controls that can be used for rotational actuating movements for two or several steps. By the removal of the key, a high degree of safety is guaranteed against unauthorized and inadvertent operation. The design measures mentioned above will have to be supplemented by notices in the form of the various coding possibilities. It is obvious that safety measures in part do not satisfy the requirements of ergonomics. Covers and recessed controls, for example, prevent quick actuation of controls, large tripping forces place greater stresses and strains on the operator, and the movements of controls in directions other than the preferred anatomical directions do not permit any favorable transfer of forces. Thus, it may be necessary to compromise, with due consideration to the priorities. Neudörfer (1982) has compiled a systematic catalog for avoiding the undesirable actuation of controls, including examples of application.

REFERENCES

Bahrick, H. P., Fitts, P. M., and Schneider, R. (1955). Reproduction of simple movements as a function of factors influencing proprioceptive feedback. *Journal of Experimental Psychology, 49*, 445–454.

Barry, J. (1993). A Review of Ergonomic Keyboards. *Work, 3 (4)*, 21–25.

Bolte, U. (1991). *Das aktive Stellteil—ein ergonomisches Bedienkonzept*, Fortschritt-Berichte, Reihe 17, No. 75. Düsseldorf: VDI-Verlag.

Brown, C. M. (1988). *Human-Computer Interface Design Guidelines.* Norwood, NJ: Ablex.

Bullinger, H.-J., and Solf, J. J. (1979). *Ergonomische Arbeitsmittelgestaltung I-III.* Bremerhaven: Wirtschaftsverlag NW.

Bullinger, H.-J., Kern, P., and Solf, J. J. (1979). *Reibung zwischen Hand und Griff.* Bremerhaven: Wirtschaftsverlag NW.

Cakir, A. (1995). Acceptance of the adjustable keyboard. *Ergonomics, 38 (9)*, 1728–1744.

Caldwell, L. S. (1959). The effect of the spatial position of a control on the strength of six linear hand movements. Report No. 411. U. S. Army Medical Research Laboratory, Fort Knox.

Chapanis, A., and Kinkade, R. G. (1972). Design of controls. In H. P. van Cott and R. G. Kinkade, Eds., *Human Engineering Guide to Equipment Design* (Revised Edition). Washington, DC.

Damon, A., Stoudt, H. W., and McFarland, R. A. (1966). *The Human Body in Equipment Design.* Cambridge: Harvard University Press.

Fitts, P. M. (1964). The information capacity of the human motor system in controlling the amplitude of movement. *Journal of Experimental Psychology, 47*, 381–391.

Foley, J. D., van Dam, A., Feiner, S. K., and Hughes J. F. (1990). *Computer Graphics: Principles and Practices.* Reading, MA: Addison-Wesley.

Gibbs, C. B. (1962). Controller Design: Interactions of controlling limbs, time-lags and gains in positional and velocity systems. *Ergonomics, 5*, 385–402.

Grandjean, E. (1991). *Physiologische Arbeitsgestaltung*, Fourth Edition. Landsberg: Ecomed.

Greenstein, J. S., and Arnaut, L. Y. (1987). Human Factors Aspects of Manual Computer Input Devices. In G. Salvendy, Ed., *Handbook of Human Factors and Ergonomics*, First Edition. New York: John Wiley.

Ilg, R. (1987). Ergonomic keyboard design. *Behaviour and Information Technology, 6 (3)*, 303–309.

ISO 9241-4 (1994). *Ergonomic requirements for office work with visual display terminals.* Keyboard requirements. International Standards Organization.

ISO 9241-12 (1994). *Ergonomic requirements for office work with visual display terminals.* Presentation of information. International Standards Organization.

Kern, P., Muntzinger, W. F., and Solf, J. J. (1984). *Entwicklung von normungsfähigen, ergonomisch richtig gestalteten Bedienteilen.* BMFT-HdA.

Kraiss, K.-F. (1970). *Beitrag zur Optimierung des Steuerkraftverlaufs von Bedienelementen*, Report No. 4. Forschungsinstitut für Anthropotechnik, Wachtberg-Werthoven.

Krüger, W. (1982). *Untersuchung der nichtlinearen Bediensignalverstärkung bei einer kontinuierlichen Trackingaufgabe.* Report No. 54. Forschungsinstitut für Anthropotechnik, Wachtberg-Werthoven.

Läubli, T., Fleischer, A. G., Krueger, H. (1989). *Gestaltung von Bildschirmarbeit*, Arbeitswissenschaftliche Erkenntnisse No. 2/79. Bundesanstalt für Arbeitsschutz, Dortmund.

Mainzer, J. (1982). *Ermittlung und Normung von Körperkräften—dargestellt am Beispiel der statischen Betätigung von Handrädern*, Fortschritt-Berichte, Reihe 17, No. 12 (VDI-Verlag, Düsseldorf).

McCormick, E. J. (1976). *Human Factors in Engineering and Design.* New York: McGraw-Hill.

Mims, F. M. (1984). A few quick pointers. *Computers and Electronics*, May, 64–117.

Muntzinger, W. F. (1986). *Ergonomische Gestaltung von Rotationsstellteilen für grob- und sensomotorische Tätigkeiten*, Dissertation. Berlin: Springer.

Neudörfer, A. (1982). Systematischer Katalog zum Vermeiden unerwünschter Betätigungen von Bedienteilen, *Werkstatt und Betrieb, 115 (12)*, 225–236.

Neumann, J., and Timpe, K. P. (1976). *Psychologische Arbeitsgestaltung.* Berlin: Deutscher Verlag der Wissenschaften.

Norman, D. A., and Fisher, D. (1982). Why alphabetic keyboards are not easy to use: keyboard layout doesn't much matter. *Human Factors, 24*, 509–519.

Ohlson, M. (1978). System design considerations for graphics input devices. *Computer, 11*, 9–18.

Plaisant, C., and Sears, A. (1993). Touchscreen interfaces for alphanumeric data entry. In B. Shneiderman, Ed., *Sparks of Innovation in Human-Computer-Interaction.* Norwood, NJ: Ablex.

Poulton, E. C. (1974). *Tracking Skill and Manual Control*. New York: Academic Press.

Price, L. A., and Cordova, C. A. (1983). Use of mouse buttons. In *Proceedings of the CHI '83 Conference on Human Factors in Computing Systems*, pp. 262–266. New York: ACM.

Röger, W. (1978). *Das Bedienelement als Informationsträger bei Bahnführungsaufgaben*, Dissertation, TU Braunschweig.

Rohmert, W. (1966). *Maximalkräfte von Männern im Bewegungsraum der Arme und Beine*, Forschungsbericht NRW No. 1616. Köln: Westdeutscher Verlag.

Rohmert, W., and Haider, E. (1981). *Leitregeln und Handbuch zur ergonomischen Gestaltung von Tastaturen*. Institut für Arbeitswissenschaft, TH Darmstadt.

Rothbauer, G. (1978). *Zum Problem des Bewegungswiderstandes bei einfachen und komplexen Stellbewegungen des Armes*, Report No. 40. Forschungsinstitut für Anthropotechnik, Wachtberg-Werthoven.

Rühmann, H. (1993). Schnittstellen im Mensch-Maschine-System. In H. Schmidtke, Ed., *Ergonomie*, (Third Edition). München: Hanser.

Rühmann, H. (1978). *Untersuchung über den Einfluß der mechanischen Eigenschaften von Bedienelementen auf die Steuerleistung des Menschen bei stochastischen Rollschwingungen*, Forschungsbericht BMVg-FBWT 78-11.

Schmidtke, H., Ed. (1991). *Handbuch der Ergonomie*. München: Hanser.

Schmidtke, H., and Rühmann, H. (1993). Betriebsmittelgestaltung. In H. Schmidtke, Ed: *Ergonomie* (Third Edition) München: Hanser.

Schmidtke, H., and Stier, F. (1960). *Der Aufbau komplexer Bewegungsabläufe aus Elementarbewegungen*, Forschungsbericht NRW No. 822. Köln: Westdeutscher Verlag.

Schulze, L. J. H., and Snyder, H. L. (1983). *A comparative evaluation of five touch entry devices.* Technical report No. HFL-83-6. Blacksburg, VA: Virginia Polytechnic Institute and State University.

Scott, J. E. (1982). *Introduction To Interactive Computer Graphics*. New York: John Wiley.

Seibt, F. (1971). *Steuerleistung in Abhängigkeit vom Übersetzungsverhältnis und von Coulombscher Reibung im Bedienelement*, Dissertation, TU München.

Shneiderman, B. (1987). *Designing the User Interface: Strategies for Effective Human-Computer Interaction.* Reading, MA: Addison-Wesley.

Taylor, R. M., and Berman, J. V. F. (1982). Ergonomic aspects of aircraft keyboard design: The effects of gloves and sensory feedback on keying performance. *Ergonomics*, *25* (*11*), 1109–1123.

Thomson, D. A. (1994). Analysis of the effect of keyboard tactile feedback on typing force. In F. Aghazadeh, Ed., *Advances in Industrial Ergonomics and Safety VI*. London: Taylor and Francis.

Woodson, W. E., Tillman, B., and Tillman, P. (1992). *Human Factors Design Handbook,* Second Edition. New York: McGraw-Hill.

CHAPTER 22

NONCONVENTIONAL CONTROLS

Grant R. McMillan
Robert G. Eggleston
***Fitts* Human Engineering Division**
Armstrong Laboratory
Wright-Patterson Air Force Base, OH 45433-7022 USA

Timothy R. Anderson
Biodynamics and Biocommunications Division
Armstrong Laboratory
Wright-Patterson Air Force Base, OH 45433-7901 USA

22.1 INTRODUCTION

In the lead chapter of the first edition of the *Handbook of Human Factors* (1987), Julien Christensen considered the origin and development of Human Factors from the perspective of the evolutionary development of human-produced artifacts used to improve the quality of life. Christensen noted that we have moved through the Age of Tools, Machines and Power and advanced to the New Age where we are designing "Machines for Minds." We now term this the Information Age, and it has begun to present formidable challenges for the design of user interfaces that allow us to harness, rather than be buried by, the full power of information. The typical workplace requires each individual to manage a more diverse and complex set of activities, many heavily dependent on the processing, creation, and dissemination of information. More routine tasks continue to be delegated to machines under the supervisory control of a human. We are just beginning to recognize that prevailing notions about the user interface in general, and controls in particular, may not be adequate for the design of the next generation of systems in the Information Age. Now more than ever we need to explore alternative or nonconventional controls and user interface paradigms. It is largely for these reasons that this chapter has been added to the second edition of the Handbook.

In this chapter, we present a glimpse into some of the nonconventional controls currently receiving attention in research laboratories. A broad range of systems from voice to gesture to brain actuated controls are reviewed. Since these technologies are all relatively immature, our goal is to provide a review of them at a functional level, identifying requirements and critical design issues where possible. Less attention is given to the current implementation of these control systems, since such factors are expected to undergo substantial change as the technologies mature into commercially usable forms. We believe this strategy of concentrating on fundamental principles and functional aspects of design will allow the material in this chapter to continue to provide valuable design guidance, at least until the next revision of the Handbook.

22.1.1 Definition of Nonconventional Controls

Simplistically, nonconventional control may be regarded as including any control concept that has not been commercialized or gained acceptance in the marketplace. From this perspective, we can anticipate that some of today's nonconventional controls will be considered conventional tomorrow, and others will simply die in the laboratory. An alternative view is to define all controls that have been tailored to meet the unique needs of individual citizens as nonconventional. Thus, sip-puff tubes, chin-operated joysticks and other specialized control devices qualify as examples, many of which are in use today to mitigate individual disabilities. We prefer another type of definition that centers around properties or characteristics of a control that set it apart from currently used technology. For the purposes of this chapter, *nonconventional controls are defined as controllers that do not require a direct mechanical linkage between the user and the input device.* With nonconventional controls the steps from intention to action will typically be shortened and more direct. Indeed, with some control devices users may not even think of their actions as control inputs, but rather regard them merely as aspects of conversation with a system.

22.1.2 Paradigms for the Design and Use of Nonconventional Controls

Nonconventional controls may be candidates for insertion into a broad spectrum of systems. To appreciate the full range of design issues that may have to be addressed by developers of these devices, it is necessary to have some understanding of the interface context and interface design paradigm or concept used to structure and guide a development. Typically the overriding design paradigm is implicit and thus taken for granted. For certain types of control insertion this does not create a problem. For others, the view of what defines a controller is itself nonconventional, and in these instances an understanding of the new interface paradigm is essential to controller design. In these situations, the potential of the controller cannot be achieved, or may not even be recognized, without formulation and understanding of the interface paradigm. This means that the control designer's task is both larger and in some ways different from the past. This should become clear after reading the immediately following sections and the closing section (22.3) of this chapter.

22.1.2.1 Simple Control Substitution

An obvious path of new control technology insertion is to replace a conventional control device with a nonconventional one. Commands to a system may be issued by voice input, for example, instead of pointing and clicking a mouse-driven cursor. Here, the goals for the control device remain unchanged from the original ones, only the mechanization of control is new.

22.1.2.2 Supplemental Controls

Another path for improving the control interface is to add a nonconventional control that will complement currently available ones. The nonconventional control may be an add-on that provides another input path for the system user. Alternatively, it may be integrated with a conventional control to increase its range of functionality. For example, a voice control device may be added to cursor control keys and a mouse as means for making menu selections, or voice and a mouse may be used together, with the mouse used to select a function or subsystem and voice used to issue the specific command. When nonconventional controls are designed as supplemental devices, more system integration issues must be considered than for simple control substitution.

22.1.2.3 Controls as Part of a "Transparent" Interface

The distinction between controls and displays is becoming increasingly artificial. The same device may have both a control and a display function, and they may have to work together synergistically to aid user performance. For example, a menu item serves as a *display* but once selected acts as a *control* for a system action. The control designer, therefore, must also engage in display design, and now is perhaps better regarded as an interface designer (Eggleston, 1988). If an aspect of the interface concept supports direct manipulation of "objects" or "resources" used by a performer to accomplish a task or job, then the interface may be taken to be, at least in some sense, transparent. Further, transparent interface design becomes equivalent to the design of a job aid. We are now beginning to stretch the concept of an interface beyond any reasonable limit; obviously a new term is needed to fully capture the design problem in the transparent interface paradigm. Even without further elaboration of this paradigm, there is little question that when control design becomes de facto job aid design, a broader range of issues must be addressed including a deeper understanding of cognitive factors involved in control.

22.1.2.4 Controls as Part of an "Intelligent" Interface

An intelligent interface is one that can interact with the user at a knowledge level. Here, control is often exercised through an abstract command that is interpreted in a context sensitive manner by the interface which issues lower level action commands (Eggleston, 1993). Control is at the "level of intentions," and control design becomes, in part, the design of a mixed-initiative dialogue that supports collaborative decision making and action taking. At this level, the interface has evolved to become a partner with human-like qualities (Eggleston, 1987, 1988). The risks of control-interface-partnership design are high, but the benefits for managing job complexity and empowering the user to achieve new levels of performance are equally high. Now the control designer must also deal with all aspects of interaction between two intelligent agents. Clearly a team of experts with diverse backgrounds is needed to design control interfaces at this level.

In these more complex paradigms the control designer must be integrated into the system development team, and not act as a vendor of a stand-alone product. This applies for conventional as well as nonconventional controls. The breadth and depth of control design issues can vary enormously depending on the type of technology insertion that is planned. Without commitment to a specific mode of insertion, adequate treatment cannot be given to identifying and addressing the scope of critical design issues. Accordingly, the following state-of-the-art reviews are necessarily incomplete in important details. Hopefully this introduction has sensitized the reader to some of the open issues and in this way provides directions for the analysis needed to execute a specific design effort.

22.2 STATE-OF-THE-ART REVIEW

To assist the reader in navigating through the control technology reviews, we have adopted a standard format. Each review begins with a brief description of the technology and a discussion of key terms and concepts. Next, the functional elements of each control technology are described using the following taxonomy:

1. ***Signal acquisition***—Elements that sense the user's input to the system, such as microphones, eye trackers and electrodes.
2. ***Signal processing***—Elements that extract meaningful patterns from the raw input, such as fast Fourier transforms, feature extraction routines and neural networks.
3. ***Control algorithms***—Elements that transform these patterns into an output capable of driving some external device. For some controllers this element is not easily separable from signal processing and the two descriptions are combined.
4. ***User feedback***—Elements that inform the user about their input, the system's interpretation thereof, and the control actions being taken. Different aspects of this feedback are more or less important for specific controllers. For example, with speech-based control, the user is typically well aware of their input. However, electroencephalographic inputs are not directly observable by the user. In both cases, the user needs feedback concerning the system's interpretation of and response to their input.

Following the technology description, applications of each control technology are reviewed. These applications are organized using the four control paradigms described above. Finally, key issues and directions for future research and development are discussed.

22.2.1 Speech-Based Control

Speech-based control is one of the most mature technologies discussed in the chapter. Although research has been ongoing for over 25 yr (Flanagan, 1972), applications have only recently become widespread and accepted by users. The slow acceptance resulted both from limitations in the technology and the very high expectations of users. Speech control must be highly accurate, robust, and reliable to meet user needs. It must be easy to use and transparent to the user; the system should adapt to the user, not force the user to adapt to the system.

22.2.1.1 Background

When discussing automatic speech recognition (ASR) systems, it is convenient to subdivide them into classes according to the problems they address. The first subdivision is based on the number of speakers they recognize. *Speaker-dependent* systems can recognize speech from only one speaker, the speaker that trained the system. *Speaker-independent* systems recognize speech from many speakers, not only the speakers that trained the system.

The next subdivision is based on the method used to handle word boundaries. *Isolated word* recognition systems require a 100–250 millisecond pause between spoken words. *Connected word* recognition systems require a very short pause between words. *Continuous speech* recognition systems require no pause between words and accept fluent speech.

An additional subdivision is based on the size of the vocabulary that the system can recognize. Vocabulary size is usually divided into *small* (less than 200 words), *large* (1000 to 5000 words), *very large* (5000 words or greater) and *unlimited* (greater than 64,000 words).

When defining a vocabulary for a specific task, a grammar may be developed that specifies which words may follow other words. This syntax, when incorporated into the recognition algorithm, reduces the total number of words that must be considered by the recognizer at any one time. This improves both the speed and accuracy of the recognizer. *Perplexity* is a common metric used to describe the complexity of a grammar. Perplexity is defined as the average branching factor of the grammar or, stated another way, the average number of words that can follow each word in the grammar. The larger the perplexity of a grammar the more difficult the recognition task.

Which combination of the above characteristics is best? The answer depends on the particular application and the characteristics of the user, task, and environment (Figure 22.1). Applications that can be accomplished with current technology are shown on the left of the dotted line, while applications to the right of the dotted line require more research and development before they are practical.

22.2.1.2 Technology Description

Automatic speech recognition can be viewed as a pattern recognition task that maps an input speech waveform to its corresponding text. Although a wide variety of specific components and processes have been used, all speech recognition systems consist of combinations of the following functional elements:

Signal Acquisition

The most commonly used microphones are the close-talking head-set microphone and the telephone hand-set, although other possibilities are lavaliere, desktop, and array. Each type of microphone presents unique challenges because of differing frequency characteristics, signal strengths, and the mode in which it is used. A desktop or array microphone, for example, allows the user to walk around the room resulting in changing signal strength as a function of the user's relative position with respect to the microphone. Such challenges are even greater for speaker-independent systems where different microphones may be used for the training and application environments.

Signal Processing and Control Algorithms

The first signal processing step in most systems is analog-to-digital conversion. The prominent peaks in the speech spectrum are in the range of 300 to 3500 Hz and sampling rates of 8000 Hz or higher are used. The speech power in specific frequency bands is

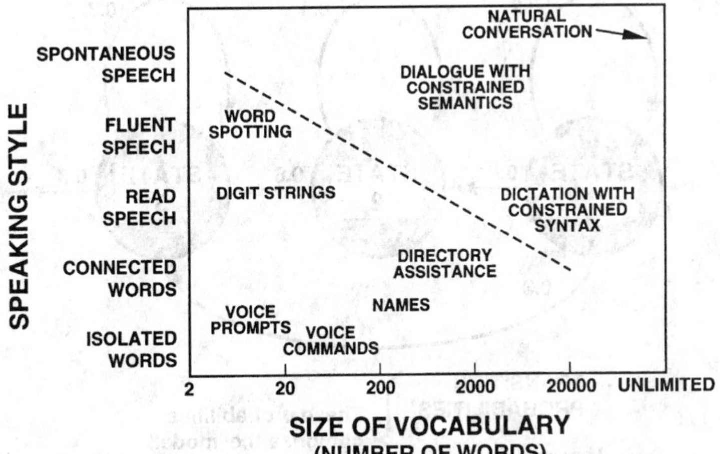

Figure 22.1 Different speech recognition tasks shown in a space of two dimensions: speaking style and size of vocabulary. Applications that can be accomplished with current technology are shown on the left of the dotted line, while applications to the right of the dotted line require more research and development before they are practical. (From Atal, 1994. Reprinted with permission from VOICE COMMUNICATIONS BETWEEN HUMANS AND MACHINES. Copyright © 1994 by the National Academy of Sciences. Courtesy of the National Academy Press, Washington, D.C.)

then estimated with fast Fourier transforms, digital band-pass filters, or auditory modeling techniques.

The analysis performed by the human cochlea takes place on a nonlinear frequency scale known as the mel scale (Parsons, 1987). This scale is linear to about 1000 Hz and is approximately logarithmic above that point. It is common to perform such a frequency warping for representations of speech. The most commonly used method of feature representation is mel-frequency cepstral coefficients or MFCCs (Davis and Mermelstein, 1980). MFCCs are generally computed every 10 milliseconds by first performing a spectral analysis using a fast Fourier transformation on a moving window of 20 milliseconds of speech. The spectrum is then shaped using the mel-frequency warping. The logarithm of this warped spectrum is taken, followed by an inverse Fourier transform. The result is called the mel-cepstrum. By keeping the first dozen or so coefficients of the cepstrum, the spectral envelope information is preserved. The resulting features are the MFCCs. This feature vector is used as the input to the pattern matching stage.

The most widely used algorithms for pattern matching are called Hidden Markov Models (HMMs). In these algorithms, a set of states is chosen for a set of phonetic or sub-word units. Three states, for example, could represent each phonetic unit (Rabiner, Levinson, and Sondhi, 1983). The states are connected left-to-right with recursive loops (Figure 22.2).

Recognition is based on a transition matrix, representing the probability of changing from one state to another, and on an output probability matrix, representing the probability that a particular set of features (e.g., MFCCs) will be observed at each state. These matrices are generated iteratively during a training process using speech from one or more speakers. The phonetic HMMs are then combined to form larger sets of states to represent words. The sets of states representing words can be combined to form the legal sentences for a particular application.

During pattern matching, each HMM model can be used to compute the probability of having generated the sequence of input spectra. This is done very effectively using the Viterbi algorithm (Forney, 1973) on the network of states used as the reference patterns. The result of the Viterbi algorithm is the total probability that the spectral sequence was generated by that series of HMMs using a specific state sequence. A different probability value results for every sequence of states.

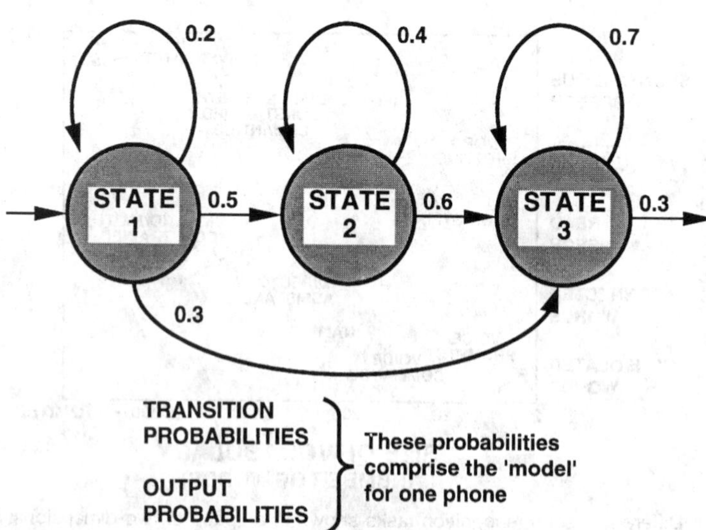

Figure 22.2 Basic structure of a three-state left-to-right phonetic Hidden Markov Model. (Adapted from Makhoul and Schwartz, 1994. Reprinted with permission from VOICE COMMUNICATIONS BETWEEN HUMANS AND MACHINES. Copyright © 1994 by the National Academy of Sciences. Courtesy of the National Academy Press, Washington, D.C.)

For recognition, the above computation is performed for all possible phoneme models and all possible state sequences. Approximate search algorithms have been used to reduce the search computation without a loss in performance. A commonly used technique known as beam search (Lowerre, 1976) is used to prune states that have low probabilities. The one sequence that results in the largest probability is declared to be the recognized phonemes/words/sentence.

User Feedback

The response provided to the user is usually application dependent. For applications over the telephone, the only viable means of feedback is a verbal response. For the control of computer menus and displays, the feedback is the selection of the appropriate button, icon, or menu item. In other applications, such as a voice-controlled wheelchair, the feedback is the action taken by the physical device. In dictation or data query tasks, text-based feedback is often provided.

Appropriate feedback is critical to the overall performance and user acceptance of the system. Feedback may enhance performance either by altering the user's speech or by allowing error correction. In the absence of feedback the user may incorrectly assume that a sequence of voice commands was executed by the system. An important human factors issue is how best to provide feedback that does not interfere with the user's primary task and maximizes system throughput.

22.2.1.3 Current Applications and Evaluations

The telecommunications market has the largest potential for speech-based control applications, with numerous examples already deployed. Many of these represent simple control substitution. American Telephone and Telegraph (AT&T) began investigating the use of limited-vocabulary, speaker-independent ASR in 1985. The goal was to reduce operator workload by automating a portion of the call handling. A seven-word vocabulary was used to automate the billing for collect, calling card, person-to-person, and bill-to-third-number calls. In 1990, AT&T developed "word spotting" technology which shifted the burden from constrained user speech to algorithms that can recognize key words spoken in an unconstrained fashion (Wilpon, Rabiner, Lee, and Goldman, 1990). Word spotting is the ability to spot key sounds in a phrase that contains both key words and non-key words. This is the first step toward understanding the essence of an utterance even when some words are not recognized. Recent field trials showed that 95% of the candidate calls were successfully automated (Wilpon, 1994).

Currently, AT&T, Sprint, and MCI are evaluating network-based voice dialing services. The user simply speaks the name of the party they wish to call. Users also speak a password or account number that is used to verify their identity before allowing the call to be placed. The AT&T system uses speaker-independent, sub-word-based ASR for the name-dialing. Other systems use speaker-dependent ASR for this function.

Speech-based control is a powerful enabling technology for people with disabilities. Here, ASR is being used as a substitute for traditional input devices. Examples include voice control of telephones, home appliances, powered hospital beds, and motorized wheelchairs (Amori, 1992; Miller, 1992). The set of voice commands needed to control such devices is small and adequate performance can be achieved using existing speech recognition and portable computer technology. Several of these applications have already found their way into the general marketplace. Examples include voice control of a cellular telephone in a moving car, remote control of a television set, or programming a VCR by voice.

ASR also is being used as a supplementary path for users to access information or control operations. In Japan, Nippon Telephone and Telegraph has combined speech recognition and synthesis in a telephone information system called Anser (Automatic Answer Network System for Electrical Requests) (Nakatsu, 1990). Anser provides information services for over 600 banks in more than 70 cities across Japan. Currently, over 360 million calls a year are processed by this system. Anser consists of a speaker-independent, isolated word speech recognition system with a vocabulary of 16 words consisting of the 10 Japanese digits and 6 control words spoken in a well-structured dialogue. Users of the system have alternative choices to interact with the system including a personal computer or a fax machine. Approximately 25% of the customers use ASR, with a recognition accuracy of 96.5% (Furui, 1992).

The U.S. Air Force, NASA, and the U.S. Navy conducted a joint program in the mid-1980s to flight test interactive voice systems in fighter aircraft. The functions being evaluated could be operated both manually and by voice. The program consisted of laboratory and simulator testing prior to flight tests. Significant improvements in recognition accuracy were made during each of the three phases of the program. Speaker-dependent, isolated word speech recognition systems were evaluated in the first two phases. Connected digit recognition capability was added in the third phase. A ten-word vocabulary was used to control Multi-Function Displays (MFDs) in the cockpit of an experimental F-16 jet aircraft. The MFDs contained programmable switches that selected pages of status information or control functions. The vocabulary words enabled the pilot to either address a particular page and then a particular function on that page, a specific function on a specific page, or to select an aircraft master mode. Recognition accuracy was approximately 90% initially, but increased to the high 90's for some pilots by the end of flight testing. For pilots with performance in the high 90's, speech was the preferred mode of MFD interaction. Pilots with performance in the low 90's preferred the manual mode of operation (Howard, 1987)

The Air Travel Information System (ATIS) is an example of an experimental system that may be deployed in the future. It employs speech control as part of a transparent interface, i.e., the user is engaging in natural dialogue to accomplish his or her job. ATIS uses voice to access travel information from the on-line Official Airline Guide (OAG). The OAG contains data on schedules and fares, hotels and car rentals, ground transportation, airport statistics, trip and travel packages, and on-time rates. Prototype systems have been developed by a number of groups (Massachusetts Institute of Technology, Carnegie-Mellon University, SRI International, Bolt Beranek and Newman Inc., AT&T, and Unisys), most under the sponsorship of the Advanced Research Projects Agency. These systems incorporate speaker-independent, continuous speech recognition and vocabularies of several hundred words. These systems run on high-end workstations in real time. Unconstrained speech is accepted including disfluencies, restarts, and corrections of queries (Figure 22.3). Some dialogue context information is employed during a session with the user. Speech recognition performance in the mid to upper 90% range is achieved by most of these systems. Feedback is in the form of a display of tables of travel information requested by the user [Figure 22.4(*a*)].

Several of the groups mentioned above have begun development of an ATIS with no visual display. These systems interface to an on-line reservation service, such as the American Airlines EAASY/SABRE system. The development of such a system requires that all information be provided to the user through spoken responses. Therefore, the system must carry on a dialogue with the user to narrow the scope of the information request and avoid long aural responses. For example, the system might ask the user to restrict the flight possibilities [Figure 22.4(*b*)]. If it becomes operational, this advanced ATIS will be an example of an intelligent interface where the system carries on a dialogue with the user in order to provide more efficient interaction.

FROM UH SSS FROM THE
PHILADELPHIA AIRPORT UM AT
OOH THE AIRLINE IS UNITED
AIRLINE AND IT IS FLIGHT
NUMBER ONE NINETY FOUR ONCE
THAT ONE LANDS I NEED GROUND
TRANSPORTATION TO UH BROAD
STREET IN PHILELD
PHILADELPHIA WHAT CAN YOU ARRANGE FOR THAT?

Figure 22.3 Example utterance with disfluencies, restarts, and corrections.

U: I WANT TO GO FROM WASHINGTON DC TO DALLAS TEXAS ON OCTOBER FIFTEEN

A: Here are the flights from Washington DC to Dallas Texas on Saturday October fifteenth.

	FLIGHT	FROM	TO	DEPARTURE	ARRIVAL
1	AA 557	IAD	DFW	6:35 A.M.	8:39 A.M.
2	AA 1831	IAD	DFW	8:03 A.M.	10:12 A.M.
3	AA 591	IAD	DFW	9:53 A.M.	11:58 A.M.
4	AA 243	IAD	DFW	12:59 P.M.	3:09 P.M.
5	AA 891	IAD	DFW	3:35 P.M.	5:43 P.M.
6	UA 1147	IAD	DFW	5:10 P.M.	7:22 P.M.

• • •

You can ask for more flights.

U: SHOW ME THE FARES FOR AMERICAN AIRLINES FLIGHT EIGHT NINE ONE

A: Please wait while I determine what fares are available for this flight. Here are the fares for American flight 891 from Washington DC to Dallas Texas on Saturday October fifteenth.

	ONE WAY	FARE CLASS	MIN-STAY	MAX-STAY	RTN REQ'D
1	$ 242.	THRIFT DISCOUNT	SUN	30	YES
2	$ 272.	COACH DISCOUNT	SUN	NO	YES
3	$ 503.	COACH CLASS	NO	NO	NO
4	$ 766.	FIRST CLASS	NO	NO	NO

• • •

(a)

U: I WANT TO GO FROM WASHINGTON DC TO DALLAS TEXAS ON OCTOBER FIFTEEN

A: What time will you be leaving?

U: I WANT TO ARRIVE JUST BEFORE DINNER

A: Hold on while I retrieve the flights from Washington DC to Dallas Texas arriving before 6:00 p.m. on Saturday October fifteenth. . . The latest flight arriving before 6:00 p.m. is American flight 891. It leaves at 3:35 p.m. and arrives at 5:43 p.m.

U: HOW MUCH DOES IT COST

A: The cheapest fare costs $242. With this fare you must stay over Saturday night and return within 30 days. It requires a round trip booking. The cheapest unrestricted fare costs $503.

• • •

(b)

Figure 22.4 (*a*) Part of a dialogue between a user and an Air Travel Information System (ATIS) when the system is operating in the combined verbal-visual dialogue mode. (*b*) Part of a dialogue between a user and an ATIS when the system is operating in the verbal-only dialogue mode. U = User and A = ATIS. (From Seneff, Zue, Polifroni, Pao, Hetherington, Goddeau, and Glass, 1995. Reprinted by permission.)

22.2.1.4 Design Methods and Principles

Because speech-based control is more mature than most of the other technologies addressed in this chapter, some general design guidelines can be identified (adapted from Simpson, McCauley, Roland, Ruth, and Williges, 1987):

1. When using speech-based control, expect the greatest payoff in task performance speed and accuracy for complex information entry tasks that must be performed in conjunction with other manual or visual tasks.
2. The selection of speech-based control should be based on an analysis of the application task requirements. When designing the vocabulary for the recognition system, use terminology that is familiar to the users and avoid the use of acoustically similar words.
3. Users should be trained to improve their pronunciation and microphone usage when possible.
4. Performance of the human–machine system should be measured in terms of operationally relevant measures such as system response time, system accuracy, and user acceptance.
5. Feedback should be provided so that the user is aware of the recognition results or the system response to the input. The more immediate the feedback, the less confusion as to source of the error. The modality of the feedback should be compatible with the demands of the task.
6. If errors occur, a correction capability should be provided that minimizes demands on the user and maximizes system throughput.

22.2.1.5 Future Research and Development

Displays that present a transcription of speech are being used increasingly by hearing impaired individuals. Current systems depend on a human to perform the transcription. A telephone relay service is an example: The speech produced by the hearing party in a telephone conversation is transcribed by a typist at a central office. The text is transmitted to the deaf party by means of a text telephone. The deaf party responds either by voice or by typing a message which is converted to speech by a synthesizer or the typist. Users of such services do not like the third-party participation. ASR is a possible solution to this problem, provided that the error rate for unconstrained continuous speech is not too high (Kanevsky, Danis, Gopalakrishan, Hodgson, Jameson, and Nahamoo, 1990; Karis and Dobroth, 1991). Real-time captioning of movies and television programs is a related application where demand is growing. Multilingual speech recognition systems will make real-time captioning and translation of foreign movies and television programs a reality.

Speech recognition performance for small and large vocabulary systems is adequate for some applications in benign environments (i.e., quiet office environments). Almost any change in the acoustic environment between recognizer training and testing causes a degradation in performance. Continued research is required to improve robustness to new speakers, new dialects, transmission channel, and microphone characteristics. Systems that have some ability to adapt to such changes have been developed (Lee, Lin, and Juang, 1991; Murveit, Butzberger, and Weintraub, 1992). Algorithms that enable ASR systems to be more robust in variable noise environments, such as airports or automobiles, have been developed (Acero and Stern, 1990; Hermansky, 1995; Hirsch, Meyer, and Ruehl, 1991), but performance is still lacking. Speech recognition performance for very large vocabularies and large perplexities is not adequate for application in any environment. Continued research to improve out-of-vocabulary word rejection, in addition to the areas mentioned above, will enable larger vocabulary ASR systems to be viable for real-world applications in the future.

Integration of automatic lip reading and speech-based control has been attempted recently as a means to improve speech recognition in noisy environments. This promising approach is addressed in Section 22.2.3.3. One key issue that must be addressed is the ability to operate speech-based controls in multitask environments. Some research has investigated the effect of task loading and other physical stressors on speech and its resultant impact on speech recognition performance (Rajasekaran and Doddington, 1986; Stanton, 1988). Continued research is needed to reduce the impact of these factors.

22.2.2 Eye- and Head-Based Control

22.2.2.1 Background

Humans naturally look at objects that they want to manipulate or use. Eye- and head-based systems attempt to harness this behavior for the purpose of control. One technical challenge is to track the head and eyes with the speed and accuracy required for natural human–machine interaction. A second challenge is control algorithm design, since exerting control by looking at something is not a natural behavior. When interacting with a computer, for example, the user will exhibit brief eye fixations connected by sudden, ballistic eye movements know as saccades. These saccades have a velocity of 400–600°/sec and can cover from 1–40° of visual angle. Movements of 15–20° are more typical with durations of 30–120 milliseconds. Fixations are typically 600 milliseconds or less in duration and have an accuracy and repeatability of about 1° of visual angle. This is driven by the fact that foveal receptor density is sufficiently high over a 1° area that one can obtain a clear view of any object within that zone. During fixations the eye exhibits microsaccades, drifts, and tremor, but they are small enough that the eye remains fixated to within about 0.5° of visual angle.

With large gaze changes, head movement is typically involved. While not normally as dynamic as the eye, head movement velocities of up to 250°/sec are commonly observed. Maximum head rates over small arcs can be as high as 800–1000°/sec. The head can be positioned with high accuracy and is more amenable to precise conscious control than the eye. Using head-based control alone, however, can lead to frequent and tiring motions which are unnatural to the user. Acceptance problems are likely if the user is required to make frequent head movements when eye movements would normally be employed.

22.2.2.2 Technology Description

Signal Acquisition and Processing

The first head tracking systems were based on direct mechanical connections to the head, but this approach is no longer common. Magnetic, ultrasonic, and electro-optical technologies are now employed. [Detailed descriptions of each approach are provided by Ferrin (1991). Kocian and Task (1995) examine the factors to be considered in selecting a particular measurement approach.] Some of the performance parameters, as well as the strengths and weaknesses of each approach, are summarized in Table 22.1. As shown there, only the electro-optical system using rotating infrared beams and the magnetic tracking systems are currently in production. Both approaches employ sensors mounted on the head and externally mounted transmitters. Magnetic trackers require only a single transmitter and receiver to measure head orientation over the ranges shown in Table 22.1. Electro-optical trackers require multiple transmitters and sensors to achieve the same coverage because of obstruction between the transmitters and sensors at large head angles. For the same reason, magnetic trackers can return accurate measurements over a wider range of head locations (head motion box) than the electro-optical trackers. Magnetic trackers do require the mapping of ferromagnetic and metal conductive surfaces in the user's environment, however.

Specifying the time delay associated with head tracking continues to be a source of misunderstanding, and a detailed discussion is beyond the scope of this chapter. [See Eggleston (1997) and Kocian and Task (1995) for a more detailed discussion of this issue.] Part of the difficulty stems from the fact that the user experiences the overall delay in the head-controlled system, not just the delay in the head tracking process itself. In addition, the magnitude of the experienced delay varies with characteristics of the movement pattern being tracked. Commonly available magnetic trackers sample at a rate of 120–240 Hz and compute new position data at about half this rate, since they require two or more update cycles to obtain good measurement convergence. This results in a minimum throughput delay of 8–12 milliseconds or greater. When graphics processing and image display times are included, system delays from 33 milliseconds to as much as 300 or 400 milliseconds are possible, depending on the movement profile being tracked. To improve overall system performance, head position prediction algorithms can be employed. If the head movements are smooth and continuous, predictive filters can effectively compensate for tracker lags of up to about 200 milliseconds (Liang, Shaw, and Green, 1991; Longinow, 1994). Other prediction techniques that can handle discontinuous profiles are under development (Maybeck, Herrera, and Evans, 1994; Kyger, 1995).

Table 22.1 Comparison of Head Tracking Technologies

Tracker Technology Category	Range of Angular Inputs (RMS)	Accuracy Range (milliradians) (RMS)	Strengths
Electro-optical using rotating IR beams or planes of light	AZ: ±180° EL: ±70° Roll: ±35°	3 to 10	Availability
Electro-optical using LED arrays	AZ: ±180° EL: ~±60° Roll: ~±45°	1 to 10	Simple installation Minimum added helmet weight High accuracy
Electro-optical using videometric techniques	AZ: ±180° EL: ~±60° Roll: ~±45°	2 to 15	No added helmet weight
Ultrasonic concepts	AZ: ±180° EL: ~±90° Roll: ~±45°	5 to 10	Minimum added helmet weight
Magnetic concepts AC or DC	AZ: ±180° EL: ~±90° Roll: ~±180°	1 to 8	Very low added helmet weight Simple mechanization Good noise immunity High accuracy Very large motion box

Tracker Technology Category	Weaknesses	Possible Interference Sources	Development Status
Electro-optical using rotating IR beams or planes of light	Helmet weight (12 oz) Reliability of moving parts No head position information	Helicopter rotor chop Sun modulation	Production (F-4 Phantom, AH-64 Apache, and A-129 Mangusta)
Electro-optical using LED arrays	Coverage Limited motion box Covertness	Reflections IR energy sources	Prototype
Electro-optical using videometric techniques	Limited motion box Helmet surface integrity	Reflections IR energy sources	Prototype
Ultrasonic concepts	Partial blockage Stray cockpit signal returns Accuracy No head position information	Air flow and turbulence Ultrasonic noise sources Multi-path signals	Prototype
Magnetic concepts AC or DC	Ferromagnetic and/or metal conductive surfaces cause field distortion (cockpit metal, moving seat, helmet CRTs)	Changing locations for metal objects Rarely—magnetic fields	AC system in low-rate production for AH-66 Comanche Several commercial variations of AC and DC designs in production

Source: Kocian and Task, 1995, from *Virtual Environments and Advanced Interface Design* edited by Woodrow Barfield and Thomas A. Furness III. Copyright © 1995 by Oxford University Press. Reprinted by permission.

A variety of technologies have been applied to the problem of eye tracking. [The reader interested in more detail should refer to reports by Young and Sheena (1988) and by Borah (1989).] Perhaps the least expensive eye-tracking technology employs the electrooculogram or EOG. The EOG is based on the 0.4–1.0 millivolt electrical potential that exists between the front and back of the eye. This dipole is approximately aligned with the optical axis of the eye and can be measured easily with skin surface electrodes and commercial biological signal amplifiers. The EOG can measure a wider range of eye movements than other technologies, exhibits high signal bandwidth, and provides an excellent means to measure eye velocity, acceleration, and the occurrence of saccades. Because of significant nonlinearities and drift problems, it is not the best technique for measuring eye line-of-sight (LOS). Knapp, Lusted, and Patmore (1995) report their continuing efforts to solve these problems using techniques such as fuzzy pattern recognition.

The most practical LOS measurement techniques involve video-based tracking of one or more features that can be optically detected on the eye (Figure 22.5). The most often used features include the limbus (boundary between the iris and sclera), the pupil, movement of the lower eyelid, the reflection of a light source from the cornea (first Purkinje image or corneal reflex), and the reflection from the rear surface of the eye lens (fourth Purkinje image).

Table-mounted and head-mounted versions of optical eye trackers are commercially available from several sources. Examples are shown in Figure 22.6.

Tracking a single feature on the eye does not permit one to discriminate between eye rotation and eye translation caused by movement of the head. This ambiguity can be resolved by tracking two features which are at different radii from the eye center of rotation. The two sets of features most commonly employed are: (1) the pupil center and first Purkinje image, and (2) the first and fourth Purkinje images. Two-feature eye trackers thus permit a small amount of head motion, assuming that the user stays within the field of view of the video camera observing the eye. Some table-mounted systems also employ

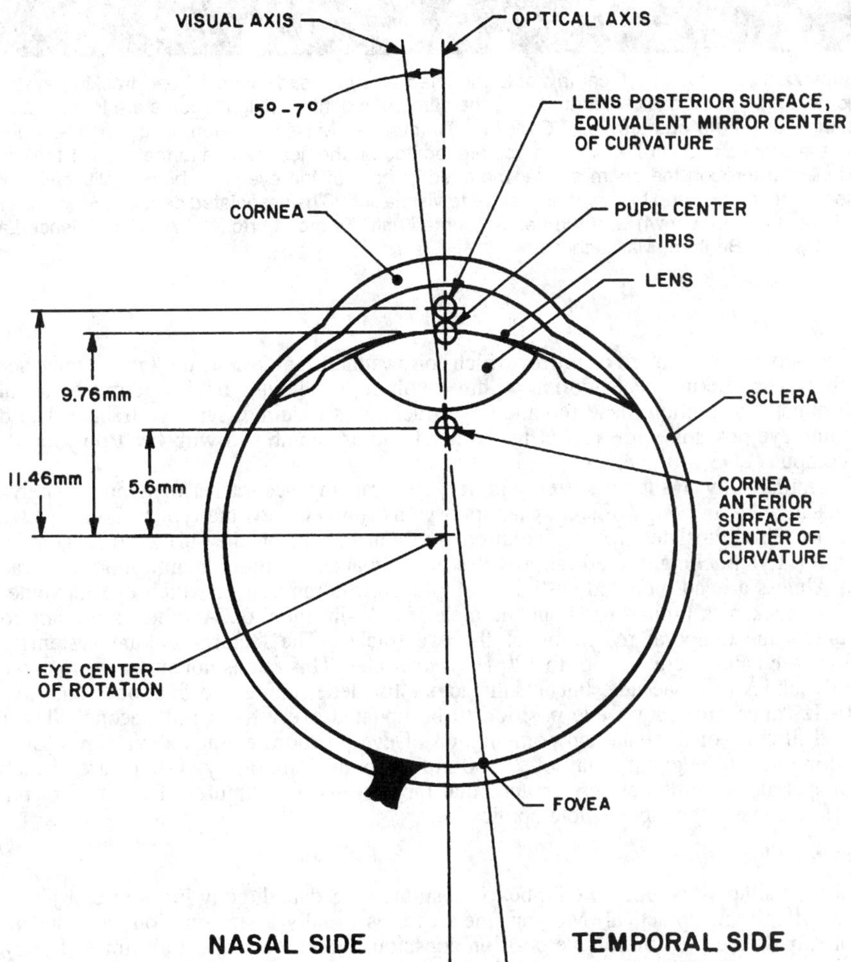

Figure 22.5 Diagram of the human eye showing features often used by video-based eye tracking systems. (From Borah, 1989.)

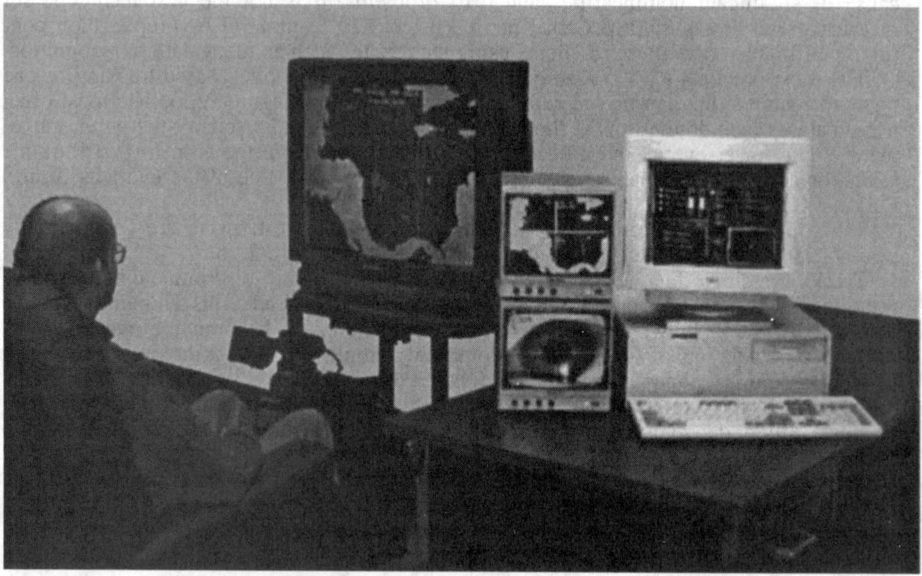

(*a*)

Figure 22.6 Photos of typical (*a*) table-mounted and (*b*) head-mounted eye tracking systems (see next pg.). In the table-mounted unit, the camera and infrared light source are located below the monitor (Photo courtesy of ISCAN Inc., Cambridge, Massachusetts). In the head-mounted unit, the camera and infrared light source are located on the headband and the eye is illuminated and viewed through the beam splitter mounted in front of the eye. This beam splitter is highly reflective to infrared and highly transmissive to visible light. The associated control and monitoring equipment are not shown with the head-mounted unit. (Photo courtesy of Applied Science Laboratories Inc., Bedford, Massachusetts.)

a servo-controlled camera or mirror which follows the user's eye as they move their head. Such systems permit head motion within a volume of about 1 ft^3. To permit essentially unlimited head motion, head-mounted eye tracking is required. The eye tracker data determine eye position with respect to the head and are combined with head position data to compute LOS to the environment.

Commercially available systems permit eye tracking accuracies of approximately 1° of visual angle in home, office, or laboratory environments. At the typical 24-in. viewing distance, this translates into a resolution of about 0.4 in. on the surface of a computer monitor. Eye tracking, therefore, provides much coarser operation than a mouse or trackball. Unless a head-mounted system is employed, the range over which eye movements can be tracked is limited to about the area of a 19-in. monitor. Another important constraint is the temporal resolution of the eye tracker. The feature-tracking systems are video-based and operate at 60- to 120-Hz frame rates. This rate is not sufficient to capture the dynamics of a saccade, but is satisfactory for determining eye fixations. The 60- to 120-Hz frame rate allows eye position to be updated every 8–17 milliseconds. The required filtering of the data and computation of eye position requires about three frames, resulting in a throughput delay of 25–50 milliseconds. Laboratory systems are available with spatial and temporal performance that far exceeds these figures. They are not practical for home, office, or vehicle applications.

Control Algorithms

While the simplest solution would be to substitute LOS data directly for mouse or joystick inputs, this is not practical. Moving one's eyes is usually a subconscious act, and it is relatively difficult to control eye position consciously and precisely at all times. The eyes tend to dart from spot to spot and it is not desirable for each such movement to initiate a control input.

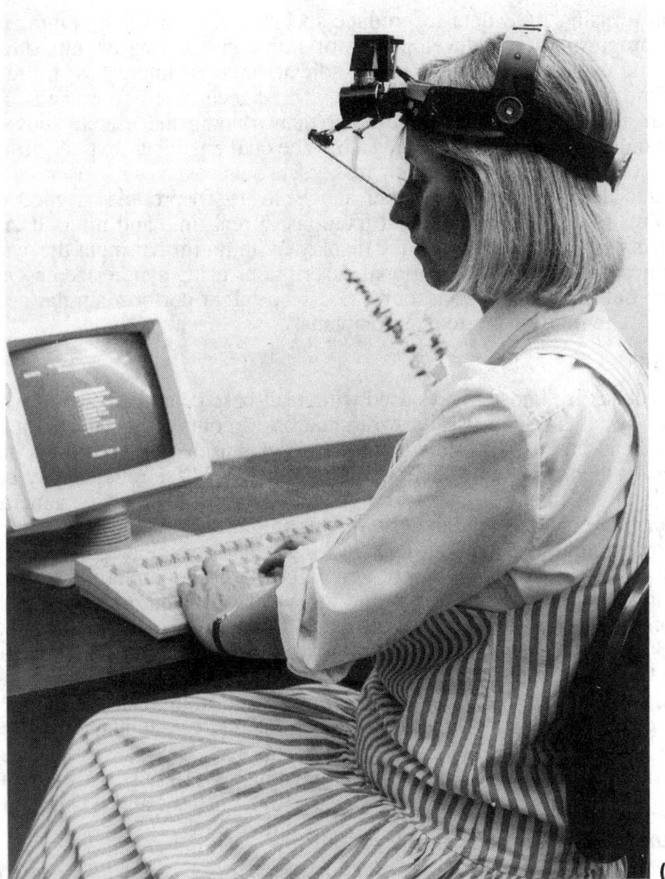

(*b*)

Figure 22.6 (*Continued*).

One common solution is to combine LOS position data with LOS dwell time criteria; the user selects an item on a display simply by looking at it for the criterion time. Dwell times in the range of 150–250 milliseconds are generally sufficient, unless an incorrect selection produces results that are hazardous or difficult to correct (Jacob, 1995). Longer dwell times tend to mitigate the speed advantage of eye-based control. To reduce the dwell time requirement and minimize the inadvertent selection of objects, a consent response made be added. Object selection can occur either by satisfying the dwell time requirement or by pressing a button, whichever occurs first. Alternatively, dwelling on an object for about 30 milliseconds can cause it to be highlighted, while a button press or voice command is used to select it. This highlighting may be maintained, without requiring constant gaze on the object, so that the user can make the consent response while performing another task (Calhoun and Janson, 1991a). Depending on the accuracy and stability of the LOS data, the selection boundaries of an object may need to be larger than its iconic representation on the screen. Some algorithms select the closest object, provided that it is "reasonably" distant from other objects on the display.

Many of the issues discussed above are less problematic when pure head-based control is employed, since there is less unintentional activity in the LOS signal. Relatively direct substitutions of head-based control for mouse or joystick inputs can be made. The helmet-mounted sights being developed for aviation environments use head position and orientation data in this manner. Head-based control is not ideal for *normal* human-computer interaction where the frequent, precise head movements would be tiring and unnatural to most users.

In addition to using LOS data to produce specific selections and commands, there is considerable progress in the development of non-command-based algorithms (Jacob, 1995). Head and eye fixations are used to indicate areas of interest, a desire for further information, and to disambiguate the referents for speech-based commands. Natural eye movements are used as implicit inputs rather than training the user to move his or her eyes in a particular way to operate the system. The challenge is to extract, from the noisy eye data, intentional components which make sense as tokens in a user–computer dialogue. A system demonstrated by Starker and Bolt (1990) provides a good example. It analyzes patterns of eye movements and fixations in real time and infers user interest in particular objects or sets of objects on the display. With no further input the system zooms in on these objects and provides additional descriptions using synthesized speech. Starker and Bolt point out the analogy to a tour guide who might perform similar actions based on the focus of attention of the tour participants.

User Feedback

Direct user feedback is almost always used with head-based control systems. The principal exceptions are systems which use head orientation to control the viewpoint of a visual display. Other head-based controllers include reticles that the user positions on the target to be acquired or the subsystem function to be activated. These reticles may be projected onto the combining element of a see-through head-mounted display, or generated as part of the imagery in a non-see-through system. One might expect that LOS feedback would be required in a system that includes eye tracking, since conscious control of the eyes is more difficult than the head. This is not generally the case. Numerous researchers have found that human performance is better and more natural when direct eye position feedback is *not* provided (Borah, 1995; Calhoun and Janson, 1991a; Jacob, 1995). Although such feedback can be used to compensate for eye tracker inaccuracies, i.e., the user can look off-target in order to place the cursor on-target, this tends to be distracting. The eye is naturally drawn to the moving cursor which can result in a positive feedback situation. If the eye tracker is working perfectly, feedback is totally redundant; the user and the system know where they are looking. Some of the eye tracking systems designed for users with disabilities employ such feedback, presumably to assist with user understanding, training, and perhaps to help compensate for limited head and postural control.

22.2.2.3 Current Applications and Evaluations

Radwin, Vanderheiden, and Lin (1990) compared head-based and mouse control on the performance of a task which required moving a cursor from the center of a computer display screen to static circular targets located in eight radial directions. Target size and distance were varied in a parametric fashion. Average movement time was 306 milliseconds (63%) greater for head input than for the mouse. In addition, head input was more affected by movement direction than the mouse. In a similar comparison of head and joystick inputs, Jagacinski and Monk (1985) found that movement time was 150–200 milliseconds greater for head control. They noted that a lighter helmet sight and additional practice should reduce this difference. Levison, Zacharias, Porterfield, Monk, and Arbak (1981) evaluated head-based tracking of moving targets and found 50% greater RMS error compared to manual joystick inputs. Taken together, these results suggest that head-based control is not the best substitute for manual inputs in human–computer interaction. Head mass, head-tracker delays, and the need for frequent, precise head movements all tend to degrade user performance. However, as reviewed below, head-based control is an excellent approach for the visually coupled systems in simulators and military aircraft. Head-based control is being applied here as a supplement or transparent interface. In addition, there are numerous demonstrations of its applicability for disabled computer users who cannot operate traditional input devices (Adams and Bentz, 1995; Radwin et al., 1990).

For applications in which high control resolution is not a requirement, eye-based control can be an effective substitute for traditional input devices. Calhoun and Janson (1991b) found no significant difference in the time required to select switches on a simulated cockpit control panel using manual and eye LOS control. The subjects performed a concurrent manual tracking task to more closely simulate flight workload. In another experiment, Calhoun and Janson (1991a) compared eye- and head-based control for the performance of target selection, weapon selection, and weapon slewing tasks designed to simulate inside- and outside-the-cockpit activities. Performance was significantly better with eye control (Figure 22.7). Ware and Mikaelian (1987) evaluated eye LOS control for computer inputs and found that object selection and cursor positioning tasks were performed approximately twice as fast with an eye tracker as with a mouse. Recall, however, that the eye tracker is resolution-limited compared to the mouse.

Jacob (1995) has argued for the use of eye LOS control in a more natural framework than that represented above. His approach contains elements of the control supplementation paradigm (the eye and mouse are used jointly) and of the transparent interface paradigm (he attempts to identify the desired system response directly from user eye movements). The control algorithms tend to focus on eye fixations and gazes rather than on saccades and the fine structure of eye activity. His application is representative of a naval command and control simulation, where an operator is viewing and managing the activities of a number of ships. When the user gazes at a ship, information about it is displayed in an attribute window located at the edge of the display. The window continues to display and update this information until a different ship is selected. Ships can be repositioned as well. Contrary to initial expectations, Jacob found that the most natural mechanism was to use the eye to select *and drag* the ship, and a mouse button to pick it up and put it down. He also identified desirable layouts for text windows which scroll when users reach the window boundary and naturally stop when users shift their gaze to read the new text.

The attempt to use eye LOS input in this non-command-based style (Nielsen, 1993) is further exemplified by systems which combine eye and voice (Glenn, Iavecchia, Ross, Stokes, Weiland, Weiss, and Zaklad, 1986) or eye, voice, and gesture inputs (Koons, Sparrell, and Thorisson, 1993). These systems typically use the eye for deictic purposes only, that is, to disambiguate the referent for a speech command or to limit the spatial area in which the system searches for referents such as: "Update the information on *that target*."

Head-based control is also well suited to non-command-based applications. For example, the head-mounted display systems used in simulated and virtual environments seamlessly update the displayed scene based on the user's head position. A similar approach has been applied in some displays that are projected on fixed screens or domes. Here a high-resolution inset is moved around a low-resolution background based on head and/or eye position (Tong and Fisher, 1984). If properly implemented, a uniform, high-resolution display can be emulated. In both of these examples, the user does not produce an explicit command to change the display viewpoint; the system senses the change in head or eye position as an implicit signal to take this action. The distinction between explicit commands and implicit signals may seem subtle, since any user action can be

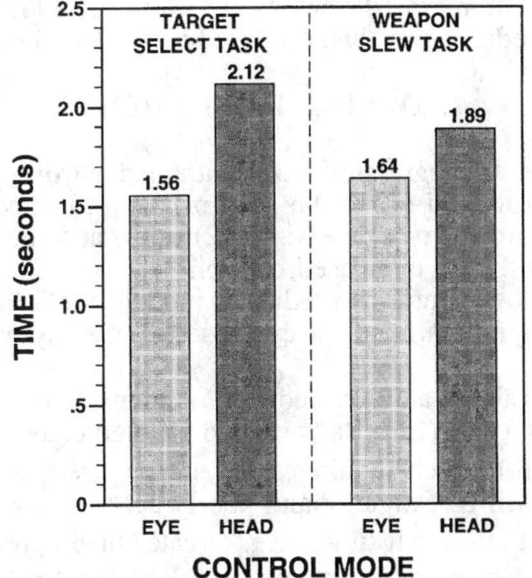

Figure 22.7 Mean times for target selection and weapon slewing using eye- and head-based control. (From Calhoun and Janson, 1991a.)

viewed as a command. The key criterion is whether the user thinks that he or she is issuing a command.

The potential of such systems to enhance human performance in demanding task environments is demonstrated by the application of "visually coupled systems" in a variety of military aircraft (Kocian and Task, 1995). Helmet-mounted display components present task-critical information, regardless of where the pilot is looking. Fully integrated helmet sights allow the pilot to lock radar and weapon systems onto targets without maneuvering the aircraft. The pilot performs these operations simply by looking at the target and executing appropriate manual consent commands. Head and eye control supplement the manual inputs in a natural, largely transparent fashion.

Despite the fact that some of the systems reviewed above may appear to be responding to a user's intentions, even before he or she expresses them, they are not true intelligent interfaces; they are not interacting with the user at a knowledge level. The work of Starker and Bolt (1990) begins to approach this level of performance. Their system "Analyzes the user's patterns of eye movements and fixations in real-time to make inferences about what item or collection of items shown holds most relative interest for the user. Material thus identified is zoomed-in for a closer look and described in more detail via synthesized speech." Although their system represents a small step, it begins to behave as an intelligent agent who is sensing and responding to the user's needs.

22.2.2.4 Design Methods and Principles

At the present time, little specific guidance can be offered concerning the design of interfaces that employ eye and head data as part of an intelligent system. There are, however, a good deal of data to assist in the design of eye- and head-based "command" systems. One set of tools that is often used to compare, and to predict performance with, various controllers are movement time models such as Fitts' Law (Fitts and Peterson, 1964). These models attempt to capture the speed-accuracy trade-off that is a necessary component of any precision control activity. They allow the designer to predict the time required to position a computer cursor on a displayed object as a function of the distance to be traveled and the size of the object to be selected. The Fitts' formulation is:

$$\mathrm{MT} = a + b \cdot \mathrm{ID}$$

where MT is movement time, ID is the index of difficulty and a and b are the intercept and slope constants, respectively. The index of difficulty proposed by Fitts is:

$$\mathrm{ID} = \mathrm{Log}_2\,[D/(W/2)]$$

where D is the distance from the starting position to the center of the target object and W is the target width. The units are in bits, reflecting an information theoretic approach. This formulation has proven to be applicable to a wide variety of controllers and tasks.

An alternative formulation was developed by Welford (1960) to correct for the tendency of Fitts' Law to predict lower than observed movement times at very low ID values:

$$\mathrm{ID} = \mathrm{Log}_2\,[D/(W + 0.5)]$$

Table 22.2 summarizes the regression model parameters from several studies of mouse, eye, and head control that used either the Fitts or Welford movement time models. As shown there, the best information-processing (IP) rates, and therefore the lowest movement times, are associated with eye-based control.

In addition to these movement time prediction models, the following guidelines should be considered when designing a head- or eye-based control system:

1. The optimal gain for head-based control of a cursor on a video monitor will be in the range of 0.3–0.6 (Lin, Radwin, and Vanderheiden, 1992).
2. Head-based control which requires frequent, precise head movements to replace eye movements will be fatiguing and poorly accepted by users.
3. The use of long LOS dwell times, e.g., greater than approximately 300 milliseconds, eliminates the speed advantage of using eye-based control (Calhoun and Janson, 1991a).

Table 22.2 Regression Model Parameters from Several Studies of Movement Time to Stationary Targets with Mouse, Eye, and Head Control

Study	Control	Intercept (sec)	Slope (sec/bit)	IP (bits/sec)	r	r^2	Model
Borah, 1995	mouse	0.474	0.149	6.71	0.97	0.94	*Welford ID*
Card et al., 1978	mouse	1.03	0.096	10.40	0.91	0.83	*Welford ID*
Epps, 1986	mouse	0.108	0.392	2.60	0.83	0.69	*Fitts ID*
Radwin et al., 1990	mouse	−0.06	0.147	6.80	0.99	0.98	*Fitts ID*
Lin et al., 1992	mouse	0.110	0.140	7.15	0.99	0.99	*Fitts ID*
Borah, 1995	eye	0.487	0.073	13.71	0.97	0.95	*Welford ID*
Ware and Mikaelian, 1987	eye	0.68	0.073	13.70	NA	NA	*Welford ID*
Borah, 1995	head	0.44	0.263	3.81	0.97	0.95	*Welford ID*
Radwin et al., 1990	head	−0.096	0.24	4.18	0.97	0.95	*Fitts ID*
Lin et al., 1992	head	0.23	0.167	5.99	0.97	0.94	*Fitts ID*
Jagacinski and Monk, 1985	head	−0.268	0.199	5.00	0.99	0.98	*Fitts ID*

Values for the Lin et al. (1992) study were computed from plots that appear in their paper. Values for the Radwin et al. (1990) study were computed by averaging the values for the different motion directions. Values for the Ware and Mikaelian (1987) study are for the button-press confirmation protocol and were derived by MacKenzie (1992). Adapted from Borah (1995).

4. The use of voice commands as a consent response may be slower than manual button presses because of delays in the speech recognition system (Calhoun and Janson, 1991b).
5. Human performance with an eye-based control system will generally be better and more natural without direct LOS feedback (Borah, 1995; Calhoun and Janson, 1991a; Jacob, 1995). With a head-based system, LOS feedback is almost always required.
6. Processing delays and lags can be significant problems in head- and eye-based systems and need to be carefully managed during the interface design process. As a rule of thumb, the delay from head/eye movement to display or system response should be kept under 100 milliseconds.

22.2.2.5 Future Research and Development

Future improvements in the accuracy and resolution of head and eye trackers are not likely to be major factors in enhancing user performance. Head trackers are already quite good in this regard and adequate for most applications. Eye trackers are approaching the 1° useful limit imposed by the size of the fovea. The primary area in which eye-tracking hardware and software needs improvement is in the stability and repeatability of the measurements. Brief dropouts and jitter that may be acceptable in research environments quickly erode user confidence in applied settings. In addition, the video-based eye-tracking systems are not particularly robust in high or variable illumination environments; infrared radiation from many sources can degrade their performance. Head and eye trackers would benefit from enhanced temporal performance. Speed improvements are needed both in the basic position measurement process and in the algorithms that transform these measurements into tokens for user–system interaction.

While some applications will continue to employ head- and eye-based control in an explicit command mode, it is clear that this mode does not fully exploit the capability of the LOS communication channel. Seamless interpretation of head and eye movements to infer user interest, needs, and desires appears to be a much more powerful, albeit challenging, path. Head and eye inputs will become just one element of multi-input interfaces. In addition to expanding the bandwidth of human–system communication, this approach may actually help to constrain overall system cost. By using eye and gesture as redundant pointing mechanisms, for example, lower precision may be acceptable in each measurement subsystem. Similarly, if speech recognition is used to issue specific commands while the referents are specified with eye and gesture, smaller vocabularies will be required.

22.2.3 Gesture-Based Control

22.2.3.1 Background

"Body Language" is an important channel for interpersonal communication. Gesture-based control seeks to exploit this channel for human–system interaction. Because traditional input devices constrain the expressive power of the human hand, scientists and engineers are developing a variety of techniques to read hand and body movements directly. Several systems for body movement tracking are commercially available. Most highly developed for human–computer interaction are systems that employ glove-based electronic input. Although the terms are sometimes used loosely, *gesture* formally refers to dynamic hand or body signs, while *posture* refers to static positions or signs. Readers interested in virtual environment applications of this technology should see Chapter 52 also.

22.2.3.2 Technology Description

Signal Acquisition

Two principal measurement systems are used in posture and gesture recognition: (1) systems which measure the *position* of limbs and body segments in three dimensional (3D) space, and (2) systems which measure the *angle* of hand joints directly [see Sturman and Zeltzer (1994) for a more detailed review of this technology].

Position measurement technologies include optical, ultrasonic, and magnetic tracking. One type of optical tracking system employs reflective markers, or infrared light emitting diodes (LEDs), located on the body to analyze the motion of body segments and limbs.

These systems use multiple cameras images to determine each marker's 3D location. Although they are highly effective for biomechanical analysis, their real-time limitations and inability to resolve closely spaced markers, e.g., on the fingers, make them less desirable for interactive applications.

Kreuger (1990), Wellner (1993), and Fukomoto, Suenaga, and Mase (1994) have demonstrated video-based systems which process silhouette images and pointing postures to provide "deviceless," but fairly constrained, interaction with a computer. Starner and Pentland (1995) have demonstrated a video-based system for American Sign Language recognition. While free-form image-based analysis is a goal that will be reached, it suffers from several current technology limitations (Sturman and Zeltzer, 1994):

1. The resolution of conventional video is not sufficient to simultaneously resolve the fingers easily and cover the field of view of broad hand motions.
2. Even 60-Hz video is insufficient to capture rapid hand movements.
3. Occlusion problems are especially difficult with finger and hand motions.
4. Computer vision techniques are limited in their ability to interpret complex visual fields in real time.

Ultrasonic trackers use high frequency sound (above the auditory range) and multiple microphones to triangulate a source within 3D space. Accuracies of a few millimeters can be achieved. However, the requirement for a clear "line-of-sight" between the source and microphones limits their applicability for high resolution interactive control.

Magnetic tracking systems are most commonly used for interactive applications since they are commercially available, operate with acceptable delays and do not rely on line-of-sight observation. They are generally accurate to better than 0.1 in. in position and 0.1° in rotation. These systems use a source element which radiates a magnetic field and a small sensor that is affixed to the body segment to be tracked. They are not commonly used to track segments as small as fingers, however.

Glove-based techniques are the current method of choice for hand and finger joint angle measurement. Glove-based systems and magnetic trackers are often combined to provide simultaneous position, orientation, and joint-angle data. Three glove systems are widely available at present: the DataGlove™, the Dexterous HandMaster™ (DHM), and the CyberGlove™ (Figure 22.8). The DataGlove is no longer produced, but is still available in many laboratories. The characteristics of these three devices are summarized in Table 22.3. As shown there, the DHM and CyberGlove have the accuracy required for complex posture and gesture recognition.

Signal Processing and Control Algorithms

The algorithms for determining positions and joint angles, based on the sensor inputs, are provided with the systems described above. With the glove-based systems, some individual user calibration is required. Magnetic trackers require the mapping of ferromagnetic and metal conductive surfaces in the user's environment, but no individual user calibration. For interactive applications that employ rapid body movements, one may need to add or modify movement prediction algorithms to compensate for system delays. Although general-purpose posture recognition software is becoming available for the glove-based systems, the development of algorithms to recognize specific postures or gestures is often left up to the user. These algorithms tend to be application specific, although some general approaches are common. For example, the recognition of a fixed set of hand postures is often based on look-up tables that contain min/max values for each position and joint measurement.

Interpreting gestures is a much more challenging problem since pattern analysis must be performed on a moving hand. Many approaches compare the motion vectors for each degree-of-freedom of the hand to reference vectors representing the target gestures. This match must be within error tolerances, and these tolerances are weighted by the significance of each hand motion vector for gesture discrimination. The weighting may be accomplished with principal components analysis (Takahashi and Kishino, 1991), with Bayesian rule-based techniques (Kramer and Leifer, 1989) or it may be performed by a neural network (Fels and Hinton, 1993). A different technique that has shown promise in several applications is a feature analysis approach developed by Rubine (1991). Originally developed for the interpretation of 2D written gestures, it has been extended to 3D hand

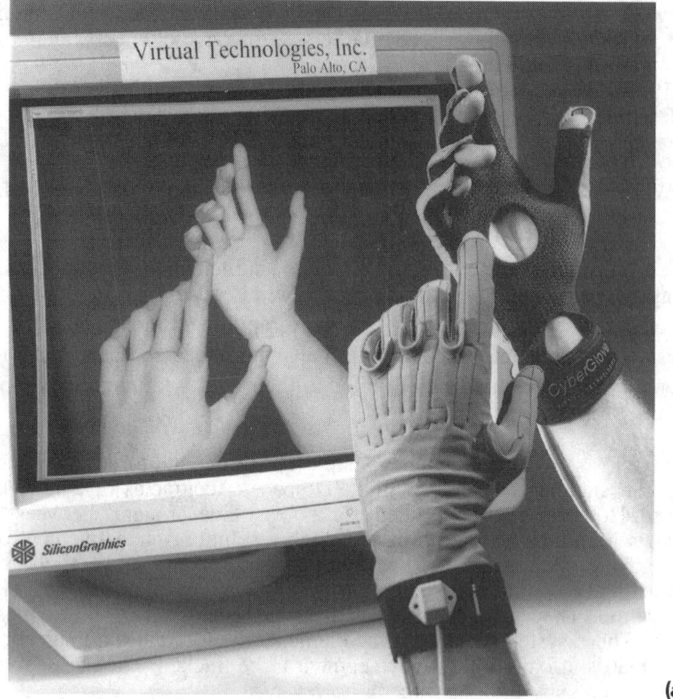

(*a*)

Figure 22.8 Photos of (*a*) the 18-sensor CyberGlove™ and VirtualHand™ software display and (*b*) the Dexterous HandMaster™ (see next pg.). (CyberGlove photo courtesy of Virtual Technologies Inc., Palo Alto, California. Dexterous HandMaster photo courtesy of Exos Inc., Woburn, Massachusetts.)

gesture recognition (Baudel and Beaudouin-Lafon, 1993). The features Rubine analyzed were prespecified measurements of the 2D movement trajectories, such as sines and cosines of the initial angle of the gesture, duration of the gesture, etc.

For telemanipulation and robot control applications resolution of the kinematic differences between the human hand and the robot hand is often required. Algebraic transformations have been employed to perform this human-to-robot mapping. Alternatively, the kinematic differences can be resolved by determining the 3D position of the user's fingertips and driving the robot's fingertips to match.

User Feedback

In many applications, the only feedback that is provided (and required) is the system's response to the recognized gesture. Examples include simulated movement in the direction toward which the user is pointing and synthesized speech following recognition of a sign language gesture. Feedback requirements for applications that involve simulated object manipulation, vehicle control, and robot operations are still the subject of research and development. In each of these cases tactile and kinesthetic feedback play an important role in normal human–system interaction. These cues are absent in most gesture-based systems. Significant progress is being made in the development of force-reflection (Hannaford and Venema, 1995) and tactile stimulation systems (Kaczmarek and Bach-y-Rita, 1995) which can provide this feedback through normal sensory modalities. In addition, there is evidence that substitute feedback can be provided with vibrotactile, auditory, and electrotactile displays (e.g., Massimino and Sheridan, 1993). However, tactile and kinesthetic feedback are not required in all cases. Massimino and Sheridan did *not* find enhanced performance of a peg insertion task when artificial or actual force cues were provided. As described in Section 22.2.4.2, users of EMG-controlled prosthetic arms can perform a grip force control task adequately with visual feedback alone, although performance is slightly enhanced when synthetic pressure cues are provided. The importance

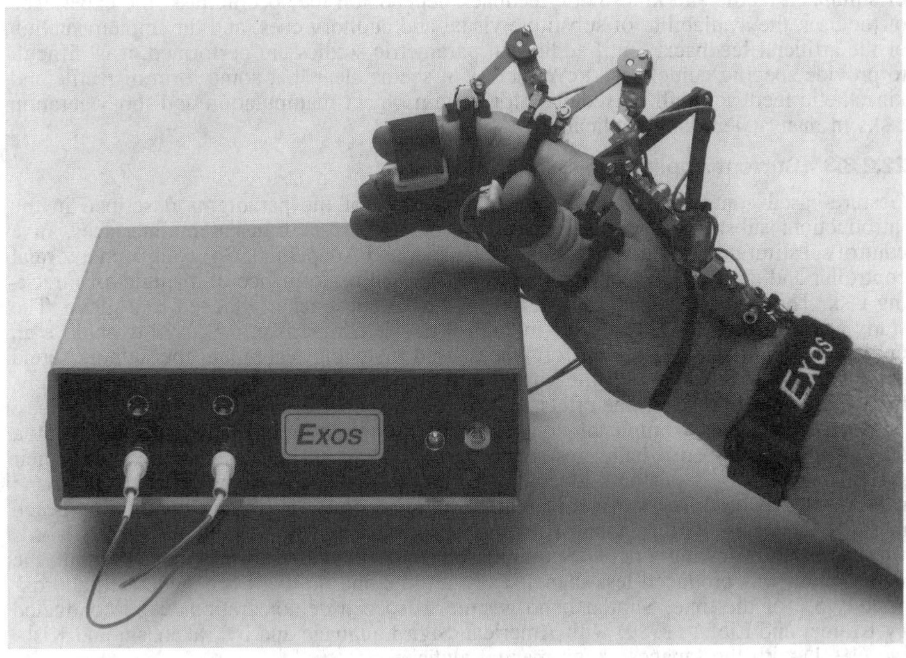

(*b*)

Figure 22.8 (*Continued*).

Table 22.3 Characteristics of Three Glove-Based Gesture Measurement Systems

	Construction	Sensing System	Accuracy
DataGlove™	Cloth glove with fiber optic bundles attached to the back	Joint bending sensed by attenuation of transmitted light. 10 flex sensors measure lower two joints of each finger and two joints of thumb. Operates at 60 Hz. Magnetic tracker can be attached to back of glove.	Rated at 1°, but 5–10° is typical. (Quam et al., 1989; Wise et al., 1990)
Dexterous HandMaster™	Exoskeleton-like device	Hall Effect sensors used as potentiometers. 20 sensors measure each finger joint, finger abduction and complex motion of thumb. Operates at up to 75 Hz with 20 sensors. Magnetic tracker can be attached.	Better than 1° (Sturman and Zeltzer, 1994)
CyberGlove™	Cloth glove with foil strain gauges sewn into the back	18 or 22 sensor models measure finger and thumb joint angles, finger and thumb abduction, palm arch and wrist bending. Operates at up to 149 Hz with 18 sensors. Magnetic tracker can be attached.	Better than 1° (Sturman and Zeltzer, 1994)

of simulated tactile and kinesthetic feedback depends on the specific task, the experience of the user, the availability of substitute visual and auditory cues, and the implementation of the artificial feedback. Until additional parametric studies are performed, it is difficult to provide specific guidelines. Nevertheless, it seems clear that some form of tactile and kinesthetic feedback will be required for certain object manipulation and tool operation tasks in many telerobotic applications.

22.2.3.3 Current Applications and Evaluations

Gesture-based control has been applied using three of the paradigms described in the introduction: substitution, supplementation, and as part of transparent interfaces. In a control substitution evaluation, Eggleston, Janson, and Adapalli (1994) compared a virtual controller and a physical displacement joystick in the performance of a single-axis tracking task. The virtual controller consisted of a magnetic tracker attached to a glove. The study also addressed a critical issue in control device comparison, i.e., how to achieve an unbiased evaluation. Using a performance-based matching technique, the authors found comparable performance with the virtual and physical controllers. In a follow-on study, they found no difference in the effects of time delay on performance with the two devices.

A more complex example of control substitution was provided by Fels and Hinton (1993) who developed a hand gesture to speech system using a neural network. Their system mapped hand postures to complete root words, followed by a directional hand movement that modified the word ending (singular, plural, etc.) and controlled speech rate and emphasis. Performance of a single "speaker" with a vocabulary of 203 words was evaluated following a network training phase. With near real-time speech output, the wrong word was produced less than 1% of the time and no word was generated approximately 5% of the time. Similar hand gesture to speech demonstrations were conducted by Kramer and Liefer (1989) with American Sign Language and by Takahashi and Kishino (1991) with the Japanese kana manual alphabet.

A compelling example of gesture as a control supplement is the application of "automatic lip reading" to enhance the performance of an acoustic speech recognizer. Both ultrasonic (Jennings and Ruck, 1995) and video-based approaches (Pentland and Mase, 1991; Petajan, Bischoff, Bodoff, and Brooke, 1988; Stork, Wolff, and Levine, 1992; Yuhas, Goldstein, and Sejnowski, 1989) have been tested to sense the lip motion, oral cavity shapes or facial movements associated with speech. Jennings and Ruck found enhanced speech recognition in a noisy environment when the ultrasonic and acoustic recognizers were combined (Figure 22.9). Yuhas et al. observed similar performance trends with their system.

An early example of a transparent interface was the "Put-that-there" demonstration described by Bolt (1980). This demonstration combined hand pointing and speech rec-

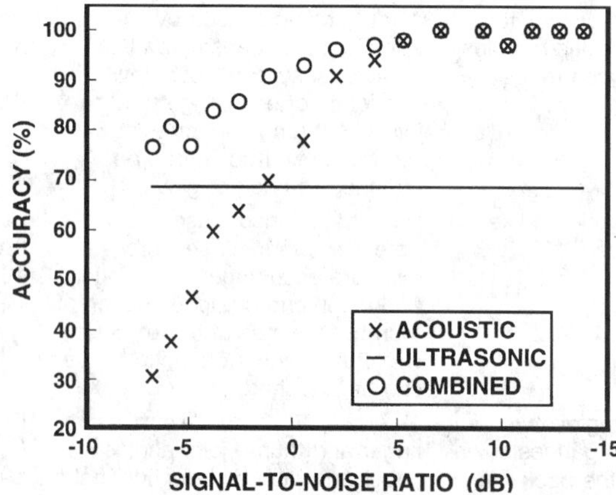

Figure 22.9 Speech recognition accuracy as a function of signal-to-noise ratio using an acoustic recognizer alone, an ultrasonic lip motion detector alone, or the combination of the two. The devices were tested in a speaker-dependent, isolated word recognition task with a vocabulary consisting of the spoken digits from zero to nine. (From Jennings and Ruck, 1995, copyright © 1995 IEEE.)

ognition to permit natural interaction with objects on a large screen display. Pointing direction was sensed with a magnetic tracker attached to the hand. The interface responded to commands such as "Name *that* X" or "Put *that* there" where *that* referred to the object being pointed at and the action was defined by voice input. This demonstration provided a compelling example of integrated nonconventional control which allowed the user to directly manipulate task objects. No visible control devices were imposed between the user and his or her task.

Additional examples are provided by 2D and 3D displays in which the user can touch, grab, and move objects by pantomiming these activities with glove-based sensors. In these applications the user actually sees a computer rendering of their hand performing the object manipulations. Researchers at NASA/Ames have used this approach in a virtual wind tunnel to explore simulations of computational fluid dynamics. Aeronautical engineers can put their hands and head into a simulated fluid flow and manipulate the patterns in real time (Bryson and Levit, 1992).

Object manipulation has been extended from virtual space to glove-based control of physical robot arms. Hale (1992) used a DataGlove to control a robot arm in a task that required retraction, slewing, and insertion of a block in a test panel. He compared his results to another study that used a conventional six degree-of-freedom hand controller as the input device. He concluded that performance with the DataGlove compared favorably with the "standard" device and that it provided a natural and intuitive user interface. Brooks (1989), on the other hand, was less optimistic about the DataGlove for robot control; his evaluation involved more complex gestures and a neural network for gesture recognition.

22.2.3.4 Design Methods and Principles

Sturman and Zeltzer (1993) have proposed a method to assist users in designing and evaluating whole-hand input for specific tasks. A flow diagram of their process is shown in Figure 22.10. In the first stage, the designer must answer questions such as: "Can existing hand signs be used to perform the task?", "Does the task require coordination of many degrees of freedom?", and "Should the absence of an intermediary control device improve performance?". If the answers to these questions support the use of whole-hand input, the designer then begins an analysis process that: (1) breaks the task down into

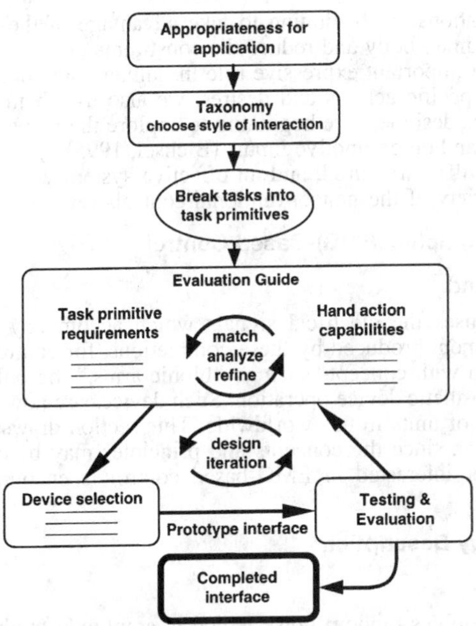

Figure 22.10 A flow diagram for designing and developing whole-hand input for specific applications and tasks. (From Sturman and Zeltzer, 1993. Included here by permission, copyright © 1993 ACM, Inc.)

primatives; (2) specifies the coordination, resolution, endurance, and other requirements for each task component; (3) determines whether hand capabilities can meet these requirements; and (4) identifies whole-hand input devices that provide the resolution, reliability, and sampling rates required to meet the task specifications. After completing these steps, the prototyping and interface evaluation process can begin.

Because of the relative immaturity of this area, detailed design principles are not available. Nevertheless, a review of the gesture literature suggests some general guidelines that are applicable to a many situations:

1. Gesture-based control should offer learning and performance advantages if the task is based on a set of already learned signs or signals. Glove-based translation of American Sign Language is an example.
2. Gesture-based control should offer learning and performance advantages if the natural coordination of the body can be employed to coordinate multiple degrees of freedom in the external device. Finger walking to control the locomotion of a legged-robot is an example (Sturman and Zeltzer, 1993).
3. Gesture-based control may be less effective than conventional control if the task requires high resolution control of a single degree of freedom. At least two factors contribute to this: (a) conventional controls often have higher resolution than gesture-based devices, and (b) conventional controls often provide support and damping that is helpful in precision control situations. This may not be true for applications in which gesture affords more natural, user-scaled control location.
4. Gesture-based control may be less effective than conventional control if tactile and kinesthetic feedback are important for task performance.
5. Gestures should be concise and quick in order to minimize fatigue. High precision over a long period of time should be avoided.
6. Since most systems capture every motion of the user's hand, the controller must provide well-defined means to detect the intention of gestures. An example is Baudel and Beaudouin-Lafon's (1993) system for controlling computer-based presentations to an audience. Gestures are acted on only when the user is gesturing within the "active zone" of the projection screen. Gestures to the audience are not recognized.

22.2.3.5 Future Research and Development

Gesture-based applications are beginning to take advantage of the dexterity and natural coordination of the human body and reduce the constraints of conventional input devices. Gesture also plays an important expressive role in human communication. While we use gestures to indicate specific actions and desires, we also use them to indicate emphasis and emotion. Interface designers are beginning to explore the recognition and application of facial expressions and other emotive inputs (Bichsel, 1995). Here deviceless, free-form gesture recognition will be required, and an effective system will undoubtedly integrate the inputs from a variety of the nonconventional controls reviewed in this chapter.

22.2.4 Electromyographic (EMG)-Based Control

22.2.4.1 Background

EMG-based control uses the electrical signals which accompany muscle contractions, rather than the movement produced by these contractions, for control. Although popularized in science fiction with concepts such as "bionic arms," the only sustained development has been for prosthetic device operation. Such devices are now of significant clinical value with thousands of units in use worldwide. This section draws heavily on the prosthetic control literature, since the concepts and principles may be extended to other applications. The reader interested in EMG-based control is encouraged to explore this literature in detail.

22.2.4.2 Technology Description

Signal Acquisition

The EMG signal resembles random noise which is amplitude modulated by changes in muscle activity (Figure 22.11). It results from the asynchronous firing of hundreds of groups of muscle fibers. The number of groups and their firing frequency controls the force produced by the muscle contraction (Parker and Scott, 1986; Scott and Parker, 1988). Prosthetic systems typically use stainless steel electrodes molded directly into the prosthesis. Because the EMG signal is most often measured at points on hairless skin, gel electrodes also represent a convenient option and are used in some general purpose EMG systems (Junker, Berg, Schneider, and McMillan, 1995; Knapp, Lusted, and Patmore, 1995). Differential recording, with two active electrodes and one reference, is typically employed to minimize common mode noise. The EMG signals of interest range in amplitude from several hundred microvolts to a few millivolts. The lower end of this range is commonly used to avoid fatigue. The EMG frequency spectrum has a peak at about 60 Hz and significant components from approximately 15 to 300 Hz. Commercial biological signal amplifiers are well suited to the amplification and bandpass filtering of the EMG signal.

Signal Processing

The raw EMG signal is a zero mean process (average value = zero), as illustrated in Figure 22.11. Mathematically, the mean absolute value of the signal is related to the strength of the muscle contraction, but the precise form of this relationship varies across muscles and individuals. In practice, the strength of the muscle contraction is commonly estimated by rectification and low-pass filtering of the EMG signal. Most current systems use this value, or its rate of change, as a control signal. For multifunction prosthetic devices (e.g., simultaneous elbow, wrist, and hand control), systems are being developed which use patterns of activity across multiple muscle groups or patterns extracted from the responses of individual muscles to control specific functions. In some cases these systems employ neural networks to recognize natural contraction patterns (Hiraiwa, Uchida, and Shimohara, 1993; Hudgins, Parker, and Scott, 1993), similar to the EEG-based approach of Pfurtscheller (see Section 22.2.5.3).

Control Algorithms

The most common prosthetic control algorithms employ simple on–off control based on the level or rate of change of EMG activity. If muscle activity at one recording site exceeds some threshold, the prosthetic hand opens. Above-threshold activity at another site causes the hand to close. Hand movement stops when the EMG at both sites is below threshold. To permit control of functions such as grip force, some on–off algorithms employ "time

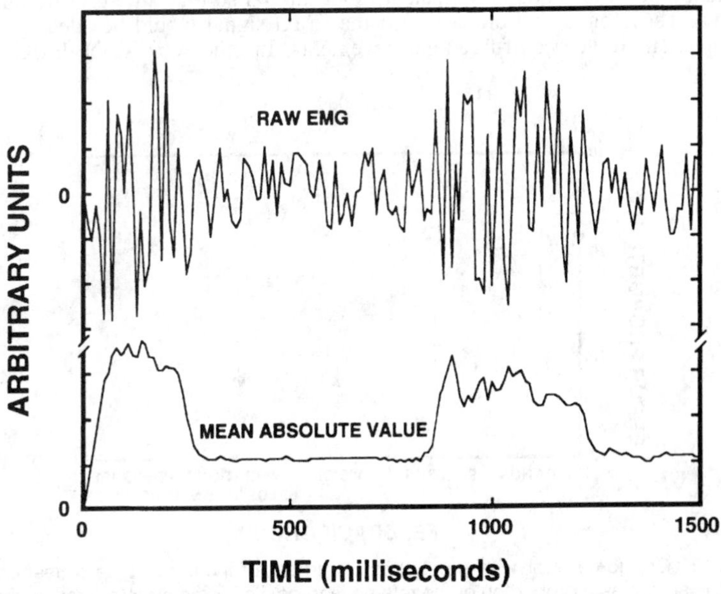

Figure 22.11 Time histories of the raw EMG signal produced by two brief muscle contractions and the same signal after rectification and smoothing with a 100-millisecond moving average filter.

proportional'' techniques, i.e., grip force increases as long as the user holds the closing signal above threshold. On–off control also can be based on EMG patterns rather than on levels or rates of change. True proportional control, in which the amplitude of the control output is directly related to the amplitude of the EMG signal, is challenging to implement in a practical fashion because of a speed-accuracy trade-off. To achieve accurate control, the EMG signal must be highly filtered. This reduces control responsiveness and is objectionable to users. Nevertheless, proportional control is being used in some single-function prostheses and in the controller developed by Junker et al. (1995).

User Feedback

Although EMG-based control provides muscle contraction feedback for users with intact sensory systems, many other feedback channels are absent. Nevertheless, most current systems rely on visual feedback, or auditory and vibration cues from prosthetic motors, to provide this information. Attempts to provide grip force feedback in prosthetic devices have most often employed vibratory or electrical cues proportional to grip force. A study by Patterson and Katz (1992) showed that pressure cues provided by an inflatable cuff permitted better grip force control than vibratory cues. However, visual cues alone appeared to be sufficient (Figure 22.12). To some extent this finding reflects limitations in the performance of current prosthetic devices, and it is generally believed that enhanced feedback will be required as the performance of prosthetic devices improves (Scott, 1990).

22.2.4.3 Current Applications and Evaluations

All applications and evaluations to date have employed EMG-based control as a substitute for conventional control, although prosthetic devices require the coordination of prosthetic and normal appendage activity. Below-elbow myoelectric prostheses (Figure 22.13) are readily available and well accepted by users despite the fact that their performance falls far short of the levels achieved with a normal hand. Simple on–off control is still most common, but proportional control systems appear to offer some performance advantages (Sears and Shaperman, 1991). Significant effort is being made to enhance multifunction prosthetic control systems and to employ natural muscle contraction patterns that require less user training (Hudgins, Parker, and Scott, 1993).

In a study sponsored by NASA, Clark and Phillips (1988) investigated the possibility of using EMG control for robotic applications. The EMG patterns produced by normal arm and hand motion were compared with the corresponding arm and hand kinematics to determine if reliable relationships existed. The objective was to use the EMG signals from human motion to produce equivalent robot motion so that little user training would be required. They found that direct use of the EMG signal would be adequate only for some simple, single degree-of-freedom movements. In other words, the EMG time his-

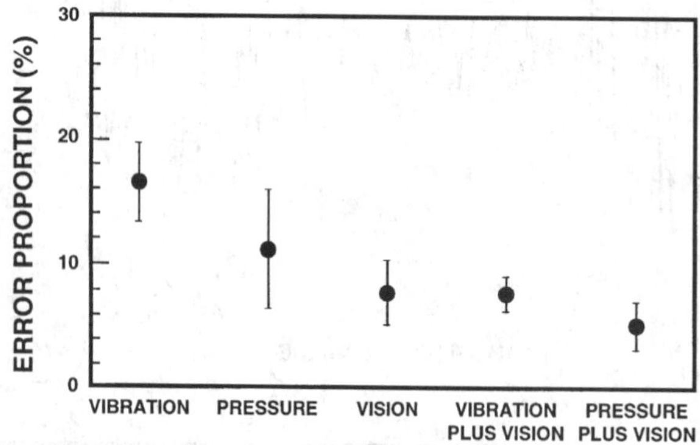

Figure 22.12 Grip force error with a prosthetic hand as a function of type of feedback. Error magnitude is shown as a proportion of the reference force, i.e., force error/reference force. (From Patterson and Katz, 1992.)

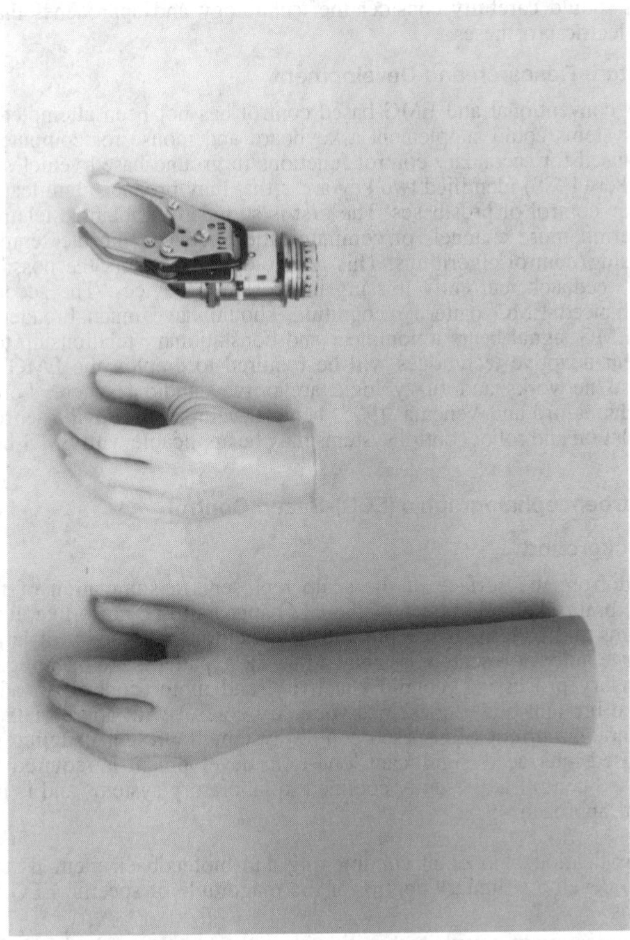

Figure 22.13 Myoelectric prosthetic hand and arm systems. (Courtesy of Otto Bock USA, Minneapolis, Minnesota.)

tories were not appropriate for controlling complex movement kinematics. This finding is not surprising and is in agreement with experience in prosthetic device development.

Junker and his colleagues (1995) have developed a hybrid system which uses a combination of forehead EMG and EEG activity for control. This system is based on proprietary algorithms which decompose the forehead signal into multiple frequency bands between 2 and 24 Hz. A variety of tasks have been controlled, but the most extensive data are available for a computer-based ping-pong task. Paddle position is directly proportional to signal amplitude in a selected frequency band. Users typically experience some control in the first few minutes of training, but precise control requires an additional time investment. In a group of 10 users, average performance ranged from 60–75% ball hits after approximately 13 hr of training. Two-axis control using one frequency band for vertical motion and another for horizontal motion has been demonstrated, but no quantitative data are available. The controller can be used also for icon selection tasks and for the control of computer-generated music and graphics.

These applications clearly demonstrate that EMG signals can be used for reliable device control, despite the fact that the control operations are limited and each of the applications requires an investment in user training. The goal of using naturally occurring EMG signals to produce equivalent movements in prosthetic devices or robots is still a distant one. Regardless of the application, readers interested in the development of EMG-based control should carefully consider the technology and approaches that have been used in myoelectric prostheses.

22.2.4.4 Future Research and Development

Integration of conventional and EMG-based control has not been attempted, but current EMG-based systems could supplement a keyboard and mouse for computer operations, and could be used for secondary control functions in ground-based vehicles and aircraft. Scott and Parker (1988) identified two key areas that may produce significant progress in the myoelectric control of prostheses. The first is surgically implanted telemetry systems that would permit more channels of communication between muscle segments and the signal processing/control algorithms. This technology also offers the possibility of artificial sensory feedback, currently missing in prosthetic devices. The second area they identified, enhanced EMG pattern recognition, should have much broader application. Because the EMG signal bears a nonlinear and nonstationary relationship to muscle activity, nonlinear adaptive techniques will be required to employ the EMG for complex control. Neural networks and fuzzy logic are prime candidates here. Force reflection technology (Hannaford and Venema, 1995) being developed to provide sensory feedback in telemanipulation and robot control systems may be applicable with EMG-based systems as well.

22.2.5 Electroencephalographic (EEG)-Based Control

22.2.5.1 Background

EEG recorded from the surface of the scalp represents a summation of the electrical activity of the brain. Although much of the EEG appears to be noise-like, it does contain specific rhythms and patterns that represent the synchronized activity of large groups of neurons. A large body of research has shown that these patterns are meaningful indicators of human sensory processing, cognitive activity, and motor control. In addition, EEG patterns can be brought under conscious voluntary control with appropriate training and feedback. Although current EEG-based control systems represent fledgling steps toward a "thought-based interface," significant long-term development is required to reach this goal. EEG-based control is presently confined to laboratory systems and is based on one of two general approaches:

1. The application of operant conditioning and biofeedback methods to enable the user to develop voluntary control of the magnitude of specific EEG responses or rhythms.
2. The application of pattern recognition algorithms to detect the EEG characteristics associated with specific body movements, eye fixations, or utterances and thereby predict a desired control action. No current algorithms attempt, or are capable of, thought or intent recognition. This approach requires no user training, but does require that the pattern recognition algorithms be trained with repetitions of the movements, fixations, or utterances. This process is directly analogous to the training of a speaker-dependent speech recognition system.

22.2.5.2 Technology Description

Signal Acquisition

The most commonly used scalp electrodes are small (approximately 0.4-in. diameter) gold or silver/silver chloride disks developed for EEG recording. Electrical contact is maintained with a conductive paste, and the electrodes are affixed with adhesive rings, tape, or the conductive paste alone. Real-world applications will benefit from convenient dry electrode systems, but these are not yet commercially available. The EEG signals of interest are in the 1–40-Hz frequency range with amplitudes ranging from 1–50 microvolts. Although these signals are very small, commercially available biological signal amplifiers are acceptable and commonly used.

Signal Processing

The first signal processing step in most systems is analog-to-digital conversion using sampling rates of 100 Hz or higher. The EEG power in specific frequency bands is then estimated with fast Fourier transforms, digital bandpass filters, or a proprietary technique. Personal computer systems of the 80486 class are sufficient to support these computations, although digital signal processing boards are sometimes used. Some systems employ neural networks to predict hand movements, joystick movements, or simple utterances based on the normally occurring EEG patterns which precede them.

Control Algorithms

Control algorithm design is particularly challenging for approaches which employ voluntary control of EEG activity rather than the recognition of normally occurring EEG patterns. Here, the system must deal with a great deal of unintentional signal variability that corrupts the intentional changes in the EEG signal. Despite this limitation, impressive control has been demonstrated, largely due to the success of the control algorithms. Low-pass filtering, signal averaging, and logic based on threshold and duration requirements have all been employed to deal with unintentional variability in the EEG signal. The work of McMillan, Calhoun, Middendorf, Schnurer, Ingle, and Nasman (1995) provides an example. They use an EEG-based system to control the roll-axis motion of a simple flight simulator. Their approach is based on self-regulation of the steady-state visual evoked response (SSVER) produced by a sinusoidally modulated light incorporated into the task display. A lock-in amplifier provides a continuous measure of the magnitude of this response. The control algorithm employs "pulsatile" logic which permits stable simulator control despite variability in the user's SSVER. This algorithm requires the user to hold the SSVER magnitude above or below set thresholds, for a set duration, before a discrete output is generated. Typical parameter settings produce an output when the SSVER remains above or below threshold for 75% of the samples in a 0.5-sec interval. These settings require the user to produce intentional changes in the SSVER; however, brief SSVER fluctuations do not interrupt simulator control. The operation of the pulsatile algorithm can be observed in Figure 22.14 as simulator motion steps separated by

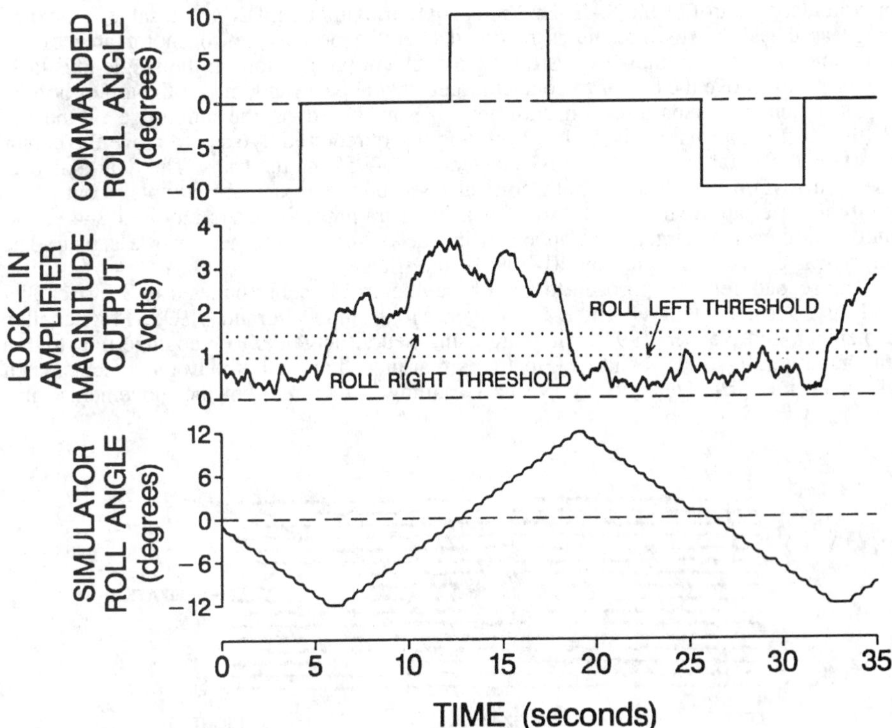

Figure 22.14 Example time history for flight simulator control using self-regulation of the steady-state visual evoked response (SSVER). The lock-in amplifier provides a continuous measure of the magnitude of this response. Responses above one threshold produce motion to the right (positive), while responses below a lower threshold produce motion to the left (negative).

0.5-sec intervals. SSVER responses above one threshold produce motion steps to the right, while responses below a lower threshold produce steps to the left. A series of roll angle commands is presented to the user to test simulator control performance.

User Feedback

Biofeedback is one of the key technologies that enabled the development of EEG systems based on neural self-regulation. User feedback has been implemented in two ways: (1) as an inherent part of the task, e.g., movement of the display element being controlled by EEG, or (2) as a separate display element when movement of the controlled element does not provide timely feedback. Biofeedback is not required in systems that are based on the recognition of EEG patterns that precede specific movements.

22.2.5.3 Current Applications and Evaluations

All applications and evaluations to date have employed EEG-based control as a substitute for conventional control. Sutter (1992) developed and extensively evaluated a system that is an EEG-based analogue to a computer light pen. It is based on the finding that a modulated stimulus in the center of visual field evokes a much larger cortical response than one in the visual periphery (Figure 22.15). A matrix of alphanumeric characters is presented on a monitor and the characters flicker in a unique sequence designed to maximize character recognition throughput. The system separately evaluates the evoked responses to each flickering character and selects the character with the largest response as the desired one, i.e., the one the user is visually fixating. No user training is required since the evoked response is involuntary; a reference evoked response template must be collected in a 10–20-min preliminary session, however. Character selection rates of 40 characters per minute with better than 90% accuracy have been achieved (Sutter, 1984). This represents a viable communication rate for persons with severe disabilities.

Wolpaw and his colleagues (Wolpaw and McFarland, 1994; Wolpaw, McFarland, Neat, and Forneris, 1991) have developed a system for computer cursor control that is based on voluntary control of the 8–12-Hz "mu" rhythm. Although it is in the same frequency range as the alpha rhythm, mu is recorded over the primary sensorimotor area of the brain and responds in known ways during movement preparation. In the single-axis task, the user must move the cursor to contact targets that appear randomly at the top or bottom of the monitor. Cursor direction and step size are based on the amplitude of the mu rhythm. After approximately 18 hr of training, users required 2–6 sec to move the cursor to a target. The target was correctly selected on 80–95% of the trials. The dual-axis task uses mu rhythm signals from both cortical hemispheres in a more complex control algorithm. After approximately 12 additional hr of training, the users required 2–4 sec to move the cursor to targets that appeared in one of the four corners of the screen. The target was correctly selected on 40–70% of the trials.

The second general approach to EEG-based control is being pursued by Pfurtscheller and his colleagues (Pfurtscheller, Flotzinger, Mohl, and Peltoranta, 1992; Pfurtscheller, Flotzinger, and Neuper, 1994). They are using neural networks to recognize mu rhythm and gamma (30–40-Hz) band EEG patterns that precede specific body movements, such as finger, toe, or tongue activity. No user training is required, but the movements must

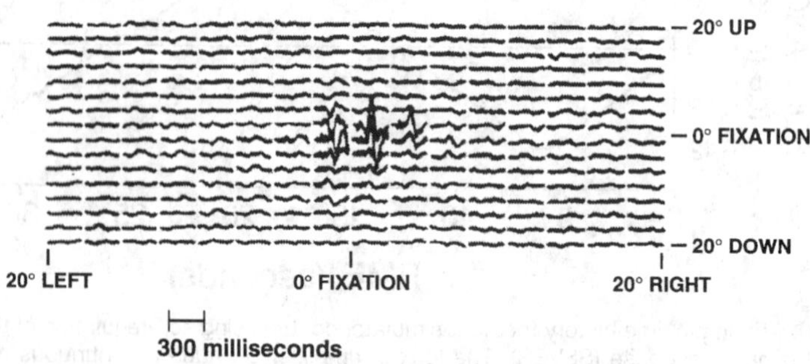

Figure 22.15 Visual evoked responses from 256 stimulus locations in the visual field. (From Sutter, 1984, copyright © 1984 IEEE.)

be repeated 100–200 times for neural network training. After training, movement prediction is possible with only 1 sec of EEG data. Pfurtscheller's off-line system achieved 89% accuracy in predicting button pushes with the left or right hand. With toe and tongue movement added, accuracy dropped to 70%. The neural network can be trained with imagined rather than actual movements, but movement prediction is slightly degraded in this case.

The use of the SSVER to control a simple flight simulator was described above. Most new subjects demonstrated some control after a single 30-min training session. Acquisition of 70–85% of the roll angle commands was typical after 5–6 hr of training. In addition to the simulator control application, Calhoun, McMillan, Morton, Middendorf, Schnurer, Ingle, Glaser, and Figoni (1995) have employed the SSVER to operate a functional electrical stimulator (FES) designed to exercise paralyzed limbs. A series of knee extension commands was presented in each trial to test user performance. One segment of such a trial is shown in Figure 22.16.

In summary, rudimentary one- and two-dimensional control and simple item selection have been demonstrated with current laboratory systems. The size, weight, and cost of these systems are not serious constraints, and learning EEG-based control does not appear to require any special skills or individual characteristics. Nevertheless, until the flexibility, precision, and reliability of EEG-based control can be increased, current applications are probably limited to assistive devices for the physically challenged and input devices for entertainment systems. EEG-based systems offer near-term potential as a communication channel for those whose profound disabilities leave them "locked in." For individuals with spared capabilities who are using alternatives such as sip-puff tubes, chin-operated joysticks, or mouth-held wands, EEG-based control offers the potential for a less awk-

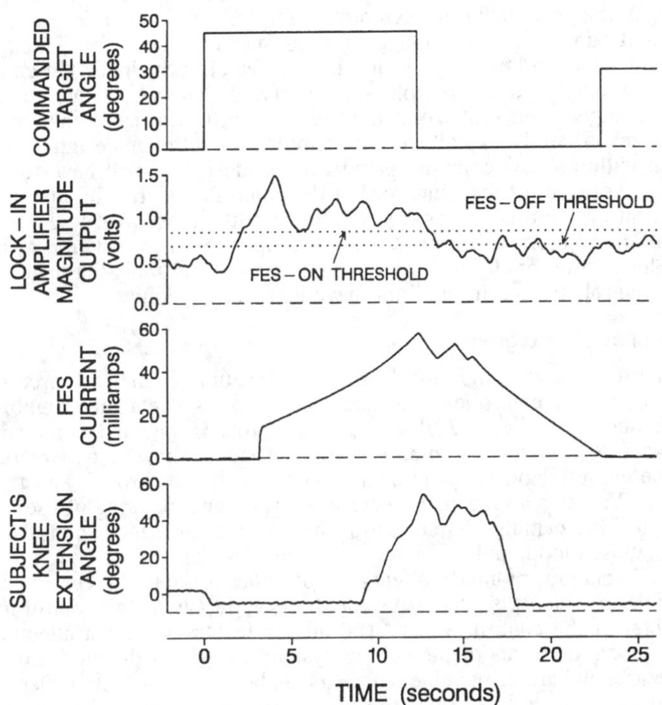

Figure 22.16 Example time history for functional electrical stimulator (FES) control of knee extension using self-regulation of the steady-state visual evoked response (SSVER). The lock-in amplifier provides a continuous measure of the magnitude of this response. Responses above one threshold turn the FES on, while responses below a lower threshold turn the FES off. Once the knee reaches the target angle, the FES current is modulated to prevent the limb from becoming hyper-extended. This is maintained until the user successfully reaches the FES-off threshold. This modulation may be seen in the figure.

ward-appearing control interface. In addition, there may be important psychological benefits from "direct willful" control of devices in the user's environment.

22.2.5.4 Future Research and Development

The use of EEG-based control as a supplement to conventional controls in high manual workload environments is the probable next step. For example, an aircraft designer could make cockpit radio frequency selection or multifunction display operation accessible with EEG-based, as well as conventional, inputs. This would require only moderate improvements in precision and reliability. The pilot could choose to use the EEG-based system when appropriate. An analogy is the availability of both keyboard and mouse functions for cursor positioning in a modern personal computer system. Users will choose one or the other depending on the nature of the task, hand location, and personal preference. Each of the approaches reviewed above could be applied in this manner. One key issue that must be addressed is the ability to operate EEG-based controls in multitask environments.

Use of EEG-based systems in transparent, intelligent interfaces will almost certainly require the extension of approaches based on the interpretation of naturally occurring EEG patterns. Here EEG-based control may achieve a utility far beyond what we now imagine. Unfortunately, this is the area where EEG-based systems are most threatening: the determination of user cognitive states, intentions, and desires. While this technology does raise the specter of "mind reading," such information is critical for truly intelligent interfaces and may be most readily available from signals recorded from the brain.

22.3 CONTROL METAMORPHOSIS

As the reviews in this chapter suggest, research and development are expanding the frontiers of control. The methods made possible by speech-, eye-, head-, gesture-, EMG-, and EEG-based technologies offer new possibilities for designing controls that are more natural to use and support a wider range of user–machine behaviors. Through effective design, including the application of sound human factors principles, the emerging technology will not only lead to controls and interfaces that can restore functionality to physically challenged users, but will also result in improved performance by all users of complex systems. To fully exploit these possibilities, controller design will have to be pursued from within a new coupling paradigm and designers will have to devote more time to the development of the "interior" of the controller itself. The final result will be the development of controls that do not look or feel like those that we know today. The controller will metamorphose into an "agent" that operates in conjunction with the human performer. This section briefly reviews emerging conceptual tools that can facilitate this transition to controllers that, in the limit, have agent-like qualities.

22.3.1 Coupling Paradigms

We use the term *Servo Paradigm* to identify the prevailing framework used to guide the development of manual controllers. The general form of user–machine coupling from this orientation is shown in Figure 22.17(*a*). The user makes an input command by direct physical linkage to a control handle. The control handle may be part of or connected directly to the end effector, but is frequently connected to it through an intervening actuator linkage. With this arrangement there is a one-to-one mapping of user input to end effector output. This paradigm is central to manual control theory, applying universally to open-loop, closed-loop, and multiloop controller designs.

The Servo Paradigm implicitly orients the designer to view the control development problem as one in which the user makes deliberative inputs to the control to "operate" the end effector of the control system. The human factors specialist attempts to ensure that the operational demands of the control system are within the capabilities and limitations of the user and are compatible with selected behavioral acts. In general, the design problem is treated as an exercise in forming a suitable actuator linkage between the performer, control system, and end effector. We may say that the controller is formed around an actuator principle.

At the beginning of the chapter, nonconventional controls were defined as the class of input devices that do not require a direct mechanical linkage between the user and system. All of the nonconventional control examples presented here establish the user–system linkage through the use of sensing techniques. When sense-based coupling is exploited from within the Servo Paradigm, we expect that the actuator principle will guide the

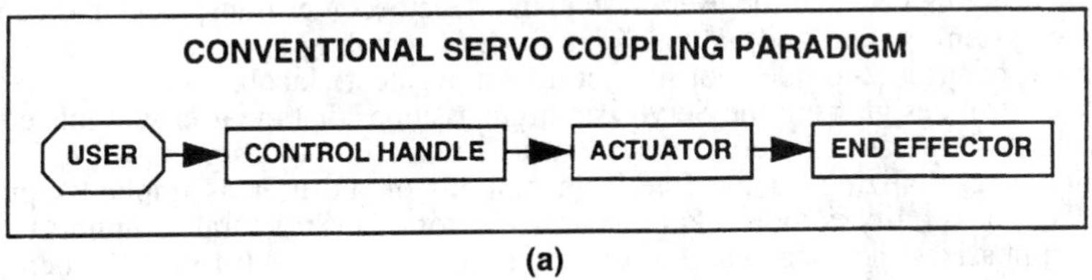

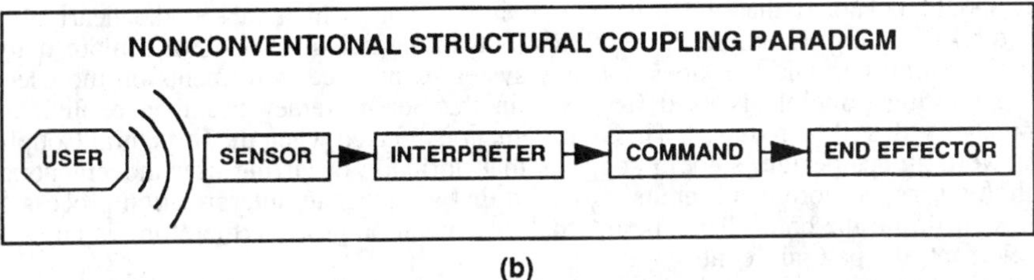

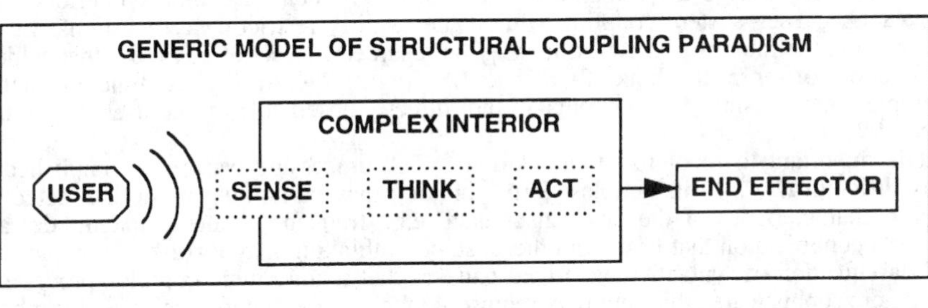

Figure 22.17 Coupling paradigms to guide the design of conventional and nonconventional controls.

development. This is most likely to lead to nonconventional controls that "look and feel" like conventional ones. The main difference will be the use of new human output channels. This will result in nonconventional controls that are most likely to be used as direct substitutes for, or supplements to, existing conventional controls. The situation changes, however, when controller design is approached from a different conceptual framework.

An alternative conceptual framework is presented in Figure 22.17(*b*). We refer to this framework as the *Structural Coupling Paradigm.* A sensor is used to monitor user actions. Further processing interprets the raw sensor data stream and results in the selection or creation of a command that is issued to the end effector. By using a sensor and the *principle of monitoring*, the need for direct mechanical linkage between the user and system is broken. As indicated in the figure, the controller itself is responsible for selecting and issuing commands to the end effector. Coupling of this form is looser than conventional approaches, since the performer does not actually manipulate the control. The two structures—the user and the controller—remain intact, physically independent, yet coupled; hence, the name of the paradigm (Eggleston, 1987, 1988, 1993; Maturana and Varela, 1988; see also Beek, 1989; Kugler and Turvey, 1987).

The shift from an actuator principle to a monitoring principle may appear subtle, but it can open new possibilities for thinking about and designing controllers. As we have noted, conventional coupling requires the performer to make explicit movements for the express purpose of manipulating the end effector. *When sense-based monitoring is employed, the performer is free to make unrestricted movements that through controller interpretation can be used implicitly to guide end effector action.* The interpreter infers user intent and issues commands to its end effector in agreement with that intent. The performer can simply "act naturally" and receive the desired support from the control system. It is in this way that the user–system interface becomes transparent and controlled use of the system is fluid and natural. This form of coupling appears to the user largely as "implicit control" and does not restrict user movement (Jacob, 1995).

Since control design from the Servo Paradigm assumes that the user will make explicit movements to "operate" the control system, it is unlikely the designer will consider the possibility of capitalizing on the free-form motions of the user as implicit input commands. This possibility is more likely to be seen from the Structural Coupling Paradigm since it emphasizes the user and control system as separate, well-formed structures that are available for coupling through monitoring or sensing. Design paradigms can have a strong influence on how emerging control technology is exploited.

It should be noted that the principle of monitoring, which lies at the heart of the Structural Coupling Paradigm, is necessary but not sufficient to achieve a more natural control system, i.e., one that does not levy system-demanded movements on the user. If the monitoring principle is used from within the Servo framework, then explicit user commands will still be required. This requirement is relaxed when the Structural Coupling framework guides design, since it suggests that normally occurring user movements can implicitly serve as command inputs, provided that an adequate interpretation process can be designed into the controller. The internal interpretation process, therefore, is an essential element of "natural" control.

Interpretation is an internal process, as reflected in the Structural Coupling Paradigm. This framework orients the designer to the task of producing "complex interiors" that allow defined user–system coupling without necessarily restricting user actions. In the general case, the interior of a controller may be regarded as a set of processes that achieve the functions of sense (S), think (T), and act (A) [Figure 22.17(*c*)]. These functions define a complex interior since their outputs are not directly observable by the user (Eggleston, 1988, 1993).

It is important to recognize that internal S-T-A processing can be accomplished in many different ways. In its simplest form, thinking may be nothing more than detecting specific characteristics of the raw sensor data (e.g., frequency content, magnitude) and issuing a command on that basis. In other instances, thinking may involve the recognition and classification of syntactically formed patterns that require more complex processing. Even more sophisticated thinking may require semantic-based interpretation and symbolic reasoning to create the controller's internal knowledge in real time, perhaps through the use of internal models. As the issuance of commands to the end effector becomes more dependent on its own ability to reason at a knowledge level, the controller begins to take on the appearance of an agent working jointly with the user. A controller in this form blurs the distinction between the control, system interface, and even the human–machine system (Eggleston, 1988). To complete the metamorphosis invited by the Structural Coupling Paradigm, the user will also have to change by adopting a new interaction style.

22.3.2 Interaction Styles

It may seem contradictory to suggest, on one hand, that the Structural Coupling Paradigm invites designers to exploit unrestricted user performance and, on the other hand, to claim a need for the user to invoke a new interaction style, since this suggests that new restrictions are being applied. In fact, the contradiction is more apparent than real. The apparent contradiction stems from the fact that the user may have to "unlearn" old restrictive interaction styles and be given an opportunity to gain comfort with a style that imposes less restrictions, but invokes a different type of human–machine coupling.

The manner in which a person makes an input to a controller or system interface may be called an interaction style. Within the Servo Paradigm, the system user is viewed as an "operator" who issues commands through manual manipulation of a control. A command to the controller is taken to be a command to the system; hence, the distinction between the controller and system is blurred. This is a general style that can be implemented with many different input devices, all of which are designed from within the Servo Paradigm. For example, a joystick, mouse, discrete switch, verbal statement, or EEG signal may be used to issue a command to a system.

Interaction style is important to consider in control design for at least two reasons. The designer may intentionally or unintentionally restrict interaction style based on the conceptual framework employed. Conversely, a user may fail to utilize capabilities made available by a new control interface if they simply carry over a style based on experience with a previous system. If the designer believes that a new interaction style is preferable, this information needs to be conveyed to the prospective user. This is especially true when constraints in the control system design are based on it. In general it is advisable for the designer to address this issue.

It is useful to consider at least three broad interaction styles that the control interface may support. These are the *Command Style*, which has already been discussed, and two others: the *Dialogue Style* and the *Transaction Style*.

The Dialogue Style has emerged over the past several years in the general domain of human-computer interaction. It is commonly associated with a "window" type interface that uses menus, icons, and dialogue boxes to support interactions. It can be implemented on a verbal level through the use of speech recognition and speech synthesis technology. The prototypical form of the Dialogue Style is a bidirectional, turn-taking conversation. For example, the user may point and click on a menu item to close a file and the system may reply by presenting a "dialogue box" to ask if changes to the file should be saved. Dialogues can degrade into one-directional commands by the user and make the control interface appear to be in a Command Style. In other words, a command interaction can be nested within the structure of a control interface that supports a more flexible Dialogue Style of interaction.

With the exception of so-called natural language interfaces, the Dialogue Style is constrained by the syntax imposed by the system. While more natural, flexible, and congenial than a command-based interface, it still imposes explicit and implicit restrictions on user actions. Unknown implicit restrictions are often the reason for a breakdown in dialogue that is not understood by the user and interferes with task performance.

Dialogues can be simple and straightforward or complex and difficult to learn. Approaching design from this style de-emphasizes the actuator principle in controller design and brings additional emphasis to the cognitive aspects of control. In general, it can be accommodated by control designs that emanate from the Servo Paradigm. However, natural language dialogues involving mixed initiative interactions, including multimodal dialogue concepts, contain elements that reflect a shift to the Structural Coupling Paradigm (Vo, Houghton, Yang, Bub, Meier, Waibel, and Duchnowski, 1995; Walker and Whittaker, 1990; Whittaker and Stenton, 1988).

The interaction style that most fully complements the Structural Coupling Paradigm is best described as the Transaction Style. This style emphasizes the agency and independence of the controller. The Transaction Style has the quality of two entities behaving implicitly as a partnership. The controller *notices* user, environment, and system states (senses), infers user intent (thinks) and issues commands (acts) in a manner expected to be consistent with user intentions. Based on experience, the user understands system actions and makes self-adjustments to stay consistent with the controller. The controller simultaneously adjusts to the user so that they can mutually achieve an implicit shared goal (Eggleston, 1987, 1988). The structure of a controller that can support this interaction style can also accommodate the more restrictive Command and Dialogue Styles. It has a complex interior similar to the top-level functional architecture shown in Figure 22.18.

A summary of some of the implications of these interaction styles for control design is shown in Table 22.4. A Command Style is essentially one-directional, from the user to the system, and invites the designer to design in terms of a master-slave relationship using the Servo Paradigm. It also invites single-control/single-display thinking in interface design. It has the advantage of being supportable by a single, one-directional data stream. Typically, physical manipulation of a control is explicit in the design. In contrast, the Dialogue Style of interaction is inherently bidirectional and imposes a "turn-taking" structure on interactions. If a task can be completed by accomplishing a connected series of user commands, turn-taking behavior is suppressed and the interaction appears command-like. This style supports queries and directives and allows interactions to be carried out at a higher cognitive level of discourse. The interface design usually contains explicit constraints on communication, while end effector manipulations may be constrained implicitly. A full duplex, bidirectional data path is required to support this style.

The Transaction Style of interaction offers the greatest range for flexible, natural interactions, since it does not place any a priori restrictions on the form of interaction. In this sense, it is "full bandwidth." Full bandwidth is achieved by having communications and manipulations be implicit rather than explicit. The controller or interface appears more like an agent behaving cooperatively with the human. The processor architecture to support this type of interaction requires parallel, asynchronous, and perhaps self-adjusting data streams.

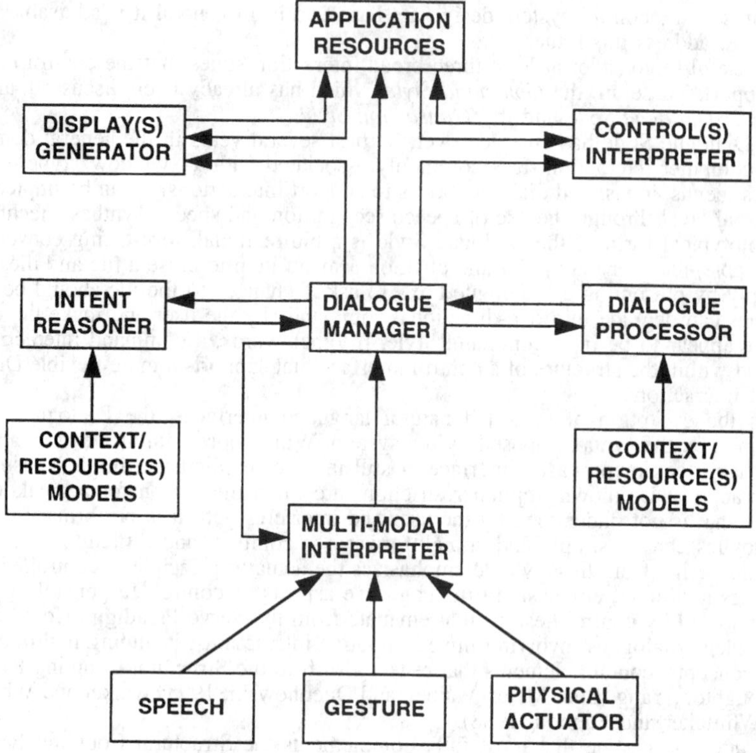

Figure 22.18 Notional adaptive intelligent interface architecture.

22.4 SUMMARY

In this brief review, we have attempted to provide an introduction to the conceptual tools, hardware, and software aspects of a range of new control concepts. Where appropriate, guidance was offered for component selection and for critical parameters relevant to nonconventional control design. We have offered the Structural Coupling Paradigm as a

Table 22.4 Characteristics of Three Styles of Human–Machine Interaction

Command Style	Dialogue Style	Transaction Style
One-directional (user to control)	Bidirectional	Independent, coordinated structures
Master-slave relationship (user to control)	Turn-taking relationship	Mutual, trust-based relationship
Explicit manipulation of control system	Explicit communication, implicit control system manipulation	Implicit communication, implicit control system manipulation
Coupling based on a single data stream	Coupling based on a single data stream with full duplex features	Parallel, asynchronous data paths
Invites *One Control–One Display* interface design	Agency (assistance)	Agency (partnership)
		Full bandwidth / non-restrictive
		Self-adjusting structures (mutual adaptation)

framework that can support the development of more flexible, natural, and transparent user-control interfaces. Other frameworks may achieve the same objective (Rouse, 1988). We hope that the material presented here will aid design engineers and human factors specialists by clarifying the admittedly fuzzy notions of natural and transparent interfaces, by identifying new design possibilities, and by partially filling the conceptual tool kit needed to design controls in a manner consistent with a user-centered philosophy.

REFERENCES

Acero, A., and Stern, R. M. (1990). Robust speech recognition by normalization of the acoustic space. *Proceedings of the International Conference on Acoustics, Speech, and Signal Processing*, Albuquerque, New Mexico, pp. 849–852.

Adams, K. D., and Bentz, B. D. (1995). Headpointing: the latest and greatest. *Proceedings of the RESNA 18th Annual Conference*, Vancouver, Canada, pp. 446–448.

Amori, R. D. (1992). Vocomotion: an intelligent voice-control system for powered wheelchairs. *Proceedings of the RESNA 15th Annual Conference*, Washington, D.C., pp. 421–423.

Atal, B. S. (1994). Speech Technology in 2001: New Research Directions. In D. R. Roe and J. G. Wilpon, Eds., *Voice Communication between Humans and Machines*. Washington, D.C.: National Academy Press.

Baudel, T., and Beaudouin-Lafon, M. (1993). Charade: remote control of objects using free-hand gestures. *Communications of the ACM*, *36*(7), 28–35.

Beek, P. J. (1989). *Juggling Dynamics*. Amsterdam: Free University Press.

Bichsel, M., Ed. (1995). *Proceedings of the International Workshop on Automatic Face- and Gesture-Recognition*. Zurich, Switzerland.

Bolt, R. A. (1980). Put-that-there: voice and gesture at the graphics interface. *Computer Graphics*, *14*(3), 262–270.

Borah, J. (1989). *Helmet-Mounted Eye Tracking for Panoramic Display Systems—Volume I: Review of Current Eye Movement Measurement Technology* (AAMRL-TR-89-019) (NTIS: AD-A273 101/6/XAB).

Borah, J. (1995). *Investigation of Eye and Head Controlled Cursor Positioning Techniques*. (AL-SR-1995-0018). Armstrong Laboratory, Wright-Patterson Air Force Base, OH.

Brooks, M. (1989). The DataGlove as a man-machine interface for robotics. *2nd IARP Workshop on Medical and Healthcare Robotics*. Newcastle upon Tyne, UK, pp. 213–225.

Bryson, S., and Levit, C. (1992). The virtual wind tunnel. *IEEE Computer Graphics and Applications*, *12*(4), 25–34.

Calhoun, G. L., and Janson, W. P. (1991a). Eye control interface considerations for aircrew station design. Paper presented at *Sixth European Conference on Eye Movements*. Leuven, Belgium.

Calhoun, G. L., and Janson, W. P. (1991b). *Eye Line-of-Sight Control Compared to Manual Selection of Discrete Switches* (AL-TR-1991-0015) (NTIS: AD-A273 019).

Calhoun, G. L., McMillan, G. R., Morton, P. E., Middendorf, M. S., Schnurer, J. H., Ingle, D. F., Glaser, R. M., and Figoni, S. F. (1995). Functional electrical stimulator control with a direct brain interface. *Proceedings of the RESNA 18th Annual Conference*, Vancouver, Canada, pp. 696–698.

Card, S. K., English, W. K., and Burr, B. J. (1978). Evaluation of mouse, rate-controlled isometric joystick, step keys, and text keys for text selection on a CRT. *Ergonomics*, *21*(8), 601–613.

Christensen, J. (1987). The Human Factors Profession. In G. Salvendy, Ed., *Handbook of Human Factors*. New York: John Wiley.

Clark, J. E., and Phillips, S. J. (1988). *The Efficacy of Using Human Myoelectric Signals to Control the Limbs of Robots In Space* (NASA-CR-182901) (NTIS: N88-25155).

Davis, S., and Mermelstein, P. (1980). Comparison of parametric representations for monosyllabic word recognition in continuously spoken sentences. *IEEE Transactions on Acoustics, Speech and Signal Processing*, *ASSP-28*(4), 357–366.

Eggleston, R. G. (1987). The Changing Nature of the Human-Machine Design Problem: Implications for System Design and Development. In W. B. Rouse, and K. R. Boff, Eds., *System Design: Behavioral Perspectives on Designers, Tools, and Organizations*. New York: North-Holland.

Eggleston, R. G. (1988). Machine Intelligence and Crew-Vehicle Interfaces. In E. Heer and H. Lum, Eds., *Machine Intelligence and Autonomy for Aerospace Systems*. Washington, D.C.: American Institute of Aeronautics and Astronautics.

Eggleston, R. G. (1993). Cognitive interface considerations for intelligent cockpits. *Combat Automation for Airborne Weapon Systems: Man/Machine Interface Trends and Technologies* (AGARD-CP-520), Edinburgh, Scotland, pp. 21/1–21/16.

Eggleston, R. G. (1997). Helmet-Mounted Display Systems and Human Performance. In J. Melzer and K. Moffitt, Eds., *Helmet-Mounted Displays: Designing for the User*. New York: McGraw-Hill.

Eggleston, R. G., Janson, W. P., and Adapalli, S. (1994). Manual tracking performance with a virtual hand controller: a comparison study. *Virtual Interfaces: Research and Applications* (AGARD-CP-541), Lisbon, Portugal, pp. 10/1–10/7.

Epps, B. W. (1986). Comparison of six cursor control devices based on Fitts' Law models. *Proceedings of the 30th Annual Meeting of the Human Factors Society*, Dayton, Ohio, pp. 327–331.

Fels, S. S., and Hinton, G. E. (1993). Glove-Talk: a neural network interface between a data-glove and a speech synthesizer. *IEEE Transactions on Neural Networks, 4(1)*, 2–8.

Ferrin, F. J. (1991). Survey of helmet tracking technologies. *Proceedings of the SPIE Conference on Large-Screen-Projection, Avionic, and Helmet-Mounted Displays.* H. M. Assenheim, R. A. Flasck, T. M. Lippert, and J. Bentz, Eds., San Jose, CA, Vol. 1456, pp. 86–94.

Fitts, P. M., and Peterson, J. R. (1964). Information capacity of discrete motor responses. *Journal of Experimental Psychology, 67*, 103–112.

Flanagan, F. L. (1972). *Speech Analysis, Synthesis and Perception.* New York: Springer Verlag.

Forney, G. D. (1973). The Viterbi algorithm. *Proceedings of the IEEE, 61*, 268–278.

Fukumoto, M., Suenaga, Y., and Mase, K. (1994). "Finger-Pointer": pointing interface by image processing. *Computers and Graphics, 18(5)*, 633–642.

Furui, S. (1992). Telephone networks in 10 years' time—technologies & services. *Proceedings of the COST-232 Speech Recognition Workshop*. Rome, Italy.

Glenn, F. A., Iavecchia, H. P., Ross, L. V., Stokes, J. M., Weiland, W. J., Weiss, D., and Zaklad, A.L. (1986). Eye-voice-controlled interface. *Proceedings of the 30th Annual Meeting of the Human Factors Society*, Dayton, OH, pp. 322–326.

Hale, J. P. (1992). *Anthropometric Teleoperation: Controlling Remote Manipulators with the DataGlove* (NASA-TM-103588) (NTIS: N92-28521/2/XAB).

Hannaford, B., and Venema, S. (1995). Kinesthetic Displays for Remote and Virtual Environments. In W. Barfield and T. A. Furness, Eds., *Virtual Environments and Advanced Interface Design.* New York: Oxford University Press.

Hermansky, H. (1995). Personal communication.

Hiraiwa, A., Uchida, N., and Shimohara, K. (1993). EMG pattern recognition by neural networks for prosthetic fingers control. *Proceedings of the IFAC Symposium on Artificial Intelligence in Real-Time Control*, Delft, Netherlands, pp. 73–79.

Hirsch, H., Meyer, P., and Ruehl, H. W. (1991). Improved speech recognition using high-pass filtering of subband envelopes. *Proceedings of Eurospeech 91*, Geneva, Switzerland, pp. 413–416.

Howard, J. D. (1987). Flight testing of the AFTI/F-16 voice interactive avionics system. *Proceedings of Military Speech Technology*, Arlington, VA, pp. 76–82.

Hudgins, B., Parker, P., and Scott, R. N. (1993). A new strategy for multifunction myoelectric control. *IEEE Transactions on Biomedical Engineering, 40(1)*, 82–94.

Jacob, R. K. (1995). Eye Tracking in Advanced Interface Design. In W. Barfield and T. A. Furness, Eds., *Virtual Environments and Advanced Interface Design.* New York: Oxford University Press.

Jagacinski, R. J., and Monk, D. L. (1985). Fitts' Law in two dimensions with hand and head movements. *Journal of Motor Behavior, 17*, 77–95.

Jennings, D. L., and Ruck, D. W. (1995). Enhancing automatic speech recognition with an ultrasonic lip motion detector. *Proceedings of the IEEE International Conference on Acoustics, Speech and Signal Processing*, Detroit, MI, pp. 868–871.

Junker, A., Berg, C., Schneider, P., and McMillan, G. R. (1995). *Evaluation of the Cyberlink Interface as an Alternative Human Operator Controller.* (AL/CF-TR-1995-0011). Armstrong Laboratory, Wright-Patterson Air Force Base, OH.

Kaczmarek, K. A., and Bach-y-Rita, P. (1995). Tactile Displays. In W. Barfield and T. A. Furness, Eds., *Virtual Environments and Advanced Interface Design.* New York: Oxford University Press.

Kanevsky, D., Danis, C. M., Gopalakrishan, P. S., Hodgson, R., Jameson D., and Nahamoo, D. (1990). A Communication Aid for the Hearing Impaired Based on an Automatic Speech Recognizer. In L. Torres, E. Masgrau, and M. A. Lagunas, Eds., *Signal Processing V: Theories and Applications.* Amsterdam: Elsevier Science.

Karis, D., and Dobroth, K. M. (1991). Automating services with speech recognition over the public switched telephone network: human factors considerations. *IEEE Journal of Selected Areas of Communication, 9(4)*, 574–585.

Knapp, R. B., Lusted, H., and Patmore, D. W. (1995). Using the electrooculogram as a means for computer control. *Proceedings of the RESNA 18th Annual Conference*, Vancouver, Canada, pp. 696–701.

Kocian, D. F., and Task, H. L. (1995). Visually Coupled Systems Hardware and the Human Interface. In W. Barfield and T. A. Furness, Eds., *Virtual Environments and Advanced Interface Design.* New York: Oxford University Press.

Koons, D. B., Sparrell, C. J., and Thorisson, K. R. (1993). Integrating Simultaneous Input from Speech, Gaze, and Hand Gestures. In M. T. Maybury, Ed., *Intelligent Multimedia Interfaces.* Menlo Park, CA: AAAI Press/The MIT Press.

Kramer, J., and Liefer, L. (1989). *The Talking Glove: An Expressive and Receptive "Verbal" Communication Aid for the Deaf, Deaf-Blind, and Nonvocal.* (Technical Report). Department of Electrical Engineering, Stanford University, Stanford, CA.

Krueger, M. W. (1990). *Artificial Reality,* Second Edition. Reading, MA: Addison-Wesley.

Kugler, P. N., and Turvey, M. T. (1987). *Information, Natural Law, and the Self-Assembly of Rhythmic Movement.* NJ: Lawrence Erlbaum Associates.

Kyger, D. W. (1995). *Reducing Lag in Virtual Displays Using Multiple Model Adaptive Estimation.* Masters Thesis (AFIT/GE/ENG/95D-11). School of Engineering, Air Force Institute of Technology, Wright-Patterson AFB, OH.

Lee, C., Lin, C-H., and Juang, B-H. (1991). A study on speaker adaptation of the parameters of continuous density hidden Markov models. *IEEE Transactions, 39(4),* 806–814.

Levision, W. H., Zacharias, G. L., Porterfield, J. L., Monk. D., and Arbak, C. (1981). A comparison of head and manual control for a position-control pursuit tracking task. *Proceedings of the 17th Annual Conference on Manual Control.* Jet Propulsion Laboratory Publication 81-95, Los Angeles, CA, pp. 641–652.

Liang, J., Shaw, C., and Green, M. (1991). On temporal-spatial realism in the virtual reality environment. *Proceedings of the ACM Symposium on User Interface Software and Technology,* Hilton Head, South Carolina, pp. 19–25.

Lin, M-L., Radwin, R. G., and Vanderheiden, G. C. (1992). Gain effects on performance using a head-controlled computer input device. *Ergonomics, 35(2),* 159–175.

Longinow, N. E. (1994). Predicting pilot look-angle with a radial basis function network. *IEEE Transactions on Systems, Man, and Cybernetics, 24(10),* 1511–1518.

Lowerre, B. T. (1976). *The Harpy Speech Recognition System,* Doctoral Thesis, Pittsburgh, PA: Carnegie-Mellon University.

MacKenzie, I. S. (1992). Fitts' Law as a research and design tool in human–computer interaction. *Human Computer Interaction, 7,* 91–139.

Makhoul, J., and Schwartz, R. (1994). State of the Art in Continuous Speech Recognition. In D. R. Roe and J. G. Wilpon, Eds., *Voice Communication between Humans and Machines.* Washington, D.C.: National Academy Press.

Massimino, M. J., and Sheridan, T. B. (1993). Sensory substitution for force feedback in teleoperation. *Presence, 2(4),* 344–352.

Maturana, H. R., and Varela, F. (1988). *The Tree of Knowledge: The Biological Roots of Understanding.* Boston: New Science Library Shambhala.

Maybeck, P. S., Herrera, T. D., and Evans, R. J. (1994). Target tracking using infrared measurements and laser illumination. *IEEE Transactions on Aerospace and Electronic Systems, 30(3),* 758–768.

McMillan, G. R., Calhoun, G. L., Middendorf, M. S., Schnurer, J. H., Ingle, D. F., and Nasman, V. T. (1995). Direct brain interface utilizing self-regulation of the steady-state visual evoked response. *Proceedings of the RESNA 18th Annual Conference,* Vancouver, Canada, pp. 693–695.

Miller, G. (1992). Voice recognition as an alternative computer mouse for the disabled. *Proceedings of the RESNA 15th Annual Conference,* Washington, D.C., pp. 55–57.

Murveit, H., Butzberger, J., and Weintraub, W. (1992). Reduced channel dependence for speech recognition. *Proceedings of the DARPA Speech and Natural Language Workshop,* Harriman, NY, pp. 280–284.

Nakatsu, R. (1990). Anser—an application of speech technology to the Japanese banking industry. *Computer, 23(8),* 43–48.

Nielsen, J. (1993). Noncommand user interfaces. *Communications of the ACM, 36,* 83–99.

Parker, P. A., and Scott, R. N. (1986). Myoelectric control of prostheses. *CRC Critical Reviews in Biomedical Engineering, 13(4),* 283–310.

Parsons, T. (1987). *Voice and Speech Processing.* New York: McGraw-Hill.

Patterson, P. E., and Katz, J. A. (1992). Design and evaluation of a sensory feedback system that provides grasping pressure in a myoelectric hand. *Journal of Rehabilitation Research and Development, 29(1),* 1–8.

Pentland, A., and Mase, K. (1991). Automatic lipreading by optical-flow analysis. *Systems and Computers in Japan, 22(6),* 67–76.

Petajan, E., Bischoff, B., Bodoff, D., and Brooke, N. M. (1988). An improved automatic lipreading system to enhance speech recognition. *CHI'88 Conference Proceedings: Human Factors in Computing Systems.* E. Soloway, D. Frye, and S. B. Sheppard, Eds., Washington, D.C., pp. 19–25.

Pfurtscheller, G., Flotzinger, D., Mohl, W., and Peltoranta, M. (1992). Prediction of the side of hand movements from single-trial multi-channel EEG data using neural networks. *Electroencephalography and Clinical Neurophysiology, 82,* 313–315.

Pfurtscheller, G., Flotzinger, D., and Neuper, C. (1994). Differentiation between finger, toe and tongue movement in man based on 40 Hz EEG. *Electroencephalography and Clinical Neurophysiology, 90,* 456–460.

Quam, D. L., Williams, G. B., Agnew, J. R., and Browne, P. C. (1989). An experimental determination of human hand accuracy with a DataGlove. *Proceedings of the Human Factors Society 33rd Annual Meeting*, Denver, CO, *1*, pp. 315–319.

Rabiner, L., Levinson, S., and Sondhi, M. (1983). On the application of vector quantization and hidden Markov models to speaker-independent isolated word recognition. *Bell System Technical Journal*, *62*, 1075–1105.

Radwin, R. G., Vanderheiden, G. C., and Lin, M-L. (1990). A method for evaluating head-controlled computer input devices using Fitts' Law. *Human Factors*, *32(4)*, 423–438.

Rajasekaran, P. K., and Doddington, G. (1986). Robust speech recognition: initial results and progress. *Proceedings of the DARPA Speech Recognition Workshop*, Palo Alto, CA, pp. 73–80.

Rouse, W. B. (1988). Adaptive aiding for human/computer control. *Human Factors*, *30(4)*, 431–443.

Rubine, D. (1991). Specifying gestures by example. *Computer Graphics*, *25(4)*, 329–337.

Scott, R. N. (1990). Feedback in myoelectric prostheses. *Clinical Orthopaedics and Related Research*, *256*, 58–63.

Scott, R. N., and Parker, P. A. (1988). Myoelectric prostheses: state of the art. *Journal of Medical Engineering and Technology*, *12*, 143–151.

Sears, H. H., and Shaperman, J. (1991). Proportional myoelectric hand control: an evaluation. *American Journal of Physical Medicine and Rehabilitation*, *70(1)*, 20–28.

Seneff, S., Zue, V., Polifroni, J., Pao, C., Hetherington, L., Goddeau, D., and Glass, J. (1995). The preliminary development of a displayless PEGASUS system. *Proceedings of the ARPA Spoken Language Systems Technology Workshop*, Austin, TX, pp. 212–217.

Simpson, C. A., McCauley, M. E., Roland, E. F., Ruth, J. C., and Williges, B. H. (1987). Speech Controls and Displays. In G. Salvendy, Ed., *Handbook of Human Factors*. New York: John Wiley.

Stanton, B. J. (1988). *Robust Recognition of Loud and Lombard Speech in the Fighter Cockpit Environment*. Doctoral Thesis, West Lafayette, IN: Purdue University.

Starker, I., and Bolt, R. A. (1990). A gaze-responsive self-disclosing display. In *Proceedings of the ACM CHI '90 Human Factors in Computing Systems Conference*, Addison Wesley/ACM Press, New York, pp. 3–9.

Starner, T., and Pentland, A. (1995). Visual recognition of American Sign Language using hidden Markov models. *Proceedings of the International Workshop on Automatic Face- and Gesture-Recognition*. M. Bichsel, Ed., Zurich, Switzerland, pp. 189–194.

Stork, D. G., Wolff, G., and Levine, E. (1992). Neural network lipreading system for improved speech recognition. *IEEE International Joint Conference on Neural Networks*, Baltimore, MD, *2*, pp. 285–295.

Sturman, D. J., and Zeltzer, D. (1993). A design method for "whole-hand" human-computer interaction. *ACM Transactions on Information Systems*, *11(3)*, 219–238.

Sturman, D. J., and Zeltzer, D. (1994). A survey of glove-based input. *IEEE Computer Graphics and Applications*, *23*, 30–39.

Sutter, E. E. (1984). The visual evoked response as a communication channel. *Proceedings: IEEE Symposium on Biosensors*, pp. 95–100.

Sutter, E. E. (1992). The brain response interface: communication through visually-induced electrical brain responses. *Journal of Microcomputer Applications*, *15*, 31–45.

Takahashi, T., and Kishino, F. (1991). Hand gesture coding based on experiments using a hand gesture interface device. *SIGCHI Bulletin*, *23(2)*, 67–74.

Tong, H. M., and Fisher, R. A. (1984). Progress report on an eye-slaved area-of-interest visual display. *Proceedings of the IMAGE III Conference* (AFHRL-TR-84-36), Phoenix, AZ.

Vo, M. T., Houghton, R., Yang, J., Bub, U., Meier, U., Waibel, A., and Duchnowski, P. (1995). Multimodal learning interfaces. *Proceedings of the ARPA Spoken Language Systems Technology Workshop*, Austin, TX, pp. 233–237.

Walker, M., and Whittaker, S. (1990). Mixed initiative in dialogue: an investigation into discourse segmentation. *Proceedings of the 28th Annual Meeting of the Association for Computational Linguistics*, Pittsburgh, PA, pp. 70–78.

Ware, C., and Mikaelian, H. T. (1987). An evaluation of an eye tracker as a device for computer input. *Proceedings of Human Factors in Computing Systems and Graphics Interface Conference*. J. M. Carroll and P. P. Tanner, Eds., Toronto, Canada, pp. 183–188.

Welford, A. T. (1960). The measurement of sensory-motor performance: survey and reappraisal of twelve years' progress. *Ergonomics*, *3*, 189–230.

Wellner, P. (1993). Interacting with paper on the DigitalDesk. *Communications of the ACM*, *36(7)*, 87–96.

Whittaker, S., and Stenton, P. (1988). Cues and control in expert-client dialogues. *Proceedings of the 26th Annual Meeting of the Association for Computational Linguistics*, Buffalo, NY, pp. 123–130.

Wilpon, J. G. (1994). Voice-processing Technology in Telecommunications. In D. R. Roe and J. G. Wilpon, Eds., *Voice Communication between Humans and Machines*. Washington, D.C.: National Academy Press.

Wilpon, J. G., Rabiner, L. R., Lee, C. H., and Goldman, E. R. (1990). Automatic recognition of keywords in unconstrained speech using hidden Markov models. *IEEE Transactions on Acoustics, Speech and Signal Processing, 38(11)*, 1870–1878.

Wise, S., Gardner, W., Sabelman, E., Valainis, E., Wong, Y., Glass, K., Drace, J., and Rosen, J. M. (1990). Evaluation of a fiber optic glove for semi-automatic goniometric measurements. *Journal of Rehabilitation Research and Development, 27(4)*, 411–424.

Wolpaw, J. R., and McFarland, D. J. (1994). Multichannel EEG-based brain-computer communication. *Electroencephalography and Clinical Neurophysiology, 90*, 444–449.

Wolpaw, J. R., McFarland, D. J., Neat, G. W., and Forneris, C. A. (1991). An EEG-based brain-computer interface for cursor control. *Electroencephalography and Clinical Neurophysiology, 78*, 252–259.

Young, L. R., and Sheena, D. (1988). Eye Movement Measurement Techniques. In J. Webster, Ed., *Encyclopedia of Medical Devices and Instrumentation*. New York: John Wiley.

Yuhas, B. P., Goldstein, M. H., and Sejnowski, T. J. (1989). Integration of acoustic and visual speech signals using neural networks. *IEEE Communications Magazine*, November, pp. 65–71.

CHAPTER 23

BIOMECHANICAL ASPECTS OF WORKPLACE DESIGN

Don B. Chaffin
Center for Ergonomics
The University of Michigan
Ann Arbor, MI 48109-2117 USA

23.1 WORKPLACE DESIGN AS A PROCESS

Most often the design of the workplace is an iterative process, wherein incremental changes are made to tools, equipment, fixtures, and chairs. Sometimes the workplace design begins with a "clean sheet of paper." Even in these latter circumstances, however, vendors will be consulted, and most often will present existing products, which then may need to be evaluated and possibly modified to meet new design criteria.

Perhaps the most important criteria to be used in any design process are derived from a fundamental understanding of the intended end user's expectations. For instance, if higher productivity is expected, then the design objective may be to minimize any human or hardware idle times. If improved process quality is needed, then the design objective may emphasize automated process control with a person supervising hardware operations. If reduction in worker injury, complaints, and/or absenteeism/turnover is desired, then the design objective would focus on reduction of work-related risk factors.

This chapter assumes that this latter objective is predominant, and that possession of a basic understanding of biomechanical stresses (as explained in Chapter 9) provides the first step in a workplace design process. The importance of understanding biomechanics is underscored in the next subsection. This is followed by a discussion of some major population attributes that must be considered in workplace design. The use of a formal workplace design process model is reviewed, followed by subsections which describe how biomechanical principles can be used to design packages, shelving, lift devices, handtools, workbenches, and workseats.

23.1.1 Why Use Biomechanical Principles for Workplace Design

Biomechanical principles provide a means by which the complex effects of specific work parameters, such as exertion related forces, awkward postures, work/rest cycles, and vibration can be more objectively understood and used to predict unwanted human outcomes (e.g., injury or pain). In this context, any workplace design that may have a person exposed to any or all of these work conditions should be evaluated to ensure that excessive physical stress (risk) is not imparted to a given population of people who are expected to function in the workplace. If the following statistics regarding work-related injuries to the musculoskeletal system are any indication, it would appear we have a major economic and social incentive to reduce physical stress in the workplace.

The most costly and prevalent work-related musculoskeletal disorder is low back pain. It is estimated by the National Council on Compensation Insurance that low back pain cases account for approximately 33% of all workers' compensation payments. When indirect costs are included, the total costs estimates range from about $27 to $56 Billion in the United States (Andersson et al., 1991). Further, about 20% of all workers' compensation claims are for low back pain, which affects over half of all working men and women during their work years (Andersson, 1981). Its occurrence and severity is associated most often with jobs involving lifting, bending, and twisting motions of the torso. Younger men and older women appear to be the most affected groups of workers.

In some industries wherein hand and arm exertions are prevalent, upper extremity cumulative trauma disorders (UECTDs) are even more costly than low back pain cases. Armstrong and Silverstein (1987) found that in plants wherein frequent hand and arm exertions predominated, more than one in 10 workers annually reported UECTDs.

In addition, standing on hard floors for prolonged periods has been associated with lower extremity and back problems. Maintaining a bent (flexed) neck posture when attempting to see small items or read a poorly designed computer generated display can cause severe neck pain and muscle spasms in the upper back. Raising the arm frequently and/or forcefully above shoulder height can cause chronic pain and mobility limitations in the shoulder (Chaffin and Andersson, 1991).

Clearly the human body is vulnerable to a large array of injurious physical stresses resulting from poor workplace design. These stresses are not often intuitively obvious, even to a trained ergonomist, until the actual worker-hardware system is set up and a person attempts to perform the required tasks comprising the job. Given this situation, how can past mistakes be avoided in the design of new workplaces? The answer relies on recognizing that the design of a workplace is a multidisciplinary process which must be continuously promoted and used. A conceptual model of this process is described in the next section.

23.1.2 A Conceptual Workplace Design Model

A typical job design, continuous improvement process cycle is graphically depicted in Figure 23.1. One often enters into this cycle at the top, by realizing through some form of surveillance scheme that the original goals for a particular job are not being met, i.e., a problem is identified. As stated at the beginning of this chapter, this is a very important step, because the statement of the problem becomes the objective for the new job design. In the context of this particular presentation, we assume that the types of problems being identified are in the realm of excessive musculoskeletal injuries and associated costs. Thus, the new job design (or redesign) goal is to reduce the risk factors believed to be causing these adverse outcomes.

Once the goal of a new design has been established, then it is necessary to turn to the fundamentals. In this context, the methods used to evaluate an existing or proposed new job will come from the biomechanical concepts described earlier in Chapter 8 (*Engineering Anthropometry*), Chapter 9 (*Biomechanics of the Human Body*), Chapter 10 (*Work Physiology*), and Chapter 13 (*Mental Workload Measurement and Analysis*) and Chapter 36 (*Work-Related Musculoskeletal Disorders of the Upper Extremity*). In reviewing these chapters, it should become very clear that there are a multitude of methods available today to assess the level of risk presented to a given population of workers. Clearly the selection of the most appropriate should be largely determined by how accurately the method predicts the type of risk factors identified as causing the problems in a particular job scenario. In other words, it is very important that the designer of a workplace formally

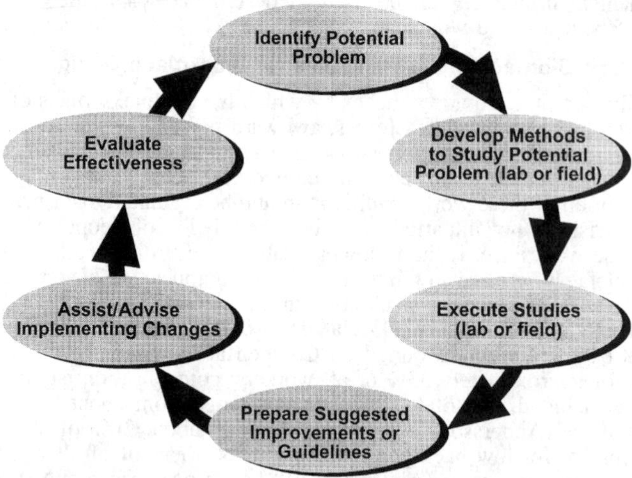

Figure 23.1 Typical job continuous improvement cycle.

review the scientific basis for a particular method of risk analysis, in that some methods being promoted today have not been well validated and peer reviewed. This situation even has legal implications. A recent federal court barred testimony of a well-regarded ergonomist in a case of alleged bad workplace design, citing Federal Rule of Evidence 702, which requires expert opinions to be based on "scientific, technical or specialized knowledge" that will assist in understanding evidence or determining a fact at issue (M. I. Rendell, 1995). Opinions regarding a workplace design that are not based on scientific studies will be questioned more and more because of the high costs of making a wrong decision when musculoskeletal injuries are involved. Thus, a designer must take the time to understand the fundamentals, and/or consult with those experts that possess such knowledge.

The next step in the job design cycle requires that a laboratory or in-plant study be performed. Many designers may question the need for this step, claiming that the general design guidelines that exist are adequate to improve present designs. Indeed this may be the case, but too often it is not. The present guidelines (some of which are described later in this chapter and elsewhere) many times are too general. The variability between tasks, special populations of workers, and hardware configurations is so great, and the combinations of conditions so complex that guidelines in the literature must be carefully interpolated and extrapolated to a specific design scenario. Often such an exercise raises many questions which can only be resolved by performing a laboratory study of a prototype design to assure the expected benefits can be achieved. Such studies are usually of a small scale, but they may be the only way to ensure that proposed changes in a workplace design have a reasonable probability of meeting the design goal. It is for this reason that a good ergonomics process will be supported by a laboratory for early evaluation of prototype designs. If this is not feasible, at least computerized simulations should be performed on the prototype designs (Evans and Chaffin, 1986; and Beck et al., 1993).

The continuous improvement cycle shown earlier in Figure 23.1 next proposes that specific design guidelines should be stated. Often this step is necessary because equipment suppliers (vendors) will become involved. These groups may not understand the basis for certain desired features in equipment (i.e., they may have very limited ergonomics expertise) and as stated earlier, they often will seek to sell off-the-shelf technologies and products to keep their costs low. It is for these reasons that purchase orders are beginning to specify ergonomic features, and the more specific these are, the better.

Next is the implementation phase, wherein new hardware is delivered and installed. At this point, many unanticipated job conditions and task requirements can affect the operations of the hardware. A good ergonomics design process requires the original designer to be present at start-up to advise (possibly train workers and supervisors) on the use of the equipment. This may require revising standard operating manuals (see Chapters

16 and 17), fine-tuning equipment designs, and setting up a monitoring system to ensure the procedure is performing as required.

The last step, which is one of the most avoided but necessary steps in a continuous design process, is to carefully evaluate the new job. Any organization that expects to improve itself over time realizes the necessity of this step. Unfortunately, even if performance data is obtained, it too often is not reported in a systematic fashion to the original designers. Because musculoskeletal disorders and complaints may not be evident for several months or even a year after a job has been initiated, there often is even less interest on the part of designers to return to an "old" design. One method to at least provide quick feedback on a design is to perform a risk factor assessment on the new design very shortly after start-up, using the same methods involved in the prototype evaluations.

In summary, a continuous design improvement process will involve a number of people and disciplines. Adapting the general improvement cycle depicted earlier (Figure 23.1) to the specifics of workplace design results in Figure 23.2. It is hoped that this type of a design process, along with the general workplace design concepts which follow will result in a real reduction in work-related musculoskeletal disorders in the future. This author is convinced that only through such a proactive approach will real benefits be sustained.

23.2 THE METHODOLOGICAL BASIS FOR ERGONOMIC WORKPLACE DESIGN

As described in Chapter 9, *Biomechanics of the Human Body*, the concepts and principles of biomechanics provide a logical means by which both limited human performance data and human tissue tolerance data can be applied to a multitude of different job conditions. The method of application most often relies on a computerized biomechanical model of the human musculoskeletal system to predict human capabilities for given task requirements. An example logic for a model used to predict whole-body exertion capabilities for a given population is depicted in Figure 23.3. In this model, specific muscle group strength data and spinal vertebrae failure data are used as the limiting values for the reactive moments at various body joints created when a person of a designated stature and body weight attempts an exertion (i.e., lifts, pushes, or pulls in a specific direction

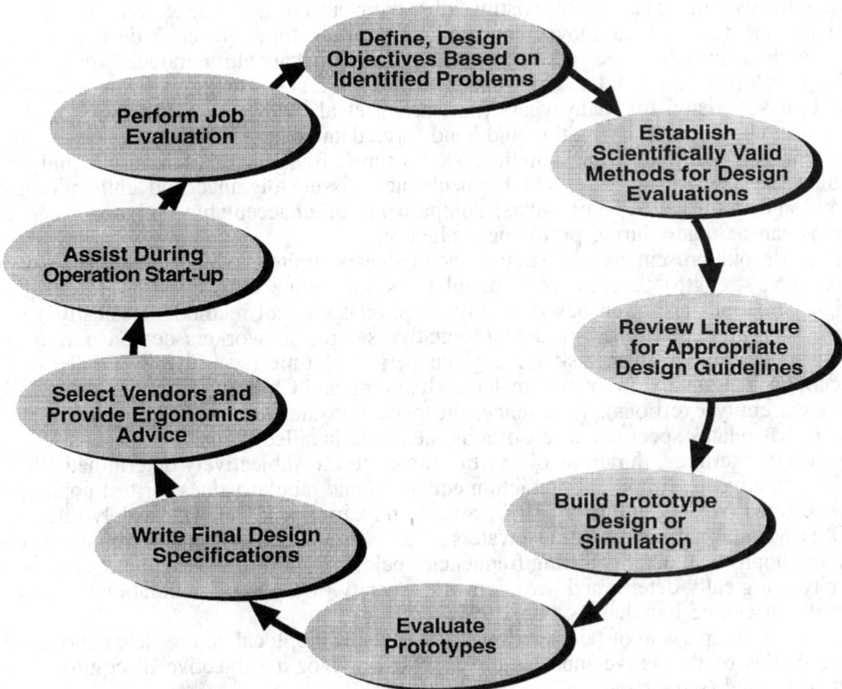

Figure 23.2 Workplace continuous improvement design process.

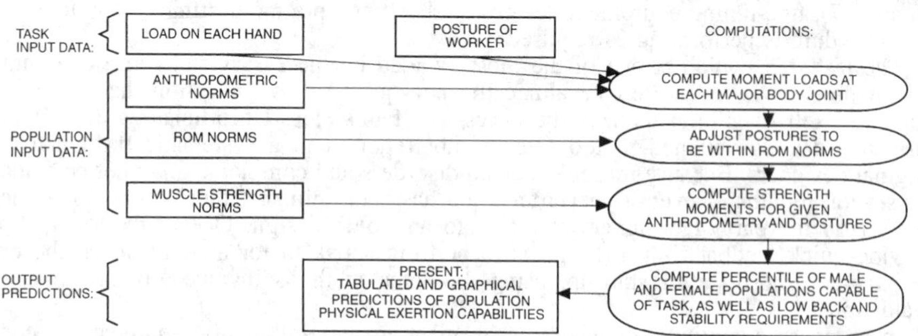

Figure 23.3 Biomechanical logic used to predict whole-body static exertion capabilities for given postures, hand force directions and anthropometric groups.

with one or both hands while maintaining a predicted posture). This logic was first developed by Chaffin (1969) and refined by Garg and Chaffin (1975), with implementation on personal computers for design purposes in 1984 at the University of Michigan. It represents one attempt to predict human capabilities over a wide range of maximum exertion requirements that may exist in a job design. Its validity to predict static exertion capabilities has been well established for men and women from 18 to 45 years old (Chaffin et al., 1987; Chaffin and Erig, 1991), with r^2 values under controlled conditions over 0.8. Studies to extend its application to older workers performing high exertion tasks are under way.

Biomechanical principles also provide the basis for comparing dynamic conditions. As described in Chapter 9, if a given manual task can be mocked up, then human motion data can be acquired by video analysis, or by goniometric devices attached to the body of a person performing the task of interest. These data can be evaluated by comparing them to various empirical norms (Marras and Mirka, 1992; McGill and Norman, 1987) to predict the risk of low back pain. Though such approaches provide realistic and valid models for dynamic analysis of existing jobs, their current dependence on complex data acquisition and 3-D human movement data limit their use for engineering design purposes.

Though quantitative risk prediction and population capability models are just now being developed and validated for upper extremity CTDs, qualitative risk assessment models have existed for many years (Armstrong et al., 1987; Keyserling et al., 1991). These use upper extremity motion and hand force data acquired from direct observations of a worker performing a task of interest. Postures, frequencies of motions, and loads handled are assessed against both biomechanical tissue tolerance and epidemiological norms, and within broad guidelines, comparisons of unacceptable and not acceptable designs can be made during prototype evaluations.

Because our present biomechanical knowledge is limited to understanding peak exertion (i.e., strength) capabilities in populations, and some even question its adequacy, design guidelines are often based strictly on psychophysical methods. Basically, a psychophysical guideline relies on a representative sample of workers demonstrating their ability to perform a given task for a given period of time (see Chapter 34 for use of psychophysical method in manual materials handling and Chapter 36 for use in evaluating upper extremity exertions). In essence, the psychophysical method is an empirical approach that relates specific test conditions, i.e., load handled, general postures used, frequencies of exertions, durations of exertions, etc., to the subjectively determined abilities of the test subjects. Empirical prediction equations and tabular values for test populations have resulted (Ayoub and Mital, 1989; Snook and Ciriello, 1991). The widely referenced NIOSH Manual Lifting Equation (Waters et al., 1993) is based on both biomechanical and psychophysical data for lifting frequencies below about 6 lifts per minute, after which the physiologically determined aerobic work capacity of a normal population dictates the weight that can be handled (Ayoub, 1992).

Last, for comparison of task exertions wherein specific, localized muscle actions exist, either EMGs of the active muscles are assessed and/or a subjective discomfort rating method is used (see Chapter 11).

In summary, for the initial design of a workplace there exists several different types of guidelines and design evaluation methods. Which one is appropriate depends on the

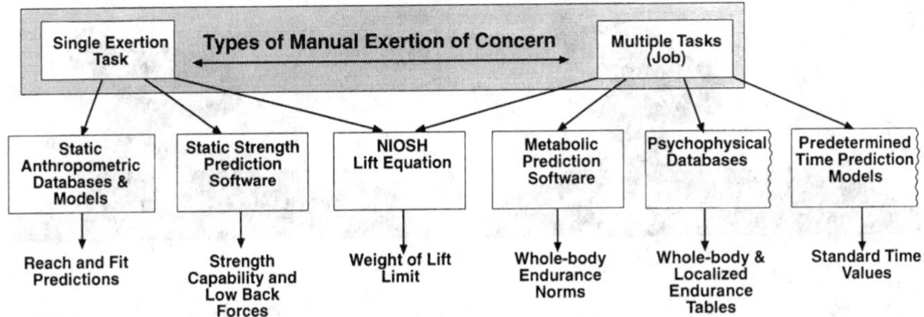

Figure 23.4 Models and databases available for proscriptive guidance in design of new workplaces with outcome predictions listed at bottom for each model or database.

types of problems one is attempting to alleviate with the new design, and whether a single exertion or multiple exertions are of concern. For initial design concept development and verification only a limited number of proscriptive ergonomic design methods exist. Their use is diagrammed in Figure 23.4.

Of course, all of these methods can (and should) be used to evaluate prototype designs of workplaces. Once a prototype is available, an experienced [representative worker(s)] can be studied to further refine the initial design. In addition to the tools in Figure 23.4, several other measurement methods can be considered; these are depicted in Figure 23.5.

It should be clear from both Figures 23.4 and 23.5 (and the concepts and principles discussed earlier in Chapters 8, 9, and 10) that there are ample methods to assure that a final workplace specification meets contemporary ergonomic guidelines. What follows is a description of some general guidelines meant to reduce the risk of common musculoskeletal problems created or aggravated by poor workplace designs.

23.3 WORKPLACE DESIGN FOR MATERIALS HANDLING

The problems created by manual materials handling tasks are well described in Chapter 34. What follows are illustrations and discussions of how workplace designs can be improved to reduce these problems.

23.3.1 Large Package Lifting and Lift Table Consideration

Package size can be problematic from a biomechanics perspective when the package is located on or near the floor. In this situation, if the object is of a size that cannot be easily straddled between the knees at the beginning of the lift, then a person must lean the torso forward. This is illustrated in Figure 23.6 for two different size boxes of the same weight using the 3-D Static Strength Prediction Program™ (Chaffin 1994). The combination of a forward torso angle and large horizontal distance between the box and low back in Figure 23.6B causes approximately a 30% to 38% increase in predicted L5/S1 disc compression force compared to small box lifting close to the body. If the object

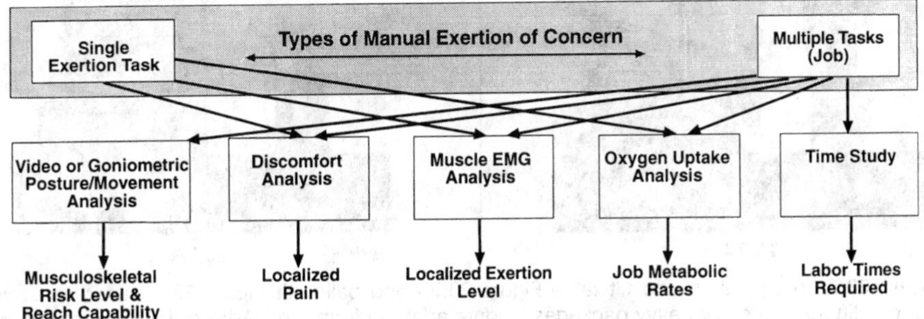

Figure 23.5 Methodologies for measuring human performance during prototype analyses, with outcome data listed for each.

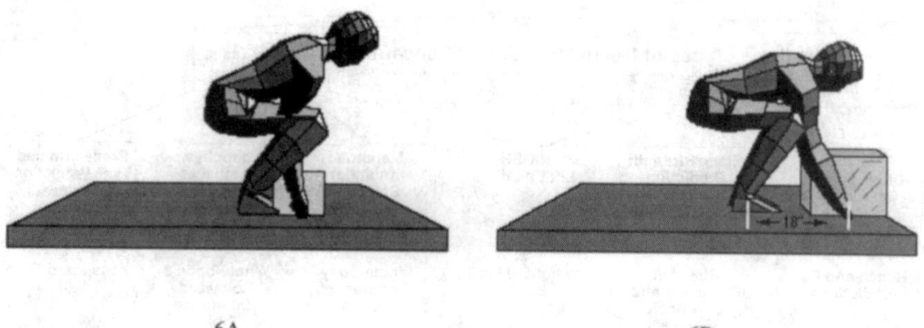

Figure 23.6 Package size creates biomechanical problems when lifted from floor. Figure 23.6A depicts small package being straddled and held close, compared to Figure 23.6B, which depicts too large a box to be easily straddled (Chaffin, 1994).

weighs over 35 lb, a significant risk of low back injury would exist in the posture shown in Figure 23.6B.

From the workplace design standpoint, when large packages are involved, they should never be presented to a worker at a height lower than about mid-thigh (or about 30 in.). Such a height allows the person to stand erect and bring the object against (or near) the torso, thus minimizing lower back bending moments and resulting spinal compression forces. The use of adjustable lift tables becomes essential when large bulky objects are handled, as depicted in Figure 23.7.

Another aspect of this particular problem is related to the orientation of the package when presented to a worker. If the package is not shaped like a cube (i.e., of nearly equal dimensions) but rather has one small dimension (e.g., like a suitcase), it should be presented in a more vertical orientation. This can allow a person to lift the object closer to the body by straddling the narrow dimension, or if a handle is provided on the top, by lifting close to the side of the body.

Figure 23.7 A typical lift and tilt table Figure 23.7A and pallet lift Figure 23.7B used to avoid stooped lifting of large or heavy packages. (Figure adapted from the UAW-Ford Job Improvement Guide, copyright © 1988, The Regents of the University of Michigan, used by permission.)

23.3.2 Package Handles

One means to assist a person who is lifting heavy objects is to provide adequate handles or hand-hold cutouts on packages. Waters et al. (1993) recommends a 10% weight reduction if packages cannot be easily grasped for lifting. When lifting boxes to shoulder height or above, the lack of a good hand-hold can force the person to either (1) abduct the arm into an extreme posture, which limits muscle strength and may injure shoulder tendons or (2) regrasp the object during the lift motion (i.e., slide the hands under the box), during which time the box may slip out of the hands. Providing a handle on the box allows reorientation of the hands during the lift without risking loss of grip, as illustrated in Figure 23.8.

In this context, NIOSH (1994) recommends the following for package handles or hand-hold cutouts:

- An optimal handle design has 0.75–1.5 in. (1.9–3.8 cm) diameter, ≥ 4.5 in. (11.5 cm) length, 2 in. (5 cm) clearance, cylindrical shape, and a smooth, nonslip surface.
- An optimal hand-hold cutout has the following approximate characteristics: ≥ 1.5 in. (3.8 cm) height, 4.5 in. (11.5 cm) length, semi-oval shape, ≥ 2 in. (5 cm) clearance, smooth, nonslip surface, and ≥ 0.25 in. (0.60 cm) container thickness.

23.3.3 Shelving Considerations

As mentioned earlier, heavy or large objects need to be presented to a person at a height that allows the person to remain erect. In addition, the person should not have to reach into the shelf to lift an object, i.e., the objects should be presented close to the torso. Both of these principles can be attained by careful planning of shelf usage and design. The former dictates that only heavy or large objects occupy shelves in the 30 to 45 in. height range. The latter suggests the use of either (1) a slanted shelf system (see Figure 23.9) or (2) a roll-out shelf or slip sheet under objects to facilitate pulling the objects to the front of the shelf before lifting.

Also, shelving and storage systems often can benefit from the use of a "lazy-Susan" rotating table. These devices can minimize the need for a worker to reach across objects to lift, thus minimizing the horizontal distance to the object being lifted.

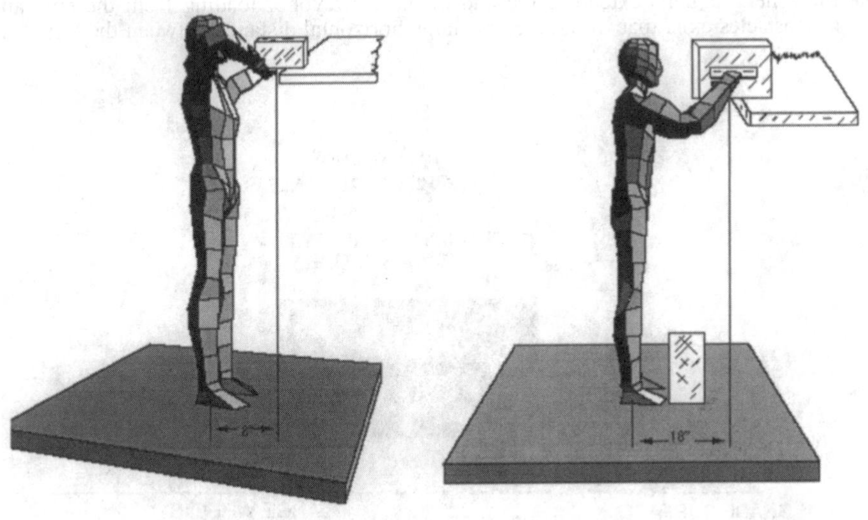

Figure 23.8 Providing handles on boxes (B) allows the hands to easily be reoriented during a high lift, thus avoiding the need for extreme arm abduction, (A). [From Chaffin (1994)].

Figure 23.9 Slanted shelf system with large objects on the middle shelves to allow erect torso lifts. (Figure adapted from the UAW-Ford Job Improvement Guide, copyright © 1988, The Regents of the University of Michigan, used by permission.)

23.3.4 Carts and Conveyors

Carrying objects requires static muscle exertions, and these can quickly cause muscle fatigue, which manifests itself as localized pain, tremors, and general loss of coordination (Chaffin, 1973). Carts and conveyors are two forms of mechanization to move objects horizontally in the workplace. Both present special design requirements.

In specifying carts it becomes most important to ensure that the carts can be easily pushed and pulled with minimum hand forces and near erect postures. Too often, carts are designed with badly placed handles and small wheels (or poorly maintained wheel bearings). These combine with heavy loads and uneven or graded floor surfaces to cause excessive peak hand forces. Studies by Snook (1978), Kroemer and Robinson (1971), Lee (1982), Chaffin, et al. (1983), Ayoub and McDaniel (1974), and Resnick et al. (1992) have provided recommendations regarding acceptable hand forces for specific push and pull conditions. These values range from about 15 lb to over 50 lb depending on handle height, floor traction, and exertion durations. In general, larger values can be achieved when good foot traction exists, the person is pushing rather than pulling, and the hands are at about hip-to-waist height, as illustrated in Figure 23.10.

Specification of conveyors needs to assure that the worker can remain with a nearly erect torso orientation while lifting objects on and off of the conveyor. This will require that the rollers or belts extend to the end of the conveyor if loading from the end, and that no obstacles exist that would create a large horizontal distance between the torso and

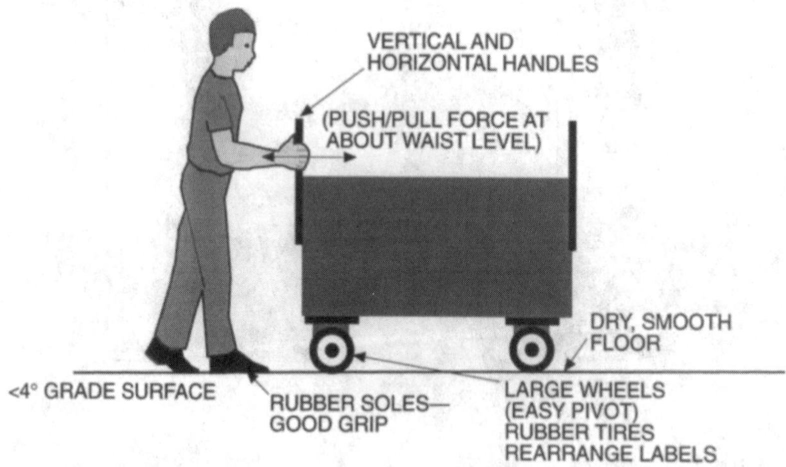

Figure 23.10 Some important specifications in cart use.

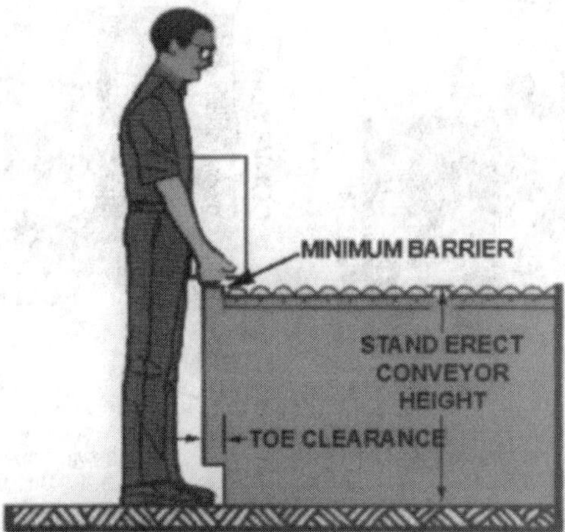

Figure 23.11 Conveyor with end load / unload station depicting desired posture provided by good design. (Figure adapted from the UAW-Ford Job Improvement Guide, copyright © 1988, The Regents of the University of Michigan, used by permission.)

object being lifted. Once again like shelving systems, the conveyor should be adjustable in height to allow an erect torso lift posture, as depicted in Figure 23.11.

If side unloading is performed from a conveyor, then it is often desirable to push objects on a wide conveyor close to the person before lifting, as depicted in Figure 23.12.

Last, by studying the flow of the materials in a plant it is often possible to present objects to a worker that minimize twisting and turning of the torso. This is illustrated in Figure 23.13, wherein a small rearrangement of the workplace eliminated the 180° torso twisting motions required by the old workstation layout.

23.3.5 Material Handling Assist Devices

In many workplaces it is desirable to reduce frequent bending, lifting and carrying of objects by providing a material handling device (MHD). These assist the worker by counterbalancing the weight of an object, using pneumatic, hydraulic, or electric power to lift and lower objects. Many different MHDs are now available. They generally can be categorized into four different types:

- **Simple Vertical Lift Device (SVLD)**—Provides antigravity lifting force and can be swung as a pendulum. Examples: chain or cable hoist, vacu-hoist, cable tool balancer.

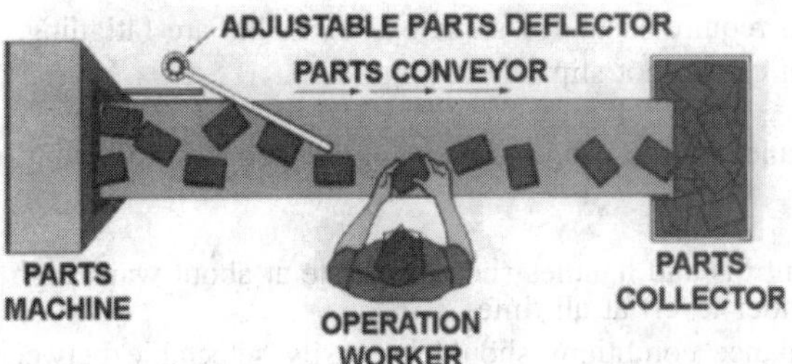

Figure 23.12 Side unloading from wide conveyor requires a parts deflector to ensure minimum horizontal lift distances. (Figure adapted from the UAW-Ford Job Improvement Guide, copyright © 1988, The Regents of the University of Michigan, used by permission.)

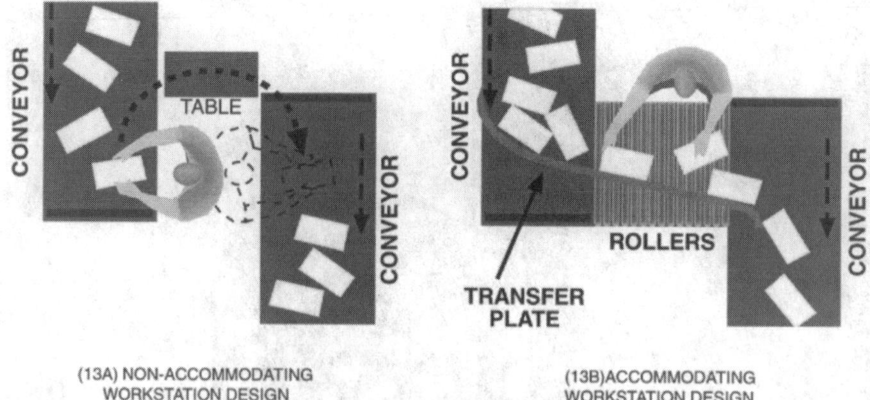

Figure 23.13 A small rearrangement in the old workstation (13A) eliminated the need for torso twisting motions and lifting parts to and from the two conveyors in 13B. (Figure adapted from the UAW-Ford Job Improvement Guide, copyright © 1988, The Regents of the University of Michigan, used by permission.)

- **Lift Device with Horizontal Transfer Movement (LDHT)**—Same as simple vertical lift device, but includes mechanism for horizontal transfer movements of the lift device. Examples: bridge crane, boom crane, trolley on overhead rails.
- **Simple Articulated Arm (SAA)**—Multiple link system with antigravity lift capability and one link attached to fixed support. Example: Congo Telis Balance Master.
- **Articulated Arm with Horizontal Transfer Movement (AAHT)**—Same as a simple articulated arm, but includes mechanism for horizontal movement of device. Example: Articulated arm on rails or overhead bridge.

Two different types of MHDs are illustrated in Figure 23.14, which also shows how horizontal forces need to be considered, as was the case with cart specifications.

The use of MHDs in industry has become very prevalent, but the following problems can be introduced into the workplace if care in MHD specification is not taken:

1. MHDs can add significant time to perform moves (compared to manual lift and carry).
2. MHDs can add significant energy requirements to job (compared to manual lift and carry).
3. MHDs can make precise object positioning difficult (loss of quality).
4. MHDs can cause dynamic push and pull hand forces which exceed strength and low-back capabilities.
5. MHDs can require sustained static exertions that are fatiguing.
6. MHDs can cause foot slip related injuries.

To reduce the chance of these problems being created, the following principles are advocated:

1. MHD controls and handles should operate at about waist high (i.e., between hip and shoulder level) at all times.
2. MHD balance conditions should be easily selectable between loaded and unloaded conditions.
3. Vertical movements should be minimized by considering this in workplace layout, (i.e., using pallet lifts, and setting conveyor heights).

Figure 23.14 Example of typical MHDs being used to lift and transport heavy objects. The top device is an LDHT and the bottom is an AAHT. (Figure adapted from the UAW-Ford Job Improvement Guide, copyright 1988, The Regents of the University of Michigan, used by permission.)

4. The number of steps needed to move the MHD should be minimized by careful workplace layout.
5. Torso twisting movements and forces should be minimized by careful workplace layout and MHD selection.
6. Movement "guides" should be used to assist in positioning MHDs when precision is needed.
7. MHD mass should be minimized by using lightweight materials and reducing component sizes.
8. Any required power or guidance controls should be installed in an attached handle to allow two-handed positioning.
9. The use of friction to assist in stopping an MHD's movements should be carefully specified to assure hand forces are not too great.
10. Object "grab" devices on the MHD should require few movements to activate.
11. High traction floors are needed to minimize foot slip potential when high push and pull forces are required.

23.4 WORKPLACE DESIGN FOR HANDTOOLS

The risk factors associated with handtool use have been presented earlier in Chapter 9 (from a biomechanics perspective) and later in Chapter 35 (from an injury prevention perspective). Most experts propose that the following job-related risk factors be controlled to prevent upper extremity cumulative trauma disorders UECTDs (Chaffin and Andersson, 1991):

- Frequency of forceful hand exertions
- Peak grip forces during manual exertions
- Awkward Postures during hand exertions
- Vibration of powered tools
- Cold temperature

What follows is a brief description of how workplace design can assist in reducing exposure to the first three risk factors.

23.4.1 Reducing Frequency of Hand Exertions

From a workplace design perspective, frequency of hand exertions is most easily reduced by introducing a powered handtool. A simple example is shown in Figure 23.15 for a fastening operation. Replacing the ordinary manual screwdriver with a powered screwdriver greatly reduces the frequency of forceful hand exertions.

What is also achieved by using a pistol style, powered screwdriver in Figure 23.15 is a less stressful wrist posture. As desirable as this design solution appears, caution must be raised, however. Powered tools add significant weight. Static exertions are necessary to support the tool weight, and if prolonged, particularly if the arm is extended horizontally, then shoulder and upper back muscle fatigue may result. This is one reason why it is always necessary to arrange the workplace so that the use of heavy tools is such that the tools are lifted close to the body, or better yet, are suspended from a tool balancer or articulated balance arm (see Figure 23.16).

23.4.2 Reducing Forceful Hand Exertions

One cause of forceful hand exertion is tool weight. Fortunately, by arranging the workplace appropriately it is often possible to use a tool balancing device, as illustrated in Figure 23.16.

Another cause of forceful hand exertions is the torque reaction caused by the sudden starting and stopping of rotating types of handtools. Torque reaction bars (as shown in Figure 16A) or articulated arm tool balancers (Figure 16B) also provide a means to control the tool rotation, and thus reduce the need to tightly hold the tool.

By providing adequate size and shape handles on boxes (discussed earlier in Section 23.3.2) and on handtools the need for stressful hand forces also can be reduced. The size of a squeeze-type tool (e.g., metal cutter, scissors, pliers, vice-grips) should be such that the tool can easily fit into the hand with the fingers wrapped securely around the handles (see Figure 23.17). To do this, the grip span should not exceed about 3 in. to accommodate most men and women (Greenberg and Chaffin, 1976), and should be long enough to distribute the grip force across the entire palm.

Hand force levels also can be reduced by orienting the handles of tools so that the primary force vectors are directed in-line with the forearm (see Figure 23.18). This minimizes the need to grip the tool hard enough to maintain a grip on the tool while applying

Figure 23.15 Frequency of forceful hand exertions is reduced by using powered handtools. (Figure adapted from the UAW-Ford Job Improvement Guide, copyright 1988, The Regents of the University of Michigan, used by permission.)

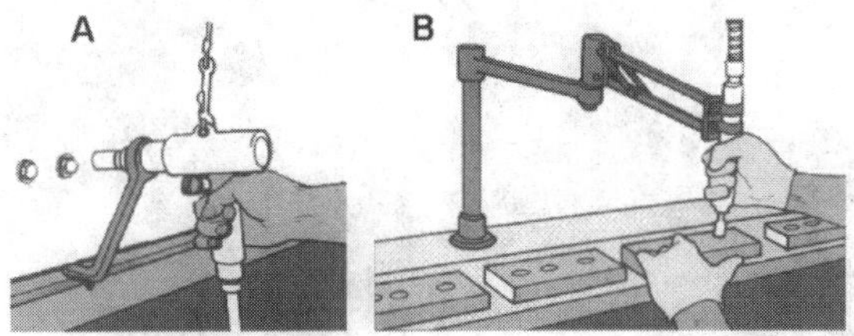

Figure 23.16 The use of a tool balancer (e.g., spring loaded cable in A or articulated arm in B provides a means to counter the effect of power tool weight. (Figure adapted from the UAW-Ford Job Improvement Guide, copyright © 1988, The Regents of the University of Michigan, used by permission.)

large push and pull forces. Providing a handle shape with a flange on the end of the handle also can allow the thumb and fingers to brace against the push and pull forces acting across the hand (Chaffin and Andersson, 1991).

23.4.3 Reducing Awkward Upper Extremity Postures

Tool shape and orientation of the workpieces must be simultaneously considered to control upper extremity postures. This is shown in Figure 23.19, wherein the choice of a pistol or in-line shaped tool depends on whether the parts to be fastened are in a near vertical or horizontal orientation.

23.5 WORKPLACE DESIGN FOR SEATED WORK

Providing a biomechanically improved seated workplace requires consideration of the size variation in the population (as discussed earlier in Chapter 8) as well as assurance that prolonged static muscle exertions will be minimized to avoid muscle fatigue (Chapter 10). In essence, the weight of human body segments must be supported (without creating high localized points of pressure), as depicted in Figure 23.20. To accommodate the variance in people's anthropometry, various international and U.S. standards organizations have promulgated recommendations on the range of adjustability in specific seat functions.

For instance, for normal office type sitting postures, the height of chairs should be easily adjustable from about 14 in. to 21 in., based on standards developed in Great Britain, Germany and Sweden (Chaffin and Andersson, 1991). In the United States, similar standards have been promulgated for use in the design of video display and keyboard data entry tasks, as described in Chapter 21.

23.5.1 Some Industrial Seating Requirements

What has become important in recent years is the awareness that an ergonomically well designed chair not only must reflect the normal variation in people's size and shape, but also will depend on the type of task being performed. For instance, in a VDT data processing task the user can often be accommodated best by a chair that allows the torso to lean back about 15 to 20° from vertical. If the chair back provides such a posture along

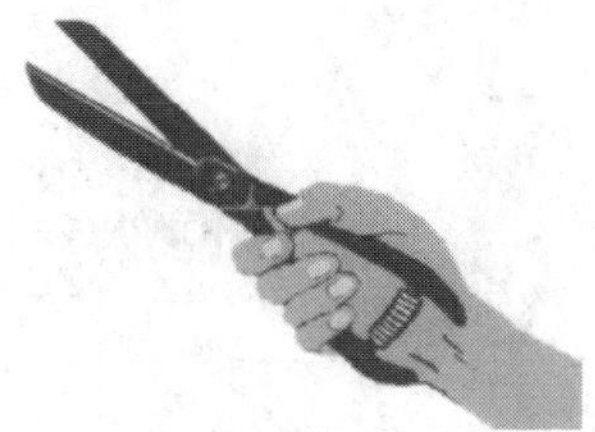

Figure 23.17 A typical squeeze tool designed to allow a full power grip and distribute forces across entire hand. (Figure adapted from the UAW-Ford Job Improvement Guide, copyright © 1988, The Regents of the University of Michigan, used by permission.)

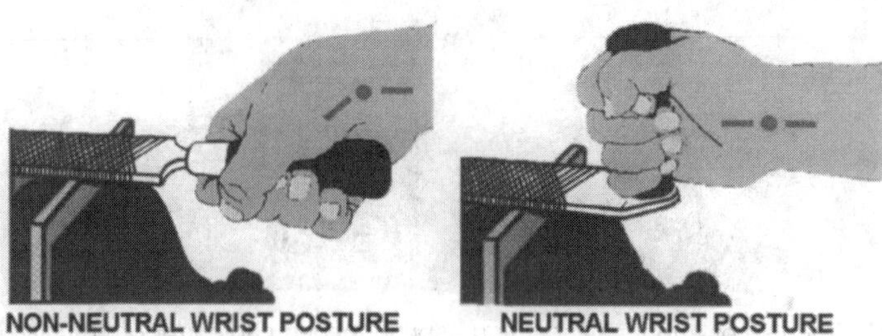

Figure 23.18 Tool handle orientation can provide both a better wrist posture and reduce grip force requirements. (Adapted from UAW-Ford, 1988.)

with a good lumbar support, most workers performing intensive keyboard entry tasks will have reduced low back muscle activity, spinal disc pressures, and fatigue-related discomfort (Andersson et al., 1974; Karwowski et al., 1994).

Similar concerns exist in manufacturing environments, but the seat design features may be quite different than those applicable to the office environment. As shown by Yu et al. (1988) and Yu and Keyserling (1989) for commercial sewing operations, an ergonomically well designed workseat would provide (1) greater height than an office chair (up to about 24 in.), (2) a seat pan shape that allowed the thighs to slope downward about 15°, (3) a narrow seat back (which avoids elbow interference) with adjustable lumbar support, and (4) an easy swivel seat. Similar features are desirable for many production line seats. To illustrate this, consider the seated workplace shown in Figure 23.21. Because the conveyor system is higher than normal table height, presumably to allow some workers to stand up, the seat must be adjustable up to about 27 in. in height. This in turn will require a footrest. The seat pan will be more shallow than an office seat, and may tilt forward (about 10° from horizontal) and/or have a drop-off on the front third of the seat pan to avoid excess pressure on the mid-to-distal end of the upper legs. The seat back may be narrower than an office chair to allow unrestricted, large arm motions needed to reach to parts and tools that often are positioned beside the workstation. Finally, instead of providing forearm supports on the chair, a padded forearm-hand support is provided along the front edge of the conveyor system. In this latter regard, the importance of providing some support for the upper extremity has been shown by Chaffin (1973). If no

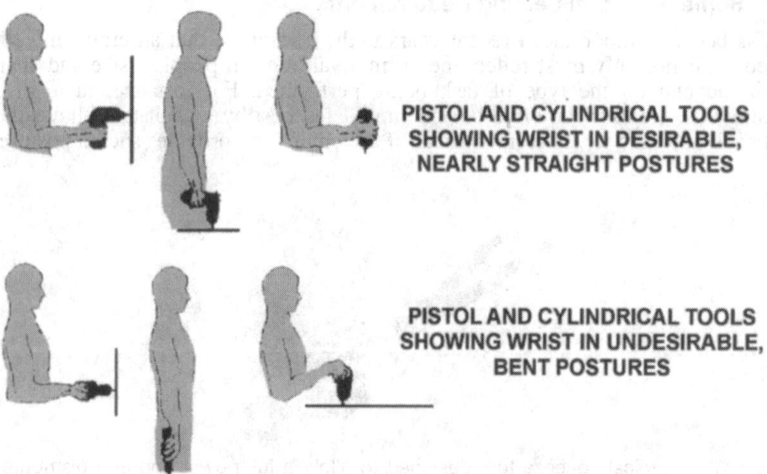

PISTOL AND CYLINDRICAL TOOLS SHOWING WRIST IN DESIRABLE, NEARLY STRAIGHT POSTURES
PISTOL AND CYLINDRICAL TOOLS SHOWING WRIST IN UNDESIRABLE, BENT POSTURES

Figure 23.19 Postures of the wrist are directly affected by tool shape and workpiece orientations. (Figure adapted from Armstrong, 1983, in Chaffin and Andersson, 1991.)

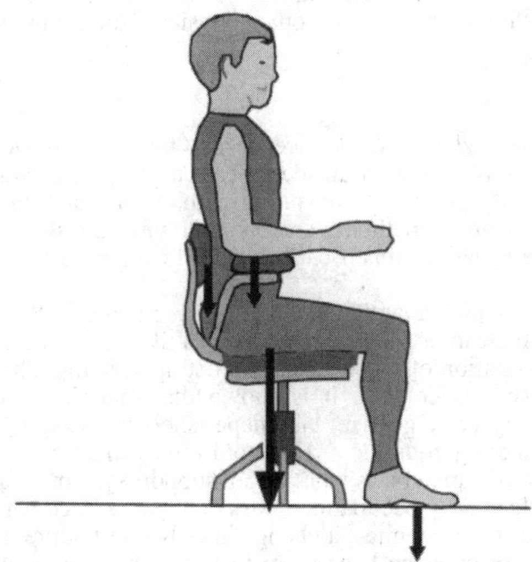

Figure 23.20 Primary body weight support vectors. (From Chaffin and Andersson, 1991.)

support is provided, and the worker's arms are extended out in front of the body, as shown in Figure 23.21, then muscle fatigue in the upper back and shoulders will develop 4 to 5 times faster than if either the work is closer to the body or the forearm or elbow is supported by an armrest or padded bar along the edge of the work surface.

Recall that to relax the back muscles it is best to have a seat back that allows the worker to lean back about 15° or more with the low back well supported. In a production operation this requires at least two workplace accommodations. First, the chair must allow the person to sit close to the point of operation. This is only possible when there is enough leg clearance under the work surface, as depicted in Figure 23.21, or the chair's armrests or base do not restrict the chair location. The second reason for a person not being able to sit in a relaxed "lean back" posture is that such a posture can impair visibility, i.e., the line-of-sight or visual acuity requirement of the task being performed requires the person to lean forward. When the head is tipped forward more than about 30° from vertical, neck muscle fatigue will develop very quickly (Chaffin, 1973). To reduce this latter problem, the primary visual task should be high enough that the head and neck are flexed less than about 30° from vertical.

In conclusion, industrial seat design is more complicated than office seating because industrial tasks vary greatly in their physical demands, and the layout of other workplace equipment and fixtures can influence specific features desired in a seat design. Because

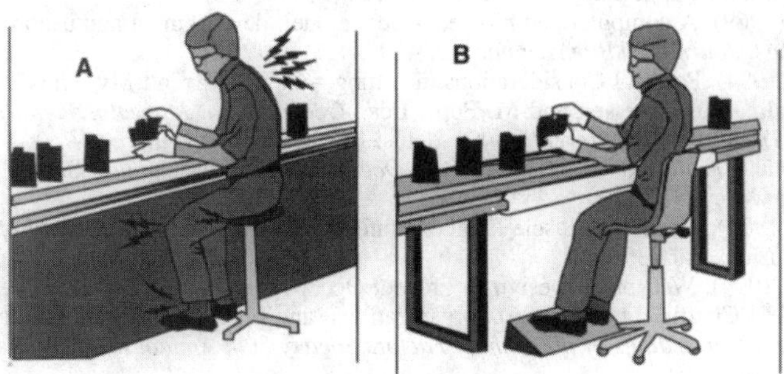

Figure 23.21 A typical conveyor-line seated workstation requires special seat design considerations shown in the improved illustration B. (Figure adapted from the UAW-Ford Job Improvement Guide, copyright 1988, The Regents of the University of Michigan, used by permission.)

of this it is imperative that the specifications of an industrial seat be done only after careful evaluation of the work tasks and other physical conditions involved in the operation of interest.

23.6 SUMMARY

This chapter has advocated that workplace design must be considered as a nine-step process. Such a process begins with an identification of specific human problems to be resolved by an improved design. By consulting with experts and the literature and using various ergonomic evaluation methods, and by following the design through mock-up, implementation, and evaluation, much can be done to improve the biomechanics of existing workplaces.

This philosophy is depicted with guidelines and graphical illustrations that display how biomechanical tools can be used in both proscriptive design of new worker-hardware systems, as well as evaluation of mock-ups and existing systems. Three workplace design issues are addressed in this context. It is shown that materials-handling tasks can be improved by considering package size and shape, shelving design, and the use of carts and other materials handling equipment. Handtool improvement can be achieved by considering shape and size parameters, as well as tool support systems. Last, industrial seating is discussed as it is affected by occupational tasks and workplace hardware requirements.

As is often the case in ergonomics, a change in only one feature of a worker-hardware system normally will not gain the benefits expected. A systematic evaluation of the existing situation must be made, and then a disciplined, design improvement process must be initiated. It is hoped that this presentation is of assistance in realizing real benefits to workers and employers in the future.

REFERENCES

Andersson, G. B. J., Örtengren, A., Nachemson, A., and Elfstöm, G. (1974). Lumbar disc pressure and myoelectric back muscle activity during sitting. I. studies on an experimental chair. *Scandinavian Journal of Rehabilitation Medicine*, *3*, 104–114.

Andersson, G. B. J. (1981). Epidemiologic aspects on low-back pain in industry. *Spine*, *6*(*1*), 53–60.

Andersson, G. B. J., Pope, M. H., Frymoyer, J. W., and Snook, S. (1991). In M. Pope, G. B. J. Andersson, J. Frymoyer, D. B. Chaffin, Eds., *Occupatonal Low Back Pain.* St. Louis: Mosby Year Book, Ch. 5.

Armstrong, T. J., and Silverstein, B. A. (1987). Upper extremity pain in the workplace—Role of usage in causality. In N. Hadler, Ed., *Clinical Concepts in Regional Musculoskeletal Illness.* Grune and Stratton, pp. 333–354.

Armstrong, T. J., Fine, L. J., Goldstein, S. A., Lifshtz, Y. R., and Silverstein, B. A. (1987). Ergonomics considerations in hand and wrist tendinitis. *J. Hand Surgery*, *12A*(*5*), 830–837.

Ayoub, M. M., and McDaniel, J. W. (1974). Effect of operator stance on pushing and pulling tasks. *AIIE Transactions*, *6*, 185–195.

Ayoub, M. M., and Mital, A. (1989). *Manual Materials Handling.* London, New York and Philadelphia: Taylor & Francis.

Ayoub, M. M. (1992). Problems and solutions in manual materials handling: the state of the art, *Ergonomics*, *35*(*7/8*), 713–728.

Beck, D. J., Chaffin, D. B., and Beck, D. J. (April, 1993). Lightening the Load, *Ergonomics in Design.*

Chaffin, D. B., Andres, R. O., and Garg A. (1983). Volitional postures during maximal push/pull exertions in the sagittal plane. *Human Factors*, *25*(*5*), 541–550.

Chaffin, D. B. (1969). A computerized biomechanical model: development and use in studying gross body actions. *Journal of Biomechanics*, *2*, 429–441.

Chaffin, D. B. (1994). Postural Considerations in Lifting—Or Why Aren't My Arms Five Feet Long? In M. Nordin, G. Andersson and M. Pope, Eds., *Occupational Musculoskeletal Disorders: Assessment, Treatment and Prevention.* St. Louis: Mosby Yearbook Inc.

Chaffin, D. B., and Andersson, G. B. J. (1991). *Occupational Biomechanics*, Second Edition, New York: John Wiley.

Chaffin, D. B. (1973). Localized muscle fatigue definition and measurement. *Journal of Occupational Medicine*, *15*(*4*), 346–354.

Chaffin, D. B. (1994). Workplace Design to Prevent Occupational Low Back Pain. Lecture note.

Evans, S. M., and Chaffin, D. B. (1986). Organizational and process differences influencing ergonomic design. *Proceedings of the Human Factors Society 30th Annual Meeting*, Dayton, OH, pp. 734–738.

Garg, A., and Chaffin, D. B. (1975). A biomechanical computerized simulation of human strength. *AIIE Transactions*, *7*(*1*).

Greenberg and Chaffin (1976). *Workers and Their Tools.* Midland, MI: Pendell Publishing Company.

Keyserling, W. M., Armstrong, T. J., and Punnett, L. (1991). Ergonomics job analysis: a structured approach for identifying risk factors associated with overexertion injuries and disorders. *Applied Occupational Environmental Hygiene*, *6*(*5*), 353–363.

Karwowski, W., Eberts, R., Salvendy, G. and Noland, S. (1994). The effects of computer interface design on human postural dynamics. *Ergonomics*, *37*(*4*), 703–724.

Kroemer, K. H. E., and Robinson, D. E. (1971). *Horizontal Static Forces Exerted by Men Standing in Common Working Postures on Surfaces of Various Tractions.* AMARL-TR-70-114, Aerospace Medical Research Laboratory, Wright-Patterson Air Force Base, Ohio.

Lee, K. (1982). *Biomechanical Modeling of Cart Pushing and Pulling*. University of Michigan, Ann Arbor, MI: unpublished doctoral dissertation.

Marras, W. S., and Mirka, G. A. (1992). A comprehensive evaluation of trunk response to asymmetric trunk motion. *Spine*, *17*, 318–326.

McGill, S. M., and Norman, R. W. (1987). Effects of an anatomically detailed erector spinae model on L4-S1 disc compression and shear. *J. Biomechanics*, *20*, 591–600.

NIOSH (1994). *Applications Manual for the Revised Lifting Equation.* (U.S. Department of Health and Human Services, Cincinnati, OH, DHHS (NIOSH) Publication No. 94-110).

Rendell, J. (1995). Juliana R. Sparks v. Consolidated Rail Corporation, in the United States District Court for the Eastern District of Pennsylvania Civil Action, No. 94-CFV-1917.

Resnick, M. L., and Chaffin, D. B. (1992). Some ergonomic considerations in the design of material handling devices. *Proceedings of the 35th Annual Meeting of the Human Factors Society*, Atlanta, GA.

Snook, S. H. (1978). The design of manual handling tasks. *Ergonomics*, *21*(*12*), 963–986.

Snook, S. H., and Ciriello, V. M. (1991). The design of manual handling tasks: revised tables of maximum acceptable weights and forces. *Ergonomics*, *34*(*9*), 1197–1213.

UAW-Ford, Chapter 4 (1988). Developing Solutions In *Fitting Jobs to People, The UAW-Ford Ergonomics Process Job Improvement Guide*, UAW-Ford National Joint Committee on Health and Safety, Regents of the University of Michigan.

Waters T. R., Putz-Anderson, V., Garg, A., and Fine, L. J. (1993). Revised NIOSH equation for the design and evaluation of manual lifting tasks. *Ergonomics*, *36*(*7*), 749–776.

Yu, C. Y., and Keyserling W. M. (1989). Evaluation of a new work seat for industrial sewing operations. *Applied Ergonomics*, *20*(*1*), 17–25.

Yu, C. Y., Keyserling W. M., and Chaffin, D. B. (1988). Development of a work seat for industrial sewing operations: Results of a laboratory study. *Ergonomics*, *31*(*12*), 1765–1786.

CHAPTER 24
NOISE*

Malcolm J. Crocker
Department of Mechanical Engineering
Auburn University
Auburn, AL 36849 USA

*Section 24.4 is based in part on Chapter 6.1 written by D. M. Jones and D. E. Broadbent in the first edition of the *Handbook of Human Factors*, edited by G. Salvendy and published by John Wiley.

24.1 INTRODUCTION

"Noise" is usually defined as unwanted sound. In the frequency range of about 15 to 16,000 Hz, sound is audible and is sensed mainly by the ear. At frequencies below about 15 Hz, sound is termed *infrasound* and, if sufficiently intense, can cause different body organs to vibrate with unpleasant sensations. Above about 16,000 Hz, sound is termed *ultrasound* and is no longer audible by people, although it can be detected by animals such as dogs and bats.

This chapter is mainly concerned with control of noise and with vibration that results in noise. Noise has several undesirable effects. In industry the main effect is that intense noise experienced for long periods throughout a working life can result in permanent deafness. Other effects, which are more difficult to attribute directly to noise, include increased accidents and reduced efficiency and productivity. Industrial noise interferes with conversation, warning signals, telephone communication, and so on at work. It can also be a source of community annoyance and complaints, and even affects sleep and other human activities in severe cases.

This chapter first reviews the way in which sound propagates, the human ear and the loudness of sound, and the ways in which it is measured. The chapter continues with the main effects of noise on people, and concludes with a discussion of methods of measuring and controlling noise and some case histories.

24.2 SOUND PROPAGATION

Sound waves propagate rather like ripples on a pond. They travel out from a source at a constant speed. Water waves on a pond are circular and two-dimensional, whereas the sound waves in air are spherical and three-dimensional, although of course they cannot be seen.

24.2.1 Sound Pressure Level and Decibels

The ear responds to both very small sound pressures and very large sound pressures, up to a million times greater. Because of this large range, acousticians normally use a logarithmic sound pressure scale. The *sound pressure level* (SPL or L_p) of a sound is defined as (see Crocker, 1997, Chapter 1).

$$\text{SPL} = L_p = 10 \log_{10} (p_{\text{rms}}^2 / p_{\text{ref}}^2), \text{ dB} \tag{1}$$

where p_{rms} is the root mean square sound pressure and p_{ref} is 0.00002 Pa, a reference sound pressure* (approximately equal to the smallest sound at 1000 Hz an average young person can hear). The units of L_p in Equation (24.1) are nondimensional and are referred to as *decibels* (dB). Figure 24.1 shows the SPL of some common sounds or noises.

24.2.2 Sound Intensity

The intensity I of a sound wave in air (in the region sufficiently far enough away from the source, known as the *far field*) is given by

$$I = p_{\text{rms}}^2 / \rho_0 c_0, \text{ W/m}^2 \tag{2}$$

where

$\rho_0 c_0$ = the characteristic impedance of air = 415 rayls (1 rayl (MKS) = 1 kg/m^{2}s)
ρ_0 = the undisturbed air density (1.21 kg/m^3)
c_0 = the speed of sound in air = 1125 ft/sec = 343 m/s
(All quantities are given at normal temperature of 20°C.)

The speed of sound in air depends only on the square root of the absolute temperature. As sound waves propagate away from a sound source at the speed of sound, so does the sound energy. *Sound intensity* is defined as the sound energy that passes through unit area in unit time (see Crocker, 1997, Chapter 1). Hence, sound intensity is sound power per unit area. Sound intensity is seen to be rather like light intensity. Further analogous behavior between sound and light waves is discussed later.

---

*0.00002 pascal = 0.00002 N/m^2 ≈ 0.0002 microbar

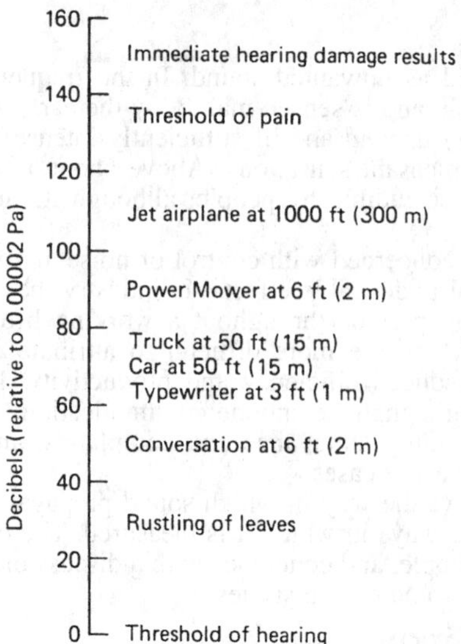

Figure 24.1 Typical sound pressure levels.

An intensity level L_I may also be defined:

$$L_I = 10 \log (I/I_{\text{ref}}), \text{ dB} \tag{3}$$

where I is the sound intensity (W/m^2) and I_{ref} is a reference intensity equal to 10^{-12} W/m^2. Note that, at 21°C, $\rho_0 c_0$ = 415 rayls (MKS), and it is easily shown that far from a source, provided that reflections are unimportant, $L_p \cong L_I$. For other temperatures in air, $L_p \approx L_I$ and slightly greater differences can result.

24.2.3 Inverse Square Law

All of the sound energy spreading out from a source has to pass through an imaginary sphere drawn around the source. Since the spherical area increases by four times each time the distance from the source is doubled, the sound power per unit area (i.e., the intensity) will reduce to one fourth of the original. Equations 24.1 and 24.2 show that the L_p correspondingly decreases by 10 $\log_{10}$ 4, or very nearly 6 dB. Hence, provided that the sound source is in free space (e.g., out-of-doors with no obstacles to cause reflections), L_p decreases by 6 dB for each doubling of distance. This is known as the *inverse square law*.

Very close to the source of sound, the sound intensity is not simply related to the sound pressure. This region close to the source is called the *near field*, and in this region there is a deviation from the inverse square law. Outside this region lies the true sound field, the *far field*. In a building, as the distance increases from a machine, the reflections become greater in magnitude than the direct sound. The distance at which the direct and reflected sound fields are equal is called the *critical distance* (r_c) and defines the boundary between the *free field* and the *reverberant field* (see Figure 24.2).

24.2.4 Reflections

If a sound wave is progressing in a medium, and the value of $\rho_0 c_0$ (the characteristic impedance of the medium), changes, then there is a reflection. Solid obstacles have very different values of $\rho_0 c_0$ from air and thus produce strong reflections. In such a case the inverse square law is violated. If strong pure tones, that is, discrete frequencies, are present in a noise spectrum, then the progressing wave and the reflected wave interfere to produce standing waves. There are regions in space of high sound pressure (known as *antinodes*) and regions of low sound pressure (known as *nodes*) for sound at these discrete frequencies. This fact is of considerable importance when the noise of a machine is measured in a room.

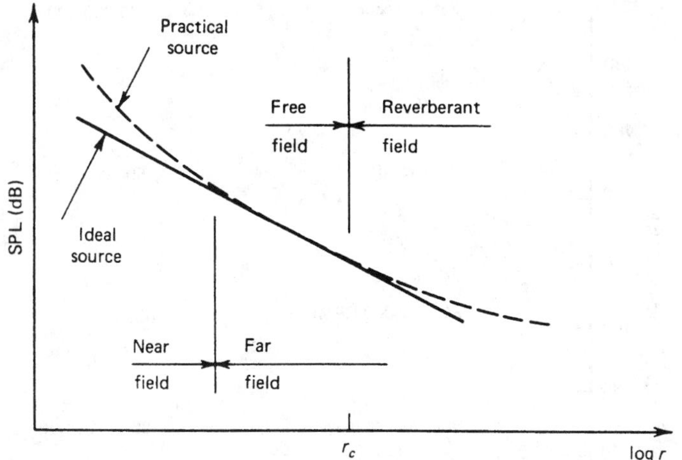

Figure 24.2 Ideal and practical sound sources shown on plot of SPL against distance r (on a logarithmic scale).

24.2.5 Plane Waves

If a source of low-frequency sound is situated in a hard-walled duct, then the sound field becomes one-dimensional, and, without absorption, there is no attenuation as the sound waves travel down the duct. Such waves are known as "plane waves."* If the wall of the duct is made absorbent to sound, then the waves are attenuated as they travel down the duct. Sound absorption is discussed later in this chapter in Section 24.7.4.

24.2.6 Diffraction

When sound waves reach an obstacle, some energy is reflected and some diffracted (or bent) around the obstacle (see Crocker, 1997, Chapter 1). The amount of sound energy that is diffracted depends on the ratio of the size of the obstacle and the sound wavelength, λ. This phenomenon is also observed with light waves. In both sound and light, if the obstacle dimension is much greater than a wavelength, a strong shadow is cast. If the dimension is much less than a wavelength, a weak shadow is cast, and much energy is diffracted around the obstacle. The wavelength, λ, is given by

$$\lambda = c_0/f \tag{4}$$

24.2.7 Sound Power

If any imaginary closed surface is drawn around a noise source, then by summing the intensity I in a direction normal to the surface, all over the closed surface, will yield the sound power of the source W, watts. For an omnidirectional source (one that radiates intensity equally in all directions), $W = 4\pi r^2 I$, where I is the intensity, W/m, at a distance r meters. The *sound power level* (PWL) of a noise source is given by Equation 24.5 where W_{ref} is the reference sound power = 10^{-12} W. Figure 24.3 shows the sound power level of some well-known sources.

$$\text{PWL} = L_w = 10 \log (W/W_{ref}), \text{ dB} \tag{5}$$

24.3 THE EAR AND THE LOUDNESS OF SOUND

24.3.1 Construction of the Ear

Hearing is probably the most highly developed human sense. Figure 24.4 shows a cross section of the ear. The ear is normally divided into three regions: *outer*, *middle*, and *inner*.

*If the frequency of the sound is low enough so the $f < 0.59\ c_0/d$, where d is the duct diameter in consistent units with the speed of sound c_0, then only plane waves can exist. At higher frequencies, cross modes can exist in addition to plane waves.

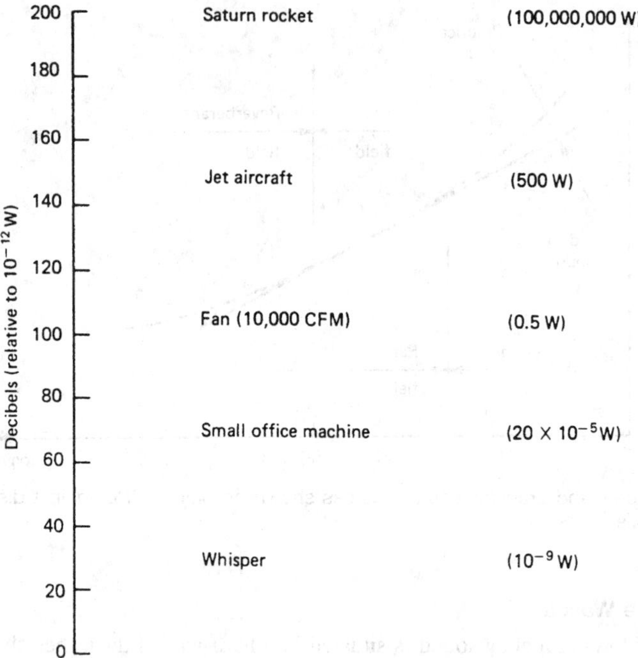

Figure 24.3 Typical sound power levels.

Sound waves are focused by the pinna into the auditory canal and travel along the canal to excite the eardrum into vibration. The eardrum (Figure 24.5) is very thin—about the thickness of paper—and under tension. The eardrum vibration is transmitted by three small bones (auditory ossicles) to the inner ear (cochlea). The outer and middle ear are filled with air, whereas the inner ear is filled with liquid (see Shaw, 1997 and Peak and Rosowski, 1997). Motion of the ossicles produces compressional waves in the fluid in the cochlea, which are sensed by thousands of microscopic hair cells. These hair cells transmit electrical signals through nerves to the brain, producing the sensation of hearing. Static pressure changes are equalized across the eardrum by swallowing, which opens the Eustachian tube to the back of the throat.

24.3.2 Loudness of Sounds

The ear is used mainly to listen to human speech, which is mostly in the range of about 250 to 4000 Hz, and is most sensitive in this frequency region. Figure 24.6 shows equal loudness contours for average young people. These contours were obtained by playing a pure tone (single frequency) sound to people at 1000 Hz and then having them adjust a tone at another frequency until it appeared equally as loud. For very intense sounds (about 100 dB), the contours are much flatter than for sounds at low levels. At low levels, the ear hears low-frequency sounds very poorly. (If it were more sensitive at low frequency, we would probably hear our own digestive processes and musculoskeletal movements!) The lowest contour shows the hearing threshold (the quietest sound that can be heard at any frequency) (see Yost and Killion, 1997). If the ear were only a little more sensitive (about 10 times) to quiet sounds, we would hear the random motion of air molecules (Brownian motion)!

Each contour in Figure 24.6 is labeled with a number in phons. The *loudness level* of a sound in phons, P, is thus the SPL of a tone at 1000 Hz that seems equally as loud. Sometimes a linear loudness scale is more desirable, and the *loudness* S of a sound in sones is internationally agreed to be given by

$$S = 2^{(P-40)/10} \tag{6}$$

Although Figure 24.6 is for pure tones, experiments with people and bands of noise have resulted in a similar set of curves for equal loudness contours of bands of noise.

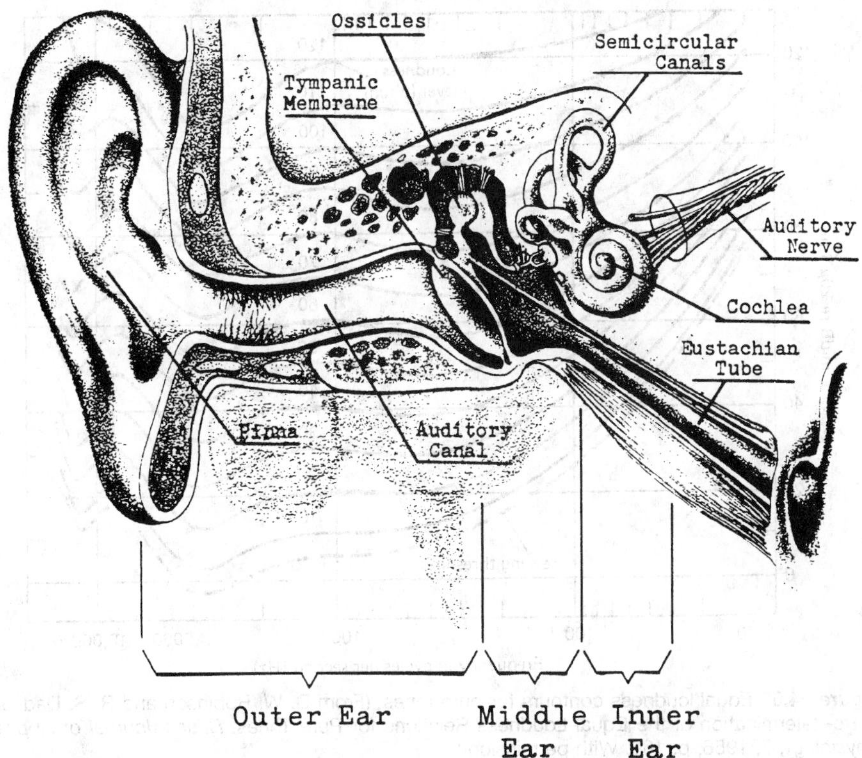

Figure 24.4 Simplified cross section through the human ear. (Based on Figure 7.1 from J. W. Palmer, *Anatomy for Speech and Hearing*, Harper and Row, New York, 1972. With permission.)

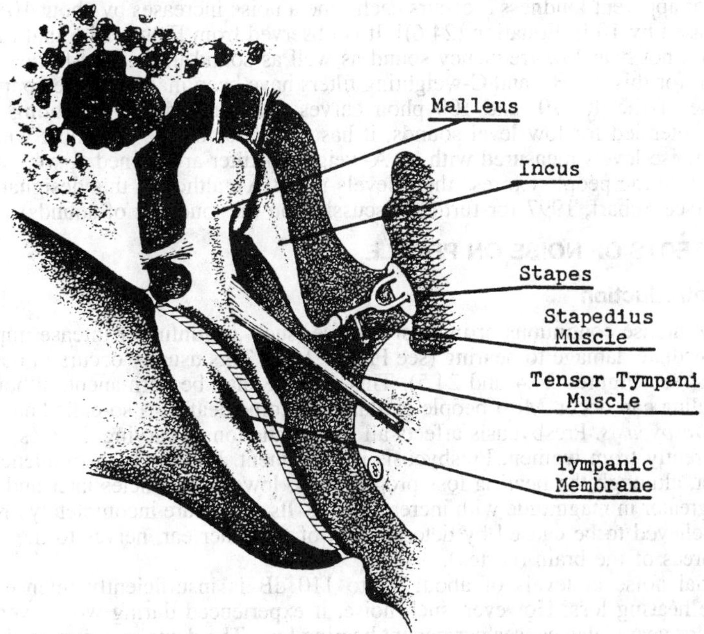

Figure 24.5 Eardrum (tympanic membrane) and auditory ossicles (malleus, incus and stapes). (Based on Figure 7.3 from J. W. Palmer, *Anatomy for Speech and Hearing*, Harper and Row, New York 1972. With permission.)

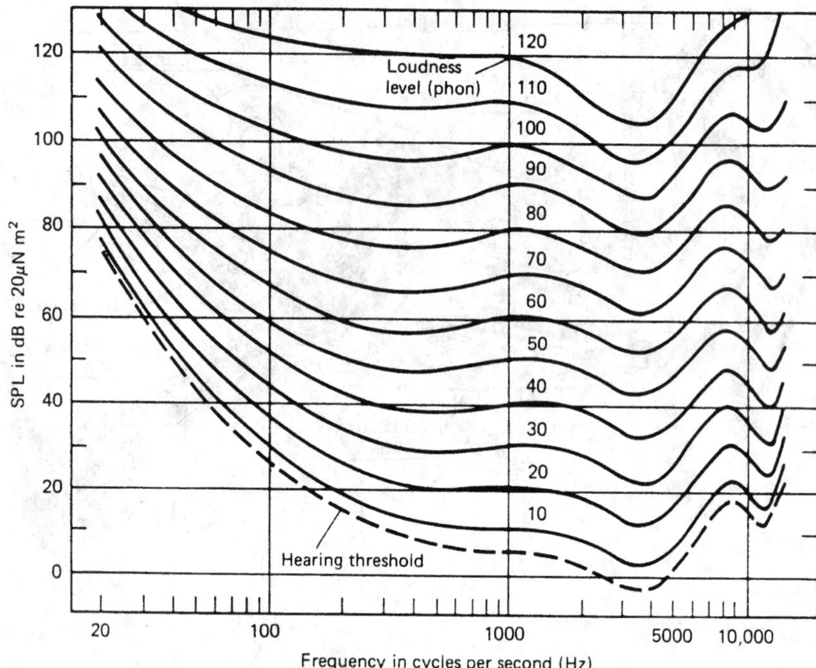

Figure 24.6 Equal loudness contours for pure tones. (From D. W. Robinson and R. S. Dadson. A Re-determination of the Equal-Loudness Relations for Pure Tones. *British Journal of Applied Physiology,* 7, 1956, p. 166. With permission.)

With noise, Equation (24.6) is again used internationally to obtain loudness. Thus, a doubling of apparent loudness S occurs each time a noise increases by about 10 dB [when P is increased by 10 in Equation (24.6)]. It is observed from Figure 24.6 that the average person does not hear low-frequency sound as well as sound in the mid-frequency range. To account for this A-, B-, and C-weighting filters have been made that are approximately the inverse of the 40, 70, and 100 phon curves. Although the A-weighting filter was originally intended for low level sounds, it has now become the one most widely used. Sound or noise levels measured with an A-weighting filter are termed A-weighted sound levels (dB). Some people express these levels in dB(A), although this designation is not preferred (see Scharf, 1997 for further discussion on the loudness of sounds).

24.4 EFFECTS OF NOISE ON PEOPLE

24.4.1 Introduction

Extremely intense continuous or impulsive noise (such as gunfire or intense impacts) can cause immediate damage to hearing (see Figure 24.1). This usually occurs in the eardrum or ossicles (see Figures 24.4 and 24.5). This damage may be permanent, although some partial healing can occur. Most people gradually become deaf by a so-called natural aging process, *presbycusis.* Presbycusis affects all the population to varying degrees and men a little differently from women. Presbycusis is permanent. Usually high-frequency hearing is lost first, although the hearing loss progresses to lower frequencies later and generally becomes greater in magnitude with increasing age. Its causes are incompletely understood, but it is believed to be caused by deterioration of the inner ear, nerves to the brain, and possibly areas of the brain (cortex).

Industrial noise at levels of about 90 to 110 dB is insufficiently intense to cause immediate hearing loss. However, such noise, if experienced during work over a period of months or years, also causes permanent hearing loss. The damage occurs in the cochlea in the inner ear, where the microscopic hair cells are gradually destroyed. The hearing loss normally called "noise-induced hearing loss" or "noise-induced permanent threshold shift" (NIPTS) usually first appears at about 4000 Hz, where the ear is most sensitive to sound. As the loss increases, it also progresses to higher and lower frequencies. Except for the characteristic loss at 4000 Hz, the hearing loss is similar to, and hard to distinguish from, presbycusis.

Noise is believed to act as a general stressor and possibly to have some effect on task concentration, efficiency, and productivity at work, and absenteeism. Other adverse health effects, sometimes attributable to noise, are increased incidence of heart attacks, miscarriages, headaches, and so on. However, these additional effects are hard to isolate and attribute directly to noise. Indeed, some authorities insist that people adapt to noise and that, provided that noise is insufficiently intense to cause a noise-induced hearing loss, these other effects will not occur. This view has been widely held in the United States, but not so much in most other countries. Noise does, of course, interfere with communication, sleep, and other human activities. Thus, it is a problem both at work and in residential communities near to industry, highways, airports and railways, etc. Speech interference is caused by the masking effect of noise—a well-known phenomenon (see Buus, 1997).

24.4.1.1 Threshold Shift

Although it is well known that exposure to intense noise results in *hearing loss*, the relationship between the two is quite complicated, and there is no simple answer to the question, "how much noise produces how much hearing loss?"

In determining hearing loss, the main parameter of interest is the *hearing threshold*: the sound pressure level at different frequencies at which sounds are just detected (see Figure 24.6). Ideally the threshold should be measured before and after exposure to loud noise and the degree of hearing loss determined from the difference between the two measurements. Three types of hearing loss resulting from exposure to loud noise are normally defined: first, *temporary threshold shift* (TTS) is a short-lived and reversible elevation of the hearing threshold; second, *noise-induced permanent threshold shift* (NIPTS) is a long-term effect of noise exposure where the hearing loss is not reversible; third, *acoustic trauma*, which is the result of a single, usually short exposure to extremely intense noise such as may arise from impulsive noise, impacts, gunfire, or explosions.

Before discussing TTS and NIPTS it should be noted that different people use different definitions of hearing loss. It may at different times be used to refer to the degree of physiological damage to the tissues in the inner ear, to the ability to detect sound, or to the ability to discriminate one sound from another. Moreover, the degree of loss, as measured by the ability to hear pure tones, for example, may not have a strong relationship to the degree of handicap. This is in part because handicap is itself difficult to define and may encompass social, domestic, and occupational aspects of life and the criterion for handicap will be different for each of these activities.

Temporary Threshold Shift

Although several features can be discerned in the recovery of hearing, that value corresponding to the threshold two minutes (2 min) after exposure is usually taken to be representative of the temporary loss (Ward, 1976). The degree of threshold shift is a function of the frequency level, and duration of noise. TTS shows a linear increase with increasing sound pressure level (SPL), at least for moderate levels of sound (80 to 105 dB) and for exposures of less than 8 hr (see Figure 24.7). Among the factors known to affect TTS are the frequency range over which the sound energy is spread and the degree of change in the spectral composition of the sound during the period of exposure. The growth of TTS to a constant level normally reaches an asymptote in 8 to 12 hr. As a general rule, if TTS at 2 min does not exceed 25 dB, recovery occurs in 16 hr. This means that although a worker at the end of an 8-hr shift may have significant impairment during the hours of rest that usually follow work, hearing improves during sleep and is normal when work commences the next day. In addition to hearing loss, TTS may be accompanied by the experience of subjective tinnitus, which is often a ringing sound heard in one or both ears (Terry, Jones, Davis, and Slater, 1983).

Noise-Induced Permanent Threshold Shift (NIPTS)

The primary concern with the study of NIPTS is the determination of the safe level of noise to which individuals can be exposed during their working lives. The specification of guidelines for such an exposure should include (1) the noise level limit for exposure

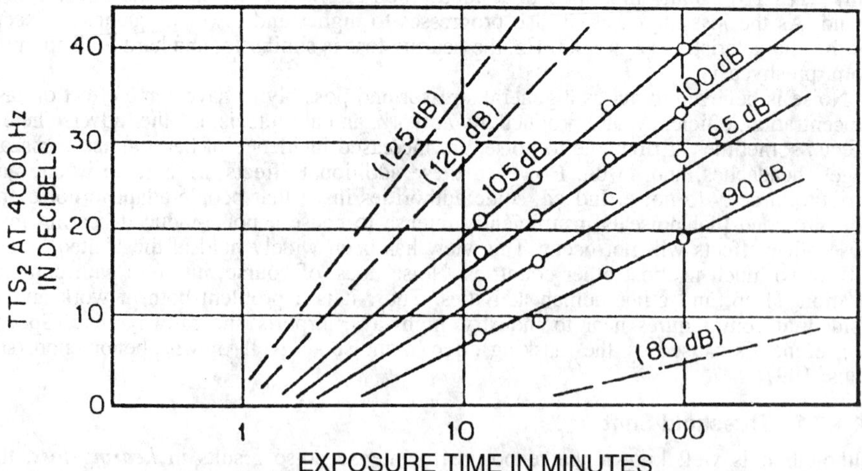

Figure 24.7 Temporary threshold shift at 4000 Hz measured 2 min after exposure to a noise band centered at 1700 Hz for a range of SPL and durations. [From Ward, Glorig, and Sklar, (1959) with the permission of W. D. Ward and the American Institute of Physics.]

to continuous noise in the typical working day; (2) some rule for trading exposure time and level, so that an increased level is compensated for by a shorter exposure in some justifiable way; and (3) some safe upper limit in both level and number of short bursts of noise.

NIPTS may arise from a variety of sources, and part of the difficulty in understanding the effects of occupational noise exposure is in assessing the amount of hearing loss contributed by these other factors. Three types of factors have been identified as contributing to the overall hearing loss: (1) effects due to age (presbycusis); (2) effects due to nonacoustic agents such as industrial chemicals, ototoxic drugs, or illness (nosoacusis); and (3) effects of noise exposure outside the work setting (socioacusis). Each of these factors may contribute in different degrees to the overall loss of hearing. Moreover, the influence of these factors may be just as great as that arising from occupational noise sources. For example, it has been estimated that the socioacoustic exposure may be as high as 80 dB(A) for 8 hr (Schori and McGatha, 1978), whereas the recommended maximum for occupational exposure may be 90 dB(A) over a similar interval.

The way in which NIPTS varies with exposure to occupational noise is fairly clear. An analysis of survey findings suggests that exposure to noise below 80 dB(A) over the working day has little effect (Passchier-Vermeer, 1974). Exposure to 85 dB(A) can produce a loss of the order of 15 dB at high frequencies but many people remain unaffected. Noise at a level of 90 dB(A) over the working day will produce a hearing loss well in excess of 20 dB in many people. At sound levels higher than 90 dB(A) the degree and scope of loss increase markedly: severe losses at high frequencies together with modest losses at low frequencies (see Figure 24.8). Such findings have served as the basis for legislation in many countries to restrict the exposure of individuals to high levels of occupational noise.

NIPTS appears to be less dependent on the type of noise to which the individual is exposed than TTS. NIPTS first appears in the region of 4000 Hz and with increasing exposure to noise, spreads to adjacent frequencies (see Figure 24.9). One reason for susceptibility to hearing loss in this frequency region is that the ear is particularly sensitive to sound in this region. Insofar as there is a dependence of hearing loss on the spectrum of noise, noise at high frequencies presents a more serious risk. This means that the A-weighted noise level is a better measure of risk to hearing than the unweighted measure.

In most countries, the trading relation between time and level is based on "the equal energy rule" so that if 90 dB(A) for 8 hr is assumed as the acceptable limit, then for every halving of exposure time, the overall level can be increased by 3 dB. Some have taken a less cautious approach by advocating 5 or 6 dB as an appropriate halving factor (e.g., Kraak, 1982). In the United States, a 5 dB trading relation is used (see Table 24.1).

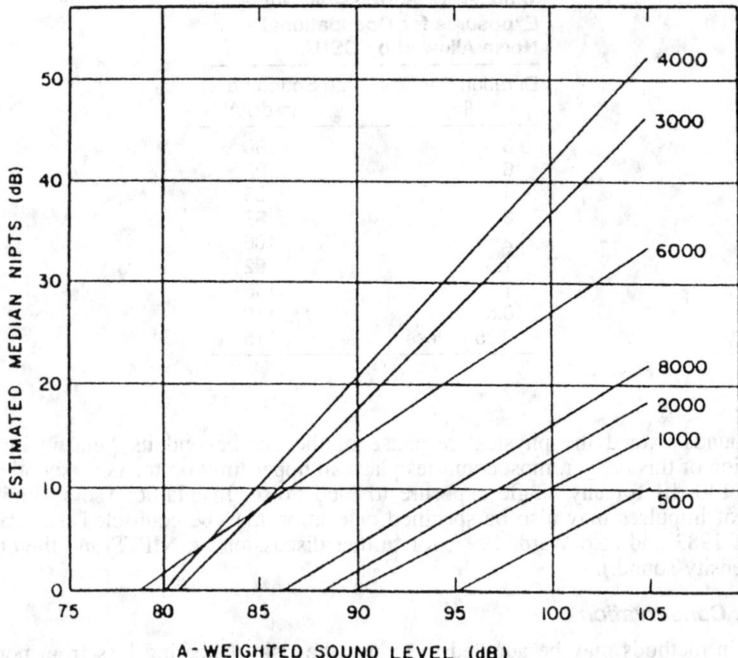

Figure 24.8 Estimated NIPTS at various frequencies produced by 10 years or more of exposure to industrial noise. The noise level is the A-weighted sound level for 8hr/day, 200 days/year. [From Ward (1983) with permission of W. D. Ward.]

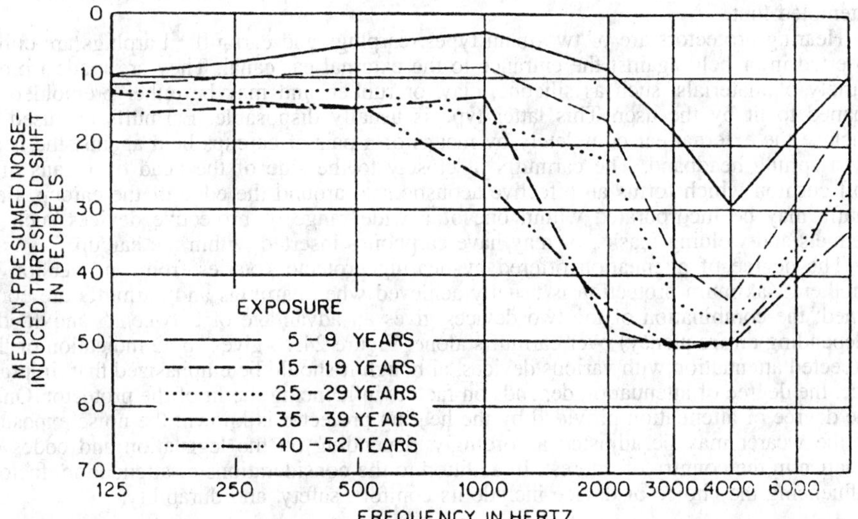

Figure 24.9 The estimated NIPTS at various frequencies as a function of exposure duration in years. [From Taylor, Pearson, Mair, and Burns (1965) with the permission of the American Institute of Physics.]

Table 24.1 Permissible Noise Exposures for Occupational Noise Allowed by OSHA

Duration Per Day (hr)	Sound Level in dB(A)
8	90
6	92
4	95
3	97
2	100
1.5	102
1	105
0.5	110
0.25 or less	115

Some sounds extend the physical response of the ear beyond its "elastic limits." In recognition of this danger, most countries place an upper limit on the peak sound pressure level of 140 dB for any single exposure to loud noise. In addition, some limit to the number of impulses may also be specified or caution may be counseled (see Passchier-Vermeer, 1983 and also Ward, 1997, for further discussion on NIPTS and the effects of high-intensity sound.).

Hearing Conservation

Four main methods may be adopted to reduce the risk of hearing loss from noise. The first is that the level of noise be reduced by the improved engineering design or modification of machines. Various engineering changes to reduce noise are described later in the chapter in Section 24.7. The second approach is to test the susceptibility of individuals to hearing loss. Unfortunately, tests or indexes of susceptibility are poor at predicting the degree of risk of hearing loss. A third method is to reduce exposure by various management or scheduling changes to rotate personnel periodically to quieter environments, to locate them where the noise levels are lower, or to reduce peak noise levels by staggering the operation of machines. A fourth approach discussed in some detail here, is to attenuate the sounds at the worker's ear by means of plugs or muffs (see Berger and Casali, 1997). This method of hearing conservation, although convenient and inexpensive, does have drawbacks, and the U.S. OSHA requires that engineering controls to reduce noise be attempted first.

Hearing protectors are of two main types: earplugs and earmuffs. Earplugs are either inserted in or held against the entrance to the external ear canal. They are made from a variety of materials, such as silicone, clay, or rubber, and may be either premolded or shaped to fit by the user. This latter type is usually disposable. Earmuffs are used to enclose the external ear completely by means of a pair of earcups held against the head by a sprung headband. The earmuffs fit closely to the side of the head by means of a soft cushion which forms an effective acoustic seal around the edge of the earcup. Earmuffs may be incorporated within one of a wide range of protective devices such as helmets and welding masks, or may have earphones inserted within the earcup.

The degree of attenuation offered by hearing protectors varies from one version to another. Maximum protection is usually achieved when earplugs and earmuffs are combined; the combination of the two devices gives an advantage of between 5 and 15 dB (depending on frequency) over earmuffs alone. Figure 24.10 gives some indication of the expected attenuation with various devices, although it should be emphasized that, in practice, the degree of attenuation depends on factors influencing the fit of the protector. Once the degree of attenuation provided by the hearing protector is known, the noise exposure of the wearer may be adjusted accordingly, depending on the legislation and codes of practice in the country of interest. In addition to the considerations of attenuation, factors influencing the choice of device include its comfort, safety, and durability.

Effects of Hearing Protectors on Talking and Listening

In continuous noise at levels between 85 and 105 dB, hearing protectors improve the *reception* of speech except for those who have hearing loss at high frequencies (Howell and Martin, 1975). This effect occurs despite the fact that noise attenuates the speech and

TYPE OF PROTECTOR	FREQUENCY RANGE IN HERTZ				
	1-20	20-100	100-800	800-8000	>8000
EARPLUGS	5-10	5-20	20-35	30-40	30-40
SEMIINSERT EARPLUGS	5-10	5-20	15-20	25-40	30-40
EARMUFFS	0-2	2-15	15-35	30-45	35-45
EARPLUGS AND EARMUFFS COMBINED	10-15	15-25	25-45	30-60	40-60
COMMUNICATION HEADSETS	0-2	2-10	10-30	25-40	30-40
HELMET	0-2	2-7	7-20	20-55	30-55
SPACE HELMET (TOTAL HEAD ENCLOSURE)	3-8	5-10	10-25	30-60	30-60

ENTRIES SHOW APPROXIMATE MINIMA AND MAXIMA OF PROTECTION AVAILABLE FROM VARIOUS TYPES OF DEVICES, IN DECIBELS.

Figure 24.10 The expected effects, in decibels, of a range of hearing protectors as a function of frequency. (From C. M. Harris, *Handbook of Noise Control*. Copyright © 1979. Reprinted with permission of McGraw-Hill Book Company.)

the noise in equal measure (the signal-to-noise ratio is the same with and without hearing protectors). In order to overcome the effects of noise, the level of speech would have to be increased. The most likely reason for the advantage of hearing protectors in the reception of speech is that speech of over 85 dB is known to produce significant degrees of aural distortion, and this effect is reduced by attenuating the overall level of sound, thus reducing speech as well as noise. At levels of continuous noise below 80 dB (with speech below 75 dB), hearing protectors may reduce the intelligibility of speech for the listener. This is because the quieter portions of speech are attenuated by the hearing protector to levels below those of the threshold of audibility.

The intelligibility of speech produced by the *talker* may be diminished when hearing protectors are worn. Wearing hearing protectors in noise causes the speaker to lower the level of the voice by 3 to 4 dB. A speaker regulates the loudness of the voice on the basis of the information about its loudness reaching the ear via bones of the skull and through the air. When hearing protectors are worn, the level of airborne speech sounds reaching the ear is reduced and speech is transmitted to the ear primarily by bone conduction. This gives the speaker the impression that there is less external noise to counteract while talking to others and thus the level of speech is adjusted to a lower level than it should be for the noisy environment.

Unfortunately, in a continuous loud noise environment the beneficial effect of wearing hearing protectors for the listener is more than outweighed by the detrimental effect of hearing protectors on the *speaker*. In intermittent noise, conventional hearing protectors severely disrupt the reception of speech when the noise temporarily stops. However, this shortcoming can be overcome partly by the use of amplitude-sensitive earplugs, which increase the degree of attenuation as the noise exceeds a certain level (Mosko and Fletcher, 1971).

Although the detectability of sounds in a noisy environment is therefore little affected by hearing protection, there is the possibility that unexpected sounds such as approaching vehicles or other hazards may be less "attention-getting." Reduction of noise at the source through improved engineering design or modifications is the preferred method of hearing protection, for this and other reasons.

24.4.1.2 Effects of Noise on Communication

Introduction

Noise may interfere with the detection of auditory signals; this effect is known as *masking*. Masking will have an effect on the detection of relatively simple auditory signals, such as warnings and alarms, and also the noise signature of a machine which may be used to assess its wear. However, by far the most pervasive effect of masking is that on speech.

Masking of Nonspeech Sounds

Figure 24.11 shows that as a pure tone is brought progressively closer to the center of a masker (a narrow band of noise), increasingly high sound pressure levels must be used for the tone to be heard. A striking feature of the family of curves shown is that at progressively higher levels of the masker, the curve describing the degree of masking becomes more asymmetric. It is observed that the effect of the masker is most pronounced for frequencies above the masker. It follows that a signal below the lower bound frequency of a masker of this sort will be a more effective alarm than one above the upper bound frequency of the masker. Only that portion of a masker close in frequency to the signal will contribute to the masking of the signal. The range of sounds that contributes to the masking of a tone is known as the *critical band*. The width of this band is roughly 50 Hz at frequencies below 1000 Hz and some 10% of the center frequency above 100 Hz. Noise falling outside these bounds will not contribute to any great degree to the masking of tones within it (see Webster, 1984, and Buus, 1997, for further details).

Intense sound signals are often used as alarms. Several different sorts of alarms may be required. Care must be taken to ensure that noise does not mask the alarm and in designing the alarm and determining the level of the alarm in relation to the background noise, the spectral and temporal features must be selected in order that alarms may be distinguished from one another.

One set of guidelines developed for use on civilian aircraft embodies a number of interesting suggestions that may be adapted to other circumstances (Patterson, 1983), as follows: (1) The alarm should be at least 15 dB above the level at which it can just be detected in the background noise. An upper limit of 25 dB above this threshold should be observed when the background levels are already high to avoid the danger of overloading the ear if the level of the alarm is excessively high. (2) If pulses of sound are used in the alarm, the length of pulse should ideally be 100 to 150 msec and should rise (and fall) in 20 to 30 msec. Pulses of shorter duration should be reserved to signify more urgent events. (3) The use of a distinctive temporal pattern (of five or more pulses) should aid identification of warnings when more than one warning is possible. (4) A sound with several harmonically related components (that is, containing component sounds that are multiples of each other in frequency) helps to give the impression of a single fused sound which again helps identification. (5) Synthesized speech may be presented during an alarm, but its prominence should depend on the purpose of the alarm.

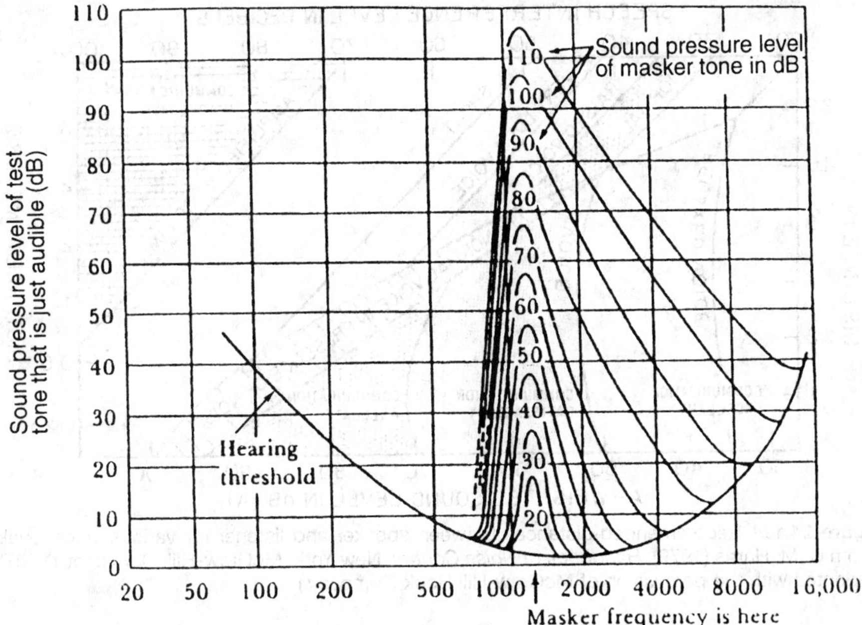

Figure 24.11 Masking curves for a narrow band of noise centered at 1200 Hz. Each curve represents the sound pressure level of a tone that is just audible in masker at the given level. [Adapted from E. Zwicker and B. Scharf (1965). A model of loudness summation. *Psychological Review, 72,* 3–26. Copyright © 1965 by the American Psychological Association. Reprinted/Adapted by permission of the author.]

Masking of Speech

Predicting the effects of noise on the production and detection of speech presents a host of complexities simply because speech is a broad-band complex acoustical signal containing periodic and aperiodic components which include impulsive sounds. Effects on the talker and on the listener may be distinguished.

Effects on the Talker Loud noise causes the overall level of speech to be raised spontaneously. In most cases this increased speech output does not match the increase in level of the noise. Values as low as a 3-dB increase have been recorded in response to a 10-dB elevation in noise level (van Heusden, Plomp, and Pols, 1979), although in other cases, values between 5 and 10 dB in response to similar elevations have been recorded (Pearsons, Bennett, and Fidell, 1977).

In the absence of masking sounds, the overall level of speech that is adopted spontaneously varies widely. When face to face with a listener 1 m away, the range in the sound pressure level of speech can be as great as 35 to 90 dB (Pickett, 1956), although there is evidence that this range is less in settings outside the laboratory (with a typical range of 48 to 70 dB(A); Pearsons et al., 1977). Idealized data for a range of distances and background noises are shown in Figure 24.12. As a general rule, voice level will be raised as the background noise level increases, as the task of communicating a message is made more demanding or when talking in front of an audience or into a microphone (Webster, 1984).

One response to the difficulties of communicating in noise is for the person to shout. Increasing vocal effort has side effects on the intelligibility of the speech. The intelligibility of speech is relatively constant at levels of speech in the range from 50 to 78 dB but intelligibility diminishes above and below this range (Pickett, 1956). The average spectral composition of speech tends to change when the level exceeds this range, which

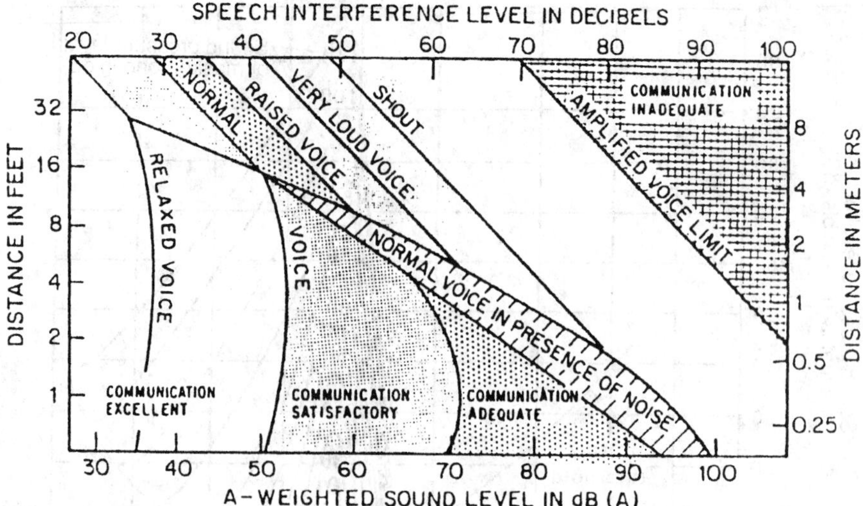

Figure 24.12 Recommended distances between speaker and listener for various voice levels. (From C. M. Harris (1979). *Handbook of Noise Control*. New York: McGraw-Hill. Copyright © 1979. Reprinted with the permission of McGraw-Hill Book Company.)

may account in part for the drop in intelligibility. As the output level increases, speech components at frequencies around 1000 Hz increases predominantly with components at frequencies below this region being less in evidence (Webster, 1979).

Effects on the Listener When speech is presented through electronic circuitry such as that on television, the effect of ambient noise may be overcome by the simple expedient of turning up the volume control. When speaking face to face, individuals may regulate the distance from one another to overcome the effect of noise. However, it is very difficult to predict these changes because the distances adopted may have as much to do with social proprieties or physical constraints as with the acoustic properties of the environment.

Although the sound pressure level of the human voice may range from 35 to 90 dB, the usable range is normally between 50 and 80 dB(A). A number of indexes of the likely effects of masking have been developed; each has some shortcomings. Part of the difficulty with the development of such indexes is that the effect of what is being said is often as great as the effect of the ratio between the signal and noise. In addition, there are effects such as that of the degree of practice at listening for sounds and familiarity with the voice (both of which influence intelligibility), which are difficult to incorporate into such indexes.

Measures of Interference with Speech by Noise

Despite its several shortcomings, one of the best ways of predicting masking properties of noise on speech intelligibility is to use the integrated A-weighted sound level. The main advantage of this type of measure (L_{eq} with A-weighting) is that it is readily available on sound level meters, with the accessibility and simplicity of measurement that this implies (Klumpp and Leonard, 1963). One can therefore use the sound level meter to take a reading of A-weighted L_{eq}, and then use Figure 24.12 with the distance of communication, to determine if communication will be adequate. Averaging methods, such as the *articulation index* and the *speech interference level* are more complicated but in general give better predictions of speech interference.

The Articulation Index (AI) This measure is based on the idea that sound in each of a range of narrow bands of frequencies makes a contribution to the total degree of interference. In its original form, 20 bands were used and the general procedure was one of calculating the difference between the speech level and the noise levels in each of the bands. In the original method there were more bands in the 2000-Hz region than in the 500-Hz region to give due recognition to the fact that speech at some frequencies contributes more than others to speech intelligibility. Variants of the method involve the use of bands (such as third-octave or octave intervals) and the use of weighting factors to achieve the same end. For a detailed analysis of the procedure consult Kryter (1995).

Speech Interference Level (SIL) This measure, unlike AI, relies on the measurement of noise alone. SIL is calculated from the arithmetic mean of the sound pressure levels for octave bands centered at 500, 1000, 2000, and 4000 Hz. Using Figure 24.12, one may assess the combination of distance and vocal effort that will give rise to satisfactory communication in the SIL in question. For example, with a SIL of 60 dB, a *normal voice in the presence of noise* may be used with a speaker-to-listener distance of 1 m (see also Crocker, 1997, Chapter 80 for further discussion on SIL and AI).

24.4.1.3 Effects of Noise on Reading

A series of studies has shown that speech at relatively low levels can disrupt memory for items presented visually, which means that other people's conversations are likely to disrupt the activities of reading and writing. This concern is in part due to the increasing prevalence of office work, the trend toward open-plan offices, and the likely negative implications on efficiency of such environments. In addition to the concern over efficiency at work, there is some suggestion that the development of reading performance may be impaired as a result of exposure to noise at home (Cohen, Glass, and Singer, 1973) and in the classroom (Crook and Langdon, 1974).

Two further features of the effect of irrelevant speech on material presented visually make the topic an important one. The first is that the effect seems to be independent of the overall level of the background speech. For example, the disruptive effect of speech on memory for visually presented material remains roughly the same over a range of levels from 40 to 76 dB(A) (Colle, 1980; Salame and Baddeley, 1983). The second important feature is that the effect is one that depends on the type of material being heard. Noise, whether it is continuous, intermittent, white, or pink, produces little or no effect on this kind of task (Murray, 1965). However, even speech in a language unfamiliar to the listener disrupts this type of performance.

Comprehension of the material being read suffers in loud noise. Not only does it take longer to read a passage, but it seems that memory for the contents of the passage is poor. Systematic studies of the type of errors produced in proofreading tasks indicate that although superficial features of the text are understood, part of the deeper meaning of the passage is lost in the presence of noise. Noise as low as 68 dB(A) impairs the detection of contextual errors (grammatical mistakes, omissions, and the presence of incorrect and inappropriate words), but both the detection of noncontextual errors (misspellings and typographical errors) and the average rate of work are unimpaired (Weinstein, 1977).

When writing in noise, efficiency may also be diminished because of the difficulty of retrieving material from that part of long-term memory concerned with the meaning of words, the so-called *semantic memory*. Noise at levels as low as 80 dB appears to facilitate the process of retrieving simple information but the retrieval of more difficult information is impaired (Eysenck, 1975). This type of effect may be produced by a shift in the confidence with which each type of material is recalled. After reading a story in loud noise, people become more cautious about recalling rare (or difficult) words and less cautious about recalling common words (Jones, Thomas, and Harding, 1982).

24.4.1.4 Effects of Noise on Task Performance

The effects of noise on performance may be divided conveniently into four categories: (1) effects of arousal, (2) effects of lack of control, (3) strategic effects, and (4) effects on attention.

Effects of Arousal

This effect takes three main forms: (1) The effect of short, often unpredictable, noise bursts that may startle the individual; (2) The effect of the variability of acoustic input, which may be increased by occasional bursts of sound and reduced by continuous unvarying noise; and (3) Loud sounds that raise the general excitability or responsiveness of a person; this may be good or bad for working efficiency, depending on the general state of the person.

Bursts of Noise Unexpected bursts of noise produce marked but transient changes in the physiological response. Three categories of response to unpredicted noises may be distinguished: startle response, orienting reflex, and defense reflex (Burns, 1979). The startle response is a potentially protective muscular response (including eye closure, facial muscle contraction, and head jerk), whereas the orienting reflex is a general alerting response (a "what is it?" reaction), and the defense reflex is a response to intense sound stimuli that are interpreted as harmful. The magnitude of the orienting response diminishes with repeated presentation of the noise burst (a phenomenon known as habituation) as the novelty of the stimulus diminishes. The startle response also habituates, although there is some evidence that it never completely disappears even after many repetitions of the stimulus (Landis and Hunt, 1939). The defense response may increase with repeated presentation of the stimulus in conditions where the significance of the noise burst becomes regarded as malignant.

The effects of bursts of noise on performance tend to occur in the period up to 30 sec following the burst. Some attempts have been made to establish the types of task that are most susceptible to such effects with generally equivocal results. Bursts of noise slow the speed of a simple motor response only when the burst arrives during the execution of the response rather than during the presentation of the stimulus (Fisher, 1973). Several studies have shown that the intake of information and other mental tasks are disrupted by noise (Salame and Wittersheim, 1978; Woodhead, 1964). It appears that the elements of the task that are susceptible to disruption by noise bursts are those that are "data limited." Activities of this kind include short perceptual or motor events in which compensatory effort on the part of the subject is not possible. Such perceptual and motor effects may be a reflection of the startle effect described above. Thus, higher mental activity such as that involved in calculation is immune to the specific effects of infrequent bursts because of the way in which short-term memory may compensate for such brief disruptions.

An important consideration is the extent to which these short-term responses diminish with repeated presentation of the burst. Although it is clear that elements of the physiological response, such as the orienting reflex, are eliminated by habituation, the startle response may resist complete elimination. Even after long periods of exposure to impulsive noise, residual effects of the eye-blink and head-jerk response are found. Laboratory studies have confirmed that repetitive impulsive noise is still capable of disrupting performance of a skilled tracing task after many repetitions (May and Rice, 1971). Performance in complex tasks is disrupted if uninterrupted vision and steady posture are necessary for the successful execution of the task.

Effects that extend to the order of 30 sec beyond the burst are likely to be ones associated with strategic effects rather than arousing effects (see Woodhead, 1966). As the effect of startle becomes less pronounced, the subject may gain tactical advantage by anticipating the appearance of a noise burst and overcome its effects by compensatory effort.

Variability of the Acoustic Background A second type of arousing effect may be found because noise, of a variable or intermittent type, may improve performance in tasks requiring vigilance after a period of long-continued work. These tasks involve the detection of small changes in the flow of information arising from one or more sources over long unbroken periods of time. Typically, the ability to detect such signals diminishes as the time at work increases. Varied and irrelevant auditory stimulation stems the decline usually found in quiet, at least when events within the task are presented at a low rate (McGrath, 1963). This suggests that in long, monotonous tasks, variable noise may serve to raise the level of arousal and hence improve efficiency (see Figure 24.13).

Similarly, even continuous noise, if loud, can reduce the effect of other conditions that produce drowsiness. For example, the effect of loud noise may counteract that of the loss of sleep. In isolation, noise and sleep loss have a deleterious effect at the end of a 40 min period of continuous work. Yet when both stressors are present together, the net effect is one of an improvement in performance (Wilkinson, 1963). Because it is generally regarded that both very high and very low levels of arousal may give rise to inefficiency, it may be argued that the de-arousing effect of sleep loss is counteracted by the arousing effect of noise (see Broadbent, 1971).

Individuals may also differ in their general level of arousal. The way in which noise interacts with other agents is very complex and details of the mechanism responsible for such effects are not well understood. However, sufficient is known to lend weight to the generalization that the effects of noise are at least in part due to the general state of the organism. In addition, these findings suggest that in industrial settings, where noise is accompanied by many other stresses, the combined effect of these stresses could be much more than is suggested from their study in isolation.

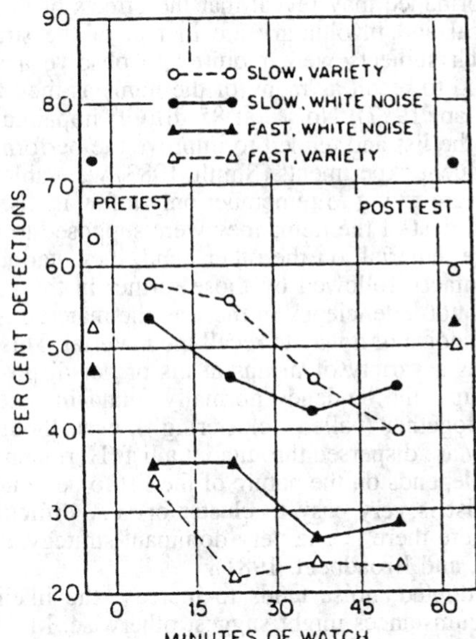

Figure 24.13 Percentage of signals detected as a function of time spent at a visual vigilance task under two different rates of stimulus presentation (slow vs. fast) and two different conditions of auditory stimulation (variable sound vs. continuous white noise). (From Buckner, D. N. and McGrath, J. J. (1963) *Vigilance: A Symposium*. New York: McGraw-Hill. Copyright © 1963. Reprinted with permission of McGraw-Hill Book Company.)

Effects of Lack of Control

Noise, in addition to being a physical entity, also occurs "in a context of cognition and social circumstances" (Glass and Singer, 1972, p. 157). Subjects exposed to bursts of random noise for 25 min were given simple verbal, numerical, and motor tasks to perform in noise. At the end of the exposure, one or a combination of the following tasks was given: a proofreading task, a task measuring tolerance to the frustration of attempting to solve problems that were in fact insoluble, and a color–word test. Those tasks undertaken during the period of exposure seemed immune to the effects of noise. However, performance of the tasks in the period following exposure was impaired by noise (Glass and Singer, 1972).

Since the original work of Glass and Singer, after effects of noise have been demonstrated in a wide range of circumstances including continuous loud noise (e.g., Jones, Auburn, and Chapman, 1982). Subsequent research has confirmed that these effects can be observed with unpredictable and uncontrollable stress of any kind and are not produced by noise alone. Furthermore, anticipation of a loud noise stressor appears to be sufficient to impair performance and the expectation of control counteracts this effect.

The issue of control over noise has a bearing on response to noise within the community. Annoyance may in part be governed by the feeling of being able to control the source of the noise as well as the acoustic properties of the noise (Graeven, 1975). Some features of performance associated with a lack of control, namely, the failure to persist at problem solving tasks, have been observed in children at schools beneath a busy approach path to an airport (Cohen, Evans, Krantz, and Stokols, 1980).

Strategic Effects

The effects of noise on performance change when details of the task, the experimental setting, or the subject population are altered. This has led some to argue that these are effects of the particular strategy adopted. The idea of a strategy suggests that a person may on different occasions perform the same task by using slightly different mental operations. According to this view, noise does not cause a straightforward reduction in efficiency, but rather causes some activities to be favored and others not.

The analysis of performance may reveal that the effects of noise that were originally thought to be mechanical and involuntary are in fact of the strategic type. One study involved a task in which subjects were required to observe a stream of items which stopped unexpectedly and to recall as many of the items as they could remember (Hamilton, Hockey, and Rejman, 1977). Noise [at 85 dB(C)] impaired memory for the items remote from the end of the list and tended to improve the performance of the last few in the list. However, subsequent experiments (Smith, 1983) were able to show that this effect was reversed if people were trying to remember only a few items. In this case, they were able to go back to the earliest of the items they were supposed to recall. When they were asked to remember more material, on the other hand, they started at the very last items (these being produced intact), followed by those further in the past (relatively less well remembered). Thus, it is not a deficiency in the way the material is stored that is changed in noise; rather, it is the choice of a way to recall the material. This effect of noise appears only if the individual has a variety of means at his or her disposal to perform the task. For instance, noise disrupts the tendency normally found in quiet for words of similar meaning to be recalled together (called "clustering"), even though when the words are originally presented, they are dispersed throughout a list (Hormann and Osterkamp, 1966). The extent of the effect depends on the nature of the list to be remembered, and the effect does not appear if the list is very easy to cluster or very difficult to cluster, these two instances being ones where there is one very dominant strategy, albeit a different one in each case (Smith, Jones, and Broadbent, 1981).

Once a strategy is adopted, noise tends to increase the likelihood of its continued adoption even when circumstances might suggest otherwise. Moreover, rapid alternation between different types of tasks is particularly damaging to performance in noise (Dornic and Fernacus, 1981). The reluctance to abandon a strategy in noise is shown by studies examining the effects of noise on the speed of response to subtle changes in the features of a task. When signals in a task are not equiprobable, noise produces faster reactions to probable signals and slower reactions to rare signals. If, without warning, the signals become equiprobable in noise, the pattern of responding previously established tends to be carried over (Smith, 1985).

Effects on Attention

Attention During Prolonged Work Early in the history of noise research, effects of very intense noise were noted on tasks involving vigilance. The way in which vigilance performance is influenced by noise has benefited from the development of sophisticated theories of signal detection (see Davies and Parasurman, 1982). These theories distinguish between effects on the efficiency of detection of response bias (which is roughly equated with the observer's readiness to make a response for a given amount of evidence) and perceptual sensitivity (the efficiency with which signal and nonsignal events can be discriminated).

Loud Noise Influences Response Bias If people are required to state the confidence of their judgment that an event is a signal, noise tends to increase the tendency to use extreme categories of judgment at the expense of intermediate categories: They are more prepared to assert that they are sure that a signal *is* or *is not* there (Broadbent and Gregory, 1965). From what is known about vigilance performance in quiet, we may predict what will happen in noise to the number of signals detected. When signals are very unlikely, people report the presence of a signal only when they have high confidence, and doubtful judgments that something is present do not produce a report. The increased certainty that results from noise then gives more correct reports. If signals are probable, however, people report them unless they are certain that no signal was present. Doubtful judgments of the absence of a signal tend to get reported as postitive detections. In that case, noise reduces the numbers of reports of signals.

The prevailing level of confidence, in addition to its effects on the reporting of signals, will have effects on the way in which observers check on the state of a display. If several sources are involved, it is possible to chart the process of interrogation by offering the observer brief glimpses of a state of each display. Typically, some displays receive more attention than others, and this tendency is exacerbated in loud noise (Hockey, 1973). The action of noise in this case seems to be one of exaggerating those biases that already prevail about where significant events requiring action are most likely to occur. Effects have also been noted on another class of vigilance tasks that are sensitive to levels of noise as low as 80 dB(C) (Jones, Smith, and Broadbent, 1979). In this case the detection of signals places a heavy reliance on memory, and the periods over which performance was assessed were relatively short. Because of these differences, the results may arise out of changes of the strategic type. Some experiments suggest that noise causes an increase in errors over time (see Figure 24.14). This experiment uses several precautions to eliminate the effects of acoustic cues (Jones, 1983).

Time Sharing Everyday tasks are made up of a range of different activities each with different priorities. The effect of noise on tasks of this sort is to swing resources away from elements of low priority and toward those that are seen to be subjectively more important (Hockey, 1970). Some variation will be observed in the patterning of response in tasks involving several different elements; this is to be expected in view of the different demands made by the tasks and by the type of instructions given for their execution (see, for example, Forster, 1978, for a detailed discussion).

Effects of Noise on Productivity

Relatively few studies have dealt with the issue of the effects of noise on efficiency at the workplace. Noise tends to be associated with accidents. The mean noise level at the place of work correlates very strongly with the frequency (but not the severity) of accidents (Kerr, 1950). Younger and less-experienced workers seem to be more susceptible to accidents, and the effects seem to be greater in levels of noise above 95 dB(A) (Cohen, 1974). It is of course possible that the effect of noise on accidents arises because noisy jobs are also difficult and dangerous. However, the incidence of accidents is reduced when ear defenders are introduced, which suggests that the greatest effect of noise is one of level rather than of exposure to risk (Cohen, 1976). Other studies have shown that productivity can be increased when the level of noise is reduced either by introducing ear defenders or by acoustical treatment of the workplace (Broadbent and Little, 1960).

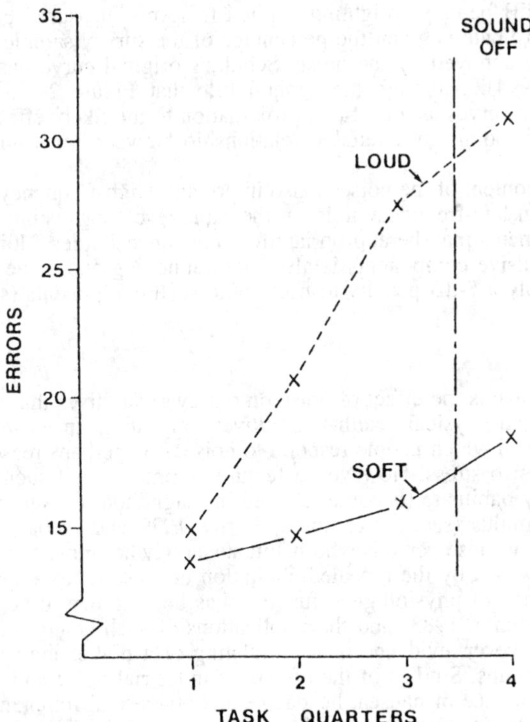

Figure 24.14 Effects of white noise on the incidence of errors in a serial reaction task lasting 40 min. In both loud (90 dB(C)) and soft (60 dB(C)) conditions the noise is switched off at the end of the third quarter.

24.4.1.5 Effects of Noise on Well-Being

This section deals with the effects of noise that are found in the community. The problem is important to human-factors engineers and ergonomists because the noise of operations may produce complaints from local residents or indeed from the work force.

Noise Annoyance

The idea of *noise annoyance* carries with it manifold suggestions of discomfort, disturbance, displeasure, unease, irritation, and complaint. From our own experience with noise we would not expect the degree of annoyance to be some simple function of the level of the noise. Certainly, it is the case that the proportion of people expressing annoyance increases as the sound level increases, but we know too that the circumstances in which a given noise arises also account in part for the annoyance.

The level of noise needed to produce the same annoyance is less when the noise is heard indoors than when it is heard outdoors (Robinson, Bowsher, and Copeland, 1963). Expectations about the noise source cloud the judgment of annoyance, as is shown by the tendency for aircraft to be judged more noisy than automobiles or rail vehicles even when they are at the same level [Robinson, Copeland, and Rennie, 1961 and Crocker, 1997 (Chapter 80)]. In addition, the difficulty of performing a task in noise may also shape attitudes to subsequent encounters with that noise (Moser and Jones, 1983).

As we would expect, the noise level plays a major role in the generation of annoyance. There are two main reasons to adopt this view: (1) If analysis is made of social surveys that determine changes in annoyance with changes in noise level in which the effects of response bias are carefully separated out (the willingness to use certain categories of annoyance) from those of the true sensitivity of noise annoyance to level, the effects on annoyance are clearly shown in the latter measure (Fidell, Horonjeff, and Green, 1981), and (2) Annoyance may be predicted from physical measurements alone. A synthesis of the results of several large-scale studies was produced by Schultz (see Schultz, 1978). Figures 24.15(*a*) and 24.15(*b*) show numbers of respondents who are highly annoyed by noise. On the abscissa is shown the metric L_{dn}, a measure based on a 24 hr L_{eq} (see above) with a 10-dB penalty weighting applied to levels measured between 22.00 and 07.00 hr. On the ordinate is given the percentage of the survey sample who report themselves to be highly annoyed by the noise. Schultz's original curve has been updated by Finegold et al. (1994). Although the general rule that Figure 24.15 represents is not without its critics, it serves as a useful approximation to the likely effects of noise within the community and points to a reliable relationship between level and annoyance (see Fidell, 1984).

The frequency content of the noise is also important. High-frequency noise is normally more annoying than low-frequency noise at the same level, suggesting that the use of A-weighted measurements may be appropriate for annoyance studies. Noise which contains pure tones or impulsive components is also more annoying and some community noise rating schemes apply a 5-dB penalty to noise with such components (see Crocker, 1997, Chapter 80).

Noise and Health

So great and pervasive is the effect of noise on our everyday lives that the idea that noise influences mental and physical health is intuitively appealing. In view of the strong and stereotyped manner in which people respond to noise, it is perhaps reasonable to suppose that this contributes to stress. However, a feature common to all such physiological responses is that they habituate (become reduced in magnitude) on successive exposure to the same noise stimulus (see, for example, Burns, 1979 and Abbas, 1997). It may be argued, however, that those persons who habituate slowly are most at risk from the likely systemic damage caused by the repeated elicitation of these reflexes (Jansen, 1961).

A very wide range of physiological function has been shown to be sensitive to noise (see Burns, 1979; Rehm, 1983) and the implications of such changes for health are still not understood. However, evidence is accumulating of a higher incidence of symptoms in noise-exposed groups. Studies of the effects of industrial noise and health have noted increases in the incidence of nausea, headache, anxiety, sexual impotence, and heart disease (for example, Jansen, 1961). In each case, however, it is difficult to distinguish the effects of the type of job and the effects of noise. Noisy jobs are usually the most physically demanding, dirty, and dangerous. The influence of these factors alone might account for the higher incidence of such symptoms.

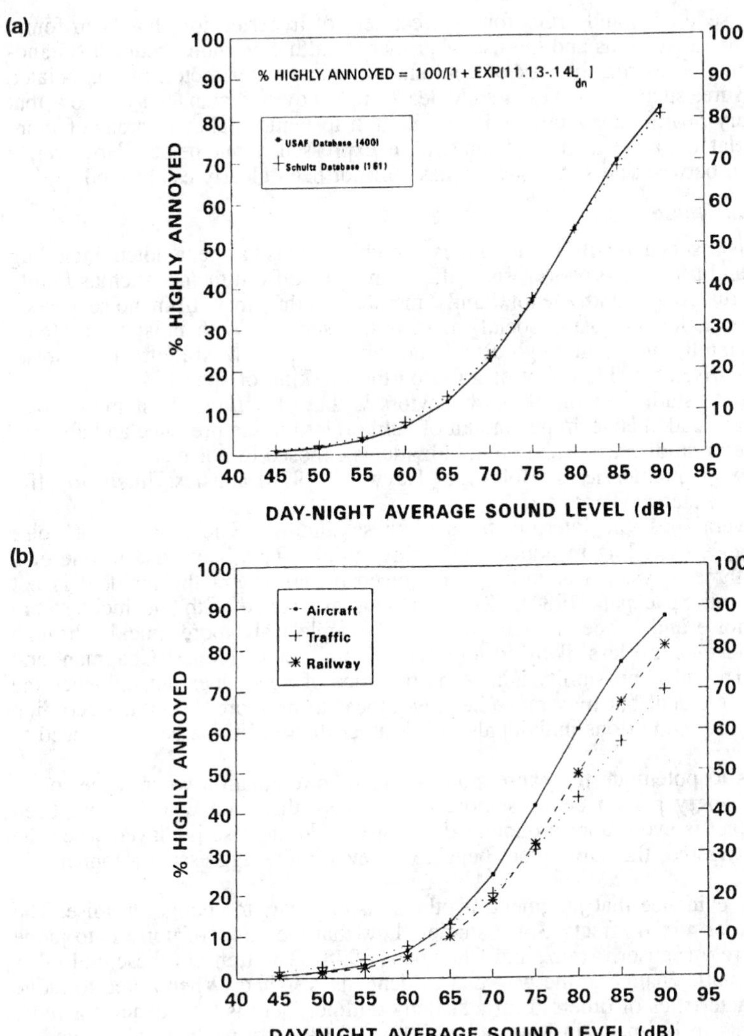

Figure 24.15 The relation between noise exposure and the percentage of survey respondents claiming to be highly annoyed. (From Finegold, Harris and von Gierke, 1994, with permission of H. E. von Gierke).

Studies of the effects of noise in the community, arising from aircraft or traffic, though not suffering the shortcomings of industrial studies, are undermined by a different class of problems. The particular context in which questions about health are asked seems to be an important variable. If no reference is made to aircraft noise while questioning respondents about their health, the overall statistical relation between aircraft noise and physical symptoms is diminished (Grandjean, 1974). One way of overcoming the effect of bias in verbal responses is to use some physiological measure instead. In these cases, changes in physiological activity are found, such as an increase in blood pressure of children living near airports (Cohen et al., 1980), but again these effects may arise owing to a number of confounding factors, such as a bias due to socioeconomic class or bias in the way in which blood pressure is recorded.

Another device for assessing the effects of noise on health, this time in an indirect way, is to compare the use of primary health care facilities in areas representing different levels of noise exposure from aircraft. In these cases the use of drugs (particularly sleeping tablets, antacids, sedatives, and drugs for the treatment of hypertension) has been found to be higher in the noisy areas and the use of primary health care more frequent (Grandjean, 1974). Such studies run the risk of confounding some other factor, not associated with noise exposure, such as social class. At least one study has managed to show that the use of primary health care facilities is, in addition to being higher in areas of more intense noise, related to the degree of annoyance expressed about health. However, a causal connection between noise and health has still not been clearly established.

Social Effects of Noise

A range of effects associated with living in noisy neighborhoods has been noted, including high crime rate and truancy. However, such effects may arise from factors such as family size, density of dwellings, and age of family members rather than from noise per se. Evidence of a reduction of casual social interaction associated with noisy areas (e.g., Appleyard and Lintell, 1972), although usually attributed to the stressful effects of noise, is more likely to arise as a side effect of noise on the masking of speech.

The most widely studied of the social behaviors is that of *helping*. Systematic comparisons have been made between the amount of help offered in the presence and absence of noise. A range of studies has shown that individuals exposed to noise above 80 dB(A) are less likely to grant interviews (Boles and Hayward, 1978) and less likely to offer various kinds of assistance (Page, 1977).

There are several possible interpretations of these findings. The first is that noise increases the chance of failing to notice the victim's plight. This may arise in one or a combination of three ways: (1) attention is narrowed by noise and the incident is not noticed (Cohen and Spacapan, 1984), (2) the sounds associated with the incident are masked by the noise and it goes unnoticed, and (3) people walk more quickly through noisy settings and thus are less likely to notice incidental details (Jones, Chapman, and Auburn, 1981). The other possibility is that the presence of noise does not influence the perception of the incident but may make helping appear to be more difficult. According to this view, in noisy conditions individuals avoid rather than fail to perceive the need to offer assistance.

Noise appears to potentiate the expression of anger. In a situation where a person is made angry, the angry person behaves more aggressively than one who has not been angered. This effect is even more pronounced in noise. If loud noise is played when the person is being angered, the subsequent behavior is even more aggressive (Donnerstein and Wilson, 1976).

There is some evidence that judgment of others also appears to change in noise. The findings show rather mixed effects. Some studies show that the effect of noise is to judge others more severely (Sauser, Arauz, and Chambers, 1978). The trend in these and other studies is a systematic change in the way that evidence is weighed. When asked to judge the general characteristics of others from a sketchy outline, there is a tendency for more extreme judgments to be made in loud noise (70 to 90 dB) irrespective of the type of attribute being judged (Siegel and Steele, 1980). This trend in noise to produce extreme judgments is one that has been observed in tasks involving memory and sustained attention and may be a general feature of the response to noise.

24.5 INSTRUMENTATION FOR NOISE MEASUREMENTS

24.5.1 Introduction

There is a large variety of instruments available for noise measurements and the instruments used will depend on the noise measurements to be made (see Section 24.6). Thus, just a few of the most basic instruments are discussed here.

24.5.2 Microphones

Just as the ear responds to sound pressure, the basic instrument available to measure sound, the microphone, produces a voltage signal proportional to the sound pressure. Microphones work on a variety of principles. The most common are the piezoelectric crystal microphone, the condenser microphone, and the electret microphone. Piezoelectric microphones have several advantages—relative immunity to damage and humidity and high capacitance—but they have a rough and limited frequency response at high frequency (5000 to 15,000 Hz). Condenser microphones have a broader, smoother frequency response, but are more expensive and susceptible to damage and humidity. They need a polarizing voltage and also have a lower capacitance, which poses problems with the associated input to the sound level meter or preamplifier with which they are used. Electret microphones are similar to condenser microphones, but are less expensive and need no polarizing voltage. They are more rugged and immune to humidity (see Busch-Visniac and Hixson, 1997 for further discussion on types of microphones).

24.5.3 Sound Level Meters

All microphones require some form of signal amplification. The sound level meter is a portable combination of microphone, amplifier, and meter. The sound level meter is used to measure the signal, and normally the root mean square pressure, p_{rms} is obtained. Most sound level meters have A-, B-, and C-weighting filters; some also have octave band filters. The A-, B-, and C-weighting filters, shown in Figure 24.16 weight the sound pressure signal measured in much the same way as the human ear. For low-level sounds, the ear responds much as the A-weighting filter. It is insensitive to low-frequency sound (below 1000 Hz), quite sensitive to middle- to high-frequency sound (between 1000 and 10,000 Hz), and then insensitive again above 10,000 Hz. The B-weighting filter more nearly represents the ear response to middle-level sounds, and the C-weighting filter to high-level sounds (where the ear frequency response is much more nearly flat). The sound level meter is sometimes used as a "loudness" meter. Most people give readings in dB(A) (decibels using the A-weighting filter). Readings in dB(B) and dB(C) are not given so frequently (see Krug, 1997 for further discussion on sound level meters).

Many people assume that microphones are nondirectional. This is true at low frequencies. However, at high frequencies (1000 Hz upwards), microphones are more sensitive to head-on sound than sound from the side. Most manufacturers alter the sound level meter design to modify this effect, and it is important to follow the manufacturer's instructions on whether the microphone should be pointed at the source or oriented sideways to the source.

The sound level meter response is affected considerably by reflections from the operator located only 2 to 3 ft (about 1 m) away. For pure tone sounds, errors of up to 5 to 10 dB can be introduced. For more accurate measurements, the microphone can be located remotely by using an extension cable to join the microphone to the sound level meter.

24.5.4 Frequency Analyzers

Another facility that may be provided on a meter (or provided as an accessory) is a series of filters that passes on for analysis and display sounds only from a limited range of frequencies (known as bandpass filters). In this way, only the sound energy in a frequency band covering an octave (an interval in the audio-frequency range whose upper bound is twice that of its lower bound), or some other convenient interval (such as one-half-, or one-third-, or one-tenth-octave), may be shown. Constant bandwidth filters (such as 10 Hz, 20 Hz, or 50 Hz) are also sometimes used. By taking readings from each filter set

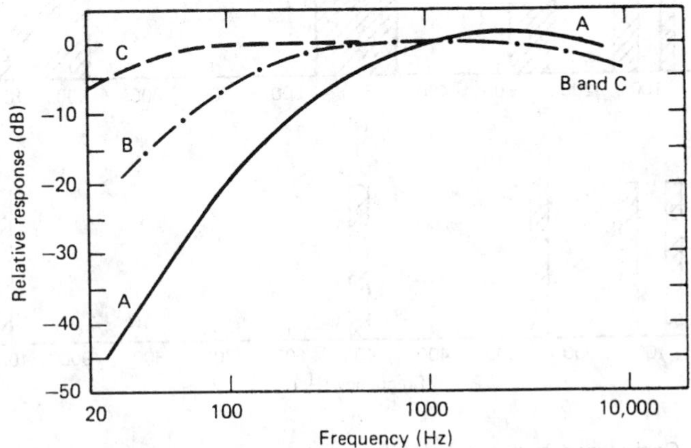

Figure 24.16 Filter response curves for A-, B-. and C-weighting filters.

in turn to cover the whole of the audio-frequency range, the distribution of the sound energy in the whole frequency spectrum may be obtained (see Figure 24.17).

24.5.5 Other Advanced Analyzers

The advent of microelectronics has allowed the range of facilities of instruments to be increased without a corresponding penalty of weight, convenience, or cost. As a result one or both of the following measures are now usually available on sound level meters or other analyzers: 1) Some instruments allow the A-weighted sound level exceeded for, respectively, 10, 50, 90, or *N*% of the total duration of the period of observation to be measured. 2) Other instruments obtain measurements of the weighted or unweighted L_{eq}. Usually, sound varies in level over time. The L_{eq} value for such a signal is the A-weighted level of a constant sound which, if continued over the same interval, would represent the same total energy as the varying sound (see Pope, 1997, for further discussion on analyzers).

24.5.6 Noise Dose Meters

Dose meters are small, simple, and inexpensive versions of the sound level meter which can be used to assess the total noise exposure of people at the workplace. They may be worn for long periods without inconvenience. The dose may be expressed as a proportion of the maximum permitted 8-hr dose. By means of relatively simple calculations, the L_{eq} may also be derived from the reading. Dose meters also signify, by means of a simple visual alarm, whether a specified maximum peak level has been exceeded. Such devices should be used with very great care to ensure that a reading is representative of the exposure. Artifacts may occur, for example, by intentional or unintentional shouting into the microphone. Wearers should be advised of the danger of false readings and if possible be supervised during the period of recording. Although noise dose meters can be designed to evaluate the noise exposure correctly where the 3-dB (energy) rule is applied, it is difficult to design a meter to evaluate exposure correctly using the 5-dB trading rule, because impulsive sounds are incorrectly assessed.

24.5.7 Sound Intensity Analyzers

Real-time and FFT analyzers, when used in conjunction with two microphones, can be used with advantage in noise source identification and machinery sound power measurements. Some real-time and FFT analyzers have the sound intensity calculation already programmed in. Although sound intensity probes can be built by individual investigators or technicians from two microphones, they are commercially available and usually incorporate phase-matched microphones. If two phase-matched microphones are not avail-

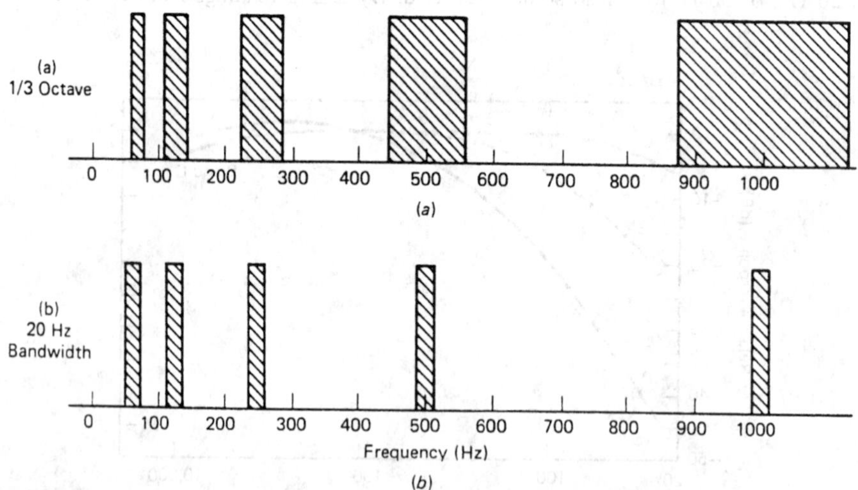

Figure 24.17 Comparison between bandwidths of (*a*) constant percentage filters (one-third octave) and (*b*) constant bandwidth filters (20 Hz bandwidth) at the same frequencies.

able, it is necessary to make corrections for phase shifts between microphones using measured phase errors and an FFT analyzer (see Crocker and Jacobsen, 1997).

24.6 NOISE MEASUREMENTS

We may want to measure noise for one or more of the following reasons:

1. To determine the noise at a certain distance from a machine (e.g., where an operator may be situated).
2. To verify that the noise produced by a machine lies within certain specified limits.
3. To determine if the noise level or level averaged over a period (such as 8 hr or 24 hr) meets industrial (e.g., OSHA) noise regulations or community noise ordinances.
4. To make a comparison between the noise produced by different production models of the same design machine or between the noise produced by different machines.
5. To try to determine the location and strength of noise sources on a machine or machines for noise control purposes.
6. To determine the sound power level of a machine for rating or labeling purposes.

Different methods will be used, depending on the reason for the noise measurement. In case 1, it may be desired simply to measure the sound level (SL) in dB(A) at the operator's head position (when he or she is absent). In many cases the machine is fixed, or a laboratory is not available, and the noise must be measured with the machine in situ. In some cases, the machine is sufficiently portable or small enough to take into a special laboratory where the noise can be measured more accurately. Two special acoustical rooms are available in most acoustical laboratories: *reverberation rooms* and *anechoic rooms*. Reverberation rooms are useful for the determination of the sound power of machines. Anechoic rooms are particularly useful when it is desired to make noise measurements without the effect of reflections.

24.6.1 Measurement of Noise from Machines

Standard methods are available for measuring machine noise. The reader should consult American National Standards (ANSI) or International Standards (ISO) depending on the type of measurement it is desired to make. The sound pressure level (SPL) or the sound level (SL) may be measured. For the SL, it is recommended that the A-weighting setting be used. For the SPL, it is suggested that octave bands be used, although for some purposes one-third octave bands may be more desirable.

24.6.2 Sound Power Level

It is desirable in some cases to determine the sound power (W) of the machine, since from the sound power it is possible to predict the SL or SPL of the machine in other surroundings. The sound power of a machine resting on a hard floor may be obtained by measuring the normal intensity I_n (W/m^2) at different points on a hemispherical surface enclosing the machine and thus computing the total sound power. Use of the new two-microphone intensity technique allows the intensity, I, to be measured directly at different points on the surface. This can then be summed over the measurement surface (see Figure 24.18) to obtain the total power, W (see American National Standard ANSI S12.12.1992 and Crocker, 1991). However, if such sophisticated equipment is unavailable, then a simplified approach must be used. In this approach, the SPL is measured on a similar hemispherical surface, and the normal intensity I_n is assumed to be given by Equation (24.2). Equation (24.2) assumes that measurements are in the far field. Background noise and reflections should be suppressed where possible (see Bruce and Moritz, 1997, for sound power of machinery).

24.7 NOISE CONTROL APPROACHES

24.7.1 Source-Path-Receiver

All noise control problems can be expressed as a source, path, and receiver, as shown in Figure 24.19. This may appear to be an obvious statement, but it is nevertheless very useful to think of each noise control problem using this concept. In many noise control problems, there are several sources, several paths of energy flow from each source, and several receivers. In any noise control problem, it is the best approach to determine both

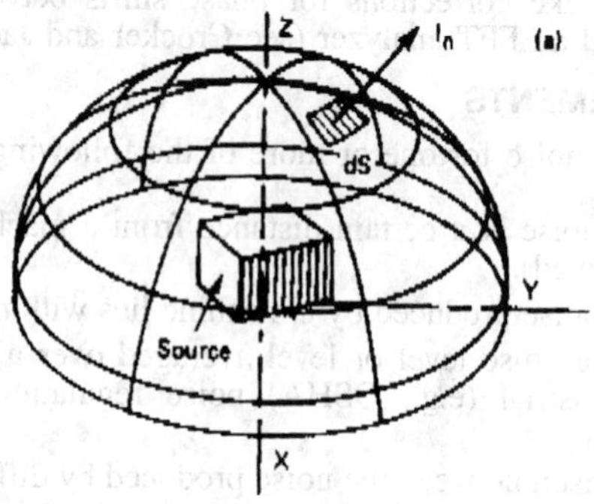

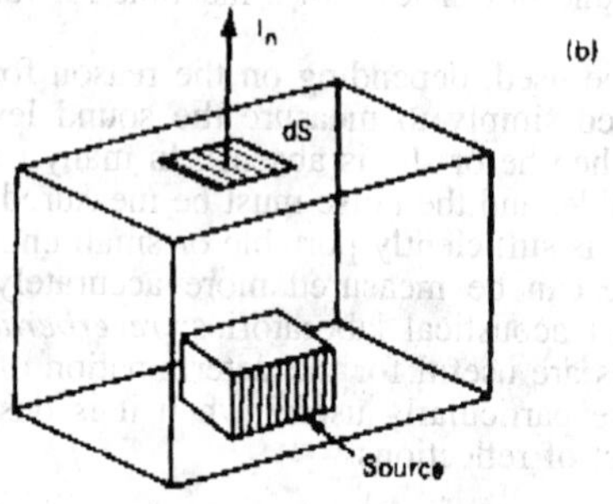

Figure 24.18 Sound intensity I_n, being measured in a direction normal to the surface on a segment dS of (a) a hemispherical enclosing surface, and (b) a rectangular box surface surrounding a source having a sound power W.

the sources and the paths of energy flow for each noise source in order of importance. The best approach is to reduce the dominant noise source if this is possible. The secondary source will now probably become more important or annoying, and it may be necessary to reduce this, and so on.

Unfortunately, in many machine noise problems, it is not always practical to reduce the strength of the source, since it either may be very expensive or time consuming or may interfere with the operation of the machine. In such cases it may be more efficient to interfere with the paths of energy transmission. Examples of this are the use of enclosures, barriers, absorbing material, vibration isolators, and vibration damping material. Some of these techniques are discussed in succeeding sections; further discussion may be found in Crocker (1997, Chapter 83).

In some cases, even the use of enclosures, barriers, absorption, and so on, is not practical since the operator may need continuous access to a machine. In this case, we must try to affect the receiver in Figure 24.4, the human ear. Using earplugs or earmuffs or making administrative changes to reduce exposure are examples of effecting changes at the receiver. Such approaches often are not completely satisfactory and should be considered as a last resort.

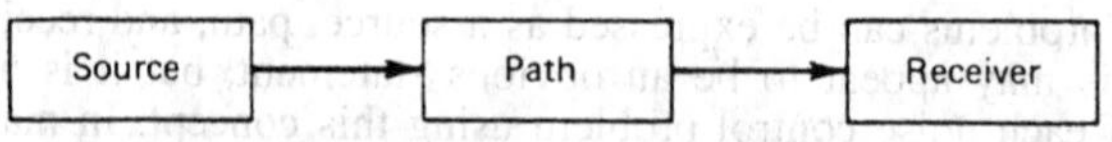

Figure 24.19 Source-path-receiver.

24.7.2 Enclosures

Enclosures are often the most effective means of reducing the acoustic energy path. Neglecting the stiffness and damping of a wall gives the transmission loss, TL, of an enclosure wall as

$$\text{TL} = 20 \log_{10} (mf) - 34, \text{ dB} \tag{7}$$

where f is the frequency (Hz) and m is the mass per unit area (lb/ft^2).

The TL predicted by Equation (24.7) agrees quite well with experiment, except at very low and high frequencies. It is seen that enclosures are effective at high frequencies and also if they are made massive. Each time the frequency (or the mass per unit area of the enclosure wall) is doubled, the TL increases by 6 dB. Life is more complicated than this, however. At low frequencies, we find that the panel and air cavity stiffness no longer can be neglected, and the TL is decreased by resonances in the enclosure walls and the air cavity (particularly if the cavity is made small). There is also a reduction in TL at high frequencies because of the coincidence effect. The critical coincidence frequency, f_c, is given by

$$f_c = 500/h, \text{ Hz} \tag{8}$$

where h is the panel thickness in inches for steel or aluminum panels. For other materials, the value 500 is replaced by another constant. The idealized TL of a typical wall is shown in Figure 24.20.

24.7.3 Vibration Isolation

Machines are often attached directly to the floor or to large metal surfaces. Such large surfaces are often efficient sound radiators at low frequencies. Careful vibration isolation can reduce this problem considerably. Vibration isolation can become particularly desirable when the air paths have been reduced by the use of enclosures. The theory of vibration isolation is well known and is discussed in most vibration textbooks. Briefly, the spring isolators must be chosen so that the natural frequency is several times smaller than the frequency of the exciting force. [For further discussion, see Crocker (1997, Chapter 83 and Crede and Ruzicka, 1996).]

24.7.4 Sound-Absorbing Materials

The use of sound-absorbing materials is very effective inside machine enclosures or in reverberant factory spaces in reducing the sound pressure caused by machines. The absorption coefficient α of a material is defined as the fraction of incident intensity I that is absorbed (see Figure 24.21). The absorption area A of a material in sabins is the product of its absorption coefficient α and its area S. Assuming that the sound field in the enclo-

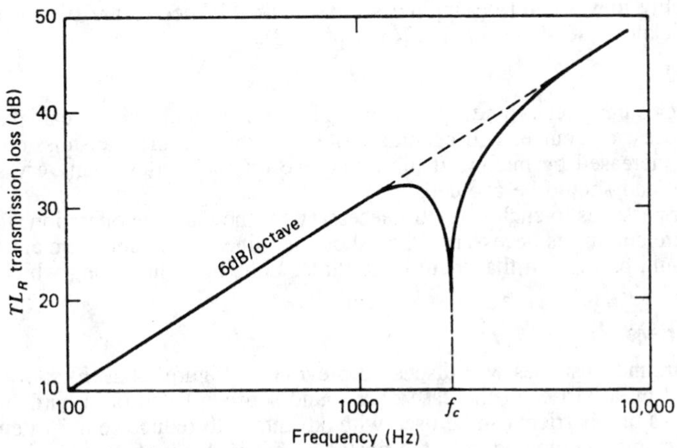

Figure 24.20 Random incidence transmission loss of a typical wall.

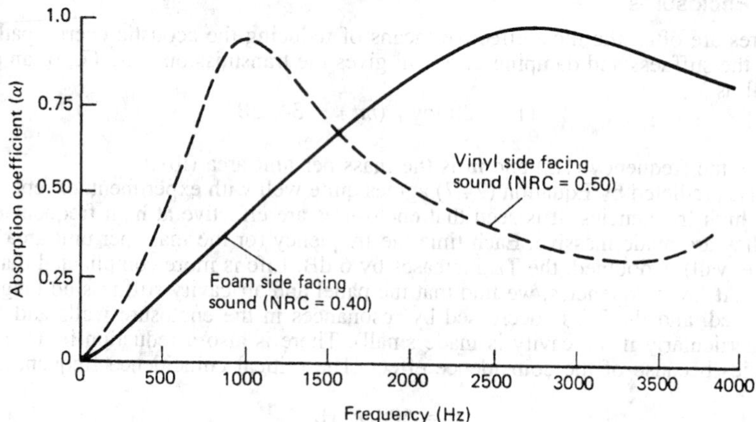

Figure 24.21 Absorption coefficient of a 3/4 in. thick layer of foam with a 1/2 lb/ft^2 vinyl coating. The noise reduction coefficient NRC is the average of the absorption coefficients α at 250, 500, 1000 and 2000 Hz.

sure or space is diffuse (which is only approximated in the higher frequency range), then the reduction in the sound pressure level, ΔSPL, caused by adding absorbing material with absorption area A_2 sabins is (see Crocker, 1997, chapter 83)

$$\Delta\text{SPL} = L_1 - L_2 = 10 \log_{10}\left(\frac{A_1 + A_2}{A_1}\right), \quad \text{dB} \tag{9}$$

where

L_I = the original SPL
L_2 = the SPL after the absorption area A_2 sabins are added
A_I = the original absorption area (sabins)

24.7.5 Damping Materials

In some cases, vibration is controlled by resonances; in these cases, the vibration amplitude is proportional to the structural damping present. Since the sound radiated is also proportional to the vibration amplitude, increasing the structural damping can reduce the sound radiated. Suitable damping materials made from viscoelastic substances are available. Preferably they should be applied so that their thicknesses are two or three times the metal thickness (see Crocker, 1997, Chapter 83).

24.7.6 Leaks

Leaks in enclosures can provide paths through which sound energy can pass with no alteration. Leaks, of course, will become more important as the transmission loss of an enclosure is increased by making it more massive. In such cases even very small leaks are important and should be avoided.

Where air passages to enclosures are necessary for the machine operation (for instance, in cases where cooling is necessary), they should be lined with acoustic material, and the passages should be bent so that there is no direct "line of sight" along which the sound can travel.

24.7.7 Barriers

Diffraction around obstacles was discussed previously. Figure 24.22(*b*) shows the attenuation caused by a barrier in the case of a sound source placed on a hard floor. Figure 24.22(*a*) shows that barriers can be used with advantage to reduce sound when the sound wavelength is small compared with the obstacle, that is, for high frequencies. Such barriers are used along roads or railroads to shield houses or apartments. Smaller barriers can be used inside factory spaces. If the factory ceiling is low, absorbing material should

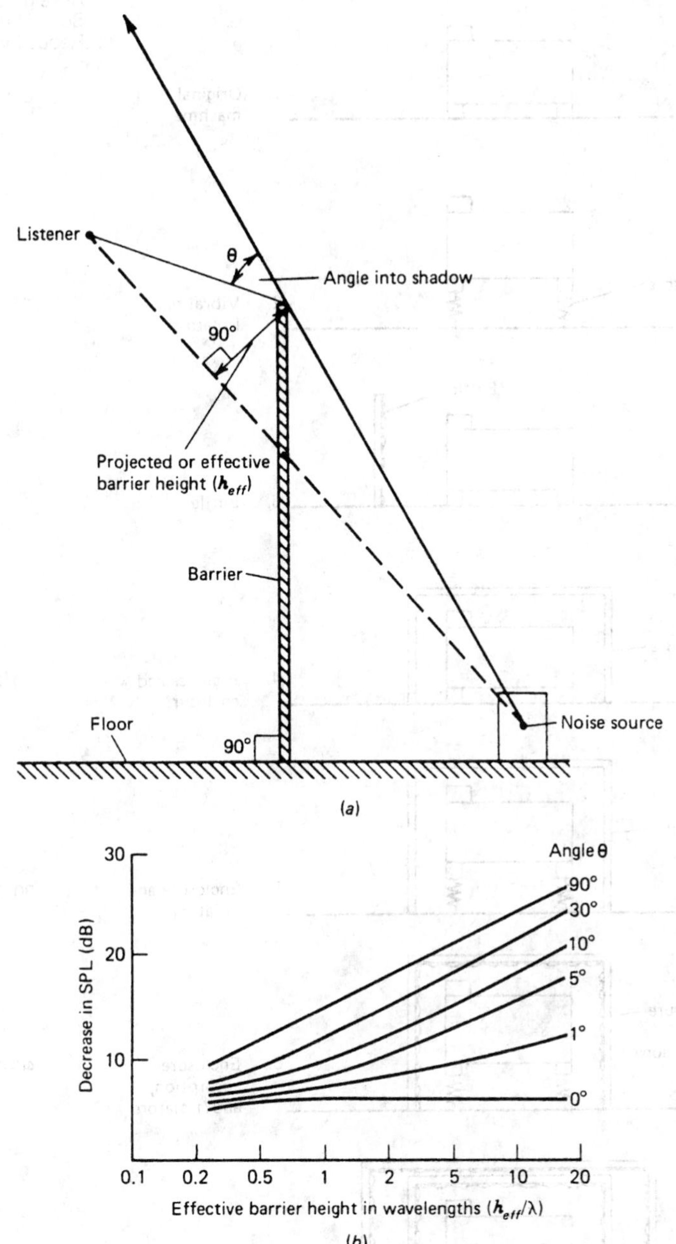

Figure 24.22 (*a*) Noise source and effective barrier height seen by listener and (*b*) noise shielding provided by a barrier.

be placed on it above the barrier to prevent reflections from the ceiling bypassing the barrier.

24.7.8 Comparison of Noise Reduction Methods

Figure 24.23 shows a comparison of noise reduction methods applied to a typical machine. Approximate A-weighted sound level reductions are given.

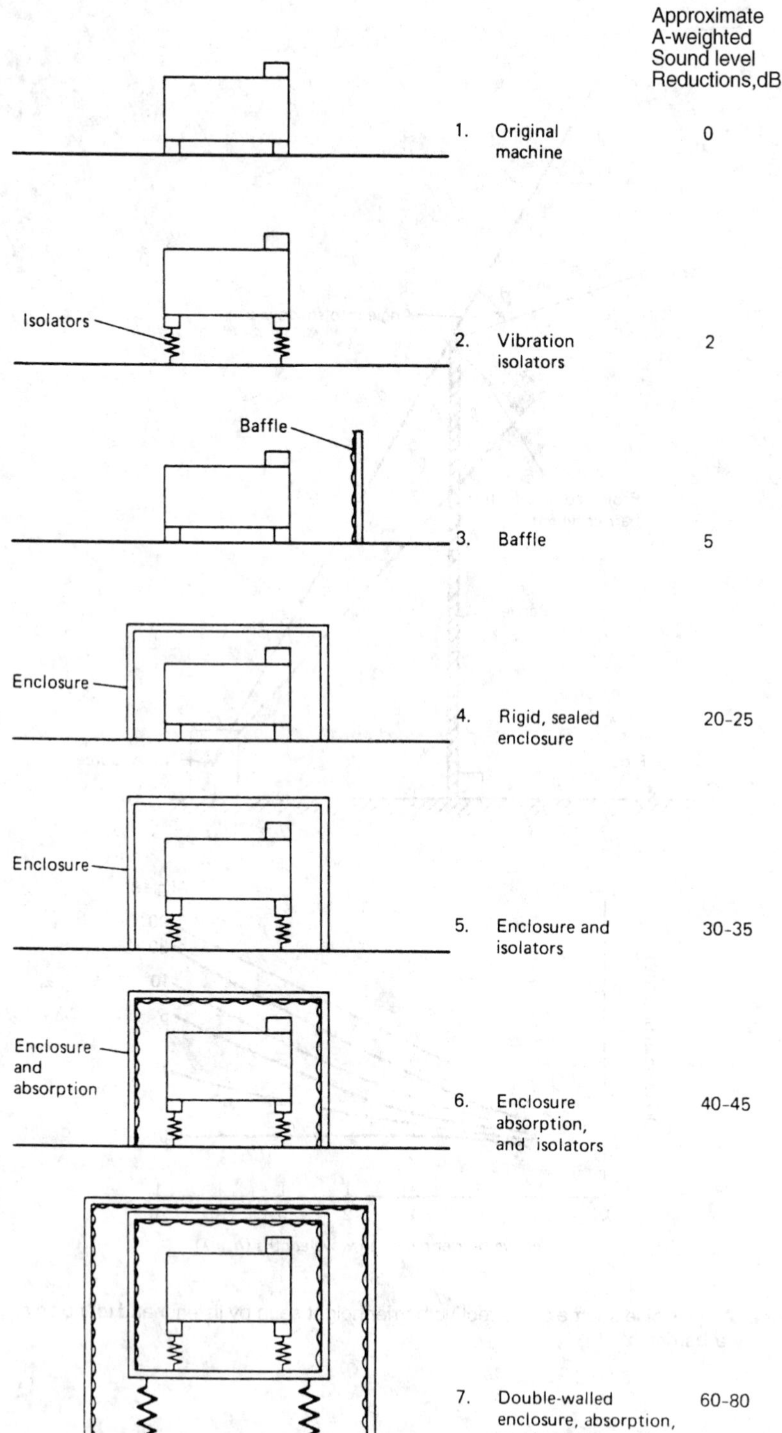

Figure 24.23 Comparison of noise reduction methods applied to a machine.

24.7.9 Personal Protective Equipment

In cases where noise reduction at the source(s) or along the path(s) is difficult or expensive, it may be necessary to consider the receiver(s) (usually the human ear). It is possible to enclose personnel completely with an acoustic booth. Such booths may need to be isolated from vibration of the floor and should have adequate transmission loss. Acoustic leaks should be minimized, and absorbing materials such as acoustic ceiling and wall tiles (and maybe carpet) should be used inside. Alternatively, earplugs, earmuffs, and even helmets may be used individually by personnel to reduce individual noise exposure. However, plugs, muffs, and helmets are uncomfortable to wear, particularly in a hot environment, and should be used as a last resort when engineering controls to reduce noise are too difficult or expensive (see section on hearing conservation).

24.8 CASE HISTORIES OF NOISE REDUCTION

24.8.1 Folding Carton Manufacture

Figures 24.24(*a*) and (*b*)show top and side views, respectively, of the cutting press and automatic strippers for removal of waste material between cartons. The main noise problem in this type of machine does not usually come from the cutting press (if it is in good mechanical adjustment), but rather from the scrap disposal system. The noise is created when pieces of paper scrap strike the sides of the intake conveyor under the press stripper, the sides of the intake hood to the fan, and the fan and outlet ducts. The noise created

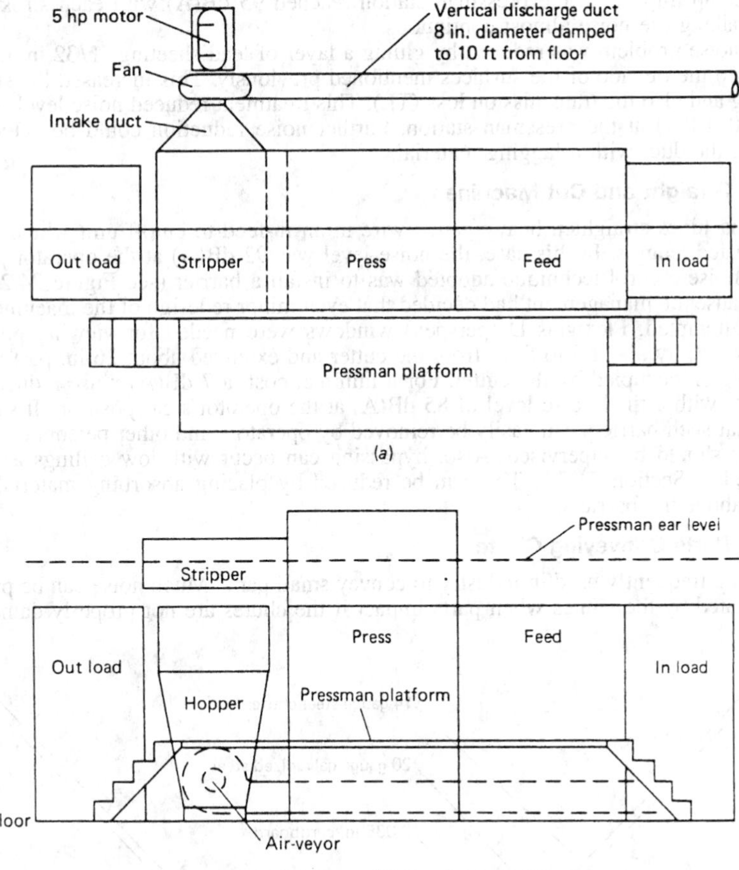

Figure 24.24 Scrap handling system for cutting press, showing (*a*) top view and (*b*) side view.

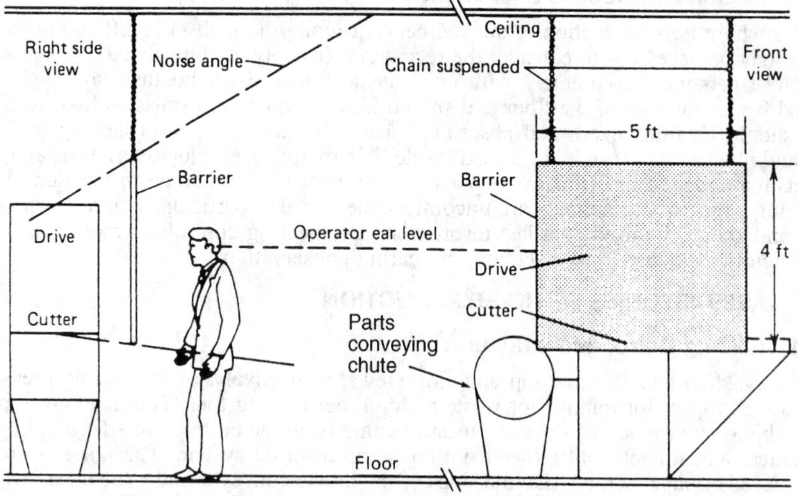

Figure 24.25 Barrier wall for straight and cut machine.

by the scrap impacts at the pressman station reached 95 dB(A) with each stroke of the press, making the noise almost continuous.

The noise problem was reduced by gluing a layer of lead sheeting (1/32 in. thick and 2 lb/ft^2 to the outside of the surfaces mentioned previously. This increased the structural damping and also the transmission loss (TL). This treatment reduced noise levels to about 88 to 90 dB(A) at the pressman station. Further noise reduction could be achieved by covering the duct with a lagging material.

24.8.2 Straight and Cut Machines

These machines straighten heavy gauge wire in an in-feed to cutoff unit which is set to cut repeated lengths. In this case, the noise level was 92 dB(A) at the operator position.

The noise control technique adopted was to install a barrier (see Figure 24.25). This was because the management had decided that even minor redesign of the machine should not be attempted. Plexiglas D (perspex) windows were needed for viewing ports. The barrier was only about 6 to 8 in. from the cutter and extended about 26 in. past the ends of the region occupied by the cutter. For a minimal cost, a 7 dB(A) noise reduction was achieved, with a final noise level of 85 dB(A) at the operator's ear position. It should be noted that such barriers can easily be removed by operators and other personnel and thus their use should be supervised. Also, bypassing can occur with low ceilings as already discussed in Section 24.7.7. This can be reduced by placing absorbing material on the ceiling above the barriers.

24.8.3 Parts Conveying Chute

Chutes are frequently used in industry to convey small parts. Much noise can be produced and radiated by the chutes when parts impact if the chutes are not properly damped.

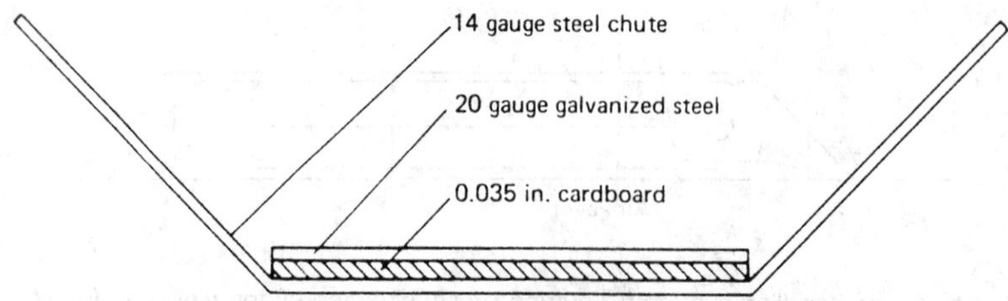

Figure 24.26 Chute for conveying cartridge cases.

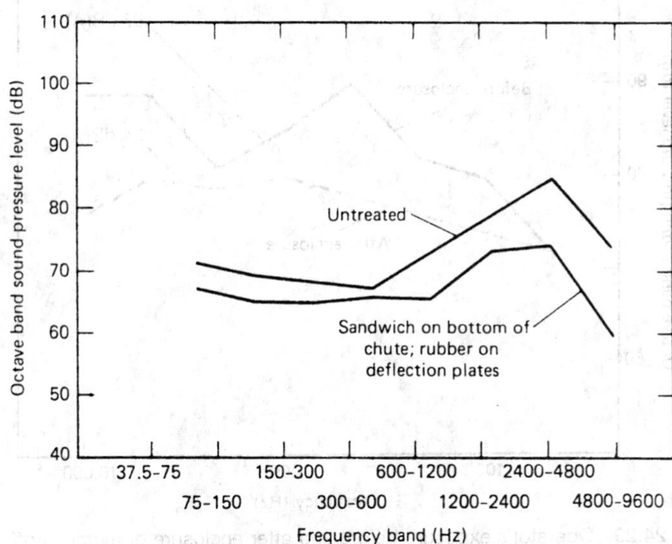

Figure 24.27 Noise spectra measured 3-ft from chute.

Constrained-layer damping applied to the chute is normally very effective. In this case history, 30-caliber cartridge cases were carried in the chute shown in Figure 24.26. The constrained layer can be placed either on the parts side or on the underside of the chute. On the parts side, the metal layer must be wear resistant to the impacting parts. The application of the cardboard and the 20-gauge galvanized sheet to the chute shown in Figure 24.26 produced enough damping to reduce the noise from 88 to 78 dB(A) at 3 ft (1 m) from the side of the chute (see Figure 24.27). Rubber deflector plates were also used to funnel parts to the center of the chute so that they did not strike the untreated sides.

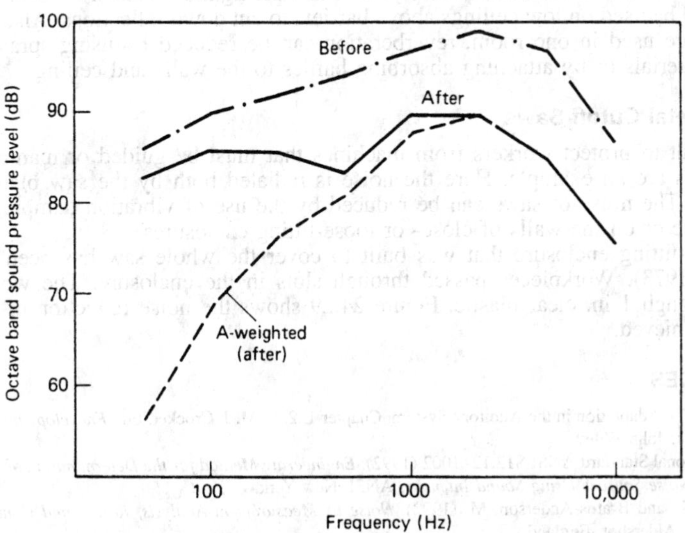

Figure 24.28 Operator position noise levels for nail-making machine.

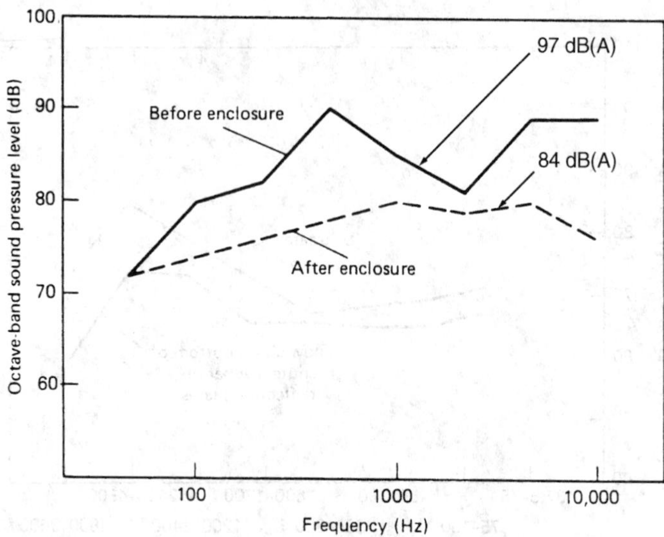

Figure 24.29 Operator's exposure before and after enclosure of metal cutoff saw.

24.8.4 Nail-Making Machine

Ten nail-making machines were mounted on a weak concrete floor. The machines were operating at 300 strokes/min. The noise level at the operator position was 103.5 dB(A).

It was believed that the vibration caused by the impacts in the machine process was being transmitted to the concrete floor and that it was being radiated as noise. An octave band analysis of the machine noise before noise reduction was attempted is shown in Figure 24.28. It was decided to use vibration isolators. It was believed that the shock pulse duration was about 10 msec. The repetition period was 1/(300/60) sec or 200 msec.

Elastomeric isolators were selected to have a natural period of 100 msec corresponding to a machine natural frequency of 10 Hz and a static deflection of 0.1 in. under load. Since 10 msec < 100 msec < 200 msec, the design conditions were fulfilled. Figure 24.28 shows the octave band levels after the isolators were installed. The A-weighted sound level was reduced by 8.5 dB(A), from 103.5 dB(A) to 95 dB(A).

Since the noise level was still high, further noise reduction could have been achieved with the use of a barrier made from plywood and plexiglass (perspex). Absorbing materials should be used on low ceilings above barriers to cut down reflections. Also, if several machines are used in one room, reverberation can be reduced by using sprayed-on absorbing materials or by attaching absorbing baffles to the walls and ceiling.

24.8.5 Metal Cutoff Saws

It is difficult to protect workers from machines that must be guided or manipulated directly. Saws are an example. Here the noise is radiated both by the saw blade and the workpiece. The noise of saws can be reduced by the use of vibration damping material on the blade or on the walls of close- or loose-fitting enclosures.

A loose-fitting enclosure that was built to cover the whole saw has been described (Handley, 1973). Workpieces passed through slots in the enclosure. The workpiece is viewed through 1 in. clear plastic. Figure 24.29 shows the noise reduction of 13 dB(A) that was achieved.

REFERENCES

Abbas, P. (1997). Adaptation in the Auditory System. Chapter 122. In M. J. Crocker, Ed., *Encyclopedia of Acoustics*, New York: John Wiley.

American National Standard ANSI S12.12-.1002 (1992). *Engineering Method for the Determination of Sound Power Level of Noise Sources Using Sound Intensity*, ANSI, New York.

Anderson, J. S., and Bratos-Anderson, M. (1992). *Noise its Measurement Analysis, Rating and Control, Avebury Technical*, Aldershot, England.

Appleyard, D., and Lintell, M. (1972). Environmental quality of city streets: The residents' viewpoint. *Journal of the American Institute of Planners, 38*, 84–101.

Bell, L. H., and Bell, D. H. (1993). *Industrial Noise Control*, Second Edition, New York: Marcel Dekker.

Beranek, L. L., Ed. (1971). *Noise and Vibration Control*, New York: McGraw-Hill.

Beranek, L. L. and Ver, I. L. (1991). *Noise and Vibration Control Engineering: Principles and Applications*, New York: John Wiley.

Berger, E. H. and Casali J. G. (1997). Hearing protection devices. Chapter 81. In M. J. Crocker, Ed., *Encyclopedia of Acoustics*, New York: John Wiley.

Boles, W. E., and Hayward, S. C. (1978). Effects of urban noise and sidewalk density upon pedestrian co-operation and tempo. *Journal of Social Psychology, 104*, 29–35.

Broadbent, D. E. (1971). *Decision and Stress.* New York: Academic.

Broadbent, D. E. (1975). Waves in the eye and ear. *Journal of Sound and Vibration, 41*, 113–125.

Broadbent, D. E. (1978). The current state of noise research: Reply to Poulton. *Psychological Bulletin, 85*, 1052–1067.

Broadbent, D. E., and Gregory, M. (1965). Effects of noise and of signal rate upon vigilance analysed by means of decision theory. *Human Factors, 7*, 155–162.

Broadbent, D. E., and Little, F. A. J. (1960). Effects of noise reduction in a work situation. *Occupational Psychology, 34*, 133–140.

Bruce, R. D. and Moritz, C. T. (1997). Sound power level predictions for industrial machinery. Chapter 86. In M. J. Crocker, Ed., *Encyclopedia of Acoustics*, New York: John Wiley.

Burns, W. (1979). Physiological effects of noise. Chapter 15. In C. M. Harris, Ed., *Handbook of noise control.* New York: McGraw-Hill.

Busch-Visniac, I. J. and Hixson, E. L. (1997). Types of microphones. Chapter 163. In M. J. Crocker, Ed., *Encyclopedia of Acoustics*, New York: John Wiley.

Buus, S. (1997). Auditory masking. Chapter 115. In M. J. Crocker, Ed., *Encyclopedia of Acoustics*, New York: John Wiley.

Cohen, A. (1974). Industrial noise and medical, absence and accident record data on exposed workers. In W. D. Ward, Ed., *Proceedings of the International Congress on Noise as a Public Health Problem.* Washington, DC: U.S. Environmental Protection Agency.

Cohen, A. (1976). The influence of a company hearing conservation program on extra-auditory problems in workers. *Journal of Safety Research, 8*, 146–162.

Cohen, S., and Spacapan, S. (1984). The social psychology of noise. In D. M. Jones and A. J. Chapman, Eds., *Noise and Society.* Chichester, Great Britain: Wiley.

Cohen, S., Glass, D. C., and Singer, J. E. (1973). Apartment noise, auditory discrimination and reading ability in children. *Journal of Experimental Social Psychology, 9*, 407–422.

Cohen, S., Evans, G. W., Krantz, D. S., and Stokols, D. (1980). Physiological, motivational, and cognitive effects of aircraft noise on children: Moving from the laboratory to the field. *American Psychologist, 35*, 231–243.

Colle, H. A. (1980). Auditory encoding in visual short-term recall: effects of noise intensity and spatial location. *Journal of Verbal Learning and Verbal Behavior, 19*, 722–735.

Colquhoun, W. P., and Edwards, R. S. (1975). Interaction of noise with alcohol on a task of sustained attention. *Ergonomics, 18*, 81–89.

Crede, C. E. and Ruzicka, J. E. (1996). Theory of vibration isolation. Chapter 30. In C. M. Harris, Ed., *Shock and Vibration Handbook.* Fourth Edition. New York: McGraw-Hill.

Crocker, M. J. (1997). General linear acoustics, Chapter 1. In M. J. Crocker, Ed., *Encyclopedia of Acoustics.* New York: John Wiley.

Crocker, M. J., Hamilton, J. F., and Price, A. J. (1971). Vibration isolation for machine noise reduction. *Sound and Vibration, 5(11)*: 30.

Crocker, M. J., and Jacobsen, F. (1997). Sound intensity, Chapter 156. In M. J. Crocker, Ed., *Encyclopedia of Acoustics*, New York: John Wiley.

Crocker, M. J. (1972). Use of anechoic and reverberant rooms for measurement of noise from machines. In M. J. Crocker, Ed., *Tutorial Papers on Noise Control, Proceedings of the Inter-Noise Conference 72* (pp. 116–123). Poughkeepsie, NY: Noise Control Foundation.

Crocker, M. J., Hamilton, J. F., and Price, A. J. (1971). Vibration isolation for machine noise reduction. *Sound and Vibration, 5*, 30.

Crocker, M. J., and Price, A. J. (1975). *Noise and Noise Control*, Vol. 1 (p. 16). Boca Raton, FL: CRC Press.

Crocker, M. J., and Kessler, F. M. (1982). *Noise and Noise Control*, Vol. 2. Boca Raton, FL: CRC Press.

Crocker, M. J. (1997). Rating measures, descriptors, criteria and procedures for determining human response to noise, Chapter 80. In M. J. Crocker, Ed., *Encyclopedia of Acoustics.* New York: John Wiley.

Crocker, M. J. (1997) The generation of noise in machinery, its control and the identification of noise sources, Chapter 83. In *Encyclopedia of Acoustics.* New York: John Wiley.

Crocker, M. J. (1991). Measurement of sound intensity. In C. M. Harris, Ed., *Handbook of Acoustical Measurements and Noise Control* (3rd ed.), New York: McGraw-Hill.

Crook, M. A., and Langdon, F. J. (1974). The effects of aircraft noise in schools around London Airport. *Journal of Sound and Vibration, 34*, 221–232.

Cudworth, A. L. (1959). Field and laboratory example of noise control. *Noise Control, 5(1)*: 39.

Davies, D. R., and Parasuraman, R. (1981). *The Psychology of Vigilance.* New York: Academic.

Dixon Ward, W. (1997). Effects of intense sounds. In M. J. Crocker, Ed., *Encyclopedia of Acoustics*, New York: John Wiley.

Donnerstein, E., and Wilson, D. W. (1976). Effects of noise and perceived control on ongoing and subsequent aggressive behavior. *Journal of Personality and Social Psychology, 34*, 774–781.

Dornic, S., and Fernaeus, S. E. (1981). Type of processing in high-load tasks: the differential effect of noise (Report No. 576). Stockholm: Department of Psychology, University of Stockholm.

Eysenck, M. W. (1975). Effects of noise, activation level, and response dominance on retrieval from semantic memory. *Journal of Experimental Psychology, 104*, 143–148.

Fahy, F. J. (1995). *Sound Intensity* (2nd ed.). London: E&FN Spon.

Faulkner, L. L., Ed. (1976). *Handbook of Industrial Noise Control* (pp. 67–112). New York: Industrial Press.

Fidell, S. (1984). Community response to noise. In D. M. Jones and A. J. Chapman, Eds., *Noise and Society*. Chichester, Great Britain: Wiley.

Fidell, S., Horonjeff, R., and Green, D. (1981). Statistical analyses of urban noise. *Noise Control Engineering, 16*, 75–80.

Finegold, L. S., Harris, S. S., and von Gierke, H. E. (1994). Community annoyance and sleep disturbance: Updated criteria for assessment of the impact of general transportation noise on people. *Noise Control Engineering Journal, 44(3)*: 25–30.

Fisher, S. (1973). The "distraction effect" and information processing complexity. *Perception*, 2, 78–89.

Forster, P. M. (1978). Attentional selectivity: A rejoinder to Hockey. *British Journal of Psychology, 69*, 505–506.

Glass, D. C., and Singer, J. E. (1972). *Urban Stress: Experiments on Noise and Social Stressor*. New York: Academic.

Graeven, D. B. (1975). Necessity control and predictability of noise annoyance. *Journal of Social Psychology, 95*, 85–90.

Grandjean, E. (1974). Sozio-psychologische Untersuchungen vor der Fluglarms. Berne: Eidgenossisches Lustant, Bundeshaus.

Hamilton, P., Hockey, G. R. J., and Rejman, R. (1977). The place of the concept of activation in human information processing theory: An integrative approach. In S. Dornic, Ed., *Attention and Performance*, Vol. 6. New York: Lawrence Erlbaum.

Handley, J. M. (1973). *Noise—The Third Pollution*. Bulletin 6.0011.0. Bronx, NY: Industrial Acoustics Company.

Harris, C. M., Ed. (1958). *Handbook of Noise Control*. New York: McGraw-Hill.

Harris, C. M., Ed. (1979). *Handbook of Noise Control* (2nd ed.). New York: McGraw-Hill.

Hockey, G. R. J. (1970). Effect of loud noise on attentional selectivity. *Quarterly Journal of Experimental Psychology*, 22, 28–36.

Hockey, G. R. J. (1973). Changes in information selection patterns in multisource monitoring as a function of induced arousal shifts. *Journal of Experimental Psychology, 101*, 35–42.

Hormann, H., and Osterkamp, J. (1966). Uber den Einfluss von Kontinvierlichem Larm auf die Organisation von Gedachtrisinhalten. *Zeitschrift fur Experimentelle und Angewandte Psychologie, 13*, 31–38.

Howell, K., and Martin, A. M. (1975). An investigation of the effects of hearing protectors on vocal communication in noise. *Journal of Sound and Vibration, 41(2)*, 181–196.

Jansen, G. (1961). Adverse effects of noise on iron and steel workers. *Stahl und Eisen, 81*, 217–220.

Jones, D. M. (1983). Loud noise and levels of control. In G. Rossi, Ed., *Noise as a Public Health Problem*. Milan: Centro Ricerche e Studi Amplifon.

Jones, D. M., Auburn, T. C., and Chapman, A. J. (1982). Perceived control in continuous loud noise. *Current Psychological Research*, 2, 111–122.

Jones, D. M., Chapman, A. J., and Auburn, J. C. (1981). Noise in the environment: A social perspective. *Journal of Environmental Psychology, 1*, 43–59.

Jones, D. M., Smith, A. P., and Broadbent, D. E. (1979). Effects of moderate intensity noise on the Bakan vigilance task. *Journal of Applied Psychology, 64*, 627–634.

Jones, D. M., Thomas, J. R., and Harding A. (1982). Recognition memory for prose items in noise. *Current Psychological Research*, 2, 33–44.

Kerr, W. A. (1950). Accident proneness and factory departments. *Journal of Applied Psychology, 34*, 167–170.

Klumpp, R. G., and Leonard, J. L. (1963). Observer variability in reading noise level with meters. *Sound*, 2, 25–29.

Kraak, W. (1982). Investigations on criterial for the risk of hearing loss due to noise. In J. V. Tobias and E. E. Schubert, Eds., *Hearing Research and Theory*, Vol. 1. New York: Academic.

Krug, R. W. (1997). Sound level meters. Chapter 155. In M. J. Crocker, Ed., *Encyclopedia of Acoustics*. New York: John Wiley.

Kryter, K. D. (1995). The Effects of Noise on Man (2nd ed.). New York: Academic.

Landis, C., and Hunt, W. A. (1939). *The Startle Pattern*. New York: Farrar and Rinehart.

Mahapatra, S. B. (1974). Psychiatric and psychosomatic illness in the deaf. *British Journal of Psychiatry, 125*, 450–451.

May, D. N., and Rice, C. G. (1971). Effects of startle due to pistol shot on control precision performance. *Journal of Sound and Vibration, 15*, 197–202.

McGrath, J. J. (1963). Irrelevant stimulation, and vigilance performance. In D. N. Buckner and J. J. McGrath, Eds., *Vigilance: A Symposium*. London: McGraw-Hill.

Moser, G., and Jones, D. M. (1983). Annoyance and performance. In G. Rossi, Ed., *Noise as a Public Health Problem*. Milan: Centro Ricerche e Studi Amplifon.

Mosko, J. F., and Fletcher, J. L. (1971). Evaluation of the gundefender earplug: Temporary threshold shift and speech intelligibility. *Journal of the Acoustical Society of America, 49*, 1732–1734.

Murray, D. J. (1965). The effects of white noise on the recall of vocalized lists. *Canadian Journal of Psychology, 19*, 333–345.

Page, R. A. (1977). Noise and helping behavior. *Environment and Behavior, 9*, 311–334.

Passchier-Vermeer, W. (1974). Hearing loss due to continuous exposure to steady state broad-band noise. *Journal of the Acoustical Society of America, 56*, 1585–1593.

Passchier-Vermeer, W. (1983). Measurement and rating of impulse noise in relation to noise-induced hearing loss. In G. Rossi, Ed., *Noise as a Public Health Problem*. Milan: Centro Ricerche e Studi Amplifon.

Patterson, R. D. (1983). Guidelines for auditory warning systems on civil aircraft: A summary and a prototype. In G. Rossi, Ed., *Noise as a Public Health Problem*. Milan: Centro Ricerche e Studi Amplifon.

Peak, W., and Rosowski, J. (1997). Acoustical properites of the middle ear, Chapter 106. In M. J. Crocker, Ed., *Handbook of Acoustics*. New York: John Wiley.

Pearsons, K. S., Bennett, R. L., and Fidell, S. (1977). Speech levels in various noise environments (EPA-600/1-77-025). Washington, DC: U.S. Environmental Protection Agency.

Pickett, J. M. (1956). Effects of vocal force on the intelligibility of speech sounds. *Journal of the Acoustical Society of America, 28*, 902–905.

Pope, J. (1997). Analyzers, Chapter 157. In M. J. Crocker, Ed., *Encyclopedia of Acoustics*. New York: John Wiley.

Rehm, S. (1983). Research on extra-aural effects of noise since 1978. In G. Rossi, Ed., *Noise as a Public Health Problem*. Milan: Centro Ricerche e Studi Amplifon.

Robinson, D. W., Bowsher, J. M., and Copeland, W. C. (1963). On judging the noise from aircraft in flight, *Acustica, 13*, 324–336.

Robinson, D. W., Copeland, W. C., and Rennie, A. J. (1961). Motor vehicle noise measurement. *Engineer, 211*, 493–497.

Salame, P., and Baddeley, A. D. (1983). Differential effect of noise and speech on short-term memory. In G. Roissi, Ed., *Noise as a Public Health Problem*. Milan: Centro Ricerche e Studi Amplifon.

Salame, P., and Wittersheim, G. (1978). Selective noise disturbance of the information input in short-term memory. *Quarterly Journal of Experimental Psychology, 30*, 693–794.

Salmon, V., Mills, S., and Peterson, A. C. (1975). *Industrial Noise Control Manual*, DHEW Publication No. (NIOSH) 75-183, U.S. Government Printing Office, Washington, DC, June.

Sauser, W.. I., Arauz, C. G., and Chambers, R. M. (1978). Exploring the relationship between level of office noise and salary recommendations: A preliminary research note. *Journal of Management, 4*, 57–63.

Scharf, B. (1997). Loudness, Chapter 118. In M. J. Crocker, Ed., *Encyclopedia of Acoustics*. New York: John Wiley.

Schonpflug, W. (1993). Continuous noise, discontinuous noise, and performance. In A. Schick, Ed., *Contributions to Psychological Acoustics*. Sixth Oldenburg Symposium on Psychological Acoustics. University of Oldenburg, 1993.

Schori, T. R., and McGatha, E. A. (1978). A real-world assessment of noise exposure. *Journal of Sound and Vibration, 12*, 24–30.

Schultz, J. (1978). Synthesis of social surveys on noise annoyance. *Journal of the Acoustical Society of America, 64*, 377–405.

Shaw, E. A. G. (1997). Acoustical properites of the outer ear. Chapter 105, In M. J. Crocker, Ed., *Encyclopedia of Acoustics*. New York: John Wiley.

Siegel, J. M., and Steele, C. M. (1980). Environmental distraction, and interpersonal judgements. *British Journal of Social and Clinical Psychology, 19*, 23–32.

Smith, A. P. (1983). The effects of noise and memory load on a running memory task. *British Journal of Psychology*.

Smith, A. P. (1985). Noise biased probability and serial reaction. *British Journal of Psychology, 76*, 89–95.

Smith, A. P., Jones, D. M., and Broadbent, D. E. (1981). The effects of noise on recall of categorized lists. *British Journal of Psychology*, 72, 299–316.

van Heusden, E., Plomp, R., and Pols, L. C. W. (1979). Effect of ambient noise on the vocal output and the preferred listening level of conversational speech. *Applied Acoustics, 12*, 31–43.

Ward, W. D. (1976). Transient changes in hearing. In G. Rossi and M. Vigone, Eds., *Man and Noise*, Milan: Edizioni Minerva Medica.

Ward, W. D. (1997). Effects of high intensity sound, Chapter 119. In M. J. Crocker, Ed., *Encyclopedia of Acoustics*. New York: John Wiley.

Webster, J. C. (1984). Noise and communication. In D. M. Jones and A. J. Chapman, Eds., *Noise and Society*. Chichester, Great Britain: Wiley.

Weinstein, N. D. (1977). Noise and intellectual performance: A confirmation and extension. *Journal of Applied Psychology, 62*, 104–107.

Wilkinson, R. T. (1963). Interaction of noise with knowledge of results, and sleep deprivation. *Journal of Experimental Psychology, 66*, 332–337.

Woodhead, M. M. (1964). The effects of bursts of noise on an arithmetic task. *American Journal of Psychology, 77*, 627–633.

Woodhead, M. M. (1966). An effect of noise on the distribution of attention. *Journal of Applied Psychology, 50*, 296–299.

Yang, S. J., and Ellison, A. J. (1985). *Machinery Noise Measurement*. Oxford: Clarendon Press.

Yost, W. A. and Killion, M. C. (1997). Hearing Thresholds, Chapter 123. In M. J. Crocker, Ed., *Encyclopedia of Acoustics*. New York: John Wiley.

CHAPTER 25

VIBRATION AND MOTION

Michael J. Griffin
Institute of Sound and Vibration Research
University of Southampton
Southampton SO17 1BJ, England

25.1 INTRODUCTION

In work and leisure activities the human body experiences movement. The motion may be voluntary (as in some sports) or involuntary (as for passengers in vehicles). Movements may occur in up to six different directions: three translational directions (fore-and-aft, lateral, and vertical) and three rotational directions (roll, pitch, and yaw). Translational movements at constant velocity (i.e., with no change of speed or direction) are mostly imperceptible, except where *exteroceptors* (e.g., the eyes or ears) detect a change of position relative to other objects. Translational motion is mainly detected when the velocity changes, causing acceleration or deceleration of the body which may be detected by *interoceptors* (e.g., the vestibular, cutaneous, kinesthetic or visceral sensory systems). Rotation of the body at constant velocity may be detected because it gives rise to translational acceleration in the body, because it re-orientates the body relative to the gravitational force of the earth, or because the changing orientation relative to other objects is perceptible through exteroceptors. Vibration is oscillatory motion: The velocity is constantly changing, and so the movement is detectable by interoceptors and exteroceptors.

Vibration of the body may be desirable or undesirable. It can be described as pleasant or unpleasant, it can interfere with the performance of various tasks and cause injury and disease. Low frequency oscillations of the body and movements of visual displays can cause motion sickness. It is convenient to consider human exposure to oscillatory motion in three categories.

Whole-body vibration occurs when the body is supported on a surface which is vibrating (e.g., sitting on a seat which vibrates, standing on a vibrating floor or lying on a vibrating surface). Whole-body vibration occurs in transport (e.g., road, off-road, rail, air and marine transport) and when near some machinery.

Motion sickness can occur when real or illusory movements of the body or the environment lead to ambiguous inferences as to the movement or orientation of the human body. The movements associated with motion sickness are always of very low frequency, usually below 1 Hz.

Hand-transmitted vibration is caused by various processes in industry, agriculture, mining, construction, and transport where vibrating tools or workpieces are grasped or pushed by the hands or fingers.

There are many different effects of oscillatory motion on the body and many variables influencing each effect. The variables may be categorized as *extrinsic* variables (those occurring outside the human body) and *intrinsic* variables (the variability that occurs between and within people), as in Table 25.1. Some variables, especially intersubject variability, have large effects but are not easily measured. Consequently, it is often not practicable to make highly accurate predictions of the discomfort, interference with activities, or health effects for an individual. However, methods exist for predicting the average effect, or the probability of an effect, for groups of people.

This chapter introduces human responses to oscillatory motion, summarizes current methods of evaluating exposures to oscillatory motion, and identifies some methods of minimizing unwanted effects of vibration.

25.2 MEASUREMENT OF VIBRATION AND MOTION

25.2.1 Vibration Magnitude

When vibrating, an object has alternately a velocity in one direction and then a velocity in the opposite direction. This change of velocity means that the object is constantly accelerating, first in one direction and then in the opposite direction. Figure 25.1 shows the displacement waveform, the velocity waveform, and acceleration waveform for a movement occurring at a single frequency (i.e., a sinusoidal oscillation). The magnitude of a vibration can be quantified by either its displacement or its velocity or its acceleration. For practical convenience, the magnitude of vibration is now usually expressed in terms of the acceleration and is measured using accelerometers. The units of acceleration are meters per second per second (i.e., ms^{-2}, or m/s^2). The acceleration due to gravity on Earth is approximately 9.81 ms^{-2}. The magnitude of an oscillation can be expressed as the distance between the extremities reached by the motion (i.e., the peak-to-peak acceleration) or the maximum deviation from some central point (i.e., the peak acceleration). Magnitudes of vibration are very often expressed in terms of an average measure of the acceleration of the oscillatory motion, usually the root-mean-square value (i.e., ms^{-2} r.m.s.

Table 25.1 Variables Influencing Human Responses to Oscillatory Motion

Extrinsic Variables
Vibration variables
vibration magnitude
vibration frequency
vibration direction
vibration input positions
vibration duration
Other variables
other stressors (noise, temperature, etc.)
seat dynamics
Intrinsic variables
Intrasubject variability
body posture
body position
body orientation (sitting, standing, recumbent)
Intersubject variability
body size and weight
body dynamic response
age
gender
experience, expectation, attitude and personality
fitness

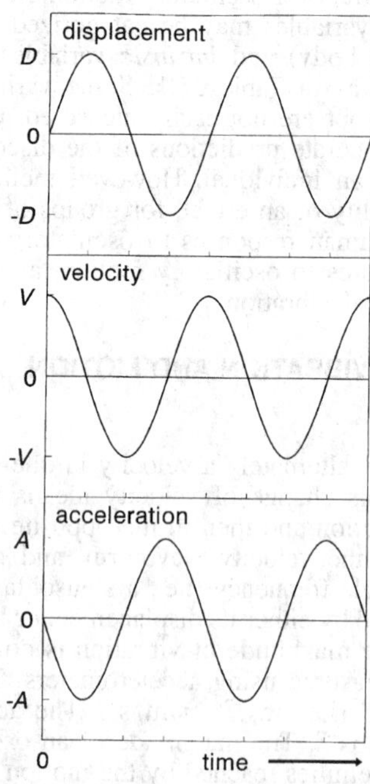

Figure 25.1 Displacement, velocity and acceleration waveforms for a sinusoidal vibration. (If the vibration has frequency, f, and peak displacement, D, the peak velocity is $V = 2\pi fD$, and the peak acceleration is $A = (2\pi f)^2 D$).

for translational acceleration, rad · s^{-2} r.m.s. for rotational acceleration). (For a sinusoidal motion, the r.m.s. value is the peak value divided by $\sqrt{2}$, i.e., approximately 1.4.)

When observing vibration, it is sometimes possible to estimate the displacement caused by the motion. For a sinusoidal motion the acceleration, a, can be calculated from the frequency, f, in Hz, and the displacement, d:

$$a = (2\pi f)^2 d.$$

For example, a sinusoidal motion with a frequency of 1 Hz and a peak-to-peak displacement of 0.1 m will have an acceleration of 3.95 ms^{-2} peak-to-peak, 1.97 ms^{-2} peak, and 1.40 ms^{-2} r.m.s. Although this expression can be used to convert acceleration measurements to corresponding displacements it is only accurate when the motion occurs at a single frequency (i.e., it has a sinusoidal waveform, as shown in Figure 25.1).

Logarithmic scales for quantifying vibration magnitudes in decibels are sometimes used. When using the reference level in International Standard 1683 (1983), the acceleration level, L_a is expressed by $L_a = 20 \log_{10}(a/a_0)$ where a is the measured acceleration (in ms^{-2}) and a_0 is the reference level of 10^{-6} ms^{-2}. With this reference, an acceleration of 1 ms^{-2} corresponds to 120 decibels (dB), an acceleration of 10 ms^{-2} corresponds to 140 dB. Other reference levels are also in use.

25.2.2 Vibration Frequency

The frequency of vibration is expressed in cycles per second using the SI unit, hertz (Hz). The frequency of vibration influences the extent to which vibration is transmitted to the surface of the body (e.g., through seating), the extent to which it is transmitted through the body (e.g., from seat to head), and the responses to vibration within the body. From Section 25.2.1 it will be seen that the relation between the displacement and the acceleration of a motion is also dependent on the frequency of oscillation: a displacement of 1 millimeter corresponds to a low acceleration at low frequencies (e.g., 0.039 ms^{-2} at 1 Hz) but a very high acceleration at high frequencies (e.g., 394 ms^{-2} at 100 Hz).

25.2.3 Vibration Direction

The responses of the body differ according to the direction of the motion. Vibration is often measured at the interfaces between the body and the vibrating surfaces in three orthogonal directions. Figure 25.2 shows a coordinate system used when measuring vibration in contact with a hand holding a tool.

The three principal directions of whole-body vibration for seated and standing persons are: fore-and-aft (x-axis), lateral (y-axis) and vertical (z-axis). The vibration is measured at the interface between the body and the surface supporting the body (e.g., on the seat

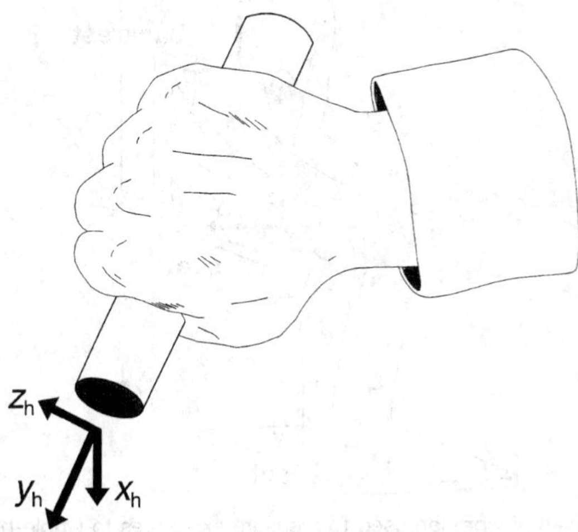

Figure 25.2 Axes of vibration used to measure exposures to hand-transmitted vibration.

beneath the ischial tuberosities for a seated person; beneath the feet for a standing person). Figure 25.3 illustrates the translational and rotational axes for an origin on a seat at the ischial tuberosities of a seated person. A similar set of axes is used for describing the directions of vibration at the backrest and feet of seated persons.

25.2.4 Vibration Duration

Some human responses to vibration depend on the duration of exposure. Additionally, the duration of measurement may affect the measured magnitude of the vibration. The root-mean-square (i.e., r.m.s.) acceleration may not provide a good indication of vibration severity if the vibration is intermittent, contains shocks, or otherwise varies in magnitude from time-to-time (see Section 25.3.3.1).

25.3 WHOLE-BODY VIBRATION

Whole-body vibration may affect health, comfort, and the performance of activities. The comments of persons exposed to vibration mostly derive from the sensations produced by vibration rather than a knowledge that the vibration is causing harm or interfering with their activities. Vibration of the whole body is produced by various types of industrial machinery and by all forms of transport (including road, off-road, rail, sea and air transport).

25.3.1 Vibration Discomfort

The relative discomfort caused by different oscillatory motions can often be predicted from measurements of the vibration. For very low magnitude motions it is possible to estimate the percentage of persons who will be able to feel vibration and the percentage who will not be able to feel the vibration. For higher vibration magnitudes, an approximate indication of the extent of subjective reactions is available in a semantic scale of discomfort.

Limits appropriate to the prevention of vibration discomfort vary between different environments (e.g., between buildings and transport) and between different types of transport (e.g., between cars and trucks) and within types of vehicle (e.g., between sports cars and limousines). The design limit depends on external factors (e.g., cost and speed) and the comfort in alternative environments (e.g., competitive vehicles).

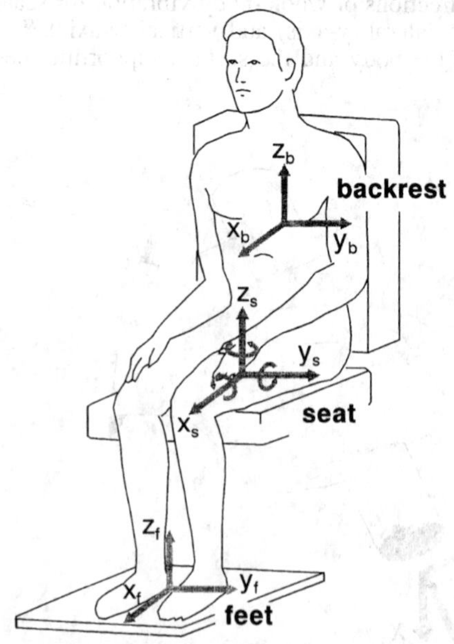

Figure 25.3 Axes of vibration used to measure exposures to whole-body vibration.

25.3.1.1 Effects of Vibration Magnitude

The absolute threshold for the perception of vertical whole-body vibration in the frequency range 1 to 100 Hz is very approximately 0.01 ms^{-2} r.m.s.; a magnitude of 0.1 ms^{-2} will be easily noticeable; magnitudes around 1 ms^{-2} r.m.s. are usually considered uncomfortable; magnitudes of 10 ms^{-2} r.m.s. are usually dangerous. The precise values depend on vibration frequency and the exposure duration, and they are different for other axes of vibration (Griffin, 1990).

A doubling of vibration magnitude (expressed in ms^{-2}) produces an approximate doubling of the sensation of discomfort. A halving of vibration magnitude can therefore produce a very considerable improvement in comfort.

25.3.1.2 Effects of Vibration Frequency and Direction

The dynamic responses of the body and the relevant physiological and psychological processes dictate that subjective reactions to vibration depend on vibration frequency and vibration direction. The extent to which a given acceleration will cause a larger or smaller effect on the body at different frequencies is reflected in 'frequency weightings': frequencies capable of causing the greatest effect are given the greatest weight, and others are attenuated in accord with their relative importance. Two different frequency weightings (one for vertical and one for horizontal vibration of seated or standing persons) were presented in International Standard 2631 (1974) and have been reproduced in other relevant standards. However, ISO 2631 is undergoing significant revision and extension (International Organization for Standardization, 1994). The draft revision of International Standard currently has a form broadly similar to British Standard 6841 (1987), which is at present the most up-to-date published standard giving general guidance concerned with vibration discomfort.

Frequency weightings for human response to vibration have been derived from laboratory experiments in which volunteer subjects have been exposed to a set of motions having different frequencies. The subjects' responses are used to determine "equivalent comfort contours." The reciprocal of such a curve forms the shape of the frequency weighting. Figure 25.4 shows frequency weightings W_b to W_f defined in British Standard 6841 (1987) as they may be implemented by analog or digital filters (Draft International Standard 2631 (1994) defines similar weightings). Table 25.2 defines simple asymptotic

Table 25.2 Asymptotic Approximations to Frequency Weightings, W(*f*), in BS 6841 (1987) for Comfort, Health, Activities, and Motion Sickness (*f* = frequency, Hz; W(*f*) = 0 Where Not Defined)

Weighting Name	Weighting Definition	
W_b	$0.5 < f < 2.0$	$W(f) = 0.4$
	$2.0 < f < 5.0$	$W(f) = f/5.0$
	$5.0 < f < 16.0$	$W(f) = 1.00$
	$16.0 < f < 80.0$	$W(f) = 16.0/f$
W_c	$0.5 < f < 8.0$	$W(f) = 1.0$
	$8.0 < f < 80.0$	$W(f) = 8.0/f$
W_d	$0.5 < f < 2.0$	$W(f) = 1.00$
	$2.0 < f < 80.0$	$W(f) = 2.0/f$
W_e	$0.5 < f < 1.0$	$W(f) = 1.00$
	$1.0 < f < 80.0$	$W(f) = 1.00/f$
W_f	$0.100 < f < 0.125$	$W(f) = f/0.125$
	$0.125 < f < 0.250$	$W(f) = 1.0$
	$0.250 < f < 0.500$	$W(f) = (0.25/f)^2$
W_g	$1.0 < f < 4.0$	$W(f) = (f/4)^{1/2}$
	$4.0 < f < 8.0$	$W(f) = 1.00$
	$8.0 < f < 80.0$	$W(f) = 8.0/f$

(i.e., straight line) approximations to these weightings and Table 25.3 shows how the weightings should be applied to the 12 axes of vibration illustrated in Figure 25.3. (The weightings W_g and W_f are not required to predict vibration discomfort: W_g has been used for assessing interference with activities and is similar to the weighting for vertical vibration in ISO 2631 (1974, 1985); W_f is used to predict motion sickness caused by vertical oscillation, see Section 25.4.)

In order to minimize the number of frequency weightings, some are used for more than one axis of vibration, with different "multiplying factors" allowing for overall differences in sensitivity between axes (see Table 25.3). The frequency-weighted acceleration should be multiplied by the multiplying factor before the component is compared with components in other axes, or included in any summation over axes. The r.m.s. value of this acceleration (i.e. after frequency weighting and after being multiplied by the multiplying factor) is sometimes called a *component ride value*.

Vibration occurring in several axes is more uncomfortable than vibration occurring in a single axis. In order to obtain an "overall ride value," the "root-sums-of-squares" of the component ride values is calculated:

$$\text{overall ride value} = (\Sigma\ (\text{component ride values})^2)^{1/2}$$

Overall ride values from different environments can be compared: A vehicle having the highest overall ride value would be expected to be the most uncomfortable with respect to vibration. The overall ride values can also be compared with the discomfort scale shown in Table 25.4. This scale indicates the approximate range of vibration magnitudes which are significant in relation to the range of vibration discomfort that might be experienced in vehicles.

25.3.1.3 Effects of Vibration Duration

Vibration discomfort tends to increase with increasing duration of exposure to vibration. The rate of increase may depend on many factors, but a simple "fourth power" time-dependency is used to approximate how discomfort varies with duration of exposure from the shortest possible shock to a full day of vibration exposure [i.e., $(\text{acceleration})^4 \times \text{duration} = \text{constant}$; see Section 25.3.3.1).

25.3.2 Interference with Activities

Vibration and motion can interfere with the acquisition of information (e.g., by the eyes), the output of information (e.g., by hand or foot movements) or the complex central processes that relate input to output (e.g., learning, memory, decision making). Effects of oscillatory motion on human performance may impair safety.

Table 25.3 Application of Frequency Weightings for the Evaluation of Vibration with Respect to Discomfort

Input Position	Axis	Frequency Weighting	Axis Multiplying Factor
Seat	x	W_d	1.0
	y	W_d	1.0
	z	W_b	1.0
	r_x (roll)	W_e	0.63
	r_y (pitch)	W_e	0.40
	r_z (yaw)	W_e	0.20
Seat back	x	W_c	0.80
	y	W_d	0.50
	z	W_d	0.40
Feet	x	W_b	0.25
	y	W_b	0.25
	z	W_b	0.40

Table 25.4 Scale of Vibration Discomfort from British Standard 6841 (1987) and Draft International Standard 2631 (1994)

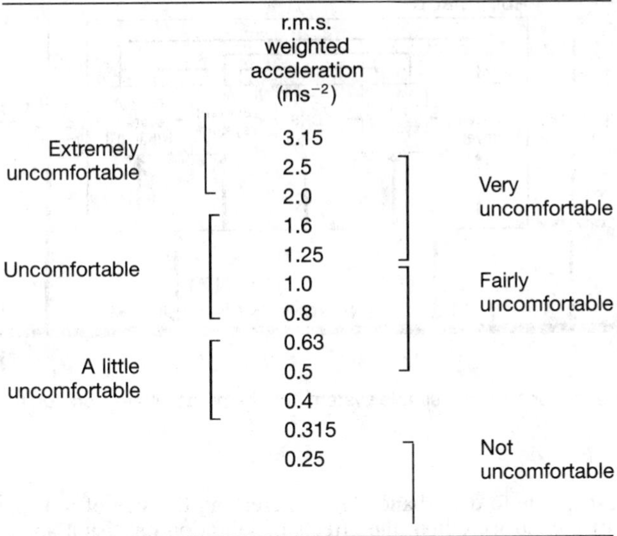

	r.m.s. weighted acceleration (ms^{-2})	
Extremely uncomfortable	3.15	
Extremely uncomfortable	2.5	Very uncomfortable
Extremely uncomfortable	2.0	Very uncomfortable
Uncomfortable	1.6	Very uncomfortable
Uncomfortable	1.25	Very uncomfortable
Uncomfortable	1.0	Fairly uncomfortable
Uncomfortable	0.8	Fairly uncomfortable
A little uncomfortable	0.63	Fairly uncomfortable
A little uncomfortable	0.5	Fairly uncomfortable
A little uncomfortable	0.4	
	0.315	
	0.25	Not uncomfortable

The greatest effects of whole-body vibration are on input processes (mainly vision) and output processes (mainly continuous hand control). In both cases there may be a disturbance occurring entirely outside the body (e.g., vibration of a viewed display or vibration of a hand-held control), a disturbance at the input or output (e.g., movement of the eye or hand), and a disturbance within the body affecting the peripheral nervous system (i.e., afferent or efferent system). Central processes may also be affected by vibration but understanding is currently too limited to make confident generalised statements (see Figure 25.5).

The effects of vibration on vision and manual control are most usually caused by the movement of the affected part of the body (i.e., eye or hand). The effects may be decreased by reducing the transmission of vibration to the eye or to the hand, or by making

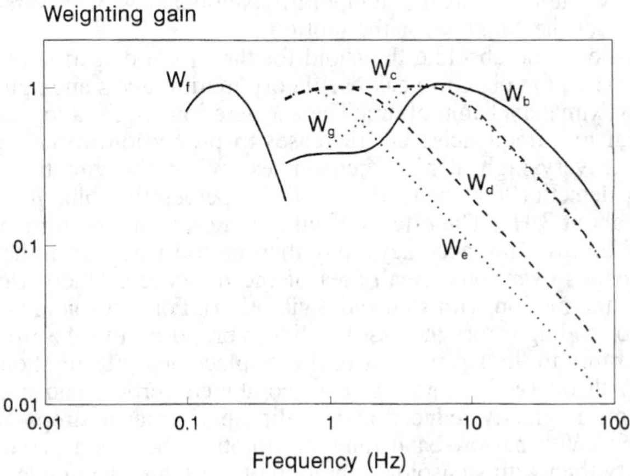

Figure 25.4 Acceleration frequency weightings for whole-body vibration and motion sickness (as defined in British Standard 6841, 1987).

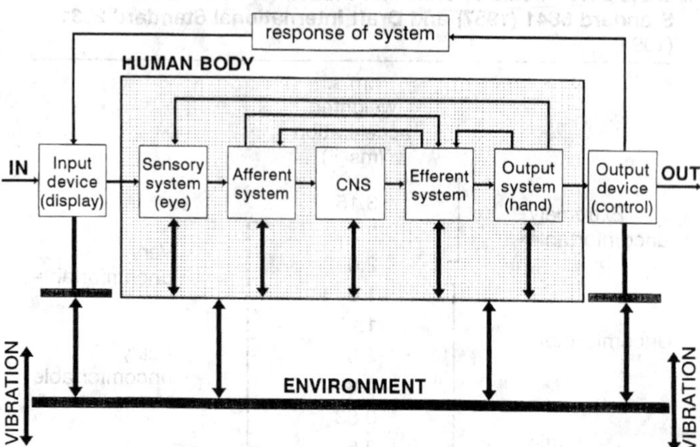

Figure 25.5 Information flow in a simple system and the areas where vibration may affect human activities.

the task less susceptible to disturbance (e.g., increasing the size of a display or reducing the sensitivity of a control). Often, the effects of vibration on vision and manual control can be much reduced by redesign of the task.

25.3.2.1 Vision

Reading a newspaper in a moving vehicle may be difficult because the paper is moving, the eye is moving, or both the paper and the eye are moving. There are many variables which affect visual performance in these conditions: It is not possible to adequately represent the effects of vibration on vision without considering the effects of these variables (Griffin and Lewis, 1978).

Stationary Observer

When a stationary observer views a moving display the eye may be able to track the position of the display using "pursuit eye movements." This closed-loop reflex will give smooth pursuit movements of the eye and clear vision if the display is moving at frequencies below about 1 Hz and with a low velocity. At slightly higher frequencies of oscillation, the precise value depending on the predictability of the motion waveform, the eye will make "saccadic" eye movements to redirect the eye with small jumps. At frequencies above about 3 Hz the eye will best be directed to one extreme of the oscillation and attempt to view the image as it is temporarily stationary while reversing the direction of movement (i.e., at the "nodes" of the motion).

In some conditions, the absolute threshold for the visual detection of the vibration of an object occurs when the peak-to-peak oscillatory motion gives an angular displacement at the eye of approximately 1 min of arc. The acceleration required to achieve this threshold is very low at low frequencies but increases in proportion to the square of the frequency to become very high at high frequencies. When the vibration displacement is above the visual detection threshold, there will be perceptible blur if the vibration frequency is above about 3 Hz. The effects of vibration on visual performance (e.g., effects on reading speed and reading accuracy) may then be estimated from the maximum time that the image spends over some small area of the retina (e.g., the period of time spent near the nodes of the motion with sinusoidal vibration). For sinusoidal vibration this time decreases (and so reading errors increase) in linear proportion to the frequency of vibration and in proportion to the square root of the displacement of vibration (O'Hanlon and Griffin, 1971). With dual-axis vibration (e.g., combined vertical and lateral vibration of a display) this time is greatly reduced and reading performance drops greatly (Meddick and Griffin, 1976). With narrow-band random vibration there is a greater probability of low image velocity than with sinusoidal vibration of the same magnitude and predominant frequency, so reading performance tends to be less affected by random vibration than sinusoidal vibration (Moseley et al., 1982). Display vibration reduces the ability to see fine detail in displays while having little effect on the clarity of larger forms.

Vibrating Observer

If an observer is sitting or standing on a vibrating surface, the effects of vibration depend on the extent to which the vibration is transmitted to the eye. The motion of the head is highly dependent on body posture but is likely to occur in both translational (i.e., in the *x*-axis, *y*-axis, and *z*-axis) and rotational (i.e., in the roll, pitch, and yaw) axes. Often, the predominant head motions affecting vision are in the vertical and pitch axes of the head. The dynamic response of the body may result in greatest head acceleration in these axes at frequencies around 5 Hz, but vibration at higher and lower frequencies can also have large effects on vision.

The pitch motion of the head is well compensated by the "vestibulo-ocular reflex" which serves to help stabilize the line of sight of the eyes at frequencies below about 10 Hz (e.g., Benson and Barnes, 1978). Although there is often pitch motion of the head at 5 Hz, there is less pitch motion of the eyes at this frequency. Pitch motion of the head, therefore, has a less than expected effect on vision—unless the display is attached to the head, as with a helmet-mounted display (see Wells and Griffin, 1984).

The effects on vision of translational motion of the head depend on viewing distance: the effects are greatest when close to a display. As the viewing distance increases the retinal image motions produced by translational displacements of the head decrease until, when viewing an object at infinite distance, there is no retinal image motion produced by translational head displacement (Griffin, 1976).

For a vibrating observer there may be little difficulty with low frequency pitch head motions when viewing a fixed display and no difficulty with translational head motions when viewing a distant display. The greatest problems occur with pitch head motion when the display is attached to the head and with translational head motion when viewing near displays. Additionally, there may be resonances of the eye within the head, but these are highly variable between individuals and often occur at high frequencies (e.g., 30 Hz and greater) where it is often possible to attenuate the vibration entering the body.

Observer and Display Vibrating

When an observer and a display oscillate together, in phase, at low frequencies, the retinal image motions (and decrements in visual performance) are less than when either the observer or the display oscillate separately (Moseley and Griffin, 1986). However, the advantage is lost as the vibration frequency is increased since there is then an increasing phase difference between the motion of the head and the motion of the display. At frequencies around 5 Hz the phase lags between seat motion and head motion may be 90° or more (depending on seating conditions) and sufficient to eliminate any advantage of moving the seat and the display together. Figure 25.6 shows an example of how reading times were affected for the three viewing conditions with sinusoidal vibration in the frequency range 0.5 to 5 Hz.

Other Variables

Some common situations in which vibration affects vision do not fall into one of the three categories in Figure 25.6. For example, when reading a newspaper on a train the motion of the arms may result in the motion of the paper being different in magnitude and phase from the motions of both the seat and the head of the observer. The dominant axis of motion of the newspaper may be different from the dominant axis of motion of the person (Griffin and Hayward, 1994).

Increasing the size of detail in a display will often greatly reduce adverse effects of vibration on vision (Lewis and Griffin, 1979). In one experiment, a 75% reduction in reading errors was achieved with only a 25% increase in the size of Landolt C targets (O'Hanlon and Griffin, 1971). Increasing the spacing between rows of letters and choosing appropriate character fonts can also be beneficial. The contrast of the display, or other reading material, also has an effect but maximum performance may not occur with maximum contrast. The known influence of all such factors are summarized in a design guide for visual displays to be used in vibration environments (Moseley and Griffin, 1986)

Optical devices may increase or decrease the effects of vibration on vision. Simple optical magnification of a vibrating object will increase both the apparent size of the object and the apparent magnitude of the vibration. Sometimes this will be beneficial since the benefits of increasing the size of the detail may more than offset the effects of increased magnitude of vibration. The effect is similar to reducing the viewing distance, which can be beneficial for stationary observers viewing vibrating displays. If the observer

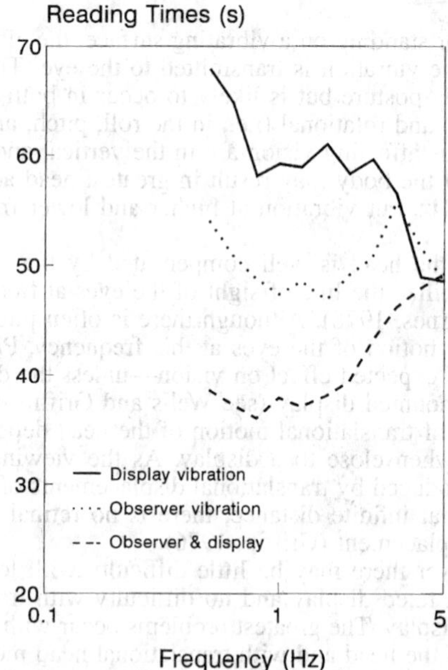

Figure 25.6 Average percentage reading times for display vibration (———), observer vibration (. . . . . . .) and simultaneous observer and display vibration (---). Sinusoidal vertical vibration at 2.0 ms^{-2} r.m.s.

is vibrating, the use of binoculars (and other magnifying devices) can be detrimental if the vibration of the device (e.g., rotation in the hand holding the binoculars) causes such an increase in the image movement that it is not sufficiently compensated by the increase in image size. The use of binoculars and telescopes in moving vehicles becomes difficult for these reasons.

25.3.2.2 Manual Control

Reading a newspaper in a moving vehicle can be impeded by the action of vibration on vision; writing, and other complex control tasks can also be impeded by vibration. Studies in which the effects of whole-body vibration on the performance of hand-tracking tasks have been reviewed elsewhere (e.g., Lewis and Griffin, 1978; McLeod and Griffin, 1989). The characteristics of the task and the characteristics of the vibration combine to determine effects of vibration on performance: A given vibration may greatly affect one type of tracking task but have little effect on another.

Effects Produced by Vibration

The most obvious consequence of vibration on a continuous manual control task is the direct mechanical jostling of the hand causing unwanted movement of the control. This is sometimes called "breakthrough" or "feedthrough" or "vibration-correlated error." The inadvertent movement of the pencil caused by "jostling" while writing in a vehicle is a form of *vibration-correlated error*. In a simple tracking task, where the operator is required to follow movements of a target, some of the error will also be correlated with the target movements. This is called *input-correlated error* and often mainly reflects the inability of an operator to follow the target without delays inherent in visual, cognitive, and motor activity. The part of the tracking error which is not correlated with either the vibration or the tracking task is called the *remnant*. This includes operator generated "noise" and any source of nonlinearity: Drawing a freehand "straight line" does not result in a perfect straight line even in the absence of environmental vibration. The effects of vibration on vision can result in increased remnant with some tracking tasks and some studies show that vibration, usually at frequencies above about 20 Hz, interferes with neuromuscular processes (e.g., Goodwin et al., 1972; Martin et al., 1984; Ribot et al.,

1986) which may be expected to result in increased remnant. The cause of the three components of the tracking error are shown in the model presented as Figure 25.7.

Effects of Task Variables

The gain (i.e., sensitivity) of a control determines the control output corresponding to a given force, or displacement, applied to the control by the operator. The optimum gain in static conditions (high enough not to cause fatigue but low enough to prevent inadvertent movement) is likely to be too great during exposure to vibration where inadvertent movement is more likely (Lewis and Griffin, 1977). First- and second-order control tasks (i.e., rate and acceleration control tasks) are more difficult than zero-order tasks (i.e., displacement control tasks) and so tend to give more errors. However, there may sometimes be advantages with such controls which are less affected by vibration breakthrough at higher vibration frequencies.

In static conditions, isometric controls (which respond to force without movement) tend to result in better tracking performance than isotonic controls (which respond to movement but require the application of no force). However, several studies show that isometric controls may suffer more from the effects of vibration (e.g., Allen et al., 1973; Levison and Harrah, 1977). The relative merits of the two types of control and the optimum characteristics of a spring-centered control will depend on control gain and control order.

The results of studies investigating the influence of the position of a control appear consistent with differences being dependent on the transmission of vibration to the hand in different positions (e.g., Shoenberger and Wilburn, 1973). Torle (1965) showed that the provision of an armrest could substantially reduce the effects of vibration on the performance of a task with a side-arm controller. The shape and orientation of controls may also be expected to affect performance—either by modifying the amount of vibration breakthrough or by altering the proprioceptive feedback to the operator.

Vibration may affect the performance of tracking tasks by reducing the visual performance of the operator. Wilson (1974) and McLeod and Griffin (1990) have shown that collimating a display by means of a lens so that it appears to be at infinity can reduce, or even eliminate, errors with some tasks. It is possible that visual disruption has played a significant part in the performance decrements reported in other experimental studies of the effects of vibration on manual control.

With some simple tasks, performance may be so easy as to be immune to disruption by vibration. At the other extreme, a task may be so difficult that any additional difficulty caused by vibration may be insignificant. Some studies suggest that, with tasks having moderate ranges of difficulty, the effects of vibration may increase as the task difficulty increases (see McLeod and Griffin, 1989).

Effects of Vibration Variables

The vibration transmissibility of the body is approximately linear (i.e., doubling the magnitude of vibration at the seat may be expected to approximately double the magnitude

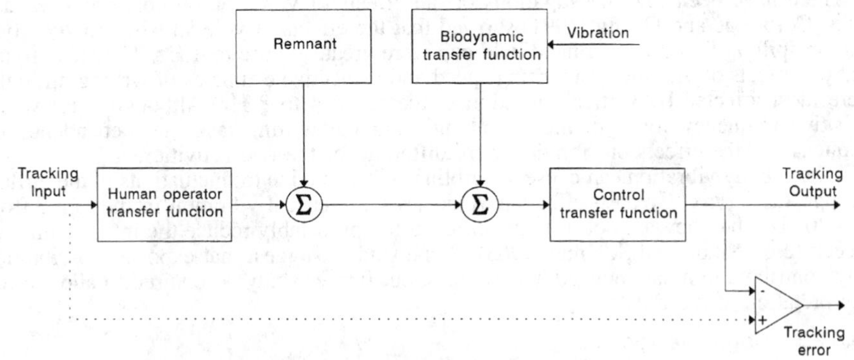

Figure 25.7 Linear model of a pursuit manual control system showing how tracking errors may be caused by the vibration (vibration-correlated error), the task (input correlated error) or some other cause (remnant).

of vibration at the head or at the hand). Vibration-correlated error may therefore increase in approximately linear proportion to vibration magnitude.

There is no simple relation between the frequency of vibration and its effects on control performance. The effects of frequency depend on the control order (which varies between tasks) and the biodynamic response of the body (which varies with posture and between operators). With zero-order tasks and the same magnitude of acceleration at each frequency, the effects of vertical seat vibration may be greatest in the range 3 to 8 Hz since transmissibility to the shoulders is greatest in this range (see McLeod and Griffin, 1989). In the horizontal axes (i.e., the *x*- and *y*-axes of the seated body) the greatest effects appear to occur at lower frequencies: around 2 Hz or below. Again, this corresponds to the frequencies at which there is greatest transmission of vibration to the shoulders. The axis of the control task most affected by vibration may not be the same axis as that in which most vibration occurs at the seat. Often, fore-and-aft movements of the control (which generally correspond to vertical movements on the display) are most affected by vertical whole-body vibration. Few controls are sensitive to vertical hand movements and these have rarely been studied.

Multiple frequency vibration causes more disruption to performance than the presentation of any one of the constituent single frequencies alone. Similarly, the effects of multiple axis vibration are greater than the effects of any one of the single axes alone.

The impression that prolonged exposure to vibration causes fatigue gave rise to the *fatigue-decreased proficiency* boundary in International Standard 2631, first published in 1974 (International Organization for Standardization, 1974, 1985). This standard proposes a complex time-dependent magnitude of vibration which is said to be "*a limit beyond which exposure to vibration can be regarded as carrying a significant risk of impaired working efficiency in many kinds of tasks, particularly those in which time-dependent effects ("fatigue") are known to worsen performance as, for example, in vehicle driving.*" Reviews of experimental studies show time-dependent effects of performance in only a few cases, with performance sometimes improving with time. It may be concluded that experimental evidence supporting the ISO fatigue-decreased proficiency boundary is very weak. There are certainly no substantial data justifying a time-dependent limit for the effects of vibration on performance with the complexity included in International Standard 2631 (1974, 1985). Any duration-dependent effects of vibration may be influenced by complex central factors including motivation, arousal, and similar concepts which depend on the form of the task: they may not lend themselves to satisfactory representation by a single time-dependent limit in an International Standard. The most common and most easily understood "direct" effects of vibration on vision and manual control are probably not intrinsically dependent on the duration of vibration exposure.

Other Variables

Repeated exposure to vibration may allow subjects to develop techniques for minimizing vibration effects by, for example, adjusting body posture to reduce the transmission of vibration to the head or the hand or by learning how to recognize images blurred by vibration. Results of experiments performed in one experimental session of vibration exposure may not necessarily apply to situations where operators have an opportunity to learn techniques to ameliorate the effects of vibration.

There have been few investigations of the effects of vibration on common everyday tasks. Corbridge and Griffin (1991) showed that the effects of vertical whole-body vibration on spilling liquid from a hand-held cup were greatest close to 4 Hz. They also found that the effects of vibration on writing speed and subjective estimates of writing difficulty were most affected by vertical vibration in the range 4 to 8 Hz. Although 4 Hz was a sensitive frequency for both the "drinking" and the writing task, the dependence on frequency of the effects of vibration were different for the two activities.

Whole-body vibration can cause a warbling of speech due to fluctuations in the airflow through the larynx. Greatest effects may occur with vertical vibration in the range from 5 to 20 Hz—but they are not usually sufficient to appreciably reduce the intelligibility of speech (e.g., Nixon and Sommer, 1963). Some studies suggest that exposure to vibration may contribute to noise-induced hearing loss, but further study is required to allow a full interpretation of these data.

25.3.2.3 Cognitive Tasks

To be useful, studies of cognitive effects of vibration must be able to show that any effects were not caused by vibration affecting input processes (e.g., vision) or output processes (e.g., hand control). Only a few investigators have addressed possible cognitive effects of vibration with care and considered such problems. For example, Shoenberger (1974) found that with the Sternberg memory-reaction-time task the time taken for subjects to recall letters presented on a display depended on the angular size of the letters. He concluded that performance was degraded by visual effects of vibration and not by cognitive effects of vibration. In most other studies, there has been little attempt to develop hypotheses to explain any significant effects of vibration in terms of the component processes involved in cognitive processing.

Simple cognitive tasks (e.g., simple reaction time) appear to be unaffected by vibration, other than by changes in arousal or motivation or by direct effects on input and output processes. This may be true also for some complex cognitive tasks. However, the scarcity and diversity of experimental studies allows the possibility of real and significant cognitive effects of vibration (see Sherwood and Griffin, 1990, 1992). Vibration may influence "fatigue" but there is little relevant scientific evidence and none which supports the complex form of the so-called fatigue-decreased proficiency limit offered in International Standard 2631 (1974, 1985).

25.3.3 Health Effects

Epidemiological studies have reported disorders among persons exposed to vibration from occupational, sport and leisure activities (see Bongers and Boshuizen, 1990; Bovenzi and Zadini, 1992; Dupuis and Zerlett, 1986; Griffin, 1990; Hulshof and van Zanten, 1987). The studies do not all agree on either the type or the extent of disorders and rarely have the findings been related to measurements of the vibration exposures. However, it is widely believed that disorders of the back (back pain, displacement of intervertebral discs, degeneration of spinal vertebrae, osteoarthritis, etc.) may be associated with vibration exposure. There may be several alternative causes of an increase in disorders of the back among persons exposed to vibration (e.g., poor sitting postures, heavy lifting). It is not always possible to conclude confidently that a back disorder is solely, or primarily, caused by vibration.

Other disorders which have been claimed to be due to occupational exposures to whole-body vibration include abdominal pain, digestive disorders, urinary frequency, prostatitis, haemorrhoids, balance and visual disorders, headaches, and sleeplessness. Further research is required to confirm whether these signs and symptoms are causally related to exposure to vibration.

25.3.3.1 Method of Vibration Evaluation

Epidemiological data alone are not sufficient to define how to evaluate whole-body vibration so as to predict the relative risks to health from the different types of vibration exposure. A consideration of such data in combination with an understanding of biodynamic responses and subjective responses is used to provide current guidance. The manner in which the health effects of oscillatory motions depend on the frequency, direction, and duration of motion is currently assumed to be similar to that for vibration discomfort (see Section 25.3.1). However, it is assumed that the "total" exposure, rather than the "average" exposure, is important and so a "dose" measure is used.

International Standard 2631 (1974, 1985) defined exposure limits (see Figure 25.8) which were ". . . *set at approximately half the level considered to be the threshold of pain (or limit of voluntary tolerance) for healthy human subjects*. . . ." Although the latest version of ISO 2631 was published in 1985, it is similar to the 1974 version, which was based on research conducted before 1970. British Standard 6841 (1987) is broadly consistent (though not identical) with the proposed revision of International Standard 2631 (International Organization for Standardization, 1994). Figure 25.8 shows an "action level" for vertical vibration based on "vibration dose values" and derived from British Standard 6841 (1987).

The *vibration dose value* can be considered to be the magnitude of a 1 sec duration of vibration, which will be equally severe to the measured vibration. The vibration dose value uses a "fourth power" time dependency to accumulate vibration severity over the exposure period from the shortest possible shock to a full day of vibration (see BS 6841, 1987; DIS 2631, 1994):

$$\text{vibration dose value} = \left[\int_{t=0}^{t=T} a^4(t)\,\mathrm{d}t\right]^{1/4}$$

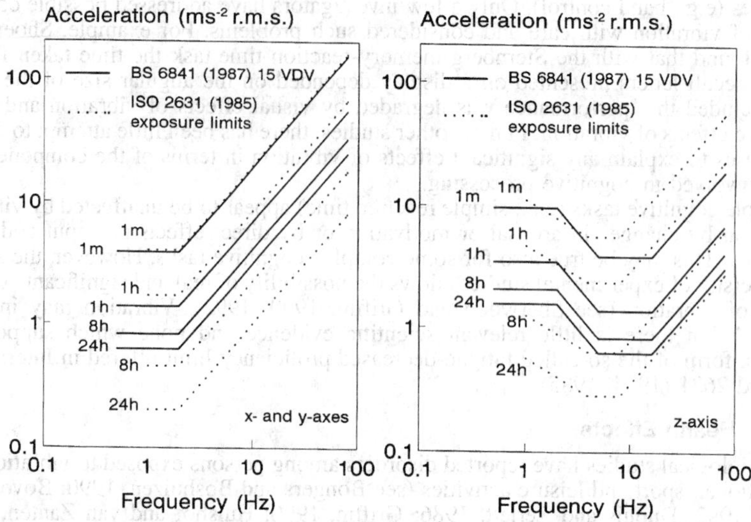

Figure 25.8 Comparison of International Standard 2631 (1985) exposure limits with an "action level" based on a vibration dose value (VDV) of 15 $ms^{-1.75}$ from British Standard 6841 (1987). When seated: *x*-axis = fore-and-aft; *y*-axis = lateral; *z*-axis = vertical). [From Griffin (1990).]

If the exposure duration (t, seconds) and the frequency-weighted r.m.s. acceleration (a_{rms}, ms^{-2} r.m.s.) are known for conditions in which the vibration characteristics are statistically stationary, it can be useful to calculate the "estimated vibration dose value," eVDV:

$$\text{estimated vibration dose value} = 1.4\, a_{rms}\, t^{1/4}$$

The eVDV is not applicable to transients, shocks and repeated shock motions in which the crest factor (peak value divided by the r.m.s. value) is high. The vibration dose value (VDV) can be used to evaluate the severity of transients, shocks and repeated shock motions.

No precise limit can be offered to prevent disorders caused by whole-body vibration, but standards define useful methods of quantifying vibration severity. British Standard 6841 (1987) offers the following guidance:

> *High vibration dose values will cause severe discomfort, pain and injury. Vibration dose values also indicate, in a general way, the severity of the vibration exposures which caused them. However there is currently no consensus of opinion on the precise relation between vibration dose values and the risk of injury. It is known that vibration magnitudes and durations which produce vibration dose values in the region of 15 $ms^{-1.75}$ will usually cause severe discomfort. It is reasonable to assume that increased exposure to vibration will be accompanied by increased risk of injury.*

An action level might be set higher or lower than 15 $ms^{-1.75}$, or two action levels may be defined corresponding to two different types of action for different severities of vibration (as in Draft International Standard 2631, 1994) At high vibration dose values, prior consideration of the fitness of the exposed persons and the design of adequate safety precautions may be required. The need for regular checks on the health of routinely exposed persons may also be considered.

The vibration dose value provides a robust measure by which highly variable and complex exposures can be compared. The tentative action level merely serves to indicate the approximate values which might be excessive. Figure 25.9 illustrates the root-mean-square accelerations corresponding to a vibration dose value of 15 $ms^{-1.75}$ for exposures between 1 sec and 24 hr.

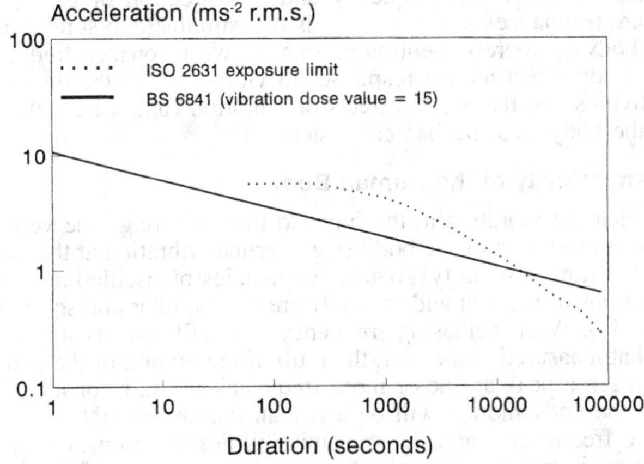

Figure 25.9 Action level corresponding to a vibration dose value (VDV) of 15 $ms^{-1.75}$ (see British Standard 6841, 1987) compared with "exposure limit" from International Standard 2631 (1974, 1985).

Unlike the "exposure limit" in the old International Standard 2631 (1974, 1985), a vibration dose value "action level" does not appear to allow very high magnitudes at short durations (a few minutes) or require very low magnitudes for long duration exposures (many hours). Any exposure to continuous vibration, or intermittent vibration, or repeated shock may be compared with the action level by calculating the vibration dose value. It would be unwise to exceed the action level without consideration of the possible health effects of an exposure to vibration or shock.

25.3.4 Disturbance in Buildings

Acceptable magnitudes of vibration in buildings are close to vibration perception thresholds. The effects of vibration in buildings are assumed to depend on the use of the building in addition to the vibration frequency, direction, and duration. Guidance is given in various standards (e.g., International Standard 2631 part 2, 1989; American National Standard S3.29, 1983; British Standard 6472, 1992). Using the guidance contained in ISO 2631 part 2 (1989) it is possible to summarize the acceptability of vibration in different types of building in a single table of vibration dose values (see Table 25.5 and British Standard 6472, 1992). The vibration dose values in Table 25.5 are applicable irrespective of whether the vibration occurs as a continuous vibration, intermittent vibration, or repeated shocks.

25.3.5 Biodynamics

The human body is a complex mechanical system which does not, in general, respond to vibration in the same manner as a rigid mass: There are relative motions between the

Table 25.5 Vibration Dose Values at Which Various Degrees of Adverse Comment May Be Expected in Buildings

Place	Low Probability of Adverse Comment	Adverse Comment Possible	Adverse Comment Probable
Critical working areas	0.1	0.2	0.4
Residential	0.2–0.4	0.4–0.8	0.8–1.6
Office	0.4	0.8	1.6
Workshops	0.8	1.6	3.2

Source: Based on International Standard 2631 Part 2 (1989) and British Standard 6472 (1992) (see Griffin, 1990).

body parts which vary with the frequency and the direction of the applied vibration. Although there are resonances in the body, it is oversimplistic to summarize the dynamic response of the body by merely mentioning one or two resonance frequencies. The biodynamics of the body affect human responses to vibration, but the discomfort, the interference with activities and the health effects of vibration cannot be well predicted solely by considering the body as a mechanical system.

25.3.5.1 Transmissibility of the Human Body

The extent to which the vibration at the input to the body (e.g., the vertical vibration at a seat) is transmitted to a part of the body (e.g., vertical vibration at the head or the hand) is described by the transmissibility. At low frequencies of oscillation (e.g., below about 1 Hz), the oscillations of the seat and the body are very similar and so the transmissibility is approximately 1.0. With increasing frequency of oscillation the motions of the body increase above that measured at the seat; the ratio of the motion of the body to the motion of the seat will reach a peak at one or more frequencies (i.e., resonance frequencies). At high frequencies the body motion will be less than that at the seat.

The resonance frequencies and the transmissibilities at resonance vary according to where the vibration is measured on the body and the posture of the body. For seated persons, there may be resonances to the head and the hand at frequencies in the range of 4 to 12 Hz for vertical vibration, below 4 Hz with x-axis vibration and below 2 Hz with lateral vibration (see Paddan and Griffin, 1988a,b). A seat back can greatly increase the transmission of x-axis vibration to the head and upper body of seated people, and bending of the legs can greatly affect the transmission of vertical vibration to the heads of standing persons.

25.3.5.2 Mechanical Impedance of the Human Body

Mechanical impedance reflects the relation between the driving force at the input to the body and the resultant movement of the body. If the human body were rigid, the ratio of force to acceleration applied to the body would be constant and indicate the mass of the subject. Because the body is not rigid, the ratio of force to acceleration is only close to the body mass at very low frequencies (below about 2 Hz with vertical vibration; below about 1 Hz with horizontal vibration).

Measures of mechanical impedance usually show a principal resonance for vertical vibration of seated subjects at about 5 Hz, and sometimes a second resonance in the range of 7 to 12 Hz (Fairley and Griffin, 1989). Unlike some of the resonances affecting the transmissibility of the body, these resonances are only influenced by movement of large masses close to the input of vibration to the body. The large difference in impedance between that of a rigid mass and that of the human body means that the body cannot usually be represented by a rigid mass when measuring the vibration transmitted through seats.

25.3.5.3 Biodynamic Models

Various mathematical models of the responses of the body to vibration have been developed. A simple model with one or two degrees-of-freedom can represent the impedance of the body, and a dummy might be constructed to represent this impedance for seat testing. Compared with impedance, the transmissibility of the body is affected by many more variables and so requires a more complex model reflecting the posture of the body and the translation and rotation associated with the various modes of vibration.

25.3.6 Protection from Whole-Body Vibration

Wherever possible, vibration should be reduced at source. This may involve reducing the undulations of the terrain, or reducing the speed of travel of vehicles, or improving the balance of rotating parts. Methods of reducing the transmission of vibration to operators require an understanding of the characteristics of the vibration environment and the route for the transmission of vibration to the body. For example, the magnitude of vibration often varies with location: lower magnitudes will be experienced in some areas adjacent to machinery or in different parts of vehicles.

25.3.6.1 Seating Dynamics

Most seats exhibit a resonance at low frequencies which results in higher magnitudes of vertical vibration occurring on the seat than on the floor! At high frequencies there is usually attenuation of vibration. The resonance frequencies of common seats are usually in the region of 4 Hz (see Figure 25.10). The amplification at resonance is partially determined by the "damping" in the seat. Increases in the damping of the seat cushioning tend to reduce the amplification at resonance but increase the transmission of vibration at high frequencies. The variations in transmissibility between seats are sufficient to result in significant differences in the vibration experienced by people supported by different seats.

A simple numerical indication of the isolation efficiency of a seat for a specific application is provided by the *seat effective amplitude transmissibility* (*SEAT*) (Griffin, 1990). A SEAT value greater than 100% indicates that, overall, the vibration on the seat is "worse" than the vibration on the floor beneath the seat:

$$\text{SEAT (\%)} = \frac{\text{ride comfort on seat}}{\text{ride comfort on floor}} \times 100$$

Values below 100% indicate that the seat has provided some useful attenuation. Seats should be designed to have the lowest SEAT value compatible with other constraints.

In practice, the SEAT value is a mathematical procedure for predicting the effect of a seat on ride comfort. The "ride comfort" that would result from sitting on the seat or on the floor can be predicted using the frequency weightings in the appropriate standard. The SEAT value may be calculated from the r.m.s. values or the vibration dose values of the frequency-weighted acceleration on the seat and the floor:

$$\text{SEAT (\%)} = \frac{\text{vibration dose value on seat}}{\text{vibration dose value on floor}} \times 100$$

The SEAT value is a characteristic of the vibration input and not merely a description of the dynamics of the seat: Different values are obtained with the same seat in different vehicles. The SEAT value indicates the suitability of a seat for a particular type of vibration.

A separate suspension mechanism is provided beneath the seat pan in "suspension seats." These seats, used in some off-road vehicles, trucks, and coaches, have low resonance frequencies (often below about 2 Hz) and so can attenuate vibration at frequencies above about 2 Hz. The transmissibilities of these seats are usually determined by the seat manufacturer, but their isolation efficiencies vary with operating conditions.

25.4 MOTION SICKNESS

Motion sickness is not an illness but a normal response to motion which is experienced by many fit and healthy people. A variety of different motions can cause sickness and reduce the comfort, impede the activities, and degrade the well-being of both those di-

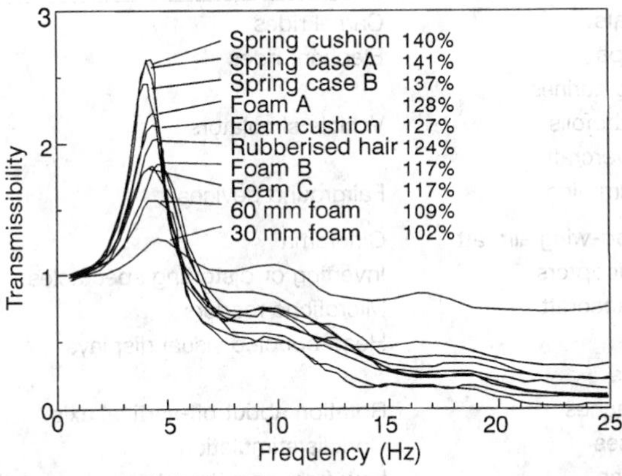

Figure 25.10 Comparison of the vertical transmissibilities and SEAT values for ten alternative cushions of passenger railway seats (data from Corbridge et al., 1989).

rectly affected and those associated with the motion sick. Although vomiting can be the most inconvenient consequence, other effects (e.g., nausea, sweating, color changes, dizziness headaches, and drowsiness) can also be unpleasant. In some cases the symptoms can be so severe as to result in reduced motivation to survive difficult situations.

25.4.1 Causes of Motion Sickness

Motion sickness can be caused by many different movements of the body (e.g., translational and rotational oscillation, constant speed rotation about an off-vertical axis, Coriolis stimulation), movements of the visual scene, and various other stimuli producing sensations associated with movement of the body (see Table 25.6 and Griffin, 1991). Motion sickness is neither explained nor predicted solely by the physical characteristics of the motion, although some motions can be reliably predicted as more nauseogenic than others.

Motions of the body may be detected by three basic sensory systems: the vestibular system, the visual system, and the somatosensory system. The vestibular system is located in the inner ear and comprises the semicircular canals, which respond to the rotation of the head, and the otoliths, which respond to translational forces (either translational acceleration or rotation of the head relative to an acceleration field, such as the force of gravity). The eyes may detect relative motion between the head and the environment, caused by either head movements (in translation or rotation) or movements of the environment or a combination of the movements of the head and the environment. The somatosensory systems respond to force and displacement of parts of the body and give rise to sensations of body movement, or force.

It is assumed that in "normal" environments the movements of the body are detected by all three sensory systems and that this leads to an unambiguous indication of the movements of the body in space. In some other environments the three sensory systems may give signals corresponding to different motions (or motions which are not realistic) and lead to some form of conflict. This leads to the idea of a "sensory conflict theory" of motion sickness in which sickness occurs when the sensory systems disagree on the motions which are occurring. However, this implies some absolute significance to sensory information, whereas the "meaning" of the information is probably learned. This led to the *sensory rearrangement theory* of motion sickness which states that: "*all situations which provoke motion sickness are characterised by a condition of sensory rearrangement in which the motion signals transmitted by the eyes, the vestibular system and the non-vestibular proprioceptors are at variance either with one another or with what is expected from previous experience*" (Reason, 1970, 1978). Reason and Brand (1975) suggest that the conflict may be sufficiently considered in two categories: intermodality (between vision and the vestibular receptors) and intramodality (between the semicircular canals and the otoliths within the vestibular system). For both categories it is possible to identify three types of situation in which conflict can occur (see Table 25.7). The theory implies

Table 25.6 Examples of Environments, Activities, and Devices Which Can Cause Symptoms of Motion Sickness

Boats	Camel rides
Ships	Elephant rides
Submarines	
Hydrofoils	Vehicle simulators
Hovercraft	
Swimming	Fairground devices
Fixed-wing aircraft	Cinerama
Helicopters	Inverting or distorting spectacles
Spacecraft	Microfiche readers
	Head-coupled visual displays
Cars	
Coaches	Rotation about off-vertical axis
Buses	Coriolis stimulation
Trains	Low frequency translational oscillation
Tanks	

Table 25.7 Type of Motion Cue Mismatch Produced by Various Provocative Stimuli

	Category of Motion Cue Mismatch	
	Visual (A)/Vestibular (B)	Canal (A)/Otolith (B)
TYPE 1 A and B simultaneously give contradictory or uncorrelated information	Watching waves from a ship	Making head movements whilst rotating (Coriolis or cross-coupled stimulation)
	Use of binoculars in a moving vehicle	Making head movements in an abnormal environment which may be constant (e.g., hyper- or hypogravity) or fluctuating (e.g., linear oscillation)
	Making head movements when vision is distorted by an optical device	Space sickness
	"Pseudo-Coriolis" stimulation	Vestibular disorders (e.g., Ménière's disease, acute labyrinthitis, trauma labyrinthectomy)
TYPE IIa A signals in absence of expected B signal	Cinerama sickness	Positional alcohol nystagmus
	Simulator sickness	Caloric stimulation of semicircular canals
	"Haunted swing"	Vestibular disorders (e.g., pressure vertigo, cupulolithiasis)
	Circularvection	
TYPE IIb B signals in absence of expected A signals	Looking inside a moving vehicle without external visual reference (e.g., below deck in a boat)	Low frequency (<0.5 Hz) translational oscillation
	Reading in a moving vehicle	Rotating linear acceleration vector (e.g., "barbecue spit" rotation, rotation about an off-vertical axis)

Source: Adapted from Benson, 1984.

that all situations which provoke motion sickness can be fitted into one of the six conditions shown in Table 25.7 (see Griffin, 1990).

There is evidence that the average susceptibility to sickness among males is less than that among females, and susceptibility decreases with increased age among both males and females (Lawther and Griffin, 1988a). However, there are larger individual differences within any group of either gender at any age: Some people are easily made ill by motions that can be endured indefinitely by others. The reasons for these differences are not properly understood.

25.4.2 Sickness Caused by Oscillatory Motion

Motion sickness is not caused by oscillation (however violent) at frequencies much above about 1 Hz: The phenomenon arises from motions at the low frequencies associated with normal postural control of the body. Various experimental investigations have explored the extent to which vertical oscillation causes sickness at different frequencies. These studies have allowed the formulation of a frequency weighting, W_f (see Figure 25.4), and the definition of a "motion sickness dose value." The frequency weighting W_f reflects greatest sensitivity to acceleration in the range of 0.125 to 0.25 Hz, with a rapid reduction in sensitivity at higher frequencies. The motion sickness dose value predicts the probability of sickness from a knowledge of the frequency and magnitude of vertical oscillation (see Lawther and Griffin, 1987; Draft International Standard 2631, 1994):

$$\text{motion sickness dose value} = a_{\text{rms}}\, t^{1/2}$$

where a_{rms} is the root-mean-square value of the frequency-weighted acceleration (ms^{-2}) and t is the exposure period (seconds). The percentage of unadapted adults who are expected to vomit is given by ⅓ MSDV. (These relationships have been derived from exposures in which up to 70% of persons vomited during exposures lasting between 20 min and 6 hr.)

The motion sickness dose value has been used for the prediction of sickness on various marine craft (ships, hovercraft, and hydrofoil) in which vertical oscillation has been shown to be the prime cause of sickness (Lawther and Griffin, 1988b). Vertical oscillation is not the principal cause of sickness in many road vehicles and some other environments: The above expression should not be assumed to be applicable to the prediction of sickness in all environments.

25.5 HAND-TRANSMITTED VIBRATION

Prolonged and regular exposure of the fingers or the hands to vibration or repeated shock can give rise to various signs and symptoms of disorder. The precise extent and interrelation between the signs and symptoms are not fully understood but five types of disorder may be identified (see Table 25.8).

The various disorders may be interconnected: More than one disorder can affect a person at the same time, and it is possible that the presence of one disorder facilitates the appearance of another. The onset of each disorder is dependent on several variables, such as the vibration characteristics, the dynamic response of the fingers or hand, individual susceptibility to damage, and other aspects of the environment. The terms *vibration syndrome*, or *hand-arm vibration syndrome*, HAVS, are sometimes used to refer to one or more of the effects listed in Table 25.8.

25.5.1 Sources of Hand-Transmitted Vibration

The vibration on tools varies greatly depending on tool design and method of use, so it is not possible to categorise individual tool types as "safe" or "dangerous." However, Table 25.9 lists tools and processes that are sometimes a cause for concern.

25.5.2 Effects of Hand-Transmitted Vibration

25.5.2.1 Vascular Disorders

The first published cases of the condition now most commonly known as *vibration-induced white finger* (*VWF*), are acknowledged to be those reported in Italy by Loriga in 1911. A few years later, cases were documented at limestone quarries in Indiana. Vibration-induced white finger has subsequently been reported to occur in many other widely varied occupations in which there is exposure of the fingers to vibration (see Gemne, 1993; Griffin, 1990; Taylor and Pelmear, 1975; Wasserman et al., 1982).

Signs and Symptoms

Vibration-induced white finger (VWF), is characterized by intermittent whitening (i.e., blanching) of the fingers. The fingertips are usually the first to blanch but the affected area may extend to all of one or more fingers with continued vibration exposure. Attacks of blanching are precipitated by cold and therefore usually occur in cold conditions or

Table 25.8 Five Types of Disorder Associated with Hand-Transmitted Vibration Exposures (some combination of these disorders is sometimes referred to as the hand-arm vibration syndrome, HAVS)

Type	Disorder
Type A	Circulatory disorders
Type B	Bone and joint disorders
Type C	Neurological disorders
Type D	Muscle disorders
Type E	Other general disorders (e.g., central nervous system)

Source: From Griffin, 1990.

Table 25.9 Tools and Processes Potentially Associated With Vibration Injuries

Type of Tool	Examples of Tool Type
Percussive metal-working tools	Riveting tools Caulking tools Chipping tools Chipping hammers Fettling tools Hammer drills Clinching and flanging tools Impact wrenches Swaging Needle guns
Grinders and other rotary tools	Pedestal grinders Handheld grinders Handheld sanders Handheld polishers Flex-driven grinders/polishers Rotary burring tools
Percussive hammers and drills used in mining, demolition and road construction	Hammers Rock drills Road drills, etc.
Forest and garden machinery	Chain saws Antivibration chain saws Brush saws Mowers and shears Barking machines
Other processes and tools	Nut runners Shoe-pounding-up machines Concrete vibro-thickeners Concrete levelling vibrotables Motorcycle handlebars

when handling cold objects. The blanching lasts until the fingers are rewarmed and vasodilation allows the return of the blood circulation.

Many years of vibration exposure often occur before the first attack of blanching is noticed. Affected persons often have other signs and symptoms, such as numbness and tingling. Cyanosis and, rarely, gangrene, have also been reported. It is not yet clear to what extent these other signs and symptoms are causes of, caused by, or unrelated to, attacks of "white finger."

Diagnosis

There are other conditions that can cause similar signs and symptoms to those associated with VWF. Vibration-induced white finger cannot be assumed to be present merely because there are attacks of blanching. It will be necessary to exclude other known causes of similar symptoms (by medical examination) and also necessary to exclude so-called Primary Raynaud's disease (also called *constitutional white finger*). This exclusion cannot yet be achieved with complete confidence but if there is no family history of the symptoms, if the symptoms did not occur before the first significant exposure to vibration, and if the symptoms and signs are confined to areas in contact with the vibration (e.g., the fingers, not the ears), they will often be assumed to indicate vibration-induced white finger.

Diagnostic tests for vibration-induced white finger can be useful but, at present, they are not infallible indicators of the disease. The measurement of finger systolic blood pressure following finger cooling and the measurement of finger rewarming times following cooling can be useful, but many others tests are in use (see Bovenzi, 1993; Nielsen and Lassen, 1977).

The severity of the effects of vibration are sometimes recorded by reference to the "stage" of the disorder. The staging of vibration-induced white finger is based on verbal statements made by the affected person. In the Stockholm Workshop staging system, the

Table 25.10 Stockholm Workshop Scale For the Classification of Vibration-Induced White Finger (If a person has stage 2 in two fingers of the left hand and stage 1 in a finger on the right hand, the condition may be reported as 2L(2)/1R(1). There is no defined means of reporting the condition of digits when this varies between digits on the same hand. The scoring system is more helpful when the extent of blanching is to be recorded.)

Stage	Grade	Description
0	—	No attacks
1	Mild	Occasional attacks affecting only the tips of one or more fingers
2	Moderate	Occasional attacks affecting distal and middle (rarely also proximal) phalanges of one or more fingers
3	Severe	Frequent attacks affecting all phalanges of most fingers
4	Very severe	As in stage 3, with trophic skin changes in the fingertips

Source: Gemne et al., 1987.

staging is influenced by both the frequency of attacks of blanching and the areas of the digits affected by blanching (see Table 25.10).

A "scoring system" is used to record the areas of the digits affected by blanching (see Figure 25.11). The scores correspond to areas of blanching on the digits commencing with the thumb. On the fingers a score of 1 is given for blanching on the distal phalanx, a score of 2 for blanching on the middle phalanx, and a score of 3 for blanching on the proximal phalanx. On the thumbs the scores are 4 for the distal phalanx and 5 for the proximal phalanx. The blanching score may be based on statements from the affected person or on the visual observations of a designated observer (e.g., a nurse).

25.5.2.2 Neurological Disorders

Neurological effects of hand-transmitted vibration (e.g., numbness, tingling, elevated sensory thresholds for touch, vibration, temperature and pain, and reduced nerve conduction velocity) are considered to be separate effects of vibration and not merely symptoms of vibration-induced white finger. A method of reporting the extent of vibration-induced neurological effects of vibration has been proposed (see Table 25.11). This staging is not currently related to the results of any specific objective test: The *sensorineural stage* is a subjective impression of a physician based on the statements of the affected person or the results of any available clinical or scientific testing. Neurological disorders are sometimes identified by screening tests using measures of sensory function, such as the thresholds for feeling vibration, heat, or coldness on the fingers.

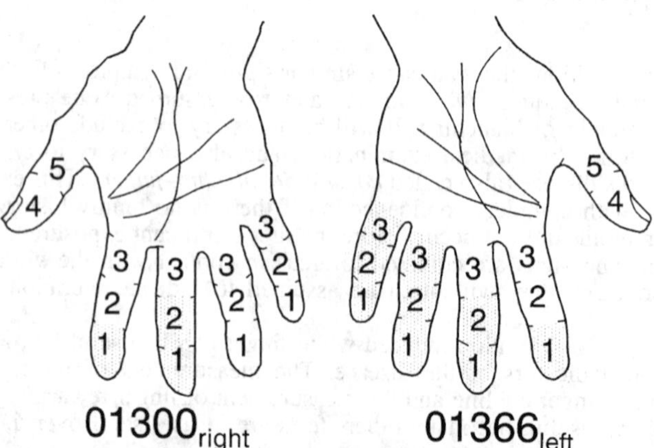

Figure 25.11 Method of scoring the areas of the digits affected by blanching. (From Griffin, 1990.) The blanching scores for the hands shown are 01300_{right}, 01366_{left}.

Table 25.11 Proposed "Sensorineural Stages" of the Effects of Hand-Transmitted Vibration

Stage	Symptoms
0_{SN}	Exposed to vibration but no symptoms
1_{SN}	Intermittent numbness with or without tingling
2_{SN}	Intermittent or persistent numbness, reduced sensory perception
3_{SN}	Intermittent or persistent numbness, reduced tactile discrimination and/or manipulative dexterity

Source: Brammer et al., 1987.

25.5.2.3 Muscular Effects

The research literature includes reports of muscle atrophy among users of vibrating tools. Workers exposed to hand-transmitted vibration sometimes report difficulty with their grip, including reduced dexterity, reduced grip strength, and locked grip. Many of the reports are derived from symptoms reported by exposed persons rather than signs detected by physicians and could be a reflection of neurological problems.

Muscle activity may be of great importance to tool users since a secure grip can be essential to the performance of the job and the safe control of the tool. The presence of vibration on a handle may encourage the adoption of a tighter grip than would otherwise occur, and a tight grip may increase the transmission of vibration to the hand. If the chronic effects of vibration result in reduced grip, this may help to protect operators from further effects of vibration, but interfere with both work and leisure activities.

25.5.2.4 Articular Disorders

Many surveys of the users of handheld tools have found evidence of bone and joint problems, most often among men operating percussive tools such as those used in metal-working jobs and mining and quarrying. It is speculated that some characteristic of such tools, possibly the low frequency shocks, is responsible. Some of the reported injuries relate to specific bones and suggest the existence of cysts, vacuoles, decalcification, or other osteolysis, degeneration, or deformity of the carpal, metacarpal, or phalangeal bones. Osteoarthrosis and olecranon spurs at the elbow, and other problems at the wrist and shoulder are also documented (Griffin, 1990).

Notwithstanding the evidence of many research publications, there is not universal acceptance that vibration is the cause of articular problems, and there is currently no dose–effect relation which predicts their occurrence. In the absence of specific information, it seems that adherence to current guidance for the prevention of vibration-induced white finger may provide reasonable protection.

25.5.2.5 Other Effects

Effects of hand-transmitted vibration may not be confined to the fingers, hands, and arms: Many studies have found a high incidence of problems such as headaches and sleeplessness among tool users and have concluded that these symptoms are caused by hand-transmitted vibration. Although these are real problems to those affected, they are "subjective" effects which are not accepted as real by all researchers. Some current research is seeking a physiological basis for such symptoms. It would appear that caution is appropriate, but it is reasonable to assume that the adoption of the modern guidance to prevent vibration-induced white finger will also provide some protection from any other effects of hand-transmitted vibration within, or distant from, the hand.

25.5.3 Preventative Measures

Protection from the effects of hand-transmitted vibration requires actions from management, tool manufacturers, technicians, and physicians at the workplace and from tool users. Table 25.12 summarizes some of the actions which may be appropriate.

When there is reason to suspect that hand-transmitted vibration may cause injury, the vibration at tool-hand interfaces should be measured. It will then be possible to predict whether the tool or process is likely to cause injury and whether any other tool or process could give a lower vibration severity. The duration of exposure to vibration should also be quantified. Reduction of exposure time may include the provision of exposure breaks

Table 25.12 Some Preventative Measures to Consider When Persons are Exposed to Hand-Transmitted Vibration

Group	Action
Management	Seek technical advice Seek medical advice Warn exposed persons Train exposed persons Review exposure times Policy on removal from work
Tool manufacturers	Measure tool vibration Design tools to minimize vibration Ergonomic design to reduce grip force, etc. Design to keep hands warm Provide guidance on tool maintenance Provide warning of dangerous vibration
Technical at workplace	Measure vibration exposure Provide appropriate tools Maintain tools Inform management
Medical	Pre-employment screening Routine medical checks Record all signs and reported symptoms Warn workers with predisposition Advise on consequences of exposure Inform management
Tool user	Use tool properly Avoid unnecessary vibration exposure Minimize grip and push forces Check condition of tool Inform supervisor of tool problems Keep warm Wear gloves when safe to do so Minimize smoking Seek medical advice if symptoms appear Inform employer of relevant disorders

Source: Adapted from Chapter 19 of *The Handbook of Human Vibration,* Griffin, 1990.

during the day and, if possible, prolonged periods away from vibration exposure. For any tool or process having a vibration magnitude sufficient to cause injury there should be a system to quantify and control the maximum daily duration of exposure of any individual.

Gloves are sometimes recommended as a means of reducing the adverse effects of vibration on the hands. When using the frequency weightings in current standards, most commonly available gloves do *not* normally provide effective attenuation of the vibration on most tools. Gloves and "cushioned" handles may reduce the transmission of high frequencies of vibration, but current standards imply that these frequencies are not usually the primary cause of disorders. Gloves may protect the hand from other forms of mechanical injury (e.g., cuts and scratches) and protect the fingers from temperature extremes. Warm hands are less likely to suffer an attack of finger blanching, and some consider that maintaining warm hands while being exposed to vibration may also lessen the damage caused by the vibration.

Workers who are exposed to vibration magnitudes sufficient to cause injury should be warned of the possibility of vibration injuries and educated on the ways of reducing the severity of their vibration exposures. They should be advised of the symptoms to look out for and told to seek medical attention if the symptoms appear. There should be pre-employment medical screening wherever a subsequent exposure to hand-transmitted vibration may reasonably be expected to cause vibration injury. Medical supervision of each exposed person should continue throughout employment at suitable intervals, possibly annually.

Table 25.13 Number of Years Before Blanching Develops in 10% to 50% of Vibration-Exposed Persons According to International Standard 5349 (1986)

Weighted Acceleration, $a_{hw(eq, 4h)}$ (ms^{-2} r.m.s.)	Percentage of Population Affected by Finger Blanching				
	10%	20%	30%	40%	50%
2	15	23	>25	>25	>25
5	6	9	11	12	14
10	3	4	5	6	7
20	1	2	2	3	3
50	<1	<1	<1	1	1

25.5.4 Standards for the Evaluation of Hand-Transmitted Vibration

International Standard 5349 (1986) and many national standards (e.g., American National Standard S3.34, 1986) use a frequency weighting (called W_h in British Standard 6842) to quantify the severity of hand-transmitted vibration over the frequency range of 8 to 1000 Hz. This weighting is applied to measurements of vibration acceleration in each of the three axes of vibration at the point of entry of vibration to the hand (see Figure 25.2). Measurements of tool vibration should be obtained with representative operating conditions. The standards imply that if two tools expose the hand to vibration for the same period of time, the tool having the lowest frequency-weighted acceleration will be least likely to cause injury or disease.

Occupational exposures to hand-transmitted vibration can have widely varying daily exposure durations—from a few seconds to many hours. Often, exposures are intermittent. To enable a daily exposure to be reported simply, the standards refer to an equivalent 4-hr or an equivalent 8-hr exposure. Table 25.13 shows a relation between years of vibration exposure, 4-hr energy-equivalent frequency-weighted acceleration and the prevalence of finger blanching as proposed in an Annex to International Standard 5349 (1986). These relationships are illustrated graphically in Figure 25.12. The values in Figure 25.12 and Table 25.13 refer to frequency-weighted acceleration referenced to the frequency range of 8 to 16 Hz. Figure 25.13 shows how the magnitudes required for a predicted prevalence of 10% vibration-induced white finger after 8 yr are assumed to depend on vibration frequency from 8 to 1000 Hz for exposure durations from 1 min to 8 hr per day.

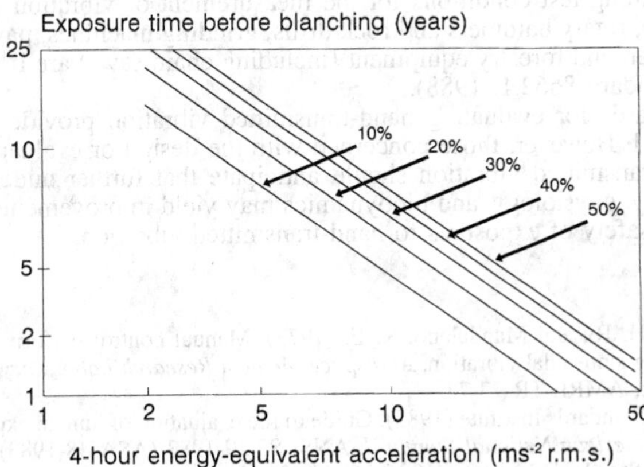

Figure 25.12 Years of exposure to various magnitudes of 4-hr energy-equivalent frequency-weighted, hand-transmitted vibration after which finger blanching in 10% to 50% of exposed persons should be expected according to International Standard 5349 (1986).

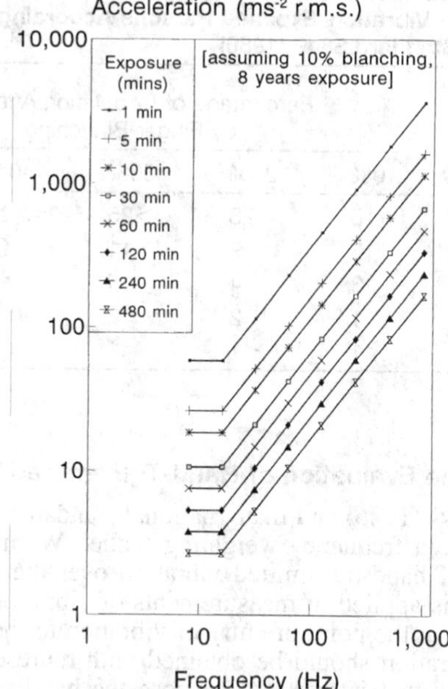

Figure 25.13 Acceleration magnitudes predicted to give 10% prevalence of vibration-induced white finger after 8 yr for daily exposure durations from 1 min to 8 hr (according to International Standard 5349, 1986).

The percentage of affected persons in any group of exposed persons will not always closely match the values shown in Table 25.13 or Figures 25.12 and 25.13. The frequency weighting, the time-dependency, and the dose-effect information are based on less than complete information, and they have been simplified for practical convenience. Additionally, the number of persons affected by vibration will depend on the rate at which persons enter and leave the exposed group. Neither the average exposure time nor the "mean latency" (i.e., the average period of vibration exposure before those with symptoms of VWF develop the condition) are appropriate measures of the exposure period for this calculation.

Standards defining test conditions for the measurement of vibration on chipping and riveting hammers, rotary hammers and rock drills, grinding machines, pavement breakers, and various garden and forestry equipment (including chain saws) are in preparation (see International Standard 8662.1, 1988).

Current standards for evaluating hand-transmitted vibration provide guidance which cannot be ignored. However, those concerned with the design or evaluation of situations involving hand-transmitted vibration should anticipate that further understanding of the relevant pathology, physiology, and biodynamics may yield improvements to the methods of assessing the safety of exposures to hand-transmitted vibration.

REFERENCES

Allen, R. W., Jex, H. R., and Magdaleno, R. E. (1973). Manual control performance and dynamic response during sinusoidal vibration. *Aerospace Medical Research Laboratory, Wright-Patterson Air Force Base*, AMRL-TR-73-78.

American National Standards Institute (1983). Guide to the evaluation of human exposure to vibration in buildings. *American National Standard*, ANSI S3.29-1983 (ASA 48-1983).

American National Standards Institute (1986). Guide for the measurement and evaluation of human exposure to vibration transmitted to the hand. *American National Standards Institute*, ANSI S3.34 (ASA 67).

Benson, A. J. (1984). Motion sickness. In: M. R. Dix and J. S. Hood, Eds., *Vertigo*. New York: John Wiley.

Benson, A. J., and Barnes, G. R. (1978). Vision during angular oscillation: The dynamic interaction of visual and vestibular mechanisms. *Aviation, Space and Environmental Medicine, 49(1)* Section II, 340–345.

Bongers, P. M., and Boshuizen, H. C. (1990). Back disorders and whole-body vibration at work. *Thesis, University of Amsterdam,* ISBN: 90-9003668-7.

Bovenzi, M. (1993). Digital arterial responsiveness to cold in healthy men, vibration white finger and primary Raynaud's phenomenon. *Scandinavian Journal of Work, Environment and Health, 19(4),* 271–276.

Bovenzi, M., and Zadini, A. (1992). Self-reported back symptoms in urban bus drivers exposed to whole-body vibration. *Spine, 17(9),* 1048–1059.

Brammer, A. J., Taylor, W., and Lundborg, G. (1987). Sensorineural stages of the hand-arm vibration syndrome. *Scandinavian Journal of Work, Environment and Health, 13(4),* 279–283.

British Standards Institution (1987). Measurement and evaluation of human exposure to whole-body mechanical vibration and repeated shock. *British Standard,* BS 6841.

British Standards Institution (1987). Measurement and evaluation of human exposure to vibration transmitted to the hand. *British Standard,* BS 6842.

British Standards Institution (1992). Evaluation of human exposure to vibration in buildings (1 Hz to 80 Hz). *British Standard,* BS 6472.

Corbridge, C., and Griffin, M. J. (1991). Effects of vertical vibration on passenger activities: Writing and drinking. *Ergonomics, 34(10),* 1313–1332.

Corbridge, C., Griffin, M. J., and Harborough, P. (1989). Seat dynamics and passenger comfort. *Proceedings of the Institution of Mechanical Engineers, 203,* 57–64.

Dupuis, H., and Zerlett, G. (1986). *The Effects of Whole-Body Vibration.* Berlin, Heidelberg, New York, Tokyo: Springer-Verlag. ISBN 0-387-16584-3.

Fairley, T. E., and Griffin, M. J. (1989). The apparent mass of the seated human body: Vertical vibration. *Journal of Biomechanics, 22(2),* 81–94.

Gemne, G., Lundström, R., and Hansson, J-E. (1993). Disorders induced by work with hand-held vibrating tools. A review of current knowledge for criteria documentation. *Arbete Och Halsa,* 1933:6, Arbets Miljo Institutet, National Institute of Occupational Health, ISBN 91-7045-209-1, ISSN 0346-7821.

Gemne, G., Pyykko, I., Taylor, W., and Pelmear, P. (1987). The Stockholm Workshop scale for the classification of cold-induced Raynaud's phenomenon in the hand-arm vibration syndrome (revision of the Taylor-Pelmear scale). *Scandinavian Journal of Work, Environment and Health, 13(4),* 275–278.

Goodwin, G. M., McCloskey, D. I., and Matthews, P. B. C. (1972). The contribution of muscle afferents to kinaesthesia shown by vibration induced illusions of movement and by the effects of paralysing joint afferents. *Brain, 95,* 705–748.

Griffin, M. J. (1976). Eye motion during whole-body vertical vibration. *Human Factors, 18(6),* 601–606.

Griffin, M. J. (1990). *Handbook of Human Vibration.* London: Academic Press. ISBN: 0-12-303040-4.

Griffin, M. J. (1991). Physical characteristics of stimuli provoking motion sickness, Paper 3. In: Motion Sickness: Significance in Aerospace Operations and Prophylaxis. *AGARD Lecture Series LS—175,* ISBN 92-835-0634-0.

Griffin, M. J., and Hayward, R. A. (1994). Effects of horizontal whole-body vibration on reading. *Applied Ergonomics, 25(3),* 165–169.

Hulshof, C., and Zanten, B. V. van (1987). Whole-body vibration and low-back pain. *International Archives of Occupational and Environmental Health, 59,* 205–220.

International Organization for Standardization (1974). Guide for the evaluation of human exposure to whole-body vibration. *International Standard,* ISO 2631 (E).

International Organization for Standardization (1983). Acoustics—Preferred reference quantities for acoustic levels. *International Standard,* ISO 1683.

International Organization for Standardization (1985). Evaluation of human exposure to whole-body vibration—Part 1: General requirements. *International Standard,* 2631/1.

International Organization for Standardization (1986). Mechanical vibration—Guidelines for the measurement and the assessment of human exposure to hand-transmitted vibration. *International Standard,* ISO 5349.

International Organization for Standardization (1988). Hand-held portable tools—Measurement of vibration at the handle—Part 1: General. *International Standard,* ISO 8662-1.

International Organization for Standardization (1989). Evaluation of human exposure to whole-body vibration—Part 2: Continuous and shock-induced vibration in buildings. *International Standard,* ISO 2631-2.

International Organization for Standardization (1994). Mechanical vibration and shock—Evaluation of human exposure to whole-body vibration—Part 1: General requirements. *Draft International Standard,* ISO/DIS 2631-1.

Lawther, A., and Griffin, M. J. (1987). Prediction of the incidence of motion sickness from the magnitude, frequency, and duration of vertical oscillation. *The Journal of the Acoustical Society of America, 82(3)*, 957–966.

Lawther, A., and Griffin, M. J. (1988a). A survey of the occurrence of motion sickness amongst passengers at sea. *Aviation, Space and Environmental Medicine, 59(5)*, 399–406.

Lawther, A., and Griffin, M. J. (1988b). Motion sickness and motion characteristics of vessels at sea. *Ergonomics, 31(10)*, 1373–1394.

Levison, W. H., and Harrah, C. B. (1977). Biomechanical and performance response of man in six different directional axis vibration environments. *Aerospace Medical Research Laboratory*, Wright-Patterson Air Force Base, OH, AMRL-TR-77-71.

Lewis, C. H., and Griffin, M. J. (1977). The interaction of control gain and vibration with continuous manual control performance. *Journal of Sound and Vibration, 55(4)*, 553–562.

Lewis, C. H., and Griffin, M. J. (1979). The effect of character size on the legibility of numeric displays during vertical whole-body vibration. *Journal of Sound and Vibration, 67(4)*, 562–565.

Loriga, G. (1911). Il lavoro con i martelli pneumatici. The use of pneumatic hammers. *Boll. Ispett. Lavoro*, 2, 35–60.

Martin, B. J., Roll, J. P., and Gauthier, G. M. (1984). Spinal reflex alterations as a function of intensity and frequency of vibration applied to the feet of seated subjects. *Aviation, Space and Environmental Medicine, 55(1)*, 8–12.

McLeod, R. W., and Griffin, M. J. (1989). A review of the effects of translational whole-body vibration on continuous manual control performance. *Journal of Sound and Vibration, 133(1)*, 55–115.

McLeod, R. W., and Griffin, M. J. (1990). Effects of whole-body vibration waveform and display collimation on the performance of a complex manual control task. *Aviation, Space and Environmental Medicine, 61(3)*, 211–219.

Meddick, R. D. L., and Griffin, M. J. (1976). The effect of two-axis vibration on the legibility of reading material. *Ergonomics, 19(1)*, 21–33.

Moseley, M. J., and Griffin, M. J. (1986). A design guide for visual displays and manual tasks in vibration environments. Part I: Visual displays. *Institute of Sound and Vibration Research, Technical Report* No. 133, University of Southampton.

Moseley, M. J., and Griffin, M. J. (1986). Effects of display vibration and whole-body vibration on visual performance. *Ergonomics, 29(8)*, 977–983.

Moseley, M. J., Lewis, C. H., and Griffin, M. J. (1982). Sinusoidal and random whole-body vibration: Comparative effects on visual performance. *Aviation, Space and Environmental Medicine, 53(10)*, 1000–1005.

Nielsen, S. L., and Lassen, N. A. (1977). Measurement of digital blood pressure after local cooling. *Journal of Applied Physiology, Respiratory Environment and Exercise Physiology*, *43*(*5*), 907–910.

Nixon, C. W., and Sommer, H. C. (1963). Influence of selected vibrations upon speech (range of 2 cps–20 cps and random). *Aerospace Medical Research Laboratories, Wright-Patterson Air Force Base*, OH, AMRL-TDR-63-49.

O'Hanlon, J. G., and Griffin, M. J. (1971). Some effects of the vibration of reading material upon visual performance. *Institute of Sound and Vibration Research, Technical Report* No. 49.

Paddan, G. S., and Griffin, M. J. (1988). The transmission of translational seat vibration to the head—1. Vertical seat vibration. *Journal of Biomechanics, 21(3)*, 191–197.

Paddan, G. S., and Griffin, M. J. (1988). The transmission of translational seat vibration to the head—II. Horizontal seat vibration. *Journal of Biomechanics, 21(3)*, 199–206.

Reason, J. T. (1970). Motion sickness: A special case of sensory rearrangement. *Advancement of Science, 26*, June, 386–393.

Reason, J. T. (1978). Motion sickness adaptation: A neural mismatch model. *Journal of the Royal Society of Medicine, 71*, 819–829.

Reason, J. T., and Brand, J. J. (1975). *Motion sickness.* London: Academic Press. ISBN 0-12-584050-0.

Ribot, E., Roll, J. P., and Gauthier, G. M. (1986). Comparative effects of whole-body vibration on sensorimotor performance achieved with a mini-stick and a macro-stick in force and position control modes. *Aviation, Space and Environmental Medicine, 57(8)*, 792–799.

Sherwood, N., and Griffin, M. J. (1990). Effects of whole-body vibration on short-term memory. *Aviation, Space and Environmental Medicine, 61(12)*, 1092–1097.

Sherwood, N., and Griffin, M. J. (1992). Evidence of impaired learning during whole-body vibration. *Journal of Sound and Vibration, 152(2)*, 219–225.

Shoenberger, R. W. (1974). An Investigation of human information processing during whole-body vibration. *Aerospace Medicine, 45(2)*, 143–153.

Shoenberger, R. W., and Wilburn, D. L. (1973). Tracking performance during whole-body vibration with side-mounted and centre-mounted control sticks. *Aerospace Medical Research Laboratory, Wright-Patterson Air Force Base*, AMRL-TR-72-120.

Taylor, W., and Pelmear, P. L., Eds. (1975). *Vibration White Finger in Industry*. New York: Academic Press, ISBN 0 12 684550 6.

Torle, G. (1965). Tracking performance under random acceleration: Effects of control dynamics. *Ergonomics*, *8*(*4*), 481–486.

Wasserman, D., Taylor, W., Behrens, V., Samueloff, S., and olds, D. (1982). Vibration white finger disease in U.S. workers using pneumatic chipping and grinding handtools. I: Epidemiology. *U.S. Department of Health and Human Services, National Institute for Occupational Safety and Health, Technical Report*, DHSS (NIOSH) No. 82-118.

Wells, M. J., and Griffin, M. J. (1984). Benefits of helmet-mounted display image stabilisation under whole-body vibration. *Aviation, Space and Environmental Medicine*, *55*(*1*), 13–18.

Wilson, R. V. (1974). Display collimation under whole-body vibration. *Human Factors*, *16*(2), 186–195.

CHAPTER 26
ILLUMINATION

Peter R. Boyce
Lighting Research Center
Rensselaer Polytechnic Institute
Troy, NY 12180-3590 USA

26.1 INTRODUCTION

Illumination is the act of placing light on an object. By providing illumination, stimuli for the human visual system are produced and the sense of sight is allowed to function. With light, we can see; without light, we cannot see. This chapter is devoted to describing how to measure and produce illumination, the effects of different lighting conditions on visual performance and visual comfort, the photobiogical and psychological effects of illumination, and the risks inherent in exposure to light.

26.2 THE MEASUREMENT OF ILLUMINATION

26.2.1 Photometric Quantities

Light is a part of the electromagnetic spectrum, lying between the wavelength limits 380 nanometers (nm) to 760 nm. What separates this wavelength region from the rest is that radiation in this region is absorbed by the photoreceptors of the human visual system, which initiates the process of seeing.

The most fundamental measure of the electromagnetic radiation emitted by a source is its radiant flux. This is a measure of the rate of flow of energy emitted and is measured in watts (*W*). The most fundamental quantity used to measure light is *luminous flux*. Luminous flux is radiant flux multiplied by the relative spectral sensitivity of the human visual system over the wavelength range of 380 nm to 760 nm.

The relative spectral sensitivity of the human visual system is based on the perception of relative brightness for each wavelength in the visual region. In fact there are two different relative spectral sensitivities, sanctified by international agreement arranged through the Commission Internationale de l'Eclairage (CIE, 1978). There are two relative spectral sensitivities because the human visual system has two classes of photoreceptor; cones, which operate primarily when light is plentiful, and rods, which operate when light is very limited. These two photoreceptor types have different spectral sensitivities, the day photoreceptor, the cones, being characterized by the CIE Standard Photopic Observer and the night photoreceptor, the rods, being characterized by the CIE Standard Scotopic Observer (Figure 26.1).

Luminous flux is used to quantify the total light output of a light source in all directions and is measured in *lumens*. While this is important, for lighting practice it is also important to be able to quantify the luminous flux emitted in a given direction. The measure that quantifies this concept is *luminous intensity*. Luminous intensity is the luminous flux emitted/unit solid angle, in a specified direction. The unit of measurement is the *candela*, which is equivalent to a lumen/steradian. Luminous intensity is used to quantify the distribution of light from a luminaire.

Both luminous flux and luminous intensity have area measures associated with them. The luminous flux falling on unit area of a surface is called the *illuminance*. The unit of measurement of illuminance is the *lumen meter*$^{-2}$ or *lux*. The luminous intensity emitted per unit projected area in a given direction is the *luminance*. The unit of measurement of luminance is the *candela meter*$^{-2}$. The illuminance incident on a surface is the most widely used electric lighting design criterion. The luminance of a surface is a correlate of its brightness. Table 26.1 summarizes these photometric quantities and the relationship between illuminance and luminance.

Unfortunately for consistency, photometry has a long history which has generated a number of different units of measurement for illuminance and luminance. Table 26.2 lists some of the alternative units, together with the multiplying factors necessary to convert

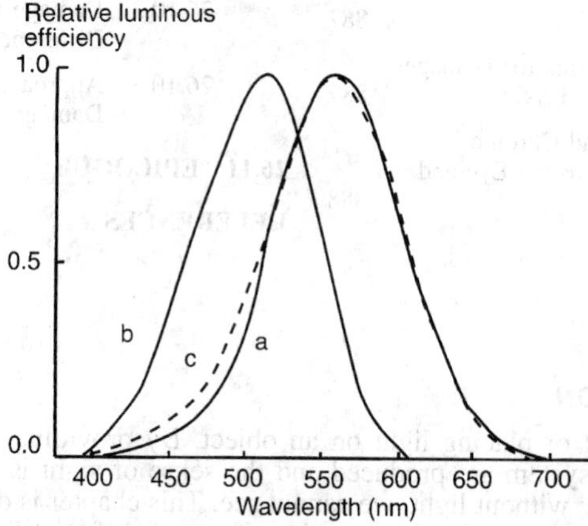

Figure 26.1 The relative luminous efficiency functions for (*a*) the CIE Standard Photopic Observer and (*b*) the CIE Standard Scotopic Observer. The CIE Standard Photopic Observer is based on a 2^0 field of view. Also shown (*c*) is the relative luminous efficiency function for a 10^0 field of view in photopic conditions.

Table 26.1 The Photometric Quantities

Quantity	Definition	Units
Luminoux flux	That quantity of radiant flux which expresses its capacity to produce visual sensation	lumen (lm)
Luminous intensity	The luminous flux emitted in a very narrow cone containing the given direction divided by the solid angle of the cone, i.e., luminous flux/unit solid angle	candela (cd)
Illuminance	The luminous flux/unit area at a point on a surface	lumen meter^{-2} (lm m^{-2})
Luminance	The luminous flux emitted in a given direction divided by the product of the projected area of the source element perpendicular to the direction and the solid angle containing that direction, i.e. luminous flux/unit solid angle/unit area	candela meter^{-2} (cd m^{-2})
Reflectance	The ratio of the luminous flux reflected from a surface to the luminous flux incident on it	
For a matte surface	$\text{Luminance} = \dfrac{\text{illuminance} \times \text{reflectance}}{\pi}$	Luminance in candelas meter^{-2}
Luminance factor	The ratio of the luminance of a reflecting surface, viewed in a given direction to that of a perfect white uniform diffusing surface identically illuminated	
For a non-matte surface for a specific viewing direction and lighting geometry	$\text{Luminance} = \dfrac{\text{illuminance} \times \text{luminance factor}}{\pi}$	Luminance in candelas meter^{-2} Illuminance in lumens meter^{-2}

Table 26.2 Some Common Photometric Units of Measurement For Illuminance and Luminance and the Factors Necessary to Change Them to the SI Units

Quantity	Unit	Dimensions	Multiplying Factor to convent to SI unit
Illuminance	lux	lumen meter^{-2}	1·00
(SI unit = lumen meter^{-2})	meter candle	lumen meter^{-2}	1·00
	phot	lumen centimeter^{-2}	10 000·00
	foot candle	lumen foot^{-2}	10·76
Luminance	nit	candela meter^{-2}	1·00
(SI unit = candela meter^{-2}	stilb	candela centimeter^{-2}	10 000·00
	—	candela inch^{-2}	1 550·00
	—	candela foot^{-2}	10·76
	apostilb°	lumen meter^{-2}	0·32
	blondel°	lumen meter^{-2}	0·32
	lambert°	lumen centimeter^{-2}	3 183·00
	foot-lambert°	lumen foot^{-2}	3·43

° These four items are based on an alternative definition of luminance. This definition is that if the surface can be considered as perfectly matte its luminance in any direction is the product of the illuminance on the surface and its reflectance. Thus the luminance is described in lumens per unit area. This definition is deprecated in the SI system.

from the alternative unit to the System Internationale (SI) units of lumens meter^{-2} for illuminance and candela meter^{-2} for luminance. The SI units will be used throughout this chapter.

Table 26.3 shows some illuminances and luminances typical of commonly occurring situations.

26.2.2 Colorimetric Quantities

The photometric quantities described above do not take into account the wavelength combination, i.e., the color, of the light being measured. There are two approaches to characterizing color, the *color atlas* and the *CIE colorimetry system.*

26.2.2.1 Color Atlases

The color atlas, as its name implies, is a physical, three-dimensional representation of color space. It is three dimensional because colors have three separate subjective attributes: *hue*, *brightness*, and *strength.* Hue tells us whether the color is primarily red,

Table 26.3 Typical Illuminance and Luminance Values

Situation	Illuminance on Horizontal Surface (lm m^{-2})	Typical Surface	Luminance (cd m^{-2})
Clear sky in summer in northern temperate zones	150 000	Grass	2 900
Overcast sky in summer in northern temperate zones	16 000	Grass	300
Textile inspection	1 500	Light grey cloth	140
Office work	500	White paper	120
Heavy engineering	300	Steel	20
Good street lighting	10	Concrete road surface	1·0
Moonlight	0·5	Asphalt road surface	0·01

yellow, green, blue, or purple. Brightness tells us to what extent the color transmits or reflects light. Strength tells us whether the color is a strong or pale.

There are several different color atlas systems used in different parts of the world (Wyszecki and Stiles, 1982). Probably the most widely used atlas is the Munsell Book of Color available from the Munsell Color Company. Figure 26.2 shows the three-dimensional color space of the Munsell atlas. The position of any color is identified by an alphanumeric code made up of three terms, Hue, Value, and Chroma, e.g., a strong red is given the alphanumeric 7.5R/4/12. Hue, Value, and Chroma are related to the three attributes of color; hue, brightness, and strength, respectively. Building materials, such as paints, plastic, and ceramics are commonly classified in terms of a color atlas.

26.2.2.2 The CIE Colorimetric System

Sometimes, it is necessary to quantify the color of a light or a surface before either exist. To meet this need and to provide a more accurate characterization of color, the CIE has developed a system of colorimetry ranging from the complex to the relatively simple (CIE 1971, 1972 and 1978a). The most fundamental characteristic of light is its spectral distribution reaching the eye. It is this spectral distribution which defines a color. Unfortunately, comparisons between spectral distributions are difficult to comprehend. The CIE has developed two three-dimensional color spaces, both based on mathematical manipulations applied to spectral distributions (Robertson 1977, CIE 1978a). These two three-dimensional color spaces, L_{ab} and L_{uv}, are the most comprehensive means of quantifying color; the L_{ab} space being used mainly for object colors, and the L_{uv} space being used mainly for self-luminous colors. If two colors have the same coordinates in one of these color spaces they will appear the same. The distance two colors are apart in color space is related to how easily they can be distinguished.

An earlier CIE color space, the 1964 uniform color space, is used in the calculation of the *CIE General Color Rendering Index*, a single-number index which is applied to light sources to indicate how accurately they render colors relative to some standard (CIE 1974). Specifically, the positions in color space of eight test colors, under a reference light source and under the light source of interest, are calculated. The separation between the two positions of each test color are calculated, the separations for all the test colors are summed and scaled to give a value of 100 when there is no separation for any of the test colors, i.e., for perfect color rendering. It should be noted that this is a very crude system. Different light sources have different reference light sources, and the summation means that light sources which render the test colors differently can have the same Color Rendering Index. Nonetheless, the Color Rendering Index is widely used as a means of classifying the color rendering capabilities of light sources.

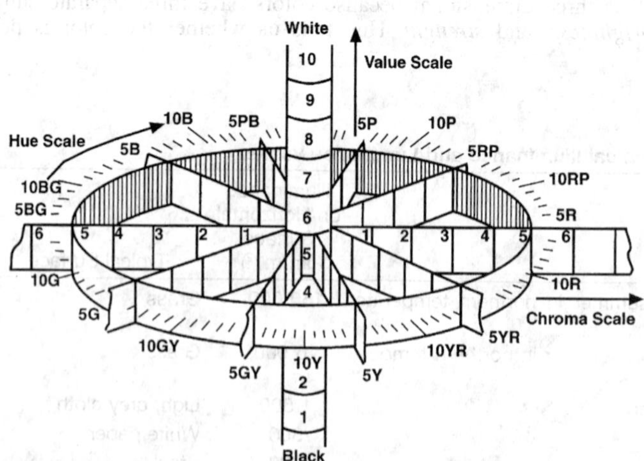

Figure 26.2 The organization of the Munsell color system. The hue letters are B = blue, PB = purple/blue, P = purple, RP = red/purple, R = red, YR = yellow/red, Y = yellow, GY = green/yellow, G = green, BG = blue/green. (From the "The Perception of Light and Colour," C.A. Padgham and J. E. Saunders, London: G. Bell and Sons, 1975).

Three, two-dimensional color surfaces are still widely used to characterize the color appearance of light sources and to define the acceptable color characteristics of light signals (CIE 1994). The most commonly used color surface is the CIE 1931 chromaticity diagram (CIE 1971). Figure 26.3 shows this diagram. Essentially it is a slice through color space at a fixed luminance. The boundary of the chromaticity diagram consists of the colors produced by single wavelengths. The equal energy point in the center of the diagram corresponds to a colorless surface. The further the coordinates of a color are from the equal energy point and the closer they are to the boundary, the greater the strength of the color. Figure 26.3 also shows several areas in which a signal light needs to fall if it is to be perceived as the specified color. The color appearance of light sources is conventionally described by their correlated color temperature. This is the temperature of the full radiator which is closest to the coordinates of the light source on the CIE 1931 chromaticity diagram (Wyszecki and Stiles 1982).

The two other two-dimensional chromaticity diagrams are the CIE 1960 and the CIE 1976 Uniform Chromaticity Scale diagrams. These are linear transformations of the CIE 1931 chromaticity diagram intended to make the surface more perceptually uniform. Whenever chromaticity coordinates are quoted, care should be taken to state the chromaticity diagram being used. A useful summary of these colorimetry systems is given in the 8th edition of the *Lighting Handbook of the Illuminating Engineering Society of North America* (IESNA 1993).

26.2.3 Instrumentation

The instrumentation for measuring photometric and colorimetric quantities can be divided into laboratory and field equipment. Laboratory equipment tends to be large and/or sophisticated and hence, expensive. Field equipment is small and portable. Luminous flux from a light source, luminous intensity distribution of a luminaire, and light source color properties are conventionally measured in the laboratory.

The two most commonly occurring field instruments are the *illuminance meter* and the *luminance meter*. Illuminance meters have three important characteristics; sensitivity, color correction, and cosine correction. Sensitivity refers to the range of illuminances

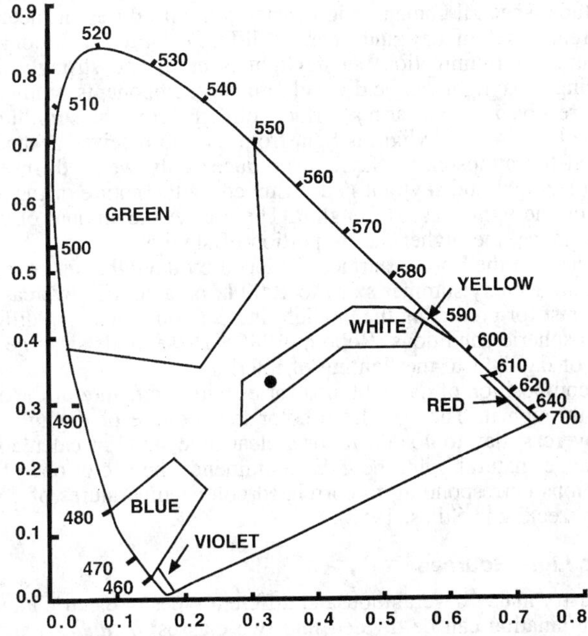

Figure 26.3 The CIE 1931 Chromaticity Diagram. The boundary curve is the spectrum locus with the wavelengths (nm) marked. The filled circle is the equal energy point. The enclosed areas indicate the chromaticity coordinates of light signals that will be identified as the specified colors.

covered, the range desired being dependent on whether the instrument is to be used to measure daylight, interior lighting or nighttime exterior lighting. *Color correction* means that the illuminance meter has a spectral sensitivity matching the CIE Standard Photopic Observer. *Cosine correction* means that the illuminance meter's response to light striking it from directions other than the normal follows a cosine law.

The luminance meter is designed to measure the average luminance over a specified area. The luminance meter has an optical system which focuses an image on a detector. Looking through the optical system allows the operator to identify the area being measured and usually displays the luminance of the area. The important characteristics of a luminance meter are its spectral response, its sensitivity, and the quality of its optical system. Again a good luminance meter has a spectral response matching the CIE standard photopic observer. The sensitivity needed depends on the conditions under which it will be used. The quality of its optical system can be measured by its sensitivity to light from outside the measurement area (CIE 1987).

Recently, imaging photometers have started to appear (Rea and Jeffrey, 1990). These instruments are based around a digitized image captured from a video camera. Such instruments are expensive but do provide a means for measuring the luminance of detailed or rapidly changing scenes.

Procedures for using illuminance or luminance meters in the field and for light measurements in the laboratory are described and referenced in the guidance published by national bodies (CIBSE, 1994; IESNA, 1993). It should be noted that virtually all commercial instrumentation used to measure illuminance and luminance uses the CIE-Standard Photopic Observer as the basis of the instrument's spectral sensitivity, even when the instrument is designed to be used in mesopic and scotopic conditions.

26.3 THE PRODUCTION OF ILLUMINATION

Illumination is naturally produced by the sun and artificially produced by electric light sources. The development and growth in use of electric light sources over the last century has fundamentally changed the pattern of life for everyone.

26.3.1 Daylight, Sunlight, and Skylight

Natural light is light received on Earth from the sun, either directly or after reflection from the moon. The prime characteristic of natural light is its variability. Natural light varies in magnitude, spectral content, and distribution with different meteorological conditions, at different times of day and year, at different latitudes. Moonlight is of little interest as a source of illumination but daylight is used, and strongly desired, for the lighting of buildings. Daylight can be divided into two components, sunlight and skylight. Sunlight is light received at the Earth's surface, directly from the sun. Sunlight produces strong, sharpedged shadows. Skylight is light from the sun received at the Earth's surface after scattering in the atmosphere. Skylight produces only weak, diffuse shadows. The balance between sunlight and skylight is determined by the nature of the atmosphere and the distance which the light passes through it. The greater the amount of water vapor and the longer the distance, the higher the proportion of skylight.

The illuminances on the Earth's surface produced by daylight can cover a large range, from 150,000 lx on a sunny summer's day to 1000 lx on a heavily overcast day in winter. Several models exist for predicting the daylight incident on a plane, at different locations, for different atmospheric conditions (Robbins 1986). These models can be used to predict the contribution of daylight to the lighting of interiors.

The spectral composition of daylight also varies with the nature of the atmosphere and the path length through it. The correlated color temperature of daylight can vary from 4000 K for an overcast day to 40,000 K for a clear blue sky. For calculating the appearance of objects under natural light, the CIE recommends the use of one of three different spectral distributions corresponding to correlated color temperatures of 5503 K, 6504 K and 7504 K (Wyszecki and Stiles, 1982).

26.3.2 Electric Light Sources

The lighting industry makes several thousand different types of electric lamps. Those used for providing illumination can be divided into two classes: *incandescent* lamps and *discharge* lamps. Incandescent lamps produce light by heating a filament. Discharge lamps produce light by an electric discharge in a gas. Incandescent lamps operate directly from mains electricity. Discharge lamps all require control gear between the lamp and the electricity supply, because different electrical conditions are required to initiate the discharge and to sustain it.

Electric light sources can be characterized on several different dimensions. They are:

- Luminous efficacy = the ratio of luminous flux produced to power supplied (lumens/watt). If the lamp needs control gear, the watts supplied should include the power demand of the control gear.
- Correlated color temperature = a measure of the color appearance of the light produced, measured in K.
- CIE General Color Rendering Index = a measure of the ability to render colors accurately
- Lamp life = the number of burning hours until either lamp failure or a stated percentage reduction in light output occurs. Lamp life can vary widely with switching cycle
- Run-up time = the time from switch-on to full light output
- Restrike time = the time delay between the lamp being switched off before it will re-ignite

Table 26.4 summarizes these characteristics for two incandescent lamp types and six discharge lamp types which are widely used, and it gives the most common applications for each lamp type. The values in the Table 26.4 should be treated as indicative only. Details about the characteristics of any specific lamp always should be obtained from the manufacturer.

26.3.3 Control of Light Distribution

Being able to produce light is only part of what is necessary to produce illumination. The other part is to control the distribution of light from the light source. For daylight, this is done by means of window shape, placement, and glass transmittance (Robbins 1986). For electric light sources, it is done by placing the light source in a luminaire. The luminaire provides electrical and mechanical support for the light source and controls the light distribution. The light distribution is controlled by using reflection, refraction, or diffusion, individually or in combination (Bean and Simons 1968). One factor in the choice of which method of light control to adopt in a luminaire is the balance desired between the reduction in the luminance of the light source and the precision required in light distribution. Highly specular reflectors can provide precise control of light distribution, but do little to reduce source luminance. Conversely, diffusers make precise control of light distribution impossible but do reduce source luminance. Refractors are an intermediate case. The light distribution provided by a specific luminaire is quantified by the luminous intensity distribution. All reputable luminaire manufacturers provide luminous intensity distributions for their luminaires. With luminaires, you tend to get what you pay for. Luminaires, well constructed, from quality materials, cost more.

26.3.4 Control of Light Output

The control of daylight admitted through a window is achieved by mechanical structures, such as light shelves, or by adjustable blinds (Littlefair 1990). Whenever the sun, or a very bright sky, is likely to be directly visible through a widow, some form of blind will be required. Blinds can take various forms; horizontal, venetian, vertical, and roller being the most common. Blinds can also be manually operated or motorized, either under manual control or under photocell control. Probably the most important feature to consider when selecting a blind is the extent to which it preserves a view of the outside. Roller blinds which can be drawn down to a position where the sun and/or sky are hidden but the lower part of the widow is still open are an attractive option. Roller blinds made of a mesh material can preserve a view through the whole window while reducing the luminance of the view out. Such blinds are an attractive option where the problem is an over-bright sky but will be of limited value when a direct view of the sun is the problem. The same applies to low transmission glass.

For electric light sources, control of light output is provided by switching or dimming systems. Switching systems can vary from the conventional manual switch to sophisticated daylight control systems which dim lamps near to windows when there is sufficient daylight. Time switches are used to switch off all or parts of a lighting installation at the

Table 26.4 Summary of the Properties of Some Widely Used Electric Light Sources

Source	Luminous Efficacy [lm/W]	Correlated Color Temperature [K]	CIE General Color Rendering Index	Lamp Life [hrs]	Run-up Time [mins]	Restrike Time [mins]	Applications
Incandescent							
Tungsten	8–19	2700	100	750–2000	instant	instant	residential, retail
Tungsten-Halogen	8–20	2900	100	2000–4000	instant	instant	display
Discharge							
Low-Pressure Mercury [fluorescent lamp]	60–110	3000–5000	50–95	9000–20,000	10	instant	commercial
Compact Fluorescent Lamp	50–70	1700–4100	80–85	9000–20,000	10	instant	commerical, retail
Hi-Pressure Mercury [Vapor]	30–60	3200-7000	15–50	16,000–24,000	4	3–10	older industrial agricultural
Hi-Pressure Mercury [Metal Halide]	50–110	3000-6500	65–95	3000–20,000	6	5–20	industrial, commercial, retail
Low-Pressure Sodium	100–180	1800	n/a	16,000–18,000	10–12	0–1	security, road
Hi-Pressure Sodium	60–140	2100–2500	20–70	10,000–24,000	4–6	1	industrial road

end of the working day. Occupancy sensors are used to switch off lighting when there is nobody in the space. Such switching systems can reduce electricity waste, but they will be irritating if they switch lighting off when it is required and they may shorten lamp life if switching occurs frequently. The factors to be considered when selecting a switching system are whether to rely on a manual or an automatic system, and if it is automatic, how to match the switching to the activities in the space. If your interest is primarily in reducing electricity consumption, a good principle is to use automatic switch off and manual switch on. This principle uses human inertia for the benefit of reducing energy consumption. If you wish to rely on voluntary manual switching of lighting, care should be taken to make the lighting being switched visible from the control panel and to label the switches so that the operator knows which lamps are being switched. Labels asking people to switch off the lighting when it is not needed can be effective.

As for dimming systems, these all reduce light output and energy consumption but a different system is required for each lamp type. The factors to consider when evaluating a dimming system are the range over which dimming can be achieved without flicker or the lamp extinguishing, the extent to which the color properties of the lamp change as the light output is reduced, and any effect dimming has on lamp life and energy consumption.

Sophisticated lighting control systems are available for some light sources which allow the user to have a number of preset scenes. These systems use dimming and switching to alter the lighting of a space. They are commonly used in rooms with multiple functions, such as conference rooms.

26.4 FUNCTIONAL CHARACTERISTICS OF THE HUMAN VISUAL SYSTEM

26.4.1 Visual System Structure

llumination is important to humans because it alters the stimuli to the visual system and the operating state of the visual system itself. Therefore, an understanding of the capabilities of the visual system and how they vary with illumination is important to an understanding of the effects of illumination. The visual system is composed of the eye and brain working together. Light entering the eye is brought to focus on the retina by the combined optical power of the air/cornea surface and the lens of the eye. The retina is really an extension of the brain, consisting of two different types of photoreceptors and numerous nerve interconnections. At the photoreceptors, the incident photons of light are absorbed and converted to electrical signals. The nerve interconnections take these signals and carry out some basic image processing. The processed image is transmitted up the optic nerve of each eye to the optic chiasma, where nerve fibers from the two eyes are combined and transmitted to the left and right parts of the visual cortex. It is in the visual cortex that the signals from the eye are interpreted in terms of past experience (Figure 26.4).

Many capabilities of the visual system can be understood from the organization of the retina. The two types of photoreceptors, called *rods* and *cones* from their anatomical appearance, have different wavelength sensitivities, different absolute sensitivities to light, and are distributed differently across the retina.

Rods are the more sensitive of the two and effectively provide a night retina. Cones are less sensitive to light and operate during daytime. In fact there are three types of cones, each with a different spectral sensitivity. These cones are commonly called *long*, *middle*, and *short* wavelength cones, from their regions of maximum spectral sensitivity. These three cone types combine together to give the perception of color. Figure 26.5 shows the distribution of rods and cones across the retina. Cones are concentrated in a small central area of the retina called the *fovea* which lies where the visual axis of the eye meets the retina, although there are cones disturbed evenly across the rest of the retina. Rods are absent from the fovea, reaching their maximum concentration about 20^0 from the fovea. This variation in concentration of rods and cones with deviation for the fovea is amplified by the number of photoreceptors connected to each optic nerve fiber. In the fovea, the ratio of photoreceptors to optic nerve fibers is close to 1 but increases rapidly as the deviation from the fovea increases. The net effect of this structure is to provide different functions for the fovea and the periphery. The fovea is the part of the retina which provides fine discrimination of detail. The rest of the retina is primarily devoted to detecting changes in the visual environment which require the attention of the fovea.

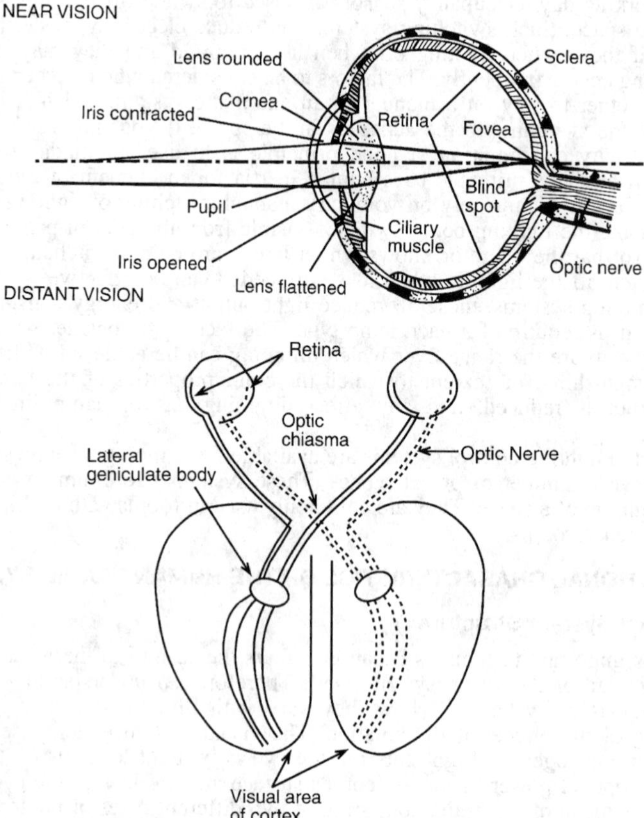

Figure 26.4 A section through the eye adjusted for near and distant vision and a schematic diagram of the binocular nerve pathways of the visual system.

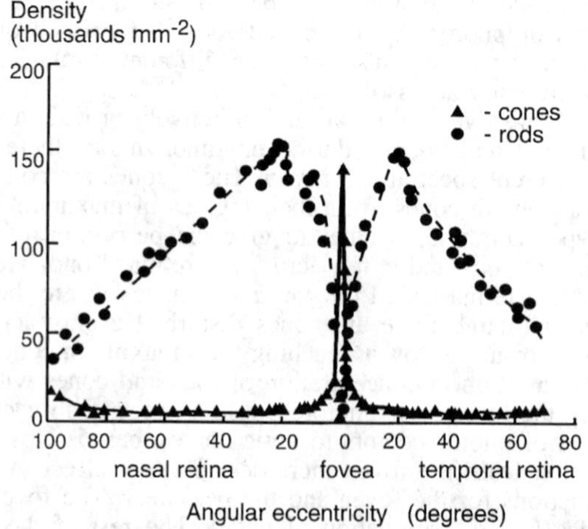

Figure 26.5 The density of rod and cone photoreceptors across the retina, on a horizontal meridian. (After Osterburg, G., *Acta Ophthalmologica, 13*, Supplement 6, 1935.)

26.4.2 Wavelength Sensitivity

The rod and cone photoreceptors have different absolute spectral sensitivities (Figure 26.6), the peak sensitivity of the rods being much greater than that of the cones. The peak sensitivity of the cones occurs at 555 nm, while that of the rods occurs at 507 nm. These spectral sensitivities form the basis of the CIE standard observers and hence, the photometric quantities discussed in Section 26.2.1. By adjusting the spectral emission of a light source to lie within the most sensitive part of the spectral response of the visual system, lamp manufacturers are able to vary the luminous efficacy of their light sources, i.e., to change the number of lumens emitted for each watt of power applied.

26.4.3 Adaptation

The visual system can operate over a range of about 12 log units of luminance, from a luminance of 10^{-6} cd m^{-2} to 10^{6} cd m^{-2}, from starlight to bright sunlight. But it cannot cover this range simultaneously. At any instant, the visual system can cover a range of 2 or 3 log units of luminance. Luminances above this limited range are seen as glaringly bright, those below as undifferentiated black. The capabilities of the visual system depend on the where in the complete range of luminances it is adapted. Three different functional ranges of luminance are conventionally identified: the *photopic*, *mesopic*, and *scotopic*. Table 26.5 summarizes the visual system capabilities in each of these functional ranges.

The visual system continuously adjusts its state of adaptation through three mechanisms: *neural*, *mechanical*, and *photochemical*. These three mechanisms differ in their speed and range of adjustment. The neural mechanism which is based in the retina, operates in milliseconds and covers a range of 2 to 3 log units in luminance. The mechanical mechanism involves the expansion and contraction of the pupil. Changes in pupil size take seconds but cover less than 1 log unit in luminance. The photochemical mechanism covers the whole range of luminance but is slow, the changes taking minutes. The exact time will depend on the starting and finishing luminance. If both starting and finishing luminances for the adaptation are greater than 3 cd m^{-2}, only cones are involved. Since the time constant for cones is of the order of 2 to 3 min adaptation takes only a few minutes. When the starting luminance is in the operating range of the cones and the finishing luminance is within the operating range of the rods, a two stage adaptation process occurs, involving both cones and rods. As rods have a time constant around 7 to

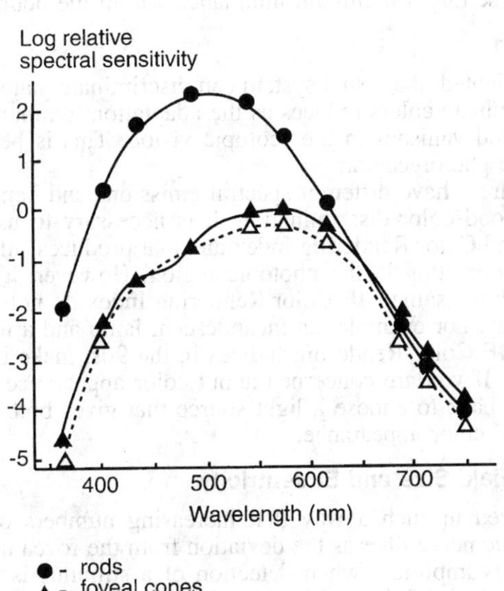

Figure 26.6 The log relative spectral sensitivity of rod and cone photoreceptors plotted against wavelength. (After Wald, G., *Science, 101*, 653, 1945.)

Table 26.5 The Functional Ranges of Visual System Capabilities

Name	Dominance Range (cd m^{-2})	Photoreceptor Active	Wavelength Range (nm)	Capabilities
Photopic	> 3	Cones	380–760	Color vision Good detail discrimination
Scotopic	< 0.001	Rods	380–760	No color vision Poor detail discrimination
Mesopic	> 0.001 and < 3	Cones and rods	380–760	Diminished color vision, reduced detail discrimination and a shift in spectral sensitivity as adaptation luminance moves from photopic to scotpic

8 min, the adaptation time is much longer. Complete adaptation from a high photopic luminance to darkness can take up to an hour.

Interior lighting is almost always sufficient for the visual system to be operating in the photopic region. Exterior lighting on roads and in urban areas is usually sufficient to keep the visual system operating in the photopic or mesopic regions. It is in very rural areas, at sea, or underground, where there is little or no exterior lighting, that the visual system reaches scotopic adaptation. The speed of adaptation is important where a large and sudden change in the luminance occurs. Examples of situations where this happens are the entrance to road tunnels during daytime (Bourdy et al., 1987) and the onset of emergency lighting during a power failure (Boyce 1985). These problems are overcome either by installing a gradual reduction in luminance which allows more time for adaptation to occur or by setting a minimum luminance within the neural adaptation range.

26.4.4 Color Vision

When photopically adapted, the visual system can discriminate many thousands of colors. This ability to discriminate colors reduces as the adaptation luminance decreases through the mesopic region and vanishes in the scotopic vision. This is because color vision is mediated by the cone photoreceptors.

Different light sources have different spectral emissions and hence, render colors differently. To ensure good color discrimination, it is necessary to use a light source that has a high CIE General Color Rendering Index and that produces sufficient light to ensure the visual system is operating in the photopic region. However, it is important to note that light sources with the same CIE Color Rendering Index do not necessarily render all colors in the same way. For example, an incandescent lamp and a fluorescent lamp, both of which can have CIE Color Rendering Indices in the 90s, make blue and green colors appear very different. If you are concerned about color appearance as well as color discrimination, you will have to choose a light source that gives both good color discrimination and the desired color appearance.

26.4.5 Receptive Field Size and Eccentricity

The retina is organized in such a way that increasing numbers of photoreceptors are connected to each optic nerve fiber as the deviation from the fovea increases. This feature of the visual system is important when detection of a stimulus is necessary and it can occur anywhere in the visual field. The visual system will normally operate by first detecting the stimulus off-axis, i.e., in the peripheral visual field, and then turning the eye so that the stimulus is brought onto the fovea for detailed examination. In order to identify a stimulus off-axis, the stimulus should be clearly different from its background, in luminance or color, and should change in space or time, i.e., it should either move or flicker. A flickering light is commonly used to draw drivers' attention to important signs placed beside or above the road.

26.4.6 Meaningful Stimulus Parameters

Any stimulus to the visual system can be described by five parameters, its visual size, luminance contrast, chromatic contrast, retinal image quality, and retinal illumination. These parameters are important in determining the extent to which the visual system can detect and identify the stimulus.

26.4.6.1 Visual Size

The visual size of a stimulus describes how big the stimulus is. The larger a stimulus, the easier it is to detect.

There are several different ways to express the size of a stimulus presented to the visual system, but all of them are angular measures. The visual size of a stimulus for detection is best given by the solid angle the stimulus subtends at the eye. The solid angle is given by the quotient of the areal extent of the object and the square of the distance from which it is viewed. The larger the solid angle, the easier the stimulus is to detect.

The visual size for resolution is usually given as the angle the critical dimension of the stimulus subtends at the eye. What the critical dimension is depends on the stimulus. For two points, the critical dimension is the distance between the two points. For two lines it is the separation between the two lines. For a Landolt ring, it is the gap size. The larger the visual size of detail in a stimulus, the easier it is to resolve the detail.

For complex stimuli, the measure used to express its dimensions is the *spatial frequency distribution.* Spatial frequency is based on the angular subtense of a critical detail and is measured in cycles per degree, the smaller the angular subtense, the greater the number of cycles/degree. Complex stimuli have many spatial frequencies and hence, a spatial frequency distribution. The match between the spatial frequency distribution of the stimulus and the contrast sensitivity function of the visual system (see Section 26.5.2) determines if the stimulus will be seen and what detail will be resolved.

Lighting can change the visual size of three-dimensional stimuli by casting shadows which extend or diminish the visual size of the stimulus.

26.4.6.2 Luminance Contrast

The luminance contrast of a stimulus quantifies its luminance relative to its background. The higher the luminance contrast, the easier it is to detect the stimulus. There are two different forms of luminance contrast. For stimuli that are seen against a uniform background, luminance contrast is conventionally defined as

$$C = (L_t - L_b)/L_b$$

where

C = luminance contrast
L_b = Luminance of the background
L_t = Luminance of the detail

This formula gives luminance contrasts that range from 0 to 1 for stimuli that have details darker than the background and from 0 to infinity for stimuli that have details bright than the background. It is widely used for the former, e.g., printed text.

For stimuli which have a periodic pattern, e.g., a grating, the luminance contrast or modulation is given by

$$C = (L_{max} - L_{min})/(L_{max} + L_{min})$$

where

C = Luminance contrast
L_{max} = maximum luminance
L_{min} = minimum luminance

This formula gives luminance contrast that ranges from 0 to 1.

Lighting can change the luminance contrast of a stimulus by producing disability glare in the eye or veiling reflections from the stimulus or by changing the incident spectral radiation when colored stimuli are involved.

26.4.6.3 Chromatic Contrast

Luminance contrast uses the total amount of light emitted from a stimulus and ignores the wavelengths of the emitted light. It is the wavelengths emitted from the stimulus that determine its color. It is possible to have a stimulus with zero luminance contrast which can still be detected because it differs from its background in color, i.e., it has chromatic contrast. There is no widely accepted measure of chromatic contrast, although various suggestions have been made (Tansley and Boynton, 1978). Fortunately, chromatic contrast only becomes important for detection when luminance contrast has reached a low level.

Lighting can alter chromatic contrast by using light sources with different spectral emission characteristics.

26.4.6.4 Retinal Image Quality

As with all image processing systems, the visual system works best when it is presented with a clear, sharp image. The sharpness of the stimulus can be quantified by the spatial frequency distribution of the stimulus—a sharp image will have high spatial frequency components present, a blurred image will not.

The sharpness of the retinal image is determined by the stimulus itself, the extent to which medium through which it is transmitted scatters light, and the ability of the visual system to focus the image on the retina. Lighting can do little to alter any of these factors, although it has been shown that light sources which are rich in the short wavelengths produce smaller pupil sizes and these tend to improve visual acuity for low contrast targets. The suggested explanation is that the smaller pupil sizes produce greater depth of field and hence, better retinal image quality (Berman et al., 1993)

26.4.6.5 Retinal Illumination

The retinal illumination determines the state of adaptation of the visual system and therefore alters its capabilities. The retinal illumination is determined by the luminances in the visual field, modified by the pupil size. Retinal illumination is measured in *trolands*, a quantity formed from the product of the luminance of the visual field and the pupil size (Wyszecki and Stiles, 1982). Illuminances and surface reflectances determine the luminance of the visual field. Luminances and light spectrum determine pupil size.

26.5 EFFECTS ON THRESHOLD VISUAL PERFORMANCE

Qualitatively, threshold visual performance is the performance of a visual task close to the limits of what is possible. Quantitatively, it is the performance of a task at a level such that it can be correctly carried out on 50% of the occasions in which it is undertaken. Threshold visual performance is affected by many different variables. For example, visual acuity is affected by the form of the target used, the luminance contrast of the target, the duration for which it is presented, where in the visual field it appears, and the luminance of the surround relative to the luminance of the immediate background. In this discussion of threshold visual performance, attention will be limited to the effects of variables that are controlled by the lighting system, i.e., the adaptation luminance and the spectral content of the light. Information on the influence of other variables can be obtained from Boff and Lincoln (1988). In the data presented, it will be assumed that the observer is fully adapted to the prevailing luminance, that the image of the target is on the fovea, that the target is presented for an unlimited time, and that the observer is correctly refracted. Again, the influence of departures from these assumptions can be estimated from the data given by Boff and Lincoln (1988).

26.5.1 Visual Acuity

Visual acuity is the limit in the ability to resolve detail. Visual acuity has been frequently measured using gratings or Landolt C's. Visual acuity can be quantified as the angle subtended at the eye by the size of detail which can be correctly detected on 50% of the occasions it is presented.

No matter what target is used, visual acuity improves, i.e., the size of detail which can be resolved decreases, as adaptation luminance increases. Figure 26.7 shows that as adaptation luminance increases from scotopic to photopic conditions, the visual acuity increases, asymptotically approaching a maximum at high luminances. The adaptation

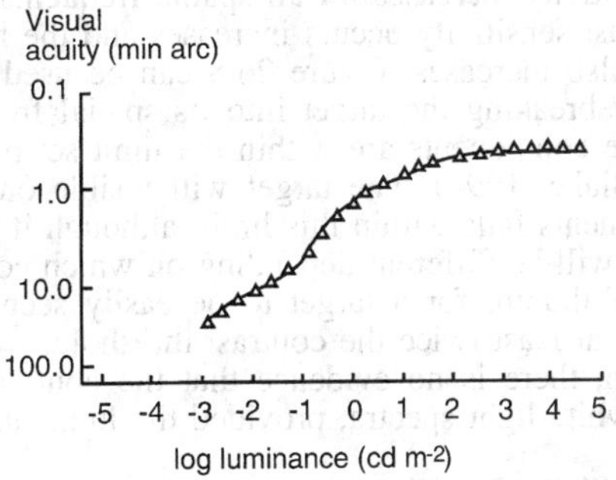

Figure 26.7 The effect of adaptation luminance on the gap size of a Landolt C target which can just be resolved. (After Shlaer, S., *Journal of General Physiology*, *21*, 165, 1937.)

luminance produced by a lighting installation will depend on the illuminances produced on different surfaces and the reflectance of those surfaces. Table 26.3 gives some luminances typically found in interior and exterior lighting installations. Given a value for the adaptation luminance, Figure 26.7 can be used to determine if detail of a given size can be resolved. A useful rule of thumb is that the detail needs to be four times bigger than the visual acuity limit if it is to be resolved sufficiently quickly to avoid affecting visual performance (Bailey, Clear, and Berman, 1993).

As for light spectrum, provided the lamp produces white light rather than an emission in a narrow spectral region, the effect on visual acuity is very small, certainly much less than the effect of adaptation luminance.

26.5.2 Contrast Sensitivity Function

Contrast sensitivity is the reciprocal of the luminance contrast which can be detected on 50% of the occasions on which it is presented. Contrast sensitivity is usually measured using a sinusoidal grating target. The contrast sensitivity function is contrast sensitivity plotted against the spatial frequency of the sinusoidal target.

Figure 26.8 shows the effect of adaptation luminance on the contrast sensitivity function. It shows that as the adaptation luminance increases from scotopic to photopic con-

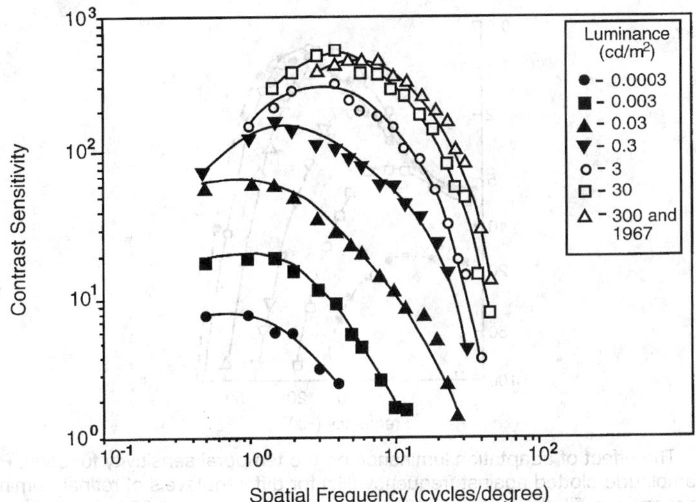

Figure 26.8 The effect of adaptation luminance on contrast sensitivity function. Contrast sensitivity plotted against spatial frequency (cycles/degree) for different adaptation luminances. (After Boff and Lincoln, 1988.)

ditions, the contrast sensitivity increases for all spatial frequencies; the spatial frequency at which the peak contrast sensitivity occurs increases and the highest spatial frequency which can be detected also increases. Figure 26.8 can be used to determine if a given target will be visible by breaking the target into its spatial frequency components and determining if any of the components are within the limit set by the contrast sensitivity function (Sekuler and Blake, 1994). The target will visible only if at least one of its spatial frequency components falls within this limit, although it should be noted that the appearance of the target will be different depending on which component or components are visible. As a rule of thumb, for a target to be easily seen, it is necessary for the luminance contrast to be at least twice the contrast threshold.

As for light spectrum, there is no evidence that the contrast sensitivity function is influenced by different white light spectra, provided the luminances are the same.

26.5.3 Temporal Sensitivity Function

The temporal sensitivity function shows percentage modulation amplitude plotted against the frequency of the modulation. Figure 26.9 shows the effect of adaptation luminance on the temporal sensitivity function. It shows that as the adaptation luminance increases from mesopic to photopic conditions, the temporal sensitivity increases for all frequencies; the frequency at which the peak temporal sensitivity occurs increases and the highest frequency which can be detected also increases. Figure 26.9 can be used to determine if a given temporal variation will be visible by breaking the waveform representing the light fluctuation into its frequency components and determining if any of the components are within the limit set by the temporal sensitivity function. The fluctuation will only be visible if at least one of its frequency components falls within this limit.

Temporal fluctuation in luminous flux, or flicker, is undesirable in lighting installations. To eliminate flicker, it is necessary to increase the frequency and/or decrease the percentage modulation sufficiently to take their combination outside the limits set by the temporal sensitivity function. In practice this is easily done. Incandescent lamps have sufficient thermal inertia to ensure that even though the frequency of the fluctuation is only twice the supply frequency (120-Hz for a 60-Hz electrical supply), the percentage modulation is small so there is little chance of seeing flicker from such a lamp. Discharge lamps, such as the fluorescent lamp, do not have thermal inertia, so their percentage modulation can be high. To ensure that fluorescent lamps do not produce visible flicker it is best to use an electronic ballast to control the lamp. Electronic ballasts typically

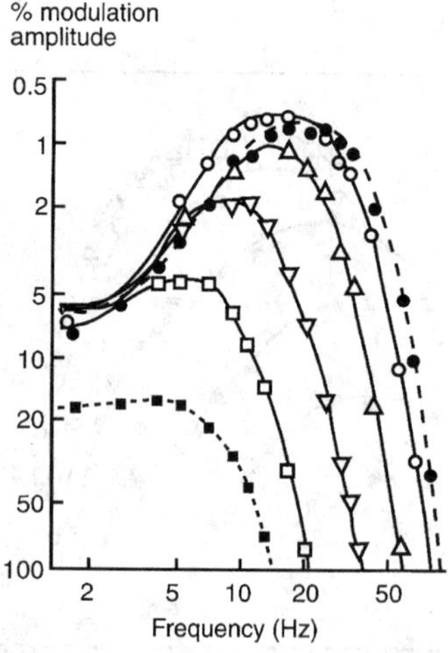

Figure 26.9 The effect of adaptation luminance on the temporal sensitivity function. Percentage modulation amplitude plotted against frequency (Hz) for different levels of retinal illumination. The retinal illuminations are; filled square = 0.06 trolands; open square = 0.65 trolands; open, inverted triangle = 7.1 trolands; open, upright triangle = 77 trolands; open circle = 850 trolands; filled circle = 9300 trolands (After Kelly, D. H., *Journal of the Optical Society of America*, *51*, 422, 1961.)

operate at frequencies in the tens of kilohertz, with small percentage modulations and, consequently, are very unlikely to produce visible flicker.

26.5.4 Color Discrimination

The ability to discriminate between two colors of the same luminance depends on the difference in spectral power distribution of the light received at the eye. Figure 26.10 shows the MacAdam ellipses, the area around a number of chromaticities, each magnified ten times, within which no discrimination of color can be made, even under side-by-side comparison conditions (Wyszecki and Stiles, 1982).

The effect of illuminance on the ability to discriminate between colors is limited in the photopic region, an illuminance of 300 lx being sufficient for good color judgment work (Cornu and Harlay, 1969). As the visual system enters the mesopic region, the ability to discriminate colors deteriorates and ultimately fails as the scotopic region is reached.

The effect of light spectrum is much more important. The position of a color on the CIE 1931 Chromaticity Diagram is determined by the spectrum of the light and, if it is reflected from or transmitted through a surface, the spectral reflectance or transmittance of that surface. Therefore, by changing the light spectrum emitted by the lamp, it is

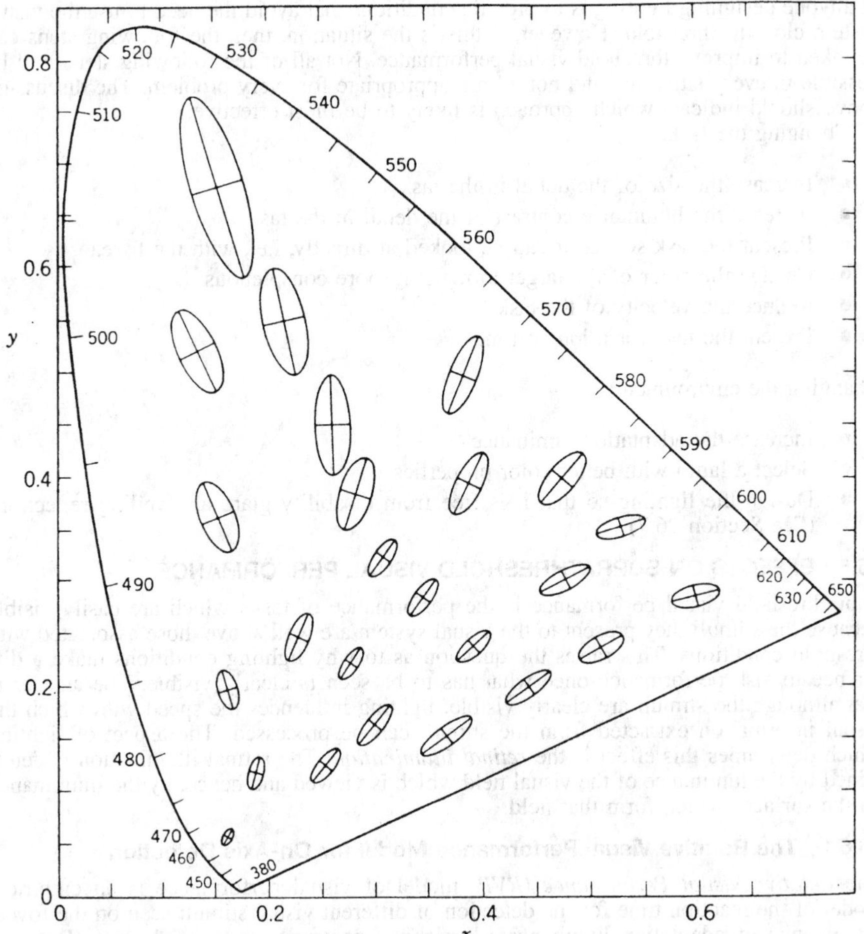

Figure 26.10 The MacAdam ellipses plotted on the CIE 1931 chromaticity diagram. The boundary of each ellipse represents ten times the standard deviation of color matches made for the indicated chromaticity (After MacAdam, D. L., *Journal of the Optical Society of America*, *32*, 247, 1942.)

possible to make colors easily discriminable or difficult to discriminate. The careful choice of light source is important wherever good color discrimination is important.

26.5.5 Interactions

The fact that there are many other variables besides adaptation luminance and light spectrum which influence threshold visual performance has been mentioned earlier. It is now necessary to introduce another complication, namely, interaction between the various components of visual system performance. As an example, consider the effect of luminance contrast on visual acuity. Visual acuity is conventionally measured using targets with a high luminance contrast. However, as the luminance contrast of the target is decreased, visual acuity also worsens. Similarly, the temporal sensitivity function as presented applies to a uniform luminance field. If the field has a pattern and hence, a distribution of spatial frequencies, the temporal sensitivity function may be changed (Koenderink and Van Doorn, 1979).

Put crudely, what this means is that as visual performance gets closer to threshold, almost everything about the stimulus presented to the visual system becomes important. Further details on some of the interactions that occur are given in Boff and Lincoln (1988).

26.5.6 Approaches to Improving Threshold Visual Performance

Working close to threshold is not easy. In fact, it can be argued that the main function of anyone designing lighting is to provide conditions that avoid the need to use the visual system close to threshold. However, if this is the situation, then the following steps can be taken to improve threshold visual performance. Not all of the following steps will be possible in every situation, and not all are appropriate for every problem. The discussion above should indicate which approach is likely to be most effective.

Changing the task:

- Increase the size of the detail in the task
- Increase the luminance contrast of the detail in the task
- Present the task so that it can be looked at directly, i.e., with the fovea
- Change the color of the target to make it more conspicuous
- Reduce the velocity of the task
- Present the task for a longer time

Changing the environment:

- Increase the adaptation luminance
- Select a lamp with better color properties
- Design the lighting so that it is free from disability glare and veiling reflections (See Section 26.7)

26.6 EFFECTS ON SUPRATHRESHOLD VISUAL PERFORMANCE

Suprathreshold visual performance is the performance of tasks which are easily visible because the stimuli they present to the visual system are well above those associated with threshold conditions. This raises the question as to why lighting conditions make a difference to task performance once what has to be seen is clearly visible. The answer is that although the stimuli are clearly visible, lighting influences the speed with which the visual information extracted from the stimuli can be processed. The aspect of lighting which determines this effect is the *retinal illumination.* The retinal illumination is determined by the luminance of the visual field which is viewed and hence, by the illuminance on the surfaces which form that field.

26.6.1 The Relative Visual Performance Model for On-Axis Detection

The *Relative Visual Performance* (*RVP*) model of visual performance is an empirical model of the reaction time for the detection of different visual stimuli seen on the fovea, for a range of adaptation luminances, luminance contrasts, and visual sizes (Rea and Ouellette, 1988, 1991). Figure 26.11 shows the form of the RVP model for four different visual size tasks, each surface being for a range of contrasts, and retinal illuminances. The overall shape of the relative visual performance surface has been described as a plateau and an escarpment (Boyce and Rea, 1987). In essence what it shows is that the

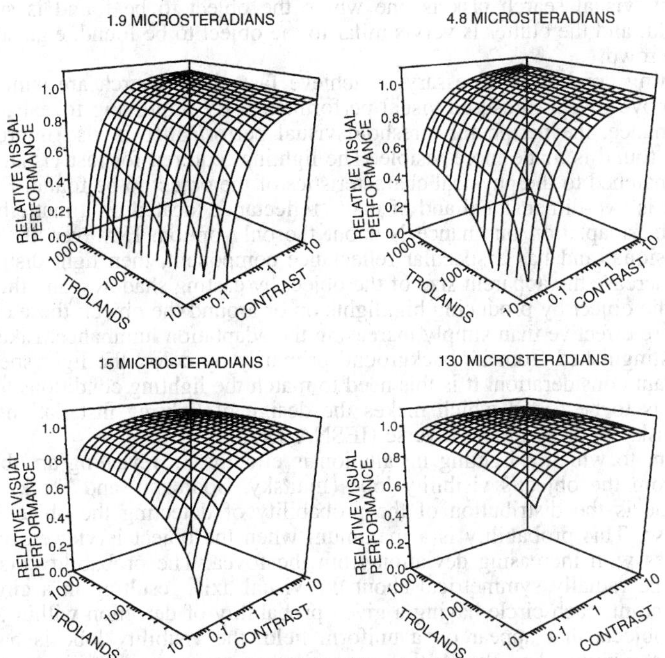

Figure 26.11 Relative visual performance surfaces plotted against retinal illumination, in trolands, and luminance contrast, for four stimuli subtending four different solid angles, measured in microsteradians. (After Rea, M. S. and Ouellette, M. J., *Lighting Research and Technology, 23,* 135, 1991.)

visual system is capable of a high level of visual performance over a wide range of visual sizes, luminance contrasts, and retinal illuminations (the plateau), but at some point, visual size, luminance contrast, or retinal illumination will become insufficient and visual performance will rapidly collapse (the escarpment) toward threshold. The existence of a plateau of visual performance, or rather a near plateau because there is really a slight improvement in visual performance across the plateau, implies that for a wide range of visual conditions, visual performance changes very little with changes in the lighting conditions. To put it bluntly, what this means is that for many visual tasks, visual performance is insensitive to lighting conditions, the visual system being flexible enough to cope equally well with a wide variety of visual stimuli.

The Relative Visual Performance model of suprathreshold visual performance provides a quantitative means of predicting the effects of changing either task size or contrast or the adaptation luminance for on-axis, suprathreshold visual performance. However, it should be noted that it has only been validated for luminances in the photopic range and does not take into consideration the effect of other variables, such as blur.

26.6.2 Visual Search

The RVP model is applicable to tasks that are seen on-axis. However, there are a whole class of tasks in which the object to be detected can appear anywhere in the visual field. These tasks involve visual search. Visual search is typically undertaken through a series of eye fixations, the fixation pattern being guided either by expectations about where the object to be seen is most likely to appear or by what part of the visual scene is most important. Typically, the object to be detected is first detected in the periphery and then confirmed or resolved by a foveal fixation. The speed with which a visual search task is completed depends on the visibility of the object to be found, the presence of other objects in the search area, and the extent to which the object to be found is different from the other objects. The simplest visual search task is one in which the object to be found appears somewhere in an otherwise empty field, e.g., paint defects on a car body. The most difficult visual search task is one where the object to be found is situated in a cluttered field, and the clutter is very similar to the object to be found, e.g., searching for a face in a crowd.

The lighting conditions necessary to achieve fast visual search are similar to those used to improve foveal threshold visual performance. By improving foveal threshold visual performance, the peripheral threshold visual performance is also improved so the object to be found is made more visible. The lighting required for fast visual search will have to be matched to the physical characteristics of the object to be found. For example, if the object is two-dimensional and of matte reflectance located on a matte background, increasing the adaptation luminance is about the only option. However, if the object is three-dimensional and has a specular reflectance component, then light distribution can be used to increase the apparent size of the object by casting shadows, and the luminance contrast of the object by producing highlights on or around the object; these changes will be much more effective than simply increasing the adaptation luminance. Likewise, if the object is distinguished from its background primarily by color, the light spectrum used is an important consideration. It is this need to match the lighting conditions to the nature of the objects to be found which makes the design of lighting installations for visual inspection tasks so difficult and diverse (IESNA 1993).

The extent to which a lighting installation is effective in revealing an object can be estimated from the object's visibility lobe (Inditsky, Bodmann, and Fleck, 1982). The visibility lobe is the distribution of the probability of detecting the object within one fixation pause. This probability is a maximum when the object is viewed on the fovea and decreases with increasing deviation from the fovea. The probability distribution is assumed to be radially symmetrical about the visual axis, resulting in a circles around the fixation point, each circle having a given probability of detection within one fixation pause. For objects that appear on a uniform field, the visibility lobe is based on the detection of the object. For objects that appear among other similar objects, the visibility lobe is based on the discriminability of the object from the others surrounding it. Visual search will be fastest for objects that have the largest visibility lobe.

26.6.3 Visual Performance, Task Performance, and Productivity

Figure 26.12 shows the relationships between the stimuli to the visual system and their impact on visual performance, task performance, and productivity. The stimuli to the visual system, including the retinal illumination, determine the operation of the visual system and hence the level of visual performance achieved. This visual performance then contributes to task performance. It is important to point out that visual performance and task performance are not necessarily the same. Task performance is the performance of the complete task. Visual performance is the performance of the visual component of the task. Task performance is what is needed in order to measure productivity and to establish cost/benefit ratios comparing the costs of providing a lighting installation with the resulting benefits in terms of better task performance. Visual performance is the only thing that changing the lighting conditions can affect directly.

Most apparently visual tasks have three components: *visual*, *cognitive*, and *motor*. The visual component refers to the process of extracting information relevant to the performance of the task using the sense of sight. The cognitive component is the process by which sensory stimuli are interpreted and the appropriate action determined. The motor component is the process by which the stimuli are manipulated to extract information and/or the actions decided on are carried out. Every task is unique in its balance between visual, cognitive, and motor components and hence, in the effect lighting conditions have on task performance. It is this uniqueness which makes it impossible to generalize from the effect of lighting on the performance of one task to the effect of lighting on the performance of another. The RVP model for on-axis tasks and the visual search models discussed above can be used to quantify the effects of lighting conditions on visual performance, but there is no general model to translate those results to task performance.

26.6.4 Approaches to Improving Suprathreshold Visual Performance

The main purpose of lighting installations is to ensure that people can perform the work they need to do quickly, easily, comfortably, and safely. To achieve this desirable aim, it is necessary to provide lighting that ensures people are working on the plateau of visual performance and not on the escarpment. The RVP model of visual performance provides a simple means of checking whether lighting is adequate for the visual performance of many on-axis tasks. The visibility lobe provides an approach to quantifying the effect of

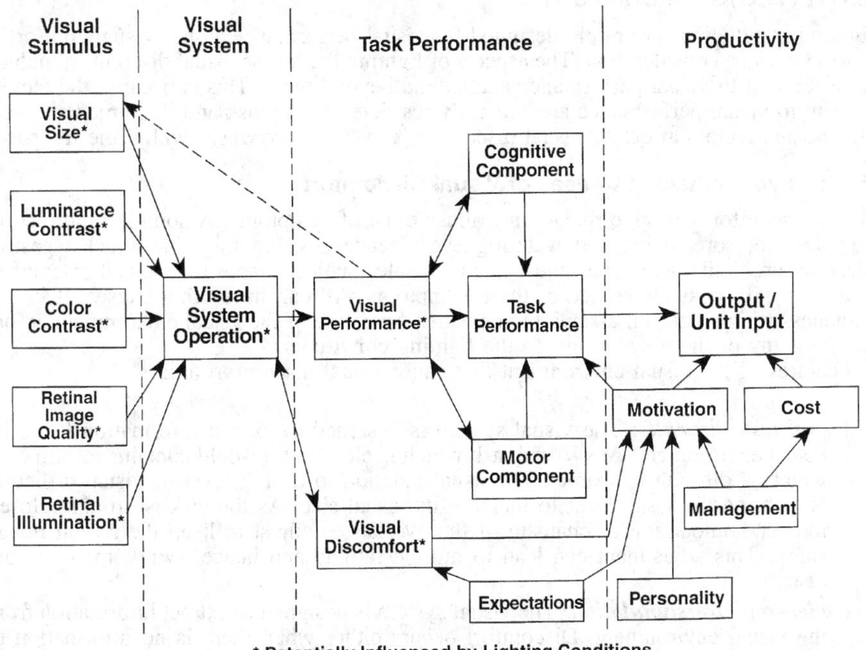

Figure 26.12 A schematic of the relationships between the stimuli to the visual system and their impact on visual performance, task performance, and productivity. The arrows indicate the direction of the effects. The dotted arrow between visual performance and visual size indicates that if visual performance is poor, a common response is to move closer to the stimulus to increase its visual size.

lighting conditions on visual search tasks. Alternatively, most countries have well-established recommendations for the illuminances to be provided for working interiors (CIBSE, 1994; IESNA, 1993). Most of these recommendations easily exceed what would be deduced as necessary from a consideration of visual performance alone.

Although the discussion above has been focused on the lighting conditions, it is important to recognize that improving suprathreshold visual performance can be achieved by changing the characteristics of the task as well as the lighting. The following list is divided into two parts, task changes and lighting changes. Not all of the following suggestions are possible in every situation, and not all are appropriate for every problem.

Changing the task:

- Increase the size of the detail in the task
- Increase the luminance contrast of the detail in the task
- For off-axis tasks in a cluttered field, make the object to be detected clearly differ from the surrounding objects on as many different dimensions as possible, e.g., size, contrast, color, and shape
- Ensure the object presents a clear, sharp image on the retina

Changing the environment:

- Increase the adaptation luminance
- Select a lamp with better color properties
- Design the lighting so that it is free from disability glare and veiling reflections (see Section 26.7)
- Design the lighting to increase the apparent size or luminance contrast of the object

26.7 EFFECTS ON COMFORT

Lighting installations are rarely designed for visual performance alone. Visual comfort is almost always a consideration. The aspects of lighting that cause visual discomfort include those relevant to visual performance and extend beyond them. This is because the factors relevant to visual performance are generally restricted to the task and its immediate area, whereas the factors affecting visual discomfort can occur anywhere within the lit space.

26.7.1 Symptoms and Causes of Visual Discomfort

Visual discomfort can give rise to an extensive list of symptoms. Among the more common are red, sore, itchy, and watering eyes; headaches and migraine attacks; gastrointestinal problems; and aches and pains associated with poor posture. Visual discomfort is not the only possible source of these symptoms. All can have other causes. It is this vagueness which makes it essential to consider the nature of the visual environment before ascribing any of these symptoms to the lighting conditions.

Features of the visual environment that cause visual discomfort are:

Visual task difficulty: The visual system is designed to extract information from the visual environment. Any visual task which is close to threshold contains information which is difficult to extract. The usual reaction to a high level of visual difficulty is to bring the task closer to increase its visual size. As the task is brought closer, the accommodation mechanism of the eye has to adjust to keep the retinal image sharp. This adjustment can lead to muscle fatigue and hence, symptoms of visual discomfort.

Under- and overstimulation: The visual system is designed to extract information from the visual environment. Discomfort occurs either when there is no information to be extracted or when there is an excessive amount of repetitive information. Examples of no information occur when driving in fog or in a "whiteout" snowstorm. In both cases, the visual system is searching for information that is hidden but may appear suddenly and require a rapid response. The stress felt while driving in these conditions is a common experience. As for overstimulation, the important point is not the total amount of visual information, but rather the presence of large areas of the same spatial frequency. Wilkins (1993) has associated the presence of large areas of specific spatial frequencies in printed text with the occurrence of headaches, migraines, and reading difficulties.

Distraction: The visual system is designed to extract information from the visual environment. To do this, it has a large peripheral field that detects the presence of objects which are then examined using the small, high-resolution fovea. For this system to work, objects in the peripheral field that are bright, moving, or flickering have to be easily detected. If, upon examination, these bright, moving, or flickering objects prove to be of little interest, they become sources of distraction because their attention-gathering power is not diminished after one examination. Ignoring objects that automatically attract attention is stressful and can lead to symptoms of visual discomfort.

Perceptual confusion: The visual system is designed to extract information from the visual environment. The visual environment consists of a pattern of luminances, developed from the differences in reflectance of the surfaces in the field of view and the distribution of illuminance on those surfaces. Perceptual confusion occurs when there is a pattern of luminances present that is solely related to the illuminance distribution and conflicts with the pattern of luminance associated with the reflectances of the surfaces.

26.7.2 Lighting Conditions That Can Cause Discomfort

There are many different aspects of lighting that can cause discomfort. Insufficient light for the performance of a task has been discussed earlier and will not be discussed again. Rather, attention will be devoted to flicker, glare, shadows, and veiling reflections. It should be noted that whether these aspects of lighting cause discomfort will depend on the context. All can be used to positive effect in some contexts.

Flicker: A lighting installation that produces visible flicker will be almost universally disliked, unless it is being used for entertainment. Individual differences, and the fact that electrical signals associated with flicker can be detected in the retina, even when there is no visible flicker (Berman et al., 1991), imply that a clear safety margin is necessary. This can be achieved by the use of high-frequency control gear for discharge lamps and/or the mixing of light from lamps powered from different phases of the electricity supply. The same approaches, which will result in a changed frequency and/or a reduced percentage modulation, can be used to diminish any stroboscopic illusions. The use of high-frequency control gear has been associated with a reduction in the prevalence of headaches (Wilkins et al., 1989).

Glare: Glare occurs in two ways. First, it is possible to have too much light. Too much light produces a simple photophobic response, in which the observer screws up his eyes, blinks, or looks away. Too much light is rare indoors but is common in full sunlight. Second, glare occurs when the range of luminance in a visual environment is too large. Glare of this sort can have two effects, a reduction in threshold visual performance and a feeling of discomfort. Glare which reduces threshold visual performance is called *disability* glare. It is due to light scattered in the eye reducing the luminance contrast of the retinal image on the fovea. The magnitude of disability glare can be estimated by calculating the equivalent veiling luminance (IESNA, 1993)

The effect of disability glare on the luminance contrast of the object being looked at can be determined by adding the equivalent veiling luminance to all elements in the formulae for luminance contrast (see Section 26.4.6.2). Disability glare is rare in interior lighting but is common on roads at night from oncoming headlights and during the day from the sun. Usually disability glare also causes discomfort, but it is possible to have disability glare without discomfort when the glare source is large in area. This can be seen by looking at a picture hung on a wall adjacent to a window. Usually the picture will be much easier to see when the eye is shielded from the window.

As for discomfort glare, this, by definition, does not cause any shift in threshold visual performance but does cause discomfort. There are many different national systems for predicting the magnitude of discomfort glare produced by interior lighting installations (CIBSE, 1994; IESNA, 1993). All these systems are based on a formula which implies that discomfort glare increases as the luminance and solid angle of the glare source increase and decreases as the luminance of the background and the deviation from the glare source increase. Lighting equipment manufacturers use these formulae to produce tabular estimates of the level of discomfort glare produced by a regular array of their luminaires for a range of standard interiors. These tables provide all the precision necessary for estimating the average level of discomfort glare likely to occur in an interior, although the precision with which they predict an individual's sense of discomfort is low (Stone and Harker, 1973).

Shadows: Shadows are cast when light coming from a particular direction is intercepted by an opaque object. If the object is big enough, the effect is to reduce the illuminance over a large area. This is typically the problem in industrial lighting where large pieces of machinery cast shadows in adjacent areas. The effect of these shadows can be overcome either by increasing the proportion of interreflected light by using high reflectance surfaces or by providing local lighting in the shadowed area. If the object is smaller, the shadow can be cast over a meaningful area, which in turn can cause perceptual confusion, particularly if the shadow moves. An example of this is the shadow of a hand cast on a blueprint. This problem can be reduced by increasing the interreflected light in the space or by providing local lighting which can be adjusted in position.

Although shadows can cause visual discomfort, it should be noted that they are also an essential element in revealing the form of three-dimensional objects. Techniques of display lighting are based around the idea of creating highlights and shadows to change the perceived form of the object being displayed.

The number and nature of shadows produced by a lighting installations depends on the size and number of light sources and the extent to which light is interreflected around the space. The strongest shadow is produced from a single point source in a black room. Weak shadows are produced when the light sources are large in area and the degree of interreflection is high.

Veiling reflections: Veiling reflections occur when a source of high luminance, usually a luminaire or a window, is reflected from a specularly reflecting surface, such as a glossy printed page or a VDT screen. The luminance of the reflected image changes the luminance contrast of the printed text or the VDT display. The extent to which this changes visual performance can be estimated using the RVP model, but the extent to which it causes discomfort is different. Bjorset and Fredericksen (1979) have shown that a 20% reduction in luminance contrast is the limit of what is acceptable, regardless of the luminance contrast without veiling reflections (Figure 26.13).

The two factors that determine the magnitude of veiling reflections are the *specularity* of the material being viewed and the *geometry* between the observer, the object, and any sources of high luminance. If the object is completely diffusely reflecting, no veiling reflections occur, but if it has a specular reflection component, veiling reflections can occur. The positions where they occur are those where the incident ray corresponding to the reflected ray which reaches the observers eye from the object comes from a source of high luminance. This means that the strength and magnitude of veiling reflections can vary dramatically within a single lighting installation (Boyce and Slater, 1981).

Like shadows, veiling reflections can also be used positively, but when they are, they are conventionally called *highlights.* Display lighting of specularly reflecting objects is all about producing highlights to reveal the specular nature of the surface.

26.7.3 Comfort, Performance, and Expectations

While lighting conditions that make it difficult to achieve good visual performance will almost always be considered uncomfortable, lighting conditions that allow a high level of visual performance may also be considered uncomfortable. Figure 26.14 shows the mean detection speed for finding a number from many laid out at random on a table, and the percentage of people considering the lighting good. As might be expected, increasing the illuminance on the table increases mean detection speed and the percentage considering the lighting good. However, as the illuminance exceeds 2000 lx, the percentage considering the lighting good declines even though the mean detection speed continues to increase. This result indicates that if you wish to achieve a satisfactory lighting installation, it is necessary to provide lighting that allows easy visual performance and avoids discomfort and that visual discomfort is more sensitive to lighting conditions than visual performance.

There is another aspect of visual comfort that distinguishes it from visual performance. Visual performance is determined solely by the capabilities of the visual system. Visual comfort is linked to peoples' expectations. Any lighting installation that does not meet expectations may be considered uncomfortable even though visual performance is adequate; and expectations can change over time. Figure 26.12 also demonstrates another potential impact of visual comfort. Lighting conditions that are considered uncomfortable may influence task performance by changing motivation even when they have no effect on the stimuli presented to the visual system and hence, on visual performance.

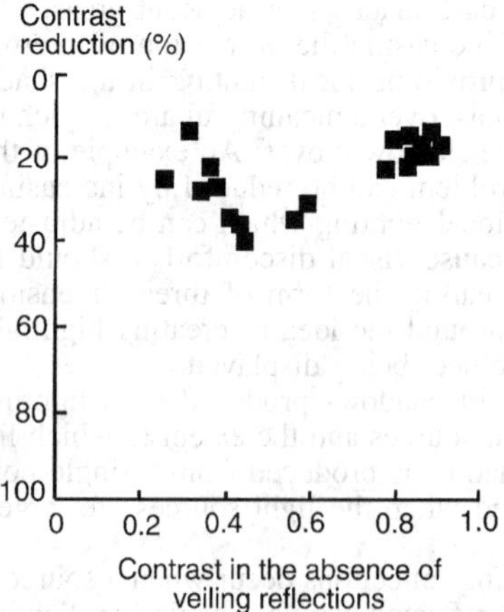

Figure 26.13 The luminance contrast reduction considered acceptable by 90% of observers plotted against the luminance contrast of the materials when no veiling reflections occurred. (After Bjorset, H. H., and Fredericksen, E. A., *Proceedings of the 19th Session of the CIE,* 1979.)

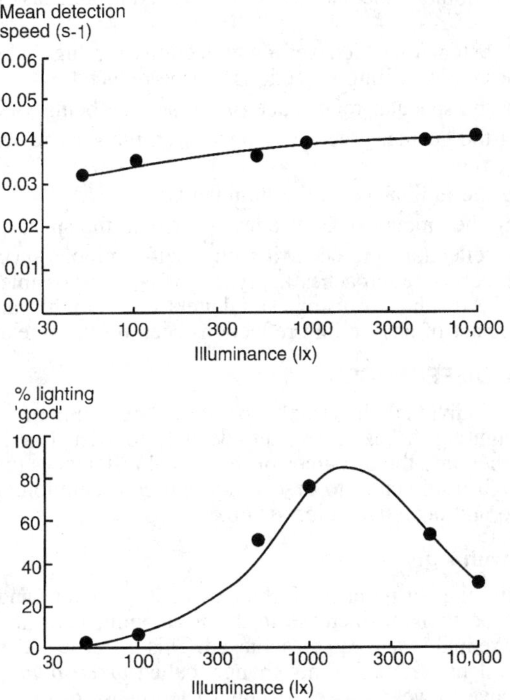

Figure 26.14 Mean detection speed for locating a specified number from amongst others at different illuminances, and the percentage of observers who consider the lighting good at each illuminance. (After Muck, E., and Bodmann, H. W., *Lichttechnik, 13*, 502, 1961.)

26.7.4 Approaches to Improving Visual Comfort

In order to ensure visual comfort, it is necessary to ensure that the lighting allows a good level of visual performance, does not cause distraction, and allows sufficient stimulation without perceptual confusion. This can be done by

- Identifying the visual tasks to be performed and then determining the characteristics of the lighting needed to allow a high level of visual performance of the tasks (see Sections 26.5 and 26.6).
- Eliminating flicker from the lighting by using appropriate control gear for discharge lamps. If this is not possible, reduce the percentage modulation of the flicker by mixing light from sources operating on different phases of the electricity supply.
- Reduce disability glare by careful selection, placing and aiming of luminaires so as to reduce the luminous intensity of the luminaires close to the common lines of sight.
- Reduce discomfort glare by careful selection and layout of luminaires. Use the appropriate national discomfort glare system to estimate the magnitude of discomfort glare. Using high reflectance surfaces in the space will help reduce discomfort glare by increasing the background luminance against which the luminaires are seen.
- Consider the density and areal extent of any shadows that are likely to occur. If shadows are undesirable and large area shadows are likely to occur, use high reflectance surfaces in the space to increase the amount of interreflected light and use more, lower-wattage lamps to supply the desired illuminance. If shadows cannot be avoided because of the extent of obstruction in the space, be prepared to provide supplementary task lighting in the shadowed areas. If dense, small area

shadows occur in the immediate work area, use adjustable task lighting to moderate their impact.

- Consider the extent to which veiling reflections (or highlights) are desirable. If they are undesirable, veiling reflections can be reduced by
  - Reducing the specular reflectance of the surface being viewed.
  - Changing the geometry between the viewer, the surface being viewed, and the offending zone.
  - Reducing the luminance of the luminaires.
  - Increasing the amount of interreflected light in the space.
- If the veiling reflections are occurring on a self-luminous surface, such as a VDT screen, all the above approaches apply, but it is also possible to increase the luminance of the display by using dark letters on a bright background. This will reduce the impact of any veiling reflections seen on the screen (Boyce, 1991).

26.8 INDIVIDUAL DIFFERENCES

Differences between individuals in visual capabilities are common and are usually dealt with by providing lighting that is more than adequate for visual performance and visual comfort. However, there are three sources of individual differences that are both common and consistent enough in direction to deserve special consideration. They are the effects of age, partial sight, and defective color vision.

26.8.1 Changes with Age

As the visual system ages, a number of changes in its structure and capabilities occur. Usually, the first to occur is an increase in the near point, i.e., the shortest distance at which a clear, sharp retinal image can be achieved. This increase occurs due to an increase in the rigidity of the lens with age. This change, called *presbyopia*, is why the majority of people over 50 have to wear glasses or contact lenses to read.

While the increasing rigidity of the lens, and other forms of focusing difficulty, can be compensated by adjusting the optical power of the eye's optical system with spectacle lenses, the other changes that occur in the eye cannot. As the visual system ages, the amount of light reaching the retina is reduced, more of the light entering the eye is scattered, and the color of the light is altered by preferential absorption of the short, visible wavelengths. The rate at which these changes occur accelerates after about age 60. The consequences of these changes with age are reduced visual acuity, reduced contrast sensitivity, reduced color discrimination, increased time taken to adapt to large and sudden changes in adaptation luminance, and increased sensitivity to glare (Sekular, Kline, and Dismukes, 1982).

Lighting can be used to compensate for these changes, to some extent. Older people benefit from higher illuminances than are needed by young people (Smith and Rea, 1979), but simply providing more light may not be enough. The light has to be provided in such a way that both disability and discomfort glare are carefully controlled and veiling reflections are avoided. Where elderly people are likely to be moving from a well-lit area to a dark area, such as from a supermarket to a parking lot, a transition zone with a gradually reducing illuminance is desirable. Such a transition zone allows their visual system more time to make the necessary changes in adaptation.

26.8.2 Helping People with Partial Sight

Partial sight is a state of vision that falls between normal vision and total blindness. Different countries use different capabilities to define the state of partial sight and blindness. The factors considered are visual acuity and the extent of the visual field. The World Health Organization accepts that a distance visual acuity of 6/18 implies that people are visually disabled and then grades the extent of disability in five steps. A visual acuity of 6/18 means that the individuals can just resolve details at 6 m that people with normal vision can resolve at 18 m.

While some people are born with partial sight, the majority of people with partial sight are elderly. Kahn and Moorhead (1973) found that among the partially sighted, 20% became partially sighted between birth and 40 yr, 21% between 41 and 60 yr and 59% after 60 yr. Surveys in the United States and the United Kingdom suggest that the proportion of the total population who are classified as partially sighted are in the range of 0.5 to 1%.

The three most common causes of partial sight are cataract, macular degeneration, and glaucoma. These causes involve different parts of the eye and have different implications for how lighting might be used to help people with partial sight.

Cataract is an opacity developing in the lens. The effect of cataract is to absorb and scatter more light as the light passes through the lens. This increased absorption results in reduced visual acuity and reduced contrast sensitivity over the entire visual field, as well as greater sensitivity to glare. The extent to which more light can help a person with cataract depends on the balance between absorption and scattering. More light will help overcome the increased absorption, but if scattering is high, the consequent deterioration in the luminance contrast of the retinal image will reduce visual capabilities. There is really little alternative to testing the effectiveness of additional light on an individual basis. What is true for everyone with cataract is that they will be very sensitive to glare from luminaires and windows. Careful selection of luminaires and window treatments to limit glare is desirable. The use of dark backgrounds against which objects are to be seen will also help.

Macular degeneration occurs when the macular of the retina, which is just in front of the fovea, becomes opaque due to bleeding or atrophy. An opacity immediately in front of the fovea implies a serious reduction in visual acuity and in contrast sensitivity at high spatial frequencies. It also implies that the ability to discriminate colors will be reduced. Typically, these changes make reading difficult if not impossible. However, peripheral vision is unaffected so the ability to orient oneself in space and to find one's way around is unchanged. Providing more light, usually by way of a task light, will help people in the early stage of macular deregulation to read, although as the deterioration progresses, additional light will be less effective. Increasing the size of the retinal image by magnification or by getting closer is helpful at all stages.

Glaucoma is shown by a progressive narrowing of the visual field. Glaucoma is due to an increase in intraocular pressure, which damages the retina and the anterior optic nerve. Glaucoma will continue until complete blindness occurs unless the intraocular pressure is reduced. As glaucoma develops, it leads to a reduction in visual field size, reduced contrast sensitivity, poor night vision, and slowed transient adaptation, but the resolution of detail seen on-axis is unaffected until the final stage. Lighting has limited value in helping people in the early stages of glaucoma, because where damage has occurred the retina has been destroyed. However, consideration should be given to providing enough light for exterior lighting at night to enable the fovea to operate.

While the extent to which providing more light is helpful will depend on the specific cause of partial sight, there is one approach which is generally useful. This approach is to simplify the visual environment and to make its salient details more visible. As an example, consider the problem of how to set a table so that a person with partial sight can eat with confidence. The plate containing the food and the associated cutlery can be made more visible by using a contrasting tablecloth, e.g., a dark tablecloth with a white plate and cutlery. The food on the plate can be made easier to identify by using an overlarge plate so that individual food items can be separated from each other. The whole scene can be simplified by using solid colors rather than patterns. This same approach of simplification and enhanced visibility can be applied to whole rooms, for example, by painting a door frame in a contrasting color to the door so that the door is easily identified.

26.8.3 The Consequences of Defective Color Vision

About 8% of males and 0.5% of females have some form of defective color vision. For most activities this causes few problems, either because the exact identification of color is unnecessary or because there are other cues by which the necessary information can be obtained. Where defective color vision does become a problem is where color is the sole means used to identify the object as, for example, in some forms of electrical wiring. People with defective color vision will have difficulty with such activities.

Where self-luminous colors are used as signals, care should be taken to restrict the range of colored lights used to those which can be distinguished by people with the most common forms of color defect. For example, the CIE has recently recommended areas on the CIE 1931 Chromaticity Diagram within which red, green, yellow, blue, and white signal lights should lie. These areas are designed so that the red signal will be named as red and the green as green by people with the most common forms of defective color vision (CIE, 1994).

26.9 OTHER EFFECTS OF LIGHT ENTERING THE EYE

Although making things visible is the most obvious effect of light entering the eye, there are two other ways in which light can affect us. The first is through *photobiology.* The second is through the psychological impact of what is visible.

26.9.1 Photobiology

Photobiology is the study of the interaction of biological systems with energy in the ultraviolet, visible, and infrared portions of the electromagnetic spectrum. Here the biological organism of interest is homo sapiens, and the responses of interest are those that follow from the absorption of radiation by the photoreceptors of the eye. The effects of ultraviolet, visible, and infrared radiation on the skin and the eye are discussed in Section 26.10.

The physiology of interest operates through the retinohypothalmic tract. This provides a route whereby signals from the retina are transmitted to the suprachiasmatic nucleus of the brain, which is believed to be the source of the biological clock in humans, and then to the many control centers of the human nervous system, including the thalamus, midbrain, brain stem and spinal cord as well as other areas of the hypothalamus (Dijk et al., 1995).

As might be expected from these widespread connections, light has been found to have an important influence on human biological rhythms (U.S. Congress, 1991). These rhythms mostly occur over one of the three geophysical rhythms found in nature: the day/night cycle, the lunar cycle, and the seasonal cycle. These cycles are important because human physiology, mood, and capabilities vary over each cycle.

The most extensively investigated cycle is the night/day (circadian) cycle. This is of considerable importance because of the prevalence of shift-work. About 20% of workers in the United States are shift workers of some sort. The short-term problems of shift work are fatigue, produced by poor quality sleep, and maintaining alertness during work. Long term, there is evidence that shift workers have a higher risk of cardiovascular disease, gastrointestinal ailments, and emotional and social problems. The short-term problems are believed to occur because of a mismatch between the demands of the work and the state of the worker's circadian rhythm. Put plainly, the workers are expected to work when their physiology is telling them to sleep and sleep when their physiology is telling them to be awake. Light is useful in alleviating this problem because it can shift human circadian rhythm so that they better match the functional requirements. Laboratory studies have shown improvements in alertness and cognitive performance following exposure to high light levels during night shift work, together with physiological changes indicative of the state of the circadian rhythm (French, Hannon, and Brainard, 1990). Aspects of light known to be important to the occurrence of these changes are the amount of retinal illumination, the duration of the exposure, and the timing of the exposure, (Moore-Ede, Sulzman, and Fuller, 1982).

Another problem associated with a bodily rhythm and that has been shown to be sensitive to light exposure, is *Seasonal Affective Disorder* (*SAD*). People with this condition experience decreased energy and stamina, depression, feelings of despair, and a greater need for sleep during the winter months. Light therapy, in which the patient is exposed to a high illuminance for a set period each day, has been shown to alleviate these symptoms in many patients. The reasons for the success of light therapy as a means of treating Seasonal Affective Disorder are still the subject of argument, the argument centering around the idea that the light treatment corrects an inappropriate phase shift in the circadian rhythm or is simply a placebo (Blehar and Lewy, 1990).

The use of light to alleviate the problems of shift work and to treat seasonal depression are just the most advanced examples of the influence of light on human well-being. Other potential applications of light therapy include treatment of sleep disorders, more general, nonseasonal depression, and jet lag. How, and if, photobiology should be brought into general lighting practice and whether it has any value for people working day jobs are questions that are just beginning to be considered.

26.9.2 Positive and Negative Affect

Psychology is a vast field, and the psychology of lighting is only a small part of it. The area relevant to lighting practice that has been most consistently studied is *perception.* Studies have been undertaken in abstract situations and have lead to quantitative relations being proposed between simple sensations such as brightness and photometric measurements such as luminance. Other studies have been undertaken in rooms with complete lighting installations and have lead to an understanding of the link between the perception of gloom and such photometric characteristics of the room as reflectance and illuminance distributions (Shepherd, Julian, and Purcell, 1989). Yet others have tried to establish if lighting generates cues by which people interpret a room, for example, does lighting the walls enhance the perception of spaciousness (Flynn et al., 1973).

While such studies have certainly influenced lighting design, they cannot be said to constitute a coherent body of knowledge. Further, they cannot form a basis for lighting practice until the impacts of specific perceptions are understood. To understand the consequences of perception of lighting, it is necessary to take a broader view. This view centers around positive affect. Positive affect, defined as pleasant feelings induced by commonplace events or circumstances, has been found to influence cognition and social behavior (Isen and Baron, 1991). Specifically, positive affect has been shown to increase efficiency in making some type of decisions, and to promote innovation and creative problem solving. It also changes the choices people make and the judgments they deliver. For example, it has been shown to alter peoples' preference for resolving conflict by collaboration rather than avoidance and also to change their opinions of the tasks they perform.

Given these usually desirable outcomes of positive affect, it is necessary to ask what can generate positive affect. The answer is both small and wide. Small, because the stimuli which have been shown to generate positive affect are low-level stimuli, ranging from receiving a small but unexpected gift from a manufacturer's representative to being given positive feedback about task performance. Wide, because positive affect can be influenced by the physical environment, the organizational structure and the organizational culture. Lighting is clearly a part of the physical environment and has been shown to influence positive affect (Baron, Rea, and Daniels 1992), but it is only one of the many factors that can do that.

As would be expected, it is also possible to generate negative affect. There is considerable information on the influence of frustration or anger on aggression and on the relationship between anxiety and performance (Baron, 1977). It seems reasonable to propose that lighting conditions which cause visual discomfort could generate negative affect.

Positive and negative affect provide a plausible route whereby the perception of the visual environment might influence the efficiency and effectiveness of organizations. As such, they represent a very different approach to identifying what is the most appropriate form of lighting for organizations to the visibility-based recommendations used in lighting practice today.

26.10 TISSUE DAMAGE

The part of the electromagnetic spectrum from 100 nm to 1 mm is called *optical radiation.* This part of the electromagnetic spectrum covers ultraviolet (100 nm–400 nm), visible (400 nm–760 nm) and infrared radiation (760 nm–1 mm). Sunlight and electric light sources all emit optical radiation. In sufficient quantities, optical radiation can cause damage to the eye and the skin. Details are given in McKinley, Harlen, and Whillock (1988).

26.10.1 Mechanisms for Damage to Eye and Skin

There are two mechanisms for tissue damage to occur: *photochemical* and *thermal.* They are not mutually exclusive; both can occur for the same incident optical radiation, but one will have a lower damage threshold than the other. Photochemical damage is related to the energy absorbed by the tissue within the repair or replacement time of the cells of the tissue. Thermal damage is determined by the magnitude and duration of the temperature rise.

The factors that determine the likelihood of tissue damage are the spectral irradiance incident on the tissue, the spectral sensitivity of the tissue, the time for which the radiation is incident, and, for thermal damage, the area over which the irradiance occurs. Spectral irradiance will be determined by the spectral radiant intensity of the source of optical radiation; the spectral reflectance and/or the spectral transmittance of materials from which the optical radiation is reflected or through which it is transmitted; and the distance from the source of optical radiation. Area is important for thermal tissue damage because the potential for dissipating heat gain is greater for a small area than for a large area.

The visual system provides an automatic protection from tissue damage in the eye, for all but the highest levels of visible radiation. This is the involuntary aversion response produced when viewing bright light. The response is to blink and look away, thereby reducing the duration of exposure. Of course, this involuntary response only works for sources that have a high visible radiation component, such as the sun. Sources that produce large amounts of ultraviolet and infrared radiation with little visible radiation are particularly dangerous because they do not trigger the aversion response.

26.10.2 Acute and Chronic Damage to the Eye and Skin

Tissue damage can be classified according to the duration of exposure it takes to produce the damage. Acute forms of damage are detectable immediately or at least within a few hours of exposure. Chronic forms of damage only become apparent after many years.

Ultraviolet radiation incident on the skin produces an immediate pigment darkening, followed a few hours later by erythema (reddening of the skin) and, ultimately, by a tan, produced by an increase in the number, size, and pigmentation of melanin granules. Excessive ultraviolet radiation incident on the eye can produce, a few hours later, an inflammation of the cornea called *photokeratitis.* This typically lasts a few days followed by recovery. As for chronic damage, prolonged exposure to ultraviolet radiation has been shown to be associated with various forms of skin cancer and cataract.

Visible radiation incident on the skin will produce erythema but not tanning, and, in sufficient quantity, skin burns. Visible radiation incident in the eye reaches the retina. This irradiance represents both an acute photochemical and an acute thermal hazard to the eye. Photochemical damage to the retina is associated with short wavelength light (blue light). The thermal damage covers retinal burns. As for chronic damage, it may be that prolonged and repeated exposure to light is involved in the retinal aging process.

Infrared radiation incident on the skin again initially produces skin reddening and, at a high enough irradiance, burns. Infrared radiation incident on the eye will cause heating of various elements of the eye, depending on the spectral content of the irradiance and the spectral transmittance of the various components of the eye. Infrared radiation from 760 nm to 1400 nm will reach the retina and can cause retinal burns. Longer wavelengths will be absorbed by other components of the eye. Prolonged heating of the lens is believed to be involved in the incidence of cataract.

26.10.3 Damage Potential of Different Light Sources

The light source with the greatest potential for tissue damage is the sun. The sun produces copious amounts of ultraviolet, visible, and infrared radiation. Voluntary staring at the sun is a common cause of retinal burns. Voluntary exposure of the skin to the sun commonly produces sunburn. However, there exist some electric light sources that can be hazardous, some being intended for lighting and others being used as a source of optical radiation for industrial processes.

The extent to which a light source represents a hazard can be evaluated by applying the recommendations of the American Conference of Governmental Industrial Hygienists (ACGIH). These recommendations take several different forms, ranging from maximum permissible exposure times to irradiance limits. Application of these standards to various commonly used electric light sources indicate that the electric light sources used for conventional interior lighting do not usually represent a hazard (Bergman, Parham, and McGowan, 1995; McKinley, Harlen, and Whillock, 1988). However, if there is any doubt about a specific light source or the way in which it is being used, the recommendations of the ACGIH should be used to assess its hazard potential.

26.10.4 Approaches to Limiting Damage

The approach to minimizing the damage caused by optical radiation is to limit the irradiance and/or the time of exposure. Whether any such action is necessary can be determined by applying the ACGIH recommendations to the situation.

For sources of optical radiation used for lighting, if the threshold limiting values are exceeded, often it will be possible to use a different light source which is less hazardous. If this is not possible, then it is necessary to filter the source to eliminate some of the hazardous wavelengths or to use some form of eye or skin protection to attenuate the optical radiation or to limit the exposure time.

For sources of optical radiation used in industrial processes, the source should be installed in an enclosure, with an interlock so that opening the enclosure extinguishes the source. If this is not possible, then appropriate forms of eye and skin protection are required.

26.11 EPILOGUE

Illumination has been a subject of study for more than 80 years. The result has been a growing understanding of how lighting conditions and the visual system can interact to facilitate visual performance and diminish visual discomfort. This knowledge has formed the framework around which many national illuminating engineering organizations have built recommendations for lighting practice (CIBSE, 1994; IESNA, 1993). These recommendations provide a firm basis for designing everyday lighting installations, provided always that the recommendations are applied with thought and not by rote.

There are three current areas of study with considerable potential. They are:

1. The impact of moderate visual discomfort on the choices people make when doing sustained, self-paced tasks.
2. The effect of light spectrum on visual performance in mesopic conditions.
3. The nonvisual effects of light, particularly the photobiological.

Knowledge gained in these areas has the potential to change the manner in which lighting is designed and the technology that is used to provide it.

REFERENCES

Bailey, I., Clear, R., and Berman, S. (1993). Size as a determinant of reading speed. *Journal of the Illuminating Engineering Society*, *22*, 102–117.

Baron, R. A. (1977). *Human Aggression*. New York: Plenum Press.

Baron, R. A., Rea, M. S., and Daniels S. G. (1992). Effects of indoor lighting (illuminance and spectral distribution) on the performance of cognitive tasks and interpersonal behaviors: The potential mediating role of positive affect. *Motivation and Emotion*, *16*, 1–33.

Bean, A. R. and Simons, R. H. (1968). *Lighting Fittings Performance and Design*. Oxford: Pergamon Press.

Bergman, R. S., Parham, T. G. and McGowan, T. K., UV emission from general lighting lamps, *Journal of the Illuminating Engineering Society*, *24*, 13–24.

Berman, S. M., Greenhouse, D. S., Bailey, I. L., Clear, R. D. and Raasch, T. W. (1991). Human electroretinogram responses to video displays, fluorescent lighting and other high frequency sources, *Optometry and Vision Science*, *68*, 645–662.

Berman, S. M., Fein, G., Jewett, D. L., and Ashford, F. (1993). Luminance-controlled pupil size affects Landolt C task performance. *Journal of the Illuminating Engineering Society*, *22*, 150–165.

Bjorset, H. H., and Frederiksen, E. (1979). A proposal for recommendations for the limitation of the contrast reduction in office lighting. *Proceedings of the 19th Session of the CIE*, Kyoto, Japan, pp. 310–314.

Blehar, M. C., and Lewy, A. J. (1990). Seasonal mood disorders: consensus and controversy. *Psychopharmacology Bulletin*, 23, 465–494.

Boff, K. R. and Lincoln J. E. (1988). *Engineering Data Compendium: Human Perception and Performance*. Harry G. Armstrong Aerospace Medical Research Laboratory, Wright-Patterson Air Force Base, OH.

Bourdy, C., Chiron, A., Cottin, C., and Monor, A. (1987). Visibility at a tunnel entrance: effect of temporal adaptation. *Lighting Research and Technology*, *19*, 35–44.

Boyce, P. R. (1985). Movement under emergency lighting: The effect of illuminance, *Lighting Research and Technology*, *17*, 51–71.

Boyce, P. R. (1991). Lighting and Lighting Conditions. In J. A. J. Roufs, Ed., *The Man-Machine Interface, Volume 15 of Visual and Visual Dysfunction*. London: Macmillan Press.

Boyce, P. R., and Rea, M. S. (1987). Plateau and escarpment: The shape of visual performance. *Proceedings of the 21st Session of the CIE*. Venice, Italy, pp. 82–85.

Boyce, P. R., and Slater, A. I. (1981). The application of contrast rendering factor to office lighting design. *Lighting Research and Technology*, *13*, 65–79.

Chartered Institution of Building Services Engineers (1994). *CIBSE Code for Interior Lighting*. London: CIBSE.

Commission Internationale de l'Eclairage (1971). CIE Publication 15 *Colorimetry*. Vienna: CIE.

Commission Internationale de l'Eclairage (1972). Supplement No. 1 to CIE Publication 15. *Special Metamerism Index: Change in Illuminant*. Vienna: CIE.

Commission Internationale de l'Eclairage (1974). CIE Publication 13.2. *Method of Measuring and Specifying Colour Rendering Properties of Light Sources*. Vienna: CIE.

Commission Internationale de l'Eclairage (1978). CIE Publication 41. *Light as a True Visual Quantity: Principles of Measurement*. Vienna: CIE.

Commission Internationale de l'Eclairage (1987). CIE Standard 69 *Methods of Characterizing Illuminance Meters and Luminance Meters: Performance, Characteristics and Specification.* Vienna: *CIE.*

Commission Internationale de l'Eclairage. (1978a). Supplement No. 2 to CIE Publication 15. *Recommendations on Uniform Color Spaces, Color-Difference Equations, Psychometric Color Terms.* Vienna: CIE.

Commission Internationale de l'Eclairage (1994). CIE Technical Report 107. *Review of the Official Recommendations of the CIE for the Colours of Signal Lights.* Vienna: *CIE.*

Cornu, L., and Harlay, F. (1969). Modifications de la discrimination chromatique en fonction de l'eclairement. *Vision Research, 9,* 1273–1280.

Dijk, D. J., Boulos, Z., Eastman, C. I., Lewy, A. J., Campbell, S. S. and Terman, M. (1995). Light treatment for sleep disorders: Consensus report, II Basic properties of circadian physiology and sleep regulation. *Journal of Biological Rhythms, 10,* 113–125.

French, J., Hannon, P., and Brainard, G. C. (1990). Effects of bright illuminance on body temperature and human performance. *Annual Review of Chronopharmacology, 7,* 37–40.

Flynn, J. E., Spencer, T. J., Martyniuck, O., and Hendrick, C. (1973). Interim study of procedures for investigating the effect of light on impression and behavior. *Journal of the Illuminating Engineering Society, 3,* 87–94.

Illuminating Engineering Society of North America (1993). *Lighting Handbook, 8th Ed.* New York: IESNA.

Inditsky, B., Bodmann, H. W., and Fleck, H. J. (1982). Elements of visual performance, contrast metric—visibility lobe—eye movements. *Lighting Research and Technology, 14,* 218–231.

Isen, A. M., and Baron R. A. (1991). Positive affect as a factor in organizational behavior. *Research in Organizational Behavior, 13,* 1–53.

Kahn, H. A., and Moorhead, H. B. (1973). *Statistics on Blindness in the Model Reporting Area 1969-70.* Publication No. (NIH) 73-427. U.S. Department of Health, Education and Welfare, Washington D.C.

Koenderink, J. J. and van Doorn, A. J. (1979). Spatiotemporal contrast detection threshold surface is bimodal. *Optics Letters, 4,* 32–34.

Littlefair, P. J. (1990). Innovative daylighting: Review of systems and evaluation methods. *Lighting Research and Technology, 22,* 1–17.

McKinlay, A. F., Harlen, F., and Whillock, M. J. (1988). *Hazards of Optical Radiation.* Bristol, U.K.: Adam Hilger.

Moore-Ede, M. C., Sulzman, F. M., and Fuller, C. A. (1982). *The Clocks That Time Us.* Cambridge, MA: Harvard University Press.

Rea, M. S., and Jeffrey, I. G. (1990). A new luminance and image analysis system for lighting and vision: 1. Equipment and calibration. *Journal of the Illuminating Engineering Society, 19,* 64–72.

Rea, M. S., and Ouellette, M. J. (1988). Visual performance using reaction times. *Lighting Research and Technology, 20,* 139–153.

Rea, M. S., and Ouellette, M. J. (1991). Relative visual performance: A basis for application. *Lighting Research and Technology, 23,* 135–144.

Robbins, C. L. (1986). *Daylighting, Design and Analysis.* New York: Van Norstrand Reinhold.

Robertson, A. R. (1977). The CIE 1976 color-difference formulae. *Color Research and Application,* 2, 7–11.

Sekular, R., and Blake, R. (1994). *Perception,* New York: McGraw-Hill.

Sekuler, R., Kline, D., and Dismukes, K. (1982). *Aging and Human Visual Function.* New York: Alan R. Liss, Inc.

Shepherd, A. J., Julian, W. G. and Purcell, A. T. (1989). Gloom as a psychophysical phenomenon. *Lighting Research and Technology, 21,* 89–97.

Smith, S. W. and Rea, M. S. (1979). Relationships between office task performance and ratings of feelings and task evaluations under different light sources and levels. *Proceedings of the 19th Session of the CIE,* Kyoto, Japan, pp. 207–211.

Stone, P. T. and Harker, S. P. D. (1973). Individual and group differences in discomfort glare responses. *Lighting Research and Technology, 5,* 41–49.

Tansley, B. W., and Boynton, R. M. (1978). Chromatic border perception: The role of red- and green-sensitive cones. *Vision Research, 18,* 683–697.

U.S. Congress, Office of Technology Assessment (1991). *Biological Rhythms: Implications for the Worker,* OTA-BA-463. U.S. Government Printing Office, Washington D.C..

Wilkins A. (1993). Reading and visual discomfort. In D. M. Willows, R. S. Kruk and E. Corcos, Eds., *Visual Process in Reading and Reading Disabilities.* Hillsdale, NJ: Lawrence Erlbaum Associates.

Wilkins, A. J., Nimmo-Smith, I., Slater, A.I., and Bedocs, L. (1989). Fluorescent lighting, headaches and eyestrain. *Lighting Research and Technology, 21,* 11–18.

Wyszecki, G., and Stiles, W. S. (1982). *Color Science: Concepts and Methods, Quantitative Data and Formulae.* New York: John Wiley.

CHAPTER 27

TOXICOLOGY AND THERMAL COMFORT*

Stephan Konz
Department of Industrial and Manufacturing Systems Engineering
Kansas State University
Manhattan, KS 66506 USA

First, we will discuss chemicals in the environment (toxicology), then air quality and volume; the emphasis is on health. Then we will discuss temperature aspects; the emphasis is on comfort. For information on extreme environments (hot and cold), see Chapter 29.

27.1 TOXICOLOGY

27.1.1 Poisons

Toxicology generally deals with the effects of foreign chemicals upon the body. Some chemicals can injure the body within hours (acute effects), while others take years

*The material in this chapter is adapted from Konz (1995).

(chronic effects). Toxicology is a very complex subject and this chapter touches the material only briefly.

Chemicals affect the body with doses producing a response.

27.1.1.1 Effect

An effect could be a permanent physical change (reduced lung function or even death). It could be reversible (pneumonia or change in dark adaption of the eye). The Threshold Limit Values (TLVs) discussed later in this chapter primarily are based on nonreversible functional changes in an organ (usually the kidney or liver); they give maximum acceptable exposure concentrations to various materials.

27.1.1.2 Body

Whose body? Giving a chemical to people and then watching them could be hard on the people! Thus, most TLVs are developed from studies on animals rather than humans. But there are marked differences between different species. For example, teratogens are substances which cause defects in fetal development. In rats and mice, thalidomide had no effect at a dose of 4000 mg/kg of body weight. Unfortunately, a dose of 0.5 mg/kg in humans was teratogenic (Mastromatteo, 1981). The TLV approach has been to consider that humans respond as the most sensitive animal species and then add a safety margin between the dose which produces the effect and the TLV. But, as you can see, the TLVs are not as precise as they seem; they are value judgments about our tolerance of different health risks.

Occasionally TLVs are based on human data. For example, some workers in polyvinyl chloride factories came down with a rare type of liver cancer after many years of exposure. The TLV for the vinyl chloride monomer gas (the raw product) then was reduced. A key point here is that it was a rare type of cancer; if it had been a common type of cancer, the danger probably never would have been noticed.

27.1.1.3 Dose/Response

With present technology, often it is possible to detect chemicals in a concentration of one part in a billion. This is roughly equivalent to measuring one sec in a period of 33 yr!

Mere detection of a chemical being present is not enough. What needs to be considered is the dose in relation to the response. Water, in excess, will kill as will salt, sugar, alcohol, sulfanilamide, or arsenic. The problem is to define "in excess."

Think of the body as a "leaky bucket" with components on shelves at various levels. (See Figure 27.1.) The problem is whether the liquid will rise high enough to cause "corrosion." Poisoning depends on the rate of poison input (liquid input), the kind of liquid, the body size (bucket size), target organ susceptibility (shelf level), and poison removal capability (hole size).

Since the same amount of poison rises much higher in a small bucket (small person), the same dose is more dangerous for small people. Although the dose should be given in proportion to body weight, for simplicity, TLVs assume a 70 kg person. Note that there is a large individual variability (5–10 fold or more) of individual susceptibility to toxins. The hole size depends primarily on the ability of the kidney and liver to transform the toxin into a less toxic compound and to eliminate it from the body. Examples of people with less elimination capacity are older people, people with hepatitis, and people who drink alcohol to excess. Another problem is that two compounds (such as alcohol and barbiturates, or cigarette smoke and cotton dust or asbestos) may both compete for the same liver enzyme (and thus slow down transformations) or both act on the same or different target organs (e.g., lungs, brain). Johns-Mansville Corporation, citing a study reporting that smokers who are occupationally exposed to asbestos have a 92 times greater chance of developing lung cancer than the general population, has banned smoking in its facilities. Alcohol and its products disturb liver function. Heavy metals (such as lead and methyl mercury) inhibit enzymes.

27.1.2 Toxin Entry Routes

Entrance to the mouth or lungs is not entrance to the body; to enter the body, a poison must enter the blood. Thus, an important characteristic of potential poisons is their ability to penetrate the body's perimeter. In general, inorganic materials (a swallowed penny) are slow to penetrate the barrier, nonlipid soluble organic compounds penetrate more quickly, and lipid soluble organic compounds (e.g., solvents) are absorbed most quickly.

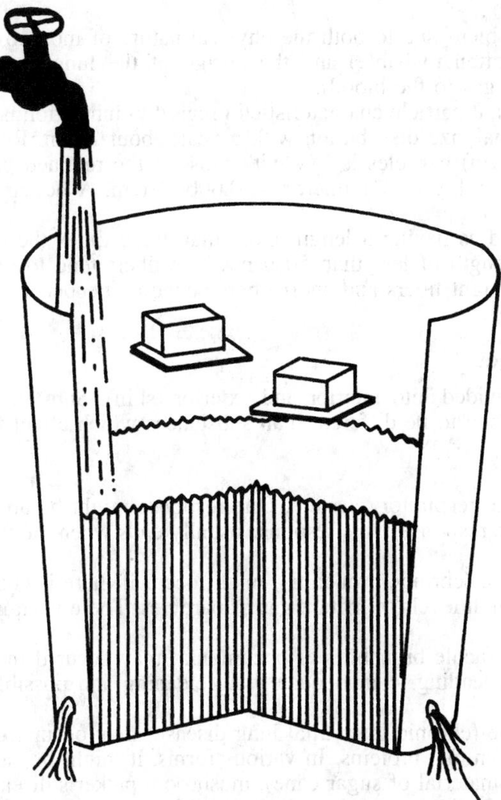

Figure 27.1 Consider the body as a leaky bucket with components on shelves at various heights. Will the liquid rise high enough to cause corrosion? (From Stephen Konz, *Work Design: Industrial Ergonomics,* 4th ed. Scottsdale, AZ: Gorsuch Scarisbrick, Publishers, 1995. Reprinted by permission of the publisher.)

The most important entrance points are the mouth, the skin, and the lungs.

27.1.2.1 Mouth

People may eat or drink the poison. Generally this is contamination of food or drink. For example, process dust may fall on sandwiches, into drink cups or onto cigarettes. The best precaution is to forbid eating, drinking, or smoking in work areas at any time. In addition to protecting the occupants from transferring toxins from their hands to food, this protects the product from food contamination. Thus clean, convenient break areas (separate from product areas) must be provided.

However, in most industries, poison entering through the mouth is not a significant problem.

27.1.2.2 Skin

The skin is a superb barrier for most compounds. Common experience shows that most compounds run off the skin rather than penetrate it. This makes the occasional compound which penetrates the skin violate the stereotype of nonpenetration and thus more dangerous. TLVs for compounds which penetrate the intact skin have the word *skin* next to the compound. Cuts and abrasions and rashes permit toxins (as well as germs) to penetrate the skin. Clothing and shoes wetted with a toxin agent increase contact time (and thus, increase danger). However, most compounds do not penetrate the intact skin. The biggest problem in poison absorption is the lungs.

27.1.2.3 Lungs

The lungs are a problem due to both the physical nature of most toxins (a gas or finely dispersed aerosol, often invisible) and the design of the lungs (very good at moving molecules from the gas to the blood).

The most important particle characteristic in regard to inhalation is their size. Airborne dust has a log-normal size distribution with a peak about 2 μm. Retention in the lungs of the larger (> 4 μm) particles is low with most of the retained particles being 0.5–2 μm. (1 micrometer = 1 μm = 1 micron = 0.000 001 m. A human hair has a 100-μm diameter.)

A fiber is defined as having a length more than three times the diameter. Almost all fibers will have a length of less than 50 μm. Since fibers tend to orient their long axis to the airstream, straight fibers and short fibers penetrate deeper than curved, U-shaped, or long fibers.

27.1.3 Toxin Targets

Toxin targets are divided into interior and exterior (skin). Substances which affect the respiratory system tend to be different than substances which affect the skin.

27.1.3.1 Interior

The first target is the respiratory system. One response might be an increase in air-flow resistance (asthmatic response); this response usually ends when the mucociliary escalator removes the irritant.

A second response (chronic bronchitis) might occur if there is continuing irritation to the mucosa due to high levels of irritant gas or particles. There is an increase in coughing and sputum.

A third response (acute bronchitis or pneumonia) may occur if the ciliary movements are paralyzed—thus leading to inadequate lung cleaning and possible improved colonization of bacteria.

A fourth response (chronic interstitial lung disease) may occur from dusts containing micro-organisms or animal proteins. In various forms, it might be called bagassosis (bagasse is the fibrous material of sugar cane), mushroom picker's disease, wheat thresher's lung, pigeon breeder's disease, feather plucker's disease, malt worker's lung, and paprika slicer's disease. Dusty spices used in food processing can be a problem.

Macrophages defend the lungs. When particles settle on the lung surfaces, the marcophages (which have the power of independent motion) engulf the particles and draw the particles through the tissues of the lung wall (1) directly into the blood, (2) to surrounding bronchioles for ciliary removal to the lymph system, or (3) simply remain permanently attached to the wall. However, inert particles present a problem. Free silica repeatedly kills the macrophases (silicosis); this leads to scar tissue and a loss of lung surface area (in effect, slow suffocation). Long asbestos fibers (over 5 μm) stick out of the macrophages and also cause scar tissue and loss of surface area. Particles of chromium, nickel, uranium and asbestos also may cause lung cancer.

Finally, a compound may pass into the blood and attack other organs. Lead primarily affects nervous tissue, cadmium affects the kidneys, carbon monoxide and cyanide affect hemoglobin, and sulfur dioxide and hydrogen sulfide affect the lungs (cough). Hydrogen sulfide also causes pulmonary edema. The French hatters in the seventeenth century used mercuric nitrate to aid fur felting; the resulting chronic mercury poisoning affected the brain and led to the expression "mad as a hatter."

A fetus is especially endangered since the baby functions as a sponge to toxins; that is, it does not have a good ability to eliminate poisons. Substances which cause defects in fetal development are called teratogens. The TLV standards are not strict enough to protect the fetus. The most critical period is the first 3 months of pregnancy. Zenz (1984) discusses the following teratogens: lead and its compounds, benzene, carbon monoxide, DBCP, chlordecone, chloroprene (2-chlorobutadiene), epichlorohydrin, ethylene dibromide, ethylene oxide, mercury, vinyl chloride, anesthetic gases, ionizing radiation, and PCBs.

There is a conflict between equal employment opportunities for women and health of the fetus. Some firms have said only sterile women or women past child-bearing age can work in fetus-harmful environments but this is not a very satisfactory solution. Brooks et al. (1994) describe the reproductive health program at Bell Labs.

27.1.3.2 Exterior (Skin)

Dermatitis, inflammation of the skin, is a common industrial disease. *Note, however, that not all dermatitis is occupationally related.*

The skin is the largest organ system of the body. Structurally the skin is composed of two layers—the dermis and epidermis. The epidermis has an outer layer of dead cells (called horny layer, keratin layer, or stratum corneum), which is a fair protection against chemicals (except alkalis and solvents) unless it has cracks (such as with eczema). Next is the Malpighian layer of living cells. The dermis (true skin) has connective tissues, nerves, hair follicles, oil and sweat glands, and blood and lymph vessels.

Atopic dermatitis (AD) tends to flare up when a person is exposed to the following triggering factors:

- Dry skin
- Low humidity
- Skin infections
- Heat, humidity, and sweating
- Emotional stress
- Irritants and allergens

Irritants include solvents, industrial chemicals, detergents, some soaps and fragrances, fumes and tobacco smoke, paints, bleach, woolens, acidic foods and astringents, and other alcohol-containing skin-care products. Allergens usually are proteins from food, pollen, or pets.

Contact dermatitis can be (1) irritant contact dermatitis (75% of cases) or (2) allergic contact dermatitis (25% of cases). Allergic contact dermatitis takes a long time (say 14 to 21 days following initial contact) for an allergy to develop. However, once sensitized, response becomes obvious within a short time (say 12 h) of exposure; it may not even occur at the contact site. Medications that are notorious for sensitizing individuals include topical anesthetics containing benzocaine, topical antibiotics containing neomycin, topical antihistamine creams, fungi creams, and the skin disinfectant thimerosal (merthiolate) (Hogan, 1986).

Causes of occupational dermatoses are:

- Mechanical and physical (abrasions or wounds, sunlight for outdoor workers, fiberglass, and asbestos)
- Chemical
  - strong irritants (e.g., chromic acid, sodium hydroxide)
  - marginal irritants (e.g., soluble cutting fluids, acetone). Marginal require prolonged contact over time. [Although coolants and cutting fluids tend to be associated with dermatitis, when applied as mists, they can cause asthma (Burge, 1989).] The normal skin is slightly acidic (pH = 6.8) (Ruane, 1989). However, typical cutting fluids tend to have pH around 9 (basic), which causes defatting of the skin.
- Plant poisons (e.g., woods such as West Indian mahogany, silver fir, spruce, when being sandpapered or polished)
- Biological agents (e.g. anthrax contracted by handlers of skins or hides from infected animals; grain or straw itch contracted by food or grain handlers from handling products infected with mites, etc.)

In addition, some metals (e.g., nickel) can sensitize the skin. Pierced ears break the skin barrier and stainless steel or white gold earrings (which contain nickel) then sensitize the entire body to nickel. Then the person gets eczema on the hands and is sensitive to faucet handles which have lost their chromium cover, to coins, jean buttons, bra buckles, clothing rivets, etc. Nickel eczema is difficult to heal and may cause early retirement.

Protect the skin with protective clothing—especially aprons and gloves. Aprons protect the chest and legs from liquids which might maintain contact with the skin for a long time.

Gloves for chemicals naturally must not be permeable to that specific chemical. However, chemicals can penetrate gloves (1) if the material degrades (change in physical property); (2) through permeation (chemical penetration on a molecular level); and (3) penetration on a nonmolecular level (seams, holes, and even pores of materials such as leather). Holes may be difficult to detect if they partially seal (i.e., a big hole becomes a small hole but still is a hole). Some gloves tear when they have a hole; another alternative is a yellow outer layer and a red inner layer so, if red shows, replace the glove.

In addition, chemical protection gloves should have long sleeves (gauntlets) to reduce chemical dripping into the glove. Rinse gloves before removal; wash and air-dry gloves between wearings. A cotton liner may reduce skin irritation due to sweating.

Gloves also can protect the skin from dry irritants and against cuts and abrasions. Machinists tend to cut their hands by wiping them with rags contaminated with metal shavings; they need to keep rags for hands separate from rags for wiping up—perhaps by using two colors of rags.

Reduce chemical exposure by good housekeeping around the workstation, by installing splashguards on machines, and by educating people concerning the danger of various compounds (so they don't expose their skin unnecessarily).

Barrier creams, a substitute for gloves, are designed to prevent dermatitis, not treat it. The two types are water-repellent and oil-repellent. Sunscreens can reduce the risk of nonmelanoma skin cancer by 90% (Hogan, 1986).

Personal cleanliness is a critical measure in preventing occupational skin disease. Lack of care in handling injurious materials and not cleaning up properly increase the amount and time of contact. Of course, engineers have a responsibility to provide adequate washing facilities near the workplace.

When cleaning the hands, the cleaner needs to be specific for the substance to be removed. When solvents are used in industrial processes, workers may clean up with solvent; this should be discouraged. Powered cleaners may be too abrasive; inorganic scrubbers (borax, silica) are more harsh than organic scrubbers (corn meal, rice hulls). When the soil is light and alkaline, liquid soaps with a neutral pH are good. Waterless cleaners tend to dry the hands. If they are used, follow with soap and a moisturizer. Avoid soaps with perfumes and coloring agents. When washing hands:

- Wet the hands
- Apply soap to the palm (the back is more sensitive)
- Rinse away soap
- Dry hands
- Apply thin layer of hand lotion

Personal clothing worn on the job, including undergarments, should be washed thoroughly before reuse. "Bystanders" (primarily the person who washes the clothes but also children in the home) can be affected by the toxins. Two solutions are (1) a double locker system with intervening shower and (2) having the clothing cleaned professionally rather than at home.

If cleaned at home (e.g., cleaning farm clothing contaminated with pesticides):

- If the clothing is saturated, throw it away.
- Wash the clothing as soon as possible.
- Wash separately from other clothing.
- Wear rubber gloves.
- Prespot the toxin.
- Use hot water and 1.5 (normal amount) of heavy-duty detergent.
- Air dry on an outside line.
- Run an empty cycle with hot water and detergent before the washer is used again.

27.1.4 Threshold Limit Values

Since the "poison is in the dose," what is an "acceptable" dose? People disagree. In my opinion, perhaps the best values are set by two private organizations, the American Conference of Government Industrial Hygienists (ACGIH) and American Industrial Hygiene Association (AIHA). The ACGIH (Kemper Woods Center, 1330 Kemper Meadow Drive,

Cincinnati, OH 45240; phone (513) 742 3355) annually issues updated Threshold Limit Value (TLV) recommendations for chemical substances, dusts, and physical agents.

The AIHA (2700 Prosperity Avenue, Fairfax, VA 22031; phone (703) 849 8888) issues Work Environment Exposure Limits (WEELs) for agents that have no current exposure guidelines established by other organizations.

The Federal Government has Recommended Exposure Limits (RELs) issued by National Institute of Occupational Safety and Health) and legal limits called Permissible Exposure Limits (PELs) (issued by Occupational Safety and Health Administration).

There are some practical problems with TLVs.

First, of the many (say 100,000) chemicals in the environment, TLVs have been established for less than 2000 (Roach, 1994). [Since the 50th percentile Lethal Concentration (LC_{50}) dose for rodents is known for about 25,000 chemicals, Roach recommends using $4(LC_{50})$ as a first approximation TLV for chemicals with no TLV.]

Second, because there often is a 20-year lag between dose and response, the scientists may not realize there is a problem with a chemical. For example, exposure to asbestos in the 1940s led to an increase in certain types of cancer in the 1970s and 1980s.

Third, because of the time lag, organizations may not realize the hazards. That is, the organizations in the 1940s did not know the problems that would occur for some of their workers 30 yr later. Related to this is that the cancer was not certain; just an increase in odds. For example, asbestos might increase the chance of a specific type of cancer (not all cancers) and might increase the odds from 1 in 1000 to 1 in 200. Even if the organization realized the problem, getting the workers to take precaution for something that might happen 30 yr later is a problem. Note the problem of convincing cigarette smokers to stop smoking!

Table 27.1 gives the TLVs for the first six values listed for chemical substances from the ACGIHs annual guide. [In addition to chemical substances, the guide gives values for respirable dusts (fibrogenic dusts which cause scar tissue in the lungs) and nuisance dusts (which cause insignificant scar tissue). The guide gives three types of threshold limit values: a time-weighted average (TLV-TWA), a short-term exposure limit (TLV-STEL), and a ceiling (TLV-C).

The TLV-TWA is the concentration, for a normal 8-h workday and 40-h work week, to which nearly all workers may be repeatedly exposed, day after day, without adverse effect. It primarily recognizes chronic (long-term) effects.

The TLV-STEL is concerned with acute (short-term) effects of (1) irritation; (2) chronic or irreversible tissue damage, and (3) narcosis of sufficient degree to increase the likelihood of accidental injury, to impair self-rescue, or to reduce work efficiency. The STEL is a 15-min time-weighted exposure, which should not be exceeded at any time during the day, even if the TLV-TWA is met. There should not be more than four STEL exposures of 15 min/day, and there should be at least 60 min between each STEL exposure.

The TLV-C is the concentration which should not be exceeded during any part of the working day.

When a recording is made of a toxin concentration in the workplace, the amount will often vary. The "high points" are called excursions or peaks. Peaks can be above the TWA and STEL but must not be above the C.

Table 27.1 TLVs For Chemical Substances for 1995–1996 (ACGIH, 1995). Only the First Six Substances Are Listed Below

	TWA		STEL		CEILING	
Substance	ppm[a]	mg/m^3[b]	ppm[a]	mg/m^3	ppm[a]	mg/m^3
Acetaldehyde	—	—	—	—	25	45
Acetic acid	10	25	15	37		
Acetic anhydrice	5	21	—	—		
Acetone	750	1780	1000	2380		
Acetone cyanohydrin as CN-Skin	—	—	—	—	4.7	5
Acetonitrile	40	67	60	101		

[a]Parts of vapor or gas per million parts of contaminated air by volume at 25°C and 760 torr.
[b]Milligrams of substance per cubic meter of air.

For example, acetic acid has a TWA of 10 ppm. An exposure of 12 ppm for 4 h can be balanced by an exposure of 8 ppm for the remaining 4 h. For TWA, you could even have an exposure of 80 for 1 hr balanced by a 0 for the remaining 7 h. This, however, is where the STEL is used. The STEL for acetic acid is 15. Thus exposure of 80 for 1 h is too high. However, an exposure of 80 for 1 min followed by 0 for the remaining 14 min (of the 15 min STEL period), gives an average exposure of 80/15 = 6, which would be below the STEL. For TWAs with no STEL listed, ACGIH (1995) recommends "Excursions in worker exposure levels may exceed 3 times the TLV-TWA for no more than a total of 30 min during a workday; under no circumstances should they exceed 5 times TLV-TWA, provided that TLV-TWA is not exceeded."

Note that, for administrative simplicity, the TLVs assume that concentration × time = a constant. That is, you would get the same effect from eight aspirins taken once in 8 h as from one taken each hour for 8 h. This is unlikely to be true, so don't push the excursions too far.

The TLVs in Table 27.1 are given both by volume and by weight. To convert (at 25 C):

$$\text{ppm} = \frac{24.45\ (\text{mg/m}^3)}{\text{molecular weight}}$$

The TLV is based on an 8-hr exposure. The value needs to be adjusted if (1) concentration varies during the day (the Time Weighted Average concept), (2) the day is not 8 h, or (3) exposure is to more than one substance.

First, assume a person was exposed to acetone for 4 h at 500 ppm, 2 h at 740 ppm, and 2 h at 1500 ppm. The equivalent exposure is:

$$\text{TWA} = \frac{C_a\, t_a + C_b\, t_b + C_c\, t_c}{8}$$

where

TWA = time-weighted average (equivalent 8-h exposure)
C = concentration of a,b,c . . . , ppm or mg/m^3
t = time of exposure to concentration a,b,c . . . , h

For the above example, TWA = (500(4) + 750(2) + 1500(2))/8 = 6500/8 = 812. Since 812 is more than the TLV of 750, the exposure is not acceptable.

Next, assume a person was exposed to 1250 ppm of acetone during the entire working day but only worked 6 h/day. Then TWA = (1250(6) + 0(2))/8 = 7500/8 = 940. Since 940 is more than 750, the exposure is not acceptable. Or, assume the person was exposed to 900 ppm for a 10–h shift. Then TWA = (900)(10)/8 = 9000/8 = 1125. Since 1125 is over 750, the exposure is not acceptable.

Assume the person was exposed to a mixture of substances. The assumption is that the effects are additive (act on the same organ system), not independent. Assume exposure for 8 h to acetone at 500 ppm, 2-butanone of 45 ppm, and toluene of 40 ppm. The TLV-TWAs are 750, 200, and 100 ppm. Then

$$\text{TWA}_{\text{mixture}} = \text{C}_1/\text{TLV}_1 + \text{C}_2/\text{TLV}_2 + \text{C}_3/\text{TLV}_3$$

where

$\text{TWA}_{\text{mixture}}$ = equivalent TWA mixture exposure (maximum of 1 permitted)
$\text{C}_{1,2,3}$ = concentration (8 h) for a specific substance
$\text{TLV}_{1,2,3}$ = TLV for a particular substance

For the above numbers, $\text{TWA}_{\text{mixture}}$ = 500/750 + 45/200 + 40/100 = 0.667 + 0.225 + 0.4 = 1.292. Since 1.292 is over 1, the exposure is not acceptable.

The above calculations have been performed mathematically. However, various potential errors and assumptions should be remembered:

- Remember the assumptions made by the scientists in determining the TLV (see Section 27.1.4).

- The TLV is not designed for "hypersensitive" people (e.g., people with kidney or liver problems, the fetus of pregnant women)
- Work concentration measurements may have errors. For example, heavy physical labor may have inhaled air volume of 20 m^3 of air/8 h versus the 5 m^3 of air/8 h for sedentary work. Another potential error is not to measure air in the worker's breathing zone. Use of a direct readout unit (where the sampling and analysis occur in one step) has the advantage that the second-by-second readout can be related to specific worker actions. This is much more useful than just knowing the total exposure over an 8–h period.

27.1.5 Controls

Controls are divided into engineering, administrative, and personal protective equipment. A key question for a manager is "Would you let your spouse work there?" If the answer is no, then the employees shouldn't either!

27.1.5.1 Engineering Controls

As Table 27.2 states, "Engineering controls are more desirable than administrative controls."

- *Substitute a less harmful material.* Use water-base cleaning compounds instead of organic-base. Use solvents with higher TLVs (toluene with TLV = 50 instead of benzene with TLV = 10). (The ACGIH (1995) has proposed to drop the TLV of benzene to .3.)

  When analyzing exposures, an instrument which gives minute by minute results (a direct reading instrument) is much more useful than one which gives only total exposure during the period. A videotape with a clock feature can show what is happening when exposures are high and low. For example, Edmonds et al. (1993) found that lead exposure from an unvented ladle took 10% of job time but gave 33% of lead exposure. For analysis, use a spreadsheet so average concentration, cumulative time, and cumulative exposure (product of average exposure and cumulative time) can be calculated easily.

  Marking a compound with a fluorescent tracer and then using video imaging is a powerful education/training tool (Archibald et al., 1995). By viewing their glowing hand, the workers could see the importance of using gloves even in "nonrisky" applications.
- *Change the material or process.* Reduce carbon monoxide by using electric-powered lift trucks instead of internal combustion engine powered trucks. Reduce dust by using low-speed oscillating sanders instead of high-speed rotary sanders. Coarse sandpaper produces less dust than fine sandpaper and dust particle diameters are larger (that's good) (Thorpe and Browne, 1995). Remove grinding particles or solder fumes with vacuum instead of redistributing them about the area with use of a compressed air hose.

  Offices also can have toxic materials. For example, consider asbestos. Asbestos has been used as a fire retardant on steel beams. Over time, vibrations shake fibers

Table 27.2 Controls for Respiratory Hazards Can Be Subdivided into Engineering Controls and Administrative Controls (Revoir, 1973). Engineeting Controls Are More Desirable Than Administrative Controls

Engineering Controls	Administrative Controls
1. Substitute a less harmful material	1. Screen potential employees
2. Change the material or process	2. Periodically examine employees (biological monitoring)
3. Enclose (isolate) the process	3. Train engineers, supervisors, and workers
4. Use wet methods	4. Reduce exposure time
5. Provide local exhaust ventilation	
6. Provide general ventilation	
7. Use good housekeeping	
8. Control waste disposal	

loose to be circulated by the ventilation systems. The problem is worst when the area between the dropped ceiling and the floor above is used as a ventilating plenum.

- *Enclose (isolate) the process.* Capture substances and vapors before they are dispersed. As a first approximation, cost of air handling is proportional to the volume moved. Three guidelines are:
  (1) Physically enclose the process or equipment.
  (2) Remove air from the enclosure (hood) fast enough so that air movement at all openings is into the enclosure (i.e., negative pressure).
  (3) Minimize the distance from an external hood to the source, as exhausting has approximately 10% of face velocity at only 1 dia. away from the face opening.

  A mechanical supply of air (i.e., a fan) usually is better than depending on air filtration. Fan operating cost is twofold: (1) the direct cost of electrical power for the fans and (2) the hidden cost of replacing the conditioned air (heated or cooled and humidified and purified to desired values).

  Heitbrink et al. (1994) reported use of a ventilated sander with a central vacuum system decreased dust exposure (versus a non-ventilated sander) from 22 mg/m^3 to 0.5 mg/m^3.

  Spray painting often is done in down-draft booths. Goyer (1995) found some booth designs were much more effective than others.

  The plastic strips used for strip doors also can be used to enclose machines or processes (such as solvent tanks) which require passage of product on conveyors. Enclosure also reduces the number of people exposed. Isolate pumps which could leak toxic compounds. A sealer coat on concrete floors reduces dust; less dust means less air changes per hour and thus lower heating costs. Maintenance workers need protection too. One automotive plant found that maintenance workers were exposed to beryllium dust when repairing copper alloy welding tools.
- *Use wet methods.* Reduce dust by using water or water with a wetting agent; dispose of the wetted particulate before it dries. Wet floors before sweeping. Other examples of wet methods are: rock drilling with hollow drills for the water, steaming cotton, moistened flint in potteries, and cleaning castings with high-pressure water jets (instead of abrasive blasting).
- *Provide local ventilation.* Local ventilation is used for "point" sources. Keep the worker upwind of the contaminant. Thus the air-flow sequence should be: (1) input (supply) air, (2) worker, (3) contaminant, and (4) exhaust air.

  Local ducts are much more efficient if they have a flange (flat plate perpendicular to the duct) at the entrance to direct the air. Disposal of "dirty air" by dumping the exhaust up the stack is not satisfactory. Clean the air with filters, cyclones, vapor traps, precipitators, etc. (Recycle the heat with rotary wheels, heat pipes, runaround coils, etc.) In some cases, this trapped waste can be sold—perhaps even for a profit. One example is spraying alfalfa dust in a cyclone with liquid lard from a rendering company. The resulting mixture is sold at a profit as cattle feed. Another example is sulfite liquor. Paper mills formerly dumped it into the river. When forbidden to do this, they found they could sell it at a profit as a dust suppressor on dirt roads.
- *Provide general (dilution) ventilation.* For non-point sources, use general ventilation to dilute the air. Forced ventilation (fans, blowers) is preferable to natural ventilation (open windows, doors) since air velocity, volume, and direction can be controlled. Inadvertent recirculation of the exhaust air is a potential problem; exhaust air should escape from the "cavity" which forms as a result of wind movement around the building.
- *Use good housekeeping.* Conventional vacuum sweepers may stir up more dust than they remove. Eliminate piles or open containers of chemicals. Fix leaking containers; immediately clean up spills of chemicals. The storage area should be fail-safe so that when a container leaks or is broken, the contents are contained.
- *Control waste disposal.* Each disposal problem should be considered separately. In addition to "conventional" waste, establish specific procedures for: unused dangerous substances, toxic residues, material containers, and containers with missing labels.

Drain systems become chemical storage systems. Be sure you know what is going to mix in your drains. In one GM plant, management had sewer covers tack-welded shut to prevent employees from dumping waste down the sewer. What do your employees do when management is not looking?

27.1.5.2 Administrative Controls

See Table 27.2.

- *Screen potential employees.* People can be (1) hypersensitive (e.g., to an allergy); (2) hypersusceptible (e.g., more likely to be affected); or (3) compromised (those who already have a problem). The goal is to restrict use of such people (just as people with bad backs have lifting restrictions). Some examples would be people with allergic contact dermatitis, people with impaired livers and kidneys, pregnant women who would be exposed to teratogens, and cigarette smokers exposed to asbestos or cotton plant bracts.
- *Use biological monitoring.* So far, the concept has been to monitor the environmental air and compare the concentration to its TLV. Biological monitoring monitors the substance concentration or a metabolite inside the person (blood, urine, hair, fingernails, or expired air). For example, the Biological Exposure Index (BEI) for carbon monoxide at the end of the shift is 20 ppm in end-expired air. The BEI for mercury is 15 μg/L of blood at the end of the shift at the end of the week.
- *Train the supervisors, engineers, and workers.* Manufacturers and distributors of hazardous chemicals must provide Material Safety Data Sheets (MSDSs) that identify the physical and health hazards of their products. Employers are responsible to (1) inform their employees about the chemicals and (2) train them in safe use of the chemicals.

  For training, emphasize costs and legal aspects to supervisors. Give details to engineers and workers. For example, tell the workers specific information such as "Change your respirator filter every four hours" instead of glittering generalities such as "Change filter when needed."
- *Reduced exposure time.* Reduced exposure time not only reduces exposure but also increases recovery time. For example, a person exposed 8 h/day has 16 h to recover or a recovery ratio of $16/8 = 2$ h recovery/h of exposure. But if exposure is cut to 4 h day (say by using two workers doing job sharing), then each person is exposed for only 4 h and the recovery ratio goes to $20/4 = 5$ h/h of exposure. Note that long shifts can cause problems. A 12-h shift has 12 h of recovery or $12/12 = 1$ h/h of exposure. Note that not all work time is necessarily exposure time.

The above assumes that both the effect and the recovery are linear versus time. However, the effect tends to increase exponentially with time and the recovery declines exponentially with time. That is, if you receive 50 ppm exposure, the first hour's exposure may be less stressful to the body than exposure when the body's defenses are being overwhelmed. In addition, most of the recovery occurs early in the break (recovery declines exponentially versus time) so, for the same total break time, short breaks often are better than long breaks occasionally. Thus, rotate jobs every couple of hours rather than rotating within days of the week. That is, it is better for Joe and Pete to alternate jobs every hour rather than Joe working on job A all day Monday and Pete working on job A all day Tuesday. Breaks within the working period (coffee, lunch) are more useful than breaks after work since the dose can be compensated for with a shorter time lag.

The concentration of a compound in the body will decline exponentially with time; the time necessary to eliminate 50% of the absorbed material is the biologic half-life. A rule of thumb is that there is a cumulative body burden (i.e., non-complete recovery to zero concentration overnight) when exposure time is greater than five half-lives (Paustenback, 1994). Holidays, weekends, and vacations help the body in these situations.

27.1.5.3 Personal Protective Equipment

Examples are safety shoes for the feet, aprons and leggings for the legs and torso (including clothing to protect against heat, cold, chemicals, welding, radiation), gloves and gauntlets (non-permeable for chemicals; tough for abrasion resistance) for the hands, respirators to protect the lungs, earmuffs and earplugs to protect the ears, safety glasses for the eyes, helmets to protect the skull, and hairnets to protect the hair from rotating machinery.

There are two problems with protective clothing. First, it is the last line of defense; if it fails, injury results. Second, much protective clothing decreases the comfort or performance of the person and thus causes a temptation not to use it.

Since workers rarely have the technical knowledge to select safety equipment properly and are tempted to purchase inferior protection to save money, organizations should purchase protective clothing and give it to the employees at no cost. Organizations also should control its maintenance.

27.2 AIR VOLUME AND QUALITY

Indoor air is removed from an environment to reduce (dilute) contamination. Environments are divided into "clean" (office) and "dirty" (shop).

27.2.1 Clean (Office) Environments

Office environments, generally but not always, have minimal contamination.

Lack of oxygen is not a problem since a sedentary person (met = 1) (1 met = 58.2 W/m^2, assuming a 70 kg person with 1.8 m^2 of surface area) has oxygen requirements of about 0.005 L/s. Assuming air is 21% oxygen and 25% of the air breathed is consumed, air requirements are only 0.12 L/s.

Carbon dioxide may be limiting. Outside air is about .03% carbon dioxide. ASHRAE (1989) uses .10% (1000 ppm) as the maximum recommended carbon dioxide level.

More commonly, air volume requirements are determined by odors. See Box 27.1 for odor perception. Most commonly, odors are from people and cigarettes. However, there may be odors from materials in the office or from contaminants in the ventilation (supply) air.

BOX 27.1 ODOR PERCEPTION

Professor P. Fanger of the Technical University of Denmark developed the two odor units (the olf and the decipol). The olf (from the Latin "olfactus" = olfactory sense) is the pollution emission rate; it is analogous to lumen (light) and watt (noise). The decipol (from the Latin "pollutio" = pollution) is the perceived level; it is analogous to lux (light) and decibel (noise).

One olf is the emission rate of air pollutants (bioeffluents) from a standard person (an average adult working in an office or similar nonindustrial workplace, sedentary, and in thermal comfort, with a hygienic standard equivalent to 0.7 bath per day). Measurement of olf values requires a panel of judges and a measurement of the supply of outside air to the space. Fanger (1988a) estimates olf values as:

Olf Value	Source
0–5	Per m^2 of materials in the office
1	Sedentary person, 1 met
5	Active person, 4 met
6	Smoker, average
11	Active person, 6 met
25	Smoker, when smoking

In a study of 15 Copenhagen offices (Fanger, 1988b), there was as average of 138 olfs, although there was an average of only 17 people. Since 1 non-smoking person gives 1 olf, the people contributed only 17/138 = 12% of the olfs. The remaining 121 olfs came from smokers (35), materials in the space (28), and the ventilation system (58). Fanger emphasized the importance of the 28 olfs from the materials and 58 olfs from the ventilation system as the typical assumption is that people contribute all the odors and the building is perfect (i.e., would have a value of 0, not 28 + 58).

One decipol is the pollution caused by 1 olf ventilated by 10 L/s of unpolluted air. That is:

$$1 \text{ decipol} = 1 \text{ olf}/10 \text{ L/s} = 0.1 \text{ olf/L/s}$$

To estimate the indoor air quality (Fanger, 1988a):

$$PD = 395 \exp(-3.25/P^{.25})$$

where

PD = people dissatisfied, percent
P = perceived air pollution, decipols

Higher humidity reduces odor intensity. For minimum odor perception and irritation, keep air water vapor pressure between 10 and 15 torr (1 torr = 1 mm Hg).

See ASHRAE (1993) for more on odors (Chapter 12) and air contaminants (Chapter 11).

ASHRAE (1989), based on odor detection by 80% or more of visitors, recommends 8 L/s (15 cfm) of outside air/person. For office space and conference rooms, the recommendation is 10 L/s (20 cfm) of outside air per person.

Recognizing the health dangers and cognitive effects of passive smoking, ASHRAE sets higher limits where smoking is permitted. For the typical office metabolic rate (1.5–2.5 met), ASHRAE recommends 17.5 L/s person. Thus banning smoking in an area or building can reduce ventilation requirements (and thus, costs) substantially.

For forced-air heating and cooling, the ventilation volume also may be determined by the room temperature. That is, to keep the temperature up in winter, warm air needs to be added. To keep the temperature down in summer, cool air needs to be added.

If the space is not occupied, odors and thermal comfort values are not a consideration. Thus ventilation requirements are drastically reduced. Since the typical office is occupied less than 50 of the week's 168 h, there is great potential for reduction of ventilation (and thus energy saving) during the time the space is not occupied.

Rather than replace the inside air completely with outside air, some of the inside air can be recirculated (if odors are removed, say by electronic precipitators). The primary advantage of the recirculated air is that it does not need to be conditioned (brought to the desired temperature and humidity levels).

The emphasis on minimizing the supply of outside air to save money has led to "tight" buildings. However, if the building air has some interior pollutants (which are diluted too slowly), the result is "sick building syndrome" and "building-related illness." Example pollutants include cigarette smoke, cleaning compounds, hydraulic elevator fluid, ozone (from copiers), and even emissions from new furniture. Carbon monoxide (from cars) can enter through supply air ducts located near busy streets or through infiltration (e.g., elevators from underground parking garages leading to the building). Micro-organisms from filters and drain parts of the HVAC system may cause bioaerosol and mold contamination.

Another potential problem is when a building is fogged with insecticides (Currie et al., 1990). Before fogging, remove or store coffee cups, stationery, and clothing. The fogging should be done Friday evening and the building well ventilated over the weekend. On new buildings, the ventilation should operate 24 h day, 7 days/week until Volitle Organic Compounds in the furniture, rugs, paint, etc. have completed their off-gassing.

Over time, fans stop working, dampers freeze shut, filters leak, occupant density in a space changes, etc. Have maintenance do periodic inspections to be sure the ventilation system works properly and is adjusted for the present occupancy.

27.2.2 Dirty (Shop) Environments

Here the problem is occupant health rather than comfort as there may be contaminants such as welding or solder fumes, solvent evaporation, smoke from ovens, etc.

The first strategy, source control, should be decreasing the concentration of the airborne contaminant (containment, isolation, substitution, and change of operating procedures). Then, through administrative control, reduce exposure duration. Finally, consider ventilation (i.e., dilution and removal).

The first strategy, source control, prevents pollutant generation. Less pollutants will reduce the need for ventilation. The capital cost of most ventilation systems over their life is relatively small compared with the operating cost. The operating cost is twofold: (1) the direct cost of electrical power for the fan and (2) the hidden cost of replacing the conditioned air (heated or cooled and humidified and purified to desired values) with new conditioned air. Thus, consider energy-recovery devices such as rotary wheels, fixed plates, heat coils, and runaround coils.

Solvents (e.g., acetone, ethanol, xylene) may cause an explosion problem. The Lower Explosive Limit (LEL) quantifies the risk. Since the Threshold Limit Values are 1 to 3%

of the LEL, ventilation to protect people's health generally protects against explosions. For example, for acetone, the TLV is 750 ppm, while the LEL is 25,500; thus, the TLV is limiting.

There are two general ventilation approaches: general (area, dilution) ventilation, and local exhaust ventilation. If the contaminant source is discrete, local exhaust ventilation usually is best.

27.2.2.1 General Ventilation

Use general ventilation when contaminant sources are diffuse and relatively non-toxic. The design rule is to keep the contaminant source between the person and the exhaust—that is, keep the source downwind from the person. In a multiple-source room, the air inlet (supply) and exhaust locations usually are fixed. The two things that can be varied easily are (1) the orientation of the equipment relative to the exhaust and (2) the location of the equipment in the room. Therefore, try to locate the workstation so the source is downwind from the operator. Try to locate the source as close to the exhaust as possible.

Clean rooms are used in the electronic and pharmaceutical industries when the process cannot be contaminated. The input air is specially filtered, the air-flow direction tends to be a vertical "shower of air" entering from the ceiling and leaving through the floor, and local contamination (such as from people's hair and hands) is carefully controlled.

27.2.2.2 Local Exhaust Ventilation

Rather than wait until the contaminant is dispersed, local exhaust ventilation attempts direct capture of the contaminant with a hood. Depending on the contaminant, the contaminated exhaust air can be dumped into the outside air or can be filtered and cleaned. The filtered/cleaned air can be exhausted or recycled into the work area. Be sure that any filter/cleaning system is fail-safe. That is, if the system fails, the exhaust air must not be recycled into inhabited space. Particles can be removed (mechanical filters and electronic air filters) as well as vapors (activated charcoal). For more on ventilation, see ASHRAE's *HVAC Applications* (1991, Chapter 25) and ASHRAE's *Handbook of Fundamentals* (1993, Chapter 23).

You can recycle office air by using the exhaust air from the office as part of the supply air for paint booths and ovens.

27.3 THERMAL COMFORT

27.3.1 Comfort Variables

ASHRAE defines comfort as "that state of mind which expresses satisfaction with the thermal environment." Comfort is influenced by seven major factors. Four are environmental (dry bulb temperature, water vapor pressure, air velocity, and radiant temperature), two are individual (metabolic rate and clothing), and one is time of exposure.

Figure 27.2, the psychrometric chart, gives the relation between adjusted dry bulb temperature (horizontal axis) and water vapor pressure (vertical axis).

27.3.2 Comfort for Standard Conditions

Figure 27.2 has a cross-hatched area giving the ASHRAE comfort zone.

The ASHRAE standard comfort zone is based on studies of 1600 subjects at Kansas State University; it has been confirmed by studies in other countries. Subjects were exposed to a variety of temperatures and humidities for 3 h; metabolic rate was standardized (sedentary sitting) and clothing standardized at 0.6 clo (see Chapter 29 for explanations of clo value). The subjects voted their comfort: TS = 1 = cold; 2 = cool; 3 = slightly cool; 4 = comfortable; 5 = slightly warm, 6 = warm; 7 = hot.

You can't satisfy all the people any of the time! For the conditions within the comfort zone, 0 voted for hot (7) or cold (1). However, 3% were warm (6) or cool (2) and 94% were slightly warm (5), slightly cool (3), or comfortable (4). Although the mean vote was 4.0, the standard deviation was .7 (Rohles and Nevins, 1973).

The comfort zone is valid for persons dressed with .5 to .7 clo with a metabolic rate of "office work." The comfort zone is different in the winter and the summer as it is assumed that people will wear more clothes in the winter than the summer. That is, clothing worn inside is affected by clothing worn outside.

The average Thermal Sensation (TS) vote can be predicted (Rohles et al., 1975):

$$TS = -1.047 + .158\ ET^* \qquad ET^* < 20.7$$

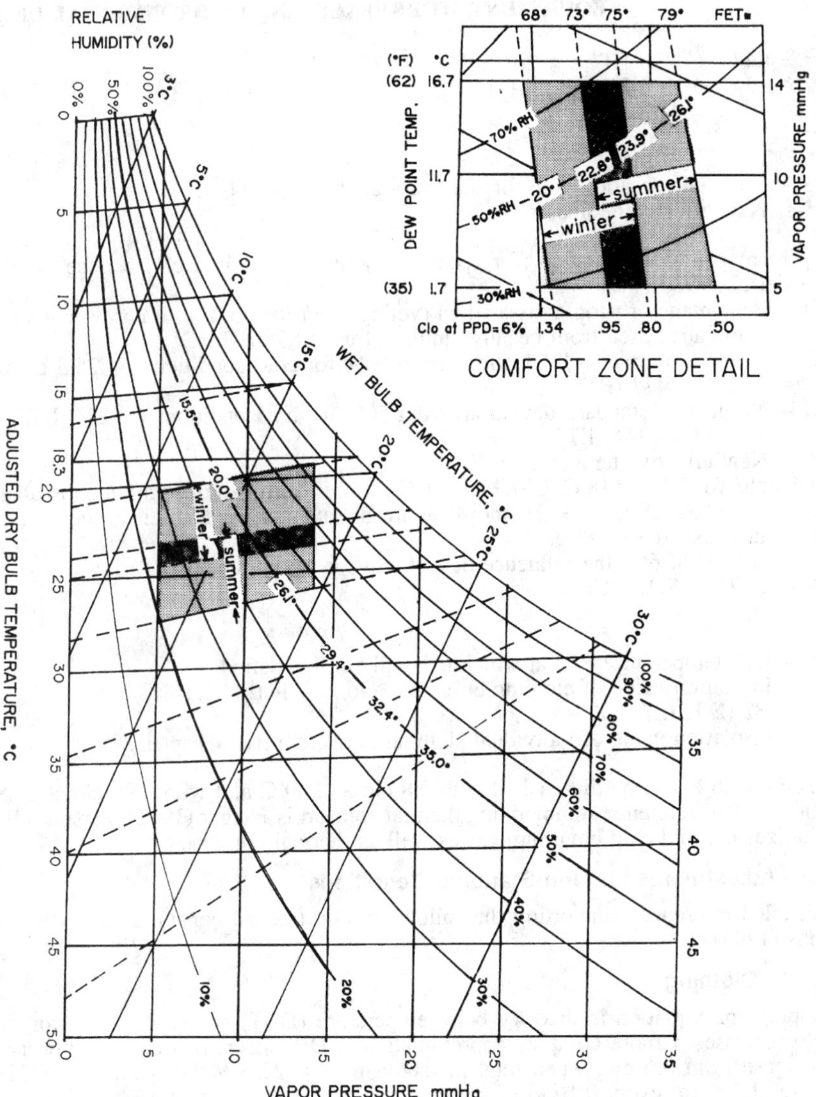

Figure 27.2 The psychrometric chart shows the relationship between temperature and humidity. The horizontal axis is adjusted dry bulb temperature (mean of air temperature and radiant temperature). The vertical axis is relative humidity (left axis) or absolute humidity (right axis).

A specific example location (point) has dry bulb temperature of 25°C and vapor pressure of 15 mm Hg (i.e., 15 torr).

- To read 64% relative humidity, use the curve coming from the left (23.5 torr is the maximum water vapor pressure in 25°C air; 15/23.5 = 64%).
- To read 17.5 dew point temperature, move horizontally to the left until you hit the 100% humidity line and then drop a perpendicular to the temperature axis.
- To read 20.2°C psychrometric wet bulb, move to the left using the solid slanting line until you hit the 100% humidity line and then drop a perpendicular to the temperature axis.
- To read 24°C effective temperature (ET), move up the dashed line until you hit the 100% humidity line and then drop a perpendicular to the temperature axis. Any point along a dashed line has approximately the same skin wettedness and gives approximately the same comfort. However, people had a hard time visualizing a 100% humidity environment. So in 1971 the number attached to the line was determined by a perpendicular from its intersection with the 50% relative humidity. ASHRAE identifies the value at ET*, new effective temperature; weather reporters call it the "temperature index."
- To read 25.5°C new effective temperature (ET*), move up the dashed line until you hit the 50% humidity line and then drop a perpendicular to the temperature axis.

(From Stephan Konz, *Work Design: Industrial Ergonomics,* 4th ed. Scottsdale, AZ: Gorsuch Scarisbrick, Publishers, 1995. Reprinted by permission of the publisher.)

$$TS = -4.444 + .326\ ET^* \qquad 20.7ET^* < 31.7$$
$$TS = -2.547 + .106\ ET^* \qquad ET^* < 31.7$$

where

TS = Thermal sensation vote for sedentary activity and .60 clo
ET* = New effective temperature, °C

The percent people dissatisfied (for sedentary activity and .5–.6 clo) can be predicted (Rohles et al., 1980):

PPD = Percentage of people dissatisfied (voting other than 3,4,5). It corresponds to the cumulative area from negative infinity for CSIG or HSIG.
CSIG = Number of standard deviations from 50% for cold conditions (< 25.3 ET*)
= 10.26 − .477 (ET*)
HSIG = Number of standard deviations from 50% for hot conditions (> 25.3 ET*)
= −10.53 + .344 (ET*)
ET* = New effective temperature, °C

For example, for ET* of 18 C, CSIG = + 1.67; from a normal table, 95% are dissatisfied. For ET* of 30°C, HSIG = −.21 and 42% are dissatisfied. At 25.3 ET*, the minimum (6%) are dissatisfied.

Given a PPD of 6%, the influence of clothing can be predicted:

$$ET^*6 = 29.75 - 7.27\ (ICL)$$

where

ET*6 = ET* temperature (°C) at which 6% will be dissatisfied
ICL = Insulation value of clothing ensemble, clo ICL < 1.1
= .82 (Σ ICLI)
ICLI = Insulation value of individual clothing items, clo (see Chapter 29)

For example, 6% will be dissatisfied with 0.8 clo at 23.9°C and with 0.95 clo at 22.8°C.

The standard reference summarizing thermal comfort is Fanger (1972). Parsons (1993) summarizes hot and cold environments as well as thermal comfort.

27.3.3 Adjustments for Non-Standard Conditions

For detailed references supporting the following adjustments, consult Konz (1995) and Parsons (1993).

27.3.3.1 Clothing

A simple approximation is that dry bulb temperature (DBT) decreases 0.6°C for every 0.1 clo increase. A more complex approximation is DBT decreases 0.6°C for every 0.1 clo increase from 0.6 clo, when total metabolism is < 225 W. For over 225 W, DBT decreases 1.2°C for every 0.1 clo increase.

27.3.3.2 Activity

Decrease DBT by 1.7°C for each 30 W increase in metabolism above 115 W. (1.0 met = seated quietly; 1.2 met = sedentary activity in office, home, school, or standing relaxed; 1 met = 58 W/m^2; a standard man = 1.8 m^2.)

27.3.3.3 Air Velocity

Up to 0.6 m/s, increase DBT by .3°C for each 0.1 m/s increase in air velocity. Between 0.6 and 1.0 m/s, increase DBT by 0.15°C for each .1 m/s increase. For a box fan (i.e., turbulent air flow) at 0.8 and 1.3 m/s, increase DBT by 0.3°C for each .1 m/s increase.

For fixed work positions with light activity, use velocity of 0.2 to 0.3 m/s. A maximum would be 0.7 to 1.0 m/s. For high metabolic rates and intermittent exposure (oscillating fan), 0.5 to 1.5 m/s is most common for spot cooling of workplaces but values of 5 to 10 are acceptable and may even go to 20 m/s. Note that air velocity from a fan drops off exponentially with distance from the fan so velocity at the fan is not the same as velocity at the person.

The effect of local cooling of the body (a "draft") can be predicted (Parsons, 1993, p. 150):

$$PPDD = 0.62\ (34 - t_a)\ (v - 0.5)\ (3.14 + 0.37\ v\ T_u)$$

where

PPDD = Percent people dissatisfied (draft)
t_a = local air temperature,°C $\quad 20 < t_a < 26$
v = local mean air velocity, m/s $\quad 0.05 < v < 0.4$
T_u = turbulence intensity (%), ratio of standard deviation of the local air velocity/local mean air velocity. If T_u is unknown, use 40%.

The air temperature difference between ankle (0.1 m above floor) and head (1.1 m above floor) should be less than 0.3°C.

27.3.3.4 Mean Radiant Temperature (MRT)

MRT is the average radiant temperature coming from all directions; it depends on the shape presented to each direction.

Change dry bulb temperature 1°C in the opposite direction of the change in MRT. For example, 25°C is a comfortable DBT when MRT also equals 25°C. But if MRT = 27, then DBT should be 23.

27.3.3.5 Time of Exposure

The comfort votes discussed above are based on the vote after 3 h of exposure. People voted about 0.3 vote warmer after 1 h than after 3 h but, for consistency, use the standard 3 h values.

27.3.3.6 Time of Day

Even though core temperature fluctuates on a 24-h cycle, thermal comfort does not vary with time of day.

27.3.3.7 Nationality/Season

Somewhat surprisingly, comfort temperature does not vary with nationality or season of the year—if clothing and metabolic rate are standardized.

However, humidity control becomes a problem in winter. Cold outside air might have only 7 torr of water vapor. When this air is brought inside and passes by a person's nose or respiratory tract (with moisture at about 45 torr), the 38 torr differential transfers water from the person to the air. This drying out of people causes sore throats, colds, and upper respiratory problems. Keep interior humidity above 12 torr to reduce colds and respiratory problems. In general, for comfort, 0.1 torr = 0.1 DBT.

27.3.3.8 Gender

If men and women have the same clothing insulation values, they have the same comfort temperature. Note, however, that women tend to wear clothing with lower insulation values.

27.3.3.9 Age

If older and younger people have the same clothing insulation values and metabolic rate, they have the same comfort temperature. Note, however, that metabolic rate tends to decline with age, so older people usually prefer warmer rooms than younger people do.

REFERENCES

ACGIH (1995). 1995–1996 Threshold Limit Values and Biological Exposure Limits, (Am. Conference of Governmental Industrial Hygienists, Cincinnati).

Archibald, B., Solomon, K., and Stephenson, G. (1995). Estimation of pesticide exposure to greenhouse applicators using video imaging and other assessment techniques. *American Industrial Hygiene Association J., 56(3)*, 226–235.

ASHRAE (1988). *HVAC Systems and Applications*. American Society of Heating, Refrigeration and Air Conditioning Engineers, Atlanta.

ASHRAE (1989). *Ventilation for Acceptable Indoor Air Quality* (Std. 62-1989). American Society of Heating, Refrigeration and Air Conditioning Engineers, Atlanta.

ASHRAE (1993). *Handbook of Fundamentals*. American Society of Heating, Refrigeration and Air Conditioning Engineers, Atlanta.

Brooks, L., Merkel, S., Glowatz, M., Comstock, M. and Shoner, L. (1994). A comprehensive reproductive health program in the workplace. *American Industrial Hygiene Association J., 55(4)*, 352–357.

Burge, P. (1989). The health effects of cutting fluids on the lungs. *Proceedings of a Conference on the Health Hazards of Cutting Oils and Their Controls*, by D. Glass, Ed., Institute of Occupational Health Birmingham, 13–28.

Currie, K., McDonald, E., Chung, I., and Higgs, A. (1990). Concentrations of diazinon, chlorpyrifos, and bendicarb after application in offices. *American Industrial Hygiene Association J., 51(1)*, 23–27.

Edmonds, M., Gressel, M., O'Brien, D., and Clark, N. (1993). Reducing exposures during the pouring operations of a brass foundry. *American Industrial Hygiene Association J., 54(5)*, 260–266.

Fanger, P. (1972). *Thermal Comfort.* New York: McGraw-Hill.

Fanger, P. (1988a). The olf and the decipol. *ASHRAE Journal, 30(10)*, 35–38.

Fanger, P. (1988b). Hidden olfs in sick buildings, *ASHRAE Journal, 30(11)*, 40–43.

Goyer, N. (1995). Performance of painting booths equipped with down-draft ventilation. *American Industrial Hygiene Association J., 56(3)*, 258–265.

Heitbrink, W., Cooper, T., and Edmonds, M. (1994). Evaluation of ventilated sanders in the autobody repair industry. *American Industrial Hygiene Association J., 55(8)*, 756–759.

Hogan, D. (1986). Skin disorders are high on the list of occupational health hazards. *Occupational Health and Safety*, 42–45, October.

Konz, S. (1995). *Work Design: Industrial Ergonomics*, Fourth Edition. Scottsdale, AZ: Publishing Horizons.

Mastromatteo, E. (1981). On the concept of threshold. *American Industrial Hygiene Association J., 42(11)*, 763–770.

Parsons, K. (1993). *Thermal Environments.* London: Taylor and Francis.

Paustenback, D. (1994). Occupational exposure limits, pharmacokinetics and unusual work schedules. In R. Harris, L. J. Cralley, and L. V. Cralley, Eds., *Patty's Industrial Hygiene and Toxicology*, Third Edition, Vol. 3A, p. 222–248, New York: John Wiley.

Revior, W. (1973). Control of respiratory hazards. *Safety Sentinel*, Southbridge, MA: American Optical Co.

Roach, S. (1994). On assessment of hazards to health at work. *American Industrial Hygiene Association J., 55(12)*, 1125–1130.

Rohles, F., Hayter, R., and Milliken, G. (1975). Effective temperature (ET*) as a predictor of thermal comfort. *ASHRAE Transactions, 81(2)*, 148–156.

Rohles, F., Konz, S., and Munson, D. (1980). Estimating occupant satisfaction from effective temperature (ET*). *Proceedings of the Human Factors Society*, 223–227.

Rohles, R., and Nevins, R. (1973). Thermal comfort: New directions and standards. *Aerospace Medicine, 44*, 730–748, July.

Ruane, P. (1989). Control of potential health hazards of formulation practice. *Proceedings of a Conference on The Health Hazards of Cutting Oils and Their Controls.* D. Glass, Ed., Institute of Occupational Health, Birmingham, 57–69.

Thorpe, A., and Brown, R. (1995). Factors influencing the production of dust during the hand sanding of wood. *American Industrial Hygiene Association J., 56(3)*, 236–242.

Zenz, C. (1984). Reproductive risks in the workplace. *National Safety News*, 38–46, September.

CHAPTER 28
CLIMATE AND CLOTHING

Carolyn K. Bensel
U.S. Army Natick Research, Development and Engineering Center
Natick, MA 01760 USA

William R. Santee
U.S. Army Research Institute of Environmental Medicine
Natick, MA 01760 USA

28.1 INTRODUCTION

Whether they work in natural or built environments, people are frequently exposed to thermal conditions that challenge the body's thermoregulatory mechanisms. The body's core temperature for normal functioning is 37°C ± 1°C. Without intervention of the thermoregulatory mechanisms, a stable core temperature could not be maintained in extreme heat or cold. Performance efficiency would deteriorate, and health and survival would be at risk.

Clothing is another intervention. It modifies the exchange of heat between the body and the thermal environment. Personal protective equipment required because of nonthermal hazards also affects heat exchange.

Minimizing the effects of high and low temperature environments on human performance, health, and safety requires an understanding of basic environmental, physiological, and clothing variables.

28.2 ENVIRONMENTAL VARIABLES

Thermoregulation is accomplished by exchange of heat between the body and the surrounding environment. The environmental parameters that determine the potential for heat exchange make up the thermal environment. Information on specific instruments to measure the environment and worksite placement of the instruments is available from a number of sources (e.g., ASHRAE Standard 55, 1992; ISO 7726, 1985; Parsons, 1993; Santee et al., 1994).

28.2.1 Air Temperature (T_a)

Air temperature and dry-bulb temperature are synonymous. Air temperature is measured using thermometers, thermistors, or thermocouples that are shaded from radiation sources.

28.2.2 Air Movement (V)

Wind speed and air velocity are common terms for air movement. Indoors, at low velocities, measurements are made with hot wire or heated-bead anemometers. Outdoor wind speeds are measured with the more rugged mechanical cup or propeller anemometers.

28.2.3 Humidity

There are a number of expressions for humidity, the amount of water vapor in a given quantity of air. These include water vapor pressure, dew-point temperature, absolute humidity, and relative humidity. Relative humidity (*R.H.*) is the ratio between the actual moisture content and the saturation value at the prevailing temperature, expressed as a percentage. When the air is saturated with water vapor, the *R.H.* is 100%. Increases in air temperature increase the capacity of the air to hold water vapor and, consequently, the amount of water vapor required for saturation. Thus, *R.H.* is dependent on air temperature. The temperature at which saturation occurs is the dew-point temperature (T_{dp}). Relative humidity, but not T_{dp}, may need to be adjusted for barometric pressure. Wet-bulb temperature (T_{wb}) varies with humidity and equals dry-bulb temperature when *R.H.* is 100%. Natural wet-bulb temperature (T_{nwb}) is obtained by exposing a thermometer fitted with a wetted cotton wick to the natural, or prevailing, air movement. The sling psychrometer is a simple instrument for measuring psychrometric wet-bulb temperature by twirling dry-bulb and wet-bulb thermometers mounted on a sling arm. The *R.H.* is determined from the difference between the two thermometer readings.

28.2.4 Nonionizing Radiation

Ultraviolet light (UV) affects human tissue, but has little impact on the body's thermal balance. Longer wavelengths of the electromagnetic spectrum, those in the visible and near infrared (IR) range, significantly affect thermal balance. Outdoors, the body may be exposed to direct rays of solar radiation, as well as to diffuse sources. Indoors, radiation emanates from hot or cold surfaces, such as large windows, or a point source like a fireplace. Radiant energy is quantified by calculating mean radiant temperature ($\overline{T}_r$). A temperature sensor inside a black copper sphere is used to measure black globe temperature (T_{bg}). Black globe temperature in °C, air velocity in m/s, and air temperature in °C, are entered into the following formula to obtain $\overline{T}_r$ in °C.

$$\overline{T}_r = [(T_{bg} + 273)^4 + 2.5\mathrm{E}8 \cdot V^{0.6} \cdot (T_{bg} - T_a)]^{1/4} - 273$$

28.3 PHYSIOLOGICAL VARIABLES

The exchange of heat between the body and the environment is accomplished through conductance, convection, radiation, and evaporation. A dual mechanism, triggered by temperature sensors in the skin and controlled centrally by the hypothalamus, regulates heat exchange to maintain thermal balance.

28.3.1 Routes of Heat Exchange

Conductance, convection, and radiation are termed dry or sensible heat exchange; they do not involve evaporation of water from the body surface and the exchanges can be measured or "sensed" as differences in temperature. Evaporation is termed wet or insensible heat exchange; it results from a phase transition of water and is not directly reflected by a change in temperature. The potential for heat exchange is determined primarily by the thermal environment and clothing.

28.3.1.1 Conductance

Conductance is heat exchange between the skin surface and a solid surface in direct contact with it. Holding a hot object in the bare hand or lying unprotected on frozen ground results in substantial heat exchange, but, normally, the amount of heat exchanged by conductance is small.

28.3.1.2 Convection

Convection is heat exchange between the skin surface and a fluid, usually air or water, that depends on movement of the fluid molecules. If fluid temperature is lower than skin temperature, the body loses heat. If fluid temperature is higher than skin temperature, the body gains heat.

28.3.1.3 Radiation

Radiation is heat exchange by nonionizing electromagnetic waves, primarily in the solar and thermal spectrum. The flow of heat by radiation does not depend on direct contact or the presence of an intervening medium, such as air; radiant heat will pass through a vacuum. The body simultaneously absorbs radiation from external sources and emits radiation. The net radiative gain or loss depends on the magnitudes and relative temperatures of the radiating sources.

28.3.1.4 Evaporation

Water released by sweat glands absorbs body heat to change from liquid to vapor. The vapor passes into the environment if the vapor pressure on the skin is higher than the vapor pressure in the surrounding environment. When not all the released water can be absorbed by the environment, liquid sweat accumulates on the skin. Further increases in sweat production do not result in heat loss because, rather than being evaporated, the sweat drips from the skin. Water vapor, and heat, is also lost by breathing. Under normal environmental conditions, in the absence of active sweating, a person evaporates about 1 L per day, losing 29 W or more of heat over 24 hr by respiration.

28.3.2 Thermal Balance

The internal temperature of the body is a summation of the surface heat exchanges and internal heat production, with the remaining energy being either work or heat storage. The relationship among these terms is expressed in the heat balance equation.

$$\pm S = M - (\pm W_k) \pm K \pm C \pm R \pm E$$

where

S = body heat storage
M = internal heat production or energy metabolism
W_k = mechanical work, where + is energy leaving the body and − is eccentric work
K = conductance
C = convective heat exchange
R = radiative heat exchange
E = evaporative heat exchange

If the body maintains thermal balance, S is zero. The heat storage term is negative when there is a net loss of heat from the body to the environment and positive when there is heat gain. Body temperature decreases with negative heat storage and increases with positive heat storage.

28.3.3 Measurement of Physiological Variables

Environmental thermal stress is a combination of factors that adversely impact the potential for thermoregulatory heat transfer between the individual and the environment. A number of physiological parameters are used to assess the resultant strain on the body's thermoregulatory mechanisms.

28.3.3.1 Core Temperature (t_c)

The temperature of deep, central areas of the body, such as the brain, is measured at sites where temperatures approximate core temperature. Rectal temperature and esophageal temperature are often used. Oral temperature is problematic because it is influenced by breathing, eating, and the like.

28.3.3.2 Skin Temperature (t_{sk})

Skin temperature is measured using thermocouples or thermistors in contact with the skin surface. Skin temperature may vary greatly over the body's surface. Measurements are taken at a number of points and the weighted points are summed to get mean skin temperature ($\bar{t}_{sk}$). Different computational schemes are used (Teichner, 1958a), but the following is a generally accepted formula.

$$\begin{aligned}\bar{t}_{sk} = {} & 0.050 \text{ instep} + 0.150 \text{ calf} + 0.125 \text{ lateral thigh} + 0.125 \text{ medial thigh} \\ & + 0.125 \text{ back} + 0.125 \text{ chest} + 0.070 \text{ upper arm} + 0.070 \text{ lower arm} \\ & + 0.060 \text{ back of hand} + 0.100 \text{ cheek}\end{aligned}$$

28.3.3.3 Sweat Rate

The simplest method for measuring the rate of water loss through the skin and the lungs is by weighing an individual at regular intervals and calculating the change in body weight after adjusting for food and water intake.

28.3.3.4 Metabolic Rate

Metabolic heat production can be quantified by measuring oxygen consumption. One liter of oxygen consumed is approximately equal to production of 20 kJ of heat. The rates of oxygen consumption and heat production are determined by measuring the oxygen concentration and volume of expired air.

Metabolic rate is often expressed in terms of body weight (W/kg) or body surface area (W/m^2). If height (Ht) in meters (m) and weight (Wt) in kilograms (kg) are known, body surface area (A) in meters squared (m^2) can be obtained from tables (Sendroy and Collison, 1960) or calculated using the DuBois height-weight formula (DuBois and DuBois, 1916).

$$A_{Du} = 0.202 \cdot \text{Wt}^{0.425} \cdot \text{Ht}^{0.725}$$

When measurement of oxygen consumption is not feasible, estimates from past research are used. Table 28.1 presents metabolic costs for some physical activities. Extensive tables are also available (Durnin and Passmore, 1967). Estimates should be employed with caution because there are large inter- and intraindividual variations in metabolic rates.

28.3.3.5 Heart Rate

Heart rate, or pulse rate, can be measured by palpation of the carotid artery at the throat or the radial artery at the wrist. For continuous monitoring, it is recorded as an electrocardiogram from electrodes affixed to the chest. Portable devices are also available that transmit a signal from a sensor on the chest to a monitor worn on the wrist.

Table 28.1 Whole Body Metabolic Rates for Selected Tasks

Activity	Metabolic Cost (W)
Resting on ground	105
Driving truck	163
Standing guard duty	137
Doing calisthenics	378
Handling pick and shovel	465–540
Walking, 1.0 m/s on hard surface	210
Walking, 1.0 m/s on hard surface with 20-kg load	255
Walking, 1.0 m/s on hard surface with 30-kg load	292
Walking, 1.56 m/s on hard surface	361
Walking, 1.56 m/s on hard surface with 20-kg load	448
Walking, 1.56 m/s on hard surface with 30-kg load	507
Walking, 2.0 m/s on hard surface	525
Walking, 2.25 m/s on hard surface	540
Walking, 1.0 m/s in loose sand	326
Walking, 1.56 m/s in loose sand	642

Source: U.S. Army Research Institute of Environmental Medicine.

28.4 CLOTHING VARIABLES

Clothing forms the body's outer shell relative to the environment. Relative to the body, clothing establishes the immediate microenvironment to which the body is exposed. There are a number of clothing parameters that affect heat exchange between the body and the environment. The theory underlying measurement of clothing properties is presented in several reference sources (ASHRAE, 1993; Gonzalez, 1988; Parsons, 1993).

28.4.1 Resistance to Passage of Dry Heat

Resistance to heat transfer by convection and radiation is combined into one general clothing property, insulation. Insulation is expressed in an arbitrary unit, the clo. In SI units, 1 clo equals 0.155°C · m^2/W. One clo unit approximates the insulation of a business suit, plus shirt, underclothing, and the air layers between them. A boundary layer of air at the clothing surface provides some resistance to heat exchange. The insulation of this air layer (I_a) is added to the intrinsic insulation provided by the clothing and air trapped between the clothing and the skin surface (I_{cl}) to obtain total insulation (I_T). The intrinsic insulation of clothing is attributable primarily to the air trapped by the fabrics.

Electronically heated devices have been developed to measure insulation. The device is placed under controlled conditions in an environmental chamber and heated to maintain a constant surface temperature. The electrical power required to maintain that temperature is measured, and the power is entered into an equation to obtain I_T. One device, the guarded flatplate, is used to test flat pieces of fabric (ASTM D 1518-85, 1985). Insulation of finished garments and clothing ensembles is measured on thermal manikins, which are full-sized human models with copper, aluminum, or plastic surfaces (ASTM F 1291-90, 1990).

Tables 28.2 and 28.3 contain clo values obtained on thermal manikins for a variety of clothing. Lists of clo values for an extensive array of clothing and ensembles are presented in a number of reference sources (e.g., ASHRAE, 1993; ISO 9920, 1993; McCullough et al., 1985). If it is not possible to find a reference containing the I_T of layered ensembles, a rough estimate can be obtained by summing the I_{cl} values of the individual components.

Figure 28.1 presents a series of curves calculated from a formula for estimating the clo units required to maintain a given average skin temperature, t_{sk} in °C, under various ambient air temperatures, T_a in °C, and metabolic rates, M in W. A t_{sk} value of 33°C is commonly used in these calculations (Burton and Edholm, 1955) and the curves in Figure 28.1 were generated using this value. The formula for the calculations is as follows.

$$I_T = [0.148(t_{sk} - T_a)]/[M \div (58.15\ A_{Du})]$$

Table 28.2 Insulation and Water Vapor Permeability Values for Selected Torso Clothing

Clothing	I_T (clo)	i_m	i_m/I_T	f_{cl}	I_{cl} (clo)
Walking shorts, short-sleeve shirt	1.02	0.42	0.41	1.10	0.36
Sweat pants, sweat shirt	1.35	0.45	0.33	1.19	0.74
Trousers, long-sleeve shirt	1.21	0.45	0.37	1.20	0.61
Knee-length skirt, long-sleeve blouse, slip, panty hose	1.22	—	—	1.29	0.67
Long-sleeve coveralls, T-shirt	1.30	—	—	1.23	0.72
Insulated coveralls, thermal underwear	1.94	0.39	0.20	1.26	1.37
Two-piece work uniform	1.34	0.41	0.31	—	—
Chemical protective ensemble, (charcoal in foam)	2.44	0.30	0.12	—	—
Temperate zone winter clothing	3.20	0.40	0.13	—	—
Arctic winter clothing	4.30	0.43	0.10	—	—

Sources: ASHRAE, 1993; Goldman, 1988a.

28.4.2 Resistance to Evaporation

Clothing slows the rate of vapor loss from the skin to the environment. If water vapor passes completely through the clothing to the external environment, heat is transferred from the body to the environment. If water vapor recondenses on the skin or within the clothing, heat is not lost to the environment. Resistance of clothing to evaporation is expressed by the water vapor permeability index (i_m), a dimensionless index (Woodcock, 1962).

Methods for measuring permeability are modifications of those for measuring insulation. Water is introduced to the surface of a guarded flatplate or manikin so that it is released into the fabric in a manner analogous to sweat evaporating from the skin. The theoretical value of i_m can range from 0 for completely moisture impermeable clothing to a maximum of 1 for completely permeable clothing.

The maximum potential for evaporative heat transfer through the clothing to the environment is a function of the ratio of the permeability index to the total insulation. This ratio approximates the percentage of the maximum evaporative potential for a given en-

Table 28.3 Insulation Values, in Clo Units, for Selected Handwear and Footwear

Clothing	I_T
Handwear	
Wool/nylon knit glove	0.6
Polyester knit glove	0.4
Uninsulated leather glove over wool/nylon knit glove	0.8
Insulated leather glove over polyester knit glove	1.3
Insulated arctic mitten	2.5
Footwear	
Unlined leather boot with one pair of socks	1.0
Insulated leather boot with one pair of socks	1.4
Shoepac with felt liners and one pair of socks	1.5
Vapor barrier arctic boot with one pair of socks	2.1

Sources: Endrusick et al., 1992; Santee and Endrusick, 1988; Santee et al., 1993.

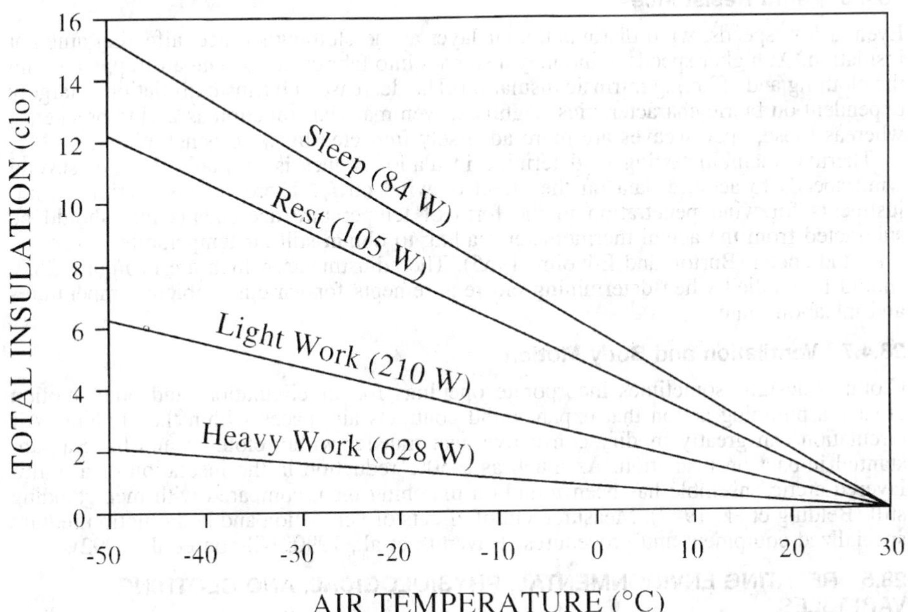

Figure 28.1 Estimated clothing insulation required for different combinations of air temperature and metabolic rate. (From *Man in a Cold Environment* (p. 108), by A. C. Burton and O. G. Edholm, 1955, London: Edward Arnold. Adapted with permission.)

vironment that may be realized when wearing specified clothing. Table 28.2 includes the values of the permeability index and the ratio i_m/I_T for a variety of clothing items.

28.4.3 Weight

Warmth with minimum weight is desirable in cold weather clothing, and a basis to compare fabrics or clothing is the insulation per unit weight. In high temperature environments, weight is a critical variable insofar as an increase in clothing weight generally represents an increase in insulation and thermal stress. There are prescribed methods for obtaining the weight of materials (ASTM D 2654-89, 1989).

28.4.4 Thickness

Clothing insulation is determined by the air trapped between fabric layers and within the fabrics themselves. Thickness is so closely correlated with insulation that measuring thickness is a way to estimate insulation. The rule of thumb is that 1 cm of thickness equals an I_T of 1.58 clo, or 1 clo equals a thickness of 0.62 cm. Fabric thickness is measured with a thickness gauge. Clothing thickness is measured on a manikin or a person where the radial thickness from outside the clothing to the skin can be obtained (Fourt and Hollies, 1970).

28.4.5 Clothing Surface Area

Heat exchange between the body and the environment takes place at the surfaces of the skin and the clothing. As the outer clothing surface area increases, the total amount of heat transferred increases. Consequently, wearing thicker and thicker clothing, which increases outer surface area, contributes proportionally less and less to protection against heat loss in a cold environment.

The clothing area factor (f_{cl}) is a dimensionless unit obtained by dividing the surface area of the clothed body (A_{cl}) by the nude body surface area (A_{Du}). Direct measurement of f_{cl} is difficult, but some references listing clo values include f_{cl} values (e.g., ASHRAE, 1993; ISO 9920, 1993).

28.4.6 Wind Resistance

Even at low speeds, wind disrupts the air layer at the clothing surface, affecting ambient insulation. At higher speeds, wind may also pass into fabrics, disturbing air trapped within the clothing and affecting intrinsic insulation. The decrease in intrinsic insulation is largely dependent on fabric characteristics. Tightly woven materials function as "wind breakers," whereas loose, open weaves are more adversely impacted by wind penetration.

Thermal manikin testing to determine insulation values is normally done at several wind speeds to acquire data on the effect of wind on I_T. There are also estimated adjustments for wind penetration in the form of temperature decrements that should be subtracted from the actual thermometer reading to obtain still air temperatures corrected for wind speed (Burton and Edholm, 1955). The adjustments, which are in Figure 28.2, should be applied when determining clo requirements for various ambient temperatures and metabolic rates.

28.4.7 Ventilation and Body Motion

Clothing designs sometimes incorporate openings for air circulation, and body motion creates a pumping action that expands and contracts air spaces within the clothing. Air circulation can greatly modify convective heat transfer within clothing and have a substantial impact on insulation. As much as a 50% reduction in the insulation of a multilayered arctic ensemble has been found on marching men, compared with men standing still (Belding et al., 1947). Measurement of effects of ventilation and body motion require specialized equipment and procedures (Havenith et al., 1990; Nilsson et al., 1992).

28.5 RELATING ENVIRONMENTAL, PHYSIOLOGICAL, AND CLOTHING VARIABLES

A number of approaches have been devised for predicting the thermal strain that an individual will experience, given the status of the key environmental, physiological, and clothing variables. The approaches include indices that reduce a number of relevant par-

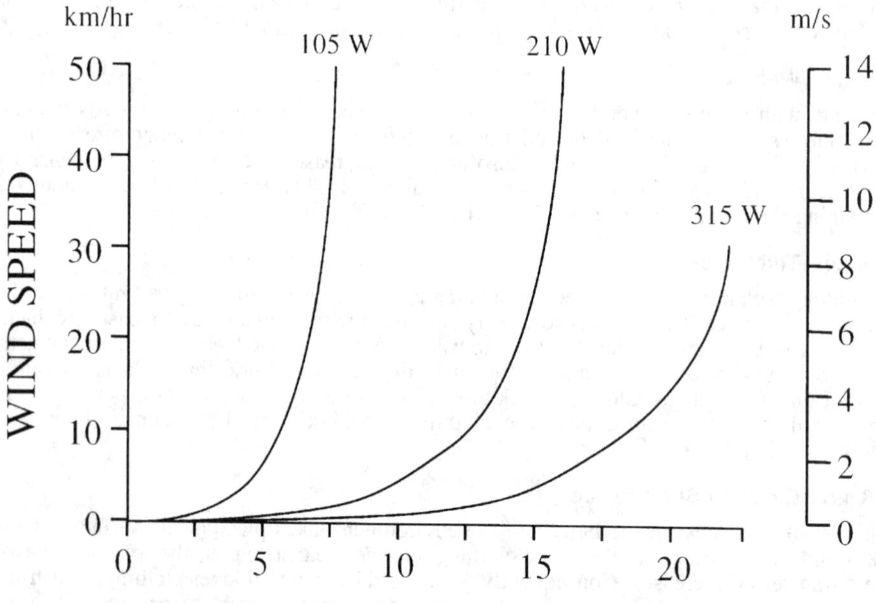

Figure 28.2 Thermal wind decrements to be subtracted from thermometer readings when determining clo requirements for various air temperatures and metabolic rates. (From *Man in a Cold Environment* (p. 116), by A. C. Burton and O. G. Edholm, 1955, London: Edward Arnold. Adapted with permission.)

ameters to a single value related to a scaled response, and more complex models that use a series of equations to calculate a number of response parameters.

28.5.1 Heat Exposure, Cold Exposure, and Effective Temperature Indices

Table 28.4 presents summaries of a number of indices, some of which are used in national and international guidelines for assessing the thermal environment (NIOSH, 1986; Parsons, 1995). The Wet-Bulb Globe Temperature (WBGT) index for hot environments was developed to provide guidance for controlling the occurrence of heat casualties among military trainees, but its use has been extended to industrial situations (NIOSH, 1986; ISO 7243, 1989). Indoors, and outdoors in the absence of a solar load, the WBGT index is computed from readings of natural wet-bulb (T_{nwb}) and black globe (T_{bg}) temperatures. The following formula is applied.

$$\text{WBGT} = 0.7T_{nwb} + 0.3T_{bg}$$

Outdoors with a solar load, air temperature (T_a) must also be measured and the formula is then as follows.

$$\text{WBGT} = 0.7T_{nwb} + 0.2T_{bg} + 0.1T_a$$

Threshold WBGT limits recommended by the National Institute for Occupational Safety and Health (NIOSH) (1986) for heat acclimatized individuals are presented in Table 28.5. The limits were set to avoid rises in core temperature above 38°C and assume that one layer of clothing is being worn. The ceiling limit set by the NIOSH recommendations is the highest WBGT to which individuals should be exposed unless they are wearing heat protective clothing. The recommendations suggest that WBGT readings be increased by 2°C when two layers of clothing are worn and by 4°C when partially impermeable ensembles are used.

The Windchill Index (WCI) for cold exposure expresses the cooling rate of wind and air temperature on bare skin and thus the danger of frostbite. Although not based on human research, the WCI is used as a guide to the severity of cold environments. Equivalent chill temperature and the expected reaction of the body are determined from wind speed and air temperature (Table 28.6). The WCI provides guidance, rather than absolute predictions. For example, Table 28.6 indicates little danger of frostbite in calm air above −26°C, but frostbite can occur whenever temperatures are below freezing.

The premise underlying equivalent temperature indices is that the net effect of the thermal environment can be characterized in temperature units. The Effective Temperature (ET) scale is illustrative of these indices. It was developed from research in which subjects compared thermal sensations experienced in still, saturated air (100% *R.H.*), set at a given temperature, with sensations produced under some other combination of temperature, humidity, and wind speed. Conditions producing the same subjective impressions of warmth or cold were assigned the same ET value. To find the ET, wind speed and air and wet-bulb temperatures are entered in a nomogram (Figure 28.3). Research indicates that the climate for long-term comfort of individuals falls between an ET of 20.0°C and 22.2°C and that an ET above 27.0°C may cause an increase in physiological strain (Schutte and Zenz, 1994).

The ET scale is widely used, but it is best suited to warm environments, where radiation effects are minimal, light clothing is being worn, and individuals are sedentary or engaging in only light activity. Furthermore, different climates with the same ET values do not result in the same tolerance times, rectal and skin temperatures, or sweat rates. The "New" Effective Temperature scale (ET*) was developed to address the limitations of the original ET (ASHRAE, 1993; Gagge et al., 1971; Gonzalez et al., 1974). More variables are entered into the calculation of ET* to obtain a more accurate prediction of thermal strain than the original ET yields.

28.5.2 Thermal Strain Models

Computer-based mathematical models for predicting physiological responses to extreme temperature environments are becoming increasingly sophisticated and are likely to be used more widely in the future than the thermal strain indices. The majority of thermal models are applicable to heat. However, a model was recently developed to predict survival time in the cold (Tikuisis, 1996). A heat strain model that has been validated for military personnel in both laboratory and field studies is the U.S. Army Research Institute

Table 28.4 Thermal Strain Indices

Index	Source(s)	Inputs	Comments
		Indices for Heat Exposure	
Wet-Bulb Globe Temperature (WBGT)	ISO 7243, 1989; NIOSH, 1986; Yaglou and Minard, 1957	T_a, T_{nwb}, T_g	WBGT requires simple input and calculations. WBGT is not recommended for conditions of high humidity.
Wet Globe Temperature (WGT)	Botsford, 1971	botsball	The botsball is an instrument consisting of a fabric-covered 60-mm black ball over a dial thermometer. WGT may be read off a scale or converted to WBGT if T_a is known.
Heat Stress Index (HSI)	Belding and Hatch, 1955; ISO 7933, 1989	T_a, T_{wb}, T_{bg} or T_r, V, M	HSI is the ratio of evaporative heat loss required to maintain a constant body temperature (E_{req}) to the maximum amount of sweat that can be evaporated under the given climatic conditions (E_{max}). Required sweat rate index (S_{req}), a further development of HSI, is used in ISO 7933.
Oxford Index (WD)	Leithead and Lind, 1964	T_a, T_{wb}	WD originated from research on men performing rescue tasks in hot underground mines. WD can be used to determine tolerance times.
		Indices for Cold Exposure	
Wind Chill Index (WCI)	Siple and Passel, 1945	T_a, V	WCI is an index for the cooling rate of wind and air temperature on bare skin. There is no adjustment for solar radiation, clothing, or activity level.
Required Clothing Insulation (IREQ)	Holmér, 1984; ISO TR 11079, 1993	T_a, $\bar{T}_r$, V, $R.H.$, M	Clothing insulation required for survival (IREQ$_{min}$) with a $\bar{t}_{sk}$ of 30°C and for maintaining thermal equilibrium (IREQ$_{neutral}$) are calculated.
		Equivalent Temperature Indices	
Effective Temperature (ET)	Houghten and Yaglou, 1923	T_a, T_{wb}, V	ET relates actual conditions to an equivalent, calm, saturated environment. ET overemphasizes effects of humidity in cool and neutral conditions and underemphasizes its effects in warm conditions.
"New" Effective Temperature (ET*)	ASHRAE, 1993; Gagge et al., 1971; Gonzalez et al., 1974	T_a, $\bar{T}_r$, V, P_a, i_m, ω, M	ET* was developed to replace ET. It includes skin-wettedness (ω) and water vapor pressure (P_a) parameters in calculating temperature of an environment at 50% *R.H.* that results in equivalent total heat loss from the skin as in the actual environment.
Operative Temperature (T_o)	Winslow et al., 1937	T_a, T_{bg} or T_r, V	T_o combines dry heat exchange parameters. There is no adjustment for work rate or effects of humidity on evaporative cooling. Formulas for approximating T_o are provided in ISO 7730 (1993).

Table 28.5 NIOSH Recommended WBGT Limits, in °C, for Heat Stress Exposure for Acclimatized Workers

	Work Load		
Hourly Work/Rest Cycle	Light (<230 W)	Moderate (230–350 W)	Heavy (>350 W)
Continuous work	<30.0	<26.7	<25.0
75% work/25% rest	30.6	27.8	25.6
50% work/50% rest	31.7	29.4	27.8
25% work/75% rest	32.2	31.1	30.0
Ceiling limit	38.9	36.7	35.0

Note Limits are for a "standard" worker weighing 70 kg with a 1.8-m^2 body surface area.
Source: NIOSH, 1986.

of Environmental Medicine (USARIEM) model (Pandolf et al., 1986). A large number of variables are input into the model (Table 28.7); the list of variables that can be outputted is likewise extensive (Table 28.8).

The heat strain model predicts how long an activity can be sustained before a preselected level of heat casualties will occur and estimates water consumption requirements. The model also calculates a work/rest cycle that will allow an activity to be sustained indefinitely while holding heat casualties below the selected level. There is an executable version of the model that can be used as a decision aid, but it includes only military clothing and militarily relevant activities (SAIC, 1993).

28.6 HIGH TEMPERATURES AND THE ROLE OF CLOTHING

When the body overheats, blood vessels dilate (vasodilation) and heart rate increases to carry warm blood to the skin where heat is lost to the environment by radiation and convection. The sweat glands are also stimulated to release water for evaporative cooling. Radiative heat loss is a direct function of surface temperature, so the higher the skin temperature, the greater the loss of thermal radiation. When air temperature is higher than

Table 28.6 Equivalent Windchill Temperatures

COOLING POWER OF WIND EXPRESSED AS "EQUIVALENT CHILL TEMPERATURE"														
		AIR TEMPERATURE (°C)												
WIND SPEED		10	5	0	-5	-10	-15	-20	-25	-30	-35	-40	-45	-50
KM/HR	M/S	EQUIVALENT CHILL TEMPERATURE (°C)												
CALM	CALM	10	5	0	-5	-10	-15	-20	-25	-30	-35	-40	-45	-50
10	2.8	8	2	-3	-9	-14	-20	-25	-31	-37	-42	-48	-53	-59
20	5.6	3	-3	-10	-16	-23	-29	-35	-42	-48	-55	-61	-68	-74
30	8.3	1	-6	-13	-20	-27	-34	-42	-49	-56	-63	-70	-77	-84
40	11.1	-1	-8	-16	-23	-31	-38	-46	-53	-60	-68	-75	-83	-90
50	13.9	-2	-10	-18	-25	-33	-41	-48	-56	-64	-71	-79	-87	-94
60	16.7	-3	-11	-19	-27	-35	-42	-50	-58	-66	-74	-82	-90	-97
70	19.4	-4	-12	-20	-28	-35	-43	-51	-59	-67	-75	-83	-91	-99
WINDS ABOVE 70 KM/HR HAVE LITTLE ADDED EFFECT		LITTLE DANGER In < 5 hr, with dry skin				INCREASING DANGER Flesh may freeze within 1 minute			GREAT DANGER Flesh may freeze within 30 seconds					
		DANGER OF FREEZING EXPOSED FLESH FOR PROPERLY CLOTHED PERSONS												

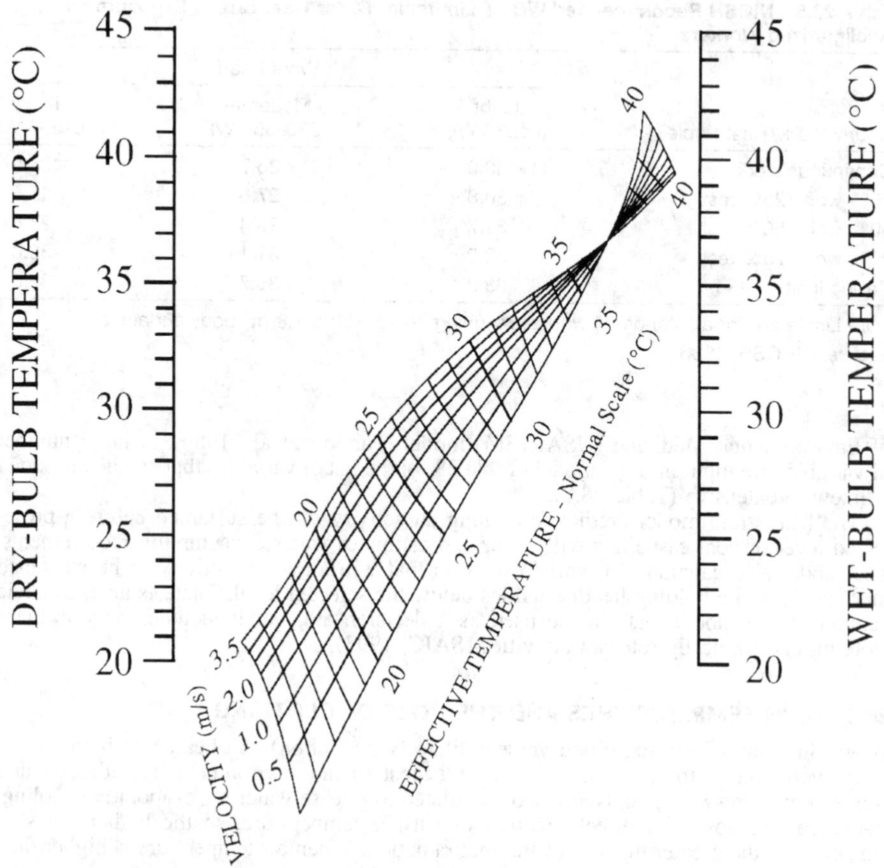

Figure 28.3 Effective temperature (ET) nomogram for lightly clothed men. (From "Physical Work and Heat Stress," by P. C. Schutte and C. Zenz. In C. Zenz, Ed., *Occupational Medicine,* Third Edition (p. 322), 1994, St. Louis: Mosby. Adapted with permission.)

skin temperature, the direction of convection is reversed and the body may gain heat. Under such conditions, the primary mechanism for heat loss is evaporation of sweat. If the water vapor is not transferred to the environment, there is no heat loss and liquid sweat accumulates on the skin. This may be followed by reduced sweat production, further increases in core temperature, and, eventually, heat illness.

A rectal temperature of 42°C is often lethal. Less extreme rectal temperatures are associated with a variety of heat illnesses (Table 28.9). Aside from the influence of environmental and clothing variables, the incidence of heat stress is affected by intra- and interindividual factors, which are listed in Table 28.10.

28.6.1 Principles of Dressing

Warm weather clothing should not impede body heat loss. Therefore, it should have low resistance to wind penetration, low thermal insulation, and high water vapor permeability, suggesting that minimal clothing be worn. The minimal approach is often impossible or impractical because of occupational and environmental hazards. Clothing, for example, does provide protection against radiant heat gain. Trade-offs must be made, and determination of the most appropriate clothing will be situation-specific. There are, however, general principles for selecting warm weather clothing. They are presented in Table 28.11.

Particularly when physical work is being performed, raising metabolic heat production, steps must be taken beyond selection of clothing if workers' health, safety, and productivity are to be maintained. Published guides for implementing heat stress control programs place priority on making engineering changes to improve environmental conditions

Table 28.7 Input Parameters for USARIEM Thermal Model

Category	Parameter	Description
Population or individual	Height	User specified or default (172 cm)
	Weight	User specified or default (70 kg)
	Level of hydration	User specified percentage or dehydrated 0–6%
	Days of acclimation	User specified in days
	Casualty level	User specified or light (<5%) to heavy
Activity	User specified	Whole body values in W
	Very light to heavy	Very light = 150 W, Heavy = 600 W
	Task menu	Specific activities (prone, walking, load carriage, etc.) in 105–642 W range
Clothing	User specified	I_T, i_m/I_T, and correction factor
	Menu of uniforms	Military clothing and ensembles, worn open or closed
Environment	Air temperature	T_a in °C or °F
	Humiditiy	*R.H.*, T_{dp}, T_{wb}, or absolute humidity
	Wind	m/s, mph, knots, or km/hr
	Solar radiation	Options include full sun, cloudy, or indoors/night

(Environmental Protection Agency, 1993; NIOSH, 1986). Where this is not feasible, other actions can be taken (Table 28.12).

As indicated in Table 28.12, acclimatization of individuals who have not worked in the heat recently is a key element in a heat stress program. Acclimatization is reflected in lower skin temperature, the onset of sweating at a lower skin temperature with increased sweat production, and reduced heart and metabolic rates. Acclimatization is achieved by exposing individuals for about 2 hr per day to temperatures and work levels approximating the conditions expected in the actual work situation. Individuals generally become fully acclimatized in 5 to 7 days (Pandolf et al., 1977).

28.6.2 Use of Personal Protective Equipment (PPE)

When PPE is worn, the heat stress imposed may negate the benefits of a worker's acclimatization, raising the risk of heat illness (Goldman, 1988b). To the extent that PPE adds

Table 28.8 Output Parameters for USARIEM Thermal Model

Category	Parameter	Units	Description
Activity duration	Maximum work time	min	Maximum value is >300 min, indicated as NL (No limit)
	Recommended work/rest cycle	min/hr	Indefinitely sustained activity below specified casualty level
Equilibrium rectal temperature (t_{re})	Final t_{re} with continuous work	°C	Based on assumption that activity could be continued until equilibrium is reached (Maximum value = 42°C)
	Final t_{re}, rest only	°C	Baseline value for combination of environmental conditions and clothing
Probability of casualties	Continuous work to final t_{re}	%	Based on USARIEM data, 5% heat casualties at 39°C, etc. Some casualties will occur below specified threshold
Water requirement	Continuous work	L/hr	Maximum consumption value is 2.1 L/hr
	Rest	L/hr	Minimum requirement is 0.2 L/hr

Table 28.9 Symptoms, Etiology, and Treatment of Heat Illnesses

Illness	Symptom	Etiology	Treatment
Heat exhaustion—In the event of collapse or unconsciousness, medical attention should be sought.	Clammy, moist skin.	Dehydration.	Rest lying down in cool, shaded area.
	Headache, dizziness, nausea, fainting, collapse.	Circulatory strain due to profusion of blood in the skin.	If individual is conscious, provide drinking water; do not use salt.
	Heart rate over 160 to 180 beats/min.	Reduced blood flow to the brain.	Loosen/remove clothing.
	t_{re}: 37.5–38.5°C.		Splash cold water on body.
			Massage arms and legs.
			If unconscious, treat for heat stroke.
			Do not return to hot environment until after overnight rest.
Dehydration exhaustion	Fatigue, nausea, headache, fainting, collapse.	Dehydration resulting from sweating.	Provide drinking water; do not use salt.
	Small volume of urine.		Do not return to hot environment until after overnight rest.
			Provide cold shower
Heat cramps	Painful spasms of arm, leg, and abdominal muscles.	Loss of body salt in sweat.	Massage affected area.
	Heavy sweating and extreme thirst.	Drinking of large volume of water dilutes electrolytes; water enters muscles resulting in spasms.	Take adequate amount of salt with meals.
			Loosen clothing.
Heat rash	Tiny raised red vesicles on skin.	Plugging of sweat gland ducts with prolonged exposure of skin to heat, humidity, sweat.	Apply mild drying lotion.
	Prickling sensation during exposure to heat.		Keep skin clean to prevent infection.
Heat stroke—Life-threatening medical emergency requiring immediate medical attention	Headache, dizziness, confusion, coma, convulsions. Onset may be sudden.	Failure of body's temperature-regulating mechanisms.	Move to shaded area, remove clothing, wrap in wet sheet, pour chilled water on and fan vigorously. Avoid overcooling.
	Hot, dry skin that is red, mottled, or cyanotic.	Reduced blood flow to brain and other vital organs.	Treat shock once temperature is lowered.
	t_{re}: 40.5°C or higher. Usually fatal if t_{re} reaches 42°C.	Dehydration and sustained exertion in heat.	Clear all vomit from nose and mouth

Sources: Environmental Protection Agency, 1993; Leithead and Lind, 1964; NIOSH, 1986.

Table 28.10 Factors Affecting the Occurrence of Heat Stress

Factor	Effect
Hydration state	Hypohydration results in lower sweat production and an increase in core temperature.
Acclimatization	Repeated heat exposure leads to earlier onset of sweating, a higher sustained sweat rate, and lower core temperature and heart rate.
Age	Sweating mechanism and circulatory system become less responsive with age, and there is high level of skin blood flow, possibly due to impaired thermoregulatory mechanism.
Physical fitness	Exercise that increases maximal aerobic capacity improves thermoregulatory responses in the heat.
Subcutaneous fat	Subcutaneous fat provides an insulative barrier, reducing transfer of heat from muscles to skin.
Gender	Although studies indicate that sweating and vasodilation occur at higher core temperatures in women than men, when controlled for fitness and menstrual phase, gender differences in the follicular phase are questionable. Women in the luteal phase have significantly higher core temperatures, which may impact thermoregulatory responses.
Body size	Leaner individuals are at an advantage because they have a larger ratio of surface area to body mass and, thus, greater capacity to dissipate heat.
Diet	Regular consumption of a balanced diet serves to replace salt and other electrolytes lost in sweat, maintaining sweating efficiency.
Previous heat illness	Previous occurrence of heat stroke increases susceptibility to subsequent heat illness.
Drugs and alcohol	Use interferes with the functioning of the central and peripheral nervous system, negatively affecting heat tolerance.

weight and bulk, higher metabolic rates also result. Furthermore, as individuals slow their work pace and take more frequent rest breaks, productivity declines.

Among the worst PPE in terms of heat stress are encapsulating ensembles, impermeable to air and water vapor, which are used to protect against toxic agents. These outfits typically include a full-face respirator and impermeable gloves and overboots. Table 28.13 is output of the USARIEM thermal model indicating suggested maximum work times, work/rest cycles, and drinking water requirements when wearing an impermeable suit and when wearing regular clothing. Even at 15°C, activities must be curtailed if heat casualties are to be avoided with the protective outfit. The NIOSH (1986) guidelines recommend that air temperature adjusted for solar heat load be used as the measure of environmental stress, rather than WBGT, when individuals are wearing impermeable ensembles, and that oral temperature and pulse rate be monitored regularly when the adjusted temperature exceeds 20°C.

Independent of the thermal burden they impose, PPE items themselves can interfere with job performance and jeopardize safety. Use of a respirator reduces speech intelligibility, and the dimensions of respirator lenses limit the visual field. Body locomotion may be impeded by bulky torso clothing and overboots. Protective gloves impair dexterity capabilities, increasing the time required to complete manual operations (Bensel, 1992).

Because of constraints on task performance and thermal strain, individuals may not wear appropriate PPE. Supervisory enforcement, although vital, is not a complete solution. This must be complemented by selection of PPE that provides the requisite level of protection and is also compatible with the work situation. Design options must be investigated and improvements in materials exploited. Furthermore, workers must be trained in their job environment while wearing the PPE. Through training, they learn to pace work output to minimize heat strain and develop skills and techniques to offset the mechanical barriers imposed by the PPE. Training also serves as familiarization with environmental distractors, such as fogged respirator lenses.

Table 28.11 Desirable Features in Hot Weather Clothing

Clothing Item	Feature
Torso clothing	Select fabrics, such as cotton, that are thin, lightweight, quick drying, and permeable to water vapor and to air.
	Select fabrics in white or light colors to minimize solar heat gain.
	Wear two-piece front-opening garment, rather than one-piece, for ease of venting and removal.
	To protect against sunburn, insects, pests, and mechanical injuries, wear long-sleeves and long trousers.
	Wear loose-fitting garments.
	Keep pockets to a minimum to avoid extra layers of fabric.
	Change into dry clothing and launder clothing frequently.
	In sunlight or around hot equipment, remain fully clothed to minimize radiant heat gain.
Headwear	To protect against solar radiation, wear a broad-brimmed hat, which also protects the skin from sunburn. Sunglasses are also needed for protection against UV injury.
	Use enclosed goggles if there is the possibility of blowing sand and dust.
Footwear	Select socks with a cushion sole and thin upper.
	Change into dry socks frequently.
	Select lightweight and low cut footwear in a material that dries quickly, such as nylon duck, unless protection from mechanical injury is required.
	If standing on hot surfaces is required, select footwear with thick, rubber outsoles.

Table 28.12 Components of a Heat Stress Control Program

Element	Components
Limit exposure time and/or temperature	Schedule heavy work for cooler times of day or year. Implement work/rest cycles. Provide cool, shaded rest areas. Add personnel. Rotate heavy work. Increase water intake. Permit worker to interrupt work. Halt work under extreme conditions.
Reduce metabolic heat load	Mechanize or redesign job. Reduce work time. Increase work force.
Enhance heat tolerance	Implement acclimatization, drinking water, and physical fitness training programs. Maintain electrolytic balance.
Train in health and safety	Monitor environmental conditions and workers. Train supervisors and workers to recognize heat stress symptoms, administer first aid, and execute contingency plans. Instruct in proper care and use of protective equipment and avoidance of drugs and alcohol. Implement buddy system.
Screen personnel for heat intolerance	Identify workers with previous heat illness, poor physical fitness levels, etc.
Establish heat alert program	Appoint and train heat alert committee, including medical personnel. Reverse winterization of work area, if required, and check drinking fountain, fan, air conditioner operation. Establish heat alert criteria and procedures. Assign responsibility for program execution.
Take additional measures	Provide auxiliary cooling garments if modifications to worker, work tasks, and environment are not adequate to alleviate risk of heat illness. Train workers in garment use and maintenance.

Sources: Environmental Protection Agency, 1993; NIOSH, 1986.

Table 28.13 USARIEM Thermal Model Outputs for a Two-Piece Work Uniform, One-Piece Coveralls With a Fire-Retardant Treatment, and a Fully Enclosed Impermeable Suit for Chemical Handling

Maximum Work Time (min)

	Work Uniform				Coveralls				Impermeable Chemical Suit			
T_a (°C)	Rest 105 W	Light 250 W	Mod. 425 W	Heavy 600 W	Rest 105 W	Light 250 W	Mod. 425 W	Heavy 600 W	Rest 105 W	Light 250 W	Mod. 425 W	Heavy 600 W
15	NL	NL	NL	NL	NL	NL	NL	NL	NL	NL	81	43
20	NL	NL	NL	171	NL	NL	NL	272	NL	NL	63	38
25	NL	NL	NL	110	NL	NL	NL	125	NL	NL	52	33
30	NL	NL	NL	78	NL	NL	NL	87	NL	96	44	28
35	NL	NL	111	55	NL	NL	138	60	NL	67	37	23
40	NL	NL	56	37	NL	NL	59	39	154	53	32	18
45	152	55	34	21	132	54	34	20	84	44	26	14

Recommended Hourly Work/Rest Cycle (min)

	Work Uniform				Coveralls				Impermeable Chemical Suit			
T_a (°C)	Rest 105 W	Light 250 W	Mod. 425 W	Heavy 600 W	Rest 105 W	Light 250 W	Mod. 425 W	Heavy 600 W	Rest 105 W	Light 250 W	Mod. 425 W	Heavy 600 W
15	60/0	60/0	60/0	60/0	60/0	60/0	60/0	60/0	60/0	60/0	30/30	18/42
20	60/0	60/0	60/0	42/18	60/0	60/0	60/0	43/17	60/0	60/0	22/38	13/47
25	60/0	60/0	60/0	36/24	60/0	60/0	60/0	38/22	60/0	60/0	11/49	5/55
30	60/0	60/0	60/0	30/30	60/0	60/0	60/0	32/28	60/0	0/60	0/60	0/60
35	60/0	60/0	33/27	21/39	60/0	60/0	36/24	22/38	60/0	0/60	0/60	0/60
40	60/0	60/0	11/49	5/55	60/0	60/0	13/47	7/53	0/60	0/60	0/60	0/60
45	0/60	0/60	0/60	0/60	0/60	0/60	0/60	0/60	0/60	0/60	0/60	0/60

Table 28.13 (*Continued*)

	Recommended Water Consumption (L/hr)											
	Work Uniform				Coveralls				Impermeable Chemical Suit			
T_a (°C)	Rest 105 W	Light 250 W	Mod. 425 W	Heavy 600 W	Rest 105 W	Light 250 W	Mod. 425 W	Heavy 600 W	Rest 105 W	Light 250 W	Mod. 425 W	Heavy 600 W
15	0.2	0.2	0.6	1.0	0.2	0.2	0.5	0.9	0.2	0.6	1.6	2.1
20	0.2	0.4	0.8	1.2	0.2	0.3	0.7	1.1	0.2	1.0	2.0	2.1
25	0.2	0.6	1.1	1.5	0.2	0.6	1.0	1.4	0.5	1.4	2.1	2.1
30	0.5	0.9	1.3	1.8	0.5	0.8	1.3	1.7	1.0	1.9	2.1	2.1
35	0.7	1.2	1.7	2.1	0.7	1.2	1.6	2.1	1.5	2.1	2.1	2.1
40	1.1	1.6	2.1	2.1	1.1	1.6	2.1	2.1	2.1	2.1	2.1	2.1
45	1.9	2.1	2.1	2.1	1.9	2.1	2.1	2.1	2.1	2.1	2.1	2.1

Note. NL = No limit. This model is a developmental analytical tool which has not been determined completely safe or suitable for use in making decisions that could affect the health and safety of personnel. Results apply only to the specified population and worker physical condition: Individuals fully hydrated, acclimatized for 15 days, average male soldier (Ht = 175.1 cm, Wt = 77.1 kg), light casualty rate (<5%). Environmental parameters: *R.H.* = 50%, wind speed = 2 m/s, full sun.

28.6.3 Auxiliary Cooling of the Body

Devices for body cooling are used to counter the stress of hot environments and the microenvironments associated with PPE. Systems include vests that hold ice or frozen gel, wetted overgarments, and more complex devices, requiring power sources and heat sinks, that deliver conditioned air or liquid to whole body garments, vests, or caps. Auxiliary cooling extends tolerance times in the heat, but devices vary in effectiveness (Speckman et al., 1988).

The selection of the most appropriate auxiliary cooling system depends on environmental parameters, workload, worker mobility requirements, and availability of resources, such as power units. The weight of portable power supplies can negate the benefits of cooling. Stationary power sources require tethers, thereby limiting mobility. Auxiliary cooling is more practical in aircraft and ground vehicles where power sources are available and operators are relatively sedentary.

28.6.4 Performance Effects

Aside from mechanical constraints of PPE and limited work durations due to heat stress, productivity can decline in heat because of poor performance of work activities. Research has focused on psychological tasks, tasks involving a minimum of physical effort and large mental or perceptual motor components. Unlike the relationship between environmental and physiological parameters in determining physical tolerance to heat, there is not clear-cut relationship on which to base predictions of the effects of heat on psychological performance. Indeed, heat may produce an increment, a decrement, or no effect on performance depending on exposure duration, acclimatization status, task skill level, and the nature of the task (Grether, 1973; Ramsey, 1995).

One interpretation of heat effects, based on dividing tasks into two categories (e.g., mental, cognitive, reaction time vs. tracking, vigilance), indicates that tasks in the former category will be minimally affected and performance may even improve, whereas a decrease in performance levels of tasks in the latter category will begin in the range of 30°C to 33°C WBGT (Ramsey, 1995). Another interpretation, drawing on the inverted-U shaped curve to relate arousal to performance, is that there is an optimal arousal level for a particular task type. Thus, heat or other extreme environmental conditions may increase arousal level, facilitating performance on one task and impairing it on another (Poulton, 1976). These interpretations are helpful in explaining extant research findings, but there is no basis on which to make specific predictions regarding the extent to which heat will affect psychological performance.

28.7 LOW TEMPERATURES AND THE ROLE OF CLOTHING

When the body is being cooled, the initial thermoregulatory response is constriction of blood vessels (vasoconstriction) to keep warm blood away from the body surface. This results in progressive cooling of the surface and a decrease in total body heat lost to the environment through convection and radiation. Blood flow to the extremities is reduced to a greater extent than to other parts of the body. The hands and feet have large surface areas relative to their mass, and decreasing blood flow is an efficient way to retard heat loss. Once cooled, resumption of blood flow to the extremities is difficult without rewarming the entire body. Body cooling is also associated with goose flesh, roughening the skin and decreasing air movement. The next response is shivering, which is initially mild and may become violent. Shivering increases metabolic heat production, but is usually not sufficient to replace heat already lost from the body.

Intra- and interindividual factors influence ability to maintain body temperatures and avoid injury in a cold environment. The principal factors are listed in Table 28.14. Depending on the status of environmental, physiological, and clothing variables, cold exposure can lead to hypothermia and, possibly, death. Critical body temperatures associated with impaired performance or cold injury are presented in Table 28.15.

28.7.1 Principles of Dressing

Cold weather clothing must meet the oftentimes incompatible requirements of providing adequate insulation, preventing overheating during exercise, and permitting the wearer to function efficiently. The layer principle is a fundamental approach to dressing for the cold. A number of thin layers are worn, rather than one or two thick layers. Insulation is adjusted to match metabolic heat production by removing clothing or opening it up for ventilation when active and putting it on or closing it when inactive. There are compro-

Table 28.14 Factors Affecting Cold Tolerance

Factor	Effect
Body size	Surface to mass ratio favors larger individuals. Smaller individuals have proportionally a greater area for heat loss. Also, larger individuals have "thermal inertia" due to greater total heat storage with greater mass.
Body shape	Because of shorter extremities and reduced surface to mass ratio, a short person with same body mass loses less heat to the environment than a tall one.
Subcutaneous fat	Subcutaneous fat provides insulation. The more subcutaneous fat, the less heat lost to the environment, particularly during cold water immersion.
Physical fitness	Regular exercise increases total heat production due to better circulation and a greater proportion of lean body mass for heat production.
Age	Younger adults are more resistant to cold injury than older because older have poorer circulation and less lean body mass for heat production. The very young, having low body mass, are not cold tolerant.
Smoking habits	Vasoconstrictive action of nicotine causes increased cooling of the extremities.
Previous cold injury	Previous injury, such as frostbite, predisposes an individual to subsequent injury.
Gender	Women have lower skin temperatures in cold than men, possibly due to body size/shape or subcutaneous fat insulation.
Drugs and alcohol	Use may modify central temperature regulation. Alcohol-induced vasodilation counters heat conservation and causes a rapid initial heat loss.

mises involved in a layered clothing system; every opening that provides for ease of venting and clothing removal also provides a route for wind and water penetration.

Layered systems have three main components: inner, intermediate, and outer layers. Table 28.16 lists the function of each component. The intermediate, insulating component may consist of more than one layer, depending on environmental conditions and activity level. Each succeeding layer of the system should fit snugly over the underlying one without compressing it, decreasing the insulation value. Layers should be added immediately when activity level is decreased and removed when exercise is increased, before the wearer begins to sweat. If the clothing becomes wet with sweat, the moisture might condense in the garments. Moisture lowers insulation value, increases weight, and promotes heat loss through evaporation from wet skin and clothing.

28.7.1.1 Torso Protection

Survival in the cold depends on keeping the torso warm and preventing the build-up of sweat within the clothing. In addition to selecting appropriate fabrics and employing a

Table 28.15 Critical Ranges of Body Temperatures During Cold Exposure

Temperature (°C)	Effect
	Body Core
35–34	Exhaustion of metabolic heat production through shivering
32	Loss of consciousness
26	Heart failure deriving from hypothermia
	Skin
18–15	Pain
5	Numbness
0	Risk of freezing

Table 28.16 Components of Layered Clothing Systems

Component	Function	Fabrics
Inner	Trap warm air next to the skin. Transfer moisture from the skin to outer layers.	Wool, polyprophylene, silk
Intermediate	Insulate, holding still air around the body.	Wool, down, synthetic batting and polyester pile
Outer	Resist wind intrusion and penetration of external moisture. Protect underlying layers from dirt and abrasion. Transfer moisture to the external environment.	Hard-finished tightly woven fabrics, polytetrafluoroethylene (PTFE) laminates and other breathable coatings

disciplined approach in managing the clothing layers, there are clothing design features that promote successful application of the layer principle to the torso. These are presented in Table 28.17.

28.7.1.2 Head, Hand, and Foot Protection

Except for the ears, there is little or no vasoconstriction in the head and face. Therefore, there can be extensive heat loss from the uncovered head. Headwear can decrease the loss substantially. Table 28.18 contains an example of the layer principle applied to headgear and Table 28.19 lists some desirable headgear features.

Protecting the hands is a particular problem in the cold. The individual fingers are very small-diameter cylinders, and the fingertips are small hemispheres. Increasing the thickness of glove fabrics in order to increase insulation greatly increases the finger surface area for heat loss. Mittens, by encasing all the fingers in one envelope of fabric, are

Table 28.17 Desirable Features in Cold Weather Clothing for the Torso

Body Area	Design Feature
Upper torso and arms	Few seams in outer layer and all seams factory-sealed
	Two-way zippered front openings in outer and intermediate layers
	Gusset backing and storm flap cover with hook-and-pile closures for front zipper in outer layer
	Zippered opening with protective cover under arms in outer layer
	All zippers made of plastic with pull tabs that can be manipulated with gloves and mittens
	Adjustable drawcord in bottom hem of outer layer
	Adjustable closures at neck and wrist of outer and intermediate layers
	Bottom hem of each layer extending to hips or lower and overlapping lower torso clothing
	Layers that stay in place when arms are raised
Lower torso and legs	Few seams in outer layer and all seams factory-sealed
	Zippered side openings in outer and intermediate layers accommodating removal of layers over boots
	Gusset backing and storm flap cover with hook-and-pile tape closures for front zipper of outer layer
	All zippers made of plastic with pull tabs that can be manipulated with gloves and mittens
	Drawcord or hook-and-pile tape adjustments at waist and ankle opening of outer layer
	Bottom hem of outer layer extending over boot tops
	Layers that stay in place during crouching and bending

Table 28.18 Layering of Headwear, Handwear, and Footwear

Layer	Clothing
	Headwear
Inner	Balaclava covering for head, face, neck
Intermediate	Cap covering ears; insulated hood attached to jacket
Outer	Wind and water resistant hood attached to jacket
	Handwear
Inner	Thin gloves covering the wrists
Intermediate	Insulated gloves or mittens covering the wrists
Outer	Wind and water resistant insulated mittens with long gauntlets extending over the wrists, onto the forearms
	Footwear
Inner	Thin socks covering the ankles
Intermediate	One or more pairs of thick socks extending above the ankles
Outer	Permeable boots with insulated insoles or impermeable, vapor barrier boots with insulated insoles; outer cover of waterproof gaiters for low-top permeable boots

superior to gloves in preventing heat loss (van Dilla et al., 1949). Application of the layer principle to the hands often involves use of both gloves and mittens (Table 28.18). The outer mittens are removed and the gloves used for some warmth and protection against contact with cold-soaked objects when dexterity is required; the outer mittens are used as handwarmers between work bouts. There is a wide array of handwear available (Gonzalez et al., 1989), and some desirable design features are presented in Table 28.19. Determination of the most appropriate type varies with dexterity requirements, the environment, and activity level.

Occurrence of cold injuries to the feet is often insidious; individuals experiencing numb feet may not realize that they are the victims of trench foot or frostbite. Applying the layer principle (Table 28.18) and keeping the feet dry (Table 28.19) are the keys to avoiding injuries.

28.7.2 Use of Personal Protective Equipment

The PPE worn to protect against nonthermal hazards can often be incorporated successfully within the cold weather clothing system, but some PPE poses problems. Lenses in

Table 28.19 Desirable Features in Cold Weather Headwear, Handwear, and Footwear

Clothing Item	Feature
Headwear	Hoods should extend beyond the face to direct warm air from within the clothing out past the face.
	Malleable wire around the hood edge allows adjustment of opening to accommodate environmental and visual requirements.
	Synthetic or fur ruff on the hood should shed frost readily.
Handwear	Handwear should stretch and recover, conforming to changing dimensions of the hand.
	Seams should not be located at fingertips where they can interfere with dexterity and serve as a conduit for heat loss.
	Handwear should not constrict the fingers, interfering with circulation and lowering hand temperatures.
Footwear	Boots should fit loosely enough to accommodate underlying sock layers and allow toe movement.
	Socks should be clean and free of worn areas.
	Damp socks and insoles should be replaced frequently during use to minimize moisture build-up and risk of cold injury.

eyewear are prone to fogging. When using a respirator and impermeable clothing for protection against toxic agents, lenses may fog, respirator valves may freeze, and the highly conductive rubber in contact with the face increases the risk of cold injury. Shrouds to prevent ice build-up on the valves are furnished with some respirators and antifogging treatments are available to retard moisture build-up on lenses. Paradoxically, because sweat is not evaporated through the impermeable garment and venting of the clothing is impossible in the presence of toxic chemicals, prolonged physical activity may lead to the occurrence of heat strain. Frequent rest breaks are required in protective shelters where the clothing can be removed and replaced.

28.7.3 Performance Effects

The performance effects associated with low temperatures are attributable to the thermal environment itself and to the clothing worn to protect against it. There is an increased energy cost of locomotion due to the weight and bulk of the clothing. Insulated footwear is relatively heavy, and it has been found that adding 1 kg to the mass of footwear increases energy cost as much as adding 5 kg to the mass of a load carried on the torso (Soule and Goldman, 1969). Furthermore, the energy costs of walking in layered clothing systems increase by 3% to 4% with each layer, an increase beyond that due to the weight change (Teitlebaum and Goldman, 1972). The increase has been attributed to frictional resistance between layers and to the hobbling effect, or the interference with joint movement, due to clothing bulk. Adding layers of cold weather clothing also restricts arm movements and limits waist flexion. Hoods interfere with head movements, hearing, and vision. Bulkier boots may be incompatible with ski and snowshoe bindings, vehicle pedals, and footholds. Even after extensive practice, dexterity capabilities with cold weather gloves are inferior to bare-hand performance (Lyman, 1957).

Productivity in a low temperature environment, and possible survival, depends on performing manual tasks. Protection of the hands from cold is often compromised in order to work without the interference of handwear. However, dexterity is impaired when hand skin temperatures fall to between 18°C and 13°C (Clark, 1961) and tactile discrimination is markedly impaired at finger temperatures below 6°C (Provins and Morton, 1960). The manner in which body core temperature, hand skin temperatures, and cooling rates interact to affect manual performance is complex. It appears that diminished blood flow to the fingers, decreased muscle temperature, increased viscosity of the synovial fluid in the finger joints, and lower responsiveness of the skin receptors all act to impair manual operations (Lockhart et al., 1975).

Cold temperatures also impair some types of complex tasks having a cognitive component (Table 28.20). Decreased performance has been attributed to distracting environmental stimuli, such as physical discomfort, that momentarily divert attention from the primary task (Teichner, 1958b).

28.7.4 Auxiliary Heating of the Body

The problem of keeping individuals warm without bulky clothing has been addressed through development of devices to heat the body. The devices range from chemically activated handwarmers to whole-body garments powered by electricity (Scott, 1988). Most emphasis has been placed on auxiliary heating of the extremities by incorporating an electrically powered heating medium in handwear. Similarly, heating elements have been

Table 28.20 Effects of Cold on Complex Tasks

Task	Finding	Reference
Simple reaction time	Unaffected down to −37°C	Teichner (1958b)
Grammatical reasoning	Unaffected during immersion in 4.7°C water	Baddeley et al. (1975)
Addition	Unaffected at 4°C	Enander (1987)
Digit classification	Increase in errors at 4°C	Enander (1987)
Pursuit tracking	Decrement at 13°C	Teichner and Wehrkamp (1954)
Vigilance	Increase in delayed responses at −3.3°C	Poulton et al. (1965)

placed in socks and boot inserts. Research indicates that provision of 3 W to each hand and 7 W to each foot is adequate to maintain skin temperatures above 4.4°C in sedentary men outfitted in arctic clothing while exposed for a 6-hr period to an ambient temperature of −40°C and a wind speed of 4.5 m/s (Goldman, 1964). Use of auxiliary heating devices is not widespread because of lack of access to power, limited durability of heating elements, high production costs, encumbrance of leads between the power supply and the clothing, and temperature regulators that cause hot spots or burning sensations on the skin (Haisman, 1988).

REFERENCES

ASHRAE (1993). *1993 ASHRAE Handbook: Fundamentals.* Atlanta: American Society of Heating, Refrigerating and Air-Conditioning Engineers.

ASHRAE Standard 55 (1992). Thermal Environment Conditions for Human Occupancy. Atlanta: American Society of Heating, Refrigerating and Air-Conditioning Engineers.

ASTM D 1518-85 (1985). Standard Test Methods for Thermal Transmittance of Textile Materials. Philadelphia: American Society for Testing and Materials.

ASTM D 2654-89 (1989). Standard Test Methods for Moisture in Textiles. Philadelphia: American Society for Testing and Materials.

ASTM F 1291-90 (1990). Standard Test Method for Measuring the Thermal Insulation of Clothing Using a Heated Manikin. Philadelphia: American Society for Testing and Materials.

Baddeley, A. D., Cuccaro, W. J., Egstrom, G. H., Weltman, G., and Willis, M. A. (1975). Cognitive efficiency of divers working in cold water. *Human Factors, 17,* 446–454.

Belding, H. S., and Hatch, T. F. (1955). Index for evaluating heat stress in terms of resulting physiological strain. *Heating, Piping and Air Conditioning, 27,* 129–136.

Belding, H. S., Russell, H. D., Darling, R. C., and Folk, G. E. (1947). Analysis of factors concerned in maintaining energy balance for dressed men in extreme cold: Effects of activity on the protective value and comfort of an arctic uniform. *American Journal of Physiology, 149,* 223–239.

Bensel, C. K. (1992). Soldiers' performance in chemical protective gear. In S. Kumar, Ed., *Advances in Industrial Ergonomics and Safety IV* (pp. 1291–1298). London: Taylor and Francis.

Botsford, J. H. (1971). A wet globe thermometer for environmental heat measurement. *American Industrial Hygiene Journal, 32,* 1–10.

Burton, A. C., and Edholm, O. G. (1955). *Man in a Cold Environment.* London: Edward Arnold.

Clark, R. E. (1961). The limiting hand skin temperature for unaffected manual performance in the cold. *Journal of Applied Psychology, 3,* 193–194.

DuBois, D., and DuBois, E. F. (1916). A formula to approximate surface area if height and weight be known. *Archives of Internal Medicine, 17,* 863–871.

Durnin, J. V. G. A., and Passmore, R. (1967). *Energy, Work and Leisure.* London: William Heinemann.

Enander, A. (1987). Effects of moderate cold on performance of psychomotor and cognitive tasks. *Ergonomics, 30,* 1431–1445.

Endrusick, T. L., Santee, W. R., DiRaimo, D. A., Blanchard, L. A., and Gonzalez, R. R. (1992). Physiological responses while wearing protective footwear in a cold-wet environment. In J. P. McBriarty and N. W. Henry, Eds., *Performance of Protective Clothing: Fourth Volume* (pp. 544–556). ASTM STP 1133. Philadelphia: American Society for Testing and Materials.

Environmental Protection Agency (1993). *A Guide to Heat Stress in Agriculture.* Document EPA-750-b-92-001. Washington, DC: Environmental Protection Agency.

Fourt, L., and Hollies, N. R. S. (1970). *Clothing: Comfort and Function.* New York: Marcel Dekker.

Gagge, A. P., Stolwijk, J. A. J., and Nishi, Y. (1971). An effective temperature scale based on a simple model of human regulatory response. *ASHRAE Transactions, 77,* 247–262.

Goldman, R. F. (1964). The arctic soldier: Possible research solutions for his protection. In G. Dahlgren, Ed., *Proceedings of the 15th AAAS Alaska Science Conference* (pp. 401–419). College, AK.

Goldman, R. F. (1988a). Biomedical effects of clothing on thermal comfort and strain. In *Handbook on Clothing: Biomedical Effects of Military Clothing and Equipment Systems.* Research Study Group 7, NATO Panel VIII. NATO, Brussels.

Goldman, R. F. (1988b). Standards for human exposure to heat. In I. B. Mekjavic, E. W. Banister, and J. B. Morrison, Eds., *Environmental Ergonomics: Sustaining Human Performance in Harsh Environments* (pp. 99–136). London: Taylor and Francis.

Gonzalez, R. R. (1988). Biophysics of heat transfer and clothing considerations. In K. B. Pandolf, M. N. Sawka, and R. R. Gonzalez, Eds., *Human Performance Physiology and Environmental Medicine at Terrestrial Extremes* (pp. 45–95). Indianapolis: Benchmark Press.

Gonzalez, R. R., Endrusick, T. L., and Santee, W. R. (1989). Thermoregulatory responses in the cold: Effect of an extended cold weather clothing system (ECWCS). *15th Commonwealth Defense Conference on Operational Clothing and Equipment.* Ottawa, Canada.

Gonzalez, R. R., Nishi, Y., and Gagge, A. P. (1974). Experimental evaluation of standard effective temperature; a new biometeorological index of man's thermal discomfort. *International Journal of Biometeorology, 18,* 1–15.

Grether, W. F. (1973). Human performance at elevated environmental temperatures, *Aerospace Medicine, 44,* 747–755.

Haisman, M F. (1988). Physiological aspects of electrically heated garments. *Ergonomics, 31,* 1049–1063.

Havenith, G., Heus, R., and Lotens, W. A. (1990). Resultant clothing insulation: A function of body movement, posture, wind, clothing fit and ensemble thickness. *Ergonomics, 33,* 67–84.

Holmér, I. (1984). Required clothing insulation (IREQ) as an analytical index of cold stress. *ASHRAE Transactions, 90,* 116–128.

Houghten, F. C., and Yaglou, C. P. (1923). Determining lines of equal comfort. *Transactions of the American Society of Heating and Ventilating Engineers, 29,* 163–176.

ISO 7243 (1989). Hot Environments—Estimation of the Heat Stress on Working Man, Based on the WBGT-Index (Web-Bulb Globe Temperature). Geneva: International Organization for Standardization.

ISO 7726 (1985). Thermal Environments—Instruments and Methods for Measuring Physical Quantities. Geneva: International Organization for Standardization.

ISO 7730 (1993). Moderate Thermal Environments—Determination of the PMV and PPD Indices and Specification of the Conditions for Thermal Comfort. Geneva: International Organization for Standardization.

ISO 7933 (1989). Hot Environments—Analytical Determination and Interpretation of Thermal Stress Using Calculation of Required Sweat Rate. Geneva: International Organization for Standardization.

ISO 9920 (1993). Estimation of the Thermal Insulation and Evaporative Resistance of a Clothing Ensemble. Geneva: International Organization for Standardization.

ISO TR 11079 (1993). Evaluation of Cold Environments—Determination of Required Clothing Insulation (IREQ). Geneva: International Organization for Standardization.

Leithead, C. S., and Lind, A. R. (1964). *Heat Stress and Heat Disorders.* Philadelphia: F. A. Davis.

Lockhart, J. M., Kiess, H. O., and Clegg, T. J. (1975). Effect of rate and level of lowered surface temperature on manual performance. *Journal of Applied Psychology, 60,* 106–113.

Lyman, J. (1957). The effects of equipment design on manual performance. In F. R. Fisher, Ed., *Protection and Functioning of the Hands in Cold Climates* (pp. 86–102). Washington, DC: National Academy of Sciences-National Research Council.

McCullough, E. A., Jones, B. W., and Huck, J. (1985). A comprehensive data base for estimating clothing insulation. *ASHRAE Transactions, 91,* 29–47.

NIOSH (1986). *Criteria for a Recommended Standard . . . Occupational Exposure to Hot Environments.* Washington, DC: U.S. Government Printing Office.

Nilsson, H. O., Gavhed, D. C. E., and Holmér, I. (1992). Effect of step rates on clothing insulation—measurement with a moveable thermal manikin. In W. A. Lotens and G. Havenith, Eds., *Proceedings of the Fifth International Conference on Environmental Ergonomics* (pp. 174–175). Maastritcht, Netherlands.

Pandolf, K. B., Burse, R. L., and Goldman, R. F. (1977). Role of physical fitness in heat acclimatization, decay and reinduction. *Ergonomics, 20,* 399–408.

Pandolf, K. B., Stroschein, L. L., Drolet, R. R., Gonzalez, R. R., and Sawka, M. N. (1986). Prediction modeling of physiological responses and human performance in the heat. *Computers in Biological Medicine, 16,* 319–329.

Parsons, K. C. (1993). *Human Thermal Environments.* London: Taylor and Francis.

Parsons, K. C. (1995). International heat stress standards: A review. *Ergonomics, 38,* 6–22.

Poulton, E. C. (1976). Arousing environmental stresses can improve performance, whatever people say. *Aviation, Space, and Environmental Medicine, 47,* 1193–1204.

Poulton, E. C., Hitchings, N. B., and Brooke, R. B. (1965). Effect of cold and rain upon the vigilance of lookouts. *Ergonomics, 8,* 163–168.

Provins, K. A., and Morton, R. (1960). Tactile discrimination and skin temperature. *Journal of Applied Physiology, 15,* 155–160.

Ramsey, J. D. (1995). Task performance in heat: A review. *Ergonomics, 38,* 154–165.

SAIC (1993). P^2NBC^2 Heat Strain Decision Aid User's Guide, Version 2.1. Joppa, MD: Scientific Applications International Corporation.

Santee, W. R., Blanchard, L. A., Chang, S. K. W., and Gonzalez, R. R. (1993). Biophysical Model for Handwear Insulation Testing. Technical Report T7/93. Natick, MA: U.S. Army Research Institute of Environmental Medicine.

Santee, W. R. and Endrusick, T. L. (1988). Biophysical evaluation of footwear for cold-wet climates. *Aviation, Space, and Environmental Medicine, 59,* 178–182.

Santee, W. R., Matthew, W. T., and Blanchard, L. A. (1994). Effects of meteorological parameters on adequate evaluation of the thermal environment. *Journal of Thermal Biology, 19,* 187–198.

Schutte, P. E., and Zenz, C. (1994). Physical work and heat stress. In C. Zenz, Ed., *Occupational Medicine,* (Third Edition) (pp. 305–333). St. Louis: Mosby.

Scott, R. A. (1988). The technology of electrically heated clothing. *Ergonomics, 31,* 1065–1081.

Sendroy, J., and Collison, H. A. (1960). Nomogram for determination of human body surface area from height and weight. *Journal of Applied Physiology, 15,* 958–959.

Siple, P. A., and Passel, C. F. (1945). Measurement of dry atmospheric cooling in subfreezing temperatures. *Proceedings of the American Philosophical Society, 89,* 177–199.

Soule, R. G., and Goldman, R. F. (1969). Energy cost of loads carried on the head, hands or feet. *Journal of Applied Physiology, 27,* 687–690.

Speckman, K. L., Allan, A. E., Sawka, M. N., Young, A. J., Muza, S. R., and Pandolf, K. B. (1988). Perspectives in microclimate cooling involving protective clothing in hot environments. *International Journal of Industrial Ergonomics, 3,* 121–147.

Teichner, W. H. (1958a). Assessment of mean body surface temperature. *Journal of Appied Physiology, 12,* 169–176.

Teichner, W. H. (1958b). Reaction time in the cold. *Journal of Applied Psychology, 42,* 54–59.

Teichner, W. H., and Wehrkamp R. F. (1954). Visual-motor performance as a function of short-duration ambient temperature. *Journal of Experimental Psychology, 47,* 447–450.

Teitlebaum, A., and Goldman, R. F. (1972). Increased energy cost with multiple clothing layers. *Journal of Applied Physiology, 32,* 743–744.

Tikuisis, P. (1996). Predicting survival time for cold exposure. *International Journal of Biometeorology, 39,* 94–102.

van Dilla, M., Day, R., and Siple, P. A. (1949). Special problem of hands. In L. H. Newburgh, Ed., *Physiology of Heat Regulation and the Science of Clothing* (pp. 374–388). Philadelphia: W. B. Saunders.

Winslow, C. E. A., Herrington, L. P., and Gagge, A. P. (1937). Physiological reactions of the human body to varying environmental temperatures. *American Journal of Philosophy, 120,* 1–22.

Woodcock, A. H. (1962). Moisture transfer in textile systems, Part I. *Textile Research Journal, 32,* 628–633.

Yaglou, C. P., and Minard, D. (1957). Control of heat casualties at military training centers. *American Medical Association Archives of Industrial Health, 16,* 302–316.

CHAPTER 29

DESIGN FOR MACROGRAVITY AND MICROGRAVITY ENVIRONMENTS

William B. Albery
Combined Stress Branch, Biodynamics and Biocommunications Division
Crew Systems Directorate, USAF Armstrong Laboratory
Wright-Patterson Air Force Base, OH 45433-7008 USA

Barbara Woolford
Flight Crew Support Division, Johnson Space Center
National Aeronautics and Space Administration
Houston, TX 77058 USA

29.1 ABSTRACT

The effects of macrogravity (G > 1) and microgravity (G < 0.001) on human factors performance are reviewed. Physiological and psychological effects are discussed as to how they relate to the human fators impact on acceleration and weightlessness. Effects on vision, memory and central processing functions, and manual control are also reviewed. Protective techniques and garments, as well as exposure limits, are presented and discussed. Included are recommendations for minimizing the impact of acceleration and weightlessness on human factors.

29.2 INTRODUCTION

29.2.1 Overview

Until this century man has not had to deal with human factors and ergonomics issues in nonterrestrial environments. Since 1903 when the Wright brothers started powered flight, man has had to work in a greater than 1 G environment. Early human factors problems involved aircraft instruments, escape at altitude from unflyable aircraft, and altitude and G protection. Most of these problems have been overcome thanks to the ability to simulate, study, and train for these environments in test facilities such as human centrifuges, modern flight simulators, and altitude chambers. Even today, however, in the modern air forces of the world, human factors related mishaps represent up to 25% of the total of all flight-related mishaps. Human factors issues in microgravity began in the United States in 1961 with the first American manned space flight. Unlike the macrogravity environment, which can be simulated on earth, weightlessness has been a phenomenon studied almost exclusively in space. Attempts to simulate weightlessness during parabolic maneuvers in KC-135 aircraft has given researchers only short glimpses of the environment. Water immersion research has provided shuttle crews the ability to practice in a simulated weightless environment and has provided human factors engineers with valuable insight for the design of tools, handholds, footholds, and restraints for extravehicular activities while in space. In this chapter, a review of human factors and ergonomics issues in these nonterrestrial environments is presented. The physiological effects of G > 1 and protective techniques and devices are summarized. The psychological effects of G > 1 are discussed including memory and central processing functions, manual control, human perception, and prolonged acceleration. The physiological effects of G < 0.001, including space motion sickness and the long-term effects of weightlessness are reviewed. Environmental factors, equipment design considerations,perceptual and cognitive factors, extravehicular activity, and combined environments are also discussed. After reading this chapter, we hope the reader has a better understanding of these nonterrestrial environments, the human factors issues of working in these environments, and figures, tables, and references that can aid the human factors engineer.

29.3 DEFINITION OF TERMS

As part of this introduction, a short explanation concerning body axes and term definitions seems warranted. The direction of acceleration forces are important with respect to the potential effects on the human operator. To uniformly describe the directions of these external forces, a biodynamic coordinate system has been standardized (American National Standards Institute, or ANSI, 1979) which references the force vectors in the x, y, and z directions to the human body (Figure 29.1). G used here is a measure of the strength of the gravitoinertial force environment. On the earth, one feels a G force equal to 1 G in magnitude directed toward the center of the earth. In flight, this 1 G force adds vectorially to accelerations in the x, y, and/or z axes and the resultant gravitoinertial force vector is generally pointing in a direction other than toward the center of the earth. Microgravity is described as G < 0.001 and not G = 0 since there is negligible gravity while orbiting the earth.

For sustained acceleration, the terminology summarized in Table 29.1 distinguishes between the direction of the acceleration vector and the resultant inertial response of the body or its components.

29.4 MACROGRAVITY (G > 1)

29.4.1 Mechanical Effects

Increased gravitational fields increase the weight of body parts. Body parts can become elongated or compressed under the G vector; this can affect the shape and function of

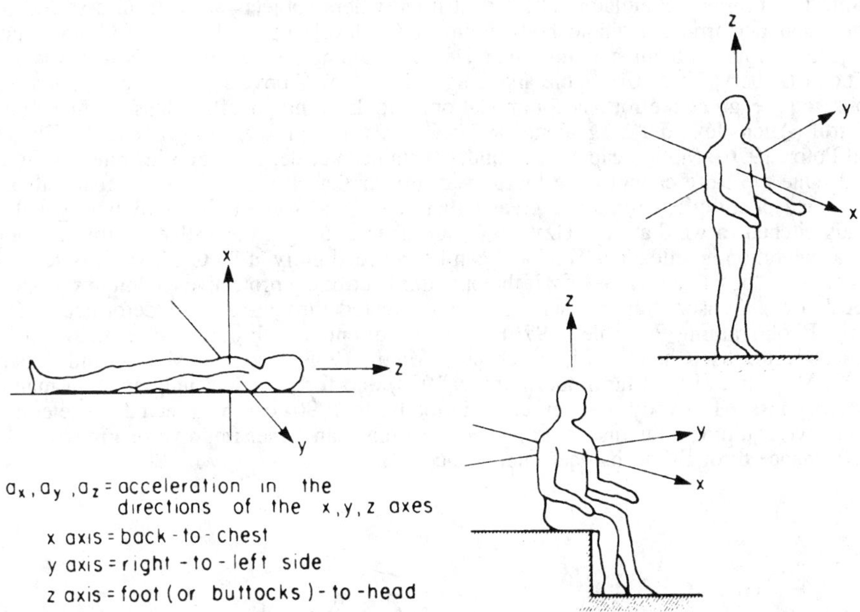

Figure 29.1 The Biodynamic Coordinate System. (ANSI, 1979; Reprinted with permission.) This is the system used by researchers to describe the directions of sustained acceleration or vibratory exposures on humans.

the soft internal organs including the heart, lungs, kidneys, liver, etc. (Gauer, 1950). Higher muscle forces are required to keep the head, torso and limbs in desired positions (Burton, 1974, 1986). At G forces of approximately +2 Gz there is increased pressure on the buttocks, drooping of the face and noticeably increased weight of all body parts; at this level of G force it is difficult to raise oneself, and at +3–4 Gz it is nearly impossible (Fraser, 1973). Experiments were conducted in the Wright-Patterson AFB (WPAFB) Dy-

Table 29.1 Nomenclature Describing Acceleration Vectors with Respect to the Operator

Direction of Acceleration

Linear Motion	Aircraft Computer Standard	Acceleration Descriptive
Forward	$+a_x$	Forward acceleration
Backward	$-a_x$	Backward acceleration
Upward	$+a_z$	Headward acceleration
Downward	$-a_z$	Footward acceleration
To right	$-a_y$	Right lateral acceleration
To left	$+a_y$	Left lateral acceleration

Inertial Resultant of Body Acceleration

Linear Motion	Physiologic Descriptive	Physiologic Standard	Vernacular Descriptive
Forward	Chest to back G	$+G_x$	Eyeballs-in
Backward	Back to chest G	$-G_x$	Eyeballs-out
Upward	Positive G	$+G_z$	Eyeballs-down
Downward	Negative G	$-G_z$	Eyeballs-up
To right	Left lateral G	$-G_y$	Eyeballs-left
To left	Right lateral G	$+G_y$	Eyeballs-right

Source: From Burton, R. R. and Whinnery, J. E. (1996). Biodynamics: Sustained acceleration. In R. DeHart, Ed., *Fundamentals of Aerospace Medicine*. 2nd Ed. Baltimore: Williams & Wilkins. Reprinted with permission.

namic Environment Simulator (DES) centrifuge where subjects arose from a seated position and performed a whole body jump at Gz levels up to 1.8 Gz. Although most subjects had no problem standing up at 1.8 Gz, jumping and leaving one's feet was very difficult (Albery, 1995; Constable and Carpenter, 1995). Above +3–4 Gz, controlled motions require greater effort, accommodation, and learning to offset loss of fine motor control (Allen, Jex, and Magdaleno, 1973; Creer, 1962; Fraser, 1973; Loose, McElreath, and Potor, 1976). Motion capabilities under sustained acceleration are illustrated in Figure 29.2. One typically cannot raise the arm at greater than 8 Gz, or legs at greater than 3 Gz. Head pitching is difficult at greater than 4 Gz and some individuals who get their heads pitched forward at high Gz (>6 G) are unable to right themselves in the seat until the acceleration is unloaded. The hand can be raised slightly at 25 Gz. Speech is severely affected, yet possible up to +9 Gz if the operator is utilizing protective techniques properly (see below). Sensory inputs such as vision affected through eyeball deformation (Air Force Pilot Training Pamphlet, 1976) vestibular orientation through the semicircular canals and otoliths (Chelette, Li, Esken, and Matin, 1995; Chelette, Martin, and Albery, 1992; Malcolm, 1987; Martin and Albery, 1993) and force/weight judgments in manual dexterity tasks (Darwood, Repperger, and Goodyear, 1990) can be affected. Acceleration protective equipment, as discussed in later sections, can either improve or limit/degrade performance through mechanical interference.

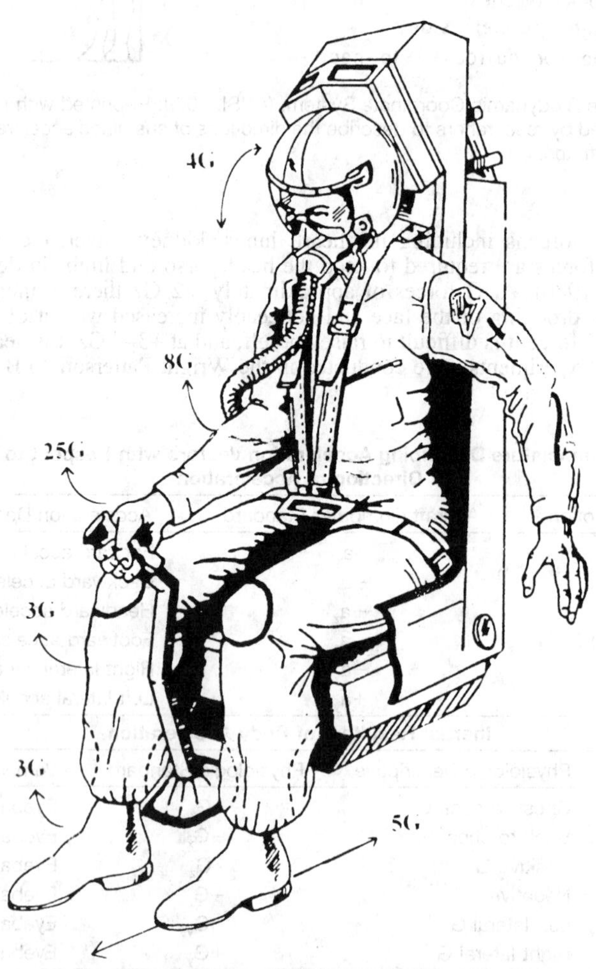

Figure 29.2 Motion capabilities under sustained acceleration. These G levels are the average maximum forces observed for human motion under sustained acceleration.

29.4.2 Physiological Effects

Besides the mechanical deformation and displacement of body parts, increased gravitational forces can result in physiological effects of severe magnitude. The most debilitating and dangerous of these physiological effects is caused by changes in the hydrostatic blood pressure column between the heart and the brain. As the level of +Gz increases, the blood pressure to the brain is decreased. This effect is illustrated in Figure 29.3.

The systolic arterial blood pressure under normal 1 G is approximately 120 mmHg. If the brain (measured at eye level) is 30 cm above the heart, approximately 98 mmHg of systolic blood pressure is available for perfusion of the brain under normal circumstances. For each additional +1 Gz increase, however, the systolic pressure measured at eye level is reduced by approximately 22 mmHg. This leads to a theoretically zero blood pressure at eye level at approximately +5.5 Gz, and a negative blood pressure at higher +Gz levels (Gauer, 1950; Gauer and Zuidema, 1961; Gillingham, 1974; Wood, 1987, 1988). Zero or negative blood pressure leads to grayout, blackout, and GLOC, or G-induced loss of consciousness (Gauer, 1950). In 1950, Lambert reported data concerning the G tolerance of male subjects as pilots controlling the onset and magnitude of acceleration and as passengers in the aircraft. Onset rates for G loads ranged from 0.3 to 1.1 G/sec, with the higher onset rates occurring in the piloting phase. For subjects piloting the aircraft, the mean G level for black-out was +5.4 Gz. As passengers, however, the same subjects experienced black-out at approximately +4.7 Gz. The average G tolerance for the pilot was +0.7 Gz higher than G tolerance as passengers. Lambert (1950) also reported data comparing pilots, passengers, and centrifuge riders. On the average, subjects had less G tolerance as centrifuge riders than as aircraft passengers, and in turn, aircraft passengers had less G tolerance than pilots. The range of +Gz values for black-out was 3.1 to 5.6, with the higher tolerance values coming from the pilots. Although the limitations of human tolerance to +Gz are modified by whether or not the subject has control over the situation (pilots vs. passengers/centrifuge riders) as well as individual differences

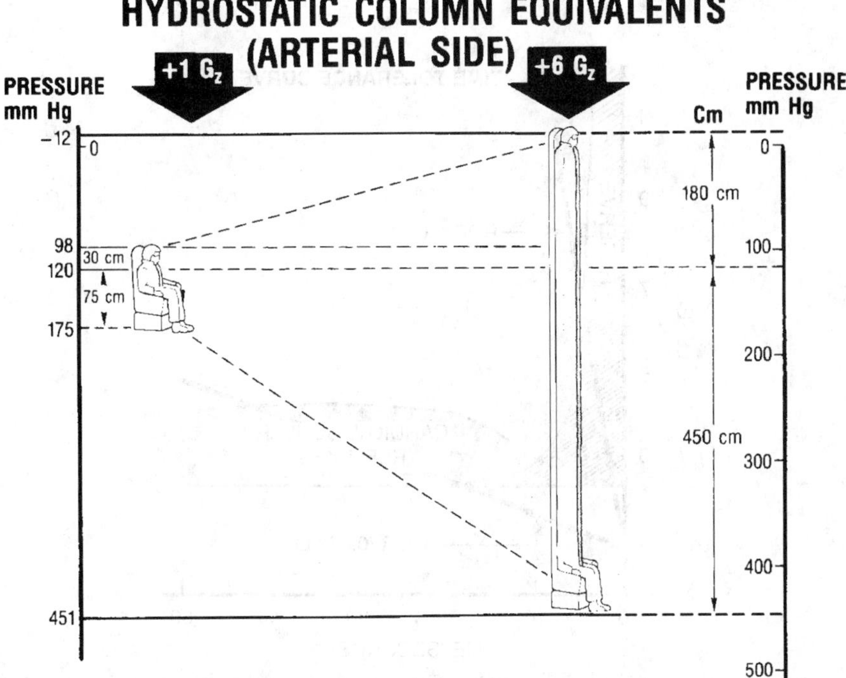

Figure 29.3 The 'hydrostatic column' effect. This shows the effect of increasing sustained acceleration on eye level blood pressure; the analogy of increasing G is lengthening the distance between the heart and head, such as stretching the neck. (From Gauer, 1950.)

in body size, structure, strength, and experience under +Gz loads, the generalized G-time tolerance curve for seated human subjects has a course similar to the one presented in Figure 29.4.

For G loads in the transverse plane (+Gx, chest to back), tolerance is greater than in the longitudinal plane (+Gz) because of reduced effects on the hydrostatic column (Cherniack, Hyde, and Zechman, 1959). This is why astronauts and cosmonauts are launched on their backs; it is the most tolerable body axis to sustained acceleration. Cherniack, Hyde, Watson, and Zechman (1961) reported human tolerances to forward acceleration up to +16.5 Gx under short durations, and up to +12 Gx with a duration of 3 min. However, inspiratory chest pain, tracheal tugging, paroxysmal coughing, and sensation of weight on the thorax were typical symptoms reported. Obviously, the greatest physiological effects of Gx acceleration are on respiratory mechanics. High sustained acceleration levels produce different physiological tolerance limits for force in other vector directions. Table 29.2 summarizes evidence obtained from the literature concerning human tolerance limits to force applied through the *x*, *y*, and *z* directions.

29.4.3 Protective Techniques and Devices

Any action or device that would increase the pressures in the hydrostatic column during increased +Gz loads should increase the level of G, and time at G, that the human operator may tolerate. This concept is the underlying basis for two commonly used G-protection principles, the *active straining maneuver* and the G suit. There are two types of straining maneuvers in use, the M-1 and the L-1 maneuvers (Howard, 1965; Leverett and Whinnery, 1985; Burton and Whinnery, 1996). The M-1 (M Mayo Clinic) consists of pulling the head down between the shoulders, slowly and forcefully exhaling through a partially closed glottis and simultaneously tensing all skeletal muscles. The L-1 (L Leverett) maneuver is similar to the M-1, except the exhalation is forcefully applied to a completely closed glottis. When properly executed, the exhalation phase of these maneuvers can result in an increase of intrathoracic pressure of approximately 50 to 100 mmHg, which raises arterial pressure at eye level. 1.5 Gz to 5.0 Gz of G-tolerance can be added, raising G-tolerance from approximately +5 Gz to +7–10 Gz. The M-1 and L-1 maneuvers, while

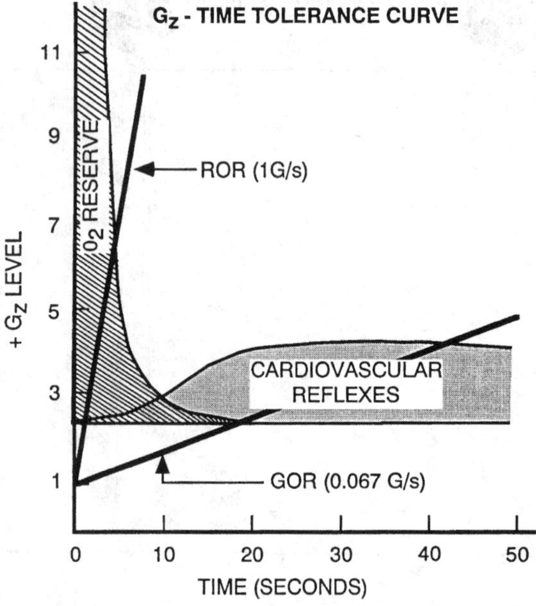

Figure 29.4 Sustained Acceleration +Gz-time Tolerance Curve. GOR Gradual Onset Rate; ROR = Rapid Onset Rate. [From Leverett, S. D., and Whinnery, J.E. (1985). Biodynamics: Sustained Acceleration. In R. DeHart, Ed., Fundamentals of Aerospace Medicine. New York: Lea and Febiger. Reprinted with permission.]

Table 29.2 G-Tolerance Limits to Sustained Acceleration in the Different Axes

Type of G and Direction of Body Movement	Aircraft Maneuver	Experimental and Operational Human Performance Exposures	Physiological Response	Facility
Positive (+G_z) Head to foot Eyeballs down	Pull out or tight turn	12.5 G for several seconds	Confusion, near loss of consciousness	F-15 spatial orientation mishap
		11.3 G for 5 sec (anti G suit)	Almost unconscious, loss of vision	NAWC
		5 to 9 G simulated aerial combat maneuvers; 10 sec plateaus; 11 min, 2 sec (with positive pressure breathing); 98 sec without PPB	Arm pain, petechiae, discomfort, musculoskeletal fatigue, pain	AL-BROOKS
		3.5 G for 1 h, 4 G for 20 min	Rotational illusions, discomfort	NAWC
Negative (−G_z) Foot to head Eyeballs up	Push over	−4.5 G for 5 sec	Pain, discomfort	NAWC
		−3 G for 32 sec (special helmet)	Facial congestion, bradycardia, throbbing headache	NAWC
		−2 G for 1 min (special suit)	Bradycardia	AL-WPAFB
Transverse (+G_x) supine Chest to back Eyeballs in	Afterburner, catapult launch	25 G_x for several seconds (unprotected)	Breathing difficulty, chest pain, loss of vision	NAWC
		31 G_x for several seconds (water immersion)	Frontal sinus pain with blood mucous	NAWC
Transverse prone (−G_x) Back to chest Eyeballs out	Arrested landing, flat spin	(−4 G, flat spin simulation)	Breathing easier than +G_x	AL-WPAFB
		−5 G for 2 sec	Distortion of vision	NAWC
		−15 G for 5 sec (special chest and leg support)	Restraint pressure	NAWC
Lateral (±G_y) Eyeballs left (−G_y) Eyeballs right (+G_y)	Side slip agility maneuver	+5 G_y for 14.5 sec	External hemorrhage, severe postrun discomfort after 10 sec	NAWC
		+3 G_y for 10 sec	Engorgement of elbow with pain	NAWC
		+2 G_y, −2 G_y for 45 sec	Neck pain, elbow pain	AL-WPAFB

Note: AL is the Armstrong Laboratory (US Air Force at Brooks AFB or at Wright-Patterson AFB (WPAFB); NAWC is the Naval Air Warfare Center (US Navy).

quite effective, are physically taxing and can interfere with speech and other communication modalities. When the maneuvers are performed for long periods (1 to 3 min), pilots can become severely fatigued. When this happens, the maneuver obviously loses its effectiveness, which in turn impacts on mission performance.

The G-suit protection garment consists of bladders inserted into a full- or partial-body coverall. The bladders inflate during the onset of acceleration. The inflation pressures usually start around 2 to 3 G and increase as G increases to a maximum pressure at +9 Gz of approximately 675 mmHg, or 13 psi in the bladders (Wood, 1988). The external increase in pressure afforded by the G suit reduces body deformation, blood pooling in the extremities, and mechanically increases internal blood pressure. The relaxed subject wearing a G suit has an increased G tolerance of approximately 2 G (Palets, Tikhonov, Popov, Popov, Arkhangelskiy, Palets, and Bondarenko, 1987; Wood, 1988). When combined with a straining maneuver, G tolerance can increase up to 3 to 5 G, allowing the human to tolerate +8 Gz to +10 Gz without losing consciousness (Burton, 1974; Palets et al., 1987; Burton and Whinnery, 1996).

Seatback angle is another method which has been shown to increase G tolerance. By decreasing the angle of the seat away from the G_z vector into the Gx vector, the hydrostatic column is reduced. Forward leaning seats can improve G tolerance by reducing the eye-to-heart distance. Li (1992) found that subjects who leaned forward 24° during +9 Gz exposures endured nearly twice as long as those seated upright (12°). Reclined seatback angles can increase G tolerance for the sitting pilot into the range of 14 to 16 Gx (Barer, Golov, Zubavin, Muranknovskiy, Rodin, Sokorina, and Tikhomivou, 1964; Burns, 1988). Deaton and Hitchcock (1991) evaluated human performance on the Navy's 50 ft radius centrifuge at Warminster PA. Subjects were required to perform both a perceptual motor task and a classification task in both reclined 27° and 67° seatback configuration while exposed to up to +10 Gz. Performance was measured pre-, during, and postexposure to high G. They found no significant differences in performance as a function of seatback angle pre- and during exposure, but did find a significant difference postexposure. Subjects' task performance was significantly better in the 67° seat postexposure. One possible explanation could be the subjects should have been less fatigued in the 67° configuration since most of the Gz would be experienced as Gx (9.2 Gx and 3.9 Gz at 10 G).

The problems with a forward leaning, or reduced seatback angle concern the entire cockpit design. Cockpit layouts are typically designed for seatbacks 12° to 30° from vertical. The pilot is sitting virtually upright, and all manual controls, visual displays, look-out capabilities and escape methods are designed around a vertical seat. A total and complete redesign of the cockpit would be needed to accommodate reduced seatback angles. In addition, flying in a supine position has been met with less than enthusiasm by the flying community; imagine trying to drive one's personal automobile in a reclined position. Such a seated position, while providing apparent safety from GLOC, introduces novel manual control, perception, and ergonomic challenges.

Another method of increasing G tolerance is assisted positive pressure breathing (APPB). Pressures of 45 to 70 mmHg have been shown to increase G-tolerance time by increasing intrathoracic pressures, reducing the mechanical effects of increased G on respiration, and reducing the effort needed to perform straining maneuvers (Albery, 1989; Burns, 1988; Chambers, Kerr, Augerson, and Morway, 1962; McCloskey, Tripp, Repperger, and Popper, 1990; Shaffstall and Burton, 1979; Shubrooks, 1973). APPB systems for G protection are now being implemented in U.S. Air Force F-16 and F-15 aircraft. Other countries including Great Britain, Sweden, France, Canada, and Germany are also considering APPB systems for their high performance aircraft. G-tolerance experiments on the Brooks AFB TX centrifuge demonstrated that subjects protected with APPB in addition to the standard anti-G suit had endurance on a 5G-9G Simulated Aerial Combat Maneuver (SACM) that was superior to those subjects protected with the standard anti-G suit, alone (Table 29.3). Also, in human performance studies involving subjects in the WPAFB DES centrifuge, those subjects protected with APPB systems were able to perform a simulated flying task to peaks of 9 G nearly twice as long as those subjects protected without APPB (Albery and Vanderbeek, 1994). The APPB raises the intrathoracic pressure of the subject/pilot, thus increasing eye level blood pressure and making it less fatiguing to breathe and to perform the straining maneuver (L-1) under high G for extended maneuvers. Although some medical concerns have been raised about APPB and high G (Jennings and Zanetti, 1988), the technique is considered completely safe (Burton and Whinnery, 1996).

Table 29.3 Acceleration Protection Method and Human Performance

G-protective Techniques (Type of G-Protection)	Task Improvement/Decrement (seatback angle)	G Level	References
Anti-G suit	Performance baseline or standard (12° or 30° seatback angle)	Up to 7 + G_z (up to 10 + G_z with straining maneuver)	Howard (1965) Burton (1974) Palets et al. (1987)
Reduced-from-vertical seatback angle	Improvement (30° seatback angle)	Improvement up to +6 G_z	Rogers (1973)
	No significant improvement (27° seatback angle)	+10 G_z	Deaton and Hitchcock (1991)
	Improvement (67° seatback angle)	Improvement up to +10 G_z	Lisher and Glaister (1978)
	Improvement noticed postexposure (67° seatback angle)	+10 G_z	Deaton and Hitchcock (1991)
Positive pressure	Improvement (supine seat—G_x exposures) (upright seat—G_z exposures)	Improvement up to 5 + G_z and 10 + G_x	Chambers et al. (1962)
	Improvement (30° seatback angle)	Improvement up to 9 + G_z	Chelette (1996) McCloskey et al. (1990) Wood (1988)

29.4.4 Psychological Effects

Very rarely does sustained acceleration affect only one factor such as speech, vision, or manual dexterity. In an effort to measure performance on complex tasks, the construct of workload has been developed. Task performance can be measured through subjective reports and physiological and behavioral measures. Usually all three types of measures are brought to bear on the multidimensional problem of workload. The subject of workload is not discussed in detail here, but has been reviewed extensively elsewhere (see Chaper 13 of this Handbook for an excellent review of the subject; Kantowitz, 1985; Navon and Gopher, 1979; O'Donnell and Eggemeier, 1986). The effect of sustained acceleration on human performance was first characterized around 1920 during the Pulitzer Air Races when a pilot nearly lost consciousness while maneuvering around a pole as part of the competition (Burton, 1986). Since that event man's intolerance to high sustained +Gz has limited his ability to control and defend his airspace. In a study involving the effects of acceleration on mental functioning, subjects were exposed to levels of acceleration high enough to produce dimming of vision and blackout (Kerr and Russell, 1944). Attendant impairment of cerebral function was also reported. This effect was observed in subjects who became confused, failed to remember parts of the experimental procedure and suffered loss of control of voluntary movement. Hallenbeck (1946) reported similar findings. He postulated that the central nervous system, and hence cognitive processes, are affected by +Gz levels below those that result in loss of consciousness (GLOC). The effects of sustained acceleration on human performance are summarized in Table 29.4.

29.4.4.1 Vision

Reaction time experiments were conducted to examine visual functioning under G stress. In one experiment, Canfield, Comrey, Wilson, and Zimmerman (1950) required subjects to gaze upon a green light in the center of a panel. Periodically, a red light would appear at different orientations to the green light. Subjects were to determine the direction of the red light in relation to the green one. Accelerations of up to +5 Gz had no effect on reaction time (RT). However, evidence for acceleration-induced increased RT had been found in a previous experiment by the same authors (Canfield, Comrey, and Wilson, 1949). In another visual discrimination experiment (Warrick and Lund, 1946), errors in dial reading increased as a function of increased +Gz level. Yet no significant effects of acceleration were found in another experiment which required subjects to match one of four similar figures to a central figure pattern (Canfield, Comrey, and Wilson, 1948). Absolute detection thresholds for both foveal and peripheral vision were shown to increase as positive +Gz increased from 1 to 4 (White, 1960; White and Monty, 1965). Threshold contrast sensitivity (detection of a target as a function of background luminance) was affected by increased acceleration in the +Gz and −Gx vectors (Braunstein and White, 1962; White, 1958). Contrast sensitivity was most affected in the +5 Gz range; acceleration in the +Gz vector produced more performance decrements than did +Gx. However, attempts to measure changes in visual color discrimination and color naming ability yielded few significant results at +3 Gz (Frankenhauser, 1945), although cognitive decrements at G were obtained with tests of simple mathematical skills. Grether (1971) reviewed the effects of acceleration on performance and concluded that both simple and choice RTs to visual signals generally increase during increased levels of +Gz. However, these effects tend to diminish or disappear as humans become more accustomed to acceleration environments.

29.4.4.2 Memory and Central Processing Functions

Numerous studies by Chambers and colleagues demonstrated memory impairment at high +Gx levels (Chambers, 1961, 1963; Chambers and Hitchcock, 1963). These studies indicated acceleration levels up to +5 Gx do not significantly affect performance on immediate memory tasks (measured as mean percent correct responses), while acceleration greater than +7 Gx does reduce memory performance. Frankenhauser (1945) found that with +3 Gz exposures 2 to 10 min in length, more time was needed to complete multiplication tests performed mentally as compared with performance during no G stress. Albery, Ward, and Gill (1985) examined performance in a maze solving task under +1.5 Gz to +6 Gz acceleration. The maze task tapped cognitive functions associated with spatial processing and problem solving. RT and error rates were collected along with subjects' opinions of how hard they were working. Performance measures (RT and error) were not

Table 29.4 Effect of Sustained Acceleration on Human Performance

Category	G Llevel	Type of Decrement	References
Vision	$>4\text{–}5 + G_z$	Visual dimming, blackout	Kerr and Russell (1944) Hallenbeck (1946)
	$>3\text{–}4 + G_z$	Detection thresholds increased for foveal and peripheral vision	White (1960), White and Monty (1965)
	$>4\text{–}5 + G_z$	Decreased contrast sensitivity	White (1958), Braunstein and White (1962)
	$>4\text{–}5 + G_z$	Increased reaction times to visual discrimination	Canfield, Comrey, and Wilson (1949)
	$>3\text{–}5 + G_z$	Increased errors in dial reading	Warrick and Lund (1946)
Memory and central processing	$>6\text{–}7 + G_z$	Increased errors in a memory task	Chambers (1961; 1963), Chambers and Hitchcock (1963)
	$>3 + G_z$	Increased reaction times to a multiplication task	Frankenhauser (1945)
	$>5\text{–}6 + G_z$	Increased subjective ratings of workload	Albery et al. (1985)
Manual control	$>4 + G_z$ $>4 \pm G_x$	Increased tracking error with lightly damped control characteristics	Creer (1962)
	$>6 + G_z$ $>14 + G_x$	Increased tracking error with heavily damped control characteristics	Creer (1962)
	$>5 + G_z$	Increased tracking error	Little et al. (1968), Piranian (1982), Loose et al. (1976)
	$+9\ G_z$	Increased tracking and communications errors (females significantly worse)	Chelette (1996), Chelette and Albery (1996)
Time and mass estimation	$>6 + G_z$	Time underestimated during long task	Frazier et al. (1990), Repperger et al. (1990)
	$>4 + G_z$	Increased error for weight estimation	Darwood et al. (1990)

affected by increasing levels of +Gz, while the subjective ratings showed an increase as +Gz increased.

29.4.4.3 Performance at G Until Exhausted

In research involving a full coverage anti-G suit (one which covered the subject from the waist down), subjects in the WPAFB DES centrifuge performed a choice reaction task involving responding to lights that were lit on a small, laptop device. Subjects rested their index and forefingers of both hands on four buttons and as a light, or lights, above the buttons were lit, the subjects pressed the corresponding button(s). Reaction times and errors were calculated as protected subjects endured a 4.5–7 G SACM until exhausted. Although, almost to the subject, they denied experiencing any cognitive effects of the SACM, subjects' reaction times increased 25% and errors (wrong buttons) increased (10% just prior to giving up on the task and exposure (Tripp et al., 1992). In an experiment involving protected centrifuge subjects performing a dual task while undergoing a 5-9G SACM, subjects who were protected with APPB and full coverage anti-G suits were able to perform the dual task twice as long as those without APPB (McCloskey, Tripp, and Martin, 1994). This dual task consisted of a primary tracking task of following a lead aircraft while performing a visual identification secondary task. The objective of the primary task was for the subject to track the simulated aircraft as accurately as they could. The secondary task involved the display of up to nine different letters, symbols, or numbers to the subject whose task was to respond to three specific symbols. Those subjects who were the most protected from G performed the dual task the longest and with fewer errors. Those subjects who had minimal protection halted the SACM exposures earlier and did not perform as well as their protected counterparts (McCloskey et al., 1994; Chelette and Albery, 1996).

29.4.4.4 Manual Control

Creer (1962) investigated the influence of various acceleration profiles as well as simulated vehicle dynamics on tracking proficiency. The maximum acceleration magnitudes were +6 Gz, +6 Gx, and −6 Gx, each 2.5 min in length. Creer (1962) found no tracking decrements at any acceleration level using a relatively easy control task involving heavily damped (no oscillation or overshoot) vehicle dynamics. In the lightly damped case, however, 20% greater errors occurred at normal 1 G, and performance deteriorated markedly above 4 G. Errors were relatively independent of the direction of acceleration. Creer's results suggest that control performance on a tracking task is no less effective at +Gz than at positive or negative Gx up to 6 G, as long as no serious visual impairment occurs. These results also indicate the importance of using a demanding task to show performance decrements, and lend support to the assertion of Chambers and Hitchcock (1963) that the more inferior the aerodynamic characteristics of the simulated vehicle, the greater likelihood of degraded human performance under G stress. In the second part of his study, Creer (1962) increased the G levels in the +Gz and +Gx directions. All tracking runs lasted 2.5 min, and vehicle motions were again damped. Between +6 Gz and +9 Gz, performance dropped rapidly, while only slight decrements were found from +1 Gx to +14 Gx. The performance differences between z and x vectors were attributed primarily to the serious visual degradation occurring above +7 Gz. Grether (1971) summarized the evidence showing that tracking and flight control exhibits progressive impairment with increasing +Gz acceleration and somewhat less impairment with +Gx acceleration. Grether also suggested that manual output functions may be more susceptible to acceleration-induced mechanical impairment than intellectual and central nervous system impairment. Loose et al. (1976) also found an increase in two-axis tracking error to be a function of increasing +Gz. This experiment involved combining +Gy motions with +1.6 Gz, +3 Gz, and +5 Gz acceleration levels. Little, Hartman, and Leverett (1968) also observed tracking decrements during accelerations of +5 Gz, +7 Gz, and +9 Gz. These authors suggested that the performance decrements resulted either from mechanical limb-loading or stress-specific factors and not from physiological insult.

Piranian (1982) investigated pilot tracking performance on a visually coupled aerial combat maneuver simulation. The task was to pursue a target aircraft into a +5 Gz circling turn and to hit the target. Tracking performance was measured in terms of pilot opinion ratings, projected miss distance from target, and time to obtain a distance of 10 miles from the target. There was an overall 20% decrement in manual performance at +5 Gz when compared to normal 1 G performance. Albery (1988) investigated manual control and reaction time using primary and secondary tasks performed simultaneously under acceleration levels of +1.4 Gz, +2.75 Gz, and +3.75 Gz. Unprotected subjects were also asked to give subjective responses as to how hard they were working. Error rates obtained from the primary tracking task increased as acceleration levels increased, whereas reaction times to the secondary target identification task did not change with acceleration level. Subjective levels of workload increased as acceleration increased. Constable and Carpenter (1995) studied subjects' muscle coordination and muscle force requirements in an altered gravitational environment. Twelve subjects were studied as they arose from a seated position and then jumped inside the cab of the WPAFB DES centrifuge at 1.2, 1.4, 1.6, and 1.8 Gz. By increasing the gravitational force, a linear decrease in jump performance was observed. Subjects were able to adapt to the macrogravity and, as the 20 ft centrifuge turned at speeds up to 17 revolutions per minute, developed a strategy to remain stable as they first stood, unassisted, and then performed a whole-body jump such as to leave their feet. These data have been used to validate a model of human locomotion in altered G environments.

Chelette (1996) found men and women protected with COMBAT EDGE, the positive pressure breathing system, were quite capable of controlling a human centrifuge, configured to simulate the F-16 aircraft, to 9 G. The eight men and eight women performed a simulated tail chase of an aircraft as a primary task and a communications secondary task. None of the trained subjects were pilots. Each subject performed four 3-minute sorties in the Dynamic Environment Simulator Centrifuge with full manual control of the device to a maximum of 9 G. Male and female performance scores were equivalent at 1 G in the simulator, however, the females' performance scores were significantly worse at high G. Twenty-four hours of sleeplessness had little or no effect on male or female performance. The females apparently could not optimize the tracking task as well as the males under dynamic conditions.

29.4.4.5 Time and Mass Estimation

Time, mass, and reaction time perception experiments have been conducted under high levels of G stress (Darwood et al., 1990; Frazier et al., 1990; Repperger et al., 1989). Time perception is affected by G levels up to and including +8 Gz (Frazier et al., 1990). The longer the time estimation, such as 16 sec, the greater the error. Subjects tended to underestimate the longer tasks, as if they wanted to finish the task early. Likewise, for mass discrimination, it was found that performance is impaired during increased +Gz (Darwood et al., 1990). The ability to discriminate between the heaviness of several objects became more difficult under G stress. In several studies involving choice reaction time in acceleration stress, it was found that reaction times actually decrease under G stress (Albery et al., 1985; Albery, 1988). This effect of improved reaction time performance under acceleration can be possibly attributed to the alerting cue phenomenon (Teichner, Arees, and Reilly, 1963).

29.4.5 Prolonged Acceleration Centrifugation

Although we live daily with prolonged G, we have adapted to this sustained acceleration and to the many problems associated with $G = 1$ (objects fall, steps are difficult to climb, jumping height is limited, etc.). The only environments where one would be exposed to $G > 1$ for any period of time (greater than 60 sec) would be in maneuvering space flight, living on a larger planet, special aircraft maneuvers, auto racing, etc. G loading in modern race cars has increased because of the increase in lap speeds and cornering speeds. It is estimated some drivers experience 2.5–3 Gy 800 times per race (Jennings and Mohler, 1988). In 1960, Carl Clark reported (1960) on his 24 hr exposure at 2 Gz in the Navy's centrifuge at Warminster PA. Clark and his colleagues set out to prove that man could easily survive a flight to Mars—one that provided a steady 2 G acceleration halfway there and then a 2 G deceleration the remainder of the trip. Such a mission would take only 24 hr to travel 45 million miles at a maximum speed of 3.8 million miles per hour! Here, 35 yr later, we are still only planning missions to Mars, but planning for 18 months each way and not 24 hr. Of interest in Clark's exposure was that he was able to live in the cab of the centrifuge, sleeping, eating, and even standing at several points. Sixteen hours into the exposure, Clark reported an anesthesia sensation in the ring and little fingers of the left hand; some tingling sensation remained for about 2 months after the experiment. A feeling of lightness lasted about 30 min after the centrifuge stopped. An abrupt head motion 30 min postexposure produced retching, but other recovery seemed uneventful (Clark, 1960).

Two experiments conducted on the WPAFB OH DES centrifuge involving shorter durations at 2 G (40 min and 90 min) were related to space applications as well (McCloskey et al., 1992; Martin and Albery, 1993). The 40 min exposure to 2 Gz was a simulation of a proposed training flight of the National Aerospace Plane (NASP) that was to be a hypersonic aircraft, capable of conventional take-off with ascent into space capability. A question was raised during the design phase of that program regarding the human performance implications of a 30+ min, 2 G turn the aircraft would make during training sessions. Many studies had been performed in the 60s and 70s regarding G exposures for space, but no one had specifically looked at human motor control and cognitive function during a 30+ min, 2 G exposure. Forty minutes at 2 G was selected to cover the entire mission. The DES cab was mocked up to resemble a partial NASP control panel. Eight subjects were exposed to +2 Gz for 40 min on the DES centrifuge, during which time they performed a choice reaction time task, a keypad entry task, and a reach task. The tasks were based on one of many preliminary NASP cockpit designs. Results indicated that with the simple reaction time tasks no performance decrements were found; in fact a slight increase in performance occurred. For the keypad task, however, performance decrements were found throughout the entire profile. The reach task showed decrements early in the exposure, but toward the end performance reached baseline levels. This suggests a "recalibration" of gross motor movement during long exposure to low-level acceleration. Results indicated that input devices which require large arm/hand movements should be mounted no farther than 21 in. from the floor and 35 in. away from the intersection of the seatback/pan of an ACES II-type seat. Keypad entry devices should have button sizes of at least 3/4 in. by 3/4 in., and should be mounted on a swivel mount to reduce uncomfortable wrist angles (McCloskey et al., 1992).

The 90 min, 2 Gz exposure experiment was related to Space Adaptation and Space Motion Sickness (SMS) research. Long duration centrifugation (3 Gx for 90 min) has been used in the Netherlands to develop SMS symptoms (Bles and de Graaf, 1993). In the WPAFB OH DES centrifuge research, eight volunteer subjects were exposed to 90 min, 2 Gz. Adaptation of the vestibular system, specifically, the otolith organs, to a nonterrestrial environment can result in space motion sickness-like symptoms when the human is re-introduced to the normal, 1 G, terrestrial environment. This premise was investigated by exposing nine subjects to 90 min of sustained 2 G acceleration in a human centrifuge and then observing and evaluating them at 1 G. Five of the subjects developed slight SMS symptoms, three developed moderate, and one developed frank sickness. Postural instabilities in two of the most affected subjects were also observed using the Equitest System postexposure. The NeuroCom Equitest System is a postural stability measurement device used in both medical and research environments. It can be used to impart both motion and visual cues to the subject via force plates, on which the subject stands, or a visual surround; reactions to these cues are then recorded and analyzed on a personal computer (Martin and Albery, 1993). Long duration exposure to a nonterrestrial G(2 G) appears to be a potential means for developing SMS-like symptoms in a ground-based human centrifuge (Martin and Albery, 1993).

29.4.6 Performance and G-Protective Techniques

Chambers et al. (1962) investigated the effects of positive pressure breathing on performance. The ability to perform a complex psychomotor task and a visual luminance discrimination task during increased +Gz and +Gx was improved while using positive pressure breathing. As was found with increased seatback angles, positive pressure breathing increased subjects' G-tolerance. Rogers (1973) measured pilot's performance in a high-G environment as a function of seatback angle. Results showed significant improvements in performance as the seatback angle was increased from 30° to 60° from vertical. This improvement in performance was attributed to the increased flow of blood to the brain as a result of minimizing the heart-to-eye hydrostatic column length. However, subject performance was found to be negatively related to accelerations above +6 Gz. Lisher and Glaister (1978) studied the effects of +Gz acceleration on the performance of a spatial orientation "manikin" task while using different seatback angles. The manikin task involved the identification of the orientation of a human figure with respect to the observer. The task seemed to tap cognitive resources associated with spatial orientation (Carter and Wolstad, 1985). Lisher and Glaister (1978) varied acceleration from +1 Gz to +10 Gz in addition to using three different seatback angles of 17°, 52°, and 67°, from vertical. Seatback angle was positively correlated with performance improvements. Using a simulated aerial combat maneuver (SACM) scenario, Burton and Shaffstall (1978) were unable to repeat the improvement in performance reported by Rogers (1973) and Lisher and Glaister (1978) using similar seatback angles with accelerations below +6 Gz. Burton and Shaffstall (1978) attributed this discrepancy to the differences in the tasks used by the three sets of researchers.

29.4.7 Human Perception/Spatial Orientation

Sustained acceleration (G > 1) can also have a profound effect on human perception (Carter and Wolstad, 1985). It is believed the otolith organs of the vestibular system are gravity transducers (graviceptors) and, as such, play an important role in human perception of motion, especially in the absence of visual cues. Sustained acceleration experienced by pilots during catapult launching of Navy aircraft off of the decks of aircraft carriers and sustained acceleration experienced by pilots in turning and looking maneuvers during flight has been implicated in contributing to aircraft mishaps. The catapult launch generates a +Gx acceleration on the pilot which can elicit a somatogravic illusion (a false sensation of body tilt that results from perceiving as vertical the direction of a nonvertical gravitoinertial force). Pilots of carrier-launched aircraft need to be especially wary of the somatogravic illusion. These pilots experience pulse accelerations lasting 2–4 sec and generating peak inertial forces of +3 to +5 Gx. Although the major acceleration is over quickly, the resulting illusion of nose-high pitch can persist for half a minute or more afterward, resulting in a particularly hazardous situation for the pilot who is unaware of this phenomenon (Cohen, Crosbie, and Blackburn, 1973). Cohen and his associates investigated this phenomenon on the Navy research centrifuge at Warminster, Pennsylvania, after several Navy pilots pitched their aircraft into the sea immediately after catapult launch. Cohen et al. were able to reproduce the pitch-up sensation experienced by pilots experiencing catapult launches, and warnings were issued to all pilots taking off from aircraft carriers.

More than a dozen air transport aircraft are believed to have crashed as a result of the somatogravic illusion occurring on takeoff (Buley and Spelina, 1970). A relatively slow aircraft, accelerating from 100 to 130 knots over a 10-second period just after takeoff, generates +0.16 Gx on the pilot. Although the resultant gravitoinertial force is only 1.01 G, barely perceptibly more than the force of gravity, it is directed 9° aft, signifying to the unwary pilot a 9° nose-up pitch attitude. Because many slower aircraft climb out at 6° or less, a 9° downward pitch correction would put such an aircraft into a descent of 3° or more—the same as a normal final-approach slope. In the absence of a distinct external visual horizon, or even worse, in the presence of a false visual horizon (e.g., a shoreline) receding under the aircraft and reinforcing the vestibular illusion, the pilot's temptation to push the nose down can be overwhelming. This type of mishap has happened at one particular civil airport so often that a notice has been placed on navigational charts cautioning pilots flying from this airport to be aware of the potential for loss of attitude reference (Gillingham and Previc, 1993).

The G-excess illusion is a false or exaggerated sensation of body tilt that can occur when the G environment is sustained at greater than 1 G (Gillingham et al., 1993). This illusion has been implicated in a number of low-level turning and looking mishaps (Lyons, Gillingham, Thomae, and Albery, 1990). It is believed the otoliths "over report" a roll or pitch when the pilot and the aircraft are in a greater than 1G maneuver. This overreporting occurs because the otoliths are "tugged" more than usual when G > 1, and this is perceived by the pilot as an increased roll or pitch attitude. If a pilot experiences the G-excess illusion, the tendency is to compensate for the illusion which can cause overbanking, loss of altitude, and even controlled flight into the terrain. Chelette and her associates duplicated the G-excess illusion on the WPAFB OH DES centrifuge (Chelette, Martin, and Albery, 1992). Rolling and/or pitching the head while in an excess G environment (>1) can cause an illusory sensation of vehicle tilt. If the pilot, or subject, is facing forward, this illusion occurs as the pilot looks up or down; if the head is turned toward the right or left shoulder, the illusion is perceived as an excessive roll of the aircraft. Twelve subjects demonstrated that true vehicle tilt up to 10° is accurately assessed without any significant illusion while head tilts in the −30° to +45° range up to 4 Gz can cause illusory tilts up to approximately 10° (Chelette et al., 1992). Chelette, Li, Esken, and Matin (1995) also investigated the effect of sustained acceleration on the perception of eye level in the WPAFB OH DES centrifuge. The pilot's perception of "horizonness," or eye level, can be influenced by other, conflicting visual cues (Chelette et al., 1995). Eight subjects estimated eye level in the darkened cab of the DES centrifuge while under the visual influence of two dimly lit vertical lines. The subjects experienced 32 combinations of seven (7) pitches of the two-line stimulus while under four different Gz conditions (1, 2, 3, 4 Gz). Sustained acceleration up to 4 Gz was found to have an effect on subjects' visually perceived eye level.

29.4.8 Macrogravity Summary

In summary, the mechanical, physiological, and psychological effects of acceleration on the human operator all act in a synergistic, interactive way to adversely impact human performance. Manual dexterity, speech, vision, and memory processes are all affected by high acceleration forces. Numerous countermeasures designed to protect man from these effects have been reviewed.

The mechanical and physiological effects of macrogravity are virtually nonexistent in microgravity. Manual dexterity, speech, vision, and memory processes are basically unaffected in the microgravity environment. As seen in the first part of this chapter, the human performs in macrogravity in terms of minutes or seconds. In microgravity, the human performs in terms of days, months, and years. This long period of performance requires adaptation and specialized workstations, tools, and protection systems for the human operator.

29.5 MICROGRAVITY (G $<$ 0.001 g)

29.5.1 Physiological Changes and Their Impacts on Human Factors

29.5.1.1 Space Motion Sickness

Space Motion Sickness (SMS) is similar in its symptoms to normal motion sickness, as experienced on earth. Its symptoms include pallor, increased body warmth, sweating, dizziness, nausea, and vomiting. It occurs to some extent in about 70% of crew members on their first flight, with about 40% of crew members suffering moderate to severe symptoms. Generally, symptoms last from 2 to 4 days (Reshke et al., 1994). Research is under way to validate a treatment which appears to prevent it (Davis et al., 1993; Bagian et al., 1994).

SMS has several implications for the human factors engineer. First, in planning activities and time lines, allowances must be made for decreased productivity on the part of at least some crew members for the first few days. Extravehicular activities, in which the astronaut is enclosed in a space suit, should not be scheduled until after the fourth day on orbit. Bags should be readily available to the crew, and equipment such as vacuum cleaners and wet wipes must be available in case cleanup is necessary. After the first few days, adaptation seems to occur in all crew members, so the impact on long-duration missions is minimal.

29.5.1.2 Anthropometrics

There are two aspects of anthropometrics which change after time in a microgravity environment. One is almost instantaneous: adaptation of the "neutral body posture." In microgravity, neither sitting nor standing straight is a comfortable position. In the absence of gravity, the freefloating body immediately adopts a posture in which the joints are flexed, effectively reducing body height from standing height. A neutral body posture was calculated from Skylab photographs, but a recent, more rigorous study (Whitmore, 1995) yielded rather different results. Table 29.5 gives the joint angles from Skylab as cited in the Man-Systems Integration Standards (MSIS) (NASA, 1989), and the actual angles measured from six subjects on a Shuttle flight. Figure 29.5 illustrates these postures with wireframe drawings of the MSIS standard and the six crewmembers.

This affects equipment design significantly. For workstations, generally only foot restraints are provided, and the crew member simply floats in place. The height of the workstation must be adjustable, to accommodate varying crew member sizes and also the personal variations in the neutral body posture. The positioning of the workstation can be achieved either through moving it up and down, or by providing foot restraints that can be easily raised, lowered, and tilted on several axes. Due to the lack of gravity, no support is needed for the buttocks. For an example, see Figure 29.6.

A second change occurs more slowly. Body height increases about 0.5 in. during the first week and stabilizes at approximately a 3% increase over longer durations, due to decompression of the spine. The distribution of body fluids shifts due to the lack of gravity, with legs losing circumference, and some upper body parts swelling, affecting the fit of clothing. There is significant fluid shift to the head, causing a puffy appearance of the face, and affecting the voice and possibly the senses of smell and taste.

Table 29.5 Neutral Body Posture Joint Angles as Presented in Man-Systems Integration Standards (MSIS NBP Column), and for Each of Six Crew Members on a Shuttle Flight. Measurements are in Degrees. Left-Right Indicates the Measurements for the Left and Right Limbs Where Applicable and Available. (MSIS Only Presented Average Numbers and Did Not Distinguish Sides. Side Dominance Data Not Available.)

Anthopometric Measurement	MSIS NBP	Crew 1	Crew 2	Crew 3	Crew 4	Crew 5	Crew 6
Joint Angles	Left-Right	Left-Right	Left-Right	Left-Right	Left-Right	Left-Right	Left-Right
Hip flexion	50	33	33–29	33	33	29	12
Hip abduction	18.5	6.5–5.5	20–16	13–17.5	15.5–16	3.5–4.5	4–9
Knee flexion	50	50	83–87	50	50	44	11–12
Ankle plantar extension	21	6–7	15–14.5	29–30	27–24	16–14	35–41
Waist flexion	0	13	0	1	0	0	2
Neck flexion	24	16	18	16	5	7	16
Left neck lateral bend	0	0	0	3	0	0	0
Shoulder flexion	36	49–46	67–64	29	33–35	60–57	36
Shoulder abduction	50	32–33	26–26.5	27–29	40.5	24–45	23–36
Medial shoulder rotation	86.6	58–61	45.5–41	71–77	74.5–74	25.5–26.5	50–48
Elbow flexion	90	78	45–53	61–57	94–91	78–80	51–64
Wrist extension	0	0	3–0	0	0	0	0
Wrist ulnar bend	0	0	0	0	0–9	0–3	0
Forearm pronation	n/a	n/a	26	20–n/a	n/a–2	16–n/a	n/a–5
Forearm supination	30	7–10	n/a	n/a–30	15–n/a	n/a–4	14–n/a
Finger flexion	0	42	60	30	21–57	55–47	25–35

29.5.1.3 Long-Term Effects and Countermeasures

Extensive research has been carried out on physiological changes in microgravity, which is well documented in Nicogossian, Huntoon, and Pool (1994). Perhaps the most significant changes for the human factors engineer are the ones in strength, as this directly affects the ability of the crew member to perform work. Jaweed (1994) reports significant (10–20%) decreases between preflight and postflight strength in the antigravity muscles (back and legs) after as few as 5–10 days on orbit. This, taken with the loss of bone mass observed (Schneider et al., 1994), indicates that countermeasures must be taken for long duration flights, and that tasks which can be performed early in flight might be more difficult or dangerous after extended time in microgravity.

The most common countermeasure for strength loss is exercise, particularly of the legs and back. Typical equipment includes bicycle ergometers, treadmills, and rowing machines. When designing a space facility, volume must be allowed for equipment storage and deployment; crew time needs to be allotted, and noise and vibration must be minimized.

29.5.2 Environmental Factors

29.5.2.1 Noise and Lighting

Noise can affect human physiology and health in a number of ways (Wheelwright et al., 1994). From the strictly human factors perspective, noise can affect performance by interfering with communications, interfering with sleep, and causing annoyance. In an assessment of the SpaceHab-1 mission (STS-57), Mount et al. (1994) found that while the measured noise levels did not generally exceed the permitted levels for the Shuttle flight deck or middeck, noise levels were substantially above design limits for the SpaceHab. This is probably because of the number and nature of experiments and equipment which were located there. However, most crew members required earplugs during sleep, even though they slept in the Shuttle; used the intercom rather than unaided voice to com-

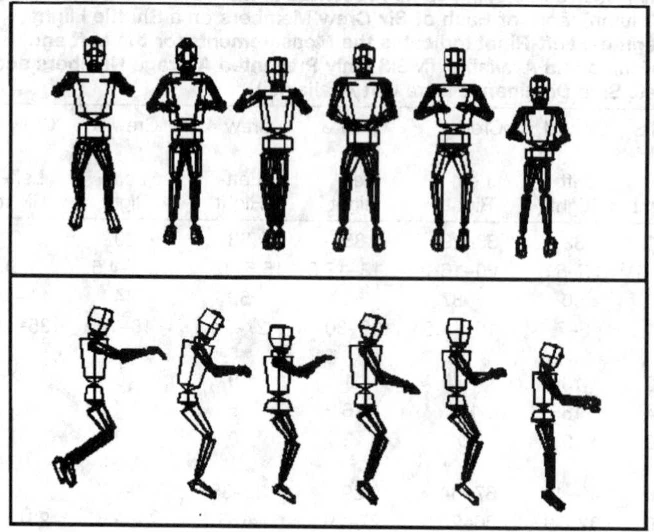

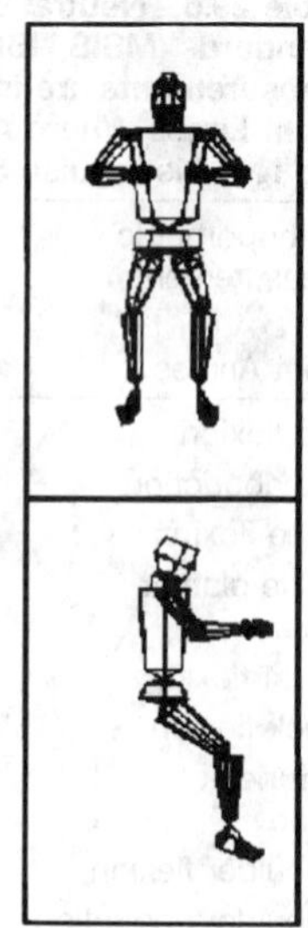

Figure 29.5 Wireframe drawings illustrating theoretical neutral body posture, as described in NASA-STD-3000 (right column) and neutral body postures as observed for six crew members on a recent shuttle flight. (Figure courtesy of Graphics Research and Analysis Facility, Johnson Space Center, NASA.)

municate, even in the same area; and reported difficulty in concentration, and noise-induced headaches and fatigue. Only one of six crew members said that such noise levels would be acceptable if endured for 6 months. A pictorial representation of the communications ranges in the various areas is shown in Figure 29.7.

Lighting is essential to performing virtually every task. When windows are present and unshuttered, the typical 90 min, low earth orbit of the Shuttle or Station causes problems with insufficient time for eyes to adapt to the rapid appearance and disappearance of sunlight. In the study by Mount et al. (1994), the most frequent report of lighting problems was from sunlight making electronic displays and TV monitors difficult or impossible to read. However, some operations require out-the-window viewing which is a favorite crew activity in any spare time. Wheelwright et al. (1994) and the Man-Systems Integration Standards (NASA, 1989) provide a number of tables and guidelines for illumination levels for various intravehicular and extravehicular tasks.

29.5.2.2 Lack of Convection and Air Flow

In the absence of gravity, warm air does not rise. That simple fact accounts for much of the noise described above, as all equipment cooling must be done by forced air (fans) or through coolant loops (pumps and compressors). Forced air circulation throughout pressurized volumes is required. In addition, it means that exercise equipment must be accompanied by a fan to cool the user, as the evaporating perspiration and radiated heat from the body have no tendency to fall or rise, but merely envelop the person.

29.5.3 Equipment Design Considerations

29.5.3.1 Restraints, Handbooks, and Foot Restraints

Restraints are needed for both personnel and equipment. The most common restraint for crew members is a foot restraint. For a location where a person will be working for

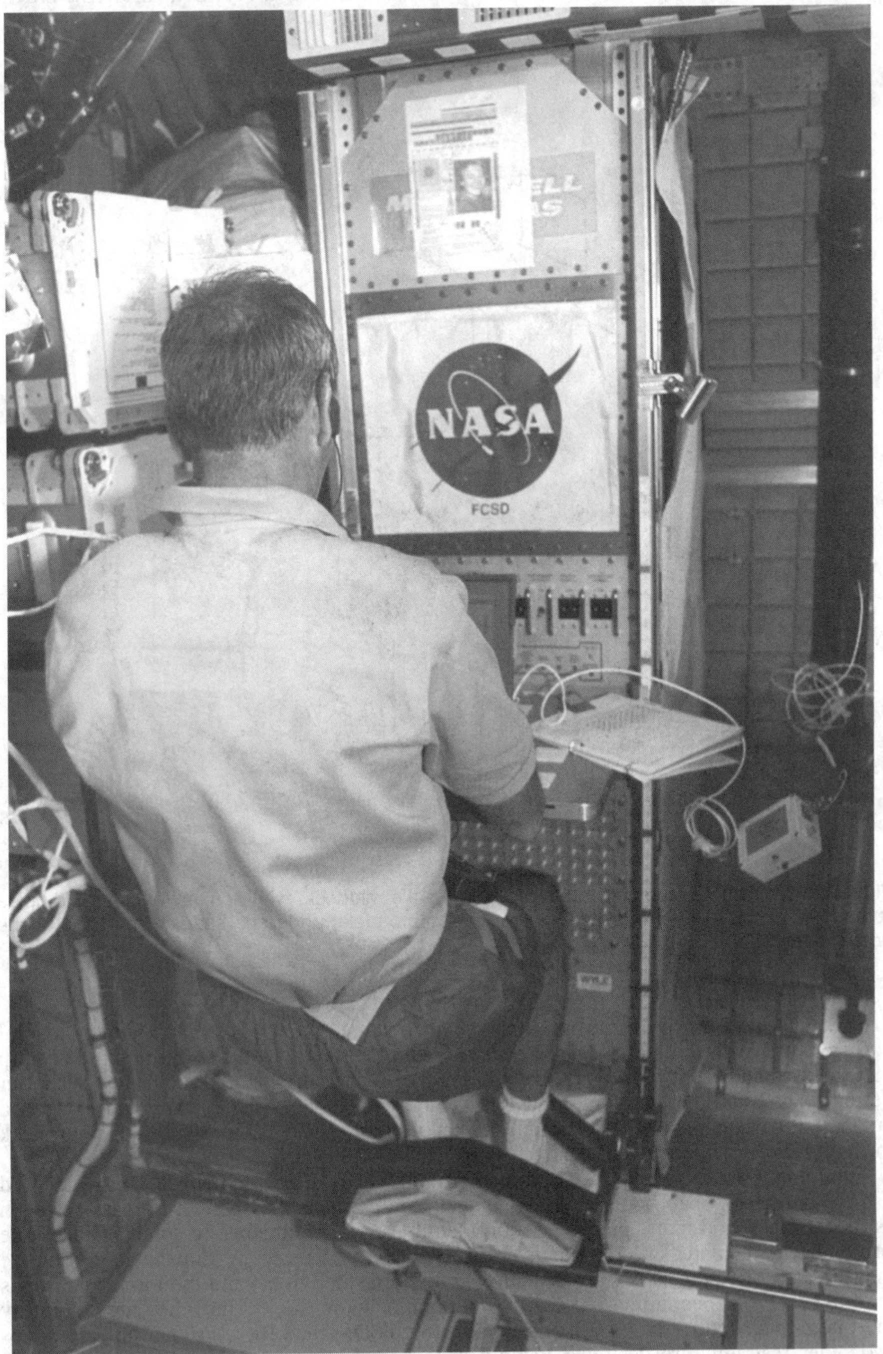

Figure 29.6 A shuttle crew member at a workstation, restrained by feet. Notice body posture and lack of support for trunk. (Photograph courtesy of National Aeronautics and Space Administration.)

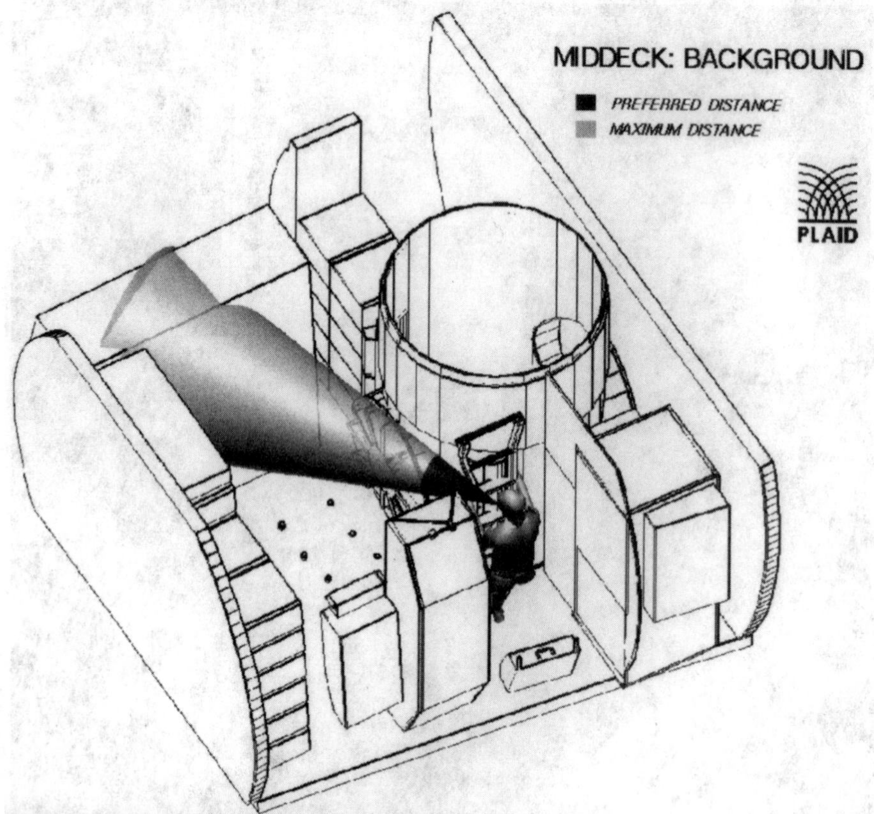

Figure 29.7a Vocal communications range in shuttle middeck, with nominal background noise levels. The darker part of the cone shows the comfortable range; the shaded cone shows the potential range.

extended periods of time, these are platforms that can be tilted to accommodate the neutral body posture, with the feet angled down, and with height adjustments. Some kind of loop that the toes and front part of the feet fit snugly under is frequently adequate for shorter periods. The Shuttle crews sometimes make their own foot restraints where one is needed but has not been installed by using gray tape stuck to the desired surface, with slack through which the foot is inserted.

Equipment also needs to be restrained (Figure 29.8). Clips or bungee cords are used to hold papers or open books. Patches of velcro are attached to almost all surfaces such as locker faces, and many pieces of equipment have velcro attached to some surface. The equipment is then stuck to the handiest piece of velcro. The major drawbacks to this are that the area of velcro is limited by flammability considerations, and the loops and hooks tend to break off, becoming free-floating contaminants in the air until they are caught by the filters in air return ducts. When a workstation is designed for a special purpose, loops and other restraints are designed to hold specific tools in convenient locations. *Tethers* are restraints that can be attached to eyelets at various locations, or to loops on the crewmember's clothing, to permit carrying equipment such as flashlights, cameras, or hand tools while leaving the hands free.

Handholds are used to initiate and to terminate locomotion. They are also sometimes used as a restraint. Grabbing a handhold with one hand leaves only one hand for performing a task, so they are generally used only for brief activities such as opening a locker, or flicking one or two switches. Crew members will push against any surface to initiate or halt translation, so in designing partitions, racks, etc., these loads must be considered. Also, switch covers are needed to prevent accidental activation by trailing feet or hands, and all three dimensions must be considered in traffic pattern analysis.

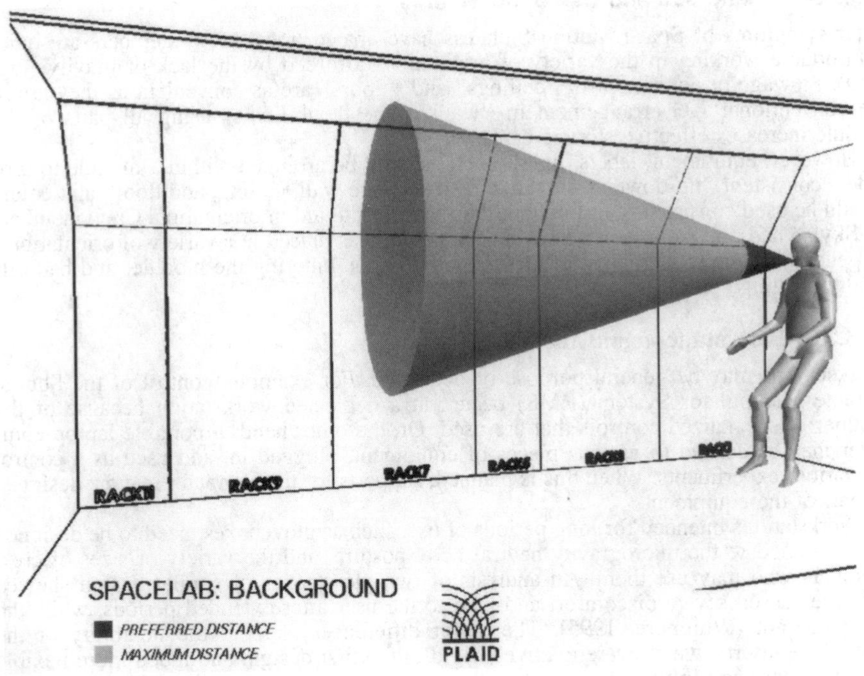

Figure 29.7b Vocal communications range in Spacelab module, illustrating that a crew member at one end of the module cannot be heard at the other end.

Figure 29.8 The need for restraints in microgravity is illustrated by the variety of free-floating manuals and cables shown. (Photograph courtesy of National Aeronautics and Space Administration.)

29.5.3.2 Orientation and Use of 3D Volume

After symptoms of Space Motion Sickness have abated, most crew members are quite comfortable working in the variety of orientations afforded by the lack of gravity (Fig. 29.9). Stowage or equipment in "ceilings" and "floors" are as convenient as they are in the conventional 1-G arrangement in "walls." The flexibility of using all parts of the module increases effective storage capacity.

However, equipment, labels, lighting, etc. should be arranged within a module to provide a consistent "up-down" orientation, and separate wall, ceiling, and floor paint colors should be used consistently throughout the spacecraft to aid in orientation. Crew members on Skylab had one module in which the equipment was placed in a variety of orientations. In general, the crew members felt "disoriented" upon entering the module, and had difficulty locating specific equipment.

29.5.3.3 Workstations and Tools

Workstations may be general purpose or dedicated. For example, control of the Shuttle Remote Manipulator System (RMS) occurs at a dedicated workstation because of the number of specialized controls that are used. On the other hand, a portable laptop computer may be carried to various pieces of equipment, plugged in, and used as a control for various experiments. When this is planned, supports for the computer may be designed as part of the equipment.

Workstations intended for long periods of use, such as gloveboxes, need to be designed to accommodate the microgravity neutral body posture, and the variety of sizes of crew members who may use them. An analysis of two gloveboxes of similar design showed that one caused severe discomfort and pain to the user after extended periods, while the other did not (Whitmore, 1995). The major differences were in the flexibility of the arm/glove inserts, which were relatively rigid in the first design, and much more flexible in the second, and in the adjustment of the foot restraints.

Tools can be essentially identical to comparable ones for earth use, with the addition of eyelets for tethering, or velcro for restraint.

Figure 29.9 The ability to work in a variety of orientations, and use of areas above normal head height for stowage and equipment, is possible in microgravity. Note the two Shuttle crew members operating perpendicular to each other. (Photograph courtesy of National Aeronautics and Space Administration.)

29.5.3.4 Liquid Containment and Hygiene

There are two major roles for liquids: human consumption and hygienic uses. Water from dispensers for drinking and to rehydrate food is generally introduced through a nozzle directly into a container. Liquids such as coffee, juices, etc., are sipped through a straw that automatically closes when suction is not present, to prevent the liquid from escaping into the atmosphere. Liquids used to prepare foods are absorbed by the dehydrated food, and when the package is opened, the food can generally be eaten with a fork or spoon, since surface tension tends to hold portions together.

Personal hygiene is accomplished on short (5–20 day) missions through the use of saturated wipes. These are premoistened with a variety of cleansers for different purposes (skin, hair, etc.) and individually packaged. Enclosed handwashers with free water have been used, with gaskets through which hands are inserted. For longer duration missions, such as Skylab or Space Station, showers are considered highly desirable. In the absence of gravity, air flow and suction are used to draw the water down the drain and preclude the user from inhaling free-floating droplets. The human factors drawback to these is the crew time required to use them. The handwasher must be wiped dry, as must the shower. Showers are typically folded and stowed rather than left standing and occupying valuable volume, so the time overhead for a shower may run as high as an hour. On Skylab, showers were generally taken only once every week or two, because of this high time cost.

29.5.4 Perceptual and Cognitive Factors

To the best of the author's knowledge, time in microgravity per se does not affect perceptual and cognitive performance. There are anecdotal reports of increased distant visual acuity, of decreased near visual acuity, and of hearing loss (caused by long-term exposure to excessive noise levels). No systematic studies of any of these effects have been found. Schiflett (1995) presented a panel discussion for the Aerospace Medical Association in which three investigations on the International Microgravity Laboratory mission (IML-2) found no consistent differences in performance between prelaunch and on-orbit tests of cognitive performance.

29.5.5 Extravehicular Activity

29.5.5.1 EVA Suit Requirements

Extravehicular activity (EVA) is work outside the spacecraft, when the crew member is wearing a "space suit," or extravehicular mobility unit (EMU). The EMU must provide pressurization against vacuum; temperature and atmosphere regulation; and radiation protection. In general, the EVA crewmember is self-propelled, although two different gas propulsion systems have been used on the Shuttle. The Shuttle EMU uses a 4.3 psi pressure of 100% oxygen, while the cabin is 10–14.7 psi oxygen/nitrogen. Therefore, an oxygen prebreathe period of about 1–2 hr is needed to remove dissolved nitrogen from tissues before EVA, to avoid decompression sickness ("bends"). Activity is limited, and eating impossible, during this period.

29.5.5.2 EVA Suits, Tools, and Aids

The EMU is illustrated in Figure 29.10. The current suit is primarily "soft"—fabrics and molded, flexible plastics are used. However, the flexibility is limited by the pressure of the suit. Advances in design of mechanical joints have produced prototype "hard" suits which are constructed of metals and hard plastics, and which can be pressurized to about 8 psia, decreasing prebreathe requirements while not significantly impairing strength performance or range of motion.

Probably the greatest EMU design challenge is the glove. Many tasks include removing bolts and fasteners, and other fine motor tasks. However, the gloves must have layers to maintain pressure, to protect from radiation and contact with very hot or very cold surfaces, and to resist penetration by micrometeoroids. This necessarily reduces tactility and dexterity. Since the hand must work against pressure to bend the glove, hand fatigue is often the limiting factor in EVA work. Excellent reviews of glove design and performance issues are provided by O'Hara, et al. (1988) and by Shepherd and Lednicky (1990).

The most important EVA aid is the restraint, since if the crew member loses contact with the shuttle, recovery may be impossible. Unless using a gas-propelled unit, the crew member must always be connected to the shuttle by a tether. In addition, when working

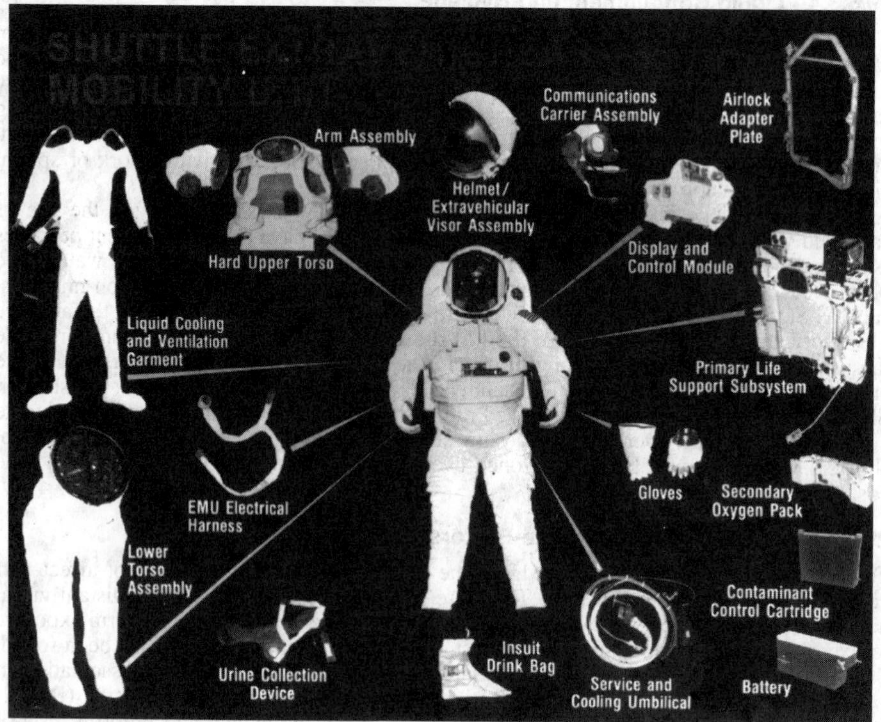

Figure 29.10 Exploded view of the shuttle Extravehicular Activity (EVA) suit. The arms, gloves, and lower torso assembly are all "soft," and provide protection from vacuum, radiation, temperature extremes, and micrometeoroids by using a number of layers of special-purpose fabrics. (Illustration courtesy of the Crew and Thermal Support Division, Johnson Space Center, NASA.)

at a fixed location, he or she may use foot restraints. The current designs require all forces to be transmitted through the ankles and/or a hand. Restraints at the hips or waist may be more advantageous from a biomechanics viewpoint.

Tools used in EVA need to be easily carried, usually by lanyards hooked to the tool and the suit; to fit the gloved hand, which is larger than the bare hand, and which cannot close as completely without excessive effort; and when possible to provide mechanical counters to torques generated, such as in an electric-powered torque wrench.

29.5.6 Microgravity Summary

The lack of gravity affects many aspects of design requirements and workload planning. All of the basics of good human factors engineering apply, but additional knowledge of the effects of microgravity is essential to creating a good environment and operable equipment (Table 29.6). To obtain much of this knowledge in a single source, the human factors practitioner designing for spacecraft is strongly urged to obtain a free copy of the Man-Systems Integration Standards, NASA-STD-3000, from the Flight Crew Support Division of the Johnson Space Center. This document contains data and requirements, guidelines, and design suggestions for designing habitable modules for space. It includes an extensive section on anthropometry and strength data, extrapolated to the year 2000; requirements for environmental aspects such as air, water, and vibration; and extensive material on design considerations for living and working spaces. One chapter is devoted to workstations, another to extravehicular activity. Besides dealing with issues of microgravity, it also contains information relevant to designs for lunar or Mars habitats. This document is regularly updated, and solicits suggestions for improvements, additions, and changes.

29.6 CONCLUSIONS AND RECOMMENDATIONS

The deleterious effects of sustained acceleration on performance can generally be attenuated by the design of protective equipment, attention to task requirements, selection and

Table 29.6 Table of Contents of Man-Systems Integration Standards, NASA-STD-3000, Illustrating Range of Issues Which Must Be Addressed in Designing For Microgravity

Man-Systems Integration Standards
NASA-STD-3000
Table of Contents

1.0	**Introduction**	8.9	Mobility aids and restraints architectural integration
1.1	Purpose	8.10	Hatches and doors
1.2	Overview	8.11	Window integration
1.3	Scope, precedence, and limitations	8.12	Interior design and decor
1.4	How to use the documents	8.13	Lighting
1.5	How to use the standards relational database	**9.0**	**Workstations**
2.0	**General requirements**	9.1	Introduction
2.1	Introduction	9.2	Workstation layout
2.2	General design considerations	9.3	Controls
2.3	General design requirements	9.4	Displays
3.0	**Anthropometry and biomechanics**	9.5	Labeling and coding
3.1	Introduction	9.6	User/computer interaction
3.2	General anthropometrics and biomechanics related design considerations	**10.0**	**Activity centers**
3.3	Anthropometric and biomechanics related design data	10.1	Introduction
4.0	**Human performance capabilities**	10.2	Personal hygiene
4.1	Introduction	10.3	Body waste management facilities
4.2	Vision	10.4	Crew quarters
4.3	Auditory system	10.5	Galley and wardroom
4.4	Olfaction and taste	10.6	Meeting facility
4.5	Vestibular system	10.7	Recreation facility
4.6	Kinesthesia	10.8	Microgravity countermeasure facility
4.7	Reaction time	10.9	Space medical facility
4.8	Motor skills (coordination)	10.10	Laundry facility
4.9	Strength	10.11	Trash management facility
4.10	Workload	10.12	Stowage facility
5.0	**Natural and induced environments**	**11.1**	**Hardware and equipment**
5.1	Atmosphere	11.1	Introduction
5.2	Microgravity	11.2	Tools
5.3	Acceleration	11.3	Drawers and racks
5.4	Acoustics	11.4	Closures and covers
5.5	Vibration	11.5	Mounting hardware
5.6	(Not used)	11.6	Handles and grasp areas (for portable items)
5.7	Radiation	11.7	Restraints
5.8	Thermal environment	11.8	Mobility aids
5.9	Combined environmental effects	11.9	Fasteners
6.0	**Crew safety**	11.10	Connectors
6.1	Introduction	11.11	Windows
6.2	General safety	11.12	Packaging
6.3	Mechanical hazards	11.13	Crew personal equipment
6.4	Electrical hazards	11.14	Cable management
6.5	Touch temperature	**12.0**	**Design for maintainability**
6.6	Fire protection and control	12.1	Introduction
6.7	Decompression hazards	12.2	Design for maintainability design considerations
7.0	**Health management**	12.3	Design for maintainability design requirements
7.1	Introduction	**13.0**	**Facility management**
7.2	Preventative care	13.1	Introduction
7.3	Medical care	13.2	Housekeeping
8.0	**Architecture**	13.3	Inventory control
8.1	Introduction	13.4	Information management
8.2	Overall architectural considerations and requirements	**14.0**	**Extracvehicular activity**
8.3	Crew station adjacenies	14.1	General EVA information
8.4	Compartment and crew station orientation	14.2	EVA physiology
8.5	Location coding	14.3	EVA anthropology
8.6	Envelope geometry for crew functions	14.4	EVA workstations and restraints
8.7	Traffic flow	14.5	EVA mobility and translation
8.8	Translation paths	14.6	EVA tools, fasteners, and connectors
		14.7	EVA enhancement systems

loading, and the ergonomic design of controls and displays compatible with environmental stress levels. Individual differences between people, which defy easy scientific measurement, must not be overlooked. Training factors and overall operator experience have a large impact on the effects of environmental influences. The research results available, reviewed, and/or referenced above should lead to appropriate guidelines with respect to safety and performance limitations.

ACKNOWLEDGMENTS

The first author expresses his appreciation to Dr. Kathy McCloskey and Dr. Henning Von Gierke of the Armstrong Laboratory who helped develop the material regarding performance in macrogravity. Roxanne Baer and Sylvia Bailey are acknowledged for typing the manuscript. The second author expresses her thanks to Mr. Robert Bond and Ms. Tandi Bagian of NASA's Johnson Space Center for the time and effort they spent in reviewing early drafts of the microgravity section, and for their insightful suggestions and comments which helped in prioritizing the material to be included. Both authors wish to thank Mr. Steven Bolia of Wright-Patterson Air Force Base for his significant contribution in preparing the manuscript and facilitating electronic communications between them.

REFERENCES

Air Force Pilot Training Pamphlet (1976). Department of the United States Air Force. AFP-160-5.

ANSI (American National Standards Institute) (1979). Guide for the evaluation of human exposure to whole-body vibration. ANSI 53.18-1979. Acoustical Society of America, New York.

Albery, W. B., Ward, S. L. and Gill, R. T. (1985). The effect of acceleration stress on human workload (AAMRL-TR-85-039). Wright-Patterson Air Force Base, OH: Armstrong Aerospace Medical Research Laboratory (DTIC-AD:156770).

Albery, W. B. (1988). The effects of biodynamic stress on workload in human operators (AAMRL-TR-88-004). Wright-Patterson Air Force Base, OH: Armstrong USAF Aerospace Medical Research Laboratory (DTIC-AD: A196720).

Albery, W. B. (1989). PBG as a means of G protection. *3rd Annual Interservice/Industry Acceleration Colloquium.* AL/CFBS, Wright-Patterson Air Force Base, OH: USA 45433-7008.

Albery, W., and Vanderbeek, R. (1994). Air Combat Command's aeromedical requirements into the 21st century. *Paper presented at the 65th Annual Scientific Meeting of the Aerospace Medical Association*, May 11, 1994, San Antonio, TX.

Albery, W. (1995). Human factors considerations of whole body movements in non-terrestrial environments. Paper No. IAF/IAA-95-G.3.02 presented at the *46th International Astronautical Federation*, Oslo, Norway, October 1995.

Allen, W. A., Jex, M. R., and Magdaleno, R. E. (1973). Manual control performance and dynamic response during sinusodal vibration (AMRL-TR-73-78). Wright-Patterson Air Force Base, OH: Aerospace Medical Research Laboratory (DTIC-AD: 773844).

Bagian, J. P., and Ward, D. F. (1994). A retrospective study of promethazine and its failure to produce the expected incidence of sedation during space flight. *Journal of Clinical Pharmacology, 34*: 649–651.

Barer, A. S., Golov, G. A., Zubavin, V. B., Muraknovskiy, S. A., Rodin, S. A., Sokorina, Y. I., and Tikhomivou, Y. P. (1964). Physiological reactions of the human organism to transverse accelerations and means of raising the resistance to such forces. *Paper presented at the XU International Astronautical Congress,* Warsaw, September 7–12, 1964. National Aeronautics and Space Administration Technical Translation F-274.

Bles, W., and de Graaf, B. (1993). Postural consequences of long duration centrifugation. *Journal of Vestibular Research, 3*, 87–95.

Braunstein, M. L., and White, W. J. (1962). The effects of acceleration on brightness discrimination. *Journal of the Optical Society of America, 52*, 931–933.

Buley, L. E., and Spelina J. (1970). Physiological and psychological factors in "The dark-night takeoff accident." *Aerospace Medicine, 41*, 553–556.

Burns, J. W. (1988). Prevention of loss of consiousness with positive pressure breathing and supinating seat. *Aviation, Space, and Environmental Medicine, 59*, 20–22.

Burton, R. R. (1974). Man at high sustained and +Gz acceleration: A review. *Aerospace Medicine, 10*, 1115–1136.

Burton, R. R. (1986). G-induced loss of consciousness: Definition, history and current status. *Aviation, Space and Environmental Medicine, 59*, 2–5.

Burton, R., and Shaffstall, R. (1978). Human tolerance to aerial combat maneuvers. *Aviation, Space, and Environmental Medicine, 51*, 641–648.

Burton, R., and Whinnery, J. (1996). Biodynamics: Sustained Acceleration pp. 201–260. In R. Dehart, Ed., *Fundamentals of Aerospace Medicine*, 2nd edition. Williams and Wilkins, Baltimore, 1996.

Canfield, A. A., Comrey, A. L., and Wilson, R. C. (1948). The effect of increased acceleration upon human abilities, Part II: perceptual speed ability (Report No. R. R. 4). Los Angeles, CA: Psychology Department, University of Southern California.

Canfield, A. A., Comrey, A. L., and Wilson, R. C. (1949). Study of reaction time to light and sound as related to increased positive radial acceleration. *Journal of Aviation Medicine*, *20*, 350–355.

Canfield, A. A., Comrey, A. L., Wilson, R. C., and Zimmerman, W. S. (1950). The effect of increased positive radial acceleration upon discrimination reaction time. *Journal of Experimental Psychology*, *40*, 733–737.

Carter, L. A., and Wolstad, J. (1985). Repeated measures of spatial ability with the Manikin Test. *Human Factors*, *27*, 209–219.

Chambers, R. M. (1961). Control performance under acceleration with side-arm attitude controllers (NADC-MA-6110). Johnsville, PA: U.S. Naval Air Development Center (DTIC-AD: 269487).

Chambers, R. M., Kerr, B. S., Augerson, W. S., and Morway, D. A. (1962). Effects of positive pressure breathing on performance during acceleration (NADC-MA-6025). Johnsville, PA: U.S. Naval Air Development Center (DTIC-AD: 298009).

Chambers, R. M. (1963). Operator performance in acceleration environments. In N.M. Burns, R. M. Chambers, and E. Hendler, Eds., *Unusual Environments and Human Behavior*, pp.193–319. New York: The Free Press of Glencoe.

Chambers, R. M., and Hitchcock, L. Jr. (1963). Effects of acceleration on pilot performance (NADC-MA-6110). Johnsville, PA: U.S. Naval Air Development Center (DTIC-AD: 408686).

Chelette, T. (1996). Female performance under high G during fatigue and after G layoff, report submitted to the Defense Women's Health Research Program, U.S. Army MCMR-PLF-DI, Ft. Detrick, MD 21 702-5024.

Chelette, T., Martin, E., and Albery, W. (1992). The nature of the G-excess illusion and its effect on spatial orientation (AL-TR-1992-0182). Wright-Patterson Air Force Base, OH: Armstrong Laboratory.

Chelette, T., Li, W., Esken, R., and Matin, L. (1995). Visual perception of eye level (VPEL) under G while viewing a pitched visual field. *Paper presented at the 66th Annual Scientific Meeting of the Aerospace Medical Association*, Anaheim, CA, May 8–12, 1995.

Chelette, T., and Albery, W. (1996). Human task performance throughout prolonged high G exposure, AL-TR-1996-0135, Wright-Patterson AFB, OH: Armstrong Laboratory.

Cherniack, N. S., Hyde, A. S., and Zechman, F. W. (1959). Effect of transverse acceleration on pulmonary function. *Journal of Applied Physiology*, *14*, 914–916.

Cherniack, N. S., Hyde, A. S., Watson, J. F., and Zechman, F. W. (1961). Some aspects of respiratory physiology during forward acceleration. *Aerospace Medicine*, *32*, 113–120.

Clark, C. (1960). Observations of a human experiencing 2G for 24 hours. *Aerospace Medicine*, April 1960.

Cohen, M. M., Crosbie, R. H., and Blackburn, L. H. (1973). Disorienting effects of aircraft catapult launchings. *Aerospace Medicine*, *44*, 37–39.

Constable, R., and Carpenter, D. (1995). Whole body movements in altered G environments. *Proceedings of the 16th Annual Gravitational Physiology Meeting*, March 19–24, 1995, Reno, NV.

Creer, B. Y. (1962) Impedance of sustained acceleration on certain pilot performance capabilities. *Aerospace Medicine*, *33*, 1086–1093.

Darwood, J. J., Repperger, D. W., and Goodyear, C. D. (1990). Mass discrimination under Gz Stress. *Aviation, Space and Environmental Medicine*, *62*, 319–324.

Davis, J. R., Jennings, R. G., Beck, B. G., and Bagian, J. P. (1993). Treatment efficacy of intramuscular prometazine for space motion sickness. *Aviation, Space, and Environmental Medicine*, *64*: 230–233.

Deaton, J. E., and Hitchcock, E. (1991). Reclined seating in advanced crewstations: human performance considerations. *Proceedings of the Human Factors Society 35th Annual Meeting*, 132–136.

Frankenhauser, M. (1945). Effects of prolonged gravitational stress on performance. *Acta Psychologica*, *104*, 10–11.

Fraser, T. M. (1973). Sustained linear acceleration. In J. J. Parker and V. R. West, Eds., *Bioastronautics Data Book.* Washington, DC: National Aeronautics and Space Administration. U.S. Government Printing Office, Washington, DC, Stock Number 3300-00474.

Frazier, J. W., Repperger, D. W., and Popper, S. E. (1990). Time estimating ability during Gz stress. *Aviation, Space, and Environmental Medicine*, *61*, A1.

Gauer, O. H. (1950). The physiological effects of prolonged acceleration. In *German aviation medicine: World War II*, pp. 554–583. Washington, DC. Department of the Air Force. U.S. Government Printing Office, Washington, DC.

Gauer, O. H., and Zuidema, G. D. (1961). Gravitational Stress in Aerospace Medicine. London: Little, Brown & Co.

Gillingham, K. (1974). Effects of the abnormal acceleratory environment of flight (*USAF SAM-TR-74-57*). Brooks Air Force Base, TX: School of Aerospace Medicine. (DTIC-AD: A009593).

Gillingham, K., and Previc, F. (1993). Spatial orientation in flight, AL-TR-1993-0022, November 1993, Armstrong Laboratory, Brooks AFB, TX.

Greenisen, M. C., and Edgerton, V. R. (1994). Human Capabilities in the Spacecraft Environment. In A. E. Nicogossian, C. L. Huntoon, and S. L. Pool, Eds., *Space Physiology and Medicine*, Third edition. Philadelphia, PA: Lea and Febiger.

Grether, W. F. (1971). Acceleration and human performance (AFAMRL-TR-71-22). Wright-Patterson Air Force Base, OH: Air Force Aerospace Medical Research Laboratory (DTIC-AD: 733814).

Hallenbeck, G. A. (1946). Design and use of anti-G suits and their activating valves in World War II, USAF (Report No. 5433). Wright-Patterson Air Force Base, OH.

Howard, P. (1965). The physiology of positive acceleration. In J. A. Gillies, Ed., *A Textbook of Aviation Physiology.* Oxford, NY: Pergamon Press.

Jaweed, M. M. (1994). Muscle Structure and Function. In A. E. Nicogossian, C. L. Huntoon, and S. L. Pool, Eds., *Space Physiology and Medicine*, Third edition. Philadelphia, PA: Lea and Febiger.

Jennings, R., and Mohler, S. (1988). Potential crashworthiness benefits to general aviation from Indianapolis motor speedway technology. *Aviation, Space, and Environmental Medicine, 59,* 67–73.

Jennings, T., and Zanetti, C. (1988). Positive pressure breathing as a G-protection device: Safety concerns. *SAFE Journal, 18,* 52–57.

Kantowitz, B. (1985). Channels and stages in human information processing: A limited analysis of theory and methodology. *Journal of Mathematical Psychology, 29,* 135–174.

Kerr, W. K., and Russell, W. A. M. (1944). Effects of positive acceleration in the centrifuge and in aircraft on functions of the central nervous system. Report No. C2719. Canada: National Research Council (DTIC-AD:494706).

Lambert, E. H. (1950). Effects of positive acceleration on pilots in flight, with a comparison of the responses of pilots and passengers in an airplane and subjects on a human centrifuge. *Aviation Medicine*, June, 195–220.

Leverett, S., and Whinnery, J. (1985). Biodynamics: Sustained acceleration. In R. DeHart, Ed., *Fundamentals of Aerospace Medicine*, pp. 202–249. New York: Lea & Febiger.

Li, W. (1992). Physiological comparisons between subjects in the forward leaning and upright postures during high Gz centrifuge tests, M.S. Thesis, Wright State University, Dayton, OH; *Aviation, Space, and Environmental Medicine, 63,* 389.

Lisher, B. J., and Glaister, D. H. (1978). The effect of acceleration and seat back angle on performance of a reaction time task. FPRC Report No. 1364. London: Ministry of Defense, Air Force Department, Flying Personnel Research Committee (DTIC-AD: B034784).

Little, V. Z., Hartman, B. O., and Leverett, S. D. (1968). Effects of acceleration on human performance and physiology, with special reference to transverse G. USAFSAM-4-68. Brooks Air Force Base, TX: USAF School of Aerospace Medicine (DTIC-AD: 676209).

Loose, D. R., McElreath, K. W., and Potor, G. Jr. (1976). Effects of direct side force control on pilot tracking performance (AFAMRL-TR-76-87). Wright-Patterson Air Force Base, OH: USAF Aerospace Medical Research Laboratory (DTIC-AD: A036083).

Lyons, T., Gillingham, K., Thomae, C., and Albery W. (1990). Low-level turning and looking mishaps. *Flying Safety*, October.

Malcolm R. (1987). How the brain and perceptual system work. Presented at the *58th Annual Scientific Meeting of the Aerospace Medicine Association.* May 10–14, 1987. Las Vegas, NV: Hilton Hotel.

Martin, E., and Albery W. (1993). Emulation of space motion sickness (SMS) on the dynamic environment simulator centrifuge (AL/CF-TR- 1993-0007), Wright-Patterson Air Force Base, OH: USAF Armstrong Laboratory.

McCloskey, K., Tripp, L., Repperger, D., and Popper, S. (1990). Subjective responses to assisted positive pressure breathing (APPB) as a means of high-G protection. Poster presented at *Human Factors Society 34th Annual Meeting*, Orlando, FL, October 8–12, 1990, also AAMRL-TR-90-056 (DTIC-AD: A230019).

McCloskey, K., Albery, W., Zehmer, G., Bolia, S., Hundt, T., Martin, E., and Black, S. (1992). NASP RE-ENTRY PROFILE: Effects of low-level +Gz on reaction time, keypad entry, and reach error (AL- TR-1992-0130), Wright-Patterson Air Force Base, OH: USAF Armstrong Laboratory.

McCloskey, K., Tripp, L., and Martin, E. (1994). Performance measurements at high G: Behavioral, subjective, and SaO2 results. Presented at the *1994 Aerospace Medical Association Meeting*, May 12, 1994.

Mount, F. E., Adam, S., McKay, T., Whitmore, M., Merced-Moore, D., Holden, T., Wheelwright, C., Koros, A., O'Neal, M., Toole, J., and Wolf, S. (1994). Human factors assessments of the STS-57 SpaceHab-1 mission, *NASA Technical Memorandum 104802.*

National Aeronautics and Space Administration (1989). Man-Systems Integration Standards: *NASA-STD-3000, Rev. A.* Johnson Space Center, Houston, TX.

Navon, D., and Gopher, D. (1979). On the economy of the human processing system. *Psychological Review, 86,* 214–255.

Nicogossian, A. E., Huntoon, C. L., and Pool, S. L., Eds. (1994). *Space Physiology and Medicine*, Third edition. PA: Lea and Febiger.

O'Donnell, R. D., and Eggemeier, F. T. (1986). Workload assessment methodology. In K. R. Boff, L. Kaufman, and J. P. Thomas, Eds., *Handbook of Perception and Human Performance*, pp. 42–1 to 42–49. New York: John Wiley.

O'Hara, J., Briganti, M., Cleland, J., and Winfield, D. (1988). Extravehicular Activities Limitations Study, Vol. II: Establishment of Physiological and Performance Criteria for EVA Gloves—Final Report, *Report No. AS-EVALS-FR-8701*, NASA Contract NAS-9-17702. Johnson Space Center, Houston, TX.

Palets, B. L., Tikhonov, M. A., Popov L., Popov, A. A., Arkhangelskiy, D. Y., Palets, L. D., and Bondarenko, R. A. (1987). Theoretical analysis of efficacy of G suits with exposure to continuously increasing acceleration. *Kosmicheskaya Biologiya I Aviakosmicheskaya Meditsina, 21*, 27–33.

Piranian, A. G. (1982). The effects of sustained acceleration, airframe buffet and aircraft flying qualities on tracking performance. *AIAA workshop on pilot workload and dynamics*. Edwards Air Force Base, CA.

Repperger, D. W., Frazier, J. W., Popper S., and Goodyear, C. (1989). Attention anomalies as measured by time estimation under acceleration stress. *NAECON Proceedings, May 1989*. Dayton, OH, Vol 2: 787–793.

Reshke, M. F., Harm, D. L., Parker, D. E., Sandoz, G. R., Homick, J. L., and Vanderploeg, J. M. (1994). Neurophysiologic Aspects: Space Motion Sickness. In A. E. Nicogossian, C. L. Huntoon, and S. L. Pool, Eds., *Space Physiology and Medicine*, Third edition. Philadelphia, PA: Lea and Febiger.

Rogers, D. (1973) Effect of modified seat angle on air-to-air weapon system performance under high acceleration (AMRL-TR-73-5). Wright-Patterson Air Force Base, OH: USAF Aerospace Medical Research Laboratory (DTIC-AD: 770271).

Schiflett, S. G. (1995). Panel: Microgravity and fatigue effects on cognitive performance of space shuttle astronauts. *Proceedings of the Aerospace Medical Association 66th Annual Scientific Meeting*. Anaheim, CA, p. A47.

Schneider, V. S., LeBlanc, A. D., and Taggart, L. C. (1994). Bone and Mineral Metabolism. In A. E. Nicogossian, C. L. Huntoon, C.L., and S. L. Pool, Eds., *Space Physiology and Medicine, Third edition*). PA: Lea and Febiger.

Shaffstall, R. M., and Burton, R. R. (1979). Evaluation of assisted positive pressure breathing on +Gz tolerance. *Aviation, Space and Environmental Medicine, 50*, 820–824.

Shepherd, C. K. Jr., and Lednicky, C. L. (1990). EVA Gloves: History, Status, and Recommendations for Future NASA Research. *NASA Report JSC-23733*. Johnson Space Center, Houston, TX.

Shubrooks, S. J. (1973). Positive-pressure breathing as a protective technique during +Gz acceleration. *Journal of Applied Physiology*, *35*, 294–298.

Teichner, W. H., Arees, E., and Reilly, R. (1963). Noise and human performance: A psychological approach. *Ergonomics*, *6*, 83–97.

Tripp, L., McCloskey, K., Repperger, D., Popper, S., and Johnston, S. (1992). Evaluation of the retrograde inflation anti-G suit (RIAGS) (AL-TR-1992-0176). Wright-Patterson Air Force Base, OH: USAF Armstrong Laboratory.

Warrick, M. J., and Lund, D. W. (1946). Effect of moderate positive acceleration (G) on the ability to read aircraft-type instrument dials, USAF. Memorandum-TSEAA-694-10. Wright-Patterson Air Force Base, OH.

Wheelwright, C. D., Lengel, R. D. Jr., and Koros, A. S. (1994). Noise, Vibration, and Illumination. In A. E. Nicogossian, S. R. Mohler, O. G. Gazenko, and A. I. Grigoryev, Eds., *Space Biology and Medicine*: Joint U.S./Russian Publication in Five Volumes, Vol. II: Life Support and Habitability, American Institute of Aeronautics and Astronautics, Washington, D.C.

White, W. J. (1958). Acceleration and Vision (WADC-TR-58-533). Warminster, PA: USAF Warminster Air Development Center (DTIC-AD: 208147).

White, W. J. (1960). Variations in Absolute Visual Thresholds During Acceleration Stress (ASD-TR-60-34). Wright-Patterson Air Force Base, OH: USAF Aerospace Systems Division (DTIC-AD: 243612).

White, W. J., and Monty, R. A. (1965). Vision and unusual gravitational forces. In C. A. Baker, Ed., *Visual Capabilities in the Space Environment*. New York: Pergamon Press, pp. 65–89.

Whitmore, M. (1995). Personal communication.

Wood, E. H. (1987). Some effects of the force environment on the heart, lungs and circulation. *Clinical and Investigative Medicine*, *10*, 401–427.

Wood, E. H. (1988). Maximum protection anti-G suits and their limitations. *SAFE Journal*, *18*, 30–40.

CHAPTER 30

ARCHITECTURE AND INTERIOR DESIGN*

John E. Harrigan
New Castle, NH 03854 USA

30.1 INTRODUCTION

Architects, interior designers, and human factors designers and engineers share many common concerns, as they plan and design for the evolving needs of clients and client organizations. Their responsibilities often force them to confront the need to identify, understand, and meet diverse and often conflicting expectations and requirements for the built environment. They recognize the potential in work, home, and community environments for helping individuals achieve a personal sense of satisfaction and effectiveness. Conversely, those responsible for the design of physical settings and planned developments also realize that the built environment can intervene in facility users' actions, fail to support important activities, and can be incompatible with existing sociocultural dynamics. Most design professionals would like to create physical settings and planned developments which meet the expectations and requirements of those for whom they are intended. This concern is seen in their attempts to be cost effective and environmentally sound in the use of resources while at the same time achieving quality at work, at home, and in the community.

Seeking to respond to this challenge, many design professionals will set out to achieve a precise, thorough, and comprehensive understanding of the human factors aspect of their areas of responsibility. On the other hand, many individuals working within the

*Adapted from *The Executive Architect: Transforming Designers into Leaders* by John E. Harrigan and Paul R. Neel, F.A.I.A., John Wiley & Sons, Copyright © 1996.

design professionals are willing to design and plan solely on the basis of personal assumptions concerning what constitutes a successful work, home, or community environment. Some may do this because they have complete faith in their own experience and insights, or because they believe the cost and effort involved in further study is not worthwhile. Others, while fully knowledgeable about building systems and recent technological innovations related to such topics as "smart buildings," "air quality systems," "safety and security," and "workstations and communication and information systems as drivers of facility design"—are not acquainted with appropriate human factors principles, practices, and methods, or distrust the results of related studies, feeling it is preferable to move into the future free of the past. Finally, some design professionals wish to avoid the collaboration and shared responsibility which is a fundamental aspect of achieving quality at work, at home, and in the community.

What is most promising for the future, is that the design professional who reads this handbook will find that human factors knowledge is readily acquired and is useful for assuring the achievement of quality of life at work, at home, and in the community. Design professionals who apply the contents of this handbook will find that expenditures in time and resources are cost effective, that information which specifies human limitations and capabilities and environmental specifications and standards can be used with confidence, that human factors applications can support the highest visionary and socially advancing goals, and that collaboration and shared responsibility with those having special preparation in human factors and ergonomics should be basic tenets of facility design and planning.

Here the focus is on managing the efforts associated with the attainment of quality of life at work, at home, and in the community where the human factors perspective must be recognized as a principal criteria for project success. Consider all the people who comprise building design and land development teams, a combination of expertise drawn from the following: client and client representatives, investors, lenders, underwriters, facility user representatives, architects, interior designers, building system consultants, facility managers, landscape architects, land planners, structural, civil, and transportation engineers, construction managers, trades and crafts workers, equipment and product manufacturers, institutional investors, marketing personnel, real estate executives, attorneys, business partners, and jurisdictional and government offices. The question facing those undertaking human factors applications is, how does the building design and land development team remain focused on the expectations and requirements of facility users? It was this challenge that led to the development of the Knowledge Base System (Figure 30.2) presented in this chapter.

30.1.1 Quality of Life at Work, at Home, and in the Community

The phrase "quality of life" is used throughout this chapter. What does it mean? Is it tangible, something that people will understand if we talk about it as a necessary consideration? Yes. Experiencing quality of life at work, at home, and in the community is the hope and demand of individuals, families, communities, and client organizations throughout the world. The idea of quality of life recognizes that a well-designed work place, a comfortable and attractive place in which to do business, a place for learning, a good home, places to rest, relax and get well, good places to entertain our friends, to visit and to shop, a place to park the car and walk, places to get away from it all and ways to get where we are going are essential aspects of life at work, at home, and in the community. At work, where people are asked to make their best possible contribution to organization success and the organization provides all that is needed to achieve skilled job performance; at home, where family members share quiet and exciting moments and try to make each day pleasant for all; in the community, where individuals and families reach out to others seeking to share experiences and develop all that is needed to foster new customs and protect treasured traditions—physical settings and planned environments must be perfected in terms of people's needs and wants.

30.2 KNOWLEDGE BASE SYSTEM STRATEGY

"Bring life to a building before there is a building" is the goal of a Knowledge Base System application; that is, develop an encompassing image of what will be a successful project outcome. This suggests that the work of building design and land development teams should be a response to nothing less than a thorough study of what people need and want.

Figure 30.1 presents in graphic form the basic strategy for applying the Knowledge Base System. It symbolizes the continual flow of information between research and de-

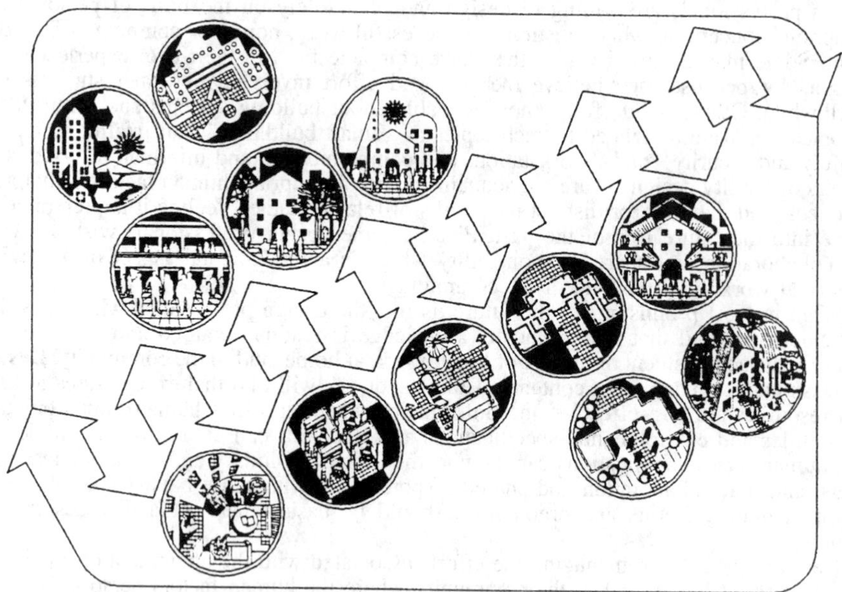

Figure 30.1 Basic strategy.

sign, showing that research and design are one process. This process is a means for accumulating ideas and sharpening perceptions, for combining design speculation and rigorous analysis. This open process strategy means that every time a new fact or finding about facility users' requirements or expectations is identified, its implications are immediately investigated. Every time a design possibility comes to mind, its potential for meeting facility users' expectations and requirements is assessed. Throughout the entire application of the Knowledge Base System, this is the way to work. Though many speculations, insights, and candidate design features will evolve, no confusion exists because the Knowledge Base System fosters a team effort, makes the process visible, and provides a format for critique.

Although the Knowledge Base System is an *open* process, it is not the least bit chaotic. The Knowledge Base System provides the "road map" that assures that the building design and land development team will realize its full potential in an organized fashion. There is a special place for every insight and analysis. In fact, the 12 sections of the Knowledge Base System and the 54 question items are mutually exclusive information categories which become unique "addresses" within the process of strategic thinking and decision making. As an open process, the question items are continually reconsidered during the entire application. The goal is to produce design concepts, schemes, forms, and features which are justified in terms of derived knowledge about expectations and requirements related to quality of life.

30.2.1 Knowledge Base System Contents

The Knowledge Base System is essentially a twelve-step process. The first step, 1.0 Quality of Life Challenge, produces a framework which specifies project concepts. This step takes into account the existing situation, anticipated project outcomes, and the archives of design, planning, and construction and engineering. The next three steps of the Knowledge Base System, 2.0 Facility Life Characteristics, 3.0 Family Life Characteristics, and 4.0 Community Life Characteristics lead to a full and accurate characterization of the expectations and requirements of facility users. Here the activities of facility users are anticipated and significant customs, lifestyles, and traditions identified. The concluding work specifies which facility user characteristics should be emphasized in design. The fifth step, 5.0 Critical Circulation Patterns, begins with a study of facility user movements and equipment and material transport. It concludes with the specification of recommended facility circulation patterns.

The sixth step, 6.0 Interior Architectural Spaces, begins the development of the facility schemes, forms, and features needed to support facility user activities. In the seventh element, 7.0 Workstations, the goal is to identify specialized activities which require customized furnishing, fixtures, equipment, and space features. The eighth step, 8.0 Communication and Information Systems, is an increasingly important design consideration. Facility design features must support communication system requirements and information development and processing activities.

When the ninth step, 9.0 Facility Space Arrangements, is reached, sufficient information exists to create floor plan schemes, options and alternatives, and associated facility management guidelines. The tenth step, 10.0 Facility Design Image, provides standards that serve to guide the development of exterior and interior design image features. The eleventh step, 11.0 Facility Site Planning, guides the achievement of compatibility between facility and site in terms of anticipated activities. The twelfth step, 12.0 Community Master Planning, considers how the area under development can be planned in order to meet quality of life objectives and neighborhood and community expectations and requirements.

30.3 KNOWLEDGE BASE SYSTEM

A notable feature of the Knowledge Base System is that each item in the system is written as a question, rather than in the form of a task description. This was done for a distinct reason. When individuals are assigned information development tasks, work becomes formalized and often inflexible. However, when asked a question, an individual begins to contribute immediately, putting what he or she knows to work. The question items in the Knowledge Base System have been specifically written to be useful and understandable by building design and land development professionals, clients, and people in general. Each question item is numerically indexed, which allows for the precise ordering and storing of data and information. Each question item is generic. The question items can be used as is, modified to meet special needs, or supplemented, with new items.

Each section of the Knowledge Base System reminds us that questions have always guided the work of building design and land development teams. As simple as any question might be, however, the resulting answers can be lengthy and complex. In quality of life research, it is essential to deal with the full complexity of people's expectations and requirements, raising the possibility that questions asked might complicate a project, rather than making it better understood. In response to this possibility, each question item in the Knowledge Base System has a selective focus that aids research organization and precisely directs inquires into the client's situation. The result is a division of research activities into basic areas of inquiry.

30.4 COMMENTARY

For purposes of this discussion, let's suppose that an electronic manufacturing corporation in Hsinchu Science Industrial Park, Taiwan, is investing in advanced memory chip applications associated with an attempt to markedly increase the features, quality, and reliability of their product line. The success of this enterprise depends on the availability of technicians with many advanced skills. For a time, the corporation trained its own people; however, the complexity of new manufacturing processes is now placing demands on the corporation's in-house training program which cannot be met without a significant improvement in their education and technical training programs. The demand is great. It takes a total of 1700 hr to move a new employee through the three training stages: Technician Levels I, II, and III. Corporate officials decide to discuss their situation with a local Junior College of Technology. They find that the present curriculum is too basic for their needs and designed for students able to attend regularly scheduled classes. The inflexibility of class schedules does not meet the needs of a corporation that uses shift changes and overtime to achieve manufacturing objectives. Another aspect of this problem is that when the corporation's technicians attend scheduled classes which take them out of the work setting, their place in the manufacturing process must be filled by someone else. This means a possible decrease in productivity and paying the replacement as well as the student, which doubles training expenditures. The corporation needs a flexible educational opportunity that accommodates the variable schedules of their manufacturing personnel. This need is so great that they offer to build a new facility for the junior college in return for the development of a more advanced and accommodating program.

The college administration and faculty suggest an "open entry, open exit" program as a solution. They say it is possible to develop and maintain an educational setting which

1.0 QUALITY OF LIFE CHALLENGE
- 1.1 PROJECT CONCEPTS. Taking into account the existing situation and anticipated project outcomes, what are the project concepts?
- 1.2 QUALITY OF LIFE OBJECTIVES. Within the context of the stated object concepts, what are the quality of life objectives?
- 1.3 DESIGN ARCHIVES. Within the archives of architecture, interior design, and facility management, what concepts and designs are significant for the specified quality of life objectives?
- 1.4 PLANNING ARCHIVES. Within the archives of planning, what concepts and site and master plans are significant for the specified quality of life objectives?
- 1.5 CONSTRUCTION AND ENGINEERING ARCHIVES. Within the archives of construction and engineering, what theories, standards, and specifications are significant for the specified quality of life objectives?

2.0 FACILITY LIFE CHARACTERISTICS
- 2.1 FACILITY USER GROUPS. Who will be active within the facility? How may these individuals be grouped by responsibilities and intentions? How many individuals does each category include?
- 2.2 ORGANIZATION STRUCTURE. What are the common and exceptional relationships between groups and organizations that use the facility and contribute to activities within the facility?
- 2.3 FACILITY ACTIVITY DESCRIPTIONS. What are the anticipated activities of facility users? What is known about the extent, time of occurrence, and duration of anticipated activities?
- 2.4 FACILITY LIFE CUSTOMS. What are the philosophies and values of the organizations and groups that will use the facility? What are the significant customs, lifestyles, norms, and traditions of facility users? Are these characteristics stable or likely to change?
- 2.5 FACILITY DESIGN OBJECTIVES. Responding to identified expectations and requirements, which facility user characteristics should be emphasized in design?

3.0 FAMILY LIFE CHARACTERISTICS
- 3.1 FAMILY LIFE GROUPS. Who will be living in the residences? How may these individuals be grouped by family life expectations and requirements? How many individuals does each category include?
- 3.2 FAMILY STRUCTURE. What are the common and exceptional relationships within and between family elements?
- 3.3 FAMILY ACTIVITY DESCRIPTIONS. What are the characteristic activities of individuals and families while at home and in their neighborhoods? What is known about the extent, time of occurrence, and duration of anticipated activities?
- 3.4 FAMILY LIFE CUSTOMS. What are the perceived roles of the family in individual and community life? What are the customs, lifestyles, norms, and traditions of individuals and families? Are these characteristics stable or likely to change?
- 3.5 HOUSING AND NEIGHBORHOOD DESIGN AND PLANNING OBJECTIVES. Responding to identified expectations and requirements, what family life characteristics should be emphasized in housing design and neighborhood planning?

Figure 30.2 Knowledge base system.

4.0 COMMUNITY LIFE CHARACTERISTICS

4.1 COMMUNITY LIFE GROUPS. Who visits and lives and works in the community? How may these people be grouped by community life expectations and requirements? How many individuals does each category include?

4.2 COMMUNITY STRUCTURE. What are the common and exceptional relationships between groups and organizations that make up the community and influence community life?

4.3 COMMUNITY ACTIVITIES. What are the activities of those who visit and live and work in the community? What is known about the extent and time of occurrence of anticipated activities?

4.4 COMMUNITY LIFE CUSTOMS. What is the perceived role of the community in individual and family life? What are the customs, lifestyles, norms, and traditions of those participating in community life? Are these characteristics stable or likely to change?

4.5 COMMUNITY PLANNING OBJECTIVES. Responding to identified expectations and requirements, what community life characteristics should be emphasized in community master plans and associated facility designs?

5.0 CRITICAL CIRCULATION PATTERNS

5.1 USER FLOW. How many people will be entering, leaving, and moving about within the facility, for what purposes, and how frequently?

5.2 EQUIPMENT AND MATERIAL TRANSPORT. What are the characteristics of the equipment and material that must be transported to and within the facility? How will these items be transported, and what is the frequency of such movements?

5.3 RECOMMENDED CIRCULATION PATTERNS. What are the recommended circulation patterns for user and equipment and material flow? In what way is this proposal a response to concerns for efficiency, convenience, safety, and security?

6.0 INTERIOR ARCHITECTURAL SPACES

6.1 SPACES. What spaces are needed to support facility users' activities?

6.2 FURNISHING, FIXTURES, AND EQUIPMENT ALLOCATIONS. What furnishing, fixtures, and equipment, fixed or mobile, does each facility space require?

6.3 CONVENIENCE, SAFETY, AND SECURITY. Will any facility user group or activity require special fixtures, furnishing, space layouts, information displays, or surface treatments? In anticipation of undesirable events, what special safety and security measures are necessary?

6.4 AMBIENT ENVIRONMENTAL CRITERIA. What provisions should be made for the effect on facility users of temperature, humidity, air quality, air movement, illumination, noise, distractions, annoyances, hazards, and climatic conditions?

6.5 INFORMATION DISPLAYS. What are the required information displays?

6.6 DURABILITY AND MAINTAINABILITY. Where do spaces require special attention to durability and maintainability of surfaces?

6.7 SPACE PLANS. What space plans best correspond to facility users' expectations and requirements?

Figure 30.2 (*Continued*).

7.0 WORKSTATIONS
- 7.1 WORKSTATION FACILITIES. Where are workstations required?
- 7.2 WORKSTATION ACTIVITIES. What are the specific workstation activities? What task sequences and timelines characterize the activities assigned to the workstation?
- 7.3 WORKSTATION FEATURES. What are the furnishing, equipment, fixture, tool, and material requirements of the subject workstation?
- 7.4 WORKSTATION LAYOUT. Since each workstation has unique activities and support requirements, how should each workstation be arranged?

8.0 COMMUNICATION AND INFORMATION SYSTEMS
- 8.1 COMMUNICATION AND INFORMATION SYSTEM FACILITIES. Where are communication and information system facilities required?
- 8.2 COMMUNICATION AND INFORMATION SYSTEM ACTIVITIES. What is the relative complexity or uniqueness of the information development and processing activities? What task sequences and timelines characterize the noted activities?
- 8.3 COMMUNICATION AND INFORMATION SYSTEM FEATURES. What are the furnishing, equipment, fixture, and software requirements of the subject communication or information system?

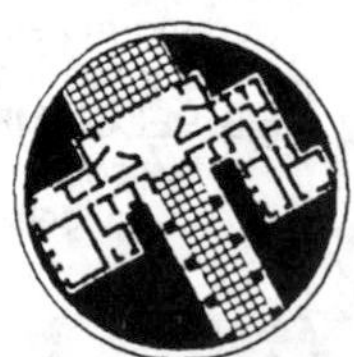

9.0 FACILITY SPACE ARRANGEMENTS
- 9.1 PROPOSED FLOOR PLAN SCHEMES. Considering all research findings and developed design concepts, what are the best schemes for achieving project concept and quality of life objectives? In terms of facility user expectations and requirements, what is the benefit and problem resolution potential of each suggested scheme?
- 9.2 SPACE REQUIREMENTS. What is the estimated square meters for each facility space and support area?
- 9.3 FACILITY MANAGEMENT SCHEME. What are the necessary guidelines and manuals that show people how to use facility operational features?
- 9.4 ALTERATION EXPECTANCIES. How soon might it be necessary to modify or expand the facility? What events would most probably lead to this requirement? How do the proposed facility schemes account for this possibility?

Figure 30.2 (*Continued*).

is open 24 hr a day all year, with the exception of national holidays and the last day of each month for maintenance. Relying on a combination of programmed learning, which permits self-instruction, and faculty supervision, a flexible and yet economical program can be established. According to their concept, all the required educational materials will be crafted for a computer–student interactive setting. Students can come in any time, work on educational assignments, and leave any time they wish. Students can be on the job when required and pursue their studies in the computer–student interactive setting

10.0 FACILITY DESIGN IMAGE

- 10.1 FORM AND STRUCTURE. What are the proposed facility form and structure design concepts?
- 10.2 EXTERIOR DESIGN IMAGES. What are the proposals for exterior facility design images, details, and accents?
- 10.3 INTRIOR DESIGN IMAGES. What are the proposals for interior spatial forms, design images, and surface colors, textures and patterns?
- 10.4 CONCEPT JUSTIFICATION. What are the facility user effects possibilities of each design image recommendation?

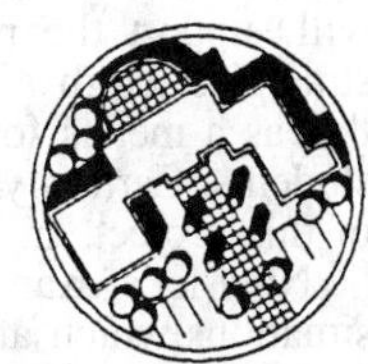

11.0 FACILITY SITE PLAN

- 11.1 SITE REQUIREMENTS. What are the site requirements and to what needs and wants do these correspond?
- 11.2 AREA IMPACT. What are the activities surrounding the site? What will be the impact of facility activities on the surrounding neighborhoods?
- 11.3 SERVICES IMPACT. What will be the impact of facility-based activities and operations on existing public and private services? Will existing services need to be improved or expanded?
- 11.4 SITE PLANS. How should the site be planned in order to achieve quality of life objectives, respond to the needs of those occupying nearby sites, and meet requirements for outdoor space in terms of amenities and landscape development?

12.0 COMMUNITY MASTER PLAN

- 12.1 COMMUNITY SERVICES. What public and private services are needed?
- 12.2 COMMUNITY FACILITIES. What facilities and planned environments are needed to support community life?
- 12.3 CHANGING COMMUNITY REQUIREMENTS. How soon might it be necessary to modify or expand community facilities and services? What events would lead to this requirement?
- 12.4 OUTDOOR SPACE. What are the requirements for outdoor space in terms of amenities, landscape development and preservation, and enhancement of existing natural features?
- 12.5 MASTER PLAN RECOMMENDATIONS. How should the community be planned in order to meet the full range of identified community life expectations and requirements?

Figure 30.2 (*Continued*).

during nonworking hours. It is also noted that, with the appropriate technology, distance learning could provide for study at the corporation's facilities. The college suggests both undertakings.

Within this project concept, the task of the building industry professional is to develop the new facility and the distance learning capability. As you might anticipate, the building design team will include college administrators and faculty, corporation managers and engineers, and the technical workforce. Each of these stakeholders will have his or her own preconceptions, some personal and some related to the corporation's objectives. Each group will have ideas about the design of the building, the interior architectural spaces, workstation and telecommunication features, and required amenities. As one stakeholder group expresses its expectations and requirements and design suggestions, the other groups will learn more about the total situation and reform their own ideas as a result. Building design professionals will work with the stakeholders to discover the unique requirements of this project. Equipment and furnishing designers and manufacturers will contribute their expertise. Government representatives from the Ministry of Education will support the project. Success in this venture may lead to new national education strategies. High school administrators and faculty may wish to participate, as they see this as a means for accelerating and enriching the educational opportunities for selected students. Certainly, other corporations facing similar needs will want to become sponsors of this project. In a short time a national showcase has been created.

Now let's examine this project from the standpoint of two scenarios that characterize situations which are not uncommon. The "open entry, open exit" project begins with an empty room and a blank blackboard. The first person to enter is an architect. What can he or she do? Given the project concept statement, the architectural professional can go to the blackboard and conceptualize the whole building. What would be the significance of this accomplishment? Very little. It expresses only a single point of view, where many are required. A second designer enters the room and helps sketch the interior architectural spaces. What has been achieved? Again, very little. These two professionals are relying on their professional training and experience without the knowledge and experience of those most familiar with what is needed. Let's have the client enter the room. He sees the schematic design presented on the blackboard, derived by a process invisible to him. Since he cannot understand what has been achieved, he rejects the proposal or, even worse, pretends enthusiasm without the intent of approving the concept. The two professional designers, not recognizing that the client's intention is to help them save face, begin to ask a series of questions about the client's requirements. As information is placed on the blackboard, the initial work of the design professionals is revised to reflect what they have learned from the client.

As a reflection of only three points of view, the design solution is still very remote from the full complexity of what is needed. The final product will be only the most meager type of building. Facility users occupying the building will find that nothing has been customized for them. Nothing has been perfected in terms of what is needed to support their activities. Certainly, there is no indication that any of their preferences have been considered in the design process. They conclude that it would have been better to buy an existing building, even if it met very few requirements, and create their own space features. This would mean discarding all of the expertise and experience of building design professionals who can assure that what is developed realizes the full potential of the capital facility investment.

A different scenario might be that the entire building design team enters the room together to work on the "open entry, open exit" project. Each takes a portion of the blackboard and begins to work toward a final problem solution. In this situation, there will be a lot of shoving and fighting for room to work. There will be no central focus. Professionals and stakeholders will go off in many directions. The result will be a blackboard full of ideas, without any synthesis or justification. In this case, there is no reliable outcome. Some of the people will then be asked to leave the room until candidate project concepts and design schemes, forms, and features have been developed. When they return, they study what is proposed. Since the process of deliberation is not visible to them, they question everything. They begin to revise, alter, amend, and redirect what was developed. The result is chaos, with no product. Finally, the team gives up and two or three people are given the responsibility for completing the project.

Whether acted out in a few days or over several weeks, the appalling aspect of the two preceding faulty scenarios is that they exaggerate only slightly what are common experiences. The best that can be said is that when the work is finished an adequate and reasonably attractive building has been produced. In fact, this is a common description of new buildings—adequate and reasonably attractive. The Knowledge Base System strategy achieves so much more on behalf of the client. First, it permits a room full of ideas, but without the chaos. Second, it forms a building design and land development team which works efficiently and with a sense of cooperation, collaboration, and shared responsibility for project success. Third, it is a visible process, where all team members can see the whole deliberation sequence unfold.

The value of the Knowledge Base System is clearly evident within deliberations associated with a project such as the "open entry, open exit" one. The commentaries that follow provide application insights and strategies for each section of the Knowledge Base System.

30.4.1 Quality of Life Challenge

The five question items in this section of the Knowledge Base System help create the boundaries of the project. A bounded problem is essential; without this framework, there is no direction or limit to the work of the building design and land development team. You cannot promote a project that is more speculative than strategic, or more assumptive than knowledgeable. The initial project description must be precise, thorough, and comprehensive.

Responding to the first question item, 1.1 Project Concepts, an image of the future is created. The second question item, 1.2 Quality of Life Objectives, establishes the emphasis that will be placed on the achievement of quality of life objectives. In question items 1.3 Design Archives, 1.4 Planning Archives, and 1.5 Construction and Engineering Archives we learn from the experiences and accomplishments of others and apply this to the project at hand. This is the first design deliberation in the Knowledge Base System application. The goal is to identify candidate design options and alternatives based on what others have achieved in projects similar to the one being undertaken.

1.1 PROJECT CONCEPTS. Taking into account the existing situation and anticipated project outcomes, what are the project concepts?

This question item establishes a reference point for all the work that follows. In some situations, project concepts are fully and clearly specified by the client. Sometimes, the degree of preparation will be incomplete. In either instance, the building design and land development team must achieve mutual agreement and understanding of project concepts and their relative importance.

This question item, as well as the ones that follow, provides an opportunity for debate and argument. No project is free of uncertainty and controversy, particularly when emphasis is placed on achieving quality of life objectives. Therefore, successful application of question item 1.1 begins to resolve aspects of the project which are complex, troublesome, unique, subtle, or unresolved. The building design and land development team continually reviews the statement of objectives, confirming or disconfirming the proposed objectives, or stating an opposing view if it exists. In every step of the Knowledge Base System process, critique and evaluation of what has been achieved or recommended is always sought. When people are given an opportunity to state their views, they become committed to the success of a project and provide insights based on their experience, identifying objectives which may have been overlooked by others.

1.2 QUALITY OF LIFE OBJECTIVES. Within the context of the stated project concepts, what are the quality of life objectives?

In the initial response to this question, it is sufficient to convey intent; that is, to identify areas within the project where quality of life is of particular concern. As you go further into the process and more information becomes available, necessary revisions may become apparent. This is true for every Knowledge Base System application. The most recent information is always used to help refine previously developed information.

There is a direct relationship between question item 1.2 and the question items contained in sections 2.0 Facility Life Characteristics, 3.0 Family Life Characteristics, and 4.0 Community Life Characteristics. As noted before, it is common to return to a question item after additional information is developed. What we learn in the process of answering the question items in sections 2.0, 3.0, and 4.0 will always lead to a revision of the answer to question item 1.2.

1.3 DESIGN ARCHIVES. Within the archives of architecture, interior design, and facility management, what concepts and designs are significant for the specified quality of life objectives?

1.4 PLANNING ARCHIVES. Within the archives of planning, what concepts and site and master plans are significant for the specified quality of life objectives?

1.5 CONSTRUCTION AND ENGINEERING ARCHIVES. Within the archives of construction and engineering, what theories, standards, and specifications are significant for the specified quality of life objectives?

The search for quality of life objectives continues with question items 1.3, 1.4, and 1.5. No one should attempt to conduct a Knowledge Base System application without examining the history of relevant building situations across the years, as well as reviewing current achievements. Even if project objectives are unique, a study of related projects is valuable. We consider the archives of planning because it is important to know which land development features have worked and which have proven unsuccessful, and under what conditions. The archives of construction and engineering are of particular importance. Failure to incorporate this data and information into design guidelines is a serious omission on the part of building design and land development teams.

30.4.2 Facility Life Characteristics

More building design and land development possibilities originate from this section of the Knowledge Base System than any other.

2.1 FACILITY USER CATEGORIES. Who will be active within the facility? How may these individuals be grouped by responsibilities and intentions? How many individuals does each category include?

This is the most important question item in the entire Knowledge Base System. Although people may be easily distinguished from one another, initial views of which people will use a building are often inaccurate. Too many people are omitted from consideration and those who will use the building are superficially characterized. The problem lies in the extensive variety of users in every type of building design and land development situation.

The first step in this question item is to distinguish one facility user category from another. To this end, question item 2.1 is first addressed by asking the client to provide the names of individuals who are most knowledgeable about each program and activity element in the client's organization. As you work together to develop a comprehensive listing of anticipated facility user groups, the close association established with these individuals will always be beneficial. A collaborative activity like this becomes a insightful experience when client's representatives reveal informally what is important to them and what their concerns are.

2.2 ORGANIZATION STRUCTURE. What are the common and exceptional relationships between groups and organizations that use the facility and contribute to activities within the facility?

When we answer this question, more is needed than a reading of existing organization charts. People must be given the opportunity to describe things as they experience them. Successful research will identify the organizational activities which are particularly important and unique to the client. From these results the building design and land development team can compose descriptions of the client's programs and activities in a manner that makes the participants feel as though they have been fairly and accurately represented.

First, people are asked to describe the principal components of the organization, identifying areas of activity and the resulting outcomes. This description is developed by those involved in the identified area of responsibility and activity. Individuals need to talk about the history of their area of responsibility, its special aspects, and their expectations for the planned facility.

2.3 FACILITY ACTIVITY DESCRIPTIONS. What are the anticipated activities of facility users? What is known about the extent, time of occurrence, and duration of anticipated activities?

As the building design and land development team works on the development of activity descriptions, they are working to keep the design effort on target and to identify the activity patterns to which a specific response must be formulated. Every facility must support a wide range of daily activities. Therefore, every square meter, each fixture, furnishing and equipment item, as well as the funds expended thereon, will eventually have to be justified in terms of the activities being supported.

People are willing to describe their activities. They want the building design and land development team to understand their day and weigh the importance of their activities as they do. Decisions about the extent of inquiry should be made on an individual basis. More important, complex, unique, or troublesome facility user groups require additional activity information.

2.4 FACILITY LIFE CUSTOMS. What are the philosophies and values of the organizations and groups that will use the facility? What are the significant customs, lifestyles, norms, and traditions of facility users? Are these characteristics stable or likely to change?

The first three question items of this section of the Knowledge Base System deal with objective descriptions of people's responsibilities and activities. This question is concerned with the ways in which people see themselves. If we are concerned with how people are seen, demographic and observationally derived distinctions can be used. On the other hand, if we are concerned with how people see themselves—the state of mind of the facility user—subjective distinctions are needed.

People with common lifestyles, traditions, norms, and customs are identified here and formed into supplementary facility user categories and become part of the answer to question items 2.1. These distinctions are critical. The American Institute of Architecture (1972) recognized this when they claimed that an understanding of differences in human needs and lifestyles is a critical first step toward understanding the circumstances surrounding professional practice. This goal poses many problems. People are suspicious about questions that delve into what they consider a highly personal domain. They will have no trouble answering question items 2.1, 2.2, and 2.3, as they believe it is appropriate to identify themselves, state where they fit into the organization, and describe what they do. In question item 2.4, they are being asked a great deal more.

Application of question item 2.4 will distinguish individuals and groups from one another in terms of lifestyle-based activities and preferences. This search for understanding begins with a careful review of the answers to questions items 2.1 and 2.2. This established spectrum of facility user groups tells us which individuals and groups are especially important. From this beginning, we begin to develop not interesting distinctions, but significant differences. This is always part of Knowledge Base System research: We need to know what is significant, not just what is interesting. The need for this strategy is evident. When designs fail, it is usually because some aspect of people's expectations did not occur to building design and land development teams, or was misunderstood. Many of the distinctions between lifestyles are difficult to understand or are so complex or subtle that they are difficult to respond to in design. Nevertheless, if an insightful level of understanding is not achieved, even the most likely opportunities for responding to people's lifestyles will be lost.

2.5 FACILITY DESIGN OBJECTIVES. Responding to identified expectations and requirements, which facility user characteristics should be emphasized in design?

This question item calls for a synthesis of the findings to this point. If this synthesis reveals remaining areas of concern, previously asked questions should be restated, to clarify and refine this information.

30.4.3 Family Life Characteristics

The application strategies developed in the preceding section are used here, with the additional consideration that objectives for quality of life at home are a response to a variety of individual and family life expectations. These expectations are related to local, regional, and national traditions, customs, and norms. Requirements are related to land and capital availability, to family income, cost of living and family size consideration, and, often, to government housing policies and subsidies. The answers to the following question items are developed in this context.

- **3.1** FAMILY LIFE GROUPS. Who will be living in the residences? How may these individuals be grouped by family life expectations and requirements? How many individuals does each category include?
- **3.2** FAMILY STRUCTURE. What are the common and exceptional relationships within and between family elements?
- **3.3** FAMILY ACTIVITY DESCRIPTIONS. What are the characteristic activities of individuals and families while at home and in the neighborhood? What is known about the extent, time of occurrence, and duration of anticipated activities?
- **3.4** FAMILY LIFE CUSTOMS. What are the perceived roles of the family in individual and community life? What are the customs, lifestyles, norms, and traditions of individuals and families? Are these characteristics stable or likely to change?
- **3.5** HOUSING AND NEIGHBORHOOD DESIGN AND PLANNING OBJECTIVES. Responding to identified expectations and requirements, what family life characteristics should be emphasized in housing design and neighborhood planning?

Attention must be paid to specific family lifestyles when establishing housing design guidelines. How does the family spend its day together? What are the child-rearing practices? How important is privacy? Concern for the quality of life at home for single people, the elderly living alone, and for those with physical limitations that require special design attention, extends the complexity of the problem of establishing objectives for a specific project.

30.4.4 Community Life Characteristics

These question items are a safeguard against the tendency of building design and land development professionals to reduce the complexity of how people live so as to fit them into what has been designed. Simplification of complex situations creates an illusion of certainty, which is a primary source of poor design and planning decisions (Boulding, 1974:8). Proshansky (1972:453), recognizing this fact, cautioned that the study of physical settings and planned environments must be done in a way that maintains the integrity of these settings, the people contained in them, and the activities occurring in them. In other words, a concern with the physical environment in all its complexity must be matched by a concern with individuals and groups of individuals in all their complexity.

- **4.1** COMMUNITY LIFE GROUPS. Who visits and lives and works in the community? How may these people be grouped by community life expectations and requirements? How many individuals does each category include?
- **4.2** COMMUNITY STRUCTURE. What are the common and exceptional relationships between groups and organizations that make up the community and influence community life?
- **4.3** COMMUNITY ACTIVITIES. What are the activities of those who visit and live and work in the community? What is known about the extent and time of occurrence of anticipated activities?
- **4.4** COMMUNITY LIFE CUSTOMS. What is the perceived role of the community in individual and family life? What are the customs, lifestyles, norms, and traditions of those participating in community life? Are these characteristics stable or likely to change?
- **4.5** COMMUNITY PLANNING OBJECTIVES. Responding to identified expectations and requirements, what community life characteristics should be emphasized in community master plans and associated facility designs?

These question items caution that it is impossible to dictate a way of life to a community. What is needed is some idea as to the extent and variety of expectations: Where do people want to experiment with lifestyles, and where do they want the comfort of the traditional? When building design and land development professionals fail to assess the sociocultural context of their projects, recognize people's preferences for one environmental scheme or feature over another, and fail to give full weight to the traditions and customs that permeate daily life, their theories and assumptions distort the reality of life in their projects.

30.4.5 Critical Circulation Patterns

For many building designers, information that describes the circulation patterns of facility users and the movement patterns of equipment and material comprises the essential insight for an appropriate facility design scheme. If facility circulation patterns are perfected, the likelihood of a successful project is markedly increased. The three question items in this section of the Knowledge Base System provide the informational basis for perfecting facility circulation, the movement and flow of people, and equipment and material. Like all question items in the Knowledge Base System, these are applied simultaneously. The objective is to determine what people do, their intentions, the importance of their actions, and how their actions vary by time and event.

5.1 USER FLOW. How many people will be entering, leaving, and moving about within the facility, for what purposes, and how frequently?

This question item is really an extension of question items 2.3, 3.3, and 4.3. Again, we need more than data; we have to discover the intentions of the facility users. In the design of a convention center, for instance, the importance of this consideration is evident. This concern for discovering intentions applies equally to hospitals, where efficient movement of visitors is one concern and what benefits can be provided for staff and patients is another. Certainly, in religious facilities, houses of state, and houses of legislature, facility user flow involves traditional ritual movement as well as the simple movement of people. Schools, student unions, government and community centers, corporate offices, housing complexes, and transportation facilities also have this two-fold aspect. This means that information should be collected along two streams: the descriptive and the interpretative.

5.2 EQUIPMENT AND MATERIAL TRANSPORT. What are the characteristics of the equipment and material that must be transported to and within the facility? How will these items be transported, and what is the frequency of such movements?

Supermarkets, department stores, and convention centers move equipment and material on a daily basis, both for normal operations and to accommodate shifts in activities. Descriptions of objectives, destinations, time of occurrence, frequency, and means of movement are the indices for determining the significance of equipment and material flow. Emergency situations require additional considerations. Descriptions of the possible emergency situations, required equipment and material, means of handling, and advance preparation, are the additional information objectives.

5.3 RECOMMENDED CIRCULATION PATTERNS. What are the recommended circulation patterns for user and equipment and material flow? In what way is this proposal a response to concerns for efficiency, convenience, safety, and security?

As with every question item in the Knowledge Base System, the answer to this question should be developed step-by-step. All the preceding information gathered from the application of the Knowledge Base System should be used to identify critical circulation patterns. After this information has been summarized, essential adjacencies are modeled into patterns of circulation, clearly indicating the nature of user and equipment and material flow.

Facility user and equipment and material flow is an hour by hour, day by day, and week by week event. A carefully arranged summary of findings, highlighting critical activities, events, periods of time, and design objectives is essential. Final summary statements of research findings must be cleverly cast, providing both research results and a basis for formulation of insights. When design decisions are being made quickly and new alternatives and options rapidly posed, there is no time to return to data summaries and begin a new evaluation. One must have an unequivocal view of what is needed, and a clear justification for each particular design recommendation.

30.4.6 Interior Architectural Spaces

In the first five sections of the Knowledge Base System, clients and facility users work to reveal their expectations and requirements. Now it is the turn of the building design and land development professionals to take the lead, working to organize and evaluate information, select design elements, and formulate final recommendations. The role of clients and facility users at this stage is to critique proposed design features, offer suggestions, and provide additional information that will speed things along and make any design more representative of individual needs and wants.

This section of the Knowledge Base System is a response to the potential of interior architectural spaces for helping individuals achieve a high standard of performance and to experience a personal sense of satisfaction. Conversely, inappropriate interior features can intervene in actions, fail to support important activities, and can be incompatible with people's preferred ways of doing things. Further, the responsibilities of building design and land development teams may force them to confront diverse and often conflicting expectations and requirements for interior architectural spaces.

The value of detailed critique is apparent as building design and land development teams deal with complex facilities and sophisticated clients. Proposed design guidelines can be adequately critiqued only if you visualize facility users as unable to move easily through the facility because allowed space restricts movement; as failing to develop a sense of orientation and direction because the visual surround is confusing and information displays are inadequate; as experiencing fatigue, stress, and frustration because the environment does not support specific activities; as attempting to adapt to light and sound levels that are intolerable; or as being exposed to an environment likely to produce accidents because of the failure to meet facility user safety needs. You must give life to your preliminary formulations by envisioning what the space means to the user. This is the sure way to avoid problems which must be corrected later, which is a burden the client certainly does not want.

6.1 SPACES. What spaces are needed to support facility users' activities?

The answer to this question provides a framework for the development of interior architectural spaces. The objective is to form a listing of facility spaces which reflects anticipated facility life characteristics as well as the building design and land development team's experience and interpretation of project and quality of life objectives. The listing of candidate spaces should be as extensive as possible, with a broad spectrum of possibilities originating, in part, from the archival research. The initial effort is concluded when the listing created seems to encompass the principal spaces appropriate to the situation. As new information is developed this listing will be reappraised, leading to revisions, additions, and deletions.

During an application of the Knowledge Base System, when do we begin to work on question item 6.1? The answer is that we do this at the same time we begin to work on question item 2.1. Question item 2.1 asks: Who will be active within the facility? How may these individuals be grouped by responsibilities and intentions? How many individuals does each category include? Question item 6.1 asks: What spaces are needed to support facility users' activities? When formed as a matrix, these two listings provide a framework for the work of the building design and land development team. The function of this matrix is to keep spaces and people before the team at all times.

6.2 FURNISHING, FIXTURES, AND EQUIPMENT ALLOCATIONS. What furnishing, fixtures, and equipment, fixed or mobile, does each facility space require?

The contents of each space is as much a determinant of facility design as any other building system component, such as structures, building materials, and site features. This is no place for arbitrary thinking and assumptions. Selections must be justified by what have been identified as facility user expectations and requirements. Those who will occupy each space must review the design formulations as well as the rationale for recommendations.

6.3 CONVENIENCE, SAFETY, AND SECURITY. Will any facility user group or activity require special fixtures, furnishing, space layouts, information displays, or surface treatments? In anticipation of undesirable events, what special safety and security measures are necessary?

These three topics are often complex, and should be given special study. The research should be guided by applying the questions in the Knowledge Base System to the case at hand.

Security is a concern requiring a complete design strategy which takes into account what must be secure and the associated threat. Threats to individual security must be described in detail; a mitigating design should then be developed. What is recommended must recognize every detail of facility users' activities. For example, if the building design and land development team anticipates a lot of movement around a site where personal security is a concern, a wide field of view is required in order to monitor movement. If individuals are located in assigned areas, limits to access and control points can be established. Patterns of anticipated behavior may lead to emphasizing exterior lighting, individually controlled locking devices, sensors and triggering mechanisms, or even the establishment of a building or neighborhood watch or escort program. The facility management strategies and design features recommended, and the associated additional cost, if this is to be approved, require crafting a justification statement that portrays the full implications of not responding to the recommendation, as well as the likelihood that the design features will work.

Accidents, fires, earthquakes, and other threats to individual safety certainly require a design response. The two elements of safety design, prevention and lessening of effects, require attention to both design details and the management of people. Detailed analyses with a special concern for safety are always warranted. Even where codes and regulations serve as standards, more study is needed, not only because of the generality of these guidelines but because they do not realize the potential that is in every specific situation for making some very thoughtful, well-directed, and effective contributions to safety. Establishing escape strategies for fires illustrates this point. The location of signs is one thing, the design of multimodality information displays that work effectively in smoke occluded spaces and when people are panicked is another. By establishing the best possible means for providing with directions and warnings ahead of time, and supporting this effort with design features, the full potential of facility design and emergency event management is realized.

There are many aspects of convenience, some related to specific expectations and requirements, and some to general preferences. This part of question item 6.3 recognizes that convenience is a major topic for design consideration and should be promoted by the results of careful analysis of daily activities. Facility user activity descriptions and circulation information should be scrutinized for opportunities to include convenience features. How significant the contribution will be to individual quality of life and whether it will be recognized and used as such determines whether the recommended design feature becomes part of the final design guidelines.

Convenience, safety, and security considerations are supported by exacting design criteria associated with the work of architects. For instance, stair design details can be developed in association the specifications and standards appearing in *The Professional Practice of Architectural Detailing,* Second Edition, by Wakita and Linde (1987); *The Architect's Studio Companion: Rules of Thumb for Preliminary Design,* Second Edition, by Allen and Iano, 1995; and on CD-ROM, the *Architectural Graphic Standard* edited by the American Institute of Architects (1996). The data and information presented in these references is highly detailed, often presented in tables and figures. The information presented indicates a range of tolerable environmental features. However, this information will mean very little unless you visualize your facility user and their associated tasks. You must give life to your source material by envisioning what the environment means to the user.

6.4 AMBIENT ENVIRONMENTAL CRITERIA. What provisions should be made for the effect on facility users of temperature, humidity, air quality, air movement, illumination, noise, distractions, annoyances, hazards, and climatic conditions?

Each interior architectural spaces generates its own micro-environment on the basis of activities, furnishing, and electrical and mechanical equipment, and each requires a selective response. In order to help the building design and land development team justify expenditures for mechanical, electrical, structural, and building materials directed toward environmental control, it is necessary to specify anticipated adverse effects on performance and satisfaction if recommended specifications are not adopted. This admonition is too often ill considered. Amal Kumar Naj (1995), reporting on the topic of "Sick Office Buildings" notes that "While workers sneeze, ache and cry from breathing indoor, air, landlords, architects and engineers are at loggerheads over how to make buildings healthier. For a dozen years, employers, workers, unions, health officials, landlords and builders have been agonizing and arguing over "sick-building syndrome"—a collection of cold-like symptoms that develop among people working in office buildings with sealed windows. Scientists have identified more than 1500 bacterial and chemical indoor-air pollutants from such sources as carpets and office machines (B1 and B8). The extent of the problem is seen in the fact that air quality is the most frequent basis for tort cases associated with building design liability.

6.5 INFORMATION DISPLAYS. What are the required information displays?

The insights needed to determine the characteristics of information displays come from all the other sections of the Knowledge Base System. Ideas should be written down as they come to mind. Toward the end of the research effort, the possibilities that were identified can be evaluated and options can be selected. The characteristics of the people using the information displays are important. What are the visual and auditory limitations and capabilities of the facility users? Will they be young or old, literate or illiterate, of one language group or several, moving at a leisurely pace or rushed? Usual conditions for the spaces involved, such as illumination and noise levels, and degree of congestion, should also be identified.

The specific proposals for information displays should include the information that will make it easier for individuals to u se the facility, under both normal and emergency conditions. The information presentations which guide and direct facility user activities include exterior and interior signs, directories, displays, logos and symbols, illuminated and nonilluminated units, tactile and auditory elements, color-coordinated surfaces and accents, and mountings and fixtures. The design features recommended should deal specifically with such detectability factors as location, placement, readability, size, and configuration. The detectability factors for auditory effects are frequency spectrum, loudness, periodicity, and directionality; for tactile effects they are texture, pattern, placement, and durability.

Information displays may also lead people through a sequence of action. Library information displays are an example of this. From the information desk to the reference computer terminals, catalogues, and reference room, onto the library's specific holdings, through the library shelving code, and to a document, the library user can be guided by directional aids. A traveler using a transportation facility is guided from the entrance road inwards, to the entrance, to check-in, to waiting lounge, to boarding, to destination, and through customs. A task sequence in a manufacturing or assembly plant is often guided by displayed information and color-coded surfaces. In a shopping or retail facility, the information displays would be in the form of signs, graphics, and surface color-coding and texture.

6.6 DURABILITY AND MAINTAINABILITY. Where do spaces require special attention to durability and maintainability of surfaces?

Whether special attention must be given to walls and ceilings, wall coverings and fixtures, floors and stairs, and doors and windows as interior and exterior building features depends on facility users' activities and regard for their surroundings. Prolonged hard use, movement of awkward material and heavy items, and vandalism are activities that must be anticipated. If there has been a history of disregard and carelessness which is unlikely to change, this should be brought to the attention of the building design and land development team. The team is capable of dealing with this area of concern if they know the likely events and situations. A space by space review by individuals familiar with these issues will identify problem areas; it will also provide valuable suggestions which deal with design features and maintenance schedules.

6.7 SPACE PLANS. What space plans best correspond to facility users' expectations and requirements?

What we know about facility life characteristics should be reflected in the work achieved here. The merit of the formulated design guideline is based on the thoroughness of the original research. This sequence of questions will define the main features of the space. With this model as a reference, the building design and land development team must then consider options and alternatives. Certainly, alternative development is a means of critique. As team members review preliminary recommendations, conflicts and omissions will be identified.

30.4.7 Workstations

7.1 WORKSTATION FACILITIES. Where are workstations required?

7.2 WORKSTATION ACTIVITIES. What are the specific workstation activities? What task sequence and time lines characterize the activities assigned to the workstation?

7.3 WORKSTATION FEATURES. What are the furnishing, equipment, fixture, tool, and material requirements of the subject workstation?

7.4 WORKSTATION LAYOUT. Since each workstation has unique activities and support requirements, how should each workstation be arranged?

The term "workstation" is used in a general sense (Harrigan and Chapman, 1991). That is, every job, home, or community activity is considered to be sited at a workstation. All types of buildings are now equipped with workstations which are designed to support specific task performance. At work, in the classroom, the manufacturing plant, the service center, or convention center; at home, where people run businesses, extend their work day, or utilize distance learning opportunities; and in the community, where people serve others in libraries, adult education programs, or with special opportunities for the disabled or handicapped, workstations are a requirement.

30.4.8 Communication and Information Systems

Complex activities within spaces, buildings, facilities, and planned developments are totally dependent on communication and information system support. The requirements for a specific situation depend on the assigned communication and information monitoring, development, exchange, and application responsibilities. The basic question items provide the framework for this research.

8.1 COMMUNICATION AND INFORMATION SYSTEM FACILITIES. Where are communication and information system facilities required?

8.2 COMMUNICATION AND INFORMATION SYSTEM ACTIVITIES. What is the relative complexity or uniqueness of the information development and processing activities? What task sequences and timelines characterize the noted activities?

8.3 COMMUNICATION AND INFORMATION SYSTEM FEATURES. What are the furnishing, equipment, fixture, and software requirements of the subject communication or information system?

It is appropriate here to reconfirm the premise that the contents of this handbook, front to back, have points of application throughout the entire Knowledge Base System. For instance, recognizing that sections 7.0 and 8.0 demand precise and thorough design criteria, find where handbook sections III. Job Design and VIII. Human-Computer Interaction offer data and analyses directly appropriate to the cited question items. Certainly, 6.0 Interior Architectural Spaces is an element of the Knowledge Base System to which apply sections IV. Equipment, Workplace, and Environmental Design and V. Design for Health and Safety.

30.4.9 Facility Space Arrangements

These question items synthesize all the information heretofore developed to produce the most promising proposals for spatial design. Emphasis is placed on critique and selective modification of proposed design schemes, forms, and features. Research findings and corresponding design recommendations are always placed in the context of such considerations as site constraints, structural systems and materials, construction methods and schedules, and cost analyses. This provides the building design and land development team with a head start on the identification of conflicts and the formulation of compromises.

9.1 PROPOSED FLOOR PLAN SCHEMES. Considering all research findings and developed design concepts, what are the best schemes for achieving project concept and quality of life objectives? In terms of facility user expectations and requirements, what is the benefit and problem resolution potential of each suggested scheme?

9.2 SPACE REQUIREMENTS. What is the estimated square meters for each facility space and support area?

9.3 FACILITY MANAGEMENT SCHEME. What are the necessary guidelines and manuals that show people how to use facility operational features?

9.4 ALTERATION EXPECTANCIES. How soon might it be necessary to modify or expand the facility? What events would most probably lead to this requirement? How do the proposed facility schemes account for this possibility?

In question item 9.1, you will synthesize your findings and provide a summary of the spatial implications for the total design. Your proposed floor plan schemes might take the form of a single-line drawing that combines information from 6.0 Interior Architectural Spaces and 5.0 Critical Circulation Patterns. Many will consider it premature to become so specific this early in design and may wish to formulate a less binding recommendation. Nevertheless, a drawing conveys more meaning with greater clarity than any other type of information statement. Alternative arrangements can be developed, evaluated, and depicted with ease. It is essential that the design concepts, schemes, forms, and features in a design guideline are presented in terms of options and alternatives. There is always more than one way to meet people's expectations and requirements.

In question item 9.2 Space Requirements, a precise calculation of total square meters for the facility is formulated. Applying a current construction cost index to this calculation provides a reliable initial cost estimate. As the building design and land development team reviews these numbers and attempts to reduce costs, deliberations can be augmented with what is known about the significance of each space. Every square meter is supported by a description of associated client and facility user expectations and requirements.

Question item 9.3 Facility Management Scheme identifies the increasing importance of facility management as a building design and land development concern. Sections 7.0 Workstations and 8.0 Communication and Information Systems illustrate the extent to which this concern is taken.

Question item 9.4 Alteration Expectancies is an attempt to guard against premature obsolescence and to extend the effective life of the facility. Few things should be of greater concern. If this understanding fails, the client will have to correct, with reorganization, additional staffing, or remodeling, problems which could have been avoided. Wherever new products, advanced technology, new activities, and new facility user groups are likely, these must be identified and the implications determined. Once likely changes in requirements are identified, they have to be evaluated in terms of design significance, relative importance to the client, and likelihood of occurrence.

Building designers often find themselves caught between the benefits of permanent structural elements and the complexity of providing a physical solution for change that is more than showing the on-site footprints of possible facility additions or the provision of a flexible furnishing and partitioning scheme. A progression of steps from the general to the detailed might be valuable here, first identifying sources of change and then evaluating these in terms of likelihood and significance. The design possibilities for both fixed spaces and managed spaces should be sketched out and a judgment made as to what to incorporate as facility features. Whether or not these deliberations are necessary depends on the client. While long-range planning is a characteristic of some organizations, other clients might find it difficult to look ahead because they are constrained by their situation, limited in view, or lacking the necessary information. The following questions need to be answered by the client and anticipated facility users: Considering what lies ahead, which of your current needs are likely to change? How extensive are the changes that you expect? When are these changes likely to occur? Are these anticipated changes part of the organization's plan? Are these changes likely to be generated by factors outside of your organization?

30.4.10 Facility Design Image

10.1 FORM AND STRUCTURE. What are the proposed facility form and structure design concepts?

10.2 EXTERIOR DESIGN IMAGES. What are the proposals for exterior facility design images, details, and accents?

10.3 INTERIOR DESIGN IMAGES. What are the proposals for interior spatial forms, design images, and surface colors, textures, and patterns?

10.4 CONCEPT JUSTIFICATION. What are the facility user effects possibilities of each design image recommendation?

Throughout this study of the Knowledge Base System there is always the concern that "design" as the personal contribution of a design professional may seem to be of secondary importance. That is not the case. Consider the design objective of achieving a building image. The work "image" can be used as an ideal. It can also be used as a pragmatic value. As an ideal, a perfect space symbolizes to the user a regard for his or her needs. Likewise, a perfect space must also be a functional success. Bruno Bettelheim (1974) has stated how symbol and function will always be the two goals of design. Bettelheim suggests that even the smallest detail of a physical setting can make a facility more useful, and at the same time convey symbolic meaning in artistic form. He would not object that design features could be the consequence of a designer's personal preferences, experiences, and convictions about the worth of what it is he or she is attempting to achieve. There are many ways to design a symbolic and functional space for human affairs. What matters is the intention with which something is done—the wish to please and arrange things to make a strong appeal while achieving all functional requirements. Providing users with the best possible physical setting—however different the specific forms and details—bespeaks of the care which has been taken to give users the best possible physical and human setting, a message the facility user will seldom fail to receive and appreciate.

Bettelheim's high regard for design is a principal foundation of the Knowledge Base System concept. All that is asked is that the information developed in the first five sections of the Knowledge Base System is always the starting point for design. Don't exclude any building design and land development team member from design endeavors. There is a role for everyone. Design is a process of growth and exploration which can be considered as a potential existing in all members of the team. A child growing and exploring in a classroom, producing demonstrations and presentations for his or her class is a designer. Likewise, the teacher, who is innovating in the direction of providing effective educational settings, is a designer. The teacher is, in this instance, neither a scientist nor a technician. He is a designer integrating knowledge and intuition, producing finely conceived and directed educational experiences and environments. Thus, although some use the term "designer" professionally, all members of the building design and land development team may share this title.

30.4.11 Facility Site Plan

11.1 SITE REQUIREMENTS. What are the site requirements and to what needs and wants do these correspond?

11.2 AREA IMPACT. What are the activities surrounding the site? What will be the impact of facility activities on the surrounding neighborhoods?

11.3 SERVICES IMPACT. What will be the impact of facility-based activities and operations on existing public and private services? Will existing services need to be improved or expanded?

11.4 SITE PLANS. How should the site be planned in order to achieve quality of life objectives, respond to the needs of those occupying nearby sites, and meet requirements for outdoor space in terms of amenities and landscape development?

The sequence of questioning for the Knowledge Base System is organized to develop design concepts, schemes, forms, and features from the "inside out." These question items run parallel with this effort. Preparation for this activity requires consideration of 2.0

Facility Life Characteristics, 3.0 Family Life Characteristics, and 4.0 Community Life Characteristics to identify the expectations and requirements that should be included in the deliberations of the building design and land development team. A thorough application of question items 2.1, 3.1, and 4.1 will identify those people who should be considered during facility site plan deliberations. Too often, facilities do not accommodate those who can benefit from the facility even though they may not be principal facility users. In many cases, insufficient attention is given to those who may threaten or disrupt facility activities. The application strategy for question items 2.1, 3.1, and 4.1 remains the same when applied to the study of site considerations. The particular situation will determine the boundaries of the study, which may extend beyond the site to include the surrounding neighborhood and community.

The client, facility users, and selected individuals representing neighborhood and community interests will contribute to the development of site design guidelines. Early in the work of the building design and land development team, these individuals will help identify existing expectations, concerns, and requirements. As site design features are formed, these individuals will critique and help revise preliminary recommendations. These findings must be summarized, indicating source and importance of considerations to various individual and groups. As in every section of the Knowledge Base System, an explanation of the way options and alternatives were developed and evaluated must always accompany final design guideline statements.

30.4.12 Community Master Plan

12.1 COMMUNITY SERVICES. What public and private services are needed?

12.2 COMMUNITY FACILITIES. What facilities and planned environments are needed to support community life?

12.3 CHANGING COMMUNITY REQUIREMENTS. How soon might it be necessary to modify or expand community facilities and services? What events would lead to this requirement?

12.4 OUTDOOR SPACE. What are the requirements for outdoor space in terms of amenities, landscape development and preservation, and enhancement of existing natural features?

12.5 MASTER PLAN RECOMMENDATIONS. How should the community be planned in order to meet the full range of identified community life expectations and requirements?

In the community, quality of life is related on one hand to change, and on the other to tradition. A child wants adventure and yet profits from a stable neighborhood experience. The young adult wants social opportunities as well as the reassurance that comes from being part of a tradition. Parents seek outside opportunities for themselves and their children and yet want the family to be the source of support at all times. The retired person wants the stimulation of new experiences as well as a place where the old is treasured. To this array of contrasting wants is added the tempo of community life with seasonal and commemorative events and the arrival of new faces.

These perspectives apply to a proposed housing development, shopping center, office building, industrial facility, or science park to be placed within an existing community. The application of the Knowledge Base System concept would also be appropriate for determining the preferred characteristics and location for a new school, hospital, community mental health center, or a halfway house. Projects such as these are probably the most complex of all Knowledge Base System applications because they tend to be highly exploratory. We do not really know the concerns and expectations of community groups until we establish that they exist.

30.5 DESIGN GUIDELINES

Knowledge Base System application results are formed as design guidelines. It is essential to note that these guidelines are only a visible record of progress in developing justified design concepts, schemes, forms, and features. A design guideline should not be considered a goal in itself, or as a document to be developed. The function of design guidelines is to bring findings and ideas to some point of synthesis. The goal is to provide a focus for discussion in which those knowledgeable about the project can employ their insights and expertise with confidence. This strategy encourages experimentation and debate.

These critical evaluations achieve something very essential. No design concept, scheme, form, or feature stands alone; it is always accompanied by a statement of justification. Therefore, every formal design guidelines statement must have two parts: the recommendation and the justification for that recommendation.

30.5.1 Unlimited Opportunity

Architects and interior designers face a world of unlimited opportunity. Cultures are evolving, economies are expanding, technology is advancing, the daily life of people is becoming more complex, and the need for perfected physical settings and planned developments is becoming more essential. In an economy where every region of the world is expanding simultaneously, the desire for quality of life at work, at home, and in the community has intensified. To make economic, social, and cultural evolution an opportunity, those responsible for the development of the built environment must look beyond the everyday aspects of professional practice and seek to learn as much as possible about human limitations and capabilities and the expectations and requirements of individuals and groups of individuals. How do architects and interior designers gain new commissions and contracts, develop the potential for future growth, and establish a market reputation for the opportunities they are in practice to attain? The answers are always found in the expectations and requirements of people and the careful translation of findings into justified design concepts, schemes, forms, and features.

How acceptable is the Knowledge Base System strategy within the professional practice of architecture and interior design? Consider the view of Charles Luckman, F.A.I.A., who created such landmarks as the Los Angeles International Airport, Cape Canaveral Missile Test Center, Madison Square Garden, and Boston's Prudential Center. In all his projects, Mr. Luckman (1988) asked his architects to find out what workers in office buildings like and want, what patients in hospitals like and want, what spectators at sporting events like and want. His buildings reflect what the people want, not what architects think they should have.

REFERENCES

Allen, E. and Iano, J. (1995). *The Architect's Studio Companion: Rules of Thumb for Preliminary Design*, Second Edition. New York: John Wiley.

American Institute of Architects (1972). *First Report of the National Task Force.* Washington, DC.

American Institute of Architects (1996). *Architectural Graphic Standards*, CD-ROM. New York: John Wiley.

Apgar, M. (1993). Uncovering your hidden occupancy costs. *Harvard Business Review.* May–June, 124–136.

Argyris, C. (1994). Good communication that blocks learning. *Harvard Business Review.* July–August, 77–86.

Ashihara, Y. (1989). *The Hidden Order: Tokyo Through the Twentieth Century.* Tokyo: Kodansha International.

Bettelheim, B. (1974). *Home for the Heart.* New York: Knopf.

Boulding, K. E. (1974). Planning may seem necessary, but it is hardly ever the essential that we thought. *Technology Review*, *77*, 8.

Cooper, L. (1994). Louis Agassiz as a teacher. In L. B. Barnes, C. R. Christensen, and A. J. Hansen, Eds., *Teaching and the Case Method.* Boston, MA: Harvard Business School Press.

Cooper, J. (1989). Science and design. *Interior Design*, August, 198–203.

Crosbie, M. J. (1994). Working in two worlds. *Progressive Architecture*, September, 78–84.

Dorsey, B. W. (1994). Project rosewood: The Mercedes-Benz siting strategy. *Site Selection*, April, 246–254.

Fisher, T. (1994). Can this profession be saved? *Progressive Architecture*, February, 45–84.

Fleenor, D. (1993). The coming and going of the global corporation. *The Columbia Journal of World Business*, Winter, 6–16.

George, S. 1992. *The Baldrige Quality System: The Do-It-Yourself Way to Transform Your Business.* New York: John Wiley.

Gercik, Patricia. (1992). *On Track with the Japanese: A Case-by-Case Approach to Building Successful Relationships.* New York: Kodansha International.

Goodman, R. (1995). *Client Assessment Knowledge System.* Unpublished Thesis: Department of Architecture, Cal Poly, San Luis Obispo, CA.

Gross, T., Pascale, R., and Athos, A. (1993). The reinvention of the roller coaster: Risking the present for a powerful future. *Harvard Business Review*, November–December, 97–108.

Hamel, G., and Prahalad, C. K. (1993). *Competing for the Future.* Boston, MA: Harvard Business School Press.

Harrigan, J. E. (1987). *Human Factors Research: Methods and Applications for Architects and Interior Designers.* Amsterdam: Elsevier Science.

Harrigan, J. E. (1987). Architecture and interior design. In G. Salvendy, Ed., *The Handbook of Human Factors.* New York: Wiley.

Harrigan, J. E., and Chapman, A. (1991). *Building a better product: A knowledge base approach to skilled job performance.* Research report to IBM Corporation.

Harrigan, J. E., and Neel, P. R. (1996). *The Executive Architect: Transforming Designers into Leaders.* New York: Wiley.

Hayes, R. H., and Pisano, G. P. (1994). Beyond world-class: The new manufacturing strategy. *Harvard Business Review*, January–February, 77–87.

Huey, J. (1993). How McKinsey Does It. *Fortune*, November 1, 56–81.

Itami, H. (1987). *Invisible Assets.* Cambridge, MA: Harvard University Press.

Keiser, T. C. (1988). Negotiating with a customer you can't afford to lose. *Harvard Business Review*, November–December, 30–37.

Kenichi, O. (1990). *The Borderless World: Power and Strategy in the Interlinked Economy.* New York: HarperBusiness.

Lazer, B. (1994). In Gross, T., Pascale, R., and Athos, A., Eds., The reinvention of the roller coaster: Risking the present for a powerful future. *Harvard Business Review*, November–December, 97–107.

Levitt, T. (1975). Marketing myopia. *Harvard Business Review*, September–October, 26–48.

Luckman, C. (1988). *Twice in a Lifetime.* New York: W. W. Norton.

Mintzberg, H. (1987). Crafting strategy. *Harvard Business Review*, July–August, 66–75.

Mintzberg, H. (1994). The rise and fall of strategic planning. *Harvard Business Review*, January–February, 107–114.

Morita, A. (1988). *Made in Japan.* New York: Penguin Books.

Naj, A. K. (1995). Squabbles delay cure of "Sick" office buildings. *Wall Street Journal*, 26 October, B1 and B8.

Nihon Keizai Shimbun (1992). Fitness clubs broaden services to ensure their fiscal health. *The Nikkei Weekly*, 18 January, 18.

Ohmae, K. (1982). *The Mind of the Strategist.* New York: McGraw-Hill.

Ohmae, K. (1990). *The Borderless World: Power and Strategy in the Interlinked Economy.* New York: HarperBusiness, 214.

Olson, C. (1992). Chrysler constructs the ultimate advantage. *Building Design & Construction*, September 23, 16.

Porter, M. (1987). The state of strategic thinking. *Economist*, May 23, 21.

Proshansky, H. M. (1972). Methodology in environmental psychology: Problems and issues. *Human Factors*, *14*, 451–460.

Rogers, E., and Chen, Y. (1990). In M. Von Glinow and S. Mohrman, Eds., *Managing Complexity in High-Technology Organizations.* New York: Oxford University Press, 15–36.

Rothstein, L. R. (1995). The Empowerment Effort That Came Undone. *Harvard Business Review*, January–February, 20–32.

Saaty, T. L. (1980). *The Analytic Hierarchy Process: Planning, Priority Setting, Resource Allocation.* New York: McGraw-Hill.

Sommer, R. (1972). *Design Awareness.* San Francisco, CA: Rinehart Press.

Spradley, J. P. (1972). Adaptive strategies of urban nomads: The ethnoscience of tramp culture. In T. Weaver and D. White, Eds., *The Anthropology of Urban Environments.* Monograph Series, Number II. Washington, D.C.: The Society for Applied Anthropology.

Spradley, J. P. (1979). *The Enthnographic Interview.* New York: Holt, Rinehart, and Winston.

Strebel, P. (1992). *Breakpoints: How Managers Exploit Radical Business Change.* Boston, MA: Harvard Business School Press.

Umlauf, E. (1991). Michigan National's headquarters reflects firm's philosophy toward its staff. *Building Design & Construction*, January, 38–43.

United States Department of Commerce. (1995). *The Malcolm Baldrige National Quality Award.* Gaithersburg, MD: National Institute of Standards and Technology.

Wakita, O., and Linde, R. (1987). *The Professional Practice of Architectural Detailing*, Second Edition. New York: John Wiley.

Wright, G. (1989). Distribution center shuns industrial image. *Building Design & Construction*, June, 42–47.

Wright, G. (1991). High-tech campus accommodates variations on a theme. *Building Design & Construction*, August, 33–37.

Zabriskie, N., and Huellmantel, A. (1994). Marketing research as a strategic tool. *Long Range Planning*, February, 107–115.

PART 5

DESIGN FOR HEALTH AND SAFETY

CHAPTER 31

OCCUPATIONAL RISK MANAGEMENT

Bernhard Zimolong
Institute of Psychology
Ruhr-Universität Bochum
Bochum D-44780 Germany

31.1 RISK MANAGEMENT PRINCIPLES

Risk management may be defined as the reduction and control of the adverse effects of the risks to which an organization is exposed. Risks include all aspects of accidental loss that may lead to any wastage of the organization's assets. These assets are personnel, materials, machinery, procedures, products, and money. The essence of risk management is to prepare, protect, and preserve the resources of the enterprise. This approach demands analyzing the current and past operating hazard, risk, and loss-producing patterns and forecasting expected hazard, risk, and loss-operating patterns. According to Bamber (1994), risk control strategies may be classified into four main areas: risk avoidance, risk retention, risk transfer, and risk reduction.

Risk avoidance means the deliberate decision on the part of the organization to avoid a particular risk. The principles of risk reduction rely on the implementation of a Health, Safety and Environment (HSE) program, whose basic aim is to protect the company's assets from wastage caused by accidental loss.

The key elements of the HSE-Management control cycle are set out below (Health and Safety Executive, 1993a):

- *Formulating a policy aimed at achieving the preservation and development of physical and human resources and reductions in financial losses and liabilities*: The policy is aligned with other human resource management policies designed to secure the commitment, involvement, and well-being of employees. It influences all activities and decisions, including those having to do with the selection of resources and information, the design and operation of working systems, the design and delivery of products and services, and the control and disposal of waste. To promote health and well-being it fosters among other things the restructuring of jobs to reduce monotony and increase flexibility, and health promotion campaigns that encourage health conduct of employees inside and outside the enterprise.

- *Designing organizational structures and formulating rules to put the HSE policy into effective practice*: This is helped by the creation of a positive culture that secures involvement and participation at all decision levels. Managers take full responsibility for controlling all those factors that could lead to ill health, injury, or loss. Designing HSE structures concerns the divisions of responsibility and distribution of formal authority, the creation of hierarchical or lean structures, the degree of self-regulation of work groups and units, and the formal relations between groups and leaders. Establishing and maintaining control is central to all management functions including HSE. The allocation of HSE responsibilities to line mangers and work groups serves as an important tool to foster the integration of HSE into the daily work activities with specialists acting as advisers.
- *Planning, devising, and formalizing the actions to be taken in respect of the whole range of hazards and risks to be expected*: The ultimate goal is to eliminate or reduce risk created by work activities, products, and services. The organizational process involves risk identification, evaluation, and control. Risk identification may be achieved by a multiplicity of techniques, including Hazard and Operability studies (HAZOP) studies, physical inspections, job safety analysis, management and worker discussions, and safety audits. The study of past accidents can also identify areas of high risk. Risk evaluation methods are used to decide priorities and set objectives for hazard elimination and risk reduction. They may be based on legal considerations, which include possible constraints from compliance with health and safety legislation, codes of practice, guidance notes and accepted standards, and other relevant legislation concerning fire prevention, pollution control, and product liability. Economic considerations may include the costs of accidents, the overall effect of the organization's profitability, the possible loss of production following the issue of improvement and prohibition notices. Social and humanitarian considerations should include the general well-being of employees and the interaction with the general public who either live near the organization's premises or come into contact with the organization's operations.
- *Measuring performance against predeterminded standards*: The techniques include a proactive examination of both hardware: premises, plant, substances; and software: people, procedures and systems, including individual behavior. Deviations from standard reveal when and where actions are needed to improve performance. Failures of control are assessed through reactive monitoring, which requires thorough investigation of any accidents, ill health, or property damage with the potential to cause harm or loss.
- *Learning from experience through performance reviews and independent audits*: This needs to be done systematically through regular reviews of performance based on data from monitoring activities and from audits of the HSE management system. It is the final step in the HSE management control cycle and part of the feedback loop needed to enable the organization to maintain and develop its ability to control successfully risks.

The above principles seem to suggest that science and industry have reasonable models of how reliable organizations work. This is not the case. As Roberts (1990) points out, the organizational literature fails to deal specifically with either hazardous organizations or high levels of performance reliability. The standard texts on safety management, e.g., Heinrich, Petersen, and Roos (1980), Bird and Loftus (1976), Bird and Germain (1987), and Ridley (1994) neither present specific models of the safety management system (SMS) nor do they provide empirical evidence of how particular aspects of the suggested frameworks contribute to the overall level of HSE. Reviews on safety research literature (Hale, Hemig, Carthey, and Kirwan, 1995) revealed a number of lines of research and isolated studies which seem to have few links with each other.[1]

Comparison of high- and low-accident companies (Cohen, 1977; Zohar, 1980) have produced lists of factors that seem to be associated with good safety performance. Anal-

[1]The term "safety" is used short for "safety, health, and environment" and refers to damage to hardware and environment as well as to people.

ysis of major accidents, among others Three Mile Island, Chernobyl (Reason, 1987a) Bhopal (Shrivastava, 1986), Zeebrugge (Wagenaar, 1992), and Piper Alpha (Department of Energy, 1990) have contributed to a deeper understanding of the interdependence of personal, technical, and organizational factors. Also the analysis of minor incidents in high risk systems has provided a catalogue of organizational factors that went wrong in those cases (Wilpert and Klumb, 1993). The main emphasis in all these investigations and reports was on the management's failure to ensure that their plant or activities were designed, operated, and maintained with sufficient reliability and safety. As a consequence, legislation establishes regulatory demands to ensure the assessment of safety management systems. Recent amendments to the European Post-Seveso Directive require major companies to have auditable safety management systems and another directive has set up a voluntary European system for environmental management auditing (European Community, 1994).

These trends have produced a variety of activities to develop and use management system audits (International Loss Control Institute, 1990). Such audit systems are largely based on the collected experience of long years of consultancy of management. As Hale et al. (1995) claim, they can give the impression of arbitrary lists of topics clustered under convenient headings that vary from one instrument to another. They do not have an explicit model of management system nor do they provide a weighing of the items or topics covered. It is not clear whether they are too detailed or not complete enough. Even the merits are unclear. As an example, Eisner and Leger (1988) very much questioned the assertion that the application of the International Loss Control Institute (ILCI) audit instrument for the mining industry was the primary cause of the accident reductions faced in South African mines.

Case studies of high-reliability organizations such as nuclear powered aircraft carriers (Roberts, 1990) revealed contradictory strategies to cope with conflicting demands. Close grouping of performers, tight coordination and control, and the existence of an extreme hierarchy seem to be factors that enhance performance reliability in this type of organization. Yet, uncertain environments require sufficient complexity and flexibility within this organization to map the uncertainty. As a consequence, organizations have developed specific strategies for avoiding the negative effects of extreme performance reliability. A vast amount of technical, informational, and task redundancy has been implemented, which deals with partial system failure, high degrees of responsibility, and accountability. Even pushing decision making to the lowest level possible, at least in the sense of vetoes, serves as a useful technique. For example, anyone on the flight deck of a carrier who sees a foreign object on deck that can be hazardous to flight operations can call a halt to the ongoing activities.

In advanced manufacturing systems (AMS), production based indicators to measure and assess the degree of system's reliability have been introduced (Zimolong and Duda, 1992a; Zimolong and Trimpop, 1994). Data on availability, particularly on organizational, personal, and hardware caused stoppage time, of flexible manufacturing systems, machining centers, and robots were used as indicators to assess the inherent risk in AMS.

Finally, descriptive studies of parts of the safety organization and management have been carried out, focusing on how these parts work, the role played by different individuals or functions in them and how effective they are. Examples of aspects covered by such studies are safety information systems (Kjellen and Larsson, 1981), safety experts (Oortman-Gerlings and Hale, 1991) safety committees and quality circles (Palmer, 1990; Ritter and Zink, 1992), management principles (Veltri, 1991), leadership principles including goal setting and feedback (McAfee and Winn, 1989), behavior modification (Sulzer-Azaroff, 1978), participative decision making (Packebusch, 1995), decision making in work groups (Brehmer, 1991) audits (Eisner and Leger, 1988) and safety culture (Hovden and Larsson, 1987).

Hale et al. (1995) conclude from literature review that there have been few attempts to produce coherent and comprehensive models of the design and management of safety organizations. There is a considerable body of knowledge on parts of the system; however, it requires some structure or framework to understand how the results might be linked to each other.

31.2 RISK MANAGEMENT FRAMEWORK

Several types of organizational models may be applied to measuring overall achievements such as productivity or safety of an organization. These measures have one characteristic in common: They are the results of the organization's achievements, which can be measured in accurate and identifiable numbers such as in productivity figures (e.g., number of manufactured goods), as level of quality standard, (e.g., number of error-free products), or level of health and safety (e.g., ill-health and accident statistics, or number of safety audits performed). The kind of measure used depends on the overall conceptual approach to be taken. Approaches of economists, managers, engineers, and social scientists, to mention but a few, primarily differ in what they are trying to learn from the measurement. To the economists, for example, safety is primarily output divided by associated inputs, for example, labor, capital, intermediate products purchased, and time. To the engineer, risk is related to the frequency of breakdowns of machines, of incidents caused by the malfunction of technology, or of deviations from technical standards. In contrast, the social scientist's approach focuses on the human subsystem of the organization, and its measures deal with health and safety and with well-being of personnel.

Pritchard, Jones, Roth, Stuebing, and Ekeberg (1988) suggested an organizational model that is based upon the goal-oriented model of Etzioni (1964), the natural systems model of Katz and Kahn (1978) and the multiple constituency approach (Connolly, Conlon, and Deutsch, 1980). The goal-oriented model states that the organization is run by a set of rational decision makers who have a manageable set of goals for the organization that can be defined well enough to be understood. Examples from the safety literature are goal setting, feedback, and behavior modification approaches (e.g., Komaki, Barwick, and Scott, 1978; Sulzer-Azaroff, Loafman, Merante, and Hlavacek, 1990). In contrast, the natural systems approach assumes that the demands on an organization are so complex and changeable that it is not possible to identify a finite set of organizational goals that are really meaningful. Instead, the model assumes that the overall goal of the organization is survival. To attain survival, different authors suggested that specific characteristics should be optimized. Examples of such characteristics include openness of communication, participation in decision making, and level of organizational trust. Thus, organizational effectiveness is thought of as the degree to which the organization is high on these key elements. Typical examples from the safety literature are studies that have looked for distinguishing factors of successful safety programs and safety performance in industry (e.g., Smith, Cohen, Cohen, and Cleveland, 1978). As an example, Cohen (1977) reviewed seven studies that dealt with critical determinants in different industrial settings. Some of the factors outlined below and in Table 31.1 seem to be more independent, controlling factors, others more interdependent and supportive ones (Cohen, 1977, p. 177):

- Strong management commitment to safety as defined by various actions reflecting management's support and involvement in safety activities
- Close contact and interaction between workers, supervisors, and management enabling open communications on safety as on other job matters
- Workforce subject to less turnover, including a large core of married, older workers with significant length of service in their jobs
- High level of housekeeping, orderly workplace conditions, and effective environmental quality control
- Well-developed selection, job placement, and advancement procedures and other employee support services
- Training practices emphasizing early indoctrination and follow-up instruction in job safety procedures
- Evidence of added features or variations in conventional safety practices serving to enhance their effectiveness

In the view of the multiple constituency approach, the organization is influenced by groups of individuals (constituencies) internal and external to the organization, such as different groups of managers, employees, customers, suppliers. These groups have different goals that are based on their own self-interest. This model of organization implies that there is no single set of goals or objectives for the organization; however, there are different goals of various groups. To reach the overall safety goals of the organization considerable efforts have to be taken at the group or unit level to establish safety commitment, and to coordinate and control the particular goals of groups and units. Examples from the safety literature are implementation of safe performance at group and department level through participative decision making, goal setting, and feedback (e.g., Misumi, 1978; Saarela, 1990). The general mechanism by which safety as well as productivity are increased is

Table 31.1 Factors of Good Safety Management. From Seven Studies in Industry[a]

GENERAL FACTOR	SAFETY PRACTICE DEFINITIONS
Management commitment	1. Safety officer holds high staff rank. 2. Top officials are personally involved in safety activities; e.g., they make personal plant safety tours and give personal attention to accidental injury reports. 3. High priority is given to safety in company meetings and in decisions on work operations. 4. Management sets clear safety policy and goals.
Safety committee and safety rules	1. The safety committee holds regular, frequent meetings. 2. Safety rules are regularly reviewed and updated in light of accident experience. 3. There is evidence of management and staff compliance with rules.
Hazard control	1. There is a high level of housekeeping. 2. There is orderly design/layout of work processes. 3. There are good environmental qualities (ventilation, lighting, noise control). 4. There is a greater number and variety of safety devices of on operating machinery.
Inspections and communications	1. There are daily worker–supervisor contacts on safety or other job matters. 2. Formal inspections are made at regular, frequent intervals. 3. There is a smaller span of supervisor control. 4. There are numerous informal contacts between workers and top officials.
Accident investigations	1. Investigations and records are kept both on disabling (lost time) and record keeping injuries and nondisabling injuries. 2. Investigations are made of property accidents and "near misses." 3. There is regular use of reports for prompting hazard control measures.
Employee support	1. There are well-established procedures for job placement and advancement. 2. There are personal counseling services. 3. There are recreational facilities and programs for off-job hours.
Safety motivation	1. A humanistic approach is used in disciplining safety violators. 2. Worker families are enlisted in safety promotions. 3. Specially designed posters or displays are used for hazard recognition. 4. Individual praise, recognition are given for safe job performance.
Safety training	1. Safety is included in new worker orientation. 2. Workers are given initial and follow-up training in safe job procedures. 3. Supervisors are given special safety training. 4. A variety of safety training techniques (lectures, films, group discussions, simulations) are used.
Makeup of workforce	1. Workers are generally older. 2. Workers generally have longer experience in their jobs. 3. There are more married workers. 4. There is less turnover and absenteeism in the workforce.

[a]Cohen (1977).

the goal-setting feedback loop (Locke and Latham, 1990). It regulates motivational and effort effects of individuals and groups. The feedback loop improves the effectiveness because the efforts of personnel would be more directly related to organizational objectives, and they would cooperate more effectively to meet those objectives.

From this point of view, risk management is a communication concept related to person, group, and unit levels. It includes the idea of a complex network of interrelationships at the intragroup, intergroup, and even interorganizational levels (Steiner, 1972). It is inappropriate to think that the goals of individuals, groups, and units are totally the product of rational decision making and are the sole determinants of organizational actions. Objectives and procedures are often the result of a process of negotiations of different constituencies with different needs and varying influences. Despite all these complexities, for the practical guidance of individuals, groups, and units goals can be identified; standards can be set and be altered; and resources can be allocated to reach these goals and standards at all levels of the organization.

Safety is not determined just by factors such as legal and social requirements, demands of the market, of strategic decision making, and of the organization's technology and safety systems. As Smith and Beringer (1987) suggested in their framework of safety performance, it is further influenced by the requirements of the physical and psychosocial elements of tasks; the design of machinery, equipment, and tools; and the hazard and danger potentials in the work environment. An effective safety management program must ensure the information, communication, and decisions related to these factors within and between tasks, decision-making levels, and phases of life cycle. These flows form the core of the SMS framework of Hale et al. (1995).

The authors proposed a model of an SMS that has certain differences from other approaches. Centrally, it is a "process-oriented" model, aimed at modeling how the safety management process is, or is not, working. Therefore, rather than looking at factors that may affect attitudes to risk management, or indicators of "good" or "poor" safety management or culture, it focuses largely on the structural and dynamic aspects of the safety management process itself. It models both the primary production processes of an organization and its management decisions. Safety management is seen as a set of problem-solving activities at different levels of abstraction in all phases of the system life cycle. "This results in a very complex structure, reflecting the fact that the management of safety in a continually changing environment, with a complex technology and with an ever changing insight into hazards is a highly demanding and complex task" (Hale, Heming, Carthey, and Kirwan, 1994, p. 37).

The coordination and control of personnel and labor are established by means of organizational structures, which, among other things, regulate the ways of performance appraisal and review process, of decision making, and of communicating, as well as through the use of goal-oriented strategies, procedures, and standards. Investigations of successful operating companies show that such a structurally embedded control is not sufficient but has to be supported and facilitated by "soft" mechanisms of control.

In that way, for example, Wollnik (1988) could show that the control effect of organizational structures which are regarded as "objective" depend on how these structures are perceived and interpreted by the members of the organization. There is a common "code" in organizations that conveys an understanding of the rules and regulations, how events are to be perceived and assessed, and what can be regarded as self-evident or "normal." This "code" constitutes the culture of a company. The idea is that safety culture reflects the attitudes, beliefs, perceptions, and values that employees share in relation to safety (Cox and Cox, 1991). It means all the basic and commonly accepted moral concepts and standards as well as patterns of thinking, of solving problems, and of behavior, which, with regard to safety, determine decision making, actions, and activities of the members of an organization. Some examples are: how different people in the organization regard their tasks and communicate about them, what kind of safety criteria are set, what priorities are chosen, and the degree of openness to admitting and learning from failures. Particularly choices are in how the tasks defined of a safety organization are mapped onto the existing company organization, and how that organization is adapted in the process. There is no one best way and there are no "best practices" that are valid for all organizations. For example, planning and procedural tasks can be allocated to experts or to work groups at the shop floor, or to task forces whose members are from different units. Monitoring tasks can be given to line or staff management or to members of the work group. Assessment of the system structure can be done by an internal audit team or an external consultant. What is an appropriate allocation of functions and tasks for a given organization will depend, among other things, upon its primary purpose and technology applied, its culture, and its existing workforce. For many authors (e.g., Schein, 1990) organizational culture is of utmost importance for the developing of an organization and for achieving goals.

31.3 ORGANIZATIONAL CONTROL OF HAZARD AND RISK

31.3.1 Planning and Controlling for Safety

Risk management and safety management are both allied to the system safety approach. Similar disciplines are total quality management and the environmental management systems, which are based on the same considerations but work with different operational processes and objectives. Quality management systems are designed to detect and correct deviations from quality standards. This principle is known as the deviation concept in safety sciences (Kjellén, 1984). Hazards are defined as deviations from a standard or ideal situation. They will lead to damage of the system elements (human, equipment, material) and/or to the system's environment (water, soil, air, people), if they are not prevented, discovered and corrected.

The overall objective of system safety is the reliability of functions: The system is to work the way it was planned and nobody should be harmed by an accident, toxic substance, or malfunction. The term "system" is used to clearly describe defined activities, processes, and equipment during the lifetime of the system. Events such as exploding spray cans, magnesium burning on dumps for weeks, or chlorofluorocarbon (CFC) that goes up into the atmosphere are examples of the need of a total system management of safety. First, systems were regarded as systems with their subcomponents; soon the systems of humans–machines and the extension of the term to organizational systems with the subcomponents human, group, organization, methods, and processes were included. According to Bird and Loftus (1976), the stages associated with system safety are as follows:

1. The preaccident (proactive) identification of potential hazards.
2. The timely incorporation of effective safety-related design and operational specification, provisions, and criteria.
3. The early evaluation of design and procedures for compliance with applicable safety requirements and criteria.
4. The continued surveillance over all safety aspects throughout the total lifespan—including disposal—of the system.

Depending on the objective of the system, each system will have a series of phases, which follows a chronological pattern. The overall lifespan of a manufactured product may include the following phases: conception and planning; design and engineering; use or operation; modification and maintenance; demolition and disposal. Each of these phases can be further subdivided. The risk management of HSE for an organization must consider all phases of the life cycle of the plant or facility it exploits and the goods and services it delivers. Each phase is an activity that must be managed safely in its own right and that must feed forward into subsequent phases and have a feedback loop in order to facilitate organizational learning processes.

This concept is illustrated in Figure 31.1 taking into consideration the lifespan of a system (Zimolong and Hale, 1989). The model emphasizes that hazards built into a technology or activity and the preventive measures to eliminate and control them are largely conditioned by the decisions made in the planning and design phase of the activity. The life-cycle phases are shown horizontally, the control and rescue measures are placed vertically: the elimination, reduction, and control of hazards as well as the limitations of the consequences of damage.

During the operational phase, the system may move from time to time outside the defined parameters or standards. The flexibility and built-in controls of the sociotechnical system, i.e., of personnel, organization, and technical protective measures, normally allow the system to return to normal operation, and no near miss, breakdown, or accident is experienced. In the overwhelming number of industrial settings, however, only a few of the deviations are discovered and effectively controlled. The system remains in an unstable situation. That state may last anywhere from seconds to months, years, and lifetimes. During this time, deviations might be detected by the operator, maintenance and repair

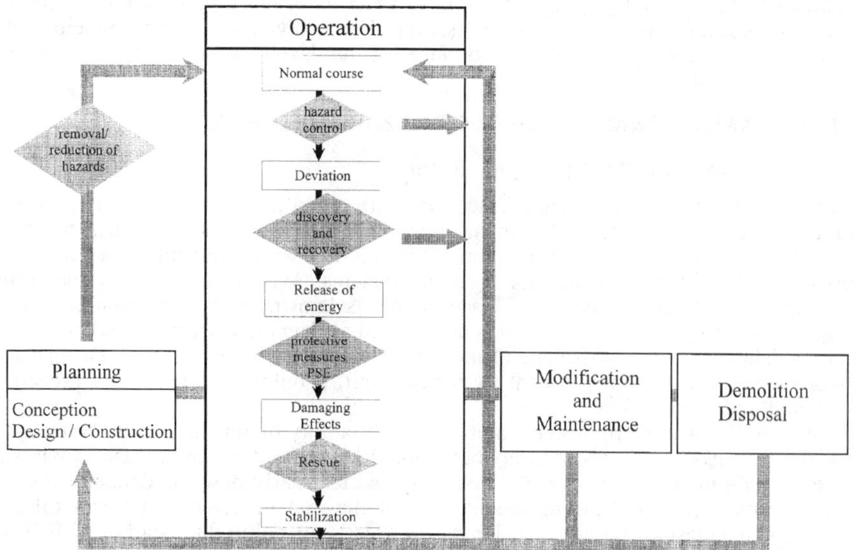

Figure 31.1 Risks as deviation from standards during a lifespan.

personnel, safety inspector, or a safety audit team. If no recovery actions are taken and the defenses are inappropriate for this very case, the latent failures eventually turn into active failures that suddenly become potent, with some that are harmless in themselves triggering conditions (Reason, 1993). Now only secondary protection measures to contain or divert the energy that was accidentally released are appropriate measures to limiting damage to less important system elements. Examples are safety goggles to guard against foreign objects, fall arrest harnesses to stop fall, and hard hats to protect against falling objects. Further damage reducing measures are rescue actions that limit the time of exposure to harmful energy, for example, disconnecting the current in electrocution cases, or washing off the caustic chemical, administering timely first aid, and proper treatment by the rescue system.

System safety essentially is planning for health and safety. The process includes establishing performance standards by which to measure and assess the HSE policy, the organizational arrangements for developing and maintaining it, the physical controls needed to meet the requirements of the performance standards (hardware control), and the systems and procedures required by the performance standards for managers, supervisors, and other employees (software control).

The British Health and Safety Executive (1993a) provides a framework for identifying key areas in the operational phase of a system for which performance standards are important. First-stage control includes the elimination and reduction of hazards and risks entering the organization. Performance standards cover the physical resources (i.e., workplaces, materials and substances, plants), the human resources (recruitment, selection of personnel and contracting organizations), and the information related to health and safety, risk control, and positive health and safety culture. Second-stage controls eliminate or minimize risks arising inside the organization. Performance standards cover hazards and risks arising from premises, plant and substances, procedures, and people. Third-stage controls minimize risks outside the organization arising from work activities, products, and services. Performance standards are required to control these risks including both organizational procedures and the control of specific risks.

31.3.2 Origins of Deviations

From a safety point of view, hazard represents a source of energy with the potential of causing immediate injury to personnel, and damage to equipment or structure. Employees are further exposed to diverse toxic substances, such as chemicals, gases or radioactivity, some of which cause health problems (Table 31.2). Unlike hazardous energies, which

Table 31.2 Some Forms of Possible Health Hazards

Category	Examples
1. Physical hazard	Noise, temperature and humidity extremes, illumination, vibration, infrared and ultraviolet radiation
2. Biological agent	Microorganisms, germs, viruses, toxins
3. Chemical hazard	Include mists, vapors, gases, fumes, dusts, liquids, pastes whose chemical composition can create health problems
4. Physical workload	Working postures, physiological load, movements and exertion of forces, manual material handling and lifting
5. Mental workload	Perceptive and cognitive workload
6. Stress	Resulting from control demands and individual control capabilities, from role conflicts and ambiguity; emotional strain resulting from aggression, loss of feedback, loss of control, helplessness

have an immediate effect on the body, toxic substances have quite different temporal characteristics, ranging from immediate effects to delays over months and years. Ozone is a good example. It causes inflammation along the entire respiratory tract. This is like a sunburn deep in the lung. For people with asthma, it increases sensitivity to allergies, frequently causing hospitalization. Often there is an accumulating effect of small doses of toxic substances that are imperceptible to humans.

The harmful effects of health hazards, such as hearing loss, cancer, liver damage, and silicoses, are regarded as illnesses. However, back pains may result from improperly designed chairs and headaches from a poor ergonomic layout of a VDT (video display terminal) workplace. This exceeds the traditional view on accidents. Consequently, controls are not always identical: the prevention of contact or its reduction to a level where no harm is done is valid only for hazardous or toxic materials, whereas illnesses resulting from poor design requires ergonomic standards, planning and sometimes complete reinstallation of the working system.

Under the total loss approach, accidents are taken as undesired events that result in harm to people, damage to property, or loss to process (Bird and Germain, 1987). They include not only those circumstances which actually cause health problems or injury, but also every event involving damage to property, plant, products or the environment, production losses, or increased liabilities. The severity of an injury that results from an accident is often a matter of chance. It depends upon many factors, such as dexterity, reflexes, physical condition, the portion of the body injured; as well as the amount of energy exchanged, what barriers were in place, whether or not protective equipment was worn. The "no injury" incident or "near miss" often has the potential to become events with more serious consequences. Analysis of the more frequently occurring property damage incidents and the near misses provides more information for guidance in the work of prevention and clearer understanding of the causes of accident problems.

Several studies have been undertaken to establish the relationship between serious and minor accidents and other dangerous events (Bird and Germain, 1987; Health and Safety Executive, 1993a). In industry, the pyramid of accident ratios is used by many companies, which is a statistical ratio between the number of fatalities, injuries, no-injury accidents and incidents. It is by no means a causal relationship; i.e., preventing all minor injuries would not result in the prevention of all serious or fatal accidents. The actual ratios in different pyramids differ significantly, indicating the problem of reliable measurement and different ratios for different locations. Traffic studies conducted by means of Traffic Conflicts Technique have clearly proved different ratios of fatalities, injuries, and conflicts (incidents) at various types of intersections (Zimolong, 1981). Not surprising, at least to the expert, is the result that apparently dangerous looking traffic light–regulated junctions have a vast amount of conflicts. The ratio of conflicts to accidents equals 1170:1. "Safe"-looking, nonsignalized urban junctions have a smaller ratio of 470:1. At "safe" junctions conflicts turn into accidents three times more often as compared to signalized junctions (Erke and Zimolong, 1978).

Most accidents happen because people commit active failures, which are called "unsafe acts." Not wearing safety glasses is one example. In terms of system safety, unsafe acts and unsafe conditions are substandard practices and substandard conditions, i.e., deviation from an accepted standard or practice. A vast number of substandard condition involve poor ergonomic design of machine, equipment and the work environment (Table 31.3).

It is essential to consider these practices and conditions only as symptoms, which point to the latent failures (Wagenaar, Groeneweg, Hudson and Reason, 1994) or basic causes behind the symptoms (Bird and Germain, 1987). The prevention of unsafe acts and conditions will be quite troublesome if their systemic nature is overlooked. They are not random events, but logical and systematic consequences of psychological states. Examples are lack of attention, haste, inexperience, reasoning errors, and misperceived risk. Psychological states are again not random events. They are caused by latent errors related to managerial and organizational failures and omissions; errors that were made long before the accident and that have been present all the time. "Haste may be caused by any one of the following: too rigorous planning, a reward system that stresses speed, lack of personnel, frequent breakdown of equipment, a motivation to complete more than the normal portion of work, exceptional emergencies that had never been foreseen" (Wagenaar, Souverijn, and Hudson, 1993, p. 159). In all these examples the cause of haste is a latent failure that has been present for a long time. Telling people not to be hasty is pointless. Haste can be prevented only by removal of latent failures that cause haste.

Various substandard practices relate to deficiencies in communication and information between functional units of the company. Equipment and materials that are inadequate or hazardous will be purchased if there are no adequate standards and if compliance with standards is not managed. Poor work process layouts and interfaces will be designed and built if there are no adequate standards and compliance for design and construction. Equipment will wear out and produce products with quality deficiencies, create waste, or break down and cause property damage, if that equipment is not properly selected, used, and maintained (Timpe, 1993).

The origins of the deviations from standard are deficiencies in management and organization. Wagenaar et al. (1994) have suggested 11 types of latent failures, which have

Table 31.3 Contents of Six Categories of Performance Standards

Category	Contents
1. Personal factors	Physical or physiological capabilities, e.g., height, strength, vision, hearing, psychomotor coordination Mental or psychological capabilities, e.g., knowledge, intelligence, emotional stability, social skill
2. Job content	Perceptive and cognitive work load Cycle times (monotony, routine, vigilance) Content demands, e.g., motivating job requirement Social contact and support Learning opportunities
3. Safe practices	Personal protective equipment Behavioral requirements of safety devices Procedures, rules, or standard operations, e.g., proper loading, placement, lifting, position for task Permission to use tools, machines, equipment
4. Ergonomics of workplace and work station	Physical layout and size Design criteria for tools, machines, equipment Design criteria for controls, information displays, user centered software
5. Working environment	Physical: illumination, noise, vibration, climate, radiation, housekeeping, order at workplace Social: work or shift schedules, work hours, payment, reward system, social support, management style, climate
6. Engineering precautions	Proper protective equipment Adequate barriers or safeguards Adequate warning system

After Bird and Germain (1987).

emerged on the basis of studies of hundreds of accidents and incidents. They are related to the work environment, to the individuals doing their job, and to management (Table 31.4). Detailed lists that cover different factors are provided in Petersen (1978) and Bird and Germain (1987).

System safety engineering involves the application of scientific and engineering principles for the timely identification of hazards and initiation of those actions necessary to prevent or control hazards within the system. It draws upon professional knowledge and specialized skills in the mathematical, physical, and related scientific disciplines, together with the principles and methods of engineering design and analysis to specify, predict, and evaluate the safety of the system. Although much of occupational safety is now being recognized as being behavioral, safety engineering still has a major role in occupational safety. General topics and methods are concerned with guarding energy sources; design and redesign of machinery, equipment, and processes; application of environmental standards; and establishment of inspection systems, such as statutory engineering inspections of pressure vessels, cranes and lifting machines, or electrical installations. Comprehensive coverage of those topics provide Bird and Germain (1987) and Ridley (1994).

In the last decades major changes have been seen in approaches to the problem of providing safer workplaces, machinery, and equipment. Machinery and equipment must conform with the standards of national laws. For the European Community (EC) market, new machinery and equipment must meet the requirements contained in the "Machinery Directive" of 1989 before it may be put on the EC market (European Community, 1989a). All existing work equipment must conform with the 'Work Equipment Directive' (1989) by 1 January 1997 (European Community, 1989b). In the United States, Department of Labor (1980) has published Concepts and Techniques of Machine Safeguarding which reflects some different practices as compared to the EC.

Workplace designs that do not take ergonomic principles into account are likely to lead to an increase in errors and accidents and a decrease in safety and efficiency. Error-prone designs place demands on performance that exceed capabilities of the user, violate the user's expectancies based on his or her past experience, and make the task unnecessarily difficult, unpleasant, or dangerous. The systems approach applied to ergonomics treats humans and machines or computers as components interacting together to bring about some desired objective. The role of the individual is characterized by his or her capabilities and limitations, mainly human sensory capabilities, perceptual or cognitive processes, and human performance abilities. For the workplace designer or safety practitioner, reliable anthropometric data, i.e., data concerning the measurement of physical

Table 31.4 List of Basic Causes of Loss and Latent Failures

Basic Causes of Loss[a]	Latent Failures[b]
1. Engineering Standards (inadequate ergonomics, missing or inappropriate standards and/or design criteria) 2. Work planning or programming 3. Purchasing (inadequate specification on requisitions, to vendors, handling of material) 4. Deficiencies or failures in preventive maintenance and repair 5. Inadequate tools and equipment 6. Work standards (inconsistent standards, procedures, rules; lack of standards, maintenance) 7. Poor housekeeping 8. Management (improper or insufficient delegation, policy, procedures, practices, guidelines, conflicting goals) 9. Deficiencies in leadership and/or supervision (objectives, goals, standards; performance feedback, consistency, instruction, orientation, and/or training).	1. Environment Hardware defects Design failures Error enforcing conditions including: Inexperience, extreme climatic conditions, low signal-to-noise ratios Poor housekeeping 2. Individuals Poor operating procedures Poor maintenance procedures Inadequate training 3. Management Incompatible goals (safety and economy left to be solved at a too low level in the organization) Organizational deficiencies Communication and information failures Deficiencies in defense planning.

[a]Bird and Germain (1987).
[b]Wagenaar et al. (1993).

features and functions of the body, are found in human factors handbooks (Van Cott and Kinkade, 1972; Woodson, 1981). Ergonomic principles for enhancing the design and safe operation of work facilities and equipment may be organized at the component, the workstation, and the work space level including the work environment. At the component level, visual displays, various types of controls, and visual or auditory warnings are considered. Workstation designs are based on anthropometric data, which are available for designing cabinets, consoles, desks, and other workstations (Corlett and Clark, 1995; Woodson, 1981). In particular, special considerations have been given to computer workplace design and to software–user interface design (Hix and Hartson, 1993).

Overall workspace design is concerned with the integration of several work areas and how to ensure that ambient environmental conditions fall within acceptable ranges. Thermal comfort, noise, and lighting are among the most important environmental factors to assess in occupational settings. The essentially multidisciplinary nature of the subject is covered in a number of useful books, among others in Grandjean (1980), Shackel (1984), and Corlett and Clarke (1995).

31.3.3 Measuring and Reviewing Performance

In the simplest case, hazards can be identified by observation, comparing the circumstances with legal standards and guidance. In more complex cases, measurements such as air sampling or examining the methods of machine operation may be necessary to identify the presence of hazards presented by chemicals or machinery. The complexity of many health risks means that the identification of health hazards and risks will generally require the measurement of exposure, calling for specific monitoring and assessment techniques and the competence to use them. Although health risks arising from the use of substances can be controlled by physical control measures, systems of work, and personal protective equipment, confirmation of the adequacy of control will often require measurements of the working environment to check that exposures are within preset limits. In special cases, surveillance of those at risk to detect excessive uptake of a substance, i.e., biological monitoring, or early signs of harm, i.e., health surveillance, may also be necessary.

Assessing risks is necessary to identify their relative importance and to obtain information about their extent and nature. This will help identify where to place the major effort in prevention and control, and to make decisions on the adequacy of control measures. The relative importance of risks may be determined by taking into consideration both the severity of the hazard and the likelihood of occurrence. There is no general formula for rating risks in relative importance but a number of simple risk estimation techniques have been developed to assist in decision making (Bird and Germain, 1987; Steel, 1990). They involve only some means of estimating the likelihood of occurrence and the severity of a hazard. An example is given in Figure 31.2.

Usually the amount of harm resulting from a critical event or accident is a matter of chance. The hazard potential can be assessed in a risk matrix. Two questions have to be answered: "What could have happened?" and "How often can the consequences occur?" Hazards — the potential to cause harm — will vary in severity. The effect is rated in terms of both injuries and damage of property. In Figure 31.2 the likelihood of harm is rated on a five-level scale ranging from once in 5 years to more than once a week.The cells of the matrix may contain some arbitrary risk numbers. The organization decides what kind of measures have to be taken by which kind of risk number. Health hazards, malfunctions, and near accidents and even accidents can be assessed in the same way. Systems of assessing relative risks can contribute not only to establishing risk control priorities but also assist in prioritizing other activities, and ranking departments and units according to their safety levels.

Hoyos and Ruppert (1993) developed a hazard potential analysis for the identification and assessment of hazards at the workplace. The Safety Diagnosis Questionnaire (SDQ) relates the hazards identified to a broad repertoire of abilities and skills that are considered the basis of hazard control by the worker.

The identification of deficiencies and deviations from standards related to managerial and organizational arrangements is usually done in an audit and review process. Some topics are for example organizational arrangements to secure the involvement of all employees in the HSE efforts, the acceptance of HSE responsibilities by line managers, the communication of policy and relevant information, and the timely consideration of HSE-related standards in the planning and design process. More topics are provided in Table 31.4 Audits are normally performed by competent people, independent of the area or

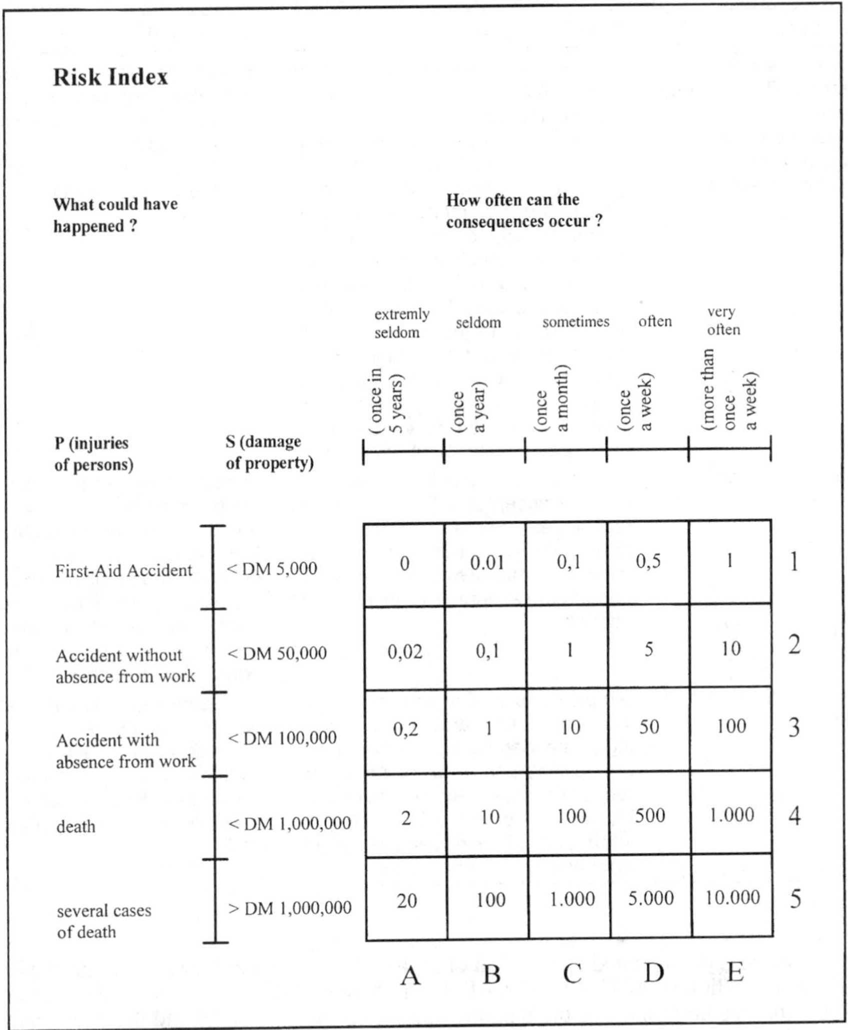

Figure 31.2 Determination of a risk index. (After Shell Company.)

activities being audited. This can be achieved either by using external consultants or by using staff from different sections, departments or sites to audit their colleagues. On the basis of structured work group discussions, Wagenaar et al. (1994) integrated the employees into the review process. The authors developed a method for predicting the causal structure of future accidents on the basis of symptoms already visible. These are the failure states that are listed in Table 31.4.

In more complex situations or in high-risk systems, qualitative or quantitative risk assessments may be required by legal standards and guidance. For example, in the chemical or nuclear industry, special techniques and systems may be applied when planning a new system or major changes of an existing system. HAZOPS and hazard analysis systems such as event- or fault-tree analysis have repeatedly been summed up and presented, among others by Hoyos and Zimolong (1988), and Ridley (1994). In Table 31.5 an overview is given on the techniques suitable for the planning and design phase of a system.

Probabilistic risk assessment (PRA) is a systematic quantitative assessment of the likelihood of the levels of damage from operating industrial settings. The assessments are derived from combining the likelihood of occurrence of hardware and human related

Table 31.5 Hazard Analysis and Risk Assessment Techniques

Instrument	Short Description	To Be Used at
Formal methods: Hazard and operability studies Failure mode and effects analysis Fault tree analysis	HAZOP and Failure mode and effect analysis (FMEA) belong to the inductive techniques: What happens if a coolant pump fails? System and its components are analyzed with regard to operability and process to identify hazards, malfunctions, weak points. Mostly complex and time-consuming, multidisciplinary brainstorming sessions. Requires high levels of expertise. Fault tree analysis (FTA) belongs to the deductive techniques: How can the failure of the coolant pump happen? Graphical presentation and analysis of logical contributions of errors and failures to one type of malfunction of a system.	Suited for plants and situations with high-risk or severe consequences. The objective is to design out risk at early stages of planning or when changes are to be made.
Standards, thresholds, limit values	Simple way of evaluating whether situation is acceptable. Powerful design technique. Some standards or threshold limits are based on epidemiological data. Gives little information about severity or probabilities of hazards.	Evaluation of health hazards, for example noise and vibration levels, exposure to chemical substances; evaluation of several aspects of workplace, especially ergonomic standards for workplace layout, space, environment, tools, equipment, machines.
Risk classification and ranking	Simple classification scheme for various kinds of hazards. A rank ordering of the identified hazards is drawn up; severity and frequency of occurrences are assessed in terms of qualitative effect classes. Outcome gives information of the risk and the priority on precautions to be taken.	Evaluation of safety and health hazards. Simple technique to be used at all levels of organization and at workplaces. Relative ranking of risks. An example is shown in Figure 31.2.

errors. One well-documented application of a procedure to obtain human error probability estimates is Technique of Human Error Rate Prediction (THERP) (Swain and Guttmann, 1983). Other methods, such as the human cognitive reliability model and the maintenance personnel performance simulation model have been discussed by Svenson (1989). A methodology still under development is the dynamic reliability technique, which identifies the origin of Human Error Probabilities (HEPs) in the dynamic interaction of the operator and plant control system (Cacciabue, Caprignano and Vivalda, 1993).

Swain and Guttmann recommend the following procedure to conduct a PRA:

1. Dividing up of the tasks in subtasks and elements
2. Analysis of possible failure with the help of a fault or event tree
3. Determining the probabilities of faults for the corresponding task element from a database or through subjective estimation
4. Computing of the reliability of tasks or frequency of failures with the rules of probability calculus

The most difficult problem still remains the availability of data. A number of attempts have been made to collect human reliability data (Miller and Swain, 1987). Sometimes expert judgment is calibrated in some way, for example, through psychophysical methods such as rating, ranking, or magnitude estimation, or through analytical methods such as simple multiattribute rating technique (SMART; Edwards, 1977) or multiattribute utility decomposition (MAUD; Humphreys and Wisudha, 1983). Research has shown that experts are liable to errors of judgment in just the same way as laypersons, especially when they are forced to go beyond their experience (Kahneman, Slovic, and Tversky, 1982). Zimolong (1992a) conducted an experiment which compared three different estimation techniques, including an expert ranking and a decision aiding technique (Success-Likelihood index methodology) (SLIM; Embrey, 1987) and THERP. A satisfactory match between estimated and empirically derived human error data was yielded only for THERP for routine tasks. The application of SLIM and ranking led to a mismatch between estimated and actual HEPs. At present, no reliable technique for the quantitative assessment of error probabilities is available.

The challenges for safety management are to devise appropriate indicators of operational and economic impact and to be able to attribute types and amounts of operational and economic impact on formulated strategies and activities. Active monitoring systems provide essential feedback on performance before an accident, ill health, or an incident happens. The methods involved in monitoring ongoing safety strategies and activities vary widely, ranging from periodic checks of compliance with performance standards, to relatively straightforward tracking of training and education services delivered, to checking on how hazardous exposures are controlled. Monitoring can include serious reexamination of whether the needs of the safety strategy and related projects as originally intended still exist, or it may suggest modification, updating, or revitalization.

Reactive systems monitor accidents, ill health, and incidents. Securing the reporting of serious injuries and ill health generally presents few problems for most organizations. However, the reporting of minor injuries, other loss events, incidents, and hazards requires special efforts and attention and creates difficulties in most organizations.

Most companies are set up to measure accountability through analysis of results. Monthly accident reports at most plants suggest that the supervisor and manager should be judged by the number and cost of accidents that occur in his or her department. Petersen (1978) strongly emphasized that they should be judged by what they do to control losses. Numerous techniques have been introduced for guiding and measuring accident prevention efforts of supervisors and line managers (Bird and Germain, 1987).

An alternative approach to measuring and evaluating the efficacy of safety management is the use of safety related financial or economic performance ratios. These ratios are performance indicators that are used to show the strategic value and operating leverage, i.e., the ratio of the percentage change in operating income to the percentage change in operating costs associated with safety management efforts (Imada, 1990). The direct costs of accidents can be determined by using one of the suggested models, for example by Veltri (1990) or Health and Safety Executive (1993b). Models disclose the direct costs of accidents and the economic impact on cost–volume–profit performance standards and profitability potential of the company. The indirect costs of accidents are difficult to calculate, chiefly because no suitable model for verifying any results in reliable and valid ways has been developed. However, Grimaldi and Simonds (1989) presented an uninsured cost model that provides a reasonable measure for determining uninsured costs of accidents.

31.4 INDIVIDUAL CONTROL OF HAZARD AND RISK

31.4.1 Hazard Perception

Most of the safety- and health-related hazards in industry cannot be eliminated, reduced, or minimized. People have to perceive, detect, and control hazards and dangers if confronted with them at work. In case of individual control of hazards, this requires careful analysis of hazards as well as what it takes to control hazards. The perception of hazards is essential to the subsequent phases of action in hazardous situations: personal risk assessment, decision making, and selection of the appropriate protective, and/or preventive action.

Saari (1976) defines the information processed during the accomplishment of a task in terms of the following two components:

- The information required to execute a task
- The information required to keep existing risks under control

For example, a construction worker perched on the top of a ladder, who is required to drill a hole in the wall, has both to keep his balance and to automatically coordinate his body–hand movements while drilling a hole; a driver searching for route information ahead simultaneously adjusts distance and speed of his own car relative to the vehicles in front of him. In both cases hazard perception is crucial to coordinate body movement to keep hazards under control, whereas conscious risk assessment plays only a minor role, if it plays one at all.

Not all hazards are directly perceptible to human senses. Examples are electricity; colorless, odorless gases such as methane and carbon monoxide; x rays and other forms of radioactivity; and oxygen-deficient atmospheres. Their very presence must be signalled by devices that translate the presence of the hazard into something which is recognizable. Most of the toxic substances are not visible at all. Ruppert (1987) found in an investigation in an iron and steel factory, in municipal garbage collecting, and in medical laboratories that from 2230 hazard indicators only 42% were perceptible to the human senses. Of the indicators 22% have to be inferred from comparisons with standards; for example, an increase in the noise level coming from the press of the garbage truck indicates the risk of being hit by particles bursting from the container. Hazard perception is based in 23% of cases on clearly perceptible events that have to be interpreted with respect to knowledge about the type of hazardousness; e.g., a glossy surface of a wet floor indicates slippery conditions. In 13% of reports hazard indicators can be retrieved only from memory; for example, current in a wall socket can only be made perceivable by the proper checking device. There are also situations where hazards exist that are not perceivable at all and cannot be made perceivable at a given time. One example is the risk of infection when opening blood samples for medical tests; another is the risk of material falling down from scaffoldings at construction sites. The knowledge that hazards exist must be deduced from one's knowledge of general principles of causality or acquired by experience.

The results demonstrate that the requirements of hazard perception range from pure detection and perception to elaborate cognitive inference processes of anticipation and assessment. Delayed or accumulating effects of health hazards; e.g., toxic substances are likely to impose additional burdens on individuals. Cause and effect relationships are sometimes unclear, scarcely detectable, or misinterpreted.

In Table 31.6 a list of requirements on perceptual processes of hazard detection and perception is presented based upon studies of 391 work sites in industry and public services (Hoyos and Ruppert, 1993). Experienced raters had to assess all hazards at a particular site, which resulted in 2373 hazards identified (Hoyos, 1995). On the average, people have to cope with six perceptual and cognitive demands per hazard in order to control the risk at work. They include visual recognition, selective attention, auditory recognition, and vigilance. As expected, visual recognition dominates auditory recognition. About three-quarters of the hazards have to be detected visually; in only 21.2% of the cases it is by auditory detection. In more than half of all hazards observed, people had to divide attention between task and hazard control, for example, observing crane

Table 31.6 Detection and Perception of Hazard Indicators in Industry. The Frequency of Demands per Hazard Is Rated in Percent[a]

Requirements	Rated Total %
1. Visual recognition	77.3
2. Selective attention	63.0
3. Division of attention	57.5
4. Rapid identification and responsiveness	56.3
5. Perception of incessant hazards	51.5
6. Observation and maintenance of distance	44.2
7. Detection of potentially dangerous objects	28.0
8. Vigilance	25.0
9. Auditory detection	21.2
10. Recognition of changing danger zones	19.9
11. Directed attention (distance)	17.9
12. Auditory recognition, e.g., of warnings	15.9
13. Visual recognition, e.g., of warnings, labels	7.3

[a]Hoyos and Ruppert (1993).

movements while working at construction sites. This is mentally strenuous and likely to be error prone. Even more alarming is the finding that in 56% of all hazards employees have to cope with rapid activities and responsiveness to avoid being hit and injured, e.g., by sudden sidestepping to avoid an oncoming vehicle.

Table 31.7 shows the cognitive demands of anticipation and assessment which are required to control hazards of the worksite. The core characteristic of all activities summarized in this table is the requirement of knowledge and experience to cope with hazards. It emphasizes the need of establishing signs and warnings and introducing personal counseling, training, and qualification efforts. As Hoyos, Bernhardt, Hirsch, and Arnhold (1991) have demonstrated, employees have little knowledge of hazards, safety rules, and proper personal protective behavior. In some cases (16.1%), perception of hazards is supported by signs and warnings; usually, however, people rely on knowledge, training, and work experience. It is without doubt mandatory to improve the indication of hazards and risks by warning signs and labels. The use of labels and warnings to combat potential hazards, however, is a controversial procedure for managing risks. Too often they are seen as a way for manufacturers to avoid responsibility for unreasonably risky products. Obviously, labels and warning signs will be successful only if the information they contain is read and understood by members of the intended group of people. Frantz and Rhoades (1993) found that 40% of clerical personnel filling a filing cabinet noticed a warning label placed on the top drawer of the cabinet; 33% read part of it, and no one read the entire label. Contrary to expectation, 20% complied completely by not placing any material in the top drawer first. Lehto and Papastavrou (1993) provided a thorough analysis of findings pertaining to warning signs and labels by examining receiver-, task-, product-, and message-related factors.

31.4.2 Personal Risk Assessment

"Personal risk assessment" refers to the decision process as to whether and to what extent the individual will be exposed to hazard; for instance, working on a high scaffolding or driving a car at high speed. It seems people must decide in the face of danger. However, people doing their jobs on a routine basis rarely consider these hazards or accidents in advance: they run risks, but they do not take them. Much of the time there will be no conscious perception or consideration of hazards as such. "The lack of safety consciousness is both a normal and a healthy state of affairs, despite what has been said in countless

Table 31.7 Prediction and Evaluation of Hazard Indicators. The Frequency of Demands per Hazard Is Rated in Percent[a]

Requirements	Rated Total %
1. Estimates of physical units (e.g., weight, force, and energy)	32.7
2. Identification (screening) of defects and inadequacies	29.6
3. Prediction of structural weaknesses	25.1
4. Expectancy of warning stimuli (e.g., railway signal lights)	19.8
5. Perception of visual cues (e.g., flags, traffic, or warning signs)	19.6
6. Subjectively perceived somatic symptoms (e.g., dizziness, breathlessness, nausea)	19.5
7. Predictions of unobvious dangers (e.g., radioactive contamination, bacterial infection)	17.7
8. Recognition of instable storage (e.g., open paint containers, excessively high piles of bricks)	16.3
9. Comprehension of warning signals (e.g., symbols, colors)	16.1
10. Evaluation of material stress (e.g., material pressures, efficacy of heat-resisting clothing)	16.0
11. Interpretation of displays and data (e.g., gauges, switches, monitors)	5.4

[a]Hoyos and Ruppert (1993).

books, articles and speeches. Being constantly conscious of danger is a reasonable definition of paranoia" (Hale and Glendon, 1987, p. 41).

There is a wide variety of interpretations of the terms "risk." On the one hand, risk is interpreted to "mean probability of an undesired event." It is an expression of the likelihood that something unpleasant will happen. A more neutral definition of risk is used by Yates (1992), who argued that risk should be perceived as a multidimensional concept, that, as a whole, refers to the prospect of loss. Technical risk assessment usually focuses on the potential for loss, which includes the probability of the loss occurring and the magnitude of the loss in terms of death, injury, or monetary costs.

Lay people's sense of risk depends on more than the probability and magnitude of loss. It may depend on such factors as potential degree of damage, as well as on dimensions such as the unfamiliarity of the consequences, the involuntary nature of exposure to risk, the incontrollability of damage, and the biased media coverage. The feeling of control in a situation may be a particularly important factor (Slovic, 1987).

A different research direction has addressed emotional reactions to risky situations. The potential for serious loss generates a variety of emotional reactions, not all of which are necessarily unpleasant. There is a fine line between fear and excitement. Again, a major determinant of perceived risk and of affective reactions to risky situations seems to be a person's feeling of control, or lack thereof. As a consequence, for many people, risk may be nothing more than a feeling (Trimpop, 1994).

The overwhelming evidence that people often make poor choices in risky situations seems also related to inappropriate risk assessment. In particular, research on judgment and choice has shown that people have methodological deficiencies such as in understanding probabilities; negligence of the effect of sample sizes; reliance on misleading personal experience; holding judgments of fact with unwarranted confidence, and misjudging risks. People are more likely to underestimate risks if they have been voluntarily exposed to risks over a longer period, such as people living in the neighborhood of reservoirs facing the risk of flooding; or in areas where earthquakes are not uncommon. Similar results have been reported from industry. In a field study (Zimolong, 1985) 153 members of 6 occupational groups mainly from the building industry and auxiliary building trade, among them carpenters, tile layers, and construction workers, rated the frequencies of falls resulting in minor injuries, and in serious or fatal injuries, respectively. Ratings were carried out for eight different work sites: ladder, scaffolding, roof, building under construction, etc. Estimates were compared with frequencies based on fall accident statistics. The results clearly indicated a job specific underestimation or overestimation of the frequencies. As can be seen from Figure 31.3 there is a mismatch between the risk as subjectively assessed and as objectively measured. Generally spoken, employees underestimate high-risk activities and overestimate low-risk activities; however, underestimation of risks is ruled by the exposure time to the specific risk. For instance, carpenters and painters frequently use ladders, and they considerably underestimate the risk of accidental fall. Tile layers typically underestimate the risk of fall from roofs, whereas scaffolding assemblers obviously seem to believe that falls from scaffoldings will never happen to them. Actually, it is the most frequent cause of minor and severe injuries. Shunters, miners, forest, and construction workers all dramatically underestimate the riskiness of their most common work activities if compared to objective accident statistics. They tend to overestimate, however, obvious dangerous activities of other fellow workers when required to rate them.

Unfortunately, expert's judgments appear to be prone to many of the same biases as those of laypersons, particularly when experts are forced to go beyond the limits of available data and rely upon their intuitions. Research further indicates that disagreements about risk will not disappear completely when sufficient evidence is available. Strong initial views are resistant to change because they influence the way that subsequent information is interpreted. New evidence appears reliable and informative if it is consistent with one's initial beliefs, contrary evidence tends to be dismissed as unreliable, erroneous, or unrepresentative (Nisbett and Ross, 1980). When people lack strong prior opinions, the opposite situation occurs—they are at the mercy of the problem formulation. Presenting the same information about risk in different ways, for example, mortality rates as opposed to survival rates, alters their perspectives and their actions (Tversky and Kahneman, 1981).

Most of the personal risk decisions in every day life are not conscious decisions at all. People are not even aware of risk. Most of our daily behavior is automatized and runs smoothly without continuous attentional control and conscious risk taking. Reason's

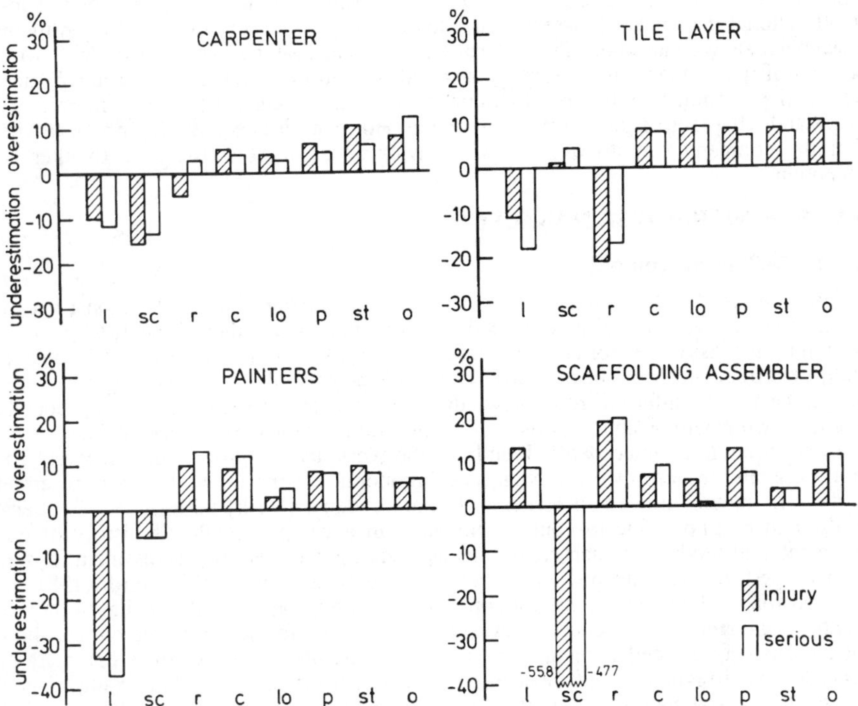

Figure 31.3 Differences of personal estimates and accident-based frequencies of work falls. (From Hoyos and Zimolong, 1988, p. 175.)

genetic error modeling system (GEMS, 1987b) describes how the transition from automatic control to conscious problem solving takes place when exceptional circumstances arise or novel situations are encountered. In normal work routines, however, conscious risk assessment and decision-making is just not present. Therefore, it cannot be argued that people's way of evaluating risk is inaccurate and needs to be improved. The notion that the acceptance of risks, identified after the occurrence of accidents, is the primary cause of the incident does not take into account that in most cases no conscious risk assessment was undertaken. In research as in practice, less attention has been paid to the conditions in which people will act automatically, follow their good feeling, or accept the first choice that is offered (Wagenaar, 1992). Contrary to these findings, there is a widespread acceptance in society and among safety and health professionals that risk taking is a prime cause of mishaps and errors. In a representative sample of Swedes aged between 18 and 70 years, 90% agreed that risk taking is the major source of accidents (Hovden and Larsson, 1987).

Preventive activities used to cope with hazards include among others: planning work procedures and steps ahead; regular checking of equipment and material for defective parts; the provision of adequate storage; selection of safe work procedures by selecting proper material and tools; setting an appropriate work pace; and inspection of facilities, equipment, machinery, and tools.

The most frequent protective measure required is the use of personal protective equipment (PPE; Hoyos and Ruppert, 1993). Together with the correct handling and maintenance it is by far the most important requirement in industry. There are major differences in the use of PPE among companies. In some of the best companies, mainly in the chemical and mineral oil industry, the use of PPE approaches 100%. In contrast, in the construction industry, safety representatives have problems even in attempts to introduce a particular PPE on a regular basis. Is risk perception and assessment the major factor that makes the difference? Again, this is doubtful. Some of the companies have successfully enforced the use of PPE, which then becomes habitualized, e.g., the wearing of safety helmets. They have established the "right safety culture" and thus have subsequently altered personal risk assessment. Slovic (1987) in his short discussion on the use of seatbelts shows that about 20% of road users wear seatbelts voluntarily, 50% would use them if it was made mandatory by law and beyond this number, only control, incentives, and punishment will serve to improve automatic usage. Thus, it is important to understand what factors govern risk perception. However, it is equally important to know what the company can do to change behavior and, subsequently, how to alter risk perception.

31.5 BEHAVIORAL RISK MANAGEMENT

31.5.1 Task Management

As Sulzer-Azaroff (1978) points out, safe and unsafe practices probably persist because they are in some way naturally reinforced. In addition, although there are natural punishers for committing risky acts, for example, injuries, pains, sicknesses, these are often delayed, weak, or infrequent. Most health hazards are not perceivable at all and negative consequences have to be inferred from knowledge, memory, or experience with a considerable risk of drawing wrong conclusions. Safety programs often have turned to heavily emphasizing behavioral antecedents directly at the workplace. Signs, guidance, threats, incentives, goals, training, work design, and ergonomic design of workplaces are among the more frequently utilized antecedents. The antecedents of risky performance frequently are difficult or impossible to control. No one can alter an individual's history of reinforcement, and modifying such conditions as hazard perception, risk assessment, or turning risky performance into preventive behavior are at best difficult. Burkardt (1981) has proposed four different strategies for superiors to modify behavior that is based on a list of antecedents and consequences of everyday's safety performance. Research has shown that emphasizing antecedents and consequences at the operating level of a company has succeeded but irregularly, and most important, not in the long run. For example, goals are antecedent statements of a particular level to be reached by a given time. Pairing goals and consequences works better than presenting goals by themselves; however, if the process is not supported and sustained by management, the gains in health and safety flattens out and subsequently returns to the baseline level (e.g., Komaki et al., 1978; Chhokar and Wallin, 1984).

Goal setting affects performance by directing the attention and actions of individuals and/or groups, mobilizing efforts, increasing persistence, and by motivating the search for appropriate performance strategies. Setting difficult, yet achievable goals, and providing performance feedback in relation to them, is widely accepted as to be one of the most powerful behavioral modification techniques. Strong support has been obtained for two of the major goal-setting propositions: setting difficult, but attainable goals and clarifying objectives to be reached (Tubbs, 1986). Goal setting has been found to have positive performance effects on tasks of varying complexity, ranging from simple brainstorming tasks, to college nurse work, to complex scientific work. Task complexity is the only variable that has been shown to have a significant and robust moderating effect on the performance gains that result from specific, difficult goals (Wood, Mento, and Locke, 1987).

Performance feedback is usually defined as information about the effectiveness of particular work behaviors and is thought to fulfill several functions. For example, it is directive, by clarifying specific behaviors that ought to be performed, it is motivational, as it stimulates greater effort, and, it is error correcting, as it provides information about the deviations from a prescribed standard or level. The relationship between goals and performance is complex, but goals have been demonstrated to mediate the effectiveness of feedback, while feedback has been shown to moderate the effectiveness of goals. The relationship between goals and feedback can be construed as the joint effects of motivation and cognition that control action. Goals, for example, inform individuals to achieve particular levels of performance in order to direct and evaluate their actions and effort, whereas performance feedback allows the individual to track how well they are doing in relation to the goal, so that if necessary, adjustments in effort, direction or possibly task strategies can be made.

Feedback has the potential to function in a variety of ways: as a reinforcer when it conveys success, as a punisher when it conveys failure, and as an antecedent when it prompts or cues the conditions under which responses will be reinforced and/or punished. Perhaps it is because it can operate in all these ways at once that feedback has been found to be an especially powerful modification technique. In a metaanalysis of psychologically based interventions Guzzo, Jette, and Katzell (1985) found goal setting to increase performance with an average effect size of .75 standard deviations (*d* statistic). Feedback was found to have an average effect size of .35 standard deviations. That does not mean, however, that feedback is less powerful than goal setting. They complement each other, and, in combination, are far more effective than either one alone, thereby providing a powerful management tool for effecting change.

Numerous studies conducted in the field of health and safety strongly support this view. To mention but a few, Alavosius and Sulzer-Azaroff (1985) introduced feedback on safe performance in patient transfer by staff in a hospital, thereby significantly improving proper lifting techniques. Goal setting in a paper mill improved performance if goals were assigned by supervisors, but not if they were set by the workers (Fellner and Sulzer-Azaroff, 1985). McCarthy (1978) used a time series design to reduce the number of "high bobbins" in a textile mill. Bobbins are spindles of thread, which, if not pushed down far enough, cause tangles. Introducing goals of increasing difficulty plus feedback led to a steady decrease in the number of high bobbins. When feedback was removed, the number of high bobbins increased and then decreased again when feedback was reintroduced.

Saari (1987) significantly improved housekeeping in a shipyard through feedback and implicit goal setting. Employees were given a written list of correct work practices, shown slides illustrating correct and incorrect practices, and given a 1-hr training seminar. The observed correct practices on the shop floor were presented with posted boards showing the performance on a quantitative index. Chhokar and Wallin (1984) used a six-stage time series design for machine shop workers: baseline, training plus goal setting, weekly feedback added, monthly feedback added, training and goal setting only, and bimonthly feedback added. Safe behavior of the employees was defined as the percentage of employees performing their jobs in a completely safe manner. They found an increase in performance that reached 95% of maximum possible safe performance, if goal setting and feedback was provided. This was an increase of 30% as compared to the baseline of 65%.

Another issue to consider with respect to the effects of goal attainment is the nature of the task. Hackman and Oldham's (1980) job characteristics theory states that the degree to which the work is seen as rewarding is dependent on the degree to which the task possesses four core attributes: personal significance, feedback, responsibility or autonomy, and identity as a whole piece of work. These core attributes are growth producing, and they fulfill important needs. A review of the literature on the relation of the core attributes, critical psychological states, and the outcomes supports the approach despite the fact that some issues have been risen concerning the method of asking the same people for the core attributes, psychological states and outcomes (Algera, 1990).

31.5.2 Participative Management

The introduction of work groups is part of a rationalization program in industry and public services that aims at improved flexibility and productivity. Self-regulation of groups, less managerial control, participation in decision making, better motivation of group members, and job inherent qualification efforts are the driving factors that may account for better quality and productivity faced in work groups (Zimolong and Windel, 1996). Problem-solving groups and quality circles are examples of the application of the self-regulation principle in work organization. In a quality circle, members from the same organizational unit meet regularly to solve work-related problems in an effort to improve the quality of the product, the production process or the delivery. Not so frequently, safety circles have been established to improve health, safety, and ergonomics. Ritter and Zink (1992) provide an overview on group-based activities to improve health and safety.

In industry as well as in science, there is little agreement on how to define work groups. Despite the common view that the diffusion of work groups is highly advanced, empirical studies reveal the contrary. In Figure 31.4 the diffusion of three different types of work groups in the mechanical engineering industry in Germany are depicted (Zimolong and Saurwein, 1995). Data are based on annual representative surveys that started in 1991. The broadest definition of a workgroup in the survey includes individuals working together at the shopfloor on a regularly time basis. This is the major group in the survey; 23% to 25% of companies have installed this type of workgroup. Most scientists would not agree upon the term "workgroup" in this case. The group as a whole has not been not given specific goals to attain; there are neither close personal contacts nor have specific structures of communication been developed between all members.

A semiautonomous workgroup as a whole has been given production goals to meet. The group is characterized by an enlarged job spectrum, and frequently the team is responsible for a certain degree of scheduling and planning of work activities in order to

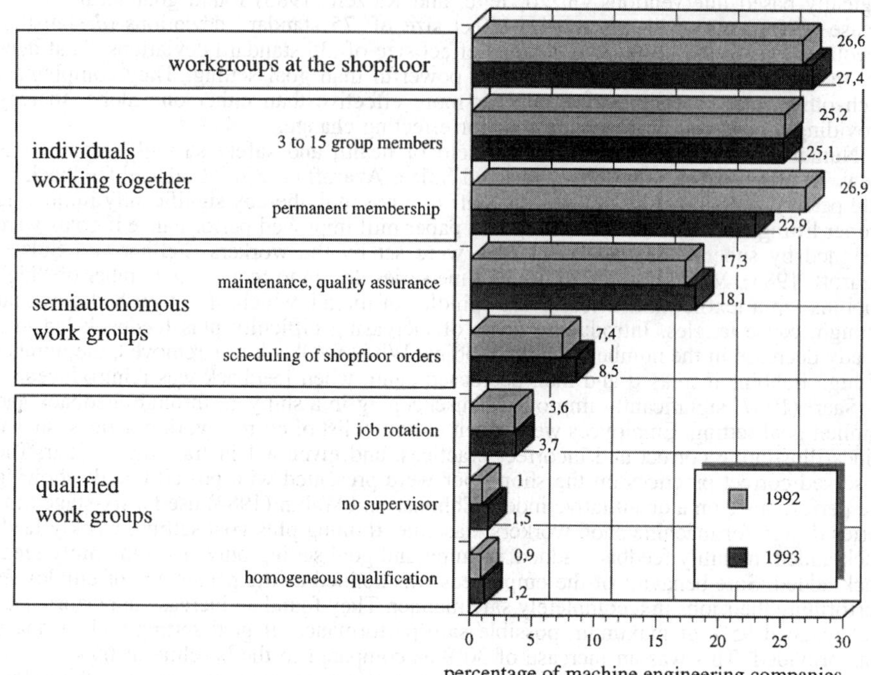

Figure 31.4 Diffusion of different types of work group in mechanical engineering industry in Germany. (After Zimolong and Saurwein, 1995.)

meet prescribed or self-assigned goals. This type of workgroup has been introduced in 9–18% of the companies. The third group is termed "qualified work group." This is the team that is often referred to in scientific discussions. Labor is self-controlled: there is a regular job rotation among the members, a rather homogeneous qualification structure, usually an elected speaker for outside communication and control, and no supervisor to the group. This type of work group has been established only in 1% of German mechanical engineering industry.

From Figure 31.4 and from data on survey results of 1991 to 1993 it can be inferred that there is no considerable growth in the diffusion of any of the workgroups so far (Zimolong and Saurwein, 1995). As a result semiautonomous and qualified workgroups in industry do not play as great a role as the frequent treatment of the subject in the literature would suggest.

Workgroups in laboratory settings have certain characteristics that distinguish them from groups in industry: They cooperate only during a short time, commitment to the goal is somewhat low, role differention is set in the beginning, and social and emotional attitudes to each other have scarcely been developed. Hence, results of performance of work groups from laboratory settings raise questions of how to generalize them for industry.

Although most of the goal setting and feedback research has focused on the impact on individual performance, a growing amount of research has been done showing positive effects for groups. Among others, Latham and Kinne (1974) and Nadler (1979) found that specific goals led to better group performance than unspecified, vague goals. Other researchers reported (Latham and Yukl, 1975) that groups performed better if their goals were difficult rather than if they were easy. Matsui, Kakuyama, and Onglacto (1987) in a laboratory experiment found that group goal setting led to higher performance than did individual goal setting. The effectiveness of task feedback in group goal setting will be maximized if feedback involves both individual and group performance information. The findings suggest that having members work as teams with a specific team goal rather than as individuals with only individual goal increases productivity. In addition, to maximize the productivity of such teams, information and control systems should contain information on team progress and individual progress.

An instructive example of the interaction of various elements of a behavioral safety management program is the study of Komaki et al. (1978). They introduced goals and feedback to improve worker safety in two departments in a food manufacturing plant. The intervention consisted of identifying safe and unsafe practices, the safe practice was introduced as the desired behavior goal, a board was posted with the baseline data of safe behavior, and the departmental goal of 90% was suggested and agreed to by the employees. Thereafter, whenever observers collected data, they posted on the graph the percentage of incidents performed safely by the group as a whole.

In addition to feedback, another planned component of the intervention program was for supervisors to recognize workers when they performed selected incidents safely. To ensure the participation of the supervisors, the president and the plant manager were asked to talk to each supervisor about the safety program at least once a week.

Following the intervention in both departments, the percentage of safe practices increased significantly. By the first week, the wrapping department of the bakery had obtained their first 100% score. During the entire intervention, no score fell below the baseline, and over half of the time, the department obtained 100% scores. Scores in the makeup department immediately rose to 100% and, with one exception, continued at this level. During the reversal phase, however, performance dropped to baseline levels (see Figure 31.5). Difficulties were encountered in implementing the recognition component of the intervention. Only 15% and 54% of the recognition checklists of the supervisors were turned in, respectively. Although management continued to give their verbal support, there were few indications that management was communicating their support to supervisory personnel.

The study reveals some of the important issues of implementing a behavioral safety program. First, a requirement analysis was performed jointly with supervisors, goal setting of 90% safe performances was suggested and agreed to by the employees, safe conduct rule was explicitly stated, and specific feedback consisting of the percentage of safe practices each week was indicated on the graph. The fact that performance returned to baseline levels during the reversal phase indicates that the program was not sufficiently supported by employees, supervisors, and management. Workers returned to their critical performances and former attitudes when feedback stopped, there was a lack of management and supervisor support of the program as indicated by the difficulties encountered

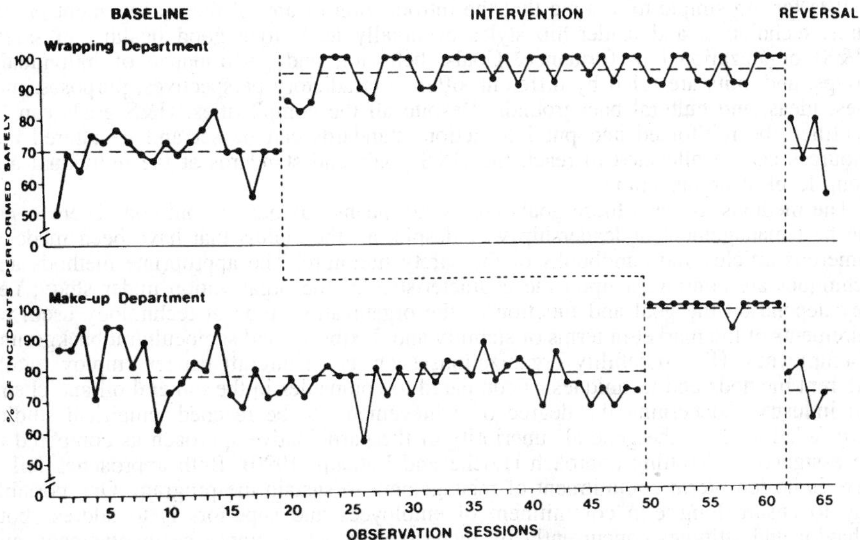

Figure 31.5 Effects of goal setting and feedback on safe performance. (From Komaki et al., 1978.)

with the recognition of workers, and the poor return of recognition checklists of supervisors and management.

Participative approaches look more promising to ensure commitment of workers and of superiors. They include: participative goal setting, establishment of problem-solving groups and use of group discussion techniques. Participative goal setting was applied to set safety improvement goals for critical behavior in a three-shift production plant (Cooper, Phillips, Sutherland, and Makin, 1994). All factory personnel, including senior management, attended their respective department's goal-setting meetings. Performance feedback was presented graphically in each department on a weekly basis. The results indicate significant improvements in safety performance, with a corresponding reduction in the plant's accident rate.

There was a lack of support, however, for an inverse relationship between actual safety performance and accident rates. It was found that the nature of the tasks mediates the relationship between safety performance and accident rates, including such factors as time pressure, manning level, and sickness absenteeism, which created a greater time pressure for workers. Nature of the task and sickness absenteeism explained 70% of the overall variance on accidents. These findings further reinforce the importance of ascertaining the effects that managerial, organizational, and job-related variables have in impacting upon workplace safety.

An intervention program using safety circles to reduce accidents with respect to housekeeping was introduced in a finnish shipyard (Saarela, 1990). Two main tasks were given to the small groups: identifying and eliminating obstacles of good order and safety and establishing new housekeeping practices in the department. Members of top management and the safety organization coordinated the program and disseminated general information concerning the program. Results indicated a 20% decrease of occupational accidents related to housekeeping during the intervention period and 1 year later after the implementation.

Brehmer (1991) reported a study using the group decision technique by Lewin to improve the commitment in safety of the drivers of a Swedish telecommunication company. Drivers met in groups of 10–15 persons to discuss general and specific problems related to safety. After the third meeting, each driver decided about his or her own future activities to solve the safety problem in question. This was only one of the conditions in the safety program, which included driver training, safety campaigns, and a bonus system for safe driving. There was a significant decrease of accident rates in the experimental group that performed under the condition of group discussion, but none in the control group. The results of this study replicate Misumi's results on bus drivers (Misumi, 1982). They show that group decisions can have considerable effects on the decrease of accidents, although it is unclear from both studies what the determining variables were.

It is far too simple to assume that the introduction of any of the management procedures, techniques, and leadership styles eventually leads to a good health and safety (H&S) organizational performance. Goals, behavior, and performance of individuals, groups, and units are ruled by different, often contradictory perspectives, purposes, motives, ideas, and cultural backgrounds. Despite all the complexities, H&S goals can be identified, be reinforced and put into action, standards can be set and be altered and resources can be allocated to reach the H&S goals and standards at the individual and group level of organizations.

The methods to reach these goals are by no means obvious or common. There is not one best management or leadership way despite all the claims that have been made in numerous articles and handbooks of the safety literature. The appropriate methods and techniques are contingent upon the characteristics of the organization under study; i.e., they depend on the goal and function of the organization, type of technology used, requirements of the market in terms of stability and flexibility, and sociocultural background of employees. High-reliability organizations such as an aircraft carrier employ totally different methods and techniques as compared to companies in the mineral oil and chemical industry. Concerning the degree of achievements to be reached, empirical studies have failed to show the general superiority of the participative approach as compared to the assigned goal-setting approach (Locke and Latham, 1990). Both approaches fail if there is no long-term commitment of management to sustain the program. One possible way to ensure long-term commitment of employees and superiors is to address both behavior and attitudes concurrently by incorporating active employee involvement, and utilizing feedback in long-term programs. This requires employees to become involved and actively participate in the identification of hazards, substandard practices and conditions, in the setting of goals, and in monitoring the safety performance of their colleagues. Management commitment and support are essential prerequisites to facilitate this type of safety intervention.

31.6 BEHAVIOR RESOURCE MANAGEMENT

Management of an organization's human resources includes such activities as job analysis, employee selection, performance appraisal, training, motivation, labor relations, decision making, and leadership. Such systems include a comprehensive job analysis to ensure the development of valid selection procedures for hiring and promotion purposes, valid performance appraisal and review systems to ensure that the person is measured on the right goals and standards and receives accurate feedback, effective training procedures to ensure that the person is adequately developed, and labor relations that are conducive to employee motivation.

A thorough job analysis is indispensable to most, if not all, human resource system because job analysis identifies the knowledge, skill, or behavior that is critical for a person to demonstrate in performing a given set of duties. With this knowledge, the performance appraisal instrument is developed. The person is then assessed in terms of the frequency with which the person demonstrates this knowledge, skills, behavior. If job analysis identifies that the person has the aptitude to do what is required but lacks the skill, a training program is needed to correct the deficiency.

A major purpose of the performance appraisal process related to health and safety is to modify safety behavior, to feed back information to the employee for counseling and development purposes so that the person will start doing or continue doing the activities critical to effectively performing on the job. The feedback must lead to the setting of and commitment to specific health and safety goals. Burke, Weitzel, and Weir (1978) showed that goal setting is a major characteristic of appraisals that are effective in bringing about a behavior change. They found that setting specific performance goals results in twice as much improvement in performance than does a discussion of general goals, or criticism without reference to specific goals. An example of an HSE review system from the mining industry is shown in Table 31.8. Only the headings of the scales are indicated, each scale included 5 to 10 items to be assessed. To some it may be surprising how little of the activities and performances addresses specific HSE goals and how much is linked to good work standards and practices. The review system strongly underline the assertion of Saari (1976) and Hoyos (1992) that personal hazard and risk control is an integral part of job activities.

A performance appraisal and review process based on behavioral observation scales (BOS) produced consistently higher levels of goal clarity, acceptance, and commitment than did a review process based on general terms and imperatives such as "Improve your sense of responsibility" or "Sharpen your leadership skills" (Tziner, Kopelman, and Livneh, 1991). Behavior based feedback pinpoints the specific desired and undesired actions

Table 31.8 Categories of a Health and Safety Review System of Safety Achievements for Managers from the Mining Industry[a]

Health and Safety Performance	Knowledge and Qualification	Leadership Skills
Planning for health and safety	Knowledge on H&S	Goal setting and reinforcement
Implementing, monitoring, and control of performance	Understanding risks within area of responsibility	Leading by example
Reviewing, auditing performance, and taking corrective actions	Understanding H&S principles	Personal safety attitude
Establishing and maintaining internal communication with managers, employees, specialists	Skills on instruction, coaching and communication	Providing rewards and sanctions
	Training of employees	Motivation of employees
	Off-the-job training	
	Providing instructions	
	On-the-job coaching and counselling	

[a]Bergwerk Rheinland, Ruhrkohle AG(1993).

to be taken or avoided. This feedback is generally more acceptable because it is seen as more factual, objective, and unbiased (Kopelman, 1986). BOS-based feedback is more conducive to setting performance goals that are specific rather than vague and job-related than nonrelated. It led to higher levels of performance and satisfaction with the review process, which includes affect toward the appraisal system, the quality of performance feedback, and the extent to which the appraisal form aids discussion of performance and how it facilitates the formulation of personal development plans.

External and internal rewards play a significant role in getting people to accept goals and motivating them to maintain goal-relevant behavior over the long term. But the key reward in organizational settings is probably not feedback but rather the consequences to which feedback leads, such as recognition, self-praise, raises, financial rewards, and promotions. It is quite clear that financial and nonfinancial incentives can indeed increase performance when the incentive system is properly designed. Guzzo et al. (1985) reported an average effect size for financial incentives of .57 standard deviations with a broad confidence interval that also included zero. They concluded that the strength of incentive effects depends heavily on the circumstances and methods of applying them. As with feedback and goal setting, literature indicates that group-based incentive systems can also be effective (Thierry, 1987).

The major findings of 24 studies that have examined the effectiveness of the use of positive reinforcement and feedback was that all studies found that incentives or feedback were successful in improving safety conditions or reducing accidents, at least on a short term (McAfee and Winn, 1989). Several studies reported situations in which safety indices did not improve. For example, Hopkins et al. (1986) found that training and praise improved the use of respirator use of only one of four sprayers (gelcoaters). Apparently, respirator use was disagreeable to the other three workers because of the discomfort and inconvenience involved. Also the question remains of whether some incentives are more effective than others.

Many consequences have been found to function effectively as reinforcers for many people. Among others, these include recognition, praise, privileges, material, or monetary rewards, which can be embedded within compensation or performance appraisal systems, and preferred activities or assignments. Unless used with care, prizes and awards soon begin to lose their appeal, whereas having one's efforts specifically and sincerely acknowledged or praised by a respected peer or supervisor will tend to keep serving as a reinforcing function. Once high performance has been demonstrated, rewards can become important as inducements to continue, but not all rewards are external. Internal, self-administered rewards that can occur following high performance include a sense of achievement based on attaining a certain level of excellence, pride in accomplishment, and feelings of success and efficacy. The experience of success will depend on reaching one's goal or level of aspiration.

In order to improve safety, organizations have traditionally relied on the use of disciplinary actions. Little evidence exists regarding the extent to which organizations use these disciplinary procedures to improve safety or whether this approach is effective. Sulzer-Azaroff (1982) reviewed the empirical evidence on the effectiveness of punishment to suppress or eliminate unsafe employee behavior. She argues that most attempts to improve safety through application of aversive consequences are not very effective. The aversive consequences are too infrequent, intermittent, delayed and often of mild intensity. The most effective application of aversive consequences requires monitoring individuals continuously to catch each unsafe act and apply aversive consequences. Obviously, in most workplaces, it would be difficult and expensive to do this. More important, this strategy is not compatible with the development and encouragement of self-responsible safety behavior.

Despite the popular assertion that training is most crucial to safety behavior, empirical research often failed support this view (Zimolong, 1992b) Training is often ineffective because it is conducted in a group setting where the members of the group are unknown to one another. Consequently, positive behavior changes that may take place during training are not reinforced by colleagues when the trainee returns to the job. In most instances specific goals for achieving or maintaining behavior change are not set, and management and supervisors do not require or support the change wanted. To overcome these limitations, an integrated management program has to be established that ensures the participation of the different departments of the company. A requirement analysis has to be conducted for the individuals in question, the requirement analysis points out the training needs, and the training goals are to be assessed after the individuals are back at their jobs. Before the requirement analysis, a hazard analysis should be conducted jointly with employees and supervisors to identify hazard control performance of incumbents. The advantage of using a joint approach is that the individual employee is involved in the hazard and job analysis that is the basis for developing a yardstick on which he or she is assessed. Moreover, the items are job related. They represent what the employees and supervisors have observed to be the critical behaviors a person must demonstrate on a given job or set of jobs to be successful. Finally, the items of the requirement analysis facilitate recognition and recall for the appraiser of what a job incumbent is does correctly or incorrectly on the job.

31.7 CONCLUDING REMARKS

Organizations that manage health and safety successfully have a number of characteristics in common. They control their health and safety risks and can display a progressive improvement in their ill health and injury record. The key elements are:

- Formulating a policy that is cost effective and aimed at achieving the preservation and development of physical and human resources and reductions in financial losses and liabilities. The policy covers all stages of the life cycle of products, processes, and services, including those to do with the selection of resources and information, the design and operation of working systems, the design and delivery of products and services, and the control and disposal of waste.
- Designing organizational structures and formulating rules to put the HSE policy into effective practice. Managers take full responsibility for controlling all those factors that could lead to ill health, injury, and loss. Organizations allocate H&S responsibilities to line managers, and specialists act as advisers. Participation in, commitment to, and involvement in H&S activities at all levels is essential, because participation complements control by encouraging responsibilities and the "ownership" of H&S policies by employees at all levels.
- Planning, measuring, and reviewing the actions to control risks means the establishment of performance standards and the measurement of performance against it. They are the basis for planning and measuring H&S achievements. Performance standards cover both organizational procedures and the control of risks. Active monitoring systems provide essential feedback on performance before an accident, ill health, or incident occurs. It involves checking compliance with performance standards and the achievement of specific objectives. Managers have the responsibilities for monitoring the achievement of those objectives and measuring compliance with those standards for which they and their workgroups or units are responsible.

The general mechanism by which H&S as well as productivity are increased is the goal-setting feedback loop. It regulates motivational and effort effects of individuals and groups to meet the standards and goals that are set by the organization. Many of the features of effective H&S management are indistinguishable from successful management practices advocated by proponents of quality and business excellence.

Organizational standards of H&S are not just determined by such factors as legal and social requirements, demands of the market, or the organization's technology and safety engineering system. Safety engineering still has a major role in creating and maintaining safe systems. An effective safety management program must ensure that the design of machinery, equipment, and tools is safe, reliable, and appropriate for the tasks to be accomplished. Workplace design that does not take ergonomic principles into account is likely to lead to an increase in errors, incidents and accidents. It eventually makes the task unnecessarily difficult, unpleasant, or dangerous.

An effective H&S program must ensure the information on performance standards and objectives and the communication and decisions related to those factors within and between groups, decision making levels, and phases of life cycle. This is only to some extent a technical problem that may be solved with the appropriate information technology. It is mainly a commitment and responsibility problem of employees at all levels of organization.

Strategies of behavioral risk management provide some guidance on this topic. Setting difficult, yet attainable goals, and providing performance feedback in relation to them, is widely accepted to be one of the most powerful behavioral modification techniques. Numerous studies conducted in the field of H&S strongly support this view not only for individuals, but also for workgroups. However, the true challenge does not relate to the short-term installation of any behavioral H&S program based on goal-setting and feedback. Programs do work on a short-term basis. If an internal member of the organization, for example the safety representative is strongly committed to the program or if there is an external consultant, the program may even survive some years. The real question is how to provide the long-term commitment of the employees and the management. There exists no one best management or leadership way to reach those goals. The appropriate methods and techniques are dependent on the characteristics of the organization, its function and goals, technology and products, market requirements, and the cultural background of the members. Participative management and equivalent leadership styles may be useful to ensure the cooperation of all members, to promote commitment and involvement in H&S activities to achieve effective risk control. Pooling knowledge and experience seems to be one of the key aspects of this approach, another is the development and maintenance of responsibilities for H&S activities at all levels of the organization. H&S management of an organization's human resources includes such activities as employee selection, performance appraisal, review process, and training based on appropriate requirement analysis. It covers decision making and the establishment of incentive systems in order to promote a positive H&S culture.

REFERENCES

Alavosius, M. P., and Sulzer-Azaroff, B. (1985). An on-the-job-method to evaluate patient lifting technique. *Applied Ergonomics, 16(4)*, 307–311.

Algera, J. A. (1990). The job characteristics model of work motivation revisited. In U. Kleinbeck, H.-H. Quast, H. Thierry, and H. Häcker, Eds., *Work Motivation* (pp. 85–104). Hillsdale NJ: Erlbaum.

Bamber, L. (1994). Risk management: Techniques and practices. In J. Ridley, Ed., *Safety at Work* (pp. 174–207). Oxford: Butterworth-Heinemann.

Bergwerk Rheinland (1993). *Beurteilungskatalog für Sicherheitsleistungen.* Moers: Ruhrkohle AG.

Bird, F. E., and Loftus, R. G. (1976). *Loss Control Management.* Loganville, GA: Institute Press.

Bird, F. E., and Germain, L. E. (1987). *Practical Loss Control Leadership.* Loganville, GA: Institute Publishing.

Brehmer, B., Gregersen, N. P., and Morén, B. (1991). *Group methods in safety work. Paper presented at Network — New Technologies and Work — Workshop, Bad Homburg, Germany 1990.*

Burkardt, F. (1981). *Information und Motivation zur Arbeitssicherheit.* Wiesbaden: Universum Verlagsanstalt.

Burke, R. J., Weitzel, W., and Weir, T. (1978). Characteristics of effective employee performance review and development interviews: Replication and extension. *Personnel Psychology, 31*, 903–919.

Cacciabue, P. C., Caprignano, A., and Vivalda, C. (1993). A dynamic reliability technique for error assessment in man-machine systems. *International Journal of Man-Machine Studies, 38*, 403–428.

Chhokar, J. S., and Wallin, J. A. (1984). Improving safety through applied behavior analysis. *Journal of Safety Research, 15(4)*, 141–151.

Cohen, A. (1977). Factors of successful occupational safety. *Journal of Safety Research, 9(4)*, 168–178.

Connolly, T., Conlon, E. J., and Deutsch, S. J. (1980). Organizational effectiveness: A multiple-constituency approach. *Academy of Management Review, 5*, 211–217.

Cooper, M. D., Philips, R. A., Sutherland, V. J., and Makin, P. J. (1994). Reducing accidents using goal setting and feedback: A field study. *Journal of Occupational and Organizational Psychology, 67*, 219–240.

Corlett, E. N., and Clark, T. S. (1995). *The Ergonomics of Workspaces and Machines.* London: Taylor and Francis.

Cox, S., and Cox, T. (1991). The structure of employee attitudes to safety: An European example. *Work and Stress, 5*, 93–106.

Department of Energy (1990). *The Public Enquiry into the Piper Alpha Disaster.* London: HMSO.

Edwards, W. (1977). How to use multiattribute utility measurement for social decision making. *IEEE Transactions on Systems, Man and Cybernetics. SMC-7*, 326–340.

Eisner, H. S., and Leger, J. P. (1988). The international safety rating system in South African mining. *Journal of Occupational Accidents, 10*, 141–160.

Embrey, D. E. (1987). SLIM-MAUD: The assessment of human error probabilities using an interactive computer based approach. In J. Hawgood and P. Humphreys, Ed. *Effective Decision Support Systems* (pp. 20–32). Aldershot (Hampshire): Technical Press.

Erke, H., and Zimolong, B. (1978). Verkehrskonflikte im Innerortsbereich. *Unfall und Sicherheitsforschung Straβenverkehr, Band 15.* Köln: Bundesanstalt für Sta βenwesen.

Etzioni, A. (1964). *Modern Organizations.* Englewood Cliffs, NJ: Prentice-Hall.

European Community (1989b). *Directive Concerning the Minimum Safety and Health Requirements For the Use of Work Equipment by Workers at Work* (No. 89/655/EEC). Luxembourg: European Community.

European Community (1989a). *Directive on the Approximation of the Laws of the Member States Relating to Machinery* (No. 89/392/EEC). Luxembourg: European Community.

European Community (1994). *EU Decree Concerning the Voluntary Participation of Companies in the Industrial Sector on a Community Environmental Management and Environmental Audit System (EMAS).* Brussels: European Community.

Fellner, D. J., and Sulzer-Azaroff, B. (1985). Occupational safety: Assessing the impact of adding assigned or participative goal-setting. *Journal of Organizational Behavior Management, 72 (1 and 2)*, 3–24.

Frantz, J. P., and Rhoades, T. P. (1993). A task analytic approach to the temporal and spacial placement of product warnings. *Human Factors, 35*, 719–730.

Grimaldi, J. V., and Simonds, R. H. (1989). *Safety Management.* Homewood, IL: Irwin.

Guzzo, R. A., Jette, R. D., and Katzell, R. A. (1985). The effects of psychologically based intervention programs on worker productivity: A meta-analysis. *Personnel Psychology, 38*, 275–291.

Grandjean, E. (1980). *Fitting the Task to the Man: An Ergonomic Approach.* London: Taylor and Francis.

Hackman, J. R., and Oldham, G. R. (1980). *Work Redesign.* London: Addison Wesley.

Hale, A. R., and Glendon, A. I. (1987). *Individual Behaviour in the Control of Danger.* Amsterdam: Elsevier.

Hale, A. R., Hemig, B., Carthey, J., and Kirwan, B. (1994). Extension of the model of behaviour in the control of danger. *Report for HSE, Volume 1 — Main Report.* Safety Science Group, Delft University of Technology, The Netherlands.

Hale, A. R., Hemig, B., Carthey, J., and Kirwan, B. (1995). *Modelling of Safety Management Systems.* Safety Science Group, Delft University of Technology, The Netherlands.

Heinrich, H. W., Petersen, D., and Roos, N. (1980). *Industrial Accident Prevention — A Safety Management Approach.* New York: McGraw-Hill.

Hix, D., and Hartson, H. R. (1993). *Developing User Interfaces.* New York: John Wiley.

Hopkins, B. L., Conrad, R. J., Dangel, R. F., Fitch, H. G., Smith, M. J. and Anger, W. K. (1986). Behavioral technology for reducing occupational exposures to styrene. *Journal of Applied Behavior Analysis, 19*, 3–11.

Hovden, J., and Larsson, T. J. (1987). Risk: Culture and concepts. In W. T. Singleton and J. Hovden, Eds., *Risk and Decisions* (pp. 47–66). New York: John Wiley.

Hoyos, C. Graf (1992). A change in perspective: Safety psychology replaces the traditional field of accident research. *The German Journal of Psychology, 16*, 1–23.

Hoyos, C., Graf (1995). Occupational safety: Progress in understanding the basic aspects of safe and unsafe behaviour. *Applied Psychology: An International Review, 44(3)*, 233–250.

Hoyos, C. Graf, Bernhardt, U., Hirsch, G. and Arnhold, T. (1991). Vorhandenes und erwünschtes Wissen in Industriebetrieben. *Zeitschrift für Arbeits- und Organisationspsychologie, 35(NF 9)*, 68–76.

Hoyos, C. Graf, and Ruppert, F. (1993). *Der Fragebogen zur Sicherheitsdiagnose (FSD).* Bern: Huber.

Hoyos, C. Graf, and Zimolong, B. (1988). *Occupational Safety and Accident Prevention. Behavioral Strategies and Methods.* Amsterdam: Elsevier Science Publisher.

Health and Safety Executive (1993a). *Successful Health and Safety Management* (Health and Safety Executive HS(G)65). London: HMSO.

Health and Safety Executive (1993b). *The Costs of Accidents at Work* (Health and Safety Executive). London: HMSO.

Humphreys, P. C., and Wisudha, A. (1983). *MAUD — An Interaction Computer Program for the Structuring Decomposition and Recomposition of Preferences Between Multiattributed Alternatives.* London: Decision Analysis Unit.

International Loss Control Institute (1990). *International Safety Rating System (ISRS).* Loganville, GA: ILCI.

Imada, A. S. (1990). Ergonomics: influencing management behavior. *Ergonomics, 33(5)*, 621–628.

Kahneman, D., Slovic, P. and Tversky, A., Eds. (1982). *Judgment Under Uncertainty.* New York: Cambridge University Press.

Katz, D., and Kahn, R. L. (1978). *The Social Psychology of Organizations* (2nd ed.). New York: John Wiley.

Kjellén, U. (1984). The deviation concept in occupational accident control — II, Data collection and assessment of significance. *Accident Analysis and Prevention, 16*, 307–323.

Kjellén, U., and Larsson, T. (1981). Investigating accidents and reducing risks — A dynamic approach. *Journal of Occupational Accidents, 3*, 129–140.

Komaki, J., Barwick, K. D., and Scott, L. R. (1978). A behavioral approach to occupational safety: Pinpointing and reinforcing safe performance in a food manufacturing plant. *Journal of Applied Psychology, 63(4)*, 434–445.

Kopelman, R. E. (1986). *Managing Productivity in Organizations.* New York: McGraw-Hill.

Latham, G. P., and Kinne, S. B. (1974). Improving job performance through training in goal setting. *Journal of Applied Psychology, 59*, 187–191.

Latham, G. P., and Yukl, G. A. (1975). Assigned versus participative goal setting with educated and uneducated wood workers. *Journal of Applied Psychology, 60*, 299–302.

Lehto, M. R., and Papastavrou, J. D. (1993). Models of the warning process: Important implications toward effectiveness. *Safety Science, 16*, 569–595.

Locke, E. A., and Latham, G. P. (1990). *A Theory of Goal Setting and Task Performance.* Englewood Cliffs NJ: Prentice-Hall.

Matsui, T., Kakuyama, T., and Onglacto, M. L. (1987). Effects of goals and feedback on performance in groups. *Journal of Applied Psychology, 72*, 407–415.

McAfee, R. B., and Winn, A. R. (1989). The use of incentives/feedback to enhance work place safety: A critique of the literature. *Journal of Safety Research, 20*, 7–19.

McCarthy, M. (1978). Decreasing the incidence of "high bobbins" in an textile spinning department through a group feedback procedure. *Journal of Organizational Behavior Management, 1*, 150–54.

Miller, D. P., and Swain, A. D. (1987). Human error and human reliability. In G. Salvendy, Ed., *Handbook of Human Factors* (pp. 219–250). New York: John Wiley.

Misumi, J. (1978). *The effects of organizational climate variables, particularly leadership variable and group decision on accident prevention.* Paper presented at the 19th International Congress of Applied Psychology, Munich, Germany.

Misumi, J. (1982). Action research on group decision making and organizational development. In H. Hiebsch, H. Brandstätter, and H.H. Kelley, Eds., *Social Psychology. Selected Revised Papers From the XXIInd International Congress of Applied Psychology.* Leipzig, GDR, July. Berlin: VEB Deutscher Verlag der Wissenschaften.

Nadler, D.A. (1979). The effects of feedback on task group behavior: A review of the experimental research. *Organizational Behavior and Human Performance, 23*, 309–338.

Nisbett, R., and Ross, L. (1980). *Human Inference: Strategies and Short Comings of Social Judgement.* Englewood Cliffs NJ: Prentice-Hall.

Oortman-Gerlings, P. D., and Hale, A. R. (1991). Certification of safety services in large Dutch industrial companies. *Safety Science, 14*, 43–59.

Packebusch, L. (1995). Gruppenbezogene Methoden in der Sicherheits- und Gesundheitsarbeit. In C. Graf Hoyos and G. Wenninger, Eds., *Arbeitssicherheit und Gesundheitsschutz in Organisationen* (pp. 197–218). Göttingen: Verlag für Angewandte Psychologie.

Palmer, T. (1990). Safety Management: Wonderland management? *Occupational Hazards, 52(4)*, 65–67.

Petersen, D. C. (1978). *Techniques of Safety Management.* New York: McGraw-Hill.

Pritchard, R. D., Jones, S. D., Roth, P. L., Stuebing, K. K., and Ekeberg, S. E. (1988). Effects of group feedback, goal setting, and incentives on organizational productivity. *Journal of Applied Psychology* (Monograph), *73*, 337–358.

Reason, J. T. (1987a). The Chernobyl errors. *Bulletin of the British Psychology Society, 40*, 201–206.

Reason, J. T. (1987b). Generic Error-Modelling System (GEMS): A cognitive framework for locating common human error forms. In K. D. Rasmussen and J. Leplat, Eds., *New Technology and Human Error* (pp. 63-83). New York: John Wiley.

Reason, J. T. (1993). Managing the management risk: New approaches to organizational safety. In B. Wilpert and T. U. Qvale, Eds., *Reliability and Safety in Hazardous Work Systems: Approaches to Analysis and Design* (pp. 7–22). Hillsdale NJ: Erlbaum.

Ridley, J., Ed. (1994). *Safety at work* (4th ed.). Oxford: Butterworth-Heinemann.

Ritter, A., and Zink, K. J. (1992). *Gruppenorientierte Ansätze zur Förderung der Arbeitssicherheit.* Berlin: Erich Schmidt.

Roberts, K. H. (1990). Some characteristics of one type of high reliability organization. *Organization Science, 1*(2), 160–176.

Ruppert, F. (1987). Gefahrenwahrnehmung — ein Modell zur Anforderungsanalyse für die verhaltensabhängige Kontrolle von Arbeitsplatzgefahren. *Zeitschrift für Arbeitswissenschaft, 41*, 84–87.

Saarela, K. L. (1990). An intervention program utilizing small groups: A comparative study. *Journal of Safety Research, 21*, 149–156.

Saari, J. (1976). Characteristics of tasks associated with the occurrence of accidents. *Journal of Occupational Accidents, 1*, 273–279.

Saari, J. (1987). Management of housekeeping by feedback. *Ergonomics, 30*(2), 313–317.

Schein, E. H. (1990). Organizational culture. *American Psychologist, 45*(2), 109–119.

Shackel, B., Ed. (1984). *Applied Ergonomics Handbook.* London: Butterworth Scientific.

Shrivastava, P. (1986). *Bhopal.* New York: Basic Books.
Smith, M. J., and Beringer, D. B. (1987). Human factors in occupational injury evaluation and control. In G. Salvendy, Ed., *Human Factors* (pp. 768-789). New York: John Wiley.
Smith, M. J., Cohen, H., Cohen, A., and Cleveland, R. (1978). Characteristics of successful safety programs. *Journal of Safety Research, 10*, 5–15.
Slovic, P. (1987). Perception of risk. *Science, 236*, 280–285.
Sulzer-Azaroff, B. (1978). The modification of occupational behavior. *Journal of Occupational Accidents, 9*, 177–197.
Sulzer-Azaroff, B. (1982). Behavioral approaches to occupational health and safety. In D. R. Frederiksen, Ed., *Handbook of organizational behavior management* (pp. 505–537). New York: John Wiley.
Sulzer-Azaroff, B., Loafman, B., Merante, R.-J., and Hlavacek, A.-C. (1990). Improving occupational safety in a large industrial plant: A systematic replication. *Journal of Organizational Behavior Management, 11(1)*, 99–120.
Steel, C. (1990). Risk estimation. *Safety Practitioner, 8(6)*, 20–21.
Steiner, I. D. (1972). *Group Processes and Productivity.* New York: Academic Press.
Svenson, O. (1989). On expert judgment in safety analyses in the process industries. *Reliablity Engineering and System Safety, 25*, 219–256.
Swain, A. D., and Guttmann, H. E. (1983). *Handbook of human reliability analysis with emphasis on nuclear power plant applications.* Washington DC: U.S. Nuclear Regulatory Commission.
Thierry, H. (1987). Payment by results systems: A review of research 1945–1985. *Applied Psychology: An International Review, 36(1)*, 91–108.
Timpe, K.-P. (1993). Psychology's contributions to the improvement of safety and reliability in the man-machine system. In B. Wilpert and T. Qvale, Eds., *Reliability and Safety in Hazardous Work Systems* (pp. 119-132). Hillsdale, NJ: Erlbaum.
Trimpop, R. M. (1994). *The Psychology of Risk Taking Behavior.* Amsterdam: Elsevier.
Tubbs, M. E. (1986). Goal setting: A meta-analytic examination of the empirical evidence. *Journal of Applied Psychology, 71(3)*, 474–483.
Tversky, A., and Kahneman, D. (1981). The framing of decisions and the psychology of choice. *Science, 211*, 453–458.
Tziner, A., Kopelman, R. E., and Livneh, N. (1991). Effects of performance appraisal format on perceived goal characteristics, appraisal process satisfaction, and changes in rated job performance: A field experiment. *Journal of Psychology, 127(3)*, 281–291.
U.S. Department of Labor. (1980). *Concept and Techniques of Machine Safeguarding* (OSHA 3067). Washington, DC: U.S. Government Printing Office.
Van Cott, H.P., and Kinkade, R. G., Eds. (1972). *Human Engineering Guide to Equipment Design.* Washington, DC: American Institutes for Research.
Veltri, A. (1990). An accident cost impact model: The direct cost component. *Journal of Safety Research, 21(2)*, 67–73.
Veltri, A. (1991). Management principles for the safety function. *Journal of Safety Research, 22(1)*, 1–10.
Wagenaar, W. A. (1992). Risk taking and accident causation. In J. F. Yates, Ed., *Risk-Taking Behavior* (pp. 257–281). Chichester: John Wiley.
Wagenaar, W. A., Groeneweg, J., Hudson, P. T., and Reason, J. T. (1994). Promoting safety in the oil industry. *Ergonomics, 37(12)*, 1999–2013.
Wagenaar, W. A., Souverijn, A. M., and Hudson, P. T. (1993). Safety management in intensive care wards. In B. Wilpert and T. Qvale, Eds., *Reliability and Safety in Hazardous Work Systems* (pp. 157–169). Hillsdale, NJ: Erlbaum.
Wilpert, B., and Klumb, P. (1993). Social dynamics, organization and management: Factors contributing to system safety. In B. Wilpert and T. Qvale, Eds., *Reliability and Safety in Hazardous Work Systems* (pp. 87–100). Hillsdale, NJ: Erlbaum.
Wollnik, M. (1988). Das Verhältnis von Organisationsstruktur und Organisationskultur. In E. Dülfer, Ed., *Organisationskultur. Phänomen-Philosophie-Technologie* (pp. 49–75). Stuttgart: Poeschel.
Wood, R. E., Mento, A. J., and Locke, E. A. (1987). Task complexity as a moderator of goal effects: A meta-analysis. *Journal of Applied Psychology, 72(3)*, 416–425.
Woodson, W. E. (1981). *Human Factors Design Handbook.* New York: McGraw-Hill.
Yates, J. F., Ed. (1992). The risk conctruct. In: J. F. Yates, Ed., *Risk-Taking Behavior* (pp. 1–25). Chichester: John Wiley.
Zimolong, B. (1981). Traffic Conflicts: A Measure of Road Safety. In H. C. Foot, A. J. Chapmann, and F. M. Wade, Eds., *Road Safety: Research and Practice.* (pp. 35–41). Eastbourne: Praeger, Holt-Saunders.
Zimolong, B. (1985). Hazard perception and risk estimation in accident causation. In R. E. Eberts and C. G. Eberts (Eds.), *Trends in ergonomics/human factors II* (pp. 463–470). Amsterdam: Elsevier.

Zimolong, B. (1992a). Empirical evaluation of THERP, SLIM and ranking to estimate HEPs. *Reliability Engineering and System Safety*, *35*, 1–11.

Zimolong, B. (1992b). Sicherheitsmanagement: Der Zusammenhang zwischen Sicherheitsorganisation, Schulung und Sicherheitsstandards. In B. Zimolong and R. Trimpop (Eds.), *6. Workshop Psychologie der Arbeitssicherheit* (pp. 85–97). Heidelberg: Asanger.

Zimolong, B., and Duda, L. (1992). Human error reduction strategies in advanced manufacturing systems. In M. Rahimi and W. Karwowski, Eds., *Human-Robot Interaction* (pp. 242–265). London: Taylor and Francis.

Zimolong, B., and Hale, A. R. (1989). Arbeitssicherheit. In S. Greif, W. Holling, and N. Nicholson, Eds., *Europäisches Handbuch der Arbeits- und Organisationspsychologie* (pp. 126–131). München: Psychologie Verlagsunion.

Zimolong, B and Saurwein, R. (1995). Maschinenbau zwischen CIM und Gruppenarbeit. *Zeitschrift für Arbeitswissenschaft*, *49*, 226–232.

Zimolong, B. and Trimpop, R. (1994). Managing human reliability in advanced manufacturing systems. In G. Salvendy and Karwowski, Eds., *Design of Work and Development of Personnel in Advanced Manufacturing Systems* (pp. 431–461). New York: John Wiley.

Zimolong, B., and Windel, A. (1996). Mit Gruppenarbeit zu höherer Leistung und humaneren Arbeitstätigkeiten? In B. Zimolong, Ed., *Kooperationsnetze, Flexible Fertigungsstrukturen und Gruppenarbeit* (pp. 140–171). Opladen: Leske and Budrich.

Zohar, D. (1980). Safety climate in industrial organizations: Theoretical and applied implications. *Journal of Applied Psychology*, *65*(*1*), 96–102.

CHAPTER 32

WORK SCHEDULES AND SUSTAINED PERFORMANCE

Donald I. Tepas
Michael J. Paley
Stephen M. Popkin
Department of Psychology
University of Connecticut
Storrs, CT 06269-1020 USA

32.1 INTRODUCTION

Although there are no current comprehensive statistics available, it is generally held that there has been a worldwide increase in the number of people who do not work regular and/or fixed diurnal hours. By one estimate, the number of individuals on shift work increased 100% between 1950 and 1974 (International Labour Office, 1978). Although night work has been practiced at least since Roman times (Scherrer, 1981), the introduction of artificial lighting has undoubtedly played an important role by making night work more practical. The current electronic revolution now appears to be refueling efforts to move workers further away from traditional daytime work. Features of the contemporary workplace that increase the need for shift work include communication and travel across multiple time zones to support the demands of international trade; development of new manufacturing processes that cannot be performed without continuous operations; the popularity of agile manufacturing and just-in-time methods; investment in automation and robotics that require around-the-clock operation for capital investment recovery; and growth in the demand for around-the-clock support services to respond to these factors. What once were optional nonusual work hours have become key and required ingredients for successful competition in the global workplace and, as such, their use will continue to increase in the future.

In the past, work schedule design was often the domain of managers who mainly worked in response to tradition and production demands. As a result, they were often ignorant of or insensitive to human ergonomic limits. Furthermore, the prevalence of shift work often varies with occupation. For example, in the United States, 64% of protective service workers but only 9% of administrative support workers are shiftworkers (U.S. Department of Labor, 1992). Data suggest that in most workplaces, the origin, justification, and impact of the work schedule used is unknown (Tepas, 1994). Finally, legislative guidelines for work schedules are either absent or limited.

Regulations governing hours of work almost always make the faulty assumption that all hours of the day can be treated as equal and interchangeable. For example, the hours of service regulations for commercial vehicle operators do not specify any differences between driving at night or during the day. In the United States, this appears to translate into millions of workers assigned to hundreds of different work schedules of unknown ergonomic merit. The magnitude of this problem is demonstrated by the Presser and Cain (1983) finding that one third of urban U.S. households include at least one shiftworker. These shiftworkers may experience a multitude of negative outcomes as well as diminished on-the-job performance associated with their work schedules. The latter has been identified as a major concern in the commercial transportation industry, where bad work schedule practices have been linked to accidents. This is a substantial problem, leading the U.S. National Transportation Safety Board to request that all appropriate government agencies review their hours of service limits in all modes of transportation (National Transportation Safety Board, 1990). Obviously, work schedule–related accidents by commercial operators place the general public at risk.

Shift work research initially focused on the problems associated with the use and unfettered spread of alternative work schedules. Often, ergonomic design efforts were simply aimed at decreasing or eliminating the practice of night work and other undesirable work scheduling practices. However, one should now view increased implementations of alternative work schedules as part of the contemporary workplace and aim to develop work arrangements that are sensitive to human limitations of working at night and for long periods of time. Despite the fact that humans are diurnal animals subject to fatigue, the demands of a global economy require schedule designs and methods that expand rather than contract work hours. It is quite clear that amateur evaluations of work schedule health and safety are often quite faulty; hence expert assessment is needed to design work schedules with consideration to both prevention and evaluation.

32.2 HISTORICAL BACKGROUND

Before 1950, the research literature on work schedule effects is limited. In the United States, most modern labor laws governing work hours have their origins before this date and were based on social policy and political factors, not scientific data. Legal constraints on night work in the United States have always been quite modest, when compared to those in other nations. By 1970, American laws covered only 4% of the workforce (Steinberg, 1982). In most cases, the studies supporting hours of service laws for commercial operators were conducted after the laws were passed and before the advent of modern chronobiological research. Unlike many industrialized nations, the U.S. government has never ratified any of the International Labour Organization conventions on night and shift work.

One might argue that the first modern study of shiftworkers in North America was published in 1965 (Mott, Mann, McLoughlin, and Warwick, 1965), and the first meeting dedicated to shift work problems was held by NIOSH in 1975 (Rentos and Shepard, 1976). European and Asian support for work schedule research appears to be more significant and long-standing. In 1957, European investigators founded what is now known as the Scientific Committee on Night and Shift Work. It was one of the first of the 25 scientific committees that make up the International Commission on Occupational Health (ICOH). In subsequent years, this ICOH committee has sponsored and published 12 international symposia and the *Shiftwork International Newsletter.* These symposia are the only dedicated and regular shiftwork meetings in the world. Selected papers from these symposia have been regularly published as books or dedicated journal issues. Clearly, the number of European and Asian publications on work schedule issues surpasses those from North America. The 1995 Foxwoods Symposium in Connecticut marked the first time the International Symposium on Night and Shiftwork was held in the Western hemisphere.

Although past history clearly shows that the primary sites for work schedule research rest in industrialized Asia and Europe, it is not appropriate to conclude that there is little need for additional research. Research has clearly demonstrated the multivariate nature of work schedule systems, and the operation of social and cultural variables. Nations and occupations vary significantly with regard to what is considered an acceptable work schedule system practice (Ong and Kogi, 1990) and their legal regulations with regard to hours of work (U.S. Congress, Office of Technology Assessment, 1991). There is much to be learned from the variety of work schedule systems practiced in other nations and the research that it has generated, but it is also clear that work schedule systems should not be imported without careful evaluation. The changes introduced by the current technological revolution make the search for better work schedule systems a global challenge. Nonetheless, it is appropriate to note that simple universal solutions to work schedule problems should not be expected, and work schedule practices will continue to vary from country to country.

32.3 TERMINOLOGY AND NOTATION

The diversity of work schedules practiced is immense. For example, U.S. fire fighters have been reported to work 150 different work schedules (Tasto and Colligan, 1977). It is reasonable to suggest that on a global basis there are thousands of different work schedules being used. This is not surprising or undesirable, given the multivariate nature of work schedule systems and the operation of social variables. The availability of a wide range of potential solutions to work schedule problems increases the options available to the ergonomic expert. Development of a good terminology and notation method is an essential ingredient to being able to apply these diverse work schedules in a reliable and valid manner.

Although the use of terminology is often consistent within a given workplace, shop-floor terms do vary from workplace to workplace, country to country, and expert to expert. Unfortunately, these differences in usage often lead to confusion and promote misinterpretation of the literature. The following three examples demonstrate these differences and the ensuing confusion. First, in the United States for some the term "swing shift" refers to any worker or group of workers who are employed on a schedule where the time of day one works changes. For others the same term is only used with regard to the afternoon–evening (second) shift. Second, U.S. workers often refer to the "night" shift as the "third" shift. Yet, this shift is sometimes called the "first," "graveyard," or "lobster" shift. Finally, most workers in the United States consider a work schedule system that changes work hours once per week to be an example of "rapid shift rotation," whereas many Europeans would reserve "rapid shift rotation" to schedules changing several times within a week.

Technical reports and journal articles often obfuscate critical issues by failing fully to define, describe, or comprehend the shiftwork terms and schedules to which they refer. Therefore, we included this section to define work scheduling terms, and demonstrate a notation method that can be used to describe most schedules. These definitions are very similar to those used in the previous edition of this handbook (Tepas and Monk, 1987), and they are used in the remainder of this chapter. Rather than assume that the terms

Table 32.1 Work Systems Definitions

Shift	The hours of a given day that an individual or a group of individuals is scheduled to be at the workplace.
Off time	The hours of a given day that an individual or a group of individuals is not normally required to be at the workplace.
Schedule	The sequence of consecutive shifts and off time assigned to a particular individual or group of individuals as their usual work assignment.
Permanent hours	A schedule for an individual or a group of individuals that does not normally require them to work more than one type of shift. That is, the time of day one works is constant.
Rotating hours	A schedule that normally requires an individual or a group of individuals to work more than one type of shift. That is, the time of day one works changes.
Basic cycle	The minimum number of days required to complete the specific sequence of shifts and off time constituting a given schedule. The number of days until a schedule begins to repeat.
Major cycle	The minimum number of days required to arrive at a point where the basic sequence of a schedule begins to repeat on the same days of the week. The number of days until the basic sequence falls on the same days when it repeats.
Work system	All of the schedule(s) implemented in a given workplace to meet the real or perceived requirements of a given plant, process or service.

used in this chapter are consistent with his or her understanding, the reader should study the terms and notation method presented to assure accurate understanding of subsequent sections of this chapter. Even more important, we approve of, recommend, and encourage use of this terminology and notation method by other authors in their future reports.

32.3.1 Work System Definitions

The work schedule definitions are presented in Table 32.1. The basic unit of these definitions is a *shift*, the time of day a worker is required to be at the workplace. By definition, all workers who are scheduled to be at a workplace on a repeated basis are shiftworkers, and it is thus proper to refer to those who work during the day as shiftworkers. Each worker has a schedule for the shifts he or she works. This schedule includes *off time*, hours (often 24 h or more for a non-workday) when the worker is not normally at the workplace. If there is no work schedule, there is no formal work schedule system. Any workplace that employs more than one individual has a formal or informal way of ensuring that all work requirements are met.

Some workers have schedules with *permanent hours*, working at the same time on each shift. Other workers have schedules with *rotating hours*, working different hours on specified days following a planned schedule. The number of days required for a worker on a given schedule to complete his or her repetitive sequence of shifts and off time is the *basic cycle* of a schedule. By definition, if a worker is employed more than 1 day, the schedule may also vary the day(s) of the week he or she works. In this case, the number of days required for the basic cycle to begin to repeat on the same days of the week is the *major cycle* of a schedule. For many schedules the basic cycle is equal to the length of the major cycle, but for others the major cycle is longer (never shorter) than the basic cycle.

The variety of work systems that might be designed and used is huge, since a work system can include many work schedules. A given work system may incorporate work schedules involving both permanent and rotating hours, as well as basic and major cycles of various lengths. Specific examples are presented in subsequent tables and text.

32.3.2 Shift Definitions

Table 32.2 provides the operational definitions for the various shift forms that are presented in this chapter. The *third shift* definition is the standard specified by the International Labour Organization (ILO) for a night shift. The remaining shift definitions use the ILO definition as a model and appear to be fairly characteristic of workplace practice in the United States. An exception to this is the definition of *irregular shifts*, which will be discussed later in this chapter. To avoid confusion with regard to overtime work and

Table 32.2 Work Shift Definitions

Three-Shift Systems:	
First shift	A work period of about 8 h in duration that generally falls between the hours of 0600 and 1700. Also known as morning or day shift.
Second shift	A work period of about 8 h in duration that generally falls between the hours of 1500 and 0100. Also known as afternoon–evening or swing shift.
Third shift	A work period of about 8 h in duration that generally falls between the hours of 2200 and 0700. Also known as night or graveyard shift.
Two-Shift Systems:	
Day shift	A work period of about 10 or more hours in duration that generally falls between the hours of 0800 and 2200.
Night shift	A work period of about 10 or more hours in duration that generally falls between the hours of 2200 and 0800.
Other Work Shift Classifications:	
Split shift	Any work period that is regularly scheduled to include two or more work periods less than 7 h, separated by more than 1 h away from work, on the same day.
Irregular shifts	Work periods that vary their shift starting time and duration in an erratic way.
Non-workday	Any calendar day in which only off time is scheduled.

to make the notation easier, definitions limit the use of the terms "day" and "night" to shifts that are significantly longer than 8 h. Users are advised to provide additional operational specifications whenever there is not an exact fit with these definitions.

It is important to note that the definitions presented do not include the term "crew." When many workers are employed on exactly the same schedule and work the same hours, they are often termed a crew. This terminology can be quite misleading, because it leads one to assume that the crew is in fact one group of people working together in an interactive and cooperative manner. In a large operation, individuals working identical schedules may never meet or interact. It is also possible that members of a crew may be assigned and scheduled to work together but for some reason they are unable to function as a team.

32.3.3 Work Schedule Characteristics

When a formal system is used to schedule workers' shifts and off time, the term *work system* is best used to describe all the shift schedules practiced in the given workplace. Four general categories of work system operations are usually noted and are defined in Table 32.3.

If a work system includes schedules for most operations 7 days a week, it is said to have *continuous* operations. A *discontinuous* operation operates less than 7 days a week, and in most cases this is a system that regularly excludes most Sunday and/or Saturday work. When a continuous or discontinuous operation regularly uses shifts longer than 8 h, they are said to have *compressed* operations. This category assumes the standard workweek is somewhere around 40 h, and that the worker can complete his or her workweek in less than 5 days. Continuous operations because of unusual events that require pro-

Table 32.3 Work Systems Operations

Discontinuous operations	A work system that does not employ around the clock and/or 7 days a week scheduling. These systems use schedules that often do not require an individual or a group of individuals to work on weekends (Saturday and/or Sunday).
Continuous operations	A work system that employs around the clock and 7 day a week scheduling. These systems use schedules that normally require an individual or a group of individuals to work some weekends.
Compressed operations	A system that uses schedules that normally include shifts of more than 8 h in length which result in a workweek of less than 5 full days of work per week.
Sustained operations	A system that allows for shifts of more than 12 h in length. These longer shifts are not usually scheduled, for they often require the worker to perform for as long as they can.

Table 32.4 Symbols for Work System Notation Tables

Symbol	Meaning
1	First shift
2	Second shift
3	Third shift
D	Day shift
N	Night shift
O	Non-workday
M	Monday
T	Tuesday
W	Wednesday
R	Thursday
F	Friday
S	Saturday
K	Sunday
S1	A designated schedule for a given work system
S2	Another schedule for the same work system
S3	Another schedule for the same work system
S4	Another schedule for the same work system
M1	The first 28 days of a given designated schedule
M2	Days 29–56 of the same schedule (follows M1)
M3	Days 57–84 of the same schedule (follows M2)
M4	Days 85–112 of the same schedule (follows M3)

longed periods of performance are sometimes termed *sustained* operations. This category often involves people ". . . performing at close to a nonstop rate for as long as they can" (Krueger, 1989, page 129).

32.3.4 Work System Notation Method

Given the definitions and terminology presented in the preceding two sections, specific work systems can be schematically presented by a notation system. Table 32.4 contains the symbols for the notation system used in this chapter. Definitions for these symbols can be found in Tables 32.1, 32.2, and 32.3. Table 32.5 is an introduction and example of how this notation system is used. This table illustrates a work system in which all workers are employed on one traditional fixed daytime work schedule. Work time for all workers is limited to 8-h daytime periods on weekdays. The table shows the work system for 16 consecutive weeks. This is a discontinuous operation with permanent hours: only one work shift is used, all workers are assigned to the same schedule, and the schedule is identical every week. In this simple but traditional example, the basic cycle and the major cycle are both 7 days because the same days of the week are worked every week.

Table 32.5 Work System with Permanent Hours and Discontinuous Workweeks Using Only One Schedule[a]

	M	T	W	R	F	S	K	M	T	W	R	F	S	K	M	T	W	R	F	S	K	M	T	W	R	F	S	K
S1M1	1	1	1	1	1	0	0	1	1	1	1	1	0	0	1	1	1	1	1	0	0	1	1	1	1	1	0	0
S1M2	1	1	1	1	1	0	0	1	1	1	1	1	0	0	1	1	1	1	1	0	0	1	1	1	1	1	0	0
S1M3	1	1	1	1	1	0	0	1	1	1	1	1	0	0	1	1	1	1	1	0	0	1	1	1	1	1	0	0
S1M4	1	1	1	1	1	0	0	1	1	1	1	1	0	0	1	1	1	1	1	0	0	1	1	1	1	1	0	0

[a]Basic cycle = major cycle = 7 days. Notation symbols for this and subsequent work system tables are found in Table 32.4.

Table 32.6 Work System with Permanent Hours and Discontinuous Workweeks Using Three Schedules[a]

	M	T	W	R	F	S	K	M	T	W	R	F	S	K	M	T	W	R	F	S	K	M	T	W	R	F	S	K
S1M1	1	1	1	1	1	0	0	1	1	1	1	1	0	0	1	1	1	1	1	0	0	1	1	1	1	1	0	0
S2M1	2	2	2	2	2	0	0	2	2	2	2	2	0	0	2	2	2	2	2	0	0	2	2	2	2	2	0	0
S3M1	3	3	3	3	3	0	0	3	3	3	3	3	0	0	3	3	3	3	3	0	0	3	3	3	3	3	0	0

[a]Basic cycle = major cycle = 7 days.

The following sections present a variety of work system shift combinations using this notation method. Readers who are familiar with the various forms work systems take may wish to skip the remainder of this section and go on to Section 32.4. Readers who are less familiar with shiftwork practices around the world are encouraged to read and study all of the succeeding sections and the related notation tables.

32.3.5 Work System Examples—Discontinuous Operations

Table 32.6 is another example of a discontinuous operation with permanent hours. In this case, three work shifts are used, workers are assigned to one of three shifts, the work hours do not vary within a given schedule, and all work shifts are on weekdays. The table shows the work schedule for four consecutive weeks for each of three schedules. Again the basic cycle and the major cycle are both 7 days because the same days are worked every week.

As in the preceding examples, the system in Table 32.7 is also limited to discontinuous operations. Unlike the preceding examples, this system has all workers on rotating hours. Three schedules are used, and the table shows the work schedule for four consecutive weeks for each of the three schedules. Each of the three schedules involves work on three different shifts, with five consecutive workdays on a specific shift, followed by two nonworkdays, then work on another shift. Thus, rotation occurs once every 7 days. In each case, rotation is from first to second to third shift. This is referred to as *forward* rotation. Again, the basic cycle and the major cycle are of equal length, but in this case they are 21 days in duration. The work system in Table 32.8 is identical to that in Table 32.7, except the direction of rotation is *backward.*

The work system in Table 32.9 provides an initial example of how rotation rate can vary. This work system is also for discontinuous operations with three rotating shift schedules. The table shows the work schedule for 16 consecutive weeks for each of the three schedules. For all three schedules, rotation is first to second to third shift (forward rotation), but rotation occurs only once every 28 days. Once again the basic cycle and the major cycle are equal, but here the length is 84 days.

32.3.6 Work System Examples—Continuous Operations

Continuous workweeks involve 7-day workplace operations. Because these work systems involve covering more hours per week, they usually include more than three work schedules. Differences in basic cycle and major cycle length are also more common. In practice, systems with rotating hours appear to be more common for continuous systems. This may be related to the fact that although permanent hours are feasible for continuous workweeks, they usually require more work schedules than a comparable rotating system.

Table 32.10 shows a European work system referred to as the *continental rota.* This continuous workweek rotating system is also known as the 2–2–3, a concise way of

Table 32.7 Work System with Rotating Hours and Discontinuous Workweeks Using Three Schedules and Forward Rotation[a]

	M	T	W	R	F	S	K	M	T	W	R	F	S	K	M	T	W	R	F	S	K	M	T	W	R	F	S	K
S1M1	1	1	1	1	1	0	0	2	2	2	2	2	0	0	3	3	3	3	3	0	0	1	1	1	1	1	0	0
S2M1	2	2	2	2	2	0	0	3	3	3	3	3	0	0	1	1	1	1	1	0	0	2	2	2	2	2	0	0
S3M1	3	3	3	3	3	0	0	1	1	1	1	1	0	0	2	2	2	2	2	0	0	3	3	3	3	3	0	0

[a]Basic cycle = major cycle = 21 days.

Table 32.8 Work System with Rotating Hours and Discontinuous Workweeks Using Three Schedules and Backward Rotation[a]

	M	T	W	R	F	S	K	M	T	W	R	F	S	K	M	T	W	R	F	S	K	M	T	W	R	F	S	K
S1M1	1	1	1	1	1	0	0	3	3	3	3	3	0	0	2	2	2	2	2	0	0	1	1	1	1	1	0	0
S2M1	2	2	2	2	2	0	0	1	1	1	1	1	0	0	3	3	3	3	3	0	0	2	2	2	2	2	0	0
S3M1	3	3	3	3	3	0	0	2	2	2	2	2	0	0	1	1	1	1	1	0	0	3	3	3	3	3	0	0

[a]Basic cycle = major cycle = 21 days.

denoting the shift and off-time sequencing. Four schedules are required for this system, and Table 32.10 shows four consecutive weeks for each of these schedules. For each schedule the basic cycle and the major cycle is 28 days.

Another European work system is shown in Table 32.11. This system is known as the *metropolitan rota*, more concisely referred to as the 2–2–2–2. The table shows eight consecutive workweeks for each of the four schedules required. For each of these schedules the basic cycle is 8 days and the major cycle is 56 days.

Shifts of 12-h duration are also used with systems involving continuous work hours. In these cases, the work system is termed a *compressed operation*, because a workweek often involves fewer than 5 days of work. Table 32.12 exhibits one of these systems, referred to as the 4–4 or the 4–4–4–4. The table shows 16 consecutive workweeks for each of the four schedules required. For each of these schedules, the basic cycle is 16 days and the major cycle is 112 days. Table 32.13 shows another compressed operations work system. This system is sometimes referred to as the 3–3 or the 3–3–3–3. The system also requires four basic schedules. Twelve consecutive workweeks for each of the schedules are presented in this table. For each of these schedules the basic cycle is 12 days and the major cycle is 84 days.

The work system shown in Table 32.14 is known as EOWEO, an acronym for "every other weekend off." This compressed operations work system is said to have originated in the United States, and it results in an equal distribution of weekend non-workdays for all workers. The system features two 3-day weekend off-time days every month for each work schedule. Again, four basic schedules with rotating work hours are used. Four workweeks for each of these schedules are shown. In this case, the basic cycle and the major cycle are both 28 days long.

32.3.7 Work System Examples—Irregular Operations

An objective definition of *irregular operations* is difficult. Some formal work systems include work schedules that vary their shift starting time and shift duration in an erratic

Table 32.9 Work System with Slowly Rotating Hours and Discontinuous Workweeks Using Three Schedules and Forward Rotation[a]

	M	T	W	R	F	S	K	M	T	W	R	F	S	K	M	T	W	R	F	S	K	M	T	W	R	F	S	K
S1M1	1	1	1	1	1	0	0	1	1	1	1	1	0	0	1	1	1	1	1	0	0	1	1	1	1	1	0	0
S2M1	2	2	2	2	2	0	0	2	2	2	2	2	0	0	2	2	2	2	2	0	0	2	2	2	2	2	0	0
S3M1	3	3	3	3	3	0	0	3	3	3	3	3	0	0	3	3	3	3	3	0	0	3	3	3	3	3	0	0
S1M2	2	2	2	2	2	0	0	2	2	2	2	2	0	0	2	2	2	2	2	0	0	2	2	2	2	2	0	0
S2M2	3	3	3	3	3	0	0	3	3	3	3	3	0	0	3	3	3	3	3	0	0	3	3	3	3	3	0	0
S3M2	1	1	1	1	1	0	0	1	1	1	1	1	0	0	1	1	1	1	1	0	0	1	1	1	1	1	0	0
S1M3	3	3	3	3	3	0	0	3	3	3	3	3	0	0	3	3	3	3	3	0	0	3	3	3	3	3	0	0
S2M3	1	1	1	1	1	0	0	1	1	1	1	1	0	0	1	1	1	1	1	0	0	1	1	1	1	1	0	0
S3M3	2	2	2	2	2	0	0	2	2	2	2	2	0	0	2	2	2	2	2	0	0	2	2	2	2	2	0	0
S1M4	1	1	1	1	1	0	0	1	1	1	1	1	0	0	1	1	1	1	1	0	0	1	1	1	1	1	0	0
S2M4	2	2	2	2	2	0	0	2	2	2	2	2	0	0	2	2	2	2	2	0	0	2	2	2	2	2	0	0
S3M4	3	3	3	3	3	0	0	3	3	3	3	3	0	0	3	3	3	3	3	0	0	3	3	3	3	3	0	0

[a]Basic cycle = major cycle = 84 days.

Table 32.10 Work System with Rotating Hours and Continuous Workweeks Using Four Schedules and Forward Rotation—The "Continental"[a]

	M	T	W	R	F	S	K	M	T	W	R	F	S	K	M	T	W	R	F	S	K	M	T	W	R	F	S	K
S1M1	1	1	2	2	3	3	3	0	0	1	1	2	2	2	3	3	0	0	1	1	1	2	2	3	3	0	0	0
S2M1	2	2	3	3	0	0	0	1	1	2	2	3	3	3	0	0	1	1	2	2	2	3	3	0	0	1	1	1
S3M1	3	3	0	0	1	1	1	2	2	3	3	0	0	0	1	1	2	2	3	3	3	0	0	1	1	2	2	2
S4M1	0	0	1	1	2	2	2	3	3	0	0	1	1	1	2	2	3	3	0	0	0	1	1	2	2	3	3	3

[a]Basic cycle = major cycle = 28 days.

way as part of an otherwise regular work system. In this case, workers are scheduled in advance to work irregular hours as part of a *mixed operation* (see the next section) in which the hours of other workers are permanent or rotate but are scheduled on a regular basis. Work schedules can also vary shift starting time and shift duration in a variable manner, which makes it quite difficult (if not impossible) for the worker to predict in advance when he or she will work and how long he or she will work. These are unpredictable irregular work hours. Table 32.15 provides an example of unpredictable irregular operations. This table presents the work schedule for a locomotive crewman whose fatal crash has been attributed to irregular work hours (National Transportation Safety Board, 1989). For Table 32.15, the actual work hours of the engineer have been adapted into the notation system of this chapter.

Although the irregular operations are more frequent than they need be and have been identified as a probable cause of accidents, there is no agreed-upon or standard way to measure the variability of a work schedule. In general, it is reasonable to conclude that highly variable schedules are easily identified by the layperson. However, it is difficult for the layperson to quantify schedule variability and predict the degree to which a given schedule constitutes a health or safety risk. For the present, risk assessment of irregular work schedules requires expert advice and analysis.

32.3.8 Work System Examples—Mixed Operations

The work systems presented thus far represent only a small number of the many work systems one might use. Within the work systems presented, the schedules are similar in that shift length is the same and all of the schedules either rotate or are permanent. More often, work schedules are mixed within one system. A work system may include work shifts of different workday lengths, both permanent and rotating work shifts, part-time and full-time workers, and other alternative work shift practices.

Table 32.16 provides an example of a mixed operation work system with continuous workweeks that includes 8- and 12-h shifts as well as full-time and part-time workers. The five shifts used in this mixed operation all have permanent hours. For all five schedules, the basic cycle and the major cycle are both 7 days in length. A variation of this schedule has 12-h shiftworkers rotating, a further mixing of operations. Another example

Table 32.11 Work System with Rotating Hours and Continuous Workweeks Using Four Schedules and Forward Rotation—The "Metropolitan"[a]

	M	T	W	R	F	S	K	M	T	W	R	F	S	K	M	T	W	R	F	S	K	M	T	W	R	F	S	K
S1M1	1	1	2	2	3	3	0	0	1	1	2	2	3	3	0	0	1	1	2	2	3	3	0	0	1	1	2	2
S2M1	2	2	3	3	0	0	1	1	2	2	3	3	0	0	1	1	2	2	3	3	0	0	1	1	2	2	3	3
S3M1	3	3	0	0	1	1	2	2	3	3	0	0	1	1	2	2	3	3	0	0	1	1	2	2	3	3	0	0
S4M1	0	0	1	1	2	2	3	3	0	0	1	1	2	2	3	3	0	0	1	1	2	2	3	3	0	0	1	1
S1M2	3	3	0	0	1	1	2	2	3	3	0	0	1	1	2	2	3	3	0	0	1	1	2	2	3	3	0	0
S2M2	0	0	1	1	2	2	3	3	0	0	1	1	2	2	3	3	0	0	1	1	2	2	3	3	0	0	1	1
S3M2	1	1	2	2	3	3	0	0	1	1	2	2	3	3	0	0	1	1	2	2	3	3	0	0	1	1	2	2
S4M2	2	2	3	3	0	0	1	1	2	2	3	3	0	0	1	1	2	2	3	3	0	0	1	1	2	2	3	3

[a]Basic cycle = 8 days; major cycle = 56 days.

Table 32.12 Work System with Rotating Hours and Continuous Workweeks Using Four Schedules and Forward Rotation—The "Compressed 4–4–4–4"[a]

	M	T	W	R	F	S	K	M	T	W	R	F	S	K	M	T	W	R	F	S	K	M	T	W	R	F	S	K
S1M1	D	D	D	D	O	O	O	O	N	N	N	N	O	O	O	O	D	D	D	D	O	O	O	O	N	N	N	N
S2M1	N	N	N	N	O	O	O	O	D	D	D	D	O	O	O	O	N	N	N	N	O	O	O	O	D	D	D	D
S3M1	O	O	O	O	D	D	D	D	O	O	O	O	N	N	N	N	O	O	O	O	D	D	D	D	O	O	O	O
S4M1	O	O	O	O	D	D	D	D	O	O	O	O	D	D	D	D	O	O	O	O	N	N	N	N	O	O	O	O
S1M2	O	O	O	O	D	D	D	D	O	O	O	O	N	N	N	N	O	O	O	O	D	D	D	D	O	O	O	O
S2M2	O	O	O	O	N	N	N	N	O	O	O	O	D	D	D	D	O	O	O	O	N	N	N	N	O	O	O	O
S3M2	N	N	N	N	O	O	O	O	D	D	D	D	O	O	O	O	N	N	N	N	O	O	O	O	D	D	D	D
S4M2	D	D	D	D	O	O	O	O	N	N	N	N	O	O	O	O	D	D	D	D	O	O	O	O	N	N	N	N
S1M3	N	N	N	N	O	O	O	O	D	D	D	D	O	O	O	O	N	N	N	N	O	O	O	O	D	D	D	D
S2M3	D	D	D	D	O	O	O	O	N	N	N	N	O	O	O	O	D	D	D	D	O	O	O	O	N	N	N	N
S3M3	O	O	O	O	N	N	N	N	O	O	O	O	D	D	D	D	O	O	O	O	N	N	N	N	O	O	O	O
S4M3	O	O	O	O	D	D	D	D	O	O	O	O	N	N	N	N	O	O	O	O	D	D	D	D	O	O	O	O
S1M4	O	O	O	O	N	N	N	N	O	O	O	O	D	D	D	D	O	O	O	O	N	N	N	N	O	O	O	O
S2M4	O	O	O	O	D	D	D	D	O	O	O	O	N	N	N	N	O	O	O	O	D	D	D	D	O	O	O	O
S3M4	D	D	D	D	O	O	O	O	N	N	N	N	O	O	O	O	D	D	D	D	O	O	O	O	N	N	N	N
S4M4	N	N	N	N	O	O	O	O	D	D	D	D	O	O	O	O	N	N	N	N	O	O	O	O	D	D	D	D

[a]Basic cycle = 16 days; major cycle = 112 days.

of a mixed schedule is shown in Table 32.17, a system for continuous operations in which all schedules have 8-h shifts.

32.3.9 Discussion

This section is included to foster a taxonomy for researchers and practitioners to communicate the makeup of various work systems. It should also impress both the practitioner and basic applied scientists with the importance of using a notation system to clearly and accurately describe the temporal characteristics of a work schedule system. Failure to use a notation system often obfuscates significant temporal variables. Further, a unified methodology allows for increased transfer of information, thereby increasing the overall knowledge base for understanding shiftwork. This system is not an evaluation methodology in that the proper evaluation of a work system includes many variables that are not obvious from the charts produced by this notation system.

Table 32.13 Work System with Rotating Hours and Continuous Workweeks Using Four Schedules and Forward Rotation—The "Compressed 3–3–3–3"[a]

	M	T	W	R	F	S	K	M	T	W	R	F	S	K	M	T	W	R	F	S	K	M	T	W	R	F	S	K
S1M1	D	D	D	O	O	O	N	N	N	O	O	O	D	D	D	O	O	O	N	N	N	O	O	O	D	D	D	O
S2M1	O	O	O	D	D	D	O	O	O	N	N	N	O	O	O	D	D	D	O	O	O	N	N	N	O	O	O	D
S3M1	N	N	N	O	O	O	D	D	D	O	O	O	N	N	N	O	O	O	D	D	D	O	O	O	N	N	N	O
S4M1	O	O	O	N	N	N	O	O	O	D	D	D	O	O	O	N	N	N	O	O	O	D	D	D	O	O	O	N
S1M2	O	O	D	D	D	O	O	O	N	N	N	O	O	O	D	D	D	O	O	O	N	N	N	O	O	O	D	D
S2M2	N	N	O	O	O	D	D	D	O	O	O	N	N	N	O	O	O	D	D	D	O	O	O	N	N	N	O	O
S3M2	O	O	N	N	N	O	O	O	D	D	D	O	O	O	N	N	N	O	O	O	D	D	D	O	O	O	N	N
S4M2	D	D	O	O	O	N	N	N	O	O	O	D	D	D	O	O	O	N	N	N	O	O	O	D	D	D	O	O
S1M3	N	O	O	O	D	D	D	O	O	O	N	N	N	O	O	O	D	D	D	O	O	O	N	N	N	O	O	O
S2M3	O	N	N	N	O	O	O	D	D	D	O	O	O	N	N	N	O	O	O	D	D	D	O	O	O	N	N	N
S3M3	D	O	O	O	N	N	N	O	O	O	D	D	D	O	O	O	N	N	N	O	O	O	D	D	D	O	O	O
S4M3	O	D	D	D	O	O	O	N	N	N	O	O	O	D	D	D	O	O	O	N	N	N	O	O	O	D	D	D

[a]Basic cycle = 12 days; major cycle = 84 days.

Table 32.14 Work System with Rotating Hours and Continuous Workweeks Using Four Schedules and Forward Rotation—The "EOWEO"[a]

	M	T	W	R	F	S	K	M	T	W	R	F	S	K	M	T	W	R	F	S	K	M	T	W	R	F	S	K
S1M1	D	D	O	O	N	N	N	O	O	D	D	O	O	O	N	N	O	O	D	D	D	O	O	N	N	O	O	O
S2M1	O	O	D	D	O	O	O	N	N	O	O	D	D	D	O	O	N	N	O	O	O	D	D	O	O	N	N	N
S3M1	N	N	O	O	D	D	D	O	O	N	N	O	O	O	D	D	O	O	N	N	N	O	O	D	D	O	O	O
S4M1	O	O	N	N	O	O	O	D	D	O	O	N	N	N	O	O	D	D	O	O	O	N	N	O	O	D	D	D

[a]Basic cycle = major cycle = 28 days.

The work systems presented in this section serve to provide examples of how the notions are applied. Equally important, these examples are points of discussion and demonstrate the variety and complexity of scheduling options. In no sense should they be viewed as a comprehensive listing of all the work systems available or used. Nor should the presented systems be viewed as ideal or even recommended approaches to the scheduling of workers. Experience and the research literature both suggest that the merit of a specific work schedule system is relative and varies as a function of human factors variables. As a result, it is appropriate to note that using the "wrong" system can diminish worker productivity, safety, and health. It also follows that a specific work schedule system may prove to be "right" in one workplace and "wrong" in another.

The complexity of the variables involved suggests that a universal and general ranking of all work schedule systems with regard to their absolute merit is not possible. In subsequent sections of this chapter we review the human factors variables that should be considered in the selection of a work schedule as well as how one might evaluate a work schedule system. The final goal is a recommended method for the selection and evaluation of work systems.

32.4 HUMAN FACTORS VARIABLES

Traditionally, night work and extended hours have been viewed as difficult and optional duty. Seniority rights and career choices were ways workers could avoid shiftwork. There were a few exceptions. As examples, nurses were needed to monitor and care for sick people 24 h a day; petroleum refineries could not be easily turned off; and search and rescue operations must proceed if victims are to survive. Within the traditional viewpoint, acute fatigue is perceived as the only significant human factors problem, and the recruitment and retention of quality workers as the major personnel problem. Research on the impact of shift work and the changes in work associated with technological developments make these traditional viewpoints an oversimplification.

In the contemporary workplace, acute fatigue is only one of many human factors concerns. Many new technologies require around-the-clock operation. Agile manufacturing methods and international operations often demand nontraditional work hours. Automation changes the typical duties of workers and often leads to production costs that are equipment intensive rather than labor intensive. In a very real sense, mastery of

Table 32.15 Irregular Work System. Actual Work Schedule of a Locomotive Engineer for 90 Days Leading to Fatal Train Accident[a,b]

	M	T	W	R	F	S	K	M	T	W	R	F	S	K	M	T	W	R	F	S	K	M	T	W	R	F	S	K
S1M1				D	1	3	O	1	2	1	O	D	2	O	2	N	N	O	D	N	O	1	O	1	1	N	3	O
S1M2	N	O	D	O	N	2	O	O	1	1	O	3	N	O	N	N	2	O	3	O	W	O	X	N	N	N	O	3
S1M3	2	1	O	D	3	O	D	2	O	Y	O	O	O	O	3	O	Z	2	O	O	O	2	O	D	O	N	N	O
S1M4	3	N	O	3																								

[a]From "Railroad accident report: Head-end collision of Consolidated Rail Corporation freight trains UBT-506 and TV-61 near Thompsontown, Pennsylvania, January 14, 1988" (Report No. NTSB/RAR-89/02); NTSB, 1989.

[b]W, Worked two shifts: one from 0000 to 0800 and another 1600 to 0030. X, Worked two shifts: one from 0100 to 0400 and another 1345 to 0400. Y, Worked two shifts: one from 0045 to 0830 and another 1630 to 0045. Z, Worked two shifts: one from 0130 to 0830 and another 1830 to 0500.

Table 32.16 Work System with Permanent Hours and Continuous Workweeks Using Five Schedules—Weekend Schedules[a]

	M	T	W	R	F	S	K	M	T	W	R	F	S	K	M	T	W	R	F	S	K	M	T	W	R	F	S	K
S1M1	1	1	1	1	1	O	O	1	1	1	1	1	O	O	1	1	1	1	1	O	O	1	1	1	1	1	O	O
S2M1	2	2	2	2	2	O	O	2	2	2	2	2	O	O	2	2	2	2	2	O	O	2	2	2	2	2	O	O
S3M1	3	3	3	3	3	O	O	3	3	3	3	3	O	O	3	3	3	3	3	O	O	3	3	3	3	3	O	O
S4M1	O	O	O	O	O	D	D	O	O	O	O	O	D	D	O	O	O	O	O	D	D	O	O	O	O	O	D	D
S5M1	O	O	O	O	O	N	N	O	O	O	O	O	N	N	O	O	O	O	O	N	N	O	O	O	O	O	N	N

[a]For all schedules, basic cycle = major cycle = 7 days.

continuous and sustained operations can increase profits if the associated human factors pitfalls are avoided. Unfortunately, many managers continue to assume that the traditional approach to shiftwork and extended operations is adequate.

Research during the past 40 years has led to major changes in how we think about work schedule issues. Studies have clearly confirmed that biology and performance vary as a function of time of day (U.S. Congress, Office of Technology Assessment, 1991). These variations are not the same as time-on-task variations (Paley and Tepas, 1994a). In addition to acute fatigue, night work has been associated with chronic effects. Significant interactions between off-the-job and on-the-job behavior have been repeatedly confirmed. The demography of shiftworkers has changed. As a result, a contemporary human factors consideration of work schedules must take a total systems approach. This section provides a review of the factors that must be considered.

32.4.1 Individual Differences

A persistent notion for many years has been the idea that some individuals are especially suited for night work. With this notion comes the idea that one might select the right workers to do shiftwork or prohibit the wrong workers from doing shiftwork. Factors considered for use in selection or prohibition include age, gender, personality, and shift preference (discussed in Section 32.5.8). In a review of individual differences in shiftwork adjustment, Monk and Folkard (1985) note that it "seems unlikely that we will ever reach a position where we can distinguish between people who are suited and those who are not suited to shift work in general" (p. 237). This conclusion continues to be appropriate.

Age

Although a criterion for selecting workers for shiftwork is not in sight, research does suggest a number of ways these variables are relevant to work system design. For example, several studies report that older shiftworkers have special problems associated with night shift work (e.g., Tepas, Duchon, and Gersten, 1993). Although this suggests that work system designs that require extensive night shift work are not appropriate when the members of the work group are older, other research is less conclusive. For example, one study showed no difference between experienced and less experienced fire fighters in their response to working on a rotating shift schedule (Paley and Tepas, 1994b). Therefore, rather than advocate age requirements for shiftwork, one should be sensitive to possible age effects. Further, this suggests that a good work schedule may become bad as a stable work force ages. Thus there is a need for evaluation of schedules over time.

Table 32.17 Work System with Permanent and Rotating Hours, Continuous Workweeks Using Four Schedules—Hybrid System[a]

	M	T	W	R	F	S	K	M	T	W	R	F	S	K	M	T	W	R	F	S	K	M	T	W	R	F	S	K
S1M1	1	1	1	1	1	1	1	0	0	1	1	1	1	1	1	1	0	0	1	1	1	1	1	1	1	0	0	0
S2M1	0	0	2	2	2	2	2	2	2	0	0	2	2	2	2	2	2	2	0	0	2	2	2	2	2	2	2	2
S3M1	3	3	0	0	3	3	3	3	3	3	3	0	0	3	3	3	3	3	3	3	0	0	3	3	3	3	3	3
S4M1	2	2	3	3	0	0	0	1	1	2	2	3	3	3	0	0	1	1	2	2	2	3	3	0	0	1	1	1

[a]S1, S2, and S3 have permanent hours, and S4 is the "continental rota." For all schedules the basic cycle and the major cycle = 28 days.

Gender

Gender difference research presents a quite different development. In the past, there have been many prohibitions against employing women on the night shift and some countries still maintain legal provisions barring women from night work. However, analysis of data from three samples of workers consistently show that the impact of the night shift on women may be less than on men (Dekker and Tepas, 1990). These findings do not suggest that women are unaffected by shiftwork or that men should not work on the night shift. Ultimately, the data point to possible differences between the sexes in their response to shiftwork, as well as to differences in the social and domestic responsibilities of the sexes, which must be considered in the design of shiftwork systems.

Morningness–eveningness

There has been considerable research on what was originally termed morningness–eveningness (Horne and Ostberg, 1976). This research is driven by the notion that individuals might be divided into two groups, evening ("owls") and morning ("larks") types, and then assigned to work shifts on the basis of this variable. In one of the more recent efforts with this venue, Smith, Reilly, and Midkiff (1989) completed a psychometric analysis of three questionnaires used to assess this dimension, and Brown (1993) has used the results to produce a scale suitable for use with industrial workers. In general, however, studies find that measures of this dimension have distributions that are nearly normal and certainly not dichotomous. Few workers are clearly identified as evening types and it would be very difficult to assemble a large number of them for night shift work (Mahan, Tepas, and Carvalhais, 1987). Therefore, the possibility of designing a work system based on morningness–eveningness principles seems inappropriate, although the practical utility may lie as a means to assess individuals who appear unable to cope with shiftwork.

32.4.2 Circadian Variation

Although there is little reason to support the proposal that workers can be selected for shiftwork on the basis of individual differences in suitability for shiftwork, one must also consider the possibility that differences in biological regulation may be related to shiftwork tolerance (Härmä, 1993). Numerous laboratory and field studies have demonstrated that people have an internal biological time keeping system. This internal system is referred to as an *endogenous clock*, oscillator, or biological clock. It has a natural cycle time of around 25 h in length (Wever, 1979). Under normal conditions, this clock is reset on a daily basis so that it has a cycle time of exactly 24 h in normal people, and it is therefore termed *circadian* (from the Latin *circa*, about, and *dies*, a day). Physical and social time cues (i.e., *exogenous clocks*), called *zeitgebers* (from the German for time giver), are responsible for these fairly modest and normal resettings of the circadian system. Research indicates that the human circadian system is normally resistant to sudden large changes in routine and is quite stable. For example, individuals placed in isolation for weeks without a clock but with total freedom to determine their activity schedule, continue to live on a self-selected routine based on a "circadian day" of about 25 h (Wever, 1979).

The circadian system produces concomitant changes in most physiological and behavioral variables. Physiological variables peak at different times of day, but under stable conditions, they are in relative synchrony with each other (Comperatore and Krueger, 1990). Both common sense and theory suggest that good health requires maintenance of this normal synchrony of the behavior rest/activity cycle (including work) with circadian biological variation. Under these conditions, circadian rhythms are said to be *entrained* or *synchronized* and body temperature is often used as a measure of this variation. Figure 32.1 is a simplified model of human circadian variation. As is shown in this figure, body temperature reaches a minimum (trough) in the early morning and a maximum (peak) about 12 h later.

Work schedules that are limited to daytime activity with nighttime rest maintain and promote this synchrony; alertness is then entrained to body temperature variation (Folkard and Monk, 1979; Monk and Embrey, 1981). That is, work can be a good *zeitgeber.* Since research indicates that many circadian biological systems are quite resilient to abrupt change in routine, synchrony of circadian systems with many work schedules, such as the third shift, may not be easily attained. As modeled in Figure 32.2, experts have proposed that exposure to some work schedules can result in a decrease in the amplitude of circadian variations (function A) or circadian rhythm *desynchronization* (function B).

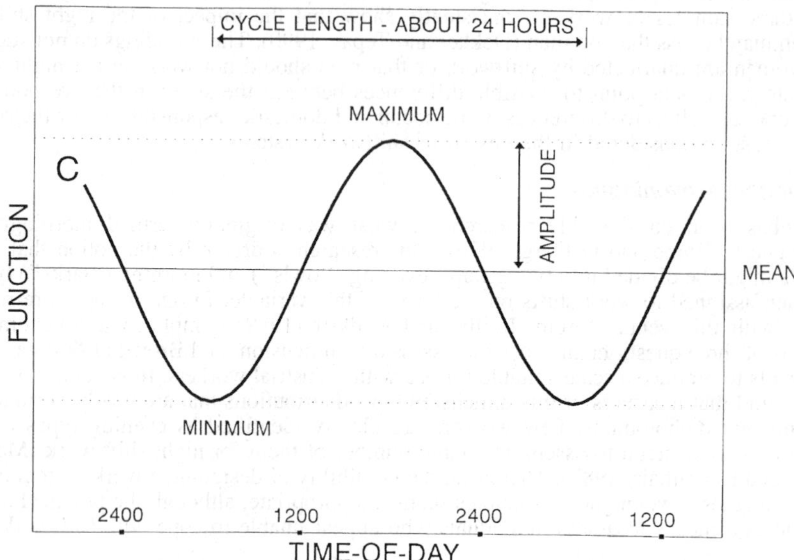

Figure 32.1 A simplified model of human circadian variation as measured by body temperature. Oral temperature is found to reach a minimum in the early morning and a maximum around 12 hours later. Alertness ratings are often entrained to these circadian variations.

Either or both of these changes in the synchronization of the circadian systems are said to show an intolerance to shift work (Reinberg et al., 1984). Both laboratory and field studies of shiftworkers suggest that over 10 days of uninterrupted nighttime work may be needed before full circadian adjustment occurs (Knauth and Rutenfranz, 1976; Åkerstedt, 1977; Dirkx, 1993; Totterdell, Spelten, Smith, Barton, and Folkard, 1995). In addition, it is also reasonable to suggest that work schedules that include work on the third shift require job performance during times around the workers' circadian trough. These times have been termed *zones of extra vulnerability*, when circadian variations make peak performance unlikely and may increase the risk of accidents (Smith, Folkard, and Macdonald, 1995). From these viewpoints, working at chronobiologically inappropriate times as well as induced alterations in circadian rhythm variation are both seen as hazards that ergonomic designers should avoid.

As mentioned, some research has shown that 10 days of night work is needed for adjustment to occur. This is not a prescription for longer periods of continuous night work, for night shift workers revert to daytime schedules on non-workdays (Tepas and Carvalhais, 1990). That is, unless a person remains on a nighttime regime on both work and non-workdays, circadian desynchronization is likely to occur. In essence, the need for 10 days suggests that full circadian adjustment is an unlikely outcome. As will be discussed in Sections 32.5.1 and 32.5.2, this is not a universally accepted position but an important issue for work schedule design.

32.4.3 Light Exposure

Given that circadian desynchronization is a common outcome of shiftwork, it should come as no surprise that during the last 15 years there has been considerable interest in the use of light to control or change circadian variation. Nonhuman animal research has clearly demonstrated that light can be a powerful *zeitgeber.* Appropriately-timed brief exposures to light can result in fairly abrupt and major changes in circadian rhythms. Recent laboratory research suggests that similar changes might be produced in humans, and thereby make adjustment to shiftwork easier. Field studies have shown that light exposure can lead to circadian rhythm change (Czeisler et al., 1990; Eastman and Miescke, 1990).

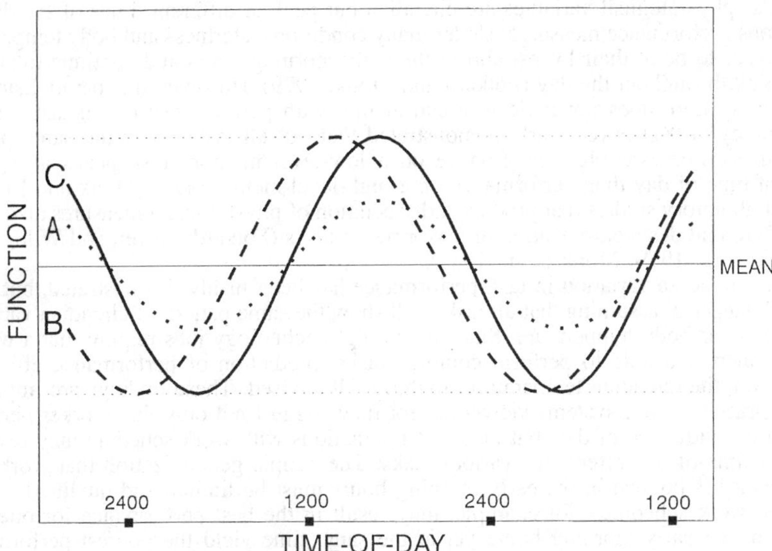

Figure 32.2 A simplified model of changes in circadian systems that may be associated with exposure to some shiftwork systems. Function C is the usual human circadian variation, as presented in Figure 32.1. Function A is an example of a decrease in the amplitude of the circadian variation, and function B is an example of circadian rhythm desynchronization.

Light exposure could be a useful easy way to produce prompt circadian entrainment to bad work schedules or sudden changes in the timing of work. If the systematic research of Eastman and her colleagues (Eastman, Stewart, Mahoney, Liu, and Fogg, 1994) is reliable and representative, circadian change requires bright light at night, dark goggles during the day, and several days of optimal exposure. In addition, this research also finds that appropriate circadian shifts are not produced in a significant proportion of the participants exposed to these treatments. Taken together, these findings suggest that the conditions required for light exposure to have a positive impact on work schedules may be low, at best. More promising results have been presented by other investigators (Czeisler et al., 1990), suggesting that further research is needed.

Assuming that light exposure may promote shiftwork adjustment, there is little agreement and even less known about the optimal light exposure parameters required for reliable impact on human circadian parameters (Rosa et al., 1990). The lighting parameters used vary from study to study with regard to intensity and exposure duration. It is even possible that light exposure, by itself, is not an adequate stimulus for shiftworker circadian change. An accurate objective interpretation of the results of these studies is impossible at the present time, since adequate and/or appropriate control groups are rarely if ever used. Since a well-lighted workplace improves visibility, there is probably little reason to suggest that light exposure schemes used in these studies have negative impact on job performance or health. In addition, the social implications of the conditions these methods may require must be fully considered before they are considered practical techniques for use by most shiftworkers.

32.4.4 Performance

Kleitman pioneered the study of circadian performance rhythms with studies of people in laboratory, cave, and submarine environments where time cues were minimal or nonexistent. Using a number of fairly simple performance tasks (such as sorting playing cards into suits) he concluded that performance varies with time of day and with body temperature (Kleitman and Jackson, 1950). Following his lead, many investigators took a position that body temperature was a reliable and simple way to index circadian changes in performance. Within this orientation, it is assumed that the body temperature model, presented in Figure 32.1, is also a model for circadian variations in job performance. As noted earlier, the contemporary view of various physiological variables is that they peak at different times of the day.

Just as physiological variables are circadian but peak at different times of the day, so do various performance measures. Under many conditions, alertness and body temperature both appear to be at their lowest ebb in the early morning hours and continue to show a strong link throughout the day (Folkard and Monk, 1979). However, this parallelism with body temperature does not indicate a causal link with performance on all tasks. Monk and Embrey (1981) have clearly demonstrated that not all tasks show the same pattern as alertness. For example, performance on a low-level memory task peaks at a much different time of day than performance on a high-level memory task (Monk and Embrey, 1981). Laboratory studies that produce a dissociation of physiological measures also result in a differential desynchronization of performance tasks (Folkard, Wever, and Wildgruber, 1983; Fröberg, 1979; Monk et al., 1983).

Thus, circadian variation in task performance has been firmly demonstrated, but there is a real danger in assuming that all tasks will show the same pattern of circadian variation as alertness or body temperature. Since many high-technology jobs require that a worker be both alert and able to perform complex tasks, prediction of performance efficiency solely from the circadian temperature rhythm is ill advised. These findings are important from a practical work systems viewpoint, for they suggest not only that worker performance varies with time of day, but also that interactions with work schedule may result in different time-of-day effects for various tasks. The simple generalization that workplace performance is poorest in the early morning hours must be limited and qualified.

Some work schedules, for example, may result in the best performance for one type of task in the early morning hours yet at the same time yield the poorest performance for another type of task at that same time of day (Folkard and Monk, 1979). In selecting or evaluating a work system, one must consider not only time-of-day variation in performance, but also the kind of work task performed. This makes the ergonomic task much more complex, emphasizing the need for specialized designers and the systematic evaluation of applications. In addition, chronic reductions of sleep length associated with permanent third shift work are associated with persistent decrements in worker performance of some tasks (Tepas, Walsh, Moss, and Armstrong, 1981). Chapter 6, *Human Error*, may provide additional relevant information on this topic.

32.4.5 Sleep

Research has clearly shown that sleep is a robust benchmark for the study and evaluation of work schedule systems. The impact of work hours on sleep is manifest in at least two ways. First, there is an acute and immediate response to a sudden change in work hours. This may occur with jet lag, changes to a new work system, or work on very irregular work schedules. It is manifest as subjective insomnia-like complaints and changes in polysomnographic sleep-stage sequencing which suggest a disturbed sleep. Second, there is a chronic response associated with extended night work exposure. In this case, the problem for the worker is getting enough sleep and his or her polysomnographic record is similar to that of a sleep-deprived person. It is this second response that troubles many experienced shiftworkers, and reductions in sleep length on workdays is a significant indicator of the problem (Tepas and Carvalhais, 1990).

The difference between acute and chronic responses is presented in the model diagrammed in Figure 32.3. This model proposes that exposure to night shiftwork is associated with what starts as a modest degree of sleep loss. Since this sleep loss is never recovered, there is a sleep loss carry-over and the debt accumulates. Within this model, it is total acute sleep loss (TASL) which increases with exposure and leads to a state of chronic sleep deprivation. It is assumed that as TASL increases there are concomitant increases in worker health and safety risks (Tepas and Mahan, 1989).

Figure 32.4 provides a good example of the sleep lengths many studies have reported for experienced shiftworkers: second-shift workers sleep the most on workdays, third-shift workers sleep the least, and first-shift workers report a sleep length that is somewhere between these two groups. For all three groups, sleep on non-workdays is significantly longer. However, the non-workday sleep lengths of these three groups do not significantly differ from each other. This, together with the finding that there is a small but positive correlation between workday and non-workday sleep length, dramatically demonstrates the impact of work and work time on sleep. At the same time these data indicate that

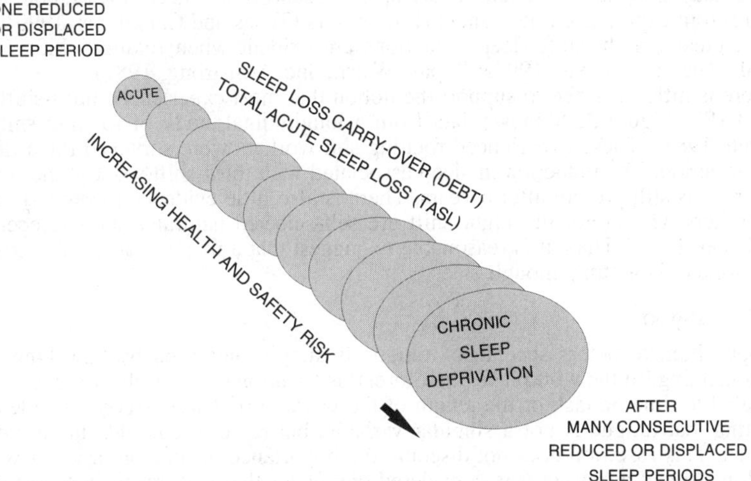

Figure 32.3 A graphical representation of the carry-over of sleep loss (debt) which increases the amount of total acute sleep loss (TASL) associated with consecutive days of reduced or displaced sleep. Chronic sleep deprivation is said to increase worker health and safety risk. (From Tepas and Mahan, 1989.)

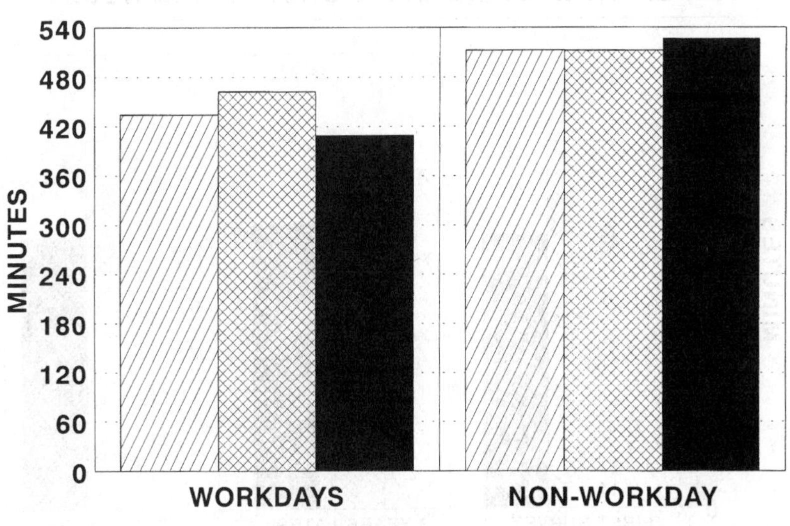

Figure 32.4 Mean workday and non-workday sleep length in minutes for permanent discontinuous hourly shiftworkers on 8-h shifts. Data graphed from a survey of 1262 workers, as reported by Tepas and Carvalhais (1990). For all three shift groups, workday sleep is significantly shorter than non-workday sleep, with third-shift workers sleeping the least. Differences between shifts for non-workday sleep are not significant.

non-workday sleep does not fully make up for lost workday sleep. Please note that these data are from experienced permanent shiftworkers (Tepas and Carvalhais, 1990) and similar but greater night shift sleep reductions are evident when rotating shiftworkers are studied (Paley and Tepas, 1994a; Tepas, Walsh, and Armstrong, 1981).

There is little evidence to support the notion that most experienced third-shift workers avoid TASL. Figure 32.5 shows data from a longitudinal study of rotating shiftworkers (Gersten, 1987). These experienced rotating shiftworkers were surveyed three times over a 6-year period. The reduction in sleep associated with third-shift work at the time of the first survey is still present after 6 years. There is also little evidence to support the notion that workers who prefer the night shift are self-selected naturally short sleepers (Tepas and Mahan, 1989). Thus, it is reasonable to suggest that a major problem for experienced night workers is getting enough sleep.

32.4.6 Fatigue

For many human factors specialists, fatigue is simply measured by how long a person does something. In the workplace, therefore, this traditional approach assumes that fatigue is equated to time on task or the length of the work shift. There is considerable evidence indicating that fatigue is not a singular variable, but rather a complex multidimensional construct. This assertion does not discount the importance of time on task, for within this multidimensional construct it is considered one factor that may result in a positive linear function such that fatigue increases as time on task increases. It is important to note that this description precludes the notion that fatigue is time on task.

A complete discussion of fatigue is beyond the scope of this chapter and the interested reader is directed to Chapter 10, *Work Physiology—Fatigue and Recovery*. However, we wish to demonstrate the multidimensional nature of fatigue. Three variables are presented as facets of fatigue: sleepiness, subjective health, and time of day. These variables were

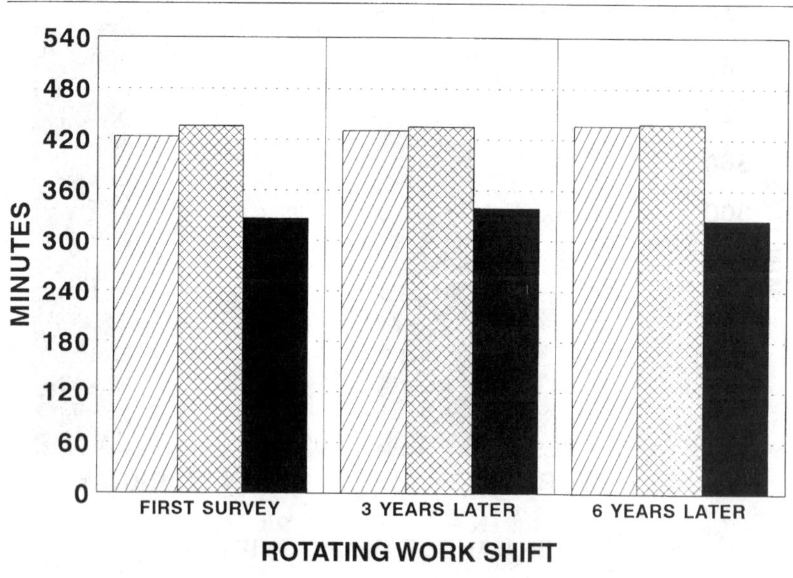

Figure 32.5 Mean workday sleep length in minutes for experienced rotating shiftworkers. Data graphed from survey reported by Gersten (1987). For these data, the same shiftworkers were surveyed three times over a 6-year period. For each survey, workday sleep length was significantly shorter on the third shift. Changes over time were not statistically significant.

chosen for each is relevant when reviewing work schedules and each helps to demonstrate that time on task alone can not describe all the processes associated with fatigue.

Sleepiness

Johnson and Naitoh (1974) suggest that the only certainty about sleep loss is an increase in sleepiness; however, they report that sleep deprivation also increases feelings of fatigue. Kribbs and Dinges (1994) point out that fatigue is a result of inadequate rest and thereby related to sleepiness and being tired. As such, one can reasonably argue that changes in sleepiness occur concomitantly with changes in fatigue. Research has shown that how physically and mentally tired a worker feels is often *not* correlated with the length of the workday (Tepas and Popkin, 1994). This indicates that time on task is not the singular causal factor. Further, as seen in the right side of Figure 32.6, a majority of workers on all shifts report feeling sleepy or tired at work, yet significantly more night workers respond in the affirmative. One explanation of this finding is that TASL, described in the previous section, is greatest among night workers and may account for these differences. Since each of the three shifts presented is 8 h long, sleep length and not time on task alone, must account for these differences.

Subjective Health

A number of health conditions are associated with fatigue, such that deteriorating health and increased fatigue have a positive correlation. Therefore, changes in subjective health ratings may be viewed as changes in fatigue. When workers are asked, "Do you think that your health would improve if you worked on a different schedule?" the results re-

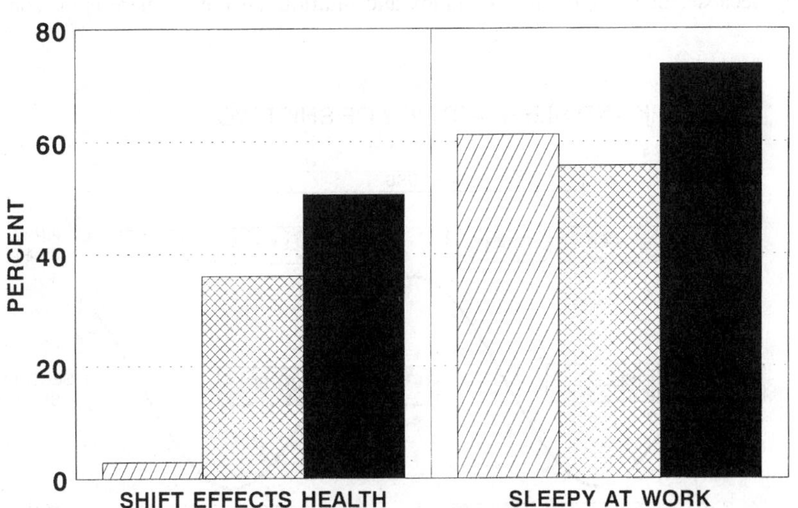

Figure 32.6 Percentage of workers responding to two survey items, one asking whether they feel that their health would improve if they worked another schedule (left graph), and the other asking about their feeling sleepy at work (right graph). These data are from a study of 1490 hourly workers on permanent discontinuous shiftwork (Tepas et al., 1985). For both variables the rate is significantly higher for third-shift workers.

semble the sleepiness question. Presented in the left side of Figure 32.6, third-shift workers respond at a significantly higher rate than those on the first or second shift. These data do not demonstrate that night workers have more health problems; rather, they indicate a link between subjective state and work schedules. Other research has shown subjective reports of well-being and mood vary with work schedule variables, further challenging the historical, singular notion of fatigue (Paley and Tepas, 1994a).

Time of Day

Unlike sleepiness or subjective health, time of day directly challenges the assumptions of a time-on-task based definition of fatigue. Figure 32.7 presents another version of the model of circadian variation presented earlier in this chapter. Superimposed on this circadian variation is a model showing sleep and work times for shiftworkers on the first, second, and third shifts, respectively. Survey data suggest that over 90% of United States' permanent and rotating shiftworkers self-select and follow this model (Tepas et al., 1985). For rotating shiftworkers, changing shifts not only changes when they work and sleep, but also the order in which these activities occur. This figure shows how biologically based circadian variation in alertness differentially affects the work and sleep behavior of shiftworkers. The way in which daily variation in alertness interacts with time-on-task changes is a function of the shift timing or, more directly, time of day. Furthermore, when discussing sleepiness, TASL was presented as a causal factor, yet time of day may also be important. One would expect sleepiness to be greater at 3:00 a.m. as compared to 11:00 a.m. The multiple influences on sleepiness also apply to fatigue. An 8-h shift starting at midnight should not and does not elicit equal amounts of fatigue as an 8-h shift starting at 9:00 a.m. This provides the work schedule designer additional areas of concern as well as the need to identify and measure interaction effects between time on task, time of day, sleep length, health, and other variables.

32.4.7 Napping

Many studies have reported that night workers have a higher incidence of napping. Despite the presence of nappers in the workforce, napping tends not to be fully understood, largely because, like fatigue, it is complex and multidimensional. There is no consensus

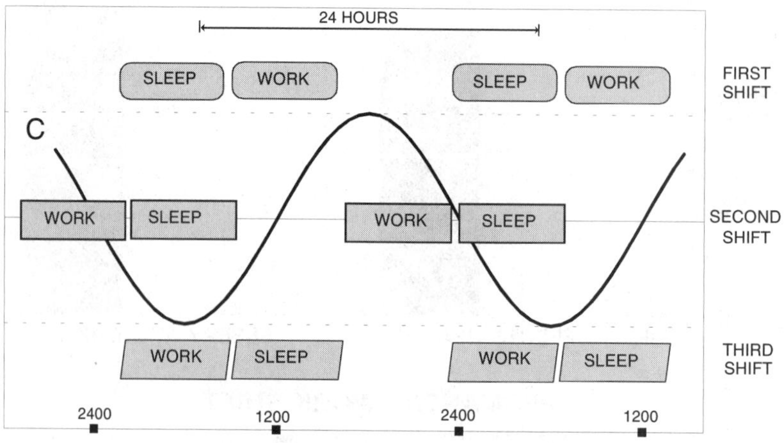

Figure 32.7 A model of when people work and sleep when on shiftwork, superimposed on the circadian model presented in Figures 32.1 and 32.2. The work and sleep times model shown in this figure fits over 90% of a total 3150 participants in two studies. These samples included workers on both rotating and permanent shiftwork, and both hourly and salary workers.

in the literature regarding the utility of napping as a coping mechanism for shiftworkers. This may rely on the fact that although napping may have some positive effects (i.e., reduction of acute sleepiness) there may also be a negative outcome (i.e., reduction of main sleep period duration). The multiple outcomes make selecting a napping strategy contingent on a systems approach that considers all elements of the work environment. It is important to note that increases in napping may be indicative of a poorly designed work schedule. In practice, strategies are often selected and combined without considering systems issues and therefore may add napping as a compensatory measure when a schedule redesign is necessary. This section will present three forms of napping strategy that merit discussion and should be considered by practitioners. These strategies are cultural, prophylactic, and replacement napping.

Cultural Napping

Cultures differ significantly in the degree to which they promote or discourage napping. Many cultures have incorporated the nap as accepted or expected behavior. The siesta of hot climates is a good example, and one investigator found that over 75% of the population he studied took naps on a regular basis (Taub, 1971). In Japan, napping on the job during work hours is a long-standing, widely practiced, and accepted tradition (Matsumoto, Matsui, Kawamori, and Kogi, 1982). In the U.S. workplace, on the other hand, sleeping on the job at anytime is often against workplace rules and can result in dismissal. Explanations for the culturalization of naps are often sermons by believers and very difficult to evaluate. Some would argue that sleep during the heat of the day is a way of avoiding the hazards of heat stress. Others would argue that napping assures quality work and avoids errors by sleepy workers. Research also suggests that some nappers are recreational nappers, who habitually nap for pleasure and differ from those who take naps for other reasons (Evans, Cook, Cohen, Orne, and Orne, 1977). In any case, a good systems approach to shiftwork should include consideration of variations in cultural napping practices.

Prophylactic Napping

Undoubtedly, some workers take naps to prepare themselves for an anticipated future period of sleep loss. This is sometimes referred to as "putting sleep in the bank." Laboratory studies of naps taken before a period of total sleep deprivation appear to yield only modest improvements in performance (Dinges, Orne, Whitehouse, and Orne, 1987). There is also some evidence that naps may have a long-term carryover (Gillberg, 1984). Casual discussions with rotating shiftworkers suggest that some of them in fact adopt a quite different strategy. That is, they sleep deprive themselves before going on the night shift to promote good sleep after their first night shift. How prevalent this strategy is remains unknown. Obviously, a late evening nap before a first night shift can have a positive and immediate affect on work performance a few hours later. However, the long-term impact of prophylactic napping on third-shift workers is open to question.

Replacement Napping

This is a napping strategy that is intended to make up previously lost sleep. When these naps are taken during a work shift, they are sometimes termed *maintenance* naps. In the United States, where sleeping on the job is most often prohibited, elevated levels of napping by shiftworkers are presumed to be efforts at replacing lost sleep. Obviously, maintenance napping is helpful when work schedules call for sustained performance or for schedules that result in chronic desynchronization. In this case, napping may improve alertness or help to anchor future sleep and thereby maintain circadian rhythm synchronization (Dinges et al., 1991; Naitoh and Angus, 1989). Just as napping is not a good way to cure insomnia, frequent napping by permanent third-shift workers appears to be a bad strategy. Third-shift workers who report that they frequently nap sleep significantly more during their main sleep period and are more likely to report problems falling or staying asleep (Tepas, 1993). Additionally, the length of *sleep inertia* (how long it takes to "wake up" and perform well) that can be tolerated by the person and TASL must be considered before concluding that napping is warranted (Naitoh, Kelly, and Babkoff, 1993). Napping, often perceived by workers as a good recovery effort, should be approached with caution since it can produce quite negative and immediate inertia performance decrements as well as lead to long-term complaints about sleep problems.

32.4.8 Social Variables

Many employers ignore the impact of off-time activity on the workplace performance of hourly workers. In recent years, a number of factors have gradually led more employers to be concerned with worker off-time activity. These factors include the increased cost of employer-paid health care; provisions to care for children and other dependents; increasing age, gender, and race diversity in the workplace; performance errors in an automated workplace, which are increasingly costly; and increases in worker turnover rate, which significantly escalate the cost of doing business. With the exception of those jobs that place the public at risk, employers cannot and should not control worker off-time activity. Employers can, however, provide a well-designed work schedule system and supporting activities that make shiftwork more usable for the majority of their workforce.

Shiftwork can have a significant positive or negative impact on worker off-time activity (Gordon, McGill, and Maltese, 1981), and the resulting off-time behavior can in turn have a significant effect on work satisfaction and performance (Thierry and Jansen, 1984). A reexamination of Figure 32.7 helps illustrate this. Many workers are job bound. They must be at work at a specific time and remain there for a given time period. Sleep is required, but workers can exercise some flexibility in determining when and how long they will sleep. The model presented in this figure is the self-selected practice of most shiftworkers, and these choices are tempered by social as well as biological factors. First-shift workers report for work shortly after they rise from their major sleep period. Workers on the second or third shift begin work after a comparatively long period of awake activity. Since third-shift workers sleep after work, their wakeup time is not job bound. Third-shift workers show the shortest sleep periods and probably experience the greatest TASL. Since an estimated 85% of these third-shift workers use an alarm or some other device to awaken (Tepas, 1982), it is reasonable to argue that social variables play a role in their reductions in sleep length. That is, the third-shift reductions in sleep length are not simply a function of endogenous circadian variables. It is reasonable to argue that most third-shift workers cut their sleep short to be with family or friends in the late afternoon and early evening, or to meet the mandatory demands of their everyday life.

Figure 32.8 provides two examples of how the home, family, and social life of workers varies dramatically with work shift schedule. These data are from permanent shiftworkers (Tepas et al., 1985). As is shown in the left half of this figure, the percentage of workers reporting that their family would prefer that they work another shift is significantly greater for second- and third-shift workers. The other side of this figure shows that fewer second- and third-shift workers report that their friends work the same shift that they do. Cross-sectional (Tepas et al., 1985) and longitudinal (Gersten, 1987) data both suggest that third-shift work may increase divorce rate. Many additional interactions between work schedule and off-time social variables have been demonstrated.

It is important to note that the interactions between work schedule and off-time variables are not all negative. For example, second- and third-shift workers in this sample appear to have less difficulty seeing a doctor about their health when not at work. Some divorced second- and third-shift workers are single parents reporting that their work schedule makes child care easier. The important thing to recognize is that worker individual differences related to social variables often determine worker suitability for a given work schedule. This in turn suggests that good work schedule system design always requires consideration of workplace, social, biological, and economic variables.

32.4.9 Diet

The timing of eating and drinking, as Figure 32.7 illustrates, cannot be the same for all work shifts. For example, third-shift workers are sleeping during the hours around noon. When compared to first-shift workers, third-shift workers report that they eat fewer meals, have poorer appetites, are less satisfied with their eating habits, and eat at different times of the day (Tepas, 1990a). These differences raise two major questions. Does nutrition vary with shift? Can eating and drinking be manipulated to improve worker adjustment to a shift?

European studies of nutrient content have failed to find any differences in the intake of shiftworkers when compared to control groups (Lennernas, 1993; Romon-Rousseaux et al., 1986). When age and other demographic variables are held constant, differences in caffeine and alcohol intake appear to be modest or nonexistent (Lennernas, 1993; Tepas, 1990a). The failure to find any differences in intake does not rule out the possibility that nutrient requirements vary with shift but are masked by other factors. It is also

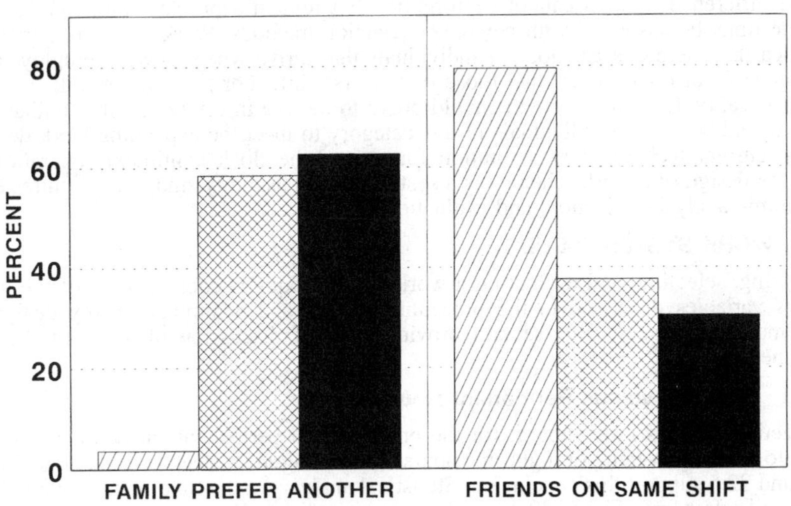

Figure 32.8 Percentage of workers responding to two survey items, one asking whether their family preferred that they work another schedule (left graph), and the other asking whether their friends worked the same shift as they do (right graph). These data are from the same study, 1490 hourly workers on permanent discontinuous shifts, as Figure 32.6. The shift difference is statistically significant for both graphs.

important to note that major cultural differences in eating habits make it difficult to arrive at any generalizable findings about the eating and drinking habits of shiftworkers (Cervinka, Kundi, Koller, Haider, and Arnhof 1984; Takagi, 1972).

Ehret (1981), Wurtman (1986), and others have suggested special diets or regimes for use by shiftworkers. These appear to be based mostly on experimentation with animals other than humans. The validity of these specific recommendations has not been satisfactorily tested with shiftworkers (Tepas, 1990a). Romon-Rousseaux, Lancry, Poulet, Frimat, and Furon, 1987) manipulated shiftworker food intake, but this isolated French study found no significant changes in mood or alertness. Although many consultants advise shiftworkers on diet issues, Popkin (1994) reported data that suggest that this sort of information is well received but does not seem to lead to any major behavioral changes. Laboratory research does indicate that meals (Smith, 1988), caffeine consumption (Walsh et al., 1990), and alcohol (Wilson, Newman, and Newman, 1956) may result in fairly immediate postconsumption positive and/or negative changes that may last for many hours. Therefore, efforts to give shiftworkers dietary advice may be meaningless if they do not include some guidance about the timing of their food intake.

32.4.10 Discussion

Advances in our understanding of work schedule systems have repeatedly confirmed the complexity of the human factors variables that must be considered in the ergonomic analysis, design, and evaluation of these systems. Circadian variations in the endogenous biological clock are closely linked with sleep behavior and variations in alertness. These three factors interact with the characteristics of the tasks performed in the workplace, the many facets of fatigue, and worker variations in napping, social activity, and diet. Given the complex nature of these interactions, what remains is the conclusion that there is no single or universal answer to the simplistic question of "What is the best work schedule?" The more appropriate question is, "What is the best work system for a specific job, for a given group of workers, at a given time, in a specific culture, and given a recognized demand for product or service?" Within this question is buried the wide range of individual differences, which cannot be forgotten but remain a problem that cannot, at the present time, be answered with personnel selection methods. Workplace lore perpetuates the idea that there is a category of individuals that thrive when they are employed on a regular basis at hours other than those of the first shift. For the present, this is a myth and not reality. Even if this myth should prove to be true in the future, it is unlikely that there would be enough individuals in this category to meet the expanding work demands of the current technological revolution for around-the-clock continuous operations. In sum, the design of a work schedule is a systems-based usability study that requires expert ergonomic analysis, selection, and evaluation.

32.5 WORK SYSTEM ISSUES

Analyzing, selecting, and evaluating a work system requires consideration of the human factors variables discussed in the preceding section and a number of key design and implementation issues. This section provides a brief discussion of some of the more prominent issues.

32.5.1 Rotating versus Permanent Hours

In practice, there is little support for the proposal that permanent shiftworkers can fully adapt to working evening or nighttime hours. Examining main sleep period length, Figures 32.4 and 32.5 support this conclusion. Statistical analysis of the time-of-day data for sleep onset and offset from Figure 32.4 is even more revealing. Like the sleep length data, non-workdays differences between the three shift schedules with regard to sleep onset and offset time are quite negligible or nonexistent. That is, on non-workdays most of these workers choose to sleep at night and be active during the day (Tepas and Carvalhais, 1990).

An accurate conclusion is that all permanent third-shift workers rotate. Their sleep/wake patterns are similar to first-shift workers when they have days off. Society is daytime oriented, and shiftworkers seemingly maintain this orientation during their non-workdays. These benchmark findings indicate that experienced shiftworkers may change their habits over time to better cope with their shift hours, but this is not the same as saying that they have typical diurnal life styles. Shiftworkers are interested in family and community activities, and they change their work/rest schedules whenever they are not bound by their work hours. A systems approach leads one to the conclusion that in the social world all night workers rotate!

If one views nighttime work hours as a hazard to be minimized, then it is appropriate to argue in favor of rotating hours. One can view rotating hours as a practical method for spreading nighttime work hours across a greater portion of the work force, thereby minimizing exposure to the individual worker. This assumes, of course, that the probable hazards associated with rotating hours are fewer than those associated with permanent third-shift work. A more moderate approach argues that rotating hours and nighttime work each have their own hazards and benefits, whose relative strength varies from job to job. This approach allows one to weigh the hazards and benefits of various schedules and select a work schedule that best matches the requirements of the job and the needs of the worker group.

32.5.2 Rotation Rate

Experts are quite divided as to what is an appropriate rotation rate, and there appear to be cultural differences in what is perceived as acceptable. In the United States, a work system like that shown in Table 32.7 is considered an acceptable but fast rotation rate, while the system in Table 32.10 is often considered to be quite unacceptable. In Europe, a work system like that shown in Table 32.7 is not considered fast rotation and is thought by many to be less acceptable than the system in Table 32.10, which is considered fast rotation. For Europeans, fast rotation usually means that a given worker rotates in some manner through first, second, and third shifts in about a week's time. In the United States, this is sometimes referred to as *swift rotation.*

Wilkinson (1992) published perhaps the most comprehensive review of the shift rotation rate literature. He concludes that "with the possible exception of personal convenience, permanent or slowly rotating night shift systems emerge superior to rapidly rotating ones as an alternative to weekly rotation" (p. 1441). Wedderburn (1992), in response to Wilkinson, argues that "a strong case for the use of permanent night shift can only be made by being selective in sources, and narrow in the range of outcomes considered" (p. 1447). He argues in favor of rapidly rotating systems. In a second response to Wilkinson, Folkard (1992) concludes that permanent shift systems may result in better performance, whereas swiftly rotating systems are perhaps more likely to make individuals happy. In his view, a work schedule like the one in Table 32.7 is worse than either of these alternatives. Williamson and Sanderson (1986) provide some data to support Folkard's contention about happy workers. They found, while holding shift length constant, improved subjective job satisfaction, health, and well-being following an increase in speed of rotation.

Both Folkard (1992) and Monk (1986) conjure positions that suggest that the search for the ideal rotation rate is somewhat like the search for one best and universal shiftwork system. That is, no general choice may ever emerge, and one should look for the best match for a given situation. A direct, controlled, and comprehensive workplace comparison of swift rotation, fast rotation, slow rotation, and permanent shift systems has never been made. This research is needed to clearly define the advantages and disadvantages that each of these approaches undoubtedly has, but this research should not be expected to lead to the identification of a single best way. The importance of worker personal satisfaction should not be ignored in considering this issue. In many countries, such as the United States, financial compensation for shiftwork duty is minimal, and an appropriate work schedule is a good way to compensate those who must work at night.

32.5.3 Direction of Rotation

Many rotating shift schedules allow the option of forward or backward rotation. Table 32.7 is an example of forward shift rotation, and Table 32.8 shows the option of backward rotation using the same basic work system form. Most field studies that change rotation direction also change other work schedule system variables at the same time. This makes it impossible to determine what impact the change in rotation has had on system operation. A study by Czeisler, Moore-Ede, and Coleman (1982) is a good example of this limitation.

Circadian rhythm laboratory research indicates that the biological clock has a forward-moving tendency that makes forward rotation better. Common sense also leads experts to argue for forward rotation, since it is easier to fall asleep when you go to bed later than when you go to bed earlier. That is, a forward rotation will push time of sleep onset forward. Workers, on the other hand, sometimes argue in favor of backward rotation since it usually provides them with a longer time period between rotation from third shift to second shift. Lavie, Tzischinsky, Epstein, and Zomer (1992) conducted a study of two groups of workers in the same factory, working on the same schedule, but rotating in opposite directions. Actigraphic data collected from these workers 6 months after one group changed to forward rotation supports the claim that forward rotation is better. This appears to be the best test of work schedule rotation direction published to date, and it continues to confirm the expert consensus that forward rotation is best. Obviously, additional research is needed, but this does appear to be one shiftwork variable that enjoys the general agreement of most investigators.

32.5.4 Length of Shift

After decades of effort by organized labor to reduce the length of the workday, there is now a popular move toward longer workdays. In most cases this is an interest in moving to 12-h-long workdays, which result in a shorter workweek and more consecutive days off. An excellent review of the literature on these compressed workweeks has been published by Duchon and Smith (1993). As one might expect, these authors conclude that compressed workweeks have both positive and negative effects. They recommend that "in industries, where accidents are a serious concern, special measures and evaluation in the use of extended workdays be considered" (p. 37).

The importance of taking a systems view, when considering the merits of a long workday, is demonstrated by two recent studies of fire fighters. Table 32.18 shows the compressed work schedule of U.S. fire fighters studied by Paley, Price, and Tepas (submitted for publication). For these workers, the day shift was 10 h long and the night shift was 14 h long; the workers were allowed to sleep at work when duty allowed; and the work schedule system was judged as acceptable for the site studied. Knauth, Keller,

Table 32.18 Work System with Rotating Hours and Continuous Workweeks Using Four Schedules—Compressed Schedule[a]

	M	T	W	R	F	S	K	M	T	W	R	F	S	K	M	T	W	R	F	S	K	M	T	W	R	F	S	K
S1M1	D	D	N	N	O	O	O	O	D	D	N	N	O	O	O	O	D	D	N	N	O	O	O	O	D	D	N	N
S2M1	N	N	O	O	O	O	D	D	N	N	O	O	O	O	D	D	N	N	O	O	O	O	D	D	N	N	O	O
S3M1	O	O	D	D	N	N	O	O	O	O	D	D	N	N	O	O	O	O	D	D	N	N	O	O	O	O	D	D
S4M1	O	O	O	O	D	D	N	N	O	O	O	O	D	D	N	N	O	O	O	O	D	D	N	N	O	O	O	O
S1M2	O	O	O	O	D	D	N	N	O	O	O	O	D	D	N	N	O	O	O	O	D	D	N	N	O	O	O	O
S2M2	O	O	D	D	N	N	O	O	O	O	D	D	N	N	O	O	O	O	D	D	N	N	O	O	O	O	D	D
S3M2	N	N	O	O	O	O	D	D	N	N	O	O	O	O	D	D	N	N	O	O	O	O	D	D	N	N	O	O
S4M2	D	D	N	N	O	O	O	O	D	D	N	N	O	O	O	O	D	D	N	N	O	O	O	O	D	D	N	N

[a]Basic cycle = 8 days; major cycle = 56 days.

Schind, and Totterdell (in press) studied German fire fighters who also worked 14 h long night shifts. All of these fire fighters were operators in a control room, answering calls, making decisions, and dispatching equipment. In this case, the work schedule was judged to be unacceptable. The recommendations of these two studies, for workers in the same occupation working 14 h night shifts, are not at odds with each other. Differences between the two groups in task assignment and work rules led to different recommendations.

Survey data reveals that workers find 12 h workdays acceptable, and at the same time they most prefer a regular workday length of 8 h (Tepas, 1990b). This inconsistency is also reflected in how field research is interpreted. For example, Rosa, Colligan, and Lewis (1989) collected extensive data showing the negative impact of 12 h shifts on the performance of control room operators in a continuous processing plant. In a second analysis of the same data, Lewis and Swain (1986) dismiss the significance of these findings and recommend adoption of extended workdays for similar operations. In a 3.5-year followup study, Rosa (1991) disagreed and has impressively revealed the persistence of these decrements in performance and alertness, which he attributes to the 12 h shifts.

Inconsistencies such as this raise a warning flag. The mix of investigators eager to adopt innovative work schedules and enthusiastic employees can easily lead some industries to overlook problems and prematurely adopt new systems. Since the interest in compressed workweeks is widespread, research on these issues continues. For organizations contemplating compressed workweeks, Duchon and Smith (1993) properly suggest that "since no *a priori* predictions from prior research can be made with certainty about the probable consequences of introducing 10- and 12-hour shifts into particular work groups, rigorous evaluation should be made on a continual basis" (p. 48).

32.5.5 Basic Cycle and Major Cycle Length

As was pointed out in the work systems presented earlier, shift schedules can vary in their basic cycle and major cycle length. As a general rule, shorter basic and major cycles are preferred over longer ones. Two concepts are relevant here. First, a short basic cycle makes it easy to track a shift schedule. That is, it makes it easy for a worker or a manager to know what shift comes next and when the change is to occur, and makes short-term off-time planning easier. The metropolitan rota shown in Table 32.11 is a good example of this. Presumably this decreases absenteeism as well as improves the quality of life of workers.

Second, a major cycle of 7 days, or a multiple of 7 days, makes it easier to plan ahead. That is, the worker can easily predict when days off will fall on a particular day of the week, making relatively long-term off-time planning easier. The work schedule system shown in Table 32.7 is a good example of such a schedule. At the other extreme, the work schedule system shown in Table 32.12 is an example of a shift schedule with a long major cycle that makes planning ahead difficult. The metropolitan rota in Table 32.11, with a short basic cycle and a long major cycle, is a good example of a work system that is said to allow good tracking but poor planning ahead.

Many work systems incorporate work schedules with relatively long basic and major cycles. Table 32.13 provides an example of a system of this sort. To overcome the problems inherent in such a system, managers sometimes issue workers a year-long pocket card or calendar with days off marked to facilitate personal planning. Although the value of such a card should not be questioned, use of a work system with a shorter cycle length is often a better solution to the problem. In most cases, it appears that a shift system that makes it difficult for a worker to track and plan ahead also makes it difficult for a manager to schedule adequate personnel on a regular basis.

In conclusion, it must be pointed out that empirical data in support of the common sense "rules" noted in this subsection are hard to come by. They may apply more to the problems associated with how to use a new system than to the ability of an individual to cope with a system once it is in practice for a period of time. It has been argued that most shiftworkers can become experts at planning around any work system (Wedderburn, 1981).

32.5.6 Hours of Service Regulations

The work schedules of commercial transportation system operators are often limited by "hours of service" regulations. In the United States most of these regulations were written decades ago in response to political practice or as manifestations of current practices. Whereas their objective is to protect the public by preventing accidents, for the most part, they are not the result of systematic research. Obviously, some regulation is needed, but the degree to which current regulations prevent accidents is not clear. The U.S. National Transportation Safety Board (1990) "Most Wanted Transportation Safety Improvements" list currently includes a request for the study and revision of these regulations in all modes of transportation. Presumably, this request has been made because current regulations are not adequate.

Most hours of service regulations share a number of faulty assumptions, since the model of the impact of work on performance was set before the discovery of biological clocks and the expansion of our knowledge about the dynamics of work, rest, and recovery. Some of these faulty assumptions are: All hours of the day can be treated as equal and interchangeable; fatigue is simply a function of the number of consecutive hours worked; recovery time is related to the number of hours worked; activities during off time are most likely to be spent resting; and acute exposure is more important than chronic exposure to bad work hours.

We now know, from research on industrial workers, that solutions to work schedule problems require a multidimensional systems approach. It is hoped that government efforts to review hours of service regulations will include a reevaluation of the historic methods used to "solve" this public problem. Until a valid approach is developed, deployment of workers in the transportation industry should be made with vigilance and caution since the failure of the current regulations has been noticed. For a further discussion of transportation issues see Chapter 59, *Human Factors in Transportation.*

32.5.7 Weekend Work

By definition, continuous workweeks require that someone work on weekends. Continuous work systems can approach this in a number of ways. Everyone can be required to work some weekends. Table 32.12 is an example of this. Another approach assigns some workers to shift schedules that only work weekends. Table 32.16 is an example of this. Other approaches limit the work of some shift schedules to weekdays in combination with other shift schedules that include both weekday and weekend shifts.

It is very important to recognize that our society is not only daytime oriented, but also weekend oriented. Important family, religious, and other social events are almost always associated with weekends, and for some workers days off on Saturdays may be more important than Sundays (Wedderburn, 1981). Thus systems that allow all workers some weekend days off are clearly preferable. A work schedule system such as EOWEO, shown in Table 32.14, is attractive to some workers because it provides all workers with 3-day weekends twice every 4 weeks on a regular and predictable basis. At the other extreme, a system such as that shown in Table 32.16, which requires workers on some shift schedules to work every weekend, may yield some special problems because it requires a commitment to work that is in conflict with community values. This can lead to increases in absenteeism and/or turnover (Tepas, Carlson, Duchon, Gersten, and Mahan, 1986).

32.5.8 Worker Preferences

The acceptability of various work schedules varies significantly from one workplace to another (Tepas et al., 1985). Further analysis of these differences supports the hypothesis that individual experience with various work schedules changes worker preferences (Tepas, 1990b). Data also indicate that workers can judge a work schedule as acceptable, and at the same time perceive that the very same schedule is having a negative impact on their quality control, workplace safety, and health (Colligan and Tepas, 1986). Numerous studies in the annals of industrial and organizational psychology have demonstrated that knowledge of how workers perceive problems and issues may be just as important as objective facts about a situation. This is clearly the case when one is evaluating or considering a change in the work schedule system installed at a given workplace.

The value of regarding the worker as an expert cannot be emphasized too much. Workers are often the best source for both demographic and preference data. Systematic surveys, interviews, record studies, and/or group discussions are all good ways of gathering information from workers. Most of these methods are also good ways of making the workforce a part of the work schedule system evaluation and selection process. As an added benefit, these methods provide a means of testing the accuracy of management and labor perceptions of worker and workplace shift preferences and problems. Viewing workers as experts minimizes the possibility that important workplace variables are overlooked by an outside expert or a zealous manager.

Care must be taken to assure that the information gathered is reliable, valid, and representative. In most cases, this requires the use of an outside third-party expert to fully and believably ensure unbiased data collection, anonymous responding, and informed consent. Finally, it must be recognized that many attitudinal measurements have little meaning unless they can be compared to equivalent norms or standards gathered from other workplaces. This is also known as benchmarking. Worker cooperation in the collection of information and the installation of changes in work schedule systems is very important. This is best ensured by informing workers in advance that they will have access to group data and project reports.

32.5.9 Education Programs

Workers have often been inserted into new work systems with little preparation. Often, workers change shifts, or workplaces change work systems, with little advance planning or notice. Instant shift schedule changes, or unilateral work schedule implementations, invite disaster and probably add to operational costs by promoting absenteeism, tardiness, and worker turnover. Educational programs may minimize the trauma and cost of change. In principle, these programs can take a variety of formats such as lectures, group discussion, videotapes, and special literature. In practice, many efforts at this are informational, but they may not be educational (Tepas, 1993).

In most cases, educational programs for shiftworkers have two objectives. First, a program explains the system used (or to be used) to the people involved in a particular workplace. This includes a full description of the work schedule system used, an explanation of why it is being used, a discussion of what is expected from the system and the workforce, and details of the implementation process. This ensures that workers know when they are to be at work, minimizes absences, and decreases complaints. Second, an educational program can provide workers with advice and information on what they, their employers, and the people they live with might do at work, in their off time, and on days off to overcome the potential problems or hazards associated with their work schedule system.

Many educational programs have been developed for shiftworkers. Their distribution and use appears to be growing. Some of these programs have good face validity, yet most of them appear to violate the principles recommended for the design of a successful training program (Tepas, 1993; see also Chapter 16, *Instructional Training Models*). Often, these programs and materials appear to be designed as substitutes for needed improvements in work scheduling practices, attempting to transfer all solutions to the problems of shiftwork within the domain of the worker. To the best of our knowledge, no systematic research has been published that demonstrates the validity of any of the commercial programs for shiftworkers. It is quite possible that none of them change behavior or improve coping.

Quality and time-intensive educational programs for shiftworkers may prove to be helpful. In an abstract, Carlson (1990) has reported on data from a noncommercial 6-week intensive training program for shiftworkers that appears to have produced modest changes in their behavior. In no sense does this isolated report justify the conduct of educational programs that do not include a solid effort to evaluate their impact on the worker. Popkin (1994) has published data that shed light on the popularity of shiftworker educational programs. Survey data collected from shiftworkers given a tailored educational program showed that over 90% of the workers would recommend the program to others, but they reported that they made little use of the coping strategies presented within that program.

32.5.10 Health Status

Many shiftwork experts are convinced that some work schedule systems lead to health problems. Evidence to support this proposal has been, in the past, quite modest. In fact, an early study by Taylor and Pocock (1972) concluded that shiftwork had no adverse effect upon mortality. Poole, Wright, and Nattrass (1992) continue to assert that there are no long-term ill effects of shiftwork. Epidemiological studies of shiftworkers are difficult, rarely done, and never perfect. Nonetheless, there is significant reason for concern (Waterhouse, Folkard, and Minors, 1993). European studies (Angersback et al., 1980; Koller, Kundi, and Cervinka, 1978) firmly demonstrate that shiftworkers are sick more than day workers and suggest that health problems are difficult to demonstrate since many of those who cannot tolerate shiftwork drop out, leaving a population of survivors.

More recent Scandinavian studies provide the most solid evidence to date that cardiovascular disease is linked to shiftwork exposure (Knutsson, 1989). This research cannot be considered conclusive, in a global sense, since cultural factors and shift design features cannot be isolated. Taken together, these studies do suggest that ergonomic designers of shiftwork systems must attend to issues related to health status. The fact that some drug susceptibility and illness symptoms have circadian variations presents another reason for considering health issues (Reinberg, 1974). A conservative precautionary position is warranted. Individuals dependent upon drugs and/or suffering from disease may be at additional risk if they are required to work during the nighttime.

In the United States, many employers appear to ignore this issue. Recent employment legislation regarding individuals with disabilities may, in fact, make it impossible to deny shiftwork to ill or diseased individuals. If this should prove true, good ergonomic work schedule system design will become even more important and may reduce employer health care costs. In Germany and other countries, special periodic health checks for employed workers doing nighttime work are recommended and practiced (Rutenfranz, 1982). At the very least, workers should be advised to inform their physician that they are shiftworkers, and physicians employed by employers should be informed when plans are being made to change or install shiftwork operations.

In the United States, permanent nighttime work is most frequently assigned to new employees by default, since the premium pay for nighttime work is small and experienced workers often use their seniority to bid out when possible. In practice, this sometimes means that those who do third-shift work are mainly the youngest, fittest, and most recently hired. In many industrialized nations, premium pay for third-shift work is high and experienced workers sometimes use their skill and tenure to obtain this work. It is quite reasonable to argue that low or no premium pay for third-shift work is a good preventive measure in that it discourages susceptible workers from exposing themselves to a hazard. One might also argue that rotating shiftwork decreases exposure time and therefore decreases the risk to workers. However, there is no definitive data, to date, that suggest that rotating shiftworkers have a lower risk for illness or disease than do permanent third-shift workers.

32.5.11 Discussion

In this section we have *briefly* discussed some of the key shiftwork issues that the human factors and ergonomic professional should address when evaluating or designing a work schedule system. Most of these issues are complex and interactive. One must approach these issues with a systems perspective. A comprehensive review of these issues is beyond the scope of this handbook, and additional issues await discovery. In many cases, experts are divided in their opinions as to how a potential hazard should be addressed. The fact remains that a good work system solution in one location may be a bad solution in another, and laboratory research recommendations may not be relevant or practical in the field. Above all, the designer is always faced with a world of compromise, where one is seeking to avoid pitfalls rather than expecting perfect solutions. Whenever possible, a quantitative evaluation should be part of a proposed solution.

32.6 WORK SYSTEM METHODOLOGY

So far, this chapter has focused on terminology and the human factors dimensions of work schedule systems. The remaining section of this chapter provides an outline of the broader domains, evaluation methods, and systems concerns that the ergonomist should consider in examining the hours of work in most organizations.

32.6.1 Evaluation of Existing Work Systems

Some of the domains which should be checked in evaluating the net value of a given work system are listed in Table 32.19. This is not intended to be a comprehensive list for all workplaces, but it does include the major factors that should be considered when present. It should be recognized that each of the items on this list has a human element

Table 32.19 Checklist of Domains To Be Evaluated

Community work system practices	What work systems are practiced by other operations in the immediate community? How well do they seem to work? How accepting is the community to these work systems?
Corporate work system practices	What work systems are practiced at other operations within the corporation? How well are they working?
Plant work system history	What work systems have been practiced in this plant? When were they practiced? How well did they work? Why were they changed?
Local job market	What is the past, current and future status of the local job market? What kinds of skills and expertise are/are not available in the workforce?
Organized labor	What does the union contract say about hours of work? Are adequate mechanisms in place to allow input and comment from union representatives?
Health and safety	What are the local and national occupational health and safety regulations that apply to this installation? Are there any environmental regulations which are relevant?
Plant personnel records	Do these records contain demographic, health and work performance information that might be helpful? Is shift assignment information in the files?
Local utility costs	Do local utility costs vary with time of day or day of the week? How do utility costs vary with volume used?
Legal requirements	What local and national work hour and wage legislation apply to this installation? What is their impact?
Maintenance requirements	How is equipment maintenance handled and does it have any restrictions? How is building maintenance handled?
Supervision and manpower requirements	Does the work system practiced interact with the manpower and supervision requirements of this installation?
Shipping and receiving	Are adequate shipping and receiving services available? Do costs and services vary with time of the day or day of the week?
Support services	Are the medical, food, transportation, and other installation and community support services adequate?
Productivity indicants	What objective measures of productivity are used at this installation, and do they allow for adequate evaluation of shift schedule differences?
Worker participation	Are adequate mechanisms in place to allow representative and cooperative input and comment from all workers?
Demand for product	What are the characteristics of the demand for the goods or services produced by this installation? What is the anticipated future market like? Is it a stable market, or is it subject to cyclic changes?

that interacts with many of the other factors associated with a given work system. Equipment maintenance is a good example of this.

A good equipment preventive maintenance program is required for the long-term satisfactory operation of a continuous work system. For some work systems, the only possible way to handle the preventive maintenance required for operation is via the purchase of duplicate equipment so that production can continue during maintenance. For other work systems, some downtime must be built into the work system so maintenance can be performed. With this solution, maintenance workers usually end up with an odd schedule and some production time is lost, but additional equipment is not required. In some cases it may be possible to change the preventive maintenance requirements so this can be done during the regular work hours with minimal production loss and little equipment duplication. A good evaluation effort requires that all reasonable maintenance alternatives be reviewed and evaluated. Because people do the actual maintenance and because maintenance tasks may have an inflexible schedule of their own, each of these alternatives has a financial cost and a human element that should be reviewed.

Each of the items in Table 32.19 should be reviewed before electing to keep or change an existing work schedule system. As stated, a human element is present for each of these items and most include a financial cost. Many of these domains require skills beyond those of the ergonomist, or data not available to an outside expert. Thus, by necessity, this review and evaluation process requires a team of relevant individuals, many of whom are employees of the workplace being evaluated. A team effort like this increases workplace participation in the evaluation and often serves as a good way to educate these participants with regard to the complexity of making work schedule system decisions.

32.6.2 Designing New Work Systems

The evaluation process described in the previous subsection is the foundation for designing a new work system. Surveys, interviews, focus groups, and other methods can be used to gather needed information. (For a more detailed examination of information gathering, see Chapter 43, *Data Collection and Evaluation of Outcome Measures.*) Examination of each of the domains listed in Table 32.19 will provide information needed to select possible arrangements for work schedule use, as the first step in an iterative design process. Once a number of potential systems have been proposed for work schedule use, a second iteration should begin, reviewing in more detail and precision the financial and human costs associated with each of the schedules under consideration. A third iteration of this process usually leads to a final design for a new work system. An iterative approach is needed to assure that the best possible selection is made. It is all too easy to select a particular work schedule system prematurely based on the financial rewards, perhaps overlooking long-term human element costs that may negate these profits.

The design and selection process should also include the development of an evaluation scenario. Table 32.20 presents a practical design evaluation methodology scenario. This methodology uses a survey for evaluation and provides a forum for discussion of significant issues. Other methods and designs may be more appropriate in some situations, and the use of control groups is desirable if practical. In any case, the primary importance of gaining individual worker cooperation in assessing human factors variables cannot be overstated. Workers and managers must understand what is being considered, agree as to how it is to be done, and be willing to participate and cooperate. A survey is not an election, but rather a fact-finding mission. One cannot provide every worker with all of the information needed to allow him or her to elect a solution. On the other hand, one can gather information from nearly every worker to provide managers with a fair and accurate view of the human element when they make their final decision. For a survey to provide this function, adequate response rates are needed. Rates of 90% or higher are realistic and to be expected when confidentiality is practiced and workers are offered participation in the survey design (Tepas et al., 1985).

32.6.3 Implementing Work Systems

The importance of evaluation, before and after implementing a new work schedule system, cannot be stressed too much. Obviously, errors in the financial evaluation of such items as utilities and maintenance can be disastrous. Of equal importance, however, are costly human factors errors that may lead to accidents, increased health care costs, high turnover rates, and labor problems (Imberman, 1983). It is easy to overlook these potential human factors losses, since they take time to develop and are sometimes masked by other developments. A preimplementation evaluation plan is a major part of a good work schedule

Table 32.20 An Evaluation Methodology Scenario

Step 1	A review of this methodology with all parties, revision of the methodology if needed, and an agreement to participate in the implementation of the agreed-upon methodology.
Step 2	Development of a survey to collect worker self-reports of their demographics, work system history, perceptions, preferences, and health status. Offering the survey developed to all workers for their voluntary, anonymous, and confidential responses.
Step 3	Analysis of the survey data and development of work system recommendations grounded in human factors and ergonomics. Recommendations presented to all parties.
Step 4	A financial and technical evaluation of all work system recommendations, and the development of a final recommendation. Acceptance by all parties of this work system recommendation and a schedule of implementation.
Step 5	If a new work system is recommended and agreed to, define criteria for use in a future evaluation of the success or failure of the new system. Agreement upon a schedule for implementation of the new work system. Selection of a time point at which an evaluation of the new work system will be made.
Step 6	Development and implementation of an instructional program for all workers. This is primarily aimed at teaching the workers how to use the new work system as well as how they might best cope with it.
Step 7	If a new work system is being installed, implement the work system following the agreed schedule.
Step 8	Following installation of the new work system, evaluation of that system is scheduled, and subsequent development of a future plan of action is initiated.

system evaluation in that these data provide the ergonomist with baseline data for later comparison. The repeated measures design is also an extremely valuable tool when faced with no control group.

As was noted in previous subsections and Table 32.20, a training program is an important part of implementing a new work system. Workers and managers alike must know what the new system is, what is expected of them, and how they might elect to cope with these changes. Many training programs offer instruction and information in the name of education. Data collected in evaluating and designing the system being implemented should provide a good needs assessment to aid in the design of a truly educational program. Since it is not clear whether this in fact happens, the evaluation process should include an evaluation of the training program used as well. Although managers often provide workers with extensive training with regard to job skills, many ignore the fact that learning how to cope with shiftwork is also a job skill. All too often, managers get caught up in a rush to implement a new work schedule system quickly and thereby ignore this important step.

32.6.4 Pitfalls To Be Avoided

The range of work schedules and systems available for potential implementation is immense. These systems include a host of interacting biological, psychological, social, and industrial variables. Given the current level of our understanding of these interactions, it should be obvious that work schedule system design is now as much art as it is science. Table 32.21 lists some of the factors within individuals that work schedule evaluators and designers should consider as potential problem areas. The factors listed are quite general and one may not assume that they are simple danger signals for all forms of shiftwork.

Age and gender provide two examples of factors that may not be the pitfalls many think they are. There is evidence to support the proposition that older workers find it difficult to tolerate third-shift work (Tepas et al., 1993), but there is recent evidence demonstrating that older workers have no more difficulty with 12-h shifts than younger workers (Keran, Duchon, and Smith, 1994). Paley and Tepas (1994b) also present data that demonstrate no differences in outcome variables between older and younger fire fighters on a rotating work schedule. For years, the International Labour Organization has had conventions limiting third-shift work by women (Kogi and Thurman, 1990), but data demonstrate that the impact of the third shift on the sleep of women may be less than the impact on men (Dekker and Tepas, 1990). Other similar examples are evident in the literature.

The importance of evaluation is clear, and the factors in Table 32.21 do provide a list of variables to be monitored in evaluating work schedule systems. As is true in all areas

Table 32.21 Factors Within an Individual That Are Likely To Cause ShiftWork Coping Problems

Age
Gender
Moonlighting
Workload at home and on the job
Circadian biological clock
Sleep disorders
Frequent napping
Physical illness and disease
Psychological characteristics
Addictions
Constipation, pain, chronic medical conditions
Social rhythms, roles, and responsibilities
Work/rest schedule experience
Religious beliefs

of ergonomic investigation, care must be taken to avoid the traditional pitfall such as investigator bias and expectancy, Hawthorne effects, halo errors, and other related effects. Since many of the hazards of shiftwork are associated with long-term chronic effects, special effort should be made to ensure that the evaluation schedule allows sufficient time for chronic symptoms to develop, appear, and be measured. Although there are no definitive data available, it is suggested that a new work system should be fully implemented for at least 6 months before any attempt is made to measure chronic, long-term effects.

The BEST network, a team of distinguished European shiftwork experts, developed what they term "guidelines for shiftworkers" (Wedderburn, 1991). Modified and tempered by an American viewpoint, Table 32.22 lists our version of the BEST list of key issues regarding the design of a shift system. For the most part, these are factors associated with work systems and work that are likely to cause shiftwork coping problems. A listing like this is a helpful guide, and this is a good list. It should not, however, be viewed as a list of absolute prohibitions. Taken together, Tables 32.21 and 32.22 provide one with helpful aids for spotting potential trouble areas and avoiding pitfalls.

32.6.5 Computer-Assisted Methods

Designing new work schedule systems and assigning people to specific slots within a work schedule system have one important dimension in common. Both take considerable time and patience. As a result there are a number of ongoing efforts to provide good computer-assisted scheduling (CAS) methods which would make these tasks easier. Since assigning people to specific work schedule slots is a large and repetitive problem, software for this appears to have developed without any serious consideration or understanding of the human factors aspects of CAS. Work schedule designers, on the other hand, are considering ergonomic criteria, but they find CAS to be a much more difficult and complicated task than it first appeared (Nachreiner, Grzech-Sukalo, Qin, Moehlmann, and Will, 1995).

Given the complexity of most shiftwork issues, this is a job that needs good CAS. It is much too early to evaluate these efforts fully. Clearly, CAS with little or no consideration of worker ergonomics is not acceptable and can lead to serious errors and problems. On the other hand, ergonomically sound CAS may not be possible at the present time due to gaps in our knowledge about shiftwork. For the present, CAS efforts should be monitored closely and used with caution. CAS may evolve as a wonderful solution to a complex and common problem or the simple-minded dispenser of bad work schedules.

32.7 CONCLUSIONS

Shiftwork is not a new idea, and it does involve a significant segment of the workforce in industrially developed countries. For decades, shiftworkers were ignored and their problems drew minimal attention. In recent years advances in chronobiology and ergonomics have generated research that allows the arguable proposal that shiftwork can place

Table 32.22 Key Issues Regarding the Design of a Shift System[a]

1. Number of consecutive night shifts
   - Permanent night shifts can be hazardous and require special consideration and limitation.
   - Although open to debate, current research seem to point in favor of fewer (2–4) consecutive night shifts worked.
2. The scheduling of work
   - The length of a work shift should be dependent on the type of tasks being performed and workload on the job.
   - Excessive numbers of consecutive workdays (more than 8) should be scheduled with caution, and screening workers for physical and mental stress may be necessary.
   - There is no firm evidence supporting either direction; however, there seems to be some advantage to rotating schedules in a forward direction.
3. The scheduling of off time
   - The purpose of off time is to allow workers to recover from the fatiguing effects of work, to sleep, and to socialize. Off time ideally should be scheduled to allow all three activities to occur.
   - The amount of off time should increase as does the number of consecutive workdays and the length of the work shifts increase.
   - A shift system should aim to maximize the amount of off time between work shifts.
   - Travel time to and from work must be considered when scheduling off time between work shifts.
   - For social reasons, shift systems should aim to include some weekend off time for all workers.
4. Considerations for time of day
   - Do not treat all hours of the day the same way. Acceptable daytime work shifts may prove to be excessive during the night.
   - Maximize the opportunity for workers to sleep during the night. Early morning shifts, for example, tend to lead to reductions in sleep length.
5. Consideration of psychosocial variables
   - Keep the rotations schedules as regular as possible. It is ideal that a worker need not check the schedule each day to figure out whether he or she must work.
   - Whenever possible allow for flexibility. This may include organizational designs such as flex time or simply allowing workers to exchange shifts with coworkers.
   - Minimize changes in shift schedules.

[a]These guidelines are an adapted and modified version of those published by the European Foundation for the Improvement of Living and Working Conditions (Wedderburn, 1991).

workers at risk. Shiftwork does change people, and many of these changes do not go away with exposure to shiftwork. This research has also demonstrated that scheduling work hours and minimizing this risk is a complex task because of a host of interacting variables.

We are now in a technological revolution where computer programs can and are scheduling work and people. Frequently this is done without fully addressing the impact of the resulting hours of work on the human element. The "misapplication hypothesis" proposes that the misapplication of computers decreases productivity (Tepas, 1994). This may become reality if the new technology simply creates more jobs that are characterized by long work hours, irregular hours of service, and/or inappropriate work schedules. Although our knowledge of work schedule systems does not as yet allow off-the-shelf delivery of systems with a proven positive impact, we do have the methodology needed to ergonomically design appropriate work schedule systems and evaluate their operation.

The time demands of international business, the development of new continuous process methods, and the use of just-in-time techniques all make it quite clear that there will be an increase in around-the-clock operations as the technological revolution continues. More shiftwork will be one of the end results of these developments. Just as contemporary research has demonstrated the risks of shiftwork, good shiftwork management has the potential of providing organizations with benefits that will protect the health of workers and make operations safe and productive. It has been said that mastery of the clock was the key to the industrial revolution. Similarly, mastery of the biological clock may be the key to the technological revolution that is now underway.

REFERENCES

Åkerstedt, T. (1977). Inversion of the sleep wakefulness pattern: Effects on circadian variations in psychophysiological activation. *Ergonomics, 20*, 459–474.

Angersbach, D., Knauth, P., Loskant, H., Karvonen, M. J., Undeutsch, K. and Rutenfranz, J. (1980). A retrospective cohort study comparing complaints and diseases in day and shiftworkers. *International Archives of Occupational and Environmental Health, 45*, 127–140.

Brown, F. M. (1993). Psychometric equivalence of an improved basic language morningness (BALM) scale using industrial populations within comparisons. *Ergonomics, 36*, 191–197.

Carlson, M. L. (1990). *Sleep-management training: An intervention program to improve the sleep of shift workers*. Unpublished doctoral dissertation, Illinois Institute of Technology, Chicago.

Cervinka, R., Kundi, M., Koller, M., Haider, M., and Arnhof, J. (1984). Shift related nutrition problems. In A. A. I. Wedderburn and P. A. Smith, Eds., *Psychological approaches to night and shift work: International research papers*. Edinburgh: Heriot-Watt University.

Colligan, M. J., and Tepas, D. I. (1986). The stress of hours of work. *American Industrial Hygiene Association Journal, 47*, 686–695.

Comperatore, C. A., and Krueger, G. P. (1990). Circadian rhythm desynchronosis, jet lag, shift lag, and coping strategies. In A. J. Scott, Ed., *Shiftwork*. Philadelphia: Hanley and Belfus.

Czeisler, C. A., Johnson, M. P., Duffy, J. F., Brown, E. N., Ronda, J. M., and Kronauer, R. E. (1990). Exposure to bright light and darkness to treat physiological maladaptation to night work. *New England Journal of Medicine, 322*, 1153–1159.

Czeisler, C. A., Moore-Ede, M. C., and Coleman, R. M. (1982). Rotating shift work schedules that disrupt sleep are improved by applying circadian principles. *Science, 217*, 460–462.

Dekker, D. K., and Tepas, D. I. (1990). Gender differences in permanent shiftworker sleep behavior. In G. Costa, G. Cesanna, K. Kogi, and A. Wedderburn, Eds., *Shiftwork: Health, Sleep and Performance*. Frankfurt-am-Main: Peter Lang.

Dinges, D. F., Connell, L. J., Rosekind, M. R., Gillen, K. A., Kribbs, N. B., and Graeber, R.C. (1991). Effects of cockpit naps and 24-hr layovers on sleep debt in long-haul transmeridian flight crews. *Sleep Research, 20*, 406.

Dinges, D. F., Orne, M. E., Whitehouse, W. G., and Orne, E. C. (1987). Temporal placement of a nap for alertness: Contributions of circadian phase and prior wakefulness. *Sleep, 10*, 313–329.

Dirkx, J. (1993). Adaptation to permanent night work: The number of consecutive work nights and motivated choice. *Ergonomics, 36*, 29–36.

Duchon, J. C., and Smith, T. J. (1993). Extended workdays and safety. *International Journal of Industrial Ergonomics, 11*, 37–49.

Eastman, C. I., and Miescke, K-J. (1990). Entrainment of circadian rhythms with 26-h bright light and sleep-wake schedules. *American Journal of Physiology, 259*, R1189–R1197.

Eastman, C. I., Stewart, K. T., Mahoney, M. P., Liu, L., and Fogg, L. F. (1994). Dark goggles and bright light improve circadian rhythm adaptation to night-shift work. *Sleep, 17*, 535–543.

Ehret, C. F. (1981). New approaches to chronohygiene for the shift worker in the nuclear power industry. In A. Reinberg, N. Vieux, and P. Andlauer, Eds., *Night and Shift Work: Biological and Social Aspects*. Oxford: Pergamon Press.

Evans, F. J., Cook, M. P., Cohen, H. D., Orne, E. C., and Orne, M. T. (1977). Appetitive and replacement naps: EEG and behavior. *Science, 197*, 687–689.

Folkard, S. (1992). Is there a "best compromise" shift system? *Ergonomics, 35*, 1453–1463.

Folkard, S. and Monk, T. H. (1979). Shiftwork and performance. *Human Factors, 21*, 483–492.

Folkard, S., Wever, R. A., and Wildgruber, C. M. (1983). Multi-oscillatory control of circadian rhythms in human performance. *Nature, 305*, 223–226.

Fröberg, J. E. (1979). *Performance in tasks differing in memory load and its relationship with habitual activity phase and body temperature* (FOA Rep. No. C-52002-H6). Stockholm: Forsvarets Forskningsanstalt, Research Institute of National Defense.

Gersten, A. H. (1987). *Adaptation in rotating shift workers: A six year follow-up study*. Unpublished doctoral dissertation, Illinois Institute of Technology, Chicago.

Gillberg, M. (1984). The effects of two alternative timings of a one-hour nap on early morning performance. *Biological Psychology, 19*, 45–54.

Gordon, G. H., McGill, W. L., and Maltese, J. W. (1981). Home and community life of a sample of shift workers. In L. C. Johnson, D. I. Tepas, W. P. Colquhoun, and M. J. Colligan, Eds., *Biological Rhythms, Sleep and Shift Work*. New York: Spectrum.

Härmä, M. (1993). Individual differences in tolerance to shiftwork: a review. *Ergonomics, 36*, 101–109.

Horne, J. A., and Ostberg, O. (1976). A self-assessment questionnaire to determine morningness-eveningness in human circadian rhythms. *International Journal of Chronobiology, 4*, 97–110.

Imberman, W. (1983). Who strikes—and why? *Harvard Business Review, 61*, 18–28.

International Labour Office. (1978). *Management of working time in industrialised countries*. Geneva: International Labour Office.

Johnson, L. C., and Naitoh, P. (1974). *The operational consequences of sleep deprivation and sleep deficit* (AGARD-AG-193, NATO). London: Technical Editing and Reproduction.

Keran, C. M., Duchon, J. C., and Smith, T. J. (1994). Older workers and longer work days: are they compatible? *International Journal of Industrial Ergonomics, 13*, 113–123.

Kleitman, N., and Jackson, D.P. (1950). Body temperature and performance under different routines. *Journal of Applied Physiology, 3*, 309–328.

Knauth, P., Keller, J., Schindele, G., and Totterdell, P. (in press). A fourteen-hour night shift in the control room of a fire-brigade.

Knauth, P., and Rutenfranz, J. (1976). Experimental studies of permanent night and rapidly rotating shift systems. *International Archives of Occupational and Environmental Health, 37*, 125–137.

Knutsson, A. (1989). Shift work and coronary heart disease. *Scandinavian Journal of Social Medicine*, Suppl. 44.

Kogi, K., and Thurman, J. E. (1990). Development of new international standards on night work. In G. Costa, G. Cesana, K. Kogi, and A. Wedderburn, Eds., *Shiftwork: Health, Sleep and Performance*. Frankfurt-am-Main: Peter Lang.

Koller, M., Kundi, M., and Cervinka, R. (1978). Field studies of shift work in an Austrian oil refinery I: Health and psychosocial wellbeing of workers who drop out of shiftwork. *Ergonomics, 21*, 835–847.

Kribbs, N. B., and Dinges, D. (1994). Vigilance decrement and sleepiness. In R. D. Ogilvie and J. R. Harsh, Eds., *Sleep onset: Normal and abnormal processes*. Washington, DC: American Psychological Association.

Krueger, G. P. (1989). Sustained work, fatigue, sleep loss and performance: A review of the issues. *Work and Stress, 3*, 129–141.

Lavie, P., Tzischinsky, O., Epstein, R., and Zomer, J. (1992). Sleep-wake cycle in shift workers on a "clockwise" and "counter-clockwise" rotation system. *Israel Journal of Medical Sciences, 28*, 636–644.

Lennernas, M. (1993). Nutrition and shift work: The effect of work hours on dietary intake, meal patterns and nutritional status parameters. *Comprehensive Summaries of Uppsala Dissertations from the Faculty of Medicien* (No. 402). Stockholm: Almqvist and Wiksell International.

Lewis, P. M., and Swain, D. J. (1986). Evaluation of a 12-hour day shift schedule. *Proceedings of the Human Factors Society 30th annual meeting* (pp. 885–889).

Mahan, R. P., Tepas, D. I., and Carvalhais, A. B. (1987). Morningness-eveningness norms for industrial shift workers. In A. Oginski, J. Pokorski, and J. Rutenfranz, Eds., *Contemporary Advances in Shiftwork Research: Theoretical and Practical Aspects in the Late Eighties*. Krakow: Medical Academy.

Matsumoto, K., Matsui, T., Kawamori, M., and Kogi, K. (1982). Effects of nighttime naps on sleep patterns of shiftworkers. *Journal of Human Ergology, 11*, Suppl., 279–289.

Monk, T. H. (1986). Advantages and disadvantages of rapidly rotating shift schedules: A circadian viewpoint. *Human Factors, 38*, 553–557.

Monk, T. H., and Embrey, D. E. (1981). A field study of circadian rhythms in actual and interpolated task performance. In A. Reinberg, N. Vieux, and P. Andlauer, Eds., *Night and Shift Work: Biological and Social Aspects*. Oxford: Pergamon.

Monk, T. H., and Folkard, S. (1985). Individual differences in shiftwork adjustment. In S. Folkard and T. H. Monk, Eds., *Hours of Work: Temporal Factors in Work-Scheduling*. New York: John Wiley.

Monk, T. H., Weitzman, E. D., Fookson, J. E., Moline, M. L., Kronauer, R. E., and Gander, P. H. (1983). Task variables determine which biological clock controls circadian rhythms in human performance. *Nature, 304*, 543–545.

Mott, P. E., Mann, F. C., McLoughlin, Q., and Warwick, D. (1965). *Shift Work*. Ann Arbor: The University of Michigan Press.

Nachreiner, F., Grzech-Sukalo, H., Qin, L., Moehlmann, D., and Will, W. (1995). Computer-aided design of shift schedules for public transport operations. *Shiftwork International Newsletter, 12*, 46.

Naitoh, P., and Angus, R. G. (1989). Napping and human functioning during prolonged work. In D. F. Dinges and R. J. Broughton, Eds., *Sleep and Alertness: Chronobiological, Behavioral, and Medical Aspects of Napping*. New York: Raven Press.

Naitoh, P., Kelly, T., and Babkoff, H. (1993). Sleep inertia: Best time not to wake up? *Chronobiology International, 10*, 109–118.

National Transportation Safety Board. (1989). Head-end collision of consolidated rail corporation freight trains UBT-506 and TV-61 near Thompsontown, Pennsylvania, January 14, 1988. *Railroad Accident Report* (NTSB/RAR-89/02). Washington: National Transportation Safety Board.

National Transportation Safety Board. (1990). NTSB adopts first list of "most wanted" safety items. *Safety Information* (SB 90 48/5299A). Washington: National Transportation Safety Board.

Ong, C. N., and Kogi, K. (1990). Shiftwork in developing countries: Current issues and trends. In A. J. Scott (Ed.), *Shiftwork*. Philadelphia: Hanley and Belfus.

Paley, M. J., Price, J. M., and Tepas, D. I. (in press). The impact of a change in rotating shift schedules: A comparison of the effects of 8, 10 and 14 hour work shifts.

Paley, M. J., and Tepas, D. I. (1994a). Fatigue and the shiftworker: Firefighters working on a rotating shift schedule. *Human Factors*, *36*, 269–284.

Paley, M. J., and Tepas, D. I. (1994b). The effect of tenure on fire fighter adjustment to shiftwork. Presented at the 23rd International Congress of Applied Psychology, Madrid, Spain.

Poole, C. J. M., Wright, A. D., and Nattrass, M. (1992). Control of diabetes mellitus in shift workers. *British Journal of Industrial Medicine*, *49*, 513–515.

Popkin, S. M. (1994). An evaluation of the impact of an educational program for freight locomotive engineers on irregular work schedules. *Proceedings of the 12th Triennial Congress of the International Ergonomics Association*, *5*, 33–35.

Presser, H. B., and Cain, V. S. (1983). Shift work among dual-earner couples with children. *Science*, *219*, 876–878.

Reinberg, A. (1974). Chronopharmacology in man. In J. Aschoff, F. Ceresa, and F. Halberg, Eds., *Chronobiological aspects of endocrinology*. Stuttgart: F.K. Schattauer.

Reinberg, A., Andlauer, P., DePrins, J., Malbecq, W., Vieux, N., and Bourdeleau, P. (1984). Desynchronization of the oral temperature circadian rhythm and intolerance to shift work. *Nature*, *308*, 272–274.

Rentos, P. G., and Shepard, R. D., Eds. (1976). *Shift work and health: A symposium* [HEW Publication No. (NIOSH) 76-203]. Washington, DC: U.S. Government Printing Office.

Romon-Rousseaux, M., Beuscart, R., Thuilliez, J. C., Frimat, P., and Furon, D. (1986). Influence of different shift schedules on eating behavior and weight gain in edible-oil refinery workers. In M. Haider, M. Koller, and R. Cervinka, Eds., *Night and Shiftwork: Long Term Effects and Their Prevention*. Frankfurt-am-Main: Peter Lang.

Romon-Rousseaux, M., Lancry, A., Poulet, I., Frimat, P, and Furon, D. (1987). Effect of protein and carbohydrate snacks on alertness during the night. In A. Oginski, J. Pokorski, and J. Rutenfranz, Eds., *Contemporary Advances in Shiftwork Research: Theoretical and Practical Aspects in the Late Eighties*. Krakow: Medical Academy.

Rosa, R. R. (1991). Performance, alertness and sleep after 3.5 years of 12 h shifts: A follow-up study. *Work and Stress*, *5*, 107–116.

Rosa, R. R., Bonnet, M. H., Bootzin, R. R., Eastman, C. I., Monk, T., Penn, P. E., Tepas, D. I., and Walsh, J. K. (1990). Intervention factors for promoting adjustment to nightwork and shiftwork. In A. J. Scott, Ed., *Shiftwork*. Philadelphia: Hanley and Belfus.

Rosa, R. R., Colligan, M. J., and Lewis, P. (1989). Extended workdays: Effects of 8-hour and 12-hour rotating shift schedules on performance, subjective alertness, sleep patterns, and psychosocial variables. *Work and Stress*, *3*, 21–32.

Rutenfranz, J. (1982). Occupational health measures of night- and shiftworkers. *Journal of Human Ergology*, *11*, Suppl., 67–86.

Scherrer, J. (1981). Man's work and circadian rhythm through the ages. In A. Reinberg, N. Vieux, and P. Andlauer, Eds., *Night and Shift Work: Biological and Social Aspects*. Oxford: Pergamon Press.

Smith, A. (1988). Effects of meals on memory and attention. In M. M. Gruneberg, P. E. Morris, and R. N. Sykes, Eds., *Practical Aspects of Memory: Current Research and Issues, Vol. 2, Clinical and Educational Implications*. Chichester: John Wiley.

Smith, C. S., Reilly, C., and Midkiff, A. (1989). Evaluation of three circadian rhythm questionnaires with suggestions for an improved measure of morningness. *Journal of Applied Psychology*, *74*, 728–738.

Smith, L., Folkard, S., and Macdonald, I. (1995). Zones of extra vulnerability. *Shiftwork International Newsletter*, *12*, 58.

Steinberg, R. (1982). *Wages and Hours: Labor and Reform in Twentieth-Century America*. New Brunswick, NJ: Rutgers University Press.

Takagi, K. (1972). Influence of shift work on time and frequency of meal taking. *Journal of Human Ergology*, *1*, 195–205.

Tasto, D. L. and Colligan, M. J. (1977). *Shift work practices in the United States* (DHEW Publ. No. 77-148). Washington DC: U.S. Government Printing Office.

Taub, J. M. (1971). The sleep-wakefulness cycle in Mexican adults. *Journal of Cross-Cultural Psychology*, *44*, 353–362.

Taylor, P. J., and Pocock, S. J. (1972). Mortality of shift and day workers 1956-68. *British Journal of Industrial Medicine*, *29*, 201–207.

Tepas, D. I. (1982). Adaptation to shiftwork: Fact or fallacy? *Journal of Human Ergology*, *11*, Suppl., 1–12.

Tepas, D. I. (1990a). Do eating and drinking habits interact with work schedule variables? *Work and Stress*, *4*, 203–211.

Tepas, D. I. (1990b). Condensed working hours: Questions and issues. In G. Costa, G. Cesana, K. Kogi, and A. Wedderburn, Eds., *Shiftwork: Health, Sleep and Performance*. Frankfurt-am-Main: Peter Lang.

Tepas, D. I. (1993). Educational program for shiftworkers, their families, and prospective shiftworkers. *Ergonomics*, *36*, 199–209.

Tepas, D. I. (1994). Technological innovation and the management of alertness and fatigue in the workplace. *Human Performance*, *7*, 165–180.

Tepas, D. I., Armstrong, D. R., Carlson, M. L., Duchon, J. C., Gersten, A. H., and Lezotte, D. V. (1985). Changing industry to continuous operations: Different strokes for different plants. *Behavior Research Methods, Instruments and Computers*, *17*, 670–676.

Tepas, D. I., Carlson, M. L., Duchon, J. C., Gersten, A., and Mahan, R. P. (1986). Moving a plant from discontinuous to continuous operation using weekend work shifts. In M. Haider, M. Koller, and R. Cervinka, Eds., *Night and Shiftwork: Long-term Effects and Their Prevention*. Frankfurt-am-Main: Peter Lang.

Tepas, D. I., and Carvalhais, A. B. (1990). Sleep patterns of shiftworkers. In A. J. Scott, Ed., *Shiftwork*. Philadelphia: Hanley and Belfus.

Tepas, D. I., Duchon, J. C., and Gersten, A. H. (1993). Shiftwork and the older worker. *Experimental Aging Research*, *19*, 295–320.

Tepas, D. I., and Mahan, R. P. (1989). The many meanings of sleep. *Work and Stress*, *3*, 93–102.

Tepas, D. I., and Monk, T. H. (1987). Work schedules. In G. Salvendy, Ed., *Handbook of Human Factors*. New York: John Wiley.

Tepas, D. I., and Popkin, S. M. (1994). Duration-of-workday and time-of-day as predictors of workday sleep length and end of workday ratings of being tired or tense. Unpublished paper presented at the 11th International Symposium on Night and Shiftwork, La Trobe University, Melbourne, Australia.

Tepas, D. I., Walsh, J. K., and Armstrong, D. (1981). Comprehensive study of the sleep of shift workers. In L. C. Johnson, D. I. Tepas, W. P. Colquhoun, and M. J. Colligan, Eds., *Biological Rhythms, Sleep, and Shift Work*. New York: Spectrum.

Tepas, D. I., Walsh, J. K., Moss, P. D., and Armstrong, D. (1981). Polysomnographic correlates of shift worker performance in the laboratory. In A. Reinberg, N. Vieux, and P. Andlauer, Eds., *Night and Shift Work: Biological and Social Aspects*. Oxford: Pergamon Press.

Thierry, H., and Jansen, B. (1984). Work and working time. In J. D. D. Drenth, H. Thierry, P. J. Willems, and C. J. deWolff, Eds., *Handbook of Work and Organizational Psychology*. New York: John Wiley.

Totterdell, P., Spelten, E., Smith, L., Barton, J., and Folkard, S. (1995). Recovery from work shifts: How long does it take? *Journal of Applied Psychology*, *80*, 43–57.

U.S. Department of Labor, Bureau of Labor Statistics. (1992). Workers on flexible and shift schedules. *News (USDL 92-491)* Washington, DC: U.S. Department of Labor.

U.S. Congress, Office of Technology Assessment. (1991). *Biological rhythms: Implications for the workers* (OTA-BA-463). Washington, DC: U.S. Government Printing Office.

Walsh, J. K., Muehlbach, M. J., Humm, T. M., Dickins, Q. S., Sugerman, J. L., and Schweitzer, P. K. (1990). Effect of caffeine on physiological sleep tendency and ability to sustain wakefulness at night. *Psychopharmacology*, *101*, 271–273.

Waterhouse, J. M., Folkard, S., and Minors, D. S. (1993). Effects of a change in shift-work on health. *Occupational Medicine*, *43*, 167.

Wedderburn, A. A. I. (1981). How important are the social effects of shiftwork? In L. C. Johnson, D. I. Tepas, W. P. Colquhoun, and M. J. Colligan, Eds., *Biological Rhythms, Sleep, and Shift Work*. New York: Spectrum.

Wedderburn, A. A. I. (1991). Guidelines for Shiftworkers. *Bulletin of European Shiftwork Topics*, No. 3. Dublin: European Foundation for the Improvement of Living and Working Conditions.

Wedderburn, A. A. I. (1992). How fast should the night shift rotate? A rejoinder. *Ergonomics*, *35*, 1447–1451.

Wever, R. A. (1979). *The Circadian System of Man*. Berlin: Springer-Verlag.

Wilkinson, R. T. (1992). How fast should the night shift rotate? *Ergonomics*, *35*, 1425–1446.

Williamson, A. M., and Sanderson, J. W. (1986). Changing the speed of shift rotation: A field study. *Ergonomics*, *29*, 1085–1096.

Wilson, R. H. L., Newman, E. J., and Newman, H. W. (1957). Diurnal variations in rate of alcohol metabolism. *Journal of Applied Physiology*, *8*, 556–558.

Wurtman, J. J. (1986). *Managing Your Mind and Mood Through Food*. New York: Rawson Associates.

CHAPTER 33

PSYCHOSOCIAL APPROACH IN OCCUPATIONAL HEALTH

Raija Kalimo
Kari Lindström
Department of Psychology
Finnish Institute of Occupational Health
Helsinki, Finland

Michael J. Smith
Department of Industrial Engineering
University of Wisconsin—Madison
Madison, WI 53706 USA

33.1 INTRODUCTION

The purpose of this chapter is to provide the ergonomic and human factors practitioner with an understanding of how psychosocial factors at work can affect health, how to recognize these conditions and their effects, and how to control and intervene in order to avoid their adverse effects and to promote health. The chapter provides information on concepts, methods, and interventions that can be used to identify, measure, and control health-related issues at work.

33.2 THEORIES AND CONCEPTS

The recognition of psychological factors as contributors of occupational health expanded along the development of the concept of the work-relatedness of occupational health since the early 1970's (Levi, 1972a). Incomplete explanatory power of the so-called traditional risk factors in the efforts to understand the causation of illnesses led to the search of possible causes in psychological factors. Stress theory was generally adopted as the conceptual framework; since then, various versions of it have remained as the main approach. In addition, the positive effects of satisfying work have gained increasing relevance (Ilgen, 1990; Smith, 1965).

33.2.1 Psychophysiological Model

The physiological approach was initially predominant when stress theory was adopted in the domain of occupational health, and it has maintained an important role since then. The basis of stress theory lies in the works of Cannon (1914) and Selye (1956), who developed the foundation for a physiological concept of stress. Stress was described as a pattern of physiological reactions called "general adaptation syndrome" (GAS) which is activated in a nonspecific, stereotyped form by any environmental demand (Selye, 1956, 1983). The syndrome is characterized by a mobilization of energy resources to fight or flight. The syndrome proceeds along three stages: alarm, resistance (adaptation), and exhaustion. The initial arousal leads to an elevated activation of bodily resources to maintain or achieve adaptation. Elevation of blood pressure and an increase in heart rate are recognized among a multitude of other changes whereby the energy stores of the organism are depleted unless a balance is achieved. This stage was characterized by an increased discharge of hormones controlled by the sympathetic nervous system and the pituitary and adrenal glands. The end results of this syndrome, if prolonged, were exhaustion and disease.

One of the characteristic features of the physiological theory was that stress was understood as a two-factorial, stimulus–response phenomenon. The concept of stress described the organism's reaction, which was activated in the same manner by any environmental demand, which would be called the stressor. General agreement was not achieved, however, on the definition of the stress concept. The response-based definition is limited because the same reaction pattern can be evoked by a wide array of stimulus conditions.

33.2.2 Cognitive Theory

Grounds for the development of the cognitive theory of stress were based on the criticism of the physiological stimulus–response theory of stress, which was first challenged most profoundly by Lazarus (1977, 1993), who claimed that stress could not be understood on the basis of a two-factor model. Support for the adoption of the cognitive theory of stress was gained also from empirical research on work, stress, and health, which indicated that within the same work conditions not all workers experienced stress.

Individual differences in cognition are thought to intervene between the stressor and the reaction (Dewe, 1991; Lazarus, 1974, 1993). Cognitive processes determine the quality and the intensity of the reactions to the environment. A stressor will not have an effect unless it is recognized and assessed by the person. Thus, an appraisal phase is added as a third component in the stress model. The appraisal phase is characterized by two consecutive steps. In the primary phase, a stressor is detected and its possible harmfulness is assessed. Three different kinds of stress situations are acknowledged: harm, threat, and challenge. Harm refers to psychological damage that has already taken place, e.g., an irrevocable loss. Threat is the anticipation of harm that has not yet taken place but may happen. Challenge results from difficult demands that one feels confident about overcoming by effective mobilization of resources. Secondary appraisal involves the assessment of the resources available to confront the stressor. Resources are then activated and manifested as coping strategies to counteract the stressors.

A three-component model with coping processes as mediators between environmental stressors and a person's reactions has been broadly adopted in psychological stress research. In spite of a great diversity of empirical findings, there is an overall agreement that effective coping reduces stress and strain. However, views on the effectiveness of various coping strategies and means differ (Latack and Havlovic, 1992).

33.2.3 Transactional and Systems Theory

Current stress models usually involve the notion of an interaction or transaction between the environment and the individual. The notion of an interaction is inherent in the coping processes of cognitive theory whereby the individual changes the environment by controlling external stressors. A more fundamental transactional point of view is involved in some other models that characterize the development of stress as an outcome of an imbalanced interaction of the environment and the person. If the environmental demands are greater than the person's capacities and/or if the person's expectations are greater than the environmental supplies, then stress will occur (Caplan, Cobb, French, Harrison, and Pinneau, 1975; Cooper and Payne, 1988; Cox and Ferguson, 1994; Ganster and Schaubroeck, 1991; Johnson and Johansson, 1991; Kalimo, 1980).

Transactional models facilitated the inclusion of a comprehensive view on health (including not only negative but also positive aspects) in the analysis of the interactions between work, coping, and health. These models do not see the environment (work) not only as a source of stressors, but also as a source of a full range of demands and supplies that may or may not match the individual. The notion of successful coping as a source of learning and mastery of new skills implies growth and development. Thus, work can be a source of life satisfaction as well as a source of distress.

33.2.4 Operationalizing Systems Theory

As a general operationalization of the approaches noted previously, the following synthesis can be made. There are two basic assumptions. First, work makes demands on the person, demands regarding, for instance, work output and participation. The person responds to these demands according to skills, motivations, and other characteristics. The other assumption is that the person has expectations regarding work, such as decision making, opportunity for development, and support. Work may present the opportunity for the achievement of such goals to a varying degree.

In the case of a dynamic balance in which environmental conditions put manageable challenge to the person, then a positive development takes place in the form of satisfaction, feelings of competence, well-being, and health. Conflict in the above transaction is the critical factor that mobilizes coping for the achievement of the expected balance. Coping processes may involve actions (1) to change the external situation, i.e., to alleviate the stressors; (2) to perceive and interpret the external situation in a new way to avoid the perception of the situation as stressful; or (3) to alleviate the consequences of stress with palliative measures such as relaxation, intake of medicines, or use of social support.

If coping at the initial appraisal stage fails, a series of negative stages begins to develop. Negative emotions including frustration, dissatisfaction, fear, and anger appear at first. These lead into the development of physiological changes via messages mediated by the pathways involving the autonomic nervous system and the hormonal and immunological systems. If coping continues to fail, psychological and behavioral context-evoked stress responses may change into a chronic state and eventual mental disorders.

If the physiological control mechanisms fail, physiological disorders may also become chronic and diseases may gradually develop. Disturbed health status again feeds back to the previous phases of the system, i.e., affects the results of stressors and the coping capacity in the future. This process is illustrated in Figure 33.1.

33.3 WORK ORGANIZATION AND JOB DESIGN

It is important to recognize those psychosocial characteristics in work, work organization, and environment that are potential contributors to the health and well-being of individual workers, groups, and the whole organization. These characteristics of the work and of the organization are overlapping and not independent. Their health effects depend on their mutual interaction and on the characteristics of the individuals. We are interested mainly in the health relatedness of the psychosocial characteristics of work and work organization, although they can be also be regarded as contributors to work motivation, or work performance. The occupational health perspective of work organization covers such outcomes as general health, mental health and well-being, competence, and aspiration, as well as integrated functioning, job satisfaction, and life satisfaction.

33.3.1 Characteristics of Work Organization

The characteristics of a work organization are often sources of occupational stress and long-term health consequences. Cooper (1986) has categorized these job stress factors to those intrinsic to the job, role in the organization, career development, and the relationships at work, as well as the organizational structure and climate. In this section factors related particularly to the physical work environment and task demands are described. Factors intrinsic to the job are: (1) physical working conditions; (2) workload, both quantitative and qualitative, and time pressure; (3) responsibilities (for lives, economic values, safety of other persons); (4) job content; (5) descision making; and (6) perceived control.

Many blue-collar jobs entail a high physical energy expenditure or difficult work movements and postures. Ergonomic factors, such as repetition, posture, and force can contribute to the development of musculoskeletal problems (e.g., musculoskeletal discomfort or fatigue), and the features of a psychosocial work environment can also contribute to their development (Lim and Carayon 1995; Lim, Rogers, Smith, and Sainfort, 1989;

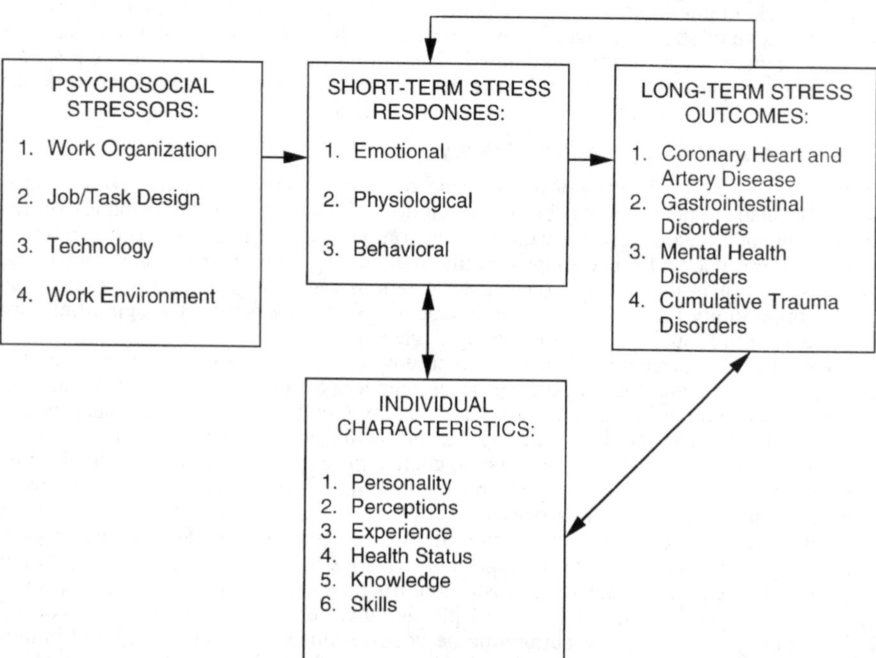

Figure 33.1 Model of psychosocial job stress process. (Adapted from Smith and Carayon, 1996.)

Smith, Cohen, Stammerjohn, and Happ, 1981; Smith and Carayon, 1996). The ergonomic aspects of the workplace are dealt with in detail in Chapters 9, 10, and 23.

The most obvious job characteristic is workload or job demands. This describes both quantitative and qualitative aspects of the task. Stress can result from overload and underload (Frankenhaeuser and Gardell, 1976). Quantitative overload means that there is too much to do or the person has to work under time pressure or meet strict deadlines. Also, demands to work excessive hours increase quantitative workload. Quantitative workload and computer-related problems have had a direct effect on psychological complaints among video display terminal (VDT) users in private and public sector organizations (Smith et al., 1981; Yang and Carayon, 1995). In nursing work, perceived stress was found to be associated with a high level of workload resulting from the level of staffing and equipment available (Hippwell, Tyler, and Wilson, 1989). Among managers and other white-collar workers, job pressure (deadlines, dealing with conflict situations) was found to be an important stressor (Turnage and Spielberger, 1991).

Qualitative overload means that a person does not feel able or capable of doing a given task, or the task demands are close to the upper limit of capability. Both quantitative and qualitative overload, as well as underload, are adverse for health and to the development of competence. Among Finnish banking and insurance employees, the difficulty of the task was the main predictor for psychological and unspecific somatic symptoms and feelings of a lack of competence (Lindström, 1991). Qualitative underload implies that a person cannot fully utilize skills and other potentials at work. Overly simple tasks lead to boredom (French and Caplan, 1973), whereas tasks that are too difficult can adversely affect the feelings of competence and self-esteem. In particular, manufacturing tasks involving paced and/or assembly operation have potential for negative effects (Salvendy and Smith, 1981).

Technological and other changes at work often temporarily bring along an increased amount of work and high demands of learning during and just after the transition to new technological systems (Lindström et al., 1995; Smith et al., 1981). In addition to high workload, these transition periods as such usually also involve uncertainty among employees.

Control at work can be defined as the ability of a person to exert some influence on his or her environment or a characteristic of the task or the work organization. In a work situation the person can have control over various job demands, such as the task itself, pacing of the work, work scheduling, the physical environment, decision making, other people, or mobility. As such, and in combination with work demands, job control has been found to be an important contributor to health and competence development at work (Karasek, 1979; Karasek and Theorell, 1990).

When job control is high, the other job demands tend to have less potential adverse effects on health. The well-known job demand–control model differentiates various kinds of jobs depending on the degree of demands and control, i.e., high- and low-strain jobs, active and passive jobs (Karasek and Theorell, 1990). In this model, job control includes both the challenges at work and the opportunities for self-development. According to this model, high workload as such is not necessarily harmful, but it becomes harmful in combination with low employee control. Under such conditions it can contribute to the development of cardiovascular disease (Karasek, 1981; Karasek, Theorell, Schwartz, Schnall, Pieper, and Michaela, 1988). Dwyer and Ganster (1991) found that among manufacturing employees high objective psychological job demands were related to a high level of tardiness and sick days only under low perceived control. Subjective workload ratings were related to job satisfaction and voluntary absence when interacting with control, but not to tardiness or sick days.

Uncertainty or lack of clarity is closely related to control because it implies a lack of needed information. Uncertainty exists when that knowledge about events requiring action is experienced as inadequate (Jackson, 1989). Uncertainty has been emphasized as one of the most predominant single sources of occupational stress (Sharit and Salvendy, 1982). It arises from job insecurity, task ambiguity, lack of knowledge of results, and unanticipated slowdown in work. With increasing use of computer technology, this kind of unpredictability has become more frequent.

Responsibility at work can be related to the life and safety of other persons and to material values. These issues are central in the social welfare and health care sector, as well as in other customer- or client-related tasks. Responsibility is also associated with occupational roles and organizational hierarchy, being greatest at the higher hierarchical levels. However, high formal responsibility in a specific occupation is not necessarily perceived as a high stressor. Among Finnish health care personnel, nurses saw it as a bigger stressor than did physicians (Huuhtanen, Nygård, Tuomi, Eskelinen, and Toikkanen, 1991).

33.3.2 Organizational Structure and Culture

The reader is referred to Chapter 18 for a more detailed discussion on organizational issues. Organizational structure and culture have been found to be important for the health and well-being of employees. Among health care professionals, an innovative climate at work was associated with higher workload (West, 1989) and higher satisfaction with the management. Organizational practices to monitor employee performance have been introduced with computer technology (Amick and Smith, 1992). Electronic performance monitoring can have both direct and indirect effects on worker stress. The research findings point to indirect effects on worker stress via the influences on job design. Such characteristics as completeness of electronic performance monitoring and comparisons of ratings among coworkers especially produce worker stress (Carayon, 1994).

Organizational commitment and career development opportunities are also essential characteristics of an organization from the individual's point of view. The relationship between organizational commitment and behavioral outcomes (absence, turnover, performance) has been found to be rather weak in general (Randall, 1990), because career stages and occupation modify the influences. Career development is related to training, skills, and security, and there usually are some organizational mechanisms affecting these. Both under- and overpromotion have been seen as stressors. During midlife, lessened job opportunities and changing work role identity and relative deprivation (i.e., how well one's own goals in one's career have been attained) have been found to accompany health complaints and job dissatisfaction (Buunk and Janssen, 1992). Recently, job insecurity is a primary issue for many workers, particularly when work life is undergoing major structural changes associated with downsizing, mergers, and business process reengineering (Smith and Carayon, 1995).

33.3.3 Management and Supervisory Systems

Supervisory style is important from the occupational health perspective. A hierarchical, authoritative management style in general has been found to have negative effects on workers' well-being (Smith, Carayon, Sanders, Lim, and LeGrande, 1992), whereas a participatory leadership style has been found advantageous (Lawler, 1986). White-collar workers in a manufacturing company attributed the highest stress to lack of opportunity for advancement and poor or inadequate supervision. Leader support and encouragement were important when aiming at the reduction of job stress (Turnage and Spielberger, 1991). Among nurses, leader support was important in predicting job satisfaction (McIntosh, 1990) and autonomy of work in decreasing anxiety. Autonomy at work can be interpreted as a message of leader confidence (McIntosh, 1990). Young and aging workers in particular have higher expectations of leader support.

33.3.4 Job Design Characteristics

The reader is referred to Chapter 14 for more detailed information about job design. Job design variables can be defined either as job characteristics, or as decision processes leading to the design of jobs (Davis and Wacker, 1982). These are factors that contribute to effective and satisfying jobs. This means that both organizational effectiveness and individual needs and goals are considered. In job design interventions, the jobs must be examined and understood in relation to the whole enterprise or organization. The job design process includes making decisions concerning the tasks to be performed, which tasks should or could be clustered into specific jobs, and how these jobs are linked together. The job design process involves technological and social choices, too.

According to quality of working life (QWL) principles the criteria in job design deal with the physical work environment, compensation systems, institutional rights and decisions, job content, internal and external social relations, and career development (Davis and Wacker, 1982). In a job design context, the following characteristics of job content are of main interest: variety of tasks and task identity; feedback from the job; perceived contribution to product or service; challenges and opportunities to use one's own skills; and individual autonomy, as well as self-regulation (which is very close to worker's self control). These are very close to the core job characteristics in the Hackman and Oldham model (Hackman and Oldham, 1980).

33.3.5 Roles and Interpersonal Relations

Work roles became the focus of occupational stress research since Kahn (1973) defined three basic types of problems in the organizational role setting: role ambiguity, role conflict, and role overload. Research on role stress has continued to concentrate on these problems, but the need for revisions has been brought up recently (King and King, 1990). Role configurations in the organizations may change considerably along with the changes in organizations, such as working in teams without a supervisor, and technology changes.

Role conflict is defined as a condition of incompatible role expectations placed on a person. Role ambiguity means a lack of clarity regarding the role expectations, and an uncertainty concerning the outcomes of one's role performance. Role overload means a scope that is too wide with too many different kinds of role expectations. Recently, role underload has been recognized as well. It is caused by too low and too few expectations placed on a person who would be capable of a greater role. Role conflict and ambiguity are very likely outcomes if a person's work goal, i.e., the purpose of the work, is poorly defined and insufficiently explained. Role dysfunctions have since been recognized as hindrances for job satisfaction (Schuler, Aldag, and Brief, 1977) and were later shown as causes of stress and disorders of well-being in a number of studies (Barling and Macintyre, 1993; Kalimo, 1980; Meyerson, 1994; Revicki, Whitley, and Gallery, 1993; Tetrick 1992).

Interpersonal conflicts are one of the major sources of stress. Recent research about interpersonal relations has been focused on aggravated conflict situations and bullying. Bullying, mobbing, or psychological harassment are a form of expression of the problems in the interaction between people at the workplace. "A person is being bullied when he or she is exposed repeatedly and over time to negative actions on the part of one or more persons at the workplace" (Vartia, 1993). It means teasing, pressure, and unfair treatment of a person in situations where the target cannot defend him or herself.

Women and men are equally often targets. Bullying may be caused by many factors: being different from the norm, the scapegoat phenomenon, intense competition between employees, greed and envy, unsatisfactory tasks (e.g., monotony), heavy work pressure, poor leadership, and insecurity caused by changes in the organization. Bullying is a serious source of stress and threat to a person's well-being. In a Finnish study (Vartia, 1993) 35% of the victims felt quite depressed and 46% had considered leaving the job. Even suicides caused by bullying have been reported (Leymann, 1990).

Social support is a resource which has been shown to influence the effects of stress (see Buunk and Peeters, 1994; House, 1981). Views on what constitutes social support have been varied. In research literature the concept has been used to cover (1) social integration, which is the size and structure of one's social network; (2) the availability of satisfying relationships characterized by love, trust, and esteem; and (3) the actual receiving of supportive acts from others when in need. House (1981) emphasized the importance of the actual receiving of social support as the key factor for the beneficial impact. He defined four forms of such support: (1) emotional support: care giving and affectionate concern; (2) appraisal support through evaluative feedback and affirmation; (3) informational support by giving suggestions or guiding; and (4) instrumental support through organizing opportunities.

Social support is a beneficial factor for well-being and job satisfaction. Two types of models have been developed to explain the mechanism of the process. The direct effect model indicates that social support is positively related to health independent of environmental circumstances, such as the presence of job stressors. On the other hand, social support is thought to exert a protective function during conditions of stress, i.e., to "buffer" a person against the harmful effects of the social environment. The general benefits derived from a supportive social environment have often been demonstrated (Nelson and Quick, 1991). The buffering effect has also been shown (Fried and Tiegs, 1993) although the evidence for it is weaker. As a whole, evidence for the beneficial effect of social support is conclusive enough to merit the promotion of efforts aimed at the improvement of interpersonal interaction at the workplace. The role of the supervisors as sources of support seems to be stronger than that of colleagues and workmates (Israel, House, Schurman, Heany, and Mero, 1989).

33.3.6 Extraorganizational Influences

Work and private life were recognized as two interdependent spheres of life in work-related stress and health. This relationship can be characterized by (1) a spillover effect of stress from one sphere to the other, (2) role conflict between work and family responsibilities, (3) a cumulative effect of work and family stressors, and (4) the buffering effect of one sphere against stress in the other.

The effect of stress and monotony caused by repetitive mechanized work has been shown to carry over to private life in the form of general passivity, as was demonstrated by Gardell (1982) in his classic work on this matter. The notion of asymmetrically permeable boundaries between the domains of work and family was introduced. Asymmetry meant that the consequences of work demands and responsibilities had a greater intrusion into the family than vice versa. The asymmetry of spreading effects has been confirmed empirically. Burke and Greenglass (1994) showed a cumulative function of work and family stress and presented a conceptual model to guide research in this sector. Role conflict between work and family roles has been shown to increase strain and lower the quality of family life. Family resource mobilization diminished symptoms of burnout in health care workers shown in a longitudinal study (Leiter, 1990).

33.4 STRESS RESPONSES

33.4.1 Attitudes and Satisfaction

Job satisfaction belongs to the most extensively studied reactions to psychosocial work conditions. It is characterized as a general positive attitude toward work. According to generally adopted views, job satisfaction has both intrinsic and external contributors. The first refer to job content, i.e., variety, skill utilization, etc., and the latter to such factors as environment and pay.

Job satisfaction can be described with the help of domain satisfactions such as satisfaction with career choice, job content, organizational climate, etc. Overall job satisfaction expressed with an answer to the question "How satisfied are you with your current job?" gives a more positive picture than answers to specific domain questions. Work-related determinants of high job satisfaction include task variety, autonomy, skill development, good interpersonal relations, fair pay, and a pleasant physical environment, which have been reviewed broadly elsewhere (Bhagat, 1982; Sekaran, 1989). Job satisfaction is higher in older people than in younger people. Satisfaction with a particular job is increased if no alternatives are available. Positive industrial relations are associated with increased job satisfaction (Kelloway, Barling, and Shah, 1993).

The importance of job satisfaction lies in the ramifications it has in work performance, organizational behavior, and health. Job dissatisfaction has shown to increase absenteeism and turnover (Curry, 1986; Dwyer and Ganster, 1991; McKee, Markham, and Scott, 1992; Sekaran 1989). It is often a component in a general job strain syndrome and burnout. The time sequence of the decrease of satisfaction and the development of strain is not clear and may not always be the same. Job dissatisfaction may be a strain preceding general attitude toward an unsatisfactory work situation (Kalimo, 1980). It has also been noted as a late reaction during the burnout process of highly job committed persons (Wolpin, Burke, and Greenglass, 1991). Job satisfaction is positively correlated with overall life satisfaction; but in multivariate analyses with extraprofessional factors, it predicted life satisfaction only to a minor degree (Rain, Lane, and Steiner, 1991; Tait, Padgett, and Baldwin, 1989).

33.4.2 Psychological Symptoms and Modifiers of Strain

Stressful experiences at work may manifest themselves in a number of psychological and behavioral reactions. A person normally copes with transactional periods of stress by either altering the situation or controlling his or her own responses. Therefore many periods of stress pass without a noticeable reaction by others. Problems arise when work conditions are in conflict with human capacities and expectations over a long period of time, and when coping fails. Cognitive appraisal of both the situation as stressful and insufficient coping evokes negative emotions. Tension, boredom, worry, anxiety, and irritability are inevitably some of the first indicators of strain in such situations. Depression and apathy are later symptoms and indicate a more harsh situation. Emotional stress reactions have been detected as a response to all kinds of stressors in work and organizations. Many reviews are available for more specific findings (Cooper and Payne, 1992; Ganster and Schaubroeck, 1991). Longitudinal studies are still quite scarce and should be made to confirm cross-sectional data (Carayon, 1993; Kinnunen and Salo, 1994).

The interaction of individual resources and individual means of coping with job stress has been another recent topic of research interest aimed at increased understanding of individual differences in the experience of stress in identical conditions (Barling and Kryl, 1990; Parkes, 1994). One of the resource constructs that has shown a relatively strong explanatory power is sense of coherence (SOC), a construct introduced by Antonovsky (1987). SOC refers to a person's general way of perceiving and controlling the environment for meaningful and appropriate action.

Direct effects of weak SOC on strain have been recognized (Kalimo and Vuori, 1991), as has the relation of strong SOC to perceived competence and life satisfaction (Kalimo and Vuori, 1990). Strain symptoms of technical designers with strong SOC decreased with greater job demands, while they increased with work demands in designers with weak SOC (Feldt, 1996). Active coping styles have been shown to be associated with improved well-being, whereas avoidance coping predicted ill health (Nowack, 1991). Situational choice is a strategy that has been emphasized (Perrez and Reicherts, 1992; Thompson and Page, 1993).

33.4.3 Burnout

The construct of burnout was introduced in the mid 1970s to describe a long-term stress syndrome among professionals in the health care sector (Maslach and Schaufeli, 1993). The syndrome develops as a result of the chronic depletion of resources and is characterized by three clusters of symptoms. These have been defined with small variations as: (1) exhaustion (emotional, intellectual, physical), (2) depersonalization (emotional detachment and cynicism), and (3) lowered professional accomplishment, self-efficacy, and self-esteem. A clear analogy can be seen between burnout and the exhaustion phase of the general adaptation syndrome (GAS) defined by Selye (1956).

An incongruity of organizational goals and professional standards with the personal and organizational competencies available for the attainment of those goals appears to be one of the crucial contributors to burnout. A number of other job stressors such as high demands and role conflicts seem to coincide with burnout as well. Other main contributors include a lack of job resources such as social support and autonomy. To indicate the multitude of organizational factors that can contribute to burnout, Cox, Kuk, and Leiter (1993) proposed the need to recognize organizational healthiness as a whole when studying the causation of burnout. Some authors also emphasize individual vulnerabilities such as over-commitment to long-term goals (Hallsten, 1993) and limitations in the person's coping strategies (Leiter, 1991) as factors increasing the probability of burnout.

33.4.4 Physiological Changes, Pain, and Discomfort

Physiological responses were the first category of stress responses noted in the early research on stress by Selye (1956). These changes were interpreted as responses to external demands for physical adaptation. Later a similar reaction pattern was recognized when persons were also exposed to psychological stressors (Levi, 1972b; Levi, Frankenhaeuser, and Gardell, 1982). The characteristic physiological response pattern includes the increased production of adrenal hormones, i.e., corticosteroids by the adrenal cortex and catecholamines by the adrenal medulla. After cognitive appraisal that the stress exceeds a person's coping capacity, these systems are activated via the neuroendocrine and the autonomic nervous system (Dantzer, 1989).

Increased activation of adrenal functions in many kinds of stressful situations has been recognized (Frankenhaeuser, 1991; Phillips, 1989). The picture of the hormonal response pattern has been differentiated according to the type of job stressors. Stress caused by work with external constraints and poor opportunity for personal control in addition to related feelings of helplessness stimulates primarily the excretion of cortisol. In opposition to this, the excretion of catecholamines is increased in situations with high demands and high effort. Frankenhaeuser and her colleagues have developed a psychobiological model of stress that specifies the psychosocial factors more likely to induce increases in cortisol and catecholamines (Frankenhaeuser, 1986; Frankenhaeuser and Johansson, 1986; Lundberg and Frankenhaeuser, 1980). There are two different neuroendocrine responses to the psychosocial environment: (1) secretion of catecholamines via the sympathetic–adrenal medullary system, and (2) secretion of corticosteroids via the pituitary–adrenal–cortical system.

Different patterns of neuroendocrine stress responses occur depending on the psychosocial characteristics of the environment. These psychosocial factors are effort and distress. The effort factor involves elements of interest, engagements, and determination, whereas the distress factor involves elements of dissatisfaction, boredom, uncertainty and anxiety (Frankenhaeuser, 1986). Effort with distress is accompanied by increases in both catecholamine and cortisol secretion. Effort without distress is characterized by increased catecholamine secretion, but no change in cortisol secretion. Distress without effort is generally accompanied by increased cortisol secretion, with a slight elevation of catecholamines.

Changes in the hormonal functions and in the activity of the autonomic nervous system lead to altered activity in the functions of most organ systems of the body, including the gastrointestinal, cardiovascular, and musculoskeletal functions. The practical importance of these changes is largely based on the feelings of pain and discomfort that they cause (Goldberger and Breznitz, 1993).

33.4.5 Psychoneuroimmunology

Since Selye's discovery of the triadic effects of stressful situations on an organism's attempts to adapt to stress, it has been recognized that stress can adversely affect the immunological system's capacity (Selye, 1956). He observed the shrinking of the thymus and the reduced capability of the lymphatic system as one aspect of the triadic response to internal disequilibrium created by stress. The effects of prolonged stress produced a general "weakening" of the body, and its ability to protect itself.

Biondi and Kotzalidis (1994) summarized research findings from animal studies of induced stress and immunological effects. The effects of stress are: (1) thymic and spleen atrophy, (2) decreased primary and secondary immune response, (3) increased susceptibility to infection, (4) decreased antiviral immunity, (5) decreased antibody forming ability, (6) changes in NK-cell activity (both decreases and increases), (7) changes in T-cell activity (mainly decreases, but some increases), and (8) decrease in B-cells.

The evidence for the same type of significant immunological system changes in humans is not as substantial as the changes defined for animals in controlled studies (Biondi and Kotzalidis, 1994). This may result from the difficulty of controlling the exposures and the confounders in the human research.

Ader, Cohen, and Felten (1995) provide a review of the field of psychoneuroimmunology. They illustrate the logic of a close connection between psychological state and immunological response. This relationship is mediated through neural connections to the primary and secondary lymphoid organs, as well as to the effects on the pituitary gland and hormonal responses. The hormones that are released when stress occurs can influence lymphatic response.

33.5 METHODS OF MEASUREMENT

33.5.1 Self-Report Measures

Emotional reactions to work belong to the crucial factors to be measured to understand and assess work-related health. The assessment should preferably cover positive and negative emotions. Different types of instruments will complement each other. Emotional reactions can be measured with questionnaires, state–trait measures, checklists, projective tests, and biological measures. It is important to use different sources of information (self-report, professional rating, family rating, and tests) (see Marsella, 1994, for a comprehensive review). Questionnaires are practical tools, but awareness of possible sources of error is important (Jex, Beehr, and Roberts, 1992).

Coping strategies are measured mostly with questionnaires (O'Driscoll and Cooper, 1994). The "ways of coping" checklist includes items covering problem-focused coping and emotion-focused coping (Folkman and Lazarus, 1980). An adaptation of the same basic type of measure to the appraisal of work stress situations is available as a 63-item checklist (Dewe, 1992). These measures are context specific and so are intended to assess coping as a form of behavior rather than coping styles over time.

Work and organizational factors can be assessed with "objective" and "subjective" methods. Objective methods usually involve expert assessment, e.g., observation. Subjective measures are questionnaires and interviews used for workers' self-assessment. Both types of measures can be used to describe how work and work organization function. Questionnaires and interviews also make it possible to assess how work and work organization are experienced, i.e., how satisfactory or stressful they are.

Measurement of work and organization has been extremely diverse, and ad hoc measures have often been used for research. Reviews are available for an overview (Cook, Hepworth, Wall, and Warr, 1981; Cox and Ferguson, 1994; Dewe, 1992; International Labour Office, 1986). A validated observation method for the assessment of psychological work stressors with a 13-dimension checklist is available for nonspecialist users (Elo,

1994). Numerous questionnaires are available and some of them have computer programs for data analysis such as the Occupational Stress Inventory (OSI), Occupational Stress Questionnaire (OSQ) (Elo, Leppänen, Lindström, and Ropponen, 1992) and the Stress Profile (Setterlind and Larsson, 1995). All of these involve items on stress responses and individual modifying factors as well.

For the assessment of burnout, specific questionnaires are available. Maslach Burnout Inventory (MBI) (Maslach and Jackson, 1981) is meant for human service employees. It assesses with 25 items the three burnout dimensions: emotional exhaustion, personal accomplishment, and depersonalization. A generally applicable burnout Inventory has been developed on that basis recently (Schaufeli, Leiter, and Kalimo, 1995).

Social support measures are numerous and are primarily questionnaires. The Daily Interaction Record in Organizations (DIRO) represents an effort to systematize the measurement with a so-called event-contingent method that requires a report every time an event meets a preestablished criterion of support (Buunk and Peeters, 1994).

33.5.2 Physiological Measures

Physiological methods can be used for the evaluation of strain caused both by physically and mentally demanding work. Recording of physiological measures such as heart rate (HR), heart rate variability (HRV), blood pressure (BP), respiratory rate (RR), and electromyography (EMG) can be made over a continuous time period. These measures are associated with a physiological system that can be considered fast acting in response to internal and external stimulation. They have the potential for conveying information that is of a momentary nature.

Biochemical measures can be obtained from various fluids of the body, such as urine, blood, sweat, and saliva. Analysis of urine has been the most popular choice. For a minimum disturbance in the work situation the collection of saliva and sweat are recommended. The most often measured biochemicals include adrenaline, noradrenaline, cortisol, steroids, glucose, uric acid, triglycerides, and lipids.

Justification for the use of these measures in the analysis of occupational stress stems from (1) clinical evidence that has demonstrated the chronic elevation of biochemicals (adrenaline, noradrenaline, triglycerides) caused by emotional reactions, which can lead to functional disturbances in various organs and organ systems, which, in turn, may lead to psychosomatic and cardiovascular diseases; (2) natural daily rhythms of these measures, which allow for evaluation of the degree to which these measures change from their normal rhythmic pattern when changes in the work requirements are initiated; (3) the availability of numerous automated processes for isolation and subsequent measure of the biochemical constituents of interest; and (4) the ability to obtain the needed fluids easily during and after work periods.

The ability to obtain indexes of stress after work hours is a distinct advantage. At times there are reasons to believe that constraints in a particular work situation may impede the individual's response to stress and produce a delayed reaction that occurs hours after the individual has left the workplace. Also, the worker may be more apt to exhibit certain responses after work hours.

In contrast to measures such as HR and RR, the practical restriction on the collection procedures for biochemical measures (and the relatively slow response they exhibit) necessarily dictates that they reflect responses to stress over fairly long periods (e.g., several hours or days). As a result, biochemical measures are relatively insensitive for identifying which components of the task are most stressful. This factor alone is most critical in selecting between these two categories of measures for the purpose of evaluation of specific situations.

Practical problems with biochemical measures concern the need to exercise rigid control in the kinds and quantity of food, liquid, and drugs ingested by the worker prior to and during the measurement process. All physiological assessments may involve procedural artifacts (Fried, 1988).

33.6 OCCUPATIONAL STRESS AND DISEASE

33.6.1 Health Risk Behavior

Behavioral changes manifested in stressful work situations may reflect efforts to cope, or they may develop as a part of the overall symptom pattern. An increase of alcohol consumption is an example of the former, while passivity for leisure may be a part of the latter. Empirically detected relationships between occupational stress and health risk behavior have been weak. Health risk behaviors often have multiple causation and, therefore, the specific role of job stress is difficult to detect from among other considerations such as socioeconomic, cultural, and gender-related factors. Another major problem is the difficulty in finding reliable measures of alcohol and drug consumption in working populations. However, some evidence is available that shows that an increased use of alcohol by a working person is related to work stress (see Kalimo and Mejman, 1987), and this evidence is probably an underestimation because of the difficulties in measurement and interpretation.

Work stress has been found to relate positively to an increase in smoking intensity (Green and Johnson, 1990) and negatively to a cessation of smoking (Caplan et al., 1975; Westman, Eden, and Shirom, 1985). Field studies have not always shown a relation between stress and smoking, whereas some stress experiments in the laboratory have led to a significant increase in smoking by persons under stress (Pomerleau and Pomerleau, 1987). Job demands have been related to smoking and sedentary behavior in Swedish working women and men when age and education were controlled. Job resources including autonomy were unrelated to smoking but predicted regular physical exercise (Johansson, Johnson, and Hall, 1991).

Assistance from others at work, considered as a form of social support, was associated with frequent physical exercise in Finnish working women (Vuori, 1994). Sedentary behavior and social passivity have been shown to result from fatigue and long recovery times among workers in machine-paced tasks with poor autonomy, and to predict self-reported physical illness (Nowack, 1991). These studies show that physical exercise which usually is considered as a buffer against stress rather than it's outcome (Cox, Gotts, Boot, and Kerr, 1988) is also subject to a stress influence.

33.6.2 Cardiovascular Diseases

Many work organization factors have been linked to short-term and long-term stress reactions. Short-term stress reactions include increased blood pressure, adverse mood states, and job dissatisfaction. In the previous sections, we have mentioned many research studies on work features and stress reactions. Several studies have shown a link between overload, lack of control and work pressure, and increased blood pressure (Matthews, Cottington, Talbott, Kuller, and Siegal, 1987; Van Ameringen, Arsenault, and Dolan, 1988; Schnall et al., 1990). Other studies have found a link between job future uncertainty, lack of social support and lack of job control, and adverse mood states and job dissatisfaction (Karasek, 1981; Sainfort, 1991; Smith et al., 1992). Long-term stress reactions include cardiovascular disease and depression. Studies have shown that job stressors are related to increased risk for cardiovascular disease (Karasek, 1981; Karasek et al., 1988).

Epidemiological studies of stress have found associations between the type of occupation and cardiovascular diseases (for reviews see Fletcher, 1988, 1991; Kasl, 1978). In a very large cohort study, Alfredsson, Spetz, and Theorell (1985) found that among men, hectic work and few possibilities to learn new things were associated with a higher incidence of hospitalization for myocardial infarction; and that among women, hectic and monotonous work was associated with a higher incidence of hospitalization for myocardial infarction. Lack of decision latitude has also been linked to increased risk of cardiovascular disease (Karasek, 1981; Pieper, LaCroix, and Karasek, 1989).

Individual factors such as personality characteristics moderate the influence of work stressors on coronary heart disease (CHD). The Framingham Heart Study showed that among young women type A behavior was associated with CHD (Haynes, Feinlab, Levine, Scotch, and Kannel, 1978; Haynes, Feinlab, and Kannel, 1980), and that type A behavior and suppressed hostility were important risk factors for CHD in both men and women (Haynes et al., 1980). Other risk factors for CHD include drinking, smoking and obesity (La Vecchia, Gentile, Negri, Parrazzini, and Franceschi, 1989; Wong, Cuppies, Ostfeld, Levy, and Kannel, 1989). High blood pressure is also a predictor of CHD (Fiebach et al., 1989; Wong et al., 1989). For a review of biological and behavioral predictors of CHD in women see Matthews, Kelsey, Meilahn, Kuller, and Wing (1989).

Several studies have shown a link between blood pressure and job stress, especially workload, work pressure, and lack of job control have been related to increase in blood pressure. Rose, Jenkins, and Hurst (1978) found that workload was associated with increased systolic and diastolic blood pressure. In a cross-sectional study of 375 female hospital workers, Van Ameringen et al. (1988) found that intrinsic pressure related to job content was related to increased standing diastolic blood pressure. Matthews et al. (1989) found that having little opportunity for participating in decisions at work was related to increased diastolic blood pressure.

Longitudinal studies of job stress and blood pressure show that blood pressure increases were related to technological change and complexity in a 1-year study of Japanese blue-collar workers (Kawakami, Haratani, Kaneko, and Arabi, 1989), being a manual worker versus a professional in a 6-year study of Australian working men (Jenner, Puddey, Beilin, and Vandongen, 1988), and various measures of work stressors in a 5-year study of Australian government employees (Chapman, Mandryk, Frommer, Edye, Ferguson, 1990). Studies have further demonstrated the link between hypertension and job stress (Schnall et al., 1990), and between emotions (e.g., anger and anxiety) and increased blood pressure (James, Yee, Harshfield, Blank, and Pickering, 1986).

33.6.3 Musculoskeletal Disorders

Studies indicate a potential link between job stress and upper extremity cumulative trauma disorders (CTDs) (Smith et al., 1992; NIOSH, 1992), and theories on these links have been proposed (Smith and Carayon, 1996). Hadler (1990, 1992) has stated that stress may be the primary cause of the symptomology associated with many upper extremity CTDs.

There are psychobiological mechanisms that make a connection between stress and CTDs plausible and likely. At the personal level, psychological stress can lead to an increased physiological susceptibility to CTDs by affecting hormonal responses and circulatory responses that exacerbate the influences of the traditional risk factors. In addition, psychological stress can affect employee attitude, motivation, and behavior that can lead to risky actions that increase CTD risk. At the organizational level, the policies and procedures of a company can affect CTD risk through the design of jobs, the length of exposure to stressors, establishing work–rest cycles, defining the extent of work pressures and establishing the psychological climate regarding socialization, career, and job security. In addition, the organization defines the nature of the task activities (work methods), employee training, availability of assistance and supervisory relations.

Stress can influence the occurrence of CTDs through its effects on a person's psychological and behavioral reactions. Thus, stress can affect psychological moods, work behavior, coping style and actions, motivation to report injury, and motivation to seek treatment for a CTD injury or symptoms of impending injury. Upper extremity CTDs involve significant pain. Stress may serve to increase the perception of pain or greater severity of pain and has been related to psychological stress among patients with low back pain (Atkinson, Slater, Grant, Patterson, and Garfin, 1988; Ryden, Lindal, Uden, and Hansson, 1985).

A few studies have shown a link between work organization and musculoskeletal disorders, such as symptoms in the back, neck, and shoulders (Linton and Kamwendo, 1989; Theorell, Ringdahl-Harms, Ahlberg-Hulten, and Westin, 1991; Ursin, Endresen, and Ursin, 1988). For a review of these studies, see Bongers and de Winter (1992). Some studies have examined work organization and upper extremity CTDs (Smith et al., 1992; NIOSH, 1990, 1992).

Figure 33.2 provides an illustration of how the exposure to psychosocial stressors at work can lead to short-term stress reactions, and if these exposures become chronic they can produce a variety of disease states such as cardiovascular disease and cumulative musculoskeletal trauma.

33.7 INDIVIDUAL AND GROUP-ORIENTED INTERVENTIONS

Table 33.1 summarizes the various categories of individual stress reduction methods that have been evaluated. These will be discussed in more detail in the following sections.

33.7.1 Stress Management and Coping Training

Specific programs have been planned and implemented for type A people, who are thought to be at risk for coronary heart disease because of their exaggerated manner of response to stressful situations. The type A behavioral pattern is characterized by impatience and irritability.

Most of these intervention programs have focused on management of stress and tension with generally applied strategies, without special measures against irritability and time urgency. For instance, a combination of physical exercise, weight control and cognitive training has been used (Roskies et al., 1989). Awareness raising of the tendency to hyperreact to daily hassles, and an acquisition of new coping patterns practiced repeatedly until they were adopted were used as methods of cognitive training. Cognitive training changed behavior and was found most efficient in achieving its purpose. Physical indices of stress did not change as a result of the program.

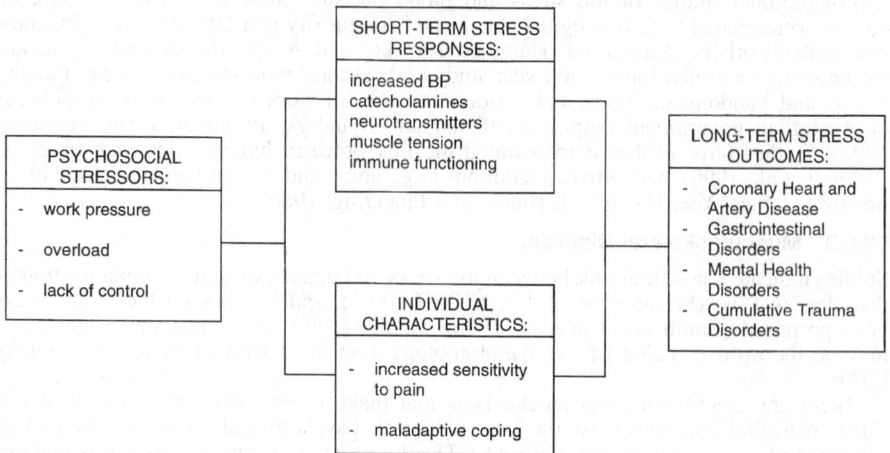

Figure 33.2 Psychobiological mechanisms of stress/disease relationship. (Adapted from Smith and Carayon, 1996.)

33.7.2 Cognitive-Behavioral Approaches

The cognitive-behavioral approach is an umbrella heading for a variety of strategies which are meant for changing the appraisal of the stressful situation and the psychological responses to the situation. Among the guiding principles of these strategies are (1) the individual's responses to the environment are a reaction to their own cognitive interpretations of it; (2) cognitions (thoughts), emotions, and behaviors are causally related to each other; and (3) a person's expectancies, beliefs, and attributions predict negative consequences (see review by Murphy, 1988).

The stress inoculation training developed by Meichenbaum (1985) involves the steps of preparation, training of skills, and training for application. In the stage of preparation the client is guided to recognize the connections of his or her maladaptive thoughts and beliefs with the maladaptive emotional reactions. The second step aims at the development of skills through giving advice about how to confront stressors and about efficient coping strategies. At the third phase the client practices the new skills. Stressful stimuli may be used at this step to increase the ecological validity of the situation and to "desensitize" the individual.

In the procedure proposed by Bramson (1985), clients are trained in coping through six phases that can be described with the following instructions: (1) assess the situation, (2) stop wishing the difficult situation would be different, (3) get some distance between you and the difficult situation, (4) formulate a coping plan, and (5) monitor the effectiveness of your coping strategy. To succeed, all coping training demands a strengthening

Table 33.1 Individual Ways To Deal with Occupational Stress

Intervention	Mechanism of Stress Control	Research and References
1. Coping training	Modify perceptions or behavior	Roskies et al. (1989)
2. Cognitive/behavioral	Modify perceptions and behavior	Murphy (1988) Meichenbaum (1985) Bramson (1985) Janis (1985)
3. Relaxation and biofeedback	Modify physiological reactions	Benson (1985) Ivancevich and Matteson (1988)
4. Debriefing	Cognitive restructuring	Braverman (1992)
5. Lifestyle change	Modify behavior	Ivancevich and Matteson (1988)

of self-confidence, hope, and perceived control. Commitment and personal responsibility must also be induced (Janis, 1985).

33.7.3 Relaxation and Biofeedback

Systematic relaxation is a means to prevent and control strain responses via a routine of procedures, which may vary. One option is the following (Benson, 1985): (1) choice of a quiet environment, (2) fixing concentration to a chosen "mental device" (a sound, a word, an object to fix gaze), (3) adoption of a passive attitude of "let it happen," and (4) maintaining of a comfortable position. Instructions can be given by a counselor or the person can think about them him- or herself. The result is a feeling of calmness and relaxation in a typical successful case. Biofeedback measures are sometimes used to help the person to recognize signs of relaxation in the body. For instance, electromyography (EMG), electroencephalography (EEG), and continuous blood pressure measurement can be used (Ivancevich and Matteson, 1988).

33.7.4 Debriefing

Debriefing is planned as a program for the immediate treatment of traumatic psychological experiences in order to avoid posttraumatic stress disorders and to prevent the development of other long-term problems as a result of an untreated trauma. The method is applicable in the situations of accidents, near-accidents and injuries, violence, catastrophes, and any situation involving acute dramatic crisis. A debriefing program comprises an initial session with the victim and the counselor within preferably 1 to 2 h after the crisis event. Rescue personnel may also be involved in this session, which is meant for an immediate exchange of the experiences of those involved. Instructions for effective coping may also be given in this session.

The main debriefing sessions are organized one to two days after the event, when people already have some recovery. The program follows a systematic plan (Braverman, 1992). Similar types of interventions may be organized in the case of organizational crises such as downsizing and mergers. In these instances a company-wide program with multiple—not only individual—approaches would be recommended.

33.7.5 Health Education and Lifestyle Change

Health education is a classic method for prevention of health problems via changes in health habits and lifestyle. Withdrawal from smoking, weight control, and physical exercise are typical topics of health education. In the more comprehensive programs, instructions are also given to improve sleeping habits, nutrition, alcohol consumption, social activities, etc. (Ivancevich and Matteson, 1988). Health education is often carried out as campaigns called wellness programs, but they also belong to the regular functions of the occupational health services.

33.7.6 Career and Skill Development

Career development interventions are designed to meet the needs of individuals in various career stages, of various ages, and both genders, and the unemployed. Interventions can take the form of using self-assessment tools, individual counseling, information services, and assessment and developmental programs (Russell, 1991). The rapid technological and structural changes in companies have illustrated the need for the development of employees' skills. At an organizational level, self-assessment tools for the personnel such as career workbooks can guide individuals in determining their strengths and weaknesses and in identifying job and career opportunities. These have been used as the basis for career planning workshops at an organization level. Other ways to support career development at the work site level are information service systems including job-posting systems. An assessment center can be used for improving the understanding of one's own skills and goals. Training programs have been initiated based on these ideas. Mentoring programs are also tools in helping to form closer relationships between junior and senior coworkers. These relationships provide both opportunities for career development and social support and friendship (Kram, 1985).

Skill development is positively related to career or job commitment (Aryee and Tan, 1992). Skill development is dependent on individual motivation or outside incentives or pressures, especially when job demands and work organization are changing. The learning organization model refers to an organization that emphasizes continuous learning and development at individual, group, and organization levels (Beck, 1989).

Traditionally, interventions to prevent career stagnation have been suggested as a part of career development programs because these interventions are either periods of frustration, or an opportunity for new challenges or reappraisal of life goals (Weiner, Remer, and Remer, 1992). The creation of new personal and organizational goals is necessary. During midlife, people become more preoccupied with decreasing job opportunities, in general, and their changing work role identity, although they still have a rather strong need for career advancement (Buunk and Janssen, 1992).

33.8 IMPROVING WORK AND ORGANIZATIONS

Improving work and work organizations is usually a practically oriented activity supported by consultants and other specialists from inside and/or outside the organization. The theoretical background is often a loose combination of several theories about organizations and human behavior. The intervention traditions and models are also strongly dependent on current social values. The goal of a specific intervention or program at work obviously influences the choice of the type of developmental intervention. Interventions focus either on the functioning of an organization, its productivity, profit, and quality of production and services, or they may aim to promote workers' health, aspiration, and competence, by making work and its characteristics favorable from the human point of view.

Table 33.2 summarizes the various categories of organizational approaches for stress reduction that have been evaluated. These will be discussed in more detail in the following sections.

33.8.1 Employee Assistance Programs

Employee assistance programs (EAP) are strategies developed in companies to help their employees to overcome problems that hamper organizational behavior and work performance. The program ideology was first developed for the intervention against alcohol abuse and to help the alcoholic. Drug abusers were soon included in the programs as well. Positive experiences were found in the sense of saved labor and money. As a result, the

Table 33.2 Organizational Ways To Deal with Occupational Stress

Intervention	Mechanism of Stress Control	Research and References
1. EAPs	Provide coping resources	Philips and Mushinski (1992)
2. Career development	Enhance technical skills and knowledge	Russell (1991) Kram (1985) Aryee and Tan (1992) Beck (1989)
3. Healthy organization	Provide a comfortable psychological work environment	Cox and Howart (1990) Cooper & Cartwright (1994) Rosen (1992) Murphy et al. (1995) Lindström (1995)
4. Ergonomic improvements	Provide a comfortable physical work environment	Konz (1990) *Handbook*, Chapter 23 *Handbook*, Chapter 34
5. Organizational change	Restructure organization for greater psychological comfort	Nadler (1988) *Handbook*, Chapter 15 *Handbook*, Chapter 18 Schweiger and Dennis (1991)
6. Job redesign	Modify tasks, responsibilities, relationships, and/or technology	Smith and Carayon (1995) *Handbook*, Chapter 14 Hackman and Oldham (1980) Lawler (1986)

same basic model was adopted for interventions with many other problems of the employees including work-related stress, personal problems, and difficulties in the family.

The programs have been modified further, and they may now also include assistance for financial and legal matters. The EAP is a referral program. The program design is for the clinician or the counselor to assess the situation and make a referral. The program is not meant to involve long-term support or therapy. Their utility has been shown in the successful prevention of problems met by the personnel administration, in saving of the occupational health care costs and in the improvement of productivity. See Philips and Mushinski (1992) for a review.

33.8.2 Healthy and Productive Work Organization

The organizational health approach (Cox and Howart, 1990) emphasizes the relationship between organizational factors and the workers' health. Usually short-sighted profit-making organizations fail in responding to market needs and in keeping the key persons in the organizations. Cox and Howarth see organizational health as the capability of an organization to function effectively in relation to various environmental factors, and to respond to environmental changes. They describe an organization at an objective and a subjective level. At the subjective level, an organization is represented in the people's understanding of it and attitude toward it and is reflected in the organizational culture.

As prerequisites for a healthy work organization, Cooper and Cartwright (1994) list factors that usually have been seen as related to occupational stress. On the intervention side they underline the central role of occupational stress and its management, at both the individual and the organizational level. A truly healthy organization is one in which secondary stress management prevention or tertiary employee assessment programs are not needed. Regular stress audits, employing the use of job and organizational stress screening, should be a tool in the maintenance of organizational health. When carrying out such audits, the companies and workplaces should have organization-directed strategies.

A value-based organization using the healthy company model involves strategies for profit making and valuation of people (Rosen, 1992). The management plays a key role in the implementation of these strategies, which are open communication and employee involvement, learning and renewal, valued diversity, institutional fairness, equitable rewards and recognition, and general economic security. People-centered technology as well as a health-enhancing environment and meaningful work are important. The balance between work and family life is also emphasized.

The occupational health approach to a healthy work organization applied by the National Institute for Occupational Health and Safety (NIOSH) (United States) and the Finnish Institute of Occupational Health (FIOH) have grounded their value base clearly in the health and well-being of people. These models for analysis and interventions are based on job and organizational stress, followed by individual- and organization-oriented interventions (Lindström, 1995; Murphy et al., 1995). This occupational health–oriented approach serves as a good basis when socially sustainable development in an organization is the goal of job design or organizational development. Although early intervention research in Finland is promising, it is too early for empirical evidence about these healthy organization approaches.

33.8.3 Improving Environmental and Ergonomic Conditions

The reader is referred to several other chapters (Chapters 20–30, 34, and 35) for explicit details on how to implement ergonomic measurements, evaluation procedures, and improvements. Ergonomics is the science of fitting the environment and activities to the capabilities, dimensions, and needs of people. In this chapter we have a specific interest in ergonomics since it is applied to adapt working conditions to the physical, psychological, and social nature of the person. Good ergonomic practice suggests that all aspects of the work system be included in job redesign improvements. From an ergonomic perspective, the design of the environment, workstations, tasks, work organization, and the tools or technology should reasonably accommodate the employees' capacities, dimensions, strengths, and skills.

Critical aspects of the success of ergonomic interventions for providing psychosocial benefits are a strong management support for the ergonomics program; the involvement of managers and supervisors in the ergonomics program; the involvement of employees in the program and their willingness to participate in defining problems, proposing solutions, and improving their work practices; and the application of engineering improvements to reduce biomechanical and physiological loads.

33.8.4 Strategy for Job Redesign

The reader is referred to the Chapters 14 and 18 for details about how to maximize job design considerations. The following section defines a strategy for redesigning jobs to reduce job stress, but does not elaborate the specific characteristics of job elements needed to achieve proper job design, because these have been defined in detail in these other chapters and earlier in this chapter.

We must recognize that there are no "perfect" jobs that provide complete psychological satisfaction for all employees and are free of all stress. In any redesign process there are tradeoffs among specific improvements and achieving the best "overall" job design solution. These tradeoffs require us to think about how to best "balance" the various needs to achieve the solution that will have the greatest positive benefit for employee health and productivity. Bohnhoff, Armin, Brandt, and Henning (1992) have called for a balance between people and technology.

Smith and Sainfort (1989) have proposed a balance among various elements of the work "system" (Figure 33.3). The essence of this approach is to reduce the negative health consequences caused by stress by "balancing" the various elements of the work system to reduce unwanted loads. It goes without saying that proper work organization and job design can best be achieved by providing those characteristics of each individual work system element that meet recognized criteria for proper physical loads, work cycles, and job content and that provide for individual physiological and psychological needs. The best designs will eliminate all sources of stress, job dissatisfaction, and discomfort.

For instance, proper workload can be established by use of appropriate methods of work analysis (Konz, 1990), whereas aspects of job control can be developed through the use of worker participation (Lawler, 1986; see Chapter 18). The negative influences of inadequate skill to use new technology can be offset by increased training of employees. Likewise, the adverse influences of low job content that creates repetition and boredom can be balanced by an organizational supervisory structure that promotes employee in-

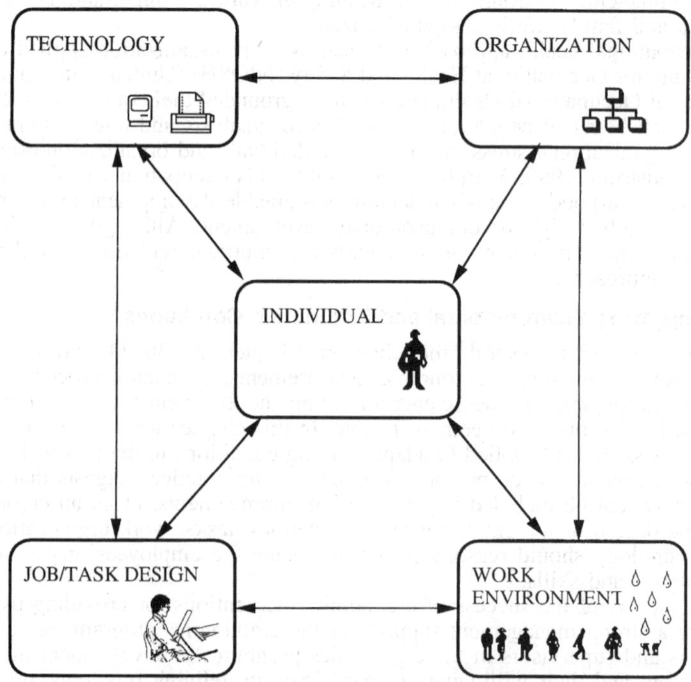

Figure 33.3 Model of the work system. (From Smith and Sainfort, 1989.)

volvement and control over tasks, and job enlargement that introduces task variety. Organizational structure can be adapted to enriched jobs in order to provide support to the individual such as increased staff or shared responsibilities or increased financial resources.

A major advantage of this model is that it does not highlight any one job factor such as job content, or a small set of factors such as autonomy and social relations. Rather, it examines the design of jobs from a holistic perspective to emphasize the potential positive elements in a job that can be used to overcome the adverse elements when such adverse elements cannot be eliminated or modified.

33.8.5 Organizational Change Management

The management of change depends on the type of change, which can be future oriented, or past oriented. The changes can be through gradual developmental or strategic actions (Nadler, 1988). The change processes are now more often structural ones, meaning that also paradigmatic and cultural changes are expected. The change can, however, cover narrow or more technical issues, e.g., the implementation of more advanced information technology, a new service or product.

All planned changes and development interventions should follow the developmental cycle. They start with problem or goal definition, and the committing and informing of all interest groups at the workplace. They then proceed to analysis of the present situation, based on survey results, and to joint planning of the intervention. The intervention itself needs clear goals, subgoals, steps, a time schedule, duties for people, and ways to overcome obstacles and problems during the process. The process seldom proceeds as planned; it needs reevaluation and checking of the methods and even the goals during the process. The intervention process and the outcome should be evaluated. This helps the organization to learn about its experiences for the next project. This part is usually left out because very often new changes and projects are already arising before the end of the former changes and projects.

A major cause of worker resistance to new work activities is the fact that changes often appear "out of the blue" without the worker knowing about an impeding change. It is of utmost importance for the successful implementation of changes in work processes, and subsequent enhancement of worker health, performance, and satisfaction, that organizations have a transition policy that includes worker participation in all stages of the change process. That is, workers should participate in the planning, then in the selection of equipment, and finally in the daily operation of the work system.

The most difficult change situations are ones involving the downsizing of activities, or the merger of earlier independent units or organizations. These usually create insecurity about the future and provoke anxiety among the personnel (Kilpatrick, Johnson, and Jones, 1991). Open communication has proved to be necessary and positive from the workers' point of view during the whole structural change process (Schweiger and Dennis, 1991).

33.9 SUMMARY

Psychosocial aspects of the workplace can have profound influences on the individual's health and well-being, satisfaction with work, and productivity. There are various characteristics of working that have generally been shown to have negative physical and/or psychological consequences, for example, machine-paced work, a lack of task control, high job demands, shiftwork, poor supervisory relations, and organizational downsizing. The extent of the negative consequences varies from individual to individual depending on the perceived threat, individual constitution, and coping mechanisms. When the person is vulnerable, substantial harm can occur, including disability and death.

Several organizational behavior approaches can be applied that will diminish or eliminate the harmful effects of poor workplace design that leads to psychological distress. Examples of these approaches include removal of the sources of stress (stressors) through job redesign, balancing the positive and negative aspects of work to provide an acceptable level of stress, developing a corporate culture that promotes self-worth and productive employment (the "healthy organization"), providing organizational resources for dealing with the consequences of stress (EAPs), and enhancing employee skills and career development.

In addition, there are personal ways for dealing with stress, for example, relaxation methods, lifestyle improvements, exercise, psychotherapy, cognitive restructuring, and social support.

If the psychosocial aspects of work design are properly developed and implemented, there can be benefits for employee's health and performance, and a "healthy corporation" will emerge. If job stress is not properly controlled, then the consequences can be reduced employee job satisfaction, health, and morale, and reduced employee and corporate performance.

REFERENCES

Ader, R., Cohen, N., and Felton, D. (1995).Psychoneuroimmunology: Interactions between the nervous system and the immune system. *Lancet*, *345*, 99–103.

Alfredsson, L., Spetz, C.-L. and Theorell, T. (1985). Type of occupation and near-future hospitalization for myocardial infarction and some other diagnoses. *International Journal of Epidemiology*, *14*, 378–388.

Amick, B. C., and Smith, M. J. (1992). Stress, computer based monitoring and measurement systems: A conceptual overview. *Applied Ergonomics*, *1*, 6–16.

Antonovsky, A. (1987). Health promoting factors at work: The sense of coherence. In R. Kalimo, M. A. El-Batawi, and C. L. Cooper, Eds., *Psychosocial factors at work and their relation to health* (pp. 153–167). Geneva: World Health Organization.

Aryee, S., and Tan, K. (1992). Antecedents and outcomes of career commitment. *Journal of Vocational Behavior*, *40*, 288–305.

Atkinson, J. H., Slater, M. A., Grant, I., Patterson, T. L., and Garfin, S. R. (1988). Depressed mood in chronic low back pain: Relationship with stressful life events. *Pain*, *35*, 47–55.

Barling, J., and Kryl, I. P. (1990). Moderators of the relationship between daily work stressors and mood. *Work & Stress*, *4*, 319–329.

Barling, J., and Macintyre, A. T. (1993). Daily work role stressors, mood and emotional exhaustion. *Work & Stress*, *7*, 315–325.

Beck, M. (1989). Learning organizations—How to create them. *Industrial and Commercial Training*, *21*(3), 21–28.

Benson, H. (1985). The relaxation response. In A. Monat and R. S. Lazarus, Eds., *Stress and Coping: An Anthology* (2nd ed., pp. 315–321). New York: Columbia University Press.

Bhagat, R. S. (1982). Conditions under which stronger job performance-job satisfaction relationships may be observed. *Academy of Management Journal*, *29*, 772–789.

Biondi, M., and Kotzalidis, G. D. (1994). Pyschoneuroimmunology today: current concepts and relevance to human disease. In C.E. Lewis, C. O'Sullivan, and J. Barraclough, Eds., *The Psychoimmunology of Cancer* (pp. 3–54). Oxford: Oxford University Press.

Bohnhoff, A., Brandt, D., and Henning, K. (1992). Dual design approach as a tool for the interdisciplinary design of human-centered systems. *International Journal of Human Factors in Manufacturing*, *2*, 289–301.

Bongers, P. M., and de Winter, C. R. (1992). *Psychosocial factors and musculoskeletal disease—A review of the literature* (NIPG Publi. No. 92.028). Leiden: Nederlands Instituut voor Praeventieve
Gezondheidszorg TNO.

Bramson, R. M. (1985). Toward effective coping: The basic steps. In A. Monat and R. S. Lazarus, Eds., *Stress and Coping: An Anthology* (2nd ed., pp. 356–370). New York: Columbia University Press.

Braverman, M. (1992). Post trauma crisis intervention in the workplace. In J. C. Quick, L. R. Murphy, and J. J. Hurrell, Jr., Eds., *Stress & Well-Being at Work: Assessments and Interventions for Occupational Mental Health* (pp. 299–316). Washington, DC: American Psychological Association.

Burke, R. J., and Greenglass, E. R. (1994). Towards an understanding of work satisfactions and emotional well-being of school-based educators. *Stress Medicine*, *10*, 177–184.

Buunk, B. P., and Janssen, P. P. M. (1992). Relative deprivation, career issues, and mental health among men in mid-life. *Journal of Vocational Behavior*, *40*, 338–350.

Buunk, B. P., and Peeters, M. C. (1994). Stress at work, social support and companionship: Towards an event-contingent recording approach. *Work & Stress*, *8*, 177–190.

Cannon, W. B. (1914). The interrelations of emotions as suggested by recent physiological researchers. *American Journal of Psychology*, *25*, 256–282.

Caplan, R. D., Cobb, S., French, J. R. P., Harrison, R. V., and Pinneau, S. R. (1975). *Job demands and worker health*. Washington, DC: U.S. Government Printing Office.

Carayon, P. (1993) Longitudinal studies of job design and VDT use: Overview and synthesis. In H. Luczak, A. Cakir, and G. Cakir, Eds., *Work With Display Units 92* (pp. 390–394). Amsterdam: North-Holland.

Carayon, P. (1994). Effects of electronic monitoring on job design and worker stress. Results of two studies. *International Journal of Human-Computer Interaction*, *6*(2), 177–190.

Carayon-Sainfort, P. (1992). The use of computer in offices: Impact on task characteristics and worker stress. *The International Journal of Human-Computer Interaction*, *4*(3), 245–261.

Chapman, A., Mandryk, J. A., Frommer, M. S., Edye, B. V., and Ferguson, D. A. (1990). Chronic perceived work stress and blood pressure among Australian government employees. *Scandinavian Journal of Work Environment and Health, 16*, 258–269.

Cherniss, C. (1990). Natural recovery from burnout: results from a 10-year follow-up study. *Journal of Health and Human Resources Administration, 13*, 132–154.

Cook, J. D., Hepworth, S. J., Wall, T. D., and Warr, P. B. (1981). *The experience of work: A compendium and review of 249 measures and their use*. London: Academic Press.

Cooper, C. L. (1986). Job distress: Recent research and the emerging role of the clinical occupational psychologist. *Bulletin of the British Psychological Society, 39*, 325–331.

Cooper, C. L., and Cartwright, S. (1994). Healthy mind; Healthy organization—A proactive approach to occupational stress. *Human Relations, 47*(4), 455–471.

Cooper, C. L., and Payne, R., Eds. (1988). *Causes, coping and consequences of stress at work*. Chichester: John Wiley & Sons.

Cooper, C. L., and Payne, R. (1992). International perspectives on research into work, well-being, and stress management. In J. C. Quick, L. R. Murphy, and J. J. Hurrell, Jr., Eds., *Stress & Well-Being at Work: Assessments and Interventions for Occupational Mental Health* (pp. 348–368). Washington, DC: American Psychological Association.

Cox, T., and Ferguson, E. (1994). Measurement of the subjective work environment. *Work & Stress, 8*, 98–109.

Cox, T., Gotts, G., Boot, N., and Kerr, J. (1988). Physical exercise, employee fitness and the management of health at work. *Work & Stress, 2*, 71–77.

Cox, T., and Howart, I. (1990). Organizational health, culture and helping. *Work & Stress, 4*(2), 107–110.

Cox, T., Kuk, G., and Leiter, M. P. (1993). Burnout, health, work stress, and organizational healthiness. In W. B. Schaufeli, C. Maslach, and T. Marek, Eds., *Professional Burnout: Recent Developments in Theory and Research* (pp. 177–193). Washington, DC: Taylor and Francis.

Curry, J. P. (1986). On the causal ordering of job satisfaction and organizational commitment. *Academy of Management Journal, 24*, 847–858.

Dantzer, R. (1989). Neuroendocrine correlates of control and coping. In A. Steptoe and A. Appels, Eds., *Stress, personal control and health* (pp. 277–294). Chichester: John Wiley & Sons.

Davis, L. E., and Wacker, G. J. (1982). Job design. In G. Salvendy, Ed., *Handbook of industrial engineering* (Sections 2.5.1.–2.5.31). New York: John Wiley & Sons.

Dewe, P. (1991). Primary appraisal, secondary appraisal and coping: Their role in stressful work encounters. *Journal of Occupational Psychology, 64*, 331–351.

Dewe, P. J. (1992). Applying the concept of appraisal to work stressors: Some exploratory analysis. *Human Relations, 45*, 143–164.

Dwyer, D. J., and Ganster, D. C. (1991). The effects of job demands and control on employee attendance and satisfaction. *Journal of Organizational Behavior, 12*, 595–608.

Elo, A-L. (1994). Assessment of mental stress factors at work. In O.B. Dickerson and E. P. Horvath, Eds., *Occupational Medicine* (pp. 945–959). St. Louis: Mosby.

Elo, A.-L., Leppänen, A., Lindström, K., and Ropponen, T. (1992). Occupational stress questionnaire: User's instructions. *Helsinki Institute of Occupational Health Reviews, 19*.

Feldt, T. (1996). The role of sense of coherence in well-being: the analysis of main and moderator effects. *Work & Stress*. (In press).

Fernandez, D. R., and Perrewè, P. L. (1995). Implicit stress theory: an experimental examination of subjective performance information on employee evaluations. *Journal of Organizational Behavior, 16*, 353–362.

Fiebach, N. H., Hebert, P. R., Stampfer, M. J., Colditz, G. A., Willett, W. C., Rosner, B., Speizer, F. E., and Hennekens, C. H. (1989). A prospective study of high blood pressure and cardiovascular disease in women. *American Journal of Epidemiology, 130*(4), 646–654.

Fletcher, B. (1991). Models of stress and disease. In B. Fletcher, Ed., *Work, Stress, Disease and Life Expectancy* (pp. 1–31). Chichester: John Wiley & Sons.

Fletcher, B. (1988). The epidemiology of occupational stress. In C. L. Cooper and R. Payne, Eds., *Causes, Coping and Consequences of Stress at Work* (pp. 3–50). Chichester: John Wiley & Sons.

Folkman, S., and Lazarus, R. S. (1980). An analysis of coping in a middle aged community sample. *Journal of Health and Social Behaviour, 21*, 219–239.

Frankenhaeuser, M. (1986). A psychobiological framework for research on human stress and coping. In M. H. Appley and R. Trumbull, Eds., *Dynamics of stress—Physiological, psychological and social perspectives* (pp. 101–116). New York: Plenum Press.

Frankenhaeuser, M. (1991). A Biopsychosocial approach to work life issues. In J. V. Johnson and G. Johansson, Eds., *The psychosocial work environment: Work organization, democratization and health* (pp. 49–60). New York: Baywood.

Frankenhaeuser, M., and Gardell, B. (1976). Underload and overload in working life: outline of a multidisciplinary approach. *Journal of Human Stress, 2*(3), 35–46.

Frankenhaeuser, M., and Johansson, G. (1986). Stress at work: Psychobiological and psychosocial aspects. *International Review of Applied Psychology, 35*, 287–299.

French, J. R. P., and Caplan, R. D. (1973). Organizational stress and individual strain. In A. J. Marrow, Ed., *The Failure of Success* (pp. 30–66). New York: AMACOM.

Fried, Y. (1988). The future of physiological assessments in work situations. In C. L. Cooper and R. Payne, Eds., *Causes, Coping and Consequences of Stress at Work* (pp. 343–373). Chichester: John Wiley & Sons.

Fried, Y., and Tiegs, R. B. (1993). The main effect model versus buffering model of shop steward social support: A study of rank-and-file auto workers in the U.S.A. *Journal of Organizational Behavior, 14*, 481–493.

Frone, M. R., Russell, M., and Cooper, M. L. (1992). Prevalence of work-family conflict: Are work and family boundaries asymmetrically permeable? *Journal of Organizational Behavior, 13*, 723–729.

Ganster, D. C., and Schaubroeck, J. (1991). Work stress and employee health. *Journal of Management, 17*(2), 235–271.

Gardell, B. (1982). Scandinavian research on stress in working life. *International Journal of Health Services, 12*, 31–41.

Goldberger, L., and Breznitz, S., Eds. (1993). *Handbook of Stress: Theoretical and Clinical Aspects*. New York: Free Press.

Green, K. L., and Johnson, J. V. (1990). The effects of psychosocial work organization on the prevalence of cigarette smoking among chemical plant employees. *American Journal of Public Health, 80*, 1368–1371.

Hackman, J. R., and Oldham, G. R. (1980). *Work Redesign*. Reading, MA: Addison-Wesley.

Hadler, N. M. (1990). Cumulative trauma disorders—An iatrogenic concept. *Journal of Occupational Medicine, 32*(1), 38–41.

Hadler, N. M. (1992, February). Arm pain in the workplace—A small area analysis. *Journal of Occupational Medicine*, 113–119.

Hallsten, L. (1993). Burning out: A framework. In W. B. Schaufeli, C. Maslach, and T. Marek, Eds., *Professional Burnout: Recent Developments in Theory and Research* (pp. 95–113). Washington, DC: Taylor and Francis.

Haynes, S. G., Feinleib, M., Levine, S., Scotch, N., and Kannel, W. B. (1978). The relationship of psychosocial factors to coronary heart disease in the Framingham study. *American Journal of Epidemiology, 107*(5), 384–402.

Haynes, S. G., Feinleib, M., and Kannel, W.B. (1980). The relationship of psychosocial factors to coronary heart disease in the Framingham study. *American Journal of Epidemiology, 111*(1), 37–58.

Hickson, D. J., and McMillan, C. J., Eds. (1981). *Organization and Nation: The Aston Programme IV*. Westmead, England: Gower.

Hippwell, A. E., Tyler, P. A., and Wilson, C. M. (1989). Sources of stress and dissatisfaction among nurses in four hospital environments. *British Journal of Medical Psychology, 62*, 71–79.

House, J. S. (1981). *Work Stress and Social Support*. Reading, MA: Addison-Wesley.

Huuhtanen, P., Nygård, C.-H., Tuomi, K., Eskelinen, L., and Toikkanen, J. (1991). Changes in the content of Finnish municipal occupations over a four-year period. *Scandinavian Journal of Work, Environment & Health, 17*(Suppl. 1), 48–57.

Ilgen, D. R. (1990). Health issues at work. *American Psychologist*, 45, 273–283.

International Labour Office. (1986). *Psychosocial factors at work: Recognition and control* (Occupational Safety and Health Series No. 56). Geneva: International Labour Office.

International Labour Office. (1986). *Introduction to work study* (3rd Ed.). Geneva: International Labour Office.

Israel, B. A., House, J. S., Schurman, S. J., Heany, C. A., and Mero, R. P. (1989). The relation of personal resources, participation, influence, interpersonal relationships and coping strategies to occupational stress, job strains and health: A multivariate analysis. *Work & Stress, 3*, 163–194.

Ivancevich, J. M., and Matteson, M. T. (1988). Promoting the individual's health and well-being. In C. L. Cooper and R. Payne, Eds., *Causes, Coping and Consequences of Stress at Work* (pp. 267–299). Chichester: John Wiley & Sons.

Jackson, S. E. (1989). Does job control control job stress? In S. L. Sauter, J. J. Hurrell, Jr., and C. L. Cooper, Eds., *Job Control and Worker Health* (pp. 25–51). New York: John Wiley & Sons.

James, G. D., Yee, L. S., Harshfield, G. A., Blank, S. G., and Pickering, T. G. (1986). The influence of happiness, anger, and anxiety on the blood pressure of borderline hypertensives. *Psychosomatic Medicine, 48*(7), 502–508.

Janis, I. L. (1985). Stress inoculation in health care: Theory and research. In A. Monat and R. S. Lazarus, Eds., *Stress and Coping: An Anthology*, Second Edition (pp. 330–355). New York: Columbia University Press.

Jenner, D. A., Puddey, I. B., Beilin, L. J., and Vandongen, R. (1988). Lifestyle- and occupation-related changes in blood pressure over a six-year period in a cohort of working men. *Journal of Hypertension, 6*(Suppl. 4), S605–S607.

Jex, S. M., Beehr, T. A., and Roberts, C. K. (1992). The meaning of occupational stress items to survey respondents. *Journal of Applied Psychology, 77*, 623–628.

Johansson, G., Johnson, J. V., and Hall, E. M. (1991). Smoking and sedentary behavior as related to work organization. *Social Science and Medicine, 32*, 837–846.

Johnson, J. V., and Johansson, G. (1991). Work organisation, occupational health, and social change: The legacy of Bertil Gardell. In J. V. Johnson and G. Johansson, Eds., *The psychosocial work environment: Work organization, democratization and health.* Amityville, NY: Baywood.

Kahn, R. L. (1973). Conflict, ambiguity, and overload: Three elements in job stress. *Occupational Health, 3*, 2–9.

Kalimo, R. (1980). Stress in work. *Scandinavian Journal of Work, Environment & Health, 6*(Suppl. 3); Institute of Occupational Health Helsinki:

Kalimo, R., and Mejman, T. (1987). Psychological and behavioural responses to stress at work. In R. Kalimo, M. A. El-Batawi, and C. L. Cooper, Eds., *Psychosocial factors at work and their relation to health* (pp. 23–36). Geneva: World Health Organization.

Kalimo, R., and Vuori, J. (1990). Work and sense of coherence—Resources for competence and life satisfaction. *Behavioral Medicine, 16*, 76–89.

Kalimo, R., and Vuori, J. (1991). Work factors and health: The predictive role of pre-employment experiences. *Journal of Occupational Psychology, 64*, 97–115.

Karasek, R. A. (1979). Job demands, job decision latitude and mental strain: implications for job redesign. *Administrative Science Quarterly, 24*, 285–306.

Karasek, R. A. (1981). Job decision latitude, job design, and coronary heart disease. In G. Salvendy and M. J. Smith, Eds., *Machine pacing and occupational stress* (pp. 45–56). London: Taylor and Francis.

Karasek, R., and Theorell, T. (1990). *Healthy work.* New York: Basic Books.

Karasek, R., Theorell, T., Schwartz, J., Schnall, P., Pieper, C. F., and Michaela, J. (1988). Job characteristics in relation to the prevalence of myocardial infarction in the U.S. HES and HANES. *American Journal of Public Health, 78*, 910–918.

Kasl, S. V. (1978). Epidemiological contributions to the study of work stress. In C. L. Cooper and R. Payne, Eds., *Stress at Work* (pp. 3–48). New York: John Wiley & Sons.

Kawakami, N., Haratani, T., Kaneko, T., and Araki, S. (1989). Perceived job-stress and blood pressure increase among Japanese blue collar workers: One-year follow-up study. *Industrial Health, 27*(2), 71–81.

Kelloway, E. K., Barling, J., and Shah, A. (1993). Industrial relations stress and job satisfaction: Concurrent effects and mediation. *Journal of Organizational Behavior, 14*, 447–457.

Kilpatrick, A. O., Johnson, J. A., and Jones, J. K. (1991). Organizational downsizing in hospitals: Consideration for management development. *Journal of Management Development, 10*, 44–52.

King, L. A., and King, D. W. (1990). Role Conflict and role ambiguity: A critical assessment of construct validity. *Psychological Bulletin, 107*, 48–64.

Kinnunen, U., and Salo, K. (1994). Teacher stress: An eight-year follow-up study on teachers' work, stress, and health. *Anxiety, Stress, and Coping, 7*, 319–337.

Konz, S. (1990). *Work Design: Industrial Ergonomics*, Third Edition. Worthington, OH: Publishing Horizons.

Kram, K. E. (1985). *Mentoring at Work.* Glenview, IL: Scott, Foresman.

Latack, J. C., and Havlovic, S. (1992). Coping with job stress: A conceptual evaluation framework for coping measures. *Journal of Organizational Behavior, 12*, 479–508.

La Vecchia, C., Gentile, A., Negri, E., Parazzini, F., and Franceschi, S. (1989). Coffee consumption and myocardial infarction in women. *American Journal of Epidemiology, 130*(3), 481–485.

Lawler, E. E., III (1986). *High Involvement Management.* San Francisco, CA: Jossey-Bass Publishers.

Lazarus, R. S. (1974). Psychological stress and coping in adaptation and illness. *International Journal of Psychiatry in Medicine, 5*, 321–333.

Lazarus, R. S. (1977). Cognitive and coping processes in emotion. In A. Monat and R. S. Lazarus, Eds., *Stress and Coping* (pp. 145–158). New York: Columbia University Press.

Lazarus, R. S. (1993). From psychological stress to the emotions: A history of changing outlooks. *Annual Reviews Psychology, 44*, 1–21.

Leiter, M. P. (1990). The impact of family resources, control coping, and skill utilization on the development of burnout: A longitudinal study. *Human Relations, 43*, 1067–1083.

Leiter, M. P. (1991). Coping patterns as predictors of burnout: The function of control and escapist coping patterns. *Journal of Organizational Behavior, 12*, 123–144.

Levi, L. (1972a). *Stress and Distress in Response to Psychosocial Stimuli.* New York: Pergamon Press.

Levi. L. (1972b). Methodological considerations in psychoendocrine research. *Acta Medica Scandinavica, 191*(Suppl. 528), 28–54.

Levi, L., Frankenhaeuser, M., and Gardell, B. (1982). Work stress related to social structures and processes. In G. Elliot and C. Eisdorfer, Eds., *Research on Stress and Human Health.* New York: Springer.

Leymann, H. (1990). Mobbing and psychological terror at workplaces. *Violence and Victims, 5,* 119–126.

Lim, S.-Y., and Carayon, P. (1995). Psychosocial work factors and upper extremity musculoskeletal discomfort among office workers. In A. Grieco, G. Molteni, E. Occhipitti, and B. Piccoli, Eds., *Work with Display Units 94* (pp. 57–62). Amsterdam:North Holland.

Lim, S.-Y., Rogers, K. J. S., Smith, M. J., and Sainfort, P. C. (1989). A study of the direct and indirect effects of office ergonomics on psychological stress outcomes. In M. J. Smith and G. Salvendy, Eds., *Work with computers: Organizational, management, stress and health aspects* (pp. 248–255). Amsterdam: Elsevier Science Publishers.

Lindström, K. (1991). Well-being and computer-mediated work of various occupational groups in banking and insurance. *International Journal of Human-Computer Interaction, 3*(4), 339–361.

Lindström, K. (1995). Searching for the model of a healthy organization. Work, Stress and Health '95: Creating Healthier Workplaces. *American Psychological Association and National Institute for Occupational Safety and Health* (September 14–16, 1995, Washington, DC, p. 37).

Lindström, K., Kaihilahti, J., and Torstila, I. (1988). *Ikäkausittaiset terveystarkastukset ja työn muutos vakuutus- ja pankkialalla* (The Finnish Work Environment Fund Publications C12). Helsinki: Finnish Work Environment Fund. (In Finnish.)

Lindström, K., Leino, T., Puhakainen, M., and Torstila, I. (1995). Follow-up of job stress and its relation to characteristics of VDT use among insurance employees. In A. Grieco, G. Molteni, E. Occhipitti, and B. Piccoli, Eds., *Work with Display Units 94* (pp. 45–50). Amsterdam: North Holland.

Linton, S. J., and Kamwendo, K. (1989). Risk factors in the psychosocial work environment for neck and shoulder pain in secretaries. *Journal of Occupational Medicine, 31*(7), 609–613.

Lundberg, U., and Frankenhaeuser, M. (1980). Pituitary-adrenal and sympathetic-adrenal correlates of distress and effort. *Journal of Psychosomatic Research, 24,* 125–130.

Marsella, A. J. (1994). The measurement of emotional reactions to work: Conceptual, methodological and research issues. *Work & Stress, 8,* 153–176.

Maslach, C., and Jackson, S. E. (1981). The measurement of experienced burnout. *Journal of Occupational Behaviour, 2,* 99–113.

Maslach, C., and Schaufeli, W. B. (1993). Historical and conceptual development of burnout. In W. B. Schaufeli, C. Maslach, and T. Marek, Eds., *Professional burnout: Recent developments in theory and research* (pp. 1–18). Washington, DC: Taylor and Francis.

Matthews, K. A., Cottington, E. M., Talbott, E., Kuller, L. H., and Siegel, J. M. (1987). Stressful work conditions and diastolic blood pressure among blue collar factory workers. *American Journal of Epidemiology, 126*(2), 280–291.

Matthews, K. A., Kelsey, S. F., Meilahn, E. N., Kuller, L. H., and Wing, R. R. (1989). Educational attainment and behavioral and biologic risk factors for coronary heart disease in middle-aged women. *American Journal of Epidemiology, 129*(6), 1132–1144.

McIntosh, N. J. (1990). Leader support and responses to work in US nurses: A test of alternative theoretical perspectives. *Work & Stress, 4*(2), 139–154.

McKee, G. H., Markham, S. E., and Scott, K. D. (1992). Job stress and employee withdrawal from work. In J. C. Quick, L. R. Murphy, and J. J. Hurrell, Jr., Eds., *Stress & Well-Being at Work: Assessments and Interventions for Occupational Mental Health* (pp. 153–163). Washington, DC: American Psychological Association.

Meichenbaum, D. (1985). *Stress Inoculation Training.* Boston, MA: Allyn & Bacon.

Meyerson, D. E. (1994). Interpretations of stress in institutions: The cultural production of ambiguity and burnout. *Administrative Science Quarterly, 39,* 628–653.

Murphy, L. R. (1988). Workplace interventions for stress reduction and prevention. In C. L. Cooper and R. Payne, Eds., *Causes, Coping and Consequences of Stress at Work* (pp. 301–339). Chichester: John Wiley & Sons.

Murphy, L. R., Rosen, R., Lindström, K., Lim, S.-Y., Kohler-Moran, S., and Cox, T. (1995). Characteristics of healthy work organizations. Work, stress and health '95: Creating healthier workplaces. In *American Psychological Association and National Institute for Occupational Safety and Health* (September 14–16, 1995, Washington, DC, p. 37).

Nadler, D. A. (1988). Organizational frame bending: types of change in the complex organization. In R. H. Kilmann and T. J. Covin, Eds., *Corporate Transformation: Revitalizing organizations for a competitive world* (pp. 66–83). San Francisco: Jossey-Bass.

Nelson, D., and Quick, J. C. (1991). Social support and newcomer adjustment in organizations: Attachment theory at work? *Journal of Organizational Behavior, 12,* 543–554.

NIOSH (1990). *Health hazard evaluation report* (HETA 89-250-2046, Newsday, Inc.). Washington, DC: U.S. Department of Health and Human Services.

NIOSH (1992). *Health hazard evaluation report* (HETA 89-299-2230, US West Communications). Washington, DC: U.S. Department of Health and Human Services.

Nowack, K. M. (1991). Psychosocial predictors of health status. *Work & Stress*, *5*, 117–131.

O'Driscoll, M. P., and Cooper, C. L. (1994). Coping with work-related stress: A critique of existing measures and proposal for an alternative methodology. *Journal of Occupational and Organizational Psychology*, *67*, 343–354.

Parkes, K. R. (1994). Personality and coping as moderators of work stress processes: Models, methods and measures. *Work & Stress*, *8*, 110–129.

Perrez, M., and Reicherts, M. (1992). *Stress, Coping, and Health: A Situational-Behavior Approach: Theory, Methods, Applications*. Seattle, WA: Hogrefe & Huber.

Philips, S. B., and Mushinski, M. H. (1992). Configuring an employee assistance program to fit the corporation's structure: One company's design. In J. C. Quick, L. R. Murphy, and J. J. Hurrell, Jr., Eds., *Stress & Well-Being at Work: Assessments and Interventions for Occupational Mental Health* (pp. 317–328). Washington, DC: American Psychological Association.

Phillips, K. (1989). Psychophysiological consequences of behavioural choice in aversive situations. In A. Steptoe and A. Appels, Eds., *Stress, personal control and health* (pp. 239–256). Chichester: John Wiley & Sons.

Pieper, C., LaCroix, A. Z., and Karasek, R. A. (1989). The relation of psychosocial dimensions of work with coronary heart disease risk factors: A meta-analysis of five United States data bases. *American Journal of Epidemiology, 129*(3), 483–494.

Pomerleau, C. S., and Pomerleau, O. F. (1987). The effects of a psychological stressor on cigarette smoking and subsequent behavioral and physiological responses. *Psychophysiology*, *24*, 278–285.

Rain, J. S., Lane, I. M., and Steiner D. D. (1991). A current look at the job satisfaction/life satisfaction relationship: Review and future considerations. *Human Relations*, *44*, 287–307.

Randall, D. M. (1990). The consequences of organizational commitment: methodological investigation. *Journal of Organizational Behavior*, *11*, 361–378.

Revicki, D. A., Whitley, T. W., and Gallery, M. E. (1993). Organizational Characteristics, perceived work stress, and depression in emergency medicine residents. *Behavioral Medicine*, *19*, 74–81.

Rogers, R. W., and Byham, W. C. (1994). Diagnosing organization cultures for realignment. In A. Howard, Ed., *Diagnosis for Organizational Change: Methods and Models* (pp. 179–209). New York: Guilford Press.

Rose, R. M., Jenkins, C. D., and Hurst, M. W. (1978). *Air Traffic Controller Health Change Study*. Washington, DC: U.S. Department of Transportation, Federal Aviation Administration, Office of Aviation Medicine.

Rosen, R. H. (1992). *The healthy company*. Los Angeles, CA: Jeremy P. Tarcher/Periguee.

Roskies, E., Seraganian, P., Oseasohn, R., Smilga, C., Martin, N., and Hanley, J. A. (1989). Treatment of psychological stress responses in healthy type a men. In R. W. J. Neufeld, Ed., *Advances in the investigation of psychological stress* (pp. 284–304). New York: John Wiley & Sons.

Russell, J. E. A. (1991). Career development interventions in organizations. *Journal of Vocational Behavior*, *38*, 237–287.

Sainfort, P. C. (1991). Stress, job control and other job elements: A study of office workers. *International Journal of Industrial Ergonomics*, *7*, 11–23.

Salvendy, G., and Smith, M. J., Eds. (1981). *Machine-Pacing and Occupational Stress*. London: Taylor and Francis.

Schaufeli, W. B., Leiter, M. P., and Kalimo, R. (1995). A self-report questionnaire to assess burnout at the workplace. Paper presented at: *Work, Stress and Health '95: Creating Healthier Workplaces*. American Psychological Association and National Institute for Occupational Safety and Health (September 14–16, 1995, Washington, DC, p. 720).

Schnall, P. L., Pieper, C., Schwartz, J. E., Karasek, R. A., Schlussel, Y., Devereux, R. B., Ganau, A., Alderman, M., Warren, K., and Pickering, T. G. (1990). The relationship between 'job strain,' workplace diastolic blood pressure, and left ventricular mass index. *Journal of the American Medical Association*, *263*, 1929–1935.

Schuler, R. S., Aldag, R. J., and Brief A. P. (1977). Role conflict and ambiguity: A scale analysis. *Organizational Behavior and Human Performance*, *20*, 111–128.

Schweiger, D. M., and Dennis, A. S. (1991). Communication with employees following the merger: A longitudinal field experiment. *Academy of Management Journal*, *34*, 110–135.

Sekaran, U. (1989). Paths to the job satisfaction of bank employees. *Journal of Organizational Behavior*, *10*, 347–359.

Selye, H. (1956). *The Stress of Life*. New York: McGraw-Hill.

Selye, H. (1983). The stress concept: Past, present, and future. In C. L. Cooper, Ed., *Stress research: Issues for the eighties* (pp. 1–20). New York: John Wiley & Sons.

Setterlind, S., and Larsson, G. (1995). The stress profile: A psychosocial approach to measuring stress. *Stress Medicine*, *11*, 85–92.

Sharit, J., and Salvendy, G. (1982). Occupational stress: Review and reappraisal. *Human Factors, 24*(2), 129–162.

Smith, K. U. (1965). *Behavior Organization and Work.* Madison, WI: College Printing & Typing Co.

Smith, M. J., and Carayon P. (1995). New technology, automation and work organization: Stress problems and improved technology implementation strategies. *International Journal of Human Factors in Manufacturing, 5,* 99–116.

Smith, M. J., and Carayon, P. (1996) Work organization, stress and cumulative trauma disorders, In S. Moon and S. Sauter, Eds., *Beyond Biomechanics: Psychosocial Aspects of Cumulative Trauma Disorders.* London: Taylor and Francis.

Smith, M. J., Carayon, P., Sanders, K. J., Lim, S.-Y., and LeGrande, D. (1992). Electronic performance monitoring, job design and worker stress. *Applied Ergonomics, 23*(1), 17–27.

Smith, M. J., Cohen, B. G. F., Stammerjohn, L. W. Jr., and Happ, A. (1981). An investigation of health complaints and job stress in video display operations. *Human Factors, 23,* 389–400.

Smith, M. J., and Sainfort, P. C. (1989). A balance theory of job design for stress reduction. *International Journal of Industrial Ergonomics, 4,* 67–79.

Summers, J. D., Rapoff, M. A., Varghese, G., Porter, K., and Palmer, R. E. (1991). Psychosocial factors in chronic spinal cord injury pain. *Pain, 47,* 183–189.

Tait, M., Padgett M. Y., and Baldwin, T. T. (1989). Job and life satisfaction: A reevaluation of the strength of the relationship and gender effects as a function of the date of the study. *Journal of Applied Psychology, 74,* 502–507.

Tetrick, L. E. (1992). Mediating effect of perceived role stress: A confirmatory analysis. In J. C. Quick, L. R. Murphy, and J. J. Hurrell, Jr., Eds., *Stress & well-being at work: Assessments and interventions for occupational mental health* (pp. 134–152). Washington, DC: American Psychological Association.

Theorell, T., Ringdahl-Harms, K., Ahlberg-Hulten, G., and Westin, B. (1991). Psychosocial job factors and symptoms from the locomotor system—A multicausal analysis. *Scandinavian Journal of Rehabilitation Medicine, 23,* 65–173.

Thompson, M. S., and Page, S. L. (1993). A test of Carver and Scheier's self-control model of stress in exploring burnout among mental health nurses. *Stress Medicine, 9,* 221–235.

Tuomi, K., Eskelinen, L., Toikkanen, J., Järvinen, E., Ilmarinen, J., and Klockars, M. (1991). Work load and individual factors affecting work ability among aging municipal employees. *Scandinavian Journal of Work, Environment & Health, 17*(Suppl. 1), 128–134.

Turnage, J. J., and Spielberger, C. D. (1991). Job stress in managers, professionals, and clerical workers. *Work & Stress, 5*(3), 165–176.

Ursin, H., Endresen, I. M., and Ursin, G. (1988). Psychological factors and self-reports of muscle pain. *European Journal of Applied Physiology, 57,* 282–290.

Van Ameringen, M. R., Arsenault, A., and Dolan, S. L. (1988). Intrinsic job stress and diastolic blood pressure among female hospital workers. *Journal of Occupational Medicine, 30*(2), 93–97.

Vartia, M. (1993). Psychological harassment (bullying, mobbing) at work. In K. Kauppinen-Toropainen, Ed., *OECD panel group on women, work and health.* (Helsinki: Publications of the Ministry of Social Affairs and Health 6, pp. 149–152).

Vuori, J. (1994). Pre-employment personal antecedents of health resources, job factors and health risk behaviour in men and women. *Work & Stress, 8,* 263–277.

Weiner, A., Remer, R., and Remer, P. (1992). Career plateauing: Implications for career development specialists. *Journal of Career Development, 19*(1), 37–48.

West, M. A. (1989). Innovation amongst health care professionals. *Social Behaviour, 4,* 173–184.

Westman, M., Eden, D., and Shirom, A. (1985). Job stress, cigarette smoking and cessation: The conditioning effects of peer support. *Social Science and Medicine, 20,* 637–644.

Wolpin, J., Burke, R. J., and Greenglass, E. R. (1991). Is job satisfaction an antecedent or a consequence of psychological burnout? *Human Relations, 44*(2), 193–209.

Wong, M. D., Cupples, L. A., Ostfeld, A. M., Levy, D., and Kannel, W.B. (1989). Risk factors for long-term coronary prognosis after initial myocardial infarction: The Framingham Study. *American Journal of Epidemiology, 130*(3), 469–480.

Yang, C.-L., and Carayon, P. (1995). Effects of job demands and social support on worker stress: a study of VDT users. *Behaviour & Information Technology, 14*(1), 32–40.

CHAPTER 34

MANUAL MATERIALS HANDLING

M. M. Ayoub
Patrick G. Dempsey
Department of Industrial Engineering and Biomedical Engineering
Texas Tech University
Lubbock, TX 79409 USA

Waldemar Karwowski
Department of Industrial Engineering
University of Louisville
Louisville, KY 40292 USA

34.1 INTRODUCTION

34.1.1 Manual Materials Handling, Low-Back Pain, and Back Disorders

Manual materials handling (MMH) tasks, which include unaided lifting, lowering, carrying, pushing, pulling, and holding activities, are the principal source of compensable work injuries affecting primarily the low back in the United States (Battié et al., 1990; Bigos et al., 1986; *Federal Register,* 1986; National Academy of Sciences, 1985; National Institute for Occupational Safety and Health, 1981). These include a large number of low-back disorders (LBDs) that are caused by either cumulative exposure to manual handling of loads over a long period of time, or isolated incidents of overexertion when handling heavy objects (Bureau of National Affairs, 1988; National Safety Council 1989; Videman, Nurminen, and Troup, 1990). According to the National Safety Council (1990), approximately 25% of all worker compensation claims are related to back injuries.

Adverse health effects on the human body caused by MMH tasks, and the financial burden of the related back injuries have long been recognized (Ayoub, Mital, Bakken, Asfour, and Bethea, 1980a; Troup and Edwards, 1985). For example, the U.S. Bureau of Labor Statistics' analysis of the 1985 data provided by 23 states revealed that about 24% of 1.2 million workers' compensation disability cases were caused by industrial back injuries (BNA Report, 1988). The overexertion injuries in the same year for the United States accounted for 32.7% of all accidents: lifting objects (15.1%); carrying, holding, etc. (7.8%); and pulling or pushing objects (3.9%). For the period of 1985–1987, back injuries accounted for 22% of all cases and for 32% of the compensation costs.

According to the National Safety Council (1989), across U.S. industries, about 28.2% of all work injuries involving disability are caused by overexertion, lifting, throwing, folding, carrying, pushing, or pulling loads that weigh under 50 lb. The analysis by industry division showed the highest percentage of such injuries occurred in service industries (31.9%), followed by manufacturing (29.4%), transportation and public utility (28.8%), and trade (28.4%). The total time lost because of disabling work injuries was 75 million workdays, whereas 35 million days were lost because of other accidents. The total work accident cost was $47.1 billion; the average cost per disabling injury was $16,800. Another study (Spengler et al., 1986), reported that although low-back injuries comprised only 19% of all injuries incurred by the workers in one of the largest U.S. aircraft companies, they were responsible for 41% of the total injury costs. Snook (1988) reported that the annual direct and indirect costs of back pain were about $16 billion. It is estimated that the economic impact of back injuries in the U.S. may be as high as 20 billion annually, with compensation costs exceeding $6 billion per year (BNA Report, 1988).

An important epidemiological study of back injuries that included worker perception of exertion with respect to task variables was conducted by the U.S. Department of Labor (Occupational Safety and Health Administration, 1982). The questionnaire survey included 906 (777 male and 129 female) blue-collar workers who sustained back injuries while lifting, placing, carrying, holding, or lowering objects. One third of the workers were in the age group of 25–34 years old, and almost 75% of the respondents were from 20 to 44 years old. The majority of back injuries were diagnosed as muscle strains and sprains. The majority of workers described their body position at a time of injury as slightly bent back (58%), legs (58%), and arms extended down (46%). The most frequent locations of the object at the start of lift were on the floor (52%), at knee height (13%) and waist height (17%). Twenty percent and 39% of the workers estimated the weight of the object lifted at a time of injury to be less than 18.1 kg (40 lb) and less than 27.2 kg (60 lb), respectively.

Fifteen percent of the workers interviewed for the study (Occupational Safety and Health Administration, 1982) stated that the heaviest object they normally lift in their jobs, without the help of a coworker or lifting equipment, was 18.1 kg (40 lb). Another 17% of the respondents reported typical weights of lift to be between 18.1 and 27.2 kg (40–60 lb). Also, 15% of the injured workers indicated that they seldomly lifted objects. For comparison purposes, Kelsey and Golden (1988) reported that the risk of lumbar disk prolapse for workers who lift more than 11.3 kg (25 lb) more than 25 times a day is over three times greater than for workers who lift lower weights. The Occupational Safety and Health Administration (OSHA) (1982) study also revealed very important information regarding workers perception of the weights lifted at the time of injury. Among the items perceived by the workers as factors contributing to their injuries were lifting too heavy objects (reported by 36% of the workers) and underestimation of weight of objects before lifting (reported by 14% of the workers).

34.1.2 Epidemiology of Low-Back Pain and Low-Back Disorders

Major components of the MMH system and the related risk factors for low-back pain (LBP) and low-back disorders (LBDs) include worker characteristics, material or container characteristics, task or workplace characteristics, and work practice characteristics (Herrin, Chaffin, and Mach, 1974). As recently discussed by Riihimäki (1991), a wide spectrum of work- and individual-related risk factors have been associated with the LBP and LBDs. However, the precise knowledge about the extent to which these factors are etiologic and the extent to which they are symptom precipitating or symptom aggravating is still limited.

An importnt review of epidemiological studies on risk factors of LBP using five comprehensive publications on LBP was made by Hildebrant (1987). A total of 24 work-related factors were found that were regarded by at least one of the reviewed sources as risk indicators of LBP. These risk factors include the following categories: (1) general: heavy physical work, work postures in general; (2) static workload: static work postures in general, prolonged sitting, standing or stooping, reaching, no variation in work posture; (3) dynamic work load: heavy manual handling, lifting (heavy or frequent, unexpected heavy, infrequent, torque), carrying, forward flexion of trunk, rotation of trunk, pushing or pulling; (4) work environment: vibration, jolt, slipping or falling; and (5) work content: monotony, repetitive work, work dissatisfaction.

According to Riihimäki (1991), this list of factors was than reduced to the "generally accepted" risk factors (i.e., referenced by at least three literature sources) as follows: general: (1) heavy physical work; static work load: (2) prolonged sitting; dynamic work load: (3) heavy manual handling; (4) heavy or (5) frequent lifting; (6) trunk rotation, (7) pushing or pulling; and work environment: (8) vibrations. Furthermore, there were also a total of 55 individual factors cited by at least one of the sources as a risk indicator of LBP. These individual factors can be grouped into the following categories: (1) constitutional factors: age, gender, weight, back muscle strength (absolute and relative), fitness, back mobility, genetic factors; (2) postural–structural factors; (3) medical factors; (4) psychosocial factors; and (5) demographic factors. Only the following six risk factors were "generally accepted," including: constitutional factors: (1) age, (2) relative muscle strength, (3) physical fitness; medical factors: (4) back complaints in the past; (5) psychosocial factors (not specified); and other factors: (6) work experience.

34.1.2.1 Work-Related Factors for LBP and LBDs

The sudden overload or fatigue caused by repeated loading on the spine may cause excessive loading of the system (Bigos et al., 1991; Hansson, 1989). As reviewed by Riihimäki (1991), many of the cross-sectional studies have shown that LBP is related to "heavy" manual work (Gyntelberg, 1974; Hult, 1954a, 1954b; Ikata, 1965; Lawrence, 1955; Lloyd, Gauld, and Soutar, 1986; Magora, 1970; Svensson and Andersson, 1983; Valkenburg and Haanen, 1983; Videman, T., Nurminen, T. and Tola, S., 1984; Wickström, Hänninen, Lehtinen, and Riihimäki, 1978). However, it was also noted that in most of the above references work was classified as "heavy" on the basis of general impression of working conditions. On the other hand, it is known that gradual increase in loading may have a training effect, as the high level of physical activity strengthens the muscles, tendons, and vertebrae, which can adapt to sustain higher loading (Porter, Adams, and Hutton, 1989). Although the evidence that links LBP and LBDs with heavy manual work is very extensive, it should be observed that some studies failed to pinpoint such a relationship (Partridge and Duthie, 1968; Sairanen, Brüshaber, and Kaskinen, 1981).

In particular, MMH activities, including those that involve sudden body motions, have been associated with LBP. Several studies indicated that the occurrence of LBP is related to lifting, carrying, pulling, pushing, or other sudden maximal efforts (Chaffin and Park, 1973; Damkot, Pope, Lord, and Frymoyer, 1984; Frymoyer et al., 1980; Frymoyer et al., 1983; Lawrence, 1955; Magora, 1972, 1973; Penttinen, 1987; Svensson and Andersson, 1983). Improper lifting was the most common self-reported cause of back injury in a manufacturing assembly work (Bigos et al., 1986). Also, among the office workers, LBP was most commonly attributed to lifting tasks (Lloyd et al., 1986). An increased risk of herniated disc was also reported in jobs involving heavy lifting combined with body twisting and forward bending (Kelsey et al., 1984).

Manual lifting tasks are often associated with adopting non-neutral trunk postures which also have been related to LBP (Frymoyer et al., 1980; Keyserling et al., 1988; Lawrence, 1955; Maeda et al., 1980; Videman et al., 1984). Riihimäki, Tola, Videman, and Hänninen (1989) have also reported a relationship between sciatic pain and working in unnatural (twisted or bent) postures among office workers who rarely engaged in heavy lifting.

34.1.2.2 Psychosocial and Individual Risk Factors for LBDs

In addition to physical factors, several studies identified a variety of psychological and psychosocial risk factors of LBP that are related to work environment (Damkot et al., 1984; Magora, 1973; Svensson and Andersson, 1983, 1989). Examples of such factors are job dissatisfaction, degree of managerial responsibility, the extent of social support, or family problems. For example, in a prospective study conducted at Boeing Company among about 3000 aircraft employees, the psychosocial factors at the workplace were better predictors of future back injuries than the physical findings (Bigos et al., 1991). It was also reported that workers with LBDs exhibited a higher frequency of psychological symptoms than those without such disorders (Donovan et al., 1981; Frymoyer et al., 1980; Gentry, Shows, and Thomas, 1974; Gilchrist, 1976; Joukamaa, 1986), and that psychological symptoms were predictive of the future incidence of LBDs (Biering-Sørensen and Thomsen, 1986; Bigos et al., 1991). However, as pointed out by Riihimäki (1991), since most of the studies have been retrospective in nature, it is difficult to determine whether these factors are antecedents or consequences of back pain (Kelsey et al., 1988), and whether these factors play a role in the etiology of LBDs or only affect the perception of symptoms and sickness behavior.

34.2 MMH CAPACITY DESIGN CRITERIA

Workers who perform heavy physical work are subjected not only to forces and stresses from the immediate physical environment, but also to mechanical forces generated from within the body. As a result of these forces and stresses, a strain is produced on the worker's musculoskeletal system, as well as on other systems such as the cardiopulmonary system. These stresses may lead to an injury. One of the most important issues in the application of ergonomics to work design is to reduce the stresses imposed on the musculoskeletal and cardiopulmonary systems (Ayoub and Mital, 1989).

The philosophy of ergonomics is to design work systems where the demands of the job are within the capacities of the work force. Figure 34.1 shows the MMH work system, its components, and the focus of ergonomics on the accommodation of a large percentage of the population. The work system identifies the risks involved in a particular work situation and the ability to control these risks through the modification of the task designs. As can be seen from this general model, several capacities, such as the biomechanical capacities of the musculoskeletal system and the physiological capacities of the cardiopulmonary system, must be contrasted with the biomechanical and physiological demands of the tasks. These capacities should be the focus when considering MMH tasks. In the context of MMH, several approaches have been used by different investigators to establish safe handling limits:

1. The psychophysical approach,
2. The physiological approach, and
3. The biomechanical approach.

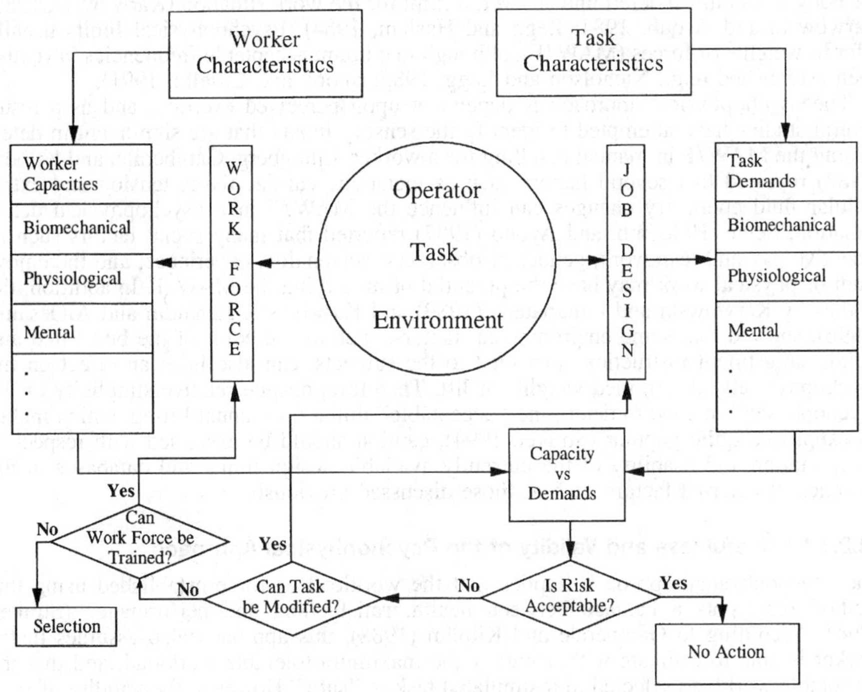

Figure 34.1 Manual materials handling model.

It is quite clear that these approaches to MMH design criteria are different and appear to be unrelated. The psychophysical approach relies on the worker's perceived exertion to quantify his or her tolerance level, thereby establishing the maximum acceptable weights or forces (MAW/F) for different MMH activities [e.g., maximum acceptable weight of lift (MAWL)]. The biomechanical approach focuses on the establishment of tissue tolerance limits of the body, especially the spine (e.g., compressive and shear force limits tolerated by the lumbar spine). The physiological approach focuses on the physiological responses of the body to physical work; therefore, this approach relates metabolic and circulatory costs of performing MMH activities to workers' physiological capacities.

34.2.1 The Psychophysical Approach

Psychophysics deals with the relationship between human sensation and their physical stimuli. Borg (1962) and Eisler (1962) found that the perception of both muscular effort and force obey the psychophysical function, where sensation magnitude (*S*) grows as a power function of the stimulus (*I*). Stevens (1975) reported the relationship between the strength of the sensation (*S*) and the intensity of its physical stimulus (*I*) by the power function:

$$S = K \times I^n \tag{1}$$

where S = strength of sensation, I = intensity of physical stimulus, K = constant, and n = slope of the line that represents the power function when plotted on log–log coordinates.

The use of psychophysics in the study of MMH tasks requires the worker to adjust the weight, force or the frequency in a handling situation until they feel it represents their MAW/F (Asfour, Genardy, Khalil, and Greco, 1984; Fernandez and Ayoub, 1988; Gamerale and Kilböm, 1988; Garg and Banaag, 1988; Legg and Myles, 1984; Mital, Genaidy, and Brown, 1989; Morrissey, Bittner, and Arcangeli, 1990; Snook and Ciriello, 1991). This approach appears to allow the individual worker to integrate the various strains on the body's systems to determine an overall limit for the work situation (Karwowski, 1983; Karwowski and Ayoub, 1984; Legg and Haslam, 1984). Psychophysical limits usually refer to weights or forces (MAW/F), although maximum acceptable frequencies have also been established (e.g., Nicholson and Legg, 1986; Snook and Ciriello, 1991).

The psychophysical approach is dependent upon perceived exertion, and as a result several studies have attempted to identify the sensory inputs that are significant in determining the MAW/F in manual handling for a worker. Ljungberg, Gamberale, and Kilböm (1982) reported that several factors such as metabolic cardiac tissue tension and extracellular fluid chemistry changes can influence the MAW/F in a psychophysical determination. Selan, Holcomb, and Ayoub (1987) reported that many social factors such as incentive pay and teamwork, gender of observers, personality, experience, and the enjoyment of physical work may have the potential of modifying the MAW/F. In addition, the studies by Karwowski and Pongpatana (1989) and Karwowski, Jamaldin and AlQesaimi (1996), showed that some environmental factors, such as the color of the box lifted and weight adjustment instructions provided to the subjects, can also have an effect on the psychophysically determined weights of lift. Therefore, despite relative simplicity of the psychophysical method to determine "acceptable" limits for manual lifting, which makes this approach quite popular (Straker, 1994), caution should be exercised with respect to interpretation and usability of the currently available design limits and databases in the presence of external factors such as those discussed previously.

34.2.1.1 Usefulness and Validity of the Psychophysical Approach

The psychophysical approach implies that the workload or force established using this method represents a balance between health, satisfaction, and performance (Straker, 1994). According to Gamberale and Kilböm (1988), this approach also assumes that a worker is able to estimate with accuracy the maximum tolerable workload, and that the acceptable workload selected in a simulated task is "safe." However, the validity of such assumptions is to some extent unknown (Chaffin and Page, 1994; Karwowski, 1989; Straker, 1994). Recently, Karwowski et al. (1996) examined the validity of the assumption that MAWL values would also be equated by subjects as "safe." A new set of instructions was developed and used to determine the maximum safe weights of lift (MSWL). It was shown that the MSWL values were significantly lower than the MAWL values, with the average weight difference between the two concepts of about 17%.

Another major concern in using the psychophysical approach is the issue of whether the MAW/F set during a short time (usually less than 30 min) represents an acceptable load or force to handle or exert for a full 8-h workday. Therefore, some of the validity studies on the psychophysical approach required subjects to work using their maximum acceptable weight for an 8-h workday. Legg and Myles (1981) had subjects perform lifting and lowering tasks from floor to knuckle height at $2\frac{1}{2}$ repetitions per minute for a full 8-h period using their MAWL. Energy expenditure and heart rate throughout the 8-h period were stable. The energy expenditure was determined to be approximately 21% of a treadmill VO_{2max} (maximum oxygen uptake) with a heart rate of 91 beats per minute. The authors concluded that MAWL can be physiologically acceptable for 8 h of work.

Mital's (1983) study involved experienced male and female industrial workers lifting at frequencies of 1, 4, 8, and 12 per minute for 8- and 12-h days. All subjects in this experiment set their MAWL during a 25-min trial and later during 8- and 12-h trials. The results showed a 35% mean reduction in weight selected by males at the end of the 8-h day, whereas the females showed only a 16% reduction. Mital concluded that a short MAWL selection time may not allow for an accurate estimation of the weight that can be handled for 8 h.

Karwowski and Yates (1986) had female students lift from the floor to knuckle height at 1, 3, 6, and 12 lifts per minute and set their MAWL during a 30-min period. Later the subjects performed the same task for a total of 4 h, during which time the weight was adjusted as often as desired. The results showed no appreciable reduction in MAWL values over the total time for the task at 1, 3, and 6 lifts per minute. However, a significant decrement of 23% in the weight lifted was observed for frequency of 12 lifts per minute. Since the oxygen consumption for both the 6 and 12 lifts/min tasks exceeded the acceptable physiological criteria for 8 h (33% of VO_{2max}), Karwowski and Yates (1986) concluded that the psychophysical method should not be used to set lifting limits for task frequencies higher than 4 lifts/min.

In the Ciriello, Snook, Blick, and Wilkenson, (1990) study, industrial subjects performed MMH tasks at frequencies 4.3 lifts/min and lower. The study used 10 male and 12 female subjects performing lifts, lowers, and push, pull, and carry tasks. Subjects performed the various tasks for 4 h using the MAW/Fs set in 40-min periods. The results showed that the MAW/Fs selected in the 40-min periods were acceptable for 4 h, as indicated by physiological measures. Fernandez, Ayoub, and Smith, (1991) investigated the validity of psychophysical determinations at 2 and 8 lifts/min in a 25-min trial for an 8-h day. For the low frequency, all subjects were able to complete 8 h of work. Only 3 of the 12 subjects completed the full day's work at the higher frequency. It is clear that in a 25-min determination, subjects could have overestimated their full 8-h capability, particularly at the higher frequency.

In view of the above discussion, it can be concluded that if MAWL is only determined for a short time of 20- or 30-min periods, this determination may not be adequate for higher frequencies because of the high physiological load over the 8-h period. Although the various studies used diffeent task conditions, different length experimental periods, and different genders, one could also conclude that the MAWL is a useful capacity measure, especially at the lower frequencies. However, results of several studies on human perception of load heaviness and the psychophysically acceptable weights, reported by Karwowski and his colleagues (Karwowski, 1989, 1991; Karwowski and Burkhardt, 1988; Karwowski, Shumate, Yates and Pongpatana, 1992), indicated that application of the psychophysical method may lead to overestimation of lifting capacity even for the infrequent tasks. It was also shown that females are better judges than males at selecting the acceptable weights of load they could "safely" lift according to the psychophysical approach (Karwowski, 1991).

A few epidemiological studies have been carried ut to investigate whether using MAW could reduce manual handling injuries. One of these studies was performed by Snook, Campanelli, and Hart (1978) during which injury data were collected from 191 companies. When compared to the Liberty Mutual database, about one quarter of the jobs involved tasks acceptable to less than 75% of the workers. Only 25% of these jobs accounted for about one half of the low-back injuries. Snook et al. (1978) interpreted these results to mean that workers performing a MMH task that is acceptable to less than 75% of the work population are three times more susceptible to low-back injury. They also stated that these data suggest that 66% of low-back injuries could be prevented if tasks were designed according to the psychophysical criteria. A large study by Herrin, Jaraiedi, and Anderson, (1986) involved both a retrospective and prospective study investigating 2934 potentially stressful manual handling tasks in 55 jobs in five major industries. In this study, several indices including psychophysically based indices were used. These include the psychophysical tables published by Snook (1978) and a modified job severity index developed by Liles, Deivanayagam, Ayoub, and Mahajan (1984). The medical experiences from this study have shown that there is a relationship between the psychophysical acceptable load and reduced injury rates.

Three methods of determining psychophysical limits have evolved:

1. Direct evaluation of the MAW/F in a specific work situation by simulation of that task
2. The use of already developed MAW/F tables such as those generated by Snook and Ciriello (1991) and Ayoub et al. (1978)
3. Predictive models that utilize task and worker variables as predictor variables

34.2.1.2 Direct Evaluation Method

The direct evaluation method is useful when establishing capacities for a unique task. In this case, a simulated task can be set up for which MAW/F can be directly determined for a particular task and population. Examples of these are given in reports by Gallagher (1991) and Ljungberg et al. (1982). Straker (1994) observed that one should utilize an appropriate sample of representative workers and with sufficient repetitions to ensure reliable results.

34.2.1.3 Existing Databases

One can use already available databases such as the one reported by Snook and Ciriello (1991). Another database was reported by Mital (1992) for symmetrical and asymmetrical lifting, and other databases include work by Ayoub et al. (1978) and Mital (1984). Using such available data replaces conducting a study for every work task and group of workers. Tables provided by the various investigators can be used to estimate the MAW/F for a range of job conditions and work populations. The databases provided in tabular format often make allowances for certain task, workplace, and/or worker characteristics. The use of these databases begins with the determination of the various characteristics with which the database is stratified. Using this information, the value in the tables that applies to the particular work situation can be used as the permissible limit.

This approach can also be used to assess the limits for combination of manual materials handling tasks. For example, Snook and Ciriello (1991) suggest calculating the MAW/F for all individual components for the combination using the same frequency, then using the lowest or the critical task maximum acceptable weight or force for the entire combination. However, it would be possible to determine the MAW/F for combination tasks as done by Jiang, Smith, and Ayoub (1986) and Taboun and Dutta (1989).

34.2.1.4 Psychophysical Models

Another method to estimate MAW/F is regression models based on the psychophysical data. Most of these models predict MAWL. Examples of models based on psychophysical data include those developed by McDaniel (1972), Dryden (1973), and Knipfer (1974). These three models were applicable in only limited work conditions and worker characteristics. Aghazadeh and Ayoub (1985) developed models based on static and dynamic strength measures. They concluded that both dynamic- and static-based models can predict the MAW with reasonable accuracy.

Jiang (1984) developed prediction models for both single and combined MMH activities. For this model, MMH activities were defined as the maximum weight a subject was willing to handle, plus body weight, for a period of 1 h under the task variables. The individual capacity prediction model was based on the isoinertial 6-ft incremental weight lift test.

34.2.2 The Physiological Approach

The physiological approach is concerned with the physiological response of the body to MMH tasks. During the performance of work, physiological changes take place within the body. Changes in work methods, performance level, or certain environmental factors are usually reflected in the stress levels of the worker and may be evaluated by physiological methods. The basis of the physiological approach to risk assessment is the comparison of the physiological responses of the body to the stress of performing a task with levels of permissible physiological limits. Many physiological studies of MMH tended to concentrate on whole body indicators of fatigue such as heart rate, energy expenditure, blood lactate, or oxygen consumption as a result of the work load. Few studies investigated the effects of MMH on local muscle fatigue, such as the study by Habes, Carlson, and Badger (1985). Several studies investigated other indicators of stress such as blood pressure (Asfour, Genaidy, Kahlil, and Muthuswamy, 1986) and the effect of heat stress (Hafez and Ayoub, 1991).

The problem for investigators in establishing recommended limits for energy expenditure at work is which value to use. Absolute values of task energy demands may be inappropriate to use because a task that demands 2 L/min of oxygen uptake may require a 50-year-old female to work at a 100% of her capacity, whereas the same task may only require 50% of the capacity of a 25-year-old-male. Similarly, the rest periods required for these workers to complete a full day's work would vary greatly (Kamon and Ayoub, 1976). To determine the energy cost limit, two basic questions must be addressed (Mital, Nicholson, and Ayoub, 1993): (1) what is the upper limit of VO_2 consumption as a percentage of aerobic capacity, and (2) which aerobic capacity should be used to express this percentage.

Based on the physiological criteria, it has been concluded that for young males, the 8-h average metabolic energy consumption rate should not exceed 5 kcal/min or 33% of the individual's maximum oxygen consumption rate (Snook and Irvine, 1969). It is generally believed that the sustained work capacity of a person for 8-h is approximately 33% to 50% of his or her aerobic capacity (Snook and Irvine, 1969). Mital (1983) reported that the average oxygen consumption associated with the psychophysically acceptable weight of load of male industrial workers was equivalent to 33% VO_2 max for the weight selected in the 25-min period and 23% VO_2 max for a 12-h lifting duration. For females, the average oxygen consumption decreased from 31% VO_{2max} for the load selected in the 25-min period to 24% VO_{2max} for the 12-h lifting period. Asfour, Genaidy, and Mital

(1988) reported that the maximum permissible limit suggested by NIOSH for manual lifting based on 33% of VO_2 max should be adjusted for the different task parameters. This means that since VO_2 max is task specific, MMH VO_2 max would be affected by the task itself and its parameters and would be expected to be different than the bicycle or treadmill VO_{2max}. Mital et al. (1993) concluded that physiological criteria for lifting activities for males should be approximately 4kcal/min and 3 kcal/min for females.

Several work- and workplace-related factors affect metabolic energy expenditure rates. For a detailed discussion on the effects of these task factors on oxygen consumption,the reader is referred to Ayoub and Mital (1989). The metabolic energy expenditure is influenced by a wide variety of factors. The effects of other factors on metabolic energy, such as posture, although too involved to be included in this table, are nevertheless significant. A primary reason for the sensitivity of metabolic energy expenditure rate to work-related factors is the fact that energy cost is dependent upon the amount of muscle groups active during task performance (Mital et al., 1993). In addition to work- and workplace-related factors, there are personal and environmental factors that also influence oxygen consumption. A detailed discussion of the effects of these factors on metabolic energy expenditure is provided by Ayoub and Mital (1989).

34.2.2.1 Validity and Use of Physiological Limits

Energy expenditure, which is the primary measure used in the physiological approach to determine MMH limits, will only capture the aspects of the physiological responses to work. The relationship between energy consumption and risk of MMH injuries is lacking.

Even though there is little experimental evidence that energy expenditure rates are related to risk of injury, there appear to be considerable consequences that energy expenditure as a measure of whole body physiological response to manual handling are important in MMH risk assessment. Asfour et al. (1988) provide a review of available physiological guidelines for lifting and lowering tasks.

There are two methods that can be used in the physiological approach:

1. A direct measurement method in which physiological measures such as energy expenditure can be directly measured for a work situation
2. Estimation of energy expenditure through available predictive energy expenditure models

34.2.2.2 Direct Measurement Method

The direct measurement method relies on the collection of physiological measures from a sample of workers performing their actual task or simulated tasks in the laboratory. Because these tasks represent the actual tasks of interest, the data collected reflect the physiological response of the body to these specific tasks, which can then be compared to permissible limits of energy expenditure. For example, Gallagher (1991) collected physiological data on coal miners performing simulated coal mining tasks.

One advantage of the direct measurement is that it allows the user to measure directly the energy expenditure requirements of a real or simulated activity. A second advantage of this method is that within the limits of equipment mobility, combinations of tasks could be measured in the same way that a single task can be assessed.

The disadvantages of this direct method lies within equipment and its limitations. Equipment to measure VO_2 accurately is expensive. Reliable energy expenditure measures through VO_2 uptake are probably possible in the laboratory setting rather than in an industrial setting.

34.2.2.3 Use of Models To Estimate Energy Expenditure Requirements

A second method for estimating energy cost of manual handling activities based on the physiological response of the body to the load is by modeling the physiological cost using work and worker characteristics. The estimates obtained from such models are then compared to the literature recommendations of permissible limits. Genaidy and Asfour (1987) and Ayoub, Mital, Asfour, and Bethea, (1980b) provided a review of available models for predicting energy expenditure and heart rate.

Garg, Chaffin, and Herrin (1978) reported metabolic cost models. Although currently in need of update, they still provide a more comprehensive and flexible set of physiological cost models as a function of the task variables. The basic form of the Garg et al. (1978) model is:

$$E_{\text{job}} = \left(\sum_{i=1}^{N_p} (E_{\text{post-}i} \times T_i) + \sum_{i=1}^{N_t} E_{\text{task}-i} \right) \div T \tag{2}$$

where E_{job} = average energy expenditure rate of the job (kcal/min), $E_{\text{post}-i}$ = metabolic energy expenditure rate due to maintenance of the ith posture (kcal/min). T_i = time duration of ith posture (min), N_p = total number of body postures employed in the job, $E_{\text{task}-i}$ = net metabolic energy expenditure of the ith task in steady state (kcal), N_t = total number of tasks in the given job, and T = time duration of the job (min).

Other models include that by Asfour (1980) developed using task variables of weight of load, frequency, container size, height of lift, range of lift, and angle of twist of the body. His results show that the load size (in the sagittal plane) and angle of twist had no significant effect on the energy cost of the lifting activity. Asfour's model and results show that for a given fixed work output, it was preferable physiologically to lift or lower heavier loads at slower paces than lighter loads at fast paces. This supports the notion that the physiological approach is more useful at high frequencies, where the loads are lighter and within the physical capability of the worker.

Each of these models have different assumptions. Perhaps the most important assumption of Garg et al.'s (1978) models are that they assume that the net metabolic cost of a series of activities can be estimated by summing the metabolic costs of the elements activities performed separately. Garg et al.'s model provides allowances for different work characteristics including different tasks, with a specific model for each of the lifting, lowering, carrying, and pushing and pulling activities. Generally different models require different input data but typically most of these models involve input information regarding task type, load weight or force, load size, height, frequency, worker characteristics, which include body weight and gender.

Physiological models in general assume that the models are generalizable to different populations. Most of the physiological models available today are based on young university students, and little validation of the models for use with typically industrial populations has been attempted (Straker, 1994). Many of the models are developed for single tasks of 1 h or less in duration. They also assume manual handling is a continuous activity, whereas in fact it is often sporadic in nature. When these physiological models are applied in realistic intermittent tasks, their predictive capacity appears to diminish, as reported by Randle (1987).

The additivity assumption in Garg et al.'s (1978) models was not verified by several subsequent studies. Genaidy, Asfour, Khalil, and Waly, (1985) evaluated this additivity assumption using their own experimental data for lifting from floor to knuckle and from knuckle to shoulder height compared with floor to shoulder heights. Based on the additivity assumption, the sum of oxygen uptakes from floor to knuckle and shoulder to knuckle lifts should equal the oxygen uptake from the floor to shoulder lift. The errors in estimating oxygen cost using the additivity assumption ranged from 19% to 36.8%. Similarly, Taboun and Dutta (1989) found that in a combination task involving lifting and carrying, the errors resulting from the additivity assumption ranged from −25.25% to 60.36%.

34.2.3 The Biomechanical Approach

Biomechanics investigates the internal and external forces acting on the body in motion and at rest. Biomechanics therefore, attempts to estimate the stresses acting on the body while performing work such as MMH activities. The levels of these stresses are then compared with levels of biomechanical stresses believed to be permissible. Some of the commonly used measures to assess these stresses include peak joint moments, peak compressive force on the lumbar spine, and peak shear forces on the lumbar spine. Other measures that have been proposed in the past include mechanical energy, average and integrated moments or forces over the lifting, and MMH activity times (Andersson, 1985; Gagnon and Smyth, 1990; Kumar, 1990). Methods used to estimate the permissible level of stress in biomechanics for MMH include strength testing, lumbar tissue failure, and the epidemiological relationship between biomechanical stress and injury.

When utilizing strength testing, investigators typically measure the static or dynamic strength of a population in order to assign personnel to jobs or to reduce task demands to a level consistent with operator's capacities. Such studies include these by Keyserling, Herrin, Chaffin, Armstrong, and Foss, (1980) and Chaffin, Herrin, Keyserling, and Garg, (1977), as well as studies discussed in more detail later.

Tissue failure studies are based on cadaver tissue strength. Generally, the research has focused on the ultimate compressive strength of the lumbar spine. Recent studies and literature reviews by Brinckmann, Biggemann, and Hilweg (1988, 1989), and Jäger and Luttmann (1991) indicate that the ultimate compressive strength of cadaver lumbar segments varies from approximately 800 N to approximately 13,000 N. Jäger and Luttman (1991) reported a mean failure for compression at 5700 N for males with a standard deviation of 2600 N. For females, this failure limit was found to be 3900 N with a standard deviation of approximately 1500 N. In addition, several factors influence the compressive strength of the spinal column. these are age, gender, specimen cross-section, lumbar level, and structure (whether disc or vertebral body). According to Jäger and Luttman (1991), the ultimate compressive strength of various components of the spine can be estimated with following regression model.

$$\text{Compressive strength (kN)} = (7.65 + 1.18G) - (0.502 + 0.382G)A/\text{decade} + (0.035 + 0.127G)C/\text{cm}^2 - 0.167L/\text{unit} - 0.89S \quad (3)$$

where: G = gender (0 for female; 1 for male), A = decades (for example 30 years = 3, 60 = 6), L = Lumbar level (0 for L5/S1; incremental values for each lumbar disc or vertebra), C = Cross-section in cm^2, S = Structure (0 for disc; 1 for vertebra).

Thus, for example, the predicted compressive force for the L2 vertebra with an area of 18 cm^2, taken from a 30-year-old man will be approximately 7 kN. Jäger and Luttman (1992a) warn that statically determined tolerances may overestimate compressive tolerances.

In contrast with compressive tolerances of the lumbar spine, there is a limited amount of literature on both sagittal and lateral shear force tolerances (Jäger and Luttman, 1989; Potvin, Norman, and McGill, 1991). Jäger and Luttman (1992b) cite studies which evaluate lateral or sagittal shear, when only one specimen was stressed to failure at 7400 N. Modeling studies by Potvin et al. (1991) suggest that erector spinae oblique elements could contribute approximately 500 N sagittal shear, to leave only 200 N sagittal shear for discs and facets to counter. He and his colleagues cite Farfan (1983) as stating that the facet joints are capable of absorbing 3100 N to 3600 N, whereas the discs support less than 900 N. Currently there are no recommendations for permissible shear limits.

Several studies support the notion that accumulated load rather than peak load in biomechanics may be the most important risk factor. A study by Magnusson et al. (1990) concluded that back pain reported by assembly line workers was more likely to result from psychological factors such as motivation and boredom. However some of their data indicate a possible cumulative mechanical etiology. Mital, Ayoub, Asfour, and Bethea (1978) found that frequency was more strongly related to injuries and days lost from lifting than weight; therefore, also suggesting a cumulative rather than a peak stress mechanism. Kumar (1990) also supports this notion. The results of this study found a significantly higher cumulative load history in those who suffered back pain compared to those who did not. Therefore, current biomechanical permissible limits, based on one repetition tolerance, may leave workers at increased risk from cumulative loading.

34.2.3.1 Biomechanical Models

The direct measurement option has rarely been used in experimental situations and has not been used in an industrial setting. As a result, modeling is a popular means to obtain estimates of stress on the body. Available biomechanical models can be divided into 2D and 3D models as well as static and dynamic. All of these models achieve the goal of estimating stresses imposed on body segments, joints, and the lumbar spine. The models developed by Chaffin (1969), Martin and Chaffin (1972), Park and Chaffin (1974), and Garg and Chaffin (1975) are all static in nature. Static models assume that the lifting action is performed quite slowly and smoothly such that forces caused by acceleration can be neglected. This means that the effects of inertia are not included. This assumption is invalid and can seriously affect the estimation of forces exerted on the body segment (Ayoub et al., 1980a). In all occupational biomechanical models, the human body is modeled as a system of rigid links of fixed length, mass, and center of gravity.

Dynamic models such as those developed by Fisher (1967), El-Bassoussi (1974), Ayoub and El-Bassoussi (1976), Chen and Ayoub (1988), and Kromodihardjo and Mital (1986, 1987) provide data for analysis in the form of the time–displacement relationships of the body segments (kinematic analysis) and the forces and torques involved (kinematic analysis). The static models have been popular and applied more widely than dynamic models for the following reasons:

1. Static models require simpler logic and task data than dynamic models.
2. Most static models compare the stresses produced by a manual task with allowable stresses (static strength data, which are readily available).

Dynamic biomechanical models serve as a valuable tool in determining stresses on a specific muscle group or body joint during an activity. For these models to be useful, the user must compare the model output with dynamic strength data. Unfortunately there is a lack of systematic data on dynamic strength for various muscle groups involved in MMH activities (Garg, Chaffin, and Freivalds, 1982).

34.2.3.2 Dynamic vs. Static Models

Because of the complexity of dynamic biomechanical models, as well as the relative lack of dynamic muscle strength data to compare with the task produced forces and moments of the body joints, assessment of the effects of lifting on the musculoskeletal system has most frequently been done with the aid of static models. Many lifting motions, which are dynamic in nature, appear to have substantial inertia components. According to Schultz and Andersson (1981), body dynamics need to be of concern only when significant inertia forces and inertia moments are produced compared with the forces and moments needed for equilibrium. When the body movement velocity is large enough to impose significant stresses on the body, dynamic modeling techniques are essential.

Leskinen, Stalhammer, and Kourinka (1983) compared the peak L5/S1 compressions in a leg lift and a back lift with the data interpolated from Garg and Herrin (1979). They reported that their data were about 70% and 100% higher, respectively. They also noted that the reasons for these differences were the dynamic effects and the inclusion of the intraabdominal pressure in their model. McGill and Norman (1985) also compared the low-back moments during lifting when determined dynamically and statically. They found that the dynamic model resulted in peak L4/L5 moments 19% higher on the average, with a maximum difference of 52%, than those determined from the static model.

34.2.3.3 Usefulness of the Biomechanical Approach

The usefulness of the biomechanical approach is best illustrated by studies that have demonstrated that once the demands of the task exceed the capacity of the musculoskeletal system, injury rates begin to rise. Particularly, epidemiological studies have shown that injury rates rise as the ratio between task demands, usually expressed as load weight, to the isometric strength of the worker increases. Several such studies are described in considerable detail in Sections 34.3.3.2 and 34.3.3.3.

Currently, the issue of whether or not lumbosacral compression is a valid MMH criterion is somewhat unclear. One particular problem with biomechanical criteria, such as L5/S1 compression, is that they appear to have only modest predictive value for preventing low-back disorders. For this reason, it may be necessary to take a more systemic approach that considers factors that are far removed from mechanical events taking place at the L5/S1 (Leamon, 1994a). It appears unwise to rely solely on L5/S1 criteria. However, it is clear that there is a relationship between the stresses placed on the musculoskeletal system and the rate of injury occurrence.

34.2.3.4 Summary of the Biomechanical Approach

The limitations of the biomechanical approach is that it tends to ignore individual differences, such as strength, over the range of motion. The approach also tends to ignore the effects on other joints besides the spine (Dutta and Taboun, 1989). Task duration and repetition are generally not accounted for in biomechanical models, nor are the cumulative nature of the stresses on the body tissues. Additionally, Leamon (1994a) stated that "there is a large potential for inaccuracies to be contained in both the criterion and the values predicted by the models." Therefore, given the complexity of the human body and the simplicity of the models as well as the disagreement among the models, values from these models can only be estimates and are best used for comparison purposes rather than suggesting absolute values (Delleman, Drost, and Huson, 1992).

34.2.4 Comparison of the Design Criteria

Both practitioners and researchers have concerns about the differences in recommended weight limits that result from the biomechanical, physiological, and psychophysical approaches. Each approach utilizes a unique criterion to generate a load limit assumed to be safe, and each approach is valid only under certain conditions; for example, the biomechanical approach is valid for low-frequency tasks, whereas the physiological approach is most applicable for high-frequency tasks (Mital et al., 1993). Although these differences are discussed here, the literature on the subject is extensive and the interested reader desiring more technical details is referred to Ayoub and Mital (1989), Garg and Ayoub (1980), Mital et al. (1993), and Nicholson (1989) for further reading.

One of the simplest approaches to determining MMH limits is to use the biomechanical approach for infrequent activities involving large loads, use the physiological approach for high-frequency tasks, and use the psychophysical approach between these extremes. Additionally, attempts have been made to develop guidelines that accommodate all three of the approaches.

Kim (1990) developed models that predict safe load limits based on biomechanical, physiological and psychophysical criteria. This study also provides good examples of how the different approaches result in different recommendations. Ultimately, Kim's suggestion was to use the lowest load provided by the three approaches as a means of dealing with the conflicting values.

Figure 34.2 illustrates biomechanical, psychophysical, and physiological load limits for a symmetrical floor to shoulder lifting task for a male, based on Kim's (1990) models. For the biomechanical criterion, 650 kg of lumbosacral compression was used, assuming an 80-kg male. For the physiological criterion, 1000 ml of oxygen consumption per minute was used, again for a 80-kg male. For the psychophysical approach, the limits represent the estimated regression model developed. Most notably, the figure illustrates how the physiological approach results in very large recommended loads at low frequencies, and how the biomechanical approach results in high recommended loads at high frequencies. The figure also illustrates how the psychophysical approach tends to result in higher recommended load limits than the biomechanical and physiological approach at low and high frequencies, respectively.

Whereas Kim (1990) suggested using the most conservative of the three design approaches to provide the load limit for a given task, Karwowski and Ayoub (1984) suggested that the acceptability of combined biomechanical and physiological stresses leads

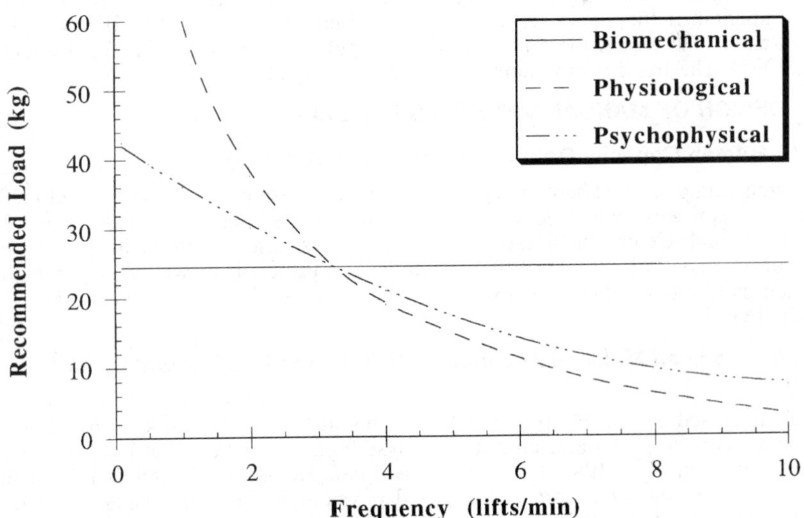

Figure 34.2 Comparison of biomechanical, physiological, and psychophysical design criteria for a floor to shoulder height symmetric lifting task. (Based on Kim's, 1990, models.)

to an overall measure of lifting task acceptability as expressed by the psychophysical values. Using fuzzy sets modeling to combine stresses, Karwowski and Ayoub (1984) found that there was a linear relationship between the acceptability of the combined biomechanical and physiological stresses and the acceptability of the psychophysical stresses.

Jung and Freivalds (1991) utilized an expert system approach to develop a means of reconciling conflicts between the different design approaches. The expert system contained biomechanics, physiology, and psychophysics modules. Unacceptability of biomechanical and physiological stresses were determined by inverting Karwowski's (1983) acceptability measures. The relative weighing of biomechanical and physiological stresses when determining acceptable loads was dependent upon frequency. The expert system generates several alternatives, and the one providing the least overall stress is the recommended solution.

The 1991 NIOSH lifting equation (Waters, Putz-Anderson, Garg, and Fine, 1993) and its predecessor (NIOSH, 1981) are widely used lifting-task design tools. Both of these equations attempt to satisfy biomechanical, physiological, and psychophysical criteria simultaneously. However, the concept of combining all these approaches into a single guideline is likely beyond the simplicity of the NIOSH equations (Dempsey, Ayoub, and Westfall, 1995). A critical point concerning the 1991 equation is that the "NIOSH perspective[s], independent of the 1991 committee" (Waters et al., 1993) were very conservative in nature; thus, it is possible that the equations do indeed satisfy all criteria for some lifting tasks.

34.2.4.1 Epidemiological Comparison and Validation of the Different Design Criteria

One off the most critical research needs in the area of MMH is the epidemiological validation and comparison of design criteria based on the different design approaches, i.e., psychophysics, biomechanics, and work physiology. MMH criteria should allow the workplace or processes to be modified to reduce the probability of LBDs. Inappropriate criteria can divert resources by prompting changes that do not reduce, or potentially increase, the probability of LBDs (Leamon, 1994b). Although there are a considerable number of MMH criteria, few have been sufficiently validated through epidemiological investigations.

The importance of validating MMH criteria is even more critical when one considers the MMH guidelines formulated with one or more criteria. For example, biomechanical, psychophysical, and physiological criteria were used in the development of the NIOSH lifting guides (NIOSH, 1981; Waters et al., 1993). Leamon's (1994b) recent review of MMH criteria casts considerable doubt on the validity of several of the criteria used in the formulation of the equations. Furthermore, although the 1981 NIOSH equation has been in use for 15 years, there is little, if any, epidemiological support for it (Dempsey et al., 1995). The trend is continuing with the 1991 equation.

34.3 DESIGN OF MANUAL MATERIALS HANDLING TASKS

34.3.1 Human Capacity Data for Manual Load Handling

The design data presented here are based upon the database of Snook and Ciriello (1991). These data represent several decades of research that has been conducted at the Liberty Mutual Research Center for Safety and Health and is based upon an industrial subject pool. For a detailed discussion of the data collection protocols as well as additional data, the reader is referred to Snook and Ciriello (1991), Smith, Ayoub, and McDaniel, (1992), and Mital et al. (1993).

34.3.1.1 Modified Maximal Acceptable Weights for Two-Handed Symmetric Lifting

As was discussed earlier, the psychophysical approach tends to result in greater recommended loads than the biomechanical and physiological approaches at low and high frequencies, respectively. Mital et al. (1993) modified Snook and Ciriello's (1991) lifting data so that they satisfied biomechanical and physiological criteria. Thus, the data presented attempt to satisfy the biomechanical, psychophysical, and physiological approaches simultaneously.

Table 34.1 provides Snook and Ciriello's (1991) two-handed lifting data for males and females, as modified by Mital et al. (1993). Those values which were modified have been

Table 34.1 Recommended Weight of Lift (kg) for Male (Female) Industrial Workers for Two-Handed Symmetrical Lifting for 8 h[a]

		Floor to 80 cm Height							
		Frequency of Lift[b,c]							
Cont. Size		1/8 h	1/30 min	1/5 min	1/min	4/min	8/min	12/min	16/min
75 cm	90	17 (12)	14 (9)	14 (8)	11 (7)	9 (7)	*7* (6)	6 (5)	4.5 (4)
	75	24 (14)	21 (11)	20 (10)	16 (9)	13 (9)	*10.5* (8)	9 (7)	7 (6)
	50	27[d] (17)	27[d] (13)	27 (12)	22 (11)	17 (10)	*14* (9)	12 (8)	9.5 (7)
	25	27[d] (20[d])	27[d] (15)	27[d] (14)	27[d] (13)	21 (12)	*17.5* (11)	15 (9)	12 (7)
	10	27[d] (20[d])	27[d] (17)	27[d] (16)	27[d] (14)	25 (14)	*20.5* (13)	18 (11)	14.5 (9)
49 cm	90	20 (13)	17 (9)	16 (8)	13 (8)	10 (8)	*7* (7)	7 (6)	6.5 (5)
	75	27[d] (16)	24 (12)	24 (10)	19 (10)	14 (9)	*10* (8)	*10* (7)	9 (6)
	50	27[d] (19)	27[d] (14)	27[d] (13)	26 (12)	19 (11)	*15* (10)	*12.5* (9)	*10* (8)
	25	27[d] (20[d])	27[d] (17)	27[d] (15)	27[d] (14)	24 (13)	18.5 (11)	*15* (10)	*12* (8)
	10	27[d] (20[d])	27[d] (19)	27[d] (17)	27[d] (15)	28 (15)	22 (13)	*17.5* (11)	*15* (9)
34 cm	90	23 (15)	19 (11)	19 (10)	15 (9)	11 (9)	*7* (8)	*7* (7)	6.5 (7)
	75	27[d] (19)	27[d] (14)	27[d] (13)	22 (12)	17 (11)	*10* (9)	*10* (8)	9.5 (7)
	50	27[d] (20[d])	27[d] (17)	27[d] (16)	27[d] (14)	22 (13)	*15* (11)	*14* (10)	*12* (8)
	25	27[d] (20[d])	27[d] (20[d])	27[d] (18)	27[d] (17)	27[d] (15)	*20* (13)	*17* (12)	*14* (10)
	10	27[d] (20[d])	27[d] (20[d])	27[d] (20[d])	27[d] (19)	27[d] (18)	*25* (15)	*21* (13)	*15* (11)

[a]Adapted from *A Guide to Manual Materials Handling,* Edited by A. Mital, A. S. Nicholson, and M. M. Ayoub, 1993, Taylor and Francis Ltd., London. Reprinted with permission.

[b]Values in parentheses are for females

[c]Italics: weight limited by physiological design criterion (4 kcal/min for males, 3 kcal/min for females).

[d]Weight limited by biomechanical design criterion (3930 N spinal compression for males, 2689 N for females).

Table 34.1 (Continued)

Cont. Size		Floor to 132 cm Height							
		Frequency of Lift[b,c]							
		1/8 h	1/30 min	1/5 min	1/min	4/min	8/min	12/min	16/min
75 cm	90	15 (10)	13 (7.5)	13 (6.5)	10 (6)	8 (6)	*6* (5)	6 (4)	4 (3)
	75	22 (12)	20 (9)	19 (8)	14.5 (7.5)	12 (7.5)	*10* (6.5)	9 (6)	7 (5)
	50	27[d] (14)	25 (11)	24 (10)	20 (9)	15 (8)	13 (7.5)	11 (6.5)	9 (6)
	25	27[d] (17)	27[d] (12.5)	27[d] (11.5)	24.5 (11)	18 (10)	15 (9)	12 (7.5)	11 (6.5)
	10	27[d] (19)	27[d] (14)	27[d] (13)	27[d] (11.5)	22 (11.5)	19 (11)	16 (9)	13 (8)
49 cm	90	18 (11)	16 (7.5)	15 (6.5)	12.5 (6.5)	9 (6.5)	*6* (6)	*6* (5)	*5* (4)
	75	27 (13)	22.5 (10)	22.5 (8)	18 (8)	14 (7.5)	*10* (6.5)	*9* (6)	*8* (5)
	50	27[d] (16)	27[d] (11.5)	27[d] (11)	24 (10)	18 (9)	*14* (8)	*12* (7.5)	*10* (6.5)
	25	27[d] (17)	27[d] (14)	27[d] (12.5)	27[d] (11.5)	22 (11)	18 (9.5)	*14* (8)	*11* (7)
	10	27[d] (19)	27[d] (16)	27[d] (14)	27[d] (12.5)	27 (12.5)	21 (11)	*17* (9)	*14* (7.5)
34 cm	90	22 (12.5)	18 (9)	18 (8)	14 (7.5)	11 (7.5)	*6* (6.5)	6 (6)	*5* (5)
	75	27[d] (16)	26 (11.5)	25 (11)	21 (10)	16 (9)	*10* (8)	*9* (6.5)	8 (5.5)
	50	27[d] (19)	27[d] (14)	27[d] (13)	27[d] (11.5)	22 (11)	*14* (9.5)	*12* (8)	*10* (7)
	25	27[d] (20[d])	27[d] (17)	27[d] (15)	27[d] (14)	27 (12.5)	*20* (11)	*14* (10)	*11* (9)
	10	27[d] (20[d])	27[d] (19)	27[d] (17)	27[d] (16)	27[d] (15)	21 (13)	*17* (11)	*14* (9)

Cont. Size		Floor to 183 cm Height							
		Frequency of Lift[b,c]							
		1/8 h	1/30 min	1/5 min	1/min	4/min	8/min	12/min	16/min
75 cm	90	15 (9)	12 (6)	12 (6)	9.5 (5)	8 (5)	*6* (4.5)	5 (4)	3 (3)
	75	21 (11)	18 (8)	17 (7)	14 (7)	11 (7)	*9* (6)	8 (5)	6 (4.5)
	50	27[d] (12.5)	24 (10)	23 (9)	19 (8)	15 (7)	12 (7)	10 (6)	8 (5.5)
	25	27[d] (15)	27[d] (11)	27[d] (10)	24 (10)	18 (9)	14 (8)	12 (7)	9 (6)
	10	27[d] (17)	27[d] (12.5)	27[d] (12)	27[d] (10)	22 (10)	18 (10)	15 (8)	12 (7)
49 cm	90	17 (10)	15 (7)	14 (6)	11 (6)	9 (6)	*6* (5.5)	*6* (4.5)	*4* (3.5)
	75	24 (12)	21 (9)	21 (7)	16 (7)	12 (7)	*9* (6)	*9* (5)	*7* (4.5)
	50	27[d] (14)	27[d] (10)	27[d] (10)	22 (9)	16 (8)	*14* (7)	*12* (7)	*10* (6)
	25	27[d] (15)	27[d] (12)	27[d] (11)	27[d] (10)	20 (10)	17 (8.5)	*14* (7)	*11* (6.5)
	10	27[d] (17)	27[d] (14)	27[d] (12)	27[d] (11)	23 (11)	20 (10)	*17* (8)	*14* (7)
34 cm	90	20 (11)	16 (8)	16 (7)	13 (7)	9 (7)	*6* (6)	*6* (5)	*4* (4.5)
	75	27[d] (14)	24 (10)	24 (10)	19 (9)	15 (8)	*9* (7)	*9* (6)	*7* (5)
	50	27[d] (17)	27[d] (12)	27[d] (12)	26 (10)	19 (10)	*14* (8.5)	*12* (7)	*10* (6)
	25	27[d] (20)	27[d] (15)	27[d] (13.5)	27[d] (12)	23 (11)	*20* (10)	*14* (9)	*11* (8)
	10	27[d] (20[d])	27[d] (17)	27[d] (15)	27[d] (14)	27[d] (13.5)	24 (12)	*17* (10)	*14* (8)

Table 34.1 (Continued)

Cont. Size		80 cm to 132 cm Height							
		Frequency of Lift[b,c]							
		1/8 h	1/30 min	1/5 min	1/min	4/min	8/min	12/min	16/min
75 cm	90	19 (13)	18 (11)	16 (10)	15 (9)	13 (8)	*7* (*6*)	*6* (6)	*5* (5)
	75	25 (15)	23 (13)	21 (12)	20 (11)	17 (9)	*8* (*7*)	*8* (7)	*7* (6)
	50	27[d] (17)	27[d] (15)	26 (14)	25 (13)	21 (11)	*12* (9)	*11* (9)	*9* (8)
	25	27[d] (20)	27[d] (17)	27[d] (16)	27[d] (14)	26 (12)	*17* (11)	*13* (10)	*12* (9)
	10	27[d] (20[d])	27[d] (19)	27[d] (17)	27[d] (16)	27[d] (14)	23 (12.5)	20 (11)	*16* (9.5)
49 cm	90	19 (13)	18 (11)	16 (10)	15 (9)	13 (8)	*7* (*6*)	*6* (6)	*5* (5)
	75	25 (15)	23 (13)	21 (12)	20 (11)	17 (9)	*8* (*7*)	*8* (7)	*7* (6)
	50	27[d] (17)	27[d] (15)	26 (14)	25 (13)	21 (11)	*12* (9)	*11* (9)	*9* (8)
	25	27[d] (20)	27[d] (17)	27[d] (16)	27[d] (14)	26 (12)	*17* (11)	*13* (10)	*12* (9)
	10	27[d] (20[d])	27[d] (19)	27[d] (17)	27[d] (16)	27[d] (14)	23 (12.5)	20 (11)	*16* (9.5)
34 cm	90	22 (14)	20 (12)	18 (11)	17 (10)	14 (9)	*7* (*7*)	*6* (*6.5*)	*5* (*6.5*)
	75	27[d] (17)	26 (14)	23 (13)	22 (12)	18 (11)	*8* (*8.5*)	*8* (*8.5*)	*7* (8)
	50	27[d] (19)	27[d] (17)	27[d] (15)	27[d] (14)	23 (13)	*12* (*11*)	*11* (10)	*9* (8.5)
	25	27[d] (20[d])	27[d] (19)	27[d] (17)	27[d] (16)	27 (14)	*17* (13.5)	*13* (*11.5*)	*12* (11)
	10	27[d] (20[d])	27[d] (20[d])	27[d] (19)	27[d] (18)	27[d] (16)	24 (14.5)	21 (13)	*16* (11.5)

Cont. Size		80 cm to 183 cm Height							
		Frequency of Lift[b,c]							
		1/8 h	1/30 min	1/5 min	1/min	4/min	8/min	12/min	16/min
75 cm	90	16 (11)	15 (9.5)	13 (9)	12 (8)	11 (7)	*7* (5)	*6* (5)	*5* (4.5)
	75	22 (13)	20 (11)	18 (10.5)	17 (9.5)	15 (8)	*8* (*6*)	*8* (6)	6 (5)
	50	27[d] (15)	25 (13)	23 (12)	21 (11)	19 (10)	*12* (8)	*11* (8)	8 (7)
	25	27[d] (17.5)	27[d] (15)	27 (14)	26 (12)	23 (10.5)	*17* (10)	*13* (9)	11 (8)
	10	27[d] (19)	27[d] (17)	27[d] (15)	27[d] (14)	27 (12)	22 (11)	18 (10)	13 (8)
49 cm	90	16 (11)	15 (9.5)	13 (9)	12 (8)	11 (7)	*7* (6)	*6* (5)	*5* (4.5)
	75	22 (13)	20 (11)	18 (10.5)	17 (9.5)	15 (8)	*8* (*6*)	*8* (6)	6 (5)
	50	27[d] (15)	25 (13)	23 (12)	21 (11)	19 (10)	*12* (8)	*11* (8)	8 (7)
	25	27[d] (17.5)	27[d] (15)	27 (14)	26 (12)	23 (10.5)	*17* (10)	*13* (9)	11 (8)
	10	27[d] (19)	27[d] (17)	27[d] (15)	27[d] (14)	27 (12)	22 (11)	18 (10)	13 (8)
34 cm	90	18 (12)	17 (10.5)	15 (10)	14 (9)	12 (8)	*7* (6)	*6* (6)	*5* (6)
	75	24 (15)	22 (12)	20 (11)	19 (10.5)	16 (10)	*8* (*7.5*)	*8* (*7.5*)	7 (7)
	50	27[d] (17)	27[d] (15)	25 (13)	24 (12)	20 (11)	*12* (*10*)	*11* (9)	9 (7.5)
	25	27[d] (19)	27[d] (17)	27[d] (15)	27[d] (14)	24 (12)	20 (11)	16 (*10*)	12 (10)
	10	27[d] (20[d])	27[d] (19)	27[d] (17)	27[d] (16)	27[d] (14)	22 (13)	18 (11)	13 (10)

Table 34.1 (Continued)

Cont. Size		132 cm to 183 cm Height							
		Frequency of Lift[b,c]							
		1/8 h	1/30 min	1/5 min	1/min	4/min	8/min	12/min	16/min
75 cm	90	15 (9)	14 (8)	12 (7)	12 (7)	9 (7)	7 (5)	6 (4)	4 (3)
	75	20 (11)	18 (9)	15 (9)	15 (8)	12 (8)	9 (6)	8 (5)	6 (4)
	50	25 (13)	23 (11)	20 (10)	19 (9)	16 (9)	12 (8)	10 (7)	7 (6)
	25	27[d] (14)	27 (12)	25 (11)	23 (10)	19 (10)	15 (9)	12 (8)	10 (7)
	10	27[d] (16)	27[d] (14)	27[d] (13)	27 (12)	22 (11)	17 (10)	13 (9)	12 (8)
49 cm	90	18 (10)	16 (9)	14 (8)	14 (7)	11 (7)	7 (5)	7 (4)	5 (3)
	75	23 (12)	21 (10)	19 (9)	18 (9)	14 (8)	9 (6)	8 (5)	6 (4)
	50	27[d] (14)	27 (12)	24 (11)	23 (10)	18 (9)	12 (8)	10 (7)	9 (6)
	25	27[d] (15)	27[d] (13)	27[d] (12)	27[d] (11)	21 (10)	15 (9)	12 (8)	10 (7)
	10	27[d] (17)	27[d] (15)	27[d] (14)	27[d] (13)	25 (11)	17 (10)	13 (9)	11 (8)
34 cm	90	20 (12)	18 (11)	17 (10)	16 (9)	13 (8)	7 (6)	6 (6)	5 (6)
	75	26 (14)	24 (12)	22 (11)	21 (11)	17 (9)	9 (7)	8 (7)	8 (7)
	50	27[d] (17)	27[d] (14)	27[d] (13)	26 (12)	21 (11)	12 (9)	11 (9)	10 (8)
	25	27[d] (19)	27[d] (16)	27[d] (15)	27[d] (14)	25 (12)	15 (11)	14 (10)	13 (9)
	10	27[d] (20[d])	27[d] (18)	27[d] (16)	27[d] (15)	27[d] (14)	17 (12)	16 (11)	15 (9.5)

identified. The data in these tables were modified so that a Job Severity Index (see Section 34.3.3.2 for a discussion of this index) value of 1.5 is not exceeded, which corresponds to 27.24 kg. Likewise, a spinal compression value that, on average, provides a margin of safety for the back of 30% was used for the biomechanical criterion, yielding a maximum load of 27.24 for males and 20 kg for females. Finally, the physiological criterion of energy expenditure was used. The limits selected were 4 kcal/min for males and 3 kcal /min for females for an 8-h working day (Mital et al., 1993).

The design data for maximal acceptable weights for two-handed pushing or pulling tasks, maximal acceptable weights for carrying tasks, and maximal acceptable holding times can be found in Snook and Cirellio (1990) and Mital et al. (1993). The maximal acceptable weights for manual handling in unusual postures are presented by Smith et al. (1992).

34.3.2 1981 NIOSH Lifting Guide

In 1981, the National Institute for Occupational Safety and Health published the *Work Practices Guide for Manual Lifting*. This guide applies to symmetrical and smooth lifting of moderate width objects in the sagittal plane only (no twisting) using good couplings (secure hand holds and low slip potential at the floor), with unrestricted posture and favorable temperature conditions. NIOSH (1981) recommendations were based on two levels of hazard: the action limit (AL), and maximum permissible limits (MPL). The AL is based on a biomechanical criterion of 3400 N spinal compression, a physiological criterion of 3.5 kcal/min, and a psychophysical criterion that the load be acceptable to 75% of women and 99% of men. The MPL is based on a biomechanical criterion of 6400 N spinal compression, a physiological criterion of 5 kcal/min, and psychophysical data indicating that the load would only be acceptable to 1% of women and 25% of men. In 1991, the equation was revised to extend the range of conditions over which the equation applies. The revised equation is discussed next.

34.3.3 Revised NIOSH (1991) Lifting Equation

The 1991 revised lifting equation expands beyond the previous guideline and can be applied to a larger percentage of lifting tasks (Waters et al., 1993). The recommended weight limit (RWL) was designed to protect 90% of the mixed (male/female) industrial working population against LBP. The 1991 equation is based on three main components, i.e., (1) standard lifting location, (2) load constant, and (3) risk factor multipliers. The standard lifting location (SLL) serves as the three-dimensional reference point for evaluating the parameters defining the worker's lifting posture. The SLL for the 1981 Guide was defined as a vertical height of 75 cm and a horizontal distance of 15 cm with respect to the midpoint between the ankles. The horizontal factor for the SLL was increased from 15-cm to 25-cm displacement for the 1991 equation. This was done in view of recent findings that showed 25 cm as the minimum horizontal distance in lifting that did not interfere with the front of the body. This distance was also found to be used most often by workers (Garg, 1989; Garg and Badger, 1986).

The load constant (*LC*) refers to a maximum weight value for the SLL. For the revised equation, the load constant was reduced from 40 kg to 23 kg. The reduction in the load constant was driven, in part, by the need to increase the 1981 horizontal displacement value from a 15 cm to a 25 cm displacement for the 1991 equation (noted above in item 1). Table 34.2 shows definitions of the relevant terms utilized by the 1991 equation.

The RWL is the product of the load constant and six multipliers:

$$\text{RWL (kg)} = LC \cdot HM \cdot VM \cdot DM \cdot AM \cdot FM \cdot CM \tag{4}$$

The multipliers (*M*) are defined in terms of the related risk factors, including the horizontal location (*HM*), vertical location (*VM*), vertical travel distance (*DM*), coupling (*CM*), frequency of lift (*FM*), and asymmetry angle (*AM*). The multipliers for frequency and coupling are defined using relevant tables. In addition to lifting frequency, the work duration and vertical distance factors are used to compute the frequency multiplier (Table 34.3). Table 34.4 shows the coupling multiplier (*CM*), whereas Table 34.5 provides information about the coupling classification.

The horizontal location (*H*) is measured from the midpoint of the line joining the inner ankle bones to a point projected on the floor directly below the midpoint of the hand grasps (i.e., load center). If significant control is required at the destination (i.e., precision placement), then *H* should be measured at both the origin and destination of the lift. This

Table 34.2 Terms of the 1991 NIOSH Equation

Multiplier	Formula (cm)
Load constant	$LC = 23\text{kg}$
Horizontal	$HM = 25/H$
Vertical	$VM = 1 - (0.003\lvert V - 75\rvert)$
Distance	$DM = 0.82 + 4.5/D$
Asymmetry	$AM = 1 - 0.0032A$
Frequency	FM (see Table 34.17)
Coupling	CM (see Table 34.18)

where:

H = The horizontal distance of the hands from the midpoint of the ankles, measured at the origin and destination of the lift (cm)

V = The vertical distance of the hands from the floor, measured at the origin and destination of the lift (cm)

D = The vertical travel distance between the origin and destination of the lift (cm)

A = The angle of asymmetry—angular displacement of the load from the sagittal plane, measured at the origin and destination of the lift (degrees)

F = Average frequency of lift (lifts/minute)

C = Load coupling, the degree that appropriate handles, devices, or lifting surfaces are present to assist lifting and reduce the possibility of dropping the load

[a]After Waters et al., 1993, *Ergonomics,* Vol. 36(7), reprinted with permission from Taylor and Francis.

procedure is required if there is a need to: (1) regrasp the load near the destination of the lift, (2) momentarily hold the object at the destination, or (3) position or guide the load at the destination. If the distance is less than 10 in. (25 cm), then H should be set to 10 in. (25 cm).

The vertical location (V) is defined as the vertical height of the hands above the floor and is measured vertically from the floor to the midpoint between the hand grasps, as defined by the large middle knuckle. The vertical location is limited by the floor surface and the upper limit of vertical reach for lifting (i.e., 70 in. or 175 cm).

The vertical travel distance variable (D) is defined as the vertical travel distance of the hands between the origin and destination of the lift. For lifting tasks, D can be computed by subtracting the vertical location (V) at the origin of the lift from the corresponding V at the destination of the lift. For lowering tasks, D is equal to V at the origin minus V at the destination. The variable (D) is assumed to be at least 10 in. (25 cm), and no greater than 70 in. (175 cm). If the vertical travel distance is less than 10 in. (25 cm), then D should be set to 10 in. (25 cm).

The asymmetry angle A is limited to the range of 0° to 135°. If $A > 135°$, then AM is set equal to zero, which results in a RWL of 0. The asymmetry multiplier (AM) is 1 − (.0032A). The AM has a maximum value of 1.0 when the load is lifted directly in front of the body and a minimum value of 0.57 at 135° of asymmetry.

The frequency multiplier (FM) is defined by (1) the number of lifts per minute (frequency), (2) the amount of time engaged in the lifting activity (duration), and (3) the vertical height of the lift from the floor. Lifting frequency (F) refers to the average number of lifts made per minute, as measured over a 15-min period. Lifting duration is classified into three categories—short duration, moderate duration, and long duration. These categories are based on the pattern of continuous *work-time* and *recovery-time* (i.e., light work) periods.

Table 34.3 Frequency Multipliers for the 1991 Lifting Equation[a]

Frequency Lifts (min)	Continuous work duration[b]					
	<8 h		<2 h		< 1 h	
	$V < 75$	$V > 75$	$V < 75$	$V > 75$	$V < 75$	$V > 75$
0.2	0.85	0.85	0.95	0.95	1.00	1.00
0.5	0.81	0.81	0.92	0.92	0.97	0.97
1	0.75	0.75	0.88	0.88	0.94	0.95
2	0.65	0.65	0.84	0.84	0.91	0.91
3	0.55	0.55	0.79	0.79	0.88	0.88
4	0.45	0.45	0.72	0.72	0.84	0.84
5	0.35	0.35	0.60	0.60	0.80	0.80
6	0.27	0.27	0.50	0.50	0.75	0.75
7	0.22	0.22	0.42	0.42	0.70	0.70
8	0.18	0.18	0.35	0.35	0.60	0.60
9	—	0.15	0.30	0.30	0.52	0.52
10	—	0.13	0.26	0.26	0.45	0.45
11	—	—	—	0.23	0.41	0.41
12	—	—	—	0.21	0.37	0.37
13	—	—	—	—	—	0.34
14	—	—	—	—	—	0.31
15	—	—	—	—	—	0.28

[a]After Waters et al., 1993, *Ergonomics,* Vol. 36(7), reprinted with permission from Taylor and Francis.
[b]Note: symbol "—" means 0.

A continuous work-time period is defined as a period of uninterrupted work. Recovery time is defined as the duration of light work activity following a period of continuous lifting. *Short duration* defines lifting tasks that have a work duration of 1 h or less, followed by a recovery time equal to 1.2 times the work time. *Moderate duration* defines lifting tasks that have a duration of more than 1 h, but not more than 2 h, followed by a recovery period of at least 0.3 times the work time. *Long duration* defines lifting tasks that have a duration of between 2 and 8 h, with standard industrial rest allowances (e.g., morning, lunch, and afternoon rest breaks).

The lifting index (LI) provides a relative estimate of the physical stress associated with a manual lifting job and is equal to the load weight divided by the RWL. According to Waters, Putz-Anderson, and Garg, (1994), the RWL and LI can be used to guide ergonomic design in several ways:

1. The individual multipliers can be used to identify specific job-related problems. The general redesign guidelines related to specific multipliers are shown in Table 34.6.
2. The RWL can be used to guide the redesign of existing manual lifting jobs or to design new manual lifting jobs.

Table 34.4 The Coupling Multipliers for the 1991 Lifting Equation[a]

Couplings	$V < 75$ cm	$V \geq 75$ cm
Good	1.00	1.00
Fair	0.95	1.00
Poor	0.90	0.90

[a]After Waters et al., 1993, *Ergonomics,* Vol. 36(7), reprinted with permission from Taylor and Francis.

Table 34.5 Coupling Classification[a]

Good coupling

1. For containers of optimal design, such as some boxes, crates, etc., a "good" hand-to-object coupling would be defined as handles or hand-hold cutouts of optimal design
2. For loose parts or irregular objects, which are not usually containerized, such as castings, stock, and supply materials, a "good" hand-to-object coupling would be defined as a comfortable grip in which the hand can be easily wrapped around the object.

Fair coupling

1. For containers of optimal design, a "fair" hand-to-object coupling would be defined as handles or hand-hold cutouts of less than optimal design.
2. For containers of optimal design with no handles or hand-hold cutouts or for loose parts or irregular objects, a "fair" hand-to-object coupling is defined as a grip in which the hand can be flexed about 90 degrees.

Poor coupling

1. Containers of less than optimal design or loose parts or irregular objects that are bulky, hard to handle, or have sharp edges.
2. Lifting nonrigid bags (i.e., bags that sag in the middle)

Notes:

1. An optimal handle design has 0.75–1.5 in (1.9–3.8 cm) diameter, ≥4.5 in. (11.5 cm) length, 2 in. (5 cm) clearance, cylindrical shape, and a smooth, nonslip surface.
2. An optimal hand-hold cutout has the following approximate characteristics: ≥1.5 in. (3.8 cm height), 4.5 in. (11.5 cm) length, semioval shape, ≥2 in. (5 cm) clearance, smooth nonslip surface, and ≥0.25 in. (0.60 cm) container thickness (e.g., double thickness cardboard).
3. An optimal container design has ≤16 in. (40 cm) frontal length, ≤12 in. (30 cm) height, and a smooth nonslip surface.
4. A worker should be capable of clamping the fingers at nearly 90° under the container, such as required when lifting a cardboard box from the floor.
5. A container is considered less than optimal if it has a frontal length >16 in. (40 cm), height >12 in. (30 cm), rough or slippery surfaces, sharp edges, an asymmetric center of mass, or unstable contents or requires the use of gloves. A loose object is considered bulky if the load cannot easily be balanced between the hand grasps.
6. A worker should be able to wrap the hand comfortably around the object without causing excessive wrist deviations or awkward postures, and the grip should not require excessive force.

[a]After Waters et al. (1994).

3. The LI can be used to estimate the relative magnitude of physical stress for a task or job. The greater the LI, the smaller the fraction of workers capable of safely sustaining the level of activity.
4. The LI can be used to prioritize ergonomic redesign. A series of suspected hazardous jobs could be rank ordered according to the LI and a control strategy could be developed according to the rank ordering (i.e., jobs with lifting indices about 1.0 or higher would benefit the most from redesign).

Finally, it should be noted that the 1991 equation should not be used if any of the following conditions occur:

1. Lifting or lowering with one hand
2. Lifting or lowering for over 8 h
3. Lifting or lowering while seated or kneeling
4. Lifting or lowering in a restricted work space
5. Lifting or lowering unstable objects
6. Lifting or lowering while carrying, pushing, or pulling
7. Lifting or lowering with wheelbarrows or shovels
8. Lifting or lowering with high-speed motion (faster than about 30 in./s)
9. Lifting or lowering with unreasonable foot to floor coupling (<0.4 coefficient of friction between the sole and the floor)

Table 34.6 General Design/Redesign Suggestions for Manual Lifting Tasks as Recommended by Waters et al. (1994)

If *HM* is less than 1.0	Bring the load closer to the worker by removing any horizontal barriers or reducing the size of the object. Lifts near the floor should be avoided; if unavoidable, the object should fit easily between the legs.
If *VM* is less than 1.0	Raise or lower the origin or destination of the lift. Avoid lifting near the floor or above the shoulders.
If *DM* is less than 1.0	Reduce the vertical distance between the origin and the destination of the lift.
If *AM* is less than 1.0	Move the origin and destination of the lift closer together to reduce the angle of twist, or move the origin and destination further apart to force the worker to turn the feet and step, rather than twist the body.
If *FM* is less than 1.0	Reduce the lifting frequency rate, reduce the lifting duration, or provide longer recovery periods (i.e., light work period).
If *CM* is less than 1.0	Improve the hand-to-object coupling by providing optimal containers with handles or hand-hold cutouts, or improve the hand-holds for irregular objects.
If the RWL at the destination is less than at the origin	Eliminate the need for significant control of the object at the destination by redesigning the job or modifying the container or object characteristics.

10. Lifting or lowering in an unfavorable environment (i.e., temperature significantly outside 66–79°F (19–26°C) range; relative humidity outside 35–50% range)

34.3.3.1 Computer Simulation of the Revised NIOSH Lifting Equation (1991)

One way to investigate the practical implications of the 1991 lifting equation for industry is to determine the likely results of the equation when applying a realistic and practical range of values for the risk factors (Karwowski, 1992). This can be done using modern computer simulation techniques in order to examine the behavior of the 1991 NIOSH equation under a broad range of conditions. Karwowski and Gaddie (1995) simulated the 1991 equation using SLAM II (Pritsker, 1986), Simulation Language for Alternative Modeling, as the product of the six independent factor multipliers represented as attributes of an entity flowing through the network. For his purpose, probability distributions for all the relevant risk factors were defined, and a digital simulation of the revised equation was performed.

As much as possible, the probability distributions for these factors were chosen to be representative of the real industrial workplace (Brokaw, 1992; Ciriello, Snook, Blick, and Wilkinson, 1990; Karwowski and Brokaw, 1992; Marras et al., 1993). Except for the vertical travel distance factor and the coupling and asymmetry multipliers, all factors were defined using either normal or log normal distributions. For all the factors defined as having log normal distributions, the procedure was developed to adjust for the required range of real values whenever necessary. The SLAM II computer simulation was run for a total of 100,000 trials, i.e., randomly selected scenarios that realistically defined the industrial tasks in terms of the 1991 equation. Descriptive statistical data were collected for all the input (lifting) factors, the respective multipliers, and the resulting recommended weight limits. The input factor distributions were examined to verify the intended distributions.

The results showed that for all lifting conditions examined, the distribution of recommended weight limit values had a mean of 7.22 kg and a standard deviation of 2.09 kg. In 95% off all cases, the RWL was at or below the value of 10.5 kg or about 23.1 lb. In 99.5% of all cases the RWL value was at or below 12.5 kg or 27.5 lb. That implies that when the LI is set to 1.0 for task design or evaluation purposes, only 0.5% of the (simulated) industrial lifting tasks would have the RWLs greater than 12.5 kg. Taking into account the lifting task duration, in the 99.5% of the simulated cases, the RWL values were equal to or were lower than 13.0 kg (or 28.6 lb) for up to 1 h of lifting task exposure, 12.5 kg (or 27.5 lb) for less than 2 h of exposure, and 10.5 kg (or 23.1 lb) for lifting over an 8-h shift.

From a practical point of view, these values define simple and straightforward lifting limits, i.e. the threshold RWL values (TRWL) that can be used by practitioners for the purpose of immediate and easy to perform risk assessment of manual lifting tasks performed in industry. Because the 1991 equation is designed to ensure that the RWL will not exceed the acceptable lifting capability of 99% of male workers and 75% of female workers, this amounts to protecting about 90% of the industrial workers if there is a 50 /50 split between males and females. The TRWL value of 27.5 lb can then be used for immediate risk assessment of manual lifting tasks performed in industry. If this value is exceeded, then a more thorough examination of the identified tasks, as well as evaluation of physical capacity of the exposed workers should be performed.

34.3.4 Prevention of LBDs in Industry: Job Design and Redesign Recommendations

34.3.4.1 General Prevention Strategies

The application of ergonomic principles to the design of MMH tasks is one of the most effective approaches to controlling the incidence and severity of LBDs. The goal of ergonomic job design is to reduce the ratio of task demands to worker capability to an acceptable level.

The application of ergonomic principles to task and workplace design permanently reduces stresses. Such changes are preferable to altering other aspects of the MMH system, such as work practices. For example, worker training may be ineffective if practices trained are not reinforced and refreshed (Kroemer, 1992), whereas altering the workplace is a lasting physical intervention.

Figure 34.3 summarizes the ergonomic approach to job (re)design. As the figure indicates, the optimal solution is to eliminate the need to handle materials manually. This can be achieved either through the implementation of mechanical aids or by redesigning the work area layout so that all materials are located on the same level. In cases where elimination of the need for MMH is physically or economically unfeasible, stresses must be reduced by decreasing job demands and minimizing stressful body movements by altering the task, workplace and/or the material being handled. Figure 34.3 provides a sample of techniques to achieve these goals.

In addition to the engineering controls discussed above, there are also techniques for job (re)design based upon the relationships between job stress and injury occurrence. Several of these methods are discussed below. Each of the techniques can be used to prioritize job interventions in terms of injury risk associated with different jobs.

34.3.4.2 Job Severity Index

The Job Severity Index (JSI) is a time- and frequency-weighted ratio of worker capacity to job demands. Worker capacity is predicted with the models developed by Ayoub et al. (1978), which use isometric strength and anthropometric data to predict psychophysical lifting capacity. JSI and each of the components are defined below.

$$\text{JSI} = \sum_{i=1}^{n} \frac{\text{hours}_i \times \text{days}_i}{\text{hours}_t \times \text{days}_t} \sum_{j=1}^{m_i} \left[\frac{F_j}{F_i} \times \frac{WT_j}{Cap_j} \right] \tag{5}$$

where: n = number of task groups, hours_i = exposure hours/day for group i, days_i = exposure days/week for group i, hours_t = total hours/day for job, days_t = total days/week for job, m_i = number of tasks in group i, WT_j = maximum required weight of lift for task j, Cap_j = the adjusted capacity of the person working at task j, F_j = lifting frequency for task j, F_i = total lifting frequency for group i. Thus,

$$F_i = \sum_{j=1}^{m_i} F_j \tag{6}$$

Table 34.7 provides the estimated regression coefficients used to predict individual worker capacity (Cap_j). The equations provide predictions of MAWL plus body weight based on the strength and anthropometric variables given in the table. Body weight is subtracted from the prediction to obtain the individual's predicted MAWL. Once MAWL

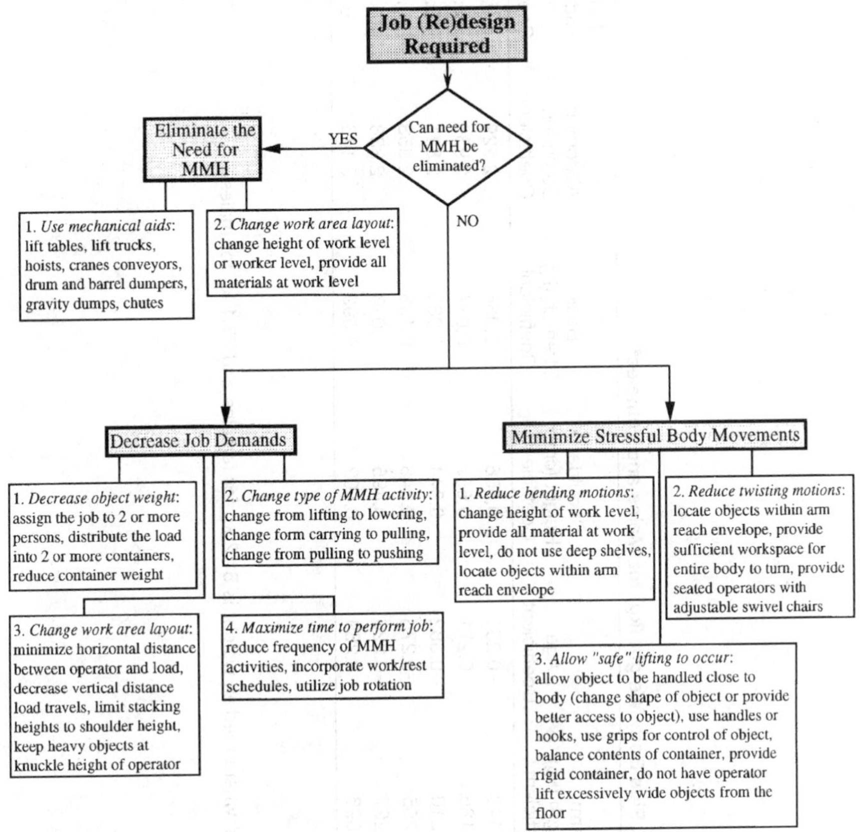

Figure 34.3 Summary of ergonomic approaches to MMH task (re)design. (Adapted from Ayoub, 1982, and Ayoub et al., 1983.)

is predicted, the capacity is adjusted for frequency using the equations in Table 34.8. Finally, the frequency-adjusted MAWL is adjusted for the dimension of the box in the sagittal plane using the equations in Table 34.9.

Liles et al. (1984) performed a field study to determine the relationship between JSI and the incidence and severity of LBDs. A total of 453 subjects were included in the study. The results off the field study indicated that both incidence and severity of recordable back injuries rose rapidly at values of JSI greater than 1.5, as illustrated in Figure 34.4. The denominator for the incidence and severity rates is 100 full-time employees, i.e., 200,000 exposure hours. JSI can be reduced to a desirable level by increasing worker capacity (e.g., selecting a worker with higher capacity) or altering task and job parameters to reduce JSI to an acceptable level.

The following example of the calculations required for the determination of JSI is based on the task parameters in Table 34.10. The job involves only one task group with two lifting tasks. The first task is a floor to knuckle height (F–K) lift and the second task is a floor to shoulder height (F–S) lift. The first step involves determining the capacity of the individual performing the job. The job is performed by a 30-year-old female weighing 59 kg with the following strength and anthropometric measurements: arm strength = 20 kg, shoulder height = 132 cm, back strength = 51 kg, abdominal depth = 20 cm, and dynamic endurance = 2.62 min. Using the coefficients in Table 34.7, the individual's MAWL plus body weight (MAWLBW) prediction for the F–K task is:

$$\begin{aligned}\text{MAWLBW} &= -32.733 - 12.852(1) + 10.996(0) + 0.143(20) - 0.251(30) \\ &\quad + 0.556(132) + 0.056(51) + 2.229(20) + 0.797(2.62) \\ &= 72.66 \text{ kg}\end{aligned} \qquad (7)$$

Table 34.7 Estimated Regression Coefficients Used To Predict MAWL Plus Body Weight (kg) for Males and Females[a]

Lifting Range[b]	Constant Term	Sex Code Coefficient[c]	Weight Code Coefficient[d]	Arm Strength (kg) Coefficient	Age Coefficient	Shoulder Height (cm) Coefficient	Back Strength (kg) Coefficient	Abdominal Depth (cm) Coefficient	Dynamic Endurance (min) Coefficient
F–K	−32.733	−12.852	10.996	0.143	−0.251	0.556	0.056	2.229	0.797
F–S	−65.958	−7.332	5.410	0.185	−0.271	0.652	0.077	2.936	1.183
F–R	−18.718	−8.824	7.337	0.210	−0.382	0.344	0.068	2.821	0.647
K–S	−25.020	−8.370	5.307	0.265	−0.275	0.348	0.105	2.853	0.642
K–R	−35.921	−8.581	7.835	0.297	−0.226	0.495	0.018	2.338	0.962
S–R	−16.982	−8.883	9.232	0.096	−0.269	0.402	0.099	2.146	0.494

[a] Adapted from Ayoub et al. (1978).

[b] F = floor, K = knuckle height, S = shoulder height, R = reach height.

[c] 0 for males, 1 for females.

[d] 0 if body weight is ≤ median weight, 1 if body weight is > median weight, where median weight is 61.2 kg for females and 77.1 kg for males.

Table 34.8 Equations To Adjust MAWL Predictions for Frequency[a]

Range of Lift	Frequency of Lift (FY) (lifts/min)[b] $0.1 < FY < 1.0$	Equation	$1.0 \leq FY \leq 12.0$	Equation
Male Capacity				
F–K, F–S, F–R	$Cap \cdot FY^{-0.184697}$	1	$Cap - 0.91(FY - 1)$	3
K–S, K–R, S–R	$Cap \cdot FY^{-0.138650}$	2	Use Equation 3	
Female Capacity				
F–K, F–S, F–R	$Cap \cdot FY^{-0.187818}$	4	$Cap - 0.5(FY - 1)$	6
K–S, K–R, S–R	$Cap \cdot FY^{-0.156150}$	5	Use Equation 6	

[a]Adapted from Ayoub et al. (1983).
[b]Capacity as determined from Table 34.7.

Thus, the subject's predicted MAWL for the F–K lift is 13.66 kg. The next step involves correcting the MAWL value for frequency using Equation 6 in Table 34.8. This correction reduces the MAWL to 11.66 kg. Finally, the frequency-adjusted MAWL is corrected for box size using Equation 4 from Table 34.9, resulting in a final value of 12.86 kg. The individual's MAWL plus body weight (MAWLBW) prediction for the F–S task is:

$$\begin{aligned}\text{MAWLBW} &= -65.958 - 7.332(1) + 5.410(0) + 0.185(20) - 0.271(30) \\ &\quad + 0.652(132) + 0.077(51) + 2.936(20) + 1.183(2.62) \\ &= 74.1 \text{ kg}\end{aligned} \tag{8}$$

Thus, the subject's predicted MAWL for the F–K lift is 15.1 kg. The frequency-adjusted MAWL value, using Equation 6 in Table 34.8, is 14.6 kg. The frequency-adjusted MAWL is corrected for box size using Equation 4 from Table 34.9, resulting in a final value of 16.8 kg.

Once all capacities have been estimated, the JSI value for the job can be calculated. Using the equation presented earlier, the JSI for the job is:

$$\begin{aligned}\text{JSI} &= \left(\frac{8}{8}\right)\left(\frac{5}{5}\right)\left[\left(\frac{5}{7}\right)\left(\frac{20.0}{12.86}\right) + \left(\frac{2}{7}\right)\left(\frac{15.0}{16.8}\right)\right] \\ &= 1.37\end{aligned} \tag{9}$$

The JSI value of 1.37 indicates that the job is a candidate for redesign. Although the value is below 1.5, the value is close enough to warrant concern. Particularly, the JSI for task 1 is 1.56, which is the problematic task. Potential solutions include eliminating the need to handle the material manually, providing the material to the operator at knuckle height, reducing the frequency of handling, reducing the weight of the load, and as a last resort, selecting an operator with higher capacity.

Table 34.9 Equations To Correct Frequency-Adjusted MAWL Predictions for Box Size[a]

Range of Lift	Box Size (BX) in Sagittal Plane (cm) $31 \leq BX \leq 46$	Equation	$BX \geq 46$	Equation
Male Capacity				
F–K, F–S, F–R	$Cap_{FA} + 0.30(46 - BX)$	1	$Cap_{FA} + 0.14(46 - BX)$	3
K–S, K–R, S–R	$Cap_{FA} + 0.20(46 - BX)$	2	Use Equation 3	
Female Capacity				
F–K, F–S, F–R	$Cap_{FA} + 0.20(46 - BX)$	4	$Cap_{FA} + 0.07(46 - BX)$	6
K–S, K–R, S–R	$Cap_{FA} + 0.10(46 - BX)$	5	$Cap_{FA} + 0.04(46 - BX)$	7

[a]Adapted from Ayoub et al. (1983).
[b]Frequency-adjusted (FA) capacity as determined from Table 34.8.

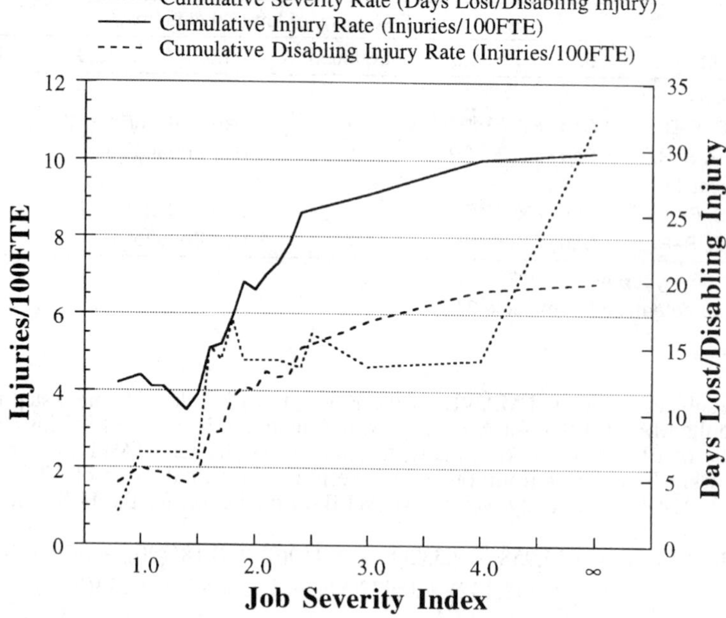

Figure 34.4 Relationships between JSI and incidence and severity of MMH-related injuries. (Adapted from Liles et al., 1984.)

34.3.4.3 Lifting Strength Rating and Related Indices

Chaffin and Park (1973) investigated the efficacy of using the Lifting Strength Rating (LSR), the ratio of maximum load lifted on a job to the maximum isometric strength of a "large/strong" man in a position similar to the task, to predict LBDs. The follow-up of 411 people lasted 1 year and indicated that an LSR value greater than 0.2 should be considered potentially harmful. Jobs with an LSR value between 0.9 and 1.0 had an overall incidence rate approximately 2 times higher than jobs with LSRs between 0.2 and 0.8, and approximately 4 times higher than jobs with LSRs below 0.2.

Chaffin (19974) reported the relationship between the ratio of the weight of load lifted for the task with the highest LSR to the mean isometric strengths of employees. Thus, instead of using the isometric strength of a large/strong man, actual strength values of employees were used. Incident rates were approximately three times greater when the stated ratio exceeded 1.0 than when it did not.

In a study similar to Chaffin (1974), Chaffin, Herrin, and Keyserling (1978) used the Job Strength Rating (JSR), the ratio of the maximum strength requirements of the job to the average task specific isometric strengths of all workers in a job. The relationship

Table 34.10 Task Parameters for JSI Example

Parameter	Task 1	Task 2
$hours_i$	8	8
$days_i$	5	5
$hours_t$	8	8
$days_t$	5	5
WT_j	20 kg	15 kg
Box size[a]	40	35
F_j	5	2

[a]Dimension in sagittal plane (cm).

between JSR and incidence rate of low-back disorders appeared to be fairly linear. The relationship between JSR and severity rate was not linear, but values greater than 1 resulted in considerably higher severity rates than values less than 1. Like the JSI, the JSR can be used either to select employees for stressful tasks or to prioritize redesign requirements.

34.3.4.4 Dynamic Model for Prediction of LBDs in Industry

Marras et al. (1993) performed a retrospective study to determine the relationships between workplace factors and trunk motion factors and LBD occurrence. A logistic regression analysis was performed to provide a model used to estimate the probability of high-risk LBD membership. High-risk jobs were defined as jobs having incidence rates of 12 or more injuries per 200,000 h of exposure. The regressors included in the model were lift rate (lifts/hour), average twisting velocity (degree/second), maximum moment Newton meter (Nm), maximum sagittal flexion (degrees), and maximum lateral velocity (degree/second).

This model can be used to guide workplace design changes, as the probability of high-risk LBD membership can be computed before and after design changes. For example, maximum moment could be reduced by decreasing the load weight or the maximum horizontal distance between the load and the lumbar spine, and the associated decrease in high-risk membership probability can be estimated. The model is considerably different from the models discussed above in that LBD risk is not assumed to be related to individual capacity.

34.4 EMPLOYEE EDUCATION AND TRAINING

34.4.1 Training Programs for Prevention of LBDs

According to Kroemer, Kroemer, and Kroemer-Elbert (1994), about 50% of injured individuals return to work within 1 week after an incidence of low-back pain. Even though as the duration of absence from work increases, the likelihood of successful return diminishes, about 90% of the affected workers attempt to return to work. The main deterrents to returning to work after the low-back injury include those related to the worker (malingering, illness behavior, etc.), the management (lack of followup, no work modification), the union (rigid work rules), the practitioner (inappropriate or ineffective treatment), or the lawyer (lump sum settlements instead of rehabilitation) (Snook, 1988). It should be pointed out that in many cases the employment conditions for these workers are subject to specific medical work restrictions.

Personnel training in "safe" lifting techniques has been advocated and practiced for many years as the means to reduce the incidence (i.e., frequency and severity rates) of LBDs. Training has also been used to promote physical fitness and manual handling skills of the workers. One of the most popular lifting techniques is the so-called "straight-back/bent knees" method, which advocates keeping the back straight, and bending the knees and lifting using the legs. The "kinetic lift" is another variation of this technique with the predefined posture of the body (Kroemer et al., 1994). The summary of guidelines for manual lifting is shown in Table 34.11.

34.4.2 The Effects of Training on LBDs

Although Miller (1977) reported success of the training program with respect to lowering the frequency of LBDs, Kroemer et al. (1994), based on the extensive review of over 100 published studies, suggested that there was strong evidence to conclude that training in "safe lifting techniques" was not an effective program for prevention of LBDs. Many of the examined studies that investigated the effectiveness of lifting training programs on prevention of LBDs at the workplace failed to show significant reductions in LBDs (Brown, 1975; Yu, Roht, Wise, Kilian, and Weir, 1984; Stubbs, Buckle, Hudson, and Rivers, 1983; Wood, 1987). For example, Snook, Campanelli, and Ford (1980) showed that the number of work-related LBDs in companies that used the training programs in "safe lifting" did not differ statistically from those that did not use such programs. Rowe (1983) concluded that "except for brief periods immediately following a training campaign when reported back injuries seem to decrease, there has been no convincing evidence that such training has effected any significant reduction in low back episodes."

34.4.3 The Effects of Education on LBDs

Another approach to training is the "back school" (Fahrini 1975; Fitzler and Berger 1983; Pope, 1987; Snook and White, 1984), mainly used for rehabilitation of patients with

Table 34.11 Guidelines for Manual Lifting[a]

Things to follow:

1. Design manual lifting (and lowering) out of the task and workplace. If it needs to be done by a person, perform it between knuckle and shoulder height.
2. Be in good physical shape. If not used to lifting and vigorous exercise, do not attempt to do difficult lifting or lowering tasks.
3. Think before acting. Place material conveniently. Make sure sufficient space is cleared. Have handling aids available.
4. Get a good grip on the load. Test the weight before trying to move it. If it is too bulky or heavy, get a mechanical lifting aid, or somebody else or help, or both.
5. Get the load close to the body. Place the feet close to the load. Stand in a stable position, have the feet point in the direction of movement.
6. Involve primarily straightening of the legs in lifting.

Things to avoid:

1. Do NOT twist the back, or bend sideways.
2. Do NOT lift or lower, push or pull, awkwardly.
3. Do NOT hesitate to get help, either mechanical or by another person.
4. Do NOT lift or lower with arms extended.
5. Do NOT continue heaving when the load is too heavy.

[a]Adapted from Kroemer, K., Kroemer, H. and Kroemer-Elbert, K. (1995). *Ergonomics, how to design for ease and efficiency,* Englewood Cliffs, NJ: Prentice-Hall, reprinted with permission.

LBDs. According to Kroemer et al. (1994), back schools focus on knowledge, awareness, and attitude change by educating the patient about anatomy, biomechanics, spine injuries, and stress management. A study by Bergquist-Ullman and Larsson (1977) showed that back schools and physical therapy treatments were equally effective in LBD abatement, but that back schools were judged as being economically more efficient. However, a study of over 3000 workers of the Boeing Company showed no significant differences in the occurrence of low-back pain between the healthy workers who attended back school and a control group of workers that did not participate in such a school (Snook and White, 1984).

As pointed out by Kroemer et al. (1994), since the material handling task characteristics and requirements differ much among different industries as well as within any one industry, depending on the specific job tasks, handling aids and equipment available, successful implementation of worker selection, and job design, it is impossible to specify a universal set of training recommendations that would be applicable across different settings. The training courses aimed to prevent the incidence of LBDs should focus on the following teaching aspects: (1) specific lifting techniques, i.e., skill improvement: (2) principles of biomechanics, awareness of and self-responsibility of back injuries; and (3) physical fitness.

REFERENCES

Aghazadeh, F., and Ayoub, M. M. (1985). A comparison of dynamic- and static-strength models for prediction of lifting capacity. *Ergonomics, 28(10),* 1409–1417.

Andersson, G. B. J. (1985). Permissible loads: Biomechanical considerations. *Ergonomics, 28(1),* 323–326.

Asfour, S. S. (1980). *Energy Cost Predicting Models for Manual Lifting and Lowering Tasks.* Ph.D. Dissertation, Texas Tech University, Lubbock, TX.

Asfour, S. S., Genaidy, A. M., Khalil, T. M., and Greco, E. C. (1984). Physiological and psychophysical determination of lifting capacity for low frequency lifting tasks. In A. Mital, Ed., *Trends in Ergonomics/Human Factors I* (pp. 149–153). Cincinnati, OH: North-Holland.

Asfour, S. S., Genaidy, A. M., Khalil, T. M., and Muthuswamy, S. (1986). Physiological responses to static, dynamic and combined work. *American Industrial Hygiene Association Journal, 47(12),* 798–802.

Asfour, S. S., Genaidy, A. M., and Mital, A. (1988). Physiological guidelines for the design of manual lifting and lowering tasks: The state of the art. *American Industrial Hygiene Association Journal, 49(4),* 150–160.

Ayoub, M. A. (1982). Control of manual lifting hazards: II. Job redesign. *Journal of Occupational Medicine, 24(9),* 668–676.

Ayoub, M. M., Bethea, N. J., Deivanayagam, S., Asfour, S. S., Bakken, G. M., Liles, D., Mital, A., and Sherif, M. (1978). *Determination and Modelling of Lifting Capacity* (Final report HEW [NIOSH] Grant No. 5R010H-000545-02). Washington, DC: National Institute of Occupational Safety and Health.

Ayoub, M. M., and El-Bassoussi, M. M. (1976). Dynamic biomechanical model for sagittal lifting activities. *Proceedings of the 6th Congress of International Ergonomics Association* (pp. 355–359).

Ayoub, M. M., and Mital, A. (1989). *Manual Materials Handling.* London: Taylor and Francis.

Ayoub, M. M., Mital, A., Asfour, S. S., and Bethea, N. J. (1980b). Review, evaluation, and comparison of models for predicting lifting capacity, *Human Factors, 22(3),* 257–269.

Ayoub, M. M., Mital, A., Bakken, G. M., Asfour, S. S., and Bethea, N. J. (1980a). Development of strength and capacity norms for manual materials handling activities: The state of the art. *Human Factors, 22(3),* 271–283.

Ayoub, M. M., Gidcumb, C. F., Hafez, H., Intaranont, K., Jiang, B. C., and Selan, J. L. (1983). *A Design Guide for Manual Lifting* (Report prepared for the Occupational Safety and Health Administration). Lubbock, TX: Institute for Ergonomics Research.

BNA Special Report. (1988). *Back Injuries: Costs, Causes, Cases and Prevention* Washington, DC: Bureau of National Affairs, Inc.

Battié, M. C., Bigos, S. J., Fisher, L. D., Spengler, D. M., Hansson, T. H., Nachemson, A. L., and Wortley, M. D. (1990). The role of spinal flexibility in back pain complaints within industry: A prospective study. *Spine, 15,* 768–773.

Bergquist-Ullman, M. and Larsson, U. (1977). Acute low back pain in industry. *Acta Orthopadica Scandinavica,* Suppl. 170.

Biering-Sørensen, F., and Thomsen, C. (1986). Medical, social and occupational history as risk indicators for low-back trouble in a general population. *Spine, 11,* 720–725.

Bigos, S. J., Battié, M. C., Spengler, D. M., Fisher, L. D., Fordyce, W. E., Hansson, T. H., Nachemson, A. L., and Wortley, M. D. (1991). A prospective study of work perceptions and psychosocial factors affecting the report of back injury. *Spine, 16,* 1–6.

Bigos, S. J., Spengler, D. M., Martin, N. A., Zeh, J., Fisher, L., Nachemson, A. and Wang, M. H. (1986). Back injuries in industry: A retrospective study: II. Injury factors. *Spine, 11,* 246–251.

Borg, G. A. V. (1962). *Physical performance and perceived exertion.* (Lund, Gleerup):

Brinckmann, P., Biggemann, M., and Hilweg, D. (1988). Fatigue fractures of human lumbar vertebrae. *Clinical Biomechanics, 3* (Suppl. 1).

Brinckmann, P., Biggemann, M., and Hilweg, D. (1989). Prediction of the compressive strength of human lumbar vertebrae. *Clinical Biomechanics, 4* (Suppl. 2).

Brokaw, N. (1992). *Implications of the Revised NIOSH Lifting Guide of 1991: A Field Study.* Unpublished M.S. Thesis, Department of Industrial Engineering, University of Louisville, Louisville, KY.

Brown, J. R. (1975). Factors contributing to the development of low-back pain in industrial workers. *American Industrial Hygiene Association Journal, 36,* 26–31.

Bureau of Labor Statistics. (1982). *Back Injuries Associated with Lifting* (Bulletin 2144). Washington, DC: U.S. Department of Labor.

Chaffin, D. B. (1969). A computerized biomechanical model: development of and use in studying gross body actions. *Journal of Biomechanics, 2,* 429–441.

Chaffin, D. B. (1974). Human strength capability and low-back pain. *Journal of Occupational Medicine, 16(4),* 248–254.

Chaffin, D. B., and Page, G. B. (1994). Postural effects on biomechanical and psychophysical weight-lifting limits. *Ergonomics, 37(4),* 663–676.

Chaffin, D. B., Herrin, G. D., and Keyserling, W. M. (1978). Preemployment strength testing: An updated position. *Journal of Occupational Medicine, 20,* 403–408.

Chaffin, D. B., Herrin, G. D., Keyserling, W. M., and Garg, A. (1977). A method for evaluating the biomechanical stresses resulting from manual materials handling jobs. *American Industrial Hygiene Association Journal, 38,* 662–675.

Chaffin, D. B., and Park, K. S. (1973). A longitudinal study of low back pain as associated with occupational weight lifting factors. *American Industrial Hygiene Association Journal, 34(12),* 513–525.

Chen, H. C., and Ayoub, M. M. (1988). Dynamic biomechanical model for asymmetrical lifting. In F. Aghazadeh, F., Ed., *Trends in Ergonomics/Human Factors V* (pp. 879–886). Amsterdam: Elsevier.

Ciriello, V. M., Snook, S. H., Blick, A. C., and Wilkinson, P. L. (1990). The effects of task duration on psychophysically-determined maximum acceptable weights and forces. *Ergonomics, 333(2),* 187–200.

Damkot, D. K., Pope, M. H., Lord, J., and Frymoyer, J. W. (1984). The relationship between work history, work environment and low-back pain in men. *Spine, 9,* 395–399.

Delleman, N. J., Drost, M. R., and Huson, A. (1992). Value of biomechanical macromodels as suitable tools for the prevention of work-related low back problems. *Clinical Biomechanics, 7,* 138–148.

Dempsey, P. G., Ayoub, M. M., and Westfall, P. H. (1995). The NIOSH lifting equations: A closer look, In A. C. Bittner and P. C. Champney, Eds., *Advances in Industrial Ergonomics and Safety VII* (pp. 705–712). London: Taylor and Francis.

Donovan, W. H., Dwyer, A. P., White, B. W. S. (1981). A multidisciplinary approach to chronic low back pain in western Australia. *Spine, 6,* 591–597.

Dryden, R. D. (1973). *A predictive model for the maximum permissible weight of lift from knuckle to shoulder height.* Ph.D. Dissertation, Texas Tech University, Lubbock, TX.

Dutta, S. P., and Taboun, S. (1989). Developing norms for manual carrying tasks using mechanical efficiency as the optimization criterion. *Ergonomics, 32(8),* 919–943.

Eisler, H. (1962). Subjective scale of force for a large muscle group. *Journal of Experimental Psychology, 64(3),* 253–257.

El-Bassoussi, M. M. (1974). *A biomechanical dynamic model for lifting in the sagittal plane.* Ph.D. Dissertation, Texas Tech University, Lubbock, TX.

Fahrini, W. H. (1975). Conservative treatment of lumbar disc degeneration, our primary responsibility. *Orthopedic Clinics of North America, 6,* 93–103.

Farfan, H. F. (1983). Biomechanics of the lumbar spine. In W. H. Kirkaldy-Willis, Ed., *Managing Low Back Pain* (pp. 9–21). New York: Churchill Livingstone.

Federal Register. (1986). October 2. *51(191),* 35–41.

Fernandez, J. E., and Ayoub, M. M. (1988). The psychophysical approach: The valid measure of lifting capacity. In F. Aghazadeh, Ed., *Trends in Ergonomics/Human Factors. V.* (pp. 837–845). Amsterdam: North-Holland.

Fernandez, J. E., Ayoub, M. M., and Smith, J. L. (1991). Psychophysical lifting capacity over extended periods. *Ergonomics, 34(1),* 23–32.

Fisher, B. O. (1967). *Analysis of spinal stresses during lifting.* M.S. Thesis. The University of Michigan, Ann Arbor, MI.

Fitzler, S. L., and Berger, R. A. (1983). Chelsea Back Program: One year later. *Occupational Health and Safety, 52,* 52–54.

Frymoyer, J. W., Pope, M. H., Constanza, M. C., Rosen, J. C., Goggin, J. E., and Wilder, D. G. (1980). Epidemiologic studies of low back pain. *Spine, 5,* 419–423.

Frymoyer, J. W., Pope, M. H., Clements, J. H., Wilder, D. G., MacPherson, B., and Ashikaga, T. (1983). Risk factors in low back pain: An epidemiological survey. *Journal of Bone and Joint Surgery, 65-A,* 213–218.

Gagnon, M., and Smyth, G. (1990). The effect of height in lowering and lifting tasks: A mechanical work evaluation. In B. Das, Ed., *Advances in Industrial Ergonomics and Safety II.* (pp. 669–672). London: Taylor and Francis.

Gallagher, S. (1991). Acceptable weights and physiological costs of performing combined manual handling tasks in restricted postures. *Ergonomics, 34(7),* 939–952.

Gamberale, F., and Kilböm, A. (1988). An experimental evaluation of psychophysically determined maximum acceptable workload for repetitive lifting work. In A. S. Adams, R. R. Hall, B. J. McPhee, and M. S. Oxenburgh, Eds., *Proceedings of the 10th congress of the International Ergonomics Association* (pp. 233–235). London: Taylor and Francis.

Garg, A. (1989). An evaluation of the NIOSH guidelines for manual lifting with special reference to horizontal distance. *American Industrial Hygiene Association Journal, 50(3)* 157–164.

Garg, A., and Ayoub, M. M. (1980). What criteria exist for determining how much load can be lifted safely? *Human Factors, 22,* 475–486.

Garg, A., and Badger, D. (1986). Maximum acceptable weights and maximum voluntary strength for asymmetric lifting. *Ergonomics, 29(7),* 879–892.

Garg, A., and Banaag, J. (1988). Psychophysical and physiological responses to asymmetric lifting. In F. Aghazadeh, Ed., *Trends in Ergonomics/Human Factors V.* (pp. 871–877). Amsterdam: North-Holland.

Garg, A., and Chaffin, D. B. (1975). A biomechanical computerized simulation of human strength. *IIE Transactions, 14(4),* 272–281.

Garg, A., Chaffin, D. B., and Freivalds, A. (1982). Biomechanical stresses from manual load lifting: Static vs. dynamic evaluation. *Institute of Industrial Engineers Transactions, 14,* 272–281.

Garg, A., Chaffin, D. B., and Herrin, G. D. (1978). Prediction of metabolic rates for manual materials handling jobs. *American Industrial Hygiene Association Journal, 39(8),* 661–675.

Garg, A., and Herrin, G. D. (1979). Stoop or squat: A biomechanical and metabolic evaluation. *AIIE Transactions, 11(4),* 293–302.

Genaidy, A. M., and Asfour, S. S. (1987). Review and evaluation of physiological cost prediction models for manual materials handling. *Human Factors, 29(4),* 465–476.

Genaidy, A. M., Asfour, S. S., Khalil, T. M., and Waly, S. M. (1985). Physiological issues in manual materials handling. In R. E. Eberts and C. G. Eberts, Eds., *Trends in Ergonomics/Human Factors II.* (pp. 571–576). Amsterdam: North-Holland.

Gentry, W. D., Shows, W. D., and Thomas, M. (1974). Chronic back pain: A psychological profile. *Psychosomatics, 15,* 174–177.

Gilchrist, I. C. (1976). Psychiatric and social factors related to low back pain in general practice, *Rheumatology and Rehabilitation, 15,* 101–107.

Gyntelberg, F. (1974). One year incidence of low back pain among male residents of Copenhagen aged 40–59. *Danish Medical Bulletin, 21,* 30–36.

Habes, D., Carlson, W., and Badger, D. (1985). Muscle fatigue associated with repetitive arm lifts: effects of height, weight and reach. *Ergonomics, 28(2),* 471–488.

Hafez, H. A., and Ayoub, M. M. (1991). A psychophysical study of manual lifting in hot environments. *International Journal of Industrial Ergonomics, 7,* 303–309.

Hansson, T. (1989). *Ländryggsbesvär och Arbete* [Low-Back Pain and Work]. Stockholm: Arbetsmiljöfonden.

Herrin, G. D., Chaffin, D. B., and Mach, R. S. (1974). Criteria for Research on the Hazards of Manual Materials Handling (Workshop Proceedings, Contract CDC-99-74-118). Cincinnati, OH: U.S. Department of Health and Human Services (NIOSH).

Herrin, G. D., Jaraiedi, M., and Anderson, C. K. (1986). Prediction of overexertion injuries using biomechanical and psychophysical models. *American Industrial Hygiene Association Journal, 47(6),* 322–330.

Hildebrandt, V. H. (1987). A review of epidemiological research on risk factors of low back pain. In P. W. Buckle, Ed., *Musculoskeletal Disorders at Work.* London: Taylor and Francis.

Hult, L. (1954a). The Munkfors investigation. *Acta Orthop. Scandinavia (Suppl.), 16,* 76.

Hult, L. (1954b). Cervical, dorsal and lumbar spinal syndromes. *Acta Orthopedica Scandinavia (Suppl.), 17,* 102.

ILO (1962). *Maximum Permissible Weight to be Carried by One Worker.* (Information Sheet No. 3). Geneva: JLO.

Ikata, T. (1965). Statistical and dynamic studies of lesions due to overloading of the spine. *Shikoku Acta Medicum 40,* 262–286.

Jäger, M., and Luttmann, A. (1989). Biomechanical analysis and assessment of lumbar stress during load lifting using a dynamic 19-segment human model. *Ergonomics, 32(1),* 93–112.

Jäger, M., and Luttmann, A. (1991). Compressive strength of lumbar spine elements related to age, gender, and other influencing factors. In P. A. Anderson, D. J. Hobart, and J. V. Danoff, Eds., *Electromyographical Kinesiology* (pp. 291–294). Amsterdam: Elsevier.

Jäger, M., and Luttmann, A. (1992a). Lumbosacral compression for uni- and bi-manual asymmetrical load lifting. In S. Kumar, Ed., *Advances in Industrial Ergonomics and Safety IV* (pp. 839–846). London: Taylor and Francis.

Jäger, M., and Luttmann, A. (1992b). The load on the lumbar spine during asymmetrical bi-manual materials handling. *Ergonomics, 35(7/8),* 783–805.

Jiang, B. C. (1984). *Psychophysical capacity modeling of individual and combined manual materials handling activities.* Ph.D. Dissertation, Texas Tech University, Lubbock, TX.

Jiang, B. C., Smith, J. L., and Ayoub, M. M. (1986). Psychophysical modelling of manual materials handling capacities using isoinertial strength variables. *Human Factors, 28(6),* 691–702.

Joukaamaa, M. (1986). *Alaselan kipu ja psykkieset tekijat: Yyoikaiseen vaestoon kohdistuva sosiaalipsykiatrinen tutkimus.* [Low-Back Pain and Psychological Factors: A Sociopsychiatric Study of the Working Age Population]. Turku, Finland: Kansanelakelaitos, Julkaisuja AL:28.

Jung, E. S., and Freivalds, A. (1991). Multiple criteria decision-making for the resolution of conflicting knowledge in manual materials handling. *Ergonomics, 34(11),* 1351–1356.

Kamon, E., and Ayoub, M. M. (1976). *Ergonomic Guide to Assessment of Physical Work Capacity.* Akron, OH: American Industrial Hygiene Association.

Karwowski, W. (1983). A pilot study of the interaction between physiological, biomechanical and psychophysical stresses involved in manual lifting activities. In *Contemporary Ergonomics 1983* (pp. 95–100). London: Taylor and Francis.

Karwowski, W. (1989). Perception of load heaviness by males. In K. H. E. Kroemer, J. D. McGlothlin and T. G. Bobick, Eds., *Manual Material Handling: Understanding and Preventing Back Trauma* (pp. 9–14). Akron, OH: American Industrial Hygiene Association.

Karwowski, W. (1991). Psychophysical acceptability and perception of load heaviness by females, *Ergonomics, 34(4),* 487–496.

Karwowski, W. (1992). Comments on the assumption of multiplicity of risk factors in the draft revisions to NIOSH Lifting Guide. In S. Kumar, Ed., *Advances in Industrial Ergonomics and Safety IV.* London: Taylor and Francis.

Karwowski, W., and Ayoub, M. M. (1984). Fuzzy modelling of stresses in manual lifting tasks. *Ergonomics, 27(6),* 641–649.

Karwowski, W., and Brokaw, N. (1992). Implications of the proposed revisions in a draft of the Revised NIOSH Lifting Guide (1991) for job redesign: A field study. In *Proceedings of the 36th Annual Meeting of the Human Factors Society.* Atlanta, GA, pp. 659–663.

Karwowski, W., and Burkhardt, A. (1988). Subjective judgement of load heaviness and psychophysical approach to manual lifting. In F. Aghazadeh, Ed., *Trends in Ergonomics/Human Factors,* (Vol. V, pp. 865–870). Amsterdam: North Holland.

Karwowski, W., and Gaddie, P. (1995). Simulation of the 1991 revised NIOSH manual lifting equation. *Proceeding of the Human Factors and Ergonomics Society Annual Meeting.* San Diego, CA, pp. 699–701.

Karwowski, W., Jamaldin, B., and AlQesami, K. K. (1996). Examination of the psychophysical approach to setting limits for manual lifting by males. Unpublished Technical Report, Center for Industrial Ergonomics, University of Louisville, Louisville, Kentucky.

Karwowski, W., and Pongpatana, N. (1989). The effect of color on human perception of load heaviness. In A. Mital (Ed.), *Advances in Industrial Ergonomics and Safety I* (pp. 673–678). London: Taylor and Francis.

Karwowski, W., Shumate, C., Yates, J. W., and Pongpatana, N. (1992). Discriminability of load heaviness in manual lifting: implications for the psychophysical approach. *Ergonomics, 37*(7–8), 729–744.

Karwowski, W., and Yates, J. W. (1986). Reliability of the psychophysical approach to manual lifting of liquids by females. *Ergonomics, 29(2),* 237–248.

Kelsey, J. L., Githens, P. B., White, A. A., Holford, R. R., Walter, S. D., O'Connor, T., Astfeld, A. M., Weil, U., Southwick, W. O., and Calogero, J. A. (1984). An epidemiologic study of lifting and twisting on the job and risk for acute prolapsed lumbar intervertebral disc. *Journal of Orthopaedic Research, 2,* 61–66.

Kelsey, J. L. and Golden, A. L. (1988). Occupational and workplace factors associated with low back pain. In R. A. Deyo, Ed., *Back Pain in Workers* (Occupational medicine: state of the art reviews 3). Philadelphia, PA: Hanley & Belfus.

Keyserling, W. M., Herrin, G. D., Chaffin, D. B., Armstrong, T. J., and Foss, M. L. (1980). Establishing an industrial strength testing program. *American Industrial Hygiene Association Journal, 41(10),* 730–736.

Keyserling, W. M., Punnett, L., and Fine, L. J. (1988). Trunk posture and back pain: Identification and control of occupational risk factors. *Applied Industrial Hygiene, 3,* 87–92.

Kim, H.-K. (1990). *Development of a Model for Combined Ergonomic Approaches in Manual Materials Handling Tasks.* PhD Dissertation, Texas Tech University, Lubbock, TX.

Knipfer, R. E. (1974). *Predictive Models for the Maximum Acceptable Weight of Lift,* Ph.D. Dissertation, Texas Tech University, Lubbock, TX.

Kroemer, K. H. E. (1992). Personnel training for safer material handling. *Ergonomics, 35(9),* 1119–1134.

Kroemer, K., Kroemer, H., and Kroemer-Elbert, K. (1994). *Ergonomics. How to Design for Ease and Efficiency.* Englewood Cliffs, NJ: Prentice-Hall.

Kromodihardjo, S., and Mital, A. (1986). Kinetic analysis of manual lifting activities: part I-development of a three dimensional computer model. *International Journal of Industrial Ergonomics, 1,* 77–90.

Kromodihardjo, S., and Mital, A. (1987). Biomechanical analysis of manual lifting tasks. *Journal of Biomechanical Engineering, 109,* 132–138.

Kumar, S. (1990). Cumulative load as a risk factor for back pain. *Spine, 15(12),* 1311–1316.

Kumar, S., and Mital, A. (1989). Safety of back! What is the margin? *Proceedings of the Human Factors Society 33rd Annual Meeting* (Vol 1, pp. 677–681). Santa Monica, CA: Human Factors Society.

Lawrence, J. S. (1955). Rheumatism in coal miners: Part III. Occupational Factors, *British Journal of Industrial Medicine, 12,* 249–261.

Leamon, T. B. (1994a). So who is counting? *International Journal of Industrial Ergonomics, 13,* 259–265.

Leamon, T. B. (1994b). Research to reality: a critical review of the validity of various criteria for the prevention of occupationally induced low back pain disability. *Ergonomics, 37(12),* 1959–1974.

Legg, S. J., and Haslam, D. R. (1984). Effect of sleep deprivation on self-selected workload. *Ergonomics, 27(4),* 389–396.

Legg, S J., and Myles, W. S. (1981). Maximum acceptable repetitive lifting workloads for an 8 hour work-day using psychophysical and subjective rating methods. *Ergonomics, 24(12),* 907–916.

Legg, S. J., and Myles, W. S. (1985). Metabolic and cardiovascular cost, and perceived effort over an 8 hour day when lifting loads selected by the psychophysical method. *Ergonomics, 28,* 337–343.

Leskinen, T. P. J., Stalhammar, H. R., and Kuorinka, I. A. A. (1983). A dynamic analysis of spinal compression with different lifting techniques. *Ergonomics, 26(6),* 595–604.

Liles, D. H., Deivanayagam, S., Ayoub, M. M., and Mahajan, P. (1984). A job severity index for the evaluation and control of lifting injury. *Human Factors, 26(6),* 683–693.

Ljungberg, A.-S., Gamberale, F., and Kilböm, Å. (1982). Horizontal lifting—Physiological and psychological responses. *Ergonomics, 25(8),* 741–757.

Lloyd, M. H., Gauld, S., and Soutar, C. A. (1986). Epidemiologic study of back pain in miners and office workers, *Spine, 11,* 136–140.

Maeda, K., Okazaki, F., Svenaga, T. (1980). Low back pain related to bowing posture of greenhouse farmers, *Journal of Human Ergology, 9,* 117–123.

Magora, A. (1970). Investigation of the relation between low back pain and occupation: II. Work history, *Industrial Medical Surgery, 30,* 504–510.

Magora, A. (1972). Investigation of the relation between low back pain and occupation: III. Physical requirements: sitting, standing and weight lifting. *Industrial Medicine and Surgery, 41,* 5–9.

Magora, A. (1973). Investigation of the relation between low back pain and occupation: IV. Physical requirements: Bending rotation, reaching and sudden maximal effort. *Scandinavian Journal of Rehabilitative Medicine, 5,* 186–190.

Magnusson, M., Granqvist, M., Jonson, R., Lindell, V., Lundberg, U., Wallin, L., and Hansson, T. (1990). The loads on the lumbar spine during work at an assembly line. *Spine, 15(8),* 774–779.

Marras, W. S., Lavender, S. A., Leurgans, S. E., Rajulu, S. L., Allred, W. G., Fathallah, F. A., and Ferguson, S. A. (1993). The role of dynamic three-dimensional trunk motion in occupationally-related low back disorders: The effects of workplace factors, trunk position and trunk motion characteristics on risk of injury. *Spine, 18(5),* 617–628.

Martin, J. B., and Chaffin, D. B. (1972). Biomechanical computerized simulation of human strength in sagittal plane activities. *AIIE Transactions, 4(1),* 19–28.

McDaniel, J. W. (1972). *Prediction of Acceptable Lift Capability,* Ph.D. Dissertation, Texas Tech University, Lubbock, TX.

McGill, S. M., and Norman, R. W. (1985). Dynamically and statically determined low back moments during lifting. *Journal of Biomechanics, 18(12),* 877–885.

Miller, R. L. (1977). Bend your knees. *National Safety News, 115,* 57–58.

Mital, A. (1983). The psychophysical approach in manual lifting—A verification study. *Human Factors, 25(5),* 485–491.

Mital, A. (1984). Comprehensive maximum acceptable weight of lift database for regular 8-hour work shifts. *Ergonomics, 27(11),* 1127–1138.

Mital, A. (1992). Psychophysical capacity of industrial workers for lifting symmetrical and asymmetrical loads symmetrically and asymmetrically for 8 hour work shifts. *Ergonomics, 35(7/8),* 745–754.

Mital, A., Ayoub, M. M., Asfour, S. S., and Bethea, N. J. (1978). Relationship between lifting capacity and injury in occupations requiring lifting. In *Proceedings of the Human Factors Society 22nd Annual Meeting* (pp. 469–473). Santa Monica, CA: Human Factors Society.

Mital, A., Genaidy, A. M., and Brown, M. L. (1989). Predicting maximum acceptable weights of symmetrical and asymmetrical loads for symmetrical and asymmetrical lifting. *Journal of Safety Research, 20(1),* 1–6.

Mital, A., Nicholson, A. S., and Ayoub, M. M. (1993). *A Guide to Manual Materials Handling.* London: Taylor and Francis.

Morrissey, S. J., Bittner, A. C., and Arcangeli, K. K. (1990). Accuracy of a ratio-estimation method to set maximum acceptable weights in complex lifting tasks. *International Journal of Industrial Ergonomics, 5,* 169–174.

National Academy of Sciences, Eds. (1985). *Injury in America.* Washington, DC: National Academy Press.

National Institute for Occupational Safety and Health. (1981). *Work Practices Guide for Manual Lifting* (DHHS (NIOSH, Publ. No. 81-122). Cincinnati, OH: U.S. Department of Health and Human Services.

National Safety Council. (1989). *Accident Facts 1989.* (National Safety Council, Chicago).

National Safety Council. (1990). *Accident Statistics.* (National Safety Council, Chicago).

Nicholson, A. S. (1989). A comparative study of methods for establishing load handling capabilities. *Ergonomics, 32,* 1125–1144.

Nicholson, L. M., and Legg, S. J. (1986). A psychophysical study of the effects of load and frequency upon selection of workload in repetitive lifting. *Ergonomics, 29(7),* 903–911.

Occupational Safety and Health Administration. (1982). *Back Injuries Associated with Lifting* (Bulletin 2144). Washington, DC: US Government Printing Office.

Park, K. S., and Chaffin, D. B. (1974). A biomechanical evaluation of two methods of manual load lifting. *AIIE Transactions, 6(2),* 105–113.

Partridge, R. E., and Duthie, J. J. R. (1968). Rheumatism in dockers and civil servants: a comparison of heavy manual and sedentary workers, *Annals of Rheumatic Disease, 27,* 559–568.

Penttinen, J. (1987). *Back pain and sciatica in Finnish Farmers* (Publ. ML; 71.) Helsinki: The Social Insurance Institution.

Pope, M. H. (1987). The biomechanical basis for early care programmes. *Ergonomics, 30,* 351–358.

Porter, R. W., Adams, M. A., and Hutton, W. C. (1989). Physical activity and the strength of the lumbar spine. *Spine, 14,* 201–203.

Potvin, J., Norman, R. W., and McGill, S. M. (1991). Reduction in anterior shear forces on the L4/L5 disc by the lumbar musculature. *Clinical Biomechanics, 6,* 88–96.

Pritsker, A. A. B. (1986). *Introduction to Simulation and SLAM II (Third Edition).* New York: John Wiley.

Randle, I. P. M. (1987). Predicting the metabolic cost of intermittent load carriage in the arms. In E. D. Megaw, Ed., *Contemporary ergonomics 1987* (pp. 286–291). London: Taylor and Francis.

Riihimäki, H. (1991). Low-back pain, its origin and risk indicators. *Scandinavian Journal of Work, Environment & Health, 17,* 81–90.

Riihimäki, H., Tola, S., Videman, T., and Hänninen, K. (1989). Low-back pain and occupation: a cross-sectional questionnaire study of men in machine operating, dynamic physical work and sedentary work. *Scandinavian Journal of Work, Environment & Health, 14,* 204–209.

Rowe, M. L. (1983). *Backache at Work.* Perinton: Fairport.

Sairanen, E., Brüshaber, L., and Kaskinen, M. (1981). Felling work, low-back pain and osteoarthritis. *Scandinavian Journal of Work Environment and Health, 7,* 18–30.

Schultz, A. B., and Andersson, G. B. J. (1981). Analysis of loads on the lumbar spine. *Spine, 6(1),* 76–82.

Selan, J. L., Halcomb, C. G., and Ayoub, M. M. (1987). Psychological variables affecting lifting capacity: A review of three studies. In S. S. Asfour, Ed., *Trends in Ergonomics Human Factors IV.* (pp. 987–993). North-Holland, Amsterdam.

Smith, J. L., Ayoub, M. M., and McDaniel, J. W. (1992). Manual materials handling capabilities in non-standard postures. *Ergonomics, 35,* 807–831.

Snook, S. H. (1978). The design of manual handling tasks. *Ergonomics, 21(12),* 963–985.

Snook, S. H. (January–March, 1988). The costs of back pain in industry. *Occupational Medicine: State of the Art Reviews, 3,* 1–5.

Snook, S. H., Campanelli, R. A., and Ford, R. J. (1980). *A Study of Back Injuries at Pratt–Whitney Aircraft.* Hopkinton, MA: Liberty Mutual Insurance Company Research Center.

Snook, S. H., Campanelli, R. A., and Hart, J. W. (1978). A study of three preventive approaches to low back injury. *Journal of Occupational Medicine, 20(7),* 478–481.

Snook, S. H., and Ciriello, V. M. (1991). The design of manual handling tasks: revised tables of maximum acceptable weights and forces. *Ergonomics, 34(9),* 1197–1213.

Snook, S. H., and Irvine, C. H. (1969). Psychophysical studies of physiological fatigue criteria. *Human Factors, 11(3),* 291–300.

Snook, S. H., and White, A. H. (1984). Education and training. In M. H. Pope, J. W. Frymoyer, and G. Andersson, Eds., *Occupational Low Back Pain* (pp. 233–244). Philadelphia: Praeger.

Spengler, D. M. J., Bigos, S. J., Martin, N. A., Zeh, J., Fisher, L., and Nachemson, A. (1986). Back injuries in industry: A retrospective study. *Spine, 11,* 241–256.

Stalhammar, H. R., Louhevaara, V., and Troup, J. D. G. (1989). Individual assessment of acceptable weights for manual sorting of postal articles. In A. Mital, Ed., *Advances in Industrial Ergonomics and Safety I.* (pp. 653–658). London: Taylor and Francis.

Stevens, S. S. (1975). *Psychophysics: Introduction to its perceptual, neural, social prospects.* New York: John Wiley.

Stubbs, D. A., Buckle, P. W., Hudson, M. P., and Rivers, P. M. (1983). Back pain in the nursing profession: II. The effectiveness of training. *Ergonomics, 26,* 767–779.

Svensson, H.-O., and Andersson, G. B. J. (1983). Low-back pain in forty to forty-seven year old men: Work history and work environment. *Scandinavian Journal of Rehabilitation Medicine, 8,* 272–276.

Svensson, H.-O., and Andersson, G. B. J. (1989). The relationship of low-back pain, work history, work environment and stress: A retrospective cross-sectional study of 38- to 64-year-old women. *Spine, 14,* 517–522.

Taboun, S. M., and Dutta, S. P. (1989). Energy cost models for combined lifting and carrying tasks. *International Journal of Industrial Ergonomics, 4(1),* 1–17.

Troup, J. D., and Edwards, F. C. (1985). *Manual Handling and Lifting.* London: HMSO.

Valkenburg, H. A. and Haanen, H. C. M. (1983). The epidemiology of low back pain. *Clin. Orthop, 179,* 9–22.

Videman, T., Nurminen, T., Tola, S., Kuorinka, I., Vanharanta, H., and Troup, D. J. (1984). Low-back pain in nurses and some loading factors of work. *Spine, 9,* 400–404.

Videman, T., Nurminen, M., and Troup, J. D. (1990). Lumbar spinal pathology in cadaveric material in relation to history of back pain, occupation, and physical loading. *Spine, 15,* 728–740.

Waters, T. R., Putz-Anderson, V., and Garg, A. (1994). *Application Manual for the Revised NIOSH Lifting Equation.* Cincinnati, OH: U.S. Department of Health and Human Services.

Waters, T. R., Putz-Anderson, V., Garg, A., and Fine, L. J. (1993). Revised NIOSH equation for the design and evaluation of manual lifting tasks. *Ergonomics, 36(7),* 749–776.

Wickström, G., Hänninen, K., Lehtinen, M., and Riihimäki, H. (1978). Previous back syndromes and present back symptoms in concrete reinforcement workers. *Scandinavian Journal of Work Environment and Health, 4*(Suppl. 1), 20–28.

Wood, D. P. (1987). Design and evaluation of a back injury prevention program within a geriatric hospital. *Spine, 12,* 77–82.

Yu, T., Roht, L. Y., Wise, R. A., Kilian, D. J., and Weir, F. W. (1984). Low-back pain in industry: An old problem revisited. *Journal of Occupational Medicine, 26,* 517–524.

CHAPTER 35

WORK-RELATED MUSCULOSKELETAL DISORDERS OF THE UPPER EXTREMITIES

Waldemar Karwowski
Department of Industrial Engineering
University of Louisville
Louisville, KY 40292 USA

William S. Marras
Department of Industrial Engineering
The Ohio State University
Columbus, OH 43210 USA

35.1 INTRODUCTION

35.1.1 Extent of the Problem

Work-related musculoskeletal disorders (WMSDs) affect several million workers in the United States, with a total cost exceeding $100 billion annually. The National Institute of Occupational Safety and Health (NIOSH, 1986) stated that musculoskeletal disorders, which include disorders of the back, trunk, upper extremity, neck, lower extremity, and traumatically induced Raynaud's phenomenon, are one of the 10 leading work-related illnesses and injuries in the United States. As reported by Praemer, Furner, and Rice (1992), the work-related upper extremity disorders (WUEDs), which are formally defined by the Bureau of Labor Statistics (BLS) as cumulative trauma *illnesses,* account for 11.0% of all work-related musculoskeletal disorders (illnesses). For comparison, the occupational low-back disorders account for more than 51.0% of all WMSDs. Significant increase in the incidence of WUEDs over the last 15 years can be linked to several occupational factors: increased production rates resulting from a worldwide competitive economic environment, a widespread use of computer technology, a higher percentage of women and older workers in the workforce, better record keeping of reportable illnesses and injuries on the job by employers, greater employee awareness of WUEDs and their relation to working conditions, and a marked shift in social policies and labor laws regarding recognition and compensation of the occupational injuries and illnesses. According to BLS (1995), in 1993 the cumulative trauma illnesses of upper extremity accounted for more than 60% of the occupational illnesses reported that year. However, it should also be noted that these work-related illnesses, which include hearing impairments caused by occupational noise exposure, represent only 6.0% of all reportable work-related injuries and illnesses (Marras, 1996).

35.1.2 Definitions

According to the most up to date and authoritative compendium on the subject matter (Kuorinka and Forcier, 1995), the umbrella term of *work-related musculoskeletal disorders* (WMSDs) defines those disorders and diseases of the musculoskeletal system that have a proven or hypothetical work-related causal component. *Musculoskeletal disorders* are pathological entities in which the functions of the musculoskeletal system are disturbed or abnormal, whereas diseases are pathological entities with observable impairments in body configuration and function. Although WUEDs are a heterogeneous group of disorders, and the current state of knowledge does not allow for a general description of the course of these disorders, it is possible nevertheless to identify a group of so-called "generic risk factors," including biomechanical factors, such as static and dynamic loading on the body and posture, cognitive demands, and organizational and psychosocial factors, for which there is an ample evidence of work relatedness and a higher risk of developing WUEDs.

According to Kuorinka and Forcier (1995), such generic risk factors, which typically interact and accumulate to form cascading cycles, are assumed to be directly responsible for the pathophysiological phenomena, which depend on location, intensity, temporal variation, duration, and repetitiveness of the generic risk factors. It was also proposed that both insufficient and excessive loading on the musculoskeletal system have deleterious effects, and that the pathophysiological process is dependent upon individual's characteristics with respect to body responses, coping mechanisms, and adaptation to risk factors. The generic risk factors, workplace design features, and the pathophysiological phenomena are parts of the generic model for WUEDs prevention proposed by Armstrong et al. (1993).

35.1.3 Cumulative Trauma Disorders of the Upper Extremity

In the United States, the term most often used for WUEDs is "cumulative trauma disorders" (CTDs) of the upper extremity. According to Putz-Anderson (1993), CTDs can be defined by combining the separate meanings for each word. "Cumulative" indicates that these disorders develop gradually over periods of time as a result of repeated stresses. The cumulative concept is based on the assumption that each repetition of an activity produces some trauma or wear and tear on the tissues and joints of the particular body part. The term "trauma" indicates bodily injury from mechanical stresses, whereas "disorders" refers to physical ailments. The above definition also stipulates a simple cause and effect model for CTD development. According to such a model, since the human body needs sufficient interval of rest time between episodes of repeated strains to repair itself, if the recovery time is insufficient, combined with a high repetition of forceful and awkward postures, the worker is at higher risk of developing a CTD. In the context of the generic model for prevention proposed by Armstrong et al. (1993), the above definition is primarily oriented toward biomechanical risk factors for WUEDs and, therefore, is incomplete.

According to Ranney, Wells, and Moore (1995), the evidence that chronic musculoskeletal disorders of the upper extremities are work related is rapidly growing. Armstrong et al. (1993) conclude that presently it is not possible to define the dose–response relationships and exposure limits for the WUED problems. This is why, for example, the Committee on the Control of Cumulative Trauma Disorders of the American National Standards Institute (ANSI, 1995), does not support the development of a specification-based standard for CTDs. In order to establish the work relatedness of these disorders, both the quantification of exposures involved in work and a determination of health outcomes, including details of the specific disorders (Hagberg, 1992, Luopajarvi, Kuorinka, Virolainen, and Holmberg, 1979; Moore, Wells, and Ranney, 1991; Stock, 1991), are needed. Also, more detailed medical diagnoses are required for choosing appropriate exposure measures as well as for structuring treatment, screening, and prevention programs (Ranney et al., 1995).

35.2 CONCEPTS AND CHARACTERISTICS OF WUEDs

35.2.1 Epidemiology of WUEDs

According to the World Health Organization (WHO, 1985), an occupational disease is a disease for which there is a direct cause–effect relationship between hazard and disease (e.g., silica–silicosis). Work-related diseases (WRDs) are defined as multifactorial when the work environment and the performance of work contribute significantly to the causation of disease (WHO, 1985). Work-related diseases can be partially caused by adverse work conditions. However, personal characteristics, environmental factors, and sociocultural factors are also recognized as risk factors for these diseases.

As reviewed by Armstrong et al. (1993) and summarized by Kuorinka and Forcier (1995), the evidence of work-relatedness of musculoskeletal disorders was firmly established by numerous epidemiologic studies conducted over the last 25 years of research in the field. A description of common musculoskeletal disorders and related job activities were summarized by Kroemer and Kroemer-Elbert (1994) and are shown in Table 35.1. It was also noted that the incidence and prevalence of musculoskeletal disorders in the reference populations were low, but not zero, most likely indicating the non–work related causes of these disorders. Such variables as cultural differences and psychosocial and economic factors, which may influence one's perception and tolerance of pain and consequently affect the willingness to report musculoskeletal problems, may have significant impact on the progressions from disorder to work disability (Leino, 1989; WHO, 1985).

35.2.2 Evidence of Work Relatedness of Upper Extremity Disorders

The WRDs of the upper extremity include, among others, carpal tunnel syndrome, tendonitis, ganglionitis, tenosynovitis, bursitis, and epicondylitis (Putz-Anderson, 1993). Workers employed in construction, food preparation, clerical and computer work, product fabrication, and mining are at a high risk of developing WUEDs. NIOSH (1977) reported that 15–20% of workers in these jobs are at a potential risk of WUEDs. Although the occurrence of WUEDs at work has been well documented (Kuorinka and Forcier, 1995), because of the high complexity of the problem, there is a lack of clear understanding of the cause–effect relationship characteristics for these disorders, which may prevent implementation of the effective control measures. The problem may be confounded by poor management–labor relationships, and lack of willingness to talk openly to each other about the potential problems and how to solve them for the fear of legal litigation, including claims of unfair labor practices.

35.2.2.1 Definition of Risk Factors

Risk factors are defined as job attributes or exposures that increase probability of the occurrence of work-related musculoskeletal disorders (ANSI, 1995). Kuorinka and Forcier (1995) classified the generic risk factors for development of WMSDs by considering their explanatory value, biological plausibility, and the relation to work environment. These generic risk factors are: (1) fit, reach, and see; (2) musculoskeletal load; (3) static load; (4) postures; (5) cold, vibration, and mechanical stresses; (6) task in invariability; (7) cognitive demands; and (8) organizational and psychosocial work characteristics. These WMSD risk factors are present at varying levels for different jobs and tasks. It should be noted that these risk factors are not necessarily causation factors of WMSDs. Also, the mere presence of a risk factor does not necessarily mean that a worker performing a job is at excessive risk of injury (Armstrong et al., 1993).

35.2.2.2 Biomechanical Risk Factors

Currently, available data have identified the following categories of biomechanical risk factors for development of WUEDs: (1) forceful exertions and motions; (2) repetitive exertions and motions; (3) extreme postures of the shoulder (elbow above midtorso reaching down and behind), forearm (inward or outward rotation with a bent wrist), wrist (palmar flexion or full extensions), and hand (pinching); (4) mechanical stress concentrations over the base of palm, on the palmar surface of the fingers, and on the sides of the fingers; (5) duration of exertions, postures, and motions; (6) effects of hand–arm vibration; (7) exposure to cold environment; (8) insufficient rest or break time; and (9) the use of gloves (Armstrong, Radurin, Hansen, and Kennedy, 1986). In addition, wrist angular (flexion–extension) acceleration was also determined to be a potential risk factor for

Table 35.1 Description of Common WMSDs[a]

Disorder Name	Description	Typical Job Activities
Carpal tunnel syndrome (writer's cramp, neuritis, median neuritis) (N)	The result of compression of the median nerve in the carpal tunnel of the wrist. This tunnel is an opening under the carpal ligament on the palmar side of the carpal bones. Through this tunnel pass the median nerve, the finger flexor tendons, and blood vessels. Swelling of the tendon sheaths reduces the size of the opening of the tunnel and pinches and the median nerve, and possibly blood vessels. The tunnel opening is also reduced if the wrist is flexed or extended, or ulnarly or radially pivoted.	Buffing, grinding, polishing, sanding, assembly work, typing, keying, cashiering, playing musical instruments, surgery, packing, housekeeping, cooking, butchering, hand washing, scrubbing, hammering
Cubital tunnel syndrome (N)	Compression of the ulnar nerve below the notch of the elbow. Tingling, numbness, or pain radiating into ring or little fingers.	Resting forearm near elbow on a hard surface and/or sharp edge, also when reaching over obstruction
DeQuervain's syndrome (or disease) (T)	A special case of tenosynovitis (see there) which occurs in the abductor and extensor tendons of the thumb, where they share a common sheath. This condition often results from combined forceful gripping and hand twisting, as in wringing cloths.	Buffing, grinding, polishing, sanding, pushing, pressing, sawing, cutting, surgery, butchering, use of pliers, "turning" control such as on motorcycle, inserting screws in holes, forceful hand wringing
Epicondylitis ("tennis elbow") (T)	Tendons attaching to the epicondyle (the lateral protrusion at the distal end of the humerus bone) become irritated. This condition is often the result of impacting of jerky throwing motions, repeated supination and pronation of the forearm, and forceful wrist extension movements. The condition is well known among tennis players, pitchers, bowlers, and people hammering. A similar irritation of the tendon attachments on the inside off the elbow is called medical epicondylitis, also known as "golfer's elbow."	Turning screws, small parts assembly, hammering, meat cutting, playing musical instruments, playing tennis, pitching, bowling
Ganglion (T)	A tendon sheath swelling that is filled with synovial fluid, or a cystic tumor at the tendon sheath or a joint membrane. The affected area swells up and causes a bump under the skin, often on the dorsal or radial side of the wrist. (Since it was in the past occasionally smashed by striking with a Bible or heavy book, it was also called a "Bible bump.")	Buffing, grinding, polishing, sanding, pushing, pressing, sawing, cutting, surgery, butchering, use of pliers, "turning" control such as on motorcycle, inserting screws in holes, forceful hand wringing

Table 35.1 (Continued)

Disorder Name	Description	Typical Job Activities
Neck tension syndrome (M)	An irritation of the levator scapulae and trapezius group of muscles of the neck, commonly occurring after repeated or sustained overhead work.	Belt conveyor assembly, typing, keying, small parts assembly, packing, load carrying in hand or on shoulder
Pronator (teres) syndrome (N)	Result of compression of the median nerve in the distal third of the forearm, often where it passes through the two heads of the pronator teres muscle in the forearm; common with strenuous flexion of elbow and wrist.	Soldering, buffing, grinding, polishing, sanding
Shoulder tendonitis (rotator cuff syndrome or tendonitis, supraspinatus tendonitis, subacromial bursitis, subdeltoid bursitis, partial tear of the rotator cuff) (T)	This is a shoulder disorder, located at the rotator cuff. The cuff consists of four tendons that fuse over the shoulder joint where they pronate and supinate the arm and help to abduct it. The rotator cuff tendons must pass through a small bony passage between the humerus and the acromion, with a bursa as cushion. Irritation and swelling of the tendon or of the bursa are often caused by continuous muscle and tendon effort to keep the arm elevated.	Punch press operations, overhead assembly, overhead welding, overhead painting, overhead auto repair, belt conveyor assembly work, packing, storing, construction work, postal "letter carrying," reaching, lifting, carrying load on shoulder
Tendonitis (tendinitis) (T)	An inflammation of a tendon. Often associated with repeated tension, motion, bending, being in contact with a hard surface, vibration. The tendon becomes thickened, bumpy, and irregular in its surface. Tendon fibers may be frayed or torn apart. In tendons without sheaths, such as within elbow and shoulder, the injured area may calcify.	Punch press operation, assembly work, wiring, packaging, core making, use of pliers
Tenosynovitis (tendosynovitis, tendovaginitis) (T)	This disorder occurs to tendons which are inside synovial sheaths. The sheath swells. Consequently movement of the tendon with the sheath is impeded and painful. The tendon surfaces can become irritated, rough, and bumpy. If the inflamed sheath presses progressively onto the tendon, the condition is called stenosing tendosynovitis. "DeQuervain's syndrome" (see there) is a special case occurring in the thumb, while the "trigger finger" (see there) condition occurs in flexors of the fingers.	Buffing, grinding, polishing, sanding, pushing, pressing, sawing, cutting, surgery, butchering, use of pliers, "turning" control such as on motorcycle, inserting screws in holes, forceful hand wringing

Table 35.1 (Continued)

Disorder Name	Description	Typical Job Activities
Thoracic outlet syndrome (neurovascular compression syndrome, cervicobrachial disorder, brachial plexus neuritis, costoclavicular syndrome, hyperabduction syndrome) (V, N)	A disorder resulting from compression of nerves and blood vessels between clavicle and first and second ribs, at the brachial plexus. If this neurovascular bundle is compressed by the pectoralis minor muscle, blood flow to and from the arm is reduced. This ischemic condition makes the arm numb and limits muscular activities.	Buffing, grinding, polishing, sanding, overhead assembly, overhead welding, overhead painting, overhead auto repair, typing, keying, cashiering, wiring, playing musical instruments, surgery, truck driving, stacking, material handling, postal "letter carrying," carrying heavy loads with extended arms
Trigger finger (or thumb) (T)	A special case of tendosynovitis (see there) where the tendon becomes nearly locked, so that its forced movement is not smooth but in a snapping or jerking manner. This is a special case of stenosing tendosynovitis crepitans, a condition usually found with digit flexors at the A1 ligament.	Operating trigger finger, using hand tools that have sharp edges pressing into the tissue or whose handles are too far apart for the user's hand so that the end segments of the fingers are flexed while the middle segments are straight.
Ulnar artery aneurysm (V, N)	Weakening of a section of the wall of the ulnar artery as it passes through the Guyon tunnel in the wrist; often from pounding or pushing with heel of the hand. The resulting "bubble" presses on the ulnar nerve in the Guyon tunnel.	Assembly work.
Ulnar nerve entrapment (Guyon tunnel syndrome) (N)	Results from the entrapment of the ulnar nerve as it passes through the Guyon tunnel in the wrist. It can occur from prolonged flexion and extension of the wrist and repeated pressure on the hypothenar eminence of the palm.	Playing musical instruments, carpentering, brick laying, use of pliers, soldering, hammering.
White finger ("dead finger," Raynaud's syndrome, vibration syndrome) (V)	Stems from insufficient blood supply bringing about noticeable blanching. Finger turns cold or numb and tingles, and sensation and control of finger movement may be lost. The condition results from closure of the digit's arteries caused by vasospasms triggered by vibrations. A common cause in continued forceful gripping of vibrating tools particularly in a cold environment.	Chainsawing, jackhammering, use of vibrating tool, sanding, paint scraping, using vibrating tool too small for the hand, often in a cold environment.

[a]Adapted from Kroemer, K., Kroemer, H., and Kroemer-Elbert, K. (1994). *Ergonomics, How to design for ease and efficiency,* Englewood Cliffs, NJ: Prentice Hall, reprinted with permission.

[b]Type of disorder: N, nerve; M, muscle; V, vessel; T, tendon.

Table 35.2 Relationship Between Physical Stresses and WMSD Risk Factors (ANSI Z-365)

Physical Stress	WMSD Factor		
	Magnitude	Repetition Rate	Duration
Force	Forceful exertions and motions	Repetitive exertions	Sustained exertions
Joint angle	Extreme postures and motions	Repetitive motions	Sustained postures
Recovery	Insufficient resting level	Insufficient pauses or breaks	Insufficient rest time
Vibration	High vibration level	Repeated vibration exposure	Long vibration exposure
Temperature	Cold temperature exposure	Repeated cold exposure	Long cold exposure

hand–wrist cumulative trauma disorders under conditions of dynamic industrial tasks (Marras and Schoenmarklin, 1993; Schoemarklin, Marras, and Leurgans, 1994). The relationship between physical stresses at work and selected risk factors for WUEDs is shown in Table 35.2. Tables 35.3 and 35.4 present a summary of work-related postural risk factors for wrist and shoulder disorders.

As discussed by Armstrong et al. (1993), poor design of tools with respect to weight, shape, and size can impose extreme wrist positions and high forces on the worker's musculoskeletal system. For example, holding a heavier object requires an increased power grip and high tension in the finger flexor tendons that causes increased pressure in the carpal tunnel. Furthermore, a task that induces hand and arm vibration cause an involuntary increase in power grip through a reflex of the strength receptors. Vibration can also cause protein leakage from the blood vessels in the nerve trunks and result in edema and increased pressure in the nerve trunks and, therefore, can also result in edema and increased pressure in the nerve (Lundborg, Dahlin, Danielson, and Kange, 1987).

Several millions of workers in occupations such as vehicle operation are intermittently exposed every year to hand–arm vibration that significantly stresses the musculoskeletal

Table 35.3 Postural Risk Factors Reported in the Literature for the Wrist[a]

Risk Factor	Results: Outcome and Details
Wrist flexion	CTS. Exposure of 20–40 h/week
	Increased median nerve stresses (pressure)
	Increased finger flexor muscle activation for grasping
	Median nerve compression by flexor tendons
Wrist extension	Median nerve compression by flexor tendons
	CTS. Exposure of 20–40 h/week
	Increased intracarpal tunnel pressure for extreme extension of 90°
Wrist ulnar deviation	Exposure response effect found: If deviation greater than 20°, increased pain and pathological findings
Deviated wrist positions	Workers with carpal tunnel syndrome used these postures more often
Hand manipulations	More than 1500–2000 manipulations per hour led to tenosynovitis
Wrist motion	1276 flexion extension motions led to fatigue
	Higher wrist accelerations and velocities in high-risk wrist WMSD jobs

[a]Adapted from Kuorinka, I. and Forcier, L. (Eds.). (1995). *Work related musculoskeletal disorders (WMSDs): A reference book for prevention,* Taylor and Francis, reprinted with permission.

Table 35.4 Postural Risk Factors Reported in the Literature for the Shoulder[a]

Risk Factor	Results: Outcome and Details
More than 60° abduction or flexion for more than 1 h/day	Acute shoulder and neck pain
Less than 15° median upper arm flexion and 10° abduction for continuous work with low loads	Increased sick leave resulting from musculoskeletal problems
Abduction greater than 30°	Rapid fatigue at greater abduction angles
Abduction greater than 45°	Rapid fatigue at 90°
Shoulder forward flexion of 30°, abduction greater than 30°	Hyperabduction syndrome with compression of blood vessels
Hands no greater than 35° above shoulder level	Impairment of blood flow in the supraspinatus muscle
Upper arm flexion or abduction of 90°	Onset of local muscle fatigue
Hands at or above shoulder height	Electromyographic signs of local muscle fatigue in less than one minute
Repetitive shoulder flexion	Tendonitis and other shoulder disorders
Repetitive shoulder abduction or flexion	Acute fatigue
Postures invoking static shoulder loads	Neck–shoulder symptoms negatively related to movement rate
Arm elevation	Tendinitis and other shoulder disorders
Shoulder elevation	Pain
Shoulder elevation and upper arm abduction	Neck–shoulder symptoms
Abduction and forward flexion invoking static shoulder loads	Neck–shoulder symptoms
	Shoulder pain and sick leave resulting from musculoskeletal problems
Overhead reaching and lifting	Pain

[a]Adapted from Kuorinka, I. and Forcier, L. (Eds.). (1995). *Work related musculoskeletal disorders (WMSDs): A reference book for prevention,* Taylor & Francis, reprinted with permission.

system (Haber, 1971). The vibration syndrome, which is characterized by intermittent numbness and blanching of the fingers with reduced sensitivity to heat, cold, and pain, affects up to 90% of workers in occupations such as shipping, grinding, and chainsawing (Taylor et al., 1976; Wasserman, Badger, Doyle, and Margolies, 1974). The vibration syndrome is primarily caused by vibration of a part or parts of the body of which the main sources are hand-held power tools such as chainsaws and jackhammers.

35.2.2.3 Work Organization Risk Factors

As mentioned before, work organization and environmental factors may also play a significant role in development of WUEDs. The mechanisms by which work organizational factors can modify the risk for WUEDs include modifying the extent of exposure to other risk factors (physical and environmental) and modifying the stress response of the individual, thereby increasing the risk associated with a given level of exposure (ANSI, 1995). Specific work organization factors that have been shown to fall into at least one of these categories include (but are not limited to) the following: (1) wage incentives, (2) machine-paced work, (3) workplace conflicts of many types, (4) absence of worker decision latitude, (5) time pressures and work overload, and (6) unaccustomed work during training periods or after returning from long-term leave. As discussed by Kuorinka and Forcier (1995), the organizational context in which work is carried out has major influences on the worker's physical and psychological stress and health. The work organization defines the level of work output required (work standards), the work process (how the work is carried out), the work cycle (work–rest regimens), the social structure, and the nature of supervision.

35.2.2.4 Basic Classification of WUEDs

Since most manual work requires the active use of the arms and hands, the structures of the upper extremity are particularly vulnerable to soft tissue injury. WUEDs are typically associated with repetitive manual tasks with forceful exertions, such as those performed at assembly lines, or when using hand tools, computer keyboards, and other devices or operating machinery. These tasks impose repeated stresses to the upper body, i.e., the muscles, tendons, ligaments, nerves tissues, and neurovascular structures. There are three basic types of WRDs to the upper extremity: (1) tendon disorder (such as tendonitis), (2) nerve disorder (such as carpal tunnel syndrome), and (3) neurovascular disorder (such as thoracic outlet syndrome or vibration-Raynaud's syndrome). Table 35.5 shows classification of different human body components at risk for WMSDs.

Generally, the greater the exposure is to a single risk factor or combination of factors, the greater is the risk of a WMSDs. Furthermore, the more risk factors that are present, the higher is the risk of injury. According to ANSI Draft Standard Z-365 (ANSI, 1995), this interaction between risk factors may have a multiplicative rather than an additive effect. However, these risk factors may pose minimal risk of injury if sufficient exposure is not present, or if sufficient recovery time is provided. It is known that changes in the levels of risk factors will result in changes in the risk of WMSDs. Therefore, a reduction in WMSDs risk factors should also reduce the risk for WMSDs.

35.2.3 Physical Assessments of Workers for WUEDs

In a recent study, Ranney, Wells, and Moore (1995) performed precise physical assessment of workers in highly repetitive jobs as part of a cross-sectional study to assess the association between musculoskeletal disorders and a set of work-related risk factors. A total of 146 female workers employed in five different industries (garment and automotive trim sewing, electronic assembly, metal parts assembly, supermarket cashiering, and packaging) were examined for the presence of potential work-related musculoskeletal disorders. The prerequisites for selection of industries and tasks within these industries were (1) existence of repetitive work using the upper limb, (2) at least 5–10 female workers performing the same repetitive job, (3) a range of jobs from light to demanding, (4) minimal job rotation, (5) no major change in the plant for at least 1 year, and (6) the support from both union or employee group and management.

The study showed that 54% of the workers had evidence of musculoskeletal disorders in the upper extremities that were judged as potentially work-related. Many workers had

Table 35.5 Body Components at Risk for WMSDs[a]

I. Effects on soft tissues
   1. Muscles (generate forces and stabilize joints)
      - strained or irritated
      - group of fibers torn apart
      - muscle atrophy (restriction in blood or nerve supply)
   2. Tendons (connect muscles to bones)
      - inflammation of a tendon sheath
      - inflammation of additional fiber tissue
   3. Ligaments (connect bones around a joint)
      - joint displacement (torn fingers)
      - ligament sprain
   4. Bursa (provides lubrication between tendons and a bone)

II. Effects on nerves and vessels
   1. Nerve fibers
      - motor (impairment of motor signals)
      - sensory (feedback) impairment or numbness
      - autonomic (control of sweat function)
   2. Blood vessels
      - vascular compression
      - reduction in blood flow delivery or ischemia

III. Effects of vibration
      - reduced diameter of blood arteries

[a]Modified from Wells, (1990).

multiple problems, and many were affected bilaterally (33% of workers). Muscle pain and tenderness was the largest problem, both in the neck–shoulder area (31%) and in the forearm–hand musculature (23%). Most forearm muscle problems were found on the extensor side. Carpal tunnel syndrome was the most common form of disorder with 16 workers affected (7 people affected bilaterally). DeQuervain's tenosynovitis and wrist flexor tendinitis were the most commonly found tendon disorders in the distal forearm (12 workers affected for each diagnosis). In view of the study results, it was concluded that muscle tissue is highly vulnerable to overuse; that the stressors that affect muscle tissue, such as static loading, should be studied in the forearm as well as in the shoulder; and that exposure should be evaluated bilaterally. Finally, the predominance of forearm muscle and epicondyle disorders on the extensor side suggested that the dual role of these muscles for supporting the hands against gravity plus postural stability during grasping.

35.3 CAUSATION MODELS FOR DEVELOPMENT OF WMSDs

35.3.1 Anatomy of the Upper Extremity

The anatomy of the upper extremity is unique compared to most components of the biomechanical system. The hand is small in dimension yet is capable of producing large amounts of force. In addition, the hand is capable of configuring itself in a variety of orientations and can generate force with either the whole hand in a power grip or with combinations of fingers in opposition to the thumb as in a pinch grip. It is this very flexibility in capability that makes the upper extremity susceptible to cumulative trauma disorders. The anatomy of the hand is illustrated in Figure 35.1.

To achieve this variety of functions, the hand is constructed so that it has itself few power-producing muscles. One of the only power producing muscles in the hand is the thenar muscle, which flexes the thumb. The fingers are flexed using the muscles of the forearm. Force is transmitted to the fingers through a network of long tendons (tendons attach muscles to bone). These tendons pass from the muscles in the forearm through the wrist (with many of them passing through the carpal canal), through the hand, and to the fingers. These tendons are secured at various points along this path with ligaments that keep the tendons in close proximity to the bones. The transverse carpal ligament forms one of the boundaries of the carpal tunnel at the proximal end of the hand. The carpal bones of the carpal canal permit the tendons to act about the bones forming a pulley system. The tendons of the upper extremity are also encased in a sheath, which assists in the sliding function of the tendon. A common sheath envelops the nine tendons passing through the carpal tunnel. In order for the fingers to generate force, a great deal of tension must be passed through these tendons. Since there are various possible combinations of tendons experiencing tension depending on the configuration of the fingers, type of grip, and the grip force required, many of these tendons experience friction. This frictional component can be exacerbated by several factors, including position of the wrist and fingers, motion of the wrist and fingers, and insufficient rest periods.

Other structures are also important to the development of cumulative trauma in the upper extremity. As shown in Figure 35.1, two major blood vessels pass through the hand. Both the radial artery and the ulnar artery provide the tissues and structures of the hand with a blood supply. One of the key structures of the hand that is often involved with cumulative trauma experiences is the nerve structure. The median nerve enters the lower arm and passes through the carpal canal. Once it passes through the carpal canal the median nerve becomes superficial at the base of the wrist and then branches off to serve the thumb, index finger, middle finger, and the radial side of the ring finger. This nerve also serves the palmar surface of the hand connected to these fingers as well as the dorsal portion up to the first two knuckles on the fingers mentioned above as well as the thumb up to the first knuckle.

35.3.2 Biomechanical Mechanisms of WRDs of the Upper Extremity

From the biomechanical viewpoint, WUEDs can be defined as an injury caused by the repeated application of force to a body structure. This repeated application of force tends to wear down the structure, thus lowering the structure tolerance to the point where the tolerance is exceeded through a process that gradually reduced the tolerance limit. Therefore, cumulative trauma represents more of a "wear and tear" on the structure. This type of trauma is becoming far more common in the workplace since more repetitive jobs are becoming common in industry and is the mechanism of concern for many ergonomics evaluations. The minimal criteria for establishing the worksite diagnosis for various WMSDs are shown after Ranney et al. (1995) in Tables 35.6 and 35.7.

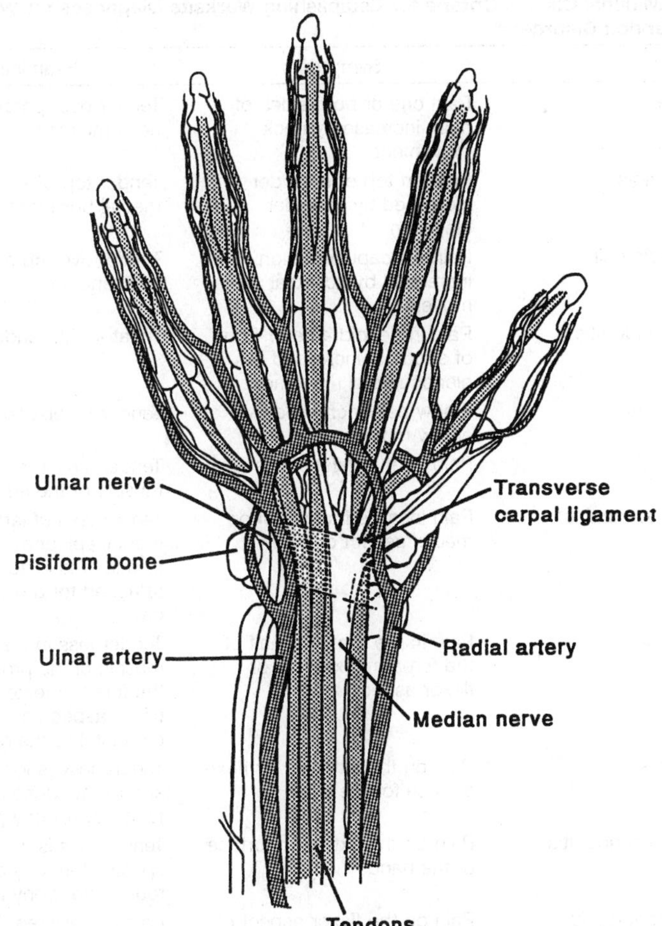

Figure 35.1 Anatomy of the hand.

Through an understanding of the unique construction of the upper extremity, the mechanism of cumulative trauma to this musculoskeletal system can be appreciated. Most problems associated with WUEDs of the upper extremity are a direct result of the increased mechanical tension and irritation among the tendons. This irritation results in friction occurring between the tendons of the wrist during the performance of work. When tension is increased in these tendons as a result of sustained or repetitive work requirements the friction between the tendons also increases. This process is accelerated if risk factors are present such as high levels of force generated by the hand, high levels of acceleration experienced by the tendons, high repetition rates, or sustained deviated postures of the wrist. Once friction occurs among these tendons the body's natural reaction is an inflammatory process. This inflammation or swelling will stimulate the activities of the nociceptors surrounding the structure and signal the central processing mechanism (brain) that there is a problem via the perception of pain.

In response to this pain the body will attempt to control the problem via two mechanism. First, the muscles surrounding the irritated area will increase their level of coactivation in an attempt to minimize the motion of the tendon since this motion will stimulate the nociceptors and result in further pain. Second, in an attempt to reduce the friction occurring within the tendon the body will increase its production of synovial fluid within the tendon sheath. However, given the limited volume of the tendon and the tendon sheath, this increased production of synovial fluid often exacerbates the problem by further increasing the volume of the tendon and in turn further stimulating the surrounding

Table 35.6 Minimal Clinical Criteria for Establishing Worksite Diagnoses for Work-Related Muscle or Tendon Disorders[a]

Disorder	Symptoms	Examination
Neck myalgia	Pain one or both sides of neck increase by neck movement	Tender over paravertebral neck muscles
Trapezius myalgia	Pain on top of shoulder increased by shoulder elevation	Tender top of shoulder or medial border of scapula
Scapulothoracic pain syndrome[b]	Pain in scapular region increased by scapular movement	Tender over rib angles 2, 3, 4, 5, and/or 6
Rotator cuff tendonitis[d]	Pain in deltoid area or front of shoulder increased by glenohumeral movement	Rotator cuff tenderness[c]
Triceps tendinitis	Elbow pain increased by elbow movement	Tender triceps tendon
Arm myalgia	Pain in muscle(s) of the arm	Tenderness in a specific muscle of the arm
Epicondylitis/tendonitis[e]	Pain localized to lateral or medial aspect of elbow	Tenderness of lateral or medial epicondyle localized to this area or to soft tissues attached for a distance of 1.5 cm
Forearm myalgia[f]	Pain in the proximal half of the forearm (extensor or flexor aspect)	Tenderness in a specific muscle in the proximal half of the forearm (extensor or flexor aspect) more than 1.5 cm distal to the condyle
Wrist tendonitis[g]	Pain on the extensor or flexor surface for the wrist	Tenderness is localized to specific tendons and is not found over bony prominences
Extensor finger tendonitis[g]	Pain on the extensor surface of the hand	Tenderness is localized to specific tendons and is not found over bony prominences
Flexor finger tendonitis[g,h]	Pain on the flexor aspect of the hand of distal forearm	Pain on resisted finger flexion localized to area of tendon
Tenosynovitis (finger/thumb)	Clicking or catching of affected digit on movement. There may be pain or a lump in the palm	Demonstration of these complaints, tenderness anterior to metacarpal of affected digit
Tenosynovitis, de Quervain's	Pain on the radial aspect of wrist	Tenderness over first tendon compartment and positive Finkelstein's test
Intrinsic hand myalgia	Pain in muscles of the hand	Tenderness in a specific muscle in the hand

[a]Adapted from Ranney, D., Wells, R., and Moore, A. (1995). *Ergonomics, 38*(7), 1408–1423, reprinted with permission.

[b]Crepitation on circumduction of the shoulder.

[c]Frozen shoulder excluded.

[d]Positive impingement test.

[e]Positive Mills's test or reverse Mills's test (lateral or medial epicondylitis).

[f]Pain localized to the muscle belly of the muscle being stressed during resisted activity.

[g]Pain localized to tendon being stressed during resisted activity.

[h]Only diagnosed moderate or severe. Classification of severity of muscle/tendon problems: *Mild,* above criteria met; *moderate,* pain persists more than 2 h after cessation or work but is gone after a night's sleep, or tenderness plus pain on resisted activity if localized in an anatomically correct manner, or see notes b, d–g; *Severe,* pain not completely relieved by a night's sleep.

Table 35.7 Minimal Clinical Criteria for Establishing Worksite Diagnoses for Work-Related Neuritis

Disorder	Symptoms	Examination
Carpal tunnel syndrome	Numbness and/or tingling in thumb, index, and/or midfinger with particular wrist postures and/or at night	Positive Phalen's test or Tinel's sign present over the median nerve at the wrist
Scalenus anticus syndrome	Numbness and/or tingling on the preaxial border of the upper lip	Tender scalene muscles with positive Adson's or Wright's test
Cervical neuritis	Pain, numbness, or tingling following a dermatomal pattern in the upper limb	Clinical evidence of intrinsic neck pathology
Lateral antebrachial neuritis	Lateral forearm pain, numbness, and tingling	Tenderness of coracobrachialis origin and reproduction of symptoms on palpation here or by resisted coracobrachialis activity
Pronator syndrome	Pain, numbness, and tingling in the median nerve distribution distal to the elbow	Tenderness of pronator teres or superficial finger flexor muscle, with tingling in the median nerve distribution on resisted activation of same
Cubital tunnel syndrome	Numbness and tingling distal to elbow in ulnar nerve distribution	Tender over ulnar nerve with positive tinel's sign and/or elbow flexion test
Ulnar tunnel syndrome	Numbness and tingling in ulnar nerve distribution in the hand distal to the wrist	Positive Tinel's sign over the ulnar nerve at the wrist
Wartenberg's syndrome	Numbness and/or tingling in distribution of the superficial radial nerve	Positive Tinel's sign on tapping over the radial sensory nerve
Digital neuritis	Numbness or tingling in the fingers	Positive Tinel's sign on tapping over digital nerves

[a]Adapted from Ranney, D., Wells, R. and Moore, A. (1995). *Ergonomics, 38(7),* 1408–1423, reprinted with permission.

nociceptors. At this point, if the affected tendon is not entrapped by other structures, the tendon sheath inflammation is the terminal mechanism of injury and will result in pain and aching at the site of the affected tendon as in tenosynovitis or DeQuervain's disease. If the process occurs in an enclosed area such as in the carpal canal, then the pressure within the canal could increase and the median nerve could be compressed. This would result in carpal tunnel syndrome. This cumulative trauma process results in chronic joint pain and a series of musculoskeletal reactions such as reduced strength, reduced tendon motion, and reduced mobility. Collectively, these reactions result in a functional disability.

A similar process occurs if the muscles are affected by cumulative trauma instead of the tendons. Muscles can become problematic when they become fatigued. Fatigue can lower the tolerance to stress and can result in muscle microtrauma. This microtrauma typically means the muscle is partially torn and the tear will cause capillaries to rupture and result in swelling, edema, or inflammation at the site of the tear. This in turn causes pain through the stimulation of nociceptors. The body reacts by cocontracting the surrounding musculature and thereby minimizing the motion of the joint. This results in the same series of musculoskeletal reactions resulting from tendon irritation (i.e., reduced strength, reduced tendon motion, and reduced mobility). The ultimate result of this process is once again a functional disability.

Even though the cumulative trauma process is somewhat similar for both tendons and muscles there is a large difference in the time required to heal from the damage to a tendon compared to a muscle. Both tendons and muscles repair themselves via blood flow to the damaged structure. Blood provides nutrients for repair as well as dissipates waste materials. However, the blood flow to a tendon is just a fraction of that supplied to a muscle. Thus, given an equivalent strain to a muscle and a tendon, the muscle will heal rather rapidly (at most about 10 days if not reinjured) whereas the tendon could take months to accomplish the same level of repair. For this reason, ergonomists must be particularly vigilant in the assessment of workplaces that could pose a danger to the tendons of the body.

35.3.3 A Conceptual Model for Development of WMSDs

Armstrong et al. (1993) developed a conceptual model for the pathogenesis of work-related musculoskeletal disorders. The model is based on the set of four cascading and interacting state variables of exposure, dose, capacity, and response, which are measures of the system state at any given time. The response at one level can act as dose at the next level (see Figure 35.2). Furthermore, it is assumed that a response to one or more doses can diminish or increase the capacity for responding to successive doses. This conceptual model for development of WMSDs reflects the multifactorial nature of work-related upper extremity disorders and the complex nature of the interactions among exposure, dose, capacity, and response variables. The model also reflects the complexity of interactions among the physiological, mechanical, individual, and psychosocial risk factors.

In the proposed model, "exposure" refers to the external factors (i.e., work requirements) that produce the internal dose (i.e., tissue loads and metabolic demands and factors). Workplace organization and hand tool design characteristics are examples of such external factors, which can determine work postures, define loads on the affected tissues, or the velocity of muscular contractions. "Dose" is defined by a set of mechanical, physiological, or psychological factors that in some way disturb an internal state of the affected worker. Mechanical disturbance factors may include tissue forces and deformations produced as a result of exertion or movement of the body. Physiological disturbances are such factors as consumption of metabolic substrates, or tissue damage, whereas psy-

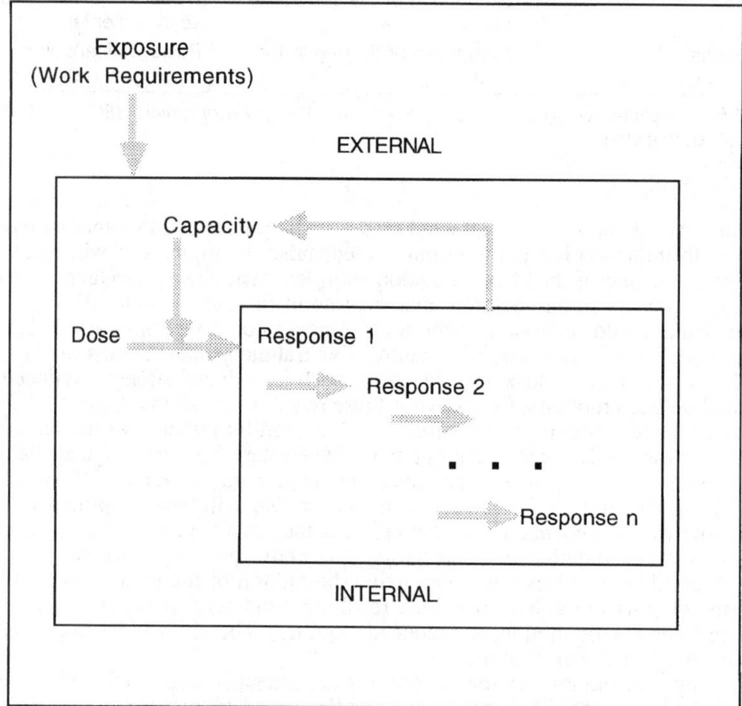

Figure 35.2 A conceptual model for development of WMSDs proposed by Armstrong et al. (1993).

chological disturbance factors are those related to, for example, anxiety about work or inadequate social support.

Changes in the states of variables of the worker are defined as "responses". A response is an effect of the dose caused by exposure. The model also allows for a given response to constitute a new dose, which then produces a secondary response (called the tertiary response). For example, hand exertion can cause elastic deformation of tendons and changes in tissue composition and/or shape, which in turn may results in hand discomfort (Armstrong et al., 1993). The dose–response time relationship implies that the effect of a dose can be immediate or the response may be delayed for a long period of time.

The proposed model stipulates that system changes (responses) can also result in either increased dose tolerance (adaptation) or reduced dose tolerance lowering the system capacity. *Capacity* is defined as the worker's ability (physical or psychological) to resist system destabilization resulting from various doses. Whereas capacity can be reduced or enhanced by previous doses and responses, it is assumed that most individuals are able to adapt to certain types and levels of physical activity. Table 35.8 shows characterization of WMSDs with respect to exposure–dose relationship, the worker's capacity, and response model proposed by Armstrong et al. (1993).

The main purpose of the *dose–response model* is to account for the factors and processes that result in WMSDs in order to specify acceptable limits with respect to work design parameters for a given individual. The exposure, dose, response, and capacity variables need to be measured and quantified. Exposure can be measured using the job title or job classification, questionnaires on possible risk factors, job checklists, or direct measurements. Dose can be measured by estimating muscle forces and joint positions. Worker capacity can be measured using anthropometry, muscle strength, and psychological characteristics. The proposed model should be useful in the design of studies on the etiology and pathomechanisms of WMSDs. The model should also complement epidemiological studies that focus on associations between the physical workload, psychological demands, and environmental risk factors of work at one end, and the manifestations of symptoms, diseases, or disabilities at the other.

35.3.4 The Postulated Causal Mechanism for Development of WUEDs

Many of the work-related muscle disorders are likely to occur when a muscle is fatigued repeatedly without sufficient allowance for recovery (Armstrong et al., 1993). An important factor in development of such disorders is motor control of the working muscle. Hägg (1991) postulated that the recruitment pattern of the motor neurons can occur according to the size principle, where the small units are activated at low forces. Given that the same units can be recruited continuously during a given work task, even if the relative load on the muscle is low, the active low-threshold motor units can work close to their maximal capacity and, consequently, may be at a high risk of being damaged. It was also reported that muscle tension caused by excessive mental load can cause an overload on some specific muscle fibers (Westgaard and Bjørkland, 1987). Karwowski, Eberts, Salvendy, and Noland (1994) showed that cognitive aspects of computer-related task design affect the postural dynamics of the operators and the related levels of perceived postural discomfort. Finally, Edwards (1988) hypothesized that occupational muscle pain might be a consequence of a conflict between motor control of the postural activity and control needed for rhythmic movement or skilled manipulations. In other words, the primary cause of work-related muscular pain and injury may be the altered motor control resulting in imbalance between harmonious motor unit recruitment and relaxation of muscles not directly invovled in the activity.

The work-related tendon disorders typically affect the wrist, forearm, elbow, and shoulder. According to the dose–response model proposed by Armstrong et al. (1993), the tendon dose is related to tensile forces from muscle contractions and to contact and shearing forces from adjacent bones and ligaments. Tendon responses can be mechanical in nature (e.g., elastic and viscous deformation) or physiological (e.g., triggering of nerve receptors or adaptation). In case of work-related nerve disorders, the dose refers to muscle contractions, joint positions, and movements that can produce pressure and deformation of the nerves and, possibly, a secondary effect of the swelling of adjacent tendons and tendon sheaths. These responses may lead to a series of physiological responses that, in turn, may impair the nerve functions. For example, disturbances in circulation in the median nerve may result in tingling, numbness, pain, and a loss of motor function in the hand. The summary of the relationships betweens WUEDs, their injury mechanisms, and the factors characterizing the soft tissue loads is shown in Table 35.9.

Table 35.8 Characterization of Work-Related Musculoskeletal Disorders in General and Muscle, Tendon, and Nerve Disorders in Particular According to Sets of Cascading Exposure and Response Variables as Conceptualized in the Model[a]

Exposure–Dose	Worker's Capacity	Response
Musculoskeletal system		
Work load	Body size and shape	Joint position
Work location	Physiological state	Muscle force
Work frequency	Psychological state	Muscle length
Work duration		Muscle velocity
		Frequency
Muscle disorders		
Muscle force	Muscle mass	Membrane permeability
Muscle velocity	Muscle anatomy	Ion flow
Frequency	Fiber type and composition	Membrane action potentials
Duration	Enzyme concentration	Energy turnover (metabolism) muscle enzymes, and energy stores
	Energy stores	Intramuscular pressure
	Capillary density	Ion imbalances
		Reduced substrates
		Increased metabolites and water
		Increase in blood pressure, heart rate, cardiac output, muscle blood flow
		Muscle fatigue
		Pain
		Free radicals
		Membrane damage
		Z-disc ruptures
		Afferent activation
Tendon disorders		
Muscle force	Anthropometry	Stress
Muscle length	Tendon anatomy	Strain (elastic and viscous)
Muscle velocity	Vascularity	Microruptures
Frequency	Synovial tissue	Necrosis
Joint position		Inflammation
Compartment pressure		Fibrosis
		Adhesions
		Swelling
		Pain
Nerve disorders		
Muscle force	Anthropometry	Stress
Muscle length	Nerve anatomy	Strain
Muscle velocity	Electrolyte status	Ruptures in perineural tissues
Frequency	Basal compartment pressure	Protein leakage
Joint position		Ruptures in perineural tissue
Compartment pressure		Protein leakage in nerve trunks
		Edema
		Increased pressure
		Impaired blood flow
		Numbness, tingling, conduction block
		Nerve action potentials

[a]Adapted from Armstrong, T. J., Buckle, P., Fine, L. J., Hagberg, M., Jonsson, B., Kilbom, A., Kuorinka, I., Silverstein, B. A., Sjogaard, G., and Viikari-Juntura, E. (1993). *Scandinavian Journal of Work and Environmental Health, 19,* 73–84, reprinted with permission.

Table 35.9 Relationships Between Four Major Work-Related Musculoskeletal Disorders, Their Injury Mechanisms, and the 12 Internal Factors Used to Characterize Soft Tissue Loads

WUEDs	Proposed Injury Mechanisms	CTD External Risk Factor	Related Internal Factor
Carpal tunnel syndrome[b]	Force of tendons on median nerve	Force Posture Time	Peak normal pressure Impulse of normal pressure during flexion
Tenosynovitis	Force of tendons on synovial sheath	Force Posture Time	Peak normal pressure Average normal pressure
	Movement of tendon with respect to tendon sheath	Motions Time	Tendon excursion Peak excursion velocity
	Friction between tendon and tendon sheath	Force Motions Posture Time	Frictional work factor Peak frictional power factor
Tendonitis	Strain in the tendons	Force Time	Peak tendon axial force Impulse of axial force
Chronic muscle strain	Muscle fatigue and overuse (constant use)	Force Posture Time	Static muscle load (10th percentile APDF)[c]
	Muscle fatigue and overuse (dynamic use)	Force Posture Motions Time	Dynamic muscle load (50th percentile APDF) Peak muscle load (90th percentile APDF)

[a]Adapted from Moore, A., Wells, R. and Ranney, D. (1991). *Ergonomics, 34(12),* 1433–1453, reprinted with permission.

[b]Carpal tunnel syndrome may also be secondary to tendosynovitis and other causes.

[c]Amplitude Probability Density Function

35.3.5 Models of Carpal Tunnel Syndrome (CTS)

35.3.5.1 Causation Mechanisms for Carpal Tunnel Syndrome

As discussed by Moore and Garg (1995), each component of the muscle–tendon unit has unique physiological and biomechanical properties, and each can be associated with unique disorders or manifestations of strain. Examples my include tendonitis, peritendonitis, muscular soreness, muscle strains, localized muscle fatigue, and stenosing tenosynovitis. Although the causation mechanism of carpal tunnel syndrome (CTS), one of the most common nerve disorders, is poorly understood, it is believed that mechanical compression of the median nerve in the carpal tunnel with persistent nerve conduction impairment is the main cause for this disorder (Moore, 1992). The surveillance case definition for CTS is given in Table 35.10.

From the epidemiological point of view, Silverstein, Fine, and Armstrong (1986a) reported that in jobs with high repetition there was a 2.8 odds ratio of CTS injury compared to the low repetitive jobs. This odd ratio increases to 30.3 if the job was both highly repetitive and required high forces (adjusted force greater than 6 kg). The highly repetitive jobs were defined as those jobs with a cycle time less than 30 s, or more than 50% of the time doing the same type of fundamental cycle. The hand–wrist posture was not found to be a significant predictor of CTS prevalence. Silverstein Fine, and Armstrong (1987) also reported that the prevalence of CTS was associated with jobs utilizing high

Table 35.10 Surveillance Case Definition for Occupational CTS[a]

A. One or more of the following symptoms suggestive of CTS is present[b]: paresthesia, hypesthesia, pain, or numbness affecting at least one part of the median nerve distribution[c] of the hand(s).

B. Objective findings consistent with CTS are present in the affected hand(s) and wrist(s):
   1. Physical examination findings—Tinel's sign[d] present or positive Phalen's test[e] or diminished or absent sensation to pin prick in the median nerve distribution of the hand.
   OR
   2. Electrodiagnostic findings indicative of median nerve dysfunction across the carpal tunnel[f]

C. Evidence of work relatedness—A history of a job involving *one or more* of the following activities before the development of symptoms[g]:
   1. Frequent, repetitive use of the same or similar movements of the hand or wrist on the affected side(s).
   2. Regular tasks requiring the generation of high force by the hand.
   3. Regular or sustained tasks requiring awkward hand positions of the affected side(s)[h].
   4. Regular use of vibrating hand-held tools.
   5. Frequent or prolonged pressure over the wrist or base of the palm on the affected side(s).

[a]After CDC (1989).

[b]Symptoms should have lasted at least 1 week or, if intermittent, have occurred on multiple occasions. Other causes of hand numbness or paresthesia, such as cervical radiculopathy, thoracic outlet syndrome, and pronator teres syndrome, should be excluded by clinical evaluation.

[c]Generally includes palmar side of thumb, index finger, middle finger, and radial half of ring finger; dorsal (back) side of same digits distal to PIP joint; and radial half of palm. Pain and paresthesia may radiate proximally into the arm.

[d]Paresthesia are elicited or accentuated by gentle percussion over the carpal tunnel.

[e]Paresthesia are elicited or accentuated by maximal passive flexion of wrist for 1 min.

[f]Criteria for abnormal electrodiagnostic finding are generally determined by the individual laboratories.

[g]A temporal relationship of symptoms to work or an association with cases of CTS in coworkers performing similar tasks is also evidence of work relatedness.

[h]Awkward hand positions predisposing to CTS include the use of pinch grip (as when holding a pencil), extreme flexion, extension, or ulnar deviation of the wrist, and use of the fingers with the wrist flexed.

forces and high task repetition, with task repetitiveness being a stronger risk factor than the task forcefulness. Armstrong et al. (1987) reported that the prevalence of hand–wrist tendonitis in workers who performed highly repetitive and forceful jobs was 29 times greater compared to those who performed jobs characterized by low task repetition and lower force levels. High task forcefulness was more significant than repetitiveness, whereas the hand–wrist posture and vibration were not associated with the prevalence of hand–wrist tendonitis. However, the wrist angular (flexion–extension) acceleration was also found to be a potential risk factor for CTS in highly dynamic, hand-intensive industrial jobs (Marras and Schoenmarklin, 1993; Schoenmarklin et al., 1994). In view of these findings, Karwowski (1993) proposed a conceptual model based on nonlinear systems dynamics, linking the risk potential for WUEDs to four work-related factors, including the wrist joint postural deviation, joint angular acceleration, joint tendon force, and motion repetition.

35.3.5.2 A Conceptual Quantitative Model for CTS

Tanaka and McGlothlin (1993) proposed a hypothetical pathogenic mechanism of carpal tunnel syndrome (CTS), and heuristic conceptual model to assess the musculoskeletal stress of manual work. The ultimate purpose of the model, which is yet to be verified through epidemiological and experimental studies, is to establish quantitative guidelines to prevent work-related CTS. The model is based on the concept that the frictional energy of manual work within the wrist initiates local tenosynovitis as the precursor of work-related CTS. Furthermore, it is assumed that the frictional energy inside the carpal tunnel is a function of the product of three biomechanical factors of manual work: internal force, repetitiveness, and wrist angles.

Specifically, it was postulated that: (1) friction between the tendons and the tendon sheaths in the carpal tunnel is the factor most responsible for initiating and/or aggravating tenosynovitis, regardless of the types of the tasks performed by the hand and fingers; and (2) that the frictional energy is proportional to the product of internal force necessary to do the work, repetitiveness of movement, and wrist angles (extension–flexion, radial–ulnar deviations, or any combination of these). Based on these postulates, it was hypothesized that work-related CTS is preceded by, or concurrent with, frictional wrist tenosynovitis, and prevention of the initial tenosynovitis would prevent work-related CTS. Assuming that compression of the median nerve is preceded by, or concurrent with, hand–wrist pain, it was also hypothesized that preventing the onset of the hand pain or discomfort would also prevent occurrence of work-related tenosynovitis and CTS. This could be accomplished by controlling the internal force required to perform the work, task repetitiveness, wrist angles, or any combination of these factors.

According to the proposed model, the product of values of internal force, task repetitiveness, and angles of hand and wrist motions must remain under certain limits to prevent inflammation. Mathematically, the exposure limit was defined as follows:

$$\mathrm{ELM} = k \times \alpha F * \beta R * e^{\gamma A} \qquad (1)$$

where

ELM = Exposure limit for manual task.
F = Internal musculoskeletal force exerted by the finger–hand–wrist–forearm complex needed to complete a typical task cycle.
R = Repetition of the task cycle per unit time period; may be substituted by the number of similar products manufactured or processed per unit time.
A = Wrist angle, expressed as the sum of the proportions of measured wrist angles while performing the task in relation to the maximum angle possible in each direction of wrist bending. [A = Flexion/(Maximum Flexion) + Extension/(Maximum Extension) + (Ulnar deviation)/(Maximum Ulnar Deviation) + (Radial deviation)/(Maximum Radial Deviation)]; $0 \leq A \leq 4$.
e = The base of the natural logarithm.
α, β, γ = Coefficients for each corresponding factor.
k = A constant to be determined for worker protection.

The proposed conceptual model was illustrated graphically in two dimensions (see Figure 35.3 for $A = 0$), and in three-dimensions (see Figure 35.4 for $A > 0$), both of which have a concavity toward the top of the figure. The practical upper limit of F is be determined by the largest internal force for safe hand exertion during a unit time. From Equation (1), if F is at the maximum, the product of R and A must be reduced to the minimum to stay below ELM, meaning that the wrist should not be deviated and only a few repetitions would be allowed. There is also a maximum limit of repetitions per unit time a human hand can make safely. If R is very high, the product of F and A must be at the minimum, i.e., the force being the weight of only the fingers and hand, and the wrist being kept in neutral position. Very low R values would represent static job or task, to which this model may not apply. If A is at the maximum, the product of F and R needs to be minimal to be safe (see upper left angle of the triangle in Figure 35.4). According to Tanaka and McClothlin (1993), after conducting surveys using standardized methods, it would be possible to define a threshold plane that, combined with the laboratory data, could also allow for determination of the coefficients of various factors (α, β, and γ) for Equation (1). Therefore, in order to prevent work-related tenosynovitis and CTS, the hypothetical value of the ELM should not exceed a limit to be derived from the population-based threshold.

35.4 QUANTITATIVE MODELS FOR CONTROL OF WUEDs

35.4.1 Problems with Modeling WUEDs

Today, only a few quantitative models that are based on the physiological, biomechanical, or psychophysical data and that relate the specific job risk factors for musculoskeletal disorders to increased risk of developing such disorders have been developed. As discussed by Moore and Garg (1995), this is mainly because: (1) the dose–response (cause–effect) relationships are not well understood; (2) measurement of some task variables, such as force, is difficult in an industrial setting; and (3) the number of task variable is very large. However, it is generally recognized that biomechanical risk factors such as force, repetition, posture, recovery time, duration of exposure, static muscular work, use

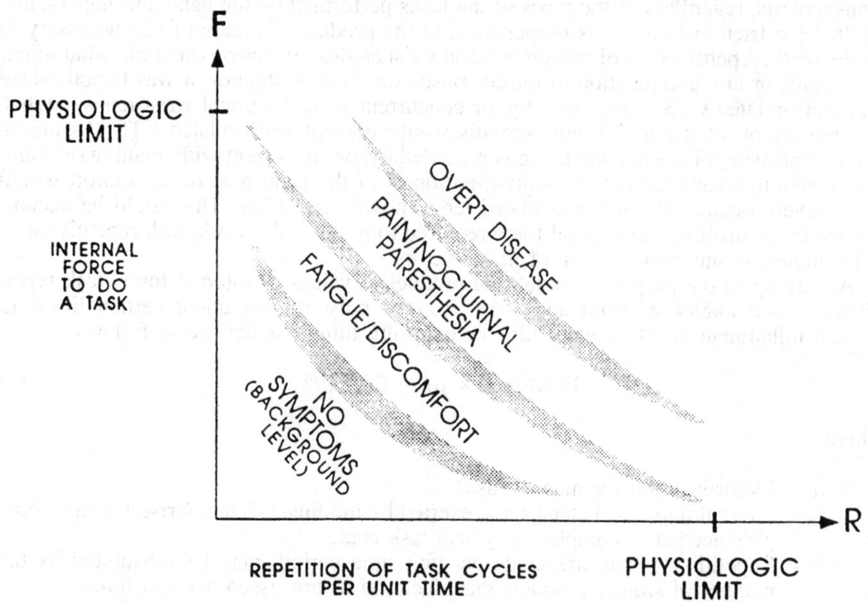

Figure 35.3 Two-dimensional illustration of the conceptual CTS model. (Adapted from Tanaka, S. and McGlothlin, J. D., 1993, *International Journal of Industrial Ergonomics,* 11, 181–193, reprinted with permission.)

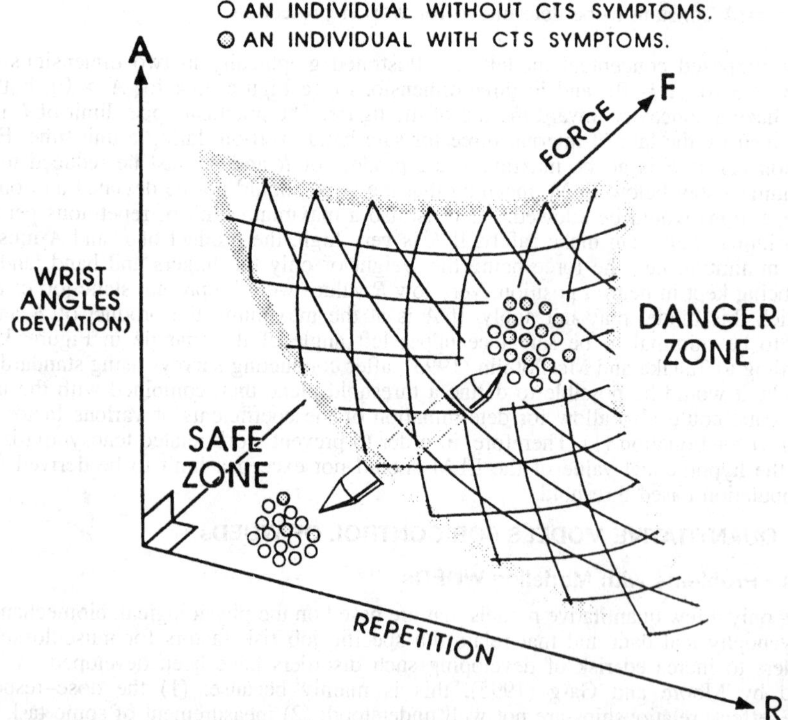

Figure 35.4 Three-dimensional illustration of the conceptual CTS model. (Adapted from Tanaka, S. and McGlothlin, J. D., 1993, *International Journal of Industrial Ergonomics,* 11, 181–193, reprinted with permission.)

of the hand as a tool, and type of grasp are important for explaining the causation mechanism of WUEDs (Armstrong et al., 1987; Keyserling, Stetson, Silverstein, and Brouver, 1993). Given the above knowledge, even though limited in scope and subject to epidemiological validation, a few methodologies that allow discrimination between safe and hazardous jobs in terms of workers being at increased risk of developing the WUEDs, were recently developed and reported in the subject literature. Some of the quantitative data and models for evaluation and prevention of WUEDs available today are described below.

35.4.2 Maximum Acceptable Forces for Repetitive Wrist Motions for Females

Recently, Snook, Vaillancourt, Ciriello, and Webster (1995) utilized the psychophysical methodology to determine the maximum acceptable forces for various types and frequencies of repetitive wrist motion, including the following: (1) flexion motion with a power grip (handle diameter, 4 mm; handle length, 135 mm); (2) flexion motion with a pinch grip (handle thickness; 5 mm; handle length, 55 mm); and (3) flexion motion with a power grip (handle diameter, 40 mm; handle length, 135 mm). Subjects were instructed to work as if they were on an incentive basis, getting paid for the amount of work that they performed. They were asked to work as hard as they could (i.e., against as much resistance as they could) without developing unusual discomfort in the hands, wrists, or forearms.

Fifteen women worked 7 h each day, 2 days per week, for 20 days in the first experiment. Repetition rates of 2, 5, 10, 15 and 20 motions per minute were used with each flexion and extension task. Maximum acceptable torque were determined for the various motions, grips, and repetition rates without dramatic changes in wrist strength, tactile sensitivity, or number of symptoms. Fourteen different women worked in the second experiment, performing a wrist flexion motion (power grip) 15 times per minute, 7 h per day, 5 days per week, for 23 days. In addition to the four dependent variables, which included the maximum acceptable torque, maximum isometric wrist strength, tactile sensitivity, and symptoms, performance errors and duration of force were also measured. The most common health symptom reported was muscle soreness (55.3%), located mostly in the hand and wrist (51.8%). Numbness in the palmar side of the fingers and thumb (69.1%) and stiffness on the dorsal (back) side of the fingers and thumb (30.6%) were also reported. The number of symptoms increased consistently as the day progressed The number of symptoms reported was 2–3 times higher after the seventh hour of work than after the first hour of work (similar to the 2–4 times higher rate in the 2 days per week exposure). Symptoms reports were 4.1 times higher at the end of the day than at the beginning of the day before testing began.

The maximum acceptable torque determined during the 5 days per week exposure was 36% lower than the task performed for only 2 days per week. Based on assumption that maximum acceptable torque decrease 36.3% for the other repetition rates used during the 2 days per week exposure, and using the adjusted means and coefficients of variation from the 2 days per week exposure, the maximum acceptable torques were estimated for different repetition of wrist flexion (power grip) and different percentages of the population. Torques were then converted into forces by dividing each torque by the average length of the handle lever (0.081 m). The estimated maximum acceptable forces for female wrist flexion (power grip) are shown in Table 35.11. Tables 35.12 and 35.13 show

Table 35.11 Maximum Acceptable Forces for Female Wrist Flexion (Power Grip) (Newtons)[a]

Percentage of Population	Repetition Rate				
	2/min	5/min	10/min	15/min	20/min
90	14.9	14.9	13.5	12.0	10.2
75	23.2	23.2	20.9	18.6	15.8
50	32.3	32.3	29.0	26.0	22.1
25	41.5	41.5	37.2	33.5	28.4
10	49.8	49.8	44.6	40.1	34.0

[a]Adapted from Snook, S. H., Vaillancourt, D. R., Ciriello, V. M., and Webster, B. S. (1995). *Ergonomics, 38*(7), 1488–1507, reprinted with permission.

Table 35.12 Maximum Acceptable Forces for Female Wrist Flexion (Pinch Grip) (Newtons)[a]

Percentage of Population	Repetition Rate				
	2/min	5/min	10/min	15/min	20/min
90	9.2	8.5	7.4	7.4	6.0
75	14.2	13.2	11.5	11.5	9.3
50	19.8	18.4	16.0	16.0	12.9
25	25.4	23.6	20.6	20.6	16.6
10	30.5	28.3	24.6	24.6	19.8

[a]Adapted from Snook, S. H., Vaillancourt, D. R., Ciriello, V. M., and Webster, B. S. (1995). *Ergonomics, 38*(7), 1488–1507, reprinted with permission.

the estimated maximum acceptable forces for female wrist flexion (pinchy grip) and wrist extension (power grip), respectively. The torques were converted into forces by dividing by 0.081 m for the power grip, and 0.123 m for the pinch grip.

35.4.3 Semiquantitative Job Analysis Methodology for Wrist–Hand Disorders

35.4.3.1 Model Structure

Recently, Moore and Garg (1995) have developed a semiquantitative job analysis methodology (SJAM) for identifying industrial jobs associated with wrist–hand disorders. An existing body of knowledge and theory of the physiology, biomechanics, and epidemiology of distal upper extremity disorders was used for that purpose. The following major principles were derived from the physiological model of localized muscle fatigue:

1. The primary task variables are intensity of exertion, duration of exertion, and duration of recovery.
2. "Intensity of exertion" refers to the force required to perform a task one time and is characterized as a percentage of maximal strength.
3. "Duration of exertion" describes how long an exertion is applied. The sum of duration of exertion and duration of recovery is the cycle time of one exertional cycle.
4. Wrist posture, type of grasp, and speed of work are considered via their effects of maximal strength.
5. The relationship between strain on the body (endurance time) and intensity of exertion is nonlinear.

The following were major principles derived form the biomechanical model of the viscoelastic properties of components of a muscle–tendon unit:

Table 35.13 Maximum Acceptable Forces for Female Wrist Extension (Power Grip) (Newtons)[a]

Percentage of Population	Repetition Rate				
	2/min	5/min	10/min	15/min	20/min
90	8.8	8.8	7.8	6.9	5.4
75	13.6	13.6	12.1	10.9	8.5
50	18.9	18.9	16.8	15.1	11.9
25	24.2	24.2	21.5	19.3	15.2
10	29.0	29.0	25.8	23.2	18.3

[a]Adapted from Snook, S. H., Vaillancourt, D. R., Ciriello, V. M., and Webster, B. S. (1995). *Ergonomics, 38*(7), 1488–1507, reprinted with permission.

1. The primary task variables for the viscoelastic properties are intensity and duration of exertion, duration of recovery, number of exertions, wrist posture, and speed of work.
2. The primary task variable for intrinsic compression are intensity of exertion and nonneutral wrist posture.
3. The relationship between strain on the body and intensity of effort is nonlinear.

Finally, the major principles derived from the epidemiological literature and used for the purpose of model development were as follows:

1. The primary task variable associated with an increased prevalence or incidence of distal upper extremity disorders are intensity of exertion (force), repetition rate, and percentage of recovery time per cycle.
2. Intensity of exertion is the most important task variable related to disorders of the muscle–tendon unit.
3. Wrist posture may not be an independent risk factor because it may contribute to an increased incidence of distal upper extremity disorders when combined with intensity of exertion.
4. The roles of other task variable have not been clearly established epidemiologically.

Moore and Garg (1995) compared exposure factors for jobs associated with WUEDs to jobs without prevalence of such disorders. They found that the intensity of exertion, estimated as a percentage of maximal strength and adjusted for wrist posture and speed of work, was the major discriminating factor. The relationship between the incidence rate for distal upper extremity disorder and the job risk factors was defined as follows:

$$IR = \frac{30 \times F^2}{RT^{0.6}} \tag{2}$$

where IR = incidence rate (per 100 workers per year); F = intensity of exertion (%MS); and RT = recovery time (percentage of cycle time).

35.4.3.2 The Strain Index

The Strain Index (SI) proposed by Moore and Garg (1995) is the product of six multipliers that correspond to six task variables, including (1) intensity of exertion, (2) duration of exertion, (3) exertions per minute, (4) hand–wrist posture, (5) speed of work, and (6) duration of task per day. An ordinal rating is assigned for each of the variables according to the exposure data. The ratings that are applied to model variables are presented in Table 35.14. The multipliers for each task variable related to these ratings are shown in Table 35.15. The Strain Index score (SI score) is defined as follows:

$$\begin{aligned}\text{Strain Index (SI)} = {} & \text{(intensity of exertion multiplier)} \times \text{(Duration of exertion multiplier)} \\ & \times \text{(Exertions per minute multiplier)} \times \text{(Posture multiplier)} \\ & \times \text{(Speed of work multiplier)} \times \text{(Duration per day multiplier)}\end{aligned} \tag{3}$$

Intensity of exertion, the most critical variable of SI, is defined as the percentage of maximum strength required to perform the task once. The intensity of exertion is estimated by an observer using verbal descriptors (see Table 35.14) and assigned corresponding rating values (1, 2, 3, 4, or 5). The multiplier values (Table 35.15) are defined based on the rating score raised to a power of 1.6 to reflect the nonlinear nature of the relationship between intensity of exertion and manifestations of strain according to the psychophysical theory. The multipliers for other task variables are modifiers to the intensity of exertion multiplier.

Duration of exertion is defined as the percentage of time an exertion is applied per cycle. The terms "cycle" and "cycle time" refer to the exertional cycle and average exertional cycle time, respectively. The duration of recovery per cycle is equal to the exertional cycle time minus the duration of exertion per cycle. The duration of exertion is the average duration of exertion per exertional cycle (calculated by dividing all durations of a series of exertions by the number of observed exertions). The percentage du-

Table 35.14 Rating Criteria for Strain Index[a]

Rating	Intensity of Exertion	Duration of Exertion (% of Cycle)	Efforts/Minute	Hand–Wrist Posture	Speed of Work	Duration per Day (h)
1	Light	<10	<4	Very good	Very slow	≤ 1
2	Somewhat hard	10–29	4–8	Good	Slow	1–2
3	Hard	30–49	9–14	Fair	Fair	2–4
4	Very hard	50–79	15–19	Bad	Fast	2–8
5	Near maximal	≥ 80	≥ 20	Very bad	Very fast	≥ 8

[a]Adapted from Moore, J. S., and Garg, A. (1995, May). *American Industrial Hygiene Association Journal, 56,* 443–458, reprinted with permission.

Table 35.15 **Multiplier Table for the Strain Index**[a]

Rating	Intensity of Exertion	Duration of Exertion (% of Cycle)	Efforts/Minute	Hand–Wrist Posture	Speed of Work	Duration per Day (h)
1	1	0.5	0.5	1.0	1.0	0.25
2	3	1.0	1.0	1.0	1.0	0.50
3	6	1.5	1.5	1.5	1.0	0.75
4	9	2.0	2.0	2.0	1.5	1.00
5	13	3.0[b]	3.0	3.0	2.0	1.50

[a]Adapted from Moore, J. S. and Garg, A., (1995, May). *American Industrial Hygiene Association Journal, 56,* 443–458, reprinted with permission.

[b]If duration of exertion is 100%, then the efforts/minute multiplier should be set to 3.0.

ration of exertion is calculated by dividing the average duration of exertion per cycle by the average exertional cycle time, then multiplying the result by 100 [see Equation (3)]. The calculated percentage duration of exertion is compared to the ranges in Table 35.14 and assigned the appropriate rating. The corresponding multipliers are identified using Table 35.15.

$$\%\text{Duration of exertion} = \frac{(\text{Average duration of exertion per cycle})}{(\text{Average exertional cycle time})} \tag{4}$$

Efforts per minute is the number of exertions per minute (e.g., repetitiveness) and is synonymous with frequency. Efforts per minute are measured by counting the number of exertions that occur during a representative observation period (as described for determining the average exertional cycle time). The measured results are compared to the ranges shown in Table 35.14 and given the corresponding ratings. The multipliers are defined in Table 35.15.

Posture refers to the anatomical position of the wrist or hand relative to neutral position and is rated qualitatively, using verbal anchors. As shown in Table 35.15, posture has four relevant ratings. Postures that are "very good" or "good" are essentially neutral and have multipliers of 1.0. As hand or wrist postures progressively deviate beyond the neutral range to extremes, they are graded as "fair," "bad," and "very bad."

Speed of work refers to perceived pace of the task or job, and is subjectively estimated. Once a verbal anchor is selected, a rating is assigned according to Table 35.14. *Duration of task per day* is defined as a total time that a task is performed per day. As such, this variable reflects the beneficial effects of task diversity such as job rotations and the adverse effects of prolonged activity such as overtime. Duration of task per day is measured in hours and assigned a rating according to Table 35.14.

Application of the Strain Index involves five steps, including (1) collecting data, (2) assigning rating values, (3) determining multipliers, (4) calculating the SI score, and (5) interpreting the results. The values of intensity of exertion, wrist posture, and speed or work are estimated using the verbal descriptors in Table 35.14. The values of percentage duration of exertion per cycle, efforts per minute, and duration per day are based on measurements and counts. These values are then compared to the appropriate column in Table 35.14 and assigned a rating. The SI multipliers are determined from Table 35.15. Table 35.16 shows the numerical example for calculating the Strain Index.

35.4.3.3 The Limitations of SJAM

The proposed Strain Index methodology aims to discriminate between jobs that expose workers to risk factors (task variables) that cause the WUEDs versus those jobs that do not. However, according to Moore and Garg (1995), the Strain Index is not designed to identify jobs associated with an increased risk of any single specific disorder. It is anticipated that jobs identified to be in the high risk category by the Strain Index will exhibit higher levels of WUEDs among workers who currently perform or historically performed those jobs which are believed to be hazardous. Finally, the authors caution that large-scale studies are needed to validate and update the proposed methodology. The Strain Index has the following limitation in terms of its application:

1. There are some disorders of the distal upper extremity that should not be predicted by the Strain Index, e.g., hand–arm vibration syndrome (HAVS) and hypothenar hammer syndrome.
2. The Strain Index has not been developed to predict increased risk for distal upper extremity disorders of uncertain etiology or relationship to work. Examples include ganglion cysts, osteoarthritis, avascular necrosis of carpal bones, and ulnar nerve entrapment at the elbow.
3. The Strain Index has not been developed to predict disorders outside of the distal upper extremity, such as disorders of the shoulder, shoulder girdle, neck, or back.

35.5 ERGONOMICS EFFORTS TO CONTROL WUEDs

35.5.1 Strategy for Prevention of Musculoskeletal Injuries

Facing the growing challenges of musculoskeletal injuries in the contemporary workplace, the *Proposed National Strategies for the Prevention of Leading Work-Related Diseases*

Table 35.16 An Example To Demonstrate the Procedure for Calculating SI Score[a]

	Intensity of Exertion	Duration of Exertion	Efforts/Minute	Posture	Speed of Work	Duration per Day
Exposure dose	Somewhat hard	60%	12	Fair	Fair	4–8
Ratings	2	4	3	3	3	4
Multipliers	3.0	2.0	1.5	1.5	1.0	1.0

SI Score = 3.0 × 2.0 × 1.5 × 1.5 × 1.0 × 1.0 = 13.5

[a]Adapted from Moore, J. S. and Garg, A. (1995, May). *American Industrial Hygiene Association Journal, 56,* 443–458, reprinted with permission.

(NIOSH, 1986) identified environmental hazards and human biological hazards, among the four main factors that contribute to human diseases. *Environmental hazards* to the musculoskeletal system associated with work were described as workplace traumatogens, i.e., a source of biomechanical stress from job demands that exceed the worker's strength or endurance, such as heavy lifting or repetitive and forceful manual exertions. Traumatogens can be measured by determining the frequency, magnitude, and direction of forces imposed upon the body in relation to posture and the point of application. *Human biological factors* include the anthropometric or innate attributes that influence a worker's capacity for safely performing the job. Examples include the worker's physical size, strength, range of motion, and work endurance. These factors partly account for variability in performance capability in the population and the potential for a mismatch between the worker and job that can be addressed by applying ergonomics principles of work design.

To reduce the extent of work-related musculoskeletal injuries, the progress in four methodological areas is expected (NIOSH, 1986): (1) identifying accurately the biomechanical hazards, (2) developing effective health-promotion and hazard-control interventions, (3) changing management concepts and operational policies with respect to expected work performance, and (4) devising strategies for disseminating knowledge on control technology and promoting their application through incentives.

35.5.2 Ergonomics Aspects of WUEDs

Ergonomic job design (and redesign) efforts focus on fitting characteristics of the job to capabilities of workers. In simple terms, this can be accomplished, for example, by reducing excessive strength requirements and exposure to vibration, improving design of hand tools and work layouts, designing out unnatural postures at work, or addressing the problem of work–rest requirements for jobs with high production rates. From the occupational safety and health perspective, the current state of ergonomics knowledge should allow for management of WUEDs in order to minimize human suffering, potential for disability, and the related workman's compensation costs. Application of ergonomics can help to: (1) identify working conditions under which the WUEDs might occur, (2) develop engineering design measures aimed at elimination or reduction of the known job risk factors, and (3) identify the affected worker population and target it for early medical and work intervention efforts.

The workplace and work design-related risk factors, which often overlap, typically involve combination of poor work methods, inadequate workstations and hand tools, and high production demands. A risk factor is defined as an attribute or exposure that increases the probability of the disease or disorder (Putz-Anderson, 1993). As discussed before, the biomechanical risk factors for WUEDs include repetitive and sustained exertions, awkward postures, and application of high mechanical forces. Vibration and cold environments may also accelerate the development of WUEDs.

Typical tools that can be used to identify the potential for development of WUEDs include conducting work-methods analyses and checklists designed to itemize undesirable worksite conditions or worker activities that contribute to injury. For example, the repetitiveness factor of the job can be defined as the number of movements that occur in a given amount of time (cycle time), or the time needed to complete the task (Silverstein et al., 1986a). A long cycle may indicate that a sequence of steps is repeated to complete the overall cycle. Jobs can be classified as *low repetitive* if the cycle time was more than 30 s, or if under 50% of the cycle time involved performing the same kind of fundamental cycle. Jobs can be classified as *high repetitive* if the cycle time was less than 30 s, or if more than 50% of the cycle time involved performing the same kind of fundamental cycle. Silverstein et al. (1986a) showed that workers in jobs that had been classified as high in repetition and force had a 31% greater risk for tendonitis than workers who had jobs classified as low in repetition and force levels.

Since job redesign decisions may require some design tradeoffs (Putz-Anderson, 1993), the ergonomic intervention should allow to: (1) perform a thorough job analysis to determine the nature of specific problems, (2) evaluate and select the most appropriate intervention(s), (3) develop and apply conservative treatment (implement the intervention), on a limited scale if possible, (4) monitor progress, and (5) adjust or refine the intervention as needed.

35.5.3 Administrative and Engineering Controls of WUEDs

The control of WUEDs requires consideration of the following aspects of this complex problem: (1) WUEDs diagnosis, (2) treatment, (3) rehabilitation and return to work; (4)

WUEDs surveillance, (5) surveillance and control of risk factors at the micro- and macro levels, (6) training and education, and (7) management and leadership with regard to WUEDs-related organizational and social aspects (Kuorinka and Forcier, 1995). The specific recommendations for prevention of WUEDs can be classified as being either primarily administrative, i.e., focusing on personnel solutions, or engineering, i.e., focusing on redesigning tools, workstations, and jobs (Putz-Anderson, 1993). In general, administrative controls are those actions to be taken by the management that limit the potentially harmful effects of a physically stressful job on individual workers. Administrative controls, which are focused on the workers, are modifications of existing personnel functions such as worker training, job rotation, and matching employees to job assignments. A summary of selected ergonomics measures that aim to control the incidence of WUEDs is shown in Table 35.17.

With respect to biomechanical risk factors, the prevention and control efforts for WUEDs should be directed toward fulfilling several recommendations based on ergonomics principles for workplace design, work methods and work organization. As discussed by Putz-Anderson (1993), these may include, for example, the following recommendations:

1. Permit several different working postures.
2. Place controls, tools, and materials between the waist and shoulder heights for ease of reach and operation.
3. Use jigs and fixtures for holding purposes.
4. Resequence jobs to reduce the repetition.
5. Automate highly repetitive operations.
6. Allow self-pacing of work whenever feasible.
7. Allow frequent (voluntary and mandatory) rest breaks.

Furthermore, with respect to hand tools used at work, the following general work design guidelines should be followed:

1. Make sure the center of gravity of the tool is located close to the body and the tool is balanced.
2. Use power tools to reduce the force and repetition required.
3. Redesign the straight tool handle; bend it as necessary to preserve the neutral posture of the wrist.
4. Use tools with pistol grips and straight grips, respectively, where the tool axis in use is horizontal and vertical (or when the direction of force is perpendicular to the workplace).
5. Avoid tools that require working with the flexed wrist and extended arm at the same time, or ones that call for the flexion of distal phalanges (last joints) of the fingers.
6. Minimize the tool weight; suspend all tools heavier than 20 N (or 2 kg of force) by a counterbalancing harness.
7. Align the tool's center of gravity with the center of the grasping hand.
8. Use special purpose tools that facilitate fitting the task to the worker (avoid standard off-the-shelf tools for specific repetitive operations).
9. Design tools so that workers can use them with either hand.
10. Use power grip where power is needed, and precision grip for precise tasks.
11. The handles and grips should be cylindrical or oval with a diameter between 3.0 and 4.5 cm (for precise operations the recommended diameter is from 0.5 to 1.2 cm).
12. The minimum handle diameter should be 10.0 cm, whereas a 11.5 cm–12.0 cm handle is preferable.
13. The handle span of 5.0 cm to 6.7 cm can be used by male and female workers.
14. Triggers on power tools should be at least 5.1 cm wide allowing their activation by two or three fingers.
15. Avoid form-fitting handles that cannot be easily adjusted.

Table 35.17 Ergonomic Measures To Control Common WMSDs[a]

Disorder	Avoid in General	Avoid in Particular	Recommendation	Design Issues
Carpal tunnel syndrome	Rapid, often repeated finger movements, wrist deviation	Dorsal and palmar flexion, pinch grip, vibrations between 10 and 60 Hz		Workplace design
Cubital tunnel syndrome	Resting forearm on sharp edge or hard surface			Workplace design
DeQuervain's syndrome	Combined forceful gripping and hard twisting			Workplace design
Epicondylitis	"Bad tennis backhand"	Dorsiflexion, pronation		Workplace design
Pronator syndrome	Forearm pronation	Rapid and forceful pronation, strong elbow and wrist flexion	Use large muscles but infrequently and for short time	Design of work object
Shoulder tendonitis, rotator cuff syndrome	Arm elevation	Arm abduction, elbow elevation	Let wrists be in line with the forearm	Design of job task
Tendonitis	Often repeated movements, particularly with force exertion; hard surface in contact with skin; vibrations	Frequent motions of digits, wrists, forearm shoulder	Let shoulder and upper arm be relaxed	Design of hand tools ("bend tool, not the wrist")
Tenosynovitis, DeQuervain's syndrome, ganglion	Finger flexion, wrist deviation	Ulnar deviation dorsal and palmar flexion, radial deviation with firm grip	Let forearms be horizontal or more declined	Design for round corners, use pad
Thoracic outlet syndrome	Arm elevation, carrying loads	Shoulder flexion, arm hyperextension		Design of work object placement
Trigger finger or thumb	Digit flexion	Flexion of distal phalanx alone		Workplace design
Ulnar artery aneurism	Pounding and pushing with heel of the hand			Workplace design
Ulnar nerve entrapment	Wrist flexion and extension	Wrist flexion and extension, pressure of hypothenar eminence		Workplace design
White finger, vibration syndrome	Vibrations, tight grip, cold exposure	Vibrations between 40 and 125 Hz		Workplace design
Neck tension syndrome	Static head posture	Prolonged static head–neck posture	Alternate head–neck postures	Workplace design

[a]Adapted from Kroemer, K., Kroemer, H. and Kroemer-Elbert, K., 1994, *Ergonomics, how to design for ease and efficiency,* Englewood Cliffs, NJ: Prentice Hall, reprinted with permission.

16. Provide handles that are nonporous, nonslip and nonconductive (thermally and electrically).

35.5.4 Standardization Efforts

The rapid growth of work-related musculoskeletal disorders in the recent past has motivated the American National Standards Institute (ANSI, 1995) and the Occupational Safety and Health Administration (OSHA, 1995) to focus on development of the voluntary standards and regulatory efforts, respectively, to control such illnesses at the workplace. OSHA has proposed the rulemaking under Section 6(b) of the Occupational Safety and Health Act of 1970, 29 U.S.C. 655. A short summary of both proposals is provided below.

35.5.4.1 Draft Ergonomic Protection Standard

In 1995, OSHA specified requirements intended for determining and implementing an appropriate strategy to control problem jobs, by reducing or preventing workers exposure to workplace risk factors. The proposed approach was based on application of engineering or administrative controls that would be considered effective in controlling the job. The key provisions of the proposed draft standard (OSHA, 1995) are summarized in Table 35.18.

The requirements of the *Draft Proposal Ergonomic Protection Standard* (OSHA, 1995) were to apply to companies where workers would be exposed daily (during the workshift) to any of the following signal risk factors: (1) performance of the same motion or motion pattern very few seconds for more than a total of 2, 3, or 4 h: (2) a fixed or awkward work posture (e.g., overhead work, twisted or bent back, bent wrist, kneeling, stooping, squatting) for more than a total of 2, 3, 4 h; (3) use of vibrating or impact tools or equipment for more than a total of 2, 3, or 4 h; (4) using forceful hand exertions for more than a total of 2, 3, or 4 h; (5) unassisted frequent or forceful manual handling (for more than a total of 1 or 2 h). It should be noted that the term "work-related musculoskeletal disorder" was used by OSHA to refer collectively to any of the following when they are caused or aggravated by exposure to workplace risk factors: signs, or persistent symptoms (at least 7 days or interfering with work), or clinically diagnosed work-related musculoskeletal disorders. An example of the workplace risk factor checklist for upper extremity is shown in Figure 35.5.

An overview of the proposed OSHA (1995) *Ergonomic Protection Standard* for employers without an existing effective job improvement process is shown in Figure 35.6. The first step would be to provide employees with information so workplace risk factors and problem jobs could be accurately identified. The second step would involve identifying the specific workplace risk factors that are present to determine whether the job is a problem job. A *problem job* was defined as one which would result in a score of more than 5 on the OSHA workplace risk factor checklist (see Figure 35.6).

According to the draft OSHA (1995) proposal, companies would be required to develop and implement a job improvement process to control each problem job. The job improvement process would be performed by an individual or team knowledgeable in identification of workplace risk factors, job analysis methods, and implementation and evaluation of control measures. Among others, the following actions would need to be taken in the job improvement process: (1) a job analysis of each problem, including description of job, including all jobs tasks performed by employees; (2) identification and description of each workplace risk factors present in the job and the task, actions, or workplace conditions that cause each workplace risk factor to be present; (3) analysis of a manual handling tasks using the most recent National Institute for Occupational Safety and Health (NIOSH) lifting equation (or accepted alternative evaluation method).

Under the considered OSHA (1995) draft provisions, each worker or ergonomic team member involved in conducting job analysis would be trained in identification of workplace risk factors, job analysis methods, and implementation and evaluation of control measures. The ergonomic awareness and job-specific training for the affected employed and their supervisors should allow them to: (1) recognize workplace risk factors and the methods for controlling them; (2) identify the signs and symptoms and the health effects of exposure to workplace risk factors, the importance of early reporting, and the employer's medical management procedures; (3) learn the procedures for reporting workplace risk factors and work-related musculoskeletal disorders, including the designated person(s) for receiving reports; (4) learn the process the employer is undertaking to address and control workplace risk factors, each employee's role in the process and how to actively participate in the process; (5) learn how to obtain a copy of this section and its appendices

Table 35.18 Key Provisions of the Proposed *Draft Ergonomic Protection Standard*[a]

APPROACH

- Organize the standard for ease of reference.
- User-friendly text wherever possible.
- Compliance assistance materials.
- Performance orientation.
- Accommodations for small employers.

SCOPE AND APPLICATION

- Problem affects broad spectrum of workers and industries.
- Tiered approach—use of signal risk factors to target jobs for further evaluation.

  Signal risk factors are:

  Performance of same motion or motion pattern every few seconds for more than (two, three, or four) hours at a time.

  Fixed or awkward postures for more than a total of (two, three, or four) hours.

  Use of vibrating or impact tools or equipment for more than a total of (two, three, or four) hours during a workshift.

  Using forceful hand exertions for more than a total of (two, three, or four) hours at a time.

  Unassisted frequent or heavy lifting (for more than a total of one or two hours).

  - Picked signal risk factors based on scientific literature review.
  - Presence of signal risk factor increases probability of potential problems.
- Additional trigger of one (or two or more) work-related musculoskeletal disorders recorded after effective date of the rule.
- Grandparenting provision: employers with existing, effective programs are exempted from certain provisions. Allows them to continue working to improve jobs rather than shifting gears to repeat activities they have already done.

IDENTIFICATION OF PROBLEM JOBS

- Information of musculoskeletal disorders and reporting procedures should be given to workers exposed to signal risk factors or recorded work-related musculoskeletal disorders.
- Risk factor checklist completed for jobs with signal risk factors or recorded WMSDs.
  - Checklist is more detailed evaluation of risk factor exposures.
  - Separates jobs that need to be controlled (i.e., problem jobs or those with a checklist score of more than 5).
  - OSHA has provided a checklist employers can use. Rule allows alternative evaluations to be used as well.

CONTROL OF RISK FACTOR EXPOSURES

- Problem jobs have to be controlled (i.e., the checklist score must be reduced to 5 or below).
  - Two approaches:
    - Quick fix (where causes of problem and appropriate controls can be readily identified and implemented).
    - Job improvement process (where causes are more complicated and additional analysis is needed before controls can be identified and implemented).
  - Engineering or administrative controls supplemented by personal protective equipment.
  - Back belts and wrist splints are not considered to be personal protective equipment under the rule.
  - Where employees routinely handle heavy packages (more than 25 lb) manually, they must be able to determine the relative weight of the package in order to lift or handle appropriately. Can mark the weight or identify contents in a way that would indicate the approximate weight.

ERGONOMIC DESIGN AND CONTROLS FOR NEW OR CHANGED JOBS

- Protect employees in the future; prevent the introduction of problem jobs started or changed after the rule is fully effective.
- Most efficient approach is to design out problems up front.

TRAINING

- Workers in problem jobs and their supervisor must be trained.
- Employers must evaluate the effectiveness of the training program.

MEDICAL MANAGEMENT

- Focuses on responding to problems and getting employees back to work. Case management rather than medical surveillance.

PHASE-IN PERIOD FOR COMPLIANCE

- Provisions are tiered with different dates for compliance.
- Small employers get an additional year to comply with all of the substantive provisions.

[a]OSHA, 1995. (Washington, D.C.)

A

Upper Extremity Risk Factors

Date: ________________________

Job: ________________________

Department: ________________________

Employee: ________________________

Analyst: ________________________

Comments: ________________________

Task	Risk Factor	Total Time

UPPER EXTREMITY RISK FACTOR SCORES					Page 1
A	B	C	D	E	F
RISK FACTOR CATEGORY	RISK FACTORS	TIME			SCORE
		2 to 4 Hours Circle the score	4+ to 8 Hours	8+ Hours Add 0.5 per hour	
Repetition (Finger, Wrist, Elbow, Shoulder, or Neck Motions)	1. **Identical or Similar Motions Performed Every Few Seconds** *Motions or motion patterns that are repeated every 15 seconds or less. (Keyboard use is scored below as a separate risk factor.)*	1	3		
	2. **Intensive Keying** *Scored separately from other repetitive tasks in the repetition category and includes steady pace as in data entry.*	1	3		
	3. **Intermittent Keying** *Scored separately from other repetitive tasks. Keyboard or other input activity is regularly alternated with other activities for 50 to 75 percent of the work.*	0	1		
Hand Force (Repetitive or Static)	1. **Grip More Than 10-Pound Load** *Holding an object weighing more than 10 pounds or squeezing hard with hand in a power grip.* Power Grip	1	3		
	2. **Pinch More Than 2 Pounds** *Pinch force of 2+ pounds as in the pinch used to open a small binder clip with the tips of fingers.* Pinch Grip	2	3		
Awkward Postures	1. **Neck: Twist/Bend** *Twisting neck to either side more than 20°, bending neck forward more than 20° as in viewing a monitor, or bending neck backward more than 5°.* 20° 5° 20° 20°	1	2		

Figure 35.5 An example of the Workplace Risk Factor Checklist for Upper Extremity (after OSHA, 1995).

from the employer (the employer shall maintain a copy of this section and its appendices and shall provide a copy to any employee within 5 days of a request); and (6) practice and demonstrate proper use of implemented control measures and safe work methods that pertain to the job.

It should be noted here that the above is a description of the proposed legislative act, which, as the date of this writing, has not been adopted by the United States Congress.

35.5.4.2 The Z-365 Standard for Control of Work-Related CTDs

The ANSI Z-365 Committee (ANSI, 1995) proposed a standard on CTD entitled *Control of Work-Related Cumulative Trauma Disorders, Part I: Upper Extremities.* The ANSI (1995) process for control of CTDs at the workplace is illustrated in Figure 35.7. The four main components of this process are: (1) management commitments, (2) a written

UPPER EXTREMITY RISK FACTOR SCORES					Page 2
A	B	C	D	E	F
RISK FACTOR CATEGORY	RISK FACTORS	TIME			SCORE
		2 to 4 Hours	4+ to 8 Hours	8+ Hours Add 0.5 per hour	
		Circle the score			
Awkward *continued*	2. **Shoulder: Unsupported Arm or Elbow Above Mid-Torso Height** *Arm is unsupported if there is not an arm rest when doing precision finger work, or when the elbow is above mid-torso height.*	2	3		
	3. **Rapid Forearm: Rotation** *Rotating the forearm or resisting rotation from a tool. An example of forearm rotation is using a manual screwdriver.*	1	2		
	4. **Wrist: Bend/Deviate** *Consider wrist bends that are more than 20 degrees flexion (bend wrist palm down) or more than 30 degrees extension (bending wrist back). Bending can occur during manual assembly and data entry.*	2	3		
	5. **Fingers** *Forceful gripping to control or hold an object, such as click-and-drag operations with a computer mouse or deboning with a knife.*	0	1		
Contact Stress	1. **Hard/Sharp Objects Press Into Skin** *Includes contact of the palm, fingers, wrist, elbow, or armpit.*	1	2		
	2. **Using the Palm of the Hand as a Hammer**	2	3		
Vibration	1. **Localized Vibration (Without Vibration Dampening)** *Vibration from contact between the hands and a vibrating object, such as a power tool.*	1	2		
	2. **Sitting/Standing on Vibrating Surface (Without Vibration Dampening)**	1	2		
Environment	1. **Lighting (Poor Illumination/Glare)** *Inability to see clearly (e.g., glare on a computer monitor).*	0	1		
	2. **Cold Temperature** *Hands exposed to air temperature of less than 60°F for sedentary work, 40°F for light work, 20°F for moderate/heavy work; cold exhaust blowing on hands.*	0	1		
Control Over Work Pace	1. **No Control Over Pace** *Machine paced, piece rate, constant monitoring, or daily deadlines. Enter 1 if one control factor is present or 2 if two or more control factors are present.*				
TOTAL UPPER EXTREMITY SCORE FOR CHECKLIST A (sum of both page 1 and page 2)					

Figure 35.5 (Continued)

program and implementation document, (3) training, and (4) employee involvement. The standard specifies requirements for surveillance, job analysis and design, medical management, and training.

According to Z-365 Committee (ANSI, 1995), the proposed technical standard specifies principles and practices for controlling work-related cumulative trauma disorders that arise from manual lifting, assembly, manipulation of tools, machinery, and other devices and other stresses to muscles, nerves, tendons, and associated soft tissues of the body. Such a standard is intended for use by those who have the responsibility for design and operation of work equipment and procedures, and for the management of medical, health, and safety programs. The proposed standard states the principles and practices in general terms only, and it requires professional judgment for application to specific work situations. It is assumed that the standard will be implemented by personnel with appropriate training. The ergonomic considerations of the standard include: (1) work postures, (2)

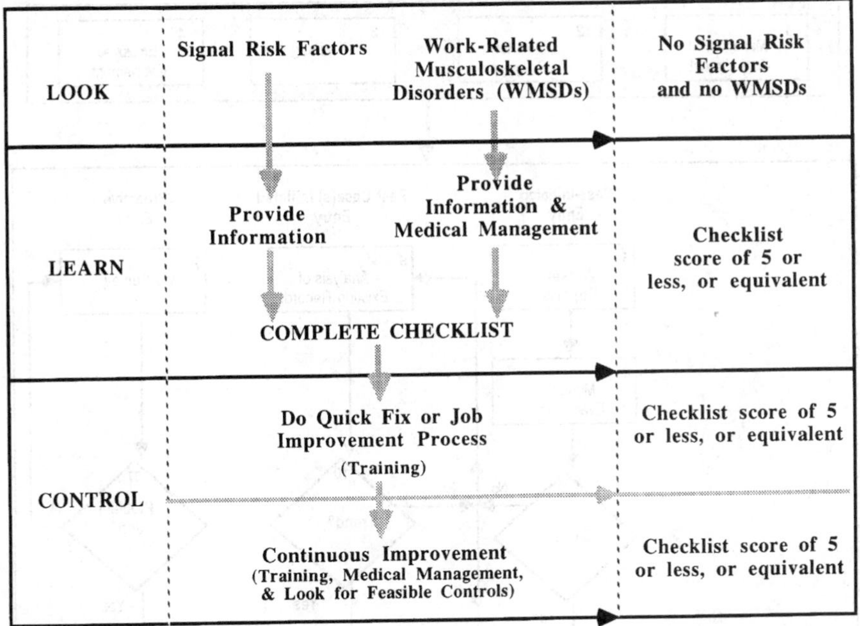

Figure 35.6 An overview of the proposed OSHA Draft Ergonomics Protection Standard (modified after OSHA, 1995).

work layout, (3) human strength requirements, (4) vibration, (5) work rates, (6) tool design and flexibility of workstations to accommodate worker variations. An example of the risk factor checklist for control of WUEDs as proposed by the Z-365 Committee (ANSI, 1995) is shown in Table 35.19.

Given the above standardization efforts, one must also recognize the significant consequences and potential impact of any performance or ergonomic design–based standards on U.S. industrial competitiveness. The effective standards and guidelines for control and prevention of the work-related hazards should significantly reduce the risk to WMSDs to employees. However, it is also generally acknowledged that the current state of the art in ergonomics science, as relevant to the WUEDs prevention, does not yet allow to specify work design parameters for a given level of risk in a given population.

35.5.5 Ergonomics Programs to Prevent WUEDs

An important component of the WUEDs management efforts is development of the well-structured and comprehensive ergonomics programs. According to Alexander and Orr (1992), the basic components of such a program should include the following: (1) health and risk factor surveillance, (2) job analysis and improvement, (3) medical management, (4) training, and (5) program evaluation. An excellent program should include participation of all levels of management; medical, safety, and health personnel; labor unions; engineering; facility planners; and workers, and contain the following elements:

1. Routine (monthly or quarterly) reviews the OSHA log for patterns of injury and illness (dedicated computer programs to be used to identify problem areas).
2. Workplace audits for ergonomic problems that are a routine part of the organization's culture (more than one audit annually for each operating area), and timely interventions as a response to the identified problems.
3. A knowledge by management and workers regarding the list of most critical problems, i.e., jobs with job title clearly identified.
4. Application of both engineering solutions and administrative controls, with engineering solution treated as the long-term solutions.

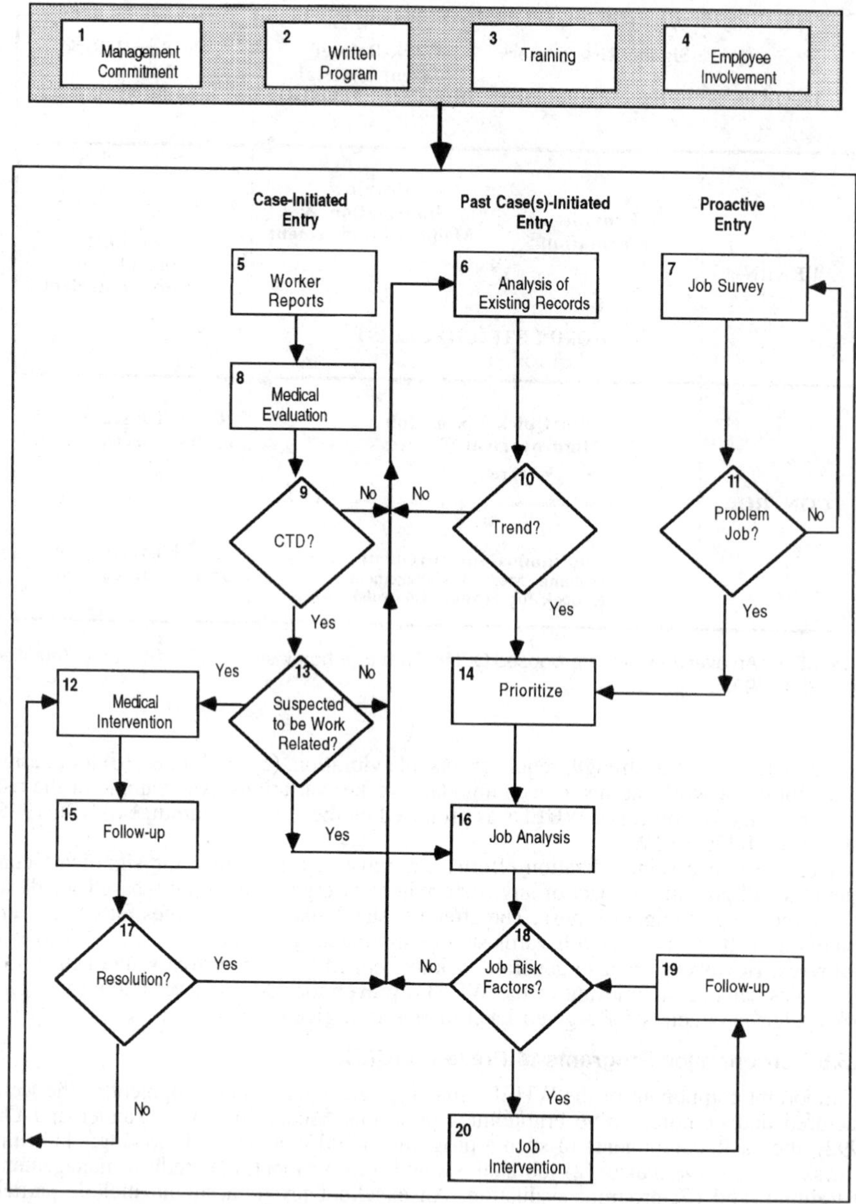

Figure 35.7 The proposed ANSI Process for Control of CTDs at the Workplace (modified after ANSI, 1995).

5. Awareness of ergonomic considerations by design engineering that utilizes them into new or reengineered designs. (People are an important design consideration.)
6. Frequent refresher training in ergonomics, including short courses and seminars for the site appointed "ergonomists."

35.6 WUEDs SURVEILLANCE AND JOB ANALYSIS AND DESIGN

35.6.1 Surveillance System

To evaluate the extent of WUEDs in the working population a surveillance system should be used. As discussed by ANSI (1995), surveillance is defined as the ongoing systematic

Table 35.19 Risk Factor Checklist for Control of WUEDs

Risk Factor	Exposure <1 h	1–4 h	>4 h	Cause of Risk Factor	Proposed Solution	Comments
REPETITION						
every few seconds	0	1	3			
every few minutes	0	0	1			
LOAD/FORCE (LIFT)						
>5–15 lb	0	0	1			
>15–30 lb	1	1	2			
>30–50 lb	2	2	3			
>50 lb	3	3	3			
LOAD/FORCE (PUSH/PULL)						
easy	0	0	1			
moderate	0	1	2			
heavy	1	2	3			
LOAD/FORCE (Carry >10 ft)						
>5–15 lb	0	0	1			
>15–30 lb	0	1	2			
>30 lb	1	2	3			
AWKWARD POSTURES:						
Neck/shoulder: overhead/bend	0	1	2			
Extended reach	0	1	2			
Elbow/forearm: twist	0	1	2			
Hand/wrist: bend/pinch	0	1	2			
Trunk: twist/bend	0	1	2			
Knee: squat/kneel	0	1	2			
USE POWER TOOLS	0	1	2			
PRESSURE POINTS	0	1	2			
SAME POSITION	0	1	2			
ENVIRONMENT: cold, hot, light/glare, vibration	0	1	2			
Continuous keyboard use	0	1	2			
Incentive work or no worker control over job pace	0	1	2			
TOTAL SCORE = 10 or more?						

[a]After ANSI (1995).

collection, analysis, and interpretation of health and exposure data in the process of describing and monitoring work-related cumulative trauma disorders. Surveillance is used to determine when and where job analysis is needed and where ergonomic interventions may be warranted. Surveillance data are used to determine the need for occupational safety and health action and to plan, implement, and evaluate ergonomic interventions and programs (Klaucke et al., 1988). Health and job risk factor surveillance provide employer and employees with a means of systematically evaluating WUEDs and workplace ergonomic risk factors by monitoring trends over time. This information can also be of benefit for planning, implementing and evaluating ergonomic interventions.

The working draft ANSI Z-365 (1995) for control of work-related cumulative trauma disorders includes two important steps of surveillance and job analysis and design. Surveillance includes: (1) worker reports, (2) analysis of existing records and surveys, and (3) job surveys and proactive entry into the process. Job analysis and design, which include description of work, definitions of risk factors, and risk factor measurement, quantification and interaction, serve two common purposes: (1) to identify potential work-related risk factors associated with musculoskeletal disorders after they are reported, and (2) to assist in identifying work-related factors associated with musculoskeletal disorders before they occur.

Detailed job analysis typically consists of analyzing the job at the element or micro level. These analyses involve breaking down the job into component actions, measuring and quantifying risk factors, and identifying the problems and conditions contributing to each risk factor. Job surveys, on the other hand, can be used for establishing work relatedness, for prioritizing jobs for further analysis, or for proactive risk factors surveillance. As illustrated in Table 35.20, the proactive job such survey methods may include facility walk-throughs, worker interviews, risk factor checklists, and team problem-solving approaches. Some of the research evaluation tools defined by the ANSI Z-365 Draft Standard for the purpose of surveillance and job analysis include the following: (1) Proactive Job Survey, (2) Quick Check Risk Factor Checklist, (3) Symptom Survey (questionnaire), (4) Posture Discomfort Survey, and (5) History of Present Illness Recording Form.

35.6.2 Procedures for Job Analysis and Design

According to ANSI (1995), the job analysis should be performed at a sufficient level of detail to identify potential work-related risk factors associated with WMSDs and include the following steps:

1. Collection of the pertinent information for all jobs and associated work methods
2. Interview of the representative sample of the affected workers
3. Breakdown of the jobs into tasks or elements
4. Description of the component actions of each task or element
5. Measurement and quantification of WRMD risk factors
6. Identification of the risk factors for each task or element
7. Identification of the problems contributing to risk factors
8. Summary of the problem areas and needs for intervention for all jobs and associated new work methods

There are three types of existing records and survey analyses: (1) initial analysis of upper limb WMSDs reported over the last 24 to 36 months, (2) ongoing trend analysis of past cases, and (3) health surveys. Analysis of existing records and surveys consists of reviewing existing databases, principally collected for other purposes to identify incidents and patterns of work-related cumulative trauma disorders and can help determine and prioritize which jobs should be further analyzed using job analysis.

35.6.3 Job Surveys

The aim of proactive job surveys is to identify specific jobs and processes that may put employees at risk of developing WMSDs before they occur. Job surveys are typically performed after the jobs identified by the previous two surveillance components have been rectified. Job surveys of all jobs or a sample of representatives should be performed. Analysis of existing records will be used to estimate the potential magnitude of the problem in the workplace. The number of employees in each job, department, or similar

Table 35.20 A List of Proactive Job Survey Methods for Control of CTDs[a]

Job survey	Description
Facility walk-throughs	A facility walk-through consists of using firsthand observations to identify exposures to recognized CTD risk factors. Knowledge of the processes, facilities, and schedules is used to conduct observations of representative activities and workers. This method may be effective at identifying the presence of some risk factors, particularly those that are most obvious. However, a single walkthrough or observation may not be sufficient to determine the magnitude, frequency, and duration of exposure to risk factors. The time required to perform a walk-through depends on the work cycle time and variability of the tasks performed, with longer work cycles and jobs with greater task variability needing more observation time. The method can be enhanced by combining it with formal or informal worker and supervisor interviews and/or risk factor checklists.
Worker and supervisor interviews	A worker and supervisor interview consists of asking questions regarding job/task attributes and associated risk factor exposures. The method relies on the firsthand observations and experiences of worker supervisor. Such interviews may be formal (written or recorded) or informal (verbal). This method may be effective at identifying the presence of risk factors and associated job attributes note easily garnered by a facility walk-through, particularly risk factor exposure duration, recovery breaks, infrequently performed tasks, or multiple methods at performing tasks. However, the method is subjective. The method is also performed as part of a more comprehensive job analysis.
Risk factor checklists	Checklists consist of a formal procedure in which CTD risk factors are enumerated on a list and checked off by the analyst through firsthand observation of a specific or representative task(s) or job(s). Checklists may have high sensitivity in identifying the presence of some risk factors, however, they may have low specificity. Checklists must be adapted and validated for each industry or occupation. Using checklists requires caution and familiarity with the job, task or process.
Team problem-solving processes	Team problem-solving processes integrate the firsthand observations and experiences of several workers and supervisors regarding job or task attributes and associated risk factor exposures and attempt to develop, test, and implement plausible solutions. Any of the above surveys tools may be incorporated into such a process.

[a]After ANSI (1995).

population needs to be determined first. Then the incidence rates can be calculated on the basis of hours worked as follows:

$$\text{Incidence (new case) rate } (IR) = \frac{\text{No. of new cases during time} \times 200{,}000}{\text{work hours during time}} \tag{5}$$

The incidence rate is equivalent to the number of new cases per 100 worker hours. Workplace-wide incidence rates (*IR*) can be calculated for all WUEDs classified by body location for each department, process or type of job. (If specific work hours are not readily available, the number of full-time equivalent employees in each area multiplied by 2000 hours can be used to obtain the denominator.) Severity rates (*SR*) traditionally use the number of lost workdays rather than number of cases in the numerator. Prevalence rates (*PR*) are the number of existing cases per 200,000 h or the percentage of workers with the condition (new cases plus old cases that are still active). It should be noted that neither

NIOSH or OSHA have defined the levels of incidence rates (*IR*) or prevalence rates (*PR*) that would be considered "excessive" in general or by any specific industry group. This is perceived as a major problem in using the general duty clause (GDC) and in promulgating any future standards.

35.6.4 Surveillance Methods for WMSDs

35.6.4.1 Surveillance Data Collection Instruments

The surveillance system aims to link the occurrence of WMSDs to work-related risk factors. Ideally, the surveillance should make it possible to identify workplace risk factors before symptoms develop. Surveillance data collection instruments can be of passive or active nature (Kuorinka and Forcier, 1995). The summary of active and passive surveillance methods are listed in Table 35.21. The passive surveillance process relies on information collected from existing databases and records (e.g., company dispensary logs, insurance records, workers' compensation records, accident reports, and absentee records) to identify the WRMD cases and patterns and potential problem jobs. Passive surveillance records are often useful in helping to determine the frequency with which active surveillance tools should be used and the required interventions or in assessing the effectiveness of ergonomics programs. In addition, brief job analysis or physical demand analysis to assess the suitability of a job for the return to work of an injured worker, can also be used for passive risk factor surveillance.

The active surveillance uses specifically designed tools and information, such as checklists and job analysis. As shown in Table 35.22, there can be both health active surveillance and risk factor active surveillance. Since most musculoskeletal disorders produce some symptoms of pain or discomfort, health questionnaires are useful in identifying new or incipient problems as well as for assessing the effectiveness of medical interventions and ergonomic controls. In addition to symptom questionnaires, medical interviews and examination can also be used to active health surveillance (Table 35.23).

35.6.4.2 Analysis and Interpretation of Surveillance Data

The surveillance data can be analyzed and interpreted to study possible associations between the WMSDs surveillance data and the risk factor surveillance data (Kuorinka and Forcier, 1995). The two principal goals of the analysis are: (1) to help identify patterns in the data that reflect large and stabler differences between jobs or departments and (2) to target and evaluate intervention strategies. This analysis can be done on the number

Table 35.21 Passive and Active Surveillance Methods[a]

Passive Surveillance[b]	Active Surveillance
Information source and method already exist and are usually designed for other administrative purposes	Information source and method specifically designed for surveillance
Relatively inexpensive	Modest to quite expensive
Usually requires additional coding of information for the purpose. For instance, surrogate(s) of exposure, e.g., job titles	Since tools are "tailor made," includes at least job title information and other data considered important by surveillance analyst. Will include data for linking of information between risk factor and WMSD data
Examples: health and safety logs, medical department logs, workers' compensation data, early retirement, medical insurance, absenteeism and transfer records, accident reports, product quality, productivity	Examples: for WMSD surveillance (confidential questionnaires without personal identifiers, questionnaire interviews, physical examinations); for risk factor surveillance (workplace walk-throughs, job checklists, postural discomfort surveys

[a]Adapted from Kuorinka, I., and Forcier, L. (Eds.). (1995). *Work related musculoskeletal disorders (WMSDs): A reference book for prevention.* Taylor and Francis, reprinted with permission.
[b]Used mostly for health surveillance since, in practice, no existing records have been used to obtain information on risk factors associated with WMSDs.

Table 35.22 Summary of Tools Used in Surveillance[a]

Approach	Tools
Health	
• Passive	• Existing records
• Active, level 1	• Symptoms surveys or questionnaires (self- or group administered)
• Active, level 2	• Health-related interviews and/or brief physical exams
Risk factors	
• Active, level 1	• Quick checklists of risk factors
• Active, level 2	• In-depth job analysis

[a]Adapted from Kuorinka, I., and Forcier, L. (Eds.). (1995). *Work related musculoskeletal disorders (WMSDs): A reference book for prevention.* Taylor and Francis, reprinted with permission.

of existing WRMD cases (cross-sectional analysis) or during a specific period of time on the number of new WRMD cases, in a retrospective and prospective fashion (retrospective and prospective analysis).

One of the simplest ways to assess the association between risk factors and WMSDs is to calculate the odds ratios (see Table 35.24). For this purpose, the prevalence data obtained in health surveillance are linked with the data obtained in risk factor surveillance. In the example shown in Table 35.24 (for more details see Kuorinka and Forcier, 1995), one risk factor is selected at a time (e.g., overhead work for more than 4 h). Using the data obtained in surveillance the following numbers of employees are counted:

- Employees with WMSDs and exposed to more than 4 h of overhead work (15 workers)
- Employees with WMSDs and not exposed to more than 4 h of overhead work (15 workers)
- Employees without WMSDs and exposed to more than 4 h of overhead work (25 workers)
- Employees without WMSDs and not exposed to more than 4 h of overhead work (85 workers)

The overall prevalence rate (*PR*), i.e., rate of existing cases, for the company is 30/140 or 21.4%. The prevalence rate for those exposed to the risk factor is 37.5% (15/40) compared with 15.0% (15/100) for those not exposed. The risk of having a WRMD

Table 35.23 Examples of Tools for WMSD Surveillance[a]

	Methods of Surveillance	
Surveillance on:	Passive	Active
•	Company dispensary logs	• Checklists
•	Insurance records	• Questionnaires
•	Workers' compensation records	• Interviews
Health (WMSDs)		
•	Accident reports	• Physical exams
•	Transfer requests	
•	Absentee records	
•	Grievances	
Workplace risk factors (associated with WMSDs)	Not really used for WMSD risk factor yet[b]	• Checklists • Questionnaires • Job Analysis

[a]Adapted from Kuorinka, I., and Forcier, L. (Eds.). (1995). *Work related musculoskeletal disorders (WMSDs): A reference book for prevention.* Taylor and Francis, reprinted with permission.
[b]The use of surrogate measures for exposure (e.g., job title or firm's department) could be viewed as "passive surveillance."

Table 35.24 Examples of Odds Ratio Calculations for a Firm of 140 Employees[a]

		WMSDs are:[b]		
		Present	Not Present	Total
Risk factor (e.g., overhead work for more than 4 h) is:	Present	15 (A)	25 (B)	40 (A + B)
	Not Present	15 (C)	85 (D)	100 (C + D)
	Total	30 (A + C)	110 (B + D)	140 (N)

[a]Adapted from Kuorinka, I., and Forcier, L. (Eds.). (1995). *Work related musculoskeletal disorders (WMSDs): A reference book for prevention.* Taylor & Francis, reprinted with permission.
[b]Number in each cell indicates the count of employees with or without WMSD and the risk factor. Odds Ratio (OR) = $(A \times D)/(B \times C) = (15 \times 85)/(25 \times 15) = 3.4$

depending on exposures to the risk factor, the odds ratio, can be calculated using the number of existing cases of WRMD (prevalence). In the above example, those exposed to the risk factor have 3.4 times the odds of having the WRMD than those not exposed to the risk factor. An odds ratio of greater than 1 indicates higher risk. Such ratios can be monitored over time to assess the effectiveness of the ergonomics program in reducing the risk of WMSDs, and a variety of statistical tests can be used to assess the patterns seen in the data.

35.7 TRAINING FOR CONTROL OF WUEDs

35.7.1 Training Program: Success Factors

As reviewed by Kroemer et al. (1994), currently it is not possible to determine the contribution made by training alone to the overall effectiveness of the intervention programs for control of WUEDs. However, training, which is often is often complementary to other interventions, can also be an important step in a global strategy to reduce or prevent the WMSDs at the workplace (OSHA, 1995; ANSI, 1995). As outlined in the ANSI (1995) proposal, a number of interventions can be introduced simultaneously, including (1) educational efforts targeting supervisors, engineers, workers, safety representatives, and occupational health and safety personnel; (2) occupational health (a medical system that aims to restrict work activities of workers with early signs of WMSDs); (3) ergonomic and engineering changes; and (4) organizational changes (including forming steering committees, comprising of management, engineers and occupational health specialists), in order to ensure effective planning, implementation, an adequate followup. A summary of the appropriate training levels for control of WUEDs as defined by ANSI (1995) is shown in Table 35.25.

Kuorinka and Forcier (1995) reviewed several case studies regarding the WRMD-related training and identified the following success factors: (1) an emphasis on active participation and learning by doing and other newer methods such as cognitive training, (2) using workers as trainers, (3) worker involvement in the design of training programs, and (4) use of audiovisual techniques and on-the-job training, as opposed to classroom teaching. The reasons for this lack of success in training was classified into the following four categories: (1) training period is too short, (2) difficult to transfer the training to the actual work context, (3) validity of the principles taught is questionable, and (4) training is not a panacea for control of work-related musculoskeletal disorders.

35.7.2 Training Methods

Training methods, which are defined by the way in which the training is actually conducted, include the following systems: train-the-trainer, cognitive training, the five-step approach, skills training, behavior modification, training by doing, and observational training. As pointed out by Silverstein (1991), the train-the-trainer approach, is particularly appropriate when a large number of workers need the WRMD-related training. In the train-the-trainer approach, a group of workers is selected to train other workers. The members of the first group are given training and then proceed to become trainers themselves. The active role of workers in providing training can be a good motivational factor. The rational of the train-the-trainer approach is that people can be trained to train other people, leading to knowledge transfer at an "exponential" rate. The number of people trained by the train-the-trainer approach is much greater than by the traditional approach, wherein one (or several persons) must train every member of the group of employees requiring training.

35.7.2.1 Application of the Train-the-Trainer Approach

The train-the-trainer approach was successfully applied at General Motors Corporation under a UAW/GM national agreement (Silverstein, 1991). At each plant, ergonomics monitors were designated to conduct ergonomic evaluations and surveillance of their work area, and to work with their supervisors to implement simple job improvements. At each plant, an ergonomics coordinator was designated as the primary in-plant trainer. These ergonomics coordinators were taught about ergonomics and WMSDs. Then, they were assigned to train the ergonomics monitors and supervisors in ergonomics. The advantages of the train-the-trainer approach include the benefits of efficiency, employee involvement, job enrichment, and increased problem-solving skills.

35.7.2.2 Cognitive Training and the Five-Step Approach

Cognitive training is a training method in which the worker is trained by mentally analyzing the tasks without actually performing the tasks. The five-step approach emphasizes the development of job skills based on the worker's mental representation of the job. The rationale is to help the worker develop mental models of tasks, activities, postures and movements before their actual performance. The cognitive training method has been used in training of assembly tasks. The five-step training approach takes into account different stages in learning a skill and combines various training methods (Vartiainen, 1987). These five steps are as follows: (1) motivate the trainee and specify the goals, (2) teach how the task should be done, (3) have the trainee rehearse mentally to learn new skills, (4) try out the skill, and (5) inspect trainee performance. Steps 3 to 5 may be repeated until the trainee achieves a previously determined level of skill.

35.7.3 Evaluation of a Training Program

Success of training programs is largely dependent on a systematic and rigorous approach to their development and implementation. Training evaluation is the systematic collection of information regarding the success of training programs (Kuorinka and Forcier, 1995). Such evaluation should include the following aspects: (1) evaluation of reactions, focusing on assessing trainees' attitudes toward the training they have received; (2) evaluation of outcomes, designed to find out whether the training actually changed something in the trainees; and (3) evaluation of impact, concerned with overall results of training. Training development and implementation consists of four phases: (1) needs assessment, (2) training development, (3) training implementation, and (4) evaluation and followup. Failure to take any of the steps into account in a rigorous way may compromise results and waste resources.

From the literature on training in manual handling Kroemer et al., 1994; Kuorinka and Forcier, 1995), one can draw a number of conclusions that could apply to WMSD-related training:

- Training aimed at improving knowledge and attitudes does not necessarily lead to changes in behavior.
- The work context must be taken into account when deciding on training content and methods.
- The broader organizational context must be considered: reinforcement, production pressures, support for proper behavior.
- There should be a systematic, rigorous approach to developing training programs: needs assessment, design of training, implementation, and evaluation.
- Time factors must be taken into account: reinforcement, refresher training, and adequate time to understand material and master new skills.
- Training is not a panacea for addressing all risk factors, but rather a complement to other interventions.

35.8 MEDICAL MANAGEMENT OF WUEDs

35.8.1 Basic Activities of Medical Management

The primary objective of medical management in occupational health and safety programs is the prevention of work-related disorders and injuries (Kuorinka and Forcier, 1995). The specific goals of occupational health programs relevant to prevention of musculoskeletal disorders were specified by the American Medical Association (1972) as follows:

1. Protecting employees against health and safety hazards in their work situation
2. Evaluating workers' physical, mental, and emotional capacity before job placement
3. Ensuring that employees can perform the work with an acceptable degree of efficiency and without endangering their own health and safety or that or others
4. Ensuring adequate medical care and rehabilitation for the occupationally ill or injured
5. Encouraging and assisting with measures for personal health maintenance, including the acquisition of a personal physician whenever possible

Medical management of WUEDs includes medical diagnosis, treatment, rehabilitation and return-to-work, and work hardening (Karwowski and Kasdan, 1988). In addition to these activities, medical management should also be involved in both passive and active health surveillance, job-skills training programs, and ergonomic task force activities (Kuorinka and Forcier, 1995). As discussed in Section 35.6.4, the use of injury reports for health surveillance purposes is a form of passive health surveillance. The effective passive health surveillance requires data that have a high sensitivity for WUEDs. Injury reports should be followed up by workplace visits and an evaluation. In a population or workers, or in a specific job category where there is a high risk or WUEDs, it may be necessary to perform the active health surveillance, i.e., the periodic medical evaluation to identify workers in the early stages of a disease, and to target these workers for early secondary prevention efforts, i.e., medical treatment.

35.8.2 Medical Treatment of WUEDs

In general, the medical treatment efforts of WUEDs in the acute phase are similar to those treatments which are used for non–work related disorders. As discussed by Kuorinka and Forcier (1995), the general therapeutic objectives for WUEDs should include the following: (1) promotion of rest for the affected anatomical structures, (2) diminish spasms and inflammation, (3) reduction of pain, (4) increase in strength and endurance, (5) increase in range of motion, (6) alteration of mechanical and neurological structures, (7) increase in functional and physical work capacity, and (8) modification of work content and social environment.

35.8.2.1 Rehabilitation, Return-to-Work, and Work Hardening Programs

The Americans with Disabilities Act (St. Clair and Shults, 1992) stipulates employers responsibilities to accommodate workers disabled by the WUEDs. A program that promotes healing and helps an injured worker to return to work, and specifies appropriate job-placement conditions based on different job tasks and work requirements, is called *occupational rehabilitation.* Since the injury may not always have only a physical basis, psychosocial (at work and outside work) and psychological disability aspects are essential parts of the rehabilitation process. According to the Commission on Accreditation of Rehabilitation Facilities (1989), a *work hardening* program is a highly structured, goal-oriented, individualized treatment program designed to maximize the individual's ability to return to work. Such a program uses a set of conditioning tasks that are graded progressively in quest to improve biomechanical, neuromuscular, cardiovascular, and psychosocial functions with real or simulated work activities.

35.8.2.2 Preemployment and Preplacement Screening for Medical Management

Although there is no scientific evidence that screening can predict the development of WUED's preemployment and preplacement screening may be an important part of the medical management activities (Kuorinka and Forcier, 1995). According to the American College of Occupational and Environmental Medicine, the Committee on Occupational Medical Practice (ACOM, 1990), screening refers to the application of at least one test (or examination) to workers in order to identify apparently healthy workers who are at high risk of developing a specific WUED from those workers who are not. Although the screening tests are not diagnostic, preemployment screening and examination are typically performed before any offer of employment can be made. On the other hand, preplacement screening process refers to an examination of an employee who has already received an offer of employment and addresses a question of employee placement in a specific job.

35.9 SUMMARY

35.9.1 Balancing the Work System for Ergonomics Benefits

As pointed out by Kuorinka and Forcier (1995), there are no perfect jobs or perfect workplaces that are free of all work-related hazards and provide ideal psychosocial conditions for complete satisfaction for all employees. Therefore, one must consider the trade-offs between competing needs for ergonomic improvements at the workplace, and establish a basis for identifying the most critical workplace characteristics for design or redesign. Such trade-offs between the biomechanical factors, personal factors, and work organizational factors, including work stress, coping strategies, and organizational practices, require one to balance various ergonomic needs to achieve the solution that will have the greatest benefits for employee health and productivity.

The balance theory–based model proposed by Smith and Sainfort (1989) takes a system's approach by focusing on the interactions between the worker, including physical characteristics, perceptions, personality and work behavior, the physical and social environments, and the organizational structure that defines the nature and level of worker involvement, interaction, control and supervision. The capabilities of technologies available to the worker to perform a specific job affect task performance and the worker's skills and knowledge needed for their effective use. Task requirements affect the required skills and knowledge of the worker. Both the tasks and technologies affect the content of the job and the physical demands. The balance theory–based model can be used to establish relationships between interacting elements such as job demands, job design factors, and ergonomic loads. Demands that are placed on the worker create loads that can be healthy or harmful. Harmful loads may lead to physical and psychological stress responses that can produce adverse health effects such as WUEDs. It should be noted that a number of personal considerations also contribute to the physical and psychological effects. These include the strength and health of the worker, previous musculoskeletal or nerve injury, personality, perceptual-motor skills and abilities, physical conditioning, prior experience and learning, motives, goals, and needs and intelligence.

35.9.2 WUEDs: Ergonomics Guidelines

The expected benefits of reduced WUEDs in industry are improved productivity and quality of work products, enhanced safety and health of the employees, higher employee morale, and accommodation of people with alternative physical abilities. Strategies for managing the WUEDs at work should focus on prevention efforts and should include, at the plant level, employee education, ergonomic job redesign, and other early intervention efforts, including engineering design technologies such as workplace reengineering and active and passive surveillance. At the macro level, management of the WUEDs should aim to provide adequate occupational health care provisions, legislation, and industry-wide standardization.

As widely recognized in Europe (Wilson, 1994), ergonomics has to be seen as a vital component of the value adding activities of the company. Even in strictly financial terms, the costs of an ergonomics management program will far outweigh the costs of not having one. A company must be prepared to accept a participative culture and to utilize participative techniques. The ergonomics-related problems and consequent intervention should go beyond engineering solutions, and must include design for manufacturability, total quality management, work organization, alongside the workplace redesign or worker training. Only then will the promise of ergonomics in managing the WUEDs at work be fulfilled.

In the absence of general applicable guidelines and criteria on minimizing and/or optimizing risk factors, two complementary approaches have merit for the prevention of WUEDs, i.e., (1) general guidelines that describe in general terms the principles and policies to be adopted in preventing WUEDs, and (2) specific guidelines that aim at the design and redesign of work and tasks that are known in detail (Kuorinka and Forcier,

1995). Since the specific guidelines draws on both scientific knowledge and the collective industrial experience, they may be much more detailed and often contain quantitative data.

Most of the current guidelines for control of the biomechanical risk factors for WUEDs at work aim to: (1) reduce the exposure to highly repetitive and stereotyped movements, (2) reduce excessive force levels, and (3) reduce the extent of movements of the joints. For example, in order to control for the extent of force required to perform a task one should (1) reduce the force required through tool and fixture redesign, (2) distribute the application of force the force, or (3) increase the mechanical advantage of the (muscle) lever system. Because of neurophysiological needs of the working muscles, adequate rest pauses (determined based on the scientific knowledge on the physiology of muscular fatigue and recovery), should be scheduled to provide relief for the most active muscles used on the job. Furthermore, reduction in task repetition can be achieved, for example, by (1) task enlargement (increasing variety of tasks to perform), (2) increase in the job cycle time, and (3) work mechanization and automation.

However, recent findings point out that the stereotyped thinking abut the risk factors for WUEDs may impair one's ability to develop effective redesign solutions. For example, the wrist posture is recognized as the risk factor for WUEDs in jobs that require primarily static exertions, such as holding tools and using a computer keyboard. It has recently been shown that in the dynamic tasks involving upper extremity, the posture of the hand itself has very little predictive power for the risk of WUEDs. Rather, it is the velocity and acceleration of the joint that significantly differentiate the WUEDs risk levels (Schoenmarklin and Maras, 1991). This is because the tendon force, which is a risk factor of WUEDs, is affected by wrist acceleration. The acceleration of the wrist in a dynamic task requires transmission of the forearm forces to the tendons. Some of this force is lost to friction against the ligaments and bones in the carpal tunnel. This frictional force can irritate the tendons' synovial membranes and cause tenosynovitis or carpal tunnel syndrome (CTS). These new research results clearly demonstrate the importance of dynamic components in assessing WUEDs risk of highly repetitive jobs.

Finally, it should be noted that many of the recommendations offered by ergonomics may be difficult to implement in practice without full understanding of the production processes, plant layouts, or quality requirements, and total commitment from all management levels and workers of the company. This is because many of the guidelines are not specific, and define what to avoid—for example, (1) Avoid high contact forces and static loading, (2) Avoid extreme or awkward joint positions, (3) Avoid repetitive finger action, and (4) Avoid tool vibration—but do not define how to avoid these risk factors. In view of the above, involvement of professional ergonomists (i.e., those who are certified by the Board of Certification in Professional Ergonomics), along with engineering personnel and production workers in a truly participative manner, is critical to the success of the ergonomic intervention efforts. Furthermore, ergonomics must be treated at the same level of attention and significance as other business functions of the plant, for example, quality management control, and be accepted as cost of doing business, rather than add-on activity caring for action only when the problems arise.

REFERENCES

ACOM (1990). Committee of Occupational Medical Practice: Preplacement/preemployment physical examinations. *Journal of Occupational Medicine, 32,* 295–9.

ANSI. (1995). *Control of work-related cumulative trauma disorders, Part I: Upper extremities,* April 17, 1995, (Z-365 Draft). Chicago: National Safety Council.

Alexander, D. C., and Orr, G. B. (1992). The evaluation of occupational ergonomics programs. In *Proceedings of the Human Factors Society 36th Annual Meeting,* pp. 697–701. Human Factors Society, Santa Monica, CA.

Ambrose, R. F., Kendall, L. G., Alarcon, G. S., Brown, S., Lipstate, J. M., Wirtschafter, D. D., Jackson, J. R., Glass, S., Rossi, K., and Margolis, C. Z. (1990). Rheumatology algorithms for primary care physicians. *Arthritis Care Research, 3,* 71–77.

American Medical Association. (1972). *Scope, objectives, and functions of occupational health programs.* Chicago: American Medical Association.

Armstrong T. (1986). Ergonomics and cumulative trauma disorders. *Hand Clinics, 2,* 553–565.

Armstrong, T. J., Buckle, P., Fine, L. J., Hagberg, M., Jonsson, B., Kilbom, A., Kuorinka, I., Silverstein, B. A., Sjogaard, G., and Viikari-Juntura, E. (1993). A conceptual model for work-related neck and upper-limb musculoskeletal disorders. *Scandinavian Journal of Work and Environmental Health, 19,* 73–84.

Armstrong, T. J., and Lifshitz, Y. (1987). Evaluation and design of jobs for control of cumulative trauma disorders. In *Ergonomic Interventions to Prevent Musculoskeletal Injuries in Industry* (pp. 73–85). Chelsea, UK: Lewis Publishers.

Armstrong, T. J., Radwin, R. G., Hansen, D. J., and Kennedy, K. W. (1986). Repetitive trauma disorders: Job evaluation and design. *Human Factors, 28,* 325–336.

BLS. (1995). *Occupational Injuries and Illness in the United States by Industry.* U.S. Department of Labor, Washington, D.C.

Bonney, R., Weisman, G., Haugh, L. D., and Finkelstein, J. (1990). Assessment of postural discomfort. *Proceedings of the Human Factors Society Annual Meeting,* pp. 684–687. Human Factors Society, Santa Monica, CA.

Caplan, R. D., Cobb, S., French, J. R. P. Jr., Harrison, R. V., and Pinneau, S. R. (1975). *Job Demands and Worker Health.* (NIOSH Pub No. 75–169). U.S. Department of Health, Education and Welfare, Washington, D.C.

Centers for Disease Control (1989). Occupational disease surveillance: carpal tunnel syndrome. *Journal of American Medical Association, 26*(2), 889.

Coburn, D. (1979). Job alienation and well-being. *International Journal of Health Services, 9(1),* 41–59.

Colombini, D., Occhipinti, E., Molteni, G., Grieco, A., Pedotti, A., Boccardi, S., Frigo, C., and Menoni, D. (1985). Posture analysis. *Ergonomics, 28(1),* 275–284.

Commission on Accreditation of Rehabilitation Facilities, 1989, *Standards Manual for Organizations Serving People with Disabilities* (report). Commission on Accreditation Organizations Rehabilitation Facilities, Tucson, AZ.

Corlett, E. N., Madeley, S. J., and Manenical, I. (1979). Posture targeting: A technique for recording working postures. *Ergonomics, 22(3),* 357–66.

Cox, T. (1985). Repetitive work: Occupational stress and health. In Cooper, C. L. and Smith, M. J., Eds., *Job Stress and Blue Collar Work.* (pp. 85–112). New York: John Wiley.

Drury, C. (1987). A biomechanical evaluation of the repetitive motion injury potential of industrial jobs. *Seminars in Occupational Medicine, 2(1),* 41–49.

Edwards, R. H. T. (1988). Hypothesis of peripheral and central mechanisms underlying occupational muscle pain and injury. *European Journal of Applied Physiology, 57,* 275–182.

Ferguson, S. A., Marras, W. S., and Waters, T. R (1992). Quantification of back motion during asymmetric lifting. *Ergonomics, 40(7/8),* 845–859.

Frankenhaeuser, M., and Gardell, B. (1976, September). Underload and overload in working life: Outline of a multidisciplinary approach, *Journal of Human Stress, 2,* 35–46.

Gruber, G. J. (1974). *Relationship Between Whole-Body Vibration and Morbidity Patterns Among Interstate Truck Drivers.* (Publ. No. 75–104). National Institute for Occupational Safety and Health, NIOSH. Cincinnati, OH.

Haber, L. D. (1971). Disabling effects of chronic disease and impairment. *Journal of Chronic Disease, 24,* 269–487.

Hackman, J. R., Oldham, G. R., Janson, R., and Purdy, K. (1975). A new strategy for job enrichment. *California Management Review, 17(4),* 57–71.

Hagberg, M. (1992). Exposure variables in ergonomic epidemiology. *American Journal of Industrial Medicine, 21,* 91–100.

Hägg, G. (1991). Static work loads and occupational myalgia—a new explanation model. In P. Anderson, D. Hobart, and J. Danoff, Eds., *Electromyographical Kinesiology.* (pp. 141–144). New York: Elsevier Science Publishers BV.

Holzmann, P. (1982). ARBAN—a new method for analysis of ergonomic effort. *Applied Ergonomics, 13,*82–86.

ISO. (1986). *Mechanical Vibration: Guidelines for the Measurement and the Assessment of Human Exposure to Hand-transmitted Vibration* (Reference no. ISO 5349:1986). (International Organisation for Standardization, Geneva).

Johansson, G., and Aronsson, G. (1984). Stress reactions in computerized administrative work. *Journal of Occupational Behavior, 5,* 159–181.

Kalimo, R., and Leppänen, A. (1985). Feedback from video display terminals, performance control and stress in text preparation in the printing industry. *Journal Occupational Psychology, 58,* 27–38.

Karhu, O., Kansi, P., and Kuorinka, I. (1977). Correcting working postures in industry: A practical method for analysis. *Applied Ergonomics, 8,* 199–201.

Karwowski, W. (1993). A catastrophe theory based model of the cumulative trauma disorders definition of space and control variables. In, R. Nielsen and K. Jørgensen, Eds., *Advances in Industrial Ergonomics and Safety V* (pp. 79–86). Taylor and Francis: London.

Karwowski, W., Eberts, R., Salvendy, G., and Noland S. (1994). The effects of computer interface design on human postural dynamics. *Ergonomics, 37(4),* 703–724.

Karwowski, W., and Kasdan, M. L. (1988). The partnership of ergonomics and medical intervention in rehabilitation of workers with cumulative trauma disorders of the hand. In *Ergonomics in Rehabilitation* (pp. 35–53). A. Mital and W. Karwowski, Eds., London: Taylor and Francis.

Kelsey, J. L., and Golden, A. L. (1988). Occupational and workplace factors associated with low back pain. *Occupational Medicine: State of the Art Reviews, 3(1),* 7–16.

Keyserling, W. M., Stetson, D. S., Silverstein, B. A., and Brouver, M. L. (1993). A checklist for evaluating risk factors associated with upper extremity cumulative trauma disorders. *Ergonomics, 36(7),* 807–831.

Kiesler, S., and Finholt, T. (1988). The mystery of RSI (repetitive strain injuries). *American Psychologist, 43,* 1004–1015.

Kilbom, A., and Persson, J. (1987). Work technique and its consequences for musculoskeletal disorders. *Ergonomics, 30,* 273–279.

Klaucke, D. N., Klaucke, D. N., Buehler, J. W., Thacker, S. B., Parrish, R. G., Trowbridge, R. L., and Berkelman, R. L. (1988). Guidelines for evaluating surveillance systems, *MMWR, 37* (Suppl. 5), 1–18.

Kroemer, K., Kroemer, H., Kroemer-Elbert, K. (1994). *Ergonomics: How to Design for Ease & Efficiency.* Englewood Cliffs, NJ: Prentice Hall.

Kuorinka, I., and Forcier, L., Eds. (1992). *Work Related Musculoskeletal Disorders (WMSDs): A Reference Book for Prevention.* London: Taylor and Francis.

Last, J. M. (1988). *A Dictionary of Epidemiology (Second Edition).* New York: Oxford University Press.

Leino, P. (1989). Symptoms of stress production and musculoskeletal disorders. *Journal of Epidemiology Community Health, 43,* 293–300.

Levi, L. (1972). *Stress and Distress in Response to Psychosocial Stimuli* New York: Pergamon Press.

Lundborg, G., Dahlin, L. B., Danielsen, N., and Kanje, M. (1990). Vibration exposure and nerve fibre damage. *Journal of Hand Surgery, 15A,* 346–51.

Luopajarvi, T., Kuorinka, I., Virolainen, M., and Holmberg, M. (1979). Prevalence of tenosynovitis and other injuries of the upper extremities in repetitive work. *Scandinavian Journal of Work Environment and Health, 5*(Suppl. 3), 48–55.

Margolis, B. L., Kroes, W. M., and Quinn, R. P. (1974). Job stress: An unlisted occupational hazard, *Journal of Occupational Medicine, 16(10),* 659–661.

Marras, W. S. (1997). Biomechanics of the human body. In *Handbook of Human Factors and Ergonomics.* 2nd Edition. G. Salvendy, Ed., New York: John Wiley.

Marras, W. S., and Schoenmarklin, R. W. (1993). Wrist motions in industry. *Ergonomics, 36(4),* 341–351.

Milner, N. P., Corlett, E. N., and O'Brien, C. (1986). A model to predict recovery from maximal and submaximal isometric exercise. In N. Corlett, J. Wilson, and I. Manenica, Eds. *Ergonomics of Working Postures.* London: Taylor and Francis.

Moore, A., Wells, R., and Ranney, D. (1991). Quantifying exposure in occupational manual tasks with cumulative trauma disorder potential. *Ergonomics, 34(12),* 1433–1453.

Moore, J. S. (1992). Carpal tunnel syndrome. *Occupational Medicine: State of the Art Reviews, 7(4),* 741–763.

Moore, J. S., and Garg, A, (1995, May). The strain index: A proposed method to analyze jobs for risk of distal upper extremity disorders. *American Industrial Hygiene Association Journal, 56,* 443–458.

NIOSH (1977). *National Occupational Hazard Survey, 1972–1974* (DHEW (NIOSH), Publ. No. 78–114). Washington, DC: National Institute for Occupational Safety and Health.

NIOSH (1986). *Proposed National Strategies for the Prevention of Leading Work-Related Diseases and Injuries, Part I* (DHEW (NIOSH), PB87-114740) Washington, DC: National Institute for Occupational Safety and Health.

OSHA (1995). *Draft Proposed Ergonomic Protection Standard.* OSHA Publications Office, Washington, D.C.

Östberg, O., and Nilsson, C. (1985). Emerging technology and stress. In *Job Stress and Blue Collar Work,* C. L. Cooper and M. J. Smith, Eds. (pp. 149–69). New York: John Wiley.

Pearcy, M. J., Gill, J. M., Hindle, J., and Johnson, G. R. (1987). Measurement of human back movements in three dimensions by opto-electronic devices. *Clinical Biomechanics 2,* 199–204.

Praemer, A., Furner, S. and Rice, D. P. (1992). *Musculoskeletal Conditions in the United States.* American Academy of Orthopaedic Surgeons, Park Ridge, IL:

Putz-Anderson, V., Ed. (1993). *Cumulative Trauma Disorders. A Manual for Musculoskeletal Diseases for the Upper Limbs.* London: Taylor and Francis.

Ranney, D., Wells, R., and Moore, A. (1995). Upper limb musculoskeletal disorders in highly repetitive industries: precise anatomical physical findings. *Ergonomics, 38(7),* 1408–1423.

Schoenmarklin, R. W., and Marras, W. S. (1991). Quantification of wrist motion and cumulative trauma disorders in industry, In *Proceedings of the Human Factors Society 35th Annual Meeting* Human Factors and Ergonomics Society, Santa Monica, CA pp. 838–842.

Schoenmarklin, R. W., Marras, W. S., and Leurgans, S. E. (1994). Industrial wrist motions and incidence of hand/wrist cumulative trauma disorders. *Ergonomics, 37*(*9*), 1449–1459.

Silverstein, B. A. (1991). Developing shop-floor ergonomics expertise using a train-the-trainer approach. In *Health work environments, health people: Participatory approaches to improving workplace health* (International Conference, June 3–5, Ann Arbor, University of Michigan, p. 74). (University of Michigan, Ann Arbor):

Silverstein, B. A., Fine, L. J., and Armstrong, T. J. (1986a). Hand, wrist cumulative trauma disorders in industry. *British Journal of Industrial Medicine, 43,* 779–784.

Silverstein, B. A., Fine, L. J., and Armstrong, T. J. (1986b). Carpal tunnel syndrome: Causes and a preventive strategy. *Seminars in Occupational Medicine, 1(3),* 213–221.

Silverstein, B. A., Fine, L. J., and Armstrong, T. J. (1987). Occupational factors and carpal tunnel syndrome. *American Journal of Industrial Medicine, 11,* 343–358.

Smith, M. J., Carayon, P., and Miezio, K. (1987). VDT technology: Psychosocial and stress concerns. In B. Knave and P.-G. Wideback (Eds.), *Work with display units 86: Selected papers from the international scientific conference on work with display units* (Stockholm, Sweden, May 12–15, 1986, pp. 695–712). Amsterdam: North-Holland.

Smith, M. J., and Sainfort, P. C. (1989). A balance theory of job design for stress reduction, *International Journal of Industrial Ergonomics, 4,* 67–79.

Snijders, C. J., van Riel, M. P. J. V., and Nordin, M. (1987). Continuous measurements of spine movements in normal working situations over periods of 8 hours or more. *Ergonomics, 30(4),* 639–653.

Snook, S. H., Vaillancourt, D. R., Ciriello, V. M., and Webster, B. S. (1995). Psychophysical studies of repetitive wrist flexion and extension. *Ergonomics, 38(7),* 1488–1507.

Spitzer, W. O., LeBlanc, F. E., Dupuis, M., Abenhaim, L., Blanger, A. Y., Bloch, R., Bombardier, C., Cruess, R. L., Drouin, G., Duval-Hesler, N., Laflamme, J., Lamureux, G., Nachemson, A., Page, J. J., Rossignol, M., Salmi, L. R., Salois-Aresenult, S., Suissa, S., and Wood-Dauphinee, S. (1987). Scientific approach to the assessment and management of activity-related spinal disorders: A monograph for clinicians: report of the Quebec Task Force on Spinal Disorders, *Spine, 12(7S),* 51–9.

St. Clair, S., and Shults, T. (1992). Americans with disabilities act: Considerations for the practice of occupational medicine. *Journal of Occupational Medicine, 34(5),* 510–517.

Stock, S. R. (1991). Workplace ergonomic factors and the development of musculoskeletal disorders of the neck and upper limbs: A meta-analysis. *American Journal of Industrial Medicine, 19,* 87–107.

Tanaka, S., and McGlothlin, J. D. (1993). A conceptual quantitative model for prevention of work-related carpal tunnel syndrome (CTS). *International Journal of Industrial Ergonomics, 11,* 181–193.

Taylor, W., and Pelmear, P. L. (1976). Raynaud's phenomenon of occupational origin: an epidemiological survey. *Acta Chiropractica Scandinavica* (Suppl. 465), 27–32.

Uhthoff, H. K., and Sarkar, K. (1990). An algorithm for shoulder pain caused by soft-tissue disorders, *Clinical Orthopaedics, 254,* 121–127.

Vartiainen, M. (1987). *The Hierarchical Development of Mental Regulation, and Training Methods.* Rep. No. 100. Helsinki University of Technology, Industrial Economics and Industrial Psychology, TKK Offset. Ataniemi.

Wangenheim, M., and Samuelson, B. (1987). Automatic ergonomic work analysis. *Applied Ergonomics, 18,* 9–15.

Wasserman, D. E., Badger, D. W., Doyle, T. E., and Margolies, L. (1974). Industrial vibration—An overview. *ASSE Journal, 19,* 38–42.

Wells, R. (1990). Unpublished workshop materials. University of Waterloo, Canada.

Westgaard, R. H., and Bjørkland, R. (1987). Generation of muscle tension additional to postural muscle load. *Ergonomics, 30(6),* 196–203.

Wilson, J. R. (1994). Devolving ergonomics: the key to ergonomics management programmes. *Ergonomics, 37(4),* 579–594.

WHO (1985). *Identification and control of work-related diseases.* Tech. Rep. No. 174, pp. 7–11. Geneva: WHO.

CHAPTER 36
WARNINGS AND RISK PERCEPTION

Kenneth R. Laughery, Sr.
Psychology Department
Rice University
Houston, TX 77251 USA

Michael S. Wogalter
Psychology Department
North Carolina State University
Raleigh, NC 27695 USA

36.1 INTRODUCTION

During the past several decades there has been an increasing concern for public safety in the United States. This concern has been manifested in many ways. One such manifestation is the much greater use of safety communications, warnings, to inform people of hazards and to provide instructions as to how to deal with them so as to avoid or minimize undesirable consequences. Warnings are used to address environmental hazards as well as hazards associated with the use of products.

In addition to the increase in general concern for safety, there is another factor that has influenced the greater use of warnings, namely, litigation. The need for and adequacy of warnings has been an increasingly prevalent issue in product liability and personal injury litigation.

As might be expected, the greater attention to and deployment of warnings has been accompanied by regulations, standards, and guidelines as to when and how to warn. Also, there has been a substantial increase in research activity on the topic. Human factors specialists, or ergonomists, have played a major role in this research and the technical literature that has resulted.

A topic that is closely associated with warnings is risk perception; that is, people's knowledge and/or understanding of hazards and their consequences. Risk perception is closely related to warnings, since when and how to warn is obviously a function of the knowledge people have about hazards and the factors that influence this knowledge.

The purpose of this chapter is to review the important principles and facts that have evolved on the topic of warnings and to discuss criteria and procedures for developing and testing warnings.

36.2 BACKGROUND

In this section several terms are defined and the role of warnings in the broader context of hazard control is discussed.

36.2.1 Definitions

It is important to establish a few definitions for terms that will be used in this chapter, particularly the concepts of hazard, danger, and risk perception. These terms are sometimes used in different ways with different meanings; hence, we want to be clear as to their meaning in this context.

Hazard is defined as a set of circumstances that can result in injury, illness, or property damage. Such circumstances may include characteristics of the environment, of equipment, and of a task someone is performing. From a human factors perspective, it is important to note that circumstances also includes characteristics of the people involved. These people characteristics encompass abilities, limitations, and knowledge.

Danger is a term that is used in a variety of ways. In this chapter it is viewed as the product of hazard and likelihood; that is, if one has quantified values of hazard and likelihood, multiplying these quantities would give a value for danger. Note, that an implication of this definition is that if either value is zero, there is no danger. If the hazard and its consequence is serious but will not occur, there is no danger. Similarly, if the probability of an event occurring is high, but there will be no resulting undesirable consequences, there is no danger.

"Risk" is a term that has had many definitions in a variety of contexts. *Risk perception* encompasses a broad notion of safety awareness. It concerns the overall awareness and knowledge regarding the hazards, likelihoods, and potential outcomes of a situation or set of circumstances.

36.2.2 Hierarchy of Hazard Control

In the field of safety there is a concept of hazard control that includes the notion of a hierarchy or priority scheme (Sanders and McCormick, 1993). This hierarchy defines a sequence of approaches to dealing with hazards in order of preference. The sequence is (1) design it out, (2) guard, and (3) warn. The notion of a design solution is that the first preference is to eliminate the hazard through alternative designs. If a nonflammable solvent can be used for some cleaning task, such a solution is preferable to wearing protective equipment or warning about the flammable hazard being near an ignition source. Of course, often it is not possible to eliminate hazards. Guarding, physical or procedural, is a second line of defense and has as its purpose preventing contact between people and the hazard. Barriers and protective equipment are examples of physical barriers, whereas designing tasks in such a way to keep people out of a hazard zone is an example of a procedural guard. However, like alternative designs, guarding is not always a feasible solution, and the third line of defense is warning. Warnings are third in the priority sequence because influencing behavior is sometimes difficult, and seldom foolproof. There is another implication of this priority scheme; namely, warnings are not a substitute for good design or adequate guarding. Indeed, warnings are properly viewed as a supplement, not a substitute, to other approaches to safety (Lehto and Salvendy, 1995).

In addition to the above three-part hierarchy, there are other steps or approaches that may be effective in dealing with hazards. Generally, they fall into the same category as warnings in that they are means of influencing the behavior of people. Training and personnel selection are examples. Another approach that includes elements similar to procedural guarding and warnings is supervisory control. These three approaches are particularly applicable to hazards in the context of job performance.

36.3 RISK PERCEPTION

This chapter does not provide a review of research and theory on risk perception. For a review of this topic see Fischhoff (1989) and Slovic, Fischoff, and Lichtenstein (1982). Our approach here is to note how risk perception considerations enter into decisions regarding the design, implementation, and effectiveness of warnings.

36.3.1 System Context

As noted earlier, an important factor in the hazards associated with any situation or product is the perception or knowledge of the people involved. Later in this chapter we discuss the purposes of warnings, but generally the goal is to influence behavior by providing information. Obviously, the information that people have from past experiences or that they glean from the existing situation or circumstances is relevant to the issue of what needs to be warned. Thus, an understanding of risk perception is important in decisions about when, where, what, and how to warn.

36.3.2 Awareness and Knowledge

The distinction between awareness and knowledge is important in understanding issues of risk perceptions and how they map on to warnings design and effectiveness. The difference is analogous to a distinction made in cognitive psychology between short-term memory (sometimes thought of as what is currently in consciousness) and long-term memory (one's permanent knowledge of the world). The point here is simply that people may have information or experience in their overall knowledge base that at a given time is not part of what they are thinking about—awareness. In the context of safety or coping with hazards, it is not enough to say that people know something. Rather, it is important that people be aware of (thinking about) the relevant information at the critical time. This distinction has important implications for the role of warnings as reminders and is further addressed later in this chapter.

There are many ways in which people can become aware and knowledgeable about hazards, consequences, and appropriate procedures or behaviors. Warnings, training, and direct supervisor inputs are among them, and it is the first of these that this chapter addresses. There are others. Experience, of course, is one way that people may acquire such safety knowledge. "Learning the hard way" by having experienced an incident or knowing about someone else who has had such experiences can certainly result in such knowledge. Such experiences, on the other hand, do not necessarily lead to *accurate* knowledge of hazards and consequences, because they may result in overestimating the degree of danger associated with some situation or product. Similarly, the lack of such experiences may lead to underestimating such dangers, or not thinking about them at all. Nevertheless, experience clearly plays an important role in risk perception.

Another source of information about dangers is the situation or product itself. In the law there is a concept of "open and obvious." The point here is that the appearance of a situation or product or the manner in which it functions may communicate the nature of the safety problem. Moving mechanical parts such as chain-driven sprockets may be an example of an open and obvious pinch point hazard. Even more obvious may be the hazard and consequence of a fall from a height in a construction setting. Of course, many safety problems are probably not open and obvious, such as some specific chemical hazards and consequences associated with solvents.

A final point to be noted regarding risk perception concerns the problem of overestimating what people know or are aware of. To the extent it is incorrectly assumed that people have information and knowledge, there may be a tendency to provide inadequate warnings. Thus, it is an important part of job, environment, and product design to take into account people's understanding and knowledge of hazards and their consequences. A further analysis and discussion of this issue can be found in a paper by Laughery (1993).

36.4 WARNINGS

In this section we discuss the purpose(s) of warnings, warnings as communications, and the concept of a warning system. Then, following a discussion of some general criteria for warnings, eight criteria for warnings design are presented and discussed.

36.4.1 Purpose of Warnings

The purpose of warnings can be stated at several levels. Most generally, warnings are intended to improve safety, that is, to decrease accidents or incidents that result in injury, illness, or property damage. At another level, warnings are intended to influence or modify people's behavior in ways that improve safety. At still another level warnings are intended to provide information that enables people to understand hazards, consequences and appropriate behaviors, which in turn enables them to make informed decisions. This latter point places warnings squarely in the category of a communication, which, of course, they are.

There are two additional points to be noted regarding the purpose of warnings, both of which are related to warnings as communications. First, warnings are a means of shifting or assigning responsibility for safety to people in the system, the product user, the worker, etc., in situations where hazards cannot be designed out or adequately guarded. This point is not to say that people do not have safety responsibilities independent of warnings; of course they do. Rather, a purpose of warnings is to provide the information necessary to enable them to carry out such responsibilities. The second point regarding the communication purpose concerns an issue that has received little attention in the technical literature; namely, people's right to know. The notion is that even in situations where the likelihood of warnings being effective may not be high, people have the right to be informed about safety problems confronting them. Obviously this aspect of warnings is more of a personal, societal and legal concern than a human factors issue, and although it is not addressed further in this chapter, it is a matter that is related to the overall purposes of warnings.

36.4.2 The Communication Model

As noted above, warnings can properly be viewed as communications. In this context it is useful to note the typical communications model or theory, because it has implications for the design and implementation of warnings. A typical and basic model is shown in Figure 36.1.

The model includes a sender, a receiver, a channel or medium through which a message is transmitted, and the message. The receiver is the user of the product, the worker, or any other person to whom the safety information must be communicated. The message, of course, is the safety information to be communicated. The medium refers to the channels or routes through which information gets there. Understanding and improving these components of a safety communication system increases the probability that the message will be successfully conveyed.

However, the communication of safety information often is not so simple as Figure 36.1 might imply. Frequently more than one medium or channel may be available and/or involved, multiple messages in different formats and/or containing different information may be called for, and the receiver or target audience may include different subgroups with varying characteristics. An example of such a warning situation would occur when a product with associated hazards is being used in a work environment. Figure 36.2 illustrates a communication model that might be applicable.

This figure reflects a much more complex situation than Figure 36.1. In addition to the sender (manufacturer) and receiver (end user), other people or entities may be involved such as distributors and employers. Further, each of these entities may be both receivers

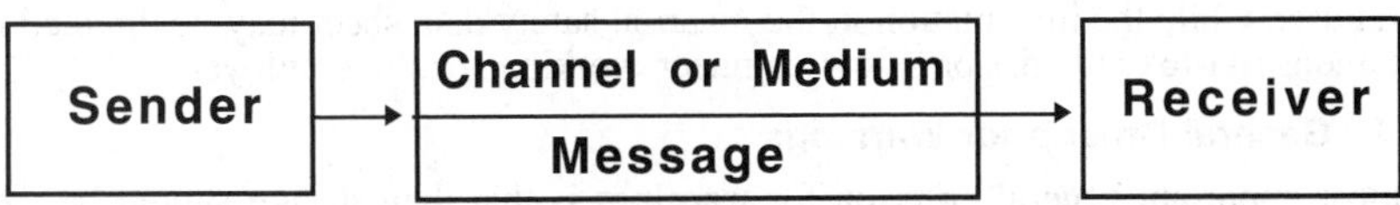

Figure 36.1 Simple communication model.

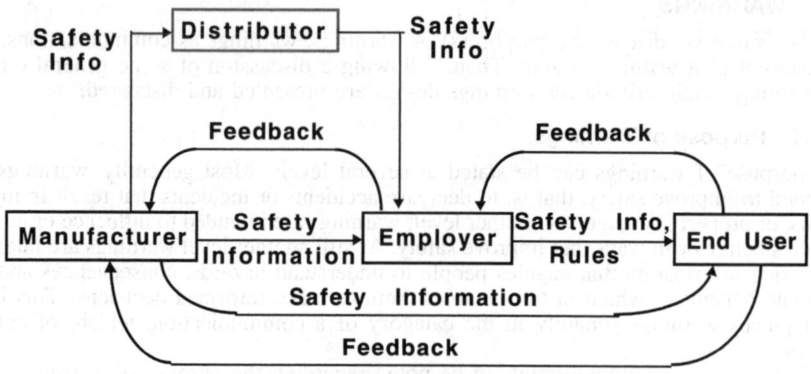

Figure 36.2 Complex communication model.

and senders of safety information. There are also more routes through which warnings may travel such as from the manufacturer to the distributor to the employer to the user, from the manufacturer to the employer to the user, or directly from the manufacturer to the user (as on a product label). The warnings may take different forms such as communications of information or rules that an employer sets to govern the behavior of employees. Thus, there are circumstances in which the warning or warning system is much more complex than a sign or label for transmitting safety information from a sender to a receiver.

36.4.3 Concept of a Warning System

The notion of warnings being a sign or a portion of a label is much too narrow a view of how such safety information gets transmitted. The concept of a warning system is that a warning communication for a particular setting or product may consist of a number of components. In the context of the communication model presented in Figure 36.2, these components may include a variety of media and messages. An example or two can help make the point.

A warning system for a product off the drug store shelf, such as NyQuil, may consist of several components: a printed statement on the box, a printed statement on the bottle, and a printed package insert. In addition, it may include verbal warnings in television advertisements about the product. A warning system for tires and rims that may be mismatched with a resulting potential explosion might consist of a number of components. Examples are: warnings in raised lettering on the sidewall of the tire, a temporary tread label on new tires, stickers or stamping on the rim, statements on wall posters in places where tires are mounted, statements in tire and rim product catalogs and manuals, statements in handouts that accompany sales of tires and rims, verbal statements by employers of people who mount tires, etc. Another example would be warnings for a solvent used in a work environment for cleaning parts. Here the components might include printed on-product labels, printed flyers that accompany the product, statements in advertisements about the product, verbal statements from the salesperson to the purchasing agent, and material safety data sheets provided to the employer.

An important point regarding warning systems is that the components may not be identical in terms of content or in terms of purpose. For example, some components may be intended to capture attention and direct the person to another component where more information is presented. Similarly, different components may be intended for different target audiences. In the example of the solvent given above, the label on the product container may be intended for everyone associated with the use of the product including the end user, while the information in the material safety data sheet may be directed more to the industrial toxicologist or safety engineer working for the employer.

36.4.4 General Criteria for Warnings

The most important general criterion for warnings is that their design should be viewed as an integral part of the overall system design process. Frantz, Rhoades, and Lehto (in press) address this issue in their excellent paper on how to go about developing product warnings. Whereas in the field of safety warnings are a third line of defense behind design and guarding, they should not be considered for the first time after the design (including guards) of the environment or product is fixed. Too many warnings are developed at this stage of design, the afterthought phenomenon, and their quality and effectiveness often reflect it. Further, warnings based on unrealistic and untested assumptions or expectations about the target audience are destined to be ineffective, and in this sense they are no substitute for good design.

In this section three general criteria for warnings will be presented: (1) when and what to warn; (2) how to prioritize warnings, and (3) whom to warn.

36.4.4.1 When and What to Warn?

There are several principles or rules that guide when a warning should be used. They include:

1. A significant hazard exists.
2. The hazard, consequences, and appropriate safe modes of behavior are not known by the people exposed to the hazard.
3. The hazards are not open and obvious; that is, the appearance and function of the environment or product do not communicate them.
4. A reminder is needed to assure awareness of the hazard at the proper time. This concern is especially important in situations of high task loading or potential distractions.

36.4.4.2 Prioritizing Warnings

In a later section we address criteria for designing a warning. Here the concern is what hazards to warn about when multiple hazards exist. How are priorities defined in deciding what to include or delete, how to sequence them, or how much relative emphasis to give them? To some extent the criteria overlap the above rules about when and what to warn. Certainly when the hazard is already known and understood or when it is open and obvious warnings may not be needed. Other considerations include:

1. *Likelihood:* The more likely an undesirable event is to occur, the greater the priority that it should be warned.
2. *Severity:* The more severe the potential consequences of a hazard, the greater priority that it should be warned. If a chemical product poses a skin contact hazard, a higher priority would be given to a severe chemical burn consequence than if it were a minor rash.
3. *Practicality:* There are occasions when limited space (a small label) or limited time (a television commercial) does not permit all hazards to be addressed in a single component of the warning system. As a general rule, unknown hazards leading to more severe consequences and/or those more likely to occur would have priority for the primary warning component, such as on the product label, whereas those hazards with lower priority would be addressed in other warning components, such as package inserts or manuals.

36.4.4.3 Whom to Warn

The general principle regarding who should be warned is that it should include everyone who may be exposed to the hazard and everyone who may be able to do something about it. There are occasions when people in the latter category may not themselves be exposed to the hazard. An example would be the industrial toxicologist who receives warning information regarding a product to be used by employees and defines job procedures and/or protective equipment to be employed in handling the material. The physician who prescribes medications that have contraindication and side effect hazards is another example.

There are, of course, situations and products where the target audience is the general public, that is, everyone. Hazards in the public environment or many products on the shelf of a drugstore or hardware store are examples. Other warnings may be directed to a very specific audience. Warnings about toxic shock syndrome in the use of tampons would be directed primarily to women of child bearing age. Warnings about contraindications associated with prescriptive medications, as noted above, may be directed primarily to physicians. If warnings are to be effective, it is imperative that the characteristics of the target audience be taken into account.

Clearly target audiences, the receivers of warnings, may differ. Laughery and Brelsford (1991) discussed several dimensions along which intended receivers may differ.

Demographic Factors

A number of studies have shown that gender and age may be factors in how people respond to warnings. With regard to gender, results indicate a tendency for women to be more likely than men to look for and read warnings (Godfrey, Allender, Laughery, and Smith 1983; LaRue and Cohen, 1987; Young, Martin, and Wogalter, 1989). Similarly, there are research results that show women are more likely to comply with warnings (Desaulniers, 1991; Goldhaber and deTurck, 1988; Viscussi , Magat, and Haber, 1986;). These findings may have implications where hazards associated with products or environments are more likely to be encountered by one of the sexes. If one is attempting to influence the safety behavior of men, the task may be more difficult.

Age has also been examined as a receiver variable in some research on warnings. Although results are mixed, there is a trend that people older than 40 are more likely to take precautions in response to warnings (Desaulniers, 1991). On the other hand, some research (Collins and Lerner, 1982; Easterby and Hakiel, 1981; Ringseis and Caird, 1995) has shown that older subjects have lower levels of comprehension for safety signs involving pictorials. Results such as these suggest that older people may be more influenced by warnings, but greater attention to issues of comprehension may be necessary.

Familiarity and Experience

Numerous studies have explored the effects of people's familiarity and experience with a product on how they respond to warnings associated with the product. Results indicate that the more familiar people are with a product the less likely they are to look for, notice or read a warning (Godfrey et al., 1983; Godfrey and Laughery, 1984; LaRue and Cohen, 1987; Otsubo, 1988; Wogalter, Allison, and McKenna, 1991). Some research has also examined the effects of familiarity on compliance (Goldhaber and deTurck, 1988; Otsubo, 1988). The results have shown that greater familiarity is associated with a lower likelihood to comply with warnings. Clearly, products that are used repetitively or used in highly familiar environments pose special warning challenges.

Competence

There are many dimensions of receiver competence that may be relevant to the design of warnings. For example, sensory deficits might be a factor in the ability of some special target audiences to be directly influenced by a warning. The blind person would not be able to receive a written warning, nor would the deaf receive an auditory warning. Further, what would be open and obvious to the normal person may not be obvious to the blind person. Opposite the sensory end of the sequence of events associated with warning effectiveness is output or behavior. If special equipment is required to comply with the warning, it must be available or obtainable. If special skills are required, they must be present in the receiver population. To some extent these sensory and behavioral limitations of receiver populations are obvious; although it is not difficult to find examples of warnings that violate such considerations—especially in the behavior domain where instructions frequently given are, at best, difficult to carry out. "Avoid breathing fumes" when using a toxic solvent in an environment where respirators are not available is an example.

Three characteristics of receivers related to cognitive competence are important in warning design: technical knowledge, language, and reading ability. The communication of hazards associated with medications, chemicals, and mechanical devices is often technical in nature. If the target audience does not have technical competence, the warning may not be successful. The level or levels of knowledge and understanding of the audience must be taken into account. This point is discussed further in a later section.

The issue of language is straightforward, and it is increasingly important. Subgroups in the American society speak and read languages other than English, such as Spanish. As trade becomes more international, requirements for warnings to be directed to non–English readers will increase. Ways of dealing with this problem include warnings stated in multiple languages and the use of pictorials.

Reading ability is another target audience characteristic whose importance is obvious. Yet, high reading levels such as a grade 12 are not uncommon for warnings intended for individuals with lower reading abilities. The usual recommendation for general target audiences is that the reading level be in the grade 4–6 range. Clearly, if comprehension of a warning is to be achieved, reading levels must be consistent with reading abilities of receivers. A discussion of reading level measures and their application to the design of instructions and warnings can be found in Duffy (1985). An additional point on reading ability concerns illiteracy. There are estimates that 16 million functionally illiterate adults exist in the American population. If so, successfully communicating warnings may require more than simply keeping reading levels to a minimum. Although simple solutions to this problem do not exist, pictorials, speech warnings, special training programs, etc., may be important ingredients of warning systems for such populations.

There are a few general principles that apply when taking receiver characteristics into account during the design of warnings:

Principle 1. Know thy receiver. Gathering information and data about relevant receiver characteristics may require time, effort, and money, but without it the warning designer and ultimately the receiver will be at a serious disadvantage.

Principle 2. When variability exists in the target audience, design warnings for the low-end extreme. Do not design for the average.

Principle 3. When the target audience consists of subgroups that differ in relevant characteristics, consider employing a warning system that includes different components for the different subgroups. Do not try to accomplish too much with a single warning.

Principle 4. Market test the warning system. Despite the designer's knowledge of receiver characteristics and efforts to apply that knowledge, warnings generally should be market tested. Such tests may consist of "trying it out" on a target audience sample to assess comprehension and behavioral intentions. This principle is addressed in a later section.

36.4.5 Criteria for Designing Warnings

In this section we present eight criteria for designing warnings. To some extent, the choice of eight such rules or guidelines, as well as the manner in which the design considerations are partitioned, is arbitrary. Others who have worked and written on the topic (Lehto and Miller, 1986; Ryan, 1991) have a somewhat different list of criteria. Although the specific terminology and/or number of criteria may differ, however, there is generally high agreement as to what factors or design issues are relevant. Indeed, a publication by the National Safety Council in 1928 outlined a set of criteria for warnings design that maps very closely onto the eight criteria presented here.

The eight criteria are attention, hazard information, consequences information, instructions, comprehension, motivation, brevity, and durability. In the sections that follow, each will be defined and discussed.

36.4.5.1 Attention

Warnings should be designed so as to attract the attention of the target audience. Except when they are in an information-seeking mode, people typically do not look for warnings; hence, "warnings have to look for people." Also, many environments and labels are cluttered and noisy, so in order for warnings to be seen or heard, they must be designed so as to stand out from the background (Wogalter, Kalsher, and Racicot, 1993a). In other words, they should be conspicuous or salient relative to the context (Sanders and McCormick, 1993). There are several factors that influence the conspicuity or salience of a warning. Standard human factors guidelines for displays are relevant here.

Contrast

Print warnings should have high contrast with the background, dark on light or vice versa. Color can also be important in achieving contrast. There is another dimension that is related to contrast that has to do with context; specifically, to the extent that warnings are separated from other information, such as on a sign or label, they may be more salient (Godfrey et al., 1991).

Size

Within some reasonable limits, bigger is generally better. However, context plays an important role with regard to size effects on salience. On a sign or label, an important factor is not just the size of the warning, but rather its size relative to other information in the display. Product labels with a bold warning where there are three other information items in larger print are not a good design if one wants the warning to be noticed.

Location

The issue of location concerns several different aspects of warnings design. Within the context of a sign or label there are a few guidelines. First, given that people tend to scan left to right and top to bottom, warnings should be located near the top or to the left, depending on the overall design of the display. Certainly, other things being equal, a warning should not be buried at the bottom. Another consideration is task related. Warnings should be located near other information that will be needed to perform a task. For example, there are warnings on sidewalls of tires regarding hazards in mounting tires on rims. One kind of information that people usually need about a tire is its size. Thus, locating the warning near the size would increase its likelihood of being noticed. Sequencing information in a label can also be important. Wogalter et al. (1987) showed warnings were more likely to be noticed and complied with if they were ahead of or above use instructions than if they followed the instructions.

Another type of location consideration concerns warning systems with multiple components. A general principle is that warnings should be located close to the hazard, both physically and in time (Frantz and Rhodes, 1993; Wogalter, Barlow, and Murphy, 1995). A warning on the battery of a car regarding a hydrogen gas explosion is much more likely to be noticed at the proper time than a warning in the car manual. A verbal warning given 2 days ago to a farm worker using a hazardous pesticide is less likely to be remembered and effective than one given immediately before the product is used. Related to the concern about warning locations, however, is the fact that at times practical considerations limit the options. A small container such as on some over-the-counter medications may simply not have room for all of the information that should go into the warning. A solution here is to capture attention to the fact that there is a hazard by putting some minimum critical information on a primary label and directing the user to additional warning information in a secondary source such an owner's manual, a package insert or (better yet) another label in another conspicuous location. Wogalter et al. (1995) have shown that such a procedure can be effective.

Signal Words

Signal words are used in warnings to capture attention. They are also intended to communicate information about the level of the hazard. The most common words used are "CAUTION," "WARNING," and "DANGER", with danger representing the most hazardous circumstance and caution the least. These three terms are also the most widely recommended for this purpose (American National Standards Institute, 1991; Chapanis, 1994; FMC Corporation, 1985; Westinghouse Electric Corporation, 1981). Further, where it is feasible to incorporate color into the warning, the different words are paired with specific colors: CAUTION (black print on a yellow background); WARNING (black print on an orange background; and DANGER (red print on a white background or vice versa). Of course, the selection of color would also be governed by the context in which the warning is presented (Young, 1991). One would not want to put a red and white warning on a red surface. Many of the guidelines or recommended design practices pair signal words with a signal icon, a triangle enclosing an exclamation point. Figure 36.3 shows an example of an icon and signal word that represents a typical portion of a warning.

Figure 36.3 Icon and signal word.

Pictorials

The role of pictorials in warnings to communicate information is discussed in a later section. However, pictorials are also very effective in attracting attention (Jaynes and Boles, 1990; Laughery, Young, Vaubel, and Brelsford, 1993a).

Habituation

An important factor regarding attention to warnings is a psychological concept called *habituation.* Repeated exposure to a warning over time may result in its attracting less attention. Even a well-designed warning incorporating the features outlined above may become habituated. Although there are no easy solutions to this problem, one approach that may have some utility is to have warnings that vary from time to time. Rotational warnings such as on cigarette packages is an example of such an approach.

Auditory Warnings

Auditory warnings usually have as their primary purpose to attract attention. One advantage of such warnings over visual warnings is that auditory signals are omnidirectional, so the receiver does not have to be looking at a particular location to be alerted. Like print warnings, their success on the attention criterion is largely a matter of salience. Auditory warnings should be more intense and distinctively different from expected background noise. Often auditory warnings are used in conjunction with visual warnings, with the auditory serving to call attention to the need to read or examine a visual or written warning that contains specific information.

36.4.5.2 Hazard Information

A warning should contain a description of the hazard(s). The point here is to tell the target audience what the safety problem is; what can go wrong. Generally this information is specific to the environment or product. Examples are:

Toxic fumes
Slippery floor
Nip point, your hand could be caught
High voltage (7200 V)

These verbal or written statements communicate hazard information. Increasingly, pictorials are also being used to communicate such information, often in conjunction with the printed verbal message. Figures 36.4a, 36.4b, 36.4c, and 36.4d show examples of pictorials whose purpose is to indicate the presence of hazards.

A general principle here is that the hazard should be spelled out in the warning. As discussed earlier, however, there are exceptions to this principle. Where a hazard is known from previous experience or general knowledge or where the hazard is open and obvious, a warning may not be needed. Where these conditions do not exist, however, hazard information is an important part of the warning (Wogalter et al., 1987).

An issue in warnings design concerns what to warn about when there are multiple hazards associated with some situation or product. This issue was addressed earlier in the section on prioritizing warnings. As noted, in addition to existing knowledge and the open and obvious concepts, other considerations in deciding what to warn about are the likelihood of an undesirable event, the severity of the potential outcomes, and practical matters such as space. There is an additional consideration that has not been mentioned; namely, "overwarning."

The concept of overwarning applies at two levels. At a general level it concerns the extent to which our world is filled with warnings to a degree that people do not attend to them or become highly selective, attending only to some. If we "put warnings on everything," do we so inundate people with such information that they tune it out? Whereas this notion has face validity, there has been little or no research on the topic to support it. The concept of habituation is not relevant here, since habituation concerns repeated exposure to the same warning. Also, cognitive overload (overloading the receiver's ability to process the information) is not the concern, because the issue is not a matter of many warnings being presented simultaneously. Perhaps it should be called the warnings ubiquity effect. Nevertheless, however we label it, overwarning may indeed be a valid concern, and unnecessary warnings should be avoided.

(a) (b)

(c) (d)

Figure 36.4 Examples of pictorials conveying hazard information. (*a*) Slippery Floor; (*b*) Electricity, (*c*) Toxic Fumes, (*d*) Pinch Point.

On another level, overwarning also applies to specific situations or products. If there are 10 hazards associated with a product, does one warn about all of them? Of course, an appropriate answer to this question is that in such circumstances the better course of action would be to redesign the product. However, when to redesign is not the primary focus of this chapter. Putting too many warnings or having a warning with too many hazards listed on a single label may discourage the product user from attending to them. A guideline here is that if there are more than three or four hazards, include the three or four having highest priority (most likely to occur, most serious consequences, least likely to be known, etc.) in the primary warning system component, such as on the product label. The remaining hazards can then be addressed in secondary components such as package inserts, manuals, etc. This approach may not always be a satisfactory solution, but it is one way of possibly addressing multiple hazard situations. Certainly if knowledge of the hazards is necessary for safety, omitting warning about some of them because there are "too many" is not an acceptable approach. "Keeping them a secret" is hardly a solution. Finally, there is another concern about omitting hazards while addressing others; namely, the presumption of safety as a result of omission. If a warning for a toxic solvent includes information about ingestion and inhalation hazards but says nothing about a skin contact hazard, the user may assume that since it is omitted, there is no skin contact problem.

36.4.5.3 Consequences Information

Consequences information concerns the nature of the injury, illness or property damage that could result from the hazard. Hazard and consequence information are usually closely linked in the sense that one leads to the other; or, stating it in the reverse, one is the outcome of the other. In warnings, statements regarding these two elements should generally be sequenced. An example would be:

Toxic Fumes Inhaling Fumes Can Lead to Severe Lung Damage

There are occasions or situations when the hazard information is presented and understood, it may not be necessary to state the consequences in the warning. This point is related to the open and obvious aspects of hazards. For example, a sign indicating "Slippery Floor" probably does not need to include a consequence statement "You Could Slip and Fall." It is reasonable to assume that people will correctly infer the appropriate consequence. Although it is desirable to keep warnings as brief as possible (the brevity criterion is discussed in a later section), there is a potential problem with omitting consequence information; specifically, people may not make the correct inference regarding injury, illness or property damage outcomes. Thus, it is important in designing warnings to assess, if necessary, whether people will correctly infer consequences (Young, Wogalter, Laughery, Magurno, and Lovvoll, 1995). If unsure on this issue, the designer should include the consequence information.

A common shortcoming of warnings is that the consequences information is not explicit; that is, it does not provide important specific details. The statement "May be hazardous to your health" in the context of a toxic fumes hazard does not tell the receiver whether he or she may develop a minor cough or suffer severe lung damage (or some other outcome). This issue will be discussed in the section on the comprehension criterion.

As a general rule, written warnings (signs and labels) are organized with an attention getting icon and signal word at the top, then hazard information, and then instructions. For purposes of getting and holding the receiver's attention, however, there are situations where it is desirable to put consequences information near the beginning of the warning (just after the icon and signal word) in larger and bolder print (Young et al., 1995). This is particularly true for severe consequences such as death, paralysis, severe lung damage, etc. Hence, the above hazard and consequence statements might be better presented as:

Inhaling Fumes Can Lead to Severe Lung Damage Toxic Fumes

The point is that knowing about severe consequences can be a motivational factor in attending to and complying with the warning message, a consideration that will be further discussed in the section on motivation.

Pictorials can also be used to communicate consequence information. Figure 36.4a actually communicates both hazard information (slippery floor) and consequence information (fall). Figure 36.5a represents an explosion (typically the explosion symbol would be in red), and Figure 36.5b shows a figure in a wheel chair indicating paralysis.

36.4.5.4 Instructions

A point to be noted at the outset of this section concerns the distinction between warnings and instructions. Our distinction is that warnings are communications about safety, whereas instructions may or may not concern safety. "Keep off the grass" is an instruction that generally has nothing to do with safety (unless the grass is infested with poisonous snakes, in which case the statement alone clearly would not be an adequate warning). Instructions on how to assemble a toy do not concern safety and have nothing to do with warnings. When instructions are concerned with safety information or safe behavior, then they can be viewed as part of a warning. In short, warnings include instructions, but not all instructions are part of a warning.

In addition to getting people's attention and telling them what the hazard and potential consequences are, warnings should instruct people about what to do or not do. Typically, but not always, instructions in a warning follow the hazard and consequence information. An example of an instructional statement that might go with the above hazard and consequence statements is:

(a)

(b)

Figure 36.5 Examples of pictorals conveying consequence information. (*a*) Explosion, (*b*) Paralysis.

Inhaling Fumes Can Lead to Severe Lung Damage Toxic Fumes Always Wear a Type 1234 Respirator When Using This Product

This instruction assumes, of course, that the receiver will know what a type 1234 respirator is and have access to it.

There are two problems that commonly occur with instructional information in warnings. One problem is that the information is not explicit; that is, sufficient detail is not provided to enable the receiver to carry out the necessary safe procedures. The classic example here is "Use with adequate ventilation." Does this statement mean open a window, use a fan, or something much more technical in terms of volume of air flow per unit time? Obviously the instruction is not clear. We address this issue in the next section, on comprehension.

The second problem commonly encountered in warnings is that instructions are given that are inconvenient, difficult, or occasionally impossible to carry out. "Do not breathe fumes" clearly cannot be accomplished by stopping breathing. "Always have two or more persons to lift" is not possible if no one else is around. "Wear rubber gloves when handling this product" may be inconvenient if the user does not have them and the hardware store is 2 miles away. The means by which people can safely function in a situation or use a product safely should be as simple, easy and convenient as possible. This issue is discussed further in a later section on cost of compliance.

Pictorials can be used to communicate instructions. Figure 36.6 shows examples of instructional information that are used in warnings. Figure 36.6a communicates that the receiver should wear goggles in this environment or in using this product. Figure 36.6b indicates something that the receiver should not do—smoke. Note that the latter pictorial uses the common negation symbol, a circle containing the pictorial with a slash through it. The circle and slash would be in red.

36.4.5.5 Comprehension

The hazard, consequence, and instruction criteria for warnings concern the kinds of information that are normally included in a warning. Comprehension is a criterion that concerns the extent to which the information in the warning is understood by the receiver. In an earlier section on whom to warn, we discussed characteristics of receivers (target populations) that need to be taken into account in designing warning systems. In this section the focus is on the design characteristics of warnings that are important for receiver comprehension.

A common but often wrong assumption of people who design warnings is that the members of the target audience will understand the hazards, consequences, and instructions as well as they do. Designers of warnings should not make such assumptions because

(a) (b)

Figure 36.6 Examples of pictorials conveying instructional information. (*a*) Wear Goggles, (*b*) Do Not Smoke.

the designers are not representative of the target audience, a population that often has a wide range of mental competence and experience. What is common knowledge to the warning designer is not necessarily common knowledge to the members of the target audience (Laughery, 1993).

Design for the Low End Principle

When there is variability in the target population for the warning, which is almost always the case (especially when the audience is the general public), the applicable design principle is to design for the low-end extreme. Safety communications should not be written at the level of the average or median percentile person in the target audience, since they will present comprehension problems for those at lower competence, experience and knowledge levels.

Reading Level

Given the information to be communicated, reading levels for written language warnings should be as low as is feasible. As noted earlier, a grade 4–6 range is usually recommended. There are readability formulas based on word frequency of use, length of words, number of words in statements, etc., that are used to estimate reading grade level (Duffy, 1985). Although these formulas have limitations, they can be useful as a preliminary guide in achieving a warning that will be understood.

Technical Information

Many hazards and consequences are technical in the sense that a full and complete understanding would require an appreciation of technical information. The chemical content of a toxic material, the maximum safe level of a substance in the atmosphere in parts per million (ppm), and the biological reaction to exposure to a substance are examples. Although there are circumstances in which it is appropriate to communicate such information (e.g., to the toxicologist on the staff of a chemical plant or the physician prescribing medicine), as a general rule it is neither necessary nor useful to communicate such information to a general target audience. Indeed, it may be counterproductive in the sense that encountering such information may result in the receiver not attending to the remainder of the message. The end user of the toxic material typically does not need to know its chemical content (such as benzene) or its density in the atmosphere. Rather, he or she needs to be informed that the substance is toxic, what it can do in the way of injury or illness, and how to use it safely. Where there are multiple groups within the target audience (the toxicologist and the employee, the physician and the patient, the parent and the child), different components of the warning system can and often should be used to communicate to the different groups.

Explicitness

An important design principle relevant to warning comprehension is explicitness (Laughery, Vaubel, Young, Brelsford, and Rowe, 1993b). Explicit messages contain information that is sufficiently clear and detailed to permit the receiver to understand at an appropriate level the nature of the hazard, the consequences, and the instructions. The key here is the word "appropriate." As noted above, technical details may not be necessary and at times may be detrimental. The bigger and more common problem, however, is that warnings are frequently not detailed or specific enough. The following two examples are warnings with hazard, consequence and instructional statements that are not sufficiently explicit.

Dangerous Environment Health Hazard Take Precautionary Measures

Mechanical Hazard You Could Be Injured Exercise Care

Alternatives to the above that would be considered more explicit and appropriate are:

Toxic Fumes Breathing Fumes Can Lead To Severe Lung Damage Always Wear Type 1234 Respirator In This Area

Moving Parts, Pinch Point Hazard Your Hand Could Be Caught In Rollers and Severely Crushed Do Not Operate Without Guard X In Place

Pictorials

Pictorials are increasingly employed in the design of warnings. Guidelines such as American National Standards Institute (ANSI) (1991) and FMC (1985) place considerable emphasis on their use. Pictorials are particularly useful in helping to increase comprehension (Boersema and Zwaga, 1989; Collins, 1983; Dewar, 1994; Laux, Mayer, and Thompson 1989; Wolff and Wogalter, 1994; Zwaga and Easterby, 1984). Obviously they can contribute to understanding warning messages for target audiences where illiterates or non–English readers are included. They can be useful where there are time constraints, such as traveling on a highway, because well-designed pictorials can cue large amounts of knowledge in a glance. Also, people who have difficulty reading print, such as the elderly, may be able to see a pictorial.

While pictorials can be very useful in the comprehension of warning information, comprehension is also a primary concern or criterion for pictorials. In some pictorials the symbol or picture directly represents the information or object being communicated and will be understood if the person recognizes the symbol or picture. Figure 36.4a is an example. In other pictorials the symbol may be recognized, but its meaning has to be learned. People may recognize a skull and crossbones, but the fact that it represents a poison hazard would have to be learned. Some pictorials are completely abstract, such as the symbols for biohazard and radioactivity hazard shown in Figure 36.7, and must be learned to be understood. As a general principle, pictorials containing symbols or pictures that directly represent the information are preferred, especially for general target audi-

(a)

(b)

Figure 36.7 Simple pictorials representing biohazard and radioactive hazard. (*a*) Biohazard; (*b*) Radioactive Hazard.

ences. Pictorials where the meaning of the symbols must be learned may be useful for special target audiences.

What is an acceptable level of comprehension for pictorials? This question has been addressed in the ANSI (1991) standard which suggests a goal of 85% comprehension by the target audience. There are two criteria that seem relevant here. The first is simply that the pictorial should be designed to accomplish the highest level of comprehension attainable. If 85% cannot be achieved, it may still be useful depending on the alternatives. A second criterion is that the pictorial not be misinterpreted or communicate incorrect information. Wogalter (1994) cites an interesting example of a misinterpretation of a pictorial that was part of a warning for the drug Acutane. This drug is used for severe acne, but causes birth defects in babies of women taking the drug during pregnancy. The pictorial shows a side-view outline shape of a pregnant woman within a circle-slash negation sign. The intended meaning of the pictorial is that women should not take the drug if they are pregnant. However, some women incorrectly interpreted the pictorial to mean that the drug might help in preventing pregnancy.

Auditory Warnings

The comprehension of auditory warnings depends on whether the signal is nonverbal (sirens, tones, bells) or verbal (speech or voice). Nonverbal auditory warnings can be further divided into simple and complex. Simple nonverbal auditory warnings are usually used as alert (attention-getting) signals after which the visual modality can then be employed to access further information (Sanders and McCormick, 1993; Sorkin, 1987). Complex nonverbal signals are composed of sounds of differing (sometimes dynamic) amplitude, frequency, and temporal patterns. Their purpose is to communicate different types or different levels of hazards. They can transmit more information than simple auditory warnings, but the listener must know what the code means. Training must be given for the meaning to be deciphered. Only a limited number of complex signals should be used, because people are limited in discriminating and remembering them (Banks and Boone, 1981; Cooper, 1977).

Complex warning messages can also be transmitted via voice (speech). In recent years voice chips and digitized sound processors have been developed making voice warnings feasible for a wide range of novel approaches and applications. Recent research indicated that voice warnings under certain circumstances can be more effective in transmitting information than printed signs (Wogalter, 1993a; Wogalter and Young, 1991). There are, however, some problems inherently associated with voice warnings. Time to transmit speech messages requires longer durations than simple auditory warnings or reading an equivalent message. Comprehension can also be a problem with complex voice messages. To be effective, voice messages should be intelligible and brief. Nevertheless, this medium for communicating safety information would appear to have considerable potential.

36.4.5.6 Motivation

The motivation criterion concerns the notion that warnings should motivate people to engage in safe behavior or not engage in unsafe behavior. There are several factors that appear to be important in the extent to which people are motivated to read and comply with warnings.

Risk Perception

One of the important factors in whether people will read and comply with warnings is their perception of the level of hazard and consequences associated with the situation or product. The greater the perceived level, the more responsive people will be to warnings (Wogalter, Brelsford, Desaulniers, and Laughery, 1991; Wogalter, Brems, and Martin, 1993b). In a sense, this factor can be viewed as a perceived cost of noncompliance; if I do not comply, what might happen to me. There are several things that can influence the risk perception or cost of not complying including familiarity and severity of consequences.

Familiarity

The "familiarity effect" states simply that the more familiar people are with a situation or product, the less they perceive associated hazards and the less likely they are to read or comply with a warning (Godfrey and Laughery, 1984; Godfrey et al., 1983; Goldhaber and deTurck, 1988; Wogalter et al., 1991). This "familiarity breeds contempt" notion, however, should not be overemphasized for at least two reasons. First, people more familiar with a situation or product may have more knowledge about the hazards and consequences as well as an understanding about how to avoid them. Second, people in situations or using products more frequently are exposed to the warnings more often, which increases the opportunity to be influenced by them. Nevertheless, where familiarity is a factor, it should be realized that stronger warnings or perhaps other efforts will be required.

Severity of Consequences

Intimately tied to risk perception or perceived cost of noncompliance are people's beliefs in how severely they might be injured. Research (Wogalter et al., 1991, 1993b) indicates that people's notions of hazardousness are almost entirely based on the seriousness of the potential outcome. Further, people do not readily consider the likelihood or probability of such events in making hazardousness judgments (Wogalter and Barlow, 1990; Young, Brelsford, and Wogalter, 1990; Young, Wogalter, and Brelsford, 1992). These findings emphasize the importance of clear, explicit consequences information in warnings. Such information can be critical to people's risk perception and thus be a major factor in driving compliance.

Cost of Compliance

The cost associated with compliance can be a strong motivator. Generally, compliance with a warning requires that people take some action. Usually there are costs associated with taking action. These costs may be in the form of convenience, time, effort, or money. Several studies have shown that such costs play a major role in whether people comply (Dingus, Hathaway, and Hunn, 1991; Wogalter et al., 1987, 1989).

Obviously in one sense the issue of compliance can be viewed as a tradeoff between the perceived cost of noncompliance and the perceived cost of compliance. The designer of the system wants to minimize the cost of noncompliance by designing a safer system and one that forgives human error or in this instance, noncompliance. But the warning designer does not want to induce noncompliance by failing to adequately warn about the hazards and consequences. Thus, it is critical that warnings contain clear explicit hazard and consequence information. Similarly, the designer wants to minimize the cost of complying with warnings so as to increase the likelihood that people will perform safely.

Social Influence

Another motivator of warning compliance is social influence. Research (Wogalter et al., 1989) has shown that if people see others comply with a warning, they are more likely to comply themselves. Similarly, seeing others not comply lessens the likelihood of complying. Social influence is a motivational variable that is an external factor with respect to warnings in that it is not part of the design. However, it does have an effect and should be kept in mind when considering motivational factors.

36.4.5.7 Brevity

Within the need to communicate required information, warnings should be as brief as possible. Two statements should not be included if one will do, such as in the slippery floor example cited earlier. Longer warnings or those with nonessential information are less likely to be read and they may be more difficult to understand. Obviously, this criterion should not be interpreted as a license to omit important information.

36.4.5.8 Durability

The durability criterion simply states that warnings should be designed to last as long as needed. There are circumstances in which durability is typically not a problem. A product off the shelf of a drug store that will be completely and immediately consumed is an example. On the other hand, products with a long life, such as cars, lawn mowers, etc., may present a challenge. Similarly, situations where warnings are exposed to weather, such as on construction sites, or extensive handling, such as on some containers, may pose durability problems.

There are several approaches to meeting the durability criterion. One solution, of course, is to make signs or labels with materials that will meet the requirements. Another is to have procedures for detecting when a replacement warning is needed and then replacing it. This approach can be useful in circumstances such as on construction sites or other work environments.

Some components of warning systems are particularly susceptible to not meeting the durability criterion. Package inserts and manuals are examples of components that get lost or discarded. Such factors should be taken into account in considering the role of such components in the overall warning system. Of course, some warning components are not intended to be durable. Tread labels on new tires that contain warnings or spoken warnings at a point of purchase are examples.

36.4.6 Criteria for Assessing Effectiveness of Warnings

In this section we will discuss issues associated with the effectiveness of warnings, and more specifically, criteria for assessing or evaluating their effectiveness. The question of effectiveness has received a great deal of attention in the technical literature in recent years, and, indeed, there has been some disagreement on issues associated with the topic. Examples of publications that contain discussions of the warnings effectiveness issues are McCarthy, Finnegan, Krumm-Scott, and McCarthy (1984), DeJoy (1989), Lehto and Papastavou (1993) and Wogalter (1994). Additional papers on the topic can be found in a collection by Laughery, Wogalter, and Young (1994).

36.4.6.1 Direct and Indirect Effects

The distinction between direct and indirect effects of warnings concerns the routes by which information gets to the target person. A direct effect occurs as a result of the person being directly exposed to the warning; he or she reads it, hears it, is instructed about it by an employer, etc. This communication route is what we usually think about when we design warnings. But warnings can also accomplish their purposes indirectly. An example is the woman who has not read the warning about toxic shock syndrome on the tampons box, but learns about it in a conversation with her neighbor. The employer or physician who reads the warnings about products with which they are concerned and then verbally communicates the information to the employee or patient are other examples. The print and broadcast news media may pick up warning information and disseminate it in ways that expose and influence people who have not seen it directly.

An example of where the concept of an indirect effect was taken into account in the design of a product warning concerned a herbicide used in agricultural settings. Given that significant numbers of farm workers in parts of the United States read Spanish but not English, there was reason to put the warning in both languages. However, there were space constraints on the product container. One aspect of the solution was to include a statement on the label in Spanish indicating that the product was hazardous and that the user should have someone translate the warning before using the product. This procedure may or may not have been the most effective way of addressing the problem, but the point here is that it was an effort to take advantage of an indirect communication to have the warning be effective.

There are situations where we rely on indirect communications to transmit warning information. Employers and physicians are examples already noted; adults who have responsibility for the safety of children are another important category. In the design of warning systems, it is important to take into account such communication routes.

36.4.6.2 An Information-Processing Model as a Context for Assessing Effectiveness

In this section a simplified model of the human information-processing system is introduced to serve as a basis for organizing the discussion of effectiveness. Its purpose is to assist in analyzing how or why warnings may fail or, conversely, what they have to accomplish to succeed. In many respects the model here is similar to, although simpler than, the information-processing model employed by Lehto and Miller (1986) and by Lehto and Papastavrou (1993) in their analyses of warnings effectiveness. A diagram of the model is shown in Figure 36.8.

The model describes the warning process in terms of human information-processing stages. Six stages are included, starting with the presentation or existence of the warning information and ending with the safe behavior. There are two basic concepts to be noted about this model. First, for the warning to be effective, it must be successful at each of these stages. This is the weak-link-in-the-chain phenomenon; if the warning is not successful at any one stage, it fails. Given that the warning information is presented, the receiver must notice and attend to it. Next, having been attended to, the message must be understood. Having been understood, the warning needs to agree with people's existing attitudes and beliefs, or if not, it must be sufficiently persuasive to change them. Next, it must motivate people to comply and perform the appropriate behaviors. Finally, the individual must be capable of carrying out the behaviors. If the warning is not noticed, or if it is not understood, or if it is rejected on the basis on existing attitudes and beliefs, or if it does not motivate one to act, or if it requires behavior that cannot be carried out, then it fails.

The second concept is that the model represents a serial processor; that is, the warning information flows through and affects the various stages sequentially. There are no feedback loops such as one from motivation to attention that would allow for a person having read the warning to be motivated to go back and read it again to gain additional information or enhance comprehension. Clearly the serial model is an oversimplification, but it is useful in considering warning effectiveness issues.

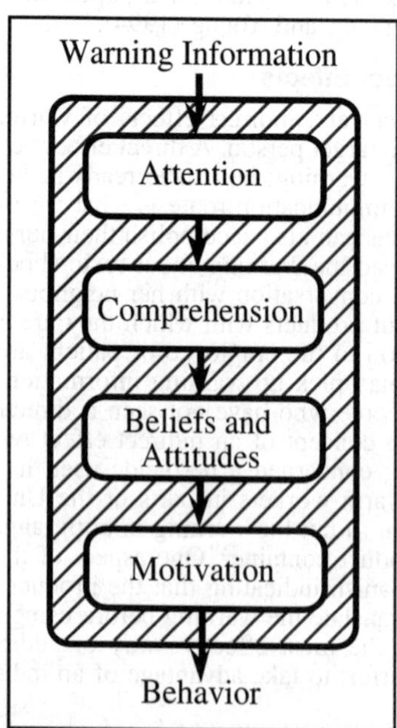

Figure 36.8 A human information-processing model of stages leading to compliance behavior.

The serial model has important implications for assessing the effectiveness of warnings. Given that the overall purpose of warnings includes influencing or modifying people's behavior, one might assume that assessing warning effects on behavior is the approach to be taken. And it is. But there are other useful methods or levels of assessing warnings. For example, if it can be determined that a particular warning is not likely to be noticed, then it is not effective. If it is not understood, then it is not effective. If it is not believed, it is not effective. Finally, if it does not provide sufficient motivation to act, it is not effective.

The implications of the above assessment logic are important for evaluating the effectiveness of a warning. Determining whether or not a warning will influence behavior is usually a difficult assignment at best. In addition to ethical problems of exposing people to hazards, actual field studies testing warnings are likely to be time consuming and costly. Certainly where feasible, such studies are desirable. Also, laboratory or other controlled simulations of warning situations can be useful in assessing behavioral effects, but such approaches, while important, leave open questions of generalizability. The implication of the model is that warnings can and should be tested at several levels. Studies that examine the effects of warnings on attention, comprehension, beliefs and attitudes, and motivation to comply can be valuable as part of the process of designing and assessing warnings. For example, such studies can help in isolating why a warning is not effective. A behavioral study that shows people do not comply with a warning may not tell us whether it failed because it was not noticed, or because it was not understood, or because it was not believed, or because it failed to motivate. Studies employing attention, comprehension, risk perception, or behaviorial intention measures can provide such information, which, in turn, can be useful in developing alternative warning designs that are effective. If a warning is noticed and understood, there may be no need to try to increase its conspicuity or lower its reading level. Instead, one may want to reconsider factors such as the cost of compliance.

Studies carried out to evaluate the potential effectiveness of a warning must, of course, incorporate appropriate principles of research design. The selection of subjects to be representative of the target population, avoiding confounding by extraneous variables, and guarding against contamination by expected outcomes are a few of the more salient factors that must be considered. For a more complete discussion of approaches to evaluating warning effectiveness, see Wogalter and Dingus (in press) and Young and Lovvoll (in press).

Warning Information

Obviously warning information has to be presented if it is to be effective. One point to be noted here is that assumptions about the target audience having preexisting knowledge or that the hazard is open and obvious should be made with care. Thus, at this level it is possible and at times important to assess the *need* for a warning by determining what knowledge people have about relevant hazards or whether the hazard is correctly recognized without a warning.

Attention

In the section on criteria for warnings design, a number of factors that influence the noticeability of warnings were presented and discussed. One means of assessing a warning with respect to attention is simply to determine the extent to which the design meets the criteria. If no signal word is used, no color employed, the print is small, the message is embedded in other types of information, etc., then the effectiveness of the warning may be questioned. More direct techniques are available for studying the attention-demanding properties of warnings such as studies employing reaction time or memory measures. While more difficult to carry out, eye movement analysis can also be a useful tool.

Comprehension

Like attention, one method of assessing comprehension of a warning is to evaluate it against the criteria discussed earlier. If the reading level is high, technical language is used, or the statements are vague and nonexplicit, then the warning is not likely to be understood. Carrying out studies to assess the extent to which a warning is understood probably has one of the best cost–benefit ratios of any procedure in the warnings design process. Relative to behavioral studies, comprehension can be assessed easily, quickly and at low cost. Well established methodologies involving memory tests, open-ended response tests, interviews, etc., are applicable. Such studies can be exceptionally valuable in determining what information in the warning was or was not understood as well as what might be done in the way of redesign to increase the level of comprehension.

Beliefs and Attitudes

Beliefs concern the extent to which information in a warning is accepted as true. Attitudes are similar to beliefs except more emotion or feeling is involved. People may understand the information in a warning, but if it is rejected as not true or irrelevant, then the warning will not be effective. This circumstance can be a problem where people's experiences with a situation or product results in their believing it is safer than it is. It can also be a problem when people believe that their own abilities or competence will enable them to overcome the hazard, such as the young adult male who believes he can safely do a shallow dive into the shallow end of a swimming pool. Here again, studies can be carried out to determine the extent to which members of the target audience accept the warning as true or valid as well as whether or not it applies to them. Negative results on these dimensions would indicate the warning is not likely to be effective.

Motivation

Some of the major factors that influence motivation to comply with warnings have been discussed in the section on criteria for warnings design. Among the most important were the cost of compliance and the cost of noncompliance (severity of the potential injury, illness, or property damage). If the warning calls for actions that are inconvenient, time consuming, or costly, there is a likelihood they will not be effective unless the consequences of noncompliance are very bad or undesirable. Motivation can be assessed by obtaining measures of behavioral intentions from members of the target audience. Although such measures will generally reflect higher levels of compliance than will actually occur, they can be useful for determining whether or not the warning is likely to be effective.

Behavior

As noted earlier, actually determining what people will do in the context of a warning is a very desirable measure of its effectiveness. Although such studies are generally difficult to execute, in situations where negative consequences of an ineffective warning are high, the effort may be warranted.

36.4.7 Warnings as Reminders

As noted earlier in this chapter, one role of a warning is to serve as a reminder. There are occasions when the target audience has knowledge of the hazards, consequences, and approproate modes of behavior, but that knowledge is not always sufficient. They must be aware of, thinking about, this knowledge or information at the proper time. No one knew better than the three-fingered punch press operators of the 1920s that their hand should not be under the piston when it stroked. Yet, such incidents occurred.

There are several circumstances in which warning reminders are useful and/or needed. Some of the more noteworthy are:

1. A hazardous situation or product (that is not open and obvious) is encountered infrequently, and forgetting may be a factor.
2. Distractions occur during the performance of a task or the use of a product.
3. Heavy task loads exceed attentional capacity.

When warnings are intended only to function as reminders, it generally is not necessary to provide the same amount of information that would normally be required. Here the emphasis should be more on noticeability, getting the person's attention. Auditory warnings can be useful, such as the buzzer in an automobile reminding occupants to fasten the seat belt. Dynamic warnings such as flashing signs are also potentially beneficial because of their ability to capture attention. The key point in considering the need for reminder warnings is to keep in mind the fact that hazard knowledge on the part of a target audience does not guarantee that that knowledge will be available when needed.

36.5 SUMMARY AND CONCLUSIONS

In recent decades warnings have become increasingly important in the field of safety. Approaches to dealing with environmental or product hazards are generally prioritized such that first one tries to solve the problem by design, then by guarding, then by warning. Thus, in the domain of safety, warnings are viewed as a third but important line of defense.

Warnings can be properly viewed as communications whose purposes include informing and influencing the behavior of people. Warnings are not simply signs or labels. They can include a variety of media through which various kinds of information get communicated to a broad spectrum of people. The use of various media or channels and an understanding of the characteristics of the receivers or target audience to whom the warning is directed are important in the design of effective warnings. The concept of a warning system with multiple components or channels for communication to a variety of receivers is useful in this regard.

The design of warnings can and should be viewed as an integral part of systems design. Too often it is carried out after the environment or product design is completed, a kind of afterthought phenomenon. Warnings cannot and should not be expected to serve as a cure for bad design.

Eight criteria can be defined that are useful in the design and assessment of warnings. They are:

Attention—Warnings should be designed so as to attract attention.

Hazard information—Warnings should contain information about the nature of the hazard.

Consequence information—Warnings should contain information about the potential outcomes.

Instructions—Warnings should instruct about appropriate and inappropriate behaviors.

Comprehension—Warnings should be understood by the target audience.

Motivation—Warnings should motivate people to comply.

Brevity—Warnings should be as brief as possible.

Durability—Warnings should last and be available as long as needed.

Of course, a specific criterion may not always be relevant. For example, a fire alarm does not have to state a consequence, and durability may not be a concern for a product off the drugstore shelf that is to be used immediately.

The issue of warning effectiveness has received a great deal of attention in recent years, especially means by which effectiveness can be assessed. Several criteria can be employed in assessing warnings, including whether they capture attention, are understood, are consistent with or capable of modifying beliefs and attitudes, motivate people to comply, and result in people behaving safely. The assessment of warning effectiveness employing approaches such as these can and should be part of the warning design process.

REFERENCES

American National Standards Institute. (1991). *Product Safety Signs and Labels, Z535.4*. (Washington, DC: National Electrical Manufacturers Association).

Banks, W. W. and Boone, M. P. (1981). *Nuclear Control Room Enunciators: Problems and Recommendations* (NUREG/CR-2147). (Springfield, VA: National Technical Information Service).

Boersema, T., and Zwaga, H. J. G. (1989). Selecting comprehensible warning symbols for swimming pool slides. In *Proceedings of the Human Factors Society 33rd Annual Meeting* Santa Monica, CA, pp. 994–998.

Chapanis, A. (1994). Hazards associated with three signal words and four colours on warning signs, *Ergonomics*, *37*, 265–275.

Collins, B. L. (1983). Evaluation of Mine-Safety Symbols. In *Proceedings of the Human Factors Society 27th Annual Meeting*. Santa Monica, CA, pp. 947–949.

Collins, B. L., and Lerner, N. D. (1982). Assessment of fire-safety symbols. *Human Factors*, *24*, 75–84.

Cooper, G. E. (1977). *A Survey of the Status and Philosophies Relating to Cockpit Warning Systems* (NASA-CR-152071). NASA Ames Research Center, CA.

DeJoy, D. M. (1989). Consumer product warnings: review and analysis of effectiveness research. In *Proceedings of the Human Factors Society 33rd Annual Meeting*. Santa Monica, CA, pp. 936–940.

Desaulniers, D. R. (1991). *An examination of consequence probability as a determinant of precautionary intent.* Unpublished doctoral dissertation, Rice University, Houston.

Dewar, R. (1994). Design and evaluation of graphic symbols. *Proceedings of Public Graphics.* University of Utrecht, the Netherlands, pp. 24.1–24.18.

Dingus, T. A., Hathaway, J. A., and Hunn, B. P. (1991). A most critical warning variable: Two demonstrations of the powerful effects of cost on warning compliance. In *Proceedings of the Human Factors Society 35th Annual Meeting.* Santa Monica, CA, pp. 1034–1038.

Duffy, T. M. (1985). Readability formulas: What's the use? In T. M. Duffy and R. Waller, Eds., *Designing Usable Texts.* Orlando, FL: Academic Press, Inc. pp. 113–143.

Easterby, R. S., and Hakiel, S. R. (1981). The comprehension of pictorially presented messages. *Applied Ergonomics, 12,* 143–152.

Fischhoff, B. (1989). Risk: A guide to controversy. In *Improving Risk Communication,* J. F. Ahearne, Ed.,. Washington DC: National Research Council.

FMC Corporation. (1985). *Product safety sign and label system.* Santa Clara, CA: FMC Corporation.

Frantz, J. P., and Rhoades, T. P. (1993). A task analytic approach to the temporal placement of product warnings. *Human Factors, 35,* 719–730.

Frantz, J. P., Rhoades, T. P., and Lehto, M. R. (In press.) Practical considerations regarding the design and evaluation of product warnings. In M. S. Wogalter, D. M. DeJoy, and K. R. Laughery, Eds., *Warnings and Risk Communication.* Taylor and Francis, London.

Godfrey, S. S., Allender, L., Laughery, K. R., and Smith, V. L. (1983). Warning messages: Will the consumer bother to look? In *Proceedings of the Human Factors Society 27th Annual Meeting.* Santa Monica, CA, pp. 950–954.

Godfrey, S. S., and Laughery, K. R. (1984). The biasing effect of familiarity on consumer's awareness of hazard. In *Proceedings of the Human Factors Society 28th Annual Meeting.* Santa Monica, CA, pp. 483–486.

Godfrey, S. S., Laughery, K. R., Young, S. L., Vaubel, K. P., Brelsford, J. W., Laughery, K. A., and Horn, E. (1991). The new alcohol warning labels: How noticeable are they? In *Proceedings of the Human Factors Society 35th Annual Meeting.* Santa Monica, CA, pp. 446–450.

Goldhaber, G. M., and deTurck, M. A. (1988). Effects of consumer's familiarity with a product on attention and compliance with warnings. *Journal of Products Liability, 11,* 29–37.

Jaynes, L. S., and Boles, D. B. (1990). The effects of symbols on warning compliance. In *Proceedings of the Human Factors Society 34th Annual Meeting.* Santa Monica, CA, pp. 984–987.

LaRue, C., and Cohen, H. (1987). Factors influencing consumer's perceptions of warning: An examination of the differences between male and female consumers. In *Proceedings of the Human Factors Society 31st Annual Meeting.* Santa Monica, CA, pp. 610–614.

Laughery, K. R. (1993, July). Everybody knows: Or do they? *Ergonomics in Design,* 8–13.

Laughery, K. R., and Brelsford, J. W. (1991). Receiver characteristics in safety communications. In *Proceedings of the Human Factors Society 35th Annual Meeting.* Santa Monica, CA, pp. 1068–1072.

Laughery, K. R., Vaubel, K. P., Young, S. L., Brelsford, J. W., and Rowe, A. L. (1993b). Explicitness of consequence information in warnings. *Safety Science, 16,* 597–613.

Laughery, K. R., Wogalter, M. S., and Young, S. L., Eds. (1994). *Human Factors Perspectives on Warnings.* Santa Monica, CA: Human Factors and Ergonomics Society.

Laughery, K. R., Young, S. L., Vaubel, K. P., and Brelsford, J. W. (1993a). The noticeability of warnings on alcoholic beverage containers. *Journal of Public Policy and Marketing, 12,* 38–56.

Laux, L. F., Mayer, D. L., and Thompson, N. B. (1989). Usefulness of symbols and pictorials to communicate hazard information. In *Proceedings of the Interface 89.* Santa Monica, CA, pp. 79–93.

Lehto, M. R., and Miller, J. M. (1986). *Warnings: Volume 1. Fundamentals, Design and Evaluation Methodologies.* Ann Arbor, MI: Fuller Technical Publications.

Lehto, M. R., and Papastavrou, J. D. (1993). Models of the warning process: Important implications towards effectiveness. *Safety Science, 16,* 569–595.

Lehto, M. R., and Salvendy, G. (1995). Warnings: A supplement not a substitute for other applications to safety. *Ergonomics, 38,* 2155–2163.

McCarthy, R. L., Finnegan, J. P., Krumm-Scott, S., and McCarthy, G. E. (1984). Product information presentation, user behavior, and safety. In *Proceedings of the Human Factors Society 28th Annual Meeting.* Santa Monica, CA, pp. 81–85.

National Safety Council. (1928). Warning Signs—Their Use and Maintenance (Safe Practices Pamphlet No. 81). Chicago: National Safety Council.

Otsubo, S. M. (1988). A behavioral study of warning labels for consumer products: Perceived danger and use of pictographs. In *Proceedings of the Human Factors Society 32nd Annual Meeting.* Santa Monica, CA, pp. 536–540.

Ringseis, E. L. and Caird, J. K. (1995). The comprehensibility and legibility of twenty pharmaceutical warning pictograms. In *Proceedings of the Human Factors and Ergonomics Society 39th Annual Meeting.* Santa Monica, CA, pp. 974–978.

Ryan, J. P. (1991). *Design of Warning Labels and Instructions.* New York: Von Nostrand Reinhold.

Sanders, M. S., and McCormick, E. J. (1993). *Human Factors in Engineering and Design.* (Seventh ed.) (New York: McGraw-Hill).

Slovic, P., Fischhoff, B., and Lichtenstein, S. (1982). In Kahneman, D., Slovic, P., and Tversky, A., Eds. *Judgment Under Uncertainty: Heuristics and Biases.* Cambridge: Cambridge University Press. pp. 3–20.

Sorkin, R. D. (1987). Design of auditory and tactile displays. In G. Salvendy, Ed., *Handbook of Human Factors.* New York: John Wiley.

Vicusi, W. K., Magat, W. A., and Huber, J. (1986). Informational regulation of consumer health risks: An empirical evaluation of hazard warnings. *Rand Journal of Economics, 17,* 351–365.

Westinghouse Electric Corporation. (1981). *Product Safety Label Handbook.* Trafford, PA: Westinghouse Printing Division.

Wogalter, M. S. (1994). Factors influencing the effectiveness of warnings. In *Proceedings of Public Graphics* (pp. 5.1–5.21), Lunteren, The Netherlands.

Wogalter, M. S., Allison, S. T., and McKenna, N. (1989). Effects of cost and social influence on warning compliance. *Human Factors, 31,* 133–140.

Wogalter, M. S., and Barlow, T. (1990). Injury likelihood and severity in warnings. In *Proceedings of the Human Factors Society 34th Annual Meeting.* Santa Monica, CA, pp. 580–583.

Wogalter, M. S., Barlow, T., and Murphy, S. (1995). Compliance to owner's manual warnings: Influence of familiarity and the task-relevant placement of a supplemental directive. *Ergonomics, 38,* 1081–1091.

Wogalter, M. S., Brelsford, J. W., Desaulniers, D. R., and Laughery, K. R. (1991). Consumer product warnings: The role of hazard perception. *Journal of Safety Research, 22,* 71–82.

Wogalter, M. S., Brems, D. J., and Martin, E. G. (1993b). Risk perception of common consumer products: Judgments of accident frequency and precautionary intent. *Journal of Safety Research, 24,* 97–106.

Wogalter, M. S., and Dingus, T. A. (In press). Evaluations of behavioral compliance/adherence. In Wogalter, M. S., DeJoy, D. M., and Laughery, K. R., Eds., *Warnings and Risk Communication.* London: Taylor and Francis.

Wogalter, M. S., Godfrey, S. S., Fontenelle, G. A., Desaulniers, D. R., Rothstein, P. R., and Laughery, K. R. (1987). Effectiveness of warnings. *Human Factors, 29,* 599–612.

Wogalter, M. S., Kalsher, M. J., and Racicot, B. (1992). Effect of presentation location and pictorials on behavioral compliance to warnings. In *Proceedings of the Human Factors Society 36th Annual Meeting.* Santa Monica, CA, pp. 1029–1033.

Wogalter, M. S., Kalsher, M. J., and Racicot, B. (1993a). Behavioral compliance with warnings: Effects of voice, context and location. *Safety Science, 16,* 637–654.

Wogalter, M. S., and Young, S. L. (1991). Enhancing warning compliance through alternative product label designs. *Applied Ergonomics, 24,* 53–57.

Wolff, J. S., and Wogalter, M. S. (1993). Test and development of pharmaceutical pictorials. In *Proceedings of Interface 93.* Santa Monica, CA, pp. 187–192.

Young, S. L. (1991). Increasing the noticeability of warnings: Effects of pictorial, color, signal icon and border. In *Proceedings of the Human Factors Society 34th Annual Meeting.* Santa Monica, CA, pp. 580–584.

Young, S. L., Brelsford, J. W., and Wogalter, M. S. (1990). Judgments of hazard, risk and danger: Do they differ? In *Proceedings of the Human Factors Society 34th Annual Meeting* Santa Monica, CA, pp. 503–507.

Young, S. L., and Lovvoll, D. R. (In press). Intermediate processing: assessment of eye movement, subjective response and memory. In Wogalter, M. S., DeJoy, D. M., and Laughery, K. R., Eds., *Warnings and Risk Communication.* London: Taylor and Francis.

Young, S. L., Martin, E. G., and Wogalter, M. S. (1989). Gender differences in consumer product hazard perceptions. In *Proceedings of Interface 89.* Santa Monica, CA, pp. 73–78.

Young, S. L., Wogalter, J. S., and Brelsford, J. D. (1992). Relative contribution of likelihood and severity of injury to risk perceptions. In *Proceedings of the Human Factors Society 36th Annual Meeting.* Santa Monica, CA, pp. 1014–1018.

Young, S. L., Wogalter, M. S., Laughery, K. R., Magurno, A., and Lovvoll, D. (1995). Relative order and space allocation of message components in hazard warning signs. In *Proceedings of the Human Factors and Ergonomics Society 39th Annual Meeting.* Santa Monica, CA, pp. 969–973.

Zwaga, H. J. G., and Easterby, R. S. (1984). Developing effective symbols or public information. In Easterby, R. S., and Zwaga, H. J. G., Eds., *Information Design: The Design and Evaluation of Signs and Printed Material.* New York: John Wiley.

PART 6

PERFORMANCE MODELING

CHAPTER 37
DECISION MAKING

Mark R. Lehto
School of Industrial Engineering
Purdue University
West Lafayette, IN 47907-1287 USA

37.1 INTRODUCTION

This chapter focuses on human decision making. Related chapters in this book include Chapter 4, Information Processing; Chapter 31, Occupational Risk Management; Chapter 40, Cognitive Modeling; and Chapter 42, Decision Support. Human decision making is a broad topic that overlaps with these chapters in many ways. Human decision making is first fundamentally related to the topic of information processing because people must gather, organize, and combine information from different sources to make many decisions. Decision making is often viewed as a stage of human information processing, but as decisions grow more complex, information processing actually becomes part of decision making and methods of decision support that help decision makers process information become of growing importance. Decision making also overlaps with problem solving. The point where decision making becomes problem solving is fuzzy, but many decisions require problem solving and the opposite is true as well. Cognitive models of problem solving are consequently relevant for describing many aspects of human decision making. They become especially relevant for describing steps taken in the early stages of a decision wherein choices are formulated and alternatives are identified.

A complete treatment of human decision making is well beyond the scope of a single book chapter.[1] The topic has its roots in economics and is currently a focus of operations research and management science, psychology, sociology, and cognitive engineering. These fields have produced numerous models and a substantial body of research on human decision making. At least three objectives have motivated this work: (1) to develop normative prescriptions that can guide decision makers, (2) to describe how people make decisions and compare the results to normative prescriptions, and (3) to determine how to help people apply their "natural" decision-making methods more successfully. The goals of this chapter are to synthesize the elements of this work into a single picture and provide some depth of coverage in particularly important areas. The integrative model presented in Section 37.1.3 focuses on the first goal. The remaining sections address the second goal.

37.1.1 Role and Utility of this Chapter

This chapter is intended to provide an overall perspective of human decision making to human factors practitioners, developers of decision tools (such as expert systems), product designers, researchers in related areas, and others who are interested in both how people make decisions and how decision making might be improved. The chapter consequently presents a broad set of prescriptive and descriptive approaches to modeling decision making and related research findings. Numerous applications are presented, and strengths and weaknesses of particular approaches are noted. Emphasis is also placed on providing useful references containing additional information on topics the reader may find of special interest.

Section 37.2 addresses topics grouped under the somewhat arbitrary heading of classical decision theory. The presented material provides a normative and prescriptive framework for making decisions. Section 37.3 summarizes decision analysis, or the application of normative decision theory to improve decisions. The discussion considers the advantages of the various approaches, how they can be applied, and what problems might arise during their application. Section 37.4 addresses topics grouped under the heading of behavioral decision theory. The material in the latter section compares human decision making to the normative models discussed earlier. Attention is given to deviations from the normative model, reasons for these deviations, and the potential of debiasing human judgments. Section 37.5 explores topics falling under the heading of dynamic and naturalistic decision theory. This material should be of interest to practitioners interested in the process followed when many "real world" decisions are made, the quality of these decisions, and why people use particular methods to make decisions. The discussion provides insight into how people perform diagnostic tasks, make decisions to take risks when using products, and develop expertise.

37.1.2 Elements of Decision Making

Decision making requires that the decision maker make a choice between two or more alternatives.[2] The selected alternative then results in some real or imaginary consequences to the decision maker. Judgment is a closely related process where a person rates or assigns values to attributes of the considered alternatives. For example, a person might judge both the safety and the attractiveness of a car being considered for purchase. Obtaining an attractive car would be a desirable consequence of the decision, whereas obtaining an unsafe car would be an undesirable consequence. The objective of a rational decision maker is to seek desirable consequences and avoid undesirable consequences.

The nature of decision making can vary greatly, depending on the decision context. Certain decisions, such as deciding where and what to eat for lunch, are routine and

---

[1]No single book covers all of the topics addressed here. More detailed sources of information are referenced throughout the chapter. Sources such as von Neuman and Morgenstern (1947), Friedman (1990), Savage (1954), Luce and Raiffa (1957), Shafer (1976), are useful texts for people desiring an introduction to normative decision theory. Raiffa (1968), Keeney and Raiffa (1976), Saaty (1988), Buck (1989), and Clemen (1991) provide applied texts on decision analysis. Kahneman, Slovic, and Tversky (1982), Winterfeldt and Edwards (1986), Svenson and Maule (1993), Payne, Bettman, and Johnson (1993), Yates (1992), and Heath et al. (1994), among numerous others, provide recent texts addressing elements of behavioral decision theory. Klein et al. (1993) provide an introduction to naturalistic decision making.

[2]Note that doing nothing can be viewed as being a choice.

repeated often. Other choices, such as purchasing a house, choosing a spouse, or selecting a form of medical treatment for a serious disease, occur seldomly, may involve much deliberation, and take place over a longer time period. Decisions may also be required under severe time pressure and involve potentially catastrophic consequences, such as when a fire chief decides whether to send firefighters into a burning building. Previous choices may constrain or otherwise influence subsequent choices (for example, a decision to enter graduate school may constrain a future employment-related decision to particular job types and locations). The outcomes of choices may be uncertain and in certain instances are determined by the actions of potentially adverse parties, such as competing manufacturers of a similar product. Decisions may be made by single individual or by a group. Within a group, there may be conflicting opinions and differing degrees of power between individuals or factions. Decision makers may also vary greatly in their knowledge and degree of aversion to risk.

Conflict occurs when a single decision maker is not sure which choice should be selected or when there is lack of consensus within a group regarding the choice. For both groups and single decision makers, conflict occurs, at the most fundamental level, because of uncertainty or conflicting objectives. Uncertainty can take many forms and is one of the primary reasons decisions can be difficult. In ill-structured decisions, decision makers may not have identified the current condition, alternatives to choose between, or their consequences. Decision makers also may be unsure what their aspirations or objectives are, or how to choose between alternatives. After a decision has been structured, at least four reasons for conflict may exist. First, when alternatives have both undesirable and desirable consequences, decision makers may experience conflict because of conflicting objectives. For example, a decision maker considering the purchase of an air bag-equipped car may experience conflict, because an air bag increases cost as well as safety. Second, decision makers may be unsure of their reaction to a consequence. For example, people considering whether to enter a raffle where the prize is a sailboat may be unsure how much they want a sailboat. Third, decision makers may not know whether a consequence will happen for sure. Even worse, they may be unsure what the probability of the consequences is, or may not have enough time to evaluate the situation carefully. They also may be uncertain about the reliability of information they have. For example, it may be difficult to determine the truth of a saleperson's claim regarding the probability of their product breaking down immediately after the warranty expires.

To resolve conflict, decision makers must deal appropriately with uncertainty, conflicting objectives, or a lack of consensus. Conflict resolution, therefore, becomes a primary focus of decision theory. The following section presents an integrative model of decision making that relates conflict resolution to the elements of decision making discussed above. This model specifically considers how decision making changes between decision contexts, if different sources of conflict are present. It also matches methods of conflict resolution to particular sources of conflict and decision rules.

37.1.3 Integrative Model of Decision Making

Human decision making can be viewed as a stage of information processing which falls between perception and response execution (Welford, 1976). The integrative model of human decision making, presented in Figure 37.1, shows how the elements of decision making discussed above fit into this perspective. From this view, decision making is the process followed when choosing a response to a perceived stimulus. The process followed depends on what decision strategy is applied and can vary greatly between decision contexts.[3] Decision strategies, in Figure 37.1, correspond to different paths between situation assessment and executing an action. The particular decision strategy followed depends upon both the decision context and whether or not the decision maker experiences conflict.[4]

At least four, sometimes overlapping categories of decision making can be distinguished. *Group decision making* occurs when multiple decision makers interact and is

[3]The notion that the best decision strategy varies between decision contexts is a fundamental assumption of the theory of contingent decision making (Payne et al., 1993), cognitive continuum theory (Hammond, 1980), and other approaches discussed in Section 37.5.1.

[4]Conflict has been recognized as an important determinant of what people will do in risky decision making contexts (Janis and Mann, 1977). Janis and Mann focus on the stressful nature of conflict and on how affective reactions in stressful situations can impact the decision strategies followed.

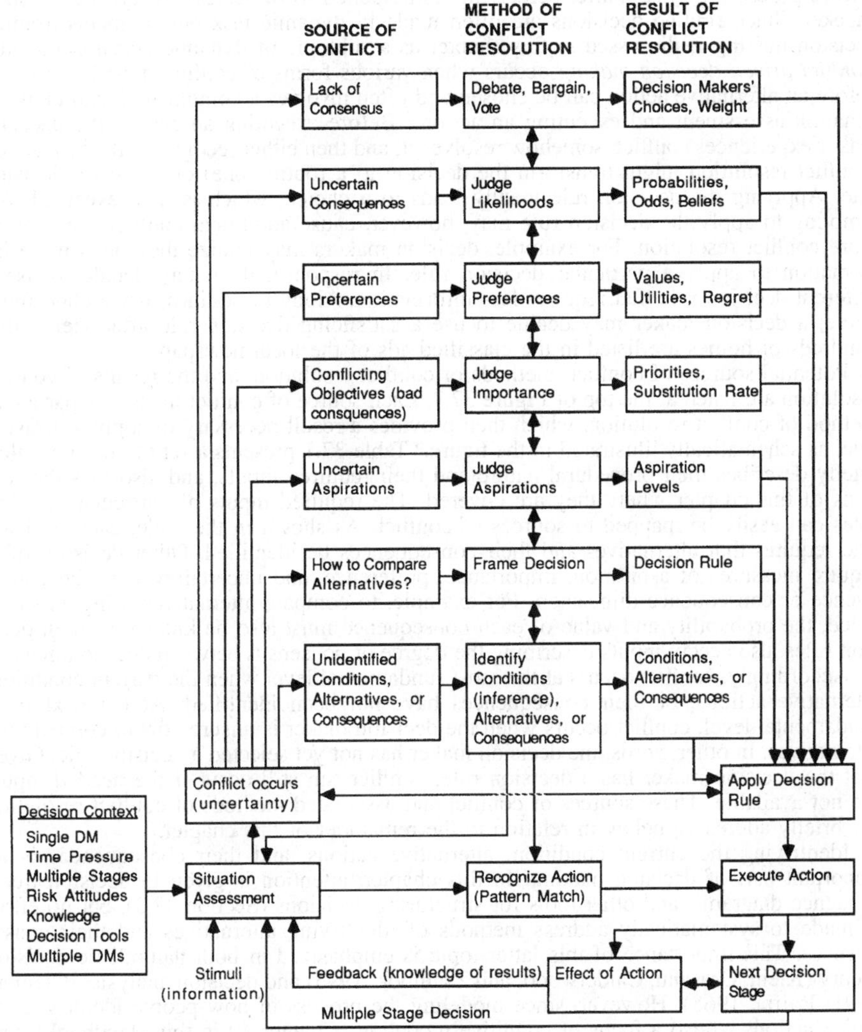

Figure 37.1 Integrative model of human decision making. DM = decision maker.

represented at the highest level of the model as a source of conflict that may be resolved through debate, bargaining, or voting. For example, members of a university faculty committee might debate and bargain before voting between candidates for a job opening. *Dynamic decision making* occurs in a changing environment, in which the results of earlier decisions impact future decisions. The decisions made in such settings often make use of feedback and are multistage in nature. For example, a decision to take a medical test almost always requires a subsequent decision regarding what to do after receiving the test results. Dynamic decision making is represented at the lowest level of the model by the presence of two feedback loops. These feedback loops show how the action taken and its effects can feed forward to the assessment of a new decision or feed back to the reassessment of the current decision.

Routine decision making occurs when decision makers use knowledge and past experience to decide quickly what to do and is especially prevalent in dynamic decision making contexts. Routine decision making is represented in Figure 37.1 as a single pattern-matching step or associative leap between situation assessment and executing an action. For example, a driver after perceiving a stop sign decides to stop; or the user of a word processing system after perceiving a misspelled word decides to activate the spell checker. Since routine decisions are often made in dynamic task environments, routine decision making is discussed in this chapter as a subtopic of dynamic decision making. *Conflict-driven decision making* occurs when various forms of conflict must be resolved before an alternative action can be chosen and often involves a complicated path between situation assessment and executing an action.[5] Before executing an action, the decision maker experiences conflict, somehow resolves it, and then either recognizes the best action (conflict resolution might transform the decision to a routine one) or applies a decision rule. Applying the decision rule ideally leads to a choice, which is then executed. Attempting to apply the decision rule may, however, cause additional conflicts, leading to more conflict resolution. For example, decision makers may realize they need more information to apply a particular decision rule. In response, they may decide to use a different decision rule that requires less information. Along these lines, when choosing a home, a decision maker may decide to use a satisficing decision rule after seeing that hundreds of homes are listed in the classified ads of the local newspaper.

Potential sources of conflict, methods of conflict resolution, and the results of conflict resolution are listed at the top of Figure 37.1. Each source of conflict maps to a particular method of conflict resolution, which then provides a result necessary to apply a decision rule, as schematically illustrated in the figure.[6] Table 37.1 presents a set of decision rules, briefly describes their procedural nature and their required inputs, and also lists the sections of this chapter where they are covered. The required inputs of particular decision rules can easily be mapped to sources of conflict. As shown in the table, each decision rule requires that alternatives and their consequences be identified. Other decision rules require measures of aspiration, importance, preference, and uncertainty for each consequence or consequence dimension. For example, to compare alternatives using expected value, the probability and value of each consequence must also be known. Certain decision rules also accept inputs describing the degree of consensus between decision makers.

Accordingly, conflict occurs at the most fundamental level when the current condition, alternative actions, or their consequences have not been identified. At the next most fundamental level, conflict occurs when the decision maker is unsure how to compare the alternatives. In other words, the decision maker has not yet selected a decision rule. Given that the decision maker has a decision rule, conflict can still occur if the needed inputs are not available. These sources of conflict and associated methods of conflict resolution are briefly addressed below in relation to the remainder of this chapter.

Identifying the current condition, alternative actions, and their consequences is an important part of decision making. In this chapter, attention is given to decision trees, influence diagrams, and other tools for structuring decisions (Section 37.3), but no effort is made to systematically address methods of identifying alternatives and their consequences. This importance of this latter topic is emphasized in both naturalistic decision theory (Klein, Orasanu, Calderwood, and Zsambok, 1993) and decision analysis[7] (Clemen, 1991; Raiffa, 1968). However, since modeling the process of how people identify alternative actions is also a focus of cognitive modeling (Chapter 40 in this Handbook), this issue is not emphasized here. This chapter does, however, consider methods of identifying the current condition falling under the topic of inference (or diagnosis). Normative approaches to inference are presented in Section 37.2.2. Section 37.4.1 presents several descriptive models of human inference and discusses potential biases.

When decision makers are unsure how to compare alternatives, they must consider what information is available and then frame the decision appropriately. The way the decision is framed then determines (1) which decision rules are appropriate, (2) what information is needed to make the decision using the given rules (as discussed earlier in reference to Table 37.1), and (3) the choices selected. As discussed in Section 37.5.1,

---

[5]The distinction between routine and conflict-driven decision making made here is similar to Rasmussen's (1983) distinction between (a) routine skill or rule-based levels of control, and (b) nonroutine knowledge-based levels of control in information-processing tasks.

[6]Note that multiple sources of conflict are possible for a given decision context. An attempt to resolve one source of conflict may also make the decision maker aware of other conflicts that must first be resolved. For example, decision makers may realize they need to know what the alternatives are before they can determine their aspiration level.

[7]Clemen (1991) includes a chapter on creativity and decision structuring. Some practitioners claim that structuring the decision is the greatest contribution of the decision analysis process.

Table 37.1 Decision Rules, Required Inputs, and Procedure Applied by the Rule

Decision Rule	Required Inputs	Procedure Applied	Section Covered
DOMINANCE	All alternatives, value of each consequence	Select alternative best on all consequences	37.2.1.2
EBA	All alternatives, value of each consequence	Select first alternative found to be best on a consequence dimension. Random order of consequences	37.2.1.3
LEXICO-GRAPHIC	All alternatives, value of each consequence, priorities	Order consequences by priority. Select first alternative found to be best on a consequence dimension	37.2.1.3
SATISFICING	At least one and up to all alternatives, aspiration level and value of each consequence	Sequentially evaluate each alternative. Stop if each consequence of an alternative equals or exceeds the aspiration level	37.2.1.4
MINIMAX COST	All alternatives, value of each consequence	Compare the worst consequence values of each alternative	37.2.1.5
MINIMAX REGRET	All alternatives, regret for each consequence	Compare largest regrets of each alternative	37.2.1.5
EV	All alternatives, probability and value of each consequence	Weight value of each consequence by its probability, for each alternative	37.2.1.6
LAPLACE	All alternatives, value or utility of each consequence	Weight value or utility of each consequence equally, for each alternative	37.2.1.7
SEU	All alternatives, probability and utility of each consequence	Weight utility of each consequence by its probability, for each alternative	37.2.1.7
MAUT	All alternatives, value or utility of each consequence, priorities	Weight value or utility of each consequence by priority, for each alternative	37.2.1.8
HOLISTIC	All alternatives and consequences	Wholistically compare the consequences of each alternative	37.2.1.9

there is reason to believe that people apply different decision-making strategies in different decision contexts. Sections 37.2.1 and 37.3.2 discuss the appropriateness of decision rules and how the particular rule used can impact choices. When the specific inputs needed by a decision rule are not available, the resulting conflict may be resolved by judging aspirations, importance, preference, or likelihood (Section 37.3). It also might be resolved by choosing a different decision rule or strategy. As noted in Section 37.5.1, there is a prevalent tendency among decision makers in naturalistic settings to minimize analysis and its required cognitive effort. In group situations, conflict resulting from a lack of consensus between multiple decision makers might be resolved through debate, bargaining, or voting (Section 37.5.2).

37.2 CLASSICAL DECISION THEORY

Classical decision theory began with the development of normative models in economics and statistics that specified optimal decisions (Savage, 1954; von Neumann and Morgenstern, 1947). Classical decision theory focuses heavily on the notion of rationality (Savage, 1954; Winterfeldt and Edwards, 1986). Emphasis is placed on the quality of the process followed when making a decision rather than on the ultimate outcome. Accordingly, a rational decision maker must think logically about the decision. To do this, the decision maker must first formally describe what is known about the decision. The decision is then made by applying principles of logic and Bayesian probability theory (Savage, 1954). This approach is therefore quantitative, and also normative or prescriptive if the numerical inputs needed are available.

The classical approach has been applied to two related problems: (1) preference and choice, and (2) statistical inference.

37.2.1 Preference and Choice

Classical decision theory represents preference and choice problems in terms of four basic elements: (1) a set of potential actions (A_i) to choose between, (2) a set of events or world states (E_j), (3) a set of consequences (C_{ij}) obtained for each combination of action and event, and (4) a set of probabilities (P_{ij}) for each combination of action and event. For example, a decision maker might be deciding whether to wear a seatbelt when traveling in an automobile. Wearing or not wearing a seatbelt corresponds to two actions, A_1 and A_2. The expected consequence of either action depends on whether an accident occurs. Having or not having an accident corresponds to two events, E_1 and E_2. Wearing a seatbelt reduces the expected consequences of having an accident (E_1). As the probability of having an accident increases, use of a belt should therefore become more attractive.

Once a decision has been represented in terms of these basic elements, the choice is then made by applying decision rules. Numerous decision rules have been developed. Decision rules are based upon basic axioms (or what are felt to be self-evident assumptions) of rational choice. Not all rules, however, make use of the same axioms. Different rules make different assumptions and can provide different preference orderings for the same basic decision. The following discussion will first present some of the most basic axioms. Then, several well-known decision rules will be briefly covered.

37.2.1.1 Basic Axioms of Rational Choice

Numerous axioms have been proposed that are essential either for a particular model of choice or for the method of eliciting numbers used for a particular model (Winterfeldt and Edwards, 1986). The best known set of axioms (Table 37.2) establishes the normative principle of subjective expected utility (SEU) as a basis for making decisions (see Luce and Raiffa, 1957; Savage, 1954 for a more rigorous description of the axioms). On an individual basis, these axioms are intuitively appealing (Stukey and Zeckhauser, 1978), but, as discussed in Section 37.4, people's preferences can deviate significantly from the SEU model in ways that conflict with certain axioms. Consequently, there has been a

Table 37.2 Basic Axioms of Subjective Expected Utility Theory

A. ORDERING OR QUANTIFICATION OF PREFERENCE

Preferences of decision makers between alternatives can be quantified and ordered using the relations:

$>$, where $A > B$ means that A is preferred to B
$=$, where $A = B$ means that A and B are equivalent
$\geq$, where $A \geq B$ means that B is not preferred to A

B. TRANSITIVITY OF PREFERENCE

If $A_1 \geq A_2$ and $A_2 \geq A_3$, then $A_1 \geq A_3$

C. QUANTIFICATION OF JUDGEMENT

The relative likelihood of each possible consequence that might result from an alternative action can be specified.

D. COMPARISON OF ALTERNATIVES

If two alternatives yield the same consequences, the alternative yielding the greater chance of the preferred consequence is preferred.

E. SUBSTITUTION

if $A1 > A_2 > A_3$, then the decision maker will be willing to accept a gamble [$p(A_1)$ and $(1 - p)(A_3)$] as a substitute for A_2 for some value of $p \geq 0$.

F. SURE-THING PRINCIPLE

If $A_1 \geq A_2$, then for all p, the gamble [$p(A_1)$ and $(1 - p)(A_3)$] $\geq$ [$p(A_2)$ and $(1 - p)(A_3)$].

movement toward developing other less restrictive standards of normative decision making (Frisch and Clemen, 1994).

Frisch and Clemen propose that: "a good decision should (a) be based on the relevant consequences of the different options (consequentialism), (b) be based on an accurate assessment of the world and a consideration of all relevant consequences (thorough structuring), and (c) make trade-offs of some form (compensatory decision rule) (p. 49)." Consequentialism and the need for thorough structuring are both assumed by all normative decision rules. Most normative rules are also compensatory. However, when people make routine habitual decisions, they often do not consider the consequences of their choices, as discussed in Section 37.5. Also, because of cognitive limitations and the difficulty of obtaining information, it becomes unrealistic in many settings for the decision maker to consider all the options and possible consequences. To make a decision under such conditions, decision makers may limit the scope of the analysis by applying principles such as satisficing and other noncompensatory decision rules discussed below. They also may apply heuristics, based on their knowledge or experience, leading to performance that can approximate the results of applying compensatory decision rules (Section 37.4).

37.2.1.2 Dominance

Dominance is perhaps the most fundamental normative decision rule. Dominance is said to occur between two alternative actions A_i and A_j when A_i is at least as good as A_j for all events E, and for at least one event E_k, A_i is preferred to A_j. For example, one investment may yield a better return than another regardless of whether the stock market goes up or down. Dominance can also be described for the case in which the consequences are multidimensional. This occurs when for all events E, the kth consequence associated with action i (C_{ik}) and action j (C_{jk}), satisfies the relation $C_{ik} \geqslant C_{jk}$ for all k, and for at least one consequence $C_{ik} > C_{jk}$. For example, a physician choosing between alternative treatments has an easy decision if one treatment is both cheaper and more effective for all patients.

Dominance is obviously a normative decision rule, since a dominated alternative can never be better than the alternative that dominates it. Dominance is also conceptually simple but can be difficult to detect when there are many alternatives to consider or many possible consequences. The use of tests for dominance by decision makers in naturalistic settings in discussed further in Section 37.5.1.5.

37.2.1.3 Lexicographic Ordering and EBA

The lexicographic ordering principle (see Fishburn, 1974) considers the case where alternatives have multiple consequences. For example, a purchasing decision might be based on both the cost and performance of the considered product. The different consequences are first ordered in terms of their importance. In terms of the above example, performance might be considered more important than cost. The decision maker then sequentially compares each alternative beginning with the most important consequence. If an alternative is found that is better than the others on the first consequence, it is immediately selected. If no alternative is best on the first dimension, the alternatives are compared for the next most important consequence. This process continues until an alternative is selected or all the consequences have been considered without making a choice. The latter situation can happen only if the alternatives have the same consequences.

The elimination by aspects (EBA) rule (Tversky, 1972) is similar to the lexicographic decision rule. It differs in that the consequences used to compare the alternatives are selected in random order, where the probability of selecting a consequence dimension is proportional to its importance. Both EBA and lexicographic ordering are noncompensatory decision rules, since the decision is made using a single consequence dimension. Returning to the above example, the lexicographic principle would result in selecting a product with slightly better performance, even if it costs much more. EBA would select either product depending on which of the consequences was first selected.

37.2.1.4 Minimum Aspiration Level and Satisficing

The minimum aspiration level or satisficing decision rule assumes that the decision maker sequentially screens the alternative actions until an action is found which is good enough. For example, a person considering the purchase of a car might stop looking once they find an attractive deal, instead of comparing every model on the market. More formally, the comparsion of alternatives stops once a choice is found that exceeds a minimum aspiration level S_{ik} for each of its consequences C_{ik} over the possible events E_k.

Satisficing can be a normative decision rule when (1) the expected benefit of exceeding the aspiration level is small, (2) the cost of evaluating alternatives is high, or (3) the cost of finding new alternatives is high. More often, however, it is viewed as an alternative to maximizing decision rules. From this view, people cope with incomplete or uncertain information and their limited rationality by satisficing in many settings instead of optimizing (Simon, 1955, 1983).

37.2.1.5 Minimax (Cost and Regret) and the Value of Information

Minimax cost selects the best alternative (A_i) by first identifying the worst possible outcome for each alternative. The worst outcomes are then compared between alternatives. The alternative with the minimum worst-case cost is selected. Formally, the preferred action A_i is the action for which over the events k, $\text{MAX}_k(C_{ik}) = \text{MIN}_i[\text{MAX}_k(C_{ik})]$. For example, in Table 37.3, the maximum cost is 5 for alternative A_1, 7 for A_2, and 8 for A_3. A_1 would be chosen since it has the smallest maximum cost. Minimax cost corresponds to assuming the worst, and therefore makes sense as a strategy where an adverse opponent is able to control the events (von Neumann and Morgenstern, 1947). Along these lines, an airline executive considering whether to reduce fares might assume that a competitor will also cut prices, leading to a no win situation.

Minimax regret involves a similar process, but the calculations are performed using regret instead of cost (Savage, 1954). Regret is calculated by first identifying which alternative is best for each possible event. The regret R_{ik}, associated with each consequence (C_{ik}) for the combination of event E_k and alternative A_i then becomes: $R_{ik} = \text{MAX}_i(C_{ik}) - C_{ik}$. Returning to earlier example, if E_1 occurs, alternative A_2 with a cost of 2 is best, resulting in a regret of 0 (2 minus 2). A_1 has a cost of 5, resulting in a regret of 3 (5 minus 2). A_3 has a cost of 6, resulting in a regret of *4* (6 minus 2). These calculations are repeated for events E_2 and E_3, resulting in regret values for each combination of events and alternative actions. The preferred action A_i is the action for which over the events k, $\text{MAX}_k(R_{ik}) = \text{MIN}_i[\text{MAX}_k(R_{ik})]$. Returning to the example, the maximum regret for A_1 (a value of 3) and A_3 (a value of 4) are both found when event E_1 occurs. The maximum regret for A_2 (a value of 2) is found when event E_2 occurs. Alternative A_2 is then selected because it has the minimum maximum regret.

Note that the minimax cost and minimax regret principles do not always suggest the same choice (Table 37.3). Minimax cost is easily interpreted as a conservative strategy. Minimax regret is more difficult to judge from an objective or normative perspective (Savage, 1954). As shown by the example, minimax regret can be less conservative than minimax cost. Alternatives that were not chosen can also impact choices made using minimax regret. For example, if alternative A_3 is removed from consideration, minimax regret and minimax cost will both select A_1. The interesting conclusion is that comparative and absolute measures of preference can result in different choices.

Bell (1982) argues persuasively that regret plays a very prominent role in decision making under uncertainty. For example, the purchaser of a new car may be happy, until finding out that a neighbor has bought the same car for $200 less from a different dealer. It is interesting to observe that regret is closely related to the value of information. This follows, since with hindsight, decision makers may regret their choice if they did not select the alternative giving the best result for the event (E_k) which actually took place. With perfect information, the decision maker would have chosen E_k. Consequently, the regret (R_{ik}) associated with having chosen alternative (A_i) is a measure of the value of having perfect information, or of knowing ahead of time that event E_k would occur. When each of the events (E_k) occur with probability P_k, it becomes possible to calculate the

Table 37.3 Example Comparison of Minimax Cost and Minimax Regret; Minimax Cost Selects A_1 and Minimax Regret Selects A_2

	E_1	E_2	E_3	MAX Cost	MAX Regret
A_1	5	5	4	5	4
A_2	3	8	2	8	3
A_3	1	10	4	10	5

expected value of perfect information [EVPI(A_i)], given that the decision maker would chose action A_i before receiving this information with the following expression:

$$\text{EVPI}(A_i) = \sum_k P_k R_{ik} \tag{1}$$

The above approach can be extended to the case of imperfect information (Raiffa, 1968), by replacing P_k, in equation (37.1), with the probability of event k (E_k) given the imperfect sample information (I). This results in an expression for the expected value of sample information [EVSI(A_i,I)], given that the decision maker would chose action A_i before receiving this information:

$$\text{EVSI}(A_i, I) = \sum_k (P_k|I) R_{ik} \tag{2}$$

The value of imperfect (or sample) information provides a normative rule for deciding whether to collect additional information. For example, a decision to perform a survey before introducing a product can be made by comparing the cost of the survey to the expected value of the information obtained. It is often assumed that decision makers are biased when they fail to seek out additional information. The above discussion shows that not obtaining information is justified when the information costs too much. From a practical perspective, the value of information can guide decisions to provide information to product users (Lehto and Papastavrou, 1991).

37.2.1.6 Maximizing Expected Value

From elementary probability theory, return is maximized by selecting the alternative with the greatest expected value. The expected value of an action A_i is calculated by weighting its consequences C_{ik} over all events k, by the probability P_{ik} the event will occur. The expected value of a given action A_i is therefore:

$$\text{EV}[A_i] = \sum_k P_{ik} C_{ik} \tag{3}$$

More generally, the decision maker's preference for a given consequence C_{ik} might be defined by a value function $v(C_{ik})$, which transforms consequences into preference values. The preference values are then weighted using the same equation. The expected value of a given action A_i becomes:

$$\text{EV}[A_i] = \sum_k P_{ik} v(C_{ik}) \tag{4}$$

Monetary value is a common value function. For example, lives lost, units sold, or air quality might all be converted into monetary values. More generally, however, value reflects preference, as illustrated by ordinary concepts such as the value of money, or the attractiveness of a work setting. Given that the decision maker has large resources and is given repeated opportunities to make the choice, choices made on the basis of expected monetary value are intuitively justifiable. A large company might make nearly all of its decisions on the basis of expected monetary value. Insurance buying and many other rational forms of behavior cannot, however, be justified on the basis of expected monetary value. Many years ago, it was already recognized that rational decision makers made choices not easily explained by expected monetary value (Bernouli, 1738). Bernouli cited the St. Petersburg paradox, in which the prize received in a lottery was 2^n, and n was the number of times (n) a flipped coin turned up heads before a tails was observed. The probability of n flips before the first tail is observed is $.5^n$. The expected value of this lottery (L) becomes:

$$\text{EV}[L] = \sum_k P_{ik} v(C_{ik}) = \sum_{n=0}^{\infty} .5^n 2^n = \sum_{n=1}^{\infty} 1 = \infty \tag{5}$$

The interesting twist is that the expected value of the above lottery is infinite. Bernouli's conclusion was that preference cannot be linear function of monetary value, since a rational decision maker would never pay more than a finite amount to play the lottery.

Furthermore, the value of the lottery can vary between decision makers. According to utility theory, this variability reflects rational differences in preference between decision makers for uncertain consequences.

37.2.1.7 Subjective Expected Utility (SEU) Theory

Expected utility theory extended expected value theory to better describe how people make uncertain economic choices (von Neumann and Morgenstern, 1947). In their approach, monetary values are first transformed into utilities, using a utility function $u(x)$. The utilities of each outcome are then weighted by their probability of occurrence to obtain an expected utility. Subjective utility theory (SEU) added the notion that uncertainty about outcomes could be represented with subjective probabilities (Savage, 1954). It was postulated that these subjective estimates could be combined with evidence using Bayes' rule to infer the probabilities of outcomes[8] (see Section 37.2.2). This group of assumptions corresponds to the Bayesian approach to statistics. Following this approach, the SEU of an alternative (A_i), given subjective probabilities (P_{ik}) and consequences (C_{ik}) over the events E_k becomes:

$$\text{SEU}[A_i] = \sum_k P_{ik} U(C_{ik}) \tag{6}$$

Note the similarity between the above formulation for SEU and the earlier equation for expected value. EV and SEU are equivalent, if the value function equals the utility function. Methods for eliciting value and utility functions differ in nature (Section 37.3). Preferences elicited for uncertain outcomes measure utility.[9] Preferences elicited for certain outcomes measure value. It accordingly has often been assumed that value functions differ from utility functions, but there are reasons to treat value and utility functions as equivalent (Winterfeldt and Edwards, 1986). The latter authors claim that the differences between elicited value and utility functions are small and that "severe limitations constrain those relationships, and only a few possibilities exist, one of which is that they are the same" (p. 213).

When people are presented choices that have uncertain outcomes, they react in different ways. In some situations, people find gambling to be pleasurable. In others, people will pay money to reduce uncertainty; for example, when people buy insurance. SEU theory distinguishes between risk-neutral, risk-averse, risk-seeking, and mixed forms of behavior. These different types of behavior are described by the shape of the utility function (Figure 37.2).

A risk-neutral decision maker will find the expected utility of a gamble to be the same as the utility of the gamble's expected value. That is, expected u(gamble) = u(gamble's expected value). For a risk-averse decision maker, expected u(gamble) < u(gamble's expected value); for a risk-seeking decision maker, expected u(gamble) > u(gamble's expected value). On any given point of a utility function, attitudes toward risk are described formally by the coefficient of risk aversion:

$$C_{\text{RA}} = \frac{u''(x)}{u'(x)} \tag{7}$$

where $u'(x)$ and $u''(x)$ are, respectively, the first and second derivatives of $u(x)$ taken with respect to x. Note that when $u(x)$ is a linear function of x, i.e., $u(x) = ax + b$, then $C_{\text{RA}} = 0$. For any point of the utility function, if $C_{\text{RA}} < 0$, the utility function depicts risk-averse behavior, and if $C_{\text{RA}} > 0$, the utility function depicts risk-seeking behavior. The coefficient of risk aversion therefore describes attitudes toward risk at each point of the utility function, given that the utility function is continuous. SEU theory consequently provides a powerful tool for describing how people may react to uncertain or risky outcomes. However, some commonly observed preferences between risky alternatives can not be explained by SEU.[10] Research on risk taking, risk perception, and risk acceptability

[8]When no evidence is available concerning the likelihood of different events, it was postulated that each consequence should be assumed to be equally likely. The Laplace decision rule makes this assumption and then compares alternatives on the basis of expected value or utility.

[9]Note that classical utility theory assumes that utilities are constant. Utilities may, of course, fluctuate. The random utility model (Bock and Jones, 1968) allows such fluctuation.

[10]Section 37.4.2 focuses on experimental findings showing deviations from the predictions of SEU.

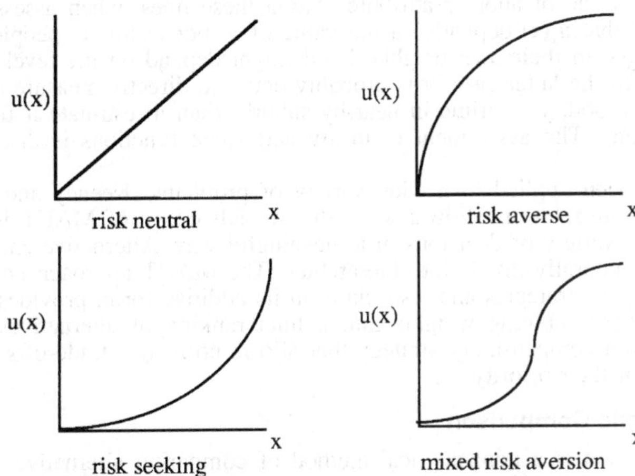

Figure 37.2 Utility functions for differing risk attitudes.

is presented in Chapter 31. Two recently published books have also focused on this topic (i.e., Singleton and Hovden, 1987; Yates, 1992).

A major contribution of SEU is that it represents differing attitudes toward risk and provides a normative model of decision making under uncertainty. The prescriptions of SEU are also clear and testable. Consequently, SEU has played a major role in fields other than economics, both as a tool for improving human decision making, and as a stepping stone for developing models that describe how people make decisions when outcomes are uncertain. As discussed further in Section 37.4, much of this work has been done in psychology.

37.2.1.8 Multiattribute Utility Theory

Multiattribute utility theory (Keeney and Raiffa, 1976) extends SEU to the case where the decision maker has multiple objectives. The approach is equally applicable for describing either utility or value functions. Following this approach, the utility (or value) of an alternative *A*, with multiple attributes *x*, is described with the multiattribute utility (or value) function $u(x_1 \ldots x_n)$, where $u(x_1 \ldots x_n)$ is some function $f(x_1 \ldots x_n)$ of the attributes *x*. In the simplest case, multiattribute utility theory (MAUT) describes the utility of an alternative as an additive function of the single-attribute utility functions $u_n(x_n)$. That is,

$$u(x_1 \ldots x_n) = \sum_n k_n u_n(x_n) \tag{8}$$

where the constants k_n are used to weigh each single attribute utility function (u_n) in terms of its importance. Assuming an alternative has three attributes, *x*, *y*, and *z*, an additive utility function is $u(x,y,z) = k_x u_x(x) + k_y u_y(y) + k_z u_z(z)$. Along these lines, a community considering building a bridge across a river, versus building a tunnel or continuing to use the existing ferry system, might consider the attractiveness of each option in terms of the attributes of economic benefits, social benefits, and environmental benefits.[11]

More complex multiattribute utility functions include multiplicative forms and functions that combine utility functions for subsets of two or more attributes (Keeney and Raiffa, 1976). An example of a simple multiplicative function would be $u(x,y) = u_x(x)*u_y(y)$. A function that combines utility functions for subsets would be $u(x,y,z) = k_{xy}u_{xy}(x,y) + k_z u_z(z)$. This latter type of function becomes useful when utility independence is violated. Utility independence is violated when the utility function for one attribute

[11]To develop the multiattribute utility function, the single-attribute utility functions (u_n) and the importance weights (k_n) are determined by assessing preferences between alternatives. Methods of doing so are discussed in Section 37.3.4.

depends on the value of another attribute. Along these lines, when assessing $u_{xy}(x,y)$ it might be found that $u_x(x)$ depends on the value of y. For example, peoples' reaction to the level of crime in their own neighborhood might depend on the level of crime in a nearby suburb. In the latter case, it is probably better to directly measure u_{xy} (x = crime in own neighborhood, y = crime in near-by suburb) than to estimate it from the single attribute functions. The assessment of utility and value functions is discussed later in Section 37.3.

MAUT has been applied to a wide variety of problems (Keeney and Raiffa, 1976; Saaty, 1988; Winterfeldt and Edwards, 1986). An advantage of MAUT is that it helps structure a large variety of decisions in a meaningful way. Alternative choices and their attributes often naturally divide into hierarchies. The MAUT approach encourages such divide and conquer strategies and, especially in its additive form, provides a straightforward means of recombining weights into a final ranking of alternatives. The MAUT approach is also a compensatory strategy that allows normative trade-offs between attributes in terms of their priority.

37.2.1.9 Holistic Comparison

Holistic comparison is a nonanalytical method of comparing alternatives. This process involves a holistic comparison of the consequences for each alternative, instead of separately measuring and then recombining measures of probability, value, or utility (Janis and Mann, 1977; Sage, 1981; Stanoulov, 1994). In so doing, a preference ordering between alternatives is obtained. For example, the decision maker might rank in order of preference a set of automobiles that vary on objectively measureable attributes, such as color, size, and price. Mathematical tools can then be used to derive the relationship between observed ordering and attribute values, and ultimately predict preferences for unevaluated alternatives, as is discussed in Section 37.3.3.4.

One advantage of holistic comparison is that it requires no formal consideration of probability or utility. Consequently, decision makers unfamiliar with these concepts may find holistic comparison to be more intuitive, and potential violations of the axioms underlying SEU and MAUT, because of their lack of understanding, become of less concern. People seem to find the holistic approach helpful when they compare complex alternatives (Janis and Mann, 1977). In fact, people may feel there is little additional benefit to be obtained from separately analyzing the probability and value attached to each attribute. This tendency becomes prevalent in naturalistic decision making, as addressed further in Section 37.5.

37.2.2 Statistical Inference

Inference is the procedure followed when a decision maker uses information to determine whether a hypothesis about the world is true. Hypotheses can specify past, present, or future states of the world or causal relationships between variables. Diagnosis is concerned with determining past and present states of the world. Prediction is concerned with determining future states. Inference or diagnosis is required in many decision contexts. For example, before deciding on a treatment, a physician must first diagnose the illness.

From the classical perspective, the decision maker is concerned with determining the likelihood that a hypothesis (H_i) is true. Bayesian inference is the best known technique, but signal detection theory, and fundamentally different approaches such as the Dempster-Schafer method, have seen application. Each of these approaches is discussed below.

37.2.2.1 Bayesian Inference

Bayesian inference is a well-defined procedure for inferring the probability (P_i) that a hypothesis (H_i) is true, from evidence (E_j) linking the hypothesis to other observed states of the world. The approach makes use of Bayes' rule to combine the various sources of evidence (Savage, 1954). Bayes' rule states that the posterior probability of hypothesis H_i given that evidence E_j is present, or $P(H_i|E_j)$, is given by the equation:

$$P(H_i|E_j) = \frac{P(E_j|H_i)P(H_i)}{P(E_j)} \tag{9}$$

where $P(H_i)$ is the probability of the hypothesis being true before the evidence E_j is obtained, and $P(E_j|H_i)$ is the probability of obtaining the evidence E_j given that the hypothesis H_i is true. For example, consider the case where a physician is attempting to determine whether a patient has a disease present in 10% of the general population. The physician has a test available that gives a positive result, 90% of the time, when administered to patients who actually have the disease. The test also gives a positive result, 20% of the time, when administered to patients who do not have the disease. If the test was administered to a member of the general population, Equation (9) predicts that the probability of having the disease given a positive test result is as:

$$P(\text{disease}|\text{positive test}) = \frac{P(\text{postive test}|\text{disease})\ P\ (\text{disease in general population})}{P(\text{positive test})}$$

also,

$$P(\text{positive test}) = P(\text{positive test}|\text{disease})P(\text{disease in general population}) + P(\text{positive test}|\text{no disease})P(\text{no disease in general population})$$

$$P(\text{disease}|\text{positive test}) = \frac{0.9*0.1}{0.9*.1 + 0.2*0.9} = 0.33$$

As discussed further in Section 37.4.1, people often fail to combine evidence consistently with the above predictions of Bayes' rule. A common finding is that people fail to consider the base rate of the hypothesis adequately. In the above example, this would correspond to focusing on $P(\text{positive test}|\text{disease}) = 0.9$, and not considering $P(\text{disease in general population}) = 0.1$. As a consequence, many people might be surprised that $P(\text{disease}|\text{positive test}) = 0.33$, rather than a number close to 0.9.

When the evidence E_j consists of multiple states $E_1 \ldots E_n$, each of which is conditionally independent, Bayes' rule can be expanded into the expression:

$$P(H_i|E_j) = \frac{\prod_{j=1}^{n} P(E_j|H_i)P(H_i)}{P(E_j)} \tag{10}$$

Calculating $P(E_j)$ can be somewhat difficult, because each piece of evidence must be dependent,[12] or else it would not be related to the hypothesis. The odds forms of Bayes' rule provide a convenient way of looking at the evidence for and against a hypothesis that does not require $P(E_j)$ to be calculated. This results in the expression:

$$\theta(H_i|E_j) = \frac{P(H_i|E_j)}{P(\sim H_i|E_j)} = \frac{\prod_{j=1}^{n} P(E_j|H_i)P(H_i)}{\prod_{j=1}^{n} P(E_j|\sim H_i)P(H_i)} \tag{11}$$

where $\theta(H_i|E_j)$ refers to the posterior odds for hypothesis H_i, $P(\sim H_i)$ is the prior probability that hypothesis H_i is not true, and $P(\sim H_i|E_j)$ is the posterior probability that hypothesis H_i is not true.

These two latter forms of Bayes' Rule provide an analytically simple way of combining multiple sources of evidence. Bayesian inference becomes much more difficult when the evidence is not certain, or when the conditional independence assumption is not met. When evidence is not certain, complex multistage forms of Bayesian analysis are required that consider the probability of the evidence being true (Winterfeldt and Edwards, 1986). When conditional independence is not true, the expanded form of Bayes' rule must be modified. For example, consider the case where the evidence consists of three events (E_1, E_2, E_3), where E_1 and E_2 are conditionally dependent, and E_3 is conditionally independent of the two other events. The posterior probability, $P(H_i|E_1,E_2,E_3)$, then becomes:

[12]Note that conditional independence between E_1 and E_2 implies that $P(E_1/H_i,E_2) = P(E_1/H_i)$ and that $P(E_2/H_i,E_1) = P(E_2/H_i)$. This is very different from simple independence, which implies that $P(E_1) = P(E_1/E_2)$ and that $P(E_2) = P(E_2/E_1)$.

$$P(H_i|E_1,E_2,E_3) = \frac{P(E_1,E_2|H_i)P(E_3|H_i)P(H_i)}{P(E_1,E_2)P(E_3|E_1,E_2)} \tag{12}$$

where $P(E_1,E_2|H_i)$ is the conditional probability of obtaining E_1 and E_2 given the hypothesis H_i; $P(E_3|H_i)$ is the conditional probability of obtaining E_3 given H_i; and $P(E_1,E_2)$ $P(E_3|E_1,E_2)$ is the probability of obtaining the evidence (E_1, E_2, E_3).

37.2.2.2 Signal Detection Theory

Bayesian inference combined with SEU leads to signal detection theory (Tanner and Swets, 1954), which has been applied in a large variety of contexts to model human performance (Wickens, 1992). In signal detection theory, the human operator is assumed to use Bayes' rule to estimate the probability that a signal actually is present from a noisy observation of the system. For example, an operator may estimate the probability a machine is going out of tolerance from a warning signal. The responses of the operator and the true state of the system together determine a set of four outcomes (Table 37.4).

The signal detection model assumes an operator receives evidence from the environment regarding the true state of the world. The relationship between the signal (S) or non-signal (N) and the evidence (E) is measured by the conditional probability [$P(E|S)$] of obtaining the observed evidence given the signal is there. The decision maker is assumed to select a criterion value (x_c) that the evidence must exceed before saying "yes." It is assumed that the value chosen will maximize utility. If the evidence is represented with a variable x, the expected utility of the operator can be described in terms of x, x_c and the four outcomes in Table 37.4. The expected utility for a given probability cutoff x_c is given by the expression:

$$\begin{aligned}\text{SEU}[x_c] &= P(x \geq x_c|S)P(S)u(h) + P(x \geq x_c|N)P(N)u(fa) \\ &\quad + P(x < x_c|S)P(S)u(m) + P(x < x_c|N)P(N)u(cr)\end{aligned} \tag{13}$$

where h is a hit, fa is a false alarm, m is a miss, and cr is a correct rejection. This expression can be maximized by first substituting $1 - P(x \geq x_c|N)$ for $P(x < x_c|N)$ and also substituting $1 - P(x \geq x_c|S)$ for $P(x < x_c|S)$ into the equation for SEU[x_c], and then setting the derivative of SEU[x_c] with respect to x_c to zero. The result at the cutoff x_c is shown below:

$$\frac{P(x = x_c|S)}{P(x = x_c|N)} = \beta^* = \frac{P(N)[u(cr) — u(fa)]}{P(S)[u(h) - u(m)]} \tag{14}$$

β^* is the optimal value of β. Substituting back the relation between $P(E|S)$ and the evidence (x), x, the optimal decision rule is to say yes if

$$\frac{P(E|S)}{P(E|N)} \geq \beta^* \tag{37.15}$$

Equation (15) can be extended to multiple operators or multiple sources of evidence (Lehto and Papastavrou, 1991). The resulting expression takes into account the probability of a false alarm and the probability of detection for the other source of information. Lehto and Papastavrou use this approach to analyze situations where the other source of information is a warning signal.

Table 37.4 Potential Outcomes Considered by Signal Detection Theory

State of the Wind			
		Noise (N)	Signal (S)
Response	yes	false alarm (fa)	hit (h)
	no	miss (m)	correct rejection (cr)

37.2.2.3 Dempster-Schafer Method

The Dempster-Shafer method (Shafer, 1976) is an alternative to Bayesian inference for accumulating evidence for or against a hypothesis. In this approach, the relation of hypotheses (H) to evidence (e) is described by a basic probability assignment (bpa) function, p. Given evidence (e), this function $p_e(n)$ assigns a value between 0 and 1 to each subset of H, such that the sum of the values assigned is 1. For example, consider the case where there are three hypotheses (A,B,C). When no evidence is available, the vacuous bpa assigns a value of 1 to the set of hypotheses H = (A,B,C), and a 0 to all subsets. That is, the subsets (A), (B), (C), (A,B), or (A,C) are each assigned a value of 0. The Bayesian approach would instead assign a probability of 0.33 to A, B, and C, respectively.

Also, given that evidence $p_e(A) = x$ supporting a specific hypothesis A is found, the Dempster-Schafer approach assigns $[1 - p_e(A)]$ to H. The Bayesian approach, of course, assigns $[1 - p_e(A)]$ to the complement of A. Returning to the above example, suppose the evidence supports hypothesis A to the degree $p_e(A) = 0.6$. Using the Dempster-Schafer approach, $p_e(\text{A,B,C}) = 0.4$. This, of course, is very different from the Bayes' interpretation, where $P(A) = 0.6$ and $P(\text{not A}) = 0.4$. The Dempster-Schafer method uses a belief function $B(n)$ to assign a total belief to n, where n is a subset of the set of possible hypotheses (H), as the sum of the beliefs assigned to m, where m is the set of possible subsets of n. In the above example, the belief in (A,B,C) after receiving evidence (e) is as given below:

$$\begin{aligned} B(\text{A,B,C}) &= p_e(\text{A,B,C}) + p_e(\text{A,B}) + p_e(\text{A,C}) + p_e(\text{B,C}) \\ &\quad + p_e(A) + p_e(B) + p_e(\text{C}) \\ &= .4 + 0 + 0 + 0 + .6 + 0 + 0 \\ &= 1.0 \end{aligned} \tag{16}$$

Similarly, the belief in (A,B) after receiving the evidence (e) is:

$$\begin{aligned} B(\text{A,B}) &= p_e(\text{A,B}) + p_e(\text{A}) + p_e(\text{B}) \\ &= 0 + .6 + 0 \\ &= .6 \\ &= B(\text{A}) \end{aligned} \tag{17}$$

To combine evidences from multiple sources e and f, Dempster-Schafer theory uses the combining function $c[p_e(X),p_f(Y)]$, where X and Y are both sets of subsets of H. For example, we might have X = [(A), (A,B,C)] and Y = [(A,B), (A,B,C)]. The combining function then assigns a value to each subset n of H. The value assigned is determined by first describing the set of subsets n' within n defined by the intersection of subsets within X and subsets within Y. A value of 0 is assigned to all subsets of n not within n'. The products $p_e(X)*p_f(Y)$ are then summed and assigned to each subset within n'. Returning to the above example, we can calculate $c(n')$ using the values given in Table 37.5. First note that the set of subsets n' for the example is defined by the inner elements of the table. Specifically, n' = [(A), (A,B), (A,B,C)]. The values used by the combining function $c(n')$ are also shown. Using these numbers, the values of $c(n')$ become: $c(\text{A}) = 0.24 + 0.36 = 0.6$; $c(\text{A,B}) = 0.16$; $c(\text{A,B,C}) = 0.24$. All remaining subsets for this evidence are assigned a value of 0.

It has been argued that the Dempster-Schafer method of assigning evidence is better suited than the Bayesian method for diagnosing medical problems (Gordon and Shortliffe, 1984). The latter researchers particularly criticize the Bayesian assumption that evidence

Table 37.5 Tableau for Dempster-Shafer Method of Combining Evidence

	Y = [(A,B),	**(A,B,C)]**
X = [(A), (A,B,C)]	p_f(A,B) = 0.4	p_f(A,B,C) = 0.6
P_e(A) = 0.6	A ; $p_e p_f$ = 0.24	A ; $p_e p_f$ = 0.36
p_e(A,B,C) = 0.4	A,B ; $p_e p_f$ =; 0.16	A,B,C ; $p_e p_f$ = 0.24

partially supporting a hypothesis should also support its negation. Gordon and Shortliffe note that the Dempster-Schafer method shows promise as a means of accumulating belief in expert diagnostic systems used in medicine.

37.3 DECISION ANALYSIS

The application of classical decision theory to improve human decision making is the goal of Decision analysis (Howard, 1968, 1988; Keeney and Raiffa, 1976; Raiffa, 1968). Decision analysis requires inputs from decision makers, such as goals, preference and importance measures, and subjective probabilities. Elicitation techniques have consequently been developed that help decision makers provide these inputs. Particular focus has been placed on methods of quantifying preferences, tradeoffs between conflicting objectives, and uncertainty (Keeney and Raiffa, 1976; Raiffa, 1968). As a first step in decision analysis, it is necessary to do some preliminary structuring of the decision, which then guides the elicitation process. The following discussion first presents methods of structuring decisions and then covers techniques for assessing subjective probabilities, utility functions, and preferences.

37.3.1 Structuring Decisions

The field of decision analysis has developed many useful frameworks for representing what is known about a decision (Clemen, 1991; Howard, 1968; Winterfelt and Edwards, 1986). In fact, these authors and others have stated that the process of structuring decisions is often the greatest contribution of going through the process of decision analysis. Among the many tools used, decision matrices and trees provide a convenient framework for comparing decisions on the basis of expected value or utility. Value trees provide a helpful method of structuring the sometimes complex relationships between objectives, attributes, goals, and values and are used extensively in multi-attribute decision-making problems. Event trees, fault trees, inference trees, and influence diagrams are useful for describing probabilistic relationships between events and decisions. Each of these approaches is briefly discussed below.

37.3.1.1 Decision Matrices and Trees

Decision matrices are often used to represent single-stage decisions (Figure 37.3). The simplicity of decision matrices is their primary advantage. They also provide a very convenient format for applying the decision rules discussed in the previous section. Decision trees are also commonly used to represent single-stage decisions (Figure 37.4), and are particularly useful for describing multistage decisions (Raiffa, 1968). Note that in a multistage decision tree, the probabilities of later events are conditioned on the result of earlier events. This leads to the important insight that the results of earlier events provide information regarding future events.[13] Following this approach, decisions may be stated in conditional form. An optimal decision, for example, may be first to do a market survey, and then to market the product only if the survey is positive.

Analysis of a single-stage or multistage decision tree involves two basic steps referred to as averaging out and folding back (Raiffa, 1968). These steps, respectively, occur at chance and decision nodes.[14] Averaging out occurs when the expected value (or utility) at each chance node is calculated. In Figure 37.4, this corresponds to calculating the

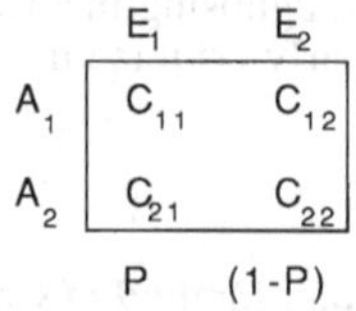

	E_1	E_2
A_1	C_{11}	C_{12}
A_2	C_{21}	C_{22}
	P	(1-P)

Figure 37.3 Decision matrix representation to a single-stage decision.

[13]For example, the first event in a decision tree might be the result of a test. The test result then provides information useful in making the final decision.

[14]Note that standard convention uses circles to denote chance nodes and squares to denote decision nodes (Raiffa, 1968).

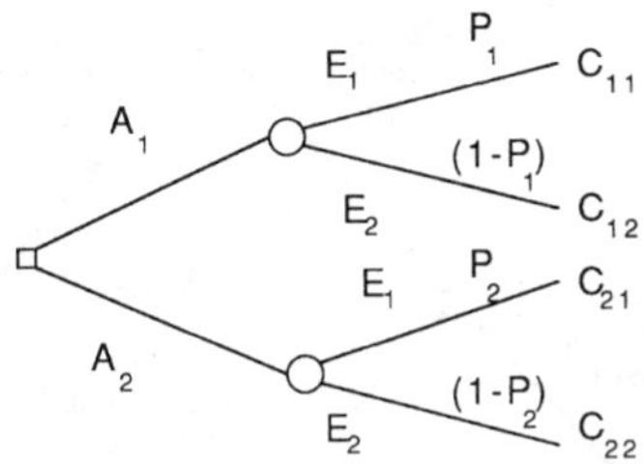

Figure 37.4 Decision tree representation of a single-stage decision.

expected value of A_1 and A_2, respectively. Folding back refers to choosing the action with the greatest expected value at each decision node.

Decision trees consequently provide a straightforward way of comparing alternatives in terms of expected value or SEU. However, their development requires significant simplification of most decisions, and the provision of numbers, such as measures of preference and subjective probabilities, that decision makers may have difficulty determining. In certain contexts, decision makers struggling with this issue may find it helpful to develop value trees, event trees, or influence diagrams, as expanded upon below.

37.3.1.2 Value trees

Value trees hierarchically organize objectives, attributes, goals, and values (Figure 37.5). From this perspective, an objective corresponds to satisficing or maximizing a goal or set of goals. When there is more than one goal, the decision maker will have multiple objectives, which may differ in importance. Objectives and goals are both measured on a set of attributes. Attributes may provide (1) objective measures of a goal, such as when fatalities and injuries are used as a measure of highway safety; (2) subjective measures of a goal, such as when people are asked to rate the quality of life in the suburbs vs. the city; or (3) proxy or indirect measures of a goal, such as when the quality of ambulance service is measured in terms of response time.

In generating objectives and attributes, it becomes important to consider their relevance, completeness, and independence. Desirable properties of attributes (Keeney and Raiffa, 1976) include: (1) completeness—the extent to which the attributes measure whether an objective is met, (2) operationality—the degree to which the attributes are meaningful and feasible to measure, (3) decomposability—whether the whole is described by its parts, (3) nonredundancy—correlated attributes give misleading results, and (4) minimum size—considering irrelevant attributes is expensive and may be misleading. Once a value tree has been generated, various methods can be used to assess preferences between the alternatives directly.

37.3.1.3 Event Trees or Networks

Event trees or networks show how a sequence of events can lead from primary events to one or more outcomes. Human reliability analysis (HRA) event trees are a classic example

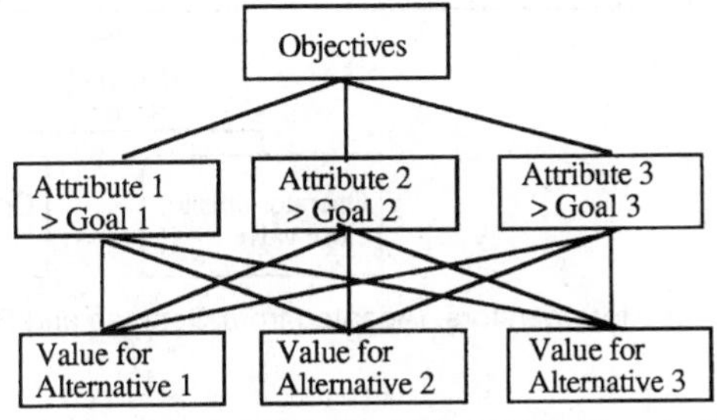

Figure 37.5 Generic value tree.

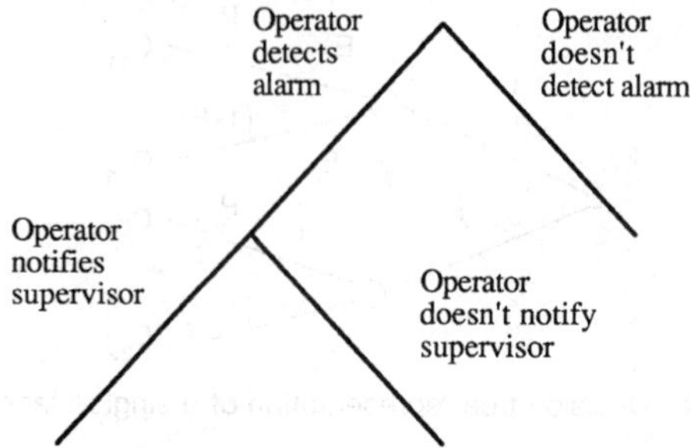

Figure 37.6 HRA event tree. (Adapted from Gertman and Blackman, 1994).

of this approach (Figure 37.6). If probabilities are attached to the primary events, it becomes possible to calculate the probability of outcomes, as illustrated in Section 37.3.2.4. This approach has been used in the field of risk assessment to estimate the reliability of human operators and other elements of complex systems (Gertman and Blackman, 1994). Chapter 31 provides additional information on human reliability analysis and other methods of risk assessment.

Fault trees work backward from a single undesired event to its causes (Figure 37.7). Fault trees are commonly used in risk assessment to help infer the chance of an accident occurring (Gertman and Blackman, 1994; Hammer, 1993). *Inference trees* relate a set of hypotheses at the top level of the tree to evidence depicted at the lower levels. This latter approach has been used by expert systems, such as PROSPECTOR (Duda, Hart, Konolige, and Reboh 1979). PROSPECTOR applies a Bayesian approach to infer the presence of a mineral deposit from uncertain evidence.

37.3.1.4 Influence Diagrams and Cognitive Mapping

Influence diagrams are often used in the early stages of a decision to show how events and actions are related. Their use in the early stages of a decision is referred to as knowledge (or cognitive) mapping (Howard, 1988). Links in an inference diagram depict causal and temporal relations between events and decision stages.[15] A link leading from an event (A) to an event (B) implies that the probability of obtaining event B depends

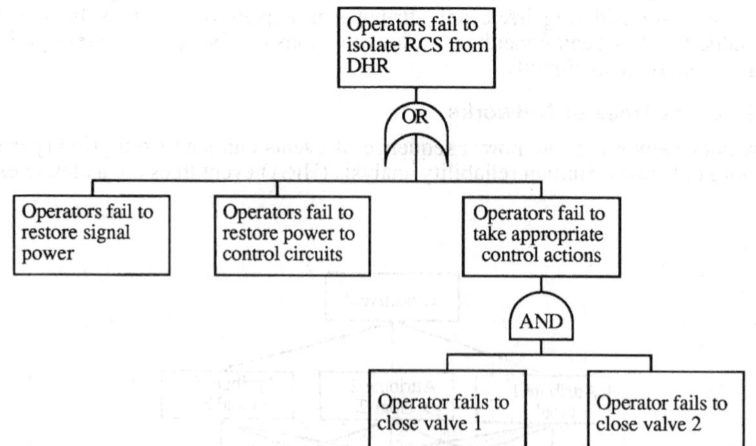

Figure 37.7 Fault tree for operators. (Adapted from Gertman and Blackman, 1994).

[15]As for decision trees, the convention for influence diagrams is to depict events with circles and decisions with squares.

on whether event A has occurred. A link leading from an decision to an event implies that the probability of the event depends on the choice made at that decision stage. A link leading from an event to a decision implies that the decision maker knows the outcome of the event at the time the decision is made.

One advantage of influence diagrams in comparison to decision trees is that influence diagrams show the relationships between events more explicitly. Consequently, influence diagrams are often used to represent complicated decisions where events interactively influence the outcomes. For example, the influence diagram in Figure 37.8 shows that the true state of the machine (i.e., whether it is in tolerance) affects both the probability of the warning signal and the consequence of the operators decision. This linkage would be hidden within a decision tree.[16] Influence diagrams have been used to structure medical decision-making problems (Holtzman, 1989) and are emphasized in modern texts on decision analysis (Clemen, 1991). Howard (1988) states that influence diagrams are the greatest advance he has seen in the communication, elicitation, and detailed representation of human knowledge. Part of the issue is that influence diagrams allow people who do not have deep knowledge of probability to describe complex conditional relationships with simple linkages between events. Once these linkages are defined, the decision becomes well defined and can be formally analyzed.

37.3.2 Probability Assessment

Several approaches have been used in decision analysis to assess subjective probabilities. In this section several of the more well-known techniques will be summarized. These techniques include: (1) direct numerical assessment, (2) fitting subjective belief forms, (3) the bisection method, (4) conditioning arguments, (5) preferences between reference gambles, and (6) scaling methods. Techniques proposed for improving the accuracy of assessed probabilities, including scoring rules, calibration, and group assessment will then be presented.

37.3.2.1 Direct Numerical Assessment

In direct numerical estimation, decision makers are asked to give a numerical estimate of how likely they think the event is to happen. These estimates can be probabilities, odds, log odds, or words (Winterfeldt and Edwards, 1986). Winterfeldt and Edwards argue that log odds have certain advantages over the other measures. Gertman and Blackman (1994) note that log odds are normally used in risk assessment for nuclear power applications because human error probabilities (HEPs) vary greatly in value. Values ranging from 1 to 0.00001 are typical.

37.3.2.2 Fitting a Subjective Belief Form

Fitting a subjective belief form requires that the questions be posed in terms of statistical parameters. That is, decision makers could be asked first to consider their uncertainty

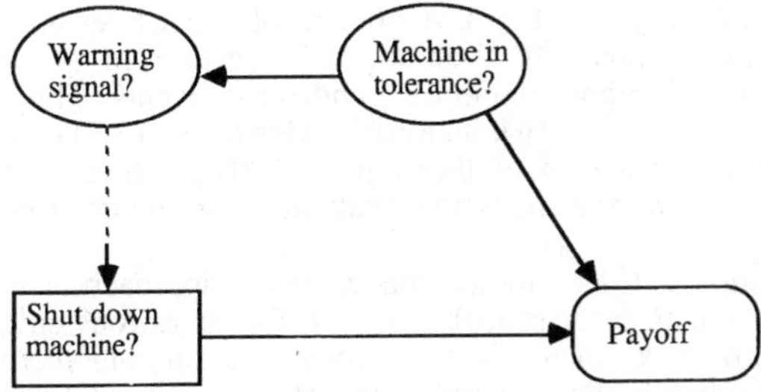

Figure 37.8 Influence diagram representation of a single-stage decision.

[16]The conditional probabilities in a decision tree would reflect this linkage, but the structure of the tree itself does not show the linkage directly. Also, the decision tree would use the flipped probability tree using *P*(warning) at the first stage and *P*(machine down|warning) at the second stage. It seems more natural for operators to think about the problem in terms of *P*(machine down) and *P*(warning|machine down), which is the way the influence diagram in Figure 37.7 depicts the relationship.

regarding the true value of a given probability, and then to estimate their mean, mode, or median belief. This approach can be further extended by asking decision makers to describe how certain they are of their estimate. For example, a worker might subjectively estimate the mean and variance of the proportion of defective circuit boards before inspecting a small sample of circuit boards. If the best estimate corresponds to a mean, mode, or median, and the estimate of certainty to a confidence interval or standard deviation, a functional form such as the Beta-1 probability density function (pdf) can then be used to fit a subjective probability distribution (Buck, 1989; Clemen, 1991).

In other words, a distribution is specified that describes the subject's belief that the the true probability equals particular values. This type of distribution can be said to express uncertainty about uncertainty (Raiffa, 1968). Given that the subject's belief can be described with a Beta-1 pdf, Bayesian methods can easily be used to combine binomially distributed evidence with the subject's prior belief (Buck, 1989; Clemen, 1991). Returning to the above example, the worker's prior subjective belief can be combined with the results of inspecting the small sample of circuit boards, using Bayes' rule. As more evidence is collected, the weight given to the subjects initial belief becomes smaller compared to the evidence collected. The use of prior belief forms also reduces the amount of sample information that must be collected to show that a proportion, such as the percentage of defective items, has changed (Buck, 1989).

37.3.2.3 Bisection Method

The bisection method (Raiffa, 1968) is another direct technique for attempting to estimate a subjective probability density function (pdf). This technique is somewhat more general than fitting the subject's belief with a functional form, such as the Beta-1, since it makes no parametric assumptions. The bisection method involves two steps that are repeated until the subject's belief is adequately described. Following this approach, the first step is to determine the median ($P_{.5}$) of the subjective pdf. This question is posed to the decision maker in a form such as "For what value of p do you feel it is equally likely the true value p^* is greater than or less than p?" This step is then repeated for subintervals to obtain the desired level of detail.

37.3.2.4 Conditioning Arguments

Statistical conditioning arguments are based on the idea that the probability of a complicated event, such as the chance of having an accident, can be determined by estimating the probability of simpler events (or subsets). From a more formal perspective, a conditioning argument determines the probability of an event A by considering the possible conditions (C_i) under which A might happen, the associated conditional probabilities [$P(A|C_i)$], and the probability of each condition [$P(C_i)$]. The probability of A can then be represented as:

$$P(A) = \sum_i P(A|C_i)P(C_i) \tag{18}$$

This approach is illustrated by the development of event trees and fault tree analysis. In fault tree analysis, the probability of an accident is estimated by considering the probability of human errors, component failures, and other events. This approach has been extensively applied in the field of risk analysis[17] (Gertman and Blackman, 1994; also see Chapter 31 for a brief discussion of this topic). THERP (Swain and Guttman, 1983) extends the conditioning approach to the evaluation of human reliability in complex systems.

SLIM-MAUD (Embrey, 1984) implements a related approach in which expert ratings are used to estimate human error probabilities (HEPs) in various environments. The experts first rate a set of tasks in terms of *performance shaping factors* (PSFs) that are present. Tasks with known HEPs are used as upper and lower anchor values. The experts also judge the importance of individual PSFs. A *subjective likelihood index* (SLI) is then calculated for each task in terms of the PSFs. A logarithmic relationship is assumed between the HEP and SLI, allowing calculation of the human error probability for task j (HEP_j) from the subjective likelihood index assigned to task j (SLI_j). More specifically:

[17]Note that it has been shown that people viewing fault trees can be insensitive to missing information (Fischhoff, Slovic, and Lichtenstein, 1978).

$$\text{Log }(1 - \text{HEP}_j) = a\text{SLI}_j + b \tag{19}$$

where

$$\text{SLI}_j = \sum_i \text{PSF}_{ij} * I(\text{PSF}_i) \tag{20}$$

and $I(\text{PSF}_i)$ is the importance of PSF_i, and PSF_{ij} is the rating given to PSF_i for task j. Gertman and Blackman (1994) provide guidelines regarding the use of this method and have generally positive conclusions. SLIM-MAUD is interesting in that it uses multiattribute utility theory as a basis for generating probability estimates.

37.3.2.5 Reference Lotteries

Reference lottery methods take a less direct approach to obtaining point estimates of the decision maker's subjective probabilities. When the objective is to measure how likely event A is to occur, the approach asks decision makers to consider a lottery where they will receive a prize x if event A occurs, and a prize y if it does not. They are then asked how much they would be willing to pay for the lottery. The amount they are willing to pay (z) is then equated to the lottery, using the relation: $z = P(A)x + [1 - P(A)]y$. From this expression it becomes possible to estimate the decision maker's subjective estimate of $P(A)$. Specifically, $P(A) = (z - y)/(x - y)$. A variant of this approach that asks decision makers to compare two lotteries over the same range of preferences may be preferable since it removes the potential effect of risk aversion (Winterfelt and Edwards, 1986).

37.3.2.6 Scaling Methods

Scaling methods ask subjects to rate or rank the probabilities to be assessed. Likert scales with verbal anchors have been used to obtain estimates of how likely people feel certain risks are (Kraus and Slovic, 1988). Another approach has been to ask subjects to do pairwise comparisons of the likelihoods of alternative events (Saaty, 1988). Pairwise comparisons of probabilities on a ratio scale correspond to relative odds and consequently have high construct validity. In fact, much of the risk assessment focuses on determining order of magnitude differences in probability. Saaty (1988), however, argues that the psychometric literature indicates that people's ability to distinguish items on the same scale is limited to 7 ± 2 categories. He consequently proposes use of a relative scale to measure differences in importance, preference, and probability that uses verbal anchors corresponding to equal, weak, strong, very strong, and absolute differences between rated items. In perhaps the most controversial aspect of his approach, these five verbal anchors are assigned the numbers 1, 3, 5, 7, and 9. Using these numbers, subjective probabilities can then be calculated from pairwise ratings on his verbal scale.

37.3.2.7 Scoring Rules, Calibration, and Group Assessment

A number of approaches have been developed for improving the accuracy of assessed probabilities (Lichtenstein, Fischhoff, and Phillips, 1982; Winterfelt and Edwards, 1986). Two desirable properties of elicited probabilities include *extremeness* and *calibration*. More extreme probabilities [for example, $P(\text{good sales}) = 0.9$ vs. $P(\text{good sales}) = 0.5$] make decisions easier since the decision maker can be more sure of what is really going to happen. Well-calibrated probability estimates match the actual frequencies of observed events. Scoring rules provide a means of evaluating assessed probabilities in terms of both extremeness and calibration. If decision makers assess probabilities on a routine basis, feedback can be provided using scoring rules. Such feedback seems to be associated with the highly calibrated subjective probabilities provided by weather forecasters (Murphy and Winkler, 1974).

Group assessment of subjective probabilities is another often followed approach, as alluded to earlier in reference to SLIM-MAUD. There is evidence that group judgments are usually more accurate than individual judgments, and that groups tend to be more confident in their estimates (Sniezek and Henry, 1989; Sniezek, 1992). Assuming that individuals within a group independently provide estimates, which are then averaged, the benefit of group judgment is easily shown to have a mathematical basis. Simply put, a mean should be more reliable than an individual observation. Group dynamics, however, can lead to a tendency toward conformity (Janis, 1972). Winterfelt and Edwards (1986) therefore recommend that members of a group be polled independently.

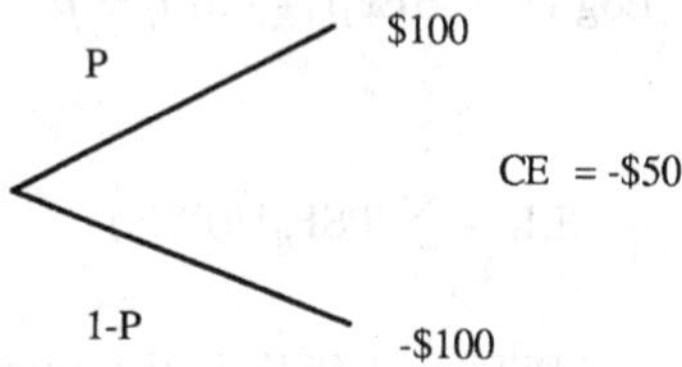

Figure 37.9 The standard gamble used in the variable probability method of eliciting utility functions.

37.3.3 Utility Function Assessment

Standard methods for assessing utility functions (Raiffa, 1968) include: (1) the variable probability method, and (2) the certainty equivalent method. In the variable probability method, the decision maker is asked to give the value for the probability of winning at which they are indifferent between a gamble and a certain outcome (Figure 37.9). A utility function is then mapped out when the value of the certainty equivalent (CE) is changed over the range of outcomes. In Figure 37.9, the value of P at which the decision maker is indifferent between the gamble and the certain loss of \$50 gives the value for $u(-\$50)$. In the utility function in Figure 37.10, the decision maker gave a value of about 0.5 in response to this question.

The certainty equivalent method uses lotteries in a similar way. The major change is that the probability of winning or losing the lottery is held constant, while the amount won or lost is changed. In most cases, the lottery provides an equal chance of winning and losing. The method begins by asking the decision maker to give a certainty equivalent for the original lottery (CE_1). The value chosen has a utility of 0.5. This follows since the utility of the best outcome is assigned a value of 1, and the worst is given a utility of 0. The utility of the original gamble is therefore:

$$u(CE_1) = pu(\text{best}) + (1 - p)u(\text{worst}) = p(1) + (1 - p)(0) = p = 0.5 \qquad (21)$$

The decision maker is then asked to give certainty equivalents for a two new lotteries. Each uses the CE from the previous lottery as one of the potential prizes. The other prizes used in the two lotteries are the best and worst outcomes from the original lottery, respectively. The utility of the certainty equivalent (CE_2) for the lottery using the best outcome and CE_1 is given by the expression:

$$u(CE_2) = pu(\text{best}) + (1 - \text{p})u(CE_1) = p(1) + (1 - p)(0.5) = 0.75 \qquad (22)$$

The utility of the certainty equivalent (CE_3) given for the lottery using the worst outcome and CE_1 is given by:

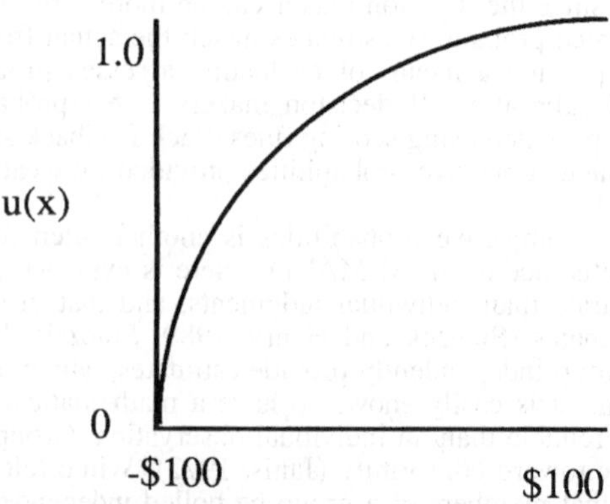

Figure 37.10 A typical utility function.

$$u(\mathrm{CE}_3) = pu(\mathrm{CE}_1) + (1 - p)u(\mathrm{worst}) = p(0.5) + (1 - p)(0) = 0.25 \quad (23)$$

This process is continued until the utility function is specified in sufficient detail. A problem with the certainty equivalent method is that errors are compounded as the analysis proceeds. This follows since the utility assigned in the first preference assessment [i.e., $u(\mathrm{CE}_1)$] is used throughout the subsequent preference assessments. A second issue is that the CE method uses different ranges in the indifference lotteries, meaning that the CEs are compared against different reference values. This may create inconsistencies, since, as discussed later in Section 37.4, attitudes toward risk usually change depending on whether outcomes are viewed as gains or losses. The use of different reference points may, of course, cause the same outcome to be viewed as either a loss or a gain.

37.3.4 Preference Assessment

Methods for measuring strength of preference include indifference methods, direct assessment, and indirect measurement (Keeney and Raiffa, 1976; Winterfeldt and Edwards, 1986). Indifference methods modify one of two sets of stimuli until subjects feel they are indifferent between the two. Direct assessment methods ask subjects to rate or otherwise assign numerical values to attributes, which are then used to obtain preferences for alternatives. Indirect measurement techniques avoid decomposition and simply ask for preference orderings between alternatives. Recently there has been some movement toward evaluating the effectiveness of particular methods for measuring preferences (Birnbaum, Coffey, Mellers, and Weiss, 1992; Huber, Wittink, Fiedler, and Miller, 1993).

37.3.4.1 Indifference Methods

Indifference methods are illustrated by the variable probability and certainty equivalent methods of eliciting utility functions presented in the previous section. There, indifference points were obtained by varying either probabilities or values of outcomes. Similar approaches have been applied to develop multiattribute utility or value functions. This approach involves four steps: (1) develop the single attribute utility or value functions, (2) assume a functional form for the multiattribute function, (3) assess the indifference point between various multiattribute alternatives, and (4) calculate the substitution rate or relative importance of one attribute compared to the other. The single attribute functions might be developed by indifference methods (i.e., the variable probability or certainty equivalent methods) or direct assessment methods as discussed later. Indifference points between multiattribute outcomes are obtained through an interactive process in which the values of attributes are systematically increased or decreased. Substitution rates are then obtained from the indifference points.

For example, consider the case for two alternative traffic safety policies, A_1 and A_2. Each policy has two attributes, x = lives lost and y = money spent. Assume the decision maker is indifferent between A_1 and A_2, meaning the decision maker feels that $v(x_1,y_1) = v(20{,}000 \text{ deaths}; \$1 \text{ trillion})$ is equivalent to $v(x_2,y_2) = v(10{,}000 \text{ deaths}; \$1.5 \text{ trillion})$. For the sake of simplicity, assume an additive value function, where $v(x,y) = kv_x(x) + (1 - k)v_y(y)$. Given this functional form, the indifference point $A_1 = A_2$ is used to derive the relation:

$$kv_x(20{,}000 \text{ deaths}) + (1 - k)v_y(\$1 \text{ trillion}) = kv_x(10{,}000 \text{ deaths}) + (1 - k)v_y(\$1.5 \text{ trillion}) \quad (24)$$

This results in the substitution rate as shown below:

$$\frac{k}{1 - k} = \frac{v_x(20{,}000 \text{ deaths}) - v_x(10{,}000 \text{ deaths})}{v_y(\$1.5 \text{ trillion}) - v_y(\$1 \text{ trillion})} \quad (25)$$

If v_x and v_y are linear functions, a value of 0.02 is obtained for k. This results in the multiattribute expression $v(x,y) = 0.02v_x(x = \text{lives lost}) + 0.98(1 - k)v_y(y = \text{money spent})$. The procedure becomes somewhat more complex when other nonadditive forms are assumed for the multiattribute function (Keeney and Raiffa, 1976).

37.3.4.2 Direct Assessment Methods

Direct assessment methods include curve fitting and various numerical rating methods (Winterfelt and Edwards, 1986). *Curve fitting* is perhaps the simplest approach. Here, the decision maker first orders the various attributes and then simply draws a curve assigning values to them. For example, an expert might draw a curve relating levels of traffic noise (measured in decibels) to their level of annoyance (on a scale of 0 to 1). *Rating methods*, as discussed earlier in reference to subjective probability assessment, include direct numerical measures on rating scales, and relative ratings.

The *analytic hierarchy process* (AHP) provides one of the more implementable methods of this type (Saaty, 1988). In this approach, the decision is first structured as a value tree (Figure 37.5). Then each of the attributes is compared in terms of importance in a pairwise rating process. When entering the ratings, decision makers can enter numerical ratios (for example, one attribute may be twice as important as another) or use the subjective verbal anchors mentioned earlier in reference to subjective probability assessment. The program then calculates a normalized eigenvector assigning importance or preference weights to each attribute. Each alternative is then compared on the separate attributes. For example, two houses might first be compared in terms of cost, and then be compared in terms of attractiveness. This results in another eigenvector describing how well each alternative satisfies each attribute. These two sets of eigenvectors are then combined into a single vector that orders alternatives in terms of preference. The subjective multiattribute rating technique (SMART) developed by Edwards (see Winterfeldt and Edwards, 1986) provides a similar easily implementable approach. Both techniques are computerized, making the assessment process relatively painless.

37.3.4.3 Indirect Measurement

Indirect measurement techniques avoid asking people to rate or rank the importance of factors that impact their preferences directly. Instead, subjects simply state or order their preferences for different alternatives. A variety of approaches can then be used to determine how individual factors influence preference. *Conjoint measurement* theory provides one such approach for separating the effects of multiple factors when only their joint effects are known. Application of the approach entails asking subjects to develop an ordered set of preferences for a set of alternatives that systematically vary attributes felt to be related to preference. The relationship between preferences and values of the attributes is then assumed to follow some functional form. The most common functional form assumed is a simple additive weighting model. Preference orderings obtained using the model are then compared to the original rankings. Example applications of conjoint measurement theory to describe preferences between multiattribute alternatives are discussed in Winterfelt and Edwards (1986). Related applications include the dichotomy-cut method used to obtain decision rules for individuals and groups from ordinal rankings of multiattribute alternatives (Stanoulov, 1994).

The *policy capturing* approach used in social judgment theory (Hammond, 1993; Hammond, Stewart, Brehmer, and Steinman, 1975) is another indirect approach for describing human judgments of both preferences and probability. The policy capturing approach uses multivariate regression, or other similar techniques, to relate preferences to attributes for one or more decision makers. The equations obtained correspond to policies followed by particular decision makers. An example equation might relate medical symptoms to a physician's diagnosis. It has been argued that the policy capturing approach measures the influence of factors on human judgments more accurately than decomposition methods. Captured weights might be more accurate because decision makers may have little insight into the factors which impact their judgments (Valenzi and Andrews, 1973). People may also weigh certain factors in ways that reflect social desirability rather than influence their judgments (Brookhouse, Guion, and Doherty, 1986). For example, people comparing jobs might rate pay as being lower in importance than intellectual challenge, while their preferences between jobs might be predicted entirely by pay. On the other hand, caution must be taken when interpreting regression weights as indicating importance, since regression coefficients are influenced by correlations between factors, their variability, and their validity (Stevenson, Busemeyer, and Naylor, 1993).

37.4 BEHAVIORAL DECISION THEORY

As a normative ideal, classical decision theory has influenced the study of decision making in a major way. Much of the earlier work in behavioral decision theory compared human behavior to the prescriptions of classical decision theory (Edwards, 1954; Einhorn and Hogarth, 1981; Slovic, Fischoff, and Lichtenstein, 1977). Numerous departures were found, including the influential finding that people use heuristics during judgment tasks (Tversky and Kahneman, 1974). On the basis of such research, pyschologists have concluded that other approaches are needed to describe the process of human decision making. Descriptive models that relax assumptions of the normative models, but still retain much of their essence, are now being evaluated in the field of judgment and decision theory (Stevenson et al., 1993).

The following discussion summarizes findings from this broad body of literature. The discussion begins by considering research on statistical estimation and inference. Attention then shifts to the topic of decision making under uncertainty and risk.

37.4.1 Statistical Estimation and Inference

The ability of people to perceive, learn, and draw inferences accurately from uncertain sources of information has been a topic of much research. The following discussion first briefly considers human abilities and limitations on such tasks. The next section introduces several heuristics people seem to use to cope with their limitations and considers how their use can cause certain biases. Attention then shifts to probabilistic information processing models and policy capturing models. These modeling approaches provide a mathematically oriented view of both how people judge probabilities, the biases that may occur, and how people learn to perform probability judgment tasks. The final section briefly summarizes findings on debiasing human judgments.

37.4.1.1 Human Abilities and Limitations

Research conducted in the early 1960s tested the notion that people behave as 'intuitive statisticians, who gather evidence and apply it in accordance with the Bayesian model of inference (Peterson and Beach, 1967). Several studies evaluated how good people are at estimating statistical parameters, such as means, variances, and proportions. Other studies have compared human inferences obtained from probabilistic evidence to the prescriptions of Bayes' rule.

A number of interesting results were obtained (Table 37.6). The research first shows that people can be fairly good at estimating means, variances, or proportions from sample data. However, this ability drops greatly when the judged events occur either rarely or very often. In particular, when people are asked to estimate the risk associated with the use of consumer products (Dorris and Tabrizi, 1978; Rethans, 1980; also see Chapter 31) or various technologies (Lichtenstein, Slovic, Fischhoff, Layman, and Coombs, 1978), estimates can be weakly related to accident data. Weather forecasters are one of the few groups of people that have been documented as being able to estimate high and low probabilities accurately (Winkler and Murphy, 1973).

Part of the issue is that risk estimates are related to factors other than likelihood, such as catastrophic potential, degree of control, or familiarity (Lichtenstein et al., 1978; Lehto, James, and Foley, 1994; Slovic, 1978, 1987). Weber (1994) provides additional evidence that subjective probabilities are related to factors other than uncertainty and argues that people will overestimate the chance of highly positive outcome because of their desire to obtain it. Weber also argues that people will overestimate the chance of a highly undesirable outcome because of their fear of receiving it. Traditional methods of decision analysis separately elicit and then recombine subjective probabilities with utilities, as discussed earlier, and assume that subjective probabilities are independent of consequences. A finding of dependency therefore casts serious doubt upon the normative validity of this commonly accepted approach.

When studies of human inference are considered, several other trends become apparent (Table 37.6). In particular, several significant deviations from the Bayesian model have been found. These include: (1) decision makers tend to be conservative in that they do not give as much weight to probabilistic evidence as Bayes' rule (Edwards, 1968); (2) they do not consider base rates or prior probabilities adequately (Tversky and Kahneman, 1974); (3) they tend to ignore the reliability of the evidence (Tversky and Kahneman, 1974); (4) they tend to overestimate the probability of conjunctive events and underestimate the probability of disjunctive events (Bar-Hillel, 1973); (5) they tend to seek out confirming evidence rather than disconfirming evidence and place more emphasis on confirming evidence when it is available (Baron, 1985; Einhorn and Hogarth, 1978), (6) they are over confident in their predictions (Fischhoff, Slovic, and Lichtenstein, 1977), especially in hindsight (Fischhoff, 1982); and (7) they show a tendency to infer illusionary causal relations (Tversky and Kahneman, 1973).

A lively literature has developed regarding these deviations and their significance (Caverni, Fabre, and Gonzales, 1990; Evans, 1989; Klein et al., 1993; Wickens, 1992). From one perspective, these deviations demonstrate inadequacies of human reason and are a source of societal problems (Hammond, 1974). From the opposite perspective, it has been

Table 37.6 Sample Findings on the Ability of People To Estimate and Infer Statistical Quantities

STATISTICAL ESTIMATION	
Accurate estimation of sample means	Peterson and Beach (1968)
Variance estimates correlated with mean Variance biases not found Variance estimates based on range	Lathrop (1967) Levin (1975) Pitz (1980)
Accurate estimates of sample proportions between 0.75 and 0.25 Severe overestimates of high probabilities; severe underestimates of low proportions Reluctance to report extreme events	Edwards (1954) Fischhoff et al. (1977); Lichtenstein et al. (1982) Du Charme (1970)
Weather forecasters provided accurate probabilities	Winkler and Murphy (1973)
Poor estimates of expected severity Correlation of 0.72 between subjective and objective measures of injury frequency	Dorris and Tabrizi (1977) Rethans (1980)
Risk estimates lower for self than for others	Weinstein (1980; 1987)
Risk estimates related to catastrophic potential, degree of control, familiarity	Lichtenstein et al. (1978)
Evaluations of outcomes and probabilities are dependent	Weber (1994)
STATISTICAL INFERENCE	
Conservative aggregation of evidence	Edwards (1966)
Failure to consider base rates Base rates considered	Tversky and Kahneman (1974) Birnbaum and Mellers (1983)
Overestimation of conjunctive events Underestimation of disjunctive events	Bar-Hillel (1973)
Tendency to seek confirming evidence Tendency to discount disconfirming evidence Tendency to ignore reliability of the evidence Subjects considered variability of data when judging probabilities People insensitive to information missing from fault trees	Einhorn and Hogarth (1978); Baron (1985) Kahneman and Tversky (1973) Evans and Pollard (1985) Fischhoff et al. (1978)
Overconfidence in estimates Hindsight bias	Fischhoff et al. (1977) Fishhoff (1982)
Illusionary correlations Gamblers fallacy Misestimation of covariance between items	Tversky and Kahneman (1974) Arkes (1981)
Misinterpretation of regression to the mean	Tversky and Kahneman (1974)

held that the above findings are more or less experimental artifacts that do not reflect the true complexity of the world (Cohen, 1993). From one such perspective, people deviate from Bayes' rule because it makes unrealistic assumptions about what is known or knowable. Simon (1955, 1983) makes a particularly compelling argument for the latter point of view. It also has been noted that researchers overreport findings of bias (Evans, 1989; Cohen, 1993).

There is an emerging body of literature that, on one hand, shows that deviations from Bayes' rule can in fact be justified in certain cases from a normative view, and, on the other hand, shows that these deviations may disappear when people are provided richer information or problems in more natural contexts. Along these lines, researchers have pointed out that (1) a tendency toward conservatism can be justified when evidence is not conditionally independent (Navon, 1979); (2) subjects do use base rate information and consider the reliability of evidence, in slightly modified experimental settings (Birnbaum and Mellers, 1983); (3) a tendency to seek out confirming evidence can offer practical advantages (Cohen, 1993) and may reflect cognitive failures, caused by a lack of understanding of how to falsify hypotheses, rather than entirely a motivational basis (Evans, 1989; Klayman and Ha, 1987); (4) subjects prefer stating subjective probabilities with vague verbal expressions, rather than precise numerical values (Wallsten, Zwick, Kemp, and Budescu, 1993), demonstrating that they are not necessarily overconfident in their predictions.

37.4.1.2 Heuristics and Biases

Tversky and Kahneman (1973, 1974) made a key contribution to the field when they showed that many of the above-mentioned discrepancies between human estimates of probability and Bayes' rule could be explained by three heuristics. The three heuristics they proposed were those of (1) representativeness, (2) availability, and (3) anchoring and adjustment.

The *representativeness* heuristic holds that the probability of an item (A) belonging to some category (B) is judged by considering how representative A is of B. For example, a person is typically judged more likely to be a librarian than a farmer when described as "A meek and tidy soul, who has a desire for order and structure, and a passion for detail." Tversky and Kahneman (1974) give several examples of biases that may occur because of the representativeness heuristic. In each case, representativeness influenced estimates more than other, more statistically oriented information. In the first study, subjects ignored base rate information (given by the experimenter) about how likely a person was to be either a lawyer or an engineer. Their judgments seemed to be based entirely on how representative the description seemed to be of either occupation. Tversky and Kahneman (1983) found people overestimated conjunctive probabilities in a similar experiment. Here, after being told that "Linda is 31 years old, single, outspoken, and very bright" most subjects said it was more likely she was both a bank teller and active as a feminist than simply a bank teller. In a third study, most subjects felt that the probability of more than 60% male births on a given day was about the same for both large and small hospitals (Tversky and Kahneman, 1974). Apparently, the subjects felt large and small hospitals were equally representative of the population.

Other behaviors explained in terms of representativeness by Tversky and Kahneman included gambler's fallacy, insensitivity to predictability, illusions of validity, and misconceptions of statistical regression to the mean. With regard to gambler's fallacy, they note that people may feel long sequences of heads or tails when flipping coins are unrepresentative of normal behavior. After a sequence of heads, a tail therefore seems more representative. Insensitivity to predictability refers to a tendency for people to predict future performance without considering the reliability of the information they base the prediction upon. For example, a person might expect an investment to be profitable solely on the basis of a favorable description without considering whether the description has any predictive value. In other words, a good description is believed to be representative of high profits, even if it states nothing about profitability. The illusion of validity occurs when people use highly correlated evidence to make a conclusion. Despite the fact that the evidence is redundant, the presence of many representative pieces of evidence increases confidence greatly. Misconception of regression to the mean occurs when people react to unusual events and then infer a causal linkage when the process returns to normality on its own. For example, a manager may incorrectly conclude that punishment works after seeing that unusually poor performance improves to normal levels following punishment. The same manager may also conclude that rewards do not work after seeing that unusually good performance drops after receiving a reward.

The *availability* heuristic holds that the probability of an event is determined by how easy it is to remember the event happening. Tversky and Kahneman state perceived probabilities will, therefore, depend on familiarity, salience, effectiveness of memory search, and imaginability. The implication is that people will judge events as more likely when the events are familiar, highly salient (such as an airplane crash), or easily imaginable. Events also will be judged more likely if there is a simple way to search memory. For example, it is much easier to search for words in memory by the first letter rather than the third letter. It is easy to see how each of the above items impacting the availability of information can result in biases. Also, these biases should increase when people lack experience or when their experiences are too focused.

Anchoring and adjustment holds that people start from some initial estimate and then adjust it to reach some final value. The point initially chosen has a major impact on the final value selected, when adjustments are insufficient. Tversky and Kahneman refer to this source of bias as an anchoring effect. They show how this effect can explain under- and overestimates of disjunctive and conjunctive events. This happens if the subject starts with a probability estimate of a single event. The probability of a single event is, of course, less than that for the disjunctive event and greater than that for the conjunctive event. If adjustment is too small, then under and overestimates, respectively, occur for the disjunctive and conjunctive events. Tversky and Kahneman also discuss how anchoring and adjustment may cause biases in subjective probability distributions. Hogarth and Einhorn (1992) present an anchoring and adjustment model of how people update beliefs that explains a number of ordering effects, such as the primacy and recency effects. This latter model holds that the degree of belief in a hypothesis after collecting k pieces of evidence can be described as follows:

$$S_k = S_k - 1 + w_k[s(x_k) - R] \tag{26}$$

where S_k is the degree of belief after collecting k pieces of evidence, S_{k-1} is the anchor or prior belief, w_k is the adjustment weight for the kth piece of evidence, $s(x_k)$ is the subjective evaluation of the kth piece of evidence, and R is the reference point against which the kth piece of evidence is compared. In evaluation tasks, $R = 0$. This corresponds to the case where evidence is either for or against a hypothesis. For estimation tasks, $R \neq 0$. The different values of R result in an additive model for evaluation tasks and an averaging model for estimation tasks. Also, if the quantity, $s(x_k) - R$, is evaluated for several pieces of evidence at a time, the model predicts primacy effects. If single pieces of evidence are individually evaluated in a step-by-step sequence, recency effects become more likely.

The notion of heuristics and biases has had a particularly formative influence on decision theory. A substantial recent body of work has emerged that focuses on applying research on heuristics and biases (Heath et al., 1994; Kahneman et al., 1982). Applications include medical judgment and decision making, affirmative action, education, personality assessment, legal decision making, mediation, and policy making. It seems clear that this approach is excellent for describing decision making in the real world at a general level. However, research on heuristics and biases has been criticized as being pre-theoretical (Slovic et al., 1977). Information-processing models provide more theoretically developed views of human judgment, as expanded upon below.

37.4.1.3 Selective Processing of Information

Evans (1989) argues that factors which cause people to process information in a selective manner or attend to irrelevant information are the major cause of biases in human judgment. Factors assumed to influence selective processing include the availability, vividness, and relevance of information, and working memory limitations. The notion of availability refers to the information actually attended to by a person while performing a task. Evan's model assumes that relevant information elements are determined during a heuristic, preattentional stage. This stage is assumed to involve unconscious processes and is influenced by stimulus salience (or vividness), and the effects of prior knowledge.

In the next stage of his model, inferences are drawn from the selected information. This is done using rules for reasoning and action developed for particular types of problems. Working memory influences performance at this stage by limiting the amount of information that can be consciously attended to while performing a task. The knowledge used during the inference process may be organized in schemas that are retrieved from memory and fit to specific problems (Cheng and Holyoak, 1985). Support for this latter conclusion is provided by studies showing that people are able to develop skills in inference tasks but may fail to transfer this skill (inference related) from one setting to another. Evans also provides evidence that prior knowledge can cause biases when it is inconsistent with provided information and that improving knowledge can reduce or eliminate biases.

Evan's model of selective processing of information is consistent with other explanations of biases. Among such explanations, information overload has been cited as a reason for impaired decision making by consumers (Jacoby, 1977). The tendency of highly salient stimuli to capture attention during inference tasks has also been noted by several researchers (Nisbett and Ross, 1980; Payne, 1980). Nisbett and Ross suggest that vividness of information is determined by its emotional content, concreteness and imageability, and temporal and spatial proximity. As noted by Evans, these factors have also been shown to affect the memorability of information. This provides a plausible explanation of both the availability heuristic and the experimental results mentioned earlier regarding biases in risk perceptions.

37.4.1.4 Probabilistic Information-Processing Models

A number of probabilistic information-processing models have been developed that mathematically describe human judgments. These approaches include social judgment theory, policy capturing, multiple cue probability learning models, information integration theory, and conjoint measurement approaches.

Social judgment theory (SJT) implements an ecological approach for explaining how environmental cues are related to psychological responses (Brehmer and Joyce, 1988; Hammond, 1993; Hammond et al., 1975). The approach can be traced back to the Brunswick lens model (Brunswick, 1952), which describes human judgments in terms of perceived environmental cues. Emphasis is placed on performing experiments where information cues reflect the statistical characteristics of the real world. Policy capturing models are also derived from the lens model, and have been applied to a wide number of real-world applications to describe expert judgments (Brehmer and Joyce, 1988). For example, policy capturing models have been applied to describe software selection by management information system managers (Martocchio, Webster, and Baker, 1993), medical decisions (Brehmer and Joyce, 1988), and highway safety (Hammond, 1993). As mentioned earlier with regard to preference assessment, linear or nonlinear forms of regression are used in this approach to relate judgments to environment cues. These equations provide surprisingly good fits to expert judgments. In fact, there is evidence, and consequently much debate, over whether the models can actually do better than experts on many judgment tasks (Brehmer, 1981; Kleinmuntz, 1984; Slovic et al., 1977).

Multiple cue probability learning models extend the lens model to the psychology of learning (Brehmer and Joyce, 1988). Research on multiple cue probability learning has provided valuable insight into factors affecting learning of inference tasks. One major finding is that providing cognitive feedback about cues and their relationship to the inferred effects leads to quicker learning than feedback about outcomes (Balzer, Doherty, and O'Connor, 1989). Stevenson et al. (1993) summarize a number of other findings, including that (1) subjects can learn to use valid cues, even when they are unreliable, (2) subjects are better able to learn linear relationships than nonlinear or inverse relationships, (3) subjects do not consider redundancy when using multiple cues, (4) source credibility and cue validity are considered, and (5) the relative effectiveness of cognitive and outcome feedback depends on the formal, substantive, and contextual characteristics of the task.

Information integration theory (Anderson, 1981) takes a somewhat different approach than SJT or the lens model to describe how cue information is used when making judgments. A major deviation is that information integration theory emphasizes the use of factorial experimental designs where cues are systematically manipulated. The goal of this approach is to determine, first, how people scale cues when determining their subjective values and, second, how these scaled values are combined to form overall judgments. Various functional forms of how information is integrated are considered, including additive and averaging functions. A substantial body of research follows this approach to test various ways people might combine probabilistic information. A primary conclusion is that people tend to integrate information using simple averaging, adding, subtracting, and multiplying models. *Conjoint measurement* approaches (Wallsten, 1972, 1976), in particular, provide a convenient way of both scaling subjective values assigned to cues and testing different functional forms describing how these values are combined to develop global judgements. By applying this approach, Wallsten (1976) was able to model primacy and recency effects.

37.4.1.5 Debiasing Human Judgments

The notion that many biases (or deviations from normative models) in statistical estimation and inference can be explained has led researchers to consider the possibility of debiasing human judgements (Keren, 1990). Part of the issue is that heuristics often work very well. It seems logical that biases based on both the availability and representativeness heuristics may be reduced if people are provided more information. Debiasing research has provided mixed results. Many biases, such as optimistic beliefs regarding health risks have been difficult to modify (Weinstein and Klein, 1995). People show a tendency to seek out information that supports their personal views (Weinstein, 1979), and are quite resistant to information that contradicts strongly held beliefs (McGuire, 1966; Nisbett and Ross, 1980). Evans (1989) concludes that preconceived notions are likely to prejudice the construction and evaluation of arguments.

Other evidence shows that experts may have difficulty providing accurate estimates of subjective probabilities even when they receive feedback. For example, most efforts to reduce both overconfidence in probability estimates and the hindsight bias have been unsuccessful (Fischhoff, 1982). One problem is that people may not pay attention to feedback (Fischhoff and MacGregor, 1982). They also may only attend to feedback which supports their hypothesis, leading to poorer performance and at the same time greater confidence (Einhorn and Hogarth, 1978). Efforts to reduce confirmation biases through training have also been in general unsuccessful (Evans, 1989).

On the positive side, there is evidence that providing feedback on the accuracy of weather forecasts may help weather forecasters (Winkler and Murphy, 1973). There is also some evidence that people can learn to perform statistical reasoning more accurately after training in statistics (Fong, Krantz, and Nisbett, 1986). Failure to consider sample size was significantly reduced after training. Another study showed that asking people to write down reasons for and against their estimates of probabilities improved calibration and reduced overconfidence (Koriat, Lichtenstein, and Fischhoff, 1980). There is evidence that overconfidence is reduced when decision makers represent subjective probabilities verbally (Wallsten et al., 1993; Zimmer, 1983). Conservatism, or the failure to adequately modify probabilities after obtaining evidence, was also reduced in Zimmer's study.

The conclusion is that debiasing human judgments is difficult, but not impossible. Some perspective can be obtained by considering that most studies showing biases have focused on statistical inference, and generally involve people not particularly knowledgeable about statistics, who are not using decision aids such as computers or calculators. It naturally may be expected that people will perform poorly on such tasks, given their lack of training and forced reliance on mental calculations (Winterfelt and Edwards, 1986). The finding that people can improve their abilities on such tasks, after training in statistics, is particularly telling, but also encouraging. Another encouraging finding is that biases are occasionally reduced when people process information verbally instead of numerically. This result may be expected given that most people are more comfortable with words than with numbers.

37.4.2 Decision Making Under Uncertainty and Risk

Much of the research evaluating human preferences under uncertainty and risk has focused on comparing observed preferences to the predictions of subjective utility theory (SEU). Initial work uncovered several paradoxes in which people's preferences violated basic axioms of SEU theory. Another violation of SEU theory was uncovered by research showing that the framing of decisions in terms of costs or benefits sometimes resulted in preference reversals. Alternative models, such as prospect theory, were consequently developed to explain these effects. Significant research has been also focused on risk acceptability and risk-taking behavior. Since this latter work is focused upon in Chapter 31, risk acceptability and risk-taking behavior are not addressed in this section.[18] Violations of SEU, framing effects, and prospect theory are briefly discussed below.

37.4.2.1 Violation of Rationality Axioms

Several studies have shown that people's preferences between uncertain alternatives can be inconsistent with the axioms underlying SEU theory. One fundamental violation of

[18]Yates (1992), and Singleton and Hovden (1987), are useful sources for the reader interested in additional details on risk perception, risk acceptability, and risk-taking behavior. Section 37.5.1.1 is also relevant to this topic.

the assumptions is that preferences can be intransitive (Budescu and Weiss, 1987; Tversky, 1969). Also, as mentioned in the previous section, subjective probabilities may depend on the values of consequences (violating the independence axiom) and, as discussed in the next section, the framing of a choice can impact preference. Another violation is given by the Myers effect (Myers, Suydam, and Gambino, 1965), where preference reversals between high (H) and low (L) variance gambles can occur when the gambles are compared to a certain outcome, depending on whether the certain outcome is positive (H preferred to L) or negative (L preferred to H). This latter effect violates the assumption of independence because the ordering of the two gambles depends on the certain outcome.

Another commonly cited violation of SEU theory is that people show a tendency toward uncertainty avoidance which can lead to behavior inconsistent with the "sure-thing" axiom. The Ellsburg and Allais paradoxes (Allais, 1953; Ellsburg, 1961) both involve violations of the sure-thing axiom and seem to be caused by people's desire to avoid uncertainty. The Allais paradox is illustrated by the following set of gambles. In the first gamble, a person is asked to choose between gambles A1 and B1, where:

Gamble A1 results in \$1 million for sure. Gamble B1 results in \$2.5 million with a probability of 0.1, \$1 million with a probability of 0.89, and \$0 with a probability of 0.01.

In the second gamble, the person is asked to choose between gambles A2 and B2, where:

A2 results in \$1 million with a probability of 0.11 and \$0 with a probability of 0.89. Gamble B2 results in \$2.5 million with a probability of 0.1 and \$0 with a probability of 0.9.

Most people prefer gamble A1 to B1, and gamble B2 to A2. It is easy to see that this set of preferences violates expected utility theory. First, if A1 > B1, then $u(\text{A1}) > u(\text{B1})$, meaning that: $u(\$1 \text{ million}) > 0.1u(\$2.5 \text{ million}) + 0.89u(\$1 \text{ million}) + 0.01u(\$0)$. If a utility of 0 is assigned to receiving \$0, and a utility of 1 to receiving \$2.5 million, then $u(\$1 \text{ million}) > 1/11$. However, from the preference A2 > B2, it follows that $u(\$1 \text{ million}) < 1/11$. Obviously, no utility function can satisfy this requirement of assigning a value both greater than and less than 1/11 to \$1 million.

As noted by Savage (1954), this set of gambles can be reframed in a way that shows this set of preferences violates the sure-thing principle. Interestingly, Savage found that his initial tendency toward choosing A1 over B1 and A2 over B2 disappeared when the problem was reframed. As noted by Stevenson et al. (1993), this is one of the first cited examples of preference reversal caused by framing of decisions, the topic discussed below.

37.4.2.2 Framing of Decisions and Preference Reversals

A substantial body of research has shown that people's preferences can shift dramatically depending on the way a decision is represented. The best known work on this topic was conducted by Tversky and Kahneman (1981). Tversky and Kahneman showed that preferences between medical intervention strategies changed dramatically, depending on whether the outcomes were posed as losses or gains. The following question, worded in terms of benefits, was presented to one set of subjects:

Imagine that the U.S. is preparing for the outbreak of an unusual Asian disease, which is expected to kill 600 people. Two alternative programs to combat the disease have been proposed. Assume that the exact scientific estimate of the consequences of the programs are as follows:

If Program A is adopted, 200 people will be saved.

If Program B is adopted, there is a 1/3 probability that 600 people will be saved, and a 2/3 probability that no people will be saved.

Which of the two programs would you favor?

The results showed that 72% of subjects preferred program A. The second set of subjects, was given the same cover story, but worded in terms of costs, as given below:

If Program C is adopted, 400 people will die.

If Program D is adopted, there is a 1/3 probability that nobody will die, and a 2/3 probability that 600 people will die.

Which of the two programs would you favor?

The results now showed that 78% of subjects preferred program D. Since program D is equivalent to B, and Program A is equivalent to C, the preferences for the two groups of subjects were strongly reversed. Tversky and Kahneman concluded that this reversal illustrated a common pattern in which choices involving gains are risk averse, and choices involving losses are risk seeking. The interesting result was that the way the alternatives were worded caused a shift in preference for identical alternatives. Tversky and Kahneman call this tendency the *reflection effect.* A body of literature has since developed, showing that the framing of decisions can have practical effects for both individual decision makers (Heath et al., 1994; Kahneman et al., 1982) and group decisions (Paese, Bieser, and Tabbs 1993). More recently, it has been shown that such framing effects can be reduced or even eliminated by changing the wording of problem statements (Kuhberger, 1995). This latter study showed that standard framing effects could be reversed by certain problem wordings and eliminated by fully describing the problems.

Other recent research has explored the theory that perceived risk and perceived attractiveness of risky outcomes are psychologically distinct constructs (Weber, Anderson, and Bernbaum, 1992). In the latter study it was concluded that perceived risk and attractiveness are "closely related, but distinct phenomena." Related research has shown weak negative correlations between the perceived risk and value of indulging in alcohol-related behavior for adolescent subjects (Lehto, James, and Foley, 1994). This latter study also showed that the rated propensity to indulge in alcohol-related behavior was strongly correlated with perceived value ($R = 0.8$), but weakly correlated with perceived risk ($R = -0.15$). Both findings are consistent with the theory that perceived risk and attractiveness are distinct constructs; but the latter finding indicates that perceived attractiveness may be the better predictor of behavior. Lehto et al. conclude that intervention methods attempting to lower preferences for alcohol-related behavior should focus on lowering perceived value rather than on increasing perceived risk.

37.4.2.3 Prospect Theory

Prospect theory (Kahneman and Tversky, 1979) attempts to account for behavior not consistent with the SEU model by including the framing of decisions as a step in the judgment of preference between risky alternatives. Prospect theory assumes that decision makers tend to be risk averse with regard to gains and risk seeking with regard to losses. This leads to a value function that disproportionately weights losses. As such, the model is still equivalent to SEU, assuming a utility function expressing mixed risk aversion and risk seeking. Prospect theory, however, assumes that the decision maker's reference point can change. With shifts in the reference point, the same returns can be viewed as either gains or losses.[19] This latter feature of prospect theory, of course, is an attempt to account for the framing effect discussed above.

A second way that prospect theory deviates significantly from SEU theory is in the way probabilities are addressed. Prospect theory weighs perceived values by a function $\pi(p)$, where p is the true probability, rather than by the probability itself, as is done in SEU. This function $\pi(p)$ is assumed generally to overweigh very low probabilities and underweigh moderate and high probabilities. It is also assumed to be discontinuous and poorly defined for probability values close to 0 or 1. Several other modeling approaches that differentially weigh utilities in risky decision making have also been proposed (Kahneman and Tversky, 1979; Stevenson et al., 1993).

Prospect theory assumes that the choice process involves an editing phase and an evaluation phase. The editing phase involves reformulation of the options to simplify

[19]The notion of a reference point against which outcomes are compared has similarities to the notion of making decisions on the basis of regret (Bell, 1982). Regret, however, assumes comparison to the best outcome. The notion of different reference points also is related to the well-known trend that buying and selling prices of assets often differ for a decision maker (Raiffa, 1968).

subsequent evaluation and choice. Much of this editing process is concerned with determining an appropriate reference point in a step called "coding." Other steps that may occur include the segregation of riskless components of the decision, combining probabilities for events with identical outcomes, simplification by rounding off probabilities and outcome measures, and search for dominance. In the evaluation phase the perceived values are then weighed by the function $\pi(p)$. The alternative with the greatest weighed value is then selected.

37.5 DYNAMIC AND NATURALISTIC DECISION MAKING

In dynamic decision making, actions taken by a decision maker are made sequentially in time. Taking actions can change the environment, resulting in a new set of decisions. The decisions might be made under time pressure and stress, by groups, or by single decision makers. This process might be performed on a routine basis or might involve severe conflict. For example, either a group of soldiers or an individual officer might routinely identify marked vehicles as friends or foes. When a vehicle has unknown or ambiguous marking, the decision changes to a conflict-driven process. Naturalistic decision theory has emerged as a new field that focuses on such decisions in real world environments (Klein et al., 1993). The notion that most decisions are made in a routine, nonanalytical way is the driving force of this approach.[20] Areas where such behavior seems prominent include juror decision making, troubleshooting of complex systems, medical diagnosis, management decisions, and numerous other examples.

The following discussion will first address models of naturalistic decision making that describe major differences in decision-making processes as a function of experience, task familiarity, context, time pressure, and stress. Research on group and team decision making in dynamic contexts is then considered. The discussion concludes by summarizing process tracing methods commonly used to document decision making in dynamic environments.

37.5.1 Models of Naturalistic Decision Making

In recent years, it has been recognized that decision making in natural environments often differs greatly between decision contexts (Beach, 1993). In addressing this topic, the involved researchers often question the relevance and validity of both classical decision theory and behavioral research not conducted in real-world settings (Cohen, 1993). Numerous naturalistic models have been proposed (Klein et al., 1993). These models assume that people rarely weigh alternatives and compare them in terms of expected value or utility. Each model is also descriptive, rather than prescriptive. Perhaps the most general conclusion that can be drawn from this work is that people use different decision strategies, depending on their experience, the task and the decision context. Several of the models also postulate that people choose between decision strategies by trading off effectiveness against the effort required.

The following discussion will briefly review models focusing upon: (1) levels of task performance (Rasmussen, 1983), (2) recognition-primed decisions (Klein, 1989), (3) image theory (Beach, 1990), (4) cognitive continuum theory (Hammond, 1980), (5) contingent decision making (Payne et al., 1993), (6) dominance structuring (Montgomery, 1989), (7) time pressure and stress (Maule and Hockey, 1993), and (8) explanation-based decision making (Pennington and Hastie, 1988).

37.5.1.1 Levels of Task Performance

There is growing recognition that most decisions are made on a routine basis in which people simply follow past behavior patterns (Beach, 1993; Rasmussen, 1983; Svenson, 1990). Rasmussen (1983) follows this approach to distinguish between skill-based, rule-based, and knowledge-based levels of task performance. Lehto (1991) further considers judgment-based behavior as a fourth level of performance.

Performance is said to be at either a skill-based or a rule-based level when tasks are routine in nature. Skill-based performance involves the smooth, automatic flow of actions

---

[20] Drucker (1985), in discussing ways of improving the effectiveness of executive decision makers, emphasizes the importance of establishing a generic principle or policy that can be applied to specific cases in a routine way. This recommendation is interesting, because it prescribes a naturalistic form of behavior.

without conscious decision points. As such, skill-based performance describes the decisions made by highly trained operators performing familiar tasks. Rule-based performance involves the conscious perception of environmental cues, which trigger the application of rules learned on the basis of experience. As such, rule-based performance corresponds closely to recognition-primed decisions (Klein, 1989). The knowledge-based level of performance is said to occur during learning or problem-solving activity during which people cognitively simulate the influence of various actions and develop plans for what to do. The judgment-based level of performance occurs when effective reactions of a decision maker cause a change in goals or priorities between goals (Etzioni, 1988; Janis and Mann, 1977; Lehto, 1991). Distinctive types of errors in decision making occur at each of the four levels (Lehto, 1991; Reason, 1989).

At the skill-based level, errors occur because of perceptual variability and when people fail to shift up to rule-based or higher levels of performance. At the rule-based level, errors occur when people apply faulty rules or fail to shift up to a knowledge-based level in unusual situations where the rules they normally use are no longer appropriate. The use of faulty rules leads to an important distinction between running and taking risks. Along these lines, Wagenaar (1992) discusses several case studies in which people following risky forms of behavior do not seem to be consciously evaluating the risk. Drivers, in particular, seem habitually to take risks. Wagenaar explains such behavior in terms of faulty rules derived on the basis of benign experience. In other words, drivers get away with providing small safety margins most of the time and consequently learn to run risks on a routine basis. Drucker (1985) points out several cases where organizational decision makers have failed to recognize that the generic principles they used to apply were no longer appropriate, resulting in catastrophic consequences.

At the knowledge-based level, errors occur because of cognitive limitations, faulty mental models, or when the testing of hypotheses cause unforseen changes to systems. At the judgment-based levels, errors (or violations) occur because of inappropriate affective reactions, such as anger or fear (Lehto, 1991).

A recent study involving drivers arrested for drinking and driving (McKnight, Langston, McKnight, and Lange, 1995) provides an interesting perspective on how the sequential nature of naturalistic decisions can lead people into "traps." The study also shows how errors can occur at multiple levels of performance. In this example, decisions made well in advance of the final decision to drive while impaired played a major role in creating situations where drivers were almost certain to drive impaired. For instance, the driver may have chosen to bring along friends, and therefore felt pressured to drive home because the friends were dependent on him or her. This initial failure by drivers to predict the future situation could be described as a failure to shift up from a rule-based level to a knowledge-based level of performance. In other words, the driver never stopped to think about what might happen if he or she drank too much. The final decision to drive, however, would correspond to an error (or violation) at the judgment-based level, if the driver's choice was influenced by an affective reaction (perceived pressure) to the presence of friends wanting a ride.

37.5.1.2 Recognition Primed Decision Making

Klein (1989) developed the theory of recognition-primed decision making on the basis of observations of firefighters and other professionals in their naturalistic environments. He found that up to 80% of the decisions made by firefighters involved some sort of situation recognition, where the decision makers simply followed a past behavior pattern once they recognized the situation.

The model he developed distinguishes between three basic conditions. In the simplest case, the decision maker recognizes the situation and takes the obvious action. A second case occurs when the decision maker consciously simulates the action to check whether it should work before taking it. In the third, and most complex case, the action is found to be deficient during the mental simulation and is consequently rejected. An important point of the model is that decision makers do not begin by comparing all the options. Instead they begin with options that seem feasible based upon their experience. This tendency, of course, differs from the SEU approach, but is comparable to applying the satisficing decision rule (Simon, 1955) discussed earlier.

Situation assessment is well recognized as an important element of decision making in naturalistic environments (Klein et al., 1993). Recent research by Klein and his colleagues has examined the possibility of enhancing situation awareness through training (Klein and Wolf, 1995). Klein and his colleagues have also applied methods of cognitive task analysis to naturalistic decision-making problems. In these efforts, they have focused on identifying (1) critical decisions, (2) the elements of situation awareness, (3) critical cues indicating changes in situations, and (4) alternative courses of action (Klein, 1995).

37.5.1.3 Image Theory

Image theory (Beach, 1990) is a descriptive theory of decision making. Beach theorizes that knowledge used to make decisions falls into three categories. The three categories are value images, trajectory images, and strategic images. The value image describes the decision makers values, and principles; the trajectory image describes goals; the strategic image describes plans to attain the goals. He also theorizes that there are two types of decisions: adoption decisions and progress decisions. Adoption decisions first involve a screening process where alternatives are eliminated from consideration. The most promising alternative is then selected from the screened set. Progress decisions involve a comparison between goals and the expected result of choosing the alternative.

Two means of evaluating decisions are applied. One test compares the compatibility of the generated alternatives to value images, trajectory images, and strategic images. The profitability test is used to further evaluate screened options in adoption decisions when more than one option survives the initial screening. Beach (1993) argues strongly for the primacy of screening as a characteristic of most real-world decision-making activity.

37.5.1.4 Cognitive Continuum Theory

Cognitive continuum theory (Hammond, 1980) distinguishes judgments on a cognitive continuum varying from highly intuitive decisions to highly analytical decisions. Hammond (1993) summarizes earlier research showing that task characteristics cause decision makers to vary on this continuum. A tendency toward analysis increases, and reliance on intuition decreases, when (1) the number of cues increase, (2) cues are measured objectively instead of subjectively, (3) cues are of low redundancy, (4) decomposition of the task is high, (5) certainty is high, (6) cues are weighted unequally in the environmental model, (7) relations are nonlinear, (8) an organizing principle is available, (9) cues are displayed sequentially instead of simultaneously, and (10) the time period for evaluation is long.

37.5.1.5 Contingent Decision Making

The theory of contingent decision making (Beach and Mitchell, 1978; Payne et al., 1993) is similar to image theory and cognitive continuum theory in that it holds that people use different decision strategies, depending on the characteristics of the task and the decision context. Payne et al. limit their modeling approach to tasks that require choices to be made (simple memory tasks are excluded from consideration). They also add the assumption that people make choices about how to make choices.[21]

Choices between decision strategies are assumed to be made rationally by comparing their cost (in terms of cognitive effort) against their benefits (in terms of accuracy). Cognitive effort and accuracy (of a decision strategy) are both assumed to depend upon task characteristics, such as task complexity, response mode, and method of information display. Cognitive effort and accuracy also are assumed to depend on contextual characteristics, such as the similarity of the compared alternatives, attribute ranges and correlations, the quality of the considered options, reference points, and decision frames. Payne et al. place much emphasis on measuring the cognitive effort of different decision strategies in terms of the number of elemental information elements that must be processed for different tasks and contexts. They relate the accuracy of different decision strategies to task characteristics and contexts, and also present research showing that people will shift decision strategies to reduce cognitive effort, increase accuracy, or respond to time pressure.

37.5.1.6 Dominance Structuring

Dominance structuring (Montgomery, 1989) holds that decision making in real contexts involves a sequence of four steps. The process begins with a preediting stage in which

[21]As such, the theory of contingent decision making directly addresses a potential source of conflict shown in the integrative model of decision making presented earlier (Figure 37.1). That is, it states that decision makers must choose between decision strategies when they are uncertain "How to compare alternatives."

alternatives are screened from further analysis. The next step involves selecting a promising alternative from the set of alternatives that survive the initial screening. A test is then made to check whether the promising alternative dominates the other surviving alternations. If dominance is not found, then the information regarding the alternatives is restructured in an attempt to force dominance. This process involves both the bolstering and de-emphasizing of information in a way that eliminates disadvantages of the promising alternative.

37.5.1.7 Time Pressure and Stress

Time pressure and stress have been shown to influence decision making in several ways. Reviews of the literature suggest that time pressure often results in poorer task performance and that it can cause shifts between the cognitive strategies used in judgment and decision making situations (Edland and Svenson, 1993; Maule and Hockey, 1993). One change is that people show a tendency to shift to noncompensatory decision rules. This finding is consistent with contingency theories of strategy selection (Section 37.5.1.5). In other words, this shift may be justified when little time is available, because a noncompensatory rule can be applied more quickly. Maule and Hockey also note that people tend to filter out low-priority types of information, omit processing information, and accelerate mental activity when they are under time pressure.

Variable state activation theory (VSAT) provides a potential explanation of the above effects in terms of a control model of stress regulation (Maule and Hockey, 1993). VSAT also proposes that disequilibriums between control processes and the demands of particular situations can lead to strong effective reactions or feelings of time pressure.

37.5.1.8 Explanation-Based Decision Making

Explanation-based decision making (Pennington and Hastie, 1986, 1988) assumes that people begin their decision-making process by constructing a mental model that explains the facts they have received. While constructing this explanatory model, people are also assumed to be generating potential alternatives to choose between. The alternatives are then compared to the explanatory model, rather than to the facts from which has been constructed.

The authors have applied this model to juror decision making and obtained experimental evidence that many of its assumptions seem to hold. They note that juror decision making requires consideration of a massive amount of data that is often presented in haphazard order over a long time period. Jurors seem to organize this information in terms of stories describing causation and intent. As part of this process, jurors are assumed to evaluate stories in terms of their uniqueness, and plausibility, completeness, or consistency. To determine a verdict, jurors then judge the fit between choices provided by the trial judge and the various stories they use to organize the information. Jurors' certainty about their verdict is assumed to be influenced by both evaluation of stories and the perceived goodness of fit between the stories and the verdict.

37.5.2 Group and Team Decision Making

Much research has been done over the past 25 years or so on decision making by groups and teams. Most of this work has focused on groups, as opposed to teams. In a team, it is assumed that the members are working toward a common goal, have some degree of interdependence, defined roles and responsibilities, and task-specific knowledge (Orasanu and Salas, 1993). Group performance has traditionally been an area of study in the fields of organizational behavior and industrial psychology. Traditional decision theory[22] has devoted some attention to group decision making (Raiffa, 1968). Obtaining group inputs

[22]Game theory, in particular, focuses on group decision making. The central notion is that the return to individual decision makers is conditional on the actions of other members of the group. In competitive games, individuals are likely to take "self-centered" actions that maximize their own return but reduce returns to other members of the group. Behavior of group members in this situation may be well described by the MINIMAX decision rule discussed in Section 37.2.1.5. In cooperative games, the members of the group take actions that maximize returns to the group as a whole. Games may differ in whether they are repeated or played only once. They also differ in terms of the number of participants and the time horizon over which the game is played. Friedman (1990) provides an excellent introduction to game theory.

when assessing probabilities and preferences, is a commonly applied approach, as discussed earlier. Group decision support systems (GDSS) have been proposed, as discussed in Chapter 42. Team performance is an area of interest in the field of naturalistic decision theory (Klein et al., 1993).

In this section some findings will be briefly reviewed regarding (1) leadership, (2) group knowledge, (3) group processes, and (4) group performance and biases. No effort will be made to consider the literature on game theory, organizational behavior and industrial psychology. Also, no attention will be given to GDSS, since they are covered in Chapter 42.

37.5.2.1 Leadership

Torrance (1953) describes retrospective accounts of military survivors lost behind enemy lines indicating that survival depended upon the leader's "leadership" skills. Important elements of leadership skills included keeping the members of the group focused on a common goal, making sure they knew what needed to be done, and keeping them informed of the current status. Related conclusions concerning the value of keeping people informed have been obtained in retrospective accounts of survivors of mining accidents (Mallet, Vaught, and Brnich, 1993). Orasanu and Salas (1993) cite research in which captains of high-performing air crews explicitly stated more plans, strategies, and intentions to the other members of the crew. They also gave more warnings and predictions to the crew members. Orasanu and Salas cite other work showing that crews performed better with captains who were task oriented and had good personal skills. Performance dropped when captains had negative expressive styles and low task orientation.

A complementary literature has been developed on leadership theory (Chemers and Ayman, 1993). Most of this research is based on leaders in organizational contexts. A sampling of factors that have been shown to be related to the effectiveness of leadership include legitimacy, charisma, individualized attention to group members, and clear definitions of goals. These results seem quite compatible with the above findings for leadership in naturalistic, dynamic contexts.

37.5.2.2 Group Knowledge

Orasanu and Salas (1993) discuss two closely related frameworks for describing the knowledge used by teams in naturalistic settings. These are referred to as *shared mental models* and the *team mind.* The common element of these two frameworks is that the members of teams hold knowledge in common and organize it in the same way. Orasanu and Salas claim that this improves and minimizes the need for communication between team members, enables team members to carry out their functions in a coordinated way, and minimizes negotiation over who should do what at what time. Under emergency conditions, Orasanu and Salas claim there is a critical need for members to develop a shared situation model. As evidence for the notion of shared mental models and the team mind, the authors cite research in which firefighting teams and individual firefighters developed the same solution strategies for situations typical of their jobs.

This notion of shared mental models and the team mind can be related to the earlier discussed notion of schemas containing problem specific rules and facts (Cheng and Holyoak, 1985). It also might be reasonable to consider other team members as a form of external memory (Newell and Simon, 1972). This approach would have similarities to Wegner's (1987) concept of transactive memory where people in a group know who has specialized information of one kind or another.

37.5.2.3 Group Processes

Many processes occur within groups during group decision making. Some of these activities include: (1) methods of attaining consensus, such as negotiation, bargaining, or voting; (2) idea generating activity, such as brainstorming; and (3) communication. Addressing these topics in any detail is well beyond the scope of this chapter. Three comments are as follows.

First, computer techniques for helping groups perform all three processes are being developed. The analytic hierarchy process (AHP) is a computer tool that has been used to help decision makers reach a consensus (Basak and Saaty, 1993; Saaty, 1988). Electronic meeting places implement convenient schemes for voting (Mockler and Dologite, 1991). Group decision support systems, and systems as simple as e-mail, are useful in the idea-generating stage of decision making and help multiple decision makers communicate (see Chapter 42). Second, approaches from behavioral decision theory are being applied to these group decision processes. Topics such as mediation and negotiation, jury decision making, and public policy setting are being evaluated from the perspective of heuristics and biases (Heath et al., 1994). There also is an emerging literature on the critical role of communication in team problem decision making (Mallet et al., 1993; Orasanu and Salas, 1993). Third, other fields have much to say about these topics. For example, Davis (1992) notes the presence of a large body of research in political science on how voting procedures can influence outcomes. Other areas of psychology also have much to say about idea generation and negotiation, as does management science.

37.5.2.4 Group Performance and Biases

Over the years, a significant body of research has compared the performance of groups to that of individual decision makers. As summarized by Davis (1992) much of the early work showed that groups were better than individuals. Later research indicated that group performance was less than the sum of its parts. Part of the issue here is the so-called phenomenon *groupthink* (Janis, 1972). The Delphi technique attempts to eliminate this effect by separating decision makers from each other. A second more recent finding was that groups tend to be more willing to select risky alternatives than individuals. One explanation of this finding is that group interactions cause people within the group to adopt more polarized opinions (Moscovici, 1976).

Duffy (1993) notes that teams can be viewed as information processes and cites team biases and errors that can be related to information-processing limitations and the use of heuristics, such as framing. Other research has studied whether groups are subject to the same biases individuals have. This research on groups has shown: (1) framing effects and preference reversals (Paese et al., 1993), (2) overconfidence (Sniezek, 1992), (3) use of heuristics in negotiation (Bazerman and Neale, 1983), and (4) increased performance with cognitive feedback (Harmon and Rohrbaugh, 1990). The conclusion is that group decisions may be better than those of individuals but are subject to some of the same problems.

37.5.3 Eliciting, Evaluating, and Applying Knowledge

The above discussion shows that knowledge is the common element that allows experts across naturalistic settings to perform tasks efficiently and effectively. A number of approaches have emerged for describing how people use and apply their knowledge when making decisions. Process tracing and behavioral protocol analysis techniques are a particularly useful example. Results obtained from applying these techniques can be compared to the results of cognitive simulations of how people use their knowledge. The obtained knowledge can also be used for other purposes, such as the development of rule-based expert systems. The following discussion first briefly summarizes process tracing and behavioral protocol analysis techniques. The topic of cognitive simulation is then briefly addressed. The final section describes a case study where expert knowledge was elicited and used to develop an expert system.

37.5.3.1 Process Tracing and Behavioral Protocols

Process tracing and the collection of behavioral protocols involves a number of procedures for documenting the sequences of problem-solving steps followed by people when they make decisions. Protocol analysis (Ericsson and Simon, 1984) focuses on asking experts to verbalize their thought processes while performing a task. The use of verbal protocols for identifying the planning, problem-solving, and decision-making activities underlying human cognitive processes has been extensively documented (for example, Bainbridge, 1974; Ericsson and Simon, 1984; Newell and Simon, 1972; Umbers, 1979). Protocol analysis has the disadvantage of being labor intensive and may interfere with the expert's ability to perform the analyzed task. However, despite the limitations associated with verbal protocols (Leplat and Hoc, 1981; Nisbett and Wilson, 1977), verbal protocols are viewed as invaluable, usually for providing insights into problem-solving and planning processes.

37.5.3.2 Cognitive Simulation

Cognitive simulation has focused on describing cognitive tasks as search within a knowledge or goal structure. The basic idea is that problem-solving activity can be described in terms of mental states and the operators that transform states into other states. The entire set of states and operators defines the problem space within which problem-solving activity takes place (Newell and Simon, 1972). The complexity of human behavior is explained in terms of simple strategies (heuristics) that guide search within the often complex problem space. Both states and operators are defined symbolically allowing the problem space to be represented as a network of connected nodes, which is often hierarchically organized. The nodes that are visited during search through the problem space define problem-solving behavior.

By analyzing the task sequences followed within the simulation, insight can be obtained regarding the factors influencing time requirements, the causes of errors, and the influences of errors in terms of: depth of reasoning, critical types of information, knowledge requirements, and memory demands [short-term memory (STM), working memory, external memory, and long-term memory (LTM)]. For example, effects such as learning can be explicitly described (Kieras, 1985) as changes in knowledge or goal structures. Errors can be traced to factors such as the similarity between the antecedent clauses of rules (Kieras, 1985), the forgetting of specific conditions, or the use of inappropriate heuristics (Johnson and Payne, 1985).

37.5.3.3 Case Study of Expert System Development

This case study considers the development of an expert system for troubleshooting the malfunctions of a 'bare chip mounter', a computer controlled device used to mount chips on integrated circuit boards (Naruo, Lehto, and Salvendy, 1990). The knowledge acquisition process followed when developing this system provides insight into the potentially valuable role of documenting the knowledge used by experts in naturalistic settings.

Malfunctions of this particular machine were mechanical, electrical, or software related. The causes of malfunctions of the machine fell into three general categories as follows: (1) human causes, such as operation mistakes; (2) external causes, such as power source or electromagnetic interference; and (3) internal causes related to both hardware and software. There were many specific causes of machine malfunction within each of these general categories. The correspondingly large number of failure modes contributed greatly to the difficulty of the troubleshooting task.

The knowledge acquisition process followed during development of the expert system focused on helping the designer build a model of how he would apply his detailed knowledge of the machine during the troubleshooting process. This required that both *declarative* knowledge (relatively static information in the form of facts and rules) and *procedural* knowledge (which describes dynamic forms of information as sequences of activities) be acquired. The first step was to develop a global view of the troubleshooting problem by hierarchically organizing the various troubleshooting related subproblems the expert system would need to solve. The second step involved developing a large design matrix that described the functional connections between each of the bare-chip mounter's components. This declarative design-related knowledge was essential for providing operators with accurate and detailed advice during troubleshooting and described information such as "which sensor is connected to which board through which line and connectors." The third and final step in knowledge acquisition was to document the designer's problem-solving strategy. To accomplish this goal, the designer developed a flow diagram, by mentally simulating his problem-solving process, for each malfunction.

Following the translation of the logic network diagrams and design matrix into a set of rules, the expert system was developed in less than a week. Of the system's 364 rules, 94 were derived from the logic network and corresponded to the expert's troubleshooting strategy. The remaining 270 rules corresponded to the declarative knowledge encoded in the design matrix. The system was capable of providing 389 different diagnoses. A post hoc analysis of malfunctions observed in the field revealed that 92% of malfunctions were successfully diagnosed by the system. The key to development was the use of appropriate mediating representations for acquiring and organizing knowledge, and the availability of simple methods for translating this knowledge into rules.

37.6 SUMMARY AND CONCLUSIONS

Beach (1993) discusses four revolutions in behavioral decision theory. The first took place when it was recognized that the evaluation of alternatives is seldom extensive and is illustrated by use of the satisficing rule rather than optimizing (Simon, 1955) and heuristics (Tversky and Kahneman, 1974). The second occurred when it was recognized that people choose between strategies to make decisions and is illustrated by contingency theory (Beach, 1990) and cognitive continuum theory (Hammond, 1980). The third is currently occurring and involves the realization that people rarely make choices and instead rely on prelearned procedures. This perspective is illustrated by the levels of processing approach (Rasmussen, 1983) and recognition-primed decisions (Klein, 1989). The fourth is just beginning, and recognizes that decision-making research must abandon a single-minded focus on the economic view of decision making. It must also consider approaches drawn from relevant developments and research on cognitive psychology, organizational behavior, and systems theory.

The discussion within this chapter parallels this view of decision making and the research that has been done. The integrative model presented at the beginning of the chapter shows how these various approaches fit together as a whole. Each path through the model is distinguished by specific sources of conflict, the methods of conflict resolution followed, and the types of decision rules used to analyze the results of conflict resolution processes. The different paths through the model correspond to fundamentally different ways of making decisions, ranging from routine situation assessment-driven decisions, to satisficing, analysis of single and multiattribute expected utility, and even obtaining consensus of multiple decision makers in group contexts. Numerous other strategies discussed in this chapter are also described by particular paths through the model.

This chapter goes beyond simply describing methods of decision making by pointing out reasons people and groups may have difficulty making good decisions. These include cognitive limitations, inadequacies of various heuristics used, biases and inadequate knowledge of decision makers, and task-related factors, such as risk, time pressure, and stress. The discussion also provides insight into the effectiveness of approaches for improving human decision making. The models of selective attention point to the value of providing only truly relevant information to decision makers. Irrelevant information might be considered simply because it is there, and especially so if it is highly salient. Methods of highlighting or emphasizing relevant information, therefore, clearly seem to be warranted. The models of selective information also indicate that methods of helping decision makers cope with working memory limitations will be of value. There also is reason to believe that providing feedback to decision makers in dynamic decision-making situations will be useful. Cognitive, rather than outcome, feedback is indicated as being particularly helpful when decision makers are learning. Training decision makers also seems to offer potentially large benefits. One reason for this conclusion is that the studies of naturalistic decision making revealed that most decisions are made on a routine, nonanalytical basis.

The studies of debiasing also partially support the potential benefits of training and feedback. On the other hand, the many failures to debias expert decision makers imply that decision aids, methods of persuasion, and other approaches intended to improve decision making are no panacea. Part of the problem is that people tend to start with preconceived notions about what they should do and show a tendency to seek out and bolster confirming evidence. Consequently, people show a tendency to develop overconfidence with experience, and strongly held beliefs become difficult to modify, even if they are hard to defend rationally.

37.7 ACKNOWLEDGMENTS

The author wishes to acknowledge Erik Eriksen and Jason Papastavrou of Purdue University for providing comments and some of the materials used in this paper.

REFERENCES

Allais, M. (1953). Le comportement de l'homme rationel devant le risque: Critique des postulates et axioms de l'ecole americaine. *Econometrica, 21*, 503–546.

Anderson, N. H. (1981). *Foundations of Information Integration Theory*. New York: Academic Press.

Bainbridge, L. (1974). Analysis of verbal protocols from a process control tasks. In E. Edward and F. B. Lees, Eds., *The Operator in Process Control.* London: Taylor and Francis.

Balzer, W. K., Doherty, M. E., and O'Connor, R. O., Jr. (1989). Effects of cognitive feedback on performance. *Psychological Bulletin, 106*, 41–433.

Bar-Hillel, M. (1973). On the subjective probability of compound events. *Organizational Behavior and Human Performance, 9*, 396–406.

Baron, J. (1985). *Rationality and Intelligence*. Cambridge: Cambridge University Press.

Basak, I., and Saaty, T. (1993). Group decision making using the analytic hierarchy process. *Methl. Comput. Modeling, 17*, 101–109.

Bazerman, M. H., and Neale, M. A. (1983). Heuristics in negotiation: Limitations to effective dispute resolution. In M. H. Bazerman, and R. Lewicki, Eds., *Negotiating in Organizations*. Beverly Hills, CA: Sage.
Beach, L. R. (1990). *Image Theory: Decision Making in Personal and Organizational Contexts*. Chichester, UK: John Wiley.
Beach, L. R. (1993). Four revoluations in behavioral decision theory. In M. M. Chemers, and R. Ayman, Eds., Leadership Theory and Reasearch. San Diego: Academic Press.
Beach, L. R., and Mitchell, T. R. (1978). A contingency model for the selection of decision strategies, *Academy of Management Journal, 3*, 439–449.
Bell, D. (1982). Regret in decision making under uncertainty. *Operations Research, 30*, 961–981.
Bernoulli, D. (1738). *Exposition of a New Theory of the Measurement of Risk*. St. Petersburg: Imperial Academy of Science.
Birnbaum, M. H., and Mellers, B. A. (1983). Bayesian inference: Combining base rates with opinions of sources who vary in credibility. *Journal of Personality and Social Psychology, 37*, 792–804.
Birnbaum, M. H., Coffey, G., Mellers, B. A., and Weiss, R. (1992). Utility measurement: Configural-weight theory and the judge's point of view. *Journal of Experimental Psychology: Human Perception and Performance, 18*, 331–346.
Bock, R. D., and Jones, L. V. (1968). *The Measurement and Prediction of Judgment and Choice*. San Francisco: Holden-Day.
Brehmer, B. (1981). Models of diagnostic judgment. In J. Rasmussen, and W. Rouse, Eds. *Human Detection and Diagnosis of System Failures*. New York: Plenum Press.
Brehmer, B., and Joyce, C. R. B. (1988). *Human Judgment: The SJT View*, Amsterdam: North-Holland.
Brookhouse, J. K., Guion, R. M., and Doherty, M. E. (1986). Social desirability response bias as one source of the discrepancy between subjective weights and regression weights. *Organizational Behavior and Human Decision Processes, 37*, 316–328.
Brunswick, E. (1952). *The Conceptual Framework of Psychology*. Chicago: University of Chicago Press.
Buck, J. R. (1989). *Economic Risk Decisions in Engineering and Management*. Ames, IA: Iowa State University Press.
Budescu, D., and Weiss, W. (1987). Reflection of transitive and intransitive preferences: A test of prospect theory. *Organizational Behavior and Human Performance, 39*, 184–202.
Caverni, J. P., Fabre, J. M., and Gonzalez, M. (1990). *Cognitive Biases*. Amsterdam: North-Holland.
Chemers, M. M., and Ayman, R., Eds. (1993). *Leadership Theory and Research*. San Diego: Academic Press.
Cheng, P. E., and Holyoak, K. J. (1985). Pragmatic reasoning schemas. *Cognitive Psychology, 17*, 391–416.
Clemen, R. T. (1991). *Making Hard Decisions: An Introduction to Decision Analysis*. Boston: PWS-Kent.
Cohen, M. S. (1993). The naturalistic basis of decision biases. In G. A. Klein, J. Orasanu, R. Calderwood, and E. Zsambok, Eds., *Decision Making in Action: Models and Methods* (pp. 51–99). Norwood, NJ: Ablex.
Davis, J. H. (1992). Some compelling intuitions about group consensus decisions, theoretical and empirical research, and interperson aggregation phenomena: Selected examples, 1950–1990. *Organizational Behavior and Human Decision Processes, 52*, 3–38.
Dorris, A. L., and Tabrizi, J. L. (1978). An empirical investigation of consumer perception of product safety. *Journal of Products Liability, 2*, 155–163.
Drucker, P. F. (1985). *The Effective Executive*. New York: Harper and Row.
Du Charme, W. (1970). Response bias explanation of conservative human inference, *Journal of Experimental Psychology, 85*, 66–74.
Duda, R. O., Hart, K., Konolige, K., and Reboh, R. (1979). *A Computer-Based Consultant for Mineral Exploration*. (Tech. Rep.), Stanford, CA: SRI International.
Duffy, L. (1993). Team decision making biases: An information processing perspective. In G. A. Klein, J. Orasanu, R. Calderwood, and E. Zsambok, Eds. *Decision Making in Action: Models and Methods*. Norwood, NJ: Ablex.
Edland, E., and Svenson, O. (1993). Judgment and decision making under time pressure. In *Time Pressure and Stress in Human Judgment and Decision Making*. O. Svenson, and A. J. Maule, Eds., 27–40, New York: Plenum.
Edwards, W. (1954). The theory of decision making. *Psychological Bulletin, 41*, 380–417.
Edwards, W. (1968). Conservatism in human information processing. In B. Kleinmuntz, Ed., *Formal Representation of Human Judgment*. (pp. 17–52). New York: John Wiley.
Einhorn, H. J., and Hogarth, R. M. (1978). Confidence in judgment: Persistence of the illusion of validity. *Psychological Review, 70*, 193–242.
Einhorn, H. J., and Hogarth, R. M. (1981). Behavioral decision theory: Processes of judgment and choice. *Annual Review of Psychology, 32*, 53–88.

Ellsberg, D. (1961). Risk, ambiguity, and the Savage axioms. *Quarterly Journal of Economics, 75*, 643–699.

Embrey, D. E. (1984). *SLIM-MAUD: An Approach to Assessing Human Error Probabilities Using Structured Expert Judgment.* (NUREG/CR-3518, Vols. 1 and 2). Washington, DC: U.S. Nuclear Regulatory Commission.

Ericsson, K. A., and Simon, H. A. (1984). *Protocol Analysis: Verbal Reports as Data.* Cambridge, MA: MIT Press.

Etzioni, A. (1988). Normative-affective factors: Toward a new decision-making model. *Journal of Economic Psychology, 9*, 125–150.

Evans, J. B. T. (1989). *Bias in Human Reasoning: Causes and Consequences.* London: Lawrence Erlbaum.

Evans, J. B. T., and Pollard, P. (1985). Intuitive statistical inferences about normally distributed data. *Acta Psychologica, 60*, 57–71.

Fischhoff, B. (1982). For those condemned to study the past: Heuristics and biases in hindsight. In D. Kahneman, P. Slovic, and A. Tversky, A., Eds., *Judgment Under Uncertainty: Heuristics and Biases.* Cambridge: Cambridge University Press.

Fischhoff, B., and MacGregor (1982). Subjective confidence in forecasts. *Journal of Forecasting, 1*, 155–172.

Fischhoff, B., Slovic, P., and Lichtenstein, S. (1977). Knowing with certainty: The appropriateness of extreme confidence. *Journal of Experimental Psychology: Human Perception and Performance, 3*, 552–564.

Fischhoff, B., Slovic, P., and Lichtenstein, S. (1978). Fault trees: Sensitivity of estimated failure probabilities to problem representation. *Journal of Experimental Psychology: Human Perception and Performance, 4*, 330–344.

Fishburn, P. C. (1974). Lexicographic orders, utilities, and decision rules: A survey. *Management Science, 20*, 1442–1471.

Fong, G. T., Krantz, D. H., and Nisbett, R. E. (1986). The effects of statistical training on thinking about everyday problems. *Cognitive Psychology, 18*, 253–292.

Friedman, J. W. (1990). *Game Theory with Applications to Economics.* New York: Oxford University Press.

Frisch, D., and Clemen, R. T. (1994). Beyond expected utility: Rethinking behavioral decision research. *Psychological Bulletin, 116(1)*, 46–54.

Gertman, D. I., and Blackman, H. S. (1994). *Human Reliability & Safety Analysis Data Handbook.* New York: Wiley.

Gordon, J., and Shortliffe, E. H. (1984). The Dempster-Schafer theory of evidence. In *Rule-Based Expert Systems: The MYCIN Experiments of the Stanford Heuristic Programming Project.* Reading, MA: Addison-Wesley.

Hammer, W. (1993). *Product Safety Management and Engineering.* (2nd ed.). Chicago: ASSE.

Hammond, K. R. (1980). Introduction to Brunswikian theory and methods. In K. R. Hammond, and N. E. Wascoe, Eds, *Realizations of Brunswick's Experimental Design.* San Francisco: Jossey-Bass.

Hammond, K. R. (1993). Naturalistic decision making from a Brunswikian viewpoint: Its past, present, future. In G. A. Klein, J. Orasanu, R. Calderwood, and E. Zsambok, Eds, *Decision Making in Action: Models and Methods.* (pp. 205–227). Norwood, NJ: Ablex.

Hammond, K. R., Stewart, T. R., Brehmer, B., and Steinmann, D.O. (1975). Social judgment theory. In M. F. Kaplan, S. and Schwartz, Eds., *Human Judgment and Decision Processes* (pp. 271–312). New York: Academic Press.

Harmon, J., and Rohrbaugh, J. (1990). Social judgement analysis and small group decision making: Cognitive feedback effects on individual and collective performance. *Organizational Behavior and Human Decision Processes, 46*, 34–54.

Heath, L., Tindale, R.S., Edwards, J., Posavac, E.J., Bryant, F.B., Henderson-King, E., Suarez-Balcazar, Y., and Myers, J. (1994). *Applications of Heuristics and Biases to Social Issues.* New York: Plenum Press.

Hogarth, R. M., and Einhorn, H. J. (1992). Order effects in belief updating: The belief-adjustment model. *Cognitive Psychology, 24*, 1–55.

Holtzman, S. (1989). *Intelligent Decision Systems.* Reading, MA: Addison-Wesley.

Howard, R. A. (1968). The foundations of decision analysis. *IEEE Transactions on Systems, Science and Cybernetics, SSC-4*, 211–219.

Howard, R.A. (1988). Decision analysis: Practice and promise. *Management Science, 34*, 679–695.

Huber, J., Wittink, D. R., Fiedler, J. A., and Miller, R. (1993). The effectiveness of alternative preference elicitation procedures in predicting choice. *Journal of Marketing Research, 30*, 105–114.

Jacoby, J. (1977). Information load and decision quality: Some contested issues. *Journal of Marketing Research, 14*, 569–573.

Janis, I. L. (1972). *Victims of Groupthink.* Boston: Houghton-Mifflin.

Janis, I. L., and Mann, L. (1977). *Decision Making: A Psychological Analysis of Conflict, Choice, and Commitment.* New York: Free Press.

Johnson, E. J., and Payne, J. W. (1985). Effort and accuracy in choice. *Management Science, 31(4)*, 395–414.

Kahneman, D., and Tversky, A. (1973). On the psychology of prediction. *Psychological Review, 80*, 251–273.

Kahneman, D., and Tversky, A. (1979). Prospect theory: An analysis of decision under risk. *Econometrica, 47*, 263–291.

Kahneman, D., Slovic, P., and Tversky, A. (1982).*Judgment Under Uncertainty: Heuristics and Biases.* Cambridge: Cambridge University Press.

Keeney, R. L., and Raiffa, H. (1976). *Decisions with Multiple Objectives: Preferences and Value Tradeoffs.* New York: Wiley.

Kieras, D. E. (1985). *The Role of Prior Knowledge in Operating Equipment from Written Instructions.* [Rep. No. 19 (FR–85/ONR–19). Ann Arbor: Department of Industrial and Operations Engineering, University of Michigan.

Keren, G. (1990). Cognitive aids and debiasing methods: Can cognitive pills cure cognitive ills? In J. P. Caverni, J. M. Fabre, and M. Gonzalez, Eds. *Cognitive Biases.* Amsterdam: North Holland.

Klayman, J., and Ha, Y. W. (1987). Confirmation, disconfirmation, and information in hypothesis testing. *Journal of Experimental Psychology: Human Learning and Memory*, 211–228.

Klein, G. A. (1989). Recognition-primed decisions. In W. Rouse, Ed., *Advances in Man-Machine System Research (pp. 5, 47–92). Greenwich, CT: JAI Press.*

Klein, G. A. (1995). The value added by cognitive analysis. In *Proceedings of the Human Factors and Ergonomics Society 39th Annual Meeting. (pp. 530–533).*

Klein, G. A., Orasanu, J., Calderwood, R., and Zsambok, E. Eds. (1993). *Decision Making in Action: Models and Methods.* Norwood, NJ: Ablex.

Klein, G. A., and Wolf, S. (1995). Decision-centered training. *Proceedings of the Human Factors and Ergonomics Society 39th Annual Meeting*, 1249–1252.

Kleinmuntz, B. (1984). The scientific study of clinical judgment in psychology and medicine. *Clinical Psychology Review, 4*, 111–126.

Koriat, A., Lichtenstein, S., and Fischhoff, B. (1980). Reasons for confidence. *Journal of Experimental Psychology: Human Learning and Memory, 6*, 107–118.

Kraus, N. N., and Slovic, P. (1988). Taxonomic analysis of perceived risk: Modeling individual and group perceptions within homogenous hazards domains. *Risk Analysis, 8*, 435–455.

Kuhberger, A. (1995). The framing of decisions: A new look at old problems. *Organizational Behavior and Human Decision Processes, 62*, 230–240.

Lathrop, R.G. (1967). Perceived variability. *Journal of Experimental Psychology, 23*, 498–502.

Lehto, M.R. (1991). A proposed conceptual model of human behavior and its implications for design of warnings. *Perceptual and Motor Skills, 73*, 595–611.

Lehto, M. R., James, D. S., Foley, J. P. (1994). Exploratory factor analysis of adolescent attitudes toward alcohol and risk. *Journal of Safety Research, 25*, 197–213.

Lehto, M. R., and Papastavrou, J. (1991). A distributed signal detection theory model: Implications to the design of warnings. In *Proceedings of the 1991 Automatic Control Conference.* (Boston, MA, 1990, pp. 2586–2590).

Leplat, J., and Hoc, J. M. (1981). Subsequent Verbalization in the Study of Cognitive Processes. *Ergonomics, 24*, 743–755.

Levin, L. P. (1975). Information integration in numerical judgements and decision processes. *Journal of Experimental Psychology: General, 104*, 39–53.

Lichtenstein, S., Fischhoff, B., and Phillips, L. D. (1982). Calibration of probabilities: The state of the art to 1980. In D. Kahneman, P. Slovic, and A. Tversky, Eds., *Judgment Under Uncertainty: Heuristics and Biases.* (pp. 306–334).

Lichtenstein, S., Slovic, P., Fischhoff, B., Layman, M., and Coombs, B. (1978). Judged frequency of lethal events. *Journal of Experimental Psychology: Human Learning & Memory, 4*, 551–578.

Luce, R. D., and Raiffa, H. (1957). *Games and Decisions.* New York: John Wiley.

Mallet, L., Vaught, C., and Brnich, M.J., Jr. (1993). Sociotechnical communication in an underground mine fire: A study of warning messages during an emergency evacuation. *Safety Science, 16*, 709–728.

Martocchio, J. J., Webster, J., and Baker, C.R. (1993). Decision-making in management information systems research: The utility of policy capturing methodology. *Behaviour and Information Technology, 12*, 238–248.

Maule, A. J., and Hockey, G. R. J. (1993). State, stress, and time pressure. In O. Svenson, A. J. and Maule, Eds., *Time Pressure and Stress in Human Judgment and Decision Making* (pp. 27–40), New York: Plenum.

McGuire, W. J. (1966). Attitudes and opinions. *Annual Review of Psychology, 17*, 475–514.

McKnight, A. J., Langston, E. A., McKnight, A. S., and Lange, J. E. (1995). The bases of decisions leading to alcohol impaired driving. In C. N. Kloeden, and A. J. McLean, Eds., *Proceedings of the 13th International Conference on Alcohol, Drugs, and Traffic Safety* (August 13th–18th, pp. 143–147). Adelaide, Australia.

Mockler, R. L., and Dologite, D. G. (1991). Using computer software to improve group decision making. *Long Range Planning*, *24*, 44–57.

Montgomery, H. (1989). From cognition to action: The search for dominance in decision making. In H. Montgomery, and O. Svenson, Eds., *Process and Structure in Human Decision Making*. Chicaster, UK: John Wiley.

Moscovici, S. (1976). *Social Influence and Social Change*. London: Academic Press.

Murphy, A. H., and Winkler, R. L. (1974). Probability forecasts: A survey of National Weather Service forecasters. *Bulletin of the American Meteorological Society*, *55*, 1449–1453.

Myers, J. L., Suydam, M. M., and Gambino, B. (1965). Contingent gains and losses in a risky decision situation. *Journal of Mathematical Psychology*, *2*, 363–370.

Naruo, N., Lehto, M., and Salvendy, G. (1990). Development of a knowledge based decision support system for diagnosing malfunctions of advanced production equipment. *International Journal of Production Research*, *1990*, *28*, 2259–2276.

Navon, D. (1979). The importance of being conservative. *British Journal of Mathematical and Statistical Psychology*, *31*, 33–48.

Newell, A., and Simon, H. A. (1972). *Human Problem Solving*. Englewood Cliffs, NJ.: Prentice-Hall.

Nisbett, R., and Ross, L. (1980). *Human Inference: Strategies and Shortcomings of Social Judgment.* Englewood Cliffs, NJ: Prentice-Hall.

Nisbett, R. E., and Wilson, T. D. (1977). Telling More Than We Can Know: Verbal Reports on Mental Processes. *Psychological Review*, *67*, 279–300.

Orasanu, J., and Salas, E. (1993). Team decision making in complex environments. In G. A. Klein, J. Orasanu, R. Calderwood, and E. Zsambok, Eds., *Decision Making in Action: Models and Methods*. Norwood, NJ: Ablex.

Paese, P. W., Bieser, M., and Tubbs, M. E. (1993). Framing effects and choice shifts in group decision making. *Organizational Behavior and Human Decision Processes*, *56*, 149–165.

Payne, J. W. (1980). Information processing theory: Some concepts and methods applied to decision research. In T. S. Wallsten, Ed., *Cognitive Processes in Choice and Decision Research.* Hillsdale, NJ: Erlbaum.

Payne, J. W., Bettman, J. R., and Johnson, E. J. (1993). *The Adaptive Decision Maker.* Cambridge: Cambridge University Press.

Pennington, N., and Hastie, R. (1986). Evidence evaluation in complex decision making. *Journal of Personality and Social Psychology*, *51*, 242–258.

Pennington, N., and Hastie, R. (1988). Explanation-based decision making: Effects of memory structure on judgment. *Journal of Experimental Psychology: Learning, Memory, and Cognition*, *14*, 521–533.

Pitz, G. F. (1980). The very guide of life: The use of probabilistic information for making decisions. In T. S. Wallsten, Ed., *Cognitive Processes in Choice and Decision Behavior.* Hillsdale, NJ: Erlbaum.

Raiffa, H. (1968). *Decision Analysis*. Reading, MA, Addison-Wesley.

Rasmussen, J. (1983). Skills, rules, knowledge: signals, signs, and symbols and other distinctions in human performance models. *IEEE Transactions on Systems, Man, and Cybernetics, SMC–13(3)*, 257–267.

Reason, J. (1990). *Human Error*. Cambridge, UK: Cambridge University Press.

Rethans, A. J. (1980). Consumer perceptions of hazards. *PLP–80 Proceedings*, 25–29.

Harmon, J., and Rohrbaugh, J. (1990). Social judgment analysis and small group decision making: Cognitive feedback effects on individual and collective performance. *Organizational Behavior and Human Decision Processes*, *46*, 34–54.

Saaty, T. L. (1988). *Multicriteria Decision Making: The Analytic Hierarchy Process*, PA: Pittsburg.

Sage, A. (1981). Behavioral and organizational considerations in the design of information systems and processes for planning and decision support. *IEEE Transactions on Systems, Man, and Cybernetics, SMC–11.*

Savage, L. J. (1954). *The Foundations of Statistics.* New York: John Wiley.

Shafer, G. (1976). *A Mathematical Theory of Evidence*. Princeton, NJ: Princeton University Press.

Simon, H. A. (1955). A behavioral model of rational choice. *Quarterly Journal of Economics*, *69*, 99–118.

Simon, H. A. (1983). Alternative visions of rationality. In H. A. Simon, Ed., *Reason in Human Affairs.* Stanford, CA: Stanford University Press.

Singleton, W. T., and Hovden, J. (1987). *Risk and Decisions*. New York: John Wiley.

Slovic, P. (1978). The psychology of protective behavior. *Journal of Safety Research*, *10*, 58–68.

Slovic, P. (1987). Perception of risk. *Science*, 236, 280–285.

Slovic, P., Fischhoff, B., and Lichtenstein, S. (1977). Behavioral decision theory. *Annual Review of Psychology*, *28*, 1–39.

Sniezek, J. A. (1992). Groups under uncertainty: An examination of confidence in group decision making. *Organizational Behavior and Human Decision Processes*, *52*, 124–155.
Sniezek, J.A., and Henry, R.A. (1989). Accuracy and confidence in group judgment. *Organizational Behavior and Human Decision Processes*, *43*, 1–28.
Stanoulov, N. (1994). Expert knowledge and computer-aided group decision making: Some pragmatic reflections. *Annals of Operations Research*, *51*, 141–162.
Stevenson, M. K., Busemeyer, J. R., and Naylor, J. C. (1993). Judgment and decision-making theory. In M. D. Dunnette, and L. M. Hough, Eds., *Handbook of Industrial and Organizational Psychology* (Vol. 1, 2nd ed.). Palo Alto, CA: Consulting Psychologists Press, Inc.
Stukey, E., and Zeckhauser, R. (1978). Decision analysis. In *A Primer for Policy Analysis*. New York: W.W. Norton, pp. 201–254.
Svenson, O. (1990). Some propositions for the classification of decision situations. In K. Borcherding, O. Larichev, and D. Messick, Eds., *Contemporary Issues in Decision Making* (pp. 17–31). Amsterdam: North Holland.
Svenson, O., and Maule, A. J. (1993). *Time Pressure and Stress in Human Judgment and Decision Making*. New York: Plenum.
Swain, A. D., and Guttman, H. (1983). *Handbook for Human Reliability Analysis with Emphasis on Nuclear Power Plant Applications* (NUREG/CR–1278). Washington, DC: U.S. Nuclear Regulatory Commission.
Tanner, W. P., and Swets, J. A. (1954). A decision making theory of visual detection. *Psychological Review*, *61*, 401–409.
Torrance, E. P. (1953). The behavior of small groups under the stress conditions of "survival." *American Sociological Review*, *19*, 751–755.
Tversky, A. (1969). Intransitivity of preferences. *Psychological Review*, *76*, 31–48.
Tversky, A. (1972). Elimination by aspects: A theory of choice, *Psychological Review*, *79*, 281–289.
Tversky, A., and Kahneman, D. (1973). Availability: A heuristic for judging frequency and probability, *Cognitive Psychology*, *5*, 207–232.
Tversky, A., and Kahneman, D. (1974). Judgment under uncertainty: Heuristics and biases. *Science*, *185*, 11241–1131.
Tversky, A., and Kahneman, D. (1981). The framing of decisions and the psychology of choice. *Science*, *211*, 453–458.
Umbers, T.G. (1979). A Study of the Control Skills of Gas Grid Control Engineers. *Ergonomics*, *22*, 557–571.
Valenzi, E., and Andrews, I.R. (1973). Individual differences in the decision processes of employment interviews. *Journal of Applied Psychology*, *58*, 49–53.
von Neumann, J., and Morgenstern, O. (1947). *Theory of Games and Economic Behavior*. Princeton, NJ: Princeton University Press.
Wagenaar, W. A. (1992). Risk taking and accident causation. In J. F. Yates, Ed., *Risk-Taking Behavior* (pp. 257–281). New York: John Wiley.
Wallsten, T. S. (1972). Conjoint-measurement framework for the study of probabilistic information processing. *Psychological Review*, *79*, 245–260.
Wallsten, T. S. (1976). Using conjoint-measurement models to investigate a theory about probabilistic information processing. *Journal of Mathematical Psychology*, *14*, 144–185.
Wallsten, T. S., Zwick, R., Kemp, S., and Budescu, D. V. (1993). Preferences and reasons for communicating probabilistic information in verbal and numerical terms. *Bulletin of the Psychonomic Society*, *31*, 135–138.
Weber, E. (1994). From subjective probabilities to decision weights: The effect of asymetric loss functions on the evaluation of uncertain outcomes and events. *Psychological Bulletin*, *115*, 228–242.
Weber, E., Anderson, C. J., and Birnbaum, M. H. (1992). A theory of perceived risk and attractiveness. *Organizational Behavior and Human Decision Processes*, *52*, 492–523.
Wegner, D. (1987). Transactive memory: A contemporary analysis of group mind. In B. Mullen, and G. R. Goethals, Eds., *Theories of Group Behavior* (pp. 185–208). New York: Springer-Verlag.
Weinstein, N. D. (1979). Seeking reassuring or threatening information about environmental cancer. *Journal of Behavioral Medicine*, *2*, 125–139.
Weinstein, N. D. (1980). Unrealistic optimism about future life events. *Journal of Personality and Social Psychology*, *39*, 806–820.
Weinstein, N. D. (1987). Unrealistic optimism about illness susceptibility: Conclusions from a community-wide sample. *Journal of Behavioral Medicine*, *10*, 481–500.
Weinstein, N. D., and Klein, W. M. (1995). Resistance of personal risk perceptions to debiasing interventions. *Health Psychology*, *14*, 132–140.
Welford, A. T. (1976). *Skilled Performance*. Glenview, IL: Scott, Foresman.
Wickens, C. D. (1992). *Engineering Psychology and Human Performance*. New York: Harper Collins.
Winkler, R. L., and Murphy, A. H. (1973). Experiments in the laboratory and the real world. *Organizational Behavior and Human Performance*, *10*, 252–270.

Winterfeldt, D. V., and Edwards, W. (1986). *Decision Analysis and Behavioral Research*. Cambridge: Cambridge University Press.

Yates, J. F. Ed. (1992). *Risk-Taking Behavior*, New York: John Wiley.

Zimmer, A. (1983). Verbal versus numerical processing of subjective probabilities. In R. W. Scholtz, Ed., *Decision Making Under Uncertainty*. Amsterdam: North Holland.

CHAPTER 38

FEEDBACK CONTROL MODELS—MANUAL CONTROL AND TRACKING

Ronald A. Hess
Department of Mechanical and Aeronautical Engineering
University of California — Davis
Davis CA 95616 USA

38.1 INTRODUCTION

38.1.1 Evolution of the Manual Control Discipline

There exist many day-to-day activities that require continuous human control for their successful and safe completion. Driving an automobile, riding a bicycle, flying an aircraft are but three examples. Each of these tasks involves the human being acting as a feedback element in a control system. The importance of such human feedback activity in the operation of many engineering systems has led to the development of a separate discipline called manual feedback control, or more simply, manual control. As a distinct discipline, manual control is approaching its fiftieth year of existence.

Most, if not all of the modern manual control research had its genesis in the work of feedback control engineers during, and immediately after, World War II. Dictated by the development of weapons, such as antiaircraft guns, that could only function in concert with human operators, pioneering studies such as those of Tustin (1947) compared the control behavior of the human to that of inanimate automatic feedback devices. Fortunately, the existing mathematical tools for feedback analysis were sufficiently mature to be applied to the problem, and the "servomechanism" model of the human operator was born. The control-theoretic representation of the human operator that evolved from this servomechanism paradigm has become the fundamental mode of representation of the human operator for most manual control practitioners (McRuer, 1980). The servomechanism paradigm, itself, may indeed have laid the groundwork for the science of cybernetics (Miller, 1982).

As a simple example of the control-theoretic approach, Figure 38.1 shows a representative human-in-the-loop control problem, here represented by a gunnery azimuth tracking task. From the reference line of sight (LOS), one can define three variables: a target LOS, $\Psi_T(t)$, a gun LOS, $\Psi_G(t)$, and a LOS error, $\Psi_E(t) = \Psi_T(t) - \Psi_G(t)$. The tracking task for the human can be defined as that of reducing $\Psi_E(t)$ to zero, if possible, through the use of the control $\delta(t)$. Figure 38.2 is a block diagram representation of the physical system of Figure 38.1. To the control system engineer, a block diagram is essentially a cause–effect representation of a physical system, where blocks denote the input–output behavior of the system element in question, e.g., the human operator or the gun, and the directed line segments represent "signals," e.g., LOS error, $\Psi_E(t)$. It is important to note that Figure 38.2 assumes that only the LOS error is being used by the human in the task at hand. Such feedback structures are referred to as being "compensatory" in nature. In reality, other information such as the target LOS, $\Psi_G(t)$, might be available to the human, given a large enough field of view. Tracking tasks in which both error and "input" information is available are typically referred to as being "pursuit" in nature.

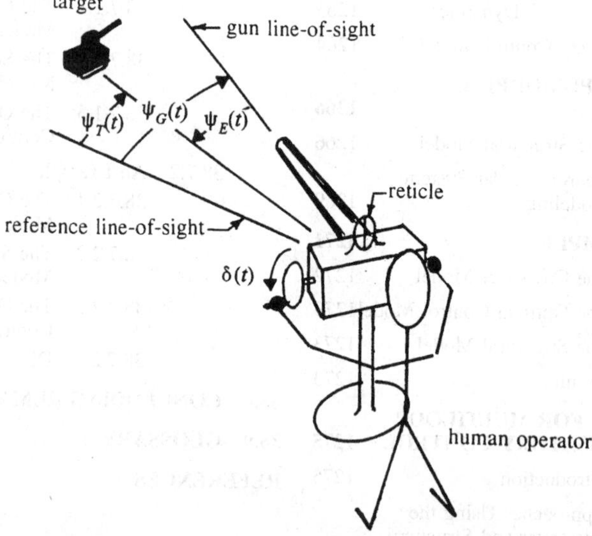

Figure 38.1 A simple manual control gunnery task.

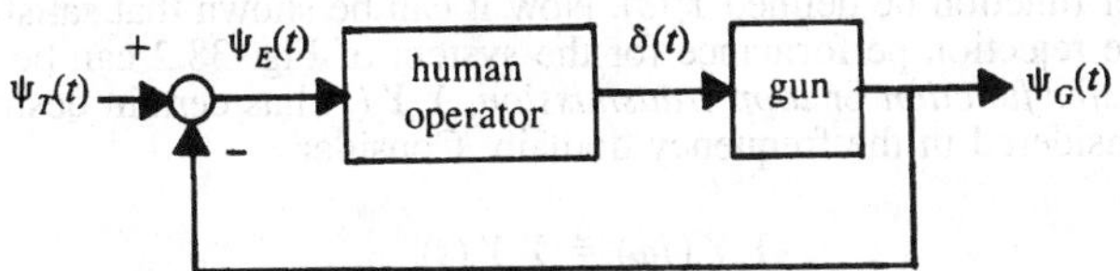

Figure 38.2 A block diagram representation of the task of Figure 38.1.

The control theoretic description of the human operator will be the basis of the modeling work discussed in this chapter. It will be assumed that the reader has had some background in elementary control theory, equivalent to an undergraduate course in the subject. For the uninitiated, it is recommended that a suitable introductory text be selected for reference, e.g. (Nise, 1995).

The majority of research in manual control theory has dealt with single-loop compensatory systems typified by Figure 38.2. It was the hope of the early manual control researchers that the human's dynamic characteristics in such systems could be described by the same types of mathematical equations used in describing linear servomechanisms, i.e. linear, constant coefficient differential equations. Indeed, in his pioneering report Tustin (1947) states:

> The object of the series of tests described in the present report was to investigate the nature of the layer's (gunner's) response to a number of particular cases and to attempt to find the laws of relationships to error. In particular, it was hoped that this relationship might be found . . . to be approximately linear and so permit the well-developed theory of "linear servomechanisms" to be applied to manual control in the same way as it is applied to automatic following.

In general, Tustin's hopes were well founded. In the intervening years, experiments have demonstrated that quasilinear describing functions could be used to describe human operator dynamics in single-loop compensatory systems with a variety of controlled elements and with random or random-appearing input signals (McRuer and Krendel, 1957) and (McRuer, Graham, Krendel, and Reisener, 1965). At the time of its publication, the report of McRuer and colleagues in 1965 represented a summation and culmination of much of the manual control research since Tustin's work.

38.1.2 Present Relevance

With increasing automation, the role of the human in many dynamics systems is moving from that of a continuous controller to that of a supervisor or systems manager. The modern airline cockpit is one notable example of this pervasive transformation. Thus the question somewhat naturally arises as to the current relevance of manual control research and human operator modeling. Although automation has made increasing and important inroads into areas that were strictly the domain of the human, there will still be instances in which at least the possibility of continuous human control must be considered. Again referring to the aircraft flight control example, although completely automatic flight is now possible from brake release to landing and rollout, a significant portion of airline flights is still done under manual control. One obvious reason for this is that pilots must retain the proficiency of their piloting skills in case the automated system does fail. With the increased performance capabilities of those systems which humans are asked to control, determining the limitations of human sensing and actuation capabilities are more critical than ever. As an example, the occurrence of a particular adverse form of pilot–aircraft coupling known as pilot-induced oscillations keeps appearing with depressing regularity on modern aircraft, from the F-22 fighter (Dornheim, 1992) to the Boeing 777 airliner (Dornheim, 1995). Thus, adequate descriptions of the human operator or controller are still relevant and, indeed, vital.

38.2 CONTROL THEORETIC MODELS

38.2.1 Single-Loop Feedback Control Principles

Consider again Figure 38.2. Assume that the element described as "gun" is represented by some transfer function, $Y_c(s)$, representing the linear dynamics of this controlled element. Also assume that one is designing an inanimate compensator to replace the human.

Let this transfer function be defined $Y_p(s)$. Now it can be shown that satisfactory tracking and disturbance rejection performance for the system of Fig. 38.2 can be obtained if the *open-loop transfer function* or *loop transmission*, $Y_pY_c(s)$ has certain desirable characteristics when considered in the frequency domain. Consider

$$Y_pY_c(j\omega) \triangleq Y_pY_c(s)|_{s=j\omega} \tag{38.1}$$

and plot a Bode plot of $Y_pY_c(j\omega)$, where the latter is a complex number that can be represented by an amplitude, $|Y_pY_c(j\omega)|$, and phase angle, $\measuredangle Y_pY_c(j\omega)$, both a continuous function of frequency ω. The Bode or frequency response plot is essentially a plot of $20\log_{10}|Y_pY_c(j\omega)|$ versus log (ω), and $\measuredangle Y_pY_c(j\omega)$ versus log (ω). Multiplying the logarithmic amplitude of $Y_pY_c(j\omega)$ by 20 defines the amplitude in decibels (dB). (When power spectral densities are involved, amplitude in decibels is defined as $10\log_{10}|\text{—}|$). From a control system design standpoint, desirable characteristics of $Y_pY_c(j\omega)$ are shown in Figure 38.3. Here one sees large amplitudes at low frequencies (small values of ω), low amplitudes at high frequencies (large values of ω), and between the two a region centered about the frequency ω_c at which $20\log_{10}|Y_pY_c(j\omega)| = 0$ dB, where the slope of the amplitude curve is approximately -20dB/decade (Maciejowski, 1989). A decade describes any frequency range ω_1 to ω_2 where $\omega_2/\omega_1 = 10$.

The desirable features of the loop transmission just described can be obtained by appropriate selection of the compensator $Y_p(s)$. As will be seen, in an attempt to achieve desirable performance, the human operator him- or herself adopts dynamic characteristics, $Y_p(s)$, similar to those that would be exhibited by an inanimate compensator designed to meet similar performance requirements.

38.2.2 The Crossover Model

Figure 38.4 shows what is termed a describing function representation of the human operator. The signal n_e shown injected with the error, e, is referred to as *remnant* and when multiplied by the transfer function $Y_p(s)$, represents that portion of the human's output that is not linearly correlated with the input *c(t)* (Graham and McRuer, 1971). By "not linearly correlated" is meant that this portion of the human's output cannot be

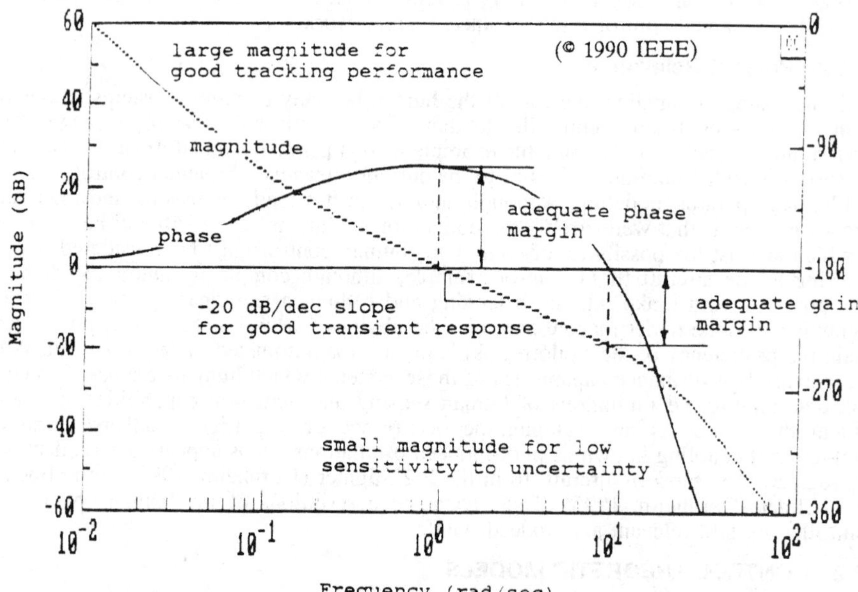

Figure 38.3 Desirable open-loop frequency domain characteristics of compensator and controlled element. (From Hess R. A., and Modjtahedzadeh, A., "A Control Theoretic Model of Driver Steering Behavior," *IEEE Control Systems Magazine*, Vol. 10, No. 5, 1990. Reprinted with permission.)

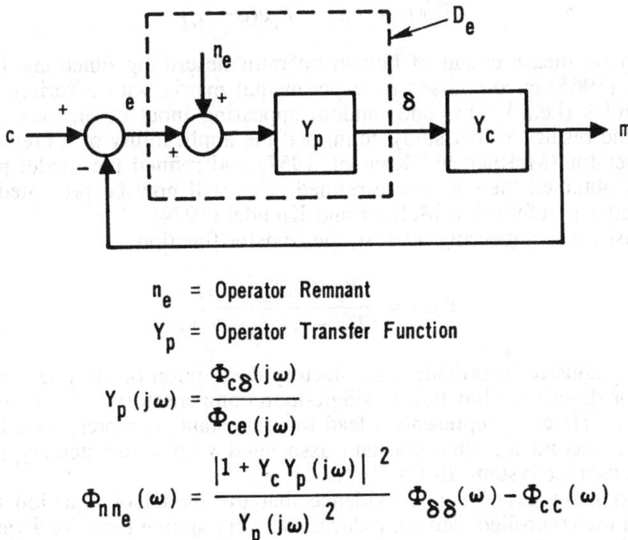

Figure 38.4 A describing function representation of the human operator.

obtained from the input signal by a linear operation. Remnant, itself, can be thought of as representing actual noise injection by the human and/or as modeling errors, i.e., errors implicit in assuming a linear, time-invariant representation of human operator dynamics. Thus, the human operator describing function, D_e, consists of a linear transfer function and a random "noise" injected with the displayed or sensed error signal.

The remnant signal in the describing function representation is typically quantified by its *power spectral density*, which can be defined as

$$\Phi_{nn_e}(\omega) = \int_{-\infty}^{\infty} \phi_{nn_e}(\tau) e^{-j\omega\tau}\, d\tau \tag{38.2}$$

where $\phi_{nn_e}(\tau)$ is called the *autocorrelation function* and is itself defined as

$$\phi_{nn_e}(\tau) = \lim_{T\to\infty} \frac{1}{2T} \int_{-T}^{T} n_e(t) n_e(t + \tau)\, dt \tag{38.3}$$

where t represents continuous time and T represents the duration of the signals in question. Figure 38.4 also indicates how $Y_p(j\omega)$ and $\Phi_{nn_e}(\omega)$ can be determined from time histories in tracking experiments.

The utility of describing function representations of the human operator derives from the rather fortuitous fact that the relative magnitude of the remnant signal tends to be rather small for most controlled elements and tasks. Thus, the transfer function part of the describing function, $Y_p(j\omega)$, characterizes most of the human operator behavior. The linear differential equation description of the human operator dynamics then stems from $Y_p(s)$, where

$$Y_p(s) = Y_p(j\omega)\big|_{j\omega=s} \tag{38.4}$$

If, for example,

$$Y_p(s) = \frac{\delta(s)}{e(s)} = \frac{K_p e^{-\tau_e s}}{T_I s + 1} \tag{38.5}$$

where K_p represents a gain and T_I represents a lag time constant, then the differential equation between the operator's input, $e(t)$, and his or her output, $\delta(t)$, is

$$T_I\dot{\delta}(t) + \delta(t) = K_p e(t - \tau_e) \tag{38.6}$$

The systematic measurement of human operator describing functions undertaken by McRuer et al. (1965) encompassed an experimental matrix with a variety of controlled element dynamics (i.e., $Y_c(s)$'s) and random appearing input commands with different bandwidths. The results of that study confirmed the applicability of a previous model of the human operator (McRuer and Krendel, 1957) and refined the model parameter adjustment rules obtained therein. These refined rules will now be presented. For further details, the reader is referred to McRuer and Krendel (1974).

In many cases of engineering interest, the transfer function

$$Y_p(s) = \frac{K_p e^{-\tau_e s}(T_L s + 1)}{(T_I s + 1)(T_N s + 1)} \tag{38.7}$$

can be suitably adjusted to provide a satisfactory description of the linear portion of the human operator describing function in single-loop compensatory systems in a frequency range around ω_c. Here, T_L represents a lead time constant, T_I represents a lag time constant, and T_N a second lag time constant associated with a rudimentary model of the human neuromuscular system. In Eq. 7, $T_N << T_I$.

There is extensive experimental evidence that the model of equation (38.7), when combined with the controlled element reduces to a very simple form for a variety of input commands and controlled elements. That is,

$$Y_p Y_c(s) = \frac{\omega_c e^{-\tau_e s}}{s} \tag{38.8}$$

The fact that, around crossover, this form for $Y_p Y_c(s)$ conforms to the dictates of a sound control system design with an inanimate compensator is worthy of note. The model of equation (38.8), along with the parameter adjustment rules to be summarized, defines what is known as the *crossover model* of the human operator. Its importance in the study of manual control cannot be overemphasized.

The model of equation (38.8) is very simple, consisting of only two parameters: the crossover frequency, ω_c and the effective time delay, τ_e. The rules for selecting these parameters are as follows:

$$\begin{aligned} \omega_c &\approx \omega_{c0} + 0.18\omega_{BW_c} \\ \tau_e &\approx \tau_0 - 0.08\omega_{BW_c} \\ \tau_0 &\approx \frac{\pi}{2\omega_{c0}} \end{aligned} \tag{38.9}$$

where ω_{BW_c} represents the bandwidth of the imput $c(t)$. The parameter ω_{c0} represents the value of the open-loop crossover frequency adopted by the human operator as the bandwidth of the command input approaches zero. Table 38.1 shows representative values of τ_0 and ω_{c0} for three different limiting forms of $Y_c(s)$ from McRuer et al. (1965). A great many of the controlled elements of practical importance in manual control systems can be closely approximated by these limiting or stereotypical forms in the important region of open-loop crossover.

Table 38.1 Approximate Parameter Values for Crossover Model of Equation (38.8).

$Y_c(s)$	τ_0 (sec)	ω_{c0} (rad/sec)
K	0.30	5.0
K/s	0.35	4.5
K/s^2	0.50	3.0

Knowledge of the form of the controlled element dynamics near crossover obviously allows one to determine $Y_p(s)$ in the crossover region. Figure 38.5 is a Bode plot for an experimental measurement for $Y_pY_c(j\omega)$ in which $Y_c(s) = K/s$ and the $\omega_{BW_c} = 1.5$ rad/sec. The solid line represents a hand-faired curve through the data which was obtained at discrete frequencies. The dashed line shows the variation caused by the approximate nature of the crossover model, here only effecting the low-frequency phase characteristics. This low-frequency phase effect is often referred to as "phase droop" and can be modeled by extensions to the simple crossover model of equation (38.8). The phase discrepancy of Figure 38.5 has almost no effect upon the quality of predicted closed-loop stability or performance.

The remaining part of the human operator describing function is the remnant which is characterized by its power spectral density, $\Phi_{nn_e}(\omega)$. Again, extensive experimental evidence suggests that this remnant power spectral density scales with the variance of the error signal to which it is added and can be represented by the following equation:

$$\Phi_{nn_e}(\omega) = \frac{R\overline{e}^2}{\omega^2 + \omega_R^2} \tag{38.10}$$

Here, $\overline{e}^2$ refers to the mean square value of the error signal, defined as

$$\overline{e}^2 \triangleq \lim_{T\to\infty} \frac{1}{2T} \int_{-T}^{T} e^2(t)\, dt \tag{38.11}$$

Table 38.2 gives approximate values of R and ω_R as a function of the limiting controlled element forms shown in Table 38.1.

Block diagram algebra e.g. (Nise, 1995) coupled with fundamental spectral analysis techniques e.g. (Bendat and Piersol, 1993) yield the following expression for the mean square error, $\overline{e}^2$ in the single-loop compensatory manual control system of Figure 38.4, assuming a stationary random input with power spectral density $\Phi_{cc}(\omega)$, and bandwidth, ω_{BW_c}

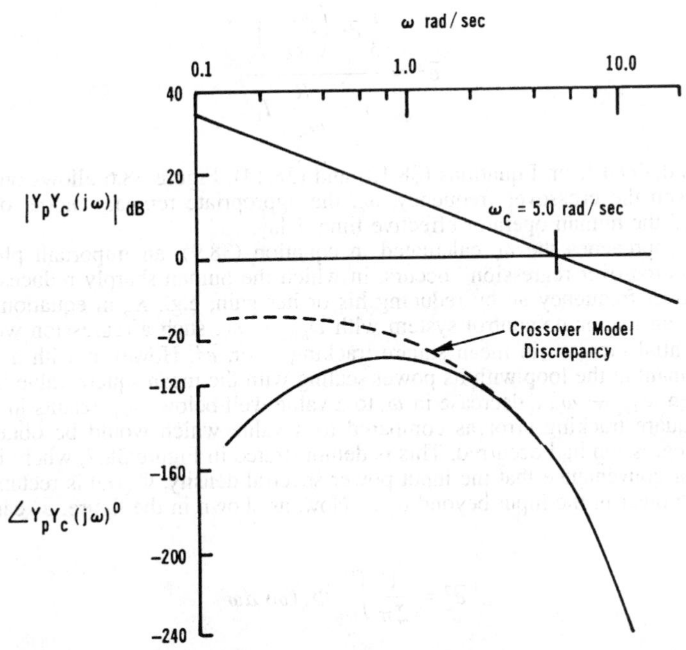

Figure 38.5 An example of an experimentally derived open-loop transfer function for $Y_c(s) = K/s$.

Table 38.2 Approximate Parameter Values for Remnant Model of Equation (38.10).

$Y_c(s)$	R	ω_R (rad/sec)
K	0.1–0.5	3.0
K/s	0.1–0.5	3.0
K/s^2	0.1–0.5	1.0

$$\bar{e}^2 = \frac{\dfrac{1}{2\pi}\displaystyle\int_{-\infty}^{\infty} \frac{\Phi_{cc}(\omega)}{|1 + Y_p Y_c(j\omega)|^2}\, d\omega}{1 - F} \tag{38.12}$$

where

$$F = \frac{1}{2\pi}\int_{-\infty}^{\infty} \Phi_{nn_e}(\omega) \left| \frac{Y_p Y_c(j\omega)}{1 + Y_p Y_c(j\omega)} \right|^2 d\omega \tag{38.13}$$

The function F can be rewritten as

$$F = \frac{R}{\omega_R^2 \tau_e} I_1 \tag{38.14}$$

For $\omega_{BW_c} < \omega_c$ which is normally the case, the numerator term in equation (38.12) can be approximated by the "1/3 power law" (McRuer and Krendel, 1974) as

$$\frac{1}{2\pi}\int_{-\infty}^{\infty} \frac{\Phi_{cc}(\omega)}{|1 + Y_p Y_c(j\omega)|^2}\, d\omega \approx \frac{1}{3}\bar{c}^2 \left(\frac{\omega_{BW_c}}{\omega_c}\right)^2 \tag{38.15}$$

Thus, equation (38.12) can be rewritten as

$$\bar{e}^2 \approx \frac{\dfrac{1}{3}\bar{c}^2 \left(\dfrac{\omega_{BW_c}}{\omega_c}\right)^2}{1 - \dfrac{R}{\omega_R^2 \tau_e} I_1} \tag{38.16}$$

where I_1 is defined from Equations (38.13) and (38.14). Figure 38.6 allows one to determine I_1 given the crossover frequency, ω_c, the appropriate remnant model of equation (38.10), and the human operator effective time delay, τ_e.

If ω_{BW_c} approaches the ω_c calculated in equation (38.9), an important phenomenon known as "crossover regression" occurs, in which the human sharply reduces the open-loop crossover frequency ω_c by reducing his or her gain, e.g., K_p in equation (38.7). In the case of an inanimate control system with $\omega_{BW_c} < \omega_c$, such a regression would result in a substantial increase in mean square tracking error, $\bar{e}^2$. However, with a human injecting remnant in the loop with its power scaling with the mean square value of the error signal, when $\omega_{BW_c} \rightarrow \omega_c$, a decrease in ω_c to a value well below ω_{BW_c} results in a decrease in mean square tracking error as compared to a value which would be obtained if no crossover regression had occurred. This is demonstrated in Figure 38.7, where it has been assumed for convenience that the input power spectral density, $\Phi_{cc}(\omega)$ is rectangular, i.e., there is no power in the input beyond ω_{BW_c}. Now, as shown in the figure, $\bar{e}^2$ can be given as

$$\bar{e}^2 = \frac{1}{2\pi}\int_{-\infty}^{\infty} \Phi_{ee}(\omega)\, d\omega \tag{38.17}$$

The solid and dashed curves in Figure 38.7 represent the error power spectral density for $\omega_{BW_c} \approx \omega_c$ and $\omega_{BW_c} \leq \omega_c$, respectively. The area under each of these curves represents

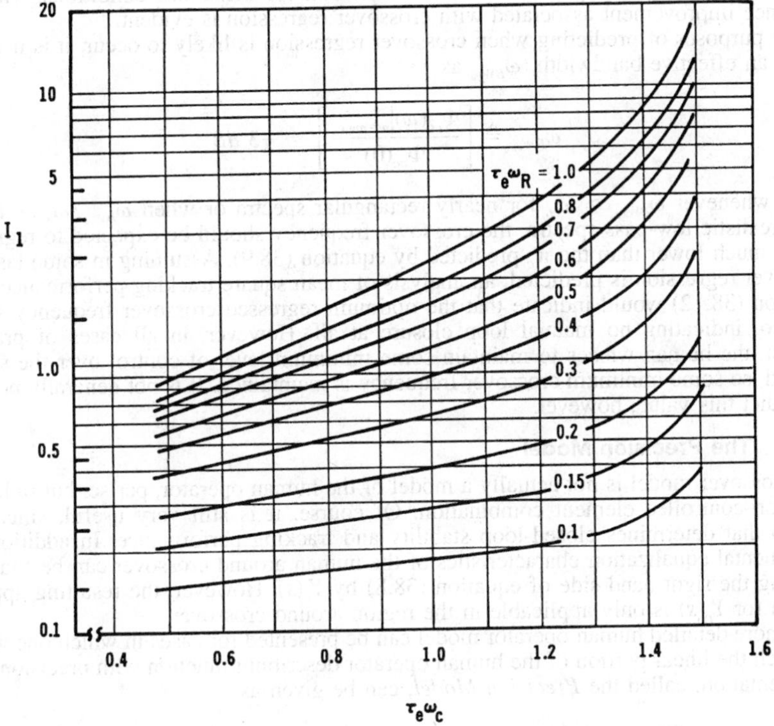

Figure 38.6 Curves for determining integral I_1 in performance calculations.

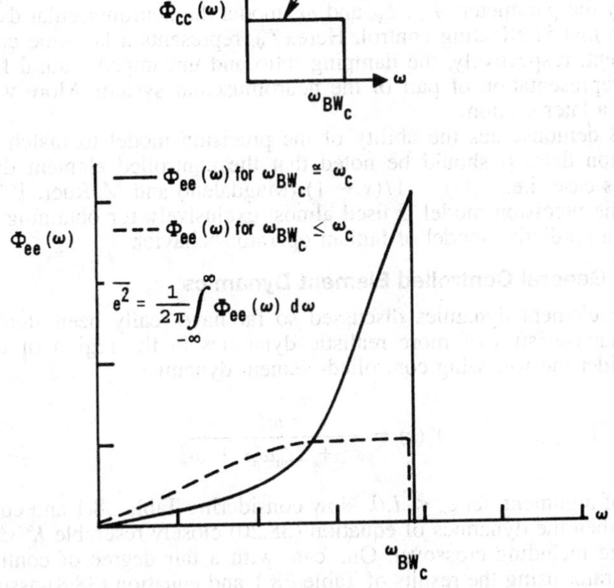

Figure 38.7 The effect of crossover regression on mean square tracking error.

the mean square tracking error which would result for these two conditions. The performance improvement associated with crossover regression is evident.

For purposes of predicting when crossover regression is likely to occur, it is useful to define an effective bandwidth, $\omega_{BW_{ce}}$ as

$$\omega_{BW_{ce}} \triangleq \left[\frac{\Phi_{cc}(\omega)|_{\omega=\omega_{BW_{ce}}}}{\Phi_{cc}(0)} \right] = -3\ dB \tag{38.18}$$

Then, whenever $\omega_{BW_{ce}} > \omega_{c0}$ for nearly rectangular spectra or when $\omega_{BW_{ce}}/\omega_c > 1.0$ for more realistic low-pass spectra, the crossover frequency should be expected to regress to values much lower than the w_c predicted by equation (38.9). Assuming in some task that crossover regression is predicted, an analysis of mean square tracking performance using equation (38.12) would indicate that the optimum regressed crossover frequency should be zero, indicating no manual loop closure at all. However, in all cases of practical interest, the human wishes to maintain some minimum level of control over the system at hand, so some minimum crossover frequency is maintained. It is not generally possible to predict this value, however.

38.2.3 The Precision Model

The crossover model is not actually a model of the human operator, per se, but rather the operator–controlled element combination. Of course, it is still very useful, since it is $Y_pY_c(s)$ that determines closed-loop stability and tracking performance. In addition, the fundamental equalization characteristics of the human around crossover can be found by dividing the right-hand side of equation (38.8) by $Y_c(s)$. However, the resulting approximation for $Y_p(s)$ is only applicable in the region around crossover.

A more detailed human operator model can be presented for cases in which one wishes to match the linear portion of the human operator describing function with precision. This representation, called the *Precision Model*, can be given as

$$Y_p(s) = K_p e^{-\tau s} \left(\frac{T_L s + 1}{T_I s + 1}\right)\left(\frac{T_K s + 1}{T'_K s + 1}\right) \left(\frac{1}{(T_{N_1}s + 1)\left[\left(\frac{s}{\omega_N}\right)^2 + \left(\frac{2\zeta_N}{\omega_N}\right)s + 1\right]} \right) \tag{38.19}$$

The parameters T_L and T_I are used to produce basic equalization capabilities of the human. The parameters T_K and T'_K are used to produce the aforementioned low frequency "phase droop." Finally the parameters T_{N_1}, ζ_N, and ω_N model the neuromuscular dynamics of the particular limb that is effecting control. Here, T_{N_1} represents a lag time constant and ζ_N and ω_N represent, respectively, the damping ratio and undamped natural frequency of a second-order representation of part of the neuromuscular system. More will be said on this subject in a later section.

Figure 38.8 demonstrates the ability of the precision model to match measured describing function data. It should be noted that the controlled element dynamics were unstable in this case; i.e., $Y_c(s) = 1/(s - 1)$ (Magdaleno and McRuer, 1971). As might be expected, the precision model is used almost exclusively for obtaining such matches rather than as a predictive model of human operator behavior.

38.2.4 More General Controlled Element Dynamics

The controlled element dynamics discussed so far have really been stereotypes representing the characteristics of more realistic dynamics in the region of crossover. For example, consider the following controlled-element dynamics:

$$Y_c(s) = \frac{\omega_n^2}{s^2 + 2\zeta_n\omega_n s + \omega_n^2} \tag{38.20}$$

For purposes of argument, let $\zeta_n < 1.0$. Now considering Table 38.1 and equation (38.8), if, $\omega_n << \omega_c$, then the dynamics of equation (38.20) closely resemble K/s^2 over a broad frequency range including crossover. One can, with a fair degree of confidence, model the human operator using the results of Table 38.1 and equation (38.8) assuming $Y_c(s) = K/s^2$. Likewise, if $\omega_n >> \omega_c$, then the dynamics of equation (38.20) closely resemble K,

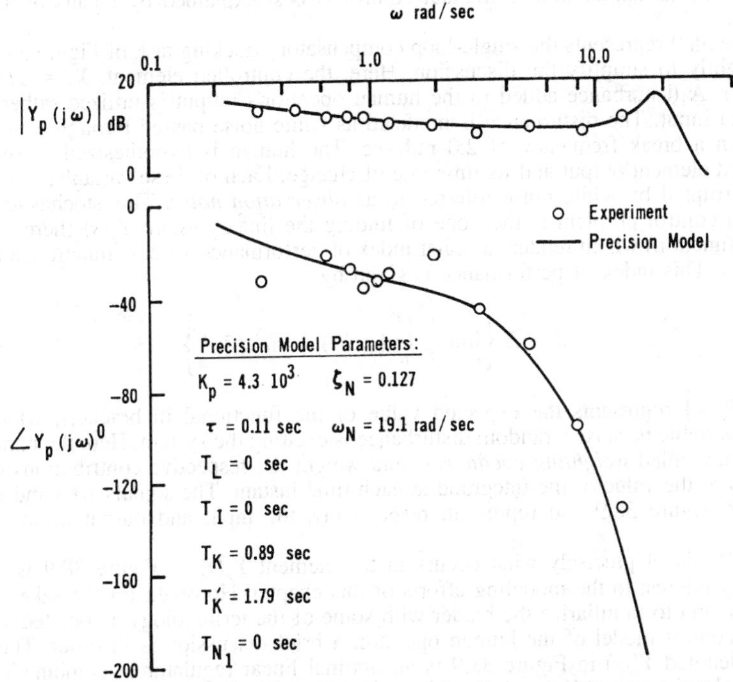

Figure 38.8 Fitting experimental data with the precision model of equation (38.19).

over a broad frequency range including crossover. Again, one can, with a fair degree of confidence, use the results of Table 38.1 and equation (38.8) assuming $Y_c(s) = K$. Now if $\omega_n \approx \omega_c$, the issue is not as clear cut. Obviously, the operator will have to generate some lead equalization to force $Y_pY_c(s)$ into the ω_c/s form around crossover. However, selection of ω_c, itself, and the effective time delay is not straightforward in this case, since the controlled element dynamics do not fall into either the K or K/s^2 class.

In the case just described, the following procedure is recommended (McRuer and Krendel, 1974). First make a reasonable estimate of ω_c. Then use equation 38.9 to yield τ_0. Next, use

$$\Delta\tau_e \approx 0.08\omega_{BW_{ce}} \tag{38.21}$$

to determine $\Delta\tau_e$, the decrement in time delay caused by forcing function bandwidth. Now the nominal crossover frequency can be determined using

$$Y_p(s) = K_p(T_L s + 1)e^{-(\tau_0 - \Delta\tau_e)s} \tag{38.22}$$

and adjusting K_p to minimize mean square error. This will yield a nominal value for ω_c. If this value differs significantly from the estimated value, repeat the process. The remnant model associated with K/s^2 dynamics (i.e., a controlled element requiring lead equalization) in Table 38.2 should be employed.

38.2.5 The Optimal Control Model

38.2.5.1 Introduction

As pointed out in Section 38.2.1, the equalization derived form the crossover model is very similar to that which would be prescribed by an experienced control system designer given the same controlled element and feedback variable, i.e., system output. This fact has led to the application of "modern" control system design techniques to human operator modeling. One such technique is based upon optimal control and estimation in what is called the stochastic linear quadratic Guassian problem (Stengel, 1986). This problem and its application to manual control is best explained by means of a simple example.

Figure 38.9 represents the single-loop compensatory tracking task of Figure 38.4 modified slightly to simplify the discussion. Here, the controlled element, $Y_c = 1/s$, is an integrator. A disturbance added to the human operator's output is utilized rather than a command input. The disturbance is modeled as white noise passed through a first-order filter with a break frequency of 2.0 rad/sec. The human is hypothesized to sense the controlled element output and its time rate of change. Each of these variables is assumed to be corrupted by white noise referred to as *observation noise*. This stochastic, linear, Guassian control problem is then one of finding the linear system $Y_p'(s)$ (here a matrix transfer function) which renders a scalar index of performance, or cost function, a relative minimum. This index of performance is given by

$$J = E\left\{\lim_{T\to\infty}\frac{1}{T}\int_0^T [qx_2^2(t) + r\dot{u}^2(t)]\,dt\right\} \tag{38.23}$$

where $E\{—\}$ represents the expected value of the functional in brackets, which is a random variable because a random disturbance is exciting the system. Here the parameters q and r are called *weighting coefficients* and weight the respective contributions of $x_2^2(t)$ and $\dot{u}^2(t)$ to the value of the integrand at each time instant. The signals $x_2(t)$ and $u(t)$ are shown in Figure 38.9 and represent, respectively, the input and output of the human operator.

The details of precisely what occurs in the element $Y_p'(s)$ of Figure 38.9 is only of secondary interest to the modeling efforts of this chapter. However, for the sake of completeness, and to familiarize the reader with some of the terminology associated with the optimal control model of the human operator, a brief discussion is in order. The linear system denoted $Y_p'(s)$ in Figure 38.9 is an optimal linear regulator in combination with an optimal estimator, or Kalman filter (Stengel, 1986). The Kalman filter produces optimal estimates of the values of the state variables [here $x_1(t)$ and $x_2(t)$] at each instant of time given noisy measurements (or observations) of linear combinations of these state variables. In Figure 38.9, these linear combinations take the form of the state variables themselves, but this need not always be the case. In fact, any combination is acceptable, as long as the resulting system possesses a property called *observability*. The linear regulator multiplies the optimal estimates of the state variables by optimal regulator gains to produce the control output $u(t)$. The regulator gains can be found if the system possesses a property called *controllability*. The properties of observability and controllability are almost always found in manual control systems, so no further discussion on these topics

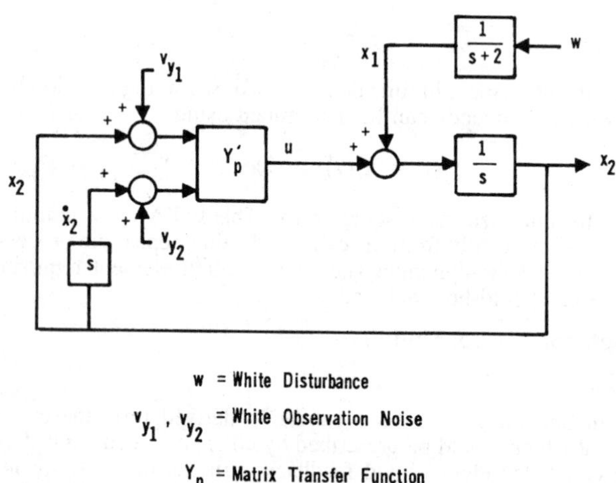

Figure 38.9 A modified form of the human–machine system of Figure 38.2.

will be pursued here. The interested reader is referred to any standard text on the subject (e.g., Nise, 1995).

One of the obvious differences between the optimal control and crossover model formulations is that the former assumes a vector input to the operator, even for "single-loop" tasks such as that of Figure 38.2. It has been shown (Levison, Kleinman, and Baron, 1969) that this vector representation leads to an equivalent single-loop remnant model similar to that given in equation (38.10). Of course, the optimal control model formulation is not limited to single-loop tasks.

The model shown in Figure 38.9 is a more simplified version of the optimal control model than is usually employed in human–machine studies (e.g., Hess and Kalteis, 1991). Figure 38.10 shows the more complete model emphasizing the state space representation of the system. Note that the operator time delay is now explicitly modeled and the Kalman estimator is followed by a predictor that produces an optimal estimate of the state vector $\hat{x}(t)$ given an optimal estimate $\hat{x}(t - \tau)$. The block labeled *neuromotor dynamics* is actually part of the optimal regulator and is a direct consequence of defining the index of performance so as to include control rate $\dot{u}^2(t)$. The noise term $\nu_u(t)$ is referred to as *motor noise* and is included to provide more realistic performance predictions. Details of the complete optimal control model are discussed by Kleinman, Baron, and Levison (1970).

38.2.5.2 Optimal Control Model Parameter Selection

The optimal control model is essentially specified by (a) the weighting coefficients q and r in the index of performance, (b) the covariances of the observation and motor noises, denoted V_y and V_u, and (c) the magnitude of the operator time delay, τ. As in the case of the precision model, these parameters can be selected to yield optimal control model transfer functions that closely match those obtained in experiment. In addition, the formulation also provides root-mean-square (RMS) performance predictions and remnant spectral densities.

As an example, consider the controlled element and disturbance given in Figure 38.9. The equations of state which described this system can be given by

$$\begin{aligned}\dot{x}_1(t) &= -2x_1(t) + w(t)\\ \dot{x}_2(t) &= x_1(t) + u(t)\\ y_1(t) &= x_2 + \nu_{y_1}(t)\\ y_2(t) &= x_1(t) + u(t) + \nu_{y_2}(t)\end{aligned} \tag{38.24}$$

The index of performance and parameter values are

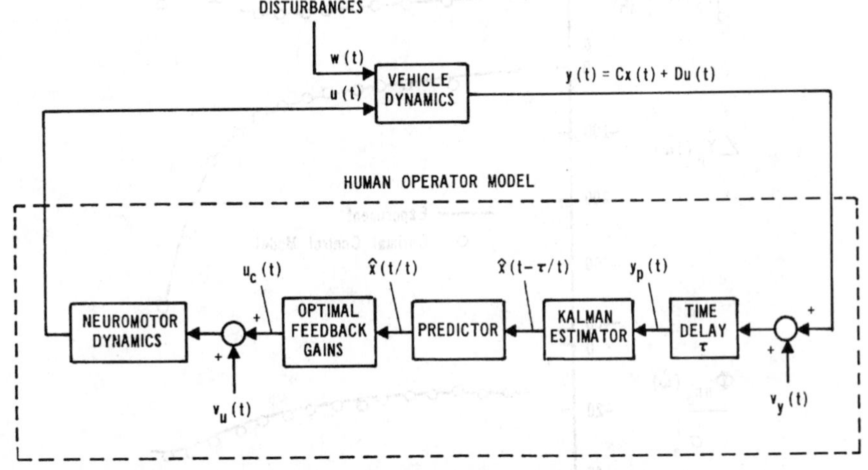

Figure 38.10 The optimal control model of the human operator.

$$J = E\left\{\lim_{T\to\infty}\frac{1}{T}\int_0^T [x_2^2(t) + 0.0017\dot{u}^2(t)]\,dt\right\}$$

$$V_{y_1} = 0.01\pi E\{y_1^2(t)\} \qquad V_u = 0.003\pi E\{u^2(t)\} \tag{38.25}$$

$$V_{y_2} = 0.01\pi E\{y_2^2(t)\} \qquad \tau = 0.15\text{s}$$

Here, the 0.01 and 0.003 factors used in defining V_{y_1} and V_{y_2} are referred to as *noise ratios*. The experimental and model generated describing functions (operator transfer function and remnant) of Figure 38.11 result (Kleinman et al., 1970).

Perhaps the most important parameters in the optimal control model formulation are the weighting coefficients q and r that appear in the index of performance definition. In many applications of the optimal control model, an approximate but very useful relationship exists between the coefficients q and r, the controlled element dynamics and the closed-loop system bandwidth. It can be shown (Hess, 1984; Kwakernaak and Sivan, 1972) that

$$\omega_{BW} \approx \left[\left(\frac{q}{r}\right)^{1/2}\right]^{1/(n-m+1)} \tag{38.26}$$

The parameters m and n are obtained from the controlled-element dynamics when expressed as

$$Y_c(s) = \frac{x_2(s)}{u(s)} = K\left[\frac{s^m + a_{m-1}s^{m-1} + \cdots + a_1 s + a_0}{s^n + b_{n-1}s^{n-1} + \cdots + b_1 s + b_0}\right] \tag{38.27}$$

The precise definition of ω_{BW} as given above is the magnitude of that closed-loop pole nearest the frequency where the amplitude of the closed-loop transfer function is 6 dB below its zero frequency value. Now the following approximate relation also allows one to approximate the open-loop crossover frequency calculated by equation (38.26):

$$\omega_c \approx 0.56\omega_{BW} \tag{38.28}$$

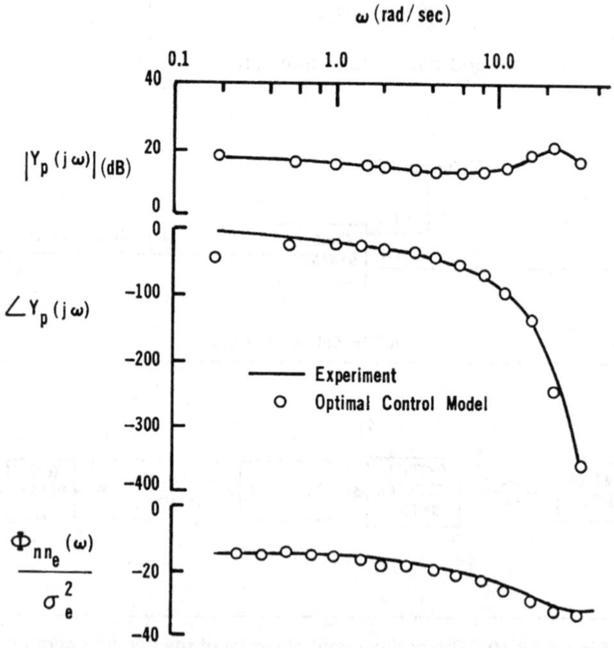

Figure 38.11 Comparison of experimental and optimal control model describing functions.

Equation (38.8) is still useful for optimal control model calculations. The time delay in the optimal control model is more representative of an actual, as opposed to an effective, delay and so a nominal value of 0.15 to 0.2 sec is often employed. This leaves the covariances of the observation and motor noises to be specified. There is fairly extensive evidence to suggest the following values for single-loop tracking tasks using ideal displays and manipulators:

$$\begin{aligned} V_{y_1} &= 0.01\pi E\{y_1^2(t)\} \qquad \text{displacement noise} \\ V_{y_2} &= 0.01\pi E\{y_1^2(t)\} \qquad \text{rate noise} \\ V_u &= 0.001\pi E\{u^2(t)\} \rightarrow .005\pi E\{u^2(t)\} \qquad \text{motor noise} \end{aligned} \tag{38.29}$$

38.2.5.3 More General Controlled Element Dynamics

In employing the optimal control model with controlled-element dynamics that differ from the stereotypical K, K/s, K/s^2, etc. one can proceed as follows: First an estimate of ω_c is made. Then equations (38.26–38.28) are used to obtain q/r. Next, using the parameters suggested by equation (38.29), (with the motor noise ratio 0.003) the ratio of q/r is varied until the mean square tracking error is minimized. This variation in q/r will, of course, result in variations in ω_c. It should be pointed out that this process is sensitive to the magnitude of the motor noise ratio. The value suggested is representative of that obtained by matching optimal control model characteristics with those of well-trained, well-motivated human operators in single-loop laboratory tracking tasks.

The author has found that a shorter procedure than that suggested above can be used in finding q/r for use in the optimal control model, either as an initial estimate or as a value to be used in preliminary analysis with the model. First, referring to the system of Figure 38.9, equation (38.23) is rewritten as

$$J = E\left\{\lim_{T\to\infty}\frac{1}{T}\int_0^T\left[\left(\frac{x_2(t)}{x_{2_M}}\right)^2 + \left(\frac{\dot{u}(t)}{\dot{u}_M}\right)^2\right]dt\right\} \tag{38.30}$$

where x_{2_M} and $\dot{u}_M$ represent "maximum allowable deviations" of $x_2(t)$ and $\dot{u}(t)$. Now an "effective time constant" T, is introduced to define these maximum allowable deviations of the integral and derivative of $\dot{x}_{2_M}$ as

$$\begin{aligned} x_{2_M} &= \dot{x}_{2_M}T \\ \dot{x}_{2_M} &= \text{specified but arbitrary} \\ \ddot{x}_{2_M} &= \frac{\dot{x}_{2_M}}{T} \\ \dddot{x}_{2_M} &= \frac{\ddot{x}_{2_M}}{T} = \frac{\dot{x}_{2_M}}{T^2} \end{aligned} \tag{38.31}$$

In a similar manner, one can write

$$\begin{aligned} u_m &= \dot{u}_M T \\ \dot{u}_m &= \text{to be selected} \\ \ddot{u}_m &= \frac{\dot{u}_M}{T} \\ \dddot{u}_M &= \frac{\ddot{u}_M}{T} = \frac{\dot{u}_M}{T^2} \end{aligned} \tag{38.32}$$

The value of $\dot{u}_M$ is not arbitrary, however, but is found using equations (38.31) and (38.32) and the vehicle dynamics as follows. Return to the controlled element dynamics as given in equation (38.27) with $Y_c(s) = \dfrac{x_2(s)}{u(s)}$. By considering the differential equation between $x_2(t)$ and $u(t)$ and replacing each derivative by its maximum allowable deviation as determined from equations (38.31) and (38.32), the following relation can be obtained:

$$\dot{u}_M = \frac{1}{K} \cdot \frac{\dfrac{1}{T^{n-1}} + \dfrac{|b_{n-1}|}{T^{n-2}} + \cdots + |b_1| + |b_0|T}{\dfrac{1}{T^{m-2}} + \dfrac{|a_{m-2}|}{T^{m-3}} + \cdots + |a_1| + |a_0|T} \cdot \dot{x}_{2_M} \tag{38.33}$$

With the introduction of equation (38.33), the selection of the relative weighting between $x_2^2(t)$ and $\dot{u}^2(t)$ becomes synonymous with the selection of the effective time constant T. Finally, the analysis of a variety of single-loop manual control tasks using the optimal control model suggests the following approximate relation for choosing T (Hess, 1984):

$$T = 0.65(\tau + \tau_D) \tag{38.34}$$

where τ is the value of the human operator time delay used in the optimal control model and τ_D is an "effective delay" obtained by approximating the actual controlled element dynamics of equation (38.27) by

$$Y_c(s) \approx K \frac{(T_L s + 1)e^{-\tau_D s}}{s\left(\dfrac{s^2}{\omega_n^2} + \dfrac{2\zeta_n}{\omega_n} s + 1\right)} \tag{38.35}$$

The approximation in equation (38.35) can be accomplished by a transfer function fitting routine.

Once T has been obtained, then q/r can be found and either used directly in the optimal control model or used as an initial estimate in the procedure to minimize the mean square error as outlined in the preceding.

Another technique has been used in the past to obtain q/r and is based upon the fact that the inclusion of control rate in the definition of the index of performance leads to the existence of a first-order lag in the optimal control model. Indeed, this is what constitutes the "neuromuscular dynamics" in Figure 38.10. By varying q/r, a "desirable" value of the lag time constant, usually denoted τ_n, can be obtained. A nominal value is $\tau_n = 0.1$ s. As q/r is increased (decreased), τ_n decreases (increases). We shall have reason to return to this method for selecting q/r in a later section.

38.2.6 Fuzzy Control Models

The models for the human operator that have been described are derived from a control theoretic viewpoint, and in particular, a viewpoint and methodology appropriate for linear, time-invariant systems. Often, the systems that humans are asked to control are nonlinear. By this is meant that these systems cannot be described by sets of linear, ordinary differential equations. In addition, it is possible that the system in question may be so complex that a precise, tractable identification of the controlled element is not possible, e.g., a national economy. The advent of what is called *fuzzy set theory* and *fuzzy control systems* offers a means of modeling human interaction with such system (e.g., Zadeh, 1973). Only the briefest of descriptions of this approach are given here.

A *set* can be defined as a collection of elements in which the members of the set are clearly and unambiguously distinguished from nonmembers. For example, one can consider the set of all real numbers, s_i, such that

$$0 \le s_i \le 1.0 \tag{38.36}$$

Clearly, the number 0.72 belongs to this set and the number 1.3 does not. Fuzzy set theory generalizes the notion of a set, so that any element can have *grades of membership* in any number of sets. These grades take on values from 0 to 1.0. In addition, fuzzy set theory makes use of (1) *linguistic variables* in place of, or in addition to, numerical variables, (2) *fuzzy conditional statements* to characterize simple relations between variables, and (3) *fuzzy algorithms* to characterize complex relations.

An example of a linguistic variable might be the word "size". The "values" of this linguistic (and fuzzy) variable might be *very small*, *small*, *somewhat small but not large*, *somewhat large*, *large*, etc. Fuzzy conditional statements can be exemplified by considering a pair of fuzzy variables x and y. IF x is *very small*, THEN y is *large*. Fuzzy algorithms are ordered sequences of instructions such as: Reduce x *slightly* if y is *very large*.

With the introduction of fuzzy algorithms, perhaps one can begin to see how fuzzy control might be related to human–machine interaction. Imagine the helmsman of a large ship explaining how he or she corrects for the ship's deviation from some desired heading. The helmsman might say, for example, "If the heading deviation is very small, I do not make a rudder correction, but if the heading deviation is small, but not large, I make a small rudder correction." Obviously, the fuzzy or linguistic variables must be quantified in some sense and this is where the membership functions come into play.

For example, consider Figure 38.12 which shows membership functions for the hypothetical heading deviation and rudder input for the helmsman problem. The shape of the membership functions have been made triangular for the sake of simplicity. Shown on Figure 38.12*a* is a heading deviation $\psi_e = -5$ degrees. As the diagram indicates, this has the following grades of membership: 0.25 (very negative), 0.75 (negative), 0 (around zero), 0 (positive), and 0 (very positive). By the same token, membership functions for the rudder input might be as shown in Figure 38.12*b*. Shown is a rudder deflection of $\delta_R = +3°$. This deflection has the following grades of membership: 0 (very negative), 0 (negative), 0 (around zero), 0.5 (positive), and 0.5 (very positive).

In terms of manual control, the activity of the human helmsman would be defined by a mapping from the input membership functions to the output ones. In fuzzy control applications this transformation is accomplished by what are termed *relational matrices* (e.g., Sutton, 1990). In the simplified example considered here, a relational matrix **R**, would be defined as

$$\mathbf{R} = M(\psi_e) \times M(\delta_R) \tag{38.27}$$

where X denotes a Cartesian product, or minimum for each element pairing of $M(\psi_e)$ and $M(\delta_R)$, and $M(—)$ is the particular grade of membership for each ψ_e and δ_R. Assume that observation of the action of a helmsman in a particular task has led to the following relational matrix:

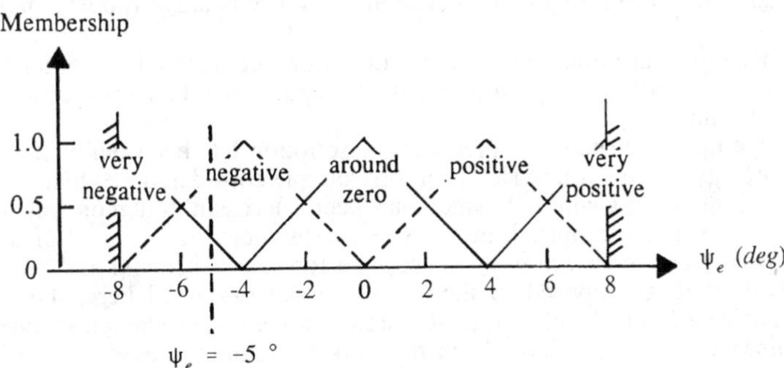

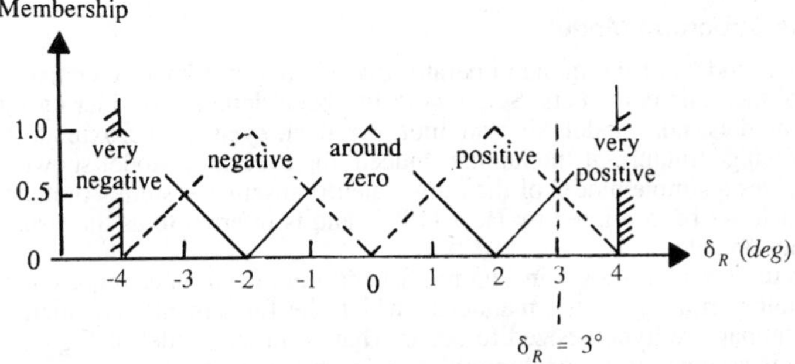

Figure 38.12 Hypothetical membership functions for the helmsman's problem.

$$\boldsymbol{R} = \begin{bmatrix} 0.6 & 0.5 & 0 & 0 & 0 \\ 0.2 & 0.75 & 0.5 & 0 & 0 \\ 0 & 0.4 & 1 & 0.4 & 0 \\ 0 & 0 & 0.5 & 0.75 & 0 \\ 0 & 0 & 0 & 0.4 & 0.6 \end{bmatrix} \tag{38.38}$$

In this rudimentary example, the R of equation (38.38) constitutes what might be called a "fuzzy control" model of the helmsman. In exercising this model, one may assume that the helmsman makes observations every ΔT seconds. For example, suppose that, at some observation instant, $\psi_e(n\Delta T) = -6°$. Figure 38.12*a* indicates that the grades of membership for this error are $M(\Psi_e) = 0.5$ (very negative), 0.5 (negative), 0 (around zero), 0 (positive), 0 (very positive). Through the use of $\boldsymbol{R}$ and $M(\psi_e)$, the membership grades of the output δ_R at this instant can be obtained as (Sutton, 1990)

$M(\delta_R)$ = max of {min[0.6,0.5], min[0.2,0.5], min[0,0], min[0,0], min[0,0]} for very negative rudder input
= max of {min[0.5,0.5], min[0.75,0.5], min[0.4,0], min[0,0], min[0,0]} for negative rudder input
= max of {min[0,0.5], min[0.5,0.5], min[1,0], min[0.5,0], min[0,0]} for around zero rudder input
= max of {min[0,0.5], min[0,0,.5], min[0.4,0], min[0.75,0], min[0.4,0]} for positive rudder input
= max of {min[0,0.5], min[0,0.5], min[0.,0], min[0,0], min[0.6,0]} for very positive rudder input

The right-hand sides of the equations above constitute the membership grades for $\delta_R(n\Delta T)$. These have been obtained by pairing the elements of $\boldsymbol{R}$ with those of $M(\psi_e)$ as would be done in the matrix multiplication $M(\psi_e)R$. Evaluating the right-hand side of the equations above yields $M(\delta_R)$ as

$M(\delta_R)$ = 0.5 (very negative rudder input), 0.5 (negative rudder input), 0.5 (around zero rudder input), 0 (positive rudder input), 0 (very positive rudder input)

These membership grades must now be quantified or "defuzzified" in order to have the model produce a control input δ_R. Note that the triangular membership functions of Figure 38.12b do not hold.

It must be emphasized that the preceding description has been deliberately oversimplified for purposes of exposition. However the example considered, i.e. that of developing a fuzzy control model of a ship helmsman, has been successfully demonstrated by Sutton and Towill (1988). For example, Figure 38.13 shows their fuzzy model of a helmsman in which two control modes are in evidence: one for course keeping and one for course changing. Note that, as opposed to the simple model discussed here, the helmsman's inputs are considered to be heading rate, $\dot{\psi}(t)$ and heading error (denoted ϵ). Fuzzy models of other human control tasks have been developed, including automobile driving, e.g., Kramer (1985).

38.3 ANTHROPOMORPHIC MODELS

38.3.1 The Structural Model

The feedback models of the human operator discussed so far have obviously been based upon control theoretic constructs. Some work has been done in considering more anthropomorphic models, i.e., models that attempt to reflect, even in approximate fashion, the signal processing structure of the human. Indeed, the Precision Model shown in equation (38.19) involves a simple model of the human neuromuscular system. Another such modeling approach has been offered by Hess (1985) and is referred to as the structural model of the human operator.

The structural model, shown in Figure 38.14, differs from other representations of the human operator primarily in the manner in which the fundamental equalization capabilities of the human are hypothesized to occur. That is, in the model of Figure 38.14, such equalization is assumed to occur through *proprioceptive* feedback from the limb effecting control. "Proprioceptive" refers to the sensory information about limb position or applied

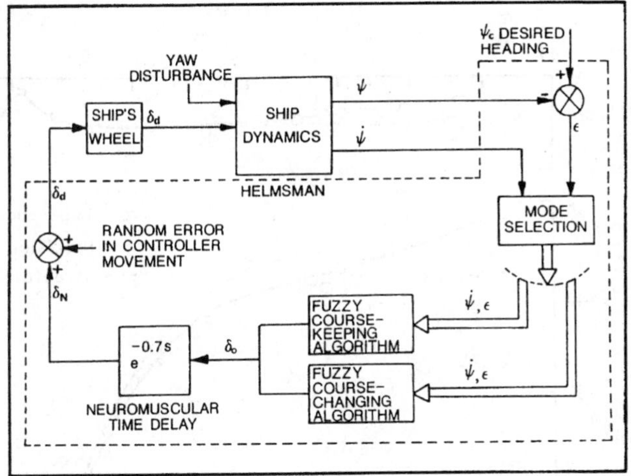

Figure 38.13 A fuzzy control model of a helmsman.

force and rate of change of position or applied force which is generated by the human neuromuscular system. The elements within the dashed portion of Figure 38.14 represent Y_p in Figure 38.4. The model has been divided into "central nervous system," "neuromuscular system," and "vestibular system," a division intended to emphasize the nature of the signal processing activity involved. The displayed system error e_d is presented to the operator and multiplied by the gain K_e. If motion cues are available, output rate, $\dot{m}$, is assumed to be sensed by the vestibular system (provided it is an angular velocity or linear or angular acceleration). The output rate is multiplied by the gain K_v and subtracted from the signal u_e. The resulting signal, u_1, is passed through a central time delay, τ_0, intended to account for latencies in the visual process sensing e_d, motor nerve conduction times, etc. The signal u_c provides a command to a closed-loop system that consists of a model of the open-loop neuromuscular dynamics, Y_{p_n}, of the particular limb driving the manipulator (e.g., control stick, steering wheel) and elements Y_f and Y_m, which form two proprioceptive feedback loops in the structural model. It is this proprioceptive feedback that provides the basic equalization capabilities of the model, and by hypothesis, the human operator. The form of Y_m is determined by the order k of the controlled element

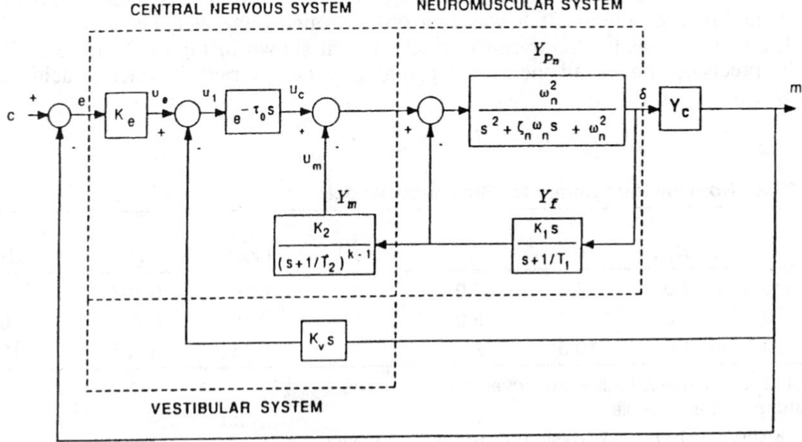

Figure 38.14 The structural model of the human operator.

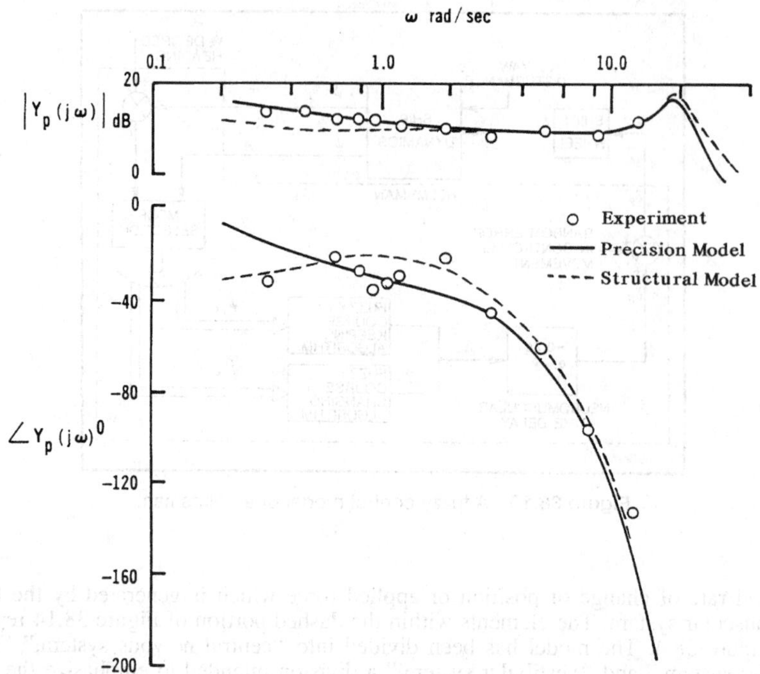

Figure 38.15 Comparison of experimental and structural model transfer functions.

dynamics, Y_c, in the region of crossover, i.e., $k = 0$ if the order is zero (e.g., $Y_c = K$), $k = 1$ if the order is 1 (e.g., $Y_c = K/s$), etc.

Compared to the rather spartan crossover model, the structural model certainly appears to be overparameterized, with 10 parameters evident in Figure 38.14. However, "nominal" values of most of these parameters can be chosen based solely upon the order of the controlled-element dynamics in the region of crossover. The remaining parameters, most notably, T_2, are selected so that the overall model obeys the dictates of the crossover model. Attention here will be focused upon cases in which no motion cues are assumed, i.e., $K_v = 0$. The selection of crossover frequency can be based upon equations (38.9). Figure 38.15 shows an operator transfer function obtained by the structural model superimposed on Figure 38.8. All the structural model parameters save K_e were chosen from the second row of Table 38.3, since the unstable controlled element used to obtain the experimental data of Figure 38.8 was first order around crossover, i.e., $k = 1$ in Table 38.3. The gain K_e was then chosen to achieve the fit shown in Figure 38.15. Recall that with the precision model, all the model parameters were varied in order to achieve the fit shown.

Table 38.3 Nominal Parameters for Structural Model

k	K_v	K_1	K_2	T_1 (sec)	T_2 (sec)	τ_0 (sec)	ζ_n	ω_n (rad/sec)
0	0	1.0	2.0	5.0	[a]	0.15	0.707	10.0
1	0	1.0	2.0	5.0	[b]	0.15	0.707	10.0
2	0	1.0	10.0	2.5	[a]	0.15	0.707	10.0

[a] Selected to achieve K/s-like crossover characteristics.
[b] Parameter not applicable.
NOTE: K_e chosen to provide desired crossover frequency.

The structural model has been employed in the past to investigate a variety of human–machine problems, including providing a rationale for human operator pulsive control behavior (Hess, 1979), describing higher levels of skill development in the human pilot (Hess, 1981), providing a theory for aircraft handling qualities (Hess and Yousefpor, 1992), and providing a basis for modeling the steering behavior of automobile drivers (Modjtahedzadeh and Hess, 1993).

38.3.2 Neuromuscular System Modeling

While the precision, optimal control, and structural models all include some rudimentary models of the neuromuscular system of the operator, there are applications in which more detailed models of this subsystem are valuable. As an example, recent problems involving adverse pilot–vehicle coupling such as a phenomenon knows as "roll ratchet" (Johnston and McRuer, 1987) have implicated the pilot's neuromuscular system as a probable contributing factor. Roll ratchet is a relatively high-frequency (approximately 2 cycles/sec) closed-loop oscillation that has been encountered in high-performance fighter aircraft engaged in roll tracking or aggressive rolling maneuvers. Although not a safety of flight issue, roll ratchet has a definite negative impact upon handling qualities and performance. Since problems such as this are likely to continue occurring as humans are asked to control systems with increasing performance capabilities (i.e., bandwidth), a brief discussion of the human neuromuscular system and associated modeling efforts is in order. This topic has already been broached in the previous discussion regarding the precision model and the structural model.

One of the early and well-documented models of the human operator including the neuromuscular system was provided by Magdaleno and McRuer (1971). The model is shown in Figure 38.16. A detailed description of the physiology behind the model of Figure 38.16 is well beyond the scope of this chapter; nonetheless, a brief description is in order. Starting from the far left, one has a visually sensed error signal, appropriate for

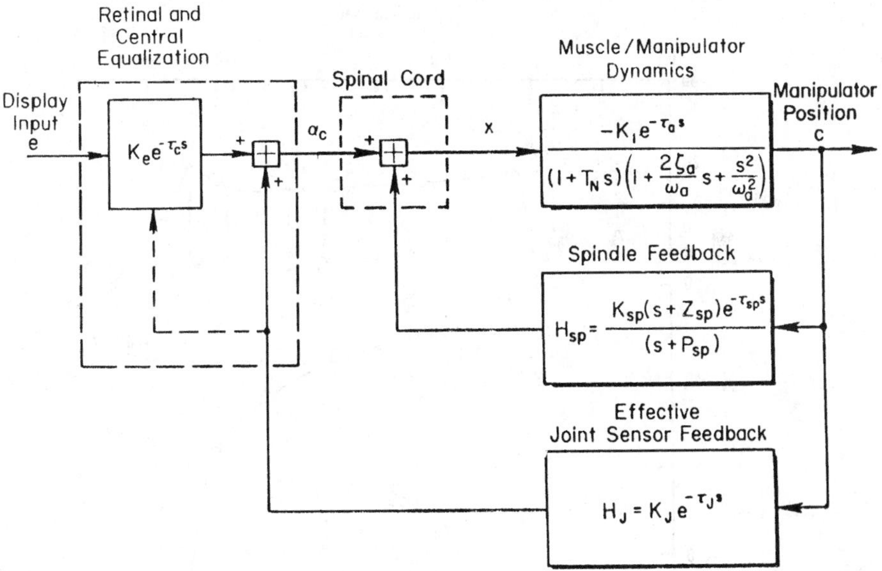

$K_{(-)}$ = gain.
$\tau_{(-)}$ = time delay.
T_N = muscle/manipulator lag time constant.
ζ_a, ω_a = muscle/manipulator damping ratio and undamped natural frequency.
Z_{SP}, P_{SP} = zero and pole of muscle spindle transfer function.

Figure 38.16 A model of the human operator neuromuscular system in tracking tasks.

a compensatory tracking task. This error is input to a block labeled "retinal and central equalization," intended to describe activity in the central nervous system, which provides the basic ability of the human to generate lead, lag, etc., appropriate for the controlled-pelement dynamics at hand. The term "retinal equalization" is intended to allow for rate sensing that may be produced directly from the retina.

One particular type of nerve cell, alpha motor neurons, innervate extrafusal muscle fibers in the limb, effecting control to produce contraction or relaxation of muscles. In the model of Figure 38.16, the alpha motor neuron command, α_c, is obtained as the difference between the output of the central equalization and that of "effective joint angle sensors" in the limb effecting control. As the name implies, these proprioceptors provide information about the angular relationship between limbs such as the forearm and upper arm, etc. The alpha motor neuron command is summed at the spinal cord level with the output of other proprioceptors called *muscle spindles.* These sensors are essentially stretch receptors within the muscles that are sensitive to changes in muscle length that accompany force application. Finally, the block labeled "muscle/manipulator" dynamics includes a pure time delay and three poles, the latter representing the dynamics of the limb and manipulator. The manipulator is assumed to be unrestrained in this model.

Given measured human operator describing function data, in particular $Y_p(j\omega)$, the parameters of the model of Figure 38.16 can be selected via a parameter identification procedure. For example, Figure 38.17 shows a model fit for a particular task when an isometric manipulator was being used. In this case, the joint-angle sensor loop is removed, and the primary proprioceptive signal is assumed to be derived from the muscle spindles. Also shown in the figure are the locations of the model poles and zeros. Note that the spindle zero at $s = -z_{sp}$ does not appear since it disappears in the spindle loop closure. Other such measurements such as those in Figure 38.17 have been made, and the contribution that the neuromuscular system can make to phenomena such as roll ratchet has been demonstrated (e.g. Johnston and McRuer, op. cit.).

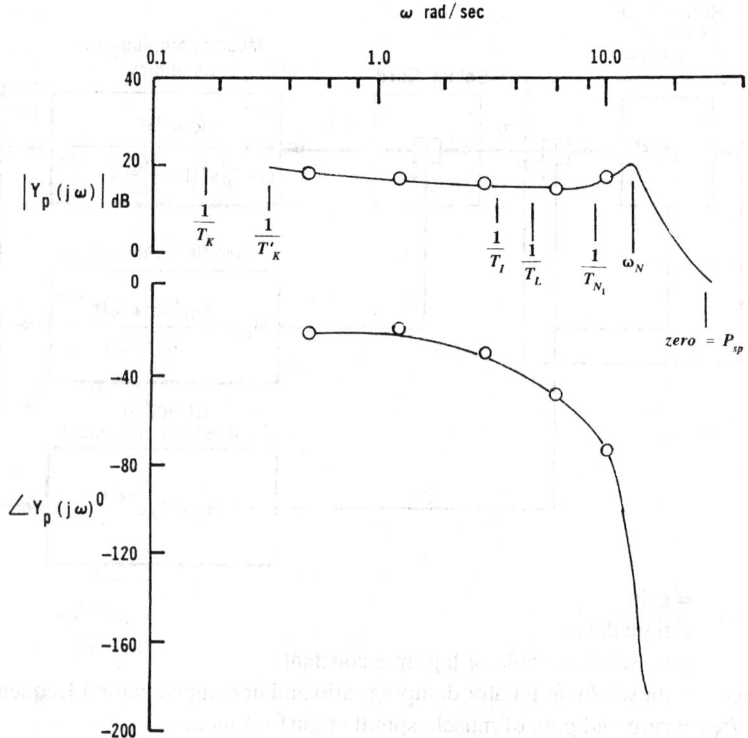

Figure 38.17 Comparison of transfer functions from experiment and from the model of Figure 38.16.

38.4 AN EXAMPLE

At this juncture it is useful to consider an example exercising the crossover, optimal control, and structural models in calculating the performance and dynamic characteristics of a well-trained, well-motivated human operator in a simple, single-loop manual control task. The fuzzy control model will not be exercised in this example, because of the significantly different manner in which the model is created. The task and controlled element are shown in Figure 38.18*a* and are based upon experiments conducted by Kleinman et al. (1970) in validating the optimal control model for single-loop applications. The frequency domain and state variable representations for the controlled element and disturbance can be given as

$$
\begin{aligned}
\dot{x}_1(t) &= -2x_1(t) + w(t) \\
\dot{x}_2(t) &= x_3(t) + x_1(t) \\
\dot{x}_3(t) &= u(t) \\
\frac{w(s)}{x_1(s)} &= \frac{1}{s+2} \\
x_2(s) &= \frac{x_1(s) + x_3(s)}{s} \\
x_3(s) &= \frac{u(s)}{s}
\end{aligned}
\tag{38.39}
$$

In the above equation, $w(t)$ represents a random, white-noise signal.

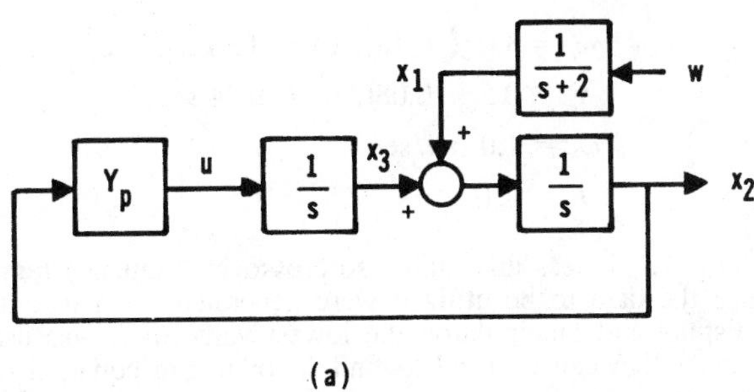

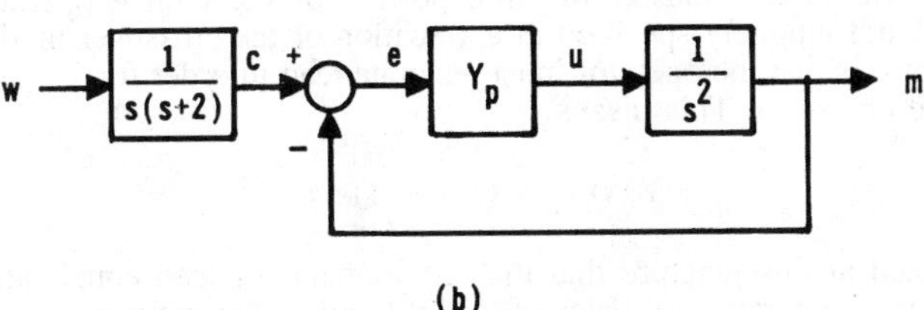

Figure 38.18 (*a*) A compensatory tracking task with a velocity disturbance, (*b*) An equivalent task with a command input.

In applying any of the human operator models, the bandwidth of the input or command signal is of obvious importance. Figure 38.18*a* utilizes a disturbance from which an equivalent input can be defined using simple block diagram algebra. This is shown in Figure 38.18*b*. The change in sign in the error signal in going from part *a* to part *b* is of no consequence here. Now the bandwidth of the equivalent input needs to be determined. The bandwidth discussed in connection with the crossover model possesses a rectangular spectrum, whereas that of the equivalent input in Figure 38.18*b* does not. In addition, equation (38.18) is not appropriate for input spectra in which the power spectral density contains free ω's in the denominator. In this case, equation (38.18) can be modified as

$$\omega_{BW_{ce}} \triangleq \frac{[\omega^p \Phi_{cc}(\omega)]_{\omega=\omega_{BW_{ce}}}}{[\omega^p \Phi_{cc}(\omega)]_{\omega=0}} = -3 \text{ dB} \tag{38.40}$$

where p is the power of the free ω in the denominator of the input power spectral density. Here, $p = 2$. Evaluating equation (38.40) for the equivalent input yields $\omega_{BW_{ce}} = 2.0$ rad/sec. The equivalent input in Figure 38.18*b* also has a mean square value that is undefined because of the free ω^2 in the denominator of its power spectral density. For purposes of using the relations derived for the crossover model, the mean square value of the equivalent input will be defined

$$\overline{c}^2 = \frac{1}{2\pi} \int_{-\infty}^{\infty} \omega^p \Phi_{cc}(\omega)\, d\omega = 0.054 \tag{38.41}$$

38.4.1 The Crossover Model

For the controlled element dynamics being considered, equations (38.9) and (38.41) together with Tables 38.1 and 38.2 suggest the following crossover model parameter values:

$$\begin{aligned} \omega_c &= 3.0 + 0.18(2.0) = 3.36 \text{ rad/sec} \\ \tau_e &= 0.5 - 0.08(2.0) = 0.34 \text{ sec} \\ \omega_R &= 1.0 \text{ rad/sec} \\ R &= 0.1 \end{aligned} \tag{38.42}$$

Also note that $\omega_{BW_{ce}}/\omega_c$ is less than unity, so crossover frequency regression will not be a problem. Since the data to be utilized were generated for a laboratory tracking task with an ideal display and manipulator, the lowest value of R was used in the remnant model. Now Figure 38.6 can be used to find I_1 for use in equation (38.16) to estimate mean square tracking error. The required abscissa value in Figure 38.6 is $\tau_e\omega_c = 1.14$ and the required curve is $\tau_e\omega_R = 0.34$. This yields $I_1 = 0.9$. Now equation (38.16) can be evaluated as follows

$$\overline{e}^2 = \frac{\frac{1}{3}(0.054)\left(\frac{2}{3.36}\right)^2}{1 - \frac{0.1}{(1)(0.34)}(0.9)} = 0.0087 \tag{38.43}$$

The actual form of the transfer function portion of the human operator describing function $Y_p(s)$ is not uniquely specified in application of the crossover model. All that is known in this case is that the operator must generate lead in order for $Y_pY_c(s)$ to resemble $\omega_c e^{-\tau_e s}/s$ around crossover. This means

$$Y_p(s) \approx K_p(T_L s + 1)e^{-\tau_e s} \tag{38.44}$$

It should be noted at this juncture that the requirement for lead equalization correlates strongly with human operator estimates of task difficulty, that is the greater the lead time constant T_L, the more difficult the task appears to the operator (McRuer and Krendel, 1974).

38.4.2 The Optimal Control Model

The index of performance for the example can be given by

$$J = E\left\{\lim_{T\to\infty} \frac{1}{T} \int_0^T [qx_2^2(t) + r\dot{u}^2(t)]\, dt\right\} \tag{38.45}$$

According to the approximate relations of equations (38.26) and (38.28), selecting $q/r = 5.0 \cdot 10^4$ would yield a crossover frequency of 3.4 rad/sec, quite close to the value suggested by the first of equations (38.42) (which are also valid for the optimal control model). The remaining optimal control model parameters were chosen to be identical to those given in equations (38.29), with the motor noise ratio set to 0.001. The operator time delay was set to the nominal value of 0.2 sec.

38.4.3 The Structural Model

In implementing the structural model, the same crossover frequency as that given in equations (38.42) was used. In the region of crossover and indeed at all frequencies, the controlled-element dynamics were second order, therefore $k = 2$ in Table 38.3. The parameter $1/T_2$ in Table 38.3 was chosen as $\omega_c/10$ to ensure a broad frequency range around crossover where $Y_pY_c(s) \approx Ke^{-\tau_e s}/s$. Finally, the gain K_e was chosen as 38.23 to give the desired crossover frequency.

38.4.4 Results

Figure 38.19 shows the operator dynamics implied by the crossover model around the crossover frequency. The experimental data are taken from the study reported by Kleinman et al. (1970). Figure 38.20 shows the same experimental data, this time with the operator transfer function generated by the optimal control model. Finally, Figure 38.21 shows a similar figure with the operator transfer function generated by the structural model. Table 38.4 compares model-generated and experimental crossover frequencies and RMS tracking performance. The data for the structural model were obtained from a computer simulation of the task. The performance scores for the optimal control model and crossover model are seen to be too small, whereas that of the structural model is seen to be too large. However, in terms of performance prediction, the values are probably ac-

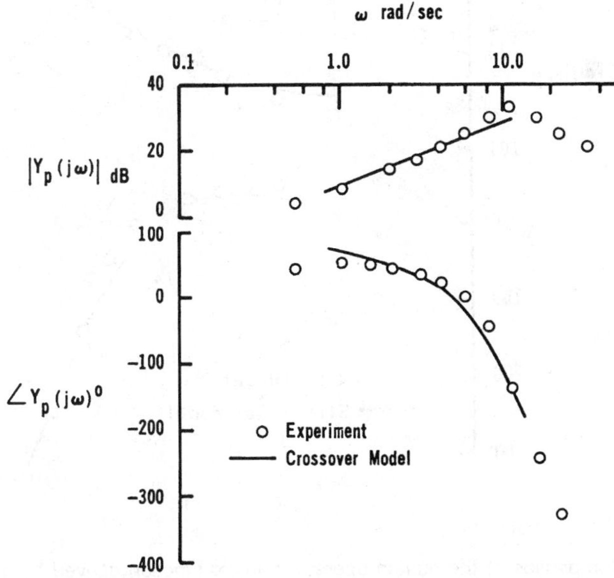

Figure 38.19 Comparison of the human operator transfer function derived from the crossover model with that from experiment.

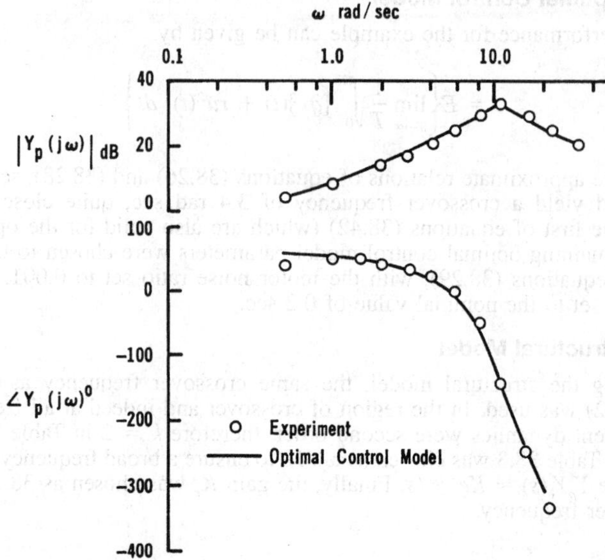

Figure 38.20 Comparison of the human operator transfer function derived from the optimal control model with that from experiment.

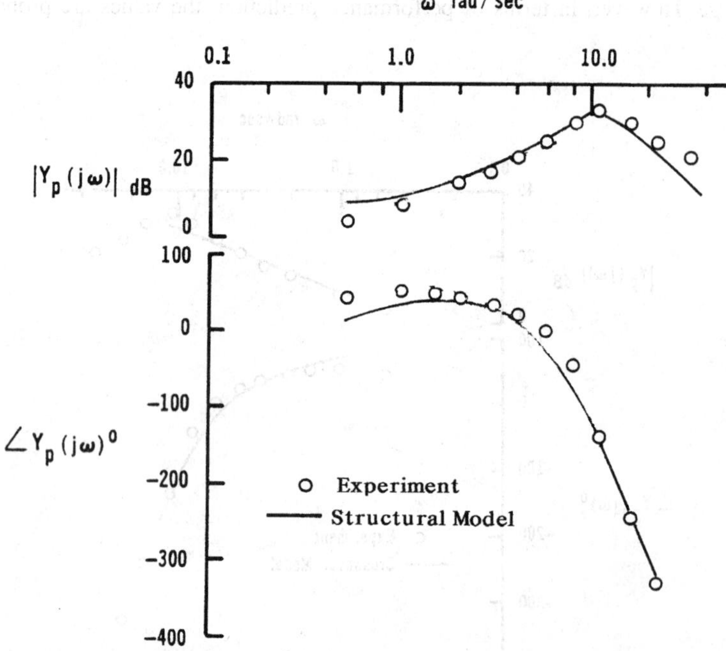

Figure 38.21 Comparison of the human operator transfer function derived from the structural model with that from experiment.

Table 38.4 Model Parameters and RMS Performance Compared with Experiment

	ω_c (rad/sec)	$\sqrt{\overline{x_2^2}} = \sqrt{\overline{e^2}}$
Experiment	3.5	0.12
Crossover model	3.4	0.093
Optimal control model	3.4	0.1
Structural model	3.4	0.14

ceptable. Of course, changes in the model parameters, particularly those associated with the remnant, could improve these matches.

38.5 MODELS FOR MULTILOOP COMPENSATORY CONTROL

38.5.1 Introduction

Consider Figure 38.22, which illustrates an automobile driving task. Figure 38.23 is a simplified block diagram representation of candidate manual control loop closures that could be used in modeling this task. In Figure 38.23 both vehicle heading error, ψ_e, and lane deviation error, y_e, are sensed by the driver and used as feedback signals in what is termed a multiloop, single-point compensatory control system. Here, "single-point" refers to the fact that one control variable, the steering wheel angle, δ_w, is utilized in controlling both vehicle heading and lane position.

Much of the success of the crossover, optimal control, and structural models of the human operator for single-loop tasks is largely due to the existence of a high-quality database for such tasks. Unfortunately, a similar database does not exist for multiloop tasks. Part of the problem lies in the fact that the classical spectral measurement techniques that have been used to determine human operator describing functions for single-loop tasks are somewhat restricted in their applicability to multiloop tasks. It can be shown, for example (Stapleford, McRuer, and Magdeleno, 1967) that the number of measurable operator transfer functions in a multiloop task is equal to the number of uncorrelated inputs times the number of operator outputs or controls. Thus, for example, in the block diagram of Figure 38.23, only a linear combination of D_ψ and D_y could be measured with the single disturbance shown.

Despite the relative scarcity of multiloop data, it has been demonstrated (Stapleford, Craig, and Tennant, 1969; Weir and McRuer, 1972) that many multiloop tasks of engineering interest can be modeled by series loop structures. In addition, in cases where display scanning is minimal, it appears that the major contributor to operator remnant is in inner-loop operation. As in the case of single-loop tasks, many of the measured human operator characteristics in multiloop tasks are coincident with those that would be selected by an experienced control system engineer in designing inanimate compensators to accomplish the same task.

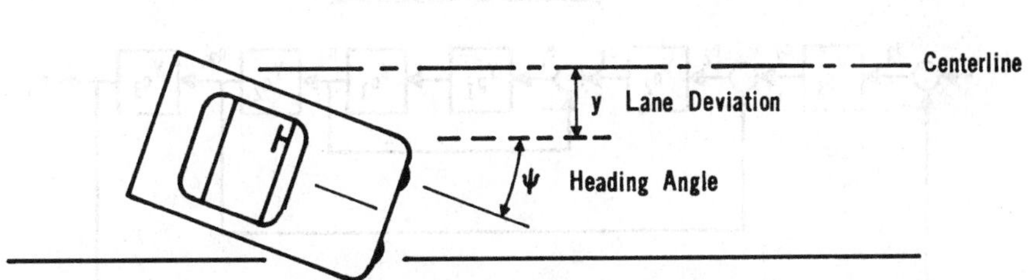

Figure 38.22 An automobile driving task.

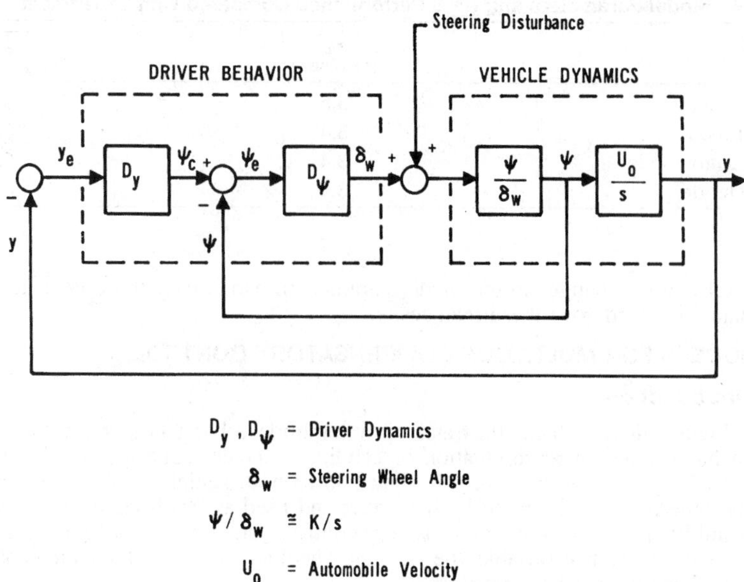

Figure 38.23 A simplified block diagram representation of the manual loop closures for the task of Figure 38.22.

38.5.2 Approaches Using the Crossover and Structural Models

A convenient stereotype of the series loop structure is shown in Figure 38.24. Here, three successive loop closures characterized by the crossover frequencies ω_{c1}, ω_{c2}, and ω_{c3}, are shown with

$$\omega_{c_1} > \omega_{c_2} > \omega_{c_3} \tag{38.46}$$

The structure of this diagram typifies many feedback systems whether under automatic or manual control. As an example Figure 38.24 might represent the flight control system

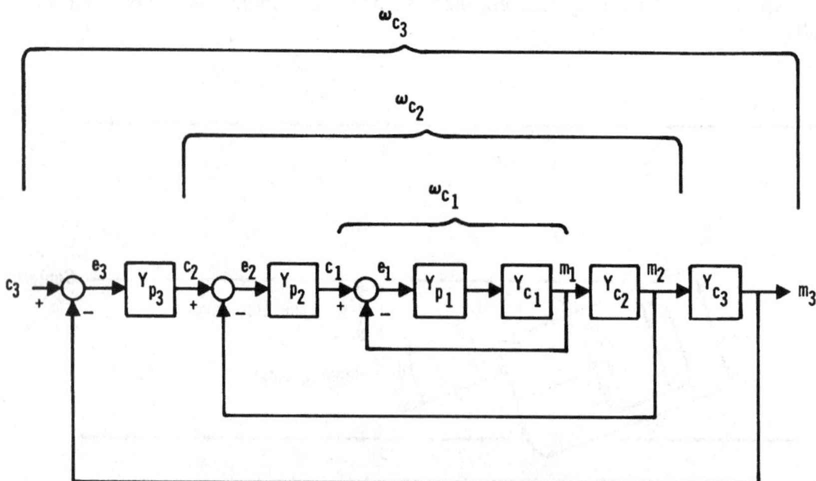

Figure 38.24 A block diagram for a single-point, multiloop manual control problem.

of a hovering vertical takeoff (VTOL) aircraft. Here, the innermost loop represents vehicle attitude control, the next loop represents vehicle translational velocity control, and the last or outermost loop represents vehicle position control. The crossover frequency (and hence, bandwidth) separation implied by equation (38.46) allows some simplification in the analysis of a system like that of Figure 38.24. For example, with the inner loop closed, the effective open-loop dynamics for the second closure, where the pertinent vehicle output would be m_2 and the "control" or input would be c_2, would take the approximate form

$$\frac{m_2(j\omega)}{e_2(j\omega)} \approx Y_{p_2}Y_{c_2}(j\omega) \text{ for } \omega < \omega_{c_1} \tag{38.47}$$

Likewise, with the first and second loops closed, the effective open-loop dynamics for the final loop closure, where the pertinent vehicle output would be m_3 and the "control" or input would be c_3, would take the approximate form

$$\frac{m_3(j\omega)}{e_3(j\omega)} \approx Y_{p_3}Y_{c_3}(j\omega) \text{ for } \omega < \omega_{c_2} \tag{38.48}$$

Equations (38.47) and (38.48) are useful, albeit approximate, relations for analysis and design. The following general guidelines can be used to formulate a single-point multiloop manual control problem, using either the crossover or structural model of the human operator, with the block diagram of Figure 38.24:

1. The operator's dynamic characteristics are distributed in serial fashion as Y_{p_1}, Y_{p_2}, and Y_{p_3}. Y_{p_1} can be assumed to be identical in form to the single-loop models that have been discussed. As has been stated, in the absence of significant scanning, operator remnant can be considered to be injected only in the inner loop. Here "significant" scanning can be interpreted as that requiring distinct shifts in the operator's eye point of view throughout the task.
2. The operator dynamics represented by Y_{p_2} and Y_{p_3} ultimately create the input signal c_1, which drives the innermost loop. It is within this loop that the actual human–machine interface occurs through the action of the manipulator. This innermost loop can be referred to as the *primary control loop* (Hess, Malsbury, and Atencio; 1993). Also note that the signals c_1 and c_2 are internally generated by the operator as opposed to being explicitly displayed or perceived from the environment. The operator's effective time delay is usually placed in the innermost loop. This is reasonable since a considerable part of the effective delay is attributable to neuromuscular effects that occur only at the human–machine interface.
3. (a) *Crossover model*: The operator dynamics are estimated by repetitive application of the crossover model, starting with the innermost loop and working out. However, the operator dynamics in the outer loops should consist of simple gains with no effective time delays included. This means, the effective open-loop dynamics given by equations (38.47) and (38.48) should appear as K/s in the appropriate crossover regions.

   (b) *Structural model*: For the innermost loop, the operator dynamics are those given in Figure 38.14 and Table 38.3. The outer-loop dynamics are treated just as with the Crossover model in step 3(a).
4. The crossover frequencies in successive loop closures should be separated by a factor of approximately 2–3. This is a rule of thumb suggested by limited experimental data, (e.g., Hess and Beckman, 1984; Ringland, Stapleford, and Magdaleno, 1971) and by sound control system design principles.
5. Estimation of loop crossover frequencies is not a simple task, even using the separation factor in item 4. Initially, one should begin at the outermost loop and estimate the bandwidth of the likely command or disturbance existing in that loop. In Figure 38.24, this would be the command c_3. The crossover frequency of the outer-loop can then be estimated based upon this bandwidth. Estimates of the remaining crossover frequencies follow using the separation factor of 2–3. If the crossover frequency of the innermost loop exceeds 0.55 of the magnitude calculated by the first of equations (38.9) and Table 38.1, suitable reductions in all

crossover frequencies should be made. The 0.55 factor is again suggested by the limited experimental data mentioned in item 4.

In the block diagram of Figure 38.24, the dynamics of the controlled element, like those of the operator, have been distributed over three loops. In reality the controlled element is multivariable in nature, as shown in Figure 38.25. In order to analyze the system as indicated in Figure 38.24, one must be able to interpret elements like Y_{c1}, etc. in terms of the multivariable dynamic system shown in Figure 38.24. This can be done, as follows:

$$\begin{aligned} Y_{c1} &= \frac{m_1}{\delta} \\ Y_{c2} &= \frac{m_2}{m_1} = \frac{m_2}{\delta Y_{c1}} = \frac{m_2}{\delta}\frac{1}{Y_{c1}} \\ Y_{c3} &= \frac{m_3}{m_2} = \frac{m_3}{\delta Y_{c1} Y_{c2}} = \frac{m_3}{\delta}\frac{1}{Y_{c1}}\frac{1}{Y_{c2}} \end{aligned} \tag{38.49}$$

38.5.3 Approach Using the Optimal Control Model

Being essentially a multivariable design technique, the optimal control model appears ideally suited to the multiloop control problem. No changes in Figure 38.10 or in the computational technique used to generate the optimal control model are needed to address the single-point, multiloop problem. Of course, with the optimal control model, no assumptions regarding loop structure are made. For each observed variable, the operator is also assumed to perceive its first time derivative, and white observation noise is added to all such variables. As in the single-loop problem, the observation noise is assumed to scale with the mean square value of the signal to which it is added. Again, motor noise and an operator pure time delay are utilized. Finally, the index of performance of equation (38.23) is modified to account for including more than two variables as follows:

$$J = E\left\{\lim_{T\to\infty} \frac{1}{T}\int_0^T \left[\sum_{i=1}^{n} q_i y_i^2(t) + r\dot{u}^2(t)\right] dt\right\} \tag{38.50}$$

As in the single-loop case, appropriate selection of the weighing coefficients, q_i, is obviously of some importance in the multivariable problem. The following general guidelines can be used to formulate a single-point multiloop manual control problem using the optimal control model, given the controlled element of Figure 38.25.

1. The operator's effective time delay is set to a nominal value of 0.2 sec.
2. The covariances of the observation and motor noises are set to values derived form single-loop applications, that is,

$$\begin{aligned} V_{y_i} &= n(0.01)\pi E\{y_i^2(t)\} \\ V_u &\geq 0.003\pi E\{u^2(t)\} \end{aligned} \tag{38.51}$$

where n is the number of explicitly displayed variables. In this case, $n = 3$. Including

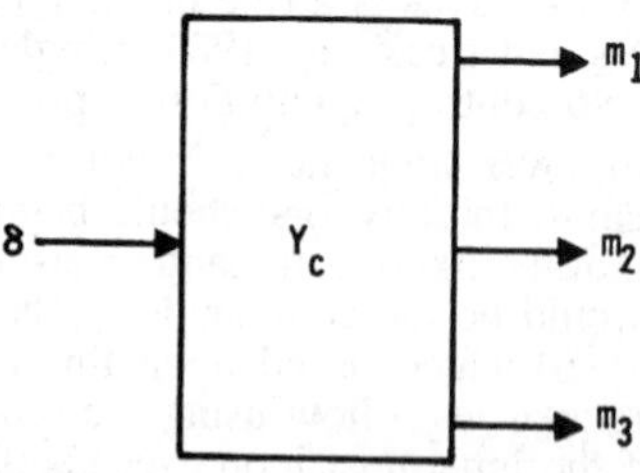

Figure 38.25 A multivariable controlled element.

the factor n is an approximate means of handing the effects of attention sharing on the part of the operator (Baron and Levison, 1973). Here,

$$y_1 = m_1 \quad y_3 = m_2 \quad y_5 = m_3$$
$$y_2 = \dot{m}_1 \quad y_4 = \dot{m}_2 \quad y_6 = \dot{m}_3 \tag{38.52}$$

3. Variables to be included in the index of performance are determined from the task definition. Unfortunately, a priori selection of the weighting coefficients is something of an art. The following procedure is recommended for preliminary analysis: The weighting coefficients of each variable appearing in the index of performance, with the exception of the control rate term, are selected as the reciprocal of the estimated "maximum allowable deviation" of that variable. This method has been offered for optimal control formulations, in general (Bryson and Ho, 1975) and has shown promise in optimal control model formulations (Baron, 1983). Finally, the weighting coefficient on control rate is selected to yield an "inner-loop" crossover frequency calculated in a manner identical to that just described for the classical approach using the crossover model. Strictly speaking, of course, no "inner-loop" exists for the optimal control model. However, equations (38.26) and (38.27) can still be used to give an approximate relationship between the weighting coefficient on the inner-loop feedback variable and that on the control rate.

38.5.4 An Example

A simple example exercising the crossover, structural and optimal control models in a multiloop task can now be discussed. The task consists of the longitudinal control of a hovering helicopter, specifically, maintaining the vehicle over a specified position on the ground in the presence of atmospheric turbulence. Figure 38.26 shows the task. The simplified vehicle equations of motion can be given by

$$\begin{aligned}\dot{x}(t) &= u(t)\\ \dot{u}(t) &= -g\theta + X_u u(t)\\ \dot{\theta}(t) &= \delta(t) + d_g(t)\\ \dot{\delta}_g &= -d_g(t) + w(t)\end{aligned} \tag{38.53}$$

Here, $g = 9.8\ m/\sec^2$, and $X_u = -0.1/\sec$. The variable d_g represents a pitch-rate disturbance caused by atmospheric turbulence. It is represented as white noise passed through a first-order filter with a break frequency of 1.0 rad/sec. The turbulence is assume to have an RMS value of 5.73 deg/sec (0.1 rad/sec).

38.5.4.1 The Crossover Model

Figure 38.27 shows the block diagram of the single-point multiloop manual control system. It is quite similar to Figure 38.24. Using the guidelines offered in the preceding

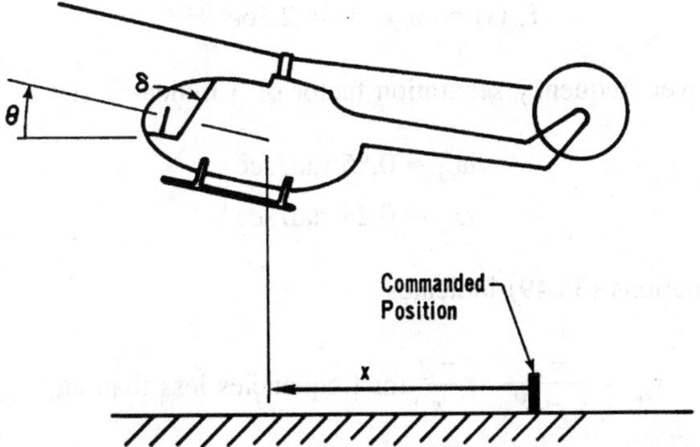

Figure 38.26 A helicopter longitudinal hover task.

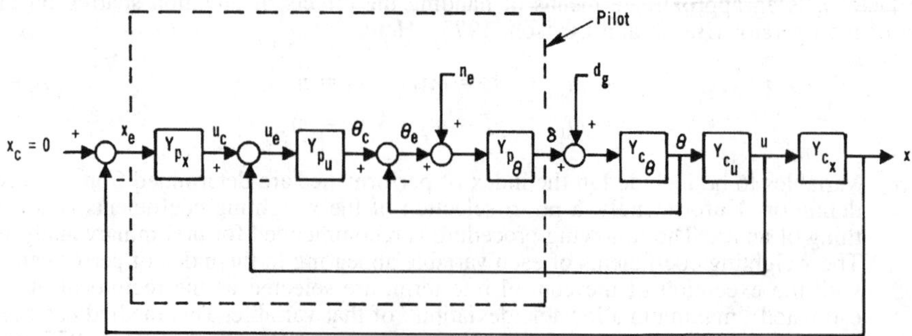

Figure 38.27 The single-point, multiloop, serial representation of the manual control task of Figure 38.26, crossover and structural models.

discussion on use of the crossover model for multiloop systems, one can proceed as follows. Crossover frequency selection begins with the inner rather than the outer loop. This is because the outer-loop command is identically zero and the primary disturbance is in the inner loop. Equation (38.40) yields

$$\omega_{BW_{ce}} = 1.0 \text{ rad/sec} \tag{38.54}$$

Now the vehicle equations of motion, equations (38.53), give

$$\frac{\theta(s)}{\delta(s)} = \frac{1}{s} \tag{38.55}$$

Equation (38.9) and Table 38.1 indicate that

$$\begin{aligned} \omega_{c_\theta} &= 0.55[\omega_{c_0} + 0.18\omega_{BW_{ce}}] = 2.56 \text{ rad/sec} \\ \tau_e &= \tau_0 - 0.08\omega_{BW_{ce}} = 0.27 \text{ sec} \\ R &= 0.5 \\ \omega_R &= 3.0 \text{ rad/sec} \end{aligned} \tag{38.56}$$

Note the 0.55 factor included in the ω_{c_θ} calculation and the fact that crossover regression is not suggested by these results. The largest R value in Table 38.2 was used since the operator is actually sensing three feedback variables. Now the crossover model indicates that the approximate form of the pilot equalization in the inner loop of Figure 38.27 is

$$Y_{p_\theta}(s) \approx \omega_{c_\theta} e^{-\tau_e s} = 2.56e^{-0.27s} \tag{38.57}$$

Applying a crossover frequency separation factor of 3 indicates that

$$\begin{aligned} \omega_{c_u} &\approx 0.85 \text{ rad/sec} \\ \omega_{c_x} &\approx 0.28 \text{ rad/sec} \end{aligned} \tag{38.58}$$

The second of equations (38.49) indicates

$$y_{c_u} = \frac{-g}{s - X_u} \approx \frac{-g}{s} \text{ for frequencies less than } \omega_{c_\theta} \tag{38.59}$$

Now, again applying the crossover model to the effective open-loop system for the second loop closure (omitting the time delay) yields

$$Y_{p_u}(s) \approx \frac{-\omega_{c_u}}{g} \tag{38.60}$$

The last of equations (38.49) indicates

$$Y_{c_x}(s) = \frac{1}{s} \tag{38.61}$$

Finally, applying the crossover model to the effective open-loop system for the third and final loop closure (omitting the time delay) yields

$$Y_{p_x} \approx \omega_{c_x} = 0.28 \tag{38.62}$$

A computer simulation of the human–machine system just defined was implemented for the purposes of calculating performance scores. These will be presented after the presentation of the structural and optimal control models.

38.5.4.2 The Structural Model

The structural model parameters are now taken from the second row of Table 38.3, since $k = 1$, i.e., the primary loop controlled element is first order at all frequencies. To achieve the desired crossover frequency of 2.56 rad/sec, the gain $K_e = 10.12$. Like the crossover frequency, the remnant model was identical to that used for the crossover model. The structural model was evaluated in a computer simulation similar to that for the crossover model.

38.5.4.3 The Optimal Control Model

Figure 38.28 shows the block diagram of a single-point, multiloop parallel representation of the manual control task of Figure 38.26. In this application the motor noise ratio was set to 0.003. The index of performance was selected as

$$J = E\left\{\lim_{T\to\infty}\frac{1}{T}\int_0^T\left[\frac{x^2(t)}{x^2_{MAX}} + \frac{u^2(t)}{u^2_{MAX}} + \frac{\theta^2(t)}{\theta^2_{MAX}} + r\dot{\delta}^2(t)\right]dt\right\} \tag{38.63}$$

In equation (38.63), the maximum allowable deviations were chosen as

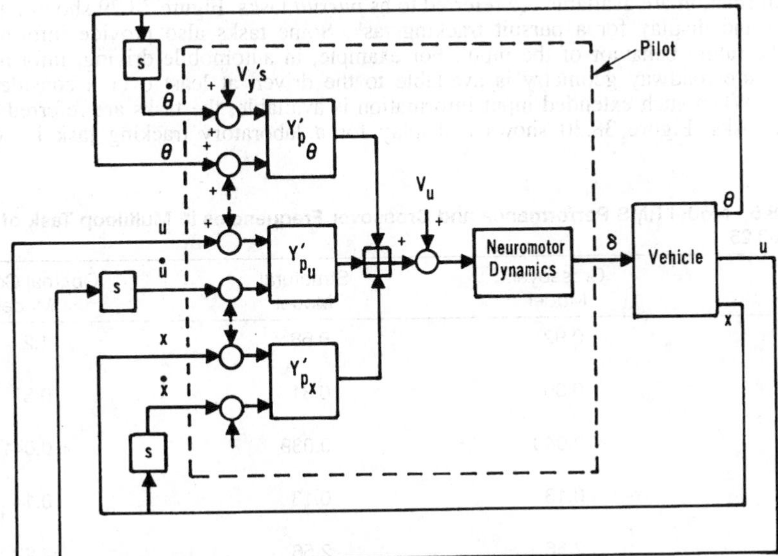

Figure 38.28 The single-point, multiloop, serial representation of the manual control task of Figure 38.26, optimal control model.

$$\theta_{MAX} = 0.088 \text{ rad (5 deg)} \qquad x_{MAX} = 7.8 \text{ m}$$
$$u_{MAX} = 2.57 \text{ m/sec} \tag{38.64}$$

These maximum allowable deviations were chosen in the following way. First, θ_{MAX} is selected. Its magnitude is immaterial. Equations (38.59) and (38.61) indicate simple integral relationships between $\theta(t)$ and $u(t)$, and $u(t)$ and $x(t)$. In a manner similar to that used in Equations (38.31)–(38.33), one can write

$$u_{MAX} = (g\theta_{MAX})T$$
$$x_{MAX} = (u_{MAX})T \tag{38.65}$$

Here, T was chosen as 3 sec simply on the basis of experience. In words, equations (38.65) state that the maximum allowable deviation on $u(t)$ is the value that would be reached if $\theta(t)$ were held at its maximum allowable value for T sec. Likewise, the maximum allowable deviation on $x(t)$ is the value that would be reached if $u(t)$ at its maximum allowable value for T sec. There is little analytical justification for this procedure. It is utilized simply because it exercises the maximum allowable deviation method often used in specifying weighting coefficients in linear quadratic regulator problems, and incorporates the kinematic equations relating $\theta(t)$, $u(t)$, and $x(t)$. Equations (38.26)–(38.28) were used to obtain an approximation for q/r, where "q" in this case is $1/(\theta^2_{MAX})$ and the ω_{c_θ} used was identical to that for the crossover and structural models. The first of equations (38.49) was used for Y_c in equation (38.27). The resulting value of r was 0.3.

38.5.4.4 Results

Table 38.5 compares the crossover, structural, and optimal control model RMS scores. Of course, there is no "correct" answer in this case, and the table is included simply to indicate the comparable results one obtains by the three modeling approaches. The largest performance difference between the models is seen to occur with the outer-loop vehicle position, x.

38.6 MODELS FOR PURSUIT AND PREVIEW CONTROL

The human operator models that have been discussed to this point share the common feature of being applied to *compensatory* tracking tasks, that is, tasks in which only error information is assumed to be used by the human. In many tasks the human operator can also perceive information about the input and output. For example, consider the gunnery task of Figure 38.1. The gunner may be able to discern target LOS, $\psi_T(t)$, and gun LOS, $\psi_G(t)$ in addition to error LOS, $\psi_E(t)$. Tasks such as these in which input, output, and error information are available are referred to as *pursuit* tasks. Figure 38.29 shows a block diagram and display for a pursuit tracking task. Some tasks also provide information about the future behavior of the input. For example, in automobile driving, information about future roadway geometry is available to the driver, at least over a considerable distance. When such extended input information is available, the tasks are referred to as *preview* tasks. Figure 38.30 shows a display for a laboratory tracking task in which

Table 38.5 Model RMS Performance and Crossover Frequencies in Multiloop Task of Figure 38.25

Variable	Crossover Model	Structural Model	Optimal Control Model
x (m)	0.92	0.68	1.2
u (m/sec)	0.30	0.31	0.37
θ (rad)	0.040	0.039	0.041
δ (rad)	0.13	0.13	0.14
ω_{c_θ} (rad/sec)	2.56	2.56	2.20

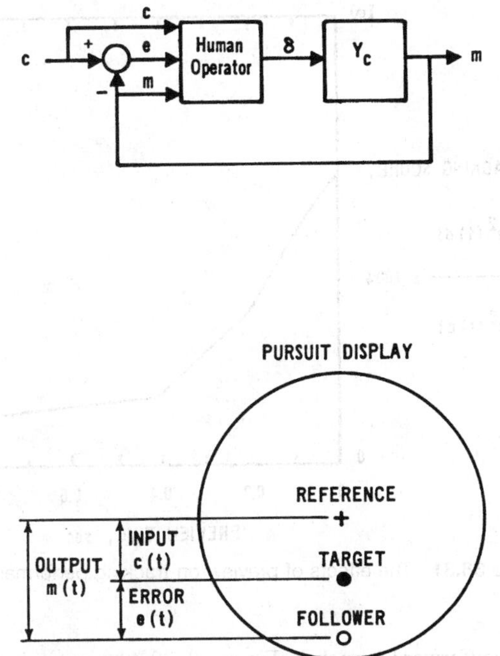

Figure 38.29 A pursuit tracking task and display.

preview information has been made available to the operator. Here the command input moves from right to left across the screen. The amount of preview defined by the visible portion of the command input is referred to as the preview time constant. Note that if the preview time constant is zero, the preview display becomes equivalent to a pursuit display. The ability of a preview display to improve tracking performance has been amply demonstrated in experiment. For example, Figure 38.31 shows the effect of preview time on tracking score from a study by Reid and Drewell (1972), where the controlled element dynamics were $Y_c(s) = K/s$.

38.6.1 The Crossover Model

A number of models for human operator pursuit and preview control have been proposed. One, based upon the crossover model representation, is shown in Figure 38.32 (Allen and McRuer, 1979). Here n is the order of $Y_c(s)$ in the region of crossover. The model implies that the operator derives or senses the nth derivative of the input. Thus, if $Y_{p_i}(s) = K_{p_i}$ (a

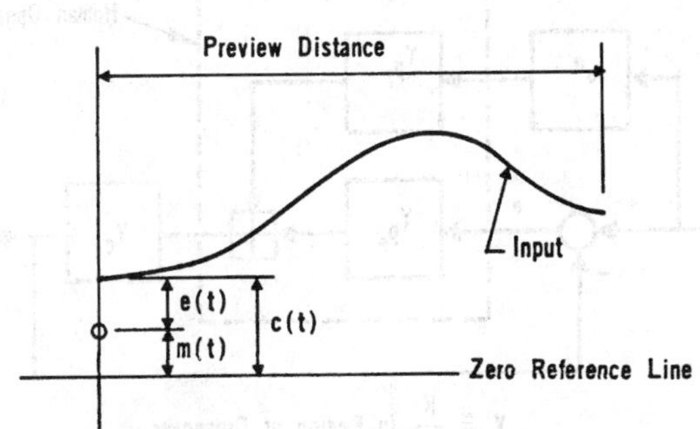

Figure 38.30 A preview display.

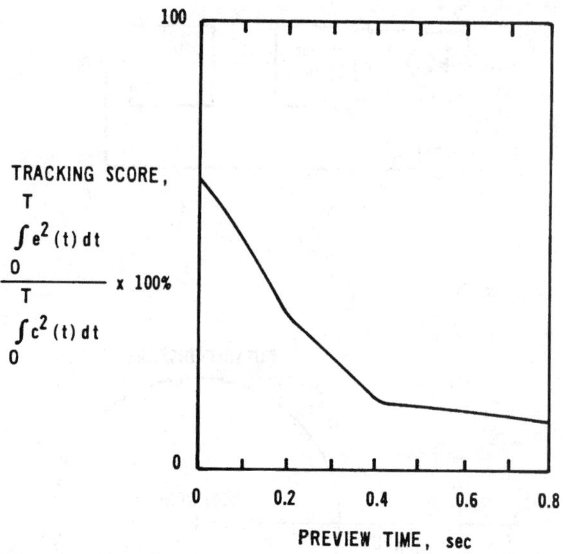

Figure 38.31 The effects of preview on tracking performance.

constant), then the feedforward branch in Figure 38.32 (the upper branch starting at the input $c(s)$ and ending in the output of the block Y_{p_i}) effectively inverts the controlled element dynamics. The element $Y_{p_e}(s)$ is that required to produce the crossover model assuming purely compensatory control. Allen and McRuer suggest that the best way to induce pursuit behavior in a particular task is to have presented in the surround or display those input quantities that will allow $Y_{p_i}(s)$ to be a constant. To indicate in approximate fashion, how tracking performance can be improved with the pursuit model of Figure 38.32, consider the following: Block diagram algebra can be used to derive the following closed-loop transfer function between input $c(s)$ and output $m(s)$:

$$\frac{m(s)}{c(s)} = \frac{K_p K s + \omega_c}{s + \omega_c} \tag{38.66}$$

Obviously, if $K_p K \approx 1.0$, then $\dfrac{m(s)}{c(s)} \approx 1.0$, and performance superior to that for compensatory tracking will result. To prove this, one can compare equation (38.66) with that which would be obtained in the absence of input information, i.e., assuming $Y_{p_i}(s) = 0$:

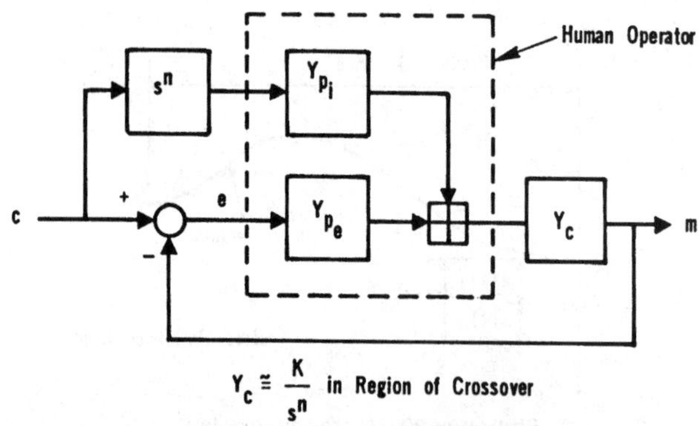

Figure 38.32 An operator model for pursuit tracking.

$$\frac{m(s)}{c(s)} \approx \frac{\omega_c}{s + \omega_c} \tag{38.67}$$

For the sake of simplicity, operator time delays were omitted from this discussion.

38.6.2 The Structural Model

Figure 38.33 shows a simplified version of the structural model of Figure 38.14, adopted for pursuit tracking. The two inner loops of Figure 38.14 have been replaced by a single loop involving the transfer function $sK_{\dot{m}}\hat{Y}_c(s)$, where $\hat{Y}_c(s)$ represents an estimate of the controlled element dynamics $Y_c(s)$. Actually, the model of Figure 38.14 evolved from the model of Fig. 38.33 (Hess, 1985). It can easily be shown that in the case of compensatory tracking ($K_m = K_{\dot{c}} = 0$), the model of Figure 38.33 will always produce the crossover model (Hess, 1985).

General guidelines can be given for selecting the model parameters in Figure 38.33 as follows: The time delay τ_e can be considered synonymous with that defined in equations (38.9). The delay τ_1 is that associated with rate sensing on the part of the human, with a value of 0.2 sec considered representative. With $K_{\dot{c}} = K_m = 0$, $K_{\dot{m}}$ and K_e are selected (one will be arbitrary) to yield a crossover frequency given by

$$\omega_c = \frac{KK_e}{K_{\dot{m}}K + 1} \tag{38.68}$$

Next, K_m is chosen so that the magnitude of $m(s)/e(s)$ resembles ω_c/s for at least a decade below ω_c (determined on a Bode plot). Finally, $K_{\dot{c}}$ is given by $K_{\dot{c}} = K_e/\omega_c$. Using Figure 38.33, one can show

$$\frac{m(s)}{c(s)} = \frac{(K_{\dot{c}}se^{-\tau_e s} + K_e)Y_c(s)e^{-\tau_e s}}{1 + sK_{\dot{m}}Y_c(s) + Y_c(s)e^{-\tau_e s}(K_m + K_e)} \tag{38.69}$$

This relation can be made considerably less formidable by considering a particular example. Let $\hat{Y}_c(s) = Y_c(s) = K/s$. Following the guidelines just given, one can show

$$\frac{m(s)}{c(s)} = \frac{(s/\omega_c)e^{-(\tau_1+\tau_e)s} + e^{-\tau_e s}}{(s/\omega_c) + e^{-\tau_e s}} \tag{38.70}$$

In deriving equation (38.70), $K_m + K_e$ was approximated by K_e. Now an analytical prediction of the utility of preview can be obtained from equation (38.70). If $c(t + \tau_1 + \tau_e)$ is sensed by the operator, rather than $c(t)$, the following relation occurs:

$$\frac{m(s)}{c(s)} = \frac{(s/\omega_c) + e^{-\tau_e s}}{(s/\omega_c) + e^{-\tau_e s}} = 1.0 \tag{38.71}$$

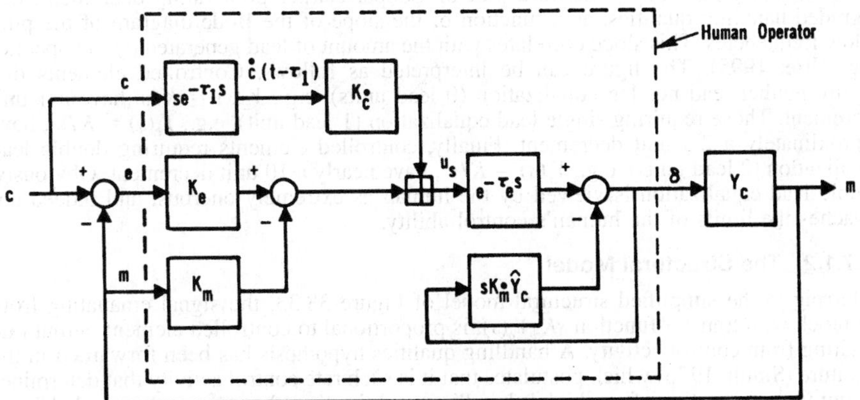

Figure 38.33 A simplified representation of the structural model for pursuit tracking.

Equation (38.71) suggests that dramatic improvements in tracking performance can occur with a preview time constant on the order of $\tau_1 + \tau_e$. With K/s controlled element dynamics, $\tau_1 + \tau_e$ would probably be on the order of 0.5 sec. Note in Figure 38.31 that this preview time agrees with that associated with near minimization of the error scores in the experiment of Reid and Drewell (1972) using K/s dynamics.

38.6.3 The Optimal Control Model

Little in the way of discussion is necessary for applying the optimal control model to pursuit tracking problems. The input and its time rate of change are simply included in the vector of variables assumed to be displayed to or perceived by the operator. Modeling preview behavior requires that the state-variable formulation be augmented to include a state or output variable proportional to the previewed input.

38.7 INTERPRETING TASK DIFFICULTY (HANDLING QUALITIES) WITH FEEDBACK CONTROL MODELS

It is desirable in the analysis of systems involving a human controller to be able to predict the "operator workload" or "task difficulty" or vehicle "handling qualities" for a particular task. Indeed, a considerable amount of research has been directed toward this goal (e.g., Hess, 1990). Space does not permit a detailed discussion here; indeed an entire chapter could easily be devoted to this topic, alone. However, a brief review is in order. This will be done by focusing upon each of the three control models discussed in previous sections, again omitting the fuzzy control model for reasons of expediency. The majority of research in this area has been directed toward aircraft handling qualities, so this will be emphasized herein. Finally, for the purposes of exposition, only the handling qualities of single-loop tasks will be considered.

38.7.1 Handling Qualities

Aircraft handling qualities are almost universally quantified using what is called the Cooper-Harper pilot rating scale, shown in Figure 38.34. This scale essentially presents the pilot with a series of dichotomous decisions regarding his or her interpretation of task difficulty, culminating in a numerical rating from 1 to 10. In addition, the scale has been divided into three handling qualities "levels," as shown in Figure 38.34, level 1 indicating "satisfactory" handling qualities, level 2 indicating "acceptable" handling qualities, and level 3 indicating "unacceptable" handling qualities.

38.7.1.1 The Crossover Model

Since the form of the crossover model of equation (38.8) is essentially invariant with task and controlled element, its use as a handling qualities predictive tool is quite limited. However, the pilot dynamics implied by the crossover model can be of some use in this area. This means appealing to the simplified operator model of equation (38.7) or the precision model of equation (38.19). In this regard, it has been found that the dominant rating-sensitive pilot parameters are the low-frequency lead equalization and the operator gain (McRuer and Krendel, 1974). Here, the lead equalization is considered "low frequency" if the reciprocal of the lead time constant is less than the crossover frequency. For example, Figure 38.35 shows a plot of Cooper-Harper pilot rating decrement, i.e., degraded handling qualities, as a function of the slope of the Bode diagram of the pilot at low frequencies. This slope correlates with the amount of lead generated by the operator (e.g. Nise, 1995). The figure can be interpreted as follows. Controlled elements that require neither lead nor lag equalization (0 lead units), e.g., $Y_c(s) = K/s$, have a 1 unit decrement. Those requiring single lead equalization (1 lead unit), e.g., $Y_c(s) = K/s^2$, have approximately a 2.5 unit decrement. Finally, controlled elements requiring double lead equalization (2 lead units), e.g., $Y_c(s) = K/s^3$, have nearly a 10 unit decrement. Obviously, double lead equalization is viewed by the human as extremely onerous, and indeed approaches the limits of the human's control ability.

38.7.1.2 The Structural Model

Referring to the simplified structural model of Figure 38.33, the signal emanating from the block with transfer function $sK_m\hat{Y}_c(s)$, is proportional to controlled element output rate resulting from control activity. A handling qualities hypothesis has been forwarded in the literature (Smith, 1976) which postulates that it is such rate control activity that determines the human's perception of a vehicle's handling qualities; i.e., the more rate control activity,

AIRCRAFT CHARACTERISTICS	DEMANDS ON PILOT	PILOT RATING	LEVEL
Excellent; highly desirable	Pilot compensation not a factor for desired performance	1	
Good; negligible deficiencies	Pilot compensation not a factor for desired performance	2	1
Fair; some mildly unpleasant deficiencies	Minimal pilot compensation required for desired performance	3	
Minor but annoying deficiencies	Desired performance requires considerable pilot compensation	4	
Moderately objectionable deficiencies	Adequate performance requires considerable pilot compensation	5	2
Very objectionable but tolerable deficiencies	Adequate performance requires extensive pilot compensation	6	
Major deficiencies	Adequate performance not attainable with maximum tolerable pilot compensation Controllability not in question	7	3
Major deficiencies	Considerable pilot compensation required for control	8	
Major deficiencies	Intense pilot compensation required to retain control	9	
Major deficiencies	Control will be lost during some portion of operation.	10	

Figure 38.34 The Cooper-Harper handling qualities rating scale.

the worse the perceived handling qualities. Hess and Yousefpor (1992) and Hess (1995) have adopted Smith's hypothesis and interpreted it in terms of the structural model through what is called the handling qualities sensitivity function (HQSF). Referring to the more complete version of the structural model in Figure 38.14, the HQSF is defined as

$$\text{HQSF} \triangleq \frac{u_m(s)}{c(s)} \tag{38.72}$$

where $u_m(s)$ is the signal proportional to output rate due to control activity in Fig. 38.14 and, of course, $c(s)$ is the input. It has been shown (e.g., Hess and Yousefpor, 1992) that level 1 handling qualities will be in evidence if

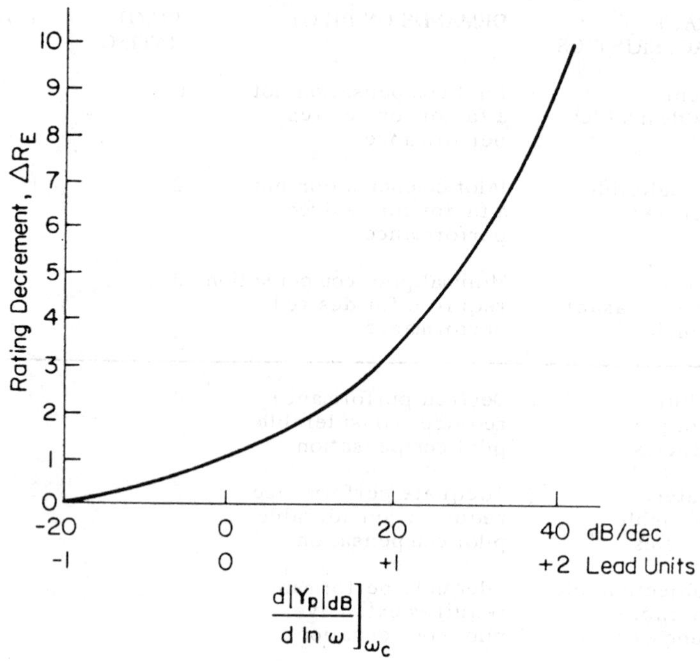

Figure 38.35 Handling qualities decrement as a function of pilot lag/lead equalization.

$$\left|\frac{u_m(j\omega)}{c(j\omega)}\right|_{\omega\approx\omega_c} < 1.0 \tag{38.73}$$

In words, equation (38.73) states that level 1 handling qualities are predicted if the amplitude of the HQSF is less that unity around the crossover frequency. One very important caveat in applying this predictive tool is that the desired crossover frequency is to be obtained by modifying the gain associated with the controlled element, rather than through the pilot model gain K_e, which is maintained at unity here. This step removes control system sensitivity as a factor in the handling qualities prediction. Finally, in applying the structural model to the handling qualities evaluation of a number of competing systems, the crossover frequency must remain invariant. This is the analysts way of demanding comparable performance from the competing systems. The selection of this crossover frequency obviously deviates from the rules outlined previously, e.g., those of the first of equations (38.9), and is typically based upon selecting the crossover frequency for desired task performance. This means ensuring that ω_c is larger than the bandwidth of the command input or disturbances. Crossover regression is ignored in this determination.

38.7.1.3 The Optimal Control Model

Some success has been achieved in relating the value of the index of performance, J, in the optimal control model formulation, to vehicle handling qualities (e.g., McRuer and Schmidt, 1990). In this approach, the weighting coefficients in the index of performance are chosen as follows: The maximum allowable deviation on displayed error is considered to be unity. The weighting coefficient on control rate is selected as that value of neuromotor time constant, τ_n, which yields minimum mean square error or $\tau_n = 0.1$ sec, whichever results in the larger value of τ_n. In addition, the work of McRuer and Schmidt suggest observation noise covariances as given in equation (38.29). However, as opposed to equation (38.29), McRuer and Schmidt employed motor noise covariances also equal to those for the observation noise, i.e., $V_u = 0.01\pi E\{u^2(t)\}$ and used an operator time delay of 0.15 sec. The derived relationship between pilot ratings and the index of performance proposed by McRuer and Schmidt is

$$\text{Handling qualities rating} \approx 5.5 + 3.7 \log_{10} \left[\frac{J}{\bar{c}^2 \omega_{BW_e}^2} \right] \tag{38.74}$$

The reader is cautioned that in applying the optimal control model to handling qualities predictions using equation (38.74), parameters must be selected in *precisely* the same way as that discussed here.

38.7.2 An Example

A single-loop tracking task, e.g., Figure 38.4, will be examined with two controlled elements $Y_c(s) = 1/s$ and $1/s^2$. The input will be filtered white noise, and the power spectral density of the input will be given by

$$\Phi_{cc}(\omega) = \frac{2\bar{c}^2}{\omega^2 + 1^2} \tag{38.75}$$

Note in this application that $\bar{c}^2 = 1.0$ and $\omega_{BW_{ce}} = 1.0$.

38.7.2.1 The Crossover Model

Little in the way of modeling is necessary here. Indeed, the expected rating decrements from "nominal" have already been outlined.

38.7.2.2 The Structural Model

In applying the structural model with $Y_c(s) = 1/s$ and $1/s^2$, the second and third rows of Table 38.3 are used. For $Y_c(s) = 1/s^2$, a value of $T_2 = 2$ sec. was used. As mentioned in Section 38.7.1.2, in applying the structural model in handling qualities investigations, the desired crossover frequency is obtained by setting $K_e = 1.0$ and artificially adjusting the "gain" of the controlled element. Here the desired crossover frequency was selected as 2.0 rad/sec, a value just beyond the bandwidth of the input command. In the two cases studied, this led to $Y_c(s) = 7.99/s$ and $22.1/s^2$. Figure 38.36 shows the resulting HQSFs. According to the structural model, the $1/s$ controlled element would be rated as level 1, while the $1/s^2$ element would be rated level 2 or 3.

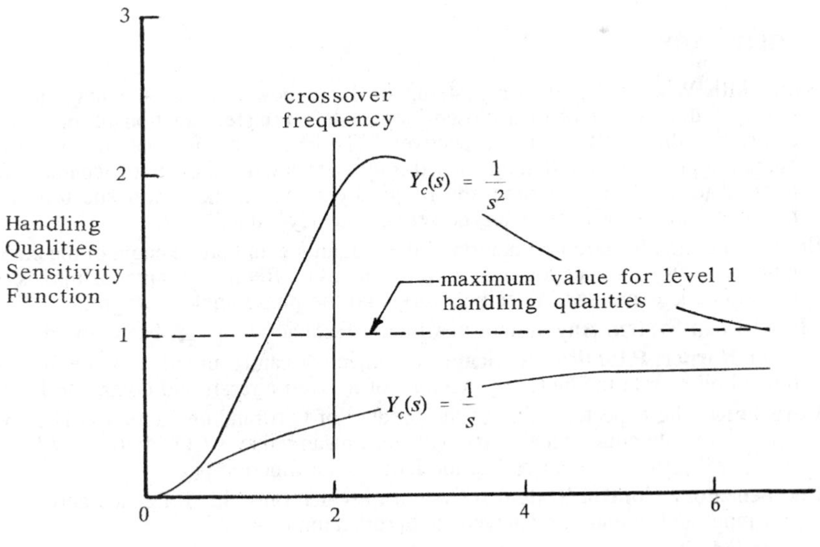

Figure 38.36 Handling qualities sensitivity functions for the structural model.

38.7.2.3 The Optimal Control Model

Employing the optimal control model in the manner outlined in Section 38.7.1.3, and using equation (38.74) yields predicted Cooper-Harper ratings of 1.8 for K/s and 4.3 for K/s^2.

38.7.2.4 Discussion

Because of the different manner in which the three modeling approaches have been adapted to handling qualities prediction, a direct comparison of results is not possible. However, a qualitative comparison can be made. The crossover model with Figure 38.35 indicates an approximate numerical rating difference between K/s and K/s^2 of 2.5 rating units. which agrees quite well with the difference in predicted ratings by the optimal control model. The structural model HQSF predicts that the K/s controlled element would be level 1 (ratings $\leqslant 3.5$) while the K/s^2 would be level 2 or worse (ratings > 3.5). Both predictions agree with those of the optimal control model. Although no experimental data are available that precisely match the conditions used in this analysis, there are certainly enough data available to corroborate level 1 ratings for K/s controlled element dynamics and level 2 ratings for K/s^2 controlled element dynamics, assuming reasonable input bandwidths and tracking performance requirements. For example, in an exhaustive pilot-in-the-loop simulation study (Mitchell, Aponso, and Hoh, 1990), K/s dynamics received Cooper-Harper ratings from 1 to 2, while K/s^2 dynamics received Cooper-Harper ratings from 4 to 7.

38.8 CONCLUDING REMARKS

The feedback models of the human operator that have been discussed in this chapter have been developed and verified in situations in which human performance and behavior is constrained by the task requirements and where the humans are highly trained and motivated. Whereas the development of the crossover, structural, and optimal control models have been based upon data in which the system disturbances and/or command inputs have been random, or at least random appearing, the fuzzy control model is not limited to these types of inputs. In addition, the structural model has been applied in situations involving discrete maneuvers, (e.g., Hess, et al., 1993).

The treatment of the manual control problem presented herein has relied heavily upon quantitative, control-theoretic representations of the human. Certainly more qualitative discussions of manual control exist and constitute recommended reading. Two pertinent examples of this more qualitative approach are provided by Jagacinski (1977) and Flach (1990).

38.9 GLOSSARY

Bandwidth When referred to a system, the frequency at which the magnitude and phase of the Bode plot of a closed-loop system transfer function become significantly less that 0 dB and 0°, respectively. The bandwidth indicates the highest frequency sinusoidal input that the closed-loop system will follow with accuracy. When referred to a function of time, the frequency at which the power spectral density becomes significantly less than the zero-frequency value.

Bode Plot A graphical representation of the magnitude and phase angle of the complex number $G(s)|_{s=j\omega}$ versus log ω as ω varies across a frequency range of interest. The magnitude is expressed in decibels (dB) and the phase angle in degrees.

Closed-Loop System Any system in which a feedback loop has been closed.

Cooper-Harper Pilot Rating Scale An adjectival and numerical scale by which trained pilots rate the handling qualities of a given aircraft and control task.

Covariance The expected value of the product of two random variables. The covariance of a white noise signal, $v_y(t)$, with zero mean value is COV $[v_y(t)] = E\{v_y(t)v_y(t + \tau)\} = V_y(\tau)\delta(\tau)$, where $\delta(\tau)$ is the Dirac delta function.

Compensatory Control System A system in which only the difference between system input and output is displayed to the operator.

Controlled Element The transfer function of the system being controlled. Simple, stereotypical controlled elements often used in experimental manual control studies are *position control* (K), velocity control (K/s) and *acceleration control* (K/s^2).

Crossover Frequency The frequency at which the magnitude portion of the Bode plot of an open-loop transfer function equals 0 dB.

Crossover Model A model of the human operator in single-loop tracking tasks which postulates that, in the frequency range around crossover frequency, ω_c, the product of the human's transfer function, $Y_p(s)$ and that of the controlled element $Y_c(s)$ can be given by

$$Y_pY_c(s) \approx \frac{\omega_c}{s} e^{-\tau_e s}$$

Crossover Regression A phenomenon in manual control systems wherein the human operator reduces the crossover frequency well below that which would be predicted by the crossover model.

Decibel (dB) $20\log_{10}|G(j\omega)|$ where $G(s)$ is a system transfer function and $10\log_{10}|\Phi(\omega)|$ where $\Phi(\omega)$ is a power spectral density.

Describing Function A representation for a nonlinear and/or time varying system consisting of a linear transfer function and a "remnant" signal injected at the input or output of the transfer function.

Differential Equation An equation involving derivatives of a function, e.g.

$$\frac{dx(t)}{dt} + x(t) = f(t)$$

Equalization The characteristics of a transfer function that, when multiplying the controlled element, will yield a closed-loop system with desired characteristics; compensation.

Fuzzy Control Model A model of the human operator that utilizes fuzzy sets and fuzzy set theory.

Fuzzy Set A set or collection of elements whose membership in the set is described by a membership function which can assume values between 0 and 1.

Gain The multiplicative constant appearing in the numerator of a transfer function.

Handling Qualities Those qualities or characteristics that determine the ease and precision with which a human operator can complete a given control task.

Handling Qualities Sensitivity Function (HQSF) A transfer function derived from the structural model of the human pilot which can be used to predict the handling qualities level of a vehicle and task.

Index of Performance In the optimal control model of the human operator, a scalar quantity defined as the weighted sum of mean square performance and control activity.

Laplace Transform The Laplace transform $F(s)$ of a time function $f(t)$ is defined as

$$F(s) = \int_0^\infty f(t)e^{-st}\, dt$$

With zero initial conditions, the Laplace transform of the nth derivative of a time function is

$$\mathcal{L}\left[\frac{d^n}{dt^n} f(t)\right] = s^nF(s)$$

Mean Square Value The mean square value of a function of time, $f(t)$, is defined as

$$\overline{f^2}(t) = \lim_{T\to\infty} \frac{1}{T} \int_0^T f^2(t)\, dt$$

Multiloop System A feedback system involving nested loop closures, each defining lower bandwidth systems as one proceeds from inner to outer loops.

Multivariable System A system whose mathematical description involves more than one differential equation.

Open-Loop System The system existing before feedback loops are closed.

Optimal Control Model A model of the human operator consisting of a combination of a linear optimal state estimator or Kalman filter and a linear regulator. The dynamic and information processing limitations of the human operator are incorporated in the model through the use of observation and motor noise, time delay, and index of performance weighting coefficients.

Power The power, P, in a continuous function of time, $f(t)$, can be defined as

$$P = \lim_{T \to \infty} \frac{1}{T} \int_0^T f^2(t)\, dt$$

Pole A value of the Laplace variable, s, which makes the denominator of a transfer function zero.

Power Spectral Density A function of frequency when integrated over all frequencies gives the total amount of power in a continuous function of time.

Preview Control System A system in which information about future values of the system input is available to the operator.

Proprioceptive Defining sensory information derived from limb position, velocity, and applied force.

Pursuit Control System A system in which information about the system input, output, and error is available to the operator.

Remnant That portion of the human operator's output which is not linearly correlated with the input.

Root Mean Square (RMS) Value The square root of the mean square value.

Servomechanism A device that measures the difference between an actual state, e.g., a position, and a desired state and uses the difference to drive the actual state to the desired.

Single-Loop System A feedback system involving only a single feedback loop closure.

Single-Point System A feedback system involving a single control variable that may have more than one feedback loop.

State Variables Those variables whose values describe all the information about a system at each instant of time. Values of the state variable at some instant of time along with the system model and future input allow one to completely describe the future behavior of the system.

Structural Model A model of the human operator that hypothesizes that the human's equalization capabilities are derived primarily from proprioceptive rather than visual feedback signals.

White Noise A fictitious continuous function of time that possesses a power spectral density that is constant over all frequencies.

Zero A value of the Laplace variable, s, which makes the numerator of a transfer function zero.

REFERENCES

Allen, R. W., and McRuer, D. T. (1979). The man/machine control interface—pursuit control. *Automatica, 15*(6), 683–686.

Baron, S. (1983). An optimal control model analysis of data from a simulated hover task. In *Proceedings of the Eighteenth Annual Conference on Manual Control.* Dayton, OH, pp. 186–206.

Baron, S., and Levison, W. H. (1973). A display evaluation methodology applied to vertical situation displays. In *Proceedings of the Ninth Annual Conference on Manual Control.* Dayton, OH, pp. 121–132.

Bendat, J. S., and Piersol, A. G. (1993). *Engineering Applications of Correlation and Spectral Analysis* (Second Edition). New York: Wiley.

Bryson, A. E., and Ho, Y. C. (1975). *Applied Optimal Control.* New York: Wiley.

Dornheim, M. A. (1992, November). Report pinpoints factors leading to YF-22 crash, *Aviation Week and Space Technology, 9,* 53–54.

Dornheim, M. A. (1995, May). Boeing corrects several 777 PIO's. *Aviation Week and Space Technology, 8,* 32.

Flach, J. M. (1990). Control with an eye for percepton: Precursors to an active psychophysics. *Ecological Psychology*, *2*(2), 83–111.

Graham, D., and McRuer, D. T. (1971). *Analysis of Nonlinear Control Systems*. New York: Dover.

Hess, R. A. (1979). A rationale for human operator pulsive control behavior, *Journal of Guidance, Control, and Dynamics*, *2(3)*, 221–227.

Hess, R. A. (1981). Pursuit tracking and higher levels of skill development in the human pilot, *IEEE Transactions on Systems, Man, and Cybernetics*, *SMC-11(4)*, 262–273.

Hess, R. A. (1984). Analysis of aircraft flight control systems prone to pilot-induced-oscillations, *Journal of Guidance, Control, and Dynamics*, *14(1)*, 198–204.

Hess, R. A. (1985). A model-based theory for analyzing human control behavior, In W. B. Rouse (Ed.), *Advances in Man-Machine Systems Research*, Vol. 2, New York: North Holland. pp. 129–175.

Hess, R. A. (1990). Methodology for the analytical assessment of aircraft handling qualities, In C. T. Leondes (Ed.), *Control and Dynamic Systems* Vol. 33, Pt. 3, San Diego: Academic Press. pp. 129–140.

Hess, R. A. (1995). Rotorcraft handling qualities in turbulence. *Journal of Guidance, Control, and Dynamics*, *10(1)*, 39–45.

Hess, R. A., and Beckman, A. (1984). An engineering approach to determining visual information requirements for flight control tasks. *IEEE Transactions on Systems, Man, and Cybernetics*, *SMC-14*(2), 286–298.

Hess, R. A., and Kalteis, R. M. (1991). Technique for predicting longitudinal pilot-induced oscillations. *Journal of Guidance, Control, and Dynamics*, *14*(*1*), 198–204.

Hess, R. A., Malsbury, T. and Atencio, A., Jr. (1993). Flight simulation fidelity assessment in a rotorcraft lateral translation maneuver, *Journal of Guidance, Control, and Dynamics*, *14*(*1*), 191–197.

Hess, R. A., and Yousefpor, M. (1992). Analyzing the flared landing task with pitch-rate flight control systems, *Journal of Guidance, Control, and Dynamics*, *15*(*3*), 768–774.

Jagacinski, R. J. (1977). A qualitative look at feedback control theory as a style of describing behavior. *Human Factors*, *19*, 331–347.

Johnston, D. E., and McRuer, D. T. (1987). Investigation of limb-side stick dynamic interaction with roll control. *Journal of Guidance, Control, and Dynamics*, *10*(2), 178–186.

Kleinman, D. L., Baron, S., and Levison (1970). An optimal control model of human response, Part I. *Automatica*, *6*(*3*), 357–369.

Kramer, U. (1985). On the application of fuzzy sets to the analysis of the system-driver-vehicle environment, *Automatica*, *21*(*1*), 101–107.

Kwakernaak, J., and Sivan, R. (1972). *Linear Optimal Control Systems*. New York: Wiley-Interscience.

Levison, W. H., Kleinman, D. L. and Baron, S. (1969). A model for human controller remnant, *IEEE Transactions on Man-Machine Systems*, *MMS-10*(*4*), 101–108.

Maciejowski, J. M. (1989). *Multivariable Feedback Design*. Wokingham, England: Addison-Wesley.

Magdaleno, R. E., and McRuer, D. T. (1971). *Experimental Validation and Analytical Elaboration for Models of the Pilot's Neuromuscular Subsystem in Tracking Tasks* (NASA CR-1757). Washington, DC: National Aeronautics and Space Administration.

McRuer, D. T. (1980). Human dynamics in man-machine systems. *Automatica*, *16*(*3*), 237–253.

McRuer, D. T., Graham, D., Krendel, E., and Reisener, W., Jr. (1965). *Human Pilot Dynamics in Compensatory Systems* (AFFDL-TR-65-15). Air Force Flight Dynamics Laboratory.

McRuer, D. T., and Krendel, E. (1957). Dynamic response of human operators. Wright Air Development Center, WADC TR 56-524.

McRuer, D. T., and Krendel, E. (1974). *Mathematical Models of Human Pilot Behavior* (AGARDograph NO. 188). NATO Advisory Group for Aerospace Research and Development.

McRuer, D. T., and Schmidt, D. K. (1990). Pilot-vehicle analysis of multiaxis tasks. *Journal of Guidance, Control, and Dynamics*, *13*(2), 348–355.

Miller, J. (1982). *The Body in Question*. New York: Random House.

Mitchell, D. G., Aponso, B. L., and Hoh, R. H. (1990). *Minimum Flying Qualities, Vol I: Piloted Simulation of Multiple Axis Flying Qualities* (AFWAL-TR-89-3125). Air Force Wright Aeronautical Laboratory, Wright-Patterson, AFB, OH.

Modjtahedzadeh, A., and Hess, R. A. (1993). A model of driver steering control behavior for use in assessing vehicle handling qualities. *Journal of Dynamic Systems, Measurement and Control*, *115*(*3*), 456–464.

Nise, N. S. (1995). *Control Systems Engineering*. Redwood City, CA: Benjamin/Cummings.

Reid, L. D., and Drewell, N. H. (1972). A pilot model for tracking with preview. In *Proceedings of the Eighth Annual Conference on Manual Control*. Ann Arbor, MI, pp. 191–204.

Ringland, R. F., Stapleford, R. L., and Magdaleno, R. E. (1971). *Motion Effects on an IFR Hovering Task* (NASA CR-1933). Washington, DC: National Aeronautics and Space Administration.

Smith, R. H. (1976). *A Theory For Handling Qualities with Application to MIL-F-8785B.* (AFFDL-TR-75-119.) Air Force Flight Dynamics Lab, Wright-Patterson, AFB, OH.
Stapleford, R. L., Craig, S. J., and Tennant, J. A. (1969). *Measurement of Pilot Describing Functions in Single-Controller Multiloop Tasks* (NASA CR-1238). Washington, DC: National Aeronautics and Space Administration.
Stapleford, R. L., McRuer, D. T., and Magdaleno, R. E. (1967). Pilot describing function measurement in a multiloop task. *IEEE Transactions on Human Factors in Electronics, HFE-8*(2), 113–125.
Stengel, R. F. (1986). *Stochastic optimal control.* New York: Wiley.
Sutton, R., and Towill, D. R. (1988). Modelling the helmsman in a ship steering system using fuzzy sets. In *Proceedings of the IFAC Conference on Man-Machine Systems: Analysis, Design and Evaluation.* Oulu, Finland.
Sutton, R. (1990). *Modelling Human Operators in Control System Design,* New York: Wiley.
Tustin, A. (1947). The nature of the operator's response in manual control and its implication for controller design, *Journal of the IEE, 94* (Part IIA, No. 2).
Weir, D. H., and McRuer, D. T. (1972). *Pilot Dynamics for Instrument Approach Tasks: Full Panel Multiloop and Flight Director Operations* (NASA CR-12019). Washington DC: National Aeronautics and Space Administration.
Zadeh, L. A. (1973). Outline of a new approach to the analysis of complex systems and decision processes, *IEEE Transactions on Systems, Man, and Cybernetics, SMC-3*(*11*), 28–44.

CHAPTER 39

SUPERVISORY CONTROL

Thomas B. Sheridan
Department of Mechanical Engineering
Massachusetts Institute of Technology
Cambridge, MA 02139 USA

This chapter is a tutorial on "supervisory control," drawing heavily on experiments done at MIT. It is not a comprehensive or even-handed review of the literature in human–robot interaction, monitoring, diagnosis of failures, human error, mental workload, or other closely related topics. Sheridan (1992) and Mouloua and Parasuraman (1994) cover these aspects more fully.

39.1 WHAT IS SUPERVISORY CONTROL?

The term *supervisory control* is derived from the close analogy between the characteristics of a supervisor's interaction with subordinate human staff members and a person's interaction with "intelligent" automated subsystems. A supervisor of people gives general directives that are understood and translated into detailed actions by staff members. In turn, staff members aggregate and transform detailed information about process results into summary form for the supervisor. The degree of intelligence of staff members determines the level of involvement of their supervisor in the process. Automated subsystems permit the same sort of interaction to occur between a human supervisor and the process (Ferrell and Sheridan, 1967; Sheridan, Fischhoff, Posner, and Pew, 1983). Supervisory control behavior is interpreted to apply broadly to vehicle control (aircraft and spacecraft, ships, and undersea vehicles), continuous process control (oil, chemicals, power generation), and robots and discrete tasks (manufacturing, space, undersea, mining).

In the strictest sense, the term supervisory control indicates that one or more human operators are setting initial conditions for, intermittently adjusting, and receiving information from a computer that itself closes a control loop (i.e., interconnects) through external sensors, effectors, and the task environment. In a broader sense, supervisory control means a computer makes complex transformations on system data to produce integrated (chunked) displays for the human, or retransforms operator commands to generate detailed control actions, but does not itself necessarily close the control loop. Figure

39.1 compares supervisory control with direct manual control and full automatic control. Figures 39.1(*c*) and 39.1(*d*) characterize supervisory control in the strict formal sense; Figure 39.1(*b*) characterizes supervisory control in the latter (broader) sense.

The essential difference between these two characterizations of supervisory control is that in the first and stricter definition the computer can act on new information independently of and with only blanket authorization and adjustment from the supervisor; that is, the computer implements discrete sets of instructions by itself, closing the loop through the environment. In the second definition the computer's detailed implementations are "open loop," that is, feedback from the task has no effect on computer control of the task except through the human operator. The two situations may appear similar to the supervisor, since he or she always sees and acts through the computer (analogous to a staff in a human organization) and therefore may not know whether it is acting open loop or closed loop in its fine behavior. In either case the computer may function principally on the efferent or motor side to implement the supervisor's commands (e.g., do some part of the task entirely and leave other parts to the human, or provide some control compensation to ease all of the task for the human). Alternatively the computer may function principally on the display side (e.g., to integrate and interpret incoming information from below, or to give advice to the supervisor as to what to do next as an "expert system"). Or it may do both.

39.2 THE EMERGENCE OF SUPERVISORY CONTROL IN TECHNOLOGICAL SYSTEMS

Supervisory control is emerging rapidly in many industrial, military, medical, and other contexts, although this form of human interaction with technology is relatively little recognized or understood in a formal way.

From the pyramid-building pharaohs of Egypt through all of the history of technology there surely has been concern about how best to extend the capabilities of human workers. Early in the present century, against the backdrop of the newly mechanichanized production line, Taylor's "scientific management" (Taylor, 1911) catalyzed a formal intellectual consciousness about the human factors involved. Taylor intended to evoke a new interest in the sensori-motor aspects of human performance. What he did not intend was the subsequent criticism that his essentially mechanistic approach was dehumanizing.

The 1940s and 1950s saw "human factors" ("ergonomics" in Europe) emerge, first in essentially empirical "knobs and dials" form, concentrating on the human–machine interface itself. This was supported over the next decade by the theoretical underpinnings

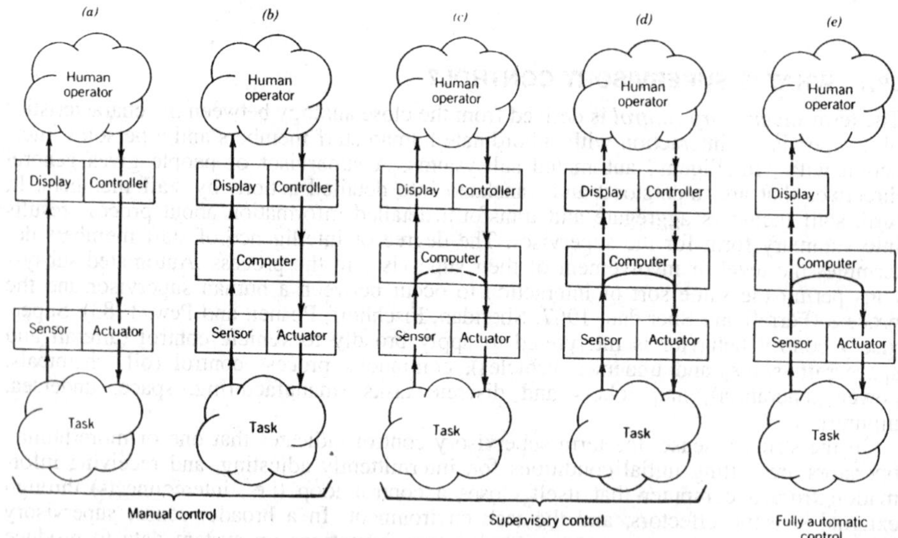

Figure 39.1 Supervisory control as related to direct manual control and full automation.

of "man–machine systems" (Sheridan and Ferrell, 1974). Such theories included control, information, signal detection and decision theories originally developed for application to physical systems, but now explicitly applied to the human operator. As contrasted with human factors engineering at the interface, human–machine systems analysis considers characteristics of the entire causal "loop" of decision, communication, control, and feedback—through the operator's physical environment and back again to the human.

From the late 1950s the computer began to intervene in the causal loop: electronic compensation and stability augmentation for control of aircraft and similar systems, electronic filtering to extract signal patterns from noise, electronic generation of simple displays. It was obvious that if vehicular or industrial systems were equipped with sensors that could be read by computers, and by motors that could be driven by computers, then—even though the overall system was still very much human controlled— control loops between those sensors and motors could be closed automatically. Thus the chemical plant operator was relieved of keeping the tank at a given level or the temperature at a reference—he or she needed only to set in that desired level or temperature signal from time to time. So, too, after the autopilot was developed for the aircraft the human pilot needed only to set in the desired altitude or heading; an automatic system would strive to achieve this reference, with the pilot monitoring to ensure that the aircraft did in fact go where desired. The automatic building elevator, of course, has been in place for many years, and is one of the first implementations of supervisory control; the passenger indicates the desired floor and the elevator goes there. Recently, developers of new systems for word processing and handling of business information (i.e. without the need to control any mechanical processes) have begun thinking along similar lines.

The full generality of the idea of supervisory control came to the author and his colleagues (Sheridan, 1960; Ferrell and Sheridan, 1967) as part of research on how people on earth might control vehicles on the moon through 3-s round-trip time delays (imposed by the limited speed of light). Under such constraint remote control of lunar roving vehicles or manipulators was shown to be possible only by performing in "move-and-wait" fashion. This means the operator can commit only to a small incremental movement "open-loop," that is, without feedback (which actually is as large a movement as is reasonable without risking collision or other error), then stopping and waiting one delay period for feedback to "catch up," then repeating the process in steps until the task is completed.[1]

It was shown that if, instead of the human operator remaining within the control loop, he or she communicates a goal state relative to the remote environment, and if the remote system incorporates the capability to measure proximity to this goal state, then the achievement of this goal state can be turned over to a remote subordinate control system for implementation. In this case there is no delay in the control loop implementing the task and thus there is no instability.

There necessarily remains, of course, a delay in the supervisory loop. This delay in the supervisor's confirmation of desired results is acceptable so long as (1) the subgoal is a sufficiently large "bite" of the task, (2) the unpredictable aspects of the remote environment are not changing too rapidly (i.e., disturbance bandwidth is low), and (3) the subordinate automatic system is trustworthy. More will be said of each of these points.

If these conditions obtain, and as computers gradually become more capable both in hardware and software (and as "machine intelligence" finally makes its real if modest appearance), it is evident that telemetry transmission delay is no way a prerequisite to the usefulness of supervisory control. The incremental goal specified by the human operator need not be simply a new steady-state reference for a servomechanism (as in resetting a thermostat) in one or even several dimensions (e.g., resetting both temperature and humidity, or commanding a manipulator endpoint to move to a new position including three translations and three rotations relative to its initial position: "put that there"). Each new goal statement can be the specification of a whole trajectory of movements (as the

[1]Attempts to drive or manipulate continuously only produce instability, as simple control theory predicts (i.e., where loop gain exceeds unity at a frequency such that the loop time delay is one half-cycle, instead of errors being nulled out they are only reinforced).
Performing remote manipulations with delayed force feedback was shown by Ferrell (1966) to be essentially impossible since forces at unexpected times act as significant disturbances to produce instability. At least the visual feedback can be ignored by the operator.

performance of a dance or a symphony) together with programmed branching conditions (what to do in case of a fall, or a broken violin string, or how to respond contingent upon audience applause).

In other words, the incremental goal statement is a program of instructions in the full sense of a computer program, which makes the human supervisor an intermittent real-time computer programmer, acting relative to the subordinate computer much the same as a teacher or parent or boss behaves relative to a student or child or subordinate worker. The size and complexity of each new program is necessarily a function of how much the computer can (be trusted to) cope with at once, which in turn depends on the computer's own sophistication (knowledge base) and the complexity (uncertainty) of the task.

Supervisory control is emerging in various forms in various industries—usually without being called as such. (More likely, each developer or vendor has its own cute acronym emphasizing how "smart" and easy it is to use the new product.) Aircraft autopilots are now "layered," meaning the pilot can select among various forms and levels. At the lowest level he or she can set in a new heading or rate of climb. Or he or she can program a sequence of heading changes at various way-points, or a sequence of climb rates initiated at various altitudes. Or he or she can program the inertial guidance system to go to (within a fraction of a mile of) a distant city. Given the existence of certain ground-based equipment, he or she can program an automatic landing on a given runway, and so on. Wiener and Curry (1980) provide a good review of how such automation is creeping into the aircraft flight deck. Modern chemical plants can similarly be programmed to perform heating, mixing, and various other processes according to a time line, but including various sensor-based conditions for shutting down or otherwise aborting the operation.

More and more a multiplicity of computers is used in a supervisory control system, as shown in Figure 39.2. One typically large computer is in the control room to generate displays and interpret commands. We call this a "human-interactive computer" (HIC), part of a "human-interactive systems" (HIS). It in turn forwards that command to various microprocessors that actually close individual control loops through their own associated sensors and effectors. We call these "task-interactive computers" (TICs), each part of its own "task-interactive system" (TIS).

The examples cited above characterize the first or stricter definition of supervisory control previously given [Figure 39.1(*c*) and 39.1(*d*)], where the computer, once programmed, makes use of its own artificial sensors to ensure completion of the assigned task. Many familiar systems such as automatic washing machines, dryers, dishwashers, or stoves, once programmed, perform their operations "open loop"; that is, there is no measurement or knowledge of results. If the task can be performed in such open-loop fashion, and if the human supervisor can anticipate the task conditions and is good at

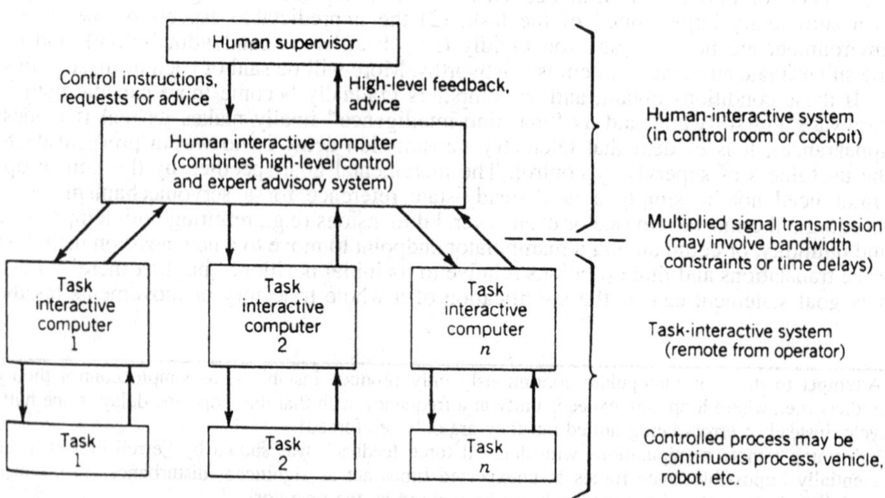

Figure 39.2 Hierarchical nature of supervisory control.

selecting the right open-loop program, there is no reason not to employ this approach. To the human supervisor, whether the lower level implementation is open loop or closed loop is often opaque and/or of no concern; the only concern is whether the goal is achieved satisfactorily. For example, a programmable microwave oven without the temperature sensor in place operates open loop, whereas the same oven with the temperature sensor operates closed loop. To the human supervisor or programmer they both look the same.

A very important aspect of supervisory control is the ability of the computer to "package" information for visual display to the human operator, including data from many sources; from the past, present, or even predicted future; and presented in words, graphs, symbols, pictures or some combination. The ubiquitous examples of such integrated displays in aircraft and air traffic control, chemical and power plants, and various other industrial or military setting are too numerous to review here.

General interest in supervisory control became evident in the mid-1970s (Edwards and Lees, 1974; Sheridan and Johannsen, 1976) and continues to grow. A report by the National Research Council (Sheridan and Hennessy, 1984) outlines problems of supervisory control, especially with regard to experimental research (which is particularly difficult because of the inherent complexity and capital cost of real supervisory control systems, both factors inhibiting simulation and experimental control).

39.3 SUPERVISORY ROLES, LOCI, AND LEVELS OF HUMAN AND COMPUTER

The human supervisor's *roles* are: (1) *planning* off-line what task to do and how to do it; (2) *teaching* (or programming) the computer what was planned; (3) *monitoring* the automatic action on-line to make sure all is going as planned and to detect failures; (4) *intervening*, which means the supervisor takes over control after the desired goal state has been reached satisfactorily, or he or she interrupts the automatic control in emergencies to specify a new goal state and reprogram a new procedure; and (5) *learning* from experience so as to do better in the future. These are usually time-sequential steps in task performance.

We may view these steps as being within two nested loops, as shown in Figure 39.3. The inner loop closes from intervening back to teaching; that is, human intervention usually leads to programming of a new goal state to the process. The outer loop closes from learning back to planning; intelligent planning for the next subtask is enhanced by learning from previous ones. Actually, an innermost loop (not drawn) might have shown monitoring closing on itself; that is, evidence of something interesting or completion of one part of a cycle of monitoring strategy leads to more investigation and monitoring. We might include minor on-line tuning of the process as part of monitoring.

The two supervisory loops operate at different time scales relative to one another. New programs are generated at somewhat longer intervals than the continual tuning which is part of monitoring. Revisions in significant task planning occur only at still longer intervals. These differences in time scale further justify Figure 39.3.

In the middle column of Figure 39.3 are shown the mental models which the human supervisor would have to have while performing each associated step, and in the righthand column is the computer model or database appropriate to decision aiding for each step, respectively.

For each of the five roles or stages of the supervisory process there are three *loci* of function in a physiological sense: *sensory* functions (accessing displays, observing, perceiving); *cognitive* functions internal to the supervisor (evaluating the situation, accessing memory, making decisions); and *response* functions. *S*, *C*, and *R* are the classic designators to differentiate these functional elements of causation through the operator.

Finally, we may appeal to the *levels* of behavior introduced by Rasmussen (1976); *skill-based* behavior (continuous, typically well-learned, sensory-motor behavior analogous to what can be expected from a servomechanism); *rule-based* behavior (what a computer programmed with a set of rules can do in recognizing a pattern of stimuli, then triggering an "if–then" algorithm to execute an appropriate response); and finally *knowledge-based* behavior ("high-level" situation assessment and evaluation, consideration of alternative actions in light of various goals, decisions and scheduling of implementation—a form of behavior machines are not now good at).

Consider the above three metacharacteristics of supervision, namely, *role*, *loci* and *level*, as three independent dimensions of supervisory behavior.

The human-interactive computer (HIC) is conceived to be a large enough computer to communicate in a human-friendly way, using near-natural language, good graphics, and

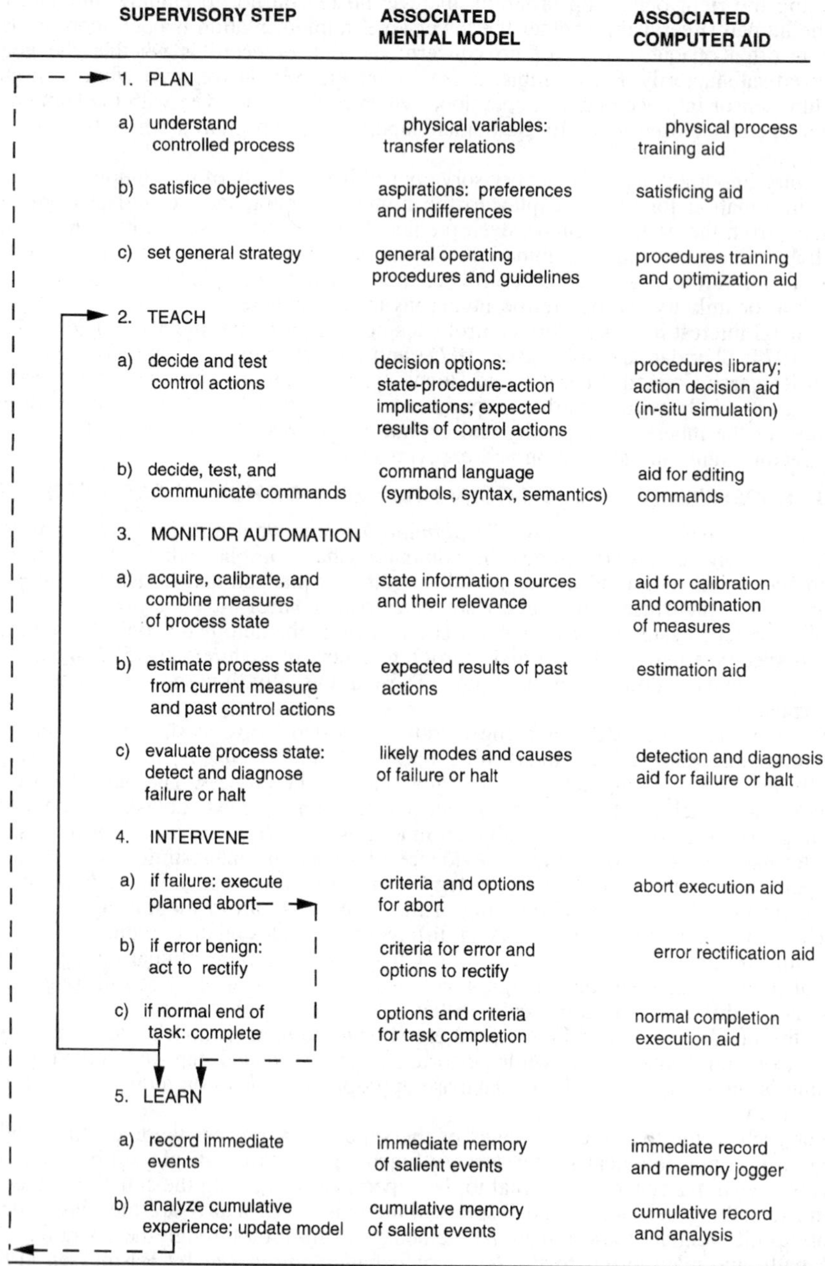

SUPERVISORY STEP	ASSOCIATED MENTAL MODEL	ASSOCIATED COMPUTER AID
1. PLAN		
a) understand controlled process	physical variables: transfer relations	physical process training aid
b) satisfice objectives	aspirations: preferences and indifferences	satisficing aid
c) set general strategy	general operating procedures and guidelines	procedures training and optimization aid
2. TEACH		
a) decide and test control actions	decision options: state-procedure-action implications; expected results of control actions	procedures library; action decision aid (in-situ simulation)
b) decide, test, and communicate commands	command language (symbols, syntax, semantics)	aid for editing commands
3. MONITIOR AUTOMATION		
a) acquire, calibrate, and combine measures of process state	state information sources and their relevance	aid for calibration and combination of measures
b) estimate process state from current measure and past control actions	expected results of past actions	estimation aid
c) evaluate process state: detect and diagnose failure or halt	likely modes and causes of failure or halt	detection and diagnosis aid for failure or halt
4. INTERVENE		
a) if failure: execute planned abort—	criteria and options for abort	abort execution aid
b) if error benign: act to rectify	criteria for error and options to rectify	error rectification aid
c) if normal end of task: complete	options and criteria for task completion	normal completion execution aid
5. LEARN		
a) record immediate events	immediate memory of salient events	immediate record and memory jogger
b) analyze cumulative experience; update model	cumulative memory of salient events	cumulative record and analysis

Figure 39.3 Functional and temporal nesting of supervisory roles. The lefthand column shows the role, the middle column the associated mental model, and the righthand column the computer model which would serve as a decision aid.

so on. This includes being able to accept and interpret commands and to give the supervisor useful feedback. The HIC should be able to recognize patterns in data sent up to it from below and to decide on appropriate algorithms for response, which it sends down as instructions. Eventually the HIC should be able to run "what would happen if" simulations and be able to give useful advice from a knowledge base, that is, serve as an "expert system."

The HIC, located near the supervisor in a control room or cockpit, may communicate across a "barrier" of time or space with a multiplicity of task-interactive computers (TICs), which probably are microprocessors distributed throughout the plant or vehicle. The latter are usually coupled intimately with artificial sensors and actuators, in order to deal in low-level language and to close relatively tight control loops with objects and events in the physical world.

The human supervisor can be expected to communicate with the HIC intermittently in information "chunks" (alphanumeric sentences, video pages, etc.) while the task communicates with the TIC continuously in computer words at the highest possible bit rates. The availability of these computer aids means that the human supervisor, while retaining the knowledge-based behavior for him- or herself, is likely to "download" some of the rule-based and almost all of the skill-based programs into the HIC. The HIC, in turn, should download a few of the rule-based programs, and most of the skill-based programs, to the appropriate TICs.

Figure 39.4 presents the functions of Figure 39.3 in the form of a flow chart. Each supervisory function is shown above, and the (usually multiple) automated subsystems of the TIC are shown below. Normally, for any given task, the planning and learning roles are performed offline relative to the online human-mediated and automatic operations of

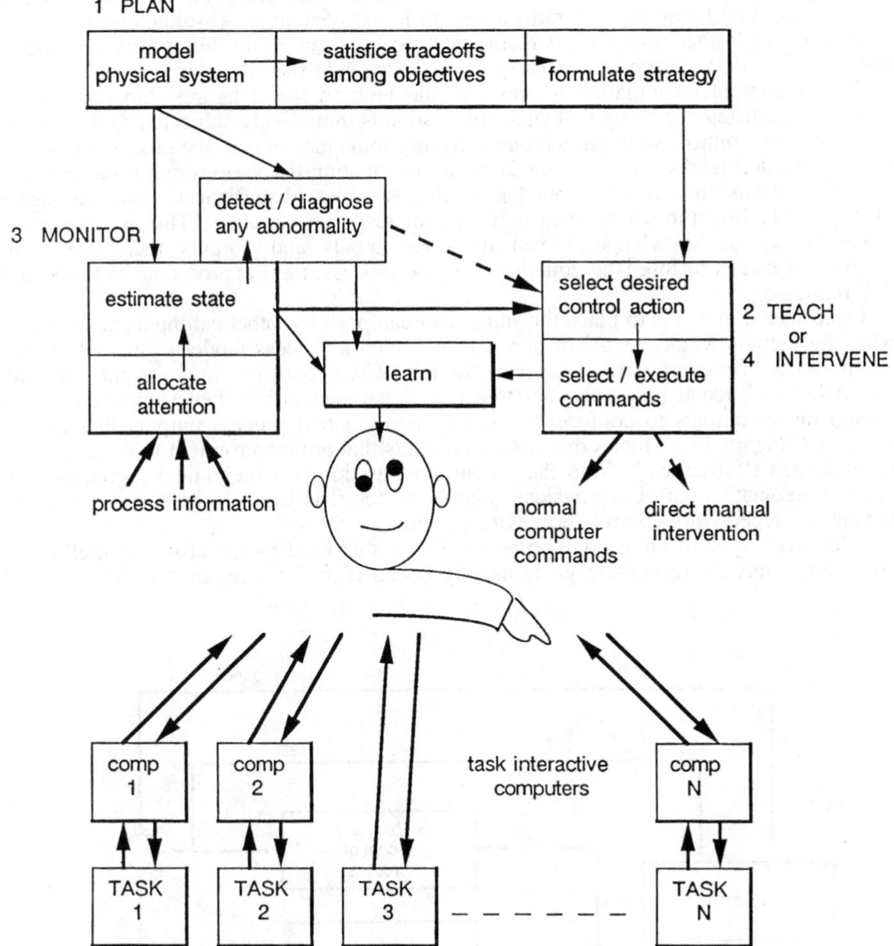

Figure 39.4 Flow chart of supervisor functions (which include both mental models and decision aids). (From Sheridan, 1992.)

the other parts of system, and therefore are shown at the top with light lines connecting them to the rest of the system. Teaching precedes monitoring on the first cycle, but thereafter follows monitoring and intervening (as necessary) within the intermediate loop. The inner loop monitoring role is carried out within the "estimate state" and "allocate attention" boxes.

In the sections that follow the various supervisory roles are discussed in more detail, bringing in examples of research problems and prototype systems that aid the supervisor in these roles.

39.4 PLANNING AND LEARNING: COMPUTER REPRESENTATION OF RELEVANT KNOWLEDGE

The first and fifth supervisory roles previously described–planning and learning–may be considered together since they are similar activities in many ways. Essentially, in planning, the supervisor asks "What would happen if?" questions of the accumulated knowledge base and considers what the implications are for hypothetical control decisions. In learning, the supervisor asks "What did happen?" questions of the database for the more recent subtasks, and considers whether the initial assumptions and final control decisions were appropriate.

The designer of an automatic control system or manual control system must ask "What variables do I wish to make do what, subject to what constraints and what criteria?" The planning role in supervisory control requires that the same kinds of questions be asked and answered, because, in a sense, the supervisor is redesigning an automatic control system each time he or she programs a new task and goal state. Absolute constraints on time, tools, and other resources available need to be clear, as do the criteria of tradeoff between time, dollars, and resources spent, accuracy, and risk of failure.

Just as computer simulation figures into planning so too it figures into supervisory control—the difference being that such stimulation is more likely subjected to time stress in supervisory control. Simulation requires having some idea of how the process or system to be controlled works, that is, a set of equations relating the various controllable variables, the various uncontrollable but measurable variables (disturbances), and the degree of unpredictability (noise) on measured system response variables. This is a common representation of knowledge. Given measured inputs and outputs there are well-established means to infer the equations—if the processes are approximately linear and differentiable.

Once such a model is in place the supervisor can posit hypothetical inputs and observe what the outputs would be. Also, one may use such a process model as an "observer" (in the sense of modern control theory). Namely, when control signals are put into both the model and actual processes, and the model parameters are then trimmed to force certain model outputs to conform to corresponding actual process outputs that can be measured (Figure 39.5), then other process outputs that are inconvenient to measure may be estimated ("observed") from the model. Just as this is a theoretical prerequisite to optimal automatic control of physical systems, so too it is likely to be useful for aiding humans in supervisory control (Sheridan, 1984a).

A different type of knowledge representation is that used by the artificial-intelligence (AI) community. Here knowledge is usually couched in the form of "if–then" logical

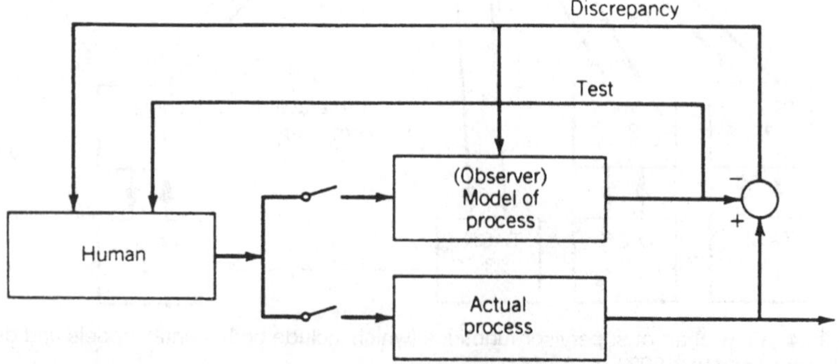

Figure 39.5 Use of computer-based observer as an aid to supervisor.

statements called "production rules," semantic association networks, and similar forms, and is usually programmed in Lisp. The input to a simulated program usually represents in cardinal numbers a hypothetical physical input to a simulated physical system. In contrast, the input to the AI knowledge base can be a question about relationships for given data or a question about data for given relationships. This can be in less restrictive ordinal form (e.g., networks of diadic relations) or in nominal form (e.g., lists).

Currently there is a great interest in how best to transfer expertise from the human brain (knowledge representation, mental model) into the corresponding representation or model within the computer, how best to transfer it back, and when to depend on each of those sources of information. This research on mental models has a lively life of its own (Falzon, 1982; Gentner and Stevens, 1983; Rouse and Morris., 1984; Sheridan, 1984a, Moray, 1997) quite independent of supervisory control.

An important aspect of planning is visualization. The now rather sophisticated tool of computer simulation, when augmented by computer graphics, enables remarkable visualization possibilities. When further augmented by human interactive devices such as head-mounted visual and auditory displays and high-bandwidth force-reflecting haptics (mechanical arms), the operator can be made to feel present in a virtual world, as has been popularized by the oxymoron *virtual reality*. Of course the idea of virtual reality is not new. The original idea of Edwin Link's first flight simulator (developed early in the 1940s) was to make the pilot trainee feel as if he or she were flying a real aircraft. The first such simulators were instrument panels only, then a realistic out-the-window view was created by flying a servo-driven video camera over a scale model, and finally computer graphics were put in to create the out-the-window images. Now all major commercial airlines and military services routinely train with computer-display, full-instrument, moving-platform flight simulators. Similar technology has been applied to ship, automobile, and spacecraft control. The salient point for the present discussion is that the new simulation capabilities now permit visualization of alternative plans, as well as better understanding of complex state information in situ, during monitoring. That same technology, of course, can be used to convey a sense of presence in an environment that is not simulated but is quite real—merely remote and communicated via closed circuit video with cameras slaved to the observer's head.

Supervisory aiding in planning the moves of a telerobot is illustrated by the work of Park (1991). His computer-graphic simulation let a supervisor try out moves of a telerobot arm before committing to the actual move. He assumed that for some obstacles the positions and orientations were already known, and he represented these in a computer model. The user commanded each straight-line move to a subgoal point in three-space by designating a point on the floor or the lowest horizontal surface (such as a table top) by moving a cursor to that point (say A in Figure 39.6*a*) and clicking, then lifting the cursor by an amount corresponding to the desired height of the subgoal point (say A) above that floor point and observing on the graphic model a vertical line being generated from the floor point to the subgoal point in space. This process was repeated for successful subgoal points (say B and C). Using the computer display, the user could view the resulting trajectory model from any desired perspective (though the "real" environment could be viewed only from the perspective provided by the video camera's location).

Either of two collision-avoidance algorithms could be invoked: a detection algorithm that indicated where on some object a collision occurred as the arm was moved from one point to another, or an automatic avoidance algorithm that found (and drew on the computer screen) a minimum-length, no collision trajectory from the starting point to the new subgoal point. Park's aiding scheme also allowed new observed objects to be added to the model by graphically "flying" them into geometric correspondence with the model display. Another aid was a means to generate "virtual objects" for any portion of the environment in the umbral region (still not visible after two video views) (Figure 39.6*b*). In this case the virtual objects were treated in the same way in the model and in the collision-avoidance algorithms as the visible objects. Experiments with this technique showed that it was easy to use and that it improved safety greatly.

At the extreme of time desynchronization is recording a whole task on a simulator, then sending it to the telerobot for reproduction. This might be workable when one is confident that the simulation matches the reality of the telerobot and its environment, or when small differences would not matter (e.g., in programming telerobots for entertainment). Doing this would certainly make it possible to edit the robot's maneuvers until one was satisfied before committing them to the actual operation. Machida, Toda, Iwata, Kawachi, and Nakamura (1988) demonstrated such a technique by which commands from a master–slave manipulator could be edited much as one edits material on a videotape

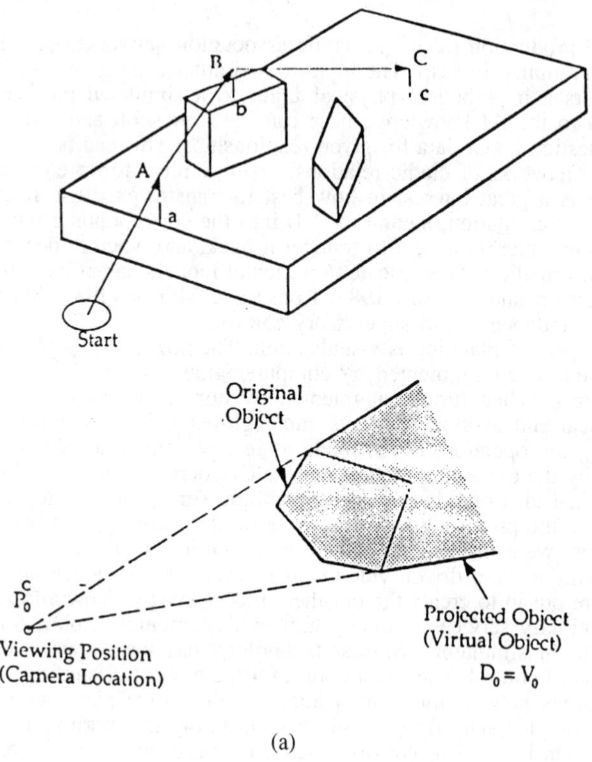

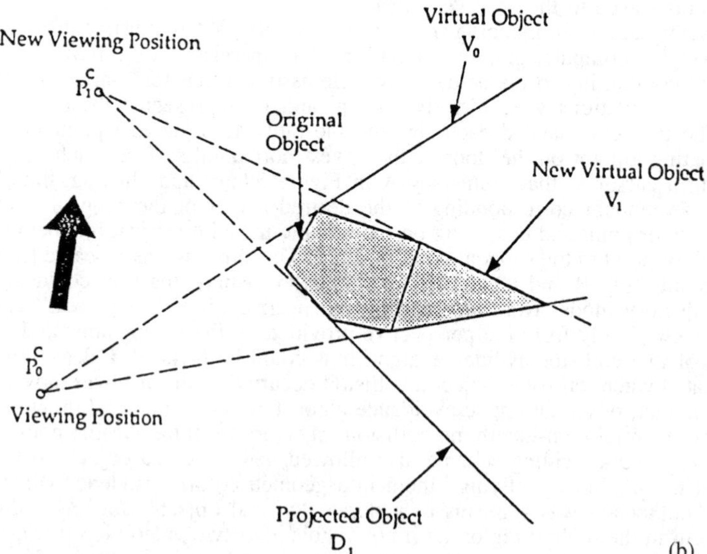

Figure 39.6 Park's display of computer aid for obstacle avoidance:
(*a*) Human specification of subgoal points on graphic model
(*b*) Generation of virtual obstacles for single viewing position (above) and pair of viewing positions (below). (From Park, 1991.)

recorder or a word processor. Once a continuous sequence of movements had been recorded, it could be played back either forward or in reverse at any time rate. It could be interrupted for overwrite or insert operations. Their experimental system also incorporated computer-based checks for mechanical interference between the robot arm and the environment.

39.5 TEACHING THE COMPUTER

Teaching or programming a task, including a goal state and a procedure for achieving it, and including constraints and criteria, can be formidable or quite easy, depending on the command hardware and software. By command hardware is meant the way in which human response—hand, foot, or voice—is converted into physical signals to the computer. Command hardware can be either *analogic* or *symbolic*. "Analogic" means that there is a spatial or temporal isomorphism among human response, semantic meaning, and/or feedback display. For example, quickly moving a control up—to rapidly increase the magnitude of a variable—which causes a display indicator to move up quickly, would be a proper analogic correspondence.

Symbolic command, by contrast, is accomplished by depressing one or a series of keys (as typing words on a typewriter), or uttering one or a series of sounds (as in speaking a sentence), each of which has a distinguishable meaning. For symbolic commands a particular series or concatenation of such responses has a different meaning from other concatenations. Spatial or temporal correspondence to the meaning or desired result is not a requisite. Sometimes analogic and symbolic commands can be combined, for example, where up–down keys are both labeled and positioned accordingly.

It is natural for people to intermix analogic and symbolic commands, or even to use them simultaneously. Typical industrial robots are taught by a combination of grabbing hold and leading the end point of the manipulator around in space relative to the workpiece, at the same time using a switch-box on a wire (a *teach pendant)* to key in codes for start, stop, speed, etc. between various reference positions. This happens, for example, when a person talks and points at the same time, or plays the piano and conducts a choir with head or free hand.

Supervisory command systems have been developed for mechanical manipulators that utilize both analogic and symbolic interfaces with the supervisor and that enable teaching to be both rapid and available in terms of high-level language. Brooks (1979) developed such a system he called SUPERMAN, which allows the supervisor to use a master arm to identify objects and demonstrate elemental motions. He showed that even without time delay for certain commands, which refer to predefined locations, supervisory control—including both teaching and execution—took less time and had fewer errors than manual control.

Yoerger (1982) developed a more extensive and robust supervisory command system that enables a variety of arm—hand motions to be demonstrated, defined, called on, and combined under other commands. In one set of experiments, Yoerger compared three different procedures for teaching a robot arm to perform a continuous seam weld along a complex curved workpiece. The end effector (welding tool) had to keep 1 in. away and retain an orientation perpendicular to the curved surface to be welded and move at constant speed. Yoerger tested his subjects in three command (teaching) modes. The first mode was for the human teacher to first move the master (with slave following in master-slave correspondence) relative to the workpiece in the desired trajectory. The computer would memorize the trajectory, and then cause the slave end effector to repeat the trajectory exactly. The second mode was for the human teacher to move the master (and slave) to each of a series of positions, pressing a key to identify each. The human would then key in additional information specifying the parameters of a curve to be fit through these points and the speed at which it was to be executed, and the computer would then be called upon for execution. The third mode was to use the master–slave manipulator to contact and trace along the workpiece itself, to provide the computer with the knowledge of the location and orientation of the surfaces to be welded. Then, using the typewriter keyboard, the human teacher would specify the positions and orientations of the end effector *relative* to the workpiece. The computer could then execute the task instructions relative to the geometric references given.

Identifying the geometry of the workpiece analogically, and then giving symbolic instructions relative to it, proved the constant winner. The reasons for this advantage apparently are the same as for Brooks' results previously described, provided of course the time spent in the teaching loop is sufficiently short.

Teaching an airplane autopilot is a good example of the teaching role in supervisory control. Modern airplanes can now automatically adjust their throttle, pitch, and yaw damping characteristics. They can take off and climb to altitude autonomously or fly to a given latitude and longitude and can maintain altitude and direction in spite of wind disturbances. They can approach and land automatically in zero-visibility conditions. To do these tasks, airplanes make use of artificial sensors, motors, and computers, programmed in supervisory fashion by pilots and ground controllers. In this sense airplanes are telerobots in the hands of their pilot teachers. In the aviation world the supervising pilot is called a "flight manager." Figure 39.7 provides a metaphoric summary of pilot information requirements for performing this task.

New aviation technology (Wiener, 1988; Billings, 1991; Hopkins, 1995; Sarter and Woods, 1994) includes TCAS (traffic collision avoidance system), ARTS (automated radar terminal system), SSR (secondary surveillance radar), and ILS/MLS (instrument or machine-aided landing systems). The "glass cockpit" came in several years ago with Boeing's 757 and 767, which integrated within computer-graphic CRT, LED, and LCD displays information previously presented on dedicated displays. The new displays have replaced the multiple independent mechanical flight instruments and have simplified the instrument panel. Autopilots have been provided with multiple control modes, e.g., for going to and holding a new altitude, flying to a set of latitude–longitude coordinates, or making an automatic landing when the airport has the supporting equipment. In the new Airbus A320 a primary flight mode is fly-by-wire through a miniature sidestick, in dramatic contrast to the old control yoke. Computer-generated and computer-based expert systems in the cockpit now give the pilot advice on engine conditions, how to save fuel, and other topics. Performance management systems are now available to optimize fuel and time.

These functions are now combined within the flight management system (FMS). The FMS is the aircraft embodiment of the human-interactive computer previously discussed. The typical FMS display has a CRT display and both generic and dedicated keysets. Multiple modules provide maps for terrain and navigational aids, procedures, and synoptic diagrams of various electrical and hydraulic subsystems. A proposed electronic map that would show flight plan route, weather, and other navigational aids is illustrated in Figure 39.8. When the pilot enters a certain flight plan, the FMS can automatically visualize the

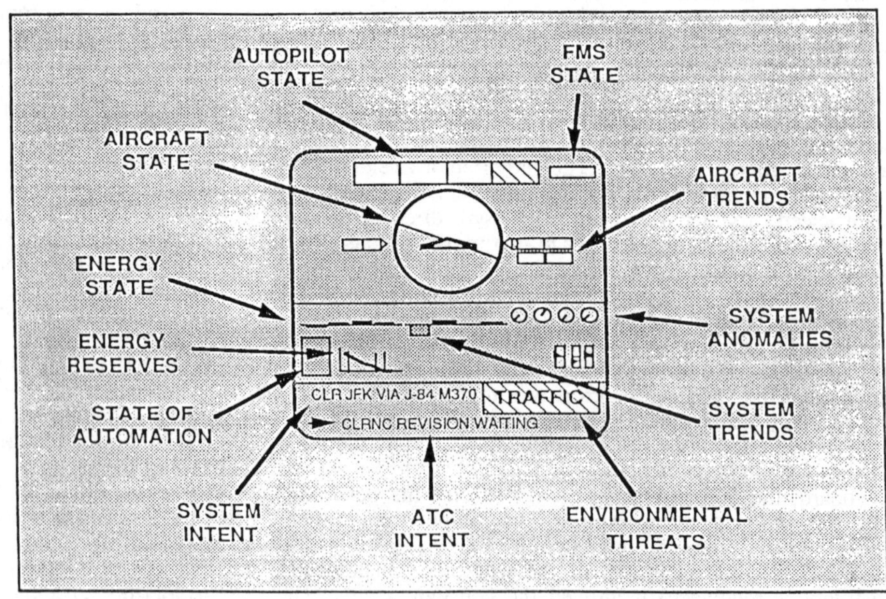

Figure 39.7 Pilot information requirements. (Courtesy C. Billings, NASA.)

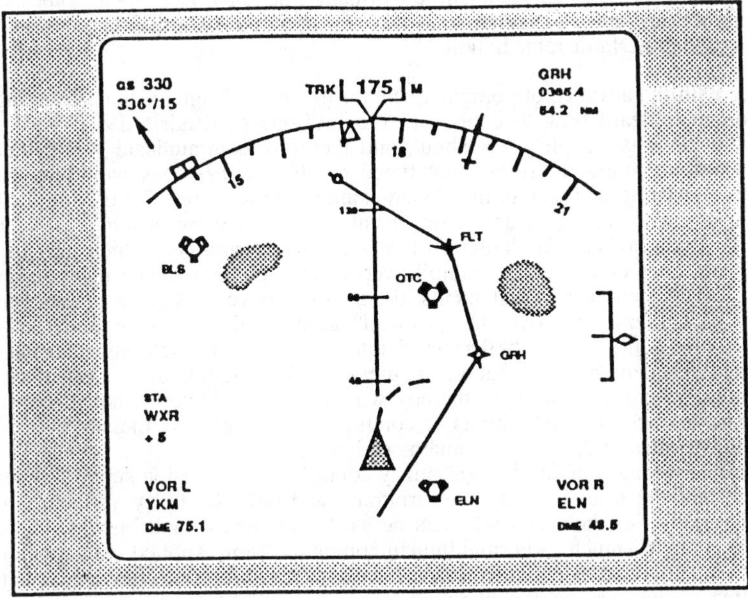

Figure 39.8 Computer-generated map display showing weather, flight plan route, and navigation aids. (Courtesy C. Billings, NASA.)

trajectory and call attention to any way-points that appear to be erroneous on the basis of a set of reasonable assumptions. (This might have prevented the programmed trajectory that allegedly took KAL 007 into Soviet territory.)

The problem of authority is one of the most difficult (Boehm-Davis, Rurry, Wiener, and Harrison, 1983). A popular myth is that the pilot is (or should be) in charge at all times. But when a human pilot turns control over to an automatic system, it is with the expectation that she or he can do something else for a while (as in the case of setting one's alarm clock and going to sleep). It is also recognized that there are limited windows of opportunity for escaping from the automation (once you get on an elevator you can get off only at discrete floor levels). People are seldom inclined to "pull the plug" unless they receive clear signals indicating that such action must be taken, and unless circumstances make it convenient for them to do so. Examples of some current debates follow:

- Should there be certain states, or a certain envelope of conditions, for which the automation will simply seize control from the pilot? In the MD11 it is impossible to exceed critical boundaries of speed and attitude which will bring the aircraft into stall or other unsafe flight regimes. The pilot can approach the boundaries of the safe flight envelope only be exerting much more than the normal force on the control stick.
- Should the computer automatically deviate from a programmed flight plan if critical unanticipated circumstances arise? The MD11 will deviate from its programmed plan if it detects windshear.
- If the pilot programs certain maneuvers ahead of time, should the aircraft automatically execute these at the designated time or location, or should the pilot be called upon to provide further concurrence or approval? The A320-340 will not initiate a programmed descent unless it is reconfirmed by the pilot at the required time.
- In the case of a subsystem abnormality, should the affected subsystem automatically be reconfigured, with an after-the-fact display of what has failed and what has been done about it? Or should the automation wait to reconfigure until after

the pilot has learned about the abnormality, perhaps been given some advice on the options, and had a chance to take the initiative? The MD11 goes a long way in automatic fault remediation.

Along with the advance of computer science in natural language understanding, it will be important to learn how to cope with the "fuzziness" (Zadeh, 1984; Kosko, 1992) inherent in the way people think about, and therefore communicate about, their tasks. That is, both memorized "rules" and typed or spoken messages would by nature be sentences consisting of fuzzy terms. As an example, a fuzzy rule for driving a car might be: "If your car is going *fast* and if the car ahead is *very* close or *moderately* close and going *slow*, then brake." The italicized terms are fuzzy variables, which may be defined with varying degrees of "membership" over a range of numerical values of speed and distance. Given a number of statements like the one above, and given membership functions for each fuzzy term over the physical variables, the "relative truth" of each of several control actions (e.g., brake, accelerate, coast) can be determined. Buharali and Sheridan (1982) demonstrated that a computer could be taught to drive a car by repeatedly giving rules, where the computer thereby would come to "know what it didn't know" with regard to various combinations of conditions and could ask the supervisor–teacher for additional rules to cover its "domains of ignorance".

The big advantages of fuzzy logic, fuzzy control, fuzzy decision support systems, and so on are that the rules can be quite arbitrary, and how the fuzzy variables map onto physical variables of the real works can be as "crisp" or as "soft" as one wishes for a particular context—much as normal human communication. Applied mathematicians and computer scientists accustomed to closed-form analysis and proof often are quite negative about "fuzzy" because so far, although it works well, "fuzzy" is not so tractable to mathematical theorems and proofs.

To complete this discourse on supervisor teaching of computer automation, it is important to emphasize that simple and ideal command-and-feedback patterns are not to be expected as systems get more complex. In interactions between a human supervisor and his or her subordinates, or a teacher and the students, it can be expected that the teaching process will not be a one-way communication. Some feedback will be necessary to indicate whether the message is understood, or to convey a request by the subordinate for clarification on some aspect of the instructions. Further, when the subordinate or student does finally act on the instruction, the supervisor may not understand from the immediate feedback what the subordinate has done and may ask for further details. This is illustrated in Figure 39.9 by the light arrows, where the heavy arrows characterize the conventional flow of information in feedback control.

39.6 MONITORING OF DISPLAYS AND DETECTION OF FAILURES

The human supervisor monitors the automated execution of the task in order to ensure proper control (Parasuraman, 1987). This includes intermittent adjustment or trimming if the process performance remains within satisfactory limits, to detect if and when it goes outside limits, and to diagnose failures or other abnormalities. The subject of failure detection in human–machine systems has received considerable attention (Rasmussen and Rouse, 1981). Moray (1986) regards such failure detection and diagnosis as the most important human supervisory role. I prefer the view that all five supervisory roles are essential and no one can be placed above the others.

The supervisory controller tends to be removed from full and immediate knowledge about the controlled process. The physical interactions he or she must monitor tend to be large in number and be distributed widely in space (e.g., around a ship or a plant). The physical variables may not be immediately sensible (e.g., steam flow and pressure) and may be computed from remote measurements on other variables. Sitting in the control room or cockpit, the supervisor is dependent on various artificial displays to give feedback of results as well as knowledge of new reference inputs or disturbances. These factors greatly affect how he or she detects and diagnoses abnormalities in the process, but whether removal from active participation in the control loop makes it harder (Ephrath and Young, 1981) or easier (Curry and Ephrath, 1977) remains an open question. Curry and Nagel (1974), Niemala and Krendel (1974), Gai and Curry (1978), and Wickens and Kessel (1979, 1981) have studied various aspects of monitoring from within the loop or from outside.

In traditional control rooms and cockpits the tendency has been to provide the human supervisor with an individual and independent display of each and every variable, and

FUNCTION OF HUMAN | FORM OF COMMUNICATION (analogic or symbolic) | FUNCTION OF COMPUTER

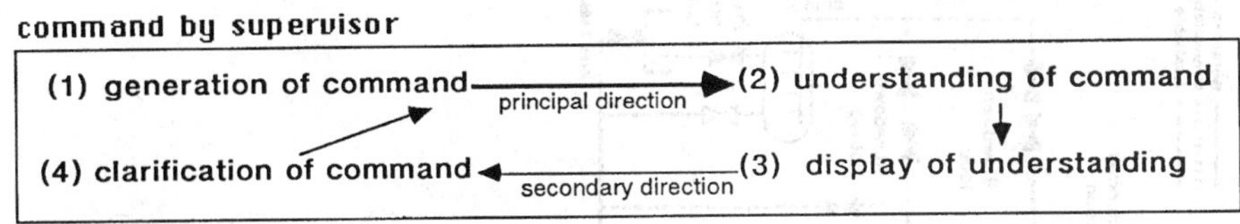

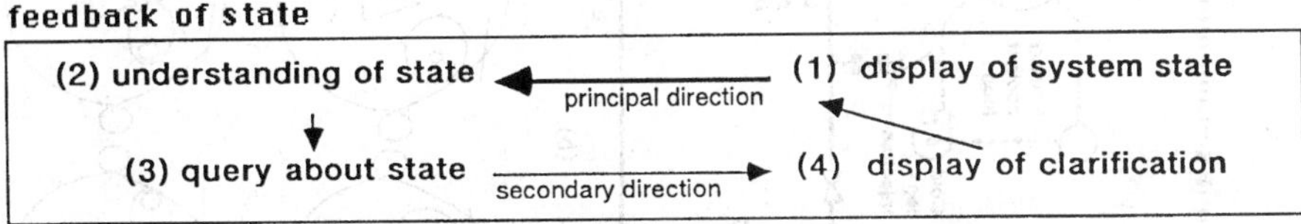

Figure 39.9 Intermediate feedback in command and display. Heavy arrows indicate the conventional understanding of functions. Light arrows indicate critical additional functions that tend to be neglected. (From Sheridan, 1992.)

for a large fraction of these to provide a separate alarm display that lights up when the corresponding variable reaches or exceeds some value. Thus modern aircraft may easily have over 1000 displays and modern chemical or power plants 5000 displays. In the writer's experience in one nuclear plant training simulator during the first minute of a "loss of coolant accident" 500 displays were shown to have changed in a significant way, and in the second minute 800 more.

Clearly no human being can cope with so much information coming simultaneously from so many seemingly disconnected sources. Just as clearly such signals in any real operating system actually are highly correlated. In every-day situations in which we move among people, animals, plants, or buildings our eyes, ears, and other senses easily take in and comprehend vast amounts of information—just as much as in the plant control room. Our genetic makeup and experience enable us to integrate the bits of information from different parts of the retina and from different senses from one instant to the next—presumably because the information is correlated. We say we "perceive patterns" but do not pretend to understand how. In any case the challenge is to design displays in technological systems to somehow integrate the information to enable the human operator to perceive patterns in time and space and across the senses. As with teaching (command), the forms of display may be either analogic (e.g., diagrams, plots) or symbolic (e.g., alphanumerics) or some combination.

In the nuclear power industry the "safety parameter display system" (SPDS) is now required of all plants in some form. The idea of the SPDS is to select a small number (e.g., 6–10) of variables that tell the most about plant safety status, and to display them in "integrated" fashion, such that by a glance the human operator can see whether something is abnormal, if so what, and to what relative degree. Figure 39.10 shows an example of an SPDS. Figure 39.10*a* gives the "high-level" or overview display (a single computer "page"). If the operator wishes more detailed information about one variable or subsystem he or she can "page down" (select lower levels), such as Figure 39.10*b*. These can be diagram having lines or symbols that change color or flash to indicate changed status, and alphanumerics to give quantitative or more detailed status. These can also be bar graphs or cross plots, or can be "integrated" in other forms. One novel technique is the "Chernoff face" (Figure 39.10*c*) in which the shapes of eyes, ears, nose, and mouth systematically differ to indicate different values of variables, the idea being that facial patterns are easily perceived. The Nuclear Regulatory Commission, fearful that some enterprising designer might employ this technique before it was proven, formally forbade it as an acceptable SPDS.

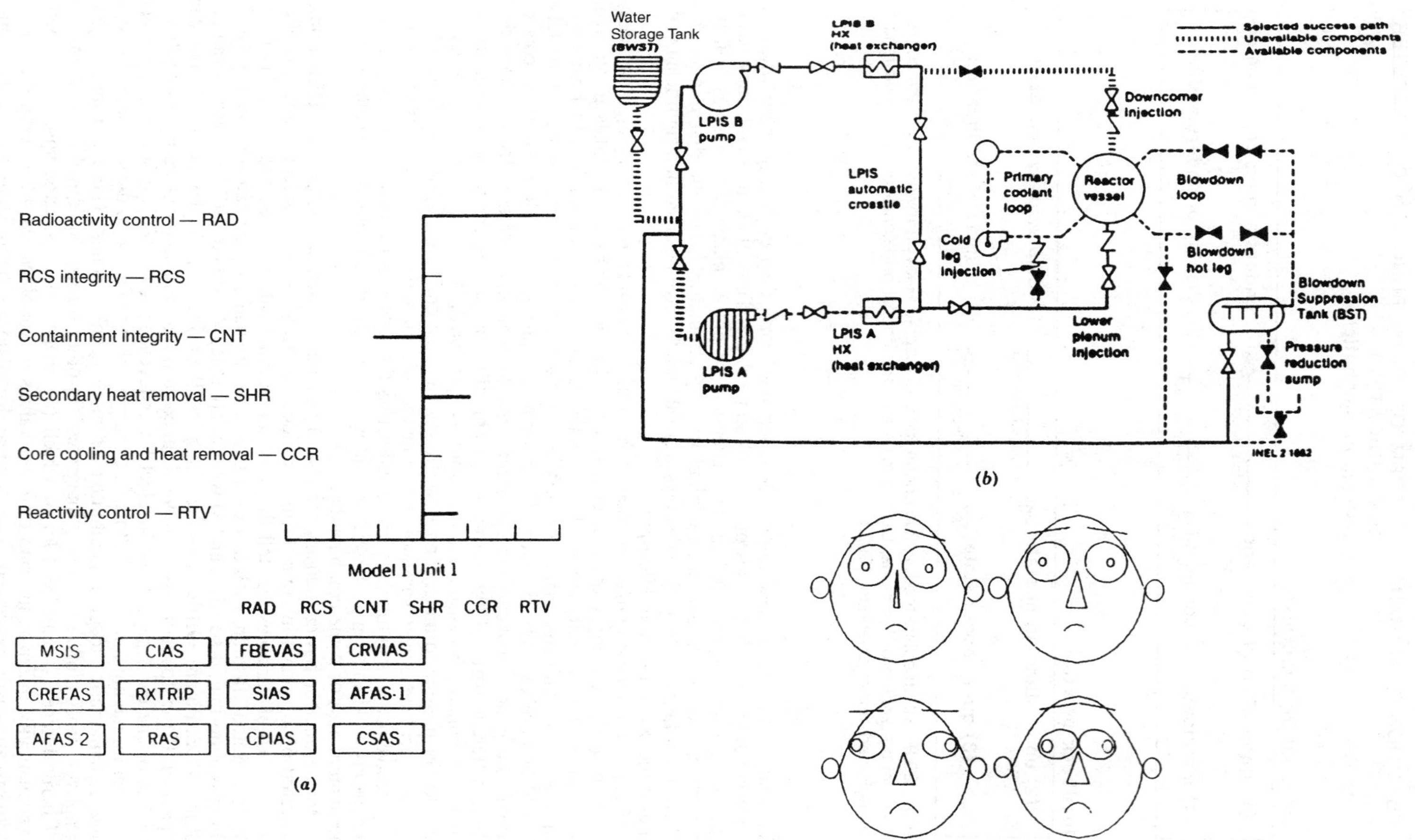

Figure 39.10 Safety parameter display system for a nuclear power plant.

Since detection and diagnosis of system failure is a critical task for the supervisor, computer aiding by the HIC in comparing, computing, and displaying has great potential. Various techniques have been proposed for doing this. One such technique (Tsach, Sheridan, and Buharali, 1983) continuously compared key measurements from the plant to corresponding variables of an online computer model; then a computer-graphic display focused the operator's attention on the discrepancies that indicate abnormality. Figure 39.11 shows one type of iconic display developed for this system—a polygon whose vertices indicate the degree to which each variable (of one subsystem in this case) is below or above a normal range (torus). The display therefore "points" to corresponding discrepancies between the measured and model variables as they evolve in time.

As previously noted (Figure 39.5) an important potential of the HIC is for modeling the controlled process. Such a model may then be used to generate a display of "observed" state variables that cannot be seen or measured directly. Another use is to run the model in fast time to predict the future, given of course that the model is calibrated to reality at the beginning of each such prediction run. A third use, now being developed for application to remote control of manipulators and vehicles in space, helps the human operator cope with telemetry time delays (as discussed in Figure 39.12) wherein video feedback is necessarily delayed by at least several seconds. By sending control signals to a computer model as a basis for superposing the corresponding graphic model on the video, the graphic model will "lead" the video picture and indicate what the video will do several seconds hence. This predictor display has been shown to speed up the execution of simple manipulation tasks by 70% (Noyes and Sheridan, 1984).

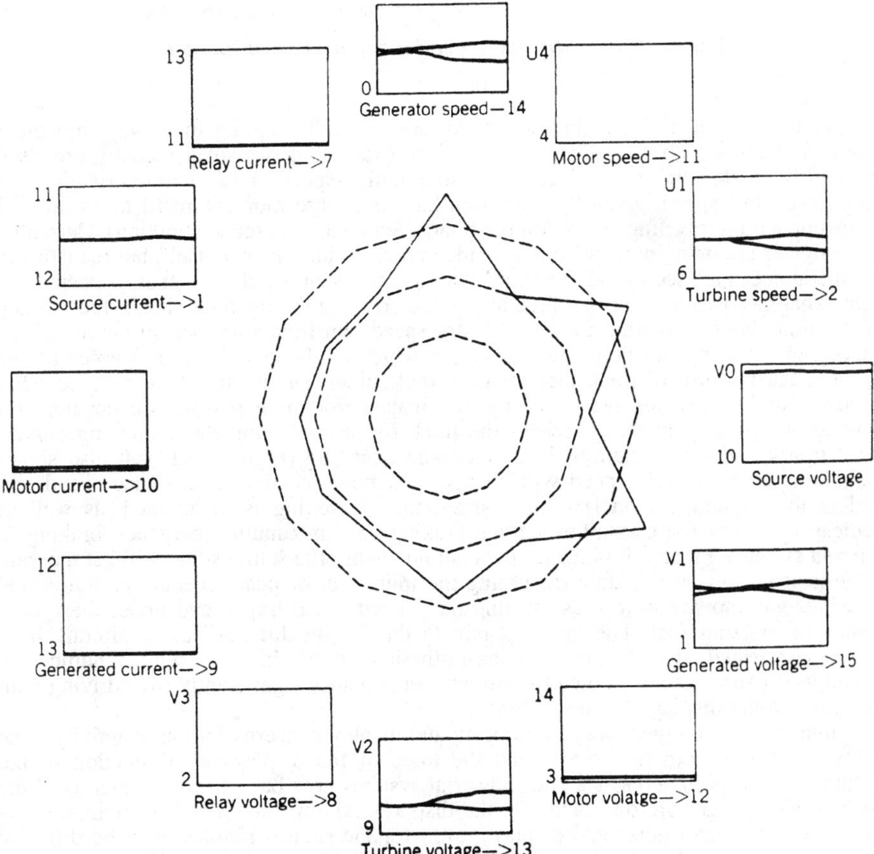

Figure 39.11 Computer display of Tsach, Sheridan, and Buharali system for failure detection and location in process control.

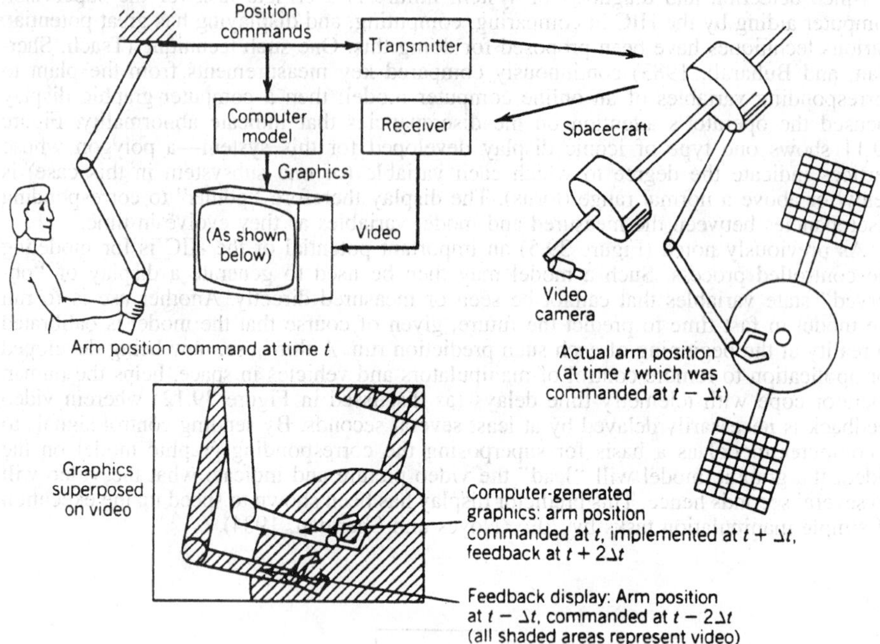

Figure 39.12 Predictor display for delayed telemanipulator.

Computer-driven displays for monitoring are gradually finding their way into older technical systems such as railway systems. One example is a display recently proposed for the driver's (locomotive engineer's) console in high-speed trains. A train driver's main job is to control speed, given that the train has very large momentum (it takes up to 3 km to stop a train traveling at 300 km per hour, even under emergency braking). Currently, speed signals are being moved from wayside indicators into the cab itself, but realistically the operator cannot see ahead for more than 1 km, usually not even that, especially at night. That means if a truck is stalled at a grade crossing the operator cannot receive any information about it in time to stop. Mostly speed constraints are set by curves in the track, grade crossings, or population densities, which are fixed and can be learned by the operator, but because of track maintenance, rock slides, snow, etc., there may be other speed constraints that are not so easily anticipated. For these reasons we developed a computer-based display that previews the track for several kilometers, showing curves, speed limits, and other features, both fixed and changing (Figure 39.13). It also shows prediction curves of how speed will change as a function of track distance ahead (according to a dynamic model) if the current throttle setting is maintained, as well as predication curves for maximum service braking and maximum emergency braking (a different system). Finally, it indicates the continuous throttle settings that will get the train to the next station on schedule (assuming the train is at or near schedule currently and the winds are known) as well as meeting the known speed limits, and under these constraints minimizing fuel. The latter is akin to the "flight director" in an aircraft. This system was tested in a dynamic human-in-the-loop train simulator with a number of trained driver subjects and shown to improve performance significantly over driving with the conventional displays (Askey, 1995).

A final aspect of supervisory monitoring and display concerns format adaptivity—the ability to change both the format and the logic of the displays as a function of the situation. Displays in aerospace and industrial systems now have fixed formats (e.g., the labels, scales, ranges are designed into the display). Alarms have fixed set points. However, future computer-generated displays, even for the same variables, may be different at various mission stages or in various conditions. Thus formats may differ for aircraft takeoff, landing, and on-route travel, and be different for plant startup, full capacity operation, and emergency shutdown. Some alarms have no meaning, or may be expected

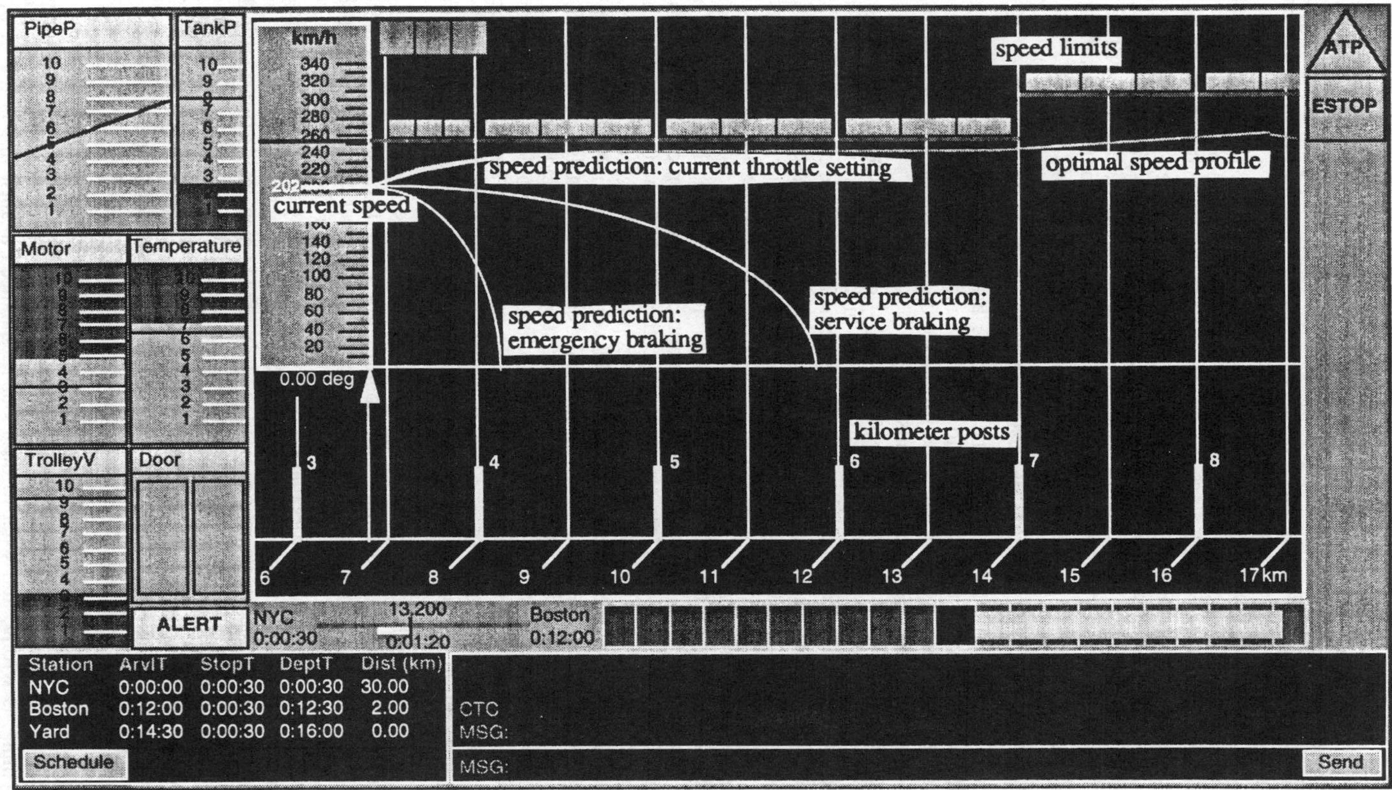

Figure 39.13 Proposed preview, predictor, and advisory display for driver of high-speed trains. (From Askey, 1995.)

to go off when certain equipment is being tested or taken out of service. In such a case adaptive formatted alarms may be suppressed, or the setpoints changed automatically to correspond to the operating mode. Future displays and alarms could also be formatted or adjusted to the personal desires of the supervisor, to provide any time scale, or degree of resolution, etc., necessary at the time. Ideally some future displays could adapt based on a running model of how the human supervisor's perception was being enhanced. There are hazards, of course, in allowing emergency displays to be too flexible, to the point where they cause errors rather than prevent them.

An example of where flexibility in monitoring displays went awry was in an aircraft accident that occurred in Europe several years ago. In this instance the pilot could ask to have either descent rate (thousands of feet per minute) or descent angle (degrees) presented, and, depending upon how the mode control panel had been set, the number was indicated by two digits displayed at the same location. In this case the pilot forgot which mode he had requested (though that information was also displayed, but at a different location). The result was a misinterpretation of a clearly displayed number and a tragic crash.

39.7 INTERVENING AND HUMAN RELIABILITY

The supervisor decides to intervene when the computer has completed its task and must be retaught, when the computer has run into difficulty and requests of the supervisor a decision as to which way to go, or when the supervisor decides to stop automatic action because system performance is judged unsatisfactory.

Intervention is a problem that really has not received as much attention as teaching and monitoring. Yet systems are being planned in which the supervisory operator is expected to receive advice from a computer-based system about remote events and within seconds decide whether to accept the computer's advice (in which case the response is automatically commanded), or reject the advice and generate his or her own commands (in effect intervene in an otherwise automatic chain of events). One such system is a state-of-the-art traffic management system currently being developed for Boston's Central Artery/Tunnel, the primary traffic system through that city, now being built at a cost of $8 billion. Magnetic, optical, infrared or smoke detectors signal sudden changes in traffic density (accident), fire and other emergencies, indicating to the operator where in the system the problem is, and displaying hopefully appropriate video pictures (from one or several of 400 cameras). The operator can accept the computer's advice (to call up fire, police or tow trucks; turn on fire extinguishers; modify electronic signals; etc.) or devise his or her own commands. The operator can add or subtract commands, or completely change the programmed execution after it has begun, but at considerable cost. An operator-interactive simulator of this system is currently being run in the author's laboratory to evaluate what one can expect of a trained operator under different conditions, and a normative intervention decision model has been devised as a basis for comparison (Kim, 1997).

It is at the intervention stage that human error most reveals itself. Errors in learning from past experience, planning, teaching, and monitoring will surely exist. Many of these are likely to be corrected as the supervisor discovers them. In operational systems such errors are relatively unlikely to be detected or counted. It is after the automatic system is functioning that those human errors make a difference, and where it is therefore critical that the human supervisor intervene in time and take appropriate action. Thus it is at the intervention stage (or where intervention is called for) where human error is most critical.

If human error is not caught by the supervisor, it is perpetuated slavishly by the computer, much as happened to the *Sorcerer's Apprentice*. For this reason supervisory control may be said to be especially sensitive to human error.

There are several factors affecting the supervisor's decision to intervene and/or his or her success in doing so:

1. *Tradeoff between collecting more data and taking action in time*. The more data collected from the more sources, the more reliable is the decision of what, if anything, is wrong, and what to do about it. Balanced against this is that if the supervisor waits too long, the situation will likely get worse, and corrective action may be too late. Formally the optimization of this decision is called the "optional stopping problem."
2. *Risk taking*. The supervisor may operate from either a risk-averse criterion such as minimax (minimize the worst outcome that could happen) or a more risk-neutral

criterion such as expected value (maximize the subjectively expected gain). Depending on the criterion the design of a supervisory control system may be very different in complexity and cost.

3. *Mental workload.* This problem is aggravated by supervisory control. When a supervisory control system is operating well in the automatic mode, the supervisor may have little concern. When there is a failure and sudden intervention is required, the mental workload may be considerably higher than in direct manual control, where in the latter case the operator is already actively participating in the control loop. In the supervisory case the supervisor may have to undergo a sudden change from initial inattention, moving physically and mentally to acquire information and learn what is going on, then making a decision on how to cope. Quite likely this will be a rapid transient from very little to very high mental workload.

Mental workload can be at issue in any human operation. The topic is reviewed by Moray (1979, 1982), Williges and Wierwille (1979), and Hart and Sheridan (1984). Ruffel-Smith (1979) studied pilot errors and fault detection under heavy cognitive workload in a realistic flight simulator and found that crews made approximately one error every 5 min.

Although the subject of human error is currently of great interest, there is no consensus on either a taxonomy or a theory of causality of errors. One common error taxonomy relates to locus of behavior; sensory, memory, decision, or motor. Another useful distinction is between errors of omission and those of commission. A third is between slips (correct intentions that inadvertently are not executed) and mistakes (intentions that are executed but that lead to failure).

In supervisory control there are several problems of human error worth particular mention. One is the type of slip called "capture". This occurs when the supervisor intends to do several steps of, but then deviates from, a well-rehearsed (behaviorally) and well-programmed (in the computer) procedure. Somehow habit, augmented by other cues from the computer, seems to "capture" behavior and drive it on to the next (unintended) step in the well-rehearsed and computer-reinforced routine.

A second supervisory error, important in both planning and failure diagnosis, results from the human tendency to seek confirmatory evidence for a single hypothesis currently being entertained (Gaines, 1976). It would be better if the supervisor could keep in mind a number of alternative hypotheses and let both positive and negative evidence contribute symmetrically in accordance with the theory of Bayesian updating (Sheridan and Ferrell, 1974). Norman (1981), Reason and Mycielska (1982), Rasmussen (1982), and Rouse and Rouse (1983) provide reviews of human error research from their different perspectives.

Theoretically anything that can be specified in an algorithm can be given over to the computer, so that the reason the human supervisor is present is to add novelty and creativity—precisely for those control demands that cannot be prespecified. This means, in effect, that the best or most correct human control behavior cannot be prespecified, and that variation from precise procedure must not always be viewed as errant noise. The human supervisor, by the nature of his or her function, must be allowed room by the system design for what may be called "trial and error" (Sheridan, 1983).

What training should the human supervisory controller receive in order to do a good job at detecting failures and intervening to avoid errors? As the supervisor's task becomes more cognitive, is the answer to provide training in theory and general principles? Curiously the literature seems to provide a negative answer (Duncan, 1981). In fact Moray (1986) in his review concludes that "There seems to be no case in the literature where training in the theory underlying a complex system has produced a dramatic change in fault detection or diagnosis." Rouse (1985) similarly concludes that the evidence (e.g., Morris and Rouse, 1985) does not support a conclusion"that diagnosis of the unfamiliar requires theory and understanding of system principles." Apparently frequent hands-on experience in a simulator (i.e., with simulated failures) is the best way to enable a supervisor to retain an accurate mental model of a process.

39.8 MODELING SUPERVISORY CONTROL

Modeling supervisory control is a challenge. For 15 years various models of supervisory control have been proposed. Mostly these have been models of particular aspects of supervisory control, not apparently claiming to model all or even very many aspects of it.

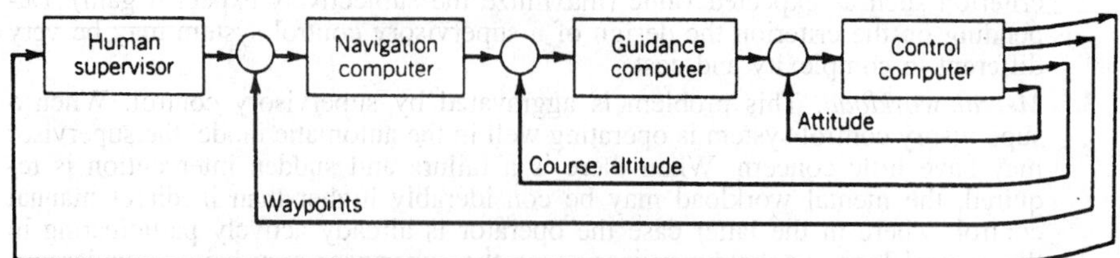

Figure 39.14 Nested control loops of aerospace vehicle.

The simplest model of supervisory control might be that of nested control loops (Figure 39.14) where one or more inner loops are automatic and the outer one is manual. In aerospace vehicles the innermost of four nested loops is typically called "control," the next "guidance," and the next "navigation," each having a set point determined by the next outer loop. Hess and McNally (1986) have shown how conventional manual control models can be extended to such multiloop situations. The outer loop in this generic aerospace vehicle includes the human operator, who, given mission goals, programs in the destination. In driving a car the functions of navigation, guidance, and control are all done by a person and can be seen to correspond roughly to knowledge-based, rule-based, and skill-based behavior, respectively.

Figure 39.15 (Sheridan, 1984b) is a qualitative functional model of supervisory control, showing the various cause–effect loops or relationships among elements of the system,

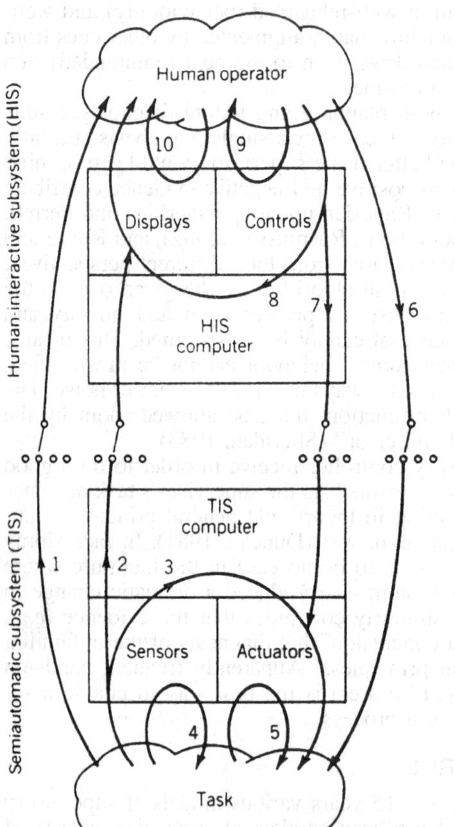

1. Task is observed directly by human operator's own senses.
2. Task is observed indirectly through artificial sensors, computers and displays. This TIS feedback interacts with that from within HIS and is filtered or modified.
3. Task is controlled within TIS automatic mode.
4. Task is affected by the process of being sensed.
5. Task affects actuators and in turn is affected.
6. Human operator directly affects task by manipulation.
7. Human operator affects task indirectly through a controls interface, HIS/TIS computers, and actuators. This control interacts with that from within TIS and is filtered or modified.
8. Human operator gets feedback from within HIS, in editing a program, running a planning model, etc.
9. Human operator orients him- or herself relative to control or adjusts control parameters.
10. Human operator orients him- or herself relative to display or adjusts display parameters.

Figure 39.15 Multiloop model of supervisory control.

and emphasizing the symmetry of the system as viewed from top and bottom (human, task) of the hierarchy.

Figure 39.16 is an abbreviated version of Rasmussen's qualitative model referred to above in Section 39.6, showing in particular the nesting of skill-based, rule-based, and knowledge-based behavioral loops. Figure 39.17 extends Rasmussen's model to show various interactions with computer aids having comparable levels of intelligence.

One problem of interest to the supervisor is how often to sample the input, how often to update a control setting, or both, particularly if there is a cost incurred each time he or she does so. Given assumptions on the magnitude distribution and autocorrelation of inputs, a utility function for value of performance resulting from a particular input and particular control action in combination, and a discrete cost of sampling, Sheridan (1970a) showed how an optimal sampling strategy could be derived to maximize expected gain. Sheridan (1976) suggested a framework for how a supervisor equipped with a variety of sensing options and a variety of motor options could try various combinations of these in "thought experiments" or simulations with an "internal model" of the controlled process and utility function. Using Bayes' theorem it is shown how expected utility can be maximized.

The supervisor faces a problem whenever allocating attention between or among different tasks, where for each switch of tasks there is a time penalty in transfer. Typically this penalty is different for different tasks, and possibly involves bodily transportation of him- or herself to different locations. Given relative worths for time spent attending to various tasks, it has been shown (Sheridan, 1970b) that dynamic programming enables the optimal allocation strategy to be established. Moray et al. (1982) applied this model to deciding whether human or computer should control various variables at each succeeding moment. For simpler experimental conditions the model fit the experimental data (subjects acted like utility maximizers), but as task conditions became complex, apparently it did not. Wood and Sheridan (1982) did a similar study where supervisors could select among alternative machines (differing in both rental cost and productivity) to do assigned tasks or do the tasks themselves. Results showed the supervisors to be suboptimal, paying too much attention to costs and too little to productivity, and in some cases using machines when they could have done the tasks more efficiently by themselves. Govindaraj and Rouse (1981) modeled the supervisor's decisions to look away from a continuous task in order to perform or monitor a discrete task.

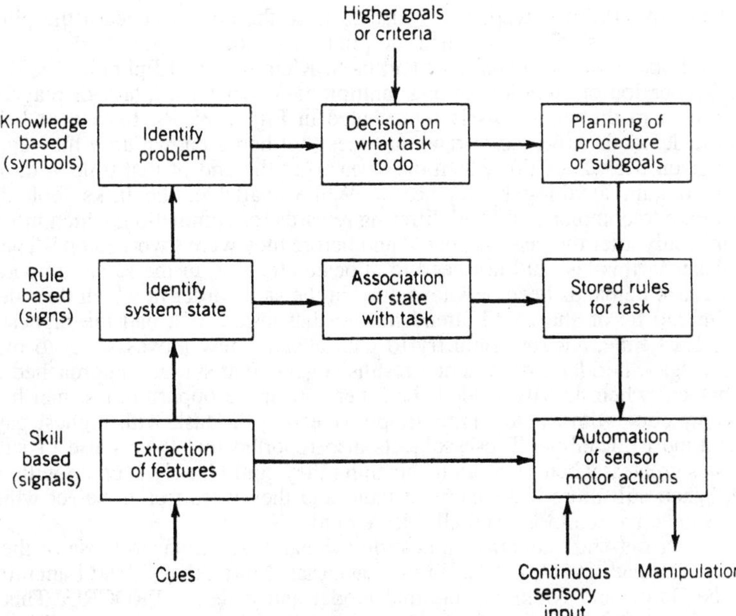

Figure 39.16 Simplification of Rasmussen's qualitative model of human behavior.

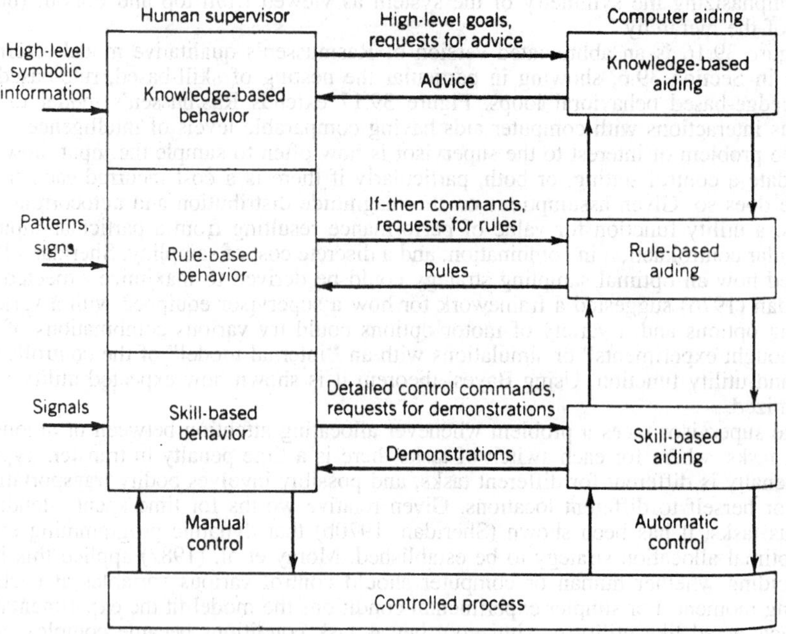

Figure 39.17 Supervisor interactions with computer decision aids at knowledge, rule, and skill level.

Rouse (1977) utilized a queuing theory approach to model whether from moment to moment a task should be assigned to a computer or to the human operator. The allocation criterion was to minimize service time under cost constraints. Results suggested that human–computer "misunderstanding" of one another degraded efficiency more than did limited computer speed. In a related flight simulation study Chu and Rouse (1979) had a computer perform those tasks that had waited in the queue beyond a certain time. Chu, Steeb, and Freedy (1980) extended this idea to have the computer learn the pilot's priorities and later make suggestions when the pilot was under stress.

Tulga and Sheridan (1980) and later Pattipatti, Kleinman, and Ephrath (1983) utilized a model of allocation of attention among multiple task demands, a task displayed on the computer screen to the subject as is represented in Figure 39.18. Instead of being stationary, these demands appeared at random times (not being known until they appeared), existed for given periods of time, then disappeared at the end of that time with no more opportunity to gain anything by an action. While available, the tasks took differing amounts of time to complete, and had differing rewards for completion, which information was available only after the tasks appeared and before they were "worked on." The human decision-maker in this task did not need to allocate attention in the same temporal order in which the task demands become known, nor in the same order in which their deadlines occurred. Instead he or she could attend first to that task which had the highest payoff or took the least time, and/or could try to plan ahead a few moves so as to maximize gains. The Tulga–Sheridan experimental results suggest that subjects approached optimal behavior, which, when heavily loaded (i.e., there are more opportunities than he or she could possibly cope with), amounts to simply selecting the task with highest payoff regardless of time to deadline. These subjects also reported that their sense of subjective workload was greatest when by arduous planning they could barely keep up with all tasks presented. When still more tasks came at them and they were free to select which they should do, subjective workload actually decreased!

Using as a "front-end" attention allocation mechanisms similar to those of the Tulga-Sheridan and Pattipatti et al. models, Baron, Zacharias, Muraldiharan, and Lancraft (1980) extended the Baron and Kleinman optimal model and called it PROCRU. This is diagrammed in Figure 39.19. It was originally built to model crew selection and implemen-

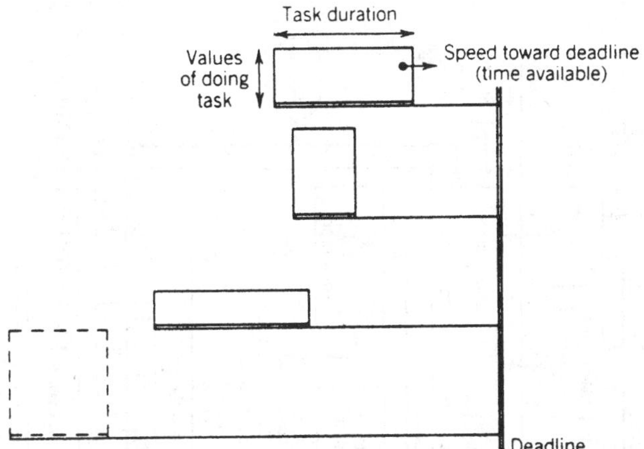

Figure 39.18 Multitask computer display used in Tulga–Sheridan experiment.

tation of procedures in aircraft approach and landing. Optimum decision and control algorithms maximize expected gain for given nominal procedure requests from the ground, aerodynamic disturbances, vehicle dynamics, and objective function.

There are a number of questions that researchers and designers of supervisory control systems must cope with. Among these are (1) how much autonomy is appropriate for the TIC, (2) how much should the TIC tell the HIC, and the HIC tell the human supervisor, and (3) how should responsibilities be allocated among the TIC, HIC, and supervisor (Johannsen, 1981)?

Rouse (1985) concludes his discussion of models of supervisory control with the interesting comments that "perhaps the most important result of the emergence and clashing of models over the past decades has been a shift away form monolithic, computationally overwhelming models to frameworks or categorizations of models, each of which may be quite simple, involving elementary control laws, a few heuristics, or pattern recognition rules. Thus, as knowledge and understanding of supervisory control has grown, researchers have come to realize that neither they nor operators need to approach tasks in such a global, brute-force manner."

The most difficult, and it might even be said impossible, aspect of supervisory control to model is that of setting in goals, conditions, and values. Even though overall goals may be given to actual systems (or given in an experiment) how those are translated into subgoals and conditional statements by human operators remains elusive. The same is true for communicating values (criteria, coefficients of utility, etc.). Although this act of evaluation remains the *sine qua non* of why human participation in system control must remain, there is little prospect for mathematical modeling of this aspect in the near future.

39.9 SOCIAL IMPLICATIONS AND THE FUTURE OF SUPERVISORY CONTROL

No one can predict the future with certainty, but it is ethically mandatory that we predict it as best we can.

One near certainty is that as technology of computers, sensors, and displays improves, supervisory control will become more prevalent. This should occur in two ways: (1) a greater number of semiautomated tasks will be controlled by a single supervisor (a greater number of TIC's will be connected to a single HIC), and (2) the sophistication of cognitive aids, including expert systems for planning, teaching, monitoring, failure detection, and learning, will increase and include more of what we now call "knowledge-based behavior" in the HIC.

Concurrently, laypeople (including those of both corporate and government operations, as well as the general public) should come to understand the potential of supervisory control much better. At the present time the layperson tends to see automation as "all-or-none," and systems as controlled either manually or automatically, with nothing in between. In robotized factories the media tend to focus on the robots, with little mention of design, installation, programming, monitoring, fault detection and diagnosis, mainte-

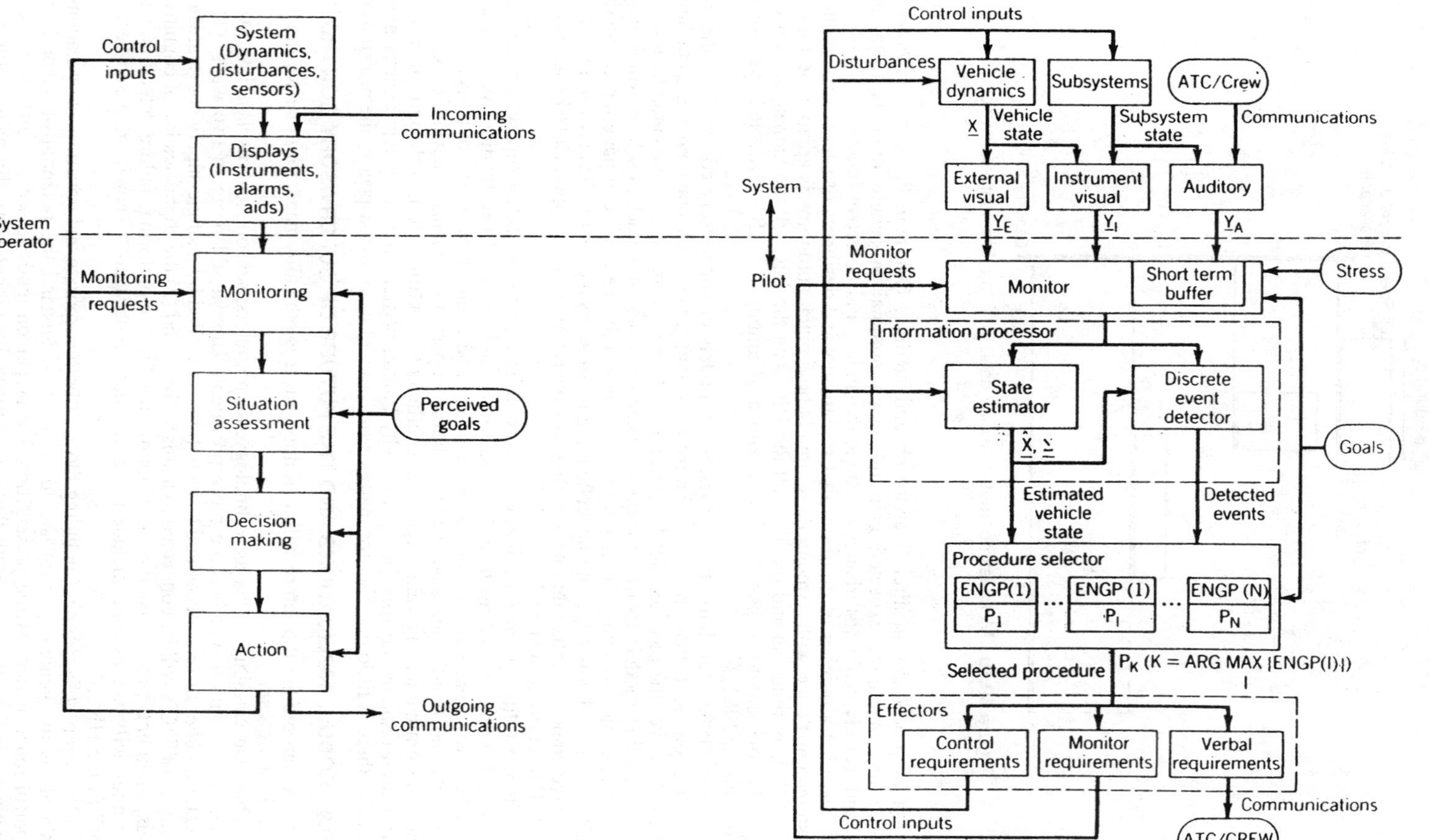

Figure 39.19 PROCRU model of Baron et al. (1980).

nance, and various learning functions that are performed by people. In the space program the same is true—options are seen to be either "automated," "astronaut in EVA," (external vehicle activity) or "astronaut or ground controlling tele-manipulator"—without much appreciation for the potential of supervisory control. These perceptions are bound to change.

In considering the future of supervisory control relative to various degrees of automation, and to the complexity or unpredictability of task situations to be dealt with, a representation such as Figure 39.20 comes to mind. The meaning of the four extremes of this rectangle are quite identifiable. Supervisory control may be considered to be a frontier (line) advancing gradually toward the upper right-hand corner.

The tendency, for obvious reasons, has been to automate what is easiest and to leave the rest to the human. This has sometimes been called the "technological imperative." From one perspective this dignifies the human contribution; from another it may lead to a hodge-podge of partial automation, making the remaining human tasks less coherent and more complex than need be, resulting in overall degradation of system performance (Bainbridge, 1983; Parsons, 1985).

As previously discussed, supervisory control may involve varying degrees of computer aiding on the afferent or incoming side, as well as on the efferent or control execution side. Table 39.1 suggests a scale of "degrees of automation" that separates the afferent (sensing) from the efferent (taking action) and breaks the afferent down into components dealing with (1) experience, (2) sensing present data, (3) interpreting present data, and (4) formulating action alternatives.

"Human centered automation" has become a popular phrase, and therefor it is important to comment on its alternative meanings. Below are 10 alternative meanings (stated in italics) that the author has gleaned form current literature. In every case the meaning must be qualified, as is done by the one or two sentences following each particular meaning of the phrase.

1. *Allocate to the human the tasks best suited to the human, allocate to the automation the tasks best suited to it.* Yes, but for some tasks that are slightly different in each case (not an exact duplication of what was done before) it really is easier to do them manually than to initialize the automation to do them. And at the other end of the spectrum are tasks that require so much skill or art or creativity that it simply is not possible to program a computer to do them.

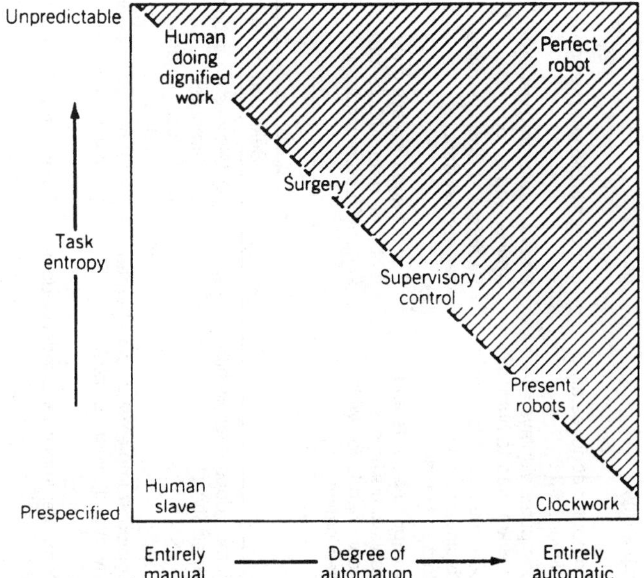

Figure 39.20 Combinations of human and computer control to achieve tasks at various difficulties.

Table 39.1 Degrees of Automation

	Degree of Compute Aiding in Acquiring and Analyzing Data			
	Past Data	Current Data	Situation Assessment	Action Alternatives
0. Human does it all (within capability).	Measures and stores relevant past data.	Observes relevnt current data.	Assess current system state or situation relative to goals.	Conjures up various action alternatives within resource constraints.
1. Computer acquires all relevant data.	Stores all relevant past data.	Measures all relevant current data.	Estimates current state or situation.	Determines all relevant action alternatives.
2. Computer displays all relevant data to human.	Displays relevant past data, trends, and so on.	Displays relevant current data, relates it to past.	Displays current state or situation.	Displays all relevant action alternatives.
3. Computer selects and displays narrow set of data to human.	Selects and displays narrow set of past data.	Selects and displays narrow set of current data.	Assesses and displays current state or situation relative to assume goals.	Determines and displays a few most salient action alternatives.
4. Computer selects and displays a single recommendation with justification.	Selects and displays only that past data sufficient to support recommendations.	Selects and displays only that current data sufficient to support recommendations.	Assesses and displays only limited information sufficient to specify current state relative to assumed goals.	Determines and displays a single recommendation for action.

	Degree of Computer Aiding in Implementation		
	Selection of Action Alternative	Conditions Accompanying Implementation	
0. Human does it all (within capability).	(a) Decides on action independent of computer. (b) Selects an action proposed by computer.	Implements action if, when, and how he or she decides.	
1. Computer implements, with human say over if, when, or how.	(a) What human selects (b) What computer recommened.	(a) Only if and when human approves (b) By a deadline if human does not stop it in time	(a) Done jointly with human participation (b) Done by computer only
2. Computer implements independent of human.	What computer recommended	(a) And necessarily tells human after the fact what it did (b) And tells human after the fact what it did only if human asks (c) And tells human after the fact what it did if it thinks he or she should be told	

2. *Keep the human operator in the decision and control loop.* That is a good idea provided the control tasks are of appropriate bandwidth, attentional demand, etc. that manual control even works.
3. *Maintain the human operator as the final authority over the automation.* Realistically, this is not always the safest solution. It depends on the task context. In nuclear plants, for example, there are safety functions that cannot be entrusted to and should never be overridden by the human operator. Examples have been given previously in the case of aircraft automation.
4. *Make the human operator's job easier, more enjoyable, or more satisfying through friendly automation.* That is fine if operator ease and enjoyment were the primary considerations, and if ease and enjoyment necessarily correlate with operator responsibility and system performance, but often these conditions are not the case.
5. *Empower the human operator to the greatest extent possible through automation.* Again one must remember that operator empowerment is not the same as system performance. Maybe the designer knows best.
6. *Support trust by the human operator.* Trust of the automation by the operator is often a good thing, but not always. Too much trust is just as bad as not enough trust.
7. *Give the operator computer-based advice about everything he or she should want to know.* We now have many examples of where too much information can overwhelm the operator, to the point where performance breaks down—even when the operator originally thought he or she wanted "all" the information.
8. *Engineer the automation to reduce human error and keep response variability to the minimum.* This, unfortunately, is a simplistic view of human error. Taken literally, it reduces the operator to an automaton. Modest, acceptable levels of error and response variability enhance learning.
9. *Make the operator a supervisor of subordinate automatic control system(s).* Though this is a chapter on supervisory control, it must be noted that for some tasks direct manual control may be best.
10. *Achieve the best combination of human and automatic control, where best is defined by explicit system objectives.* In some ideal case, where objectives can reliably be reduced to mathematics, this would be just fine. Unfortunately, prior judgment of what is good and bad in each particular situation is seldom possible for human or machine. Fortunately, judgment of what good and bad in a particular situation *in situ* is almost the essence of what it is to be human.

The bottom line is that proper use of automation depends upon context, which in turn depends upon designer and operator judgment.

I have written elsewhere about the long-term social implications of supervisory control (Sheridan, 1980; Sheridan, Vamos, and Aida, 1983). My concerns might be reviewed here very briefly:

1. *Unemployment.* This is the factor most often considered. More supervisory control means more efficiency, less direct control, fewer jobs.
2. *Desocialization.* Although cockpits and control rooms now require two- to three-person teams, the trend is toward fewer people per team, and eventually one person will be adequate in most installations. Thus cognitive interaction with computers will replace that with other people. As supervisory control systems are interconnected, the computer will mediate more and more interpersonal contact.
3. *Remoteness from the product.* Supervisory control removes people from hands-on interaction with the workpiece or other product. The people become not only separated in space but also desynchronized in time. Their functions or actions no longer correspond to how the product itself is being handled or processed mechanically.
4. *Deskilling.* Skilled workers "promoted" to supervisory controller may resent the transition because of fear that when and if called on to take over and do the job manually, they may not be able to. They may also feel loss of professional identity in losing a skill built up over an entire working life.

5. *Intimidation by higher stakes.* Supervisory control will encourage larger aggregations of equipment, higher speeds, greater complexity, higher costs of capital, and probably greater economic risk if something goes wrong and the supervisor does not take the appropriate action.
6. *Discomfort in the assumption of power.* The human supervisor will be forced to assume more and more ultimate responsibility. Depending on one's personality, this could lead to insensitivity to detail, anxiety about being up to the job requirements, or arrogance.
7. *Technological illiteracy.* Supervisory controllers may lack the technological understanding of how the computer does what it does. They may come to resent this and resent the elite class who do understand.
8. *Mystification.* Human supervisors of computer-based systems could become mystified about the power of the computer, even seeing it as a kind of magic or "big brother" authority figure.
9. *Sense of not being productive.* Since the efficiency and mechanical productivity of a new supervisory control system (perceived as "automated") may far exceed that of an earlier manually controlled system a given person has experienced, that person may come to feel no longer productive as a human being.
10. *Eventual abandonment of responsibility.* As a result of the factors previously described, supervisors may eventually feel they are no longer responsible for what happens; the computers are.

These ten potential negatives may be summarized with a singe word: *alienation.* In short, if human supervisors of the new breed of computer-based systems are not given sufficient familiarization with and feedback from the task, sufficient sense of retaining their old skills, or ways of finding identity in new ones, they may well come to feel alienated. They must be trained to feel comfortable with their new responsibility, must come to understand what the computer does and not be mystified, and must realize that they are ultimately in charge of setting the goals and criteria by which the system operates. If these principles of human factors are incorporated into the design, selection, training, and management, supervisory control has a positive future.

39.10 CONCLUSION

Computer technology, both hard and soft, is driving the human operator to become a supervisor (planner, teacher, monitor, and learner) of automation and an intervener within the automated control loop only for abnormal situations. A number of definitions, models, and problems have been discussed. There is little or no present consensus that any one model characterizes in a satisfactory way all or even very much of supervisory control with sufficient predictive capability to entrust to the designer of such systems. It seem that for the immediate future we are destined to run breathless behind the lead of technology, trying our best to catch up.

REFERENCES

Askey, S. (1995). *Design and Evaluation of Decision Aids for Control of High Speed Trains: Experiments and Model.* Ph.D. Thesis. Cambridge, MA: MIT.

Bainbridge, L. (1983). Ironies of automation. *Automatica, 19,* pp. 775–779.

Baron, S., Zacharias, G., Muraldiharan, R., and Lancraft, R. (1980). PROCRU; a model for analyzing flight crew procedures in approach to landing. *Proceedings of 16th Annual Conference on Manual Control.* Cambridge, MA: MIT, pp. 488–520.

Billings, C. S. (1991). Human-centered aircraft automation: a concept and guideline. NASA Technical Memorandum 103885. Moffett Field, CA: NASA Ames Research Center.

Boehm-Davis, D., Rurry, R., Wiener, E., and Harrison, R. (1983). Human factors of flight deck automation: report on a NASA-industry workshop. *Ergonomics, 267,* pp. 953–961.

Brooks, T. L. (1979) *SUPERMAN: A System for Supervisory Manipulation and the Study of Human Computer Interactions* (SM Thesis). Cambridge, MA: MIT.

Buharali, A., Sheridan, T. B. (1982). Fuzzy set aids for telling a computer how to decide. *Proceedings on IEEE International Conference on Cybernetics and Society.* Seattle, WA, pp. 643–647.

Chu, Y. Y., and Rouse, W. B. (1979). Adaptive allocation of decision making responsibility between human and computer in multi-task situations. *IEEE Transaction in Systems, Man and Cybernetics.* SMC-9, pp. 769–778.

Chu, Y. Y., Steeb, R., and Freedy, A. (1980). Analysis and Modeling of Information Handling Tasks in Supervisory Control of Advanced Aircraft (PATR-1080-80-6). Woodlands, CA: Perceptronics.

Curry, R. E., and Ephrath, A. R. (1977). Monitoring and control of unreliable systems. In T. B. Sheridan and G. Johannsen, Eds., *Monitoring Behavior and Supervisory Control.* New York: Plenum, pp. 193–203.

Curry, R. E., and Nagel, D. C. (1974). Decision behavior changing signal strengths. *Journal of Mathematical Psychology, 14*, pp. 1–24.

Duncan, K. D. (1981). Training for fault diagnosis in industrial process plants. In J. Rasmussen and W. B. Rouse, Eds., *Human Detection and Diagnosis of System Failures.* New York, Plenum, pp. 553–524.

Edwards, E., and Lees, F. (1974). *The Human Operator in Process Control.* London: Taylor and Francis.

Ephrath, A. R., and Young, L. R. (1981). Monitoring vs. Man-in-the-loop detection of aircraft control failures. In J. Rasmussen and W. B. Rouse, Eds., *Human Detection and Diagnosis of System Failures.* New York: Plenum, pp. 143–154.

Falzon, P. (1982). Display structures: computability with the operator's mental representation and reasoning processes. *Proceedings of the Second Annual Conference on Human Decision Making and Manual Control*, Delft, Netherlands. pp. 297–305.

Ferrell, W. R. (1966, October). Delayed force feedback. *Human Factors*, pp. 449–455.

Ferrell, W. R., and Sheridan, T. B. (1967). Supervisory control of remote manipulation. *IEEE Spectrum, 4(10)*, pp. 81–88.

Gai, E. G., and Curry R. E. (1978). Preservation effects in detection tasks with correlated decision intervals. *IEEE Transactions on Systems, Man. And Cybernetics.* SMC-8, pp. 93–110.

Gaines, B. R. (1976). On the complexity of casual models. *IEEE Transactions of Systems, Man and Cybernetics.* SMC-6, pp. 56–59.

Gentner, D., and Stevens, A. L., Eds. (1983). *Mental Models.* Hillsdale: NJ: Erlbaum.

Govindaraj, T., and Rouse, W. B. (1981). Modeling the human controller in environments that include continuous and discrete tasks. *IEEE Transactions on Systems, Man and Cybernetics*, SMC-I, pp. 411–417.

Hart, S. G., and Sheridan, T. B. (1984). Pilot workload, performance, and aircraft control automation. *Proceedings of AGARD Symposium on Human Factors Considerations in High Performance Aircraft.*

Hess, R. A., and McNally B. D. (1986). Automation effect in a multi-loop manual control system. *IEEE Transactions on Systems, Man, Cybernetics* SmC-16. no. 1. New York: IEEE, pp. 111–121.

Hopkin, V. D. (1995). *Human Factors in Air Traffic Control.* London: Taylor and Francis.

Johannsen, G. (1981). Fault management and supervisory control of decentralized systems. In Rasmussen, J. and Rouse, W. B., Eds., *Human Detection and Diagnosis of System Failures.* New York: Plenum Press.

Kim, S. (1997). *Theory of Human Intervention and Design of Human-Computer Interfaces in Supervisory Control: Application to Traffic Incident Management.* PhD Thesis, Cambridge, MA: MIT.

Kosko, B. (1992). *Neural Networks and Fuzzy Systems.* Englewood Ciffs, NJ: Prentice Hall.

Machida, K., Toda, Y., Iwata, T., Kawachi, M., and Nakamura, T. (1988). Development of a graphic simulator augmented teleoperator system for space applications. In *Proceedings of 1988 AIAA Conference on Guidance, Navigation, and Control*, Part I: pp. 358–364.

Moray, N. (1986). Monitoring behavior and supervisory control. In K. Boff, L. Kaufman and J. P. Thomas, Eds., *Handbook of Perception and Human Performance*, vol. 2, New York: John Wiley.

Moray, N., Ed. (1979). *Mental Workload: Its Theory and Measurement.* New York: Plenum Press.

Moray, N. (1982). Subjective mental workload. *Human Factors, 24*, pp. 25–40.

Moray, N. (1986). Monitoring Behavior and supervisory control. In K. Boff, Ed., *Handbook of Perception.* New York: John Wiley.

Moray, N. (1997). Models of models of . . . mental models. In T. Sheridan and T. van Lunteren, T., Eds., *Perspectives on the Human Controller*, Hillsdale: NJ: Erblaum.

Moray, N., Sanderson, P., Shiff, B., Jackson, R., Kennedy, S. and Ting, L. (1982). A model and experiment for the allocation of man and computer in supervisory control. In *Proceedings of IEEE International Conference on Cybernetics and Society*, pp. 354–358.

Morris, N. M., and Rouse, W. B. (1985). The effects of type of knowledge upon human problem solving in a process control task. *IEEE Trans. Systems, Man and Cybernetics*, SMC–15: pp. 698–707.

Mouloua, M., and Parasuaman, R. Eds. (1994). *Human Performance in Automated Systems; Recent Research and Trends.* Hillsdale, NJ: Erlbaum.

Niemala, R., and Krendel, E. S. (1974). Detection of a change in plant dynamics of a man-machine system. In *Proceedings of 10th Annual Conference on Manual Control*, pp. 97–112.

Norman, D. A. (1981). Categorization of action slips. *Psychological Review, 88*: pp. 1–15.

Noyes, M. V., and Sheridan, T. B. (1984). A novel predictor for telemanipulation through a time delay. In *Proceedings of Annual Conference on Manual Control*, Moffett Field, CA, NASA Ames Research Center.

Parasuraman, R. (1987). Human-computer monitoring. *Human Factors, 29*, pp. 695–706.

Park, J. H. (1991). *Supervisory Control of Robot Manipulators for Gross Motions.* PhD Thesis. MIT, August.

Parsons, H. M. (1985). Automation and the individual: comprehensive and comparative views. *Human Factors*, Vol.27, No. 1, ppl 99–111.

Pattipatti, K. R., Kleinman, D. L. and Ephrath, A. R. (1983). A dynamic decision model of human task selection performance. *IEEE Transactions on Systems, Man and Cybernetics.* SMC-13; pp. 145–166.

Rasmussen, J. (1976). Outlines of a hybrid model of the process plant operator. In T.B. Sheridan and G. Johannsen, Eds., *Monitoring Behavior and Supervisory Control*, New York: Plenum.

Rasmussen, J. (1982). Human errors: A taxonomy for describing human malfunction in industrial installations. *J. Occupational Accidents* 4: pp. 311–335.

Rassmussen, J., and Rouse, W. B., Eds. (1981). *Human Detection and Diagnosis of System Failures.* New York: Plenum.

Reason, J. T., and Mycielska, K. (1982). *Absent Minded? The Psychology of Mental Lapses and Everyday Errors.* Eaglewood Cliffs, NJ: Prentice Hall.

Rouse, W. B. (1974). The effect of display format on the human paerception of statistics. *Proceedings of the 10th Annual Conference on Manual Control.* Wright Patterson AFB, OH.

Rouse, W. B. (1977). Human–computer interaction in multi–task situations. *IEEE Trans. System, Man and Cybernetics.* SMC–7. *5*, pp. 384–392.

Rouse, W. B. (1985). Supervisory control and display systems, In J. Zeidner, Ed., *Human Productivity Enhancement.* New York: Praeger.

Rouse, W. B., and Morris, N. M. (1984). *On Looking into the Black Box: Prospects and Limits in the Search for Mental Models.* Norcross, GA: Search Technology.

Rouse, W. B., and Rouse, S. H. (1983). Analysis and classification of human error. *IEEE Trans. System, Man and Cybernetics.* SMC–13, *4*, pp. 539–599.

Ruffel–Smith, H. P. A. (1979). *A Simulator Study of the Interaction of Pilot Workload with Error, Vigilance and Decisions.* NASA TM–78482. Moffett Field, CA, NASA Ames Research Center.

Sarter, N., and Woods, D. D. (1994). Decomposing automation: autonomy, authority, observability and perceived animacy. In M. Mouloua and R. Parasuaman, Eds., *Human Performance in Automated Systems; Recent Research and Trends.* Hillsdale, NJ: Erlbaum.

Sheridan, T. B. (1992). *Telerobotics, Automation, and Human Supervisory Control.* Cambridge, MA: MIT Press.

Sheridan, T. B. (1960). Human metacontrol. In *Proceedings of Annual Conference on Manual Control.* Wright Patterson AFB, OH.

Sheridan, T. B. (1970a). On how often the supervisor should sample. *IEEE Trans. Systems Science and Cybernetics.* SSC–6, pp. 140–145.

Sheridan, T. B. (1970b). Optimum allocation of personal presence. *IEEE Trans. Human Factors in Electronics* HFE–10, pp. 242–249.

Sheridan, T. B. (1976a). Toward a general model of supervisory control. In T.B. Sheridan and G. Johannsen, Eds., *Monitoring Behavior and Supervisory Control.* New York: Plenum.

Sheridan, T. B. (1980). Computer control and human alienation. *Technology Review*, *83*, October, pp. 60–73.

Sheridan, T. B. (1983). Measuring, modeling and augmenting reliability of man–machine systems. *Automatica, 19.*

Sheridan, T. B. (1984a). *Interaction of Human Cognitive Models and Computer-based Models in Supervisory Control.* Cambridge, MA: MIT Man-Machne Systems Lab.

Sheridan, T. B. (1984b). Supervisory control of remote manipulators, vehicles and dynamic processes. In W. B. Rouse, ed., *Advances in Man–Machine Systems Research*, Vol. 1, New York: JAI Press.

Sheridan, T. B., and Ferrell, W. R. (1974). *Man–Machine Systems.* Cambridge, MA: MIT Press.

Sheridan, T. B., Fischoff, B., Posner, M., and Pew, R.W. (1983). (Eds.) *Supervisory control systems. In Research Needs in Human Factors.* Washington, D.C.: National Academy Press.

Sheridan, T. B. and Hennessy, R. T., Eds. (1984). *Research and Modeling of Supervisory Control Behavior.* National Research Council, Committee on Human Factors, Washington, DC: National Academy Press.

Sheridan, T. B., and Johannsen, G., Eds. (1976). *Monitoring Behavior and Supervisory Control.* New York: Plenum.

Sheridan, T. B., Vamos, T., and Aida, S. (1983). Adapting automation to man, culture and society. *Automatica, 19(6)*, pp. 605–612.

Taylor, F. W. (1922). *Principles of Scientific Management.* New York: Harper Brothers.

Tsach, U., Sheridan,T. B., and Buharali, A. (1983). Failure detection and location in process control: integrating a new model-based technique with other methods. In *Proceedings of American Control Conference.* San Francisco, June, pp. 22–24.

Tulga, M. K. and Sheridan, T. B. (1980). Dynamic decisions and workload in multi–task supervisory control. *IEEE Trans. on Systems, Man and Cybernetics.* SMC–10, *5*, pp. 217–231.

Wickens, C. D., and Kessell, C. (1979). The effects of participatory model and task workload on the detection of dynamic system failures. *IEEE Trans. on Systems, Man and Cybernetics*. SMC–9, pp. 24–34.

Wickens, C. D., and Kessell, C. (1981). Failure detection in dynamic systems, In J. Rassmussen and W.B. Rouse (Eds.) 1981. *Human Detection and Diagnosis of System Failures*. New York: Plenum.

Wiener, E. L. (1988). Cockpit automation. In E. L. Wiener and D. Nagel, Eds. *Human Factors in Aviation*. San Antonio, TX: Academic Press pp. 433–461.

Wiener, E. L., and Curry, R. E. (1980). Flight deck automatic: promises and problems. *Ergonomics*, *23*, 995–1011.

Williges, R. C., and Wierwille, W. W. (1979). Behavioral measures of aircrew mental workload. *Human Factors*, *21*, 549–574.

Wood, W. and Sheridan, T. B. (1982). The use of machine aids in dynamic multi-task environments: a comparison of an optimal model to human behavior. *Proceedings of IEEE International Conference on Cybernetics and Society*. Seattle, pp. 668–672.

Yoerger, D. (1982). *Supervisory Control of Underwater Telemanipulators: Design and Experiment* (Ph.D. Thesis). Cambridge, MA: MIT.

Zadeh, L. A. (1984). Making computers think like people. *IEEE Spectrum*, *21*, 26–32.

CHAPTER 40
COGNITIVE MODELING

Ray Eberts
School of Industrial Engineering
Purdue University
West Lafayette, IN 47907-1287 USA

The cognitive modeling of products has represented an important advance in human factors research and laboratory techniques. Similar to other engineering models, cognitive modeling can be used to make predictions about the usability of products. As a laboratory technique in the development of products, cognitive modeling can be used in conjunction with or as an alternative to other laboratory procedures such as informal and formal experimentation.

Cognitive modeling has some distinct advantages when compared to experimentation. Experimentation is limited in that at least a mock-up of the product is needed before experiments can be run. The data collected would be more valid with the actual product or a working prototype. For cognitive modeling, the task analyst does not even need a mock-up but can perform the modeling with only a sketchy description of the product. This can save time and money because important aspects about usability can be decided before incurring the expense of mock-ups, prototypes, or the actual product.

Experiments are often limited to the use of novice subjects. The time needed and expense to run an experiment from first use to expert use is prohibitive, especially for complicated products such as computer software. Cognitive modeling can be used to model the usability of products with different expertise levels for the users; the same modeling technique can be used to predict usability for novices and experts. Once again, using these techniques can save money. Other advantages of cognitive modeling will be considered throughout this chapter.

Although many cognitive modeling techniques are available, this chapter concentrates on procedural models. These models assume that people learn to use products by generating rules for use and then "run" their models, when interacting with the product, by sequencing through the set of rules. Although the specifics of other modeling techniques not considered in this chapter may be different from those discussed, the general approach for the others will be similar. The models discussed in this chapter are the Model Human Processor; goals, operations, methods, and selection rules (GOMS); (basic model, keystrokes-level model, and Model-UT); natural GOMS language (NGOMSL); production systems; and grammars (Backus-Naur Form, BNF, and task-action grammar, TAG). To integrate the discussion of these different models, a single task, modeling computer menu systems, will be modeled for each of the models. This chapter provides enough detail to understand the modeling techniques and do some simple models. The original references, provided throughout this chapter, should be consulted for more detail. Also, Eberts (1994) provides more detail on all these modeling techniques.

40.1 MODEL HUMAN PROCESSOR

The Model Human Processor of Card, Moran, and Newell (1983, 1986) follows a long line of human information processing models developed in cognitive psychology. These models specified the discrete stages that information would flow through in order to be used and processed. Card et al. enhanced these models by quantifying the memory and processing parameters, which could be used to make quantitative predictions for people interacting with machines. Use of the model indicates the implicit assumptions of the discrete stages models of human information processing. Basically, the user of this model must accept the assumptions that processing occurs in stages, that processing in a stage is completed before information is passed to the next stage, and that information flows in a sequential manner from one stage to the next. If we know the stages that the information must pass through and we know the timing characteristics of those individual stages, then we can add together the timing values to determine an estimate for the total task time. When times need to be determined for the complicated tasks, the tasks need only be broken down into the individual components, of which we have the relevant timing characteristics. The total time is determined by adding together the times of the individual components.

40.1.1 Parameters of the Model Human Processor

Following the tradition of human information processing models, the Model Human Processor is composed of three systems or stages: the perceptual, cognitive, and motor systems. The perceptual and cognitive systems have memories associated with them; the motor system does not have a memory. Several parameters can be used to quantitatively describe these three systems and, thus, make predictions about the timing of information through these systems. The Model Human Processor and the parameters are shown in Figure 40.1.

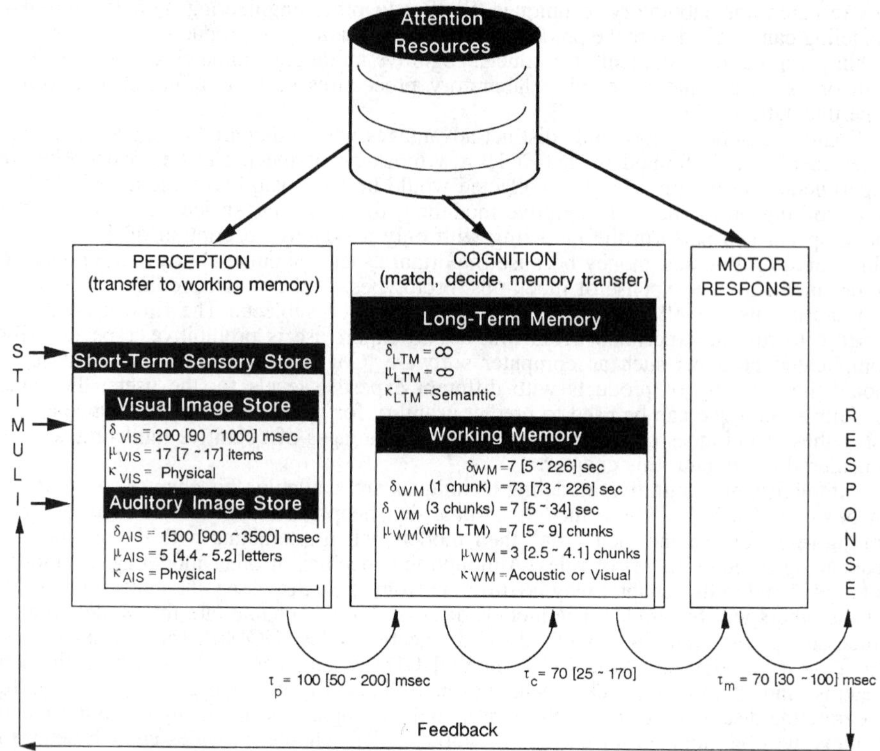

Figure 40.1 Illustration of the human information-processing model with the parameters of the Model Human Processor.

40.1.1.1 Cycle Time Parameters

One of the most important parameters for determining the timing characteristics of a task is the cycle time, τ, for the three systems. This is the time that it takes for the information to be processed through the stages. While a stage is processing the information, it cannot process any other information. The values for the cycle times for the three systems—perceptual, cognitive, and motor—are as follows:

$$\tau_p = 100\ [50 \sim 200]$$
$$\tau_c = 70\ [25 \sim 170]$$
$$\tau_m = 70\ [30 \sim 100]$$

The first number represents the most plausible value for the parameter, or the average value. The numbers in brackets represent low and high values for these parameters. These are not really ranges, in the statistical sense, but represent plausible values. Card et al. (1983) refer to the first value as the "Middle Man" value, the low value as the "Fast Man" value, and the high value as the "Slow Man" value.

40.1.1.2 Memory Parameters

The other parameters of the Model Human Processor are those associated with the memories of the perceptual and cognitive systems. Three parameters describe the processing of information for these memories: the code (κ), the decay time (δ), and the capacity (μ). The code is a nonquantifiable variable referring to the form in which the information is stored in the memory. The decay time is the half-life, which is the amount of time that the information will remain in memory with the probability of retrieval greater than 50%. The capacity of memory is the number of items that can be stored in memory.

For human–computer interaction tasks, the perceptual system has two memories relevant to these tasks: the visual image store (VIS) and the auditory image store (AIS). A definition of these perceptual processors is that information is unprocessed or it is not transformed in any way. Therefore, the code, κ, of both memories is a physical code wherein information is an unidentified, nonsymbolic analog to the actual stimulus. This is represented as follows:

$$\kappa_{VIS} = \text{physical}$$
$$\kappa_{AIS} = \text{physical}$$

The decay time of these perceptual memories is:

$$\delta_{VIS} = 200\ [90 \sim 1000]\ \text{msec}$$
$$\delta_{AIS} = 1500\ [900 \sim 3500]\ \text{msec}$$

Again, the numbers in the brackets represent the plausible ranges of values and the number outside the brackets represents the most likely value. All of these values for the Model Human Processor are taken from experimental studies and the numbers represent summaries of experiments to determine these values (e.g., Averbach and Coriell, 1961, was used to determine the capacity of VIS).

The capacity of the perceptual memories are:

$$\mu_{VIS} = 17\ [7 \sim 17]\ \text{letters}$$
$$\mu_{AIS} = 5\ [4.4 \sim 6.2]\ \text{letters}$$

The "letters" units are used because this is the form in which the data were collected for the experiments. Letters can refer to chunks of information. With these parameters, the characteristics of the perceptual memories are completely specified.

The cognitive memories consist of a working memory and a long-term memory. The working memory (WM) holds information under current consideration and constitutes the general registers of the cognitive processor. Information is coded in two ways in working memory:

$$\kappa_{WM} = \text{acoustic or visual}$$

The decay time (seconds) for working memory is dependent on the number of chunks that must be retained.

$$\delta_{WM}\ (1\ \text{chunk}) = 73\ [73 \sim 226]\ \text{s}$$
$$\delta_{WM}\ (3\ \text{chunks}) = 7\ [5 \sim 34]\ \text{s}$$

Information can be retained in working memory indefinitely if the person can rehearse the items; the decay will occur for unrehearsed items.

Finally, the capacity of working memory can be described as follows:

$$\mu_{WM} = 7\ [5 \sim 9]\ \text{chunks}$$

This value corresponds to Miller's (1956) famous paper on the "magic number plus or minus 2." Subsequent work has found that the capacity of working memory may not be as large as was considered at that time. Working memory only has this capacity if it can be augmented by long-term memory; which may be the case in only some of the applications. Without long-term memory augmentation, the capacity of working memory can be described as follows:

$$\mu_{WM} = 3[2.5 \sim 4]\ \text{chunks}$$

Long-term memory (LTM) is the mass of available memory and is, in most cases conceptualized as a network of related chunks. It is information that retains very little from what is originally perceived, and so it has a coding according to its meaning, or semantics, instead of any physical characteristics:

$$\kappa_{LTM} = \text{semantic}$$

The information does not decay from long-term memory and so the decay time is:

$$\delta_{LTM} = \text{infinite}$$

Finally, the capacity is unlimited:

$$\mu_{LTM} = \text{infinite}$$

The capacity of long-term memory is considered to be infinite. In other words, you do not have to worry about learning too much information. Still, people know that trying to learn too much all at once is difficult, so how can capacity be unlimited? The problem occurs again with retrieval cues. When learning something, the person must be careful to develop retrieval cues with the right kinds of links so that the information can be retrieved later. If an item is learned, and then a new item is learned with the same link, then it will be difficult to retrieve the earlier learned item.

40.1.2 Examples of Applying the Model Human Processor

Consider an example for the design of menu displays for interactive computer programs. An important issue in this area is how to organize the items on the screen. Should all the items be displayed on one screen or should only a few items be displayed on one and then, depending on the choice, more screens be accessed? This issue is considered in terms of the depth or breadth of the menu items. With all items on one screen the display has high breadth; with items embedded on several screens, the display has depth.

For this example, consider two kinds of organizations. For the high-breadth layout, all 16 menu items are on one screen (see Figure 40.2). For the high-depth layout, four items are on one screen and for any menu choice that is made, the user would then have another choice of four items (see Figure 40.3). In both layouts, the menu item has a number or letter associated with it so that the user would enter a number or letter for the menu choice. Assume that the user knows the menu item needed and is experienced in using this kind of display. We also assume that the user is an expert typist so that key searches on the keyboard are not needed.

First, consider the example for the high-breadth display. The following steps would be needed:

```
 1.  Menu item a
 2.  Menu item b
 3.  Menu item c
 4.  Menu item d
 5.  Menu item e
 6.  Menu item f
 7.  Menu item g
 8.  Menu item h
 9.  Menu item i
10.  Menu item j
11.  Menu item k
12.  Menu item l
13.  Menu item m
14.  Menu item n
15.  Menu item o
16.  Menu item p

Enter the number of the menu item:  ______
```

Figure 40.2 An example of the high-breadth display. In this case, all 16 items are displayed on the one screen. To choose a menu item, the user must read through the items until the desired one is found, and then enter the number of that item.

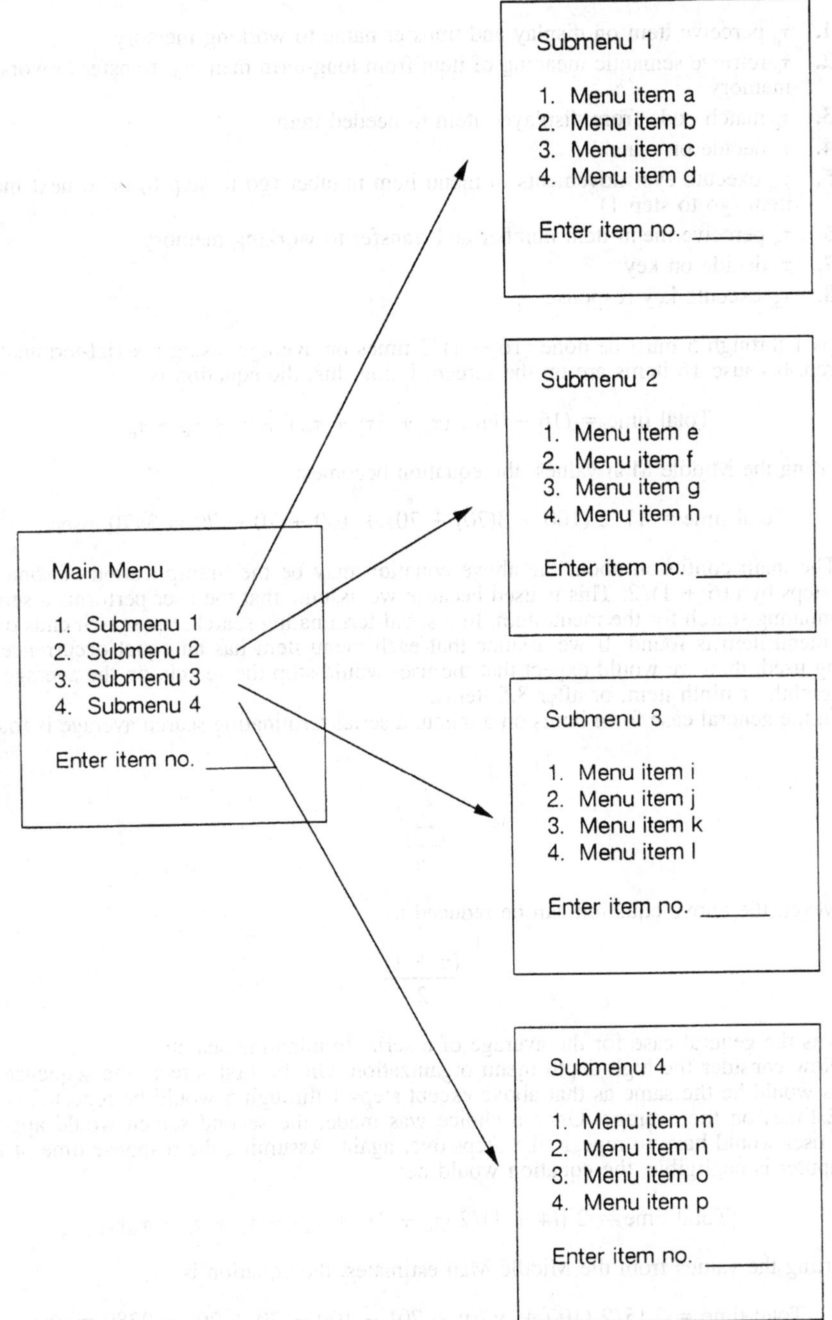

Figure 40.3 An example of the high-depth display. In this case, all 16 items are divided between two levels. On the first screen, four items will appear. The user must make a choice by typing the number of the menu item. Depending on the choice for the first screen, next one of the four screens will appear. If the user chose item 1 on the first screen, then the menu with items a–d will appear on the second screen. If the user chose item 2 on the first screen, then items e–h will appear on the second screen, and so forth. For the high-depth display, the user only has to search four items on a screen, but two screens have to be searched.

1. τ_p perceive item on display and transfer name to working memory
2. τ_c retrieve semantic meaning of item from long-term memory, transfer to working memory
3. τ_c match code from displayed item to needed item
4. τ_c decide on match
5. τ_m execute eye movements to menu item number (go to step 6) or to next menu item (go to step 1)
6. τ_p perceive menu item number and transfer to working memory
7. τ_c decide on key
8. τ_m execute key response

Steps 1 through 5 must be done (16 + 1)/2 times on average, using a serial-terminating search, because 16 items are on the screen. Using this, the equation is:

$$\text{Total time} = (16 + 1)/2\ (\tau_p + 3\tau_c + \tau_m) + \tau_p + \tau_c + \tau_m$$

Inserting the Middle Man values, the equation becomes:

$$\text{Total time} = 17/2\ (100 + 3(70) + 70) + 100 + 70 + 70 = 3470 \text{ msec}$$

The main confusion about the above equation may be the multiplication of some of the steps by (16 + 1)/2. This is used because we assume that the user performs a serial-terminating search for the menu item. In a serial-terminating search, the search ends once the menu item is found. If we assume that each menu item has an equal occurrence of being used, then we would expect that the user would stop the search, on the average, at the eighth or ninth item, or after 8.5 items.

In the general case for n items on a menu, a serial-terminating search average is found by

$$\frac{\sum_{i=1}^{n} i}{n}$$

However, the above equation can be reduced to

$$\frac{(n + 1)}{2}$$

This is the general case for the average of a serial terminating search.

Now consider the high-depth menu organization. On the first screen, the sequence of steps would be the same as that above except steps 1 through 5 would be repeated (4 + 1)/2 times on the average. Once a choice was made, the second screen would appear. The user would have to repeat all 8 steps over again. Assuming the response time of the computer is negligible, the equation would be:

$$\text{Total time} = 2\ [(4 + 1)/2\ (\tau_p + 3\tau_c + \tau_m) + \tau_p + \tau_c + \tau_m]$$

Inserting the values from the Middle Man estimates, the equation is:

$$\text{Total time} = 2\ [5/2\ (100 + 3(70) + 70] + 100 + 70 + 70) = 2380 \text{ msec}$$

Using these calculations, we would predict that the high-depth organization would be faster than the high-breadth organization by more than a second.

40.1.3 Summary

Card et al. show that the Model Human Processor is very accurate at making estimates, especially for simple tasks. As with all of the models discussed in this chapter, different researchers may analyze the task differently, resulting in different parameterizations of the same task. The assumptions, such as the skill level of the operators, are a very important consideration also, which may result in very different analyses. Any such application of the Model Human Processor must be preceded by a careful task analysis and possibly field interviews of the operators or users to determine which of the assumptions are valid.

Several analysis techniques are considered in the next sections. The complexity of the task should be matched to the scope of the analysis problem. The Model Human Processor is good for the simple problems; the analysis techniques that follow are good for the more complicated problem. In the following models, the individual components of the task are not at the very atomistic level of cognitive processes, but are at the higher level of keystrokes or goals and intentions.

40.2 GOMS

The class of models known as GOMS (goals, operators, methods, and selection rules) is based on the assumption that interacting with machines is similar to solving a problem. The problem must be decomposed into subproblems, the subproblems decomposed into subsubproblems, and so on. The problem solver must determine the goals and subgoals for attacking the problems. Usually, the means of solving the problem are reduced to a certain set of methods or operators that can be used in the solution.

Human–machine interaction tasks have often been described as problem-solving tasks. In such a model, Norman (1986) would say that the user would sequence through the following seven stages in order to solve the problem posed by a human–computer interaction task. To provide an example for these seven stages, consider the problem of a person editing a memo to make it more readable. The seven stages are:

- Establishing the goal (to reorder the paragraphs in the memo to make it more readable)
- Forming the intention (the user may intend to move paragraph 1 behind paragraph 2)
- Specifying the action sequence (this particular word processor may require that the user highlight paragraph 1, use a menu to "cut" the paragraph, move the cursor behind paragraph 2, and then use the menu to "paste" paragraph 1)
- Executing the action sequence (execute the above steps)
- Perceiving the system state (if the word processor shows the changes as they occur, then the user perceives the screen and reads the changed memo)
- Interpreting the state (the user determines the consequences of the changes)
- Evaluating the system state with respect to the goals and intentions (the user evaluates whether the reordered memo makes it more readable)

These stages do not have to be performed sequentially. For example, multiple intentions may be formed before an action sequence is determined. If the evaluation determines that the goal has not been satisfied, then new intentions are constructed.

Formal predictive modeling techniques have also been devised based upon cognitive problem-solving behavior. One of these models, GOMS (Card, Moran, and Newell, 1983), will be discussed in this section. A derivative of the GOMS model, NGOMSL, which is described by Kieras (1988), will be discussed in the next section.

GOMS was developed by Card et al. (1983) based upon previous work on problem solving by Newell and Simon (1972). In the Newell and Simon book, the concepts of goals and a goal stack are used to describe how humans solve problems. Their model of problem solving consists of decomposing a primary goal into a hierarchical tree of subgoals with branches of lengths that depend on the degree of subgoal decomposition. At the end nodes of the tree are subgoals to which elementary information processes can be applied. Newell and Simon define elementary information processes to be elemental components of a problem-solving system from which all problem-solving methods are constructed. They use a stack to store the subgoal tree. Two stack operations, "pushing" and "popping," control goal decomposition and application of elementary information processes. The tree is decomposed in a depth-first fashion; thus only parts of the tree are on the stack at one time. The stack lengthens when a goal or subgoal is split and shortens when elementary information processes are performed. If every subgoal at every level can be split into other subgoals or elementary information processes, then eventually the stack will become empty, at which time the attainment of the primary goal is achieved.

The GOMS model is actually three models: the basic model, a keystroke model, and the Model-UT. All are discussed in this chapter. These models can be used to explain user–machine interaction cognitive behavior at several levels of detail. The basic GOMS model outlines the cognitive behavior of a user by decomposing the problem into subgoal and goal stacks. A variant of the basic GOMS model, called the Keystrokes-Level model, was designed to identify observable behavior such as keystrokes and mouse manipulations. This level of the model is especially useful for making specific time predictions based upon simplifying assumptions. Finally, the Model-UT is used in task analyses for more complicated tasks.

40.2.1 The Basic GOMS Model

Card et al. (1983) based the logic of their GOMS model upon what they called the "rationality principle." The rationality principle states that a person acts so as to attain goals through rational action, given the structure of the task and inputs of information and bounded by limitations on knowledge and processing ability:

$$\text{Goals} + \text{Task} + \text{Operators} + \text{Inputs} + \text{Knowledge Process Limits} \rightarrow \text{Behavior}$$

This states that users rationally attain their goals. It also implies that, similar to the Model Human Processor, a small number of information-processing operators underlie the detailed behavior of a particular user. The user's behavior can then be described through a sequence of these operators. If each of these operators can be assigned a time value, then the time the user requires to act is the sum of the times of the individual operators.

40.2.1.1 Components

The components of the GOMS model will be discussed through an analysis of each letter of its name. The G of GOMS represents the goals of the task. Card et al. (1983) state that a goal is a symbolic structure defining a state of affairs to be achieved and determining a set of possible methods to accomplish it. The goals can serve many purposes. A goal can serve as a memory point that can be returned to upon failure or error. A goal contains information about what is desired, about the methods available, and about what has already been tried.

The O of GOMS represents the operators. Card et al. (1983) define "operators" as elementary perceptual, motor, or cognitive acts whose execution is necessary to change any aspects of the user's mental state or to affect the task environment. The grain of analysis is dependent on the particular task analysis desired. The perceptual, cognitive, and motor cycles of the Model Human Processor can be considered a very low level or fine grain of analysis, which usually would be too fine a grain of analysis to be considered for the GOMS analysis. Usually, the grain of analysis corresponds to observable behaviors such as keystrokes, mouse movements, or user movements (e.g., looking at the screen). User behavior consists of the serial execution of operators.

The M of GOMS represents the methods. The methods describe procedures for accomplishing a goal in terms of operators and other goals. The user usually has a choice of several different methods. A good example is positioning the cursor on the screen. In most word processors and text editors, the user would have several choices for cursor control: arrow keys, a scroll bar on the side of the screen, a search or find function, mouse operations, or commands such as "u" or "d." The user should choose the method that is the most efficient for performing the task. If the desired position in the text is far away from the current position, then the user would choose a different method than if the desired position was close to the current position.

The final letter, S, of the GOMS model represents selection rules. To completely specify the user behavior, the choice of methods must be represented in the model. Selection rules are essentially the control structure of the model. This control structure is specified in terms of if–then rules. To use the previous example of cursor control, the methods would be specified through if–then rules. As an example, IF the desired position and the current position are both on the screen, THEN the arrow key method would be used. IF the desired position is on a different screen than the current position, THEN the search command method would be used. These selection rules could be refined based upon the task.

40.2.1.2 Example

The GOMS model attempts to capture the goals of the user, how these goals are decomposed into subgoals and subsubgoals, and how the observable behaviors, such as keystrokes, are used to satisfy these goals. To further explain the operation of the GOMS model, a high-breadth/high-depth menu display example will be used. Consider first the high-breadth display, where all menu items are displayed on one screen. The overall goal of this task would be to carry out some kind of command sequence; the command considered with the Model Human Processor example may be only one unit of this command sequence. We will consider this example in the context of its larger goal. For instance, the user may be in a word processing program and wish to choose the size (e.g., 12 points), font (e.g., Times Roman), and style of the characters (e.g., bold) thus requiring three sequential interactions with the menu displays to satisfy the overall goal of setting the type style. Therefore, the highest level goal implicit to the particular task is to perform a command sequence. This can be represented as follows:

GOAL: PERFORM-COMMAND-SEQUENCE

The single menu item choice from the menu example is one unit of this high-level goal and each of these units must be considered separately.

The next goal is to perform the unit task of the command, or:

GOAL: PERFORM-COMMAND-UNIT-TASK

To perform this unit task, the user must first determine which command needs to be performed, this is referred to as acquiring the unit task. Then the user must execute the responses by interacting with the computer that carries out the actions needed; this is referred to as executing the task. The latest goal can be decomposed into the following two subgoals:

GOAL: ACQUIRE-UNIT-TASK
GOAL: EXECUTE-UNIT-TASK

Each of these goals must now be satisfied.

To satisfy the acquisition of the goal, the user must determine the next command. This is done by looking at the screen and finding the location of the command on the screen. These actions no longer constitute a subgoal, but they are a series of specific operators used to satisfy the last goal on the stack. The following operators would be needed:

GET-NEXT-COMMAND

The operators are usually characterized by some kind of external action by the user that can be observed; sometimes the operators are internal cognitive actions that cannot be observed. Most of the execution of the operator of getting the command would be unobserved cognitive activities. These activities would correspond to the cognitive processes outlined for this task Model Human Processor: reading the menu items on the screen, matching the menu item to the desired command, and then deciding on the response. For the GOMS model, however, we do not analyze this task at such an atomistic level. All of these cognitive activities can be combined into just the one operator of getting the command. Although this internal cognitive activity cannot be observed, the external action of looking at the screen can be observed and so would be associated with this operator.

Once the command is obtained, the acquire task subgoal is satisfied and the execute task subgoal can be attended to. The execution of the task requires that the user enter the one or two digits corresponding to the desired command and then verify that the correct menu item has been entered. Since the menu items number from 1 through 16 for this particular example, the user has the possibility for using two different methods for accomplishing this: entering a single key or entering a string of keys. The selection of the appropriate method depends on the menu item that is to be entered. If the menu item number is between 1 and 9, then the key method will be used. If the menu item number is between 10 and 16, then the string method will be used. As can be seen, the previous two sentences easily could be converted into selection rules as follows:

IF menu item number is between 1 and 9, THEN use KEY METHOD
IF menu item number is between 10 and 16, THEN use STRING METHOD

The operators to complete the task can be represented as follows for the key method:

USE KEY METHOD
VERIFY ENTRY

For the string method, the operators to complete the task can be represented as follows:

USE STRING METHOD
VERIFY ENTRY

When this execution of the task is completed, the user is finished with the menu interaction task. The steps for the high-breadth example is shown in Figure 40.4. The high-depth procedure is similar and is shown in Figure 40.5.

40.2.1.3 Execution Time Predictions

Relative time differences in executing the task using the two different menu displays can be considered by counting the number of steps for each display (the step number is shown in the first column of Figures 40.4 and 40.5). A step occurs when a goal is "pushed" on the stack or "popped" off a stack or an operator is applied.

As with the Model Human Processor on the menu displays, the number of steps needed for performing the task can be determined in the general case. Assuming a serial-terminating search and using n as the number of menu items on the display, then the number of menu items which must be searched is $(n + 1)/2$. In the high-breadth display, two steps (8 and 9 displayed in Figure 40.4) are needed for this search process. Therefore, on average, $(2(16 + 1))/2$ steps would be needed for this search, or 17 steps. Since the last step before the search began was step 7, the step after the search is concluded will be step 25. On step 25, either the key method or the string method can be used to enter the menu item number, depending on the number of digits in this menu item. The task is finished after 33 steps.

The analysis for the number of steps, in the general case, can also be performed for the high-depth menu display of Figure 40.5. The whole process would be finished on step 24.

For the general case, the high-breadth display should require 33 steps and the high-depth display should require 24 steps. Card et al. (1983) do not make specific predictions for how long each step should take, but we can assume that each step would require about an equal amount of time. Just by comparing the relative numbers of steps and

Steps	Contents of Goal Stack	Operator Executed	Action
1	DO-CS		
2	DO-CS,DO COMM-UT		
3	DO-CS,DO COMM-UT,ACQ-UT		
4	DO-CS,DO COMM-UT,ACQ-UT	GET-NEXT-COMM	access LTM
5	DO-CS,DO COMM-UT		
6	DO-CS,DO COMM-UT,EXE-UT		
7	DO-CS,DO COMM-UT,EXE-UT,MENU-COMM		
8	DO-CS,DO COMM-UT,EXE-UT,MENU-COMM	READ-SCREEN	look at screen
9	DO-CS,DO COMM-UT,EXE-UT,MENU-COMM	USE MATCH METHOD	WM match
. . .	Steps 8 and 9 are repeated, on the average, (n + 1)/2 times, or 8.5 times		
25	DO-CS,DO COMM-UT,EXE-UT,MENU-COMM	USE KEY METHOD or USE STRING METHOD	enter digits from keys
26	DO-CS,DO COMM-UT,EXE-UT,		
27	DO-CS,DO COMM-UT,EXE-UT,VERIFY COMM		
28	DO-CS,DO COMM-UT,EXE-UT,VERIFY COMM	READ-SCREEN	look at screen, hit ENTER
29	DO-CS,DO COMM-UT,EXE-UT,VERIFY COMM	USE DECIDE METHOD	WM decide
30	DO-CS,DO COMM-UT,EXE-UT,VERIFY COMM	USE KEY METHOD	hit ENTER key
31	DO-CS,DO COMM-UT,EXE-UT		
32	DO-CS,DO COMM-UT		
33	DO-CS		

Figure 40.4 A representation of the goal stack operations for the high-breadth menu display. (CS = command sequence; COMM = command; UT = unit task; ACQ = acquire and EXE = execute).

Steps	Contents of Goal Stack	Operator Executed	Action
1	DO-CS		
2	DO-CS,DO COMM-UT		
3	DO-CS,DO COMM-UT,ACQ-UT		
4	DO-CS,DO COMM-UT,ACQ-UT	GET-NEXT-COMM	look at screen
5	DO-CS,DO COMM-UT		
6	DO-CS,DO COMM-UT,EXE-UT		
7	DO-CS,DO COMM-UT,EXE-UT,MENU-COMM		
8	DO-CS,DO COMM-UT,EXE-UT,MENU-COMM	READ-SCREEN	look at screen
9	DO-CS,DO COMM-UT,EXE-UT,MENU-COMM	USE MATCH METHOD	WM match
...	Steps 8 and 9 are repeated, on the average, (n + 1)/2 times, or 2.5 times		
13	DO-CS,DO COMM-UT,EXE-UT,MENU-COMM	USE KEY METHOD	enter digit from keys
14	DO-CS,DO COMM-UT,EXE-UT		
15	DO-CS,DO COMM-UT,EXE-UT,MENU-CAT		
16	DO-CS,DO COMM-UT,EXE-UT,MENU-CAT	READ-SCREEN	look at screen
17	DO-CS,DO COMM-UT,EXE-UT,MANU-CAT	USE MATCH METHOD	WM match
...	Steps 16 and 17 are repeated, on the average, (n + 1)/2 times, or 2.5 times		
21	DO-CS,DO COMM-UT,EXE-UT,MENU-CAT	USE KEY METHOD	enter digit from keys
22	DO-CS,DO COMM-UT,EXE-UT		
23	DO-CS,DO COMM-UT		
24	DO-CS		

Figure 40.5 A representation of the goal stack operations for the high-depth menu display. (CS = command sequence; COMM = command; UT = unit task; ACQ = acquire; EXE = execute; and CAT = category).

assuming that each step requires about the same amount of time, the high-depth display should be performed faster than the high-breadth display. Using a 100-msec estimate for each step, we could say, however, that the high-depth would be about 0.9 s faster than the high-breadth, a conclusion that is similar to that of the Model Human Processor. In that analysis, the high-depth display was predicted to be faster by about half a second.

The basic GOMS model, considered above, is best for making qualitative predictions about differences between the tasks. The two other variants of the GOMS model, Keystrokes-Level model and the Model-UT (for unit task), considered later in this chapter, were designed by Card et al. to make quantitative predictions for the tasks. The Keystrokes-Level model makes these predictions by quantifying values for observable operators; the Model-UT predicts the times by establishing average times for each unit task through experimentation.

40.2.1.4 Error Predictions

The GOMS model can be used to determine the methods and operators needed for performing a task. One could assume that the more methods and operators which must be learned, the more chances for making errors. Thus, the number of errors should be positively related to the number of methods and operators to be learned.

The number of methods and operators to be learned is not equivalent to the number of steps in the process as used for the execution time prediction. The methods and operators which had to be learned are listed in Table 40.1 for the high-breadth display and in Table 40.2 for the high-depth display. The high-breadth display requires that six methods and operators be learned; the high-depth display requires that only four methods and operators be learned. Based on this, we would predict the high-depth display to exhibit fewer errors.

40.2.1.5 Learnability

The GOMS model was not designed to measure learnability explicitly, but one can see how the model could be applied to predict the learning time. The more methods and operators to learn, the longer the learning time. Like errors, we would expect a relationship between the number of methods and operators with learnability. By examining Tables

Table 40.1 Methods and Operators That Have To Be Learned for the High-Breadth Display

List of Methods and Operators
GET-NEXT-COMMAND
READ-SCREEN
USE MATCH METHOD
USE KEY METHOD
USE STRING METHOD
USE DECIDE METHOD

40.1 and 40.2, we can see that the high-depth display should be faster to learn. The NGOMSL and production system models, considered in later sections, have developed formal techniques for predicting learning time.

40.2.2 Keystroke-Level Model of GOMS

One of the problems of using the GOMS model is that obtaining time predictions for the model could be difficult and time consuming. For the basic model, no general technique of making time predictions was used. As seen in the next section for the Model-UT, the time predictions are obtained after taking verbal protocols of users, collecting keystroke data, and then determining average times for the unit tasks. This is a detailed procedure that may overwhelm any time savings of GOMS, because of the modeling of performance, over the usual method of experimentally determining the differences between the system. The Keystroke-Level model of GOMS was designed by Card et al. (1983) to obtain time predictions easier. Card et al. (1983) referred to this class of models as engineering models because it represents more of a practical design tool.

The Keystroke-Level model analyzes only the observable behavior such as keystrokes and mouse movements. It does not analyze the nonobservable behavior such as the time for acquiring goals. Not estimating the acquisition time could remove a large chunk of the actual time to perform a task. As an example, Card et al. (1983) estimate, based on experimental studies, that acquisition times can be 2 to 3 s per unit task if the task is already defined (such as having a marked-up manuscript). Acquisition times could be 5 to 30 s if the task needs to be generated in the user's mind. In addition, the Keystroke-Level model assumes error-free performance because of the difficulty of predicting the errors and determining the times of errors. The Keystroke-Level model is used to predict the times given a method; it does not predict the methods.

Card et al. base this level of the model on six operators: *K*, a keystroke or mouse button push; *P*, pointing to a target on a display with a mouse or some other pointing device; *D*, moving the mouse to draw a set of straight-line segments; *H*, moving the hands from the mouse to the keyboard; *M*, mental preparation for doing an operation; and *R*, system response time. Based upon empirical work, Card et al. chose values for operators (shown in Table 40.3). *K* depends on the experience of the typist. *P* and *H* are invariant. *D* depends on the number of line segments needed (N_D) and the length of the line segments (l_D). *M* is the only operator that is not observable and a useful estimate is 1.35 s.

Table 40.2 Methods and Operators That Have To Be Learned for the High-Depth Display

List of Methods and Operators
GET-NEXT-COMMAND
READ-SCREEN
USE MATCH METHOD
USE KEY METHOD

Table 40.3 Time Estimates for the Operators from the Keystroke-Level Model of GOMS

Operator	Description	Time (s)
K	Time varies with typing skill	
	Best typist (135 wpm)	0.08
	Good typist (90 wpm)	0.12
	Average skilled (55 wpm)	0.20
	Average unskilled (40 wpm)	0.28
	Typing random letters	0.50
	Typing complex codes	0.75
	Worst typist	1.20
P	Point with mouse	1.10
H	Home hands on keyboard	0.40
$D(n_D, l_D)$	Draw segment	$0.9n_D + 0.16l_D$
M	Mentally prepare	1.35

R is not listed because it depends on the particular computer system and the load on the system. Since M is the only operator that is questionable as to when to apply it, Card et al. (1983) formulated rules for its application (see Table 40.4).

The execution time for a task is found by stringing the operators together for the task, assigning the parameter values to the operator, and then summing the times. Different systems can be compared by examining the final summation of the times. The system with the least amount of time needed for the task is assumed to be the best. Card et al. (1983) validated these times with experiments on expert users and found the predictions to be close to the actual values. The model was especially accurate in predicting the qualitative relationships among the different systems tested.

40.2.2.1 Examples

As an example of the application of the Keystroke-Level model of GOMS, consider the high-breadth and high-depth menu interaction tasks that have been used throughout this chapter. In the high-breadth display, the user would be required to search the 16 items on the screen to find the one desired, to enter the one- or two-digit number corresponding to the desired menu item, and then to hit the return key. Using the operators from the Keystroke-Level model, this would require an M operator for searching the display, one or two K operators for entering the menu item number, and then another K operator for the enter key. To apply the model, a few assumptions must be made. First, we will assume that the user is an average skilled typist. Next, we will assume that the 16 menu items have an equal probability of use. The operators needed for the task are:

$$M + K(\text{first digit}) + 0.44K(\text{second digit}) + K(\text{Enter})$$

Table 40.4 Rules For When to Apply the M Operator For the Keystroke Level Model

Rule	Example
When an operation is fully anticipated in another, M can be chunked with the other operator.	Pointing with the mouse and then hitting the mouse button.
An obvious syntactic unit constitutes a chunk when it must be typed out in full.	When using the DIR command (for a directory), the M operator follows the typing of DIR instead of having an M operator after each keystroke.
User will bundle redundant terminators into a single chunk.	If a command is followed by ESC ENTER, the M operator will occur after ENTER.
A terminator of a constant-string chunk will be assimilated into a chunk.	If an ENTER is needed after the command DIR, then the M operator will be after ENTER.

Since all menu items have the same probability of use, the second digit will be used for items 10 through 16 and these 7 items would occur with a probability of 7/16 or 0.44. Applying the time values for the operators, the predicted time for this task is:

$$1.35 + 0.20 + 0.44(0.20) + 0.20 = 1.84 \text{ s}$$

This time can be compared to the time prediction of this task using the Model Human Processor, which was 3.47 s. The discrepancy in the predictions occurs because the Model Human Processor explicitly predicts the search time but the Keystroke-Level model only provides an estimate of 1.35 s for this and other mental tasks. In addition, the time of the Keystroke-Level analysis is shorter because it only predicts the execution time and not the acquisition time.

A time can also be predicted for the high-depth display. For this task, the user must think about the menu item to be entered, enter the single digit from the first menu, and then enter the single digit from the second menu. In this example, we will assume that the expert is an expert user so that an M operator is not needed between the first and second keystrokes. In other words, the user would know where to look for the menu item without searching for it on the screen. We assume that the computer is very fast and that the response time, R, between the two menus is negligible or, essentially zero. As in the previous example, we also assume that the user is an average skilled typist. The operators for this task are:

$$M + K(\text{digit}) + K(\text{digit})$$

Substituting the values of the operators, the time to perform the task is:

$$1.35 + 0.2 + 0.2 = 1.75 \text{ s}$$

This time can also be compared to the time prediction of this task using the Model Human Processor which was 2.975 s. Once again, the Keystroke-Level predicted a time less than the Model Human Processor.

It is interesting to note that the two models make the same qualitative prediction: The high-depth display should be faster than the high-breadth display. The particular time predictions are different as are the differences between the predicted times for the two menu operations.

40.2.2.2 Design Revisions

The GOMS models can be used to revise a computer design to make it easier to execute, faster to perform, and easier to learn for a user. Card et al. (1983) provided some suggestions for how their models could be used. Some of the important suggestions follow.

1. A basic unit of analysis is the method. When performing a task analysis, all the methods should be determined and written down. The methods should then be examined to eliminate particularly long or awkward methods.
2. An emphasis in these models has been placed on the expert. Make the performance of experts more efficient by eliminating operators or combining operators.
3. Each interface design will have different methods for performing the same task. Design the set of alternative methods for a task so that the rule for selecting each alternative is clear for the user and easy to apply.
4. Finally, the interface should be easy to learn. A good technique for learnability is to first design a few general-purpose methods which are easy to learn. A novice would learn these to begin with. Then design alternatives for the user based upon the general purpose methods. Only when the user has learned the general purpose methods can knowledge then be extended to the alternatives. In this way, novices can use the system quickly without knowing too much, and they can continue to learn and become more expert by building on this basic knowledge.

40.2.3 Model-UT

In some cases, the Keystroke-Level model may provide an inappropriate level of analysis. It is a good model if interface designs can be specified in terms of the keystrokes. At the conceptual levels of designs, however, specifying such detail would be inappropriate, and probably not needed. A more appropriate level would be at the unit task level. When discussing the basic GOMS models, unit tasks were examined. Thus, the Model-UT can be used estimate task times by specifying the unit tasks.

The unit task may be an appropriate level of analysis because of the assumption for a unit task that the user's behavior is highly integrated and the dependencies between the unit tasks are minimal. Recall that a major assumption of the GOMS model was that different segments of behavior can be seen to be composed of the same few units differently combined. It may be possible to define these units sufficiently independently of each other that the time required by a unit in isolation is a good approximation to the time it requires as part of a sequence. The time prediction would consist of first finding the times for the units (t_{UT}), analyzing the task to determine which unit tasks would occur and how many times within the total task each would occur (n_{UT}), and then summing all of these products together. Thus, for each unit task, the product $t_{UT}\ n_{UT}$ would be determined and all these products for each unit task would be totaled to determine the time of the total task. If the system response time is large, then this factor would have to be included in the total.

Determining what constitutes a unit task could be difficult. It must be a meaningful unit, it must be highly integrated, and it must be independent of other unit tasks. Many of these decisions are subjective, and so the choice of unit tasks is a subjective matter. Usually a unit task is typified by keystrokes or other types of behavior that are chunked together. If an experiment can be performed measuring the pauses between behaviors, then the choice of unit tasks can be confirmed or disconfirmed (e.g., Robertson and Black, 1986).

The time estimates for a unit task can come from several sources; the source used depends on the availability of information and the amount of time available to perform the analysis. The most exact method to determine the times would be to determine experimentally times for a particular system. This may be impossible especially if such an analysis is being used for a conceptualized system, one that has not yet been designed or implemented. In this case, the times for unit tasks can be determined from other designs that may correspond closely to the design being conceptualized.

40.2.4 Conclusions

The GOMS models are a family of models that can be used for very detailed analyses, such as when applying the Model-UT, or for quick global analyses for predicting execution times, such as when applying the Keystroke-Level model. The applications of these models depends greatly on the skill of the task analyzer. Each person analyzing a task may come up with a slightly different variant of the model. This is not too much of a problem when comparing different systems, as long as the task analyzer has been consistent in how the model has been applied to the different systems. Some of the later modeling techniques, such as the NGOMSL model or the production system models, attempt to solve this problem. The model has been very influential in human–computer interaction research. Some of the other applications of the GOMS models, and the tests of the model, are considered below.

Several examples of validation of the GOMS models were presented in the Card et al. (1983) book ranging from text editing to CAD design tasks. They found that the model could predict the methods used by the user 80–90% of the time on a text-editing task and it predicted the operators used 80–100% of the time for the same task. For a CAD task, the model predicted an execution time of 1192 s, whereas the observed execution time was 1028 s.

The predictions from the model were very close to the actual times. These should be read with caution, though, because they depend somewhat on the skill of the task analyzer. For predicting the methods and operators, the task analyzer would have to examine the task and formulate the methods and operators for the task. The model only provides very limited procedural information for mapping the task to the methods and operators. When the high predictions for the methods and operators are espoused, this prediction means that most of the times the subjects in the experiments have been rational and have not randomly chosen methods and operators. This would probably only hold true for novices; an expert would have a difficult time predicting all the methods and operators for a novice.

The prediction for the execution times only considered the times to execute the unit tasks; not the time to acquire the unit tasks. The execution time could only be a small part of the total time, but some might argue that this would be the important component of the task. The acquisition times could vary widely. Without the acquisition times, the execution time forms kind of ceiling for how fast the task can be performed under optimal conditions. The close correspondence of the predictions to the actual times is due, in part, to the fact that the unit tasks were determined from experimental data, and then the times for each individual unit task were strung together to find a total execution time.

The GOMS models have been very important in trying to understand how a user interacts cognitively with interactive software and in quantifying aspects of the interaction even before the software is prototyped. It has been criticized as being too vague in how to apply it (Kieras, 1988) and only applicable to error-free performance (Carroll and Olson, 1988). GOMS has provided a good start; more work is being performed to refine and proceduralize these techniques.

40.3 NGOMSL

NGOMSL, the Natural GOMS Language, was developed by Kieras (1988) to enable the GOMS-like model and the resulting task analysis to be more specific. NGOMSL is a proposed procedural method. To analyze a task using the NGOMSL procedure, the interaction of the user with a computer is described in a computer programming–like language. The activities of the user are described in terms similar to the subroutines of computer programming languages. The subroutines, in turn, have decision statements (if–then statements), flow of control statements (goto), and memory storage and retrieval. Having experience writing computer programs is advantageous for using the NGOMSL task analysis.

Just like GOMS, NGOMSL decomposes a task into goals, operators, methods, and selection rules. A goal is something that a user tries to accomplish and is described as an action–object pair in the form <verb noun>. As an example, the goal of "delete word" would be such an action-object pair.

Operators are actions that a user executes and can be external operators, which are observable, or can be mental operators, which are not observable. Examples of external operators are to press a key or to move a mouse. External operators can also include perceptual operations such as scanning a page. Mental operators must be inferred by the task analyst; examples include recalling information from long-term memory, actively retaining information in memory, and making decisions. In many ways, operators are very similar to goals; the distinction is subjective on the part of the task analyst. The main difference is that a goal is something to be accomplished, whereas an operator is something that is executed.

A method is a sequence of steps that accomplishes a goal. A step in a method consists typically of either external or mental operators. Each of the statements and the steps correspond to a specific production rule (see the next section) so that cognitive complexity can be determined directly from these steps.

Finally, the selection rules are needed if more than one method is available to accomplish a goal. Selection rules are the same as that for the GOMS model. The information needed for the selection can be formulated in if–then rules for selecting the appropriate method based upon the context.

40.3.1 Structures and Statements of NGOMSL

The basic structure of the NGOMSL analysis is the method. A method must be specified for accomplishing each of the goals of the task. Accomplishing the method is specified through the steps needed. Three possibilities for this method structure are shown in Figure 40.6 (from Kieras, 1988). In this notation, the word in brackets refers to variables that can change depending on the situation. The first statement always states the name of the method in terms of the goal or subgoal that it accomplishes. The steps are operators, goals that must be accomplished through other methods, or the "report goal accomplished." If a step only has an operator, then this represents a low-level operator or primitive at the lowest possible level corresponding to an internal operator or an external operator. This operator is executed, and the step cannot be decomposed any further. Acceptable operators will be explained later. If a step is a goal that must be accomplished, then the method for doing this must be specified in the steps of another method somewhere in the task analysis. The final step in the method is to report the goal accomplished. This does not have to be the step at the very bottom; depending on the flow of control in the method this step could be located physically elsewhere.

The other structure of the NGOMSL model is in terms of selection rule sets. Figure 40.7, from Kieras (1988), shows the general structure for a selection rule set. The selection rules are used when the user has a choice of methods to accomplish the goal. The "if,"

Method Structures

Method to accomplish goal of <goal description>

Step 1. <operator>. . .
Step 2. <operator>. . .
. . .
Step n. Report goal accomplished

Method to accomplish goal of <goal description>

Step 1. <operator>
Step 2. <operator>
. . .
Step k. Accomplish the goal of <subgoal description>
. . .
Step n. Report goal accomplished

Method to accomplish goal of <subgoal description>

Step 1. <operator>
Step 2. <operator>
. . .
Step k. Accomplish the goal of <subsubgoal description>
. . .
Step n. Report goal accomplished
. . .

Figure 40.6 The three structures for a method. (From Kieras, 1988.)

or conditional, part of the rules specify the conditions that must be satisfied for applying the method to accomplish the goal specified in the "then" part of the rule. The selection rules must be written so that only one of the conditions of the ifs is true. The selection rule set can consist of as many rules as necessary for specifying the options. A selection rule always ends with "Report goal accomplished."

The acceptable operators are shown in Figure 40.8. Any of these operators could be substituted in the positions labeled <operator>. The mental primitives for flow of control are analogous to flow of control statements in computer programming languages. "Accomplish the goal of <goal description>" is analogous to the CALL statement for accessing a subroutine. The "Report goal accomplished" operator is analogous to a RETURN statement that returns the control back to the calling method or, analogously, the calling subroutine. The Decide operator determines the truth of the operator(s) after the "if" and performs the operator after the "then" part if the conditional has been true. In the second Decide operator, if the conditional was false, then the Else operator is performed. Finally, a goto command can transfer control to another step within the method. As in structured programming, this kind of control should be avoided as much as possible because the flow of control is difficult to determine with too many gotos.

A very important part of the NGOMSL description of a task is the specification of the memory requirements for performing it. The memory storage and retrieval operators spec-

Selection Rules

Selection rule set for goal of <general goal description>

If <condition> Then accomplish goal of <specific goal description>

If <condition> Then accomplish goal of <specific goal description>

Report goal accomplished

Figure 40.7 General sturcture for a selection rule set. (From Kieras, 1988.)

Operators

Mental primitives for Flow of Control

Accomplish goal of <goal description>
Report goal accomplished
Decide: If <operator. . .> Then <operator>
Decide: If <operator. . .> Then <operator> Else <operator>
Goto Step <number>

Memory Storage and Retrieval

Recall that <WM-object-description>
Retain that <WM-object-description>
Forget that <WM-object-description>
Retrieve-LTM that <LTM-object-description>

Primitive External Operators

Home-hand to mouse
Press-key <key name>
Type-in <string of characters>
Move-cursor to <target coordinates>
Find-cursor-is-at <returned cursor coordinates>
Find-menu-item <menu-item-description>

Analyst Defined Mental Operators

Get-from-task <name>
Verify-result
Get-next-edit-location

Figure 40.8 Acceptable operators for NGOMSL. (From Kieras, 1988.)

ify the operations of working memory (WM) and long-term memory (LTM). WM and LTM have the same cognitive parameters as that explained by the Model Human Processor. From WM, the user can recall, retain, and forget items in this memory. Recall means to fetch from WM and retain means to store the item in WM. Forget means that the item is no longer needed and so can be forgotten. These memory operators are very important for determining the mental workload of the task. Any kind of cognitive processing on information can only occur in WM, so the user may first have to retrieve some information from LTM before it can be manipulated.

The primitive external operators refer to operators that can be observed. This set of operators may change depending on the particular system being modeled. As an example, some systems may not have a mouse so the first operator, homing the hand to the mouse, would not be needed. Some systems may require more or different kinds of operators to define their capabilities. As an example, a system may incorporate a touchscreen, in which case the operators for this device could be specified.

Finally, the last set of operators refers to mental operators defined by the task analyst. These are mental operations too complicated to be defined in the NGOMSL description. The task analyst has much discretion in the kinds of operators that can be included in this set. Kieras (1988) has suggested the set in the figure. In some cases, this set may not be specific enough. In particular, Verify result may be defined as a goal instead of an operator, so a method would need to be described for doing this.

40.3.2 Example

Kieras (1988) suggests that NGOMSL can be applied in the following steps. The technique should be applied top-down by first describing the top-level goal. The top-level goal is accomplished by a method that includes a series of steps of high-level operators (these can be goals themselves). Each step, or operator, in the high-level goal needs a method in which to accomplish it. The method for accomplishing each step or operator is then specified. The process is continued in this manner until the operators are composed of primitives. A primitive is usually some elementary process such as a button press or a cognitive process. The primitive level can change depending on the needs for the task analysis. For example, the following task analyses of high-breadth and high-depth menus carries the task analysis to a low level for the primitives by specifying the memory, match, and decision operators for choosing a menu item. This need not be done in all task analyses. Kieras (1988) provides an operator, called Find-menu-item (see Figure 40.8) which could subsume all the primitive menu operators listed previously. Since two kinds of menus are being compared to determine which is best, a detailed analysis, with low-level or fine-grain primitives, is required. The differences between the menus may only be captured in that detailed analysis. If the focus of the task was not on the menus, but a menu method was a relatively insignificant component of the task, then the higher level primitive of Find-menu-item could be used to describe the menu operations. The grain of analysis, and the level of the primitives, depends upon the purpose of the task analysis.

The NGOMSL representation of the high-breadth display is shown in Figure 40.9 and the high-depth display is shown in Figure 40.10. For the high-breadth display, the first goal is to perform all the steps in the command sequence. To accomplish this, five steps must be performed. If a step is a primitive, then no methods need be specified for that step. If the step is a nonprimitive, then the method for the step must be described fully. In the first method, accomplishing the goal of performing the command sequence, step 1 is a primitive, step 3 is a decision step, step 4 is a control step, and step 2 is a nonprimitive

NGOMSL: High-Breadth Display

(1a) Method to accomplish goal of executing command sequence

Step 1. Retrieve-LTM that current item in command sequence is MENU-COMMAND
Step 2. Accomplish goal of performing the command unit task
Step 3. Decide: If more commands in command sequence, then Goto 1
Step 4. Report goal accomplished

(2a) Method to accomplish goal of performing the command unit task

Step 1. Recall MENU-COMMAND and accomplish goal of locating MENU-COMMAND
Step 2. Recall N and accomplish goal of enter the menu item number
Step 3. Recall N and N' and accomplish goal of verify the entry
Step 4. Forget MENU-COMMAND and report goal accomplished

(3a) Method to accomplish goal of locating MENU-COMMAND

Step 1. ReadScreen next command line and retain as MENU-ITEM
Step 2. Decide: if MENU-COMMAND is different from MENU-ITEM then forget MENU-ITEM and Goto 1
Step 3. Retain that N is number of MENU-ITEM
Step 4. Forget MENU-ITEM and report goal accomplished

Selection rule set for goal of enter the menu item number

If N is between 1 and 9, then accomplish goal of do key-method
If N is 10 or above, then accomplish goal of do string-method
Report goal accomplished

(4a) Method to accomplish goal of do key-method

Step 1. Press-key N
Step 2. ReadScreen and retain feedback as N'
Step 3. Report goal accomplished

(5a) Method to accomplish goal of do string-method

Step 1. Type-in N
Step 2. ReadScreen and retain feedback as N'
Step 3. Report goal accomplished

Figure 40.9 A representation of the high-breadth menu display using NGOMSL.

(6a) Method to accomplish the goal of verify the entry

Step 1. Decide: if N and N' do not match, then forget N' and accomplish goal of deleting entry
Step 2. Press-key ENTER
Step 3. Forget N and N'
Step 4. Report goal accomplished

(7a) Method for accomplishing goal of deleting entry

Step 1. Press-key DELETE
Step 2. Decide: if more digits in entry space on screen, Goto 1
Step 3. Forget N', recall N, and accomplish the goal of enter the menu item number
Step 4. Report goal accomplished

Figure 40.9 *(Continued)*.

NGOMSL: High-Depth Display

(1b) Method to accomplish goal of executing command sequence

Step 1. Retrieve-LTM that current item in command sequence is MENU-COMMAND
Step 2. Accomplish goal of performing the command unit task
Step 3. Decide: If more commands in command sequence, then Goto 1
Step 4. Report goal accomplished

(2b) Method to accomplish goal of performing the command unit task

Step 1. Accomplish goal of issuing MENU-CATEGORY
Step 2. Accomplish goal of issuing MENU-COMMAND
Step 3. Report goal accomplished

(3b) Method to accomplish goal of issuing MENU-CATEGORY

Step 1. Recall MENU-COMMAND and accomplish goal of locating MENU-CATEGORY
Step 2. Recall N and Press-key N
Step 3. Forget N and MENU-CATEGORY
Step 4. Report goal accomplished

(4b) Method to accomplish goal of issuing MENU-COMMAND

Step 1. Recall MENU-COMMAND and accomplish goal of locating MENU-COMMAND
Step 2. Recall N and Press-key N
Step 3. Forget N and MENU-COMMAND
Step 4. Report goal accomplished

(5b) Method to accomplish goal of locating MENU-CATEGORY

Step 1. Retrieve-LTM that the category of MENU-COMMAND is MENU-CATEGORY
Step 2. ReadScreen next command line and retain as MENU-ITEM
Step 3. Decide: if MENU-ITEM is different from MENU-CATEGORY then forget MENU-ITEM and goto 2
Step 4. Retain that N is number of MENU-ITEM
Step 5. Forget MENU-ITEM and report goal accomplished

(6b) Method to accomplish goal of locating MENU-COMMAND

Step 1. ReadScreen next command line and retain as MENU-ITEM
Step 2. Decide: if MENU-ITEM is different from MENU-COMMAND then forget MENU-ITEM and goto 1
Step 3. Retain that N is number of MENU-ITEM
Step 4. Forget MENU-ITEM and report goal accomplished

Figure 40.10 A representation of the high-depth menu display using NGOMSL.

step. Step 2 is the only one in which methods must be described. The step 2 method is described by the second method in the figure. Each of the steps of this method must be primitives, decision steps, control steps, or described by other methods. This continues until all the steps end in primitives.

Kieras (1988) suggests that even after the NGOMSL representation has been completed, the representation should be examined to make sure that it is complete and the division of steps to methods is plausible. After completing a draft of the NGOMSL analysis, the complete task analysis should be checked for consistency and conformance to guidelines. Kieras (1988) states that the following should be checked:

1. Check on method detail and length. The length of the method should probably be no more than five steps. If more than five steps, it should be divided into other operators. The method detail should have only one accomplish-goal operator to a step.
2. Check on consistent assumptions about user's expertise with regard to the number of operators in a step. The rules for the relationship of operators to user expertise is not very clear. Some broad guidelines can be specified. For ordinary or novice users, no more than one external primitive operator should occur in a step. For expert or experienced users, several external primitive operators can be used in a single step if the sequence of external primitive operators is often performed without any decisions or subgoals involved.
3. Identify high-level operators and check that each high-level operator corresponds to a natural goal.
4. Check for consistency of terminology and usage with already defined operators.

The NGOMSL task analyses shown in Figures 40.9 and 40.10 have already gone through several revisions before inclusion in this chapter. The models shown in the figures can now be used to make predictions about user performance.

40.3.3 Applications of NGOMSL

Kieras (1988) presents several ways in which the NGOMSL model can be applied: estimating execution time; estimating learning times; design revisions; estimating the mental workload of a task; and analyzing the documentation. The procedures for each of these will be reviewed.

40.3.3.1 Execution Times

Execution time could be dependent on the time to execute the operators and the number of cognitive steps involved. The formula used by Kieras (1988), based upon experimentation and modeling of the results, is:

Execution time =
  NGOMSL statement time
  + Primitive external operator time
  + Analyst-defined mental operator time
  + System response time
NGOMSL statement time =
  Number of NGOMSL statements executed · 0.1 s
Primitive external operator time =
  Total of times for external operators
Analyst-defined mental operator time =
  Total of times for mental operators defined by the analyst
System response time = Total time when user is idle

To determine the estimated times, a specific task instance must be used because each specific task may use different statements. Let's assume that the user is searching for a menu item found in the 14th position.

To estimate the times, we need to determine the number of NGOMSL statements, the Primitive external operator time, the Analyst-defined mental operator time, and the System response time. A detailed description of the 62 statements that must be processed for this task can be found in Eberts (1994).

From the execution time equation, the NGOMSL statement time is determined by taking the number of NGOMSL statements executed times 0.1 s. For the 62 statements executed, the time taken is 6.2 s. The 0.1 s time is similar to the time needed for a cognitive processor (τ_c) from the Model Human Processor.

Kieras (1988) says that good estimators for the operators are the primitive operator values used by Card et al. (1983) in their Keystroke-Level model. In this particular example, only three external operators occur: typing the two digits (14) of the menu item along with the ENTER key. Recalling the values for keystrokes from Table 40.3 for an average skilled typist (0.20 sec), the total Primitive external operator time is 0.60 s.

If Analyst-defined mental operators are included, then these times would have to be determined individually through experimentation or, in the absence of experimentation, values from the Keystroke-Level model of GOMS or the Model Human Processor values can be used. In this particular case, one Analyst-defined mental operator, ReadScreen, is used 15 times to read the 14 menu items and then the menu item number (14) that is typed in. Since this is a relatively primitive process, the perceptual processor (τ_p) from the Model Human Processor can provide a good estimate (100 msec from the Middle Man Model) for this time instead of having to go through the trouble of running an experiment. Since the ReadScreen occurs 15 times, the Analyst-defined mental operator time is 1.5 s.

As done in the previous sections, we will assume that the System response time is negligible or equal to 0. Since we are not analyzing the task for any particular system, it is difficult to predict the system response time. If the menu displays were being compared for a particular system, then these times could be determined exactly.

Totaling the execution time, it is:

$$\text{Execution time} = 62\,(0.1)\text{ s} + 0.6\text{ s} + 15\,(0.1)\text{ s} + 0 = 8.3\text{ s}$$

Using the NGOMSL model, we would estimate that the time needed to choose the 14th menu item is 8.3 s.

The estimated execution time for the high-depth display can be determined in a similar way (see Eberts, 1994, for a more detailed analysis of this). The execution of this task requires 36 statements. The user has two external primitive operators: the single-digit keystroke for the main menu and the single-digit keystroke for the secondary menu resulting in a Primitive external operator time of 0.4 s. ReadScreen is the only Analyst-defined mental operator and it occurs four times for the category menu and twice for the command menu. Using the 0.1 estimate, the total time for Analyst-defined mental operators is 0.6 s. Once again, we assume that the system response time is negligible. The equation for estimating the time is:

$$\text{Execution time} = 36\,(0.1)\text{ s} + 0.4\text{ s} + 0.6\text{ s} + 0 = 4.6\text{ s}$$

Comparing this time of 4.6 s for the high-depth display to the time of 8.3 s for the high-breadth display, we would predict that the high-depth display would be faster to execute by over 3 s for choosing menu item 14. These qualitative results are similar to that found for the Model Human Processor and the GOMS models.

In the above example, the execution time was solved for a particular case for finding the 14th item; it can also be solved for the general case. The general case is a little bit more difficult to determine. Without going into great detail (see Eberts, 1994, for more detail), the high-breadth display execution time is

$$\text{Execution time} = 4.55\text{ s} + 0.49\text{ s} + 1.35\text{ s} + 0 = 6.39\text{ s}$$

The equation for the high-depth display is

$$\text{Execution time} = 33\,(0.1)\text{ s} + 0.4\text{ s} + 0.6\text{ s} + 0 = 4.3\text{ s}$$

Once again, this model predicts that the high-depth display will be faster on average than the high-breadth.

40.3.3.2 Learning Times

Kieras (1988) also reports some experimentation on estimating the learning time for a novice user to learn a task that has been analyzed by the NGOMSL task analysis. He indicates that this is an estimate of "pure" learning time that can be taken as an estimate of minimum learning time. The estimate is determined as follows:

Learning time = (30–60) min + 30 s per number of NGOMSL statements

For estimating execution time, a specific task, finding menu item *m*, had to be executed. In this case, the number of NGOMSL statements refers to the total number of statements in the task analysis independent of any specific task. The learning statements are counted as follows:

1. Method title or selection rule title = 1 statement
2. Each if–then step or selection rule = 1 statement
3. All other steps = 1 statement

The number of statements used for learning time estimation can be determined simply by counting the number of steps of Figures 40.9 and 40.10 and adding one more statement for the method title. For the high-breadth display, the number of statements can be found as follows:

Method 1a: title + 4 steps = 5 learning statements
Method 2a: title + 4 steps = 5 learning statements
Method 3a: title + 4 steps = 5 learning statements
Method 4a: title + 3 steps = 4 learning statements
Method 5a: title + 3 steps = 4 learning statements
Method 6a: title + 4 steps = 5 learning statements
Method 7a: title + 4 steps = 5 learning statements
Selection rules title + 2 rules + 1 step = 4 learning statements

The grand total is 37 statements which must be learned. For the high-depth display, the number of statements to be learned, calculated in the same way, is 30. We would predict that the high-breadth display would take the following amount of time to learn:

Learning time = 30–60 min + 0.5 (37) min = 48.5–78.5 min

The high-depth display would take the following amount of time to learn:

Learning time = 30–60 min + 0.5 (30) min = 45–75 min

Because of the 30–60 minute baseline value, nonoverlap in learning times between displays will occur only when the number of NGOMSL statements is very high, which may be the case for many interface designs. When comparing the high-breadth and high-depth displays, the overlap in estimated times is too high so that it is difficult to determine which one would be easiest to learn with a great degree of certainty.

40.3.3.3 Gains from Consistency

Kieras (1988) argues that if the methods are designed to be consistent or similar then the learning time will not be as long because the user will be able to determine the pattern of the methods. This is called a "gain due to consistency" and it can be calculated by evaluating the similarities between the methods. Determining similarities depends on the goal description of the method. Recall that the goal description is in terms of a verb–noun pair. Kieras (1988) suggests that to use this technique on methods A and B, either the verb or the noun term of the goal descriptions of A and B must be the same. If this is the case, then the similar statements in methods A and B are determined by changing the statements of a copy of A to get method B after making a simple substitution of the different goal term in the statements. Kieras (1988) specifies the following steps:

1. Make a copy of A
2. Identify the single term in the goal specifications of the two methods that is different, change it to a "parameter," and, change this term throughout the copy of A.

3. Count how many NGOMSL statements are identical in these methods A and B; these statements are the ones classified as "similar" and need to be charged learning time only once.

Let's now apply this technique to determining any gains from consistency for either the high-breadth or the high-depth displays. Examining the verbs and nouns of the goal descriptions of the methods for the high-breadth display shows a possible match between methods 4a and 5a (both have the same verb of "do"). For determining gains due to consistency, the first step is to make a copy of A or, in this case, method 4a (see Figure 40.9). For the second step, the nonmatching verb or noun is parameterized and changed throughout the copy. Since the verbs of these two match, then the noun of method 4a, string-method, will be parameterized. This parameter does not occur elsewhere in the method steps so that no further changes need to be made. In the final step, count how many statements are the same in methods 4a and 5a. As can be seen, two of the statements are the same. Similar procedures can be performed with the other methods that match in terms of verbs or nouns.

Performing this analysis will show that both the high-breadth and high-depth display have a gain due to consistency. For the high-breadth display, the original number of statements that had to be learned from the previous section was 37. This results in a learning time of 37 × 0.5 min or 18.5 min above the baseline. If method 4a were learned before method 5a, then the number of statements to be learned could be reduced to 34 (reductions of 1 method title and 2 statements) resulting in a learning time of 17 min above the baseline.

For the high-depth display, the original number of statements which had to be learned was 30. This resulted in a learning time of 30 × 0.5 min or 15 min of learning time above the baseline. With the consistency between 3b and 4b (reduce by 5 statements) and 5b and 6b (reduce by 5 statements), 20 statements would need to be learned resulting in a learning time of 20 × 0.5 min or 10 min of learning time above the baseline.

The estimation for learning time showed that both could be expected to be learned in similar amounts of time. The analysis for the gain from consistency must be performed to obtain a better estimation of learning time. Because of the consistencies in the design of the high-depth display, the estimated learning time will be 7.5 min less for this display as compared to the high-breadth display.

40.3.3.4 Design Revisions

Kieras (1988) has extended many of the design revision guidelines suggested by Card et al. (1983) for the GOMS model discussed previously. His extensions are:

1. Try to reduce learning time by eliminating, rewriting, or combining methods, especially to get consistency.
2. If a selection rule cannot be stated clearly and easily, then consider eliminating one or more of the alternative methods.
3. Eliminate the need for operators that retrieve information from LTM which require that the user memorize information, and may be slow until heavily practiced.
4. If there are WM load problems, see if the design can be changed so the user needs to remember less.
5. Eliminate high-level bypassing operators if possible, especially if they involve a slow and difficult cognitive process.
6. The basic way to speed up execution time is to eliminate operators by shortening the methods. But notice that complex mental operators are usually much more time consuming than simple motor actions, and so it can be more important to reduce the need for thinking than to save a few keystrokes. So do not reduce the number of keystrokes if an increase in mental operators is the result.

40.3.3.5 Mental Workload

Kieras (1988) indicates that mental workload can be considered in several ways. One aspect of mental workload, the working memory load, is the technique that is the easiest to apply and the technique most easily grounded in empirical validation. Recall from the Model Human Processor that the capacity of working memory was about five chunks with a maximum of nine chunks if long-term memory augmented it.

Mental workload can be determined from the NGOMSL model by counting the number of items in working memory for each step. Two kinds of items are stored in working memory: variable names or values and the goal that is being performed. The variable names are placed into working memory through the operators RETAIN, RECALL, and RETRIEVE-LTM. They are removed from memory using the FORGET operator. The goal is placed in memory by the "Accomplish goal" operator and removed by "Report goal accomplished." When analyzing workload, the task analyst must list each of the steps in the task (the same steps used to calculate execution time), count the number of items in working memory on each step, and then sum all those numbers together.

Two measures can be taken from this calculation. The first is the peak load on WM. This is found by determining the statements in which the highest number of items are stored in WM. For the high-breadth display to find the 14th item, the peak load is 7 items and for the high-depth display on the corresponding task, the peak is 8 items (see Eberts, 1994).

The other measure of workload, the workload density, is the average amount of items that have to be stored for each statement. This measure can be found by counting the total number of items in memory at each step and then dividing through by the number of steps. For the high-breadth display, the total number of items across all steps is 284, 62 steps are needed to accomplish the task, and so the workload density is 4.58. The high-depth display has 185 items to be remembered across 36 steps for a workload density of 5.14. The higher workload of the high-depth display could be expected because the high-breadth display is meant to be a "flat" display and the high-depth is meant to have more layers and thus more things to remember.

40.3.3.6 Documentation

Kieras suggests that NGOMSL can be used for determining completeness and accuracy of the documentation by comparing the documentation to the procedures of NGOMSL. If the documentation is incomplete, then any omissions should stand out in the comparison. Elkerton (1988) also suggests that the index, table of contents, and headings for the documentation should be organized according to the natural goals of the computer system rather than the specific name provided by the particular system.

40.3.3.7 Expert–Novice Differences Analysis

An important test of a modeling technique is whether the model can be used to differentiate between novices and experts. It is clear that experts perform tasks differently from novices. Not only are experts faster than the novices, they also apparently perform the task with less effort. The main mechanism for differentiating novices and experts with the NGOMSL analysis is through the combination of operators in steps. As an example of this, one of the statements from the high-depth display is:

Recall MENU-COMMAND and accomplish goal of locating MENU-CATEGORY

A novice would be likely to have two steps for this task, recalling the MENU-COMMAND and then accomplishing the goal. Experts are likely to combine several operators in one step. When analyzing a task using NGOMSL, there are no good rules to follow for what level of expertise to combine the statements and what level of expertise not to combine the statements. This is left to the discretion of the individual task analyst. More research is needed for determining learning characteristics of computer users. It is also likely that experts will have different goal structures from novices (Lang, Eberts, Gabel, and Barash, 1991), but changing the goal structure using NGOMSL is difficult and has not been specified.

40.3.4 Conclusions

Since the GOMS and NGOMSL models both originated from the same source, they are very similar. Comparing the GOMS model to the NGOMSL analysis shows that many similarities exist. The number of cycles should be about the same in both analyses. However, GOMS does not always have as much detail as the statements from the NGOMSL analysis. In particular, the WM details, such as when to recall and forget memory items, are not very specific so that many of these cycles are not present in GOMS. One other change is in the decision statements. NGOMSL counts an IF and THEN part of a statement as separate cycles. The GOMS analysis does not necessarily do this. Therefore,

NGOMSL, being more specific than a GOMS analysis usually is, will have more cycles than the corresponding goal stack model.

NGOMSL is clearly an extension of the GOMS models of Card et al. (1983) and many of the same analyses can be performed using either analysis. Both GOMS and NGOMSL are based upon goals, operators, methods, and selection rules. These are defined in the same way using both models. Both models have mechanisms for making time estimates for tasks that which may only be in the prototype stage.

NGOMSL goes beyond the GOMS analysis by combining several of the GOMS models into one integrated model. But NGOMSL goes beyond the goal stack model in several ways. First, NGOMSL places a significance to the cycles needed to complete a task using these cycles as a part of the time estimation equation. NGOMSL integrates the Keystroke-Level model and the goal stack model of GOMS by specifying clearly the primitive operators which are the same as the Keystroke-Level operators. An important difference, however, is that NGOMSL has clear mechanisms, through the time estimate of the cycles, to determine exact estimates for the M operator. In GOMS, M is estimated to be 1.35. In NGOMSL, M would be equivalent to the number of cycles needed to accomplish the operation.

NGOMSL also incorporates the Model Human Processor of Card et al. (1983) through the use of the Analyst-defined operators and the specification of the cognitive cycle time to be equal to the statement cycle time. For the Analyst-defined operators, estimators in some cases can be found by using cycle times from the Model Human Processor. In the examples provided in this section, the perceptual processor cycle time was used for determining the time to read the screen. An alternative method for estimating times is through experimentation. The Model-UT of GOMS is not clearly incorporated in NGOMSL except for this idea of using experimentation to find time estimates for some of the units of the model.

The experimentation associated with the NGOMSL analysis allows many different kinds of estimations not easily possible with the GOMS models. In particular, using NGOMSL one can determine estimates for learning time and gains due to consistency. These are not possible using the GOMS models. Another difference between the two is that GOMS is only applicable to experts, whereas NGOMSL has some techniques, although they are rather crude at this time, to specify different models for experts and novices.

40.4 PRODUCTION SYSTEMS

Production systems have been used to describe how people process and store information. The concept of a production system was developed mostly in the late 1960s and early 1970s culminating in a book called *Human Problem Solving* by Newell and Simon (1972). This conceptualization has been very influential in many areas: cognitive psychology, cognitive science, computer science, and artificial intelligence. As indicated by the book's name, the purpose of production systems was to determine how people solved problems and to specify these steps in a system resembling, but slightly different from, the computer programming languages of the time, which were used to specify the steps for machine problem solving. If the steps could be specific enough, then only a short jump is needed to specify the steps for a machine to solve problems much like a human does. This was an impetus for much of the work in artificial intelligence and the application area of expert systems.

40.4.1 Basic Production Systems

A production system contains declarative knowledge (the facts) and procedural knowledge (how to process the facts). The declarative knowledge is contained in the production rules, which have the form of:

IF <condition> THEN <action>

The procedural knowledge is how the interpreter, that part of the computer program that interprets the programming statements, sequences through the rules. The sequencing occurs through a recognize/act cycle. In the recognize stage, all of the conditions of the rules are matched against the contents of working memory (WM). The act stage occurs when all the rules that match all the conditions are fired and the interpreter executes the actions. When this is completed, one cycle of the process has been completed.

40.4.2 Production System Modeling

Kieras and Polson (1985) developed a production system model for human–computer interaction tasks. This was based upon the GOMS model of Card et al. (1983) and was a precursor to Kieras' NGOMSL model. It was designed, for the most part, to show the relationships between NGOMSL and the underlying rules. Also, since the representation is very similar to a computer program, it could be run on a computer. It is doubtful that very many people would model an interface directly using this kind of production system. Since the production system was designed to be equivalent to NGOMSL, NGOMSL would be used to model the interface, NGOMSL would be converted into the product system rules, and the production system would be used to test the model on a computer.

40.4.2.1 The Parsimonious Production System Rule Notation

Bovair, Kieras, and Polson (1990) found that it was most convenient to represent the production system in terms of the PPS (Parsimonious Production System) rule notation which was developed earlier by Kieras. This notation was designed with the following constraints: (1) On any cycle, any rule whose conditions are currently satisfied will fire; (2) rules must be written so that a single rule will not fire repeatedly; (3) only one rule will fire on a cycle; and (4) all procedural knowledge is explicit in the rules rather than being explicit in the interpreter. The basic structure of a rule using this notation is:

<name> IF <condition> THEN <action>

The name is not functional and is used to assist the programmer in reading the code. The condition is a list of clauses that must all be matched for the rule to be true. The notation for clauses is described below. The actions are sequences of operators similar to that for the NGOMSL model. These operators can be used to modify working memory usually so that other rules can fire on the next cycle.

40.4.2.2 Clauses

The notation conventions for clauses are as follows. Condition clauses have the following form:

$$(\text{Tag Term}_1 \text{ Term}_2 \ldots \text{Term}_k)$$

The tag can be one of the following:

GOAL—This corresponds directly to the goals from GOMS or NGOMSL.
NOTE—This is information kept in working memory over several firings.
STEP—This is a control execution sequence within a method.
DEVICE—This is information provided directly from the device.
LTM—This is sometimes used to indicate information stored in LTM.

The terms are used to specify the tags. Each tag has a predefined number of terms associated with it. As in NGOMSL, each GOAL tag has two terms associated with it: the verb or action of the goal and the object of the verb or action.

The NOTE term has several functions. One function of the NOTE is to specify the ordering of goals and actions. In some situations, the action of a rule may be to set a goal and to specify the next step. Since both of these are set, the rule condition containing the goal and the rule condition containing the step will both fire unless a NOTE is used to limit the firing to only one. This is usually done by including, in the condition with the step, a NOTE that the goal has been performed so that the rule with the step will only fire after completion of the goal. In the example of the menu displays, the NOTE is also used to indicate variable values such as for the menu-item number. In this case, the first term of the NOTE will always be VARIABLE. Finally, the NOTE is used to set the context of some of the rules and could contain various kinds of information.

The STEP term has a structure very similar to the GOAL term in that it contains two tags: a verb and an action. It is usually used in conjunction with a GOAL to specify how the user is stepping through the different methods steps in the goal.

The DEVICE term is not specified clearly by Bovair et al. (1990) and can probably be left somewhat to the discretion of the task analyzer. In the examples that follow, DEVICE has three tags: the object of the information displayed (usually the USER), the location of the relevant information on the screen, and the action required of the information. In some situations, the action is to remember what has been displayed on the screen, in which case the user is to add a NOTE to store in working memory. In one situation, the action is merely to indicate whether there are any characters in the location specified.

The LTM term is also not specified clearly by Bovair et al. (1990) and is also left to the discretion of the task analyzer. In the following menu examples, LTM is used to access the command sequence from LTM or to determine the menu category for the command. It has only one tag, which refers to a general description of the information being accessed.

40.4.3 Applying Production Systems to the Menu Displays

In this section, a full detail of applying production systems to modeling is not provided. Instead, the production system model for the high-breadth and high-depth displays can be found in Figures 40.11 and 40.12, which can then be compared to the NGOMSL representations. A complete description of how to model human–computer interaction tasks using production systems can be found in Bovair et al. (1990) and the complete high-breadth and high-depth menu examples can be found in Eberts (1994). To follow how the production system works in Figures 40.11 and 40.12 (see p. 1357 and p. 1361), determine which rules will "fire" by determining which conditions are true. Only one condition should be true at any time in a method. The firing of a rule will cause a new action to occur, which will change the conditions for a new rule to fire. Remember that only one rule can fire at a time.

For the NGOMSL model, the techniques were specified for estimating execution time, estimating learning times, performing design revisions, estimating the mental workload of a task, and analyzing the documentation. Since NGOMSL and production systems provide equivalent representations of a task, in slightly different terms, then the same estimations and predictions can be used for production systems. In some ways, however, the representation of NGOMSL has been set up to make these kinds of estimations and predictions easy to perform. These predictions are not discussed in this section. To perform these estimations, one could first transform the production rules into the methods and operators of NGOMSL or convert some of the equations to fit with the production rule parameters. However, the production systems seem to be more appropriate for some estimations than the NGOMSL model, especially expert–novice differences. Expert–novice differences are discussed; the other predictions can be taken from the NGOMSL analysis.

Differences between experts and novices are more clearly stated using the production system models than using the NGOMSL model. The style rules for the production system are listed in Table 40.5 (see p. 1364). These apply only to someone who is neither really an expert nor a novice; someone in between. The style rules for novices and experts are slightly different from these original ones. The additions needed for novice behavior will be considered first.

The first addition to the style rules for novice users is:

Novice Rule 1: Each overt action requires a separate production rule.

An overt action can include a key press, a mouse manipulation, looking at a manuscript, or looking at a screen. To model novices, each one of these would have to be placed in a separate rule. Other actions manipulating working memory, such as adding or deleting NOTEs, do not have to be enclosed within separate steps.

The second addition to the style rules is:

Novice rule 2: Novices explicitly check all feedback.

The checking of feedback occurs for the high-breadth display. If this were a model for an expert, this step would not have been included because we would expect that the RETURN key would be hit automatically without the explicit check.

By examining the production systems for the high-breadth and high-depth displays, it can be seen that modeling was performed with a novice user in mind. It is interesting to

High-Breadth Display

```
(Top.Start      IF    ((GOAL PERFORM COMMAND-SEQUENCE)
                      (NOT (NOTE PERFORMING COMMAND-SEQUENCE)))
              THEN    ((Add STEP GET COMMAND-UNIT-TASK)
                      (Add NOTE PERFORMING COMMAND-SEQUENCE)))
(Top.P1         IF    ((GOAL PERFORM COMMAND-SEQUENCE)
                      (STEP GET COMMAND-UNIT-TASK))
              THEN    ((RetainLTM LIST-ITEM as (NOTE VARIABLE MENU-COMMAND))
                      (Add STEP EXECUTE COMMAND-UNIT-TASK)
                      (Delete STEP GET COMMAND-UNIT-TASK)))
(Top.P2         IF    ((GOAL PERFORM COMMAND-SEQUENCE)
                      (STEP EXECUTE COMMAND-UNIT-TASK))
              THEN    (Add GOAL PERFORM COMMAND-UNIT-TASK)
                      (Add STEP CHECK TASKS-DONE)
                      (Delete STEP EXECUTE COMMAND-UNIT-TASK)))
(Top.P3         IF    ((GOAL PERFORM COMMAND-SEQUENCE)
                      (STEP CHECK TASKS-DONE)
                      (NOTE PERFORM COMMAND-UNIT-TASK FINISH)
                      (LTM (More commands in command sequence)))
              THEN    ((Add STEP GET COMMAND-UNIT-TASK)
                      (Delete STEP CHECK TASKS-DONE)
                      (Delete NOTE PERFORM-COMMAND-UNIT-TASK FINISH)))
(Top-Finish     IF    ((GOAL PERFORM COMMAND-SEQUENCE)
                      (STEP CHECK TASKS-DONE)
                      (NOTE PERFORM-COMMAND-UNIT-TASK FINISH)
                      (LTM (No more commands in command sequence)))
              THEN    ((Delete GOAL PERFORM COMMAND-SEQUENCE)
                      (Delete STEP CHECK TASKS-DONE)
                      (Delete NOTE PERFORMING COMMAND-SEQUENCE)
                      (Delete NOTE PERFORM-COMMAND-UNIT-TASK FINISH)
                      (Stop Now)))

    (StartUnitTask    IF    ((GOAL PERFORM COMMAND-UNIT-TASK)
                            (NOT (NOTE EXECUTING COMMAND-UNIT-TASK)))
                    THEN    ((Add NOTE EXECUTING COMMAND-UNIT-TASK)
                            (Add STEP LOOK SCREEN)))
    (UnitTask.P1      IF    ((GOAL PERFORM COMMAND-UNIT-TASK)
                            (STEP LOOK SCREEN))
                    THEN    ((Add GOAL LOCATE MENU-COMMAND)
                            (Add STEP ITEM ENTRY)
                            (Delete STEP LOOK SCREEN)))
    (UnitTask.P2      IF    ((GOAL PERFORM COMMAND-UNIT-TASK)
                            (STEP ITEM ENTRY)
                            (NOTE LOCATE-MENU-COMMAND FINISH))
                    THEN    ((Add GOAL ENTER N)
                            (Add STEP DO VERIFY)
                            (Delete STEP ITEM ENTRY)
                            (Delete NOTE LOCATE-MENU-COMMAND FINISH)))
    (UnitTask.P3      IF    ((GOAL PERFORM COMMAND-UNIT-TASK)
                            (STEP DO VERIFY)
                            (NOTE ENTER-N PERFORMED))
                    THEN    ((Add GOAL VERIFY ENTRY)
                            (Add STEP FINISH TASK)
                            (Delete NOTE ENTER-N PERFORMED)
                            (Delete STEP DO VERIFY)))
    (FinishUnitTask   IF    ((GOAL PERFORM COMMAND UNIT-TASK)
                            (STEP FINISH TASK)
                            (NOTE VERIFY-ENTRY FINISH))
                    THEN    ((Add NOTE PERFORM-COMMAND-UNIT-TASK FINISH)
                            (Delete GOAL PERFORM COMMAND-UNIT-TASK)
                            (Delete STEP FINISH TASK)
                            (Delete NOTE VERIFY-ENTRY FINISH)
                            (Delete NOTE VARIABLE MENU-COMMAND)
                            (Delete NOTE EXECUTING COMMAND-UNIT-TASK)))
```

Figure 40.11 The production system model for the high-breadth display.

```
(StartLocate       IF    ((GOAL LOCATE MENU-COMMAND)
                          (NOT (NOTE EXECUTING LOCATE MENU-COMMAND)))
                 THEN    ((Add NOTE EXECUTING LOCATE MENU-COMMAND)
                          (Add STEP READ ITEMS)))
(ReadRule          IF    ((GOAL LOCATE MENU-COMMAND)
                          (STEP READ ITEMS))
                 THEN    ((DEVICE USER MENUline (Add NOTE VARIABLE MENU-ITEM))
                          (Add STEP MATCH ITEMS)
                          (Delete STEP READ ITEMS)))
(MatchFalse        IF    ((GOAL LOCATE MENU-COMMAND)
                          (STEP MATCH ITEMS)
                          (MatchWM (NOT (MENU-COMMAND MENU-ITEM))))
                 THEN    ((Add STEP READ ITEMS)
                          (Delete STEP MATCH ITEMS)
                          (Delete NOTE VARIABLE MENU ITEMS)))
(MatchTrue         IF    ((GOAL LOCATE MENU-COMMAND)
                          (STEP MATCH ITEMS)
                          (MatchWM (MENU-COMMAND MENU-ITEM)))
                 THEN    ((Add STEP UPDATE ITEMS)
                          (Delete STEP MATCH ITEMS)))
(UpdateRule        IF    ((GOAL LOCATE MENU-COMMAND)
                          (STEP UPDATE ITEMS))
                 THEN    ((DEVICE USER MENU-ITEMnumber (Add NOTE VARIABLE N))
                          (Add STEP FINISH LOCATE)
                          (Delete STEP UPDATE ITEMS)))
(FinishLocate      IF    ((GOAL LOCATE MENU-COMMAND)
                          (STEP FINISH LOCATE))
                 THEN    ((Add NOTE LOCATE-MENU-COMMAND FINISH)
                          (Delete GOAL LOCATE MENU-COMMAND)
                          (Delete STEP FINISH LOCATE)
                          (Delete NOTE EXECUTING LOCATE-MENU-COMMAND)
                          (Delete NOTE VARIABLE MENU-ITEM)))

  (SelectKeyMethod    IF  ((GOAL ENTER N)
                           (NOTE CHECK (N between 1 and 9))
                           (NOT (NOTE EXECUTING ENTER-N)))
                    THEN  ((Add GOAL DO KEY-METHOD)
                           (Add NOTE EXECUTING-ENTER N)))
  (FinishSelectKey    IF  ((GOAL ENTER N)
                           (NOTE ENTER-N FINISH))
                    THEN  ((Add NOTE ENTER-N PERFORMED)
                           (Delete GOAL ENTER N)
                           (Delete NOTE ENTER-N FINISH)
                           (Delete NOTE EXECUTING ENTER-N)))
  (StartKeyMethod     IF  ((GOAL DO KEY-METHOD)
                           (NOT (NOTE EXECUTING KEY-METHOD)))
                    THEN  ((Add NOTE EXECUTING KEY-METHOD)
                           (Add STEP DO SINGLE-KEY)))
  (KeyMethod          IF  ((GOAL DO KEY-METHOD)
                           (STEP DO SINGLE-KEY))
                    THEN  ((Press-key N)
                           (Add STEP GET FEEDBACK)
                           (Delete STEP DO SINGLE-KEY)))
  (KeyFeedback        IF  ((GOAL DO KEY-METHOD)
                           (STEP GET FEEDBACK))
                    THEN  ((DEVICE USER EntryFeedbackLine (Add NOTE VARIABLE N'))
                           (Add STEP FINISH KEY)
                           (Delete STEP GET FEEDBACK)))
  (KeyMethodFinish    IF  ((GOAL DO KEY-METHOD)
                           (STEP FINISH KEY))
                    THEN  ((Add NOTE ENTER-N FINISH)
                           (Delete GOAL DO KEY-METHOD)
                           (Delete STEP FINISH KEY)
                           (Delete NOTE EXECUTING KEY-METHOD)))
```

Figure 40.11 (*Continued*).

```
(SelectStringMethod  IF    ((GOAL ENTER N)
                            (NOTE CHECK (N greater than 9))
                            (NOT (NOTE EXECUTING ENTER-N)))
                     THEN  ((Add GOAL DO STRING-METHOD)
                            (Add NOTE EXECUTING ENTER-N)))
(FinishSelectString  IF    ((GOAL ENTER N)
                            (NOTE ENTER-N FINISH))
                     THEN  ((Add NOTE ENTER-N PERFORMED)
                            (Delete GOAL ENTER N)
                            (Delete NOTE ENTER-N FINISH)
                            (Delete NOTE EXECUTING ENTER-N)))
(StartStringMethod   IF    ((GOAL DO STRING-METHOD)
                            (NOT (NOTE EXECUTING STRING-METHOD))
                     THEN  ((Add NOTE EXECUTING STRING-METHOD)
                            (Add STEP DO STRING)))
(StringMethod        IF    ((GOAL DO STRING-METHOD)
                            (STEP DO STRING))
                     THEN  ((Type-In N)
                            (Add STEP GET FEEDBACK)
                            (Delete STEP DO STRING)))
(StringFeedback      IF    ((GOAL DO STRING-METHOD)
                            (STEP GET FEEDBACK))
                     THEN  ((DEVICE USER EntryFeedbackLine (Add NOTE VARIABLE N')
                            (Add STEP FINISH STRING)
                            (Delete STEP GET FEEDBACK)))
(StringMethodFinish  IF    ((GOAL DO STRING-METHOD)
                            (STEP FINISH STRING))
                     THEN  ((Add NOTE ENTER-N FINISH)
                            (Delete GOAL DO STRING-METHOD)
                            (Delete STEP FINISH STRING)
                            (Delete NOTE EXECUTING STRING-METHOD)))

(StartVerifyMethod   IF    ((GOAL VERIFY ENTRY)
                            (NOT (NOTE EXECUTING VERIFY ENTRY)))
                     THEN  ((Add NOTE EXECUTING VERIFY ENTRY)
                            (Add STEP DO MATCH)))
(MatchInputFalse     IF    ((GOAL VERIFY ENTRY)
                            (STEP DO MATCH)
                            (MatchWM (NOT (N N'))))
                     THEN  ((Add GOAL DELETE ENTRY)
                            (Add NOTE MORE-DIGITS)
                            (Delete NOTE VARIABLE N')
                            (Add STEP DO TERMINATE)
                            (Delete STEP DO MATCH)))
(MatchInputTrue      IF    ((GOAL VERIFY ENTRY)
                            (STEP DO MATCH)
                            (MatchWM (N N')))
                     THEN  ((Add STEP DO TERMINATE)
                            (Delete STEP DO MATCH)))
(TerminateRule       IF    ((GOAL VERIFY ENTRY)
                            (STEP DO TERMINATE)
                            (NOT (NOTE EXECUTING DELETION)
                            (NOT (NOTE MORE DIGITS)))
                     THEN  ((Press-Key ENTER)
                            (Add STEP VERIFY-UPDATE)
                            (Delete STEP DO TERMINATE)))
(VerifyUpdate        IF    ((GOAL VERIFY ENTRY)
                            (STEP VERIFY UPDATE))
                     THEN  ((Add STEP VERIFY-DONE)
                            (Delete NOTE VARIABLE N)
                            (Delete NOTE VARIABLE N'
                            (Delete STEP VERIFY-UPDATE)))
(FinishVerifyMethod  IF    ((GOAL VERIFY ENTRY)
                            (STEP VERIFY DONE))
                     THEN  ((Add NOTE VERIFY-ENTRY FINISH)
                            (Delete GOAL VERIFY ENTRY)
                            (Delete STEP VERIFY DONE)
                            (Delete NOTE EXECUTING VERIFY-ENTRY)))
```

Figure 40.11 *(Continued).*

```
(StartMethod           IF   ((GOAL DELETE ENTRY)
                             (NOT (NOTE EXECUTING DELETION)))
                     THEN   ((Add NOTE EXECUTING DELETION)
                             (Add STEP DO KEY-DELETE)))
(DeleteKey             IF   ((GOAL DELETE ENTRY)
                             (STEP DO KEY-DELETE))
                     THEN   ((Press-Key DELETE)
                             (Add STEP DECIDE MORE-DIGITS)
                             (Delete STEP DO KEY-DELETE)))
(DecideMoreDelete      IF   ((GOAL DELETE ENTRY)
                             (STEP DECIDE MORE-DIGITS)
                             (DEVICE USER EntrySpace DigitsPresent))
                     THEN   ((Add STEP DO KEY-DELETE)
                             (Delete STEP DECIDE MORE-DIGITS)))
(DecideNoMoreDelete    IF   ((GOAL DELETE ENTRY)
                             (STEP DECIDE MORE-DIGITS)
                             (DEVICE USER EntrySpace NoDigitsPresent))
                     THEN   ((Add GOAL ENTER N)
                             (Add STEP DO DELETE-TERMINATE)
                             (Delete NOTE MORE DIGITS)
                             (Delete STEP DECIDE MORE-DIGITS)))
(FinishMethod          IF   ((GOAL DELETE ENTRY)
                             (STEP DO DELETE-TERMINATE)
                             (NOTE ENTER-N PERFORMED))
                     THEN   ((Add NOTE DELETE-ENTRY FINISH)
                             (Delete GOAL DELETE ENTRY)
                             (Delete STEP DO DELETE-TERMINATE)
                             (Delete NOTE VARIABLE N')
                             (Delete NOTE EXECUTING DELETION)
                             (Delete NOTE ENTER-N PERFORMED)))
```

Figure 40.11 *(Continued)*.

note that the production system model is probably easiest to apply to novice users because only having one overt action in a step is a very specific rule to follow, thus reducing the variability of modeling between different task analysts.

The opposites of the above two novice rules can be formulated to reflect expert performance. From the second novice rule, we would assume that experts would not explicitly check all the feedback. Any rules associated with this could be removed, which would also have the effect of reducing the number of rules. The production systems for the high-breadth and high-depth displays can be changed to remove these prompts.

For the first novice rule, we would expect that the rules would be reduced by compacting the rules into a more efficient rule structure. For the production system modeling, emphasis is placed on making the rule sets more compact rather than generating new specialized methods.

40.4.4 Conclusions

The production system studied in this section is very similar to the other models studied in the previous sections. In fact, Bovair et al. (1990) claim that the production system is exactly equivalent to the NGOMSL model or various combinations of the GOMS models. We have seen in this section that some of the statements from NGOMSL, especially the Decide statements, are difficult to incorporate in a production system and this can change the estimates for the number of cycles needed for performing a human–computer interaction task. For the most part, though, the production system model is highly equivalent to the NGOMSL model and the two could be used interchangeably. This is especially the case when estimating execution times, learning times, gains due to consistency, and the workload measurements.

With the two models being so highly similar, what are the reasons for using one over the other? NGOMSL has a fairly simple structure so that the model can be formulated and it could be easily understood when shown to someone else. The statements are in a very concise form although much of the processing and control structure is implicit to the model instead of being explicitly shown. The production system has much more

High-Depth Display

```
(Top.Start      IF    ((GOAL PERFORM COMMAND-SEQUENCE)
                       (NOT (NOTE PERFORMING COMMAND-SEQUENCE)))
              THEN    ((Add STEP GET COMMAND-UNIT-TASK)
                       (Add NOTE PERFORMING COMMAND-SEQUENCE)))
(Top.P1         IF    ((GOAL PERFORM COMMAND-SEQUENCE)
                       (STEP GET COMMAND-UNIT-TASK))
              THEN    ((RetainLTM LIST-ITEM as (NOTE VARIABLE MENU-COMMAND))
                       (Add STEP EXECUTE COMMAND-UNIT-TASK)
                       (Delete STEP GET COMMAND-UNIT-TASK)))
(Top.P2         IF    ((GOAL PERFORM COMMAND-SEQUENCE)
                       (STEP EXECUTE COMMAND-UNIT-TASK))
              THEN    (Add GOAL PERFORM COMMAND-UNIT-TASK)
                       (Add STEP CHECK TASKS-DONE)
                       (Delete STEP EXECUTE COMMAND-UNIT-TASK)))
(Top.P3         IF    ((GOAL PERFORM COMMAND-SEQUENCE)
                       (STEP CHECK TASKS-DONE)
                       (NOTE PERFORM COMMAND-UNIT-TASK FINISH)
                       (LTM (More commands in command sequence)))
              THEN    ((Add STEP GET COMMAND-UNIT-TASK)
                       (Delete STEP CHECK TASKS-DONE)
                       (Delete NOTE PERFORM-COMMAND-UNIT-TASK FINISH)))
(Top-Finish     IF    ((GOAL PERFORM COMMAND-SEQUENCE)
                       (STEP CHECK TASKS-DONE)
                       (NOTE PERFORM-COMMAND-UNIT-TASK FINISH)
                       (LTM (No more commands in command sequence)))
              THEN    ((Delete GOAL PERFORM COMMAND-SEQUENCE)
                       (Delete STEP CHECK TASKS-DONE)
                       (Delete NOTE PERFORMING COMMAND-SEQUENCE)
                       (Delete NOTE PERFORM-COMMAND-UNIT-TASK FINISH)
                       (Stop Now)))

(StartUnitTask        IF    ((GOAL PERFORM COMMAND-UNIT-TASK)
                             (NOT (NOTE EXECUTING COMMAND-UNIT-TASK)))
                    THEN    ((Add NOTE EXECUTING COMMAND-UNIT-TASK)
                             (Add STEP FIND CATEGORY)))
(UnitTask.P1          IF    ((GOAL PERFORM COMMAND-UNIT-TASK)
                             (STEP FIND CATEGORY))
                    THEN    ((Add GOAL ISSUE MENU-CATEGORY)
                             (Add STEP FIND COMMAND)
                             (Delete STEP ISSUE MENU-CATEGORY)))
(UnitTask.P2          IF    ((GOAL PERFORM COMMAND-UNIT-TASK)
                             (STEP FIND COMMAND)
                             (NOTE ISSUE-MENU-CATEGORY FINISH))
                    THEN    ((Add GOAL ISSUE MENU-COMMAND)
                             (Add STEP FINISH TASK)
                             (Delete STEP FIND COMMAND)
                             (Delete NOTE ISSUE-MENU-COMMAND FINISH)))
(FinishUnitTask       IF    ((GOAL PERFORM COMMAND-UNIT-TASK)
                             (STEP FINISH TASK)
                             (NOTE ISSUE-MENU-COMMAND FINISH))
                    THEN    ((Add NOTE PERFORM-COMMAND-UNIT-TASK FINISH)
                             (Delete GOAL PERFORM COMMAND-UNIT-TASK)
                             (Delete STEP FINISH TASK)
                             (Delete NOTE ISSUE-MENU-COMMAND FINISH)
                             (Delete NOTE EXECUTING COMMAND-UNIT-TASK)))
```

Figure 40.12 The production system model for the high-depth display.

```
(StartCategory        IF  ((GOAL ISSUE MENU-CATEGORY)
                          (NOT (NOTE EXECUTING ISSUE-MENU-CATEGORY)))
                    THEN  ((Add NOTE EXECUTING ISSUE-MENU-CATEGORY)
                          (Add STEP LOCATE CATEGORY)))
(CatLocateRule        IF  ((GOAL ISSUE MENU-CATEGORY)
                          (STEP LOCATE CATEGORY))
                    THEN  ((Add GOAL LOCATE MENU-CATEGORY)
                          (Add STEP CAT KEY-PRESS)
                          (Delete STEP LOCATE CATEGORY)))
(CatKeyRule           IF  ((GOAL ISSUE MENU-CATEGORY)
                          (STEP CAT KEY-PRESS)
                          (NOTE LOCATE-MENU-CATEGORY FINISH))
                    THEN  ((Press-Key N)
                          (Add STEP UPDATE CAT-ITEMS)
                          (Delete STEP CAT KEY-PRESS)
                          (Delete NOTE LOCATE-MENU-CATEGORY FINISH)))
(CatUpdateRule        IF  ((GOAL ISSUE MENU-CATEGORY)
                          (STEP UPDATE CAT-ITEMS))
                    THEN  ((Delete NOTE VARIABLE N)
                          (Delete NOTE VARIABLE MENU-CATEGORY)
                          (Add STEP FINISH CATEGORY)
                          (Delete STEP UPDATE CAT-ITEMS)))
(FinishCategory       IF  ((GOAL ISSUE MENU-CATEGORY)
                          (STEP FINISH CATEGORY))
                    THEN  ((Add NOTE ISSUE-MENU-CATEGORY FINISH)
                          (Delete GOAL ISSUE MENU-CATEGORY)
                          (Delete STEP FINISH CATEGORY)
                          (Delete NOTE EXECUTING ISSUE-MENU-CATEGORY)))

  (StartCommand         IF  ((GOAL ISSUE MENU-COMMAND)
                            (NOT (NOTE EXECUTING ISSUE-MENU-COMMAND)))
                      THEN  ((Add NOTE EXECUTING ISSUE-MENU-COMMAND)
                            (Add STEP LOCATE COMMAND)))
  (ComLocateRule        IF  ((GOAL ISSUE MENU-COMMAND)
                            (STEP LOCATE COMMAND))
                      THEN  ((ADD GOAL LOCATE MENU-COMMAND)
                            (Add STEP COM KEY-PRESS)
                            (Delete STEP LOCATE COMMAND)))
  (ComKeyRule           IF  ((GOAL ISSUE MENU-COMMAND)
                            (STEP COM KEY-PRESS)
                            (NOTE LOCATE-MENU-COMMAND FINISH))
                      THEN  ((Press-Key N)
                            (Add STEP UPDATE COM-ITEMS)
                            (Delete STEP COM KEY-PRESS)
                            (Delete NOTE LOCATE-MENU-COMMAND FINISH)))
  (ComUpdateRule        IF  ((GOAL ISSUE MENU-COMMAND)
                            (STEP UPDATE COM-ITEMS))
                      THEN  ((Delete NOTE VARIABLE N)
                            (Delete NOTE VARIABLE MENU-COMMAND)
                            (Add STEP FINISH COMMAND)
                            (Delete STEP UPDATE COM-ITEMS)))
  (FinishCommand        IF  ((GOAL ISSUE MENU-COMMAND)
                            (STEP FINISH COMMAND))
                      THEN  ((Add NOTE ISSUE-MENU-COMMAND FINISH)
                            (Delete GOAL ISSUE MENU-COMMAND)
                            (Delete STEP FINISH COMMAND)
                            (Delete NOTE EXECUTING ISSUE-MENU-COMMAND)))
```

Figure 40.12 (*Continued*).

```
(StartCatLocate        IF  ((GOAL LOCATE MENU-CATEGORY)
                            (NOT (NOTE EXECUTING LOCATE-MENU-CATEGORY)))
                     THEN  ((Add NOTE EXECUTING LOCATE-MENU-CATEGORY)
                            (Add STEP ACCESS LTM)))
(AccessLTMRule         IF  ((GOAL LOCATE MENU-CATEGORY)
                            (STEP ACCESS LTM)
                            (LTM (MENU-COMMAND category)))
                     THEN  ((Add NOTE VARIABLE MENU-CATEGORY)
                            (Add STEP MATCH CAT-ITEMS)
                            (Delete STEP READ CAT-ITEMS)))
(ReadCatRule           IF  ((GOAL LOCATE MENU-CATEGORY)
                            (STEP READ CAT-ITEMS))
                     THEN  ((DEVICE USER MENUline (Add NOTE VARIABLE MENU-ITEM))
                            (Add STEP MATCH CAT-ITEMS)
                            (Delete STEP READ CAT-ITEMS)))
(MatchCatFalse         IF  ((GOAL LOCATE MENU-CATEGORY)
                            (STEP MATCH CAT-ITEMS)
                            (MatchWM (NOT (MENU-CATEGORY MENU-ITEM))))
                     THEN  ((Add STEP READ CAT-ITEMS)
                            (Delete NOTE VARIABLE MENU-ITEM)
                            (Delete STEP MATCH CAT-ITEMS)))
(MatchCatTrue          IF  ((GOAL LOCATE MENU-CATEGORY)
                            (STEP MATCH CAT-ITEMS)
                            (MatchWM (MENU-CATEGORY MENU-ITEM)))
                     THEN  ((Add STEP CATEGORY RETENTION)
                            (Delete STEP MATCH CAT-ITEMS)))
(CatRetentionRule      IF  ((GOAL LOCATE MENU-CATEGORY)
                            (STEP CATEGORY RETENTION))
                     THEN  ((DEVICE USER MENU-ITEMnumber (Add NOTE VARIABLE N))
                            (Add STEP UPDATE CAT-ITEMS)
                            (Delete STEP CATEGORY RETENTION)))
(UpdateCatRule         IF  ((GOAL LOCATE MENU-CATEGORY)
                            (STEP UPDATE CAT-ITEMS))
                     THEN  ((Delete NOTE VARIABLE MENU-ITEM)
                            (Add STEP FINISH CAT-LOCATE)
                            (Delete STEP UPDATE CAT-ITEMS)))
(FinishCatLocate       IF  ((GOAL LOCATE MENU-CATEGORY)
                            (STEP FINISH CAT-LOCATE))
                     THEN  ((Add NOTE LOCATE-MENU-CATEGORY FINISH)
                            (Delete GOAL LOCATE MENU-CATEGORY)
                            (Delete STEP FINISH CAT-LOCATE)
                            (Delete NOTE EXECUTING LOCATE-MENU-CATEGORY)))
```

Figure 40.12 *(Continued).*

overhead to it. Developing this kind of model is very difficult because it forces the task analyst to be very precise for when the NOTEs are to be added and deleted and the kinds of condition statements accompanying each rule. As can be seen when comparing the production system of Figures 40.11 and 40.12 to the NGOMSL models of Figures 40.9 and 40.10, the production system requires many more statements.

It may be useful to use both models during the course of a task analysis. The NGOMSL model is useful for designing the task analysis in the initial stages. It provides the task analyst with a good overall structure, it can be shown to others in the group to get comments or changes, and it is very concise. With less detail, however, the task analyst could possibly become sloppy and make mistakes or not be specific enough. This is why production system modeling would be useful in the next stage. The structure of NGOMSL would provide the task analyst with an overall structure, a specification of the goals, determination of the selection rules, and determination of the primitive operators. This useful NGOMSL structure can then be used as a guide to the overall production system structure. In particular, the goals, operators, methods, and selection rules would already be specified and the task analyst would only have to use the templates to fill in the appropriate statements. In so doing, the task analyst is also forced to be very specific about each of the statements, especially the condition statements of the rules, and the adding and deleting of NOTEs. Many of these kinds of details could have been forgotten through the NGOMSL analysis.

```
(StartComLocate        IF  ((GOAL LOCATE MENU-COMMAND)
                            (NOT (NOTE EXECUTING LOCATE-MENU-COMMAND)))
                     THEN  ((Add NOTE EXECUTING LOCATE-MENU-COMMAND)
                            (Add STEP MATCH COM-ITEMS)))
(ReadComRule           IF  ((GOAL LOCATE MENU-COMMAND)
                            (STEP READ COM-ITEMS))
                     THEN  ((DEVICE USER MENUline (Add NOTE VARIABLE MENU-ITEM))
                            (Add STEP MATCH COM-ITEMS)
                            (Delete STEP READ COM-ITEMS)))
(MatchComFalse         IF  ((GOAL LOCATE MENU-COMMAND)
                            (STEP MATCH COM-ITEMS)
                            (MatchWM (NOT (MENU-COMMAND MENU-ITEM))))
                     THEN  ((Add STEP READ COM-ITEMS)
                            (Delete NOTE VARIABLE MENU-ITEM)
                            (Delete STEP MATCH COM-ITEMS)))
(MatchComTrue          IF  ((GOAL LOCATE MENU-COMMAND)
                            (STEP MATCH COM-ITEMS)
                            (MatchWM (MENU-COMMAND MENU-ITEM)))
                     THEN  ((Add STEP COMMAND RETENTION)
                            (Delete STEP MATCH COM-ITEMS)))
(ComRetentionRule      IF  ((GOAL LOCATE MENU-COMMAND)
                            (STEP COMMAND RETENTION))
                     THEN  ((DEVICE USER MENU-ITEMnumber (Add NOTE VARIABLE N))
                            (Add STEP UPDATE COM-ITEMS)
                            (Delete STEP COMMAND RETENTION)))
(UpdateComRule         IF  ((GOAL LOCATE MENU-COMMAND)
                            (STEP UPDATE COM-ITEMS))
                     THEN  ((Delete NOTE VARIABLE MENU-ITEM)
                            (Add STEP FINISH COM-LOCATE)
                            (Delete STEP UPDATE COM-ITEMS)))
(FinishComLocate       IF  ((GOAL LOCATE MENU-COMMAND)
                            (STEP FINISH COM-LOCATE))
                     THEN  ((Add NOTE LOCATE-MENU-COMMAND FINISH)
                            (Delete GOAL LOCATE MENU-COMMAND)
                            (Delete STEP FINISH COM-LOCATE)
                            (Delete NOTE EXECUTING LOCATE-MENU-COMMAND)))
```

Figure 40.12 (*Continued*).

Two advantages of the production system approach over the NGOMSL approach are that it has a good theoretical basis and that it could be used easily in a computerized simulation. Production systems had been shown in the past to be reasonable models of human performance (Anderson, 1976; Newell and Simon, 1972) for tasks other than human–computer interaction. In addition, the production system is, essentially, a computer program that could be run easily in a simulation to obtain measurements and to test the completeness of the analysis.

40.5 GRAMMAR REPRESENTATIONS

An interesting research problem in human–computer interaction has been the investigation of techniques to represent the interface languages under a common organizational scheme.

Table 40.5 The Nine Rules That Should Be Followed in the Task Analysis When Using a Production System Model

1. The production rules must generate the correct sequence of user actions.
2. The representation should conform to structure programming principles.
3. The top-level representation for most tasks should have a unit task structure.
4. The top-level unit task structure should be based on the top-level goal.
5. Selection rules select which method to apply.
6. Information needed by a method should be supplied through working memory.
7. Sequencing within a method is maintained by chaining steps.
8. Labels for steps should be based on the action of the rules and should be used consistently.
9. Use a lockout NOTE to prevent a rule from firing repeatedly.

In linguistics, this has been done through a production rule grammar (Chomsky, 1964). In computer programming, compilers are designed to translate programming statements into machine code (Lewis, Rosenkranz, and Stearns, 1976). Several researchers—Moran (1981), Reisner (1981), Young (1981), and Payne and Green (1986, 1989)—have developed grammars to describe interactive languages with a goal toward defining a metric for the cognitive complexity and ease of use of these programs.

Grammars were one of the earliest methods used to model human-computer interaction languages. The Reisner (1981) model borrowed many concepts from linguistics by incorporating the BNF (Backus–Naur Form) method. This technique, and others like it, when applied to human–computer interaction tasks can describe complex languages and operating systems in a relatively small number of statements. In many ways, the statements of the grammars are the same as or similar to the production rules discussed previously. The BNF grammars emphasized the relationships between the syntax and the actions needed to perform the commands. The Payne and Green grammar, called TAG for Task-Action Grammar, was designed to emphasize the family resemblances among the language elements. In the following discussion, the Reisner grammar will be discussed first, followed by a discussion of TAG.

40.5.1 Reisner's Grammar

Generally, a production rule grammar describes a language as a set of rules for specifying correct strings in the language. Linguists have struggled for some time to understand the grammars that underlie languages. Following Reisner (1981), a grammar for a language such as English consists of the following features:

1. A set of terminal symbols (e.g., the words in the language)
2. A set of nonterminal symbols (e.g., invented constructs used to show the structure of the language such as the "noun phrase")
3. A starting symbol (e.g., "S" for sentence)
4. The metasymbols "+", "|", and "::=" (some common meanings for these are "and," "or," and "is composed of," respectively)
5. Rules constructed from the above (e.g., S::= noun phase + verb phrase)

The "cognitive grammar," developed by Reisner, uses a similar terminology for describing human–computer interaction languages. The various parts of the above grammar, such as words and sentences, can be replaced with human–computer interaction features. In particular, terminal symbols are the actions that a user has to learn and remember, such as hitting a key or positioning a mouse. Nonterminal symbols represent sets of similar actions that can be grouped together, such as sets of drawing actions. The starting symbol is a high-level task to be performed by the user, such as "edit text" for word processing. The metasymbols are the same as in the above grammar. The rules must be constructed for the interaction grammar used; Reisner constructed 101 rules for her particular application.

40.5.1.1 Calculations from the Grammar Model

Reisner (1981) used the formal grammar for the interaction language to test two interactive drawing programs to predict which one would be easier to use. This grammar has been used specifically to determine the consistency of the design, defined as the number of rules necessary to describe the structure of some set of terminal strings, and to determine the simplicity of the interaction, defined as the lengths of the terminal strings for particular tasks. Empirical studies on two alternative drawing programs showed that the predictions of the model were accurate; the users found it easier to select the correct actions for the program predicted to be simpler, and the users found it easier to learn and remember the program that was predicted to be consistent. Examples of performing these calculations for the high-breadth and high-depth displays will be shown in a later section.

Although not stated explicitly by Reisner, one could assume that execution time for a task could also be determined. Similar to the other models in the previous sections, execution time would be most likely related to the number of steps needed to perform the task. As each grammar rule is used, this constitutes a particular step.

40.5.1.2 The BNF Representation

The BNF representation for the high-breadth and high-depth displays is shown in Figures 40.13 and 40.14. Each display can be described by four grammar rules. For the high-

```
TASK FEATURES

digits                  one, two

SIMPLE TASKS

Choose a menu item below 10 {digits = one}
Choose a menu item 10 or above {digits = two}

RULE SCHEMAS

Task [digits = one] --> keypress + complete entry
Task [digits = two] --> keypress + keypress + complete entry

keypress --> "digit"
complete entry --> ENTER

SIMPLE TASKS

Choose a menu item

TASK-ACTION RULE SCHEMAS

Task --> Menu1 + Menu2

Menu1 --> keypress
Menu2 --> keypress
keypress --> "digit"
```

Figure 40.13 BNF grammar representation for high-breadth menu.

breadth display, the overall task is to choose a menu item. This can be composed of a digit operation and a complete entry. Both of these are nonterminal and so must be decomposed further. A digit operation is composed of a digit and a digit operation, providing it with a recursive structure so that it can accept multiple digits or a single digit. If the menu item number is less than 10, then a single digit is needed. If the menu item is 10 or more, then another digit is needed and the digit operation is repeated. On the repeat of the digit operation, the remaining single digit is entered. The specification of a digit is a nonterminal symbol, and the last rule indicates the OR'ed choices for the digits. From the first rule, a complete entry is also a nonterminal symbol. The third rule shows that a complete entry is described by the ENTER key.

The representation for the high-depth display is slightly different. In this case, choosing a menu item involves selecting from the main menu and selecting from the secondary menu. Each selection requires a single digit. A digit is a nonterminal symbol; the last rule shows the terminal symbol possibilities for a digit.

Prediction of Execution Time

Execution time will be predicted by the number of steps needed to complete the task. For the high-breadth display, the following nonterminal steps must be completed to perform the task:

Choose menu item
Digit operation
Digit

```
choose menu item ::= select main menu + select secondary menu
select main menu ::= digit
select secondary menu ::= digit
digit ::= 0 | 1 | 2 | 3 | 4 | 5 | 6 | 7 | 8 | 9
```

Figure 40.14 BNF grammar representation for high-depth menu.

Complete entry

This is the case when the user only has to enter one digit for the menu item. If two digits are needed, then the following steps would be performed:

Choose menu item
Digit operation
Digit
Digit operation
Digit
Complete entry

In the one-digit situation, four steps are needed. In the two-digit situation, six steps are needed. If each menu item has an equal probability of being chosen, then the average number of steps needed to perform the task would be:

$$4 \text{ steps} + 10/16\ (2 \text{ steps}) = 5.25 \text{ steps}$$

For the high-depth display, the following nonterminal steps would be needed to perform the task:

Choose menu item
Select main menu
Digit
Select secondary menu
Digit

This task requires five steps total. From this analysis, we would predict that the high-depth menu display could be performed slightly faster than the high-breadth menu display. Notice that this analysis only accounts for the executable parts of the task, similar to the Keystroke-Level model of GOMS, and not for the cognitive processing of the task.

Consistency

The grammar can be used to determine the consistency of the design, and Reisner (1981) defined this as the number of rules necessary to describe the structure of some set of terminal strings. In this case, the analysis is very simple. One has only to look at Figures 40.13 and 40.14 and count the number of rules. Both menu displays have four rules, so the model would predict no differences in consistency between the two menu displays. This analysis is slightly different from the consistency analyses for NGOMSL and the production system model, called consistency gains in those cases, in that the other models predicted that the high-depth display would have higher consistency.

Simplicity

Reisner (1981) also states that the grammar can be used to determine the simplicity of the interaction. This is defined as the lengths of the terminal strings for particular tasks. In Figures 40.13 and 40.14, both the high-breadth and the high-depth displays have the same terminal string, so we would expect no differences in terms of consistency between the two displays. What is meant by simplicity? It could mean that a simpler display is faster to execute, and so it is related to the execution time. It could also mean that a simpler display is easy to learn. This is most probably the meaning of the definition. Simplicity can be equated with learnability.

40.5.2 Task-Action Grammars

Payne and Green (1986, 1989) used a very similar grammar, called the Task-Action Grammar (TAG), to investigate consistency in more detail, concentrating on the overall structure of the language rather than individual rules. To accomplish this, TAG represents family resemblances (the overall sentence structure of commands), the degree to which a task language relies on well-learned world knowledge, and the organization of tasks and subtasks. Ease of learning and use depends on the number of simple-task rule schemas, which is based upon the overall structure of the language. Unlike the other predictive models discussed in this section, TAG was designed to make predictions about the relative complexity of designs rather than to provide quantitative measures and predictions of performance. In other words, the focus of this model is to generate experimental hypotheses, based upon the model, and then test those hypotheses experimentally.

Payne and Green (1989) describe the important aspects of TAG as follows:

1. Identify "simple tasks" that the user can perform without problem solving, and that contain no control structure.
2. Describe simple tasks in a dictionary by sets of semantic components reflecting categorizations of the task world.
3. Rewrite rules that map simple tasks onto action specifications, in which grammatical tokens can be tagged with semantic features derived from the simple-task dictionary. The featural tagging supplies selection restrictions and allows the grammar to capture generalizations.

The most important aspect of TAG is that it can determine well-defined categories of tasks. The tasks with the categories that are well defined are those with the most structural consistency. Arbitrary collections of tasks have poorly defined categories. More detail on how to build a TAG model can be found in Payne and Green (1989) and in Eberts (1994).

40.5.2.1 The Use of TAG in Human–Computer Interaction

Design revisions, consistency, and learnability are all evaluated in terms of the top-level rule schema. An interface is considered to be highly consistent if the whole task can be described using one top-level rule schema. The task becomes less and less consistent as more top-level rule schemas are needed to describe the task. A consequence of this is that the total number of rules is unimportant; only the number of top-level rule schemas is important for consistency.

Learnability is closely related to the consistency evaluation. An assumption is that most of the learning time is involved in learning the top-level rule schemas and, therefore, learning time is a function of the number of these rules to be learned. Once again, having only the one top-level rule should make an interface easy to learn.

If the interface design contains more than one top-level rule schema, then a revision of the design should be considered. The structure of the commands, if commands are being used, should be evaluated to determine whether they can be made more consistent by reducing the number of top-level rule schemas.

Payne and Green (1989) conducted an experiment to test the hypothesis that the number of top-level rules is more important in learnability than the total number of rules. They designed one interface, which they hypothesized to be consistent, that could be described by TAG with 28 rules, one of which was a top-level rule. A different interface design, hypothesized to be inconsistent, could be described by 12 rules, 8 of which were top-level rules. If the number of total rules was important, then learning should take longer for the 28-rule interface than for the 12-rule interface. If the number of top-level rules was important, then learning should take longer for the 12-rule interface than for the 28-rule interface. The results of an experiment supported the predictions from the TAG analysis: learning was faster for the 28-rule interface as compared to the 12-rule interface. Although the rules used in TAG are not equivalent to the rules used in the production system model considered previously, the other models would probably have a difficult time explaining this result for consistency.

In a variation on TAG, Howes and Young (1991) have developed a model in which learnability and consistency can be determined not from specifying all the task-action mappings, as in TAG, but by providing the Programmable User Model (PUM) with examples for how to perform the task. As an example, the PUM can learn how to perform a menu pull-down task when provided with one example. Learnability is assessed by how quickly PUM can acquire the rule for the learning. Consistent interfaces can be learned faster than inconsistent interfaces.

40.5.2.2 Example

The TAG representation of the high-breadth and high-depth displays is shown in Figures 40.15 and 40.16. Because this menu operation is so simple, the BNF and TAG representations are very similar: For both models, both displays can be described in four rules.

SIMPLE TASKS

Choose a menu item

TASK-ACTION RULE SCHEMAS

Task --> Menu1 + Menu2

Menu1 --> keypress
Menu2 --> keypress
keypress --> "digit"

Figure 40.15 TAG representation of the high-breadth display.

For the high-depth display, the user would have a choice of simple tasks based upon the menu-item number: If below 10 then the variable digits has a value of one; for 10 and above the variable digits will have a value of two. The rule schemas show that two top-level tasks are needed, depending on the number of the menu item. All parts of the top-level task are nonterminal symbols that can be further decomposed. A key press can be decomposed into a terminal digit, and the complete entry can be decomposed into a terminal ENTER key.

The TAG representation for a high-depth display is shown in Figure 40.16. In this case, there is only one simple task, which is to choose the menu item. Because of this, we do not need to specify variables and features of the variables. Both Menu1 and Menu2 are decomposed into a single key press. A key press must be a digit.

Execution Time

The main focus of the TAG analysis of the task is not on execution time but rather it is on consistency of the task. Thus, Payne and Green (1989) make no predictions about execution time. However, as with the other models, we could make the simple assumption that the model that showed the fewest number of steps needed could be executed fastest. This may be too simple of an assumption, because consistency should also have some effect on execution time. As an example, a task that is consistent could come to be performed with little cognitive processing, and thus very fast, whereas an inconsistent task would need much cognitive processing and so would be performed slowly.

Just looking at the number of steps needed, we see that the high-breadth display would require the following steps for the nonterminal rules:

Task
Key press
Complete entry

TASK FEATURES

digits one, two

SIMPLE TASKS

Choose a menu item below 10 {digits = one}
Choose a menu item 10 or above {digits = two}

RULE SCHEMAS

Task [digits = one] --> keypress + complete entry
Task [digits = two] --> keypress + keypress + complete entry

keypress --> "digit"
complete entry --> ENTER

Figure 40.16 TAG representation of the high-depth display.

The preceding steps would hold for entering one digit. For two digits, the following steps would be needed:

Task
Key press
Key press
Complete entry

With equal probability of choosing any menu item, the number of steps needed on average would be

$$3 \text{ steps} + 10/16 \text{ step} = 3.63 \text{ steps}$$

The high-depth display would require the following steps:

Task
Menu1
Menu2
Key press

This would require four steps for its operation. Since the high-depth display requires more steps, we would say that this display should be slightly slower to operate when compared to the high-breadth display. This is the only model that makes the prediction in this direction.

Consistency

The TAG representation primarily is designed to capture the consistency differences between the two displays. For TAG, consistency is determined by the number of top-level rules. The TAG analysis shows that the high-depth display is more consistent than the high-breadth display because only one top-level task is needed for the high-depth display and two top-level tasks are needed for the high-breadth display. Even though they both have the same number of rules, the high-depth display should be more consistent. Payne and Green (1989) equate learnability with consistency.

One problem with the TAG representation, and a problem with all the other modeling techniques, is that alternative representations making slightly different predictions are also logical. In particular, the high-breadth display could be modeled using one top-level task if the second key press could take a value of NULL when the menu item number is less than 10. With only one top-level task, the consistencies and learnability of the high-depth and high-breadth displays should be the same.

40.5.3 Summary

The grammars are especially useful for evaluating consistency of the interface design and for offering design revisions based upon consistency. For BNF grammars, consistency is related to the number of grammar rules needed to describe the interface. For the TAG model, consistency is related to the number of top-level rules. In some cases, BNF and TAG may make dissimilar predictions about the consistency of an interface. For the high-breadth display, the BNF grammar predicted no difference between the interfaces, whereas the TAG model predicted a difference. The TAG model corresponded more closely to the predictions of the other models discussed in this chapter.

Payne and Green (1989) equate consistency with the learnability of the interface. Those displays which are consistent, through a TAG analysis, should be easier to learn than those displays which are inconsistent according to TAG. Experimental results on interfaces hypothesized to be consistent and inconsistent confirmed their predictions.

Out of all the models considered in this chapter, the grammar models are probably most appropriate for complex systems. The menu interaction task considered in this and the other sections was probably too simple to model effectively with the grammars. Out of all the models considered, the grammars required the fewest number of statements. One characteristic of the grammars is that they could not be taken to a low level of detail such as eye movements, matching items in memory, or making decisions. Many of the fine-level cognitive activities could not be easily modeled or have not been traditionally modeled using these techniques. This means that if a task has a large component devoted to these kinds of cognitive activities, then some of the other models would be appropriate, especially the Model Human Processor, NGOMSL, or production systems. If the task is fairly complex and the fine-detail cognitive activities are minimal or nonessential to the task, then the grammars can be used.

The grammars also make no predictions about execution time or mental workload. Performance for a particular task could be unimportant if one is just analyzing a large-scale interface design without considering particular tasks or "benchmarks." If mental workload is considered along the lines of consistency, then perhaps the grammars could be altered to consider this concept. The other models did not provide a complete analysis of mental workload.

40.6 COMPARISON OF THE MODELS

Table 40.6 compares all the models considered in this section across execution time, learnability, consistency, workload, and expert–novice differences. Some of the models can be used to make these kinds of predictions, and others cannot.

For execution time, the Model Human Processor, the Keystroke-Level GOMS model, NGOMSL, and production systems can all be used to make quantitative predictions. All of them predict that the high-breadth display will be slower than the high-depth display. The actual estimates vary depending on the model. The Model Human Processor places most of the emphasis on the cognitive processing of the material and little emphasis on its execution. The Keystroke-Level model, on the other hand, places most of the emphasis on the execution, not the cognitive processing. NGOMSL and production systems emphasize both aspects of the task about equally.

The GOMS goal stack model and the two grammar models do not make specific time estimates for executing tasks. Execution time depends on the number of statements or steps needed to execute the task. For the GOMS goal stack, the high-breadth display should be slower because it requires 29 steps as compared to the 24 steps of the high-depth display. For the grammars, the modelers do not state explicitly the relationship between number of steps and the execution time; one can only assume the implicit relationship. The BNF grammar predicts that the high-breadth display has more steps and, thus, would require more execution time. The TAG grammar makes the opposite prediction. The high-depth display would require more steps. This is the only model that makes this prediction.

The second section of Table 40.6 displays the predictions for learnability of the displays. The GOMS goal stack, NGOMSL, production system, and BNF grammar models can all predict learning time by determining the number of rules or methods that must be learned. For NGOMSL and production systems, specific learning time predictions can be made from the number of methods to be learned. The others are not so explicit in the prediction. Three of the models (GOMS goal stack, NGOMSL, and production systems) predict that the high-breadth display will require more methods and, thus, will take longer to learn. The BNF grammar model predicts no differences in the number of methods.

The exception to the technique for estimating learnability is the TAG grammar model. TAG equates learnability with consistency of the task. Consistency can be determined through the number of top-level rules for performing the task. Experimentation showed that the number of rules was not a determinant of consistency. From this analysis, the high-breadth display has two top-level rules and the high-depth display has one top-level rule. The high-depth display should be easier to learn. The two other models (Model Human Processor and GOMS Keystroke-Level) make no predictions about learnability.

The third part of Table 40.6 examines predictions for consistency. NGOMSL and production system models determine consistency through the gains due to consistency. Some methods or rules are highly similar so that they do not need to be learned again. The estimated gain due to consistency can be determined exactly from these two models, and both predict that the high-depth display will have a higher consistency gain. For TAG, because learnability and consistency are the same, the top-level rule is considered. TAG also predicts that the high-depth display is more consistent than the high-breadth display. The BNF grammar predicts no difference between the two displays.

Only the NGOMSL and production system models are designed to predict mental workload of the task. Workload is measured according to the peak workload and the density, or average, workload over all steps. Both predict that the high-breadth display will have less workload.

Table 40.6 Comparison of High-Breadth and High-Depth Menu Displays for All Models

Model	High-Breadth Display	High-Depth Display
Execution Time		
Model Human Processor	3.47 s	2.38 s
GOMS goal stack	33 steps	24 steps
GOMS Keystroke-Level	1.84 s	1.75 s
NGOMSL	6.39 s	4.30 s
Production system	5.38 s	5.00 s
BNF grammar	5.25 steps	5 steps
TAG grammar	3.63 steps	4 steps
Learnability		
Model Human Processor	No prediction	No prediction
GOMS goal stack	6 methods	4 methods
GOMS Keystroke-Level	No prediction	No prediction
NGOMSL	48.5–78.5 min	45–75 min
Production system	33.9 min	24.1 min
BNF grammar	No difference (4 rules)	No difference (4 rules)
TAG grammar	2 top-level rules	1 top-level rule (easier)
Consistency		
Model Human Processor	No prediction	No prediction
GOMS goal stack	No prediction	No prediction
GOMS Keystroke-Level	No prediction	No prediction
NGOMSL	1.5-min gain	5-min gain
Production system	Reduce 43 rules to 39	Reduce 34 rules to 22
BNF grammar	4 rules	4 rules
TAG grammar	2 top-level rules	1 top-level rule (more consistent)
Workload		
Model Human Processor	No prediction	No prediction
GOMS goal stack	No prediction	No prediction
GOMS Keystroke-Level	No prediction	No prediction
NGOMSL	Peak of 7 items Density of 4.58[a]	Peak of 8 items Density of 5.14[a]
Production system	Peak of 7 items Density of 4.45[a]	Peak of 8 items Density of 4.87[a]
BNF grammar	No prediction	No prediction
TAG grammar	No prediction	No prediction
Expert–Novice Differences		
Model Human Processor	Use Fast Man model	Use Fast Man model
GOMS goal stack	No prediction	No prediction
GOMS Keystroke-Level	Only typing skill	Only typing skill
NGOMSL	Combine steps	Combine steps
Production system	Combine rules	Combine rules
BNF grammar	No prediction	No prediction
TAG grammar	No prediction	No prediction

[a]Not the general case, for item 14.

Finally, some models can be used to determine differences between experts and novices. Only one model, the production system model, has been designed explicitly to determine expert–novice differences. Bovair et al. (1990) worked out several techniques for combining and compacting rules to capture these differences. To a lesser extent, some of the steps in the NGOMSL model can be combined for the experts. These procedures are not stated explicitly in Kieras (1988), however. Finally, some aspects of the other models capture some of the expert–novice differences. For the Model Human Processor, the Fast Man model could be used for the experts and the Slow Man model could be used for the novices. For the GOMS Keystrokes-Level model, the only skill differences would be in typing speed. The tables provide estimates for keystroke times based upon typing skill.

Each of the models considered in this chapter can be used for different aspects of the analysis of human–computer interaction tasks. The NGOMSL and production system models are the most comprehensive of all the models; but one could also argue that because of the comprehensiveness of these models, they may only consider surface characteristics of the aspect they are trying to measure. As an example, the workload measures only capture one small part of workload, the load on working memory, because this is most easily determined from the model. Other workload measures, such as the amount of attentional resources required for a task, are not considered. The Keystroke-Level model can be used for a quick determination of execution times. The grammars are probably best for complex tasks in which consistency of the task is an important characteristic. Placing all the emphasis on consistency, instead of being a comprehensive model, may not be a bad approach because of the importance of consistency in usability of the interface (e.g., Nielsen, 1989).

REFERENCES

Anderson, J. R. (1976). *Language, Memory, and Thought.* Hillsdale, NJ: Erlbaum.

Averbach, E., and Coriell, A. S. (1961). Short-term memory in vision. *Bell System Technical Journal, 40,* 309–328.

Bovair, S., Kieras, D. E., and Polson, P. G. (1990). The acquisition and performance of text-editing skill: A cognitive complexity analysis. *Human Computer Interaction, 5,* 1–48.

Card, S. K., Moran, T. P., and Newell, A. L. (1983). *The Psychology of Human Computer Interaction.* Hillsdale, NJ: Erlbaum.

Card, S. K., Moran, T. P., and Newell, A. L. (1986). The Model Human Processor. In K. R. Boff, L. Kaufman, and J. P. Thomas, (Eds.), *Handbook of Perception and Human Performance* (Vol. II). New York: Wiley.

Carroll, J. M., and Olson, J. R. (1988). Mental models in human-computer interaction. In M. Helander, Ed., *Handbook of Human–Computer Interaction.* Amsterdam: Elsevier.

Chomsky, N. (1964). *Syntactic Sructures.* The Hague, Netherlands: Mouton.

Eberts, R. E. (1994). *User Interface Design.* Englewood Cliffs, NJ: Prentice-Hall.

Elkerton, J. (1988). Online aiding for human-computer interfaces. In M. Helander, (Ed.), *Handbook of Human–Computer Interaction.* Amsterdam: Elsevier.

Howes, A., and Young, R. M. (1991). Predicting the learnability of task-action mappings. *Human Factors in Computing Systems Conference Proceedings, CHI'91* (pp. 113–118). New York: Association of Computing Machinery (ACM).

Kieras, D. E. (1988). Towards a practical GOMS model methodology for user interface design. In M. Helander, Ed., *Handbook of Human–Computer Interaction.* Amsterdam: Elsevier.

Kieras, D. E., and Polson, P. (1985). An approach to the formal analysis of user complexity. *International Journal of Man–Machine Studies, 22,* 365–394.

Lang, G. T., Eberts, R. E., Gabel, M., and Barash, M. (1991). Extracting and using procedural knowledge in a CAD task. *IEEE Transactions on Engineering Management, 38,* 257–268.

Lewis, P. M., Rosenkranz, D. J., and Stearns, R. E. (1976). *Compiler Design Theory.* Reading, MA: Addison-Wesley.

Miller, G. (1956). The magical number 7, plus or minus two: Some limits on our capacity for processing information. *Psychological Review, 63,* 81–97.

Moran, T. P. (1981). The command language grammar: A representation for the user interface of interactive computer systems. *International Journal of Man-Machine Studies, 15,* 3–50.

Newell, A., and Simon, H. (1972). *Human Problem Solving.* Englewood Cliffs, NJ: Prentice Hall.

Nielsen, J., Ed. (1989). *Coordinating User Interfaces for Consistency.* Boston: Academic Press.

Norman, D. A. (1986). Cognitive engineering. In D. A. Norman and S. W. Draper, (Eds.), *User Centered System Design.* Hillsdale, NJ: Erlbaum.

Payne, S. J., and Green, T. R. G. (1986). Task-action grammars: A model of the mental representation of task languages. *Human–Computer Interaction, 2,* 93–133.

Payne, S. J., and Green, T. R. G. (1989). The structure of command languages: An experiment on task-action grammar. *International Journal of Man-Machine Studies, 30,* 213–234.

Reisner, P. (1981). Formal grammar and human factors design of an interactive graphics system. *IEEE Transactions on Software Engineering*, *SE-7*, 229–240.

Robertson, S. P., and Black, J. B. (1986). Structure and development of plans in computer text editing. *Human-Computer Interaction*, *2*, 201–226.

Young, R. M. (1981). The machine inside the machine: Users' models of pocket calculators. *International Journal of Man–Machine Studies*, *15*, 51–85.

CHAPTER 41

COMPUTER MODELING AND SIMULATION

K. Ronald Laughery, Jr.
Micro Analysis and Design, Inc.
Boulder, CO 80301 USA

Kevin Corker
NASA—Ames Research Center
MS-262-3 Moffet Field, CA 94305 USA

41.1 INTRODUCTION

Over the past few decades, the human factors and ergonomics practitioner has increasingly been called upon early in the system design and development process. Early input from all disciplines results in better and more integrated designs as well as lower costs than if one or more disciplines finds that changes are required later. Our goal as human factors and ergonomics practitioners should be to provide substantive and well supported input regarding the human(s), his or her interaction(s) with the system, and the resulting human–system performance. Furthermore, we should be prepared to provide this input from the earliest stages of system concept development and then throughout the entire system or product life cycle.

To meet this challenge, many human factors and ergonomics tools and technologies have evolved over the years to support early analysis and design. Many of these technologies have taken the form of design guidance (e.g., Boff, Kaufman, and Thomas, 1986; O'Hara, Brown, Stubler, Wachtel, and Persensky, 1995) and user high-fidelity rapid interface prototyping (e.g., Dahl, Allender, Kelley, and Adkins, 1995). Design guidance technologies, in the form of either handbooks or computerized decision support systems, put selected portions of the human factors and ergonomics knowledge base at the fingertips of the designer, often in a form tailored to a particular design problem such as nuclear power plant design or UNIX computer interface design. However, design guides have the shortcoming that they do not often provide methods for making tradeoffs in *system* performance as a function of design. For example, design guides may tell us that a high-resolution color display will be better than a black and white display, and they may even tell us the value in terms of increased response time and reduced error rates. However, this type of guidance will rarely provide good insights into the value of this improved element of the human's performance to the *overall system's* performance. As such, design guidance is less valuable for providing concrete input to *system-level performance prediction.*

Rapid prototyping, on the other hand, provides system-level analysis of how a specific design and task allocation will affect human and system level performance. The disadvantage of prototyping, as will all human subjects experimentation, is that it is costly. Also, in contrast to user–computer interface prototypes, hardware-based systems, such as aircraft and machinery, are relatively expensive to prototype at the system level, particularly at early design stages when there are many widely divergent design concepts. To be sure, hardware and software prototyping has been an important tool to the human factors practitioner, and its use is growing in virtually every application area. What is often needed is a methodology to extrapolate from the general knowledge base of human factors and ergonomics, as reflected in design guides and the literature, to make system-level performance predictions as a function of human factors design alternatives. This methodology should also bind with rapid prototyping and experimentation in a mutually supportive way. As has become the case in many engineering disciplines, one of these methodologies is computer modeling and simulation.

Computer modeling of human behavior and performance is hardly a new endeavor. Computer models of complex cognitive behavior have been around for over 20 years (e.g., Newell and Simon, 1972) and tools for computer modeling of task level performance have been available since the 1970s (e.g., Wortman, Duket, Seifert, Hann, and Chubb, 1978). However, two things have changed appreciably in the past decade that promote the use of computer modeling and simulation of human performance as a standard tool for the practitioner. First is the rapid increase in computer power and the associated development of modeling tools that are easier to use. The individual with an interest in predicting human performance through simulation has a variety of compute-based tools available (for a comprehensive list of these tools, see McMillan et al., 1989). Second is the increased focus by the research community on the development of *predictive* model (Gray, John, and Atwood, 1993) represents the integration of research into a model for making predictions of how humans will perform in a realistic task environment. Another example is that much of the research in cognitive workload that has been represented as computer algorithms (e.g., McCracken and Aldich, 1984; Farmer, 1995). Given a description of the tasks and equipment that humans are engaged in, these algorithms allow the assessment of when workload-related performance problems are likely to occur. These are just two examples of the continued improvements in the knowledge base of first principles of human cognition and performance.

From another perspective, simulation allows the human factors and ergonomics team to "step up to the table" with the other engineering disciplines who also rely increasingly on computer models of the phenomena of interest. What we will discuss in this chapter is the human factors and ergonomics contribution to computer-aided system design.

41.2 OBJECTIVES OF THIS CHAPTER

This chapter will discuss some of the current computer tools for modeling and simulating human/system performance. It is intended to provide the reader with

1. An understanding of the types of human factors and ergonomics issues that can be addressed with modeling and simulation
2. An understanding of some of the specific tools that are now available to assist the human factors and ergonomics specialist in conducting model-based analyses
3. An appreciation of the level of expertise and work that will be required to use these technologies

We cannot begin this chapter without two caveats. The first is that we are not yet at a point where computer modeling of human behavior allows sufficiently accurate predictions that no other analysis method (e.g., prototyping) is likely to be needed. In early stages of system or product concept development, modeling human–system behavior may be all that is possible. As the system or product moves into the design process, human factors and ergonomics designers will often want to confirm and enhance modeling and simulation predictions with prototyping and experimentation. In addition to providing high-fidelity system performance data, these data can be used to enhance and refine the models. This concept of human performance modeling supporting and being supported by experimentation with human subjects is represented in Figure 41.1. In essence, simulation provides the human factors and ergonomics practitioner with a means of extending the knowledge base of human factors and of amplifying the effectiveness of limited experimentation.

The second caveat is that the technologies discussed herein are evolving rapidly. One can be certain that (1) every tool discussed is undergoing constant change, and (2) new modeling tools are being developed. We are discussing computer-based tools and the reader should expect the pace of change in these tools to mirror the pace in other computer tools such as word processors and spreadsheets. These detailed discussions of several of

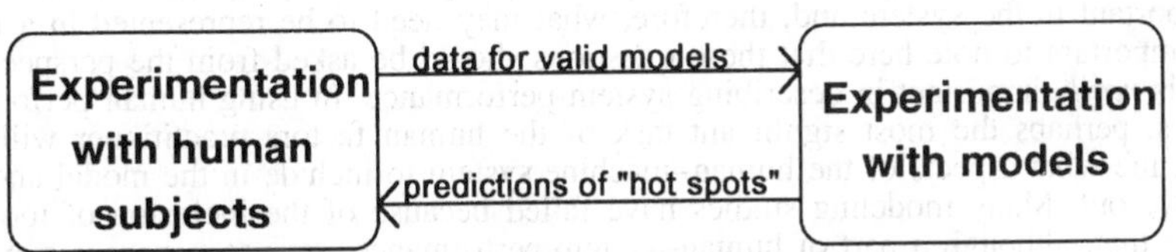

Figure 41.1 The synergy between modeling and experimentation.

the modeling tools are included to facilitate a better understanding of human performance modeling tools. We encourage the reader to contact citations in this chapter to assess the state of any tool at that time.

41.3 THE TYPES OF QUESTIONS THAT ARE BEING ADDRESSED BY HUMAN PERFORMANCE MODELS

Below is a list of some classes of problems to which human–system modeling has been applied:

- How long will it take a human or team of human to perform a set of tasks as a function of system design, task allocation, and individual capabilities? Additionally, what are the tradeoffs in performance for different design, task allocation, and individual capabilities?
- What would the expected error rates be for a human or team of humans as a function of system design, task allocation, and individual capabilities and styles?
- What are the workload demands on the human as a function of system design, and task allocation and automation? How will human performance and resulting system performance change as the demands of the environment change? How many individuals are required on a team to ensure safe, successful performance?
- How will human–system performance be affected by environmental stressors such as heat, cold, or the use of drugs?

This list should be seen as a sampling rather than an exhaustive list. The tools we discuss in this chapter are inherently flexible and one of our consistent discoveries is that these tools can be used to solve problems that the tool developers never conceived.

To assess the potential of simulation to answer questions, every potential human performance modeling project should first conduct a critical assessment of what is important in the human–machine system being modeled and, therefore, what should be included in the model. First, what elements of human system performance are believed to be important to system behavior and, therefore, should be represented in the model? The questions that should be considered about the system include:

- *Human performance representation:* What time or duration of performance is important? What time or initiation of human performance can be defined and what resolution of behavior is required? What time profiles of human performance including task management, load management, and goal management structure are expected?
- *Equipment representation:* With what equipment are the tasks to be accomplished? To what level of functional or physical description can and should the equipment be represented? Is it operable by more than one human or system component?
- *Interface requirements:* What information meeds to be conveyed to the humans and when? Is there required transformation of information?
- *Control requirements:* What processes need to be controlled by the human and to what level of resolution?
- *Logical and physical constraints:* What allowable performance is supported on equipment operability and on procedural sequences? What alarms and alerts should be represented?
- *Simulation driver:* What make the system function? The occurrence of well-defined events (e.g., a procedure), the movement of time (e.g., the control of a vehicle), or a hybrid of both?

By answering these questions, the human factors practitioner will get a sense of what is important in the system and, therefore, what may need to be represented in a model. It is important to note here that these questions should be asked from the perspective of what is *really* important in describing system performance. In using human performance models, perhaps the most significant task of the human factors practitioner will be to determine what aspects of the human–machine system to include in the model and what to leave out. Many modeling studies have failed because of the inclusion of too many factors that, although a part of human–system performance, are not system performance drivers. Consequently, the models become overly complex and untenable. In our experience, it is better to begin with a model with too few aspects of the system represented and then add to it than to begin a modeling project by trying to model everything. The first approach may succeed, whereas the second is virtually doomed.

Second, the human factors practitioner should consider the measures of effectiveness of the system that the model should be designed to predict. In building the model, it is important to always remember that the goal of the model will be to predict system performance or measures of human performance that will clearly impact system performance. Therefore, a clear definition of what is important to performance is necessary. The following aspect of performance measures should be considered:

- *Success criteria:* What operational success measures are important to the system?
- *Range of performance to be studied:* What are the experimental variables that are to be explored by the model? How important is it to establish a range of performance for each experimental condition as a function of the stochastic (i.e., random) behavior of the system?

By asking these questions before beginning a modeling project, the human factors practitioner can develop a better sense of what is important in the system, both in terms of aspects that drive system performance as well as the measures of effectiveness that are truly of interest. Then, and only then, can a human performance modeling project begin with a reasonable hope of success.

The remainder of this chapter will discuss the two classes of modeling tools for human performance simulation. After discussion each class of modeling tool, we shall provide specific examples of a modeling tool and then provide case studies about how these tools have been used in answering real human performance questions.

41.4 THE CLASSES OF SIMULATION MODELS OF HUMAN–SYSTEM PERFORMANCE

Human performance can be highly complex and involve many types of processes and behavior. Over the years many models have been developed that predict sensory processes (e.g., Gawron, Laughery, Jorgensen, and Polito, 1983), aspects of human cognition (e.g., Newell, 1990), and human motor response (e.g., Fitt's law). The current literature in the areas of cognitive engineering, error analysis, and human–computer interaction contains many models, descriptions, methodologies, metaphors, and functional analogies. However, in this chapter, we are focusing not on the models of these individual elements of human behavior, but on models that can be used to describe human performance in systems. These human–system performance models typically include some of these elemental behavioral models as components but provide a structural framework that allow them to be put in the context of human performance of tasks in systems.

We separate the world of human–system performance models into two general categories that can be described as *reductionist* models and *first-principle* models. *Reductionist* models use human–system task sequence as the primary organizing structure as shown in Figure 41.2. The individual models of human behavior for each task or task element are connected to this task sequencing structure. We refer to it as reductionist because the process of modeling human behavior involves taking the larger aspects of human system behavior (e.g., "Perform the mission") and then successively reducing them to smaller elements of behavior (e.g., "Perform the function, perform the tasks") until a level of decomposition is reached at which reasonable estimates of human performance for the task elements can be made. One can also think of this as a top-down approach to modeling human–system performance. The example of this type of modeling that we will use in this chapter will be *task network modeling*, where the basis of the human–system model is a task analysis.

First-principle models of human behavior are structured around an organizing framework that represents the underlying goals and principles of human performance. Tools that support first-principle modeling of human behavior have structures embedded in them that represent elemental aspects of the human. For example, these models might directly represent processes such as goal-seeking behavior, task scheduling, sensation and perception, cognition, and motor output. To use tools that support first principle modeling, one must describe how the system and environment interacts with the modeled human process.

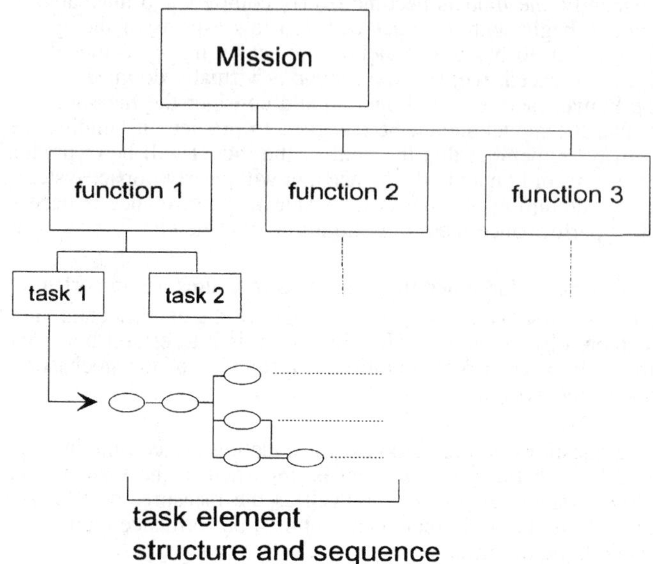

Figure 41.2 The concept of reductionist models of human performance.

An example of a very simple structure to support this type of modeling environment is presented in Figure 41.3. The example we will use of a tool designed to support this type of modeling is the Man Machine Integrated Design and Analysis System (MIDAS).

It is worth noting that these two modeling strategies are not mutually exclusive and, in fact, can be mutually supportive in any given modeling project. Often, when one is modeling using a reductionist approach, one needs models of basic human behavior to represent behavioral phenomena accurately and, therefore, must draw on elements of first-principle models. Alternately, when one is modeling human–system performance using a

Behavior

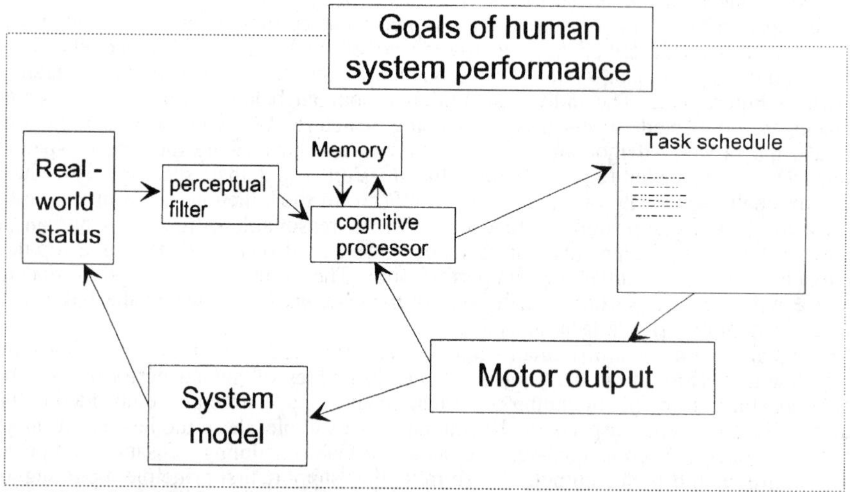

Figure 41.3 An example of the concept of first-principle models of human performance where elemental models of basic human behaviors interact to predict complex human behavior.

first-principled approach, some aspects of human–system performance may be more easily defined using a reductionist approach.

41.5 AN EXAMPLE OF A REDUCTIONIST APPROACH — TASK NETWORK MODELING

One technology that has proved useful for predicting human–system performance is *task network modeling*. In a task network model, human performance of an individual performing a function (e.g., performing a procedure) is decomposed into a series of subfunctions, which are then decomposed into tasks. This is, in human factors engineering terms, the task analysis. The sequence of tasks is defined by constructing a *task network*. This concept is illustrated in Figure 41.4, which presents a sample task network for dialing a telephone.

Task network modeling is an approach to modeling human performance in complex systems that has evolved for several reasons. First, it is a reasonable means for extending the human factors staple—the task analysis. Task analyses organized by task sequence are the basis for the task network model. Second, in addition to complex operator models, task network models can include sophisticated submodels of the plant hardware and software to create a closed-loop representation of the human–machine system. Third, task network modeling is relatively easy to use and understand. Recent advancements in task network modeling technology have made this technology more accessible to human factors practitioners. Finally, task network modeling can provide reasonable input to many types of issues. With a task network model, the human factors engineer can examine a design (e.g., control panel redesign) and address questions such as "How much longer will it take to perform this procedure?" and "Will there be an increase in the error rate?" Generally, task network models can be developed in less time and with substantially less effort than would be required if a prototype were developed and human subjects used.

Task network models of human performance have been subjected to validation studies with favorable results (e.g., Lawless, Laughery, and Persensky, 1995). However, as with any modeling approach, the real level as which validation must be considered is with respect to a particular model, not with respect to the general approach.

41.5.1 What Goes into a Task Network Model?

To represent complex, dynamic human–system behavior, many aspects of the system may need to be modeled in addition to simply task lists and sequence. In this subsection, we will use as an example the task network modeling tool *Micro Saint*.

The basic ingredient of a Micro Saint task network model is the task analysis as represented by a network or series of networks. The level of system decomposition (i.e., how finely we decompose the tasks) and the amount of the system that is simulated depends on the particular problem. For example, in a power plant model, one can create separate networks for each of the operators and one for the power plant itself. Whereas the networks may be independent, performance of the tasks can be interrelated through shared variables. The relationships among different components of the system (which are represented by different segments of the network) can then communicate through changes in these shared variables. For example, when an operator manipulates a control, this may initiate an "open valve" task in a network representing the plant. This could ripple through to a network representing other operators and subsystems.

This basic task network is built in Micro Saint via a point and click drawing palette. Through this environment, the user creates a network as shown in Figure 41.5. Networks can be embedded within networks allowing for hierarchical construction.

Figure 41.4 An example of a task network model representing a human dialing a telephone.

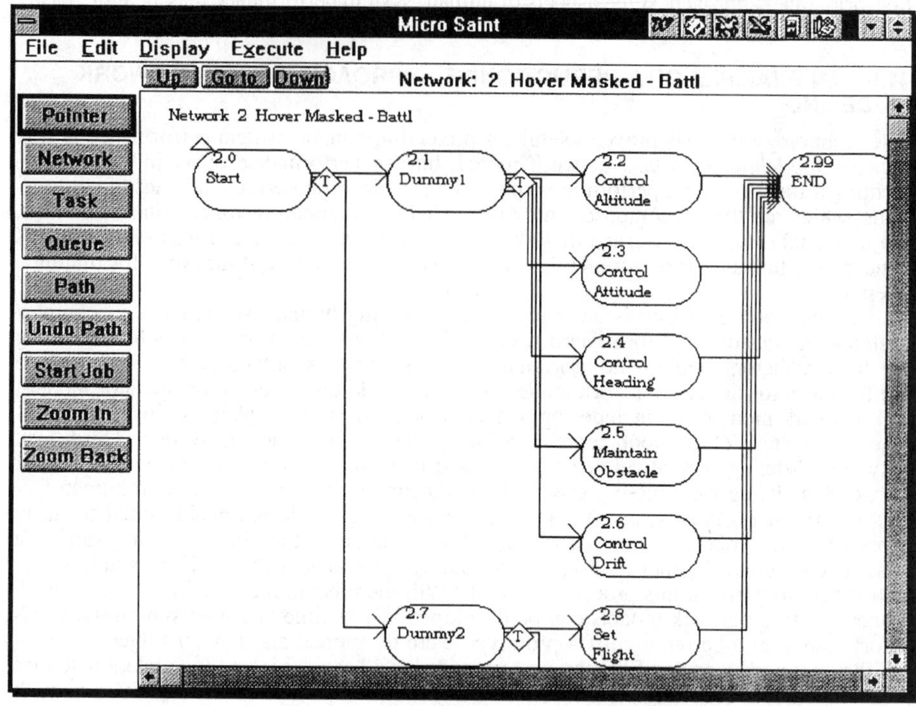

Figure 41.5 The main window in Micro Saint for task network construction and viewing.

To reflect complex task behavior and interrelationships, more detailed characteristics of the tasks need to be defined. By pointing and double clicking on a task, the user opens up the Task Description window as shown in Figure 41.6. Below are descriptions of each of the items on this window.

Task Number — An arbitrary number for task referencing.

Task Name — Any name used to identify the task.

Time distribution — Micro Saint will conduct Monte Carlo simulations with task performance times sampled from a distribution as defined by this option (e.g., normal, beta, exponential).

Mean time — This parameter defines average task performance time for this task. This can be a number, equation, or algorithm, as can all values in the fields described below.

Standard deviation — Standard deviation of task performance time.

Release condition — This condition (if one is defined) will hold up task execution until the condition is met. For example, a condition stating that task will not start before an operator is available may be represented by a release condition such as the following:

Operator = = 1;

In other words, there must be an operator available for the task to commence. If all operators were busy, the value of the variable "operator" would equal zero until a task is completed at which time an operator becomes available. This task would remain suspended until the condition was true before beginning execution, which would probably occur as a result of the operator completing the task he or she is currently performing.

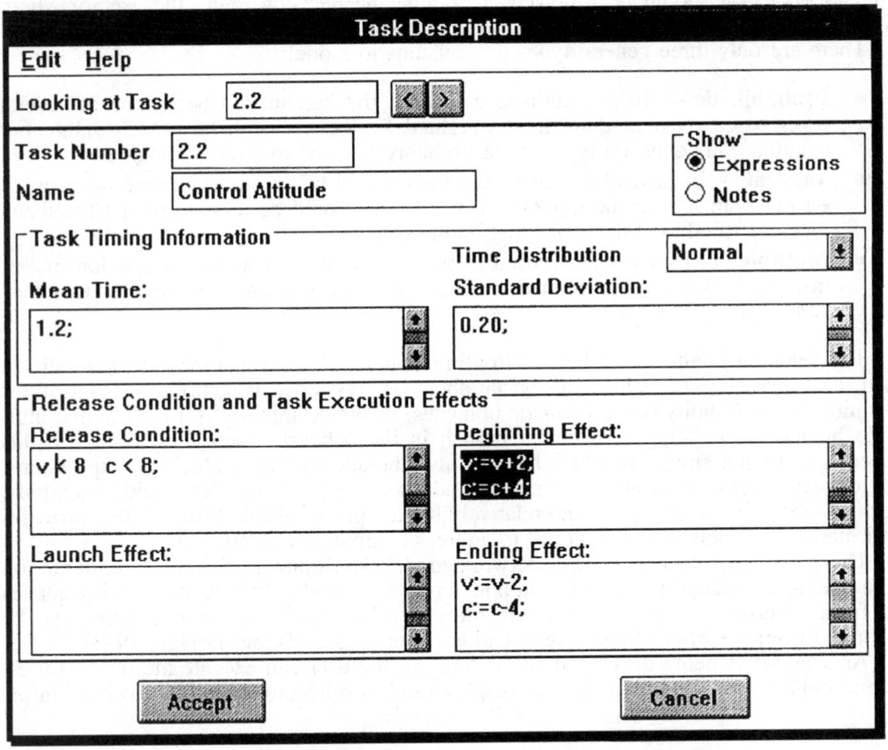

Figure 41.6 The user interface in Micro Saint for providing input on a task.

Beginning effect — This field permits the user to define how the system will change as a result of the commencement of this task. For example, if this task used an operator that other tasks might need, we could set the following condition to show that the operator is unavailable while he performed this task:

Operator := 0;

Beginning effects are one key way in which tasks are interrelated.

Launch effect — Similar to a task beginning effect but used to launch high-resolution animation of the task.

Ending effect — This field permits the definition of how the system will change as a result of the completion of this task. From the previous example, when this task was complete and the operator became available, we could set the ending effect as follows:

Operator := 1;

at which point the task using this as a release condition could begin. Ending effects are another key way in which tasks can be interrelated.

Another notable aspect of the Task Network Diagram window shown in Figure 41.5 is the diamond-shaped icons that follow some tasks. These are present every time more than one path out of a task is defined. In a task network model, this means that there are several tasks that may commence at the completion of this task. Implicitly, this means that a decision must be made by the human to select which of the following potential courses of action should be followed. To define the decision logic, the user of Micro Saint would double-click on the diamond to open up a window as shown in Figure 41.7.

There are only three general types of decisions to model:

- **Probabilistic** — In probabilistic decisions, the human will begin one of several tasks based on a random draw weighted by the probabilistic branch value. For example, this decision type would probably be used to model human error.
- **Tactical** — In tactical decisions, the human will begin one of several tasks based on the branch with the highest "value." This would be used to model the many types of rule-based decisions that humans make.
- **Multiple** — This would be used to begin several tasks at the completion of this task, such as when one human issues a command that begins other crewmembers activities.

The fields on Figure 41.7 labeled "Routing Condition" represent the values associated with each branch. The values can be numbers, expressions, or complicated algorithms defining the probability (for probalistic branches) or the desirability (for tactical and multiple branches) of taking a particular branch in the network. Again, any value on this screen can be not simply numbers but can also include variables, algebraic expressions, logical expressions, or groups of algebraic and logical expressions that would, essentially, form a subroutine. Indeed, it is the relatively hidden power of this "parser" that provides task network models with the ability to address complex problems.

There are other aspects of task network model development including the definition of a simulation scenario, defining continuous processes within the model, defining queues in front of tasks, and several other features. Further details of these features can be obtained from the Micro Saint User's Guide (Micro Analysis and Design, 1993).

As a model is being developed and debugged, the user can execute the model to test it and collect data. There are several display models reflecting differing levels of infor-

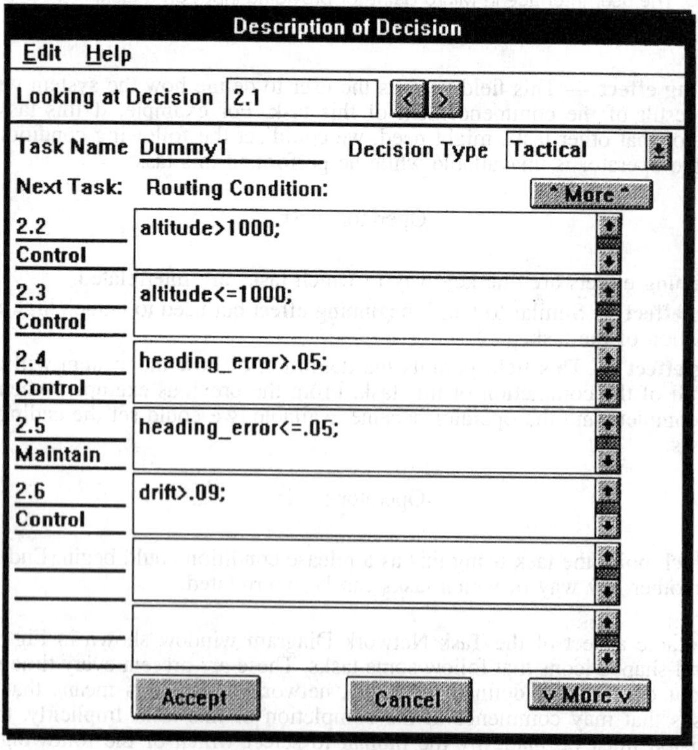

Description of Decision
Edit Help
Looking at Decision 2.1
Task Name Dummy1 Decision Type Tactical
Next Task: Routing Condition: ^More^
2.2 Control altitude>1000;
2.3 Control altitude<=1000;
2.4 Control heading_error>.05;
2.5 Maintain heading_error<=.05;
2.6 Control drift>.09;
Accept Cancel v More v

Figure 41.7 The user interface in Micro Saint for defining task branching decision logic.

mation provided to the user during execution. In the most detailed mode, the simulation pauses after every simulated task and will not continue until the user depresses the spacebar. Another mode shows the user nothing about the simulation except when it is completed. There is also a model animation mode whereby the task network is drawn on the screen and, as tasks are executed, they are highlighted. In the model animation mode, the analyst can get a very clear picture of what events are occurring in what sequence in the model. Figure 41.8 presents a sample display during model animation.

Once a model is executed and data are collected, the analyst has a number of alternatives for data analysis. The data created during a model execution can be reviewed in the model analysis environment or exported to statistical and graphics packages.

The above discussion should indicate that task network modeling is a relatively straightforward concept that is a logical expression of task and system analysis. Task network modeling is an evolution, not a revolution to the human factor practitioner. The basis for task network models of human performance is the mainstay of human engineering analysis, the task analysis. Much of the information discussed is generally included in the task analysis. Task network modeling, however, greatly increases the power of task analysis since the ability to simulate a task network with a computer permits *prediction* of human performance rather than simply the *description* of human performance that a task analysis provides. What may not be as apparent, however, is the power of task network modeling as a means for modeling human performance in *systems.* Simply by describing the systems activities in this step-by-step manner, complex models of the system can be developed where the human's interaction with the system can be represented in a closed-loop manner.

The previous discussion, in addition to being an introduction to the concepts, is also intended to support the argument that task network modeling is not simply an idea but is supported by technology. Task network modeling is a mature technology in search of applicable problem domains.

41.5.1.1 An Example of a Task Network Model of a Process Control Operator

This simple hypothetical example illustrates how many of the basic concepts of task network modeling can be applied to studying human performance in a process control environment. It is intended to practically illustrate many of the concepts described above.

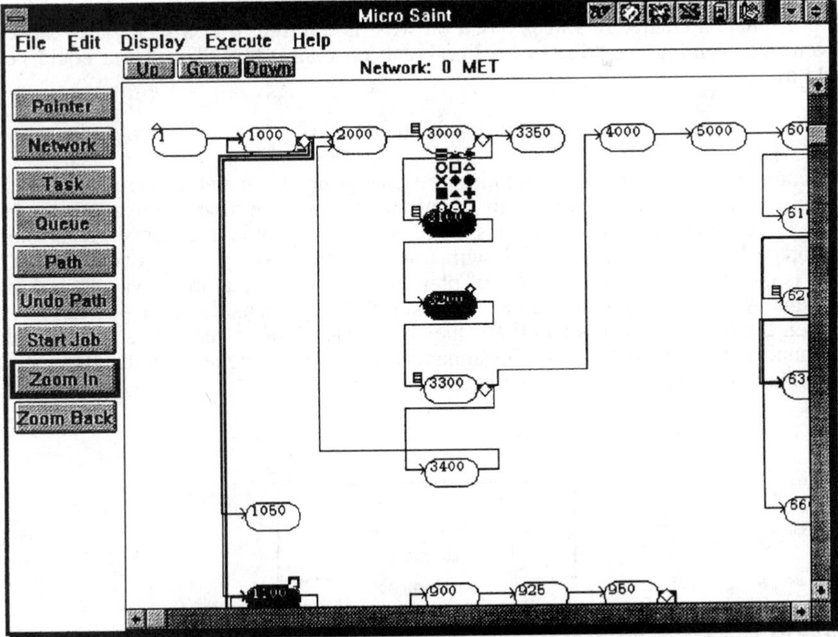

Figure 41.8 An example of a task network animation during model execution in Micro Saint.

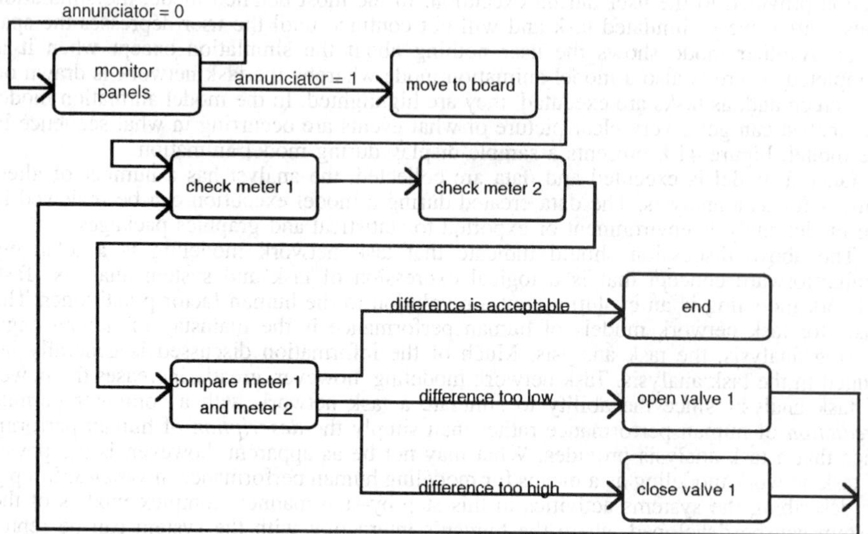

Figure 41.9 Sample task network model of a process control operator responding to an annunciator.

The simple human task that we want to model is of an operator responding to an annunciator. The procedure requires that he or she compare two meter readings. Based on the relative values of these readings, the operator must either open or close a valve until the two meter values are nearly the same. The operator activities for this model are represented by the task network in Figure 41.9. Also, to allow the study of the effects of different plant dynamics (e.g., control lags), a simple one-node model of the line in which the valve is being opened is included in Figure 41.10. The operator portion of the model will run the "monitor panels" task until the values of the variables "meter1" and "meter2" are different. The simulation could begin with these values being equal and then precipitate a change in values based on what is referred to as a scenario event (e.g., an event representing the effects of a line break on plant state). This event could be as simple as:

$$\text{meter1} = \text{meter1} + 2.0$$

or as complex as an expression defining the change in the meter as a function of line break size, flow rates, etc. An issue that consistently arises in model construction is how complex the plant or system model should be. If the problem under study is purely operator performance, simple models will usually suffice. However, if overall plant behavior is of interest, then the models of plant dynamics, such as meter values, are more important. Again, we recommend the "start simple" approach whenever possible.

When the transient occurs and the values of "meter1" and "meter2" start to divulge, the annunciator signal will go on. This annunciator would be triggered in the plant portion of the model by a task ending effect such as:

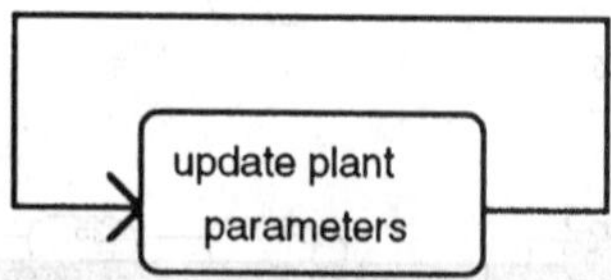

Figure 41.10 A simple one-node model of the plant that is integrated with the detailed operator model.

if meter1 <> meter2 then annunciator = 1

Once the plant model sets the value of the variable "annunciator" to 1, the operator will begin his or her activities by moving to the appropriate board. Then, the operator will continue through a loop where he or she checks the values for "meter1" and "meter2" and either opens "valve1," closes "valve1," or makes no change. The determination of whether to make a control input is determined by the difference in values between the two meters. If the value is less than the acceptable threshold, then the operator would open the valve further. If the value is greater than the threshold, then the operator would close the valve. This opening and closing of the valve would be represented by changes in the value of the variable "valve1" as a task ending effect of the tasks "open valve1" and "close valve1." In this simple model, operators do not consider rates of change in values for "meter1" and, therefore, would get into an operator induced oscillation if there was any response lag. A more sophisticated operator model could use rates of change in the value for "meter1" in deciding whether to open or close valves.

Again, this is a very small model reflecting simple operator activity on one control via a review of two displays. However, it illustrates how large models of operator teams looking at numerous controls and manipulating many displays could be built via the same building blocks used in this model. The central concepts of a task network and shared variable reflecting human–system dynamics remains the same.

Given a task network model of a process control operator in a "current" control room, how might the model be modified to address human centered design questions? Some examples are:

1. Modifying task times based on changes in the time required to access a new display
2. Modifying task times and accuracies based upon changes in the content and format of displays
3. Changing task sequence, eliminating tasks, and/or adding tasks based upon changes in plant procedures
4. Changing allocation of tasks and ensuing task sequence based upon reallocation of tasks among operators
5. Changing task time and accuracies based upon stressors such as sleep loss or drug effects

This list is not intended as a definitive list of all the ways that these models may be used to study design or operations concepts but should illustrate how these models can be used to address design and operational issues.

41.5.2 Case Studies in the Use of Task Network Modeling to Address Specific Design Issues

In this section, we examine two case studies in the use of task network simulation for studying human performance issues. The first case study explores how task network modeling can be used to assess workload issues in a cognitively demanding environment. The second example will explore how task network modeling has been used to extend laboratory and field research on human performance under stress to new task environments.

We should state clearly that these examples are intended to be representative of the types of issues that task network modeling can address as well as approaches to modeling human performance with respect to these issues. *They are not intended to be comprehensive with respect to either the issues that might be addressed or the possible techniques that the human factors practitioner might apply.* Simulation in modeling is a technology whose application leaves much room for creativity on the part of the human factors practitioner with respect to application areas and methods. These two case studies are representative.

41.5.2.1 Using Task Network Modeling To Evaluate Crew Mental Workload

Perhaps the greatest contributor to human error in many systems is the extensive workload placed upon the human operator. The inability of the operator to cope effectively with all of his or her information and responsibilities contributes to many accidents and inefficiencies. In recognition of this problem, new control or display technologies have been introduced to reduce mental workload during periods of high stress. Unfortunately, these technical solutions often introduce new tasks to be performed that affect the visual, auditory, and/or psychomotor workload of the operators. Recently, new concepts in crew coordination have focused on better management of the human workload. This area shows tremendous promise and is benefiting from efforts of human factors researchers. However, these efforts are hindered because the identification of human operator activities where excessive workload will occur is rarely straightforward. For example, high workload is not typically caused by a single task but by situations in which multiple tasks must be performed or managed simultaneously. These situations will not typically be discovered through normal human engineering task analysis or subjective workload analysis until there is a system to be tested. That is often too late to influence design.

To rectify this problem, there has been a significant amount of recent research and development aimed at human mental workload *prediction* models. Predictive models allow the designers of a system to estimate the net effect on operator workload *without human subjects experimentation.* From this and other research, a solid theoretical basis for human workload prediction has evolved as is described in Wickens, Sandry, and Vidulich (1989).

This section discusses a study using task network modeling to predict human workload. Although these examples are posed in the context of the design of a military system, the same techniques have been used in nonmilitary applications such as process control and user–computer interface design.

Modeling Crew Workload for the LHX Helicopter

In 1985, a number of projects were underway to assess the feasibility of a one-person cockpit for a complex Army helicopter that had traditionally been manned by two operators. The central design issue was workload: Could one individual reasonably be expected to perform all of the tasks required within the available time? Many studies were undertaken to assess this matter, including the one described below, which used task network modeling of human performance.

The basis of this technique is an assumption that excessive human workload is not usually caused by one particular task required of the operator. Rather, it is the human having to perform several tasks simultaneously that leads to overload. Since the factors that cause this type of workload are intricately linked to these dynamic aspects of the human's task requirements, task network modeling provides a good basis for studying how task allocation and sequencing can affect operator workload.

However, task network modeling is not inherently a model of human workload. The only relevant output common to all task network models is the time required to perform a set of tasks and the sequence in which the tasks are performed. Time information alone would suffice for some workload evaluation techniques such as Siegel and Wolf (1969) whereby workload is estimated by comparing the time available to perform a group of tasks to the time required to perform the group of tasks. Time available is driven by system performance needs and time required could be computed with a task network model. However, it has long been recognized that this simplistic analysis misses many aspects of the human's tasks that influence both perceived workload as well as ensuring performance. At the very least, this approach misses the fact that some pairs of tasks can be performed in combination better than other pairs of tasks.

Based on a review of the literature, Drews, Laughery, Kramme, and Archer (1985) concluded that the most promising theory of operator workload that was consistent with task network modeling was the multiple resource theory proposed by Wickens (e.g., Wickens, Sandry, and Vidulich, 1983). Simply stated, the multiple resource theory suggests that humans have not one information-processing resource that can only be tapped singly but several different resources that can be tapped simultaneously. Depending upon the nature of the information processing tasks required of a human, these resources would have to process information sequentially (if different tasks require the same types of resources) or possibly in parallel (if different tasks required different types of resources).

Whereas the multiple resource theory provided a theoretical basis for addressing the issues, the particular approach that we selected to include into the task network models was developed by McCracken and Aldrich (1984). This approach, although by no means a perfect manifestation of the multiple resource theory, does provide a fairly straightforward and manageable representation of the concept of multiple human information processing resources. Using McCracken and Aldrich's technique, each operator activity in a task network is characterized by the workload demand required in each of four channels, the auditory channel, the visual channel, the cognitive processing channel, and the psychomotor output channel. McCracken and Aldrich present benchmark scales for determining demand for each channel. As an example, the scale for visual attentional demands is presented below:

Value:	Activity:
1	Monitor, scan, survey
2	Detect movement, change in size, brightness
3	Trace, follow, track
4	Align, aim, orient
5	Discriminate symbols, numbers, words
6	Discriminate based on multiple aspects
7	Read, decipher text, decode

Similar scales have been developed for the auditory, cognitive, and psychomotor channels. Using this approach, each operator task can be characterized as requiring some amount of each of the four kinds of attentional demand, as represented by a value between one and seven. All operator tasks can be analyzed with respect to these demands and values assigned accordingly.

In performing a set of tasks pursuant to a common goal (e.g., engage an enemy tank), an operator frequently must perform several tasks simultaneously, or at least nearly so. For example, he or she may be required to monitor his or her hover position while receiving a communication. Given this, the workload literature indicates that the operator may either accept the increased workload (with some risk of performance degrading) or begin dumping tasks he or she perceives as less important. To factor these two issues into task network simulations, two approaches can be incorporated: (1) Evaluate combined operator workload demands for tasks that are being performed concurrently and/or (2) determine when the operator would begin dumping tasks because of overload. The method followed by Drews et al. focuses on the first approach.

During a task network simulation, the model of the operator may indicate that he or she is required to perform several tasks simultaneously as defined by the task network. The task network model evaluates total attentional demands for each of the four channels (visual, auditory, psychomotor, and cognitive) by simply summing the attentional demands across all tasks that are being performed simultaneously. For example, let us assume that at some point in the mission, the operator is simultaneously monitoring the altimeter while he or she is looking at the multifunction display to evaluate weapon status.

Let us assume that the attentional demands of these tasks are as follows:

Channel	Check Altitude	Evaluate Weapons	Combined Tasks
Visual	3	2	5
Auditory	1	0	1
Cognitive	2	5	7
Psychomotor	0	0	0

The last column above indicates what his combined attentional demands would be for each of the four channels according to the McCracken–Aldrich method.

In the task network models for helicopter operation, Drews et al. assigned values of workload to each task for each of the four channels (visual, auditory, cognitive, and psychomotor). For example, a subnetwork that describes procedures for acquiring a target is presented in Figure 41.11. As shown in the Task Description window for the "Acq Tgt, Laser, Auto" task (acquire target automatically using the laser), in Figure 41.12, when the task begins, the Task Beginning Effect increments four variables, V, A, C, and P, corresponding to visual, auditory, cognitive, and psychomotor attentional demands. Then, while the tasks are being performed, these four variables track attentional demands. When the tasks are completed, the Task Ending Effect reduces the values of these variables accordingly.

In addition to the humans' tasks, the overall model used by Drews et al. included a simple submodel of the helicopter and a submodel of the threat environment. All of these submodels interacted with the human model. This task network model was then used to simulate the performance of the pilot's tasks under different scenarios. The models produced estimates of the total attentional demands on the pilot across all tasks throughout the simulation. These workload values were characterized graphically, an example of

Micro Saint
File Edit Display Execute Help
Up Go to Down
Network: 12 Acquire Target
Pointer
Network
Task
Queue
Path
Undo Path
Start Job
Zoom In
Zoom Back
arget
12.1 Select Sensor
M
12.2 Prepare DTV
12.5 Acq Tgt, Laser,
12.6 Prepare DVO
12.9 Acq Tgt, Laser,
12.10 Prepare FLIR
12.13 Acq Tgt, FLIR, Auto
12.99 END

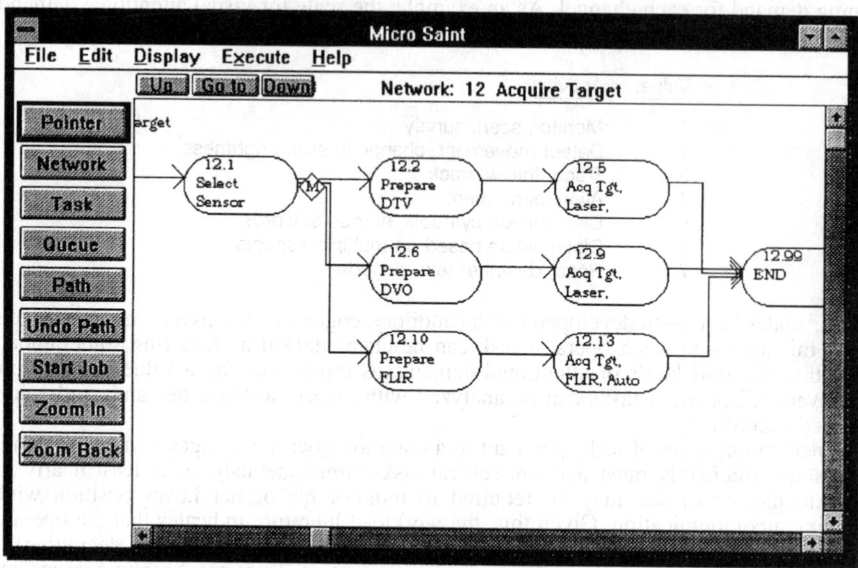

Figure 41.11 Task network for target acquisition procedures for the designed helicopter.

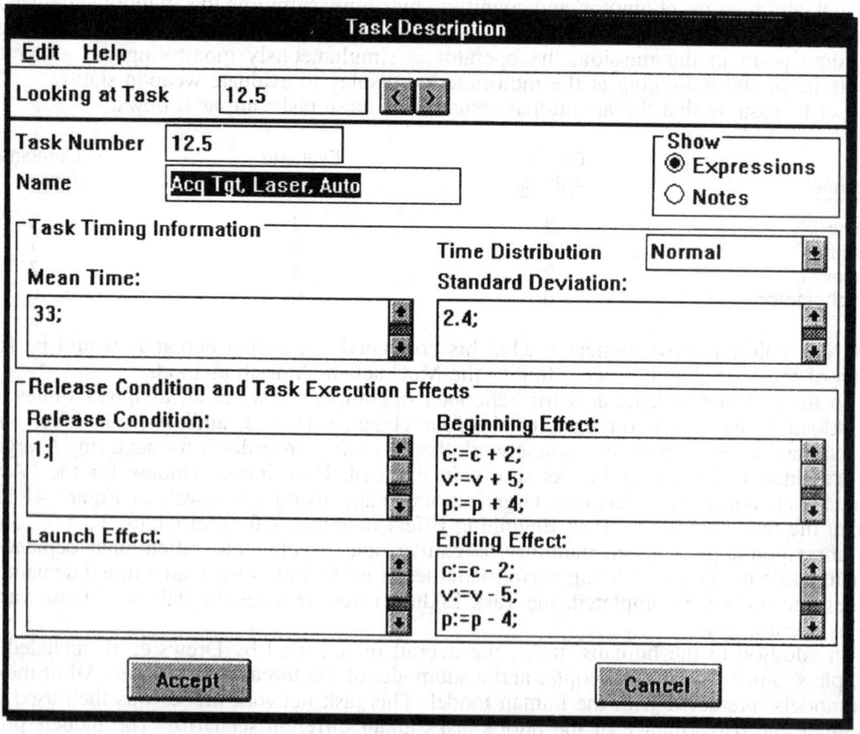

Task Description
Edit Help
Looking at Task 12.5
Task Number 12.5
Name Acq Tgt, Laser, Auto
Show
Expressions
Notes
Task Timing Information
Time Distribution Normal
Mean Time:
33;
Standard Deviation:
2.4;
Release Condition and Task Execution Effects
Release Condition:
1;
Beginning Effect:
c:=c + 2;
v:=v + 5;
p:=p + 4;
Launch Effect:
Ending Effect:
c:=c - 2;
v:=v - 5;
p:=p - 4;
Accept
Cancel

Figure 41.12 Task Description window for "Acq Tgt, Laser, Auto" task that includes workload channel variable manipulations used to predict human mental workload.

which is shown in Figure 41.13. Using the scales developed by McCracken and Aldrich, any value of workload over seven was cause for concern and, when these were observed, further reviews were conducted. By examining the points in the mission at which these attentional demands are high, Drews et al. were able to assess the mission segments for which operator workload would be excessive.

At the end of this study, it was determined that a one-person cockpit was going to be quite different to achieve, even with the most optimistic projections of aiding, automation, and pilot support that could be achieved. These findings were supported by other research and the helicopter was built as a two-person system.

Extensions of This Approach To Simulating Crew Mental Workload to Other Environments

After this study was completed, computer software to specifically support the above simulation methodology was developed by the Army as part of the HARDMAN III human engineering tool set. This tool, described in Allender et al. (1995) and Fontenelle and Laughery (1988), integrates task network modeling software with features to specifically support the workload analysis method discussed above. It provides the human factors practitioner with an environment that supports the analysis of task assignment to crewmembers based on three factors:

1. *Workload of crewmembers* — Tasks should be assigned to minimize the amount of time crewmembers will spend in situations of excessive workload.
2. *Time performance requirements* — Tasks must be assigned and/or sequenced so that they are completed within the available time. This consideration is essential since time constraints often drive the need to perform several tasks simultaneously.
3. *Access to controls and displays* — Tasks cannot be assigned to crewmembers that do not have access to the necessary controls and displays.

The method for computing workload used by HARDMAN III is nearly identical to that used by Drews et al. The time to perform a series of tasks is a natural output of task

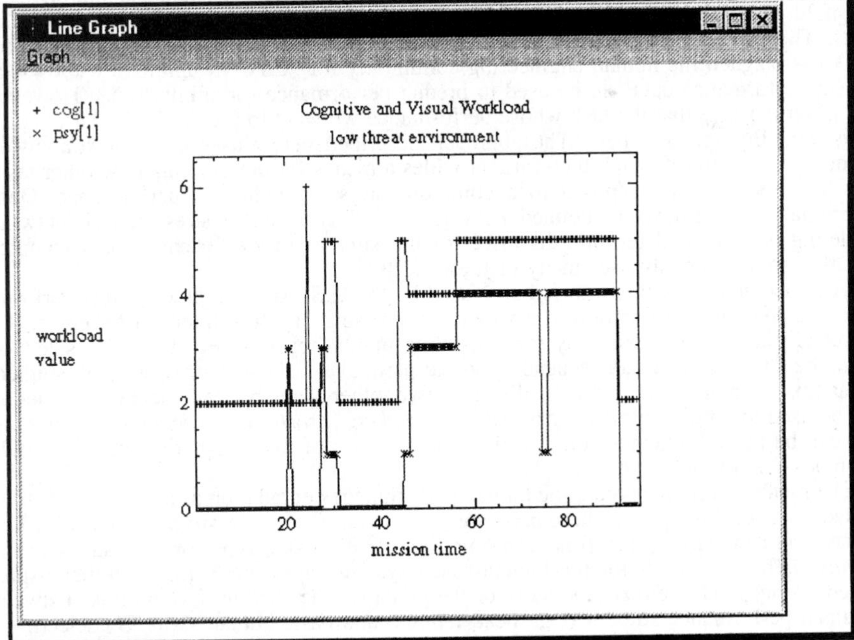

Figure 41.13 Sample predicted workload profile for the designed helicopter that was developed with a Micro Saint model.

network simulations. The access to controls and displays is information that the analyst must provide. HARDMAN III provides a convenient means for simulating human system performance based upon these three factors using task network simulation.

Of course there are numerous theoretical questions regarding this simplistic approach to assessing workload in an operational environment. However, even the use of this simple approach has been shown to provide useful insight during design. For example, in a study conducted by the Army (Allender, 1995), a three-man crew design was evaluated using HARDMAN III model. The three-person model was constructed using data from a prototype four-person system. From this model-based analysis, the three-person design was found to be unworkable. Later, human subjects experimentation verified that the model's workload predictions were sufficiently accurate to point the design team in an essential direction.

Finally, the Army has supported the additional development of more refined methods of predicting workload. This approach as manifested in another custom software package WinCrew, is described in Plott (1995). WinCrew overlays the W/INDEX manifestation of the multiple resource theory of workload (Boettcher, North, and Riley, 1989) into a task network–based environment. In addition to a better estimate of workload, WinCrew is unique in that it has built-in constructs for simulating *workload management* strategies that operators would employ to accommodate points of high operator workload. The ultimate result of simulating the workload management strategies is that the operator task network being modeled is *dynamic*. In other words, the task sequence and assignments to operators may change in response to excessive operator workload as the task network model executes. These changes may be as simple as one operator handing tasks off to another operator to reduce workload to an acceptable level or as complex as the operator beginning to timeshare tasks in order to complete all the assigned tasks, potentially requiring more time for the lower priority ones. This innovation in modeling provides greater fidelity in efforts that model human behavior in the context of system performance, particularly in high-workload environments such as complex system control and management.

41.5.2.2 Using Task Network Modeling as a Means of Extending Research Findings on Human Performance Under Stress to New Task Environments

In this example of the use of task network modeling, we demonstrate how network modeling has been used by LaVine, Laughery, and Peters (1995) to extend laboratory data and field data collected on one set of human tasks to predicting performance on similar tasks. The problem of extending laboratory or field human performance data to other tasks has plagued the human engineering community for years. We intuitively know that human performance data can be used to predict performance for similar tasks. However, it is often the case that the task whose performance we want to predict is similar in some ways, but different in others. The approach described below uses a skill taxonomy to quantify task similarity, and, therefore, provides a means for determining how other tasks will be affected when exposed to a commons stressor on human performance. Once functional relationships are defined between a skill type and a stressor, task network modeling is used to determine the effect of the stressor on performance of a complex task that simultaneously uses many of these skills.

The specific approach below is being used by the U.S. Army to predict crew performance degradation as a function of a variety of stressors. It is not intended to represent a universally acceptable taxonomy for simulating human response to stress. The selection of the best taxonomy would depend upon the particular tasks and stressor being studied. What this example is intended to illustrate is another way that task network modeling can be used to predict human performance by *making a series of reasonable assumptions* that can be played together in a model for the purpose of making predictions that would be impossible to make otherwise.

The methodology for predicting human performance degradation as a function of stressors consists of three parts. These parts are (1) a *taxonomy* for classifying tasks according to basic human skills, (2) *degradation functions* for each skill type for each stressor, and (3) *task network models* for the human-based system whose performance is being predicted. Conceptually, either laboratory or field data can be used to develop links between a human performance stressor (e.g., heat, fatigue) and basic human skills. By selecting a skills taxonomy that is sufficiently discriminating to make this assumption reasonable, one can assume that the effects of the stressor on all tasks involving the skill will be approximately the same. The links between the level of a stressor (e.g., fatigue) and resulting skill performance (e.g., the expected task time increase from fatigue) are defined mathematically as the degradation function. The task network model is the means for linking these back to complex human–system performance.

The Taxonomy

The basic premise behind the taxonomy is that the tasks that humans perform can be broken down into basic human skills or atomic tasks (Roth, 1992). The taxonomy that was used by Roth consists of five skill types: (1) attention, (2) perception, (3) psychomotor, (4) physical, and (5) cognitive skills. These taxonomic skills are described by Roth as follows:

- *Attention* — The ability to attend actively to a stimulus complex for extended periods of time in order to detect specified changes or classes of changes that indicate the occurrence of some phenomenon that is critical to task performance
- *Perception* — The ability to detect and categorize specific stimulus patterns embedded in a stimulus complex
- *Psychomotor* — The ability to maintain one or more characteristics of a situation within a set of defined conditions over a period of time, either by direct manipulation, or by manipulating controls that cause changes in the characteristics
- *Physical* — The ability to accomplish sustained, effortful muscular work
- *Cognitive* — The ability to apply concepts and rules to information from the environment and from memory in order to select or generate a course of action or a plan. This includes communicating the course of action or plan to others

These five skills covered most of the tasks that were of interest to the Army, and still provided a manageable number of categories for an analyst to use.

Degradation Functions

The degradation functions quantitatively link skill performance to the level of a stressor. The degradation functions can be developed from any data source including standard test batteries or actual human tasks. Through statistical analysis, one can build skill degradation functions for each taxon. These functions map the performance decrement expected on a skill based on the parameters of the performance shaping factor (e.g., time since sleep). An example of these functions is presented in Figure 41.14.

Incorporating the Degradation Functions into Task Network Models To Predict Overall Human–System Performance Degradation

The key to making this approach useful to predicting complex human performance is the task network model of the new task. In the task network model of the human's activities,

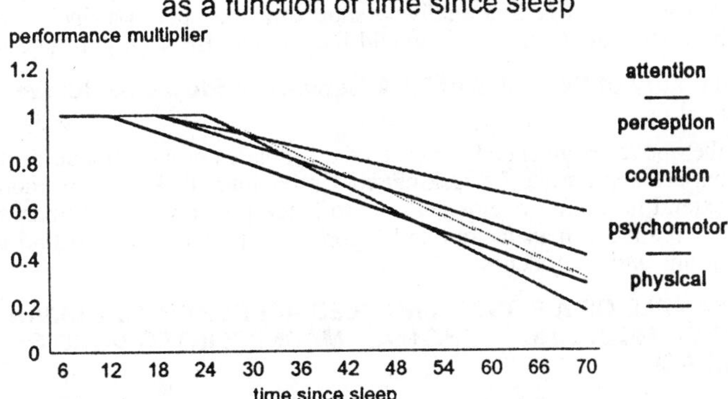

Figure 41.14 An example of the performance degradation functions associated with each of the human skills from the taxonomy.

all tasks in the task network models are defined with respect to the percentage of each skill required from the taxonomy. For example, the following are ratings for tasks faced by a console operator responding to telephone contacts:

- Detect ring — 50% attention, 50% perception
- Select menu item using a mouse — 40% attention, 60% psychomotor
- Interpret customer's request for information — 100% cognitive

In building the task network model, one can build functions to degrade a specific task's performance through an arithmetic weighing of skill degradation multipliers that are derived from the degradation functions. For example if the fatigue parameter was "time since sleep" and the value of that parameter was "36 hours since sleep," then the multipliers would be as follows in the example above:

- Attention performance multiplier =0.82
- Perception performance multiplier = 0.808
- Cognition performance multiplier = 0.856
- Psychomotor performance multiplier = 0.784
- Physical performance multiplier = 0.727

Based upon these multipliers and the above task weightings, the specific task effects would be:

- Detect ring — (50% attention, 50% perception)
  task multiplier = $0.5 \times .82 + 0.5 \times 0.808 = 0.814$
- Select menu item using a mouse — (40% attention, 60% psychomotor)
  task multiplier = $0.4 \times 0.82 + 0.6 \times 0.784 = 0.7984$
- Interpret customer's request for information — 100% cognitive
  task multiplier = 0.856

In a model of the complex tasks being examined by LaVine et al. (1995), the task networks consisted of several dozen, or even several hundred tasks. Through the approach described above, each task in a model exhibited a unique response to a stressor depending upon the particular skills that is required. The task network model then provided the means for relating these individual tasks' performance to overall human–system performance as a function of stressors level (e.g., the time to perform a complex series of tasks involving decision making and error correction). Through this type of analysis, LaVine et al. were able to develop curves such as that shown in Figure 41.15 relating human performance to a stressor. These relationships would have been virtually impossible to develop experimentally.

Again, there were a number of simplifying assumptions that were made in this research. However, by being willing to accept these assumptions, LaVine et al. were able to characterize how complex human–system performance would be affected by a variety of stressors over a wide range in a relatively short time. As such, LaVine et al. were able to estimate the effects of stressors that would have otherwise been pure guesswork.

41.5.2.3 Summary of Examples of Task Network Modeling of Human–System Performance

Once again, the above are intended to serve as examples, not a catalogue of problems or approaches that are appropriate for task network modeling. Task network modeling is an approach to extend task and systems analysis to make predictions of human system performance. The creative human factors and ergonomics practitioner will find many other useful applications and approaches.

41.6 AN EXAMPLE OF A FIRST-PRINCIPLED APPROACH TO HUMAN–SYSTEM PERFORMANCE MODELING — THE MAN–MACHINE INTEGRATED DESIGN SYSTEM (MIDAS)

The other functional approach to modeling human performance is based upon the mechanisms that underlie and cause human behavior. Since this approach is based on fundamental principles of the human and his or her interaction with the system and

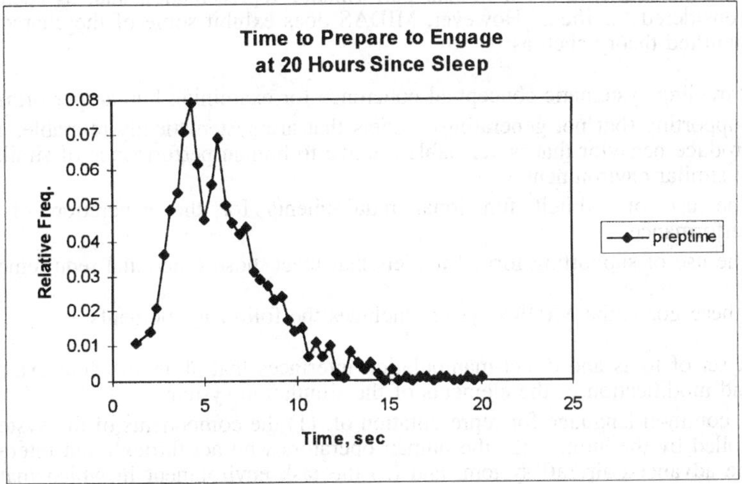

Figure 41.15 Frequency distribution of expected human performance as a function of time since sleep that was derived using task network modeling.

environment, we have designated them as *first-principle* models. By integrating these models with models of the system and environment, the human factors specialist can predict the full behavior of large-scale interactive human–machine systems. The Man Machine Integrated Design and Analysis System (MIDAS) follows in the tradition of integrated, first principled models of human performance such as PROCRU (Baron, Muralidharan, Lancraft, and Zacharias, 1980) in that the modeling framework provides models of emergent human behavior based on elementary models of human behaviors such as perception, attention, working memory, and decision making. In the operation of these elementary models, MIDAS shares some of the characteristics of the task network approach. However, MIDAS is focused around an integrated architecture whereby micromodels of human performance feed forward and feedback to the other constituent models in the *human system* rather than being linked primarily to the human's *activities* as in task network models.

41.6.1 Background

The joint Army–NASA Aircrew/Aircraft Integration (A3I) Program, was initiated in 1985 to explore of computational representations of human–machine performance to aid crew system designers. The major product of this effort is a human factors computer-aided-engineering system called MIDAS (Man-Machine Integration Design and Analysis System). MIDAS's goal is to revise the system design process in order to place more accurate information into the hands of the designers early in the process of human engineering design. It is also intended to identify and model human–automation interactions with flexible representations of human–system function. With MIDAS, designers can work with computational representations of the crew station and human operators to discover problems and ask "what-if" questions regarding the projected mission, equipment, and environment.

In addition to its use in development and design, MIDAS offers a framework in which models of human cognition and performance can be tested with respect to their ability to predict complex human–system behavior. The MIDAS framework systemizes the interaction of computational representations of the elemental aspects of human performance. Models of human performance from perception through cognition and action can be implemented within this framework. Then, in MIDAS interplay of the models produces simulations of behavior that can be compared to empirical data, thereby providing a test of the theoretical perceptual and cognitive models.

Whereas the MIDAS framework supports models that produce simulations of human behavior, the structure is not intended to be a "unified theory of cognition," perception, or action in the terms specified by Newell (1990). The MIDAS system does not make "structural" assumptions about human performance representation and, as such, it can not be considered "unified." However, MIDAS does exhibit some of the characteristics of good unified theory such as:

- Providing systematic conceptual coherence for examining human performance
- Supporting (but not generating) models that are systematically refinable, and that produce behavior that is verifiable relative to human performance of similar tasks in similar environment
- The use of explicit functional requirements for the generation of human performance
- The use of supporting formal models that meet those functional requirements.

Toward these goals, the MIDAS system includes the following elements:

- A set of tools and direct-manipulation interfaces that allow efficient examination and modification of the elements of the simulation system
- A common language for representation of: (1) the components of the system controlled by the human, (2) the human operators who act through and interact with the advanced aircraft system, and (3) the task environment in which that action takes place
- A mechanism for propagating the impact of manipulation in the scenario, equipment, or operator profiles to ensuing mission performance

MIDAS supports the aforementioned elements with the following characteristics:

- *Modifiability and manipulability.* The basic mode of operation for MIDAS users is to explore the impact of changes to the baseline design. Thus, the capability for systematic change is critical. Of equal important is system extensibility. To be generally useful, the modeling environment should be applicable to many types of design changes, and to many operational domains. The MIDAS architecture is designed to allow extensions of this type with minimal disruption to the existing core MIDAS system.
- *Transparency.* The analysis system must provide designers with explicit and transparent reference to the rules, decision making strategies, heuristics, and assumptions under which the human–machine system is assumed to be operating, as well as predicted performance. For example, at any point in the simulation a designer should be able to examine the cognitive state of the human operators, the rules that are being used to guide their behavior, and their nominal workload. The designer should also be able to perform sensitivity analyses on critical parameters of the human–machine system. Similarly, the state of equipment or mission progress should be able to be probed in order to relate the system state to the operator's performance.
- *Dynamic analysis capability.* The simulation system must produce a stream of behavior in the form of dynamic timelines describing not only its state and structure, but also sequences of action over time and contingent responses of the human–system behavior. The system must support testable hypotheses. Designers must be able to analyze the events occurring in a simulation scenario and relate this performance to "man-in-the-loop" simulation data. In MIDAS each action taken, decision made, and communication event is logged by the analysis system.

41.6.2 System Architecture

There are two perspectives on the MIDAS system architecture that describe the system to support these modes of analysis. These are the functional architecture and structural bases of the system.

41.6.2.1 MIDAS Functional Architecture

MIDAS is intended to function in three modes. First, there is a *specification mode* in which users, i.e., designers of prototype systems, are provided with a set of tools to specify operator characteristics, mission characteristics, and characteristics of the physical plant or crew station. Editors are provided for activities, equipment, mission, the rule base, and the human performance models. Once the system to be examined is specified, MIDAS supports two paths for analysis. The first path is termed *interactive mode.* This mode uses scenario-independent layout and animation of displays and controls for assessments of visibility and legibility, examination of anthropometric characteristics, and analyses of cockpit topology and configuration. The output of interactive mode analyses are useful for comparison to design guidelines, such as MIL-STD-1472. The other analysis path for MIDAS users to take is that the *simulation mode.* This mode provides facilities so that the specification of the human operator, the equipment, and the mission goals can be run in a dynamic simulation. The simulation mode exercises all of the system and the elemental human performance models in an integral process that results in activity traces, task load timelines, information flow analyses, as well as mission performance measures. Operator measures are generated that enable identification of significant human performance variables such as potential resource conflicts, task loading as a function of configuration or mission, and information requirements. Also in the simulation mode, mission performance measures are generated that include operational timelines, routes, contingencies, and mission effectiveness metrics.

The functional view of the full MIDAS system and its components is provided in Figure 41.16.

41.6.2.2 MIDAS Structural Architecture

The MIDAS system is built around object-oriented software structures. An object-oriented simulation is composed of objects and software entities that maintain and manipulate values representing human, equipment, and environmental states. An object's state (the description of its parameters and defining values) is stored as a set of numerical and symbolic values in structures called slots. The slot values of an object comprise the specification of its attributes. For example, in the case of a physical object, shape, size, and position values are typical attributes, MIDAS's object-oriented software also provides for procedures or methods that can operate on some, or all, of an object's slot values.

Objects in MIDAS interact with each other by exchanging messages. The way an object responds to messages and the way it alters its slots depends on an object's "class." In an object-oriented simulation system, it is frequently the case that the description of a class of objects inherits the description of other classes. For example, in a MIDAS model of air traffic control, the human-operator class and the flight-maintenance–computer class include the characteristics of the class ingredient agent. The behavior of an intelligent agent, once described, can then be incorporated every time a class of object embodying the same type of intelligent behavior is described.

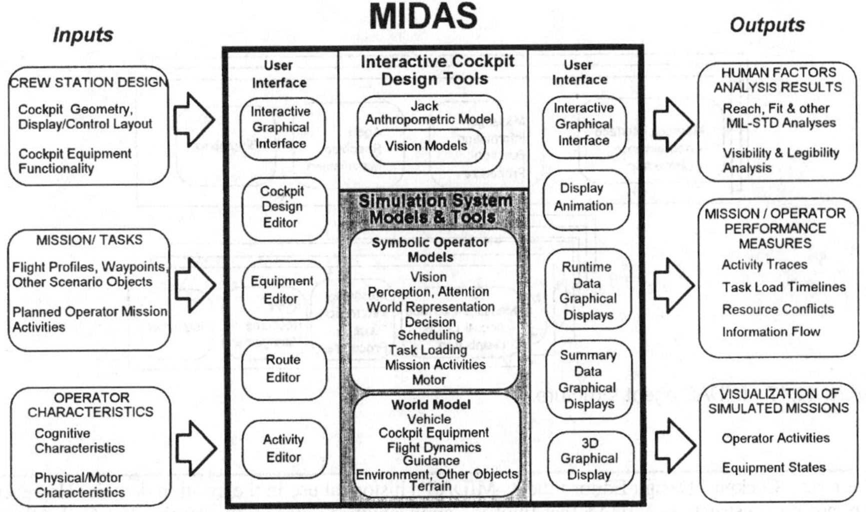

Figure 41.16 Functional view of the MIDAS system.

Agent Architecture

The simulation is further organized by the imposition of an architecture that provides for the interaction among all of the objects in a MIDAS simulation. The interaction of these elements of MIDAS is provided through an *agent* architecture. This architecture is designed to provide a basis for multiple, concurrent, and independent intelligent agents as described in Agha (1985). The agent architecture supports the MIDAS modeling system goals of modularity, extensibility, perspicuity, and analytic tractability. Further, the agent architecture provides a syntax for communication among system components as well as a uniform description of the objects that comprise the simulation system. All agents in MIDAS share a common structure, as illustrated in Figure 41.17.

Each agent consists of a message interface that determines the other agents to and from which messages can be passed. The agent has a function, method, or procedure it performs either (1) as a consequence of messages received from other agents or (2) at specified simulation times. Agents can also have the capability to manage resources associated with their own action (e.g., they can determine the sufficiency or availability of information necessary to support a required activity). Finally, each agent has a biographer that tracks all of the messages received, operations performed and messages sent.

MIDAS Agent Content. The MIDAS system includes the classes of agents described below. This description is the current set that has been implemented to support the simulation and analyses conducted to date. The agent structure encourages incremental development of new capabilities and extension of base capabilities into other application domains.

Physical Component Agents.

- *Equipment:* MIDAS uses current technologies and capabilities of commercially available Computer-aided design (CAD) databases. The MIDAS graphical representation for the physical entities in an environment are created and held in a system called the Cockpit Design Editor (CDE).[1] The MIDAS system not only represents the graphical or physical aspects of projected equipment, but also supports definition of the ways in which equipment components in a MIDAS simulation operator. It associates standard operating procedures and functional activities with each component. The equipment agent is designed to allow the MIDAS user to develop specific components from more generic ones.

The behavior of an equipment component may be expressed in four different formats. These four formats are not mutually exclusive — a single component can use a maximum of any or all of them. The different formats are: a finite state machine (FSM), a time script, a stimulus–response script, or a LISP method and associated functions.

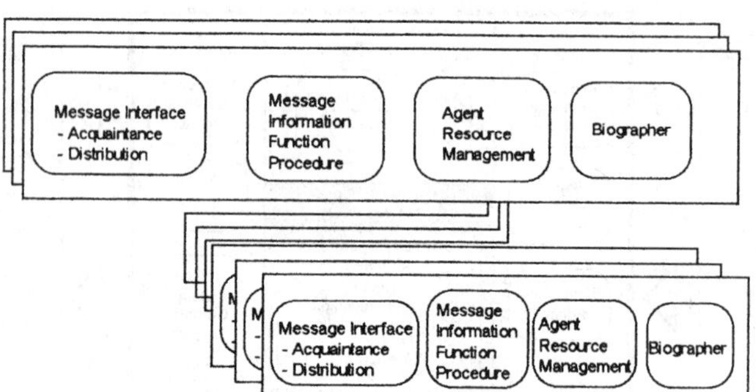

Figure 41.17 MIDAS agent structure.

[1]The term "Cockpit" Design Editor reflects MIDAS's historical use in the aviation domain. However, there are no constructs in MIDAS that limit it's applicability to other task domain. The CDE allows rapid prototyping and animation of virtually any 3D physical entity and has been used for modeling spacecraft, ground vehicles, and commercial aircraft components.

- *Physical World Agents:* In the application that are currently the focus of MIDAS, the physical world is represented by terrain and aeronautical equipment. The MIDAS system also provides the designer with tools to specify the area in which the operations take place. Physical components or environments can also be represented within MIDAS to support a system or simulated operator's reasoning about physical properties such as location, distance, routes, and visibility. For example, the MIDAS system uses a simple aerodynamics and guidance model to fly a prototype helicopter. Given a predicted reference point in space, a control component computes the collective, cyclic, and pedal control movement required to effect the desired trajectory to the reference point.

Human Operator Agents. The next, and more challenging task, is to provide system designers with an examinable, consistent, and valid representation of the human operator(s) who will interact with the equipment and respond to a scenario. The particular interest of this modeling is characterization of the process of perception, decision making, activity selection, timing, and task loading as the operator interacts with the system under study. In addition to the behavioral components of an operator's interaction, MIDAS is capable of identifying and tracking the various information requirements and transformations that occur as the scenario behaviors unfold.

- *Human Performance Representations:* In order to populate MIDAS with simulated operators to carry out and supervise automated processes and vehicular control, MIDAS includes representations of human cognitive, perceptual, and motor operations. These models describe (within their limits of accuracy) the responses that can be expected of human operators in several areas that are critical to safe and reliable operation of advanced automated systems. The fundamental human performance elements of these representations can be applied to any human–machine interaction environment. Each of the human operators modeled by MIDAS contains the following models and structures, the interaction of which will produce a stream of activities in response to mission requirements, equipment requirements, and models of human performance capabilities and limits. These models are undergoing constant revision as more is known about the first principles of these elemental human behaviors.
- *Physical Representation:* An anthropometric model of human has been developed in conjunction with the University of Pennsylvania (Badler, Phillips, and Webber, 1993). The model used is called Jack™, and is an agent in MIDAS. The Jack™ agent's purpose is to represent human figure data (e.g., size and joint limits) in the form of an animated mannequin that moves through various postures representing the physical activities of the simulated human operator. The graphic representation of the Jack assists designers in questions of workstation geometry, reach accommodation, restraint, and occlusion.
- *Perception and Attention:* The simulated human operator in MIDAS constantly receives sensory input. MIDAS presently focuses on modeling visual perception. During each simulation cycle, the perception agent computes what environment or cockpit objects are imaged on the operator's retina, tagging them as in or out of the peripheral and foveal fields of view (90° and 5°, respectively), in or out of the attention field of view (variable depending on the task), and in or out of focus, relative to the fixation plane. An environmental object can be in one of several states of perceptual attention. Objects in peripheral visual fields are perceived and salient changes in their state, (e.g., change in color or flashing) will be passed to the internal world representation. For detailed information to be fully perceived (e.g., reading of textual messages) the data of interest must be in focus, attended, and within the foveal field of view for 200 mse.

The perception agent also controls the simulation of commanded eye movements via defined scan, search, fixate, and track modes. Differing stimuli salience and pertinence are also accommodated through a model of preattention in which specific attributes (e.g., color or flashing) are monitored to signal an attentional shift. MIDAS's models of attention and preattention are patterned after the work of Remmington, Johnston, and Yantes (1992).

- *Updatable World Representation (UWR):* In MIDAS, the UWR provides a structure whereby each of the multiple, independent human agents, representing individuals and cooperating teams of pilots and flight crews, accesses its own tailored or personalized information about the operational world. The contents of an UWR are determined, first, by presimulation loading of required mission, procedural, and equipment information. Then, data are constantly updated in each operator's UWR as a function of the perceptual mechanisms previously described. The data of each operator's UWR are operated on by *daemons and rules* (discussed below) to guide behavior and are the sole basis for a given operator's activity. Providing each operator with his or her own UWR accounts for the significant operational, reality that differing members of a cooperating control team have different information about the world in which they operate. Further, the individual operator may, or may not, receive a piece of information available to the sensory apparatus as a function of perceptual focus as the relevant point in the mission. Although ideally the human operators' representation of the world is consonant with the state of the world, in fact, this is rarely the case. The capability for both systematic and random deviation from the "ground truth" of the simulated world is an important component of any system that is designed to represent and analyze complex human performance.

As described above, input to the UWR is mediated by attention and perception. These functions are activation filters that allow more or less of the stimuli in the environment to enter the memory structure.

The organization of perceptual data and knowledge about the world in a UWR is accomplished with a structure called a semantic net. A semantic net is a structure containing objects called nodes that represent concepts. Relationships among nodes in a semantic network are expressed as links. The types of links can be described by an analyst or designer to represent relationships that the analyst or designer considers critical. Hierarchies in which system–subsystem relationships are known can also be implemented. For example, "part-of" links can be used to relate a physical subsystem to the system it is part of among the physical elements of the crewstation. An altimeter node could be linked to a pilot-primary flight instruments node using a part-of link. This link would support critical reasoning in an emergency, if it were important to simulate an operator's reasoning that a malfunction in the primary flight instruments could be causing an anomalous altimeter reading. Other uses of part-of links include "team–base" linking among the human elements of the simulation (e.g., a pilot is part-of a flight crew, and a flight crew is part-of an aircraft squad) and mission-based temporal relationships) e.g., an ingress phase is part-of a mission, and a way-point is part-of a flight route). Information in the UWR is subject to decay, representing a forgetting function. The decay function for working memory is encoded as an exponential that is calculated on the time since an activity or object entered the store.

- *Activity Representation:* Activities are MIDAS objects that simulate actions that may be performed by agents in the system. Representations of activities available to an operator are contained in that operator's UWR. Activities are characterized by:
  - Preconditions, that define the allowable conditions for their spawning and decomposition
  - Satisfaction conditions, which define their successful completion
  - Spawning specifications, which detail the temporal and logical constraints on any "child activities" that might be needed for activity performance
  - Decomposition methods, that describe in a context-sensitive way what children should be spawned to accomplish a higher level activity's goals
  - Interruption status and interruption specifications, which detail the interruption and resumption methods for that activity
  - Loads, which indicate task performance requirements in terms of visual, auditory, cognitive, and psychomotor resource requirements
  - Duration, either estimated or calculated by an activity-specific function
  - Priority, which in this implementation is provided as a fixed table of relative priorities assigned to action and constrained by operator type

Activities can also be "forgotten" through a mechanism that defers activities through interruption. A user-defined number of interruptions of an activity may result in its permanent removal from the active goal queue and its consequent forgetting.

Activities are performed by all of the dynamic entities (agents) in the simulation. However, activities performed by human operators and other intelligent agents in the simulation are of particular interest to the analysts of human–machine systems. An aggregation of MIDAS's many separate human performance elements, which we term a *symbolic operator model* (SOM) agent, organizes and mediates the performance of activities through the operations of scheduling, task loading and prioritization.

- *Rule-based and Decision Activities:* This method of introducing activities into the simulation world provides an SOM with the ability to respond to contingencies in the simulation world. Contingent activities involve the application of daemon, rule, and decision theoretic models that act on data incoming to the SOM.
  - *Daemons.* Informaton about the simulation world is available to a simulated operator's UWR through the action of objects that model perceptual processes. Changes to the value of any information currently held in the UWR are monitored by "daemons" that may notify other objects in the SOM when a significant change in state has occurred.
  - *Rules.* If conditions are appropriate, a rule may spawn activities in response to changes in the simulation world. If rules are not available to guide behavior then a more computationally intensive decision process is invoked.
  - *Decisions.* In general, decisions are made when the SOM notices anomalous situations triggered by deviations from the expected state. Human decision making has been shown to be context sensitive and variable (Tversky and Kahneman, 1974). MIDAS provides for prescriptive decision methods to be applied according to the amount of time available to make the decision (the "decision horizon"). The decision horizon is calculated as a function of vehicle and operator state and is used in the selection and application of particular decision strategies. The decision-making process as currently simulated in MIDAS is a computationally deterministic process. The prescriptive decision making agent includes six algorithms, each of which can be used for selecting among alternative options. As adapted from Payne, Bettman, and Johnson (1988), these include:
    - Weighted additive
    - Equal weighted additive
    - Lexicographic
    - Elimination by aspect
    - Satisficing conjunctive
    - Majority of confirming dimensions

    Each of these algorithms uses a different combination of attribute values, attribute weights, and attribute cutoff values for calculating the "goodness" value of the options.
- *Scheduler:* Activities derived from the goals of the mission by the SOM and its subagents are queued and passed to the schedular as reported in Shankar (1991). The scheduler then interacts with the task loading model. The task loading model computes task loads on the human operators of the system to determine an estimated load for the activities to be scheduled and to determine an order of activity performance based on a set of operator strategies for scheduling around the available resources. Strategies such as "balanced loading over all resources dimensions" and "task time minimization" are implemented in MIDAS.
- *Task Loading Model:* The task loading model in MIDAS is calculated on the basis of an assumed multiple resource representation of capacity in the dimensions of vision, audition, physical, and cognitive components. The model further assumes that the operator has fixed amounts of these resources available, and that the loads imposed in these four categories vary as a function of the task. This is very similar to the technique developed by McCracken and Aldrich (1984) discussed earlier. The task loading methods applies as the simulation runs are responsive to the task ensemble and to the context in which the tasks are to be performed. The resulting loads are returned to the scheduler, which uses them as human resource constraints in scheduling the activities.

To produce a stream of human-system behavior, the MIDAS agents described above execute collectively during a mission simulation as depicted in Figure 41.17. As previously mentioned, declarative and procedural information about the designated mission and vehicle equipment is held in the simulated operator's updatable world representation (UWR). Then, during each simulation time cycle (presently 100 mse), information from the world is filtered by perception and attention models and passed to the UWR. The operator uses this sensed information as required by the mission goals to select appropriate lower level activities. These activities are then scheduled and passed to Jack for execution, where they generally affect the cockpit equipment models, allowing the cycle to then repeat.

As this discussion illustrates, MIDAS is a model of human performance that is based upon the underlying principles of what drives human behavior in a goal oriented, and resource constrained complex environment. Although fundamentally different from a reductionism (e.g., task network) approach, the objectives of modeling and predicting human performance in real, complex systems is identical.

41.6.3 Case Studies in MIDAS Applications

41.6.3.1 Midas Case Study 1—Predicting Flight Crew Performance in the Advanced Air Traffic Management System

With the goals of providing safer, more flexible, more fuel-efficient routing, and increased capacity in the United States air traffic system, NASA and the Federal Aeronautics Administration (FAA) have undertaken programs to exploit advances in flight management systems, communication and automation in air traffic control (ATC) aiding. These goals are to be met by the development of aiding and communications technologies and the redefinition of roles and procedures in the National Airspace System. The Advanced Air Transportation Technologies (AATT) initiative and the Terminal Area Productivity (TAP) program have research elements explicitly focused on the development of technologies for effective integration of flight deck automation, flight crew, ATC, and ATC-aiding systems. This integration should produce optimized routing, sequencing, and scheduling in the terminal area while relaxing constraints in en route environments to accommodate user-preferred routing and schedules.

NASA used MIDAS to predict flight crew and ATC performance in the advanced air traffic management system. The model-predicted behavior was then compared to human performance in full mission simulation of the descent scenario. NASA followed the paradigm in the application of human performance modeling shown in Figure 41.1. First the aiding technologies under consideration were modeled to the extent possible in their conceptual design state. The MIDAS human performance model was then exercised under a set of scenarios that make use of the advanced technology across a range of operating conditions. Then, to verify the accuracy of the model's predicted behavior, NASA examined an instance of the advanced technology and ran a full-mission simulation with human operators in the loop following the same scenario that was used in the MIDAS model. A comparison of the model predictions to the human performance provided the basis for generalization of the model's data to technology developments not explicitly tested in the "man-in-the-loop" simulation environment. The model analysis then was used to provide design guidance for a range of conditions and technological variations that would be difficult or expensive to simulate.

Experiment 1: Exploring an Optimum Time Range for Issuing a CTAS Descent Clearance

The MIDAS simulation and editing tools were used to develop a model of the top of descent procedures enacted with Center Tracon Automation System (CTAS) aiding in a Boeing 757. CTAS is a decision aid that provides a descent profile that is optimized for multiple aircraft descending into a terminal area. However CTAS currently does not have exact wind and weight information from the aircraft. As a result the flight path that CTAS proposes may differ from the flight path generated by aircraft's flight management system. Therefore, the flight crew must sometimes interpret and incorporate this difference vie the aircraft flight controls (Erzberger, Davis, and Green, 1993; den Braven, 1992).

The issue to be addressed by the MIDAS model was to define the ideal range of time for the issuance of a CTAS descent clearance so that the air crew would be likely to accept the clearance and enact it using flight deck automation (as opposed to manually commanding the descent). A full range of communication (voice and datalink) and automation implementation modalities [Autoload, Control Display Unit (CDU), and Mode Control Processor (MCP)] were modeled (Corker, Lozito, and Pisanich, 1995).

Development of the Activity Set. A nominal activity set and sequence was developed to describe the flight crew tasks associated with servicing the top of descent CTAS clearance. These activities included:

- Receive CTAS clearance from ATC.
- Determine if time is available to decide on the clearance.
- Determine whether the descent can safely be initiated.
- Communicate intent to appear or reject the clearance to ATC.
- Determine automation level for the descent procedure.
- Implement descent procedure.

Interruptions (such as responding to traffic calls) were another class of activities represented within the simulation. These activities can suspend the current clearance activity, delaying its completion while the interrupt is serviced.

Development of the Decision Rules. The rule set used to make decisions at top of descent was developed from an expert opinion survey focused on standard procedures. This rule set defines categories of decisions based on time available to top of descent, safe descent, and automation usage. For example, the rules associated with a safe descent involve weather, equipment limitations, and passenger comfort considerations. Weather-specific rules consult such conditions as reported icing, high cloud tops (visible convection) ahead, and reported turbulence.

Development of the Updatable World Representation. The UWR structure was developed to represent the information that the flight crew understands about the operations around the top of descent. Categories of UWR information included the aircraft state and configuration, meteorology, descent information, traffic, and pilot operational knowledge.

Model Runs. The factors manipulated in the MIDAS experiments were:

- How the clearance was able to be implemented on the flight deck: "automation implementation mode" (Autoload, CDU, or MCP)
- How the clearance was delivered: "communications mode" (data link or voice)
- When the clearance was provided relative to the top of descent (TOD) point: "time provided to implement the clearance" (five times corresponding to increasing distances from the top of descent).

A full factorial experiment was run with the MIDAS simulation.

Analysis. In examining a portion of the data output from the MIDAS model (Figure 41.18) we see that as the TOD point is approached the air crew model tends to select a flight mode that involves less automation (MCP operations). Further, the time from TOD that the crew selects that less automated operation is sensitive to the medium through which the clearance is presented; voice presentation of the data provides a more pronounced use of nonautomated modes.

The interaction between the communication medium and the flight mode with which the flight crew responds to the clearance was also replicated in the human performance data. As the aircraft approaches within 5 to 8 miles from the CTAS required TOD point, the number of successes in any clearance compliance is reduced significantly.

Neither of the above two findings was intuitive. Based on these results, design guidance was provided to the CTAS development process that details a tradeoff between level of successful automation use in enacting a clearance and the distance before TOD at which that clearance is provided. Further guidance can be provided to suggest that the mode of clearance enactment is interactive with the medium of presentation.

Experiment 2: Evaluating Model Predictions vs. Human Performance

The results of the first experiment confirmed that the pattern of behavior that the modeled flight crew exhibited in responding to a CTAS (Center Tracon Automation System)-generated TOD is similar to that exhibited by the actual flight crews. A follow-on experiment was performed to evaluate the predictive nature of the model. The CTAS/Denscent Advisor (DA) study was performed in a full mission simulator (Boeing 747-400). In this study, four two-pilot crews participated, with each pilot making 2 approaches

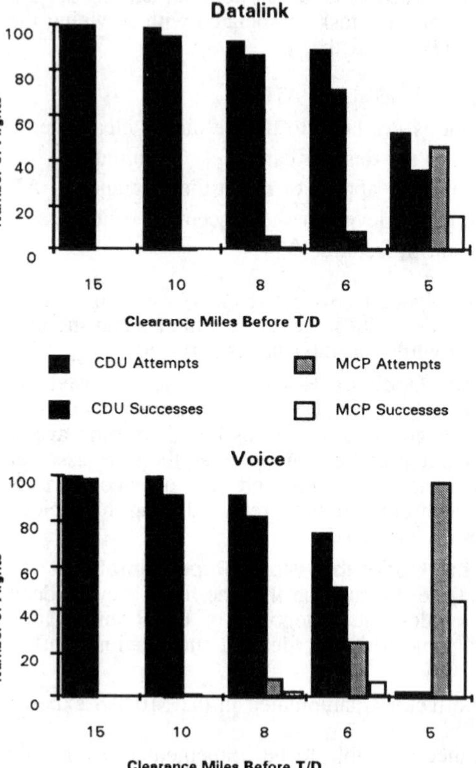

Figure 41.18 The effect of communication medium on automation selection and success rate.

and 1 landing. The result was 4 runs per flight crew across the experimental manipulations and 16 runs for the human pilots overall. To compare the MIDAS predictions of crew performance to the human flight crew, the model was initialized with the same information as the human flight crew. The modeled flight crew was then directed to fly descent profiles with the same conditions of speed, crossing restriction, and distance to top of descent as experienced by the human flight crews. A split-halves method was used in which the model's activity times (e.g., FMS operation and "button-push" time) were derived from one half of the human performance data. The model was then tested to see whether it could predict the behavior of the remaining half of the human performance data.

Model Runs. The model generated activity data comparable to that exhibited by the human operators. For example, data collected for the human flight crew included time to complete access and to enter the clearance, time to receive the clearance, time to decide on clearance acceptability, ATC communication times, and FMS operation times. The model similarly generated activity times for that sequence of behaviors that can be used in direct comparison.

Analysis. The hypothesis tested in this study was that the model performance across the manipulations of CTAS would not significantly differ from that of the flight crew in the same performance regimen. Three statistical comparisons were performed. First, a comparison was made between the behavior of the model across the experimental conditions and the performance of the flight crew across those conditions (split-halves comparison). Second, a comparison was made between one model run and cumulative data from the four simulation model runs to check for internal consistency in the model cumulative model data versus single run. Finally, the accumulated human data were compared to a single model run, chosen at random, to see whether there was an effecft on the model variance encountered by summing across model runs (flight crews versus simulation run). In all cases the statistical test revealed no significant differences between the data sets compared. The comparisons are presented in Table 41.1.

Table 41.1 Idealized Choice-Entry Time as Function of Task Parameters *N*, *A*, and *W*

N	Decision H^T (bits)	*D* (in.)	*W* (in.) (Width of Target)	Index of movement difficulty $\log^2 (2D/W)$	Total Time (s)
4	2	12	0.25	6.5	2.19
4	2	12	0.5	5.5	1.99
4	2	12	1.0	4.5	1.79
4	2	18	0.25	7.2	2.31
4	2	18	0.5	6.2	2.11
4	2	18	1.0	5.2	1.91
4	2	24	0.25	7.6	2.39
4	2	24	0.5	6.6	2.19
4	2	24	1.0	5.6	1.99
8	3	12	0.25	6.6	2.41
8	3	12	0.5	5.6	2.21
8	3	12	1.0	4.6	2.01
8	3	18	0.25	7.2	2.53
8	3	18	0.5	6.2	2.33
8	3	18	1.0	5.2	2.13
8	3	24	0.25	7.6	2.61
8	3	24	0.5	6.6	2.41
8	3	24	1.0	5.6	2.21

The data suggest that the model performance is predictive of the flight crew performance across the conditions of this experiment. The lack of significant difference between the model and the cumulative human performance and the check for internal consistency in this model supported the validity of MIDAS.

This case study represents an excellent example of how models and experimentation can be used synergistically to address a complex human performance issues.

41.6.3.2 MIDAS Case Study 2—Predicting Air Traffic Control Crew Performance in the Advanced Air Traffic Management System

The issues associated with the advanced air traffic management process require an ability to describe the procedures, equipment interaction, information state and control flow among the multiple agents within the airspace management process. To that end, MIDAS was used to study *differing individual behaviors* of air traffic control operators as well as pilots and copilots with whom they interact. Separate pilot, copilot, and ATC specialists are represented, with individual symbolic operator models (UWR, activities, and rule set) for each operator.

Developing the Symbolic Operator Models (SOMs) and Simulating Multiple Operator Differences. To represent the action of multiple independent and interactive operators in the simulation, certain characteristics had to be modeled differently for each of the intelligent entities participating in the scenario. The differences between operators (physical and cognitive) and how the interaction of these differences may affect the overall system that were modeled included:

- World representation, decision rules
- Activity sequences and scheduling
- Physical and cognitive response rates
- Communication activities and performance
- Expectation processing and satisfiers

Development of the Multiple SOM representation required duplicating the major functions (UWR and scheduler) for each operator. Also, major changes to the MIDAS UWR involved building provisions to support the modeling of *communications and expectations* as described below.

Communications Model. MIDAS uses a model of information communication between SOMs that can examine the modalities, channels, and impediments in existing and future communications. The SOMs included three specific modes of airspace communications including voice, aircraft datalink, and electronic (such as those through networked ATC display stations). Specific communication channels that were defined included ATC to flight crew member), and ATC to ATC (simulating handoffs of responsibility between sectors).

MIDAS also allows the representation of features that can impede or degrade efficient communications. Interfering environmental effects specific to the communications modes (such as noise, frequency and traffic, or even glare on a display) were included to interact with the SOM-based functions (forgetting or decay of content). An information processing model limits the processing of incoming data to no more than three informational segments (for example, a clearance with no more than three commands).

Expectations. Central to this study was the modeling of human "expectations." In the MIDAS model, an expectation activity is generated by an operator based on the knowledge of activities that he or she has initiated (for example, requesting information) or would normally expect to occur based on outside input (as is expecting to receive a clearance by some point). Modeling expectations permit the study of the interactions and timing of dependent actions within multiple human–machine systems. For example, expectation activities are based on the appearance of specific information (for example, an acknowledgment) in the operator's UWR within some specified time. The information is judged as to whether it meets the expectation using success, failure, or time-out satisfaction conditions. The time-out condition also has a specified time that automatically satisfies if the other conditions have not been met within that time. As modeled, the time between the establishment of a expectation and its conclusion is an interval of potential error, either through an interference with the expectation or the insertion of erroneous data into its satisfaction conditions.

Testing and Evaluation of Model

Communications Scenario. The combination of expectations and communication loops provided a structure that was sufficient to capture known communication errors. The model predicted that two types of errors would be common. The first type is seen when the communications link integrity is broken by an interruption and the expectation is aborted. In such cases, erroneous behavior results in the uptake of earlier or inappropriate expectations, or by a lack of appropriate expectations. In cases where the internal representation of the expected value biases the perceptual process, the expectation content itself is changed. This model also has the ability to simulate these error types by the manipulation of the communication loop structure to "forget" an expectation in the fact of an interruption.

Results. The goal of this case study was to provide an analytic tool to answer questions about human–automation interaction in advanced airspace management. Specific questions as to procedure sequences and performance timing were analyzed and model performance verified. Expansion of the basic model to support multiple independent and interactive human operators was also undertaken. A limited simulation was run of multiple ATC and flight crew operators interacting in profile descent clearance exchange. In addition, an examination of incident data for ATC and Flight crew clearance exchange errors indicated that the MIDAS model was sufficient to simulate certain types of error that are expressed as interruption of the communication cycle and corruption of the contents of activity expectations. The model was also applied to issues in the definition of free flight operating concepts and will be expanded to incorporate human error theory in an attempt to predict performance errors.

41.7 SUMMARY

This chapter has provided the need for simulating performance of complex human-based systems as an integral part of system design, development, testing, and life-cycle support. It has also defined two fundamentally different approaches to modeling human performance, a reductionist approach and a first-principled approach. Additionally, we have provided detailed examples of two modeling environments that typify these two approaches along with representative case studies.

As we have stated and demonstrated repeatedly throughout this chapter, the technology for modeling human performance in systems is evolving rapidly. Furthermore, the breadth of questions begin addressed by models is constantly expanding. We encourage the human factors practitioner with a little creativity and computer savvy to consider how computer simulation can provide a better and more cost effective basis for human factors analysis.

REFERENCES

Agha, G. (1985). *Actors: A model of concurrent computation in distributed systems* (Tech. Rep. 844). Cambridge, MA: *MIT Artificial Intelligence Laboratory.*

Allender, L. (1995, December) Personal Communication.

Allender, L., Kelley, T., Salvi, L., Headley, D. B., Promisel, D., Mitchell, D., Richer, C., and Feng, T. (1995). Verification, validation, and accreditation of a soldier-system modeling tool. In *The Proceedings of the 39th Human Factors and Ergonomics Society meeting*, October 9–13, San Diego, CA. Available from the Human Factors and Ergonomics Society, Santa Monica, CA.

Archer, R., Drews, C. W., Laughery, K. R., Dahl, S. G. (1986). Data on the usability of Micro Saint. *Proceedings of NAECON Meeting*, Dayton, Ohio, May.

Badler, N., Phillips, C., and Webber, B. (1993). *Simulating Humans: Computer Graphics, Animation, and Control.* Oxford University Press.

Baron, S., Muralidharan, R., Lancraft, R., and Zacharias, G. (1980). *PROCRU: A Model for Analyzing Crew Procedures in Approach to Landing.* Contractor Report NASA—Ames Research Center CR-152397. Moffet Field, CA.

Boetcher, K., North, R., and Riley, V. (1989). On Developing Theory-Based Functions to Moderate Human Performance Models in the Context of Systems Analysis. In *Proceedings of the Human Factors Society 33rd Annual Meeting.* Denver, Colorado, October.

Boff, K. R., Kaufman, L., Thomas, J. P. (1986). *Handbook of perception and cognition*, New York: John Wiley and Sons.

Corker, K. M., Lozito, S., and Pisainch, G. (1995). Flight crew performance in automated air traffic management. In Fuller, Johnston, and McDonald, Eds., *Human factors in aviation operation* (Vol. 3). Hants, UK: Avebury Aviation.

Corker, K. M., and Pisanich, G. M. (1995). When reasonable expert systems disagree. Paper presented at the Topical Meeting of the American Nuclear Society, Computer-Based Human Support Systems: Technology, Methods, and Future. Philadelphia, PA.

Corker, K. M., and Pisanich, G. M. (1995). Analysis and modeling of flight crew performance in automated air traffic management systems. Presented at 6th IFAC/IFIP/IFORS/IEA symposium: Analysis, design, and evaluation of man-machine systems, Boston MA.

Corker, K. M., and Smith, B. (1993). An architecture and model for cognitive engineering simulation analysis: Application to advanced aviation analysis. Presented at AIAA conference on computing in aerospace, San Diego, CA.

Dahl, S. G., Allender, L., Kelley, T., and Adkins, R. (1995). Transitioning softwarre to the Windows environment — Challenges and innovations. *Proceedings of the 1995 Human Factors and Ergonomics Society meeting.* Santa Monica, CA: Human Factors and Ergonomics Society.

den Braven, W. (1992). *Design and Evaluation of an Advanced Air-Ground Data-Link System for Air Traffic Control.* NASA Technical Memorandum (TM-103899). January.

Drews, C., and Laughery, K. R. (1985). A modeling system designed around the user interface. Presented at the Summer Computer Simulation Conference, Chicago, IL, July.

Drews, C., Laughery, R. R., Kramme, K., and Archer R. (1985). *LHX cockpits: Micro SAINT simulation study and results.* Report prepared for Texas Instruments — Equipment Group, Dallas, Texas.

Erzberger, H., Davis, T. J., and Green S. M. (1993). Design of Center—TRACON Automation System. AGARD Guidance and Control Symposium. May. Berlin, Germany.

Fontenelle, G. A., and Laughery, K. R. (1988). A Workload Assesment Aid for Human Engineering Design. In *Proceedings of the 32nd Annual Human Factors Society Meeting.* Anneheim, CA, October.

Gawron, V. J., Laughery, K. R., Jorgensen, C. C., and Polito, J. (1983). A computer simulation of visual detection performance derived from published data. In *Proceedings of the Ohio State University aviation psychology symposium*, Columbus, Ohio, April.

Gray, W. D., John, B. E., and Atwood, M. E. (1993). Project Ernestine: Validating a GOMS analysis for predicting and explaining real-world task performance. *Human-Computer Interaction, 8*, 237–309.

Farmer, E. W., Belyavin, A. J., Jordan, C. S., Bunting, A. J., Tattershall, A. J. and Jones, D. M. (1995). *Predictive Workload Assessment: Final Report* (Report No. DRA/AS/MMI/CR95100). Farnborough, UK: Defence Research Agency.

LaVine, N. D., Laughery, K. R., and Peters, S. D. (1995). *A methodology for predicting and applying human response to environmental stressors.* Boulder, CO: Micro Analysis & Design, Inc.

Lawless, M. L., Laughery, K. R., and Persensky, J. J. (1995). *Micro Saint to predict performance in a nuclear power plant control room: A test of validity and feasibility* (NUREG/CR-65). Washington, DC: Nuclear Regulatory Commission.

McCracken, J. H., and Aldrich, T. B. (1984). Analysis of Selected LHX Mission Functions: Implications for Operator Workload and System Automation Goals. Technical Note ASI 479-024-84(b) prepared by Anacapa Sciences, Inc., June 1984.

McMillan, G. R., Beevis, D., Salas, E., Strub, M. H., Sutton, R., and Van Breda, L. (1989). *Applications of human performance models to system design.* New York: Plenum Press.

Micro Analysis and Design. (1993). *Micro Saint for Windows user's guide.* Boulder, CO: Micro Analysis and Design.

Newell, A. (1990). *Unified theories of cognition.* Cambridge MA: Harvard University Press.

Newell, A., and Simon, H. A. (1972). *Human problem solving.* Englewood Cliffs, NJ: Prentice-Hall.

O'Hara, J. M., Brown, W. S., Stubler, W. F., Wachtel, J. A., and Persensky, J. J. (1995). *Human-system interface design review guideline: Draft report for comment.* (NUREG-0700 Rev. 1). Washington, DC: U.S. Nuclear Regulatory Commission.

Payne, W., Bettman, J. R. and Johnson, E. J. (1988). Adaptive strategy in decision making. *Journal of Experimental Psychology: Learning, Memory and Cognition, 14(3)*, 534–552.

Plott, B. (1995). *Software user's manual for WinCrew, the Windows-based workload and task analysis tool.* Aberdeen Proving Ground, MD: U.S. Army Research Laboratory.

Remington, R. W., Johnston, J. C. and Yantis, S. (1992). Involuntary attentional capture by abrubt onsets. *Perception & Psychophysics, 51(3)*, 279–290.

Roth, J. T. (1992). *Reliability and validity assessment of a taxonomy for predicting relative stressor effects on human task performance* (Tech. Rep. 5060-1 prepared under contract DNA001-90-C-0139). Boulder, CO: Micro Analysis and Design, Inc.

Shankar, R. (1991). "Z-Scheduler: Integrating theories of scheduling behavior into a computational model. In *Proceedings of the IEEE international conference on systems, man, and cybernetics* (pp. 1219–1223).

Siegel, A. I., and Wolf, J. A. (1969). *Man-machine simulation models.* New York: Wiley-Interscience.

Tversky, A., and Kahneman, D. (1974). Judgment under uncertainty: Heuristics and biases. *Science, 185*, 1124–1131.

Wickens, C. D., Sandry, D. L., and Vidulich, M. (1983). Compatibility and resource competition between modalities of input, central processing and output. *Human Factors, 25*.

Wortman, D. B., Duket, S. D., Seifert, D. J., Hann, R. I., and Chubb, A. P. (1978). *Simulation using Saint: A user-oriented instruction manual. Wright-Patterson AFB, OH: Aerospace Medical Research Laboratory. (AMRL-TR-77-61).*

CHAPTER 42
DECISION SUPPORT SYSTEMS

Andrew P. Sage
School of Information Technology and Engineering
George Mason University
Fairfax, VA 22030-4444 USA

42.1 INTRODUCTION

In very general terms, a decision support system (DSS) is a system that supports technological and managerial decision making by assisting in the organization of knowledge about ill-structured, semistructured, or unstructured issues. For our purposes, a structured issue is one that has a framework with elements and relations between them that are understood. The primary components of a decision support system are generally described as:

- a database management system (DBMS)
- a model-base management system (MBMS)
- a dialogue generation and management system (DGMS)

Emphasis in the use of a DSS is upon provision of support to decision makers in terms of increasing the effectiveness of the decision-making effort and this need should be a major factor in the definition of requirements for, and subsequent development of, a DSS.

A DSS is generally used to support humans in the formal steps of problem solving, or systems engineering (Sage, 1992, 1995), that involve

- The *formulation* of alternatives
- The *analysis* of their impacts
- *Interpretation* and selection of appropriate options for implementation

Efficiency in terms of time required to evolve the decision, although important, is usually secondary to effectiveness. DSSs are intended for use in strategic and tactical situations, and less in operational situations. In operational situations, which are often well structured, an expert system may often be gainfully employed to assist novices. Those very proficient in operational tasks generally do not require support, except perhaps for automation of some routine and repetitive chores. There are many application areas in which the use of a decision support system is potentially promising. These include management and planning, command and control, system design, health care, operations management, and essentially any area in which management has to cope with decision situations that have an unfamiliar structure, and where there are information uncertainty and imprecision.

42.1.1 Taxonomies for Decision Support

Numerous disciplinary areas have contributed to the development of decision support systems. One area is computer science, which provides the hardware and software tools necessary to implement decision support system design constructs. In particular, computer science provides the database design and programming support tools that are needed in a decision support system. The field of management science and operations research provides the theoretical framework in decision analysis that is necessary to design useful and relevant normative approaches to choicemaking, especially those that are concerned with systems analysis and model base management. The areas of organizational behavior and behavioral and cognitive science provide rich sources of information concerning how humans and organizations process information and make judgments in a descriptive fashion. Background information from these areas is needed for the design of effective systems for dialogue generation and management systems. The area of systems engineering is concerned with the process and technology management issues associated with the definition, development, and deployment of large systems of hardware and software, including systems for decision support.

There have been many attempts to classify different types of decisions. Among the classifications of particular interest here is the decision type taxonomy of Anthony (Anthony, 1965; Anthony, Dearden, and Govindarajan, 1992) which describes four types of decisions:

1. *Strategic planning decisions*, which are decisions related to choosing highest level policies and objectives and associated resource allocations
2. *Management control decisions*, which are decisions made for the purpose of assuring effectiveness in the acquisition and use of resources
3. *Operational control decisions*, which are the decisions made for the purpose of assuring effectiveness in the performance of operations
4. *Operational performance decisions*, which are the day-to-day decisions made while performing operations

Figure 42.1 illustrates the way in which these decisions are related and the way in which they normatively influence organizational learning. Generally, low-consequence decisions are made more frequently than high-consequence decisions. Also, strategic decisions are associated with higher consequences, are likely to involve more significant risk, and must be made on the basis of considerably less perfect information than most operational control decisions.

There are a number of abilities that a decision support system should support. It should support the decision maker in the *formulation* or framing or assessment of the decision situation in the sense of recognizing needs, identifying appropriate objectives by which to measure successful resolution of an issue, and generating alternative courses of action that will resolve the needs and satisfy objectives. It should also provide support in enhancing the abilities of the decision maker to obtain the possible impacts on needs of the alternative courses of action. This *analysis* capability must be associated with provision

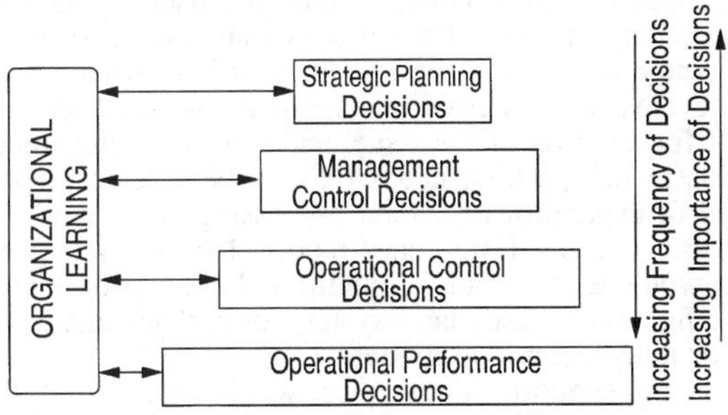

Figure 42.1 Organizational information and decision flow.

of capability to enhance the ability of the decision maker to provide an *interpretation* of these impacts in terms of objectives. This interpretation capability will lead to evaluation of the alternatives and selection of a preferred alternative option. These three steps of formulation, analysis, and interpretation are very fundamental ones for formal analysis of difficult issues. They are the fundamental steps of systems engineering and are discussed at some length in Sage (1991, 1992, 1995); Sage (1991) is the source from much which of this chapter is derived. It is very important to note that the purpose of a decision support system is to support humans in the performance of primarily cognitive information-processing tasks that involve decisions, judgments, and choice. Thus, the enhancement of information processing in systems and organizations (Sage, 1990) is a major feature of a DSS. Even though there may be some human supervisory control of a physical system (Sheridan, 1992) through use of these decisions, the primary purpose of a DSS is support for cognitive activities that involve human information processing and associated judgment and choice. Associated with these three steps must be the ability to acquire, represent, and utilize information or knowledge, and the ability to implement the chosen alternative course of action.

The extent to which a support system possesses the capacity to assist a person or a group to formulate, analyze, and interpret issues will depend upon whether the resulting system should be called a management information system (MIS), a predictive management information system (PMIS), or a decision support system (DSS). We can provide support to the decision maker at any of these several levels, as suggested by Figure 42.2. Whether we have a MIS, a PMIS, or a DSS depends upon the type of automated computer-based support that is provided to the decision maker to assist in reaching the decision. Fundamental to the notion of a decision support system is assistance provided in assessing the situation, identifying alternative courses of action and formulating the decision situation, structuring and analyzing the decision situation, and then interpreting the results of analysis of the alternatives in terms of the value system of the decision maker.

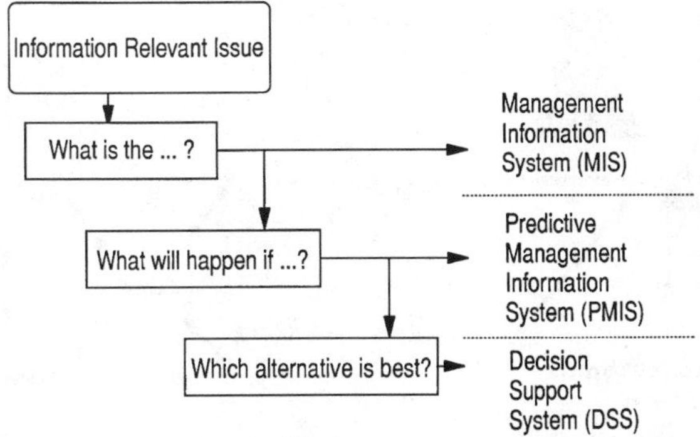

Figure 42.2 Conceptual differences between MIS, PMIS, and DSS.

In a classical management information system, the user inputs a request for a report concerning some question, and the MIS supplies that report. When the user is able to pose a "what if?"–type question and the system is able to respond with an if–then type of response, then we have a predictive management information system. In each case there is some sort of formulation of the issue, and this is accompanied by some capacity for analysis. The classic MIS, which needs only to be able to respond to queries with reports, is comprised of capabilities for data processing, structured data flows at an operational level, and preparation of summary reports for the system user. The predictive management system would also include an additional amount of analysis capability. This might require an intelligent database query system, or perhaps just the simple use of some sort of spreadsheet or macroeconomic model.

To obtain a decision support system, we would need to add the capability of model–base management to a MIS. But much more is needed, for example, than just the simple addition of a set of decision trees and procedures to elicit examination of decision analysis–based paradigms. We also need a system that is flexible and adaptable to changing user requirements such as to provide support for the decision styles of the decision maker as these change with task, environment, and experiential familiarity of the support system users with task and environment. We need to provide analytical support in a variety of complex situations. Most decision situations are fragmented in that there are multiple decision makers and their staffs, rather than just a single individual decision maker. There are also temporal and spatial separation elements involved. Further, as Mintzberg (1973) has indicated so very well, managers have many more activities than decision making to occupy themselves with, and it will be necessary for appropriate DSS to support many of these other information related functions as well. Thus, the principal goal of a DSS is *improvement in the effectiveness of organizational knowledge users through use of information technology*. This is not a simple objective to achieve as has been learned in the process of past DSS design efforts.

42.1.2 Frameworks for Designing Decision Support Systems

As we have discussed, there are three principal components of a decision support system: a database management system (DBMS), a model base management system (MBMS), and a dialogue generation and management system (DGMS). An appropriate decision support system design framework will consider each of these three component systems and their interrelations and interactions. Figure 42.3 illustrates the interconnection of these three generic components and illustrates the interaction of the decision maker with the system through the DGMS. We will describe some of the other components in this figure soon.

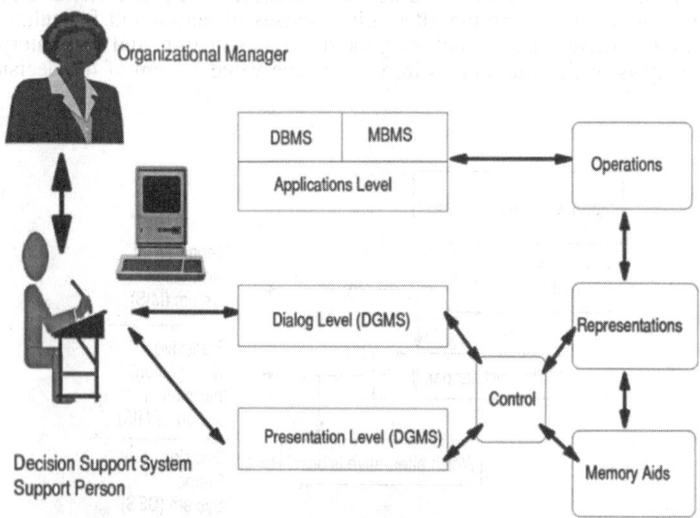

Figure 42.3 Generic components in decision support systems.

Sprague and Carlson (1982), authors of an early and seminal book on decision support systems, have indicated that there are three technology levels at which a DSS may be considered. The first of these is the level of DSS tools themselves. This level contains the hardware and software elements that enable use of system analysis and operations research models for the model base of the DSS and the database elements that comprise the database management system. The purpose of these DSS tools is to design a specific DSS that is responsive to a particular task or issue. The second level is that of a decision support system generator. The third level is the specific DSS itself. The specific DSS may be designed through the use of the DSS tools only, or through use of the DSS generator only that may call upon some elements in the generic MBMS and DBMS tool repository, or through combined use of these.

Often the best designers of a decision support system are not the specialists primarily familiar with DSS development tools. The principal reason for this is that it is difficult for one person or small group to be very familiar with a great variety of tools, as well as to have the ability to identify the requirements needed for a specific DSS and the systems management skills needed to design a support process. This suggests that the decision support generator is a potentially very useful tool, in fact a design level, for DSS system design. The DSS generator is a set of software products, similar to a very advanced generation system development language, that enables construction of a specific DSS without the need formally to use microlevel tools from computer science and operations research and systems analysis in the initial construction of the specific DSS. A DSS generator contains an integrated set of features, such as inquiry capabilities, modeling language capabilities, financial and statistical (and perhaps other) analysis capabilities, and graphic display and report preparation capabilities. The major support provided by a DSS generator is that it allows the rapid construction of a prototype of the decision situation and allows the decision maker to experiment with this prototype and to refine it such that it is more representative of the decision situation and more useful to the decision maker. This generally reduces, often to a considerable degree, the time required to design and build a DSS. This notion is not unlike that of software prototyping, one of the principal macroenhancement software productivity tools (Sage and Palmer, 1990) in which the process of constructing the prototype DSS through use of the DSS generator leads to a set of requirements specifications for a DSS that are then realized in efficient form using DSS tools directly.

The primary advantage of the DSS generator is that it is something that the DSS designer can use for direct interaction with the DSS user group. This eliminates, or at least minimizes, the need for DSS user interaction with the content specialists most familiar with microlevel tools of computer science, systems analysis, and operations research. Generally, a DSS user will seldom be able to identify or specify the requirements for a DSS initially. In such a situation, it is very advantageous to have a DSS generator that may be used by the DSS engineer, or developer, in order to obtain prototypes of the DSS. The user may then be encouraged to interact with the prototype in order to assist in identifying appropriate requirements specifications for the evolving DSS design.

The third level in this DSS design and development effort results from adding a decision support systems management capability. Often, this will take the form of the dialogue generation and management subsystem referred to earlier, except perhaps at a more general level since this is a DGMS for DSS design rather than a DGMS for a specific DSS. This DSS design approach is not unlike that advocated for the systems engineering of large-scale systems in general and DSS in particular. There are many potential difficulties. Among these are inconsistent, incomplete, and otherwise imperfect system requirements specifications; system requirements that do not provide for change as user needs evolve over time; and poorly defined management structures. The major difficulties associated with the production of trustworthy systems have more to do with the organization and management of complexity than with direct technological concerns. Thus, whereas it is necessary to have an appropriate set of quality methods and tools, it is also very important that they be used within a well-chosen life-cycle process and set of systems management strategies.

Since a decision support system is intended to be used by decision makers with varying experiential familiarity and expertise with respect to a particular task, it is especially important that a DSS design consider the variety of issue *representations* or frames that decision makers may use to describe issues, the *operations* that may be performed on these representations to enable formulation analysis and interpretation of the decision situation, the automated *memory aids* that support retention of the various results of operations on the representations, and the *control mechanisms* that assist decision makers in using these representations, operations, and memory aids. A very useful control mechanism results in the construction of heuristic procedures, perhaps in the form of a set of production rules, to enable development of efficient and effective standard operating policies to be issued as staff directives. Other control mechanisms are intended to encourage the decision maker to control and direct use of the DSS personally, and also to acquire new skills and rules based on the formal reasoning based knowledge that is called forth through use of a decision support system. This process independent approach toward development of the necessary capabilities of a specific DSS is due to Sprague and Carlson (1982) and is known as the ROMC approach, where ROMC is an acronym for representations, operations, memory aids, and control mechanisms. Figure 42.3 illustrates the ROMC elements, together with the three principal components (DBMS, MBMS, and DGMS) of a DSS.

42.2 DATABASE MANAGEMENT SYSTEMS

As we have noted, a Database Management System (DBMS) is one of the three fundamental technological components in a decision support system. Figure 42.3 indicates the generic relationship among these components, or subsystems. We can consider a database management system as comprised of a database (DB) and a management system (MS). This sort of expansion holds for the model base management system also. Figure 42.4 indicates the resulting expanded view of a DSS. The DBMS block itself can be expanded considerably, as can the other two subsystems of a DSS. Figure 42.5 indicates one possible expansion to indicate many of the very useful components of a database management system that we examine briefly later in this section.

Three major objectives for a DBMS are *data independence*, *data redundancy reduction*, and *data resource control*, such that software to enable applications to use the DBMS and the data processed are independent. If this is not accomplished, then such simple changes to the data structure as adding four digits to the ZIP code number in an address might require rewriting many applications programs. The major advantage to data independence is that DSS developers do not need to be explicitly concerned with the details of data organization in the computer, or how to access it explicitly for information processing purposes, such as query processing. Elimination or reduction of data redundancy will assist in lessening the effort required to make changes in the data in a database. It may also assist, generally greatly, in eliminating the inconsistent data problem that often results from updating data items in one part of a database but (unintentionally) not up-

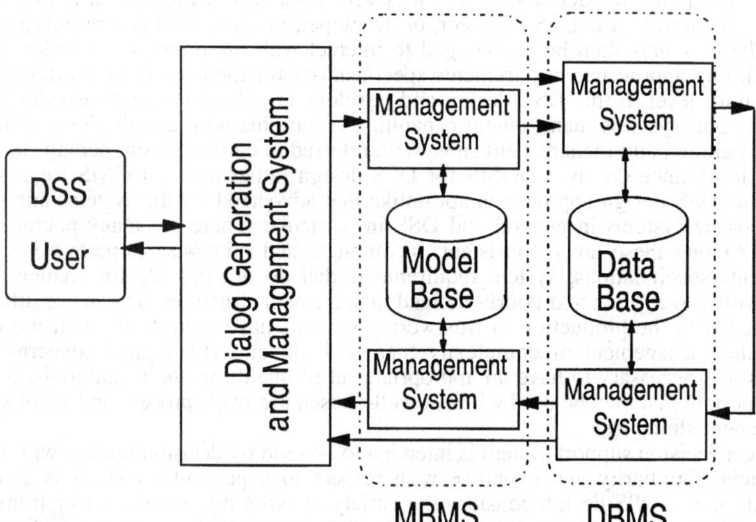

Figure 42.4 Expanded view of DSS indicating database and management system for DBMS, and model base and management system for MBMS.

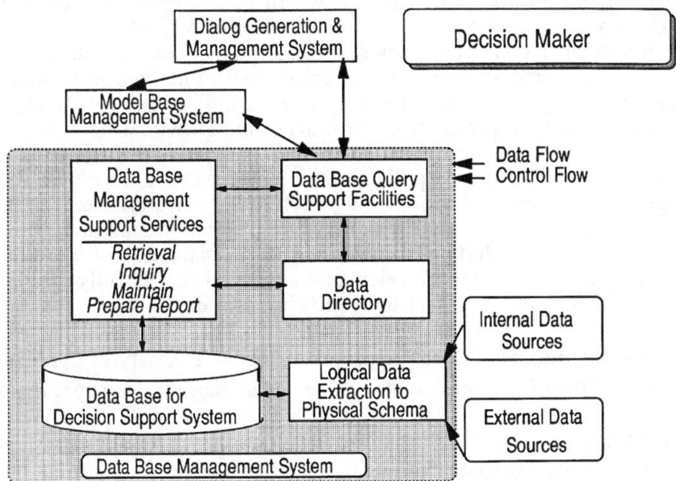

Figure 42.5 Generic components in decision support system—with expansion of database management system (DBMS).

dating this same data that exist elsewhere in the database because of data redundancy. With many people potentially using the same database, resource control is essential. The presence of these three features in a DBMS is the major factor that differentiates it from a file management system.

The management system (MS) for a database is comprised of the software that is beneficial to the creation, access, maintenance, and updating of a database. A database contains data that are of value to an individual or to an organization and that an individual or an organization desires to maintain. Maintenance should be such that the data survives even when the DBMS system hardware and/or software fail. Maintenance of data is one of the important functions provided for in DBMS design.

There are many tasks that we desire to perform using a database. These include five very important tasks:

1. *Capturing* relevant data for use in the database
2. *Selection* of relevant data from the database
3. *Aggregation* of data to form totals, averages, moments, and other items that support decision making
4. *Estimation, forecasting, and prediction* in order to obtain extrapolations of events into the future, and such other activities as
5. *Optimization* in order to enable selection of a "best" alternative

We note that these database use notions raise issues that are associated with the model base management system. In a similar way, the dialogue generation and management system determines how data is viewed and is, therefore, also important for use of a DBMS.

42.2.1 DBMS Design, Selection, and Systems Integration

As with any information systems engineering based activity, DBMS design should start with an identification of the DBMS design situation and the user requirements. From this, identification of the logical or conceptual data requirements follows specification of a logical database structure in accordance with what is known as a *data definition language* (DDL). After this, the physical database structure is identified. This structure must be very concerned with specific computer hardware characteristics and efficiency considerations. Given these design specifications, an operational DBMS is constructed, and DBMS operation and maintenance efforts begin. The logical database design and physical database design efforts can each be disaggregated into a number of related activities. The typical database management system life cycle will generally follow one of the systems engineering development life cycles discussed in Sage (1992, 1995) and elsewhere.

The three important DBMS requirements—data independence, redundancy reduction, and increased data resource control—are generally applicable both to logical data and to physical data It is highly desirable, for example, to be able to change the structure of the physical database without affecting other portions of the DBMS. This is denoted *physical data independence.* In a similar way, *logical data independence* denotes the ability of software to function using a given applications-oriented perspective on the database even though changes in other parts of the logical structure have been made. The requirements specification, conceptual design, logical design, and physical design phases of the DBMS development life cycle are specifically concerned with satisfaction of these requirements.

A number of questions need to be asked and answered successfully in order to design an effective DBMS. Among these are the following (Arden, 1980):

1. Are there data models that are appropriate across a variety of applications?
2. What are DBMS designs that enable data models to support logical data independence?
3. What DBMS designs enable data models to support logical data independence, and what are the associated physical data transformations and manipulations?
4. What features of a data description language will enable a DBMS designer to control both the logical and physical properties of data independently?
5. What features need to be incorporated into a data description language in order to enable errors to be detected at the earliest possible time such that users will not be affected by errors that occur at a time prior to their personal use of the DBMS?
6. What are the relationships between data models and database security?
7. What are the relationships between data models and errors that may possibly be caused by concurrent use of the database by many users?
8. What are design principles that will enable a DBMS to support a number of users having diverse and changing perspectives?
9. What are the appropriate design questions such that applications programmers, technical users, and DBMS operational users are each able to function effectively and efficiently?

The bottom line question that summarizes all of these is, How does one design a data model and data description language to enable efficient and effective data acquisition, storage, and use? There are many related questions; one of them concerns the design of what are called standard query languages (SQLs) such that it is possible to design a specific DBMS for a given application.

It is very important that, whenever possible, a specific DBMS be selected before design of the rest of the DSS. There are a variety of reasons that this is quite desirable. The collection and maintenance of the data through the DSS is simplified if there is a specified single DBMS structure and architecture. The simplest situation of all occurs when all data collection (and maintenance) is accomplished before use of the DSS. The DBMS is not then used in an interactive manner as part of DSS operation. The set of datebase functions that the DSS needs to support is controlled when we have the freedom to select a single DBMS structure and architecture before design of the rest of the DSS. The resulting design of the DSS is therefore simplified. Further, the opportunities for data sharing among potentially distributed databases are increased when the interoperability of databases is guaranteed. Many difficulties result when this is not possible, and sometimes it is not possible. Data in a DBMS may be classified as *internal data*, stored in an internal database, and *external data*, stored in an external database. Every individual and organization will necessarily have an internal database. Although there may exist no problem in insuring DBMS structure and architecture compatibility for internal data, this may be very difficult to do for external data. If both of these data can be collected and maintained before to use of a DSS, then there will generally be no data integration needs. Often, however, this will not be possible. Because we will often be unable to control the data structure and architecture of data obtained externally, the difficulties we cite will often be real, especially in what are commonly called real-time, interactive environments. When we have DBMSs that are different in structure and architecture, data sharing across databases is generally difficult, and it is often then necessary to maintain redundant data. This can lead to a number of difficulties in the systems integration that is undertaken to ensure compatibility across different databases.

In this chapter, as well as in much DSS and information system design in practice, we may generally assume that the DBMS is preselected prior to design of the rest of the DSS, and that the same DBMS structure and architecture are used for multiple databases that may potentially be used in the DSS. This is often appropriate, for purposes of DSS design, since DBMS design technology (Date, 1983, 1986) is now relatively mature in contrast to MBMS and DGMS design. This does not suggest that there are not continuing evolution and improvements to DBMS designs. It is usually possible, and generally desirable, to select a DBMS, based on criteria we will soon discuss, and then design the MBMS and DGMS based on the existing DBMS. This will usually, but surely not always, allow selection of off-the-shelf DBMS software. The alternate approach of designing a MBMS and DGMS first and then specifying the requirements for a DBMS based on these is possible and may, in some cases, be needed. Given the comparatively developed state of DBMS software development, as contrasted with MBMS and DGMS software, this approach will usually be less desirable from the point of view of design economy.

42.2.2 Data Models and Database Architectures

We will now expand very briefly on our introductory comments concerning data model representations and associated architectures for database management systems. Some definitions are appropriate. A data model defines the types of data objects that may be manipulated or referenced within a DBMS. The concept of a logical record is a central one in all DBMSs. Some DBMS designs are based on mathematical relations, and the associated physical database is a collection of consistent tables in which every row is a record of a given type. This is an informal description of a *relational* database management system, a DBMS type that is a clear outgrowth of the file management system philosophy. Other DBMS may be based on *hierarchical* or *network* structures that resemble the appearance of the data in the user's world and may contain extensive graphical and interactive support characteristics.

There are several approaches that we might take to describing data models. Date (1983, 1986), for example, discusses the three level representation in which the three levels of data models are:

1. An external model, which represents a data model at the level of the user's application and is the data model level closest and most familiar to users of a DBMS or DSS
2. A conceptual model, which is an aggregation model that envelopes several external models
3. An internal model, which is a technical level model that describes how the conceptual model is actually represented in computer storage.

Figure 42.6 is a generic diagram that indicates the mappings and data translations needed to accommodate the different levels of data models and architectures. The relations be-

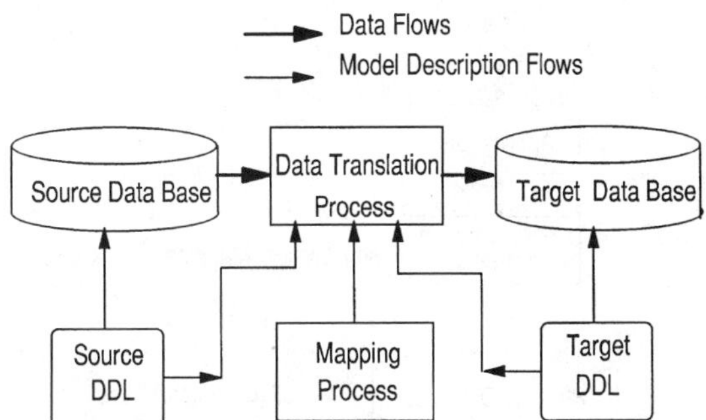

Figure 42.6 Data transforms and model mappings.

tween the various levels are called *mappings*. These mappings specify and describe the transformations that are needed in order to obtain one model from another. The user supplies specifications for the source and target data structures in a Data Definition Language (DDL) and also describes the mapping that is desired between source and target data. Figure 42.7 represents this general notion of data transformation. This could, for example, represent target data in the form of a table that is obtained from a source model comprised of a set of lists.

What we have shown in Figure 42.7 is just a very simple illustration of the more general problem of mapping between various schemas. Simply stated, a *schema* is an image used for comparison purposes. A schema can also be described as a data structure for representing generic concepts. Thus, schemas represent knowledge about concepts and are structurally organized about some theme. In terms appropriate here, the user of a database must interpret the real world that is outside of the database in terms of real-world objects, or entities and relationships between entities, and activities that exist and that involve these objects. The database user will interact with the database in order to obtain needed data for use or, alternately, to store obtained data for possible later use. But, a DBMS cannot function in terms of real objects and operations. Instead, a DBMS must use data objects, comprised of data elements and relations between them, and operations on data objects. Thus, a DBMS user must perform some sort of mapping or transformation from perceived real objects and actions to those objects and actions representations that will be used in the physical database.

The single-level data model, which is conceptually illustrated in Figure 42.7, represents the nature of the data objects and operations that the user understands when receiving data from the database. It is in this fashion that the DBMS user models the perceived real world. To model some action sequence, or impact of an action sequence on the world, the user maps these actions to a sequence of operations that are allowed by the specific data model. It is the data manipulation language, or DML, that provides the basis for the operations submitted to the DBMS as a sequence of queries or programs. The development of these schemas, which represent logical data, results in a DBMS architecture or DBMS framework. This architecture or framework describes the types of schemas that are allowable and the way in which these schemas are related through various mappings. We could, for example, have a two-schema framework or a three-schema framework, which appears to be the most popular representation at this time. Here, the external schemas define the logical subset of the database that may be presented to a specific DBMS user. The conceptual schemas define conceptual models as represented in the database. The internal schema describes physical data structures in the database. These data structures provide support to the conceptual models. The mappings used establish correspondence between various objects and operations at the three levels. If consistency of the conceptual models is to be assured, we must be able to map the internal, or physical, and external, or logical, schemas to the conceptual schemas.

Useful documentation about the database structure and architecture is provided through the various schemas, which represent explicit data declarations. These declarations represent data about data. The central repository in which these declarations are kept is called a *data dictionary* or *data directory* and is often represented by the symbol DD. The data

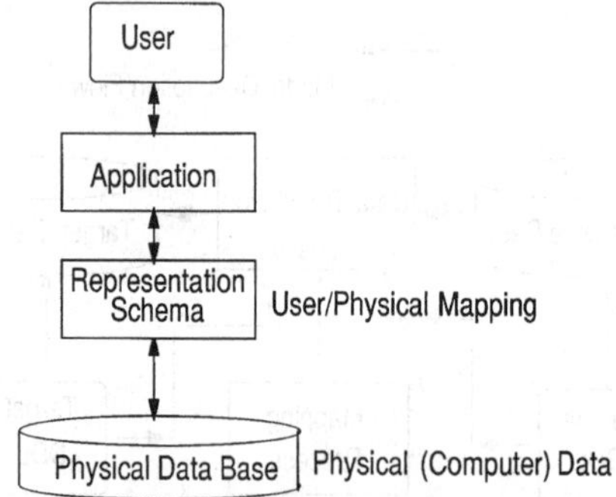

Figure 42.7 Single-level model of data schema.

directory is a central repository for the definitions of the schemas and the mappings between schemas. Generally, a data dictionary can be queried in the same manner as can the database, thereby enhancing the ability of the DBMS user to pose questions about the availability and structure of data. It is often possible to query a data directory with a high-level, or fourth generation, query language.

A data dictionary is able to tell us what the records in the data dictionary consist of. It also contains information about the logical relationships that pertain to each of the particular elements in the database. The development of a data dictionary generally begins with formation of lists of desired data items or fields that have been grouped in accordance with the entities that they are to represent. A name is assigned for each of these lists, and a brief statement of the meaning of each is provided for later use. Next, the relationships between data elements should be described and any index keys or pointers should be determined. Finally, the data dictionary is implemented within the DBMS. In many ways, a data dictionary is the central portion of a DBMS. It performs the critical role of retaining high-level information relative to the various applications of the DBMS and thereby enables specification, control, review, and management of the data of value for decision making relative to specific applications.

For very large system designs, it is necessary that the data dictionary development process be automated. A typical data dictionary for a large system may include several thousand entries. It is physically impossible to manually maintain a dictionary of this size or to retain consistent and unambiguous terms for each data element or composite of data elements. Therefore, automated tools are needed for efficient and effective development and maintenance of a data dictionary. These are provided in contemporary DBMSs.

This ability to specify database structures through use of schemas enables efficient and effective management of data resources. This is necessary for access control. The use of one or more subschemas generally simplifies access to databases. It may also provide for database security and integrity for authorized users of the DBMS. Since complex physical (computer) structures for data may be specified, independent of the logical structure of the situation that is perceived by DBMS users, it is thereby possible to improve the performance of an existing and operational database without altering the user interface to the database. This provides for simplicity in use of the DBMS.

The data model gives us the constructs that provide a foundation for the development and use of a database management system. It also provides the framework for such tools and techniques as user interface languages. There are three types of user interface languages:

1. Data definition languages (DDL) provide the basis for definition of schemas and subschemas.
2. Data manipulation languages (DML) are used to develop database applications.
3. Data query languages (DQL or simply query languages (QL) are used to write queries and reports. Of course, it is possible to combine a DDL, a DML, and a DQL into a single database language. The user of these languages is often called a database administrator (DBA).

In general, a data model is a paradigm for representation, storing, organizing, and otherwise managing data in a database. There are three component sets in most data models. These, and their functions, are:

1. A set of *data structures* that define the fields and records which are allowed in the database. Examples of data structures include lists, tables, hierarchies, and networks.
2. A set of *operations* that define the admissible manipulations which are applied to the fields and records that comprise the data structures. Examples of operations include retrieve, combine, subtract, add, and update.
3. A set of *integrity rules* that define or constrain allowable or legal states or changes of state for the data structures that must be protected by the operations.

There are three generically different types of data model representations, and several variants within these three representations. Each of these is applicable as an internal, external, or conceptual model. We will be primarily concerned with these three modeling representations as they affect the external, or logical, data model. This model is the one with which the user of a specific decision support system interfaces. For use in a decision support system generator, the conceptual model is of importance since it is the model that influences the specific external model that various users of a DSS will interface with after an operational DSS is realized. This does not mean that the internal data model is unimportant. It is very important for the design of a DBMS. Through its data structures, operations, and integrity constraints, the data model controls the operation of the DBMS portion of the DSS.

The three fundamental data models for use in the external model portion of a DSS are record-based models, structurally based models, and expert system-based models. Each of these will now be briefly described.

42.2.2.1 Record-Based Models

We are very accustomed to using forms and reports, often prepared in a standard fashion for a particular application. Record based models are computer implementations of these spreadsheet-like forms. Two types can be identified. The first of these, common in the early days of file processing systems (FPS), or file management system, is the *individual record model.* This is little more than an electronic file drawer in which records are stored. It is useful for a great many applications. More sophisticated, however, is the relational database data model in which mathematical relations are used to electronically "cut and paste" reports from a variety of files. Relational database systems have been developed to a considerable degree of sophistication and many commercial products are available. dBase and Oracle are two leading examples.

The individual record model is surely the oldest data model representation. Whereas the simple single record tables characteristic of this model may appear quite appealing, the logic operations and integrity constraints that need to be associated with the data structure are often undefined, and are perhaps not easily defined. Here, the data structure is simply a set of records with each record consisting of a set of fields. When there is more than a single type of record, one field contains a value that indicates what the other fields in the record are named.

The *relational model* is a modification of the individual record model that limits its data structures and which thereby provides a mathematical basis for operation on records. Data structures in a relational database may consist only of relations, or field sets that are related. Every relation may be considered as a table. Each row in the table is a record or tuple. Every column in each table or row is a field or attribute. Each field, or attribute, has a domain which defines the admissible values for that field.

Often, there is only a modest difference in structure between this relational model and the individual record model. The major difference is that relationships are represented by fields in the various records of the individual record model, whereas relationships among fields or attributes in a relation, are denoted by the name of the relation.

Although structural differences between the relational model and the individual record model are minimal, there are major differences in the way in which the integrity constraints and operations may affect the database. The operations in a relational database form a set of operations that are defined mathematically. The operations in a relational model must operate on entire relations, or tuples, rather than only on individual records. The operations in a relational database are independent of the data structures and, therefore, do not depend on the specific order of the records or the fields. There is often controversy about whether or not a DBMS is truly relational. Whereas there are very formal definitions of a relational database, a rather informal one is sufficient here. A relational database is one described by the following statements.

- **a.** Data are presented in tabular fashion without the need for navigation links, or pointer structures, between various tables.
- **b.** A relational algebra exists and can be used to prepare joins of logical record files automatically.
- **c.** New fields can be added to the database, without the necessity to rewrite any programs that used previous versions of the database.

If a DBMS does not satisfy these three criteria, then it is almost surely NOT a relational database.

42.2.2.2 Structural Models

In many instances, data are intended to be associated with a natural structural representation. A typical hierarchical structure is that of an organization. A more general representation than a hierarchy, or tree, is known as a network. It is often possible to represent a logical data model with a hierarchical data structure. In a hierarchical model, we have a number of nodes that are connected by links. All links are directed to point from "child" to "parent," and the basic operation in a hierarchy is that of searching a tree to find items of value. When a query is posed with a hierarchical database, all branches of the hierarchy are searched and those nodes that meet the conditions posed in the query are noted and then returned to the DBMS system user in the form of a report.

Some comparisons of a hierarchical data model with a relational data model are of interest here. The structures in the hierarchical model represent the information that is contained in the fields of the relational model. In a hierarchical model, certain records must exist before other records can exist. The hierarchical model is generally required to have only one key field. In a hierarchical data model, it is necessary to repeat some data in a descendant record that need be stored only once in a relational database regardless of the number of relations. This is so since it is not possible for one record to be a descendent of more than one parent record. There are some unfortunate consequences of the mathematics involved in creating a hierarchical tree, as contrasted with relations among records. Descendants cannot be added without a root leading to them, for example. This leads to a number of undesirable characteristic properties of hierarchical models that may affect our ability to add, delete, and update or edit records easily.

A network model is quite similar to but more general than the hierarchical model. In a hierarchy, data have to be arranged such that one child has only one parent, and there are many instances when this is unrealistic. If we force the use of a hierarchical representation in such cases, data will have to be repeated at more than one location in the hierarchical model. This redundancy can create a number of problems. A record in a network model can participate in several relationships. This leads to two primary structural differences between hierarchical and network models. Some fields in the hierarchical model will become relationships in a network model. Further, the relationships in a network model are explicit and may be bidirectional. The navigation problem in a network data model can become severe. Since search of the database can start at several places in the network, there is added complexity in searching, as well.

Although spreadsheet-type relational records are very useful for many purposes, it has been observed (Kent, 1979) that not all views of a situation, or human cognitive maps, can be represented by relational data models. This has led to interest in entity- and object-oriented data models, and to data models based on artificial intelligence techniques. The basic notion in use of an entity–relationship (ER) model is to accomplish database design at the conceptual level, rather than at the logical and/or physical levels. ER models (Chen, 1976) are relatively simple and easy to understand and use, in large part because of the easy graphical visualization of the database model structure. Also, ER modeling capability is provided by many computer aided software engineering (CASE) tools. Whereas there are such limitations as lack of a very useful query language (Atzeni and Chen, 1983), much interest in ER data models, especially at the conceptual level, exists at this time. Figure 42.8 illustrates this conceptual orientation of the ER modeling approach.

ER models are based on two premises. Information concerning entities and relationships exist as a cognitive reality. This information may be structured using entities and relationships among them as data. An entity relationship data model is a generalization of the hierarchical and network data models. It is based on well-established graph theoretic developments and is a form of structured modeling. The major advantage of the ER approach is that it provides a realistic view of the structure and form for the DBMS design and development effort. This should naturally support the subsequent development of appropriate software. In addition, the approach readily leads to the development of the data dictionary. The primary difficulties that may impede easy use of the ER method are in the need for selection of appropriate verbs for use as contextual relationships, elimination of redundancy of entities and assurances that all of the needed entity–relationships have been determined and properly used. Often, it may take a considerable amount of time to identify an appropriate set of entities and relations and to obtain an appropriate structural model in the form of an ER diagram.

42.2.2.3 Expert Database Models

An expert database model, as a specialized expert database system, involves the use of artificial intelligence technology. The goal in this is to provide for database functionality in more complex environments that require at least some limited form of intelligent capability. To allow this may require adaptive identification of system characteristics, or learning over time as experience with an issue and the environment into which it is

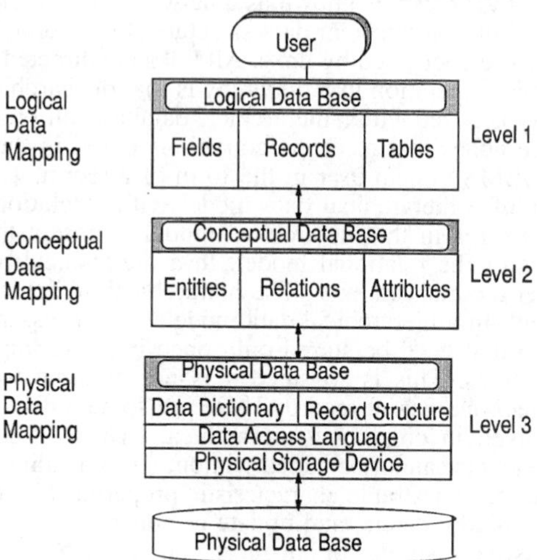

Figure 42.8 The potential role of entity—relationship data models at the conceptual data level.

imbedded increases. The principal approach that has been developed to date comprises an object-oriented database (OODB) design with any of the several knowledge representation approaches useful in expert system development. The field is relatively new, and we can anticipate new and potentially exciting developments in the future. It is described much more fully in Kerschberg (1987, 1989); Myloupoulos and Bridie (1989); Parsaye, Chignell, Khoshafian, and Wong (1989); and Zdonik and Maiser (1989).

The efforts to date in the object-oriented area concern what might be more appropriately named object-oriented database management systems design (OODBMS). The objective in OODBMS design is to cope with database complexity, potentially for large distributed databases, through a combination of object-oriented programming and expert systems technology. Object-oriented design approaches generally involve notions of separating internal computer representations of elements from the external realities that lead to the elements. A primary reason for using object-oriented language is that it naturally enables semantic representations of knowledge through use of 10 very useful characteristics of object-oriented approaches (Parsaye et al., 1989): information hiding, data abstraction, dynamic binding and object identity, inheritance, message handling, object-oriented graphical interfaces, transaction management, reusability, partitioning or dividing or disaggregating an issue into parts, and projecting or describing a system from multiple perspectives or viewpoints. These could include, for example, social, political, legal, and technoeconomic perspectives (Sage, 1992).

There are at least two approaches that we might use in modeling a complex large-scale system: functional decomposition and structuring, and purposeful or object decomposition and structuring. Both approaches are appropriate and may potentially result in useful models. Most models of real-world phenomena tend to be purposeful. Most conventional high-level programming languages are functionally, or procedurally, oriented. To use them, we must write statements that correspond to the functions that we wish to provide in order to solve the problem that has been posed. An advantage to object decomposition and structuring is that it enables us to relate the structure of a database model more easily to the structure of the real system. This is the case if we accomplish our decomposition and structuring such that each module in the system or issue model represents an object or a class of objects in the real issue or problem space. Objects in object-oriented methodology are not unlike elements or nodes in graph theory and structural modeling. It is possible to use one or more contextual relations to relate elements together in a structural model. An object may be defined as a collection of information and those operations that can be performed upon it. We request an object to perform one of its allowable operations by instructing it with a message.

Figure 42.9 illustrates the major conceptual difference between using a conventional programming approach and using an object-oriented approach. In the conventional ap-

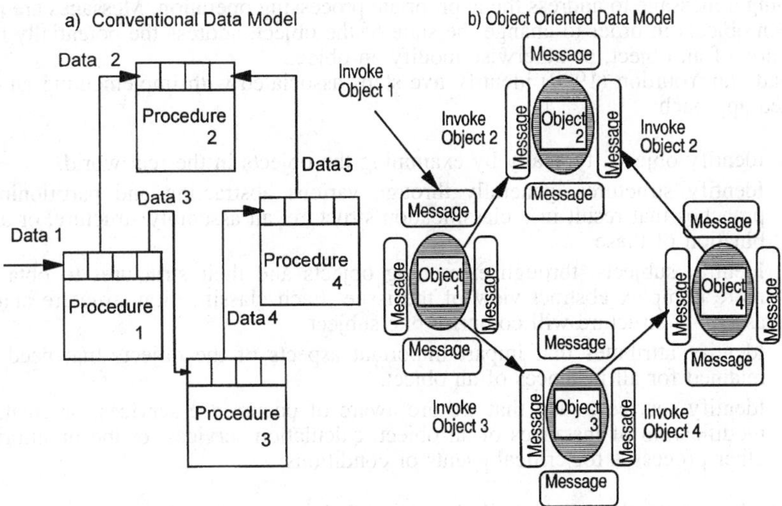

Figure 42.9 Conventional and object-oriented data models.

proach, procedures are at the nexus and procedures update and otherwise manipulate data and return values. In the object-oriented approach, the collection of independent objects are at the nexus and communicate with each other through messages or procedures. Objects investigate requests and behave according to these messages. Object-oriented design often provides a clear and concise interface to the problem domain in that the only way to interact with an object is through the operations or messages to the object. These messages call for operations that may result in a change of state in the object in question. This message will affect the particular object called and only that one, because no other object is affected. This provides a high degree of modularity and increased ability to verify and validate outcomes, and thus provides an increased sense of reliability in the resulting DBMS design. Object-oriented languages are, for the most part, very high level languages used to accomplish precisely the same results as high level languages. By focusing upon the entities of objects and the relationships between objects, they often provide a simpler way to describe precisely those elements needed in detailed design procedures.

Object-oriented DBMS design is based on appropriate linking of objects, or data items or entities, and the operations on these, such that the information and processing is concentrated on the object classes, attributes, and messages that transfer between them. The features that set object-oriented approaches from other approaches are the capability of object-oriented languages to support abstraction, information hiding, and modularity. The items used in object-oriented DBMS design are objects, abstracts or physical entities that pertain to the domain of the system; attributes, or characteristics of objects; operations, or dynamic entities that may change with time; and messages, or requests to an object to perform an operation. Objects represent such real-world entities as machines, files, products, signals, persons, things, or places. They may be either physical entities or abstract representations. The attributes are characteristics that describe the properties ascribed to objects. Each attribute may be dynamically associated with a numerical value and the combination of these values together with the description of the object in the problem domain presents the state of the object. Thus, the attributes are related to the object as subobjects. Attributes include such things as the name of the object, the number of specific objects in a file, or the name for the place, and the like. The basic attribute, one that may not be further decomposed, is the primitive type of object or subobject. Attributes may also be nonnumeric. Operations consist of processes and data structures that apply to the object to which it is directed. These are generally dynamic entities whose value may change over time. Each object may be subjected to a number of operations that provide information relative to control and procedural constructs that are applied to the object. Information hiding is achieved by defining an object to have a private part and then assigning a message to address the appropriate processing operation. Messages are passed between objects in order to change the state of the object, address the potentially hidden data parts of an object, or otherwise modify an object.

Coad and Yourdon (1990) identify five steps associated with implementing an object oriented approach:

1. Identify objects, typically by examining the objects in the real world.
2. Identify structures, generally through various abstracting and partitioning approaches that result in a classification structure, an assembly structure, or a combination of these.
3. Identify subjects, through examining objects and their structures to obtain this more complex abstract view of the issue. Each classification structure and each assembly structure will comprise one subject.
4. Identify attributes that impart important aspects of the objects that need to be retained for all instances of an object.
5. Identify services, such that we are aware of occurrence services, or creation or modification of instances of an object, calculation services, or the monitoring of other processes for critical events or conditions.

Steps such as these should generally be accomplished in an iterative manner, since there is no unique way to accomplish specification of a set of objects, or structural relations between objects.

A major result from using object-oriented approaches is that we obtain the sort of knowledge representation structures that are commonly found in many expert systems. Thus, object-oriented design techniques provide a natural interface to expert system based techniques for DBMS design. This is the case since objects, in object-oriented design, are provided with the ability to: store information such that learning over time can occur; process information by initiating actions in response to messages; compute and communicate by sending messages between objects; and generate new information as a result of communications and computations. Object-oriented design also lends itself to parallel processing and distributed environments.

42.2.3 Distributed and Cooperative Databases

We have identified the three major objectives for a DBMS as data independence, data redundancy reduction, and data resource control. These objectives are important for a single database. When there are multiple databases potentially located in a distributed geographical fashion, and potentially many users of one or more databases, additional objectives arise. These include:

1. Location independence or transparency, to enable DBMS users to access applications across distributed information bases without the need to be explicitly concerned with where specific data are located
2. Advanced data models, to enable DBMS users to access potentially nonconventional forms of data such as multidimensional data, graphical data, spatial data, and imprecise data
3. Extensible data models that will allow new data types to be added to the DBMS, perhaps in an interactive real-time manner, as required by specific applications.

An important feature of a distributed database management systems (DDBMS) is that provisions for database management are distributed. Only then can we obtain the needed "fail-safe" reliability and "availability" even when a portion of the system breaks down. There are a number of reasons that distributed databases may be desirable. These include and are primarily related to distributed users and cost savings. There are also some additional costs involved and these must be justified.

A distributed database management system will generally look much like replicated versions of a more conventional single location database management system. We can thus imagine replicated versions of Figure 42.7. These, taken together, represent a possible conceptual architecture of a distributed database in which there are two requests for data being simultaneously submitted. Through use of protocols, all requests for data are entered as if that data were stored in the database of the requesting user. The data dictionaries of the distributed system are responsible for finding the requested data elements and delivering them to the requesting user. All of these auxiliary requests are accomplished in a manner that is transparent to the requesting database user. Data transmission rate concerns must be resolved in order for this to be done satisfactorily. Potentially, however, real-time access to all of the data in the entire distributed system can be provided to any local user.

It is important to note that completion of the above steps does not transfer the requested data into the database of the requesting user, except perhaps in some temporary storage fashion. The very good reason that this is not done is that doing so would result in database redundancy. Doing this would also increase database security difficulties.

An alternate approach to this distributing of data is to use what is often called cooperative processing or cooperative database management. This involves central database and distributed database concepts, blended to produce better results than can be had with either approach alone. Central to this are various approaches for the distributing and coordinating of data. We could use function, geography, or organizational level as the basis for distributing and cooperating. Various perspectives concerning these issues have been taken (Ceri and Pelagatti, 1984; Lorin, 1988; Mullender, 1989). Possible system configurations may be based upon background communications, microcomputer-based front-end processing for host applications, and peer-to-peer cooperative processing (Altman, 1989).

The database component of a DSS provides the knowledge repository that are needed for decision support. There must be some sort of management system associated with a database in order to provide the intelligent access to data that are needed. We have, in this section, examined a number of existing constructs for database management systems.

42.3 MODEL BASE MANAGEMENT SYSTEMS

There are four primary objectives of a DBMS:

1. To manage a large quantity of data in physical storage
2. To provide logical data structures, which interact with humans, that are independent of the structure used for physical data storage
3. To reduce data redundancy and maintenance needs, and to increase flexibility of use of the data, by provision of independence between the data and the applications programs that use it
4. To provide effective and efficient access to data by users who are not necessarily very sophisticated in the microlevel details of computer science.

Many support facilities will typically be provided with a DBMS to enable achievement of these purposes. These include data dictionaries to aid in internal housekeeping, and information query, retrieval, and report generation facilities to support external use needs.

The function of a model base management system (MBMS), a structure for which is illustrated in Figure 42.10, is quite analogous to those of a DBMS. The primary functions of a DBMS are separation of system users, in terms of independence of the application, from the physical aspects of database structure and processing. In a similar way, a MBMS is intended to provide independence between the specific models that are used in a DSS and the applications that use them. The purpose of a MBMS is to transform data from the DBMS into information that is useful for decision making. An auxiliary purpose might also include representation of information as data such that it can later be recalled and used.

The term *model management system* was apparently first used over 20 years ago (Will, 1975). Soon thereafter, the MBMS usage was adopted in Sprague and Carlson (1982) and the purposes of an MBMS were defined to include creation, storage, access, and manipulation of models. Objectives for a MBMS include:

1. To provide for effective and efficient creation of new models for use in specific applications
2. To support maintenance of a wide range of models that support the formulation, analysis, and interpretation stages of issue resolution
3. To provide for model access and integration, within models themselves as well as with the DBMS
4. To centralize model base management in a manner analogous and compatible with database management
5. To insure integrity, currency, consistency, and security of models

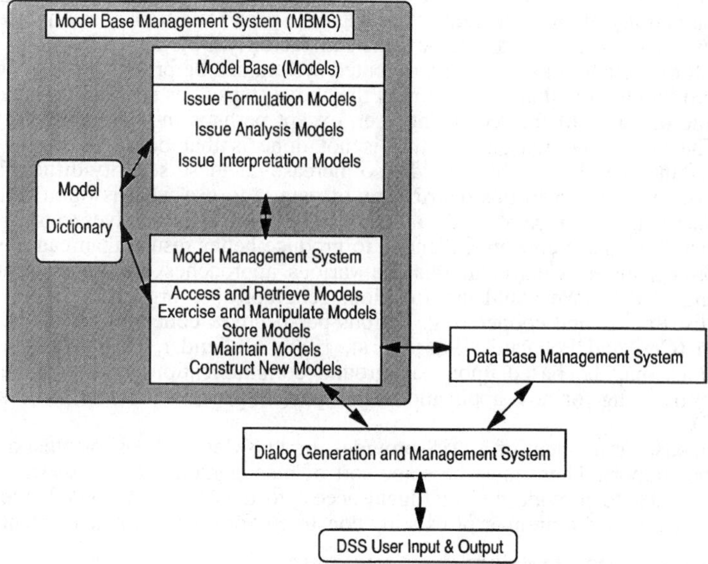

Figure 42.10 Prototypical structure of model management system.

Just as we have physical data and logical data processing in a DBMS, so also do we have two types of processing efforts in a MBMS: model processing and decision processing MBMS (Applegate, Konsynski, and Nunamaker, 1986). A DSS user would interact directly with a decision-processing MBMS, whereas the model processing MBMS would be more concerned with provision of consistency, security, currency, and other technical modeling issues. Each of these support the notion of appropriate formal use of models that support relevant aspects of human judgment and choice.

There are several necessary ingredients for a study of MBMS. The first of these is a study of formal analytical methods of operations research and systems engineering that supports the construction of models that are useful in issue formulation, analysis, and interpretation. To present even a small fraction of the analytical methods and associated models that are in current use would be a mammoth undertaking. Rather than do this, we will discuss models in a somewhat general context.

42.3.1 Models and Modeling

In this section we present a brief description of a number of models and methods that can be used as part of a systems engineering based approach to problem solution, or issue resolution. Systems engineering (Sage, 1992, 1995) involves the application of a general set of guidelines and methods useful to assist clients in the resolution of issues and problems, often through the definition, development, and deployment of a trustworthy system. Three fundamental steps may be distinguished in a formal systems based approach that is associated with each of these three phases of system fielding or problem solving:

1. Problem or issue formulation
2. Problem or issue analysis
3. Interpretation of analysis results, including evaluation and selection of alternatives, and implementation of the chosen alternatives

These steps are conducted at a number of phases throughout a systems life cycle. As we have indicated in earlier, this life cycle begins with definition of requirements for a system through a phase where the system is developed to a final phase wherein deployment, or installation, and ultimate system maintenance and retrofit occur.

42.3.1.1 Issue, or Problem, Formulation Models

The first part of a systems effort for problem or issue resolution is typically concerned with problem or issue formulation, including identification of problem elements and characteristics. The first step in issue formulation is generally that of definition of the problem or issue to be resolved. Problem definition is generally an outscoping activity, because it enlarges the scope of what was originally thought to be the problem. Problem or issue definition will ordinarily be a group activity involving those familiar with or impacted by the issue or the problem. It seeks to determine the needs, constraints, alterables, and social or organizational sectors affecting a particular problem and relationships among these elements.

Of particular importance are the identification and structuring of objectives for the policy or alternative that will ultimately be chosen. This is often referred to as *value system design*, a term apparently first used by Hall (1969) in one of his seminal works in systems engineering. Option generation (Keller and Ho, 1990), or alternative identification, is a very important and often neglected portion of a problem-solving effort. This option generation, or system or alternative synthesis, step of issue formulation is concerned primarily with the answers to three questions:

1. What are the alternative approaches for attaining objectives?
2. How is each alternative approach described?
3. How do we measure attainment of each alternative approach?

The answers to these three questions lead to a series of alternative activities or policies and a set of activities measures.

Several of the methods that are particularly helpful in the identification of issue formulation elements are based on principles of collective inquiry (McGrath, 1984). The term "collective inquiry" refers to the fact that a group of interested and motivated people is brought together in the hope that they will stimulate each other's creativity in generating elements. We may distinguish two groups of collective inquiry methods here depending upon whether or not the group is physically present at the same physical location.

Brainstorming, Synectics, and Nominal Group Techniques

These approaches typically require a few hours of time, a group of knowledgeable people gathered in one place, and a group leader or facilitator. The nominal group technique is typically better than brainstorming in reducing the influence of dominant individuals. Both methods can be very productive: 50–150 ideas or elements might be generated in less than 1 hr. Synectics, based on problem analogies, might be very appropriate if there is a need for truly unconventional, innovative ideas. Considerable experience with the method is a requirement, however, particularly for the group leader. The nominal group technique is based on a sequence of idea generation, discussion, and prioritization. It can be very useful when an initial screening of a large number of ideas or elements is needed. Synectics and brainstorming are directly interactive group methods, whereas nominal group efforts are "nominally" interactive in that the members of the group do not directly communicate.

Questionnaires, Survey, and Delphi Techniques

These three methods of collective inquiry do not require the group of participants to gather at one place and time, but they typically take more time to achieve results than the methods above. In most questionnaires and surveys a large number of participants is asked, on an individual basis, for ideas or opinions, which are then processed to achieve an overall result. There is no interaction among participants. Delphi usually provides for written anonymous interaction among participants in several rounds. Results of previous rounds are fed back to participants, and they are asked to comment, revise their views as desired, etc. A Delphi can be very instructive, but usually takes several weeks or months to complete.

Use of some of the many structuring methods, in addition to leading to greater clarity of the problem formulation elements, will typically lead also to identification of new elements and revision of element definitions. Most structuring methods contain an analytical component; and they may, therefore, be more properly labeled as analysis methods. The following element structuring aids are among the many modeling aids available that are particularly suitable for the issue formulation step.

There are many approaches to problem formulation (Volkema, 1990). In general, these approaches assume that "asking" will be the predominant approach used to obtain issue formulation elements. Asking is often the simplest approach; but often valuable information can be obtained from observation of an existing and evolving system, or from study of plans and other prescriptive documents. When these three approaches fail, it may be necessary to construct a "trial" system and determine issue formulation elements through experimentation and iteration with the trial system. These four methods (asking, study of an existing system, study of a normative systems, experimentation with a prototype system) are each very useful for information requirements determination.

42.3.1.2 Models and Analysis

The analysis portion of a DSS effort typically consists of two steps. First, the options or alternatives defined in issue formulation are analyzed to assess their expected impacts on needs and objectives. This is often called impact assessment. Second, a refinement or optimization effort is often desirable. This is directed toward refinement or fine tuning a viable alternative through adjustment of the parameters within an alternative, such as to obtain maximum performance in terms of needs satisfaction, subject to the given constraints.

Simulation and modeling methods are based on the conceptualization and use of an abstraction, or model, which is hoped to behave in a way similar to the real system. Impacts of alternative courses of action are studied through use of the model, something that often cannot easily be done through experimentation with the real system. Models are, of necessity, dependent on the value system and the purpose behind utilization of a model. We want to be able to determine the correctness of predictions based on use of a model and thus be able to validate the model. There are three essential steps in constructing a model:

1. Determine those issue formulation elements which are most relevant to a particular problem.
2. Determine the structural relationships among these elements.
3. Determine parametric coefficients within the structure.

We should interpret the word "model" here as an abstract generalization of an object or system. Any set of rules and relationships that describes something is a model of that thing. The MBMS of a DSS will typically contain formal models that have been stored into the model base of the support system.

Gaming methods are modeling methods in which the real system is simulated by people taking on the roles of real-world actors. The approach is very appropriate for studying situations in which people's reactions to each other's actions are of great importance, such as competition between individuals or groups for limited resources. It is also a very appropriate learning method. Conflict analysis (Fang, Hipel, and Kilgour, 1993; Fraser and Hipel, 1984) is an interesting and appropriate game theoretic based approach that may result in models that are particularly suitable for inclusion into the model base of a MBMS.

Trend extrapolation or time series forecasting models or methods are particularly useful when sufficient data about past and present developments are available, but there is little theory about underlying mechanisms causing change. The method is based on the identification of a mathematical description or structure that will be capable of reproducing the data. Then, this description is used to extend the data series into the future, typically over the short to medium term. The primary concern is with input–output matching of observed input data and results of model use. Often, little attention is devoted to assuring process realism, and this may create difficulties affecting model validity. Whereas such models may be functionally valid, they may not be purposefully or structurally valid.

Continuous-time dynamic-simulation models or methods are generally based on postulation and qualification of a causal structure underlying change over time. A computer is used to explore long-range behavior as it follows from the postulated causal structure. The method can be very useful as a learning and qualitative forecasting device. Often, it is expensive and time consuming to create realistic dynamic-simulation models. Continuous-time dynamic models are quite common in the physical sciences and in much of engineering.

Input–output analysis models are especially designed for study of equilibrium situations and requirements in economic systems in which many industries are interdependent. Many economic data formats are directly suited for the method. It is, relatively simple, conceptually, and can cope with many details. Often, input-output models are very large.

Econometrics or macroeconomic models are primarily applied to economic description and forecasting problems. They are based on both theory and data. Emphasis is placed an specification of structural relations, based upon economic theory, and the identification of unknown parameters, using available data, in the behavioral equations. The method requires expertise in economics, statistics, and computer use. It can be quite expensive and time consuming. It has been widely used for short- to medium-term economic analysis and forecasting.

Queuing theory and discrete-event simulation models are often used to study, analyze, and forecast the behavior of systems in which probabilistic phenomena, such as waiting lines, are of importance. Queuing theory is a mathematical approach, and discrete-event simulation generally refers to computer simulation of queuing theory-type models. The two methods are widely used in the analysis and design of systems such as toll booths, communication networks, service facilities, shipping terminals, scheduling, etc.

Regression analysis models and estimation theory models are very useful for the identification of mathematical relations and parameter values in these relations from sets of data or measurements. Regression and estimation methods are used frequently in conjunction with mathematical modeling, in particular with trend extrapolation and time series forecasting, and with econometrics. These methods are often also used to validate models. Often these approaches are called "system identification approaches" when the goal is to identify the parameters of a system, within an assumed structure, such as to minimize a function of the error between observed data and the model response.

Mathematical programming models are used extensively in operations research systems analysis, and in management science practice for resource allocation under constraints, planning or scheduling, and similar applications. It is particularly useful when the best equilibrium or one-time setting has to be determined for a given policy or system. Many analysis issues can be cast as mathematical programming problems. There have been a very significant number of mathematical programming models developed, including linear programming, nonlinear programming, integer programming, and dynamic programming. There are many appropriate reference texts, including Hillier and Liberman (1986) that discuss this important class of modeling and analysis tools.

42.3.1.3 Interpretation Models

The third step in a decision support systems effort starts with evaluation and comparison of alternatives, using the information gained by analysis. Subsequently, one or more alternatives are selected, and a plan for their implementation is designed. Thus, an MBMS must provide models for interpretation, including evaluation, of alternatives.

It is important to note that there is a clear and distinct difference between the refinement of individual alternatives, or optimization step of analysis, and the evaluation of sets of refined alternatives. In some cases refinement or optimization of individual alternative policies is not needed in the analysis step. But evaluation of alternatives is always needed; for if there is but a single policy alternative, then there really is no alternative at all. It is especially important to avoid a large number of cognitive biases in evaluation and decision making. Clearly, the efforts involved in the interpretation step of evaluation and decision making interact most strongly with the efforts in the other steps of the systems process.

There are a number of methods for evaluation and choice making that are of importance. A few are described briefly here. Chapter 37 discusses decision analysis and decision making in some detail.

Decision analysis (Raiffa, 1968) is a very general approach to option evaluation and selection. It involves identification of action alternatives and possible consequences, identification of the probabilities of these consequences, identification of the valuation placed by the decision maker upon these consequences, computation of the expected value of the consequences, and aggregating or summarizing these values for all consequences of each action. In doing this we obtain an evaluation of each alternative act; and the one with the highest value is the most preferred action or option.

Multiple attribute utility theory (Keeney and Raiffa, 1976) has been designed to facilitate comparison and ranking of alternatives with many attributes or characteristics. The relevant attributes are identified and structured, and a weight or relative utility is assigned by the decision maker to each basic attribute. The attribute measurements for each alternative are used to compute an overall worth or utility for each alternative. Multiple attribute utility theory allows for various types of worth structures, and for the explicit recognition and incorporation of the decision makers attitude towards risk in the utility computation.

Policy capture (or *social judgment theory)* (Hammond, McClelland, and Mumpower, 1980) has also been designed to assist decision makers in making values explicit and known. It is basically a descriptive approach toward identification of values and attribute weights. By knowing these, one can generally make decisions that are consistent with values. In policy capture, the decision maker is asked to rank order a set of alternatives. Then, alternative attributes and their attribute measures or scores are determined by elicitation from the decision maker for each alternative. A mathematical procedure involving regression analysis is used to determine that relative importance, or weight, of each attribute that will lead to a ranking as specified by the decision maker. The result is fed back to the decision maker who, typically, will express the view that some of his or her values, in terms of the weights associated with the attributes, are different. In an iterative learning process, preference weights and/or overall rankings are modified until the decision maker is satisfied with both the weights and the overall alternative ranking.

42.3.2 Model Base Management

As we have noted, an effective model base management system [MBMS] will make the structural and algorithmic aspects of model organization and associated data processing transparent from users of the MBMS. Such tasks as specifying explicit relationships between models to indicate formats for models and which model outputs are input to other models are not placed directly on the user of a MBMS but handled directly by the system. Figure 42.10 presents a generic illustration of a MBMS. It shows a collection of models or model base, a model base manager, a model dictionary, and connections to the DBMS and the DGMS.

There are a number of capabilities that should be provided by an integrated and shared MBMS of a DSS (Barbosa and Herko, 1980; Liang, 1985) construction, model maintenance, model storage, model manipulation, and model access (Applegate et al., 1986). These involve control, flexibility, feedback, interface, redundancy reduction, and increased consistency:

1. *Control*—The DSS user should be provided with a spectrum of control. The system should support both fully automated and manual selection of models that seem most useful to the user for an intended application. This will enable the user to proceed at the problem-solving pace that is most comfortable for the user's experiential familiarity with the task at hand. It should be possible for the user to introduce subjective information and to not have to provide full information. Also, the control mechanism should be such that the DSS user can obtain a recommendation for action with this partial information at essentially any point in the problem solution process.
2. *Flexibility*—The DSS user should be able to develop part of the solution to the task at hand using one approach and then be able to switch to another modeling approach, if this appears preferable. Any change or modification in the model base will be made available to all DSS users.
3. *Feedback*—The MBMS of the DSS should provide sufficient feedback to enable the user to be aware of the state of the problem-solving process at any point in time.
4. *Interface*—The DSS user should feel comfortable with the specific model from the MBMS that is in use at any given time. The user should not have to laboriously supply inputs when the user does not wish to do this.
5. *Redundancy reduction*—Through use of shared models and associated elimination of redundant storage that would otherwise be needed.
6. *Increased consistency*—Through the ability of multiple decision makers to use the same model and the associated reduction of inconsistency that may result from use of different data or different versions of a model.

In order to provide these capabilities, it appears that a MBMS design must allow the DSS user to:

1. Access and retrieve existing models
2. Exercise and manipulate existing models, including model instantiation, model selection, and model synthesis, and the provision of appropriate model outputs
3. Store existing models, including model representation, model abstraction, and physical and logical model storage
4. Maintain existing models as appropriate for changing conditions
5. Construct new models with reasonable effort when they are needed, and usually by building new models by using existing models as building blocks

There are a number of auxiliary requirements that must be achieved in order to provide these five capabilities. There must be, for example, appropriate communication and data changes among models that have been combined. It must also be possible to locate appropriate data from the DBMS and transmit it to the models that will use it.

It must also be possible to analyze and interpret the results obtained from using a model. There are a number of ways in which this can be accomplished. In this section, we will examine two of them: relational MBMS, and expert system control of a MBMS. The objective is to provide an appropriate set of models for the model base and appropriate software to manage the models in the model base; integration of the MBMS with the DBMS; and integration of the MBMS with the DGMS. We can expand further on each of these needs. Many of the technical capabilities needed for a MBMS will be analogous to those needed for a DBMS. These include model generators that will allow rapid building of specific models, model modification tools that will enable a model to be restructured easily on the basis of changes in the task to be accomplished, update capability that will enable changes in data to be input to the model, and report generators that will enable rapid preparation of results from using the system in a form appropriate for human use.

As is the case with a relational view of data, a relational view of models is based on a mathematical theory of relations. Thus, a model is viewed as a virtual file or virtual relation. It is a subset of the Cartesian product of the domain set that corresponds to these input and output attributes. This virtual file is, in principle, created by exercising the model with a wide spectrum of inputs. These values of inputs and the associated outputs become records in the virtual file. The input data become key attributes and the model output data become content attributes.

Model base structuring and organization is very important for appropriate relational model management. Records in the virtual file of a model base are not individually updated, however, as they are in a relational database. When a model change is made, all of the records that comprise the virtual file are changed. Nevertheless, there are possible processing anomalies in relational model management. Transitive dependencies in a relation, in the form of functional dependencies that affect only the output attributes, do occur and are eliminated by projecting into an appropriate normal form.

Another issue of considerable importance relates to the contemporary need for usable model base query languages, and needs within such languages for relational completeness. The implementation of joins is of concern in relational model base management, just as it is in relational database management. A relational model join is simply the result of using the output of one model as the input to another model. Thus, joins will normally be implemented as part of the normal operation of software and a MBMS user will often not be aware that they are occurring. However, there can be cycles, since the output from a first model may be the input to a second model, and this may become the input to the first model. Cycles such as this do not occur in relational DBMS.

Expert system applications in MBMS represent another attractive possibility. Four different potential opportunities exist. It might be possible to use expert system technology to considerable advantage in the construction of models (Murphy and Stohr, 1986), including decisions with respect to whether or not to construct models in terms of the cost and benefits associated with this decision. Artificial intelligence (AI) and expert system technology may potentially be used to integrate models. This model integration is needed to join models. AI and expert system technology might be potentially useful in the validation of models. Finally, this technology might find potential use in the interpretation of the output of models. This would especially seem to be needed for large-scale models, such as large linear programming models (Greenberg, 1987). Whereas MBMS approaches based on a relational theory of models and expert systems technology are new as of this writing, they offer much potential for implementing model management notions in an effective manner. As has been noted, they offer the prospect of data as models (Dolk, 1986) that may well prove much more useful than the conventional information systems perspective of "models as data."

42.4 DIALOGUE GENERATION AND MANAGEMENT SYSTEMS

In our efforts in this chapter, we envision a basic DSS structure of the form shown in Figure 42.11. This figure also shows many of the operational functions of the database management system (DBMS) and the model base management system (MBMS). The primary purpose of the dialog generation and management system (DGMS) is to enhance the propensity and ability of the system user to use and benefit from the DSS. There are doubtlessly few users of a DSS who use it because of necessity. Most uses of a DSS are optional. There are a variety of ways in which this use can occur. In all uses of a DGMS, it is the DGMS that the user interacts with. In an early seminal text, Bennett (1983) posed three questions to indicate this centrality:

1. What presentations is the user able to *see* at the DSS display terminal?
2. What must the user *know* about what is seen at the display terminal in order to use the DSS?
3. What can the DSS user *do* with the systems that will aid in accomplishing the intended purpose?

Bennett refers to these elements as the *presentation language*, the *knowledge base*, and the *action language*.

It is generally felt that there are three types of languages or modes of communications: *words, mathematics*, and *graphics*. The presentation and action languages, and the knowledge base in general, many contain any or all three language types. The mix of these that is appropriate for any given DSS task will be a function of the task itself, the environment into which the task is embedded, and the nature of the experiential familiarity of the person performing the task with it and the environment. The DSS, when one is used, becomes a fourth ingredient, although it is really much more of a vehicle supporting effective use of words, mathematics, and graphics.

Notions of DGMS design are relatively new, especially as a separately identified portion of the overall design effort. To be sure, user interface design is not at all new. However, the usual practice has been to assign the task of user interface design to the design engineers responsible for the entire system. In the past, user interfaces have not been given special attention. They were merely viewed as another hardware and software component in the system. Often, system designers were not particularly familiar with, and perhaps not even especially interested in, the user-oriented design perspectives necessary to produce a successful interface design. As a result, many user interface designs have provided more of what the designer wanted, rather than what the user wanted and needed. Notions of dialogue generation and dialogue management, the architecture for which is illustrated in Figure 42.12, extend far beyond that of interface issues, although

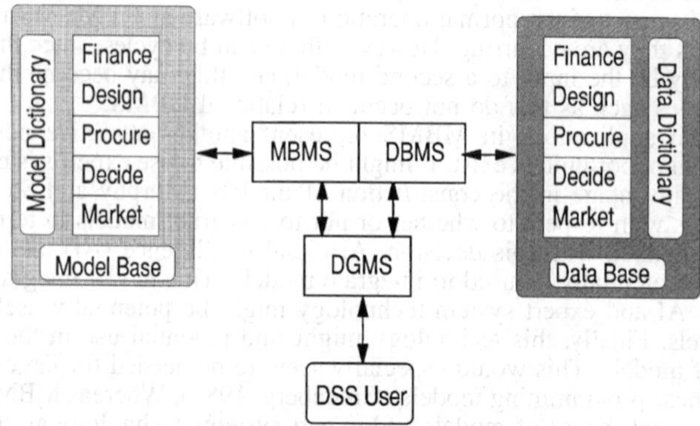

Figure 42.11 Organizational information and decision flow.

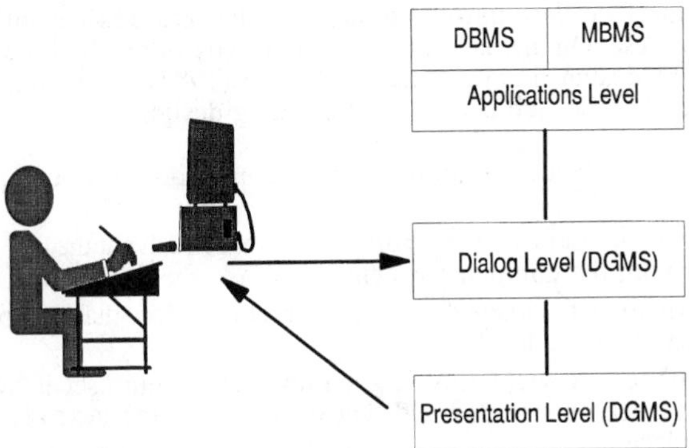

Figure 42.12 Dialogue generation and management system architecture.

the interface is a central concern in dialogue generation and dialogue management. A number of discussions of user interface issues are contained in Part VIII of this handbook, especially in Chapters 50 and 51.

Figure 42.13 illustrates an attribute tree for interface design based on the work of Smith and Mosier (1986). This attribute tree can be used, in conjunction with the evaluation methods of decision analysis, to evaluate the effectiveness of interface designs.

- Quality of Interface
  - Data Entry
    - Data Entry Transaction Consistency
    - Minimal User Input Actions
    - Minimal User Memory Load
    - Data Entry and Display Compatibility
    - Flexible User Control of Data Entry
  - Information Display
    - Consistent Data Displays
    - Efficient Information Assimilation
    - Minimal Human Memory Burden
    - Data Display and Entry Compatibility
    - Flexible User Control of Data Display
  - Sequence Control
    - Consistency of Control Actions
    - Minimal Control Actions by User
    - Minimal Memory Load on User
    - Compatibility with Task Requirements
    - Sequence Control Flexibility
  - User Guidance
    - Consistency of Operational Procedures
    - Efficient Use of Full System Capabilities
    - Minimum Memory Load on User
    - Minimal Learning Time
    - Flexibility in Support of Different Users
  - Data Transmission
    - Consistency of Data Transmission
    - Minimal User Actions
    - Minimal Memory Load on User
    - Compatibility with Other Inf. Handling Elements
    - User Control Flexibility in Data Transmission
  - Data Protection
    - Efficient Data Security
    - Minimal Entry of Bad Data
    - Minimal Erroneous Changes to Stored Data
    - Minimal Loss of Needed Data
    - Minimal Interference with Normal Inf. Processing

Figure 42.13 Attribute tree of Smith and Mosier elements for interface design evaluation.

There are other interface descriptions, some of which are less capable of instrumental measurement than these. On the basis of a thorough study of much of the human-computer interface and dialog design literature, Schneiderman (1987) has identified eight primary objectives, often called the "golden rules" for dialog design.

1. Strive for consistency of terminology, menus, prompts, commands, and help screens.
2. Enable frequent users to use shortcuts that take advantage of their experiential familiarity with the computer system.
3. Offer informative feedback for every operator action that is proportional to the significance of the action.
4. Design dialogues to yield closure such that the system user is aware that specific actions have been concluded and that planning for the next set of activities may now take place.
5. Offer simple error handling such that, to the extent possible, the user is unable to make a mistake. Even when mistakes are made, the user should not have to (for example) retype an entire command entry line but, rather, just edit the portion that is incorrect.
6. Permit easy reversal of action such that the user is able to interrupt and then cancel wrong commands rather than having to wait for them to be fully executed.
7. Support internal locus of control such that users are always the initiators of actions rather than the reactors to computer actions.
8. Reduce short-term memory load such that users are able to master the dialogue activities that they perform.

Clearly, all of these will have specific interpretations in different DGMS environments and need to be sustained in and capable of extension for a variety of environments.

Because of their major contemporary importance, it is not surprising that there have been a number of recent approaches to human–computer interaction and associated interface design issues. Some of these are almost totally empirical. Some involve almost totally theoretical and formal models (Harrison and Thimbleby, 1990). Others attempt approximate predictive models that are potentially useful for design purposes (Card, Moran, and Newell, 1983). One word that has appeared often in these discussions is "consistency." This is Schneiderman's first "golden rule of dialogue design," and many other authors advocate this as well. A notable exception to this advocacy of consistency appears in the writing of Grudin (1989), who argues that issues associated with consistency should be placed in a very broad context. Three types of consistency are defined.

1. *Internal consistency*—The physical and graphic layout of the computer system, including such characteristics as those associated with command naming and use, and dialogue forms, are consistent if these internal features of the interface are the same across applications.
2. *External consistency*—If an interface has unchanging use features when compared to another interface with which the user is familiar, it is said to be externally consistent.
3. *Applications consistency*—If the user interface uses metaphors or analogous representations of objects that correspond to those of the real-world application, then the interface may be said to correspond to experientially familiar features of the world and to be applications consistent.

Grudin makes several relevant observations. The two most relevant to interface consistency are that ease of learning can conflict with ease of use, especially as experiential familiarity with the interface grows, and that consistency can work against both ease of use and learning. On the basis of some experiments illustrating these hypotheses, he establishes these three appropriate dimensions for consistency.

There are a number of desirable characteristics for user interfaces. Roberts and Moran (1982) identify functionality of the editor, learning time required, time required to perform tasks, and errors committed, as the most important attributes of text editors. To this might be added the cost of evaluation. Harrison and Hix (1989) identify usability, completeness, extensibility, escapability, integration, locality of definition, structured guidance, and direct manipulation as well in their more general study of user interfaces. They also note a number of tools useful for interface development, as does Lee (1990).

Evaluating usability of human computer interfaces is the subject of a monograph by Ravden and Johnson (1989). Nine top-level attributes were identified in this effort: visual clarity, consistency, compatibility, informative feedback, explicitness, appropriate functionality, flexibility and control, error prevention and correction, and user guidance and support. These are each disaggregated into a number of more measurable attributes. These attributes can be used as part of a standard multiple attribute evaluation.

A goal in DGMS design is to define an abstract user interface that can be implemented on specific operating systems in different ways. The purpose of this is to allow for device independence such that, for example, switching from a command line interface to a mouse-driven pull-down menu interface can be easily accomplished. Separating the application from the user interface should do much toward ensuring portability across operating systems and hardware platforms without modifying the MBMS and the DBMS that, together, comprise the applications software portions of the DSS.

42.5 GROUP AND ORGANIZATIONAL DECISION SUPPORT SYSTEMS

The term group decision support system (GDSS) refers to an information technology–based support system designed to provide decision-making support to groups and/or organizations. This could refer to a group meeting occurring at one physical location at which judgments and decisions are made that affect an organization or group. Alternately, it could refer to a spectrum of meetings of one or more individuals that occur at distributed locations, or distributed in time, or distributed spatially and temporally. Often, GDSSs are called *organizational decision support systems.* There are other terms often used. These include *executive support systems* (ESS), which are information technology based systems designed to support executives and managers, and *command and control systems,* which is a term often used in the military for a decision support system. We will generally use the term GDSS to describe all of these.

Managers and other knowledge workers spend much time in meetings. Much research into meeting effectiveness suggests that it is low, and there have been proposals to increase this through information technology support (Johansen, 1988). Specific components of this information technology–based support might include computer hardware and software, audio and video technology, and communications media. There are three fundamental ingredients in this support concept: technological support facilities, the support processes provided, and the environment in which they are embedded. A noteworthy commentary concerning the need for group efforts was provided by Kraemer and King (1988) in their overview of GDSS efforts. They indicated that group activities are economically necessary, efficient as a means of production, and reinforcing of democratic values.

There are a number of predecessors for group decision support technology. Certainly, decision rooms, or situation rooms, where managers and boards meet to select from among alternative plans or courses of action, are very common. The first computer-based decision support facility for group use is reported to be one created by Douglas C. Engelbart, the inventor of the (computer) mouse, at Stanford in the 1960s. A discussion of this and other early support facilities is contained in Johansen (1988).

The electronic boardroom type design of Engelbart is acknowledged to be the first type of information technology based GDSS. The electronic format was, however, preceded by a number of nonelectronic formats. The cabinet war room of Winston Churchill is perhaps the most famous of these. Maps placed on the wall and tables for military decision makers were the primary ingredients of this room. The early 1970s saw the introduction of a number of simple computer-based support aids into situation rooms. The first system that resembles that of the GDSS in use today is often attributed to Gerald Wagner, the chief executive officer of Execucom, who implemented a planning laboratory that was comprised of a U-shaped table around which people sat, a projection TV system for use as a public viewing screen, individual small terminals and keyboards available to participants, and a minicomputer to which the terminals and keyboards were connected. This enabled participants to vote and to conduct simple spreadsheet-like exercises. Figure 42.14 illustrates the essential features of this concept. Most present-day GDSS centralized facilities appear much like the conceptual illustration of a support room, or situation, room, shown in this figure.

As with a single-user DSS, appropriate questions for a GDSS that have major implications for design are questions that concern the perceptions and insights that the group

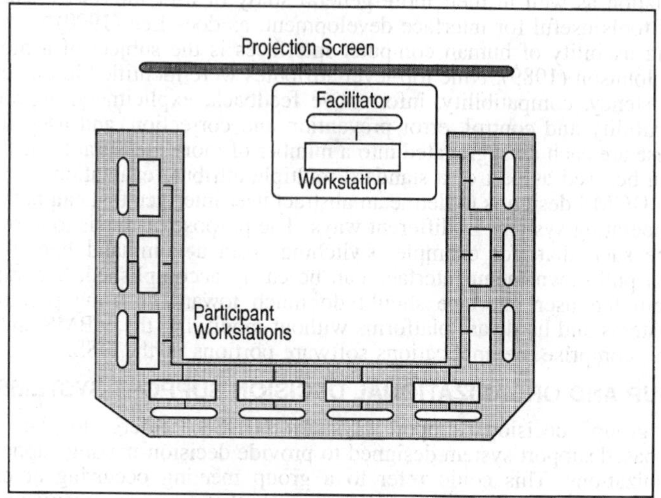

Figure 42.14 Early group decision support system (GDSS) situation room.

obtains through use of the GDSS, and the activities that can be carried out through its use. Also, additional concerns arise regarding the public screen, interactions between the public screen and individual screens, the characteristics of individual work screens, and contingency task structural variables associated with the individuals in the group using the GDSS (Gray and Olfman, 1989).

Huber (1982) has indicated both the needs for GDSS and how an appropriately designed GDSS can meet these needs. He identifies four interacting and complicating concerns.

1. Effective decision making requires not only obtaining an appropriate decision, but also assuring that participants are happy with the process used to reach the decision and that they will be willing to meet and work cooperatively in the future.
2. Productivity losses occur because of dominant individuals and group pressures that lead to conformity of thought, or groupthink.
3. Miscommunications are common in group situations.
4. Insufficient time is often spent in situation assessment, problem exploration, and generation of alternative courses of action.

Huber further indicates that a GDSS can help improve the unaided decision situation, which often suffers from imperfect information processing and suboptimal decision selection.

A GDSS is composed of:

1. *Technological components*—in terms of computer hardware and software, and communication equipment
2. *Environmental components*—in terms of the people involved, their locations in time and space, and their experiential familiarity with the task at hand
3. *Process components*—or variables that are composed of the conventions used to support task performance, and to enable the other components of decision making to function appropriately.

We have already described the technological design features of DSSs. Thus, our commentary on this can be brief here. We do need to provide a perspective on groups and organizations and we do this in our next two sections. Then, we turn to some architectural considerations specifically relevant to GDSS.

42.5.1 Information Needs for Group and Organizational Decision Making

The nature of the decisions, and the type of information that is required, differs across each of these four levels identified in Figure 42.1. Generally, operational activities occur much more frequently than strategic planning activities. Also, there is a difference in the degree to which the knowledge required for each of these levels is structured. In 1960, Herbert Simon (Simon, 1960) described decisions as structured or unstructured depending upon whether or not the decision making process can be explicitly described before the time when it is necessary to make a decision. This taxonomy would seem to lead directly to that in which expert skills (wholistic reasoning), rules (heuristics), or formal reasoning (holistic analysis) are normatively used for judgment. Generally, operational performance decisions are much more likely than strategic planning decisions to be prestructured. This gives rise to a number of questions concerning efficiency and effectiveness tradeoffs between training and aiding (Rouse, 1991) that occur at these levels.

There are a number of human abilities that a GDSS, which is a generic term that describes a computerized system that supports knowledge workers in performing cognitive tasks, should augment.

1. It should help the decision maker to formulate, frame, or assess the decision situation. This includes identifying the salient features of the environment, recognizing needs, identifying appropriate objectives by which we are able to measure successful resolution of an issue, and generating alternative courses of action that will resolve the needs and satisfy objectives.
2. It should also provide support in enhancing the abilities of the decision maker to obtain and analyze the possible impacts of the alternative courses of action.
3. There should also exist the capability to enhance the ability of the decision maker to interpret these impacts in terms of objectives. This interpretation capability will lead to evaluation of the alternatives and selection of a preferred alternative option.

Associated with each of these three formal steps of formulation, analysis, and interpretation, must be the ability to acquire, represent, and utilize information and associated knowledge, and the ability to implement the chosen alternative course of action.

There are many attributes that will affect the quality and usefulness of the information that is obtained, or which should be obtained, relative to any given decision situation. These variables are very clearly contingency task dependent. Included among these attributes are the following (Keen and Scott Morton, 1978).

- *Inherent and required accuracy of available information*—Operational control and performance situations will often deal with information that is relatively accurate. The information in strategic planning and management control situations is often inaccurate.
- *Inherent precision of available information*—Generally, information available for operational control and operational performance decisions is very imprecise.
- *Inherent relevancy of available information*—Operational control and performance situations will often deal with information that is relatively relevant to the task at hand because it has been prepared that way by management. The information in strategic planning and management control situations is often obtained from the external environment and may be irrelevant to the strategic tasks at hand, although it may not initially appear this way.
- *Inherent and required completeness of available information*—Operational control and performance situations will often deal with information that is relatively complete and sufficient for operational performance. The information in strategic planning and management control situations is often very incomplete and insufficient to enable great confidence in strategic planning and management control.
- *Inherent and required verifiability of available information*–Operational control and performance situations will often deal with information that is relatively verifiable to determine correctness for the intended purpose. The information in strategic planning and management control situations is often unverifiable, or relatively so, and this gives rise to a potential lack of confidence in strategic planning and management control.

- *Inherent and required consistency and coherency of available information*—Operational control and performance situations will often deal with information that is relatively consistent and coherent. The information in strategic planning and management control situations is often inconsistent and perhaps even contradictory or incoherent, especially when it comes from multiple external sources.
- *Information scope*—Generally [but not always] operational decisions are made on the basis of narrow scope information related to well-defined events that are internal to the organization. Strategic decisions are generally based upon broad scope information and a wide range of factors that often cannot be fully anticipated before the need for the decision.
- *Information quantifiability*—In strategic planning, information is very likely to be highly qualitative, at least initially. For operational decisions, the available information is often highly quantified.
- *Information currency*—In strategic planning, information is often rather old, and it is often difficult to obtain current information about the external environment. For operational control decisions, very current information is often needed and present.
- *Needed level of detail*—Often very detailed information is needed for operational type decisions. Highly aggregated information is often desired for strategic decisions. There are many difficulties associated with information summarization that need attention.
- *Time horizon for information needed*—Operational decisions are typically based on information over a short time horizon, and the nature of the control may be changed very frequently. Strategic decisions are based on information and predictions based on a long time horizon.
- *Frequency of use*—Strategic decisions are made infrequently, although they are perhaps refined fairly often. Operational decisions are made quite frequently and are relatively easily changed.
- *Internal or external information source*—Operational decisions are often based upon information that is available internal to the organization, whereas strategic decisions are much more likely to be dependent on information content that can only be obtained external to the organization.

These attributes, and others, could be used to form the basis for an evaluation of information quality in a decision support system.

Information is used in a DSS for a variety of purposes. In general, information is equivalent to, or may be used as, evidence in situations in which it is relevant. Often information is used directly as a basis for testing a hypothesis. Sometimes, it is used indirectly for this purpose. There are three different conditions for describing hypotheses according to Schum (1987, 1994).

1. Different alternative hypotheses or assessments are possible if evidence is imperfect in any way. A hypothesis may be imperfect if it is based on imperfect information. Imperfect information refers to information that is incomplete, inconclusive, unreliable, inconsistent, or uncertain. Any or all of these alternate hypotheses may or may not be true.
2. Hypotheses may refer to past, present, or future events.
3. Hypotheses may be sharp (specific) or diffuse (unspecified). Sharp hypotheses are usually based on specific evidence rather than earlier diffuse hypotheses. An overly sharp hypothesis may contain irrelevant detail and invite invalidation by disconfirming evidence on a single issue in the hypothesis. An overly diffuse hypothesis may be judged too vague and uninteresting by those who must make a decision based upon the hypothesis.

The support for any hypothesis can always be improved by either revising a portion of the hypothesis to accommodate new evidence or by gathering more evidence that infers the hypothesis. Hypotheses can arise be of potentially significant for four uses.

1. *Explanations*—An explanation usually involves a model, which can be elaborate or simple. The explanation consists of the rationale for why certain events have occurred.
2. *Event predictions*—In this case the hypothesis is proposed for a possible future event. It may include the date or period when the possible event is to occur.
3. *Forecasting and estimation*—This involves the generation of a hypothesis based on data that does not exist, or which is inaccessible.
4. *Categorization*—Sometimes it is useful to place persons, objects, or events into certain categories based upon inconclusive evidence linking the persons, objects, or events to these categories. In this case the categories represent hypotheses about category membership.

Assessment of the validity of a given hypothesis is inductive in nature. The generation of hypotheses and the determination of evidence relevant to these hypotheses involves deductive and abductive reasoning. Hypotheses may be generated on the basis of the experience and prior knowledge that lead to analogous representations and to recognitional decision making, as has been noted by Klein (1990).

Although no theory has been developed that is widely accepted on how to quantify the value of evidence, it is important to be able to support a hypothesis in some logical manner. Usually there is a major hypothesis that is inferred by supporting hypotheses, and each of these supporting hypotheses is inferred by its supporting hypothesis, and so on. Evidence is relevant to the extent that it causes one to increase or decrease the likeliness of an existing hypothesis, or causes one to modify an existing hypothesis, or causes one to create a new hypothesis. Evidence is direct if it has a straightforward bearing on the validity of the main hypothesis. Evidence is indirect if its effect on the main hypothesis is inferred through at least one other level of supporting hypothesis.

There are a number of human information-processing capabilities and limitations that interact with organizational arrangements and task requirements to strongly influence resource allocations for organizational problem solving. Needs in this area have led to the development of group decision support systems (GDSS). The purpose of these GDSS as computerized aids to planning, problem solving, and decision making include:

1. Removing a number of common communication barriers in groups and organizations
2. Providing techniques for the formulation, analysis, and interpretation of decisions
3. Systematically directing the discussion process, and associated problem solving and decision making, in terms of the patterns, timing, and content of the information that influences the actions which follow from decisions that have been taken

There are a number of variations and permutations possible in the provision of group and organizational decision support. These are associated with specific realization or architectural format for a GDSS to support a set of GDSS performance objectives for a particular task in a particular environment.

The same maladies that affect individual decision making and problem solving behavior, as well as many others, can result from group and organizational limitations. There exists a considerable body of knowledge, generally qualitative, relative to organizational structure, effectiveness, and decision making in organizations. The majority of these studies suggest that a bounded rationality or satisficing perspective, often heavily influenced by bureaucratic political considerations, will generally be the decision perspective adopted in actual decision making practice in organizations. To cope with this effectively requires the ability to concurrently deal with technological, environmental, and process concerns as they each, separately and collectively, motivate group and organizational problem-solving issues.

The influencers of decision and decision process quality are particularly important in this. At this point, we should sound a note of caution relative to some possibly overly simplistic notions relative to this. Welch (1989) identifies a number of potential imperfections in organizational decision making and discusses their relationship to decision process quality. In part, these are based on an application of seven symptoms identified in Herek, Janis, and Hurth (1987) to the Cuban missile crisis of 1962. These potential imperfections include:

1. Omissions in surveying alternative courses of action
2. Omissions in surveying objectives
3. Failure to examine major costs and risks of the selected course of action (COA)
4. Poor information search, resulting in imperfect information
5. Selective bias in processing available information
6. Failure to reconsider alternatives initially rejected, potentially by discounting favorable information and overweighing unfavorable information
7. Failure to work out detailed implementation, monitoring, and contingency plans

The central thrust of this study is that the relationship between the quality of the decision-making process and the quality of the outcome is difficult to establish. This strongly suggests the usefulness of the contingency task structural model construct, and the need for approaches that evaluate the quality of processes, and not just decisions.

Organizational ambiguity is a major reason why much observed "bounded rationality" behavior is asserted as being so pervasive. March (1983) and March and Wessinger-Baylon (1986) show that this is very often the case, even in situations when formal rational thought or "vigilant information processing" might be thought to be a preferred decision style. March (1983) indicates that there are at least four kinds of opaqueness or equivocality in organizations: *ambiguity of intention*, *ambiguity of understanding*, *ambiguity of history*, and *ambiguity of human participation.* These four ambiguities relate to an organization's structure, function, and purpose, as well as to the perception of these decision-making agents in an organization. They influence the information that is communicated in an organization, and generally introduce one or more forms of information imperfection. The notions of organizational management and organizational information processing are, indeed, inseparable. In the context of human information processing, it would not be incorrect to define the central purpose of management as development of a consensual grammar to ameliorate the effects of equivocality or ambiguity. This is the perspective taken by Karl Weick (1979, 1985) in his noteworthy efforts concerning organizations.

Starbuck (1985) notes that much direct action is a form of deliberation. He indicates that action should often be introduced earlier in the process of deliberation than is usually the case and that action and thought should be integrated and interspersed with one another. The rational behind this argument is that probative actions generate information and tangible results, which modify potential thoughts. Of course, any approach that involves act-now, think-later–type behavior should be applied with considerable caution.

Much of the discussion to be found in the judgment, choice, and decision literature concentrates on what may be called formal reasoning and decision selection efforts that involve the issue resolution efforts that follow as part of the problem-solving efforts of issue formulation, analysis, and interpretation that we have often discussed. There are other decision-making activities, of decision associated activities, as well. Very important among these are activities that allow perception, framing, editing, and interpretation of the effects of actions upon the internal and external environments of a decision situation. These might be called *information selection activities.* There will also exist *information retention activities* that allow admission, rejection, and modification of the set of selected information or knowledge such as to result in short-term learning and long-term learning. Short-term learning results from reduction of incongruities, and long-term learning results from acquisition of new information that reflects enhanced understanding of an issue. Although the basic GDSS design effort may well be concerned with the short-term effects of various problem-solving, decision-making, and information presentation formats, the actual knowledge that a person brings to bear on a given problem is a function of the accumulated experience that the person possesses, and thus long-term effects need to be considered, at least as a matter of secondary importance.

It was remarked earlier that a major purpose of a GDSS is to enhance the value of information and, through this, to also enhance group and organizational decision making. Three attributes of information appear dominant in the discussion so far relative to value for problem-solving purposes and in the literature in general. These are:

1. *Task relevance*—Of course, information must be relevant to the task at hand. It must allow the decision maker to know what needs to be known in order to make an effective and efficient decision. This is not as trivial a statement as might initially be suspected. Relevance varies considerably across individuals, as a function of the contingency task structure, and in time as well.
2. *Representational appropriateness*—In addition to the need that information be relevant to the task at hand, it must be represented in a form that is appropriate for use by the person who needs the information.
3. *Equivocality reduction*—It is generally accepted that high-quality information may reduce imperfection or equivocality. This equivocality generally takes the form of uncertainty, imprecision, inconsistency, or incompleteness. It is very important to note that it is neither necessary nor desirable to obtain decision information that is unequivocal or totally "perfect." Information need be only sufficiently unequivocal or unambiguous for the task at hand. To make it better may well be a waste of resources!

Each of these top-level attributes may be decomposed into those attributes at a lower level which are needed as fundamental metrics for valuation of information quality. We have indicated that some of the components of equivocality or imperfection are uncertainty, imprecision, inconsistency, and incompleteness. A few of the attributes of representational appropriateness include naturalness, transformability to naturalness, and conciseness. These attributes of information presentation system effectiveness relate strongly to overall value of information concerns and should be measured as a part of the DSS and GDSS evaluation effort, even though any one of them may appear to be a secondary theme.

There are many ways in which we can characterize information. Among attributes that we noted earlier and that we might use are accuracy, precision, completeness, sufficiency, understandability, relevancy, reliability, redundancy, verifiability, consistency, freedom from bias, frequency of use, age, timeliness, and uncertainty. Our concerns with information involve at least five desiderata (Sage, 1987):

1. Information should be presented in very clear and very familiar ways, such as to enable rapid comprehension.
2. Information should be such as to improve the precision of understanding of the task situation.
3. Information that contains an advice or decision recommendation component should contain an explication facility that enables the user to determine how and why results and advice are obtained.
4. Information needs should be based upon identification of the information requirements for the particular situation.
5. Information presentations and all other associated management control aspects of the support process should be such that the decision maker, rather than a computerized support system, guides the process of judgment and choice.

It will generally be necessary to evaluate a GDSS to determine the extent to which these information quality relevant characteristics are present.

42.5.2 Group Decision Support System Design Constructs

In an organization, there are two fundamental types of decision making—individual, and group or organizational. Individual decisions are decisions made by a single person, and group or organizational decisions are those made by a collection of two or more people. It is, of course, possible to disaggregate this still further. An individual decision may, for example, be based on the value system of one or more people, and the individual making the decision may or may not have his or her values included. In a multistage decision process, the various decisions may be made by different people. Some authors differentiate between group and organizational decisions (King and Star, 1992), but we see no need for this here even though this may be warranted in some contexts. There can be no doubt at all, however, of the fact that a GDSS needs to be carefully matched to an organization that may use it.

Often, groups make decisions differently from the way an individual does. Groups need protocols that allow effective inputs by individuals in the group or organization, a method for mediating a discussion of issues and inputs, and algorithms for resolution of disagreements and reaching a group consensus. Acquisition and elicitation of inputs and the mediation of issues are usually local to the specific group, informed of personalities, status, and contingencies of the members of the group. Members of the group are usually desirous of cooperating in reaching a consensus on conflicting issues or preferences. The support for individual versus group decisions is different and hence DSSs and GDSSs may require different designs. Because members of a group have different personalities, motivations, and experiential familiarities with the situation at hand, a GDSS must assist in supporting a wide range of judgment and choice perspectives.

It is important to note that the group of people may be centralized at one spot, or decentralized in space and/or time. Also, the decision considered by each individual in a decision making group may or may not be the ultimate decision. The decision being considered may be sequential over time and may involve many component decisions. Alternately, or in addition, many members in a decision making group may be formulating and/or analyzing options, and preparing a short list of these for review by a person with greater authority or responsibility over a different portion of the decision making effort.

Thus, the number of possible types of GDSS may be relatively extensive. Johansen (1988) has identified no less than 17 approaches for computer support in groups in his discussion of groupware. It is of value to describe these here.

1. *Face-to-face meeting facilitation services*—This is little more than office automation support in the preparation of reports, overheads, videos, and the like that will be used in a group meeting. The person making the presentation is called a "facilitator" or "chauffeur."
2. *Group decision support systems*—By this, Johansen essentially infers the GDSS structure shown in Figure 42.14 with the exception that there is but a single video monitor under the control of a facilitator or chauffeur.
3. *Computer-based extensions of telephony for use by Workgroups*—This involves use either of commercial telephone services, or private branch exchanges (PBX). These services exist now and Northern Telecom Meridian is an example of a conference calling service.
4. *Presentation support software*—This approach is not unlike that of approach one, except that computer software is used to enable the presentation to be contained within a computer. Often, the presentation material is prepared by those who will present it, and this may be done in an interactive manner to the group receiving the presentation.
5. *Project management software*—This is software that is receptive to presentation team input over time, and which has capabilities to organize and structure the tasks associated with the group, often in the form of a Gantt chart. This is very specialized software and would be potentially useful for a team interested primarily in obtaining typical project management results in terms of program evaluation and review technique (PERT) charts and the like.
6. *Calendar management for groups*—Often, individuals in a group need to coordinate times with one another. They indicate times that are available, potentially with weights to indicate schedule adjustment flexibility in the event that it is not possible to determine an acceptable meeting time.
7. *Group authoring software*—This allows members of a group to suggest changes in a document stored in the system, without changing the original. A lead person can then make document revisions. It is also possible for the group to view alternative revisions to drafts. The overall objective is to encourage group writing, and to improve the quality and efficiency of group writing. It seems very clear that there needs to be overall structuring and format guidance which, while possibly group determined, must be agreed upon before the structure is filled out with report details.
8. *Computer supported face-to-face meetings*—Here, individual members of the group work directly with a workstation and monitor, rather than having just a single computer system and monitor. A large-screen video may, however, be

included. This is the sort of DSS envisioned in Figure 42.14. Although there are a number of such systems in existence, the Colab system at Xerox Palo Alto Research Center (Stefik et al., 1987) is probably the most sophisticated of these. A simple sketch of these would appear somewhat as in Figure 42.14. Generally, there is both public and private information contained in these systems. The public information is shared, and the private information, or a portion of it, may be converted to public programs. The private screens normally start with a menu screen from which participants can select activities in which they engage, potentially under the direction of a facilitator.

9. *Screen sharing software*—This software enables one member of a group to share screens selectively with other group members. There are clearly advantages and pitfalls in this. The primary advantage to this approach is that of sharing information with those who have a reason to know specific information and not having to bother others who do not need it. The disadvantage is just this also, and it may lead to a feeling of ganging up by one subgroup on another subgroup.
10. *Computer conferencing systems*—This is the group version of electronic mail. Basically, what we have is a collection of DSSs with some means of communication among the individuals that comprise the group. This form of communication might be regarded as a product hierarchy in which people communicate.
11. *Text filtering software*—This allows system users to search normal or semistructured text through the specification of search criteria that are used by the filtering software to select relevant portions of text. The name of the original system to accomplish this was Electronic Mail Filter (Malone, Grant, Turbak, Brobst, and Cohen, 1987) although there is now also emphasis on an information lens that will enable the system to obtain information matching rules that are specified by the system user.
12. *Computer supported audio or video conferences*—This is simply the standard telephone or video conferencing, as augmented by each participant having access to a computer and appropriate software.
13. *Conversational structuring*—This involves identification and use of a structure for conversations that is presumably in close relationship to the task, environment, and experiential familiarity of the group with the issues under consideration (Winograd and Flores, 1986). For these group participants, structured conversations should often provide for enhanced efficiency and effectiveness or there may be a perception of unwarranted intrusions that may defeat the possible advantages of conversational structuring.
14. *Group memory management*—This refers to the provision of support between group meetings such that individual members of a group can search a computer memory in personally preferred ways through the use of very flexible indexing structures. The term "hypertext" (Nielson, 1989) is generally given to this flexible information storage and retrieval. One potential difficulty with hypertext is the need for a good theory of how to prepare the text and associated index such that it can be indexed and used as we now use a thesaurus. An extension of hypertext to include other than textual material is known as hypermedia.
15. *Computer supported spontaneous interaction*—The purpose of these systems is to encourage the sort of impromptu and extemporaneous interaction that often occurs at unscheduled meetings between colleagues in informal setting, such as a hallway. The need for this could occur for example when it is necessary for two physically separated groups to communicate relative to some detailed design issue (Goodman and Abel, 1987).
16. *Comprehensive work team support*—This refers to integrated and comprehensive support, such as perhaps might be achieved through use of the comprehensive DSS design philosophy described earlier.
17. *Nonhuman participants in team meetings*—This essentially refers to the use of unfacilitated DSS and expert systems that automate some aspects of the process of decision making.

According to Johansen (1988), the order in which these are described above also represents the order of increasing difficulty of implementation and successful use. These scenarios of support for decision making are also characterized in terms of support for: face-to-face meetings (1, 2, 4, 8), support for electronic meetings (3, 9, 10, 11, 12), and support between meetings (5, 6, 7, 13, 14, 15, 16).

A GDSS may and doubtlessly will influence the process of group decision making, perhaps strongly. A GDSS has the potential for changing the information-processing characteristics of individuals in the group. This is one reason that organizational structure and authority concerns are important ingredients in GDSS designs.

In one study of the use of a GDSS to facilitate group consensus (Watson, DeSanctis, and Poole, 1988), it was found that:

1. GDSS use tended to reduce face-to-face interpersonal communication in the decision-making group.
2. GDSS use posed an intellectual challenge to the group and made accomplishment of the purpose of their decision-making activity more difficult than for groups without the GDSS.
3. The groups using the GDSS became more process oriented and less specific issue oriented than the groups not using the GDSS.

Support may be accomplished at any, or all, of the three levels for group decision support identified by DeSanctis and Gallupe (1987) in their definitive study of GDSS. A GDSS provides a mechanism for group interaction. A GDSS may impose any of various structured processes on individuals in the group such as, for example, a particular voting scheme. A GDSS may impose any of several management control processes on the individuals on the group, such as that of imposing or removing the effects of a dominant personality. The design of the GDSS and the way in which it is used are the primary determinants of these.

DeSanctis and Gallupe (1987) have developed a taxonomy of GDSS. A level I GDSS would simply be a medium for enhanced information interchange that might lead ultimately to a decision. Electronic mail, large video screen displays that can be viewed by a group, or a decision room that contains these features, could represent a level I GDSS. A level I GDSS provides only a mechanism for group interaction. It might contain such facilities as a group scratchpad, support for meeting agenda development, idea generation, and voting software.

A level II GDSS would provide various decision structuring and other analytic tools that could act to reduce information imperfection. A decision room that contained software that could be used for problem solution would represent a level II GDSS. Thus, spreadsheets would primarily represent a level II DSS. To become a level II GDSS, there would also have to be some means of enabling group communication. Figure 42.15 represents a level II GDSS. It is simply a communications medium that has been augmented with some tools for problem structuring and solution with no prescribed management control of the use of these tools.

A level III GDSS also includes the notion of management control of the decision process. Thus, there is a notion of facilitation of the process, either through the direct intervention of a human in the process, or through some rule-based specifications of the management control process that is inherent in level III GDSS. Clearly, there is no sharp transition line between one level and the next and it may not always be easy to identify at what level a GDSS is operating. The DSS generator, such as discussed in our preceding section, would generally appear to produce a form of level III DSS. In fact, most of the DSSs that we have been discussing in this book are either level II or level III DSSs or GDSSs. The GDSS of Figure 42.15, for example, becomes a level III GDSS if is supported by a facilitator.

DeSanctis and Gallupe identify four recommended approaches:

1. Decision room for small group face-to-face meetings
2. Legislative sessions for large group face-to-face meetings
3. Local area decision networks for small dispersed groups
4. Computer mediated conferencing for large groups that are dispersed

They discuss the design of facilities to enable this, as well as techniques whereby the quality of efforts such as generation of ideas and actions, choosing from among alternative courses of action, and negotiating conflicts, may be enhanced. On the basis of this, these

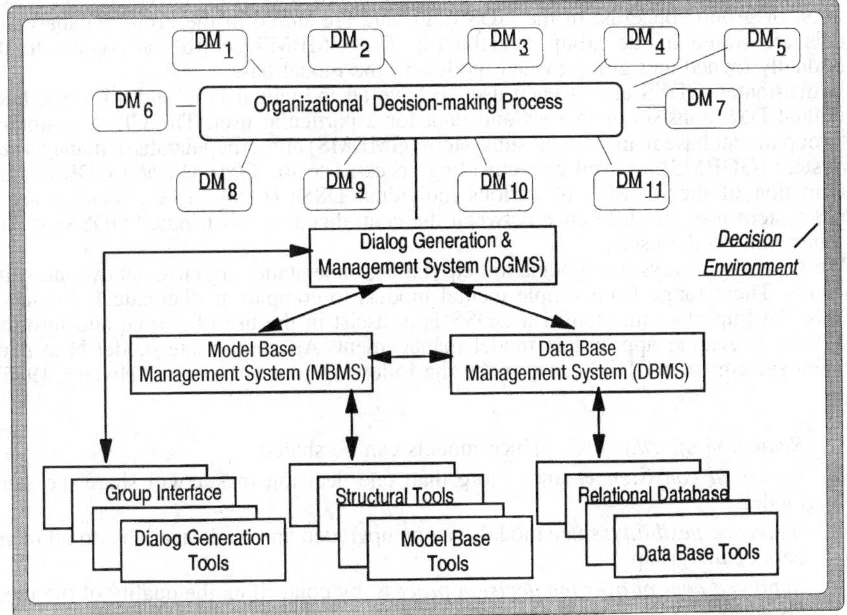

Figure 42.15 Level II (no facilitator) and level III (with facilitator) GDSS.

authors recommend six areas as very promising for additional study: GDSS design methodologies; patterns of information exchange; mediation of the effects of participation; effects of (the presence or absence of) physical proximity, interpersonal attraction, and group cohesion; effects on power and influence; and performance–satisfaction tradeoffs. Each of these supports the purpose of computerized aids to planning, problem solving, and decision making: removing a number of common communication barriers; providing techniques for structuring decisions; and systematically directing group discussion, and associated problem solving and decision making, in terms of the patterns, timing, and content of the information that influences these actions.

In our discussions, we have mentioned the need for a GDSS model base management system (MBMS). They are a great variety of MBMS tools. Some of the least understood are group tools that aid in the issue formulation effort. Since these may represent an integral part of a GDSS effort, it is of interest to describe GDSS issue formation, as one component of a MBMS, here.

Other relevant efforts and interest areas involving GDSS include the group processes in computer-mediated communications study, the computer support for collaboration and problem solving in meetings study of Stefik et al. (1987); the organizational planning study of Applegate, Chen, Konsynski, and Nunamaker (1987), and the knowledge management and intelligent information sharing systems study of Malone et al. (1987). Particularly interesting current issues surround the extent to which cognitive science and engineering studies that involve potential human information processing flaws can be effectively dealt with, in the sense of design of debiasing aids, in GDSS design.

42.5.3 Distributed GDSS

A single user DSS must provide single-user–single-model and single-user–multiple-model support whereas a GDSS model base management system (MBMS) must support multiple-user–single-model and multiple-user–multiple-model support. A centralized GDSS induces three basic components:

1. A group model base management subsystem (GMBMS)
2. A group database management subsystem (GDBMS)
3. A group dialogue generation and management subsystem (GDGMS).

These are precisely the components needed in a single-user DSS, except for the incorporation of group concerns. In the GDSS, all data are stored in the group database and models are stored in the group model base. The GMBMS controls all access to the individually owned and group-owned models in the model base.

A distributed GDSS allows each user to have an individual DSS and a GDSS. Each individual DSS consists of models and data for a particular user. The GDSS maintains the group model base management subsystem (GMBMS) and group database management subsystem (GDBMS), as well as controlling accesses to the GMBMS and GDBMS and coordination of the MBMSs of various individual DSSs (Liang, 1988). During actual GDSS system use, the difference between the centralized and distributed GDSSs should be transparent to the users.

We generally always use models to help define, understand, organize, study, and solve problems. These range from simple mental models to complex mathematical simulation models. An important mission of a GDSS is to assist in the use of formal and informal models by providing appropriate model management. An appropriate model base management system for a GDSS can provide the following four advantages (Hwang, 1985):

1. *Reduction of redundancy*, since models can be shared
2. *Increased consistency*, since more than one decision maker will share the same model
3. *Increased flexibility*, since models can be upgraded and made available to all members of the group
4. *Improved control over the decision process*, by controlling the quality of the models adopted

A model base management system (Blanning and King, 1993) provides for at least the following five basic functions: construction of new models, storage of existing and new models, access and retrieval of existing models, execution of existing models, and maintenance of existing models. MBMSs should also provide for model integration and selection. Model integration by using the existing model base as building blocks in the construction of new, or integrated, models is very useful when ad hoc or prototype models are desired. Model integration is needed in the production of operational MBMSs.

In this section, we have provided a very broad overview of group decision support systems that potentially support group and organizational decision making functions. Rather than concentrate on one or two specific systems, we have painted a picture of the many requirements that must be satisfied to produce an acceptable architecture and design for these systems. This provides much fertile ground for research in many GDSS relevant cognitive systems engineering areas (Rasmussen, Pejtersen, and Goodstein, 1995; Andriole and Adelman, 1995).

REFERENCES

Altman, R. (1989). An assessment of current cooperative processing architectures. In P. C. Tinnirello, P. C., Ed., *Systems management: Development and Support,* pp. 287–311. Boston: Auerbach Publishers.

Andriole, S., and Adelman, L. (1995). *Cognitive Systems Engineering for User-Computer Interface Design, Prototyping, and Evaluation.* Hillsdale, NJ: Erlbaum.

Anthony, R. N. (1965). *Planning and control systems: A framework for analysis.* Cambridge, MA: Harvard University Press.

Anthony, R. N., Dearden, N. J., and Govindarajan, V. (1992). *Management Control Systems.* Homewood, IL: Richard D. Irwin.

Applegate, L. M., Chen, T. T., Konsynski, B. R., and Nunamaker, J. F. (1987). Knowledge management in organizational planning. *Journal of Management Information Systems, 3(4)*, 20–38.

Applegate, L. M., Konsynski, B. R., and Nunamaker, J. F. (1986). Model management systems: Design for decision support. *Decision Support Systems, 2(1)*, 81–91.

Arden, B. W., Ed. (1980). *What Can Be Automated? The Computer Science and Engineering Research Study* (Chapter 10). Cambridge, MA: MIT Press.

Atzeni, P., and Chen, P. P. (1983). Completeness of query languages for the entity-relationship model. In P.P. Chen, Ed., *Entity-Relationship Approach to Information Modeling and Analysis.* Amsterdam: North Holland.

Barbosa, L. C., and Herko, R. G. (1980). Integration of algorithmic aids into decision support systems. *MIS Quarterly. 4(3)*, 1–12.

Bennet, J. L., Ed. (1983). *Building Decision Support Systems.* Reading, MA: Addison Wesley.

Blanning, R. W., and King, D. R. (1993). *Current Research in Decision Support Technology*. Los Altos, CA: IEEE Computer Society Press.

Card, S. K., Moran, T. P., and Newell, A. (1993). *The Psychology of Human Computer Interaction*. Hillsdale, NJ: Erlbaum.

Ceri, S., and Pelagatti, G. (1984). *Distributed Databases: Principles and Systems*. New York: McGraw Hill.

Chen, P. P. S. (1976). The entity-relationship model: Towards a unified view of data. *ACM Transactions on Database Systems, 1*, 9–36.

Coad, P., and Yourdon, E. (1990). *Object Oriented Analysis*. Englewood Cliffs NJ: Prentice-Hall.

Date, C. J. (1983). *Database: A Primer.* Reading, MA: Addison Wesley.

Date, C. J. (1986). *An Introduction to Database Systems*. Reading, MA: Addison Wesley.

DeSanctis, G., and Gallupe, R. B. (1987). A foundation for the study of group decision support systems. *Management Science, 33*, 547–588.

Dolk, D. R. (1986). Data as models: An approach to implementing model management. *Decision Support Systems, 2(1)*, 73–80.

Fang, L., Hipel, K. W. and Kilgour, D. M. (1993). *Interactive Decision Making: The Graph Model for Conflict Resolution.* New York: John Wiley.

Fraser, N. M., and Hipel, K. W. (1984). *Conflict Analysis: Models and Resolution.* New York: North Holland.

Goodman, G. O., and Abel, M. J. (1987). Communications and collaboration: Facilitating cooperative work through communications. *Office Technology and People, 3*(2), 129–146.

Gray, P., and Olfman, L. (1989). The user interface in group decision support systems. *Decision Support Systems, 5*(2), 119–137.

Greenberg, H. J. (1987). A Natural Language Model to Explain Linear Programming Models and Solutions. *Decision Support Systems, 3(4)*, 333–342.

Grudin, J. (1989). The case against user interface consistency. *Communications of the ACM, 32*, 1164–1173.

Hall, A. D. (1969). Three dimensional morphology of systems engineering. *IEEE Transactions on Systems Science and Cybernetics, 5*, 156–160.

Hammond, K. R., McClelland, and Mumpower, J. (1980). *Human Judgment and Decision Making: Theories, Methods, and Procedures.* New York: Praeger Publishers.

Harrison, H. R., and Hix, D. (1989). Human computer interface development. *ACM Computing Surveys, 21(1)*, 5–92.

Harrison, M., and Thimbleby, Eds. (1990). *Formal Methods in Human Computer Interaction.* Cambridge, UK: Cambridge University Press.

Herek, G. M., Janis, I. L., and Hurth, P. (1987). Decision Making During International Crises: Is Quality of Process Related to Outcome. *Journal of Conflict Resolution, 31*(2), May, 203–226.

Hillier, F. S., and Lieberman, G. J. (1986). *Operations Research* (Fourth Edition). San Francisco: Holden Day.

Huber, G. P. (1982). Group decision support systems as aids in the use of structured group management techniques. *Proceedings 2nd International Conference on Decision Support Systems,* (pp. 96–108) San Francisco, June 1982.

Hwang, S. (1985). Automatic model building systems: A survey. *Proceedings DSS-85*, 22–32, Atlanta, GA, 1985, pp. 22–32.

Johansen, R. (1988). *Groupware: Computer Support for Business Teams.* New York: Free Press.

Keen, P. G. W., and Scott Morton, M.S. (1978). *Decision Support Systems: An Organizational Perspective.* Reading, MA: Addison-Wesley.

Keeney, R. L., and Raiffa, H. (1976). *Decisions with Multiple Objectives.* New York: John Wiley and Sons.

Keller, L. R., and Ho, J. L. (1990). Decision problem structuring. In Sage, A. P., Ed., *Concise Encycopedia of Information Processing in Systems and Organizations* (pp. 103–110). Oxford: Pergamon Press.

Kent, W. (1979). Limitation of record based information models. *ACM Transations on Database Systems, 4*, 107–131.

Kerschberg, L., Ed. (Vol. 1 1987, Vol. 2 1989). *Proceedings of the International Conference on Expert Database Systems.* Menlo Park, CA: Benjamin-Cummings.

King, J.L., and Star, S.L. (1992). Organizational decision support processes as an open systems problem. In T. Stohr, and B. Konsynski, Eds., *Information Systems and Decision Processes* (pp. 150–154). Los Altos, CA: IEEE Press.

Klein, G. A. (1990). Information requirements for recognitional decision making. In A.P. Sage, Ed., *Concise Encyclopedia of Information Processing in Systems and Organizations* (pp. 414–418). Oxford, UK: Pergamon Press.

Kraemer, K. L., and King, J. L. (1988). Computer based systems for cooperative work and group decision making. *ACM Computing Surveys, 20*(2), 115–146.

Lee, E. (1990). User-Interface development tools. *IEEE Software, 7(3)*, 31–36.

Liang, B.T. (1988). Model management for group decision support. *MIS Quarterly, 12*, 667–680.

Liang, T. P. (1985). Integrating model management with data management in decision support systems. *Decision Support Systems, 1(3)*, 221–232.

Lorin, H. (1988). *Aspects of Distributed Computer Systems.* New York: John Wiley and Sons.

Malone, T. W., Grant, K. R., Turbak, F. A., Brobst, S. A., and Cohen, M. D. (1987). Intelligent information sharing systems. *Communications of the ACM, 30*, 390–402.

March, J., and Wessinger-Baylon, T., Eds. (1986). *Ambiguity and Command: Organizational Perspectives on Military Decision Making.* Boston: Pitman.

March, J. G. (1983). Bounded rationality, ambiguity, and the engineering of choice. *Bell Journal of Economics, 9,* 587–608.

McGrath, J. E. (1984). *Groups: Interaction and Performance.* Englewood Cliffs, NJ: Prentice-Hall.

Mintzberg, H. (1973). *The Nature of Managerial Work.* New York: Harper and Row.

Mullender, S., Ed. (1989). *Distributed Systems.* Reading, MA: Addison Wesley.

Murphy, F. H., and Stohr, E. A. (1986). An intelligent systems for formulating linear programs. *Decision Support Systems, 2(1),* 39–47.

Mylopoulos, J., and Brodie, M. L., Eds. (1989). *Artificial Intelligence and Databases.* San Mateo, CA: Morgan Kaufman.

Nielson, J. (1989). *Hypertext and Hypermedia.* San Diego: Academic Press.

Parsaye, K., Chignell, M., Khoshafian, S., and Wong, H. (1989). *Intelligent databases: Object-Oriented, Deductive, Hypermedia Technologies.* New York: John Wiley.

Raiffa, H. (1968). *Decision Analysis.* Reading, MA: Addison Wesley.

Rasmussen, J., Pejtersen, A. M., and Goodstein, L. P. (1995). *Cognitive Systems Engineering.* New York: John Wiley.

Ravden, S., and Johnson, G. (1989). *Evaluating Usability of Human-Computer Interfaces: A Practical Method.* Chichester, UK: John Wiley.

Roberts, T. L., and Moran, T. P. (1982). A Methodology for evaluateing text editors. In B. Curtis, Ed., *Proceedings of the IEEE Conference on Human Factors in Software Development.* Gaithersburg, MD.

Rouse, W. B. (1991). Conceptual design of a computational environment for analyzing tradeoffs between training and aiding. *Information and Decision Technologies, 17,* 143–152.

Sage, A. P. (1987). Information systems engineering for distributed decision making. *IEEE Transactions on Systems, Man and Cybernetics, 17*(6), 920–936.

Sage, A. P., Ed. (1990). *Concise Encyclopedia of Information Processing in Systems and Organizations.* Oxford: Pergamon Press.

Sage, A. P. (1991). *Decision Support Systems Software Engineering.* New York: John Wiley.

Sage, A. P. (1992). *Systems Engineering.* New York: John Wiley.

Sage, A. P. (1995). *Systems Management for Information Technology and Software Engineering.* New York: John Wiley.

Sage, A. P., and Palmer, J. D. (1990). *Software Systems Engineering.* New York: Wiley Interscience.

Schneiderman, B. (1987). *Designing the User Interface: Strategies for Effective Human Computer Interaction.* Reading, MA: Addison Wesley.

Schum, D. A. (1987). *Evidence and Interface for the Intelligent Analyst (Vols. I and II).* Lanham, MD: University Press of America.

Schum, D. A. (1994). *Evidential Foundations of Probabilistic Reasoning.* New York: John Wiley.

Sheridan, T. B. (1992). *Telerobotics, Automation, and Human Supervisory Control.* Cambridge, MA: MIT Press.

Simon, H. A. (1960). *The New Science of Management Decisions.* New York: Harper.

Smith, S. L., and Mosier, J. N. (1986). *Guidelines for designing user interface software (Tech. Rep. MTR-10090, ESD-TR-86-278).* Bedford, MA: MITRE Corporation.

Sprague, R. H. Jr., and Carlson, E. D. (1982). *Building Effective Decision Support Systems.* Englewood Cliffs, NJ: Prentice-Hall.

Starbuck, W. E. (1985). Acting first and thinking later. In J. Pennings, Ed., *Organizational Strategy and Change* (pp. 336-372). New York: Jossey Bass.

Stefik, M., Foster, G., Bobrow, D. G., Kahn, K. Lanning, S., and Suchman, L. (1987). Beyond the chalkboard: Computer support for collaboration and problem solving in meetings. *Communications of the ACM, 30(1),* 32–47.

Volkema, R. J. (1990). Problem formulation. In A.P., Sage, Ed., *Concise Encyclopedia of Information Processing in Systems and Organizations* (pp. 377–382). Oxford: Pergamon Press.

Watson, R. T., DeSanctis, G., and Poole, M. S. (1988). Using a GDSS to facilitate group consensus: Some intended and unintended consequences. *MIS Quarterly, 12,* 463–477.

Weick, K. E. (1979). *The Social Psychology of Organizing.* Reading, MA: Addison Wesley.

Weick, K. E. (1985). Cosmos vs. chaos: Sense and nonsense in electronic context. *Organizational Dynamics, 14,* 50–64.

Welch, D. A. (1989). Group decision making reconsidered. *Journal of Conflict Resolution, 33(3),* 430–445.

Will, H. J. (1975). Model management system. In E. Grochia, and N. Szyperski, Eds., *Information Systems and Organizational Structure* (pp. 468–482). Berlin: deGruyter.

Winograd, T., and Flores, F. (1986). *Understanding Computers and Cognition.* Los Altos, CA: Ablex Press.

Zdonik, S. B., and Maiser, D. (1989). *Readings in Object Oriented Databases.* Los Altos, CA: Morgan Kaufman.

PART 7

EVALUATION

CHAPTER 43

DATA COLLECTION AND EVALUATION OF OUTCOME MEASURES

Gavriel Salvendy
School of Industrial Engineering
Purdue University
West Lafayette, IN 47907-1287 USA

Pascale Carayon
Ecole des Mines de Nancy
Parc de Saurupt
54042 Nancy Cedex
France

43.1 INTRODUCTION

Human factors and ergonomics (HFE) professionals study, change, and/or optimize work systems in order to improve the interface and the relationship between people and their work environment. The system, comprised of the person and his or her work environment, can be conceptualized as a system composed of five factors: (1) the individual, (2) the tasks performed by the individual, (3) tools and technologies used by the individual in performing the tasks, (4) the physical environment in which the work is being done, and (5) the organizational environment (Smith and Carayon-Sainfort, 1989). These factors form a work system that can produce adverse working conditions that increase the risk of negative effects for the individual, as well as for organizations and society at large. According to Grandjean (1985), the prime objective of HFE is "to contribute to human needs in a work environment, including promotion of health and well-being" (page ix). The objective of HFE is therefore to design work systems that not only do not "harm" people, but also contribute to their health and well-being. It thus becomes important to measure the "outcomes" of work systems. Outcome measures are good indicators of how well or poorly work systems are designed. Outcomes can be defined as the resulting effects of work systems on the individual, as well as on the organizations and society at large. This chapter focuses on individual-level outcomes.

Outcome measures of interest to HFE professionals include comfort, physical and mental health, performance, and attitudes. Data collection and evaluation of outcome measures resulting from HFE intervention represent an important contribution of HFE to our society. HFE interventions have resulted in increased productivity, quality, health, and satisfaction associated with the use of products or processes. This is an indicator of the impact that the profession of HFE has. Therefore, it is important to choose the right outcome measures, to use them effectively, and to measure them properly. In this chapter, we first examine the nature of criteria and then review the different HFE outcome measures, their use, the methods used to gather data on them, and examples of actual measures.

43.2 CRITERIA

A criterion is an evaluation standard that may be used as a surrogate or correlate of outcome measures such as system effectiveness, human performance, and attitudes. There are a number of important aspects to be considered when selecting criteria measures, including the relative strengths and weaknesses of the various data-collection methods, and the balance between the costs of the methods and the available resources. The balance between the costs and the resources is measured not only in terms of money, but also in terms of time, personnel, and expertise. The important aspects of criteria include relevance, linearity, and homogeneity. Each of these is discussed below.

43.2.1 Relevance

In order for a criterion to be used as an evaluation standard, it has to be relevant and to contribute significantly to the objective and mission of the overall system. Because of this, frequently a single-criterion measure is inadequate to evaluate the outcome of a given work system.

For instance, effectiveness of the work system is affected by speed of performance and speed of response time of the workers. Errors committed by individuals affect the system's effectiveness. In addition, the individuals' skills and abilities needed to operate the system influence the length of the learning process, that is, the length of the time that the individuals need to become skillful in operating the system. Because of this it becomes apparent that evaluations of the system effectiveness require a variety of criteria for evaluating its own effectiveness by taking into account variables that contribute to the system's effective functioning. This example shows that the evaluation of any HFE outcome via some criterion requires taking into account other variables indirectly related to the outcome of interest.

43.2.2 Nonlinearity

Criterion measures are frequently assumed to be linear functions, when in effect they are not. For example, in an evaluation of the impact of intelligence of industrial workers on industrial performance for 181 operators, Salvendy (1973) initially concluded that there was no relationship between scores derived on intelligent tests and performance effectiveness ($R = 0.15$). However, a more detailed evaluation of the data revealed that intellectual abilities up to an IQ of 90 contributed significantly to the success of outcome measures such as performance ($R = 0.46$), whereas an IQ level between 90 and 110 had no impact on performance ($R = 0.04$). But being too intelligent for the job (that is, IQs above 110) had a significant negative impact on performance ($R = -0.52$). This may be explained by the fact that being too intelligent for the job can cause boredom and, therefore, affect performance. This example illustrates that using the linear function for criteria measure resulted in misleading information on the relationship between the criteria measures and the outcome measures that were evaluated.

Linearity of measures can also be viewed in a variety of other settings. For example, if the maximum performance is 100, such as in test-taking, for the individual who scored 20 on the first test, it would be much easier to improve to 50 by 30% than for an individual who had 69 score in the first test to improve by 30% to a score of 99 score.

We suggest that, in all evaluation measures, the relationship between the criterion measure and the outcome must be specifically evaluated and the linearity relationship should not be assumed. When one looks at criterion measures in terms of human ability required to perform the task, we have a multicriterion measure. As Fleishman and Quaintance (1984) have documented (Table 43.1), there are 52 distinct human abilities associated with all types of human performance. The notion that there are 52 abilities implies that an individual who has a high performance on one ability may or may not have a high performance on another ability since the correlation between the abilities measures is very low. Hence if one individual is good on one performance criterion, this does not guarantee his or her performance on another ability. This multidimensionality of human attributes has implications in determining the ability requirements associated with human performance to different systems. It has implications for the selection and training of individuals for task performance and for job rotation, in particular. For example, let us assume that tasks consist of one set of abilities in which an individual had a high performance level. If that individual is rotated to another type of task (i.e. because of organizational and social needs of job rotation), there is no guarantee that the individual's performance on that other set of tasks that require different human abilities will be as good as before the rotation.

Table 43.1 A Taxonomy of Cognitive (1–14), Perceptual (15–19), Psychomotor (20–31), and Physical; (32–52), Abilities. (Adapted from Fleishman and Quaintance, 1984.)

1. Oral comprehension	27. Finger dexterity
2. Written comprehension	28. Wirst–finger speed
3. Oral expression	29. Speed of limb movement
4. Written expression	30. Selective attention
5. Fluency of ideas	31. Time sharing
6. Originality	32. Static strength
7. Memorization	33. Explosive strength
8. Problem sensitivity	34. Dynamic strength
9. Mathematical reasoning	35. Trunk strength
10. Number facility	36. Extend flexibility
11. Deductive reasoning	37. Dynamic flexibility
12. Inductive reasoning	38. Gross body coordination
13. Information ordering	39. Gross body equilibrium
14. Category flexibility	40. Stamina
15. Speed of closure	41. Near vision
16. Flexibility of closure	42. Far vision
17. Spatial orientation	43. Visual color discrimination
18. Visualization	44. Night vision
19. Perceptual speed	45. Perceptual vision
20. Control precision	46. Depth perception
21. Multilimb coordination	47. Glare sensitivity
22. Response orientation	48. General hearing
23. Rate control	49. Auditory attention
24. Reaction time	50. Sound localization
25. Arm-Hand steadiness	51. Speech hearing
26. Manual dexterity	52. Speech clarity

43.2.3 Nonhomogeneity

Another important dimension of criteria is nonhomogeneity. For instance, Figure 43.1 shows that human performance is nonhomogeneous (these data were derived from Dudley, 1968). It can be seen that there is a warmup period at the beginning of the workday and a slowdown period at the end of the day, and that performance, in effect, is not the same during the entire workday. Hence when a sample of the performance is taken for workers during the workday, caution has to be exercised in interpreting its applicability to the whole workday because of the warmup and slowdown periods during the workday and the work week. There have been extensive studies done over a period of three decades, in the 1940s, 1950s, and 1960s, on the reliability of production performance of industrial workers. This results as a whole indicate that production performance has reliability values varying from 0.7 to 0.9, and that there is a momentary fluctuation in performance output of 5–7%. A detailed review of studies in this area has been conducted by Salvendy and Seymour (1973).

Nonhomogeneity can be relevant for a range of criteria, including quantity of output (performance) as seen above. This nonhomogeneity has important consequences for the data-collection process: collecting data at the right time, and/or over an adequate period of time is critical to obtaining 'good' representative data.

43.3 DEPENDABILITY OF MEASURES

To ensure the appropriateness of conclusions drawn from an HFE intervention or some other process where HFE outcomes are important (e.g., experimental study in the laboratory), it is critical to ensure the "goodness" of the outcome measures. In particular, it is important to evaluate the reliability, constant error, and validity for each measure. Each of these evaluations is discussed below. Much of the information presented below is based on Salvendy and Seymour (1973), Standards for Educational and Psychological Testing (1985), Anastasi (1988), and Bohrnstedt (1983).

43.3.1 Reliability

Reliability refers to the consistency of measures obtained by individuals when reexamined with the same criterion measure on different occasions or with different sets of equivalent tasks. Reliability provides information on error of measurement of a single criterion and indicates the extent to which individual differences in criterion measured are attributable to "true" differences in characteristics under consideration and the extent to which they are attributable to random errors. Reliability is usually expressed in terms of a correlation. Thus for a measure with a reliability of 0.80, 64% of the variance is explained by the individual's data and 36% of the variance associated with this measure cannot be explained and is, in effect, a random error.

Reliability measures can be divided into two major classes: measures of stability and measures of equivalence. Reliability evaluated by correlating a measure across time is called a measure of stability or test-retest reliability. Reliability measures of equivalence include: split-half, alternate form and Cronbach alpha. The four different methods for measuring reliability are described below.

- *Test-retest reliability:* Test-retest reliability is the correlation between the repetition of identical measures. This measure can be obtained, for example, by correlating the performance of two successful days, or by correlating anthropometric data

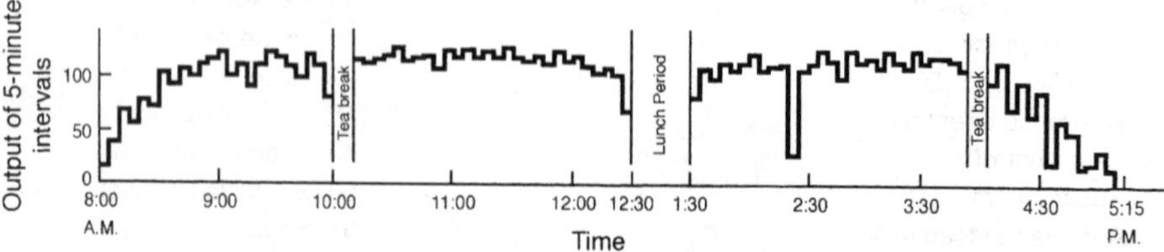

Figure 43.1 Production output curve of one British operator performing a repetitive manual operation during a work day. Similar results were obtained for other repetitive tasks and for other operators (after Dudley 1968).

taken on a population before and after a short tutorial. The higher the reliability, the less susceptible the measures are to the random changes in the condition of the individual affecting the measure or the measuring environment.

- *Split-half reliability:* Split-half reliability is obtained by correlating measures derived from one half of the data on which the measure is based with the other half. This method of reliability testing may be useful when test-retest reliability is either not practical or not feasible. The problem with split-half reliability is how to split the data to ensure that the two parts are equal. A way of solving the problem may be to compute the correlation between the first half with the second half or the odd questions with the even questions. In the context of HFE outcome measures, split-half reliability raises the problems as to how to account for learning, fatigue, and boredom since these variables may change within a single session.
- *Alternate form:* It can be informative to correlate alternate forms of a measure. Two alternate forms of a measure, designed to be as similar as possible, are administered to the same group of persons. The correlation between the two measures on the two forms indicates the degree of reliability of either form taken separately. The two administrations can be done at the same time, successively, or separated by a certain period of time. The difficulty with this approach is to construct two truly equivalent measures of the same HFE outcome.
- *Cronbach alpha:* A common method for assessing the reliability in survey research is to determine the internal consistency using the Cronbach alpha calculation (Cronbach 1951, 1990). Reliability is typically measured for sections or groups of questions that are determined (or assumed) to measure one variable or factor.

  The Cronbach alpha reliability measure for checking internal consistency considers the variance of individual items in the group over all respondents, the variance of the group of items over all respondents and the number of items in the group. It indirectly considers the number of respondents since the variance of individual items would be expected to decrease with the increase in the number of respondents in the calculation. The formula for calculating the Cronbach alpha coefficient, a_k, is:

$$a_k = \frac{k}{k-1} * \left[1 - \frac{\Sigma s^2_{\text{items}}}{s^2_{\text{total}}}\right]$$

where: k is the number of items in the group, s^2_{item} is the variance of an item in the group over all respondents, Σs^2_{item} = the sum of item variances, and s^2_{total} is the variance of a group of items over all respondents.

For more information on the different types of reliability, see Carmines and Zeller (1979).

43.3.2 Constant Error

A constant error is an error that appears consistently in repeated measurements. Suppose, for example, that the time used for a car has a 10% larger circumference than that recommended for use by the car manufacturer; thus each revolution of the wheel will cover a distance of 10% more than what the car odometer will actually show; since at each revolution of the wheel the distance covered will be 10% more.

The implications of constant error of measure for ergonomics research and practice is evident from both the physical and the psychological ergonomics point of view. This points to the repeated need to calibrate all instruments of measurement against known standards.

43.3.3 Validity

Whereas the concept of reliability addresses whether or not the results of a measurement are the same when the operation is repeated or when similar measurement operations are done, the concept of validity concerns whether a measurement operation measures what it intends to measure. A measure that is reliable is not necessarily valid. It is important to examine the reliability and the validity of a measure. There are several types of validity: content validity, construct validity, and criterion-related validity.

- *Content validity:* When examining the content validity of an HFE outcome measure, one tries to ensure that the *universe of content* of the construct is adequately represented. The "universe of content" of the construct (in our case, the construct is the outcome measure) can be thought of as having various facets. A measure that has content validity is one that has sampled adequately these various facets. In some cases, an argument to ensure content validity can be based on correlations between the submeasures of a measurement test (internal consistency). Some general guidelines can be provided to evaluate content validity (Bohrnstedt, 1983):

1. Determine how the construct of interest has been used by others (e.g., conduct a literature search and review).
2. List the different facets within the universe of content of the construct. In order to ensure the representation of the various facets within the universe of content, the universe can be stratified into its major facets. Then, one needs to ensure that all major facets are represented in the measurement operation.
3. Pretest the measure on a sample of people.

- *Construct validity:* The issue of construct validity concerns two issues: (1) appropriateness of the measurement operation for the construct, and (2) boundaries around the construct. In our case, the construct can be any HFE outcome.

There are three issues in attempting to ensure the construct validity of the outcome measure of interest:

1. Need to specify the universe of content of the construct.
2. Need to examine the relationship among the different facets of the construct. The techniques available to perform this task include: internal consistency, factor analysis, convergent and discriminant validity, and the multitrait–multimethod matrix.
3. Need to examine relations among the construct of interest and some other construct.

- *Criterion-related validity: Concurrent and predictive validity:* There are two types of criterion-related validity: concurrent validity and predictive validity. The concurrent validity of an HFE outcome measure can be assessed by examining the association between this measure and some other measure that is known to be valid and reliable (Bohrnstedt, 1983). Techniques to examine the concurrent validity of a measure include correlating the measure with some other concept and the "known-group" technique.

  For example, the concurrent validity of measures of psychosocial work factors among office workers has been examined by Carayon-Sainfort (1990) by using the known group technique. She showed that the measures of psychosocial work factors varied among three job categories, i.e., clerical workers, professionals, and supervisory and managerial jobs. The known group technique is used to examine whether the HFE outcome measure varies among certain groups, as expected.

  The other type of criterion-related validity, predictive validity, examines whether the HFE outcome measure predicts some other measure. The question to be answered is whether or not one can generalize from the HFE outcome measure to that other measure. One of the most difficult problems in examining criterion-related validity is to find a "good" measure to use that has been shown to be valid and reliable.

43.4 USES OF OUTCOME MEASURES

Why do we need to collect data on HFE outcomes? Data collection and evaluation of outcome measures represent an important contribution of HFE. Ultimately, data on HFE outcomes are used to ensure that the desired objective is achieved. In a problem-solving approach, data on HFE outcomes are gathered before the intervention takes place in order (1) to collect baseline information, and (2) to diagnose the HFE problems. Data on HFE outcomes are also gathered after the implementation of the HFE intervention to assess the overall effectiveness of the intervention and the process used to implement the intervention.

43.4.1 Baseline Measures

Baseline measures of HFE outcomes are often collected before the intervention is designed and implemented. These data can be used to make comparisons between different groups and to identify the groups most "at risk." For instance, in a surveillance program for controlling and preventing cumulative trauma disorders (CTDs), data on musculoskeletal discomfort and injuries are gathered and used to compare different groups of workers (Hagberg et al., 1995). These different groups of workers can be, for instance, workers performing different jobs or workers in different departments of a single organization. The data collected can be used to design interventions that address the groups most at risk, for instance. Baseline measures of different groups of workers can be used to prioritize the groups most in need. This is particularly useful when resources are limited: the groups most in need will be the one to which the limited resources are allocated. When additional information is available on (potential) risk factors for CTD's, then analyses can be conducted to examine the relationship between those risk factors and the outcome (e.g., discomfort and/or recorded injuries).

Baseline data are also useful because they provide information for establishing a "diagnosis" of the situation, in particular to determine HFE problems and concerns (see Section 43.2.2.2) and to establish baseline data, which are compared to data collected after the intervention (see Section 43.2.2.3).

43.4.2 Diagnosis of Problems

Whenever possible, it is important to collect data not only on HFE outcomes, but also on potential contributors to the outcomes. As discussed in section 43.1, the potential contributors to HFE outcomes can be conceptualized as a system with five factors: (1) the individual, (2) tasks, (3) tools and technology, (4) physical environment, and (5) the organizational environment (Smith and Carayon-Sainfort, 1989). Various methods can be used for collecting data on work-related risk factors. See, for example, Chapter 35 for a description of measurement of risk factors of work-related musculoskeletal disorders, and Chapter 33 for a discussion of psychosocial work factors and their effects on health.

A study of the relationship between the outcome measures and the potential work-related risk factors can provide useful information for identifying the facets of the work environment that need to be changed in an ergonomic intervention. Carayon (1994) has developed a set of tools for evaluating office work environments and their impact on employee health. This set of tools and the associated implementation process can help in the diagnosis of ergonomic deficiencies existing in the work environment under study. Hagberg et al. (1995) proposed guidelines for designing surveillance systems for CTDs. An important element in the surveillance system is the collection of valid and reliable data on CTDs or precursors of CTDs. This information can be used to diagnose ergonomic deficiencies and their effect on people.

43.4.3 Effectiveness of Ergonomic Interventions

HFE interventions are aimed at preventing or reducing some ergonomic deficiency and at improving outcomes, such as comfort, health, performance, and attitudes. The ultimate use of HFE outcome measures is to assess the effectiveness of ergonomic interventions. Ideally, data on outcome measures are gathered before <u>and</u> after the intervention is implemented, and the comparison of these data provides useful information on how effective the intervention is. In some instances, several outcomes are targeted by the intervention; therefore, data should be gathered on all of these outcomes.

Showing the positive effects of HFE interventions is an important challenge to HFE professionals. Collecting valid and reliable data on outcome measures and using the adequate evaluation mechanism(s) are critical elements in this process. The close examination of a particular ergonomic intervention will demonstrate this. Chatterjee (1992) conducted a prospective longitudinal study at an electromechanical plant between 1980 and 1988. Several methods were used to identify the causes of work-related musculoskeletal disorders. Simultaneously, several interventions were put in place: (1) educational interventions for supervisors, engineers, workers, safety representatives, and occupational and safety personnel; (2) occupational health surveillance; (3) ergonomic and engineering changes (e.g., new, adjustable workstations); and (4) implementation of a steering committee with management, engineers, and occupational health specialists to ensure effective planning and implementation and adequate followup. The result of this multidisciplinary intervention was a dramatic reduction in incident rates for new cases that fell from 2.1 in 1987 (when major engineering and ergonomic modifications were put in place) to 0.1 in 1990. The results of the intervention were very positive, as the comparison between the outcome measures (i.e., incident rates for new CTD cases) taken before and after the intervention shows.

43.4.4 Temporal Dimensions

Outcome measures are temporally bounded. First, they often contain temporal features (e.g., number of forms processed during a certain period of time). Second, their collection occurs over time. The collection and evaluation of outcome measures is a process that takes time. Therefore, it is important to consider the temporal dimensions of this process, in particular the temporal dimensions of the data-collection phase.

43.4.4.1 Outcome Measures

Many HFE measures contain important temporal features, such as rates or frequencies of responses, latencies, and reaction times. A rate or frequency is a record of the number of some kind of occurrence per some predetermined unit of time; it ignores the distribution of events throughout the interval. Other HFE outcome measures are measures of temporal duration, that is, the temporal interval between the onset and the cessation of some event (for example, reaction time).

Another temporal dimension of the outcome measures is the time period used to measure the outcome. For instance, which time period is used to measure performance, such as number of units produced? The time period of the outcome measures can be measured in seconds, minutes, hours, days, weeks, months, or years.

Another temporal dimension concerns the differentiation between frequency, duration and severity. For instance, musculoskeletal discomfort can be measured in terms of frequency of discomfort or pain experienced, duration of uncomfortable or painful events, and severity of discomfort or pain. The questionnaire used to measure musculoskeletal discomfort and pain in several National Institute for Occupational Safety and Health (NIOSH) studies differentiates between these three dimensions and asks three different questions for each of these three dimensions for each body part (neck, shoulders, back, elbow, hands/wrists) (NIOSH, 1990, 1992). For each body part, an indicator is then built that combines the three dimensions of frequency, duration, and severity.

Finally, the method used to collect data on HFE outcomes poses certain challenges with regard to time. For instance, self-reports can be collected (almost) at any time, before, during, or after an event (e.g., the HFE intervention). Observational measures can provide online, direct data on the outcome(s) of interest. For a complete discussion of the temporal challenges of various data-collection methods, such as self-report measures, observational measures, trace measures and archival records, see Kelly and McGrath (1988).

It is important to specify clearly the temporal features of the outcome of interest. Then, the most appropriate (and feasible) method for collecting data on the outcome can be chosen. This decision needs to take into account the weaknesses and strengths of the various data-collection methods, in particular with regard to time.

43.4.4.2 Data-Collection Process

The concept of "cause" is closely related to time. First, a cause must precede an effect in time. Second, all causal processes take some finite amount of time before complete implementation. During an HFE intervention, we manipulate one variable (the "cause," that is, some work-related factors) and then, after a period of time, we measure the effects of that manipulation on the other (the "effect," that is, the HFE outcome measures). There are several temporal dimensions of the data-collection process: periodicity of measurement, duration of measurement, and temporal window between measurements and events (e.g., before and after the ergonomic intervention) (Kelly and McGrath, 1988).

The temporal interval is the amount of time necessary and sufficient for the cause (the HFE intervention) to have the effect (positive effect on the HFE outcomes). If the interval is set too short, then the cause may not yet have had time to yield the intended effects that are to be observed. If the interval is set too long, the effects of interest may have come and gone, or they may already have been altered by counterforces in the system. During the time between the cause and the effect, several other events could occur. This temporal window is the interval through which extraneous factors can enter and bias results. It is important to optimize the length of the temporal window, that is, the various times of collection of HFE outcome measures:

- There should be just enough time for the causal processes to operate to their fullest.
- There should not be so much time that some of it is empty of the operation of the causal processes.
- There should not be so much time that some other events have begun to operate so as to counter the causal forces under study.

- There should not be too much time so that the effects of the cause will have begun to wane (Kelly and McGrath, 1988).

43.5 DATA-COLLECTION METHODS

Data collection can be achieved in a variety of ways. The major considerations in deciding which method of data collection to use are:

- Cost of collecting data
- Convenience—duration and time,
- Dependability of the measures, including constant error, reliability, and construct validity
- Knowledge and skills possessed by the investigator in specific data-collection methods.

Each of the major methods of data collection, including questionnaires, interviews, observations and videotaping, records, documents, and archival data, and online measurement methods is discussed separately below.

43.5.1 Interview

Interviews are often used for measuring and evaluating HFE outcomes. One of the best way to begin addressing a problem is sometimes by asking the people what they think. The major advantage of interviews is the collection of qualitative, rich data. The interviewer can elicit a fuller, more complete response than will a questionnaire requiring respondents to write out answers. The interviewer can ask respondents to clarify their answers. The major weaknesses of interviews are a small number of respondents, data that are difficult to quantify, and expense of the method (time and personnel). Interviews can be done either face to face or over the telephone. Both methods require trained interviewers to ensure that interviewers do not induce additional bias in the data collection.

So-called "exit interviews" are often performed when individuals are exposed to a "treatment" or an HFE intervention. During these interviews, information is gathered from the individuals on their perceived effectiveness of the intervention. Such interviews generally ask open-ended questions about the individuals' perception of and feelings toward the HFE intervention. This information is important to assess the global effect of the intervention on the individuals for which this intervention has been designed.

The degree of structure of an interview varies. Unstructured interviews are very open: an outline with a limited set of very broad questions is used as a guide for conducting the interview. Semistructured interviews are more structured: a set of open-ended questions is designed before the interview is conducted and is included in the interview guide along with instructions for asking followup questions for clarification. A structured interview is similar to a questionnaire, except that the mode of administration is different. In the case of a questionnaire, the individual respondent is by him or herself to answer the questions. In the case of a structured interview, the individual respondent is asked the questions by an interviewer. At the opposite of a questionnaire, the structured interview allows the respondent to raise his or her opinions and explain his or her answers, which can be recorded by the interviewer. However, this mode of administration is much more expensive because it requires the presence of interviewers.

43.5.2 Observations and Videotaping

Direct observations or observations via videotaping allow the collection of data on HFE outcomes, such as performance and behaviors. Some behaviors can be indicators of other HFE outcomes. For instance, rubbing one's neck or fidgeting are behaviors that can reflect the individual's state of musculoskeletal discomfort. The main strengths of observations and videotaping are objectivity (direct collection of data), face validity, and richness of data. The weaknesses of observations and videotaping are small sample size, biases (e.g., interpretation of the observer, effect of the observer on the person being observed), sampling of events and time, difficulty of analyzing and interpreting data, and cost in time and personnel.

Observational methods vary in several ways (Proctor and Van Zandt, 1994):

- The observations can be recorded at the time of the observation or later.
- The content and amount of detail in the observations can vary.

- The duration of observations can be short or long.
- Observations can vary in terms of inference.

In order to adequately conduct observations, the HFE professional must develop a tool for recording the events or behaviors of interest. In some cases, a checklist is developed in advance and is used by the observer to record the presence or absence of some events or behaviors. When only certain events or behaviors are of interest, the observer will record only the events or behaviors that correspond to predetermined categories. A videotaped observation can be examined subsequently for evaluating the phenomenon of interest.

43.5.3 Records, Documents, and Archival Data

Another method for measuring and evaluating HFE outcomes consists in the analysis of records, documents, and archives. Records of interest to HFE professionals include medical records, accidents and injuries, workers' compensation claims, absenteeism, turnover, and performance. The main strengths of this set of methods is that it allows a historical evaluation of the phenomenon under study. This historical analysis can link certain events (e.g., an ergonomic intervention) to outcome measures (e.g., accidents) to examine the effectiveness of these events over time. An example of the use of such data-collection methods is given in Section 43.2.2.3. The main weaknesses of these methods are limited information and lack of reactivity.

43.5.4 Online Measurement and Instrumentation

Another method for collecting information on HFE outcomes is the use of online measurement and instrumentation. Many bodily functions can be evaluated with online measurement and instrumentation, such as heart rate, blood pressure, and muscular activity. For instance, Chapter 35 discusses the collection of muscular activity via electromyogram (EMG). One of the strengths of online measurement and instrumentation is the connection between particular events in the work environment and HFE outcomes. Hennigan and Wortham (1975) measured heart rate data from managers throughout the workday. These data were related to the activities experienced by the managers during the day and illustrate the use of online measures for evaluating white-collar work stress and the relationship between particular work activities and stress. Salvendy and Knight (1983) have monitored blood pressure, heart rate, and rate of breathing for the full working day for a one year duration with specially designed equipment for this purpose (Knight et al., 1979, 1980). Other strengths of online measurement and instrumentation include objective, rich data; face validity; and high reactivity. The main weaknesses are cost (personnel, time, instruments) and small sample size.

43.5.5 Performance

HFE professionals are also interested in performance, not only in terms of quantity, that is, the amount or output produced or processed during a specific time period, but also in terms of quality. The design of the work environment can enhance and foster performance, but can also hinder performance. For instance, Chapters 28 and 29 describe performance in extreme environment. Chapter 14 describes job design theories and approaches for optimizing performance, both in terms of quantity and quality, but also for improving attitudes such as job satisfaction.

The goal of ergonomics is to design safe and healthy work systems, in particular the interactions between the individual and various factors of the work system, in order to improve the "functioning" of the overall system. An important facet of the overall system functioning is its performance. Examples of measures of performance include more output (quantitative aspect of performance), more output from fewer inputs to the system (productivity), and increased reliability (qualitative aspect of performance).

Two concepts are closely related to performance, that is effectiveness and efficiency. Effectiveness is "doing the right things," whereas efficiency is "doing things the right way." Effectiveness and efficiency are two important aspects of performance that go hand in hand. A person may be doing his or her tasks efficiently, but may be doing the wrong things (poor effectiveness). Therefore, it is important to examine the effectiveness and efficiency aspects of performance, as well as the qualitative and quantitative aspects of performance.

Methods for measuring performance objectively are discussed in Salvendy (1992), and an example of a questionnaire used to measure overall performance is presented in Figure

43.2. The questionnaire includes 15 items that ask about discomfort in the back, neck, shoulders, arms and wrists, and legs. The questionnaire items can be grouped together to compute various indicators of discomfort. A global indicator of musculoskeletal discomfort can be computed by adding all 15 items, and indicators of discomfort in specific parts of the body can be computed by adding some specific questionnaire items. The computations for creating these discomfort indicators are explained below:

Total musculoskeletal discomfort
MUSCUL = (Q#1 + Q#2 + Q#3 + Q#4 + Q#5 + Q#6 + Q#7 + Q#8 + Q#9 + Q#10 + Q#11 + Q#12 + Q#13 + Q#14 + Q#15)/15
Hand-arm discomfort
HAND = (Q#6 + Q#11 + Q#13 + Q#14 + Q#15)/5
Legs discomfort
LEG = (Q#7 + Q#9)/2
Upper body discomfort
UPPER = (Q#2 + Q#4 + Q#8 + Q#10 + Q#12)/5

This questionnaire has been used in many studies, in particular in studies of VDT workers (Smith et al., 1981; Sauter et al., 1983). Data from these studies can be used as comparison basis. The reliability of the various sub-scales of the musculoskeletal discomfort questionnaire varies between .60 and .88 (Sainfort and Carayon, 1994).

There are two types of performance measures of interest to HFE professionals: task performance and overall performance. The main conceptual difference between task performance and overall performance is the temporal boundary. The concept of task performance is at a lower level than overall performance. In addition, typically, overall performance is measured on a longer period of time than task performance.

43.5.5.1 Task Performance

There are several indicators of task performance that can provide useful information to the HFE professional. Typically, quantity and quality of task performance are of direct interest. However, other aspects of task performance, such as performance deterioration, performance on secondary tasks, and performance variability, can provide important information on other HFE outcomes.

QUESTIONNAIRE ON COMFORT (MUSCULOSKELETAL DISCOMFORT)

The following questions concern your body and the way you and it function. Please try to answer each question by circling a number to indicate how often you have experienced each of the following items within the past year in general.

	Never	Occasionally	Frequently	Constantly
Q#1. Swollen or painful muscles and joints...	1	2	3	4
Q#2. Back pain....................................	1	2	3	4
Q#3. Pain or stiffness in your arms or legs....	1	2	3	4
Q#4. Pain or stiffness in your neck and shoulders..................................	1	2	3	4
Q#5. Persistent numbness or tingling in any part of your body.......................	1	2	3	4
Q#6. Pain down your arms.....................	1	2	3	4
Q#7. Leg cramps.................................	1	2	3	4
Q#8. Feeling of pressure in the neck..........	1	2	3	4
Q#9. Difficulty with feet and legs when standing for long periods..............	1	2	3	4
Q#10. Shoulder soreness..........................	1	2	3	4
Q#11. Loss of feeling in the fingers or wrist....	1	2	3	4
Q#12. Neck pain that radiates into shoulder, arm or hand....................................	1	2	3	4
Q#13. Cramps in hands/fingers relieved only when not working........................	1	2	3	4
Q#14. Loss of strength in arms or hands........	1	2	3	4
Q#15. Stiff or sore wrists........................	1	2	3	4

Figure 43.2 A sample questionnaire used to measure musculoskeletal discomfort. (After Smith et al., 1981.)

Deterioration of performance over time can be a good indicator of HFE deficiencies in the work system. In a continuous, paced inspection task, a 2% miss level in detection performance was noted in the first 10 min. of the task. Performance rapidly deteriorated over time: The miss level increased to 3% after 30 min and to 5% after 40 min (Eastman Kodak Company, 1983).

Measures of secondary task performance can be used as an index of primary task load. This measurement method is based on the assumption that the individual has a limited capacity to process information. If the main (primary) task is not too difficult, processing capacity will be available for performing the secondary task. If the primary task is difficult, less capacity will be left for the secondary task and performance on the secondary task will decrease. A decrease in performance on the secondary task is assumed to be an indicator of high mental workload on the primary task. A variety of secondary tasks have been developed. See, for example, Bridger (1995) for a list of the most common secondary tasks.

Performance variability can also be an important facet of performance to measure. It has been suggested that performance variability can be an indicator of fatigue and boredom. The lack of regularity in performance seems to be an indicator of the accumulation of fatigue and boredom over time (Henning et al., 1989). Pan, Shell, and Schleifer (1994) conducted a laboratory study of 24 data-entry workers under two conditions: (1) electronic performance monitoring, and (2) no electronic monitoring. Self-ratings of fatigue and boredom, and speed and accuracy of task performance were measured during the course of the study. Increases in fatigue and boredom over the course of the study were accompanied by an increase in variability of speed performance, but were not related to accuracy variability.

43.5.5.2 Overall Performance

The overall performance of an employee in an organization can be evaluated both informally and formally. Informal performance evaluations usually occur on a continuous basis as an exchange of information between the employee, on one hand, and his or her supervisor, colleagues, and clients or customers, on the other hand. Formal performance evaluation is called performance appraisal and is typically done on a yearly basis. A performance appraisal can have positive results, such as increased employee motivation, clarification of work expectations, and increased feedback; but can also have negative results, such as feelings of unfairness, worsening of employee–supervisor relationships, and reduced motivation.

The development of an effective, useful performance appraisal is very difficult. One important element of an effective performance appraisal is the development of valid and reliable measures of individual performance. The two most popular rating scales of individual performance are the Behaviorally Anchored Rating Scale (BARS) and the Behavior Observation Scale (BOS) (Landy and Farr, 1980). The BARS lists specific anchor behaviors, whereas the BOS requires the rater to give a frequency associated with the performance of each behavior. The most important type of appraisal error is due to the rater, usually the supervisor. Rater bias cannot be eliminated; it can be reduced by training (Landy and Farr, 1980).

43.5.6 Questionnaire

Another data-collection method for measuring and evaluating HFE outcomes is the questionnaire. The major advantages of questionnaires are ease of use, quantifiable data, relatively inexpensive method, and large number of respondents from a number of organizations or geographical locations. The major weaknesses of questionnaires are lower response rate, biases (e.g., response selectivity, unanswered questions), limited information (e.g., no opportunity to ask followup questions), and subjectivity. In Chapter 18, Hendrick describes the use of organizational surveys for measuring macro-ergonomic variables.

Many different questionnaires already exist to measure different types of HFE outcomes. However, if one needs to develop a new set of questions, several steps need to be followed to ensure its reliability and validity. Table 43.2 displays the process to be used for developing a new set of questions. It is important to not only ask the right questions, but also to ask questions the right way. For more information on principles for writing questions, see Converse and Presser (1986).

A questionnaire must be well constructed. As with any other measurement methods, the data obtained is as good as the measurement tool. Important issues for constructing

Table 43.2 Steps for Developing a Questionnaire

1. Outline analysis	Define the domain of content of the construct. Determine the different facets of the domain.
2. Select items	Select the items about which questions are going to be written.
3. Literature review	Identify items that can be used from the literature. Evaluate problems using or analyzing existing items and questionnaires.
4. Develop or redevelop questions	Initial development: begin with very open-ended interviews, individual interviews, or focus groups, then more focused interviews or discussions. Techniques: "concurrent think aloud," paraphrasing, "retrospective think aloud."
5. Write and re-write questions	Decide: open versus closed. Using familiar words. No parentheses in a question. No double-barrelled questions. Balanced questions.
6. Response categories	Define the response categories. Temporal dimensions: occurrence, frequency, timing, interval. Avoid adding, percentages or proportions. Offering or not a middle category. Offering or not a "Don't Know" category.
7. Formal and final draft	Writing instructions. Order of questions. Cover letter. Consistency in the format of the questionnaire.
8. Pretesting	Several pretests can be done if budget and/or time allows.
9. Data processing review	Format questionnaire for data processing.
10. Pilot study	Goal: to make sure to have all bugs out of instrument.

a good questionnaire include not only the content and format of the questions, but also the flow of questions, the instructions, and the administration of the questionnaire (e.g., perceived confidentiality, motivation of the respondents). Questions to evaluate outcome measures are frequently used for comfort, physical and mental health, and attitude. Each of these are discussed below.

43.5.6.1 Comfort

Comfort can be defined as a subjective, perceptual reaction of the employee to his or her work environment. Comfort (or discomfort) has been conceptualized as either a reaction specific to a factor of the work system (e.g., thermal comfort) or a general reaction to the entire work system (e.g., generalized muscular discomfort).

HFE professionals are interested in comfort relative to various factors of the work system, for example, thermal comfort and chair comfort. Thermal comfort refers to comfort with regard with one factor of the work system, that is the physical environment, and more specifically the temperature, humidity and air velocity. Fanger (1977) defines *comfort zones* as a range of temperatures and humidities (combined with air speed) that most people would find comfortable while performing tasks that do not require heavy physical activity. Therefore, thermal comfort depends on several factors, among them temperature, humidity, and air velocity, and there are individual preferences about what constitutes a comfortable indoor temperature. See Eastman Kodak Company (1983) and Chapter 27 for more details on thermal comfort. Another type of comfort of interest to HFE professionals is chair comfort. For a discussion of the concept of chair comfort, see for example Lueder (1983).

The concept of comfort also applies to general reactions of the person to his or her work system. These general reactions can be conceptualized as subjective, perceptual effects on the body. In this case, the term "discomfort" is more often used and is a synonym of inconvenience. Discomfort is often considered as being related to symptoms of pain. However, the concept of discomfort is more fuzzy than the concept of pain.

One of the most widely used methods for measuring musculoskeletal discomfort is Corlett's body part diagram (Corlett and Bishop, 1976). It has been used in various populations, including various industrial groups (Wilson and Corlett, 1990) and Video

Display Terminal (VDT) data-entry workers (Sauter, Schleifer, and Knutson, 1991). Other methods for measuring musculoskeletal discomfort include questionnaires, such as the NIOSH health checklist which has been used in many studies of office workers and VDT operators (Sainfort and Carayon, 1994; Smith, Cohen, Stammerjohn, and Happy 1981). Hagberg et al. (1995) present a variety of tools for measuring musculoskeletal discomfort in different body parts. An important (unresolved) issue is to know whether the discomfort symptoms are precursors of future illness. There is insufficient research to solve this issue. However, discomfort data are important for evaluating the employees' perception of their well-being. In addition, followup studies have shown that discomfort data are useful in reducing fatigue (Corlett and Bishop, 1976). Discomfort data are also useful to evaluate the optimality of a job (Hagberg et al., 1995).

For most of these questionnaire-based methods, data exist for a variety of jobs in a variety of work environments in different countries. These data can be used as a basis to compare the sample under study to other groups for which data are available. Such comparisons can provide very useful information to assess the extent of the problem in the sample under study. Data collected via questionnaires are often easy to obtain, especially when questionnaires are short and questions are easy to answer. In this instance, questionnaire data can be collected on a regular basis, and trends over time can be examined. Questionnaire data can, for instance, be collected a few months before and after an ergonomic intervention. Such data can also be collected on a regular basis, for example, every year, as part of a surveillance program. Analysis of trends of comfort data over time can be a useful tool for preventing severe health problems from actually occurring.

It has been suggested that musculoskeletal discomfort can result in a need to move (Branton, 1969). Consequently, one approach to assess discomfort would be to measure these moves by, for instance, structured observations using a checklist.

Questionnaires are a very popular method for measuring musculoskeletal discomfort. However, other methods, such as interviews, have been used to measure musculoskeletal discomfort. Westlander (1994) has developed an interview to examine musculoskeletal discomfort, and its effect on VDT workers. The interview asks closed questions on musculoskeletal discomfort in the neck, shoulders, back, and arms and open-ended questions on the individual's subjective experiences of work-related musculoskeletal discomfort (e.g., activities related to discomfort, coping with discomfort, and circumstances of the experience of discomfort). This in-depth interview-based method provides very rich data on the link between musculoskeletal discomfort and life situation.

An example of a questionnaire used to measure musculoskeletal discomfort is presented in Figure 43.3. The questionnaire on overall performance was developed by Carayon (1992). It is comprised of 4 items, and its Cronbach-alpha score was .87 in a group of 171 office workers (Carayon, 1992). An indicator of overall self-reported performance is the mean score derived from the four questions.

In practice, the concept of discomfort is often defined by the methodology used to measure it (for example, measurement of musculoskeletal discomfort; Corlett and Bishop, 1976). Visual discomfort and musculoskeletal discomfort are two examples of this type

---

QUESTIONNAIRE ON OVERALL PERFORMANCE

A short questionnaire on overall performance was developed by Carayon (1992). It is comprised of 4 items, and its Cronbach-alpha score was .87 in a group of 171 office workers (Carayon, 1992). The following table gives a detailed description of this questionnaire.

The following question deals with how you think you do your job in comparison to others. Rate your performance, as compared to others who do similar work, on the following dimensions (circle the appropriate number):

	LOW									HIGH
Q#1. quantity of "output"	1	2	3	4	5	6	7	8	9	10
Q#2. efficiency	1	2	3	4	5	6	7	8	9	10
Q#3. work quality	1	2	3	4	5	6	7	8	9	10
Q#4. dependability	1	2	3	4	5	6	7	8	9	10

Figure 43.3 A sample questionnaire used to measure overall performance.

of HFE outcomes that have received much attention. The concept of visual discomfort has received attention in the research literature on VDT work and its health consequences. Visual discomfort can be defined as a set of symptoms experienced by the individual; the symptoms include pain in the eyes, burning or itching eyes, blurred or double vision, and other troubles (Grandjean, 1987; Sauter, Chapman, and Knutson, 1985). Recently, the concept of musculoskeletal discomfort has received attention in relation to the phenomenon of cumulative trauma disorders or work-related musculoskeletal disorders. Discomfort, along with fatigue and pain, is considered the most common first symptoms associated with work-related musculoskeletal disorders (Hagberg et al., 1995).

There are two types of comfort of interest to HFE professionals: comfort with regard to specific facets of the work environment, such as thermal comfort and chair comfort, and general comfort reactions to the entire work system, such as visual comfort and musculoskeletal discomfort. In this section, various methods for measuring musculoskeletal discomfort are reviewed.

43.5.6.2 Physical and Mental Health

Ultimately, the goal of HFE is to prevent or reduce the negative effects of work on the individual, in particular the negative effects on worker health, and to enhance health and well-being. It is now accepted and well established that poorly designed work environments can induce physical and psychological stress that can affect physical and/or mental health. For instance, Chapter 24 describes how noise can affect hearing. Chapter 33 examines the role of psychosocial work factors in mental and physical health.

As was discussed earlier, there are multiple factors of the work system that can affect people and their mental and physical health. These factors can be categorized as individual, task, tools and technologies, physical environment, and the organization. They can be described as either physical or psychosocial or organizational (Cox and Ferguson, 1994). Cox and Ferguson (1994) developed a model of the effects of physical and psychosocial or organizational factors on health. According to this model, the effects of work factors on health are mediated by two pathways: (1) a direct physicochemical pathway, and (2) an indirect psychophysiological pathway. These pathways are present at the same time and interact in different ways to affect health. Physical work factors can have direct effects on health via the physicochemical pathway, and indirect effects on health via the psychophysiological pathway, but can also moderate the effect of psychosocial or organizational work factors on health via the psychophysiological pathway. Psychosocial or organizational work factors have indirect effects on health and moderate the effects of physical factors via the psychophysiological pathway. Cox and Ferguson's model is useful to describe and explain the various effects of work on physical and mental health.

A recent health issue that has challenged the profession of HFE is the increasing number of work-related musculoskeletal disorders in a range of jobs and work environments. Work-related musculoskeletal disorders include a range of trauma of the back, neck, shoulders, and arms (Putz-Anderson, 1988; Hagberg et al., 1995). Work-related musculoskeletal disorders represent a very serious threat to the physical and mental health of workers all over the world.

Two main types of techniques have been developed and used to directly measure health: self-administered instruments (questionnaires) and expert evaluations. Self-assessment is generally more efficient in time and cost, but reliability and validity are usually not as good as expert assessments. Self-assessment of health can be based on single-item or multiitem scales. In general, multiitem scales are more satisfactory than single-item scales because single items are often less reliable and valid. Consequently, several instruments with multiitem scales have been developed to measure physical and mental health. The NORDIC questionnaire focuses on the measurement of musculoskeletal symptoms (Kuorinka et al., 1987). Kasl (1992) reviews indicators of psychological health and functioning that have been used in the occupational stress field. The NIOSH health checklist covers many different aspects of physical and mental health of concerns for VDT operators (Sainfort and Carayon, 1994; Smith et al., 1981).

Questionnaire-based measures of overall health are much less reactive than physiological measures of precursors of health problems. The questions asked in a questionnaire have a variety of temporal dimensions. For example, the NIOSH health checklist and the NORDIC questionnaire ask about discomfort and pain experienced within the past year. Other temporal dimensions include the present moment and the past week. However, questionnaire data are not useful to measure effects over very short periods of time. In this regard, physiological measures are much more reactive than questionnaire data.

Physiological measures can provide an indirect evaluation of health by measuring precursors of health problems. In particular, a variety of methodologies have been developed to measure the physiological effects of stress. Physiological measures of stress can be used for evaluating both physically and mentally demanding work. They include heart rate, heart rate variability, blood pressure, respiratory parameters, electromyography, and biochemical measures, such as catecholamines and cholesterol. Many of these measures can be considered as fast responses to internal and external stimuli (Sharit and Salvendy, 1982). They can be used to isolate sources of stress and to evaluate potential health consequences.

Aasman, Mulder, and Mulder (1987) and Vicente, Thornton, and Moray (1987) have shown that sinus arrhythmia can be used as an index of mental effort, and therefore is an important HFE outcome measure. In a laboratory study, Hwang and Salvendy (1988) used measures of heart rate and sinus arrhythmia (heart rate variability) along with performance and subjective measures of stress (via questionnaire) to study the effects of allocation of functions and number of machines in a flexible manufacturing system.

Respiratory parameters have been used to measure the stress effects of work. End-tidal PCO^2 seems to be a respiratory parameter of value to measure the stress effects of VDT work, in particular (Schleifer and Ley, 1994). End-tidal PCO^2 is the peak concentration of carbon dioxide in a single breath of exhaled air at the end of the expiratory phase. Under stressful conditions, hyperventilation occurs; this leads to an excessive loss of CO^2, which is reflected in a reduction of CO^2 in end-tidal PCO^2 concentrations. Therefore, stress effects can be measured as a reduction in end-tidal PCO^2. Schleifer and Ley (1994) conducted a laboratory study to examine the usefulness of end-tidal PCO^2 as an index of psychophysiological stress among VDT workers. A group of 11 VDT workers were examined under the following conditions: (1) during a self-relaxation baseline period, (2) during a progressive muscle relaxation period, and (3) during a period of computer-based data-entry work. Measures of end-tidal PCO^2, respiration frequency, cardiac interbeat interval, and self-ratings of relaxation and tension were taken during the course of the study. Results show that end-tidal PCO^2, cardiac interbeat interval, and relaxation ratings during data-entry work were lower than during either baseline relaxation and progressive muscular relaxation, whereas respiration frequency and tension ratings were higher. This study demonstrates that end-tidal PCO^2 may be useful in measuring the psychophysiological stress effects of VDT work.

Biochemical measures can be obtained from various fluids of the body, such as urine, blood, and saliva. They include cholesterol, adrenaline, noradrenaline, steroids, glucose, uric acid, triglycerides, and lipids. Johansson (1981) has shown that machine-paced work as compared to self-paced work produced increased levels of adrenaline and noradrenaline. Frankenhaeuser and Gardell (1976) studied acute stress reactions of 24 sawmill workers by measuring catecholamine (adrenaline and noradrenaline) excretions during and after work. The results pointed to systematic relations between specific job characteristics, such as monotony and machine control, and symptoms of stress as reflected in high catecholamine levels. In the high-risk group (machine-paced, repetitive job) catecholamine increased toward the end of the day, whereas it decreased in the control group. Johansson and Aronsson (1984) linked the occurrence of unpredicted computer breakdowns to increase in blood pressure and adrenaline excretion.

Many precautions should be taken when using physiological measures. Various confounding factors, such as age, gender, time of the day, posture, and diet, have to be taken into consideration (Fried, Rowland, and Ferris, 1984; Kak, 1981; Levi, 1972; Sharit and Salvendy, 1982). Physiological measures require also reliable and valid instrumentation that is well calibrated.

43.5.6.3 Attitudes

HFE professionals are interested in negative consequences or effects of work on people, but they are also interested in fostering positive outcomes, such as performance and attitudes. Attitudes of interest to HFE include satisfaction, resistance to change, and involvement. HFE is concerned with either satisfaction with one's job in general or satisfaction with particular job facets (for example, satisfaction with physical working conditions, satisfaction with pay).

An example of a questionnaire used to measure attitude is presented in Figure 43.4. The facet-free job satisfaction questionnaire (Quinn et al., 1971) has been used in many studies, and there exists 'benchmark' data for a variety of jobs (see for example, Quinn et al., 1971; Caplan et al., 1975). This questionnaire has been used in many studies and

JOB SATISFACTION QUESTIONNAIRE

Q#1. All in all, how satisfied would you say you are with your job? (circle one number)
1. Very satisfied
2. Somewhat satisfied
3. Not too satisfied
4. Not at all satisfied

Q#2. If you were free to go into any type of job you wanted, what would your choice be? (circle one number)
1. I would want the job I have now.
2. I would want to retire and not work at all.
3. I would prefer some other job to the job I have now.

Q#3. Knowing what you know now, if you had to decide all over again whether to take the job you now have, what would you decide? (circle one number)
1. I would decide without hesitation to take the same job.
2. I would have some second thoughts.
3. I would decide definitely not to take the same job.

Q#4. In general how well would you say that your job measures up to the sort of job you wanted when you took it? (circle one number)
1. Very much like the job I wanted.
2. Somewhat like the job I wanted.
3. Not very much like the job I wanted.

Q#5. If a good friend of yours told you he or she was interested in working in a job like yours for your employer, what would you tell him or her? (circle one number)
1. I would strongly recommend it.
2. I would have doubts about recommending it.
3. I would advise the friend against it.

Figure 43.4 A sample questionnaire used to measure attitudes and job satisfaction.

has been found reliable and valid. The Cronbach-alpha score for this questionnaire was .85 in a study of 2,010 men in 23 occupations (Caplan et al., 1975) and .82 in a study of 113 computer users (Carayon et al., 1995). The job satisfaction score is computed by using the mean weight score for the 5 questions. The weight for each question is as follows:

Q1: 5, 3, 1, 1; Q2: 5, 1, 1; Q3: 5, 3, 1; Q4: 5, 3, 1; Q5: 5, 3, 1

Improving workers' attitudes is, directly or indirectly, an outcome of importance in any HFE intervention or program. Directly, an HFE intervention may be aimed, for instance, at improving the job content in order to foster satisfaction (see for example Chapter 14 on job design). Indirectly, whenever an HFE intervention is implemented (for instance, improved design of workstations), satisfaction of the workers can be affected either positively or negatively, and therefore the success of the intervention can be either fostered or limited. Recently, methods have been developed to improve the process by which ergonomic interventions are introduced, and consequently to ensure the 'complete' success of the intervention. Nagamachi and Imada (1992) have shown that macroergonomic interventions (e.g., using participatory ergonomics) conducted in Japan and in the United States can reduce accident rates by as much as 72–90%. For a more complete discussion of participatory ergonomics, see Noro and Imada (1991) and Chapter 15 of this handbook.

There are numerous questionnaires to measure attitudes, such as job satisfaction, motivation, involvement, and tension. Cook, Hepworth, Wall, and Warr, (1981) reviewed numerous questionnaire-based measures of attitudes.

An important set of attitudes is the individual's psychological state. The use of mood checklists is a popular method to measure affective states related to mental health and psychological stress. Mackay, Cox, Burrows, and Lazzerini (1978) developed a mood adjective checklist that differentiates "stress" from "arousal". The Profile of Mood States

(POMS) is a commonly used checklist comprised of 64 adjectives grouped in 6 scales: tension-anxiety, depression-dejection, anger-hostility, vigor-activity, fatigue-inertia, and confusion-bewilderment (McNair, Lorr, and Droppleman, 1971). It has been used in recent studies of office automation to show that, in general, clerical computer users reported more stress than clerical noncomputer users and professional computer users (Smith et al., 1981). A shorter version of the POMS has been developed and used by NIOSH researchers (see, for example, Schleifer and Amick, 1989).

Attitudes can also be indirectly evaluated by examining organizational records of absenteeism and turnover. Folger and Belew (1985) reviewed the literature on absenteeism measurement. Measures of absenteeism include: (1) time lost index (total number of days of worker absences), (2) frequency index (total number of absences), (3) short-term index (total number of one- and two-day absences), and (4) percentage short-term index (number of short-term absences relative to the total number of absences). Care should be taken in the reporting or recording process of absenteeism to avoid problems such as underreporting (Folger and Belew, 1985).

43.6 CONCLUSION

This chapter has provided information on various HFE outcomes, on methods for measuring and evaluating HFE outcomes, and on measures of HFE outcomes. We have also reviewed the criteria for assessing the "goodness" of different measures, and the strengths and weaknesses of various methods and measures.

The temporal factors in the measurement and in the data-collection process need to be taken account. Temporal dimensions of the HFE outcome itself are important. For instance, one needs to specify whether the measure is a rate or a frequency measure. In addition, it is critical to take into account such temporal dimensions of the data-collection process, such as temporal window, time of measurement, and periodicity of measurement.

Finally, it is obvious that no one measure of an HFE outcome can satisfy all the "goodness" criteria. Therefore, it is important, whenever possible, to use multiple methods and multiple sources of information for evaluating and measuring HFE outcomes. When deciding about HFE outcome measures, one should attempt to use multiple methods with complementary strengths and with strengths that can compensate for weaknesses.

REFERENCES

Aasman, J., Mulder, G., and Mulder, L. J. M. (1987). Operator effort and the measurement of heart-rate variability. *Human Factors, 29*(2), 161–170.

Anastasi, A. (1988). *Psychological Testing* (sixth ed.). New York: Macmillan.

Bohrnstedt, G. W. (1983). Measurement. In P. H. Rossi, J. D. Wright, and A. B. Anderson, Eds., *Handbook of survey research* New York: Academic Press. (pp. 69-–121).

Branton, P., 1969, Behavior, body mechanics and discomfort, *Ergonomics*, *12*, 316–327.

Bridger, R. S. (1995). *Introduction to Ergonomics* (New York: McGraw-Hill.

Caplan, R. D., Cobb, S., French, J. R. P. Jr., Van Harrison, R., and Pinneau, S. R. Jr. (1975). *Job Demands and Worker Health* (HEW Publication No. (NIOSH) 75–160) U.S. Department of Health, Education, and Welfare, NIOSH.

Carayon, P. (1992), *Preliminary Report–The Use of Computers in Offices: Impact on Job Stress and Quality of Working Life* (Project funded by the National Science Foundation, No: IRI-9109566). Madison, WI: University of Wisconsin.

Carayon, P. (1994). A systems approach to reducing physical and psychological stress: Application in automated offices. In G. E. Bradley and H. W Hendrick (Eds.), *Human Factors in Organizational Design and Management—IV*, Amsterdam: Elsevier. (pp. 733–738).

Carayon-Sainfort, P. (1990). Perceptions of work environment and psychological strain across categories of office jobs. In *Proceedings of the Human Factors Society 34th annual meeting* (pp. 849–853). Santa Monica, CA: The Human Factors Society.

Carayon, P., Yang, C.-L., and Lim, S.-Y. (1995). Examining the relationship between job design and worker strain over time: A longitudinal study of office workers. *Ergonomics, 38(6)*, 1199–1211.

Carmines, E. G., and Zeller, R. A. (1979). *Reliability and Validity Assessment*, Newbury Park, CA: Sage Publications.

Chatterjee, D. S. (1992). Workplace upper limb disorders: A prospective study with intervention. *Occupational Medicine, 42(3)*: 129–136.

Converse, J. M., and Presser, S. (1986). *Survey Questions.* Newbury Park, CA: Sage Publications.

Cook, J. D., Hepworth, S. J., Wall, T. D., and Warr, P. B. (1981). *The experience of work—A Compendium and Review of 249 Measures and their Use.* London: Academic Press.

Corlett, E. N., and Bishop, R. P. (1976). A technique for assessing postural discomfort. *Ergonomics, 19*(2), 175–182.

Cox, T., and Ferguson, E. (1994). Measurement of the subjective work environment. *Work and Stress, 8(2)*, 98–109.

Cronbach, L. J. (1951). Coefficient Alpha and the internal structure of tests, *Psychometrika, 16*, 297–334.

Cronbach, L. J. (1990). *Essentials of Psychological Testing*, (Harper & Row, New York).

Dudley, N. A. (1968). *Work Measurement: Some Research Studies*. New York: Macmillan.

Eastman Kodak Company. (1983). *Ergonomic Design for People at Work* (vol. 1). Belmont, CA: Lifetime Learning Publications.

Fanger, P. O. (1977). Local discomfort to the human body caused by nonuniform thermal environments. *Annals of Occupational Hygiene, 20*, 285–291.

Fleishman, E. A., and Quaintance, M. K. (1984). *Taxonomies of Human Performance*. Orlando, FL: Academic Press.

Folger, R., and Belew, J. (1985). Nonreactive measurement: A focus for research on absenteeism and occupational stress. *Research in Organizational Behavior, 7*, 129–170.

Frankenhaeuser, M., and Gardell, B. (1976). Underload and overload in working life: Outline of a multidisciplinary approach. *Journal of Human Stress, 2(3)*, 35–46.

Fried, Y., Rowland, K. M., and Ferris, G. R. (1984). The physiological measurement of work stress: A critique. *Personnel Psychology, 37*, 585–615.

Grandjean, E. (1985). *Fitting the Task to the Man*. London: Taylor and Francis.

Grandjean, E. (1987). *Ergonomics in Computerized Offices*. London: Taylor and Francis.

Hagberg, M., Silverstein, B., Wells, R., Smith, M. J., Hendrick, H. W., Carayon, P., and Perusse, M. (1995). In I. Kuorinka and L. Forcier Scientific Eds., *Work-Related Musculoskeletal Disorders (WMSDs): A Reference Book for Prevention*. London: Taylor and Francis.

Hennigan, J. K., and Wortham, A. W. (1975). Analysis of workday stresses on industrial managers using heart rate as a criterion. *Ergonomics, 18*, 675–681.

Henning, R. A., Sauter, S. L., Salvendy, G. and Krieg, E. F. (1989). Microbreak length, performance, and stress in a data entry task. *Ergonomics, 32*, 855–864.

Hwang S.-L., and Salvendy, G. (1988). Operator performance and subjective response in control of flexible manufacturing systems. *Work and Stress, 2(1)*, 27–39.

Johansson, G. (1981). Psychoneuroendocrine correlates of unpaced and paced performance. In G. Salvendy and M. J. Smith, Eds., *Machine Pacing and Occupational* Stress. London: Taylor and Francis (pp. 277–286.).

Johansson, G., and Aronsson, G. (1984). Stress reactions in computerized administrative work. *Journal of Occupational Behaviour, 5*, 159–181.

Kak, A. V. (1981). Stress: An analysis of physiological assessment devices. In G. Salvendy and M. J. Smith, Eds., *Machine Pacing and Occupational Stress*. London: Taylor and Francis. (pp. 135–142).

Kasl, S. V. (1992). Surveillance of psychological disorders in the workplace, In G. P. Keita, and S. L. Sauter, Eds. *Work and Well-Being—An Agenda for the 1990s*, Washington, DC: American Psychological Association. (pp. 73–95).

Kelly, J. R., and McGrath, J. E. (1988). *On Time and Method*. Newbury Park, CA: Sage Publications.

Knight, J. L., Salvendy, G., and Geddes, L. A. (1979). *A Microcomputer System for Long-Term Automatic Blood Pressure Monitoring. The Annals of Biomedical Engineering, Vol. 7*, pp. 369–374.

Knight, J. L., Geddes, L. A., and Salvendy, G. (1980). *Monitoring the Respiratory and Heart Rate of Assembly Line Factory Workers. Medical and Biological Engineering and Computing, Vol. 18*, pp. 797–798.

Kuorinka, I., Jonsson, B., Kilbom, A., Vinterberg, H., Biering-Sorensen, F., Andersson, G., and Jorgensen, K. (1987). Standardised Nordic questionnaires for the analysis of musculoskeletal symptoms. *Applied Ergonomics, 1(3)*, 233–237.

Landy, F. J., and Farr, J. L. (1980). Performance rating. *Psychological Bulletin, 87*, 72–107.

Levi, L. (1972). *Stress and Distress in Response to Psychosocial Stimuli*. New York: Pergamon Press.

Lueder, R. K. (1983). Seat comfort: A review of the construct int the office environment. Human Factors, *25(6)*: 701–771.

Mackay, C., Cox, T., Burrows, G., and Lazzerini, T. (1978). An inventory for the measurement of self-reported stress and arousal. *British Journal of Social Clinical Psychology, 17*, 283–284.

McNair, D. M., Lorr, M. and Droppleman, L. F. (1971). *Profile of Mood States: Manual*. San Diego, CA: Educational and Industrial Testing Service.

Nagamachi, M., and Imada, A. S. (1992). A macroergonomic approach for improving safety and work design. In *Proceedings of the Human Factors Society 36th annual meeting*. (pp. 859–861). Santa Monica: CA: Human Factors Society.

National Insitute for Occupational Safety and Health. (1990). *Health hazard evaluation report—HETA 89-250-2046—Newsday, Inc.* Washington, DC: U.S. Department of Health and Human Services.

National Insitute for Occupational Safety and Health. (1992). *Health Hazard evaluation report—HETA 89-299-2230—US West Communications*. Washington, DC: U.S. Department of Health and Human Services.

Noro, K., and Imada, A. (1991). *Participatory Ergonomics*. London: Taylor and Francis.

Pan, C. S., Shell, R. L. and Schleifer, L. M. (1994). Performance variability as an indicator of fatigue and boredom effects in a VDT data-entry task. *International Journal of Human-Computer Interaction, 6(1)*, 37–45.

Proctor, R. W., and Van Zandt, T. (1994). *Human Factors in Simple and Complex Systems*. Boston: Allyn & Bacon.

Putz-Anderson, V. (1988). *Cumulative Trauma Disorders—A Manual for Musculoskeletal Diseases of the Upper Limbs*. London: Taylor and Francis.

Quinn, R., Seashore, S., Kahn, R., Mangione, T., Campbell, D., Staines, G., and McCullough, M. (1971). *Survey of Working Conditions: Final Report on Univariate and Bivariate Tables* (Document No. 2916-0001). Washington, DC: U.S. Government Printing Office.

Sainfort, F. and Carayon, P. (1994). Self-assessment of VDT operator health: Hierarchical structure and validity analysis of a health checklist. *International Journal of Human-Computer Interaction, 6(3)*, 235–252.

Salvendy, G. (1973). The Non-Linearity Model of Intellectual Abilities for the Selection of Operators: The G. T. 70/23 Non-Verbal Intelligence Test-Norms, Reliabilities & Validities. *Studia Psychologica, 15*, pp. 79–83.

Salvendy, G. Ed. (1992). *Handbook of Industrial Engineering* (2nd Ed.). New York: John Wiley & Sons.

Salvendy, G., and Knight, J. L. (1983). Circulatory response to machine-paced and self-paced work! An industrial study. *Ergonomics, 26(7)*, 713–717.

Salvendy, G. and Seymour, W. D. (1973). *Prediction and development of industrial work performance*. New York: John Wiley.

Sauter, S.L., Gottlieb, M.S., Rohrer, K. M., and Dodson, V. N. (1983). *The Well-Being of Video Display Terminal Users*. Madison, WI: Department of Preventive Medicine, University of Wisconsin-Madison.

Sauter, S. L., Chapman, L. J. and Knutson, S.J. (1985). *Improving VDT Work—Causes and Control of Health Concerns in VDT Use. Madison, Wisconsin: The University of Wisconsin Board of Regents.*

Sauter, S. L., Schleifer, L. M. and Knutson, S. J. (1991). Work posture, workstation design, and musculoskeletal discomfort in a VDT data entry task. *Human Factors, 33(2)*, 151–167.

Schleifer, L. M., and Amick, B. C. III. (1989). System response time and method of pay: Stress effects in computer-based tasks. *International Journal of Human-Computer Interaction, 1*, 23–29.

Schleifer, L. M., and Ley, R. (1994). End-tidal PCO^2 as an index of psychophysiological activity during VDT data-entry work and relaxation. *Ergonomics, 37*(2), 245–254.

Sharit, J. and Salvendy, G. (1982). Occupational stress: Review and appraisal. *Human Factors, 24(2)*, 129–162.

Smith, M. J. and Carayon-Sainfort, P. (1989). A balance theory of job design for stress reduction. *International Journal of Industrial Ergonomics, 4*, 67–79.

Smith, M. J., Cohen, B. G. F., Stammerjohn, L. W., and Happ J. (1981). An investigation of health complaints and job stress in video display operations. *Human Factors, 23*, 387–400.

Standards for educational and psychological testing. (1985). Washington, DC: American Psychological Association.

Vicente, K. J., Thornton, D. G. and Moray, N. (1987). Spectral analysis of sinus arrhythmia: a measure of mental effort, *Human Factors, 29*(2), 171–182.

Wechsler, D. (1952). *Range of human capacities*. Baltimore: Williams & Wilkins.

Westlander, G. (1994). The full-time VDT operator as a working person: Musculoskeletal work discomfort and life situation. *International Journal of Human-Computer Interaction, 6(4)*, 339–364.

Wilson, J. R., and Corlett, E., Eds. (1990). *Evaluation of human work. London: Taylor and Francis.*

CHAPTER 44

EXPLORATORY SEQUENTIAL DATA ANALYSIS: QUALITATIVE AND QUANTITATIVE HANDLING OF CONTINUOUS OBSERVATIONAL DATA

Penelope M. Sanderson
Swinburne Computer–Human Interaction Laboratory
Swinburne University of Technology
Hawthorn, Victoria, Australia 3122

Carolanne Fisher
U S West
Advanced Technologies
Denver, CO 80202 USA

44.1 USE OF EXPLORATORY SEQUENTIAL DATA ANALYSIS TECHNIQUES IN HUMAN FACTORS

In many human factors investigations we need to observe individuals or groups while they work. For example, in cockpit resource management we may need to observe how crew interactions are managed; in usability studies we may be interested whether an interface affords trouble-free progress toward goals; in product evaluation our concern may be whether instructions are easy to follow and use is unproblematic; and in a videoconferencing environment we may want to judge whether cues to normal conversational turn-taking have been successfully preserved. Such human factors issues lead us increasingly to observational studies in which we must maintain the temporal coherence of people's activities as they work (Grudin, 1990; Olson, Olson, and Kragt, 1992).

However, the analysis of observational data has never been easy. There are many possible techniques that human factors professionals can use, but very few principles for deciding which technique is appropriate in a specific situation. In addition, observational data analysis always threatens to be impractically time consuming and adopting an inappropriate technique only adds to that cost. As a result, human practitioners have sometimes been wary of formal approaches to observational data analysis, whereas they could have benefited from the results.

The enterprise that has been dubbed Exploratory Sequential Data Analysis, or ESDA for short, is a recently proposed integrated framework for thinking about observational data analysis (Sanderson and Fisher, 1994). ESDA is defined as follows:

> *Exploratory sequential data analysis (ESDA) is any empirical undertaking seeking to analyze systems, environmental, and/or behavioral data (usually recorded) in which the sequential integrity of events has been preserved. The analysis of such data (a) represents a quest for their meaning in relation to some research or design question, (b) is guided methodologically by one or more traditions of practice, and (c) is approached (at least at the outset) in an exploratory mode. (Sanderson and Fisher, 1994, p. 255)*

Therefore ESDA is a term meant to synthesize a family of observational and sequential data analysis techniques that already exist, and to help to conceptualize that family more effectively, rather than to propose a new technique. The agenda is to map out the conceptual and methodological possibilities across a wide range of techniques and to provide researchers and designers—including human factors practitioners—with guidelines as to how best to analyze their observational data and how to avoid common traps along the way. Human factors practitioners typically use ESDA in two ways: first, to describe the sequential or temporal patterning of events as people interact with devices to achieve goals (for instance in descriptive task analysis) and, second, to evaluate the impact of human factors interventions on how goals are achieved over time (for instance, when assessing changes in system design, product design, procedures, or training).

44.1.1 Philosophical Relation to Tukey's Exploratory Data Analysis

It is important to explain exactly how ESDA is exploratory and to distinguish exploratory from confirmatory sequential data analysis. The term "exploratory" in ESDA is used in much the same way as it is in Tukey's (1977) Exploratory Data Analysis (EDA). The connection of EDA to observational data was first noted by MacKay (1989) in her work with exploratory video annotation. EDA is now a highly technical area of statistics, but for ESDA we borrow the philosophy of EDA rather than its techniques and we focus on the analysis of data where the ordinal, and possibly also temporal, relations between events are critical.

Tukey's (1977) EDA is distinct from more familiar confirmatory forms of data analysis because of its skepticism of numerical summaries, its use of strong visualizations, its emphasis on hypothesis generation more than hypothesis testing, and its encouragement to the analyst to remain in contact with both the raw data and higher order transformations of it at the same time and to continually explore reexpressions of the data. Tukey talks of data having "smooth" parts—those already accounted for or modeled—and "rough" parts—those unaccounted for. In contrast with this, confirmatory data analysis relies upon numerical summaries and focuses on hypothesis testing using inferential statistics. Exploration and confirmation are complementary: Exploration yields hypotheses that are later worth testing in a confirmatory manner, but always on data sets independent of those on which the hypotheses were originally generated. All the above concerns are just as relevant for exploring sequential data as they are for exploring nonsequential data. Researchers who rely heavily upon sequential data—even those researchers most inclined to use objective confirmatory techniques—all emphasize the importance of first exploring data, especially in a new area of investigation (for example, see Bakeman and Gottman, 1986; Suen and Ary, 1989).

44.1.2 Varieties of ESDA

Many forms of ESDA have become traditions of practice and are already familiar to human factors professionals by name. They include verbal and nonverbal protocol analysis, video analysis, statistical sequential data analysis, process tracing, certain kinds of usability testing, conversation analysis, interaction analysis, classical observational data analysis, some task analysis techniques, and so on. (Some of these techniques are discussed in other chapters of this handbook, such as Chapter 46 on usability testing and Chapter 45 on system effectiveness testing.) For the most part, the forms of ESDA listed above are an amalgamation of specific ESDA techniques and have been applied to the following human factors activities:

- Usability testing
- Analyzing group work phenomena
- Understanding people's use of consumer products
- Understanding how people follow instructions, heed warnings, etc.
- Performing requirements analysis and certain forms of task analysis
- Evaluating operating procedures
- Studying physical activity at workstations
- Analyzing eye movements to study information gathering, attentional deployment, etc.
- Analyzing decision-making and problem-solving behavior in individuals or teams.

Table 44.1 presents the principal ESDA techniques to be discussed in this chapter. The first column itemizes the techniques, categorizing them as either quantitative or qualitative and subclassifying then within each category. The second column provides a pivotal reference for each technique. The third column identifies properties that are needed to use a particular technique. The fourth column is a brief description of the kind of information the technique provides, and the fifth column gives one or more representative

Table 44.1 Principal ESDA Techniques Discussed in This Chapter

Technique	Key reference(s) for technique	Properties needed to use technique[a]	Kind of result or information provided	Sample use(s) in human factors
Quantitative (numerical figures of merit)				
Statistical sequential data analysis				
Markov analysis	Kemeny and Snell (1960), Gottman and Roy (1990)	O C ST N SS	Dependencies between two or more adjacent events	Analyzing motion sequences, eye movements, allocation of attention
Lag sequential analysis	Bakeman and Gottman (1986), Gottman and Roy (1990)	O C ST N SS	Dependencies between two isolated events at specified distance	Finding causes for actions, patterns of multiagent interactions
Grammatical Techniques				
Probabilistic finite state automata	Gaines (1976)	O C ST N SS	Whole system of legal transitions between events	Identifying rules underlying motor sequences
Fisher's cycles	Fisher (1988)	O C ST	All possible sequential paths between two specified events	Identifying and evaluating pivotal events that organize behavior
Maximal repeating patterns	Siochi and Hix (1991)	O C	Longest subsequence that is repeated at least once	Finding keystroke problems in HCI, identifying behavioral units
Regular expressions	Unix *grep* manual	O C (I)	All subsequences matching the sequential pattern specified	Finding tactical modules of behavior and their frequency
Rewriting	Olson et al. (1994)	O C I	Recoding certain subsequences with one summary term	Noting and distinguishing design flaws and their effects
Comparison Techniques				
Sequence comparisons	Ritter and Larkin (1994)	O C P MC	Degree of ordinal similarity between two sequences	Assessing match of behavior to formal operating procedures or model predictions
Item comparisons	Cohen (1960), Suen and Ary (1989)	O/U C P MC N	Degree of similarity in set of paired events	Assessing inter- or intraobserver coding reliability

Qualitative techniques (systematic nonnumerical inferences)				
Cognitive				
Protocol analysis	Ericsson and Simon (1993)	O R/C I	Cognitive processes induced from observable cognitive products	Assessing mental workload; effects of design or training; decision-making processes
Social				
Qualitative classification	Miles and Huberman (1994)	O/U R/C I TS	Rigorous process of simplification exposing critical properties	Classifying critical incidents; identifying causes of error
Constant comparisons (grounded theory)	Strauss (1987)	O/U R/C I TS	Rich qualitative theory fully grounded in data	Identifying social or organizational factors shaping professional decision making
Interaction analysis	Jordan and Henderson (1995)	O R I TS	Account of how agents make sense to each other in context	Identifying effects of information technology on collaborative work
Visualization techniques				
Identations	Bainbridge and Sanderson (1995), Rasmussen et al. (1994)	O R/C (I)	Structure (if any) at coarser grain than data elements	Analyzing task structure, exploring reasoning processes
Tree structures	Sanderson et al. (1994)	O C ST	Family of paths through partly specified sequence of events	Evaluating behavioral consistency and strength of preferred tactics
Timeline displays	Harrison (1995), Kirwan and Ainsworth (1992)	O C	Linear patterning (if any) of actions over time	Seeing temporal coordination, interagent communication patterns
Link diagrams	Kirwan and Ainsworth (1992)	O C ST N	Visualization of set of first-order transitions	Evaluating display and control layouts
Quasi-realism	Moll van Charante et al (1992), Kirwan and Ainsworth (1992)	O R/C I	Reexpression of data partly in physical or functional context	Describing human activity in space or with equipment

[a]O = technique requires ordered data, U = technique can handle unordered data, C = input is coded data, R = input is raw data, ST = sequence must have stationarity, N = need adequate sample of data for statistical analysis, SS = statistical sampling assumptions must be met, TS = thematic sampling assumptions used, MC = technique requires matching codes, NC = technique can handle nonmatching codes, P = technique requires events to be paired, I = interpretive framework required to apply technique, ( ) = technique sometimes needs, or can be helped with, the property indicated, / = either property can be present.

examples of the technique's use in human factors. The entries in Table 44.1 will be more fully explained as each technique is covered.

In the remainder of this chapter we cover the following topics. First, we discuss the generic ESDA process and describe three research traditions—behavioral, cognitive, and social—that underlie different ESDA techniques. Second, we outline some important considerations and challenges that human factors practitioners often face when using ESDA techniques. Third, we present the ESDA techniques listed in Table 44.1—first the quantitative and then the qualitative techniques. Finally, we discuss software support for ESDA and indicate features that make some ESDA software systems particularly successful.

44.2 THEORETICAL AND METHODOLOGICAL TRADITIONS BEHIND ESDA PRACTICE IN HUMAN FACTORS

44.2.1 The Generic ESDA Process

Whatever the technique used and whatever the question, ESDA always involves the basic elements shown in the "generic ESDA diagram" in Figure 44.1. There is a fundamental research or design question. Observations are made while the work of interest is under way; electronic or paper-and-pencil logs may be collected and audiovisual recordings

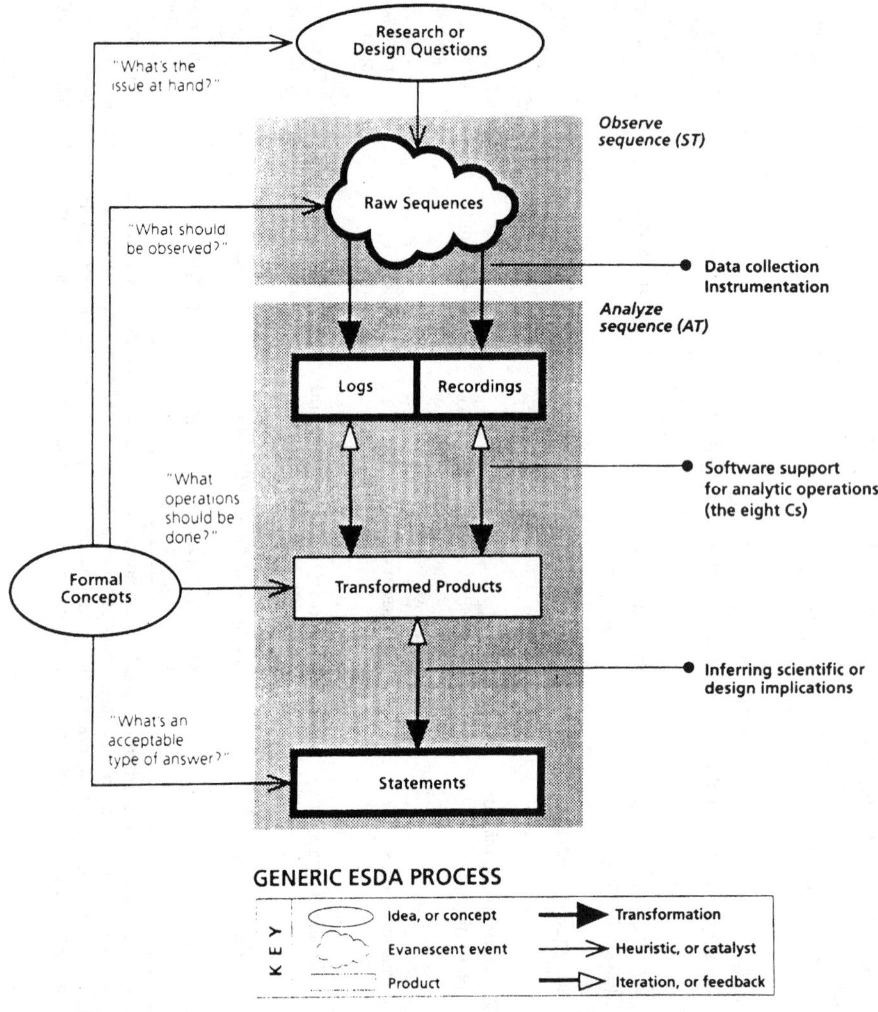

Figure 44.1 The generic ESDA diagram. (From Sanderson and Fisher, 1994.)

may be made. The analyst then transforms the logs and recordings into notes, summaries, coded sequences of events or video segments, etc. The analyst infers answers to the initial research or design question and makes a final statement in the form most relevant for the audience involved, whether scientific colleagues, product designers, software engineers, or managers. Each step is motivated by a question: "What's the issue at hand?" "What should be observed?" "What operations should be done?" and "What's an acceptable type of answer?" How each question is answered is motivated by formal concepts from a particular tradition of performing observational data analysis.

44.2.2 Effect of Theoretical Orientation on Transformations in the Generic ESDA Process

The way each of the four questions in the generic ESDA diagram is answered depends, first, on the question being asked in a specific investigation and, second, on the general theoretical tradition—behavioral, cognitive, or social—that is brought to the investigation. Unfortunately for the human factors practitioner using ESDA for the first time, many descriptions of observational and sequential data analysis techniques are strongly rooted in one theoretical tradition, presenting the methodology associated with that tradition as self-evident rather than simply as one methodological option among many. Yet when choosing an ESDA technique, human factors practitioners should understand the pervasive influence of theoretical tradition at all stages of analysis and should be keenly aware of the wide range of options that exist. It is important to be able to distinguish the concerns, analytic approaches, and forms of verification used in the different ESDA traditions and to adopt the most appropriate approach for the investigation at hand. In addition, adopting useful features of ESDA techniques from other traditions can be a powerful tool for resolving methodological or conceptual impasses during analysis.

Table 44.2 lays out the different methodological options that exist at each point in an investigation in which ESDA techniques are being used (Sanderson and Fischer, 1994). The first column lists the four ESDA questions listed at the end of Section 44.2.1 and the second column identifies ESDA dimensions that correspond to each question. The three rightmost columns represent the three general traditions of ESDA—behavioral, cognitive, and social—that usually suggest different answers to the questions posed. Therefore, each cell entry is a typical (though by no means universal) way that the ESDA question or dimension is handled within a certain tradition. The three traditions are described below. Fuller descriptions can be found in Sanderson and Fisher (1994) and in the references mentioned within each of the following sections.

44.2.2.1 Behavioral

This is the oldest tradition in the ESDA family, having its roots in early ethological studies. Its practitioners have a commitment to the scientific method, often working in the context of experimental designs, and have developed objective, statistically based approaches that often result in confirmatory analyses. The hallmark is the encoding of observables with carefully developed coding schemes, establishing intercoder reliability statistically, and analyzing encoded data using sequential statistics or grammatically based approaches. Researchers in the behavioral tradition have generated a rich array of statistical sequential data analysis techniques and there is now considerable experience in their use. Overviews of the behavioral tradition and its ESDA techniques can be found in Suen and Ary (1990), Bakeman and Gottman (1986), Sackett (1978), Scherer and Ekman (1982), Gottman and Roy (1990), and van Hooff (1982).

In human factors, studies of workplace methods and certain kinds of task analysis have borrowed many concepts from this tradition (Kirwan and Ainsworth, 1992). Further examples of the behavioral tradition being used in human factors investigations are Olson, Herbsleb, and Reuter (1994); Vortac Edwards, and Manning (1994); Cooke, Neville, and Rowe (1996); and Kanki and Foushee (1989).

44.2.2.2 Cognitive

The cognitive tradition of ESDA borrows important elements from the behavioral tradition, such as use of objective coding methods and establishing intercoder reliability statistically. However, its roots are in laboratory studies of problem solving (Newell and Simon, 1982) and its focus is on inferring how performance is mediated by cognitive processes that are observable only indirectly and that may be symbolic. Since the dominant cognitive paradigm has been to model cognition in the individual, there is more

Table 44.2 Methodological Options at Different Points in the ESDA Process.

ESDA Question	ESDA Dimension	Tradition		
		Behavioral	Cognitive	Social
"What's the issue at hand?"	Investigative approach	Use scientific method, often hypothetico-deductive tests, and achieve objective results	Model cognitive processes within the individual over time	Provide accounts of social phenomena; empirical, ethnographic, strong in inductive methods
"What should be observed?"	Setting	Field settings sought to ensure ecological validity; laboratory sometimes used	Systematic laboratory investigations, but also applied field settings	Field observations, emphasize value of observer's participation
	Sampling	Sampling theory and measurement theory used for subject and code selection	Ideally as for behavioral; detail needed often constrains sampling adequacy	Cover representative situations, pursue themes, artifacts or agents; seldom statistical sampling
	Focus of analysis	Objectively identified behavioral observables	Individuals' verbalization and action as evidence for cognitive processes and structures	Social interactions, communicative devices; utterance, gesture, action
"What operations should be done?"	Coding and description	Formal encoding using standardized terms; concern with reliability and objectivity	As for behavioral; debate whether use of context constitutes bias; induction often needed	Interpretation parallels encoding; participant involvement; use of multiple interpretations
	Means of analysis	Sequential, nonsequential statistical methods	Intensive assessment of goodness of model fit for a few subjects	Emphasis on qualitative rather than quantitative analysis; ethnographic
"What's an acceptable type of answer?"	Sources of rigor	Quantitative; concern with replicability and generality of results	Relation of account to extant models of cognition, computational adequacy	Qualitative; plausibility and robustness of account; finding one sound interpretation from many

Source: Sanderson and Fisher, 1994.

emphasis on building an adequate account of individual performance from one or only a few subjects' data, rather than analyzing a full sample of data files, although the latter is valued if it can be managed. Overviews of the cognitive tradition and its ESDA techniques can be found in Ericsson and Simon (1993), Simon and Kaplan (1989), Anderson (1993), and Ritter and Larkin (1994).

In human factors the cognitive approach to ESDA is seen in cognitive task analysis, in GOMS (goals, operators, methods, and selection rules) approaches to human–computer interaction (HCI) (Card, Moran, and Newell, 1983), in the use of models such as ACT-R to model cognitive performance on work-related tasks (Anderson, 1993), and in process-tracing methods used in decision-making studies (Woods, 1993). Examples of the cognitive tradition being used in human factors are Ritter and Larkin (1994), Sanderson, Verhage, and Fuld (1989a); Gray, John, and Atwood (1993); and Woods (1993). Further information about cognitive modeling can be found in Chapter 40 of this handbook.

44.2.2.3 Social

The naturalistic tradition of the social sciences has always emphasized field-based participant observation (Blumer, 1954) and the qualitative rather than quantitative analysis of data. The growth of interpretive sociology and anthropology (Schwandt, 1994) and of ethnomethodology (Garfinkel, 1967) in the last couple of decades has given rise to many new qualitative data analysis techniques, many of which focus on the temporal organization of work (Jordan and Henderson, 1995). In the social tradition of ESDA the focus is on the group—the human and machine agents—rather than on the individual alone. Instead of using formal scientific methods and statistics, naturalistic social methods usually proceed by applying a rich body of theory to data to aid in an inductive process of interpretation.

Overviews of qualitative data analysis methods used in the naturalistic social tradition can be found in Denzin and Lincoln (1994) and Miles and Huberman (1994). Interaction analysis, a qualitative technique increasingly used in computer-supported cooperative work settings, is described more fully in section of this chapter (Jordan and Henderson, 1995). Finally, Strauss (1987) provides a helpful overview of problems and solutions encountered in qualitative data analysis that apply equally well to many forms of ESDA.

The naturalistic social tradition is unfamiliar for many human factors practitioners, who usually have been trained in the behavioral sciences or engineering, but it is increasingly influential in the study of humans in the workplace. Examples of the naturalistic social tradition of ESDA being used to examine issues of interest to human factors are Suchman (1987), Frohlich, Drew, and Monk (1994), Heath and Luff (1992a; 1992b), and Douglas (1995).

44.2.3 Fundamental ESDA Operations (the Eight Cs)

The previous sections distinguish different traditions of ESDA but do not reveal what they have in common, apart from what is shown on the generic ESDA diagram. In fact, there appears to be a set of eight fundamental operations—the eight "Cs"—that help analysts find the appropriate grain of analysis, develop an appropriate coding scheme (if needed), and find meaningful regularities in the data. Not all eight fundamental operations are always used, but an examination of a wide range of studies using ESDA reveals that at least some of them are always used (Sanderson and Fisher, 1994). The eight Cs are, in effect, "smoothing" techniques and are simply an application of Tukey's (1977) EDA principles (see Section 44.1) to sequential data.

The eight Cs are listed below, together with examples of their use. Because these operations are usually performed on raw data, they are usually adequately described only in working notes or technical reports. However, Rasmussen, Pejtersen, and Goodstein (1994) provide an excellent example reproduced in Figure 44.2 and Figure 44.3 (described more fully in Sanderson, 1994) and Tang's (1991) video analysis example using NoteCards (Figure 44.4) provides a further good example.

- *Commenting:* Making unstructured annotations to clarify or interpret some segment of data. Comments can be simple reminders and clarifications or rich theoretical interpretations. When an investigator uses an ESDA technique that requires codes, comments often precede code development and help the investigator develop a coding scheme. However in other ESDA techniques comments are sufficiently rich and informative that they remove the need for encoding. See Figures 44.2 and 44.3 for an examples of comments connected with raw data.

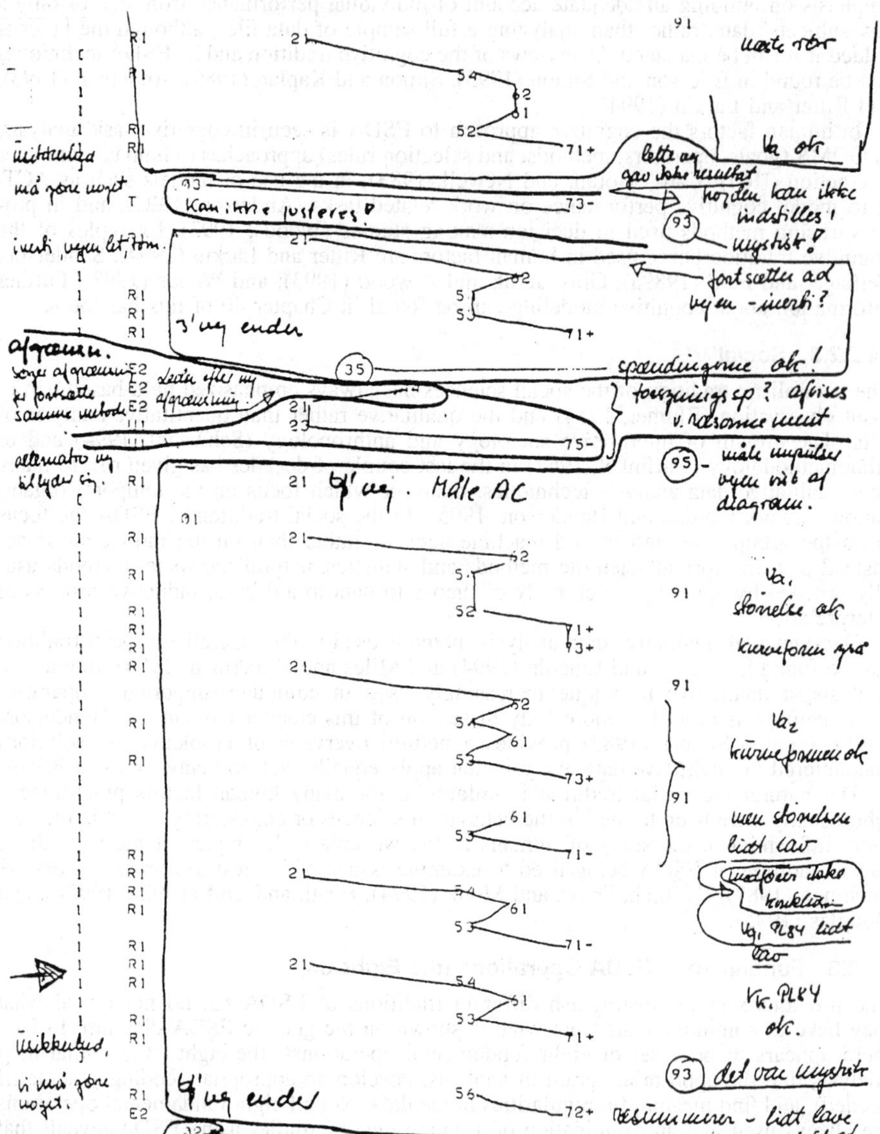

Figure 44.2 Coded and converted verbal protocol based on an electronic troubleshooters' verbalizations. Codes are represented by numbers, which have been collected into numerical decades and printed in columns. (From Rasmussen, Pejtersen, and Goodstein, 1994.)

- *Chunking:* Identifying a set of contiguous events or actions as belonging with each other. Chunking is an important way of finding more coarsely grained structure in observational data. For an outstanding example of visual chunking see Figure 44.2 and for algorithmic approaches to chunking see van Hooff (1982) and Olson et al.'s (1994) discussion of rewriting.
- *Coding:* Reexpressing the data in a smaller, theoretically meaningful set of terms. Coding removes distinctions considered unimportant and systematizes distinctions considered important for the research or design question. Coding either can involve redescription of the raw data in a systematic and theoretically rich language, as seen in Frohlich et al. (1994; their Figures 6 through 10) or can be a set of simple

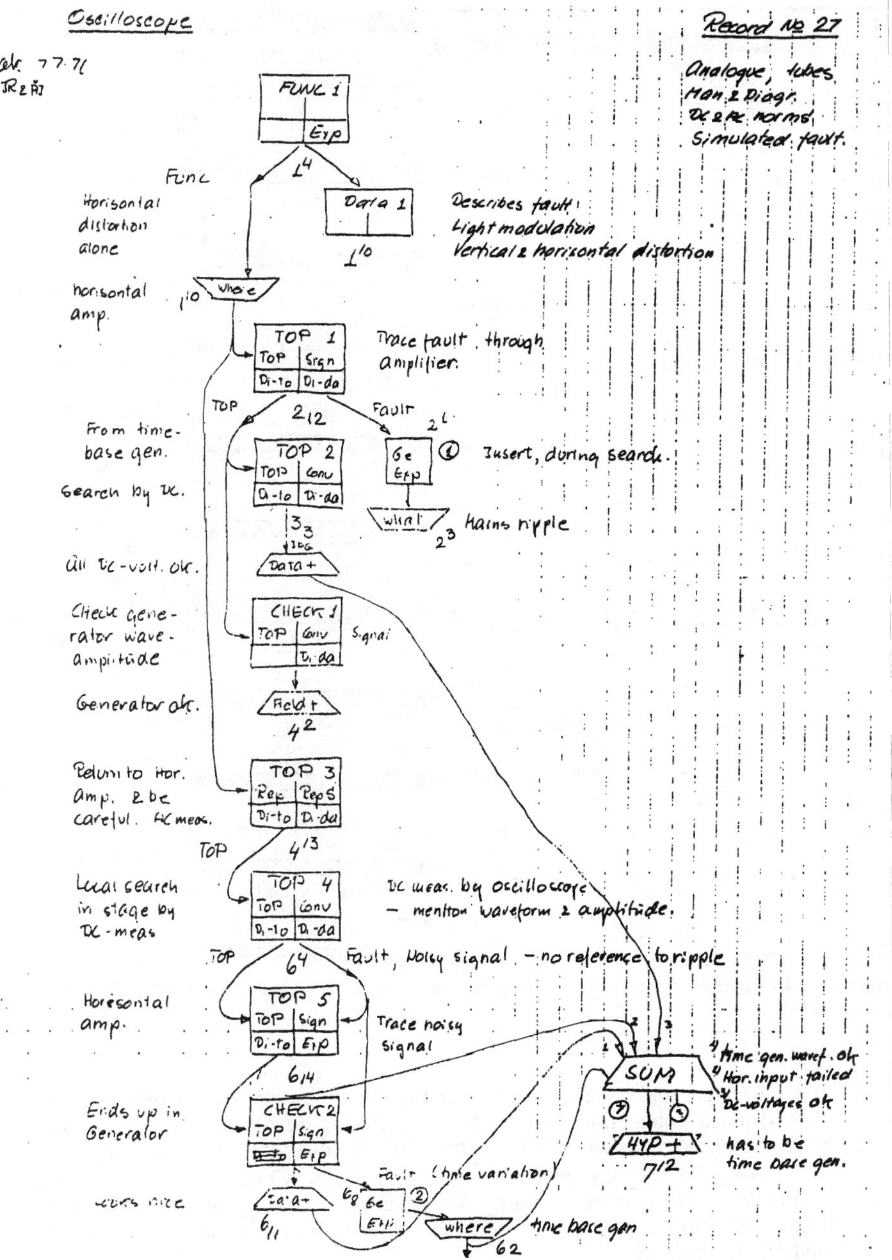

Figure 44.3 Reworking of an electronic troubleshooter's protocol into a format that embodies an emerging theory of fault diagnosis. (From Rasmussen, Pejtersen, and Goodstein, 1994.)

terms on which computations are performed, as in Vortac et al. (1994; their Figure 2), various figures in Cuomo (1994), and Figures 44.2 and 44.4.

- *Connecting:* Noting the relation between noncontiguous events or actions, or the relation between two forms of the data. In the first sense, connections highlight the "threaded" and often discontinuous nature of sequential activity—activities may be suspended while a new issue is dealt with, and resumed afterwards. See

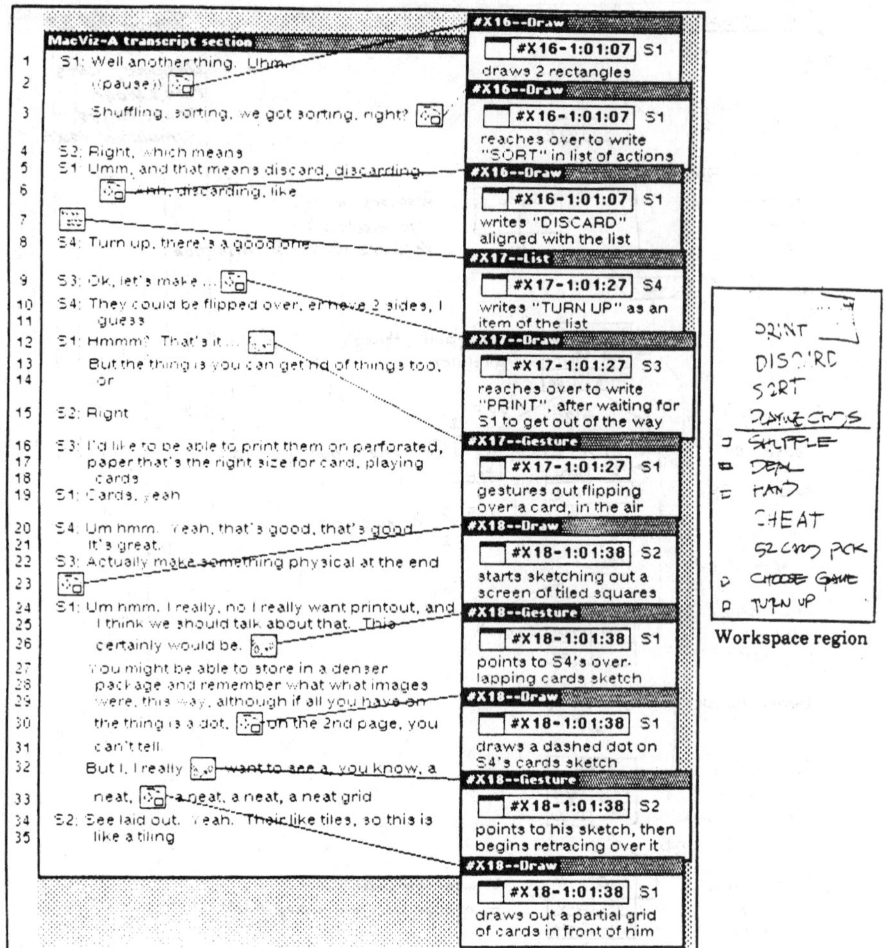

Figure 44.4 Record of interaction during a design meeting, analyzed with NoteCards. (From Tang, 1991.)

Roth, Bennett, and Woods (1988; their Figures 1 through 18) for examples. In the second sense, connections point to the same data in different forms during analysis, as in Figure 44.3, where small numbers refer to location in the original transcript.

- *Comparing:* Making comparisons between two sets of events or actions to test for similarity of ordering or similarity of content. Comparing can be important when testing the predictions of a model. See Ritter and Larkin (1994; their Figures 7 and 10) for examples.
- *Constraining:* Temporarily bringing forward a subset of the data for display or analysis and filtering out the remainder of the data. Constraining can help the analyst discover threads or discern higher order structure in subsets of the data.
- *Converting:* Reexpressing the data in an alternative modality or format to highlight different properties. Conversion to graphical form often helps analysts discover patterns, as shown in Figures 44.2 and 44.3, and manipulating visualizations can be a powerful pattern-detection tool: see Sanderson et al. (1994; their Figure 12) and Olson and Olson (1991; their Figure 3) for examples. Conversions are often final presentation formats, as in van Hooff (1982; his Figure 7.6) and Olson et al. (1994; their Figure 8).

- *Computing:* Performing algorithmic operations on the data that involve either statistical calculations or symbolic inferences. See Vortac et al. (1994), Olson et al. (1994), Cooke, Neville, and Rowe (1996) for discussions of typical procedures.

An awareness of this full range of manipulations can help human factors practitioners make fuller use of their data and take new routes if one approach to analysis does not work. For example, if an investigator carries out detailed encoding without having the kind of overview that chunking and commenting can provide, it may be very difficult to identify less detailed temporal or thematic structuring in the data. If data are already coded, or are sufficiently objective to be manipulated, then trying some conversions to different modalities or formats can reveal structure that may be missed by some sequential data analysis routines but picked up nicely by others. Constraining data displays or analyses to certain events or people may reveal connections between disparate parts of the data that would otherwise have gone unnoticed.

44.3 CONSIDERATIONS AND CHALLENGES WHEN CONTEMPLATING ESDA

It requires experience for an investigator to navigate his or her way confidently through the different traditions of ESDA and the fundamental operations identified in the eight Cs. Moreover, discovering how to apply these traditions and operations in the context of new research or design questions takes time even for an experienced investigator. A further concern is the amount of time it takes to review and analyze the data. Therefore, investigators have often been wary of using ESDA techniques. It is much easier and quicker, for example, to conduct interviews or to administer subjective assessment instruments such as questionnaires and surveys. As Sanderson and Fisher (1994) have proposed, ESDA should not be undertaken lightly but instead only when:

- The data needed to answer a question cannot be gathered any other way—for example, when the sequencing and temporal organization of activity are central to a question.
- The analyst has identified or adapted a viable technique for moving from questions to statements in the context of available logs and recordings (see Figure 44.1) and is competent to implement the technique.
- It is clear that the practical or conceptual benefit of using an ESDA technique is worth the extra time involved and that time is available. (Sanderson and Fisher, 1994, p. 258)

In the next sections we survey some of the principal considerations human factors practitioners must take into account when contemplating ESDA and we outline ways to face some of the challenges.

44.3.1 Qualitative vs. Quantitative ESDA Techniques

When facing the analysis of observational data for the first time, it can be unclear whether qualitative or quantitative ESDA techniques should be used. For example, human factors practitioners are often drawn to the quantitative techniques of the behavioral tradition, but the data may be too sparse or too idiosyncratic to yield to such approaches. Trying to use statistics to analyze the data as a linear sequence of events will fail (see Rasmussen and Jensen, 1973, for such an account). Investigators with a cognitive training may lean toward developing a cognitive process model, but the events observed may depend so much on interpersonal interactions and reveal so few cognitive products that the appropriate data are unavailable. Sometimes the truly informative parts of an observation are errors or misunderstandings that happen only once (Hutchins and Klausen, 1992) and so must be either interpreted qualitatively in the light of available theory or accounted for with a cognitive model, rather than dealt with quantitatively. The latter approaches can be thought of as nonlinear because relations are inferred that go beyond the linear ordering of events (Roth et al., 1988). However, when there are enough data for quantitative estimates to be reliable, the resulting information about frequencies and durations of critical events and about sequential patterns of activity can be powerful input to design decision making.

Overall, an investigator using ESDA needs to recognize when qualitative vs. quantitative methods are needed and to combine them adaptively to answer the research or design questions at hand (Denzin and Lincoln, 1994). Being aware of the different ESDA traditions summarized in Table 44.2 and the techniques summarized in Table 44.1 lets an investigator more freely consider alternative techniques if the first one does not work. Some helpful discussions of the advantages and disadvantages of different techniques—especially cognitive vs. social approaches—can be found in Suchman (1987), Tang (1989), Douglas (1995) and Lawrence, Atwood, Dews, and Turner (1995).

44.3.2 Handling the Volume of Data

As soon as electronic or audiovisual recording of observed events becomes possible, the amount of data available threatens to become overwhelming. It is tempting to try to analyze it all in detail. However, the fact that the data have to be experienced over time to be analyzed means that analysis can be extremely time consuming. Practitioners talk of analysis time to sequence time (AT:ST) ratios to describe how many time units it takes to analyze one unit of real recorded time. Ratios range from 100:1 and even more for detailed scientific studies to around 2:1 for the most time-pressured kinds of commercial usability studies (Smith, Smith, and Kuptsas, 1993; Weiler, 1993; Roschelle and Goldman, 1991).

One factor that significantly increases AT:ST ratios is the need to transcribe verbalization or detailed nonverbal activity; rough transcripts can be produced with AT:ST ratios of 8:1 (Swarts, Flower, and Hayes, 1984) but more detailed transcripts may go up to 20: 1 or higher (Jordan and Henderson, 1995). One solution is to avoid transcription by using ESDA software environments that provide instant access to digitized audiovisual segments (MacWhinney, 1991; Sellen, 1992, 1995; Sanderson and Mainzer, 1996); another is to reflect whether transcription is strictly scientifically necessary. Whenever encoding is undertaken, a second factor that increases AT:ST ratios is coding at an unnecessarily fine grain. The encoding effort may be abandoned or the resulting analyses may not be meaningful. Finding the right grain of analysis is discussed in the next section.

44.3.3 Finding an Appropriate Descriptive Language and Grain of Analysis

The most important conceptual tools for finding an appropriate descriptive language and grain of analysis are having a well-defined research or design question and having a notion of the general class of statements that would answer it (statistical results, logical argument, or critical anecdotes).

The transformational operations identified as the eight Cs can then help the investigator identify the most appropriate grain of analysis and the most useful descriptive language that will answer the question persuasively. ESDA practitioners from diverse traditions have insights to offer here. Bakeman and Gottman (1986) have a helpful discussion on choosing and applying coding schemes, noting that no amount of clever analysis will "wrest understanding" from an inadequate coding scheme. To this end, they warn against borrowing coding schemes unless there is a very good reason for doing so:

> *sometimes the development of coding schemes is approached almost casually and so we sometimes hear people ask: Do you have a coding scheme I can borrow? This seems to us a little like wearing someone else's underwear. Developing a coding scheme is very much a theoretical act, one that should begin in the privacy of one's own study, and the coding scheme itself represents an hypothesis, even if it is rarely treated as such. After all, it embodies the behaviors and distinctions that the investigator thinks important for exploring the problem at hand. (Bakeman and Gottman, 1986, p. 19)*

Even though their interaction analysis technique does not involve coding, Jordan and Henderson (1994) have discussed the kinds of activities and events that are significant in workplace interaction and that an analyst might attend to as markers for important interactions. In the context of cognitive engineering studies of cockpits, control rooms, and operating theaters, Woods (1993) discusses how descriptive languages can embody theoretical concepts. Finally, Strauss (1987) provides a detailed description of developing an interpretive coding scheme in the context of grounded theory (see Section 44.5).

A coding scheme intended to reveal sequential relations is usually focused on events at a certain grain of analysis, or "frequency band" as Sanderson and Fisher (1994) call it. Rasmussen et al. (1994) discuss the process of finding the most effective grain of analysis in an electronic troubleshooting example that is more fully described in Sanderson (1994). Once a coding scheme has been established and encoding performed, then a wide array of analysis options exist, which are covered in the next sections.

44.4 ESDA AND QUANTITATIVE ANALYSIS

When ESDA is influenced strongly by the behavioral and cognitive traditions, events are usually objectified and given codes, as discussed above. If the events are well ordered and if there are enough of them, then there are many quantitative or computational techniques that can be used to find structure in the observational data and so help infer its meaning. This section reviews the principal quantitative and computational analysis routines in use for such data—of the eight Cs, these are computations on codes that have often been constrained in some way. Although many routines include procedures for confirmatory analysis, they can be very powerful tools for exploration as well. The basics of how these routines work on individual data sequences are presented, followed by an example of their use in exploring the structure of a very simple data set. For further reading, a seminal on authoritative account of these and other related methods can be found in van Hooff (1982). Gottman and Roy (1990) provide a thorough survey, providing full statistical details. Shorter descriptions with examples can be found in Bakeman and Gottman (1986) and Olson et al. (1994).

In many of the data analysis examples that follow we have used letters of the alphabet to stand for codes that an investigator might have developed. Using simple letter codes gives clarity to the examples and saves space. However, we have supplemented each example with references in the human factors literature that show how the technique in question has been used.

44.4.1 Sequential Data Analysis Techniques

Among the many statistical sequential data analysis techniques available, Markov analysis and lag sequential analysis are presented in detail here because they are most widely used. However, other statistical sequential techniques that have been used successfully in both exploratory and confirmatory ways include time series analysis (Box and Jenkins, 1976) and log-linear modeling (Fienberg, 1980). For example, time series analysis has been used by Lee and Moray (1992) to model how operators' trust of automation is sustained or eroded depending upon prior events. Log-linear modeling has been used to study differences in communication patterns between effective and ineffective flight crew (Bowers, Jentsch, Salas, and Braun, 1995) and between sequences of discussion topics in design meetings in which participants either were or were not supported by electronic meeting technology (Olson et al., 1994).

44.4.1.1 Markov Analysis

Markov analysis (Kemeny and Snell, 1960) is a family of techniques for determining whether transitions from one or more strictly ordered antecedent (previous) events to a consequent (following) event happen more often or less often than by chance. For example, an investigator using eye movement data to study visual scanning patterns in the cockpit may be interested in which instrument the pilot looks at after looking at weather radar. In a first-order Markov model one examines the dependence of the consequent code, which we will call B, on a single antecedent code, which we will call A. The conditional probability of finding B after A is expressed as $p(\mathrm{B}|\mathrm{A})$. In a second-order Markov model one examines the dependence of a consequent code C on two strictly ordered antecedent codes, AB, which would be expressed as $p(\mathrm{C}|\mathrm{AB})$. Higher order transitions can also be calculated.

Individual Transitions

Gottman and Roy (1990) propose the following formula for testing sequential dependencies in specific code sequences. If interested in the sequence ABCD, one can test the conditional dependency of D on C (first order), D on BC (second order), and D on ABC (third order). The simplest case is to test for first-order dependency, here expressed with a general formula:

$$z = \frac{p(x_i|y_{i-1}) - p(x_i)}{\sqrt{\dfrac{p(x_i|y_{i-1})(1 - p(x_i|y_{i-1}))(1 - p(y_{i-1}))}{Np(y_{i-1})}}}$$

where $p(x_i)$ is the probability of finding item x at location i in a sequence, $p(x_i|y_{i-1})$ is the conditional probability of finding item x at location i given that item y is found immediately beforehand at location $i - 1$, and $p(y_{i-1})$ is the probability of finding item y in location $i - 1$. This formula can be generalized to individual higher-order transitions by substituting the appropriate higher-order terms. For example, to test whether the second-order transition $p(x_i|z_{i-2}y_{i-1})$ adds further dependency over $p(x_i|y_{i-1})$, in the above equation one would substitute $p(x_i|z_{i-2}y_{i-1})$ for $p(x_i|y_{i-1})$, $p(x_i|y_{i-1})$ for $p(x_i)$, and $p(z_{i-2}y_{i-1})$ for $p(y_{i-1})$. Equivalent substitutions would be made for higher-order tests.

Mean First Passage Time

One figure of merit that Markov analysis provides is the mean first passage time, or MFPT. In this case we view a sequence of codes as a sequence of states. If a system is currently in state i (which might be code A, for example) then the mean first passage time from state i to state j (MFPT_{ij}) is the mean number of states through which the system will pass in order to get from state i to state j for the first time. A variance is associated with each MFPT. Details of the calculations can be found in Kemeny and Snell (1960).

Moray, Richards, and Low (1980) applied MFPT to measure the workload of radar operators. They measured empirically the rate at which operators forgot the location of each radar echo. They then measured (and modeled) visual attention in terms of eye movement transitions. They used the MFPT and its variance to predict the probability that an interval would elapse between successive fixations on each echo that would be so long that significant forgetting would occur. Moray et al. were able to predict when radar operators would lose track of aircraft on their displays.

Overall Sequential Dependency

It is possible to test the overall strength of sequential dependency at different orders in the data—first, second, third, etc.—rather than focusing on individual transitions. Gottman and Roy (1990) describe an information-theoretic technique for identifying the transition order that captures the most sequential structure in a data sequence. The information, H, at a number of different transition orders from single symbols to third- or fourth-order transitions is calculated using the formula:

$$H_{\text{order}} = -\Sigma\, p_i \log_2 p_i$$

where p_i is the probability of each transition at the order indicated. The larger H_{order} is, the more information (uncertainty) there is in the sequence, and thus the less sequential dependency (redundancy) there is at that order. Then, the difference in information between adjacent orders is calculated:

$$H_n = H_{\text{order}+1} - H_{\text{order}}$$

The larger H_n is, then the less added sequential dependency there is at $H_{\text{order}+1}$ than at H_{order}. Finally, the difference between the amount of information gained between one set of levels and the next is calculated:

$$T_{\text{gained}} = H_{\text{n}} - H_{n+1}$$

T_{gained} and the total number of observations, K, are multiplied by a constant to provide a X^2 sample statistic to test against a criterion population chi-squared value, χ^2:

$$X^2 = 1.3863KT_{\text{gained}}$$

The pattern of results for X^2 across different values of T_{gained} tells us how many events into the past we can go and still be getting significantly more information about the sequence. Even when one is not trying to determine the order of sequential constraint in a sequence, Markov analysis is useful for finding individual higher order sequences that can provide clues as to the structure in a data set at given points.

44.4.1.2 Lag Sequential Analysis

As Gottman and Roy (1990) have commented, "lag [sequential] analysis is a trick to get around the problem of not having enough data for a complete Markov analysis of second or third order" (p. 100). Lag sequential analysis can be used, for example, to see whether a pilot tends to look at the flight plan approximately three steps after looking at the weather radar, regardless of what he or she has looked at directly after the flight plan. Therefore, instead of seeking completely defined subsequences in the data, such as $p(\text{B}|\text{CA})$ for the sequence CAB, with lag sequential analysis one can examine the data to see whether B occurs three steps after C, $p(B_{t+3}|\text{C}_t)$ without specifying what happens one step after C. Therefore the probability $p(\text{B}_{t+3}|\text{C}_t)$ aggregates over everything that might happen one step after C. Full treatments of lag sequential analysis (LSA) can be found in Bakeman and Gottman (1980), Sackett (1978) and Gottman and Roy (1990) as well as in further articles mentioned herein.

Usually when performing LSA the transitional frequencies or probabilities for a range of lags are calculated, as shown in Figure 44.5. The central code around which events at various lags are examined is the "key" or "from" term and the event being plotted at different lags around the key term is the "target" or "to" term. As Figure 44.6 makes clear, if the sequential dependency happens at approximately, rather than precisely, the lag in question, it will nonetheless be seen. An additional advantage is that analyses can seek dependencies across parallel data streams, such as between a pilot's verbal commands and a first officer's actions.

To test whether an individual transition is more or less likely to occur than would be expected by chance, one needs to take into account the base probability of the antecedent code or code sequence and the base probability of the consequent code.

The expected probability must take into account the base probabilities of the antecedent and consequent events and so is the product of the marginal probabilities. The way it is calculated depends on whether a code can be seen more than once in sequence (e.g., ABBBC) (see Bakeman and Gottman, 1986, for the formulae).

A z-statistic can be used to test whether an observed transitional probability is greater or less than would be expected by chance (see Allison and Liker, 1982; Bakeman and Gottman, 1986; Gottman, 1980; and Gottman and Roy, 1990). A z-score often referred to in the context of LSA as Allison and Liker's z is widely used and is calculated with the following formula:

$$z = \frac{p(\text{target}|\text{key}) - p(\text{target})}{\sqrt{\dfrac{p(\text{target})(1 - p(\text{target}))(1 - p(\text{key}))}{N \times p(\text{key})}}}$$

where $p(\text{target}|\text{key})$ is the condition probability of finding the target term given the key term, $p(\text{target})$ is the overall probability of the target term and $p(\text{key})$ is the overall probability of the key term. However, also available is Markov z (Faraone and Dorfman, 1987). This is an alternative measure of lagged dependency useful when the sequences from which the key and the target terms are taken are separate data sequences (for example, when the key term is an action of an airplane pilot and the target is an action of a first

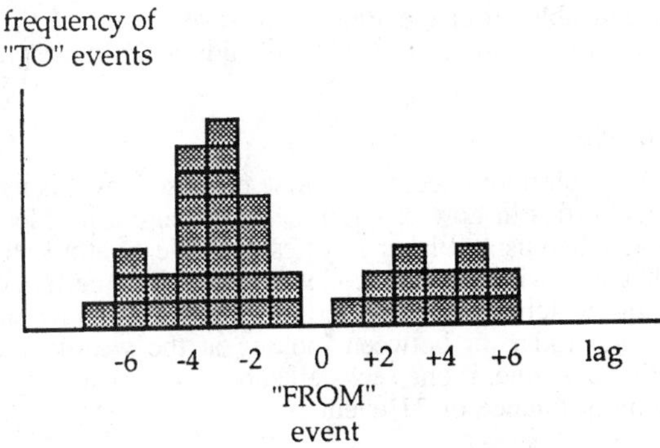

Figure 44.5 Histogram integrating the descriptive results of several lag sequential analyses of the relation between a key term and a target term at several lags.

officer). Markov z takes into account the degree of first-order autocorrelation in the each of the sequences. For more details about the relation between Allison and Liker's z (1982) and Faraone and Dorfman's Markov z (1987), see Faraone and Dorfman (1987).

Continuous sequences suggested by LSA can be verified without having to use Markov analysis by finding (a) significance at the lags of interest and (b) significant first-order transitions between the items. Techniques for this have been suggested by Gottman and Roy (1990) and other authors.

LSA has been used by Olson et al. (1994) in a study of human factors in collaborative design to distinguish two different clusters of sequential dependencies in the way discussion topics unfold during design meetings.

44.4.2 Grammatical Techniques

Grammatical techniques have proved useful for discovering structure in sequences and are effectively a form of chunking. Several approaches are possible.

- Build up a high-level (coarser grain) description of the data through successive rewritings of meaningful subsequences found in the data. This approach requires generating a set of rewrite rules and successively applying them to the data.
- Propose smaller grammars that account for subsequences within the larger data sequence. Regular expressions have proved to be very useful for this, particularly as they are ready at hand in many operating systems and are easy to implement.
- Use grammatical inference techniques that infer a full grammar for a data sequence. With all the variability in behavioral data this can be extremely difficult to do. However, some researchers have taken the approach of building probabilistic finite state automata to describe behavioral sequences.

44.4.2.1 Probabilistic Finite State Automata

One grammmatical approach is to try to summarize a data sequence succinctly as a probabilistic finite state automation (Gaines, 1976; Hingston and Lees, 1994; Patrick and Chong, 1991). A finite state automation (FSA) is a formal description of a fixed set of transitions between states, as for example, one might generate when describing the physical location of the actions a manual worker performs in the course of a shift. A probabilistic finite state automation (PFSA) is usually presented graphically: acts or events are shown on the arcs and states on the nodes, and the transitions show the legal sequence of actions. A PFSA includes information about the probability with which each arc is traveled (see Hingston and Lees, 1994, for details).

Given a series of strings of data, a PFSA can be induced algorithmically. The resulting graph can be simplified yet still adequately describe all the sequences. However, there are usually many possible simplifications and some way is needed to choose the best one. The Wallace Information Measure (Wallace and Boulton, 1968) has been used by Hingston and Lees (1994) and Patrick and Chong (1991) to find the PFSA with the minimum message length—an information-theoretic figure of merit. Patrick and Chong have developed a heuristically based variant to find good rather than optimal PFSAs in real time and have used it to analyze tactics in Australian rules football games on the fly. Patrick and Chong's system also allows the retrieval of recorded video segments on the basis of specifying strings producable from the PFSA. However, none of these approaches addresses the issue raised by Olson et al. (1994) of finding the most practically and theoretically useful result.

44.4.2.2 Fisher's Cycles

Fisher's cycles are an exploratory technique first proposed by Fisher (1988, 1991) for finding sequential regularities in how programmers structure a problem in the process of software development. When using Fisher's cycles, the investigator seeks all subsequences in a data sequence that start with a certain code and end with a certain code. For example, we might wish to know which instrument the pilot looks at between each occasion he or she looks at the weather radar, or between looking at the weather radar and the flight plan. Using a simplified example, if one seeks all subsequences starting with A and ending with D in the following sequence of 31 events:

AABCDCCCDXYABCCDDDXAABDDDCDYXADC

the following subsequences are returned if we assume that A and D cannot be included in the middle of the sequence:

ABCD
ABCCD
ABD
AD

From this we learn that B usually follows A, that C never follows A, and that where C does occur, it can repeat. Such information is usually only a small part of a much larger picture, but it can help an analyst start to build up an understanding of the regularities in a data sequence. For example, Fisher (1988, 1991) used cycles to discover the structure of programmers' thinking as they represented a programming problem before starting to code it.

The regularities found in a Fisher's cycles analysis may suggest rewritings (see Section 44.4.2.5) that could be performed or even a regular expression (see Section 44.4.2.4) that could capture the sequential relations seen. The relation between Fisher's cycles and regular expressions is outlined in the next section and graphical methods for displaying cycles will be discussed in Section 44.5.3.2.

44.4.2.3 Maximal Repeating Patterns

A related approach to uncovering higher order structure in sequential data is to use the maximal repeating pattern technique developed by Siochi and Hix (1991). With this technique one seeks the longest pattern in the data that repeats itself at least once elsewhere in the data.

Depending upon context, such repeating patterns can either indicate problems in system design that are leading to overly long series of actions (Siochi and Hix, 1991) or indicate constellations of activity suggesting habitual tactics that should not be disrupted when functions are reallocated or when interface systems are redesigned (Vortac et al., 1994).

44.4.2.4 Regular Expressions

Regular expressions are small grammars that describe rules governing the ordering of events in a data sequence. For example, there may be several times in the data where, after looking at the weather radar, the pilot tends to look at either the rate of climb or the attitude indicator several times, then the radio frequency setting once, and then the flight plan. Regular expressions can be applied to a data sequence to see whether any subset of the observed sequence can be matched by the rules in the regular expression. So, for example, in the following sequence of 31 events:

AABCDCCCDXYABCCDDDXAABDDDCDYXADC

the subsequences AABCDCCCD, ABCCDDD, and AABDDDCD would all be matched by the regular expression:

$$A^{+}B(C|D)^{+}$$

and the subsequences XY, X, and YX by the regular expression:

$$(X|Y)^{+}$$

where $^{+}$ means one or more instances of the preceding expression, | is an "or" operator, and parentheses enclose areas to disambiguate the range of other operators.

Together these two regular expressions describe the whole sample data sequence except for the ADC at the end. A well-known implementation of regular expressions is the UNIX "grep" function, and readers are referred to that environment for further information and to try it out. A human factors example is provided by Moray, Sabadosh, and Kijowski (1992) who used a simplified form of regular expressions in a simulated helicopter rescue task to summarize the order in which subjects scanned their basic flight instruments.

Note that the Fisher's cycles analysis is a special case of applying a regular expression to a sequence. If "?" indicates a wildcard character, * indicates "zero or more of," S is the starting term, and E is the ending term, then all Fisher's cycles can be expressed in the regular expression:

S ?* E

44.4.2.5 Rewriting

Rewriting "smooths" a data sequence by successively removing low-level detail by finding a simpler description for it. For example, when an aircraft pilot shows a sequence of eye movements between the rate of climb and the attitude indicator, we could summarize these in the single term "checks primary instruments." In the letter code previously used, event sequences can be rewritten until the whole data sequence is summarized in nine terms. (Regular expression notation is used here for sake of clarity, but usually recursive rewrite rules are used.)

rewrite rule	*data sequence after rewriting(s) at left*
	AABCDCCCDXYABCDDDXAABDDDCDYXADC
A ← A+	rewrite repeated As as single instances
C ← C+	rewrite repeated Cs as single instances
D ← D+	rewrite repeated Ds as single instances
	ABCDCDXYABCDXABDCDYXADC
E ← AB	rewrite AB sequences as "E"
	ECDCDXYECDXEDCDYXADC
F ← (C|D)+	rewrite sequences of C or D as "F"
	EFXYEFXEFYXAF
G ← EF	rewrite sequences of EF as "G"
	GXYGXGYXAF
Z ← (X|Y)+	rewrite sequences of X or Y as "Z"
	GZGZGZAF

In the above case, the analyst could have used quite different rewriting rules, which could have led to a different result. Such rewritings can continue until the analyst has a description of the data at the grain of analysis that is most meaningful and informative, given the research questions at hand. Olson et al. (1994) provide an informative discussion of the use of rewriting techniques in ESDA, using the Definite Clause Grammar formalism (Pereira and Schieber, 1987) implemented in Prolog and LISP and they illustrate this kind of ESDA with design meeting data. They also suggest statistical figures of merit to judge the effectiveness of rewriting, such as the percentage reduction achieved through a series of rewritings.

44.4.3 Comparison Techniques

44.4.3.1 Sequence Comparisons

Recently there has been considerable interest in finding effective procedures for comparing the similarity of two data sequences (John, 1994; Johnston, 1994; Ritter and Larkin, 1994). The motivation for comparing two sequences can be any of the following:

- Comparing different encodings of the same event sequence (intraobserver reliability, interobserver reliability). We might compare two coders' encodings of a pilot's eye-scanning pattern to ensure they are the same.
- Judging the effect of some manipulation on a person or group's performance on the same task (training, practice, interface redesign, etc.). We might test the effect on scanning of a conventional cockpit as opposed to a glass cockpit.
- Comparing the performance of different people or groups on the same task (experts vs. novices, cultural differences, etc.). We might compare the eye-scanning pattern of novice pilots with that of expert pilots.
- Evaluating how closely a person or group's performance conforms to sequence predicted by a model or a performance standard, such as an ACT-R (Anderson, 1993) or Soar-based (Newell, 1990) model of some cognitive task, or a standard

operating procedure such as a cockpit checklist. We might build a statistical model (e.g., time series) or a cognitive model to predict eye scanning and compare it with observed eye scanning patterns.

There are two problems to be dealt with when comparing two sequences: event alignment and event match. Alignment is the process of determining which events in one data sequence should be paired with which events in another, and event match is whether the contents of paired events are the same. Event alignment may proceed in the absence of event matches, as in cases 1 and 2 below, or event matches may be part of the condition for event alignment, as in cases 3 and 4 below.

1. *Alignment by serial position:* Events are aligned simply in order of occurrence with no regard to their timestamps (if any) or contents. This kind of alignment yields valid results only if the events at a certain serial position in the two sequences are, in principle, logically or temporally related. Reliability statistics can be performed validly on the results if the events are codes.
2. *Alignment by event timestamp:* Events are aligned if their timestamps are the same or within some time tolerance, with no regard to their contents. This approach provides robustness when events may have been sampled at a different grain and at slightly different times in the two sequences. Reliability statistics can be performed validly on the results if the events are codes, but allowance must be made for unmatched events.
3. *Alignment by event contents:* Events are aligned by their contents, often in such a way as to maximize the number of events matched and with no regard to timestamps. This approach is used if both sequences involve the same task, but the task can be carried out at any speed. This kind of alignment would be used for testing the predictions of behavioral models or adherence to standard procedures. Various algorithms have been proposed for this kind of alignment (see review by Kruskal, 1983) but those used for ESDA have been longest common subsequence (LCS) algorithms (Card et al., 1983; Hirschberg, 1975).
4. *Alignment by contents and timestamp:* Events are aligned by their contents only if the two events' timestamps are within some time tolerance. This approach is used if it is expected that the sequences should unfold at the same speed, as when external events pace work or when a model such as the Model Human Processor (Card et al., 1983) provides time estimates.
5. *Manual alignment by recognizing relatedness:* The analyst may recognize that certain unaligned events refer to the same real events and so force an alignment between them. This procedure can be used to avoid completely wrong alignments in the latter two cases. John's (1994) Trace and Transcribe program provided a flexible environment for manual alignments—including several kinds of alignment disallowed by algorithmic approaches such as the alignment of "crossed pairs" of events.

Ritter and Larkin (1994) have developed a powerful software environment (Spa-mode in the Soar/MT environment) for determining best alignments of sequences by event contents, with human and alignment algorithms working interactively. They used their software to evaluate Soar-based models of how humans browse an online help system.

Using a slightly different alignment algorithm, Johnston (1994) has implemented a simplified version of Ritter and Larkin's (1994) environment within MacSHAPA (Sanderson et al., 1994) and has extended it to handle timestamps and a variety of item comparison statistics. Figure 44.6*b* shows an alignment per Johnston (1994) which has adjusted the sequence alignments to maximize the number of matches regardless of the events' timestamps. In Figure 44.6*c* timestamp information has been used so that events match only if they contain the same code and are within 1 s of each other.

Alignments of type 1 and 2 in the list above are usually performed as a first step in performing item comparison and are discussed in the next section. Alignments of type 3, 4, and 5, which have already included contents as a criterion for alignment, are usually performed to judge sequential similarity. Sequential similarity may be measured simply as the proportion of events in each sequence that have a match in the other sequence or the degree to which the alignment has lengthened the data sequence over the length of the longer of the two sequences.

subject 1	subject 2
A (00:04)	A (00:02)
B (00:10)	A (00:15)
C (00:20)	B (00:22)
B (00:32)	C (00:36)
A (00:57)	A (00:55)
D (01:16)	D (01:20)
E (01:59)	E (01:50)
F (02:30)	F (01:58)

(*a*)

subject 1		subject 2
A (00:04)	-----	A (00:02)
		A (00:15)
B (00:10)	-----	B (00:22)
C (00:20)	-----	C (00:36)
B (00:32)		
A (00:57)	-----	A (00:55)
D (01:16)	-----	D (01:20)
E (01:59)	-----	E (01:50)
F (02:30)	-----	F (01:58)

(*b*)

subject 1		subject 2
A (00:04)	-----	A (00:02)
		A (00:15)
B (00:10)		
C (00:20)		
B (00:32)	-----	B (00:22)
		C (00:36)
A (00:57)	-----	A (00:55)
D (01:16)	-----	D (01:20)
E (01:59)	-----	E (01:50)
		F (01:58)
F (02:30)		

(*c*)

Figure 44.6 Different alignments of two sequences; letters are coded events and numbers are timestamps for each event. (*a*) Alignment by serial position, (*b*) alignment to maximize the number of pairs with matching values, (*c*) alignment to maximize the number of pairs with matching values whose timestamps are also within 10 seconds of each other.

44.4.3.2 Item Comparisons

A perennial concern in sequential data analysis is the reliability of encoding. Reliability analyses assume a sequence alignment of type 1 or 2 in the list above. Strictly speaking, item comparison techniques are not sequential analyses, despite assuming a valid alignment of the two sequences to be compared, but they are important in ESDA because reliability sets a limit on the validity of any true sequential analyses that follow.

The simplest approach is to calculate the percentage of paired events that have been coded the same way, but this fails to take into account the probability that two events may match by chance. Cohen's (1960) kappa is a well-known statistic that takes chance matches into account. A table is constructed where rows and columns represent the codes used in the two sequences. In the cells are tallied the number of times a code from sequence X is aired with a code from sequence Y. Cohen's kappa is calculated with the following equation:

$$\text{kappa} = \Sigma\Sigma\ (p_o - p_c)/(1 - p_c)$$

where p_o is the observed probability of any cell in the matrix (cell frequency divided by overall N) and p_c is the probability expected by chance. In addition, the maximum attainable value with the present data and the significance of Cohen's kappa can also be calculated (see Cohen, 1960 for details).

Cohen's kappa is really only relevant when the same codes are used in sequences X and Y. However even if two analyses use completely different codes there may be some structural similarity in the way they apply codes to events. Any such structure can be detected by calculating the information transmitted (Edwards, 1964) between sequence X and Y (Johnston, 1994). The information content of a set of data is calculated as:

$$H(\text{rows}) = -\Sigma\ p_i \log_2 p_i$$
$$H(\text{columns}) = -\Sigma p_j \log_2 p_j$$
$$H(\text{cells}) = -\Sigma\Sigma\ p_{ij} \log_2 p_{ij}$$

where p_i is the probability associated with events in each row i, p_j is the probability associated with events in each column j, and p_{ij} is the probability associated with events in each cell ij. Information transmitted is then calculated as:

$$\text{Information transmitted} = H(\text{rows}) + H(\text{columns}) - H(\text{cells})$$

A further discussion of reliability and item comparison techniques can be found in Suen and Ary (1989). In particular, they discuss Cronbach's generalizability theory, which is a way of seeing whether encodings distinguish the variables that should be distinguished even if maximum possible reliability is not achieved.

44.4.4 General Issues

Perennial issues with sequential data analysis are stationarity and homogeneity. "Stationarity" refers to whether the sequential properties of a sequence stay the same across the whole sequence or undergo shifts within it. For example, the sequence ABCABCA-BCXBAXBAXB has different properties in its first and second halves and sequential analyses across the whole sequence will be typical of neither half. "Homogeneity" refers to whether the same sequential properties are found in data sequences collected on separate occasions on different subjects. For example, one subject's sequence may run ABCABCABCABCABCABC, whereas another subject's sequence may run AXBAXBAXBAXBAXBAXB, leading to a completely different pattern of transitions. Combining transition matrices produced by the two sequences would produce results typical of neither.

44.5 ESDA AND QUALITATIVE DATA ANALYSIS

Qualitative data analysis can refer broadly to any data analysis that is not quantitative, instead involving systematic nonnumerical inferences, or it can refer more narrowly to a rich family of techniques in the social sciences that stand in contrast to statistically based hypotheticodeductive techniques. Many qualitative methods of analyzing observational data are useful for human factors practitioners when a research or design question requires an interpretive final statement rather than a numerical result or the result of an experimental contrast. Much cognitively oriented modeling is qualitative in nature when it is not supported by computational models (Bainbridge, 1985; Sanderson et al., 1989a; Woods 1993). Within the social sciences, particularly with the arrival of audiovisual recording, approaches to studying interaction have emerged that are being applied very effectively to the study of the workplace (Heath and Luff, 1992a, 1992b; Hughes, Randall, and Shapiro, 1992; Tang, 1991). In addition, qualitative methods are forced upon us when quantitative methods either fail to yield informative results or when data are inadequate for quantitative techniques.

In this section we survey some qualitative approaches to ESDA, focusing on aspects that are of most use to human factors practitioners. In the first part we present a qualitative cognitive approach. In the second part we outline methodological themes common to most qualitative research methods and highlight three particular approaches: Miles and Huberman's (1994) qualitative classification approach, Strauss' (1987) grounded theory, and finally interaction analysis, which has gained currency in HCI and CSCW (Jordan and Henderson, 1995; Suchman and Trigg, 1991). In the third part, we discuss visualization as a powerful qualitative technique that cuts across many theoretical orientations.

44.5.1 Cognitive Approaches: Verbal and Nonverbal Protocol Analysis

The term "protocol analysis" indicates a cognitive approach to the analysis of sequential data, where "protocol" refers to a continuous unfolding sequence of verbalizations and/or actions. The term is usually qualified so that we speak of verbal protocols, in which we ask a person to think out loud as he or she performs a task, or of nonverbal protocols, in which we capture the succession of actions, gestures, expressions, body motions, etc., as a person performs a task. In either case, the verbalizations or activities captured are viewed as observable cognitive products that, given the correct framework of interpretation, can indicate the unobservable cognitive processes that may have produced them.

Verbal protocol analysis is particularly useful for human factors practitioners when trying to understand the mental processes of humans in work contexts where there is little natural talk and action is relatively rare, such as supervisory control environments. In such cases, people have to be asked to think aloud for the investigator really to understand what is going on. Verbal protocol analysis, as described here, should therefore be distinguished from the analysis of naturally occurring work-related verbalization that happens when people discuss current work conditions and future events, communicate information, or negotiate plans of action. The analysis of such socially produced talk usually requires as much interpretive input from the communication and social sciences as from cognitive science in order to be fully understood. Human factors–oriented papers contrasting cognitive approaches to verbalization with social approaches that deal more with naturally occurring talk are Suchman (1987), Douglas (1995), and Lawrence et al. (1995).

The most thorough recent treatment of verbal protocol analysis is provided by Ericsson and Simon (1993). As Ericsson and Simon argue, verbal protocols provide valid evidence for cognitive processes only when the protocols are produced under the following conditions:

- Subjects must simply report the contents of their short-term memory—in other words, the cognitive products available to them—rather than speculating about their own cognitive processes.
- The cognitive processes involved must generate short-term memory products that are inherently verbal or propositional in nature so they can be verbalized directly without the need for mental translation from another modality. For example, reporting the contents of short-term memory during a verbal reasoning task is probably more valid than trying to report short-term memory contents during a highly visuospatial task.
- The rate at which the cognitive task being examined unfolds must not be greater than the rate at which the subject can verbally report the contents of short-term memory while doing the task.

Human factors practitioners contemplating verbal protocol analysis should take pains to ensure that the above conditions obtain. Protocol analysis usually involves the following steps.

1. Collection of verbal and/or nonverbal data through audio or video recordings. Subjects are asked to think aloud continuously as they perform a task, simply reporting what they are thinking rather than "theorizing" about what they are doing or saying.
2. Transcription of the recordings (although specialized software that provides rapid visually guided access to locations in digitized recordings often removes the strict necessity for this step: see Section 44.6).
3. Division of the stream of verbal and/or nonverbal behavior into segments or chunks of a size that promises to be meaningful for the current investigation.
4. Establishment of an encoding vocabulary or classification system to simplify the data at an appropriate grain of analysis by highlighting important distinctions, removing redundancies, and eliminating detail considered unimportant.

5. Application of the vocabulary or classification system to some or all segments to produce an encoded (simplified) version of the protocol.
6. Transformations of the protocol encoding to reveal patterns, relationships, knowledge structures, and sequences that suggest the nature of the cognitive processes that might have produced the protocol. Some relevant techniques are described in Section 44.5.3 of this chapter. Further transformations include representing cognitive processes as a succession of information processing stages organized within a means-ends framework (Card, Moran, and Newell, 1983), as steps in a decision flowchart, as a grammar (Hollnagel, 1979), or as symbolic manipulations on mental contents (Peck and John, 1992).
7. Crystallization and presentation of the results of the previous step in a descriptive or computational model of the cognitive processes involved in the task.

General techniques for performing protocol analysis have been outlined in Ericsson and Simon (1993) and treatments focusing on human factors uses of protocol analysis are provided by Bainbridge (1979), Woods (1993), Sanderson et al. (1989a), Cuomo (1994), Kessler and Anderson (1986), McNeese and Roe (1995), and Bainbridge and Sanderson (1995).

44.5.2 Qualitative Data Analysis (QDA) in the Social Sciences

Qualitative data analysis (QDA) is an intricate network of methods that have evolved over decades in response to changing theoretical positions in the history of the social sciences. Therefore there is no one form of QDA. Miles and Huberman (1994; their Figures 1.1 and 1.2) present two family trees of qualitative techniques, revealing how complex their relations are, and chapters in Denzin and Lincoln's (1994) recent handbook of qualitative research provide glimpses into the history, theoretical preoccupations, methods, and uses of qualitative research.

Miles and Huberman (1994) provide a useful list of the characteristics of "naturalistic" or qualitative research, some of which are paraphrased below:

- Research is conducted in a field setting that constitutes normality for the participants being observed.
- The researcher tries to gain a holistic, systemic, view of the context being studied.
- The researcher tries to gain evidence for how the context being studied is experienced by the normal participants themselves.
- The researcher may review some material with informants but keeps material in its original form.
- The material may be interpreted in many ways, but some interpretations are better either because they have strong theoretical implications or because they maintain better internal consistency.
- Standardized instrumentation (measures, surveys, etc.) tends not to be used.
- Most analysis is done with words, which can be arranged, clustered, etc., to help the analyst see patterns in them.

An all-important theme in qualitative research is the rejection of objective, positivist, "scientific" methods for the social sciences, even though some practitioners feel that quantitative methods can work productively in support of qualitative methods (Strauss and Corbin, 1994; Miles and Huberman, 1994). Another broad theme is the importance of interpretation as the goal of qualitative research: the investigator focuses on the meanings that are shared between people in the social world, and tries to construct a theoretically rich interpretation of these meanings, while recognizing that there may be other valid interpretations. Most qualitative data analysis techniques do not focus specifically on the temporal order of events (conversation analysis and interaction analysis are important exceptions) but as video recordings become easier to navigate, manipulate, and annotate, a focus on temporal aspects is becoming more frequent.

The techniques that qualitative data analysts bring to bear on large, complex datasets have a great deal to offer the human factors practitioner. Interestingly, qualitative researchers deal far more frankly in their writings with handling difficulties that emerge in analysis than quantitative researchers do, probably because the process of finding an analytic approach is a significant first step in the formal process of understanding the data. This openness is very helpful to investigators handling qualitative data for the first time. An awareness of qualitative techniques can help investigators move quickly to a systematic approach, so decreasing AT:ST ratios and increasing the quality of analysis.

44.5.2.1 Qualitative Classification: Miles and Huberman's Approach

Miles and Huberman (1994: see their Figure 13.1) present a systematic approach to qualitative data analysis that is an attempt to make it more rigorous and replicable and for its methods to be more evident (Tesch, 1990). The procedure is essentially an elaboration of the generic ESDA diagram in Figure 44.1. Briefly, a conceptual framework and research questions together help determine what will be sampled or observed. As data start to be collected there are usually some adjustments in the sampling technique and the means of collecting data and possibly also in the conceptual framework and research questions themselves. A coding scheme evolves, but in this case codes are not so much entities to be analyzed (as with sequential data analysis) as they are entities reflecting the result of analysis.

Once a coding system is reasonably stable the analyst works within cases to build diagrams that describe the data. Miles and Huberman often use tabular or "matrix" displays that show relations between codes, but they also use concept maps, influence diagrams, hierarchical structures, etc., to show causality or relatedness. Such qualitative manipulations are increasingly supported by software; for example Nud•ist can display the data in some of the ways mentioned above and it connects the data to the original video segments using the CVideo program (Roschelle, 1992). Further qualitative software tools are discussed in Weitzman and Miles (1995).

From such displays the analyst tries to induce explanations that speak to the research question and to represent these in explanatory diagrams. If there are multiple cases this cycle is repeated for each case. Miles and Huberman's (1994) approach is distinguished by the amount of report preparation and report writing that goes on before, during, and after the coding takes place and by the continual feedback between and adjustment of research questions, conceptual framework, coding system, illustrative displays, and report contents. Overall, Miles and Huberman seek well-grounded causal explanations. Their approach is rigorous, systematic, and sufficiently well-articulated that the human factors practitioner trained in quantitative techniques will not find it alien and can readily grasp it.

44.5.2.2 Constant Comparisons: Grounded Theory

Grounded theory (Strauss, 1987; Strauss and Corbin, 1990, 1994) is a very widely used approach to qualitative data analysis that has been defined as follows:

> *Grounded theory is a general methodology for developing theory that is grounded in data systematically gathered and analyzed. Theory evolves during actual research, and it does this through continuous interplay between analysis and data collection. A central feature of this analytic approach is "a general method of . . . comparative analysis" . . . hence the approach is often referred to as the* constant comparative method *(Strauss and Corbin, 1994, p. 273)*

The relevance of grounded theory for ESDA has been noted elsewhere (Pidgeon, 1992; Sanderson and Fisher, 1994). As was the case for Miles and Huberman's (1994) approach, grounded theory is a systematic way of inducing well-substantiated conclusions from data, but it places particular emphasis on effectively organizing the multiplicity of ideas that emerge and on developing "conceptually dense" theories based on concepts that can be demonstrated to account for large amounts of variation in the data corpus. Theory is developed alongside intensive comparative analysis of data and its fit to emerging concepts. In this process, richer theory is sought and at a more solid empirical founding for the developing theory is sought.

From the very beginning of data collection, sampling is guided by a nascent theory and so is termed "theoretical sampling." Events, utterances, actions, etc., are seen as indicators of a concept hypothesized by the analyst. Indicators of a hypothesized concept are examined and compared and then coded—grounded theory offers rich tactics for coding. Along the way, theoretical memos are written that keep track of the research questions, hypotheses currently being tested, and the coding system currently being used. Eventually, core categories start to appear that account for most of the variation in the observed events. The analysts then tries to "theoretically saturate" these core categories by establishing that they are tightly connected to all aspects of the data and to other categories, and that further analysis is not contributing anything new to the category.

Strauss (1987) includes a chapter in which typical questions and problems that arise during analysis are listed, and practical suggestions for solutions are provided. The questions include such things as:

- "Is it necessary to transcribe all your interviews or fieldnotes?"
- "What is a category, an indicator, a dimension, a code? How do I recognize one?"
- "How do I rid myself of thinking in terms of quantitative methods?"
- "What do I do when I have a whole bunch of integrative diagrams, all rather different?"
- "How much can I do in a day? How do I pace myself in doing these analyses?"
- "How do I present grounded theory results to quantitatively trained people?"

In summary, grounded theory is an approach that places great emphasis on developing empirically justified theory is a systematic way. Its principles apply equally to the analysis of observational data in which temporal relations are important, as is the case for ESDA.

44.5.2.3 Interaction Analysis

A particularly influential qualitative method being used in the study of cooperative work settings is interaction analysis (Heath and Luff, 1992a, 1992b; Jordan and Henderson, 1995; Suchman and Trigg, 1991; Tang, 1989). A definition of interaction analysis follows:

> *Interaction Analysis . . . is an interdisciplinary method for the empirical investigation of the interaction of human beings with each other and with objects in their environment. It investigates human activities such as talk, nonverbal interaction, and the use of artifacts and technologies, identifying routine practices and problems and the resources for their solution. Its roots lie in ethnography (especially participation observation), sociolinguistics, ethnomethodology, conversation analysis, kinesics, proxemics, and ethology. (Jordan and Henderson, 1995, p. 39)*

The basic premise of interaction analysis, which distinguishes it markedly from the cognitive tradition, is that expertise and competence are not to be found in mental representations and plans, but instead in the interactions between members of a particular community engaged with the material world. Meaning is to be found in the details of the particular, rather than in codes and models abstracted from actual events. Therefore interaction analysis stands in marked contrast to the previous two qualitative techniques. Because of the focus on naturally unfolding real-world interactions, interaction analysts rely heavily on video recordings and have developed strong ways of managing, manipulating, and interpreting observational data (Roschelle and Goldman, 1991; Suchman and Trigg, 1991).

One of the most important contributions of interaction analysis to the kind of ESDA that human factors practitioners increasingly perform is that it focuses attention on the mechanics of social and communicative interactions between people—mechanics that may be altered when workplace collaboration is mediated by communication technology. The investigator seeks regularities in the following phenomena:

- How interactive episodes are structured—how they are started and how they are ended.
- How activity is organized over time—coarse-grain structuring, rhythms of work, and "entrainment" (temporal capture of one kind of activity or event by another).
- How participants take turns when communicating.
- What kinds of participation structures exist—who is included, and when.
- How misunderstandings are detected and resolved.
- How activity is organized in space and between people.
- What kinds of artifacts, tools and documents are used, how they are used, and by whom.

These phenomena can help an investigator judge the effect of workplace layout, communications technology, team composition, authority, etc. Such factors are important in the study of team performance in human factors; for example, in group design activity (Tang, 1989, 1991), cockpit resource management (Wiener, Kanki, and Helmreich, 1993), multimedia technology for remote communication (Sellen, 1995), and navigation (Hutchins, 1995; Hutchins and Klausen, 1992). A further important practical contribution of interaction analysis to ESDA is that it shows how analysis can be a collaborative achievement of a group of researchers, so broadening the range and richness of possible interpretations to be considered, rather than exclusively the responsibility of a single researcher.

As mentioned, in interaction analysis coding schemes are not used and regularities are instead induced by working through examples. Figure 44.7 shows a rich qualitative reworking of airport ground operations observations (Suchman and Trigg, 1991) performed with the help of the VideoNoter software environment (Roschelle and Goldman, 1991). We see a transcription of conversation (transcript), a content log noting principal events artifacts used (content log), and a diagrammatic representation of what is achieved in the interaction (three-way aircraft switch). The video segment associated with an event can be cued by selecting the event. This is a "connection" to the raw data themselves from the context of higher level interpretation, in the spirit of EDA. As another example, Figure 44.8 shows a diagram from Suchman (1987) illustrating users' problematic interactions with the expert help system of a photocopier. The temporal structuring of events is preserved. The two center columns show the information available to the machine from the user, and to the user from the machine—a format that proves to be effective for exposing mismatches. The sources of analytic rigor in interaction analysis come from the extent to which interpretations make reference to and contribute to a body of theory, and the extent to which they are appropriately grounded in the data. A further very lucid exposition of interaction analysis, and detailed description of how the method was applied in a study of group design, can be found in Tang (1989).

Further examples of interaction analysis relevant for human factors are Heath and Luff's (1992a,b) studies of the control room of the London Underground and of video-based office communications systems, and Frohlich et al.'s (1994) study of users' interactions with a prototype query-by-format database interface. Developing coherent

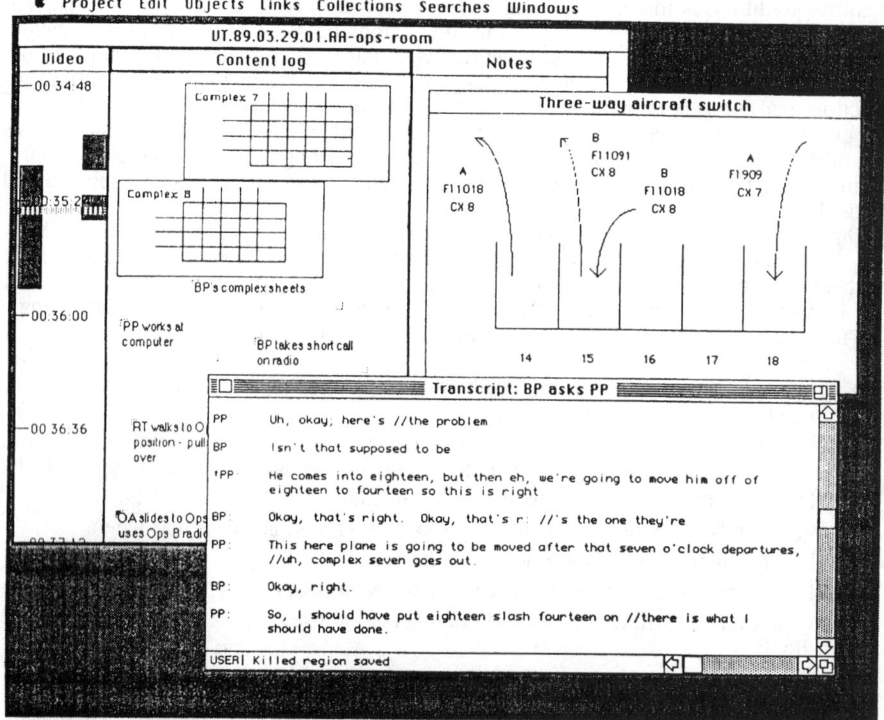

Figure 44.7 Interaction analysis of airport ground operations, supported by VideoNoter software. (From Suchman and Trigg, 1991.)

Sequence 11. *Again A and B are making two-sided copies of a bound document, this time with reduction. (The document is still on the copier glass, the document cover is closed.)*

THE USERS		THE MACHINE	
Not available to the machine	Available to the machine	Available to the user	Design rationale
		DISPLAY 1	Selecting the proceedure
B: It's supposed to– it'll tell "Start," in a minute.			
A: Oh. It will?			
B: Well it did: in the past. (pause) A little start: box will:			
		DISPLAY 4	Ready to print
B: There it goes.			
A: "Press the Start button"			
	SELECTS START		
		STARTS PRINTING	
Okay.			

Figure 44.8 Diagram showing the activities of a human and computer agent (outside columns) and the information from each available to the other (inside columns). (From Suchman, 1987.)

symbolic representations of nonverbal interaction such as body positioning, gesture, mouse and keyboard input, and computer output is challenging. Various notational schemes have been developed (for an example see Frohlich et al., 1994) but they are difficult to read. Consequently, such transcripts are starting to be supplemented with a series of video screen dumps, as in Heath, Jirotka, Luff, and Hindmarsh (1994), and when fully electronic journal publishing becomes available, we can expect to see full video segments.

44.5.3 Modality Shifts and Visualization Techniques

One of the most powerful of the eight Cs is converting—reexpressing the data in a different format, in a different modality, or on a different timescale. Successful conversions reexpress the data in a way that reveals their meaning for the analyst and so are often used to support statements made at the conclusion of an investigation (see Figure 44.1). In the sections above we have seen the importance of graphical displays for qualitative data analysis, where displays not only present the data but very often present a higher level interpretation of the data with great theoretical or practical importance.

44.5.3.1 Indentations

Data conversions need not be complex to be effective: Figure 44.2 shows how well a simple conversion, supplemented by a constraint, can serve the analyst. In the study discussed in Rasmussen et al. (1994) verbal protocols taken from electronics troubleshooters were coded, with similar activities being coded within the same numerical decade (e.g., code 71, code 72, etc). The data sequence was then printed out so that codes within a decade were collected and arranged in columns, so constraining them. This format conversion let the analysts see regularities that had not been visible in transition matrices and it led to further structuring, as the hand annotations on the figure indicate (see Figure 44.8).

Another simple kind of format change is to use indentations to indicate hypothesized hierarchical structuring in activity as it unfolds: Figure 44.9 is from Bainbridge's classic studies of decision making in the steel industry (see Bainbridge and Sanderson, 1995). Cuomo (1994) has also used indentations to good effect in a HCI usability study focusing on uncovering behavioral structure.

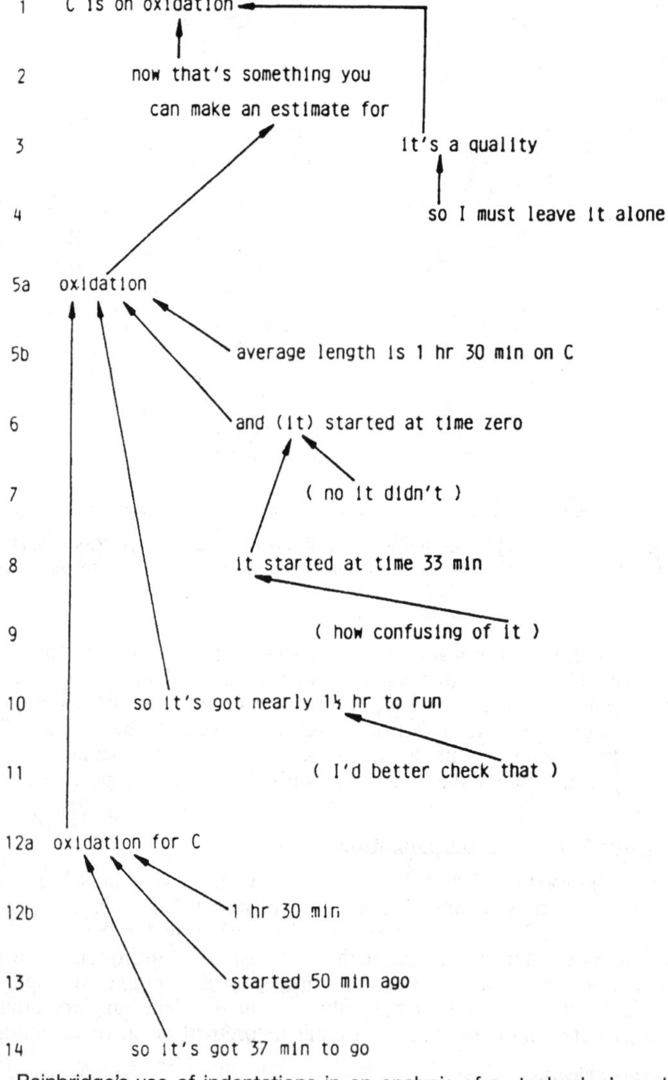

Figure 44.9 Bainbridge's use of indentations in an analysis of a steelmaker's problem-solving process. (From Bainbridge and Sanderson, 1995.)

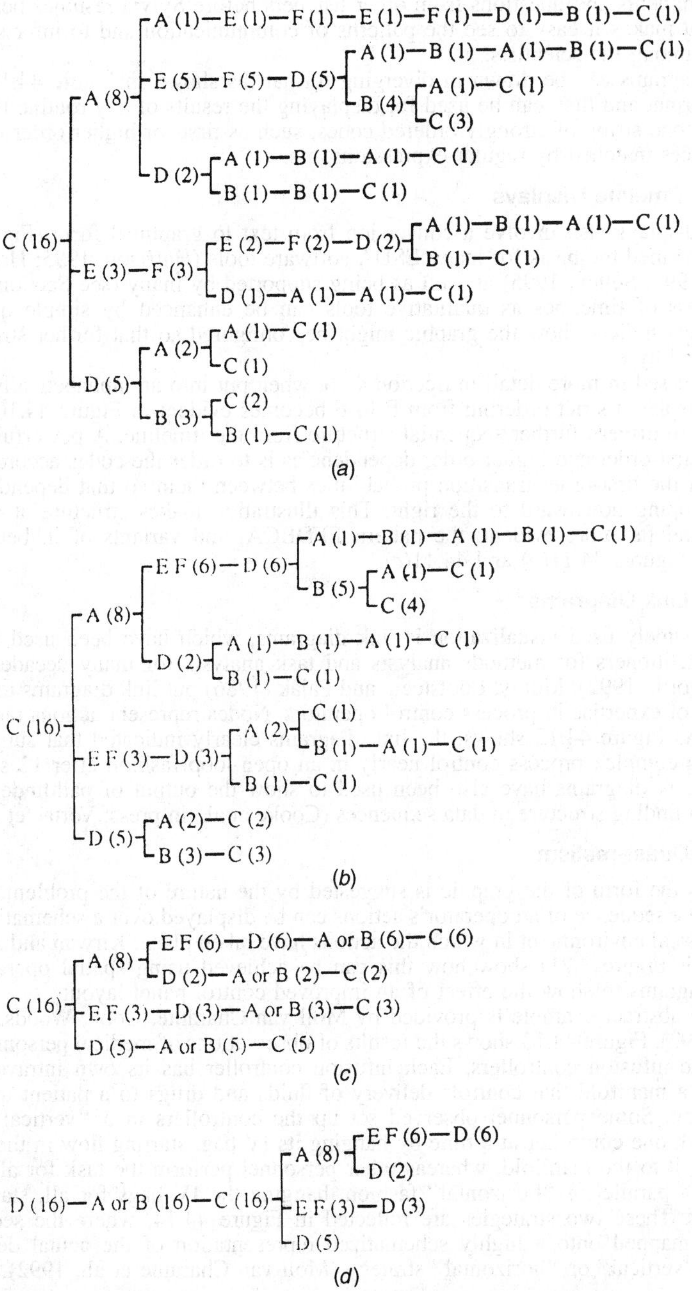

```
C (16)
├─ A (8)
│   ├─ A (1)—E (1)—F (1)—E (1)—F (1)—D (1)—B (1)—C (1)
│   ├─ E (5)—F (5)—D (5)
│   │   ├─ A (1)—B (1)—A (1)—B (1)—C (1)
│   │   └─ B (4)
│   │       ├─ A (1)—C (1)
│   │       └─ C (3)
│   └─ D (2)
│       ├─ A (1)—B (1)—A (1)—C (1)
│       └─ B (1)—B (1)—C (1)
├─ E (3)—F (3)
│   ├─ E (2)—F (2)—D (2)
│   │   ├─ A (1)—B (1)—A (1)—C (1)
│   │   └─ B (1)—C (1)
│   └─ D (1)—A (1)—A (1)—C (1)
└─ D (5)
    ├─ A (2)
    │   ├─ A (1)—C (1)
    │   └─ C (1)
    └─ B (3)
        ├─ C (2)
        └─ B (1)—C (1)
```

(*a*)

```
C (16)
├─ A (8)
│   ├─ E F (6)—D (6)
│   │   ├─ A (1)—B (1)—A (1)—B (1)—C (1)
│   │   └─ B (5)
│   │       ├─ A (1)—C (1)
│   │       └─ C (4)
│   └─ D (2)
│       ├─ A (1)—B (1)—A (1)—C (1)
│       └─ B (1)—C (1)
├─ E F (3)—D (3)
│   ├─ A (2)
│   │   ├─ C (1)
│   │   └─ B (1)—A (1)—C (1)
│   └─ B (1)—C (1)
└─ D (5)
    ├─ A (2)—C (2)
    └─ B (3)—C (3)
```

(*b*)

```
C (16)
├─ A (8)
│   ├─ E F (6)—D (6)—A or B (6)—C (6)
│   └─ D (2)—A or B (2)—C (2)
├─ E F (3)—D (3)—A or B (3)—C (3)
└─ D (5)—A or B (5)—C (5)
```

(*c*)

```
D (16)—A or B (16)—C (16)
├─ A (8)
│   ├─ E F (6)—D (6)
│   └─ D (2)
├─ E F (3)—D (3)
└─ D (5)
```

(*d*)

Figure 44.10 Fisher's cycles procedure applied to the sample data and displayed in tree format. (*a*) Tree summary of cycles around code A, reduced successively in subfigures (*b*) and (*c*) using rewriting techniques, and (*d*) displayed cycling around a different key term, D.

44.5.3.2 Tree Structures

Tree diagrams move data from text format to graphical format. Figure 44.10 shows a Fisher's cycles output reformatted into tree diagrams[1] and will be discussed in Section 4.6. A further example owing to Dunbar (in press; shown in Sanderson et al., 1994 as their Figure 12) shows the patterning of exchanges between participants at a scientific seminar where one person, Sylvia, is giving a talk. The tree diagrams makes it clear that one other person, Chris, is Sylvia's principal interlocutor and that Chris's comments attract further comments and questions from other listeners before Sylvia resumes her talk. The tree format makes it easy to see the patterns of communication and to infer status relationships among the scientists.

Tree diagrams can be drawn in diverging format (as shown in Figure 44.10) or converging format and they can be used for displaying the results of any routine that returns more than one string of strongly ordered codes, such as first- or higher order transitions, subsequences matched by regular expressions, etc.

44.5.3.3 Timeline Displays

Timeline displays also involve a conversion from text to graphical form (Figure 44.11) and have formed the basis for many ESDA software tools (Harrison, 1995; Harrison and Baecker, 1992; Sellen, 1995) as well as being supported by many (see Section 4.6). The effectiveness of timelines as qualitative tools can be enhanced by simple quantitative analyses that indicate how the graphic might be configured so that further structure can be discerned by eye.

As discussed in more detail in Section 44.6, when put into an alphabetically arranged timeline display a strict ordering from E to F becomes evident in Figure 44.10(*a*), but it is difficult to discern further sequential structure from the timeline. A powerful tactic for revealing first-order and higher order dependencies is to order the codes according to the strength of the first-order transition probabilities between them so that dependencies are patterns sloping downward to the right. This illustration makes structure at a low frequency band far more evident; the pattern EFDBCA, and variants of it, become very evident in Figures 44.11(*b*) and 44.11(*c*).

44.5.3.4 Link Diagrams

A further widely used visualization is link diagrams, which have been used by human factors practitioners for methods analysis and task analysis for many decades (Kirwan and Ainsworth, 1992). Moray, Lootsteen, and Pajak (1986) put link diagrams to good use in a study of expertise in process control operators. Nodes represent actions taken by the subjects. As Figure 44.12 shows, the link diagrams clearly indicated that subjects were performing complex process control nearly in an open-loop fashion after 12 sessions of practice. Link diagrams have also been used to show the output of pathfinder analyses focused on finding structure in data sequences (Cooke et al., in press; Vortac et al., 1994).

44.5.3.5 Quasi-realism

Sometimes the form of the graphic is suggested by the nature of the problem itself. For example, the sequence of an operator's actions can be displayed over a schematic diagram of the physical environment in which the actions have taken place. Kirwan and Ainsworth (1992; their Figure 3.21) show how this can be achieved using spatial operational sequence diagrams to show the effect of an improved control panel layout.

A more abstract example is provided by Moll van Charante, Cook, Woods, Yue, and Howie (1992). Figure 44.13 shows the results of observations of medical personnel setting up multiple infusion controllers. Each infusion controller has its own intravenous line leading to a manifold that controls delivery of fluids and drugs to a patient undergoing an operation. Some personnel observed set up the controllers in a "vertical" fashion, dealing with one controller at a time by hanging its IV bag, starting flow in the line, and connecting it to the manifold, whereas other personnel perform the task for all the controllers in a parallel or "horizontal" fashion, hanging the IV bags for all, starting flow for all, etc. These two strategies are reflected in Figure 44.14, where the sequence of actions is mapped onto a highly schematized representation of the actual devices, revealing a "vertical" or "horizontal" strategy (Moll van Charante et al., 1992).

---

[1]The tree output for Fisher's cycles report was conceived and developed by Jeff James in March, 1991, for incorporation into the MacSHAPA software package.

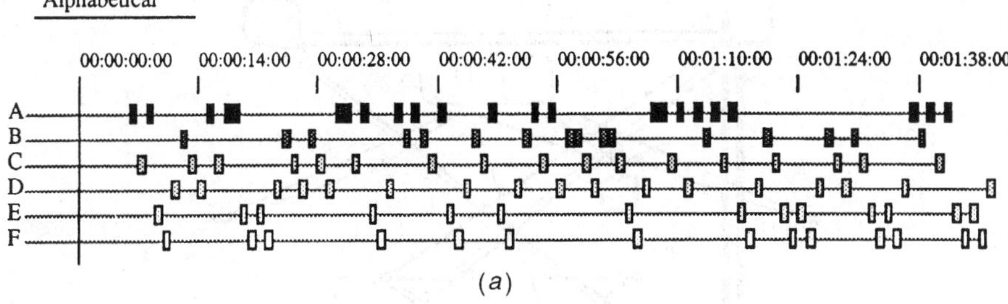

(*a*)

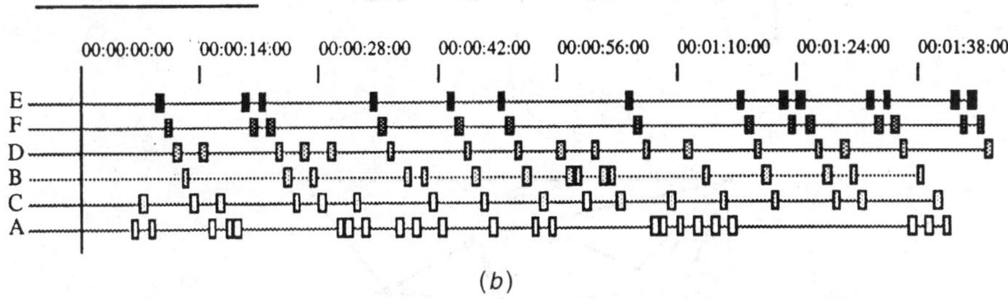

(*b*)

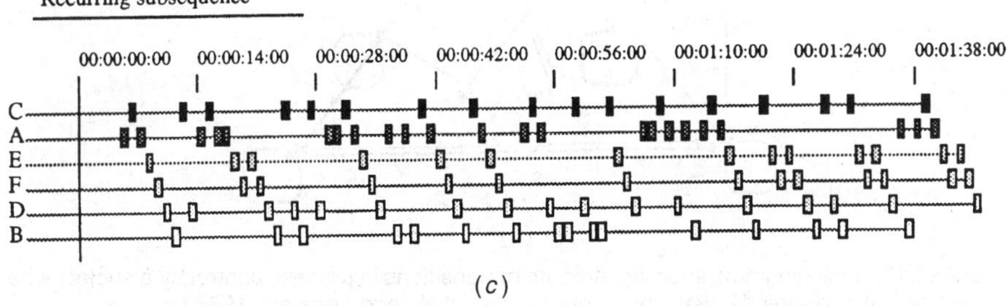

(*c*)

Figure 44.11 Timelines representing the data in the sample data set in different organizations. (*a*) Data are represented in alphabetical order. (*b*) Data are ordered according to the strength of first-order transitions between codes. (*c*) Data are ordered according to strength of recurring subsequences.

44.6 QUANTITATIVE AND QUALITATIVE TECHNIQUES WORKING TOGETHER

To illustrate the different ways that qualitative and quantitative techniques can be used in an exploratory fashion, a very simple data set will be used that consists of 101 tokens taken from the set {A, B, C, D, E, F}:

A C A E F D B C D A C A A E F E F D B C D B C D A A C A E F D A B A B C
A E F D B C A E F D B A C A D B B C D B B C E F D A A C A D A B
A C A E F D B C E F E F D B C D B C E F E F D A B A C A E F E F D

Note that because these data do not have any context, the social techniques cannot be used. However, this example will show how far one can progress with a completely "bottom-up" analysis of sequential data. A further example of ESDA that uses a variety of quantitative and computational techniques is Olson et al.'s (1994) analysis of design meeting discussion topics.

At first glance at the sample data file, it is evident that there are many CA and EF sequences. A good initial tactic often is to throw the data into a timeline display. In the timeline shown in Figure 44.11*a* the ordering of codes from top to bottom is alphabetical. Fortuitously, codes E and F are adjacent, and a strict ordering from E to F becomes

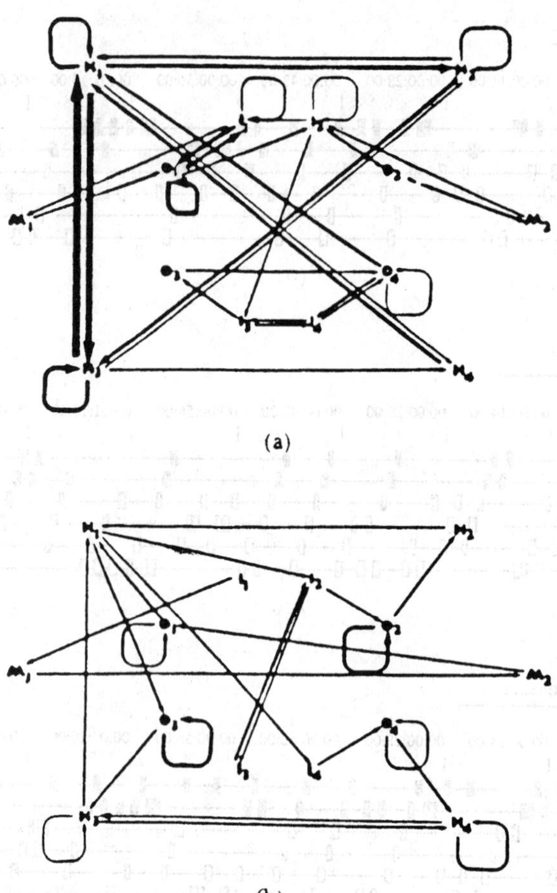

Figure 44.12 Link diagrams showing strength of transitions in process control by a subject when unpractised (*a*) and after 12 trials (*b*). (From Moray, Pajak, and Loostein, 1986.)

evident. However, C and A are not adjacent and any relation between them has been obscured. It is difficult to discern further sequential structure from the timeline.

A powerful tactic for revealing first-order and higher order dependencies in data is to order the codes in the timeline display according to the strength of the first-order transition probabilities between them. In the present case, the strongest conditional transition probability is from E to F, $p(F|E) = 1.0$, so E is assigned to the first line of the timeline display and F to the second line. From F, then, the strongest transition is D, $p(D|F) = 0.714$, and so D is assigned to the third line. The process is continued until all codes are accounted for. Any dependencies in the data will be seen as patterns sloping downward to the right.

Visual inspection of Figure 44.11*b* now reveals much greater patterning in the data. In fact, the patterning is even more evident if one squints one's eyes, so forcing the visual system to become a low-bandpass filter! This makes structure at a low-frequency band far more evident. The pattern EFDBCA, and variants of it, are now very evident. However, there are many departures from this strict sequence that still need to be accounted for.

For our sample data file, the maximal repeating pattern (Siochi & Hix, 1991) is EFEFDBCDBC (which we will refer to as pattern 1), which appears twice in the data. We have already established the strong links between E and F, F and D, D and B, and B and C from the first-order transitions, so at first sight this maximal repeating pattern appears not to be serendipitous. The next longest repeating pattern in the data that is not merely a subset of EFEFDBCDBC is ACAEFDBC (pattern 2), which appears twice. The

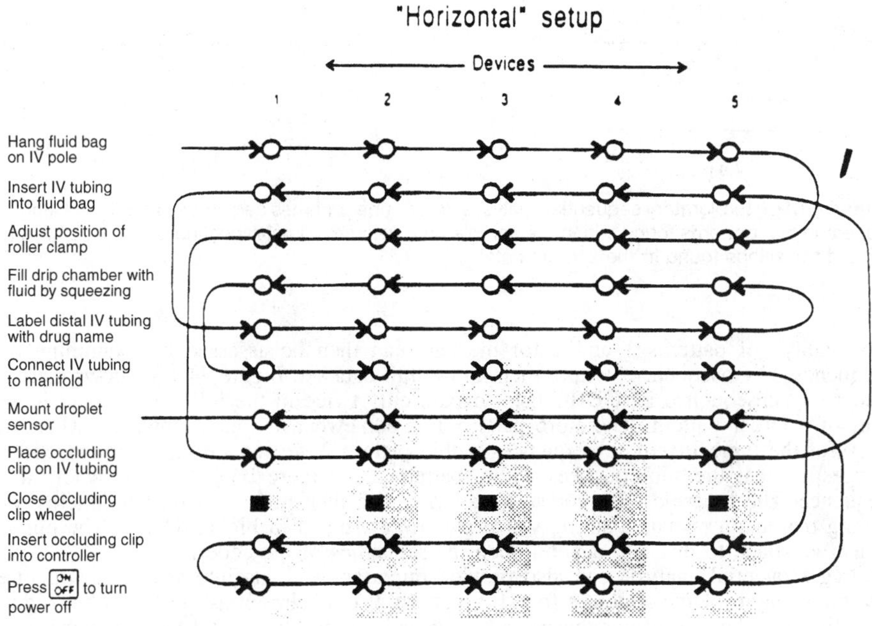

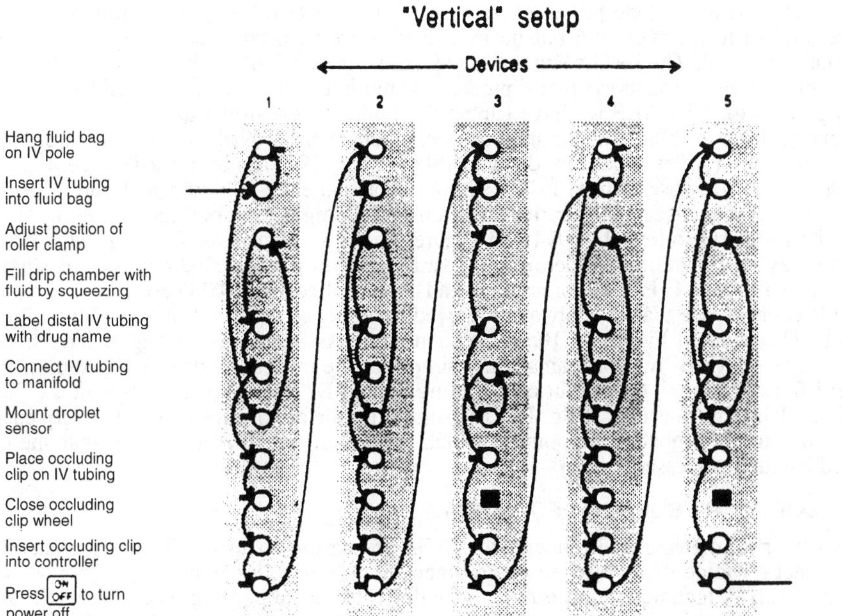

Figure 44.13 Quasi-realistic reexpression of data showing vertical vs. horizontal patterns of mounting infusion devices. (From Moll van Charante, Cook, Woods, Yue, and Howie, 1992.)

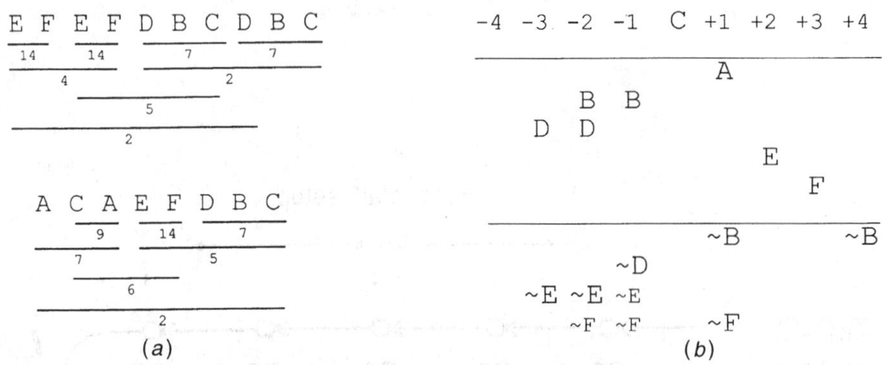

Figure 44.14 Exploratory sequential analyses. (*a*) Number of times sequences of different lengths appear in two patterns found within the sample data using maximal repeating patterns. (*b*) Strong lagged transitions found in the sample data.

"typicality" of patterns 1 and 2 for the data can then be assessed by examining the frequency of subsequences within them in the full data set. Figure 44.14*a* shows that the pattern 1 subsequence DBCDBC itself occurs only twice in the full data set and is not part of a larger pattern. Therefore pattern 1 is somewhat atypical. However, when we look for the typicality of subsequences within pattern 2 we see strong interconnections, suggesting that the subsequences within pattern 2 are more typical of the whole data sequence. The innermost six codes of pattern 2 can then provide a further timeline ordering that again reveals a strong visual pattern, as shown in Figure 44.11*c*. The importance we attach to this will depend upon the semantics of the codes.

Lag sequential analysis can also provide important insights into sequence structure. On the strength of the structure found in pattern 2, C is chosen as the key term. Figure 44.14*b* shows lag transitions that reached $z = \pm 1.96$ on Allison and Liker's (1982) test. The analysis is exploratory rather than confirmatory: the type I error rate would normally be unacceptably high with such a low z-value. Evidence that a code is not seen at a given lag is shown with the code preceded by a tilde, such as "~B." In smaller font are cases where z failed to reach ± 1.96 but no transitions were seen and a Yule's Q measure of associativity is 1.0. The results suggest a sequence running DBCAEF. Further LSA analysis with D as key term shows that F precedes D more than would be expected by chance, hinting that with DBCAEF we have captured a typical cycle in the data.

Finally, different kinds of grammatical techniques can be used to find structure. One technique is the Cycles procedure developed by Fisher (1988) as described above. Figure 44.10*a* shows all subsequences in the sample data sequence that start and end with C. Repeating subsequences can then be rewritten into a single term or "metacode" to simplify the result (Olson et al., 1994). In Figure 44.10*b* sequences of E and F have been rewritten as "E F" and then repetitions of "E F" or of any other code rewritten as single instances so that E F E F becomes E F and A A A becomes A. These manipulations make it easier to see that D is always followed by either A or B, but in an undetermined pattern. Therefore in Figure 44.10*c* all sequences of As and Bs following D have been rewritten as "A or B" which dramatically simplifies the cycles. Figure 44.10*d* shows the same data as in Figure 44.10*c* but now cycling around D rather than C to exploit the fact that D is always followed by "A or B." This simplifies the description of the data sequence even more and is probably the point at which an analyst would feel satisfied that she or he had captured its essence.

44.7 SOFTWARE SUPPORT FOR ESDA

Many powerful software environments for ESDA are starting to appear. The human factors practitioner's choice of a software environment for ESDA is as important as his or her choice of ESDA technique, and often comes down to the same thing because all ESDA software environments assume a certain approach to ESDA in the analyses they make available. A comprehensive review of general and specialized software being used for ESDA is provided in Sanderson (1994) and Ritter (1992) and shorter versions can be found in Bainbridge and Sanderson (1995), Sanderson et al. (1994), and Ritter and Larkin (1994). Harrison (1995) surveyed some influential video-based ESDA software tools. An excellent critical survey of 24 software environments being used in qualitative research, many of which are suitable for ESDA, is provided in Weitzman and Miles (1995).

ESDA software support can in principle reduce AT:ST ratios and increase the quality of analyses performed and there has been much written about these goals (MacKay, 1989; Ritter, 1992; Roschelle and Goldman, 1991; Sanderson and Fisher, 1994). However, as Sanderson (1994) has noted, ESDA software can be a Faustian bargain because each tool allows certain kinds of regularities in data to be discovered and obscures others. ESDA software tools can constrain how data can be represented and transformed and they embody quite distinct models of research practices and data management. If there is a match between the regularities sought by an investigator and the regularities discernible by the software, then the limitations actually can become advantageous. However, if an ESDA software tool channels an investigator into analyses that are poorly matched to his or her research or design question or poorly matched to the kind of data that have been collected, then a great deal of time is wasted and the results of analysis are uninformative.

Five general classes of ESDA software support are described below, classified according to the type of ESDA they have been designed to support. Software references for the most part are limited to tools that are generally available and supported at the time of writing.

- *General productivity software:* These include word processors (Word, WordPerfect, MacWrite), smart editors (emacs), spreadsheets (Excel), databases (Oracle, Double Helix), flowcharting applications (MacFlow), statistical programs (SPSS, S, SAS, Systat, Data Desk), and video editing and multimedia presentation software (Director, Media Tool, PowerPoint). Further information about multimedia tools can be found in Chapter 55 of this Handbook.
- *Multimedia ethnographic tools:* These have usually developed within the naturalistic social tradition of ESDA and are often hypermedia based, so provide strong support for operations such as chunking and connecting. Broadly available tools are CVideo (Roschelle, 1992), HyperResearch (Hesse-Biber, Dupuis, and Kinder, 1991), and Nud•ist (Richards and Richards, 1994). Although the latter two are not specialized for ESDA, they can be effectively adapted to it. For information about these and many more software tools for qualitative data analysis, Weitzman and Miles (1994) provide an excellent and exhaustive survey.
- *Online coding and review tools:* These have usually evolved within the behavioral tradition of ESDA where efficient electronic or symbolic capture of observable events is important. Generally available tools of this kind are A.C.T. (Segal and Andre, 1993), Kronos (Kerguelen, 1992), OCS Tools (Randle and Szostak, 1993), The Observer (Noldus, 1991), MacSHAPA (Sanderson et al., 1994), and DRUM (Bevan and MacLeod, 1994). There are also many commercially available systems for usability testing that provide hardware as well as software.
- *Sequential statistical tools:* These have usually also emerged from the behavioral tradition and are the only implementations of many of the statistical routines discussed in this chapter. They include MOOSES (Tapp, Wehby, and Ellis, 1995), The Observer, MacSHAPA, and ELAG (Bakeman, 1993). CHILDES-CLAN (MacWhinney, 1991) is specialized for verbalization but provides good support for sequential querying.
- *Cognitively oriented tools:* ESDA tools to support cognitive modeling have been developed by researchers and include tools such as PAW (Fisher, 1988, 1991), WE tools (Smith et al., 1993), and Trace & Transcribe (John, 1994) but the latter are not generally available. However, modules of the Soar/MT environment can be obtained (Ritter and Larkin, 1994).

ESDA software continues to be developed. Human factors practitioners interested in staying informed about such developments should consult journals such as *Behavior and Information Technology; Behavior Research Methods, Instruments, and Computers; International Journal of Human-Computer Studies;* and *Human-Computer Interaction* as well as publications such as *Ergonomics in Design,* the ACL SIGCHI's *interactions,* and *SIGCHI Bulletin.* In addition, the Dutch organization iec ProGAMMA provides information about a great many software packages—including ESDA software environments—in its SIByl social science software databank on the WWW at the following URL:

http://www.gamma.run.nl/home.html.

Human factors practitioners should judge ESDA software by the following criteria:

- Does it help the analyst avoid lengthy reworkings of the data before analysis can be undertaken?
- Does it encourage exploration of the data?
- Does it allow multiple linked views of the data?
- Does it provide flexible tools for querying and reexpressing the data?
- Does it support an appropriate syntax for the kind of descriptions and encodings contemplated (see Sanderson, 1994, for details of alternative syntaxes)?
- Is it easy to use the software alongside other software applications, sharing data or importing and exporting data very easily?
- Are the outputs of analyses and statistics provided in a form that can be seamlessly moved to presentation environments?
- Does it provide tools to manage large collections of data and to support the activities of multiple researchers working on the same data for similar or different purposes?

The following quotation from Bainbridge and Sanderson (1995) relates to protocol analysis, but its message is critical for choosing software support for any form of ESDA:

It cannot be emphasized too strongly that researchers should exercise critical judgment when contemplating the use of software to support protocol analysis. Each research should find a tool—or combination of tools—that supports how he or she wants to analyze protocols, rather than adopt a tool because someone else has found it useful, or because it purports to be designed for protocol analysis. The software designer's intentions and the tool's success elsewhere do not guarantee that it will be suitable for the needs of a given research programme. If a tool is not helpful this does not mean that the researcher has failed to follow some canon for performing protocol analysis, but instead simply that the tool was helpful. (Bainbridge and Sanderson, 1995, pp. 192–193)

44.8 SUMMARY AND CONCLUSIONS

Human factors practitioners very often collect observational data but very seldom extract full value from them. Many obstacles stand in the way, such as lack of appreciation of the richness that can emerge from observational data once one moves beyond simple content analysis, ignorance of an appropriate method for analyzing the data, and perceived lack of time to explore the data. Exploratory sequential data analysis is the name of a framework for organizing, comparing, and contrasting all the different observational data analysis techniques so that investigators can choose a technique suitable for their research question, their data, and the kind of statement that will be effective. We distinguished between three traditions of ESDA—the behavioral, cognitive, and social traditions—and showed how they led to different values and analytic operations at different points in analysis. However, we noted that there is a set of fundamental data manipulations—the eight Cs—that seem to cut across the different ESDA traditions.

Some important challenges that have to be met when using ESDA are, first, finding the appropriate balance between quantitative and qualitative techniques, finding ways to handle the volume of data that typically comes with electronic data capture and audiovisual recording, and finding an appropriate descriptive language and grain of analysis. Qualitative data analysis techniques can be particularly helpful with the latter.

There are a great many quantitative sequential data analysis techniques that can be used when exploring observational data and also to make confirmatory statements. Sequential data analysis techniques covered in this chapter include Markov analysis, lag sequential analysis, various grammatically based techniques, techniques for comparing the match between sequences, and item comparison techniques. Other techniques exist: time series analysis is a strong sequential data analysis technique (Box and Jenkins, 1976;

see Lee and Moray, 1992, for a human factors example) and log-linear analysis is increasingly used to discover sequential relations in data and to distinguish the effects of independent variables on sequential structure (Fienberg, 1980; Gottman and Roy, 1990; see Olson et al., 1994 for a human factors example).

Quantitative sequential data analysis has the charm of seeming very straightforward to carry out. However, an investigator's research question or the nature of the data collected may preclude quantitative analysis and instead point to the need for a more qualitative interpretation. Most human factors practitioners have not been trained in qualitative data analysis of any kind and so are at a loss as to how to proceed beyond mere anecdote. Moreover, qualitative researchers have for the most part not developed specific methods for analyzing sequential data, in the way that there is a clear distinction in quantitative methods between sequential and nonsequential data analysis techniques. Three widely used qualitative data analysis techniques were presented: Miles and Huberman's approach, Strauss's grounded theory, and interaction analysis. The application of these techniques to sequential data is relatively unproblematic because the same notions of coding interpreting, and verifying hold for qualitative analysis regardless of whether the data are inherently sequential, as in recordings of work activities, or nonsequential, as with interview data, field notes, examination of artifacts, etc.

ESDA will become an increasingly viable data analysis option for human factors as our conceptual understanding of what ESDA offers us increases, as we develop better methods for doing ESDA, and as we develop software to support ESDA effectively. As Ritter (1992) has noted, there is a tight interdependence between the above three factors. Software that lets investigators easily manipulate their data in the ways captured in the eight Cs lets investigators learn quickly which methods work best under which conditions and even to establish new methods. Therefore, appropriate software can lead to better methods. Better methods, in their turn, lead investigators to richer and more effective conceptualizations of their data with less wasted time. The different approaches outlined in this chapter capture some of the methodological options in a broad array of possibilities and point to some of the experiences and achievements of human factors practitioners who have used ESDA to good effect.

ACKNOWLEDGMENTS

The authors would like to thank the following people who contributed motivation, ideas, and material to this chapter: Ken Boff, Josiane Caron-Pargue, Donna Cuomo, Wayne Gray, Phillip Hingston, Catherine Lees, Michael McNeese, Neville Moray, John Tang, and Peter Thomas. Many thanks also to Neville Moray and Gavriel Salvendy for helpful commentaries on an earlier draft of the paper.

REFERENCES

Allison, P. D., and Liker, J. K. (1982). Analyzing sequential categorical data on dyadic interaction: A comment on Gottman. *Psychological Bulletin, 2,* 392–403.

Anderson, J. R. (1993). *Rules of the Mind.* Hillsdale, NJ: Erlbaum.

Bainbridge, L. (1979). Verbal reports as evidence of the process operator's knowledge. *International Journal of Man-Machine Studies, 11,* 411–436.

Bainbridge, L. (1985). Inferring from verbal reports to cognitive processes. In M. Brenner, J. Brown, and D. Canter, Eds., *The Research Interview.* London: Academic Press.

Bainbridge, L., and Sanderson, P. M. (1995). Verbal protocol analysis. In J. Wilson and E. N. Corlett (Eds.), *Evaluation of Human Work* (Second Edition). London: Taylor and Francis.

Bakeman, R. (1983). Computing lag sequential statistics: The ELAG program. *Behavior Research Methods and Instrumentation, 15,* 530–535.

Bakeman, R., and Gottman, J. (1986). *Observing Interaction: An Introduction to Sequential Analysis.* Cambridge, UK: Cambridge University Press.

Bevan, N., and Macleod, M. (1994). Usability measurement in context. Behavior and Information Technology, *13(1–2)*, 132–145.

Blumer, H. (1954). What is wrong with social theory? *American Sociological Review, 19,* 3–10.

Bowers, C. A., Jentsch, F., Salas, E., and Braun, C. C. (1995). *Performance differences among aircrews: Analysis of communication patterns.* Manuscript submitted for publication.

Box, G. E. P., and Jenkins, G. M. (1976). *Time Series Analysis: Forecasting and Control.* San Francisco, CA: Holden-Day.

Card, S., Moran, T., and Newell, A. (1983). *The Psychology of Human-Computer Interaction.* Hillsdale, NJ: Erlbaum.

Cohen, J. (1960). A coefficient of agreement for nominal scales. *Educational and Psychological Measurement, 20,* 37–46.

Cooke, N. J., Neville, K. J., and Rowe, A. L. (1996). Procedural network representations of sequential data. *Human-Computer Interaction, 11,* 29–68.

Cuomo, D. L. (1994). Understanding the applicability of sequential data analysis techniques for analyzing usability data. Behavior and Information Technology, *13(1–2),* 171–182.

Denzin, N. K., and Lincoln, Y. S. (1994). *Handbook of qualitative research.* Thousand Oaks, CA: Sage.

Dunbar, K. (in press). How scientists think: online creativity and conceptual change in science. In T. B. Ward, S. M. Smith, and S. Vaid, Eds., *Conceptual Structures and Processes: Emergence, Discovery and Change.* Washington, D.C.: APA Press.

Douglas, S. A. (1995). Conversation analysis and human-computer interaction design. In P. Thomas, Ed., *The social and interactional dimensions of human-computer interfaces.* Cambridge, UK: Cambridge University Press.

Edwards, E. (1964). *Information Transmission.* London: Chapman and Hall.

Ericsson, K. A., and Simon, H. A. (1993). *Protocol analysis: Verbal Reports as Data* (Second Ed.). Cambridge, MA: MIT Press.

Faraone, S. V., and Dorfman, D. D. (1987). Lag sequential analysis: Robust statistical methods. *Psychological Bulletin, 101,* 312–323.

Fienberg, S. (1980). *The Analysis of Cross-Classified Categorical Data.* Cambridge, MA: MIT Press.

Fisher, C. (1988). Advancing the study of programming with computer-aided protocol analysis. In G. Olson, E. Soloway, and S. Sheppard, Eds., *Empirical Studies of Programmers, 1987 Workshop* (pp. 198–216). Norwood, NJ: Ablex Publishing Corporation.

Fisher, C. (1991). *Protocol Analyst's Workbench: Design and evaluation of computer-aided protocol analysis.* Unpublished Ph.D. thesis. Department of Psychology, Carnegie-Mellon University, Pittsburgh, PA.

Frohlich, D., Drew, P., and Monk, A. (1994). The management of repair in human-computer interaction. *Human-Computer Interaction, 9,* 385–425.

Gaines, B. R. (1976). Behavior/structure transformations under uncertainty. *International Journal of Man-Machine Studies 8,* 337–365.

Garfinkel, H. (1967). *Studies in Ethnomethodology.* Englewood Cliffs, NJ: Prentice-Hall.

Gottman, J. (1980). On analyzing for sequential connection and assessing interobserver reliability for the sequential analysis of observational data. *Behavioral Assessment, 2,* 361–368.

Gottman, J. M., and Roy, A. K. (1990). *Sequential Analysis: A Guide for Behavioral Researchers.* Cambridge, UK: Cambridge University Press.

Gray, W. D., John, B. E., and Atwood, M. E. (1993). Project Ernestine: Validating a GOMS analysis for predicting and explaining real-world task performance. *Human-Computer Interaction, 8,* 237–308.

Grudin, J. (1990). The computer reaches out: The historical continuity of interface design. In *Proceedings of the CHI '90 Conference on Human Factors in Computer Systems.* New York: ACM. J. Chew and J. Whiteside, Eds., April 1–5 Seattle, WA, pp. 261–268.

Harrison, B. L. (1995). Multimedia tools for social and interactional data collection and analysis. In P. J. Thomas, Ed., *The Social and Interactional Dimensions of Human-Computer Interfaces.* Cambridge, UK: Cambridge University Press. Canadian Information Processing Society, Toronto. San Mateo, CA: Morgan Kaufman.

Harrison, B. L., and Baecker, R. M. (1992). Designing video annotation and analysis systems. In *Proceedings of the Graphic Interface '92 Conference.* Vancouver, BC, May 11–15.

Heath, C., Jirotka, M., Luff, P., and Hindmarsh, J. (1994). Unpacking collaboration: The interactional organization of trading in a city dealing room. *Computer Supported Cooperative Work: An International Journal, 3,* 147–165.

Heath, C., and Luff, P. (1992a). Media space and communicative asymmetries: Preliminary observations of video-mediated interaction. *Human-Computer Interaction, 7,* 315–346.

Heath, C., and Luff, P. (1992b). Collaboration and control: Crisis management and multimedia technology in London Underground line control rooms. *Computer Supported Cooperative Work (CSCW), 1,* 69–94.

Hesse-Biber, S., Dupuis, P., and Kinder, T. S. (1991). HyperRESEARCH: A computer program for the analysis of qualitative data with an emphasis on multimedia analysis and hypothesis testing. *Social Science Computer Review, 9,* 452–460.

Hingston, P. C., and Lees, C. (1994). Sequential analysis with probabilistic finite state automata (Tech. Rep. 94/12). Department of Computer Science, University of Western Australia, Perth.

Hirschberg, D. S. (1975). A linear space algorithm for computing maximal common subsequences. *Communications of the ACM, 18,* 341–343.

Hollnagel, E. (1979). *A framework for the description of operator behavior* (Tech. Rep. N-35-79). Roskilde, Denmark: Electronics Department, Riso National Laboratory.

Hughes, J. A., Randall, D., and Shapiro, D. (1992). Faltering from ethnography to design. In *Proceedings of the CSCW '92 Conference on Computer-Supported Cooperative Work* (pp. 115–122). New York: ACM.

Hutchins, E. (1995). *Cognition in the Wild.* Cambridge, MA: MIT Press.

Hutchins, E., and Klausen, T. (1992). Distributed cognition in an airline cockpit. In D. Middleton and Y. Engestrom, Eds., *Cognition and communication at work.* Cambridge, UK: Cambridge University Press.

John, B. E. (1994). *A database for analyzing sequential behavioral data and their associated cognitive models.* (Tech Rep CMU-CS-94-127). Pittsburgh, PA: School of Computer Science, Carnegie-Mellon University.

Johnston, T. P. (1994). *A General Tool for Sequence Comparison for Use in Exploratory Sequential Data Analysis.* Unpublished M.S.C.S. thesis. Department of Computer Science, University of Illinois at Urbana-Champaign.

Jordan, B., and Henderson, A. (1995). Interaction analysis: Foundations and practice. *Journal of the Learning Sciences, 4,* 39–103.

Kanki, B., and Foushee, C. (1989). Communication as group process mediator of aircrew performance. *Aviation, Space, and Environmental Medicine, 60,* 402–410.

Kemeny, J. G., and Snell, J. L. (1960). *Finite Markov Chains.* Princeton, NJ: Van Nostrand Co.

Kerguelen, A. (1992). *Utilisation du module DRONOS sur PSION ORGANISER LZ Version 1.4.* Technical manual, Laboratoire d'Ergonomie Physiologique et Cognitive de l'EPHE, Paris.

Kessler, C. M., and Anderson, J. R. (1986). Learning flow of control in recursive and iterative procedures. *Human-Computer Interaction, 2,* 135–166.

Kirwan, B., and Ainsworth, L. K. (1992). *A Guide to Task Analysis.* London, UK: Taylor and Francis.

Kruskal, J. B. (1983). An overview of sequence comparison: Time warps, string edits, and macromolecules. *SIAM Review, 25,* 201–237.

Lawrence, D., Atwood, M. E., Dews, S., and Turner, T. (1995). In P. Thomas, Ed. *The Social and Interactional Dimensions of Human-Computer Interfaces.* Cambridge, UK: Cambridge University Press.

Lee, J. D., and Moray, N. (1992). Trust, control strategies and allocation of function in human-machine systems. *Ergonomics, 35(10),* 1243–1270.

MacKay, W. E. (1989). EVA: An experimental video annotator for symbolic analysis of video data. *SIGCHI Bulletin, 21,* 68–71.

MacWhinney, B. (1991). *The CHILDES Project: Tools for Analyzing Talk.* Hillsdale, NJ: Erlbaum.

McNeese, M. D., and Roe, M. (1993). *Putting Knowledge to Use: The Acquisition and Problem Solving Transfer of Knowledge in Situated Problem Solving Environments.* AL/CF-TR-1993-0052. Armstrong Laboratory, Wright-Patterson Air Force Base, OH.

Miles, M. B., and Huberman, A. M. (1994). *Qualitative data analysis* (Second Edition). Beverly Hills, CA: Sage.

Moll van Charante, E., Cook, R. I., Woods, D. D., Yue, L., and Howie, M. B. (1992). *Human-computer interaction in context: Physician interaction with automated intravenous controllers in the heart room* Cognitive Systems Engineering laboratory (CSEL) Tech. Rep. Columbus, OH: Department of Industrial and Systems Engineering, The Ohio State University.

Moray, N., Lootsteen, P., and Pajak, J. (1986). Acquisition of process control skills. *IEEE Transactions on Systems, Man, & Cybernetics, SMC-16,* 497–504.

Moray, N., Richards, M., and Low, J. (1980). *The behaviour of fighter controllers.* Contract Report for Ministry of Defence, United Kingdom. London, UK: Ministry of Defence.

Moray, N., Sabadosh, N., and Kijowski, B. (1992). STRESSFUL: An Experimental Environment for the Study of Strategic Behavior. *Proceedings of the 10th European Conference on Manual Control.* Valenciennes, November.

Newell, A. (1990). *Unified theories of cognition.* Cambridge, MA: Harvard University Press.

Newell, A., and Simon, H. A. (1972). *Human Problem Solving.* Englewood Cliffs, NJ: Prentice-Hall.

Noldus, L. P. J. (1991). The Observer: A software system for collection and analysis of observational data. *Behavior Research Methods, Instruments, and Computers, 23,* 415–429.

Olson, G. M., Herbsleb, J., and Rueter, H. (1994). Characterizing the sequential structure of interactive behaviors through statistical and grammatical techniques. *Human-Computer Interaction, 9,* 427–472.

Olson, G. M., and Olson, J. S. (1991). User-centered design of collaboration technology. *Journal of Organizational Computing, 1,* 61–83.

Olson, G. M., Olson, J. S., and Kragt, R. E. (1992). Introduction to this special issue on computer-supported cooperative work. *Human-Computer Interaction, 7* 251–256.

Patrick, J. D., and Chong, K. E. (1991). Real time inductive inference for analysing human behavior. Paper presented in workshop on Integrating Artificial Intelligence into Databases. In *Proceedings of the International Joint Conference on Artificial Intelligence (IJCAI '91),* Sydney, Australia, 24–30 August. San Mateo, CA: Morgan Kaufman.

Peck, V., and John, B. E. (1992). Browser-Soar: A computational model of a highly interactive task. In *Proceedings of the CHI'92 Conference on Human Factors in Computing Systems* (pp. 165–172). New York: ACM. P. Bauersfeld, J. Bennett, and G. Lynch, Eds. May 3–7, Monterey, CA.

Pereira, F. C. N., and Shieber, S. M. (1987). *Prolog and natural-language analysis* (CSLI Lecture Note 10). Menlo Park, CA: Center for the Study of Language and Information / SRI International.

Pidgeon, N. (1990). Grounded theory and protocol analysis. *Ergonomics,*

Randle, J. D., and Szostak, T. K. (1993). *The Observational Coding System of Tools—A Modular System Integrating Observational Research and Computer Analysis.* Unpublished manuscript. Research Triangle Park, NC: Triangle Research Collaborative, Inc.

Rasmussen, J., and Jensen, A. (1973). A study of mental procedures in electronic troubleshooting. Riso Tech. Rep. No. Riso-M-1582. Roskilde, Denmark: Riso National Laboratory.

Rasmussen, J., Pejtersen, A. M., and Goodstein, L. P. (1994). *Cognitive Engineering: Concepts and Applications.* New York: John Wiley.

Richards, T., and Richards, L. (1994). Using computers in qualitative analysis. In N. Denzin and T. Lincoln, Eds. *Handbook of Qualitative Research.* Thousand Oaks, CA: Sage.

Ritter, F. E. (1992). *A Methodology and Software Environment for Testing Process Models' Sequential Predictions with Protocols.* Unpublished Ph.D. thesis. Department of Psychology, Carnegie-Mellon University, Pittsburgh, PA.

Ritter, F. E., and Larkin, J. (1994). Process models for understanding human action sequences in HCI. *Human-Computer Interaction, 9,* 345–383.

Roschelle, J. (1992). *CVideo Manual.* San Francisco, CA: Knowledge Revolution, Inc.

Roschelle, J., and Goldman, S. (1991). VideoNoter: A productivity tool for video data analysis. *Behavior Research Methods, Instruments, & Computers, 23,* 219–224.

Roth, E., Bennett, K., and Woods, D. D. (1988). Human interaction with an 'intelligent' machine. *International Journal of Man-Machine Studies, 27,* 479–525.

Sackett, G. P. (1978). *Observing behavior* (Vol. 2). Baltimore: University Park Press.

Sanderson, P. M. (1991). *ESDA: Exploratory Sequential Data Analysis.* EPRL Tech. Rep. EPRL-91-04. Engineering Psychology Research Laboratory, Department of Mechanical and Industrial Engineering, University of Illinois at Urbana-Champaign.

Sanderson, P. M. (1994). *Exploratory Sequential Data Analysis: Software* Tech. Rep. EPRL-94-01. Engineering Psychology Research Laboratory, Department of Mechanical and Industrial Engineering, University of Illinois at Urbana-Champaign.

Sanderson, P. M., and Fisher, C. (1994). Exploratory sequential data analysis: foundations. *Human-Computer Interaction, 9(3),* 251–317.

Sanderson, P. M., James, J. M., and Seidler, K. S. (1989b). SHAPA: An interactive software environment for protocol analysis. *Ergonomics, 32,* 1271–1302.

Sanderson, P. M., and Mainzer, J. (1996). *MacSHAPA's Digital Video Analysis: Documentation.* Engineering Psychology Research Laboratory, Department of Mechanical and Industrial Engineering, University of Illinois at Urbana-Champaign.

Sanderson, P. M., Scott, J. J. P., Mainzer, J., Johnston, T., and James, J. M. (1994). MacSHAPA: A software environment for ESDA. *International Journal of Human-Computer Studies 41(5),* 633–681.

Sanderson, P. M., Verhage, A. G., and Fuld, R. B. (1989a). Verbal protocol and state space approaches to continuous process control. *Ergonomics, 32,* 1343–1372.

Scherer, K. R., and Ekman, P. (1982). Methodological issues in studying nonverbal behavior. In K. R. Schere and P. Ekman, Eds. *Handbook of Methods in Nonverbal Research* (pp. 1–44). Cambridge, UK: Cambridge University Press.

Schwandt, T. A. (1994). Constructivist, interpretivist approaches to human inquiry. In N. Denzin and Y. S. Lincoln, eds. *Handbook of Qualitative Research* (pp. 118–137). Thousand Oaks: Sage.

Segal, L. D., and Andre, A. D. (1993). *Activity Catalog Tool (A.C.T.) v2.0 user manual* (NASA Contractor Rep. CR-177634. Moffett Field, CA: NASA Ames Research Center.

Sellen, A. (1992). Speech Patterns in Video-Mediated Conversations. In *Proceedings of the CHI'92 Conference of Human Factors in Computer Systems* (pp. 49–59). New York: ACM.

Sellen, A. J. (1995). Remote conversations: The effects of mediating talk with technology. *Human-Computer Interaction, 10,* 401–444.

Simon, H. A., and Kaplan, C. (1989). Foundations of cognitive science. In M. Posner, Ed. *Foundations of Cognitive Science* (pp. 1–47). Cambridge, MA: MIT Press.

Siochi, A.C., and Hix, D. (1991) A Study of Computer-Supported User Interface Evaluation Using Maximal Repeating Pattern Analysis. *Proceedings of the CHI '91 Conference on Human Factors in Computer Systems.* New York: ACM Press. S. Robertson, G. Olson, and J. Olson, Eds. New Orleans, LA, April 27–May 2. pp. 301–305.

Smith, J. B., Smith, D. K., and Kuptsas, E. (1993). Automated protocol analysis. *Human-Computer Interaction, 8,* 101–145.

Strauss, A. (1987). *Qualitative Analysis for Social Scientists.* Cambridge, UK: Cambridge University Press.

Strauss, A., and Corbin, J. (1990). *Basics of qualitative research: Grounded theory procedures and techniques.* Newbury Park, CA: Sage.

Strauss, A., and Corbin, J. (1994). Grounded theory methodology: An overview. In N. K. Denzin and Y. S. Lincoln, Eds. *Handbook of Qualitative Research.* Thousand Oaks: Sage.

Suchman, L. (1987). *Plans and Situated Actions: The Problem of Human-Machine Communications.* New York: Cambridge University Press.

Suchman, L., and Trigg, R. (1991). Understanding practice: Video as a medium for reflection and design. In J. Greenbaum and M. Kyng, Eds. *Design at Work* (pp. 65–89). Hillsdale, NJ: Erlbaum.

Suen, H. K., and Ary, D. (1989). *Analyzing Quantitative Behavioral Observation Data.* Hillsdale, NJ: Erlbaum.

Swarts, H., Flower, L. S., and Hayes, J. R. (1984). Designing protocol studies of the writing process: An introduction. In R. Beach and L. S. Bridwell, Eds. *New Directions in Composition Research* (pp. 53–71). New York: Guilford.

Tang, J. (1989). *Listing, drawing, and gesturing in design: A study of the use of shared workspaces by design teams.* (Tech. Rep. SSL-89-3). Palo Alto, CA: Xerox Corporation, Palo Alto Research Center.

Tang, J. (1991). Findings from observational studies of collaborative work. *International Journal of Man-Machines Studies, 34,* 134–160.

Tapp, J., Wehby, J., and Ellis, D. (1995). A multiple option observation system for experimental studies: MOOSES. *Behavior Research Methods, Instruments, and Computers, 27,* 25–31.

Tesch, R. (1990). *Qualitative Research.* New York: Falmer.

Tukey, J. W. (1977). *Exploratory Data Analysis.* Reading, MA: Addison-Wesley.

van Hooff, J. A. R. A. M. (1982). Categories and sequences of behavior: methods of description and analysis. In K. R. Scherer and P. Ekman, Eds. *Handbook of methods in nonverbal behavior research* (pp. 362–439). Cambridge, UK: Cambridge University Press.

Vortac, O. U., Edwards, M. B., and Manning, C. A. (1994). Sequences of actions for individual and teams of air traffic controllers. *Human-Computer Interaction, 9,* 319–343.

Wallace, C. S., and Boulton, D. M. (1968). An information measure for classification. *Computer Journal, 2,* 185–194.

Weiler, P. (1993). Software for the usability lab: A sampling of current tools. In S. Ashlund, Ed. *Proceedings of the ACM Conference on Human Factors in Computing Systems (InterCHI'93).* Amsterdam, April 24–29.

Weiner, E., Kanki, B., and Helmreich, R. (1993). *Cockpit Research Management.* San Diego: Academic Press.

Weitzman, E. A., and Miles, M. B. (1995). *Computer Programs for Qualitative Data Analysis.* Thousand Oaks: Sage.

Woods, D. D. (1993). Process tracing methods for the study of cognition outside of the experimental psychology laboratory. In G. Klein, R. Calderwood, and J. Orasanu, Eds. *Decision Making in Action: Models and Methods* (pp. 29–51). New York: Ablex.

CHAPTER 45

EFFECTIVENESS TESTING OF COMPLEX SYSTEMS

Annelise Mark Pejtersen
Center for Cognitive Research
Risø
4000 Roskilde, Denmark

Jens Rasmussen
HURECON
Smorum DK 2765 Denmark

45.1 INTRODUCTION

Experienced system designers are deeply embedded in their local context and very often will treat their current design as an update or modification of prior solutions. They will have many preconceived ideas and solutions at several goal–function–form levels. In other words, they will not approach design as an orderly top-down synthesis but instead consider the process as a sideways modification of prior solutions to similar problems. They will implicitly or explicitly have to conform with constraints from many different sources. These include the ultimate user needs, the competitors' products, the company policy and image, marketing strategies and product style, the financial policies of the company and its preferred part suppliers, the hot issues of their own profession, and their subjective preferences, personal styles, and creative images. All of this influences their design choices and cannot be ignored from a rational design point of view. Many possibilities for choice are present within the space of acceptable product solutions—especially with the variety of new technologies.

In this situation, it is more realistic to ask designers to pause at suitable points in the design process, and then to review and evaluate their present results consciously and systematically, than it will be to try to formalize the design process itself. That is, instead of suggesting guidelines in term of a well-ordered normative design process, efforts to formalize should be spent on evaluations after the fact of design concepts, prototypes, and products as they emerge. This process of evaluation will serve to make the conceptual basis for a design explicit and available later for reference (e.g., during maintenance and modifications) and to test the match between system capabilities and potential users' needs.

For a creative design process, we have suggested considering the support for design and evaluation in terms of *maps of the relevant knowledge territories* involved in design, rather than route instructions in terms of guidelines (this approach is discussed in detail in Rasmussen, Pejtersen, and Goodstein, 1994).

45.1.1 Outline of the Chapter

It is a general feature of many evaluation experiments that it is difficult explicitly to define what functional features have actually been tested, and how comprehensive the test has been. The main purpose of this chapter is to discuss how a framework for work analysis and system design can be used to structure different evaluation approaches by offering a set of compatible boundaries for planning experiments in the laboratory as well as at the user's workplace. The chapter is organized as follows: Reasons for an increased need for evaluation are suggested, followed by some examples of methodological issues that need to be considered, including a distinction between verification and validation of complex systems. The organization of the chapter is then controlled by a distinction between two major approaches to evaluation: the analytical approach and the empirical approach. A short description of a cognitive framework for work analysis is introduced before a more detailed discussion of analytical and empirical approaches to evaluation, since this framework is proposed as a general support tool for a systematic design of evaluation experiments. The next two paragraphs on analytical and empirical evaluation both suggest complementary analytical and empirical methods such as usability testing and cognitive walkthrough. The use of the framework for empirical evaluation is illustrated by discussions of experimental setup and examples of actual evaluation experiments and techniques. Finally, the need to consider evaluation as a complex and dynamic process of design and evaluation is discussed.

45.2 NEEDS FOR EVALUATION

An evaluation can be carried out for several reasons, such as to assess whether a system conforms to some standard or goal or to compare alternative approaches and to identify eventual improvements or changes in a system to be considered for a redesign.

The present fast pace of change of technology immediately increase the need for careful and systematic evaluation of new designs, because rapid changes disturb established practices and traditions. During stable periods, designers and users to a certain degree share a common tradition as the basis for discussions about the merits and difficulties connected with proposed designs, and "expert judgment" can then play an important role.

However, the present situation can be characterized by the following features:

- The rapid pace of technological development makes a smooth empirical and incremental development of systems difficult.
- New technology upsets the traditions and practices that encourage mutual understanding between designers and users or consumers during periods of technological stability.
- The new multimedia technology gives new means for recording and analyzing data, a development that has resulted in a great number of usability laboratories in industry (Nielsen, 1994).
- An increased concern with large, centralized systems having potential high hazards—explosions, releases, loss of property, money, information, or, at the worst, life.
- The advent of more rigorous guidelines and recommendations as a result of the increase of the number of products on the market having similar qualities, which aim at common international standards, such as the International Standard Organization's guidelines for menus and measuring the usability of products (ISO, 1993).

45.3 EVALUATION OBJECTIVES, MEASURES, AND METHODS

Evaluation is a broad topic and the following are examples of relevant questions that need to be answered before an evaluation is started.

- What is to be evaluated: A product, a concept, a partial solution, a prototype with surface levels of the total functionality of the system, or a prototype with full functionality of only a part of the system?
- Should several system solutions be chosen for evaluation? Evaluations can be *comparative* if several systems are to be checked for a differential result to support choice among design alternatives. Or they can be *absolute* in testing whether a single system will be able to or does in fact achieve a given goal and level of performance. Comparative evaluation, for example, can be useful when several systems are to be integrated during use, and consistency in representation is required to reduce the amount of resources needed for learning the systems. In order to be useful, comparative evaluation should be based on an exhaustive list of clearly defined and compatible qualities and functions of the different systems to be compared, which are then correlated with identical performance measures. The cognitive framework described in this chapter is suggested to aid the performance of

a systematic comparative evaluation and generalization among different design solutions.

- What constitutes an unambiguous definition of goals and objectives that can be transferred to the evaluation level (what is the evaluation supposed to establish)? An example of definitions of effectiveness, efficiency, and satisfaction can be found in section 45.9.4 on empirical usability evaluation methods.
- What are the (categories of) situations to be evaluated?
- How will performance be defined and how will it be measured? What will be the linking between evaluation objectives and measurable performance variables?
- What are the effects of the intermediate variables (training, experience, task, environment, etc.)?
- Who will participate in the evaluation? Real end users, test subjects, design team members, colleagues? In iterative design, a distinction should be made between use of subjects from the workplace in a work situation, test users in a laboratory, and the testing done among the design team members and colleagues in the project group. In the first case, when users from the workplace participate as test subjects, the system design will have to be advanced and close to a finished version of the system, which is not the case when design team members and colleagues evaluate.
- Where to perform the evaluation? In a laboratory or at the users' workplace?
- What evaluation data should be collected? Evaluations can be subjective and qualitative if they are based mainly on user and/or expert judgment, or objective if attempts are made to measure (quantitatively) the degree to which objective criteria are met.
- What quantitative data should be collected? Quantitative measurements can be performed as objective measures or as subjective measures. For example, the cognitive workload or mental effort required to perform a task can be measured objectively by equipment recording heart rate variability and respiration, and subjectively by users' rating on a scale the amount of effort they perceive to have invested in task performance (Bevan and Macleod, 1994).
- What quantifiable performance measures are relevant? Quantifiable measurements may include time to do a task, error rate, number of features actually used, number of features never used, number of features the user can remember after the test, frequency of use of help system, number of positive or negative user statements, number of users that prefer one system to another, number of times the user is sidetracked from focus on the task, frequency of users' choice of work strategies, etc.
- What methods to use to capture data? Synchronized audio recording and videotaping, questionnaires, interviews, logging of observational data, of the actual use of the varied functionality of a product. Automatic data logging written into the software is frequently used, but in public domains the method may involve ethical and privacy concerns with respect to user anonymity. Many data loggings have been concerned with logging errors, error messages received, or errors made by users during use of interface commands (Bradford, Murray, and Carey, 1990).
- What methods to select for data analysis and data encoding to obtain reliability? What statistical methods and what data integration method for coding, sampling, and analysis? What methods to choose for qualitative analysis of case studies? What methods to select for presentation of results for customers or test subjects? Will a summary of videotapes be effective?

Which objectives, methods, and measures to choose depends on the context of the system design, and on the subsequent utilized approach to the experimental design of the evaluation. It is outside the scope of the present chapter to suggest particular experimental designs and objectives for evaluation. Usually, a combination of methods, measurements, objectives, and evaluation techniques is likely to be required.

45.4 VERIFICATION AND VALIDATION

Evaluation in itself is a rather general term and a distinction is often made between verification and validation. *Verification* refers to a test of the correspondence of a concept or product with the specifications as intended by the designer, whereas *validation* is concerned with the correspondence of the design product with the actual needs of the end user (Figure 45.1).

45.4.1 Verification

Verification is an assessment of the degree to which the design product meets the design specifications as they are formulated as a basis for the design. Verification is thus supposed to answer the question "Is the design right?" Does the product meet the design intentions?

The verification process will often raise a number of questions about the intentions behind and the rationality of the design specifications. Specifications of a complex product cannot eliminate all alternatives and many decisions must be made during the design and development process that are not covered by the specifications. Such open questions are then resolved by designers and developers from their perception of the users' situation and needs, and the intentions expressed in design teams. Thus verification very likely will result in identification of issues that need to be reconsidered in the design specification by the design team. Records of open questions coming up along the trajectories of the design process, and the decisions then taken, will be important information for the subsequent verification of the design product. Such a documentation of the design intentions, the decisions made and their rationale, tradeoffs, constraints, and solutions embedded in the product is also necessary for later comparison and generalization of problems and their solutions, when new (re)designs are initiated. One approach proposed for such design documentation is the design rationale, which refers to records of design alternatives and the arguments used to choose among different options, called QOC: questions, options, criteria, (MacLean, Young, Bellotti, and Moran, 1991; Carroll and Moran, 1991).

Verification of the match of a proposed design concept or product with the design specifications may be based on a proposed concept or product's compliance with a work domain analysis, with accepted practice within a company or a professional community, with heuristic rules of thumb advocated by experts, and with guidelines and international standards.

45.4.2 Validation

Validation is an assessment of the degree to which the design product actually serves the needs of the end user. Thus validation is supposed to answer the question "Is it the right design?"

The validation process will include the customers' requirements to the functionality of the product as well as its impact on the users' general work situation. A validation is performed when a work analysis and/or user requirements are used as references for

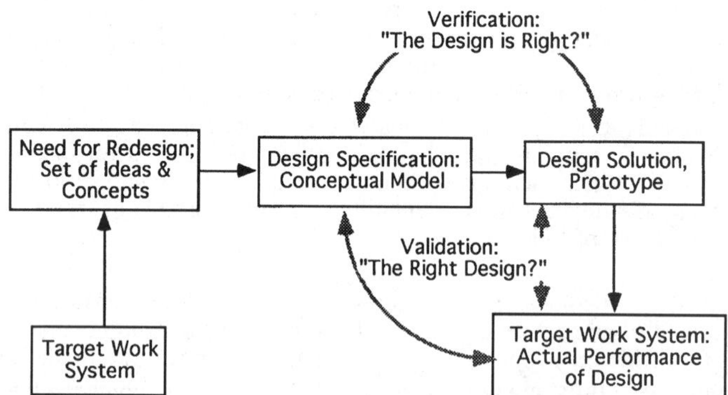

Figure 45.1 In evaluation, a distinction is drawn between *verification,* a test of correspondence between design specifications and the product, and *validation,* a test of the correspondence between the design specifications and the actual needs of the users. Both verification and validation can be performed empirically or analytically, depending on the reference for judgment used, whether the reference is a conceptual model of the actual performance, based on work analysis, or the reference is a set of experimental tests.

experimental performance, and when empirical testing of the product is done with the actual end users.

A successful validation of the performance of a design solution will have validity relative to the correspondence of the experimental conditions with the conditions in a real work environment. For example, a validation of a product is more valid if it includes performance measures in the evaluation test that reflect the performance measures adopted by end users in a in real-life work situation.

Generally speaking, the relationship between verification and validation (Hopkin, 1993), can be viewed as complementary:

- Verification and validation tend to be serial rather than parallel processes.
- Verification normally precedes validation, when both processes are applied.
- Both processes should usually occur, although there could be circumstances when just one would suffice.
- Verification and validation should be treated as complementary and mutually supportive rather than alternatives.

45.5 ANALYTICAL AND EMPIRICAL EVALUATION

It is suggested that a systematic evaluation of complex systems should be well structured and performed at several well-defined levels of user–workplace interaction. At each of these levels, evaluation should be performed either analytically or empirically, or both approaches should be applied. Issues related to the *contents* of the information and the functionality of the system can be evaluated analytically, whereas issues related to its *form* involve context, user experience, and preferences and, therefore, very likely will need an empirical approach. An analytical evaluation depends on a structured comparison of work requirements as defined by a work analysis to the characteristics of the design specifications and/or a system prototype. In contrast, empirical evaluation involves experimental tests of the performance of system characteristics with reference to design objectives (verification) or its actual performance in a laboratory with test users or in the ultimate context of use in a real workplace (validation).

Empirical evaluation pose special problems with respect to validity and generalization of results. To enable *generalization* and the *transfer of findings* among actual work analyses and different experimental designs, a consistent framework is necessary. To structure an analytical evaluation of the match between a new design and the work domain including the characteristics of its users, we need a comprehensive framework for description of work systems. Section 45.6 is a short description of a framework that can be used as a reference for analytical and empirical evaluation.

45.6 FRAMEWORK FOR WORK DOMAIN ANALYSIS AND EVALUATION

In modern work, stable work procedures are not the norm. Many tasks are discretionary. Explicit consideration of goals and constraints and exploration of the boundaries of acceptable performance are often required to optimize effectiveness. For this reason, the object of modeling can no longer be the "task" but must include all the features of the work environment and the interpretation of these features by the actors. The interaction of work environment and actors' resource constraints creates the task ad hoc. A wide variety of options is found with respect to when and how to approach a given task. Therefore, to understand why a particular piece of behavior is preferred instead of another possible pattern, we have to understand how the action alternatives in a particular situation are eliminated so that one unique sequence of behavior can manifest itself. Only then can we hope to predict how a new tool will change a present work practice and the users' interaction with a new system design.

In other words, we have to identify the constraints within a work environment that shape the behavior of users together with the subjective performance criteria they apply to optimize performance within the remaining action possibilities.

45.6.1 Levels of Work Analysis

A framework for analysis and evaluation must serve to represent the characteristics of both the physical work environment and the "situational" interpretation of this environment by the actors involved, depending on their physical, perceptual properties and their skills, strategies, and values. The analysis of the work domain activity includes the identification of behavior-shaping constraints represented by the boundaries of Figure 45.2.

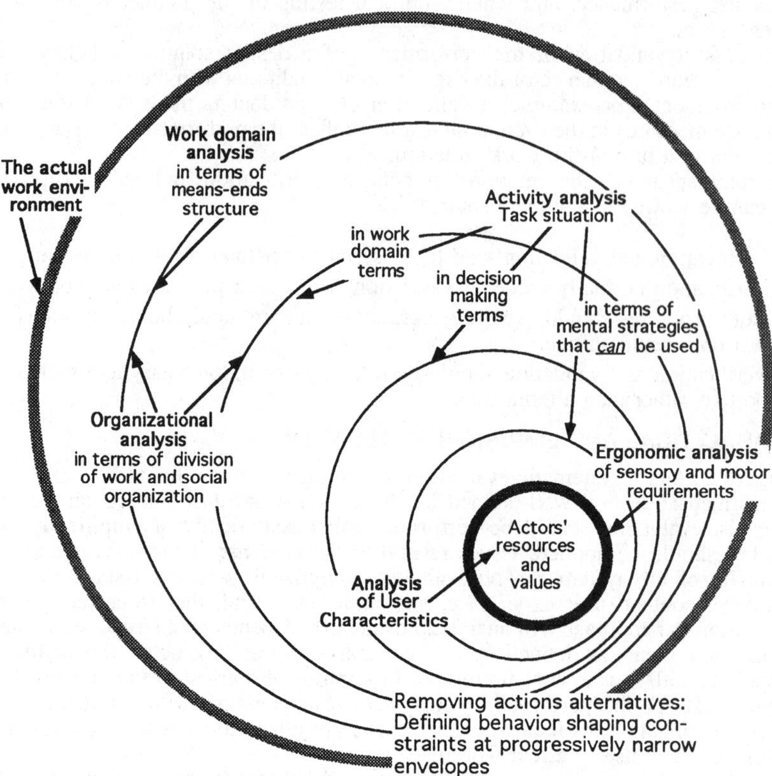

Figure 45.2 Several levels of analysis of work activities are necessary, with corresponding shifts in representation language, in order to relate properties of a work environment to the cognitive resource profiles of the actors. In the framework the work space dimensions used for the design input are compatible with those work space dimensions used for evaluation of the design output. (Reproduced from Rasmussen, Pejtersen, and Goodstein (1994), by permission from John Wiley and Sons.)

The following section presents an overview of the dimensions of analysis that have been identified from field studies in several different work places as a useful, comprehensive representation of user work interaction (for more details, see Rasmussen et al., 1994; Pejtersen, Sonnenwald, Buur, Govindarej, and Vincente, 1995).

45.6.1.1 Work Domain, Means–Ends Space

This dimension represents the landscape within which the work takes place and it serves to make explicit its goals, constraints, and productive resources. The representation presents an inventory of system elements and it is, in the short perspective, independent of particular situations and tasks. It identifies the functional elements and their means–ends relations or, in other words, the productive resources that are available for the actors to "design" their local activity. The analysis is structured at several levels of functional abstraction and, in this way, include representations of physical configuration and anatomy, of physical work processes, of general functions, of priority measures, and, finally, of system goals and constraints with reference to the environment, see Table 45.1

An analysis within this dimension of the framework will identify the structure and general content of the global knowledge base of the work system that must be considered for design of work support systems.

45.6.1.2 Activity Analysis: Task Situation in Domain Terms

This dimension instantiates that subset of the basic means–ends network which is relevant for a particular task. Analysis should not be made in terms of work procedures but in

Table 45.1 The Classes within the Means–Ends Hierarchy

MEANS-ENDS RELATIONS	PROPERTIES REPRESENTED
Purposes and Constraints	Properties necessary and sufficient to establish relations between the performance of the system and the reasons for its design, i.e., the purposes and constraints of its coupling to the environment. *Categories are in terms referring to properties of environment.*
Abstract Functions and Priority measures	Properties necessary and sufficient to establish priorities according to the intention behind design and operation: Topology of flow and accumulation of mass, energy, information, people, monetary value. *Categories in abstract terms, referring neither to system nor environment.*
General Functions	Properties necessary and sufficient to identify the 'functions' which are to be coordinated irrespective of their underlying physical processes. *Categories according to recurrent, familiar input-output relationships.*
Physical Processes and Activities	Properties necessary and sufficient for control of physical work activities and use of equipment: To adjust operation to match specifications or limits; to predict response to control actions; to maintain and repair equipment. *Categories according to underlying physical processes and equipment.*
Physical Form and Configuration	Properties necessary and sufficient for classification, identification and recognition of particular material objects and their configuration; for navigation in the system. *Categories in terms of objects, their appearance and location.*

Source: Reproduced from Rasmussen, Pejtersen, and Goodstein (1994), by permission from John Wiley and Sons.

terms of the objectives, functions and resources active in prototypical work situations, and the related information requirements. A set of such prototypical work situations can be used in various combinations to characterize a set of task situations to be considered for information system design.

45.6.1.3 Activity Analysis: Task Situation in Decision Terms

For the next dimension of analysis, a shift in representational language is made. For each of the activities defined, the relevant tasks are identified in terms of decision-making functions, such as situation analysis and goal evaluation, planning, or actual execution. This representation breaks down work activities into subroutines that can be related to the cognitive activities of the involved people and that serves to identify the cognitive tasks that are the targets for support systems. The information gained in this analysis will identify the knowledge items from the work domain representation that are relevant in a particular situation. In addition, it assists in identifying the queries that are likely to be made by decision makers for retrieving information.

45.6.1.4 Activity Analysis in Terms of Mental Strategies

A further analysis of the decision task requires another shift in language in order to be able to compare task requirements with the cognitive resources and subjective preferences of the individual actors. For this purpose, the mental strategies that can be used for each of the decision functions are identified by detailed analyses of the actual work performance (e.g., by protocol analysis). Each strategy is based on a particular kind of mental model, a set of tactical rules and a related mode of interpretation of observations. The characteristics of the various strategies are identified with reference to subjective performance criteria such as time needed, cognitive strain, amount of information required, or cost of failure. Knowledge about the available effective strategies is important for the user-interface design, because it supplies the designer with several coherent sets of mental models, data formats, and tactical rule sets that can be used by actors with varying expertise and competence and, therefore, should be supported by an interface.

45.6.1.5 Analysis of Users' Cognitive Resources and Values

At this stage, the action possibilities in work performance of the individual have been delimited through an identification of the work-dependent behavior-shaping constraints down to the level of mental strategies that can be employed for the decision functions allocated to each individual actor. In order to judge which strategy will actually be used, the resource requirements of the various strategies have to be compared to the cognitive resource profiles of the actors. Therefore, this perspective of analysis is focused on the background of the relevant user category and on the level of expertise and the performance criteria of the individual actors.

45.6.1.6 Analysis of Social, Organization, Role Allocation, and Coordination of Work

In order to identify the actors actually involved in the prototypical task situations, it is necessary to find the principles and criteria governing the allocation of roles among the groups and individuals involved. This allocation of roles to actors is governed by the social organization and management structure and is dynamically dependent upon the circumstances and criteria such as actor competency, access to information, minimizing the communication needed for coordination, sharing of work load, and complying with regulations (e.g., union agreements).

A work analysis will not proceed as an orderly top-down through these perspectives as described above, starting with the work domain analysis and finishing with the users' cognitive resources and value criteria. The broader context of the entire work environment as shown in Figure 45.2 will be activated during the analysis of both task activity and user characteristics. In particular, the analysis of division and coordination of work and social organization will be closely related to the analysis of the work domain and the task situation and frequent iterations among the perspectives will be necessary.

45.7 USE OF THE FRAMEWORK FOR ANALYTICAL AND EMPIRICAL EVALUATION

Ideally, for a structured, consistent system design, each of these perspectives of analysis should be considered explicitly in a top-down design process. However, as discussed above, actual design sessions are creative processes trading-off ideas against constraints, a process that should not be constrained further by "rational" guidelines. Instead, at suitable times during the design process, the result of the creative effort should be evaluated systematically with reference to a comprehensive analysis of the particular work system for which new support systems are designed or existing systems are redesigned. In other words, the framework described above for work analysis is as important as a reference for a systematic evaluation of a design concept or a system prototype. For each of the perspectives of analysis, explicit evaluation questions can be posed (Figure 45.3). In the following sections, the discussion of analytical and empirical evaluation techniques will be structured according to the framework of this figure.

Both analytical and empirical evaluation should be considered for all the levels of Figure 45.3, but the sequence of the levels considered will be different for an analytical and an empirical approach. For analytical evaluation of design objectives and requirements, a natural approach will be top-down from global system properties to detailed task functions. For empirical evaluation, one should go from details to global features. Complex experiments involving, e.g., entire task situations will be meaningless if the system does not match user characteristics at the elementary level, for example, if the readability of the interface information is poor.

45.8 ANALYTICAL EVALUATION APPROACHES

Analytical evaluation involves the comparison of design concepts or prototype characteristics with a reference model, which may be derived from work analysis within the particular organization for which the design is aimed or based on generalization from similar work systems.

It follows that a systematic analytical evaluation according to the scheme shown in Figure 45.3 will proceed systematically inward from global to local features. At the innermost boundaries, evaluation is focused on the match of the form of the displayed information to the users preferred strategies and tactics for coping with work requirements in an effective and for them acceptable way.

The actual structure and focus of an analytical evaluation depends on the phase of the design process when an evaluation is performed. Frequently, an iteration between design and evaluation throughout the process will be used. In addition, an analytical evaluation will not be a well-formed top-down progression, iteration between several levels will often be convenient, without losing the systematic use of a comprehensive work analysis.

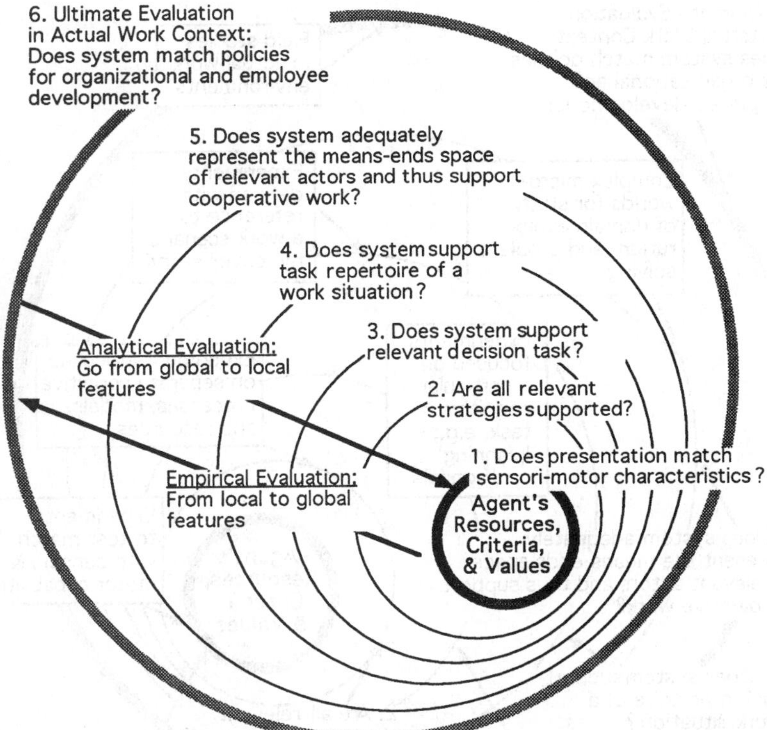
6. Ultimate Evaluation
in Actual Work Context:
Does system match policies
for organizational and employee
development?
5. Does system adequately
represent the means-ends space
of relevant actors and thus support
cooperative work?
4. Does system support
task repertoire of a
work situation?
Analytical Evaluation:
Go from global to local
features
3. Does system support
relevant decision task?
2. Are all relevant
strategies supported?
1. Does presentation match
sensori-motor characteristics?
Empirical Evaluation:
From local to global
features
Agent's
Resources,
Criteria,
& Values

Figure 45.3 The figure demonstrates how different evaluation questions can and should be asked at the various levels described in the framework for work analysis. In addition, it is shown that a different ordering of the evaluation questions should be considered for an analytical and for an empirical approach to evaluation. (Reproduced from Rasmussen, Pejtersen, and Goodstein (1994), by permission from John Wiley and Sons.)

Several other systematic approaches to analytical design evaluation than the framework proposed here have been suggested, which refer to different boundaries of Figure 45.3. One of the first methods was the computational GOMS (goals, operators, methods, and selection rules) model, a predictive model of user performance, developed to constrain the design space and increase prediction of usability requirements (Card, Moran, and Newell, 1983). It addresses several hierarchical abstraction levels of goals, tasks, operators, and selection rules. The most recent analytical approaches are the "cognitive walk-through" and "heuristic evaluation" methods, which are both based on expert judgments and a pencil-and-paper comparison of system characteristics with established guidelines, models, and cognitive theories.

45.8.1 Usability Inspection and Heuristic Evaluation

Usability inspection is an informal, analytical evaluation of a user interface and it is applied as a substitute for empirical end-user testing. It is based on subjective expert judgments of how well an interface is likely to perform with reference to criteria derived from usability guidelines or human factors guides. Systematic usability inspection methods imply evaluation sessions during the design phases after an interface design has been generated. An open-ended evaluation is conducted during a couple of hours by a small number of usability experts. These experts try to "recognize" the most common and severe interface usability problems—sometimes aided by a typical task scenario. They are followed by observers, that is, note-taking domain experts, who capture their comments and communicate these and the related domain knowledge to the evaluator.

An example of a systematic usability inspection method is "heuristic evaluation," also called "discount usability." The approach advocated is to apply the 10 most important

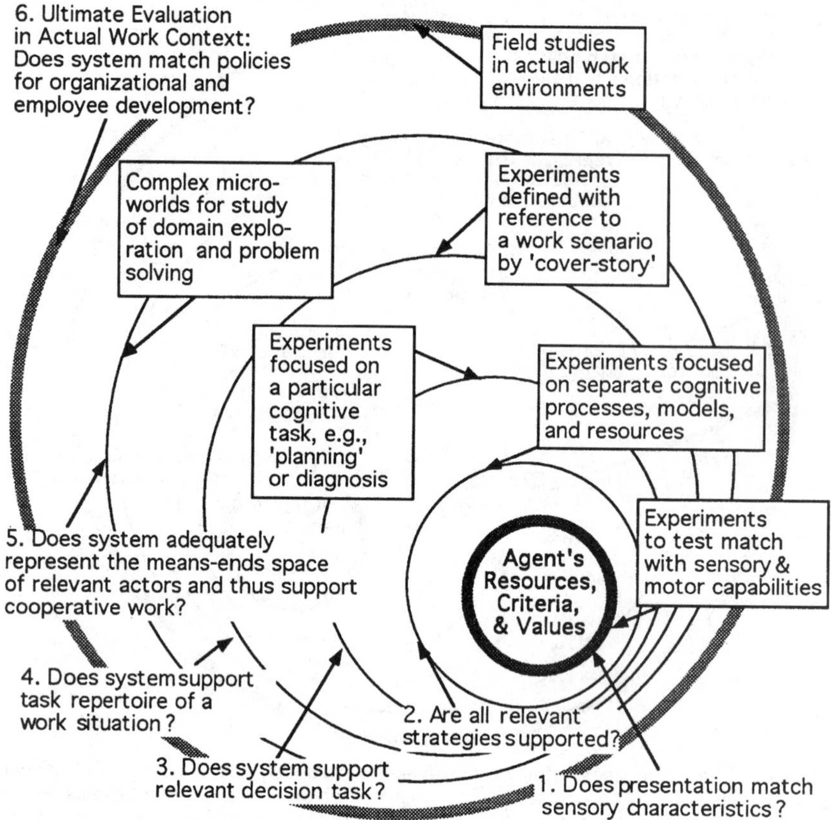

Figure 45.4 The boundary conditions of evaluation analysis and experiments in field studies and laboratory tests. The framework can be utilized to support system designers and evaluators in performance of analytical evaluation and in specifying appropriate experiments for empirical evaluation and/or accessing the experimental results available in the literature. It illustrates the fact that boundaries defined by the dimensions of the framework can be made compatible with the constraint envelopes used for defining experiments. (Reproduced from Rasmussen, Pejtersen, and Goodstein (1994), by permission from John Wiley and Sons.)

usability rules found in guidelines for a "quick and dirty" evaluation of an interface (Table 45.2). Guideline principles used in usability heuristics are features such as:

- Use simple and natural dialogue.
- Use as few steps as possible.
- Match to task.
- Match system concepts and user concepts.
- Display only needed information at the right time and place.
- Classify information used together.
- Match sequence of information to users' most effective way of acting.
- Let user control the dialogue.
- Adjust sequence to preference; a sequence chosen by designer will indicate the suggested sequence.

Debriefing sessions are then used to brainstorm about ideas to remedy usability problems. Recommended by usability evaluators who focus on usability defects of an interface is alternation between heuristic evaluation and empirical user tests in an iterative design

Table 45.2 Examples of Criteria Involved in Heuristic Evaluation

Visibility of system state,
Simple and natural dialogue,
User control and freedom,
Recognize rather than recall,
Speak the user's language,
Minimize the user's memory load,
Consistency and standards,
Feedback,
Clearly marked exits,
Shortcuts,
Good error messages,
Prevent errors and support error recovery,
Help and documentation.

process (Nielsen, 1993). Usability inspection includes *severity ratings,* which are the determination of how serious a problem with an interface is, such as: How much does the problem impede user progress; Quantitatively, how many users will be bothered; how much delay in task performance; and qualitatively, how annoying is the problem; will it only be a first-time problem for the first-time user?

These are all factors that are used to judge how much a problem is in need of repair and such judgments reflect the experience and knowledge of the evaluator. Usability inspection of interface defects results in a prioritized list of problems with respect to severity of the problems (severity judged by frequency of problem, its user impact and persistency of occurrence) as well as recommendations and new suggestions for redesign. This involves analysis of the larger user context and a deeper understanding of the task domain, for which the usability inspection approach is not applicable (Conklin, 1991).

45.8.2 Cognitive Walkthrough

Cognitive skill acquisition theories suggest that exploration of effective actions in relation to current goals and task contexts is a problem solving process that will facilitate learning. Based on this, a basic assumption of *cognitive walkthrough* is that a design should support learning by exploration. Cognitive walkthrough is an analytical evaluation of the ease of learning a design with reference to accepted attributes correlated with ease of learning. Focus of the evaluation process is the identification of mismatches between designers' and users' conceptualization of a task by simulating the user's problem-solving process and tracking the correct actions that a user should be able to take without problems imposed by the interface (Lewis, Polson, Wharton, and Rieman, 1990; Wharton, Rieman, Lewis, and Polson, 1994).

The method involves three major stages of analysis, and it is performed by individual experts or groups of experts at any design phase using a mock-up, a prototype or a complete interface:

1. The input to a "walkthrough" is defined in terms of identification of the users and their background knowledge, the sample tasks for evaluation, the action sequences needed for task completion and the implementation of task actions in the interface with focus on links between task actions and system feedback.
2. The walkthrough the action sequences for each task includes telling a credible story to create a task scenario and speculate whether the user will try to achieve the right effect, notice the available correct actions, associate correct actions with the desired effect and notice that progress has been made.
3. The recorded evaluation session is analyzed and used to suggest design revisions. The information source for redesign is the critical information captured about the system's requirement to user knowledge and user learning as well as other discovered issues that need to be fixed.

A description of the cognitive walkthrough process is found in Table 45.3.

Table 45.3 An Overview of the Steps to be Taken in a Cognitive Walkthrough Evaluation Process

1. Define inputs to the walkthrough
Identification of the users
Sample tasks for evaluation
Action sequences for completing the tasks
Description or implementation of the interface
2. Convene the analysis
3. Walk through the action sequences for each task
Tell a credible story considering..
Will the user try to achieve the right effect?
Will the user notice that the correct action is available?
Will the user associate the correct action with the effect that the user is trying to achieve
If the correct action is performed, will the user see that progress is being made toward solution of the task?
4. Record critical information
User knowledge requirements
Assumptions about the user population
Notes about the side issues and design changes
The credible success story
5. Revise the interface to fix problems

Source: Reproduced from Nielsen and Mack (1994), by permission from John Wiley and Sons.

The analyst tells a "credible story," or presents a scenario that involves assumptions about the information processing required by a given user with a certain task and an assumed background knowledge. From here, the effective action sequence required by the interface is inferred. This simulation implies that links are interface features that are more critical than other features, such as links between a user's task description and a correct action, and links that provide feedback about the progress of a previous action. This theory-based evaluation approach is derived from several cognitive theories and models of the problem solving process, which in some cases are derived from laboratory experiments. The work on cognitive walkthrough has been developed during several years through experiments with the objective of finding out how the underlying theories can be implemented in operational procedures that are not too time consuming or too complicated for practitioners. Examples of such theories are general models of skill acquisition and learning mechanisms that demonstrate how representations of correct actions are stored with current goals and task context, and how problem solving is a process of discovering effective actions (Anderson, 1987; Newell, 1990). A comparison of the overview of the walkthrough items in Table 45.2 with the levels of Figure 45.3 demonstrates that the proposed framework for cognitive work analysis will be useful to ensure consistency of a walkthrough session.

Comparison of the cognitive the walkthrough approach, human factors guidelines, and usability testing have been done within dimensions such as:

- **a.** Total number of problems captured
- **b.** Severity of usability problems
- **c.** Scope of problems captured
- **d.** Level of task action and task content that the methods cover

Comparisons of the interface problems found by human factors experts using different analytical usability methods such as heuristic evaluation and cognitive walkthrough for verification of the interface quality have shown differences among these methods, for instance, in cost effectiveness or number, type, and severity of defects that are discovered by a single method. Similarly, these methods have been compared with empirical validations during experimental user testing in laboratories. Such comparison shows, for example, that user testing is generally better at discovering more serious problems than heuristic evaluation inspections performed with guidelines.

Expert inspection and empirical end-user testing seem to lead to the identification of different usability problems and therefore supplement each other (Nielsen, 1993). For a more detailed discussion of comparison of these methods and their strong and weak points in system evaluation see Desurvire (1994) and Nielsen and Mack (1994).

Several different, and more practical analytical techniques can be applied during analytical design–evaluation iterations. For example, in the *parallel design approach,* design alternatives are explored and evaluated during brain storming sessions by making participants in a design team come up with their independent suggestions through rough drafts and paper sketches (Nielsen, 1993). Another analytical approach is *forward scenario simulation,* by asking "what-if" questions about the use of the system in a simulated task situation, and the action possibilities the system offers. After evaluation sessions, and before a new prototype can be around, discussions in design teams with *hypothesis testing* and *brainstorming sessions* are often used to decide about possible features to change (Cordingly, 1989). Hypothesis testing and brainstorming are actually commonplace methods that can be used without any prototype, and in particular at the early stages of the design evaluation. In *retrospective testing* all evaluation records are collected, and users review this material and comments on the course of a session of system use.

45.9 EMPIRICAL EVALUATION APPROACHES

An empirical evaluation of a design concept or a prototype may be necessary for several reasons. Evaluation experiments can be necessary at many levels of functional detail, from very focused experiments to test interface features to more global experiments to test overall system functions. Experiments for empirical system evaluation can take on many different forms—from small-scale experiments for testing local design features or particular human attributes, to complex microworld setups for evaluating hypotheses about user adaptation and preferences for certain strategies in a complex environment.

Very often, evaluation questions are zapping among several levels of analysis with very different requirements for control of the boundary conditions. This makes it very difficult to generalize the evaluation results to other, similar work domains and support systems. Based on our experience from analysis of actual work systems and from system evaluation sessions, we find it necessary to define the boundary conditions of empirical evaluation experiments very carefully with reference to a consistent framework, as for instance the one shown in Figure 45.3.

In laboratory experiments, the processes and anatomy of the experimental setup replace those of the real work domain. A subject will normally be well aware of this fact, in particular when "in trouble." It is often found that a subject will try to explain unexpected performance by peculiarities in the experimental design or bugs in a computer program. In an actual work environment, there are very subtle many-to-many relationships between the goals, the functions, and their possible implementations at the physical process level, and it will be very difficult for a subject to judge what is, and what is not, included in the simulation. Therefore, the subject will have to infer the scope of the experiment, as well as the goals, constraints and functions included, on the basis of more or less intuitive assumptions about the work domain or the "source world" as conceived by the designer of the experiment. In this situation, it is important to be able to analyze and describe explicitly how the source domain, e.g., the actual work domain and its users is treated in the research hypothesis, and to make explicit how the behavior-shaping constraints are transferred to the research domain. It is only possible to draw conclusions from selective experiments and simulations if we can demonstrate unambiguously that the constraints from the experimental source domain and the experimental conditions, results in a similar space around the subject as does the constraints from the actual work conditions.

For empirical evaluation experiments, it is necessary to establish an experimental work situation that creates a well-defined space of constraints around the subject, and to study whether subjects' responses to this boundary leads to the mode of behavior that was assumed as the design basis. For a comprehensive experimental evaluation of a system, it is necessary to define a suitable sequence consisting of a set of boundary conditions (Figure 45.4) with increasing distances from the actor to be able to evaluate more encompassing features of the system. For proper integration of the results, such experimental evaluation scenarios should be compatible with the structure of the work analysis underlying design and with the design specifications and concepts used for analytical evaluation (Figures 45.2 and 45.3). Usually a section of a work environment is separated and transferred to an experimental setup. Proper experimental control requires an explicit definition of the boundary conditions at the particular cut that separates the experiment from the influences of the rest of the environment.

It should be noted that complex experiments for evaluating advanced information systems can be wasted, if the evaluation is carried out too early at too "high" a boundary. For example, if experiments are planned to evaluate the functionality of a prototype in advance of a test of the interface readability.

The boundary along which the cut is made varies considerably, depending on the aim of the experiment. Therefore, a suitable framework for identifying the experimental boundary conditions must be a subset of the framework for the analysis of work in general. Figure 45.4 links the various dimensions of the framework for work analysis to various boundaries useful to define different types of evaluation experiments. The innermost boundaries correspond most closely to the traditions of experimental psychology. The remaining boundaries "move" the context successively further from the actor to encompass more and more of the total work content in some kind of increasingly complete simulation.

The various categories of evaluative experiments will now be discussed with reference to the use of the framework for specifying the boundary conditions in order to evaluate how well a system match the constraints defined at each boundary. The categories are labeled with reference to the boundaries shown in Figure 45.4: Evaluation of how well a system matches the users' sensory-motor characteristics, the users' cognitive capabilities and mental processes, the cognitive decision task and strategies, the work task situation, and the work environment and its social organization of work.

45.9.1 Boundary 1: Does the System Match Sensory–Motor Characteristics

Experiments with the ergonomics of the system at this boundary will examine whether the physical configuration of a system, the equipment, and its arrangement at the users' workplace correspond to the anthropometric and sensory characteristics of the user group. Evaluation of, for instance, a computer system will focus on the size of letters, the readability of the typography, and the graphics of the displays. Checks will be made on whether the keys can be clearly distinguished and whether the size and form of prompts and controls are functional. Experiments will serve to determine for example whether the sound is audible at a proper distance, and whether the use of colors is in agreement with current standards for visual perception.

This level will normally be covered by a great number of ergonomic standards, and it can be evaluated with conventional ergonomic guidelines and human factors checklists (see, e.g., Boff and Lincoln, 1988; Boff and Lincoln, 1995; Mayhew, 1992; and other chapters in this Handbook).

45.9.1.1 Experimental Setup

Suggestions for choice of test methods and the related experimental setup can be found in standard ergonomic handbooks. At this level a typical group of experiments takes place in usability laboratories for the comparison of competitive products such as mechatronics, HI-FI technology and computer products, etc. Such tests typically involve the use of questionnaires, interviews, and video or audiotaped task scenarios (Kennedy, 1989). Frequently used performance criteria are user satisfaction, number of steps for actions, percentage of user population capable of performing tasks using specific controls etc.

However, evaluation of the next generation of multimedia interfaces will be more complex as they include new features such as speech recognition, voice communication, and coupling of interfaces to a variety of communication network programs, for example the various graphical browsers and home pages in the Internet. In addition, when interfaces become more adaptive, they are likely to be based on intelligent agents and computer control of a changing interface composition. The computer may observe the users' actions and adapt its displays to the user situation through eye tracking, gesture recognition, and other observation techniques. Virtual reality technology will allow user interfaces to support exploration and manipulation of a complex, multidimensional representation of the work domain. In that case additional empirical data and new methods and techniques will be necessary to guide the evaluation and design process. This includes both the evaluation of anatomical and perceptual properties of the user population, and the ergonomic quality of the system, as well as the evaluation of the match of the interface with the users' cognitive capabilities.

45.9.2 Boundary 2: Does the Presentation Match the Users' Cognitive Processes and Mental Models

This level addresses evaluation principles that are important for the understandability of the information flow in the communication between the system and the user. Numerous experiments and empirical data have been published during the last decade in human factors handbooks, checklists, and guidelines leading to general design principles for display composition, consistency, coherence, use of colors, icons, WYSIWYG interfaces, WIMP (windows, icon, menu, pointer) systems, overlapping windows, etc. Experimental evaluation and tests have treated questions such as: How do mental strategies and cognitive criteria influence the understandability of the display of information, the single messages, and the provided functions and action buttons? Are the sequences of the interface messages sufficiently coherent, their number acceptable, and the labels of identical functions consistent? Is the understandability of the communication supported adequately by uniform prompts of same functions in different contexts? Is the flow and speed of information appropriate to follow the natural sequences of acts and movements? Are the relevant action possibilities immediately recognizable by the user? Is the user's memory supported by the display of identical categories of messages in uniform units, and are similar functions placed in the same screen position?

Many of these aspects may not need a new empirical evaluation if appropriate experience and human factors handbook data, checklists, and guidelines were available during design. However, some answers to questions related to the understandability of the communication may be very context sensitive and thus demand empirical evaluation, such as: Does the linguistic level and the use of "interface language" correspond to the language level and vocabulary of the assumed categories of users in a specific work domain? Is it familiar to such an extent that a novice user understands the communication and can get started without getting lost or stuck in the process? How is the best combination of different media such as voice, animation, graphics, video, and text, compared to the task situation and the characteristics of the users in a specific domain?

45.9.2.1 Experimental Setup

Experiments at this level are concerned with very basic human traits and often with very general (standardized) interface features. Experiments in this category may therefore be carried out by system evaluators both in connection with a specific application, and frequently also as general efforts to establish common knowledge, data banks, and accumulate knowledge to be used for instance in handbooks and guidelines.

In this category of experiments, the formulation of the subject's instruction is at the procedural level and is very explicit. Methods employed are generally established methods from experimental psychology. They serve to define the constraint boundary around the experimental situation, and isolate it from (a) the general, personal knowledge background and performance criteria of the subject, and (b) any eventual higher level considerations within the experimental domain itself. The use of only one experimental method for each study is commonplace.

In addition to identifying the boundary conditions with reference to the constraints shown on Figure 45.4, it is necessary to clarify the links between the experimental work domain as seen by a subject and her or his private world of experiences. It is obvious that although an experiment is embedded in the world of the designer or evaluator, it interacts at the same time in some way with the private world of the subject. In traditional psychological laboratories, the setup of the experiment and the instructions to the subject are intended to decouple the experimental conditions from this private world.

Thinking aloud is an example of an experimental, psychological evaluation technique. It has become a popular evaluation method used in laboratories usually for the purpose of getting access to users' thought process about what they do, how they do it, and why they do it. It aims at the examination of users' mental capabilities and cognitive decision making at boundary 2, as well as at boundary 3 of the framework described in the following (Lewis, 1982; Ericsson and Simon, 1984; Dening, Holem, Simpson, and Sullivan, 1990).

45.9.2.2 Examples

This is the typical experimental level for most work in human–computer interaction, from where these examples are drawn. For example, the many tests of users' perception of icons to be used in the user interface.

In a laboratory experiment, a number of tests were conducted with four different sets of icons, each containing 17 icons designed for a graphical interface, in order to compare their ease of learning, efficiency of use, and the subjective user satisfaction with the icons. A number of laboratory experiments were conducted and different measurements adopted (Bewly, Roberts, Schroit, and Verplank, 1983). Rogers (1986) studied comprehensibility of icons in relation to the complexity of icons, which was defined as the number of elements contained in each icon. Users compared sets of icons with different complexity with a textual description of each icon, and they were asked to match text and icon. Icons with many elements were more difficult to understand than were icons with only one element, Best understood were icons combining elements of concrete objects and an abstract representation of an action.

Evaluation can be dedicated to user characteristics such as age (Czaja, 1988), gender (Fowler and Murray, 1987), and to assumed differences between groups of users, which have proved to be quite general such as spatial memory and reasoning capabilities (Gomez, Egan, and Bowers, 1986).

A large amount of experiments have been dedicated to evaluation of hierarchical menus. Using a hierarchical menu system will require good spatial memory abilities, and a laboratory experiment with subjects using a hierarchical file system was conducted to test how spatial memory ability would effect the use of the system. The experiment showed that people with this ability will be able to remember information within a hierarchy without any overview map, but for people without good spatial ability, an overview of the hierarchy helps remembering location of information (Vicente and Willeges, 1988).

An indirect empirical evaluation of a display for diagnostic performance has been suggested by Vicente, (1991). In his experiments, a direct diagnostic evaluation is replaced by a test of a match of an interface to the mental models of subjects by means of a memory test (based on DeGroot's (1965) experiments with chess experts). This consisted of a test of whether a group of experts, on the basis of a short presentation, could accommodate the data presented by a display in their mental model, and regenerate the individual data on request—and do it better than novices. Later experiments with this memory recall test (Moray et al., 1991) have shown that great care should be taken to identify reliably the cognitive mechanisms that are to be considered for an empirical test.

45.9.3 Boundary 3: Does System Support Cognitive Decisions and Task Strategies

Going from boundary 2 to 3 results in a focus on studies of problem solving and decision situations. Here, the cognitive processes are more complex, and more constrained by the task environment, but less constrained by the experimental tool. A basic question to be asked at this level is: Does the system effectively support the cognitive decisions that have to be made during task performance? Does the system support the actor's decision making—Are exploration, situation analysis, goal evaluation, and planning supported for familiar as well as less familiar situations? Can decisions be made smoothly without any interruptions, without constraints on decision sequences, and without unnecessary delays? Does the system support the user's attempt to learn the task through exploration, experiments, and trial and error? Is the effect of errors visible and can they be corrected, for example, by shift of strategy?

Experiments can be designed to test the correspondence of the semantic content and the functions provided by the system with the end users' information needs during decision making. This implies that work domain information is organized to map the decision-related information needs as means and ends at any current state of task execution.

For this, the system's support of users' goal formulations and situation analysis in complex task situations should be evaluated together with its support of their choice among action alternatives. The users' choice of strategies in task decisions that call upon the relevant strategies should be observed to see if the effective work strategies for the task goal are supported.

At this level few adequate guidelines and standards are available (apart from the guidelines for error recovery). The requirements are very likely to vary with the work domain, the tasks and the users, and thus a systematic work analysis is necessary to serve as a reference for evaluation.

45.9.3.1 Experimental Setup

This category moves one step away from the actor toward the total work situation by focusing on the study of separate decision functions such as diagnosis, goal evaluation, planning, and/or the execution of planned actions. Therefore, more complex task situations must be simulated, which also serve to evaluate the subject's goal formulations, and which allow for a faithful interaction between the action alternatives presented. Experiments designed for the evaluation at the mental strategy level will be concerned with the question: Are the effective work strategies supported? In this case, classical, well-controlled experiments can be run. Task situations that call upon the relevant strategies can be designed and the subjects instructed with respect to the goal of the task and the use of the strategies. For such experiments, tasks can very well be "artificial" without relation to the work situation for which the system is intended. The question is: Given that the user adopts a particular strategy, does the interface enforce the relevant mental model, and is the information necessary for the strategy present in an understandable form?

Many different experimental evaluation techniques can be applied, for example, a variation of the thinking aloud method mentioned under boundary 2, which is called *constructive interaction or codiscovery learning* (Kennedy, 1989). The method is characterized by the verbalizing of joint problem solving, when two test users verbalize their reactions, and thus, implicitly, seem to probe each others' attitudes. This leads to more comments than a single-user thinking aloud experiment and may be well suited for more complex evaluation experiments (Berry and Broadbent, 1990; Hackman and Biers, 1992; Wright and Converse, 1992).

45.9.3.2 Examples

A typical group of experiments evaluating performance during use of particular mental strategies is concerned with diagnosis and fault management in technical systems (see, e.g., Rouse, 1981; Sanderson, 1990). Such experiments are often labeled by a cover story, a task label, such as "fault-finding" or "troubleshooting." However it seems clear that the diagnostic conditions in an actual work situation are different from those in an investigation requiring a subject to locate a fault in a constructed logic network. Therefore, the results of the experiments can only be generalized when the strategy imposed on the subjects by the experimental setup is properly identified.

A decision activity study of conceptual representations of task strategies adopted a case study method to evaluate "on the job" online searching in large databases. Observations about the professional searchers were recorded by the experimenter, and their "thinking aloud" during the three task stages of preparation, proper search, and quality controls and assessments was taped in order to identify individual characteristics of searching styles as well as general regularities of cognitive styles. The iterative, conceptual moves during cognitive decisions of analysis, evaluation, and revision typically fell into two groups: The operationalist search using system features to change the search without change of the meaning of the original conceptual representation. The conceptualist search would use different domain sources to arrive at new concepts that would change the initial conceptual representation of the problem (Fidel, 1984).

45.9.4 Boundary 4: Does the System Support Relevant Task Situations

Does the system adequately support the actual work task situation? Is its capacity adequate? The question here is whether the system supports the entire task repertoire—Are the tools adequate and their functionality sufficient, and does the information cover the complete work task space?

Experiments may serve to evaluate whether information is available about the basic concepts of the system and its overall architecture. Is it possible to navigate among tasks, and to pursue several, different task-related goals? Is the content and form of different screen displays and navigation among these controlled by a coherent task design? Are the relevant action possibilities available for different task situations, and do they help the user to formulate goals and keep track of the task execution?

Another important, but highly domain-dependent aspect, is the evaluation of the safety of the system and its error tolerance, and how well unacceptable consequences such as loss of data and resource demanding repetitions are avoided.

This level requires an extensive evaluation of the degree of effectiveness of the system functions in relation to the individual work tasks, which will require evaluator domain expertise and detailed knowledge of the work tasks that the system should support.

45.9.4.1 Experimental Setup

At this level, experiments serve to test well-defined task scenarios. The development of advanced information technology, and the need for models of cognitive processes in work performance have brought about a significant interest in more complex experiments and usability tests with focus on actual task situations. Consequently, the evaluation and simulation of task situations controlled at boundary 4 in Figure 45.4 has become popular.

The heuristic usability inspection method mentioned previously is an analytical method that defines a few number of the attributes of a system. The approach of usability measurement in context of use (Bevan and Macleod, 1994) is an empirical approach that covers a complex set of task attributes to be evaluated in task scenarios. The method defines the aspects of the context of use of the product that are most likely to have an important impact on the quality of use of the overall system, and that therefore should be included in the usability measurements. The choice of evaluation context should be representative of attributes of the real use context. The quality of use is for most systems suggested to be its usability measured quantitatively as percentage by:

a. *Effectiveness,* which relates to the accuracy and completeness with which goals of using the system can be achieved—measured for instance by number of errors and by number of completed units. For instance, task effectiveness can be measured by a quantitative measure of the amount of the task goals completed by a user in the output of the task. Qualitatively, task effectiveness can be measured as the degree to which the output achieves the task goals.

b. *Efficiency,* which relates to the utilization of resources required to accomplish a goal such as amount of mental resources, physical resources, time, and financial costs. For instance, efficiency can be measured as productive time defined as the task time spent on the progressing toward the goal, after help, search, and snag times have been removed.

c. *Satisfaction,* which relates to the acceptability and usability of the system as perceived by its users, and by people who are in some way affected by the system. Satisfaction is measured by number of positive or negative comments during use, or by attitude rating scales. For instance, user satisfaction can be based on 50 questions in a questionnaire asking the user to answer whether it was difficult to learn new functions, and whether the software was frustrating to use, etc.

The method of usability in context advocates empirical tests based on user participation and video recording of representative users in typical work tasks. It operates with a relative importance of usability measurements depending on the empirical evidence gathered from an analysis of the context. In order to create an evaluation context that is compatible with the real use or workplace context, it is suggested to gather information by use of *The Usability Context Analysis Guide* developed in the MUSiC project (Metrics for Usability Standards in Computing by Bevan and Macleod (1994); see Table 45.4). The methods and usability measurements are supported by various software tools. The final result is a global assessment of usability given by a single numerical figure with a breakdown of the overall assessment into affect, efficiency, helpfulness, control, and learnability.

Setting up the experiments of task situations can of course combine several techniques such as a workplace field study to identify critical tasks, and the *scenario technique,* which is a frequently used approach to anticipate how users will interact with a system. A task scenario can be set up to evaluate a variety of implementations, such as a prototype with minimal functionality and limited scope, or premature design ideas. A scenario can also be an experimental setting recorded over a certain time period that involves the individual user, who, with a specific set of computer facilities, is asked to achieve a specific outcome under specified task-oriented circumstances (Pejtersen and Austin, 1984; Pejtersen, 1990; Carroll and Rosson, 1990, 1992). A task scenario can be extended with the character scenario technique of making up a "user character" to identify prototypical user characteristics, which are then used in laboratory task scenarios (Buur, 1995; Campbell, 1992; Carroll, 1994, 1995; Verplank, 1993).

Table 45.4 Examples of the Parameters Involved in the Decomposition of the Context of System Use or Intended System Use to Be Identified for an Adequate Description of the Context, Which Should Be Covered by the Experimental Evaluation. Naturally, What Attributes to Include in the Usability Test Will Depend on the Intended Use of the Product

Equipment
- *Basic description*
  - Product identification
  - Product description
  - Main application areas
  - Major functions
- *Specification*
  - Hardware
  - Software
  - Materials
  - Other items

Users
- *Personal details*
  - User Types
  - Audience and secondary users
- *Skills and knowledge*
  - product experience
  - System Knowledge
  - Task experience
  - Organizational experience
  - training
  - keyboard and input skills
  - Qualifications
  - Linguistic ability
  - General knowledge
- *Personal attributes*
  - Age
  - Gender
  - Physical capabilities
  - Physical limitations and disabilities
  - Intellectual ability
  - Attitude
  - Motivation

Task
- Task breakdown
- Task name
- Task goal
- Task frequency
- Task duration
- Frequency of events
- Task flexibility
- Physical and mental demands
- Task dependencies
- Task output
- Risk resulting from error

Environment
- *Organizational environment*
- *Structure*
  - Hours of work
  - Group working
  - Job function
  - Work Practices
  - Assistance
  - Interruption
  - Management structure
  - Communications structure
  - Remuneration
- *Attitudes and culture*
  - Policy on use of computers
  - Organizational aims
  - Industrial relations
- *Job design*
  - Job flexibility
  - Performance monitoring
  - Performance feedback
  - pacing
  - Autonomy
  - Discretion
- *Technical environment*
  - Configuration
  - Hardware
  - Software
  - reference materials
- *Physical environment*
  - *Workplace conditions*
    - Atmospheric conditions
    - Auditory environment
    - Thermal environment
    - visual environment
    - Environmental instability
  - *Workplace design*
    - Space and furniture
    - User Posture
    - Location
  - *Workplace safety*
    - Health hazards
    - Protective clothing and equipment

Source: Reproduced from Bevan and Macleod (1994), by permission from Taylor and Francis.

Evaluation at the task situation level has been used to evaluate decision-support systems for process plant control rooms, for instance, "disturbance analysis systems" for hazardous systems, such as nuclear power plants. A discussion of such a task evaluation and the related methodological issues is found in Rouse (1984). He distinguishes among the following aspects of the human–work interaction:

- Compatibility with human sensory and homomorphic characteristics—That is, can the displays be read and the controls operated?
- Understandability—Can the user understand the text and graphics of the interface? Does the user understand the provided functions?
- Effectiveness—Do the functions provided meet the task requirements of the end user? Can the system be used?

In this way, his work task evaluation includes questions matching the three inner boundaries of Figure 45.4.

45.9.4.2 Examples

Experiments span many different work domains, and the performance measures are often defined by the problem of the task domain instead of using general performance measures such as error frequency or reaction time. Therefore, work analysis and identification and evaluation of performance measures become an intertwingled process, as illustrated by an experiment made by Woods and Sarter (1993). A scenario experiment was designed to study the possible effects of fatigue on cognitive processes during the task situation of pilots' long-haul transport operations, such as the effect of narrowing the perceptual focus, loss of cohesive perception, increased distractibility and inaccurate recall of operational events. In cooperation with domain experts, a model of expected crew behavior was developed for each episode of a scenario, with operational tasks and events that would probe these tasks and their related cognitive processes that might be effected by fatigue. The model of the task context specified a set of expected problems that could occur, such as how an episode and envelopes of acceptable trajectories could evolve over time, how actual pilot behavior might deviate from standard procedures and still be acceptable, and what behavior would lead to significant operational problems.

A set of evaluation experiments at this level study the performance of information retrieval systems at the task situation level, either in a realistic environment with realistic users, or with subjects who perform searches on test questions devised by the experimenters usually to evaluate and compare the precision of retrieved items, and the number of recalled items compared to what is estimated to be relevant for a search question (Beaulieu, 1991; Saracevic, 1995). The ongoing TREC (Text Retrieval Conference) experiments were designed to allow large-scale laboratory testing of information retrieval techniques. Some investigations have been concerned with the evaluation of automatic and interactive query expansion facilities implemented in different interface environments. The retrieval task is viewed as two distinct subtasks: One concerned with query construction and the other with viewing or browsing results. Laboratory experiments address interface features that cater for each subtask (Beaulieu, Robertson and Rasmussen, 1996; Robertson, Walker and Beaulieu, 1997).

In this category, we also find studies of decision making and diagnostic behavior in complex work situations using simulators of high functional fidelity for aviation, chemical process plants, power plants, etc. (Woods, 1984).

45.9.5 Boundary 5: Does the System Adequately Represent Work Environment

A bottom-up approach will be necessary at this level beginning with the user characteristics, and closing with the top level of the final work context including all the boundaries as shown in Figure 45.2, and discussed above in relation with experiments as illustrated in Figure 45.3. Evaluation experiments may investigate the relationship between the actual use of the system and intellectual and emotional aspects of users' working style, the way they like to work, and their personal problem-solving habits in the total workplace context. A typical example is the evaluation of the impact of users' subjective value criteria and preferences on their acceptance of a new system. Users' acceptance can relate to the ability to circumvent difficulties without error messages and utilization of help texts and documentation manuals. Does the system support several task strategies, and can the user shift goals and tasks concurrently without loosing support from the system? Does the system provide all the necessary functions, and is the user's mental model of the work domain supported by the interface—also during distributed decision making? Does the system improve the cooperation and task coordination among users, and does it improve their competence and, hence, the quality of their work situation?

For this boundary, the evaluation must be based on actual work scenarios generated from an actual work analysis, when professional users are observed using the system in a specific work domain and work organization. The aim is not a task simulation but a workplace simulation. The methods will be similar to the evaluation at the work task situation level, but the complexity and scope of the simulation will be greater.

This is the ultimate level of simulation complexity and must include not only the system's effect on a complex work place situation, but also facilities for simulating different interpersonal communication modes among cooperating users.

45.9.5.1 Experimental Setup

Experiments at this level depend on techniques similar to field studies. A more recent category of experiments has been focused on human problem-solving behavior in complex simulated work environments, in which the entire decision process is activated including value formation and goal evaluation as well as emotional factors. In addition, the match of the system with the different levels of human work characteristics and its possible impact on the influence from the work environment as described in Figure 45.2.

Evaluation of real-life task situations can be made at various boundaries which increase stepwise in their "distance" from the actor. When planning experiments, the evaluator can get inspiration from the examples of the concepts allocated the different dimensions of the framework for work analysis as illustrated in Figure 45.4.

Experiments in this category aim at a study of the performance of the subjects in their explorations of an entire problem domain, including their formulations of the problem present, the goal to adopt, and the solution to choose. Because of their very nature, the instructions for these experiments become very open. Instructions given by a "cover story" may relate the experiment to the private world and general experience of the subject, and the performance will then be strongly influenced by the particular background of the individual subject. In addition, it is very difficult to control reliably the actual goals and performance criteria adopted by naive subjects (e.g., students). At this level, the use of subjects who are very familiar with the actual work domain will be necessary, and the instructions and the experiment will have to simulate a realistic task situation.

In addition to the scenario techniques mentioned at boundary 4, evaluation techniques such as *participatory design, focus group interviews,* and *interactive and rapid prototyping* are frequently used methods, which enable designers to evaluate and discuss on the spot with users, how well their design ideas represent the total work situation, as well as their proposed system changes covering the total work situation (Kyng, 1991; Greenbaum and Kyng, 1994).

45.9.5.2 Example

A large-scale study of the information-seeking situation was conducted in a real environment under as real-life conditions as possible with participation of users and professional searchers. The objectives were to evaluate the interaction among users and information retrieval systems through a series of observations and experiments, including a considerable amount of contextual work situation factors. In particular, five classes of variables that would impact an information retrieval task were included in the experimental setup: *User:* effects of the context of real questions and constraints coming from genuine users. *Searcher:* effects of cognitive traits of professional and real life searchers. *Question:* effect of categories of questions based on classification of real questions and their structures. *The search:* described by the effect of different types of searches and their overlap, effectiveness, and efficiency. *Items retrieved:* magnitude of retrieval of relevant and non-relevant items. For a large amount of combination of variables explored no statistical significant result was found, and contributions to changes in the design of the user–system interaction was difficult to find (Saracevic, Kantor, Chamis, and Trivison, 1988; Saracevic, 1995).

45.9.6 Boundary 6: Field Evaluation in Actual Work Environment

As the ultimate step evaluation at boundary 6 in the actual work context will address the question: Does the system match policies for organizational and employee acceptance and development? How is its impact on the work context and the quality of the work situation? Does the system support several coherent work task activities and the cooperative coordination of activities among several users, maybe in different departments of the organization, and does it support interaction and coordination with institutions outside the organization? In other words, will the system be accepted by the social work organization and objective company goals and strategies? Whether the system will be utilized according to the designers' intention may depend on many organizational conditions such as changes in the organization and work routines caused by the introduction of new tasks, role allocations, and task procedures.

This evaluation may also answer the question whether the design approach and the assumed work organization does match the performance criteria and preferences of the users. Will the system be used? And do the users like to use the system, and do they actually use it over a longer time span in the daily task situations, for which it was designed and to the degree as was expected? How will the system impact the role allocation and the coordination of work tasks? What task functions will be changed and what functions will be superfluous as the users adapt to the new system? How will the system change users' behavior, their goal formulation, value criteria, and preferences?

Field studies at the workplace are very resource demanding, and although they provide new useful insights into organizational issues that are very hard to simulate experimentally, they also introduce new methodological problems.

45.9.6.1 Experimental Setup

Real-life studies of daily work routines when cooperating professional users are observed using the system in a specific work domain and work organization is suggested as the ultimate test. As was the case in boundary 5, all the boundaries as shown in Figure 45.2, and all the levels of user characteristics, can be included in real workplace evaluations. For example, field evaluation of boundary 3 to 6 in Figure 45.2 can be used to validate at the actual workplace the correspondence of the design implementation with experimental results performed to test and evaluate hypotheses. The closer the levels are to the boundaries of the actor, such as boundaries 1 and 2, reflecting cognitive, perceptual, and sensory-motor user characteristics, the more cost effective it is to perform evaluation experiments as simulations of task situations in laboratories.

The empirical evaluation is best suited for separate tasks or functions for which a reasonable level of operational skill can be developed, and appropriate performance criteria found. For more complex task situations, and for evaluation carried out in the context of the users' workplace, the empirical approach is much more difficult to use convincingly and realistically because of all the uncontrolled (uncontrollable) variables. This leads to the danger that when the results of an evaluation are not in accordance with the intuition or expectations of the designer, then conditions will be readjusted until they match (confirmation bias). No explicit stop rule exists for the termination of these adjustments of the experimental conditions. When the designer strives to explain negative outcomes, he or she may very likely discover situational factors that need tighter control. This is a natural consequence of causal explanations and does not imply a manipulative experimenter.

In a realistic work situation, it is difficult to keep complete check of the complex set of often uncontrollable variables affecting users' behavior, and it is difficult to record users' responses and every detail in their use of various data-collection tools, because they are busy doing their tasks. Use of several, different methods for observation of behavior can provide more reliable results—especially if every data-collection method gives the same result. The use of multiple methods will often result in large amounts of data with some redundancy. But it can be worthwhile, as redundancy offers some possibilities for checking what users do against what they say they do.

Examples of methods that can be applied in field studies for elicitation of users' acceptance of a system, and evaluation of a system's performance and impact on the work situation, are observation of users' behavior and elaborate recording of task situations, for instance, registration of training sessions in system use, in a logbook or audio- and videotaped. Interviews, open-ended, structured, critical incidents interviews, and focus group interviews are well suited for many purposes (Caplan, 1990), for instance, to get user feedback from a variety of sources and to clear up successful and poor performance caused by tradeoffs. For collection of large quantitative data about the functionality of the system, software with data logging and online questionnaires can be a very valuable source. It can be used to compare actual system use with qualitative, verbal user statements in unstructured interviews about how well the system supports their goals, tasks, preferences, and problem-solving behavior.

A study of the ratings made by 13 usability engineers of 33 design and evaluation techniques showed that among the four most important were the *iterative design* and *task analysis,* which were the most important, followed by *empirical tests with real users, participatory design,* and last, but still highly rated, *visit to workplace or customer location* before start of design, and *field study* used for evaluation after the installation of the system (Nielsen, 1993).

45.9.6.2 Examples

This category is illustrated by a selection of evaluation tests from the design and evaluation of a full scale library system, which was based on a work analysis (see framework in Figure 45.2), tested in laboratory experiments (Figure 45.4), and evaluated less constrained at the workplace within the framework boundaries of Figure 45.3. The library system was developed as an iterative design process, and it was subject to an exhaustive functional verification to ascertain that the system could meet the functionality requirements that had been imposed, i.e., that the design was right in that the results met the specifications. Extensive experimental validations in the laboratory were performed to ascertain that the system could meet the work requirements before the evaluation of the system took place in the actual work context in a library. The subsequent evaluation using the general public was thus an attempt to validate whether or not the system actually were the right design for supporting library users in the task situation of information retrieval. This evaluation with unconstrained conditions was necessary to confirm or disprove the experimental data and analysis from earlier evaluations of separate parts of system design and aspects of the user–system interaction, which were constrained and performed within the boundary 1 to 5 as described above. Each evaluation sequence was planned to focus heavily on data collection related to design concepts developed from field studies, and on the implementation of these in the final system based on results from controlled experimental laboratory studies. The overall goal of this phase, which took place over 6 months in a public library, was to evaluate whether the information system:

1. Could be accessed and was accepted by the general public and professional librarians. This included a quantitative and qualitative evaluation by public library users of the functionality of the system (the interface, search dialogue, retrieval mechanisms, and database content) within each of the five boundaries of Figure 45.4, and with use of all the different methods described above.
2. Provided the books asked for to the users' satisfaction. This evaluation took place at the user's workplace or in their homes; i.e., their satisfaction and the relevance of the books was evaluated after their use.
3. Would impact the library work in a way that was satisfying to the public and the professionals, and would be cost effective to the organization.

The evaluation experiments are described in Pejtersen (1991, 1992, 1994), Goodstein and Pejtersen (1989), and Rasmussen et al. (1994).

45.9.6.3 Boundary of Users' Resources

One of the tests conducted to pursue the first goal was an experimental laboratory and workplace evaluation at boundary 1 and 2 of the iconic interface. Field studies of users' information-seeking behavior showed that pictures on book covers were an important information source for a quick judgment of book content. It was hypothesized that icons in the interface could be used to browse, associate, and search for book content. A controlled, multiple choice association test of icons was conducted in a laboratory with different groups of users (children and adults). The purpose was to validate the comprehensibility of the meaning of icons, and to test how users associated words with icons intended to be used in the interface. Words and icons were then paired together in a many to many mapping that followed the associative relationships that most frequently occurred in the user test. The efficiency of use, the comprehensibility of icons and the subjective user satisfaction was then evaluated at the workplace in a full-scale prototype system by 1030 users, who responded to online questionnaires that appeared automatically on the screen when the user ended his or her session with the system. The questionnaire adapted to the individual user's navigation trajectory and displayed those icons, which the user had met at the interface and actually employed during a search. It contained questions about the understandability of icons, which were used both as action buttons, and as a means to express the topics contained in books. Fifteen different icons used as action buttons were displayed together with a textual list of action possibilities, and users were then asked to select the action that would match the icon. Evaluation of the associative relationship between the message of the icons and the contents of the books in the database was measured on a scale that expressed the users' perception of degree of match. Finally, users' subjective satisfaction with an icon based interface was evaluated relative to a similar text-based interface. The result of the quantitative test at the workplace was then tested qualitatively by 75 observations and interviews with library users after they had used the system.

45.9.6.4 Boundary of Strategies

Field studies of task strategies before the design showed that several different strategies were employed such as analytical search by attributes, search by analogy and similarities with previous examples, browsing strategies, etc. Laboratory simulation at boundary 3 of the work situation with prototype databases to be searched by several different strategies pointed at the requirements for individual retrieval functionalities and display content for each strategy and led to design of search dialogues along each strategy. This change of total system functionality led to a drastic shift in task strategies: Analytical strategies were very rarely encountered at the workplace before the system was introduced because of its high demands on knowledge, time, and memory resources etc. At the workplace, 7100 online logging of all dialogue events (mouse clicks, etc.) tracked the users' strategy choice, and 220 questionnaires gave answers to their reason for choice of strategy, its ease of use, their strategy preference, etc. During use of the system over a longer time span, the analytical strategy became the most popular strategy, as users and librarians adapted their strategy choice to the most effective strategy in the new environment.

45.9.6.5 Boundary of Decision Task

The second goal was pursued by an evaluation at boundary 3 in Figure 45.3 of users' subjective satisfaction with the books they had retrieved from the database by use of the classification scheme. Field studies in public libraries of users' decision making queries during information retrieval tasks clearly indicated that information needs could be expressed within several dimensions, which referred to different means and ends at different abstraction levels. Based on users' need formulations, a multifaceted classification scheme was developed and tested in a laboratory by professional intermediaries, who searched an experimental database on a large set of test questions. The support of the classification scheme, its keywords and book descriptions in retrieval of relevant books was evaluated in structured questionnaires by 120 end users based on their reading of books. The most important performance measure was the precision of retrieved books based on users' comparison of the database classification of book contents with their own estimation of the book content and its relevance in a use situation.

45.9.6.6 Boundary of Work Space

Another type of experiment was used to pursue the third goal aimed at an evaluation of the impact of a new retrieval system on user behavior and preferences, on the means and ends required, and on the total work situation. Professional intermediaries working with a new computer system in information retrieval and cultural mediation tasks reported in questionnaires and focus group interviews at boundary 5 how the system changed their roles and left more resources for cooperation and a thorough dialogue with the users. The system supported their cultural mediation strategies and allowed a shift to the role of a consultant analyzing task problems, evaluating the quality of alternative proposals, and assisting in choice of solutions. Second, they reported how important it was for the professional image and pleasure of use that errors did not occur as the system supported exploration of alternatives and no error messages occurred.

45.9.6.7 Boundary of Organizational Work Context

The possible positive or negative impact on quality of work and the system's potential deterioration of professional skills during changes in role allocation among users and librarians was evaluated at boundary 6. Whether the new system would lead to a simplistic interpretation of users' needs, an impoverishment of their reading experiences and, as well, an impoverishment of the librarian's domain knowledge. During the experimental design phase, user participatory design was a continuous activity that primarily focused on elicitation of domain knowledge from librarians which was not present in the field data. This was done to ensure that database content was superior in completeness of book descriptions compared to the descriptions that were communicated in user–librarian negotiations. A computer logging of librarians' and users' use of the system was designed at boundary 3 and combined with focus group interviews with the staff and user groups. Both types of data were compared with records of librarians' and users' book descriptions from earlier field studies, to make sure that the database information exceeded in number and breath their book knowledge, so that both users and librarians through the use of the system would increase their competence and knowledge about the document collection.

The impact on cost effectiveness was measured by the increase in number and distribution in loan of high-quality books, because the ultimate institutional goal for public libraries is to promote education and cultural values. A more even distribution of book loans means more effective use of the book stock, which has economic implications for a library's costs for book acquisition.

45.10 EVALUATION AND DESIGN, A DYNAMIC PROCESS

If a match during analytical or empirical assessments between requirements or hypotheses and results is not seen to be possible, then changes will be required—either in the system or in the actual work environment. This leads in turn to the realization that evaluation actually should be a dynamic process—i.e., a continuing design refinement throughout the design process itself, in the transition from design to operations and during the subsequent operational period. In this way, a formalized evaluation of a design can serve as a link between design and operations to support a continuing review process of actual operations vs. assumptions and conditions.

Trade-off in designs cannot always be anticipated and has the effect that changes in the design may introduce new problems, and prototyping and iterative design will always be driven by evaluation. The depth in evaluation will be dependent on the stages of the design, an increasingly elaborate number of evaluation tests will take place as the system design progresses, which will lead to increase in quantitative data. Evaluation as a link between design and system use is particularly important for complex decision support systems which are very costly to develop and which possess a rich functionality with a potential for changing the organization of work tasks in a way that is very difficult to predict.

Experiments, simple and complex, can fill many roles during the design and evaluation of information systems to meet users' requirements for compatibility, understandability, effectivity, and acceptance. This is particularly the case when new versions cannot be based to any great extent on incremental changes of prior designs. The design of a new information system depends to a certain extent on the transfer of reliable models, concepts, and solutions from a current (or a similar) application to the new context. However, as discussed earlier, given an existing system as the "starting point," the identification of behavior-shaping constraints and subjective preferences, some of which may no longer be active and vocalizable because of the formation of habits and established practice, can be difficult. Therefore a continuing experimental evaluation—either of hypotheses and models emerging from field studies or of system features and user preferences arising during system design—will often be necessary. In addition, experiments may be required to test hypotheses about the effectiveness of new tools, about users' preferences for certain mental strategies, when new interfaces and tools change the demand–resource match, etc. Following on after the conceptual design, experiments based on prototypes of varying complexity can be used to validate the design ahead of expensive system manufacture. Thus the experimental type and design can be matched to the scope and nature of the problem.

It seems obvious that some optimum combination of the analytical and empirical approaches to evaluation is required, although a specific recommendation is difficult to give. A natural approach will be to perform the analytical evaluation more or less concurrently with the design and, in the case that prior experience or analyses of work in existing systems do not support the necessary design decisions, to test the new concepts empirically during design. In general, this can be called a form of hybrid evaluation.

In conclusion, the framework suggested here has been developed to be used concurrently for analysis of user–work interaction in system design and evaluation both in the laboratory and in field studies of real work environments. The framework can serve to resolve much of the standing controversy about the value of laboratory experiments for understanding "real-life" work—a controversy caused by designers' problems with the lack of explicit descriptions of the boundary conditions of experiments in terms that could facilitate the transfer of results to their work contexts. It can—and should—be used in conjunction with other evaluation methods as illustrated in this chapter.

ACKNOWLEDGMENTS

This work has been supported by a grant from SHF, the Danish National Research Council for the Humanities as a preparatory work required for a systematic evaluation of information support systems for design in industry under the project of "Semantic Information Retrieval in Communication Networks supported by Multimedia Techniques."

REFERENCES

Anderson, J. R. (1987). Skill acquisition: Compilation of weak-method solution. *Psychological Review, 94,* 192–211.

Bealulieu, M. (1991). Evaluation of on-line catalogues: Elicitation of information from the user. *Information Processing and Management, 27*(*5*), 22–24.

Beaulieu, M., Robertson, S. E., Rasmussen, E. (1996). Evaluating interactive systems in TREC. *Journal of the American Society for Information Science. 47*(*1*), 85–94.

Berry, D. C., and Broadbent, D. E. (1990, May–June). The role of instruction and verbalization in improving performance on complex search tasks. *Behavior and Information Technology, 9*(*3*), 175–190.

Bevan, N., and Macleod, M. (1994). Usability measurement in context. *Behavior and Information Technology, 13*(*1* and 2), 132–145.

Bewly, W. L., Roberts, T. L., Schroit, D., and Verplank, W. L. (1983). Human factors testing in the design of Xerox's 8010 Star office workstation. In Janda, A., Ed., *Human Factors in Computing Systems. Proceedings of the ACM CHI'83 Conference.* Boston, MA, December 12–15, Amsterdam: North Holland (pp. 72–77).

Boff, K. R., and Lincoln, J. E., Eds. (1988). *User's guide, engineering data compendium, human perception and performance.* Dayton, OH: Armstrong Aerospace Research Laboratory, Wright-Patterson Air Force Base.

Boff, K. R., and Lincoln, J. E., Eds. (1995). *Computer-Aided Systems Human Engineering.* Performance Visualisation System. Engineering Data Compendium. Compact Disc. Version 1. Machintosch.

Bradford, J. H., Murray, W. D., and Carey, T. T. (1990). What kind of errors do Unics users make? In *Proceedings of the IFIP Third International Conference on Human-Computer Interaction* (Cambridge, UK, August 27–31, pp. 43–46).

Buur, J., and Nielsen, P. (1995). Design for usability: Adapting human computer interaction methods for the design of mechatronical products. In *Proceedings of the 10th International Conference in Engineeering Design: Design Science for and in Design Practice (ICED).* August 22–24, Praha, Czech (pp. 952–957).

Campbell, R. L. (1992, April). Will the real scenario please stand up? *ACM SIGCHI Bulletin, 24*(2), 6–8.

Caplan, S. (1990). Using focus groups methodology for ergonomic design. *Ergonomics, 33*(5), 527–533.

Card, S. K., Moran, T., and Newell, A. (1983). *The Psychology of Human-Computer Interaction.* Hillsdale, NJ: Earlbaum.

Carroll, J. M., and Rosson, M. B. (1992). Getting around the task-artifact cycle: How to make claims and design by scenario. *ACM Transditions on Information Systems, 10*(2), 181–212.

Carroll, J. M., and Moran, Th. P. (1991). Introduction to this special issue on design rationale. *Human-Computer Interaction, 6,* 197–200.

Carroll, J. M., and Rosson, M. B. (1990). Human-computer interaction scenarios as a design representation. In *Proceedings of the Twenty-Third Annual Hawaii International Conference on System Sciences IEEE HICSS-23, 23rd Hawaii International Conference on System Sciences,* January 2–5, Vol. II. IEEE Comp. Soc. Press, Los Alamitos, CA (pp. 555–561).

Carroll, J. M., Ed. (1995). Scenario-Based Design: Envisioning Work and Technology in System Development. New York: John Wiley.

Conklin, J. (1991). Bringing usability effectively into product. *Workshop on development.* In Rudisill, Ed., *Human-computer interface design: Success stories, emerging methods, and real-world context.* (Boulder, CO, July 24–26), Morgan Kaufmam.

Cordingly, E. (1989). Knowledge elicitation techniques for knowledge based systems. In D. Diaper, Ed. *Knowledge Elicitation: Principles, Techniques, and Applications* (pp. 89–172). Chichester, UK: Ellis Horwood.

Czaja, S. J. (1988). Microcomputers and the elderly. In Helander, M., Ed., *Handbook of human-computer interaction* (pp. 581–598). Amsterdam: Elsevier.

De Groot, A. D. (1965). Thought and Choice in Chess. The Hague: Mouton.

Denning, S., Holem, D., Simpson, M., and Sullivan, K. (1990). The value of thinking-aloud protocols in industry: A case study at Microsoft Corporation. In Countdown to the 21st Century *Proceedings of the Human Factors Society 34th Annual Meeting, October 1990, Orlando, FL* (pp. 1285–1289).

Desurvire, H. W. (1994). Faster, cheaper!! Are usability inspection methods as effective as empirical testing? In J. Nielsen and R. L. Mack, Eds., Usability Inspection Methods. New York: John Wiley, pp. 173–199.

Ericsson, K. A., and Simon, H. A. (1984). *Protocol analysis: Verbal reports as data.* Cambridge, MA: MIT Press.

Fidel, R. (1984). On-line searching styles: A case-study-based model of searching behavior. *Journal of the American Society of Information Science, 35*(4), 211–221.

Fowler, C. J. H., and Murray, D. (1987). Gender and cognitive style differences at the human-computer interface. *Proceedings of the IFIP Interact'87 Second Human-Computer Interaction, Stuttgart, Germany,* September 1–4 (pp. 709–714).

Gomez, L. M., Egan, D. E., and Bowers, C. (1986). Learning to use a text editor: Some learner characteristics that predict success. *Human-Computer Interaction, 2*(*1*), 1–23.

Goodstein, L. P., and Pejtersen, A. M. (1989). *The BOOK HOUSE. System functionality and evaluation.* (Tech. Rep. Risø-M-2793) Roskilde Denmark: Risø National Laboratory.

Greenbaum, J., and Kyng, M., Eds. (1991). *Design at Work: Cooperative Design of Computer Systems.* Hillsdale, NJ: Lawrence Earlbaum.

Hackman, G. S., and Biers, D. W. (1992). Team usability testing: Are two heads better than one? *Proceedings of the Human Factors Society 36th annual meeting* (pp. 1205–1209).

Hopkin, D. V. (1993). Verification and validation: Concepts, issues and applications. In J. A. Wise, D. V. Hopkin, and P. Stager, Eds., *Verification and validation of complex systems: Human factors issues.* (Nato ASI Series. Springer Verlag).

ISO (1993). Ergonomic Requirements for office work with visual display terminals. Guidance on usability. ISO DIS 9241-11. Geneve.

ISO (1993). Ergonomic Requirements for office work with visual display terminals Menu Dialogues. ISO DIS 9241-14. Geneve.

Jørgensen, H. A., and Aboulafia, A. (1989). Perceptions of design rationale. In K. Nordby, P. Helmersen, A. D. Gilmore, and S. Arnesen, Eds. *Human-Computer Interaction: Interact '95* (pp. 61–66). New York: Chapman Hall.

Kennedy, S. (1989, October). Using video in the BNR usability lab. *ACM SIGCHI Bulletin, 21*(2), 92–95.

Kyng, M. (1994). Scandinavian Design: Users in Product Development. In Adelson, B. Dumais, and S. Olson, J., Eds., *Human Factors in Computing Systems, CHI'94, Celebrating Interdependence.* Boston, MA.

Lewis, C. Polson, P., Wharton, C., and Rieman, J. (1990). Testing a walkthrough methodology for theory-based design of walk-up—and use interfaces. In *Empowering People: CHI '90 Conference Proceedings.* Seattle, WA, April 1–5. New York: ACM Press, pp. 235–242.

Lewis, C. (1982). *Using the 'thinking aloud' method in cognitive interface design* (Research Report RC9265). Yorktown Heights, NY: Watson Research Center.

MacLean, A., Young, R. M., Bellotti, V. M. E., and Moran, Th. P. (1991). Questions, options, and criteria: Elements of design space analysis. *Human-Computer Interaction, 6,* 201–250.

Mayhew, D. J. (1992). *Principles and Guidelines in Software User Interface Design.* Englewood Cliffs, NJ: Prentice Hall.

Moray, N., Jones, G. J., Rasmussen, J., Lee, J. D., Vicente, K. J., Brock, R., and Djemil, D. (1991). *A Performance indicator of the effectiveness of human-machine interfaces for nuclear power plants* (Tech. Rep. NUREG/CR-5977). (Nuclear Regulatory Commission, Washington, DC).

Newell, A. (1990). *Unified theories of cognition.* Cambridge, MA: Harvard University Press.

Nielsen, J. (1993). *Usability Engineering*. New York: Academic Press.

Nielsen, J., Ed. (1994). Behavior and information technology. *International Journal on the Human Aspects of Computing. Special Issue on Usability Laboratories, 13*(*1* and *2*).

Nielsen, J., Mack, R. L., Eds. (1994). *Usability inspection methods*. New York: John Wiley.

Pejtersen, A. M., Sonnenwald, D. H., Buur, J., Govindarej, T., and Vincente, K. (1995). *Using Cognitive Engineering Theory to Support Knowledge Exploration in design. Proceedings of the 10th International Conference in Engineering Design: Design Science for and in Design Practice (ICED)* August 22–24, Praha, Czech, pp. 419–427.

Pejtersen, A. M. (1991). *Interfaces Based on Associative Semantics for Browsing in Information Retrieval* (Tech. Rep. Risø M-2794). Roskilde, Denmark: Risø National Laboratory.

Pejtersen, A. M. (1992a). The Book house. An icon based database system for fiction retrieval in public libraries. In B. Cronin, Ed., *The Marketing of Library and Information Services 2* (ASLIB, London), pp. 572–591.

Pejtersen, A. M. (1992b). New model for multimedia interfaces to online public access catalogues. The electronic library. *International Journal for Minicomputer, Microcomputer and Software Applications in Libraries, 10*(*6*), 359–366.

Pejtersen, A. M. (1994). Designing hypermedia representations from work domain properties. *CBT Forum, No. 2,* 17–34.

Pejtersen, A. M., and Austin, J. (1984). Fiction retrieval: Experimental design and evaluation of a search system based on users' value criteria. Part 1. *Journal of Documentation, 39*(*4*), 230–246.

Pejtersen, A. M., and Austin, J. (1984). Fiction retrieval: Experimental design and evaluation of a search system based on users' value criteria. Part 2. *Journal of Documentation, 40*(*1*), 25–35.

Rasmussen, J., Pejtersen, A. M., and Goodstein, L. P. (1994). *Cognitive systems engineering*. London: John Wiley (especially Chapters 7–13).

Robertson, S. E., Walker, S., and Beaulieu, M. (1997). To appear in *Journal of Documentation, 53*(*1*) January.

Rogers, Y. (1986, April). Icons at the interface: Their usefulness. *Interacting with Computers, 1*(*1*), 115–117.

Rouse, W. B. (1981). Experimental studies and mathematical models of human problem solving performance in fault diagnosis tasks. In J. Rasmussen and W. B. Rouse, Eds., *Human Detection and Diagnosis of System Failures*. New York: Plenum.

Rouse, W. B. (1984). *Developing an Evaluation Plan: Computer-generated Display System Guidelines* EPRI-NP-3701 (vol. 2). Search Technology, Atlanta, GA.

Sanderson, P. M. (1990). Knowledge acquisition and fault diagnosis: Experiments with PLAULT. *IEEE Transactions of Systems, Man and Cybernetics, SMC-20,* 225–242.

Saracevic, T. (1995). Evaluation of evaluation in information retrieval. In E. Fox, P. Ingwersen, and R. Fidel, Eds., *SIGIR '95 Proceedings of the 18th Annual International ACM SIGIR Conference on Research and Development in Information Retrieval.* Seattle, WA July 9–13, 1995. New York: ACM, pp. 138–146.

Saracevic, T., Kantor, P. Chamis, A. Y., and Trivison, D. (1988). A study of information seeking and retrieving. (Three parts). *Journal of the American Society of Information Science, 39(3)*, 161–216.

Verplank, B. (1993). Observation and invention: The use of scenarios in interaction design. Tutorial notes. In *INTERCHI '93: Bridges between worlds. Adjunct Proceedings of the International Conference on Human Factors in computing systems.* (INTERACT '93 and CHI'93), Amsterdam, Netherlands, April 24–29, 1993. New York: ACM.

Vicente, K. (1991). *Supporting knowledge-based behavior through ecological interfaces.* Unpublished Dissertation. Urbana, IL: University of Illinois at Urbana-Champaign.

Vicente, K. J., and Willeges, R. C. (1988). Accommodating individual differences in searching a hierarchical file system. *International Journal of Man-Machine Studies, 29(6)*, 647–668.

Wharton, C., Rieman, J., Lewis, C., Polson, P. (1994). The cognitive walkthrough method: A practitioner's guide. In J. Nielson and R. L. Mack, Eds., *Usability Inspection Methods.* New York: John Wiley.

Woods, D. D. (1984). Some results on operator performance in emergency events. In D. Whitfield, Ed., *Ergonomic Problems in Process Operations* (Institute of Chemical Engineering Symposium Serial 90).

Woods, D. D., and Sarter, N. B. (1993). Evaluation of the impact of new technology on human-machine cooperation. In J. Wise, V. D. Hopkin, and P. Stager, Eds., *Verification and validation of complex and integrated human-machine systems.* Berlin: Springer-Verlag.

Wright, R. B., and Converse, S. A. (1992). Method Bias and Concurrent Verbal Protocol in Software Usability Testing. *Proceedings of the Human Factors Society 36th Annual Meeting.* Santa Monica, CA: Human Factors Society (pp. 1220–1224).

CHAPTER 46

USABILITY TESTING

Jakob Nielsen
Strategic Technology
Sun Microsystems
Mountain View, CA 94043 USA

46.1 INTRODUCTION

User testing with real users is the most fundamental usability method and is in some sense irreplaceable, since it provides direct information about how people use computers and what their exact problems are with the concrete interface being tested. Even so, other usability engineering methods (Nielsen, 1994a) can serve as good supplements to gather additional information or to gain usability insights at a lower cost. In particular, user testing can often be combined with heuristic evaluation (Nielsen, 1994b) or other usability inspection methods (Nielsen and Mack, 1994) for greater efficiency. Briefly explained, heuristic evaluation is a way of finding usability problems in a design by contrasting it with a list of established usability principles. Even though this method works very well and provides fast evaluation results, there are always some surprises left that require user testing.

There are several methodological pitfalls in usability testing (Holleran, 1991; Landauer, 1988a), and as in all kinds of testing one needs to pay attention to the issues of reliability and validity. Reliability is the question of whether one would get the same result if the test were to be repeated, and validity is the question of whether the result actually reflects the usability issues one wants to test.

46.1.1 Reliability

Reliability of usability tests is a problem because of the huge individual differences between test users. It is not uncommon to find that the best user is 10 times as fast as the slowest user, and the best 25% of the users are normally about twice as fast as the slowest 25% of the users (Egan, 1988). Because of this well-established phenomenon, one cannot conclude much from, say, observing that user A using interface X could perform a certain task 40% faster than user B using interface Y; it could very well be the case that user B just happened to be slower in general than user A. If the test were repeated with users C and D, the result could easily be the opposite. For usability engineering purposes, one often needs to make decisions on the basis of fairly unreliable data, and one should certainly do so since some data is better than no data. For example, if a company had to choose between interfaces X and Y as just discussed, it should obviously choose interface X since it has at least a little bit of evidence in its favor. If several users have been tested, one could use standard statistical tests[1] to estimate the significance of the difference between the systems (Brigham, 1989). Assume, for example, that the statistics package states that the difference between the systems is significant at the level $p = .20$. This means that there is a 20% chance that Y was actually the best interface, but again one should obviously choose X since the odds are 4 to 1 that it is best.

Standard statistical tests can also be used to estimate the confidence intervals of test results and thus indicate the reliability of the size of the effects. For example, a statistical claim that the 95% confidence interval for the time needed to perform a certain test task is 4.5 ± 0.2 minutes means that there is a 95% probability that the true value is between 4.3 and 4.7 (and thus a 5% probability that it is really smaller than 4.3 or larger than 4.7). Such confidence intervals are important if the choice between two options is dependent not just on which one is best but also on how much better it is (Landauer, 1988a, 1988b). For example, a usability problem that is very expensive to fix should be fixed only if one has a reasonably tight confidence interval showing that the problem is indeed sufficiently bothersome to the users to justify the cost.

In a survey of 36 published usability studies, I found that the mean standard deviation was 33% for measures of expert-user performance (measured in 17 studies), 46% for measures of novice-user learning (measured in 12 studies), and 59% for error rates (measured in 13 studies). In all cases, the standard deviations are expressed as a percentage of the measured mean value of the usability attribute in question. These numbers can be used to derive early approximations of the number of test users needed to achieve a desired confidence interval. Of course, since standard deviations vary a great deal between studies, any particular usability test might well have a higher or lower standard deviation than the values given here, and one should perform further statistical tests on the actual data measured in the study.

In all, the results show that error rates tend to have the highest variability, meaning that they will generally require more test users to achieve the same level of confidence as can be achieved with fewer test users for measures of learnability and even fewer for measures of expert user performance. A confidence level of 95% is often used for research studies, but for practical development purposes, it may be enough to aim for an 80% level of confidence.

46.1.2 Validity

Validity is a question of whether the usability test in fact measures something of relevance to usability of real products in real use outside the laboratory. Whereas reliability can be addressed with statistical tests, a high level of validity requires methodological understanding of the test method one is using as well as some common sense.

Typical validity problems involve using the wrong users or giving them the wrong tasks or not including time constraints and social influences. For example, a management information system might be tested with business school students as test users, but it is

[1] Statistics is of course a topic worthy of several full books in its own right. See, for example, (Pedhazur and Schmelkin, 1991) for basic methods, and (Lehmann and D'Abrera, 1975) for more specialized statistical tests.

likely that the results would have been somewhat different if it had been tested with real managers. Even so, at least the business school students are people who likely will become managers, so they are probably more valid test users than, say, chemistry students. Similarly, results from testing a hypertext system with a toy task involving a few pages of text may not always be relevant for the use of the system in an application with hundreds of megabytes of information.

Confounding effects may also lower the validity of a usability test. For example, assume that you want to investigate whether it would be worthwhile to move from a character-based user interface to a graphical user interface for a certain application. You test this by comparing two versions of the system: one running on a 24×80 alphanumeric screen and one running on a 1024×1024 pixel graphics display. At a first glance, this may seem a reasonable test to answer the question, but more careful consideration shows that the comparison between the two screens is as much a comparison between large and small screens as it is between character-based and graphical user interfaces.

46.2 TEST GOALS AND TEST PLANS

Before any testing is conducted, one should clarify the purpose of the test since it will have significant impact on the kind of testing to be done. A major distinction is whether the test is intended as a formative or summative evaluation of the user interface. *Formative evaluation* is done to help improve the interface as part of an iterative design process. The main goal of formative evaluation is thus to learn which detailed aspects of the interface are good and bad, and how the design can be improved. A typical method to use for formative evaluation is a thinking-aloud test. In contrast, *summative evaluation* aims at assessing the overall quality of an interface, for example, for use in deciding between two alternatives or as a part of competitive analysis to learn how good the competition really is.[2] A typical method to use for summative evaluation is a measurement test.

46.2.1 Test Plans

A test plan should be written down before the start of the test and should address the following issues:

- The goal of the test: What do you want to achieve?
- Where and when will the test take place?
- How long is each test session expected to take?
- What computer support will be needed for the test?
- What software needs to be ready for the test?
- What should the state of the system be at the start of the test?
- What should the system or network load and response times be? If possible, the system should not be unrealistically slow, but neither should it be unrealistically fast because the experimental system or network has no other users. One may have to slow down the system artificially to simulate realistic response times.
- Who will serve as experimenters for the test?
- Who are the test users going to be, and how are you going to get hold of them?
- How many test users are needed?
- What test tasks will the users be asked to perform?
- What criteria will be used to determine when the users have finished each of the test tasks correctly?
- What user aids (manuals, online help, etc.) will be made available to the test users?
- To what extent will the experimenter be allowed to help the users during the test?
- What data are going to be collected, and how will they be analyzed once they have been collected?
- What will the criterion be for pronouncing the interface a success? Often, this will be the "planned" level for previously specified usability goals, but it could also

[2] Remember, by the way, that manual or paper-based solutions that do not involve computers at all are also in the running and should be studied as well.

be a looser criterion such as "no new usability problems found with severity higher than 3."

46.2.2 Test Budget

The test plan should also include a budget for the test. Some costs will be out of pocket, meaning that they have to be paid cash. Other costs are in the nature of using company staff and resources that are already paid for. Such indirect costs may or may not be formally charged to the usability budget for the specific project, depending on how the company's accounting mechanisms are set up, but they should be included in the usability manager's internal budget for the test in any case. Typical cost elements of a user test budget are

- Usability specialists to plan, run, and analyze the test: out-of-pocket expense if consultants are used
- Administrative assistants to schedule test users, enter data, etc.
- Software developers to modify the code to include data collection or other desired test customization
- The test users' time: out-of-pocket expense if outside people are hired for the test
- Computers used during testing and during analysis
- The usability laboratory or other room used for the test
- Videotapes and other consumables: out-of-pocket expense

The cost estimates for the various staff members should be based on their loaded salary and not on their nominal salary. A loaded salary is the total cost to the company of having a person employed and includes such elements as benefits, vacation pay, employment taxes or fees, and general corporate overhead.

The test budget should be split into fixed and variable costs, where fixed costs are those required to plan and set up the test no matter how many test users are run, and variable costs are the additional costs needed for each test user. Splitting the cost estimates in this way allows for better planning of the number of test users to include for each test. Obviously, both fixed and variable costs vary immensely between projects, depending on multiple factors such as the size of the interface and the salary level of the intended users. Based on several published budgets, estimates for a representative, medium-sized usability test can be derived, with fixed costs of $3000 and variable costs of $1000 per test user (Nielsen and Landauer, 1993). Note that any specific project is likely to have different costs than these estimates.

Given estimates for fixed and variable costs, it then becomes possible to calculate the optimal number of test users if further assumptions are made about the financial impact of finding usability problems and the probability of finding each problem with a single test user. Unfortunately, these latter two numbers are much harder to estimate than the costs of testing, but any given organization should be able to build up a database of typical values over time. Again based on values from published studies, a representative value of finding a usability problem in a medium-sized project can be taken as $15,000.

Nielsen and Landauer (1993) showed that the following formula gives a good approximation of the finding of usability problems:

$$\text{Usability\_Problems\_Found}(i) = N[1 - (1 - \lambda)^i]$$

where i is the number of test users, N is the total number of usability problems in the interface, and λ is the probability for finding any single problem with any single test user. The values of N and λ vary considerably between projects and should be estimated by curve fitting as data become available for each project. It is also recommended that one keep track of these numbers for one's own projects such that one can estimate "common" values of these parameters for use in the planning of future tests.

For several projects we studied, the mean number of problems in the interface, N, was 41 and the mean probability for finding any problem with a single user, λ, was 31% (Nielsen and Landauer, 1993). The following discussion uses these mean values to illustrate the use of the mathematical model in the budgeting of usability activities. Of course, one should really use the particular N and λ values that have been measured or estimated for the particular project one wants to analyze.

Given the assumptions mentioned above, Figure 46.1 shows how the payoff ratio between the benefits and the costs changed in our average example with various numbers of test users. The highest ratio was achieved with three test users, where the projected benefits were $413,000, and the costs were $6000. However, in principle, one should keep testing as long as the benefit from one additional test user is greater than the cost of running that user. Under the above model, this would imply running 15 test users at a cost of $18,000 to get benefits worth $613,000. If iterative design is used, it will be better to conduct more, smaller, tests (one for each iteration) than to spend everything on a single test.

46.2.3 Pilot Tests

No usability testing should be performed without first having tried out the test procedure on a few pilot subjects. Often, one or two pilot subjects will be enough, but more may be needed for large tests or when the initial pilot tests show severe deficiencies in the test plan. The first few pilot subjects may be chosen for convenience among people who are easily available to the experimenter even if they are not representative of the actual users, since some mistakes in the experimental design can be found even with subjects such as one's colleagues. Even so, at least one pilot subject should be taken from the same pool as the other test users.

During pilot testing, one will typically find that the instructions for some of the test tasks are incomprehensible to the users or that they misinterpret them. Similarly, any questionnaires used for subjective satisfaction rating or other debriefing will often need to be changed based on pilot testing. Also, one very often finds a mismatch between the test tasks and the time planned for each test session. Most commonly, the tasks are more difficult than one expected, but of course it may also be the case that some tasks are too easy. Depending on the circumstances of the individual project, one will either have to revise the tasks or make more time available for each test session.

Pilot testing can also be used to refine the experimental procedure and to clarify the definitions of various things that are to be measured. For example, it is often difficult to decide exactly what constitutes a user error or exactly when the user can be said to have completed a given test task, and the pilot test may reveal inconsistencies or weaknesses in the definitions contained in the test plan. Pilot testing is worth the added investment in time since it draws out the problems in the test plan, reduces surprises, saves the cost of running an experiment only to find that that it was flawed. In some cases pilot testing may even reduce the final number of test participants since it may show that some parts of the system do not need further testing.

46.3 GETTING TEST USERS

The main rule regarding test users is that they should be as representative as possible of the intended users of the system. If the test plan calls for a "discount usability" approach

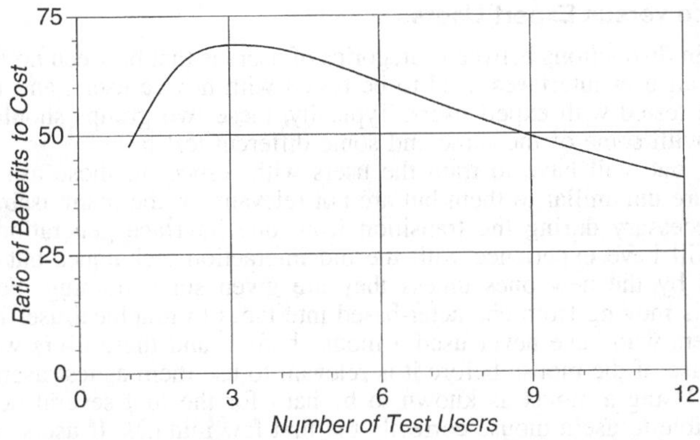

Figure 46.1 The payoff ratio (how much larger the benefits are than the costs) for user tests with various numbers of test users under the assumptions for a "typical" medium-sized project described in the text.

with very few test users, one should not choose users from outlier groups but should take additional care to involve average users. If more test users are to be used, one should select users from several different subpopulations to cover the main different categories of expected users.

The main exception from the rule that test users should be representative of the end users is testing with sales people. For many products, sales staff is used to give demonstrations to prospective customers, and the ease with which they can give these demos may be a major selling point. Often, sales people handle multiple products and do not get extensive experience from actual use of any individual product. The experience of demonstrating a user interface is very different from that of actually using it for a real task, and even though most usability efforts should aim at making the system easier to use for the users, it may increase sales significantly if "demoability" is also considered as a usability attribute.

Sometimes, the exact individuals who will be using a system can be identified. This is typically the case for systems that are being developed internally in a company for use in a given department of that company. In this case, representative users are easy to find, even though it may present some difficulties to get them to spend time on user testing instead of doing their primary job. Internal test users are often recruited through the users' management who agrees to provide a certain number of people. Unfortunately, managers often tend to select their most able staff members for such tests (either to make their department look good or because these staff members have the most interest in new technology), so one should explicitly ask managers to choose a broad sample with respect to salient user characteristics such as experience and seniority.

In other cases, the system is targeted at a certain type of users, such as lawyers, the secretaries in a dental clinic, or warehouse managers in small manufacturing companies. These user groups can be more or less homogeneous, and it may be desirable to involve test users from several different customer locations. Sometimes, existing customers are willing to help out with the test since it will get them an early look at new software as well as improving the quality of the resulting product, which they will be using. In other cases, no existing customers are available, and it may be more difficult to gain access to representative users. Sometimes, test users can be recruited from temporary employment agencies, or it may be possible to get students in the domain of interest from a local university or trade school. It may also be possible to enter a classified advertisement under job openings in order to recruit users who are currently unemployed. Of course, it will be necessary to pay all users recruited with these latter methods.

Yet other software is intended for the general population, and one can in principle use anybody as a test user, again using employment agencies, students, or classified advertising to recruit test users. Especially when testing students, one should consider whether the system is also intended to be used by older users, since they may have somewhat different characteristics (Czaja, Hammond, Blascovich, and Swede, 1989; Nielsen and Schaefer, 1993) and may therefore need to be included as an additional test group. A good source of older test users is retirees, who may also serve as a pool of talent with specific domain expertise.

46.3.1 Novice versus Expert Users

One of the main distinctions between categories of users is that between novice and expert users. Almost all user interfaces need to be tested with novice users, and many systems should also be tested with expert users. Typically, these two groups should be tested in separate tests with some of the same and some different test tasks.

Sometimes, one will have to train the users with respect to those aspects of a user interface that are unfamiliar to them but are not relevant for the main usability test. This is typically necessary during the transition from one interface generation to the next, where users will have experience with the old interaction techniques but will be completely baffled by the new ones unless they are given some training. For example, a company that is moving from character-based interfaces to graphical user interfaces will have many users who have never used a mouse before, and these users will have to be trained in the use of the mouse before it is relevant to use them as test users of a mouse-based system. Using a mouse is known to be hard for the first several hours, and it is almost impossible to use a mouse correctly the first few minutes. If users are not trained in the use of the mouse and other standard interaction techniques before they are asked to test a new interface, the test will be completely dominated by the effects of the users' struggle with the interaction devices and techniques, and no information will be gained as to the usability of the dialogue.

One example of the potentially devastating effect of not training users before a test was a study of the use of a single window versus multiple windows for an electronic book (Tombaugh, Lickorish, and Wright, 1987). When novice users without any specific training were used as test subjects, the single-window interface was best for reading the electronic book. The time needed to answer questions about the text was 85 s when using the multiwindow interface and 72 s when using the single-window interface. In a second test, the test users were first given 30 min of training in the use of a mouse to control multiple windows, and the time to answer the questions about the text was now 66 s for the single-window interface and only 56 s for the multiwindow interface. Thus, both interfaces benefitted from having more experienced users, but the most interesting result is that the overall conclusion with respect to determining the "winner" of the test came out the opposite. The single-window solution would be best for a "walk-up-and-use" system for users without prior mouse experience. On the other hand, the multiwindow solution would be best in the more common case where the electronic book was to be used in an office or school environment where people would be using the same computer extensively. For such environments, the wrong conclusion would have been drawn if a test with untrained users had been the only one.

46.3.2 Between-Subjects versus Within-Subjects Testing

Often, usability testing is conducted in order to compare the usability of two or more systems. If so, there are two basic ways of employing the test users: *between-subject testing* and *within-subject testing.*

Between-subject testing is in some sense the simplest and most valid since it involves using different test users for the different systems. Thus, each test user participates in only a single test session. A major problem with between-subject designs is the huge individual variation in the skills of different users. Therefore, it can be necessary to run a very large number of test users in each condition in order to smooth over random differences between test users in each of the groups.

Between-subject testing also risks a bias resulting from the assignment of test users to the various groups. For example, one might decide to test 20 users, call for volunteers, and assign the first 10 users to one system and the next 10 to the other. Even though this approach may seem reasonable, it in fact introduces a bias, since users who volunteer early are likely to be different from users who volunteer late. For example, early volunteers may be more conscientious in reading announcements, or they may be more interested in new technology, and thus more likely to be superusers. There are two methodologically sound ways to assign test users to groups: The simplest and best is random assignment, which minimizes the risk of any bias but requires a large number of test users because of individual variability. The second method is matched assignment, which involves making sure that each group has equally many users from each of those categories that have been defined as being of interest to the test. For example, users from different departments might be considered different categories, as might old versus young users, men versus women, and users with different computer experience or different educational backgrounds.

Alternatively, one may conduct the test as a within-subject design, meaning that all the test users get to use all the systems that are being tested. This method automatically controls for individual variability since any user who is particularly fast or talented will presumably be about equally superior in each test condition. Within-subject testing does have the major disadvantage that the test users cannot be considered as novice users anymore when they approach the other systems after having learned how to use the first system. Often, some transfer of skill will take place between systems, and the users will be better at using the second system than they were at using the first. In order to control for this effect, users are normally divided into groups, with one group using one system first and the other group using the other system first. The issues discussed above regarding the assignment of users to groups also apply to this aspect of within-subject testing.

46.4 CHOOSING EXPERIMENTERS

No matter what test method is chosen, somebody has to serve as the experimenter and be in charge of running the test. In general, it is of course preferable to use good experimenters who have previous experience in using whatever method is chosen. For example, a study where 20 different groups of experimenters tested the same interface, there was a correlation of $r = .76$ between the rated quality of the methodology used by a group and the number of usability problems they discovered in the interface (Nielsen, 1992c). When running 3 test subjects, experimenters using very good methodology found about

5–6 of the 8 usability problems in the interface, and experimenters using very poor methodology only found about 2–3 of the problems.

This result does not mean that one should abandon user testing if no experienced usability specialist is available to serve as the experimenter. First, it is obviously better to find a few usability problems than not to find any, and second, even inexperienced experimenters can use a decent (if not perfect) methodology if they are careful. It is possible for computer scientists to learn user test methods and apply them with good results (Nielsen, 1992c; Wright and Monk, 1991).

In addition to knowledge of the test method, the experimenter must have extensive knowledge of the application and its user interface. System knowledge is necessary for the experimenter to understand what the users are doing as they perform tasks with the system, and to make reasonable inferences about the users' probable intentions at various stages of the dialogue. Often, users' actions will go by too fast for experimenters, who are trying to understand what the system is doing at the same time as they are analyzing the users.

The experimenter does not necessarily need to know how the system is implemented, even though such knowledge can come in handy during tests of preliminary prototypes with a tendency to crash. If the experimenter does not known how to handle system crashes, it is a good idea to arrange to have a programmer with the necessary skills stand by in a nearby office.

One way to get experimenters with a high degree of system knowledge is to use the system's designers themselves as evaluators (Wright and Monk, 1991). In addition to the practical advantages, there are also motivational reasons for doing so, since the experience of seeing users struggle with their system always has a very powerful impact on designers (Jørgensen, 1989). There are some problems with having people run tests of their own systems, though, including a possible lack of objectivity that may lead them to help the users too much. A common weakness is the tendency for a designer to explain away user problems rather than acknowledging them as real issues. To avoid these problems, developers can serve as one part of the usability testing team while usability specialists handle relations with the users (Ehrlich, Butler, and Pernice, 1994).

46.5 ETHICAL ASPECTS OF TESTS WITH HUMAN SUBJECTS

Users are human, too. Therefore, one cannot subject them to the kind of "destructive testing" that is popular in the components industry. Instead, tests should be conducted with deep respect for the users' emotions and well-being (Allen, 1984; American Psychological Association, 1982).

At first, it might seem that usability testing does not represent the same potential dangers to the users as would, say, participation in a test of a new drug. Even though it is true that usability test subjects are not normally bodily harmed, even by irate developers resenting the users' mistreatment of their beloved software, test participation can still be quite a distressful experience for the users (Schrier, 1992). Users feel a tremendous pressure to perform, even when they are told that the purpose of the study is to test the system and not the user. Also, users will inevitably make errors and be slow at learning the system (especially during tests of early designs with many severe usability problems), and they can easily get to feel inadequate or stupid as they experience these difficulties. Knowing that they are observed, and possibly recorded, makes the feeling of performing inadequately even more unpleasant to the users. Test users have been known to break down and cry during usability testing, even though this only happens in a small minority of cases.

At first, one might think that highly educated and intelligent users would have enough self-confidence to make fear of inadequacy less of a problem. On the contrary, high-level managers and highly specialized professionals are often especially concerned about exhibiting ignorance during a test. Therefore, experimenters should be especially careful to acknowledge the professional skills of such users up front and emphasize the need to involve people with these users' particular knowledge in the test.

The experimenter has a responsibility to make the users feel as comfortable as possible during and after the test. Specifically, the experimenter must never laugh at the users or in any way indicate that they are slow at discovering how to operate the system. During the introduction to the test, the experimenter should make clear that it is the system that is being tested and not the user. To emphasize this point, test users should never be referred to as "subjects," "guinea pigs," or other such terms. I personally prefer the term "test user," but some usability specialists like to use terms such as "usability evaluator" or "participant," which emphasize even more that it is the system that is being tested. Since the term "evaluator" technically speaking refers to an inspection-oriented role where usability specialists judge a system instead of using it to perform a task, I normally do not use this term myself when referring to test users.

The users should be told that no information about the performance of any individual users will be revealed and specifically that their manager will not be informed about their performance. The test itself should be conducted in a relaxed atmosphere, and the experimenter should take the necessary time for small talk to calm down the user before the start of the experiment, as well as during any breaks. It might also be a good idea to serve coffee, soft drinks, or other refreshments—especially if the test takes more than an hour or so. Furthermore, to bolster the users' confidence and make them at ease, the very first test task should be so easy that they are virtually guaranteed an early success experience.

The experimenter should ensure that the test room, test computer, and test software are ready before the test user arrives in order to avoid the confusion that would otherwise arise because of last-minute adjustments. Also, of course, copies of the test tasks, any questionnaires, and other test materials should be checked before the arrival of the user so that they are ready to be handed out at the appropriate time. The test session should be conducted without disruptions: typically, one should place a sign saying, "User test in progress—Do not disturb" outside the (closed) door and disable any telephone sets in the test room.

The test results should be kept confidential, and reports from the test should be written in such a way that individual test users cannot be identified. For example, users can be referred to by numbers (User1, User2, etc.) and not by names or even initials. The test should be conducted with as few observers as possible, since the size of the "audience" also has a detrimental effect on the test user: It is less embarrassing to make a fool of yourself in front of 1 person than in front of 10. And remember that users *will* think that they are making fools of themselves as they struggle with the interface and overlook "obvious" options, even if they only make the same mistakes as everybody else. For similar reasons, videotapes of a user test session should not be shown publicly without explicit permission from the user. Also, the users' manager should never be allowed to observe the test for any reason and should not be given performance data for individual users.

During testing, the experimenter should normally not interfere with the user but should let the user discover the solutions to the problems on his or her own. Not only does this lead to more valid and interesting test results,[3] it also prevents the users from feeling that they are so stupid that the experimenter had to solve the problems for them. On the other hand, the experimenter should not let a user struggle endlessly with a task if the user is clearly bogged down and getting desperate. In such cases, the experimenter can gently provide a hint or two to the user in order to get on with the test. Also, the experimenter may have to terminate the test if the user is clearly unhappy and unable to do anything with the system. Such action should be reserved for the most desperate cases only. Furthermore, test users should be informed before the start of the test that they can always stop the test at any time, and any such requests should obviously be honored.

After the test, the user should be debriefed and allowed to make comments about the system. After the administration of the questionnaire (if used), any deception employed in the experiment should be disclosed in order not to have the user leave the test with an erroneous understanding of the system. An example of a deception that should be disclosed is the use of the Wizard of Oz method (Maulsby, Greenberg, and Mander, 1993) to simulate nonexisting computer capabilities. Also, the experimenter can answer any additional user questions that could not be answered until after the user had filled in the questionnaire for fear of causing bias. The experimenter should end the debriefing by thanking the user for participating in the test and explicitly state that the test helped to identify areas of possible improvement in the product.[4] This part of the debriefing helps

[3] It is a common mistake to help users too early. Since users normally do not get help when they have to learn a computer system on their own, there is highly relevant information to be gained from seeing what further difficulties users get into as they try to solve the problem on their own.

[4] However, it may also be necessary to mention that the development team will not necessarily be able to correct all identified problems. Users can get very disappointed if they find that the system is released with one of "their" problems still in the interface in spite of a promise to correct it.

users recover their self-respect after the many errors they probably felt they made during the test itself. Also, the experimenter should endeavor to end the session on a positive and relaxed note, repeating that the results are going to be kept confidential and also engaging in some general conservation and small talk as the user is being escorted out of the building or laboratory area.

In addition to following the rules outlined here, it is a good idea for the experimenters to have tried the role of test subjects themselves a few times, so that they know from personal experience how stupid and vulnerable subjects may feel.

46.6 TEST TASKS

The basic rule for test tasks is that they should be chosen to be as representative as possible of the uses to which the system will eventually be put in the field. Also, the tasks should provide reasonable coverage of the most important parts of the user interface. The test tasks can be designed based on a task analysis or based on a product identity statement listing the intended uses for the product. Information from logging frequencies of use of commands in running systems and other ways of learning how users actually use systems, such as field observation, can also be used to construct more representative sets of test tasks for user testing of similar systems (Gaylin, 1986).

The tasks need to be small enough to be completed within the time limits of the user test, but they should not be so small that they become trivial. The test tasks should specify precisely what result the user is being asked to produce, since the process of using a computer to achieve a goal is considerably different from just playing around. For example, a test task for a spreadsheet could be to enter sales figures for six regions for each of four quarters, with some sample numbers given in the task description. A second test task could then be to obtain totals and percentages, and a third might be to construct a bar chart showing trends across regions. Test tasks should normally be given to the users in writing. Not only does this ensure that all users get the tasks described the same way, but having written tasks also allows the user to refer to the task description during the experiment instead of having to remember all the details of the task. After the user has been given the task and has had a chance to read it, the experimenter should allow the user to ask questions about the task description, in order to minimize the risk that the user has misinterpreted the task. Normally, task descriptions are handed to the user on a piece of paper, but they can also be shown in a window on the computer. This latter approach is usually chosen in computer-paced tests where users have to perform a very large number of tasks.

Test tasks should never be frivolous, humorous, or offensive, such as testing a paint program by asking the user to draw a mustache on a scanned photo of the President. First, there is no guarantee that everybody will find the same thing funny, and second, the nonserious nature of such tasks distracts from the test of the system and may even demean the users. Instead, all test tasks should be business oriented (except, of course, for tests of entertainment software and such) and as realistic as possible. To increase both the users' understanding of the tasks and their sense of being realistic usage of the software, the tasks can be related to an overall scenario. For example, the scenario for the spreadsheet example mentioned above could be that the user had just been hired as sales manager for a company and had been asked to give a presentation the next day.

The test tasks can also be used to increase the user's confidence. The very first test task should always be extremely simple in order to guarantee the user an early success experience to boost morale. Similarly, the last test task should be designed to make users feel that they have accomplished something. For example, a test of a word processor could end with having the user print out a document. Since users will feel inadequate if they do not complete all the given tasks, one should never give the users a complete listing of all the test tasks in advance. Rather, the tasks should be given to the users one at a time such that it is always possible to stop the test without letting the user feel incompetent.

46.7 STAGES OF A TEST

A usability test typically has four stages:

1. Preparation
2. Introduction
3. The test itself
4. Debriefing

46.7.1 Preparation

In preparation for the experiment, the experimenter should make sure that the test room is ready for the experiment, that the computer system is in the start state that was specified in the test plan, and that all test materials, instructions, and questionnaires are available. For example, all files needed for the test tasks should be restored to their original content, and any files created during earlier tests should be moved to another computer or at least another directory. To minimize the user's discomfort and confusion, this preparation should be completed before the arrival of the user. Also, any screen savers should be switched off, as should any other system components, such as e-mail notifiers, that might otherwise interrupt the experiment.

46.7.2 Introduction

During the introduction, the experimenter welcomes the test user and gives a brief explanation of the purpose of the test. The experimenter may also explain the computer setup to users if it is likely to be unfamiliar to them. The experimenter then proceeds with introducing the test procedure. Especially for inexperienced experimenters, it may be a good idea to have a checklist at hand with the most important points to be covered, but care should be taken not to make the introduction seem mechanical, as could easily be the case if the experimenter were simply to read from the checklist.

Typical elements to cover in a test introduction include the following:

- The purpose of the test is to evaluate the software and not the user.
- Unless the experimenter is actually the system designer, the experimenter should mention that he or she has no personal stake in the system being evaluated, so that the test user can speak freely without being afraid of hurting the experimenter's feelings. If the experimenter did design the system, this fact is probably better left unsaid in order to avoid the opposite effect.
- The test results will be used to improve the user interface, so the system that will eventually be released will likely be different from the one seen in the test.
- A reminder that the system is confidential and should not be discussed with others. Even if the system is not confidential, it may still be a good idea to ask the test user to refrain from discussing it with colleagues who may be participating in future tests in order not to bias them.
- A statement that participation in the test is voluntary and that the user may stop at any time.[5]
- A reassurance that the results of the test will be kept confidential and not shown to anybody in a form by which the individual test user can be identified.
- An explanation of any video or audio recording that may be taking place. In cases where the video record will not be showing the user's face anyway, but only the screen and keyboard and the user's back, it is a good idea to mention this explicitly to alleviate the user's worries about being recorded.
- An explanation that the user is welcome to ask questions since the experimenters want to know what the users find unclear in the interface, but that the experimenter will not answer most questions during the test itself, since the goal of the test is to see whether the system can be used without outside help.
- Any specific instructions for the kind of experiment that is being conducted, such as instructions to think out loud or to work as fast as possible while minimizing mistakes.
- An invitation to the user to ask any clarifying questions before the start of the experiment.

Many people have the test users sign an informed consent form that repeats the most important instructions and experimental conditions and states that they have understood them. I do not like these forms since they can increase the user's anxiety level by making

---

[5] Even though this may not need to be mentioned explicitly in the experimenter's introduction, users who do elect to stop the experiment should still get whatever payment was promised for the time they have spent, even if they did not complete the experiment, and even if the data from their participation cannot be used.

the test seem more foreboding than it really is. Sometimes, consent forms may be required for legal reasons, and they should certainly be used in cases where videotapes or other records or results from the test will be shown to others. In any case, it is recommended to keep any such forms short, to the point, and written in everyday language rather than legalese, so that the users do not fear that they are being entrapped to sign away more rights than they actually are.

During the introduction phase, the experimenter should also ensure that the physical setup of the computer is ergonomically suited for the individual test user. A common problem is the position of the mouse for left-handed users, but it may also be necessary to adjust the chair or other parts of the room such that the user feels comfortable. If the actual computer model is unfamiliar to the user, it may be a good idea to let the user practice using some other software before the start of the test itself, to avoid contaminating the test results with the user's initial adjustments to the hardware.

After the introduction, the user is given any written instructions for the test, including the first test task, and asked to read them. The experimenter should explicitly ask the test user whether he or she has any questions regarding the experimental procedure, the test instructions, or the tasks before the start of the test.

A special step at the end of the introduction may be to train the users on any part of the interface that they will need to know in order to use the part of the system that is being tested. A classic example was the need to train virtually all users in the use of a mouse before they could be used as test participants in early tests of graphical user interfaces. If separate mouse training had not been provided, we would have found nothing but mouse problems in the test. These days, of course, most users do know how to use a mouse, but there may be other specialized parts of a system that are not relevant for the test and that users should be taught. If in doubt, then do not train the users: it is often better to observe as many usability problems as possible.

46.7.3 Running the Test

During the test itself, the experimenter should normally refrain from interacting with the user, and should certainly not express any personal opinions or indicate whether the user is doing well or poorly. The experimenter may make uncommitted sounds like "uh-huh" to acknowledge comments from the user and to keep the user going, but again, care should be taken not to let the tone of such sounds indicate whether the user is on the right track or has just made a ridiculous comment. Also, the experimenter should refrain from helping the test user, even if the user gets into quite severe difficulties.

The main exception from the rule that users should not be helped is when the user is clearly stuck and is getting unhappy with the situation. The experimenter may also decide to help a user who is encountering a problem that has been observed several times before with previous test users. The experimenter should only do so if it is clear beyond any doubt from the previous tests what the problem is and what different kinds of subsequent problems users may encounter as a result of the problem in question. It is tempting to help too early and too much, so experimenters should exercise caution in deciding when to help. Also, of course, no help can be given during experiments aiming to time users' performance on a task.

In case several people are observing the experiment, it is important to have appointed one of them as the official experimenter in advance and only have that one person provide instructions and speak during the experiment. In order not to confuse the user, all other observers should keep completely quiet, even if they do not agree with the way the experimenter is running the experiment. If they absolutely need to make comments, they can do so by unobtrusively passing the experimenter a note or talking with the experimenter during a break.

46.7.4 Debriefing

After the test, the user is debrief and is asked to fill in any subjective satisfaction questionnaires. To avoid any bias from comments by the experimenter, questionnaires should be administered before any other discussion of the system. During debriefing, users are asked for any comments they might have about the system and for any suggestions they may have for improvement. Such suggestions may not always lead to specific design changes, and one will often find that different users make completely contradictory suggestions, but this type of user suggestion can serve as a rich source of additional ideas to consider in the redesign.

The experimenter can also use the debriefing to ask users for further comments about events during the test that were hard for the experimenter to understand. Even though users may not always remember why they did certain things, they are sometimes able to clarify some of their presumptions and goals.

Finally, as soon as possible after the user has left, the experimenter should check that all results from the test have been labeled with the test user's number, including any files recorded by the computer, all questionnaires and other forms, as well as the experimenter's own notes. Also, the experimenter should write up a brief report on the experiment as soon as possible, while the events are still fresh in the experimenter's mind and the notes still make sense. A full report on the complete sequence of experiments may be written later, but the work of writing such a report is made considerably simpler by having well-organized notes and preliminary reports from the individual tests.

46.8 PERFORMANCE MEASUREMENT

Measurement studies form the basis of much traditional research on human factors and are also important in the usability engineering lifecycle for assessing whether usability goals have been met and for comparing competing products. User performance is almost always measured by having a group of test users perform a predefined set of test tasks while collecting time and error data.

A major pitfall with respect to measurement is the potential for measuring something that is poorly related to the property one is really interested in assessing. Figure 46.2 shows a simple model relating the true goal of a measurement study (the usability of the system) to the actual data-collection activities that may sometimes erroneously be thought of as the core of measurement. As indicated by the model, one starts out by making clear the goal of the exercise. Here, we will assume that "usability" as an abstract concept is the goal, but it could also be, e.g., improved customer perceptions of the quality of a company's user interfaces.

Goals are typically quite abstract, so one then breaks them down into components, such as usability attributes. Figure 46.2 shows two such components, learnability and efficiency of use. One then needs to balance the various components of the goal and decide on their relative importance. Once the components of the goal have been defined, it becomes necessary to quantify them precisely. For example, the component "efficiency of use" can be quantified as the average time it takes users to perform a certain number of specified tasks. Even if these tasks are chosen to be representative of the users' normal task mix, it is important to keep in mind that the test tasks are only that: test tasks and not all possible tasks. In interpreting the results from the measurement study, it is necessary to keep in mind the difference between the principled component that one is aiming for, that is, efficiency of use in general, and the specific quantification that is used as a proxy for that component (i.e., the test tasks). As an obvious example, an iterative design process should not aim at improving efficiency of use for a system just by optimizing

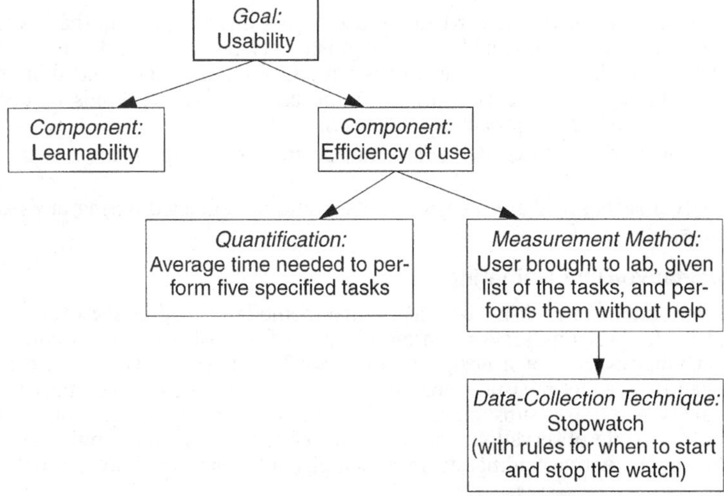

Figure 46.2 Model of usability measurement.

the interface for the execution of the five test tasks and nothing else (unless the tasks truly represent all of what the user ever will do with the system).

Given the quantification of a component, one needs to define a method for measuring the users' performance. Two obvious alternatives come to mind for the example in Figure 46.2: Either bring some test users into the laboratory and give them a list of the test tasks to perform, or observe a group of users at work in their own environment and measure them whenever a task like the specified test tasks occurs. Finally, one needs to define the actual activities that are to be carried out to collect the data from the study. Some alternatives for the present example could be to have the computer measure the time from start to end of each task, to have an experimenter measure it by a stopwatch, and to have users report the time themselves in a diary. In either case it is important to have a clear definition of when a task starts and when it stops.

Typical quantifiable usability measurements include

- The time users take to complete a specific task.
- The number of tasks (or the proportion of a larger task) of various kinds that can be completed within a given time limit.
- The ratio between successful interactions and errors.
- The time spent recovering from errors.
- The number of user errors.
- The number of immediately subsequent erroneous actions.
- The number of commands or other features that were utilized by the user (either the absolute number of commands issued or the number of different commands and features used).
- The number of commands or other features that were never used by the user.
- The number of system features the user can remember during a debriefing after the test.
- The frequency of use of the manuals and/or the help system, and the time spent using these system elements.
- How frequently the manual and/or help system solved the user's problem.
- The proportion of user statements during the test that were positive versus critical toward the system.
- The number of times the user expresses clear frustration (or clear joy).
- The proportion of users who say that they would prefer using the system over some specified competitor.
- The number of times the user had to work around an unsolvable problem.
- The proportion of users using efficient working strategies compared to the users who use inefficient strategies (in case there are multiple ways of performing the tasks).
- The amount of "dead" time when the user is not interacting with the system. The system can be instrumented to distinguish between two kinds of dead time: response-time delays where the user is waiting for the system, and thinking-time delays where the system is waiting for the user. These two kinds of dead time should obviously be approached in different ways.
- The number of times the user is sidetracked from focusing on the real task.

Of course, only a subset of these measurements would be collected during any particular measurement study.

46.8.1 Qualitative Data Gathering

Many usability studies in industry use qualitative methods instead of the exact measurements listed in the previous section. Often, the main focus of a test is to collect information about what aspects of a design seem to work and what aspects cause usability problems, even if no exact measures are available to discriminate between the two. Qualitative data analysis requires substantial usability engineering experience if one is to mine a study for all possible observations, but even inexperienced usability engineers can discover the major qualitative findings from a usability test since they are normally pretty glaring.

In our lab, most studies do indeed use qualitative methods and we often have several observers present behind the one-way mirror to take notes about the usability events experienced during the test. Most such notes concern usability problems where the user had difficulties using the product, but the observers are also encouraged to note any positive elements of the design. Notes are typically made on sticky notes, using different colored paper for each test participant. It is also possible to use personal digital assistants (PDAs) for note taking, the benefit being that the PDA can automatically timestamp each observation, making it easier to coordinate the observations with the videotape of the test.

After a usability test with maybe three or four users, we typically have 100 or 200 sticky notes with observations. Many of the observations will be the same, either because several observers made the same comment or because several users encountered the same usability problems. Furthermore, many notes will concern similar aspects of the system, even if they are not the same. For example, one often finds many notes related to the user's navigation through the user interface. We often use an affinity diagramming technique as illustrated in Figure 46.3: All the sticky notes are pasted on a large wall, and several people move them around, placing notes that are identical over each other and notes that are similar near each other. After a while, the major groupings of observations appear, and the last step is to provide names for each category. Typically, observers from the usability test are used to produce the affinity diagram since they are better capable of understanding the notes from the test.

46.9 THINKING ALOUD

Thinking aloud may be the single most valuable usability engineering method. Basically, a thinking-aloud test involves having a test subject use the system while continuously thinking out loud (Lewis, 1982). By verbalizing their thoughts, the test users enable us to understand how they view the computer system, and this again makes it easy to identify the users' major misconceptions. One gets a very direct understanding of what parts of the dialogue cause the most problems, because the thinking-aloud method shows how users interpret each individual interface item.

The thinking-aloud method has traditionally been used as a psychological research method (Ericsson and Simon, 1984), but it is increasingly being used for the practical

Figure 46.3 Building an affinity diagram by moving sticky notes with observations from usability test. (Photo courtesy of SunSoft.)

evaluation of human–computer interfaces (Denning, Hoiem, Simpson, and Sullivan, 1990). The main disadvantage of the method is that it does not lend itself very well to most types of performance measurement. On the contrary, its strength is the wealth of qualitative data it can collect from a fairly small number of users. Also, the users' comments often contain vivid and explicit quotes that can be used to make the test report more readable and memorable.

At the same time, thinking aloud may also give a false impression of the cause of usability problems if too much weight is given to the users' own "theories" of what caused trouble and what would help. For example, users may be observed to overlook a certain field in a dialogue box during the first part of a test. After they finally find the field, they may claim that they would have seen it immediately if it had been in some other part of the dialogue box. It is important not to rely on such statements. Instead, the experimenter should make notes of what the users were doing during the part of the experiment where they overlooked the critical field. Data showing where users actually looked has much higher validity than the users' claim that they would have seen the field if it had been somewhere else. The strength of the thinking-aloud method is to show what the users are doing and why they are doing it while they are doing it in order to avoid later rationalizations.

Thinking out loud seems very unnatural to most people, and some test users have great difficulties in keeping up a steady stream of utterances as they use a system.[6] Not only can the unnaturalness of the thinking-aloud situation make the test harder to conduct, but it can also have an impact on the results. First, the need to verbalize can slow users down, thus making any performance measurements less representative of the users' regular working speed. Second, users' problem solving behavior can be influenced by the very fact that they are verbalizing their thoughts. The users might notice inconsistencies in their own models of the system, or they might concentrate more on critical task components (Bainbridge, 1979), and these changes may cause them to learn some user interfaces faster or differently than they otherwise would have done. For example, Berry and Broadbent (1990) provided users with written instructions on how to perform a certain task and found that users performed 9% faster if they were asked to think aloud while doing the task. Berry and Broadbent argue that the verbalization reinforced those aspects of the instructions which the users needed for the task, thus helping them become more efficient. In another study (Wright and Converse, 1992), users who were thinking aloud while performing various file system operations were found to make only about 20% of the errors made by users who were working silently. Furthermore, the users in the thinking-aloud study finished their tasks about twice as fast as the users in the silent condition.

The experimenter will often need to prompt the user continuously to think out loud by asking such questions as, "What are you thinking now?" and "What do you think this message means?" (after the user has noticed the message and is clearly spending time looking at it and thinking about it). If the user asks a question such as, "Can I do such-and-such?" the experimenter should not answer, but instead keep the user talking with a counterquestion such as, "What do you think will happen if you do it?" If the user acts surprised after a system action but does not otherwise say anything, the experimenter may prompt the user with a question such as, "Is that what you expected would happen?" Of course, following the general principle of not interfering in the user's use of the system, the experimenter should not use such prompts as, "What do you think the message on the bottom of the screen means?" if the user has not noticed that message yet.

Since thinking aloud seems strange to many people, it may help to give the test users a role model by letting them observe a short thinking-aloud test before the start of their own experiment. One possibility is for the experimenter to enact a small test, where the experimenter performs some everyday task such as looking up a term in a dictionary while thinking out loud. Alternatively, users can be shown a short video of a test that was made with the sole purpose of instructing users. Showing users how a test videotape looks may also help alleviate their own fears of any videotaping that will be done during the test.

Users will often make comments regarding aspects of the user interface that they like or do not like. To some extent, it is one of the great advantages of the thinking-aloud

---

[6] Verbalization seems to be hardest for expert users who may perform many operations so quickly that they have nothing to say. They may not even consciously know what they are doing in cases where they have completely automated certain common procedures.

method that one can collect such informal comments about small irritants that would not show up in other forms of testing. They may not impact measurable usability, but they may as well be fixed. Unfortunately, users will often disagree about such irritants, so one should take care not to change an interface just because of a comment by a single user. Also, user comments will often be inappropriate when seen in a larger interface design perspective, so it is the responsibility of the experimenter to interpret the user's comments and not just accept them indiscriminately. For example, users who are using a mouse for the first time will often direct a large proportion of their comments toward aspects of moving the mouse and pointing and clicking, which might be interesting for a designer of more intuitive input hardware but are of limited use to a software designer. In such a test, the experimenter would need to abstract from the users' mouse problems and try to identify the underlying usability problems in the dialogue and estimate how the users would have used the interface if they had been better at using the pointing device.

46.9.1 Constructive Interaction

A variation of the thinking-aloud method is called *constructive interaction* and involves having two test users use a system together (O'Malley, Draper, and Riley, 1984). This method is sometimes also called *codiscovery learning* (Kennedy, 1989) or *paired-user testing* (Wildman, 1995). The main advantage of constructive interaction is that the test situation is much more natural than standard thinking-aloud tests with single users, since people are used to verbalizing when they are trying to solve a problem together. Therefore, users may make more comments when engaged in constructive interaction than when simply thinking aloud for the benefit of an experimenter (Hackman and Biers, 1992). The method does have the disadvantage that the users may have different strategies for learning and using computers. Therefore, the test session may switch back and forth between disparate ways of using the interface, and one may also occasionally find that the two test users simply cannot work together.

Constructive interaction is especially suited for usability testing of user interfaces for children since it may be difficult to get them to follow the instructions for a standard thinking-aloud test.

Constructive interaction is most suited for projects where it is easy to get large numbers of users into the lab, and where these users are comparatively cheap, since it requires the use of twice as many test users as single-user thinking aloud.

46.9.2 Retrospective Testing

If a videotape has been made of a user test session, it becomes possible to collect additional information by having the user review the recording (Hewett and Scott, 1987). This method is sometimes called *retrospective testing*. The user's comments while reviewing the tape are sometimes more extensive than comments made under the (at least perceived) duress of working on the test task, and it is of course possible for the experimenter to stop the tape and question the user in more detail without fearing to interfere with the test, which has essentially already been completed.

Retrospective testing is especially valuable in cases where representative test users are difficult to get hold of, since it becomes possible to gain more information from each test user. The obvious downside is that each test takes at least two times as long, so the method is not suited if the users are highly paid or perform critical work from which they cannot be spared for long. Unfortunately, those users who are difficult to get to participate in user testing are often exactly those who are also very expensive, but there are still some cases where retrospective testing is beneficial.

46.9.3 Coaching Method

The coaching method (Mack and Burdett, 1992) is somewhat different from other usability test methods in having an explicit interaction between the test subject and the experimenter (or "coach"). In most other methods, the experimenter tries to interfere as little as possible with the subject's use of the computer, but the coaching method actually involves steering the user in the right direction while using the system.

During a coaching study, the test user is allowed to ask any system-related question of an expert coach who will answer to the best of his or her ability.[7] Usually, the exper-

[7] One variant of the coaching method would be to restrict the answers to certain predetermined information. In an extensive series of experiments, one could then vary the rules for the coach's answers in order to learn what types of answers helped users the most. Unfortunately, this variant

imenter or a research assistant serves as the coach. One variant of the method involves a separate coach chosen from a population of expert users. Having an independent coach lets the experimenter study how the coach answers the user's questions. This variant can be used to analyze the expert coach's model of the interface. Normally, though, coaching focuses on the novice user and is aimed at discovering the information needs of such users in order to provide better training and documentation, as well as possibly redesigning the interface to avoid the need for the questions.

The coaching method has proved helpful in getting Japanese users to externalize their problems while using computers (Kato, 1986). Other, more traditional methods are sometimes difficult to use in Japan, where cultural norms make some people reluctant to verbalize disagreement with an interface design.[8]

The coaching situation is more natural than the thinking-aloud situation. It also has an advantage in cases where test users are hard to come by because the intended user population is small, specialized, and highly paid. Coaching provides the test users with tangible benefits in return for participating in the test by giving them instruction on a one-to-one basis by a highly skilled coach.

Finally, the coaching method may be used in cases where one wants to conduct tests with expert users without having any experts available. Coaching can bring novice users up to speed fairly rapidly and can then be followed by more traditional tests of the users' performance once they have reached the desired level of expertise.

46.10 USABILITY LABORATORIES

Many user tests take place in specially equipped usability laboratories (Nielsen, 1994c). Figures 46.4 and 46.5 show the two main rooms in one of Sun Microsystem's usability laboratories: the participant room and the control room. I should stress from the beginning

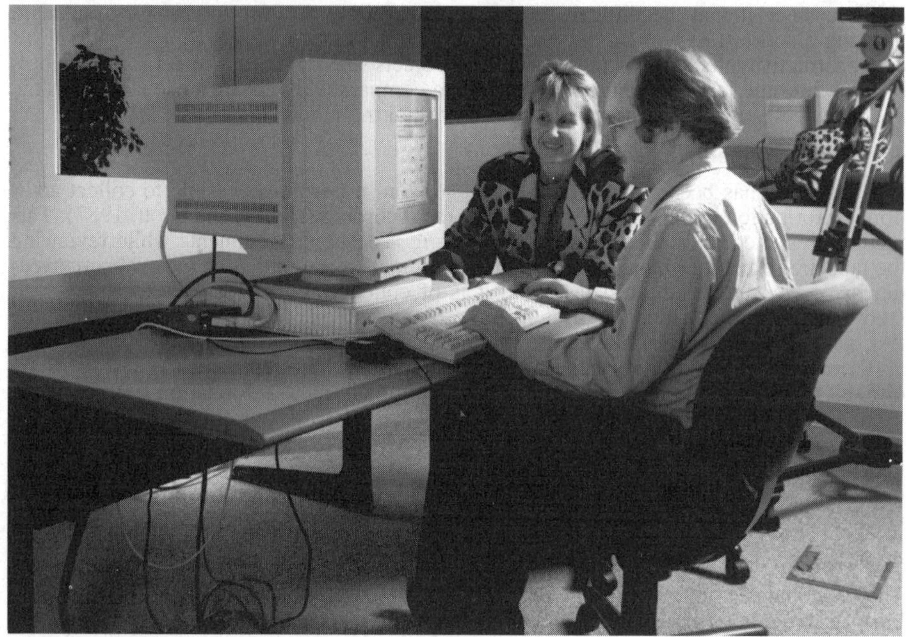

Figure 46.4 View of a usability laboratory showing the participant room: An experimenter is in the room with the user to take notes and manage the test while other team members may observe from the control room (see Figure 46.5). (Photo courtesy of SunSoft.)

requires extremely skilled and careful coaches, since they need to compose answers on the fly to unpredictable user questions.

[8] In general, usability methods sometimes have to be adjusted for use in other countries, and there is an emerging discipline of how to do international usability studies (Nielsen, 1996).

Figure 46.5 View of the control room in a usability laboratory. The participant room (see Figure 46.4) is visible through the one-way mirror. (Photo courtesy of SunSoft.)

that special laboratories are a convenience but not an absolute necessity for usability testing. It is possible to convert a regular office temporarily into a usability laboratory, and it is possible to perform usability testing with no more equipment than a notepad.

In September 1993, I surveyed 13 usability laboratories from a variety of companies (Nielsen, 1994c). The median floor space of the laboratories was 63 m^2 (678 ft^2), and the median size of the test rooms was 13 m^2 (144 ft^2). The smallest laboratory was 35 m^2 (377 ft^2) with only 9 m^2 (97 ft^2) for the test user. The largest laboratory was 237 m^2 and had 7 rooms, allowing a variety of tests to take place simultaneously (Lund, 1994). The largest single test room was 40 m^2 (430 ft^2) and was found in a telephone company with a need to test groupware interfaces with many users. See Table 46.1 for further details of the survey. Even though the survey was conducted in the end of 1993, it seems that more recently built usability laboratories have about the same characteristics as the ones in the survey.

Having a permanent usability laboratory decreases the overhead of usability testing (once it is set up, that is!) and may thus encourage increased usability testing in an organization. Having a special room and special equipment dedicated to usability testing means that there will be fewer scheduling problems associated with each test and also makes it possible to run tests without disturbing other groups.

Usability laboratories typically have soundproof, one-way mirrors separating the observation room from the test room to allow the experimenters, other usability specialists, and the developers to discuss user actions without disturbing the user. Users are not so stupid that they do not know that there are observers behind a wall with a large mirror in a test room, so one might as well briefly show the users the observation room before the start of the test. Knowing who and what are behind the mirror is much less stressful for the users than having to imagine it. People usually come to ignore unseen observers during the test, even though they know they are there.

Having an executive observation area in the back of the main observation area allows a third group of observers (e.g., the development team) to discuss the test without disturbing the primary experimenters and the usability specialists.

Typically, a usability laboratory is equipped with several video cameras under remote control from the observation room: The average number of cameras in each test room was 2.2 in my survey, with 2 cameras being the typical number and a few labs using 1

Table 46.1 Overview of the usability laboratories I surveyed in September 1993. For each company, only a single usability laboratory was surveyed, even if the company had multiple labs[a]

Company Name	Main Product	Other Labs in Company?	Date of First Usability Lab in Company	Floor Space of Typical Subject Room in Sq. Meters	Floor Space of Total Lab Area in Sq. Meters	Number of Rooms (Subject Rooms, Control Rooms, etc.)	Number of Cameras in Typical Subject Room	Scan Converter Used to Directly Tap Screen Image?	One-Way Mirror?	Usability Staff Supporting/Utilizing Lab
Ameritech	Communications service	No	1989	12.5	237	7	2	Yes	Yes	1/10
Bellcore	Telco software	Yes	1985	12.3	121	7	2	No	Yes	0.3/30
BT (British Telecom)	Telephone service	No	1988	40	96	3	3	Yes	Yes	0.5/70
IBM	Computer systems	Yes	1981	11.7	165	14	2	No	Yes	0.1/4
MAYA Design Group	Design consultants	No	1990	8.8	42.9	3	2	No	Yes	3/12
Microsoft Corp.	PC software	No	1989	10.8	181.3	19	2	Yes	Yes	4/22
NCR	Computer systems	Yes	1966	13.4	31.2	3	2	Yes	Yes	2/15
Philips, Corp. Design	Consumer electronics	Yes	1990	30	40	2	3	No	Yes	0.25/10
Philips, IPO	Consumer electronics	Yes	1990	9	35	3	2	No	No	1/25
SAP	Standard business apps	No	1992	37	63	2	1	No	Yes	2/12
SunSoft	Workstation software	No	1988	25.1	202.3	8	3	Yes	Yes	3.5/8
Symantec	PC software	No	1992	23.8	47.6	2	2	Yes	Yes	1/1
Taligent	PC operating systems	No	1992	13.4	26.8	2	2	No	Yes	1/2
Mean		38%	1987	19.1	99.2	5.8	2.2	46%	92%	1.5/17.0
Median		No	1989	13.4	63.8	3	2	No	Yes	1/12

[a] Notes: The first number in the entry for usability staff supporting and using the lab indicates the head count of the staff that is dedicated to keeping the lab up and running. The second number in this entry indicates the number of usability specialists who share the lab for their work, even though they may not all be using it full time. One square meter is 10.76 square feet. As used in this table, the abbreviation "PC" refers to the concept "personal computers."

or 3. These cameras can be used to show an overview of the test situation and to focus in on the user's face, the keyboard, the manual and the documentation, and the screen. A producer in the observation room then typically mixes the signal from these cameras to a single video stream that is recorded, and possibly timestamped for later synchronization with an observation log entered into a computer during the experiment. Such synchronization makes it possible to later find the video segment corresponding to a certain interesting user event without having to review the entire videotape.

More rarely, usability laboratories include other equipment to monitor users and study their detailed behavior. For example, an eyetracker can be used to collect data on what parts of the screens the user looks at (Benel, Ottens, and Horst, 1991).

Figure 46.6 shows some equipment from Sun Microsystem's usability laboratory. In many ways, the most important equipment in the figure is the "Do Not Enter" sign (Figure 45.6*a*), since it makes it possible to conduct the usability test without interruptions. As long as one has a room with a Do Not Disturb sign, one can conduct usability tests without any further equipment (one will not even need a computer if one is doing paper prototyping!). The second most important piece of equipment may be high-quality microphones (Figure 46.6*b*); since there is normally a good deal of background noise from the computer, it will be impossible to hear what the user is saying unless professional microphones are used and unless the user is actually wearing the microphone. The figure also shows a typical movable camera (Figure 46.6*c*) and the camera controls, but as further discussed below, videotaping is not nearly as critical as is the ability to hear what the user is saying.

46.10.1 To Videotape or Not

Having videotapes of a user test is essential for many research purposes where one needs to study the interaction in minute detail (Mackay and Tatar, 1989). For practical usability engineering purposes, however, there is normally no need to review a user test on videotape since one is mostly interested in finding the major "usability catastrophes" anyway. These usability problems tend to be so glaring that they are obvious the first time they

Figure 46.6 Typical equipment from a usability laboratory: Do Not Enter sign on the door (*a*), track-mounted camera (*b*), wireless lavaliere microphones for the participant and the experimenter (*c*), and monitors and remote controls for operating the cameras from the control room (*d*). (Photos courtesy of SunSoft.)

Figure 46.6(b) *(Continued)*

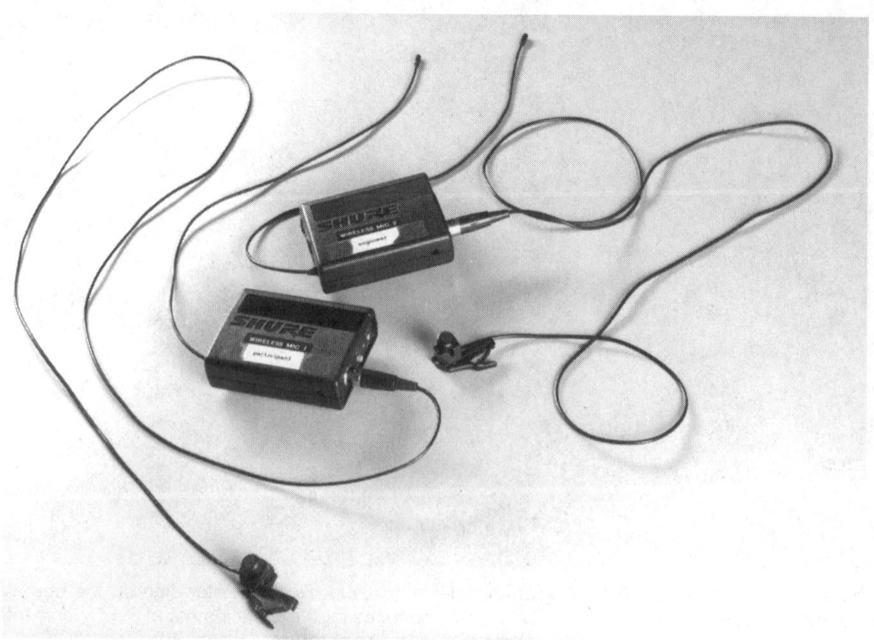

Figure 46.6(c) *(Continued)*

Figure 46.6(d) *(Continued)*

are observed and therefore do not require repeated perusal of a record of the test session. This is especially true considering estimates that the time needed to analyze a videotape is between 3 and 10 times the duration of the original user test. In most cases, this extra time is better spent running more test subjects or testing more iterations of the design.

Videotape does have several uses in usability engineering, however. For example, a complete record of a series of user tests is a way to perform formal impact analysis of usability problems (Good, Spine, and Whiteside, 1986). Impact analysis involves first finding the usability problems and then going back to the videotapes to investigate exactly how many users had each usability problem and how much they were delayed by each problem. Since these estimates can be made only after one knows what usability problems to look for, an impact analysis requires a videotape or other detailed record of the test sessions. Alternatively, one can run more tests and count the known problems as they occur. Impact analyses can then be used to prioritize the fixing of the usability problems in a redesign such that the most effort is spent on those problems that are faced by many users and impact them severely.

Videotape also serves as an essential communications medium in many organizations, where it may otherwise be difficult for human factors professionals to persuade developers and managers that a certain usability problem is in fact a problem. Seeing a video of a user struggling with the problem often convinces these people. This goal can also be achieved by simpler means, however, since it is normally even more effective to have the doubter observe a user test in person.[9]

A final argument in favor of videotaping and equipment-extensive usability laboratories is the need to impress upper management and research funding agencies with the unique aspects of usability work. Some usability specialists feel that simpler techniques may not

[9] Doing so requires strict adherence to the "shut up" rule: The developers should be advised in advance that they are not supposed to interfere with the user during the experiment. Doing so can be extremely hard for a person who normally has quite strong defensive feelings toward the design. Developers have been known to forcibly interrupt a test user's "maltreatment" of their beloved system and shout, "Why don't you press that function key!" This, of course, totally destroys the test.

be sufficiently impressive to outsiders, whereas having an expensive laboratory will result in increased funding and respect because of its "advertising value" (Lindgaard, 1991).

46.10.2 Cameraless Videotaping

The main aspects of a test session can be captured on videotape without the use of cameras. Many computers provide a video output that either is directly compatible with video recording or can be made so fairly cheaply by a scan converter.[10] This video signal can be fed directly into the "video in" jack of the video recorder and will thus allow the recording of the exact image the user sees on the monitor. This technique will normally result in better image quality than filming the monitor with a camera, but the video resolution will still be poorer than that of most computer monitors. Furthermore, an audio signal can be fed into the video recorder's "audio in" jack from a microphone, thus creating a composite recording of the screen and the user's comments (Connally and Tullis, 1986).

Cameraless videotaping has the obvious disadvantages of not including the user in the picture and not making it possible for a camera operator to zoom in on interesting parts of the screen or the manual page being studied in vain by the user. Unless a high-definition television standard is used, one will also suffer a loss of resolution since current television standards use a poorer quality signal than that used by almost all computer monitors. These limitations may make the resulting videotape less appealing and convincing in some cases. At the same time, cameraless videotaping is considerably cheaper because neither cameras nor operators are needed, and the users are normally less intimidated by a microphone than by a camera.

46.10.3 Slave Monitors

Whether or not videotaping is used, it is critical to allow the observers in the observation room to follow what is happening on the user's screen. Often, this is done by having one or more cameras pointing at the user's screen and feeding the images from these cameras to monitors in the observation room. The advantage of the video shots is that the observers can see not just the computer display but also what elements of the display the user may be pointing to.

In my experience, the best image quality is derived not from a video shot of the user's screen but from a slave monitor driven directly by the video port on the user's computer. Most computers generate a video signal that can be fed through a splitter, with one signal fed through to the user's monitor and the other signal used to drive a slave monitor in the neighboring room. With a slave monitor, the observers see exactly the same screen image as the user, and there is no risk of the user's head obscuring part of the video if the user gets too eager and leans forward toward the screen. On the other hand, a slave monitor does not allow the observers to see where the user is pointing, so an optimal lap setup combines the slave monitor with one or more video images.

46.10.4 Portable Usability Laboratories

In addition to permanent usability laboratories, it is possible to use portable usability laboratories for more flexible testing and for field studies. With a portable usability laboratory, any office can be rapidly converted to a test room, and user testing can be conducted where the users are rather than having to bring the users to a fixed location.

A true discount portable usability laboratory need consist of no more than a notepad and possibly a laptop computer to run the software that is being tested. Normally, a portable usability laboratory will include slightly more equipment, however. Typical equipment includes a camcorder (possibly just home video equipment, but preferably of professional quality since the filming of user interfaces requires as high resolution as possible) and a lavaliere microphone (two microphones are preferred so that the experimenter can also get one). The regular directional microphone built into many camcorders is normally not sufficient because of the noise of the computer. Also, a tripod helps steady the image and carry the camera during the hour-long test sessions.

46.10.5 Usability Kiosks

A final approach to the collection of usability data is the usability kiosk, which really is a self-served usability laboratory for use as part of a hallway methodology (Gould, Boies,

---

[10] Scan converters were used by 46% of the labs in my survey.

Levy, Richards, and Schoonard, 1987). In general, the hallway method involves putting a user interface on display in a heavily trafficked area such as outside a company cafeteria in order to collect comments from users and other passersby. A usability kiosk can conduct automated usability testing with self-selected users in such a setting by providing access to a computer running a test interface, suggesting various test tasks to the users, and recording their task times and any comments they might have.

REFERENCES

Usability testing is the main topic of the annual meetings of the Usability Professionals' Association. For further information contact its office: Usability Professionals' Association, 10875 Plano Road, Suite 115, Dallas, TX 75238, USA. Tel. +1-214-349-8841, fax +1-214-349-7946, e-mail upadallas@aol.com.

Allen, R. B. (1984, July). Working paper on ethical issues for research on the use of computer services and interfaces. *ACM SIGCHI Bulletin 16*(1), 12–16.

American Psychological Association. (1982). *Ethical Principles in the Conduct of Research with Human Participants.* Washington, DC: American Psychological Association.

Bainbridge, L. (1979). Verbal reports as evidence of the process operator's knowledge. *International Journal of Man–Machine Studies, 11,* 411–436.

Benel, D. C. R., Ottens, D. Jr., and Horst, R. (1991). Use of an eyetracking system in the usability laboratory. In *Proceedings of the Human Factors Society 35th Annual Meeting* (pp. 461–465). Santa Monica, CA: Human Factors and Ergonomics Society.

Berry, D. C., and Broadbent, D. E. (1990, May–June). The role of instruction and verbalization in improving performance on complex search tasks. *Behaviour and Information Technology, 9*(3), 175–190.

Brigham, F. R. (1989). Statistical methods for testing the conformance of products to user performance standards. *Behaviour and Information Technology, 8*(4), 279–283.

Connally, C. S., and Tullis, T. S. (1986). Evaluating the user interface: Videotaping without a camera. In *Proceedings of the Human Factors Society 30th Annual Meeting* (pp. 1029–1033).

Czaja, S. J., Hammond, K., Blascovich, J. J., and Swede, H. (1989). Age related differences in learning to use a text-editing system. *Behaviour and Information Technology, 8*(4), 309–319.

Denning, S., Hoiem, D., Simpson, M., and Sullivan, K. (1990). The value of thinking-aloud protocols in industry: A case study at Microsoft Corporation. *Proceedings of the Human Factors Society 34th Annual Meeting* (pp. 1285–1289). Santa Monica, CA: Human Factors and Ergonomics Society.

Egan, D. E. (1988). Individual differences in human–computer interaction. In Helander, M. (Ed.), *Handbook of Human–Computer Interaction* (pp. 543–568). Amsterdam: North-Holland.

Ehrlich, K., Butler, M. B., and Pernice, K. (1994). Getting the whole team into usability testing. *IEEE Software, 11*(1), 89–91.

Ericsson, K. A., and Simon, H. A. (1984). *Protocol Analysis: Verbal Reports as Data.* Cambridge, MA: MIT Press.

Gaylin, K. B. (1986). How are windows used? Some notes on creating an empirically-based windowing benchmark task. *Proceedings of the ACM CHI'86 Conference* (Boston, MA, April 13–17, pp. 96–100). New York: Association for Computing Machinery.

Good, M., Spine, T. M., Whiteside, J., and George, P. (1986). User-derived impact analysis as a tool for usability engineering. *Proceedings of the ACM CHI'86 Conference* (Boston, MA, April 13–17, pp. 241–246). New York: Association for Computing Machinery.

Gould, J. D., Boies, S. J., Levy, S., Richard, J. T., and Schoonard, J. (1987). The 1984 Olympic Message System: A test of behavioral principles of system design. *Communications of the ACM, 30*(9), 758–769.

Hackman, G. S., and Biers, D. W. (1992). Team usability testing: Are two heads better than one? *Proceedings of the Human Factors Society 36th Annual Meeting* (pp. 1205–1209). Santa Monica, CA: Human Factors and Ergonomics Society.

Hewett, T. T., and Scott, S. (1987). The use of thinking-out-loud and protocol analysis in development of a process model of interactive database searching. *Proceedings of the IFIP INTERACT'87 Second International Conference on Human–Computer Interaction* (Stuttgart, Germany, September 1–4, pp. 51–56). Amsterdam: Elsevier Science Publishers.

Holleran, P. A. (1991, September–October). A methodological note on pitfalls in usability testing. *Behaviour and Information Technology, 10*(5), 345–357.

Jørgensen, A. H. (1989). Using the thinking-aloud method in system development. In G. Salvendy and M. J. Smith (Eds.), *Designing and Using Human–Computer Interfaces and Knowledge Based Systems* (pp. 743–750). Amsterdam: Elsevier Science Publishers.

Kato, T. (1986, December). What 'question-asking protocols' can say about the user interface. *International Journal of Man–Machine Studies, 25*(6), 659–673.

Kennedy, S. (1989, October). Using video in the BNR usability lab. *ACM SIGCHI Bulletin, 21*(2), 92–95.

Landauer, T. K. (1988a). Research methods in human–computer interaction. In Helander, M. (Ed.), *Handbook of Human–Computer Interaction* (pp. 905–928). Amsterdam: North-Holland.

Landauer, T. K. (1988b). Relations between cognitive psychology and computer system design. In Carroll, J. M. (Ed.), *Interfacing Thought: Cognitive Aspects of Human–Computer Interaction* (pp. 1–25). Cambridge, MA: MIT Press.

Lehmann, E. L., and D'Abrera, H. J. M. (1975). *Nonparametrics: Statistical Methods Based on Ranks.* San Francisco: Holden-Day.

Lewis, C. (1982). *Using the 'thinking-aloud' method in cognitive interface design (Research Rep. RC9265).* Yorktown Heights, NY: IBM T.J. Watson Research Center.

Lindgaard, G. (1991, August). Impressions from HUSAT. *CHISIG Newsletter* (Computer-Human Interaction Special Interest Group of the Ergonomics Society of Australia), 1–2.

Lund, A. M. (1994, January–April). Ameritech's usability laboratory: From prototype to final design. *Behaviour and Information Technology, 13*(1 and 2), 67–80.

Mack, R. L., and Burdett, J. M. (1992). When novices elicit knowledge: Question-asking in designing, evaluating and learning to use software. In R. Hoffman (Ed.), *The Psychology of Expertise: Cognitive Research and Empirical AI* (pp. 245–268). New York: Springer-Verlag.

Mackay, W. E., and Tatar, D. G. (1989, October). Introduction to this special issue on video as a research and design tool. *ACM SIGCHI Bulletin, 21*(2), 48–50.

Maulsby, D., Greenberg, S., and Mander, R. (1993). Prototyping an intelligent agent through Wizard of Oz. In *Proceedings of the ACM INTERCHI'93 Conference* (Amsterdam, The Netherlands, April 24–29, pp. 277–284). Association for Computing Machinery, New York.

Nielsen, J. (1994a). *Usability Engineering.* Boston: AP Professional.

Nielsen, J. (1994b). Heuristic evaluation. In J. Nielsen and R. L. Mack (Eds.), *Usability Inspection Methods* (pp. 25–62). New York: John Wiley & Sons.

Nielsen, J. (1994c, January–April). Usability laboratories. *Behaviour and Information Technology, 13*(1 and 2), 3–8.

Nielsen, J. (1996). International usability engineering. In E. del Galdo and J. Nielsen (Eds.), *International User Interfaces,* New York: John Wiley & Sons.

Nielsen, J., and Landauer, T. K. (1993). A mathematical model of the finding of usability problems. In *Proceedings of the ACM INTERCHI'93 Conference* (Amsterdam, The Netherlands, April 24–29, pp. 206–213). Association for Computing Machinery, New York.

Nielsen, J., and Mack, R. L. (1994). *Usability Inspection Methods.* New York: John Wiley & Sons.

Nielsen, J., and Schaefer, L. (1993, July–August). Sound effects as an interface element for older users. *Behaviour and Information Technology, 12*(4), 208–215.

O'Malley, C. E., Draper, S. W., and Riley, M. S. (1984). Constructive interaction: A method for studying human–computer–human interaction. In *Proceedings of the IFIP INTERACT'84 First International Conference Human–Computer Interaction* (London, September 4–7, pp. 269–274). International Federation for Information Processing, Geneva, Switzerland.

Pedhazur, E. J., and Schmelkin, L. P. (1991). *Measurement, Design, and Analysis: An Integrated Approach.* Hillsdale, NJ: Erlbaum.

Schrier, J. R. (1992). Reducing stress associated with participating in a usability test. In *Proceedings of the Human Factors Society 36th Annual Meeting* (pp. 1210–1214). Human Factors and Ergonomics Society, Santa Monica, CA.

Tombaugh, J., Lickorish, A., and Wright, P. (1987, May). Multi-window displays for readers of lengthy texts. *International Journal of Man–Machine Studies, 26*(5), 597–615.

Wildman, D. (1995, July). Getting the most from paired-user testing. *ACM Interactions, 2*(3), 21–27.

Wright, P. C., and Monk, A. F. (1991, December). A cost-effective evaluation method for use by designers. *International Journal of Man–Machine Studies, 35*(6), 891–912.

Wright, R. B., and Converse, S. A. (1992). Method bias and concurrent verbal protocol in software usability testing. In *Proceedings of the Human Factors Society 36th Annual Meeting* (pp. 1220–1224). Human Factors and Ergonomics Society, Santa Monica, CA.

CHAPTER 47
MAINTAINABILITY

Anthony E. Majoros
McDonnell Douglas Aerospace
Long Beach, CA 90807-4418 USA

Edward Boyle
Air Force Armstrong Laboratory
Wright-Patterson Air Force Base, OH 45433-7604 USA

47.1 THE REALITIES OF MAINTENANCE

Complex equipment is costly to maintain, and when it is "down," it has, as one writer put it, "no more value than a child's pull toy" (Martin, 1981, p. 163). Military equipment is a prime example, where maintenance accounts for 30–50% of life cycle cost (e.g., Bond, 1987; U.S. Army Materiel Command, 1976). Other industries, of course, see significant costs for maintenance. The expense of repairing and servicing equipment produces a steady pressure favoring systems that operate longer without maintenance and are easier to fix when they fail. Manufacturers in turn can satisfy customers and improve market share with maintainable products (Lynch, 1995).

Equipment design characteristics can obviously interact with human capacities and limitations to foil or promote maintenance efforts, although technicians' skill, quality of maintenance instructions, and work environment also account for much variability in maintenance. The reader can get a sense of the many opportunities for human factors engineering to help reduce time and expense in maintenance by studying the "problems" in Table 47.1.

Table 47.1 Sequence of Events in Corrective Maintenance and Examples of Typical Setbacks Requiring Extra Time

Maintenance Event	Potential "Problem" Conditions or Extra Time Elements
Failure occurs	Delay before trouble is noticed
Failure is detected, operator produces "squawk"	No failure occurred, operator misjudged equipment condition
Technician notified	Gather tools, supplies, and materials Special tools required—more gathering time Cannot locate technical data (maintenance/fault isolation instructions), spend time tracking it down, or "wing it"
Technician prepared and work begins	Set up for task (work stand, crawler, lighting, etc.) Wait for equipment to become available for maintenance Technical data is incomplete—get advice or "wing it" No failure as squawked, cannot duplicate (CND) fault or failure mode Failure is real but BIT is unreliable, cannot match symptoms to squawk (CND) Systematic troubleshooting perceived as not working, begin "shotgunning" Technician called away to support another task or for administrative functions Technician passes incomplete task to next shift, second technician covers same ground
Failed LRU located	BIT is uninterpretable, get help
Access gained	Remove good parts to gain access, good parts suffer damage in the process
Fault located and correction started	Technical data is outdated, discover problem after trying as-written procedure Disconnect failed LRU, fasteners or connectors have problems (corrosion, etc.) Cannot recall procedure, meticulously study technical data
Obtain replacement unit	Replacement LRU not in stock, try to fix component in LRU, or cannibalize, or deadline equipment Confusion arises about part numbers and effectivity
Install new item	Re-install all other parts removed to gain access to LRU (Test cannot commence unless all parts are re-installed) Calibrate, add shims, torque, install safety wire, or do other subsidiary tasks
Replacement complete and test started	Do more adjusting and calibrating Failure indication won't go away Repeat LRU removal and replacement
Adjustment complete, equipment closed	Re-assemble miscellaneous support hardware
Checkout	Wrap up/clean up area Discover that a tool is missing, possibly left in equipment
Repair complete	Return technical data and special tools Document repair
Parts counter returns LRU to factory	LRU retests okay (RTOK) at the factory

Ease of maintenance is sometimes achieved by a breakthrough idea. For example, U. S. Army technicians in the past checked infrared night-flight lights on helicopters by observing the lights through expensive goggles; a handheld sensor costing a few dollars now does the job (John Ledoux's better idea, 1987). Much more commonly though, equipment is designed with certain features that make for easier maintenance. This concentration on features and attributes comes through in references to the field: according to the *Maintainability Engineering Guide*, the discipline is "concerned with *incorporating required maintainability features* in system/equipment design" (U.S. Army Materiel Command, 1976) (italics added). In Blanchard's words (1986, p. 32), "Maintainability is an *inherent design characteristic* dealing with the ease, accuracy, safety, and economy in the performance of maintenance functions" (italics added). A few of the hundreds of documented maintainability characteristics are illustrated in Figure 47.1.

47.1.1 Maintainability Defined

A system, device, or equipment is effective if it operates when needed (it is available), and it operates up to its performance parameters (it is capable). Maintainability is one of the pieces in system effectiveness: Systems that require little maintenance or, when main-

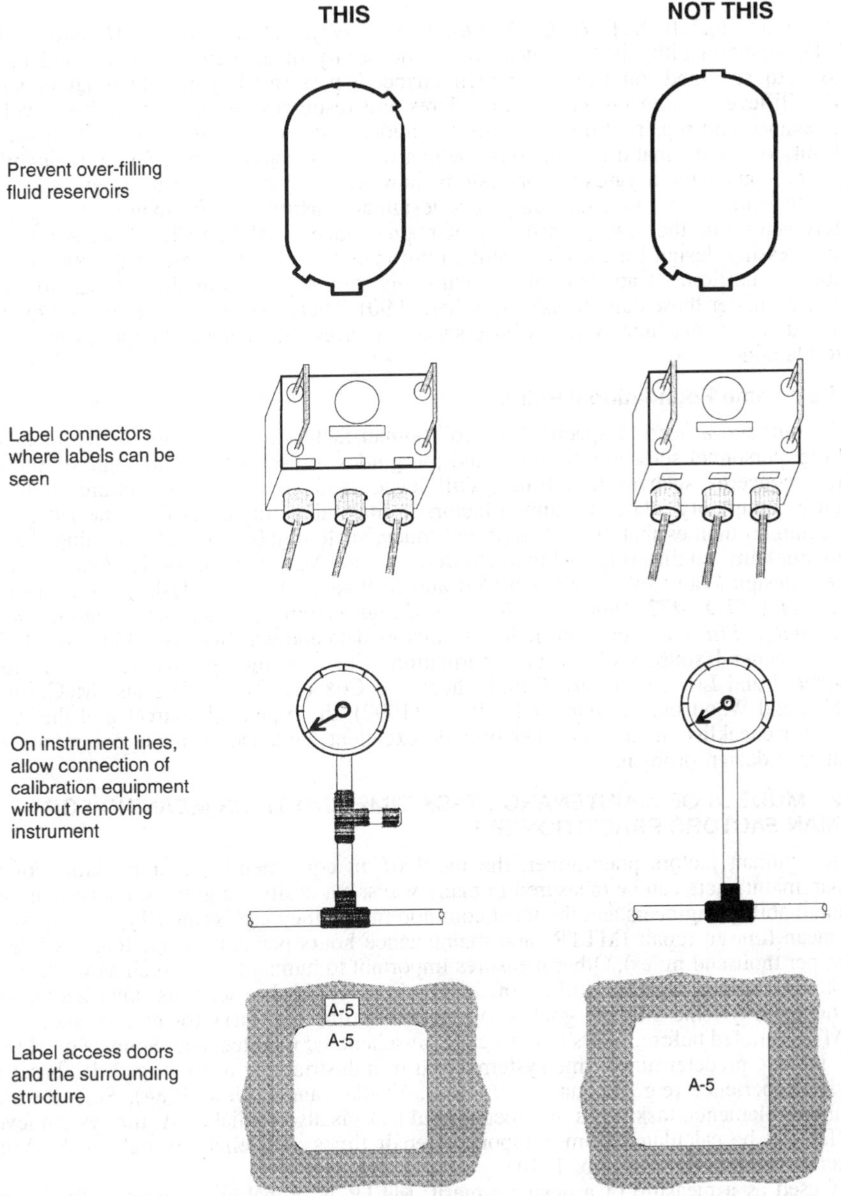

Figure 47.1 Examples of design characteristics to improve maintainability.

tenance is needed, are easy to service or repair, tend to be ready for use when they are needed; that is, they have high availability.

The degree to which ease of maintenance increases availability is measurable; the impact of savings in maintenance time is especially noticeable when equipment has a relatively low reliability, according to the following relationship:

$$A = R/(R + M) \tag{47.1}$$

where

A = availability term
R = reliability term (time between failures)
M = maintainability term (time to repair)

According to MIL-STD-721C, *Definitions of Terms for Reliability and Maintainability* (1981), maintainability is "the measure of the ability of an item to be retained in, or restored to, specified condition when maintenance is performed by personnel having specified skill levels, using prescribed procedures and resources, at each prescribed level of maintenance and repair." Stramler (1993) includes "probability" in his definition—a direct link to "maintainability functions," which we will discuss shortly. So maintainability might be considered a yardstick or scale of how well a system is set up to allow people to service and repair a system. Equipment design and instructions for maintenance people largely determine the ease or difficulty of maintenance. MANPRINT, a U.S. Army program stressing design for maintainability, reinforces attention to human capacities with a multipart question: "Can this soldier, with this training, perform these tasks, to these standards, under these conditions?" (Booher, 1990). Therefore maintainability is a quality of a maintainer–machine system whose success (correct maintenance) depends on deliberate planning.

47.1.2 Some Foundational Points

In the military, a need to specify "macro" human factors requirements for maintenance staffing, personnel selection, training, and job guides forced attention to "micro" human factors concerns such as task times, skill levels, and crew sizes. Literature from the defense community related to human factors in maintainability dates from the 1950s and has common themes that are still repeated today, such as job aiding (Lumsdaine, 1957), maintainability engineering (Altman, Marchese, and Marchiando, 1961; Folley, 1960), system design (Van Cott, 1960; Van Cott and Kinkade, 1972), and task analysis (Miller, 1953). *MIL-STD-1472, Human engineering design criteria for military systems, equipment, and facilities*, compiled basic human factors data and was first issued in about 1970. Current general sources of design information with relevance to maintenance include Blanchard and Lowery (1969), Cunningham and Cox (1972), Sanders and McCormick (1987), and Woodson, Tillman, and Tillman (1992); these general sources and the references for checklists mentioned later provide excellent guidance on many commonly encountered design problems.

47.2 MODELS OF MAINTENANCE TASK TIME AND THEIR MEANING FOR HUMAN FACTORS PRACTITIONERS

To the human factors practitioner, the merit of an equipment's accommodation of its human maintainers can be measured in many ways, but easily the greatest concern among maintainability engineers and the most common metric they use is time. Typical measures are mean time to repair (MTTR) and maintenance hours per unit of use (e.g., per flight hour, per thousand miles). Other measures important to human factors such as workspace, workload, and comprehensibility of instructions could be seen as intermediary in maintainability—the ultimate goal is to reduce the time necessary for maintenance.

MTTR, noted before, refers to corrective, unscheduled maintenance. Sources of MTTR data include predetermined time systems used in industrial engineering, and databases of service experience (e.g., Fuqua, 1980; Rose, Voytko, and Davolt, 1984). Software that combines elemental task times into meaningful tasks is also available. At the system level, MTTR can be calculated from component repair times and reliability data (U.S. Army Materiel Command, 1976, pp. 1–14).

If used as a measure of a design's merit, MTTR is probably too general for human factors engineers. The metric mixes the elements that make up maintenance actions, it may hide maintenance crew size, and MTTR predictions tend to underestimate. However, MTTR warrants attention: It is ubiquitous in maintainability engineering, and its distribution is dependent on the nature of the maintenance task. MTTR distributions suggest that many maintenance tasks are cluttered with subtasks that are not part of the core repair job on an equipment—"nit noids," as an acquaintance of the first author called them.

In the case of a single task, MTTR is the expected time for performance in a randomly selected occurrence of that task. If the task is an extremely simple one with no subtasks, MTTR will be normally distributed (maintenance tasks like this are not common). If the task contains significant subtasks, MTTR will form a distribution that is skewed to the right (Blanchard, 1986, pp. 36–37). In the case of equipment requiring a variety of tasks, MTTR is the expected time for repair of a randomly selected failure. If the equipment requires only tasks that are extremely simple, the distribution of these task times will be normal. Much more common is the situation where equipment will require tasks that vary on complexity and frequency dimensions—MTTR for such equipment will form a distribution skewed to the right. Figure 47.2 plots the frequency of repair times for an infrared transmitter/receiver reported by Towne, Johnson, and Corwin (1983). Bond (1987) presents a similar distribution of repair times.

The easiest way to make use of the fact that MTTR distributions are skewed to the right is to regard the underlying data as exponentially distributed, assuming that a goodness-of-fit test supports this assumption. A plot of an exponential probability density function (pdf) is conceptually similar to the plot in Figure 47.2, but the function gives the probability that some instance of the variable lies between specified limits, which is not too practical a calculation in maintainability. Instead, a cumulative distribution function (cdf) of an exponentially distributed variable gives the probability that a given instance of the variable is less than or equal to some value. In maintainability terms, a cdf is produced by

$$M(t) = 1 - e^{-t/\mathrm{MTTR}} \tag{47.2}$$

where

$M(t)$ = probability that repair will be successfully completed in time t when it starts at $t = 0$
t = variable repair time
MTTR = mean time to repair
e = base of the natural logarithm ($e = 2.71828\ldots$)

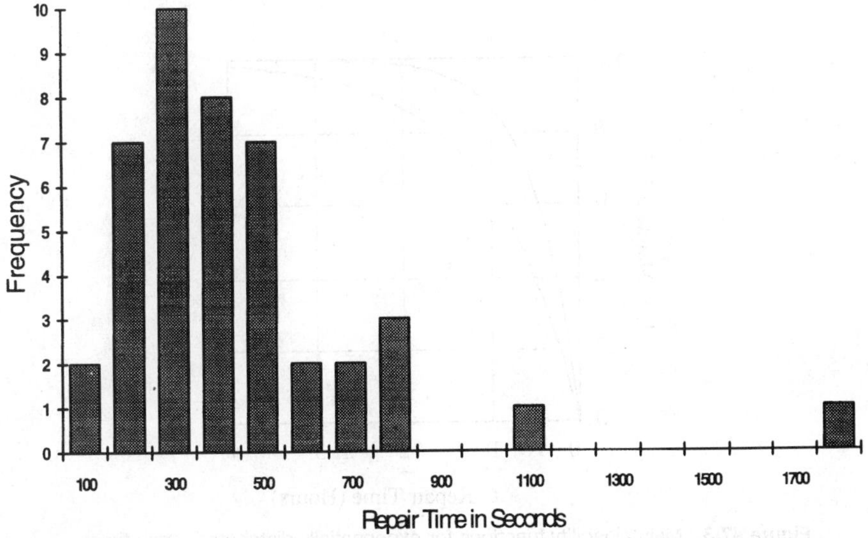

Figure 47.2 Frequency histogram of repair times for an infrared transmitter/receiver.

MTTR will produce a cdf approximately like those shown in Figure 47.3, which presents one function for an equipment with an MTTR of 0.5 hr and another for an equipment with an MTTR of 1 hr. For the equipment in Figure 47.3 with an MTTR of 0.5 hr, the probability of accomplishing repair in 0.5 hr is approximately .63 (the probability of completing a repair in a time equal to the mean of any exponential distribution is .632), and the probability of accomplishing repair in 0.25 hr (15 min) is only about .40. Similar computations for the equipment with an MTTR of 1.0 hr are possible. It is clear that the function in Figure 47.3 indicates that the likelihood of large values progressively decreases, just like an empirical distribution of maintenance task times in Figure 47.2. Such a cumulative function plot, with probability of completion on the ordinate and time on the abscissa (regardless of the shape of the distribution being presented), is a "maintainability function."

Use of the exponential distribution is appropriate when repair involves more than simple responses but is still easy. As we pointed out though, many tasks in maintenance are not quick and easy. Therefore, a distribution of MTTR might be more appropriately modeled as a lognormal distribution. We can plot the pdf of a lognormal variable for the sake of observing its shape. The equation is

$$f(x) = \frac{1}{x \cdot \sqrt{2 \cdot \pi \cdot \sigma^2}} \cdot \exp - \frac{(\ln(x) - \mu)^2}{2 \cdot \sigma^2} \tag{47.3}$$

where

x = random variable
π = the constant (3.14159 . . .)
σ = shape parameter
μ = scale parameter
exp = base of natural logarithms (2.71828 . . .)

The parameters μ and σ determine the basic form of a lognormal distribution. With $\mu = 3$ and $\sigma = .5$, the distribution of 100 values of x is shown in Figure 47.4. Small changes in μ and σ would change the distribution shape, but with the selected parameters the density function visually reveals why it can be a model of time to repair—it appears to describe the data in Figure 47.2. The very useful aspect of lognormally distributed variables is that the natural logarithm of the variable will be normally distributed. So probability estimates are seldom made with a lognormal cdf; instead, analysts transform

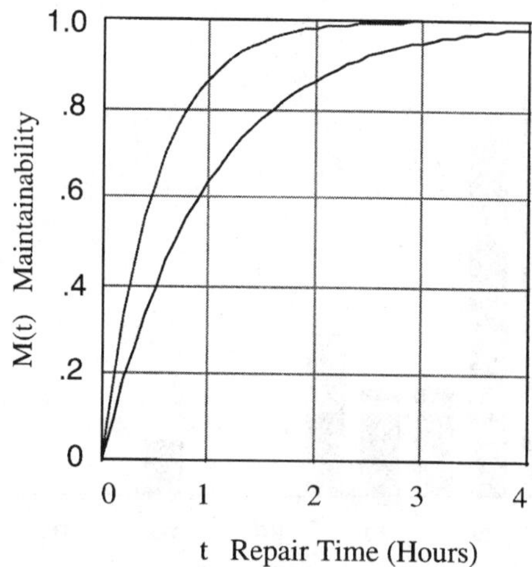

Figure 47.3 Maintainability functions for exponentially distributed repair times.

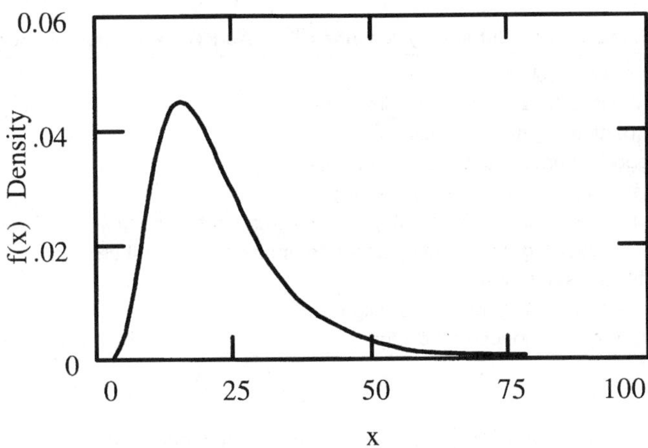

Figure 47.4 Lognormal distribution of x.

lognormally distributed repair time values into normally distributed ones and make probability estimates using normal tables.

Lognormality is the most frequently cited model of MTTR. There are several reasons why empirical distributions of MTTR are skewed to the right in a way that makes the theoretical lognormal distribution a good model. First, in the single task case, it is logical to expect that faster-than-optimum performance is much less likely than slower-than-optimum performance. In theory, a task could go on indefinitely. In the case of equipment requiring a variety of tasks, lognormality is tantamount to saying that a significant proportion of the equipment's tasks take more time to perform than typical tasks for the equipment. What characteristics of maintenance tasks exist to make this so? The demand for fault location (AMCP 706–133, 1976, pp. 8–13 to 8–17), the presence of mostly complex tasks (Blanchard, 1986, pp. 36–37), the presence of irregularly required subtasks (NAVAIR 01-1A-33,1977, pp. 2–11), or the presence of tasks with a mean greater than 1 hr and with numerous subtasks (MIL-STD-472, 1966, pp. 1–12 to 1–14) can all contribute. Equipment requiring a small proportion of tasks (i.e., a percentage of the total tasks) that take an excessively long time to complete can actually produce bimodal distributions of task times (Towne, Fehling, and Bond, 1981).

Therefore, the shape of the distribution of maintenance task times for most complex equipment suggests the relatively common presence of tasks that are complicated or protracted by time-consuming subtasks. For example, a manufacturer of a consumer apparel product with identical European and North American assembly lines obtained a difference in repair time for the lines after binding errors were detected (a brief but frequent problem). The difference was attributable to "safeing" the line (setting locks to prevent accidental startup) at a location many meters from the binding machine. In Europe, two workers coordinated their efforts—one immediately entered the machinery to repair it after he was signaled to do so by another worker who set the safety locks. In North America, one worker did both actions, first setting the locks and then running to the repair location for the actual repair activity. In this case, setting safety locks is a time-consuming subsidiary (albeit a very important one) of the actual repair. One possible maintainability improvement is designing safety lock actuation closer to the failing machinery. So while MTTR is not specific for type of task, its distribution may lead to a search for subsidiary activities in maintenance that are difficult to improve on once they have been designed in. A design requiring maintenance tasks with relatively many steps may offer more opportunities to improve maintainability than a design requiring maintenance tasks with relatively few steps. Table 47.2 lists some of these subsidiary activities.

47.2.1 Constraints and Allocations

The amount of time necessary to maintain a system, regardless of the type of measurement, is generally recognized to be established during design, so designers make con-

Table 47.2 Subtasks and Subsidiary Activities That Add Time to Maintenance Tasks

- Setup/cleanup for a task
- Dealing with overly difficult access requirements
- Use of tools, particularly special tools
- Removal of good components to reach faulty ones
- Removing/installing many different types of fasteners
- Following convoluted test procedures (e.g., poorly planned test points)
- Searching for or resolving uncertainties about component part numbers
- Securing parts with safety wire
- On-equipment calibration, shimming, or alignment
- Recovering from errors induced by design

scious, deliberate, design decisions intending to reduce maintenance requirements. However, many factors influence decisions for maintainability, as the following constraints suggest.

- If the intended users (i.e., maintainers) already have the skills to perform maintenance, some maintenance complexity may be okay. (Demographic issue.)
- If it will be too costly to train users to perform maintenance, plan for little or no field maintenance. (Human resources issue.)
- If special test apparatus or unusual tools needed for maintenance will not be readily available to users, avoid maintenance requirements that demand these items. (Logistics issue.)
- If the equipment being designed must be returned to service after a breakdown more quickly than the equipment it is replacing, invest more to make maintenance easy. (Contract issue.)
- If a device will be obsolete before it fails, plan for no maintenance. (Economic issue.)

47.2.2 Trades

How easy to make maintenance is based on questions like the ones above. However, design work to ensure a level of maintainability will add cost to a program. Designs that are already robust for maintainability may offer little practical gain for more design work. For ground electronic systems, the Air Force estimates that program cost will increase 7% for every 50% reduction in equipment repair downtime (U.S. Air Force, 1988); similar costs would not be a surprise in other types of equipment. So producers perform trade studies or trade-offs—referring to the concept of trading some characteristic for another—to find the least costly alternative for a desired level of effectiveness. For example, all types of engine coolant (glycol or "antifreeze") recycling equipment filter the used coolant. Basic filtering (e.g., pleated filters only, no prefilter or filtration membranes) will reduce the cost of the equipment but may increase servicing steps (more frequent filter changes, adding treatment chemicals, disposing of spent filters) and thereby increase coolant treatment cost on a per gallon basis (American Trucking Associations, 1995). Obviously, the maintenance burden imposed by basic filtering methods would control the outcome of a trade study comparing basic with more sophisticated methods. Trades on various dimensions are common, including reliability versus maintainability, repair level, repair/replace/discard, corrective versus preventive maintenance, level of automation in equipment, and others. Blanchard (1986) provides a good treatment of trade studies. One of the pleasing interactions between maintainability and reliability is that equipment that is easy to service tends to receive regular servicing and therefore suffers fewer breakdowns.

Constraints and trades dictate a target level of maintainability. Maintenance time requirements can be allocated among subsystems and components within subsystems to achieve the target level. The goal is an overall, system-level maintainability, and identifying elements with good candidates for eliminating setup steps, special tools, on-equipment calibration, and so on may allow relaxation of maintainability goals for another element. An overview of allocation is contained in Majoros and Chen (1992).

47.3 BEST PRACTICES

Anyone who has changed automobile oil and filters may have wished for captive drain plugs (to prevent plugs from falling into oil pans) and easily accessible filters. These and hundreds of other ideas have been used to ease maintenance. Some commonly recurring "themes" are described below. Human factors methods described later can help to evaluate whether a design exhibits these and other characteristics.

Easy access. Poor access to components and service points may be the most frequent complaint of maintenance people regarding equipment. Many solutions are possible. To improve access to air conditioners mounted below the crew cabins of its switching yard engines, a locomotive manufacturer mounted the units on telescoping slides (Locomotive heat, 1995). Now, rather than a forklift and extra personnel, one person can open a hinged door and pull the heavy (840 lb) unit out for servicing or checking the coolant system. The setup is shown in Figure 47.5.

Modularity. A very successful small turbine for aircraft propulsion was designed with two basic modules: a power section and a gas generator. For maintenance on either the turbine or combustion sections, the modules can be easily separated by unbolting a single flange (Most, 1988).

Effective labeling. Equipment may have hundreds or even thousands of components subject to maintenance, and the components may not be grouped or organized in any way that aids in their identification. A study of power plant maintainability found scores of instances where maintenance personnel made crude labels on pipe runs and valves to overcome this type of design neglect (Electric Power Research Institute, 1980).

Only common tools needed, no special tools. User-level engine maintenance for the U.S. Army's new light utility helicopter requires no special tools and only six common hand tools (Weand and Thagard, 1989).

Figure 47.5 Locomotive air conditioner mounted on telescoping slides below cabin can be pulled to extended position (shown) for easy access.

No delicate shimming, adjustment, or other "shop" work needed in field. A pin elevator is a device that picks up dead pins from a bowling lane and loads them into a pinspotter for resetting. One manufacturer has eliminated the need for repetitive, manual re-tensioning of the elevator drive belt by incorporating spring-loaded tensioners (Tensioners keep pins, 1995).

Sensible fault isolation. The heating system of a popular public transit bus is a closed loop of pneumatic and water lines, valves, components, and controls without a single indicator or test point. Troubleshooting such a system is a grim task of item-by-item testing. One municipal owner of the vehicles built a custom test fixture to get around the design problem (Fuchs, 1982).

47.4 FAULT ISOLATION/TROUBLESHOOTING

Fault isolation is a significant and persistent problem in equipment maintenance. On complex equipment, most corrective maintenance time is spent on fault isolation and, unfortunately, fault isolation is not necessarily effective. In commercial air transport, for example, nearly half of all avionics equipment returned to the manufacturer for repair has, after testing, no fault found (Ott, 1995). Even with accurate built-in-test equipment (BITE), fault isolation can be frustrating if the BITE is difficult to use or comprehend. In a recent case involving a passenger transport aircraft, a centralized fault display system was believed to be so capable that the manufacturer provided only a small amount of supplemental information beyond that provided by the automated system. However, in service, maintenance people did not incorporate the system into their diagnostic work, prompting the manufacturer to create a comprehensive troubleshooting manual covering aircraft faults *and* use of the fault display system (Cheaper to run, 1995). On top of an all-too-common uncertainty in fault isolation, removals of modules can result in loose and bent pins in connectors, damage due to handling, and other induced problems that become fault isolation headaches in their own right.

Identifying the faulty component in a degraded or failed system is obviously related to repair or correct removal and replacement. When testing is easy (checking status at test points and following diagnostic logic), maintainers will probably try to pinpoint a faulty module before any removals rather than skip over the fault isolation step and simply begin removing any and all suspect modules—a practice called "shotgunning." When testing is easy *and* effective, maintainers will probably pinpoint and remove only the truly defective module. Thus, the most basic objective in designing for effective fault isolation is that faults are readily traceable to discrete components. Figure 47.6 illustrates the logic necessary for simple systems; system simulators, computer-aided software engineering (CASE), and graph theory are used in more complex cases. However, despite the logical design of a system, technicians' attempts to track down faults, an evolving propagation of faults, and a mosaic of fault indicators can interact to greatly complicate the sequence of events in troubleshooting (Woods, 1988).

The human factors goal is to influence the design of equipment and troubleshooting aids based on an awareness of troubleshooters' proclivities and limitations. In this regard, Bond's (1987) review of the knowledge on troubleshooting behavior is still current. In general, troubleshooters' behavior is geared to minimize time, but may not necessarily be systematic or logical. Troubleshooting difficulties in the field are substantial enough that maintainers' likelihood of tracking down failures in a reasonable amount of time is routinely overestimated. Several reasons for the overestimation and potential corrections are shown in Table 47.3; Yoon and Hammer (1985) list others.

47.5 SELECTED METHODS

47.5.1 Task Analysis

Task analysis is ". . . a method of describing what an operator is required to do, in terms of actions and/or cognitive processes, to achieve a system goal. It is a method of describing how an operator interacts with a system, and with the personnel in that system" (Kirwan and Ainsworth, 1992). Task analysis is, or should be, the starting point for human factors in maintainability. A thorough analysis of maintenance tasks improves equipment availability by identifying resource requirements in time for effective planning, and it is useful, and usually necessary, for finding opportunities for improving the flow of events and for estimating the time requirements and workload in a maintenance task so that errors can be minimized.

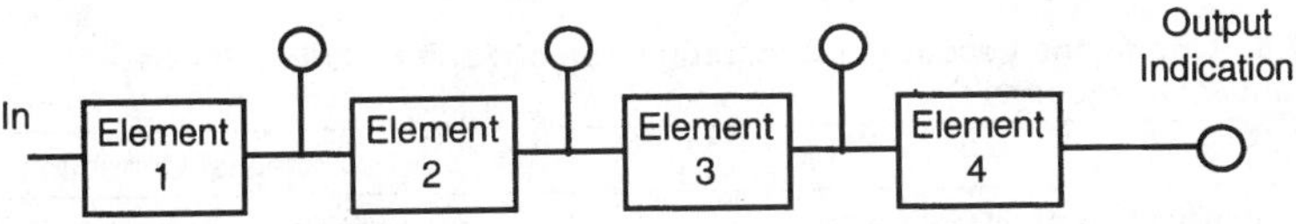

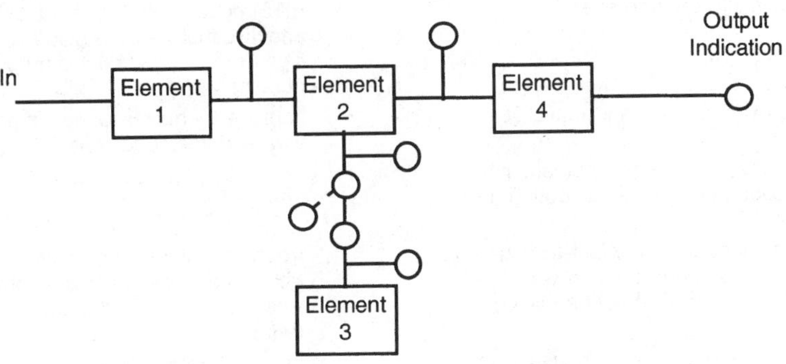

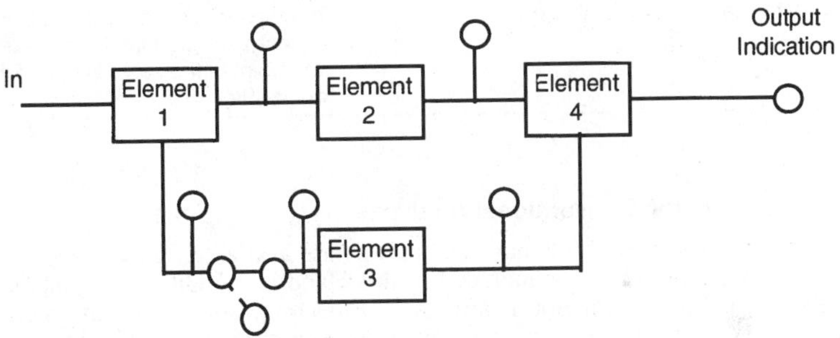

= Test Point

Figure 47.6 Energy flow patterns encountered in electrical circuits.

A task is often described as a set of discrete behaviors (i.e., logical units of behavior with clear starting and ending points) stated in an action–object pairing (e.g., remove fasteners). Common maintenance task analysis objectives include the following:

- Decompose tasks into logical subtasks and sequences.
- Describe task and/or subtask characteristics in terms of required knowledge, skills, and abilities.
- Identify hazards and sources of error and their consequences.
- Identify necessary tools and support equipment.
- Document proper task performance for training and maintenance manuals.
- Establish timelines for task performance.

Table 47.3 Conditions Leading to Overestimation of Fault Isolation Success and Potential Corrections

Condition	Potential Correction
Test and qualification of BITE in a laboratory by inserting known faults is conservative	Maintainability demonstrations performed in the intended operational environment
Maintenance manuals are incomplete, superficial, inaccessible, or confusing	Simplified English and a recognized, effective format
BITE is difficult to comprehend	BITE documentation is comparable to effective maintenance manuals; BITE designed to display meaningful words and phrases
Pressure exits to finish a job quickly	Improved maintenance management and resource conditions
Maintenance people, being human, may exhibit suboptimal troubleshooting in several ways:	————
• Reasoning poorly in effect-to-cause terms if equipment function was encoded as cause-to-effect in training	Troubleshooting checklists that align with system descriptions taught in maintenance training (Goldberg and Gibson, 1986)
• Choosing a diagnosis because it is easily recalled from memory	Troubleshooting aids (e.g., computers) that display a full set of hypotheses
• Resorting to favorite, inefficient, troubleshooting techniques	Periodic hands-on refresher training
• Favoring data that is easy to obtain	Troubleshooting aids that allow any diagnostic procedure to be executed at any point in the session (e.g., Flight Control, 1994)

47.5.1.1 Methods for Conventional Analyses

Numerous techniques in conventional task analysis are available (Kirwan and Ainsworth, 1992). (A chapter on the topic is included in this Handbook.) One, *task simulation*, will be described here. Task simulation is a method of decomposing a task and modeling the relationships among its elements in such a way that experimentation with task parameters can be performed. One tool for this method is discrete event and continuous state simulation software. A user of this software enters (1) the elements of a task; and (2) parameters such as the expected time for the element, the type of distribution characterizing samples of the element's time, probabilistic effects, waiting/delay/queue instructions for each element, the effects of completing elements, and so on. When a task is fully modeled, the simulation software builds task modeling rules based on the entered parameters and probabilities. Modeled tasks are "run" repeatedly to generate a sample of performances; the effects of changes in parameters are examined by comparing performance samples under baseline and revised parameters.

In an application of Microsaint™ simulation software to electrical assembly testing, analysts broke the test into about 20 steps. Figure 47.7 shows part of the task network for the test. Analysts manipulated parameters of the elements "Test Team ID Line Items" and "Get Next Line Item" to create two conditions: a baseline condition and a modified condition where one electrical test technician could be added to the job. Specifically, elapsed time and delay parameters in "Test Team . . ." and "Get Next . . ." were expected to differ under the two conditions, so these steps' parameters were experimentally manipulated while all other steps were held constant. Measured output variables were number of assemblies tested per unit time and total time assemblies remained in the testing process. This experimental comparison of baseline and proposed staffing demonstrated that the cost of one additional technician would be paid back in just a matter of months.

Often, task analysis for maintenance will not follow a standard method. One type of "nonstandard" analysis occurs as manufacturers prepare maintenance instructions for an

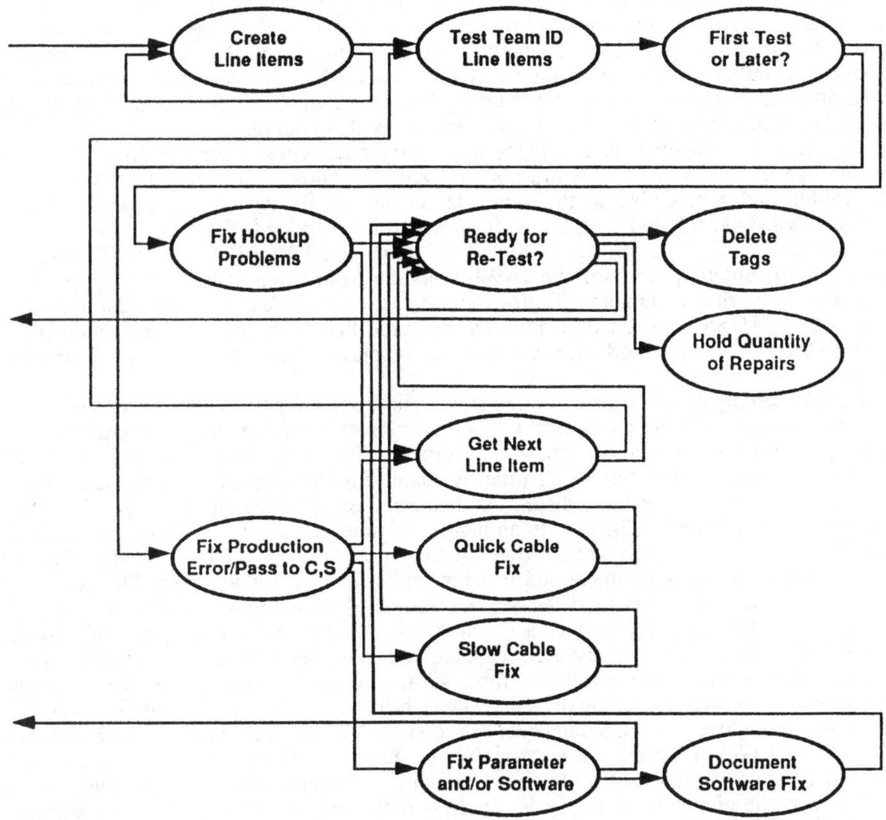

Figure 47.7 Portion of an electrical assembly task modeled with Microsaint™ simulation software.

equipment, plan for maintenance training, and describe maintenance staffing requirements. Many details about maintenance tasks emerge in these processes, although feedback to design during any of them is not customary. Another type of informal analysis occurs when some stakeholder in a project discovers what he or she believes is a condition that will create a very difficult maintenance requirement and someone begins to scrutinize the specific task elements involved. Detailed simulations or models of the task elements in question might be constructed to resolve the concern. The logic appears to be that of assigning a detailed level of representation to the parts of a task that prompted the scrutiny in the first place, and very coarse-grain representation, if any, to all other portions of the task.

47.5.1.2 Cognitive Task Analysis

Analyzing Cognitive Requirements Expected for Maintenance. Cognitive requirements of maintainers should not be trivialized. Ever-increasing equipment complexity, vague instructions, stressful work environments, and tasks that may be attention-demanding yet boring or practice-dependent yet infrequent can combine to create significant needs for evaluation, problem solving, attention, and recall. After a jetliner flying among the Hawaiian islands lost part of its upper fuselage due to structural failure, the National Transportation Safety Board found that the probable cause was the "failure (of mechanics) to detect the presence of significant disbonding and fatigue damage . . ." (NTSB, 1989, p. 73). The report also concluded that "There are human factors issues associated with visual and nondestructive inspection which can degrade inspector performance to the extent that theoretically detectable damage is overlooked" (p. 72). This conclusion, widely noticed in the industry, pointed to task-related conditions affecting attention, evaluation, and comparison skills of maintainers. In fault isolation particularly, where cognitive demands can be characterized at the highest level of Rasmussen's (1983) skill-rule-knowledge taxonomy (Swain and Weston, 1988), intellect can be frequently challenged. Fault isolation takes up the bulk of corrective maintenance time and the fact that successful fault isolation is so difficult (e.g., see statistics reported by Ott, 1995; Ruffner, 1990) strongly suggests that this major aspect of maintenance is cognitively demanding.

Cognitive task analysis (CTA) helps to identify requirements for cognitive skills, although use of the technique should not imply that cognition and related constructs are discernible and predictable to the same degree as are the observable aspects of task performance. One method in CTA is Task Analysis/Workload (TAWL), a program developed for the U. S. Army (Hamilton, Bierbaum, and Fulford, 1991). TAWL and its companion simulation program, the TAWL Operator Simulation System (TOSS), produce measures of workload (amount of attentional demand) that vary across the duration of a task. TAWL/TOSS is based on a multiple resource theory of human information processing (Wickens, 1984) and allows up to six resources (e.g., cognitive, psychomotor, visual) to be modeled.

A TAWL analysis progresses in three stages. In the first stage, task/workload analysis, a user organizes a task into a hierarchy (TAWL organization is oriented to aircraft piloting and uses the terms "mission," "phase," "segment, "function," and "task"). Then, at the "task" (i.e., most elemental) level, time estimates for the elements and resource types associated with the elements are listed. At this point, even before the first stage of TAWL is complete, this listing is essentially an analog of time line analysis. We might conclude that CTA has been accomplished. However, the expectation in CTA is that the effect of workload imposed by equipment and mission on performance is assessed. Therefore, the remainder of the first stage and the second and third stages exist for that purpose. To complete the first stage, wherever a demand for a resource occurs in the task listing, ratings of workload for that resource are obtained, possibly from subject matter experts (SMEs), and entered into worksheets. Table 47.4, adapted from the TAWL User's Guide, provides a taxonomy of cognitive and related abilities. The table contains five resource types and seven levels, arranged in order of increasing attention demand, for each resource type. In a TAWL analysis, each level would have a numeric value associated with it—often simple ordinal values like those shown in the upper portion of the table. As the analyst lists the elements of a task, he or she—with the help of an SME—compares the similarity of elements to resource types and assigns a rating to the element corresponding to the perceived level of attention demand.

The second stage is model construction. Here, the analyst describes how the elements are combined to form higher levels. The necessary decisions involve the nature (discrete vs. continuous) and timing (fixed vs. random) of elements. When the descriptions are entered into TOSS program pages, they constitute rules to be followed in a simulation of the task. The third stage is simulation. Randomization, workload summation, and overload computation algorithms in TOSS produce outputs for the actor (maintainer, operator) showing estimates of resource workload over time, overloads, the frequency of randomly occurring events, and other values. A CTA tool like TAWL/TOSS forces rigor into the possibly speculative aspects of projecting cognitive demands. CTA can also distinguish types of tasks on the basis of cognitive requirements.

When applied to fault isolation, task analysis has the goal of predicting whether an equipment system will be quick or time-consuming to diagnose. The work in this area usually relates a measure of equipment complexity to the likely ease or difficulty of troubleshooting (Rouse and Rouse, 1979). For example, Towne (1985) successfully related his measure of complexity—the number of indicators for fault conditions—to fault isolation performance.

Opportunities exist on many fronts—in design, technical media, and logistics—to improve fault isolation. Maintainers' cognitive effort in these areas can be anticipated to some degree by an awareness of human expectations, perception, memory, decision making, and attention, and, of course, by analysis of the tasks that will place demands on these human abilities. Many sources are available that discuss human information-processing capabilities. Sanders and McCormick (1987) present an overview from a human factors standpoint. Some aspects of information processing capabilities are contained in Table 47.5, with examples of intended effects in design.

An Emerging Approach in Cognitive Task Analysis. Recent developments in cognitive research, such as the use of multidimensional scaling (MDS), could reveal more about

Table 47.4 Five Types of Cognitive Abilities with Levels within Tasks Arranged in Order of Increasing Attention Demand

Type of Task	Levels within Task
Cognitive	Automatic (Simple association) (Sample ordinal value = 1) Alternative selection (2) Sign/signal recognition (3) Evaluation/judgment (Consider single aspect) (4) Encoding/decoding, recall (5) Evaluation/judgment (Consider several aspects) (6) Estimation, calculation, conversion (7)
Visual	Visually register/detect (Detect occurrence of image) Visually discriminate (Detect visual differences) Visually inspect/check (Discrete inspection/static condition) Visually locate/align (Selective orientation) Visually track/follow (Maintain orientation) Visually read (Symbol) Visually scan/search/monitor (Continuous inspection, multiple conditions)
Auditory	Detect/register sound (Detect occurrence of sound) Orient to sound (General orientation/attention) Orient to sound (Selective orientation/attention) Verify auditory feedback (Detect occurrence of anticipated sound) Interpret semantic content (Speech) Discriminate sound characteristics (Detect auditory differences) Interpret sound patterns
Kinesthetic	Detect discrete activation of switch Detect preset position or status of object Detect discrete adjustment of switch Detect serial movements (Keyboard entries) Detect kinesthetic cues conflicting with visual cues Detect continuous adjustment of switches Detect continuous adjustment of controls
Psychomotor	Speech Discrete actuation (Push button) Continuous adjustive (Adjust calibration screw) Manipulative (Handle parts) Discrete adjustive (Change rotary switch setting) Symbolic production (Writing) Serial discrete manipulation

the cognitive parts of maintenance. MDS is a statistical procedure that calculates a spatial representation of items based on some measure of how the items are related to one another (Ashby, 1992; Shepard, 1987). The technique presents items in an *n*-dimensional space, with items similar to one another lying close together, while dissimilar items lie farther apart in the space. For example, if a representative sample of American cities (in terms of their distance from one another) is subjected to MDS, the resulting spatial representation of the cities will reveal a plot approximating the shape of the United States. If, on the other hand, measures of item relatedness are based on human judgment, the representation is said to be cognitive or perceptual (depending on the type of task and stimuli used), and the dimensions may reveal the underlying cognitive factors used by individuals in their information processing. For example, an MDS analysis of cognition-related words by Jonsson (1994) produced two dimensions. In Figure 47.8, the two-dimensional (2-D) space revealed by the technique is occupied by cognitive tasks spaced according to their relatedness; the vertical axis was interpreted as a dimension of complexity, and the horizontal axis was interpreted as a dimension reflecting cognitive activities involved in planning versus those involved in diagnosis.

In a recent application, Jonsson and Ricks (1995) used MDS to determine how airline pilots represented and processed flight deck information. MDS indicated that three dimensions were underlying pilots' judgments. Experts interpreted the dimensions as cor-

Table 47.5 Design-Revelant Aspects of Information Processing Capabilities

Information-processing capability	Design goal	Examples
Expectations	Relationship between objects and responses is compatible with users' expectations.	• Movement of displays is consistent with movement of controls. • Controls are physically grouped with their associated displays. • Symbols are congruent with the object or condition being symbolized.
Perception	Meaningful stimuli are detectable.	• A dial with white face and black letters has a bright orange pointer. • Equipment near potential hydraulic leaks is painted white. • Cautions and warnings are audible/visible against background. • Service points are painted yellow.
Memory	Reliance on short- and long-term memory is minimized.	• Part numbers are sectioned into groups: 2138-7221-LH, not 21387221LH. • The number of instructions in a single sentence is minimized.
Decision making	Mitigate maintainers' tendency toward bias in decision making.	• The need for calculation and interpretation in maintenance instructions is avoided. • Key words are used consistently. • Input–output test points are provided at every link in linearly arranged systems.

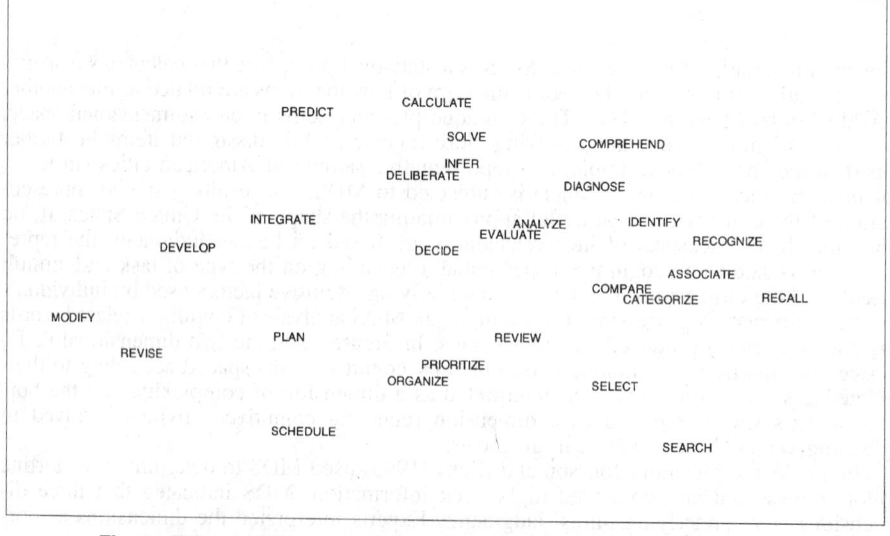

Figure 47.8 Multidimensional scaling analysis of cognition-related words.

responding to (1) flight function, (2) flight action, and (3) the frequency with which pilots sampled the information. The authors proposed using these dimensions to assess the types of cognitive activities pilots were engaged in for typical flight deck tasks.

The methodology behind this approach could be applied to the cognitive analysis of maintenance tasks. The first requirement is to identify a low-level condition or element of task performance. For the experiment dealing with pilots above, flight deck "information" was used; for maintenance fault isolation, "resources" might be selected, as in "What cognitive *resource* is needed or what problem-solving needs to happen for this type of maintenance task element to be successfully accomplished?"

The resources are then listed, as in Table 47.6, and a sample of these items is drawn at random and presented in all possible pairs to maintenance SMEs for rating the perceived similarity of item pairs (1 = very similar, 5 = very dissimilar, or some comparable scale). MDS analysis is applied to the ratings, and the dimensions along which the resources fall are revealed. In interpreting these dimensions, it is most useful to have another group of SMEs (who did not participate in the original rating exercise) interpret them. In our current example, dimensions emerging might be "information comprehensibility," "material availability," or others.

These dimensions can then be applied to tasks expected for the design so that elements of the tasks are assigned dimensional values. In other words, a task decomposition can be constructed for various maintenance situations/scenarios and the dimensions revealed by MDS can be used by SMEs to rate task elements: Element 1 might be rated low on one dimension but high on another, Element 2 might be rated high on both dimensions, and so on. The ratings can be used to reveal the pattern of cognitive demands imposed by fault isolation for a new design. It is also possible to develop methods, based on conjoint measurement theory (Krantz and Tversky, 1971), that reduce the task analysis to a single number representing "cognitive effort" or some similar metric.

In 1985, Towne concluded that "we [currently] have neither the ability to predict the cognitive content of a complex diagnosis activity nor the means for quantifying the time requirements of those mental processes" (p. 11). It is still difficult to predict the degree of cognitive "burden" in fault isolation or even in maintenance tasks considered predominantly physical, but steady inroads in cognitive task analytic methods look promising.

47.5.2 Checklists

In maintainability, checklists are used to remind designers to consider the needs of maintainers or to evaluate designs for ease of maintenance. The formal intent of any checklist is to call attention to or call for an action with respect to some item and, when the item is complete, release attention from the completed item and direct it to the next item. Maintainability checklists do not intend the pacing or controlled progression though steps in the way cockpit lists do, but it is still their purpose to guide behavior. Checklists are popular because they are obvious memory aids, the criteria they list is statistically related to maintenance time (Crawford and Altman, 1972), and as a group, they assemble design ideas for improved maintenance that engineers have collected over the years. Good checklists are contained in Blanchard (1986); Woodson (1981); and *MIL-STD-1472C, Human engineering design criteria for military systems, equipment and facilities* (1981).

Despite the popularity of maintainability checklists, the devices have several potential faults. First, they normally group items into gross categories such as displays, labeling, access panels, connectors, fasteners, and so on, and the items themselves are not specific. Furthermore, many items in the typical maintainability checklist will not be applicable to

Table 47.6 Sample Resources for Task Element Accomplishment

Job card is accessible.
Job card is interpretable.
Job card is current.
Equipment has true fault.
Design of subsystem structure is identifiable.
Design is related to physical components.
Components of subsystem are identifiable.

a given design, either because some items are outdated by the design's technology or because the checklist is too broad, so the designer is required to find meaningful items in a checklist. Another possible shortcoming of checklists is that some of the interpretations required by items are complex; in fact, they can amount to evaluations that require far more time than a checklist format implies. In summary, maintainability checklist items are often unclear, of uncertain relevance for any given design, and call for actions of varying complexity.

So why are checklists useful? What makes them useful is careful tailoring to a specific class of equipment. Tailoring here means to revise vague and unclear items so they use the language of and refer to the usual characteristics of the design under development. Also, grouping of items should not be in categories such as connectors, displays, and the like; rather grouping should coincide with the systems and assemblies that make up logical units of the equipment; items that are not relevant for a particular group should be omitted. Goldberg and Gibson (1986) observed more accurate circuit board inspection when inspectors used logically organized checklists (i.e., consistent with the organization presented when inspectors were trained) than when they used unorganized checklists. The authors believed the organized list's superiority was due to its congruence with inspectors' mental representation of circuit board defects. [Gentner and Stevens (1983) discuss mental models of system functioning.]

47.5.3 CAD and Virtual Reality Technologies

For the human factors engineer, CAD and virtual reality (VR) technologies, like scientific visualization ("sci-vi"), transform data into pictures to help users comprehend meaning and patterns. For example, posture is vastly easier to interpret with an electronic human form model than with a table of human joint angles. In VR, designers and maintainers can interact with image representations of equipment.

47.5.3.1 Approaches Designed for CAD

Electronic mockups and virtual prototyping based on CAD graphics are beginning to replace physical mockups and hardware testing for many types of engineering evaluation (Boyle, 1991). These developments have stimulated growth of what can be called a "computational" approach to human factors evaluation.

Human Form Modeling. Certain aspects of maintainability analysis can now be accurately and easily performed using CAD human form models. In this approach, computer-graphics mannequins are placed in the CAD scene to represent human interactions with equipment, tools, and other objects in the workspace (Society of Automotive Engineers, 1994). Most of the human form modeling systems on the market today display mannequins that are simplified models of the human body composed of skeletal link segments and joints; depending on a given system's features, they may also exploit CAD capabilities to provide collision detection, vision cones, and tool sweep paths. Recent advances in modeling capabilities include animation and control of body movements (e.g., walking), strength-based motion, and detailed hand motion. The U.S. Air Force "Crew Chief" model can calculate some task performance times. Badler, Phillips, and Webber (1993) discuss current capabilities of and prospects for this technology. Several case studies of human form model use are described in McDaniel and Hofmann (1990).

To date, the most common use of human form modeling has been in the evaluation of physical accommodation, such as space for body positioning or movement, reaching, seeing, and manipulating tools. For example, Figure 47.9 shows a human form removing an avionics box from a rack in a transport aircraft. In order to evaluate accommodation in this application, the analyst defined a reasonable posture, set the model in position to manipulate the box, executed the grasp and removal, and examined clearance during various elements of the task. Capabilities such as mannequin options (to represent various populations) and joint angle limits can be invoked in most models to improve the generalizability or validity of an analysis.

Despite their visual appeal and demonstrated value in design evaluation, human form models require considerable skill and judgment to create serious applications and to interpret results. The user needs some skill in CAD, or support from someone with that skill; sound and supportable decisions must be made regarding what structure/human form model scenario best explores a design question; and the user needs to make assumptions about the margin of error when representing a task. Task visualization is also an essential part of analysis with human form models, and should meet definable objectives (Majoros, 1991).

Figure 47.9 Electronic human form model used to demonstrate removal of an avionics box.

Work Envelopes. Envelopes, or representations of the total area occupied during human motion have been used for years to prescribe maximum dimensions for reach or minimum dimensions for clearance. One type of envelope, the *work* envelope, defines the maximum excursions of human anatomy, tools, and components during the performance of some task, In design practice, work envelopes are used to define "keep-out" space in equipment interiors so that enough clearance for hand, tool, and component movement is provided (Majoros and Taylor, in press). Figure 47.10 presents a work envelope for hand and tool access to a hinge fitting in a transport aircraft rudder.

The most capable CAD approach to model work envelopes is currently in development (Society of Automotive Engineers, 1994). In this "numerical" method, the shape of an object is abstracted from CAD geometry and a path generation program follows rules to progressively twist, turn, and advance the component (in CAD) past obstructions from an origin to a defined destination. Other programs produce a trail or trace of the component's movement, connect trail segments together in a lattice-like structure, and form a surface on the lattice. This numerical method shows only the component removal portion of an envelope; that is, it does not yet include human anatomy and tool sweep, but capability to add those remaining parts in this method will very likely appear soon.

A practical, "manual" technique for creating the component removal portion of work envelopes is available to anyone with CAD and is presented in Table 47.7. Although its accuracy is limited, this approach communicates the work envelope concept to designers.

47.5.3.2 Virtual Reality and Augmented Reality

Virtual reality refers to computer-generated, usually visual, representations of real-world objects which a user can navigate or manipulate. The elements comprising a VR system can be assembled along several lines which vary in cost and complexity. The most well-known approach is "immersive," where the real world is opaque to the user and he or

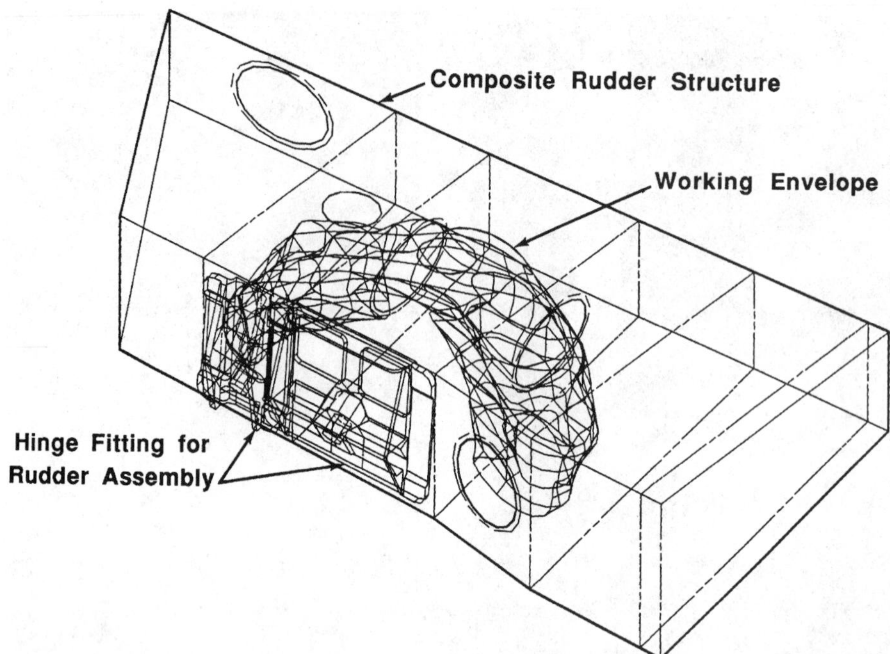

Figure 47.10 Work envelope for tool and hand access to parts in an aircraft rudder. (From Majoros, A. E., and Chen, H. C. (1992). Techniques in the design of aircraft for maintainability. In C. T. Leondes (Ed.), *Control and Dynamic Systems, Advances in Theory and Applications, Volume 52: Integrated Technology Methods and Applications in Aerospace Systems Design.* San Diego: Academic Press. Used with permission).

Table 47.7 Construction Procedure for Removal Portion of a Work Envelope

1. Lay out the relevant equipment zone in three dimension CAD, including all elements (e.g., components, tubes, structure) that could obstruct movement.
2. Define the entity to be moved as an object and select points on the outer surfaces of the object.
3. Establish the object in an initial position; define a destination position.
4. Make a copy of the object.
5. Translate the copy on a straight path from the initial position as far as possible toward the destination; stop short of any obstruction.
6. Fix the copy in the stopped position.
7. Connect corresponding points with lines between original and copied objects.
8. Use the lines to estimate the outer surface of a volume and create planes based on the estimate.
9. Delete the original object and regard the copy as a new original.
10. Translate, and rotate if necessary, a second copy to a new position on a path that will go around the obstruction and toward the destination or an interim position; stop short of any new obstruction.
11. Repeat steps 6 through 10 until a copy is fixed in the destination location.

she is provided the sensation of interacting directly with computer-generated objects. In other approaches, VR shares certain attributes of three-dimension CAD (Stone and Angus, 1995).

Methods in Virtual Reality. In immersive VR, a head-mounted display (HMD) is worn and its position in space is tracked. As the user moves his or her head, aspects of the computer-generated object appropriate to the HMD position are displayed. The user navigates by pointing and grasping while wearing a position-sensitive glove. One application—installation and removal of a fighter aircraft engine (including the operation of tools and a hydraulic dolly)—was executed in virtual experience (Phillips, 1995). Developers concluded that the experience involved "many of the (same) maintenance procedures and challenges" as does a physical mockup. Other variations of immersive VR have been developed, such as mounting the display on a boom, which frees the user from the limits of an HMD tracking system. One immersive VR configuration designed for maintainability evaluation is shown in Figure 47.11. The "cherry picker" in the figure

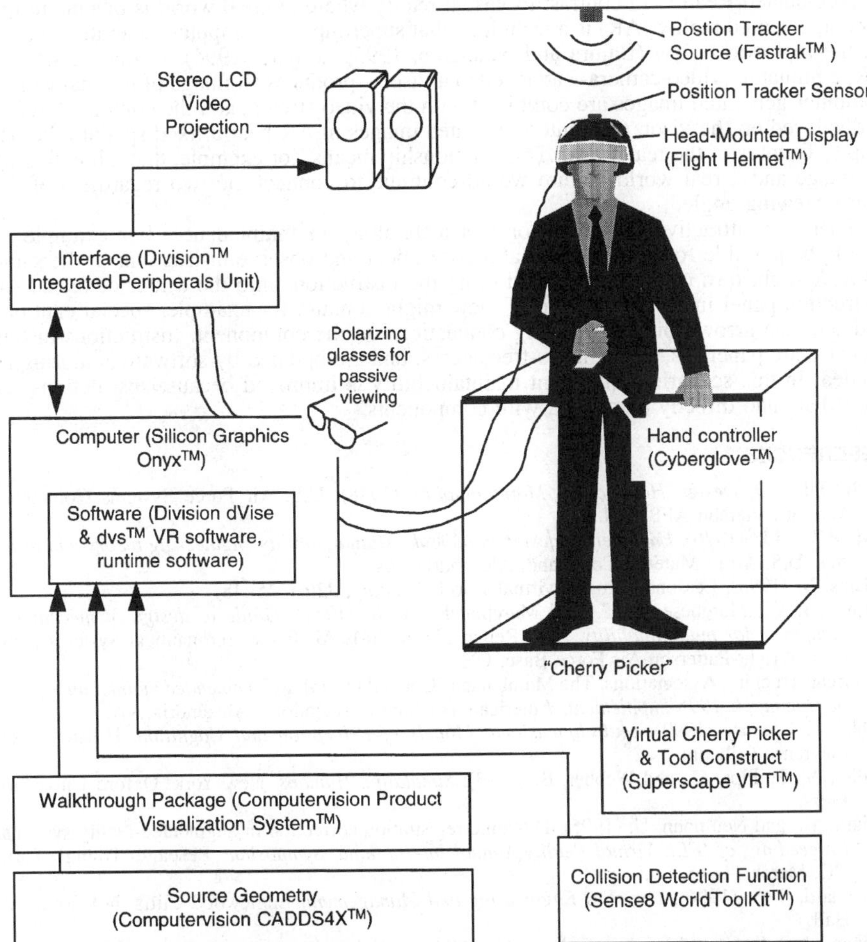

Figure 47.11 Elements of an applied immersion/projection test bed for virtual maintenance evaluation. (Adapted from Stone, R. J. (1995). Unpublished figure: "InSys—Rolls-Royce—VSEL immersion/projection text bed for virtual maintenance evaluations." Used with permission. Trademarks belong to the following: CADDS4X & Product Visualization System - Computervision; Cyberglove - Virtual Technologies, Inc.; dVise & dvs - Division, Ltd.; Fastrak - Polhemus, Inc.; Flight Helment - Virtual Research; Onyx - Silicon Graphics, Inc.; Sense8 WorldToolKit - Sense8 Corp.; Superscape VRT - Superscape Ltd.).

refers to a physical platform on which subjects stand; the platform is reproduced in virtual form, creating the experience of moving on a lift or gantry (Stone and Angus, 1995).

Stereo viewing is a workstation or personal computer approach that relies on specialized glasses worn by the user to achieve depth in the monitor display. Navigation and manipulation of objects are achieved with an input device such as a glove or three-dimension (3-D) mouse. One company recently developed such a personal computer-based VR system for the National Guard that allows reservists to practice maintenance on "virtual tanks" (Machlis, 1995). Stereo viewing and the desktop VR described below have been called "fly-through" VR (Teschler, 1995).

Desktop or "window-on-world" (Hayward, 1993) VR is an approach that produces a type of virtual world without an HMD and where stereo viewing glasses are optional. An object is viewed on a desktop computer monitor, and navigation and manipulation of the object is achieved with a mouse. In a preliminary study of assembly training methods, Motorola created virtual assembly lines for a desktop VR group and an HMD VR group, and compared their training progress with a conventional classroom group. The desktop and classroom groups were equal in learning performance (measured by the number of errors produced); the HMD group was superior to the other methods (Adams, 1995).

Augmented Reality. In contrast to virtual reality where the real world is opaque to the user, augmented reality (AR) is a technique that superimposes computer-generated images on the real-world view (Bajura and Neumann, 1995; Larijani, 1994). In one version of AR, a miniature video camera—attached to a visor—produces an image of the real world, computer-generated images are combined with the video stream, and the combined image is displayed on the visor. Computer-generated images in AR bear a fixed spatial relationship to features of the real-world. This relationship means, for example, that a line linking an image and a real world feature would continue to connect the two regardless of the user's viewing angle.

There are attractive prospects for this technology in maintenance. For example, it should be possible for users to look at a component and observe instructions at the same time. A slight turn of the head would bring the instructions into the center of focus. An instruction panel in the user's field of view might contain, for example, special cautions and show an arrow pointed to tubing connections on the component. Instructions in this scenario are paperless, allow hands-free access, and are updated by software authoring as needed. In this scenario, equipment maintainability is improved because instructions are immediate and directly associated with components.

REFERENCES

AFSC DH 1-9, *Design Handbook - Maintainability* (1988). U.S. Air Force Systems Command, Wright-Patterson AFB, OH.

AMCP 706-133 (1976). *Engineering design handbook: Maintainability engineering theory and practice.* U.S. Army Materiel Command, Alexandria, VA.

Adams, N. (1995). Lessons from the virtual world, *Training, 32(6)*, 45–48.

Altman, J. W., Marchese, A. C., and Marchiando, B. W. (1961). *Guide to design of mechanical equipment for maintainability.* ASD Report No. 61-381. Air Force Aeronautical Systems Division, Wright-Patterson Air Force Base, OH.

American Trucking Associations, The Maintenance Council (1995). *Recommended maintenance practices manual: 1995 Supplement.* American Trucking Associations, Alexandria, VA.

Ashby, F. Gregory (1992). *Multidimensional Models of Perception and Cognition.* Hillsdale, NJ: Lawrence Erlbaum.

Badler, N., Phillips, C., and Webber, B. (1993). *Simulating Humans.* New York: Oxford University Press.

Bajura, M., and Neumann, U. (1995). Dynamic registration correction in augmented-reality systems. *Proceedings of IEEE Virtual Reality Annual International Symposium.* Research Triangle Park, NC, March 11–15.

Blanchard, B. S. (1986). *Logistics Engineering and Management.* Englewood Cliffs, N.J: Prentice-Hall.

Blanchard, B. S. Jr., and Lowery, E. E. (1969). *Maintainability: Principles and practices.* New York: McGraw-Hill.

Bond, N. A. (1987). Maintainability. In G. Salvendy, Ed. *Handbook of Human Factors.* New York: John Wiley.

Booher, E., Ed. (1990). *MANPRINT: An approach to systems integration.* New York: Van Nostrand.

Boyle, E., Ed. (1991). *Design for Maintainability: Workshop Proceedings.* Report No. AL-TP-91-0010. Armstrong Laboratory, Logistics Research Division, Wright-Patterson Air Force Base, OH.

Cheaper to run, but at a cost. (1995). *Aircraft Maintenance International*, April, pp. 22–23.
Crawford, B. M., and Altman, J. W. (1972). Designing for maintainability. In H. P. Van Cott and R. G. Kinkade, Eds. *Human Engineering Guide to Equipment Design.* American Institutes for Research, Washington, DC.
Cunningham, C. E., and Cox, W. (1972). *Applied Maintainability Engineering. New York: John Wiley.*
Electric Power Research Institute, 1980, *Human Factors Review of Power Plant Maintainability, Report No. EPRI NP-1567-SY* (EPRI, Palo Alto, CA).
Flight control maintenance diagnostic system testing. (1994). *Aerospace Engineering, 14(7)*, July, pp. 13–16.
Folley, J. D. (1960). Design for maintenance. In H. Van Cott, Ed. *Human Factors Methods for System Design.* American Institutes for Research, Pittsburgh.
Fuchs, F. (1982). A current application of design-for-maintainers (DFM) technology. *Proceedings of the Human Factors Design for Maintainers Conference*, D. K. McBride, J. V. Lambert, L. Murray, and L. Hitchcock, Eds. pp. 11–42. Naval Air Development Center, Pensacola Beach, FL.
Fuqua, N. B. (1980). *Electronic equipment maintainability data, Report No. RAC-EEMD-1.* U.S. Air Force Reliability Analysis Center, Griffiss AFB, New York.
Gentner, D., and Stevens, A. L. (1983). *Mental Models.* Hillsdale, NJ: Lawrence Erlbaum.
Goldberg, J. H., and Gibson, D. C. (1986). The effects of training method and type of checklist upon visual inspection accuracy. In W. Karwowski, Ed., Trends in Ergonomics/Human Factors, *3*, North-Holland, Amsterdam.
Hamilton, D. B., Bierbaum, C. R., and Fulford, L. A. (1991). *Task Analysis/Workload (TAWL) user's guide: Version 4. Report No. ARI Research Product 91–11.* U. S. Army Research Institute for the Behavioral and Social Sciences, Alexandria, VA.
Hayward, T. (1993). *Adventures in Virtual Reality.* Que, Carmel, IN.
John Ledoux's better idea. (1987). *Time*, *130(21)*, November 23, p. 31.
Jonsson, J. (1994). *Cognitive categories for the empirical analysis of verbal protocol data.* (Unpublished manuscript.)
Jonsson, J. E., and Ricks, W. R. (1995). *Cognitive models of pilot categorization and prioritization of flight-deck information.* NASA Technical Paper 3528. NASA Langley Reseach Center, Hampton, VA.
Kirwan, B., and Ainsworth, L. K., Eds. (1992). *A Guide to Task Analysis.* London: Taylor and Francis.
Krantz, D. H., and Tversky, A. (1971). Conjoint-measurement analysis of composition rules in psychology. *Psychological Review*, *78*(2), pp. 151–169.
Larijani, L. C. (1994). *The Virtual Reality Primer.* New York: McGraw-Hill.
Locomotive heat problem slip-slides away. (1995). *Machine Design*, *67(12)*, July 13, p. 20.
Lumsdaine, A. (1957). Behavioral guides for improving utilization and performance of technical personnel. *American Psychologist, 12(6).*
Lynch, T. (1995). Design's next step: Serviceability. *Design News*, *50(12)*, June 26, pp. 78–82.
Machlis, S. (1995). Innovation shines in North Carolina. *Design News*, *50(12)*, June 26, pp. 70–75.
Majoros, A. E. (1991). Aircraft design for maintainability. *Journal of Aircraft*, *28(3)*, pp. 187–192.
Majoros, A. E., and Chen, H. C. (1992). Techniques in the design of aircraft for maintainability. In C. T. Leondes, Ed., *Control and dynamic systems, advances in theory and applications; Vol. 52: Integrated technology methods and applications in aerospace systems design.* San Diego: Academic Press.
Majoros, A. E., and Taylor, S. A. (In press). *Work envelopes in equipment design.*
Martin, J. (1981). *Avionics component standardization: The key to maintainability.* American Institute of Aeronautics and Astronautics (AIAA). Paper No. 81–2252.
McDaniel, J., and Hofmann, M. (1990). Computer-aided ergonomic design tools. In H. Booher, Ed., *MANPRINT: An approach to systems integration.* New York: Van Nostrand.
MIL-STD-472, *Maintainability prediction.* (1966).; Naval Air Systems Command, Washington, DC.
MIL-STD-721C, *Definitions of terms for reliabiltiy and maintainability.* (1981). Department of Defense, Washington, DC.
MIL-STD-1472C, *Human Engineering Design Criteria for Military Systems, Equipment and Facilities.* (1981). Department of Defense, Washington, DC.
Miller, R. (1953). *A Method for Man-Machine Task Analysis.* Pittsburgh: American Institutes for Research.
Most, M. (1988). 30 years of success for Pratt & Whitney's PT6. *Aviation Mechanics Journal, 19(1)*, September, pp. 37–41.
National Transportation Safety Board (1989). *Aircraft Accident Report. Report No. NTSB/AAR-89/03.* NTSB, Washington, DC.
NAVAIR 01-1A-33, *Maintainability engineering handbook.* (1977). Naval Air Systems Command, Washington, DC.
Ott, J. (1995). Maintenance executives seek greater efficiency. *Aviation Week and Space Technology, 142(20)*, May 15.

Phillips, D. (1995). Virtual reality offers real savings at MDA. *McDonnell Douglas Team Talk, 10(11)*, p. 8.

Rasmussen, J. (1983). Skills, rules, and knowledge; signals, signs, and symbols, and other distinctions in human performance models. *IEEE Transactions on Systems, Man, and Cybernetics, SMC-13(3)*, May/June, pp. 257–266.

Rose, J., Voytko, J. J., and Davolt, J. A. (1984). *Maintainability time standards for electronic equipment.* Report No. RADC-TR-84-165. U.S. Air Force Rome Air Development Center, Griffiss AFB, New York.

Rouse, W. B., and Rouse, S. H. (1979). Measures of complexity of fault diagnosis tasks. *IEEE Transactions on Systems, Man and Cybernetics, SMC-9(11)*, pp. 720–727.

Ruffner, J. W. (1990). *A survey of human factors methodologies and models for improving the maintainability of emerging Army aviation systems.* U.S. Army Research Institute for the Behavioral and Social Sciences, Alexandria, VA.

Sanders, M. S., and McCormick, E. J. (1987). *Human factors in engineering and design.* New York: McGraw-Hill.

Shepard, R. N. (1987). Toward a universal law of generalization for psychological science. *Science, 237*, pp. 1317–1323.

Society of Automotive Engineers (1994). *Human modeling technology and applications.* Videotape, Special Publications No. SP-1100. Society of Automotive Engineers, Warrendale, PA.

Stone, R. J. and Angus J. (1995). Virtual Maintenance. *Aerospace*, May, pp. 17–21.

Stramler, Jr., J. H. (1993). *The dictionary for human factors/ergonomics.* Boca Raton, FL: CRC Press.

Swain, A. D., and Weston, L. M. (1988). An approach to the diagnosis and misdiagnosis of abnormal conditions in post-accident sequences in complex man-machine systems. In L. P. Goodstein, H. B. Andersen, and S. E. Olsen, Eds., *Task, Errors and Mental Models.* London: Taykor and Francis.

Tensioners keep pins picked up (1995). *Machine Design, 67(12)*, p. 68.

Teschler, L. (1995). Walk-through realism slashes development time, *Machine Design, 67(10)*, pp. 60–70.

Towne, D. M. (1985). *Cognitive workload in fault diagnosis.* Office of Naval Research Contract No. N00014-80-C-0493. Report No. ONR-107. University of Southern California Behavioral Technology Laboratories, Los Angeles.

Towne, D. M., Fehling, M. R., and Bond, N. A. (1981). *Design for the maintainer: Projecting maintenance performance from design characteristics.* Office of Naval Research/Naval Air Development Center Contract No. N00014-80-C-0493, Report No. 1. (University of Southern California, Behavioral Technology Laboratories, Los Angeles.

Towne, D. M., Johnson, M. C., and Corwin, W. H. (1983). *A performance-based technique for assessing equipment maintainability.* Office of Naval Research Contract No. N00014-80-C-0493, Report No. ONR NR196-165. University of Southern California, Behavioral Technology Laboratories, Los Angeles.

Van Cott, H. P., Ed. (1960). *Human factors methods for system design.* American Institutes for Research, Pittsburgh.

Van Cott, H. P., and Kinkade, R. G., Eds. (1972). *Human engineering guide to equipment design.* American Institutes for Research, Washington, DC.

Weand, S., and Thagard, P. (1989). T800 remembers the soldier. *MANPRINT Bulletin, 4(3)*, November/December, pp. 6–8.

Wickens, C. D. (1984). *Engineering psychology and human performance.* Columbus, OH: Merrill.

Woods, D. D. (1988). Coping with complexity: The psychology of human behaviour in complex systems. In L. P. Goodstein, H. B. Anderson, and S. E. Olsen, Eds., *Tasks, Errors and Mental Models.* London: Taylor & Francis.

Woodson, W. E. (1981). *Human Factors Design Handbook.* New York: McGraw-Hill.

Woodson, W. E., Tillman, B., and Tillman, P. (1992). *Human Factors Design Handbook (Second Edition).* New York: McGraw-Hill.

Yoon, W. C., and Hammer, J. M. (1985). Aiding the operator during novel fault diagnosis. In *Proceedings of the IEEE International Conference on Systems, Man, and Cybernetics*, Tucson, AZ, November 12–15, pp. 362–365.

CHAPTER 48
HUMAN FACTORS AUDITS

Colin G. Drury
Department of Industrial Engineering
State University of New York—Buffalo
Amherst, NY 14260 USA

When we audit an entity, we perform an examination of it. Dictionaries typically emphasize official examinations of (financial) accounts, reflecting the accounting origin of the term. Accounting texts go further, e.g., ". . . testing and checking the records of an enterprise to be certain that acceptable policies and practices have been consistently followed" (Carson and Carlson, 1977, p. 2). In the human factors field, the term is broadened to include nonfinancial entities, but remains faithful to the concepts of checking, acceptable policies/practices, and consistency.

Human factors audits can be applied, as can human factors itself, to both products and processes. Both applications have much in common, as any *process* can be considered a *product* of a design procedure, but this chapter emphasizes process audits because product evaluation is covered in detail in Chapter 46, on usability testing. Product usability audits have their own history (e.g., Malde, 1992) which is best accessed through the product design and evaluation literature (e.g., McClelland, 1990).

A second point needs to be made about the scope of this chapter: the role of checklists. As will be seen, checklists have assumed importance as techniques for conducting human factors audits. They can also be used alone as evaluation devices, in applications as diverse as VDT workplaces (Cakir, Hart, and Stewart, 1980) and risk factor assessment (Keyserling, Brouwer, and Silverstein, 1992). Hence, the structure and use of checklists will be covered in some detail independently of their use as an auditing technique.

48.1 THE NEED FOR AUDITING HUMAN FACTORS

Human factors or ergonomics programs have become a permanent feature of many companies, with typical examples shown in Alexander and Pulat (1985). As with any other function, human factors/ ergonomics needs tools to measure its effectiveness. Earlier, when human factors operated through individual projects, evaluation could take place on a project-by-project basis. Thus, the interventions to improve apparel-sewing workplaces described by Drury and Wick (1984), could be evaluated to show changes in productivity and reductions in cumulative trauma disorder causal factors. Similarly, Hasslequist (1981) showed productivity, quality, safety, and job satisfaction improvements following human factors interventions in a computer-component assembly line. In both cases, the objectives of the intervention were used to establish appropriate measures for the evaluation.

However, ergonomics/human factors is no longer confined to operating in a project mode. Increasingly, the establishment of a permanent function within an industry has meant that ergonomics is more closely related to the strategic objectives of the company. As Drury, Kleiner, and Zahorjan (1989) have observed, this development requires measurement methodologies which also operate at the strategic level. For example, as a human factors group becomes more involved in strategic decisions about identifying and choosing the projects it performs, evaluation of the individual projects is less revealing. All projects performed could have a positive impact, but the group could still have achieved more with a more astute choice of projects. It could conceivably have had a more beneficial impact on the company's strategic objectives by stopping all projects for a period to concentrate on training the management, workforce, and engineering staff to make more use of ergonomics.

Such changes in the structure of the ergonomics/human factors profession indeed demand different evaluation methodologies. For example, a powerful network of individuals who can, and do, call for human factors input in a timely manner can help an enterprise more than a number of individually successful project outcomes. Audit programs are one of the ways in which such evaluations can be made, allowing a company to focus its human factors resources most effectively. They can also be used in a prospective, rather than retrospective, manner to help quantify the needs of the company for ergonomics/human factors. Finally, they can be used to determine which divisions, plants, departments, or even product lines are in most need of ergonomics input.

48.2 DESIGN REQUIREMENTS FOR AUDIT SYSTEMS

Returning to the definition of an audit, the emphasis is on checking, acceptable policies, and consistency. The aim is to provide a fair representation of the business for use by third parties. A typical audit by a certified public accountant would comprise the following steps (adapted from Koli, 1994):

1. ***Diagnostic investigation:*** Description of the business and high lighting of areas requiring increased care and high risk.
2. ***Test for transaction:*** Trace samples of transactions grouped by major area and evaluate.
3. ***Test of balances:*** Analyze content.
4. ***Formation of opinion:*** Communicate judgment in an audit report.

Such a procedure can also form a logical basis for human factors audits. The first step chooses the areas of study, the second samples the system, the third analyses these samples while the final step produces an audit report. These define the broad issues in human factors audit design:

1. ***How to sample the system:*** How many samples and how these are distributed across the system?
2. ***What to sample:*** Specific factors to be measured, from biomechanical to organizational.
3. ***How to evaluate the sample:*** What standards, good practices or ergonomic principles to use for comparison?
4. ***How to communicate the results:*** Techniques for summarizing the findings, how far can separate findings be combined?

A suitable audit system needs to address all of these issues (see Section 48.3), but some overriding design requirements must first be specified.

48.2.1 Breadth, Depth, and Application Time

Ideally, an audit system would be broad enough to cover any task in any industry, it would provide highly detailed analysis and recommendations, and would be applied rapidly. Unfortunately, the three variables of breadth, depth and application time are likely to trade off in a practical system. Thus, a thermal audit (Parsons, 1992) sacrifices breadth to provide considerable depth based on the heat balance equation, but requires measurement of seven variables. Some can be obtained rapidly (air temperature, relative humidity), but some take longer (clothing insulation value, metabolic rate). Conversely, structured interviews with participants in an ergonomics program (Drury, 1990a) can be broad and rapid, but quite deficient in depth.

At the level of audit instruments such as questionnaires or checklists, there are comprehensive surveys such as the Position Analysis Questionnaire (McCormick, 1979), the Arbeitswissenschaftliche Erhebungsverfahren zur Tátikgkeitsanalyse (AET) (Rohmert and Landau, 1989) which takes 2–3 hr to complete, or the simpler Work Analysis Checklist (Pulat, 1992). Alternatively, there are simple single page checklists such as the Ergonomics-Working Position-Sitting Checklist (SHARE, 1990) which can be completed in a few minutes.

Analysis and reporting can range in depth from merely tabulating the number of ergonomic standards violated, to expert systems which provide prescriptive interventions (Ayoub and Mital, 1989).

Most methodologies fall between the various extremes given above, but the goal of an audit system with an optimum trade-off between breadth, depth, and time is probably not realizable. A better practical course would be to select several instruments and use them together to provide the specific breadth and depth required for a particular application.

48.2.2 Use of Standards

The human factors/ergonomics profession has many standards and good practice recommendations. These differ by country (ANSI, BSI, DIN) although commonality is increasing through joint standards such as those of the International Standards Organization (ISO). Some standards are quantitative, such as heights for school furniture (BSI, 1965), sizes of characters or a VDT screen (ANSI/HFS-100), and occupational exposure to noise. Other standards are more general in nature, particularly those which involve management actions to prevent or alleviate problems, such as the OSHA guidelines for meat packing plants (OSHA, 1990). Generally, standards are more likely to exist for simple tasks and environmental stressors, and are hardly to be expected for the complex cognitive activities with which human factors predictions increasingly deal. Where standards exist, they can represent unequivocal elements of audit procedures, as a workplace which does not meet these standards is in a position of legal violation. A human factors program which tolerates such legal exposure should clearly be held accountable in any audit.

However, merely meeting legal requirements is an insufficient test of the quality of ergonomics/human factors efforts. Many legal requirements are arbitrary or outdated, for example weight limits for manual materials handling in some countries. Additionally, other aspects of a job with high ergonomic importance may not be covered by standards, for example, presence of multiple stressors, work in restricted spaces resulting in awkward postures, or highly repetitive upper extremity motions. Finally, there are many "human factors good practices," which are not the subject of legal standards. Examples are the NIOSH lifting equation (Waters, Putz-Anderson, Garg, and Fine, 1993), the Illuminating Engineering Society (IES) codes (1993), or the zones of thermal comfort defined by ASHRAE (1989) or Fanger (1970). In some cases, standards are available in a different jurisdiction from that being audited. As an example, the military standard MIL-1472D (DOD, 1989) provides detailed standards for control and display design which are equally appropriate to process controls in manufacturing industry, but have no legal weight there.

Standards, in the legal sense, are a particularly reactive phenomenon. It may take many years (any many injuries and accidents) before a standard is found necessary and agreed upon. The NIOSH lifting equation referenced above addresses a back injury problem which is far from new, yet it still has no legal force. Standards for upper extremity cumulative trauma disorder prevention have lagged disease incidence by many years. Perhaps because of busy legislative agendas, we cannot expect rapid legal reaction, unless a highly visible major disaster occurs. Human factors problems are both chronic and acute, so that legislation based on acute problems as the sole basis for auditing is unlikely ever to be effective.

Despite the lack of legislation covering many human factors concerns, standards and other instantiations of good practice do have a place in ergonomics audits. Where they exist, they can be incorporated into an audit system, without becoming the only criterion.

Thus, noise levels in the United States have a legal limit for hearing protection purposes of 90 dBA. But at levels far below this, noise can disrupt communications (Jones and Broadbent, 1987) and distract from task performance. An audit procedure can assess the noise on multiple criteria, i.e., on hearing protection and on communication interruptions, with the former criterion used on all jobs and the latter only where verbal communication is an issue.

If standards and other good practices are used in a human factors audit, they provide a quantitative basis for decision making. Measurement reliability can be high and validity self-evident for legal standards. However, it is good practice in auditing to record only the measurement used, and not its relationship to the standard, which can be established later. This removes any temptation by the analyst to "bend" the measurement to reach a predetermined conclusion. Illumination measurements, for example, can vary considerably over a work space, so that an audit question:

Work surface illumination >750 Lux ☐ yes ☐ no

could be legitimately answered either way for some work spaces by choice of sampling point. Such temptation can be removed, for example, by an audit question.

Illumination at four points on workstation:

☐ ☐ ☐ ☐ Lux

Later analysis can establish whether, for example, the mean exceeds 750 Lux, or whether any of the four points fall below this level.

It is also possible to provide later analyses which combine the effects of several simple checklist responses, as in Parsons's (1992) thermal audit where no single measure would exceed good practice even though the overall result would be cumulative heat stress.

48.2.3 Evaluation of an Audit System

For a methodology to be of value it must demonstrate validity, reliability, sensitivity, and usability. Most texts which cover measurement theory treat these aspects in detail, e.g., Kerlinger, 1964. Shorter treatments are found in human factors methodology texts, e.g., Drury in Wilson and Corlett (1990b) or Osburn in Salvendy (1987).

Validity is the extent to which a methodology measures the phenomenon of interest. Does our ergonomics audit program indeed measure the quality of ergonomics in the plant? It is possible to measure validity in a number of ways, but ultimately all are open to argument. For example, if we do not know the "true" value of the "quality of ergonomics" in a plant, then how can we validate our ergonomics audit program? Broadly, there are three ways in which validation can be tested.

Content validity is perhaps the simplest but least convincing measure. If each item of our measurement device displays the correct content, then validity is established. Theoretically, if we could list all of the possible measures of a phenomenon, content validity would describe how well our measurement device samples these possible measures. In practice it is assessed by having experts in the field judge each item for how well its content represents the phenomenon studied. Thus, the heat balance equation would be judged by most thermal physiologists to have a content which well represents the thermal load on an operator. Not all aspects are as easily validated!

Concurrent (or predictive) validity has the most immediate practical impact. It measures empirically how well the output of the measurement device correlates with the phenomenon of interest. Of course, we must have an independent measure of the phenomenon of interest, which raises difficulties. To continue our example, if we used the heat balance equation to assess the thermal load on operators, then there should be a high correlation between this and other measures of the effects of thermal load. Perhaps measures such as frequency of temperature complaints or of heat disorders (e.g., heat stroke, hyperthermia, hypothermia). In practice, however, measuring such correlations would be contaminated by, for example, propensity to report temperature problems or individual acclimatization to heat. Overall outputs from a human factors audit (if such overall outputs have any useful meaning) should correlate with other measures of ergonomic inadequacy, such as injuries, turnover, quality measures, or productivity. Alternatively, we can ask how well the audit findings agree with independent assessments of qualified human factors engineers (Keyserling et al., 1992; Koli, Drury, Cuneo, and Lofgren, 1993) and thus validate against one interpretation of current good practice.

Finally, there is construct validity. This is concerned with inferences made from scores, evaluated by considering all empirical evidence and models. Thus, a model may predict that one of the variables being measured should have a particular relationship to another variable not in the measurement device. Confirming this relationship empirically would help validate the particular construct underlying our measured variable. Note that different parts of an overall measurement device can have their construct validity tested in different ways. Thus, in a board human factors audit, the thermal load could differentiate between groups of operators who do and do not suffer from thermal complaints. In the same audit a measure of difficulty in a target aiming task could be validated against Fitts' Law. Other ways to assess construct validity are those which analyze clusters or factors within a group of measures. Different workplaces are audited on a variety of measures and the scores, which are then subjected to factor analysis, should show an interpretable, logical structure in the factors derived. This method has been used on large databases for job-evaluation-oriented systems such as McCormick's Position Analysis Questionnaire, PAQ (McCormick, 1979).

Reliability refers to how well a measurement device can repeat a measurement on the same sample unit. Classically, if a measurement X is assumed to be composed of a true value X_t and a random measurement error X_e, then

$$X = X_t + X_e$$

For uncorrelated X_t and X_e, taking variances gives

$$\text{Variance }(X) = \text{Variance }(X_t) + \text{Variance }(X_e)$$

or

$$V(X) = V(X_t) + V(X_e)$$

We can define the reliability of the measurement as the fraction of measurement variance accounted for by true measurement variance.

$$\text{Reliability} = \frac{V(X_t)}{V(X_t) + V(X_e)}$$

Typically, reliability is measured by correlating the scores obtained through repeated measurements. In an audit instrument, this is often done by having two (or more) auditors use the instrument on the same set of workplaces. The square of the correlation coefficient between the scores (either overall scores, or separately for each logical construct) is then the reliability. Thus, PAQ was found to have an overall reliability of 0.79, tested using 62 jobs and two trained analysts (McCormick, 1979).

Sensitivity defines how well a measurement device differentiates between different entities. Does an audit system for human computer interaction find a difference between software generally acknowledged to be "good" and "bad"? If not, perhaps the audit system lacks sensitivity, although of course there may truly be no difference between the systems except blind prejudice. Sensitivity can be adversely affected by poor reliability, which increases the variability in a measurement relative to a fixed difference between entities, i.e., gives a poor signal-to-noise ratio. Low sensitivity can also come from a floor or ceiling effect. These arise where almost all of the measurements cluster at a high or low limit. For example, if an audit question on the visual environment was

Does illumination exceed 10 Lux? ☐ yes ☐ no

then almost all workplaces could answer "yes" (although the author has found a number that could not meet even this low criterion). Conversely, a floor effect would be a very high threshold for illuminance. Sensitivity can arise too when validity is in question. Thus, heart rate is a valid indicator of heat stress, but not of cold stress. Hence, exposure to different degrees of cold stress would be only insensitively measured by heart rate.

Usability refers to the auditor's ease of use of the audit system. Good human factors principles should be followed (for example, document design guidelines in constructing checklists, Patel, Drury, and Prabhu, 1993; Wright and Barnard, 1975). If the instrument does not have good usability, it will be used less often, and may even show reduced reliability due to auditors' errors.

48.3 AUDIT SYSTEM DESIGN

As outlined in Section 48.2, the audit system must choose a sample, measure that sample, evaluate it and communicate the results. In Section 48.3, we approach these issues systematically.

An audit system is not just a checklist; it is a methodology which often includes the technique of a checklist. The distinction needs to be made between methodology and technique. Almost three decades ago Easterby (1967) used Bainbridge and Beishon's (1964) definitions:

Methodology: a principle for defining the necessary procedures
Technique: a means to execute a procedural step.

Easterby notes that a technique may be applicable in more than one methodology.

48.3.1 The Sampling Scheme

In any sampling, we must define the unit of sampling, the sampling frame, and the sample choice technique. For a human factors audit the unit of sampling is not as self-evident as it appears. From a job evaluation viewpoint, e.g., McCormick (1979), the natural unit is the job which is composed of a number of tasks. From a medical viewpoint the unit would be the individual. Human factors studies focus on the Task/Operator/Machine/Environment (TOME) system (Drury, 1992) or equivalently the Software/Hardware/Environment/Liveware (SHEL) system (ICAO, 1989). Thus, from a strictly human factors viewpoint the specific combination of TOME can become the sampling unit for an audit program.

Unfortunately, this simple view does not cover all of the situations for which an audit program may be needed. While it works well for the rather repetitive tasks performed at a single workplace typical of much manufacturing and service industry, it cannot suffice when these conditions do not hold. One relaxation is to remove the stipulation of a particular incumbent, allowing for jobs which require frequent rotation of tasks. This means that the results for one task will depend on the incumbent chosen, or that several tasks will need to be combined if an individual operator is of interest. A second relaxation is that the same operator may move to different workplaces, thus changing environment as well as task. This is typical of maintenance activities, where a mechanic may perform any one of a repertoire of hundreds of tasks, rarely repeating the same task. Here, the rational sampling unit is the task, which is observed for a particular operator at a particular machine in a particular environment. Examples of audits of repetitive tasks (Drury, 1990a; Mir, 1982) and maintenance tasks (Chervak and Drury, 1995) are given later to illustrate these different approaches.

Definition of the sampling frame, once the sampling unit is settled, is more straightforward. Whether the frame covers a department, a plant, a division, or a whole company, enumeration of all sampling units is at least theoretically possible. All workplaces, or jobs, or individuals can in principle be listed, although in practice the list may never be up to date in an agile industry where change is the normal state of affairs. Individuals can be listed from personnel records, tasks from work orders or planning documents, workplaces from plant layout plans. A greater challenge, perhaps, is to decide whether indeed the whole plant really is the focus of the audit. Do we include office jobs or just production? What about managers, chargehands, part-time janitors, etc.? A good human factors program would see all of these tasks or people as worthy of study, but in practice they may have had different levels of ergonomic effort expended upon them. Should some tasks or groups be excluded from the audit merely because most participants agree that they have few pressing human factors problems? These are issues which need to be decided explicitly before the audit sampling begins.

Choice of the sample from the sampling frame is well covered in sociology texts. Within human factors it typically arises in the context of survey design (Sinclair, 1990). To make statistical inferences from the sample to the population (specifically to the sampling frame), our sampling procedure must allow the laws of probability to be applied. The most often used sampling methods are as follows:

Random Sampling: Each unit within the sampling frame is equally likely to be chosen for the sample. This is the simplest and most robust method, but it may not be the most efficient. Where subgroups of interest (strata) exist and these subgroups are not equally represented in the sampling frame, one collects unnecessary information on the most populous subgroups and insufficient information on the least populous. This is because our ability to estimate a population statistic from a sample depends on the absolute sample size and not, in most practical cases, on the population size. As a corollary, if subgroups are of no interest, then random sampling loses nothing in efficiency.

Stratified Random Sampling: Each unit within a particular stratum of the sampling frame is equally likely to be chosen for the sample. With stratified random sampling we can make valid inferences about each strata. By weighting the statistics to reflect the size of the strata within the sampling frame, we can also obtain population inferences. This is often the preferred auditing sampling method as, for example, we would wish to distinguish between different classes of tasks in our audits: production, warehouse, office, management, maintenance, security, etc. In this way our audit interpretation could give more useful information concerning where ergonomics is being used appropriately.

Cluster Sampling: Clusters of units within the sampling frame are selected, followed by random or nonrandom selection within clusters. Examples of clusters would be the selection of particular production lines within a plant (Drury, 1990a), or selection of "representative" plants within a company or division. The difference between cluster and stratified sampling is that in cluster sampling only a subset of possible units within the sampling frame is selected, whereas in stratified sampling, all of the sampling frame is used as each unit must belong to one stratum. Because clusters are not randomly selected, the overall sample results will not reflect population values so that statistical inference is not possible. If units are chosen randomly within each cluster, then statistical inference within each cluster is possible. For example, if three production lines are chosen as clusters, and workplaces sampled randomly within each, the clusters can be regarded as fixed levels of a factor and the data subjected to Analysis of Variance to determine whether there are significant differences between levels of that factor. What is sacrificed in cluster sampling is the ability to make *population* statements. Continuing this example, we could state that the lighting in line A is better than in lines B or C, but still not be able to make statistically-valid statements about the plant as a whole.

48.3.2 The Data Collection Instrument

So far we have assumed that the instrument used to collect the data from the sample is based upon measured data where appropriate. While this is true of many audit instruments, this is not the only way to collect audit data. There have been interviews with participants (Drury, 1990a), interviews and group meetings to locate potential errors (Fox, 1992), and use of archival data such as injury of quality records (Mir, 1982). All have potential uses with, as remarked earlier, a judicious range of methods often providing the appropriate composite audit system.

One consideration on audit technique design and use is the extent of computer involvement. Computers are now inexpensive, portable, and powerful so that they can be used to assist data collection, data verification, data reduction, and data analysis (Drury, 1990a). With the advent of more intelligent interfaces, checklist questions can be answered from mouse-clicks on buttons, or selection from menus, as well as the more usual keyboard entry. Data verification can take place at entry time by checking for out-of-limits data, or odd data, such as the ratio of lluminance to illuminance implying a reflectivity greater than 100%. In addition, branching in checklists can be made easier, with only valid follow-on questions highlighted. The "checklist user's manual" can be built into the checklist software using context-sensitive help facilities, as in the EEAM checklist (Chervak and Drury, 1995). Comptuers can, of course, be used for data reduction (for example, finding the insulation value of clothing from a clothing inventory), data analysis, and results presentation.

Having made the case for computer use, some cautions are in order. Computers are still bulkier than simple pencil-and-paper checklists. Computer reliability is not perfect, so that inadvertant data loss is still a real possibility. Finally, software and hardware dates much more rapidly than hard copy, so that results safely stored on the latest media may be unreadable 10 yr later. How many of us can still read punched cards or 8-in. floppy disks? In contrast, hard copy records are still available from before the start of the common era.

48.3.2.1 Checklists and Surveys

For many practitioners the proof of the effectiveness of an ergonomics effort lies in the ergonomic quality of the TOME systems it produces. A plant or office with appropriate human–machine function allocation, well-designed workplaces, comfortable environment, adequate placement/training, and inherently satisfying jobs almost by definition has been well served by human factors. Such a facility may not have human factors specialists, just good designers of environment, training, organization, etc. working independently, but this would generally be a rare occurrence. Thus, a checklist to measure such inherently ergonomic qualities has great appeal as part of an audit system.

Such checklists are almost as old as the discipline. An early paper by Burger and deJong (1964) lists four earlier checklists for ergonomic job analysis before going on to develop their own. Theirs was commissioned by the International Ergonomics Association in 1961 and is usually known as the IEA checklist. It was based in part on one developed at the Philips Health Centre by G. J. Fortuin and provided in detail in Burger and deJong's paper.

Checklists have their limitations, though. The cogent arguments put forward by Easterby (1967) provide a good early summary of these limitations and most are still valid today. Checklists are only of use as an aid to designers of systems at the earliest stages of the process. By concentrating on simple questions, often requiring yes/no answers, some checklists may reduce human factors to a simple stimulus-response system, rather than encouraging conceptual thinking. Easterby quotes Miller (1967): "I still find that many people who should know better seem to expect magic from analytic and descriptive procedures. They expect that formats can be filled in by dunces and lead to inspired insights . . . We should find opportunity to exorcise this nonsense" (Easterby, p. 554). Easterby finds checklists can have a helpful structure, but often have vague questions, make nonspecified assumptions, and lack quantitative detail. Checklists are seen as appropriate for some parts of ergonomics analysis (as opposed to synthesis) and even more appropriate to aid operators (not ergonomists) in following procedural steps. This latter use has been well covered by Delgani and Wiener (1990) and will not be further presented here.

Clearly, we should be careful, even 30 yr on, to heed these warnings. Many checklists are developed, and many of these published, which contain design elements fully justifying such criticisms.

A checklist, like any other questionnaire, needs to have both a helpful overall structure and well-constructed questions. It should also be proven reliable, valid, sensitive, and usable, although there are precious few which meet all of these criteria. In the remainder of this section, a selection of checklists will be presented as typical of (reasonably) good practice. Emphasis will be on objective, structure, and question design.

The IEA Checklist. The IEA Checklist (Burger and deJong, 1964) was designed for ergonomic job analysis over a wide range of jobs. It uses the concept of functional load to give a logical framework relating the Physical Load, Perceptual Load and Mental Load to the Worker, the Environment, and the Working Methods/Tools/Machines. Within each cell (or subcell, e.g., Physical Load could be static or dynamic) the load was assessed on different criteria such as force, time, distance, occupational medical, and psychological criteria. Table 48.1 shows the structure and typical questions. Dirken (1969) modified the IEA checklist to improve the questions and methods of recording. He found that it could be applied in a median time of 60 min/workstation. No data are given on evaluation of the IEA checklist, but its structure has been so influential that it is included here for more than historical interest.

Position Analysis Questionnaire. The PAQ is a structured job analysis questionnaire using worker-oriented elements (187 of them) to characterize the human behaviors involved in jobs (McCormick, Mecham, and Jeanneret, 1969). The PAQ is structured into six divisions, with the first three representing the classic experimental psychology approach (information input, mental process, work output) and the next a broader sociotechnical view (relationships with other persons, job context, other job characteristics). Table 48.2 shows these major divisions, examples of job elements in each and the rating scales employed for response (McCormick, 1979).

Construct validity was tested by factor analyses of data bases containing 3700 and 2200 jobs, which established 45 factors. Thirty-two of these fit neatly into the original six-division framework, with the remaining 13 classified as "overall dimensions." Further proof of construct validity was based on 76 human attributes derived from the PAQ, rated by industrial psychologists and the ratings subjected to principal components analysis to develop dimensions "which had reasonably similar attribute profiles" (McCormick, 1979, p. 204). Inter-rater reliability, as noted earlier, was 0.79 based on another sample of 62 jobs.

Table 48.1 IEA Checklist: Structure and Typical Questions

A: Structure of the checklist

		A	B	C
Load 1. Mean 2. Peaks Intensity, Frequency, Duration		Worker	Environment	Working Method, Tool—s, Machines
I. Physica—l load	1. Dynamic 2. Static			
II. Perceptual load	1. Perception 2. Selection, decision 3. Control of movement			
III. Mental load	1. Individual 2. Group			

B: Typical Question

I B. Physical Load/Enviroment	2.1 Physiological Criteria
	1. Climate: high and low temperatures 1. Are these extreme enough to affect comfort or efficiency? 2. If so, is there any remedy? 3. To what extent is working capacity adversely affected? 4. Do personnel have to be specially selected for work in this particular environment?

The PAQ covers many of the elements of concern to human factors engineers, and has indeed much influenced subsequent instruments such as AET. With good reliability and useful (though perhaps dated) construct validity, it is still a viable instrument if the natural unit of sampling is the job. The exclusive reliance on rating scales applied by the analyst goes rather against current practice of comparison of measurements against standards or good practices.

AET (Arbeitswissenschaftliche Erhebungsverfahren zur Tätikgkeitsanalyse). The AET has been published in German (Landau and Rohmert, 1981) and was published later in English (Rohmert and Landau, 1983). It is the job analysis subsystem of a comprehensive system of work studies. It covers "the analysis of individual components of man-at-work systems as well as the description and scaling of their interdependencies" (Rohmert and Landau, 1983, pp. 9–10). As with all good techniques, it starts from a model of the system (REFA, 1971; referenced in Wagner, 1989), to which is added Rohmert's stress/strain concept. This latter sees strain as being caused by the intensity and duration of stresses impinging upon the operator's individual characteristics. It is seen as useful in the analysis of requirements and work design, organization in industry, personnel management, and vocational counseling and research.

AET itself was developed over many years, using PAQ as an initial starting point. Table 48.3 shows the structure of the survey instrument with typical questions and rating scales. Note the similarity between AET's Job Demands Analysis and the first three categories of the PAQ, and between the scales used in AET and PAQ (Table 48.2).

Measurements of validity and reliability of AET are discussed by H. Luczak in an appendix to Landau and Rohmert, although no numerical values are given. Cluster analysis of 99 AET records produced groupings which supported the AET constructs. Seeber, Schmidt, Kierswelter, and Rutenfranz (1989) used AET along with two other work analysis methods on 170 workplaces. They found that AET provided the most differentiating aspects (suggesting sensitivity). They also measured postural complaints and showed that only the AET groupings for 152 female workers found significant differences between complaint levels, thus helping establish construct validity.

AET, like PAQ before it, has been used on many thousands of jobs, mainly in Europe. A sizable database is maintained which can be used for both norming of new jobs analyzed, and analysis to test research hypotheses. It remains a most useful instrument for work analysis.

Table 48.2 PAQ: Structure and Typical Questions

A: Structure of the Checklist

Division	Definition	Examples of Questions
1. Information input	Where and how does the worker get the information he uses in performing his job?	1. Use of written materials 2. Near-visual differentiation
2. Mental processes	What reasoning, decision making, planning, and information processing activities are involved in performing the job?	1. Level of reasoning in problem solving 2. Coding - decoding
3. Work output	What physical activities does the worker perform and what tools or devices does he use?	1. Use of keyboard devices 2. Assembling - unassembling
4. Relationships with other persons	What relationships with Other People are required in performing the job?	1. Instructing 2. Contacts with public or customers
5. Job context	In what physical or social contexts is the work performed?	1. High temperature 2. Interpersonal; conflict situations
6. Other job characteristics	What activities, conditions, or characteristics other than those described above are relevant to the job?	1. Specified work pace 2. Amount of job structure

B: Scales use to rate elements

Types of scale	
Identification	**Type of Rating**
U	Extent of Use
I	Importance of the job
T	Amount of Time
P	Possibility of Occurence
A	Applicability (yes/no only)
S	Special Code

Scale values	
Rating	**Definition**
N	Does not apply
1	Very minor
2	Low
3	Average
4	High
5	Extreme

Ergonomics Audit Program (Mir, 1982; Drury, 1990a). This program was developed at the request of a multinational corporation to be able to audit its various divisions and plants as ergonomics programs were being instituted. The system developed was a methodology of which the Workplace Survey was one technique. Overall, the methodology used archival data or outcome measures (injury reports, personnel records, productivity) and critical incidents to rank order departments within a plant. A cluster sampling of these departments gives either the ones with highest need (if the aim is to focus ergonomic effort) or a sample representative of the plant (if the objective is an audit). The Workplace Survey is then performed on the sampled departments.

The Workplace Survey was designed based on ergonomic aspects derived from a Task/Operator/Machine/Environment model of the person at work. Each aspect formed a section of the audit, and sections could be omitted if there were clearly not relevant, e.g., manual materials handling aspects for data entry clerks. Questions within each section were based on standards, guidelines and models, such as the NIOSH (1981) lifting equation, ASHRAE Handbook of Fundamentals for thermal aspects, and Givoni and Goldman's (1972) model for predicting heart rate. Table 48.4 shows the major sections and typical questions.

Data were entered into the computer program, and a rule-based logic evaluated each section to provide messages to the user in the form of either a "section shows no ergonomic problems" message:

MESSAGE
Results from analysis of auditory aspects:

Table 48.3 AET: Structure and Typical Questions

A: Structure of the Checklist

Part	Major Division	Section
A. Work systems analysis	1. Work objects	1.1 Material work objects 1.2 Energy as work object 1.3 Information as work object 1.4 Man, animals, plants as work objects
	2. Equipment	2.1 Working equipment 2.2 Other equipment
	3. Work environment	3.1 Physical environment 3.2 Organizational and social environment 3.3 Principles and methods of remuneration
B: Task analysis	1. Tasks relating to material work objects 2. Taks relating to abstract work objects 3. Man-related tasks 4. Number and repetitiveness of tasks	
C: Job demand analysis	1. Demands on perception	1.1 Mode of perception 1.2 Absolute/relative evaluation of perceived information 1.3 Accuracy of perception
	2. Demands for decision	2.1 Complexity of decisions 2.2 Pressure of time 2.3 Required knowledge
	3. Demands for response/activity	3.1 Body postures 3.2 Static work 3.3 Heavy muscular work 3.4 Light muscular work, active light work 3.5 Strenuousness and frequency of movements

B: Scales used to rate the elements

Types of scale	
Code	Type of Rating
A	Does this apply?
F	Frequency
S	Significance
D	Duration

Typical Scale values	
Duration Value	Definition
0	Very infrequent
1	Less than 10% of shift time
2	Less than 30% of shift time
3	30% to 60% of shift time
4	More than 60% of shift time
5	Almost continuously during whole shift

Everything OK in this section

or discrepancies from a single input:

MESSAGE
Seats should be padded, covered with non-slip materials and have front edge rounded

or discrepancies based on the integration of several inputs:

MESSAGE
The total metabolic workload is 174 W
Intrinsic clothing insulation is 0.56 clo
Initial rectal temperature is predicted to be 36.0°C
Final rectal temperature is predicted to be 37.1°C

Table 48.4 Workplace Survey: Structure and Typical Questions

Section	Major Classification	Examples of Questions
1. Visual aspects		Nature of task Measure illuminance at task midfield outer field
2. Auditory aspects		Noise level, dBA Main source of noise
3. Thermal aspects		Strong radiant sources present? Wet bulb temperature (Clothing inventory)
4. Instruments, controls, displays	Standing vs. Seated Displays Labeling Coding Scales, dials, counters Control/display relationships Controls	Are controls mounted between 30″-70″ Signals for crucial visual checks Are trade names deleted? Color codes same for control & display? All numbers upright on fixed scales? Grouping by sequence or subsystem? Emergency button diameter > 0.75″?
5. Design of workplaces	Desks Chairs Posture	Seat to underside of desk > 6.7″? Height easily adjustable 15″ - 21″? Upper arms vertical?
6. Manual materials handling	(NIOSH Lifting Guide, 1981)	Task, H, V, D, F
7. Energy expenditure		Cycle time Object weight Type of work
8. Assembly/repetitive aspects		Seated, standing, or both? If heavy work, is bench 6″–16″ below elbow height?
9. Inspection aspects		Training time until unsupervised? Number of fault types?

Counts of discrepancies were used to evaluate departments by ergonomics aspect, while the messages were used to alert company personnel to potential design changes. This latter use of the output as a training device for nonergonomic personnel was seen as desirable in a multinational company rapidly expanding its ergonomics program.

Reliability and validity have not been assessed, although the checklist has been used in a number of industries (Drury, 1990a). The workplace survey has been included here as, despite its lack of measured reliability and validity, it shows the relationship between audit as methodology and checklist as technique.

ERGO, EEAM and ERNAP. (Koli, Drury, Cuneo, and Lofgren, 1993; Chervak and Drury, 1995). These checklists are both part of complete audit systems for different aspects of civil aircraft hangar activities. They were developed for the Federal Aviation Administration to provide tools for assessing human factors in aircraft inspection (ERGO) and maintenance (EEAM) activities respectively. Inspection and maintenance activities are non-repetitive in nature, controlled by task cards issued to technicians at the start of each shift. Thus, the sampling unit is the task card, not the workplace, which is highly variable between task cards. Their structure was based on extensive task analyses of inspection and maintenance tasks, which led to generic function descriptions of both types of work (Drury, Prabhu, and Gramopadhye, 1990). Both systems have sampling schemes and checklists. Both are computer-based with initial data collection on either hard copy or direct into a portable computer. Recently, both have been combined into a single program (ERNAP) distributed by the FAA's Office of Aviation Medicine. The structure of ERNAP and typical questions are given in Table 48.5.

As in Mir's Ergonomics Audit Program, the ERNAP checklist is again modular, and the software allows formation of data files, selection of required modules, analysis after data entry is completed, and printing of audit reports. Similarly, the ERGO, EEAM and ERNAP instruments use quantitative or Yes/No questions comparing the entered value

Table 48.5 ERNAP Structure and Typical Questions

Audit Phase	Major Classification	Examples of Questions
I. Premaintenance	Documentation	Is feedforward information on faults given?
	Communication	Is shift change documented?
	Visual Characteristics	If fluorescent bulbs are used, does flicker exist?
	Electric/Pneumatic Equipment	Do push buttons prevent slipping of fingers?
	Access Equipment	Do ladders lave nonskid surfaces on landings?
II. Maintenance	Documentation (M)	Does inspector sign off workcard after each task?
	Communication (M)	Explicit verbal instructions from supervisor?
	Task Lighting	Light levels in four zones during task, fc.
	Thermal Issues	Wet bulb temperature in hanger bay, °C
	Operator Perception	Satisfied with summer thermal environment?
	Auditory Issues	Noise levels at five times during task, dBA
	Electrical and Pneumatic	Are controls easily differentiated by touch?
	Access Equipment (M)	Is correct access equipment available?
	Hand Tools	Does the tool handle end in the palm?
	Force Measurements	What force is being applied, kg?
	Manual Material Handling	Does task require pushing or pulling forces?
	Vibration	What is total duration of exposure on this shift?
	Repetitive Motion	Does the task require flexion of the wrist?
	Access	How often was access equip. repositioned?
	Posture	How often were following postures adopted?
	Safety	Is inspection area adequately cleaned for inspect?
	Hazardous Material	Were hazardous materials signed out and in?
III. Postmaintenance	Buy Back	Are discrepancy worksheets readable?

with standards and good practice guides. Each takes about 30 min/task. Output is in the form of an audit report for each workplace, similar to the messages given by Mir's Workplace Survey, but in narrative form. Output in this form was chosen for compatibility with existing performance and compliance audits used by the aviation maintenance community.

Reliability of a first version of ERGO was measured by comparing the output of two auditors on three tasks. Significant differences were found at $P<0.05$ on all three tasks showing a lack of interrater reliability. Analysis of these differences showed them to be largely due to errors on questions requiring auditor judgment. When such questions were replaced with more quantitative questions the two auditors had no significant disagreements on a later test. Validity was measured using concurrent validation against six Ph.D. human factors engineers who were asked to list all ergonomic issues on a power plant inspection task. The checklist found more ergonomic issues than the human factors engineers. Only a small number of issues were raised by the engineers which were missed by the checklist. For the EEAM checklist, again an initial version was tested for reliability with two auditors, and only achieved the same outcome for 85% of the questions. A modified version was tested, and the reliability was considered satisfactory with 93% agreement. Validity was again tested against four human factors engineers, this time the checklist found significantly more ergonomic issues than the engineers, without missing any issues they raised.

The ERNAP audits have been included here to provide examples of a checklist embedded in an audit system where the workplace is *not* the sampling unit. They show that nonrepetitive tasks can be audited in a valid and reliable manner. In addition, they demonstrate how domain-specific audits can be designed to take advantage of human factors analyses already made in the domain.

Upper Extremity Checklist. (Keyserling, Stetson, Silverstein, and Brouwer, 1993). As its name suggests, this checklist is narrowly focused on biomechanical stresses to the upper extremities which could lead to cumulative trauma disorders (CTDs). It does not claim to be a full-spectrum analysis tool, but is included here as a good example of a special-purpose checklist which has been carefully constructed and validated. The checklist (Table 48.6) was designed for use by management and labor to fulfill a requirement in the OSHA guidelines for meat-packing plants. The aim is to screen jobs rapidly for harmful exposures, rather than to provide a diagnostic tool. Questions were designed based upon the biomechanical literature, structured into six sections. Scoring was based on simple presence or absence of a condition, or on a three-level duration score. As shown in Table 48.6, the two or three levels were scored as o, √ or * depending on the stress-rating built into the questionnaire. These symbols represented insignificant, moderate, or substantial exposures. A total score could be obtained by summing moderate and substantial exposures.

The Upper Extremity Checklist was designed to be biased toward false positives, i.e. to be very sensitive. It was validated against detailed analyses of 51 jobs by an ergonomics

Table 48.6 Upper Extremity Check List: Structure, Questions, and Scoring

A: Structure of the Checklist

Major Section	Examples of Questions
Worker Information	Which hand is dominant?
Repetitiveness	Repetitive use of the hands and wrists? If "yes" then: Is cycle < 30 s? Repeated for > 50% cycle?
Mechanical stress	Do hard or sharp objects put pressure localized pressure on: back or side of fingers? Palm or base of hand . . .
Force	Lift, carry, push or pull objects > 4.5 kg? If gloves worn, do they hinder gripping? . . .
Posture	Is pinch grip used? Is there wrist deviation? . . .
Tools, Handheld objects and equipment	Is vibration transmitted to the operator's hand? Does cold exhaust air blow on the hand or wrist? . . .

B: Scoring Scheme

Question	Scoring		
Is there wrist deviation?	No	Some	> 33% cycle
	O	√	*
C. Overall Evluation			
Total score	Number of √ + Number of *		

expert. Each section (except the first which only recorded dominant hand) was considered as giving a positive screening if at least one * rating was recorded. Across the various sections, there was reasonable agreement between checklist users and the expert analysis, with the checklist being generally more sensitive, as was its aim. The original reference shows the findings of the checklist when applied to 335 manufacturing and warehouse jobs.

As a special purpose technique in an area of high current visibility for human factors, the Upper Extremity Checklist has proven validity, can be used by those with minimal ergonomics training for screening jobs, and takes only a few minutes per workstation. The same team has also developed and validated a legs, trunk, and neck job screening procedure along similar lines (Keyserling, Brouwer, and Silverstein, 1992).

Ergonomic Checkpoints. The Workplace Improvement in Small Enterprises (WISE) methodology (Kogi, 1994) was developed by the International Ergonomics Association (IEA) and the International Labour Office (ILO) to provide cost-effective solutions for smaller organizations. It consists of a training program and a checklist of potential low-cost improvements. This checklist, called *Ergonomics Checkpoints*, can be used both as an aid to discovery of solutions and as an audit tool for workplaces within an enterprise.

The 128-point checklist has now been published (Kogi and Kuorinka, 1995). It covers the nine areas shown in Table 48.7. Each item is a statement rather than a question, and is called a "checkpoint." For each checkpoint there are four sections, also shown in Table 48.7. There is no scoring system as such; rather each checkpoint becomes a point of evaluation of each workplace for which it is appropriate. Note that each checkpoint also covers why that improvement is important, and a description of the core issues underlying it. Both of these help the move from rule-based reasoning to knowledge-based reasoning as nonergonomists continue to use the checklist. A similar idea was embodied in the Mir (1982) Ergonomic Checklist.

Other Checklists. The above sample of successful audit checklists has been presented in some detail to provide the reader with their philosophy, structure, and sample questions. Rather than continue in the same vein, other interesting checklists are outlined in Table 48.8. Each entry shows the domain, the types of issues addressed, the size or time taken in use, and whether validity and reliability have been measured. Most textbooks now provide checklists, and a few of these are cited. No claim is made that Table 48.8 is

Table 48.7 Ergonomic Checkpoints: Structure, Typical Checkpoints, and Checkpoint Structure

A: Structure of the Checklist

Major Section	Typical Checkpoints
Materials handling	• Clear and mark transport ways.
Handtools	• Provide handholds, grips or good holding points for all packages and containers.
Productive machine safety	• Use jigs and fixtures to make machine operations stable, safe and efficient.
Improving workstation design	• Adjust working height for each worker at elbow level or slightly below it.
Lighting	• Provide local lights for precision or inspection work.
Premises	• Ensure safe wiring connections for equipment and lights.
Control of hazards	• Use feeding and ejection devices to keep the hands away from dangerous parts of machinery
Welfare facilities	• Provide and maintain good changing, washing and sanitary facilities to keep good hygiene and tidiness.
Work organization	• Inform workers frequently about the results of their work.

B: Structure of each checkpoint

WHY?	Reasons why improvements are important.
HOW?	Description of several actions each of which can contribute to improvement.
SOME MORE HINTS	Additional points which are useful for attaining the improvement.
POINTS TO REMEMBER	Brief description of the core element of the checkpoint.

Source: (from Kogi, private communication, November 13, 1995).

Table 48.8 A Selection of Published Checklists. (First nine from Landau and Rohmert, 1989; next three from Berchem-Simon, 1993)

Name	Authors	Coverage	Reliab.	Validity
TBS	Hacker et al. (1983)	Mainly mental work		vs. AET
VERA	Volpert et al. (1983)	Mainly mental work		vs. AET
RNUR	RNUR. (1976)	Mainly physical work		
LEST	Guèlaud. (1975)	Mainly physical work		
AVISEM	AVISEM. (1977)	Mainly physical work		
GESIM	GESIM. (1988)	Mainly physical work		
RHIA	Leitner et al. (1987)	Task hindrances, stress	0.53–0.79	vs. many
MAS	Groth. (1989)	Open structure, derived from AET		vs. AET
JL and HA	Mattila and Kivi. (1989)	Mental, physical work, hazards	0.87–0.95	
	Bolijn. (1993)	Physical work checklist for women	tested	
	Panter. (1993)	Checklist for load handling		
	Portillo Sosa. (1993)	Checklist for VDT standards		
Work analy.	Pulat. (1992)	Mental and physical work		
Thermal aud.	Parsons. (1992)	Thermal audit from heat balance		content
WAS	Yoshida and Ogawa (1991)	Workplace and environment	tested	vs. expert
Ergonomics	SHARE (1990)	Short workplace checklists		
	Cakir et al. (1980)	VDT checklist		

comprehensive, but it is rather a sampling with references so that readers can find a suitable match to their needs. The first nine entries in the table are conveniently colocated in Landau and Rohmert (1989). Many of their reliability and validity studies are reported in this publication. The next entries are results of the Commission of European Communities fifth ECSC programme, reported in Berchem-Simon (1993). Others are from texts and original references. The author has not personally used all of these checklists so cannot specifically endorse them. Also, omission of a checklist from this table implies nothing about its usefulness.

48.3.2.2 Other Data Collection Methods

Not all data comes from checklists and questionnaires. We can audit a human factors program using outcome measures alone (e.g., Chapter 44). However, outcome measures such as injuries, quality, and productivity are nonspecific to human factors: Many other external variables can affect them. An obvious example is changes in the reporting thes-hold for injuries, which can lead to sudden apparent increases and decreases in the safety of a department or plant. Additionally, injuries are (or should be) extremely rare events. Thus, to obtain enough data to perform meaningful statistical analysis may require aggregation over many disparate locations and/or time periods. In ergonomics audits, such outcome measures are perhaps best left for long-term validation, or for use in selecting cluster samples.

Besides outcome measures, interviews represent a possible data collection method. Whether directed or not (e.g., Sinclair, 1990) they can produce critical incidents, human factors examples, or networks of communication (e.g., Drury, 1990a) which have value as part of an audit procedure. Interviews are routinely used as part of design audit procedures in large-scale operations such as nuclear power plants (Kirwan, 1989) or naval systems (Malone, Baker, and Permenter, 1988).

A novel interview-based audit system was proposed by Fox (1992) based on methods developed in British Coal (reported by Simpson, 1994). Here an error-based approach was taken, using interviews and archival records to obtain a sampling of actual and possible errors. These were then classified using Reason's (1990) active/latent failure scheme and orthogonally by Rasmussen's (1987) skill-, rule-, knowledge-based framework. Each active error is thus a conjunction of skill/mistake/violation with skill/rule/

knowledge. Within each conjunction, performance shaping factors can be deduced, and sources of management intervention listed. This methodology has been used in a number of mining-related studies: Examples will be presented in 48.4.

48.3.3 Data Analysis and Presentation

Human factors as a discipline covers a wide range of topics from workbench height to function allocation in automated systems. An audit program can only hope to abstract and present a part of this range. With our consideration of sampling systems and data collection devices we have seen different ways in which an unbiased abstraction can be aided. At this stage the data consist of large numbers of responses to large numbers of checklist items, or detailed interview findings. How can, or should, these data be treated for best interpretation?

Here there are two opposing viewpoints: One is that the data are best summarized across sample units, but not across topics. This is typically the way the human factors professional community treats the data, giving summaries in published papers of the distribution of responses to individual items on the checklist. In this way, findings can be more explicit, for example that the lighting is an area which needs ergonomics effort, or that the seating is generally poor. Adding together lighting and seating discrepancies is seen as perhaps obscuring the findings rather than assisting in their interpretation.

The opposite viewpoint, in many ways, is taken by the business community. For some, an overall figure of merit is a natural outcome of a human factors audit. With such a figure in hand, the relative needs of different divisions, plants, or departments can be assessed in terms of ergonomic and engineering effort required. Thus, resources can be distributed rationally from a management level. This view is heard by those who work for manufacturing and service industries who ask after an audit "How did we do?" and expect a very brief answer. The proliferation of the spreadsheet, with its ability to sum and average rows and columns of data, has encouraged people to do just that with audit results. Repeated audits fit naturally into this view as they can become the basis for monthly, quarterly, or annual graphs of ergonomic performance.

Neither view alone is entirely defensible. Of course, summing lighting and seating needs produces a result which is logically indefensible, and which does not help diagnosis. But equally, decisions must be made concerning optimum use of limited resources. The human factors auditor, having chosen an unbiased sampling scheme and collected data on (presumably) the correct issues, is perhaps in an excellent position to assist in such management decisions. But so too are other stakeholders, primarily the workforce.

However, audits are not the only use of some of the data collection tools. For example, the Keyserling, et al. (1993) Upper Extremity Checklist was developed specifically as a screening tool. Its objective was to find which jobs/workplaces are in need of detailed ergonomic study. In such cases, summing across issues for a total score has an operational meaning, i.e., that a particular workplace needs ergonomic help.

Where interpretation is made at a deeper level than just a single number, a variety of presentation devices have been used. These must show scores (percent of workplaces, distribution of sound pressure levels, etc.) separately but so as to highlight broader patterns. Much is now known about separate versus integrated displays and emergent features (e.g., Wickens, 1992, pp. 121–122), but the traditional profiles and spider's web charts are still the most usual presentation forms. Thus, Wagner (1989) shows the AVISEM profile for a steel industry job before and after automation. The nine different issues ("Rating Factors") are connected by lines to show emergent shapes for the old and the new jobs. Landau and Rohmert's (1981) original book on AET shows many other examples of profiles. Klimer, Kylian, Schmidt, and Rutenfranz (1989) present a spider web diagram to show how three work structures influenced ten issues from the AET analysis. Mattila and Kivi (1989) present their data on the job load and hazard analysis system applied to the building industry in the form of a table. For six occupations, the rating on five different loads/hazards is presented as symbols of different sizes within the cells of the table.

There is little which is novel in the presentation of audit results: Practitioners tend to use the standard tabular or graphical tools. But audit results are inherently multidimensional so that some thought is needed if the reader is to be helped toward an informed comprehension of the audit's outcome.

48.4 AUDIT SYSTEMS IN PRACTICE

Almost any of the audit programs and checklists referenced in previous sections give examples of their use in practice. Only two examples will be given here as others are readily accessible. These examples were chosen as they represent quite different approaches to auditing.

48.4.1 Auditing a Multiplant Company

Using the checklist developed for another multiplant company, Drury (1990a) audited a company with three plants making automotive subassemblies. The audit was part of a larger assessment of the company's competitiveness, prior to interventions through a regional economic development agency. Management and unions cooperated on this larger assessment, which covered strategic plans, quality, manufacturing methods, labor relations, and workforce training in addition to ergonomics. This use of ergonomics within an economic development context has been described by Drury, Kleiner, and Zahorjan (1989). The following section covers the use of the Workplace Survey (quoted from Drury, 1990a).

At the three plants, 76 workplaces were chosen for the Workplace Survey. The selection was based on observation and photographs of all jobs in the plants. True randomness was not used, but rather workplaces were chosen to represent the various activities (e.g., assembly, machine operating, materials handling, inspection) within clusters, typically different production lines within each plant. In this way questions of interest to the plant could be answered from survey data (e.g., "has new technology led to better ergonomics?" or "is line 15 different from line 27, which is the model for future lines?"). No attempt was made to include or exclude jobs suspected of high injury rates. Surveys were conducted using the current workplace incumbents, and included operators from all three shifts.

The results can only be summarized here, and will be presented by section of the Survey. Overall, there were few significant differences between the three plants, or between lines within the plants, indicating both a uniform level of ergonomics application (or lack of it!) and no measurable improvement in new lines.

Visual aspects: The distribution of lighting levels at the task, at midfield, and outer field showed satisfactory mean values, but high variance. Level of luminance was too low in 37% of workplaces and only 36% of workplaces had progressively decreasing luminance from task to mid- to outer field.

Auditory aspects: While the mean noise level was less than the OSHA limit, 42% of workstations were above the 8 hr limit of 85 dBA. All were above 75 dBA so that communications interference could be expected.

Thermal aspects: For the level of energy expenditure in the tasks, even for sedentary work, the thermal environment was too hot and too humid for comfort, but below the levels of potential heat stress.

Instruments/controls/displays: Apart from a new workplaces lacking control shape coding and adequate label sizes, the survey showed that human–machine communication was well up to ergonomics standards.

Design of workplaces: Very few workplaces met ergonomics standards. Chairs were of poor design, footrests absent, and knee room under most conveyors was nonexistent. The results were seen in excessive bending, twisting, reaching, and awkward hand–arm postures. In the Repetitive Tasks section, most jobs were found to have high repetition rates associated with these awkward postures, thus predisposing operators to repetitive trauma injuries (Putz-Anderson, 1988).

Manual materials handling (MMH) and energy expenditure: Energy expenditures were within typical recommendations, but forces on the body, estimated from static biomechanical models, were high. Ten percent of the 21 jobs where MMH was evaluated gave disc compressive forces exceeding 2500N and 85% exceeded 1000N. In terms of the NIOSH Action Limit concept (NIOSH, 1981), 19% had actual loads exceeding the Action Limit.

Inspection tasks: Training for inspection was found to be inadequate in all plants.

Conclusions from this study were that relatively simple changes to the visual environment were possible and desirable, but less noisy equipment would have to wait for new designs meeting ergonomics criteria. A major effort in workplace design was recommended, and demonstration projects were started in all plants as a final part of the assessment. MMH analyses were recommended as standard procedure for all new jobs and equipment.

While the Workplace Survey was being completed, two parallel surveys were conducted to determine how and where ergonomics was being used in the company. Interviews were conducted with providers of ergonomics and users of ergonomics: Results from these latter will be presented here. At both corporate and plant levels, two types of users were interviewed: line managers and support staff. They were asked the same set of open-ended questions, shown as table headings 1 to 5 in Table 48.9. Question 4 asked for details of their most recent ergonomics project, so that data from question 4 is omitted from Table 48.9. Not all respondents answered all questions, and some gave multiple answers. All comments and answers were classified into the subheadings shown in Table 48.9.

These interviews assisted the audit process by assessing knowledge and utilization of ergonomics at both corporate and plant levels. For example, at the plant level, most users

Table 48.9 Responses to Ergonomics Users

Question and Issue	Corporate		Plant	
	Mgt	Staff	Mgt	Staff
1. What is Ergonomics?				
1.1 Fitting job to operator	1	6	10	5
1.2 Fitting operator to job	0	6	0	0
2. Who do you call on to get ergonomics work done?				
2.1 Plant Ergonomics people	0	3	3	2
2.2 Division Ergonomics people	0	4	5	2
2.3 Personnel Department	3	0	0	0
2.4 Engineering Department	1	8	6	11
2.5 We do it ourselves	0	2	1	0
2.6 College Interns	0	0	4	2
2.7 Vendors	0	0	0	1
2.8 Everyone	0	1	0	0
2.9 Operators	0	1	0	0
2.10 University faculty	0	0	1	0
2.11 Safety	0	1	0	0
3. When did you last ask them for help?				
3.1 Never	0	4	2	0
3.2 Sometimes/Infrequently	2	0	1	0
3.3 1 year or more ago	0	1	4	0
3.4 one month or more ago	0	0	2	0
3.5 less than 1 month ago	1	0	3	4
5. Who else should we talk to about ergonomics?				
5.1 Engineers	0	0	3	2
5.2 Operators	1	1	2	0
5.3 Everyone	0	0	2	0
6. General Ergonomics Comments				
6.1 Ergonomics Concerns				
6.11 Workplace design for safety/ ease/stress/fatigue	2	5	13	5
6.12 Workplace design for cost savings/productivity	1	0	2	1
6.13 Workplace design for worker satisfaction	1	1	0	1
6.14 Environment design	2	1	3	0
6.15 The problem of finishing early	0	0	1	1
6.16 The Seniority/Bumping problem	0	3	1	0
6.2 Ergonomics Program Concerns				
6.21 Level of reporting of ergonomics	0	1	7	0
6.22 Communication/who does erognomics	7	1	4	0
6.23 Stability/staffing of ergonomics	0	0	10	4
6.24 General evaluation of ergonomics				
Positive	1	3	3	4
Negative	4	10	10	3
6.25 Lack of financial support for ergonomics	0	0	1	0
6.26 Lack of priority for ergonomics	2	2	1	4
6.27 Lack of awareness of ergonomics	2	1	6	1

saw ergonomics as fitting the job to the operator (1.1), whereas at the corporate level, the staff at least were equally divided between this definition and a more personnel-oriented one (1.2). To get ergonomics work done, users went both to ergonomics specialists and to the engineering department, although the latter had in fact little ergonomics knowledge. Most revealing was the distribution of time since the last ergonomics involvement (question 3). Corporate people had rarely used ergonomics, whereas at the plant level, there was much more frequent involvement.

The general comments told much about both ergonomics itself and the company's program. Most users saw ergonomics as design for human well-being (6.11) rather than design for performance (6.12). Although the ergonomics program gave generally positive comments (6.24), it was still seen as having low priority and awareness (6.26, 6.27). At the plant level, staffing was a major problem (6.23) as people were rapidly rotated through ergonomics positions. Finally, communications were seen as poor, particularly at the corporate level (6.22).

Clearly, although users were aware of ergonomics, it was not being fully utilized. Specific recommendations were given to the company to improve the use of ergonomics at corporate and plant level. As a result of the overall competitiveness evaluation, many of these organizational changes have been implemented. The company has the Workplace Survey and user–provider interview methodologies, so that the future status of ergonomics within the plant can be tracked.

48.4.2 Error Reduction at a Colliery

In a 2-yr project, reported by Simpson (1994) and Fox (1992), the Human Error Audit described in Section 48.3.2 was applied to two colliery haulage systems. The results of the first study are presented here. In both systems, data collection focused on potential errors and the performance shaping factors (PSFs) which can influence these errors. Data was collected by "observation, discussion and measurement within the framework of the broader man-machine systems and checklist of PSFs," taking some 30–40 shifts at each site. The whole haulage system from surface operations to delivery at the coal-face was covered.

The first study found 40 Active Failures (i.e., direct error precursors) and nine Latent Failures (i.e., dormant states predisposing the system to later errors). Four broad classes of active failures were as follows:

1. Errors associated with loco-maintenance (7 errors) (e.g., fitting incorrect thermal cut-offs).
2. Errors associated with loco-operation (10 errors) (e.g., locos not returned to service bay for 24-hr check).
3. Errors associated with loads and load security (7 errors) (e.g., failure to use spacer wagons between overhanging loads).
4. Errors associated with the design/operation of the haulage route (10 errors) (e.g., continued use despite potentially unsafe track).
5. (plus a small miscellaneous category)

The Latent Failures were (Fox, 1992):

1. Quality assurance in supplying companies
2. Supplies ordering procedures within the colliery
3. Locomotive design
4. Surface "Make up" of supplies
5. Lack of equipment at specific points
6. Training
7. Attitudes to safety
8. The safety inspection/reporting/action procedures

As an example from 3: Locomotive Design, the control positions were not consistent across the locomotives fleet, despite all originating from the same manufacturer.

Using the slip/mistake/violation categorization, each potential error could be classified so that the preferred source of action (intervention) could be specified.

This audit led to the formation of two teams, one to tackle locomotive design issues and the other for safety reporting and action. As a result of team activities many ergonomic actions were implemented. These included management actions to ensure a uniform wagon fleet, autonomous inspection/repair teams for tracks, and multifunctional teams for safety initiatives.

The outcome was that the accident rate dropped from 35.40 per 100,000 person-shifts to 8.03 in 1 yr. This brought the colliery from worst in the regional group of 15 collieries to best in the group, and indeed in the United Kingdom. In addition, personnel indicators, such as industrial relations climate and absence rates, improved.

48.5 FINAL THOUGHTS ON HUMAN FACTORS AUDITS

An audit system is a specialized methodology for evaluating the ergonomic status of an organization at a point in time. In the form presented here, it follows auditing practices in the accounting field, and indeed in such other fields as safety. Data is collected (typically with a checklist), analyzed, and presented to the organization for action. In the final analysis, it is the action which is important to human factors engineers, as the colliery example above shows. Such actions could be taken using other methodologies, such as active redesign by job incumbents (Wilson, 1994): Audits are only one method of tackling the problems of manufacturing and service industries. But as Drury (1991) points out, industry's moves toward quality are making it more measurement driven. Audits fit naturally into modern management practice as measurement, feedback, and benchmarking systems for the human factors function.

REFERENCES

Alexander, D. C., and Pulat, B. M. (1985). *Industrial Ergonomics: A Practitioner's Guide.* Atlanta: GA: Atlanta, GA. Industrial Engineering and Management Press.

ASHRAE (1989). Physiological principles, comfort and health. *Fundamentals Handbook.* Chapter 8. Atlanta GA: American Society of Heating, Refrigerating and Air-Conditioning Engineers.

AVISEM (1977). Techniques d'Amélioration des Conditions de Travail dans l'Industrie. Suresnes-France: Editions Hommes et Techniques.

Ayoub, M. M., and Mital, A. (1989). *Manual Materials Handling.* London: Taylor and Francis.

Bainbridge, L., and Beishon, R. J. (1964). The place of checklists in ergonomic job analysis. *Proceedings of the 2nd I.E.A. Congress, Dortmund. Ergonomics Congress Proceedings Supplement.*

Berchem-Simon, O., Ed. (1993). *Ergonomics Action in the Steel Industry.* EUR 14832 EN. Commission of the European Communities, Luxembourg.

Bolijn, A. J. (1993). Research into the employability of women in production and maintenance jobs in steelworks. In O. Berchem-Simon., Ed., *Ergonomics Action in the Steel Industry.* EUR 14832 EN. Commission of the European Communities, Luxembourg, pp. 201–208.

British Standards Institution (1965). *Office Desks, Tables and Seating.* London: British Standard 3893.

Burger, G. C. E., and de Jong, J. R. (1964). Evaluation of work and working environment in ergonomic terms. *Aspects of Ergonomic Job Analysis*, pp. 185–201.

Carson, A. B. and Carlson, A. E. (1977). *Secretarial Accounting* (10th Edition). Cincinnati, OH: South Western Publishing Company.

Cakir, A., Hart, D. M. and Stewart, T. F. M. (1980). *Visual Display Terminals* (John Wiley and Sons, New York) 144-152, 159-190, App. I.

Chervak, S. and Drury, C. G., 1995. Simplified English validation, *Human Factors in Aviation Maintenance - Phase 6 Progess Report*, DOT/FAA/AM-95/xx, Federal Aviation Administration (Office of Aviation Medicine) (National Technical Information Service, Springfield, VA).

Degani, A., and Wiener, E. L. (1990). *Human Factors of Flight-Deck Checklists. The Normal Checklist*, NASA Contractor Report 177549. Ames Research Center, CA.

Department of Defense (1989). *Human engineering design criteria for military systems, equipment and facilities.* MIL-STD-1472D, Washington, D.C.

Dirken, J. M. (1969). An ergonomics checklist analysis of printing machines. *ILO (Geneve), 2,* 903–913.

Drury, C. G. (1990a). The ergonomics audit. In E. J. Lovesey, Ed., *Contemporary Ergonomics.* London: Taylor and Francis, pp. 400–405.

Drury, C. G. (1990b). Computerized data collection in ergonomics. In J. R. Wilson and E. N. Corlett, Eds., *Evaluation of Human Work. 8.1*, London: Taylor and Francis, pp. 200–214.

Drury, C. G. (1991). Errors in aviation maintenance: taxonomy and control. *Proceedings of the 35th Annual Meeting of the Human Factors Society*, San Francisco, CA, pp. 42–46.

Drury, C. G. (1992). Inspection performance. In G. Salvendy, Ed., *Handbook of Industrial Engineering. 88*, New York: John Wiley 2282–2314.

Drury, C. G., Kleiner, B. M., and Zahorjan, J. (1989). How can manufacturing human factors help save a company: intervention at high and low levels. *Proceedings of the Human Factors Society 33rd Annual Meeting*, Denver, CO, pp. 687–689.

Drury, C.G., Prabhu, P., and Gramopadhye, A. (1990). Task analysis of aircraft inspection activities: methods and findings. *Proceedings of the Human Factors Society 34th Annual Conference*, Santa Monica, CA, pp. 1181–1185.

Drury, C. G., and Wick, J. (1984). Ergonomic applications in the shoe industry. *Proceedings of the International Conference on Occupational Ergonomics, 1*, pp. 489–493.

Easterby, R. S. (1967). Ergonomics checklists: An appraisal. *Ergonomics, 1967, 10(5)*, 549–556.

Fanger, P. O. (1970). *Thermal Comfort, Analyses and Applications in Environmental Engineering.* Copenhagen: Danish Technical Press.

Fox, J. G. (1992). The ergonomics audit as an everyday factor in safe and efficient working. *Progress in Coal, Steel and Related Social Research*, 10–14.

GESIM (Groupement des Entreprises Sidérurgiques et Minières) (1988). Connaissance du poste de travail, II conditions de l'activitC. Metz: GESIM.

Givoni, B., and Goldman, R. F. (1972). Predicting rectal temperature response to work, environment, and clothing. *Journal of Applied Physiology, 32(6)*, 812–822.

Groth, K. M. (1989). The modular work analysis system (MAS), *Recent Developments in Job Analysis, Proceedings of the International Symposium on Job Analysis*, University of Hohenheim, March 14–15, New York: Taylor and Francis, pp. 253–261.

Guélaud, F., Beauchesne, M.-N., Gautrat, J., and Roustang, G. (1975). Pour une analyse des conditions de travail ouvrier dans l'entreprise. Recherche du Laboratoire d'Economie et de Sociologie du Travail C.M.R.S. Third Édition, Aix-en-Provence, Paris: Librairie Armand Colin.

Hacker, W., Iwanowa, A., and Richter, P. (1983). *Tätigkeitsbewertungssystem.* Berlin (OST): Psychodiagnostisches Zentrum.

Hasslequist, R. J. (1981). Increasing manufacturing productivity using human factors principles. *Proceedings of the Human Factors Society 25th Annual Conference*, Santa Monica, CA, pp. 204–206.

Illuminating Engineering Society (1993). *Lighting Handbook, Reference and Application*, Eighth Edition. New York: The Illuminating Engineering Society of North America.

ICAO (1989). *Human Factors Digest No. 1 Fundamental Human Factors Concepts.* Circular 216-AN/131. Canada: International Civil Aviation Organization.

ISO Publication (1987). *Assessment of Noise-Exposure during Work for Hearing Conservation Purposes*, Geneva.

Jones, D. M., and Broadbent, D. E. (1987). Noise. In E. Salvendy, Ed., *Handbook of Human Factors Engineering*, New York: John Wiley and Sons.

Kerlinger, F. N. (1964). *Foundations of Behavioral Research.* New York: Holt, Rinehart and Winston.

Keyserling, W. M., Stetson, D. S., Silverstein, B. A., and Brouwer, M. L. (1993). A checklist for evaluating ergonomic risk factors associated with upper extremity cumulative trauma disorders. *Ergonomics, 36*(*7*), 807–831.

Keyserling, W. M., Brouwer, M., and Silverstein, B. A. (1992). A checklist for evaluation ergonomic risk factors resulting from awkward postures of the legs, truck and neck. *International Journal of Industrial Ergonomics, 9(4)*, 283–301.

Kirwan, B. (1989). A human factors and human reliability programme for the design of a large UK nuclear chemical plant. *Proceedings of the Human Factors Society 33rd Annual Meeting--1989*, Denver, CO, pp. 1009–1013.

Klimer, F., Kylian, H., Schmidt, K.-H., and Rutenfranz, J. (1989). Work analysis and load components in an automobile plant after the implementation of new technologies. In K. Landau and W. Rohmert, Eds., *Recent Developments in Job Analysis.* New York: Taylor and Francis, pp. 331–340.

Kogi, K. (1994). Introduction to WISE (work improvement in small enterprises) methodology and workplace improvements achieved by the methodology in Asis, *Proceedings of the 12th Triennial Congress of the International Ergonomics Association*, Vol. 5, Human Factors Association of Canada, Toronto, CA, pp. 141–143.

Kogi, K., and Kuorinka, I., Eds. (1995 (in press)). *Ergonomic Checkpoints.* Switzerland: ILO Publications.

Koli, S. T. (1994). *Ergonomic Audit for Non-Repetitive Task.* Unpublished M.S. Thesis. State University of New York at Buffalo.

Koli, S., Drury, C. G., Cuneo, J., and Lofgren, J. (1993). Ergonomic audit for visual inspection of aircraft, *Human Factors in Aviation Maintenance - Phase Four, Progress Report, DOT/FAA/AM-93/xx.* Springfield, VA: National Technical Information Service.

Landau, K., and Rohmert, W. (1981). *Fallbeispiele zurArbeitsanalyse.* Switzerland: Verlag Hans Huber Bern Stuttgart Wien.

Landau, K., and Rohmert, W., Eds. (1989). *Recent Developments in Job Analysis. Proceedings of the International Symposium on Job Analysis*, University of Hohenheim, March 14–15, New York: Taylor and Francis.

Leitner, K. and Greiner, B. (1987). Assessment of job stress: the RHIA instrument. In *Recent Developments in Job Analysis. Proceedings of the International Symposium on Job Analysis*, K.

Landau and W. Rohmert (eds.) (1989), University of Hohenheim, March 14–15, New York: Taylor and Francis.

Malde, B. (1992). What price usability audits? The introduction of electronic mail into a user organization, *Behaviour & Information Technology, 11*(6), 345–353.

Malone, T. B., Baker, C. C., and Permenter, K. E. (1988). Human engineering in the naval sea systems command. *Proceedings of the Human Factors Society-32nd Annual Meeting—1988.* Anaheim, CA, Vol. 2, pp. 1104–1107.

Mattila, M., and Kivi, P. (1989). Job load and hazard analysis: A method for hazard screening and evaluation. In K. Landau and W. Rohmert, Eds., *Recent Developments in Job Analysis. Proceedings of the International Symposium on Job Analysis.* University of Hohenheim, March 14–15, New York: Taylor and Francis, pp. 179–186.

McClelland, I. (1990). Product assessment and user trials. In *Evaluation of Human Work*, J. R. Wilson and E. N. Corlett, Eds. New York: Taylor and Francis, 218–247.

McCormick, E. J. (1979). *Job Analysis: Methods and Applications.* New York: AMACO .

McCormick, W. T., Mecham, R. C. and Jeanneret, P. R. (1969). *The Development and Background of the Position Analysis Questionnaire*, Occupational Research Center (Purdue University, Lafayette).

Mir, A. H. (1982). *Development of Ergonomic Audit System and Training Scheme.* Unpublished M.S. Thesis, State University of New York at Buffalo.

Muller-Schwenn, H. B. (1985). Product design for transportation. *Ergonomics International, 85*, 643–645.

NIOSH (1981). *Work Practices Guide for Manual Lifting*, DHEW-NIOSH publication, 81–122, OH: Cincinnati.

Osburn, H. G. (1987). Personnel selection. In G. Salvendy, Ed., *Handbook of Human Factors.* New York: John Wiley and Sons, pp. 911–933.

Occupational Safety and Health Administration (OSHA) (1990). *Ergonomics Program Management Guidelines for Meatpacking Plants.* Publication No. OSHA-3121. U.S. Department of Labor, Washington, D.C..

Panter, W. (1993). Biomechanical damage risk in the handling of working materials and tools - analysis, possible approaches and model schemes. In O. Berchem-Simon, Ed., *Ergonomics Action in the Steel Industry.* EUR 14832 EN. Commission of the European Communities, Luxembourg.

Parsons, K.C. (1992). The thermal audit. A fundamental stage in the ergonomics assessment of thermal environment. In E. J. Lovesey, Ed., *Contemporary Ergonomics 1992.* London: Taylor and Francis, pp. 85–90.

Patel, S., Drury, C. G., and Prabhu, P. (1993). Design and usability evaluation of work control documentation. *Proceedings of the Human Factors and Ergonomics Society 37th Annual Meeting*, Seattle, WA, pp. 1156–1160.

Portillo Sosa, J. (1993). Design of a computer programme for the detection and treatment of ergonomic factors at workplaces in the steel industry. In O. Berchem-Simon, Ed, *Ergonomics Action in the Steel Industry.* EUR 14832 EN, (Commission of the European Communities, Luxembourg) 421–427.

Pulat, B. M. (1992). *Fundamentals of Industrial Ergonomics.* New Jersey: Prentice Hall.

Putz-Anderson, V. (1988). *Cumulative Trauma Disorders: A Manual for Musculo-Skeletal Diseases of the Upper Limbs. London: Taylor and Francis.*

Rasmussen, J. (1987). Reasons, causes and human error. In J. Rasmussen, K. Duncan, and J. Leplat, Eds., *New Technology and Human Error*, New York: John Wiley, 293–301.

Reason, J. (1990). *Human Error.* New York: Cambridge University Press.

Régie Nationale des Usines Renault (RNUR) (1976). Les profils de ostes. Méthode d'analyse des conditions de travail. Collection Hommes et Savoir. Masson, Sirtès, Paris.

Rohmert, W. and Landau, K. (1989). Introduction to job analysis. *A New Technique for Job Analysis, Part 1.* London: Taylor and Francis Ltd, pp. 7–22.

Rohmert, W., and Landau, K. (1983). *A New Technique for Job Analysis.* London: Taylor and Francis Ltd.

Seeber, A., Schmidt, K.-H., Kierswelter, E., and Rutenfranz, J. (1989). On the application of AET, TBS and VERA to discriminate between work demands at repetitive short cycle tasks. In K. Landau and W. Rohmert, Eds., *Recent Developments in Job Analysis*, New York: Taylor & Francis, pp. 25–32.

SHARE (1990). *Inspecting the Workplace.* SHARE Information Booklet. Australia: Occupational Health and Safety Authority.

Simpson, G. C. (1994). Ergonomic aspects in improvement of safe and efficient work in shafts. *Ergonomics Action in Mining, Eur 14831* (Commission of the European Communities, Luxembourg) 245–256.

Sinclair, M. A. (1990). Subjective assessment. In J. R.Wilson and E. N. Corlett, Eds., *Evaluation of Human Work*, London: Taylor and Francis, pp. 58–88.

Volpert, W., Oesterreich, R., Gablenz-Kolakovic, S., Krogoll, T., and Resch, M. (1983). *Verfahren zur Ermittlung von Regulationserfordernissen in der Arbeitsttigkeit (VERA).* Verlag TÜV Rheinland: Köhn.

Wagner, R. (1989). Standard methods used in French-speaking countries for workplace analysis. In K. Landau, K. and W. Rohmert, Eds., *Recent Developments in Job Analysis.* New York: Taylor and Francis, pp. 33–42.

Wagner, R. (1993). Ergonomic study of a flexible machining cell involving advanced technology. *Ergonomics Action in the Steel Industry*, EUR 14832. Luxembourg: Commission of the European Communities, pp. 157–170.

Waters, T. R., Putz-Anderson, V., Garg, A., and Fine, L. J. (1993). Revised NIOSH equation for the design and evaluation of manual lifting tasks. *Rapid Communications, Ergonomics, 1993, 36*(*7*), 749–776.

Wickens, C. D. (1992). *Engineering Psychology and Human Performance*, Second Edition. New York: Harper-Collins.

Wilson, J. R. (1994). A starting point for human-centered systems. *Proceedings of the 12th Triennial Congress of the International Ergonomics Association*, Toronto, Canada, *6*(*1*), 141–143.

Wright, P., and Barnard, P. (1975). Just fill in this form--a review for designers. *Applied Ergonomics*, *6*, 213–220.

Yoshida, H., and Ogawa, K. (1991). Workplace assessment guideline - checking your workplace. In W. Karwowski and J. W. Yates, Eds., *Advances in Industrial Ergonomics and Safety III.* London: Taylor and Francis. pp. 23–28.

CHAPTER 49

ASSESSING COST/BENEFITS OF HUMAN FACTORS

William B. Rouse
Enterprise Support Systems
Atlanta, GA 30071-4707 USA

Kenneth R. Boff
Armstrong Laboratory, AL/CFH
Wright-Patterson AFB, OH 45433-6573 USA

49.1 INTRODUCTION

This chapter is concerned with the problems of assessing the returns on investments in human factors. More specifically, both economic and noneconomic returns due to investments of monies in human factors research, development, and applications are of interest. The term *assessing* is meant to include both estimating past or current returns and predicting future returns.

It is essential at the outset to emphasize the importance of knowing why one is concerned with assessing costs/benefits. For example, are the results to be used as part of a resource allocation activity? Or, is such an assessment being used to evaluate past investments? As is later discussed, the problem is less one of assessing cost/benefit in general than it is one of supporting formulation and resolution of specific decision-making problems.

More technically, the purpose of the analysis should drive the representation of the problem in terms of:

- What attributes will be measured
- How these attributes will be mapped to commensurate metrics
- Whose preferences regarding these metrics are to be taken into account.

Once a representation is chosen, the calculation of cost/benefits is relatively straightforward. Put simply, the most difficult aspect of cost/benefit analysis is representation, not calculation.

Representation issues are particularly difficult for several reasons:

- Investments or costs almost always occur long before returns are realized—thus, benefits are delayed or denied because they are obscured by time or no longer relevant.
- The value of benefits received in the future usually is discounted due to the cost of the capital, e.g., interest paid or foregone, during the period prior to receiving these benefits, as well as uncertainty about benefits being realized.
- Benefits are seldom purely monetary—hence, value or utility has to be attributed to benefits that are not readily converted to monetary units.
- Costs and benefits often present accounting and measurement problems, and especially prediction problems—therefore, agreement on how costs and benefits are to be calculated in principle often leaves substantial problems in practice.
- Costs and benefits are often distributed among a wide range of stakeholders with differing values, concerns, and perceptions—consequently, means of comparing the relative utilities of gains and losses across stakeholders are needed.

These problems are certainly not unique to the discipline of human factors. In fact, these problems are ubiquitous. For example, Gibson and Rogers (1994) discuss the difficulties of assessing the benefits of investments in electronics R&D. Harris (1994) and Watterson (1995) discuss the problems of assessing the benefits of investments in information technology.

There are several reasons why these problems should be addressed. All disciplines need to communicate—present their case—in the context of design and evaluation of products and systems. This context includes business plans, product plans, and economic considerations in general. The ability to communicate successfully in this context enables disciplines to gain organizational roles and associated resources. This also builds "buy in" for the results of the discipline's efforts.

Beyond the difficulties experienced by all disciplines, human factors faces additional problems. Numerous people have addressed these problems in the context of the economic value added of human factors (e.g., Beevis, 1994; Corlett, 1988; Lane, 1987; Rouse & Cacioppo, 1989). Lane provides a detailed summary of the problems faced by human factors, especially as they relate to human factors engineering within military systems. He primarily focuses on the reasons why determination of costs and benefits, or cost-effectiveness, of human factors applications is very difficult:

- Benefits can be difficult to link to specific human factors activities independent of other activities.
- Benefits may be most closely related to events that do not happen, such as accidents that are avoided by good human factors design.
- Benefits may primarily be the freeing up of resources that can be used to add value elsewhere and consequently are not directly linked to human factors contributions.
- Benefits may be quite localized and obscured relative to more global measures of return.
- Benefits of good systems design are difficult to assess in general, and consequently, human factors contributions are methodologically difficult to assess.

To a great extent, the difficulties summarized by Lane reflect the role that human factors has often played in the design of products and systems. Rouse (1986) reports on this role based on interviews of aerospace crew system designers. In this study, which eventually involved a population sample of 240 people, designers in several aerospace companies were asked what role human factors played in their design efforts.

The designers interviewed termed human factors professionals "the cold water crowd" due to their inclinations to point out why concepts would not work from a human factors point of view. However, when human factors professionals were asked what should be done, designers reported that these people had "empty guns." They seldom could provide guidance on how something should be designed. Hence, Rouse's report was entitled "Cold Water and Empty Guns."

This conclusion implies less about the inherent value of human factors than it does about the usual organizational role of human factors. It is quite common for human factors people to be called in after a design concept has been detailed to assess the extent of any "people problems." If they find problems, the recipients of this information perceive the cold water effect. If they do not find any problems, they are perceived as adding no value. Human factors inputs need to be much earlier if they are to be perceived as adding value to the design—usually in terms of design features attributable to human factors efforts—and hence have potential for contributing auditable benefits for the costs that they incur.

Human factors is also strongly affected by its roots in the defense industry. This industry has little tradition of economically justifying its efforts. Usually the focus has been on meeting requirements rather than providing auditable value. In this context, human factors in its after-the-fact role has often focused solely on whether or not requirements will be compromised by human-related issues and problems. However, with the rest of the economy, the defense industry is changing and economic justifications are becoming increasingly crucial.

The foregoing implies that human factors benefits primarily occur via features of designed products and systems. However, many human factors activities are much broader and not always directly related to design. For example, human factors R&D produces human performance data and standards, as well as concepts for training and aiding personnel in a variety of systems. Applied R&D produces design guidance and tools. In these cases, eventual benefits may be far removed from initial human factors activities.

Thus, any cost/benefit accounting scheme has to be able to deal with both direct and indirect linkages between investments and return. To illustrate how such a wide range of accounting problems can be handled, this chapter later addresses four archetypical problems:

- Relating investments in health and safety improvements to cost savings and productivity increases.
- Linking features of individual products—for example, "widgets" bought and used by consumers—to benefits experienced by end users.
- Relating human factors activities associated with developing complex systems—for example, military weapon systems—to mission-related benefits.
- Demonstrating how the products of R&D—for example, understanding perceptual processes underlying virtual reality—provide accountable benefits.

These four examples serve to illustrate use of the methodology advanced in this chapter, as well as to show the nuances and difficulties of assessing cost/benefits of human factors applications.

The foregoing discussion begs the question of why cost/benefit assessments are needed. Why is human factors being challenged in terms of costs and benefits? The answer is straightforward. The 1990s have emerged as a period when all activities of all enterprises are being scrutinized. Everything is a target for delayering, downsizing, and rightsizing. Human factors is not being singled out. Instead, it is facing the same challenge as all disciplines. Cost/benefit analysis can provide the basis for addressing this challenge.

49.2 OVERVIEW OF COST/BENEFIT ANALYSIS

There is an immense literature related to cost/benefit analysis. Sage (1995) provides an overview of this literature. He differentiates cost/benefit and cost-effectiveness assessments. The concept of "effectiveness" is used when benefits cannot be expressed in purely economic metrics. While considerable attention is devoted to noneconomic returns in this chapter, the phrase cost/benefit is retained to keep things simple.

Sage notes the difficulty of assessing benefits for public sector activities. Hence, cost/benefit analyses in these activities usually focus on choosing the best program rather than justifying that benefits exceed costs. The goal in these types of analyses involves determining the degree to which objectives are achieved rather than questioning the economic value of achieving these objectives. He also notes the difficulty of assessing the costs/benefits of R&D investments due to great uncertainty about the nature of outcomes—never mind the benefits of these outcomes. A related problem concerns the assessment of costs because it often is not clear when investments started.

Most, and perhaps all, of the issues raised by Sage are relevant to assessing costs/benefits of human factors activities. For example, human factors investments are often associated with improving public safety in, for instance, air transportation. As another illustration, human factors investments are often made to improve broad knowledge of human abilities and limitations and, in this way, contribute to better system and product design for yet to be determined domains of application. Of particular interest, virtually all of Sage's issues apply to human factors despite the fact that human factors was not one of the topics of his treatment of cost/benefit analysis. Clearly, many endeavors suffer from the same assessment problems that are endemic to human factors.

It is essential that we differentiate the overall problems of assessment from the specific issue of calculation. Equation (1) summarizes the basic calculations. Given projections of costs, c_i, $i = 0, 1, \ldots N$, and returns, r_i, $i = 0, 1, \ldots N$, the calculations of Net Present Value (NPV), Internal Rate of Return (IRR), or Cost/Benefit Ratio (CBR) are quite straightforward elements of financial management (Brigham and Gapenski, 1988). The only subtlety is choosing a discount rate to reflect the expected cost of capital, CC.

$$\text{NPV} = \sum_{i=0}^{N} (r_i - c_i)/(1 + \text{CC})^i$$

$$\text{IRR} = \text{CC such that} \sum_{i=0}^{N} (r_i - c_i)/(1 + \text{CC})^i = 0$$

$$\text{CBR} = [\sum_{i=0}^{N} c_i/(1 + \text{CC})^i]/[\sum_{i=0}^{N} r_i/(1 + \text{CC})^i]$$

$$\text{CC} = \text{cost of capital} \tag{1}$$

NPV reflects the amount one should be willing to pay now for benefits received in the future. These future benefits are discounted by the cost of the capital, CC, invested now to receive these later benefits. IRR, in contrast, is the value of CC if NPV is zero. This metric enables comparing alternative investments by forcing the NPV of each investment to zero. The CBR simply reflects the discounted cash outflows divided by the discounted cash inflows.

These calculations become more complicated when returns are not readily transformable to economic terms. Benefits such as safety, quality of life, and aesthetic value are very difficult to translate into strictly monetary values. Multiattribute utility models provide a means for dealing with such situations. Noneconomic costs and returns are transformed to common utility scales using $u(c_i)$ and $u(r_i)$, $i = 0, 1, \ldots N$. These utility functions serve as inputs to the overall utility calculations shown in Equation (2) (Keeney and Raiffa, 1976).

$$U(c,r) = U[u(c_1),u(c_2), \ldots u(c_N), u(r_1), u(r_2), \ldots u(r_N)] \tag{2}$$

Note that the time value of benefits depicted in Equation (1) can be included in the equations on Equation (2) in two ways. One approach involves assessing utility functions for discounted benefits, possibly discounted as shown in Equation (1). Another approach is to include time explicitly in the assessment of utility functions. The benefit of the latter approach is that people are forced to deal with capital costs and uncertainty separately and explicitly, while the use of a discount rate, CC, tends to confound these two factors.

These mappings from c_i and r_i to $u(c_i)$ and $u(r_i)$, respectively, enable dealing with the subjectivity of preferences for noneconomic returns. In other words, utility theory enables one to quantify and compare things that are often perceived as difficult to objectify. Unfortunately, models based on utility theory do not always reflect the ways in which human decision making actually works. One needs to draw upon psychology to help this construct better depict human decision making.

Feather (1982) discusses the psychology that is needed. In particular, one needs to deal with the fact that people's perceptions of probabilities usually differ from what an outside observer might characterize as objectively "correct" probabilities. People have expectations. They attribute cause to the events they observe. They have mental models of the ways in which the world works.

Subjective expected utility (SEU) theory reflects these human tendencies. Thus, to the extent that one accepts that perceptions are reality, one needs to consider the SEU point of view when one makes expected utility calculations. In fact, one probably should make these calculations using both objective and subjective probabilities to gain an understanding of the sensitivity of the results to perceptual differences.

Once one admits the subjective, one needs to address the issue of whose perceptions are considered. Most decisions involve multiple stakeholders—in other words, people who hold a stake in the outcome of a decision. It is, therefore, common for multiple stakeholders to influence a decision. Consequently, the cost/benefit calculation needs to take into account multiple sets of preferences. The result is a group utility model, as shown in Equation (3) (Keeney and Raiffa, 1976; Kirkwood, 1979).

$$U = U[U_1(c,r), U_2(c,r), \ldots U_M(c,r)] \tag{3}$$

Formulation of such a model requires that two important issues be resolved. First, mappings from attributes to utilities must enable comparisons across stakeholders. In other words, one has to assume that $u = 0.8$, for example, implies the same value gained or lost for all stakeholders, although the mapping from attribute to utility may vary for each stakeholder. Thus, all stakeholders may, for instance, have different needs or desires for safety and, hence, different utility functions. They also may have different time horizons within which they expect benefits. However, once the mapping from attribute to utility is performed and utility metrics are determined, one has to assume that these metrics can be compared quantitatively.

The second important issue concerns the relative importance of stakeholders. Equation (3) implies that the overall utility attached to each stakeholder's utility can differ. For example, it is often the case that primary stakeholders' preferences receive more weight than the preferences of secondary stakeholders. The difficulty of this issue is quite straightforward. Who decides? Is there a super stakeholder, for instance. Do the groups of stakeholders, or their representatives, simply vote on who gets how much weight? Such a procedure has its own theoretical problems which cannot be addressed here.

It should be very clear from the foregoing that the context of the decision to be made has a huge impact on how cost/benefit calculations should be pursued. What attributes over what time periods are important? Whose preferences are important? What is the relative importance of each set of preferences? Answers to these questions are very much context dependent.

Beyond formulating how cost/benefit calculations should be made, there is the central problem of determining the model parameters and attribute values—the inputs to the calculations. There are well-developed procedures for identifying parameters of multi-attribute utility models (Keeney and Raiffa, 1976). There are also a wide variety of methods for forecasting levels of attributes, at least for attributes that exhibit some pattern of continuity. It is common to employ sensitivity and "what if" analyses to assess the impact of uncertainties on overall cost/benefit calculations, as well as the impact of these uncertainties on alternative decisions.

These procedures are very useful when there is uncertainty about the exact attribute levels that will result if a particular decision is made. Indeed, the origins of utility theory are based on dealing with such uncertainties. It is very different, however, when one is uncertain about what attributes will be affected. Such is the case, for example, when one considers the returns of investing in research on human errors in complex systems. The results of such research are likely to affect a variety of unknown future systems in unknown ways. Accounting for such returns is very difficult. In many domains, consequently, such investments have often been made based on "faith" in R&D as a process. In an era of downsizing, rightsizing, etc., such faith is stretched pretty thin and R&D must show how it provides strategic business advantages (Roussel, Saad, and Erickson, 1991).

Summarizing this overview of cost/benefit analysis, the necessary calculations are quite straightforward. There can be some difficulty in obtaining estimates of parameters and attribute levels for these calculations. Further, the real need is for forecasted estimates, not just assessed estimates of current or past parameters and levels.

The potential for differing, and perhaps conflicting, preferences among stakeholders complicates the situation substantially. Available means for dealing with these differences

Table 49.1 Types of Human Factors Products

- Solutions (very tangible)
- Tools
- Methods
- Guidance (e.g., principles and standards)
- Information (e.g., data, technology base)
- Advice
- Influence (less tangible)

are assumption laden, but nevertheless useful. Probably the most important aspect of dealing with these types of problems involves understanding the decision-making context within which cost/benefit analysis is to play a role. Seldom is it the case that the resulting cost/benefit metrics will dictate decisions independent of other factors.

49.3 PRODUCTS, BENEFITS, AND COSTS

In Table 49.1, types of human factors products are listed. As indicated, they range from very tangible to much less tangible. A specific problem solved—an unreachable input device made reachable—is very tangible. Tools that enable assessment of reachability are also tangible, but less direct in that the problems that will be solved with these tools are at least partially unknown. Methods for assessing reachability are even less direct in that there is likely to be considerable variation in how users of these methods implement them.

Guidance in terms of principles and standards, as well as information in the form of data and the technology base in general, are the typical goals of much of human factors R&D. These types of products tend to be more abstract—relative to specific problems solved—and their impact is also much more indirect. Consequently, as might be imagined, accounting for benefits becomes more difficult.

Advice can be a fairly amorphous product, and influence more so. This is especially the case in terms of measuring benefits. These types of products typically are packaged as the services of experts who can advise and advocate human-related issues in system and product design, development, and evaluation. The idea that the availability of services from knowledgeable and skilled people constitute products is a fairly abstract notion. Not surprisingly, explicitly accounting for the benefits of having such people available can be rather difficult.

The notion that availability of expertise provides benefits is similar to the idea that libraries provide benefits. Most of the potential benefits of libraries are never realized—for example, many books are never consulted or checked out. Further, the benefits received by those who do consult or check books out are often quite different than might have been imagined by, for instance, the authors who wrote the books or the librarians who acquired them. Nevertheless, many people would readily agree to the notion that such benefits are real.

The products in Table 49.1 tend to provide one or more of the types of benefits listed in Table 49.2 Activities that result in solving a problem that could not otherwise be solved provide clear benefits. For example, if the reachability problem had not been solved and the airplane could not be flown, the benefit is obvious.

Table 49.2 Types of Human Factors Benefits

- Solution not otherwise possible (very tangible)
- Acceptable performance/cost not otherwise possible
- Performance/cost improvements
- Enhanced customer/user perceptions and willingness to pay
- Cost avoidance
- Mishap avoidance
- Increased confidence (less tangible)

Less obvious, however, is the economic value of that benefit. Does, for example, the activity that resulted in solving the reachability problem get the credit for the total economic value of the airplane which could not otherwise be used? If the answer is "yes," then what value is attributed to other elements of the aircraft—the tires, for example, without which the airplane could not take off?

The difficulty of answering these questions becomes compounded for the much larger number of situations where investments are made to ensure that acceptable performance /cost results. Investments are also made to gain performance/cost improvements beyond the level of minimal acceptability. In some contexts, increments of performance and cost can be converted to the economic terms of the equations in Equation (1) and used to calculate cost/benefit. However, in a surprisingly large number of cases, it is difficult to convert performance gains to economic returns, and the equations on Equations (2) and (3) are appropriate.

Another type of benefit is enhanced customer/user perceptions and willingness to pay. Investments in improving usability of consumer products tend to provide these types of benefits. While accounting for cost/benefits in such cases would seem straightforward, a problem quickly emerges of attributing benefits to human factors activities versus marketing and sales efforts, or product support activities. As much of the foregoing illustrates, the problem of attribution of benefits is pervasive and affects all disciplines. However, once certain activities are accepted as the standard way of doing things, we usually no longer try to justify them. Thus, attribution of benefits is often not a concern.

A fairly clear type of benefit is cost avoidance. The ability to account for this benefit is determined by the ability to account for costs that would have occurred without human factors activities. While one could claim that the solution would have been, for instance, twice as expensive as what it was without human factors involvement, in many cases the economics of the situation would not have allowed this to occur. People, for example, simply would not buy such an expensive product and, hence, it would not have been developed.

Even more complicated is accounting for benefits due to mishaps avoided. Do the redesigned control room panels get full credit for avoiding a nuclear power plant meltdown? Does the touchpad in the new Boeing 777 get full credit for avoiding a crash? In this situation, we encounter the dual problems of estimating the frequencies with which things do *not* happen, as well as attributing benefits among the many investments that helped ensure that these things did *not* happen.

Perhaps the least tangible benefit produced by human factors activities is increased confidence in airplanes, power plants, medical devices, and so on. As a consequence, people fly in airplanes, let power plants get commissioned, and go to hospitals. While the phenomenon is clear, it certainly is not clear how one should account for the benefits received by the airlines, utilities, hospitals, and of course, consumers of these enterprises' products and services.

It is very important to emphasize that the lists in Tables 49.1 and 49.2 are not at all specific to human factors activities. Every discipline has problems demonstrating benefits and, in particular, isolating those benefits that are solely attributable to the discipline's activities. Thus, while this chapter focuses on human factors, the concepts and methods discussed are widely applicable.

As one moves from more tangible to less intangible in both Tables 49.1 and 49.2, the investment that one is willing to make is quite likely to substantially decrease. This begs the question of the nature of investments or, more generally, costs. While costs are usually more straightforward than benefits—that is, easier to account for—there are a variety of complications.

As discussed by Orlansky and String (1981) in the context of training simulators, there are issues of both scope and time. Scope concerns accounting for both direct and indirect costs which raises the usually complicated issue of attributing indirect costs. For example, what proportion of the maintenance costs for the roads on a military base should be attributed to the costs of performing R&D in a laboratory housed in a building on that base?

The issue of time concerns when the cost clock is started and stopped. Should the costs of a human factors solution include a portion of the R&D costs that produced the principles and data—usually many years earlier—that enabled creating the solution? From a practical point of view, such costs are usually not counted. However, such long-term payoffs are likely to be one of the primary reasons why R&D investments are made. If such benefits are to be counted, it seems reasonable to also count the costs. Clearly, there is substantial potential for accounting nightmares.

There are further complications in capturing costs and linking them to benefits. For example, how should one account for use of sunk costs—for instance, the aforementioned costs of the roads on a military base? Such investments were often made in the distant past. However, many ongoing costs are also treated as sunk costs. For instance, the marginal costs of assigning personnel to an effort are often discounted—or not counted—because such personnel are viewed as sunk costs. If such sunk costs are not counted, cost/benefit assessments will yield much more attractive results than a true accounting would provide.

Another issue concerns possible synergies among multiple investments. Costs may be higher, and the benefits lower, if only a subset of desired investments are made. Further, it is quite possible that overall costs may have to be increased somewhat to increase benefits substantially and, thereby, minimize cost/benefits—or maximize benefits per unit cost. Here again, the issue of scope arises when accounting for both costs and benefits.

In light of the foregoing discussion of issues and difficulties, it should be clear that cost/benefit analysis can become very complicated. How can one cope with this complexity? How can one keep track of the various primary, secondary, tertiary, etc. costs and benefits? The simple answer is that such is not necessary.

If one looks at the decision-making context within which a cost/benefit analysis is being pursued, it quickly becomes clear that few decision makers can possibly digest the range of issues and interactions portrayed thus far. An analysis, therefore, only must deal with the level of complexity that decision makers, or in general stakeholders, want to pursue. In other words, an understanding of the abilities, limitations, and preferences of the users of results of cost/benefit analyses can enable important and substantial simplifications.

The ability to simplify the analysis is the key to making cost/benefit analysis tractable. Understanding the decision-making process and decision makers involved can enable such simplifications. Moreover, as Arthur (1995) and Levison (1995) emphasize, understanding the underlying psychology and sociology of the decision-making process greatly increases the possibility of affecting it.

49.4 COST/BENEFIT METHODOLOGY FOR HUMAN FACTORS

Thus, as emphasized earlier, a very central aspect of successful cost/benefit analysis involves knowing why such an analysis is being performed. Put simply, since value is context dependent and in the eye of the beholder, it is essential that one knows who is interested in the results of such an analysis and to what purpose they are likely to put the results. Understanding these issues can make an analysis tractable as well as useful (Boff, 1988).

A valuable construct for facilitating this understanding is the value chain from investments to returns. More specifically, it is quite helpful to consider the value chain from: (1) investments (or costs) to (2) products to (3) benefits to (4) stakeholders to (5) utility of benefits to (6) willingness to pay to (7) returns. This value chain is summarized in Table 49.3.

The process starts with investments which result—or will result—in particular products over time. These products yield benefits, also over time. A variety of people—or stakeholders—have a stake in these benefits. These benefits provide some level of utility to each stakeholder. The utility perceived—or anticipated—by each stakeholder affects their willingness to pay for these benefits. Their willingness to pay affects their "purchase" behaviors which result in returns for investors.

The central methodological question concerns how one can predict the inputs and outputs of each element of this value chain. One approach to answering this question

Table 49.3 Value Chain From Investments to Returns.

Investments (costs)	to	Resulting products over time
Products over time	to	Benefits of products over time
Benefits over time	to	Range of stakeholders in benefits
Range of stakeholders	to	Utility of benefits to each stakeholder
Utility to stakeholders	to	Willingness to pay for utility gained
Willingness to pay	to	Returns to investors

involves the use of computational models and related tools. A variety of examples, listed chronologically, include:

- Mason and Sassone (1978) provide a model of the value of information based on an economic-oriented analysis that distinguishes private and social costs. Benefits are defined in terms of willingness to pay for information. The model does not, however, handle people's uncertainty about the usefulness of the information obtained.
- Eggleston and Kulwicki (1984) discuss the Department of Defense system development and acquisition cycle and present a value analysis methodology for assessing likely impacts of emerging technologies. They model value using linear combinations of attributes such as survivability, effectiveness, workload, technical risk, cost, weight, aircraft performance, utility, and vulnerability.
- Rouse (1985) provides a model that relates human error reduction and/or error tolerance to investments in selection, training, equipment design, job design, and aiding. He uses sensitivity analysis to show how the optimal allocation of development resources among these five investments depends on various parameters.
- Rouse (1987) provides a model for determining online, in real time, whether personnel need tutoring, explanations, or aiding. The choice is found to depend on task frequency, aptitude of personnel, performance changes possible, and delivery time. Sensitivity analysis is used to illustrate various trade-offs.
- Corlett (1988) provides a process model for cost/benefit analysis of ergonomic and work design changes. This model relates areas of ergonomic change to intermediate effects, intermediate results, and long-term results. He illustrates use of the model with several examples related to quality of working life.
- Rouse and Cacioppo (1989) present a model that represents the impact on economic return of investments in job requirements; manpower, personnel, and training; safety; human factors engineering; and technology. They focus on the availability of data to validate their model, as well as how this model can be used for making investment decisions.
- Kaplan and Holman (1989) discuss the HARDMAN III model which focuses on projecting manpower, personnel, and training requirements to ensure acceptable system performance and availability. The model is based on comparability analysis using baseline systems, jobs, and tasks.
- Alexander and Getty (1995) focus on cost analysis and approaches to justification of costs including economic analysis, required expenditures, allocation of funds (budgeting), and other techniques. They present examples from office and industrial settings, with emphasis on relatively straightforward health and safety improvements as translated to costs of improvements versus direct and measurable monetary benefits of these improvements.

While these models are very interesting and offer much potential, they all suffer from a central shortcoming. Except perhaps for health and safety issues, there is an almost overwhelming lack of data for estimating model parameters, as well as a frequent lack of adequate input data (Rouse, 1990). Use of baselines can help, but the validity of baselines depends on new systems and products being very much like their predecessors. Overall, the paucity of data dictates development of a more qualitative methodology whose usefulness is not totally determined by availability of hard data. The remainder of this section outlines such a methodology.

The overall cost/benefit methodology is summarized in Table 49.4. The first step involves identifying the stakeholders who are of concern relative to the investments being entertained. Usually this includes all of the people in the value chain summarized in Table 49.3. This might include, for example, those who will provide the resources that will enable a solution, those who will create the solution, those who will implement the solution, and those who will benefit from the solution.

The next step involves defining the benefits and costs involved from the perspective of each stakeholder. These benefits and costs define the attributes of interest to the stakeholders. The value that stakeholders attach to these attributes are defined by stakeholders' utility functions, which are determined in the next step. The utility functions enable mapping disparate benefits and costs to a common scale.

Table 49.4 Overall Cost/Benefit Methodology

- o Identify stakeholders in alternative investments
- o Define benefits and costs of alternatives in terms of attributes
- o Determine utility functions for attributes (benefits and costs)
- o Decide how utility functions should be combined across stakeholders
- o Assess parameters within utility models
- o Forecast levels of attributes (benefits and costs)
- o Calculate expected utility of alternative investments

Next, one determines how utility functions should be combined across stakeholders. At the very least, this involves assigning relative weights to different stakeholders' utilities. Other considerations such as desires for parity can make more complicated the ways in which utilities are combined by, for example, requiring interaction terms in the equation depicted in Equation (3).

The next step focuses on assessing parameters within the utility models. For example, utility functions that include diminishing or accelerating increments of utility for each increment of benefit or cost involve rate parameters that must be estimated. As another instance, estimates of the weights for multistakeholder utility functions have to be estimated. Fortunately, there are a variety of standard methods for making such estimates.

With the cost/benefit model fully defined, one next must forecast levels of attributes or, in other words, benefits and costs. Thus, for each alternative investment, one must forecast the stream of benefits and costs that will result if this investment is made. Quite often, these forecasts involve probability density functions rather than point forecasts. Utility theory models can easily incorporate the impact of such uncertainties on stakeholders' risk aversions.

The final step involves calculating the expected utility of each alternative investment. These calculations are performed using a specific form of the equations shown in Equations (1)–(3). This step also involves using sensitivity analysis to assess, for example, the extent to which the rank ordering of alternatives, by overall utility, changes as parameters and attribute levels of the model are varied.

Some elements of the cost/benefit methodology just outlined are more difficult than others. The overall calculations, summarized in Equations (1)–(3), are quite straightforward. The validity of the resulting numbers depends, of course, on stakeholders and attributes having been identified appropriately. It further depends on the quality of the inputs to the calculations.

These inputs include estimates of model parameters and forecasts of attribute levels. The quality of these estimates is often compromised by lack of available data, which is discussed in detail by Rouse and Cacioppo (1989). They found, for example, that data which one would think would be readily available is, in fact, seldom ever collected.

Perhaps the most difficult data collection problems relate to situations where the impacts of investments are both uncertain and very much delayed. In such situations, it is not clear what data should be collected and when it should be collected. The types of situations where these problems are most prevalent are discussed later in this chapter.

49.5 EXAMPLE APPLICATIONS

This section illustrates use of the cost/benefit methodology in two ways. First, nine case studies/vignettes are briefly reviewed and then recast in terms of the cost/benefit methodology. Second, the four examples noted in the Introduction are discussed in light of the insights gained from the nine case studies.

49.5.1 Nine Case Studies/Vignettes

Booher and Rouse (1990) summarize five cost/benefit studies conducted in military systems. These studies are quite useful for illustrating how the cost/benefit methodology presented in this chapter could have been used to predict the results actually observed. These results included:

1. Without increasing the cost of the T-800 engine, the 134 tools for organizational maintenance of the predecessor engine were reduced to only six tools for the new engine.

2. A $300,000 investment in decreasing human factors errors for the Pedestal Mounted Stinger resulted in a $61,000,000 cost avoidance by increasing probability of success in using this weapon from 0.82 to 0.92.
3. An $800,000 investment in decreasing the human factors errors for the Line of Sight Forward Heavy resulted in an $80,000,000 cost avoidance.
4. A $1,500,000 investment in removing a no-longer-needed bundle of wires from the Airborne Target Hand Over System resulted in a $1,000,000 annual cost avoidance due to decreased personnel requirements.
5. A $20,000,000 investment in a Jeep Rollover Protection System resulted in a greater than 90% decrease in annual damage costs and a greater than 98% reduction in nonfatal injuries.

Note that stakeholders for these five cases included both operations and maintenance personnel. Also, both immediate cost savings and savings over time were considered. However, results do not reflect costs to another type of stakeholder: industry. While the nature of this environment is such that costs are passed through to the government, it is still very possible that the investments noted might have negatively impacted profits. Of course, it might also have been that industry would have gladly compromised profits in order to become players in the competition for much larger contracts.

If the equations in Equation (1) were employed, the NPV for these examples would undoubtedly be attractive for any reasonable cost of capital. IRR and CBR would also be attractive. Considering the equations in Equations (2) and (3), the analysis could have included nonmonetary attributes such as ease of maintenance for the T-800 and reduction of injuries for the Jeep, both of which could be partially translated to economic returns. Thus, a more robust analysis was possible, but in these cases not apparently needed.

Rouse and Cacioppo (1989) also provide several examples of measured costs and benefits. The subtleties of interpreting the results of these case studies provide additional insights into the methodology presented here.

6. Investments over a 14-yr period of under $50,000/yr in eye protection equipment in a naval shipyard resulted in a savings of over $100,000/yr in compensation and medical expenses.
7. A 100% increase in investment in safety equipment in manufacturing resulted in a 2% decrease in loss of work days—return to work was highly related to the workmen's' compensation benefits in the state where the injury occurred.
8. Investments in driver training reduced the number of traffic violations but did not reduce the number of accidents.
9. Increases in training time for National Guardsmen resulted in a decrease in retention due to conflicts with work and family commitments.

Example 6 shows a clear and substantial benefit. Example 7 illustrates a small benefit, clouded by a somewhat external factor. For example 8, the primary effect is not on the expected variables. Finally, for example 9, the effect was the opposite of what was intended. It is easy to imagine this happening frequently, as is illustrated in a later example for repetitive stress injuries.

These examples serve to illustrate two points not evident in examples 1–5. First, the effects of any particular investment may be quite small unless other factors are also manipulated. Second, effects of investments can be quite subtle and possibly counterintuitive.

The cost/benefit analysis methodology discussed in this section could have helped to anticipate these results. By expanding the stakeholders considered, additional attributes would inevitably have been included in the analysis. Further, by considering noneconomic attributes such as additional time off and impact on family life, results might have been predicted. Succinctly, the apparent problem representation or framing was inappropriate for several of these cases, not the calculation method.

49.5.2 Four Examples

49.5.2.1 Health and Safety Improvements

The first example concerns relating investments in health and safety improvements to cost savings and productivity increases. Put simply, do changes of job design, workplace layout, seating, lighting, safety devices, and so on provide payoffs in excess of their costs? This example has the potential to yield a purely economic answer.

Estimating the costs of the improvements should be straightforward, especially if one ignores sunk costs and other than primary overhead costs. To determine likely benefits, one has to estimate impacts on job performance, errors, accidents, lateness, absenteeism, turnover, etc. One then has to translate these impacts into economic terms.

What are the economic consequences of improved job performance, fewer errors, and reduced chances of accidents? The answers to this question are, obviously, context dependent. However, to the extent that overall business processes have been characterized, it should be possible to estimate the impact of human performance improvements. Thus, for example, for processes such as automobile assembly and semiconductor fabrication, this question is reasonably tractable. In contrast, for jobs where tasks are not standardized and repetitive, it is likely to be much more difficult to estimate impacts.

Nevertheless, this example provides a straightforward illustration of use of the cost/benefit methodology. Stakeholders can be limited to investors and workers. The general nature of benefits and costs is fairly clear. A purely economic analysis is likely to be possible.

49.5.2.2 Consumer Products

The second example involves linking features of individual products—for example, "widgets" bought and used by consumers—to benefits experienced by end users. The value chain, depicted in Table 49.3, is relatively straightforward for this example. Investors want to invest in features that customers are very likely to buy.

The primary stakeholders in this example include customers, who are often also the end users, and the investors in product development. Attributes of interest to customers include the benefits provided by the features of interest, e.g., ease of use or speed of response, and the purchase price of these features. From the investors' point of view, central attributes are the flow of monies invested and the resulting flow of returns from product sales.

If features can be mapped to willingness to pay, and hence, sales over time, this example can be viewed purely economically. However, if there are trade-offs among features, it may be that a multiattribute utility model is needed. This is especially the case if these trade-offs involve multiple stakeholders such as investors, product development, manufacturing, marketing and sales, product support, and end users. Nevertheless, this example illustrates a fairly straightforward application of the cost/benefit methodology.

49.5.2.3 Weapon Systems

The third example involves relating human factors activities associated with developing complex systems—for example, military weapon systems—to mission-related benefits. The stakeholders in such efforts are substantially more diffuse than those for the previous example. The procurers and users of the weapon systems are certainly primary stakeholders. In fact, these two broad classes of stakeholders are composed of many different stakeholders—for example, pilots, maintainers, unit and force commanders, logistics organizations, procurement organizations, technology developers, and so on. Further, there are contractors and subcontractors, investors in these companies, and employees of these companies. There are also those who provide products and services to these employees.

The complexity of this web of relationships is usually limited by focusing on metrics related to the performance of the weapon system relative to specified targets and threats. It is assumed that achieving particular levels of performance will result in increased chances of military victory, which will lead to political solutions, which will precipitate economic gain, and which will justify substantial investments in the development of these weapon systems. As might be imagined, there is substantial uncertainty associated with this causal chain. For this reason, attributes are usually limited to performance requirements and costs of acquisition and operation of the weapon system.

Limiting the analysis in this way results in the cost/benefit analysis being tractable. Often, however, the limitations imposed are too crisp. While crisp requirements make procurement easier and also decrease engineering uncertainties, the fact is that end users seldom have strict requirements such as: any response time less than 1 sec is perfect and any response time greater than 1 sec is useless. In reality, needs are usually much "softer." The utility theory-based approach summarized in Equations (2) and (3) can readily incorporate such soft requirements.

The limiting of such analyses to mission performance versus costs also tends to ignore larger issues that are, in fact, not really ignored by the web of stakeholders. For example, weapon systems procurements are often affected by how many jobs are affected in which congressional districts. These jobs are among the benefits of weapon systems procurements and, ideally at least, should be taken into account in a cost/benefits analysis. Simplifying the analysis by assuming that such factors do not play an important role causes the results of the analysis to play a much lesser role in the actual decision making.

49.5.2.4 Products of R&D

The fourth example involves demonstrating how the products of R&D—for example, understanding perceptual processes underlying virtual reality—provide accountable benefits. In this case, one may be able to identify the stakeholders in general, but it is likely to prove difficult in particular. One may know that the company's product lines will benefit 10 yr from now. However, what specific products and what specific benefits are likely? One could easily pose similar questions for military R&D and its impact on future weapon systems.

Moving from stakeholders to attributes, R&D faces the inherent problem that the ways is which its products are used tend to be unpredictable. Knowledge in general, and particularly new technologies, often provide benefits that were not anticipated by their originators. Nevertheless, metrics such as speed, accuracy, and cost are fairly pervasive attributes. Volume, weight, and power consumption are also ubiquitous attributes. Thus, one is more likely to define a reasonable set of attributes than one is to identify the stakeholders that will eventually benefit from these attributes.

This results in one knowing what utility functions are needed, but not knowing whose preferences to assess to determine specific forms of these functions. One is, therefore, compelled to use known stakeholders—for example, product managers for existing product lines—as surrogates for eventual stakeholders. The primary limitation resulting from this practice is a strong tendency of these stakeholders to emphasize focusing R&D efforts on near-term incremental improvements of current product lines.

Breakthroughs, then, tend to be adopted by stakeholders unknown at the time of the investment decision. An excellent example is provided by the history of graphical user interfaces for computers that employ windows as the central metaphor. Levy (1994) discusses how Douglas Englebart developed this concept with R&D funds from DoD's Advanced Research Projects Agency (ARPA). Xerox attempted to exploit the concept, but got cold feet. Apple took advantage of Xerox's efforts to create the Mac. And, finally, Microsoft is the biggest beneficiary by far with its Windows operating system. It would have been very difficult, indeed, to have projected these consequences of ARPA's decision to fund Englebart's research 30 yr ago.

Does this imply that cost/benefit analysis should not be used for evaluating alternative R&D investments? The answer is categorically, "No." The framework provided by the cost/benefit methodology in Table 49.4 provides a nominal line of reasoning that is valuable for all R&D efforts. However, any numbers obtained as a result of employing the equations in Equations (1)–(3) should be used cautiously, especially for basic and long-term R&D.

At most, these numbers should be employed on a relative basis. For example, for organizations who have a history of investments in R&D and the subsequent returns, it should be possible to develop benchmarks. Some of these benchmarks can be quantitative, such as projected versus realized cost/benefit ratios. Others can be more qualitative—for example, related to projected stakeholders and attributes versus actual stakeholders and attributes identified after the fact.

49.5.3 Summary

In this section, 13 examples were considered—nine quite specific and quantitative and four fairly broad and qualitative. The primary goal of this discussion was to show how the methodology summarized in Table 49.4 could be applied to a wide variety of problems, ranging from the quite straightforward use of the equations in Equations (1)–(3), to problems where representation of stakeholders, attributes, and so on is much more complicated.

49.6 MORE ON COST/BENEFITS OF R&D

What guidance does the foregoing provide? At one extreme, if one's concern is the cost/benefit of adding a high-tech whatzit to one's current widget product line, the analysis is straightforward. Data may be expensive to collect, but there are no fundamental barriers to the cost/benefit assessment.

In contrast, if one is trying to appropriately allocate resources among alternative human factors R&D activities, everything is much more complicated. The primary product of R&D is information whose payoff is usually very much delayed and difficult to identify. Substantial efforts have been invested in trying to understand the value of information, both in general (Ozog, 1979) and for science and technology in particular (Chase and Samios, 1991; King, 1993). Such an endeavor is further complicated by the need for secondary information (e.g., handbooks and other reference works) to facilitate transferring primary information to use. In general, the more that products are intangible and the more that the concern is with accounting for benefits that are both byproducts and delayed, the more that the value of information comes into play. This is especially true in situations where baseline cost/benefit results for direct and immediate products are not compelling, as is usually the case for R&D.

One of the primary paths whereby by R&D investments eventually yield benefits involves the impact of R&D generated information on system and product design. Boff (1990) discusses the nature of design and cost/value considerations in the use of information by designers. He considers the cost/benefit of investing in information generation—that is, R&D—and how to improve the returns via design support systems. In a similar vein, Tushman and Nadler (1980) conceptualize an R&D laboratory as an information-processing system which must attend to work-related uncertainty through patterns of technical communication. They conclude that R&D managers must ensure that patterns of technical communication and the distribution and characteristics of special technical roles fit the information-processing requirements of their unit's work.

A variety of commentators have emphasized the role of technical communications in technology transfer. Harrison and Debs (1988) suggest that technical communicators, as boundary spanners, both disseminate and make sense of information required for coordination between organizational groups and for effective responses to the environment. Leonard-Barton (1990) discusses the difference between point-to-point and diffusion modes in the communications underlying technology transfer. She notes that the key skills underlying each mode differ, with negotiation skills being central to the point-to-point mode and marketing skills being key to the diffusion mode.

The centrality of communication to technology transfer does not only apply to information flow from R&D to applications. Moenaert and Souder (1990) provide a conceptual framework and integration mechanisms for interfunctional information transfer, uncertainty reduction, and new product innovations within business organizations. Doheny-Farina (1992) discusses several case studies of the roles of business plans, instructional texts, and communications specialists in technology transfer in product development. Rogers (1982) presents a case study of technological-information exchange among private firms in the solar flatplate collector industry and microprocessor industry in Silicon Valley.

Thus, the subtlety of information—both in its value and its transfer—is intertwined with all products of human factors activities, as well as those of other disciplines. Benefits often do not accrue without additional communications processes operating. One is then faced with the problem of attributing benefits to these communications. If, on the other hand, one insists on only attributing benefits to end results, the immediate question concerns identifying the "end" because one stakeholder's end is usually only another stakeholder's means.

A recent treatise by Scalia (1994) illustrates many of the points made in this section. Scalia questions the validity of current knowledge of Repetitive Stress Injuries (RSI), especially in terms of the extent to which OSHA regulations can be based on this knowledge. In particular, he criticizes the lack of "dose-response" relationships that predict what combination of loadings and frequencies will cause RSI. He also indicts those who would proceed with regulations for imposing substantial costs on private industry with few if any defensible benefits.

Scalia argues, and many ergonomists agree, that much more R&D is needed in the area of RSI before there is sufficient evidence to warrant OSHA regulations. While considerable research has been done already, the phenomenon of RSI is highly multidimensional and, consequently, the dose-response relationships are expected to be much more complex than such relationships for many other phenomena.

Consider the costs/benefits of potential investments in this R&D. First of all, it is not clear that these investments will result in useful dose–response relationships. Further, it may be that factors over which employers have little if any control contribute very significantly to RSI. Finally, it is certainly possible that such R&D will illuminate the phenomenon of RSI but provide few practical clues for avoiding such injuries.

Thus, there is considerable uncertainty about what will be learned, the value of learning it, and whether the returns on this R&D will justify the investments needed to proceed. Does this mean that RSI studies should not be performed? To answer this question, a broader perspective is needed.

Back, arm, and wrist injuries that are likely to be attributable to RSI account for billions of dollars of lost productivity and medical expenses. Thus, clearly something should be done to determine if and how these costs might be substantially reduced. Work redesign, training, and selection are possible solutions, all of which depend on gaining better understanding of the conditions under which RSI occurs.

However, Scalia and others of similar persuasion might still object to such investments, claiming perhaps that the costs/benefits are not attractive. This line of reasoning would likely result in terminating investments in virtually all R&D. This is due to the fact that almost all R&D is not targeted at problems as tangible and substantial as RSI. For example, study areas such as workload, stress, leadership, to name a few, are much less tangible and the costs of deficiencies in these areas are much less clear.

The conclusion of this discussion of the costs/benefits of R&D has to be that investments in R&D cannot be made solely on the basis of quantitative cost/benefit metrics. The framework provided by the cost/benefit methodology in this chapter can be very useful to qualitatively compare alternative R&D investments. In some cases, quantitative comparisons may also be possible. However, diffuse and unidentified stakeholders and attributes make it unlikely that quantitative closure can be gained on the cost/benefits of R&D.

49.7 IMPLICATIONS FOR DECISION MAKING

How should the cost/benefit methodology presented in this chapter be incorporated into decision making? First of all, it helps tremendously if one knows what decisions are to be made far enough in advance to be able to collect data as they are generated rather than in retrospect. The pervasive lack of data to serve as inputs to cost/benefit analyses often reflects the fact that data were not collected when it would have been easy and are much too costly to collect after the fact. For example, it is much easier to capture data regarding stakeholders, attributes, and so on when solutions are being designed than it is to reverse engineer a solution and track down how decisions were made.

Nevertheless, data are likely to remain problematic. This is not unique to human factors—all disciplines share this problem. A very powerful way to ease this problem is through the use of models and sensitivity analysis. While a large number of data items might intuitively seem essential for making a particular decision, it is almost always the case that a few items are dominant. Modeling and sensitivity analysis can help to identify such items. Data collection can then be targeted on these few data items.

The efficiency and effectiveness of modeling, sensitivity analyses, data collection, and cost/benefit assessments can be greatly enhanced with appropriate methods and tools. The alternatives range from off-the-shelf spreadsheet and database packages, to specialized tools for decision analysis and quality function deployment, to creation of tailored cost/benefit analysis software. For all of these alternatives, the goal is to make attractive the cost/benefit of performing cost/benefit analyses. Without such methods and tools, the cost of analysis—in terms of both money and time—can easily exceed the benefit of having the results.

Well-planned and efficient data collection, modeling, and sensitivity analysis, as well as methods and tools that directly support cost/benefit assessments, can greatly enhance the transfer of human factors products into use. This should, as a consequence, enable human factors to play a more central organizational role, earlier rather than later. The result will be that the indictment of "cold water and empty guns" will no longer be relevant.

Human factors professionals should be able to combine this leverage with another factor to gain substantial advantage in influencing decision making. More than most—but not all—other disciplines, human factors is in a position to understand the decision making processes that are to be influenced. Understanding the psychology, sociology, and politics of decision processes can greatly facilitate technology transfer and, thereby, enhance costs/benefits.

An example can best illustrate this point. In the 1980s, the notion of pilot-centered design—a version of user-centered design—became quite popular. Those involved with human factors R&D invested considerable effort to ensure that pilots were heavily involved in their research into, for example, cockpit automation. The idea was that pilot involvement would result in better solutions and much greater chances for potential innovations being adopted.

In the late 1980s and early 1990s, an important discovery was made. While pilots may fly 'em, they don't build 'em or buy 'em. The reality of decision making is that the airlines have to want to buy a solution, the airframe manufacturers have to want to sell a solution, the avionics companies have to want to supply a solution, the regulatory authorities have to be willing to approve a solution, and pilots have to want to use a solution. If one understands these stakeholders and the attributes that reflect their needs and desires, then and only then will a potential innovation become an innovation in reality.

Human factors, with its rich multidisciplinary mix of engineering, behavioral science, and social science, is a position to understand the often complex web of relationships and issues that underlie decision processes. For this reason, human factors professionals have the prerequisites for moving beyond "cold water and empty guns." It is only a matter of employing the methods discussed in this chapter in the context of knowledge and skills that are ubiquitous within the profession.

49.8 CONCLUSIONS

This chapter began with a discussion of the substantial difficulties associated with cost/benefit assessment of human factors activities. It was also emphasized that these difficulties are, by no mean, unique to human factors. Every discipline complains about these problems.

The cost/benefit methodology presented in this chapter does not eliminate these problems. However, it does bring a degree of order and rigor to the topic. In particular, it places the emphasis where it should be—representation rather than calculation. Representing cost/benefit decisions within an appropriate context-specific framework is essential to the resulting calculations being relevant and useful.

At the same time, the quantitative results of such calculations are often far from definitive. They may be only useful for comparative purposes. It may be that the primary benefit of an analysis is the thinking that went into it, rather than any specific metrics that resulted. Thus, cost/benefit analysis does not provide a panacea. Decision makers still have to make judgments, often intuitive judgments.

Another distinct possibility is that many things, when forced into such an analysis, may not be "worth" it. This may be due, in part, to the great difficulty of accounting for all direct and indirect benefits. However, it may also be, for example, that a particular experiment, model, method, or tool does not provide enough value added to justify the money and time necessary to produce it. Less accurate answers, for instance, may be good enough for the decision at hand.

Nevertheless, it is always worthwhile to think in terms of costs and benefits. It is always useful to think about stakeholders and the attributes that concern them. It is always valuable to think about who benefits, how these benefits will get to them, and what these benefits will be worth. It is better yet when these questions can be answered quantitatively.

REFERENCES

Alexander, D. C., & Getty, R. L. (1995). *Cost Justification for Ergonomics.* (James Publishing, Santa Ana, CA).

Arthur, W. B. (1995). Complexity in economics and financial markets. *Complexity, 1(1)*, 20–25.

Beevis, D. (1994). System Ergonomics. *Proceedings of the 35th DRG Seminar in Improving Military Performance Through Ergonomics*, Brussels, Belguim, pp. 61–80.

Boff, K.R. (1988). The value of research is in the eye of the beholder. *Human Factors Bulletin, 1(6)*, 1–4.

Boff, K. R. (1990). Meeting the Challenge: Factors in the Design and Acquisition of Human-Engineered Systems. *MANPRINT: An Approach to Systems Integration.* H. R. Booher, Ed., New York: Van Nostrand Reinhold.

Booher, H. R., and Rouse, W. B. (1990). MANPRINT as the Competitive Edge. *MANPRINT: An Approach to Systems Integration.* In H. R. Booher, Ed., New York: Van Nostrand Reinhold.

Brigham, E. F., and Gapenski, L. C. (1988). *Financial Management: Theory and Practice.* Chicago, IL: The Dryden Press.

Chase, R. P., and Samios, N. P. (1993). Managing the unmanageable. *The Atlantic Monthly*, p. 82.

Corlett, E. N. (1988). Cost/benefit analysis of ergonomic and work design changes. *International Reviews of Ergonomics*, 2, 85–104.

Doheny-Farina, S. (1992). *Rhetoric, Innovation, Technology: Case Studies of Technical Communication in Technology Transfers*. Cambridge, MA: MIT Press.

Eggleston, R.G., and Kulwicki, P.V. (1984). Estimating the Value of Emerging Fighter/Attack System Technologies. *Proceedings of the NATO Defense Research Group Panel VIII Workshop*, Shrivenham, England.

Feather, N. T., Ed. (1982). *Expectations and Actions: Expectancy-Value Models in Psychology*, Hillsdale, NJ: Erlbaum.

Gibson, D. V., and Rogers, E. M. (1994). *R&D Collaboration on Trial: The Microelectronics and Computer Technology Corporation*. Boston, MA: Harvard Business School Press.

Harris, D., Ed. (1994). *Organizational Linkages: Understanding the Productivity Paradox* (National Academy Press, Washington, DC).

Harrison, T. M., & Debs, M. B. (1988). Conceptualizing the organizational role of technical communications: A systems approach. *Journal of Business and Technical Communication, 2*, 5–21.

Kaplan, J. D., & Holman, C. (1981). *HARDMAN III Decision Support System*. (U.S. Army Research Institute, Alexandria, VA).

Keeney, R. L., and Raiffa, H. (1976). *Decisions With Multiple Objectives: Preferences and Value Tradeoffs*. (Wiley, New York).

King, E. (1993). Topics in science: Leaving room for failure. *Scientific Computing and Automation Magazine*, pp. 8–10.

Kirkwood, C.W. (1979). Pareto optimality and equity in social decision analysis. *IEEE Transactions on Systems, Man, and Cybernetics, 9(2)*, 89–91.

Lane, N. E. (1987). *Evaluating the Cost-Effectiveness of Human Factors Engineering*. (Institute for Defense Analyses, Arlington, VA).

Leonard-Barton, D. (1990). The Intraorganizational Environment: Point-to-Point Versus Diffusion. *Technology Transfer: A Communication Perspective*, F. Williams & D.V. Gibson, (eds.) (Sage, Newbury Park, CA).

Levison, M. (1995). Dismal science grabs a couch. *Newsweek*. April 10, 41–42.

Levy, S. (1994). *Insanely Great: The Life and Times of Macintosh, the Computer That Changed Everything*. (Viking, New York).

Mason, R.M., and Sassone, P.G. (1978). A lower bound cost benefit model for information services. *Information Processing & Management, 14*, 71–83.

Moenaert, R. K. and Souder, W. E. (1990). An information transfer model for integrating marketing and R&D personnel in new product development projects. *Journal of Product Innovation Management, 7*, 91–107.

Orlansky, J., and String, J. (1981). *Cost-Effectiveness of Maintenance Simulators for Military Training*. Arlington , VA: Institute for Defense Analyses.

Ozog, S. (1979). On the value of information. *Journal of the American Society for Information Science*, 310–315.

Rogers, E. M. (1982). Information exchange and technological innovation. In D. Shahal, Ed., *The Transfer and Utilization of Technical Information*. Lexington MA: Lexington Books.

Rouse, W. B. (1985). Optimal allocation of system development resources to reduce and/or tolerate human error. *IEEE Transactions on Systems, Man, & Cybernetics, 15*, 620–630.

Rouse, W. B. (1986). *Cold Water and Empty Guns: A Report From the Front*. Presentation to the Department of Defense Human Factors Engineering Technical Advisory Group. Goleta, California.

Rouse, W. B. (1987). Model-based evaluation of an integrated support system concept. *Large-Scale Systems, 13*, 33–42.

Rouse, W. B. (1990). Human resource issues in system design. In N. Moray, W. R. Ferrell, and W. B. Rouse, Eds. *Robotics, Control, and Society*. London: Taylor and Francis.

Rouse, W. B., and Cacioppo, G. M. (1989). *Prospects for Modeling the Impact of Human Resource Investments on Economic Return*. (Department of the Army, Office of the Deputy Chief of Staff for Personnel, Washington, DC).

Roussel, P. A., Saad, K. N., and Erickson, T. J. (1991). *Third Generation R&D: Managing the Link to Corporate Strategy*. (Arthur D. Little, Cambridge, MA).

Sage, A. P. (1995). *Systems Management*. (Wiley, New York).

Scalia, E. (1994). *Ergonomics: OSHA's Strange Campaign to Run American Business*. (The National Legal Center, Washington, DC).

Tushman, M. L., and Nadler, D. A. (1980). Communication and technical roles in R&D laboratories: An information processing approach. *TIMS Studies in the Management Sciences, 15*, 91–112.

Watterson, K. L. (1995). *Client/Server Technology for Managers*. Reading, MA: Addison-Wesley, Reading.

PART 8

HUMAN–COMPUTER INTERACTION

CHAPTER 50

DESIGN OF COMPUTER TERMINAL WORKSTATIONS

Michael J. Smith
William J. Cohen
Department of Industrial Engineering
University of Wisconsin—Madison
Madison, WI 53706 USA

50.1 INTRODUCTION

Professor Etienne Grandjean produced a monumental chapter on the Design of VDT Workstations for the *Handbook of Human Factors* in 1987. Professor Grandjean was a giant in the field of ergonomics, with decades of research and practical experience examining the important considerations in the design of computer workstations as they influenced employee health and performance. With the recent death of Professor Grandjean, a major force in the field of office ergonomics was lost. The authors of this chapter were asked to follow in the footsteps of this great person by updating the material in his chapter and providing some additional flavor, with greater emphasis on the implementation of physical ergonomic design and the influence of psychosocial factors on ergonomic design. We have accepted this challenge with humility, and admiration, and respect for Professor Grandjean's genius; we are aware that no one can replace Etienne Grandjean's knowledge. In this chapter, we will provide many of Professor Grandjean's ideas which have been taken from the 1987 edition of the *Handbook of Human Factors* (Salvendy, 1987), as well as our own unique perspectives. We are hopeful that Professor Grandjean's family, colleagues, and students will recognize much of his influence in this chapter, and will forgive any inadvertent misuse or misinterpretation of his work. The reader is encouraged to examine Professor Grandjean's landmark chapter in the *Handbook of Human Factors* (Salvendy, 1987), and the chapter entitled "Human-Computer Interaction" by Smith, Carayon, Eberts, and Salvendy in the *Handbook of Industrial Engineering* (Salvendy, 1992).

Millions of employees around the world use computer terminals (CTs) daily, primarily in office settings using desks or tables. Portable CTs are increasingly being used during commutes between work and home, at home, or in unstructured environments where there is no access to desks (for instance, in an airport). Some of these terminals are referred to as "laptops" since they are often placed on the person's lap when being used. The number of CTs in use has been steadily increasing over the last decade and by the turn of the century nearly 65–75% of the workforce will be using them. CTs are a tool of uncommonly widespread usage in a variety of settings for a wide range of work tasks and jobs, from scientist to teacher to field repair technician to store clerk. In essence, they are being used everywhere. This poses special problems when devising principles of workstation design for CTs, since task characteristics and environmental conditions have been recognized as critical aspects in defining the health and performance effects of their use (Cakir et al., 1979; Grandjean, 1979, 1987; Smith, 1984; Smith et al., 1981; Smith et al., 1992). This chapter will concentrate on defining the problems associated with CT use due to workstation factors in traditional office settings. Implications for less common uses of the computer are not discussed.

A fundamental perspective in this chapter is that workstation design influences employee musculoskeletal fatigue, discomfort, and pain. These conditions may have an adverse influence on employee performance and/or health, for example the risk for work-related musculoskeletal disorders (WMSDs) of a cumulative nature (Hagberg et al., 1995). In addition, workstation design will affect visual fatigue and discomfort that can lead to performance deficits. The research about these relationships is extensive, going back some 20 yr.

50.2 FIELD STUDIES ON THE COMPLAINTS OF VDT OPERATORS

In a traditional "paper-and-pencil" office job, an employee carries out a number of physical activities and has much space for various postures and movements: The person might look for documents, take notes, use the telephone, read a text, exchange information with colleagues, type, and perform many other tasks during the course of the working day. In such a physically diverse job, a desk or chair that is too low or too high, or that has other ergonomic shortcomings, is not likely to cause annoyance or physical discomfort to the employee. The situation is, however, entirely different for a modern office job where computers are used. Employees work with a visual display terminal (VDT), often for several hours without interruption. Such a VDT operator is tied to a human-machine system where physical movements are restricted, attention is primarily directed toward the monitor, and both hands are fixed to the keyboard. Such operators are more susceptible to ergonomic shortcomings such as inadequate lighting conditions and uncomfortable furniture due to these prolonged exposures. They are more sensitive to visual strain and to unsuitable desk levels that cause constrained postures. Such circumstances mandate the need for the science of ergonomics to be directed at the design of the modern office environment.

As long as engineers and other highly motivated experts operated VDTs, nobody complained about adverse effects. However, the situation changed drastically with the expansion of VDTs in workplaces where traditional working methods had formerly been applied. Moreover, complaints from VDT operators about visual strain and physical discomfort in the neck–shoulder area became more and more frequent. This provoked differing reactions: Some believed that the complaints were highly exaggerated and mainly a pretext for social and political claims; others considered the health hazards to be a serious problem requiring immediate measures to protect operators from injuries.

Since the initial work of Hultgren and Knave (1973), Ostberg (1975), and Gunnarsson and Ostberg (1977) in Sweden, there have been hundreds of research studies from every corner of the globe examining the working conditions of VDT and visual display unit (VDU) operators and their associated health complaints. There have been several international conferences devoted to these issues starting in 1980, and there are already programs specified for the year 2000 and beyond. The findings from this research and meetings have generally indicated that poor ergonomic conditions have been associated with large numbers of VDT operators complaining about visual discomfort, musculoskeletal discomfort and pain, and psychological distress (Bergqvist, 1984; Berlinguet and Berthelette, 1990; Bullinger, 1991; Cakir et al., 1979; Cakir and Cakir, 1993; Grandjean 1979, 1987; Grandjean and Vigliani, 1980; Grieco et al., 1995; Knave and Wideback, 1987; Luczak, Smith and Salvendy, 1993; Smith, 1984, 1987; Smith and Salvendy, 1989).

50.2.1 Early VDT Field Studies

This section addresses research about VDT work and its effects on health as presented in the first version of this chapter developed by Professor Grandjean in 1987. The most frequent complaints of VDT operators in the early field studies were visual discomfort and fatigue. These complaints typically occurred after continuous viewing of a computer terminal screen for one or more hours. VDT operators reported that their eyes were sore or hurt, tired, or failed to focus properly. Many of these field studies have shown that more VDT users report visual complaints than comparison employees who do not use VDTs (see Bergqvist, 1984; Smith, 1987 for reviews). However, some experts (Läubli and Grandjean, 1984; NAS, 1983) have questioned the appropriateness of the comparison groups, and dispute the contention that using VDTs increased the risk of visual complaints. Rather, they have proposed that adverse visual task demands and environmental conditions in both VDT and non-VDT work are the causes of visual discomfort, and not the VDT specifically (Grandjean, 1987).

Gunnarsson and Ostberg (1977) conducted one of the first large-scale field evaluations of VDT use in industry that achieved worldwide attention. This was a study of the reservation office at SAS Airlines that had extensive use of VDTs. The study encompassed interviews of operators, photographic documentation of operator working postures, measurements of geometric dimensions of operator "fit" with the workstation, and measurements of lighting conditions. The results demonstrated that the nature of job tasks had a major influence on the postural and viewing behavior of the operators. Job task requirements also influenced how the workstations and technology were positioned for convenience of use. A substantial percentage of operators reported visual disturbances; 75% eye problems, 60% visual fatigue, 48% burning or itching eyes. In addition, environmental conditions, such as screen glare and reflections, as well as improper viewing distances, were identified by the researchers as possible sources of these complaints.

Many operators also reported musculoskeletal complaints; 65% muscle problems, 54% shoulder problems, and 32% lower back problems. Only 6% reported arm or hand problems. The musculoskeletal complaints were believed to be related to long periods of static seated posture and work postures that produced unacceptable muscle loading. The percentage of operators reporting visual and muscular complaints was directly tied to the nature of the job tasks performed. Gunnarsson and Ostberg (1977) established critical considerations for VDT operator comfort and health. These included design of environmental conditions, workstation "fit" with the operator and task requirements, and demands of the task in producing specific operator body postures.

Several other studies of VDT operators have been conducted in Sweden (see Bergqvist, 1984 for a review of early studies). Knave et al. (1985) conducted a questionnaire survey to examine subjective symptoms and discomfort in VDT operators compared to referents. The purpose of the study was to compare employees who used VDTs for at least 5 hr day with referents whose tasks were as similar as possible to the VDT operators but who did not use VDTs themselves. Participants were selected from an insurance company, an airline, a post office and three daily newspapers. The findings indicated that more VDT operators than referents had eye discomfort for all worksites, except at the newspapers where many of the referents and VDT operators experienced similar eye problems. For the entire population, females reported more eye discomfort, musculoskeletal complaints, headaches, and skin disorders than males. Special measurements with a device to examine gaze direction indicated that the amount of time looking at the screen and longer work hours were related to eye and musculoskeletal discomfort. There were also correlations among the various health complaints, such that employees with eye discomfort also reported musculoskeletal and headache complaints.

Hagberg and Sundelin (1986) conducted a field evaluation of VDT operator postures and muscle contraction in the shoulder. Surface electrodes were placed on the shoulders of six female VDT operators doing word processing tasks to measure shoulder muscle tension while they worked. EMG measures of the descending part of the trapezius muscle were taken. Immediately before and after VDT work, the operators rated their discomfort levels using a Borg comfort rating scale. Working conditions were manipulated so that exposures of 5 hr of continuous work interrupted only by lunch and coffee breaks were compared to 3 hr of continuous work broken only by coffee breaks compared to 3 hr of work with short pauses taken 10 times per hour. VDT operators reported their discomfort in the eyes, neck, shoulders, elbows, hands, and back. The overall finding was that VDT work increased musculoskeletal discomfort for VDT operators the longer they worked without rest breaks. When short work pauses were introduced, the rated discomfort decreased. The various working conditions did not have an influence on trapezius muscle tension as measured by EMG.

In a series of Swiss studies in the early 1980s, Grandjean and his colleagues established critical relationships between VDT technology characteristics, environmental conditions, workstation design characteristics, job task requirements, and VDT operator health complaints. In 1980, these researchers reported the results of a study of accounting machine operators who worked 8.5 hr day entering coupon data via a numerical keyboard (Hünting, Grandjean, and Maeda, 1980; Maeda, Hünting, and Grandjean, 1980). The operators were compared to shop clerks who had no data entry duties to examine differences in reported muscular complaints. They found that the accounting machine operators reported more tiredness and pain in both hands than the shop clerks, while the shop clerks reported more tiredness and pain in both legs than the accounting machine operators. These complaints could be tied to specific task requirements of each job.

Detailed measurements were made of the postures of the accounting machine operators while they worked. Neck angles of greater than 55° from a forward gaze were related to a greater percentage of operators reporting neck pain, but not to neck stiffness. Elbow angles of greater than 75° were related to shoulder stiffness, while there was no relation between shoulder pains and elbow angle. Ulnar deviation of the wrist was related to a higher percentage of operators reporting righthand tiredness, but not to right-hand pain or cramps. The authors concluded that: (1) adverse body postures would not be important unless the exposure was for many hours and led to constrained postures; (2) the keyboard should be rotated counterclockwise by 15 to 20° to eliminate ulnar deviation; (3) the keyboard level from the floor should be 70–72 cm; and (4) source documents should be properly placed.

In a second study (Hünting, Läubli, and Grandjean, 1981; Läubli, Hünting, and Grandjean, 1981), the visual and postural demands placed on VDT operators were examined. This field evaluation examined 162 VDT operators and 133 control subjects engaged in office work. The results indicated that the percentage of employees reporting health complaints was related to postural and visual demands due to task requirements, workstation design, VDT screen characteristics, and environmental design considerations. The authors examined specific aspects of the workplace and the results of medical examinations of the employees. Their general conclusions were that employees in all jobs reported pain in the neck, shoulders, and arms, but that job specific requirements and ergonomic conditions led to greater problems for some of the jobs. Overall, one-third of the employees had medical findings showing pathology. Measurements at workstations indicated that "unnatural" body postures were common and were due to unsuitable workplace layout and workstation design. Extreme body postures were associated with increased health complaints. There were connections between poor workstation dimensions and conditions and poor body postures and health complaints.

Grandjean, Hünting, and Pidermann (1983) and Grandjean, Hünting, and Nishiyama (1984) conducted a field experiment with experimental adjustable workstations to determine the preferred settings of workstations of 68 VDT operators. Five experimental workstations were tested by employees at four different companies while they were engaged in their usual job activities over a 5-day period. First, ergonomic measurements of VDT operators' regular workstations were conducted, and then the adjustable workstations were installed at the typical settings of the regular workstations. Employees were then instructed to use adjustment mechanisms to establish preferred settings during the course of the workday. On some of the days, but not on others, operators were given forearm–wrist supports. Postural angles were measured during normal working activities, and a questionnaire was used to examine postural comfort.

The results indicated that once subjects established their preferred workstation settings, they did not change them over the course of the experiment. Furthermore, the preferred workstation settings and VDT operator posture did not comport with what would be expected from the research literature and from expert recommendations. VDT operators did not maintain upright torso postures, and in fact many leaned backward between 97 and 121°. In addition, the VDT operators preferred higher keyboard levels to those recommended in the German DIN standard No. 4549. VDT operator musculoskeletal comfort ratings for neck, shoulder, and back improved with the use of the experimental workstations as compared to the conventional workstations. The adjustable workstations allowed the VDT operators to attain more easily their preferred postures. Eighty percent of the VDT operators favored the forearm–wrist supports.

Grandjean (1984) summarized the findings of these early studies. He asserted that when repeated daily over a long period, poor postural conditions can lead to permanent muscle aches and may also adversely effect other tissues such as the tendons, joints and discs. He proposed that designers should include the following considerations: (1) the furniture should be as flexible as possible with adjustment ranges to accommodate users; (2) controls for workstation adjustment should be easy to use; (3) there should be sufficient knee space for seated operators; (4) the chair should have an elongated backrest with an adjustable inclination and a lumbar support; and (5) the keyboard must be moveable on the desk surface.

In Singapore, Ong, Hoong, and Phoon (1981) examined 62 VDT operators to evaluate subjective health symptoms as compared to 41 traditional office employees. The study participants completed a questionnaire about health symptoms, working conditions, job stress, and aspects of VDT equipment. Site visits were made to each workstation where dimensions, illumination, and glare were measured. VDT operators, in comparison to the traditional office employees, reported more eye strain, back pain, and shoulder aches. Insufficient lighting and poor quality source documents were identified as sources of eye strain. VDT operator muscular complaints varied depending on the task requirements. Data entry operators had greater problems with their left side related to neck posture when using the keyboard. Both data entry and conversational VDT operators had back pain due to static seated posture for prolonged periods. Upper limb fatigue in VDT operators was believed to be due to prolonged holding of the arms in a fixed position.

In the United States, the National Institute for Occupational Safety and Health (NIOSH) has conducted many studies to assess the potential health risks of working with a VDT. In an insurance company and four newspaper operations, Smith et al. (1980, 1981) conducted a questionnaire survey of working conditions and health complaints in VDT operators and to employees who did not use VDTs (non-VDT). Two hundred fifty four (254) VDT operators and 158 non-VDT employees participated. For a wide range of visual, musculoskeletal, and emotional complaints, more clerical VDT operators reported problems than non-VDT employees; but for professionals who used VDTs, the range of complaints was much less than for the clerical VDT users, even though some complaints were higher than those of the non-VDT employees. This illustrated the importance of job task factors in defining which VDT operators had health complaints. In addition, there was a relationship between the reported levels of stress due to working conditions and the extent of health complaints reported by VDT operators.

In a companion study, Stammerjohn et al. (1981), conducted ergonomic evaluations of workstations to examine how the design of the workstations and environmental conditions related to VDT operator health complaints. Several characteristics of the readability of the VDT screen were related to visual discomfort. These were the clarity of the characters, screen glare, and screen brightness. In addition, the height of keyboards for more than half of the workstations exceeded maximum height recommendations of the US Military standards. VDT operator complaints about such physical working conditions correlated well with objective ergonomic measurements.

Studies by Bell Laboratories of American Telephone and Telegraph did not confirm the findings from Sweden, Switzerland, or the NIOSH studies in the United States that showed that VDT operators had more problems than non-VDT employees. Starr et al. (1982) conducted a questionnaire survey to assess working conditions and health complaints for 145 directory assistance operators using VDTs as compared to 105 control subjects doing the same job without VDTs. The findings indicated that there was only one difference between VDT and non-VDT operators out of 15 health complaints (which suggests no effects of VDT use). Interestingly, the percentage of VDT users reporting visual and musculoskeletal health complaints in this study was about the same level as the European and NIOSH studies. The primary difference was the high percentage of employees who did not use VDTs who reported these health complaints. This may indicate that job requirements and ergonomic conditions were as important for the development of health complaints for the non-VDT employees as for the VDT operators.

In a second study (Starr et al., 1984), the same questionnaire methodology was used to study business office employees to determine if the results held for jobs with different tasks. Two hundred eleven (211) VDT users were compared to 148 comparison subjects, who did not use VDTs but did similar tasks to the VDT users. The findings were similar to the first study (Starr et al., 1982) in that only 3 of 14 health complaints indicated differences between VDT users and nonusers, and all were higher for the nonusers. There were also some slight differences in employee perceptions of job satisfaction and job security that were more positive for the VDT users.

A third study (Starr et al., 1985) was a field ergonomic evaluation of VDT operator postures using photographic methods. In addition, VDT users reported health complaints using a questionnaire. One hundred (100) of the 145 directory assistance operators using a VDT in the prior questionnaire survey (Starr et al., 1982) were photographed at their workstation 2–4 weeks after the questionnaire survey data from the first study were collected. The postures and geometric angles of body parts with the workstation and input devices were compared to VDT operator reports of health complaints. Correlational analysis indicated only one significant correlation from among 28 possible relationships examined. The authors concluded that they were unsuccessful in relating postural parameters to predict health complaints.

Shute and Starr (1984) studied the effects of improved ergonomic furniture on directory assistance operator comfort in a telephone company. Twenty-nine VDT operators participated in an evaluation of enhanced workstations, while 28 other VDT operators participated in an evaluation of enhanced chairs. A dual adjustable work-surface table replaced a nonadjustable work table. With the dual table, the height of the keyboard and VDT screen were independently adjusted. (Note: This was the same table used in the Dainoff (1982) study that will be described in the Laboratory Studies section of this chapter.) The enhanced chair had easier to use adjustments for seat pan height and back rest depth.

The table study lasted 8 weeks, with the last 2 weeks including both the improved table and improved chair. The chair study lasted 5 weeks. In each study, there were 2 weeks of baseline comfort measurements followed by 2 weeks of familiarization and then 1–3 weeks of evaluation. There was improvement in VDT operator comfort in the lower back for the improved table and the improved chair both when they were used separately and when combined. There was less shoulder discomfort when using the improved chair. The combination of improved table and improved chair produced a reduced incidence and reduced intensity of reported pain for lower and upper back, buttocks, thigh, shoulder, neck, and sore eyes.

Smith (1987) reviewed the findings from a portion of the scientific literature on the relationship between VDT use and health complaints. He concluded that a large number of VDT users experienced pain and discomfort of the eyes and musculoskeleture, while some also experienced job stress and related psychological symptoms. Most of these conditions were not considered pathological, but for a small proportion of VDT users serious health problems of the musculoskeleture were documented. Stress was defined as a critical factor for employees using VDTs, and the design of work tasks was seen as a central determinant of potential health problems.

50.2.2 Recent VDT Field Studies

The studies described in the previous section represent research conducted before the first version of this chapter was published in 1987. Dozens of field studies of VDT operators have been done since then. This section will examine a cross section of these studies to illustrate the major findings. Generally, these studies shifted their focus from visual difficulties to a focus on musculoskeletal disorders and psychological stress.

There was continued research into the health effects of VDT work at the Swedish National Institute for Occupational Health (SIOH). Building on the work of Knave et al. (1985), Bergqvist et al. (1992) reported on a longitudinal study of VDT operator health. This research compared the responses of 341 employees (VDT operators and nonoperators) who completed questionnaires about working conditions and health in 1981 and again in 1987. In 1981, approximately 70% of the group worked with VDTs and this increased to 83% by 1987.

The findings indicated that there was an increased prevalence of vision, neck–shoulder and hand–wrist discomfort for VDT users from 1981 to 1987. For the nonusers, the prevalence of eye discomfort decreased as did the prevalence of hand–wrist discomfort, but the prevalence of neck–shoulder discomfort increased. In 1981, more nonusers than VDT operators reported neck–shoulder and hand–wrist discomfort than the VDT operators; this was only consistent for the neck–shoulder discomfort observed in 1987. Furthermore, slightly more VDT operators reported hand/wrist discomfort in 1987 than the nonusers. The authors concluded that VDT use could be linked to eye discomfort and potentially to wrist discomfort, but not to neck–shoulder discomfort.

In another study at SIOH, Westlander (1994) examined 25 data-entry task VDT users and 12 dialogue task VDT users to review sources of musculoskeletal discomfort. In-depth interviews were conducted with each VDT user concerning working conditions, nonwork conditions and life situation. The results indicated that 58% of the VDT users reported shoulder blade complaints, while 44% reported shoulder complaints, 36% arm complaints, and 11% back complaints. All but one of the VDT users felt that the VDT work had a major role in their muscular complaints. However, the perceptions of the specific working and nonworking conditions that were felt to influence musculoskeletal discomfort varied widely across VDT users, even those in the same job category. Essentially, very few agreed on the specific sources of their discomfort. In addition, off-the-job activities were reported as important contributors to muscular discomfort for about one-half of the VDT users.

A third Swedish study by Kamwendo, Linton, and Moritz (1991) examined 420 medical secretaries using VDTs through a questionnaire. They found that 63% of the secretaries reported neck–shoulder pain, 51% low back pain, 30% wrist–hand pain, and 15% elbow pain sometime in the previous year. Neck and shoulder pain increased with age and length of present employment. There was no relationship between the extent of time sitting each day (greater than 5 hr) and neck and shoulder pain. Working with office machines more than 5 hr day was associated with more shoulder pain. Of specific interest, poor psychosocial factors of the job were highly predictive of musculoskeletal pain.

In the Netherlands, Verbeek (1991) studied the use of VDT operator training in how to gain optimal adjustment benefits from an adjustable chair in combination with a fixed height workstation. If VDT operators wanted to increase the workstation height, they could order wooden blocks to put under the desk legs. Sixty-eight administrative VDT operators in a municipal government office, who were not trained typists, were taught how to adjust their chairs. The chairs had multiple adjustments for the seat pan height, seat pan depth, backrest height, and arm rest height. The chairs were delivered by the manufacturer without any instructions.

The program consisted of working individually with each VDT operator to take: (1) anthropometric measurements of body dimensions, and (2) furniture measurements. After the measurements, the data were noted on a form and each VDT operator was instructed in how to make proper chair adjustments. In addition, the VDT operators were given a leaflet explaining proper adjustments. At this time, VDT operators could request wooden blocks to raise the desk and/or a foot rest. Prior to instruction the mean deviation from the ideal seat pan height was 71 mm and 70 mm for desk height. After instruction, the deviation was reduced by 11 mm for seat pan height and 18 mm for desk height. However, only 7% of the VDT operators adjusted the seat pan height and only 13% adjusted the desk height as advised. This suggests that the usefulness of instruction is questionable.

In the United States, several field studies were undertaken. Cornell and Kokot (1988) carried out a study similar to Grandjean et al. (1983), except they examined the dimensional qualities of adjustable workstations being used by VDT operators in the workplace. The workstation measurements of 91 VDT operators were observed at night after the VDT operators had left work for the day. Measurements were made of keyboard surface height, and angle, keyboard angle, display surface height and angle, display angle, seat pan height, and pitch. These measurements were repeated twice for 73 VDT operators, and three times for 50 VDT operators. Anthropometric measurements were taken for 40 of the 91 VDT operators.

The results showed that seat pan height settings ranged from 16.8 to 21.8 in., but 88% of the settings were between 19 and 21 in. The ranges in keyboard home row height was 26.5 to 33.2 in., with 79% from 27 to 30 in. When comparing body dimensions with preferred chair and keyboard settings, the preferred settings were higher than expected. Cornell and Kokot (1988) questioned the validity of the recommendations for the adjustability ranges in the ANSI/HFS-100 (1988) standard and the CSA (1989) standard for office ergonomics.

Several studies examined health complaints among VDT users. Rossignol et al. (1987) examined 1061 clerical workers who used VDTs and 359 nonusers with a questionnaire survey. The results indicated that 13% of the nonusers had vision complaints. In addition, vision complaints were reported by VDT operators with different VDT daily use time; 13% of the VDT operators with 3 or less hr of VDT daily, 22% of VDT operators with 4–6 hr of daily use, and 31% of VDT operators with 7 or more hr of daily use. The most significant vision problem was eyestrain or sore eyes.

In terms of musculoskeletal complaints, 17% of the nonusers reported musculoskeletal discomfort, but 15% of VDT operators with 3 or fewer hr of VDT daily use, 20% of VDT operators with 4–6 hr of daily use, and 31% with 7 or more hr of daily use reported musculoskeletal discomfort. The neck, shoulders, and back were the anatomical sites with the greatest number of complaints of discomfort. Interestingly, more nonusers (29%) complained of pain in hands, fingers, or wrists than the VDT operators (16%). This study showed a higher prevalence of vision complaints and neck, shoulder, and back pain in VDT operators with 7 or more hr of daily exposure to VDT work, but less arm pain and stiffness and less pain in the hands, fingers, or wrists.

Researchers at the U.S. National Institute for Occupational Safety and Health (NIOSH) continued their work examining health effects of VDTs. Sauter, Schleifer, and Knutson (1991) used a questionnaire survey to examine 905 VDT operators in two government agencies. A subsample of 40 data entry VDT operators was also selected for detailed ergonomic evaluations of their workstations. Charts were completed by VDT operators to identify body areas where they had almost constant discomfort. The results indicated that 33% reported almost constant discomfort for the low back, 27% for the neck, 27% for the buttocks, 15% for the right shoulder, 13% for the right hand, 12% for the right wrist, 6% for the left hand, and 5% for the left wrist. The differences in complaint levels for the right versus left arm led the researchers to examine a subsample of VDT operators where tasks required more bilateral arm use.

Factor analysis indicated that arm discomfort was a composite indicator of pain. For the right arm, discomfort included the fingers, wrist, forearm, and biceps area. The left arm discomfort also included the left shoulder. No differences were observed between the right and left arms in this group, leading to the conclusion that task demands defined the loading characteristics that produced arm musculoskeletal discomfort. The type of chair (swivel/tilt versus secretarial), difference in seat pan height versus popliteal height, and difference in pan compression all predicted leg discomfort.

VDT operators with swivel-tilt chairs reported less leg discomfort. In addition, when popliteal height was less than pan height, in combination with a less compressible seat, VDT operators reported less leg discomfort. The nature of the VDT operator's seated posture (erect, stooped, reclining) and seat backrest height relative to the seventh cervical vertebra predicted back discomfort. An erect seated posture was associated with less back discomfort. As the ratio of the distances between the seventh cervical vertebra and the seat backrest height decreased, more VDT operator back discomfort was observed. The keyboard height relative to elbow height predicted both right and left arm discomfort. Discomfort was less frequent as the keyboard was lowered to elbow level.

Cohen, Piotrkowski, and Coray (1987) and Piotrkowski, Cohen, and Coray (1992) examined 625 female office workers (mostly VDT operators) with a questionnaire. The purpose of the study was to determine the relationships between physical and psychosocial aspects of work, employee stress, and health and well-being. The findings indicated that working conditions influenced the level of job stress, as well as employee reports of health complaints. Poor organizational climate, lack of job control, a poor physical environment, excessive workload, and interpersonal tension were all related to greater psychological and somatic health complaints. The pattern of associations revealed that a poor organizational climate was related to higher levels of psychological distress, while a poor physical environment was related to more frequent somatic complaints.

In another NIOSH study, Hales et al. (1994) examined 533 telecommunications workers who used VDTs to determine the extent and possible causes of musculoskeletal disorders. Employee medical records were analyzed, a questionnaire survey about working conditions and health complaints was completed, and medical examinations were conducted. Of specific interest was the role of work organization factors in upper extremity musculoskeletal disorders. For a subsample of 174 of the VDT operators, explicit counts of keystrokes were available through an electronic monitoring system

The findings from the questionnaire indicated that 22% of the VDT operators had symptoms consistent with a definition of an upper extremity musculoskeletal disorder. There were 15% who reported tendon-related disorders, 8% muscle-related disorders, and 4% potential nerve entrapment disorders. Several psychosocial factors were predictive of upper extremity musculoskeletal disorders, especially neck pain. These psychosocial factors included fear of being replaced by computer technology, high work pressure, high information processing demands, low job variety, and lack of production standards. There was no association between the number of keystrokes and upper extremity musculoskeletal disorders.

Several studies have been conducted at the University of Wisconsin—Madison examining ergonomic, health, and stress considerations in VDT work. Carayon, Swanson, and Smith (1987) conducted a field study of 25 VDT users in a purchasing department. An ergonomic checklist was used by engineers to evaluate workstation and chair dimensions, environmental characteristics, and VDT features. In addition, photographs of VDT operator postures were taken; and a questionnaire survey was completed by VDT operators about working conditions, ergonomic features of their workstations and VDT, and health complaints. VDT operators averaged 5 hr of VDT use per day.

Eighty percent reported back pain, 72% shoulder stiffness, 24% sore wrists, 80% eye strain, and 72% burning eyes. Working surface height averaged 73 cm with ranges of 59–79 cm. Separate keyboard surface height averaged 67 cm with a range from 64 to 73 cm. VDT operator self-reports of chair comfort indicated a preference for a chair with a full back rest rather than one with a split back rest (secretarial style). However, VDT operator reports of back and shoulder pain were higher for the full back chair than for the split back, an unexpected finding with no clear explanation. Glare on VDT screens was correlated with higher vision complaints, as were poor screen characters (blurred, distorted edges).

In a second study (Lim, Rogers, Smith, and Sainfort, 1989), the direct and indirect effects of ergonomic design factors were related to VDT operator stress and somatic health complaints. A model was proposed in which an ergonomic feature (chair design) would influence back pain and have effects on perceived psychological stress. Engineers conducted ergonomic evaluations on the workstations, chairs, and environmental features of a group of 125 VDT operators in a government office. In addition, the VDT operators completed a questionnaire regarding their perceptions of ergonomic features, working conditions, and health complaints (physical and psychological).

The results indicated that VDT operator subjective perceptions of chair comfort were predictive of both back pain and psychological stress. The objective ergonomic characteristics of the chair were not predictive of back pain or psychological stress. However, the objective measures were correlated with the subjective perceptions of the VDT operators. The results show the importance of how objective ergonomic features are perceived by the VDT operators in predicting back pain and psychological stress, and a possible indirect path for affecting health symptoms.

In a third study (Carayon, 1994; Smith et al., 1992), the effects of job conditions related to stress were examined to see the effects on physical and psychological health complaints of VDT operators. A group of VDT operators in telecommunications jobs, who had their work performance electronically monitored, were compared to VDT operators in the same class of jobs who were not monitored. A questionnaire survey was used to gather information about working conditions, ergonomic features, and health complaints. The results indicated that those VDT operators who were monitored reported greater job stress, more psychological complaints, and more musculoskeletal health complaints of the back, shoulders, and wrists.

In a fourth study (Lim and Carayon, 1993), VDT operator risk of upper extremity musculoskeletal health complaints was evaluated for psychosocial factors and ergonomic features. A total of 171 VDT operators were studied using videotape analysis of upper extremity postures. Additionally, an engineer took measurements of ergonomic features and administered a questionnaire survey of working conditions, ergonomic features, job stressors, and health complaints (physical and psychological).

The results indicated that employee perceptions of work pressure and work pace were correlated to awkward postures of the upper extremities. The requirement to meet production standards was correlated with increased upper extremity repetition and awkward postures. Regression analysis indicated that psychosocial factors were predictive of employee behavior that in turn influenced the "ergonomic fit," and this "fit" was predictive of reported fatigue. The psychosocial factors were not directly predictive of upper extremity musculoskeletal pain, but were indirect predictors through their influence on behavioral risk factors. Lim and Carayon (1995) verified these results in a longitudinal questionnaire study of 129 VDT operators. Their findings showed that psychosocial work factors were related to upper extremity musculoskeletal discomfort in an indirect way as mediated by their influence on awkward postures and the frequency of motions.

Carayon (1993) reported on the results of a longitudinal study of about 150 VDT operators over a 3-yr year period. A questionnaire survey of working conditions, ergonomic features, and health complaints was given to VDT operators on three separate occasions (baseline, 18-mon follow-up, and 36-month follow-up). The results indicated that lack of job control and lack of social support from supervisors were chronic job stressors that affected psychological distress. In addition, lack of job control was correlated with symptoms of musculoskeletal pain and discomfort for all three rounds of measures. Interestingly, work pressure was not correlated with musculoskeletal symptoms for any of the three rounds, but was correlated with life distress in all three rounds.

Smith and Carayon (1995) and Lim and Carayon (1995) summarized the relationships among work organization factors, psychosocial perceptions of working conditions, job stress, and musculoskeletal disorders. They concluded that the risk of musculoskeletal disorders was based on a multifactorial relationship among all of these factors which work in concert to produce adverse symptoms. They further conclude that efforts to solve workplace musculoskeletal disorders need to take a systems approach to encompass work organization, psychosocial stress, and biomechanical factors simultaneously.

Faucett and Rempel (1994) examined 150 VDT operators in the editorial department of a large newspaper using a questionnaire and an ergonomic analysis of 70 workstations. VDT operator work postures were recorded during the ergonomic evaluations. Fifty-nine percent (59%) of the VDT operators reported musculoskeletal pain, 38% eye strain, and 21% psychological anxiety. No ergonomic (postural) or psychosocial factors were related to upper extremity pain severity, but psychosocial factors were related to upper extremity numbness.

Head rotation away from midline and keyboard height exceeding elbow level contributed to upper torso musculoskeletal pain and stiffness. Several interaction effects showed psychosocial factors that moderated the effects of ergonomic factors. For instance, employees who reported low decision latitude had more severe upper extremity numbness and upper torso pain when the keyboard position was above the elbow than those reporting high decision latitude. For employees with better supervisory relations, greater symptoms' severity was related to higher relative keyboard height and lower seat back height. The opposite relationship held for employees with poor supervisory relations.

Martin and Dain (1988) examined VDT operators who wore bifocal lenses due to presbyopia. They observed that these operators often assumed awkward postures when viewing the CRT screen. The authors tested the difference in motion patterns and postures when using the bifocal glasses versus experimental monofocal glasses made especially for viewing the CRT focal distance. When wearing the experimental glasses, the number of head movements looking from the keyboard to the screen was reduced by two-thirds. In addition, the number of movements of flexion of the wrists and elbows was reduced when wearing the monofocal glasses. Sauter et al. (1983) have observed that VDT operators who wore bifocal lenses reported more visual and neck disturbances.

Studies of VDT operator ergonomics provided an understanding that use of computers in offices created different working conditions than traditional office work. Large numbers of VDT operators reported visual and musculoskeletal discomfort that led to concerns that the equipment and/or office ergonomic conditions could lead to serious health problems. As interest in these issues grew, the number of studies about VDT operator health rapidly grew so that hundreds were completed between 1980 and 1995. In these studies, working conditions and technology design features were identified as contributors to VDT operator discomfort. As more research was completed, an understanding developed that ergonomic problems were due to multiple factors including work organization and psychosocial conditions as well as biomechanics and work physiology.

50.3 LABORATORY STUDIES

Several laboratory studies have been undertaken to define specific risk factors due to ergonomic design considerations that may influence musculoskeletal symptoms. Descriptions of the experimental arrangements of several studies are presented, followed by a summary of the general findings. Grandjean, Nishiyama, Hünting, and Pidermann (1982) studied 30 trained female typists. The group had a normal distribution of body stature, exceeding the mean body height of European women by 5 cm (mean = 166 cm); 13 subjects wore glasses. Only two subjects reported pains in the neck and shoulders during the last few weeks. The home row of the keyboard was 8 cm above desk level, a support for forearms and wrists was provided, and the chair had a high backrest with an adjustable inclination. The subjects typed a text of five lines on the screen and afterward recopied the same text for 10 min. The preferred dimensions were assessed before, during, and after the 10-min typing test. Afterwards, subjects had to repeat the typing tasks with imposed settings.

Cushman (1984) tested 20 experienced female VDT operators who entered text from paper copy for 50 min. Their average stature was 164 cm with a standard deviation of ±8 cm. The subjects performed the task for 10 min for each of five keyboard heights (from 70 to 86 cm above floor). Keying rate and error data, as well as subjective judgments, were obtained for all five test conditions. The keyboard was 7 cm high and movable. An adjustable chair with a fixed backrest inclination was provided. There was no handrest in front of the keyboard.

Rubin and Marshall (1982) tested 25 men and 25 women aged between 17 and 73 with three different positions of a VDT workstation. Five groups were formed, each consisting of five males and five females, corresponding to the 5th, 25th, 50th, 75th, and 95th percentile of the British civilian population. All subjects were naive users of keyboards and VDTs, so they had to glance frequently from the screen to the keyboard to ensure correct key selection. The three positions are defined as follows: A "standard position," corresponding to dimensions that might be found in a typical office; a "user-preferred position," taking into account the preferred settings of the subjects; and an "ergonomist determined position" that meets the recognized human factors recommendations. Each experiment lasted 10–15 min.

Weber, Sancin, and Grandjean (1984) recorded the EMG of the trapezius muscle at the preferred keyboard height, as well as 5 cm above and below it. Each condition was tested with and without forearm–wrist support. Furthermore, the pressure load of forearms, wrists, and hands on the support and the keyboard was recorded. Twenty trained subjects had to imitate a VDT job by operating the keyboard and looking alternately at source document and screen. Each experiment lasted 10 min.

The preferred settings obtained from these six laboratory studies are presented in Table 50.1. As previously mentioned, the experimental conditions of the six studies differed greatly from each other. For instance, simulated VDT work versus other test activities; different stature distribution; "with" versus "without" wrist support; and trained versus naive subjects. The disagreement in the results of Table 50.1 is therefore not surprising.

In all of these studies, the range of preferred settings is rather wide. Taking into account all extreme settings, the following ranges can be observed:

- Keyboard heights: 64–84 cm.
- Screen heights: 78–118 cm.
- Screen distances from table edge: 44–96 cm.
- Screen angles: 0–21°.
- Seat heights: 32–55 cm.

The heights of the keyboard are not in accordance with the usual ergonomic recommendations, which are mainly based on anthropometric considerations. The ranges reveal that

Table 50.1 Preferred Settings of Adjustable VDT Workstations of Six Laboratory Experiments

		Miller and Suther (1981)	Brown and Schaum (1980)	Grandjean et al. (1982)	Cushman (1984)	Rubin and Marshall (1982)	Weber et al. (1984)
Keyboard height[b]	$\bar{x}$[a] (cm)	71	74	77	74–78	70.5	78[g]
	range	64–80	72–84	71–84	—	—	74–84
Screen height[c]	$\bar{x}$ (cm)	92	100	109	—	86.7	97
	range	78–106	88–108	94–118	—	—	85–108
Screen angles[d]	degrees	3°	10°	0°	—	—	11°
	range	0–7°	3–17°	0–16°	—	—	0–21°
Screen distance[e]	$\bar{x}$ (cm)	—	52	65	—	—	71
	range	—	44–66	47–94	—	—	60–96
Seat height	$\bar{x}$ (cm)	41	50	47	—	41.8	47
	range	32–49	44–52	43–51	—	—	43–55

[a] $\bar{X}$ = mean values.
[b] Home row height above floor.
[c] Center of the screen above floor.
[d] Upward titled screens related to a vertical line.
[e] Screen center to table edge.
[f] Settings with best subjective ratings, highest keying performances, and lowest error rates.
[g] With wrist support.

a large number of operators prefer keyboard heights above the recommended levels. Cushman (1984) obtained preferred heights that were 5–10 cm above elbow level, as opposed to ergonomic textbooks that recommend elbow height. The preferred values for the screen level are, in general, also higher than recommended. About 50% of the subjects fixed the screen center at levels exceeding 95 cm above the floor. This means that many operators prefer a nearly horizontal line of sight or a slightly downward visual angle when looking at the screen, as opposed to the recommended downward visual angle.

All six laboratory studies have one common, important drawback: The experiments were carried out only over a short period of time (10 min or even less). It is very doubtful whether subjects engaged in a short-term experiment will have the same postures prefer the same settings, or experience the same level of discomfort as those working at a VDT workstation for much longer periods of time. Despite these shortcomings, the laboratory studies disclosed some interesting results that shall be discussed here briefly. The experiments with preferred and imposed settings (Grandjean et al., 1982) revealed an increase of physical discomfort in the neck–shoulder–arm area under the conditions of imposed keyboard heights and screen distances. These results, compared with those of Cushman (1984), lead to the conclusion that subjects were guided by a feeling of relaxation when using the preferred workstation settings, which were, in turn, associated with high keying performance and low error rates.

In one study, 20 out of 30 subjects preferred a keyboard with wrist support, and 24 of these subjects claimed that the wrist support did not impede typing activities (Grandjean et al., 1982). Weber et al. (1984) examined the effects of wrist support in a more systematic way. The pressure load exerted on the support remained surprisingly constant over the 10-min typing periods. Without wrist support, the pressure load on the keys was nearly zero; when working with support, the pressure load ranged between 15 and 35 N on average and increased significantly with higher keyboards. In each working condition with wrist support there was a significant negative correlation between EMG level and exerted pressure on the wrist rest.

From this, it can be inferred that the more the arms and hands rest on the support, the lower the electrical activity in the trapezius muscle and the higher the force on the keys. A comparison of the experiments with and without wrist support showed that, in the former, the EMG activity of the trapezius muscle was always lower, independent of keyboard height. At the end of the experiment, 12 out of 20 subjects preferred a keyboard with wrist support.

Dainoff (1982,1983) reported on a series of studies to examine the effect of VDT workstation design characteristics and rest pauses on operator performance, subjective symptoms, and objective measures of visual function. An experimental VDT workstation was set up at the NIOSH research laboratories in a room that had environmental controls (noise, climate, lighting, and glare). The working surfaces for the keyboard and screen were independently adjustable; the task chair was height adjustable; an in-line document holder was used for documents; a wrist rest was provided; a parabolic lighting fixture provided illumination at 550 Lux on the hard copy; and the VDT was fitted with a circular polarized screen filter to control glare.

Simulated data entry and data retrieval tasks were carried out by 13 trained typists who had no prior experience using a VDT. Subjects were videotaped during each 4-hr work session. Pre- and postmeasures of vision, spatial, and temporal contrast, acuity, and lateral phoria were taken. A financial incentive system for production was used to maintain subject motivation over the course of 4 days of 4-hr work sessions at the VDT.

Ergonomic design was manipulated by changing the adjustments of the workstation, chair, and glare on the VDT screen, and use of the wrist rest and copy holder. Subjects themselves were not informed of nor allowed to make workstation adjustments. Subjects were tested for 2 days for 4 hr under poor ergonomic conditions and for 2 days for 4 hr under proper ergonomic adjustments, in a counterbalanced format. The good ergonomic conditions provided chair adjustment to the proper height, lumbar support, VDT screen, and keyboard adjustment to provide a 90° horizontal forearm angle, a viewing angle to screen center of 20°, wrist rest, document holder, and glare control. In the poor ergonomic condition, the workstation was adjusted to provide forearm angles of 45°, a screen viewing angle of 30° at center screen, and the wrist rest and document holder were removed. In addition, the glare filter was removed, and illumination was directed on the screen that produced a character contrast ratio below 3:1.

The results indicated a 25% performance enhancement with the good ergonomic conditions. There were higher ratings of pain and discomfort for the arms, neck, shoulders, and back for the poor ergonomic conditions, ranging from 10% to 49%. There were no differences in objective vision measures.

Life and Pheasant (1984) carried out a laboratory experiment to examine posture in keyboard operation because in their earlier field research they found that VDT operators worked with their keyboards higher than normally recommended. Twelve skilled female typists performed a typing task at four keyboard elevations that affected arm, wrist, elbow, and shoulder postures. These keyboard elevations were at elbow height, 5 cm above elbow, 10 cm above elbow, and 20 cm above elbow; the order of exposure was counterbalanced. Subjects typed a passage of prose, and performance was measured for the first 2750 keystrokes. Six subjects read the copy script from a stand, whereas for the other six, the copy was laid flat on the desk. The subjects did not look at the screen during the task.

Performance was virtually unaffected when the keyboard was raised from elbow level to 20 cm above elbow level. There was, however, a consistent increase of the torque in the shoulder as the working height was increased. This means that a higher keyboard level required more static activity by the shoulder muscles to support the weight of the upper limb. The torque at the neck (C7 articulation) was slightly decreased when the keyboard level was raised. The six subjects with script in stand complained about a significant increase of discomfort as the keyboard level was raised. This effect was less pronounced with the subjects working with the laid-out script. Overall, the strongest discomfort levels were felt in the forearms, arms, and shoulders. The authors concluded that the home row of the keyboard should be approximately at the elbow height of the operator to reduce shoulder load.

This study demonstrates how delicate laboratory experiments can be and how dangerous it is to draw general conclusions. The subjects did not have chairs with proper backrests, suitable supports for the wrists, or a task requiring visual contact with the screen. Furthermore, the typists adopted an upright trunk posture and did not rest wrists on a support. These conditions might well be the reason for the disagreement of recommended keyboard height with all the previously cited studies. The conclusions of Life and Pheasant (1984) may only be valid for those under the specific laboratory setting conditions.

Waersted, Bjorklund, and Westgaard (1991) examined shoulder tension in a limited VDT task using electromyographic measurements. Eighteen subjects were seated at VDT screens approximately 70 cm from the eyes, with a line of sight approximately 10° below the horizontal. The lumbar region of the back was supported and the forearms were on chair arm rests. Subjects responded to stimuli on the screen by pushing either of two push-buttons on the right armrest. The tasks were a simple reaction time response or a two-choice reaction time response. Speed and accuracy of response were measured.

Subjects used a Borg scale to indicate muscular discomfort before and after each experimental session. Electromyographic measures of the right and left descending parts of the right trapezius muscle were taken using surface electrodes.

Most subjects generated low-level static muscle tension during the tests. The two tasks did not have different levels of muscle tension. There were, however, eight subjects who displayed a higher overall muscle tension, and this was highest for the two-choice reaction time task. The authors suggest that "psychogenic" mediation may be responsible for the higher muscle tension.

Fernstrom, Ericson, and Malker (1994) examined electromyographic activity in the forearm and shoulders when typing on a typewriter and a VDT keyboard. There were eight subjects who typed at a typewriter and at a keyboard that could be either flat or positioned at a 20° vertical angle. Each of eight experimental conditions of typing was tested for a period of 5 min, after subjects became accustomed to each of the keyboards for 10–20 min. The conditions were: (1) mechanical typewriter; (2) electromechanical typewriter; (3) electromechanical typewriter with palm rest; (4) electronic typewriter; (5) flat computer keyboard; (6) flat computer keyboard with a palm rest; (7) 20° angled computer keyboard; and (8) 20° angled computer keyboard with a palmrest.

There were small differences in the muscular force in the mechanical typewriter conditions. Using the angled computer keyboard produced lower extensor muscle activity in the left forearm as compared to the electronic typewriter. No muscle force differences were observed between the angled and flat computer keyboard. The right shoulder muscle activity was higher when using the electronic typewriter than when using the mechanical one and the palmrests had no effect on muscle activity.

These laboratory studies provided much needed objective verification that working conditions of VDT operators could create musculoskeletal strain. In addition, the results provided some direction for ergonomic improvements based on testing of specific workstation and technology features. The next sections of the chapter address our understanding of how working conditions affect VDT operator sensory/motor capabilities and the potential impact on visual and musculoskeletal discomfort, fatigue, and health. When possible, recommendations about improvements in working conditions are provided.

50.4 VISUAL FUNCTION, SCREEN TECHNOLOGY, AND LIGHTING

The cathode ray tube (CRT) remains the most extensively used display device for VDTs, although greater use of flat panels is being seen as this technology becomes less expensive. The primary advantages of the CRT are its low cost compared to other technologies and its capability to create sharp images. Laptop portable computers are almost exclusively flat panel, due to size and weight restrictions that limit the usefulness of the CRT. The CRT display has characteristics that tend to lead to problems from environmental influences. For instance, the luminance from the screen compared to overall lighting levels may cause excessive visual contrast, the reflections from the glass surface of the screen may create glare, and the accumulation of dust particles on the screen may distort images. These conditions affect the ability to read the screen and can lead to visual fatigue and dysfunction. Specific characteristics of the environment, such as illumination and glare, have been related to VDT operator vision problems. Others such as noise, air quality, and facility layout have been related to other VDT operator health complaints and psychological distress. See Chapter 20 on Visual Displays for details on how various display technologies function and for proper design and application criteria.

50.4.1 Visual Functions

The main visual functions involved in VDT work are accommodation, convergence, and adaptation.

Accommodation and convergence are the abilities of the eye to bring objects at varying distances from infinity to the nearest point of distinct vision (the so-called "near point of accommodation") into sharp focus. Focusing on near objects is achieved by adapting the curvature of the lens through contraction of the ciliary muscles (muscles of accommodation). The ciliary muscles change the curvature of the lens so that it bulges just the right amount to throw the sharp image back onto the plane of the retina.

The movements of each eyeball are controlled by six external muscles. Convergence is regulated by eye movements and achieved through focusing. The optical axis of both eyes is brought to the observed object, thus permitting both eyes to converge on an object. If convergence is not well regulated, double images occur. The proper level of illumination and the sharpness of contrasts between characters and background are essential for accommodation and convergence, and therefore legibility of images on the screen. If lighting is poor and contrasts are low, both speed and precision of accommodation and convergence are reduced.

The alignment of lighting in relation to the VDT workstation, as well as levels of illumination in the area surrounding a VDT workstation, have been shown to influence the ability of the VDT operator to read hardcopy and the VDT screen (Cakir et al., 1979; Dainoff, 1983; Grandjean, 1987; Stammerjohn et al., 1981). Readability is also affected by the differences in luminance contrast in the work area. The level of illumination affects the extent of reflections from working surfaces and from the VDT screen surface. Mismatches in these characteristics and the nature of the job tasks have been postulated to cause the visual system to overwork and lead to visual fatigue and discomfort (Cakir et al., 1979; NAS, 1983).

Generally, it has been shown that excessive illumination leads to increased screen and environmental glare, and poorer luminance contrast (Ghiringhelli, 1980; Gunnarsson and Ostberg, 1977; Läubli et al., 1981). Several studies have shown that screen and/or working surface glare are problematic for visual disturbances (Cakir et al., 1979; Gunnarsson and Ostberg, 1977; Läubli et al., 1981; Stammerjohn, 1981). Research by van der Heiden (1984) has shown that VDT users spend a considerable amount of their viewing time looking at objects other than the VDT screen. Bright luminance sources in the environment can produce reflections and/or excessive luminance contrasts that create excessive pupillary response.

Adaptation is the capacity of the eye to adapt its sensitivity to incident light flow. In fact, the aperture of the pupil, as well as the sensitivity of the retina, are continuously adapted to the prevailing lighting conditions. Both the aperture size of pupil and retinal adaptation, prevent over-or underlighting of the retina. The adjustment of the pupil aperture takes a measurable time between a few tenths of a second to one second. The adaptation of the sensitivity of the retina to total dark takes a comparatively longer time of about 30 min.

At VDT workstations the visual field of an operator often contains dark and bright areas. In such situations only a partial adaptation of the retina takes place, such that the bright area projected on the retina reduces the retinal sensitivity, whereas the dark area increases it. This form of disturbance is called *relative glare*; it reduces the general visual capacities such as visual acuity and visual sensitivity to contrasts. Adaptation conflicts are a frequent phenomenon at VDT workstations where the dark background of the screen lies near bright source documents or other bright or reflective surfaces.

An important conclusion can be drawn from these considerations. All surfaces within the visual field of an operator should be of a similar order of brightness. The temporal uniformity of the surface luminance is as important as the static spatial uniformity. Rhythmically fluctuating surface luminances in the visual field are distracting and reduce visual performance. Such unfavorable conditions prevail if the work requires the operator to alternately glance at a bright and a dark surface, or if the light source generates an oscillating light. As mentioned previously, the pupil and retina can cope with changes in brightness only after a certain delay, so that oscillating brightness leaves the eyes either under or overexposed most of the time.

Since fluorescent tubes operate from alternating current, the light is generated with 100 light oscillations per second in Europe and with 120 per second in the United States. Below a certain frequency, these oscillations are perceived as a flickering light; above this level, the oscillation is not seen. The threshold of perception is called the *critical fusion frequency* (*CFF*). The 100- and 120-Hz oscillations of fluorescent tubes are usually not perceived as flicker. It can, however, become noticeable as a stroboscopic effect on moving reflective objects, and sometimes when reflected from the VDT screen (CRT).

When fluorescent tubes wear out or are defective, they develop a slow, easily perceptible flicker, especially at the periphery. Earlier studies have revealed that many tubes show a small 50-Hz oscillation superimposed on the main 100-Hz one. This 50-Hz oscillation apparently comes from asymmetrical emissions of the electrodes, is perceptible, and is likely to cause visual discomfort. Visible flicker has adverse effects on the eye mainly because of the repetitive overexposure of the retina to light. Flickering light is extremely annoying and causes visual discomfort.

When fluorescent lighting was first introduced on a large scale in European offices, several complaints of irritated eyes and eye strain were reported. On the assumption that the oscillating character of fluorescent light was the cause of visual discomfort, the lighting technology developed phase-shifting equipment that produced an almost constant light. Complaints have decreased in offices where phase-shifted fluorescent tubes were installed. But the question of which degree of oscillation can be tolerated and which is likely to produce visual discomfort remains open. In places where 120 Hz is the standard (e.g., Europe), it is generally concluded that offices should never be lighted with single fluorescent tubes, but always with two or more phase-shifted tubes inside one luminary.

50.4.1.1 Lighting

Lighting is an important aspect of the visual environment that influences CRT screen and hardcopy readability, glare on the VDT screen (CRT), and viewing in the general environment. Figure 50.1 illustrates the various sources of light in an office environment and potential impact on the VDT user.

There are four types of general workplace illumination of interest to the computer user's environment. These are direct radiants, indirect lighting, mixed direct radiants and indirect lighting, and opalescent globes:

1. ***Direct radiants.*** Direct radiants are the main source of office lighting. These can be incandescent or fluorescent lights. Fluorescent lights are more prevalent in workplaces and stores. Direct radiants direct 90% or more of their light toward the object(s) to be illuminated in the form of a cone of light. They have a tendency to produce glare.
2. ***Indirect lighting.*** This approach uses reflected light to illuminate work areas. Indirect lighting directs 90% or more of the light onto the ceiling and walls, which then reflects back into the room. Indirect lighting has the advantage of reducing glare, but supplemental lighting is often necessary for hardcopy tasks when indirect lighting is used.
3. ***Mixed direct radiants and indirect lighting.*** With this approach, part of the light (about 40%) radiates in all directions, while the rest is thrown directly onto objects to be illuminated or indirectly onto the ceiling and walls.

Figure 50.1 Various sources of light in an office environment and the potential impact on the VDT user.

4. ***Opalescent globes.*** These lights give illumination equally in all directions. Because they are bright, they often cause glare.

Modern light sources used in the four general approaches to workplace illumination are typically of two kinds: Electric filament lamps and fluorescent tubes. The following are advantages and drawbacks of these two light sources:

1. ***Filament lamps.*** On the one hand, the light from filament lamps is relatively rich in red and yellow rays. It changes the apparent colors of objects and thus is unsuitable when correct assessment of color is essential. Filament lamps have the additional drawback of emitting heat. On the other hand, employees like the warm glow of filament lamps that is associated with evening light and a more natural, cozy atmosphere. Filament lamps are less energy efficient than fluorescent lights, and hence, are more expensive to operate.
2. ***Fluorescent tubes.*** Fluorescent lighting is produced by passing electricity through a gas. Fluorescent tubes usually have a low luminance and thus are less of a source of glare. They have the ability to match their lighting spectrum to daylight, which many employees find preferable. They may also be matched to other spectrums of light that can fit office decor or employee preferences. Standard spectrum fluorescent tubes, however, are often perceived as a cold, pale light and may create an unfriendly atmosphere. Fluorescent tubes may also produce flicker, especially when they become old or defective.

The intensity of illumination or the illuminance being measured is the amount of light falling on a surface. In practice, this level depends on both, the direction of flow of the light and on the spatial position of the surface being illuminated in relation to the light flow. Illuminance is measured in both the horizontal and vertical planes. At computer workplaces, both the horizontal and vertical illuminances are important. A document lying on a desk is illuminated by the horizontal illuminance, whereas the computer screen is illuminated by the vertical illuminance. In an office that is illuminated from overhead luminaries, the ratio between the horizontal and vertical illuminances is usually between 0.3 and 0.5. So, if the illuminance in a room is said to be 500 lux, this implies that the horizontal illuminance is 500 lux while the vertical illuminance is between 150 and 250 lux (0.3–0.5 of the horizontal illuminance).

The illumination required for a particular task is determined by the visual requirements of the task and the visual ability of the employees concerned. The illuminance in workplaces that use computer screens should not be as high as in workplaces that exclusively use hardcopy. Lower levels of illumination will provide better computer screen image quality and reduced screen glare. Illuminance in the range of 300–700 lux measured on the horizontal working surface (not the computer screen) is normally preferable. The lighting level should be set up according to the visual demands of the tasks performed. For instance, higher illumination levels are necessary to read hardcopy and lower illumination levels are better for work that just uses the computer screen. Thus, a job in which a hardcopy and a computer screen are both used should have a general work area illumination level of about 500–700 lux; and a job that only requires reading the computer screen would have a general work area illumination of 300–500 lux.

Conflicts can arise when both hardcopy and computer screens are used by different employees who have differing job task requirements or differing visual capabilities and are working in the same room. As a compromise, room lighting can be set at the lower level (300 lux) or intermediate level (500 lux) and additional task lighting for hard copy tasks can be provided at each workstation as needed. Such additional lighting must be carefully shielded and properly placed to avoid glare and reflections on the computer screens and adjacent working surfaces of other employees. Furthermore, task lighting should not be too bright in comparison to the general work area lighting, since the contrast between these two different light levels may produce eyestrain.

Task lighting refers to localized lighting at the workstation to replace or supplement ambient lighting systems used for more generalized lighting of the workplace. Task lighting assists in illuminating hardcopy when the room lighting is set at a low level, which can hinder document visibility.

50.4.1.2 Luminance

Luminance is a measure of the brightness of a surface; the amount of light leaving the surface of an object, reflected by the surface (as from a wall or ceiling), emitted by the surface (as from the CRT characters), or transmitted (as light from the sun that passes through translucent curtains). Luminance is expressed in units of candelas per square meter (cd/m^2). High-intensity luminance sources (such as windows) in the peripheral field of view should be avoided. In addition, a balance among luminance levels within the computer user's field of view should be maintained. The ratio of the luminance of a given surface or object to another surface or object in the central field of vision should be around 3:1, while the luminance ratio in the peripheral field of vision should not exceed 10:1.

50.4.2 The Generation of Characters on the Screen

Most VDTs in use today are based on the cathode ray tube (CRT) technology. On the inner surface of the screen there is a phosphor layer that, stimulated by the electron beam, generates a light emission. A number of VDTs use luminous characters on a rather dark screen background. In Europe this is called *negative presentation*, while in the United States it is called *positive contrast.* However, there is an increasing tendency on the market to offer VDTs with dark characters on a bright background. This is called *positive presentation* in Europe and *negative contrast* in the United States.

The most common techniques of generating characters on the screen are the dot matrix method or, less frequently, the continuous stroke method. The dot matrix system has proven superior in legibility and is generally preferred by operators. With this technique, characters are generated on an ideal grid of dots covering the entire surface of the screen. The CRT draws horizontal lines (scanlines) on the screen. The electron beam is turned on or off as required to produce line segments of symbols and characters. The dot spacing depends on the size of the scanning spot and on the raster pitch. Raster pitch is caused by the fact that the horizontal scan lines setting up the raster are not quite horizontal but slightly curved. If the scan line spacing is equal to the spot size of the scanning beam, then the spots composing the characters will partially overlap, producing almost stroke-like characters. The more scan lines that are used to form a character, the better the legibility.

A 525-line raster display presents visible spaces between raster lines, which cause dot visibility. A well-designed display consisting of 729 or 1029 lines is likely to have raster lines that are barely visible. A visible raster structure should be avoided as it is detrimental to legibility (Beamon and Snyder, 1975). Even more important is the visibility of the matrix structure of the individual characters that is caused by dot spacing being greater than dot diameter. Snyder and Maddox (1978) showed that an increase of the spaces between dots leads to a prolonged reading time. The more a dot matrix character resembles a stroke character, the more readable is the text. The same authors demonstrated that character font and matrix size can have a significant effect on legibility and readability. A 5×7 dot matrix font was less legible than a 7×9 dot matrix font, which, in turn, was less legible than a 9×11 matrix font (see also Snyder, 1984).

50.4.3 Colors

The eye is more sensitive to the central part of the visible spectrum, which appears as a yellow-green color and seems to be brighter than other colors. Many VDT models have characters of this yellow-green color. Some operators prefer green colors, although they cannot provide a rationale for this preference. It is possible that green characters are more easily distinguished if disturbing reflections appear on the glass surface of the screen.

A few VDT models have an amber color screen background and characters of a shining yellow phosphor that operators seem to like as well. There is no scientific reason for recommending one color over another, and the color of characters is mainly a matter of personal preference. Many of today's displays are multicolor. Different colors can emphasize certain parts of the text; they function as codes and facilitate identification processes. Differing colors may cause problems for the accommodation mechanism through chromatic aberration. In fact, red colors are focused behind the retina, blue colors are focused in front of it, and yellow-green is focused directly on the retina. Krueger and Mader (1982) showed that the colors used in VDTs are not associated with noticeable chromatic aberration.

50.4.4 Display Characteristics

The first question that arises concerning visual comfort at VDTs is: What may be the difference between reading a printed text and reading a text displayed on a VDT? Compared with a printed text, the main and often observed differences in VDTs are as follows:

- The characters are luminous on a dark background.
- The face may be flickering; its luminance is of an oscillating kind.
- The text often has low sharpness.
- The contrasts between characters and background may be low.
- The characters may be moving and unstable.
- The text sometimes shows an insufficient geometric design of characters and background.

It must be assumed that sharpness, contrasts, and poor design of characters influence the speed and precision of accommodation, and that contrasts of surface luminances as well as a flickering background may disturb the adaptation of the retina. One important consequence of disturbed accommodation and retinal adaptation might be lower legibility and occasional visual fatigue.

50.4.4.1 Techniques of Display Measurement

Measuring display characteristics involves two photometric parameters: *luminance* and *illumination.* Luminance measures the brightness in cd/m^2. An older unit is the footlambert (fL); 1 fL = 3.426 cd/m^2. Illumination measures the stream of light falling onto a surface in lux units (lx). An older unit of illumination is the footcandle (fc); 1 fc = 10.76 lx. Since luminance is a function of the light that is emitted or reflected from the surface of walls, furniture, and other objects, it is greatly affected by the reflective power of the respective surface. If the luminances of various surfaces are compared, they can be expressed also as reflectance (percentage of the reflected luminous flux).

The luminance in cd/m^2 and the illumination in lux are related as follows:

$$\text{reflectance } (\%) = \frac{0.32\text{cd/m}^2}{\text{lx}}$$

Fellman, Bräuninger, Gierer, and Grandjean (1982), Bräuninger et al. (1982, 1983, and 1984) designed the following equipment to measure the different lighting characteristics of displays: A microscope picked up the luminance of a small dot of 0.1 mm inside a bar of a character, led it to a photomultiplier that amplified the signals and transferred them to an oscilloscope, a DC voltmeter, an AC voltmeter, a Fourier analyzer, and a linearcorder. The bandwidth of the system was 1 MHz. In order to establish the degree of oscillation, the luminance of a 5 × 7-cm display surface was measured by means of a camera. The luminances of the various surfaces at the VDT workstations were measured with a Tektronix instrument (Types J 16 and J 6523). All measurements were carried out under standardized lighting conditions: indirect constant light with 400 lux vertical and 160 lux horizontal. Many of the measurements were conducted with an adjusted luminance of the characters, the so-called "preferred" luminance. These preferred figures were between 20 and 50 cd/m^2 and were assessed by the experimenters. In practice, operators adjust luminances between 9 and 77 cd/m^2, the mean value being 33 cd/m^2 (Läubli et al., 1981).

Another procedure to measure image quality is the modulation transfer function (MTF). Snyder (1980) applied the MTF to quantify the quality of the displayed image of VDTs. This measurement is based on a Fourier transformation of a luminance contrast. Snyder explains the MTF as "the contrast (modulation) expressed as a function of the size of the bars on a sine-wave grating, with increasing spatial frequency (e.g., cycles per unit visual angle) denoting decreasing bar width. More modulation per unit spatial frequency indicates greater contrast and perceived sharpness to the displayed image" (Snyder, 1984). With the MTF, sharpness of characters, reflected glare, and character contrast were determined.

50.4.4.2 Oscillating Luminances of Characters

The light of characters, generated by the stimulation of the phosphor through the electron beam, is composed of light flashes with the frequency of the refresh rate of the CRT. For that reason the light of the characters is not constant but oscillating. The *critical fusion frequency* (*CFF*), which is the threshold of perceived flicker, depends not only on the refresh rate but also on the phosphor persistence that determines how long the phosphor remains illuminated after the electron beam has excited it. This is also called a "ghost image", and it appears when scrolling procedures are used. If the phosphor persistence is too slow, a "smearing" of the image may occur. If the phosphor persistence is too fast, characters appear as flickering. There are important individual differences in one's sensitivity to see flicker. Furthermore, the size and brightness of the target, as well as the waveform, influence the threshold of perceived flicker. As already mentioned, visible flicker is extremely annoying and causes strong visual discomfort.

Some VDTs have refresh rates of 50 or 60 Hz. These rates seem to be in the critical range where some operators might already begin to perceive flicker. Gyr, Nishiyama, Läubli, and Grandjean (1984) used simulated VDT equipment with bright characters on a dark background and measured CFF levels between 45 and 55 Hz on 28 subjects. Bauer (1984) measured the CFF of a bright screen background (80 cd/m^2) with a rather fast phosphor on 30 subjects and observed a range between 73 and 93 Hz. At present, there exists no scientific basis to assess the tolerance limit for the oscillation degree of character luminances. Fellman et al. (1982) recommend an oscillation degree that does not exceed that of a phase-shifted fluorescent light.

The oscillation degree of a light source can be determined by recording the amplitude of the oscillation over the mean luminance. Bräuninger et al. (1982, 1983, and 1984) adopted a procedure based on the following formula:

$$a = \frac{1}{\text{Lm}} \sqrt{\sum_{n=1}^{20} \text{A}_n^2 \text{ eff}}$$

where

- a = oscillation degree
- Lm = mean luminance
- A_n eff = amplitude of the groundware and of 20 first harmonies of a Fourier transformation.

Preference is to be given to CRTs with a degree of oscillation of character luminances comparable to figures shown by phase-shifted fluorescent tubes: a should be lower than 0.2. Refreshing rates of 80–100 Hz with phosphor decay times of about 10 msec for the 10% luminance level should be suitable.

50.4.4.3 Sharpness of Characters

An important characteristic of image quality is the sharpness of characters or image resolution. It is generally accepted that characters with sharp outlines guarantee a comfortable legibility, whereas characters with blurred edges offer lower visual comfort. Gomer and Bish (1978) studied the effects of image resolution on evoked potentials of the brain. An image of higher resolution produced a stronger and more clearly defined evoked potential than an image of lower resolution. Rupp, McVey, and Taylor (1984), using five subjects, investigated the effects of a sharp and blurred text on a display terminal. The focused display had a blurred border zone of about 0.25 mm, whereas the defocused characters had one of approximately 0.35 mm. The mean values for accommodation and stability of accommodation were the same order for both conditions. These results do not explain why operators prefer characters with sharp edges.

Printed texts of good quality have sharp outlines, whereas VDT characters have a relatively blurred border area. The extent of the blurred border area determines the sharpness of characters. Snyder (1980) and Snyder and Maddox (1978) used the MTF procedure to assess the degree of sharpness of displayed characters. Bräuninger et al. (1982, 1983, and 1984) also examined the sharpness of VDT characters using the methods described earlier and also determined the blurred border area of characters of 33 different VDT models. The evaluation is, to some extent, arbitrary and chiefly based on the observation that a border zone r of less than 0.3 mm is not perceived, whereas higher values for r reveal visible blurred edges. Only five models out of 33 examined showed good sharpness with a blurred border zone of less than 0.3 mm; 19 models had blurred border areas of more than 0.4 mm. The reason for poor character sharpness is often found in an insufficient focusing device of the CRT. In some cases antireflective equipment, such as a micromesh filter, substantially reduces the sharpness of characters.

50.4.4.4 Character Contrasts

Various parameters have been proposed to describe the luminance difference between the image and its background. Among them are contrast ratio, the MTF, and several formulas to define contrast and percent contrast. Contrast ratio is certainly the easiest figure to be assessed. Although the MTF is a more appropriate contrast specification, contrast ratio will be used here because this parameter is easy to understand. Printed texts of good quality usually have high contrast ratios of 1:20 and more, while ratios of 1:10 are considered to be suitable.

Shurtleff (1980), Snyder and Maddox (1978), and other authors have studied the influence of character contrasts on the legibility of display symbols. Shurtleff concludes that the minimum contrast ratio acceptable for general VDT display conditions is within the range of 10:1 to 18:1. Field studies (Läubli et al., 1981) have revealed that operators in conversational VDT jobs adjust the character luminances in such a way as to keep contrasts in the range between 2:1 and 31:1, the mean contrast ratio being 9:1. The ratio of 10:1 has become a generally accepted industrial standard for display design. This recommendation is valid for character sizes between 16 and 25 arcmin. The smaller the size of the characters, the higher the contrast ratio should be.

There is also an important reciprocal relationship between sharpness and contrasts of characters: Poor sharpness requires higher character contrasts if good legibility is to be maintained (Snyder and Maddox, 1980). This applies in particular if the contrast ratio between bar luminance and interspace luminance (space between two characters) is considered. In fact, CRTs with an insufficiently focused electron beam indicate an increased luminance of the space between characters, and the corresponding contrast ratios can be as low as 1:2. The luminance in the space between two characters is called *rest luminance* and is expressed as percentage of the luminance of the bars. It is recommended that the rest luminance of the space between two characters should not exceed 14%. This corresponds to a contrast ratio of about 1:7.

Snyder and Maddox (1980) conclude that any symbol luminance above roughly 65 cd/m^2 is adequate as long as a sufficient contrast is maintained. Field studies (Läubli et al., 1981) have revealed that operators engaged in conversational jobs prefer character luminances in the range of 9–77 cd/m^2, with a median value of 33 cd/m^2.

The luminance of the screen background depends on the luminous flux in the room and on the reflection characteristics of the display screen. It is therefore hardly possible to recommend a precise luminance of the screen background. In the previously cited field study (Läubli et al., 1981), the measured background luminances range between 1 and 11 cd/m^2 with a median figure of 4 cd/m^2. This gives a mean contrast ratio for characters of 9:1. A Swedish study at a telephone information center showed screen background luminances between 0.2 and 5.6 cd/m^2, with a mean luminance of 1.3 cd/m^2 during the day, and of 2.0 cd/m^2 during the night shift (Shahnavaz, 1982).

Some authors (Cakir, Hart, and Stewart, 1979) recommend a rather high screen background luminance of between 15 and 20 cd/m^2. Such background luminances will mislead operators into adjusting high character luminances of more than 100 cd/m^2 with the risk of obtaining poor sharpness and visible flicker. Thus, it is advisable to keep the screen background luminances below 8 cd/m^2.

Bräuninger et al. (1982, 1983, and 1984) studied the character contrasts of 33 VDT models under standardized conditions. The results reveal that 18 models out of 33 had good or acceptable character contrasts with rest luminances of less than 25%. On the other hand, nine models showed character contrasts of less than 3:1, which must be considered a very poor contrast. The previously mentioned relationship between sharpness and contrast of characters led the authors to conclude that a sharpness r of less than 0.3 mm combined with a character contrast of 1:5 should guarantee fairly good legibility. If one of these parameters does not meet the recommended level, then the other one should show an optimum figure.

50.4.4.5 Stability of Characters

If the electron beam is well regulated, the background appears stable. If the regulation is insufficient, the characters show a poor stability. This phenomenon occurs as drift, jitter, or disturbances of linearity. Drift is a change in the position of a symbol and can cause a merging of characters. These movements are rather slow. Jitter is a brief, small, abrupt, and repetitive change in the position of a symbol. Disturbances of linearity refer to bends in the displayed lines. Such electronic interferences may produce annoying sensations.

There may be two main reasons for jitter. First, the noise in the electronic line and image-deflection circuits may cause irregular displacements of the single dot. Second, jitter can be produced by ac fields of external sources that may be superimposed on the deflection field. The operators report irregular movements or additional blurring of characters. Bauer (1984) studied the phenomenon of jitter under the conditions of reversed presentation (dark characters on bright background). For 10 subjects the mean threshold value for jitter movements at 10 Hz was 25.4 mm, which corresponded to a visual angle of 17.5 in. of arc. The author concluded that for VDT operators using 80 cd/m^2 bright screens, the physical jitter at 10 Hz must be less than 15 in. of arc. At frequencies above 30 Hz, movements became blurred.

Even if the jitter-induced movements are eliminated, other flicker-like interferences may occur on the bright background. Fellman et al. (1982) as well as Bräuninger et al. (1982, 1983, and 1984) assessed the stability of characters. In order to compare and to evaluate the stability of characters, the recorded variations were determined and expressed as variance in percentage of the maximal luminance. It was observed that a variance of less than 5% is not perceived. The results revealed that eight models out of 34 disclosed a high instability of characters, that resulted in letters merging into each other.

Ideally, the display should be completely free of perceptible movements, such as flicker or jitter. CRT screens are refreshed a number of times each second so that the characters of the screen appear to be solid images. When this refresh rate is too low, users perceive screen flicker. Furthermore, the perceptibility of screen flicker depends on illumination, screen brightness, polarity, contrast, and individual sensitivity. For instance, as we get older and our visual acuity diminishes, so too does our ability to detect flicker.

Screens with a dark background and light characters show less flicker than screens with dark lettering on a light background. However, the light characters on a dark screen also show more glare. In practice, flicker should not be observable. To achieve this, a screen refresh rate of 80–100 Hz for each line on the CRT screen is recommended. With such a refresh rate, flicker should not be a problem for either type of screen polarity. It is a good idea to test a CRT screen for image stability. This can be done as follows: Turn the lights down, increase the screen brightness and contrast to its highest settings, and fill the screen with letters. Flickering of the entire screen or jitter of individual characters should not be perceptible when viewed with peripheral vision or from the side of the CRT.

50.4.4.6 Reflections on Screen Surfaces

The surface of the screen is made of glass that reflects about 4% of incident light, and this suffices to reflect clear images of the office surroundings such as lights, the keyboard, or the operator. The luminance of the reflections decreases character contrasts and disturbs legibility; it can be so strong that it produces a glare. Image reflections are annoying, especially since they also interfere with focusing mechanisms; the eye is induced into focusing alternately the text and the reflected image. Thus, reflections are also a source of distraction. Stammerjohn, Smith, and Cohen (1981) as well as Elias and Cail (1983) observed that bright reflections on the screen are often the principal complaint of operators. The reflected luminances reached values between 3 and 50 cd/m^2.

It is certainly not always possible to completely avoid disturbing bright reflections, and that is why many manufacturers have developed anti-reflective technologies and devices. Some of these systems are described and evaluated below:

1. Micromesh filters are fine fabrics placed directly onto the surface or in front of the display screen. They give the screen a black appearance and are certainly efficient in reducing reflections. They have one main drawback: The sharpness of characters and their luminance are reduced, since part of the emitted light is absorbed and diffused. These effects induce operators to increase the character luminance, which causes even poorer sharpness and rest luminance between the characters. When the filter becomes dusty, the quality of the image is further reduced.
2. Etching or roughening the screen glass is a frequently applied procedure. This is a chemical or mechanical treatment of the outer surface of the front glass which produces an optically irregular surface. It does not reduce much of the total amount

of reflected light, but it breaks up the reflected image and makes it more diffuse. The more the surface is roughened, the more the reflection is broken up. The reflected image becomes softer because it is dispersed over a large area. The light rays generated by the excited phosphor are also dispersed when they pass through the front glass. This produces a blurring of the display characters. The roughening procedure is moderately efficient in reducing reflections, and the effects on character sharpness can be kept at a minimum if the surface is not roughened too much. Bräuninger et al. (1982, 1983, and 1984) tested several VDT models with roughened front glass, finding reduced reflections but, nevertheless, good character sharpness.

3. The coating of the screen with a thin antireflective film is a very efficient means of reducing reflections. This film layer, usually a Lambda/4 layer, has the thickness of a quarter of the wavelength of light and does not diminish the sharpness of characters. The only drawbacks are the sharp outlines of the remaining image reflections and the fact that the surface is easily soiled by fingerprints.
4. Polarization filters polarize the incident light and partially reduce reflections. The main drawback of this device is the occurrence of double images. Moreover, the outer surface of the polarization filter is easily soiled.

All antireflective technologies have serious drawbacks. If efficiency is weighed against drawbacks, the Lambda/4 coatings and the etching-roughening procedures are preferable over the micromesh and polarization filters. The correct placement of lights and an appropriate positioning of the screen with respect to windows and lights remain the most efficient preventive measures.

No unscreened light should appear in the visual field of any working employee. The line from eye to light must show an angle of more than 30° with the horizontal plane. If a smaller angle cannot be avoided (for example, in large offices), the lights must be effectively shaded. To avoid glare from the reflection of the desk, the lights should be arranged on either side of the workplace in order to avoid a coincidence of the line of sight with the line of the reflected light.

50.4.4.7 Luminances of Surfaces

The surfaces in the visual field of a VDT operator are the screen, the CRT frame, the surfaces of the VDT set, the desk, the keyboard, the source documents, and some elements of the surroundings, such as walls and windows. If the ergonomic recommendations concerning brightness contrast ratios in the visual field are applied to a VDT workstation, the luminance contrasts between the dark screen and the neighboring surfaces (including source documents, parts of the VDT set, and keyboard) should ideally not exceed a ratio of 1:3. Some ergonomists object to this and claim that the eyes looking at the screen are focusing the bright characters only and not the dark background. These experts recommend surface contrasts between screen and source documents that do not exceed the ratio 1:10. Conversely, if the eyes are directed toward the source document, they adapt to the brightness of that surface and the screen is in the periphery of the visual field; this contrast should not exceed 10:1.

Although nowadays not all problems of spatial and temporal differences of luminance in the visual field of VDT operators are solved, it is reasonable and realistic to make the following propositions: The luminance contrast between dark screen and source document should not exceed the ratio 1:10. All other surfaces in the visual field should have a luminance of an average value between that of screen and source document. The designer of a VDT workstation should select colors of similar brightness for the different surfaces, replace eye-catching effects with black and white contrasts, avoid reflecting materials, and give preference to dim colors.

50.4.4.8 Design of Screen Characters

Image quality is a major factor for reducing eye strain and visual fatigue, and good character design can help improve image quality. The proper size of a character is dependent on the task, the display parameters (brightness, contrast, glare treatment, etc.), and the viewing distance. Character size that is too small can make reading difficult and causes the visual focusing mechanism to overwork. This produces eye strain and fatigue. Character heights should preferably be 20 to 22 min of visual arc, while character widths should be between 70 and 80% of the character height. This approximately translates into a minimum lowercase character height of 3.5 mm with a width of 2.5 mm at a normal viewing distance of 50 cm, and uppercase characters of 4 mm high with a width 75% of the height.

Good character design and proper horizontal and vertical spacing of characters can help improve image quality. To ensure adequate discrimination between characters and good screen readability, the character spacing should be in the range of 20–50% of the character height [25% preferred by Grandjean (1987)]. The interline spacing should be between 100 and 150% of the character height (Grandjean, 1987). The geometric design of the characters influences their readability. Some characters are hard to decipher, such as the lowercase "g" that looks like the numeral "9." A good font design minimizes character confusion and enhances the speed at which characters can be distinguished and read. Two excellent fonts are the Huddleston and the Lincoln-Mitre.

50.4.4.9 Viewing Distance

Experts have traditionally recommended a viewing distance between the screen and the operator's eye of 45–50 cm, but no more than 70 cm. However, experience in field studies has shown that users may adopt a viewing distance greater than 70 cm and are still able to work efficiently and not develop visual problems. Thus, viewing distance should be determined in context with other considerations. It will vary depending on the task requirements, CRT screen characteristics, and individual visual capabilities. For instance, with poor screen or hardcopy quality, it may be necessary to reduce viewing distance for easier character recognition. Typically, the viewing distance should be 50 cm or less due to the small size of characters on the VDT screen.

50.4.4.10 The Screen and Viewing

As we have indicated earlier, poor screen images, fluctuating and flickering screen luminances, and screen glare cause user visual discomfort and fatigue. Screens with glass surfaces have a tendency to pick up glare sources in the environment and to reflect them. This can diminish the contrast of images on the screen. To reduce environmental glare, the luminance ratio within the user's near field of vision should be approximately 1:3 and approximately 1:10 within the far field of vision. For luminance on the screen itself, the character-to-screen background luminance contrast ratio should be at least 7:1. To give the best readability for each operator, it is important to provide VDTs with adjustments for character contrast and brightness. These adjustments should have controls that are obvious and easily accessible from the normal working position (e.g., located at the front of the screen).

50.5 INPUT TECHNOLOGY ISSUES

The reader should consult Chapters 20 (Design of Displays), 21 (Design of Controls), and 22 (Design of Nonconventional Controls) of this volume, which provide extensive descriptions of the design characteristics, features, and benefits of various components that are used in operating a VDT.

Computer input devices are the means by which users provide instructions to the computer. There are a wide variety of devices for interfacing, including the keyboard, mouse, touch panel, light pen, pointer, tablet, and speech recognition. Any mechanical or electronic device that can be tied to a motor response can serve as a computer interface. The most common interface still in use today is the keyboard, and it will be used as an example to illustrate how to conceive the design for better human–computer interfaces.

50.5.1 The Keyboard

The keyboard for typing letters was invented in 1868. It was a mechanical device that required a design of four parallel rows of keys. To operate these keys quickly the typist had to keep their hands parallel to the rows. With the development of electronics, the mechanical typewriter was replaced by the electric one, which was then replaced by the electronic keyboard. The mechanical resistance of keys was much reduced and the operation of the keyboard was made easier with the electronic keyboard.

At VDT workstations the typing activity is similar to the traditional operation of typewriters. There are some slight differences, though: First, the number of keys has increased with specially arranged numerical keys and several function keys for operating the computer. Second, in conversational tasks, VDT operators frequently wait for a response from the computer, and thus are not typing continuously. According to the Swedish study of Johansson and Aronsson (1980), computer response times of more than 5 sec were experienced as annoying and stressful. During these unwanted pauses, operators often looked for suitable supports to rest their forearms and wrists.

This has induced some VDT designers to develop flatter keyboards that allow operators to rest their forearms and wrists on the desk. In addition, some newer keyboard designs provide a supporting surface at the front of the keyboard for resting the hands and wrists. Another alternative keyboard design attempts to deal with the constrained posture of hands when keying by reducing the inward turning and ulnar abduction of hands. Several such keyboards have been developed in Europe and the United States.

A number of keyboard features can influence an employee's comfort, health, and performance. The ANSI/HFS-100 (1988) standard provides guidance in the design and use of keyboards. This standard is currently undergoing revision, and new requirements should be available soon. The keyboard should be detachable and movable from the screen, thus providing flexibility for independently positioning the keyboard and screen. The keyboard should be stable and have nonslip materials on its bottom surface to ensure that it does not slide when placed on a tabletop and being used. To assist in achieving a favorable arm height when keying, the keyboard should be as thin as practical. The slope or angle of the keyboard should be between 0° and 25° measured from the horizontal. Adjustability of keyboard angle is useful to help in achieving good wrist posture.

The shape of the key top must satisfy several ergonomic requirements, such as minimizing reflections, aiding the accurate location of the operator's finger, providing a suitable surface for the key legends, and being neither sharp nor uncomfortable when depressed. The surface of the key tops, as well as the keyboard itself should have a matte finish to reduce reflections. The key top should be large enough to be easily struck without striking the adjacent keys (for instance, ANSI/HFS-100 recommends a minimum horizontal width of 12 mm). ANSI-HFS-100 also suggests that the spacing between the key centers should be about 18–19 mm horizontally and 18–20 mm vertically for effective keying.

The ANSI/HFS-100 (1988) standard indicates that feedback of key actuation is important, since it indicates to the operator that the keystroke has been successfully completed. There are three types of keyboard feedback: Visual, tactile, and auditory. Visual feedback provides an indication on a display that the key has been depressed. It shows whether the key has been successfully depressed, because the content of that key is displayed on the screen. Tactile feedback can be provided by a collapsing spring that increases in tension as the key is depressed, or by a snap-action mechanism when key actuation occurs. The snap-action indicates to the operator that the key has been actuated. Auditory feedback (e.g., a "click" or "beep") can also be used to indicate that the key has been actuated. There is debate about the usefulness of feedback with experienced typists.

Güggenbuhl and Krueger (1990, 1991) and Rempel and Gerson (1991) have shown that finger movements of experienced typists are "preprogrammed" and operating without benefit of feedback of response. In fact, for experienced typists the finger movements are too fast to be affected by feedback, since by the time the feedback is received the fingers have moved on to the subsequent movements. Thus, the need for feedback of actuation for purposes, other than initially learning to use the keyboard, is questionable.

The keyboard layout has also generated debate. Many feel that the standard QWERTY layout makes the weakest fingers work the hardest. However, there is no research evidence that the extent of work done by the fingers using a QWERTY layout, as opposed to others, is excessive or increases the risk of finger fatigue or injury (especially because alternate keyboard layouts are rarely used). Thus, the conventional QWERTY layout seems to be a reasonable design, as does the DVORAK layout. It should be recognized that it is very difficult for experienced VDT operators to switch between keyboards with different layouts.

The use of a wrist rest when keying can help minimize extension (backward bending) of the hand/wrist, but the use of a wrist rest for operator comfort and health has generated some debate. When the hand or wrist is resting on the wrist rest there is compression of the tissue that may create increased intercarpal canal pressure or local tissue ischemia. On the other hand, the wrist rest allows the hands and shoulders to be supported with less muscular tension which is beneficial to VDT operator comfort. At this time, there is no scientific evidence that the use of a wrist rest either causes or prevents serious musculoskeletal disorders of the hands, wrists or shoulders. Thus, the choice to use a wrist rest should be based on employee comfort and performance considerations until scientific evidence suggests otherwise.

If used, the wrist rest should have a fairly broad surface (5 cm minimum) with a rounded front edge to prevent cutting pressure on the wrist and hand. Padding further minimizes skin compression and irritation. Height adjustability is important so that the wrist rest can be set to a preferred level in concert with the keyboard height and slope. Arm holders are also available to provide support for the hands, wrists, and arms while keyboarding and have shown to be useful for shoulder comfort.

50.5.2 Alternative Keyboards

Several alternative keyboard designs have been developed to address VDT operators' shoulder, arm, and hand discomfort, pain, and fatigue. Nakaseko, Grandjean, Hünting, and Gierer (1985) developed an experimental keyboard that split the keyboard in half. This allowed the opening angle to be manipulated to reduce the extent of ulnar deviation. Based on earlier field study findings (Hünting, Läubli, and Grandjean, 1980), three different opening angles (15°, 25°, and 35°) and two lateral displacements (0° and 10°) were compared with a traditional keyboard (with and without forearm supports) for user preference. Subjects preferred the keyboard with an opening angle of 25° and 10° of lateral displacement. A second experiment was conducted to compare an experimental keyboard with the preferred settings and a large forearm support, to the experimental keyboard with a small forearm support, to a traditional keyboard with a large forearm support. The results showed that the experimental keyboards produced less ulnar abduction than the traditional keyboard.

Ilg (1987) reported on a 5-yr evaluation project to develop the optimal criteria for keyboard design. This research was conducted at the Fraunhofer Institute in Stuttgart. Fifteen design parameters of keyboards were investigated to define the optimal characteristics for each parameter. The method of investigation was sequential and in succession. That is, as each parameter was optimized, the next parameter was evaluated in a keyboard configuration that contained the previously optimized parameters. Thirty subjects with varying keyboard experience were used to test the various parameters. Four levels of each parameter were sequentially tested for about ten minutes each in a laboratory evaluation using an experimental keyboard. Subject performance was examined regarding typing speed and accuracy, and subject preferences were recorded. The following results were found to be noteworthy:

- ***Keyboard geometry***—The alpha/numeric field was divided in two. The existing row layout was replaced with a column key arrangement. It was found that the splaying of columns according to finger positioning had only a slight influence on performance.
- ***Key spacing***—A key spacing of approximately 19 mm both horizontally and vertically was found to be preferable.
- ***Inclination of the keyboard***—Lateral inclination of about 8° was optimal. The optimal frontal inclination was between 5 and 10°.
- ***Curvature***—The results of longitudinal and transverse curvature for fitting the fingers was inclusive.
- ***Depth of key depression***—4 mm of key travel was advantageous.
- ***Force of key actuation***—The result was not definite. The optimal was felt to be between 0.20 N and 0.70 N.
- ***Key design***—The key surface should be concave with a minimum radius of 30 mm. The key should be square with a side length of 14 mm.
- ***Hand-rest***—All subjects found the hand rest desirable.

In order to evaluate hand posture, Hedge and Powers (1995) examined a negative slope keyboard and forearm supports using video-motion analysis. Twelve experienced female typists participated in a laboratory experiment where they typed on a computer keyboard for approximately 50 min under different arm support and keyboard slope conditions. The keyboard was a standard IBM keyboard which was fitted into a fixture that provided adjustable negative angling. Subjects were tested in different conditions comparing the conventional keyboard flat on the desk (CK), a condition in which the forearms were supported in moveable arm supports (FMFS) with the conventional keyboard flat on the desk, and the negative sloping keyboard with wrist support (NSKS).

The findings indicated that there were no differences in subject postures between the forearms supported condition (FMFS) and the conventional keyboard flat on the desk condition (CK). The FMFS condition did not yield improved wrist postures. The negative slope condition (NSKS) produced improved wrist extension over the conventional keyboard (CK). Five of the subjects had greater than 15° wrist extension using the conventional keyboard, and none had this much extension using the negative slope keyboard. However, the NSKS condition did not show improvements for ulnar deviation or elbow angle.

New concepts for computer keyboards have been developed based on the research of Grandjean and his colleagues and the work done at the Fraunhofer Institute (see Ilg, 1987). Several alternative keyboards are now commercially available. The primary change from a conventional keyboard is that of providing geometric alterations (such as split half alpha/numeric keys and angling of the rows and/or columns). This changes the keyboard user's wrist angles of extension, pronation, and ulnar deviation. The success of these keyboards in real-life settings can be assessed as their use increases.

50.5.3 Other VDT Input Devices

The reader is referred to Chapter 21 for a review of other input devices including the mouse, joystick, trackball, touch screen, graphic tablet, and light pen. Each of these devices can create ergonomic problems at the workstation. The primary ergonomic issue for any of these input devices is how the need of the VDT user interfaces with the input medium.

As with any tool used in the workplace, it is important to understand the nature of the VDT task prior to selecting an input device. The necessary ergonomic consideration is that the input device matches the need for its use. However, as with the keyboard, it is necessary to provide a workstation that has enough space for the proper and intended use of any input device. It is inadvisable to use a mouse in a cramped environment since a trackball device can function similarly within smaller space requirements. However, if the choice of the input device is mediated by the lack of room for a mouse, it is likely that the workstation is too small or cramped. Space savings is probably better achieved by reorganizing the work flow at the station.

The device must not cause extended reaching, but at the same time, the device cannot inhibit the free movement by the VDT operator. In addition, the input device that is appropriate for some job tasks may not be appropriate for others. Thus, it cannot be overstated that the nature of the job task itself must be understood before settling on the choice for any input device. Because it is unlikely that the typical VDT workstation will have more than the keyboard and another peripheral input device, the choice will often be a compromise as to which device is best. Chapter 21 provides the pros and cons of each input device and the environment for which it is best suited.

As with any tool, the input device should not require undo force to operate. It should be designed according to known spatial and force compatibility in that the reaction of movement of the device should correspond to the movement upon the screen in direction and speed. Trankle and Deutschmann (1991) investigated the problem of cursor positioning using a mouse and determined that compatibility principles were necessary in design of the mouse especially when the mouse movements were for short distances. The maintenance of the input device is also a potential ergonomic problem given that mechanisms can become dirty and respond less accurately than the user expects. For instance, rollers that respond to the roller ball inside a mechanical mouse can become so dirty as to require forceful and repetitive movements in order to move a cursor on the screen. To avoid having the VDT operator struggle with using the input device as intended, regular maintenance is required for all input devices with mechanical actuators.

Adjustable gain controls should be available as well. High response gains can require the operator to make frequent corrections to desired screen positions, whereas too low gains can be frustrating and may cause the operator to actuate the input device with more force than is necessary. The device should also provide feedback to the user not unlike that provided by key actuation in a keyboard. The feedback can vary depending on the device used, but should not interfere with the task.

The placement of the device should not induce awkward postures in its use. The device should be placed within easy reach of the operator, especially when it will be used frequently during work. Keyboard trays should allow for placement of other input devices directly on the tray instead of on other working surfaces.

The use of touch screens and light pens present new ergonomic considerations to the design of VDT workstations. These devices may inherently create extended reaching and thus, are not appropriate for use in the typical, seated VDT environment. Brocklehurst (1991) investigated the use of an input pen for a flat panel display interface that used a light pen for drawing and pointing. He reported that a potential ergonomic problem existed because users were required to hold the pen in a more upright posture than was found when using a standard hardcopy writing pen. He explained that the design of the pen was still based on the need to point rather than write. However, the use of a touch screen or light pen is certainly warranted in other situations such as at standing workstations or when specialized drawing is needed. Shneiderman (1990) outlined some important considerations in the use of touch screens which included arm fatigue, optical interference, or increased glare on the screen and the design issue that a finger can obscure parts of the screen.

The ergonomic problems associated with alternative input devices will become more evident as their use increases and is more systematically studied. As voice-activated input devices become more refined and inexpensive, what new cumulative trauma might we expect to the voice box of the VDT operator? At this time, it can only be postulated as to the impact these input interfaces will have on employee safety and health.

50.6 ERGONOMIC DESIGN OF VDT WORKSTATIONS

50.6.1 Medical Aspects

As already mentioned earlier, the introduction of VDTs into the workplace has produced an integration of employees in a person–machine system. The relationship between the operator and the machine is reciprocal; it is acting as a closed system. One of the consequences of this is a restriction of space for physical activities that leads to constrained postures, together with long-lasting static contractions of several muscles of the back and shoulders. Such static effort reduces blood irrigation of the muscles and creates local fatigue (see Figure 50.2). This leads to symptoms of tiredness, pain, and even cramps. In addition, heavy workload, chronic repetition, and other biomechanical strains can cause the same problems.

Figure 50.2 An older style VDT workstation showing several ergonomic risk factors such as a twisted sitting posture, sharp desk edges, improper work surface height for VDT operation, and insufficient work surface dimensions.

These postural, repetition, and workload problems lead to reduced performance and productivity, and in the long run, they may also affect employee well-being and health. In fact, if these adverse ergonomic conditions are repeated daily over a long period, more or less permanent aches and pains may affect the upper extremities and back and may involve not only muscles but also other soft tissues, such as tendons and nerves. Thus, long-lasting, adverse ergonomic conditions may lead to a deterioration of joints, ligaments, and tendons. Field studies (Grandjean, 1987; Hagberg et al., 1995), as well as general experience, have shown that these conditions may be associated with a higher risk of:

- Inflammation of the joints
- Inflammation of the tendon-sheaths
- Inflammation of the attachment-points of tendons
- Symptoms of degeneration of the joints in the form of chronic arthroses
- Painful induration of the muscles
- Disc troubles
- Peripheral nerve disorders

Workstation design is a major element in ergonomic strategies for improving user comfort and particularly for reducing musculoskeletal problems. Figure 50.3 illustrates the relationships between the working surface, VDT, chair, documents, and various parts of the body. The task requirements will determine critical layout and dimensional characteristics of the workstation. The relative importance of the screen, keyboard, and hardcopy (e.g., source documents) depends primarily on the task, and this defines the design considerations necessary to improve operator performance, comfort, and health. Data entry jobs, for example, are typically hardcopy oriented. The operator spends little time looking at the screen, and tasks are characterized by high rates of keying. For this type of task it is logical for the layout to emphasize the keyboard and the hardcopy, as these are the primary tools used in the task, while the screen is of lesser importance. On the other hand, data acquisition operators spend most of their time looking at the screen and seldom use hardcopy. For this type of task, the screen and the keyboard layout should be emphasized.

Cohen et al. (1995) used a case study method to determine the effects of task demands, customer needs, and organizational environment on the recommendations for ergonomic redesign in a large pension and insurance organization. The employee union hired the

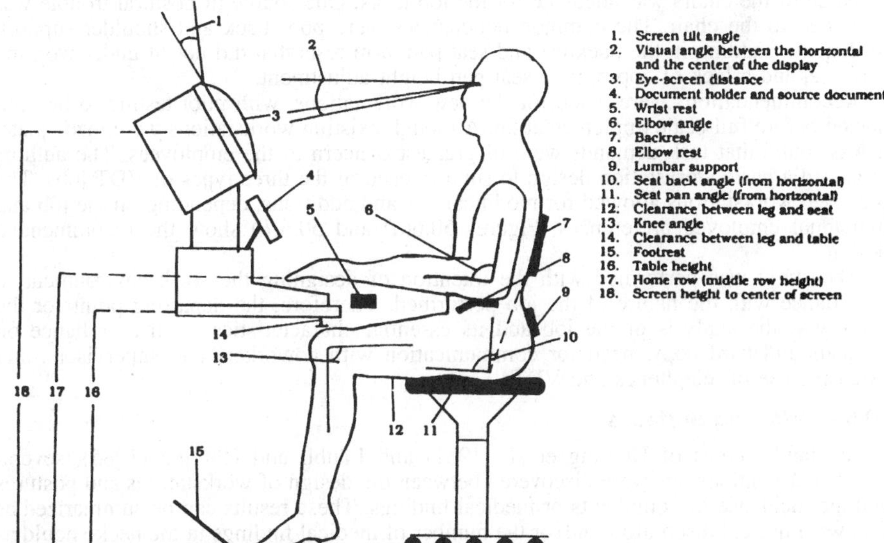

Figure 50.3 Relationships between the working surface, VDT, chair, source document, and various parts of the body. (From Helander, 1982.)

authors to investigate computer workstation needs of its employees throughout their U.S. nationwide office environments. The purposes of the study were to provide recommendations regarding future purchases of system furniture, to suggest retrofitting options for those offices already using newer workstations, and to complete a thorough job analysis in order to ensure that workstation recommendations were appropriate for job requirements. At the time of the study, the organization was planning a major renovation of 40,000 workstations and chairs, but employees were concerned that their opinions and recommendations for the design of the new furniture were not solicited or valued by management.

The researchers studied three jobs using semistructured interviews, job observations, and measurements of workstations and the office environment. The clearest finding from the ergonomic analysis was that the current workstations were not adequate to support the task demands and high workload associated with each of the three jobs. The job analyses revealed very high workloads and backlogs of work. Thus, it was advised, in light of new workstation design, to address the entire process of work for each of these jobs because furniture replacement can be considered only as one element of the work system.

Potential health risk factors were identified, including the risks associated with static postures of the trunk, neck, and arms; awkward twisting and reaching motions; and eyestrain. For some very old workstations [see Figures 50.4(*a*) and 50.4(*b*)], ergonomic risk factors were caused by the placement of the VDT and keyboard on different surfaces, insufficient work surface dimensions, glare on the CRT, insufficient knee and toe space, sharp desk edges, noise sources, and the inability for the chair armrests to fit under any of the work surfaces. The arrangement resulted in awkward postures, with excessive twisting of the trunk and neck flexion. This problem was worsened by the lack of space and clutter on working surfaces from hard copies. These factors led to static and rigid working postures.

The most troublesome problem at the newer workstations was the need to share VDTs between two employees. The keyboard was placed on a swivel base that could be turned 180° from one employee to an adjacent person. This caused extended reaches when using the VDT. Frequently, the base became stuck or was blocked by desk clutter. The swivel base was unstable and required continuous adjustment while typing to keep it aligned with the employee. If the keyboard was removed from the swivel (to bring it closer to the desk edge), there was no room for wrist placement in front of the keyboard. In addition, the use of shared VDTs resulted in eyestrain and poor postures because of extended reaches and extended viewing distances. Inadequate storage space and poor housekeeping further prevented the VDT from being easily accessed Figure 50.5 illustrates the shared VDT workstation.

None of the chairs was adequate for the job tasks. One source of postural trouble was attributed to the chair. The common deficiencies were poor back and shoulder support, inadequate padding in the backrest and seat pan, arm rests that did not fit under working surfaces, and a lack of appropriate seat pan height adjustment.

Recommendations were made for the new workstations, with pilot testing to be conducted before full-scale implementation. Although existing workstations were inadequate, it was found that task demands were of greatest concern to the employees. The authors proposed a basic workstation design to support each of the three types of VDT jobs. The proposed workstations allowed for modifications and additions, depending on the job and individual employee preferences. Figures 50.6(*a*) and 50.6(*b*) show the recommended design.

The study was performed with the intention of designing the work environment in accordance with the nature of the job performed. Therefore, the departing point for the study was the analysis of the job and its essential characteristics, such as reliance on manuals and hard copy, need for communication with coworkers and supervisors, and extensive use of telephones and VDTs.

50.6.2 Working Surfaces

In the field studies of Hünting et al. (1981) and Läubli and Hünting (1985), several significant relationships were discovered between the design of workstations and postures and the incidence of complaints or medical findings. These results can be summarized as follows: Physical discomfort and/or the number of medical findings in the neck-shoulder-arm-hand area are likely to increase when:

- The keyboard level above the floor is too low

(a)

(b)

Figure 50.4*a* and Figure 50.4*b* Numberous ergonomic risk factors impact VDT operators in these old workstations. Note the VDT and keyboard placement, insufficient work surface dimensions, glare on the CRT, insufficient knee and toe space, sharp desk edges, noise sources, and the inability for the chair armrests to fit under any of the worksurfaces.

- Forearms and wrists cannot rest on an adequate support.
- The keyboard level above the desk is too high.
- Operators have a marked head inclination.
- Operators adopt a slanting position of the thighs under the table due to insufficient space for the legs.
- Operators disclose a marked sideward twisting (*ulnar abduction*) of hands while operating the keyboard.

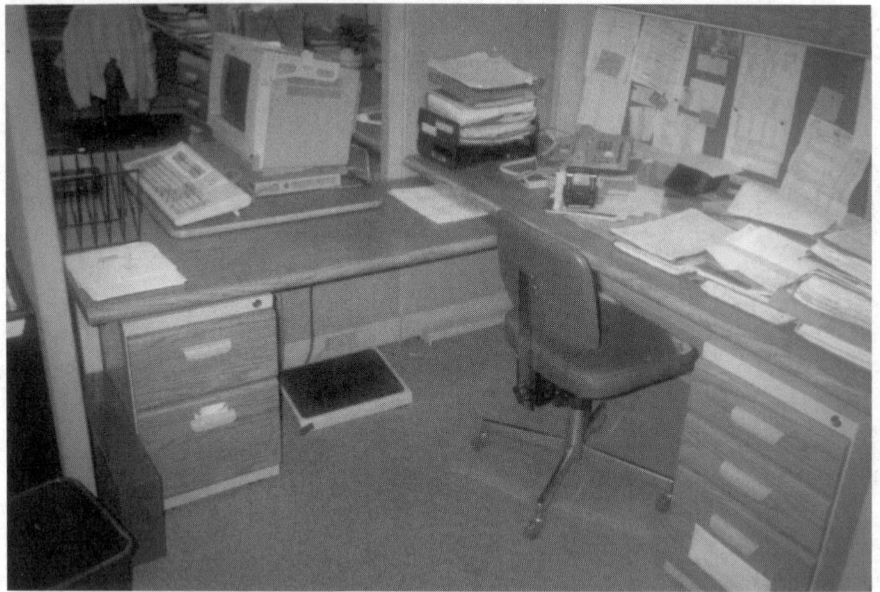

Figure 50.5 A shared VDT workstation using a lazy susan style swivel base. Note the poor chair design.

The recommended size of the work surface is dependent on the task(s), the documents, and the technology. The primary working surface (e.g., those supporting the keyboard, the display, and documents) should be sufficient to: (1) allow the screen to be moved forward or backward, (2) allow a detachable keyboard to be placed in several locations, and (3) allow source documents to be properly positioned. Additional working surfaces (e.g., secondary working surfaces) may be required in order to store, lay out, read and/or write on documents or materials. Sometimes computer workstations are configured so that multiple pieces of equipment and source materials can be equally accessible to the user. In this case, additional working surfaces are necessary to support these additional tools and should be arranged to allow for easy movement from one surface to another.

The tabletop should be as thin as possible to provide clearance for the operator's thighs and knees. Moreover, it is important to provide unobstructed room under the working surface for the feet and legs so that operators can easily shift their posture. Knee-space height and width and toe depth are the three key factors for the design of clearance space under the working surfaces. The recommended minimum width for leg clearance is 51 cm, while the preferred minimum width is 61 cm (ANSI/HFS-100, 1988). The minimum depth under the work surface from the operator edge of the work surface should be 38 cm for clearance at the knee level and 59 cm at the toe level (ANSI/HFS-100, 1988). A good workstation design accounts for individual body sizes and often exceeds minimum clearances to allow for free postural movement.

Table height has been shown to be an important contributor to computer user musculoskeletal problems (Grandjean et al., 1983; Hünting et al., 1981). In particular, tables that are too high (normal desk height of 30 in./76 cm is too high for most people) cause the keyboard to be too high for many operators. This puts undue pressure on the hands, wrists, arms, shoulders, and neck. It is desirable for table heights to vary with the height of the operator, particularly if the chair is not height adjustable. Height-adjustable working surfaces are effective for this. Adjustable multisurface tables enable good posture by allowing the keyboard and display to be independently adjusted to appropriate keying and viewing heights for each individual and each task. Tables that cannot be adjusted easily are a problem when used by several individuals of differing sizes, especially if the chair is not height adjustable either. When adjustable tables are used, the ease of adjustment is essential. Adjustments should be easy to make, and operators should be instructed how to adjust the workstation to be comfortable.

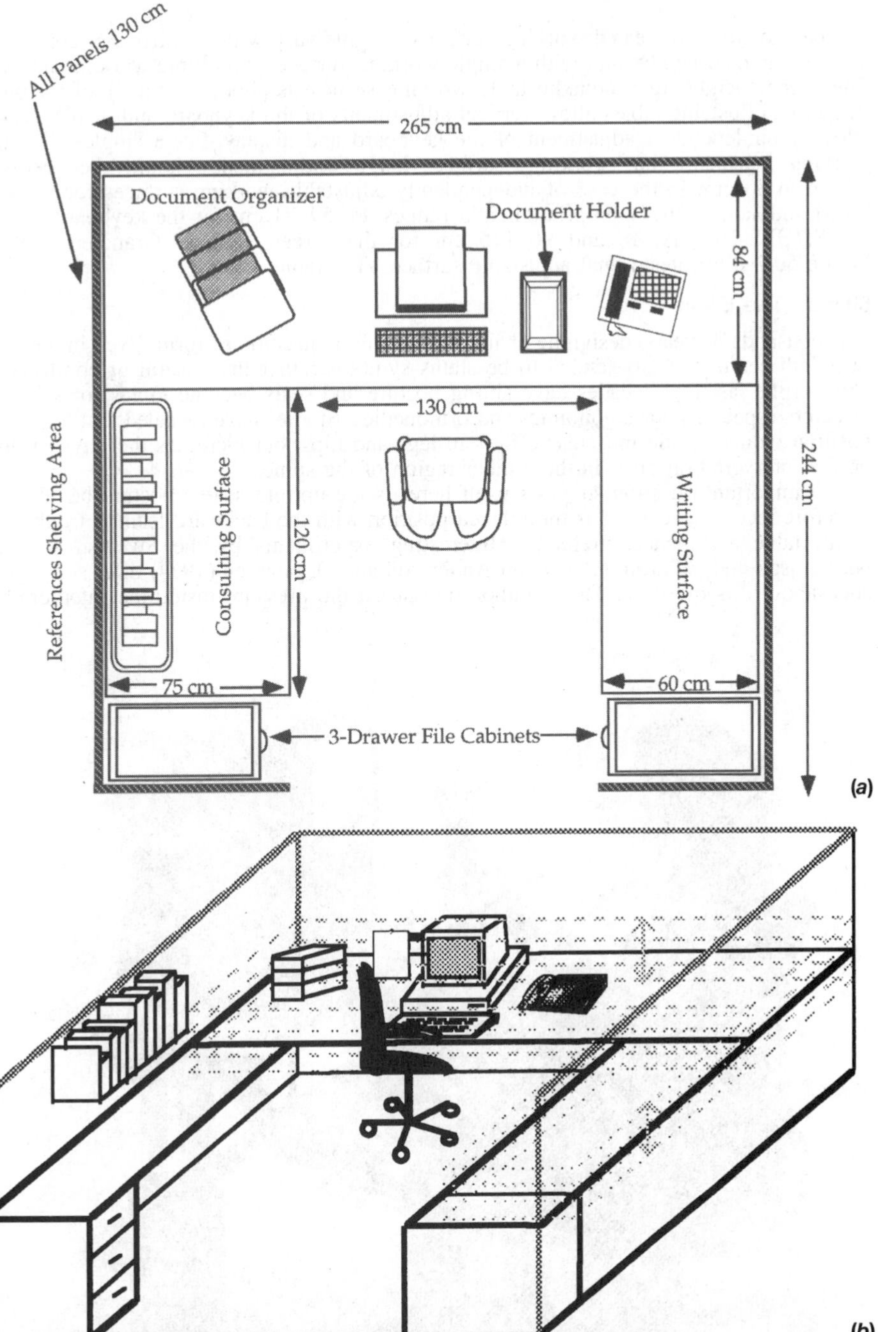

Figure 50.6*a* and Figure 50.6*b* Bird's eye view (*a*) and 3-dimensional view (*b*) of a recommended workstation layout and dimensions based on ergonomic measurements and task analysis for VDT operators. (Adapted from Cohen et al., 1995.)

Specifications for seated working surfaces' heights vary with whether the table is adjustable or fixed height, and with a single working surface or multiple working surfaces. The proper height for a nonadjustable working surface is about 70 cm (floor to top of surface). Adjustable tables allow vertical adjustments of the keyboard and display. Some allow for independent adjustment of the keyboard and display. For a single adjustable working surface, the working surface height adjustment should be between 60 cm to 80 cm. In the case of independently adjustable working surfaces for the keyboard and screen, the appropriate height ranges are 59–71 cm for the keyboard surface (ANSI/HFS-100, 1988), and 90–115 cm for the screen surface (Grandjean, 1987). Figure 50.7 illustrates a dual adjustable surface VDT table.

50.6.3 The Chair

For thousands of years, designing chairs was mainly a question of form. Even in the early twentieth century, chairs tended to be status symbols rather than useful or comfortable. Only in the last few decades have sitting posture and seats become topics for scientific research, especially for ergonomics and orthopedics. Studies have revealed that the sitting position reduces static muscular efforts in legs and hips, but increases the physical load on the intervertebral discs in the lumbar region of the spine.

An important question suggests itself here. Is the upright seated posture healthy and therefore recommended, or is the relaxed position with the backward-leaning trunk more comfortable and hence, preferred? Interesting experiments by the Swedish surgeons Nachemson and Elfstrom (1970) and Andersson and Ortengreen (1974) offer some guidance about this question. These authors measured the pressure inside the intervertebral

Figure 50.7 A dual adjustable surface VDT table. Though not an ideal design, this VDT workstation illustrates flexibility and adjustability. (Source: Steelcase Corp.)

discs as well as the electrical activity of the back muscles in relation to different sitting postures. When the backrest angle of the seat was increased from 90 to 120°, subjects exhibited an important decrease of the intervertebral disc pressure and of the electromyographic activity of the back. Since heightened pressure inside intervertebral discs means that they have more stress, it is concluded that a sitting posture with reduced disc pressure is more healthy and desirable.

The results of the Swedish studies indicate that by leaning the back against an inclined backrest some of the weight of the upper part of the body is transferred to the backrest. This reduces considerably the physical load on the intervertebral discs and the static strain of the back and shoulder muscles. Thus, VDT operators instinctively do the right thing when they prefer a backward-leaning posture and ignore the traditionally recommended upright trunk position.

Most ergonomic standards for VDT workstations are based on traditional views on healthy sitting postures. Mandal (1982a,b) reports that the "correct seated position" goes back to 1884 when the German surgeon Staffel recommended the well-known upright position (see Figure 50.8) Mandal (1982a) stated with exaggerated subtlety:

> But no normal person has ever been able to sit in this peculiar position (upright trunk, inward curve of the spine in the lumbar region and thighs in a right angle to the trunk) for more than 1-2 minutes, and one can hardly do any work as the axis of vision is horizontal. Staffel never gave any real explanation why this particular posture should be better than any other posture. Nevertheless, this posture has been accepted ever since quite uncritically by all experts all over the world as the only correct one. (pp. 520–524)

It is indeed a fact that the sitting posture of students in the lecture hall or of any other audience is very seldom a "correct upright position of the trunk." On the contrary, most people lean backward (even with unsuitable chairs) or in some cases lean forward with elbows resting on the desk. It is most probable that these two preferred trunk positions are associated with a substantial decrease of intervertebral disc pressure, as well as lessened tension of muscles and other tissues in the lumbar and thoracic spine (Andersson and Ortengreen, 1974; Nachemson and Elfstrom, 1970).

One restriction is suggested here: Some special work situations (such as manual work requiring freedom of movement or physical effort) might call for an upright trunk position with elbows down and forearms in a horizontal plane. Presumably the old mechanical typewriters requiring key forces of several hundred grams (g) were more easily operated in such a posture. But the advances in electronic keyboard technology permit very rapid keying with low key forces of 40–80 g and key displacements of 3–5 mm. The new keyboard is mainly operated by finger movements with hardly any assistance of forearms. These conditions might, to some extent, explain why VDT operators in offices prefer to

Figure 50.8 Recommended and typical postures at VDT workstations. The left drawing illustrates "wishful thinking" where the trunk is in an upright posture with the elbows down and forearms nearly horizontal. The right drawing is most commonly observed for VDT work and represents the preferred body posture.

lean backward, keep the upper arms slightly forward with the wrists on a support (which can be the desk itself), and adjust the keyboard height to a rather high level.

Poorly designed chairs (see Figure 50.9 for an example) can contribute to computer user discomfort. Chair adjustability in terms of height, seat angle, and lumbar support helps to provide trunk, shoulder, neck, and leg postures that reduce strain on the muscles, tendons, and discs. The postural support and action of the chair help maintain proper seated posture and encourage good movement patterns. A chair that provides swivel action encourages movement, whereas backward tilting increases the number of postures that can be assumed. The chair height should be adjustable so that the VDT operator's feet can rest firmly on the floor with minimal pressure beneath the thighs. The minimum range of adjustment for seat pan height should be between 38 and 52 cm.

To enable very small users to sit with their feet on the floor without compressing their thighs, it may be necessary to add a footrest. A well-designed footrest has the following features: (1) it is inclined upwards slightly (about 5–15°); (2) it has a nonskid surface; (3) it is heavy enough that it does not slide easily across the floor; (4) it is large enough for the feet to be firmly planted; (5) it is portable; and (6) it is height adjustable.

The seat "pan" is where the person sits on the chair. It is the part of the chair that directly supports the weight of the buttocks. The seat pan should be wide enough to permit operators to make slight shifts in posture from side to side. This not only helps

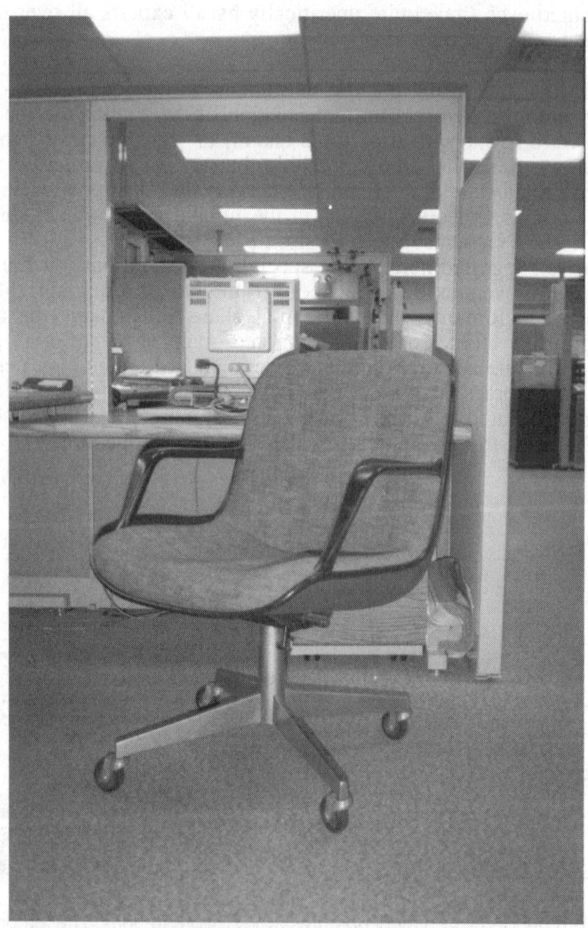

Figure 50.9 An example of poor chair design. Note the wide and deep seat pan, wide back rest, lack of lumbar support, thin armrests too far apart, lack of adjustability, and four legged base. This is a chair to avoid.

to avoid static postures, but also accommodates a large range of individual buttocks sizes. The seat pan should not be too U-shaped because this can lead to static sitting postures. The minimum seat pan width should be 45 cm, and the depth between 38 and 43 cm (ANSI/HFS-100, 1988). The front edge of the seat pan should be well rounded downward to reduce pressure on the underside of the thighs that can affect blood flow to the legs and feet. This feature is often referred to as a "waterfall" design chair. The seat needs to be padded to the proper firmness that ensures an even distribution of pressure on the thighs and buttocks. A properly padded seat should compress about ½ to 1-in. when a person sits on it. Some experts feel that the seat front should be elevated slightly (up to 7°), while others feel it should be lowered slightly (about 5°). There is little agreement among the experts about which is correct. Many chairs allow for both front and backward angling of the seat pan front edge. Seat pan adjustments should be accessible and easy to use (from a seated position).

The tension and tilt angle of the backrest should be adjustable. Inclination of chair backrest is important for operators to be able to lean forward or back in a comfortable manner, while maintaining a correct relationship between the seat pan angle and the backrest inclination. A backrest inclination of about 110° is considered an appropriate posture by many experts. However studies have shown that operators may incline backwards as much as 125°. Backrests that tilt to allow an inclination of up to 125–130° are therefore a good idea. Backrest tilt adjustments should be accessible and easy to use. The advantage of having an independent tilt angle adjustment is that the backrest tilt will then have little or no effect on the front seat height or angle. This also allows operators to shift postures readily.

Chairs with high backrests are preferred since they provide both lower back and upper back (shoulder) support. This allows employees to lean backward or forward, adopting a relaxed posture and resting the back and shoulder muscles. A full backrest with a height of around 45–51 cm is recommended. To prevent back strain, it is also recommended that chairs have lumbar (midback) support, because the lumbar region is one of the most highly strained parts of the spine when sitting.

For most computer workstations, chairs with rolling castors are desirable: They are easy to move and facilitate postural adjustment, particularly when the operator has to reach for equipment or materials that are on the secondary working surfaces. Chairs should have five supporting legs.

Armrests are another important chair feature. Both pros and cons to the use of armrests at computer workstations have been advanced. On the one hand, some chair armrests can present problems of restricted arm movement, interference with keyboard operation, pinching of fingers between the armrest and table, restriction of chair movement, such as under the work table, irritation of the arm or elbows, and adoption of awkward postures. On the other hand, well-designed armrests or elbowrests can provide support for resting the arms to prevent or reduce arm, shoulder, and neck fatigue. Properly designed armrests can overcome the problems mentioned above. Removable armrests are an advantage because they provide greater flexibility for individual operator preference. For specific tasks such as using a numeric keypad, a full armrest can be beneficial in supporting the arms. Many chairs have height adjustable armrests that are helpful for operator comfort. Figure 50.10 is an example of an adjustable chair specifically designed for VDT work.

50.6.4 Screen Swivel and Tilt

Reorientation of the screen around its vertical and horizontal axes can help to position a screen to reduce screen reflections and glare. Reflections can be reduced by simply tilting the display slightly back or down, or to the left or right away from the source of glare. The perception of screen reflections depends not only on screen tilt, but also on the operator's line of sight.

50.6.5 Other Workstation Considerations

An important component of the workstation that can help reduce musculoskeletal loading is a document holder. When properly designed, proportioned, and placed, document holders reduce awkward inclinations of the head and neck and frequent movements up and down and back and forth. They permit source documents to be placed in a central location at the same viewing distance as the computer screen. This eliminates needless head and neck movements and reduces eyestrain. In practice, some flexibility about the location, adjustment, and position of the document holder should be maintained to accommodate both task requirements and operator preferences. Dainoff (1982) showed the effectiveness

Figure 50.10 A highly adjustable chair, specifically designed for the VDT workstation. This is not the recommended position for the armrests, but rather, a demonstration of adjustability. (Source: Steelcase Corp.).

of an in-line document holder. The document holder should have a matte finish so that it does not reflect light.

Privacy requirements include both visual and acoustical control of the workplace. Visual control prevents physical intrusions, contributes to confidential/private conversations, and prevents the individual from feeling constantly watched. Acoustical control prevents distracting and unwanted noise (from machine or conversation) and permits speech privacy. While certain acoustical methods and materials such as free standing panels are used to control general office noise level, they can also be used for privacy. In addition, they also provide workstation privacy. Planning for privacy should not be made at the expense of visual interest or spatial clarity. For instance, providing wide visual views can prevent the individual from feeling isolated. Thus, a balance between privacy and openness enhances user comfort, work effectiveness, and office communications. Involving the employee in decisions of privacy can help in deciding the compromises between privacy and openness.

50.7 THE ENVIRONMENT

50.7.1 The Auditory Environment

A major advantage of computer technology over the typewriter is less noise at the office workstation. Even so, it is not unusual for computer users to complain of bothersome noise that interferes with job tasks. While noise levels encountered in offices are below established limits for hearing loss (e.g., below 85 dba), employees have expectations that their work areas will be quiet. Task demands often require concentration that makes noise an annoyance. In this case, noise can disrupt the ability to concentrate; this may lead to stress. Problems of noise may be exacerbated in open-plan offices in which noise is harder for the individual employee to control than in enclosed offices.

Noise control can be achieved through the use of acoustical materials on the ceiling, floor and walls, furniture, and equipment to absorb sound rather than reflect it. Ceilings that scatter, absorb, and minimize the reflection of sound waves are desirable to promote speech privacy and reduce general office noise levels. The most common means of blocking a sound path is to build a wall between the source and the receiver. Walls are not only sound barriers, but are also a place to mount sound-absorbent materials. In open-plan offices, freestanding acoustical panels can be used to act as walls that reduce the ambient noise level and also to separate an individual from the noise source. The full effectiveness of acoustical panels is best achieved in concert with the soundabsorbent materials and finishes applied to the walls, ceiling, floor, and other surfaces. For instance, carpets not only cover the floor, but also serve to reduce noise. This is achieved in two ways: (1) carpets absorb the incident sound energy, and (2) gliding and shuffling movements on carpets produce less noise than on bare floors. Furniture coverings and draperies are also important for noise reduction. The use of acoustical covers for line printers can be very effective in lowering the noise level associated with this equipment although the increasing use of higher quality printers achieves a better result.

Acoustical control can also be achieved by proper space planning. For instance, workstations that are positioned too closely do not provide suitable speech privacy and can be a source of disturbing conversational noise. As a general rule, there should be a minimum of 8–10 ft between employees, separated by acoustical panels or partitions, to provide normal speech privacy.

50.7.2 Heating, Ventilating, and Air Conditioning (HVAC)

Temperature, humidity, air flow, and air exchanges are important parameters for employee performance and comfort. True, it is unlikely that offices will produce either excessively hot or cold temperatures that could be physically harmful to employees. However, thermal comfort is an important consideration in employee satisfaction and one that can influence job performance. Satisfaction is not based on the ability to tolerate and experience temperature extremes, but on moderation. Many studies have shown that most office employees are not satisfied with their thermal comfort (Smith, 1984; Stammerjohn et al., 1981). The definition of a comfortable temperature is usually a matter of personal preference. Opinions as to what is a comfortable temperature vary within an individual from time to time and certainly across individuals. Seasonal variations of ambient temperature influence perceptions of thermal comfort. Office employees sitting close to a window may experience the temperature as being too cold or hot depending on the outside weather, while those in the center of the room are comfortable. It is virtually impossible to generate one room temperature in which all employees are equally satisfied over a long period of time.

As a general rule, it is recommended that the temperature should be maintained in the range 20–24° C (68–75° F) in winter and in the range 23–27° C (73–81° F) in summer (NIOSH, 1981). Air flows across a person's neck, head, shoulders, arms, ankles, and knees should be kept low (below 0.15 m/sec in winter and below 0.25 m/sec in summer). It is important that ventilation not produce currents of air that blow directly on employees or their work materials. This is best handled by proper placement of the workstation in relation to fans and other sources of ventilation.

Relative humidity is also an important component of office climate and influences employee comfort and well-being. Air that is too dry tends to dry out the mucous membranes of the eyes, nose, and throat. Individuals who wear contact lenses and work at VDTs may be especially uncomfortable because of dry air. In instances where intense, continuous near-vision work at the computer is required, very dry air has been shown to irritate the eyes. As a general rule, it is recommended that the relative humidity in office environments be at least 50%, and less than 60%. Air that is too wet enhances the growth of unhealthy organisms that can cause disease (such as Legionnaire's disease).

50.8 GENERAL DESIGN ISSUES

50.8.1 Just What Is Ergonomics and What Does It Mean for Computer Workstation Design?

Ergonomics is the science of fitting the environment and activities to the capabilities, dimensions, and needs of people. Ergonomic knowledge and principles are applied to adapt working conditions to the physical, psychological, and social nature of the person. The goal of ergonomics is to improve performance while enhancing comfort, health, and safety. Computer workstation design is more than just making the computer interfaces easier to use or making furniture adjustable in various dimensions. It also involves integrating design considerations with the work environment, the task requirements, social aspects of work, and job design.

50.8.2 Implementing Ergonomic Changes and the Use of Participatory Ergonomics

The reader is referred to Chapter 18 *Organizational Design and Macroergonomics* for more details about organizational strategies for implementing ergonomic improvements and participatory ergonomics. Implementing a change in technology (workstation) or work methods is a complex process, because it impacts various components of the work system and organizations (Smith and Carayon, 1995; Smith and Sainfort, 1989). It is a serious mistake to think that technological enhancements are easy to make and that performance and health improvement results of such changes will be substantial and immediate. Proper technological implementation involves changes in work organization, job content, socialization at work workstation redesign, task improvements, and changes in job demands. In fact, the complexity of the change process may hamper early success. Planning for change can help the success of implementation and reduce the stress generated by the changes. But the success of implementing change depends heavily on the involvement and commitment of the concerned parties: top management, middle management, technical staff, support staff, first-line supervision, and employees.

Planning for change in an organization requires the coordination of the various system components. This creates the need to have interaction among these subsystems and to gather information from each for proper coordination. This duty falls to the administrative or technical functions in the organization, because there are sometimes differences of opinion among the subsystem managers about the relative importance of the strategic aspects of change. Organizational change must be directed by the chief executive officer of the company to ensure agreement (and compliance) among the subsystems' managers. Without the involvement of top management it is virtually impossible to gain the necessary coordination among subsystems, and lack of coordination can lead to serious implementation problems. Top management involvement also demonstrates the importance of the proposed changes to managers and employees. Top management needs to share and support a common vision of the changes (Burnes and James, 1992).

There is agreement among change experts that the most successful strategies for workplace change include aspects of involvement by all subsystems that will be affected by the change (Lawler, 1986). Involvement assumes that there is an active role in, not only providing strategic information, but also providing opinions about the implications of the change(s) for the subsystem. There are different aspects of involvement that are important, such as timing, nature, and method. In the planning phase, it is important for top management to consider the timing of involvement, for instance, how soon to involve employees. Effective involvement requires the sharing of critical information among the subsystems to identify the best possible strategies and potential problems. Active participation generates greater motivation and better acceptance of solutions than passively providing information and taking orders. Active participation is achieved by soliciting opinions and sharing authority to make decisions about solutions. However, one drawback of active participation is the need to develop consensus among participants who have differing opinions and motives. This usually takes more time than more traditional management decisionmaking, and can bring about conflict among subsystems.

The *participatory ergonomics* approach specifies that the "end users of ergonomics" (namely, the employees) should be heavily involved in planning and implementing the ergonomic change (Hendrick, 1986). Participatory ergonomics is a technique that has proven successful to solve ergonomic problems and to implement ergonomic changes. It can be particularly useful because employees can have a say in the identification and analysis of ergonomic problems. Participative ergonomics can take various forms, such as design decision groups, quality circles, and worker-management committees. Some of the common characteristics of these various programs are: Employee involvement in developing and implementing ergonomic solutions, dissemination and exchange of information, pushing ergonomics expertise down to lower levels, cooperation between experts and nonexperts, and consideration of employees' opinions.

One of the characteristics of participatory ergonomics is the dissemination of information (Noro, 1991). Participative ergonomics can be beneficial to reduce or prevent resistance to change because it encourages employee involvement, but also because of the information provided to the various members of the organization concerned with new technology. Uncertainty and lack of information are two major causes of resistance to change and have been linked to increased employee stress. If employees are informed about potential ergonomics changes in advance, they are less likely to actively resist the change.

The role of the "technical expert" varies considerably in the various forms of participatory ergonomics. At the beginning of the change process, the expert may play a very active role in informing, teaching, and directing; then move to a coaching role; and finally become an advisor or mentor. The training and information received by the employees is important to the success of participatory ergonomics. Employees have much knowledge and expertise about their job, but not necessarily about the new technology or organizational strategy. It is possible that employees may never become complete "technical experts" and that well-trained technicians will still be needed as implementation proceeds. However, it is important to involve employees early in the change process to reduce resistance to change, give employees a chance to start learning about the technology, and to demonstrate commitment to them and their importance to the company. Most applications of participatory ergonomics have addressed physical ergonomics problems. Participatory ergonomics should be expanded to examine not only physical stress, but also psychological stress. Given the influence of automation on both physical and psychological stress (Landau and Rohmert, 1992; Smith, 1986; Smith and Carayon, 1995), participatory ergonomics can be a useful approach to foster employee involvement and reduce the negative aspects of VDT automation.

50.8.3 Work Practices

Good ergonomic design of computer workstations has the potential to reduce visual and musculoskeletal complaints, as well as to increase employee performance. However, regardless of how well a workstation is designed, if operators must adopt static postures for a long time, performance, comfort, and health problems may still exist. Thus, designing tasks that induce employee movement in addition to work breaks can contribute to comfort and help relieve employee fatigue.

50.8.3.1 Work Breaks

As a minimum, a 15-min break should be taken after 2 hr of continuous computer work. Breaks should be more frequent as visual, muscular, and mental loads increase. VDT operators should be encouraged to take frequent "informal" rest breaks when they feel fatigued or tired (Henning et al., 1993, 1994).

50.8.3.2 Workload and Work Standards

The determination of appropriate workloads also contributes to employee comfort and performance. Workload for computer operators should be set using scientific methods of time and motion evaluation and not be set by other criteria, such as economic need to recover capital investment or by the limits or capacity of the computer system. Human cognitive, physical, and perceptual motor capabilities have to be taken into consideration. Work standards should take into consideration task loads, individual needs (e.g., rest breaks), and equipment requirements (e.g., breakdowns, slow responsiveness). This is especially true when work standards are enforced by monitoring of employee performance.

50.8.3.3 Management Style

In terms of management style, a supportive type of supervision can be very helpful and contribute to the successful implementation and use of computers. Supervisors should use positive motivational and employee support approaches to enhance performance and use of ergonomic features of the workstation. The training of supervisors in supportive approaches can help to buffer the effects of stress and promote good work practices. It is also important to have technically skilled supervisors who can assist those operators having technical difficulties.

Total quality management (*TQM*) is a customer-driven approach that emphasizes the need for companies to design production processes and organizational structures that can produce high-quality products. ISO 9000 is an international standard designed to encourage companies to set up these processes and structures. The principal concepts of ISO 9000 are outlined in the following quote (ISO, 1987).

> An organization should seek to accomplish the following three objectives with regard to quality:
> a) The organization should achieve and sustain the quality of the products or service produced so as to meet continually the purchaser's stated or implied needs.
> b) The organization should provide confidence to its own management that the intended quality is being achieved and sustained.
> c) The organization should provide confidence to the purchaser that the intended quality is being, or will be, achieved in the delivered product or service provided. When contractually required, the provision of confidence may involve agreed upon demonstration requirements.

To achieve these objectives the organization has to have the following elements: (1) a quality policy expressed by top management; (2) management that determines and implements the quality policy; (3) an organizational structure and process for implementing quality management; (4) quality control or measurement and monitoring techniques and activities for fulfilling quality management requirements; and (5) quality assurance to ensure that products will satisfy quality requirements.

To put all of these activities in place requires the active involvement of management, technical specialists, and employees. It also requires a corporate climate of trust and confidence among the members of the organization, and a willingness by all to participate. In addition, it must be recognized that product quality improvement goes hand-in-hand with improvements in the quality of jobs and the overall quality of working life (Smith, Sainfort, Carayon-Sainfort, and Fung, 1989). Concern for employee health and safety is a requisite condition for successful quality management efforts.

The application of ISO 9000 and the implementation of TQM concepts have three major implications for technology implementation and ergonomic redesign. First, ISO 9000 emphasizes employee involvement as a tool necessary for quality. Second, TQM can be an essential element of management philosophy that can facilitate technology redesign and implementation. Third, ISO 9000 emphasizes the need for TQM organizations to be concerned with the health and safety of their employees.

50.8.3.4 Training

Training computer users on how the technology and the workstation function and operate has often been a neglected element in office automation. Many times the extent of operator training is limited to reading the manual and to learning by trial and error. In some cases, operators may go to classes to review the material in the manual and give hands-on practice with the new equipment for limited periods of time. The problem with these approaches is that there is usually insufficient time for users to develop the skills and confidence to adequately use the new equipment. Experience in implementing new workstations indicates that training of VDT operators in how to use the workstation properly is important especially if the adjustment controls are neither obvious nor intuitive.

For discussion of general approaches for training, please refer to the chapter on training issues in this volume. Hagberg et al. (1995) have indicated that employee training is a necessary component to any ergonomic program for reducing work-related musculoskeletal disorders. Green and Briggs (1989) found that adjustable workstations are not always effective without appropriate information about benefits of adjustments and training in how to use the equipment. Hagberg et al. (1995) discuss some basic problems that can reduce the effectiveness of training for the reduction of musculoskeletal strain. These are: (1) the training period is too short; (2) the training does not transfer to real-world activities; (3) the principles taught may not be valid; (4) the training is not accompanied by other ergonomic improvements; (5) the organizational context is ignored; and (6) training to improve knowledge and attitudes may not lead to behavior changes. Based on these problems, they suggest the following considerations for ergonomics training programs:

1. Have employees involved in the development and process of training. Using employee work experiences can be helpful in illustrating principles to be learned during training. In addition, using employees as instructors can be motivational for the instructors and learners.
2. Use active learning processes where learners participate in the process and apply "hands-on" methods of knowledge and skill acquisition. This approach to learning enhances acquisition of inputs and motivation to participate.
3. Apply technology to illustrate principles such as audiovisual equipment and computers. Much like active processes, technology provides opportunities for learners to "visualize" the course materials and to test their knowledge dynamically and immediately.
4. Use of on-the-job training is preferred over classroom training. Both can be effective in combination.

50.8.4 Achieving "Balance" in a Work System

As we have previously indicated, there is a need for addressing the entire work system when designing workstations that reduce the risk of musculoskeletal fatigue, strain, and disease, and that enhance employee performance. It is essential to look at the work organization aspects of the human–workstation interface to achieve a proper "balancing" of system components. This requires some consideration of the job task and organizational management elements of the work system in addition to the workstation design requirements (see Figure 50.11).

There are two aspects of "balance" when addressing job and organizational design. These are (1) system balance and (2) compensatory balance. System balance is based on the idea that a workplace or process or job is more than the sum of the individual components of the system. The interplay among the various components of the system produces results that are greater (or lesser) than the additive aspects of the individual parts. It is the way in which the system components relate to each other that determines the potential for the system to produce positive or negative results. If the designer concentrates solely on the technological component of the system, then there is an "imbalance" because the personal, psychosocial, and task aspects are neglected.

The second type of balance is "compensatory" in nature. It is seldom possible to maximize all aspects of a work system due to financial considerations, or inherent aspects of job tasks. In such a case, good system elements can be used to balance the negative effects of poor system elements. For instance, changes in workstation design can reduce biomechanical stress when changes in work methods are not possible. This type of com-

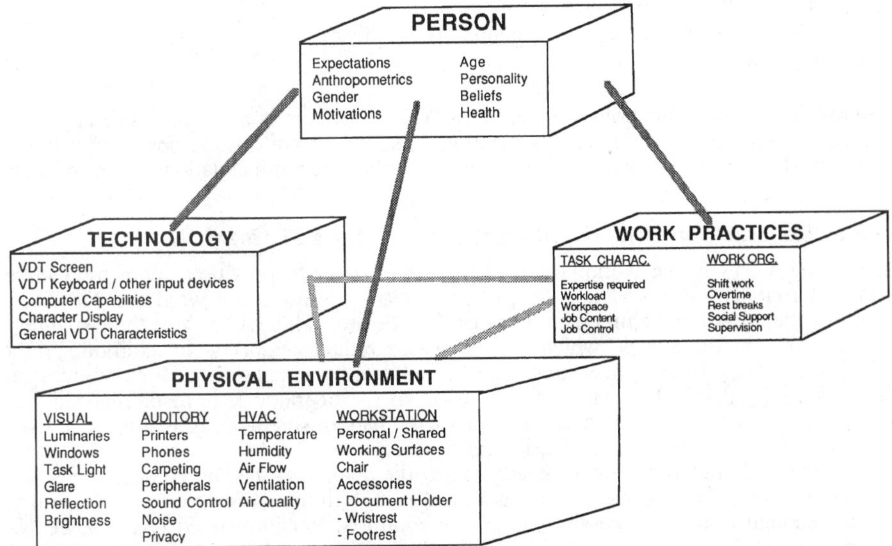

Figure 50.11 The work system model of the VDT office environment.

pensation has been used very successfully to reduce biomechanical stressors due to the use of video display terminals (OTA, 1985). In this same example, it would be important to provide proper workstation design to enhance employee shoulder, arm, wrist, hand, and finger postures. This would provide positive system balance because research has shown that postural factors are related to employee health and comfort (Grandjean, 1984; Silverstein et al., 1987). The ergonomic design to reduce biomechanical strain also serves to enhance the psychosocial environment by making employees more comfortable and demonstrating their importance to the company by investing in better workstations.

The essence of this approach is to reduce physical and psychological stress and the negative health consequences caused by VDT automation by "balancing" the various elements of the work system. Proper work organization and job design can best be achieved by meeting recognized criteria for physical loads, work cycles, and job content such that individual physiological and psychological needs are met. The best designs will eliminate all sources of stress, job dissatisfaction, and discomfort. However, such a perfect job can seldom be achieved. Using positive work elements to compensate for negative work elements can serve to diminish the adverse performance and health effects. This, in turn, moderates negative factors and reduces the total negative load on the system.

A major feature of the ergonomics approach is that the job task characteristics will help to define the ergonomic conditions and the workplace design requirements. Specific recommendations and guidelines which result from using the ergonomic approach may have to be modified to account for varying technology, personal, situational, or organizational needs, as well as improved knowledge about human–computer interaction. Furthermore, it cannot be overstated that such recommendations and guidelines are dynamic and should not be viewed as fixed specifications.

50.9 SUMMARY OF RECOMMENDATIONS

A VDT workstation is composed of the VDT, the furniture where the VDT is placed, and the physical environment in which the VDT is used. The design of these elements and how they fit together play a crucial role in improving VDT workstation design to allow the operator to minimize potential adverse loads and optimize positive ones. This can promote a healthier and more productive workplace environment. The recommendations presented in this section address the physical environment. It is important to note, however, that this is only one component of the larger system to be improved. Hence, important consideration should also be given to organizational factors and task related factors as they affect and/or depend on the individual. It is also important that unique situational factors be considered when an ergonomic intervention is being implemented.

With regard to the physical environment, there are four major participants to consider:

- VDT manufacturers
- Office furniture designers
- VDT workstation designers/ergonomics engineers
- Operator

Generalizing recommendations relevant to all VDT operators is a mistake. The approach presented in this chapter emphasizes meeting operators' needs so that positive loads are emphasized and negative ones are minimized. For these recommendations, refer to Table 50.2.

50.9.1 General Workstation Recommendations for VDT Users

Positioning VDTs in the workplace is important in order to provide a more productive work environment. VDTs should be placed at right angles to the windows. However, windows should not be behind or in front of the operator. This will reduce the possibility of reflections on the screen, which can otherwise reduce legibility. In addition, bright reflections coming from light sources can be reduced by placing these light sources on either side of and parallel to the line of vision of the operator. It is recommended that only phase-shifted fluorescent tubes with prismatic pattern shields or grid shields be used in the office environment where VDT work is done.

Moreover, illumination should be adapted to the quality of the source documents and the task required. This is done by having a high enough illumination to enhance legibility, yet low enough to avoid excessive luminance contrasts. Recommended levels range between 300 and 700 lux.

Table 50.2 Summary of Recommendations for VDT Workstation Design

DESIGN COMPONENT	RECOMMENDATIONS
VDT PLACEMENT	**VDT WORKSTATION DESIGNERS** ● VDT must be arranged at a right angle to the window front. This is to avoid risk of reflection on the screen. • Windows should be neither behind nor in front of the operator • If possible, place intermediate screen against windows or bright surfaces. • Viewing distance: 45–50 cm between screen and operator's eye (no more than 70 cm)
VDT WORKSTATION	**OFFICE FURNITURE DESIGNERS** • Controls for adjustment of workstation dimensions should be easy to use and easily accessible. **VDT WORKSTATION DESIGNERS** • Adjustment ranges: • keyboard height (home row to floor): 59–71 cm (up to 80 cm preferred) • Screen center above floor: 90–115 cm • Screen backward inclination to horizontal plane: 88° to 105° • Keyboard (home row) to table edge: 10–26 cm • Screen distance to table edge: 50–75 cm
WORKSTATION CHAIR	**OFFICE FURNITURE DESIGNERS** • Allow for backward leaning posture, so that relaxation of back muscles is possible and load on intervertebral discs is decreased. • Seat pan height: easily adjustable between 38 and 52 cm. • Backrest length: 45 to 51 cm above seat surface. • Adjustable inclination backward: up to 130°. • Backrest shape: slightly concave form on the thoracic level, and lumbar support.
SPACE FOR LEGS	**VDT WORKSTATION DESIGNERS** Space depth at the level of the knees: • at least 38 cm from table edge • at least 59 cm at feet level • Space width for leg clearance: • at least 51 cm.
LIGHTING	**VDT WORKSTATION DESIGNERS** • Use light sources with phase-shifted fluorescent tubes • Do not place light sources behind operator • Lights should be placed on either side of and parallel to the line of vision • Fluorescent light sources should be provided with prismatic pattern shields or with grid shields • Angle should not exceed 45° to a vertical line • Illumination : Between 300 and 700 lux. High enough for good legibility, but low enough to avoid excessive luminance contrast • Adapt illumination levels to quality of source documents and task requirements.
LUMINANCE	**VDT MANUFACTURERS** • Within user's near field of vision: 1:3 • Within user's far field of vision: 1:10 • Character-to-screen background: 7:1
KEYBOARD	**VDT MANUFACTURERS** • Keying force action (i.e. keying force and displacement): should promote VDT operator comfort and performance. **VDT WORKSTATION DESIGNERS** • Wrist rest or support surface of 15 cm is recommended to rest wrist and forearms. • Keyboard must be movable on the desk, but stable when in use.

Table 50.2 (Continued)

DESIGN COMPONENT	RECOMMENDATIONS
SCREEN	**VDT MANUFACTURERS** • Contrast: 1:7 between background and characters to guarantee good legibility • Background luminance: 6 to 8 cd/m^2 for character luminance of 40 to 50 cd/m^2 • Reflection glare on screen should be reduced by 5–10 times without decreasing sharpness of characters nor darkening the screen background too much. • Background of screen should not be too dark or too bright
OSCILIATION DEGREE OF CHARACTER LUMINANCE	**VDT MANUFACTURERS** • Lower degree of oscillation of characters to a level comparable value for phase-shifted fluorescent tubes • Character refresh rates: up to 80 Hz or 100 Hz with a phosphor decay time of about 10 msec for the 10% luminance level.
CHARACTER SHARPNESS	**VDT MANUFACTURERS** • Blurred border area of characters: should not exceed 0.3 mm
CHARACTER CONSTRAST	**VDT MANUFACTURERS** • Contrast of luminances in the spaces between characters and the bars of characters: ratio of 1:7 • Brightness between characters: should not exceed 14% of brightness of characters
CHARACTER STABILITY	**VDT MANUFACTURERS** • Electronic control of the electron beam must secure good stability • Light dot recorded from middle of character should not exceed a variance luminance of 20%.
CHARACTER SIZE/FACE	**VDT MANUFACTURERS** • Height of capital letters: 3 to 4.2 mm • Height of lower case letters: 3.5 mm • Width of characters: 75% of height • Distance between characters: 25% of height • Space between lines:100-150% of height

50.9.2 General Design Recommendations for VDT Manufacturers

As designers of screen and keyboard interface devices, VDT manufacturers need to be aware of the following guidelines for VDT operator performance issues.

50.9.2.1 Screen/Character Display Characteristics

The background of the screen should be neither too dark nor too bright. A luminance of 1:7 between background and characters provides good legibility. The degree of oscillation of characters should be lowered to a comparable value for phase-shifted fluorescent tubes. Character refresh rates up to 80 or 100 Hz with a phosphor decay time of approximately 10 msec for the 10% luminance level are recommended.

The following specifications should be met regarding the character display. For character sharpness, the blurred border area of characters should not exceed 0.3 mm. For better contrast enhancing legibility, the brightness between characters should not exceed 14% of the brightness of characters. For character stability, a light dot recorded from the middle of a character bar should not exceed a variance of luminance of 20%. In addition to these, some more specific character size specifications are:

- Height of capital letters: 4 mm
- Height of lowercase letters: 3.5 mm
- Width of the characters: 75% of height
- Distance between characters: 25% of height
- Space between lines: 100–150 % of height

50.9.2.2 Keyboard Characteristics

The keyboard must be movable on the work surface, but stable when in use. Characteristics of keying action (for example, keying force and displacement) should promote VDT operator comfort and performance. A wristrest or support surface of 15 cm in depth is recommended to rest the wrist and forearms.

50.9.3 General Design Recommendations for Office Furniture Designers

In order to accommodate user variability and preference, task demands variability, and technological demands variability, office furniture designers need to provide products that allow for adjustability and flexible use. This will promote VDT operator comfort and can contribute to a healthier and more productive workplace. VDT workstations should allow for the following adjustments:

- Keyboard height (home row to floor): 59–71 cm (up to 80 cm preferred)
- Screen center above floor: 90–115 cm
- Screen backward inclination to horizontal plane: 88–105°
- Keyboard (home row) to table edge: 10–26
- Screen distance to table edge: 50–75 cm

A VDT workstation without adjustable keyboard height and without adjustable height and distance of the screen is not reasonable for continuous work with a VDT. The controls for adjusting the dimensions of a workstation and chair should be easy to use. Such adjustability is particularly important at workstations and chairs used by more than one employee.

Furthermore, sufficient space for the legs should be provided to allow for comfort and to avoid unnatural or constrained postures. The space depth at the level of the knees should be at least 38 cm from table edge and at least 59 cm at the level of the feet.

With regard to chair design for VDT workstations, designers and manufacturers should be aware of the fact that a backward-leaning posture allows for relaxation of the back muscles and decreases the load on the intervertebral discs. In order to allow for this, height should be easily adjustable from 38 to 52 cm. The chair should also have a minimum backrest height (above the seat surface) range from 45 to 51 cm, and adjustable inclination backward of up to 130°. In addition, the backrest should have a lumbar support and a slightly concave form on the thoracic level.

These recommendations are not exhaustively inclusive of all aspects of VDT workstation design. The chapter described other important workstation criteria in Sections 50.4–50.7.

ACKNOWLEDGMENTS

The authors would like to thank Antoinette Derjani Bayeh for her invaluable insights and technical contributions to the chapter. Marla C. Haims, Francisco Moro, and Myrna Kasdorf provided excellent critical reviews and editorial assistance. Inadvertent mistakes are the sole responsibility of the authors.

REFERENCES

Andersson, B. J. G., and Ortengreen, R. (1974). Lumbar disc pressure and myoeletric back muscle activity. *Scandinavian Journal of Rehabilitation Medicine*, *3*, 115–121.

ANSI (1988). *American national standard for human factors engineering of visual display terminal workstations.* ANSI/HFS Standard no. 100-1988. Santa Monica, CA: The Human Factors Society.

Bauer, D. (1984). What causes flicker in bright-background VDUs and how to cure it. In E. Grandjean, Ed., *Ergonomic and Health Aspects in Modern Offices*. London, England: Taylor and Francis, Ltd.

Beamon, R. S., and Snyder, H. L. (1975). *An experimental evaluation of the spot wobble method of suppressing raster structure visibility.* Technical Report AMRL-TR-75-63: Wright-Patterson Air Force Base.

Bergqvist, U., Knave, B., Voss, M., and Wibom, R. (1992). A longitudinal study of VDT work and health. *International Journal of Human-Computer Interaction*, *4*(2), 197–219.

Bergqvist, U. O. (1984). Video display terminals and health: A technical and medical appraisal of the state of the art. *Scandinavian Journal of Work, Environment and Health, 10* (Supplement 2), 87 pp.

Berlinguet, L., and Berthelette, D., Eds., (1990). *Work With Display Units 89.* Amsterdam: Elsevier.

Bräuninger, U. (1983). *Lichttechnische eigenschaften der bildschirmgerate aus ergonomischer sicht.*, Zurich: Swiss Federal Institute of Technology, Department of Ergonomics and Hygiene.

Bräuninger, U., Grandjean, E., Fellman, T., and Gierer, R. (1982). Lighting characteristics of VDTs. In *Proceedings of the Zurich Seminar on Digital Communication*, Zurich.

Bräuninger, U., Grandjean, E., van der Heiden, G., Nishiyama, K., and Gierer, R. (1984). Lighting characteristics of VDTs from an ergonomic point of view. In E. Grandjean, Ed., *Ergonomic and Health Aspects in Modern Offices.* London, England: Taylor and Francis, Ltd.

Brocklehurst, E. R. (1991). The NPL Electronic Paper project. *International Journal of Man Machine Studies, 34(1)*, 69–95.

Bullinger, H. J. (1991). *Human Aspects in Computing: Design and Use of Interactive Systems and Information Management.* (Vol. 18B). Amsterdam: Elsevier Science Publishers.

Bullinger, H. J. (1991). *Human Aspects in Computing: Design and Use of Interactive Systems and Work with Terminals.* (Vol. 18A). Amsterdam: Elsevier Science Publishers.

Burnes, B., and James, H. (1992). Human factors in manufacturing: The need for a consistent strategy. *The International Journal of Human Factors in Manufacturing, 2(1)*, 67–79.

Cakir, A., Hart, D. J., and Stewart, T. F. M. (1979). *The VDT Manual.* Darmstadt: Inca-Fiej Research Association.

Carayon, P. (1993). Chronic effects of job control, supervisor social support and work pressure on office worker stress. To be published in Job Stress 2000: Emerging Issues.

Carayon, P. (1994). A longitudinal study of quality of working life among computer users: Preliminary results. In A. Grieco, G. Molteni, E. Occhipinti, and B. Piccoli, Eds., *Proceedings of the Fourth International Scientific Conference on Work with Display Units*, pp. 39–44. Milan, Italy: Elsevier.

Carayon, P., Swanson, N., and Smith, M. J. (1987). Objective and subjective ergonomic evaluations of automated offices. In J. M. Flach, Ed., *Proceedings of the Fourth Midcentral Ergonomics/Human Factors Conference*, pp. 358–366. Urbana, ILL: University of Illinois.

Cohen, B. G. F., Piotrkowski, C. S., and Coray, K. E. (1987). Working conditions and health complaints of women office workers. In G. Salvendy, S. L. Sauter, and J. J. Hurrel, Eds., *Social, Ergonomic and Stress Aspects of Work with Computers*, Vol. 10A, pp. 365–372 Amsterdam: Elsevier.

Cohen, W. J., James, C. A., Taveira, A. D., Karsh, B., Scholz, J., and Smith, M. J. (1995). Analysis and design recommendations for workstations: A case study in an insurance company. In *Proceedings of the Human Factors and Ergonomics Society 39th Annual Meeting, 1*, pp. 412–416. San Diego, CA: Human Factors and Ergonomics Society.

Cornell, P., and Kokot, D. (1988). Naturalistic observation of adjustable VDT stand usage. In *Proceedings of the Human Factors Society 32nd Annual Meeting*, pp. 496–500. San Diego, CA: Human Factors and Ergonomics Society.

CSA. (1989). *A Guideline on Office Ergonomics: A National Standard of Canada.* Document no. CAN/CSA-Z412-m89. Rexdale, Ontario: Canadian Standards Association.

Cushman, W. H. (1984). Data-entry performance and operator preferences for various keyboard heights. In E. Grandjean, Ed., *Ergonomic and Health Aspects in Modern Offices.* London, England: Taylor and Francis, Ltd.

Dainoff, M. J. (1982). Occupational stress factors in visual display terminal (VDT) operation: A review of empirical research. *Behaviour and Information Technology, 1*(2), 141–176.

Dainoff, M. J. (1983). Video Display Terminals: The relationship between ergonomic design, health complaints and operator performance. *Occupational Health Nursing, December*, 29–33.

Elias, R., and Cail, F. (1983). *Constraints et astreints devant les terminaux a ecran cathodique, 1109.* Institut National de Recherche et de Securite.

Faucett, J., and Rempel, D. (1994). VDT-related musculoskeletal symptoms: Interactions between work posture and psychosocial work factors. *American Journal of Industrial Medicine, 26*, 597–612.

Fellman, T., Bräuninger, U., Gierer, R., and Grandjean, E. (1982). An ergonomic evaluation of VDTs. *Behavior and Information Technology, 1*, 69–80.

Fernstrom, E., Ericson, M. O., and Malker, H. (1994). Electromyography activity during typewriter and keyboard use. *Ergonomics, 37(3)*, 477–484.

Ghiringhelli, L. (1980). Collection of subjective opinions on use of VDUs. In E. Grandjean and E. Vigliani, Eds., *Ergonomic Aspects of Visual Display Terminals*, pp. 227–232. London, England: Taylor and Francis, Ltd.

Gomer, F. E., and Bish, K. G. (1978). Evoked potential correlates of display image quality. *Human Factors, 20*, 589–596.

Grandjean, E. (1979). *Ergonomical and medical aspects of cathode ray tube displays* . Zurich, Switzerland: Federal Institute of Technology.

Grandjean, E. (1984). Postural problems at office machine work stations. In E. Grandjean, Ed., *Ergonomics and Health in Modern Offices*, pp. 445–455. London, England: Taylor and Francis, Ltd.

Grandjean, E. (1987). Design of VDT workstations. In G. Salvendy, Ed., *Handbook of Human Factors*, pp. 1359–1397: John Wiley.

Grandjean, E., Hünting, W., and Nishiyama, K. (1982). Preferred VDT workstation settings, body posture and physical impairments. *Journal of Human Ergology, 11(1)*, 45–53.

Grandjean, E., Hünting, W., and Nishiyama, K. (1984). Preferred VDT workstation settings, body posture and physical impairments. *Applied Ergonomics, 15*(2), 99–104.

Grandjean, E., Hünting, W., and Pidermann, M. (1983). VDT workstation design: Preferred settings and their effects. *Human Factors, 25*(2), 161–175.

Grandjean, E., Nishiyama, K., Hünting, W., and Pidermann, M. (1982). A laboratory study on preferred and imposed settings of a VDT workstation. *Behaviour and Information Technology, 1*, 289–304.

Grandjean, E., and Vigliani, E., Eds. (1980). *Ergonomic Aspects of Visual Display Terminals.* London, England: Taylor and Francis, Ltd.

Green, R. A., and Briggs, C. A. (1989). Effect of overuse injury and the importance of training on the use of adjustable workstations by keyboard operators. *Journal of Occupational Medicine, 31*, 557–562.

Grieco, A., Molteni, G., Occhipinti, E., and Piccoli, B. (1995). *Work with Display Units 94*. Amsterdam: Elsevier.

Güggenbuhl, U., and Krueger, H. (1990). Musculoskeletal strain resulting from keyboard use. In L. Berlinguet and D. Berthelette, Eds., *Proceedings of Work With Display Units 89*. Elsevier.

Güggenbuhl, U., and Krueger, H. (1991). Ergonomic characteristics of flat keyboards. In Y. Queinnec and F. Daniellou, Eds., *Proceedings of the Eleventh Congress of the International Ergonomics Association, 1* pp. 730–732. Paris, France: Taylor and Francis, Ltd.

Gunnarsson, E., and Ostberg, O. (1977). *Physical and emotional job environment in a terminal-based data system, 1977:35*: Department of Occupational Safety, Occupational Medical Division, Section for Physical Occupational Hygiene.

Gyr, S., Nishiyama, K., Läubli, T., and Grandjean, E. (1984). The affects of various refresh rates in positive and negative displays. In E. Grandjean, Ed., *Ergonomic and Health Aspects in Modern Offices*. London, England: Taylor and Francis, Ltd.

Hagberg, M., Silverstein, B., Wells, R., Smith, M. J., Hendrick, H., Carayon, P., and Peruse, M. (1995). *Work Related Musculoskeletal Disorders (WRMSDs): A Reference Book for Prevention.* London, England: Taylor and Francis, Ltd.

Hagberg, M., and Sundelin, G. (1986). Discomfort and load on the upper trapezius muscle when operating a wordprocessor. *Ergonomics, 29(12)*, 1637–1645.

Hales, T. R., Sauter, S. L., Peterson, M. R., and Fine, L. J. (1994). Musculoskeletal disorders among visual display terminal users in a telecommunications company. Special Issue: Telecommunications. *Ergonomics, 37(10)*, 1603–1621.

Hedge, A., and Powers, J. R. (1995). Wrist posture while keyboarding: Effects of a negative slope keyboard system and full-motion forearm supports. *Ergonomics, 38*, 508–517.

Hendrick, H. (1986). Macroergonomics: A conceptual model for integrating human factors with organizational design. In O. Brown and H. Hendrick, Eds., *Human Factors in Organizational Design and Management*, pp. 467–477. Amsterdam: Elsevier.

Helander, M. G. (1982). *Ergonomic Design of Office Environments for Visual Display Terminals* (DTMD), Cincinnati, OH: National Institute for Occupational Safety and Health.

Henning, R. A., Ortega, A., Callaghan, E., and Kissel, G. (1994). Self management of rest breaks by VDT users. In *Proceedings of the Human Factors and Ergonomics Society 38th Annual Meeting*, 2 pp. 754–758. Nashville, TN: Human Factors and Ergonomics Society.

Henning, R. A., Alteras-Webb, S. M., Jacques, P., Kissel, G., and Sullivan, A. (1993). Frequent, short breaks during computer work: The effects on productivity and well-being in a field study. In H. Luczak, A. Cakir, and G. Cakir, Eds., *Work With Display Units 92*. Berlin, Germany: Elsevier.

Hultgren, G., and Knave, B. (1973). Contrast blinding and reflection disturbances in the office environment with display terminals. *Arbete Och Halsa.*

Hünting, W., Grandjean, E., and Maeda, K. (1980). Constrained postures in accounting machine operators. *Applied Ergonomics, 11(3)*, 145–149.

Hünting, W., Läubli, T., and Grandjean, E. (1980). Constrained postures of VDU operators. In: Grandjean, E. & Viglinai, E. (Eds.), *Ergonomic Aspects of Video Display Terminals.* London, Philadelphia: Taylor & Francis, Ltd. pp. 175–184.

Hünting, W., Läubli, T., and Grandjean, E. (1981). Postural and visual loads at VDT workplaces: I. Constrained postures. *Ergonomics, 24(12)*, 917–931.

Ilg, R. (1987). Ergonomic keyboard design. *Behaviour and Information Technology, 6(3)*, 303–309.

ISO. (1987). *Quality management and quality assurance standards: Guidelines for selection and use.* Geneva: International Organization for Standardization.

Johansson, G., and Aronsson, G. (1980). *Stress reactions in computerized administrative work.* (Supplement 50): Department of Psychology, University of Stockholm.

Kamwendo, K., Linton, S. J., and Moritz, U. (1991). Neck and shoulder disorders in medical secretaries. *Scandinavian Journal of Rehabilitation Medicine, 23*, 135–142.

Knave, B., and Wideback, P. G. (1987). *Work with Display Units*. Amsterdam: Elseviers.

Knave, B. G., Wibom, R. I., Voss, M., Hedstrom, L. D., and Bergqvist, O. V. (1985). Work with video display terminals among office employees: I. Subjective symptoms and discomfort. *Scandinavian Journal of Work, Environment and Health, 11(6)*, 457–466.

Krueger, H., and Mader, R. (1982). Der einfluss der farbsattingung auf den chromatischen fehler der akkomodation des menschlichen auges. *Fortschritte der Ophthalmologie, 79*, 171–173.

Landau, K., and Rohmert, W. (1992). Evaluation of worker workload in flexible manufacturing industry. *The International Journal of Human Factors in Manufacturing, 2(4)*, 369–388.

Läubli, T., and Grandjean, E. (1984). The magic of control groups in VDT field studies. In E. Grandjean, Ed., *Ergonomic and Health Aspects in Modern Offices*. London, England: Taylor and Francis, Ltd.

Läubli, T., and Hünting, W. (1985). Gesundheitsprobleme bei ganztagiger bildschirmarbeit am beispiel der flugreservationskontrolle. *Zeitschrift fur Arbeitswissenschaft.*

Läubli, T., Hünting, W., and Grandjean, E. (1981). Postural and visual loads at VDT workplaces: II. Lighting conditions and visual impairments. *Ergonomics, 24(12)*, 933–944.

Lawler, E. E. (1986). *High-Involvement Management*. San Francisco, CA: Jossey-Bass.

Life, M. A., and Pheasant, S. T. (1984). An integrated approach to the study of posture in keyboard operation. *Applied Ergonomics, 15(2)*, 83–90.

Lim, S. Y., and Carayon, P. (1993). An integrated approach to cumulative trauma disorders in computerized offices: The role of psychosocial factors, psychological stress and ergonomic risk factors. In M. J. Smith and G. Salvendy, Eds., *Proceedings of the Fifth International Conference on Human-Computer Interaction, 19A* pp. 880–885. Orlando, FL: Elsevier.

Lim, S. Y., and Carayon, P. (1995). Psychosocial and work stress perspectives on musculoskeletal discomfort. In *Proceedings of PREMUS 95*. Montreal, Canada.

Lim, S. Y., Rogers, K. J. S., Smith, M. J., and Sainfort, P. C. (1989). A study of the direct and indirect effects of office ergonomics on psychological stress outcomes. In M. J. Smith and G. Salvendy, Eds., *Work with Computers: Organizational, Management, Stress and Health Aspects*, (pp. 248-255). Amsterdam: Elsevier.

Luczak, H., Cakir, A., and Cakir, G. (1993). *Work with Display Units 92*. Amsterdam: Elsevier.

Maeda, K., Hünting, W., and Grandjean, E. (1980). Localized fatigue in accounting-machine operators. *Journal of Occupational Medicine, 22(12)*, 810–816.

Mandal, A. C. (1982[a]). The Seated Man: Theories and Realities Proceedings of the Human Factors Society, Santa Monica, CA., pp. 520–524.

Mandal, A. C. (1982b). The correct height of school furniture. *Human Factors, 24(3)*, 257–269.

Martin, D. K., and Dain, S. J. (1988). Postural modifications of VDU operators wearing bifocal spectacles. *Applied Ergonomics, 19(4)*, 293–300.

Nachemson, A., and Elfstrom, G. (1970). Intravital dynamic pressure measurements in lumbar discs. *Scandinavian Journal of Rehabilitation Medicine* (Supplement 1), 1–40.

Nakaseko, M., Grandjean, E., Hünting, W., and Gierer, R. (1985). Studies on ergonomically designed alphanumeric keyboards. *Human Factors, 27(2)*, 175–187.

NAS. (1983). *Video Terminals, Work and Vision* . Washington, DC: National Academy Press.

NIOSH. (1981). *Potential health hazards of video display terminals* (Publication no. 81–129). Cincinnati, OH: National Institute of Occupational Safety and Health.

Noro, K. (1991). Concepts, methods and people. In K. Noro and A. Imada, Eds., *Participatory Ergonomics*, (pp. 3–29). London, England: Taylor and Francis.

Ong, C. N., Hoong, B. T., and Phoon, W. O. (1981). Visual and muscular fatigue in operators using visual display terminals. *Journal of Human Ergology, 10(2)*, 161–171.

Ostberg, O. (1975). Health problems for operators working with CRT displays. *International Journal of Occupational Health and Safety, November/December*, 24–52.

OTA. (1985). *Automation of America's Offices* (OTA-CIT-333). Washington, DC: Office of Technology Assessment, Congress of the United States.

Piotrkowski, C. S., Cohen, B. G., and Coray, K. E. (1992). Working conditions and well-being among women office workers. Special Issue: Occupational stress in human-computer interaction: II. *International Journal of Human Computer Interaction, 4(3)*, 263–281.

Rempel, D., and Gerson, J. (1991). Fingertip forces while using three different keyboards. In *Proceedings of the Human Factors and Ergonomics Society 35th Annual Meeting, 1* pp. 253–255. San Francisco, CA: Human Factors Society.

Rossignol, A. M., Morse, E. P., Summers, V. M., and Pagnotto, L. D. (1987). Video display terminal use and reported health symptoms among Massachusetts clerical workers. *Journal of Occupational Medicine, 29(2)*, 112–118.

Rubin, T., and Marshall, C. J. (1982). Adjustable VDT workstations: Can naive users achieve a human factors solution? In *Proceedings of the International Conference on Man-Machine Systems*. Manchester: IEE Conference Publication No. 212.

Rupp, B. A., McVey, B. W., and Taylor, S. E. (1984). Image quality and the accommodation response. In E. Grandjean, Ed., *Ergonomic and Health Aspects in Modern Offices*. London: Taylor and Francis, Ltd.

Salvendy, G., Ed.. (1987). *Handbook of Human Factors*. New York: John Wiley.

Salvendy, G., Ed. (1992). *Handbook of Industrial Engineering*. New York: John Wiley.

Sauter, S. L., Gottlieb, M. S., Jones, K. C., Dodson, V. N., and Rohrer, K. M. (1983). Job and health implications of VDT use: Initial results of the Wisconsin-NIOSH study. *Communications of the ACM, 26(4)*, 284–294.

Sauter, S. L., Schleifer, L. M., and Knutson, S. J. (1991). Work posture, workstation design, and musculoskeletal discomfort in a VDT data entry task. *Human Factors, 33(2)*, 151–167.

Shahnavaz, H. (1982). Lighting conditions and workplace dimensions of VDU-operators. *Ergonomics, 25(12)*, 1165–1173.

Shurtleff, D. A. (1980). *How to Make Displays Legible.* La Mirada, CA: Human Interface Design.

Shute, S. J., and Starr, S. J. (1984). Effects of adjustable furniture on VDT users. *Human Factors, 26(2)*, 157–170.

Silverstein, B., Fine, L., and Armstrong, T. (1987). Occupational factors and the carpal tunnel syndrome. *American Journal of Industrial Medicine, 11*, 343–358.

Smith, M. J. (1984). Health issues in VDT work. In J. Bennet, D. Case, J. Sandlin, and M. J. Smith, Eds., *Visual Display Terminals*, pp. 193–228. New Jersey: Prentice Hall.

Smith, M. J. (1984). Human factors issues in VDT use: Environmental and workstation design considerations. *IEEE CGandA* (November), 56–63.

Smith, M. J. (1986). Job Stress and VDUs: Is technology a problem? In *Proceedings of the International Scientific Conference: Work with Display Units*, (pp. 189–195). Stockholm, Sweden: Swedish National Board of Occupational Safety and Health.

Smith, M. J. (1987). Mental and physical strain at VDT workstations. *Behaviour and Information Technology, 6(3)*, 243–255.

Smith, M. J., Carayon, P., Eberts, R., and Salvendy, G. (1992). Human-Computer Interaction. In G. Salvendy, Ed., *Handbook of Industrial Engineering*, pp. 1107–1144. New York: John Wiley.

Smith, M. J., and Carayon, P. C. (1995). New technology, automation and work organization: Stress problems and improved technology implementation strategies. *International Journal of Human Factors in Manufacturing, 5*, 99–116.

Smith, M. J., Cohen, B. G., Stammerjohn, L. W., and Happ, A. (1981). An investigation of health complaints and job stress in video display operations. *Human Factors, 23(4)*, 387–400.

Smith, M. J., Sainfort, F., Carayon-Sainfort, P., and Fung, C. (1989). Efforts to solve quality problems, *Investing in People-- A Strategy to Address America's Workforce Crisis*, pp. 1949–2002. Washington, D.C.: (Commission on Workforce Quality and Labor Market Efficiency) U.S. Department of Labor.

Smith, M. J., and Sainfort, P. C. (1989). A balance theory of job design for stress reduction. *International Journal of Industrial Ergonomics, 4*, 67–79.

Smith, M. J., and Salvendy, G. (1989). *Designing and Using Human-Computer Interfaces and Knowledge Based Systems* Vol. 12B. Amsterdam: Elsevier.

Smith, M. J., and Salvendy, G. (1989). *Work with Computers: Organizational, Management, Stress and Health Aspects* Vol. 12A. Amsterdam: Elsevier.

Smith, M. J., and Salvendy, G. (1993). *Human-Computer Interaction: Applications and Case Studies* Vol. 19A. Amsterdam: Elsevier.

Smith, M. J., and Salvendy, G. (1993). *Human-Computer Interaction: Software and Hardware Interfaces* Vol. 19B. Amsterdam: Elsevier.

Smith, M. J., Stammerjohn, L., Cohen, B., and Lalich, N. (1980). Video display operator stress. In E. Grandjean and E. Vigliani, Eds., *Ergonomic Aspects of Visual Display Terminals*, pp. 201–210. London, England: Taylor and Francis.

Shneiderman, B. (1990). Future directions for human-computer interaction. *International Journal of Human Computer Interaction, 2(1)*, 73–90.

Snyder, H. L. (1980). *Human visual performance and flat panel display image quality.* Technical Report HFL-80-1. Virginia Polytechnic Institute and State University.

Snyder, H. L. (1984). Lighting characteristics, legibility and visual comfort at VDTs. In E. Grandjean, Ed., *Ergonomic and Health Aspects in Modern Offices*. London: Taylor and Francis.

Snyder, H. L., and Maddox, M. E. (1978). *Information transfer from computer-generated dot matrix displays.* Technical Report HFL-78-3. Virginia Polytechnic Institute and State University.

Snyder, H. L., and Maddox, M. E. (1980). On the image quality of dot-matrix displays. In *SID, 21*, pp. 3–7.

Stammerjohn, L. W., Smith, M. J., and Cohen, B. G. F. (1981). Evaluation of work station design factors in VDT operations. *Human Factors, 23(4)*, 401–412.

Starr, S. J. (1984). Effects of video display terminals in a business office. *Human Factors, 26(3)*, 347–356.

Starr, S. J., Shute, S. J., and Thompson, C. R. (1985). Relating posture to discomfort in VDT use. *Journal of Occupational Medicine, 27(4)*, 269–271.

Starr, S. J., Thompson, C. R., and Shute, S. J. (1982). Effects of video display terminals on telephone operators. *Human Factors, 24(6)*, 699–711.

Trankle, U., and Deutschmann, D. (1991). Factors influencing speed and precision of cursor positioning using a mouse. *Ergonomics*, *34*(2), 161–174.

van der Heiden, G. H., Braeuninger, U., and Grandjean, E. (1984). Ergonomic studies on computer aided design. In E. Grandjean, Ed., *Ergonomic and Health Aspects in Modern Offices*. London, England: Taylor and Francis.

Verbeek, J. (1991). The use of adjustable furniture: Evaluation of an instruction programme for office workers. *Applied Ergonomics*, *22*(*3*), 179–184.

Waersted, M., Bjorklund, R. A., and Westgaard, R. H. (1991). Shoulder muscle tension induced by two VDU-based tasks of different complexity. *Ergonomics*, *34*(2), 137–150.

Weber, A., Sancin, E., and Grandjean, E. (1984). The effects of various keyboard heights on EMG and physical discomfort. In E. Grandjean, Ed., *Ergonomic and Health Aspects in Modern Offices*. London, England: Taylor and Francis.

Westlander, G. (1994). The full-time VDT operator as a working person: Musculoskeletal work discomfort and life situation. *International Journal of Human-Computer Interaction*, *6*(*4*), 339–364.

CHAPTER 51

SOFTWARE–USER INTERFACE DESIGN

Yili Liu
Dept. of Industrial and Operations Engineering
University of Michigan
Ann Arbor, MI 48109 USA

As computer systems pervade more and more environments, the need for effective and easy to use software-user interfaces becomes more pressing. It has become clear that the interface is often the most expensive part of software development and the most important factor in determining the success or failure of a software application.

This chapter discusses human factors considerations for the design of software-user interfaces. To present a comprehensive view of the interface design problem, this chapter starts with a conceptual model for interface design, which helps organize the contents of this chapter. This description is followed by a discussion of the design objectives, the design process, and the major methods for interface design. Various aspects of interface design are then described in subsequent sections as important issues that need to be considered in the design process.

51.1 A CONCEPTUAL MODEL FOR INTERFACE DESIGN

Software–user interfaces appear in remarkably diverse applications. In many areas of application, interfaces play an intermediary role of connecting users with a target system. Examples include computer-aided process control, library information systems, air traffic control, and road network management. In this class of systems, users use interfaces to query, influence or control a target system. However, for such applications as word processing, desktop publishing, and data analysis, software interfaces allow users to accomplish their tasks without connecting with a target system. Furthermore, interfaces for communications and networking such as electronic mail and computer supported cooperative work provide support for user communication and cooperation, which may or may not involve direct communication links with a target system. For instance, a distributed air traffic control system allows a network of software users to communicate with each other in controlling the air traffic system. Distributed word editors, however, allow a group of users to work on the same document, which may describe a particular target system but does not have direct communication links with the system.

Apparently, the design of different types of interfaces requires the consideration of different issues, which have been the subjects of discussion of a large body of literature (e.g., Baecker and Buxton, 1987; Eberts, 1994; Helander, 1988; Myers, 1989; Norman, 1983, 1988; Rasmussen, Pejtersen, and Goldstein, 1994; Shneiderman, 1992; Vicente and Rasmussen, 1992). For example, Norman (1983, 1988) analyzed the role of the interface in bridging the gap between the designer's understanding of a target system and the user's understanding of the same system. Vicente and Rasmussen (1992) and Barfield (1993) described information abstraction and information presentation as two critical aspects of the interface design problem. Smith (1987) discussed the environmental considerations of interface design. Shneiderman (1992) analyzed the needs of networked interfaces.

To organize the discussion of these diverse issues and the related research work in a coherent manner, this section proposes a conceptual model of interface design as shown in Figure 51.1, which serves as the background for the discussions throughout this chapter. The model depicts the target system, the software interface, and the user as three abstract entities and their relationships as two pairs of unidirectional information transformations. Since separate software interfaces are simply separate instances of the same abstract entity called interface, communications between separate interfaces are shown in Figure 51.1 as a two-directional link starting and ending at the same abstract interface entity. Users interact with a software interface in a certain social, organizational, and physical environment, which is shown in the background of Figure 51.1 as "task environment." The three entities, the five relationships, and the task environment define nine important aspects of the interface design problem. Of course, users may communicate with each other or with the target system directly without using a software interface. But Figure 51.1 does not depict explicitly these possible direct links between individual users, between target systems, or between users and target systems. Instead, these types of activities are treated as part of the task environment because they do not involve the use, but may influence the use, of software interfaces. Task environment is not discussed in this chapter, but the other eight aspects are discussed in detail in later sections.

Each entity is shown in Figure 51.1 as a box, and each relationship as a line connecting a pair of entities with an arrow indicating the direction of information transformation. Two of the entities (software interface and user) and the associated pair of transformations (interface presentation and user manipulation) are involved in the design and use of all software interfaces, and shown as solid boxes and lines. The other information transformations and the target system are not involved in all applications and so are drawn with dashed lines. Although all interface activities take place in a task environment, interface

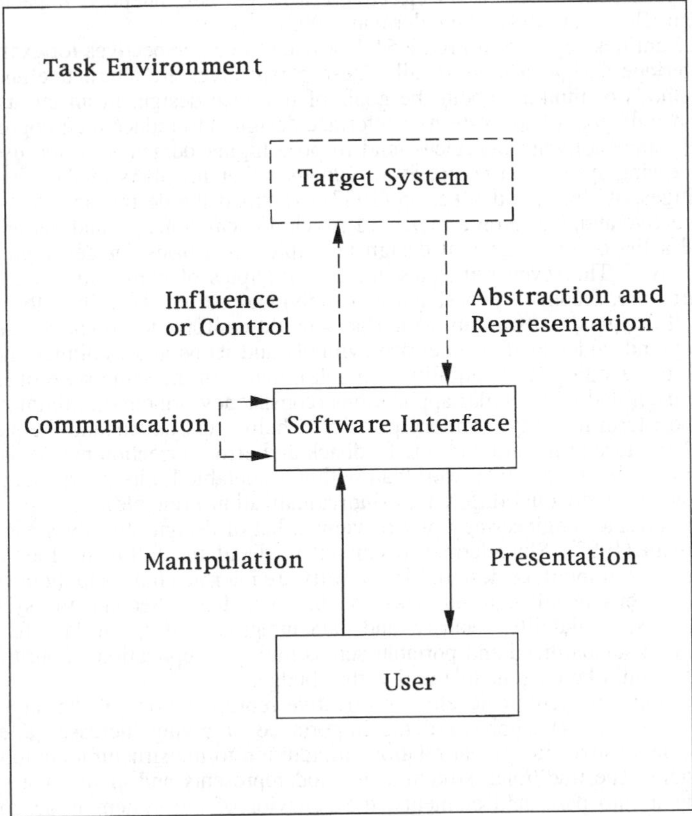

Figure 51.1 A model for interface design. The model defines nine aspects of the interface design problem, some of which are involved in all interface designs (shown with solid lines) and others of which are involved in some, but not all, design activities (dashed lines).

designers may not have to consider it explicitly in all their design activities, particularly when the task environment is simple and static and has been considered in the design of the earlier versions of the same application. For example, interface design for a distributed computer-aided process control system would require the consideration of all the nine aspects of the interface design problem, whereas the design of a single-user word processor may not require a detailed analysis of all the nine aspects.

51.2 GOALS OF INTERFACE DESIGN

Software-user interface is a special member of the user-interface family, which includes interfaces for mechanical, electrical, and electronic devices, as well as computer interfaces. The design of a software-user interface shares many of the general user-interface design objectives such as effectiveness, efficiency, comfort, and safety. A well-designed software-user interface that meets these objectives can improve the quality of work, increase the satisfaction of users, improve the productivity of the workforce, and enhance the safety of the system that the software program controls.

The conceptual model presented in Figure 51.1 establishes a background for achieving a comprehensive view of the various aspects of the goals of interface design. The content of an interface should abstract the critical features of the target system (Norman, 1983; Rasmussen et al., 1994; Vicente and Rasmussen, 1992); the form of interface presentation should be consistent with human perceptual and cognitive characteristics (Carroll, Mack, and Kellogg, 1988; Tullis, 1988; Wickens, 1992a); the means for interface manipulation should be compatible with human response tendencies (Myers, 1991; Shneiderman, 1983);

interface actions should help users effectively achieve their intended impact upon the target system (Sheridan, 1984; Shneiderman, 1992).

The three entities depicted in Figure 51.1 provide three perspectives for examining the goals of interface design in more detail. These perspectives are alternative and complementary methods of thinking about the goals of interface design. From the user's point of view, a specific goal of software-user interface design is to reduce their cognitive loads and stresses, since software interfaces tend to pose higher demands on the users' information processing systems in comparison with the other members of the user interface family. Williges, Williges, and Elkerton (1987) examined the design objectives from the user's perspective and, based on a review of psychological concepts and research results, summarized a list of seven general design principles as a basis for developing specific design objectives. The seven principles are the principles of compatibility, consistency, memory, structure, feedback, workload, and individualization. According to these principles, a well-designed software-user interface should conform to population stereotypes in perceiving and understanding interface symbols and icons and minimize the demand for information recoding (compatibility principle), maintain the same style of interaction both within itself and with similar applications (consistency principle), minimize demand on user's short-term memory (memory principle), help users understand the structure of the system (structure principle), provide feedback and error-correction mechanisms (feedback principle), keep user mental workload within acceptable limits (workload principle), and accommodate individual differences (individualization principle).

From the interface engineering point of view, a list of design objectives can be found in Shneiderman (1992). Shneiderman discussed in detail the following four classes of engineering goals of interface design. First, a software interface must supply the necessary functionality to ensure all required tasks can be carried out. Second, the system must ensure reliability, availability, security, and data integrity. Third, interface features and styles should be standardized and portable across multiple applications. Fourth, interface development should be on schedule and within budget.

From the point of view of developing effective representations of the target system, Rasmussen et al. (1994) emphasized the importance of giving increased attention for representing the abstract functional relations in addition to the structural compositions of a target system. The traditional structural method represents and analyzes a system by decomposing it into parts and elements; the behavior of the system is described as a causal chain of events occurring at the component level. Although this approach is an important tool for modeling human interaction with complex systems, the approach is limited in its ability to model high-level goal-oriented behavior. The functional method complements the structural method by looking at the entire system and representing it by abstraction to a proper functional level and separation of relevant functional relations.

It should be pointed out here that an interface should not only help users understand the target system, but also assist them to achieve their intended impact or influences upon it. In different domains of applications users have different relations with the target system and have different goals and intentions. In supervisory control applications such as air traffic control, users need to exert authoritative, immediate, and effective control over the target system (Sheridan, 1984). In hortatory operations environments such as highway traffic administration, however, the operator's role is often that of influencing the target system by providing information, advice, or warnings to it (Murray and Liu, 1995a, 1995b, 1996). A goal of interface design is certainly to help users accomplish their task of controlling or influencing a target system, in addition to understanding it.

51.3 PROCESS OF INTERFACE DESIGN

Interface design is a creative, iterative, and cooperative activity. The design of high-quality interfaces requires scientific reasoning, artistic imagination, and aesthetic judgment; involves many rounds of revisions; and usually takes places in a multidisciplinary team environment. Both the process and the final product of design are marked with uncertainty.

Many theories and ideas have been proposed to characterize the design process, and virtually all of them describe design as a process comprised of a number of stages. These stages include identification of goals and constraints, requirements analysis and task analysis, design generation and synthesis, prototyping and iterative redesign, and testing and evaluation. These stages are shown in Figure 51.2 and are described in this section. The major techniques and tools that might be used in each stage of the process are described in the next section.

Design is a goal-oriented activity and is subjected to various constraints. The first stage of the interface design process is to identify the goals and constraints. A general descrip-

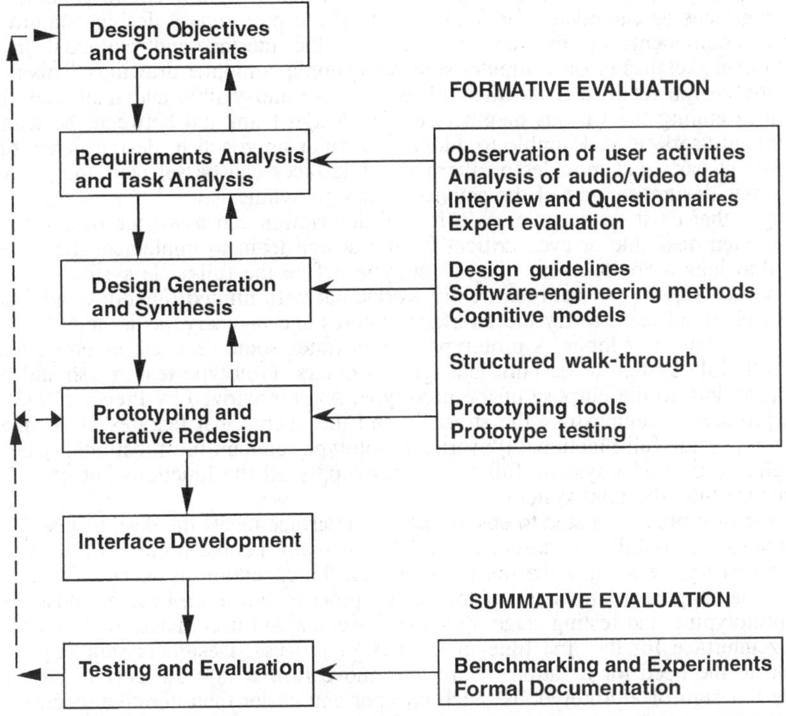

Figure 51.2 Process of interface design and human factors evaluation methods.

tion of the design objectives is provided in the preceding section. A list of constraints would include factors such as budget and cost considerations, schedule requirements, technical feasibility, and performance limitations of special types of users. Because of the existence of various constraints and the potential interactions between the various aspects of the design objectives, it is often not feasible or is impractical to attempt to meet all the objectives that can be listed. It is important that an interface be neither underdesigned nor overdesigned (Shneiderman, 1992; Williges et al., 1987). An underdesigned interface will not meet the design objectives, whereas an overdesigned interface will provide unnecessary functions and features, thereby wasting time and effort. Furthermore, excessive features may distract the users from the important ones and jeopardize their performance and increase their workload. As Shneiderman (1992) pointed out, excessive functionality is probably the more common mistake of designers. The rapid development of hardware and software capabilities is offering more options for interface development, but at the same time also making it easier to build excessive features on the interface.

Requirements analysis identifies the required functions of the interface and specifies what the interface should do. There has been a strong emphasis in the user interface design literature on early focus on users and user needs in establishing interface requirements. Gould and Lewis (1985) emphasized that the design team should focus on users and their needs at early stages of design by establishing active communication with potential users before an interface is developed. Numerous authors have emphasized the importance of user participation in the design process, as reflected in the articles collected in a recent book titled *Participatory Design* (Schuler and Namioka, 1993). The design of interfaces for complex target systems requires not only the consideration of user needs, but also a thorough analysis of the characteristics of the target system. In addition to working with representative users, interface designers often need to communicate with the designers of the target system, particularly if the complexity of the system exceeds the scope of knowledge of the interface designers. Requirements analysis should identify not only what the users want, but also what the users should have to interact with the target system effectively. It is also desirable at this stage to establish evaluation criteria that can be used later to verify the functions of the developed interface.

After specifying interface requirements, a design team often needs to generate alternative interfaces as candidates for final design. These preliminary designs should show the major components and the control structure of the interface and may exist on paper as illustrative sketches or on computer screen as simple computer drawings. This process allows the design team to make their ideas concrete and visible and facilitates further discussions among the various members of the design team and between the team and the potential users. It is desirable to adopt a modular approach in design generation so that different parts and modules of alternative interfaces can be selected and assembled to form new designs as part of the activity of design synthesis.

Designs that exist on paper or as a formal description can assist the design process, but it is often desirable or even critical for the design team to implement their ideas or design drawings more concretely as a prototype before the full-scale system is built. A prototype is a software system that really works, but with limited functions and features; it allows potential users to try the interface features and can serve as a tangible tool for further collecting user input. A prototype demonstrates some selected, essential features of the full-scale system to be built and ignores others. Prototype features should be selected according to the purpose of the prototype. As summarized by Preece (1993), horizontal prototypes demonstrate the structure and the operational features of an interface but do not provide full functionality; vertical prototypes contain full functionality but only for a selected part of a system; full prototypes supply all the functions but provide less performance than the final system.

An interface must be tested to ensure that the interface meets the design specifications and satisfies the usability requirements. Although many people prefer to "get it right" with the first try, iteration is the norm rather than the exception in design. The need for iteration may appear at any stage of the design process, but often become most obvious at the prototyping and testing stage after the designers and users have tried and tested a working interface for the first time in the design process. Designers should pay close attention to the need for iteration before they move further into the development stage: iterative redesign of a prototype is much cheaper and easier than iterative reconstruction of a full-scale system.

Evaluation means more than testing; it can take place at various points in the design and development process and may be conducted with a variety of methods. Depending on when it is conducted, evaluation can be classified as formative evaluation and summative evaluation (Preece, 1994; Williges et al., 1987). Formative evaluation can occur at the requirements analysis and task analysis stage, at the design generation and synthesis stage, or at the rapid prototyping stage, but before the interface is constructed. The purpose is to accommodate user inputs early in the design process, to identify the needs for iterative redesign, and to select the best interface configuration from among the candidates. In contrast, summative evaluation is conducted after an interface has been developed with the aim to test its functionality and usability.

51.4 METHODS AND TOOLS FOR INTERFACE DESIGN

51.4.1 Goal Identification

This section describes a number of techniques and tools that might be used in each stage of the design process. A simple but useful method that might be used at the first stage of the design process is to prioritize the various design goals by either assigning numerical rankings or categorizing each goal as "essential," "desirable," or "optional." This process can help the designers focus on the most critical issues and, at the same time, allow them to strive to meet as many goals as possible.

In prioritizing various goals, the designers should realize that specific goals of interface design vary for different designers and different intended users and applications. For instance, for applications such as word processing, data analysis, and video games, designers need to focus on the goal of satisfying a large and diverse population of users—common users with diverse backgrounds, rather than a special group of well-trained users. For these types of applications, market competition is tough, and user satisfaction is critical. Low cost, ease of learning or no need for learning, and attractive and effective interface features are some of the ways to meet the needs of a large and diverse population of common users.

On the other hand, design objectives related to system reliability are most important for software interfaces used by specially trained and well-motivated users to control complex, and often hazardous, target systems such as nuclear power plant and chemical processing. The design team needs to examine available interface technology for interface development; reducing cost is an objective but should be ranked lower in importance than system reliability and safety. The design team should also take into account the characteristics of future users, identify and reduce their training needs, and examine ways to keep their workloads within optimum boundaries. When all the objectives cannot be met at once, user performance should outrank some other user related objectives such as ease of training (Shneiderman, 1992).

The designers should also realize that goal identification is an iterative process. This is because consumer requirements and design objectives and constraints may not be very well understood at the early stages of software design and development. Further information about consumer requirements identified in later stages can provide feedback to the designers and help them achieve a better understanding of their design goals and constraints (Goel and Pirolli, 1992).

51.4.2 Requirements Analysis and Task Analysis

Throughout the interface design process, a number of methods can be employed to obtain user requirements, among which the most commonly used include observation of current user activities and workplaces, and use of techniques such as interviews, focus groups, and questionnaires. Observing and analyzing user activities and workplaces can often help designers detect user errors and difficulties, discover their use patterns and frustrations in using current interfaces, and identify possible areas for improvement. This process often help designers become better prepared for organizing interviews and designing questionnaires.

Interviews and group discussions are usually conducted as in-depth sessions involving a small number of users at a time. Interviews and discussions can be either structured with a fixed format and similar sequences of questions for all sessions, or unstructured with a great degree of flexibility in the content and the style of interview. Participants may also be asked to try to work with an interface, in addition to answering interviewer's questions. With the consent of the participants, interview and discussion sessions may also be audio- or videotaped for detailed analysis. Session participants are usually selected according to some criteria established according to the purpose of the interview. For example, they may all come from the same company and perform the same job function, or they may all belong to the same age and gender group. A special type of group discussion called a *focus group* is used in market analysis for identifying customer needs, whose participants are usually the current users of a particular system (Mantei and Teorey, 1988).

Use of questionnaires is desirable if the design team needs to obtain the opinions of a large number of individuals on some relatively well-defined issues. Thus, this method is particularly useful when the design team either does not have enough time or does not see a need to conduct many in-depth interview sessions. Questionnaires usually consist of a collection of multiple choice questions, bipolar scales with end point descriptors, and open-ended questions. Multiple-choice questions require the users to select the answer that best describes their situation from a list of given choices for each question. For example, a user may be given a list of three choices "never, seldom, often" for the question "How often do you use this system?" and the user may select any one of the three choices that best describes the frequency of use. The most commonly used bipolar scales for interface evaluation are unidimensional 1-to-7 or 1-to-9 scales with descriptors for the two end points representing the lowest and the highest levels. For example, a 1-to-7 scale may be used for the question "Were the error messages easy to understand?" The end points 1 and 7 could be labeled as "very difficult" and "very easy," respectively. Open-ended questions encourage the survey participants to make comments on issues that are not covered in the questionnaire and to express their thoughts freely. The questions could be either completely open-ended like "Please write down your comments" or asked with some focus or guidance, which often help the participants express more specific ideas. An example of this type of open-ended question is "Which part of the interface do you feel is most difficult to use? How would you like it changed?"

The outcome of requirements analysis is usually an interface specification describing the functionality of the designed interface. Specifications of interface requirements can be documented with informal methods such as text descriptions, design sketches, and diagrams. Recently, user interface researchers have advocated the use of software engineering methods as formal techniques for specifying interface requirements, particularly when the requirements are concerned with interface organization and control structure, interface language definition, or interface response to user commands.

Two notable classes of the formal methods that have been developed for specifying interface requirements are those based on formal grammars and those based on transition diagrams. Formal grammar-based models take the form of production rules for producing correct language statements. Examples of this type of models include Reisner's "cognitive grammar" for describing human–computer interaction language (Reisner, 1981), and Payne and Green's task-action grammar (TAG) for representing interface structure and evaluating interface consistency (Payne and Green, 1986). Transition diagram–based models represent system status and user actions in the form of graphs composed of labeled nodes and arcs. Each node represents a state of the system in which the interface awaits user input, and each arc—starting at one node and ending at another—shows an action that users are allowed to take when the system is at the state represented by the "starting node" to change the system to the state represented by the "ending node." An example of a transition diagram–based model is that of Wasserman (1985).

While requirement analysis identifies what the interface should do, task analysis provides information about how it would be used by a typical user. Over the past two decades a number of techniques for task analysis have been developed. Central to most of the techniques is a systematic decomposition of a user's task into a number of smaller task components. Different techniques organize the task components in different ways: Some organize them as a sequence, some as a hierarchy, and others as a network. The fundamental challenge is to identify what the task components are and how they are organized in accomplishing the task. Some of the task components such as typing a key are observable behaviors and can be measured objectively. But many others are internal mental processes that are not amenable to open inspection; their possible existence and characteristics can only be inferred from analyzing observable behavior. In this regard, theories and models of human perception and cognition become of extreme importance for cognitive task analysis. Table 51.1 includes a list of computational models of cognitive performance.

Although cognitive psychologists have developed numerous theoretical models of human information processing, most of them are not developed specifically for the purpose of direct application. A prominent example of a design-oriented model is the Model Human Processor (MHP), which is an engineering model of human information processing developed by Card, Moran, and Newell(1983). MHP models human information processing as the activities of three interacting subsystems: the perceptual system, the cognitive system, and the motor system. The purpose of the model is to make quantitative and approximative predictions about user performance time with interactive systems, based on a set of principles and assumptions about how the three subsystems operate in a task situation.

Table 51.1 Samples of Computational Models of Cognitive Performance

Specification Language or Grammar-Based Models	Models Based on Operations Research Methods
MHP (Model Human Processor) and GOMS (goals, operators, methods, and selections) Models (Card, Moran, and Newell, 1983)	Queueing theoretic models (Moray, 1986; Rouse, 1980)
Natural GOMS Language (NGOMSL) (Kieras, 1988)	Control theoretic modesl (Baron and Levinson, 1977; Pattipati and Kleinman, 1992)
Cognitive complexity theory (CCT) (Kieras and Polson, 1985)	SAINT (systems analysis of integrated networks of tasks) models (Laughery, 1989; Siegal and Wolf, 1969)
CPM-GOMS (critical path method-GOMS) (John, 1988)	Critical path network models (John, 1988; Schweikert, 1978)
Command Language Grammar (Moran, 1981)	Queueing network models (Liu, 1994, 1996a, 1997)
Cognitive Grammar (Reisner, 1981)	
Task-action grammar (Payne and Green, 1986)	

Closely related to MHP is the GOMS (goals, operators, methods, and selection rules) family of models (Card et al., 1983), which analyze cognitive tasks in terms of operators and methods for achieving the goals of the task and rules for choosing among different possible methods. The model can be used to describe the methods that a user employs to carry out a set of tasks and to estimate the time it takes to complete those tasks. In order to make the GOMS approach more usable for practical applications, Kieras (1988) developed the Natural GOMS Language (NGOMSL), which allows the modeler to describe user-computer interaction in a specification language similar to computer programming languages. Cognitive complexity theory (CCT) represents another extension of the basic GOMS model (Kieras and Polson, 1985). John (1988) extended the GOMS approach to the analysis of parallel activities. Reisner's "cognitive grammar" and Payne and Green's (TAG) mentioned earlier for requirements specification can also be used for task analysis. The Command Language grammar (CLG) developed by Moran (1981) is also a grammar-based task analysis method. For a review, see Olson and Olson (1990). For a detailed discussion of the GOMS approach to cognitive modeling and interface analysis, the readers are referred to Chapter 40 (*Cognitive Modeling*) of this Handbook.

In addition to the models mentioned above that are based on specification languages or formal grammars, numerous models based on operations research and industrial engineering methods have also been developed for analyzing human performance in complex interactive systems. Queueing theoretic models, control theoretic models, discrete network models, and queueing network models are four classes of these models. Queueing theoretic models and control theory–based models share an assumption that describes human information processing system as a single server system or a single central-processing-unit (CPU) processor. These models assume that the cognitive system rapidly switches its processing capacity among concurrent competing task demands according to some scheduling or resource allocation policy (see, e.g., Rouse, 1980).

The task network modeling methodology started with SAINT (systems analysis of integrated networks of tasks) models human interaction with the task environment as a sequence of tasks. Alternative sequences to accomplish a goal may exist and form a network; parallel sequences represent alternatives rather than concurrence of processing (Laughery, 1989; Siegal and Wolf, 1969). The critical path network model of psychological processes developed by Schweikert (1978) also represents an information processing activity as a task network, but parallel paths in a critical path network represent concurrent rather than alternative processes. John (1988) integrated the critical path method (CPM) and the GOMS approach and developed a CPM–GOMS method for analyzing concurrent interface activities. In both the SAINT network and the critical path network, a process cannot start until the preceding process on the same path of the network has been completed. Thus, at any instant, only one process on a path can be active. Because of this characteristic, they both belong to the class of discrete networks.

Queueing network models integrate the considerations of queueing theoretic and discrete network models and support the modeling of a broader range of possible mental structures that can be subjected to empirical testing (Liu, 1994, 1996a, 1997). The models describe the cognitive system as a network of distinct servers, each of which is responsible for a specific task function and has a waiting space for task demands to wait if they cannot immediately receive their requested service. Different aspects of the task demands are serviced by separate servers in parallel and concurrently; multiple queues of task demands may exist in front of the various servers when the task demands are high. The models allow processes on the same path to be active at the same time and support the modeling of both alternative and concurrent processes, as well as sequential and parallel processes (Liu, 1994, 1996a, 1997).

It has become increasingly clear that in order to reduce the design cycle time and to serve a proactive role in the early design process, rather than a passive role in the late testing phase, human factors input to system design needs to be quantitative and model driven (Elkind, Card, Hochberg, and Huey, 1989; McMillan et al., 1989). Computational models of cognitive performance provide an important means for cognitive task analysis and can help designers make quantitative and approximative predictions about user performance in the design process before an interface or a system is built.

A useful method at the design generation and synthesis stage is one called structured walk-through. If the preliminary design exists in a paper form, users and designers can perform a paper-and-pencil exercise to "walk through" the various interface states by following the "directions" indicated on paper. If the design exists in the form of a formal language or a transition diagram, users and designer can navigate the interface according to the structure defined by the more abstract language or diagram descriptions. This activity is not unlike that of evaluating the design of a house when it exists only as paper drawings or as a detailed language description. A potential buyer and the designer of the house can "walk-through" the house even though it has not been built.

51.4.3 Design Guidelines

A large number of interface design guidelines have been developed and can be of great value to designers. Although guidelines cannot guarantee the success of a design, they may prevent the designers from developing interfaces that will be clearly undesirable. Some guidelines are relatively general, but many are developed for specific aspects of the interface design. For example, at a more general level, Shneiderman (1992) stated the following eight important guidelines, which should be considered and adapted by designers for each design task: Strive for consistency; enable frequent knowledgeable users to use shortcuts; provide information feedback; organize sequences of actions into groups; offer simple error handling mechanisms; allow easy reversal of actions; enable users to be in control of the system; and reduce short-term memory load. At a more specific level, an example is the list of 162 guidelines developed by Smith and Mosier (1986) specifically for the design of data displays. It is beyond the scope of this section to present the large body of design guidelines, many of which can be found in Helander (1988), Smith and Mosier (1986), and books on specific topics. It should be emphasized here, however, that interface designers should realize that many design guidelines have been developed and can be of great value in interface design. Samples of interface design guidelines are presented in Table 51.2.

51.4.4 Prototyping, Testing, and Evaluation

A prototype should be built quickly and cheaply; it should also be easy to modify in order to support redesign. Some prototypes are discarded after they have completed their mission, whereas others evolve into the final systems. Tools are available to support prototyping, a partial list of which would include Motif for the XWindows environment, HyperCard and MacApp for the Macintosh, Toolbox and Visual Basic for the IBM-PC, and the widget sets for the DECWindows and Microsoft Windows, and other GUI (graphical user interface) builders such as Powerbuilder, Visual C++, and other Hypermedia- or multimedia-based interface prototyping tools. These tools can greatly reduce the time and effort that are needed for prototype development, and there is no doubt that new tools will continue to emerge to offer more advanced prototyping supports. For more detailed discussions of prototyping toolkits, see e.g., Hartson and Hix, 1989; Myers, 1988, 1989; and Shneiderman, 1992.

Usability testing is a method that is widely used to test the usability of software systems. Effective testing requires the test team to have clearly defined test goals, well-prepared test plans, and carefully selected test procedures. Test participants (often called subjects) should be representative of the intended user population. Testing data should be collected and analyzed in a systematic way according to preplanned experimental methods. Testing data may include performance data such as response time and error rate, subjective data such as workload and satisfaction rating scores, video data such as videotapes of testing sessions, and audio data such as subject's verbal reports recorded when they are asked to "think aloud" about what they are doing.

As mentioned earlier, evaluation includes formative evaluation and summative evaluation. In fact, the activities mentioned above for analyzing user characteristics such as user observation, interview, group discussion, user survey, cognitive modeling, prototype testing, and "cognitive walkthrough" constitute different aspects of the formative evaluation process, which is conducted before an interface is constructed. After an interface has been developed, summative evaluation can be conducted to compare this interface with alternative designs and to test its usability. Lab or field experiments are the most commonly adopted method. Some summative evaluations use standard tests and tasks (often called benchmark tests) for the purpose of standardizing the evaluations. It is desirable to use benchmark tests, because they can serve as metrics for evaluating and comparing user performance on different systems. Unfortunately, few benchmark tests are available for software interface evaluation (Williges et al., 1987), and one of the challenges for interface researchers is to develop valid and reliable benchmark tests that can be used in interface evaluation. Chapter 46 is devoted to the topic of usability testing; interested readers are referred to that chapter for detailed information.

Table 51.2 Samples of Interface Design Guidelines

Samples of General Design Guidelines (Adapted from Shneiderman, 1992)

- Strive for consistency.
- Enable knowledgeable frequent users to use shortcuts.
- Provide information feedback.
- Organize sequence of actions into groups.
- Offer simple error handling mechanisms.
- Allow easy reversal of actions.
- Enable users to be in control of the system.
- Reduce short-term memory load.

Samples of Data Display Guidelines (Adapted from Smith and Mosier, 1986)

- Left justify columns of alphabetic data to allow rapid scanning.
- Label each page to show its relation to other pages.
- Maintain consistent format from one display to another.
- Display data in directly usable forms.
- Use short, simple sentences.
- Use affirmative rather then negative statements.
- Provide an informative header or title for every display.
- When blink coding is used, the blink rate should be 2 to 5 Hz.

Samples of Screen Design Guidelines (Adapted from Tullis, 1988)

- Make appropriate use of abbreviations.
- Avoid unnecessary details.
- Use concise wording.
- Use familiar data formats.
- Use tabular formats with column headings.
- Arrange related items as groups.
- Use highlighting to attract user attention to certain elements.
- Present information in a proper sequence.

Samples of Color Usage Guidelines (Adapted from Murch, 1987)

- Avoid pure blue for text, thin lines, and small shapes.
- Avoid red and green in the periphery of large-scale displays.
- Not all colors are equally discernible.
- Do not overuse colors.
- Use similar colors to convey similar meanings.
- Use a common background color to group related elements.
- Use brightness and saturation to draw viewer attention.
- For color-deficient viewers, avoid single-color distinctions.

Samples of Error Message Guidelines (Adapted from Shneiderman, 1992)

- Be as specific and precise as possible.
- Be positive: Avoid condemnation.
- Be constructive: Tell user what needs to be done.
- Be consistent in grammar, terminology, and abbreviations.
- Use user-centered phrasing.
- Use consistent display format.
- Test the usability of error messages.
- Try to reduce or eliminate the need for error messages.

51.5 EIGHT ASPECTS OF THE INTERFACE DESIGN PROBLEM

The above discussions indicate that software interface design is an iterative process; techniques and tools are available that can be used in each stage of the design process. As mentioned earlier, the design of different types of software-user interfaces requires the consideration of different issues throughout the design process. The conceptual model depicted in Figure 51.1 organizes the diverse issues of interface design into a coherent structure and defines nine aspects of the interface design problem, eight of which are discussed in turn in this section.

51.5.1 Interface Capabilities

One of the important issues that interface designers must consider is the available interface technology, which is shown as the middle box in Figure 51.1. Interface hardware and software are changing on a daily basis; the following description is only a snapshot of the vast and fast-changing technological landscape. Interface designers must keep track of the changes and stay informed about the options available so that they can select the best options to serve their needs.

51.5.1.1 Hardware Devices

Hardware devices are the physical input–output devices used for information communication between the user and the computer. Although the most commonly used hardware devices nowadays are still keyboard, mouse, and CRT display, other input devices such as touchscreen, stylus, trackball, trackpad, and touch-sensitive panel are replacing or augmenting the role of conventional keyboard and mouse as input devices. The rapid development of computer technology and the increased concern for human factors have led to the appearance of a large variety of new devices. Devices for tracking eye movement, hand gesture, or whole-body movement and those for speech recognition provide more novel ways of sending input to the computer. Output devices such as high-resolution graphic displays and multimedia displays with sophisticated video and audio capabilities are also becoming more and more prevalent in workplaces, and they provide more options for displaying information to the users. Many human factors issues are involved in the design of conventional and advanced input–output devices; many of these issues are related to human sensory and motor characteristics as well as physiological and biomechanical factors. For detailed discussions of some of these issues, the readers are referred to Chapter 20 (*Visual Displays*), Chapter 21 (*Controls*), Chapter 22 (*Nonconventional Controls*), and Ch 50 (*Design of Computer Terminal Workstations*) of this Handbook.

51.5.1.2 Software Widgets

In parallel with the rapid development of hardware capabilities, there has also been a fundamental change in the way software interfaces are constructed and used. From the interface developer's point of view, there is no longer the need to build an interface from "scratch" with conventional programming languages; with the advent of interface programming toolkits, interfaces can now be constructed with "prefabricated building blocks," often called interface widgets (or widgets for short). From the users' point of view, widget-based interfaces allow them to interact with the interface through physical manipulation of widget objects, which is often more natural and easier to perform than typing complex text commands as required by traditional command-based interfaces.

The current style of interface design is to design interfaces that are composed of a number of windows, each of which occupies a section of the computer screen and is responsible for a specific part of a user's task by communicating with the user within the space provided by the window. A window can be built with a window widget contained in the widget set of a programming toolkit. In order to facilitate the communication with users, a number of widget items can also be constructed and attached to a window. For example, a label names a window and a scroll-bar allows users to scroll the contents of the window. In this case, the label and the scroll-bar exist only in the context of the window to which they are attached. This type of relationship between two widgets is called a parent–child relationship. The child widget exists only in the context of its parent widget, moves on the screen with its parent, and is deleted from the screen automatically when its parent is deleted. This parent–child relationship thus reflects the hierarchical organization of various interface components. A clear understanding of widget functions and their parent–child relationships can greatly facilitate interface design and programming.

The most commonly used widgets include labels, forms, lists, scroll-bars, pushbuttons, toggle switches, radio boxes, windows, dialog boxes, text widgets, menu-bars, and pull-down menus. Different interface toolkits may contain different sets of widgets, but many of them supply these basic widgets. Toolkit manuals accompanying a toolkit describe in detail what widgets are provided and how to use them in interface programming. For a general discussion of interface widgets, see Eberts (1994).

Figure 51.3 is an example of an interface constructed with software widgets. It is an interface for the Three Dimensional Static Strength Prediction Program™ (3DSSPP™), a software program developed at the Center for Ergonomics at the University of Michigan as a job analysis and evaluation tool. The interface shown in Figure 51.3 illustrates widget-based interface features such as windows, dialog boxes, labels, menu-bars, and pushbuttons. Figure 51.3 is also used in later sections of this chapter to illustrate a number of other interface design concepts.

51.5.1.3 Graphics Programming Tools

Whereas the interface widgets help establish the control structure and "open the windows" of an interface, graphics libraries help construct and render images in the space provided by the windows. To a great extent, we can say that widgets are used for making frames and graphics libraries are for painting the pictures.

A graphics library contains a collection of subroutines for drawing and animating two-dimensional (2D) and three-dimensional (3D) color graphics images. More specifically, the subroutines can be called from a program to display characters that form texts, to draw 2D and 3D geometric primitives that form complex objects, to show color and lighting effects, to create animation, to provide coordinate transformations such as viewing and projection transformations, to render curves and surfaces, to create and edit objects, to provide depth cueing and hidden-surface removal for 3D scenes, to collect user inputs,

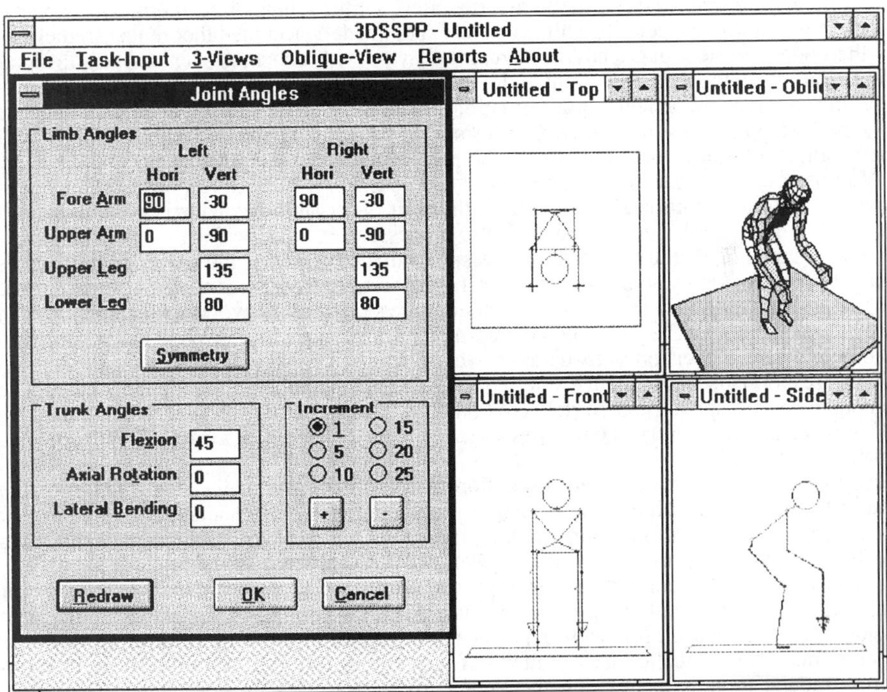

Figure 51.3 Interface for a computer-aided ergonomics analysis software program—3DSSPP™ (University of Michigan, Ann Arbor, MI). This display is used to illustrate a number of concepts discussed in the text, including interface widgets, abstraction and selective representation, object display and emergent features, visualization, codes and modalities, direct manipulation, and control–display compatibility.

and so on. Examples of graphics programming libraries include Xlib of the XWindow system, graphics libraries on platforms such as Sun and HP, the IRIS Graphics Library of SGI (Silicon Graphics), and its publicly available, vendor-neutral version called OpenGL. For detailed discussions about graphics libraries and graphics programming, see, e.g., Foley, Van Dam, Feiner, and Hughes, 1990; and Neider, 1993.

51.5.1.4 Multimedia and Virtual Reality

Multimedia interfaces and virtual reality (VR) are two new comers in the field of interface technologies. Multimedia, as its name suggests, attempts to improve the "look and feel" of interfaces and change the mode of user-computer interaction by providing high-resolution computer graphics, digital full-motion interactive video, stereo sound, and CD-ROM all on a personal computer. Multimedia technology distinguishes itself from current audio and video technologies such as television by the interactivity it provides users. However, the current technological trends suggest that the distinction between the two technologies will undoubtedly become blurred as they merge with each other (Hoffert and Gretsch, 1991). The readers are referred to Chapter 55 of this Handbook for a detailed discussion of multimedia interfaces.

Virtual reality is an emerging computer technology that attempts to eradicate the barriers between users and computers by creating a "virtual" environment where users "live" and "navigate" while performing their tasks. Head-mounted displays, 3D goggles, and data-gloves are some of the enabling technologies for virtual reality, and new concepts and devices are emerging rapidly. Chapter 52 is devoted to the topic of virtual reality.

51.5.2 User Characteristics

No matter how sophisticated computer interface technology may become, the ultimate judges of the technology are the human users (shown in Figure 51.1 as the bottom box), with their information-processing capabilities and limitations. Interface designers should take into account the characteristics of user behavior and preferences and be knowledgeable about related psychological theories and explanations.

Although competing schools of theories are emerging that offer alternative explanations to psychological activities, the currently most widely accepted theoretical framework is the cognitive view that analyzes human behavior with concepts such as mental representations, symbolic computations, and information-processing modules or stages. According to this view, the human information-processing system—also called the cognitive system—responds to environmental information by forming mental representations and performing symbolic computations on the representations. This activity is carried out by a number of interconnected modules or stages, each of which is responsible for a specialized information-processing function; thus, an observable behavior is composed of a collection of internal, unobservable mental processes.

More specifically, the cognitive view assumes that the sensory and perceptual processes are responsible for receiving information from the task environment and comparing the input information with the existing mental representations in working memory and in long-term memory. The outcome of this matching process is delivered to the decision module, where a decision is made as to what response to make, but the actual execution of the response is carried out by a separate module, called the response module. The outcomes of a response can be delivered back to the perceptual module, which compares the desired and the observed outcomes and uses their difference as a basis for performance improvement.

Chapter 4 presents a detailed discussion of the functions and the organization of the various modules in the human information-processing system. Readers are referred to that chapter for specific information. In the context of software-user interface design, this section selects the concept of mental models for a more detailed elaboration.

The concept of mental models has proved to be a useful conceptual device in user interface design for explaining user behavior. According to Norman (1983, 1986), when using a device or working with a physical system, users tend to form their views or mental models (also called user models) of how the device or system works. Users construct mental models by relating the system to some similar systems they are already familiar with; thus, past experience and training play an important role in the formation of mental models.

Although users tend to form mental models of devices or systems, there is no guarantee that these models are always adequate. In fact, people's mental models are often deficient in a number of ways. As summarized in Norman (1983), mental models are incomplete, unstable, lack of firm boundaries, unscientific, and parsimonious. Furthermore, people have limited ability to run them. People tend to forget the details of the system they are using, to confuse similar systems, and to maintain erroneous behavior or action patterns if they help reduce their mental efforts.

An important task of a designer is to help users form adequate mental models, which need not be technically accurate but must be functional to allow users to accomplish their tasks effectively and efficiently. To do so, designers need to understand the relationship between the designer's "conceptual model" of the system, the "system image" presented to the user, and the user's "mental model" formed on the basis of that image (Norman, 1983). The designers' conceptual model is an accurate, consistent, and complete abstraction of the system, expressed in designer's terminology or some formal modeling language and used in designing and developing the system. The system image is what can be seen by the user about the system. The designers expect that the user's mental model matches the designer's conceptual model, but the system image is the only channel through which the designers can communicate with the users. Therefore, the designers should carefully construct the system image so as to minimize potential differences between their model of the system and the user's mental model (Figure 51.4).

51.5.3 The Target System

As mentioned earlier in this chapter, some software applications such as word processing do not need to connect the software users with a target system. The text file that a user is working on can describe a system, but does not establish a direct communication link with the system. But in many other areas of applications, users use the interface to influence or control a target system (shown as the box at the top of Figure 51.1), and thus the interface plays an intermediary role between the users and the target system. The

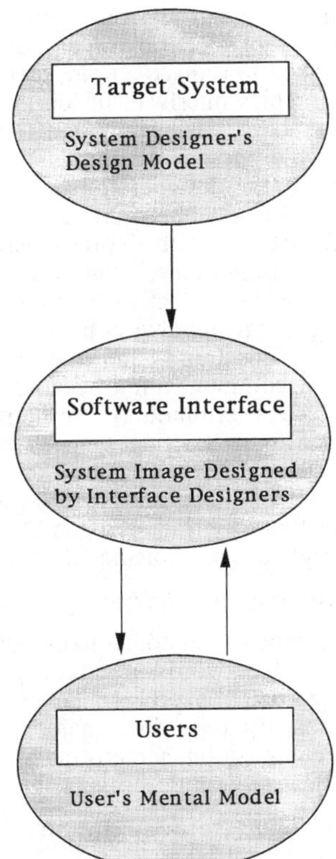

Figure 51.4 Distinction between design model, system image and user's mental model. Interface designers need to use well-designed system image to bridge the gap between system designer's design model and user's mental model of the same system.

effective design of such interfaces requires not only the consideration of available interface technology and user characteristics, but also a clear understanding of the target system, including its purpose, its function, the concrete and abstract objects that comprise the system, the relationships among these objects, the system variables that define the state of the system, and the factors that determine the values of these variables.

A target system can be analyzed and described in different ways and at different levels of detail and rigor. One way to describe a system is to categorize it into a particular class according to one of its attributes. Since a target system has a number of attributes, there are a number of dimensions along which one can categorize a system. For instance, a target system can be classified as either dynamic or static, depending on whether the status of the system changes over time or not; as discrete or continuous, depending on whether the system variables can take a finite or an infinite number of possible values; as tightly or loosely connected, depending on whether the objects that comprise the system have extensive interactions with each other or not; or as tree or network structured, depending on whether the organization of system components is hierarchical (like a tree) or flat (like a 2D network). Categorization is a method that is used frequently not only in the design process but also in everyday life.

Another commonly used method for system analysis is modeling, which allows designers to analyze and describe a system at a desired level of detail and rigor by selecting a proper modeling methodology. A model is an abstraction of a system, expressed in some modeling language; it not only helps designers analyze and understand a system, but also establishes a framework for communication and documentation. A multitude of mathematical and engineering modeling methodologies is available, each of which is most useful for modeling a certain type of system but ill-suited for others. For instance, discrete network models can be used to model systems comprised of discrete objects and their relationships, but continuous flowlike system behavior is best described with differential equations.

The target system and the software-user interface are usually designed by separate individuals, who are often from different groups or organizations. The two types of design activities may take place at different locations or at different times. For example, an interface design team may be asked to replace an old, text-based interface of an existing complex system with a color graphics interface. In this type of situations, the model of the target system—the "design model," using Norman's (1983) terminology introduced in the previous section—initially resides in the minds or the documents of the system designers but are not possessed by the designers of the interface (the "system image"). Therefore, interface designers need to establish early and direct contact not only with the future users (system operators), but also with the designers of the target system so that both the "user model" and the "design model" can be correctly and clearly identified.

It should be noted that establishing effective communication between the interface designers and the system designers may not always be an easy task. The two design teams may have different objectives and constraints and may encounter conceptual and "language" barriers in the process of communication because of the possible differences in their conceptual bases and their use of terminology. Furthermore, if the interface designers are asked to add a new interface to an existing system, the time gap between the new interface design and the target system design may add more difficulties to the establishment of effective communication. Interface designers must realize and overcome these difficulties so that they can successfully bridge the gap between the user's mental model and the system designer's conceptual model of the same target system.

51.5.4 Interface Representation of the Target System

Much of interface design is concerned with constructing abstract models of systems and presenting them to the users. In other words, interface designers need to solve both a representation problem and a presentation problem to decide both the content and the form of the interface. This section discusses three major methods (abstraction and selective representation, structural decomposition, and abstraction hierarchy) that might be used to represent a system, and the next section describes some of the presentation methods.

51.5.4.1 Abstraction and Selective Representation

Different aspects of a target system are not of equal importance to the users; some are essential, some are optional, whereas others can be distracting to the users in carrying out their tasks. An interface should represent the essential aspects of the target system and filter out the potentially distracting aspects.

Examples of abstraction and selective representation abound. Maps do not resemble the real world but contain critical geographical information for navigation. Graphical illustrations used in textbooks and instructional manuals are often drawn as simple but informative schematic diagrams. Another example can be found in Figure 51.3. There is no doubt that everyone can recognize that the graphic forms shown in Figure 51.3 represent a human figure, although nobody resembles the figure in the real world. The graphic human forms contain the critical information for biomechanical computations, such as the joint angles and the lengths of limbs, but have left out the nonessential and potentially distracting information such as the flesh and the organs.

Different users have different needs in their interactions with different target systems, and apparently they should be served with different ways of abstraction and selective representation. A map of city roads and highways is useful for automobile drivers; an aviation map is useful for aircraft pilots; however, a map showing both the road and the air navigation information can be distracting to both types of users. The goals and the requirements of the user and the target system should determine the particular form of abstraction and selective representation for a specific application.

51.5.4.2 Structural Decomposition

One of the methods for developing representations of target systems is the method of structural decomposition, which represents a system by decomposing it into structural components and elements. The behavior of the system is described as a causal chaining of events occurring at the component level. For instance, an automobile can be represented as an aggregation of automobile components such as brakes, steering wheel, throttle, and so on; automobile driving behavior can be analyzed as a causal chain of events occurring at these components.

Structural decomposition is a useful method, because humans often need to know the parts and elements that comprise a system, the functions and characteristics of these components, and the relationships among them. Structural information is often particularly important for the training of new users and for the diagnosis of system failures. New users often learn how to operate a system by remembering the functions of each component and the operational procedure about when and how to use the various components. Similarly, failure diagnosis often relies on a reasoning process that links one component to another.

When using the structural decomposition method to represent a system, interface designers need to consider such questions as the following: What are the components and relationships that are relevant to the user's task? At what level of detail or granularity should the decomposition be made? When and how to change from one level of decomposition to another?

51.5.4.3 The Abstraction Hierarchy

Although the structural decomposition approach is an important tool for modeling human interaction with complex systems, the approach is limited in its ability to model high-level, goal-oriented behavior (Rasmussen et al., 1994). Rasmussen (1985) proposed a theoretical framework called abstraction hierarchy that complements the structural approach and emphasizes the importance of looking at the entire system and representing it by abstraction to a proper functional level and separation of relevant functional relations. Furthermore, the abstraction hierarchy emphasizes the need to include several levels of abstraction and decomposition in representing a work system.

In the work of Rasmussen and his colleagues, five levels of abstraction are found to be useful (Figure 51.5): the purposes of the system and its environmental constraints (the highest level); the priority measures established according to the design intention and the abstract functions expressed in terms of accumulation and flow of mass, energy, information, people, and monetary value (the second level); the general functions that the system is designed to perform irrespective of the underlying physical properties (the third level); the physical processes and activities (the fourth level); and the physical form, location and configuration of material objects (the lowest level). The five levels are further linked by a means–end relation, according to which changes in higher level objectives will propagate downward as reasons for proper functions, whereas changes in the lower level properties will propagate upward as causes of malfunction. The terms "means–ends space" and "means–ends network" have also been used to refer to the overall structure of the five abstraction levels and their means–ends relations.

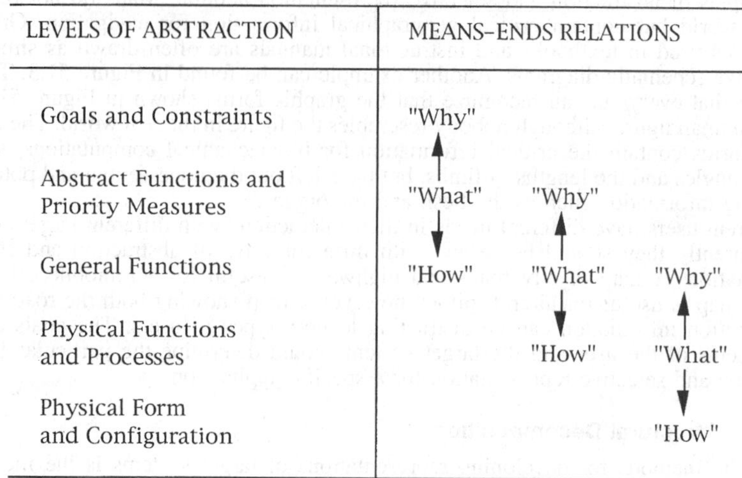

Figure 51.5 Levels of abstraction in system representation. (Adapted from Rasmussen, Pejtersen, and Goodstein, 1994.)

An important property of the abstraction hierarchy is that adjacent levels in the hierarchy are linked by a "WHAT, WHY, and HOW interrelationship"—any work function at a certain level (WHAT needs to be done) can be considered as a goal (WHY) for another function at the lower level, as well as a means (HOW) for another function at the higher level. For instance, if one needs to know WHY a general function (at the third level) has to be performed and HOW to accomplish it, then the description at the adjacent higher level (the second level in this case) gives answers to the question WHY, and the adjacent lower level (the fourth level) answers the question HOW. Thus a representation of a target system in the form of an abstraction hierarchy not only allows one to work with the system at a proper level under regular operating conditions, the means–ends relation among the various levels can also be used to support system planning and development as well as emergency troubleshooting.

Based on Rasmussen's constructs, Rouse, Cody, Boff and Frey (1990) presented a two-dimensional design-space model. The first dimension of the model, called level of aggregation, define four levels of system decomposition—system, subsystem, assembly and component. The second dimension is called level of abstraction, and has three levels—purpose, function, and form. Rouse et al. described "purpose" as the representation of design requirements and objectives (via scenarios, documents, simulations, etc.); "function" as the representation of various relationship such as physical and computational relationships (via diagrams, equations, simulations, etc.); and "form" as the representation of appearance such as geometry and assembly of objects (via drawings, pictures, mockups, etc.).

In the context of interface design, the abstraction hierarchy establishes the foundation for the ecological interface design approach proposed by Vicente and Rasmussen (1992; Rasmussen and Vicente, 1989). This approach suggests that in order to support problem-solving behavior, the interface should represent the work domain in the form of an abstraction hierarchy. Two experiments conducted by Vicente, Christofferson, and Pereklita (1995) using a simulated process control task demonstrated that a display containing both physical information and information about functional variables supported subject's performance better than a display containing only physical information.

Murray and Liu (1995a) applied the abstraction hierarchy in their analysis of operator behavior in a highway traffic control center. They found that the more experienced operators tended to move up and down through the various levels of the hierarchy more fluidly and rapidly in their way of seeking information and reasoning. The more junior operators, on the other hand, were more inclined to focus their attention on a tighter range at the lower part of the abstraction hierarchy. This finding suggests that an interface represented in the form of the abstraction hierarchy may not only enhance operator performance, but also can serve as a useful device for user training and testing.

Although a small number of formal methods of developing system representations for interface design have been developed, the data and knowledge base in this area remains sparse. It is hoped that user interface researchers will pay more attention to this issue and strengthen the knowledge base from which effective system representations can be developed.

51.5.5 Presentation of Information to the User

51.5.5.1 Overview

Whereas the representational problem of interface design discussed in the previous section is concerned with the substance or the content of an interface, the presentational problem discussed in this section is concerned with the style or the format of presenting information to the user. Although the topic of user manipulation of the interface is discussed in a separate section, it should be noted that information presentation and user manipulation are intertwined with each other in user activities and are two interdependent aspects of the interface design problem. A pull-down menu, for example, is a both a display containing information about available choices and a control for users to select an item. Very often it is the perspective from which one examines an interface feature that decides whether the feature is more of a display or more of a control. Therefore, many issues discussed in this section on information presentation are fundamentally linked with the discussions in the next section on interface manipulation. Similarly, the topics covered in the next section are intrinsically tied with the discussions in this section.

This section does not attempt to offer a comprehensive review of the vast body of literature on how to present information effectively through a software interface. Instead, five topics are selected for discussion, each involving an interface technique and a fundamental psychological or human factors concept: The methods of screen design are discussed from the perspective of visual search; the value of object displays is attributed to the creation of emergent features; the benefits of visualization are explained with the concept of metaphor; the effects of 3D graphics are linked with the depth cues; and the selection of multimedia options is examined in terms of codes and modalities of information presentation. The orientation of the discussion is from human factors, rather than interface techniques, and the purpose is to illustrate the relevance and importance of human factors concepts and theories in selecting the proper format of information presentation.

It should be emphasized that effective interface design requires the consideration of a large number of issues that are not covered in this section. A sample of these issues would include typography (Baecker, 1990; Sassoon, 1993), feedback and error messages (Shneiderman, 1992), online help (Kearsley, 1988), documentation (Boehm-Davis and Fregly, 1985; Carroll, 1984; Crown, 1992), multiwindow strategy (Billingsley, 1988; Henderson and Card, 1986; Hopgood, Duce, Fielding, Robinson, and Williams, 1985; Marcus, 1992), benefits and costs of interface consistency (Grudin, 1989; Nielsen, 1989), hypertext and hypermedia (Rivlin, Botafogo, and Shneiderman, 1994; Rizk, Streitz, and Andre, 1990), and voice interfaces (Baber and Noyes, 1993; Strathmeyer, 1990) (Table 51.3). Designers should realize the importance of these issues, which are discussed in detail in the references mentioned above and in a number of books on interface design, including Baecker and Buxton (1987), Helander (1988), and Shneiderman (1992).

51.5.5.2 Screen Design and Visual Search

When computer screens are cluttered with homogeneous items, users often need to spend a long time to find the item they are looking for. The psychological process involved in finding a target in a crowded visual field is called visual search or visual selective attention. It is both consistent with intuition and demonstrated by numerous psychological studies that distinct items attract visual attention and can be found more quickly and easily than nondistinct items. An item is distinct if it differs significantly from the other items on at least one physical attribute—size, brightness, color (more precisely, hue), rate of blinking, and so on.

Richard Christ (1975) reviewed the results of 42 studies on the effects of color on visual search and showed that color aids searching if the color of the target is unique and known to the users in advance. Although few of the studies reviewed by Christ were conducted on computer displays, some recent studies on computer graphic displays have provided evidence that color coding is an effective way to increase user satisfaction (e.g., Tullis, 1981) or to improve user performance (e.g., Benbasat, Dexter, and Todd, 1986; Hoadley, 1990; Liu and Wickens, 1992; Murray and Liu, 1993). Shneiderman (1992)

Table 51.3 Major Display Design Issues

Screen design—Color, highlighting, layout, and information density
(Marcus, 1992; Shneiderman, 1992; Tullis, 1981, 1988).

Typography
(Baecker, 1990; Sassoon, 1993).

Data display format
(Smith and Mosier, 1986; Tullis, 1988).

Object displays
(Chernoff, 1973; Coekin, 1969; Woods, Wise, and Hanes, 1981).

Information visualization
(Pickover and Tewksbury, 1994; Rosenblum, 1994; Tufte, 1983, 1990).

Three-dimensional displays
(Ellis, 1991; Okoshi, 1980; Wickens, Todd, and Seidler, 1989)

Multiwindow strategy
(Billingsley, 1988; Henderson and Card, 1986; Marcus, 1992)

Selection of display modalities
(Wickens, Sandry, and Vidulich, 1983; Wickens and Liu, 1988).

Hypertext and hypermedia
(Nielsen, 1990; Rivlin, Botafogo, and Shneiderman, 1994; Rizk, Streitz, and Andre, 1990; Shneiderman and Kearsley, 1989).

Design of voice interfaces
(Baber and Noyes, 1993; Strathmeyer, 1990).

Design of error messages and online help
(Isa, Boyle, Neal, and Simone, 1983; Kearsley, 1988; Shneiderman, 1992).

Documentation
(Boehm-Davis and Fregly, 1985; Carroll, 1984; Crown, 1992).

discussed 14 guidelines for the use of color in user interfaces. Other authors have also provided important suggestions about how to create effective color displays (Brown, 1988; Doney and Seton, 1988; Durrett, 1987; Galitz, 1989; Marcus, 1992; Thorell and Smith, 1990). For example, it is suggested that color can be used to group related items; changes in color can be used to indicate status changes. However, the designers should avoid the use of too many colors on the same screen and should be consistent in color coding by using the same color-coding rules throughout the application. Another suggestion is that the designers should be attentive to user expectations about color codes. The same color could convey different information in different task contexts (Shneiderman, 1992).

Tullis (1981) showed that the effectiveness of a screen design is not only determined by the total number of items on a display, but also by how these items are arranged. Use of indention and space to group related items can make the screens much easier to read. Other methods such as the use of target highlighting (Fisher, Coury, Tengs, and Duffy, 1989), size coding (Liu and Wickens, 1992a), and target blinking (Smith and Goodwin, 1971) have all been shown to be effective ways of reducing search time. A list of 162 data-display guidelines developed by Smith and Mosier (1986) contains specific suggestions about how different coding methods should be used. For instance, one of the guidelines suggests that when size coding is used, a larger symbol should be at least 1.5 times the height of the next smaller symbol.

Tullis (1988) developed a comprehensive list of guidelines for screen design, which includes specific guidelines with regard to the amount of information on a screen, grouping of information, display of text, use of highlighting, standardization of displays, and graphic representations. For example, it is suggested that the designers should minimize the amount of information on a screen by displaying only the necessary information and by using concise wording. Highlighting can be achieved by using flashing, underlining, reverse video, unique color, or bolder letters. It is also suggested that a consistent format should be used for all the screens of the same application so that the users will know where a given piece of information is located at. For text displays, one of the guidelines states that upper case characters are effective in attracting user attention, but cannot be read as quickly as regular upper and lower case texts.

51.5.5.3 Object Displays and Emergent Features

Although the screen design methods described in the previous section are mainly concerned with designing displays that allow users to find and process a given piece of information quickly, the idea of object displays was developed to help users process multiple pieces of information in parallel. The idea of object displays is based on psychological theories of preattentive perceptual processes in processing visual information. A large number of studies have shown that visual processing of the multiple dimensions or attributes of a single object is a parallel and obligatory process. It is parallel because people pay attention to all the dimensions or attributes at once, rather than one of them at a time. It is obligatory because people have no control over this process—one cannot notice one dimension or attribute of an object but ignore the others. For instance, people notice both the height and the width of a rectangle at the same time, rather than one dimension after another or one dimension only. Similarly, people notice both the color and the shape of an apple at once, rather than one attribute at a time.

When users need to process and integrate multidimensional information in order to accomplish their tasks, it is desirable for user interface designers to capitalize on this human perceptual characteristic by creating multidimensional object displays. In fact, numerous examples of object displays have appeared, including polygon–polar diagrams (Coekin, 1969; Woods, Wise, and Hanes, 1981), Chernoff's faces (Chernoff, 1973), box displays (Barnett and Wickens, 1988), and other multidimensional object displays (Carswell and Wickens, 1987; Wickens and Andre, 1990). A polygon–polar display represents the normalized magnitudes of n system variables as a regular polygon with n sides. A change in the magnitudes of the variables causes a corresponding change in the shape of the polygon; certain shapes signal the occurrence of certain types of system problems. Chernoff's face display represents the multiple attributes of a system as various features of a human face; different facial expressions are associated with different status of the represented system. A rectangle is an example of a box display, whose height and width can be used to represent the magnitude of two variables, and the area of the rectangle makes visible the multiplicative relation between the two variables.

The shape of a polygon, the facial expression of a human face, and the area of a rectangle are all called emergent features in the psychology literature—global perceptual features that are created via interactions among individual parts or local features (e.g., primitive geometric units such as line segments and curves) (Pomerantz, 1981). The benefits of an object display are due to its creation of emergent features, but it should be noted that emergent features need not be created only by single objects. Sanderson, Flach, Buttigieg, and Casey (1989) showed that a row of parallel bar graphs can also create emergent features—a horizontal line connecting the tops of the bars. Liu and Wickens (1992a) showed that clustered ordering of different colors, different-sized squares, or 3D contours can all create emergent features and facilitate information integration tasks. Other authors have also demonstrated the importance of using emergent features to display relationships among system variables (e.g., Bennett and Flach, 1992; Bennett, Toms, and Woods, 1993).

As pointed out by Bennett et al. (1993), it is not an easy task to represent variables as emergent features that highlight the critical data relationships. For example, the same set of data can be mapped into different versions of a facial display (Kleiner and Hartigan, 1981), and some features of the face become more salient than others (MacGregor and Slovic, 1986). Bennett et al. (1993) suggested that the designers need to decide which variables should be included in the graphic form and how they should be assigned to the dimensions of the object. Furthermore, decision must also be made about whether all variables should be converted to a common scale, and about how to represent the task context. It is also suggested that some of the graphical features should be made more salient if the variables they represent are of special importance to the users.

51.5.5.4 Visualization and Metaphor

Object displays discussed above represent abstract data and variables as the geometric dimensions and visible features of objects that allow users to visualize the relationships among the variables. In fact, graphical presentation of information has been used in diverse areas of application and has demonstrated its value as a method of displaying information (Tufte, 1983, 1990; Tukey, 1977). The advancement of computer graphics technology is providing more options for portraying abstract concepts and data relations in various graphics forms. Over the past few years, visualization has emerged as a major field of computer graphics research; visualization techniques have also been successfully applied in diverse disciplines (see, e.g., Frenkel, 1988; McCormick, Defanti, and Brown, 1987; Pickover and Tewksbury, 1994; Rosenblum, 1994).

Proper use of metaphors is critical for creating effective graphical representations of abstract concepts and variables. Metaphor is an important emerging concept in software-user interface design; interfaces incorporating proper metaphors can make the strange familiar, the invisible visible, and the abstract concrete. Metaphors can assist users to relate a new, unfamiliar situation to an old, familiar one so that they can apply their existing knowledge to the new situation. Abstract concepts can be depicted as concrete objects; the hidden structure of a database can be revealed as layouts of spatial arrays; the complex relationships among a number of variables can be illustrated as changes of visible features of spatially arranged objects.

Metaphors have been used extensively in user-interface design, and many of these applications are summarized in Carroll et al. (1988) and Eberts (1994). For instance, the desktop metaphor represents files and folders as file- and folderlike icons, lists of choices as menu icons, and file deletion as the action of moving a file icon to a wastebasket icon. Some spreadsheet applications use the ledger sheet metaphor to show the matrix structure of the database, whereas some drawing and painting applications use the metaphor of drawing and painting on paper with pen and paintbrush. The room or office metaphor store different applications in different rooms or areas of the office and allow users to "walk" through doors to navigate between the rooms, each of which is used to store a certain kind of application.

Metaphors and icons have their limitations, too. Icons and graphic representations take up excessive screen space (Shneiderman, 1992) and may discourage new ways of thinking and interacting with a domain (Hutchins, Hollan, and Norman, 1986). A poorly selected metaphor or icon could be confusing if it does not convey its meaning clearly, and it could be misleading if it supports alternative interpretations. Furthermore, metaphors are limited in the ability to express ideas, so a perfect match may not exist for some operations. Interface designers need to be clear about the benefits and costs of using metaphors for their specific application.

A systematic approach to the use of metaphors in interface design is suggested in the four-steps approach of Carroll et al. (1988) (Figure 51.6). The first step of this approach requires the designers to identify candidate metaphors that match the desired attitude of the user, share a similar theme, and are as interesting as possible. The second step involves the detailing of the metaphor–software matches in the context of representative user scenarios. The third step is to identify the likely mismatches between the metaphors and the objects or actions they intend to represent, and the final step is to develop effective design strategies to help the users manage possible mismatches.

51.5.5.5 3D Graphics and Depth Perception

The development of computer graphics technology has made it possible to develop 3D computer graphic displays that provide the viewers a sense of three dimensionality—"realistic" and "natural" viewing conditions that closely resemble perceiving 3D objects in a 3D space.

As summarized by Wickens, Todd, and Seidler (1989), there are both benefits and costs associated with using 3D displays. One benefit is that a 3D visual scene is a more natural, "ecological," or compatible representation of a 3D world than that provided by 2D displays. Another benefit is that a single integrated display of a 3D object or scene reduces the need for mental integration of two or three 2D displays. A cost of using 3D displays is that any projection of a 3D world inevitably produces an inherent perceptual ambiguity, in addition to the potential costs associated with additional hardware, software, and programming requirements for implementing 3D displays. Although a 3D display makes it easier to achieve a holistic perception of the whole object or scene, it makes it difficult to make accurate judgments on relations or values along any one particular axis, particularly when one attempts to judge the distance between two points along the line of sight. These benefits and costs can be illustrated with Figure 51.3. The 3D human form shown in the upper-right window of the figure provides the viewers with a holistic view of a human posture, but the three 2D displays provide a better means for making accurate judgments on the values of particular joint angles (Liu, Zhang, and Chaffin, 1993).

In order to create a sense of three dimensionality, designers of 3D interfaces should be familiar with the perceptual cues that the human perceptual system uses to form 3D

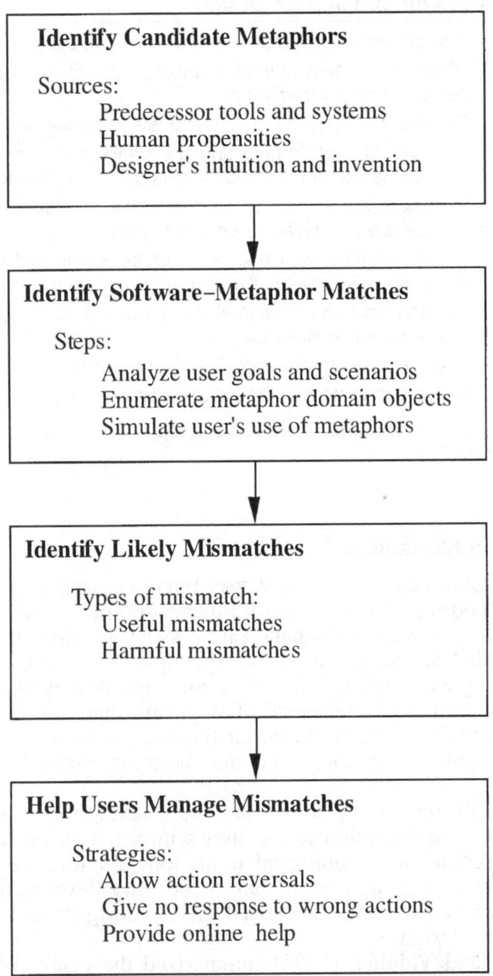

Identify Candidate Metaphors
Sources:
Predecessor tools and systems
Human propensities
Designer's intuition and invention
Identify Software–Metaphor Matches
Steps:
Analyze user goals and scenarios
Enumerate metaphor domain objects
Simulate user's use of metaphors
Identify Likely Mismatches
Types of mismatch:
Useful mismatches
Harmful mismatches
Help Users Manage Mismatches
Strategies:
Allow action reversals
Give no response to wrong actions
Provide online help

Figure 51.6 A four-step approach to the use of metaphors in software interface design. (Adapted from Carroll, Mack, and Kellogg, 1988.)

perception. These perceptual cues can be classified as observer-centered cues and world-centered, pictorial cues. Observer-centered cues include binocular disparity, convergence, and accommodation. Binocular disparity refers to the difference between the images received by the two eyes. Convergence refers to the proprioceptive messages from the eye muscles that rotate the eyes inward or outward when looking at objects at varying distances. Accommodation is based on the messages from the eye muscles that adjust the shape of the lens to bring the image into focus on the retina. Wickens (1992a) summarized nine pictorial cues, which are listed in Table 51.4. Pictorial cues can be implemented on a computer display with computer graphics methods to create a sense of three dimensionality.

Based on an extensive review of the role of the various depth cues in 3D perception and 3D display design, Wickens et al. (1989) concluded that stereopsis, motion, and occlusion are particularly salient cues. They further emphasized the important role of motion in creating the sense of three dimensionality, since it appears that the usefulness of stereopsis is diminished and may vanish altogether when displays are dynamic. Interested readers are referred to Wickens (1992) and Wickens et al. (1989) for detailed discussions of the various depth cues and their applications in the design of flight deck displays, air traffic control, meteorology, teleoperation, and computer graphics.

Table 51.4 Pictorial Cues for 3D Interface Design[a]

Linear perspective	Two converging lines are perceived as two parallel lines receding in depth.
Interposition	When one object obscures another object, the obscured object is perceived to be more distant.
Height in the plane	Objects higher in the visual field are regarded as more distant.
Light and shadow	The direction of lighting and the characteristics of an object's shadow offer clues about the object's orientation relative to the perceiver.
Relative size	If objects are known to be the same true size, those look smaller on display are perceived to be more distant.
Textual gradient	When the plane of a textured surface is oriented toward to line of sight, the grain will grow finer at a greater distance.
Proximity–luminance covariance	Continuous reductions in illumination and intensity are perceived to represent receding distance.
Aerial perspective	"Hazier" objects are perceived to be farther away.
Relative motion gradient	Objects that move faster tend to be perceived to be closer.

[a]Based on Wickens (1992a).

51.5.5.6 Codes and Modalities

When the advent of multimedia technology, user interface designers have more and more options for selecting codes and modalities for information presentation. The term "codes" refers to the distinction between information processing activities that are analog-spatial in nature and those that are verbal-linguistic. Examples of spatial codes include spatial orientations, analog representations, velocity vectors, and flow fields. Examples of verbal codes include mental arithmetic, rehearsal of linguistic material, text, and speech processing. The term "modalities" refers to the distinction between visual and auditory perception (called perceptual modalities) and that between manual and voice responses (response modalities).

Although the rapidly increasing power and sophistication of multimedia technology make the assignment of information to the user's information processing channels less constrained by the traditional technological limitations, appropriate theory-based guidelines for selecting the various options remain to be fully developed. This section summarizes two theories that have been developed to help designers select proper codes and modalities in interface design.

Wickens, Sandry, and Vidulich (1983) summarized the results of a large number of studies and proposed a stimulus–central processing–response (S–C–R) compatibility principle, which states that the visual and the manual modalities are better suited for processing spatial-analog information, whereas the auditory and the voice modalities are better for processing verbal-categorical information, but not the converse. This principle can be illustrated with the human posture display shown in Figure 51.3. One can perceive the spatial configuration of a human posture easily with a visual, rather than an auditory, display; one can change the displayed posture more easily by using a mouse to "click and drag" the body joints than by uttering voice commands.

The potential role of codes and modalities in interface design for multitask environments, where operators are performing several tasks at once, is described in the multiple resources theory (Wickens, 1984). According to this theory, spatial and verbal codes define two separate resources of human information processing, so do the visual and auditory perceptual channels and the manual and voice response modalities. The theory predicts that concurrent tasks competing for the same resource will interfere with each other more than tasks employing separate resources. In a driving environment, for instance, the theory predicts that the use of visual displays or manual control devices such as electronic maps or push buttons will produce greater interferences with the driving task than the use of voice displays and speech-activated controls, because driving is primarily visual and manual in nature.

The S–C–R principle and the multiple resources theory offer critical insights into the complex and important issue of selecting codes and modalities in interface design, but the issue is far from resolved. It has become increasingly clear that other important behavioral factors should also be considered, and they often make the role of codes and modalities not as clear-cut and straightforward as discussed above. A partial list of these factors would include the alerting and preemptive characteristics of auditory presentation (Posner, Nissen, and Klein, 1976; Wickens and Liu, 1988), the visual scanning requirements of visual interfaces (Liu, 1993, 1996b, 1996c; Liu and Wickens, 1992b), and the similarity and integrality of concurrent tasks (Polson, Wickens, Klapp, and Colle, 1989; Wickens, 1992b). Since software-user interfaces are being used in more and more multitask situations, as demonstrated in the increasing use of electronic maps in automobiles, interface design for multitask environments is a great challenge for interface researchers and developers in the years to come.

51.5.6 User Manipulation of the Interface

51.5.6.1 Overview

Based on the way that users provide inputs to the computer, interfaces can be classified into four major classes (Table 51.5): command language–based, menu-selection, direct manipulation, and anthropomorphic interfaces. This section selects direct manipulation interfaces and the important concept of control–display compatibility for a detailed discussion; the other three styles of interaction are briefly described below.

Among the four classes of interfaces, command language–based interfaces are the oldest and often the most difficult to use. Users are required to remember and type command sequences that are often cryptic and composed by complex syntax. User frustration in front of a command-based interface is often observed because of the great

Table 51.5 Four Classes of User Interfaces Classified in Terms of the Way in Which Users Provide Inputs to the Computer. Samples of Related Design Issues and Recent Studies Are Included in the Table.

Command Language Interfaces

(Benbasat and Wand, 1984; Black and Moran, 1981; Carroll and Thomas, 1982; Green and Payne, 1984; Laudauer, Colotti, and Hartwell, 1983; Norman, 1981)

- Command organization strategies
- Selection of command names
- Command abbreviation
- Formal models of command languages

Menu Selection and Form-fill-in Interfaces

(Brown, 1982; Card, 1982; Kiger, 1984; Koved and Shneiderman, 1986; Laverson, Norman, and Shneiderman, 1987; Norman, 1991; Perlman, 1984)

- Menu structure
- Sequence and phrasing of menu items
- Menu layout and graphical menu features
- Display rate and response time
- Menu selection mechanisms
- Embedded menus
- Menu shortcuts

Direct Manipulation Interfaces

(Carroll and Thomas, 1982; Hutchins, Hollan, and Norman, 1986; Myers, 1991; Rubin, Golin, and Reiss, 1985; Shneiderman, 1983, 1992; Van de Vegte, Milgram, and Kwong, 1990; Ziegler and Fahnrich, 1988)

- Metaphors and icons
- Direct manipulation methods
- Direct manipulation programming
- Remote direct manipulation
- Explanations of direct manipulation

Anthropomorphic Interfaces

(Don, Brennan, Laurel, and Shneiderman, 1992; Eberts, 1994; Fels and Hinton, 1995; Miller and Walker, 1990; Schmandt, 1994; Shneiderman, 1992; Smith, 1994; Takeuchi and Nagao, 1993; Walker, Sproull, and Subramani, 1994)

- Natural language interfaces
- Gesture interfaces
- Facial expression interfaces
- Eye movement interfaces

memory and typing demands and the low tolerance of errors shown by this type of interfaces—a confusing error message is often the consequence of a single typo in a long command. Important issues that have been considered by command language researchers include command abbreviation behavior (Benbasat and Wand, 1984), command organization strategies (Carroll and Thomas, 1982; Norman, 1981), and selection of command names (Black and Moran, 1982; Laudauer, Calotti, and Hartwell, 1983).

Menu selection avoids many of the problems associated with command language interfaces. The interface displays the choices as menu items and users can select an item easily by pointing and clicking or with one or two keypresses. Menu selection can thus greatly reduce the need for memorization and typing of complex command sequences. The design of effective menu selection interfaces requires the consideration of numerous issues, including menu structure, sequence and phrasing of menu items, shortcuts through the menus for frequent users, menu layout and graphical menu features, display rates and response time, and selection mechanisms. Detailed discussions about these issues can be found in Norman (1991) and Shneiderman (1992).

Anthropomorphic interfaces interact with users in a way that is analogous to the way humans interact with each other. Natural language interfaces and interfaces that recognize gestures, facial expressions, or eye movements all belong to this type of interfaces. The design and development of anthropomorphic interfaces require not only hardware and software supports but also a clear understanding of how humans communicate with each other with natural languages and through gestures, facial expressions, and eye contacts. Signal must be separated from noise, meaning must be identified from ambiguous messages, and intentions must be recognized from vague expressions. The design and development of anthropomorphic interfaces represent one of the frontiers of interface research and development. A number of books and articles have been published covering related topics, such as Don, Brennan, Laurel, and Shneiderman (1992), Eberts (1994), Fels and Hinton (1995), Miller and Walker (1990), Schmandt (1994), Shneiderman (1992), Smith (1994), Takeuchi and Nagao (1993), Walker, Sproull, and Subramani (1994).

51.5.6.2 Direct Manipulation

Direct manipulation interfaces (DMI) are based on the idea that users should be allowed to manipulate computer interfaces in a way that is analogous to the way they manipulate objects in space. Since humans interact with objects in space everyday, the argument goes, DMIs would represent a more natural and familiar environment for user action than command-based interfaces.

Shneiderman (1983) coined the term "direct manipulation interface" and described DMI as containing the following three characteristics:

1. There is a visible and continuous representation of the objects and actions of interest.
2. The objects of interest are manipulated by physical actions or labeled button presses rather than complex syntax.
3. Operations are rapid, incremental, and reversible; their effect on the objects of interest is immediately visible.

Examples of DMI can be found in many areas of applications, including wordprocessing, desktop publishing, computer-aided design (CAD), flight simulations, and video games. Most word processors in use nowadays are DMI word processors, also called WYSIWYG (what you see is what you get) word processors. DMI word processors display the document in the form that it will appear when printing is done, allow users to make changes directly to the displayed document, and display the changes immediately. DMI employs familiar icons and associated actions to help users accomplish their tasks without the need to remember abstract commands. For example, a user may delete a file by dragging the corresponding file icon into a wastebasket icon.

DMI-based CAD systems allow designers to draw and change objects directly on the screen and see the effects immediately. Figure 51.3 contains an example of a DMI. A user can change the posture of the displayed human figure by selecting the joints to be moved and clicking the screen to reposition the selected joints to the desired positions. This DMI-based method is much more natural and easier to use than the text entry method, which requires the user to enter numerical values into the joint angle dialog box shown on the left part of Figure 51.3.

Two theories have been proposed to explain the observed benefits of DMI. The first of these is the syntactic-semantic model of objects and actions (SSOA model) of Shneiderman (1983). The model suggests that users have syntactic knowledge about device-dependent details and semantic knowledge about task concepts and computer concepts. The syntactic knowledge is acquired by rote memory and easily forgotten, whereas the semantic knowledge is acquired by meaningful learning and stable in memory. DMI-based interfaces display the objects of interest and the results of action directly so that the users can be greatly relieved from the distraction of dealing with complex commands and syntax that are at the level of syntactic knowledge.

The second theory is the explanation based on the gulfs of evaluation and execution (Hutchins, Hollan, and Norman, 1986). The gulf of execution refers to the gap between a user's intentions and the inputs that a computer recognizes, whereas the gulf of evaluation refers to the gap between the computer's output and the user's expectations and mental model of the task. The two gulfs can be bridged either by better computer interface design or by greater user efforts. Traditional command language–based interfaces require the users to bridge the gulf. DMI-based interfaces, according to this theory, let the computer bridge the gulfs, making the users feel as if they are interacting with the domain rather than an interface.

DMI has both benefits and problems (Shneiderman, 1992). One of the benefits is that DMI reduces user anxiety and encourages learning. Learning is by analogy; there is little syntax to remember; thus, operational knowledge is better retained. Another benefit is that there is little need for error messages because the results of actions are visible and obvious. Since DMIs often rely on metaphors and icons, many of the problems of DMI are similar to those of metaphors and icons discussed in Section 51.5.5.4, such as possible user confusion and waste of screen space. Another problem of DMI can be illustrated with Figure 51.3. If a users intends to adjust the posture accurately according to a set of predetermined joint angle values, entering these values into the dialog box from a keyboard is easier and faster than the DMI method.

51.5.6.3 Control–Display Compatibility

One of the fundamental principles in human factors is the principle of control–display compatibility, which suggests that displays and controls should be designed in a way that is consistent with human expectations. The following is a summary of five specific principles of compatibility that should be considered in software-user interface design (Table 51.6): the principles of location compatibility, control movement compatibility, conceptual compatibility, compatibility of display orientation, and compatibility of display movement.

According to the principle of location compatibility, controls should be located close to the corresponding displays and should be arranged in a way that allows users to tell easily which control should be used for a particular display. Touchscreen displays overlay controls on their respective displays and thus achieve the maximum degree of location

Table 51.6 Five Principles of Control–Display Compatibility That Should Be Considered in User Interface Design[a]

Principle of Location Compatibility
Controls should be located to the corresponding display, and their spatial arrangement should allow users to tell easily which control is used for a particular display.

Principle of Movement Compatibility
The indicator of a display should move in the same direction as its control.

Principle of Conceptual Compatibility
The layout and the operational methods of controls should be consistent with expectations of the intended user population.

Principle of Compatibility of Display Orientation
For analog displays, the orientation and ordering of the display should be consistent with those of the mental representation.

Principle of Compatibility of Display Movement
The direction of movement of the moving part of a display should be consistent with user expectation.

[a]Based on Wickens 1992a.

compatibility. When it is impossible or undesirable to use touchscreens, this principle suggests that an interface designer should maximize the similarity between the control panel layout and the configuration of the displays.

A complete description of the principle of movement compatibility would require a detailed discussion of the diverse types of control devices, which is beyond the scope of this chapter. In the context of software-user interface design, however, the main suggestion of this principle is that the indicator of a display should move in the same direction as its control. For instance, the mouse cursor displayed on a computer screen should move in the same direction as the mouse—a leftward movement of a computer mouse should move the mouse cursor to the left, rather than some other direction.

The principle of conceptual compatibility asserts that the layout and the operational methods of controls should be consistent with user expectations, also called population stereotypes. For example, the layout of a QWERTY keyboard is what users expect for a keyboard—imagine what would happen if a user is asked to use a keyboard with a different layout. With regard to the expected operational method of a control, a simple example is that in order to increase the value of a variable, users would expect to press the up-arrow rather than the down-arrow key.

The principle of compatibility of display orientation, also called the principle of pictorial realism (Roscoe, 1968), suggests that for analog displays (presentations of continuously varying quantity), the orientation and ordering of the display should be consistent with those of the mental representation. For instance, altitude information should be displayed with a linear, vertically oriented display, rather than with a horizontally oriented linear display or with a circular display. Furthermore, the linear vertical display should present high altitudes at the top section of the display and low ones at the bottom, rather than the converse.

According to the principle of compatibility of display movement, also called the principle of the moving part (Roscoe, 1968, 1981), the direction of movement of the moving part of a display should be consistent with the user expectation. In the context of user interface design, this principle can be interpreted as suggesting that the moving part of a display should move in the direction that the user command requests. The implications of this principle can be illustrated with the help of the upper-right window in Figure 51.3. In computer-aided design (CAD), designers often need to move a 3D object such as the displayed 3D human figure along the left–right direction. A "Left" command could be used to produce a leftward movement of the object (the moving-object method), or a leftward movement of the "viewing window" (the "moving-window" method), which would result in a rightward movement of the object in its visual effect. For example, suppose software users control the display by using the left-arrow and the right-arrow keys of a keyboard. Pressing the left-arrow would move the object to the left, if the moving-object method were used, but would move the object to the right if the "moving-window" method were used. The principle of moving parts suggests that the moving-object method should be used—in the eyes of the users it is the displayed human figure rather than the window that is the moving part, and the moving part should move in the direction that the user command requests.

Although the principle of moving parts suggests the use of the moving-object method, the "moving-window" method is preferred when a software-user interface is used by the users to navigate in a large information space, only a small part of which is displayed and viewed at any time by the user. This method is not only necessary if the information space is too large to be displayed on one screen, but also more consistent with user expectations in some task situations. An example of an application of the "moving-window" method can be found in the design of computer-based driving simulators: the direction of movement of the displayed objects is opposite to that of the control movement, accurately reflecting a user's experience in a driving environment.

51.5.7 Effects of Interface Activities on the Target System

In many areas of application software-user interfaces do not stand alone; they play an intermediary role of connecting users with a target system. In different domains of application users have different relations with the target system and have different goals and intentions. This section discusses two classes of systems in which interface users have different goals and intended impact or influence upon the target system while performing interface activities such as typing computer commands, selecting menu items, or manipulating displayed graphics. The two classes of systems also define two types of relations between users and a target system.

51.5.7.1 Supervisory Control

The first type of relation between users and a target system can be described with the term "supervisory control," which was proposed by Sheridan (1984) to describe a broad class of human–machine systems in which the human operator plays the role of a supervisor who gives directives and commands, while the computers serve as subordinates by generating and carrying out detailed control actions. Supervisory control characterizes the behavior of systems such as vehicle control (including automobile, aircraft, and ships), discrete and continuous process control (including manufacturing, chemical, and power plant operations), and robot control in remote or hazard environments. Broadly speaking, supervisory control also includes manual controls that are aided by computers; the specific computer support may come in the form of computerized integrated data displays or computer facilitated control actions.

Because of the prevalence and the importance of supervisory control systems, the last 20 years have seen a great increase in research attention to this class of application. Chapter 39 of this Handbook discusses supervisory control in detail. In the context of software-user interface design, user-interface issues have been investigated in two types of supervisory control systems, each having its own user-interface problems.

In the first type of supervisory control systems the operators not only assume the supervisory role with an authoritative power but also possess sufficient technological mechanisms to receive timely information from and exert effective control upon the target system. For this type of system the research focus has been on operator monitoring behavior and support for failure diagnosis. A list of the important interface design questions that have been addressed would include: How should the system data be formatted on the display so that operators' error detection performance can be improved and their workload reduced? How should the interface be designed so that the operators can maintain an adequate mental model of the system even after they have been removed from active participation in the control loop for an extended period of time? What cognitive biases do operators tend to demonstrate in a supervisory system (Ephrath and Young, 1981; Gilmore, Gertman, and Blackman, 1989; Liu, Fuld, and Wickens, 1993; Moray, 1986; Parasuraman, 1987).

In the second type of supervisory control systems, operators still assume the authoritative role, but they have to deal with some complicating factors in exercising their control. For example, imperfect communications technology and long distance of transmission could both cause time delays, so could slow-responding devices or systems. Time delays, incomplete feedback, feedback from multiple sources, and unanticipated interferences are some of the complicating factors discussed in Sheridan (1984) and Shneiderman (1992). In terms of user-interface design, Sheridan (1984) suggested the use of predictor displays to help operators deal with time delays—a predictor display shows not only the current information but also the predicted future status of the system on the basis of a computer model of the controlled system. One solution suggested by Shneiderman (1992) is to make explicit the time delays by showing the users the changes of the system status as time progresses; moreover, spatially defined commands may be better than temporal commands—users should use commands that specify a destination rather than a motion.

51.5.7.2 Hortatory Operations

Murray and Liu (1995a, 1995b, 1996) used the term "hortatory operations" to describe another type of human interaction with complex systems. In this class of systems, operators in an operations center such as those associated with freeway traffic management have quite limited influence or control over the target system, compared to the level of authoritative control in supervisory environments. Operators handling the management of a freeway network have responsibilities that typically involve monitoring traffic levels, detecting and resolving problems that cause congestion and incidents, and taking proactive measures to prevent the occurrence of certain problems. They can help "inform" travelers about current or expected traffic conditions and can "advise," "encourage," or even "urge" them to take certain actions. However, unlike supervisory control systems such as air traffic or power plant process control, a freeway operations center has limited ability to "control" the actions of travelers and the status of the system. Furthermore, the operations center may have limited ability to monitor the target system—time delays and incomplete or even conflicting information become more severe than those in remote supervisory control because of the nature of the target system.

Hortatory operations systems pose some challenges to human factors researchers. Clearly, the control–response relationships in hortatory environments are defined much less clearly than in supervisory control systems. The target system may not respond to an operator's action at all, or it may respond after a significant time delay, or in a way different from that which the operator expects. All these scenarios can make it difficult to identify the effects of each operator action and thus complicate the operator's task of monitoring and controlling the spontaneous changes in system status, whose existence constitutes the main reason for the operator's presence.

With regard to software-user interface design, researchers and designers need to identify effective interface mechanisms that can help operators monitor and influence the target system effectively under the constraints of a hortatory environment. More specifically, since time delay and incomplete or potentially conflicting information become more severe and more prevalent, there is a great need for software interfaces that make explicit the time delays and integrate the diverse sources of information on a single display. For instance, since the target system responds to the operator's advisory information only after a significant time delay and it may respond in a way that is different from what the operator expects, interfaces may need to make explicit the time and the type of each operator action and the time and type of each status change in the various parts of the system. Effective graphical displays integrating these information can help operators identify the likely effects of various actions and develop the most effective strategies of interacting with the target system. Predictor displays also have great potential value if the behavior of the target system can be modeled with an acceptable level of accuracy.

51.5.8 Networked Interfaces

The design of networked interfaces—interfaces for user communication and cooperation through a computer and communications network—is a challenging new area of investigation for software-user interface researchers and practitioners. In addition to the seven aspects of the interface design problem discussed above, the design of networked interfaces requires the consideration of a multitude of other factors, some of which have been identified but many more of which remain hidden and elusive. Readers are referred to Chapter 53 for a broad discussion of the emerging domain of computer supported cooperative work (CSCW). In the context of software-user interface design, this section summarizes some recent studies on the design of CSCW interfaces for the purpose of illustrating some of the main design issues.

A useful framework for analyzing different types of collaborative applications is the time–space matrix of Ellis, Gibbs, and Rein (1991), which classifies different applications into four classes: asynchronous local activities that take place at the same place but at different times (e.g., project scheduling); asynchronous distributed activities that occur at different places and at different times (e.g., e-mail); face-to-face interactions that require the participants to be at the same place at the same time (e.g., meetings and classes); and synchronous distributed interactions that take place at different places but at the same time (e.g., use of shared editors).

Different types of distributed applications have different characteristics and interface requirements. Research on e-mail interfaces has been focusing on organization of messages, automatic filtering of incoming messages, automatic routing and replies (Malone, Grant, Turbak, Brobst, and Cohen, 1987; Flores, Graves, Hartfield, and Winograd, 1988), and development of multimedia e-mail (Borenstein, 1991). Research on bulletin boards and electronic conferences has been focusing on access policy, browsing and searching methods, and filtering tools (Hiltz and Turoff, 1985; Shneiderman, 1992). The need for addressing these issues becomes more pressing with the emergence of the World Wide Web, which represents a new era in computer support for distributed asynchronous activities (December and Randall, 1994; Reiss and Radin, 1995). For face-to-face applications such as meetings and lectures, a number of systems have been developed and researchers have been addressing interface design issues such as the characteristics of computer supported meetings and classroom activities, the use of shared and private workspaces, and tools for meeting activities (brainstorming, ranking, voting, and decision making) (e.g., Greif, 1988; Malone and Crowston, 1990; Olson, Olson, Storrosten, and Carter, 1993; Valacich, Dennis, and Nunamaker, 1991)

The design of interfaces for remote synchronous interactions poses a special challenge to interface researchers and developers. In addition to the functions and features of face-to-face applications, an interface for remote synchronous work needs to provide effective mechanisms to support general awareness among remotely located users so that they can feel as if they are working at the same place. It has been shown that high-quality audio is important to remote synchronous work (Fish, Kraut, and Chalfonte, 1990; Tang and Isaacs, 1993). Recently, Olson, Olson, and Meader (1995) demonstrated that video adds significant value to groups of people working at a distance. In their study, the groups used a CSCW software program called ShrEdit to design an automated post office, and were connected to each other either with a high-quality stereo audio or the same audio plus high-quality video. The results showed that the quality of work was as good as that face-to-face when video was used, but the quality suffered when audio only was provided. Advanced media for collaboration is an area of active research and development (e.g., Benford, Bowers, Fahlen, Greenhalgh, and Snowdon, 1995; Bly, Harrison, and Irwin, 1993; Ishii, Kobayashi, and Arita, 1994; Mantei et al., 1991).

51.6 SUMMARY

This chapter has described the goals, the process, and the methods of software-user interface design, and discussed the different design issues that need to be considered. The conceptual model proposed at the beginning of the chapter helps organize the diverse design issues as eight aspects of the interface design problem and emphasizes the intermediary role of an interface in a user's work environment. Design is a creative process; there are no formulas to ensure a successful design. The conceptual model and the discussions presented in this chapter offer a way to achieve a comprehensive view of the software-user interface design problem. Interested readers are encouraged to read the books and articles referenced in this chapter for more detailed discussions of the various aspects of interface design.

REFERENCES

Baber, C., and Noyes, J. M., Eds. (1993). *Interactive speech technology: Human factors issues in the application of speech input/output to computers.* Bristol, PA: Taylor and Francis.

Baecker, R. M. (1990). *Human factors and typography for more readable programs.* Reading, MA: Addison-Wesley.

Baecker, R. M., and Buxton, W. A. S., Eds. (1987). *Readings in human-computer interaction: A multidisciplinary approach.* San Mateo, CA: Morgan Kaufmann.

Barfield, L. (1993). *The user interface concepts and design.* Wokingham, England: Addison-Wesley.

Barnett, B. J., and Wickens, C. D. (1988). Display proximity in multicue information integration: The benefit of boxes. *Human Factors, 30,* 15–24.

Baron, D., and Levison, W. H. (1977). Display analysis with the optimal control model of the human operator, *Human Factors, 19,* 437–457.

Benbasat, I., Dexter, A. S., and Todd, P. (1986). The influence of color and graphical information presentation in a managerial decision simulation. *Human-Computer Interaction, 2,* 65–92.

Benbasat, I., and Wand, Y. (1984). Command abbreviation behavior in human-computer interaction. *Communications of the ACM, 27,* 376-383.

Benford, S., Bowers, J., Fahlen, L. E., Greenhalgh, C., and Snowdon, D. (1995). User embodiment in collaborative virtual environments. In *Proceedings of CHI' 95* (pp. 242–249). New York, ACM.

Bennett, K. B., and Flach, J. M. (1992). Graphical displays: Implications for divided attention, focused attention, and problem solving. *Human Factors, 34,* 513–533.

Bennett, K. B., Toms, M. L., and Woods, D. D. (1993). Emergent features and graphical elements: Designing more effective configural displays. *Human Factors, 35,* 71–97.

Billingsley, P. A. (1988). Taking panes: Issues in the design of windowing systems. In M. Helander, Ed., *Handbook of Human-Computer Interaction.* Amsterdam: Elsevier.

Black, J., and Moran, T. (1982). Learning and remembering command names. In *Proceedings of the conference on computer human interaction* (CHI '82) (pp. 8–11). New York, ACM.

Bly, S., Harrison, S., and Irwin, S. (1993). Media spaces: Bringing people together in a video, audio, and computing environment. *Communications of the ACM, 36*(1), 28–47.

Boehm-Davis, D. A., and Fregly, A. (1985). Documentation of concurrent programs. *Human Factors, 27*(4), 423–432.

Borenstein, N. S. (1991). Multimedia electronic mail: Will the dream become a reality? *Communications of the ACM, 34*(4), 117–119.

Brown, C. M. (1988). *Human-computer interface design guidelines.* Norwood, NJ: Ablex.

Brown, J. W. (1982). Controlling the complexity of menu networks. *Communications of the ACM, 25*(7), 412–418.

Card, S. K. (1982). User perceptual mechanisms in the search of computer command menus. In *Proceedings of the conference on computer human interaction* (*CHI '82*) (pp. 190–196). New York, ACM.

Card, S. K., Moran, T. P., and Newell, A. (1983). *The psychology of human-computer interaction.* Hillsdale, NJ: Erlbaum.

Carroll, J. M. (1984). Minimalist training. *Datamation, 30,* 125–136.

Carroll, J. M., Mack, R. L., and Kellogg, W. A. (1988). Interface metaphors and the user interface design. In M. Helander, Ed., *Handbook of human-computer interaction* (pp. 67–85). Amsterdam: Elsevier.

Carroll, J. M., and Thomas, J. (1982). Metaphor and the cognitive representation of computing systems. *IEEE Transactions on Systems, Man, and Cybernetics, SMC-12*, 107–115.

Carswell, C. M., and Wickens, C. D. (1987). Information integration and the object display. *Ergonomics, 30*, 511–527.

Chernoff, H. (1973). The use of faces to represent points in k-dimensional space graphically. *Journal of the American Statistical Association, 68*, 361–368.

Christ, R. E. (1975). Review and analysis of color coding research for visual displays. *Human Factors, 17*, 542–570.

Coekin, J. A. (1969). A versatile presentation of parameters for rapid recognition of system state. In *International symposium on man-machine systems*, IEE Conference Record No. 69, 58-MMS, Vol. 4, Cambridge, IEE.

Crown, J. (1992). Effective computer user documentation. New York: Van Nostrand Reinhold.

December, J., and Randall, N. (1994). *The world wide web unleashed*. Indianapolis, IN: SAMS Publishing.

Don, A., Brennan, S., Laurel, B., and Shneiderman, B. (1992). Anthropomorphism: from Eliza to Terminator 2. In *Proceedings of the conference on computer human interaction* (*CHI '92*) (pp. 67–70). New York, ACM.

Doney, A., and Seton, J. (1988). Using color. In T. Rubin (Ed.), *User interface design for computer systems*, Chichester: Ellis Horwood.

Durrett, H. J., Ed. (1987). *Color and the computer*. New York: Academic Press.

Eberts, R. (1994). *User interface design*. Englewood Cliffs, NJ: Prentice Hall.

Elkind, J. I., Card, S. K., Hochberg, J., and Huey, B. M., Eds. (1989). *Human performance models for computer-aided engineering*. Washington, DC: National Academy Press.

Ellis, C. A., Gibbs, S. J., and Rein, G. L. (1991). Groupware: Some issues and experiences. *Communications of the ACM, 34*(1), 680–689.

Ellis, S. R. (Ed.) (1991). *Pictorial communication in virtual and real environments*. London: Taylor and Francis.

Ephrath, A. R., and Young, L. R. (1981). Monitoring versus man-in-the-loop detection of aircraft control failures. In J. Rasmussen and W. B. Rouse, Eds., *Human detection and diagnosis of system failure* (pp. 143–154). New York: Plenum.

Fels, S., and Hinton, G. (1995). Glove-TalkII: An adaptive gesture-to-formant interface. In *Proceedings of the conference on computer human interaction* (*CHI '95*) (pp. 456–463). New York, ACM.

Fish, R. S., Kraut, R. E., and Chalfonte, B. (1990). The VideoWindow system in informal communication. In *Proceedings of CSCW '90*, (pp. 1–11). New York, ACM.

Fisher, D. L., Coury, B. G., Tengs, T. O., and Duffy, S. (1989). Minimizing the time to search visual displays: The role of highlighting. *Human Factors, 31*, 167–182.

Flores, F., Graves, M., Hartfield, B., and Winograd, T. (1988). Computer systems and the design of organizational interactions. *ACM Transactions on Office Information Systems, 6*(2), 153–172.

Foley, J. D., Van Dam, A., Feiner, S. K., and Hughes, J. F. (1990). *Computer graphics: Principles and practice*. Reading, MA: Addison-Wesley.

Frenkel, K. A. (1988). The art and science of visualizing data. *Communications of the ACM, 31*, 111–121.

Galitz, W. O. (1989). *Handbook of screen format design* (Third Edition). Wellesley, MA: Q. E. D. Information Sciences.

Gilmore, W. E., Gertman, D. I., and Blackman, H. S. (1989). *User-computer interface in process control: A human factors engineering handbook*. San Diego, CA: Academic Press.

Goel, V., and Pirolli, P. (1992). The structure of design problem spaces. *Cognitive Science, 16*(3), 395–429.

Gould, J. D., and Lewis, C. (1985). Designing for usability: Key principles and what designers think. *Communications of the ACM, 28*(3), 300–311.

Green, T. R. G., and Payne, S. J. (1984). Organization and learnability in computer languages. *International Journal of Man-Machine Studies, 21*, 7–18.

Greif, I. (1988). *Computer-supported cooperative work: A book of readings*. San Mateo, CA: Morgan Kaufmann.

Grudin, J. (1989). The case against user interface consistency. *Communications of the ACM, 32*, 1164–1173.

Hartson, H. R., and Hix, D. (1989). Human-computer interface development: Concepts and systems for its management. *ACM Computing Surveys, 21*(1), 5–93.

Helander, M. (Ed.) (1988). *Handbook of human-computer interaction*. Amsterdam: Elsevier Science Publishers.

Henderson, A., and Card, S. K. (1986). Rooms: The use of multiple virtual workspaces to reduce space contention in a window-based graphical user interface. *ACM Transactions on Graphics*, *5*(3), 211–243.

Hiltz, S. R., and Turoff, M. (1985). Structuring computer-mediated communication systems to avoid information overload. *Communications of the ACM*, *28*(7), 680–689.

Hoadley, E. D. (1990). Investigating the effects of color. *Communications of the ACM*, *33*(2), 120–139.

Hoffert, E. M., and Gretsch, G. (1991). The digital news system at EDUCOM: A convergence of interactive computing, newspapers, television and high-speed networks. *Communications of the ACM*, *34*, 113–116.

Hopgood, F. R. A., Duce, D. A., Fielding, E. V. C., Robinson, K., and Williams, A. S., Eds. (1985). *Methodology of window management*. Berlin: Springer-Verlag.

Hutchins, E. L., Hollan, J. D., and Norman, D. A. (1986). Direct manipulation interfaces. In D. A. Norman and S. W. Draper (Eds.), *User centered system design*. Hillsdale, NJ: Erlbaum.

Isa, B. S., Boyle, J. M., Neal, A. S., and Simons, R. M. (1983). A methodology for objectively evaluating error messages. In *Proceedings of the conference on computer human interaction (CHI '83)* (pp. 68–71). New York, ACM.

Ishii, H., Kobayashi, M., and Arita, K. (1994). Iterative design of seamless collaboration media. *Communications of the ACM*, *37*(8), 83–97.

John, B. E. (1988). *Contributions to engineering models of human-computer interaction*. Unpublished doctoral dissertation, Carnegie Mellon University, Pittsburgh, PA.

Kearsley, G. (1988). *Online help systems: Design and implementation*. Norwood, NJ: Ablex.

Kieras, D. E. (1988). Towards a practical GOMS model methodology for user interface design. In M. Helander, Ed., *Handbook of human-computer interaction* (pp. 135–157). Amsterdam: North Holland.

Kieras, D. E., and Polson, P. G. (1985). An approach to the formal analysis of user complexity. *International Journal of Man-Machine Studies*, *22*, 365–394.

Kiger, J. I. (1984). The depth/breadth tradeoff in the design of menu-driven user interfaces. *International Journal of Man-Machine Studies*, *20*, 201–213.

Kleiner, B., and Hartigan, J. A. (1981). Representing points in many dimensions by trees and castles. *Journal of the American Statistical Association*, *76*, 260–269.

Koved, L., and Shneiderman, B. (1986). Embedded menus: Menu selection in context. *Communications of the ACM*, *29*, 312–318.

Laudauer, T. K., Calotti, K. M., and Hartwell, S. (1983). Natural command names and initial learning. *Communications of the ACM*, *23*, 556–563.

Laughery, K. (1989). MicroSAINT: A tool for modeling human performance in systems. In G. McMillan, D. Beevis, E. Salas, M. Strub, R. Sutton, and L. Van Breda, Eds., *Applications of human performance models to system design* (pp. 219–230). New York: Plenum.

Laverson, A., Norman, K, and Shneiderman, B. (1987). An evaluation of jump-ahead techniques for frequent menu users. *Behavior and Information Technology*, *6*, 97–108.

Liu, Y. (1993). Visual scanning, memory scanning, and computational human performance modeling, In *Proceedings of the 37th annual meeting of the Human Factors Society* (pp. 142–146). Santa Monica, CA: Human Factors and Ergonomics Society.

Liu, Y. (1994). A queueing network model of human performance of concurrent spatial and verbal tasks. In *Proceedings of the 1994 international conference on systems, man, and cybernetics* (pp. 2761–2766). New York, IEEE.

Liu, Y. (1996a). Queueing network modeling of elementary mental processes. *Psychological Review*, *103*, 116–136.

Liu, Y. (1996b). Quantitative assessment of effects of visual scanning on concurrent task performance. *Ergonomics*, *39*, 382–399.

Liu, Y. (1996c). Interactions between memory scanning and visual scanning in process monitoring. *Ergonomics*, *39*, 1038–1053.

Liu, Y. (1997). Queueing network modeling of human performance of concurrent spatial and verbal tasks. *IEEE Transactions on Systems, Man, and Cybernetics*, *27*.

Liu, Y., Fuld, R., and Wickens, C. D. (1993). Monitoring behavior in manual and automated scheduling systems. *International Journal of Man-Machine Studies*, *39*, 1015–1029.

Liu, Y., and Wickens, C. D. (1992a). Use of computer graphics and cluster analysis in aiding relational judgment. *Human Factors*, *34*, 165–178.

Liu, Y., and Wickens, C. D. (1992b). Visual scanning with or without spatial uncertainty and selective and divided attention. *Acta Psychologica*, *79*, 131–153.

Liu, Y., Zhang, X., and Chaffin, D. (1993). *Visualization and perception of human posture information for computer aided ergonomic analysis*. University of Michigan Center for Ergonomics Technical Report, Ann Arbor, Michigan.

MacGregor, D., and Slovic, P. (1986). Graphic representation of judgmental information. *Human-Computer Interaction*, *2*, 179–200.

Malone, T. W., and Crowston, K. (1990). What is coordination theory and how can it help design cooperative work systems? In *Proceedings of the third conference on computer-supported-cooperative work* (pp. 206–215). New York, ACM.

Malone, T. W., Grant, K. R., Turbak, F. A., Brobst, S. A., and Cohen, M. D. (1987). Intelligent information-sharing systems. *Communications of the ACM, 30*, 390–402.

Mantei, M. M., Baecker, R., Sellen, A., Buxton, W., Milligan, T., and Wellman, B. (1991). Experiences in the use of a media space. In *Proceedings of the conference on computer human interaction (CHI '91)* (pp. 203–208). New York, ACM.

Mantei, M. M., and Teory, T. J. (1988). Cost/benefit analysis for incorporating human factors in the software lifecycle. *Communications of the ACM, 31*, 428–439.

Marcus, A. (1992). *Graphic design for electronic documents and user interfaces.* New York: ACM Press.

McCormick, B. H., Defanti, T. A., and Brown, M. D., Eds. (1987). Visualization in scientific computing. *Computer Graphics, 21*, 1–87.

McMillan, G. R., Beevis, D., Salas, E., Strub, M. H., Sutton, R., and Van Breda, L. (Eds.) (1989). *Applications of human performance models to system design.* New York: Plenum.

Miller, R. K., and Walker, T. (1990). *Natural language and voice processing: An assessment of technology and applications.* Englewood Cliffs, NJ: Fairmont Press.

Moran, T. P. (1981). The command language grammar. *International Journal of Man-Machine Studies, 15*, 3–50.

Moray, N. (1986). Monitoring behavior and supervisory control. In K. Boff, Ed., *Handbook of perception and human performance* (pp. 40/1–40/51). New York: John Wiley.

Murch, G. M. (1987). Color graphics: Blessing or ballyhoo? In Baecker, R. M., and Buxton, W. A. S., Eds., *Readings in human-computer interaction: A multidisciplinary approach* (pp. 333–341). San Mateo, CA: Morgan Kaufmann.

Murray, J., and Liu, Y. (1993). Going with the flow: Computer visualization of road traffic information. In *Proceedings of the 37th annual meeting of the human factors society* (pp. 353–357). Santa Monica, CA: Human Factors Society.

Murray, J., and Liu, Y. (1995a). *Towards a distributed intelligent agent architecture for human-machine systems in hortatory operations.* (Tech. Rep. 95–17). Ann Arbor: Dept. of Industrial and Operations Engineering, University of Michigan.

Murray, J., and Liu, Y. (1995b). *The colloquium: Ontological support for hortatory operations* (Tech. Rep. 95–18). Ann Arbor: Dept. of Industrial and Operations Engineering, University of Michigan.

Murray, J., and Liu, Y. (1996). Hortatory operations in highway traffic management. To appear in *IEEE Transactions on Systems, Man, and Cybernetics.*

Myers, B. A. (1988). *Creating user interfaces by demonstration.* Boston, MA: Academic Press.

Myers, B. A. (1989). User interface tools: Introduction and survey, *IEEE Software, 6*(1), 15–23.

Myers, B. A. (1991). Demonstrational programming: A step beyond direct manipulation. In D. Diaper and N. Hammond (Eds.), *People and computer*, VI (pp. 11–30). Cambridge: Cambridge University Press.

Neider, J. (1993). *OpenGL programming guide: The official guide to learning openGL.* Reading, MA: Addison-Wesley.

Nielsen, J., Ed. (1989). *Coordinating user interfaces for consistency.* Boston: Academic Press.

Nielsen, J. (1990). *Hypertext and Hypermedia.* New York: Academic Press.

Norman, D. A. (1981). The trouble with UNIX. *Datamation, 27*, 139–150.

Norman, D. A. (1983). Some observations on mental modes. In D. Gentner and A. L. Stevens, Eds., *Mental models.* Hillsdale, NJ: Erlbaum.

Norman, D. A. (1986). Cognitive engineering. In D. A. Norman and S. W. Draper, Eds., *User centered system design.* Hillsdale, NJ: Erlbaum.

Norman, D. A. (1988). *The psychology of everyday things.* New York: Basic Books.

Norman, K. L. (1991). *The psychology of menu selection: Designing cognitive control of the human-computer interface.* Norwood, NJ: Ablex Publishing.

Okoshi, T. (1980). Three-dimensional displays. *Proceedings of the IEEE, 68*(5), 548–564.

Olson, J. R., and Olson, G. M. (1990). The growth of cognitive modeling in human-computer interaction since GOMS. *Human-Computer Interaction, 5*, 221–265.

Olson, J. R., Olson, G. M., and Merder, D. K. (1995). What mix of video and audio is useful for small groups doing remote real-time design work? In *Proceedings of the conference on computer human interaction (CHI '95)* (pp. 362–368). New York, ACM.

Olson, J. R., Olson, G. M., Storrosten, M., and Carter, M. (1993). Groupwork close up: A comparison of the group design process with and without a simple group editor. *ACM Transactions on Information Systems, 11*, 321–348.

Parasuraman, R. (1987). Human-computer monitoring. *Human Factors, 29*, 695–706.

Pattipati, K. R., and Kleinman, D. L. (1992). A review of the engineering models of information processing and decision-making in multi-task supervisory control. In D. Damos, Ed., *Multiple task performance* (pp. 35–68). London: Taylor and Francis.

Payne, S. J., and Green, T. R. G. (1986). Task-action grammars: A model of the mental representation of task languages. *Human-Computer Interaction*, *2*(2), 93–133.

Perlman, G. (1984). Making the right choices with menus. In *Proceedings of '84 international conference on human-computer interaction (INTERACT '84)* (pp. 291–295).

Pickover, C. A., and Tewksbury, S. K. (1994). *Frontiers of scientific visualization*. New York: John Wiley.

Polson, M. C., Wickens, C. D., Klapp, S. T., and Colle, H. A. (1989). Human interactive informational processes. In P. A. Hancock and M. H. Chignell, Eds., *Intelligent interfaces: Theory, research and design* (pp. 129–164). Amsterdam: Elsevier.

Pomerantz, J. R. (1981). Perceptual organization in information processing. In M. Kubovy and J. R. Pomerantz, Eds., *Perceptual organization* (pp. 141–180). Hillsdale, NJ: Erlbaum.

Posner, M. I., Nissen, J. M., and Klein, R. (1976). Visual dominance: An information processing account of its origins and significance. *Psychological Review*, *83*, 157–171.

Preece, J., Ed. (1993). *A guide to usability*. Reading, MA: Addison-Wesley.

Preece, J. (1994). *Human-computer interaction*. Reading, MA: Addison-Wesley.

Rasmussen, J. (1985). The role of hierarchical knowledge representation in decision making and system management. *IEEE Transactions on Systems, Man, and Cybernetics*, *SMC-15*(2), 234–243.

Rasmussen, J., Pejtersen, A. M., and Goodstein, L. P. (1994). *Cognitive systems engineering*. New York: John Wiley.

Rasmussen, J., and Vicente, K. J. (1989). Coping with human error through system design: Implications for ecological interface design. *International Journal of Man-Machine Studies*, *31*, 517–534.

Reisner, P. (1981). Formal grammar and human factors design of an interactive system. *IEEE Transactions on Software Engineering*, *SE-7*(2), 229–240.

Reiss, L., and Radin, J. (1995). *Open computing guide to mosaic*. Berkeley, CA: Osborne McGraw-Hill.

Rivlin, E., Botafogo, R., and Shneiderman, B. (1994). Navigating in hyperspace: Designing a structure-based toolbox. *Communications of the ACM*, *37*(2), 87–96.

Rizk, A., Streitz, N., and Andre, J., Eds. (1990). *Hypertext: Concepts, systems and applications*. Cambridge: Cambridge University Press.

Roscoe, S. N. (1968). Airborne displays for flight and navigation. *Human Factors*, *10*, 321–332.

Roscoe, S. N. (1981). *Aviation psychology*. Ames, IA: Iowa State University Press.

Rosenblum, L. J., Ed. (1994). *Scientific visualization: Advances and challenges*. London: Academic Press.

Rouse, W. B. (1980). *Systems engineering models of human-machine interaction*. Amsterdam: Elsevier.

Rouse, W. B., Cody, W. J., Boff, K. R., and Frey, P. R. (1990). Information systems for supporting design of complex human-machine systems. In C. T. Leondes, Ed., *Advances in control and dynamic systems*. Orlando, FL: Academic Press.

Rubin, R. V., Golin, E. J., and Reiss, S. P. (1985). *Thinkpad: A graphics system for programming by demonstrations*. *IEEE Software*, *2*(2), 73–39.

Sage, A. P. (1991). *Decision support systems engineering*. New York: John Wiley.

Sanderson, P. M., Flach, J. M., Buttigieg, M. A., and Casey, E. J. (1989). Object displays do not always support better integrated task performance. *Human Factors*, *31*, 183–198.

Sassoon, R. (1993). *Computers and typography*. Oxford: Intellect.

Schmandt, C. (1994). *Voice communication with computers: Conversational systems*. New York: Van Nostrand Reinhold.

Schuler, D., and Namioka, A., Eds. (1993). *Participatory design: Principles and practices*. Mahwah, NJ: Erlbaum.

Schweikert, R. (1978). A critical path generalization of the additive factor method: Analysis of a Stroop task. *Journal of Mathematical Psychology*, *18*, 105–139.

Sheridan, T. (1984). Supervisory control of remote manipulations, vehicles and dynamic processes. In W. B. Rouse, Ed., *Advances in man-machine systems research* (Vol. 1, pp. 49–137). New York: JAI Press.

Shneiderman, B. (1983). Direct manipulation: A step beyond programming languages. *IEEE Computer*, *16*(8), 57–69.

Shneiderman, B. (1992). *Designing the user interface: Strategies for effective human-computer interaction* (Second Edition). Reading, MA: Addison-Wesley.

Shneiderman, B., and Kearsley, G. (1989). *Hypertext Hands-On! An introduction to a new way of organizing and accessing information*. Reading, MA: Addison-Wesley.

Siegal, A. I., and Wolf, J. J. (1969). *Man-machine simulation models*. New York: John Wiley.

Smith, M. J. (1987). Human factors issues in VDT use: Environmental and workstation design considerations, In R. M. Baecker, and W. A. S. Buxton, Eds., *Readings in Human-Computer interaction: A multidisciplinary approach* (pp. 109–116). San Mateo, CA: Morgan Kaufmann.

Smith, R. W. (1994). *Spoken natural language dialog systems: A practical approach.* New York: Oxford University Press.

Smith, S. L., and Goodwin, N. C. (1971). Blink coding for information display. *Human Factors, 13,* 283–290.

Smith, S. L., and Mosier, J. N. (1986). *Guidelines for designing user interface software* (Report ESD-TR-86-278). Bedford, MA: Electronic Systems Division, the MITRE Corporation. Available from National Technical Information Service, Springfield, VA.

Strathmeyer, C. R. (1990). Voice in computing: An overview of available technologies. *IEEE Computer, 23*(8), 10–16.

Takeuchi, A., and Nagao, K. (1993). Communicative facial displays as a new conversation modality. In *Proceedings of the conference on computer human interaction* (CHI '93) (pp. 187–193). New York, ACM.

Tang, J. C., and Isaacs, E. (1993). Why do users like video: Studies of multi-media supported collaboration. *Computer Supported Cooperative Work, 1,* 163–196.

Thimbleby, H. (1990). *User interface design.* New York: ACM Press.

Thorell, L. G., and Smith, W. J. (1990). *Using computer color effectively.* Englewood Cliffs, NJ: Prentice-Hall.

Tufte, E. R. (1983). *The visual display of quantitative information.* Cheshire, CT: Graphics Press.

Tufte, E. R. (1990). *Envisioning information.* Cheshire, CT: Graphics Press.

Tukey, J. W. (1977). *Exploratory data analysis.* Reading, MA: Addison-Wesley.

Tullis, T. S. (1981). An evaluation of alphanumeric, graphic and color information displays. *Human Factors, 23,* 541–550.

Tullis, T. S. (1988). Screen design. In M. Helander, Ed., *Handbook of human-computer interaction* (pp. 377–411). Amsterdam: Elsevier Science Publishers.

Valacich, J. S., Dennis, A. R., and Nunamaker, J., Jr. (1991). Electronic meeting support: The GroupSystems concept. *International Journal of Man-Machine Studies, 34*(2), 261–282.

Van de Vegte, J. M. E., Milgram, P., and Kwong, R. H. (1990). Teleoperator control models: Effects of time delay and imperfect system knowledge. *IEEE Transactions on Systems, Man, Cybernetics, 20*(6), 1258–1272.

Vicente, K. J., Christopherson, K., and Pereklita, A. (1995). Supporting operator problem solving through ecological interface design. *IEEE Transactions on Systems, Man, and Cybernetics, SMC-25*(4), 529–545.

Vicente, K. J., and Rasmussen, J. (1992). Ecological interface design: Theoretical foundations. *IEEE Transactions on Systems, Man, and Cybernetics, SMC-22*(4), 589–606.

Walker, J., Sproull, L., and Subramani, R. (1994). Using a human face in an interface. In *Proceedings of the conference on computer human interaction* (*CHI '94*) (pp. 85–91). New York, ACM.

Wasserman, A. I. (1985). Extending state transition diagrams for the specification of human-computer interaction. *IEEE Transactions on Software Engineering, 11,* 699–713..

Wickens, C. D. (1984). Processing resources in attention. In R. Parasuraman and R. Davies, Eds., *Varieties of attention* (pp. 63–101). New York: Academic Press.

Wickens, C. D. (1992a). *Engineering psychology and human performance,* (2nd ed.). New York: HarperCollins.

Wickens, C. D. (1992b). Processing resources and attention. In D. Damos, Ed., *Multiple task performance* (pp. 3–34). London: Taylor and Francis.

Wickens, C. D., and Andre, A. D. (1990). Proximity compatibility and information display: Effects of color, space, and objectness on information integration. *Human Factors, 32,* 61–78.

Wickens, C., and Liu, Y. (1988). Codes and modalities in multiple resources: A success and a qualification. *Human Factors, 30*(5), 599–616.

Wickens, C. D., Sandry, D., and Vidulich, M. (1983). Compatibility and resource competition between modalities of input, output, and central processing. *Human Factors, 25,* 227–248.

Wickens, C. D., Todd, S., and Seidler, K. (1989). *Three-dimensional displays: Perception, implementation, and applications* (CSERIAC Rep. CSERIAC-SOAR-89-001). Wright-Patterson Air Force Base, OH: CSERIAC AAMRL.

Williges, R. C., Williges, B. H., and Elkerton, J. (1987). Software interface design. In G. Salvendy, Ed., *Handbook of Human Factors* (pp. 1416–1449). New York: John Wiley.

Woods, D. D., Wise, J., and Hanes, L. (1981). An evaluation of nuclear power plant safety parameter display system. In *Proceedings of the 25th annual meeting of the Human Factors Society.* Santa Monica, CA: Human Factors Society.

Ziegler, J. E., and Fahnrich, K.-P. (1988). Direct manipulation. In M. Helander, Ed., *Handbook of human-computer interaction* (pp. 123–133). Amsterdam: Elsevier.

CHAPTER 52

VIRTUAL ENVIRONMENTS

Hans-Jörg Bullinger
Wilhelm Bauer
Martin Braun
Fraunhofer Institute for Industrial Engineering (IAO), Stuttgart, and Institute for Human Factors and Technology Management (IAT)
University of Stuttgart, Germany

52.1 INTRODUCTION AND BASICS

Increasing requirements concerning quality, economy, and flexibility have been characterizing the ergonomic design of work systems, work places, tools, and products for many years. In this connection, the requirements are not exclusively restricted to the design results, but they include the design process equally.

To put the ergonomic knowledge into practice, efficient computer-supported methods of integrated work system design are being investigated, developed, and used, which make new dimensions of application possible. Essential characteristics of these methods, which are based on virtual environment technologies, are the three-dimensional modeling and simulation of virtual objects and situations, where the users are intensively and multisensorily integrated by means of intuitive, real time–oriented interaction modes.

Above all, the term *virtual environment* (VE) describes a computer-based generation of an intuitive perceivable and experienceable scene of a natural or abstract environment. VE applications will contribute to enhancing the qualities of human–computer interaction, the importance of which, in view of increasing complex information and communication applications, is constantly rising. VE technologies are more able than conventional computer applications to influence the thinking and behavior of people and to come to grips with social processes. Consequently, VE applications are challenging technical and social concepts, but also philosophical ideas.

In this context, VE technologies are on the one hand the methodical basis for ergonomic work system design, for which applications must be identified and developed. On the other hand, VE systems themselves are objects of ergonomic analysis and design. Problems in the fields of naturalness, comfort, safety, health, and task appropriateness, but also individual and social concerns as well as practical experiences will be sources of theoretical and experimental research input by ergonomists. For some time, this input will contribute to the design of advanced devices, displays, and software.

This chapter presents potential application areas as well as basic technical and physiological knowledge of VE systems and discusses requested dimensions. These remarks are intended to give advice and help with decisions regarding the application and design of VE systems for the purpose of ergonomic work system design.

52.1.1 Historical Overview

The development of virtual environments can be split into three phases.

The idea of later development of virtual environments was anticipated in literary circles during the mid 1960s. Science fiction works such as *Summa Technologiae* by Lem (1964) and *Sirius Transit* by Franke (1967) foresaw the expansion of new forms of human communication and experience. Objects of such literary work are on the whole a naturalistic, artificially generated habitat, in which a number of mentally influenced people intensively take part in a mutually experienced computer simulation. The information given to the human is exclusively determined by the machine (Glitz, 1994).

Simultaneously to the expansion of VE as a science fiction vision, research began aiming to make human–computer interaction more intensive. Sutherland (1965) presented his ideas of ultimate display and immersion into computer-generated environments by means of new types of input and output devices. At that time, however, because of limited computer capacity, one was only able to create primitive geometric objects and environments. Decades later, when powerful graphic computers for real time–oriented rendering were available, Sutherland's concept was seen in a new light. First VE applications were employed under the U.S. military and at NASA, who tested telepresence for the purpose of remote control tasks in space. In the third phase, the availability of powerful graphic computers led toward a technological push and the research in diverse application fields within science, industry, and entertainment. As a result, the commercial development of VE applications started during the 1980s. Today, both, science and industry commit themselves worldwide to the further development and expansion of VE systems.

52.1.2 Definition of Virtual Environments

An application-oriented definition, which might place the most accurate minimal demand, describes a VE system as a combination of computer-based display and interaction techniques.

Ellis (1995) defines a virtual environment as a synthetic, interactive, illusory environment, perceived when a user wears or inhabits appropriate apparatus, providing a coordinated presentation of sensory information, imitating a physical environment.

All VE applications are founded on the generation, perception, and manipulation of naturalistic or abstract virtual worlds, without any physical equivalent. Objects existing within virtual worlds can possess various qualities and behaviors. Examples of these object representations are graphics, sound, force feedback, etc. It is being attempted to generate the most possible intuitive virtual environments through the multifactoral addressing of the human sensoric. Generating is being attempted multifactorily; i.e., VEs can be experienced through a visualization which is marked out by three-dimensional object representations, and realtime–oriented interaction modes.

To distinguish VE from the multitude of computer-based visual simulation techniques, the following minimal requirements must be fulfilled:

- Graphical display has to be dependent on the position and orientation of the user.
- 360° visualization must comply with all three coordinate axes.
- The modeled objects must display realistic three-dimensional behavior.
- The most intuitive interaction modes possible with objects, adapting to human experiences and behavior, must occur.
- Object manipulation must be possible in all real or requested degrees of freedom (DOF).
- Object response must be in quasi–real time.

VE systems can be defined as computer-based information technologies for interactive, real time–oriented simulation, and multisensory representation of objects, processes, and their results. With the surmounting of conventional input and output interaction forms, the user should be given the impression that he or she is situated in a virtual environment. By doing this, the human's central cognitive abilities, such as the interpretation of visual data and the recognition of complex structures, are supported.

Other terms have been established to describe computer-generated worlds apart from virtual environment. One speaks for instance of *virtual reality, artificial reality, telepresence* and *cyberspace*. The difference is among virtual reality as a superior term for computer-generated, interactive, three-dimensional scenes; cyberspace as a place of communication connecting several users; and telepresence, which overcomes space and time, are accepted in most points. With the term "virtual reality," one tendentiously associates the vision of a meticulously detailed and perfect computer-generated duplicate of natural environment. In contrast to this, the term virtual environment particularly emphasizes the self-contained character of the application, whose abstact representations are oriented more to the user's ideas than on real objects. The term "virtual reality," which is used in many contexts, often resulted in utopic visions and the arousal of unfulfillable expectations; hence, for reasons of precise description, experts prefer the term "virtual environment." In this context it is to be said that a perfect simulation of a natural environment, with its endless variety, is impossible to create, because it would presume a consistent mental concept of the world.

The degree of *immersion* indicates to what extent a user is physically tied in a virtual environment. The impact of immersion can be described as an opportunity for real time–oriented perception and interaction. The characteristic of immersion is the believability of the environment. Immersion possesses a technical dimension, a content-representative dimension, and an individual-psychic dimension. These dimensions contribute correspondingly to an immersive experience (see Section 52.3.1.4).

52.1.3 Classification of VE Applications

Each VE application can be assigned to spatial-interactive, virtual presentation techniques.

The simulation of objects and processes is the main application area regarding VE. This suggests that VE can be used as a presentational simulation technique. Alongside the presentation of perceivable sensory objects, a special use lies within the expression of nonmaterialistic and invisible objects, or in saving of physical products. The presentation can rely on a static scene as well as on a dynamic process. With a pure presentational simulation the degree of interaction is limited, because the emphasis is laid on the presentation of knowledge. However, with the control simulation by means of VE, the user takes a more intensive part in the events, while the qualities of presentation have been preserved. Apart from the use of control simulation for entertainment purposes, its uses lie with the handling of critical situations, the control of devices in inaccessible areas, and working with meager resources. The third application type is particularly characterized by its interaction forms, where the virtual objects are not just moved, but are also used for design simulation. With regards to these applications, decision making in product develop process should be supported.

Aside from the process simulation, VE systems can be used as mediators for other, generally real, objects. The purpose of presentation of any object serves communication. Communication between humans can occur by means of a technical medium, which is subsumed under the term "cyberspace." Furthermore, VE can be used as a medium for human–machine interaction. In this context, communication with something irreal, for instance, a further virtual representation, might be possible. Finally, the possibility exists to integrate virtual information within a real environment.

A third dimension of a potential VE application is defined as *teleapplications.* This term combines simulation, which does not take place at the application location, and communication. These applications are founded on real spatial distances. Application areas are differentiated into two areas, that of *telepresence,* which is a spatially distanced presentational simulation, and *teleoperation,* which is a type of simulation for manipulation or design purpose.

52.1.4 Methods for Human–Computer Interaction

Foley and Silbert (1989) define the human-computer interface as the determination of all user inputs into a computer, the determination of all computer outputs to the user, and the determination of sequences of the inputs–outputs made accessible to the user.

In the context of VEs, generic modes of human–computer interaction allow themselves to be classified in *formal language interaction, direct manipulative interaction, natural language interaction,* and *gestic interaction.* These interaction modes are complemented by *combined interaction.*

52.1.4.1 Formal Language Interaction

Formal interaction languages are classified into programming languages, command languages, and formal interrogation languages. With command languages the meanings are predominantly laid down in the vocabulary. An accurate, defined volume of available commands and parameters exists. The meaning of a command arises itself from the inbuilt sequence and the relative position within an expression. Interactions based on formal languages are technically effective, however, unfamiliar to most users. As formal languages have to be learned before use, they are only suitable to a limited number of users. Programming languages are established for system implementation and form the basis of all other interaction languages.

52.1.4.2 Direct Manipulative Interaction

Direct manipulative interaction techniques, which are applied with graphical user interfaces, make use of familiar metaphors of daily life. The dialogue of direct manipulative interaction techniques is based on a permanent visual presentation of all relevant objects, and the function execution with single-stage reversible operations. The impacts of actions on the relevant objects are received on a direct visual feedback. Direct manipulative systems make easy learning, use, and extension of system functions possible. The steps of dialogue within the direct manipulation are slight in their complexity and their range, as with natural or command languages (Hanne, 1993). Disadvantages of direct manipulative systems are high-implementation expenditures and the impossibility of activating objects that are nonvisible on a graphical user interface. In direct manipulative systems effects, such as quantification, vague expressions, etc., are not or are insufficiently representable.

52.1.4.3 Natural Language Interaction

Natural language systems can use conventional methods of electronic data processing or interfere with knowledge-based systems. In the framework of human–computer interaction, natural language is an adequate means to express references to objects, actions, and abstract facts. Natural language offers possibilities of expression, which can be represented by other forms of interaction, only with an incomplete or large expenditure. The use of natural language in user interfaces raises the number of possible users considerably, in particular the unpracticed. Sounds, however, exist only in the present and cannot be called up in format as written language can. Analyzed, natural language proves inefficient, imprecise, manipulable, and only applicable for selected dialogue (Shneiderman, 1993). Natural speech interaction occurs by means of appropriate input–output speech devices.

52.1.4.4 Gestic Interaction

Gestic interaction can be defined as a command-based toolkit that allows the user to interact through nonverbal, nonsymbolic commands and instructions, using gestics, hand signals, and movements. A special kind of gestic interaction is *manual interaction.* Manual interaction forms are marked out by grips and movements of virtual objects, corresponding to a physical environment. The application of natural and intuitive gestic interaction modes aimes at an enhancement of efficiency in human–computer interaction.

The naturalism and intuitivism of gestic interaction forms result from the inclusion of human's sensorimotoric characteristics and abilities, as well as the integration of hand signs, mimics, and gestures, which have been culturally conditioned and ingrained by daily life. The quality of a gestic, in particular manual interaction, is oriented toward a maximal usability of the adaptability and dexterity of the human hand. A consequent allocation of hand movements and accompanying actions contribute to the understanding of gestic interaction.

Gestic interaction allows specification of certain commands and parameters with high expressive properties, as, for example, the pointing out of directions, the gripping of objects, the control of complex kinematic movements, or the parametrization of object qualities. Trivial gestic interaction is easy to learn and to apply and does not assume lingual knowledge. A further advantage lies with the directness of gestic interaction; i.e., the hand serves as an immediate medium.

With gestic interaction, every command must be represented by a certain hand gesture, clearly distinguishable from other gestures. In particular for complex commands and control processes not enough gestures are available, that results in the fact that gestic interactions, based on arbitrary gestures, are both unmethodical and inefficient. Discrete values and vague expressions are scarcely or not at all representable. Complex gestic interaction, e.g., for shaping and design applications, has partly proved as imprecise and is applicable only with selected commands.

Furthermore, with regards to gestic interaction within the control process, sensory feedback is an essential decision-making criteria. For detailed information on gesture based interaction see Section 52.2.3.

52.1.4.5 Combined Interaction

Combined forms of interaction do not represent generic interaction modes. Rather, they can be defined as an application-oriented combination of existing interaction modes, i.e., primarily gestic, direct manipulative, and speech input.

The isolated use of some interaction modes often leads to a one-sided application and performance profile. With a symbiosis of gesture-based, natural language and direct manipulative forms of interaction, the scope of human–computer interaction, as well as the functionality of the information input can be increased, and therefore it is possible to maximize the usability and efficiency.

If one examines the interaction modes applied within virtual environments within the historical context of human communication, one will notice that they do not represent anything new. In fact, only the reduction of communication to screen and keyboard, resulting from the computer technology of the past decades, will be reversed and communication will be shifted back to a human standard (Figure 52.1).

52.1.5 Design Concepts of VE Interfaces

The fundamental design principle of VE interfaces is to support the human's mental processes through an extensive communication and interaction environment, and therefore increase the scopes of human handling and decision making (Glitz, 1994). Resulting from these principles are, for example, the reduction of the degree of enforced handling sequences or the use of the computer for the establishment and expression of relations. From these principles the following differentiated design criteria for the development of user interfaces and computer-based tools can be derived which comply with ergonomic system design (Brooks, 1988):

- Three dimensionality of the modeled objects

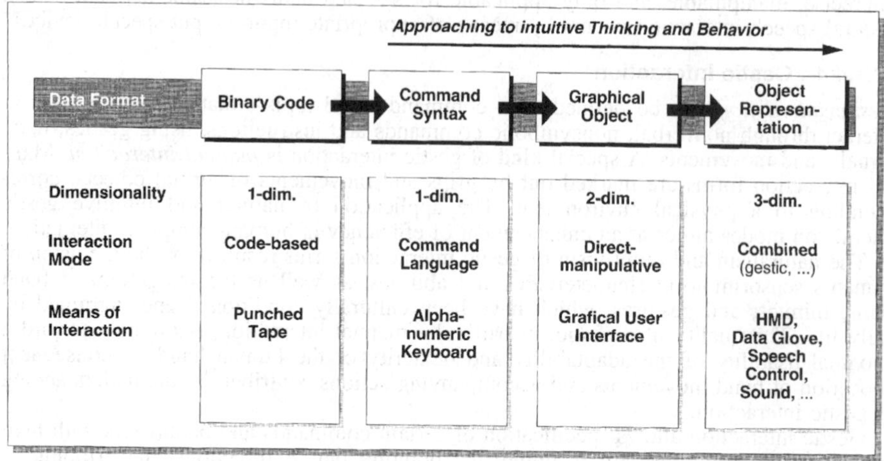

Figure 52.1 Developments in the basis of information of the human–computer interface.

- Direct manipulative and intuitive interaction, instead of formal interaction
- Interactivity rather than sequence professed routines
- Multisensory addresses rather than pure visual perception

These requirements are characterized by the basic idea that human–computer interaction will be comparable with the communication of humans, by means of speech, gestures, mimics, and body movements. In order to gain a human-oriented access to a computer, all of the human sensory channels should be included in the interaction, if possible. Apart from effective interaction modes, the quickest and most definite conception and interpretation of mediated information is strived for. The principle technical challenge regarding the design of interactive VE systems lies within the development of human–computer interfaces, which provide a possibility to convert internal data structures into sensory perceptible representations that possess a clear and consistent behavior. In the same way, development of devices that translate the human's movements into computer commands is required.

52.2 TECHNICAL DESCRIPTION OF VE SYSTEMS

To be able to design and use effective VE systems it is necessary to understand the technical concepts related to virtual environments, to be aware of the limitations of the available technology, and to know the design approaches that lead to the creation of successful virtual environments.

Although there are many VE systems in use, VE system and device design, as well as the basic use concepts, are fields open for research.

52.2.1 Functioning of a VE System

VE systems are used for computer-based real time–oriented interactive simulation and multisensory display of virtual objects, structures, and processes. The technical components of a virtual environment consist of hardware and software systems, which can be divided into sensors, effectors, and interlinkage. The construction and the functioning of a VE system is shown in Figure 52.2.

52.2.2 Hardware

To create immersive experiences within virtual environments, VE systems integrate a combination of several hardware technologies. These technologies can be grouped under the following categories:

- Display systems: Present the virtual environment to the user.

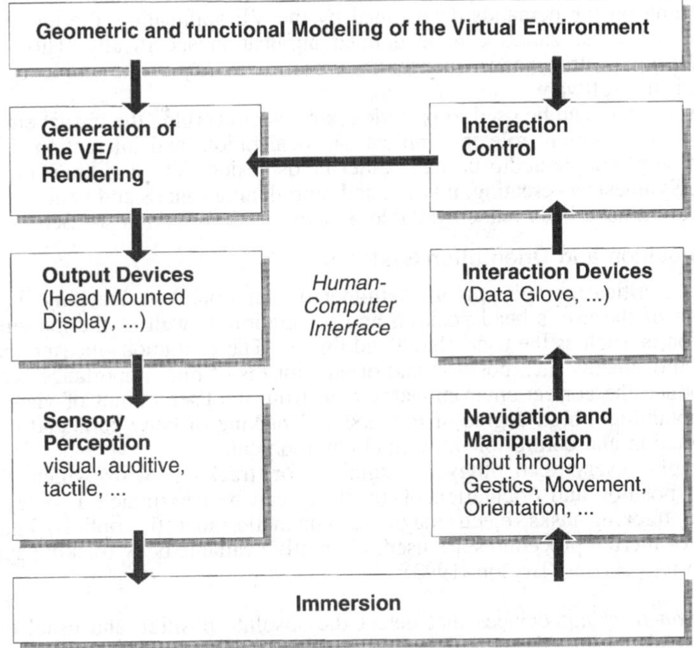

Figure 52.2 Principles of construction and functioning of a VE system.

- Position and orientation systems: Track the user's position and orientation in the virtual environment. They are also used for interaction purposes.
- Interaction and manipulation systems: Provide manipulation of the virtual environment.
- Computation systems: Perform the computations required for the generation of a virtual environment.
- Networks: Allow integration of several distributed user systems in one common environment.

52.2.2.1 Display Systems

Main elements of VE-interfaces are displays that mediate visual, auditory, and tactile, as well as proprioceptive sense stimuli. The visual and auditory displays, which are described below, are of essential significance.

Visual displays are devices that present virtual environments to the user. The degree of immersion given by a particular VE system mainly depends on the visual interface display. There are several kinds of visual displays currently available: monitors, head mounted displays (HMD), binocular omnioriented monitors (BOOMs), and projection systems. These systems are capable of producing wide-angle stereoscopic views of the scene. Although in some cases monoscopic vision is used.

HMDs (that are the most often used visual displays in immersive VE systems), and BOOMs place a pair of display screens directly in front of the user's eyes. With HMDs, the screens are mounted on a helmet which the user wears while staying in the virtual environment. The BOOM is like a pair of binoculars that is freely attached to a flexible swivel arm construction that can be easily maneuvered by the hand through open space. Both HMDs and BOOMs are coupled with a tracking system to determine the viewer's position in space. The virtual environment is displayed in stereo from the user's point of view, which leads to an enhancement of the degree of immersion; users are completely surrounded by the virtual environment. Other kinds of visual displays are projection-based systems. In such systems the user's position and actions are tracked, and the corresponding virtual scene is projected onto large screens.

With the visualization, what must be considered above all is its purposed use. Therefore, depending on the demands determined by the VE application, the required degree of detail of the visualization can be defined application specifically. This *concept of selected fidelity* has strong consequences on type and quality of the employed hardware as well as on the software.

Auditory displays can be used to provide feedback concerning the virtual environment. Sound plays an important role in sonification, localization, and interaction. Because of the application of simple audio in many other fields besides VE, it is known as a mature technology. Synthesizers creating, mixing, and reproducing sounds and systems for speech input and output have been made available. For further information see Section 52.3.1.2.

52.2.2.2 Position and Orientation Systems

Tracking is a critical component of an immersive environment. Tracking includes the measurement of the user's head position and orientation as well as the measurement of other body parts, such as the user's hand and fingers. The continuous measurement of the user's head movements (i.e., position and orientation) is of high importance because it is used to produce the correct environmental view from the user's point of view, which is critical to obtaining a high degree of immersion. Tracking of body parts and movements allows interaction and control of the virtual environment.

In principle, every technology is suitable for tracking, with which the three-dimensional position and orientation of an object can be determined as delay free and reliable. For tracking tasks, electromagnetic, kinematic, acoustic, optical, image-work-processing, or inertial procedures are used. Currently available body tracking systems are divided into two classes (Bryson, 1993):

- *Position tracking:* devices that detect the absolute position and usual orientation of a tracker in the three-dimensional space.
- *Angle measurement:* devices that detect the angle of bend of a body part. It is usually found in data gloves to measure the angle of finger joints.

52.2.2.3 Interaction and Manipulation Systems

Interaction and *manipulation systems* are devices that allow manual exploration of the virtual environment and the manipulation of virtual objects. These devices, functionally based on a tracking system, measure the position and forces of the user's hand and other body parts and can apply forces to the user to produce the sensation that the virtual objects are "real" (Cruz-Neira, 1993).

Interaction and manipulation devices have been classified as

- Pointing and selection devices
- Force and torque feedback devices
- Tactile devices
- Devices to produce stimulus such as temperature, etc.

The most well-known representative of interaction devices is the *data glove.* After having been calibrated specifically to the user, the data glove allows the computer to determine the user's hand gesture, e.g., fist, open hand, pointing, etc. Often the hand's physical position in a three-dimensional space is also determined. The glove's coordinates are overlain into the virtual environment, where the physical glove position controls a three-dimensional virtual hand in this environment. The virtual hand can be used to pick up virtual objects, push virtual buttons, etc. Furthermore, numerous interface devices exist, such as the *free flying joystick* and *spaceball,* which have been developed further in consideration of three-dimensionality, deriving from conventional interface devices, as well as systems for tactile and force feedback. For more information see Section 52.3.1.3.

Another interface technology for effective control and interaction in VEs is speech control. Speech control in virtual environments facilitates tasks and exploration. Not being tied to tapping onto a keyboard or moving, a speech control device frees the user's hands and adds to the feeling of immersion (Iovine, 1994).

52.2.2.4 Computation Systems

Computation systems are the computer hardware, which is used to control the overall operation of a virtual environment. Therefore, computer hardware has to handle several tasks: the generation of the graphics for the scene, the computation of the state of the environment, and the control of input and output devices. In a VE system, all these tasks have to be integrated and synchronized.

52.2.2.5 Networks

Networks serve the exchange of data between different virtual environments and lay therefore the foundations for distributed VE systems. Distributed VE systems make it possible to connect several persons which are located in different places in order to take part in a joint communication and design process. With the appropriate tools provided, such systems might be powerful platforms for interdisciplinary work forms, independent of location, submitted accordingly to the sense of *computer supported cooperative work* (CSCW).

52.2.3 Software

For the processing of information inputs and for the generating of an appropriate output, an internal computer functional hierarchy exists. Essential functions of this hierarchy are represented by the modules *modeling*, *simulation control* and *rendering*. Built onto appropriate hardware structures, these functionalities are realized by software technology.

52.2.3.1 Modeling

Above all, the virtual environment of an application is conceived and constructed in the modeling. As a rule, object data are converted from the modeling data and are completed through color and surface data. Besides the geometric description, the modeling also contains the function modeling, which exceeds the CAD applications. In function modeling the total number of functions and the geometric free degrees of a respective object are determined.

The structure of geometric objects can be described through parameters (object data). The totality in a computer stored object data gives the internal computer representation of real objects. The mathematical model of a real object is composed not only from data structures but also from algorithms, which intervene in these structures. It forms the base for all built-up modules such as simulation control and rendering.

52.2.3.2 Simulation Control

During the running of a VE application, the arrangement of objects is constantly calculated by putting the contents of the data bases into relation with the interaction instructions of the user—respectively, the behavioral parameters of autonomous objects. The simulation control is closely linked to the communication of data processing. The basis for a coordinated running is real-time management.

Simulation requires the determination of certain functions. Therefore, interaction parameters or collision identification have to be defined. Complex simulations require very high computing performance, which may lead to bottlenecks within the real-time management.

52.2.3.3 Communication

Communication includes data transformation, i.e., the coordination of data transfer between input–output devices and simulation control, as well as data exchange between users and units, which are involved in a multiuser system.

During the data transformation the transformation software interprets the interaction instructions of the user and passes the input commands to the simulation control, where the real time–oriented computation of the virtual environment happens. Additionally, the data representing the virtual environment are reconverted by the transformation software into appropriate interface signals and offered as simulation output. With that, the data transformation primarily serves to the data transfer between the user–interfaces and the simulation control.

Within a multiuser application the communication coordinates the data exchange between the active participants and units involved in a network. The real time–oriented data exchange serves to an information of all participants about relevant interaction processes as well as the actualization of the databases.

52.2.3.4 Rendering

The term *rendering* subsumes different procedures of the transformation of parametric data models into discrete images and sounds. Therefore, rendering is a part of communication within data processing, but because of its complexity and importance in VE it is specially dealt with. The requirements of real-time rendering concerning the computative performance are high. With VE applications the performance bottleneck lies mostly within the field of rendering, often effecting a field of tension between computating speed and quality of presentation, for visual as well as for acoustic rendering.

The visual presentation usually occurs with a pair of images, which must be computated with a minimum frame rate of 20 pictures per second in order to avoid picture disturbance.

Image generation is based on lighting models. One differentiates local and global lighting models. Local models determine the reflections of the single surfaces, independent from the environmental lighting situation. Global models consider the independence of lighting and reflections, and seem therefore more realistic; they are, however, unequally harder to generate (Göbel, 1992).

To keep the system update-rate running as high as possible as well as providing a high image quality, it is appropriate to minimize the complexity of the VE database, without undermining the system's task. One such strategy involves storing within the database different *levels of detail* (LOD) for specific objects. The operating systems automatically selects the model description that matches the current view and operation mode. Higher levels of detail can then be accessed on a range basis. As the user approaches different objects, they might be introduced into the user's viewpoint for the first time, or their geometric description may be improved by making a model substitution. As the user withdraws from an object, the reverse occurs (Vince, 1995).

Acoustic presentation within a virtual environment orientates itself toward the geometry and surface formation of the modeled room. For acoustic simulation two methods are suitable: The first is the *image-source method,* which calculates virtual sound waves, inclusive of their reflection behaviour at the modelled surfaces. From this, the received sound energy can be determined. With the *particle-tracing method,* the sent out sound impulses are registrated at placed detectors, from which the impulse response can be gained.

52.3 HUMAN FACTOR ISSUES

VE systems can be defined as the use of computers to create an illusion of immersion in a computer-generated environment (Bryson, 1993). As of this time, the human factors issues involved in creating this illusion are only partly known. One of the greatest challenges of VE-research is the understanding of human factor issues involved in the creation of an immersive computer-generated environment that the user takes seriously as a reality.

There were several failure modes in attempted early applications, including poor interface quality, poor system performance, and poorly designed environments, which were in violation of the human factor requirements. The lack of maturity of user interfaces has also proved to be an impediment to successful applications. For example, the data glove was thought to be an intuitive, easy use device. Its actual use, however, has proved to be somewhat inaccurate and more difficult than anticipated (Bryson, 1993). Among technical failures, the human factor impacts of these inaccuracies are only partially understood.

Basic ergonomic knowledge is represented below, and further requested dimensions are discussed. These remarks aim to give hints and help with decisions regarding the design and application of VE-systems.

52.3.1 Physiology in Virtual Environments

Sensory perception relates to external events as well as to internal processes of the human body. The existence of sensory phenomena requires definite physical and physiological events. VE-technologies take advantage of aspects concerning physiological perceptions in which an artificial working stimulation can be produced on the basis of human sense organs. The sensory perceptions of computer-generated representations can extend themselves over all physiological-sense areas in which a transmission of artificially produced sense stimuli is possible. This concerns, above all, the visual sense and the auditive sense as well as the haptic, tactile, and thermoreceptive senses. The sense channels ideally put under stimulation depend on the application. Integrated within a function structure, Figure 52.3 illustrates potential receptory channels and human–computer interface devices.

With perception forms, variations must be made among the technical reproduction of a perceptible environment, the sensory perception of the human, and the human's experience, which is set up. The impression of reality within virtual environments is based on the tendency of the human psyche to close the gaps in sensory perception and, thereby, to produce the impression of a consistent reality. If the human brain interprets a virtual

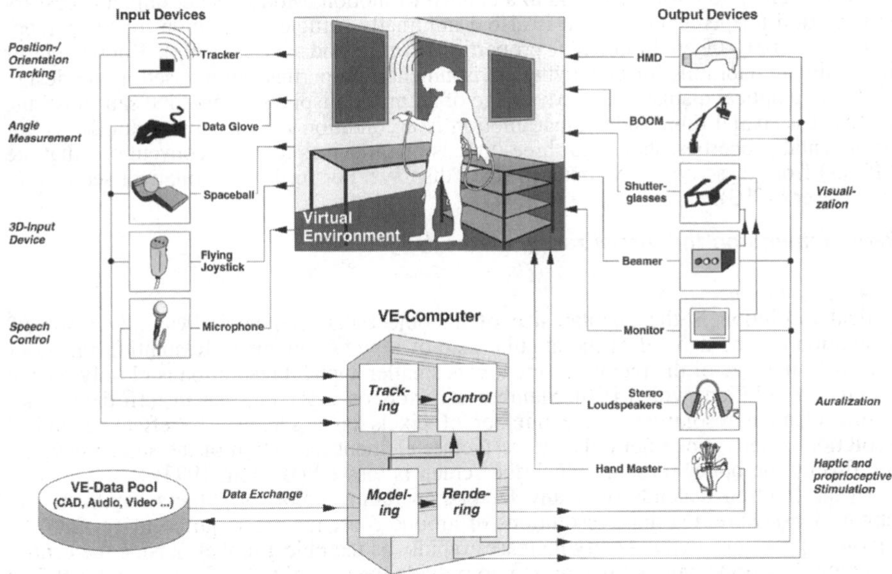

Figure 52.3 Structure and components of a VE system.

environment as reality after having had an adaptation phase, technical shortcomings can be compensated.

Nevertheless, it must be taken into account that by no means can VE technical stimulated impressions be compared with the richness of sensory impressions of a natural environment. In order to gain full control of sensory impressions, direct links between the human nerve system and the VE system have to be established, which at that moment is not deemed possible (Rötzer, 1993).

52.3.1.1 Visual Perception

Physiological Basics

Among all of the senses, the visual sense dominates. The light that is sent out from a source of light, or one that is reflected by an object, is sent into the eyes. By refraction and collection of the light, it is united on the retina, where the photo receptors are embedded. Following this, there is an arousal of the nerves, which is escorted along the optical track and is analyzed by the brain. With accommodation, the eye sets itself to the distance of the object being looked at. With the adaptation to dark and light, the eyes are able to change their acuity within wide ranges. The eye's absolute threshold with scotopic vision is at 2 to 6×10^{-17} J. Seeing in color occurs when there is sensory impression from the stimulation of the retina through light with wavelengths in the range of 400–780 nm (Müller-Limmroth, 1993).

When a three-dimensional object is looked at, there is a projection of a pair of two-dimensional images, one on each retina. As a result, each eye receives a different image. Both images are converted as impulse patterns in the brain, where they are centrally united, creating a three-dimensional object. In order to determine the distance of an object from the eye, the convergence angle between the eyes' axis and the accommodation is used. For detailed information see Chapter 3.

Visualization

Visual perception plays an important role in feedback in virtual environments. The optical information is presented by the use of optical displays. Depending on the application, display devices with various inserted performance profiles are used, making it possible to create anything from simple monochrome graphics and line models to a photorealistic rendering. Specifications of visualization technologies are discussed in the next subsection.

Visual perception always occurs in a context of motion. Motion is not only a necessary precondition for perception, but it is also a minimally sufficient condition for the perception of a variety of environmental properties (Proffitt and Kaiser, 1993). Knowledge of the minimal conditions for perceiving environmental properties can be used in the design of effective optical displays. The advantage of animated displays is that time can substitute for the lost spatial dimension. Since motion is a condition for perceiving numerous environmental properties, its use in three-dimensional devices requires conventions that are different from these typically found in static displays. For further information, see Proffitt and Kaiser (1993).

Requirements on the Visualization Methods

- *Resolution*

Optical resolution is the angular size of an object that can be individually resolved. Resolution can be defined as the angular size of a picture element. Resolution increases as the angular size of the picture element gets smaller. Optical resolution is closely related to screen resolution, which is the number of pixels on a screen, that in turn determines picture element resolution or the number of pixels on a screen. The effective optical resolution is determined not only by the picture element resolution of the screen devices, but also by the optics through which the screen is viewed (Bryson, 1993).

Eye resolution depends on many factors, including color, brightness, contrast, and length of exposure. On axis, resolutions of around 5 arcmin are required to get into the region of peak sensitivity. Acuity increases rapidly as the object moves outside the central 2° region. It is principally sufficient only to provide the central field of vision with detailed pictures, because the natural resolution of the eye in the periphery strongly decreases. At 10° off-axis eccentricity, acuity drops around 10 arcmin (Helman, 1993).

However, the marginal areas are particularly sensitive for low light intensities and movements. To achieve natural resolution capabilities, screens with 6000 picture lines are required. For a convincing visual display with a field of view of 150°/60° a resolution of ideally 9000 $\times$ 3600 pixels is recommended.

- *Stereoscopsis and Depth*

The stereoscopic visual presentation usually occurs through a pair of pictures with a slightly shifting convergence angle. The limit of stereo vision typically occurs for a binocular disparity of 12 arcsec. With the methods of stereoscopic visual presentation, one must differentiate between *time-parallel* (both eyes have simultaneous visual presentation) and *time-multiplexed systems* (nonperceivable alternating visual presentations). The time-parallel stereoscopic systems produce the displacement of the optical convergence axis either by different monitors for each eye, or by two half-pictures differentiating in color and perspective, which are looked at by toned or polarized 3D glasses. On the other hand, the time-multiplexed systems insert electrooptic or mechanic (so called) shutter systems for the viewing of alternating pictures, displayed on a monitor (Göbel, 1992).

Three-dimensional visual perception through object-differentiated accommodation can not occur with a HMD, because the monitor consists of a flat surface. When looking at a stereoscopic computer-generated imagery, the eye's accommodation and convergence often do not match because they must focus at the screen or the image plane defined by the optics of a HMD, but directed at an angle dictated by the rendered images. Without proper calibration (or monoscopic system), neither focus nor convergence may reflect the actual position of the virtual object relative to the viewer. Because of these inconsistencies many users of stereoscopic systems have trouble fusing stereo images. To place the images at infinity, thereby making convergence and focus match closely for distance, collimated optics can be used. Computer graphics applications that do not need to depict accurately the scale of depth can artificially adjust the parallax to allow more comfortable viewing. For virtual environments requiring closeup manipulation of objects or accurate registration of virtual objects with the physical world, all variables affecting stereo viewing should be considered. Size, image distance, and overlap of the system must be chosen carefully to match the task and operation distance (Helman, 1993). Alternative approaches propose for mediation of a visual depth impression, instead of convergent image display, the use of multiplanary visualization systems, where the accommodation is taken into account. Because of the bad visual quality and hardware expense, such systems have found only little acceptance until now (Aukstakalnis and Blatner, 1992).

However, other solution approaches try to realize the adaptation of depth with the changeability of the focus. In this connection the eye's focus is measured by means of a laser beam reflected by the retina, and in consequence the depth of focus is followed by image visualization (Aukstakalnis and Blatner, 1992).

With monocular picture presentation by means of a monitor or big-screen projection, depth effects methods such as object moving, solid modeling, shadowing, perspective and covering can be used to sharpen a three-dimensional pictural impression. For additional information see Profitt and Kaiser (1993).

- *Field of View (FOV)*

Each eye has approximately a 150° FOV horizontally and 120° FOV vertically. The binocular overlap, when focused at infinity, is approximately 120°. With VE display devices, a wide field of view is very desirable to convey a feeling of immersion. For HMDs a minimum of 120° horizontal and 60° vertical FOV is recommended. The tradeoffs for higher FOV are lower effective resolution and usually more distortion in the periphery of the picture. Distortion can be avoided by using complex optics. However, distortion is a problem with see-through HMDs because in this case the natural environment provides a reference for "straightness." With the relatively small FOVs of HMDs, large overlaps of more than 50% have been found useful (Helman, 1993).

- *Brightness and Luminance*

Because of the bad visual quality and hardware expense, including dark adaptation, the eye has a dynamic range of around seven orders of magnitude, far greater than any current display device. The eye is sensitive to ratios of intensities rather than to absolute differences. At high illuminations the eye can detect differences in luminance as small as 1%. Brightness is a basic precondition for the perception of objects.

As brightness is the available luminance for each picture element, a typical HMD based on a cathode ray tube (CRT) can display no more than about 400 perceptible luminance levels. Sufficient brightness is particularly a problem for liquid crystal displays (LCD).

To minimize disturbing lighting conditions and differences of contrasts between the physical environment and the virtual simulation, high brightness is especially important for see-through HMDs.

- *Contrast*

Contrast is the dynamic range of the luminance that the display supports. Contrast is important to the perception of structure in an image. Low-contrast systems are difficult to interpret. Therefore low-contrast and low-brightness displays do not serve a high degree of immersion. For screen displays a 5:1 contrast ratio for scenery and 25:1 for light points is recommended (Helman, 1993).

In LCDs the display brightness and contrast depend greatly on the viewing angle. The display viewing angle is characterized as the angle between the normal to the display surface and the line between the center of the display and the user's eye. For high image quality, the viewing angle should be as small as possible.

- *Color Saturation*

Color is a significant feature of visual images. The quality of color can be characterized in terms of the decomposition of a color signal into three primary components. Usually these components will be red, green, and blue primary colors measured by their luminance. In another categorization, color components may include the hue, luminance, and value components.

In display systems, color is usually attained by grouping pixels into sets of color components. For example, a red pixel, a blue pixel, and a green pixel are grouped into a triple, which comprises one full-color picture element. Although this method works when the optical pixel resolution is very high, in wide-field displays the individual component pixels are easily individually visible. Another method of attaining color, called time-multiplexed color, is to use a monochromatic display screen with rotating colored filters. There are three rotating color filters, one for each primary color. According to the activated filter, the monochromatic image with the lumination of the correspondent colored filter is displayed.

The quality of the color signal depends on several aspects, such as quality of color component pixels, etc. The dynamic range of the luminance value for each component will determine the full range of colors that can be achieved.

Apart from color displays, for several purposes monochrome displays are in use. In general, monochrome displays offer high resolution at equivalent price.

- *Masking*

A display quality issue closely related to color is the problem of masking, or nonabutting pixels, which occurs when the pixels are separated by a blank space. Masking is a problem because the wide-field optics magnify the display screen and so magnify the space between the pixel elements. To avoid the problem of visible pixel elements and masking, the image in wide-field displays is often intentionally degraded, usually done by a diffusion screen which blurs the pixel.

- *Performance*

To avoid picture disturbance, the frame rate must at the least be 20 pictures per second. The frequency at which modulation is no longer perceptible varies from 15 Hz up to around 70 Hz for high illumination levels. Bright displays with large FOVs can require frame rates up to 85 Hz. In order to minimize movement dizziness, the perceptional latency must lie below 0.1 s.

- *Refresh Rate*

Refresh rate is time required for the picture elements to change state (*on* or *off*). These times need not be the same. Typically, a pixel takes longer to go off than to go on. If the refresh time of a pixel is too long, ghosting effects will occur in rapidly changing images.

Visualization Methods

In dependency on the grade of immersion and the extent of the interactions, graded concepts from fully immersive VEs down to a nonimmersive monitor VE can be realized. VE applications for entertainment purposes mainly aim for the simulation of self-contained fantasy worlds, whereas in scientific applications above all the sensory acquirement of a reduced reality is strived for. Although the tendency of early development was almost exclusively toward fully immersive concepts, because of their shortcomings, nowadays one finds an increasing number of partly immersive VE system applications.

- *Fully Immersive VE*

With fully immersive VE the user wears a head mounted display (HMD), equipped with a visual stereo display and headphones. Thereby, the user is visually and auditively sealed off from the physical environment and primarily perceives impressions of the virtual environment. The HMD, a well-known fully immersive display system, is made up of two monitors, which project the image directly before the eyes, and a lens system, which widens the image to the natural field of view. On the basis of the convergence angle of the screen arrangement, the image has a certain appearance of depth.

An alternative interface for HMD is the BOOM (binocular omnioriented monitor). Here in contrast to the HMD, the monitor is not mounted on the head of the user, but fixed onto a flexible swivel arm construction, so that it can be freely moved in space by hand. During the application, the monitor is always held before the eyes. Because all gravitational forces are compensated by the device, no force expenditure is required for the guidance of the BOOM.

For the use of HMDs and BOOM systems, on the one hand, only display technologies with small build sizes and light weight are suitable. On the other hand, a high-quality replay should be guaranteed by a high-resolution capability. At the moment, high-resolution cathode ray tubes (CRT) or liquid crystal displays (LCD) are used. LCDs are distinguished by their light weight and extreme flatness. The disadvantages of LCDs are low resolution, the lesser contrast, and the lack of color saturation. CRTs on the other hand allow a high visual resolution to be made possible, with high light densities, however with larger build sizes and heavier weight.

By means of a laser microscanner (*virtual retinal display,* VRD), actual developments intend to project a high-resolution picture of approximately 6000 × 8000 pixels directly on the user's retina. The VRD is not a screen-based technology. The long-range development goal is to build a very small and lightweight display, mountable on glasses, with high performance qualities, concerning high resolution, large field of view, superior colour resolution, and image depth modulation. VRDs will be capable of fully inclusive or see-through display modes. For detailed information see, for example, Tidwell, Johnston, Melville, and Furness (1995), and Holmgren and Robinett (1993).

Ideally, contact lenses would be suitable for visual presentation, because they move with the eye, thus always covering the whole field of view. Until now, however, it has not been possible to develop electronic contact lenses, which cover the whole of the pupils with several million light emitting elements.

- *Projection-VE*

With projection-VE, stereo pictures are thrown up on surrounding walls by means of a special projection system. Systems for projection-VE allow the user a higher degree of movement, because, besides 3D glasses, no other devices have to be worn. Such systems make it possible to link up several users within a VE application. However, the user's orientation and interaction within such a virtual environment is harder because of its lower grade of immersion.

As an example of projection–VE, see Cruz-Neira, Sandin, Defanti, Kenyon, and Hart (1992), who describe the CAVE system (audio-visual experience automatic virtual environment, recursive acronym).

- *Augmented Reality*

With an augmented reality semipervious data glasses are used, which allow computer-generated objects and information to be linked by superimposing them into the perception process of the physical environment. This lets parts of the physical environment remain perceivable, at the same time as virtual elements contribute to an enrichment of information. Semipervious HMDs prove their worth in every case, where the superimposure of computer-generated pictures onto a real backdrop will facilitate the visualization.

- *Monitor VE*

In contrast to immersive VE systems, monitor VE portrays a low-price alternative for a three-dimensional, nonimmersive visualization of virtual worlds. As with projection VE, the user wears 3D glasses, which allows him or her stereoscopic sights of the virtual world on a monitor. By a tracking system, head movements of the user are measured. The tracking data are raised by the computation of the orientation-dependent displayed sight. Monitor VE makes the viewing of virtual objects from different directions and distances possible. A decided advantage of the monitor VE in opposition to fully immersive VE is the possibility of simultaneous interaction with the physical environment. Besides, in contrast to head mounted displays, present monitors make substantial higher visual resolution available. As monitor-VE is limited to a visual frame, only a slight feeling of immersion is possible.

- *Responsive Workbench*

Responsive workbenches enable direct interaction with virtual objects, appearing in physical environments. At a so-called responsive work place, several users move around a table equipped with a glass desktop. By means of a stereoscopic visual projection from the under side of the table, three-dimensional virtual objects are generated, which appear to the users as if they were found on the workbench. The users can perceive as well the virtual object and the workbench as the own person and the colleagues. Examples of responsive workbench applications are employed in the areas of medicine and architecture (Krüger and Fröhlich, 1994).

52.3.1.2 Auditory Perception

Physiological Basics

The human ear consists of two different sense organs. Beneath the organ for perception of sound there is the so-called vestibule apparatus, an organ for perception of acceleration. The ear is subdivided into the outer, the middle, and the inner ear.

The sense organ for perception of sound, the cochlea, is found in the inner ear. The outer ear (with pinna) and middle ear serve for the transmittance of airborne sound on the receptors. The ear drum vibrations, caused by sound waves, are transmitted through the auditory ossicles, which are the hammer, anvil, and stirrup. The stirrup makes the connection to the inner ear. There one finds the cortex sense organ. Within this organ lie the hair cells, which as mechanicoreceptors express the sense cells. The auditory nerves come out of the hair cells (Müller-Limmroth, 1993). The auditive system makes possible perception and analysis of tones, noises, and sounds from frequency and amplitude. The tone phase is not perceptible.

For the strength of a sound perception, the sound pressure is decisive. Based on a 1 kHz tone, the audibility limit amounts to 2×10^{-5} N/m^2 for the maximum frequency limit. With regards to auditive perception, the pain limit where sound becomes uncomfortable lies at 2×10^2 N/m^2 for a sound with a frequency of 1 kHz. The minimum audible field, and thereby the volume-related hearing sensation, is strongly dependent on frequency. The human audible frequency scope of younger people ranges from about 16 Hz to 20 kHz. With increasing age, the upper maximum frequency limit falls.

Hearing with both ears makes spatial orientation possible. Directional hearing is based above all on the coverage of running times and intensity differences between the two ears, as well as on the frequency-dependent directional characteristic of the hearing. The double point threshold for the directional hearing lies around 1°–5°. In contrast to directional hearing, distance-dependent hearing is only insufficiently formed (Müller-Limmroth, 1993).

Acoustic Simulation

The generation of sounds, their synchronization to motion, and the modeling of environmental effects are important aspects because sound provides a relevant feedback in virtual environments. Sound can be segregated between speech and nonspeech audio. Speech audio is used in voice recognition and speech generation for the purpose of interaction. The most important criterion for speech transmission is maintaining intelligibility throughout the entire communication system chain. With nonspeech audio, the functions and means for producing sound are more varied. Nonspeech audio can be divided into three categories: alarms and warning messages, status and monitoring messages, and encoded messages (Begault, 1994). Sound is applied in four areas (Cruz-Neira, 1993):

- *Interaction:* Using audio for input and output, e.g., voice recognition and speech synthesis.
- *Sonification:* Presenting (numerical) information as sound.
- *Localization:* Associating a sound with an object or a location in space.
- *Navigation:* Using audio as an additional cue to orient users in the VE.

Auditive displays exist temporally and independent of place. Because sound fades away, presented information is only available during a limited period. For auditory perception, the user does not need to face the source. Sound suits itself to sequential display of information, which can occur simultaneously to other information channels.

Requirements on the Acoustic Simulation Methods

With the application of sound, it must be considered above all, what the proposed use of sound will be, how sound will motivate, alert, guide, or immerse the user. This proposes questions as to how detailed the envisioned virtual audio world will be.

Subsequently, essential demands on the acoustic simulation are placed together, which have to be fulfilled according to the application:

- *Adaptation to the Physiological Performance Characteristics of Hearing*

Sound may be characterized as having an amplitude component and a frequency component, which translate from hearing to loudness, noisiness, pitch, and tone color or timbre. The hearing possesses temporal and spatially resolution capabilities. The standards for acoustic simulation must be oriented toward the human hearing. These demands also apply to some extent to the means of speech interaction, i.e., synthesized speech must be clear cut, perceivable and understandable. In addition to this, acoustic mediation of information should be sufficient for the mental information-processing capacity of the user; i.e., simultaneous signals have to be distinguishable and interpretable.

Individual head and ear forms lead to different individual auditory perceptions. With the creation of sounds, these influences of anatomy must be taken into consideration.

- *Realism (Modulation, Real-Time Response)*

Electronic sound generators using conventional modes of audio signals produce steady periodic waveforms, which offer little irregularity or complexity to the listeners. Natural sounds are almost never periodic, and the ones that are truly periodic lead to uninteresting listening. A great challenge of sound synthesis is to recreate a natural complexity, which is done by involving combinations of different wave forms (Bargar, 1993).

An immersive experience within virtual environments requires among others a real time–oriented acoustic simulation. Dependent on the application it is to decide, whether spatialized sounds need to be generated in real time, or whether the sounds can be prespatialized.

- *Three-Dimensional, Specific-Environment Auditory Impressions*

The human auditory sense makes it possible to localize sound sources and contributes to the obtaining of information concerning the specific environmental qualities.

Each environment has specific acoustic cues, such as reflections, echos, and reverberation. Spatial hearing is substantially dependent on physical transformations and psychoacoustic cues. With the simulation of spatial sounds, it is to simulate the specific acoustic environment and make it audible.

- *Movement-Dependent Modeling of Sound Fields*

Sound synchronized to motion provides additional cues that are helpful in defining spatial and temporal relationships (Hahn, Gritz, Darken, Geigel, and Won Lee, 1993).

The auditory perception of a situation- and environment-specific sound field is above all dependent on relative movements between the hearing and the sound source. This effect has to be considered with regard to the simulation of spatial sound, i.e., the sound field must be modeled in dependency on the user's spatial position (position and orientation changes).

Only with a dependency on the modeled sound field from the user's spatial position can auditory perception be guaranteed, which contributes to a spatial orientation.

- *Undistorted Sound Reproduction*

Sound can be reproduced even by loudspeakers or headphones. With binaural hearing, i.e., listening through headphones, the left and right ears hear the sounds only from the individual transducers. Additionally, there is some sound transmission through the skull from one ear to another. With the use of stereo loudspeakers, each ear hears the sound emanating from both loudspeakers. In the case of binaural sound, containing phase and intensity cues, the sound stage appears to be located within the listener's head (Vince, 1995).

The problem with using loudspeakers for three-dimensional sound is that control over perceived spatial imagery is greatly sacrificed, since the sound will be reproduced in an unknown environment. The room and the loudspeakers will impose unknown transformations that usually cannot be compensated by the designer of the three-dimensional audio system.

Additionally, the use of a reasonable sound reproduction system should provide a spatial sound perception to a group of people rather than to an individual.

Methods of Acoustic Simulation

Computer-generated sound display is a very large field. It can be given a cursory discussion here only.

- *Direction-Independent Sound*

Direction-independent sound can be classified in monaural sound and in stereo sound.

Simple *monaural sounds* can be used for interaction or sonification purposes to indicate events or provide feedback to the user. Both events and feedback must be global in nature, because monaural sound cannot provide a sense of location by itself. In certain applications, there may be many opportunities to use sound to indicate global events and states, located outside of the visual field, or not immediately perceivable. In combination with graphics, a monaural sound may be used to associate with a particular object or location. Monaural sounds are also suitable for speech input and output within the framework of natural speech human–computer interaction.

Conventional *stereo sound* enhances the realism of the sound. Stereo provides sounds with a position along a one-dimensional line between the ears. The use of conventional stereo sound in virtual environments gives individual sounds something of a spatial location, which is usually head stable. Thus it is difficult to give conventional stereo sounds a sense of presence in a location which is independent of the user (Bryson, 1993).

Monaural sound as well as stereo sounds in VEs are typically generated by electronic synthesizers. Parameters such as amplitude, frequency, and phase can be controlled nearly continuously.

- *Direction-Dependent (Spatially Localized) Sound*

Giving sound a spatial location provides it with a considerable sense of presense. Because sound has the quality of a three-dimensional position, it can be used to assign qualities to objects within virtual environments. Alternatively, sound can be a system-stable object without being associated to any graphical object. Sound can be used to display data or, as an independent element, for localization within VE. Besides this, three-dimensional acoustic simulation can be used for the evaluation of spatial sound qualities on behalf of planning and prognosis.

With acoustics, the totality of all physical parameters that influence the sound in the audible frequency range are described as an *acoustical environment.* This is perceived by the listener as *auditory environment;* i.e., acoustic phenomena (source of sound, reflection surfaces, etc.) are transformed into auditory phenomena (spatial impression, location of auditive impression).

One approach to create a spatially localized sound is through the use of *head-related transfer functions* (HRTF) measured in the human's ear canals. The HRTF data encode how sound waves interact with the human's hearing and characterizes how humans exploit signal propagation delays between the two ears to localize sound sources. When anechoic signals are processed by the HRTFs and are heard through headphones, a realistic spatial sound stage arises (Vince, 1995).

The spectral shaping of a sound is significant depending upon the pinnae's influence as well as the spatial origin of the sound source. Thus, the brain learns to extract spatial information from the unique earprint the pinnae impress upon the incoming pressure waves. The shaping of incoming sound by the pinnae is responsible for creating an external rather than an internal sound stage.

Because headphones, especially the small ones plugged directly in the middle ear, ignore or destroy the action of the pinna, the perceived sound stage is internal. With head-related transfer functions, the left and right channels driving the headphone are artifically shaped electronically, so that the brain gets the impression that the sounds have an external reality. To achieve this, the spectral shaping (or transfer function) introduced by the pinnae must be discovered. This can be done by measuring the spectral difference between an external sound and that incident on the eardrums (Vince, 1995).

Head-related transfer functions relate to an individual user's head. HRTFs are influenced by the head's and ears' sizes and shapes, as well as the shoulders and body. Indeed, they are often derived from a population of individuals, resulting in an averaged set of characteristics. Generalized HRTF may cause perceptional errors in directional localization, or make it difficult to imagine the sound stage external from one's head.

Research continues into the simulation of spatial localization cues and the real-time generation of synthetic audio cues, as a part of multisensory interface within virtual environments. For detailed information on acoustic simulation, see, for example, Begault (1994) and Stephenson (1994).

52.3.1.3 Haptic, Kinesthetic, and Other Perceptions

Physiological Basics

Through the skin, a variety of sensations such as touch, pressure, vibration, warmth, cold, and pain can be aroused. Tactile sensations are aroused through *mechanicoreceptors, thermoreceptors,* and *nocireceptors.*

Mechanosensitivity gives information by means of touch or pressure, with periodical disruption by means of vibration, on an object's form, size, weight, consistency, and surface qualities. An adequate stimulus is a mechanical deformation of the skin. The regions of the human body with the highest density of mechanicoreceptors have the most acute sensitivity to tactile stimulation. The localization of mechanical stimulus on skin is very exact (Müller-Limmroth, 1993).

Stimulus parameters for the sensation of temperature are the absolute skin temperature, its changing speed, and the size of the stimulus surface. The sensation of temperature is polar, structured into warm and cold. In between there exists a thermic area of indifference. Arousal releases physiological processes of temperature regulation, which create new sensations. Thermoreceptors can be aroused only chemically.

The tasks of the nocireceptors are to protect the human organism from damaging influences. Each skin stimulus leads from a crossing of a certain intensity to a feeling of pain. There is differentiation between light surface pain sensation and dull body pain sensation.

The proprioceptive sensitivity offers mediation for sensation of orientation, movement, and force. Muscle spindles (reflectoric control of motoric), tendon receptors (force sensation), and the nerve endings in joint capsules and ligaments are efficient kinesthetic proprioceptive receptors.

The truth of the matter is that there is almost no way to really separate mechanicoreceptive from proprioceptive cues. Tactile and force feedback are closely connected.

Tactile and Force Feedback

For numerous manipulation processes—for instance, with the design of complex geometric and freeform surfaces—as well as the evaluation of object properties, the implementation of extensive haptic and kinesthetic functionalities is indispensable. Thereby object properties are made assessable and experiencable in a clear way for the user. However, the inclusion of haptic sense is much more difficult than the visual and auditive senses, here the mechanico- and thermoreceptors are distributed all over the body. They are extremely sensitive and small. Even harder is proprioceptive stimulation; here sensors lying deep beneath the skin have to be stimulated. Figure 52.4 shows the basics of haptic, respectively kinesthetic, perception and shows forms of device-based generation of corresponding sensory feedback.

In order to create haptic and kinesthetic sensations, several tactile and force feedback devices have been developed. However, the task of developing technologies capable of generating haptic sensations is one of the most difficult challenges VE researchers are encountering.

With artificial stimulation it must always be taken into consideration that the human sensorium is of such high complexity, that from a technical point of view the creation of perfect stimuli nowadays and probably in future will never be realized.

Requirements on the Feedback Simulation Methods

Feedback systems have to provide feedback information from VE, in order for the user to feel the sensation that he or she is physically manipulating the virtual objects. The following basic demands can be placed on the variety of different feedback systems:

- Body endangering and damaging effects caused by mechanical or other further influences of the VE system must safely be ruled out.
- Feedback systems should provide maximum user's freedom of motion and reduced fatigue so natural interactions can occur. Finger, hand, and arm movements having a certain degree of accuracy should be possible in the interaction space.
- Feedback to independent body parts, especially fingers, should be provided.
- The user must be able to impart movements and forces to virtual objects.
- For applications with (dextrous) tasks, feedback systems should provide high-level forces as well as discretized forces.

Methods of Feedback Simulation

For generation of a sensory feedback, various methods exist that can be classified in both *portable* and *nonportable systems*. Nonportable device systems limit the freedom of motion of the user's hand to a small volume next to the device. Portable systems represent an improvement in user's freedom of motion and a corresponding enlargement in simulation volume. Essential methods are described below.

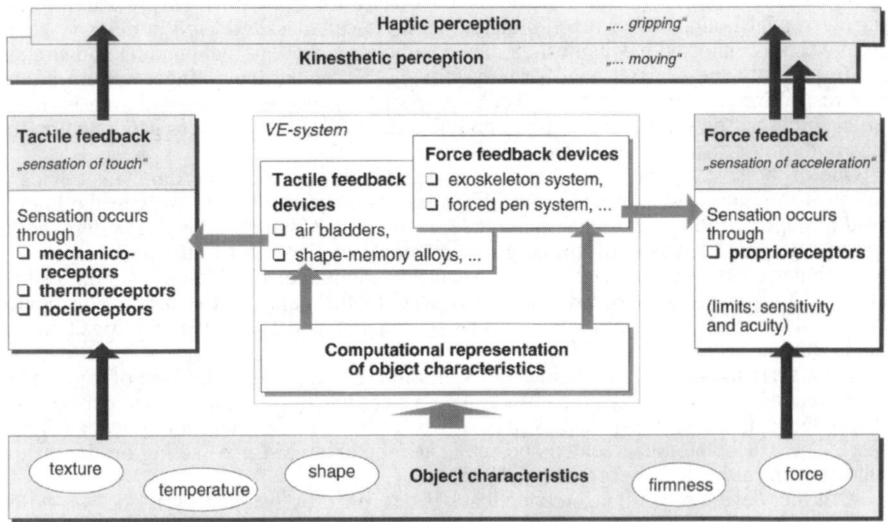

Figure 52.4 Haptic and kinesthetic perception and synthetic stimulation (scheme).

- *Tactile Feedback*

The simulation of a tactile feedback should enable virtual objects to be "touched." For this, different technical approaches exist.

One type of device for generating tactile feedback uses *air bladders* that can be rapidly inflated and deflated. Such systems consist of a glove containing arrays of air bladders distributed about the underside of the fingers, thumb, and palm. The differing shapes and sizes of the bladders accommodate the varying levels of sensitivity of the corresponding regions of the palm, fingers, and thumb (Aukstakalnis and Blatner, 1992). During use, as when a virtual hand comes in contact with a virtual object, force patterns are retrieved from the systems memory. A computer controls the amount of air pressure that enters or is released from the bladders through two miniature solenoid valves. By appropriate inflation or deflation, a particular tactile sensation is created. With this method, the generation of a sensation of rudimentary pressure is considered realistic, however a different force simulation is impossible.

Another means of providing tactile stimulation involves the use of *shape-memory alloys.* These alloys assume one shape when initially cast but remain flexible enough to be formed into other shapes. When heated again, the alloy reassumes the shape in which it has been originally cast. This feature of shape-memory alloys has been used for a tactile display: Here several tactors, consisting of thin semistiff metal beams, anchored at one end, are arranged in a matrix pattern. When the tactors are pulsed with an electrical charge, which effectively heats the metal, the capped end of the metal beam projects out through a small hole in the top of the array. By pulsing various combinations of these tactors at different times, simple textures and physical features can be created (Aukstakalnis and Blatner, 1992).

Certain approaches try to stimulate tactile feedback through the application of *contact elements.* Thereby location-dependent directional forces are generated, which correspond with the penetration resistance of the virtual object. In principle, such systems deal with kinematic driven "pipes," which wrap around the finger, or respectively the hand. The finger and the pipe are coaxially aligned and hold a large tolerance. In tracking mode, the position of the finger located in the pipe is taken up by the system, which conducts the pipe according to the movements of the finger, so that a contact between the finger and the pipe is avoided. Stopping the pipe's movement in a defined position causes a collision between finger and pipe. The resistance forces that are transmitted to the finger at the same time lead to a rudimentary sensation of touch. However, with this method the sensation of elastic surface qualities is only limitationaly possible.

- *Force Feedback*

Force feedback systems are applied for force-related manipulation of virtual objects. Therefore several of these systems are built around gestural input devices such as the data glove. Force feedback is also used for boundary detection, i.e., to protect from running into a virtual wall.

Some prototypes allow force feedback on certain fingers with high precision within narrow spatial bounds. The functioning of these so-called *exoskeleton systems* is based on moveable connected mechanical elements, which transfer external bending and stretching forces onto the skeletal hand–arm system (HAS), or the hand–finger system of the user respectively. Kinesthetic force feedback must be generated by considering both external forces acting on the virtual body and also internal grasping forces generated on the fingers during grasping operations in VEs.

One of the best known force feedback systems is the *dexterous master.* This device is designed to be used with a data glove. The master conveys force feedback to the human hand through piston like cylinders mounted on ball joints (Figure 52.5). The system works by passing force feedback information from a robot hand's sensor to the portable master. A computer controls the amount of air pressure that enters or is released from the piston. These pistons press against ball joints mounted in the palm of the hand. The device imparts the realistic sensations that virtual or remote objects are being gripped in the hand (Burdea, 1995).

Other systems integrate the whole HAS into interaction. The replication of kinesthetic force feedback at the levels of the HAS is achieved by means of a mechanical exoskeleton wrapping up the user's arm and connected to a glove at the metacarpus level (Figure 52.5). The exoskeleton is actuated and replicates the external forces acting on the virtual hand or arm in the VE (Bergascamo, 1993).

A further device providing force feedback is described by Iwata (1993). The *pen-based force display* allows six degrees of freedom (DOF) and can be used for the sensory

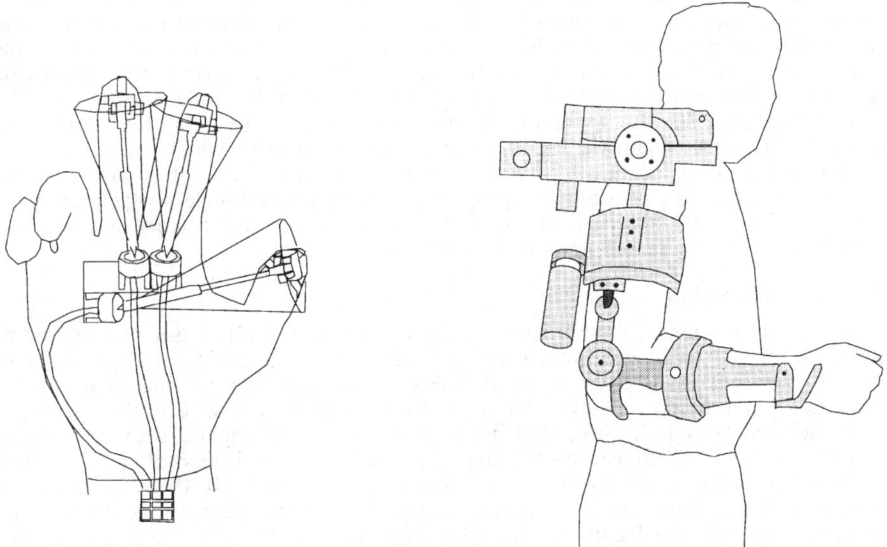

Figure 52.5 Left: Dextrous master. Right: exoskeleton system. (From Bergascamo, 1993; Burdea, 1995.)

manipulation of virtual geometric models and free form surfaces (Figure 52.6). The force display is attached by a three-axis manipulator at both ends. The manipulators generate the requested reaction forces which are needed for simulation, simultaneously compensating all gravity forces. The handling of the force display is comparable to a conventional pen. The force display does not have to be specifically calibrated for the user.

Other Perception

The human can differentiate among several thousand qualities of smell. Olfactory sensations, which are mostly influenced by the gustatory sense, are based on the stimulation of nerve endings in the nostrils by substances that emit odors.

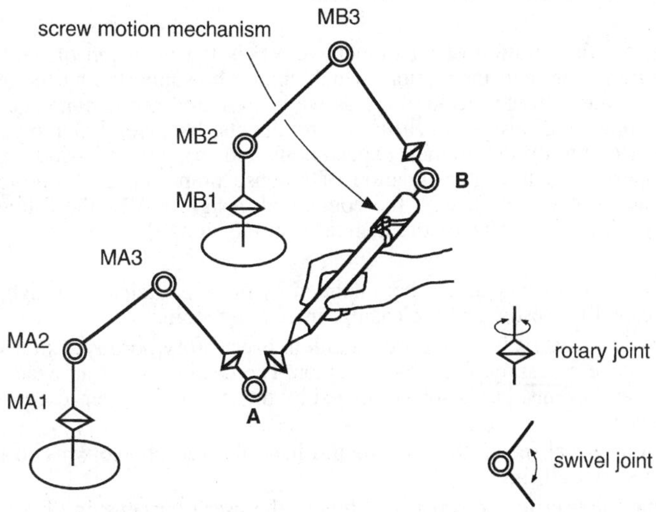

Figure 52.6 Pen-based force display. (From Iwata, 1993.)

The taste of a substance is based on a combined stimulation from various senses under essential participation of the sense of smell. One defines the qualities of taste, sweet, sour, salty and bitter, which are normally combined. The sensation of taste comes from the tongue, palate, back of the throat, and larynx. The intensity of a sensation depends on the concentration density, the exposure time, and the size of the stimulus surface.

The simulation of olfactory and gustatory events are currently being researched. The simulation of smell requires a variety of individual substances, which are made up of endless molecular connections. For the synthetic generation of gustatory and olfactory feedback, it is therefore possible to apply preselected physical substances, which are given to the user by means of a taste shower. For an extensive use of gustatory and olfactory devices much more research work will lead the way.

52.3.1.4 Immersion

As previously discussed, the purpose of immersion is to make the user feel that he or she is part of the virtual environment and can interact with the environment just as he or she can interact with the physical world. The sense of immersion (or presence) in a VE is reached when the user believes that he or she is in a world other than the one where his or her real body is located. With immersion, one can distinguish between *external factors,* which are determined by the employed technology, and *internal factors,* which is how inputs to the human are processed internally (Slater and Usoh, 1993). As discussed in Section 52.1.2, these factors themselves can be divided into technical, contentual, and individual aspects. The factors leading to immersion seem to have completing effects, i.e., a specific deficit of performance can be compensated by complementary performances. However, so far there is no definite science concerning the concept of immersion.

The external factors are determined by the hardware and software, which drive the displays and interaction devices. According to Slater and Usoh (1993), the following requirements have to be fulfilled:

- High-quality, high-resolution information must be presented to the human's sensory organs, in a manner that does not indicate the existence of the devices or displays.
- The sensory features of the environment presented to the human should be consistent across all displays.
- The environment should be one with which the user can interact, including objects and other participants that spontaneously react to the human.
- The self-representation of the user ("virtual body") should be similar in appearance to the user's own body, should respond correctly, and should be seen to correlate with the movements of the user.
- The connection between the user's actions and their effects should be simple enough for the user to model over time.

One basis for the creation of an immersive VE is the inclusion of natural forms of perception, orientation, and interaction, which should be supported by the use of appropriate hardware and software technology as well as environmental contents.

With the contentual design of virtual environments, behavioral stereotypes—i.e., individually and culturally determined expectations and experiences—and objects' familiarity are important to the intensification of a sense of presence. According to Kelso, Weyhrauch, and Bates (1993), with the contentual design of VEs the following aspects have to be considered to enhance the degree of immersion:

- Virtual environments should be inhabited by other participants or dynamic objects and shaped by aesthetically pleasing, interesting stories.
- To provide a user's behavioral orientation, every story portrayed in a virtual scene should have a destiny, fixed by a creator. The destiny is not an exact sequence of actions and events, but is subtly shaped by the system in order to create a cathartic experience.
- Within a virtual scene, the user should have the full choice what to think, to do, and to say at all times.
- Pushing the user to a specific goal breaks the user's suspension of belief. The user feels manipulated and unsatisfied.

- Little inconsistencies in either the story's contents or the behavior of the involved objects are acceptable, because the user mostly is caught up in the story, and often does not notice such inconsistencies. With an ephemeral and immersive activity, sensory immediacy might be more important than the structural elegance of event sequences (Laurel, 1991).

A virtual scene should have a structure that provides unity and cohesion and yet still permits rich interaction. Providing a satisfying experience for a user relies on maintaining a balance between user's freedom and creator's control (Kelso et al., 1993).

A sense of presence, however, is not exclusively determined from the extent and the quality of the technical equipment employed, as well as the mediated contents. For example, it is known, that photorealistic rendering does not seem to be a critical aspect to enhance the degree of immersion. Virtual objects do not need to look like anything in the physical world to have a sense of presence (Bryson, 1993).

Immersion is also dependent on internal factors. Internal factors are those dimensions of the human's individual-psychic capacity relating to how sensory inputs received in the immersive VE are processed. Internal factors are strongly dependent on subjective experience, which is encoded in internal visual, auditory, and kinesthetic representations. These representations are always seen from an individual perceptual position (Slater and Usoh, 1993).

For the creation of immersive VEs, the awareness of the human's cognitive processes is required, i.e., how the brain receives sensory impressions from the environment, and how, from these impressions, an extensive and consistent mental picture of the world is constructed. Many of the conditions leading to satisfactory degree of immersion have yet not been explored. It must be examined which aspects of real-world perception as well as internal representations are important for creating the illusion of an immersive computer-generated environment.

As illustrated in Section 52.3.3, natural sensory perception is based on the inclusion of several sense channels, whose coordinated and simultaneous stimulus contributes to a consistent impression. This aspect also has to be taken into consideration with the technical generation of a virtual environment. A basic precondition for the consistent perception of a virtual environment is the simultaneous and coordinated presentation of its total sensory characteristics. Above all, this consistence includes a real time–oriented object responding with the interactions. Inconsistencies of sensations of particular perception channels lead to dissonances.

Another critial aspect to obtaining an immersive feeling is the provision of appropriate forms for spatial orientation and navigation. Therefore, an essential requirement is an understanding of a human's orientation and navigation in the physical world. One characteristic of an environment is motion (see Section 52.3.1.1). With the design of VEs the motion either of the user or of the environmental objects have to be considered (Proffitt and Kaiser, 1993). Immersion is enhanced by allowing the user's head movements to control the direction of gaze of the computer-generated images; this provides the brain with motion parallax information to complement other visual cues (Vince, 1995).

Whether a high degree of psychic immersion, however, is needed to improve human performance efficiency is still uncertain. In the context of actual VE technologies, the occasion should arise, the marking of new behavioral stereotypes as part of human capacity may be requested. Consequently, a fitting of the human to the technically determined circumstances may take place.

It should be noted that providing a sense of immersion does not necessarily mean to be isolated from the physical world. In fact, isolation of the physical world can be highly intrusive and disorienting because, even though users cannot see the physical world, they are still aware of the physical surroundings, fearing events such as tripping over a cable or walking into a wall. This isolation and awareness of the physical environment can destroy the immersive experience (Cruz-Neira, 1993).

52.3.2 Human Performance Efficiency

The designing of complex systems requires integral methods that comply with the knowledge, experience, working methods, and intuitive behavior of all partakers. Only then the methods can be applied efficiently, cooperatively, and to achieve an optimization of results.

To fulfill the based claim of an application-related rise in efficiency, VE-systems must correspond to the requirements of *user friendliness* as well as *task appropriateness.* In this context a user-friendly design above all relates itself to the existing perception and interaction forms.

With the designing of a user-friendly system, it should be made particularly possible for nonexperts to deal intuitively with the computer systems. Until now, established criteria of user friendliness within VE systems have been only vaguely defined. Technical efforts at user friendliness are found in the area between system and dialogue. This is so as to avoid shifting performances into the system, presumed to be user-friendly characteristics, actually identified as counterproductive. So, for example, dialogue capability is naturally a human property, which relates itself to every human sense and includes several levels of communication. To use dialogue capability in the sense of user friendliness and enhancement of efficiency, it may have to be completely formalized and implemented, which is impossible according to empirical examinations (Degele, 1994).

With a user-friendly human–computer interaction, a well-balanced relationship between human and computer as well as their specific contental, technical, and social-communicative abilities and properties has to be strived for. Hereby the interaction with tasks and not with the computer always must be to the fore. A consequent alignment on the expectations and competences of the users as well as their early participation contributes decisively to the successful application of VE systems (Degele, 1994).

In the sense of task appropriateness, the following criteria and minimal preconditions for the application of VE-methods should be fulfilled:

- The method must be efficient; i.e., the application system has to be adapted to the application task, and the costs connected with the application stand in justifiable relation to the use.
- The installed technology must be reliable, robust, and maintainable, and it must satisfy the relevant safety and health regulations (see Section 52.3.3).
- The application of VE must find wide social acceptance, contributing to social benefit and complying with ethnic norms of all involved persons (see Section 52.3.4).

Human orientation and navigation is one essential precondition for high performance efficiency in VE. When a VE exhibits similarities to the physical world, orientation principles from the physical world can be extended to apply in the virtual environment. As VEs become more abstract and complex, the development of new paradigms for movement and navigation will become essential for performance of general tasks in these spaces. Broadly examined, in terms of a rise in efficiency, a great need for a stronger user- and task-oriented system design exists, in particular with the development of new immersion and interaction concepts.

52.3.3 Safety and Health Issues (Especially Sickness)

So far, existing references on health dangers resulting from the use of VE systems have concentrated on the appearance of simulator sickness as well as the strain of radiation from display CRTs.

Because the user is simultaneously within a physical and a virtual environment, considerable sensory dissonances can result, which are felt as disturbances within the person's physical and cognitive areas. Such appearances are subsumed under the term *simulator sickness.* Symptoms of simulator sickness include vasomotor dysfunctions, mental disorientation, and nausea. Drowsiness, fatigue, eyestrain, and headache are among the more common symptoms.

There is a general agreement that simulator sickness has two prerequisites, a functioning vestibular system and a sense of (visual) motion. The vestibular system gives the human sense of orientation and acceleration. It is believed that simulator sickness arises from mismatch between visual motion cues and the vestibular system. This would commonly occur when visual motion does not match physical motion either because no motion platform is used or because the motion platform lags behind the visualization and cannot match all the accelerations and orientations.

Studies have proved, that bright imagery is more likely to induce sickness than nighttime scenes, and that wide fields of view cause more problems than near ones (Helman, 1993). Furthermore, it has been pointed out that great individual differences exist, in proness to simulator sickness, adaptability to virtual environments, and times of acclimatization and reorientation (Kalawsky, 1993). The occurring dissonances are attributed to system deficits such as lower screen resolution, tracking delays, and lower feedback. But since problems occur even in systems with good optical quality and no tracking issues, one will always be confronted with the effect of simulator sickness in some types of VE systems, particularly those with large visually induced motions. To prevent some symptoms of simulator sickness it is necessary to fulfill, as a minimum, the requirements of high, constant frame rates, and low latencies. For further information on motion sickness see, for example, Oman (1993).

Health dangers, such as *radiation strain*, that result from the use of HMDs placed directly in front of the eyes, are yet not sufficiently studied. In particular, the application of CRTs is to be judged critically. Strong electromagnetic fields can heat body tissue. Heavy tissue and eye damage caused by the use of HMDs cannot be ruled out. From an incorrect system-determined accommodation, higher *eye strain* can result. Examinations showed a higher rate of headaches, sickness, and disturbance of perception after even a short use of HMDs (N.N., 1993).

In order not to lose sight of the cursor's position, mainly inexperienced users keep their hands at eye level when interacting and manipulating. The resulting *static muscle strain* from this unnatural arm position produces tiredness and aching.

A decided aspect of user acceptance is involved in putting on and taking off the HMD. Part of *wearability* is affected by the basic design of the HMD, part by the aspects of adjustment. Adjustment is of importance if various people use the same HMD. If an application requires use by many people or for short periods of time, special attention has to be paid to the time it takes to enter and leave the display. Other major issues concerning wearability of HMDs is weight (especially for long-term use), balance (make sure the HMD does not tip one's head forward or back), and comfort (ventilation), which is an overall, subjective criterion. An HMD should be usable without any restriction by people wearing glasses, too.

The same criteria regarding wearability likewise needs to be used with the data glove, which, in order to function, requires a specific adjustment and calibration for each user.

Actual VE systems require users to wear encumbering devices and attachments, such as headgear or cables from trackers and other devices. Normally, there attachments restrict users from walking more than a limited distance, and from making fast changes in position. Interface hardware often limits the user's well-being and comfort. In general, all current VE systems suffer from intrusive features. In the field of the equipment's wearability and comfort, a great ergonomic research work is yet to be done.

52.3.4 Potential Individual and Social Issues

The development, implementation, and use of VE systems, has enormous effects on individual and social life, which contains opportunities as well as risks (Table 52.1). Individual and social effects may influence professional life as well as the private world of leisure and entertainment. In particular, VE-games, such as role plays, adventure games, war games, strategy games, and sports, in view of their considerable market penetration and their high intensity of application, have a strong influence on society. The effect VE will have on individual and social life, however, depends on the form the technology takes, and how it will be used by people.

The potential social effects of VE systems appear so far to be ambivalent because virtual environments may enrich human contacts by new communication forms; however,

Table 52.1 Opportunities and Risks from the Use of VE

Opportunities	Risks
Increased Communication	**Anonymization**
❑ Broadened Communication Techniques	❑ Lack of contact, isolation
❑ Cooperation	❑ Stimulus overload
Expansion of Experiences	**Loss of Reality**
❑ Greater Horizons of Experience	❑ Unfixed values, trivialization
❑ Expanding of knowledge	❑ Addiction
Development of Personality	**Manipulation**
❑ Development of creativity	❑ Demagogy
❑ Independency	❑ Surveillance

at the same time, they may isolate people from the surrounding world. Virtual environments open up new forms of *communicational possibilities*, and consequently may contribute to the enlargement of interhuman contacts, either by means of common VE experiences or by an exchange of experience (Glitz, 1994). The drawback of the ubiquitous presence and availability at the virtual may possibly lie in the *anonymity* and the lack of communicational depth, i.e., that one can be present at any location without really being there. If technical isolation of actual virtual environments cannot be overcome, there is a danger that it will be supplemented by a social isolation. That means the users may retreat into virtual worlds, equipped according to each of their individual needs, but still standardized.

With their new modes of learning and practicing, virtual environments open up the possibility of creating new thinking structures, and of contributing to the *enlargement of experience horizons*. This increase of experience stands in contrast to the loss of natural sensuality, for the enrichment of sensory perception in virtual environments, compared with the natural environment, is always limited. Additionally, there is a *loss of experience*, caused more by the replacement of noninfluenceable and marking events in individual and social life than by the lesser display qualities within VEs. This can lead to a loss of social skills and an emptying of sense. When everything appears to be an illusion, people might lose their sense of life's rhythm and orientation. As with no other instrument, virtual environments offer the possibility to run away from reality. If one feels threatened by the physical world around one, one can hide in a virtual universe offering unlimited changes. Moreover, VE offers great potential to enable people to perform *violent acts*.

The established danger is, that information spread through VE will lose its significance to the individual, because he or she cannot judge the varied information and its respective relevance any more.

Creativity and independence, which are potentially supported by VE-systems, are principally regarded as positive characteristics of the personality. The possibilities of *manipulation* through VEs range from a general influence of mentality to demagogy. In building up the delusion of ideal and illusory worlds, which may only be slightly different from reality, the individual is distanced from his or her natural environment and influenced by an illusory reality. It is also to be expected that VE-systems will be used to raise additional demand, which will lead to economic dependencies.

Apart from consequences to the individual, central areas influenced by virtual environments concern aspects regarding social integration and organization of the professional life, which place new social challenges.

At present, the degree of penetration of VE systems in future professional and private life is yet unknown. What remains open is, whether the use of VE systems in future will be limited on a small elite, or whether VE will develop to a mass phenomenon. In the first case it can be estimated that existing education and power differences will become stronger within society, whereas in the second case these differences according to tendency will decrease.

With the installation of extensive and common, accessible communication networks the possibility exists of making the process of arriving at opinions and decisions more transparent, according to democratic and rational principles. On the other hand, there is the possibility of perfecting methods of *surveillance* in both professional and private life.

Within virtual environments, goods and values are in principle indifferent and interchangeable, because the possibilities of creating these goods and values are the same for all participants. Thereby social differences based on status will eventually be raised, leading to the alignment of living conditions.

For economic reasons, human action in professional life is being replaced by automated processes to an increasing extent. The motivation for the use of VE systems may lie less in the substitution of human action, but more in the strived for rise of production efficiency created by extended forms of interaction. Such working methods may lead to a considerable increase in performance.

VE systems support the promotion of a "high-tech culture," which includes a technification and effectuation of all areas of life, in the form of unlimited approval as well as a provocative refusal. Protagonists and critics are equally united by the fascinating vision of a VE culture, marked by the desire for imagination and new outlook upon life. "Artificial" is judged as innovative and intelligent, "hyperreal" deemed as better than reality. For further information on social issues see, for example, Stone (1993) and Whitbeck (1993).

52.4 APPLICATIONS OF VIRTUAL ENVIRONMENTS

In numerous research and application projects the potential uses of VE have been examined and appropriate applications have been developed. The main work has been done in the fields of architecture and infrastructure planning (Schmitt, Wenz, Kurmann, and van der Mark, 1995), manufacturing (Wilson, D'Cruz, Cobb, and Eastgate, 1995), product design, health care and medicine, telerobotics (Sheridan, 1992), entertainment (Aukstakalnis and Blatner, 1992), education, and training (Brown, Cobb, and Eastgate, 1995).

In the following pages, some exemplary application areas are discussed, which are of special interest, considering aspects of simulation, utilization, and visualization. Although some of the applications presented in the following sections are still at a stage of conception, others are already in use as prototypes or demonstration models. From examples of applications, methodological demands, results, and benefits shall be demonstrated as well as possible deficiencies of VE technologies.

52.4.1 Virtual Prototyping

Under the term *prototyping,* numerous technical, methodical and organizational measures have been grouped, extending from a concept's formulation to its finished draft. In the face of increasing task complexity and shorter innovation cycles, the various requirements, design areas, methods and planning participants can, however, only be integrated in a common development-, design- and communication process by considerable computer support. For these tasks, computer-based planning and design tools are requested, which possess a high degree of vividness at their human-machine interface. VE-technologies are suited to a high degree as methodical foundations for the prototyping.

52.4.1.1 Virtual Product Design

Bauer, Breining, and Rößler (1995) developed the following summary concept of *virtual product design:* The goals for virtual product design can largely be derived from the requirements. Cooperative, virtual product design can offer a high degree of illustration, of self-management and of efficient team work. As far as market-oriented goals are concerned, the following areas can be mentioned: increase of innovation rate, reduction of development cycles, and wide utilization of the knowledge available in a planning team. The concept of virtual product design can briefly be defined as follows: Based on a virtual cooperation platform, which is realized as a multiuser system by a VE technology, several persons plan and design a product (including a service product) in parallel. The virtual working environment replaces all activities relevant to planning and design. Cooperative, virtual planning and design will make a decisive contribution to an innovative product development.

Computers have largely made their entrance into design departments. Normally two-dimensional CAD systems are used, which are essentially employed for the preparation of drawings for simple components. If objects with complex, three-dimensional geometrics are designed or if the CAD-generated information on shape is further processed, it will be meaningful to use CAD systems that include a three-dimensional, computer-internal representation of model data. In the field of three-dimensional CAD a trend toward feature-based modeling systems has been recognized. Such a CAD system makes high demands on the user's ability to analyze and abstract in terms of geometry. However, particularly for the design of freeform surface bodies, it is desirable to be able to describe the shape without having to deal with aspects of mathematics and geometry. Conventional human–machine interfaces often make very exacting demands on the imaginative faculty of their operators. Mainly the mental translation of three-dimensional inputs into computer instructions via low-dimensional input devices such as a mouse or keyboard are frequently very difficult to manage. The consistent translation of the whole process chain from the product idea to the finished product in the three-dimensional space will bring a number of advantages.

Apart from the three-dimensional interaction with complex geometrics, the efficient handling of complex action chains together with a large number of influencing variables (e.g., quality, technology, cost) is an essential prerequisite for innovative and speedy product development. For this purpose intuitive interaction within the computer-generated working environment is necessary. The intuitive interaction forms used are based on the system integration of application-optimized functions with advanced input and output devices. Figure 52.7 shows the basic structure for a virtual design system.

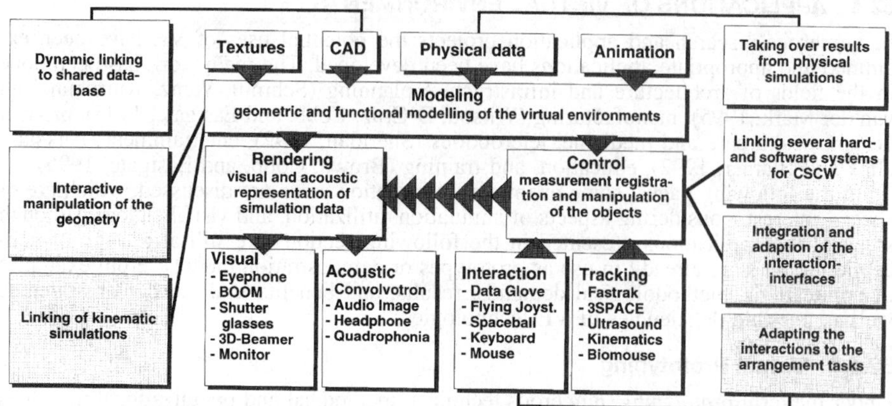

Figure 52.7 Function structure for a virtual product design system.

Based on VE technology, a virtual working environment will be created in which all work steps necessary for the product development process are feasible. This means, that developers—by means of appropriate input and output devices—enter the virtual working environment. In this environment, the objects to be developed can be designed, altered, or evaluated. Moreover, the persons participating in the development process can communicate with each other. The participating persons will also be represented in the virtual working environment. The long-term goal is the communication in decentral structures of worldwide distribution.

With basic functions, such as the real time–oriented, view-dependent 3D visualization and intuitive interaction forms, VE-systems applicated for virtual product design will have additional minimum functions for their disposal:

- Imports and exports from basic objects based on standard CAD-geometrics
- Free position, orientation, and scaling of the objects
- Free design of objects' shape, surface, and color
- Simulation of various lighting conditions
- Generation of geometric and object hierarchy, i.e., links and detachment of objects
- Rejection of objects
- Visual, auditory, and haptic representations

Product designers take great interest in the currently known methods for direct manipulation by means of data gloves or similar input devices. It is more than imaginable that formal changes of shape can be made on a virtual product by means of virtual design tools, such as a virtual saw, virtual grinding paper, and an virtual airbrush gun.

Advantageous with virtual working environments is the three-dimensional visualization resulting from the individual line of sight and the most varied, real time–oriented interaction forms. As a result of the free positioning and scaling of the user inside and outside the object to be planned, it will be possible that a large number of persons can work simultaneously on an object. In doing so, this object will not have to be broken down into its elements.

52.4.1.2 Virtual Cockpit Design

Conventional planning and design methods are greatly enhanced when merged with VE systems. For example, car designs modeled in three dimensions can be placed in a virtual environment where the operator can enter the virtual car. Designers can apply VE systems to studying the interior ergonomics of new cars. This creative design process allows comparison of various models at an early stage before the actual production.

With car development the use of VE systems for the purpose of virtual cockpit design opens up better design qualities for planners and designers in the following areas:

- Sight investigations
- Anthropometric design of vehicle interior especially the drivers cabin
- Interface design (controls, displays, etc.)
- Access systems (doors, steps, ladders, etc.)
- Aesthetics and design

Brückmann and Gottlieb (1995) describe an approach using VE-systems for virtual design of vehicle interior. The authors' aim is to cover all aspects of people's behavior relating to driving a car, ranging from the customer situation of buying a car, to the passenger riding smoothly and comfortably in a car. The VE system is a useful aid in studying and comparing car designs for operability and maintainability while the design still exists only in digital form. Not only is the VE interface an aid to design process, it also means that performance simulations can be carried out without having laboriously to build physical models.

To judge the vehicle interior, a computer-based three-dimensional inner vehicle model is built. From this vehicle model, on the basis of subjective judging tests, criteria relevant to customer-oriented vehicle design have been established and evaluated. These criteria concern characteristics such as the inner room size, color design, materials, surfaces, forms, and geometry. Additionally, potential influences from driving on the customer's judgement of the vehicle interior design are assessed.

To drive a virtual car, the subject has to sit down in a real driver's seat fixed on a simple wooden platform. A control juts out of an instrument dashboard. Putting on the head mounted display, the driver is located just inside the virtual car. Figure 52.8 represents the cockpit of a virtual car. Depending on the direction in which the head is turned, the interior of a car is seen: The instrument board, seats with colored seat covers, back seats, etc. Pressing a button on the screen causes the car to change its color; pressing another modifies the size of the car. These techniques are used to find out if and how subjects perceive modifications of the color and the volume.

Figure 52.8 Inner room of a virtual car. (Provided by Daimler-Benz.)

52.4.2 Education, Training, and Health Care

By means of virtual environments a learning context can be created, where the multiple aspects of human intelligence are addressed. Under the precondition of sufficient perception and interaction qualities, which mediate to the learner a feeling of intuitive handling with virtual objects, the following listed principle VE applications unfolds in the area of apprenticeship.

- Interactive demonstration of objects and events
- Mediation of (abstract) knowledge based on one's own experiences (e.g., exploring a molecular structure)
- Mediation of skills and training of behavior in dangerous situations and environments
- Knowledge mediation beyond human physical limitations (e.g., the exploring of the way a petrol engine functions by opening up the combustion chamber)
- Exploring distant places or past epochs
- Advancement of creative abilities by free design and impart of imagined objects

The use of VE-systems in teaching, education, and training can be very effective, because it complies with the psychological learning and pedagogic knowledge so that the largest learning success can result from the practical application of an educational program. The success of learning can be raised by parallelly addressing, that is, frequently changing the sense channels, and by the inclusion of one's own actions. The ability of concentration and learning success increase, when thinking structures are fixed into the imagination, or when ideas of a third person are made intelligible.

Particularly in the artistic field, VE can contribute support to the students' creativity. In addition to this, VE can be installed into language lab education. By the simulation of a fitting environment, the learning of a foreign language will be made more intensive, and therefore more effective.

52.4.2.1 Language Instruction

Cleal, Giles, and Schroeder (1994) report on a VE system that has been developed at the University of Nottingham for language instruction with educationally disadvantaged individuals.

The bases for the VE application in language instruction are the knowledge and methods of the cognitive psychology. Therefore, the child's ability to speak develops according to concepts that are formed by interaction with the environment and how these concepts can be verbally expressed. The development of speech can be linked directly with human ability to understand both object-related and abstract concepts, and to use them in a meaningful way. Abstraction is important, because language portrays itself as an abstract instrument for the description of the environment. Abstract thinking capabilities hold the ability to take up information and to generalize and use it in other areas. The inability to generalize information is one of the biggest impediments in the acquirement of speech.

Using VE-technology, whether delays in cognitive development could be counteracted by initiating intensive human experiences within a virtual environment was investigated. The VE system setup, which was used to complement the existing educational methods and principles, offered large communicative and kinesthetic potential for students with speech impediments or physical disabilities. During 2 weeks of investigations, 25 students between the ages of 5 and 19 years were observed while using a VE system. Along with the use of the VE system a number of difficulties arose, especially problems concerning of the use input devices. The results of the studies showed distinct differences from the original expectations: It was concluded that students who had learning disabilities had mostly profited from the use of VE systems, because here the promotion and support potentials were at the best. Indeed for students with higher grades of ability, and those who could best handle the system, the experience with virtual environments was mostly animating. The supposition is obvious that fundamental cognitive and mental hurdles exist within the use of VE systems.

52.4.2.2 Driving and Flight Training

At present, there are various training systems in development and use that are meant to support the learning of skills, for instance with an assembly process, or the training of certain behavioral modes in dangerous working conditions and environments.

Flight simulation was one of the first applications of virtual environments. The technology originated for training commercial and military aircraft pilots, without risk to pilot or aircraft. Flight simulation allows pilots to train for emergency situations that occur only rarely. By repeatedly training in flight simulation, the pilot learns to handle the emergency should it ever occur during a real flight. Additionally, the cost of flight training in virtual simulation is less than conventional flight training in aircraft (Iovine, 1994).

Apart from flight simulations, various successful efforts toward a *vehicle driving school* exist. Learner drivers are given the possibility of driving in the first couple of lessons, which are commonly not entirely safe, completing these lessons, and then later switching back to a street-based vehicle.

Schmidt, Müller, and Trost (1993) compiled an HGV (heavy good vehicle) driving simulator with a five-channeled outside view presentation (Figure 52.9). The driving simulator includes a driver's cabin, a visualization system, a movement system, a sound simulator, as well as a graphic and simulation computer. The visualization system consists of a three-channeled front view system, with a view angle of 170° horizontally and approximately 40° vertically. As further means of view, there are two monitors, acting as side-wing mirrors. Each image channel is calculated on a separate graphic computer and is projected as a large picture onto screens, which are arranged around the driver's cabin. The drivers cabin, which is attached to a hydraulically driven moving platform, is complete with all displays and controls holding full function capability, also in relation to the required actuating forces. The dynamic model, based to the simulator, contains acceleration and deceleration movements, steering geometrics and uneven road surfaces. As a result, besides the cabin vibration also situation related, realistical roll and pitch movements can be simulated.

The scenes that are projected onto the walls result from the applied landscape data, the chosen environmental conditions, and the driver's actions, such as steering, accelerating, and braking. With the graphical and data synchronization, always jerk-free, consistent, and delay-free front view presentation is achieved. In order to gain a realistical impression while training a student in driving a HGV, scenarios of realistic town traffic with their total complexity are reproduced.

The practical testing of the driving simulator was carried out by eight learner drivers, who had no previous driving experience. The objective regarding the testing series was to compare HGV training with both a real vehicle and the use of a simulator, where in every case eight driving lessons comprising 30 min each were completed in both the HGV and the simulator. The aims of training were condensed to 12 aspects, such as turning left, driving up gradients and down gradients, and maneuvering with trailers. Apart from the learning successes and failures of the learner drivers, physiological parameters, such as strain, fatigue, and state of health, were investigated.

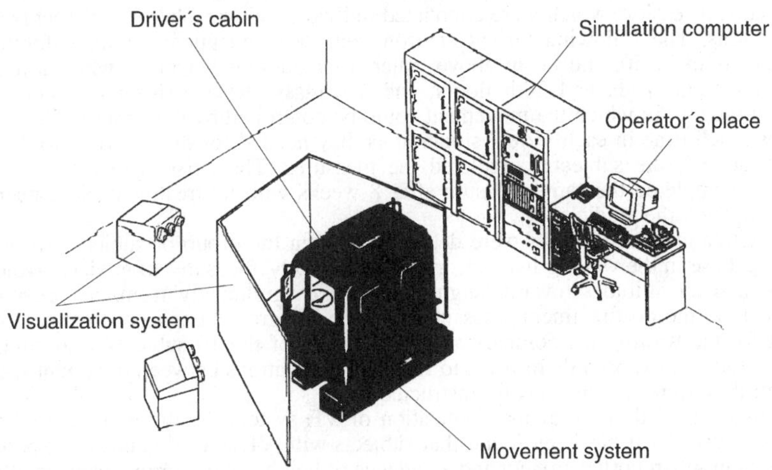

Figure 52.9 Schematic construction of an HGV driving simulator. (From Schmidt, Müller, and Trost, 1993.)

The scheduled learning program was completed by the drivers without any significant problems. With regard to the health of the drivers, there were only few problems for a certain number of persons, who also suffered under other conditions from travel sickness.

The evaluation of the tests showed a generally high acceptance of the driving simulator on the part of the drivers. While carrying out the tests they were highly motivated, which surely the exclusiveness of the employed technology also reinforced. The ride in the simulator was considered ambitious and demanding, but not stressful. In contrast to conventional training, the compact design of the virtual landscape passed through permitted an increase of complexity and variety of training programs as well as a considerably lessening of wasted time. With the use of the driving simulator, great learning success with vehicle control was achieved, as well as driving with trailers. Learning such processes as passing over curves was handled without any problems, which points to the realism of the total training environment, in particular to the quality of the recorded view information.

This assessment found its confirmation at a concluding training drive through real town traffic, to which all of the drivers were subjected. Regardless to most difficult driving conditions, the drivers were certified as having good control of the vehicle. In comparison to drivers who had been purely conventionally trained, drivers who were trained with the use of the simulator often gained even better results (Schmidt et al., 1993).

52.4.2.3 Health Care

VE applications can also be found within the areas of medicine and psychological therapy. Here concepts and applications in the fields of diagnosis, surgery operation planning, rehabilitation, prevention, and care are known. (For further information, see, for example, Satava, Morgan, Sieburg, Mattheus, and Christensen, 1995.) Success has also been achieved within the treatment of phobias using VE systems. The term *phobia* paraphrases an unfounded fear from a harmless situation, and the resulting avoidance of this situation at any price. With the therapy by means of a VE system, patients slowly can be led to their threshold of fear by simulating a comparable situation, always knowing about the risklessness of the attempt. Through confrontation with the fearful situation and a developed growing confidence, the phobia can finally be overcome.

Rothbaum et al. (1995) describe a therapy approach, using VE as a means to overcome *acrophobia* (fear of heights). The advantages of psychological treatment using a VE system lie in the locally independence of treatment and in the potentially offering of more control over exposure stimuli. Thus, it may offer a time- and cost-effective way to conduct exposure therapy.

The study's goal was to examine the efficiency of computer-generated graded exposure in the treatment of acrophobia. For the methodical tests 20 subjects with an average age of 20 years were drawn, who indicated substantial fear and avoidance of heights. Twelve subjects were assigned to VE treatment and eight subjects were assigned to the waiting list. After a pretreatment assessment, the subjects in the treatment group received their first treatment session, which were conducted individually over 7 weeks in weekly 35–45 min sessions. The following virtual environments were encountered: three footbridges with heights of 7, 50, and 80 m above water; four outdoor balconies with railings that were on ground, 10th, and 20th floors; and one glass elevator rising 49 floors, up to 147 m at the top, in which the movement could be controlled by the subject. The subjects spent as much time in each virtual situation as they needed for their anxiety to decrease. After treatment, the subjects completed the measures. The persons in the waiting list condition completed the same measure after 8 weeks with no treatment. Seventeen subjects completed the study.

No pretreatment differences were detected between the group of subjects given treatment and those in the waiting list. According to the study, measures of anxiety, avoidance, distress, and all attitudes toward heights decreased significantly from the pretreatment assessment to the posttreatment assessment for the VE graded exposure treatment group, but not for the waiting list comparison group. Seven of the 10 subjects who completed the VE treatment exposed themselves to real height situations between treatment sessions, although they were not specifically instructed to do so.

In this controlled study of the application of VE systems to the treatment of a psychological disorder it has been found that subjects with VE graded exposure experienced reductions in self-reported anxiety and avoidance of heights and improvements in attitudes toward heights. Subjects without VE graded exposure did not evidence any change. Despite some limitations—such as absence of a standard treatment comparison group, absence of followup data, and no formal assessment of phobic avoidance—the findings provide support for the use of virtual environments exposure in the treatment of acrophobia. VE also appears applicable in the treatment of other anxiety disorders in which exposure-based treatments are recommended (Rothbaum et al., 1995).

52.5 CONCLUSIONS

In the future, the importance of virtual environments as an integrating human–machine interface will increase enormously in many fields of application. However, VEs have to be considered not only as highly developed user interfaces, but as information technologies, where it is not the human's dealing with the computer that is to the fore, but the direct interaction between the human and computer-generated visions and objects, events, and other persons.

At the moment, VE-technologies are used and investigated particularly in technical and scientific applications as well as in the fields of entertainment and leisure. At present, it is not possible to identify completely the potentials of application of VE systems. New applications will arise mainly in fields with high degrees of interaction, which have to be designed with regard to humanity and efficiency.

In this chapter it has been demonstrated that the VE technologies with their specific visualization and interaction techniques ideally can serve as a basis for tasks concerning ergonomic work system design. Furthermore, VE will become an integrating, human-oriented technology for communicational and technical processes in fields of creativity, learning, service, production, etc.

In spite of the current immaturity and limitations of VE technology it does offer great potential in ergonomic planning and design processes, which will likely be implemented in industry within the next years. However, it is essential that the ongoing research and development of VE technology be directed toward providing what is actually required. In this context, VE systems themselves have to be regarded as objects of ergonomic analysis and design.

Virtual environments have great potential to do ergonomic applications far better than at present, but they will not be a panacea even when further developed (Wilson et al., 1995). With regard to an expected growth market in the future, it is necessary to realize fields of application, potentials, and deficits in order to design human- and application-related VE technologies as well as to use them appropriately and efficiently.

REFERENCES

Aukstakalnis, S., and Blatner, D (1992). *Silicon Mirage: The Art and Science of Virtual Reality.* Berkeley, CA: Peachpit Press.

Bargar, R. (1993). Sound for virtual immersive environments. In *Course notes 23* (ACM Siggraph '93, Chicago, 4.1–4.18). New York: ACM.

Bauer, W., Breining, R., and Rößler, A. (1995). Co-operative, virtual planning and design. In *Virtual Reality World '95* (Conference documentation, Stuttgart, 21.-23.02.1995, 213–223). München: IDG.

Begault, D. (1994). *3D Sound for Virtual Reality and Multimedia.* Boston: Academic Press.

Bergascamo, M. (1993). The GLAD-IN-ART project. In *Virtual reality '93.* (IPA-/IAO-Forum, Stuttgart, 4.–5. February 1993, 251–258). Berlin: Springer.

Brooks, F. P. (1988). Grasping reality through illusion-interactive graphics serving science. In *Proceedings of the Conference on Computer-Human Interaction (CHI '88),* Washington, D.C.

Brown, D. J., Cobb, S. V., and Eastgate, R. M. (1995). Learning in virtual environments (LIVE). In J. Vince and R. Barnshaw, Eds. *Virtual Reality Applications.* London: Academic Press.

Brückmann, R., and Gottlieb, W. (1995). Spatial perception of vehicle interior. In *Virtual reality world '95* (Conference documentation, Stuttgart, 21.-23.02.1995, 459–461) München: IDG.

Bryson, S. (1993). Implementing Virtual Reality. In *Course notes 43* (ACM Siggraph '93, Chicago). New York: ACM.

Burdea, G. (1995). Research on portable force feedback masters for virtual reality. In *Virtual Reality World '95* (Conference documentation, Stuttgart, 21.-23.02.1995, 317–324) München: IDG.

Cleal, B., Giles, W., and Schroeder, R. (1994). Virtual Reality im Sprachunterricht für Lernbehinderte. In *Virtual reality '94* (IPA-/IAO-Forum, Stuttgart, 9–10. Febr. 1994, 227–240). Berlin: Springer.

Cruz-Neira, C. (1993). Virtual reality overview. In *Course Notes 23* (ACM Siggraph '93, Anaheim, 1.1-1.18). New York: ACM.

Cruz-Neira, C., Sandin, D., Defanti, T., Kenyon, R., and Hart, J. (1992). The cave. Audio visual experience automatic virtual environment. *Communications of the ACM, 6*(35), 64–72.

Degele, N. (1994). Zur sozialen Dimension nutzerfreundlicher VR-Systeme. In *Virtual reality '94* (IPA-/IAO-Forum, Stuttgart, 9–10. Febr. 1994, 365–375). Berlin: Springer.

Ellis, S. (1995). Human engineering in virtual environments. In *Virtual Reality World '95* (Conference documentation, Stuttgart, 21.-23.02.1995, 295–301). München: IDG.

Foley, J., and Silbert, J. (1989). User-computer interface design. In *Lecture Notes* (No. 1, CHI '89 Conference, Austin, Texas).

Franke, H. W. (1967). *Sirius Transit.* Frankfurt: Suhrkamp.

Glitz, R. (1994). *Technikfolgenabschätzung Virtuelle Realität.* Düsseldorf: VDI-Verlag.

Göbel, M. (1992). *Virtuelle Realität—Technologie und Anwendungen* (ZGDV-Documentation). Darmstadt: Zentrum für Graphische Datenverarbeitung.

Hahn, J. K., Gritz, L., Darken, R., Geigel, J., and Won Lee, J. (1993). An integrated virtual environment system. *Presence, 2(4)*, 353–360.

Hanne, K.-H. (1993). *Systeme Kombinierter Multimodaler Mensch-Rechner-Interaktionen.* Dissertation. Berlin: Springer.

Helman, J. (1993). Designing virtual reality systems to meet physio- and psychological requirements. In *Course notes 23* (ACM Siggraph '93, Anaheim, 5.1-5.17). New York: ACM.

Holmgren, D. E., and Robinett, W. (1993). Scanned laser displays for virtual reality: A feasibility study. *Presence 2(3)*, 171–184.

Iovine, J. (1994). *Step into Virtual Reality.* New York: McGraw-Hill.

Iwata, H. (1993). A six degree-of-freedom pen-based force display. In *Proceedings of the Fifth International Conference on Human-Computer Interaction,* Orlando (pp. 651–656).

Kalawsky, R. (1993). *The Science of Virtual Reality and Virtual Environments.* Reading MA: Addison-Wesley.

Kelso, M. T., Weyhrauch, P., and Bates, J. (1993). Dramatic presence. *Presence, 2(1)*, 1–15.

Krüger, W., and Fröhlich, B. (1994). Responsive Workbench. Towards a user-centered, application-driven multimedia human-computer interface. In *Virtual reality '94* (IPA-/IAO-Forum, Stuttgart, 9.-10. Febr. 1994, 73–80). Berlin: Springer.

Laurel, B. (1991). *Computers as Theatre* (Second Edition). Reading, MA: Addison-Wesley.

Lem, S. (1964). *Summa technologiae.* Frankfurt: Insel. (German ed. 1976).

Müller-Limmroth, W. (1993). Sinnesorgane. In H. Schmidtke, Ed., *Ergonomie* (3rd ed).

N. N. (1993). Britische Forscher warnen. Krank durch Virtual Reality. *VDI-Nachrichten,* 24.9.93.

Oman, C. M. (1993). Sensory conflict in motion sickness: An observer theory approach. In S. R. Ellis, Ed., *Pictoral Communication in Virtual and Real Environments* (2nd ed., pp. 362–376). London: Taylor and Francis.

Proffitt, D. R., and Kaiser, M. K. (1993). Perceiving environmental properties from motion information: minimal conditions. In S. R. Ellis (Ed.), *Pictoral Communication in Virtual and Real Environments* (2nd ed., pp. 47–60). London: Taylor and Francis.

Rothbaum, B., Hodges, L., Kooper, R., Opdyke, D., Williford, J., and North, M. (1995). Effectiveness of computer-generated (virtual reality) graded exposure in the treatment of acrophobia. *American Journal of Psychiatry, 152*(4), 626–628.

Rötzer, F., 1993, Virtuelle und reale Welten. In F. Rötzer and P. Weibel, Eds., *Cyberspace. Zum medialen Gesamtkunstwerk* (pp. 81–113). München: Hanser.

Satava, R. M., Morgan, K., Sieburg, H. B., Mattheus, R., and Christensen, J., Eds. (1995). Interactive technology and the new paradigm for healthcare. In *Medicine Meets Virtual Reality III Proceedings*, San Diego, 19–22 January 1995. Amsterdam: IOS Press.

Schmidt, R., Müller, W., and Trost, N. (1993). Virtual Reality am Beispiel einer fünf-kanaligen LKW-Fahrsimulation. In *Virtual Reality '93* (IPA-IAO-Forum, Stuttgart, 4–5 February 1993, 271–280). Berlin: Springer.

Schmitt, G., Wenz, F., Kurmann, D., and van der Mark, E. (1995). Toward virtual reality in architecture: Concepts and scenarios from the Architectural Space Laboratory. *Presence 4(3)*, 267–285.

Schneiderman, B. (Ed.). (1993). *Sparks of Innovation in Human-Computer-Interaction.* Norwood, NJ: Ablex.

Sheridan, T. B. (1992). *Telerobotics, Automation, and Human Supervisory Control.* Cambridge, MA: MIT Press.

Slater, M., and Usoh, M. (1993). Presentation systems, perceptual position, and presence in immersive virtual environments. *Presence, 2(3)*, 221–233.

Stephenson, U. (1994). Raumakustische Simulation und Auralisation—Methoden und Anwendungen. In *Virtual Reality '94* (IPA-/IAO-Forum, Stuttgart, 9–10. February 1994, 181–212). Berlin: Springer.

Stone, V. E. (1993). Social interaction and social development in virtual environments. *Presence, 2(2)*, 153–161.

Sutherland, I. E. (1965). The ultimate display. In A. Kalvenich, Ed., *Information Processing. Proceedings of the IFIP Concress 3* (pp. 506–508). Washington: Spartan Books.

Tidwell, M., Johnston, R., Melville, D., and Furness T. (1995). The virtual retinal display—A retinal scanning imaging system. In *Virtual Reality World '95* (Conference documentation, Stuttgart, 21.-23.02.1995, 325–333) München: IDG.

Vince, J. (1995). *Virtual Reality Systems.* Reading, MA: Addison-Wesley.
Whitbeck, C. (1993). Virtual environments: Ethical issues and significant confusions. *Presence, 2*(2), 147–152.
Wilson, J. R., D'Cruz, M., Cobb, S. V., and Eastgate, R. M. (1995). *Virtual Reality for Industrial Application: Opportunities and Limitations.* Nottingham: University Press.

CHAPTER 53

SOCIAL COMPUTING: COMPUTER SUPPORTED COOPERATIVE WORK AND GROUPWARE

Anitesh Barua
Ramnath Chellappa
Andrew B. Whinston
Center for Information Systems Management
Department of MSIS
Graduate School of Business
The University of Texas at Austin
Austin, TX 78712 USA

53.1 INTRODUCTION

Computers in society today have undergone a rapid transformation from being personal tools to their present role as the gateway to the information superhighway that spans not only the nation but the globe. Computers are no longer used just by one individual to satisfy occasional number crunching needs. Instead, they are used on a daily basis to facilitate communication among groups of people.

Humans have always been social animals, and communication has been their key to survival in society. Advancements in technology have made it possible for the computer to be used as a key tool to fill our communication needs. The one technological innovation primarily responsible for this is networking, which is quite simply a way for computing systems to communicate—to "socialize"—with each other. Hence, the birth of the concept of social computing.

Social computing can be defined as the use of computers and computing systems for the purpose of interaction between people for work, entertainment, or even simple communication. Whereas use of computing systems outside the workplace is growing, traditionally they have been used for organizational purposes. Here too, social computing thrives since almost every organization is composed of divisions or departments where people work together in groups toward a common goal. This need to bring together the process of collaboration and computing has led to the development of various collaborative systems. Over the past two decades, this area of computing has undergone tremendous changes, spurred on by rapid development of the underlying technologies. Social computing systems are now able to simulate the face-to-face collaborative environment better than ever, and systems that maximize user value by combining real-time, interactive, and multimedia capabilities are no longer only a vision.

This chapter provides a brief look at the history of the various types of collaborative systems in light of the technological advancements that have made progress possible. In fact, collaborative systems and their enabling technologies are changing so rapidly that it is no longer feasible to describe various systems that are currently available since they may well be outdated by the time the information reaches the reader. Rather, we here describe the various types of social computing systems by looking behind their static forms. By examining instead the factors and features that increase user value in collaborative computing, we present a generic framework by which all systems can be analysed—past, present, and future. It is the desire to better incorporate and integrate these factors and features that has driven system developments in the past and that continue to spur new efforts in the social computing field. In an effort to retain as much relevance as possible, we have weighted our chapter heavily toward the most recent—albeit the fastest growing—medium for collaborative systems: the Internet, and in particular, its multimedia World Wide Web. We provide a detailed description of the multitude of factors and features that must be taken into consideration in designing and implementing collaborative systems for this medium that highlights the challenges that designers and users of present and new systems will face. Although the possible applications will change rapidly, we hope that the underlying framework will continue to provide guidance.

53.2 THE SHIFTING PARADIGM OF GROUP SUPPORT SYSTEMS

Ever since they broke loose from their early constraining role as purely computational aids, computers and computing systems have been pervasively used by individuals as tools in the decision-making process. We now take it for granted that individuals rely on computers each and every step of the way toward the final decision: We garner relevant information from extensive corporate databases, analyze it using spreadsheet software, and present it in impressive reports complete with charts and graphs produced on user-friendly word processing and computer graphics software.

Often, however, we do not go through this alone. Developing software that could be used by groups of individuals involved in a decision making process was clearly a natural progression in organizational computing given the frequency of group interactions. A significant amount of work in business is achieved through a collaborative process by groups rather than individuals (Bair, 1985). Managers and other professionals in particular spend a large part of their time working in groups. Although exact numbers vary among jobs and individuals, Information Systems (IS) managers, for example, have been found to spend between 60% and 70% of their time in group activities (Ives and Olson, 1981; Hymowitz, 1988; Mintzberg, 1983; Mosvick and Nelson, 1986).

Organizational computing has not been blind to the importance of group interactions. The perceived need "to compensate for the limitations of human decision making and group process losses" (Huber, 1984) has spurred the development of systems that support group interactions. In the past couple of decades, information technologies have emerged that achieve this by creating a collaborative environment or by problem structuring or both.

Although the need may be generally accepted, the terminology used is often confusing. Group support systems (GSS) have in general been defined as "software, hardware and language components and procedures that support a group of people engaged in decision-related meeting" (Huber 1984). Information technology systems used to support group activities can be further classified into two categories; namely, group decision support systems (GDSS) and computer-based systems for cooperative work, or computer supported cooperative work (CSCW) (Chen, Lynch, Himler, and Goodman, 1992). In recent years, the term "groupware" has also become widely popularized by commercial products such as LotusNotes.[1] And collaboratories based on the Internet's World Wide Web (WWW), now in the development stage, will surely be a reality by the time this *Handbook* reaches the reader. Although the common objective of all these is the facilitation of group interaction, as we shall see, they each do so at various levels and with varying degrees of interaction. The development of the various types of collaborative systems can best be illustrated by placing them along a continuum of increasing technological sophistication over time. As we move along the continuum seen in Figure 53.1, technological advancements in computing and communications have made it possible to fulfill a wider set of user needs, desires, and expectations in the facilitation of group interactions. Whereas some early group decision support systems essentially used a single computer system with multiple terminals, all located in the same room to assist in problem structuring, collaboratories allow users across the globe to conference via real-time video, to access shared databases, and to collaborate on the drafting of a single decision. The following historical review of group support systems shows the increasingly advanced features and ease of use that combine to simulate a progressively more realistic environment.

53.3 TRADITIONAL GROUP DECISION SUPPORT SYSTEMS

The earliest objective of group decision support systems was to aid professionals to conduct group meetings that would otherwise have been handled using paper and pencil.

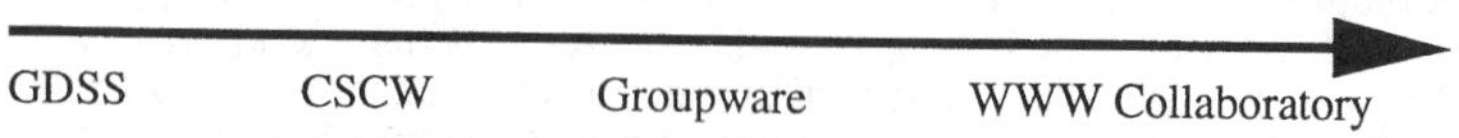

GDSS CSCW Groupware WWW Collaboratory

Figure 53.1 Evolution of group support systems.

[1]© Trademark of Lotus Corporation.

Since decision makers were typically confronted with unstructured problems (DeSanctis and Gallupe, 1985) and were often viewed as rushing to a decision before adequately defining a problem (Lewis and Keleman, 1990), early group decision support systems focused on problem definition and documentation. Preliminary systems typically involved use of a single computer with multiple terminals. These systems were essentially restricted to automating most manual actions of a group meeting and were primarily used to document the proceedings. Further improvements in GDSS led to the addition of various structuring tools. The most common tool was a voting system that allowed users not only to discuss the issues among themselves, but also to rank or rate them. The role of these systems became more analogous to facilitators or moderators of group meetings, and in many cases the apparent objective seemed to be to automate nominal group techniques.

This narrow focus of early GDSS can be attributed at least in part to the restrictions stemming from the technology of the time. One key factor was the geographical limitation. Users were confined to a single room with no possibility of communicating with the outside world since computers could not yet "talk" to each other. Another major drawback during this period was the restriction to one medium only. All group interactions on the computer were limited to text-based exchanges—there was no means of illustrating a chart or sketch other than the age-old use of overhead projectors. The question of voice exchanges over the computer did not even arise since all the participants were in the same room. Technological advancements soon allowed developers to push to overcome these limitations.

53.4 COMPUTER SUPPORTED COOPERATIVE WORK AND GROUPWARE

The single most important technological advancement fueling the growth of collaborative systems is undoubtedly the computer network. The very concept of connecting computers together is in itself highly analogous to social networking or grouping. It only stands to reason that since the aim of collaborative systems is to bring together groups working toward a common goal, the tools used by the individuals in these groups must also have the ability to communicate with each other. In fact CSCW has been defined as "the use of computer and electronic communication tools as a media for communicating" (Johnson, Weaver, Olson and Dunham 1986).

The emergence of networking in the late 1980s resulted in greater focus being placed on the ability of groups—even those geographically dispersed—to interact with each other. As a result, four main types of collaborative systems were identified, each with distinct functions: messaging systems, conferencing systems, meeting systems and coauthoring systems (Rodden, 1991). Clearly, collaborations generally encompass more than one type of group activity or process, and most products have attempted to integrate these functions to varying degrees in order to develop systems that function as needed by users. The ultimate goal, of course, was to simulate as best as possible the processes found in face-to-face interactions. If we dig deeper into the underlying elements of group collaboration that cut across all of the functional uses of CSCW, three characteristics can be found to span all of these functional groups: media richness, geographical scope, and real-time communication.

53.4.1 Media Richness

Media richness, from a social perspective, defines a medium's ability to convey expressions, pitch, tone, etc. In contrast with face-to-face meetings, many have argued that computer-based information systems restrict feedback and are not "rich" (Daft and Lengel, 1986). Put in somewhat more technical terms, this can also be called the "bandwidth of communication" (Barua, Chellappa, and Whinston, 1994a), which refers to the number of different data types that can be used for interaction. In the case of CSCW, a system can be rated according to its ability to support not only textual material but also pictures, charts, voice, and video images.

53.4.2 Geographical Scope

With the emergence of networking in the late 1980s, the geographical scope of group interactions burst out of the confines of a single room and became restricted only by the size of either the users' local area network or wide area network.

1. **Local area network:** A local area network—also known as a LAN—is a network of machines 'physically' connected to each other by some medium—coaxial cable, telephone wire, or now even radio waves. LANs are typically restricted in size by the limitations of this medium.

The concept of networking in a local environment first emerged with the introduction in the late 60s of the UNIX[2] operating system developed at AT&T Bell Laboratories. One of the key goals of this project was to devise a means to share files among computers. This was originally accomplished by hooking all machines up to a single file server running an operating system such as UNIX that enabled the exchange of files and using LAN management software to support the process. With the spread of LANs and the increase in power of personal computers, network software was developed such as Novell NetWare that would run not only on UNIX but on the most widespread operating systems in use such as DOS, Windows, and the Macintosh.

2. **Wide area networking:** A wide area network, or WAN, is a network of machines with the ability to communicate with each other without regard to the geographical distance between them. Networks such as IBM's BITNET and TYMNET fall into this category, as do the vast number of other proprietary WANs operated by large corporations. The largest of them all is the Internet—essentially a system of interconnecting networks around the globe that communicate with each other on the basis of the TCP/IP protocol.[3] Although LANs and WANs that do not use the TCP/IP protocol can link up with the Internet through the use of a router or gateway that bridges this gap, the spread of TCP/IP has become so pervasive that new operating systems such as Windows '95 come already bundled with it.

53.4.3 Real-Time Communication

Real-time communication defines the ability of a CSCW tool to work synchronously. Ever since the beginning of collaborative systems, this feature has been granted great significance because collaboration in a group more often than not requires immediate input and feedback to be effective. One example of a common situation requiring real-time communication is when group members need simultaneously to view and correct a document. CSCW tools initially concentrated on providing a platform for this—which led to the development of a wide array of "shared editors" or coauthoring tools, as described in the following section.

53.4.4 Semi-integrated CSCW Tools

Although the value of media richness, unbounded geographical scope, and real-time communication ability is quickly recognized by users, tools that fully integrate all three have taken a long time to develop. Barriers to earlier development include level of technological sophistication, lack of standardization, and prohibitive costs. In the meantime, tools have been developed that incorporate various subsets of these features. Past trends—and the costs of deviating from them—have also played a substantial role in the continued development of CSCW tools for both the electronic meeting room environment and the desktop environment.

As noted earlier, because of the technological limitations of the time, GDSS were traditionally used in a single room that eventually came to be known as an "electronic meeting room." In this room, alternatively known as a "telemeeting environment," collaboration took place through the use of individual private desktop computers and a large public display controlled by the meeting facilitator. When required by the task at hand, geographical scope could be increased through the use of LAN technology and real-time communication ability could be ensured by the use of CSCW tools such as shared editors. Some CSCW applications for the desktop environment concentrated on the collaborative process of creating, editing, and presenting ideas. Apple's SharedEdit, for example, was one of the first to allow the simultaneous viewing and editing of a document by one or many people on the Macintosh personal computer.

While this application featured a limited real-time communication ability and the potential for great geographical scope via networks, in this early stage it lacked even the basics of media richness and was unable to display or edit graphs or charts. Media richness

---

[2]© Trademark of AT&T Bell Laboratories.

[3]A protocol refers to "language" that computers use to speak to each other. Transmission control protocol/Internet protocol (TCP/IP) defines a set of rules or guidelines that allow machines with a unique IP (Internet protocol) number to recognize each other. Any application that requires communication with other machines on such a network needs to have this protocol implemented in it.

was enhanced beyond this only by the use of audio devices such as the telephone or live satellite TV connection.

Because of the often crude integration of sets of tools with different capabilities, at this time, computer-mediated communication generally received low ratings in terms of its ability to support interactive, expressive communication (Zmud, Lind, and Young, 1990). From a social perspective, this may well stem from a mismatch between the properties of the medium used and the nature of the tasks involved. Still others feel the fault lies in the technology itself (Barua et al., 1994a). For example, the use of a telephone as an audio device could prove disruptive when it is one more tool that needs to be handled in addition to the CSCW tool itself. As we will see later in Section 53.4.6 on groupware, technological advancements, including increased physical bandwidth capability, have since led to the development of live audio/video conferencing facilities that add considerably greater media richness to the CSCW environment.

In the area of geographical scope, local and wide area networking has also made it possible to introduce CSCW outside of a common room, in essence bringing it to the desktop—and beyond. When desktop computers began to be hooked up to the Internet starting in the late 1980s, the CSCW environment moved beyond the LAN to the outer reaches of the world in just one small step.

53.4.5 Collaboration on Wide Area Networks: The Internet Tools

In the late 1980s unprecedented expansion in the potential geographical scope of collaboration was achieved by the global growth of the Internet, the ultimate in wide area networks. Far from being a monolithic system, the Internet is a conglomeration of interlinked smaller regional, national, and international networks. It evolved out of the United States' Advanced Research Projects Agency Network, or ARPANET as it is commonly known. Launched in 1969, the goals of the original project were to facilitate research among academic and government institutions spread around the country by sharing computer resources and facilitating information retrieval and communication through a government-funded network of interlinked computers (Kalakota and Whinston, 1995). In order to accomplish this, however, groundbreaking work had to be done on developing new means for the host computers to communicate, including the first packet switching technology and store and forward mechanisms.

As the network grew, the logistics of internetworking heterogeneous computers and network protocols and point-to-point communications required the development of a new protocol that could manage this efficiently. In the mid 1970s TCP/IP was developed and within a decade was chosen by the U.S. government as the standard protocol for the entire Internet (Quarterman and Carl-Mitchell, 1994a). Collaboration on the Internet quickly took on a new face: It was informal and global, as evidenced by its various tools.

53.4.5.1 e-mail

One of the Internet's most popular applications that was introduced early on was electronic mail (e-mail) (Quarterman and Carl-Mitchell, 1994b). First "invented" in 1973, e-mail was originally intended to be a communication tool between two users and is probably the first form of asynchronous computer-mediated communication. But e-mail had no collaborative overtones and was thought of more as a personal messaging system.

53.4.5.2 Usenet

Instead, the Internet community, consisting of like-minded professionals, saw the need for a platform-independent forum to discuss work-related issues. This grew into what is now called Usenet,[4] a virtual network on top of the Internet that is perhaps the first of its kind to function as a global conferencing and discussion system. In essence, Usenet constitutes a global asynchronous CSCW environment. It is made up of a wide variety of "newsgroups" grouped under various topics to which readers can post relevant "articles." Although the topics initially focused on computer hardware and software issues, newsgroups today span the gamut of imaginable topics.

Since it was never intended to function as a task-specific bulletin board, apart from the subject-based hierarchies, Usenet lacks the formalism and structure of later, more

[4]Refer to *http://www.cis.ohio-state.edu/hypertext/faq/usenet/what-is-usenet/part1/faq.html* for the history of Usenet.

specialized CSCW tools available on the Internet. Still today, it remains entirely text based and can accommodate pictures or other binary files only after they have been converted to ASCII documents by special software or software agents. Other Internet CSCW tools have added additional features to serve one or more special needs. One example is the Worm Community System (WCS). This customized piece of software, which serves biologists studying genetics in geographically dispersed areas, has been enhanced with numerous graphical features, increasing the media richness of the tool.

53.4.5.3 Internet Relay Chat

The Internet offers numerous tools that allow users to engage in real-time discussions. Although they can accommodate only text-based interchanges at this time, Internet relay chat (IRC) and "talk" applications allow synchronous or real-time discussion and collaboration. Much like the subgroupings of the newsgroups under Usenet, IRC consists of various "channels" dedicated to discussion of different topics. Participants contribute to the ongoing open discussion one or two lines at a time and even have the ability to communicate privately with each other. The separate channels for hardware, software, sports, etc., eliminate unnecessary noise and provide a structure that facilitates meaningful interactions, while still remaining general purpose tools accessible to every user of the Internet.

53.4.5.4 Bulletin Board Systems

The once sharp distinction between commercial online service providers such as CompuServe, America OnLine, and Prodigy and true access providers to the originally government-sponsored Internet is now quite blurred. In many instances, online service providers have also patterned tools after those available on the Internet. One common tool is the bulletin board system, which allows users to post and read messages as if on a normal office bulletin board similar to the Internet-based newsgroups. With its growing popularity, more and more features have been added to provide new functionalities such as real-time communication, interactive games, etc.

53.4.5.5 Client–Server Architecture and Collaborative Internet Tools

A key feature of wide area networks, including the Internet, is the reliance on client–server architecture. This technology, along with the TCP/IP communication protocol, enables users to access and apply features that would otherwise be impossible even in an open system such as the Internet. For example, in order to accommodate a "simple" text-based IRC session, two software programs are needed. On the one end, the client need only be able to send information to and receive information from the server. On the other end, the IRC server, the more dynamic of the two, must be able to listen to this and other clients requests and to manage the process of responding to them. Whether we are talking about real-time discussion groups, interactive role-playing games called MUDs (multiuser dungeons), or bulletin board systems, the client–server format allows multiple users from theoretically anywhere in the world to link up to a server easily and seamlessly to access a particular tool, an application, or data. As we will see later, the client–server architecture was a fundamental technological prerequisite to the recent development of the World Wide Web, the multimedia offspring of the Internet.

53.4.6 Groupware

Although the above advances in wide area network technology have enlarged the geographical scope of numerous tools for group interaction, advancements in CSCW products for the desktop environment—enabled by technological improvements in local area networks—have created a new class of collaborative system products that have come to be known as groupware. Definitions of "groupware" range from being essentially an umbrella term for "the technologies that support person-to-person collaboration" including anything from "E-mail to Electronic Meeting Systems to workflow" (Coleman, 1995) to tying it in somewhat more specifically with communication, document management, and database access. In essence, groupware can be thought of as a superset of desktop environments and messaging systems on LANs. Whereas CSCW desktop environments concentrate on the editing, creation, and presentation of ideas, groupware products elevate to a new level of media richness. User interfaces, group processes, and concurrency control are key features of groupware products that aid group editing, distributing, and sharing of documents (Ellis, Gibbs, and Rein, 1991). All this places high demands for robustness and responsiveness as well as data replication on groupware products. One of the common barriers to providing the desired media richness—lack of sufficient bandwidth—was never a problem in the LAN environment. LANs provide adequate speed, e.g., 10 Mbps on an ethernet to 100 Mbps on fiber distributed data interface (FDDI) rings.

However, the problem of integrating disparate functions such as document processors, messaging systems, and group activity monitors remained a challenge. As the sharing of critical information among workgroups has gained importance in the business world, a wide array of applications have become available on the market. Although the common ground among all these commercial products is their focus on communication, collaboration, and coordination (Ellis et al., 1991), the products range from those intended for small, narrowly focused groups to those serving enterprise-wide strategic programs. Some groupware products, such as cc:Mail and LotusNotes from Lotus Corporation, Microsoft Mail and Microsoft Schedule from Microsoft Corporation and OpenMail from Hewlett-Packard address a particular subsection of needs—in this case communication or messaging. Microsoft's Windows for Workgroups, on the other hand, focuses on communication and shared access to documents, reflecting the document centric nature of office work. With emerging document management structures, documents no longer tend to be entirely text based but include images, spreadsheets, and other "embedded" objects as well. Microsoft's MS Office is a classic example of such an integrated application environment in the area of desktop publishing.

Other groupware products are devoted primarily to database management. Lotus Corporation's SmartSuite, for example, claims to incorporate bidirectional replication; i.e., once two replicas of a database are synchronized, workgroup members on different networks and servers begin to make changes, deletions, and additions to them. Clearly, many of the tools that form groupware applications such as these have existed for a long time. It is the increasing level of integration and media richness that differentiates groupware from its predecessors. This revolution has been made possible by the changes in the underlying supporting network technologies. Groupware indeed lies at the "convergence of technical, economics, social and organizational trends" (Coleman, 1995). Groupware has remained at the focal point of current trends by continuing to develop; this time by expanding into the hitherto unexplored waters of the Internet. The historical path that groupware development has followed—concentrating on the closed, well-defined environment of local area networks—has left a legacy that in many cases makes the leap to use on nonproprietary WANs highly complicated. Many groupware products such as LotusNotes that are popular today are based entirely on proprietary standards. Their ability to work with nonproprietary, open networks such as the Internet was originally quite limited.

Nevertheless, the phenomenal geographical scope of the Internet has made it an irresistibly attractive messaging channel. To overcome the limitations of the proprietary systems, certain customized "gateways" were developed to enable them to communicate over the Internet. At first glance it sounds like the best of both worlds: Gateways integrate powerful LAN-based features of groupware products with TCP/IP for wider geographical reach.

However, the use of gateways still poses serious limitations and drawbacks. For example, while LotusNotes can work over wide area networks like the Internet through the use of Lotus's Internotes gateway, the end users must all have LotusNotes to be able to communicate with each other. In addition, customization is traditionally very expensive and potential incompatibilities often render it a short-term approach only. When a vendor other than the original develops an additional component for groupware applications, it is often necessary to redesign the customized gateway to interface with the new components. With every new technology introduction or product release, additional patchwork is necessary.

In addition to increasing its geographical scope, groupware is also capitalizing on new technological advances to enhance its real-time capability. Since most groupware products are highly document centric, they have traditionally lacked synchronous communication ability. Unlike electronic meeting rooms, which are designed for specific meetings, groupware is more of a desktop, day-to-day collaboration technology. When occasional conferencing is necessary, a separate telemeeting facility with video and audio conferencing capabilities is normally required. However, rapid advances in communication technology are now enabling "desktop conferencing" to become a reality.

With a video camera and microphone attached to a multimedia desktop computer, groupware users could possibly work and discuss with each other on a real-time basis.

Although the basic technology to accomplish this is already available,[5] questions remain about its application and use. For example, work still needs to be done on integrating such conferencing systems with other groupware components, i.e., storing the conference session and linking it as a multimedia document for future reference. Whereas integration of this type is undoubtedly possible to a certain extent with vendor-based products, the costs of doing so may be prohibitive because of a lack of common standards among vendors.

53.5 WIDE AREA COLLABORATIVE ENVIRONMENTS: THE CASE OF "SHARED"

As discussed in earlier sections, many tasks involve group interaction and coordination. Examples range from CAD design of an engineering problem to molecular biologists discussing a certain strain of a virus. As do the tasks, the tools required for each environment may vary considerably. Consider, for example, a collaborative environment for an engineering design problem.

Engineering tasks typically involve acquiring and manipulating data or information to produce a final design of a product. Therefore, one main objective is to archive data in relevant formats for future access. Such data could include CAD diagrams, specially formatted documents and other types of information ranging from audio to video files. Also, in order to leverage the Internet as the physical medium for communication, nonproperietary communication protocols must be used in the archiving and managing of these media-rich data formats. In this section we take a look at an example of a project that has taken on this technological challenge.

"SHARE," a collaborative environment for mechanical engineers developed by Enterprise Integration Technologies,[6] concerns itself primarily with how information technology can help engineers develop products. The need for such an environment is obvious as, increasingly, product development involves teams of engineers from multiple organizations or from a single organization that is geographically dispersed.

The SHARE environment is a CSCW tool for mechanical engineers. It consists of synchronous and asynchronous tools for communication. SHARE follows a layered architecture in defining the tools it uses for collaboration. In essence, it is a four-layered architecture consisting of an application layer, a services layer, a transport layer, and a storage layer.

The storage layer is comprised of both a traditional file system and a shared repository technology. The transport layer could be a local area network or a wide area network. The services layer includes technology that facilitates the sharing of application specific data and control information. And finally, the application layer provides the front end for generic applications and CAD tools.

The key component of this environment is clearly the service layer, which enables sharing of nontextual data. The service layer consists of a number of applications such as NoteMail, Xshare and Service Mail and uses MIME data types to define specific document formats for applications. In essence, this setup not only allows for engineers to communicate with each other by mail but also allows for simultaneous use of Xwindows applications on multiple machines. By providing the right Multi-Purpose Internet Mail Extensions (MIME) tags, it is further possible for users to retrieve or transfer specific application files or multimedia documents themselves.

Clearly, SHARED is an example of a sophisticated, flexible collaborative environment where documents are stored and retrieved and applications can be shared over the Internet. Limitations do still exist, however. The use of Xwindows, which typically operates only on a UNIX™ or a VMS™ platform, may limit the users' choice of platform. Also, each participant of this collaborative environment needs to obtain the specific layers before being able to take advantage of SHARED.

---

[5]Numerous desktop video-conferencing products exist today. AT&T's Vistium and Intel's ProShare are examples of proprietary systems in this field. There are also other systems such as MBONE and Cu-Seeme, which operate on the Internet and are based entirely on published protocols. More information on these can be obtained from *http://www2.ncsu.edu/eos/service/ece/project/succeed_info/dtvc_survey/products.html*

[6]Refer to URL: *http://www.eit.com/projects/share/share/share-home.html*

53.6 COLLABORATIVE ENVIRONMENTS ON THE WEB

To recap briefly, the above evaluation of collaborative environments focused on the three main characteristics of media richness, geographical scope, and real-time communication ability, which until recently were developed largely independently of each other. This is understandable since these developments in the GDSS and CSCW fields took place during a period when the world of computing and communications was characterized by proprietary systems and standards. It has, however, left a legacy of its own. As we have seen, most collaborative systems are not only platform specific but also restrictive in their ability to communicate with other systems. Although these systems cater well to the needs of specific organizations or workgroups, they appear less suitable for supporting true global interactions.

As we saw in the case of groupware, it is possible to bring together existing developments to suit emerging needs through the use of customized gateways. The cost of doing so, however, can be not only prohibitive but also inhibitive in its ability to integrate with future technologies. We also described a concurrent engineering environment that represents a groundbreaking advance in fulfilling the needs of CSCW users for media-rich interaction in a global environment. However, these systems are still not "open" in the sense that it is not possible for users with different applications to interact seamlessly on a global basis.

More and more, users perceive the need for creating an open platform where they can engage in information exchange and discussions without being forced to use proprietary applications and data formats. Clearly, optimal collaborative computing solutions need to do more than integrate the three dimensions that have affected development to date, they also need to be nonproprietary in nature. The bottom-up approach of collaboration that we've seen so far, which builds upon existing components, is fast being replaced by a top-down view that encompasses all these characteristics—and customizes according to need. The platform for the development of such systems is already in existence: the Internet's World Wide Web (WWW).

As we have seen, the Internet is a platform that is ideally suited for collaboration not only because of its geographical expanse, but also because of its open nature, which allows users from around the globe to communicate and interact with each other without the need to have specific, proprietary end-user applications. This is made possible by the TCP/IP protocol and a suite of other communication protocols built on top of this stack. This holds true for the Internet's WWW as well. For example, Web browsers like Mosaic and Netscape make it possible to invoke other applications, such as word processing and spreadsheets, seamlessly. One user could make available a spreadsheet file that can then be accessed, analyzed, and possibly even changed by others using Netscape.

In addition to open access, the Web provides an extremely high degree of media richness. It is the first of the Internet applications to offer multimedia and hypertext capability on a seamless basis, made possible by its client–server architecture. Even though the Internet offers other systems such as Gopher that are primarily text based, the Web is the first to allow multimedia documents to be shared across the Internet. The "document" in this case includes hypertext links to other information resources and inline images that are part of the document itself. Another feature that distinguishes the Web from Usenet or Gopher systems is its unique client–server implementation that allows actions or "scripts" to be executed on a remote server. One illustration of the use of such scripting is provided later in the chapter.

Indeed, the advancements the Web embodies—global geographical scope, high degree of media richness, nonproprietary open system, client–server architecture—all combine to open up a vista of opportunities for collaboration on a global basis. Whereas Gopher has integrated diverse capabilities such as content and location search for information resources, document transfer, and remote access, the Web takes a revolutionary step in the direction of the ultimate goal of worldwide information dissemination and interaction with the ability to link information resources anywhere in the world.

Although the World Wide Web brings us closer than ever to this goal, which is theoretically possible already today, advancements still need to be made in two particular areas: interactivity and information organization. The Web was primarily designed for dissemination of information rather than as an interactive medium and so lacks the real-time communication ability essential to group interactions. For the WWW to be an electronic forum for productive interaction, interactive capabilities will need to be integrated with the information repositories distributed over the Internet.

Also, the Web's phenomenal success may actually be a hindering factor. There is no doubt that if properly organized conceptually and technically, the Web can serve as a highly effective foundation for a collaborative system. On the other hand, isolated and idiosyncratic developments of Web servers will quickly lead to information overload, inefficiency, and chaos. The growth to date has been staggering.[7] In the very near future, in the absence of shared conceptual foundations for organization of information, a serious problem of "infoglut" will arise. Both increased interactive ability and improved search capabilities are the focus of innumerable development efforts that will undoubtedly succeed in making the Web even more conducive to successful group collaboration.

Despite these issues, the enormous potential of the Web is incontestable, and pioneering open collaborative systems are now being developed. Designing such a system is no small task. Unlike a proprietary system or application, which provides only a few customization choices to users through limited application programming interfaces, an open system such as the Internet provides a wide range of design choices, all of which require careful consideration on the part of the developers. Although specific applications and solutions will vary in each instance, the general principles used in designing open collaborative systems, which we call collaboratories, hold true throughout. Just as we presented a framework to use in evaluating the historical development trends of CSCW, in the following sections we will describe in some detail the factors involved in designing a collaboratory. We will use the example of a collaboratory developed for the Management Information Systems (MIS) field to illustrate the issues relevant to both the design phase and the actual implementation phase.

53.6.1 Designing a Collaboratory Based on a Shared Conceptual Foundation

A collaborative system, or collaboratory, can be defined as an "open electronic platform for individuals or groups with common interests to efficiently exchange, disseminate and create issues, ideas and knowledge" (Barua, Chellappa, and Whinston, 1995a).

Applying this viewpoint to the domain of MIS, researchers at the University of Texas at Austin's Center for Information Systems Management (CISM) identified two main objectives of their MIS collaboratory[8]: (1) efficient dissemination of information through organization of information resources, search, and resource linking capabilities, and (2) a global forum for asynchronous and synchronous interactions involving MIS issues, ideas, and research articles. The collaboratory, which is a Web server with information organized according to the developers' vision of the MIS field, also incorporates an online electronic forum and announcement and search features. In sum, the collaboratory supports multimedia document-based interactions among MIS professionals and academics. In designing the MIS collaboratory, the developers selected specific design characteristics from the large number of alternatives in such a way that net user value would be maximized. Quite briefly it was assumed that certain factors, objects, people, or processes have a value synergy among themselves, i.e., they are complementary. Accordingly, when they are chosen in a coordinated fashion, the combination yields greater value than the sum of the values created by individual factors in isolation. It is similar to the popular saying "the whole is worth more than the sum of the parts."[9] The first step in the design process is to identify the key determinants of the net value to the ultimate user. In the case of the MIS collaboratory, three key factors were identified based on the primary goals of usage: level of information access, interaction richness and, information–interaction cost. The first two determine the gross benefit that accrues to users from spending time on the collaboratory, whereas the third relates to the cost of this. Essentially a standard cost–benefit analysis is conducted that will determine the user value of the overall system. Each of these three factors is influenced by a number of different considerations as illustrated in Figure 53.2.

---

[7]Annual rate of growth for WWW traffic is put at 341,634% and Gopher traffic at 997%. For more information refer to *http://www.openmarket.com/diversions/internet-index/93-12.html*

[8]© A. Barua, Chellappa, and A. B. Whinston, CISM, 1994a.

[9]Whereas this concept of synergy is not new to the business world, recent academic research in economics (Milgrom and Roberts, 1990) and information systems (Barua, Lee, and Whinston, 1995b, Barua and Whinston, 1994) has focused on the ideal combinations of organizational and technology-related factors to maximize organizational payoff.

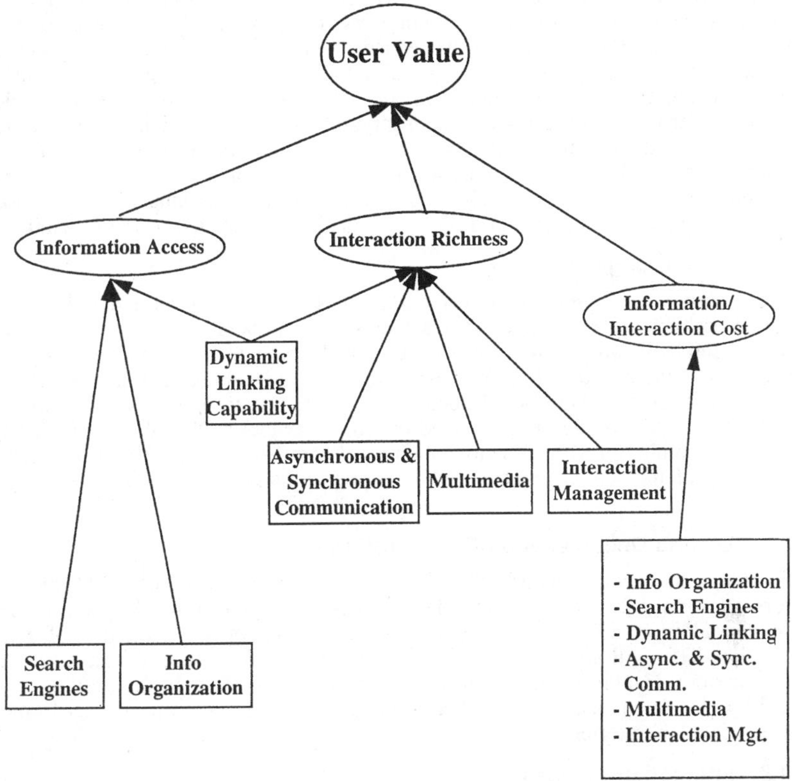

Figure 53.2 Complementary drivers of collaboratory users' values.

53.6.1.1 Information Access

One of the major objectives of using a collaboratory is information gathering, whereby a user can access relevant information dispersed throughout the Internet without having to travel extensively in Gopherspace or Webspace. This does not imply that there should be only one collaboratory for each discipline to simplify the search process. This would go against the very spirit of the Internet and would pose an overwhelming administrative load on the site maintaining the collaboratory. Rather, a set of central points distributed throughout the globe are envisioned that serve as gateways to a specific field of interest and related information. The search among these collaboratories would be made more efficient and effective through the appropriate organization of information, the use of context-specific search mechanisms, and the Web's dynamic linking capability.

53.6.1.2 Information Organization

As discussed earlier, the Internet is already struggling with information overload, as exemplified by the results of queries with popular search engines. While this is definitely a positive indicator of Internet usage and its popularity, it raises the issue of "infoglut." Clearly, it will become increasingly important for information on the Internet to be organized in some structured form that enables easy user access.

Although the physical accessibility of data on the Internet has been well defined in terms of addresses and protocols for various files and formats, the process of how to identify relevant information still needs improvement. The traditional way has been keyword-based searches, such as WAIS (wide area information servers), which index words in documents based on efficient algorithms. The problem of sheer numbers remains, however, when large volumes of documents need to be searched. To narrow the search further, subject- or author-based indexing is commonly used.

A third option, which was chosen by the MIS collaboratory, is to prepare an intellectual indexing of documents and other information resources that links a set of related documents to each other. For example, in the world of MIS, an issue such as end-user computing has many related facets involving technological, social, and economic factors. A collaboratory that organizes information on end-user computing without grouping relevant documents under these categories with appropriate links between them is likely to be less useful in the search for information on, say, social issues in end-user support. Such a categorization and linking scheme helps a user efficiently locate a set of related documents on a given topic. A detailed description of the intellectual indexing of information for this collaboratory is given in Section 53.6.3.1, for those interested in more detail.

53.6.1.3 Search Engines

The search process is central to information access and retrieval on the WWW. In a distributed architecture consisting of a set of core collaboratories across the globe, a document search would involve both location and content. At individual sites typical Internet search engines maintain WAIS-based (keyword) indexing schemes. However, with the proliferation of Web-based documents, location now plays a central role in the search process. This problem is addressed by Web crawlers or WWW worms that travel in Webspace searching for documents which meet certain user-specified criteria. Clearly, the efficiency of the search process can be further enhanced by making these searches context specific.

53.6.1.4 Linking Documents on Collaboratories

Travel through the virtual network of collaboratories can also be improved by the linking of information on different collaboratories. This is supported by the Hypertext links on the Web. In addition, collaboratories will also have to support the capability of dynamic link creation. This will allow a user who finds a relevant document to link it to a collaboratory, thereby making it immediately available to other users. In the absence of such a capability, the potential exists for creating pockets of information that is hard to access on different collaboratories.

53.6.1.5 Interaction Richness

Apart from information access, another major purpose of using a collaboratory is to have meaningful interactions with other members of the same profession, to debate issues and problems of importance and interest, as well as to gain new insights into research, practice, and education. Interaction richness can be defined as the extent to which the barriers of space, time, and media/document formats can be overcome in interacting with others. This would include:

- Ability to talk, see, write, and draw in both synchronous and asynchronous manner
- Access to relevant reference information
- Archiving of interactions for future review
- Debate of issues, problems, ideas, articles, etc., in open public forums on a global basis

This list raises the question of whether all these features are actually necessary. In fact, some may question whether the ability to conduct face-to-face meetings through a collaboratory actually adds to or detracts from user value. For example, a considerable section of the literature on GDSS supports the notion that anonymity of interactions can help shy reticent group members participate more freely (Nunamaker, Dennis, Valacich, Vogel, and George, 1991) and that face-to-face interaction may even lead to undesirable outcomes.

In any case, whether a feature such as anonymity has a positive or negative impact on the interaction outcomes is an empirical issue, the answer to which depends on the context of use. And to some extent the issues is only important when specific instances are examined since the ability exists in the top-down design of the collaboratory to customize the application depending on the context of the interaction—in other words only a subset of all the available features need be chosen.

The design features that provide the highest level of interaction richness can be grouped under the following categories: dynamic linking capability, synchronous and asynchronous communication, multimedia support, and interaction management.

53.6.1.6 Dynamic Linking Capability

As in the case of efficient information access, dynamic linking capability is also critical to interactions based on sound arguments, rational, and empirical or anecdotal evidence. For example, in a discussion of client–server effectiveness on an electronic forum, a user may take a position that this architecture allows better integration of diverse computing platforms than does a mainframe environment. In an ideal electronic forum, the user is able to create links to documents located elsewhere on the Internet that support his or her stand on the issue or that counter someone else's claim. In order to do this, the user will embed a Uniform Resource Locator (URL), gopher, or File Transfer Protocol (FTP) link to relevant articles, comments, or cases within his or her posting about the issue of integration in a client–server environment.

53.6.1.7 Asynchronous and Synchronous Communication and Multimedia Support

The ability to interact in both asynchronous and synchronous modes is crucial to group interactions and should clearly be an integral part of a collaboratory. As we saw earlier, groupware applications have achieved this capability within the small geographical regions defined by local area networks. And other Internet applications such as newsgroups and Internet relay chat also feature these abilities. At this time, however, the Web does not typically support real-time interactions since it was primarily designed for the dissemination of information stored in the form of documents on Web servers. However, in contrast to the other Internet tools, the Web does support a wide variety of multimedia documents, another key element in simulating a realistic collaborative environment. Once again, this particular feature may vary in importance depending on the context of use. For example, the ability to handle high-quality images may be more critical in fields such as astronomy and biology than it is in MIS. Nevertheless, we will discuss in Section 53.6.3 how the MIS collaboratory overcame this dichotomy to provide both multimedia support and real-time interactions in its design. Other solutions are currently under development and may soon be available for use.

53.6.1.8 Interaction Management

Finally, collaboratories must provide mechanisms to manage interactions in an orderly fashion. For example, if an idea is posted for discussion, the input of users will be less meaningful if organization is lacking. In contrast, if the comments are organized according to their position on the question under discussion, a new reader will find it relatively easy to know what others think. Similarly, the contribution itself could have been written in a structured form where the writers clearly state their position and provide empirical or analytical support for it. Without such explicit mechanisms for managing discussions and argumentation, it may be impossible to obtain the full benefits that a collaboratory can offer. To achieve this, a database environment is necessary that is capable of linking related discussions and dynamically reorganizing such links based on certain semantic criteria.

53.6.1.9 Information and Interaction Cost

The "cost" and "the benefit" side of design issues are kept distinct from each other and in different scenarios may vary greatly. For example, different search engines or different approaches to organizing information may lead to the same set of documents being retrieved, resulting in the same benefit to the user. However, one may be more user friendly or more efficient than the other, allowing the user to perform the search faster and with greater ease. In this case, two different designs can lead to the same benefit, but at far different costs. In designing a collaboratory, the term "cost" does not usually refer to a real dollar cost but rather to the opportunity cost associated with the time and effort put in by the user.

53.6.2 Complementarity Between Collaboratory Design Factors

In addition to determining the individual design factors that impact user value, choosing appropriate combinations of design factors is critical to maximizing user value. As mentioned earlier by combining features and capabilities in a coordinated fashion within a single integrated system that we call a collaboratory, the level of information access and interaction richness can be increased while reducing the users' cost at the same time. In essence, the sum of the whole is greater than the sum of its parts.

Theoretically, it is possible to enjoy all or most of the above features on the Internet in an isolated manner. For example, Web and Gopher sites already have a large volume of relevant information on virtually every topic. The Web also supports multimedia documents. Usenet bulletin boards allow asynchronous interactions on a global basis. Internet relay chat provides synchronous interaction capabilities. And WAIS, Archie, and their derivative search applications allow users to search for documents and their locations. However, from a user's perspective, the cost of the time and effort involved will be too high if separate applications are needed for each aspect of an interaction. Incorporating these capabilities within one system can significantly reduce the users' cost of spending time on a collaboratory. We have already seen this in the case of interactive ability and multimedia. A traditional Web, which has multimedia capabilities, cannot usually support interactions as can newsgroups. On the other hand, newsgroups lack the multimedia capabilities of Web-based applications. Clearly, the benefit of having both features together is very high from the users' standpoint. This has been recognized and new interactive newsgroups for the Web are now being developed.

Similarly, the ability to link dynamically to an information resource to support one's stand on an issue is a key feature of a constructive interaction. Whereas the Web allows hypertext links, it does not come with a platform for asynchronous or synchronous communication. Once again, the benefit of having the platform and the linking capability together is higher than the sum of the values derived from the two separate systems, one with the linking capability and the other with the forum.

Or consider the interplay between multimedia support and dynamic linking capability. Being able to link dynamically to, say, FTP sites, but not retrieve diagrams or pictures associated with a document, will be much more tedious (and hence costly) than using a system that allows both dynamic links and multimedia documents.

For those interested in more detail, the following two sections provide a more involved description of the design considerations discussed above and the resulting implementation challenges.

53.6.3 Case Study: Complementarity-based Design and Implementation of an MIS Collaboratory

53.6.3.1 Information Organization

In December 1994, the Center for Information Systems Management of the University of Texas at Austin took the first step in the process of creating a collaboratory and organized information according to their conceptualization of the MIS field. Namely, that MIS creates new knowledge and value for both academics and professionals by combining technological and managerial viewpoints.

The managerial viewpoints come through multiple disciplines such as cognitive psychology, economics, organizational behavior, marketing, and technical developments. Their relationship is visualized in Figure 53.3. Information technology (IT) refers to computing and communications hardware and software. The reference disciplines enable those in MIS to understand how managers invest and use IT, and how IT impacts various

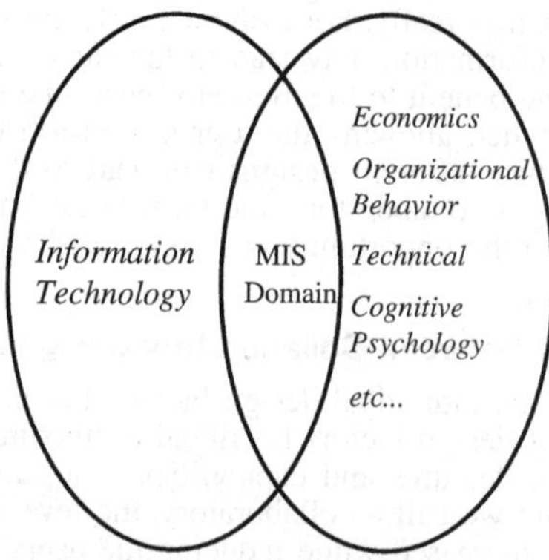

Figure 53.3 The domain of MIS.

cognitive, social, and economic aspects of the organization. In other words, for the world of MIS, the area represented by the intersection in the diagram is more valuable than the sum of the values obtained from pure IT and from pure reference disciplines. Clearly, other fields may have different perspectives, which will govern the way information should be intellectually organized in a collaboratory for their discipline.

Almost any topic of current interest can be used to illustrate the benefits of intellectual organization of information. Suppose a user is exploring the possibility of migrating to a client–server environment. A multitude of related questions arise: What technology is needed (database servers, network operating systems, front-end tools, etc.)? What are the organizational implications (user acceptance, learning, etc.)? What are the economic factors involved? Thus, one single issue spans a variety of different fields that can be searched most efficiently if they are organized according to their relation to the MIS field.

Take another issue that has received widespread attention recently: commercialization of the Internet. This topic involves issues ranging from advertising to legal issues, from technical challenges to economic analysis of pricing and resource allocation. Academics and professionals interested in this area could possess highly diverse backgrounds and could publish their thoughts in widely varied journals. If every aspect of the commercialization of the Internet were documented, it would be impossible to search through various documents to find one of interest and relevance within a reasonable amount of time. However, if the same issues are organized on the basis of the developers' conceptualization, then commercialization of the Internet falls under the MIS domain with documents annotated with links to commercial sites and academic articles on pricing and resource allocation on the Internet. In fact, it was found that almost all MIS issues could be accommodated under the suggested conceptual design. The value to a user is far greater than that from traditional indexing of pages by authors' names or organizations.

53.6.3.2 Supporting Dynamic Interactions on the MIS Collaboratory

The other major objective in developing the collaboratory was to advance the relatively young field of MIS through an electronic forum for generating research ideas and stimulating debates involving researchers and practitioners. An ideal public CSCW system was conceptualized for the Internet which would allow person-to-person and group-to-group communication in a multimedia format. The relevant tasks involved in such a communication would include mailing, posting, and viewing rich documents as shown in Figure 53.4.

An online forum for supporting interactions between MIS academics and professionals was added to the collaboratory. Although the prototype has only a subset of the features ultimately envisioned, it does allow global dynamic interactions on the Web. Not only does the forum display issues and comments, a traditional Web function, it allows users to post comments on these issues and discussions or post a new issue for discussion by others.

A typical scenario will illustrate the benefits of this integrated interactive application. Suppose a visiting professional is interested in commenting on a particular research article and wishes to follow up on others' observations. In a traditional, noninteractive Web environment, this involves at least two parties—the user and the Web administrator.

This user would:

Read the document.
Write comments about the document.
Mail back the comments.

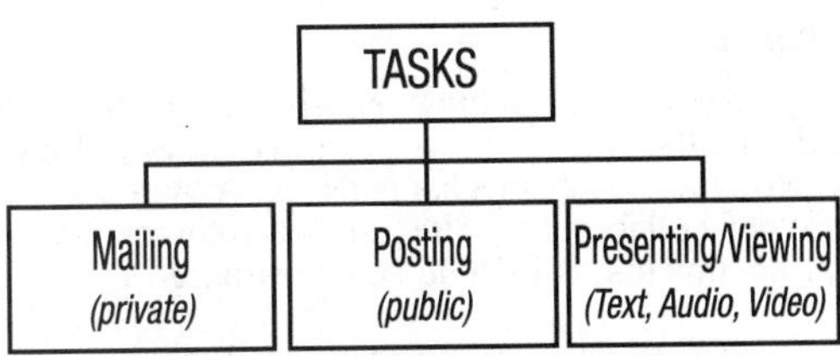

Figure 53.4 Complementary task blocks.

The system administrator would then:

Save the mail as a file.
Convert it to a document in the desired form.
Attach it to the annotation system.

All of the above can be completed with tools available on the Internet. But whereas they constitute one related group of actions for the user, they are typically handled as separate tasks. In contrast, the collaboratory's forum transforms this into one seamless process:

The user:

Looks at a Web page.
Reads the article.
Clicks on a hypertext link that leads to a form.
Fills out the comments.
Clicks on a button that automatically executes a common gateway interface (CGI) script to immediately post the article in the relevant form on the Web page.

The system administrator does nothing.

In this manner, the MIS collaboratory has achieved the interactive nature of newsgroups, but with the multimedia and linking capabilities of the WWW. In other words, this forum, which can be seen in Figures 53.5 and 53.6, simulates the features of Usenet on the WWW, while still allowing users to embed a variety of different URL (e.g., http, Gopher, FTP, etc.) links in their comments. These features are of course in addition to incorporating inline images, audio or even video clips.

53.6.3.3 Behind the Scenes: Server Processes

In the implementation described above, the collaboratory server handles user requests initiated through browsers such as Mosaic or Netscape and provides a "back end" process for information organization and interaction. Three key processes at the server end provide the necessary functionality: CGI processes, search engines, and organized information repositories.

53.6.3.4 CGI Processes

Information dissemination and interaction are handled on the Web through the use of common gateway interface (CGI) mechanisms that allow http requests to execute processes on a remote server through the use of a normal browser and form inputs. The CGI processes have two core duties: document display and request handling. Consider a user who visits the MIS collaboratory for the first time. The document displayer responds to the browser's request and sends back the main document with embedded links to various other relevant documents and forums. If the user chooses a link to any kind of interaction, a form will be displayed with a button to activate a backend CGI script. For example, the user may wish to add a conference announcement. She or he enters the requisite input and submits the document. The request handler receives this submission and evaluates it to determine the appropriate repository. Once the document is placed there, the repository sends a message to the document displayer unit to update the index document on conference announcements. Thus, both the information positioning and dynamic indexing are handled by this unit. In the course of their work, CGI processes rely heavily on search engines.

53.6.3.5 Search Engines

Search engines are responsible for searching repositories for relevant information. In its present stage, the MIS Collaboratory uses a simple keyword-based mechanism that searches only locally. However, other mechanisms like WAIS and Archie can be used to expand the search to other collaboratory sites on the Internet. Its capability will extend beyond simple textual documents to include searches based on URLs, document types, etc.

Enhancing existing search engines is important so that additional qualifications such as document type, location, etc., can be used in addition to specific keyword search. Any

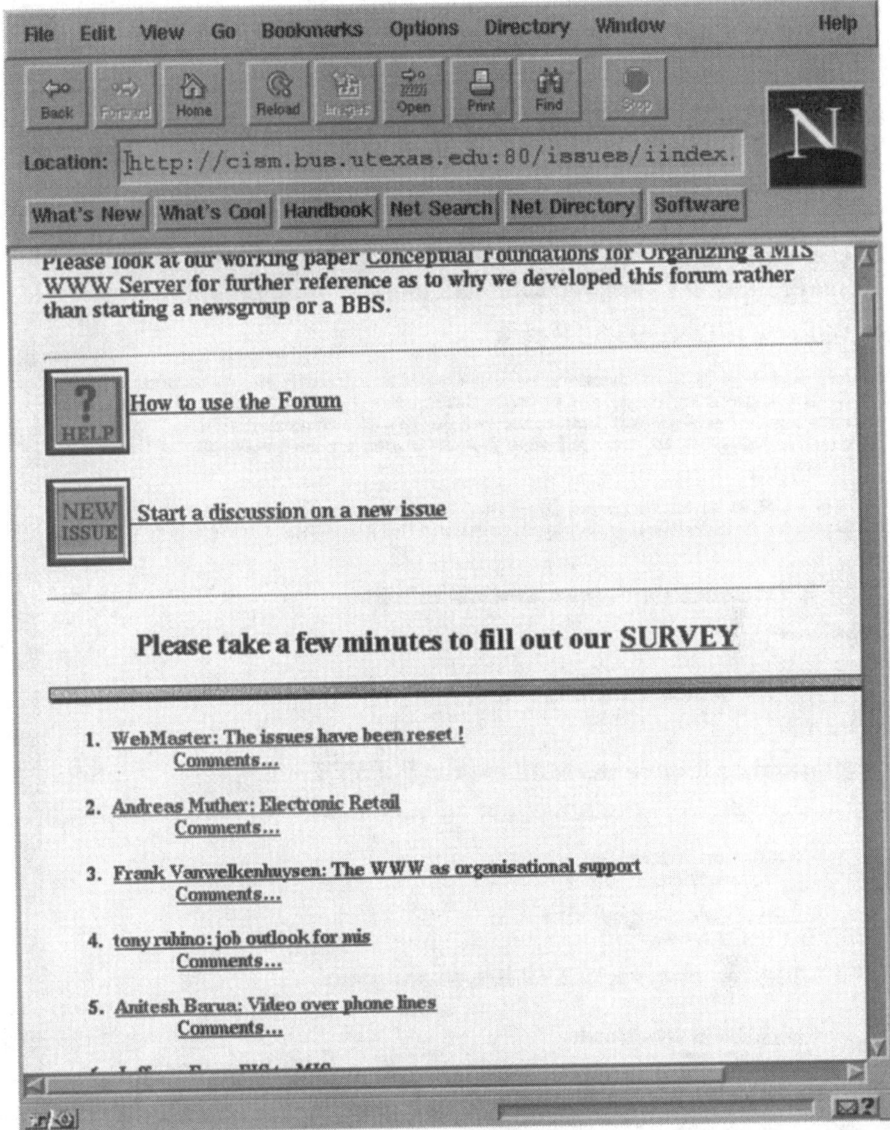

Figure 53.5 The MIS forum.

request for a search will be handled by the CGI process, which invokes the search engine by passing relevant input parameters. While the user enters the input in a normal fashion, a CGI script parses it to provide a suitable string to the search engine and further redisplays results in a formatted fashion. The pool from which search engines cull their information is the information repository.

53.6.3.6 Organized Information Repositories

The information repository is the unit that deals with actual storage of "documents," which include text, audio, video, references to links, etc. All core information as well as information added by users and logs of interactions are stored here. Although the MIS Collaboratory uses a simple flat file system with UNIX directory structures to store documents, it is possible to use a commercial database server as well. Such a database would

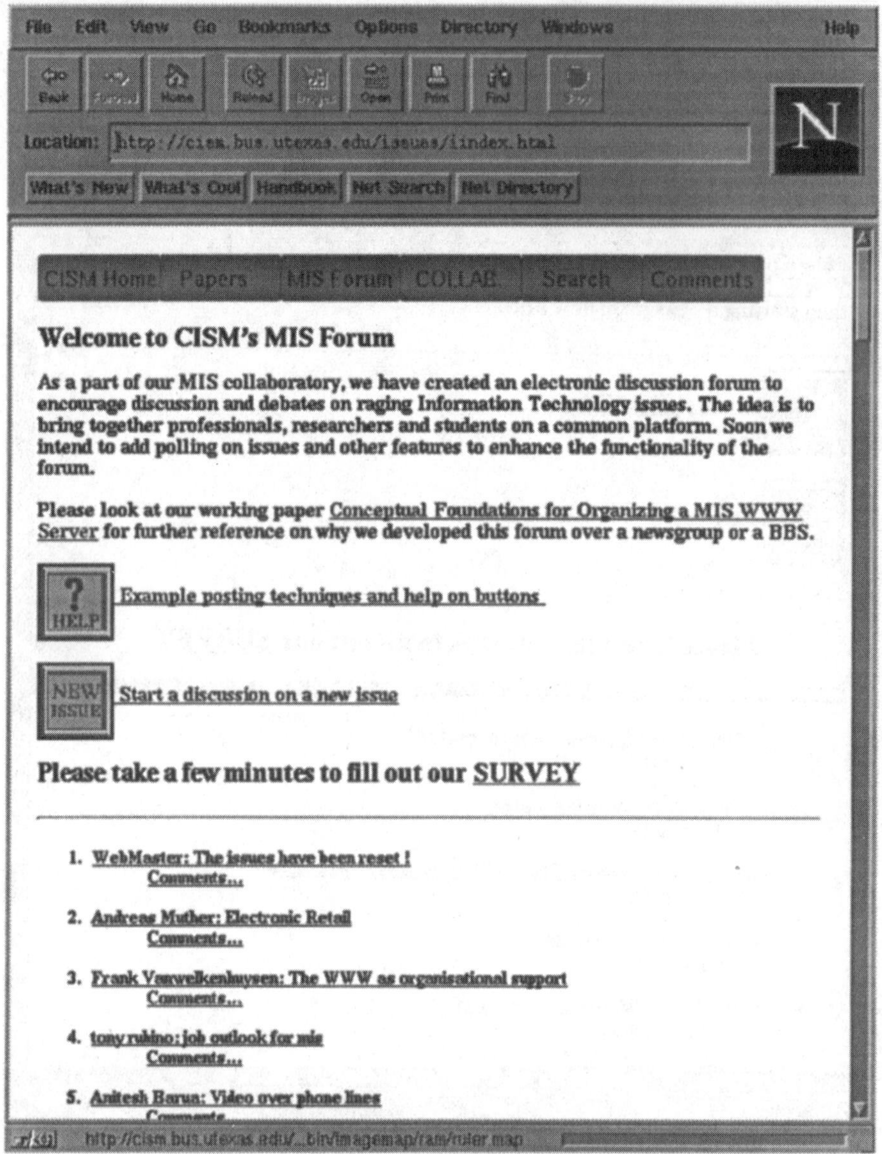

Figure 53.5 Continued.

provide better indexing capabilities and would also allow construction of optimal query scripts for the CGI processing unit. Dynamic modifications are needed in this repository not only to add new information but also to reorganize from time to time the "intellectual index" and the referencing information.

53.6.4 Collaboration: A Holistic View for Future Collaboratories

The collaboratory at the Center for Information Systems Management represents only one small island in the MIS ocean. Thinking in more holistic terms, one vision supposes that there will be centers of collaboration that will act as gateways to linked repositories of MIS-related information (Figure 53.7). The key idea is to maintain the distributed nature

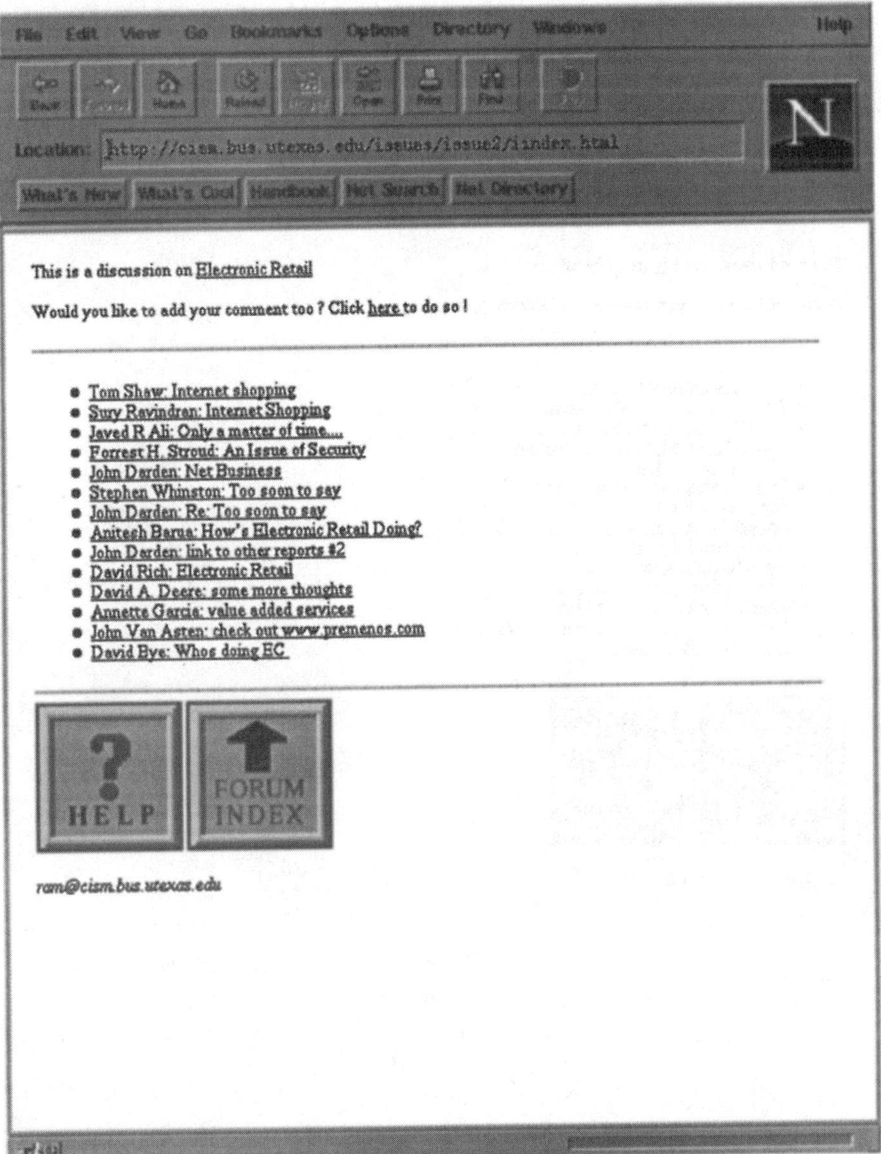

Figure 53.6 A sample discussion in the MIS forum.

of the Web, while avoiding excessive duplication of repositories. Since this is a collaborative environment, there is bound to be some amount of duplication of material by servers that wish to maintain a local copy of a certain repository.

Although it may be difficult to determine an optimal repository distribution strategy, the entire MIS Collaborative environment could be made up of five to seven core collaboratories distributed across the continents. Depending upon usage load factors, "mirror sites" could be used to replicate the same repositories and the ability to execute similar search agents. Because of the seamless nature of the Internet, users would not even be aware of the geographical location of these servers as they access collaboratories all over the globe.

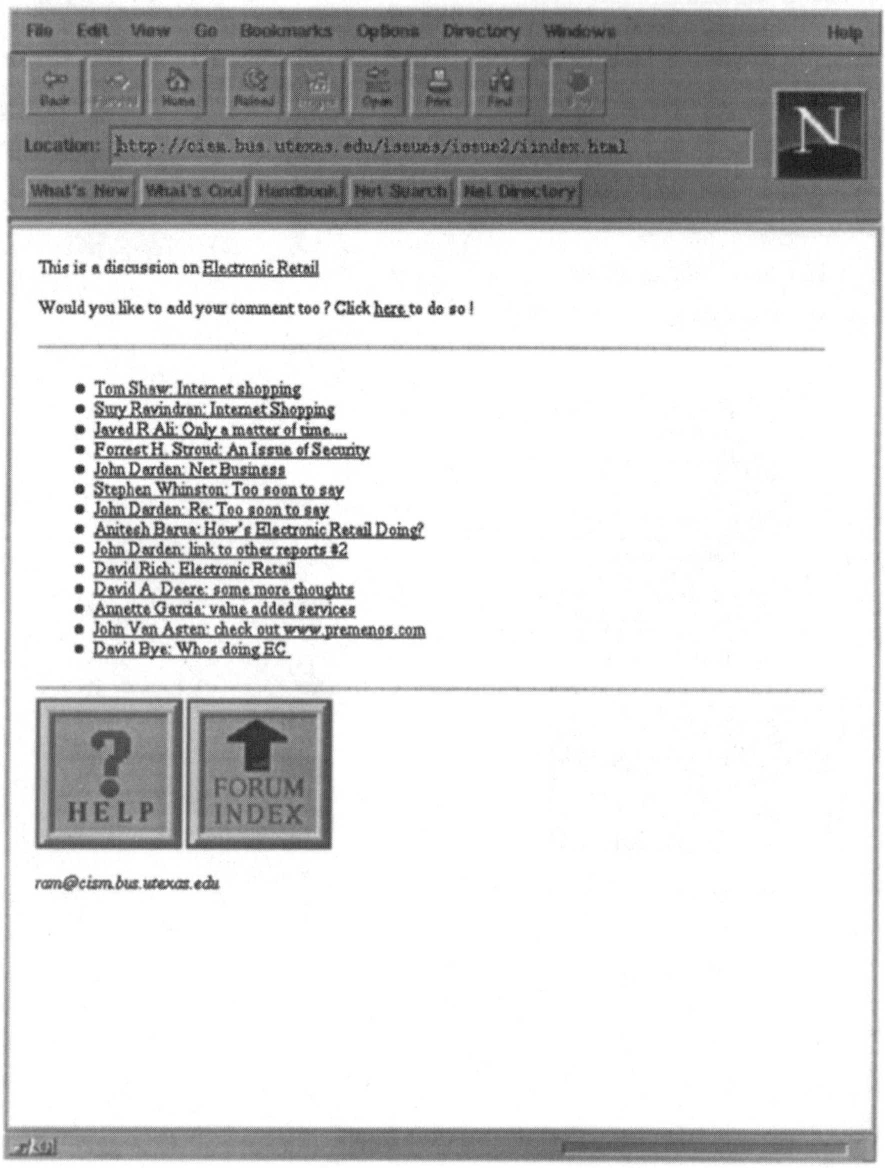

Figure 53.6 Continued.

53.6.5 Future Research

This detailed example of a collaboratory focused on creating a platform for global interactions is made possible to an unprecedented degree by the geographical scope of the Internet. However, geographical scope is not the only revolutionary advantage. The "openness" of the Web also has far-reaching implications within a single organization. The Web makes it possible to customize a system to meet specific organizational requirements or business needs without the restraints of proprietary standards and protocols. In fact, a highly promising area of future research is the development of a collaborative system on the Web to handle the integration of workflow with messaging systems and

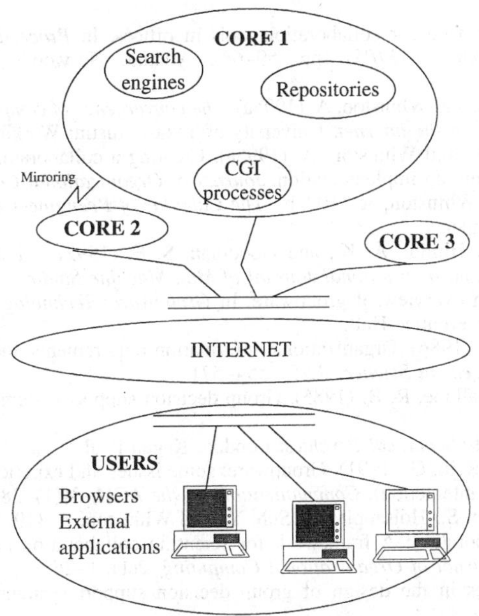

Figure 53.7 Proposed architecture for linking collaboratories.

databases through the use of "normative scripting." A script in the context of the Web refers to shell programs or CGI processes, which can be triggered by different user actions. By scripting or documenting these actions on the basis of certain rules, or "norms," an entire process could be automated.

The possible applications within organizations are multitudinous. For example, consider a Web-based system for a marketing group that is collaborating on a marketing plan for a product. The play has two facets, one involving a comparison of similar products on the market, and the other an internal decision making process which approves or rejects the plan. A normative script can be helpful in both facets. First, it will send out "smart agents" such as Web crawlers to search for references in the plan to existing products and to retrieve the information for the marketing group. Second, if a business rule of the firm requires, for example, that at least two thirds of the group must agree on a certain plan, a CGI process can be created to verify the group members' positions and to message the document automatically to a senior manager if there is at least a two-thirds approval.

Whereas commercial groupware can handle messaging and workflow automation, they also require adherence to proprietary standards. A Web-based collaboratory can achieve the same functionality on a world-wide basis through normative scripting of processes. These and similarly revolutionary changes in business processes will become possible through the Internet and the tools now under development.

53.7 CONCLUSIONS

The history of collaborative computing has been peppered with enormous leaps in system development made possible by technological advancements. As we have seen, the cutting edge of technology today—the World Wide Web—once again has created a new platform for group interactions that only a few years ago was an almost unattainable vision. Today, collaboratories support multimedia interactions for the exchange and creation of new ideas and concepts on a global scale.

Throughout this endless sea of change, the one constant is the increasing ability of cooperative systems to meet the needs of their users. Technological advancements have made it possible to break down the barriers of geography and time. And as a result, collaboration through computers resembles face-to-face group interaction more than ever.

REFERENCES

Bair, J. H. (1985). The need for collaboration tools in offices, In *Proceedings of the 1985 office automation conference (AFIPS)* (pp. 59–68). Atlanta. Norwood, NJ: Ablex Publishing Corporation.

Barua, A., Chellappa, R., and Whinston, A. (1994a). *The convergence of complementary technologies: The case of mbone on the internet.* University of Texas, Austin: Working Paper. 35 pages.

Barua, A., Chellappa, R., and Whinston, A. (1995a). Creating a collaboratory in cyberspace: Theoretical foundation and an implementation. *Journal of Organizational Computing.*

Barua, A., Lee, C., and Whinston, A. (1995b). *The Calculus of Reengineering.* University of Texas, Austin: Working Paper.

Chen, H., Lynch, K. J., Himler, A. K., and Goodman, S. E. (1992). Information management in research collaboration. *International Journal of Man-Machine Studies, 36*(3), 419–445.

Coleman, D. (1995). An overview of groupware. In *Groupware: Technology and applications.* Englewood Cliffs, NJ: Prentice Hall.

Daft, R., and Lengel, R. (1986). Organizational information requirements, media richness and structural design. *Management Science, 32*(5), 554–571.

DeSanctis, G. L., and Gallupe, R. B. (1985). Group decision support systems: A new frontier. *Data Base, 16,* 3–9.

Edgeworth, F. (1881). *Mathematical Psychics.* London: Kegan Paul.

Ellis, C., Gibbs, S., and Rein, G. (1991). Groupware: some issues and experiences. (Using computers to facilitate human interaction). *Communications of the ACM, 34*(1), 38–58.

Hamalainen, M., Hashim, S., Holsapple, C., Suh, Y., and Whinston, A. (1992). Structured discourse for scientific collaboration: A framework for scientific collaboration based on structured discourse analysis. *Journal of Organizational Computing, 2*(1), 1–26.

Huber, G. (1984). Issues in the design of group decision support systems. *MIS Quarterly, 8*(3), 195–204.

Hymowitz, C. (Jan.8, 1988). A survival guide to the office meeting. *Wall Street Journal,* p. 35.

Ives, B., and Olson, M. H. (1981). Manager or technician? The nature of the information systems manager's job. *The Management Information Systems Quarterly, 5,* 49–62.

Johnson, B., Weaver, G., Olson, M. H., and Dunham, R. (1986). Using a computer-based tool to support collaboration: A field experiment. In *Proceedings of the conference on computer-supported cooperative work (CSCW'86)* (pp. 343–353).

Kalakota, R., and Whinston, A. (1996). *Frontiers of electronic commerce.* Reading, MA: Addison-Wesley.

Lewis, L., and Keleman, K. (1990). Experiences with gdss development: Lab and field studies. *Journal of Information Science, 16,* 195–205.

Milgrom, P., and Roberts, J. (1990). The economics of modern manufacturing: Technology, strategy and organization. *American Economic Review, 80(3)*, 511–528.

Mintzberg, H. (1983). *The Nature of Managerial Work.* New York: Harper and Row.

Mosvick, R. K., and Nelson, R. B. (1986). *We've Got to Start Meeting Like This: A Guide to Successful Business Meeting Management.* New York: Scott Foresman.

Nunamaker, J. F., Dennis, A., Valacich, J., Vogel, D., and George, J. (1991). Electronic meeting systems to support group work. *Communications of the ACM, 34*(7), 40–61.

Quarterman, J. S., and Carl-Mitchell, S. (1944a). *The Internet Connection: System Connectivity and Configuration.* Reading, MA: Addison-Wesley.

Quarterman, J. S., and Carl-Mitchell, S. (1944b). *The E-mail companion: Communicating effectively via the internet and other global networks.* Reading, MA: Addison-Wesley.

Rodden, T. (1991). A survey of cscw systems. *Interacting with Computers, 3*(3), 319–353.

Zmud, R., Lind, M., and Young, F. (1990). An attribute space for organizational communication channels. *Information Systems Research, 1*(4), 440–457.

CHAPTER 54

HUMAN FACTORS IN INFORMATION ACCESS OF DISTRIBUTED SYSTEMS

Ray A. Reaux
John M. Carroll
Department of Computer Science
Virginia Polytechnic Institute and State University
Blacksburg, VA 24060 USA

Anyone who has a computer can access an ocean of information as well as add their own information to this ocean. The user population of this universal information system encompasses not only scientists and engineers, but also elementary school students, private business owners, and people in their living rooms. This accessibility has changed the nature of computing from primarily scientific data generation and analysis to communication, education, and entertainment. The revolution in computer-mediated information access is reflected by the proliferation of new ways of accessing and presenting information such as news groups, bulletin-boards, and shared virtual experience systems such as multiuser domains (MUDs).

Information access raises issues that go far beyond the technical challenges that it poses to researchers and technicians. Distributed information access is now pervasive in society and will become a basic foundation of our civilization. How we use information is also changing. The broadcast–receiver model which television and radio brought us is being supplanted by an author model. In this model, every user can be a creator as well as a consumer of information.

Freedom of access and electronic publishing of information has tremendous potential for business, education, entertainment, research, and communication. It also poses tremendous challenges to our model of information access and usage. The familiar data access model from traditional database systems is inadequate to characterize how people will use information with a complex system such as the Internet. Because so many users of information systems are inexperienced or infrequent users, new ways of designing information and information access systems are required. Information systems designers can no longer count on having a dedicated user group trained in formal database query languages who have accrued enough experience with an information system to be able to understand complex information structure. The fact that anyone can be an electronic publisher raises issues of information glut, making it harder for people to find information of relevance, and once found, determine its relevance. For information authors, the challenge is to ensure that published information is accessed and used by its intended audiences.

54.1 RECENT EVOLUTION OF INFORMATION ACCESS

Distributed information systems used to be homogeneous networked systems, multiple computers of the same kind sharing a common software environment and connected to form distributed databases. Today, the Internet brings us to the most recent phase of distributed computing, a self-evolving heterogeneous distributed system with a dynamic and ill-defined boundary. Developed in the 1970s Cold War era, the Internet at first connected only major government, business, and academic institutions. Now, commercial access providers such as America Online, CompuServe, and Prodigy serve as gateways for home users and small businesses. Originally designed to transmit text and numeric data, the Internet now carries graphics, video, and audio. In fact, the most rapidly growing arena of the Internet is the World Wide Web (WWW), perhaps the most heterogeneous distributed hypermedia system.

New information system technologies have transformed the basic information tasks of querying, browsing, and authoring. In part, this was just a matter of new technologies enabling potentially better solutions to old problems, for instance, faster and better graphics to help scientists visually analyze data more efficiently. But this transformation also came about because of the development of a fundamental societal need, a need for mechanisms to manage increasingly complex and evolving information. Ten years ago, the dominant sources of information were the public library and the print media; today, a major source for a large segment of the population is the Internet.

Any new information technology we produce must address two fundamental goals of users: finding desired information in a complex information space and providing users with effective and easy methods of authoring information.

54.1.1 Searching the Information Space

Basic to information use is finding relevant information. Traditional information systems supported two kinds of information search tasks: querying and browsing. Querying is a direct access mechanism in which users explicitly specify a criteria for the desired information. Browsing is a more casual or opportunistic search mechanism. A person browsing for information may not necessarily know the exact criteria of desired information. Instead, he or she begins a search in some known part of an information space and navigates from there until suitable information sites are found.

Querying and browsing are not necessarily mutually exclusive, and people often switch from one mode of search to another. This is most common in browsing systems that supplement a browsing mechanism with querying or string search mechanisms. Similarly, traditional querying systems such as relational database systems often have browsing mechanisms built on top of the databases. Recent work suggests systems can be implemented that recognize the search mode in which a user operates and detect transitions between modes (Loeb, 1992; Sanchez and Leggett, 1994). If so, a computer may take a more active role in suggesting interesting information to be examined, particularly when users are browsing.

An open-architecture information space such as the Internet is highly asynchronous. As Johnson (1995) pointed out, it is not always possible to isolate users from the underlying mechanisms that support network communication, especially with a high-bandwidth medium such as the WWW. Because of variation in net traffic load, retrieval time of a document from a remote site can be almost instantaneous during periods of low traffic, and agonizingly slow during periods of high traffic. If a server containing the information fails, the requested information may not be available for some time. If the server does not go down, but gateway machines (those machines through which the information has to pass to get from the server to the requesting client) or connections fail or become congested, the information may have to be rerouted, with differing transmission times.

Users find it difficult to complete within a set time interval retrieval tasks which suffer from unpredictable access times (Pejtersen, 1989). Users also find unpredictable system delays frustrating, and this frustration can lead to user errors (Kuhmann, 1989). Providing a user with feedback on the status of a query, or even the fact that a query is still being executed, is important within this unpredictable environment (Johnson, 1995; Nielsen and Sano, 1994). Users can benefit from even more informative system responses which tell how much of a document is left to be transferred, how much of a query is left to be executed, how long a file transfer will take to complete, or how long before a query terminates. To provide a user-centered control mechanism, the user should be given the opportunity to cancel an executing query or information access function at any time and suffer no serious consequences from canceling the execution.

54.1.2 Authoring Information

For information to be accessed, it must first be authored by someone. Since the goal of an information author is to create information that other people will access and use, it benefits an author to make data salient to intended users and easy to comprehend and use. Management and distribution of meaningful data require that they be made available in a consistent and reliable format to a wide range of users for multiple purposes. Since the audience is diverse, designing information for the users is much more complex than designing information for a single user or user group. An information accessor can have widely differing educational and cultural backgrounds and technical skills and may have different reasons for accessing the data.

In a traditional database system, authoring data is relatively simple, particularly since the user population and the data are relatively homogeneous. The design of traditional databases is highly structured with database schemas determined by designers before the databases are implemented. Even additions to a database is tightly controlled by the database administrator. Users must have write privileges to enter new records in the database, and most users cannot add or modify the database structure, only add data in prescribed formats using standard data entry forms. Authoring information in distributed open-environment information systems such as the Internet is much more complex. Users can author and present to other users much more than simple text files or database field values. Modern information systems can include scientific data, graphics, video, and digitized sound.

54.2 THE QUERY PARADIGM

Query systems use a relational data model, where data consists of entities connected by relationships, and these entity-relationships are modeled as tables. Query languages have varying degrees of usability. Vassiliou and Jarke (1984) evaluated several query language types based on three interaction parameters: query formulation, language power, and output presentation.

With a graphic query language, such as Shneiderman's (1994) filter flow, the user manipulates visual symbols, which represent entities and relationships in the database, to formulate a query. The restricted natural language allows a user to interact with the database using a subset of native language. Linear keyword languages such as Structured Query Language (SQL) use a limited set of English commands with a definite syntax and make up the majority of query languages in use today. Record-at-a-time logic for data retrieval such as a file transfer protocol (FTP) archive is typified by file management systems. Mathematical query language uses precise mathematical formalisms for short and succinct, but powerful expressions.

Vassiliou and Jarke's (1984) evaluation based on query formulation is shown in Table 54.1. Query formulation describes the overall effort the user employs to work with the system. They also evaluated the query languages based on language power and output

Table 54.1 Taxonomy of Query Languages[a]

	Thinking			ERRORS				TRAINING		
				Probability of Errors						
Query Language	Syntactic Constructs	Model Complexity	INPUT	Clerical	Syntactic	Semantic	Correction Handling	User Type Level	Composition	Comprehension
Function-key use	Low	Low	Low	Low	Low	Low	Low	Low	Low	Med
Menu selection	Low	Low	Low	Low	Low	Low-med	Low-med	Low-med	Low	Med-med
Line-by-line prompt	Med	Low	Med	Med	Low	Low-med	High	Med	Med	Med
Graphic	Med	Med	Low-med	Low-med	Med	Low-med	Med	Med	Med-high	Med
Restricted natural	Low-high	Low	High	High	Low-med	High	High	Low-med	Low-med	Low
Linear keyword	High	Med-high	High	High	Med-high	Med-high	High	Med	Med-high	Low-med
Record at a time	High	High	High	High	High	High	High	High	High	Med
Mathematical	High	High	Med	High	High	Med-high	High	Med	Med-high	High
Optimal	Low	Low	Low	Low	Low	Low	Low	N/A	Low	Low

[a]Modified from Vassiliou and Jarke (1984).

presentation, as shown in Table 54.2. Language power is a representation of how much a user can do with the language and what factors determine this capability. Output presentation, or what the user sees, depends primarily on the system, not on the user.

Function keys, menu-driven dialogues, and line-by-line prompting (also called parameterized interaction) are not commonly used for distributed information systems except for specialized applications such as automated teller machines (ATMs), or information kiosks. Similarly, graphical user interfaces (GUIs) are generally not practical in distributed systems since they are computationally expensive and require high bandwidth. The exception being a client-server model in which most processing is done at the local machine based on some transferred protocol. Linear keywords and mathematical queries demand lots of user experience and skill. A summary rule of thumb might be to use function keys and menu selection for novice users with limited querying options; use line-by-line prompts, graphics, and restricted natural language for moderately skilled users; and use linear keyword, record at a time, and mathematical queries for skilled users.

Most query languages are command driven, and SQL is the industry standard. Several attempts have been made to use other querying strategies such as the filter-flow graphic query builder (Shneiderman, 1994), a menu-driven query builder, a mathematical query builder, and restricted natural language query builders. These strategies have failed to replace command-driven query languages for various reasons. The expressive powers of graphic query builders and menu-driven query builders do not approach what can be achieved with command-driven query languages. A graphic query builder also has significant hardware and software costs. A mathematics-based query language has limited applications and, like formal query languages, requires extensive user training. Finally, natural language processing technology has, on the most part, failed to meet expectations.

Querying is a labor intensive task. People trained to use queries generally have well-defined and immediate information needs. They are also willing to expend quite a bit of energy in formulating their queries and have lengthy discourse with the information system to retrieve specific information items of interest. To support this complex task, a user must know the syntax required to manipulate the system, the limitations of the querying mechanism, and have a reasonable understanding of the information available to the user. Thus, traditional querying systems were primarily of use for people with extensive working knowledge of the information system. Users gained the expertise to work effectively with these querying systems only with training or extensive experience.

The relational structure is naturally suited to a querying paradigm but is too multidimensional (each relationship between two attributes is a dimension) to easily fit within the concept of information space. Users cannot easily build an effective and complete model of a relational structure. In fact, any mental model a user builds of a relational system is incomplete and generally based on table schema. The human cognitive process is not adept at tabular processing, whereas it is adept at spatial processing, an advantage that can be applied to understanding information space such as hierarchical data structures, or simple hyperlinked structures.

The assumption that information users are experienced may be valid in a dedicated database system, however, it is not valid in a public information system such as the Internet. Within this huge, dynamic environment, attempting to understand the information space comprehensively is impossible and, for most cases, may not be necessary, especially if we assume that the results of a query cannot be exhaustive. In a tightly controlled database system, the query results would be exhaustive. The question then becomes, at what level of understanding can users effectively retrieve information using a query paradigm.

Methods are needed to guide users in formulating queries that maximize hits and minimize misses and false alarms. Since misses are failures to identify relevant information, they have adverse impact on system efficiency. False alarms, however, burden a user more by forcing the user to do additional cognitive filtering of the information to decide that the presented information is irrelevant. With false alarms, the user is performing a function that the query mechanism was intended to provide. Excessive false alarms are not acceptable, for they can lead to one of two results: The user may inadvertently accept false data as correct data, or the user may distrust the database. In either case, the effectiveness of the database is compromised.

54.2.1 Command Line Query Languages

Command line query languages have either been formal languages such as SQL or limited natural languages. Formal languages use an explicitly defined dictionary and syntax. Nat-

Table 54.2 Power and Output Presentation on Query Languages[a]

	LANGUAGE POWER				OUTPUT PRESENTATION			
Query Languages	Application Dependence	Database Dependence	Selectivity	Functionality	Control	Variation	Responsiveness	Customization
Function-key Use	High	Low	Low	Low	Low	Low	Low-medium	High
Menu selection	High	Low-medium	Low-medium	Low-medium	Low-medium	Low	High	High
Line-by-line prompt	High	Low-medium	Low-medium	Low-high	Low	Low	Medium-high	Medium-high
Graphic	Low	Medium-high	Medium-high	Low-medium	Low-high	Low-medium	High	Medium
Restricted natural	Medium-high	Medium	Medium	Medium	Low	Low	Low	Medium
Linear keyword	Low	High	Low	High	Medium	Medium	Low	Low-medium
Record at a time	Low	High	Low	High	Medium	Medium	Low	Low
Mathematical	Low	High	Low	High	Medium	Medium	Low	Low
Optimal	Low	Low	High	High	High	High	High	High

[a]Modified from Vassiliou and Jarke, 1984.

ural language systems attempt to resolve commands expressed in common language format.

Formal languages benefit users by requiring them to learn a constrained language with a concise and unambiguous way of communicating with the computer. However, formal languages are not the first choice for use by casual users since they require extensive training and explicit knowledge of the database and are hard to use. Because formal languages are highly constrained, often overly verbose, and complicated, users often make spelling, synonym, syntax, punctuation, and database structure errors.

Natural language systems tend to be more forgiving of command ambiguity. The pitfall with natural language systems is that users often fail to understand the hidden constraints of the natural language dialogue, particularly since the limitations of these systems are not explicitly stated to the user. Several studies (Vassiliou and Jarke, 1984; Shneiderman, 1978) have shown that novice users have trouble with natural language interfaces when they have limited knowledge of the database or its structure, and in general, they generate more invalid queries. Chechile, Fleishman, and Sadoski (1986) argued that what users need is not English-like syntax but uncomplicated syntax. Roop, Eike, and Heasly (1987) found that users did not rate English-like syntax as an important feature of a database, instead, what seemed to be important to users were the size of the command language set. Users preferred a set limited to a few commands, although combinations of these commands could be complex. Malhotra (1985) found that a large portion of natural language queries can be classified into a fairly small number of syntactic types. Of the commands used in a simulated database retrieval application, 78% of sentences could be parsed into 10 sentence types, and 81% of the parsed sentences could be accounted by three of these sentences types.

Whether a formal language or natural language interface is used, the system should minimize the cognitive load on the user. For instance, it should support automatic root form translation (Roop et al., 1987). Translating keywords or syntactical units into their root forms before executing a search allows the user to verify that the query was correctly expressed and reduces a user's cognitive load since the human user does not have to perform the transformation. A variant strategy is to provide automatic substitution of synonyms for keywords. In all cases, the users should have control over whether these aids should be used.

54.2.2 Other Models of Query Formation

In addition to command line query languages, two other query formation methods should at least be considered. These are menu-based query languages and graphics query languages. Intuitively, we might assume that novice users prefer and perform better with context-rich interaction paradigms such as a menu-driven query system, however, that is not necessarily true. Retrieval performance with menu-driven query systems, even for novice users, generally fall below performance with command-driven systems. Experienced users prefer command-driven query systems because of their expressive power and the speed they provide for query generation. With industry standards, such as SQL, an experienced database user can also transfer knowledge, at least about syntax, from one database system to another.

An example of a graphical query language is Shneiderman's (1994) filter flow. Users build a query by selecting filter widgets, such as sequential flow lines indicating ANDs and parallel flow lines indicating ORs, from a set of attributes and combining them to create Boolean queries. Users found this graphical approach of query composition was significantly more effective than SQL for composing and comprehending queries. However, although the filter flow provides full Boolean functionality, the Boolean functions are at too basic a level to provide the power of SQL command sets. Working with Boolean functions, a person designing the query has to specify not only what data need to be accessed, but how the data are accessed. Working with SQL, a person specifies what data are accessed, not how.

For applications in which users are not dedicated full-time database searchers, graphics-based query formation may be effective. A study by Catarci and Santucci (1995), in which they compared user performances with a graphical query language and a text-based language, showed that in general, people were more effective with a graphics-based query language such as Query By Diagram (QBD) than with a text-based language such as SQL. They contend that graphical query languages give users higher levels of abstraction, have simpler syntax with less memory requirements, have shorter interaction time since graphical interaction generally takes less time than writing SQL, and engage a user's interest. They also contend that naive and intermediate users are more effective with graphical queries since they have trouble learning a declarative language such as SQL. Eberts and Bittianda (1993) also found that people had faster performance on complex query tasks using direct manipulation interfaces. They showed that users preferred the direct manipulation mental model of database interaction, particularly if the solution was concrete and had a graphics base for visualization.

54.2.3 Relevance Feedback

Once a user executes a query, he or she must judge the relevance of the query results. Judgment of relevance assumes that a user understands the retrieved object. However, without contextual information, a user may not be able to assess its relevance. Fischer and Stevens (1991) believe that some of the burden of query construction and relevance assessment must be shifted to the system by supplying domain knowledge that supports the cognitive tasks of location and comprehension.

Methods addressing the relevance issue for queries have been explored, most notably, relevance feedback techniques. With relevance feedback, a system rates the results of each query against a criterion and rank orders the returned information objects. Ranking of query output based on some known criterion, such as frequency of keyword occurrences, allows a user to formulate broader queries without having to survey the retrieved information in depth (Roop et al., 1987). It also provides cues useful in determining the success of a query and can increase a user's confidence in the query mechanism. Many WWW query mechanisms, such as Lycos, score search results using the number of hits for specified words and the weights assigned to the found words. The Wide-Area Information Service (WAIS) is another Internet resource that weighs and ranks query output for the user.

A problem with relevance feedback may be undesired generalization, especially for document matching systems. That is, as a user modifies a query by adding more relevant text, the focus of the search may drift or expand to cover areas of information the user does not intend. This places a heavier cognitive load on the user who has to monitor the possibility that the query may not be narrowing the search space but may instead be increasing or shifting it.

Generalization may not always be undesirable. An approach currently under exploration is the use of "approximate answers." The approximation approach drops or relaxes a selection predicate when a user-specified query fails and uses statistical approximations or intentional responses instead of full enumerations when query results are very large. At issue is when generalization should be used, and how results of generalization, particularly if they are based on statistical approximations, should be explained to the user.

With multiple database environments, another problem that may arise is how to merge the results of a distributed search. The results must be merged in order to produce an overall ranking of retrieved items. However, this merging of relevance is difficult since individual rankings may be incompatible in the sense that the numbers used to produce these rankings may not be directly comparable (Croft, 1995).

54.2.4 Query by Reformulation

Simple query techniques effective for feature selection of a small collection of data items do not scale well; that is, they are not sufficient to discriminate data items in a much larger database with thousands of features. People initially think about ideas in terms of prototypical examples as opposed to formal or abstract attributes. Often at the beginning of a search, users don't know how to express what they are looking for. They refine an initial guess by continuously elaborating upon a prototype (Fischer, Henninger, Redmiles, 1991; Rao, Card, Jellinek, Mackinlay, and Robertson, 1992).

Salton and Buckley (1991) found that query by reformulation, a process of feedback and iterative refinement of queries, created significant improvements in search. Although applicable for all class of users, reformulation systems may be particularly beneficial for novice or intermediate users. In fact, in Henninger's (1994) application, users had faster retrieval using reformulation than using simple relevance queries, but only for ill-defined searches initiated from prototypes.

Retrieval by reformulation involves two steps. In the first step, a user builds an initial "best guess" query by specifying keywords or a template document. In the second step, once the system returns the results of the query, the user tells the system which returned items are relevant. With user-provided information and a system-defined algorithm, the system modifies its query parameters and initiates a new query. Each successive query is iteratively refined until the user finds the desired data item (Erickson and Salomon, 1991). The X-WAIS and INQUERY (Broglio, Callan, and Croft, 1995) systems have search engines which implement retrieval by reformulation.

Many query by reformulation systems work on entire documents. Roop et al. (1987) found that users liked using entire documents as queries. This approach may model the traditional method of literature search used by most people. Users find appropriate documents and expand the search by adding the documents to their set or prototypes, and from that point the characteristics of the selected documents are used to retrieve other documents.

The central problems in relevance feedback are selecting "features" such as words or phrases from relevant documents and calculating weights for these features in the context of a new query. These problems are substantially more difficult in environments with large databases of full-text documents. Feedback techniques assume that the searcher would provide a few documents from which to start a search. In many real interactions, however, users specify only a single relevant document, and in some cases the relevant document may not be strongly related to the initial query. Although intended to improve an initial query, feedback techniques often lead users to browse the material using feedback and exploratory information retrieval (Croft, 1995).

Henninger (1994) suggests that intelligent software systems might be able to assist users in better defining feedback queries. He suggests that intelligent reformulations that go beyond simple relevance feedback mechanisms by justifying relevance ratings may assist a user in building a cognitive model of a system's behavior. An intelligent reformulation system thus trains users on how to create effective queries.

54.3 THE BROWSING PARADIGM

Browsing is a more casual search approach than querying and is often practiced in activities such as exploratory learning. It follows a navigation and spatial metaphor and supports an opportunistic model of information access. People who browse data are generally not willing to expend the same amount of resource or time as they would if they were querying for specific information, and they often treat information access as much entertainment as task or work directed. They tend to skim the information they receive and rely on the layout of the information to give them a quick overview.

Traditional information systems, especially text-based systems, do not reflect or support the way people browse information. Most information systems store textual files, with little to help the reader to distinguish one item of information from another. That is not to say that eye-catching flashing icons or graphics have to be displayed on each page of information, for that strategy just introduces distractions that produce their own cognitive loads. What is needed are guidelines for effective use of visual cues that support scanning.

54.3.1 Structure and Browsing

Traditional browsing systems are based on either a linear or a hierarchical structure. A linear structure treats nodes as elements of a single-level list, and to navigate through the linear space, users generally page or scroll. A simple example of a linear structured information is a scrollable text file. Mohageg (1992) found in a test of user search skills that users had lower performance with linear structures than with hierarchical structures. Users had consistently worst performance on linear searches primarily because they limit spatial cues.

A hierarchical structure has two dimensions of navigation: breadth of possible siblings of a node and depth of children lower in the tree. Users navigate from node to node, going up, down, and sideways in the tree. The UNIX file management system typifies a hierarchical information system. Shafrir and Nabkel (1994) suggest that hierarchical structures facilitate search and navigation for three reasons. First, they encourage authors to progressively disclose information. Branches higher in a tree describe general information and group subsequent details of lower level branches and leaf topics. Second, the vertical dimension dramatically expands user's cognitive spatial model (Mohageg, 1992). Finally, hierarchical structures allow users to learn and recall information groups and discrete parent–child relationships for individual topics.

In some cases where a hierarchy is not explicitly defined for an information system, Robertson, Mackinlay, and Card (1991) suggest imposing a hierarchy by creating auxiliary links. The strategy of imposing a hierarchical structure on a nonhierarchical space has been used with several applications. Robertson et al. simplified a network information structure by clipping some links and adding others to form a hierarchy. Some newsgroup browsers take what is ordinarily a flat information space and create a hierarchical information space based on threads and subthreads in the subject heading. Rennison (1994) takes the idea one step further in his Galaxy of News application which uses a user-defined database of relations as a filter to group articles in the news information space into hierarchical subject groups. Even if a hierarchical structure is not explicitly implemented, with some linear information space, a symbolic hierarchical grouping of sections and subsections may be useful even though the symbolic hierarchy does not reflect the actual structure of the document (Shafrir and Nabkel, 1994). For instance, a long Web page document may have up front a table of contents with internal links to specific locations later within the same document.

54.3.2 Navigation Issues

In a location-centric metaphor, such as directories and folders, keeping path history is important and useful for maintaining a user's cognitive map of the information. Orienting users is nontrivial, especially when users have difficulty knowing the boundaries and determining the quantity of online information. Whereas book readers intuitively use the spatial cues of thickness, shape, and layout to locate oneself and to navigate, no similar intuitive cues are available in browsable systems. The cues must be built into the displayed information. Take, for instance, a linear information space such as a text file. A location cue commonly used in graphic interfaces is a draggable icon in a scroll bar on a display window. The position of the scroll bar reflects the "depth" of the current displayed page relative to the entirety of the information space of the document. Although this provides some cues, it is still not as rich as a physical book. A book has thickness, but the length of a scroll bar is a function of the length of a display window, not the length of the information space.

Spatial maps have been suggested as a method of supporting navigation in a complex information space. However, spatial maps lose their cognitive advantages when applied to large and complex data structures. Creation of spatial maps also demands large computing overhead with little benefit to the user, especially if the analogy between the real situation and the modeled situation is not very clear (Webb and Kramer, 1987).

Gentner (1983) theorized in her structure mapping theory of analogy that a good analogy between a database structure and some well-known structure is more important than a spatial map. In her analogy approach, user's knowledge of a base domain anchors an unknown target domain by a statement similar in form to "A (target) is like a (base)." In the process of mapping the analogy, a user gains inferential power useful in manipulating the newly learned system. Webb and Kramer (1987) found that database users retained more knowledge of the database over time when using analogy mapping than when using spatial maps. With an analogy, only the relevant information needed to map the known to the unknown need to be specified, and superficial aspects of the domain which make up much of the information in a spatial map are irrelevant.

54.4 AUTHORING INFORMATION

In modern information systems, authoring involves far more than preparing text. Preparing presentable text is hard enough, but now, authors may have to work with scientific data, graphics, video, and digitized sound. Authors must also deal with the fact that a significant barrier to information access and understanding is the lack of contextual cues in the information system. They must not only design information but also the contextual information that support users in understanding the content or the structure of the information system.

54.4.1 Use of Media

Laurel, Oren, and Don (1990) argued that the greater the variety of form, style, media, and point of view, the greater the potential for a person to explore, penetrate, and understand the topic. Multimedia information provides viewers with different perspectives to the same problem and flexibility in learning the subject material. However, multimedia must be used carefully, with the intended subject and audience in mind. Currently, only preliminary issues have addressed use of different media in information authoring and information access.

Historically, written text has symbolized learning, especially since until recently, only the highly educated and the social elite could read or write. This has resulted in the phenomena that users often perceive text articles as a more reliable source of "truthful" or "academic" information than other media types. In our print-biased culture, a statement such as "the book says" is practically equivalent to "this is true."

Graphics is an information dense medium, a justification often used by people who apply visualization techniques to data (Tufte, 1991). However, although Peters, Yastrop, and Boehm-Davis (1988) found that users were quicker to respond to spatial data, they more accurately responded to textual data. Graphics can also distract, particularly when the graphics are "aesthetical" rich but information poor, in other words, flashy but with little substance.

Graphical data exploit human perceptual ability to perceive and process visual differences. Thus, Shneiderman (1994) suggests that with scientific or inherently graphical data, textural representation should be used as a last resort. When designers cannot identify natural two-dimensional representations of data, they should resort to such displays as scatter plots, or bar charts.

Although video is accepted as a useful medium, there is relatively little experience with video or audio data in information systems. Research is needed to develop guidelines that aid an information author to use them effectively. What is known is that media such as video or audio do not fare as well as traditional text in perceived veracity (Laurel et al., 1990). This may be because their metaphors come primarily from television and motion pictures, both products of entertainment industries. Users also have high expectations of the presentation quality of video and audio, a result of people's exposure to a polished film and recording industry, and an audience may reject an information-rich video if its presentation is poor. Laurel et al. also found that the video medium may even distract an audience from the intended information content.

Digitized data require tremendous bandwidth to transmit, which makes the use of multimedia information problematic for users who access the information through slow data feeds such as phone lines connected to an Internet gateway. Thus, downloading multimedia information should be optional, and never forced on a user. However, this suggests that multimedia information can be used only as supplemental information. Emergence of new technology such as the Virtual Reality Markup Language (VRML), a standard for transmitting three-dimensional graphics and object rendering specification instead of bit-mapped files over the Internet, and high-capacity transmission lines, such as fiber optic cables, may alleviate much of the bandwidth issue.

Integration of multiple media in information systems raises usability issues. One issue is mixing control metaphors across media types. Typically, developers provide different user interface metaphors for different media, for instance a tape player for video, and a scrollable window for text. The advantage of this approach is that each interface serves the exact needs of a particular medium; the disadvantage is a lack of consistency.

Another issue is granularity of media data. For instance, whereas a page can be a meaningful unit of information, it is unclear how to define meaningful units of video or animation data. Instead, a group of related frames, such as a scene or film clip, more commonly make up a unit of video data. What are the meaningful units below this level from which media links can be established? For text these units are words or phrases, for video, they might be objects displayed over a span of frames.

Cataloging media data is also difficult, and methods for indexing multimedia are not well understood. Where they exist, the solutions are of limited utility. An example of this is indexing images by their color distribution. This technique can be used effectively in some applications, such as retrieving pictures of fabric in specified color shades, but is difficult in most other applications, such as indexing entertainment, news, or video clips.

54.4.2 Contextual Information

Users need to understand the database system and its data model in order to understand particular data. Authors can help them understand by enriching the data they create with contextual information. Contextual information can be provided with a metaphoric model that frames the information system, by metadata that describe objects in the database, and by the display of nearest-neighbor data. The first deals with providing the user with an anchor point on which to hang his or her understanding of the information system; the second deals with understanding the nature of the information; and the last deals with understanding the relationship of data to other data. Display of nearest-neighbor data is necessary for understanding trends in scientific data, something scientific visualization is intended to address. We discuss more about scientific visualization in Section 54.7.1.

In a casual use information system, what people know and bring to the interaction must be taken into account and put to work if possible. Metaphors provide a mechanism for leveraging effective conceptual models (Carroll, Mack, and Kellogg, 1988). However, as recent work on interaction icons and metaphors used in a global context have shown, metaphors are often specific to a culture (Nielsen, 1990, 1994; Russo and Boor, 1993).

Information systems have gone beyond the goal of just providing data. It now has the goal of providing expertise and insight, as well as data. According to Matheus, Chan, and Piatetsky-Shapiro (1993), the most important information that experienced users employ when they apply the data found in databases to problem solving is "domain knowledge." Domain knowledge provides users with the knowledge to search the database efficiently and process and interpret the obtained data (Gurd and Jones, 1994; Register and Gerone, 1994) effectively. Information systems must also provide multiple forms of access to the data and tie data to additional information resources such as overviews which describe the origins and contexts of the data, conditions for use, and the availability or distribution of the data.

Domain knowledge and overviews are examples of metadata. With scientific or technical data, meta data should provide information useful for characterizing the collection and analysis of the data and provide background and descriptive information to enhance public understanding of the contexts and benefits of research projects, thereby helping to assure their continued support. Data published on the Internet should provide metadata by embedding links within HTML pages that provide the context for interpreting the data. FTP sites should provide separate explanation files or "readme" files within directories that contain data files identified only by cryptic file names.

Meta data may also include user annotations. Such annotations can serve as memory cues about which aspects of the information is of importance and can also help users search for information (Erickson and Salomon, 1991). Metadata may also be automatically generated. The Indexer robot (Yuwono, Lam, Ying, and Lee, 1995) is an example of an Internet agent that traverses hyperlinks and builds metadata analysis by analyzing keywords in HTML pages, focusing on words in HTML format tags, e-mail addresses, directory paths, file names, and network host names.

54.4.3 Additional Issues in Information Authoring

An author must design not only the information content, but the structure of the information space, particularly for information intended for a browsing system. The author must consider that information may be accessed by a diverse group of people. With the huge difference in capabilities and technical skill levels of users of open distributed information systems, people should have complete control of how detailed or deep they want to get into the information. Thus an author should design information to support multiple paths into and through the information structure and multiple learning models.

Frenkel (1991) suggests that the most daunting problem of integrating heterogeneous databases may not be the diversity of data models, but the different underlying concepts and definitions used to build these distributed databases. Furnas, Landauer, Gomez, and Dumais (1987) showed that people named even common objects very differently. The problem then is a basic, semantic modeling problem of getting the concepts to have the same meaning, and at the same time, retain their original intent. Interdisciplinary and interorganizational attempts to develop common data dictionaries represent efforts to incorporate aspects of information gathering and analysis practices into electronic systems. Register and Gerone (1994) suggest that these efforts at developing common dictionaries extend the consistency that characterizes scientific or disciplinary process models to the handling of digital data and computer technologies.

When authoring information, displaying time information is critical. In a study of Web pages on the Internet, Nielson (1994) found that users liked the use of a "new" indicator to highlight new information in a list. They were irritated by information that was not up to date, but liked information that had dates to indicate how current the information was. Since users also show little tolerance for sites with long transfer time, high bandwidth data should be used sparingly. When a link to long transfer time information is displayed, the information author should provide some cue to prepare the user for a longer transfer time. However, testing is required whether cues such as displaying the length of the information item with the link will overcome a user's impatient tendency to "time-out" an access.

54.5 HYPERLINKS

Introduction of Hypertext in the 1970s introduced a new paradigm of information access based on network structures. Hypertext is a series of documents, each document displaying links to other documents in the set. A user navigates through the set by selecting a link. The link takes the user to the selected document, and from there the user can select other links. A hypertext system follows a network structure metaphor in which information items are nodes in a network space, and hyperlinks are directed links in the network. Originally, hypertext consisted exclusively of linked text documents, but recent innovations have added multimedia information. Thus, the term "hypertext" has given way to "hypermedia" and "hyperlink system." The best known hyperlink system is the WWW which uses the Hypertext Markup Language (HTML). If there is one factor that has led to the explosive popularity of the Internet, it is its incorporation of hyperlinks.

54.5.1 Issues in Navigation

Mohageg (1992) showed that users had better search performance with hierarchical structures than network structures. This may be because a network structure is more complex and allows more paths than a hierarchical structure (Laurel et al., 1990). Users may be reluctant to traverse network links since it is harder to predict where the links will take them, and they often get disoriented from a rapid succession of jumps. Maps of the information space can ameliorate these problems, but maps easily become too complex, and the user must learn the map. Web browsers provide hot lists and global history files as a solution. Hot lists provide user-specified anchor points in the information space that users can always go to if they get lost. Global history files keep a trail of the user's passage through Web pages. With it, a user can go directly to any point on the trail, rather than back tracking link by link.

The highly textual content of most hypertext screens also hinders navigation by making all screens look alike. Although hyperlinked graphics is becoming more common, links embedded in text are typically underlined words or phrases which rely heavily on recognizing typography. However, word links have no inherent dimension (Shafrir and Nabkel, 1994). A Web browser application has no knowledge of remote information structure and cannot provide users with infrastructure navigation cues other than a trace of the hyperlinks they followed. One approach is to embed movement and location cues into the document. With graphics, an information author can provide recognizable metaphoric landmark symbols which visually identify modules, helping users create cognitive spatial maps of the information (Nielson, 1994; Shafrir and Nabkel, 1994).

54.5.2 When the Information Space Is Too Large

In using distributed databases, a general problem is locating the best databases to search in an environment that may have hundreds or even thousands of databases. A network such as the WWW is much too complicated and large for purely a browsing access paradigm. Robots such as Lycos and webcrawlers, which search the distributed databases and keep catalogues of references, have been developed to solve this problem. They periodically update their catalogues, and a user's query searches these databases. A query engine produces a list of links to relevant sites in the information space, and from them, a user can again resume a browsing search. At present, most of these query engines allow the user to construct only primitive AND and OR queries, typically only one of these relationships per query. The user has to develop effective query strategies, with little or no help from the query engines. Although the power and complexity of formal query languages such as SQL may be too much for the majority of inexperienced and casual users of the WWW, the limitations of current search engines indicate the need for better query mechanisms and software that support users in creating better queries. A practical balance needs to be reached in the power and ease of use of these query engines.

54.6 INFORMATION FILTERING AND AGENTS

Potentially relevant information accumulates constantly throughout the Internet, creating a need for new information management methods that go beyond traditional querying and browsing methods. A person may want only a fraction of the information in the news groups, bulletin boards, or messages that he or she receives through list serves or e-mail, but may have to expend significant resources of time and effort to extract the few information items of interest.

54.6.1 Information Filtering

Querying and browsing is an active information seeking process, information filtering is passive. Table 54.3 provides an overview of the differences between information retrieval and filtering. With information retrieval, a mechanism is used to go out and collect rel-

Table 54.3 Differences Between Retrieval and Filtering

Retrieval	Filtering
Active seeking of information	Passive receiving of information
Collects data	Removes unwanted data
One-time query to meet immediate goal	Enduring screening to meet long-term goal
Process data with long life time	Process data with short life time
Proactive, user willing to expend large amounts of time and resources	Casual user unwilling to expend large amounts of time and resources

evant information in the system. With information filtering, a mechanism screens a data stream, either broadcast by a news service or sent by other sources such as e-mail, and only passes information that meets a specified criteria. The Eudora Pro communication software built by Qualcomm is an example of a popular e-mail application that implements a filter. The personal computer application compares e-mail against a user-specified profile of keywords, and either blocks or redirects incoming mail to user-specified mail boxes.

Filters offset the distributed and continuous effort of scanning and classifying information by the massed and perhaps one-time effort of creating a filter profile. Fischer and Stevens (1991) found that many newsgroup readers were interested in reading about more topics, but the large volume of information prohibited them from effectively utilizing the available information. This suggests that users need help finding relevant information and supports the conclusion that the critical resource in information access on the Internet may not be the availability of information but the time and effort necessary to find and utilize that information.

Loeb (1992) defined four major dimensions of filtering applications: user disposition, time scale, information delivery, and information content, that can describe filtering applications scenarios (Table 54.4). Loeb suggests that a primary consideration in designing an effective filtering system is to encourage users to be as specific as possible without limiting them to a system-defined list. Relevance feedback may provide just such a capability, especially if coupled with the idea of a self-adapting profile. A software filter, based on the user's usage patterns, may alleviate the need for precise profile definition at the filter's inception. The filter agent would decide whether to actually deliver an information item which matches a user-specified profile.

The timeliness of filtered data is often of overriding significance. Whereas data on a Web page may stay relevant for months or years, articles posted on a newsgroup or

Table 54.4 Characterization of Filtering Applications[a]

Dimensions	Characteristics
User Disposition	
User type	Proactive or casual
Privacy protection	Protected profiles, usage history, information
Time Scales	
Information lifetime	Minutes (stock market), days (news)
Source availability pattern	Stored or live information
Filter delivery pattern	Continuous, synchronous, or asynchronous
User usage pattern	Single session, irregular, regular, continuous
User feedback delivery mode	Real-time, off-line
User information modes	
Information delivery	
Information media	Media composition, size
Information transport architecture	Broadcast, narrowcast, switched
User equipment	Television, personal computers
Information content	Full-context indexing, descriptors

[a]From Loeb (1992).

bulletin board service may be relevant for only a few days. Data in e-mail may have shorter relevance lives of a few hours. A method is needed to represent temporal constraints, that is, how to understand when a text is likely to be timely for a particular user, and what timeliness means in specific contexts.

54.6.2 Agents

Agents are processes that execute autonomously or semiautonomously, carrying out functions delegated by the user. The idea of agents is to free users from the drudgery of administrative tasks to perform more cognitive demanding tasks, such as information analysis or search strategy formulation. It is important to note that it has been the field of information processing, and most notably information filtering that has been the breeding ground for the concept of software agents (Laurel, 1990; Laurel et al., 1990; Norman, 1994; Recken, 1994; Sheth and Maes, 1993). A few possible applications of specialized agents are background processes that perform knowledge domain mapping to aid the user in navigating through data, synonym vocabulary building to aid users in using command driven systems, and tutoring information retrieval system.

There has been significant usage of the term agents in recent human–computer interaction (HCI) literature, and in many cases, it has been overused. A user-defined static filter by itself is not an agent, although an intelligent, adaptive filter is. Nor is a clever, nonadaptive query mechanism an agent, although one that learns based on a user's information access pattern is. Foner (1995) lists some properties of software agents: autonomy, personalizablity, discourse, cooperation, risk and trust, domain, expectations, and anthropomorphism. Sanchez and Leggett (1994) also discusses an agent's scope.

An agent should have a measure of autonomy from its user so that it can perform an assigned task independent of its user. On the surface, this need for autonomy may conflict with the HCI notion of user control, however, what the user relinquishes with an agent is not control as much as immediate supervision. The user still maintains control over an agent since he or she sets the goals and conditions for its behavior. After setting up an agent, the user is free to pursue other work. The concept of autonomous software can be both liberating and daunting. It is liberating since it frees users from mundane tasks. However, it is frightening since to achieve autonomous software, major issues need to be addressed concerning security, level of control, and how best to convey the necessary system knowledge that users may require to make full use of the autonomy, especially for naive or infrequent users who do not have the interest or the time to cultivate the system knowledge.

Since people don't necessarily do the same tasks in the same way with the same measure of skill, agents must be adaptable to the task and the characteristics of individual users. Since the purpose of an agent is to free the human user's resources, a user should not be required to program the agent explicitly. Maes and Kozierok (1993) showed that agents that learn achieve a level of personalization impossible with knowledge engineering and without direct user intervention. Fully dynamic agents can expand their knowledge by observation or by explanation. An agent that learns by observation infers behavior patterns by sampling user behaviors and formulating predictions about upcoming actions. An agent that learns by explanation uses an analytic method and a domain knowledge base to formulate generalizations from a single example (Sanchez and Leggett, 1994). How an agent infers or analyzes human behavior, what an agent should learn, and how fast the agent should learn are issues that have yet to be satisfactorily addressed, especially since cognitive scientists have yet to answer these questions for human performance.

An agent must also provide a service that the user cannot manage alone or does not have the time to manage alone. In a study that used user–interface agents as guides to historical information, Laurel et al. (1990), found that although users liked an agent's level of engagement and other experiential qualities, they often saw agents as an obstruction rather than as an aid on information access tasks in which they knew exactly what they were looking. This suggests that unless agents can learn and grow with its user, a user may eventually outgrow the need for certain agents.

For an agent to be effective, it must carry on a discourse with the user or another agent. A discourse is a two-way feedback in which both parties make their intentions and abilities known, and negotiate an implicit contract that details responsibilities and goals. It is the foundation for building cooperation between user and software. Proponents of agents suggest that an interaction style based on cooperation leads to a more symbiotic interaction between a user and a computer (Kay, 1990; Laurel, 1990).

An agent can also talk to other software agents, possibly belonging to other users, to learn how to deal with unknown situations (Lashkari, Metral, and Maes, 1994). This cooperation among agents brings up issues of risk and trust among agents. In some cases, an agent may rely on advice from other trusted agents to inform them whether external agents can be trusted. But fundamentally, collaboration among agents makes issues of security and privacy more difficult to resolve.

Agents are ideally suited toward some domain of work, but are poorly suited towards others. Within a social or entertainment context, an agent failure can be recovered from fairly easily. However, failures in safety or precision critical systems, such as a stock market investment software, might result in significant and irrecoverable loss. Domain is closely tied with the notion of risk and trust, and the domain set suitable for agent applications has as yet not been identified.

The effectiveness of agents is founded on realistic expectations between user and agent. A user has expectations of an agent's capabilities and behaviors, and an agent has expectations of its users. If expectations are mismatched, distrust results and a user will be reluctant to delegate to an agent or will spend more time verifying the agent's action. In either case, the effectiveness of an agent is drastically reduced. Expectations must be explicitly developed within the discourse between a user and agents. Some critics claim that the concept of agents leads users to have unrealistic expectations, particularly if human traits are attributed to agents. Thus, agents must be based on appropriate metaphors that convey agent capabilities accurately but that do not create unrealistic expectations.

In some quarters, anthropormorphism has been equated with agents; however, anthropomorphism is not a necessary, let alone sufficient, condition for agency. Negroponte (1990) showed that an agent does not have to be anthropomorphic to be useful. Anthropomorphosizing agents places a high demand on modeling a complex, humanlike personality. Laurel et al. (1990) believe that a notion of dramatic character is both more appropriate and more tractable than human personality as a model for interface agents. A dramatic character can simply be a set of behavior predispositions and perceptible actions and behaviors that allow users to infer those predispositions.

An agent has scope, either local or global. Local agents generally work on behalf of one user and uses individual-specific knowledge. Typical local agents may observe how a particular user uses a system and then provide a user context-sensitive help or orientation. Global interface agents have a system perspective and watch a user community. Although a global agent assists individual users, the type of help it provides and when it assists is based on knowledge of a user class, not knowledge of a specific user.

The question of whether to use local or global scoped agents may be answered by noting that global scoped agents may be inherently server (information provider) side and local scoped agents may be inherently client (information user) side. From an information provider perspective, creating and maintaining unique agents, such as a help agent, for each user that accesses the information is prohibitive. Thus, providing the help agent as local agents on the client side may be more appropriate (Sanchez and Leggett, 1994).

54.6.3 Issues in Using Agents and Filters

The use of filters and agents introduces serious issues of privacy, security, and reliability. A filter defines a user's interests, and as such, reveals a lot about the user, some of which a user may not want known to other users. This has serious consequences for both personal and business privacy. An example where a filter may reveal private information about a person is Loeb's (1992) implementation of mood-sensitive filters for accessing a library of digitized songs. A filter used to screen business information can betray a company's new business strategies or interests.

Associated with the need for privacy is the need for security. Filters have to be secure from unauthorized users, something that may be difficult to ensure in a distributed environment. The security issue is further complicated by the fact that an agent that implements a filter learns by modifying its own knowledge. This has two implications for security. One is that if an external source is aware of the agents learning mechanisms, it can influence how the agent will adapt. The other implication is that anomalous behavior in an adaptive agent may be harder to detect, especially if the user trusts the agent.

In order for a user to trust an agent, it must be reliable. However, software reliability is a serious consideration that has still not been fully addressed. Designing complete reliability is impossible within a dynamic agent paradigm that works within a complex and unpredictable environment such as the Internet. Testing for reliability is equally impossible. Thus, what designers are left with are comfort levels of reliability, but in the context of trust and discourse, what are the comfort levels? From a system standpoint, how does a system detect its own unreliability, and how does it recover from the situation? From a user's perspective, how does a person detect unreliable behavior.

54.7 VISUALIZING INFORMATION

Low-cost displays and high-performance graphics–capable workstations allow a richer, graphic-oriented presentation of large amounts of data instead of forcing a text-based tabular presentation. Gains in computing power have spurred development of new methods of data visualization and highly interactive information access, particularly for graphical data. Emerging technologies for three dimensional visualization and interactive animation offer approaches to displaying complex data structures of large databases, and hold promise for solving the large information space but small viewing window problem.

54.7.1 Scientific Visualization

Leung and Apperley (1994) developed a taxonomy of presentation techniques for large graphical data spaces which is shown in Figure 54.1. According to them, some data are inherently nongraphical, those generally found in the fields and records of relational databases. Others, such as scientific data, lend themselves to graphic presentations. "Nongraphical" data does not mean that data items cannot be represented in an abstract graphical form, just that there are no inherent spatial relationships among the data items. In many cases, visualization can be applied to nongraphical data, although as a mechanism to show the structure in such information as hierarchically organized text files. Graphical data have implicit spatial relationships among its data items, for instance, a topographical map, and lend themselves to visualization techniques.

Visualization techniques make use of the advantages a human visual system has over a computer system in efficiently analyzing midsize amounts of data and recognizing patterns (Keim and Kriegel, 1994). It has as its goal, the effective representation of data such that the human observer can make inferences and deduce relationships about the data.

Scientific visualization is particularly useful for visualizing multivariate information, and generally up to five variables can be easily handled following real-world coding techniques. These coding techniques are the three dimensions that constitute volume, another dimension which can be represented as time, and a fifth dimension that can be represented as color. Although additional coding techniques may be possible, for the most part, having more than five parameters significantly clutters up a display and creates an unintended cognitive load of dealing with the clutter and distractions. In these cases, the obvious solution is to separate multivariate display into several displays, in the process often collapsing several variables into one. Separating displays is difficult for two reasons. From a systematic perspective, it requires that the relationships between the variables be

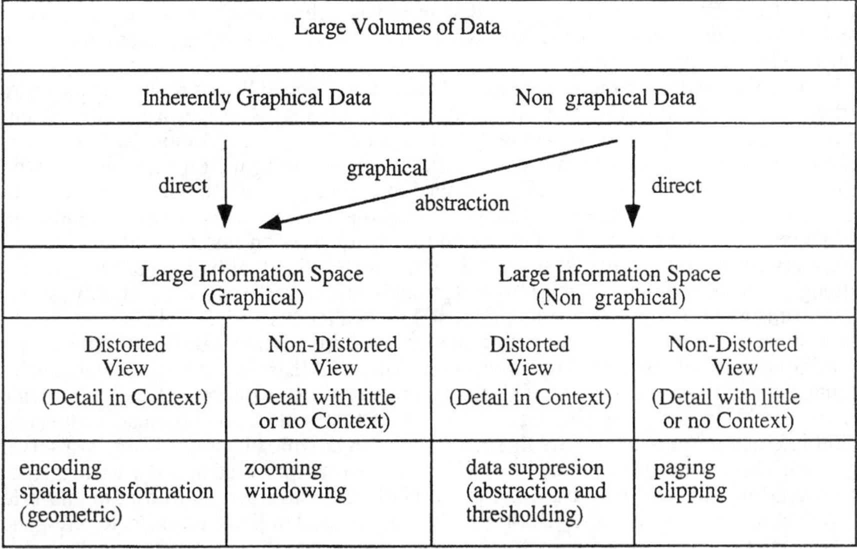

Figure 54.1 Taxonomy of presentation techniques for large graphical data. (From Leung and Apperley, 1994.)

explicitly known, something that is not assured since the purpose of scientific visualization is to identify these relationships. Second, it requires the scientist to cognitively reintegrate separate displays. This latter chore may not be as negative as it sounds. Given control over what variables are collapsed, a user may benefit from the exploratory learning opportunities.

54.7.2 Highly Interactive Querying

Highly interactive systems make use of dynamic mechanisms which apply the principles of direct manipulation to the database: visual presentation of the query components; visual presentation of results; rapid, incremental, and reversible control of the query; selection by pointing; and immediate and continuous feedback. The inherent strengths of highly interactive systems such as direct manipulation systems are ease of query formulation, low training requirements, high system feedback, and gamesmanship. The last is an interaction's capacity to challenge a user and capture the person's interest.

Novices learning to formulate queries in a command language may take several hours with many errors in syntax and semantics. Dynamic queries shorten learning time but often at the risk of increased input error. Most interactive systems make up for the risk of increased errors by providing fast and easy means of correcting for error. Catarci and Santucci found that in some cases, naive users performed as well as intermediate users when using dynamic queries. A limitation, however, of highly interactive systems is that they generally constrain the user in what queries they can form (Catarci and Santucci, 1995), and they place a greater burden on the application designer to develop useful iconic and spatial representations (Vassiliou and Jarke, 1984).

With a dynamic query, a user controls a visual query parameter that generates a rapid, animated visual display of database search results. Search results are continuously and immediately, within 100 msec, updated as users adjust graphical interfaces such as sliders or select buttons to ask simple questions of fact or to find patterns or exceptions. In many cases, manipulation of a parameter may also effect the display or constraints of other parameters. Ahlberg and Truevé (1995) suggest that constraining the meaningful queries and displaying this constraint greatly simplify a user's cognitive load and reduce the chance of getting empty queries, that is queries that generate no hits.

According to Shneiderman (1994), dynamic queries can empower users to perform more complex searches by using visual search strategies. People have a highly developed visual system which they can use to grasp the content of a picture much faster than they can scan and understand text. Interface designers can capitalize on this by shifting some of the cognitive load of information retrieval to the perceptual. In a test of dynamic queries, Shneiderman (1994) found that people could quickly perceive patterns in data and rapidly generate new queries based on what they discovered through incidental learning. He also found that users quickly discovered which sections of a multidimensional search space were densely populated; where there were clusters, gaps, or outliers; and what trends were in the test data.

Although the dynamic query approach is attractive, especially for novice users, current hardware, software systems, and networks cannot support its use in a distributed environment. Widespread use of dynamic queries assume standardized input and output and tool kits that easily integrate into existing database and information systems. Currently, application-specific programming is necessary to capture the advantage of dynamic queries. To be more widely useful, dynamic queries have to be expanded to include more than simple conjunctions or disjunctions and range queries on numeric values.

Immersive virtual reality is at the extreme end of the highly interactive spectrum. Although this technology holds promise for applications requiring high fidelity, such as remote sensing, training, and simulation, it has limited use in distributed information access since the demands on computing power and transmission bandwidth outweigh the potential benefits of total sensory immersion. The VRML is an emerging standard and language for creating desktop virtual reality content on the Internet. Java, (Sun Microsystems, 1995) a scripting language for creating distributed applets, also promises to extend interactive spectrum on the Internet. Users can download applets from Java servers, and these applets or miniapplications will run Java scripts that specify user–interface behavior. Thus, the user can use a mouse to click on a Java scripted desktop and effect an application at a server site. Both VRML and Java standards promise to bring new levels of interactivity to the Internet, an interactivity approaching that of standalone desktop workstations.

54.7.3 Visualizing Structure

Case studies performed by Laurel et al. (1990), Mackinlay, Robertson, and Card (1991), and Johnson and Shneiderman (1991) suggest that spanning properties, such as time and space, often reveal the structure in an underlying document collection. Mackinlay et al. (1991) differentiated between space and time techniques for displaying information. A space technique uses layout and graphic design to pack information into one view, while a time technique uses view transitions to spread information over multiple views. Although the space–time distinction is useful, large information spaces often require simultaneous use of both strategies, which can lead to complex interactions among the techniques (Mackinlay et al., 1991; Rennison, 1994; Robertson et al., 1991). Take a windowing system as an example. The system can use windows as a space technique to group related information. It can also provide some contextual information through the proximity of windows on the display. The system can also use windows as a time technique for switching among views (for example, overlapped windows or foreground/background shading).

Virtual desktops which implement windows quickly become cluttered and unusable during work with large amounts of information. The Rooms window manager (Mackinlay et al., 1991) is a time technique that increases the effective working size of a window system by allowing users to switch among window "working sets" as they switch tasks. Rooms also supports navigation with various techniques including an "Overview Room" that allow users to see and work with the entire workspace.

Laurel et al. (1990) believe users experience information as a series of events unfolding over time. If so, a temporal metaphor such as that embodied by narrative may be a good match for user's experience. When they presented a video with information from a database, Laurel et al. found that users noticed no functional distinction between the video information and the contents of a database. Instead, users perceived the stories presented in a video as part of the flow of information. This collapse of content frame suggests that a temporal metaphor can reduce cognitive load, at least in experiental and browsing activities.

Robertson et al. (1991) illustrated the use of temporal spanning in their implementation of the Cone Tree (Figure 54.2). They used animation to shift the cognitive load to the human perceptual system, engage the user, and help the user understand the information structure. In their application, a user selects new areas of interest by rotating a cone to bring the area of interest into the brightest light and closest angle. A user sees parts of the structure pass by the focus area in the rotation animation of the Cone Tree. Robertson et al. found that without animation, users took seconds or tens of seconds to cognitively reassimilate structural relationships after a tree transformation. Animation reduced the required time to generally less than a second. Rennison (1994) also used animation, in this case zooming and panning in information space, to reduce cognitive load in viewing large news information space.

A three-dimensional representation maximizes the effective use of screen space, and since most hierarchies tend to be broad and shallow, the use of depth cues allows the display of breath without reducing the view of the entire structure. Depth cues may include subtle perceptual tricks such as lighting and angle of view to focus a user's interest. Shadows enhance the perception of depth, as well as focus the user's attention on the more relevant information areas. Although users do not seem to focus directly on the information in shadow, shadowing does appear to help users understand the structure (Robertson et al., 1991). Another method of using a three-dimensional space cue to maximize information presentation in a limited display space is to use translucent display objects (Kramer, 1994). A translucent object provides the user a method of overlaying an information object on a related information object without totally obscuring the object forced into the background.

Robertson et al. (1991) found that since visual processing is heavily dependent on the perception of visual differences, the uniform appearance of substructures in a balanced tree makes perceptual differentiation difficult. This is a prime example where systematic performance is diametrically opposed to usability. Although from a systematic performance perspective, a balanced tree is highly desirable, from a visualization perspective, a balanced tree is highly undesirable since it provides less distinctive perceptual cues than an unbalanced tree.

With a complex hierarchy, a user may often need to hide selected parts to focus on a particular substructure. Therefore, a user should be able to prune or grow views of a

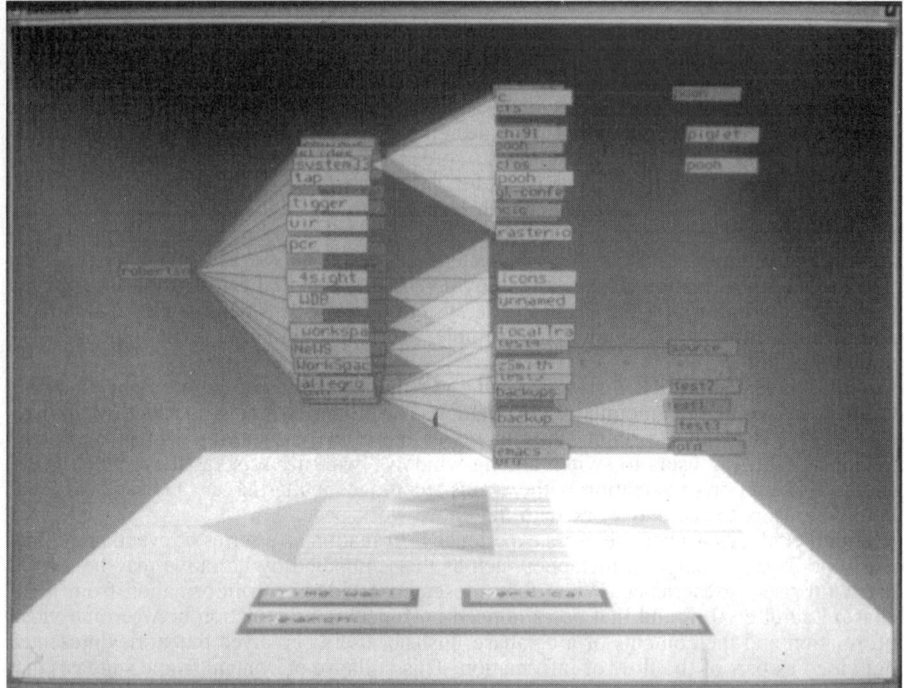

Figure 54.2 Cone Tree for visualizing structure. (From Robertson, Mackinlay, and Card, 1991.)

structure. A simple example is using overlapping windows. A more sophisticated example is demonstrated by the animation in the Cone Tree (Robertson et al., 1991). A user can perform information retrieval by directly selecting a node or executing a query. Since the Cone Tree treats a node and its subnodes as an object, a user can rearrange the structure by grasping nodes and dragging its substructure to another anchor node. A user can also hide subnodes by collapsing the subnodes into a node, making the node look as if it was the leaf of the tree). Other sophisticated examples of using space to combat information overload is the pyramidal encoding in the Galaxy of News (Rennison, 1994), and the geometrically vanishing perspective in the Perspective Wall (Mackinlay et al., 1991).

54.7.4 Views on a Limited Screen

Because information spaces can be large, data often exceed the limited viewing window of a display. Interface designers have to face two problems associated with presentation of complex information in a confined viewing area: the limited spatial area of the viewing window and the suppression of irrelevant data. Having unwanted and unnecessary data on display can distract users and load their processing resources, resources that they could apply to more useful work. However, with limited viewing windows, users have difficulty locating and interpreting specific items of information and relating them in context to other items not seen within the viewing area.

Leung and Apperley (1994) broadly categorized two techniques for accessing large volumes of data in a limited viewing area as distortion oriented and nondistortion oriented. The most familiar nondistortion technique for presenting textual data is scrolling or paging of data, in which a portion of the information is displayed at a time. This technique assumes a flat information space.

The principal obstacles to visualizing linear information structures are the large amount of information that must be displayed and the difficulty of accommodating the extreme aspect ratio of a linear structure on the screen. A common technique for integrating detail and context is to have two simultaneous views: an overview with a scale reduced version of a workspace, and a detailed view into the workspace. Nielson and Sano (1994) found in a usability study of Internet home pages that users consistently praised screens that provided overviews of large information spaces. Another common technique is to divide the total information space into portions which can be displayed, and providing hierarchical access to these pages. As the user moves down a hierarchy, the information becomes more detailed within a smaller area of the information space. Two of these techniques are the two-dimensional Tree Map (Johnson and Shneiderman, 1991), which exploits tree structures in the data, and the three-dimensional Cone Tree (Robertson et al., 1991).

Often the results of a query can be displayed as thumbnail sketches, a reduced view of the item. However, a uniform scale reduction of a workspace may cause it to appear very small, and although suitable for graphic files, thumbnail sketches provide little distinguishing information for text files. Even with graphic files, a thumbnail is generally a reduced view, not a filtered view of the most distinguishing characteristic of the displayed item. Therefore, the distinguishing characteristic may be lost or obscured in the reduction (Rao et al., 1992). Nielson and Sano (1994) found that users complained about thumbnail pictures in which too much detail was provided in too little space, and they appreciated thumbnails which show simple pictures with fewer but more salient details. Another problem with uniform scale reduction is that important contextual information, such as items in the immediate neighborhood of the viewing region, is just as small as unimportant details.

Nondistortion-oriented techniques may be adequate for small text-based applications but they do not provide users with adequate context to support navigation of large-scale information spaces. Instead of a uniform overview of a workspace, an effective strategy may be to distort the view so that details and context are integrated. Distortion techniques make use of the human capacity for selective omission of information and aggregation of experienced information into more abstract forms. This inherent capability makes it possible for the human eye and the other biological systems to process vast amounts of information by smoothly integrating a focused view for the detail with a general unfocused view for the context.

The main feature of a distortion-oriented technique is that it allows the user to examine a local area in greater detail on a section of the screen, but still provides a user with an overall context, generally at reduced magnification (Leung and Apperley, 1994; Rao et al., 1992). The diffused information of contextual data acts as an aid in spatial orientation. The view is generally divided into several distorted regions around a main undistorted viewing region. A mathematical transformation function calculates for each detail in the distorted regions a relevance factor based on the distance between the information item and the point of focus in the structure (Furnas, 1986; Mackinlay et al., 1991). Thresholding is often used with distortion techniques. If the function value of a data item exceeds a certain threshold value, the data item is displayed; otherwise it is suppressed or shown in lesser detail (Furnas, 1986).

A fisheye view (Furnas, 1986; Hollands, Carrey, Mathews, and McCann, 1989; Mitta, 1990; Sarkar and Brown, 1992) is a distortion method that uses functions which determine degrees of interest and match them against thresholds to determine the relevance of displayed information. However, thresholding results in gaps of displayed information. The Bifocal Display (Spence and Apperley, 1982) is a combination of a detailed view flanked by two horizontally distorted views arranged in narrow vertical strips. Because it is two dimensional, a Bifocal Display does not integrate a detail and its context smoothly or intuitively, and it must maintain both the detailed and distorted views of an object.

The Perspective Wall (Figure 54.3) technique (Mackinlay et al., 1991) takes advantage of hardware support for three-dimensional interactive animation to imitate the architecture of the eye. It folds a two-dimensional layout into a three-dimensional wall that smoothly integrates a region of viewing details with perspective regions for viewing context. The distortion provides efficient space utilization and allows smooth transitions of views. A view transition, which is accompanied by animation, is equivalent to scrolling along the perspective wall. This animation provides the user with object constancy, thereby making obvious the relationship between a data object in the detail panel and its context. It also shifts the cognitive load of assimilating the changed view to the perceptual system which can more readily handle the task. To further optimize the use of spatial information in the display, the Perspective Wall also uses the vertical dimension as a means of codifying layers of information.

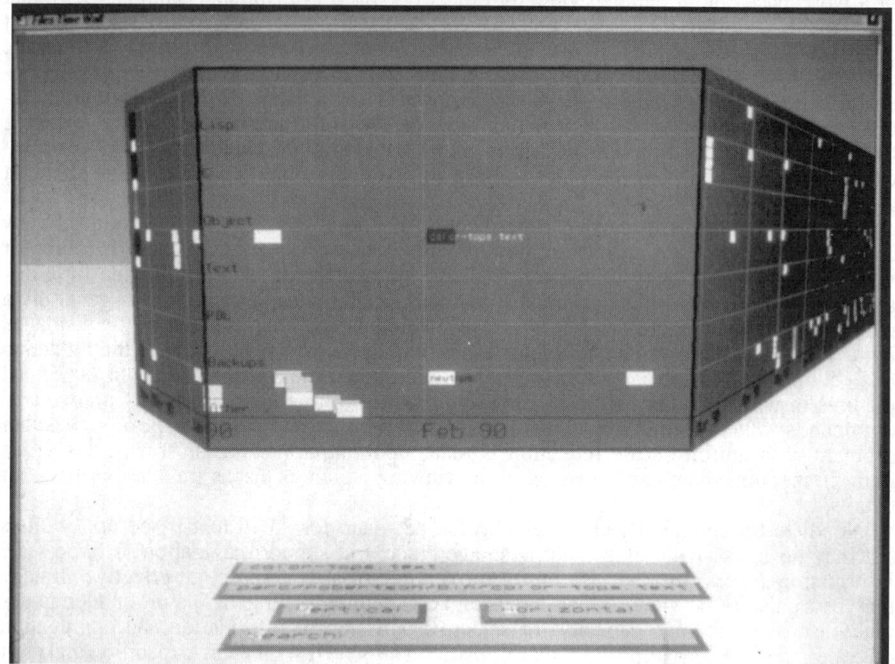

Figure 54.3 Perspective Wall for visualizing information space. (Mackinlay, Robertson, and Card, 1991.)

54.8 ORGANIC INFORMATION SPACE

Distributed information spaces such as the Internet are becoming more and more organic. By organic, we mean that an information space evolves not based on the design of any one designer or group of designers, but as a result of the combined environmental forces and the interactions of all users of the information space. Although nodes in an information space are locally designed, how that node interacts with other nodes in the information space and users that travel the space is unpredictable within the larger context of a global system such as the Internet.

54.8.1 Information Exchange as Computer-Mediated Communication

Computer-mediated communication (CMC) magnifies existing filtering and retrieval problems, as well as introducing new ones. Authors do not create information in isolation, especially in the dynamic environment of the Internet and the current shift in computing into the new paradigm of CMC. The CMC paradigm has bearing on distributed information system since many of the mechanisms for CMC such as newsgroups, bulletin board services, and MUDs fit the definition of information systems. These communication mechanisms and online publication media such as Web pages are converging to support greater communications and user design of the information space. Almeida, Roque, and Figueiredo (1995) argue that CMC-based information systems has greater relevancy for information exchange in the real world than traditional distributed database systems.

Traditional database systems view data themselves as the entire entity, with no explicit consideration of who or what has produced the data or how it is exchanged. A goal of traditional database systems is to make other database users transparent to a user. Consider as an example an airline reservationist who works with a reservation booking database. From the reservationist's perspective, he or she is interacting with only the reservation booking database. However, fundamentally, even a traditional distributed database system is a mechanism for communicating, sharing, and coordinating information among multiple users. For the previous example, the reservationist is in actuality carrying on a dialogue with other remotely located reservationists who are also using the database. Within CMC-based information systems, the user-centric concept of user may no longer be appropriate. User transparency is not a desired characteristic and may actually be detrimental to information processing and understanding. Instead, a less centric term such as "discussant" may better fit and reflect the individual's role in the process of information exchange.

CMC opens up new communication opportunities, but also usability challenges. CMC brings with it the promise of using multiple channels of data exchange, and new models of collaborative work. For instance, video or text conferencing may be used side by side with common access of HTML pages. Some existing MUDs provide users with a common virtual environment for text-based user communication and have, within the virtual environments, objects that access HTML pages. The user-interaction implications of these multiple channels of data exchange has yet to be explored.

Embedded within the CMC data will be the information the user needs. Therefore, a user needs agents or software tools with which to extract the relevant information from the irrelevant data. Archiving all CMC data is prohibitive, especially if the CMC medium is high-bandwidth data, for instance video capture of televideo conferencing. Even archiving e-mail can be prohibitive since it is not uncommon for a person to get over 100 e-mail messages in a day.

54.8.2 Collaborative Design of Information Space

An open-architecture information system such as the Internet is a collective design created by people who use the system. People can design segments of the information space, but the entire information space is organic. It grows as people add new information sites, and contracts as others are removed. For instance, a hobbyist interested in electric trains may put up a Web site containing a collection of photographs and facts about trains. The hobbyist may also decide to put in links to other Web sites created by people with similar interests, or to the Web sites of a transportation museum, a train manufacturing company, a FTP or Gopher site containing archived train-related information, or even a newsgroup dedicated to trains. Every time a user puts up a Web page with links to other sites or creates an Internet node and plugs that node into the Internet, the information space grows.

The organic nature of the information space also extends to information nodes. Some information nodes are not so much user designed as they are group designed. Consider two examples of a group designed information node: a newsgroup and a MUD. A newsgroup may have a theme and original charter, but the contents of that newsgroup is determined by the people who post to that newsgroup. With time, the theme of the news group may drift as the interests of its members change. Newsgroups may also spawn off other newsgroups, and a subset of the user group may find themselves with an interest different from the original theme of the newsgroup and may fission off to create a new one. A MUD has greater initial design than a news group, but is also subject to design by its users. As people manipulate objects within that MUD environment, they create small changes in the virtual commons. They can make larger changes by constructing additional objects or adding new regions within the virtual space of the MUD (Curtis and Nichols, 1993; Meyer, Blair, and Hadar, 1994).

Group design of information space can be an effective mechanism for organizing and processing the huge volume of information within even a local node. The thread is a commonly used mechanism for structuring news articles in a newsgroup information space. Another example of group design of information space is the recent work in collaborative filtering systems (Goldberg, Nichols, Oki, and Terry, 1992; Resnick, Iacovou, Suchak, Bergstrom, and Reidl, 1994). With collaborative filtering, users help one another filter documents by recording their reactions to what they read. A user may use other people's annotations to a document to determine whether the document is of interest, and then, after reading the document, annotate his or her reactions to the document so that others can use that information in their document selection. The product of collaborative filtering is a collective knowledge that reduces the information load on all members of the people who participate.

54.8.3 Constructionist Model for Scientific Databases

Even scientific databases may benefit from a more organic perspective of information space. No database technology currently exists that supports a fluid scientific environment. Instead, the common model for a database is one in which the underlying structure is preconceived and evolves slowly with great difficulty. However, a scientific database cannot be just a snapshot of what everybody thinks today but should be something that will last indefinitely. Scientific databases should support a constructionist process in which the scientist/database user uses the database to model an unknown universe, and in the process, learn more about that universe. Researchers also need support to interact with the data to develop new relationships and insights not explicitly stored in the information system.

REFERENCES

Ahlberg, C., and Truevé, S. (1995). Tight coupling: Guiding user actions in a direct manipulation retrieval system. In M. Kirby, A. Dix, and J. Finley (Eds.), *Proceedings of the Human-Computer Interaction '95 conference*) (pp. 305–321). New York: ACM Press.

Almeida, A., Roque, L., and Figueiredo, A. (1995). Cyberspace: The HCI frontier? In *Proceedings of the Human-Computer Interaction '95 conference* (pp. 51–62).

Broglio, J., Callan, J. P., and Croft, W. B. (1995). An overview of he INQUERY system as used by the TIPSTER project. At http://ciir.cs.umass.edu/info/inquery.html

Carroll, J., Mack, R., and Kellogg, W. (1988). Interface metaphors and user interface design. In M. Helander (Ed.), *Handbook of human computer interaction* (2nd ed., pp. 67–85).

Catarci, T., and Santucci, G. (1995). Are visual query languages easier to use than traditional ones? An experimental proof. In M. Kirby, A. Dix, and J. Finley (Eds.), *Proceedings of the HCI '95 conference* (pp. 323–338).

Croft, W. B. (1995). What do people want from information retrieval? In *Digital-Library Magazine* or at http://www.dlib.org/dlib/november95/11/croft.html

Croft, W. B., and Das, R. (1990). Experiments with query acquisition and use in document retrieval systems. In *SIGIR '90, proceedings of the thirteenth annual ACM SIGIR conference on research and development in information retrieval* (pp. 349–368). New York: ACM Press.

Curtis, P., and Nichols, D. (193). MUDs grow up. Paper presented at the Third International Conference on Cyberspace, May 1993. At ftp://parcftp/xerox.com/pub/MOO/papers/MUDsGrowUp.txt

Eberts, R., and Bittianda, K. (1993). Preferred mental models for direct manipulation and command-based interfaces. *International Journal of Man-Machine Studies, 38,* 769–785.

Erickson, T., and Salomon, G. (1991). Designing a desktop information system: Observations and issues. In *Proceedings of the CHI '91 conference on human factors in computing systems* (pp. 49–54).

Fischer, G., and Stevens, C. (1991). Information access in complex, poorly structured information space. In *CHI '91 Proceedings* (pp. 63–70).

Frenkel, K. A. (1991). The human genome project and informatics. *Communications of the ACM, 34*(11), 41–51.

Furnas, G. (1986). Generalized fisheye views. In *Proceedings of the CHI '86 conference on human factors in computing systems,* New York: Association for Computing Machinery. (pp. 16–23).

Furnas, G., Landauer, T., Gomez, L., and Dumais, S. (1987). The vocabulary problem in human-system communication. *Communications of the Association for Computing Machinery, 30*(11), 964–971.

Goldberg, D., Nichols, D., Oki, B. M., and Terry, D. (1992). Using collaborative filtering to weave an information tapestry. *Communications of the Association for Computing Machinery, 35*(12), 61–70.

Gurd, J. R., and Jones, C. B. (1994). The global-yet-personal information system. In Ian Wand and Robin Milner (Eds.), *Computing tomorrow.* Cambridge: Cambridge University Press.

Hollands, J., Carey, T., Matthews, M., and McCann, C. (1989). Presenting a graphical network: A comparison of performance using fisheye and scrolling views. In G. Salvendy and M. Smith (Eds.), *Designing and using human-computer interfaces and knowledge based systems* (pp. 313–320) Amsterdam: Elsevier.

Johnson, B., and Shneiderman, B. (1991). Tree maps: A space-filling approach to the visualization of hierarchical information structures. In *Proceedings of the 2nd international IEEE visualization conference* (pp. 284–291) New York: IEEE.

Johnson, C. (1995). Time and the web: Representing and reasoning about temporal properties of interaction with distributed systems. *Proceedings of the HCI '95 Conference* (pp. 39–50).

Keim, D., and Kriegel, H. (1994). VisDB: database exploration using multidimensional visualization. *IEEE Computer Graphics and Applications, 14*(5) 40–47.

Kramer, A. (1994). Translucent patches: Dissolving windows, In *Proceedings of the seventh annual symposium on user interface software and technology* (pp. 121–130).

Kuhmann, W. (1989). The stress inducing properties of system response times. *Ergonomics, 32*(3), 271–280.

Laurel, B. (1990). Interface agents. In B. Laurel (Ed.), *The art of human-computer interface design* (pp. 355–366).

Laurel, B., Oren, T., and Don, A. (1990). Issues in multimedia interface design: Media integration and interface agents. In *Proceedings of the CHI '90 conference on human factors in computing systems* (pp. 133–139).

Leung, Y. K., and Apperley, M. D. (1994). A review and taxonomy of distortion-oriented presentation techniques. *ACM Transactions on Computer-Human Interaction, 1*(2), 126–160.

Loeb, S. (1992). Architecting personalized delivery of multimedia information. *Communications of the ACM, 35*(12), 39–48.

Mackinlay, J. D., Robertson, G. G., and Card, S. K. (1991). The perspective wall: Detail and context smoothly integrated. In *Proceedings of the CHI '91 conference on human factors in computing systems* (pp. 55–62).

Malhotra, A. (1985). *Design criteria for a knowledge-based English language system for a management: An experimental analysis* (Project MAC Report TR-146). Cambridge, MA: MIT Press.

Meyer, T., Blair, B., and Hadar, S. (1994). *A MOO-based collaborative hypermedia system for WWW.* At http://www.cs.brown.edu/people/twm/wwwmoo.html

Mohageg, M. F. (1992). The influence of hypertext linking structures on the efficiency of information retrieval. *Human Factors, 34*(3), 351–367.

Norman, D. (1994). How might people interact with agents. *Communications of the ACM, 37*(7), 68–71.

Nielsen, J. (1990). *Designing user interfaces for international use.* Amsterdam: Elsevier Science Publishers (North-Holland).

Nielsen, J., and Sano, D. (1994). SunWeb: User interface design for Sun Microsystem's internal web. At http://www.sun.com/sun-on-net/uidesign/sunweb/sunweb.html

Pejtersen, A. (1989). A library system for information retrieval based on a cognitive task analysis and supported by an icon based interface. In *SIGIR '89, proceedings of the twelfth annual ACM SIGIR conference on research and development in information retrieval* (pp. 40–47).

Rao, R., Card, S., Jellinek, H., Mackinlay, J., and Robertson, G. (1992). A framework for information retrieval and retrieval-centered applications. In *Proceedings of the fifth annual symposium on user interface software and technology* (pp. 23–32).

Recken, D. (1994). A conversation with Marvin Minsky about agents. *Communications of the ACM, 37*(7), 23–29.

Rennison, E. (1994). Galaxy of news: An approach to visualizing and understanding expansive news landscapes. In *Proceedings of the seventh annual symposium on User Interface Software and Technology* (pp. 3–12).

Resnick, P., Iacovou, N., Suchak, M., Borgstrom, P., and Riedl, J. (1994). GroupLens: An open architecture for collaborative filtering of Netnews. In *Proceedings of the ACM conference on computer supported cooperative work* (pp. 175–186).

Robertson, G. G., Mackinlay, J. D., and Card, S. K. (1991). Cone Trees: Animated 3D visualizations of hierarchical informaton. In *Proceedings of the CHI '91 conference on human factors in computing systems* (pp. 189–194).

Roop, E. A., Eike, S. R., and Heasly, C. C. (1987). User-computer interface requirements for remote access data bases. *Proceedings of the Human Factors Society—31st annual meeting* (pp. 973–977).

Russo, P., and Boor, S. (1993). How fluent is your interface? Designing for international users. In *Proceedings of ACM INTERCHI '93 conference on human factors in computing systems* (pp. 342–347).

Salton, G., and Buckley, C. (1991). Improving retrieval performance by relevance feedback. *Journal of the American Society for Information Science, 41*(4), 288–297.

Sanchez, J. A., and Leggett, J. J. (1994). Hyperactive: Extending an open hypermedia architecture to support agency. *ACM Transactions on Computer-Human Interaction, 1*(14), 357–382.

Shneiderman, B. (1978). Improving the human factors aspect of database interactions. *ACM Transactions on Database Systems, 3*(4), 417–439.

Shneiderman, B. (1994). Dynamic queries for visual information seeking. *IEEE Software,* 70–77.

Shafrir, E., and Nabkel, J. (1994). Visual access to hyper-information: Using metaphors with graphic affordance. ACM SIGCHI '94 Interactive Poster, 1994.

Sheth, B., and Maes, P. (1993). Evolving agents for personalized information retrieval. In *Proceedings of the 9th IEEE conference on artificial intelligence and applications* Orlando, FL (pp. 1–11) New York: IEEE.

Sun Microsystems, Inc. (1995). Hot Java. At http://java.sun.com/.Sun Microsystems Inc., Mountain View, CA.

Vassiliou, Y., and Jarke, M. (1984). Query languages—a taxonomy. In Y. Vassilou (Ed.), *Human factors and interactive systems* (pp. 47–82).

Webb, J. M., and Kramer, A. F. (1987). Learning hierarchical menu systems: A comparative investigation of analogical and pictoral formats. In *Proceedings of the Human Factors Society—31st annual meeting* (pp. 978–982).

Yuwono, B., Lam, S. L. Y., Ying, J. H., and Lee, D. L. (1995). A World Wide Web resource discovery system. *Fourth international World Wide Web conference.* At http://www.w3.org/pub/Conferences/WWW4/Papers/66/

CHAPTER 55
MULTIMEDIA

Mark Chignell
Department of Mechanical and Industrial Engineering
University of Toronto
Toronto, Ontario, M5S1A4 Canada

John Waterworth
Department of Informatiks
Umeå University
Umeå, 901 87 Sweden

55.1 INTRODUCTION

Modern computers began as large calculating machines, in spite of much broader early concepts of general computing (e.g., McCorduck, 1979). Computers continued to serve mainly as calculating and simple text processing machines until the early 1980s. However, experimental models and prototypes developed at SRI in the 1960s, and Xerox Parc in the 1970s were pointing toward a new kind of computer system that could also handle images and other media. The successor to these experiments was the Apple Macintosh, released in 1984. The Macintosh brought the graphical user interface and the desktop metaphor originally developed at Xerox Parc into widespread usage. The graphical user interface subsequently became standard across all major platforms (e.g., as X-Windows in Unix, and Windows on the PC platform).

The graphical user interface used the principle of direct manipulation (Shneiderman, 1983) to allow the person to operate on visual objects to perform tasks instead of typing out command strings. At the same time that the user interface was becoming media rich, computers and media were converging. As computing power on the desktop increased throughout the 1980s, more and more media elements entered mainstream computing, first using analogue media such as laser disks, and later using digital coding of media into binary data that the computer could process directly (e.g., on CD-ROM).

The convergence of computing and media presentation has led to "multimedia", a catchall term that covers the vast domain of computer-controlled presentations using a variety of media. In our view, multimedia represents much of the future of both computing and electronic publishing. Fundamentally, multimedia is concerned with communication, that most pervasive of human activities.

In this chapter we survey the field of multimedia from a human factors and ergonomics perspective. We focus in particular on issues of multimedia development, interface design, and ergonomic guidelines for multimedia design. The results and recommendations provided in this chapter are necessarily tentative, given that multimedia technology is still in its infancy.

This chapter will emphasize the document model of multimedia, which is becoming the dominant approach in a variety of task domains, including computer-based training, electronic publishing, online documentation, digital advertising, and communication in general. The chapter begins by highlighting multimedia and its use. More extensive discussions of multimedia theory may be found in Waterworth (1992) and in Waterworth and Chignell (1996).

After setting the theoretical context, we describe the multimedia life cycle and approaches to multimedia authoring. Later sections of the chapter deal with the related issues of design and evaluation. We present several ergonomic strategies for multimedia design and outline the main approaches to evaluating multimedia in terms of utility, integrity, usability, and aesthetics. This is followed by an account of design guidelines. We then outline several unresolved problems in the development of ergonomically sound multimedia and indicate the research challenges that must be met to address these issues. We conclude with summary remarks and pointers toward the future of multimedia interaction design.

55.2 MULTIMEDIA AND ITS USE

Multimedia began with the vision that information, in all its diverse forms, could be accessed in a flexible and interactive way from a common user interface. Perhaps the earliest coherent expression of this vision was Bush's classic paper on the Memex, published in 1945. Vannevar Bush served as Director of the Office of Scientific Research and Development, coordinating the activities of American scientists in World War II. Near the end of the war, Bush urged scientists to make the world's huge and growing store of knowledge more accessible and usable. Bush argued that earlier methods for transmitting and reviewing information had become outdated. As an example of this, he described how the concepts of Mendelian genetics lay buried in an obscure publication for years until they were rediscovered.

In the following excerpt, Bush proposed a mechanism (now referred to as Hypertext, or Hyperlinking) for loosely organizing heterogeneous information in a flexible fashion.

> *Our ineptitude in getting at the record is largely caused by the artificiality of systems of indexing. When data of any sort are placed in storage, they are filed alphabetically or numerically, and information is found (when it is) by tracing it down from subclass to subclass. It can be in only one place, unless duplicates are used; one has to have rules as to which path will locate it, and the rules are cumbersome. Having found one item, moreover, one has to emerge from the system and re-enter on a new path. The human mind does not work that way. It operates by association. With one item in its grasp, it snaps instantly to the next that is suggested by the association of thoughts, in accordance with some intricate web of trails carried by the cells of the brain. It has other characteristics, of course; trails that are not frequently followed are prone to fade. . . . (Bush, 1945, p. 106).*

Bush went on to outline the associative network (hypertext) of information in more detail, and proposed an implementation using the technologies available at that time (e.g., microfiche). The interactivity that Bush envisaged, plus the heterogeneity of data, and the nonlinear forms of access constitute major components of multimedia as it is recognized today.

55.2.1 What Is Multimedia?

Elsewhere (Waterworth and Chignell, 1996) we have defined multimedia as computer-controlled interactive presentations. This definition is extremely broad. However, multimedia also involves the notion that the presentations are designed, crafted, or authored in some way. Even in those cases where multimedia is constructed dynamically (on the fly) from available databases (e.g., by using templates and heuristics) choices must still be made in terms of how to present and coordinate multimedia elements within the user interface.

The almost ubiquitous use of the term "multimedia" for media-rich interactive computing obscures the fact that "multimedia" refers to a broad range of computer applications or information presentations. The essence of multimedia is interactivity, and the skillful use of different modalities and forms of presentation to communicate essential ideas and concepts. Thus multimedia necessarily includes the use of multiple media, yet accidental or inappropriate mixtures involving multiple media should not be considered multimedia. For instance, sequences of readouts from a SQL database or a spreadsheet are not multimedia unless they are coordinated effectively within an interactive presentation. In our view, the multiplicity of media is only one of a number of factors in the definition of what multimedia is. In addition, the information within multimedia should be coherent, the user interface should be interactive, and the overall effect should be pleasing to the senses, and should promote, rather than obscure, the assimilation of information.

From the user interface and ergonomic perspective, multimedia is a conceptual structure that may be used to navigate among information. Multimedia also acts as a kind of visual menuing system (cf. Shneiderman, 1987), and as a bridge (or interface glue) between different media, database environments, and text environments. In addition, multimedia can provide a structural (often spatial) metaphor within which information can be embedded. Thus multimedia is a broad topic that shares, with general user interface design, issues such as navigation, information structuring, and information metaphors.

55.2.2 Three Types of Multimedia

There are three main approaches to multimedia, which are referred to below as performance multimedia; presentation multimedia; and document multimedia. Each of these approaches to multimedia is based on a different metaphor, as summarized in Table 55.1.

Table 55.1 Three Metaphors of Multimedia

Multimedia Type	Metaphor
Performance	A film or stage play
Presentation	A slide show
Document	A book

These different approaches to multimedia can be distinguished in terms of both their goals and the methods that they use.

Performance multimedia is concerned with creating a sensory experience through the orchestration of various actors and events. This is multimedia presented like a stage play, and so it is not surprising that one of the principal authoring tools for this type of multimedia (Macromedia Director) used the metaphor of a stage with actors on it. Performance multimedia has been used frequently in the area of "edutainment," i.e., in applications that straddle the boundary between education and entertainment. Examples of such applications are travelogues, e.g., the walking tour of Amsterdam constructed by Hardman, van Rossum, and Bunkman (1993) to demonstrate their structured approach to authoring of performance multimedia. In performance multimedia, timing is very important. Media elements are synchronized, and the playing of media elements must be coordinated with user actions.

Presentation multimedia is an extension of the traditional business presentation which was formerly conducted as a slide show with 35-mm slides or overheads, and more recently as a computer presentation using tools such as PowerPoint and Persuasion. In presentation multimedia, the linear sequence of slides can be supplemented both by the addition of multimedia elements (e.g., video clips or animations) to slides, and by creating links between slides. Since presentations are usually fairly focused, the branching structure created by links between slides tends to be simple, consisting of short detours from the main backbone or sequence of the presentation. The distinguishing characteristic of presentation multimedia is the existence of a main path, or core set of ideas, that form the focus and intention of the presentation. The goal of the presentation is to communicate these core ideas to the user or reader.

In performance multimedia, the sensory experiences associated with media elements tend to dominate. In presentation multimedia, the focus is on a particular set of ideas organized into an approximately linear structure. These ideas are typically represented by "bullets" (i.e., brief textual statements) and are supported by media elements (often charts and graphs in the case of business presentations).

The third type of multimedia is based on the metaphor of the book and is termed here "document multimedia." Document multimedia focuses on text and ideas, which are typically discussed in more detail than they are in performance and presentation multimedia. The media elements enrich the text, but they do not completely replace it. Sometimes, the text forms a narrative for the media elements, whereas in other cases the media elements embellish the ideas discussed in the text. Thus document multimedia consists of pages or nodes that are organized around a textual framework. The result is a multimedia document, which may not look like a conventional document, but which still retains many of the familiar properties of documents (including the ability to index and search based on text content). Text continues to organize content within multimedia documents. These documents also contain media-rich presentations (e.g., videos, animations, etc.) that are annotated in some way by the text.

Document multimedia was neglected to a large extent in the early days of multimedia. Performance multimedia seemed to be more innovative technologically and thus attracted more attention, both in research and practice. However, a number of researchers pointed out the advantages of linear documents over nonlinear hypertext and multimedia (e.g., McKnight, Dillon, and Richardson, 1991), and others demonstrated that document-styled applications could be combined with hypertext (e.g., Remde, Gomez, and Landauer, 1987; Valdez and Chignell, 1992). There were also problems with early performance multimedia applications because their media displays were not always well integrated. For instance, there might be relatively little text, but a lot of images, sounds, and videos hyperlinked together. In such a situation it was easy for users to get disoriented and lose sense of the overall structure of the application. Experience with these media-rich, but text-deficient applications demonstrated to some, at least, a need for textual frameworks to organize and annotate multimedia.

By the early 1990s it was clear that there were compelling theoretical and practical reasons to develop a document model of multimedia that would supplement the performance and presentation approaches. The development of HTML (the hypertext markup language) and the rise of the World Wide Web (or simply Web) to prominence in 1993 gave a huge boost to document multimedia. Web documents were a combination of text, pictures, sound, and video that could be served across the Internet anywhere that had Internet and Web connections. Web documents could be linked to other Web documents around the world (using http, the hypertext transfer protocol), creating a global hypertext, one that was at least a rough approximation of the world brain idea envisaged by H.G. Wells 60 years earlier.

By 1995, "surfing the Net," i.e., navigating through the Web, had become a pastime for millions of people. In fact, although not generally acknowledged, the Web had become the preeminent publishing medium for multimedia. This meant that the majority of multimedia applications actually being read and used were multimedia documents. Thus, by the mid 1990s, a stage had been reached where performance multimedia was being distributed on CD-ROM, document multimedia was being distributed on the Web, and multimedia presentations were being presented directly to audiences, typically using a notebook computer connected to a large-screen presentation device.

The current trend is for each type of multimedia to invade alternative distribution channels. For instance, Hot Java is a tool for distributing orchestrated multimedia across the Web. VRML (the virtual reality markup language) is an alternative to HTML that also enables a more performance-oriented version of multimedia on the Web. Meanwhile, document multimedia is increasingly being distributed on CD-ROM, in the form of electronic books and plays (e.g., Shakespeare's *Macbeth* distributed by the Voyager company, and children's books distributed by companies such as Discis).

The invasion of HTML from the Web to the CD-ROM publishing medium is also exemplified by CD-ROMs that contain a collection of HTML documents and a Web browser. Two early such CD-ROMs were the 1994 collection of *BYTE* magazine articles issued in 1995, and the Proceedings of the CHI '95 conference (the Association for Computing Machinery's, or ACM's annual conference on human-computer interaction). As corporations become more aware of the advantages of multimedia on the Internet and on intranets (equivalent networks within companies and organizations), multimedia presentations also began to be delivered more frequently across networks to spatially (and temporally) distributed audiences.

The document model of multimedia is preeminent on the Internet and Web, and is becoming an established alternative for multimedia on CD-ROM. This has made it possible for multimedia to be published simultaneously on the Web and on CD-ROM without significant additional authoring effort in porting the multimedia document from one distribution medium to the other. An advantage of the HTML approach to document multimedia authoring is that HTML is platform independent so that Unix machines, Macs, PCs, etc., can read the multimedia, providing that they have an appropriate HTML browser. Various "helper" applications are also needed to play the sound, video, and other dynamic presentations inside, or referenced by, the multimedia documents. It is our expectation that the document model of multimedia will continue to grow in importance, and that it will subsume much of the role of information dissemination currently assigned to printed texts.

55.2.3 Why Is Multimedia Important?

Multimedia is important because, with the related technologies of cyberspace (global information networking) and virtual reality, it represents much of the future of computing, and of human communicating, collaboration, and information exploration. The waning years of the twentieth century have seen an unprecedented migration of information storage and presentation from print to electronic (and computer-controlled) media. This is happening because of lower costs (including storage and shipping), less pollution (much of the material dumped in landfills has been printed matter), and the greater functionality provided by electronic information presentation (e.g., it is easier to copy, can be distributed almost instantaneously over global networks, and can be "hyperlinked" to other information and to active data across live feeds).

Multimedia is an embryonic technology for communication using electronic information in domains such as education, entertainment, and business, or other applications.

However, multimedia is evolving rapidly and is not a passing fad. It will likely be the dominant method of experiencing, organizing, and communicating information in the twenty-first century.

55.2.4 When Should Multimedia Be Used?

Our view of multimedia is extremely broad. Multimedia, in the literal sense of the term, means the use of multiple media. Thus multimedia presentations can be rich sensory displays. Multimedia also implies integrated information, but this integration works best when there is an overall coherence to the organization and structure of the information. Incoherent multimedia is like a collection of statistical tables and graphs that have no explanatory text and no obvious sequencing.

Multimedia should be used whenever rich media, hyperlinking, and interactivity can add significant value to an information presentation. For instance, multimedia can bring a Shakespearean play to life, while annotating the presentation with critical information that might be suitable for an essay on Shakespearean drama. The multimedia version of *Macbeth* (Voyager, 1994) contains the complete text of the play, an audio performance of the play by the Royal Shakespeare Company, video excerpts from films of the play directed by prominent filmmakers (Kurosawa, Polanski, and Welles), a picture gallery of relevant images (including maps of Scotland and Shakespeare's London), and various summaries and commentaries. In this application, several media are exploited to enrich the presentation, and yet the core of the application and its main focus is still the text. One interesting part of the multimedia version of *Macbeth* is a kind of theatrical karaoke, where the user can read the part of one of the characters, while the others are played by members of the Royal Shakespeare Company.

The example above shows how multimedia can enrich a play (surprisingly, perhaps, the result in this case is more of a multimedia document than a performance multimedia application). With imagination, multimedia can enrich almost any information presentation. Thus perhaps the question in the title of this section should have been "When should multimedia not be used?" There are at least two answers to this question:

1. When the multimedia content is superfluous or distracting
2. When the effort expended in creating multimedia add-ons could be more profitably focused on development of basic information content

55.2.5 Ergonomics and Multimedia

Multimedia has typically been viewed as a combination of display technologies. However, for the ergonomist it may be more usefully viewed as a user interface. From this perspective, the problems that attracted the attention of multimedia technologists early on (e.g., compression, multimedia databases, multimedia indexing, and content retrieval) can be fixed with "technological plumbing." The enduring problems in multimedia are defined by the limitations and characteristics of human intellect, skills, and preferences.

The ergonomics of reading or browsing multimedia can be divided into micro and macro aspects. At the micro level, the concern is with individual presentations consisting of screenfuls of information. At the macro level the concern is with the connectedness, coherence, and navigability of a set of individual presentations that jointly constitute the multimedia application or document. General principles of ergonomic screen design apply at the multimedia micro level, i.e., at the level of individual screens. These include highlighting of important information, organization of blocks of information within screen layouts, and use of appropriate fonts, font sizes, and colors. Most micro level issues in multimedia apply to screen and layout design in human–computer interaction, and are covered elsewhere in the ergonomics literature, and in this *Handbook*.

A major macrolevel issue in multimedia is the degree of nonlinearity. The nonlinearity of some multimedia applications makes it difficult for people to find their way around them (e.g., Nielsen, 1990a). Although long-term memory is associative and somewhat nonlinear, learning seems to be most efficient for material that is structured linearly (cf. McKnight et al., 1991). One can get a sense of the need for linearity by navigating the World Wide Web. Clicking from one document to another on the Web is easy and intriguing, but often one gets a sense of not knowing where one is going, and of being disoriented. This may be why a number of alternative interfaces for navigation on the Web have developed, including hotlists that may either be stored locally on the client machine, or added to pages on the Web to help orient people visiting those pages and to suggest other interesting places to visit. Search engines have also become popular on the Web, with prominent engines in early 1996 including Lycos, Webcrawler, and Alta Vista. Another type of navigation interface is the hierarchical menu or catalogue, typified by Yahoo, an online catalogue of the Web originally developed at Stanford University.

At the micro level, one of the salient characteristics of multimedia is that it provides multisensory output (and may accept multimodal input). Thus at the level of the individual information node or screen, a rich sensory experience can be used to convey information. This appears to increase the motivation and interest of the reader or viewer, but also allows for multiple channels of information to reinforce the message, and for integrated information presentation (Wickens, 1992).

The rich sensory experience of multimedia represents a considerable challenge to ergonomists and psychologists, who have tended to focus either on the details of screen design (e.g., color, Murch, 1984; complexity: Tullis, 1984; legibility: Jorna and Snyder, 1991) or on cognitive aspects of information presentation (e.g., Norman, 1986). However, one area of ergonomics that seems well suited to the analysis and improvement of multimedia is that of ecological interface design (EID) that promotes perceptual pick up of information (Vicente and Rasmussen, 1992). Ecological displays are carefully crafted to convey goal-relevant information in an easily assimilated perceptual form, based on ideas concerning perceptual invariants and affordances originally developed by J. J. Gibson (1966, 1979). For instance, complex interactions between different parameters and processes in a nuclear powerplant can be captured in a relatively simple graphical display where key information and trends are shown as emergent properties of the display (e.g., Dinadis and Vicente, in press). EID has been used by Toshiba as the basis for designing its advanced control room for a next generation boiling water reactor plant (Monta et al., 1991).

Whereas multimedia will benefit from ergonomic input, ergonomics itself can also use multimedia to its advantage. The techniques of information design and media presentation utilized in multimedia can be applied in areas such as display design. For instance, techniques for highlighting different portions of a screen might prove useful in the design of a large overview display for process control. Alternatively, the highly engaging and interactive style of communication developed for multimedia might also be usefully applied in designing human–machine interactions.

Multimedia can also assist ergonomics as a strategy for implementing training and documentation. Although formal research results are as yet sparse, there does appear to be a general consensus that well-implemented multimedia can be effective. Ergonomists have long known the importance of using different modalities in areas such as display design (e.g., Sanders and McCormick, 1993), and thus it would seem reasonable for ergonomics as a discipline to pay more attention to the use of multiple media and interactive styles of presentation in the design of displays and information presentations.

55.3 THEORETICAL ISSUES

Multimedia raises issues concerning navigation methods and organization principles that are important generally in human–computer interaction. In this section we explore some of the theoretical issues specifically relevant to multimedia.

55.3.1 Organization and Overviews

One of the first ergonomic issues in multimedia is when and how to use hyperlinking. It is often claimed that the nonlinearity of hypertext is especially well suited to effective browsing, suggesting that hyperlinking should be used whenever possible. However, a study by Monk, Walsh, and Dix, (1988) illustrates that, in some circumstances at least, a scrolling browser is preferable to a hypertext browser (i.e., that hyperlinking can make performance worse rather than better). Monk et al. found that performance with the hypertext browser was significantly improved when an overview map of the information structure was provided. This led them to conclude that "finding your way about a hypertext structure may distract from the primary task" (Monk et al. 1988, p. 433).

Maps, tables of contents, and paths are all tools ("cognitive prostheses," cf. Wright, 1991) or "cognitive organizers" for reducing cognitive overhead. These organizers reduce the load on human long-term and working memory (Baddeley, 1975; Norman, 1977) by summarizing information about structure and organization. In books, typically, the organization is carried out by the author. Often the author's organization, as well as the rest of the text, is subjected to considerable editorial review, frequently resulting in organizations that are judged to be of high quality. Increasingly, though, large information structures and databases are not authored by a single person, or even by a group of people with a shared editorial vision and policy. In many cases the sheer volume of information makes it impractical for it to be organized by people. This has led to development of a number of automated methods for organizing information, including automated indexing (e.g., Salton and McGill, 1983), clustering (Everitt, 1990), and various forms of factor analysis (e.g., Deerwester, Dumais, Landauer, Furnas, & Harshman, 1990).

55.3.2 Navigation

One of the paradoxes of multimedia is that while the predominant metaphors of the play, slide show, and book are more conceptual than spatial in nature, the overall metaphor of multimedia and hypertext is generally assumed to be spatial. In this spatially oriented view of multimedia, browsing is a process of navigating through some information structure. This contrasts with the collection metaphor typically used with online retrieval systems. The navigational aspect of hypertext can be reinforced by the use of maps (typically 2D box and arrow representations of nodes and their relative position) to assist in browsing. However, the navigation metaphor is not a necessary part of hypertext or multimedia, as systems such as SuperBook have shown.

When a person clicks on a hypertext hotspot, what appears to happen next depends on the interaction style that is used. In the collection interaction style, a new page is brought to the user and replaces or overlaps the previous page. In the navigation interaction style the users themselves "move" to the new piece of information. Differentiating between these two metaphors in the user interface presumably requires some cues relating to command names, screen visual effects, and the use of related tools such as maps (navigation) or querying (collection).

Multimedia is increasingly associated with the navigation metaphor, but navigating through networks can be disorienting (Campagnoni and Ehrlich, 1989; Conklin, 1987; Nielsen, 1990a). Navigation is difficult when one cannot see the structure through which one is navigating. Users of large Hypertext systems have to deal with a huge structure of interconnected nodes, but the interface lets them display only a restricted view of this structure. Thus using some multimedia applications is a bit like a person with tunnel vision or a very limited field of view trying to navigate through a spatial world. Not surprisingly, this tunnel vision makes it difficult for multimedia users to learn about information structure and aquire cognitive maps of multimedia documents. The processes of navigation in the real world discussed by Wickens (1992) and others seem to be lacking in much of current multimedia. Study of physical navigation indicates the importance of landmarks, and the need for an evolution from route knowledge (knowing where to go when) to "survey knowledge" which corresponds to a fairly detailed mental map of the environment. Currently, multimedia seems to have coopted the metaphor of physical navigation, without incorporating the techniques and principles of successful navigation.

55.3.3 Spatial Representation

Navigation generally implies movement through some sort of space. In the case of a menu hierarchy or network, users may not be aware of what information structure they are moving through, or of how large that structure actually is. Thus the question arises as to when and how spatial representation should be used in making information structure apparent to users. Spatial representation may increase the memorability of multimedia, e.g., through the use of the method of loci to enhance memory (c.f., Yates, 1984). In that mnemonic system, a familiar spatial structure (e.g, the inside of a house) is used to organize arbitrary sets of objects, concepts, or events, which can then be more easily remembered. Recall then consists of moving through the remembered space and picking up the objects stored there.

Spatial models of information structure can serve as a convenient external memory for their users, thus functioning as cognitive prostheses (Wright, 1991). These spatial models can utilize computer graphics and user interface design techniques to enhance interaction with the spatial model. However, ergonomic rules or guidelines for how to enhance spatial models of information to make them more usable and effective have yet to be developed.

One recent example of using spatial maps to provide overviews of multimedia was provided by Hasan, Mendelzon, and Vista (1995b), who visualized the structure within the world's largest multimedia document database, the World Wide Web. In their method, when a user browses through a set of nodes (documents), the links within each visited node form a subgraph which is displayed as a two-dimensional network. An example of a graph produced by their system is shown in Figure 55.1. In this example the highlighted

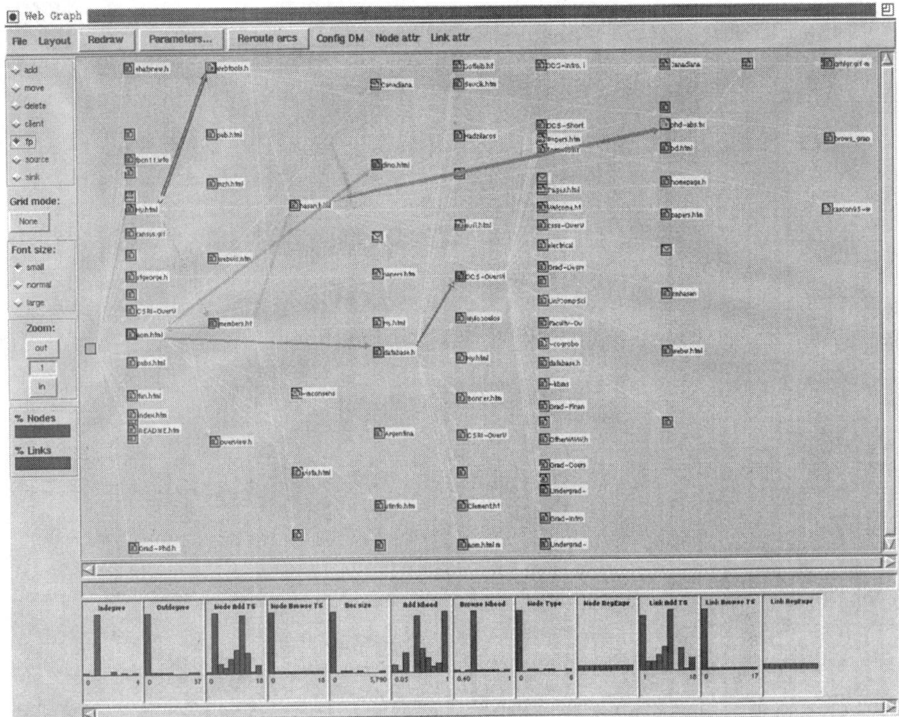

Figure 55.1 A subgraph of the World Wide Web constructed from a browsing interaction.

links reflect the nodes actually visited by the user. Links to nodes not on the highlighted path represent links that appeared in the documents (nodes) visited but that were not traversed.

Charoenkitkarn, Chignell, and Golovchinsky (1996) recently found evidence that simple maps can be helpful in information retrieval. They constructed concept maps based on a history of queries, which Charoenkitkarn et al. (1996) referred to as "query graphs." These query graphs visualize the cumulative query term history. Edges in the graph join pairs of terms that have been linked by an AND operator in a query. Different layouts could then be used to cluster terms visually. An example of a query graph produced in this way is shown in Figure 55.2. Line thickness and color is used in a query graph to indicate how frequently the AND relationship was used for the two terms connected by each link in the query graph.

55.3.4 The Sensory Ergonomics of Multimedia Design

Multimedia has been criticized for its tendency to produce vivid experience at the expense of reflective thought (e.g., Norman, 1993). The application of "cognitive ergonomics" has been presented as an antidote to these poisonous developments. However, multimedia is intrinsically experiential in nature, and that experience is overwhelmingly sensory. Designers of multimedia are involved in sensory, rather than cognitive, ergonomics. They are designing the sensory experiences of users (see Waterworth, 1995). An essential aspect of multimedia design is the selection of media and modalities. With multimedia, we can choose the medium in which ideas and information are presented. We can take information in one medium and choose to have it presented in another (either directly as designers, or by designing in suitable controls for users).

There are many examples of research that exploits the capability to take information in one modality, and presents it to users in one or more other modalities (such systems are dubbed "synaesthetic media" in Waterworth, 1995). A relevant case, the PianoFORTE project (Smoliar, Waterworth, & Kellook, 1995), developed a computer system that recorded electric piano performances as MIDI data. The obvious way to display these data

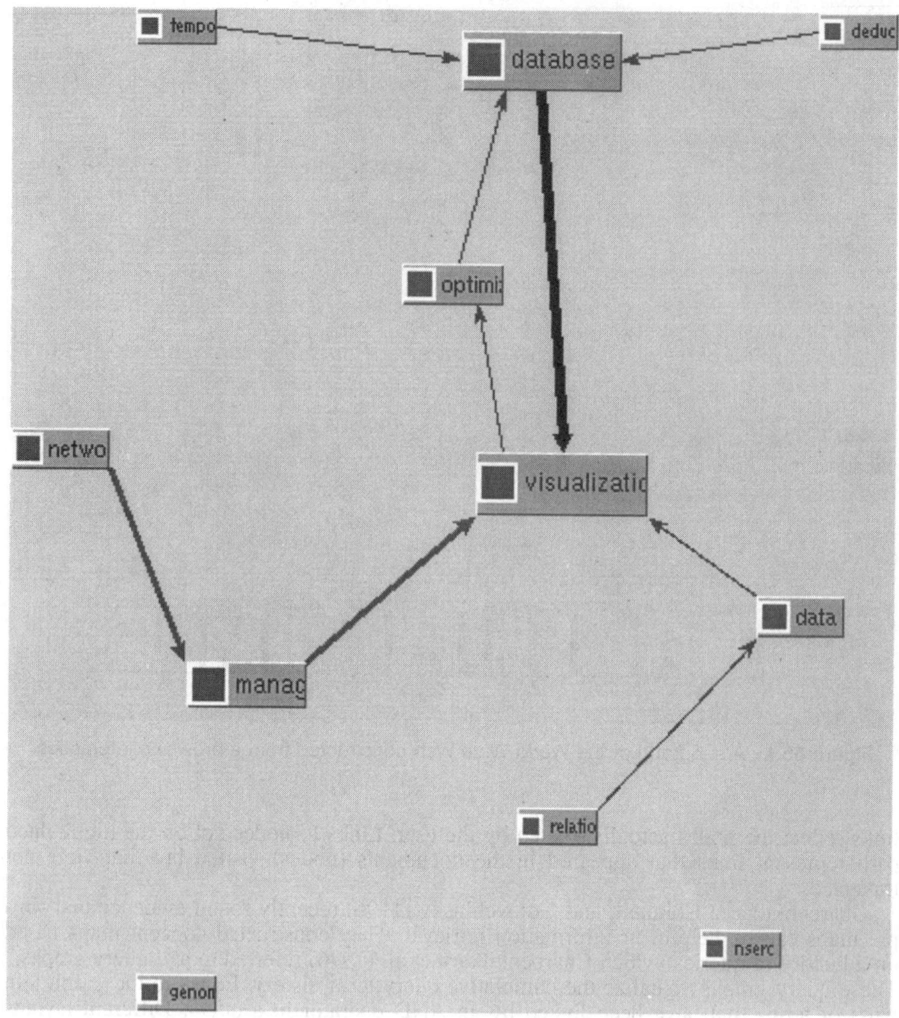

Figure 55.2 An example of a query graph.

would be as sound, to be experienced as a repeat of the acoustic events that were produced during a performance. But pianoFORTE aims to be a medium of communication between the piano student and his teacher. It does this by allowing piano performances to be displayed visually, as graphical annotations to the score that is played. The display can be chosen to reflect different aspects of the performance, such as dynamics, tempo, and articulation. The displays comprise a synaesthetic medium of communication between teacher and student that, unlike the original performance, is not ephemeral. It is significant that the *visual* displays were found to enhance *listening* ability; both students and teachers heard things in a performance after viewing the displays that they would otherwise not have detected.

Other examples of the use of synaesthetic media include:

- Visualization of numerical data
- Auralization of information
- Graphical display of motor performance
- Use of gestures for musical composition
- Associating information with color and/or spatial location

These can all be seen as attempts to support direct sensory experience and de-emphasise cognitive interpretations. Rather than being seen as a problem for cognitive ergonomics, recent developments in multimedia present great opportunities for the ergonomics and information design communities. They enable us to become more aware that what we are doing with multimedia is essentially sensory, not primarily 'cognitive' in nature, and to start to explicitly study and enhance sensory ergonomics.

55.4 THE MULTIMEDIA LIFE CYCLE

The multimedia life cycle is still evolving as tools for multimedia development evolve. The multimedia life cycle of the late 1980s, for instance, involved a lot of handcrafting. At the most general level, the multimedia life cycle involves the activities shown in Figure 55.3. This view of multimedia development focuses on the information and content, and on how it is built into a multimedia application. Another point of view on multimedia development is provided by Balasubramaniam and Turoff (1995), who recommended a series of procedural steps to follow in constructing multimedia. Some of their recommendations are incorporated into the description of the multimedia life cycle given below.

In addition to the elements shown in Figure 55.3, activities such as requirements and tasks analyses (including analyses of user requirements) should be carried out to determine what the requirements of multimedia are. The task analysis should include an analysis of the needs and abilities of likely users.

Figure 55.3 shows the multimedia life cycle as a linear flow. In practice, there may be a number of cycles leading back from later stages to earlier stages of the life cycle. The life cycle itself will tend to vary somewhat from project to project. However, the stages shown in Figure 55.3 are fairly representative of a large number of projects. Each of those stages can now be described briefly.

55.4.1 Content Development

Content development is concerned with creating or selecting the content information (nodes). Content is the information that will be viewed by the user. For instance, in one application it may be the text of a newspaper article, in another it may be a video sequence. Content development may involve both authoring or writing of original material,

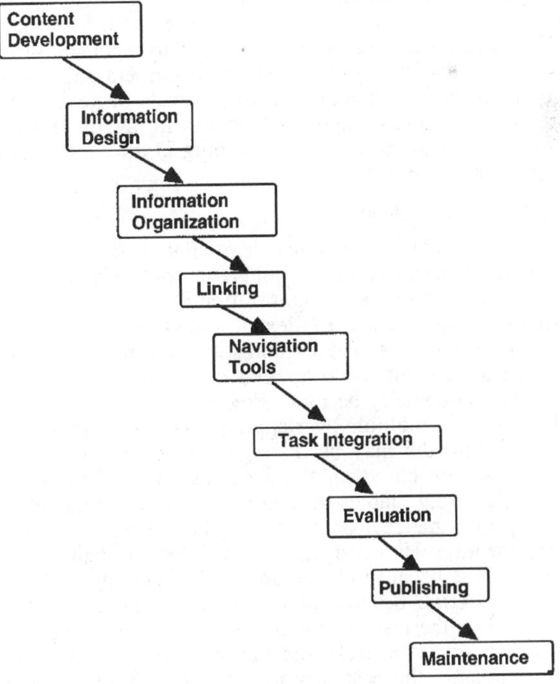

Figure 55.3 The multimedia life cycle.

and selection of existing material. Content development in multimedia, as in other applications, should be responsive to user and task requirements.

Content development frequently involves "repurposing" of existing material. For instance, images from existing photographs may be scanned, filtered, and transformed, and then combined with computer graphics to create new composite images. In a typical multimedia development project, content development is one of the most difficult and time consuming activities. One of the problems in content development is that of getting appropriate copyright clearances and permissions. In the era of electronic information, content material is becoming "intertwingled" and "transcluded" (Nelson, 1981). In practice, this means that it is becoming more and more difficult to know where the original source of material came from. The whole notion of copyright is challenged by large scale hypertext and multimedia, and legislators and lawyers are struggling to redefine it in the light of new media and distribution mechanisms.

Even without the problem of copyright, finding good information content can be a major problem. For instance, an innovative recent use of multimedia is for electronic navigation using global positioning systems and detailed maps of the surrounding environment. One company is mapping large parts of North America and Europe to create a detailed electronic navigation system (Barkow, 1995). The main problem is the huge effort in building the necessary database containing all the maps and local data that is needed: "One section of downtown Philadelphia had 22,000 'links', a term used to denote a road length or an address range along a street." (Barkow, 1995, p. 101). Often, the available information in maps, surveys, aerial photos, etc., must be supplemented by going out to locations and filming them. This is an extremely laborious process. Furthermore, with an estimated 10% of the data changing per year, the task of content development in this case never stops.

55.4.2 Information Design

As in other design disciplines, form should follow function in multimedia, rather than vice versa. Information design involves the "look and feel" of multimedia content. It is the place where content development and graphic design meet. The look and feel of the multimedia should include a consistent framework for presenting information in. Consistent frameworks or templates are often good for both author and reader. For authors, they cut down on the amount of work required, and for readers they provide consistency and make it easier to learn how to interact with the multimedia effectively.

Information design is concerned with how to lay out and signpost information (graphically and aurally) so that it is both entertaining and easy to assimilate. Issues in information design include the style and format of text, the arrangement of text fields (blocks), whitespace, and margins; the style and placement of borders and frames; and the use of visual cues and icons to signal different topics and levels of nesting. Information design is an emerging area that combines graphic art, document layout, and visual design. Some insights into information design have been developing in the context of developing online documentation (e.g., Horton, 1991).

55.4.3 Information Organization

It is often said when organizing objects that there should be a place for everything, and everything should be in its place. The same holds true for information. Information should be organized. One of the best forms of organization is a hierarchy (or table of contents). In addition, information objects can be indexed, allowing them to be searched by using index terms as content descriptors. The organization step in the multimedia life cycle is often difficult because there is either no obvious organization or else there are too many organizations that could potentially be used. Frequently, organizations may be partial or incomplete and there may be multiple overlapping versions of these partial organizations.

Organizations in multimedia may be distinguished between those that are based on metaphors and those that are based on standard containers. Examples of standard containers include hierarchies (e.g., tables of contents) and lists. Examples of metaphors are a bookshelf or a map of some real or imagined place. Early multimedia tended to use metaphor based organization. However, designers should probably begin by considering standard containers. This is because people are used to dealing with standard containers, which thus have some of the benefits of consistency and familiarity. Metaphor-based containers work best when the task or content domain has a strong and memorable metaphor associated with it. In some tasks the metaphor is so closely associated with the task that it is unquestionably the best way to organize the information. For instance, in dealing with geographical data it will generally make sense to use maps to organize the data.

55.4.4 Linking

The nodes that represent the various information objects need to be linked into the network structure of multimedia. Many different types of links can be used (e.g., Tam and Chignell, 1992). The formation of linking structure is something of a black art. A sufficient number of links is needed to facilitate navigation, but having too many links may be confusing. In addition, linear links as well as hyperlinks should be used. For instance, "next" and "previous" links may be used to provide access to an overall sequencing of the nodes, or to a series of paths or trails through the multimedia. It is one of the ironies of multimedia that the linear concept of a path or trail, which was so strongly emphasized by Bush in his original formulation, should have been largely ignored in many early multimedia applications.

Manual linking can be supplemented with automated linking techniques based on heuristics such as similarity in wording used between nodes (e.g., Salton, Allen, & Buckley, 1994). The results of automated linking should then be subjected to editing and review. Methods of network restructuring (described later in this chapter) may also be used to add or remove links so as to ensure that all nodes are sufficiently reachable, without the whole network being cluttered with too many links.

55.4.5 Navigation Cues

There are a number of ways in which multimedia authors can help readers and users. Maps and overviews should be used to orient readers. Consistent visual layout and formatting of information screens should be used to highlight navigational cues. Development of navigation tools often goes hand in hand with changes in information design, which should provide graphical support to the navigation techniques and tools being used. Search tools, hierarchical menus, etc., can also provide alternatives to browsing by links. In addition, landmarks can be used to create convenient jumping off points, and landmarks can also serve as familiar locations to return to if the user gets disoriented.

Lynch (1960) argued that navigability could be built into the design of a city by developing structures and connections that created clearly defined paths, nodes, edges, districts, and landmarks. We would expect that use of these structuring principles should also enhance the navigability of multimedia.

Aside from this built-in navigability through use of appropriate structuring, navigability may be enhanced by explicit navigation tools and organizers such as maps and overviews. One problem with current multimedia is that maps and overviews are sometimes used as bandaids to deal with flaws in the inherent navigability of multimedia based on its structure. In our view it is preferable to build the navigable structure right into the multimedia and then use maps and overviews to enhance, rather than correct for, this structure.

55.4.6 Task Integration

The task that motivates the use of multimedia has to be considered throughout its development. Multimedia has sometimes been viewed as a stand-alone structure that includes tools for browsing and visualization. In practice, though, multimedia is an artifact used as part of a task. If it is not task oriented, then multimedia becomes part of a rather aimless activity such as flipping through the pages of the Encyclopedia Britannica or back issues of National Geographic on a wet Sunday afternoon (which will generally be of less use or interest to ergonomic practitioners). In a task-oriented educational application, for instance, multimedia may be used as an information resource that is called up in response to different instructional goals, which in turn imply different tasks and uses.

Task integration may include the development of training and instructional materials to help intended users utilize the multimedia application effectively. It may also include behavioral goals that are embedded in the multimedia. For instance, portions of the multimedia network may be accessible only after certain competency or understanding has been demonstrated.

Task integration may also involve pilot testing and incorporation of additional task-related content into the application. In practice, task integration will be closely associated with evaluation. Although task integration is listed as a fairly late stage in the multimedia life cycle shown in Figure 55.3, it is good development practice to consider task-related factors at all stages of multimedia design and development. For instance, the types of graphics and layouts chosen in information design should depend on the types of users and tasks. "Playful" graphics may work well for some users and situations, but not for others.

55.4.7 Evaluation

Evaluation of multimedia is difficult, but necessary. Evaluation may be carried out with respect to the overall comprehensibility and effectiveness of a multimedia application, or with respect to the user interface and its navigability. The latter type of evaluation tends to be easier because it can utilize established methods of software usability testing (e.g., Mark and Nielsen, 1994; Nielsen, 1997). Evaluation of comprehensibility tends to be more subjective.

The perceptual content of multimedia should be checked to ensure that it meets perceptual requirements of the user population (e.g., will colorblind users, or those with reduced visual acuity, such as the elderly, have problems in reading the information screens or viewing the presentations effectively?) In addition to the evaluation of multimedia content and use, there should also be evaluation of the overall information design (Balasubramaniam and Turoff, 1995). Multimedia developers should check that the information design does not confuse, distract, or annoy significant numbers of users. The learnability and usability of navigation tools should also be tested, particularly those that are innovative or unusual.

Frequently, the style of evaluation will need to be tailored to the type of multimedia. For instance, educational multimedia should be evaluated with respect to how well its users achieve appropriate learning objectives. Similarly, multimedia documentation can be evaluated in terms of how easily users can find and understand the information they need to perform a particular task. Evaluation of multimedia usability should be done early and often in the development process. Multimedia evaluation is considered in more detail in a later section of this chapter.

55.4.8 Publishing

There are a number of publishing formats for multimedia including CD-ROM, the Web, and multimedia kiosks (e.g., in museum exhibits). Each of these formats have different properties. For instance, publishing on the Web means that the multimedia document is instantly available to millions of people worldwide. However, there is still the problem of making people aware of the availability of the document in a network that already contains millions of documents. In contrast, publishing on CD-ROM is more restrictive, but with good distribution channels CD-ROMs can be targeted to the people who want them. Multimedia also coexists with print media in some situations. For instance, some books may contain a multimedia CD-ROM. The question then is whether the print or multimedia format takes precedence. Sometimes, multimedia is used to embellish the primary print product, in other situations, the print product provides an introduction and overview to the multimedia product.

55.4.9 Maintenance

The final stage in the simple multimedia life cycle described here is maintenance. Multimedia is a form of software application and thus maintenance is an important issue. In principle, since multimedia contains electronic information it should be easier to update than print documents. In practice, it is difficult to update a complex network of nodes and links, where apparently simple changes may propagate over a number of nodes. Expectations for maintenance of multimedia are higher on the Web, where a multimedia application consists of a set of files stored on the hard disk at a Web site. Here files may be continuously updated as new information becomes available. However, whereas there are high expectations for maintenance of content on the Web, there are fewer expectations concerning training and technical support. This also because multimedia on the Web is generally less of a commercial activity (as of this writing). In contrast, maintenance of multimedia CD-ROMs may include technical support and reasonably frequent updates.

55.4.10 Localization

Users of applications are typically not homogeneous. Different versions of a word processor are needed for speakers of different languages, and an engineer's requirements in a mathematical and statistical package may be very different from an educator's. With multimedia too, there is a need to customize or "localize" the document or application to the needs of different user groups. The interactivity and flexibility of multimedia provides considerable scope for carrying out this localization without sacrificing the overall integrity and coherence of the application.

For instance, multimedia can be implemented in different languages. Each text element and icon could depending on the language selected by the user. Video and sound clips could also reflect the choice of language. Orchestration of multilingual speech with video clips can be difficult, though, as is sometimes seen when movies are dubbed into different languages. Difficulties may arise if speech in different languages has to be orchestrated with other precisely timed events within the multimedia application.

For some multimedia applications, localization is not an issue, and thus we have chosen not to include it in the representation of the multimedia life cycle shown in Figure 55.3. When it does occur, localization will have an impact on a range of activities within the life cycle, including task integration, content development, information design, and evaluation.

55.5 MULTIMEDIA AUTHORING

In our view, more attention should be paid to the authoring process, since it is difficult to correct poorly authored multimedia by using sophisticated browsing tools. Usability needs to be built into multimedia during authoring, not after it. In contrast to multimedia reading and multimedia technology development, there has been relatively little discussion and research on the multimedia authoring (development) process. The role of the author in multimedia is poorly defined at present and there are no standard means of evaluating the effectiveness of authoring. As explained below, it is possible that authoring (by adding links) may be harmful, rather than helpful, in some situations.

In addition to providing a critical evaluation of multimedia authoring, this section also discusses manual authoring of multimedia links and introduces two techniques that can assist multimedia authoring. The first of these techniques uses various semiautomated methods to convert existing content material into multimedia. These methods were originally developed for hypertext conversion and have met with mixed success. The second technique involves simplified drag and drop authoring of multimedia. It has become predominant in the authoring of multimedia documents for the World Wide Web. Techniques for improving the author's awareness of how the document being created will look and feel to readers and users are also considered. Figure 55.4 provides a pictorial overview of this section. The section begins by considering a set of preliminary questions associated with multimedia authoring, before reviewing available strategies for the linking portion of the multimedia development task are reviewed.

55.5.1 What Is Authoring and Who Is the Author?

"Authoring" is a term commonly used to describe the task of creating or developing multimedia. However, it is not always clear who the author is, or what the authoring task consists of. In writing a book or article, the author is assumed to be the person forming the words, either by directly inputing them, or by dictating them to someone else who transcribes them into a text file or onto paper. In some cases, the authorship of a book

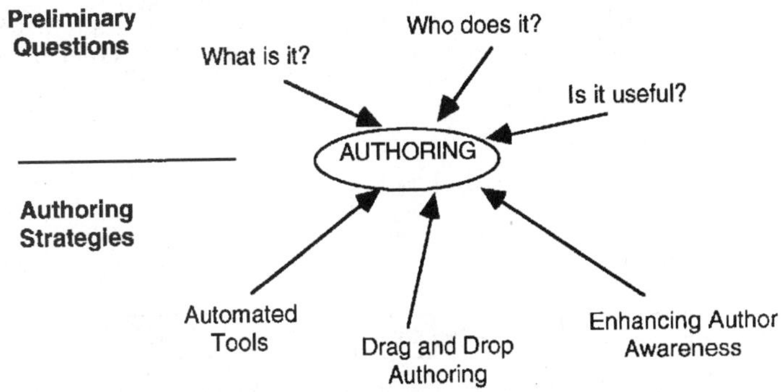

Figure 55.4 Authoring issues.

may be unclear, e.g., when a famous person collaborates with a journalist or "ghost writer" to publish a volume of memoirs. In this type of situation the "author" may provide ideas, rather than construct text.

The concept of authorship is much more complex in multimedia development than in writing. As outlined in Figure 55.3, the multimedia authoring process contains a number of steps including developing or selecting content material, repurposing and editing that material, creating links, and building an appropriate user interface for a particular task. All of these steps may be considered to be authoring of some sort.

For large-scale multimedia projects, the authoring task is distributed among a number of specialists, typically led by a project leader or "system architect" (cf. Rechtin, 1991). In such cases, there is no one author. Other approaches to multimedia authoring (e.g., Cotton and Oliver, 1993, p. 82) have identified the multimedia designer as an information architect "responsible for coordinating all the components of the media matrix into a media environment that the user can explore at will." According to Cotton and Oliver "The designer's job is to develop a structure that shows in schematic form the relationships between every part of the programme." Thus in their view storyboarding is a major task of the multimedia designer. This storyboard (or flowchart) then provides a "program map" for the production team.

Multimedia authoring is often taken to be the act of linking information together into a network. This is the usage that we will apply in this chapter, recognizing that other aspects of multimedia development also constitute significant authoring activity. One argument in favor of this approach is that linking is the activity which sets apart multimedia development from content development generally. Thus linking is the aspect of authoring that is unique to multimedia (and hypertext).

If linking is the main authoring activity in multimedia, who is the author? Multimedia links can be constructed by people, or by software programs, or by some combination thereof. Furthermore, there is no training procedure that we know of that will enable a person to create good multimedia links, and there are no generally accepted methodologies for evaluating the effectiveness of multimedia links once they have been created. Thus the linking portion of multimedia authoring is not well understood, and systematic methods for designing and evaluating links are urgently needed.

55.5.2 Is Authoring Harmful?

It is generally assumed that authoring is a useful activity that adds value to information. However, informal observation in our research laboratories has raised questions about the value of multimedia authoring, at least in some situations.

Sometimes there is absolutely no doubt about the value of authoring. For instance, the *MYST* "mystery game" is a beautifully crafted application that represents a landmark in multimedia usage. *MYST* represents a handcrafted world that was constructed out of the minds of its authors.

In other situations the value of authoring (and linking in particular) can be called into question. For instance, in document multimedia, there may already exist a large amount of content material and the question is how to link it together effectively. In such situations we have found that human authors don't necessarily achieve good results. For instance, in the Jefferson project (Chignell and Lacy, 1991), human authors created sparsely linked hypertexts that often contained inconsistencies (e.g, there was a link from A to B, but no corresponding link from B to A even though it was clearly needed).

In document multimedia, the value of a particular link will generally depend on the task being carried out and the interests of the user (Figure 55.5). Problems will occur

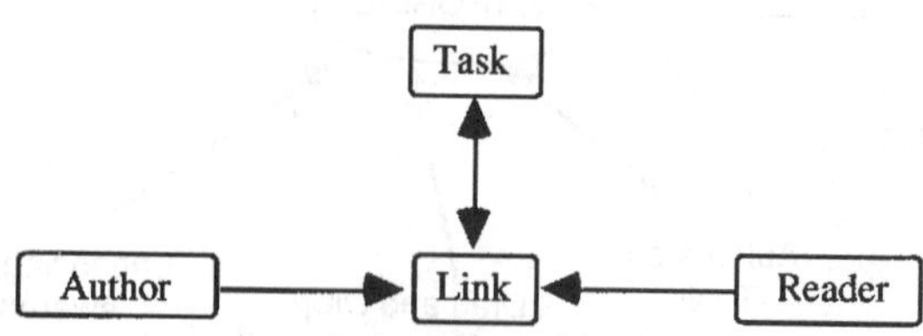

Figure 55.5 The context of linking.

when the author does not correctly anticipate the interests and needs of the user. Since the usefulness of a link is subjective and dependent on the particular situation (e.g., who the user is and what the task is), the notion of a formal and correct authoring process is lost, because now the value of the multimedia varies, depending on the situation or context in which it is used. Thus multimedia that works well for one combination of user and task may be inappropriate for a different user or task.

Research in our laboratory has confirmed the subjective nature of multimedia authoring (Baron, Brown, and Chignell, 1993). In one experiment, subjects were asked to author multimedia by creating links of different types. They had a variety of links available to them, some of which were semantic (e.g., similar to, contrasted with, a kind of) while others were rhetorical (e.g., explanation, illustration, continuation, summary). For the same content material, different authors created very different sets of links. In addition, each author used either a predominance of semantic links or a predominance of rhetorical links. The behavior of readers was then studied. Like authors, readers (as indicated by their selection of which links to follow) preferred to either focus on semantic or rhetorical links (but not both). Thus there appear to be strong individual differences or cognitive styles in terms of creating and using multimedia (hypertext) links.

This led us to consider the somewhat radical view that for some content domains and tasks at least, authoring might actually get in the way of the reader, particularly if the cognitive or authoring style of the developer clashes with the reading style of the user.

Perhaps this should not be surprising if one takes the standpoint of semiotics. Use of multimedia involves a communication process. In this process, the multimedia artifact has to be internalized inside the mind of the reader (cf. Chignell, 1993). Thus there is the externally defined multimedia, and then there is the version of the multimedia actually constructed inside of the head of the reader or user.

Presumably it is the way in which multimedia is interpreted or utilized that really matters. This suggests that the focus of multimedia developers should be on creating multimedia that will be interpreted or utilized effectively, rather than on multimedia as an attractive external artifact (i.e., something with a good appearance). One way to increase the usefulness and interpretability of multimedia is to let readers themselves construct it to some extent. This can be done by giving readers flexible tools for collecting and organizing content material. In one such approach (Golovchinsky and Chignell, 1993), users markup concepts they are interested in, on text. This conceptual markup then launches queries in a text retrieval system with matching documents then being returned to the user in a list. One interpretation of this type of interaction is that the conceptual markup defines a type of link. In this case, links are not preauthored but constructed through interaction with the user.

In our view, multimedia authoring can sometimes be harmful. When different interpretations can and should be put on content material, predefined authoring stands in the way of readers understanding or interpreting multimedia in accordance with their needs or interests.

55.5.3 Enhancing Usability Through Author Awareness

In the act of assimilating and organizing content material, authors need to explore and browse that material, thereby taking on the role of readers and users. Thus techniques for improving the usability of multimedia may also be used to advantage in helping authors carry out their task.

One useful approach to this problem is to create authoring tools that put the author in the reader's position. To some extent this is what WYSIWYG (what you see is what you get) word processors already do. The author sees the document from the readers' point of view. However, word processed documents are not interactive, in the way that multimedia documents are. Thus the author needs more insight into the reader's experience. Ideally, the author gets a sense of WIFIWYF (what I feel is what you feel), where "feeling" refers to the complete multimedia experience of exploring links, reading text and graphics, and playing back multimedia elements.

The first requirement in the development of such tools is that the authoring tools include a playback option. This can be done simply by having a special keypress indicating that an input is a playback action. For instance, clicking the mouse button on a point would normally move the insertion point there, whereas clicking with the option key held down would indicate a playback action where the link at that point is traversed. Simple interface techniques such as this create a relatively modeless environment where the author is also a browser. Encouraging authors to browse their own multimedia documents as they create them should increase the usability and effectiveness of the resulting multimedia documents.

55.5.4 Multimedia and Hypertext Conversion

Glushko (1990) challenged hypermedia (multimedia) application developers to create tools for automatic conversion of text into hypertext since in many cases it is not possible to discard vast amounts of information already existing in printed form. He also coined the phrase "hypertext engineering," to point out that there is a need to use systematic methods for the creation of hypertexts. In this section we will consider how this challenge of automated multimedia creation is being met.

The multimedia life cycle described earlier was concerned with creating multimedia from scratch. However, in many situations there is a considerable amount of existing content material and the question is how to convert earlier presentations into a multimedia format. One of the main challenges in this regard is the conversion of (linear) printed text into (nonlinear) hyperlinked material. Some of the issues and difficulties that arise in such conversion projects have been discussed by Glushko (1990). Research carried out at the Institute of Systems Science in Singapore (Chignell, Nordhausen, Valdez, and Waterworth, 1991) demonstrated the feasibility of developing semiautomated methods of converting text into hypertext. In one of their experiments, Chignell et al. converted a book into a hypertext document within the space of a working day using software tools developed for hypertext conversion.

Although a number of different conversion methods have been developed, hypertext (and multimedia) conversion has been used relatively infrequently. One reason may be the doubts that remain about the effectiveness of hypertext conversion methods. For instance, a study by Nordhausen, Chignell, and Waterworth (1991) that compared manually authored and automatically converted hypertext yielded mixed results. In one experiment the automatically converted hypertext did quite well, while in another experiment it fared poorly in comparison to a human author.

An alternative to automated hypertext conversion is "dynamic" hypertext/multimedia. The advantage of dynamic multimedia is that links can be constructed on the fly based on the profile and activity of the current user. However, there are few examples of dynamic multimedia at present. One approach that has been tried in addressing this problem is to use a search engine (information retrieval) interface that builds queries based on markup (Golovchinsky and Chignell, 1993). This combines the point and click style of interaction typically associated with hypertext, with the dynamism provided by a search engine (where users are free to build their own queries, rather than rely on preauthored links).

Dynamic visualization is another promising way of reducing authoring effort. In a very large multimedia development project it can be prohibitively expensive and time-consuming to develop all the possible charts and graphs that might be useful to readers. In such cases automated or interactive visualization of charts and graphs is an attractive option (Roth, Kolojejchuk, Mattis, & Goldstein, 1994). Golovchinsky et al. 1995; and Reichenberger, Kamps, and Golovchinsky, 1995) developed methods for automatically creating charts and graphs based on a structured representation (semantic network) of the content material within the *Macmillan Dictionary of Art.* These projects demonstrate the potential of automated visualization in multimedia. It is likely that an increasing number of large-scale multimedia projects will employ automated visualization in the future.

55.5.5 Drag and Drop Authoring of Multimedia Documents

The existence of the Web has lowered the barriers to entry for multimedia publishing. Given access to an account on a Web server, almost anyone can create a credible document (or home page) that has potentially as much visibility as the home page of giant corporations such as General Motors or Sony. Simpler authoring tools are needed that will enable nontechnical people to publish multimedia, whether it be as part of corporate training, or to communicate daily events in a household to friends and family around the world.

Multimedia authoring tools such as Director and SuperCard typically require at least some programming. However, a new generation of authoring tools are emerging that rely on a much simpler drag and drop functionality. On the Silicon Graphics platform, WebMagic is an HTML authoring tool that allows users to drop media and links onto a text document to create multimedia. Personal Page is a similar drag and drop authoring tool available on the Macintosh (NewSoft, Inc., 1996). Figures 55.6 and 55.7 show two

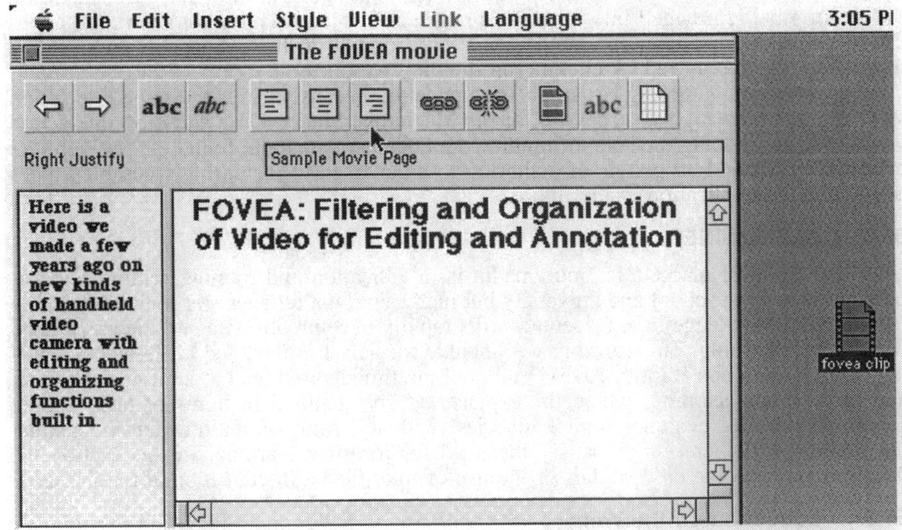

Figure 55.6 A movie file added to a Presto! Personal Page document.

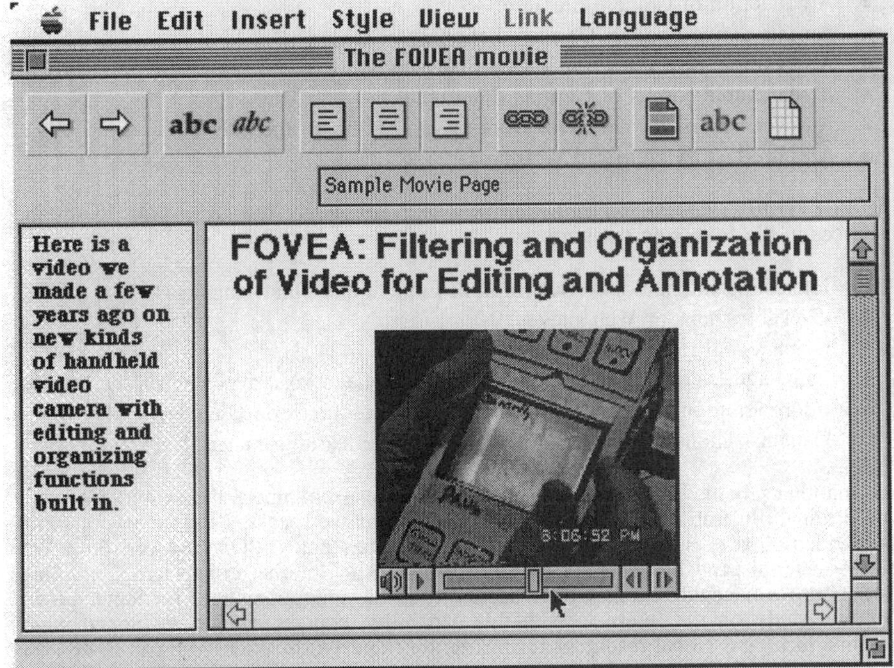

Figure 55.7 The Personal Page document after the movie file has been dragged and dropped into it.

screen shots from Personal Page. The first screen shot shows a file icon for a movie on the desktop being dragged into an HTML page. The second screenshot shows the application moments later after the movie has been dropped in. It then appears as an inline movie that can be viewed by clicking on the play button.

Drag and drop authoring is particularly appealing because it uses the same interaction style that has been highly successful for the Macintosh Finder. Drag and drop interaction works well in defining the sorts of multimedia documents that are found on the Web. We expect the drag and drop style of authoring to be adapted to increasingly complex types of multimedia application in the future.

55.6 CASE STUDIES

One of the largest markets for multimedia is in education and training. Training is important not only at school and university but increasingly at all ages. As societies become increasingly knowledge based, specific skills rapidly become obsolete and there is a need for life-long learning. This learning will include formal, informal, and professional learning (Marchionini and Haurer, 1995). Authored multimedia will tend to be used for formal and professional learning, where the expense can be justified in terms of task-specific goals. Meanwhile, evolving digital libraries, with a wealth of multimedia and textual information will increasingly satisfy the need for informal learning. In this section we briefly describe some case studies of the use of multimedia in education and training.

55.6.1 Education on the Web

Perhaps the largest market for multimedia is education and training. An example of the use of multimedia for education is the Web server implemented by The Virginia Commonwealth University Department of Foreign Languages (Godwin-Jones, 1995), which provides information on foreign language and international resources on the Web. This case is interesting in that it demonstrates how multimedia can be published successfully on the World Wide Web. The site described by Godwin-Jones contains annotated lists of language resources available on the internet, grouped by language. Listed resources include links to:

- An audio-based introduction to Arabic
- A collection of Chinese literature
- An interactive English–German dictionary
- An example of a French medieval manuscript
- A searchable corpus of German literature
- An overview of resources for working in Cyrillic on the internet
- An introductory course in Welsh

The *VCU Trail Guide to International Sites and Language Resources* links to Internet resources in the following categories:

1. Interesting international sites (in both English and other languages)
2. Clickable maps of Web sites
3. Country and regional information
4. Language learning materials and foreign language texts available on the Internet
5. Information on discussion groups and other resources on language studies
6. Demonstrations of locally developed language learning materials

Information available includes several children's stories from nineteenth-century Germany, with original illustrations, recordings of the texts by native speakers and, in several cases, short video clips of selected scenes. As Godwin-Jones notes, "Once placed on a Web server, materials can be accessed by any user with an Internet connection. . . . Once learners are connected . . . they have access to all the many resources for language and literature study on the Internet. With this capability, courses can be developed which combine local and global resources for instruction. The World Wide Web can in this way serve as a virtual classroom."

This is but one example of many sites for multimedia in education that are springing up on the Web and the Internet. For instance, a site in Texas has implemented a "Virtual

Lecture Hall," where courses for many institutions are listed. Syllabi and lecture notes can be placed online (with the amount of information available varying greatly between the different courses) and are then available as global information resources.

55.6.2 Multimedia CD-ROMs: Children's Books

Aside from the Internet, CD-ROM is the major medium for publishing multimedia in education. Examples of CD-ROMs designed to build interest and skills in early reading are *Peter Rabbit* and *Thomas' Snowsuit* from Discis books. Multimedia is particularly well suited for early reading, because when done well, the multimedia content supplements the text material and also helps to motivate young readers. Some of the ways in which sound is used to supplement the text in *Thomas' Snowsuit* include:

- Pronunciation—Speaking words out loud
- Sound effects
- Syllables—Sounding out each syllable
- Explanation—Explaining the meaning of each word
- Part of speech—Indicating the part of speech (noun, verb, etc.) for each word

Multimedia is particularly attractive in this type of application because it enables the reader to link the sounds of words to their visual appearances. Illustrations can also be linked to animations and pop-up information, both increasing the motivation of the child and providing supplementary information. Each of the items in an illustration can be annotated with a pop-up name. When the child moves the cursor over an object in a picture, its name is either shown or spoken.

Currently, entertainment and instruction for young children appears to be one of the main successes for multimedia. By 1995, many parents seemed to have accepted that multimedia was a good medium of instruction for children, and surveys indicated that the most frequently given reason for buying home computers was to help educate children.

55.6.3 The Visual Almanac: An Early Development Project

From a historical perspective the Visual Almanac is an extremely interesting and instructive project. It was developed at a time when handcrafted multimedia was preeminent. It demonstrated the difficulties of trying to scale up the handcrafted approach to deal with large amounts of multimedia content.

The Visual Almanac was a project carried out by Apple's multimedia group between November 1987 and January 1990 (Apple Computer, 1991). The Visual Almanac "was designed to give people a glimpse of multimedia's potential for learning, teaching and communicating" (Apple Computer, 1991, p. 5). As stated in that overview:

> *we thought our previous experiences would stand us in good stead; after all, the difference between creating our design examples and developing the Visual Almanac was basically just a matter of scale. That was the understatement of the year. Scale turned out to be the difference between building a log cabin and constructing a skyscraper—in an environment that had never seen a skyscraper, much less built one. It meant proceeding without precedents, and it proved to be one of the most frustrating, exhausting and exhilarating challenges we had ever known.*

In hindsight, the difficulties in constructing the Visual Almanac were not surprising. By the mid-1990s people had learned that it cost in the hundreds of thousands of dollars to produce a high-quality multimedia CD-ROM. However, as of 1987, few if any people had any idea of the expense and effort that large-scale multimedia projects would require.

Based on the published description, the development of the Visual Almanac appears to have been an exploration, without a clear concept of what exactly the application was at the outset. One of the lessons learned that was cited in the project report was the need to plan ahead, with more research up front leading to less work later.

If the experiences reported by the Visual Almanac team are any guide, it appears that project management issues are particularly important in multimedia development. This need to incorporate effective project management into a creative process is a recurring theme in multimedia development. It is something that the entertainment industry has had to deal with for a number of years. However, even in Hollywood, where film making styles itself as an "industry," there are still project management disasters where bloated movies come in late, well over budget, and often with little apparent artistic merit.

In the Visual Almanac project, Woolsey and her colleagues identified five different stages in the production, including three design phases (initial design, software development, and final design), and two production phases (prerelease and final production). The initial design was concerned with choosing the right audience for the multimedia and selecting the content material, which was primarily images and video clips in the Visual Almanac. In 1987 image libraries for multimedia were just beginning and thus a great deal of effort went into finding images and clips that would be suitable for the project. One of the lessons learned in the Visual Almanac project was that "revisions and refinements to the final videodisc cost as much as all the production preceding it" (Apple, 1991, p. 31). This is not surprising to book authors who have to labor through multiple revisions and brushes with editors, but it was quite unexpected for many participants in the fledgling multimedia industry.

Overall, the Visual Almanac project was unique in being a research project that was designed both to showcase multimedia and to document the activity of a multimedia development team. The members of the Visual Almanac made a number of recommendations on how multimedia projects should be carried out. Some of these recommendations are included in the guidelines for multimedia development listed later in this chapter.

55.7 ERGONOMIC STRATEGIES

This section discusses the use of ergonomic strategies in multimedia development. The most obvious need is for more task analysis and evaluation. Since general techniques for task analysis and evaluation are well understood by ergonomists, our focus here will be on ergonomic strategies that address the particular properties of multimedia.

55.7.1 Provide Orientation

One of the major goals of multimedia developers should be to promote a sense of orientation for those reading or using the multimedia application. A well-oriented person should be able to answer the following questions:

- Where am I?
- Where did I come from?
- Where do I want to go?

If one has trouble answering these questions then one has become disoriented. Disorientation is one of the major ergonomic problems in multimedia (e.g., Nielsen, 1990a, 1990b). Perhaps the best means of reducing disorientation is to create a straightforward and usable organization of the content material that makes it easier for users to figure out where things are and how to reach them.

Additional solutions to the problem of disorientation in multimedia include:

- Making structure apparent (e.g., through the use of titles that indicate the place of the current node within a structure of topics and subtopics)
- Providing maps and overviews
- Use of annotation to increase rhetorical structure
- Creation of paths and hotlists
- Addition of search tools to provide alternate means of "getting around"

One of the main strategies for dealing with disorientation has been to provide maps of various sorts. A map functions as a kind of template. Other templates are more abstract such as the grid model used in a spreadsheet. The layout of a typical newspaper also provides a well-organized template for assimilating the news. That template includes headlines, information arranged in columns, and the partitioning of different sections of the newspaper according to topics. The use of templates is intended to speed up the process of assimilating content material. The idea that templates can both simplify authoring and reading has been exploited in the information mapping approach to technical documentation (Horn, 1989).

The mapping of nonspatial aspects of information onto 3D coordinates is both problematic and (potentially) powerful. One simple way of achieving this is shown in Figure 55.8 (from Waterworth and Serra, 1994), where a 3D "SuperCube" illustrates a general-purpose way of mapping informational dimensions onto those of space. In this particular implementation, what is displayed on the front face of the cube depends on how the cube is rotated (in a virtual reality environment).

Rotating upward about the horizontal causes more general material to be displayed, whereas rotating downward results in the display of more specific information. Rotating to the left about the vertical is equivalent to moving back in time (perhaps following back a history trail), whereas rotating to the right results in the display of more recent material. Finally, turning the front face clockwise or anticlockwise (about the z-axis) selects the way in which the user wants material to be presented (magnifying the view, for example, or showing only images or only text).

The same three-dimensional mapping can be used in a more directly navigational setting, by locating information accordingly in a virtual space for user exploration. And, of course, many other plausible mapping schemes could be developed. For example, closeness (near the user) could signal currency or importance, with distance increasing with declining relevance or elapsed duration. The main point to be made is that the mapping scheme must be simple and easily understood by users. Ideally, users could develop their own idiosyncratic scheme to suit their own needs. Whatever scheme is applied, as long as it is applied consistently it should aid memorability and serve as a useful tool for informal information organization.

Theoretical discussion of visual information processing and maps has been provided by Chase (1986). A more applied discussion of maps and navigation has been provided by Wickens (1992). Maps are generally considered to be useful in multimedia. However, there are no definitive findings as yet on what types of maps should be used when.

55.7.2 Provide Multithreaded Navigation

Many current systems limit users to following only one active navigation thread. Obviously, it is advantageous to have processes that may take some time going on in the background, so that the user can get on with other things. To be successful in following more than one navigational thread, users need to be able to maintain an awareness of the context within each thread and to be alerted to the contextual links between threads.

Currently, the navigational model with which most Web "explorers" are familiar is not rich enough to support much more than the fairly haphazard collection of interesting Uniform resource locators (URLs). The navigational approach with typical Web browsers is depth first, with the user following a path consisting of a series of nodes to his or her current "place" somewhere in the Web. At any particular node, the user has the option to choose a link onward in his or her journey, or follow the path back from whence he or she came. After going back the user can choose to go forward again to the next node on the most recent path taken. But he or she cannot go back to revisit an earlier path, and this is a major weakness with current browsers (the "back–forward" problem). If the

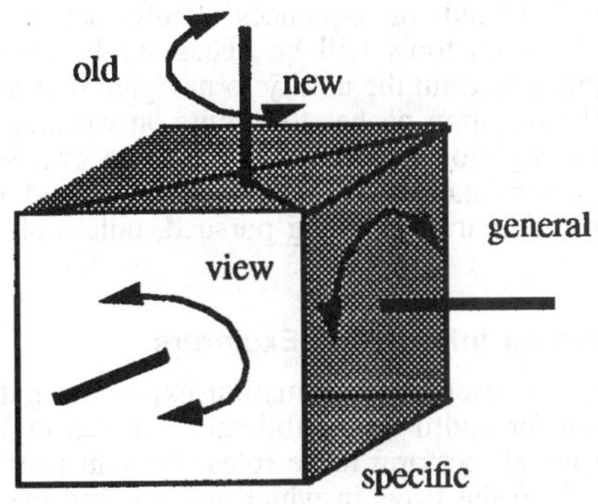

Figure 55.8 An example of a SuperCube.

browser provides multiple active threads (most do not as of this writing), the user may have several active exploration paths in progress. But no relationships between the items in the multiple threads are shown. It is as if each thread were a completely separate entity that might as well be running on a different machine with a different user.

Deckscape (Brown and Shillner, 1995) is a system developed to circumvent these limitations. Deckscape is based on a stack metaphor, where Web pages are stacked on top of each other with only the top one visible. When the user follows a link to a Web page, the new page appears on top of the stack. As he or she follows a single path, the stack gets bigger, with the current page visible on top. The user can revisit the pages in a stack, by leafing through the entire stack, by jumping to the top or bottom of the stack, and by choosing a particular page from a separate list of contents. All new pages are fetched in the background, in a separate thread, so that neither following a slow link nor downloading a large file will freeze the entire applicationl—a common problem when navigating the Web.

Deckscape keeps all pages until the user chooses to discard them, a facility that can be used to get around the back–forward problem. For example, consider a user who navigates from page A to several other pages including page B and stops at page C. If he or she then backtracks to B and moves forward to a previously unvisited page, Deckscape inserts this page (and subsequent ones) after B, and retains the previously visited pages between B and C. Existing browsers, in contrast, would discard all pages between B and C (and including C itself). Users can also choose to have particular pages displayed separately from the stack and retained there, even if a link on such a page is followed (the path continues on the stack, and the chosen page can be retained for reference while exploration continues). Users can also form their own stacks comprised of particular collections, such as hotlists.

A weakness with Deckscape is the way the concepts have been visualized. Having only the topmost page on a stack visible means that it looks just like a single page. The separate list of contents does not reinforce the metaphor and selection from it is indirect. The use of real or pseudo-3D to spatialize stacks would add considerable realism and would tap human spatial abilities.

An alternative approach to Web surfing is to intermix browsing and querying. This can be done by extending the graphical querying approach (Golovchinsky and Chignell, 1993) to information exploration on the Web. In the Multisurf system (Hasan et al., 1995a), for instance, Web documents can be marked up with a specially modified Motif client to create queries that are then executed through an index server. The matching documents are then shown in a hit list within the client/browser window. The end result is that Multisurf acts as an integrated front end for searching and browsing on the Web. Figure 55.9 shows a screenshot from the Multisurf system. In Figure 55.9, the user has marked up a query (which may be considered as a user-defined link) in the document shown on the right hand side. Titles of documents retrieved by that query are shown in the list of hits on the left hand side of the screen. Note that the HTML links are also active on the documents that have been marked up and can be clicked on. Thus the same document can be both marked up for querying or used as the starting pointing for browsing via links.

Deckscape and MultiSurf provide examples of how software functionality may assist users in following multiple threads or sequences of information at the same time. From an ergonomic perspective, such tools will be necessary because of the way in which cognitive limitations likely constrain the ability to navigate in a multithreaded fashion. In particular, multithreaded navigation pushes the limits on working memory and attention. Furthermore since each navigation thread may require its own sense of orientation, the user's overall sense of disorientation will likely be multiplied in accordance with the number of navigation threads currently being pursued, unless effective orienting support is provided.

55.7.3 Provide Support for Information Explorers

We can think of multimedia users as information explorers, gatherers, and organizers. The Web, the main forum for multimedia publication, increasingly functions as a virtual world-at-large in which we all perform these roles. We can learn lessons for the design of usable virtual spaces from the ways in which we use and explore physical spaces in the real world. We also need assistance with navigation by having the system answer our questions and help us find those things in which we are interested.

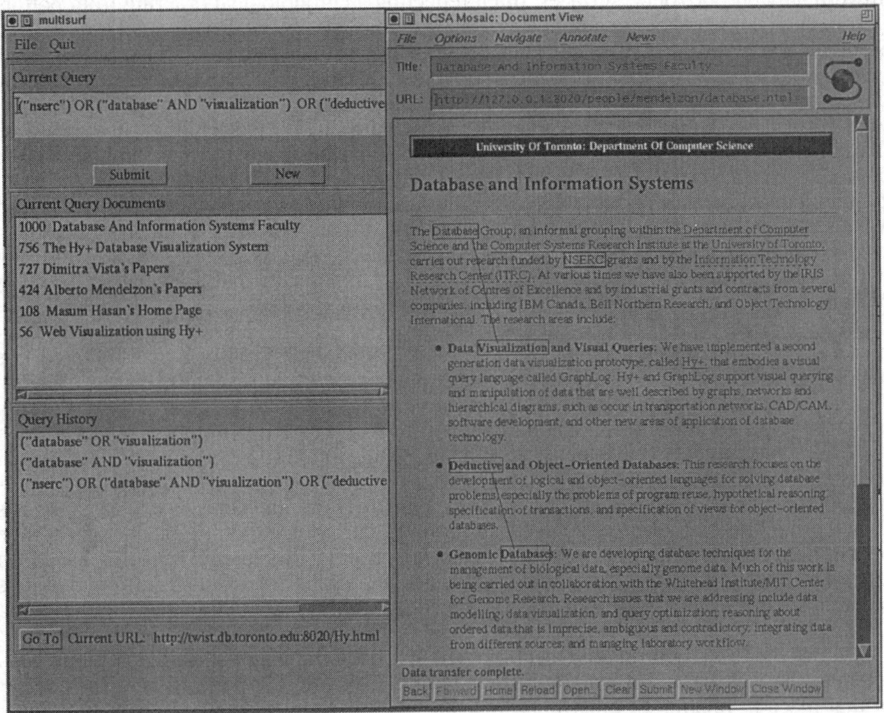

Figure 55.9 A marked up query in MultiSurf.

55.7.3.1 Users as Travelers

Like all travelers in a strange land, users in a virtual world or information space have three main problems: making sense of what they come across, finding things of interest, and communicating with others and sharing what they have found. Travelers in cyberspace can benefit from the experiences of other travelers. They can share commented itineraries, travelers' tales, and travelogues. It is actually relatively easy to communicate about spatial exploration, since people are used to memorizing and talking about locations, describing routes and so on (Dieberger and Tromp, 1993; Yates, 1984). Such records of journeys would need some editable record of the route taken and facilities for user annotation. There are serious problems with evaluating the success (or otherwise) of information explorations, given the absence of objective measures. Analysis of travelers' tales may provide one route to assessing information exploration environments and tools.

As collectors of interesting finds, travelers need bags, bins, or other containers in which to collect things during the process of exploration. Bates (1989) has referred to the act of collecting interesting information finds as "berry picking" and has used the analogy of a basket as a container in which a person can place finds during information seeking. These finds can be examined in detail later and passed to others. One model of the steps involved in this process of exploration and collection is known by the acronym GROPE—gather, review, organize, publicize, and edit. First, we gather what appears, at a cursory glance, to be relevant to our interests. When we have finished collecting, we stop and review what we have found, going through the things we have collected at our leisure. We then seek to organize the materials in a way that makes sense to us, and to communicate our findings with others (the publicize phase). Finally, on reflection and perhaps in response to feedback from others, we edit what we have organized, discarding items that no longer seem very relevant or important.

55.7.3.2 Capitalizing on How the World Works and Looks

The traveler has to interact with the environment. One important aspect of our experience of terrestrial space is the presence of gravity. We make use of this when we pile things up vertically on horizontal surfaces, often reflecting a chronological ordering from bottom to top. A pile metaphor for temporary organizations of data was suggested by Mander, Saloman, and Wong (1992). A further consequence of evolving in a world with gravity is that we tend to be happier exploring horizontal surfaces than vertical, which somewhat offsets our tendency to stack things in piles. We cope better with horizontal layouts than with vertical, as witnessed by our failure to socialize with others living or working on floors below or above us, as compared with those located nearby on the same level. The consequence for spatial design is that most navigation should be conducted on the horizontal plane and stacks should be vertical, but not too high. In personal workspaces, such as our offices, we make use of walls and other vertical surfaces to display calendars, posters and other reminders of future events or tasks, such as "to do" lists. These not only remind ourselves of information, they often communicate that information to others or help us to do so.

Many aspects of real space planning and development are relevant to designing multimedia. For an excellent account of these topics, see Lynch (1960). Lynch outlined five major properties of cities that improve their navigability: paths, edges, districts, nodes, and landmarks. The principles that determine how easy it is to find one's way in a city may also apply to multimedia. Habitable towns and cities are not simply planned. Rather, there is a meeting between top-down planning and bottom-up "evolution." Successful paths and landmarks, for example, develop with use. Navigation is also assisted by "writing on the world" (Dieberger and Tromp, 1993) of various kinds, such as signposts showing the direction of places and labels showing current location (the name on the side of a building, for example). Another aspect of reading the spatial world is based on knowing that as things get older they look older. Similarly, as collected information gets older it should look its age, aiding identification and later selection (Hill and Hollan, 1992).

One navigationally difficult aspect of built spaces is the inside–outside problem. This refers to the fact that when we are inside a building we lose our perception of the context of that place in relation to other places in the vicinity. From the balcony of a skyscraper we might get a magnificent panoramic overview of the city. But once we are deep inside such a building, it is very easy to lose our sense of orientation to the world at large. The inside–outside problem also applies to three-dimensional information space. If we construct three-dimensional structures that are explorable from the inside as well as the outside, we are likely to lose the contextual overview once we enter into buildings. This is the case with several multimedia systems that use the metaphor of buildings and cities (e.g. Andrews and Pichler, 1994: Dieberger and Tromp, 1993: Musil and Pigel, 1993). The structures serve well as identifiers of categories, and from a high altitude provide a good overview of what is available. But once a user zooms down and actually enters a building, she is likely to become disorientated rapidly.

The easiest way to avoid this problem is to have three-dimensional objects in a three-dimensional space but allow objects to be viewed and explored only from the outside. This, of course limits the complexity of organization that is possible. Transparency can also be used to circumvent the inside–outside problem. Consider a mole constructing a complex network of tunnels. Although we are standing above the whole system, we can no more get an overview than can the mole deep within a particular tunnel, because we are outside and cannot see past the surface. But if the rendering is reversed with the earth shown as transparent and the tunnels as black, then we can get a tunnel overview by virtue of being elevated above the network.

55.7.3.3 Using Agents

Agents are one way of dealing with situations where it is preferable to have the system search for material, rather than have the user navigate. We can distinguish between private agents and public agents (Waterworth and Singh, 1994). Public agents are the same for everyone who uses them, they are not customized to the needs of particular users. Public agents are suitable for providing information about the information exploration environment itself, and for carrying out standard searches in the background while the user does other things. Private agents, on the other hand, are in some sense familiar with an individual user (or possibly group of users). They may be used to track the kind of information in which a user is interested over a period of time, so that suggestions can be made about new material as it comes in. They may also be used to carry out specific tasks for a given user, based on his or her pattern of exploration. Examples of this might be noting times spent at particular locations as an aid to reminding the user about past navigation, or to plan a route to a particular location taking account of the user's current location.

The currently prevalent distinction between system-mediated searches (in response to a query from the user) and user navigation (while browsing) is misleading and limits interaction design in inappropriate ways. For some applications, it is more useful to blend the two, so that a user request for system-mediated search is used as an initial step in user navigation. Conversely, users will sometimes want to navigate to a general region and then request a system-mediated search of that locality (through a private agent).

There is currently a vogue for the use of agents as slaves to users' information exploration needs. One or more agents can be dispatched to find information on a particular topic, which amounts to no more than placing a traditional information retrieval query (except that the agent usually operates in the background, freeing the user to do other things). More interestingly (though potentially dangerous), agents can be designed to build up a picture of the kinds of topic in which a user is interested. Items on this topic can be collected, users can be alerted when new information appears on that topic, and so on. They may also leave marks to indicate where they have been, creating paths for users (who then assume a role somewhat analogous to that of dung beetles following a trail of agents' droppings).

One drawback with agents, as with other attempts to embody intelligence in the information system ('user modeling,' for example), is that the extent to which users' information needs can be anticipated is very limited. At a simple level, a user cannot specify, nor can the system infer, the new material that the user might be interested in—new material, that is, that is not obviously related to the items in which that user has been interested in the past. More generally, information exploration is an activity that is little understood, but is at the heart of human cognitive functioning. Until it is better understood, it is unlikely to be successfully automated through the use of agents or other autonomous search mechanisms.

55.7.3.4 Navigating in Time

Information exploration can be viewed as a journey, consisting of a series of "time slices" which we can revisit (and do other things with). Stacking is an obvious way of doing this, with the most recent items on the top. Transparency and blur have been shown to be very effective in enhancing the perception of layers of graphic display and in isolating objects of interest (Colby and Scholl, 1991). One can show history by allowing users to look through from the present into the past, by providing transparent or semitransparent overlaying of more recent items (e.g. Genau and Kramer, 1995).

Some 3D representations of information have used relative movement to convey a sense of the depth of items in the display, capitalizing on our perceptual tendency to use motion parallax as the dominant depth cue (e.g., SemNet: Fairchild, Poltrock, & Furnas, 1988). Generally, either the viewpoint of the observer is moved from side to side (or up and down)—the "camera position"—or the display is joggled about in an attempt to create a stronger impression of depth. Moving the camera position is much more successful than moving the display, presumably because this is closer to natural depth perception and because it is hard to concentrate on a jittery display. If head motion detection is available, then depth can be conveyed even more naturally, without having to explicitly manipulate "camera" position (this is the typical approach with virtual reality applications). When depth is used in this way, relative movement of layers can also be useful, particularly when there are many time slices with which to deal. This allows users to grab a handle attached to a time slice and move that slice around while the other slices remain fixed. Combined with transparency and viewpoint manipulation, this can be a powerful way of using depth to convey overlaid information (e.g., Silvers, 1995).

55.7.3.5 Information Currency and Overload

The amount of information that is available is expanding at an ever-increasing rate, to the point where we feel overwhelmed by the almost infinite number of information nuggets we might discover during exploration. This is the crisis of information provision since, if we define information as that which reduces uncertainty, when we have a practically infinite number of items of information we effectively have no information. However, if we shift the focus of design from *access to* information to *selection from* information, we start to see the user again as a purposive explorer of information space making sense (literally) by the items chosen but also by those ignored or rejected. This orientation plays down the importance of exhaustive, precise searching (by machine or human) and stresses satisfying immediate needs and serendipity. What follows from this is that it is not necessarily desirable to retain all found items. As with human memory, some forgetting by the system might allow us to get on with the current situation with less distraction (Jones, 1989). This idea has been revived by Norman (1993). If we have not used an item of information for some time, we might want it to become less salient. And as time passes and the item remains unaccessed, perhaps it should fade away altogether. Although the prospect of losing information will alarm some users and will not be appropriate for all tasks or situations, temporal pruning will often become essential to avoid multimedia information overload.

Related to the issue of old material is that of new. As old information becomes progressively less important, so new information becomes ever more important. Currency is extremely valuable, and should be marked to alert the user to the fact that he or she is accessing new information (a process sometimes referred to as monitoring). One popular way of marking new material is with flashing markers. Making text and objects flash on the screen (e.g., with large brightness changes) is one small example of an increasing trend to use animation at the interface.

55.7.3.6 Importance and Relevance

Lynch (1960, p. 78) noted in his study of how people navigated in cities that "There seemed to be a tendency for those more familiar with a city to rely increasingly on systems of landmarks for their guides." Lynch also pointed out that a key property of landmarks is singularity, because landmarks are going to be singled out from other objects in the environment. Landmarks, Lynch proposes, are easily identified if (1) they have a clear form, (2) they contrast with their background, and (3) there is some prominence of spatial location.

Waterworth and Chignell (1989) point out the need for importance and relevance to be indicated in (or form the basis of) multimedia overviews. Importance generally refers to how well-connected a particular node is, whereas relevance is related to the user's current (and/or previous) task and is therefore harder to quantify. Importance can be seen as global relevance—how relevant an item of information is to users in general. See Sperber and Wilson (1986) for a detailed account of the role of relevance in cognition and communication.

One approach to relevance is to track the items a user is interested in (as indicated by relevance feedback) and build up an interest profile of the individual user (often on a particular task). An example of this approach was recently found on the Web (described as a "collaborative filtering service") where users were invited to send URLs of pages they find of interest. Interest profiles were then compiled for each user and compared with those of other users. When there was a good match between a pair of users their submitted URLs were exchanged, on the assumption that the URLs of the closely matching user were also likely to be of interest (relevant).

55.7.3.7 Topography and Geography

At the heart of designing spatial metaphors for multimedia information is the notion of topography—the physical features of an object or objects, and their structural relationships—and how this relates to the information content. One problematic aspect of this is the representation of geographical information about the physical world. If, when developing a multimedia application, we use a geographically spatial model, such as "Information Islands" (Waterworth and Singh, 1994) where a landscape of geographical features serves as the world model to convey the structure of a large set of information, how do we represent geographical information? There are two problems here. One is the question of "Where in the world is the world?" and should there be a 'Geography Island,' where all geographical information is located? The other is the question of how to show the geographical location of sources of information represented as entities in a spatial interface world. If users want to know where items of information are physically located in the real world, how best to convey this?

One approach to solving this problem is to use a geographical structure, and represent this as an explorable map world. Information entities (the I-world) would be attached to their geographical locations (the G-world) (Figure 55.10). This approach seems satisfactory on the face of it, but once we consider that the geographical location of information is often irrelevant to our interests and that the design shown in Figure 55.10 takes no account of any relationships between items of information (other than geographical), it soon comes to look rather inadequate. Adding in just a few links between items and representing these on this hypothetical design brings out this inadequacy (Figure 55.11). Unless the focus of system use is entirely geographical, the I-world is bigger than the G-world and contains it. In other words, a useful design orientation is to treat the physical world as a place in the user's spatial world model of the multimedia application.

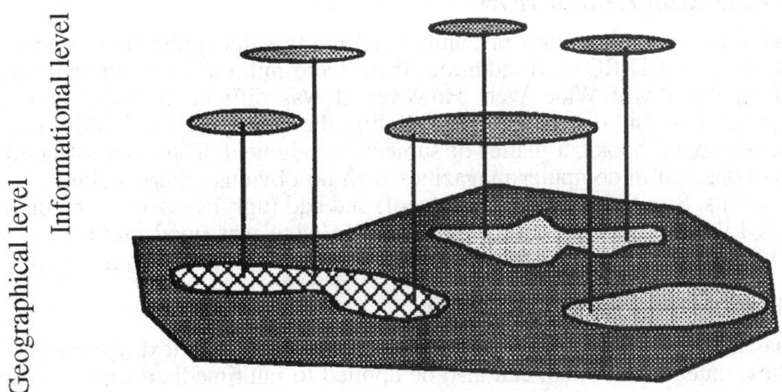

Figure 55.10 Connecting information entities (the I-world) to their geographical locations (the G-world).

55.7.4 Provide Choice of Medium

As Alty (1991) points out, we know only a little about the advantages and problems of using particular media for different types of task. Auditory (instead of visual) presentation seems better for later recall of dialogue, whereas events are better recalled from a presentation that includes visual material. Diagrams are often successful in conveying ideas, but text is better for detail. On the whole, however, knowledge on this topic is rudimentary and incomplete. The study of detailed ergonomic aspects of multimedia design and use is still in its infancy.

The value of users being able to choose the medium in which material is presented is increasingly recognized. The power of digital media resides, to a large extent, in the fact that information can (at least in principle) be readily converted between media—as a graphic diagram, as audio, as an animation, as full video. Negroponte (1995) refers to this as "mediumless." Waterworth (1995) suggests that the more different media forms and modalities are experienced, the more likely the user is to gain creative insight into an issue. He coins the term "synaesthetic media" to emphasize both how multimedia implies cross-modal translation and presentation of information and the creative possibilities raised (since natural synaesthetes are often very creative individuals). This view can be set against others (e.g., Norman, 1993) that suggest multimedia inevitably leads to less reflective thought because of its predominantly experiential nature.

In our view, providing a means for users to choose the medium or media of presentation that best suits their needs may well be more successful than trying to develop a typology of media and uses.

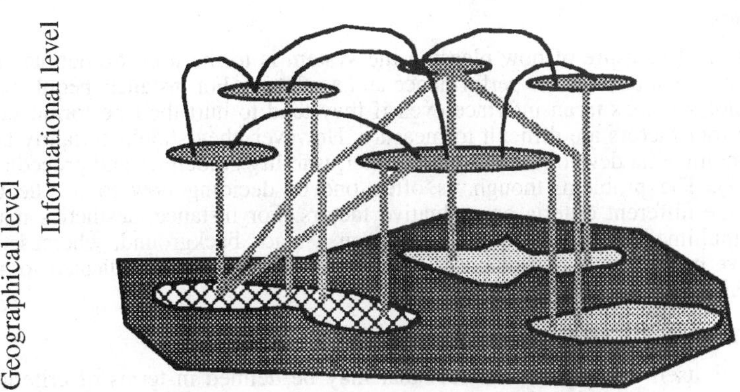

Figure 55.11 A Visual demonstration of how information relationships (as well as geograhpical relationships) need to be represented.

55.8 MULTIMEDIA EVALUATION

As of 1995 there were thousands of commercial multimedia applications, many of which were published on CD-ROM. In addition, there were millions of multimedia documents published on the World Wide Web. However, it was difficult to judge how good the various applications and documents were. Rating the quality of CD-ROMs seemed to be much like reviewing books, a matter of subjective judgment. Thus, lists of good and bad CD-ROMs appeared in computer magazines with no obvious criteria being used to make those judgments. Similarly, lists of good (cool) and bad (ugly) sites appeared on the World Wide Web. Like art, people could often agree on what was good multimedia and what was not, without being able to give rules that separated the good from the bad.

55.8.1 Four Categories of Evaluation

Nielsen has discussed four categories for the evaluation of hypertext documents (Nielsen, 1991). These categories (which can also be applied to multimedia documents) are utility, integrity, usability, and aesthetics.

Utility

Utility is a measure of whether the hypertext document actually helps a user perform the intended task. This has to be compared with performing the same tasks with linear text. Much of the available research on this topic (some of which is summarized by Waterworth and Chignell, 1997) has shown that readers exhibit poorer performance with hypertext than with paper documents.

Integrity

Integrity is a measure of the completeness of the document—whether it is up to date and not misleading.

Usability

Evaluation of multimedia usability is inherently more difficult than evaluation of software usability in general. In multimedia, usability must be measured in terms of the effectiveness of communication. Thus, it is the transmission of the content material that is important. In contrast, software usability is determined by what problems, if any, users have in carrying out a standard set of actions using the software. Usability tests can be constructed with a set of standard tasks, and performance on those tasks can be measured objectively. These objective measures can then be augmented with subjective ratings concerning the ease of use of the software, plus think-aloud protocols of users describing the problems they are having while using the software to perform the tasks.

It is less obvious how to assess multimedia usability because there is no notion of a standard task for many multimedia applications, nor are there obvious measures of performance on tasks using multimedia. Instead, one generally has to rely on subjective ratings of the quality and usability of the multimedia, along with tests of comprehension (at least in educational applications) of its content after the multimedia has been used. Methods for assessing multimedia usability are outlined further in a following section.

Aesthetics

Aesthetics is a measure of how pleasing the system is to the user. Sometimes there is a trade-off between usability or performance and aesthetics. For instance, people may prefer gaudy color schemes in an interface even if they tend to hurt their performance.

These four factors are difficult to measure. However, they should certainly be considered by multimedia developers, as should other usability guidelines and procedures (Nielsen, 1997). The problem, though, is often one of deciding how to handle trade-offs between the different criteria or evaluative factors. For instance, aesthetics may dictate that the multimedia application be overlaid on a black background, whereas the utility perspective may suggest that some of that background space be dedicated to additional navigation tools.

55.8.2 Diagnostic Measures of Usability

Usability is itself a complex construct that may be defined in terms of criteria such as ease of learning and ease of use, or in terms of the structural properties of multimedia that promote usability.

Ease of learning—How fast a reader can start learning the content material and navigating through a hypertext document. Good presentation structure and graphic design are key determinants of ease of learning.

Ease of use—This measure refers to how well a user can navigate through the multimedia. Appropriately defined links, search mechanisms, landmarks, and other navigational aids can greatly increase ease of use.

Error handling—This is a measure of how many errors a reader makes and how easy it is for him or her to recover from such errors. Ideally, multimedia applications should be "forgiving" and errors, even with relatively unskilled users, should be few and far between.

There are also a number of structural measures of multimedia that may be defined, but which (if any) of these is predictive of usability? To our knowledge, there is currently no standard method of evaluating multimedia usability, although there are a number of ways of assessing usability in user interfaces that may possibly be applicable to multimedia documents as well (Jeffries, Miller, Wharton, and Ugeda, 1991; Karat, Campbell, and Fiegal, 1992; Nielsen and Philips, 1993). Methods that may be used include questionnaires and measures of frequency of use for systems that are readily available to a group of users that have free access to the multimedia being evaluated. It is desirable for such global measures to be supplemented with diagnostic measures of multimedia usability that can be used to guide creation and revision of links.

We propose four empirical measures of usability that may be diagnostic of structural properties, recognizing that multimedia usability is probably a heterogeneous concept that may well include a number of different components.

- Node accessability
- Link recognizability
- Landmark recognition
- Convergence of conceptual structure

The idea behind node accessibility is that if one is at a particular node in the multimedia and wants to get to another (target) node, it shouldn't be too difficult to find a path from the current node to the target node. Note here that we are not assuming the presence of browse tools, maps, or indices that may facilitate the access process. Our hypothesis is that node accessibility based on the ability of users to find paths of associative links between pairs of nodes will be correlated with an important aspect of multimedia usability. To put it another way, even if good browse tools are available, it may be desirable to have a network structure where associative paths between nodes are generally easy to find (i.e., browse tools should not be used as bandaids for otherwise poorly connected network structures).

The availability of paths is one aspect of usability; the recognizability of links is another. In usable multimedia, the links that are available for each node should generally make sense to the user. One empirical test of link recognizability is to make up a list containing the links that actually exist for a node, randomly intermixed with distractor links (which may in turn be selected in a number of different ways, each leading to a somewhat different measure of link recognizability) and see whether or not the user can discriminate between the actual links and the distractor links. The user's performance could be quantified in a number of ways, including the use of signal detection theory and a measure of discriminability such as d' (e.g., Wickens, 1992, Chapter 2). A similar strategy could be adopted for landmark recognition where hypothesized landmarks are grouped with distractors and the user is asked to identify the "important" nodes. Different variants of this measure could be constructed by operationalizing the definition of "importance" in different ways.

The final measure of usability to be discussed here was introduced earlier by Teshiba and Chignell (1988). The basic idea behind this measure is that for usable hypertext or multimedia, the user's model of the structure will converge toward the system's representation of structure after a certain period of usage. This measure can be derived using a pre–post test strategy. First users construct a representation of the way they think the information should be structured. Then, they use the multimedia over a period of time. Finally, they are again asked to represent the information structure. Convergence is indicated if the proximity between the user representation and system representation of structure increases between the pretest and the posttest.

Once diagnostic structural measures are identified and usability measures collected, the question is, Which structural measures are predictive of multimedia usability, and what profiles of these structural values are associated with usable multimedia? Perhaps the development of usability profiles may provide a useful basis for predictive assessment of multimedia usability. If so, then it may also provide a basis for restructuring networks to be more usable, thereby providing an important tool for iterative multimedia creation.

55.9 ERGONOMIC DESIGN GUIDELINES

Multimedia "inherits" many of the design rules and principles that govern user interface design in general. For instance, the following interface design heuristics (proposed by Nielsen and Molich, 1990) also apply to multimedia:

Use simple and natural dialogue.
Speak the user's language.
Minimize the memory load.
Be consistent.
Provide feedback.
Provide clearly marked exits.
Provide shortcuts.
Prevent errors.
Provide good error messages (when necessary).

In this section we discuss authoring and design guidelines for enhancing the usability and ergonomics of multimedia. There appear to be relatively few detailed multimedia design guidelines available in the existing literature. The guidelines presented here should be treated as provisional and subject to considerable revision and refinement. These guidelines have been collected from a number of sources and from our own experience.

55.9.1 First Steps in Multimedia Design

The following suggestions provide an initial orientation for multimedia design.

55.9.1.1 State Design Goals as Early as Possible

Set out well-defined goals for the multimedia, in the form of a product specification. This specification should then be referred to in subsequent design reviews.

55.9.1.2 Sketch Out Early Prototypes of the Interaction

Storyboards should be shown to a typical user to test comprehension of the basic interface. If it is difficult to explain to a potential user, revisions need to be made. Writing and testing user documentation can serve a similar function in early evaluation. This guideline echoes the emphasis on rapid prototyping and early evaluation of design concepts that is used in user interface design generally.

55.9.1.3 Avoid Early Commitment

Multimedia design is a special case of design in general. Design generally goes through a divergent phase where different design concepts are explored, followed by a convergent phase where there is commitment to a particular design concept, along with detailing and implementation of that concept. Commitment to a design concept should not be made too early, as it will be difficult and expensive to change to a different concept later on.

55.9.1.4 Be Aware of Interdependencies

The "ripple effect" of changes is dangerous. An example of this is found in the following observation made by the first author when studying the design process in a large aerospace company. During an interview, where an engineer was describing the process of working on part of a satellite, he started discussing the ripple effect. In one project, the engineer found that a particular mechanical structure was too weak. To strengthen it, he increased its volume slightly. This in turn reduced the distance to the surrounding parts, resulting in a heating problem. At this point, a number of engineers had to deal with this problem, trading off issues of size, strength, and thermal properties. In multimedia too, there will be interdependencies. Change in nodes affect the viability of links, and so on.

55.9.2 General Guidelines

Multimedia is a field that borrows from the arts and sciences, in addition to utilizing the latest technologies. As a result, there are many different perspectives on multimedia, and many different views on how it should be designed, constructed, and used. In this section we provide a number of guidelines that have been drawn from different areas and from user interface design in particular.

55.9.2.1 Find a Strong Image or Metaphor

A strong image or metaphor can provide a focal point for the entire project. This helps the design team, and it will ultimately benefit the user. A good metaphor provides a framework for icon design, development of interaction styles, and even overall product packaging. A good metaphor will also increase the predictability of the multimedia for users, providing them with a sense that the interface works as they would expect on the basis of the familiar metaphor. Metaphors are a good starting point for interface design, because they provide building blocks for concepts and language (Lakoff and Johnson, 1980). Since metaphors also serve as models, it is important that interface metaphors imply appropriate actions, and that they provide support in areas where the user's understanding is weakest (Erickson, 1990).

Identification of an appropriate metaphor is a good starting point for hypertext (multimedia) development, but it is difficult to get any one metaphor to adequately cover an entire multimedia application (e.g., Waterworth and Chignell, 1989). More frequently one can identify a collection of metaphors that cover different aspects of the application. The challenge in this case is to make sure that the different metaphors do not conflict with each other. Some metaphors naturally go together, e.g., books and bookshelves.

Sometimes a user interface may succeed in spite of an incongruous mixing of metaphors. The classic example of this is the familiar desktop metaphor in the graphical user interface. In the Macintosh finder for instance, the only way to view the screen as a desktop is if one is looking down on the surface of the desktop. However, the trashcan, and windows are also lying flat on the surface of the desktop. Clearly this is an unusual desktop. Yet the graphical user interface has been remarkably successful, even with these incongruities. It seems that the principles of consistency and direct manipulation overrule or compensate for any inadequacies in the underlying metaphor, at least in this case.

Madsen (1994) provided guidelines for generating, evaluating, and developing metaphors. Madsen's guidelines are summarized and adapted in the remainder of this section, with modifications intended to customize the guidelines for use in multimedia development.

Generating Metaphors

- **A.** Build on existing metaphors.
- **B.** Use familiar artifacts and tools as metaphors.
- **C.** Capitalize on metaphors or structures already existing in the problem or task domain.

The major task in generating metaphors is to provide the benefits of familiarity. The three guidelines above all capitalize on familiarity in some way. As the example of the desktop metaphor shows, a metaphor may "work" even though it contains inconsistencies (e.g., windows on desks). Thus people can learn to handle incongruities in metaphors (e.g., scrolling windows with "elevators" or "thumbs"). However, metaphors should help users decide how to carry out their intentions in terms of actions at the user interface. In this case the ultimate test of a metaphor is in whether it assists or supports appropriate action.

Evaluating Metaphors

- **A.** Check the mapping between metaphor and action.
- **B.** Keep the user in mind.
- **C.** Ensure that the metaphor is well understood.

Metaphors should probably be evaluated in action, since there are examples of incongruous metaphors that nevertheless work. Thus testing of the usability of a user interface can also be a test of the underlying metaphors. However, it is useful to have principles that can screen out potentially unrewarding metaphors if a number of different alternatives are available. The principle of mapping between metaphor and action is particularly important. The performance of a metaphor is ultimately judged on whether it facilitates appropriate action, and if there is not a strong mapping between the metaphor and associated actions, then it is unlikely to be successful in use.

The second and third guidelines above focus on the understandability and meaningfulness of metaphors. A metaphor that is good for one type of user (e.g., a sports metaphor) may not work for another group of users. Thus, metaphors should be meaningful and familiar to the target users of the multimedia.

Developing Metaphors

A. Identify triggering concepts.
B. Adapt the metaphor appropriately.
C. Check the assumptions.

Useful metaphors contain key concepts, which in turn imply actions that are appropriate within the context of the metaphor. For instance, a library metaphor implies actions such as cataloguing and circulation (loaning) of books. There should be a one to one mapping between triggering concepts in the metaphor and key actions in the user interface.

The desktop metaphor and the graphical user interface demonstrate that metaphors can be adapted considerably and yet still work. It appears that metaphors can be adapted to create new metaphors. Consistency and rich feedback appear to be important determinants of success in adapting metaphors.

A metaphor also carries with it a set of assumptions about what is important and what is not. For instance, Madsen (1994) gives the example of how a warehouse metaphor tends to highlight the objects being borrowed (e.g., books), while a meeting place metaphor emphasizes the borrowers rather than the books. In multimedia, for instance, the metaphor of a guide has been utilized. The assumption in this case is that the guide knows more than the user about the topic, or at least about where to find information about the topic within the multimedia document or application. This assumption is violated if the guide is not a credible source of knowledge. For instance, multimedia researchers at Apple Computer implemented a multimedia guide system (Salomon, Oren, and Kreitman, 1989; Oren, Salomon, and Don, 1990). They found that they had to include an introduction from each of the guide characters explaining their background and expertise, so that users would find them more credible.

55.9.2.2 Avoid Excessive Complexity

Multimedia gives one the capability to design rich and complex environments, but there is a lot to be said for simplicity in communication, as the Visual Almanac team noted (Apple Computer, 1991):

> *[O]ur evaluation experience with teachers taught us to keep things as simple and familiar as possible; just giving them images and sounds to retrieve and put in HyperCard stacks was sufficiently breathtaking . . . And above all, the improved simplicity of the collections' structure made it easier for the beauty of the content to show through, and for it to be appropriated and used by our novice users. (p. 44)*

Avoidance of complexity is another way of stating the KISS (keep it simple, stupid) principle that is a cornerstone of applied ergonomics. The KISS principle should be applied to most areas of multimedia design and development, including content development and information design, and authoring (linking).

55.9.2.3 Provide Multiple Entrances

Getting into the right location in a large multimedia document can be critical. Starting off from the wrong place can be distracting and disorienting. Good access points or entrances into multimedia increase the chances that users will "start off on the right foot." The Visual Almanac, for instance, was designed with multiple entries to maintain overall coherence:

> *People typically enter houses through front doors, and books from the first page. In electronic environments, people should be able to enter materials from a very wide range of different directions, depending on their interests and their needs. . . . For example, one can read a composition, open an object and use it as a door to the collections. or one can play with an activity, retrieve the source information for the video and learn more about the image provider. (Apple Computer, 1991, p. 143)*

55.9.2.4 Practice Good Information Design

Utilize the screen layout (workspace, message areas, etc.) to good effect in presenting information, improving orientation, and maintaining consistency. Web browsers are starting to create expectations about consistent layout. In the same way that graphical user interfaces have converged on the desktop metaphor, we can expect multimedia interfaces to converge over time toward a few well-known models or templates that are known to work well for particular families of applications.

Since consistency is so important, it has to be taken into account along with good information design. Sometimes information design that is good in absolute terms may be bad in practice, because it conflicts with prevailing expectations and standards.

On a larger scale, a certain amount of innovation is both desirable and necessary. However, too much innovation within a single project will create problems for users who have grown used to earlier standards and usages.

55.9.2.5 Test and Enhance Internal Consistency

Although it is difficult to provide an overall evaluation of multimedia, it can be quite easy to identify inconsistencies within multimedia. As multimedia is being developed, the following questions should be asked.

- Can all nodes and screens within the multimedia be reached from each other?
- Is there a reasonable entry point?
- Is information design reasonably consistent across the different screens and nodes?
- What actions can be performed on the nodes?
- Are the interactions and dialogs necessary to perform equivalent actions consistent across the various nodes in the application?

Software quality assurance should be carried out on multimedia to answer these questions. Some of these issues can be addressed objectively. For instance, the ability to reach all the nodes in the network from each other can be calculated using graph theoretic measures. Other issues, such as the consistency of information design, will need to be assessed subjectively, both by typical users and by experts in the field.

55.9.2.6 Review User Interaction States

In "vanilla" hypertext or multimedia, and at a general level of description, there are basically only two interaction states, i.e., visiting a node or traversing a link. However, at a greater level of detail there can be many states, such as playing a video, scrolling down a page of information, operating a simulation, building a query, accessing a menu, etc. As multimedia applications mature, they are likely to include more and more computer functionality. Thus the identification and design of user interaction states will become increasingly important. Problems with interaction may stem from information design, content, use of media elements, or some combination thereof.

For each state that exists in the multimedia, authors and designers should ask the following questions:

- What is the goal of the current interaction (why is it being included)?
- Is the interaction achieving this goal effectively?

55.9.2.7 Implement Effective Help

Task-oriented multimedia needs help just like other computer applications. There should be both interface help (i.e., how to use the tools provided in the interface) and content help (i.e., what is available, how it is organized, and where different pieces of information may be found). Ideally, this help should be context sensitive. At a minimum, context-sensitive help should be available for the various navigational tools and buttons. This may include the use of "bubble help," where the user can move the cursor over an interface object and get a brief preview or description of what clicking on that object will do.

The help system for a multimedia application should be coordinated with overviews and maps. In addition, these overviews and maps should be annotated with summaries of the different neighborhoods that exist in the multimedia application.

A neighborhood (or district) is a region of closely related nodes. In terms of structure, a neighborhood can be defined as a collection of nodes that are all easily reachable (have short path distances) from each other. Thus a neighborhood can be analytically identified as a cluster of nodes that have short distances between each other, and longer distances to nodes outside the cluster (neighborhood). Neighborhoods may also be defined in terms of related concepts. For instance, nodes that refer to a particular country or location may be part of the same conceptual neighborhood even though they are not closely related in terms of linking structure. Thus conceptual neighborhoods may not necessarily correspond to the structural neighborhoods defined by linking patterns.

55.9.2.8 Use Composite Nodes

A composite node consists of a cluster of individual nodes and links that belong together (e.g., Chua and Lai, 1991). Composite nodes can reduce the structural complexity of multimedia if they are designed so that there is only one incoming link (gateway) into the composite node (Thüring, Haake, and Hannemann, 1991).

Composite nodes represent a layer of abstraction in multimedia. Instead of having to deal with all the nodes, the reader can deal with a reduced number of nodes, with some of the detail being nested within the composite nodes.

55.9.2.9 Use Terminology and Labels Appropriately

Nodes and terminology within nodes should be labeled clearly. Efforts should also be made to use terminology consistently within the multimedia document or application. In a large application, text analysis tools can be used to support the author in rooting out terminological inconsistencies. For instance, latent semantic indexing (Deerwester et al., 1990) or lexical chaining (Morris and Hirst, 1991) can be used to identify terms associated with topic clusters. Terms within topic clusters can then be examined to see whether the number or diversity of terms can be reduced (e.g., by restricting the unnecessary use of synonyms or different phrases with equivalent meaning).

Terminology and labeling is also important for links. Use meaningful link labels to make link semantics clear to readers, and name nodes and composites clearly, making sure that names represent what they are naming. As Thüring et al. (1991) point out, this helps to increase the overall coherence of the multimedia.

Link labels should generally be verbs. It should then be possible to construct comprehensible sentences of the form source–link–destination. Thüring et al. (1991, p. 167) give the example: "robot reply–criticizes–Chinese room argument," where "criticizes" is the link label (verb).

55.9.2.10 Design for Specific Types of User

Thüring et al. (1991) recommended tailoring the information and its organization to particular types of reader.

> *For example, instead of writing completely new documents for novices and experts of AI the author can define two diverse organization structures by referencing different elements of the content net thus creating reader specific variants of his document. These variants can differ from each other with respect to several features: they may contain different parts of the content net, they may use different content nodes as the reader's starting point, and they may arrange the selected elements of the content net in different orders. (p. 168)*

Authors should also choose an appropriate starting point for each type of user. In reality, not every user will visit every node in a multimedia document or application. Thus a large multimedia application may actually be thought of as a large number of different multimedia documents, with each document being defined by a different usage. Ideally, different users should utilize different subgraphs within the overall multimedia network depending on their interests and needs. If the needs of a particular class of users can be identified, they should have a separate starting point that launches them into the appropriate region of the multimedia network.

55.9.2.11 Provide Navigation Alternatives

People should have different ways of navigating through multimedia: "providing a handful of simple ways to travel from one object to another (by keyword, object search, text type or random choice) lets people create many paths through a rich territory, without getting lost or hitting a dead-end." (Apple Computer, 1991, p. 46).

This principle was exploited in the SuperBook project (Remde et al., 1987; Egan, Rembde, Landauer, Lochbaurn, and Gomery, 1989), where an electronic book was created that allowed people to conduct searches, follow hypertext links, move via a table of contents, etc. Valdez (1992) found that people tended to use a mixture of different navigation tools when the different tools were all made available.

55.9.3 Linking

Linking is often regarded as the most important element of multimedia authoring, yet there are few guidelines on how it should be carried out. Clearly, links should be authored with the user in mind and should represent meaningful associations and relations between different nodes. Links should also be supplemented (as discussed elsewhere in this chapter) with effective overviews, maps, and other prosthetics.

55.9.3.1 Check Links and Link Anchors

Linking involves relations between nodes. Every link is a traversal from one presentation to the next. The resulting transitions should be smooth, or at least understandable. Ideally, links should be labeled, so that readers have a good sense of the type of relation or link that exists and that can be traversed.

In addition, the choice of link anchor is important, as the anchor may serve as a prediction of the information content that exists at the end point of the link. Generally, one should avoid including the same link (anchor) multiple times within the same node, since this will inflate the apparent number of items attached to that node, and users find it annoying to go back to a place already visited when they expect to go somewhere different.

55.9.3.2 Use Dynamism

Static multimedia is limited in that a constant set of links is used for all readers. Some flexibility is provided by the fact that readers can choose different paths through the multimedia. However, the author may not always be able to anticipate and construct all the possible links that will be of interest to different types of user. Furthermore, putting all links that might possibly be useful in multimedia would be counterproductive, since many of the links would be irrelevant or distracting to any particular reader.

Creating links dynamically is a way of solving both these problems. Dynamic links can be more responsive to the interests of a particular reader, and dynamic links can be constructed as needed, reducing the overall number of links that have to be considered when navigating through the multimedia.

Golovchinsky and Chignell (1993) provide an early example of more autonomous user navigation, in their Queries-R-Links system. Users start by making queries (i.e., by requesting the system to search for something) that are expressed as markup on the text of a document. Documents that match the query are then returned in a hit list (a menu or selectable list of one line excerpts from each document). Documents can then be selected from the hit list and displayed in a text window for further markup. In this approach the user implicitly defines links by markup of text followed by selection of link end points. This type of system can then be extended to a link authoring environment, by saving graphical markup as link anchors, and saving selected documents as link end points (with an option to have all documents for the query serve as a larger, but secondary set of link end points).

Figure 55.12 shows a screen from a recent system based on the Queries-R-Links approach, where a newspaper metaphor is used to display multiple documents at the same time. This approach could be extended to authoring by "freezing" the graphical markup to create a permanent link anchor on a document or set of documents. In this way, user-defined links can be created from graphical queries with little authoring effort.

The graphic representation of system-constructed links from user queries satisfies an important general need when computer-initiated action is taken, that of enactment (Bernstein, 1993). Enactment refers to the requirement that users be informed by the system about the action that has been taken. In other words, whoever has structural responsibility (system or user) also has responsibility for enactment and must communicate their actions to the other in some way (either descriptively or referentially; see Waterworth and Chignell, 1991).

Currently, dynamism is implemented in multimedia using information retrieval methods, such as those described in the preceding paragraphs. Fortunately, dynamic linking makes the most sense in situations in which information retrieval is appropriate, i.e., where the multimedia consists of many documents, only some of which will be useful for a user carrying out a particular task.

Thus, if it is possible to carry out meaningful information retrieval and search on the multimedia content, then it is probably better to let users build multimedia interactively through query and concept markup and similar heuristics and techniques, than it is to try and rely entirely on static links. Note that our recommendation is for dynamic links to supplement, rather than replace, static links.

55.9.4 Prosthetics

A number of tools and techniques have been developed for navigating through multimedia. In this section we consider some guidelines associated with these techniques.

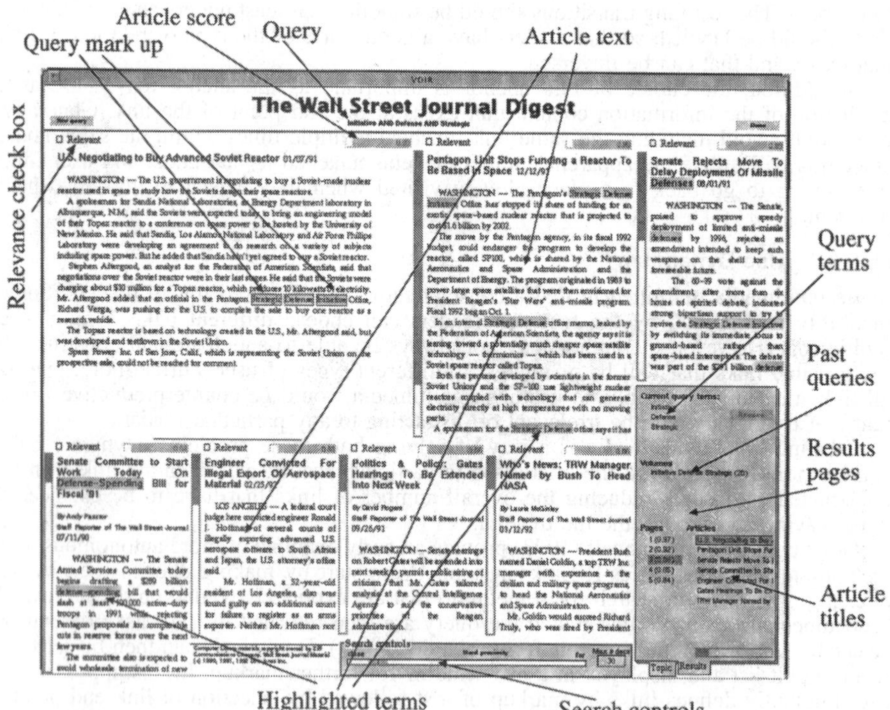

Figure 55.12 A screenshot from the VOIR system showing query markup combined with the Newspaper Metaphor.

55.9.4.1 Use Overviews and Organizers

Orientation is one of the key ergonomic issues in multimedia. Users need to have a sense of where they are and where they can go. Overviews and organizers help provide the necessary sense of orientation in multimedia. Maps, tables of contents, and other tools help users to plan their interactions more effectively. However, they should be designed into the multimedia application in such a way that users can ignore or bypass them if they want to.

General methods for constructing overviews and multimedia are becoming available. An example of this approach is the development of query graphs out of query histories by Charoenkitkarn et al. (1996). Another promising approach is to create map overviews automatically from the multimedia structure (Hasan et al., 1995a, 1995b).

For historical and geographical material, maps are frequently available. For instance, it is natural to include a map to document Napoleon's march on Russia, and this has been used to good effect in graphic design (Tufte, 1983). Sometimes, however, maps developed by original authors are inadequate or unavailable. In this case multimedia authors are faced with the task of creating a meaningful overview of the information.

55.9.4.2 Provide Tools to Assist Authors and Developers

Multimedia authors need specialized tools for multimedia development that will support them in building multimedia efficiently while preventing the inclusion of unnecessary problems such as dangling references.

Perez (1991) recommends use of the following tools to support authoring of hypertexts:

1. Tools for listing links and nodes
2. Tools that provide information about nodes
3. Tools for outlining
4. Tools for indexing
5. Tools for word frequency and analysis reporting

These tools help an author maintain awareness of the organization and content of material to be added to multimedia, and of the material currently contained within a multimedia document currently under construction.

Authors should also find tools for automatically generating maps and overviews particularly useful. Mukherjea and Foley (1995a, 1995b) discussed a number of these techniques, including graphical layout methods for representing graphs as networks (see also Noik, 1994), building hierarchies out of graphs, clustering based on link data, and creation of landmarks. Much of the discussion of such tools has focused on how they can assist readers or users of multimedia. However, these tools are likely to be equally useful for authors.

Other tools that are notably lacking but sorely needed are tools for cooperative authoring of multimedia. These tools will likely parallel other tools for collaborative editing in computer-supported cooperative work (see Baecker, 1993, for a compilation of research in this area), including the ability for multiple authors to work at a distance, edit each other's work, and maintain satisfactory version control.

Annotation is particularly important in collaborative authoring of multimedia. Annotation allows readers to communicate with each other through multimedia, and thus increases the value and usability of multimedia, by allowing users to share their experience. However, it is very hard to control the process of annotation. Thus the value of annotation depends to a large extent on the ability and motivation of the user community of a multimedia document. In the Web for instance, hotlists placed at different Web sites are used as tools that assist navigation by others. Some sites become prominent not for their intrinsic content, but for the lists of other sites that they contain.

In large stand-alone projects, such as an encyclopedia or a dictionary of art, the finished multimedia document results from the collaboration of many people, including writers, editors, and artists. Annotation is a way of communicating asynchronously between other collaborators, through the document itself. Furthermore, annotation may form the basis for further content material. The issue of how to create tools that encourage effective annotation have yet to be addressed, but represents a fruitful research area for study in the ergonomics of multimedia.

55.9.5 Structural Guidelines

In addition to the general guidelines listed above the creation of appropriate structures and representations can generally improve the usability of multimedia. Useful structures include: maps, hierarchies, neighborhoods, paths, and landmarks. Each of these structures will be considered in the remainder of this section.

Multimedia typically has a network structure, where the network is constructed as a set of interconnected nodes. The structural properties of multimedia networks may provide useful diagnostic tools for evaluating and improving multimedia (see also Parunak, 1991). The structural analysis of networks that has been developed in disciplines as diverse as mathematical sociology, and the analysis and control of switching in telecommunications networks, can be utilized in developing these tools. All of the structural measures mentioned in this section have well-known algorithms (e.g., Burt, 1982) associated with their calculation and may be estimated using existing and readily available software packages.

The structural properties of networks can generally be divided into two general types, i.e., node properties that refer to individual nodes, and network properties that refer to the network as a whole. The importance and reachability of a node may both have usability consequences, but quantitative measures are needed that correspond to these basic concepts. In the following paragraphs we introduce three measures of node importance, each of which is based on how well the nodes are connected to the rest of the network.

One measure of node importance involves the direct connectivity of that node to other nodes in the network. Connectivity may be defined in a number of different ways. One way is to select a path distance (such as 2) and count the number of other nodes that are no more than two links away from the node in question. Another method is to calculate the average (outgoing) path distance from the node in question to all the other nodes in the network. Note that the path distance in this case between two nodes is calculated as the minimum number of links necessary to get from one node to the other.

A second measure of an individual node is its reachability. In the extreme case, a node is unreachable if there are no incoming links to it. In practice, nodes may be theoretically reachable, but nevertheless difficult to get to from most places in the network. Thus we may estimate the reachability of a node as the average incoming path distance. For instance, if we wanted to know the reachability of node A we would ask what is the average (e.g., mean) of the path distances from B to A, C to A, and so on. Using this definition, reachability is the conceptual inverse of connectivity, since reachability is concerned with incoming path distances while connectivity is concerned with outgoing path distances. This distinction arises in multimedia networks where asymmetric or unidirectional links may be defined.

A third measure associated with an individual node is its centrality. Centrality refers to the position of the node within the network. In general, measures of centrality will be correlated with measures of connectivity. One measure of centrality (i.e., the average path distance of the node in question to other nodes) coincides with measures of connectivity and/or reachability that were defined earlier. An alternative measure is based on the proportion of links in which a node is involved. Thus, if a node were involved in 10% of the links (both ingoing and outgoing) in the network and a second node were involved in only 5% of the links, then the first node would have higher centrality according to this measure.

In addition to the structural properties of individual nodes, one can also measure the overall structural properties of networks. Concepts that describe the overall properties of a network include:

Density
Compactness
Clustering

Density can be simply defined as the proportion of possible links that actually exist. It can be expressed either as a decimal fraction or as a percentage. It is instructive to calculate the density of links in hypermedia applications that are in use. We have found density measures of a fraction of 1% to be typical in manually authored hypermedia networks. If, as seems likely, human authors will tend to connect a roughly constant number of links to each node (on average) then it can be seen that the density of links for authored hypermedia will generally decrease with larger network size since the number of authored links would be increasing linearly, while the number of possible links would be increasing in proportion to the square of the number of nodes[1] (as shown in Table 55.2).

Compactness may be defined as the average path distance between pairs of nodes in the network. The precise measure of compactness will depend on which measure of central tendency is used to compute this average (e.g., mean or median).

Clustering is a global property of multimedia networks that is more difficult to operationalize than density and compactness. The basic concept of clustering is expressed in the question "To what extent do the nodes within the network fall into well-connected subgroups?" In sociological analysis of networks these subgroups are sometimes referred to as cliques. However, the term "clique" carries with it the notion that each subgroup is fairly disconnected from other subgroups, which may not generally be true for multimedia networks. One strategy for identifying clusters is to develop a correlation matrix by correlating rows and columns of the path distance matrix. Each correlation would then be the extent to which two nodes share the same pattern of path distances (i.e., if one node is close to a third node, so is the other; if one node is distant from an outside node, so is the other). This correlation matrix could then be clustered using one of a number of clustering techniques that are available (e.g., Everitt, 1990). Alternatively, the path distance matrix could be clustered directly. One hypothesis is that multimedia networks that contain well-defined clusters may be more usable. In addition, clustering information may be used in building browse tools. For instance, a browser could be created that allowed the user to travel between different clusters or "zoom in" for a detailed look at a particular cluster.

55.9.5.1 Build In Good Structure

The navigation task tends to be simpler when users can explore within collections or neighborhoods, thereby focusing on coherent subsets of nodes. In large multimedia databases, such neighborhoods may be hard to define ahead of time, since potentially there can be very many of them, and they may overlap. In such cases, filters can be used to construct neighborhoods at runtime, as they are needed.

The principles of urban architecture outlined by Lynch (1960) are highly relevant here. For instant, clustering may be used to define districts or neighborhoods within multimedia applications. Calculation of minimal spanning trees, combined with analysis of node content may be used to define interesting paths through the multimedia.

Paths should also be defined on the basis of what sequence of interactions is appropriate for people carrying out specific tasks. Development of paths is particularly relevant for online documentation, where they may be many paths, each corresponding to a particular problem or information need.

Table 55.2 The Relationship Between Number of Nodes and Density When a Consistent Number of 5 Links per Node Is Used

Number of Nodes	Number of Possible Links	Number of Links (5 per node)	Density (%)
20	380	100	26.3
50	2,450	250	10.2
100	9,900	500	5.1
500	249,500	2,500	1
1,000	999,000	5,000	0.5

[1]The density is equal to the number of links divided by the total number of possible links ($N*[N-1]$). If we assume a constant number k of links per node, then the number of links becomes $k*N$, and the density is then equal to $k*N/(N*[N-1])$, or $k/N-1$ after dividing top and bottom by N. Thus, the density will generally be inversely proportional to the number of nodes, unless authors, or automated linking tools, increase the number of links per node as the number of nodes increases.

55.9.5.2 Restructure Networks

Creation of paths, landmarks, and neighborhoods adds value to multimedia with superordinate structures that complement the basic nodes and links. Multimedia can also be enhanced by restructuring the basic network so that only the most useful links are used. In the following discussion we suggest some simple methods for adding and removing links in multimedia. This suggestions are offered with the disclaimer that they have not been subjected to much empirical testing and that they address the structural, rather than conceptual or semantic, features of links.

We anticipate that there may be an inverted U-shaped relation between network density and usability. That is, networks with too many or too few links may be difficult to use, with some moderate level of density proving to be more usable. In some cases, links may be added or removed to provide an appropriate network density. Even when density is at an appropriate level, ineffective links may be removed to make way for more useful links.

One heuristic for adding links is to connect nodes that have low path distances, but are not currently connected (a transitivity heuristic). Another approach is to use the correlation between path distances as the criterion for establishing new links. Thus if Nodes A and B both tend to be at short distances from some nodes, and large distances from other nodes, then they should be linked together, on the grounds that they are similar in their relationships to other nodes. Naturally, one can evaluate whether or not link revisions using such heuristics or criteria are effective. For instance, one could test link recognizability before and after link revision to determine whether the revision led to improvements in estimated usability.

A second heuristic for adding links is to emphasize links to important nodes. This reflects the expectation that important nodes are likely to have more incoming links. This expectation has a theoretical basis in terms of asymmetric models of similarity (e.g., Tversky and Gati, 1978) where the similarity from A to B is stronger than the similarity from B to A if B is a stronger referent (i.e., better landmark) than A. Thus, one might want to create links between landmarks (important nodes) and to create additional links from regular nodes to the landmarks they are close to.

In general, given the low link density of manually authored networks, link addition is likely to be used more frequently than link removal. However, when link removal is necessary, it could also be based on an adjacency criterion. That is, consider links as candidates for removal if they connect nodes that otherwise appear to be very dissimilar. Once, again the correlation between path distances could be used as a criterion. Since it is easy to skip a poor or irrelevant link, but impossible to traverse one that is not there, link removal should be done cautiously and sparingly.

One of the most important criteria for link removal is that it should not significantly reduce the overall connectivity and (compactness) of the network (since we expect connectivity-compactness to be an important determinant of usability). For instance, one could examine the compactness of the matrix before and after the link deletion to determine whether or not the loss of compactness associated with the removal of the link was too great.

Voting procedures based on actual usage can also be used as a basis for adding links. If links are traversed frequently by users, this can be taken as a sign that they are useful. In contrast those links that are traversed infrequently, or not at all, are candidates for removal. However, voting procedures should be used for link removal with caution since some links could be used relatively infrequently, but still be very important for a particular group of users.

55.9.5.3 Create a Hierarchical Organization of All Nodes

Thüring et al. (1991) recommended that composite nodes be used to form hierarchical structures within multimedia, thus increasing the coherence of the resulting structure. Hierarchical structures provide overviews of multimedia and may also serve as useful access mechanisms. Hierarchies provide users with familiar organizational structures (e.g., tables of contents and menus).

A number of methods are available for creating hierarchies in multimedia. These include hierarchical clustering (e.g., Everitt, 1990) and techniques described by Mukherjea and Foley (1995a). However, our recommendation is that multimedia authors should be directly involved in the clustering process. Clustering techniques tend to be fairly ad hoc, and can be usefully supplemented by human judgment.

One approach to organization multimedia is to use a clustering tool, followed by reassignment of the resulting clusters based on the judgement of multimedia authors (who could be content developers or multimedia developers). Another approach is to let the authors define high-level clusters, and then use partitioning techniques (e.g., K-means clustering) to obtain more detailed clusters within the hierarchical structure.

55.9.5.4 Build Neighborhoods

The neighborhood or district was one of the key architectural concepts that Lynch (1960) identified as enhancing imageability. Hierarchical clustering can be used to organize multimedia nodes into layers representing different levels of abstraction. Each cluster then represents a neighborhood within the multimedia application and could be represented by one or more landmarks within a higher level of abstraction (which might be associated with a browsing map). Tam and Chignell (1992) characterized a neighborhood as a region consisting of combinations of physically, relationally, syntactically, and/or semantically "close" nodes. This suggests that a single node may exist within a number of different neighborhoods, each of which is defined by the notion of distance or closeness that is used in the clustering criterion that defines the neighborhood.

Once neighborhoods are constructed they should be highlighted within the user interface. For instance, an overview map might be created showing the neighborhoods and their conceptual relationships to each other. Neighborhoods could then be used as access points, with users being able to select a neighborhood and then enter the corresponding subgraph or region represented by that neighborhood.

55.9.5.5 Create Meaningful Sequences and Paths

The multimedia developer or author should create sequences (paths) of links to assist the reader (for instance a multimedia document could consist of a sequence of three composite nodes such as introduction, overview, and summary).

Bush (1945) referred to paths as "trails" and saw them as one of the key tools in his memex model of hypertext/multimedia. He envisioned that these paths or trails would be manually authored by a profession of trailblazing. Most modern multimedia and hypertext applications do not include built-in paths to any great extent. This may be one of the reasons why multimedia tends to be difficult to navigate. Lynch (1960) in his seminal work on design of cities argued that the existence of well-defined paths was a key component contributing to the imageability of a city.

Manual authoring of paths in large multimedia applications and documents (of which the Web is the largest) is a daunting task. However, there are a number of tools and techniques which assist in the development of paths, as discussed in the following paragraphs.

A path is a kind of thread running through multimedia. Threads can be identified by term or by concept. The simplest way to recognize a thread is to base the recognition on a single term. For instance, a product called SmartText (marketed by Lotus before they were bought by IBM in 1995) built threads on the basis of a selected term. If one selected a term such as "Alaska," all the instances of Alaska were then linked together in sequence (this method requires that the underlying text documents be linear, i.e., sequenced).

There are drawbacks to relying on a single term in this way, because a single term may not be sufficiently diagnostic of what information will be relevant to a particular thread. On the one hand, the term "Alaska" may appear in many different contexts, only some of which are of interest to a thread or path. On the other hand, some sections of content material may be relevant to the thread even though they never mention the word Alaska (for instance, discussions on the city of Anchorage, the Exxon Valdez oil spill, or oil drilling on Prudhoe Bay). Concept-based thread recognition can be used to address these problems. In this case the thread is described and identified in terms of concepts. Three general approaches to concept identification are passage recognition, lexical chaining, and authoring.

Passage recognition relies on a textual description of the concept as a paragraph of text. The thread or path is then built up as the successive instances of this concept within the sequence of text that underlies a multimedia document. Instances of the concept are recognized based on their similarity to the passage used to define the concept. This similarity can be calculated in a number of ways. One of the simplest is to extract the content bearing words from the passage (ignoring stop words such as "and," "or," "the," etc.) and then use a vector space (Salton, 1989) or similar model to identify matching sections of text based on similarity of word usage.

Lexical chaining (Morris and Hirst, 1991) is a technique that uses word associations to convert words into contexts. Lexical chaining can be applied to the text by using a thesaurus to identify clusters of related terms. Instances of terms belonging to each conceptual cluster are then linked together to form lexical chains. Each lexical chain running through a sequence of text is then a potential path.

The final approach considered here is authoring, which is similar to the "trailblazing" originally envisaged by Bush (1945). In the most extreme case, authoring consists of manually selecting each node in a path and then sequencing all such nodes. This process might then be repeated many times, once for each of the paths created.

With currently available tools, path creation is probably best done with a judicious mixture of automated tools and manual authoring. Techniques such as passage recognition and lexical chaining can be used to:

1. Identify possible paths
2. Identify possible nodes on a given path

The human author can then edit the paths and nodes, add in new nodes where necessary, and sequence the nodes within paths. The paths should then be signposted and highlighted in the user interface in some way, so that users know of their existence and how to follow them. One technique is to add a path icon or marker to each link. Each link would then function like a highway offramp, and, as with highways, different paths could be numbered. Clicking on a link with the number "11" next to it might take the user to the next node in path number 11. Moving the mouse pointer over the link anchor without clicking the mouse button might bring up a help balloon with a description of what path 11 represents.

55.9.5.6 Develop Landmarks

It has generally been assumed that knowledge of spatial structures (and by analogy, knowledge of networks) is organized by landmarks. Fisheye views (e.g., Furnas, 1986) extend this idea by developing maps from a particular vantage point where many close landmarks are shown and correspondingly fewer distant landmarks.

Valdez, Chignell, and Glenn (1988) studied the use of landmarks in hypermedia navigation. They hypothesized that landmarks will (1) be better remembered, (2) tend to represent either basic level or high level categories, and (3) will have a higher connectivity value.

Valdez et al., (1988) also carried out a study that investigated how well different nodes served as landmarks within a prototype software application. They found a strong correlation ($r > .6$) between a structural measure of node connectivity and an empirically obtained measure of landmark quality (subjective connectivity) where subjects estimated whether or not one node was on a path between two other nodes (the more paths a node was judged to be on, the better its assumed importance or landmark quality). This small study (with eight subjects) suggested that users may be sensitive to the structural properties of multimedia networks and that measurable structural properties may be determinants of usability. Thus, the study of landmarks offered the first clue that structural properties of a network might be used to enhance usability.

Valdez et al. suggested several methods for identifying landmarks in hypertext and developed a method for empirically assessing the quality of a landmark. The basis of that method is the intuitive notion that landmarks often function as waypoints on the path from one location to another (e.g., "Go down Pico Boulevard until you see the bowling alley and then turn right").

Valdez et al. defined an operational measure of landmark quality based on path inclusion (PI) by asking people to think about getting from point A to point B and then judging whether a point C (potential landmark) is on the path between A and B. On average, C should be more likely to be selected as being on the path if it acts as a landmark. Valdez et al. used an experimental procedure where in pairs of points were displayed and a participant was asked to judge whether or not a third point lay on a path between the first two points. This PI (path inclusion) landmark judgment task was implemented as a HyperCard stack on the Apple Macintosh computer.

The landmark quality of a node was calculated as the proportion of times in which it is judged to be on the path between two other nodes. Valdez et al., (1988) found that subjects could make these judgments and that the resulting empirical measures of landmark quality correlated fairly well with theoretical predictions based on the second-order connectedness of the nodes in question and their memorability or psychological salience.

The second-order connectedness of a node is the number of other nodes that can be reached from that node by following no more than two links. The suggestion is that this measure can be used to calculate important nodes which might serve as landmarks in an overview. In a related approach, Botafogo, Pivlin, and Shneiderman, (1992) suggested that nodes with high back second-order connectedness also make good landmarks. Back second-order connectedness is the number of nodes that can reach the candidate node in no more than two steps. In addition, outdegree and indegree (the number of nodes that can be reached *from* the node by following one link, and the number of nodes that can reach *to* the node by following one link, respectively) also provide evidence of relative importance.

Mukherjea and Foley (1995b) suggest that when importance is calculated as a weighted sum of second-order connectedness (SOC), back second-order connectedness (BSOC), outdegree (O) and indegree (I), a landmark can be defined as a node whose importance is greater than a threshold, and they used a threshold of 10% of the total number of nodes in the information space. Their procedure is:

1. Calculate: importance = (I + O) * wt1 + (SOC + BSOC) *wt2 where wt1 + wt2 = 1.0 (they found best results with 0.4 and 0.6)
2. If importance > 10% of total n of nodes, then the node is a landmark.

Once identified, information landmarks should be visually and organizationally highlighted. Landmarks can be visually highlighted through the use of color, animation and distinctive shape. Organizationally, landmarks can be highlighted by placing them near the top of hierarchical overviews or tables of contents, and by ensuring that they are well connected to other landmarks. Organizational highlighting increases the connectedness of landmarks and thus inflates their importance as estimated by the equation shown above.

In the real world, landmarks are defined not only by connectedness, but also by attributes such as distinctiveness and incongruousness. For instance, a nondescript building might become a landmark if it is badly burnt in a fire or if it is painted with a colorful mural.

In multimedia, landmarks often emerge from structure, rather than by design. For instance, in the World Wide Web, the independent actions of millions of authors create a single (rather confusing and overwhelming) structure. This structure is dynamic and continuously evolving, as new nodes are added and links get made, broken, and made again. Landmarks arise in the Web, often by design, but sometimes not. In the relative anarchy of the Web, there is little control over what connects to what. In particular, if back second-order connectedness is the main driver of what is a landmark, as Mukherjea and Foley (1995b) suggest, then the establishment of landmarks on the Web is largely out of the control of the authors of the original node. Instead, they must rely on others to point or link to their node before it can become a landmark.

55.10 UNSOLVED ERGONOMIC PROBLEMS

In this section we discuss some of the criticisms that have been leveled at uses of multimedia (cf, Parsaye and Chignell, 1993, pp. 273–278), most of which follow from an ergonomic critique of the current state of the art.

55.10.1 Multimedia Is Often Used Inappropriately

Multimedia is sometimes accused of being "shallow," of having too much "glitz" and not enough content. One of the enduring problems of multimedia is that there is no single task that motivates the development of the technology. Concepts such as "docuverse" and "world brain" are simply too broad and remote to provide a direct impetus to research and development. It can be argued that currently the development of multimedia is being driven by the World Wide Web, a vast global experiment in mass authoring of multimedia. As of 1997, any time spent surfing the Web should convince the user that with some exceptions, the Web does not seem to be a particularly efficient way of communicating information. In the present authors' view, the popularity of the Web comes from the content that it contains, not from the way that content is organized or presented.

Aside from the basic content, there is a great deal of variation in the implicit models of multimedia that are used. There are many different categories of literature (e.g., plays, poems, biographies, novels, etc.), likewise there are many different types of multimedia. In literature, methods have developed for assessing good form in the various styles that are used. Multimedia has typically been much less disciplined.

Research Challenge: Practical methods of multimedia engineering are needed that take into account the different styles of multimedia that are appropriate to different tasks. Multimedia engineering should be based on a meaningful taxonomy of different types of multimedia, where each of these types is linked to one or more types of communication task for which it is useful or appropriate.

55.10.2 A More Effective Model of Browsing Is Needed

People get lost in large multibranching multimedia applications. Successful browsing generally requires at least a minimal awareness of structure. Multimedia systems, as currently implemented, generally give the user little assistance in figuring out the structure of the information. This is in contrast to books, where the structure is provided by a hierarchical grouping of material into sections and subsections, which are then listed in the table of contents. Cues about information structure are generally very helpful to the user (e.g., Egan et al., 1989; Valdez, 1992).

Currently, browsing seems to be conceived of as an all-or-nothing process. Either one is browsing by going from link to link, or else one explicitly enters a search mode in order to track down a particular document or piece of information. However, people often want to browse in a particular topic area. They generally do not want to follow links and then find themselves off in an unrelated topic. Even if it is interesting, the new information may distract them from the task at hand. Thus, current browsers seem to be suited for open-ended exploration rather than for task-oriented information gathering.

It seems natural to mix retrieval and browsing. For instance, one can identify a set of documents that are on the topic of interest and then browse among those documents. This tends to be what people do when they are browsing in the library. People find an area of the library that they are interested in (e.g., psychology or business) and then browse through the stacks where books on the topic of interest are located.

Research Challenge: New models of browsing are needed that allow constrained exploration within a neighborhood or topic.

55.10.3 Screen Real Estate Needs To Be Better Managed

One problem with early browsers in large hypertexts was a difficulty in showing all the information on a single screen. The browsers got so large that they had to be split up into a number of screens and this led to navigation problems within the browser, which is ironic, since the whole idea of browsers was to alleviate navigation problems in the first place. However, a number of researchers are working on technological solutions to this problem. Such solutions include the use of three-dimensional hierarchical browsers (e.g., Card et al, 1991), fisheye views (e.g., Sarkar and Brown, 1992), and outlining tools. The use of transparent layers of information (Harrison, Ishii, Vicente, and Buxton, 1995) to arrange windows in depth rather than around the screen is a particularly interesting way of reducing screen real estate requirements.

Research challenge: Find ways to reduce screen real estate requirements without reducing functionality.

55.10.4 Usable Multimedia Is Needed on a Large Scale

Early multimedia systems were about as common as vintage Rolls Royces, but a family sedan type of multimedia is needed for large-scale use. More realistic and heuristic methods of hypertext/multimedia engineering are needed to develop multimedia on a large scale (cf. Glushko, 1990). Manual authoring of a relatively small size hypertext can be time consuming and prone to errors. Manual authoring in Project Jefferson (Chignell and Lacy, 1991), for instance, was found to lead to errors in the linking process because of fatigue and forgetting. For large-volume hypertexts, the common approach today is to rely on indexing documents in order to then use Boolean operators and information retrieval techniques based on word frequency techniques (Salton and McGill, 1983). However, Boolean searches with simple queries yield too many entries to be useful, including many that are irrelevant (Glushko, 1990).

One of the current problems in manually authoring hypertext is that there are few good tools available to support the task, so that it is even more labor intensive than it needs to be. Size and complexity are difficult for any database or information system to handle.

The problem is particularly severe with multimedia because the number of possible binary (leave alone n-ary) links in a graph (hypertext structure) increase in proportion to the square of the number of the nodes in the structure.

The importance of a tool and publishing medium is amply demonstrated by the dramatic growth in the World Wide Web, where a relatively simple markup language and communication protocol led to the development of millions of documents in a couple of years. Now methods are needed to coordinate these largely individual efforts into large-scale multimedia documents and applications.

Research Challenge: Tools are needed for large-scale multimedia engineering.

55.10.5 Linear and Nonlinear Structures Should Be Integrated

Linearity and nonlinearity are both useful structuring principles. Many early proponents of hypermedia celebrated its liberating nonlinearity (e.g., Nelson, 1981). However, experience with highly nonlinear hypertexts has shown that they can be disorienting and confusing. Multimedia systems are now evolving to the point where they contain both linear and nonlinear features (e.g., the SuperBook system, Remde et al., 1987).

The notion of nonlinearity in hypertext is sometimes confused with the general responsiveness of electronic text (Parsaye and Chignell, 1993). People have compared hypertext vs. hardcopy but this really confounds two issues: linear vs. nonlinear, and hardcopy vs. electronic. Linear features are functions that can be found in linearly designed documents (whether in electronic or paper form). Examples of linear functions are: going to the table of contents, going to the index, flipping to the next or previous page, etc. Nonlinear functions have a different use: e.g., the users can "jump" to a different page in the document where more information on some topic of interest is presented to him or her.

Valdez (1992) developed methods for testing the use of linear and nonlinear functions in books and hypertexts. His research showed that nonlinear functions (jumping between sections and papers) are sometimes used just as frequently in printed documentation as they are in hypertext documentation. He also found individual differences, with some people tending to rely more on nonlinear functions, whereas others relied on linear functions. These results suggest that the intuitive idea that hypertexts are nonlinear and books are linear is incorrect. Linearity vs. nonlinearity is as much a function of the strategy a person uses in reading a material as it is a result of the particular structure or presentation of the material in the document. Since people tend to have different navigational styles, and they use whatever tools or cognitive prosthetics (Wright, 1991) are available, in most applications a judicious mixture of linear and nonlinear structuring will work best.

Research Challenge: Determine the different styles of navigation that are used and how they may best be accommodated by a multimedia application. In particular, the relationship between linear and nonlinear structuring and navigation strategy needs to be examined.

55.10.6 Associative Structures in Multimedia Should Be Compatible

It has been argued that associative multimedia should be easy to use because human memory is associative. This argument ignores the fact that the actual structure and content of associative memories may differ widely between people. Thus for many applications there may not be a single associative structure that will be compatible with everyone's associative memory.

Parsaye and Chignell (1993) argued that the psychological justification for the large scale use of nonlinearity in multimedia and hypertext is ill-founded. They cited a number of psychological experiments that demonstrate that linearity in processing new information is a central component in the dominant model of human information processing and cognition (e.g., Anderson, 1985; Lindsay and Norman, 1977; Neisser, 1967).

Research Challenge: Methods are needed for identifying associative structures that are maximally compatible for a wide range of users. Usability studies are needed to evaluate the various maps and overviews that have been developed for visualizing multimedia structure.

55.10.7 Text Should Be Exploited More Effectively

Multimedia applications often emphasize nontextual media. However, in task-oriented multimedia more structuring is needed, and text frequently provides that structure. Text can be exploited with a number of processing techniques to create links, summaries, etc.

Full-scale natural language understanding is still an unsolved problem. However, progress has been made in computational linguistics using heuristic analyses of the surface features of text. For instance, lexical chaining (Morris and Hirst, 1991) can be used to identify lexical chains (topics) with the aid of a thesaurus.

Tables of contents, indices, and search engines are other text-oriented techniques that should be incorporated into most multimedia applications.

Research Challenge: Learn to exploit the meaning in text so as to coordinate collections of information nodes based on the text that they contain.

55.11 CONCLUSIONS

This chapter has focused on the role of ergonomics in multimedia. It should be clear from the foregoing discussion that ergonomics has a major role to play in the design and development of effective multimedia.

One of the chief roles of ergonomics in multimedia will be in the development of further guidelines for designing and evaluating multimedia. More ergonomic input is needed to address the following deficiencies in current multimedia:

- A paucity of research on multimedia usage and usability
- A lack of standardized models of multimedia for different tasks
- A dearth of guidelines for multimedia development
- Relatively primitive tools for constructing multimedia
- A shortage of research findings pertaining to the psychology of multimedia and the effectiveness of multimedia in a variety of tasks, particularly in education and training.

Ergonomics has a key role to play in addressing these deficiencies. Methods of task analysis should also be applied more frequently to multimedia. What are the key tasks in which multimedia should be applied? Who are the users and what are their needs? What tools, interaction techniques, metaphors, and prosthetics will be most compatible for different types of user?

Multimedia is more than CD-ROMs that are fun to play with and learn from. Multimedia is a set of forms of highly interactive electronic information communication. In its first decade, multimedia became a preeminent publishing medium, and a global computer publishing medium, largely through the explosive growth of the World Wide Web. The first decade of multimedia has been marked by considerable technical progress. Increasingly, however, the key problems are not technical, but behavioral and social. In this chapter we have touched on some of the key ergonomic problems in multimedia. We have also outlined some tentative design guidelines.

In spite of the problems noted in this chapter, there is no turning back to earlier modes of communication and publishing. The development of multimedia systems (including virtual realities and cyberspace), is an opportunity to create and experience more richly compelling interactive environments than ever before. Multimedia will continue to grow in importance, requiring ergonomists and designers to develop tools and techniques that can assist in the coordination and presentation of "media that are instant, global, and multisensory" (Cotton and Oliver, 1993, p. 21).

ACKNOWLEDGMENTS

Thanks are due to the many colleagues who have contributed to our understanding of the nature and significance of multimedia and related topics. These include Luis Serra, Stephen Smoliar, Pete Kellock, Ron Baecker, Beverly Harrison, Gene Golovchinsky, Jim Tam, Manny Noik, David Modjeska, Torsten Nilsson, and Erik Stolterman.

The image showing the representation of the G-world and the I-world was suggested by Professor Guriev, Head of the Information and Technology Centre for Analytical Research, Administration of the President of the Russian Federation.

The research described in this chapter was supported by funding from the Information Technology Research Centre of Excellence of the Province of Ontario to the first author, and by a BT Short Term Research Fellowship to the second author.

REFERENCES

Alty, J. L. (1991). Multimedia—What is It and How do we Exploit It? In D. Diaper, and N. Hammond, Eds., *People and Computers, VI*. Cambridge: Cambridge University Press.

Anderson, J. R. (1985). *Cognitive Psychology and its Implications* (2nd ed.) San Francisco: Freeman.

Andrews, K., and Pichler, M. (1994). Hooking up 3-D space: Three-dimensional models as fully-fledged hypermedia documents. In *Proceedings of MHVR'94 east-west conference on multimedia, hypermedia and virtual reality*, Moscow, September 1994.

Apple Computer. (1991). *Visual Almanac Technical Report*. Cupertino, CA: Multimedia Lab, Apple Computer.

Baddeley, A. (1975). *Human Memory*. New York: Basic Books.

Baecker, R. M. (1993). *Readings in groupware and computer-supported cooperative work: facilitating human-human collaboration*. Los Altos, CA: Morgan Kaufmann.

Balasubramaniam, V., and Turoff, M. (1995). A systematic approach to user interface design for hypertext systems. In *Proceedings of the twenty-eighth Hawaii international conference on system sciences*, Wailea, Hawaii, January 3-6, 1995 (pp. 241–250). IEEE Computer Society Press.

Barkow, T. (December 1995). Ground truth. *Wired*, 96–106.

Baron, L., Brown, E., and Chignell, M. H. (1996). Unpublished manuscript. Department of Industrial Engineering, University of Toronto. A study of the Relationship between Hypertext Author and Reader Working Paper #96-13. Department of Industrial Engineeting, University of Toronto.

Bates, M. J. (1989). The design of browsing and berrypicking techniques for the online search interface. *Online Review, 13(5)*, 407–424.

Bernstein, M. (1993). Enactment in information farming. In *Proceedings of the ACM hypertext '93 conference* (pp. 242–249). New York: ACM.

Botafogo, R., Rivlin, E., and Shneiderman, B. (1992). Structural analysis of hypertexts: Identifying hierarchies and useful metrics. *ACM Transactions on Office Information Systems, 10*(2), 142–180.

Brown, M., and Shillner (1995). A New Paradigm for Browsing the web. In conference companion, CHI '95; Conference on Human Factors in Computing Systems (Denver, May 1995), pp. 320–321. New York: ACM.

Burt, R. S. (1982). *Toward a structural theory of action: Network models of social structure, perception, and action*. New York: Academic Press.

Bush, V. (1945). As we may think. *Atlantic Monthly*, July 1945. See also Bush, V. (1967). memex revisited. In V. Bush (Ed.), *Science is not enough*. New York: William Morrow.

Campagnoni, J., and Ehrlich, K. (1989). Information retrieval using a hypertext-based help system. *ACM Transactions on Office Information Systems*.

Card, S. K., Robertson, G. G., and Mackinlay, J. D. (1991). The Information Visualizer, an Information Workspace. In *Proceedings of CHI '91 Conference on Human Factors in Computing Systems*, New Orleans, April 1991. (pp. 1811–188). New York: ACM.

Charoenkitkarn, N., Chignell, M. H., and Golovchinsky, G. (1996). *Proceedings of TREC-4*. Gaithersburg, MD: National Institute of Standards and Technology.

Chase, W. G. (1986). Visual information processing. Chapter 28 in K.R. Boff, L. Kaufman, and J.P. Thomas, Eds., *Handbook of perception and human performance. Vol. II. Cognitive processes and performance*. New York: John Wiley.

Chignell, M. H. (1993). Cooperative human-machine reasoning: Communication through the user interface. In R.J. Jorna, B. van Heusden, and R. Posner, Eds., *Signs, search and communication: Semiotic aspects of artificial intelligence*, (pp. 348–368). Berlin: Walter de Gruyter.

Chignell, M. H., and Lacy, R. (1991). Instructional resources for researching and writing: the Jefferson notebook. *Journal of Computing in Higher Education, 2(2)*, 18–43.

Chignell, M. H., Nordhausen, B., Valdez, J. F., and Waterworth, J. A. (1991). The HEFTI model of text to hypertext conversion. *Hypermedia, 3(3)*, 187–205.

Chua, T-S., and Lai, E. (1991). Composition editor for a hypermedia environment. In J. A. Waterworth, Ed., *Multimedia: Technology and Applications*. Chichester, UK: Ellis Horwood, Simon and Schuster International.

Colby, G., and Scholl, L. (1991). Transparency and blur as selective cues for complex visual information. *Proceedings of the ISOE, 1460*, 114–124.

Conklin, J. (1987). Hypertext: An introduction and survey. *IEEE Computer*, September 1987.

Cotton, R., and Oliver, R. (1991). Understanding Hypermedia. From Multimedia to Virtual Really. London: Rhaidon Press Ltd. p. 82.

Deerwester, S., Dumais, S. T., Landauer, T. K., Furnas, G. W., and Harshman, R. A. (1990). Indexing by latent semantic analysis. *Journal of the Society for Information Science*, *41*(*6*), 391–407.

Dieberger, A., and Tromp, J. G. (1993). The information city project—A virtual reality user interface for navigation in information spaces. In *Proceedings of the Virtual Reality Symposium*, Vienna, December 1–3, 1993.

Dinadis, N., and Vincente, K. J. "Ecological interface design for a power plant feedwater subsystem, "IEEE Transactions on Nuclear Science, vol. 43, pp. 266–277, 1996.

Egan, D., Remde, J., Landauer, T., Lochbaum, C., and Gomez, L. (1989). Behavioral evaluation and analysis of a hypertext browser. In *Proceedings of CHI '89* pp. 205–210). New York: ACM.

Erickson, T. (1990). Working with interface metaphors. In B. Laurel, Ed., The art of Human-Computer Interface Design. Reading, MA: Addison-Wesley.

Everitt, B. S. (1990). *Cluster Analysis* (3rd ed.). London: Heinemann.

Fairchild, K., Poltrock, S. E., and Furnas, G. W. (1988). SemNet: Three-dimensional graphic representations of large knowledge bases. In R. Guindon, Ed., *Cognitive Science and its Applications for Human-Computer Interaction.* (pp. 201–233). Hillsdale, NJ: Erlbaum.

Furnas, G. W. (1986). Generalized fisheye views. In *Proceedings of the ACM CHI '86 Conference.* (pp. 16–23). New York: ACM.

Garber, S. R. and Grunes, M. B. (1992). The Art of Search: A Study of Art Directors. In *Proceedings of the ACM CHI '92 Conference.* (pp. 157–164). New York: ACM.

Genau, A., and Kramer, A. (1995). Translucent history. In *Conference companion, CHI '95 Conference on Human Factors in Computing Systems.* New York: ACM.

Gibson, J. J. (1966). *The Senses Considered As Perceptual Systems.* Boston: Houghton-Mifflin.

Gibson, J. J. (1979). *The Ecological Approach to Visual Perception.* Boston: Houghton-Mifflin.

Glushko, R. J. (1990). Designing "electronic encyclopedias" with hypertext software. *Human Factors Society Bulletin, 33(1)*, 6–9.

Golovchinsky, G., and Chignell, M. H. (1993). Queries-R- Links: Graphical markup for text navigation. In *Proceedings of the ACM CHI '93 Conference* (pp. 454–460). New York: ACM.

Golovchinsky, G., Kamps, T., and Reichenberger, K. (1995). Subverting Structure—Data-driven Dynamic Diagram Generation. *Proceedings of IEEE Visualization '95* (pp. 217–223). Atlanta, GA: October 29–November 3, 1995.

Godwin-Jones, R. (1994). Language Learning and the World Wide Web. In *Electronic Proceedings of the Second World Wide Web Conference '94. Mosaic and the Web.* http://141.142.3.30/SDG/IT94/Proceedings/Arts/godwin-jones/godwin. html.

Gray, S. (1994). *Hypertext and the technology of conversation.* Westport, CT:: Greenwood Press.

Hardman, L., van Rossum, G., and Brinkman, D. (1993). Structured multimedia suthoring. In *Proceedings of the ACM Multimedia '93.*

Harrison, B. L., Ishii, H., Vicente, K. J., and Buxton, W. A. S. (1995). Transparent layered user interfaces: An evaluation of display design to enhance focused and divided attention. In *Proceedings of CHI '95* (pp. 317–324) New York: ACM.

Hasan, M., Golovchinsky, G., Noik, E., Charoenkitkarn, N., Chignell, M. H., Mendelzon, A., and Modjeska, D. (1995a). Browsing local and global information. In *Proceedings of CASCON '95.* Toronto, Canada, November 7–9, 1995.

Hasan, M., Mendelzon, A., and Vista, D. (1995b). Visual web surfing with Hy+. In *Proceedings of CASCON '95.* Toronto, Canada, November 7–9, 1995.

Hill, W. C., Hollan, J. S., Wroblewski, D., and McCandless, T. (1992). Edit wear and read wear. In *Proceedings of the ACM CHI '92 Conference on Human Factors in Computing Systems.* New York: ACM.

Horn, R. E. (1989). *Mapping hypertext.* Lexington, MA: The Lexington Institute.

Horton, W. (1991). *Illustrating computer documentation: The art of presenting information graphically on paper and online.* New York: John Wiley.

Jackson, W. Y, Lee, J. D., & Sanquist, T. F. (1995). *Review of ecological interface design research: Applications of the design philosophy and results of empirical evaluations* (Battelle Tech. Rep.). Seattle, WA: Battelle.

Jeffries, R., Miller, J. R., Wharton, C., and Uyeda, K. M. (1991). User interface evaluation in the real world: A comparison of four techniques. In *CHI '91 Proceedings.* New Orleans, LA (pp. 119–124). New York: ACM.

Jones, W. P. (1989). As we may think?—Psychological considerations in the design of a personal filing system. In R. Guindon, Ed., *Cognitive Science and its Applications for Human-Computer Interaction.* Hillsdale, NJ: Erlbaum.

Jorna, G. C., and Snyder, H. L. (1991). Image quality determines differences in reading performance and perceived quality with CRT and hard-copy displays. *Human Factors, 33*, 459–470.

Karat, C.-M., Campbell, R., and Fiegel, T. (1992). Comparison of empirical testing and walkthrough methods in user interface evaluation. In *CHI '91 proceedings* (pp. 397–404). New York: ACM.

Lakoff, G., and Johnson, M. (1980). *Metaphors We Live By.* Chicago: The University of Chicago.

Lindsay, P., and Norman, D. A. (1977). *Human Information Processing: An Introduction to Psychology (2nd ed.).* New York: Academic Press.

Lynch, K. (1960). *The Image of the City.* Cambridge, MA: MIT Press.

Madsen, K. H. (1994). A guide to metaphorical design. *Communications of the ACM, 37(12)*, 57–62.

Mander, R., Salomon, G., and Wong, Y. (1992). A 'Pile' metaphor for supporting casual organisation of information. In *Proceedings of ACM CHI '92 Conference on Human Factors in Computing Systems* (pp. 627–634). New York: ACM.

Marchionini, G., and Haurer, H. (1995). The role of digital libraries in teaching and learning. *Communications of the ACM, 38(4)*, 68–75.

Mark, R. L., and Nielsen, J. (1994). *Usability Inspection Methods.* New York: John Wiley.

McCorduck, P. (1979). *Machines who think, a personal inquiry into the history and prospects of artificial intelligence.* San Francisco, CA: W.H. Freeman.

McKnight, C., Dillon, A., and Richardson, J. (1991). *Hypertext in Context.* Cambridge: Cambridge University Press.

Monk, A. F., Walsh, P., and Dix, A. J. (1988). A comparison of hypertext, scrolling, and folding. In R. Winder, Ed., *People and Computers, IV.* Proceedings of HCI '88. Cambridge: Cambridge University Press.

Monta, K., Takizawa, Y., Hattori, Y., Hayashi, T., Sato, N., Itoh, J., Sakuma, A., and Yoshikawa, E. (1991). An intelligent man-machine system for BWR nuclear power plants. In *Proceedings of AI 91: Frontiers in innovative computing for the nuclear industry* (pp. 383–392).

Morris, J., and Hirst, G. (1991). Lexical cohesion computed by thesaural relations as an indicator of the structure of text. *Computational Linguistics, 17*, 21–48.

Mukherjea, S., and Foley, J. D. (1995a). Showing the context of nodes in the world-wide web. In *Conference companion, CHI '95 conference on human factors in computing systems*, Denver, May 1995. New York: ACM.

Mukherjea, S. and Foley, J. D. (1995b). Visualizing complex hypermedia networks through multiple hierarchical views. In *Proceedings of CHI '95 conference on human factors in computing systems*, Denver, May 1995. New York: ACM.

Murch, G. M. (1984). Physiological principles for the effective use of color. *IEEE Computer Graphics and Applications, 4(11)*, 49–54.

Musil, S. and Pigel, G. (1993). Virgets: Elements for building 3D user interfaces. In *Proceedings of the ssymposium on virtual reality, Vienna*, December 1993. Available as TR 93/13, Vienna User Interface Group, Lenaugasse 2/8, A-1080 Vienna.

Negroponte, N. (1989). A personal perspective: An iconoclastic view beyond the desktop metaphor. *International Journal of Human-Computer Interaction, 1(1)*, 109–113.

Negroponte, N. (1995). *On Being Digital.* New York: Knopf.

Neisser, U. (1967). *Cognitive psychology.* New York: Appleton-Century-Crofts.

Nelson, T. H. (1981). *Literary Machines.* Swarthmore, PA: T. H. Nelson.

NewSoft, Inc. (1997). *Presto! Personal Page*, Fremont,CA. http://www.tophat.com.

Nielsen, J. (1990a). The art of navigating through hypertext. *Communications of the ACM, 33*, 297–310.

Nielsen, J. (1990b). *Hypertext and Hypermedia.* New York: Academic Press.

Nielsen, J. (1991). Panel discussion—The Nielsen ratings: Hypertext reviews. In *Proceedings of Hypertext '91.* New York: ACM.

Nielsen, J. (1997). Usability. In M. Helander and T. K. Landauer, Eds., *Handbook of human-computer interaction.* Amsterdam: North-Holland.

Nielsen, J., and Molich, R. (1990). Heuristic evaluation of user interfaces. In *Proceedings of CHI '90* (pp. 249–256). New York: ACM.

Nielsen, J., and Phillips, V. L. (1993). Estimating the relative usability of two interfaces: Heuristic, formal, and empirical methods compared. In *Proceedings of INTERCHI '93* (pp. 214–221). New York: ACM.

Noik, E. G. (1994). A space of presentation emphasis techniques for visualizing graphs. In *GI '94: Graphics interface 1994* (pp. 225–234). Alberta: Banff.

Nordhausen, B., Chignell, M. H., and Waterworth, J. A. (1991). The missing link? Comparison of Manual and Automated Linking in Hypertext Engineering. In *Proceedings of the 35th Annual Meeting of the Human Factors Society*, San Francisco, September 1991.

Norman, D. A. (1977). *Memory and Attention.* New York: Academic Press.

Norman, D. A. (1986). Cognitive Engineering. In D. A. Norman and S. Draper, Eds., *User-Centred System Design.* Hillsdale, NJ: Erlbaum.

Norman, D. A. (1993). *Things That Make Us Smart.* Reading, MA: Addison-Wesley.

Oren, T., Salomon, G., and Don, A. (1990). Guides: Characterizing the Interface. In B. Laurel, Ed., *The Art of Human-Computer Interface Design.* Reading, MA: Addison-Wesley.

Parsaye, K., and Chignell, M. H. (1993). *Intelligent Database Tools and Applications.* New York: Wiley.

Parunak, H. van D. (1991). Ordering the information graph. Chapter 20 in E. Berk and J. Devlin (Eds.), *Hypertext/Hypermedia Handbook.* New York: Intertext Publications, McGraw-Hill.

Perez, E. (1991). Tools for Authoring Hypertext. In E. Berk and J. Devlin, Eds., *Hypertext/Hypermedia Handbook.* New York: Intertext Publications, McGraw-Hill.

Rechtin, E. (1991). *Systems Architecting*. Englewood Cliffs, NJ: Prentice-Hall. Towards a generative theory of diagram design. *Proceedings of the Information Visualisation Symposium '95*, pp. 11–18. *Atlanta, GA, October 29–November 3, 1995.*

Remde, J. R., Gomez, L. M., and Landauer, T. K. (1987). SuperBook: An automatic tool for information exploration-hypertext? In *Proceedings of Hypertext '87*, Chapel Hill, North Carolina. New York: ACM.

Roth, S. F., Kolojejchuk, J., Mattis, J., and Goldstein, J. (1994). Interactive graphic design using automatic presentation knowledge. In *Proceedings of CHI '94*. New York: ACM.

Salomon, G., Oren, T., and Kreitman, K. (1989). Using guides to explore multimedia databases. In *Proceedings of the 22nd Annual Hawaii International Conference on Systems Sciences* (Vol. III, pp. 3-12).

Salton, G. (1989). *Automatic text processing: The Transformation, Analysis, and Retrieval of Information by Computer.* Reading, MA: Addison-Wesley.

Salton, G., Allen, J., and Buckley, C. (1994). Automatic structuring and retrieval of large text files. *Communications of the ACM, 37(12)*.

Salton, G., and McGill, M. J. (1983). *Introduction to Modern Information Retrieval.* New York: McGraw-Hill.

Sanders, M. and McCormick, E. (1993). *Human factors in engineering design (7th ed.). New York: McGraw-Hill.*

Sarkar, M., and Brown, M.H. (1992). Graphical fisheye views of graphs. In *Proceedings of the CHI '92* (pp. 83–91). New York: ACM.

Shneiderman, B. (1983). Direct manipulation: A step beyond programming languages. *Computer, 16(8)*, 57–69.

Shneiderman, B. (1987). User interface design for the HyperTies electronic encyclopedia. *Proceedings of Hypertext '87*, Chapel Hill, North Carolina. New York: ACM.

Silvers, R. (1995). Livemap: A system for viewing multiple transparent and time-varying planes in three dimensional space. In *Conference Companion, CHI '95 Conference on human factors in computing systems*, Denver, May 1995. New York: ACM Press.

Smoliar, S. W., Waterworth, J. A., and Kellock, P. R. (1995). pianoFORTE: A system for piano education beyond notation literacy. In *ACM multimedia '95 Conference*, San Francisco, November 1995. New York: ACM.

Sperber, D., and Wilson, D. (1986). *Relevance.* Oxford: Blackwell.

Tam, J. C., and Chignell, M. H. (1992). Neighborhoods and links: Examining the assumptions of hypermedia. In *Proceedings of the 25th annual Conference of the Human Factors Assocation of Canada.*

Teshiba, K., and Chignell, M.H. (1988). Development of a user evaluation technique for hypermedia based interfaces. *Proceedings of the Human Factors Society.*

Thüring, M., Haake, J. M., and Hannemann, J. (1991). What's Eliza doing in the Chinese room? Incoherent hyperdocuments and how to avoid them. In *Proceedings of Hypertext '91* (pp. 161–177). New York: ACM.

Tufte, E. R. (1983). *The Visual Display of Quantitative Information.* Chesire, CT: Graphics Press.

Tullis, T. S. (1984). *Predicting the usability of alphanumeric displays.* Ph.D dissertation. Available from the Report Store, Lawrence, Kansas.

Tversky, A., and Gati, I. (1978). Studies of similarity. In E. Rosch and B.B. Lloyd, Eds., *Cognition and Categorization.* Hillsdale, NJ: Erlbaum.

Valdez, J. F. (1992). *Navigational strategies in using documentation.* Unpublished Ph.D dissertation. Department of Industrial and Systems Engineering, University of Southern California, Los Angeles.

Valdez, J. F., and Chignell, M. H. (1992). Methods for assessing the usage and usability of documentation. In *Proceedings of the Third Conference on Quality in Documentation*, Waterloo, Ontario, November 1992.

Valdez, J. F., Chignell, M. H., and Glenn, B. (1988). Browsing models for hypermedia databases. In *Proceedings of the Annual Meeting of the Human Factors Society. Santa Monica, CA: Human Factors Society.*

Vicente, K. J., and Rasmussen, J. (1992). Ecological interface design: Theoretical foundations. IEEE Transactions on Systems, Man, and Cybernetics, *SMC-22*, 589–606.

Voyager Company. (1994). *Macbeth CD-ROM Version.* Los Angeles: Voyager Company.

Waterworth, J. A. (1992). *Multimedia Interaction: Human Factors Aspects.* Chichester, England: Simon and Schuster International.

Waterworth, J. A. (1995). HCI design as sensory ergonomics: Designing synaesthetic media. In D. Dahlbom, F. Kämmerer, F. Ljungberg, J. Stage, and C. Sorensen, Eds., In *Proceedings of IRIS 18 Conference*, Denmark, August 1995. *Gothenburg Studies in Informatics*, Report 7, 1995, 744–753.

Waterworth, J. A., and Chignell, M. H. (1991). A model of information exploration. *Hypermedia, 3(1)*, 35–58.

Waterworth, J. A., and Serra, L. (1994). VR management tools: Beyond spatial presence. In *Conference Companion to CHI '94 Conference on Human Factors in Computing Systems*, Boston, April 1994. New York: ACM.

Waterworth, J. A., and Singh, G. (1994). Information islands: Private views of public places. In *Proceedings of MHVR '94 East-West International Conference on Multimedia, Hypermedia and Virtual Reality*, Moscow, September 14–16, 1994.

Waterworth, J. A., and Chignell, M. H. (1989). A manifesto for hypermedia usability research. *Hypermedia, 1*(*3*), 205–234.

Waterworth, J. A., and Chignell, M. H. (1997). Multimedia interaction. In M. H. Landauer and Helander, M., Eds., *The Handbook of HCI.* Amsterdam: North Holland.

Wickens, C. D. (1992). *Engineering Psychology and Human Performance* (2nd Ed.) Columbus, OH: Charles E. Merrill.

Wright, P. (1991). Cognitive overheads and prostheses: Some issues in evaluating hypertexts. In *Proceedings of Hypertext '91* (pp. 1–12).

Yates, F. A. (1984). *The Art of Memory*. London: Routledge and Kegan Paul. First published in 1966.

Yoder, E., Akscyn, R., and McCracken, D. (1989). Collaboration in KMS, a shared hypermedia system. In *Proceedings of CHI '89, Conference on Human Factors in Computing Systems*. New York: ACM Press.

PART 9

SELECTED APPLICATIONS OF HUMAN FACTORS

CHAPTER 56

HUMAN FACTORS IN MANUFACTURING

Waldemar Karwowski
Department of Industrial Engineering
University of Louisville
Louisville, KY 40292, USA

Hans J. Warnecke
Fraunhofer Society (FhG)
80636 Munich, Germany

Manfred Hueser
Fraunhofer Institute for Manufacturing Engineering and Automation (IPA)
70569 Stuttgart, Germany

Gavriel Salvendy
School of Industrial Engineering
Purdue University
West Lafayette, IN 47907, USA

56.1 INTRODUCTION

Contemporary manufacturing industry faces several challenges of increasing worldwide competition in rapidly changing global market demands. The traditional approach of mass production and reducing costs to satisfy customer needs is no longer an effective corporate strategy in world markets (Goldman, Nagel, and Preiss, 1995; Jaikumar, 1986; Karwowski, Parsaei and Wilhelm, 1988; Office of Technology Assessment, 1984; Stalk and Hout, 1990; Wall, Clegg, and Kemp, 1987; Johnson and Wilson, 1988; Warnecke, 1993;

Womack, Jones, and Roos, 1990). It is predicted that by the year 2000 manufacturers will need technologies that promote rapid product development and significant variety of parts (Majchrzak and Paris, 1995). For example, the critical technologies of importance may include sophisticated computer-aided engineering tools, flexible assembly system, reconfigurable equipment, computer-aided design and manufacturing (CAD/CAM) links, and online programming capabilities, and automated inventory systems. Customers of today often require high-quality products with competitive price and value that match their particular needs and are delivered in a timely manner. Such diverse market demands have put significant pressure on organizations, resulting in a wave of massive restructuring of manufacturing entities.

With a variety of dynamic changes taking place in the contemporary manufacturing at both the technological and organizational levels, a consideration of human factors issues can be beneficial to manufacturing industry, especially at the design stages (Wilson et al., 1994). Adaptation of technology to a work system and vice versa involves human factors in the wider sense of organizational structures, incentive schemes, and management strategies (Rasmussen, 1993). In a period of rapid technological change, the human factors problems call for cross-disciplinary cooperation in a study of the mechanisms governing the behavior of large sociotechnical systems in dynamic environments. Since the contemporary manufacturing systems are both social and technical systems, the appropriate methodological basis for their design and implementation needs to be developed and implemented in practice (Corbett, 1988; Goldman et al., 1995; Jaikumar, 1986; Karwowski and Salvendy, 1994; Kidd, 1987, 1994; Majchrzak, 1988).

This chapter discusses the human factors–related issues that affect design and operation of contemporary manufacturing systems, and are not covered in other chapters of this Handbook. Table 56.1 provides insights on the topics of different chapters of this Handbook in generically addressing key human aspects which are directly applicable to the manufacturing area.

56.1.1 Importance of Human Factors in Manufacturing

Contemporary manufacturing companies are exposed to tensions resulting from market demands, rapid progress in technology, legal provisions, and social changes. These demands influence the organization and management of human work in various respects and lead to complex and dynamic changes in business environment. Because of the need to produce with shorter life cycles and a greater variety of models, companies are forced to adapt their manufacturing program to customer demands of short delivery time and smaller batch sizes in order to keep the finished stock as little as possible (Stalk and Hout, 1990).

A study by the Office of Technology Assessment (1984) on the impact of modern manufacturing automation on employment and education identified the following human factors areas of concern: (1) changes in skill levels and occupational structure, (2) operator training requirements, (3) interdependence between workers, (4) boredom and stress at work, (5) system downtime and maintenance, (6) safety issues, and (7) labor–management relations. The Manufacturing Studies Board (1986a) in the United States concluded that realizing the full benefits of modern manufacturing will require interrelated changes in seven areas of human resource practices: planning; plant culture; plant organization; job design; compensation and appraisal; selection, training, and education; and labor–management relations.

Sinclair (1986) discussed implications for the ergonomics and human factors profession in context of the trend toward automated factories in which dominant human tasks are of a supervisory nature. The implications of manufacturing automation were classified under the categories of *fitness for use* based on the concepts of quality, including locus of control and accountability; organization, staffing and training issues; decision aids; the transfer of knowledge among multiple users; and knowledge capture. It was pointed out that although much of the human factors knowledge is relevant to the above areas of concern, significantly more research is needed to address the broad scope of emerging human issues. Such areas in human factors research include the following:

1. An understanding of how operators generate and maintain cognitive models of complex dynamic system, the resulting structures, and the functional relationships of these structures to the real system.

Table 56.1 Human Aspects in Manufacturing Covered in Other Chapters of This Handbook

Human Issue	Some Possible Solutions	For a Thorough Treatment of Subject, Please Read Chapter No.
Material handling	Minimize load size; optimize material movement direction; mechanize when possible; provide personnel training.	34
Maintainability	Isolate faults and use cognitive task analysis and virtual reality technology to improve maintainability.	47
Human factors audit	Use checklist to identify potential ergonomics problem areas in the workplace.	48
Supervisory control	Computer technology is driving the human operator to become a supervisor (planner and monitor) of factory automation and an intervener within the automated control loop only for abnormal situations.	39
Scheduling work	Use the proposed step method to evaluate current shift work and use the 16-item checklist to design new shift work.	32
Lighting	Provides specific information as to which lighting system to use for which applications in order to maximize comfort and productivity and accommodate individual differences.	26
Noise and vibration	The best way to cope with noise and vibration is to reduce it at the source by either engineering design changes or isolation and absorption. When that is not possible, the human must wear protective gear and limit duration of exposure to both noise and vibration.	24, 25
Training	Operator training will increase productivity, morale, and agility of manufacturing operations. For this purpose, a systematic long-term training method must be developed for each job function and each individual.	16
Job design	In order to optimize job design, the hybridization of mechanistic motivational, perceptual-motor, and biological approaches is proposed. In order to optimize team work, approaches for combining tasks and selecting team members is outlined.	14
Workplace design	Ensure that the workplace design is adjustable for the range of human variations in body size and strength, cognitive capabilities, and with regard to individual's likes and dislikes.	8, 9, 21, 50
Human error	Reduce human error by designing equipment, workplace and jobs that avoid ambiguity and is compatible with human capabilities and limitations. Individuals doing the job must be carefully selected to ensure that they have the capability to be trained to perform the task.	6
Human–machine complementarity	Before the human–machine system is designed, the relative capabilities and limitations of the human must be considered and the manual load demand considered. Too high or too low a manual load for a job will result in suboptimal performance.	11, 13

2. The effect of cognitive models and states of the system on the behavior of human operators, including consideration of such factors as operator motivation, confidence, and trust.
3. A rationale for developing desirable relationships between human operators and computers regarding the responsibility for and control of complex systems.
4. The design of interfaces that will allow communication between humans and computers of their respective goals and plans.
5. Determination of what constitutes appropriate support to enable humans to maintain control of the system.
6. An understanding of organizational structures appropriate to computer-based control, which involves knowledgeable and skilled people and identification and communication of explicit knowledge.

The studies on the fusion of flexible manufacturing systems and new information technologies in Europe (Brödner, 1987) concluded that organization is the key element in economic success of modern production systems, and should be valued and appreciated at the level equal to new technology. It was also pointed out that the concept of production culture is important for use and the kind of technology applied, organization of work and demand for skill. Martin (1993) reported that many manufacturing automation projects have failed because of insufficient automatibility (automation flexibility), inadequate user–system interfaces, and incompatibility between the human needs and system requirements.

Contemporary manufacturing systems that utilize sophisticated tools of information technology must be designed with due consideration to the human operators who monitor, program, supervise, and maintain these systems. In the recent past, designers of such systems perceived the need for human skill as a problem and the human operator as the source of errors. Therefore, a predominant design goal would be to eliminate human intervention within the system control loop as much as possible. Such approach limited human roles to those tasks which the designers did not know how to automate (Bainbridge, 1983; Sanderson, 1989; Wilson et al., 1994). It is widely acknowledged today, however, that successful operation of modern manufacturing systems depends upon abilities of the human resources to compensate for limitations of computer-based technology. According to Bi and Salvendy (1994), the roles of the human operators in contemporary manufacturing systems include: (1) monitoring the automated systems, fine-tuning the processes, overriding of the automatic system when necessary; (2) detection, diagnosis, and compensation for scheduling system failures or infeasible routings; (3) communication with other departments, making some necessary tradeoffs or negotiate among the alternative solutions to some tasks, and (4) dealing with unplanned events and unexpected contingencies within the system.

As discussed by Wilson (1991a), there are a number of ironies or paradoxes implicit to the involvement of people in modern manufacturing systems. These include the following issues:

1. The nature of human roles is more important in less labor-intensive, highly automated systems, and the more advanced a control system the more crucial may be the contribution of the human operator (Bainbridge, 1983).
2. The typical designer's view of human operators is often that they are unreliable and inefficient system elements. The designer who attempts to eliminate people by automating a system will often still leave, for the operators, an arbitrary and mainly unplanned collection of tasks that cannot be automated.
3. It is only the very fact that people are flexible and intelligent that permits automated systems to be operated at all (Clegg and Corbett, 1987).
4. Important human ability lies in the complex knowledge-based skills for problem solving in novel situations, which is gained through experiences with low-level system monitoring interactions.
5. Operators may acquire little knowledge in the more complex work situations representative of the emergencies where human skills are required.
6. The human problem-solving skills are at least effective in those situations which are stressful to operators.

Table 56.2 Examples of Problems Related to Lack of Understanding About the Organizational and Human Resource Changes That Affect Implementation of New Technology

- Design of process and product technologies affect how productive people are on their jobs, how effectively they communicate with others, and how they feel about their work.
- Effective implementation of a new technology demands sufficient advanced planning so that job design, training, and organizational structural changes can occur in conjunction with the technology change (rather than after the implementation).
- Involvement of technology users in the design of the technology and the concomitant organizational changes is an essential ingredient to ensuring adequate user motivation and commitment
- Firms that effectively integrate the design of new technology with the design of organizational and human resource systems are more successful than those that have incompatible system.

Source: After Majchrzak and Paris (1995).

The above discussed ironies of automation (Bainbridge, 1983) are summarized in Table 56.2.

A recent review by Majchrzak and Paris (1995) concluded that technology implementation is more likely to be successful when the technology, organization, and human resource or people issues have been designed to complement and integrate with each other. It was suggested that high failure rates in advanced technology implementation efforts can be attributed to managers and technology designers lacking an understanding about the organizational and human resource changes that are often needed with new technology. Examples of such problems are listed in Table 56.3.

Seppälä, Touminen, and Koskinen (1992) investigated the impact of the introduction of advanced manufacturing technology and flexible production philosophy on the organization of work, job contents, work demands, and employees' well-being in nine Finnish companies. The components of modern manufacturing included the flexible manufacturing systems (FMS), computer numerical control (CNC) machining centers, and robotized

Table 56.3 Examples of Design Criteria for Contemporaty Manufacturing

Area	Examples of Criteria
Design process	*Participation*: the design process should be undertaken participatively; e.g., system users, supporters and managers participate at all stages in the design process.
Allocation of function	*Complementarity*: decisions on allocation of function should recognize that humans an machines may complement one another in synergistic ways: e.g., humans and machines both contribute to production scheduling.
Job design	*Control*: the operators should have some control over their jobs, which should included distinct areas of responsibility and decision making: e.g., operators are responsible for deciding the order in which tasks are undertaken.
Organization structure	*Boundary management*: supervisors and managers of the system should be responsible for managing its boundaries; e.g., supervisors provide operators with warning of alterations to work schedules.
Hardware ergonomics	*Safety and prevention of accidental operation*: the workstation and equipment should be safe and the workspace and the components within it should be designed/chosen and arranged so that accidental (possibly dangerous) operation is avoided; e.g. controls are guarded by recessing of buttons and switches.
Software ergonomics	*Informative feedback*: users should be given clear, informative feedback regarding where they are in the system, what actions they have taken, whether these actions have been successful, and what actions should be taken next; e.g., any error message or warning explains what it is, and how it can be corrected.

Source: After Clegg and Corbett (1987).

machining cells. The results showed that the advanced manufacturing does not inevitably result in impoverishment of job contents and deskilling if the corporations design the production systems based on a flexible and multiskilled workforce.

56.1.2 Historical Account of Manufacturing Concepts

Design of work in manufacturing settings originated with Adam Smith and Charles Babbage, who were first to advocate the use of narrow and precise definitions of tasks leading to job specification and tasks grouping (or departmentalization) of the organizational structure. Such *horizontal differentiation* was measured through the number of required specialties and the level of training to be designed into the jobs (Hendrick, 1987). Based on the division of work and allocation of responsibility between the management and workers, Taylor (1911) developed the structural form for the vertical specialization called the *principles of scientific management.* The creation of a new management science, dependent upon one optimal method for definition of work task and determining all of its details, formed the basis of Taylor's system which was characterized by division and simplification of work and work standards, and led to a separation of planning, controlling, and monitoring activities. Other scientists, such as Maslow and Herzberg, analyzed the individual needs of people and planned new forms of work, being better adapted to the abilities and endeavors of the individual. Some 30 years later the concept of *ideal bureaucracy,* introduced by Weber (1946), reinforced and extended the Tayloristic design principles. Such bureaucratic design was based on organizational structure that focused on efficiency, stability and control, and division of labor through well defined, simple, and routine tasks to be performed by the workers.

Even though the widespread adoption of Taylor's principles of scientific management in the manufacturing area led to drastic job fragmentation, from the technological standpoint it was considered as progress toward creation of an effective factory organization (Hayes and Jaikumar, 1988). The weaknesses of this approach, however, lay in its social effects (MacKenzie and Wajcman, 1985; Rosenbrock, 1983). Separation of managers and staff from workers, disregard for the need for continual improvements of workers skills and organizational capabilities, and job fragmentation proved to be serious deficiencies of many production systems. This technology-driven or technocentric approach to design and management of contemporary manufacturing systems was in principle no different from the Tayloristic method (for examples see Brödner, 1987; Cooley, 1986; Corbett, 1985; Kidd, 1987). As such, the quest for computer-based manufacturing as practiced in the recent past further enhanced vertical division of work and aimed to reduce human involvement in the manufacturing processes. Therefore, the human resource utilization is considered today as the key to meet the needs of modern organizations for efficient manufacturing and high-quality product under technological uncertainty (Jaikumar, 1986; Kember and Murray, 1988; Majchrzak, 1988; Manufacturing Studies Board, 1986a,b; Wall et al., 1987).

The changing economic conditions that influence the contemporary manufacturing environment, as well as technical developments in automation for flexible production, required introduction of different ways of thinking, and a paradigm change from previous manufacturing philosophy. This upheaval in the manufacturing area can be characterized by the following assumptions (Warnecke 1993): (1) manufacturing is a service, (2) flexibility ensures productivity, (3) productivity in information management increasingly influences the productivity of material processing (utilization of work, capital, material, energy), (4) more productivity by uniting information processing, decision making and realization in small, fast control loops or work groups, (5) the work force becomes more important, and (6) manufacturing systems have to be designed as more use-oriented than cost-oriented units. In order to realize these objectives and requirements, a set of new ideas and new ways of thinking have been developed.

As pointed out by Martin (1990), the source of all productivity in a contemporary (automated) manufacturing are human-embodied skills that can provide the basis for future learning, experimentation, and other insights. Therefore, the dominant factor of contemporary manufacturing paradigms is the necessity of an integrated approach to technology, people and the organization (Goldman et al., 1995; Karwowski and Salvendy, 1994; Karwowski et al., 1994; Kidd, 1990, 1994; Kidd and Karwowski, 1994; Majchrzak, 1988; Majchrzak and Gasser, 1992; Majchrzak et al., 1994; Salvendy and Karwowski, 1994). In the framework of this integration, the appropriate manufacturing technology must be selected and implemented, along with requirements for selection and training of human resources, and providing control to the people charged with the use of this technology for meeting corporate objectives. In addition, appropriate organizational structures must be developed and implemented in order to manage the available resources and adapt quickly to a changing marketplace.

The concept of *lean manufacturing,* first introduced by Womack et al. (1990), considers the organizational structures and human resources, in addition to advanced manufacturing technology, as important factors for generating expected productivity and quality results, and replaces traditional mass production paradigm. Lean production combines the advantages of both craft and mass production, including high volume and diversified production using flexible automation, multiskilled teamwork, and an overall corporate culture of continuous improvement and high quality (Womack et al., 1990).

A study by Wobbe and Charles (1994) has identified five general principles that tend to be evident when pursuing the goals of lean production, including the following:

1. The more complex products become, the more quality is dependent upon upgrading of all stages of manufacturing and demands the full dedication of employees at all levels,
2. The more sophisticated manufacturing technology becomes, the more it is vulnerable and dependent upon human skills for control and maintenance,
3. The more customized productions are, the more human intervention is necessary with regard to change over, setting up machines, adaptation, adjustment, and control,
4. If products demand a high service input and after sales service and maintenance, skilled people are required to deal with this,
5. The shorter the life cycle of products becomes, the more innovativeness comes into play; take off phases occur more frequently and their mastery is dependent upon experienced personnel with formal knowledge to overcome new challenges connected with the start of a new product.

A more recent manufacturing philosophy and business approach that encompasses the above attributes and provides much more focused emphasis on system flexibility and agility is the concept of *agile manufacturing* (Goldman et al., 1995; Kidd, 1994; Kidd and Karwowski, 1994). An agile company aims to manage the change and uncertainty through the entrepreneurial approach, combined with flexible organizational structure and distributed managerial decision-making authority, which allows for rapid reconfiguration of the human and technological resources in a quest for meeting globally changing market requirements.

56.2 COMPUTER-INTEGRATED MANUFACTURING

56.2.1 Background

Advanced manufacturing can be defined as the application of information technology to automate and integrate different functions in the manufacturing system. However, as pointed out by Wilson (1991a), because of technological limitations, full manufacturing automation with only a very limited role or even no role for people is not feasible; even if it were feasible then requirements of good job design would mean it would not be desirable; even if it were feasible and suitable, then the abilities of people to optimize and intervene and their adaptability and flexibility would make it not optimal.

Although contemporary manufacturing is typically identified with the introduction of computer-based systems into design, planning, and manufacturing, it also involves many other aspects of advanced technology, including, for example, management of human resources, marketing, purchasing, engineering, planning, financing, and production planning and control. All these functions must be integrated in order to assure successful implementation of contemporary manufacturing tools on the basis of increasing competitive advantages and market opportunities.

At present, most of the computer-integrated manufacturing systems are incompatible with the inherent capabilities of the human operators as expressed by their skills and knowledge necessary for the effective control and monitoring of these systems (Karwowski, 1991; Kidd, 1994). Such incompatibility arises at all levels of the human, machine, and human–machine functioning and can be defined within a framework of the individual and the whole factory. Problems with integrating people and technology occur early at the contemporary manufacturing design stage. These problems can be conceptualized using the following model of the complexity of interactions (I) between contemporary manufacturing designers (D) and users (U) of contemporary manufacturing technology, and the technology (T) itself:

$$I(U, T) = F[I(U, D), I(D, T)]$$

where I stands for relevant interactions and F indicates functional relationships between designers (D), users (U), and technology (T).

This model points out that the interactions between the users and contemporary manufacturing technology are determined by the outcome of the integration of the two earlier interactions, i.e., (1) those between the designers and potential users, and (2) those between the designers and manufacturing technology (at the level of machines and system integration). Although strong interactions typically exist between the designers and technology, only very few examples of strong connections between the designers and the human operators can be found. Designers of contemporary advanced manufacturing systems focus primarily on computer-aided integration of sophisticated machines and other equipment as parts of technology, rarely paying much attention to the paramount needs for the effective human integration within such systems (Kidd, 1994; Majchrzak, 1988).

Modern manufacturing seeks to design a technology that enhances the human communication mode, and supports the cooperative work between people. According to Wall et al. (1987), there are two kinds of human personal needs that affect work design in contemporary manufacturing: (1) social needs, or the desire for significant social relationships; and (2) growth needs, or the desire for personal accomplishment, learning, and development. People in modern manufacturing have to be treated as resources and knowledge contributors so that they reinforce each other's work in the organization. Examples of different categories of design criteria for modern manufacturing systems are shown in Table 56.4.

To assure full benefits from the contemporary manufacturing concepts, a practical implementation of a much broader vision of the system integration, which includes people, organization and technology, is needed (Kidd, 1991). For example, even though the idea of *human and computer-integrated manufacturing systems* (HCIMS) remain at the

Table 56.4 The Ironies of Automation

1. *Operators "out of the control loop."* The human operators are present in the system to exercise control when needed, but by being "out of the control loop" they lose manual skills and long-term system knowledge that is required in case of an emergency.
2. *Outdated "mental picture."* The human operators may not be able to respond quickly to changes in the system if they have not been following the last few minutes of the operation closely. The operators' up-to-date knowledge or mental picture of the manufacturing system functioning may be, therefore, severely limited making it impossible to be able to respond as needed.
3. *Disappearing generations of skills.* New operators may not be able to adequately learn about the computerized manufacturing system, and therefore, will be unable to exercise effective control when needed.
4. *Expert system limitations.* Rapidly changing market requirements may induce changes in informational needs in the system supervision which will have to be expressed through changes in the knowledge bases of the expert systems.
5. *Authority of automatics.* If the computerized manufacturing system has been implemented because it can perform the required tasks better than the human operator, the question arises "On what basis does the operator decide that correct or incorrect decisions are being made by the automatic systems?"
6. *Solving the wrong problem.* Computerized manufacturing may not offer the best solutions to production or market problems, as these could be better solved through the conventional methods.
7. *Emergence of the new types of errors due to automation.* Computerized manufacturing leads to new types of errors and accidents which cannot be analyzed within the framework of traditional techniques utilized in the manual production environments.

Source: After Sanderson (1989).

conceptual stage, the desirability of such systems seems unquestionable (Kidd, 1994). The design and integration of enabling technologies should address the system's organizational architecture, including the following: (1) considerations of the network of groups, (2) the structure of each group, (3) the interaction between groups, (4) the nature of the supporting software, and (5) technical communication and integration needs between supporting software modules.

56.2.2 Manufacturing as an Open Manufacturing System

Kidd (1991) discussed the following requirements for human and computer-integrated manufacturing systems: (1) flexibility, (2) adaptation, (3) improved responsiveness, and (4) the need to motivate people and make better use of their skills, judgment, and experience. The above also requires that organizational structures, work practices, and technologies be developed as to allow people at all levels in the company to adapt their work strategies to the variety of systems control situations (Kidd, 1989a). Therefore, the organizations, work practices, and technologies will have to be designed and developed as open systems. The term open system is used to describe a system that receives inputs from and sends outputs to the systems environment (Kidd, 1991). The term was associated in manufacturing with system architectures based on the *International Standards Organization Open Systems Interconnection model.* The idea can be applied not only to system architectures and organizational structures, but also to work practices, human–computer interfaces, and the relationship between people and technologies, such as scheduling, control systems, and decision support systems.

An open manufacturing system allows people a large degree of freedom to define the mode of operating the system and should adapt to its environment. Kidd (1990) has demonstrated the concept of an open manufacturing system for the human–computer interface (HCI) in the workshop-oriented CNC systems. Such HCI system allows the human operator to customize the interface to his or her own personal preferences by changing the dialogue, the screen layout, etc. In comparison, the closed manufacturing system typically restricts the user freedom of action or forces the user to use the manufacturing system in a particular way. A closed system also attempts to automate a particular task, but when the manufacturing system fails, it leaves the user without the necessary computer-based decision support. In the open, adaptable manufacturing system, the relationship between the user and the computer is determined by the user and not by the designer. The role of the designer of an open manufacturing system is to create a system that will satisfy the user's personal preferences, and allow the users to work in a way that they find most appropriate.

The open systems philosophy prevents the designers of a modern manufacturing system from imposing on the system user their own views on what the computer should do and what the user should do, by minimizing the effect of the value judgments made by designers. For example, Wall, Corbett, Martin, Clegg, and Jackson (1990) examined two main work design styles used to manage and operate the CNC stand-alone systems. The work designs of interest were: (1) the specialist control, and (2) the operator-centered control. In the specialist control mode, engineers and computer specialists maintain, repair, write and fine tune the programs, while the operator has minimal involvement. In the operator-centered control mode, the operator is responsible for monitoring and maintenance and programming of problems as they occur. A sociotechnical criterion for predicting performance through the concept of production variance was used. It was hypothesized that increasing operator control improves performance more for high-variance manufacturing systems than for low-variance ones, and that increasing operator control enhances operator well-being. The study results revealed that introduction of enhanced operator control over CNC assembly machines led to reduction in downtime for high-variance machines and suggested that work redesign improved intrinsic job satisfaction and reduced feelings of job pressure among operators.

56.3 COGNITIVE ENGINEERING AND MACROERGONOMICS

56.3.1 Human Roles in Manufacturing and Design Issues

The application of information technology in contemporary manufacturing systems affects the human roles both hierarchically and horizontally (Sanderson, 1989). The hierarchical effect of information technology manifests itself through the process of automation and leads to human operators acting as supervisors of the artificially intelligent manufacturing processes. The horizontal effect of information technology can be described in information-processing terms and illustrates the situation in which the human operators have access to information about all aspects of the manufacturing system. Traditional approaches to design and management of advanced manufacturing systems consider workers as deterministic input–output systems and tend to disregard teleological nature of human behavior, i.e., the goal-oriented behavior focused on seeking relevant information and active selection of goals (Rasmussen, 1983). However, to be successful, the design and management of advanced manufacturing systems must be based on the description of human mental functions needed for a specific task. With increasing automation efforts seen in industry today, the nature of tasks in modern manufacturing systems shifts from those that require perceptual-motor skills to cognitive activities of problem solving and decision making in monitoring and supervisory control tasks (Goodstein, Anderson, and Olsen, 1988). Human operators in contemporary manufacturing systems act as monitors, problem solvers, and decision makers.

Sanderson (1989) identified two complementary approaches to defining the role of people in contemporary manufacturing systems: (1) cognitive engineering, based on the cognitive and motivational psychology and focusing on the individual, and (2) macroergonomics, based on combination of industrial or organizational psychology, social psychology, and systems theory, which focuses on the entire organization. These two approaches are complementary since the actions of the individual operators are restricted by the policies and culture of the organization. On the other hand, the functioning of the entire manufacturing organization is reflected in the actions of the people who supervise the system operation. The cognitive activities of the human supervisor are: (1) planning what should be done in the manufacturing plant for a period of time and determining how it should be done, (2) teaching or setting up plans through human-interactive computers, (3) monitoring the manufacturing process to ensure all is going well, (4) intervening if abnormalities arise or if priorities change, and (5) learning through feedback from the plant about the impact of the above four activities (Sheridan, 1994).

Cognitive engineering approach (Hollangel and Woods, 1983) emphasizes appropriate goals for the human roles, a need to understand details of the particular system, and system complexity. Human knowledge and skill are considered as an inherent part of system design requirements. The interaction between technical and human parts of the system is to be designed to make the best use of the knowledge and skills of both parts of the systems. Sobol and Lei (1994) discussed the relationship between different types of organization-based knowledge (explicit vs. tacit) and its application to utilizing various forms of modern manufacturing technology. Whereas explicit knowledge allows corporations with a comparable skill and technology base to apply the knowledge in a similar manner, the tacit knowledge is based more on insight, practice, and cumulative learning that is deeply embedded within the organization's dynamic routines and operating systems. The effective learning and application of new types of contemporary manufacturing skills and techniques is primarily based on the firm's cultivation of tacit, embedded knowledge to build competitive advantage.

Cognitive engineering proposes that man–machine (hybrid) systems need to be conceived, designed, analyzed, and evaluated in terms of human mental processes (operator mental model of the adaptive systems). According to Sanderson (1989), both the cognitive engineering and organizational approaches, such as macroergonomics, offer important frameworks for thinking about emerging human roles in advanced manufacturing systems. These frameworks were defined as follows:

1. The human is an active agent of organizational goals who has specialized, often hard-to-capture, skills and expertise.
2. Human roles in manufacturing systems are becoming more difficult and responsible rather than less.
3. Information systems help to eliminate boundaries of traditional job descriptions.
4. "Unsafe acts" are viewed as the result of system design flaws and organizational inadequacies just as much as the result of systematic human fallibility.
5. Error, when contained within the system, is viewed as necessary for the development of skills.

The cognitive systems approach focuses on the human operator and development of an internal model that describes the operation and function of the system using the operator's experience and training with relation to the nature of man–machine interfaces.

System design should allow us to build into the machines or other artificial systems a model of the user's characteristics (a system's image of the operator) (Hollnagel and Woods, 1983). The system's image of the worker should contain not only human physical characteristics, but also the cognitive functioning capacity of the worker.

The human operators functioning in modern manufacturing systems can also be classified according to Rasmussen's (1983) three major categories, i.e., (1) skill-based behavior, (2) rule-based behavior, and (3) knowledge-based behavior. "Skill-based behavior" refers to sensorimotor performance during acts or activities that take place without conscious control as smooth, automated, and highly integrated patterns of behavior. In this view, the human activities are considered as a sequence of skilled acts composed for the actual situation. The term "rule-based behavior" refers to goal-oriented performance structured by feedforward control through a stored rule or procedure allowing composition of a sequence of subroutines in a familiar work situation. Rule-based performance is based on the explicit knowhow of employing the relevant rules, whereas the reference data consist of references for recognition and identification of states, events, or situations (Rasmussen, 1983).

"Knowledge-based behavior" refers to the goal-controlled performance, where the goal is explicitly formulated based on the knowledge of environment and aims of the person (Johannsen, 1988). The internal structure of the system is represented by a *mental mode.* This kind of behavior allows us to develop and test different plans under unfamiliar and, therefore, uncertain control conditions and is needed when skills or rules are either unavailable or inadequate so that conscious problem solving and planning are called for to meet the demands of unfamiliar situations (Goodstein, 1981). The most important information to use for planning human interactions for unfamiliar occasions in complex control systems is knowledge of the beliefs and value structures of the work environment related to the abstraction level of functional purpose.

56.3.2 Job Design and Development of Human Skills in Manufacturing

The necessary reduction of the internal complexity of contemporary manufacturing enterprises can be achieved by comprehensive use of the unique worker skills, and on the basis of human-oriented organizational schemes that reintegrate all necessary tasks to process classes of similar objects (Kidd, 1994). Human skills can be developed and comprehensively utilized by taking into consideration all the factors that define the working tasks and management procedures. These procedures, in turn, have to be appropriately designed with respect to the organization, use of technology, and corporate policies. An appropriate framework needs to be developed to deal with change, risk, and complexity. This requires involvement from production workers and management learning to trust their abilities. Since change takes a long time, one should think in terms of long time scales (Savage, 1990).

Smith and Carayon (1995) noted that in order to assure the most effective application of contemporary manufacturing technology, the following critical job design considerations should be incorporated into the work systems: (1) job content, (2) control of the work process, (3) job demands, (4) performance monitoring and feedback, (5) socialization on the job, and (6) supervision style. In addition, the methods that manufacturing corporations can use to enhance effective use of new technology by employees, which are also important for facilitating the management of technological change, include: (1) involving employees in the decision process of selection and implementation of the new technology, (2) maintaining an ongoing process of employee participation in the production process, (3) developing ways of increasing employee control over their own work activities, (4) establishing a supervision process that is helpful and problem solving rather than monitoring and coercive, (5) providing employee learning and growth throughout a career, and (6) designing jobs that provide employee self-esteem and self-worth (Smith and Carayon-Sainfort, 1989).

An integration of the enterprise personnel across departments and at different hierarchical levels in developing the strategic company vision should be followed by the planning strategy to realize this vision within a new organizational structure. The support for a change process can be established through such actions as: (1) development of skills and training for new forms of teamwork, (2) implementation of the reward systems for participation in teams, (3) design of a comprehensive communication system that incorporates a knowledge of change, and (4) establishment of procedures that allow for experimentation, creating a culture of change, providing facilities to support change, i.e., meeting rooms, tools etc. The knowledge and tools of the human factors discipline can be very useful in this process. The process of learning and long-term collaborative approach to working with human factors researchers needs to be established, e.g., in the area of ergonomics, human–machine interface, new systems design, job design, teamwork, management of the change process, and organizational restructuring principles.

56.4 HUMAN AND COMPUTER-INTEGRATED MANUFACTURING

As mentioned above, the human- and computer-integrated manufacturing refers to developing integrated systems of people, organization, and technology (Kidd, 1991). Three main types of system integration must be applied here, i.e.: (1) integration of people, by assuring the effective communication between people, (2) human–computer integration, by designing suitable interfaces and interaction between people and computers, and (3) technological integration, by assuring effective interfacing and interactions between machines. One of the main ingredients of an integrated manufacturing is the concept of skill-based technology, a strategy that involves people who work within appropriate organizational structures, and are supported by advanced technology. The above defined concept of manufacturing takes it origins from the early ideas of the human-centered approach to design of advanced manufacturing technology (Rosenbrock, 1989). A comparison of design choices for modern manufacturing is shown in Table 56.5.

56.4.1 Basic Requirements of Human-Centered Manufacturing

The human-centered approach in manufacturing focuses on five basic principles of work design (Corbett, 1990; Murphy, 1989), i.e., (1) retention and enhancement of existing human skills, (2) extension of operator choice and control (human controls the technology), (3) minimal subdivision of work (unites planning, execution, and monitoring), (4) maximization of the human operator knowledge of the whole production process (encourages social communication and interaction), and (5) consideration of the ergonomic factors (layout of equipment, design of keyboards, software, etc.). According to Corbett (1988), the human-centered, computer-integrated manufacturing systems should satisfy the following basic requirements (see Table 56.6):

1. Human-centered manufacturing system accepts the present skill of the user and allows it to develop.
2. The more degrees of freedom open to users to shape their own working behavior and objectives, the more human centered the manufacturing system.
3. Human-centered manufacturing system unites the planning, execution, and monitoring component of work.
4. Human-centered manufacturing encourages social communication (both formal and informal) between users, preserving the face-to-face interaction in favor of electronically transmitted data exchange.
5. Generally, human-centered manufacturing systems should provide a healthy, safe, and efficient work environment.

Badham and Schallock (1991) outlined the human-centered work-oriented model of advanced manufacturing in Germany based on the strategy of integrated group manufacturing (Badham and Schallock, 1991; Brödner, 1986, 1991; Corbett, 1990). The model was adapted to conditions of smaller corporations and batch sizes, high quality and customized markets, and relatively high level of skills, unionization, and "high-trust" production cultures (Badham and Schallock, 1991). This approach has been developed within the ESPRIT research program (Kidd, 1990). By splitting orders, instead of dividing labor, job shop manufacturing was changed into a group manufacturing where part families were manufactured in their entirety (Brödner, 1986).

Based on the Maslow (1954) theory of human needs, Märtensson (1995) derived the human-centered requirements of work and used them as guidelines for design of the human–machine interactions and evaluation of job content in flexible manufacturing systems (FMS) in Sweden. These following general requirements were identified: (1) a versatile work content, (2) responsibility and participation, (3) information processing, (4) influence on the physical work performance of the task, (5) contact and cooperation with colleagues, and (6) competence development. These requirements (Table 56.7), described in system design criteria, can also be used as a checklist when designing the work organization and a decision support for the human operators. The applicability of these

Table 56.5 Comparison of Technology-Based Manufacturing Concepts with Modern Concepts

Criteria	Technocratic Approaches	Lean Production	Anthropocentric Approach
Labor market	Numerical flexibility core/periphery	Core of high-skill segmented core/periphery	Core of high skill and skill
Employment relationship	High substitutability of labor	High dependency on skilled labor	High dependency but labor mobility
	Low trust	High trust	High trust
Industrial relations	Adversarial	Corporatist company unions	Corporatist or pluralist participative social contract
Qualifications	Low skill, minimal training	High training, horizontal/multiskilling company based	High education and training, multiskilled transferrable skills (craft origins), vertical/horizontal skilling
Design paradigm	Taylorist mechanist metaphor	Organic system metaphor	Organic tool metaphor
Organization structure	Hierarchic divisionalized/functional reports	Leadership, less hierarchy, collaborative networks, and teamwork	Flat hierarchies decentralized semiautonomous work groups
	Bureaucratic	Post bureaucratic	Postbureaucratic
Organizational culture	"One best way" scientific management	Teamwork collectivist "one best way" harmony consensus	Group based plus occupational identity, pluralist, partnership

Source: After Wobbe and Charles (1991).

Table 56.6 The Requirements of the Human-Centered Design Approach to CIM

1. Compatibility	Operation should not require skills unrelated to existing skills but should allow existing skills to evolve. The operator should input and receive information that is compatible with conventional ship floor practice. In this way, the interface will conform to the user's prior knowledge and skill.
2. Transparency	One cannot control a system without understanding it. Therefore, the operator must be able to "see" the internal processes of the software in order to facilitate learning. A transparent system makes it easy for users to build up an internal model of the decision-making and control functions that the system can perform.
3. Minimum shock	The system should not do anything that operators find unexpected in the light of the information, detailing the present state of the system, available to them.
4. Disturbance control	Uncertain tasks (as defined by the choice structure analysis) should be under operator control with computer decision-making support.
5. Fallibility	Operators' tacit skills and knowledge should not be designed out of the system. They should never be put in a position where they helplessly watch the software direct an incorrect operation.
6. Error reversibility	Software should supply sufficient feedforward of information to inform the operator of the likely consequences of a particular operation or strategy.
7. Operating flexibility	The system should offer operators the freedom to trade off requirements and resource limits by shifting operation strategies, without losing software support.

Source: After Corbett (1988).

requirements was tested in attitude surveys of workers in a Swedish manufacturing industry (Märtensson, 1995).

Alasoini (1996) discussed the Finnish Research Program on Work, Culture, and Technology, carried out in four Finnish metal manufacturing shops. The purpose of the two projects was to test and consolidate a new mode of operation for adaptable manufacturing within the companies involved. It was assumed that the shift to adaptable production called for changes in four dimensions: (1) group and network relations, (2) tools and procedures for developmental activities, (3) management and employee skills, and (4) work and enterprise cultures. The goal of the first project, entitled *Techno-Organizational Change and the Adaptable Product Shop,* was to create and test in practice a model for an adaptable product shop capable of short response and continuous development. The second project, entitled *Development of an Adaptable Production Systems as a Social and Cultural Process,* aims to examine the prerequisites for advanced modes of manufacturing in Finnish industry. These two projects are carried out in close cooperation with four Finnish engineering workshops (a steel foundry, a bus coach assembly plant, a manufacturer of paper and board machines, and a manufacturer of special machine tools). The overall goal of these studies was to identify a set of requirements for an adaptable mode of operation in the four target companies in four dimensions (Table 56.8), including the following issues (Alasoini, 1996):

1. Group and network relations. Cooperation within manufacturing and between manufacturing and its support functions. Collaborative group and network relations

Table 56.7 Requirements and Criteria for Work To Be Used in the Design of the Man–Machine System (e.g. the FMS Operator)

Requirements	Criteria
Versatile job content	This individual should *plan, perform and monitor* his or her job, which will become a well-defined part of the process (e.g., scheduling, programming, machining, monitoring, disturbance handling).
Responsibility and participation	Individual/group *responsible* for the *whole task* (e.g., quality control carried out by the group leads to replanning).
Information processing	*Participation in design* process (e.g., of new machines).
Influence on the physical work performance	Cognitive activity in *new situations including decision making* (e.g., problem solving at disturbances).
	Operator's *working pace controlled* by process *only temporarily* (e.g., start and finish of machine process).
	Operator *chooses* working *method* (e.g., manual or automatic performance).
Contact and cooperation	Verbal and visual contact with at least one person.
	Contact with colleagues in other "process steps" (e.g., planning or assembly department).
	Cooperation in a team.
Competence development	Competence of the individual is being used in *more qualified tasks* (e.g., learning new machines).
	Continuous training (e.g., updating knowledge on programming).

Source: Märtensson (1996).

required to satisfy the company's relations to its customers as well as its materials and component suppliers.

2. Tools and procedures for developmental activities. Networklike work patterns are not possible without new tools and procedures that support problem solving and decision making and their implementation and evaluation.
3. Skills. Besides multiskilling, an adaptable mode of operation calls for ability to work as a member of a team and to take part in production-improvement activities.
4. Work and enterprise culture. Adoption of the new mode of operation calls for involvement on the part of management and the different groups of employees in long-term collaboration in line with jointly agreed goals and procedures.

56.4.2 Work Decentralization Principles in Contemporary Manufacturing

The decentralizing principles for human-centered manufacturing were described by Badham and Schallock (1991) as follows. First, operators tasks are widened as far as possible. Second, computer-aided planning facilities are located at the shop floor level rather than the planning department level. Third, as much as possible, planning and scheduling functions are supported at the production island rather than the foreman or area control level. An example of an automated manufacturing cell designed using the human-centered concept was discussed by Cooley (1989). The operator runs the manufacturing cell with the aid of powerful software tools. The operator's job includes: (1) creation of the machine programs using high-level software tools, (2) optimization of these programs using his or her skills and experience to minimize the cutting time, (3) machine scheduling, (4) programming the work handler to load and unload, and (5) doing the leftover tasks. For example, Table 56.9 compares two plausible design scenarios for modern manufacturing systems.

A review of concepts about the human computer-integrated manufacturing systems (Kidd and Karwowski, 1994) indicates that full human and machine integration may be

Table 56.8 Comparison of Rationalized Craft Production and Adaptable Production

Dimension	Rationalized Craft Production	Adaptable Production
Group and network relations	Within a company	Within a company
	Little cooperation between different groups of employees and functions. Cooperation is situational, reactive, and unsystematic	Comprehensive cooperations • within production teams • between production teams • between manufacturing and staff functions
	Company–supplier bargaining	Company–customer cooperation
	Company–customer bargaining	Company–customer cooperation
Tools and procedures for developmental activities	Developmental activities are uncoordinated and situational, resulting in group- and function-bound suboptimization	Developmental activities are steered by jointly agreed goals, models, tools, and procedures
Skills	Craft-bound and specialized technical skills	Multiskilling Ability and readiness to team work Mastering of tools and methods for developmental activities
Work and enterprise culture	Hierarchical system of production production management	Lean system of production management
	Rigid boundaries between jobs and functions	Flexible boundaries between jobs and functions
	Remuneration focuses on individual performances (amount, quality)	Remuneration focuses on skills and knowledge, and collective performance.
	Low-trust workplace industrial relations	High-trust workplace industrial relations.

Source: Alasoini (1996).

possible through a fully adaptive system architecture. A prerequisite for the above is development of the adaptation design methodology, and the enabling, computer-supported technology for its implementation in the design process. Such adaptation methodology should include the following phases of intelligent system design and implementation: (1) requirements analysis: variability, acceptability and usability of manufacturing requirements; (2) viability analysis: practicality, adaptability and usability of contemporary manufacturing design; (3) design: organizational system architectures, user interfaces, and adaptive system behavior modeling; (4) adaptive human supervisory control methodology and models; and (5) evaluation of system adaptation capabilities during operation.

56.4.3 Organizational Design Concepts for Contemporary Manufacturing

Characteristics of contemporary manufacturing depend on the choice of hardware and software, associated allocation of functions at different levels of control, the form of system development, and the human–machine interface (Badham and Wilson, 1989). In recent past, the attention of human factors engineers and hardware or software ergonomists has typically been restricted to design of human–machine interfaces, ignoring the organizational framework. However, at the core of contemporary manufacturing systems is the computer-supported organizational change (Badham and Schallock, 1991; Majchrzak, 1990). For example, the concept of human-centered manufacturing allows us to extend the traditional human factors considerations through the design of manufacturing architecture within adaptive organizational structures. Under this concept, modern manufacturing systems are viewed as sociotechnical systems where complementary technical and social components are designed in parallel (Badham and Schallock, 1991), with focus on the following issues: (1) parallel design—a cyclical and iterative process of simultaneous planning of both human and technical aspects of the new systems; (2) the

Table 56.9 Two Scenarios Illustrating Different Choices for the Design of Jobs within an Automated Flexible Assembly Cell

Scenario 1: Control by Specialists (Computer Integration)	Scenario 2: Control by Operators (Human–Computer Integration)
LOGIC	LOGIC
Expensive equipment: therefore needs managing and controlling by "experts"	Expensive equipment, needs to be controlled by an expert who can solve problems as they arise
EMPHASIS:	EMPHASIS:
Specialization Centralized control	Self-control and flexibility Local control with support specialists
ROLES:	ROLES:
Specialist machine setters	Operators set machines
More programmers to write, prove out, and edit programs	Fewer programmers (writing only)
More engineers to maintain and repair electrical, electronic, and mechanical components of cell	Fewer engineers as operators carry out routine maintenance and error recovery
Operators who mind machines and call for help when required	Operators who set machines, prove-out and edit programs, carry out routine problem solving
More quality inspectors	Fewer inspectors as operators are responsible for wider areas of quality control
COST/BENEFITS:	COST/BENEFITS:
Low direct labor costs	High direct labor costs
High indirect costs	Low indirect costs
Low motivation and commitment of operators	High motivation and commitment of operators
Waiting for experts; therefore, poorer utilization	Speedy resolution of problems, therefore, better utilization
Experts deal with simple problems, which is wasteful of their expertise and distracts them from other, more specialized problems	Operators deal with simple problems, leaving experts free to deal with more specialized problems

Source: After Clegg (1988).

complementarity of technical and social components of the new systems—design of equipment hardware and software as a set of tools to perform routine transfer control tasks and enhance the ability of humans to make choices in situations of uncertainty and deal with unpredictable events; and (3) new system design techniques created to overcome designers lack of understanding of contemporary manufacturing operating requirements.

56.4.4 Enabling Design Concepts for Modern Manufacturing

Womack et al. (1990) introduced the term *lean production,* which was used to describe application of new collaboration concepts in advanced manufacturing systems and upgrading of the human operator roles. In the concept of lean manufacturing organization, the organization and human, rather than the technological factors, are considered most important for generating expected results. The concept of lean production attempts to define the key elements of a new production paradigm replacing traditional mass production. Contrary to the flexible specialization, lean production combines the advantages of both craft and mass production, including high volume but diversified production using flexible automation, multiskilled team work, and an overall corporate culture of continuous improvement and high quality. Lean production stipulates more rewarding human roles in manufacturing, which involves more control over work, allows for problem solving and learning in the workplace, and leads to higher employee motivation (Womack et al., 1990).

Another concept of contemporary manufacturing integration is based on the idea of the anthropocentric production system (APS). According to Wobbe and Charles (1994), the anthropocentric approach assumes that people play a central role in manufacturing and relates to work organization of the management structure and organizational culture. To achieve these principles, technology has to allow for the collaborative production structures. The particular measures needed to establish APS and to organize different levels of change toward APS principles are shown in Table 56.10. The above referenced production concepts represent a departure from the central organizational principles of traditional mass production: increased division of labor, semi- or unskilled work, and tall, bureaucratic hierarchies. Table 56.10 also compares important features of the ideal types of lean production and anthropocentric production systems.

The APS systems aim to combine both the humanization of work and economic efficiency under conditions of volatile markets leading to increased requirements for flexible specialization (Brödner, 1990). APS are called anthropocentric because they are focused on skilled human labor instead of technology as the main resource for highly flexible, customer-oriented, and quality-based production. The defining characteristics of the anthropocentric production system are: Work in semiautonomous groups; holistic task assignments to the groups, including both the horizontal integration of task (e.g., integration of technical maintenance and quality assurance into manufacturing groups) and the vertical integration of tasks (e.g., integration of numerical control (NC) programming, planning, and scheduling into manufacturing groups); a decentralized factory organization with a comprehensive delegation of planning and controlling functions to the semiautonomous units; internal rotation of tasks, leading to job enlargement and job enrichment for the group members; and high and polyvalent skills and continuous upskilling at work.

For example, Eichener (1996) described viable approaches to APS in the model of the production island, with the variation manufacturing island, assembly island, office island, or the service island. Production islands are spatial, organizational, and social units dedicated to the production of complete products and thus integrating all machinery, equipment, and workplaces necessary for the production of a family of similar products (e.g., a family of similar parts in manufacturing islands, of similar aggregates in assembly islands). The islands have a high degree of autonomy and are responsible for logistics, technical maintenance (with the opportunity to call for external specialists), quality assurance, and, very important, workshop control, planning, scheduling, and programming.

Table 56.10 Anthropocentric Production System: Organization Design on Different Plant Levels

Levels	Measures and Principles
Factory	• Small decentralized production units • Product shops • Companies within a company • Delegation of responsibility to lower levels
Interdepartmental relations	• Cooperation between design and manufacturing • Interactivity between workshop and engineering departments concerning programming of machine tools • Integration of business department, technical department and shop floor concerning planning and scheduling
Group	• Installation of production islands • Group work in flexible manufacturing systems (FMS) • Semiautonomous assembly groups
Workplace	• Workshop programming • Integration of intellectual and manual functions • In highly automated areas: integration of programming, planning, maintenance, and processing tasks • In low automated assembly areas: work enrichment with decision space on execution sequence and performance

Source: After Wobbe and Charles (1994).

Manufacturing islands receive job shop orders from the production planning department, and are self-responsible for the fine scheduling and machine and human task assignments.

56.5 MANUFACTURING SYSTEM INTEGRATION

56.5.1 Work Design and Human Networking

Development of modern manufacturing systems has a profound impact on work design, i.e., on how the tasks are grouped into jobs and allocated among different work groups. Two key technological features can affect work design success: (1) technical interdependence, or the extent to which the technology requires cooperation among employees to produce a product or service, and (2) technical uncertainty, or the amount of information processing and decision-making workers must exercise during task execution (Wall et al., 1987). The work design is very much affected by the task environment. When task environments are relatively stable the job can be programmed and standardized. However, when the environment is relatively dynamic the job must be managed adaptively. Implementation of modern manufacturing requires nontraditional work designs and significant modifications in the work context. Specifically, the new forms of technology tend to increase technical interdependence, uncertainty, and environmental dynamics.

In addition to assessment of skills, training, and human–machine interfaces, effective design of modern manufacturing systems should include examination of organizational structures that have traditionally determined the system of reward of skills, direction of technological innovation, and the form of human operator control over production operations. In particular, emphasis should be placed on the new forms of cooperation, establishment of flat organizational hierarchies, and development of the human cooperation strategies. Savage (1990) conceptualized a transition from the steep organizational hierarchies, and formulated the principles of *human networking* in which people can work in cross-functional teams with their knowledge effectively utilized. The principles behind human networking, treated here as the enabling organizational design and management technology for modern manufacturing, include the following components: (1) peer-to-peer networking, (2) integrative process, (3) work as a dialogue, (4) human time and timing, and (5) virtual teams. These concepts proposed by Savage (1990) are briefly discussed below.

The *peer-to-peer networking* type of organizational structure allows people or teams to communicate directly with other people or teams without going through a hierarchical arrangement. This assumes that people have ready access to information and to one another's knowledge. Superior–subordinate relations will no longer exist. Instead there is one-to-one interaction on an equal basis. Peer-to-peer networking will also allow multiple teams to work on the same project that the quality of work is better.

The *integrative process* considers people and resources as virtual resources available at any time. It involves continuous development, training, and coordination within the organization and requires that the organization be in constant touch with its environment.

In the era of modern manufacturing, work is seen as a process or a series of acts. The *work process* in modern manufacturing combines the knowledge and vision of the worker to define a product and working toward that product by going through a process or a series of acts.

In modern manufacturing, the *human time* is not easily separated into distinct parts. People can anticipate the future while at the same time reviewing the past. Likewise, organizations can anticipate the future while at the same time reviewing customer response patterns. The organization can draw conclusions from past knowledge and take steps to improve its competitiveness in the market (Savage, 1990).

An alternative to hiring or training shop floor personnel who possess numerous skills and a variety of knowledge required by a particular problem, is to create *virtual teams.* "Virtual" means "being in force though not actually real" (Goldman et al., 1995; Savage, 1990). *Virtual organizations* set up cross-functional teams (CFT) to work on various projects and to rely on the talent and knowledge of their members. These CFT's are not permanent; they are defined and redefined according to needs. The greatest advantage of virtual teams is their flexibility. Implementation of these highly flexible, self-configurable, and autonomous organizational structures require a *shift of responsibility and enabling scheme.* This means that workers should be given more knowledge about the whole production process and the company's goals. Virtual teams allow for dynamic cooperation of the team members in combining the areas of product design and testing, production, and marketing. The virtual team members stay together for as long as the problem-solving process or the project continues, and then the team can be reorganized to deal with other projects necessitated by new market demands. A network of such virtual teams can also be formed. The decentralized, autonomous working groups can be interconnected by an adaptive information system providing tools for each of the groups to organize and perform their work and have unrestricted access to the information about company's goals, policies, and performance measures at all times.

56.5.2 System Integration and Organizational Effectiveness

In general, the manufacturing organization is effective only when it accomplishes its stated goals, makes maximum use of its resources and when its strategic constituencies are satisfied (Savage, 1990). The goal attainment approach seeks to satisfy the goals of an organization. However, there can be long- and short-term goals, and individual and organization goals. By making use of the principles of human networking in modern manufacturing it is possible to build cross-functional teams with a common agreement of goals. Peer-to-peer interaction and an integrated structure can ensure timely flow of information between teams and efficient sharing of knowledge and skills.

The integration of technology, organization, and people has become a cornerstone for success in advanced manufacturing environments (Karwowski et al., 1994; Majchrzak, 1988, 1995). This is partly because of the complexity and variety of issues that need to be considered, including those listed in Table 56.11. The four main categories of relevant issues can be identified as: (1) skill-based design of automated manufacturing; (2) job design and design of work, including sociotechnical and organizational design, designing the human infrastructure for advanced technology, managing for continuous improvement or total quality management, and managing the change to automated manufacturing; (3) human reliability, health, and safety, worker attention and fatigue, and personnel selection and training; and (4) human roles, function allocation, human system control, and human issues in planning and scheduling. These and other related issues are briefly discussed next.

56.5.3 Organizational Design Issues

The fast-paced technological developments in many fields require manufacturing engineering equipment that makes it possible to keep up with technological innovations. These innovations should be integrated into existing work structures without disturbing production sequences. Manufacturing companies can fulfill these requirements by utilizing work structures whose technical features meet the dimensions of flexibility concerning the following: (1) production quantity, (2) use of personnel resources, (3) variety of types or variants and new or changed products, and (4) new or changed manufacturing methods. Furthermore, the organizational work structures should be organized to minimize disturbing effects and assure the best manufacturing quality.

In addition, relevant social demands, which are reflected by national labor laws, labor standards, and trade tariffs, influence the organizational structures in the same way as markets or technological developments do. As a result, the organizational measures in the manufacturing companies must be adapted to the changing needs of the human resources within the company. In contrast to conventional structures, such new structures and especially organizational measures, aim at assigning members of the workforce according to their abilities and needs. The need for work enrichment and reduction of task monotony overlies today's social basic factors, such as work safety, reduction of physical strain, public health welfare, social security, and environmental conditions, which take place within the conditions of higher educational standards of the contemporary society. Jones (1995) discussed current research in sociotechnical systems, theory computer-supported cooperative work, and cognitive systems engineering and proposed a framework for integrating operational concerns into the concurrent engineering process.

As any other organizations, the contemporary manufacturing systems must use the process of *differentiation* in order to divide its staff, functions, and processes into the distinguishable units (Majchrzak, 1988). At the same time, however, these units must also be closely coordinated through the process of (computer) integration. Maintaining the balance between the integration and differentiation is a central dilemma for organizing manufacturing systems. Majchrzak (1988) also proposed a model describing six major components of the human infrastructure in contemporary manufacturing systems. This model is arranged into four components as stages to signify the fact that earlier decisions affect later options. These stages are:

1. The consideration of equipment features (parameters) and selection of equipment

Table 56.11 Impact of Short-Term Human-Centered Activities on the Four Competitive Factors of Advanced Manufacturing

	Impact on Competitive Factors			
Issues	Cost	Quality	Flexibility	Time
Human Issues				
View workforce skills as an asset	Changes view of labor costs	N/A[b]	N/A[b]	N/A[b]
Select personnel with ability to learn and adapt	Initial selection costs may rise, but training costs lower.	More qualified workforce producing fewer mistakes	Workforce capability to adapt is increased.	Capability to learn quickly can decrease setup and turnaround time.
Select personnel with communication skills	Reduce defect rate through communication among workers.	Interaction among workers to solve quality problems among themselves		Rapid, effective communication decreases time.
Implement long-term training programs	Initially increase cost, returns from higher in-house skills and less turnover.	Higher skills lead to fewer quality problems and provide worker with capability to solve quality problems.	Workers have wider range of skills to be more flexible.	Higher skill levels should reduce cycle time.
Specify specific training objectives to meet corporate objectives	Focus of training yields efficiency.	Can focus training on quality.	Can focus training on multifunctional worker.	Can focus training on cycle time reduction.
Implement virtual teams	Can embody necessary skills among several individuals.	Multiskilled teams to address quality issues	Multiskilled teams provide flexibility.	Time may increase slightly in team environment.
Design communication networks for teaming	Reduce need for unproductive time due to transportation to meetings.	Quicken sharing of quality information	Decisions made more quickly and more people can be included.	Rapid information transmission can reduce time.
Organization				
Consider sociotechnical view of congruence	Focus on costs of congruence rather than on isolated components.	Address quality problems resulting from interactions among components.	Enabling interactions allows greater flexibility.	Reduce time to market through more efficient design process.

Derive productivity measures for knowledge workers	New basis for costing job.	N/A[b]	Better understanding of work procedure can allow design of job flexibility.	Can decrease time by increasing productivity.
Identify areas of productivity improvement for knowledge workers and implement program.	Improve productivity, reduce costs.	New work methods can eliminate errors.	Better understanding of work procedure can allow design of job flexibility.	Can decrease time by increasing productivity.
Decentralize organizational structure where appropriate.	Initial increase in cost, but communication efficiencies gained.	More direct decision-making for quality problems.	Autonomous business unit enhances adaptability to market conditions.	Allows more rapid decision making.
Consider tacit knowledge in organizational design.	Assigning value to tacit knowledge may increase valuation of personnel.	Can apply tacit knowledge to solve quality problems.	N/A[b]	N/A[b]
Develop organizational structures for change and organizational learning.	Reduce long-term costs by adapting and removing inefficiencies.	Organization can adapt to address quality issues.	Adapting organization provides flexibility to adjust to competition.	Allows more rapid adjustment to changing conditions and product design decisions.
Technology				
Design ergonomic checklists that account for advanced manufacturing environmental features.	Direct cost reduction through better ergonomic design.	Improve product quality through more effective work via ergonomic design.	N/A[b]	Enables faster equipment design.
Develop cognitive task analysis program.	Consideration of cognitive work may change job standards.	Improving decision making process can reduce errors.	Can build flexibility into decision-making process of workers.	N/A[b]
Design equipment not to exceed operator skill and ability limitations.	Initial design costs may increase, but accident and error rates will decrease.	Reduction in errors will improve quality.	Operator limitations on adaptability understood and accounted for in design.	Reduce time as a result of less rework.

[a]After Karwowski et al. (1994).

[b]N/A = Not Applicable

2. Effects that decisions about equipment features could have on jobs (first-order effects)
3. Effects that the decisions about jobs can have on personnel training, selection, and other related policies, and on organizational structure (second-order effects)
4. The ultimate significance of the equipment for the production process, human infrastructure benefits, and organizational survival.

The two remaining components of the model proposed by Majchrzak (1988) are the *planned change process* for implementing the decisions, and constraints on the human infrastructure decisions, i.e., work force, control over resources, predictability of marketplace, and management of human resources priorities. The choices about human infrastructure must be compatible with equipment parameters to achieve optimal use of the new technology.

One of the most important issues of implementing and managing contemporary manufacturing technology is the ability to integrate successfully the required set of changes, which include the skills, habits, values of people, the work process (i.e., the tasks, procedures, and technology), workplace design (i.e., ergonomics, safety, and plant organization, including workgroup design) communication, decision-making processes, human resources, wage system, etc., into the organization (Salvendy and Karwowski, 1994).

56.5.4 Sociotechnical View of the Organization

Within the last few years many corporations have made an effort to develop and introduce new work structures that would be compatible with different branches of modern manufacturing systems (Womack et al., 1990). Significant improvements were made by structuring manufacturing into functional blocks (products in assemblies, production in work groups and manufacturing cells), leading to greater system transparency of internal processes, as well as higher system flexibility. In a decentralized system, with people who are adequately skilled, production planning and control can be shifted to the operational (factory floor) level, which leads to shorter and faster control loops and smaller organizational hierarchy. The organizational flexibility was also enhanced by including in such a change other support activities, including: programming, maintenance, servicing and tasks connected with quality assurance. The improved use of the system (availability) compensated for additional costs for personnel caused by higher qualification. Finally, utilization of the teamwork concept has allowed us to increase flexibility concerning quantity and personnel. Recently, planning and controlling tasks are being increasingly transferred to task groups, so that there is an increase not only in autonomy but also in responsibility.

In order to account for the multitude of factors encompassed in human, organization, and technology issues, a sociotechnical approach has been promoted (Jones, 1995; Majchrzak, 1988). In this view, the organization is not designed to optimize any one component, such as technology or people, but it should focus on the interactions or congruence of these factors. The manner in which such factors match one another is more important than any single factor alone, and it is the degree of appropriate match that the organization should promote. Some of the important tools designed to assure congruence between different manufacturing factors were developed by Majchrzak and Gasser (1992) and Majchrzak and Paris (1995). These tools, such as HITOP and ACTION, incorporate expert system techniques and knowledge bases in order to provide managers with structured information about organizational structures given the technology and human attributes of the manufacturing facility.

56.6 TOOLS FOR MANUFACTURING INTEGRATION

According to Smith and Carayon (1995), the integration of manufacturing technology and related issues of work design, organization, and management lags with respect to application of workplace automation and workplace design which are needed to accommodate the new technology and workforce needs. New theories of work organization and design have emphasized the need for more workforce involvement in the planning for manufacturing automation and during the implementation of new technology and for better workplace design to enhance human–machine interfaces. Smith and Carayon (1995) noted that theories such as sociotechnical systems, macroergonomics, and high-involvement management provide some insight into the problems of automation and solutions. The sociotechnical systems theory (STS) focuses on joint optimization of both human and technological consideration in system design and operation (Majchrzak, 1990). According to Jones (1995), a focus on design teams necessitates the study of the relationship between group work and technology as studied in the field of computer-supported cooperative work (CSCW). The cognitive system engineering approach extends the principles of STS to design of manufacturing systems in which machines serve as flexible, context-sensitive resources for human problem solving. The balance theory of Smith and Carayon-Sainfort (1989) provides a framework for successful implementation of manufacturing automation.

As discussed by Majchrzak and Paris (1995), the computer-integrated manufacturing differs from the nonintegrated systems with respect to technical complexity and the scope of operations. Complexity means that with integrated manufacturing the equipment often serves multiple and flexibly interchangeable functions, and significant information is typically needed by the human operator to interpret, evaluate, and diagnose events. Such a situation may lead to technical problems that are more difficult to diagnose than those which can occur in the nonintegrated manufacturing systems. The scope of operations means that under integrated manufacturing systems the equipment performs many more operations than before, making disturbance removal more difficult; i.e., a solution to a problem at one machine needs to be considered in relation to other machines. Therefore, appropriate management strategies must be developed to allow workers to cope effectively with the greater complexity and scope of integrated manufacturing.

Different ways to support workers in managing the enlarged scope and complexity include the following (Liker, Fleischer, and Arnsdorf, 1993; Majchrzak, 1988, 1992):

1. Broadening manufacturing operators' job responsibilities to include machine repair, process improvements, and inspection
2. Enlarging maintenance workers' job responsibilities to include teaching, ordering parts, scheduling, and machine operations
3. Extending supervisory job responsibilities to include working with other departments to resolve problems
4. More maintenance people to compensate for increasing equipment unpredictability
5. Increased use of work teams to provide a coordinated response to broad problems
6. Operator selection based more on human relations skills than seniority to ensure necessary communication and coordination capabilities
7. Increased training in problem solving and how the various manufacturing processes function to handle the increased scope of problems

Majchrzak (1995) postulated that plants which successfully utilize integrated manufacturing technology exhibit more of the management practices discussed above than the less successful plants. In addition to implementing integrated manufacturing, the successful plants also have these management practices in place. For corporations with nonintegrated manufacturing, management practices needed for integrated manufacturing are not nearly as critical for the plant to perform successfully. With flexible manufacturing automation workers need to be able to respond to a greater number of issues than with manual operations, and thus they need to have the appropriate authority, technical skills, latitude in work procedures, and motivating rewards for quick and effective response.

As discussed by Majchrzak (1995) an effective design of technology for the purpose of organizational-technology integration requires the following. First, there should be broadly diffused expertise among workers, engineers, technicians, and managers about different technology, organization, people (TOP) integration options and their impacts on each other. Second, different disciplines and stakeholders should work together as a team to create a comprehensive picture of the as-is and to-be technology, organization, people (TOP) aspects of the enterprise. Third, modeling techniques should allow a design team to explore and assess the consequences of different integration options so that intuitive predictions about effects of technological or organizational change can be tested against best practice models. Finally, the designs should be documented with the expectation that they will evolve over time in response to ongoing learning and adjustment so that as workers gain experience with new technical and organizational designs, they can improve their designs. The following is the discussion of the contemporary system tools for integrating manufacturing technology, organization, and people.

56.6.1 The HITOP System

HITOP, or *h*igh *i*ntegration of *t*echnology, *o*rganization, and *p*eople (Majchrzak, Fleisher, M., Roitman, D., and Mokray, J., 1990), is a methodology that allows to conduct an analysis of a TOP system. A HITOP analysis involves the design team completing a series of checklists and forms that describe their organization and current technology plans, and then helps the design team to identify the implications of those plans on organizational and people issues (Table 56.12). However, the HITOP analysis contains no knowledge base, nor does it recommend specific design options.

HITOP-A, a highly integrated technology, organization, and people—automated system reported by Majchrzak and Gasser (1992), is a knowledge-based decision support and simulation tool that supports the design and planning of advanced manufacturing systems. This integration tool takes into consideration the factors of organizational readiness to change, environmental constraints, strategic business goals, production variances, management values, characteristics of the planned technology, human capabilities, and motivational needs of the workforce. HITOP-A can be used to assess the impact of various manufacturing technologies on the human infrastructure of an organization, and allows to generate the required human infrastructure defined in terms of work design, skills, performance management, and organizational structure, including behavioral norms, coordination mechanism, work procedures, and centralization of authority.

56.6.2 The GRIPS Model

The Swiss Federal Institute of Technology (ETH) has developed the *work-oriented* approach, called GRIPS, to facilitate implementation of new technologies based on diagnoses of technology, organization, and people integration and to develop recommendations for system redesign. The GRIPS model, which assumes that effective organizations focus not only on technical aspect but also on the design of work organization and the use of skills and qualifications of its human resources, aims to diffuse the knowledge base about effective integration options and various design choices for the small and medium-sized manufacturing companies. From the organizational viewpoint, such an approach is characterized by decentralization at the enterprise level, functional integration at the organizational unit level, collective regulation at the group level, and complete and challenging tasks at the individual level.

Table 56.12 The Framework of HITOP

1. Organizational readiness
   How ready is the organization to make changes recommended by a HITOP analysis?
2. Critical technical features
   Critical technical features are features of the planned technology that are most likely to impact the integration of the technology with organization and people.
3. Essential role requirements
   For the four primary functions in a manufacturing workforce (core, support, supervision, and management), HITOP identifies eight role requirements, including degree and type of interdependence, information exchange, decision authority, and involvement and complexity of strategic goal setting.
4. Job designs
   The HITOP analysis requires the team to develop a set of job design values, such as "workers should have control over resources for those areas over which they are responsible."
5. Skill requirements (including selection and training)
   The minimal skill requirements (categorized by perceptual, conceptual, manual dexterity, problem solving, technical, and human relations) for each role requirement are determined in this step and a determination of which skills will be trained versus selected.
6. Reward systems
   Forms are provided to help the design team make three decisions about rewards: basis for pay (e.g., merit, hours, performance), basis for nonfinancially recognizing and rewarding performance, and future career paths.
7. Organization design
   Forms are provided to help the design team work through five organizational design changes typically seen with the implementation of new technology: changes in reporting lines, procedural formality, unit grouping, cross-unit coordination mechanisms, and organizational culture.

Source: After Majchrzak et al. (1991).

56.6.3 The COSAT System

COSAT, or Cross-Organizational STEP Adoption Tool, is a hypermedia-based software tool designed to assist managers to make the organizational and human resource changes necessary to gain the maximum benefit from technologies which implement the Standard for the Exchange of Product Model Data (STEP) as part of concurrent engineering (Industrial Technology Institute, 1995). COSAT provides online guidance to concurrent engineering (CE) by walking the user through the cases of best practice in CE. The system presents a relatively traditional planning process for organizational changes and offers design principles for optimizing work and information flow and organizational structure. Using matrices and worksheets, the COSAT system guides a team through the assessment and design process with structured activities related to the planning and changes processes.

56.6.4 The OPTISS System

Salo and Karwowski (1995) reported development of the fuzzy logic-based system for simulating the integration of new technology, organizational design, and human resources. The OPTISS (Organization, People, and Technology Integration Simulation System) is designed to facilitate an evaluation, coordination, and decision-making aid to optimize the integration of technology, organization, and people (TOP) in advanced manufacturing environments, in order to address the needs of the decision makers and designers of contemporary manufacturing systems. OPTISS, which was built using the fuzzy logic control software, TILShell 3.0, also allows for simulation of different TOP integration design solutions with the indication of final output design values. OPTISS was developed to address the needs of managers, system planners, and decision makers who aim to integrate and coordinate the critical components of modern manufacturing.

The main goal in OPTISS development was to develop a tool with quantitative outputs in order to help in design and implementation of the advanced manufacturing technology in companies. Another goal of this project was to develop a system that could be used to evaluate the present solutions for integration of new technology, organizational design, and human resources. OPTISS also includes a consultation level where cautionary and advisory suggestions are provided to implicate the effects of different design solutions for integration efforts. OPTISS is based on three interrelated modules that, if necessary, can be used independently. These modules showed great development potential. Each module can be easily modified. Input variables, as well as output variables and rulebases, can be added, changed, or modified in every module without difficulties. These characteristics will be valuable for future OPTISS development efforts.

56.6.5 The ACTION System

ACTION is an interactive software system and a methodology that embodies an extensive knowledge base about relationship among technical, organizational, and strategic features of the manufacturing corporations (Majchrzak, 1995). The system allows us to model the impacts of different organizational, technology, and strategy choices. This can be done, for example, by specifying characteristics of technology, best practice skills, information, performance measures, rewards and norms, and empowerment needs for the comprehensive data sets related to different activities, process variance control strategies, and business strategies (including new product development flexibility, minimizing throughput time, and maximizing process quality). A profile of the ideal organization can then be invoked from the knowledge base in terms of the required activities, information, skills, technologies, etc.

According to Majchrzak (1995), the identified system integration-related problems (gaps) and alternative priorities for solving such problems that are based on different prioritization criteria can be in analyzed ACTION and presented to the user. The prioritization is calculated based on a probability model in which a high probability of accomplishing a particular business strategy is calculated based on the number of problem areas meeting particular criteria of successful organizational designs; minimizing coordination needs, maximizing unit capabilities, and motivating workers through appropriate performance metrics and rewards. ACTION promotes the modeling of alternative TOP integration designs by a computer-accessible knowledge depository of best practices. Furthermore, the system uses a simple mode of changing inputs to describe alternative ideas, and allows for arbitrary order for input of data, design scenarios, information processing, and output.

The following questions need to be answered in preparation for using the above discussed manufacturing integration tools in practice (Majchrzak and Gasser, 1992):

1. Is the management and engineering staff aware of the technology failure rates internal to the organizations?
2. Have there been any efforts in the organization to document learnings from the failures to confirm that the failure have been attributable to lack of planning how the technology would be integrated with the people and organization? Have these learnings been disseminated to managers and engineers?
3. Have there been any efforts to examine in the organization how technology design, especially as it related to organizational and people issues, is conducted today? Has there been any effort to benchmark how the organization does technology design with how other companies do it?
4. Have there been any efforts to examine how a tool could be used to help facilitate the technology, organization, people integrative design process? Has a list of criteria for a worthwhile tool been generated?
5. Have there been any efforts to search for available tools that have been or could be used to help facilitate a TOP integrative design process?
6. Have there been any efforts, albeit small steps, to involve shop floor workers in continuous process improvements efforts in which they work collaboratively with engineers?

56.6.6 A Framework for Integrating Humans and Technology

Karwowski et al. (1994) proposed a conceptual framework to address the long-term issues related to competitiveness, complexity and uncertainty issues relevant to the human side of contemporary manufacturing enterprises. The GOPRIST framework (Figure 56.1) starts with the overall company *go*als, the set of design *pri*nciples as a basis to fulfill these goals, a set of management and organizational *st*ructures that correspond to the given principles, and the specific *t*echniques to implement these principles. The goals refer to achieving the desired state of the manufacturing enterprise, reflected in its responsiveness and the level of organizational performance. The highly responsive organization can be defined as flexible, adaptable, having a rapid response capacity, and having rapid product innovation. The high-performance organization can be identified by being productive,

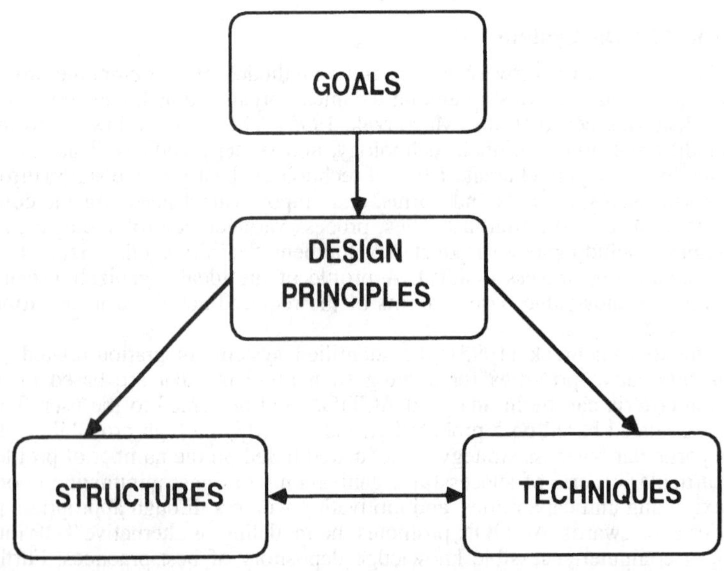

Figure 56.1 A framework for competitive advanced manufacturing enterprise. (After Karwowski et al., 1994.)

delivering high-quality products at low cost, using effective management of trade-offs techniques, and being environmentally conscious.

A change in the world view of manufacturing, i.e., a paradigm shift, requires development of a new set of principles. More self-configurable and adaptive organizational structures, and a much more people-oriented approach rather than a technology-centered one are needed. The human factors discipline can offer specialized knowledge on such aspects of the paradigm change, and specifically the management of change as follows: (1) work organization; (2) job design, new forms of organizing manufacturing processes; (3) skill-oriented control and responsibility; (4) ideas for managing the change process by assessing the critical change factors and developing systems, procedures, and tactics to address them; (5) evaluating change by aiming to make problems visible and create energy for change; (6) determining the costs–benefits of solution alternatives based on the degree and type of change; and (7) specific tools, techniques, and methods in the above listed and other areas.

The organizational structures of the GOPRIST framework (see Figure 56.1) correspond to the given design principles. Such structures translate these principles into specific actions by utilizing a set of available organizational design techniques. In general the organization subsystem focuses on work processes (tasks, procedures), workgroup design, communication and decision-making processes, and it includes: (1) the learning organization and (2) a set of integrated organization design principles. The learning organization principle allows for a high level of cooperation, open communication, and continuous improvement and leads to an integrated organization design principle. Finally, the technology subsystem focuses on the "technology as a tool" design principle.

Existing functional organizations of the manufacturing enterprises are too rigid to cope with external complexity and dynamics of the markets, and with rapid changes in products and processes suitable to meet the market demands. Such functional organizational structures waste specific skills of the human operators, i.e., workers' ability to make fine judgments, to cope with the system's uncertainties by modifying rules when appropriate, etc. Contemporary manufacturing companies should create self-configurable and highly adaptive organizational structures and procedures which enhance communication and cooperation between different organizational units and enable people to do what they can do best. The process of enabling people includes: (1) professional or technical qualifications (including social competence); (2) a shift of decision-making authority; (3) advanced, computer-aided tools for information handling; and (4) financial feedback and budgeting.

The knowledge of change implementation should be translated into a comprehensive and practically usable methodology (Karwowski et al., 1994). Elements to be included into such a tool are: (1) risk assessment, (2) cost–benefit analysis, (3) predictive models, (4) methods to constantly assess and modify and adjust the end state, and (4) specific application of change management theory at each stage of the change process. Examples of organizational design techniques that can be used to implement specific organizational structures include: (1) joint technical and organizational design, (2) job design principles, (3) user participation (participatory design), (4) organizationally appropriate technologies, and (5) comprehensive training and professionalization of personnel.

56.7 HUMAN ROLES IN PLANNING AND SCHEDULING

According to Savage (1990), the organizational functioning of modern manufacturing is based on the strategic constituencies approach, which holds that any organization must satisfy its strategic constituencies, owners, employees, customers, suppliers, unions, and government to survive. To satisfy all these groups the organization must develop a network connecting all groups. Employees, both blue and white collar, are critical to the success of an organization. In the modern manufacturing system, work is seen as a process, and the worker is intimately connected with the product through this process. In this view, the workers have the vision about the product and the knowledge of the process to work towards the envisioned product.

56.7.1 Cognitive Task Design and Human Workload

The recognition of human roles in contemporary manufacturing raises questions on how the system should be modeled and how much demand can be imposed on human supervisors (Bi and Salvendy, 1994). Typically, human supervisory control is characterized by dynamic, discrete, and random decision-making tasks and the functions allocated to human supervisors are highly correlated with human subjective stress and system performance. It was reported that humans appeared better at controlling a FMS under task conditions characterized by an externally induced workload compared to conditions where the workload was more internally induced (Sharit, 1984).

With human controllers in the system loop, there is a concern about cognitive task allocation based on human mental workload and performance, and the traditional task analysis techniques based on observable actions are no longer appropriate. As tasks in manufacturing become more cognitive in nature, traditional tools for analysis become ineffectual. Analysis of cognitive tasks is required for providing input to equipment design. It is recommended that attention in equipment design be given to the process in which cognitive tasks are executed and errors in decision making are committed. Given the complexity of these systems, the ramification of human errors in decision making becomes increasingly costly (Reason, 1990). Therefore, a practical method is needed for assessing the operator's cognitive ability to deal with information processing requirements imposed by the design of the different tasks that are likely to be performed in the automated manufacturing environment. Management and control systems in contemporary manufacturing should be designed so that individual operators can get the information that is compatible with their level of skills, knowledge, competence, and capacity.

Bi and Salvendy (1994) developed an analytical model to predict task load and human workload based on system engineering parameters, including task arrival rate, task complexity, task uncertainty, and task performance requirement. An experimental study, in which the sensitivity of the proposed engineering parameters on human workload and the validity of the analytical model were studied, was also conducted. A real team scheduling simulator of an advanced manufacturing system was implemented and configured with experimental data. The results support strongly the analytical model and, further, the conceptual model of human workload prediction. It was suggested that the derived cognitive task analysis method and prediction models could be used for dynamic decision-making system design with random task arrivals.

56.7.2 Human Planner and Scheduler

Much of the available literature on the human planner and scheduler have been reviewed by Sanderson (1989, 1991) and Nakamura and Salvendy (1994). Sanderson developed the Model Human Scheduler (MHS) in a quest to build a complete model of a human scheduler that would encompass the detailed mental activities in the human scheduling process. A framework for understanding and modeling the human operator performing scheduling functions was also presented by Papantonopoulos and Salvendy (1991). A general quantitative methodology for cognitive task allocation was proposed and validated in the control of manufacturing systems.

There are five major steps constituting the effective design of the human planner and scheduler, three of which follow:

1. Determining the relative capabilities of computer and human with regard to each subtask (Table 56.13).
2. Allocating the respective tasks to operational research models, knowledge-based systems and humans from which it is concluded that a hybrid model integrating operations and research knowledge-based systems and human provides the best planning and scheduling results in contemporary manufacturing setting (Figure 56.2).
3. For the work assigned to the human designer, special attention will have to be given to ensure that the human–computer interaction is well designed (Chapter 51), and that the operator has appropriate support where needed (Chapter 42), and that the job is well designed (Chapter 14) incorporating the best available knowledge in workplace design (Chapters 8 and 23) since the human planner and scheduler in manufacturing setting is a special case of a supervisory job.

56.7.3 Diagnosing Machine Failure

With increased utilization of flexible automation in the workplace and increased use of manufacturing cells, both the complexity and the importance of prompt diagnosis of machine status is needed. Basically, there are three classes of methods available to diagnose machine failure:

1. Fault diagnosis made by human (e.g., Morrison, 1988). The advantage of this approach is that it does not require investment in technology; no additional time

Table 56.13 Capabilities of Human and Computer in Planning and Scheduling Tasks of Advanced Manufacturing Systems (AMS)

Subtask	Description	Computer	Human
1. Detection	Detect information and data for jobs and machines.	Computer can easily detect information and data.	Humans take a long time to detect the presence of information and data.
2. Identification of system status	Identify the present state of the system.	If the identified pattern was predetermined, computer can quickly identify it.	Humans can recognize the important features in the planning, and scheduling environment. (But this is nonlinguistic knowledge.)
3. Interpretation	Interpret performance criteria and set the final goal for planning and scheduling.	Computer can decide if the program connecting the present state with the final goal is stored.	Humans can set the reasonable goal from among many criteria which conflict with each other.
4. Order selection	Select an order to be scheduled according to a priority.	Heuristic algorithm can provide a "good" solution, but no guarantee on optimal one.	Human intuition makes the best feasible solution.
5. Time assignment	Determine the start time and finish time for each operation of the selected order.	It is difficult to take balance between job waiting time and machine idle time.	Coordinating human with computer helps to determine efficient time assignment.
6. Resource allocation	Select the resources (machines, tools, fixtures, NC program, etc.) to produce an order.	Computer program can easily check whether machines tools, fixtures, and NC program are available.	Humans select many alternative solutions.
7. Evaluation and modification	Evaluate the plan or schedule and if not satisfied, modify it.	Poor, but updates the overall plan or schedule at least once every minute.	Humans modify overall plan or schedule with flexible decision making abilities.
8. Generation	Generate the plan or schedule sheet and issue it to the floor.	Computer can do it very easily.	Slow, not suitable.
9. Control	Check the difference between the plan or schedule and the practice.	Computer can do it easily under normal conditions.	Humans can adapt at abnormal conditions.

Source: After Nakamura and Salvendy (1994).

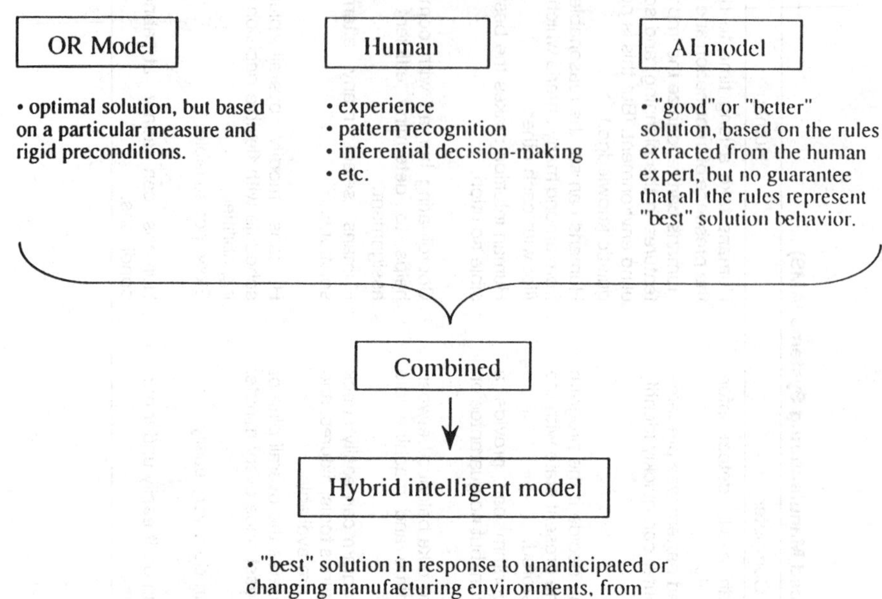

Figure 56.2 Hybrid intelligent model for planning and scheduling in AMS. (After Nakamura and Salvendy, 1994.)

needed to develop a method since the human expertise is instantaneous. However, the disadvantages of this approach are: (1) high cost of human labor; (2) limited availability of skilled personnel; and (3) frequently it takes a long time to diagnose machine failure, which increases system downtime and increases cost of production.

2. Fault diagnosis made by knowledge-based system (e.g., Naruo et al., 1990). The disadvantage of this system is that it takes time and costs to develop the system. However, once the system is developed, it can diagnose faults much faster than a human, which reduces downtime and increases productivity, especially when there is a need to wait for the approval of human expertise to diagnose the faults. In the Naruo et al. (1990) study, a knowledge-based system using 364 rules was utilized to diagnose the faults for a specific type of computer-controlled electromechanical equipment used in the electronics industry. This knowledge-based system diagnosed 92% of the machines malfunctions correctly in a 2-year period. The only failures it was not able to diagnose were ones that had no sensors to monitor their status. This included loose screws and stress factors.
3. Fault diagnosis made by neural networks (e.g., Lin et al., 1995; Ye et al., 1993). The main advantage of using a neural network instead of a knowledge-based system is that the neural network has the potential to learn from its performance, which traditional knowledge-based systems typically do not have. In the Lin et al. (1995) study, neural network was utilized for fault diagnosis of hydraulic forging presses in China. Neural networks with 30,000 iteration training provided a 99% correct identification of the causes of failure of hydraulic forging presses in China.

The above illustrates the dual role of human factors in manufacturing. It demonstrates that understanding the human thought processes can aid in both designing the task for the human (Akita et al., 1995) and evaluating the basis for taking the human out of the system by building a knowledge-based flexible automation system.

56.8 SAFETY, SYSTEM DESIGN, AND HUMAN ERRORS

Although modern manufacturing systems often remove the human operators from traditional energy flows during normal operating conditions, many tasks in automated manufacturing systems still need to be carried out in direct contact with some types of energy flow (Kjellén, 1990). As pointed out by Sugimoto (1987), as manufacturing automation progresses, workers may no longer be able to keep up with the sophistication and complexity of automation. Since the number of computer-controlled system components is growing rapidly, special emphasis must be placed on workers' comfort and safety through effective injury control and prevention strategies.

56.8.1 Safety and Workplace Design Issues

Worker safety and health hazards in contemporary manufacturing are critical factors for successful system design, implementation, and operation. Both productivity and safety may decrease if the principles of human factors engineering and ergonomics applicable to work and process design are not followed (Helander and Nagamachi, 1990; Karwowski et al., 1994; Salvendy and Karwowski, 1994). The contemporary manufacturing environment brings with it a new set of circumstances that require modification of traditional safety and ergonomic practices. Kjellén (1984) outlined five characteristics of environments:

1. Computer-integrated technology is associated with an increased complexity; i.e., systems are made up of networks of closely related subsystems and their components.
2. Several accident problems emerge at interfaces between subsystems and when disturbances progress from one subsystem to the other.
3. Traditional safety measures such as guards are in many cases unfeasible.
4. It is difficult for the worker who moves inside the physical system to anticipate the possible energy flows, and this difficulty in anticipating the performance and effects of the system is enhanced in systems that utilize exotic physical processes.
5. Many contemporary manufacturing systems are characterized by the existence of substantial amounts of energy and, consequently, of a high potential severity of accidents.

According to the International Standards Organization (ISO, 1991), ISO/TC 184/WG4, the risks associated with hazards caused by manufacturing automation vary with the types of machine tools incorporated into specific manufacturing system and the application of such a system as to how it is installed, programmed, operated, maintained and repaired. Easy to use, computer-based checklists are needed to evaluate whether the proposed system design meets the relevant human factors requirements and applicable design principles. Such tools can also be useful in helping the contemporary manufacturing industry to redesign existing environments through automation efforts.

56.8.2 System Safety, Operator Stress, and Human Error

As reviewed by Smith and Carayon (1995), manufacturing automation has traditionally brought about psychological stress problems for workers caused by issues of deskilling, job uncertainty, increased job demands, and organizational indifference to human resources (OTA, 1985; Smith, Carayon, Sanders, Lim, and LeGrande, 1992). The effects of these stresses may lead to poorer workers' health, decreased work productivity and safety performance, lower product quality, increased human error and work absenteeism, and greater employee resistance to organizational change (Keita and Sauter, 1992; Smith, 1987). The common sources of potential system errors that may occur in the flexible manufacturing systems include (Majchrzak, 1988): (1) operator errors, (2) errors in the cell or shop level computer, (3) errors in the part or program that runs the specific manufacturing process at the machine level, and (4) faults in the electronic logic that relays the mechanical movements of the machine and/or signals to the central computer to terminate operations upon completion. Morrison (1993) pointed out that system error and the sources of inefficiency may result from problems at several levels in a manufacturing system, not just the shop floor.

According to Reason (1990) there exist three basic human error types, i.e., skill-based slips and lapses, rule-based mistakes, and knowledge-based mistakes. This taxonomy is based on the modified Rasmussen's skill–rule–knowledge classification of human performance and allows to define three basic human error types in advanced manufacturing: (1) skill-based slips (and lapses), (2) rule-based mistakes, and (3) knowledge-based mistakes. The generic error-modeling systems (GEMS), which attempt to locate the origins of the basic human error types, can be used to derive the overall taxonomy of human behavior in advanced manufacturing. GEMS seeks to integrate two distinct areas of error research (1) slips and lapses, in which actions deviate from current intention due to execution failures and/or storage failures, and (2) mistakes, in which the actions may run according to plan but where the plan is inadequate to achieve its desired outcome.

At the skill-based level, human performance is governed by stored patterns of preprogrammed instructions represented as analogue structures in a time–space domain level are related to the intrinsic variability of force, space, or time coordination. The rule-based level is applicable to tackling familiar problems in which solutions are governed by stored rules (productions) of the type if (state) then (diagnosis) or if (state) then (remedial action). Here, errors are typically associated with the misclassification of situations leading to the application of the wrong rule or with the incorrect recall of procedures. The knowledge-based level comes into play in novel situations for which actions must be planned online, using conscious analytical processes and stored knowledge. Errors at this level arise from resource limitations ("bounded rationality") and incomplete or incorrect knowledge. With increasing expertise, the primary focus of control moves from the knowledge-based toward the skill-based levels, but all three levels can coexist.

56.8.3 Safety Aspects in Modern Manufacturing Systems

An accident is only one of the several outcomes of a man–machine interactions under hazardous conditions; near accidents and damage incidents are much more common (Zimolong and Hale, 1989). Generally speaking the occurrence of an error can lead to one of these consequences: (1) the error remains unnoticed; (2) the error can be compensated by the system; (3) the error leads to a machine breakdown and/or system stoppage; or (4) the error leads to an accident. Since not every human error that results in a critical incident will cause an actual accident, the further distinction of the following outcome categories is appropriate as follows (Swain, 1985): (1) unsafe incidents, i.e., any unintentional occurrence that may or may not result in injury, damage, or loss; (2) an accident, i.e., unsafe event resulting in injury, damage, or loss; (3) damage incident, i.e., unsafe event, which only resulted in some kind of material damage; (4) near accident, i.e., unsafe event in which injury, damage, or loss was fortuitously avoided despite a close call; and (5) accident potential, i.e., unsafe events that could have resulted in injury, damage, or loss, but in which, owing to circumstances, not even a close call was experienced.

Zimolong and Duda (1992) concluded that available data of reported incidents on advanced manufacturing components, in general, and robot installations in particular, are very poor, and that only a few original reports on accidents, critical incidents, and abnormal stoppage cases have been reported so far. There is also a lack of information on supplementary hazard data such as number of employees exposed, work hours of personnel. As a consequence, it is unclear how dangerous work with advanced manufacturing really is with respect to such basic concepts as number of accidents/injuries per 1000 employees, per 1 million work hours, or per number of robots.

Järvinen and Karwowski (1995) examined 103 self-reported accident cases attributed to advanced manufacturing environments, based on the results of an anonymous questionnaire requesting information about one serious accident that occurred in the respondents' manufacturing facility. A majority of the accidents occurred within the context of the stand-alone automated equipment, followed by flexible manufacturing systems or cells. The injured personnel were operators of automated equipment (67%), followed by maintenance or repair personnel (20%). Typical activities performed at the time of the accident were related to production disturbances, such as clearing a blockage, fault finding or rectification, and adjustment of part or machine. In 55% of the analyzed accidents and 75% of the robot accidents, the manufacturing equipment was in the automatic operating mode. In 74% of the reported accidents, the safeguarding applied to the involved equipment was claimed to be inadequate. These results suggest a need for the design of more effective safety systems that do not interfere with the human operator's work.

56.8.4 Disturbance Control in Automated Manufacturing Systems

Because of the relatively large investment of capital required for installation and operation of modern manufacturing systems, a high utilization of manufacturing capacity and low downtime resulting from system disturbances are critical to system productivity (Jarvinen et al., 1996). Production disturbances can also affect safety of personnel involved in operating and maintaining the FMS installations. According to Märtensson (1996), many of the automated systems in Swedish manufacturing industry exhibit a downtime of as much as 30%, with handling of unexpected events the main task of the human operators.

In advanced manufacturing systems, the human operators are needed for the purpose of controlling, programming, maintaining, presetting, servicing, or troubleshooting tasks. Disturbances in the system lead to situations that make it necessary for workers to enter hazardous areas. In this respect, it can be assumed that disturbances remain the most important reason for human interference in advanced manufacturing, because the systems will more often than not be programmed from outside the restricted areas (Anon, 1984). One of the most important issues for advanced manufacturing safety is to prevent disturbances, since most risks occur in the troubleshooting phase of the system. The avoidance of disturbances is the common aim for both safety and cost effectiveness.

The disturbance in advanced manufacturing is a state or function of a system that deviates from the planned or desired state. In addition to productivity, the disturbances during the operation of an FMS have direct effect on the safety of the people involved in operating the system. The study reported by Kuivanen (1990) showed that about one half of the disturbances in automated manufacturing decrease the safety of the workers. The main causes for disturbances were errors in system design (34%), system component failures (31%), human error (20%), and external factors (15%). Most machine failures were caused by the control system, and, in the control system, most failures occurred in sensors. An effective way to increase the safety of FMS is to reduce the number of disturbances. This requires that the critical system components and functions be identified (Kuivanen, 1990) by using analysis methods, such as Failure Mode and Effect Analysis (FMEA) at the design stage. Risks and hazards can be identified by using, for example, the energy analysis method. Even better results can be obtained, if these two methods are combined to identify critical system failures and disturbances.

The main research issues in manufacturing disturbance prevention are (Kuivanen, 1990): (1) the major causes of the disturbances, (2) unreliable components and functions, (3) the impact of the disturbances on safety, (4) the impact of the disturbances on the function of the system, (5) material damage, and (6) repairs. Toikka et al. (1991) described the orientation model for evaluation of the operator's role in advanced manufacturing disturbance control. The model includes five stages as follows: (1) withdrawal (somebody else solves the problems), (2) routine disturbance handling (habitual, not concerned about the causes of disturbances), (3) unofficial developing activities (private log books for ideas and inventions), (4) official optimizing of the system (organizational forms to develop disturbance control), and (5) systematic way of working (developing the functional principles and organization of contemporary manufacturing).

A classical approach in safety research has been to study the effects of the characteristics of the production system on human information processing and the probability of human errors. A current approach is to study human responses to deviations or disturbances in the system (Kjellén, 1990). Human actions in disturbed systems both prevent and contribute to the occurrence of accidents in the advanced manufacturing environment. For example, a study of accidents related to malfunctions of technical control systems showed that about one-third of the accident sequences included a human intervention in the control loop of the disturbed system (Backström and Harms-Ringdahl, 1984). Järvinen, Vannas, Mattila, and Karwowski (1996) reported a study aimed to: (1) determine the causes of production disturbances and their effects on the operation and safety performance of the FMS installations, (2) analyze and compare disturbance data for the FMS installations in the United States and Finland, and (3) identify ways to improve safety and efficiency in FMS implementations. The survey included 14 FMS installations in the United States and 31 FMS installations in Finland. The field survey questionnaire was used in data collection, with six sections: (1) background information, (2) planning, (3) performance, (4) organization, (5) training, and (6) production disturbance situations. About 12% of all the disturbances observed in the United States and 35% of the disturbances in Finland caused hazardous situations or accidents. The FMS disturbances were classified into design errors, component failures, human errors, and external factors. More than one third (34%) of the disturbances in the United States and in Finland (35%) were mainly caused by system design-based errors.

56.8.5 Safety Planning Procedures for Integrated Manufacturing Systems

According to the ISO (1991), an integrated manufacturing system is a group of industrial machines working together in a coordinated manner normally interconnected with and operated by a supervisory controller or controllers capable of being reprogrammed for the manufacturing of discrete parts or assemblies. The design phase of the proposed ISO (1991) safety strategy includes: (1) specification of the limits of parameters of the system, (2) application of a safety strategy, (3) identification of the hazards, (4) assessment of the associated risks, and (5) removal of the hazards or limitation of the risks as much as practicable. The advanced manufacturing safety specification (ISO, 1991) should include: (1) description of function; (2) layout and/or model; (3) survey of the interaction of different working processes and manual activities; (4) analysis of process sequences, including manual interaction; (5) description of the interfaces with conveyor or transport lines; (6) process flow charts; (7) foundation plans; (8) plans for supply and disposal devices; (9) determination of the space required for supply and disposal of material; and (10) available accidents records.

The work hazards of computer-integrated manufacturing can be characterized as follows: (1) The human operator must enter the danger zone during disturbance recovery, service, and maintenance tasks; (2) the danger zone is difficult to determine, to perceive, and to control; (3) the work may be monotonous; and (4) the accidents occurring within the system are often serious. The safety requirements of advanced manufacturing and its components must be adapted to the new production conditions. All necessary requirements granting a safe operation need to be considered in the design of systematic safety planning procedure (ISO, 1991). This includes all protective measures to reduce hazards effectively and requires: (1) integration of the man–machine interface, (2) early definition of the position of those working on the system (in time and space), (3) early consideration of ways of cutting down on isolated work, and (4) consideration of environmental aspects. The safety planning procedure should address among other aspects, the following safety related issues (ISO, 1991):

1. Selection of the operating modes of the system. The control equipment should have provisions for at least the following operating modes: (1) normal or production mode: all normal safeguards connected and operating; (2) operation with some of the normal safeguards suspended; and (3) operation in which system or remote manual initiation of hazardous situations is prevented (e.g., local operation, isolation of power to or mechanical blockage of hazardous conditions).
2. Training, installation, commission, and functional testing. When personnel are required to be in the hazard zone, the following safety measures should be provided in the control system: (1) hold to run, (2) enabling device, (3) reduced speed, (4) reduced power, and (5) moveable emergency stop.
3. Safety in advanced manufacturing programming, maintenance, and repair. During programming, only the programmer should be allowed in the safeguarded space. The system should have an inspection and maintenance procedures to ensure continued intended operation of the system. The inspection and maintenance program should take into account the recommendations of the system supplier and those of suppliers of various elements of the systems. Personnel who perform maintenance or repairs on the system should be trained in the procedures necessary to perform the required tasks.
4. Fault elimination. Where fault elimination is necessary from inside the safeguarded space, it should be performed after safe disconnection (if possible lockout). Additional measures against erroneous initiation of hazardous situations should be taken. Where hazards can occur during fault elimination at sections of the system or at the machines of adjoining system or machines, these should also be taken out of operation and protected against unexpected starting. By means of instruction and warning signs, attention should be drawn to fault elimination at system that cannot be observed completely.

According to the ISO (ISO, 1991), the risk assessment in computer-integrated manufacturing systems should be performed to minimize all risks, and to serve as a basis for determining safety objectives and measures in the development of programs or plans in order to create a safe working environment, and to ensure safety and health to personnel. Each identified hazard should be assessed for its risk and appropriate safety measures should be determined and implemented to minimize that risk. Hazards should also be ascertained for the single units, and interaction between single units, the operating sections of the system, and operation of the complete system for all intended operating modes and conditions including conditions in which normal safeguarding means are suspended for such operations as programming, verification, troubleshooting, maintenance, or repair.

56.9 HUMAN RESOURCE PRACTICES

56.9.1 A Manufacturing Paradigm Shift

The contemporary manufacturing industry is undergoing a major paradigm shift that requires innovation in several areas, including management processes, organizational and management structures, human resources, job design, skill development, and performance evaluation (Karwowski et al., 1994). The manufacturing industry embraces such a paradigm shift even though it involves a great deal of risk, uncertainty, and challenge to its existing structure and order. One of the basic elements of agile manufacturing environments (Goldman et al., 1995; Kidd, 1994) is the recognition that fulfilling demands for greater operational flexibility, critical to eventual success of the contemporary manufacturing enterprises, cannot be obtained without a qualified workforce. Importance is placed upon workforce skills and these skills are viewed as a corporate asset. This approach leads to investment in personnel selection and training and better utilization of worker skills. Since almost all organizational activities are based on human resources, the manufacturing industry should select people with the capability to learn and perform to a satisfactory level, provide sufficient opportunity for them to learn and acquire technical skills, and produce an environment in which employees are motivated to do their best work. Given the asset-based view of personnel that coincides with the agile-manufacturing perspective, selection and training can be considered investments.

Regarding selection of a qualified workforce, the identification of workers who can perform the current job is no longer sufficient. Given the exponential growth of technology in the workplace, personnel must be constantly learning and updating their skills. Therefore, corporations must attend to selecting people with the capability to adapt to new environments and learn new skills. Selecting workers exclusively for the jobs they will perform today may lead to an ill-equipped workforce tomorrow. In addition, as more cognitive-oriented and human interaction skills are required in the workplace, traditional selection techniques must be revised to account for these changes.

A variety of training techniques are available to produce a qualified workforce. Required regular classroom training has been used successfully by a number of companies. Others have used pilot projects, mentoring and teamwork as forms of on-the-job training. Regardless of the method used to deliver training, a careful evaluation of training goals in order to meet corporate objectives is critical because of rapidly changing workplace requirements. Readily available resources, such as HITOP (Majchrzak et al., 1992), Netmap, and design principles, can be used for this purpose. Although an organization may select promising candidates and provide adequate training, without a motivated worker overall system performance will be less than optimal. Attention must be paid to providing an environment where workers are satisfied and motivated. Control over the workplace and work pace contributes significantly to worker motivation and satisfaction (Wall, 1986). Whereas contemporary manufacturing technology can be used for workplace control, this may not be the optimal strategy when considering overall system productivity. Rather, providing workers a degree of control contributes by providing the motivational attributes and allowing workers the flexibility necessary to adapt to changing fluctuations in daily production requirements. As such, design of the workplace to allow for appropriate amount of worker control (Majchrzak, 1988; Rasmussen, 1993; Wilson, 1991a).

56.9.2 Planning the Workforce

According to Wilson (1991), as the roles of human resources in manufacturing corporations are changing as different forms of advanced technology and different methods of organizing production are introduced, there is an increasing recognition in industry of the input that can be made by the human factors profession with respect to potential benefits of developing systems around the concept of the skilled, responsible operator with local control. Hörte et al. (1994) analyzed the performance effects of new manufacturing technology and the related issues of human resource management between 1987 and 1992 based on the sample of 70 Swedish companies. Three different paths of development were identified, including: (1) technological upgrading, (2) development of the human resources of the company and organizational changes, and (3) a combined approach aimed at a synchronous development of technology and human resource management. The results showed significant correlations between development of human resource management and the performance measures, with only few performance increases correlated with upgraded manufacturing technology.

The rapid progress in technical developments leads to new challenges for planning of manpower resources. Manufacturing enterprises must be prepared for future developments on the job market, and proposals for ensuring the quantitative and qualitative manpower resources must be worked out according to technical and economical plans. From the perspective of society at large, the important human resource issues include avoidance of social strain, resulting from insufficiently planned decisions in regard to personnel (dismissals that can be avoided, claiming in labor courts, etc.), responsible planning and management of labor needs and reductions, and (3) realization and fulfilling of legal regulations and sociopolitical objectives.

Apart from the corporate-political requirements, an importance of personnel planning as a sociopolitical instrument, which can serve as additional aid for job market policies, also increases. Therefore, planning of personnel should serve the purpose or reducing negative effects of structural changes on employees and ensure preservation of their social status as well as improve their opportunities for future development. The following human resource management issues are of interest to employers: (1) availability of the skillful workforce, (2) use of personnel according to requirements and suitability, (3) improvement of the qualification level of employees, (4) avoidance of costs for personnel recruitment by using company-internal employees, (5) motivation of employees, (6) keeping track of the development of costs for personnel, and (7) workforce availability.

From the employee viewpoint, the important human resource issues include: (1) workplace guarantee or avoidance of cases of hardship if personnel are moved or released, (2) diminishing of risks that can result from technical and economical changes, (3) earned income that is secure and appropriate to demands and performance, (4) humane and healthy working conditions, (5) opportunities for professional and further training, (6) promotional prospects within the company, and (7) protection of special groups of employees (older, handicapped, and young persons).

The central difficulty in personnel planning results from the fact that people can hardly be planned. Although human resources planning cannot undo wrong management decisions of the past, it is increasingly turned into a strategic and essential instrument of company management practices. The results from planning manpower resources are the input for planning the personnel recruitment and personnel cutbacks. For a short period of time, the quantitative planning of manpower resources can deliver relatively sure predictions, because the production programs as well as the sales programs have usually been set up for the same time. On the medium- and long-term horizon, however, the majority of input factors for the quantitative and qualitative planning of requirements (work productivity, future demands on the work place) must be considered variable. Information about these future developments is often imprecise and incomplete. To assure effective human resource plans, one needs to consider characteristics of the particular industrial sector, type of production, stability of the environmental situation, a planning horizon, current situation on the job market, and the considered group qualifications.

In practice, different ways of estimating the future manpower resources are used, including the following: (1) estimation method (questioning of experts); (2) global requirement predictions (extrapolation of trends, calculations concerning regressions and correlations, on the basis of past developments); (3) index of code number method, e.g., the development of work productivity indices; (4) time-unit based personnel assessment method, (5) job planning method, where current and future organizational structures and workplaces are treated as variables and are used in the organizational chart that shows the manpower resources; and (6) qualification research based on the work tasks and methods that are developed to show how a company can determine the manpower qualification requirements.

The task of planning personnel recruitment can be defined as analyzing the internal and external job markets. Specific decision criteria must be established before hand to determine whether the manpower resources are to be covered internally or externally. The most important advantage of the internal recruitment process is the fact that it is easier and less risky and requires less expenditure. The disadvantage is the danger of internal inbreeding. Having a broader spectrum of human resources to choose from is the advantage of an external recruitment; however, there is a disadvantage concerning the cost and time expenditure and the higher risk. These requirements can be overcome by job ads, analysis of job applications, personnel leasing (personnel for a short period of time), fixed-term contracts, employment agencies, conclusion of contracts for work, as well as employment of part-time workers.

56.9.3 Function and Work Task Allocation Practices

To optimize manufacturing system performance, appropriate methods for function allocation between humans and computers need to be developed, with due emphasis on the consideration of human cognitive abilities (Bi and Salvendy, 1994). Since both an overload and underload affect human performance in a supervisory control, a predictive model is needed to evaluate the human workload with a specific manufacturing system design. Allocation of work tasks means assigning the workforce to the specific workplaces and vice versa. The assignment of work tasks means mastering quantitative as well as qualitative adaptation problems and covers two main areas of responsibility. In the medium and long terms, the assignment of work tasks aims for adapting the workforce to the requirements of the workplace, and for adapting the workplaces and working conditions to the workforce. In the short term, task allocation should allow for the time- and capacity-related integration of the workforce into the work process. Assigning work tasks requires intensive cooperation between various departments of a company, including workplace analysis and workplace design and ergonomics and planning of personnel development. The basis for assigning worktasks are process plans and job descriptions. The current process plans that have been established for the future show the number and kind of workplaces that have to be filled and, furthermore, job profiles concerning future employees. These job profiles must be compared with the ability profiles of the future employees.

56.9.4 Planning Personnel Training

Planning of personnel training, which is also called qualitative personnel planning, is responsible for determining and meeting requirements of new workplaces caused by technical and organizational changes or by promotions and hiring of new workers. Planning the personnel training also means keeping the discrepancy between the requirements of a workplace and the qualifications of employees as small as possible. There are different methods for personnel training. Job and workplace descriptions, the assessment of the performance of workers, as well as their development potential are the basis for planning the personnel training.

56.9.5 Planning Personnel Cutbacks

Planning personnel cutbacks means avoiding or diminishing cutting down personnel over capacities, and therefore reducing the labor force. There are various reasons for personnel cutbacks, such as slow economic growth, work-saving technologies, business fluctuations, and seasonal fluctuations. Because personnel cutbacks often conflict with the interests of the employees who want to have and keep their workplaces, there are regulations and laws guarding against unlawful employee dismissal. For example, in Europe, apart from legal regulations, collective labor agreements (agreement on rationalization protection, prohibition of dismissing older employees, as well as agreements on severance pay) are becoming more and more important. Other forms and possibilities for cutting back personnel or preventing such a cutback may include: (1) acceptance of orders from other companies, (2) taking back outsourcing contracts (however, this only results in a shift of the workplace risk), (3) taking advantage of fluctuation and at the same time stopping recruitment, (4) reducing extra work and subcontracted labor (equal use of all workplace capacities), and (5) shortening the regular working hours. The problem combined with indirectly reducing personnel is the fact that although it is better for the employees, it cannot solve the economic problems of unemployment.

56.9.6 Training the Workforce

In describing the training system of a company, a difference must be made between basic training and specialized training. For example, in Germany, in comparison to basic training, the system of specialized training has been subordinated. At the moment, most of the medium-sized and large companies in Germany spend 2.5% to 3.5% of the yearly gross pay and the salary sum for training and specialized training. However, educating the trainees already costs about 1.8% of the yearly gross pay. Considering the fact that the share of trainees is relatively small compared to the total number of employees, it becomes obvious that there is not much money left in the corporations to be allocated for continuous specialized training.

Internal specialized training refers to job-accompanying educational measures which are meant to provide additional knowledge and skills for working adults. Because internally carrying out and organizing specialized training would cause disproportionately high costs for small companies, these prefer to use outside institutions that offer specialized training. The target groups of these external institutions reach from semiskilled and unskilled workers to members of the top management. The main emphasis of internal educational measures is on imparting technical specialist knowledge, which is necessary for the use of new technologies; on the introduction into electronic data processing; and on the training of new working techniques. In specialized training activities, activities on the level of qualified personnel and specialists as well as workers are usually limited to imparting specialist knowledge and skills. The courses are offered during working hours as well as following them. On the level of semiskilled and unskilled workers, one cannot talk about specialized training in the true sense of the meaning; in this case methods for teaching basic skills predominate.

56.10 HUMAN FACTORS IN MANUFACTURING MANAGEMENT

Effective implementation of computer-integrated technologies in manufacturing depends upon a considerable degree of organizational adaptation. Bessant, Levy, Ley, Smith, and Tranfield (1992), reviewed evidence for the emergence of a paradigm shift that is changing the rules governing best practice in modern manufacturing and stipulates that organizational design needs to involve a complete system reappraisal rather than a minor adjustment. It was concluded the organizational design principles appropriate for use by managers in designing effective manufacturing organization beyond 1990s may be as follows:

1. An increase in the automation of standardized tasks. The specific role of integrated technologies is vital given that many tasks, although standardized, will require integrated treatment to produce flexible responses.
2. An increased sense of corporate focus through a sense of shared purpose. Coordination through the sharing of common purpose is an emerging feature of organization design.
3. An emphasis on intrinsic rather than extrinsic motivation. Because jobs are both fewer and more complex, and they include the nonprogrammable elements that remain outside of the formal system, individual commitment, personal understanding, and identification with the overall purpose requires intrinsic motivation.
4. A greater empowerment of individuals brought about by a decrease in emphasis on external control. A corollary of (3) above is the changing role of management to that of providing direction and support rather than control and externally mediated coordination.
5. A greater flexibility and fluidity of structure—a loose structure held together by a tight culture.
6. Design on the principle of redundancy of function and not on redundancy of parts. Because individuals and work groups will have a variety of skills and both the capability and requirement to undertake multiple roles within fluid structures, job functions are expected to become outdated, or temporarily unnecessary.

56.10.1 Management Issues in Manufacturing

In general, two basic management styles, i.e., the authoritative style and the cooperative style, have been utilized in the past by the manufacturing industry (Figure 56.3). The differences between these two styles results from the degree of involvement of employees with respect to task- and person-related decisions within the company. If the management is characterized by extreme authority, decisions are made only by the superior and the employees do not have any influence. Cooperative management is based on the principle of delegating responsibilities, authority to issue directives, and competence in as well as responsibility for one's actions. Gerwin and Kolodny (1992) advocate that different management practices are needed at different levels of manufacturing flexibility. Furthermore, to maximize the process innovation, different organizational structures should be used at different levels of manufacturing automation (Collins et al., 1988).

Contemporary management styles adopted different cooperative forms of organization, and several *management by* techniques were developed. *Management by delegation* is

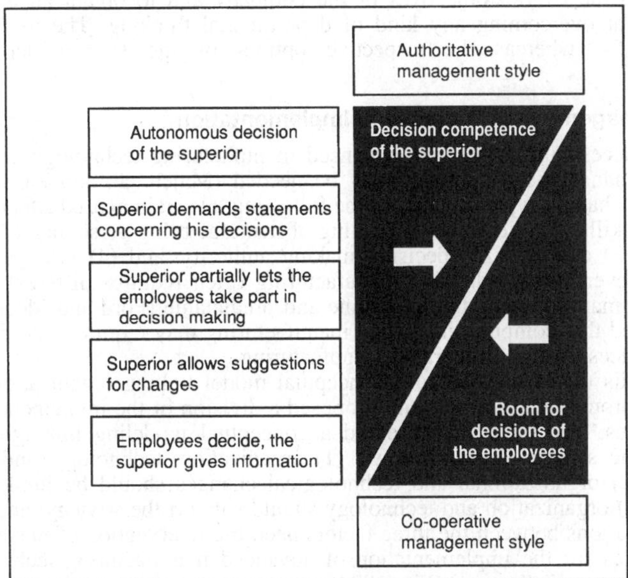

Figure 56.3 Characteristics of authoritative and cooperative management styles.

characterized by division of labor in the management. Delegation means transferring tasks, authority (competencies), and responsibility. The degree of independence and professional qualification determines the maximal extent of delegation. The delegation principle of the "management with employee relationship" is the separation of management responsibility as well as of responsibility for action.

In *management by objectives,* employees are involved at different management levels and are given concrete (partial) objectives, clearly showing them the direction for work. However, the measures for fulfilling these objectives are largely up to the employees. The degree to which the objectives have been fulfilled serves as a criterion for assessing the performance. From the point of view of methodology, objectivizing the instructions and performance is not easy.

Management by results means that management concentrates on the results produced in the several sections. Special emphasis is given not the solution of problems, but the use of the most successful possibilities is given. The produced results serve as criterion for evaluating the priorities of management measures. For evaluating the results, the time horizon is essential: Short-term success does not necessarily lead to long-term success ("job hoppers").

In *management by exception,* the top management level is to be relieved of routine jobs. Its work concentrates on "exceptional situations" while the lower management level deals with "normal situations." Concrete standards as well as a continuous comparison between the existing situation and the situation-to-be by different persons are the prerequisites. Deviations of the situation-to-be from the existing situation that exceed a certain limit characterize a situation as "exception" and trigger off interventions of the higher management level. However, there are problems if the management style depends on the quality of instructions. Furthermore exceptional management involves the problem of interference.

Management by motivation concentrates on human motive power (human needs) and focuses on motivating the employees to give their best for the company. Motivation means activating and increasing the willingness of the employees to do a good job. It also means giving them the feeling of identity with the company objectives. However, the concrete and individual needs must be considered, and the way for obtaining lasting motivation.

In the *management by systems* concept, a company is considered a dynamic unit, having an active (influence) as well as a passive (adaptation) relationship with the environment. This style is an attempt to connect subsystems of the company, on order to direct them to the overall objectives of the company and to obtain an overview. This method aims at overcoming any kind of departmental thinking. The total optimum is given emphasis, whereas the respective optima of the several departments are subordinated.

56.10.2 Management of Technology Implementation

In order to successfully implement advanced manufacturing technology, major changes in corporate management strategies may be needed (Majchrzak and Paris, 1995). The recommended changes typically include the following: (1) an increased attention to human relations and skills for selection and training of manufacturing personnel, (2) broader job descriptions, (3) decentralized decision making authority, and (4) new pay and regard policies. However, less predictable manufacturing often requires different management practices than manufacturing that is routine and predictable (Ford and Slocum, 1977). It should be noted that computer-integrated manufacturing may require quite different management practices than nonintegrated manufacturing.

Sun and Riis (1994) developed a conceptual model linking organization technology and strategy through management and proposed a division of the implementation process into four stages. The basic ideas behind a conceptual modeling linking organization, technology, and strategy are as follows: (1) organization, technology, and strategy are interrelated; (2) organizational and technological changes should be implemented synchronously; (3) organization and technology should support the strategy; and (4) changes of and interrelations between the three factors need the intervention of management. The following stages for the implementation of advanced manufacturing technology (Table 56.14) were proposed (Sun and Riis, 1994):

1. Initiation and justification stage. In this stage, a champion starts to specify the problems and to recommend the potential contemporary manufacturing to the top management for approval.
2. Preparation and design stage. At this stage, the design and planning of both physical and organizational issues related to the contemporary manufacturing are conducted.
3. Installation and training stage. The tasks at this stage include the selection of an installation place, the acquisition of physical equipment, the installation, and the selection and training of staff.
4. Routinization and learning stage. After the equipment has been there on the shop floor, it still needs time for both the managers and the workers to gain enough technical and organization experiences for the normal and continuous operation of the contemporary manufacturing. The organizational designs and changes are implemented and modified as needs.

Majchrzak and Paris (1995) showed that different management strategies are needed for successful plant performance with nonintegrated versus integrated manufacturing. Also, different management practices may correspond to different performance measures within the same corporation. Table 56.15 illustrates that management practices related to effective performance for one dimension of performance may not be related to effective performance for another dimension. Furthermore, managing technical complexity and the broader scope of operations of integrated manufacturing using these management strategies, which include broader job responsibilities, enhanced job training, more support staff, decentralization, and less formal procedures, may not be sufficient to achieve high performance. Although these strategies accounted for over 50% of the variance in performance for plants with stand-alone equipment, a much lower amount of variance was accounted for in plants with the integrated manufacturing equipment.

56.11 HUMAN FACTORS IN CONCURRENT ENGINEERING

56.11.1 Principles of Concurrent Engineering

Concurrent engineering (CE) involves the near simultaneous design of new products and processes (Sinclair, Siemieniuch, Cooper, and Waddell, 1995). The critical issues that also need to be taken into account in such a process are operational requirements for human workers in the manufacturing system, including the structure of human tasks, their organization and coordination, sharing of information and resources, and a coherent set of

Table 56.14 The General Framework of the Implementation Guide

Issues and Stages	Initiation and Justification	Preparation and Design	Acquisition and Installation	Routinization and Learning
Strategic	• Goal specification	• Elaboration of goals	• Human resource strategy	• (Goal change)
Technological	• Objective of AMT	• Preparation • Design	• Acquisition • Installation	• Learning
Organizational	• Conditions • Labor union • Staff	• Work organization • Integration	• (Recruit) • Training	• Learning • Organizational changes
Managerial	• Champion • Management support • Communication • Justification	• Project team • Cooperation with vendor	• Site selection • Participation • Staff selection	• Management of changes • Uncertainty • Auditing

Source: Sun and Riss (1994).

Table 56.15 Selected Survey Questions of Management Practices

Questions	Mean (SD)
Job responsibilities of AMT personnel	
Proportion of weekly time, on the average, that AMT machine operator:	
A. Identifies process improvements	7% (0.08)
B. Monitors machine	51% (0.26)
C. Performs preventive maintenance	5% (0.08)
D. Diagnoses minor maintenance problems	7% (0.08)
E. Moves materials and supplies	10% (0.13)
F. Inspects quality	19% (0.18)
Proportion of weekly time, on the average, that AMT maintenance worker:	
A. Performs scheduled preventive maintenance	18% (0.16)
B. Repairs and troubleshoots	51% (0.29)
C. Informally teaches others	9% (0.09)
D. Orders parts	6% (0.06)
E. Schedules maintenance work	10% (0.14)
F. Operates equipment	6% (0.13)
Proportion of weekly time, on the average, that the first-line supervisor spends:	
A. Scheduling production	23% (0.14)
B. Assigning jobs	30% (0.17)
C. Motivating and evaluating subordinates	21% (0.13)
D. Working with other departments to resolve problems	26% (0.15)
AMT operator job design	
Which best describes the way the jobs of operators of automated equipment are designed:	
A. Work teams	2 yes
B. Each operator given set of tasks, regardless of the machine	10 yes
C. Operators rotate between jobs	10 yes
D. Each operator responsible for ≥1 machine	21 yes
Selection criteria for AMT personnel	
How important is each criterion for selection and promotion of AMT operators and maintenance?	
A. Seniority	2.1 (0.9)
B. Dependability	2.7 (0.8)
C. Relevant machine skills	2.9 (0.6)
D. Human relations skills	1.8 (0.8)
E. Past performance on other jobs	2.7 (0.8)
F. Interest in the position	2.7 (0.8)
Organizational design factors	
Characteristics of the organization:	
A. How are your departments in the plant divided: by manufacturing process? by functions?	12 yes 24 yes
B. Centralization of decisions (three questions on level in plant with authority over supplier contracts, personnel, and purchasing with higher levels meaning more centralization)	5.4 (1.5)
C. Number of levels of authority	4.1 (1.1)
D. Index of Formalization (three questions on use of organizational chart, process manuals, and formal job descriptions)	8.6 (1.1)
E. Proportion of job titles relative to number of hourly workers	11% (0.12)
F. Proportion of supervisors to number of hourly workers	11% (0.13)
Criteria on which hourly pay is based	
To what extent is hourly pay based on:	
A. Quality of work	1.8 (0.9)
B. Seniority	1.9 (0.8)
C. Innovativeness	1.5 (0.6)
D. Degree of responsibility	2.6 (0.6)
E. Delivery of finished work	1.9 (0.8)
F. Absenteeism	1.7 (0.8)
Criteria for evaluating departmental performance	
How important is each criterion for evaluating the performance of a department in this plant?	
A. Quantity of work performed	1.6 (0.6)
B. Costs per unit produced	2.6 (0.8)
C. Machine uptime	2.5 (0.9)
D. Met delivery schedules	1.8 (0.7)
E. Customer satisfaction	2.6 (0.8)

Source: After Majchrzak and Paris (1995).

job responsibilities. A key feature in CE is having a way to identify effects of a range of possible design features on each other as well as on intended performance criteria. Included in this range of design features are such management practices as organizational design, job design, and training.

Under the concurrent engineering process, the manufacturability, assembly and cost are the factors that are considered early in the product design (Jones, 1995). According to Duffy, Danck, and Salvendy (1995), manufacturers that use concurrent engineering perform the design of the product and process in parallel and may utilize teams as the primary work group. To implement concurrent engineering, manufacturers rely on a variety of new technologies. King and Majchrzak (1995) investigated different management practices needed given a key design decision regarding the stand-alone versus integrated manufacturing systems and application of advanced management technology. In particular, the following questions was addressed: *Is there an empirically supported relationship between degree of technical integration and types of management practices adopted that is tied to plant performance?* Since different management practices may be needed for varying degrees of technical integration, by identifying which management practices can be implemented in parallel with or independent of the CE design process can be very helpful to the design team to improve the technical design decision process. In addition, an understanding of management practices associated with different technical decisions is useful to assess the likely impact on management practices of technical design decision changes during the design process.

Duffy et al. (1995) proposed that methods developed for the effective management of technological and organizational change could be used to predict the levels of success of the corporation attributed to concurrent engineering practices. An integrated model that considers the human and organizational variables important to implementation of new technology, was developed and validated at 25 manufacturing companies. Results showed that the time spent by management in determining the task, work structure, and technology implementation accounted for about 50% of the variance attributed to the effectiveness of organizations. Duffy et al. (1995) examined key elements of the three significant indicators of success attributed to concurrent engineering, i.e., (1) the task and work structure, (2) skills and knowledge, and (3) success in implementing new technology. These elements are shown in Table 56.16.

56.11.2 CE Assumptions and Their Plausibility

Adachi, Shih, and Enbawa (1994) discussed a framework for supporting the organization and structuring of product development teams under concurrent engineering environment, which was validated based on real cases in product development projects. The concepts of team structuring dimensions (TSD) and development characteristics and factors (DCF) were defined with respect to team organization and structuring. The results have shown that more appropriate team organization facilitates the communication and information sharing between the team members, and the quality of this communication is essential for product development effectiveness and efficiency.

King and Majchrzak (1995) examined nine different assumptions typically made about concurrent engineering and grouped them into the three sets: (1) communication and (lateral) coordination between disciplines, (2) data representation for designers, and (3) human–computer interaction of individuals (Tables 56.17 and 56.18). The following four assumptions from the first set related to communication and coordination between disciplines were examined:

1. An integrated computing environment is sufficient for information sharing to occur.
2. There is a single representation of key design features that is understandable to the entire design community.
3. CE tools can support the interaction of multidisciplinary teams.
4. A designer needs and wants access to all information relevant to a design. The primary type of information used in design is digitally representable and context free.

The second set of assumptions relates to data representation for designers. The following three assumptions were examined:

5. A designer needs and wants access to all information relevant to a design.

Table 56.16 Research Issues Related to Key Indicators of Success in Implementing New Technology

Indicator	Research Issues
Task and work structure	• What are the individual and team tasks defined to be within successful organizations? • What work structures exist for concurrent engineering in successful organizations? • What are the functions performed by teams and who do the teams interact with in successful organizations? Where does the authority lie? • What are the roles and responsibilities of individuals within the teams in successful organizations? Roles and responsibilities of the teams? • What are the potential changes in jobs and work structure when implementing concurrent engineering and related technologies in successful organizations?
Skills and knowledge	• How are the required skills and knowledge determined to effectively utilize technologies associated with concurrent engineering in successful organization? • What skills or personalities are found to enhance learning in successful organizations? • What motivates concurrent engineering team members to learn in successful organizations? • How do successful organizations provide opportunities for learning?
Technology implementation	• How do successful organizations integrate new technologies with existing equipment and processes? • How do successful organizations estimate the reliability and flexibility of new technologies prior to implementation? • What types of feedback are useful is assessing the benefit of new equipment and technologies associated with concurrrent engineering?

Source: After Duffy et al. (1995).

6. The primary type of information used in design is digitally representable and context free.
7. The process of doing design work can be codified resulting in opportunities to intelligently aid design (synthesis) activities.

The last set of assumptions relates to human–computer interaction of individuals. The following two assumptions were examined:

8. Well understood principles of enhancing human-computer interaction (HCI) exist.
9. Existing user interface technology is sufficient to adapt to individual differences in tool use and designer styles.

A summary of the plausibility of these assumptions is shown in Table 56.17. All of the nine CE tool developers' assumptions listed above were found to be inconsistent with the documented behavior identified from related technology developments. King and Majchrzak (1995) pointed to a need to develop strategies for creating tools that would allow refocussing their development work on the restated assumptions (see Table 56.18) and recommended the following three complementary development strategies: (1) CE tools based on principles of user-centered design, (2) components for CE tools (such as wrappers) that capture information on the contextual realities in which the CE tool is being used and use this contextual information to modify the tool to the needs of the user; and

Table 56.17 Assumptions About Concurrent Engineering

Assumption	CE Tool Technologies	Results from Other Technologies
Lateral coordination		
1. An integrated computing environment is sufficient for information sharing to occur.	Communicatin networks; linked databases	Unexpected changes in communication patterns and organizational interaction
2. There is a single representation of key design features that is understandable to the entire design community.	Data standards; knowledge translation	A physical (geometric) representation represents the needs of only selected members of the design community
3. CE tools can support the interaction of multidisciplinary teams.	Group decisions support; negotiation and conflict management	Unpredictable team interaction; inability to reconcile multiple-dimension views; organizational resistance
4. Linking remote designers using a CE tool can substitute for physical colocation.	Communication; CSCW	Proximity aids cooperation; visible interaction needed in design activities
Data needs of designers		
5. A designer needs and wants access to all information relevant to a design.	Interoperable databases and tools; translators; knowledge representation	Experts search for the minimum amount of information; good coordination requires less information
6. The primary type of information used in design is digitally representable and context free.	Data standards; knowledge representation; multimedia	Lack of agreement on standard representation; difficulty in representing tacit knowledge from many unique design domains
7. Design process can be codified resulting in opportunities to aid design (synthesis) activities intelligently.	Models of design process, cognition collaboration	Design process is poorly understood; many of processes are nonanalytical and ill defined
Human–computer interaction		
8. Well-understood principles of enhancing HCI exists in cooperative problem solving.	User interfaces; CSCW	Unexpected group dynamics and social issues
9. Existing user interface technology is sufficient to adapt to individual differences in tool use and designer in styles	Adaptive interface, decision-support tools, HCI	User interfaces unable to accommodate wide variety of designer styles and differences in problem-solving approaches

Source: After King and Majchrzak (1995).

(3) using CE tools differently when they are used for early conceptual design than when they are used for later detailed design.

Under the first strategy proposed by King and Majchrzak (1995), the user-centered design approach was postulated as well suited to organizations that commit to change over several design cycles. Such a strategy would promote the effective application of many of the restated assumptions by providing the context-specific understanding for the particular setting in which the tool would be used. A second strategy states that CE tools should integrate human aspects of design with the technical elements and allow sharing of cognitive models as well as technologies to accommodate variety of perspectives that team members bring to a CE design process. Finally, a tool for conceptual design should encourage information sharing by suggesting sharing of ideas to appropriate people at the closeout of each session, rather than demanding it using standardized project management techniques.

56.12 AGILE MANUFACTURING

The term "agile manufacturing" was universally accepted after publication of the *21st Century Manufacturing Enterprise Strategy* (Iacocca Institute, 1991), which makes three important points in an effort to define agile manufacturing. These key points are as follows:

Table 56.18 Restated Assumptions Based upon Literature

Assumption	Restated Assumption	Applications to CE Tool Design
I Coordination		
1. An integrated computing environment is sufficient for information sharing to occur.	Information sharing will occur with a computing environment that integrates technology (data), architecture, semantics, and allows user to focus on tasks—not on computing technology.	Semantics and users are better integrated into a CE tool when design processes are understood and congruent organizational design features exist. Need clear goals and guidance on which interactions occur in design process.
2. There is a single representation of key design features that is understandable to the entire design community.	Representations based on explicit cognitive models are needed so an entire design community can understand key design features vis-à-vis the explicit model.	A single representation is sufficient when cognitively similar activities are involved (e.g., metal fabrication). Products that include mechanical, electrical, and (computer) processing elements require multiple representations. These should have explicit cognitive models drawn and overlaps between them described.
3. CE tools can support the interaction of multidisiplinary teams.	Tools will support interaction of multidisciplinary teams only when interaction is supported by organizational structure, culture, training, and experience.	Tool design incorporates social and organizational constraints (e.g., e-mail with rules of engagement and management check-offs) coupled with an implementation plan that outlines needed changes to organization.
4. Linkage of remote designers with a CE tool substitutes for their physical colocation.	Once designers have established communication norms, remote linkage of CE tools can substitute for context-free data exchanges	Team building and negotiation of critical design issues in face-to-face setting is prerequisite for utilizing remote communication methods. Consider checklist of prerequisites before remote usage.
Data needs of designers		
5. A designer needs and wants access to all information relevant to a design.	Designers need to forcus their search quickly to relevant design information and be guided through the proper interpretation of relevant design information not within their discipline.	Designers have to be shown crucial boundaries where external information is needed so that cognitive overload is avoided; guidance for interpreting information; and guidance on reducing information search efforts.
6. The primary type of information used in design is digitally representable and context free.	Design information is a mixture of prescriptive, tacit and context-specific data.	Tools are suitable for prescriptive analysis (mathematical and algorithmic aspects). Information should be presented in a way to facilitate user's tacit and context-specific data (such as through queries).
7. The design process of doing design work can be codified resulting in opportunities to intelligently aid design (synthesis) activities.	Design process is most easily codified when tasks have minimum interdependencies.	Design process must be thoroughly understood before sequencing of process is automated. Many sources of variations in processes have to be considered as task interdependence (e.g., multiple disciplines) and product complexity increase.
Human–computer interaction		
8. Well-understood principles of enhancing HCI exist in cooperative problem solving.	HCI principles of effective collaborative problem solving are highly contingent upon task type and social factors.	Treat human–computer and computer-mediated collaboration separately. For collaborative problem solving, design interfaces to enhance reasoning process, participation according to knowledge, and reinforcing informational influence.
9. Existing user interface technology is sufficient to adapt to individual diferences in tool use and designer styles.	Balance needs to be struck between degrees of standardization and individual preferences of users.	Identify minimal critical specifications that can be standardized; allow user to modify all others.

Source: After King and Majchrzak, (1995).

1. A new competitive environment is emerging, which is acting as a driving force for change in manufacturing.
2. Competitive advantage will accrue to those enterprises that develop the capability to respond rapidly to the demand for high-quality, highly customized products.
3. To achieve the agility that is required to respond to these driving forces and to develop the required capability, it is necessary to integrate flexible technologies with a highly skilled knowledgeable, motivated and empowered workforce functioning within the organization and management structures that stimulate cooperation both within and between firms.

In view of the above, the concept of agility, which implies a quickly moving, nimble, and active system, is not the same as manufacturing flexibility, which means adaptability and versatility. Whereas that flexibility is a necessary requirement for the competitive corporations, it does not constitute agility. Neither does the concept of *lean manufacturing,* i.e., doing everything with less (Jones, 1992). According to Kidd (1994), manufacturing leanness and agility are two different concepts. Although lean manufacturing is necessary for agility, it is not sufficient. Similarly, computer-integrated manufacturing and the computer-integrated enterprises, even though they assure rapid communication and exchange and reuse of data, are not necessarily agile. However, computers that link across applications, functions, and enterprises may provide for the necessary condition of agile manufacturing (Goldman et al., 1995).

Agile manufacturing can be considered a structure within which every company can develop its own business strategies and products (Kidd, 1994). Such a structure is supported by three primary resources: (1) innovative management structures and organization, (2) a skill base of knowledgeable and empowered people, and (3) flexible and intelligent technologies. Agility can be achieved through the integration of these three resources into a coordinated and interdependent system that aims to achieve cooperation and innovation in response to the need to supply customers with high-quality customized products. The basic principle of agile manufacturing is that an enterprise should be built on the competitive foundations of continuous change, rapid response, quality improvement and social responsibility in terms of environment and employees (Kidd, 1994). According to the Iacocca Institute (1991), the agile manufacturing enterprise exhibits the following characteristics: (1) concurrency in all activities, (2) continuing education for all employees, (3) customer responsiveness, (4) dynamic multiventuring capabilities, (5) employees valued as vital assets, (6) empowered individuals working in teams, (7) environmental concern and proactive approach, (8) accessible and usable information, (9) skilled and knowledgeable employees, (10) open-system architectures, (11) right first time designs, (12) total quality philosophy, (13) short cycle times, (14) technology awareness and leadership, (15) enterprise integration, and (16) vision-based management.

56.12.1 A Conceptual Framework for Agile Manufacturing

Agile manufacturing is primarily a business concept that bring together many ideas in order to develop an appropriate manufacturing response to global market opportunities (Goldman et al., 1995; Kidd 1994) and that can be realized through virtual enterprises. The agility comes from integration of organization, people, and advanced technology into a coordinated interdependent system. Agile manufacturing can also be defined in terms of a conceptual framework of the enterprise built on the competitive foundations of continuous change, rapid response, quality improvement, social responsibility, and total customer focus (Table 56.19). According to Kidd (1994), the core concepts of agility are as follows: (1) a strategy to become an agile manufacturing enterprise; (2) a strategy to exploit agility to achieve competitive advantage; (3) integration of organization, people, and technology into a coordinated interdependent system which is the competitive weapon; and (4) an interdisciplinary design methodology to achieve the integration of organization, people, and technology.

A framework of generic features of an agile manufacturing enterprise is comprised of the following elements (Kidd, 1994): (1) integrated enterprises; (2) human networking organization; (3) enterprises based on natural groups; (4) increased competencies of all people; (5) focus on core enterprise competencies; (6) virtual corporations; (7) an environment supportive of experimentation, learning, and innovation; (8) multiskilled and flexible people; (9) team working; (10) empowering of all the people in the enterprise; (11) knowledge management; (12) skill and knowledge enhancing technologies; (13) con-

Table 56.19 Conceptual Framework for Agile Manufacturing

Generic Features Model
- Integrated enterprises
- Human networking organization
- Enterprises based on natural groups
- Increased competencies of all people
- Focus of core competencies
- Virtual corporations
- An environment supportive of experimentation, learning, and innovation
- Multiskilled and flexible people
- Teamworking
- Empowerment of all the people in the enterprise
- Knowledge management
- Skill and knowledge enhancing technologies
- Continuous improvement
- Change and risk management

Core Concepts
- Strategy to achieve agility
- Strategy to exploit agility
- Integration of organization, people, and technology
- Interdisciplinary design methodology

Competitive Foundations
- Continuous change
- Rapid response
- Quality improvement
- Social responsibility
- Total customer focus

Source: After Kidd, 1994.

tinuing improvement involving all people; and (14) change and risk management. The two important building blocks of manufacturing agility are human networking and employee empowerment. Under the concept of human networking (Naisbitt, 1984; Savage, 1990) everybody is networked together by interactions that occur among all people in the organization. Traditional, computer-based networks can be used to support the human networking.

56.13 CONCLUSIONS

Although a growing body of human factors research in comtemporary manufacturing suggests the need for integration of various human-embodied skills and participation with automated production, many corporations are still confronted with a productivity paradox, regardless of the amount of time and money invested with computer-based technology (Sobol and Lei, 1994). Successful human factors initiatives in advanced manufacturing technology is dependent upon integration of cognitive ergonomics as applied to human–machine systems and psychosocial ergonomics as applied to work organization and job design (Wilson, 1991). According to Helander (1994), the three critical problem areas where greater emphasis must be spent on human factors issues in contemporary manufacturing include incorporation of human factors evaluation criteria in concurrent engineering, the use of computer aids to support the cognitive requirements of the design process, and organizational impact of the introduction of computer systems for manufacturing, planning, and production.

The involvement of the human factors profession in facilitating the implementation of contemporary manufacturing should be guided by the following prescriptions (Majchrzak, 1993):

1. Technology is best conceived not as a generic model or type of equipment, but rather as a configuration of technical options, workplace decisions, human factors, and sociopolitical factors.

2. As an interdisciplinary concept, the human factors profession must be concerned not simply with physiologically oriented equipment design, but with macrolevel factors such as organizational structure, professional associations, suppliers, and key assumptions underlying the technology.
3. Technological innovation process occurs continually, feeding to and from between design and implementation, from one generation of a technology on the next, and it is subject to contingencies and is highly uncertain.
4. The design of technology is not conceived as a single point in time that is influenceable by human factors professionals; rather, the design of technology is conceived as a process that is extended in time to embrace postadoption incremental innovation.
5. Involvement of users in manufacturing technology design needs to be more broadly conceptualized to not only involve end users but the involvement of all potential designers of system configurations, ranging from employees to suppliers, from customers to decision makers.

Rasmussen, Pjetersen, and Goodstein (1992) discussed several research issues related to human factors of contemporary manufacturing systems. These issues can be summarized as follows. First, since modern manufacturing systems often operate in a turbulent environment and are subject to dynamic changes in system goals, requirements, and opportunities existing in the outside environment, the means and tools to pursue corporate goals and adapt to changes vary. Therefore, work organization structures depend heavily on adaptive mechanisms that allow for changes in system properties to maintain a match with current needs when the internal conditions and the environment change. Second, the control of system adaptation is typically distributed across all individuals, terms, and organizations, and affects the structure of cooperative organizational patterns, the role allocation among people, and performance of individuals. The usefulness of any theoretical framework for human factors in modern manufacturing in analysis of work and prediction of responses to changes in working conditions depends upon mechanisms underlying the evolution of work practices. Third, the manufacturing system adaptation is an evolutionary process, difficult to plan and to explain in terms of linear cause and effect relations. The structure and performance of the system emerge from a survival of the fittest of the structures and performance and depend to a large degree on the outcomes of trial and error experiments, planned and unplanned, conscious and unconscious.

56.13.1 Short-Term Strategies for Contemporary Manufacturing

Human factors research needs to be aimed at helping industry to improve and develop its capabilities in the areas of (1) management of change (2) technology deployment; (3) integration of organization, people, and technology into a coordinated system aimed at delivering competitive advantaged; (4) development of computer-based and knowledge-based tools to support analysis of complex organizations, based on holistic methodologies; (5) development of skills and competencies in the area of systems strategy, systems architecting, and systems integration, focusing on both soft systems and hard systems; and (6) development of technologies that leverage the skills and knowledge of users (Karwowski et al., 1994). In the above context, the contemporary manufacturing industry should reevaluate how organizational responsibilities are assigned, how power and decision making are shared, and how human skills are developed at all levels. The degree to which the manufacturing enterprises decide to promote a professional model for new job structures at all organizational levels should also be considered. The short-term focus should be on specific areas for improvements, for example, promotion of the skill-based design concept, or breaking down functional distinctions of the traditional organizational structures.

Particular emphasis of the human factors profession should be directed toward facilitating development of agile manufacturing enterprises. There is a need for new methods and tools to support an interdisciplinary approach, with special emphasis on combining top-down and bottom-up methods, holistic approaches, organizational simulation, appropriate and selective use of technology, increased user involvement, rapid prototyping, etc. All of these are relevant tools that will be required to support a new approach to the design of agile manufacturing enterprises. More emphasis should also be placed on developing decisions-support systems that expand the ranged of possible decision alternatives considered by people. More effort is needed to address the human factors research needs in manufacturing as a whole, and not just research into new manufacturing technologies. There is a need to develop new management practices, technology deployment techniques, new reward systems, more sophisticated financial justification techniques, new management accounting methods, etc., to support agile manufacturing.

56.13.2 Long-Term Agenda for Human Factors in Manufacturing

Contemporary manufacturing industries position themselves for the increasingly competitive global markets of the twenty-first century by adopting the concepts of agility, including the ability to rapidly develop and produce new products (Forsythe and Karwowski, 1995). Agility manifests itself in many different forms, with the agile manufacturing paradigm proposed by the Iacocca Institute offering a generally accepted, long-term vision (Kovac, 1993; Nagel and Dove, 1992). The manufacturing industry has never before experienced such a dramatic infusion of new technologies, or such an extensive changes in work culture and work practices as today. The human factors profession has an opportunity to play an important role in accomplishing the technical and social objectives of agile manufacturing, as well as to participate in shaping the future manufacturing paradigms. According to Jones (1995), the challenges to human factors profession with respect to contemporary manufacturing focus on the issues of design of manufacturing processes and environments in order to support effective work practices and appropriate organizational structures and takes into account the quality of working life.

There are several gaps in the existing capabilities of the human factors discipline and the requirements of contemporary manufacturing that modify practical realization of a new paradigm of agility (Karwowski et al., 1994). The major categories of these gaps are as follows: (1) human factors in business and manufacturing strategies, (2) educational and vocational training, (3) technological issues, (4) cultural issues (including social norms and habits), (5) usability of technology and people, (6) management and design philosophies, (7) communication abilities, (8) publicity campaign and promotion, (9) dissemination and marketing, (10) proof of profitability, (11) acceptance of the human factors discipline as a viable design tool, and (12) coping with complexities and uncertainties of competitive manufacturing environment. To fill the knowledge gaps listed above, Karwowski et al. (1994) proposed that several research and development themes be undertaken through cooperative agreements among the manufacturing industries, academia, and governments. These themes are listed in Table 56.20.

Siemieniuch and Sinclair (1995) reviewed changes in strategic approaches and philosophies of many contemporary manufacturing organizations and outlined the potential areas of human factors input. It was postulated that the fundamental principles of human factors contributions to development of modern manufacturing systems should be based on the user-centered design principles and allow for the input of ergonomics expertise at all stages of system design. The following steps were proposed to be incorporated into the human factors strategy:

1. Implementation of the user-centered or user-led design concepts, where project development teams have a coherent strategy for involving the future users of manufacturing technology at all stages of system design. Project development teams will also need to ensure a strong commitment from a selected group of users over the duration of the project.
2. Stakeholder analysis to identify those individuals and groups who have a legitimate stake in the new manufacturing system, to specify the requirements for and nature of the user interaction with the system, and to identify the scenarios under which the new manufacturing system is designed to cope.
3. Analysis of the user requirements specification which aims to gain an in-depth understanding of the main requirements, constraints and variables of both current and proposed organizational structures, allowing for early identification of the crucial human roles and tasks.
4. Detailed task analysis based on the in-depth interviews with the key stakeholders, to identify in more detail the components of those tasks which are considered to be critical, to gain a description of the activities and data, control and information flows considered to be critical, and to represent these graphically.
5. Allocation of function between technology and humans to ensure that the jobs left over for the human operators to do are within recognized human capabilities, and make the best use of human skills and knowledge in combination with the technology available, based on the creation of a functional view of an organization and identification of competencies required to carry out the recognized functions.

6. Definition of roles, responsibilities, corporate policies, and organizational structures for the new manufacturing systems.
7. Human–computer interaction (HCI) input to system design that allows for utilization of the human factors expertise and HCI principles, including design of user interfaces to information technology (IT) support tools, information presentation and formatting, use of multimedia techniques, knowledge of human cognitive processing limitation, or equipment and workstation design.
8. Validation and evaluation of the collected data is to ensure that any options selected are worthwhile, representative and relevant to the needs of a future system conducted in an iterative fashion throughout the design cycle. A representative range of end users must be closely involved at all points and both formal and information procedures need to be established to facilitate this.
9. A coherent implementation strategy developed prior to system implementation, which considers aspects such as integration with existing systems, training requirements, changes in working practices or policy, or information dissemination.

56.13.3 Human Factors in Agile Manufacturing

As discussed by Forsythe and Karwowski (1995), the potential human factors contributions to agile manufacturing reflect the variety of human components of an industrial enterprise and include the following broad areas of application: (1) development of business practices, (2) design of enabling technologies, and (3) management of the introduction and fielding of new technologies and business practices. Implementation of the agile manufacturing principles may often require extensive changes in the existing business practices, or even a complete overhaul of existing business practices (Goldman et al., 1995; Greiwss, 1993). Since the success may depend on rapid development and delivery of quality products, modern corporations must maximize their ability to capture and utilize corporate experience in product design through concurrent engineering practices (Goldman and Priess, 1992). Also, an implementation of collaborative design efforts is required to assure fast paced design decisions in a competitive environment with no error tolerance (Forsythe and Ashby, 1994).

Consideration of human factors affecting decision making within the dynamic markets is very much needed for development of manufacturing agility. This include the knowledge of team dynamics, individual information requirements, information management and utilization, and monitoring and assessment of the status of complex and dynamic systems (Forsythe and Ashby, 1996). The appropriate organizational structure and infrastructure support for the corporate communication and information transfer are also needed to accomplish the goals of agile manufacturing, including system and software compatibility. It should be noted that software compatibility cannot be maintained without the coordination and empowerment of administrative and support staff (Forsythe and Ashby, 1996). As pointed out by Haney, Reece, Wilhelmsen, and Romero (1994), in agile manufacturing enterprises, the integration and networking or information technologies occurs at all levels of the organizational structure. Therefore, the support needs of a complex infrastructure and the numerous human points of failure in supporting such an infrastructure must be addressed.

56.13.4 Human Factors in Manufacturing: Challenges of the Future

Computer-controlled systems, such as machining centers with automated change of tools and workpieces, constitute an integral part of the daily routine in the contemporary manufacturing industry. Although the manufacturing technologies are advancing at a fast pace, they often lack desirable usability characteristics and lead to user dissatisfaction. For example, the computer-aided product design and manufacturing, which are fundamental to agile manufacturing, are currently underutilized, partially because they lack full integration with the existing work practices (Wiebe, 1995; Forsythe and Ashby, 1996). The human factors discipline has an important role to play in both advanced manufacturing technology development and in defining various technology systems and their usage (Karwowski et al., 1994; Forsythe and Karwowski, 1995).

As discussed above, many of the challenges posed by agile manufacturing concepts are sociotechnical in nature. Agile manufacturing imposes new demands on managers and floor workers. An empowerment of product development teams and the increased openness of information sharing leads to substantial decrement of power on the part of management. Much control exercised by designers is lost in the collaborative development of

Table 56.20 Long-term Research and Development Activities and Their Impact on Fundamental Corporate Activities

Long-Term Research and Development Items[b]	Fundamental Corporate Activities[a]			
	Develop Design Principles	Develop and Implement Organizational Structures	Develop and Implement Techniques and Tools	Focus on Process of Change
1. Perform an intercultural comparison of agile manufacturing implementation.	X			
2. Develop a vauation methods/hybrid approach normative tool.			X	
3. Increase software usability by adaptive technologies.			X	
4. Develop methods for information browsing and filtering.			X	
5. Perform a comparative study of "work attractiveness" under different self-configurable organizational structures.	X	X		
6. Document successes and failures in implementation of AMT.	X			
7. Provide a theory of human factors and manufacturing systems.	X			
8. Integrate human factors data into the manufacturing design process.	X			
9. Explore adaptabiity of technology to organizational structures.	X	X	X	X
10. Provide virtual organizational structures and human networking principles.	X	X		
11. Develop a concept of contingencies and appropriateness of technology.	X			
12. Establish direct links betwen the design of work and design of technology.	X			
13. Define effective tools for implementing organizational changes.			X	X
14. Identify specific problems that can be addressed by the human factors and skill-oriented approaches.	X			
15. Evaluate effectiveness of the skill-based principles of design.	X			
16. Explore participative design approaches.	X			
17. Investigate the relationship between adaptive organizational structures and potential stresses on workers in automated environments because of task complexity, time pressure, neurophysiological effects of aging population, etc.		X	X	
18. Develop planning documents and instruments, i.e., project management tools that can be used for integrating human factors, technology, and organizational changes into a comprehensive implementation plan			X	X
19. Develop cost–benefit analysis instruments to be used to help management make decisions about what to change, in what order, and at what speed, including human resource elements as well as traditional hard dollar numbers.			X	X
20. Establish performance measures for teamwork in advanced manufacturing systems.		X	X	
21. Investigate the relation between reliability and system productivity.	X			
22. Compare concepts of hybrid automated manufacturing and fully integrated manufacturing systems.	X			

23. Investigate the specific features that make a technical artifact a tool (e.g., task appropriateness, transparency, etc.), including the development and evaluation of prototype applications.			X	
24. Develop methods for implementing adaptable information technology systems according to the specific organizational needs.			X	X
25. Develop new training methods for action competence.			X	
26. Apply the advanced communication technology in manufacturing plants for the shop floor applications.			X	
27. Consider the relationships between system designers, system operators, maintenance personnel, etc.	X	X		
28. Develop a new paradigm of self-organization.	X	X		
29. Develop a system for self-organizing knowledge engineers.		X	X	
30. Develop solution strategies for integration of technology, organization, and people.	X			
31. Estabish a set of pilot courses for interdisciplinary engineering and human factors education.	X			
32. Fund an international workshop on human factors to establish an alliance to Intelligent Manufacturing Systems (IMS) project in the human factors area	X			
33. Develop methods/tools for an integrated modeling of work flow and information systems with respect to task design, motivation, human skill, and advanced manufacturing system usability.			X	
34. Develop a computer-supported conference/or meeting room system for engineers to improve the design phase of a new product (including 3D capabilities, virtual reality, information processing, multimedia capabilities).			X	
35. Develop autonomous intelligent work units.		X		
36. Develop artificially intelligent systems to prevent accidents.			X	
37. Organize a joint research group to deal with the research and development related human aspects of advanced manufacturing technology.	X			
38. Develop technology deployment techniques.			X	
39. Develop organizational rules in order to outline the rapid prototyping approach.			X	
40. Apply proven techniques to solving complex engineering processes.			X	
41. Develop more active and more intelligent tools and interfaces for small groups.			X	
42. Develop routing strategies and tools for a rapid information retrieval.			X	
43. Build an effective change management system to ensure the implementation of the quantum leaps.			X	

Source: After Karwowski et al. (1994).

[a]Process of change is impacted by all long-term research and development items.

[b]Items not presented in any particular order.

designs. At the factory floor level, there is an increased responsibility on the part of the workers who are brought into the product development and decision-making processes, and the threat posed by computerization and automation of fabrication and assembly tasks. These typically occur within the stressful and fast-paced manufacturing environment that places high cognitive demands on the personnel.

Cognitive task demands and cognitive characteristics of the workers need be considered in the design and management of contemporary manufacturing systems in order to assure their compatibility with the worker's internal that describes the operations and functions of these systems. Consequently, the system's description level should be shifted from the skill-based to the rule-based and knowledge-based human functioning. For the above requirements, new methods for cognitive task analysis need to be developed to identify the operator's model of a system. The challenge to human factors profession is to assist in design of contemporary manufacturing systems which incorporate the characteristics of human cognition and explicitly built into design process both physical and cognitive images of the knowledge workers. Furthermore, new forms of organizational structures that are self-adaptive to the requirements and changes in technology must be developed in order to realize full economic benefits of the contemporary manufacturing technology.

Helander (1994) described a systems approach for implementation of human factors in contemporary manufacturing, including the technologies of computer-integrated manufacturing (CIM) and concurrent engineering, design of appropriate human–computer interactions. The design of a manufacturing organization was presented (Figure 56.4) as an iterative process with top-down planning and bottom-up evaluation. According to this model, management makes decisions that are implemented by engineers working with CIM and other design tools. As an outcome of design, resources will be tentatively allocated between various agents, and the implications for productivity and job satisfaction must be assessed. Since in contemporary manufacturing the various production responsibilities are often delegated to the workers on the shop floor, the organization must be built bottom-up and contingencies must be made at the higher levels of the organization to support the lower levels.

The human factors profession can contribute to development of skill and knowledge enhancing technologies along with the insights from the organizational and psychological

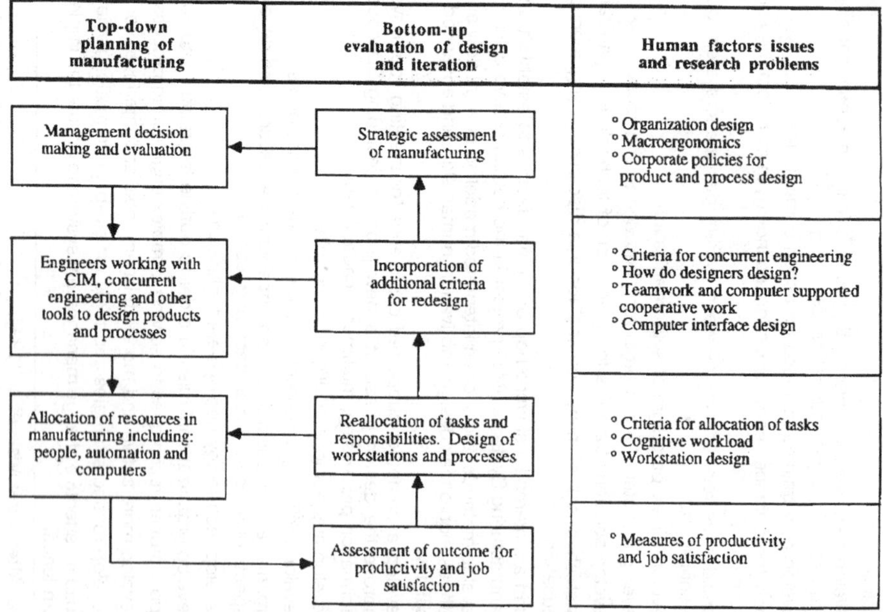

Figure 56.4 Top-down planning and bottom-up evaluation of manufacturing and associated human factors issues and research problems. (After Helander, 1994.)

sciences. The important areas for further research with the emphasis on seeking skill and knowledge enhancement should include computer-supported cooperative working teams; system monitoring; database management and networking; applications of genetic algorithms, neural networks to support decision making; vision systems; workshop oriented programming of robots and machine tools; etc. Since new technologies are required to support the group work, computer-supported cooperative work is an area of research that needs to be addressed (Forsythe and Karwowski, 1995). Finally, human factors knowledge can be useful in development of technologies that support the learning corporation, as an important component of agile manufacturing.

REFERENCES

Adachi, T., Shih, L-C., and Enkawa, T. (1994). Strategy for supporting organization and structuring of development teams in concurrent engineering. *The International Journal of Human Factors in Manufacturing, 4(2)*, 101–120.

Alasoini, T. (1996). A learning factory: Experimenting with adaptable production in Finnish engineering workshops. *The International Journal of Human Factors in Manufacturing, 6(1)*, 3–19.

Badham, R., and Schallock, B. (1991). Human factors in CIM: A human-centered perspective from Europe. *International Journal of Human Factors in Manufacturing, 1*(2), 121–141.

Bainbridge, L. (1983). Ironies of automation. *Automatica, 19, 775–779.*

Bessant, J., Levy, P., Ley, C., Smith, S., and Tranfield, D. (1992). Organization design for factory 2000. *The International Journal of Human Factors in Manufacturing, 2(2)*, 95–125.

Bi, S., and Salvendy, G., (1994). Analytical modeling and experimental study of humann workload in scheduling of advanced manufacturing systems. *The International Journal of Human Factors in Manufacturing, 4*(2), 205–234.

Brödner, P. (1986). Skill based manufacturing vs. "unmanned factory"—Which is superior? *International Journal of Industrial Ergonomics, 1,* 145–153.

Brödner, P., Ed. (1987). *Strategic Options for New Production Systems—CHIM: Computer and Human Integrated Manufacturin.* FAST Occasional Papers. Brussels, Belgium: Directorate General for Science, Research and Development, Commission of the European Communities.

Brödner, P. (1990). *The Shape of Future Technology. London: Springer.*

Brödner, P., (1991). Design of work and technology in manufacturing. *International Journal of Human Factors in Manufacturing, 1(1)*, 1–16.

Bullinger, M. J. (1986). Technology trends—a challenge to research and industry. In T. Lupton, Ed., *Human Factors: Man, Machine and New Technology.* IFS. Berlin: Springer-Verlag.

Clegg, C. (1988). Appropriate technology for manufacturing: Some management issues. *Applied Ergonomics, 19(1)*, 25–34.

Clegg, C., and Corbett, H. M. (1987). Research and development into "humanizing" advanced manufacturing technology. In T. D. Wall, C. W. Clegg, and N. J. Kemp, Eds., *The Human Side of Advanced Manufacturing Technology* (pp. 173–195). New York: John Wiley.

Clegg, C., Ravden, S., Corbett, M., and Johnson, G. (1989). Allocating functions in computer integrated manufacturing: A review and a new method. *Behaviour and Information Technology, 8(3)*, 175–190.

Cooley, M. J. E. (1986). Problems of automation. In T. Lupton, Ed., *Human Factors: Man, Machine and New Technology.* Berlin: IFS, Springer-Verlag.

Cooley, M. (1989). Human-centred systems. In H. H. Rosenbrock, Ed., *Designing Human-Centred Technology* (pp. 133–143). New York: Springer-Verlag.

Corbett, J. M. (1985). Prospective work design of a human-centered CNC lathe. *Behaviour and Information Technology, 4(3)*, 201–214.

Corbett, J. M. (1988). Ergonomics in the development of human-centred HAS, *Applied Ergonomics, 19,* 35–39.

Corbett, J. M. (1990). Human centred advanced manufacturing systems: From theoretic to reality. *International Journal of Industrial Ergonomics, 5,* 83–90.

Duffy, V., Danek, A., and Salvendy, G. (1995). A predictive model for the successful integration of concurrent engineering with people and organizational factors: Based on data of 25 companies. *The International Journal of Human Factors in Manufacturing, 5(4)*, 429–445.

Eichener, V. (1996). The impact of technical standards on the diffusion of anthropocentric production systems. *The International Journal of Human Factors in Manufacturing, 6*(2), 131–145.

Forsythe, C., and Ashby, M. R. (1994). *Developing Communications Requirements for Agile Product Realization,* Report No. SAND–94-0481C. Sandia National Laboratories, Sante Fe.

Forsythe, C. and Ashby, M. R. (1996), *Human factors in agile manufacturing. Human Factors and Ergonomics in Manufacturing,* 7(2)1), 11–20.

Forsythe, C., and Karwowski, W. (1995). Human factors in agile manufacturing. In *Proceedings of the Human Factors and Ergonomics Society 39th Annual Meeting, Human Factors and Ergonomics Society,* Santa Monica, CA, pp. 538–540.

Goldman, S. L., Nagel, R. N., and Preiss, K. (1995). *Agile Competitors and Virtual Organizations.* New York: Van Nostrand Reinhold.

Goldman, S., and Preiss, K. (1992). *21st Century Manufacturing Enterprise Strategy, Vol. 2, Infrastructure.* Bethlehem, PA: Lehigh University Press).

Goodstein, L. P., Anderson, H. B., Olsen, S. E., Eds. (1988). *Tasks, Errors and Mental Models.* London: Taylor and Francis.

Greiss, H. A. (1993, June 1). American industrial dominance will depend upon agility to management Change. *Focus.*

Haney, L. N., Reece, W. J., Wilhelmsen, C. J., and Romero, H. A. (1994). Modeling cognitive aspects of human error. In P. T. Kidd and W. Karwowski, Eds., *Advances in Agile Manufacturing: Integrating Technology, Organization and People* (pp. 335–338). Amsterdam: IOS Press.

Helander, M. G. (1994). Cognitive and sociotechnical issues in Design for manufacturability. *The International Journal of Human Factors in Manufacturing, 4(4)*, 375–390.

Hendrick, H. (1987). Organizational design. In G. Salvendy, Ed., *Handbook of Human Factors* (pp. 470–494). New York: John Wiley.

Hollnagel, E., and Woods, D. (1983). Cognitive systems engineering: new wine in new bottles. *International Journal of Man-Machine Studies, 18,* 583–600.

Hwang, S. L., Barfield, W., Chang, T. C., and Salvendy, G. (1984). Integration of humans and computers in the operation and control of flexible manufacturing systems. *International Journal of Production Research, 22(5)*, 841–856.

Iacocca Institute. (1991). *21st Century Manufacturing Enterprise Strategy. An Industry-Led View (Vols. 1 & 2).* Bethlehem, PA: Iacocca Institute.

ISO (1991). *Industrial Automation Systems—Safety of Integrated Manufacturing Systems—Basic Requirements* (CD 11161). Geneva, Switzerland: ISO.

Jaikumar, R. (1986 November-December). Post-industrial manufacturing. *Harvard Business Review,* 69–76.

Järvinen, J., and Karwowski, W. (1995). Analysis of self-reported accidents attributed to advanced manufacturing systems. *The International Journal of Human Factors in Manufacturing, 5(3)*, 251–266.

Järvinen, J., Vannas, V., Mattila, M., and Karwowski, W. (1996). Causes and safety effects of production disturbances in FMS installations: A comparison of field survey studies in the USA and Finland. *The International Journal of Human Factors in Manufacturing, 6(1)*, 57–72.

Johannsen, G. (1988). Categories of human operator behavior in fault behavior situations. In L. P. Goodstain, h. B. Andersoon, and S. E. Olsen, Eds., *Tasks, Errors, and Mental Models.* (pp. 251–277). London: Taylor and Francis.

Johnson, G. I., and Wilson, J. R., Eds. (1988). *Ergonomics Matters in Advanced Manufacturing Technology.* London: Butterworths.

Jones, D. T. (1992). Beyond the Toyota production system: The era of lean production. In C. A. Voss, Ed. *Manufacturing Strategy. Process Content* (pp. 189–210). London: Chapman and Hall.

Jones, P. M. (1995). Designing for operations: towards a sociotechnical systems and cognitive engineering approach to concurrent engineering. *International Journal of Industrial Ergonomics, 16, 283–292.*

Karwowski, W. (1991). New perspectives on human factors in the design and management of advanced manufacturing systems: A Review. *The Japanese Journal of Ergonomics* (in Japanese), *27(6)*, 301–312.

Karwowski, W., Parsaei, H. R., and Wilhelm, M. R., Eds. (1988). *Ergonomics of Hybrid Automated Systems I.* Amsterdam: Elsevier.

Karwowski, W., and Rahimi, M., Eds. (1990). *Ergonomics of Hybrid Automated Systems II,* Amsterdam: Elsevier.

Karwowski, W., and Salvendy, G., Ed. (1994). *Organization and Management of Advanced Manufacturing.* New York: John Wiley.

Karwowski, W., Salvendy, G., Badham, R., Brödner, P., Clegg, C., Hwang, S. L., Iwasawa, J., Kidd, P. T., Kobayashi, N., Koubek, R., LaMarsh, J., Nagamachi, M., Naniwada, M., Salzman, H., Seppala, P., Schallock, B., Sheridan, T., and Warschat, J. (1994). Integrating people, organizations, and technology in advanced manufacturing: A position paper based on the joint view of industrial managers, engineers, consultants, and researchers. *International Journal of Human Factors in Manufacturing, 4(1)*, 1–19.

Keita, G. P. and Sauter, S. L. (1992). *Work and Well Being. Washington, DC: American Psychological Association.*

Kember, P., and Murray, H. (1988). Towards socio-technical prototyping of work systems. *International Journal of Production Research, 26(1)*, 133–142.

Kidd, P. T. (1990). An open systems human-computer interface for a workshop oriented CNC lates. In W. Karwowski and M. Rahimi, Eds., *Human Aspects of Hybrid Automated Systems II* (pp. 537–544). Amsterdam: Elsevier.

Kidd, P. T. (1991). Human and computer integrated manufacturing: A manufacturing strategy based on organization, people and technology. *International Journal of Human Factors in Manufacturing, 1(1)*, 17–32.

Kidd, P. T. (1994). *Agile Manufacturing: Forging New Frontiers.* Reading, MA: Addison Wesley.

Kidd, P. T., and Karwowski, W., Ed. (1994) *Advances in Agile Manufacturing: Integrating Technology, Organization and People.* Amsterdam: IOS Press.

King, N., and Majchrzak, A. (1995). Concurrent engineering tools: Are the human issues being ignored. *IEEE Transactions on Engineering Management (in press).*

Kjellén, U. (1984). The role of deviations in accident causation and control. *Journal of Occupational Accidents, 6,* 117–126.

Koubek, R., and Karwowski, W., Ed. (1996). *Human Factors in Agile Manufacturing.* Amsterdam: IEA Press.

Kovac, F. J. (1993). Agility focuses on collaboration. *Focus,* June 2.

Kuivanen, R. (1990). The Impact on Safety of Disturbances in Flexible Manufacturing Systems. In W.Karwowski and M. Rahimi, Eds., *Ergonomics of Hybrid Automated Systems II* (pp. 951–956). Amsterdam: Elsevier.

Liker, J. K., Fleischer, M., and Arnsdorf, D. (1992). Fulfilling the promises of CAD. *Sloan Management Review, 33(3)*, Spring, 74–86.

MacKenzie, D., and Wajcman, J., Eds. (1985). *The Social Shaping of Technology.* Philadelphia, PA: University Press.

Majchrzak, A. (1988). *The Human Side of Factory Automation* San Francisco: Jossey-Bass Publishers.

Majchrzak, A. (1993). Commentary. *The International Journal of Human Factors in Manufacturing, 3(1)*, 89–90.

Majchrzak, A., Fleischer, M., Roitman, D., and Mokray, J. (1991). *Reference Manual for Performing the HITOP Analysis.* Ann Arbor, MI: Industrial Technology Institute.

Majchrzak, A. and Gasser, L. (1992). HITOP-A: A tool to facilitate interdisciplinary manufacturing systems design. *International Journal of Human Factors in Manufacturing, 2(3)*, 255–276.

Majchrzak, A. and Paris, M. L. (1995). High-performing organizations match technology and management strategies: Results of a survey. *International Journal of Industrial Ergonomics, 16(4–6)*, 309–326.

Mann, F. C., and Hoffman, L. R. (1960). *Automation and the Worker.* New York: Henry Holt and Co.

Manufacturing Studies Board (1986a). *Toward a New Era in U.S. Manufacturing: The Need for National Vision* Washington, D.C.: National Academy Press.

Manufacturing Studies Board (1986b). *Human Resource Practices for Implementing Advanced Manufacturing Technology.* Washington, D.C.: National Academy Press.

Märtensson, L. (1995). *Requirements on work organization—From work environment design of "Steelworks 80" to human-machine analysis of the aircraft accident at Gottr ra.* Doctoral dissertation, Department of Work Science, The Royal Institute of Technology, Stockholm, Sweden.

Märtensson, L. (1996). The operator's requirements for working with automated systems. *The International Journal of Human Factors in Manufacturing, 6(1)*, 29–39.

Martin, T. (Ed.) (1984). *Design of Work in Automated Manufacturing Systems.* Oxford: Pergamon Press.

Martin, T. (1990). The need for human skills in production: The case of CIM. *Computers in Industry, 4,* 205–211.

Martin, T. (1993). Considering social effects in control system design-a summary. In *Pre-prints of IFAC 12th World Congress,* Sydney, Australia, July 18–23 (Vol. 7, pp. 325–330).

Maslow, A. H, (1954). *Motivation and Personality.* New York: Harper.

McGregor, D. (1960). *The Human Side of Enterprise.* New York: McGraw-Hill.

Morrison, D. L. (1993). Organizational aspects of manufacturing technology management (guest editorial). *The International Journal of Human Factors in Manufacturing, 3(2)*, 111–115.

Murphy, S. (1989). The ESPRIT Project. In H. H. Rosenbrock, Ed., *Designing human-centred technology* (pp. 145–168). New York: Springer-Verlag.

Nagel, R. N. and Dove, R. (1992). *21st Century Manufacturing Enterprise Strategy.* Bethlehem, PA: Iacocca Institute, Lehigh University.

Naisbitt, J. (1984). *Megatrends. Ten New Directions Transforming Our Lives.* New York: Warner.

National Research Council. (1988). *A Research Agenda for IMS—Information Technology,* Manufacturing Studies Board, Washington, D.C., National Academy Press.

Nakamura, N., and Salvendy, G. (1988). Human planner and scheduler. In G. Salvendy and W. Karwowski, Eds., *Design of Work and Development of Personnel in Advanced Manufacturing Systems* (pp. 331–354). New York: John Wiley.

Office of Technology Assessment. (1984). *Computerized Manufacturing Automation: Employment, Education and the Workplace (OTA-CIT-235).* Washington, DC: U.S. Government Printing Office.

Office of Technology Assessment. (1985). *Automation of America's Offices.* Washington, DC: U.S. Congress.

Rasmussen, J. (1983). Skills, rules, and knowledge: signals, signs, and symbols, and other distinctions in human performance models. *IEEE Transactions on Systems, Man and Cybernetics, SMC 13(3)*, 257–266.

Rasmussen, J. (1993). Commentary: What are we looking for in the black box? *The International Journal of Human Factors in Manufacturing, 3(1)*, 89–90.

Rasmussen, J., Pejtersen, A. M., and Goodstein, L. P. (1992). *Cognitive Engineering Concepts and Applications.* London: John Wiley.

Rosenbrock, H. H. (1983). Designing automated systems-need skill be lost *Science and Public Policy, 10,* 247–277.

Rosenbrock, H. H., Ed. (1989). *Designing Human-Centred Technology.* New York: Springer-Verlag.

Salvendy, G., and Karwowski, W., Eds. (1994). *Design of Work and Development of Personnel in Advanced Manufacturing.* New York: John Wiley.

Sanderson, P. M. (1991). Towards the model human scheduler. *International Journal of Human Factors in Manufacturing, 1,* 195–219.

Sanderson, P. M. (1989). The human planning and scheduling role in advanced manufacturing system: An emerging human factors domain. *Human Factors, 31,* 635–667.

Savage, C. M. (1990). *Fifth Generation Management. Integrating Enterprises Through Human Networking.* Bedford, MA: Digital Press.

Savage, C. M. (1991). *Fifth Generation Management.* Bedford, MA: Digital Press.

Seppälä, P, Tuominen, E., and Koskinen, P. (1992). Impact of flexible production philosophy and advanced manufacturing technology on organizations and jobs. *The International Journal of Human Factors in Manufacturing, 2(2)*, 177–192.

Siemieniuch, C. E., and Sinclair, M. A. (1995). Information technology and global developments in manufacturing: The implications for human factors inputs. *International Journal of Industrial Ergonomics, 16,* 245–262.

Sinclair, M. A. (1986). Ergonomics aspects of the automated factory. *Ergonomics, 29(12)*, 1507–1523.

Sinclair, M. A., Siemieniuch, C. E., Cooper, K. A., and Waddell, N. (1995). A discussion of simultaneous engineering and the manufacturing supply chain, from an ergonomic perspective. *International Journal of Industrial Ergonomics, 16(4–6)*, 263–281.

Smith, M. J., (1987). Occupational stress. In G. Salvendy, Ed., *Handbook of Ergonomics/Human Factors* (pp. 844–860). New York: John Wiley.

Smith, M. J., and Carayon, P. (1995). New technology, automation, and work organization: stress problems and improved technology implementation strategies. *The International Journal of Human Factors in Manufacturing, 5(1)*, 99–116.

Smith, M. J., and Carayon-Sainfort, P. (1989). A balance theory of job design for stress reduction. *International Journal of Industrial Ergonomics, 4,* 67–79.

Smith, M. J., Carayon, P., Sanders, K. J., Lim, S-Y., and LeGrande, D. (1992). Employee stress and health complaints in jobs with the without electronic performance monitoring. *Applied Ergonomics, 23,* 17–27.

Sobol, M. G. and Lei, D. (1994). Environment, manufacturing technology, and embedded knowledge. *The International Journal of Human Factors in Manufacturing, 4(2)*, 167–189.

Stalk, G., and Hout, T. M. (1990). *Competing against Time.* New York: The Free Press.

Sugimoto, N. (1987). Subjects and problems of robot safety technology. In K. Noro, Ed., *Occupational Safety and Health in Automation and Robotics* (p. 175). London: Taylor and Francis.

Sun, H., and Riis, J. O. (1994). Organizational, technical, strategic, and managerial issues along the implementation process of advanced manufacturing technology-a general framework of implementation guide. *The International Journal of Human Factors in Manufacturing, 4(1)*, 23–36.

Taylor, F. W. (1911). *The Principles of Scientific Management.* New York: W.W. Norton.

Wall, T., Corbett, J. M., Martin, R., Clegg, C., and Jackson, P. (1990). Advanced manufacturing technology, work design, and performance: A change study. *Journal of Applied Psychology, 75(6)*, 691–697.

Wall, T. D., Clegg, C. W. and Kemp, N.J. (1987). *The Human Side of Advanced Manufacturing.* New York: John Wiley.

Warnecke, H. J. (1993). *The Fractal Company—A Revolution in Corporate Culture.* Berlin, London, and New York: Springer.

Weber, M. (1946). *Essays in Sociology.* New York: Oxford.

Wilson, J. R. (1991). Personal perspective: Critical human factors contributions in modern manufacturing. *International Journal of Human Factors in Manufacturing, 1(3)*, 281–297.

Wilson, J. R., Koubek, R., Salvendy, G., Sharit, J. and Karwowski, W. (1994). *Human Factors in Advanced Manufacturing: A Review and Reappraisal.* In W. Karwowski and G. Salvendy, Eds., *Organization and Management of Advanced Manufacturing* (pp. 379–415). New York: John Wiley.

Wobbe, W., and Charles, T. (1994). Human roles in advanced manufacturing technology. In W. Karwowski and G. Salvendy, Eds., *Organization and Management of Advanced Manufacturing* (pp. 61–80). New York: John Wiley.

Womack, J., Jones, D., and Roos, D. (1990). *The Machine that Changed the World.* New York: Rawson Associates.

Zimolong, B., and Duda, L. (1992). Human error reduction satrtegies in advanced manufacturing systems. In M. Rahimi and W. Karwowski, Eds., *Human-Robot Interaction.* (pp. 242–285). London: Taylor and Francis.

CHAPTER 57
AUTOMATION SURPRISES

N. B. Sarter
D. D. Woods
C. E. Billings
Cognitive Systems Engineering Laboratory
The Ohio State University
Columbus, OH 43210 USA

"The road to technology-centered systems is paved with user-centered intentions." (*Woods, 1994*)

57.1 INTRODUCTION

In a variety of domains, the development and introduction of automated systems has been successful in terms of improving the precision and economy of operations. At the same time, however, a considerable number of unanticipated problems and failures have been observed. These new and sometimes serious problems are related for the most part to breakdowns in the interaction between human operators and automated systems. It is sometimes difficult for the human operator to track the activities of their automated partners. As a result, the operator is surprised by the behavior of the automation and is asking questions like, what is it doing now, why did it do that, or what is it going to do next (Wiener, 1989). Thus, automation has created surprises for practitioners who are confronted with unpredictable and difficult to understand system behavior in the context of ongoing operations. The introduction of new automation has also produced surprises for system designers or purchasers who experience unexpected consequences because their automated systems fail to work as team players.

This chapter describes the nature of unanticipated difficulties with automation and explains them in terms of myths, false hopes, and misguided intentions associated with modern technology. Principles and benefits of a human-centered rather than technology-centered approach to the design of automated systems are explained. The chapter points out the need to design cooperative teams of human and machine agents in the context of future operational environments.

Automation technology was originally developed in hope of increasing the precision and economy of operations while, at the same time, reducing operator workload and training requirements. It was considered possible to create an autonomous system that required little if any human involvement and therefore reduced or eliminated the opportunity for human error. The assumption was that new automation can be substituted for human action without any larger impact on the system in which that action or task occurs, except on output. This view is predicated on the notion that a complex system is decomposable into a set of essentially independent tasks. Thus, automated systems could be designed without much consideration for the human element in the overall system.

However, investigations of the impact of new technology have shown that these assumptions are not tenable (they are what could be termed the substitution myth). Tasks and activities are highly interdependent or coupled in real complex systems. Introduction of new automation has shifted the human role to one of monitor, exception handler, and manager of automated resources.

As a consequence, only some of these anticipated benefits of automation have, in fact, materialized—primarily those related to the improved precision and economy of operations, i.e., those aspects of system operation that do not involve much interaction between human and machine. Other expectations were not met, and unanticipated difficulties were observed. These problems are primarily associated with the fact that even highly automated systems still require operator involvement and therefore communication and coordination between human and machine. This need is not supported by most systems, which are designed to be precise and powerful agents but are not equipped with communicative skills, with comprehensive access to the outside world, or with complete knowledge about the tasks in which it is engaged. Automated systems do not know when to initiate communication with the human about their intentions and activities or when to request additional information from the human. They do not always provide adequate feedback to the human who, in turn, has difficulties tracking automation status and behavior and realizing there is a need to intervene to avoid undesirable actions by the automation. The failure to design human–machine interaction to exhibit the basic competencies of human–human interaction is at the heart of problems with modern automated systems.

Another reason that observed difficulties with automation were not anticipated was the initial focus on quantitative aspects of the impact of modern technology. Expected benefits included reduced workload, reduced operational costs, increased precision, and fewer errors. Anticipated problems included the need for more training, less pilot proficiency, too much reliance on automation, or the presentation of too much information (for a more comprehensive list of automation-related questions, see Wiener and Curry, 1980). Instead, it turned out that many of consequences of introducing modern automation technology were of a qualitative nature, as will be illustrated in later sections of this chapter. For example, task demands were not simply reduced but changed in nature. New cognitive demands were created, and the distribution of load changed over time. Some types of errors and failures declined, whereas new error forms and paths to system breakdown were introduced.

Some expected benefits of automation did not materialize because they were postulated based on designers' assumptions about intended rather than actual use of automation. The two can differ considerably if the future operating environment of a system is not sufficiently considered during the design process. In the case of cockpit automation, for example, the actual and intended use of automation are not the same because air traffic control procedures do not match the abilities and limitations designed into modern flight deck systems, and the various operators of highly advanced aircraft have different philosophies and preferences for how and when to use different automated resources.

Finally, design projects tend to experience severe resource pressure, which almost invariably narrows the focus so that the automation is regarded as only an object or device that needs to possess certain features and perform certain functions under a narrowed range of conditions. The need to support interaction and coordination between the machine and its human user(s) in the interest of building a joint human–machine system becomes secondary. At this stage, potential benefits of a system may be lost, gaps begin to appear, oversimplifications arise, and boundaries are narrowed. The consequences are challenges to human performance.

Actual experiences with advanced automated systems confirm that automation does, in fact, have an effect on areas such as workload, error, or training. However, its impact turns out to be different and far more complex than anticipated. Workload and errors are not simply reduced, but changed. Modified procedures, data filtering, or more-of-the-same training are not effective solutions to observed problems. Instead, the introduction of advanced automation seems to result in changes that are qualitative and context dependent rather than quantitative and uniform in nature. In the following sections, some unexpected effects of automation will be discussed. They result from the introduction of automated systems that need to engage in, but were not designed for, cooperative activities with humans.

57.2 UNEXPECTED PROBLEMS WITH HUMAN–AUTOMATION INTERACTION

57.2.1 Workload—Unevently Distributed, Not Reduced

The introduction of modern technology was expected to result in reduced workload. It turned out, however, that automation does not have a uniform effect on workload. As first discussed by Wiener (1989) in the context of modern technology for aviation applications, many automated systems support pilots most in traditionally low-workload phases of flight but are of no use or even get in their way when help is needed most, namely in time-critical highly dynamic circumstances. One reason for this effect is the automation's lack of comprehensive access to all flight-relevant data in the outside world. This leads to the requirement for pilots to provide automation with information about target parameters, to decide how automation should go about achieving these targets (e.g., selecting levels and modes of automation), to communicate appropriate instructions to the automation, and to monitor the automation closely to ensure that commands have been received and are carried out as intended. These task requirements do not create a problem during low-workload phases of flight but once the descent and approach phases of flight are initiated, the situation changes drastically. Air traffic control (ATC) is likely to request frequent changes in the flight trajectory, and given that there is not (at this stage) a direct link between ATC controllers and automated systems, the pilot has the role of translator and mediator. He or she needs to communicate every new clearance to the machine and needs to (know how to) invoke system actions. It is during these traditionally high-workload, highly dynamic phases of flight that pilots report an additional increase in workload. Wiener (1989) coined the term "clumsy automation" to refer to this effect of automation on workload—a redistribution of workload over time rather than an overall decrease or increase because the automation creates new communication and coordination demands without supporting them well.

Workload is not only unevenly distributed over time but sometimes also between operators working as a team. For example, the pilot-not-flying on many advanced flight decks can be much busier than the pilot-flying as she or he is responsible for most of the interaction with the automation interface, which can turn a simple task (such as changing a route or an approach) into a "programming nightmare."

The effect on workload was also unexpected in the sense that the quality rather than the quantity of workload is affected. For example, the operator's task has shifted from active control to supervisory control by the introduction of automated systems. Humans are no longer continuously controlling a process themselves (although they still sometimes need to revert to manual control), but instead they monitor the performance of highly autonomous machine agents. This imposes new attentional demands, and it requires that the operator know more about systems in order to be able to understand, predict, and manipulate their behavior.

57.2.2 New Attentional and Knowledge Demands

The introduction of modern technology has created new knowledge and attentional requirements. Operators need to learn about the many different elements of highly complex systems and about the interaction of these elements. They need to understand input–output relationships to be able to anticipate effects of their own entries. In addition to knowing how the system works, they need to explore "how to work the system," i.e., operators must learn about available options, learn and remember how to deploy them across a variety of operational circumstances, and learn the interface manipulations required to invoke different modes and actions. Finally, it is not only the capabilities but also the limitations of systems that need to be considered.

Empirical research on human–automation interaction (e.g., Sarter and Woods, 1994a) has shown that operators sometimes have gaps and misconceptions in their model of a system. Sometimes operators possess adequate knowledge about a system in the sense of being able to recite facts, but they are unable to apply the knowledge successfully in an actual task context. This is called the problem of "inert" knowledge. One way to eliminate this problem is through training that conditionalizes knowledge to the contexts in which it is utilized.

Since the complexity of many modern systems cannot be fully covered in the amount of time and with the resources available in most training programs, operators learn only a subset of techniques or "recipes" to be able to make the system work under routine conditions. As a consequence, ongoing learning needs to take place during actual operations and has to be supported to help operators discover and correct bugs in their model of the automation. Recurrent training events can be used to elaborate their understanding of how the automation works in a risk-free environment.

Another problem related to knowledge requirements imposed by complex automation technology is that operators are sometimes miscalibrated with respect to their understanding of these systems. Experts are considered well calibrated if they are aware of the areas and circumstances for which they have correct knowledge and those in which their knowledge is limited or incomplete. In contrast, if experts are overconfident and wrongly believe that they understand all aspects of a system, then they are said to be miscalibrated (e.g., Wagenaar and Keren, 1986).

A case of operator miscalibration was revealed in a study on pilot–automation interaction where pilots were asked questions such as, "Are there modes and features of the Flight Management System that you still don't understand?" (Sarger and Woods, 1994a; these kinds of questions were asked in an earlier study by Wiener, 1989). When their responses to this question are compared with behavioral data in a subsequent simulator study, there is some indication that these "glass cockpit" pilots were overconfident and miscalibrated about how well they understood the Flight Management System. The number and severity of pilots' problems during the simulated flight was higher than was to be expected from the survey. Similar results have been obtained in studies of physician interaction with computer-based automated devices in the surgical operating room (Cook, Potter, Woods, and MacDonald, 1991; Moll van Charante, Cook, Woods, Yue, and Howie, 1993).

Several factors contribute to miscalibration. First, areas of incomplete or inaccurate knowledge can remain hidden from operators because they have the capability to work around these areas by limiting themselves to a few well-practiced and well-understood methods. In addition, situations that force operators into areas where their knowledge is limited and miscalibrated may arise infrequently. Empirical studies have indicated that ineffective feedback on the state and behavior of automated systems can be a factor that contributes to poor calibration (e.g., Cook et al., 1991; Norman, 1990; Wagenaar and Keren, 1986).

The need for adequate feedback design is related not only to the issue of knowledge calibration but also to the attentional demands imposed by the increased autonomy exhibited by modern systems (Moray, 1986). Operators need to know when to look where for information concerning (changes in) the status and behavior of the automation and of the system or process being managed or controlled by the automation. Knowledge and attentional demands are closely related, because the above mentioned mental model of the functional structure of the system provides the basis for internally guided attention allocation. In other words, knowing about inputs to the automation and about ways in which the automation processes these inputs permits the prediction of automation behavior which, in turn, allows the operator to anticipate the need for monitoring certain parameters (Gopher, 1991). This form of attentional guidance is particularly important in the context of normal operations.

In case of anomalies or apparently inconsistent system behavior, it can be difficult or impossible for the user to form expectations. Therefore, under those circumstances, the system needs to provide some form of external attentional guidance to the user to help detect and locate problems (for example, Jonides and Yantis, 1988). The system interface needs to serve as an external memory for the operator by providing cues that help realize the need to monitor a particular piece of information or to activate certain aspects of knowledge about the system.

Two frequently observed ways in which attention allocation can fail is (1) a breakdown in the "mental bookkeeping" required to keep track of the multiple interleaved activities and events that arise in the operation of highly complex technology, and (2) a failure to revise a situation assessment in the presence of new conflicting information. In the latter case, called "fixation error," evidence that is not in agreement with an operator's assessment of his or her situation is missed, dismissed, or rationalized as not really being discrepant.

The above problems—gaps and misconceptions in an operator's mental model of a system as well as inadequate feedback design—can result in breakdowns in attention allocation which, in turn, can contribute to a loss of situation, or more specifically, system and mode awareness.

57.2.3 Breakdowns in Mode Awareness and "Automation Surprises"

Norman (1988, p. 179) explains device modes and mode error quite simply by suggesting that one way to increase the possibilities for error is to "change the rules. Let something be done one way in one mode and another way in another mode." Mode errors, in this case, occur when an intention is executed in a way appropriate for one mode when, in fact, the system is in a different mode.

With more advanced systems, each mode itself is an automated function that, once activated, is capable of carrying out long sequences of tasks autonomously in the absence of additional commands from human supervisors. This increased autonomy produces situations in which mode changes can occur based on situational and system factors. This capability for "indirect" mode changes, independent of direct and immediate instructions from the human supervisor, creates the potential for mode errors of omission. Both forms of mode error—errors of commission and errors of omission—are symptoms of a breakdown in mode awareness, i.e., the ability of a supervisor to track and to anticipate the behavior of automated systems (Sarter and Woods, 1995a,b).

Breakdowns in mode awareness result in "automation surprises." These automation surprises have been observed and reported in various domains (most notably flight deck and operating room automation, e.g., Sarter and Woods, 1994a; Moll van Charante et al., 1993) and have contributed to a considerable number of incidents and accidents. Breakdowns in mode awareness can lead to mode errors of omission in which the operator fails to observe and intervene with uncommanded and/or undesirable system behavior.

Early automated systems tended to involve only a small number of modes that were independent of each other. These modes represented the background on which the operator would act by entering target data and by requesting system functions. Most functions were associated with only one overall mode setting. Consequently, mode annunciations (indications of the currently active as well as planned modes and of transitions between mode configurations) were few and simple and could be shown in one central location. The consequences of a breakdown in an operator's awareness of the system configuration tended to be small, in part because of the short time–constant feedback loops involved in these systems. Operators were able to detect and recover from erroneous input relatively quickly.

The flexibility of more advanced technology allows and tempts automation designers to develop much more complex mode-rich systems. Modes proliferate as designers provide multiple levels of automation and various methods for accomplishing individual functions. The result is a large number of indications of the status and behavior of the automated system(s), distributed over several displays in different locations. Not only the number of modes but also, and even more importantly, the complexity of their interactions has increased dramatically.

The increased autonomy of modern automated systems leads to an increase in the delay between user input and feedback about system behavior. These longer time–constant feedback loops make it more difficult to detect and recover from errors and challenge the human's ability to maintain awareness of the active and armed modes, the contingent interactions between environmental status and mode behavior, and the contingent interactions across modes.

Another contributing factor to problems with mode awareness relates to the number and nature of sources of input that can evoke changes in system status and behavior. Early systems would change their mode status and behavior only in response to operator input. More advanced technology, on the other hand, may change modes based on sensor information concerning environment and system variables as well from input by one or multiple human operators. Mode transitions can now occur in the absence of any immediately preceding user input. In the case of highly automated cockpits, for example, a mode transition can occur when a preprogrammed intermediate target (e.g., a target altitude) is reached or when the system changes its mode to prevent the pilot from putting the aircraft into an unsafe configuration.

Indirect mode transitions can arise as side effects of direct operator input to an automated system. This potential is created by the fact that the effects of operator input depend on the status of the system and of the environment at the time of input. The user intends one effect but the coupling between different automated subsystems and modes may automatically result in other unintended changes. Thus, an action intended to have one particular effect can have a different effect or additional unintended side effects. Missing these side effects is a predictable error form that is accentuated because of weak feedback concerning mode status and transitions and when there are gaps or misconceptions in the user's mental model of the system. Incidents and accidents have shown that missing these side effects can be disastrous in some circumstances.

57.2.4 New Coordination Demands

When new automation is introduced into a system or when there is an increase in the autonomy of automated systems, developers often assume that adding "automation" is a simple substitution of a machine activity for human activity (the substitution myth). Empirical data on the relationship of people and technology suggest that is not the case. Instead, adding or expanding the machine's role changes the cooperative architecture, changing the human's role, often in profound ways. Creating partially autonomous machine agents is, in part, like adding a new team member. One result is the introduction of new coordination demands. When it is hard to direct the machine agents and hard to see their activities and intentions, it is difficult for human supervisors to coordinate activities. This is one factor that may explain why people "escape" from clumsy automation as task demands escalate. Designing for coordination is a postcondition of more capable machine agents. However, because of the substitution myth, development projects rarely include specific consideration of how to make automation an effective team player or of how to evaluate possible systems along this dimension.

One concrete example of coordination occurs when automation compensates for a fault but only up to a point. When the automation can no longer handle the situation, it gives up and turns the problem back over to the human crew. The problem is that this transfer of control can easily be far from bumpless. Human operators may not be aware of the problem or may not fully understand its evolution if the automation compensates for the fault silently. Suddenly, when the effects of the fault are already substantial, they are forced to take over control. This is a challenging situation for them that has contributed to accidents in aviation and to incidents in anesthesiology.

The above case is an example of a decompensation incident (Woods, 1994). Decompensation incidents in managing highly automated processes are one kind of complication that can arise when automatic systems respond to compensate for abnormal influences generated by a fault. As the abnormal influences produced by the fault persist or grow over time, the capacity of the automation's counterinfluences to compensate becomes exhausted. When the automation's capacity to counteract is exhausted, it hands control back to human team members. However, they may not be prepared to deal with the situation (they may not appreciate the seriousness of the situation or may misunderstand the trouble), or the situation may have progressed too far for them to contribute constructively.

The presence of automatic counterinfluences leads to a two-phase signature. In phase 1 there is a gradual falling off from desired states over a period of time. Eventually, if the practitioner does not intervene in appropriate and timely ways, phase 2 occurs—a relatively rapid collapse when the capacities of the automatic systems are exceeded or exhausted. During the first phase of a decompensation incident, symptoms may not be present or may be small, which makes it difficult for human supervisors to understand what is occurring. This can lead to great surprise when the second phase occurs.

In those situations, the critical information for the human operator is not the symptoms per se, but the force with which they must be resisted. An effective human team member would notice and communicate the need to exert unusual control effort (Norman, 1990).

Thus, lack of information about automatic system activities can contribute to the failure to recognize the seriousness of the situation and the failure of the supervisory controller to act to invoke stronger counteractions early enough to avoid the decompensation.

This example illustrates the need for effective feedback in human–machine cooperation. But what kind of feedback will prove effective? We could address each specific example of a need for better feedback about automation activities individually, one at a time. But this piecemeal approach will generate more displays, more symbolic codings on displays, more sounds, more alarms. More data will be available, but these will not be effective as feedback because they challenge the crew's ability to focus on and digest what is relevant in a particular situation. Instead, we need to look at the set of problems that all point to the need for improved feedback to devise an integrated solution. For example, the decompensation signature shows us a class of feedback problems that arise when automation is working at the extreme of its envelope or authority. The automation doesn't clearly tell the human supervisor that this is the case; when control passes back to people, they are behind the developing situation and the transfer is rather bumpy or worse.

For this class of cases new feedback and communication between the human and machine agents is needed to indicate:

- When I (the automation) am having trouble handling the situation
- When I (the automation) am taking extreme action or moving toward the extreme part of my authority

This specifies a performance target. The design question is how to make the system smart enough to communicate this intelligently and how to define what are "extreme" regions of authority in a context-sensitive way. When is an agent having trouble in performing a function (but not yet failing to perform)? How does one effectively communicate moving toward a limit rather than just invoking a threshold crossing alarm?

From experience and research, we know some constraints on the answers to these questions. Threshold crossing indications (simple alarms) are not smart enough—thresholds are either set too late or too early. We need a more gradual escalation or staged shift in level or kind of feedback. We know that providing an indication whenever there is any automaton activity (e.g., auditory signal) says too much, too soon. We want to indicate trouble in performing the function or extreme action to accomplish the function, not simply any action.

We know that there are certain errors that can occur in designing feedback that should be avoided in this case. We must avoid

- Nuisance communication such as voice alerts that talk to you too much in the wrong situations
- Excessive false alarms
- Distracting indications when more serious tasks are being handled (e.g., a warning on constantly or at a high noise level during a difficult situation—"Silence that thing!")

In other words, misdesigned feedback can talk too much, too soon or it can be too silent, speaking up too little, too late as automation moves toward authority limits.

Should the feedback occur visually or through the auditory channel or through multiple indications? Should this be a separate new indication or integrated into existing displays? Should the indication be of very high perceptual salience? In other words, how strongly should the signal capture user attention? Working out these design decisions requires developing prototypes and adusting the indications in terms of perceptual salience, along a temporal dimension (when to communicate), and along a strength dimension (how gabby) based on human performance in context. All this requires thinking about these new indications in the context of other possible signals.

57.2.5 The Need for New Approaches to Training

With the introduction of highly advanced automation technology, traditional approaches to training no longer seem adequate to prepare operators for their new task of supervisory control of highly dynamic and complex systems. The aviation domain is one area where this problem became highly noticeable in the 1980s. Failure rates in transition training for "glass cockpit" aircraft were at an all-time high (Wiener, 1993), and even today the introduction of a new highly advanced airplane to an airline's fleet tends to create challenges and problems. Whereas today most pilots successfully complete their training, they still report that the first transition to one of these new airplanes is considerably more difficult and demanding than the transition between two conventional aircraft.

The observed problems are interpreted by many as indications of a need for more training time to acquire more knowledge about these complex systems. However, training-oriented research in other similarly complex domains (e.g., Feltovich, Spiro, and Coulson, 1991) and a better understanding of the kinds of problems experienced by "glass cockpit" pilots suggest that it is the nature of training that needs to be reconsidered rather than its duration. Cockpit technology has changed in fundamental ways that require new ways of learning and practice. It is no longer possible to learn about these systems by accumulating compartmentalized knowledge about individual components and simple input–output relations. Instead, pilots need to form a mental model of the overall functional structure of the system to understand its contingencies and interactions. Such a model is a prerequisite for being able to monitor and cordinate activities with cockpit automation as was explained in the earlier section on attentional demands imposed by modern automated systems.

A mental model helps build expectations of system behavior, and it contributes to the adequate allocaton of attention across and within numerous information-rich cockpit displays. It also supports pilots in dealing with novel situations by allowing them to derive possible actions and solutions based on their general understanding of how the system works (Carroll and Olson, 1988). These affordances of a mental model make it the desirable objective of training for advanced cockpit systems, which no longer allow for exposure to all aspects of their operation during training—even with more training time.

To support the formation of an effective mental model, training needs to encourage pilots to explore actively the available options and dynamics of the automation. Inventing a model based on experimentation has been shown to be preferable to the explicit teaching of a system model (Carroll and Olson, 1988). As Spiro, Coulson, Feltovich, and Anderson (1988) point out, "knowledge that will be used in many ways has to be learned, represented, and tried out in many ways." In contrast, rote memorization is antithetical to the development of applicable knowledge, i.e., knowledge that can be activated in context. Learning recipes results in "inert" knowledge, whereby the user can recite facts but fails to apply this knowledge effectively in actual line operations (see Feltovich et al., 1991).

In summary, part of the solution to observed problems with flying highly automated aircraft may be new approaches to and objectives for training rather than simply more training time. But it is important to realize that even such improved training cannot solve all observed problems—training cannot and should not be a fix for bad design.

57.2.6 New Opportunities for New Kinds of Error

Another anticipated benefit of automation was a reduction in human error—but again, operatonal experience and systematic empirical research proved otherwise. Instead of reducing the overall amount of errors, automation provided new opportunities for different kinds of error (Woods, Johannesen, Cook, and Sarter, 1994). One example that has recently gained considerable interest is the case of mode awareness and its relationship to automation surprises, especially in the context of new flight decks.

Interest in research on mode error on "glass cockpit" aircraft was triggered by the results of a study by Wiener (1989) who conducted a survey of B-757 pilots where about 55% of all respondents said that they were still being surprised by the automation after more than 1 year of line experience on the aircraft. In a followup study, Sarter and Woods (1992) sampled a different group of pilots at a different airline who were flying a different glass cockpit aircraft (the B737-300/400). They replicated Wiener's results and, more importantly, they gathered detailed information concerning the nature of and underlying reasons for "automation surprises," which can be seen as symptoms of a loss of mode awareness, i.e., awareness of the status and behavior of the automation. In addition, an experimental simulator study was carried out to assess pilots' mode awareness in a more systematic way (Sarter and Woods, 1994a). Overall, this research confirmed that "automation surprises" are experienced even by pilots with a considerable amount of line experience on highly automated aircraft. Problems with mode awareness were shown to occur most frequently in nonnormal and time critical situations.

Mode errors seem to occur because of a combination of gaps and misconceptions in operators' model of the automated systems and the failure of the automation interface to provide users with salient indications of its status and behavior (Sarter and Woods, 1994a, 1995b. During normal operations, bugs in operators' mental model can make it difficult or impossible to form accurate expectations of system behavior that provide guidance for effective allocation of attention across and within displays. In the case of nonnormal events that cannot be anticipated by the crew, the automation interface needs to attract the user's attention to the relevant piece(s) of information—many current systems fail to do so.

Mode errors of omission are particularly disturbing because of their implications for error recovery. Mode errors of omission occur in the absence of an immediately preceding directly related pilot action or input. Therefore, operators are less likely to look for and /or notice a change in system status or behavior. The trend toward errors of omission also has implications for the development of countermeasures to mode error. One proposal for dealing with mode errors of commission has been to introduce forcing functions that prevent the user from carrying out certain actions or inputs. In the case of errors of omission, however, such measures would not be useful as there is no specific action to prevent.

It seems that the detection of unexpected and possibly undesired changes in automation behavior is one of the major difficulties with the coordination between humans and advanced automated systems. The problem is difficult because the task is not simply to detect an abnormal state or behavior. Instead the task is to detect that a certain system behavior, which may be normal and acceptable in some circumstances, requires operator intervention in a particular context.

57.2.7 Complacency and Trust in Automation

Complacency has been proposed as another factor contributing to operators' failures to detect and intervent with system failures or undesirable system behavior. Complacency refers to the development of a false sense of security as operators come to rely on automation which is, in fact, highly reliable but can still fail without warning (Billings, 1991, 1996).

This view has raised concerns as it seems to blame the human by implying a lack of motivation and concentration on the task at hand. It seems to suggest that if the human would only try harder, it would be possible for him or her to perform his or her duties successfully. It fails to acknowledge that people will come to rely on systems that appear to be reliable at least for situations they encounter with high frequency. The design of the joint human–machine system has created a role where people must monitor for rare events—a sustained attention task. Complacency may therefore be more indicative of the need to rethink the human–machine architecture.

Trust in automation is a related issue that has received considerable consideration in the literature on human–automation interaction (e.g., Muir, 1987). Trust miscalibration is a problem associated with current increasingly autonomous and powerful systems that can create the image of a highly intelligent and proficient partner. What do we mean by "trust" in the context of human–machine interaction? One recent definition of trust (Barber, 1983) includes two important dimensions, namely, the expectation of technically competent role performance and the expectation that a partner will carry out his or her fiduciary obligations and responsibilities, i.e., his or her duty to place, in certain situations, other's interests before his or her own. This latter form of trust is important in situations where it is not possible for the user to evaluate the technical competence of another agent. In those cases, the user has to rely on the moral obligation o the other agent not to misuse the power given to him/her/it. Muir (1987, p. 530) has predicted that "the issue of machine responsibility will become more important in human–machine relationsihps to the extent that we choose to delegate autonomy and authority to 'intelligent,' but prosthetic machines. The more power they are given, the greater will be the need for them to effectively communicate the intent of their actions, so that people who use them can have an appropriate expectation of their responsibility and interact with them efficiently."

Finally, trust in automation is related to the issue of responsibility. Jordan was one of the first to point out that "we can never assign them [i.e., the machines] any responsibility for getting the task done; responsibility can be assigned to man only" (Jordan, 1963 p. 164). Therefore, he required that responsibilities be clearly assigned to each human in the system and that each human be provided with the means necessary to effectively control the tasks and systems for which he or she is responsible. Jordan's point that responsibility cannot be assigned to a machine has been expanded on by Winograd and Flores (1986) who point out that one of the major differences between human–human interaction and human–machine interaction is that we may treat an intelligent machine as a rational being but not as a responsible being. "An essential part of being human is the ability to enter into commitments and to be responsible for the courses of action that they anticipate. A computer can never enter into a commitment (although it can be a medium in which the commitments of the designers are conveyed)" (Winograd and Flores, 1986, p. 106). The last sentence of this statement is important because it suggests that designers may have to be held responsible for the machines they build. Just because machines cannot be responsible may not mean that their operators have to bear all responsibility alone. Designers need to ensure that their systems comply with requirements of human-centered automation such as being predictable, accountable, and dependable (see Billings, 1996), all of which are aspects of responsibility.

In the preceding paragraphs, unanticipated problems associated with the design of and training for modern automated systems have been discussed. It was shown that anticipated benefits and disadvantages of automation were often conceived of in quantitative terms—less workload, more precision, less errors, more training requirements—but that observed problems tended to be qualitative in nature—new temporal workload patterns, different kinds of errors and failure patterns, the need for different approaches to training. Observed problems with human–automation interaction are related to the need for, but lack of support for, communication and coordination between human operators and machine agents. This, in turn, is the consequence of increasingly high levels of system complexity, coupling, autonomy, and authority in combination with low system observability (Sarter and Woods, 1994b; Woods, 1996). In the following section, a number of accidents involving modern technology are presented that illustrate how these system properties can lead to breakdowns in overall system performance.

57.3 ACCIDENTS INVOLVING BREAKDOWNS IN HUMAN–MACHINE COORDINATION

In the following section, two accidents involving breakdowns in the communication and coordination between human and machine agents in various domains are described. They illustrate how a combination of several of the above discussed factors, as well as additional factors such as organizational pressures, can contribute to the evolution of disastrous events.

57.3.1 Radiation Accidents with THERAC-25

A series of accidents related to human–machine interaction occurred in the medical world with a system called THERAC-25, a computerized radiation therapy machine for cancer therapy. The machine can be used to deliver a high-energy electron beam to destroy tumors in relatively shallow tissue areas; deeper tissue can be reached by converting the electron beam into x-ray photons.

In the mid-1980s, several accidents happened with this device, all of which involved a lack of feedback about the status and activities of the machine. One of those accidents, which has been analyzed in quite some detail, occurred in 1986 at the East Texas Cancer Center. In this case, a patient was undergoing his ninth treatment as a followup to the removal of a tumor from his back. During this procedure, the technician operating the device would normally be in contact with the patient via a video camera and the intercom in the treatment room. On this particular day, however, both these sources of feedback were temporarily inoperative. Another thing happened that day that had not happened before. The technician made a mistake when setting up the device for the treatment. She entered and "x" (the entry for x-ray treatment) instead of an "e" (for electron mode), realized her mistake after a few more entries, and tried to correct it quickly by selecting the "edit" function of the display and by entering an "e" in place of the "x." Then she hit the return key several times to leave all other entries unchanged.

What the technician did not know and could not see was that the machine did not accept her corrections despite the fact that the screen displayed the correct therapy entries. The particular sequence of keystrokes she entered plus the speed she entered them (less than 8 s because she was proficient with the interface) had revealed a software design problem. When the machine was activated by the technician, the machine paused and a display indicated to her "Malfunction 54" and "treatment pause." This general feeback was familiar to her; it indicated that the treatment had not been initiated because of some minor glitch. On a separate documentation sheet, she found the error number explained as a "dose input 2" error but the technician did not understand the meaning of this message. In general the machine paused when a minor glitch occurred; when there were more significant problems, in her experience the machine abandoned the treatment plan and reverted to an initial state. Being used to many quirks of the machine, she decided simply to activate the beam again. When she reactivated the device, the same sequence of events occurred.

Meanwhile, inside the treatment room, unbeknown to the technician, the patient was hit by a massive radiation overdose first in the back and then on his arm as he tried to get up from the treatment table after the first painful episode.

After the patient reported intense pain during and after the treatment, the equipment was examined but no malfunction was diagnosed and nothing physically wrong was found with the patient. As no problems could be identified and because other patients were waiting in line, they started using the equipment again the same day. The patient died a few months later from complications related to the overdose he received on this day (for a detailed description and analysis of this accident, see Leveson and Turner, 1993).

This case illustrates how a lack of feedback concerning both the effected process (the display of dosage could not go high enough to indicate the dosage received) and the status and behavior of the automated system (the gap between what the machine said it would do and what instructions it was actually following; the pause type alarm was a familiar occurrence and minor issue) make it impossible for the operator to detect and intervene with undesirable system behavior. It also shows how organizational pressures such as the large number of patients waiting to be treated with the THERAC-25 device added to the problem faced by the operator and prevented a timely investigation of the device. And finally, it shows that virtually all complex software can be made to behave in an unexpected fashion under some conditions. (Leveson and Turner, 1993).

57.3.2 A Fatal Test Flight

This accident occurred in the context of a test flight of one of the most advanced automated aircraft in operation. It involved a simulated engine failure at low altitude under extreme flight conditions. A number of things were out of the ordinary on this flight and would later be presented as contributing factors in the accident investigation report (for a detailed account of this accident see *Aviation Week and Space Technology*, April 3, April 10, and April 17, 1995). During takeoff, the copilot rotated the aircraft rather rapidly, which resulted in a pitch angle of slightly more than 25° within 6 sec after takeoff. At that point, the autopilot was engaged as planned for this test. Immediately following the autopilot engagement, the captain brought the left engine to idle power and cut off one hydraulic system to simulate an engine failure situation. This particular combination of actions and circumstances led up to a crash killing everyone aboard the aircraft, which was totally destroyed. How did this happen?

When the autopilot was selected, it immediately engaged in an altitude capture mode because of the high rate of climb and the rather low pilot-selected level-off altitiude of 2000 ft. At the same time, because the pitch angle exceeded 25° at that point, the declutter mode of the primary flight display activated. This means that all indications of the active mode configuration of the automation (including the indication of the altitude capture mode) were hidden from the crew because they had been removed from the display for simplification.

Another important factor in this accident is the inconsistency of the automation design in terms of protection functions, which are intended to prevent or recover from unsafe flight attitudes and configurations. One of these protection functions guards against excessive pitch, which results in too low an airspeed. This protection is provided in all automation configurations except one—the very altitude acquisition mode in which the autopilot was operating. As a consequence, the automation continued to try to follow an altitude acquisition path even when it became impossible to achieve it (after the captain had brought the let engine to idle power). Ultimately, the automation flew the aircraft into a stall, and the crew was not able to recover given the low altitude.

Clearly, a combination of factors contributed to this accident and was cited in the report of the accident investigation. Included in the list of factors are the extreme conditions under which the test was planned to be executed, the lack of pitch protection in the altitude adquisition mode, and the inability of the crew to determine that the automation had entered that particular mode because of the decluttering of the primary flight display. The time available for the captain to react to the abnormal situation (12 s) was also cited as a factor in this accident.

A more generic contributing factor in this accident was the behavior of the automation, which was highly complex, inconsistent, and difficult to understand. These characteristics made it hard for the crew to anticipate the outcome of the maneuver. In addition, the observability of the system was practically nonexistent when the declutter mode of the primary flight display activated upon reaching a pitch angle of more than 25° up.

These problems and accidents are clearly related to and call for changes in the design of highly automated systems in the interest of making them more observable and cooperative agents. To achieve this goal, various approaches to system design have been proposed and are discussed in the following section.

57.4 DIFFERENT APPROACHES TO AUTOMATION DESIGN

The problems and accidents with automated systems described in the previous sections are to a large extent the result of "technology-centered" design that does not consider the need for supporting communication and cooperation between human and machine agents. To counteract and prevent the reoccurrence of such difficulties, a new approach, called "human-centered" automation, has been developed. The following sections provide an overview of the basic principles underlying this new perspective. We will also discuss extentions of the basic human-centered view to consider supporting not just one but a group of human users and their interaction and to consider integrating humans, machines, and their task environments in an effort to improve overall system performance.

57.4.1 Human- versus Technology-Centered Automation

The unanticipated problems associated with clumsy automation has led to the call for human-centered rather than technology-centered automation. What do we mean by these labels? Norman illustrated the difference by quoting and rewriting the motto of the Chicago World's Fair (1933). The original was, "Science finds, industry applies, man conforms." In contrast, Norman (1933) suggested, "people propose, science studies, technology conforms," as a human-centered alternative.

In a technology-centered approach the primary focus is technological feasibility—what is needed to create machines that can function more autonomously. Human-centered automation, on the other hand, is oriented toward operational needs and practitioner requirements. Its objective is to support, not supplant or replace the human operator. The primary focus becomes how to make automated systems team players.

Basically, in a user-centered approach designers consider, up front, the impact of introducing new technology and automation on the role of people in the system and on the structure of the larger system of which the automation is a part. This approach is needed because of technological success: The dominant question is rarely what can be automated but what should be automated in support of human operators and of human–machine cooperation.

It is very important to be clear that human-centered automation is *not* a call for less technology. In contrast, it calls for developing high technology that is adapted to the pressures of the operational world. People have always developed and skillfully wielded technology as a tool to transform and amplify their work, whether physical work or cognitive work. As Norman (1988) puts it, "technology can make us *smart* and technology can make us *dumb* (emphasis added)." The central problem is not less or more technology, but rather skillful or clumsy technology.

Guiding principles for a human-centered approach to design have been put forward by Billings (1991), who suggests that as long as human operators bear ultimate responsibility for operational goals, they must be in command. To be in command effectively, operators need to be involved in and informed about ongoing activities and system states and behaviors. In other words, the automation must be observable, and it needs to act in predictable ways (see Billings, 1991 and 1996 for a comprehensive discussion of human-centered automation).

Let us illustrate the difference between human- and technology-centered perspectives by examining some recent incidents and accidents in the aviation domain. Some of these events involved pilots who were trying to take control of an airplane upon observing unexpected or undesirable automation behavior. They attempted but failed to disengage or overpower the automation, thus creating a "fight" between man and machine over the control of the aircraft—in some cases with fatal consequences (for examples see Dornheim, 1995).

The technology-centered view of these events puts people and machines into opposition. Technologists assert that the problem was caused by the inappropriate behavior of a pilot who interfered with the activities of an automated system that acted as designed and as instructed earlier by the same pilot. The only alternative, from a technology-centered point of view is to say that the machine was to blame. The problem must be either in the people or in the machine.

In contrast, a human-centered approach shifts the boundaries. Human and machine agents are together part of one system. The kinds of accidents mentioned earlier reveal a breakdown in coordination between the human and machine portions of a team. Solutions involve developing the mechanisms to produce improved team play between machine and human agents just as we have recognized that effective team play is critical for the success of crews of human operators.

The two perspectives also try to address the problem in very different ways. The technology-centered approach suggests efforts to modify people—remedial training, new procedures, more technology intended to eliminate human activity. In contrast, the human-centered view focuses on changing the human–machine system in ways to support better coordination. Should an automated subordinate continue to act in opposition to the pilot's latest input, or is this an act of insubordination? Should an effective subordinate communicate with his or her supervisor to point out conflicting inputs and to ask for clarification?

Being in command is possible only if one is provided with or has access to all information necessary to assess the status and behavior of the automation and to make decisions about future courses of action based on this assessment. A major objective of keeping the operator informed is to avoid undermining his or her authority (for a broader discussion of these issues, see Billings, 1996). If information is hidden from the operator or not provided except in a hazardous situation, he or she is forced into a reactive mode.

Authority also implies that the operator has means to instruct, redirect, and if need be "escape" from the automation when deemed necessary. Having available the nominal means to instruct or escape is not sufficient. These mechanisms have to be usable under actual task conditions of multiple tasks, a dynamic world, and multiple data sources competing for attention. If, for example, control over the automation can be achieved only by means of a sequence of rarely executed actions that may require the diversion of attention away from critical system parameters in an escalating problematic situation, the automation is only a burden and not a resource or support.

57.4.2 Automation Design—User-Centered, Crew-Centered, Practice-Centered?

The original label for the perspective outlined in the previous section and introduced by several authors was "human-centered" (or "user-centered"). This label has some limits. Someone may interpret it as saying that the human to be supported is whoever does the task today. But technology change does transform the roles people play; a human-centered process focuses on supporting the new roles of people as supervisory controllers, exception handlers and monitors and managers of automated resources. Also, the label "human-centered" seems to suggest that there is an individual human who should be supported by new designs. It turns out, of course, that the issue is rarely a single individual. Perhaps the label should be "team-centered." Still others may object that our goal is to improve system performance and not merely to "enrich" an individual's or a team's job. Emphasizing the integration across many practitioners, instruments, and tasks leads some to prefer such labels as "use-centered" or "practice-centered" (e.g., Flach and Dominguez, 1995). "Use-centered" means that (1) we are trying to make new technology sensitive to the constraints and pressures acting in the actual operational world, and (2) we are focused on multiple actors at different levels with different scopes of responsibility embedded in a larger operational system. Some of these agents are human and some machines. We need to think of new automation as part of this control and management system rather than simply divide the world into machine and human parts. We need to design this system of interacting control and management agents to perform effectively given the demands of the domain.

Human-, team-, practice-centered; all of the labels center design on someone or something. But all point to an overall systems perspective that implies that no single element of the system should be at the center of designers' considerations. Instead, all system components are viewed as mutually dependent and interacting with one another. All of them involve constraints as well as the potential for change within certain limits. The integration of those constraints is the design objective.

57.4.3 The Gap Between User-Centered Intentions and Actual Practice

Interestingly, the need for a "human-centered" approach to automation design is accepted by many people involved in the design and evaluation of modern technology. For example, almost all parties in the aviation industry agree that human-centered automation is an appropriate goal. Yet, as we have discussed, flight deck automation is a well-studied case in which many automation designs exhibit problems with human–automation coordination. Hence the epigraph that opens this chapter. This gap between human-centered intentions and human-centered development practices indicates that there is either a misunderstanding about the concept or human-centeredness or an inability to translate its underlying ideas into actual designs.

Gaps between user-centered intentions and actual design practices arise in part because the developers of new technology and the people who must use it in actual work environments have two different perspectives. The developers' eye view is *apparent simplicity*. They justify their new technology in terms of potential benefits, often benefits derived from their claims about how the new technology will affect human performance. The practitioners' eye view is *real complexity*. They see all of the complicating factors that can arise in the operational world at the margins of normality and in more exceptional circumstances. They experience the new burdens created by clumsy use of technology. They must, as responsible agents, make up for the gaps between the developer's dreams and the actual complexities of practice.

The gap between user-centered intentions and technology-centered development occurs when designers:

- Oversimplify the pressures and task demands from the users' perspective.
- Assume that people can and will call to mind all relevant knowledge.
- Are overconfident that they have taken into account all meaningful circumstances and scenarios.
- Assume that machines never err.
- Make assumptions about how technology impacts on human performance without checking for empirical support or despite contrary evidence.
- Define design decisions in terms of what it takes to get the technology to work.
- Sacrifice user-oriented aspects first when tradeoffs arise.
- Focus on building the sytem first, then trying to integrate the results with users.

57.5 AUTOMATION—A WIDE RANGE OF TOOLS AND AGENTS

Taking a human-centered and system-oriented approach to the design and evaluation of modern technology requires that one aim at identifying and integrating the constraints associated with the human user(s), the automation, and the task (environment). These constraints are changing over time. Therefore, mismatches may disappear and new ones may be created.

One important system element that is undergoing considerable changes is the machine element. Increasingly sophisticated automated systems are being introduced to a wide range of domains where they can serve a variety of purposes. Automated systems can support or take over control of subtasks, they can process an present information, or they can carry out system management tasks (Billings, 1996).

Because of the different demands associated with these tasks and functions, and also because of the continuing evolution of computational power and of automation philosophies, automated systems differ to a considerable extent. It may therefore not be appropriate to use the term "automation" as though it referred to one class of homogenous systems. Instead, these systems differ with respect to many important properties that affect the relationship between the system and its human user(s).

Examples of such properties are a system's level of authority and autonomy as well as its complexity, coupling, and observability (Woods, 1996). The term "authority" refers to the power to control a process. The level of authority of an automate control system has implications for the role and responsibility assigned to its human operator. "Autonomy" denotes a system's capability to carry out sequences of actions without requiring (immediately preceding) operator input. In other words, autonomy refers to a system's level of independence from the human user for some specific task. System complexity is determined by the number of system components and especially by the extent and nature of their interactions. Coupling refers to the potential for an event, fault, or action to have multiple cascading effects. The higher the level of autonomy, complexity, and/or coupling of a system, the greater is the need for communication and coordination between human and machine to support the operator's awareness of the state and behavior of automation.

System awareness is the prerequisite for realizing the need for intervention with system activities that are not desirable or may even be dangerous. The key to supporting human–machine communication and system awareness is a high level of system observability. Observability is the technical term that refers to the cognitive work needed to extract meaning from available data (Rasmussen, 1985). This term captures the fundamental relationship among data, observer, and context of observation that is fundamental to effective feedback. Observability is distinct from data availability, which refers to the mere presence of data in some form in some location. Observability refers to processes involved in extracting useful information. It results from the interplay between a human user knowing when to look for what information at what point in time and a system that structures data to support attentional guidance.

To summarize, automated systems are not homogeneous but rather differ and continue to change along a number of important dimensions that can affect the interaction between humans and machines. One prerequisite for making progress in understanding and supporting the coordination between these two agents is to define them both in terms of their abilities, strategies, and limitations. Just as we consider differences between human operators such as the different abilities and strategies of novices versus experts, we need to better define the nature of automated systems because it is the degree to which human and machine properties match and support each other that determines to a large extent the overall system performamce.

57.6 AUTOMATION DESIGN—CURRENT TRENDS AND FUTURE NEEDS

57.6.1 Ongoing Trends in Automation Design

Automated systems differ significantly as the result of a continuous evolution of technological capabilities in combination with the different automation philosophies that determine how these capabilities are utilized and implemented. One trend in automation design is toward higher levels of system autonomy, authority, complexity, and coupling. These properties create an increased need for communication and coordination between humans and machines. To support this need, the design of system feedback would have to be improved in the interest of providing the automation with communicative skills. This need has not been sufficiently addressed, however. The amount of information that is potentially available to the operator has increased; but its quality does not match the mechanisms and limitations of human information processing. As a consequence, the gap between available and required feedback is growing. This has been shown to have a negative impact on the ability of operators to maintain awareness of automation status and behavior and to coordinate their activities with these advanced systems (e.g., Sarter and Woods, 1995a,b).

57.6.2 The Need for Cooperative Systems

Unexpected problems with new automated systems have been explained in two different ways. One group of commmentators blamed problems on the human element in the system. They stated that the machines worked as intended, and that it was the variability in the operator's performance or "human error" that caused the problem. This argument was and is still being made by those who think that the solution to automation-related difficulties is even more automation. The very same problems have been explained by others as evidence of "overautomation."

Research that has examined the effects of new automation on human performance indicates that both of these reactions are too simple (e.g., Norman, 1990). The data indicate that the problems are associated with breakdowns in the coordination between human and machine agents. These coordination breakdowns follow a general template. An event occurs or a set of circumstances come together that appears to be of minor importance, at least in principle. This initial event or action triggers an evolving situation from which it is possible to recover. But through a series of commissions and omissions, misassessments and miscommunications, the human and automation team manage the situation into a much more serious and risky incident or even accident. The mismanagement hinges on the misassessments and miscommmunications between the human and machine agents. It is results of this kind that have led to the recognition that machine agents cannot be designed merely as strong individual agents but need to be designed to support coordinated activity across the team. Similarly, the human supervisory role requires new knowlege and skills to manage automated resources effectively across a variety of potential circumstances.

Thus, the critical unit of analysis is the joint human–machine system. Success depends on supporting effective communication and coordination between humans and increasingly autonomous and powerful systems that can act independently of operator commands. This emphasizes the need to identify and integrate the constraints associated with machines, humans, tasks, and the environment in which they cooperate in order to increase the efficiency and safety of operations.

As other domains begin to introduce new levels of automation, they can avoid the problems that have occurred in the past by considering how to make automated systems team players early in the development process. One domain where the new levels of automation are likely to be introduced in the near future is air traffic control. Plans for future air traffic management are based on the idea to provide airspace users with more flexibility and authority in order to increase the efficiency of air carrier operations and the capacity of the air traffic system. This increased flexibility will reduce the potential for long-term planning, which forms the basis for current controller decisions and interventions. New airborne and groundbased automated systems will have to be introduced to support a more short-term approach to the safe handling of traffic.

The first challenge created by the introduction of more and more complex and powerful systems will be to ensure that these systems not only perform their assigned tasks of traffic separation and guidance but that they communicate to their users about their status, reasoning, and behavior. For some of the systems that will still exist in the future environment, it has already been shown that breakdowns in the interaction between humans and these machines occur because it can become very difficult for the user to keep track of his "strong but silent" counterpart. It is not known how the addition of more systems, the large number of human and machine agents, and the potential for machine–machine communication may affect the overall safety and efficiency of this system.

57.6.3 Networks of Automated Systems

Increasingly, networks of automated systems are envisioned and created to handle highly complex tasks and situations by engaging in negotiations and coordination among themselves without a need for direct involvement of the human operator. Such machine networks may become part of future air traffic management where, for example, ground-based computers may talk to airborne flight management computers to negotiate and communicate in flight plans.

The challenge will be to develop technology that contributes to the communication and coordination among all human and machine players involved in this distributed network of decision makers. It will be critical to avoid additional management, monitoring, and coordination tasks for the human operator resulting from a "clumsy" and "silent" implementation of automation.

57.7 CONCLUSION

The introduction of advanced technology has created automation surprises: System operators are surprised by the behavior of their strong but silent machine partners; system designers or purchasers are surprised to find new problems that concern the coordination of people and automated systems. Practitioners have to cope with resulting breakdowns in user–system coordination and with uncommanded, unanticipated, and sometimes undesirable activities of their machine counterparts. Designers are faced with unexpected consequences of the failure to support communication adequately between humans and machines. These surprises are not simply the result of overautomation or human error. Instead, they represent a failure to design for a coordinated team effort across human and machine agents as one cooperative system.

This design failure arises because there is a difference between commonly held beliefs about the impact of new automation on human and system performance and the actual impact of new technology on the people who must use it in actual work. The developers try to justify their new technology in terms of potential benefits, often benefits derived from their assumptions about how the new technology will affect human performance. In contrast, the evidence from research investigations on the actual impact of new technology has shown that many of these assumptions are not tenable.

Table 57.1 summarizes this contrast by juxtaposing certain beliefs prevalent in developer communities (column 1: putative benefits or effects of new technology) with the results from investigations of the impact of new technology on human performance (column 2: the real complexity of how technology affects performance in challenging fields of practice).

Table 57.1 Designer's Eye View of Apparent Benefits of New Automation Contrasted with the Real Experience of Operational Personnel

Putative Benefits	Real Complexities
Improves results with only simple substitution.	Transforms practice and creates new roles for human operators.
Frees up resources:	
I. Offloads work.	I. Creates new cognitive demands, often at busy or critical times.
II. Offloads requirements for attention.	II. Requires actively tracking and user integrating multiple activities and changes.
Requires less knowledge.	Requries different knowledge and new skills.
Operates autonomously.	Requires but does not support teamplay between humans and machines.
Integrates all necessary data.	Requires new levels and types of feedback to recognize what's informative in context.
Provides generic flexibility.	Explosion of features/options/mode creates new demands, error opportunities, and paths toward failure.
Reduces human error.	Creates new kinds of failure related to breakdowns in human–machine coordination.

New user- and practice-oriented design philosophies and concepts are being developed to address deficiencies in human–machine coordination. Their common goal is to provide the basis to design integrated human–machine teams that cooperate and communicate effectively as situations escalate in tempo, demands, and difficulty. Another goal is to help developers identify where problems can arise when new automation projects are considered and therefore help moblilize the design resources to prevent them.

ACKNOWLEDGMENT

The preparation of this manuscript was supported in part under a Cooperative Agreement (NCC 2-592) with the Aerospace Human Factors Research Division of the NASA-Ames Research Center (Technical Monitor: Dr. Everett Palmer).

REFERENCES

Aviation Week and Space Technology. (1995, April 3). A 330 crashed in Cat.3 test flight, 72–73.

Aviation Week and Space Technology. (1995, April 10). A 330 test order included study of autopilot behavior, 60.

Aviation Week and Space Technology. (1995, April 17). Toulouse A330 flight swiftly turned critical, 44.

Barber, B. (1983). *The logic and limits of trust*. New Brunswick, NJ: Rutgers University Press.

Billings, C. E. (1991). *Human-centered aircraft automation: A concept and guidelines* (NASA Technical Memorandum 103885). Moffett Field, CA: NASA-Ames Research Center.

Billings, C. E. (1996). *Aviation automation: The search for a human-centered approach*. Hillsdale, NJ: Erlbaum.

Carroll, J. M., and Olson, J. R. (1988). Mental models in human-computer interaction. In M. Helander (Ed.), *Handbook of human-computer interaction* (pp. 45–65). New York: Elsevier Science Publishers.

Cook, R. I., Potter, S. S., Woods, D. D., and McDonald, J. M. (1991). Evaluating the human engineering of microprocessor-controlled operating room devices. *Journal of Clinical Monitoring*, 7, 217–226.

Dornheim, M. A. (1995). Dramatic incidents highlight mode problems in cockpits. *Aviation Week and Space Technology*, January 30, 6–8.

Feltovich, P. J., Spiro, R. J., and Coulson, R. L. (1991). *Learning, teaching, and testing for complex conceptual understanding* (Tech. Rep. No. 6). Springfield, IL: Southern Illinois University School of Medicine—Conceptual Knowledge Research Project.

Flach, J. M., and Dominguez, C. O. (1995, July). Use-centered design: Integrating the user, instrument, and goal. *Ergonomics in Design*, pp. 19–24.

Gopher, D. (1991). The skill of attention control: Acquisition and execution of attention strategies. In D. Meyer and S. Kornblum, Eds., *Attention and Performance XIV*. Hillsdale, NJ: Erlbaum.

Jonides, J., and Yantis, S. (1988). Uniqueness of abrupt visual onset in capturing attention. *Perception and Psychophysics, 43*(4), 346–354.

Jordan, N. (1963). Allocation of functions between man and machines in automated systems. *Journal of Applied Psychology, 47*(3), 161–165.

Leveson, N. G., and Turner, C. S. (1993). An investigation of the THERAC-25 accidents. *Computer,* July, 18–41.

Moll van Charante, E., Cook, R. I., Woods, D. D., Yue, L., and Howie, M. B. (1993). Human-computer interaction in context: Physician interaction with automated intravenous controllers in the heart room. In H. G. Stassen, Ed., *Analysis, Design and Evaluation of Man-Machine Systems 1992*. Pergamen Press, pp. 263–274.

Moray, N. (1986). Monitoring behavior and supervisory control. In K. R. Boff, L. Kaufman, and J. P. Thomas, Eds., *Handbook of perception and human performance* (Vol. 2, Chapter 40). New York: Wiley.

Muir, B. M. (1987). Trust between humans and machines, and the design of decision aids. *International Journal of Man–Machine Studies*, 527–539.

Norman, D. A. (1988). *The Psychology of Everyday Things*. New York: Basic Books.

Norman, D. A. (1990). The problem with automation: Inappropriate feedback and interaction, not 'over-automation.' *Philosophical Transactions of the Royal Society of London, B327*, 585–593.

Norman, D. A. (1993). *Things That Make Us Smart: Defending Human Attributes In the Age of the Machine*. Reading, MA: Addison-Wesley.

Rasmussen, J. (1985). Trends in human reliability analysis. *Ergonomics, 28*(8), 1185–1196.

Sarter, N. B. (1996). Cockpit automation: From quantity to quality, from individual pilot to multiple agents. In R. Parasuraman and M. Mouloua, Eds., *Automation and human performance: Theory and applications*. Hillsdale, NJ: Erlbaum.

Sarter, N. B., and Woods, D. D. (1992). Pilot interaction with cockpit automation: Operational experiences with the flight management system. *International Journal of Aviation Psychology, 2*(4), 303–321.

Sarter, N. B., and Woods, D. D. (1994a). Pilot interaction with cockpit automation II: An experimental study of pilots' model and awareness of the flight management and guidance system. *International Journal of Aviation Psychology, 4*(1), 1–28.

Sarter, N. B., and Woods, D. D. (1994b). Autonomy, authority, and observability: The evolution of critical automation properties and their impact on man-machine coordination and cooperation. Paper presented at the 6th IFAC/IFIP/IFORS/IEA Symposium on Analysis, Design, and Evaluation of Man–Machine Systems. Cambridge, MA, June 1995.

Sarter, N. B., and Woods, D. D. (1995a). *"Strong, silent, and out-of-the-loop": Properites of advanced, cockpit) automation and their impact on human-machine coordination* (Tech. Rep. No. 95-TR-01). Cognitive Systems Engineering Laboratory, The Ohio State University.

Sarter, N. B., and Woods, D. D. (1995b). How in the world did we ever get into that mode? Mode error and awareness in supervisory control. *Human Factors, 37*(1), 5–19.

Spiro, R. J., Coulson, R. L., Feltovich, P. J., and Anderson, D. K. (1988). Cognitive flexibility theory: Advanced knowledge acquisition in ill-structured domains. In *Proceedings of the Tenth Annual Conference of the Cognitive Science Society*, August 17–19. Hillsdale, NJ: Erlbaum.

Wagenaar, W. A., and Keren, G. B. (1986). Does the expert know? The reliability of predictions and confidence ratings of experts. In E. Hollnagel, G. Mancini, and D. D. Woods, Eds., *Intelligent decision support in process environments* (pp. 87–103). New York: Springer.

Wiener, E. L. (1982). Human Factors of Advanced Technology, ("Glass Cockpit") Transport Aircraft (NASA Contractor Report No. 177528). Moffett Field, CA: NASA-Ames Research Center.

Wiener, E. L. (1993). Crew coordination and training in the advanced-technology cockpit. In E. L. Wiener, B. G. Kanki, and R. L. Helmreich, Eds., *Cockpit resource management* (pp. 199–223). San Diego: Academic Press.

Wiener, E. L., and Curry, R. E. (1980). Flight-deck automation: Promises and problems. *Ergonomics, 23*(10), 995–1011.

Winograd, T., and Flores, F. (1986). *Understanding Computers and Cognition*. Reading, MA: Addison-Wesley.

Woods, D. D. (1994). Cognitive demands and activities in dynamic fault management: Abduction and disturbance management. In N. Stanton, Ed., *Human factors of alarm design*. London: Taylor and Francis.

Woods, D. D. (1996). Decomposing automation: Apparent simplicity, real complexity. In R. Parasuraman and M. Mouloua, Eds., *Automation and human performance: Theory and applications*. Hillsdale, NJ: Erlbaum.

Woods, D. D., Johannesen, L., Cook, R. I., and Sarter, N. B. (1994). *Behind human error: Cognitive systems, computers, and hindsight* (State-of-the-Art Report). Dayton, OH: Crew Systems Ergonomic Information and Analysis Center.

CHAPTER 58
HUMAN FACTORS IN PROCESS CONTROL

Neville Moray
Université de Valenciennes
Laboratoire d'Automatique et Mecanique Industrielles et Humaines
Valenciennes, Cedex 59304, France

58.1 INTRODUCTION

58.1.1 What Is a "Process Control Industry"?

Process control plants provide the greatest challenge to the application of human factors to the design of human–machine systems. In this chapter, they are taken to include industries involving continuous or batch processing of materials and energy to produce a product by means of physical or chemical transformations. They are characterized usually by complexity, great size (both physical and conceptual), the presence of risk and high levels of hazard, high levels of operator workload, complex real-time dynamics, and a need to integrate the activities of many people at many levels of a plant, from management to maintenance workers. There can be very high levels of automation, and very high implicit or explicit costs in terms of performance and safety. In this chapter, we specifically exclude systems such as aircraft, ships, and terrestrial surface transportation, and also discrete manufacturing, such as assembly line work or computer-integrated manufacturing.[1]

As Woods and Hanes (1986) described the process industries,

> *The process industry involves systems where material and energy flows are made to interact and to transform each other. Examples include the generation of electricity in conventional fuel and nuclear power plants; the separation of petroleum by fractional distillation in refineries into gas, gasoline, oil, and residue; hot-strip rolling in steel production; chemical pulping in the production of paper; pasteurization of milk; and high pressure synthesis of ammonia. (Woods and Hanes (1986), p. 1726)*

Thus, process industries at their simplest include such operations as baking and pasteurization, whereas at the other extreme lie the most complex industrial systems ever built, including many where sometimes the physics and chemistry are only imperfectly understood, and hence where quite unforeseen events can occur under special conditions of abnormal operation, with the risk of potentially catastrophic releases of toxic materials or energy (Haynes and Bojcun, 1988; Hazarika, 1986).

58.1.2 Examples Of Process Control Systems

Many examples of process control systems are discussed in the classic reference by Edwards and Lees (1974), although all the papers in that collection date from a period before the massive introduction of automation that has occurred in recent decades. There are characteristics in common across both scale and type of plant throughout the process control industries, but there are also important differences that may have a significant impact on how to support human–machine integration. For that reason, this chapter adopts some different emphases from that in an earlier edition of this Handbook, where Woods and Hanes (1986) used nuclear power plants as an example of typical high-end process control. There was a good justification for that choice in that more attention has been paid to the role of the human operator and in particular the design of control rooms in that industry than in almost any other. (There are more than 200 publications relating to the human factors of the nuclear industry published by the U.S. Nuclear Regulatory Commission alone.) But, at the same time, the particular characteristics of nuclear power make it in some respects atypical.

Because of the peculiar hazards associated with high levels of radiation and the potential consequences of even small accidents, let alone catastrophic accidents such as that at Chernobyl (Haynes and Bojcun, 1988), all personnel in nuclear plants are more remote from the physical processes in which matter and energy are transformed than in most process control plants. In many process industries, for example steel making, at least certain parts of the physical processes are directly observable, and indeed audible and accessible to other senses. This is clearly very different from the case of nuclear power, where there is a series of barriers between the physical events of the process and the workers. This has important consequences for the way in which the process is controlled, especially in regard to the use of what De Keyser et al. (1987) calls "informal information" and for the importance of human–human communication among members of the team. Furthermore, the degree to which the role of the operator is constrained by standard operating procedures is extreme in the nuclear industry, matched only by some cases of

[1]Although such industries have some characteristics in common with process control, particularly real-time complex dynamics, there are sufficient differences to require that those industries be considered as special cases on their own, even though much of what is discussed in this chapter, and much of what was discussed by Woods and Hanes (1986), will be found applicable.

the chemical industry. In many process industries, as we shall see, much of the process is physically visible to operators and other members of the teams that run the plant, and indeed direct physical manipulation of equipment and even of product may be common even in automated plants, with the result that there are important differences in the human factors requirements. The chapter by Woods and Hanes (1986) should be regarded as complementary to the present chapter, and, along with the Edwards and Lees's (1974) review, should be regarded as basic reading for anyone interested in this area. Petrochemical refineries lie somewhere between nuclear power and steel making.

58.1.3 Challenges for Human Factors in Process Control

Woods and Hanes (1986) also noted that historically human factors have often been neglected in the design of process plants, and there are many well-documented cases (of which Three Mile Island is probably the most famous) in which the lack of integration between civil and mechanical engineering on the one hand, and human factors and ergonomics on the other, has directly led to costly and dangerous incidents. Woods and Hanes identified three challenges for human factors in modern process control plants. These, together with three additional challenges, are shown in Table 58.1.

58.2 CHARACTERISTICS OF PROCESS CONTROL SYSTEMS

Continuous process systems are complex and large—often involving hundreds of variables and many degrees of freedom. Because of their massive flows of energy and matter, and the dynamics of physical and chemical reactions, transfer functions are frequently very complex and include large time delays and lags, making it extremely difficult for workers[2] to identify the state of the plant.

A second characteristic is that plants are frequently physically very large, covering many hectares. The process is geographically widely distributed, with subsystems and components spread over great distances in three dimensions. This implies significant delays if a worker must be sent to inspect or to operate a subsystem in a remote part of the plant.

A third important characteristic is the increasing use of automation. Whereas this has many advantages, both for safety and efficiency (productivity), it also has the effect of making the physical processes increasingly remote from the workers in many process industries. It is not uncommon to hear workers complain that they are worried about losing contact with the real plant, and feel that they are really only in control of the interface. Often control room operators walk about the plant to remind themselves of its layout—where pumps are located, where particular pipes run, etc.

In some cases dynamics and time delays are extreme: Electricity generating plants and petrochemical plants, even those which are highly automated, may take many hours or even several days to start up. During that period the dynamics of the plant may be quite

Table 58.1 Human Factors Challenges in Process Control

- The design of control rooms that will support safe and economic operation.
- The shift from physical skills to cognitive skills as a result of increasing automation, with a subsequent need to support decision making in complex real-time situations rather than to support manual control.
- The design and evaluation of advanced (computer driven) human–machine interfaces (HMIs).
- The integration of behavior between members of teams who have very different responsibilities (such as operators, maintenance, "plant floor" operation, and even management).
- The identification of a good balance between prescribed activities and creative intervention.
- The management of the psychological relation between humans and the equipment they use—the role of trust and confidence in the tools provided.

[2]The term "worker" will be used as a generic term to include control room operators and plant floor workers such as fitters, maintenance workers, and foremen. Where the word "operator" is used it will have a more specific reference to someone performing a particular task, such as control room operator.

abnormal in comparison with those seen during full power operation. On the other hand, once running, plants may run for months or even years without a full shutdown.

It is characteristic of traditional plants that control rooms also tend to be large and complicated (see Figure 58.1) In traditional control rooms there may be hundreds or even thousands of displays and controls, spread over an area of many square meters and requiring operators to walk distances of more than ten meters from instruments representing one subsystem to those representing another. In part this is because of the "single-sensor-single-instrument" or "one-measurement-one-indicator" philosophy of control room design (Goodstein, 1981). Just as the components of the plant are geographically distributed widely in the environment, so are data widely distributed around the control room.

Woods and Hanes (1986) provide a guide to the identification of significant ergonomic and human factors problems in control room design, including examples of bad design and guides to improvements. Several general guides exist for the application of human factors and ergonomics to process industries (Essex Corporation, 1984; General Physics Corporation, 1985; Gilmore, Gertman, and Blackman, 1989; Ivergard, T., 1989; Mitchell, Stewart, Bocast, and Murphy, 1982) and there is no reason for bad design where traditional displays and controls are used unless designers do not consult these sources.

The most important problem here arises from the distinction between *data* and *information.* The single-sensor–single-display philosophy provides the operator with immense amounts of data. But what is required for effective plant operation is *appropriate information*, that is, data that have been transformed so as to be meaningful, not merely a list of values of individual variables. One of the improvements that can in principle result from the introduction of computers into process control is to provide a solution to this problem. No longer need a worker look at the values of several instruments and try to remember them and work out the relation between them. Such relationships can be calculated by the process computer and displayed directly.

58.2.1 The Tasks of Process Control

The most important impact of computers in the process industries, however, has been an enormous increase in automation during the last 20 years. The still frenetic pace of innovation in hardware and software at all levels of computer technology guarantees that

Figure 58.1 A typical control room. (From Kragt, 1992.)

there will be equally dramatic changes in the impact of computers, more generally microprocessors, in the coming decade. We can expect to see increasingly sophisticated sensors, actuators, communication systems, artificial intelligence, artificial pattern recognition, machine learning, etc., applied to process control. Is there still a role for human workers, and what is the need for human factors?

The most direct answer is that none of the advanced computer technology is mature. Indeed it is possible that the extremely rapid development of computer technology will mean that at no time in the future will that technology as applied to process control reach a steady state. The result of continuing and extremely rapid technical innovation will be an industry where the technology is always only partly evaluated, and where there is considerable uncertainty about many of the properties of advanced systems. As fast as one system is complete it will tend to be superseded by another using more powerful technology. It is precisely in the flexibility and creativity of human workers that protection against failures and inefficiencies of such technology is to be found. Rather than technology guaranteeing increased reliability compared with humans, only in an effective symbiosis of technology and human intelligence can technical solutions be satisfactorily implemented in a period of continuous innovation.

Table 58.2 provides a list found in one form or another in many major studies of process control tasks involving human workers (see, for example, Woods and Hanes, 1986)

Much of this list resembles the description of supervisory control functions given by Sheridan (see Chapter 39, this volume). This is not surprising, since supervisory control was developed in continuous process industries. The list of typical supervisory control activities (monitoring, intervention, learning, etc.) is effectively a superset of the activities listed in Table 58.2. It is important, however, to note that the latter contains a mixture of tasks that are typically ascribed to the control room operator, and others that implicate a far wider range of people. One of the biggest problems of research on the role of the human operator in process control has been that both basic research and applied research has tended to concentrate almost exclusively on the role of the control room operator. Worse than that, most of the research has been centered on the behavior or role of a single operator. Thus, one finds many papers on the theory of display design, but almost

Table 58.2 The Tasks of Process Controllers

- During normal operation the process must be monitored.
- Disturbances and their consequences must be detected.
- Any such disturbances must be counteracted.
- If abnormalities (faults) occur they must be detected.
- The causes of faults must be diagnosed.
- Appropriate countermeasures to control the effects of faults must be applied.
- Operating procedures must be consulted as needed.
- Databases of information about possible options may need to be consulted.
- Appropriate strategies must be adopted to support both safety and productivity.
- Changes may be made to the system either during normal or abnormal operations in the light of observations of the system state in order to prevent or compensate for drifts or faults.
- Such changes may be made manually or by changing the program of automated controllers.
- A record (log) must be kept of significant events.
- Significant events must be communicated to other members of the crew and, where necessary, to management, maintenance, etc., so that operations may be coordinated and required maintenance operations undertaken at appropriate times.
- From time to time special actions may be needed during the handover at the end of a shift, or during special conditions such as startup or shutdown.
- Introduce long-term changes and adjustments to the system so that it will tend to evolve toward a more efficient system.
- Undertake training and retraining to ensure the retention and improvement of skills.
- Perform emergency shutdown or other control actions to avoid dangerous accidents, or cooperate with automated systems for this purpose.

all of them assume that there is a single observer monitoring the process. The research problem has been seen as how to support a single operator who is faced with a very large amount of dynamic information so that he or she can assess the state of the system, make appropriate decisions, and choose and implement appropriate actions; but the reality of many process industries is quite different.

Process industries usually involve several operators, who may or may not be simultaneously present in the same room. In addition, foremen or supervisors, maintenance personnel, and fitters, "rondiers," or "plant floor" personnel may all be active at the same time. The plant floor personnel are members of the crew who move around the plant from place to place checking functions and manually intervene in the flow of energy and materials. Thus, if a skip full of materials fails to discharge its content into a blast furnace, typically a fitter will be sent to examine the situation visually, and perhaps to intervene directly on the spot. Or if the control room shows an alarm indicating that a valve has not closed, a fitter may be sent to inspect the valve at close quarters and perhaps to manipulate it by a hand wheel.

It is extremely important to take such factors into consideration. If process control is seen as the work of a single isolated operator, then clearly, as the systems become more and more complex, there will be an ever-increasing tendency to rely on massive application of computer technology and artificial intelligence to support the work of that operator, since the complexity of the process becomes overwhelming for a single human. If, however, the operation of a process control system is seen not as the work of an isolated individual, but as the combined operation of an integrated crew, the picture becomes quite different. We now have a case of distributed intelligence and distributed knowledge over humans as well as between humans and machines and over machines. The design of the system from a human factors viewpoint should be centered not on the centralization of information, display and control, but on the integration of the abilities of the entire crew. The quality of the communication systems that link the members of the crew to permit a flow of instructions and information among them is at least as important, perhaps moreso, than the provision of the latest high-technology control room.

It is instructive in this regard to compare the description by Woods and Hanes (1986) of a typical United States nuclear power control room, with the description by De Keyser (1981) and De Keyser et al. (1987) of a continuous casting steel plant in Belgium.

58.2.1.1 Nuclear Power Process Control

The typical nuclear power control room is quiet, well lit, and completely isolated from the physical processes that are being controlled. All information arrives in the room through the hardwired connections that drive the huge arrays of annunciators, alarms, strip chart recorders, meters, etc. Control is exercised by an equally massive array of switches, buttons, etc. In addition there are several telephones that are used to communicate with the plant, and one or more intercoms or telephone links using loudspeakers rather than hand sets. In U.S. plants there are usually three operators, of whom one is the senior shift supervisor and two are directly in charge of controlling the moment-to-moment evolution of the process of power generation, and a shift technical advisor who is highly qualified in nuclear engineering. In more modern plants with advanced computer control the level of manning may be even lower, and human intervention even less. In some very recent plants desks with computer displays are to be found, and information is displayed on the screens rather than on the walls. Communication does take place between the operators and the maintenance or plant floor personnel, but that communication is usually between locations remote from the control room, and it is quite possible for days to pass without the control room crew seeing plant floor or maintenance personnel.

58.2.1.2 Process Control in Steel Making

In contrast, the following is a description of continuous casting in the steel industry (De Keyser, 1981). Molten steel arrives in huge ladles holding many tons of metal. It is poured into a smaller vessel, which in turn distributes it in a continuous pour so that it falls vertically in a molten stream and is cooled by jets of water, which cause it to solidify as its temperature drops. It is caught between rollers that guide it into a horizontal position, where it is progressively rolled into the required thickness for the billets and is cut with oxygen torches into pieces of the required sizes, after which it is passed through further rolling mill machines. The whole process is extremely hot, noisy, and dangerous. If the strip of cooling metal ruptures, or if, at the downstream end of the process, the ribbon sticks or jams, the result can be a ribbon of red or white hot metal flailing around in the air up to a height of many meters.

The continuous casting is heavily automated; that is, the rate of pouring, the speed and pressure of the rollers, etc., are not under moment-to-moment manual control. On the other hand, from time to time some of the components must be realigned or even changed, so that the operator must intervene to stop and then restart the process. In addition, the operator must monitor the rate of arrival of the molten steel, take note of its chemical composition, and look for abnormal incidents such as the rupture of the ribbon being cast or blockages in the pipes through which the molten steel is poured. Maintenance is extremely important and is an almost continuous process. Because of the massive quantity of materials involved, the time taken to transport the ladles by overhead cranes is considerable, and a great deal of forward planning in real time is involved.

The plant is organized as follows. Along the length of the continuous casting machine there is a series of local control stations, each for a different task. One monitors the quality of the initial stage of the pour, one the oxygen cutting operation, one the final emergence of the product at the end of the machine, etc. There is a superordinate control room some 30 meters from the head of the continuous casting machine, and situated considerably above the floor of the plant so that the controller has a direct view of the entire plant through its windows. In addition the operator has two telephones that connect both within the plant and with the world outside, and an intercom with direct pushbutton call-up facilities for 32 different locations in the plant, including the control stations, maintenance, hydraulic systems, water supply, the "planning man," and the sectors responsible for the transportation of the ladles, chemical tests, etc. The control room also has a black and white alphanumeric computer screen with data input and output, which is particularly concerned with three kinds of data: mandatory records of certain events such as the cause of a pause in the process, records of steps taken to remedy situations if they are not recorded automatically by the system, and optional records such as comments on the progress of the casting. There is a second computer screen that provides alphanumeric information from different parts of the process such as chemical analysis of steel, planning of the next stages, times at which steel is expected to arrive, etc. A third computer screen is dedicated to alarms, including some measure of alarm filtering, and a primitive indication of the part of the plant in which the problem may have occurred, together with detailed information about plant variables of the alarmed process. There are two color screens with 15 pages of images representing plant state variables and including some forms of graphics in addition to alphanumeric information, and two more color screens with bar graph information primarily used for maintenance. There is a mimic diagram with binary luminous indicators, and five mimic diagrams each dedicated to a particular part of the casting process (level of molten steel supply, cooling process, etc.). Finally, there is a giant display on the plant floor, in a position such that both the control room operator and the plant floor personnel can see it, that displays major state variable values. It displays the tonnage of steel in the ladle, the tonnage in the tank that feeds the casting system, the temperature of the steel, the predicted remaining time for this casting, the speed of casting, and the length of the casting. As De Keyser summarizes the situation,

> *In summary, the information system available to the operator includes a direct view of the casting process, two telephones, an intercom, seven screens, six mimic diagrams, and a giant display. (De Keyser, 1981, p. 8)*

De Keyser also notes the importance of what she calls "informal information" such as the sounds of the machinery, the heat, the smell of the process, the color of the metal, vibration, etc. To experienced operators these are sources of extremely important information that may indeed allow a more accurate assessment of plant state on some occasions than the formally sensed and displayed information.

It is clear that taking the nuclear power plant operation as a paradigm of process control can be very misleading. In nuclear power the physical contact of the crew with the process is of course minimal. In addition, the extent of "informal" information is almost negligible, except perhaps in the generating hall. Furthermore, since nuclear power plants are generally used for base load generation, the process dynamics are much reduced during normal operation: the aim is to run the plant for many months between startup and shutdown, whereas in steel making there is a cycle of startup and shutdown on the scale of hours or even minutes rather than months. Furthermore, the distributed nature of control in the steel process means that situations frequently arise where a member of the crew other than the control room operator may have critical information at his or her work station which he or she can communicate to the operator before any changes appear directly on the screens or mimic diagrams of the latter, and such members of the crew may intervene directly in the control of the process. The steel industry is in some senses a less sophisticated technology than that of nuclear power, but it is a much better example of typical process control.

58.2.2 Manual Control versus Automation

In recent years there has been a considerable discussion of the merits of increased automation (Rasmussen, Pejtersen, and Goodstein, 1995; Wiener and Curry, 1980; Woods, Chapter 57 in this book; Woods and Roth, 1988). Two main reasons have been advanced for increasing automation: safety and productivity, with the latter often linked to the idea that with automation manning levels can be reduced. But it is now clear that although initially there was the possibility of a greatly reduced labor force resulting from the introduction of automation, in many industries we may have reached the point of diminishing returns. It is impossible to achieve a completely autonomous automated system even for most simple, let alone complex, systems. This is because such systems are designed by humans, and in a sense the properties of the designer remain in the system, which bears the marks of any weaknesses in design, and also because all but trivial systems require maintenance, and maintenance requires human intervention. In addition, as Bainbridge (1983) noted in an oft-cited paper, engineers automate those things they can well understand; the remaining tasks, left to human operators, are therefore even harder than before automation. Situations can arise that are "beyond design basis" and are thus unforeseen, and these situations require human intervention. Chapter 57 in this Handbook by Woods should be consulted for a discussion of these points.

What is actually required is a system that makes the best synthesis of human abilities and those of automation and that provides the best kind of work situation for the operating crew, not just for the control room operator. It makes no sense to argue the case of "manual control vs. automation." One must try to see how to integrate both methods of process control. Indeed it is misleading to put the question in the form of a balance between manual and automatic control. From a systems design viewpoint one should rather ask the question, "What are typical tasks which humans are required to perform in the operation of complex process control systems, and how can human skill be integrated with technically advanced components so that they may be best carried out?". Typical situations requiring human intervention are shown in Table 58.3.

58.2.2.1 Typical Situations for Manual Intervention

Table 58.3 Typical Situations for Manual Intervention

- Start-up and shutdown
- Intervention to control drift beyond the point where automation can cope
- Abnormal conditions:
  - Initiation of emergency operating procedures
  - Initiation of beyond design basis conditions
- Manipulation of plant to discover properties not described in procedures
- Manipulation of plant to make use of windows of opportunity not seen by algorithms
- Fault and catastrophe management

58.2.2.2 Implications for "Control Room" Design

Typically all of the tasks listed in Table 58.3 require a mixture of human intervention and automated control. In designing such systems we therefore need to consider what kind of information, assistance, communication and control should be available to the human crew. We put the phrase "control room" in quotes to emphasize what is apparent from De Keyser's description of continuous steel casting: we must consider the entire crew, not just control room operators.

58.3 THE PSYCHOLOGY OF THE PROCESS OPERATOR

All members of a process control crew are faced with common problems of information processing, although each will have certain special problems associated with the particular tasks for which he or she is responsible. To understand the background against which the design of the human–machine interfaces takes place one needs a general understanding of human information processing, decision making, cognition, and motivation. Accounts can be readily found at a general level in Wickens (1992), Grandjean (1980), and many other textbooks of ergonomics and human factors sources, as well as relevant chapters in this Handbook. More technical and specialized reviews of many relevant issues can be found in Boff, Kaufmann, and Thomas (1986). These sources that are specially slanted toward an understanding of the psychology of process control crews are Bainbridge (1978a, 1996), and Hoc (1996).

Figure 58.2 provides a broad outline of the information flow in human decision making. Any or all of the components could be examined in detail, but it is worth mentioning a few characteristics of the psychology of the process control worker that are particularly pertinent to the problem of designing human–machine interfaces and support systems.

58.3.1 Detection and Perception

The perception of information that reaches the senses of an operator can be modeled by a two-parameter quantitative model called the theory of signal detection. By signal we mean any pertinent event that is processed by the nervous system in such a way that the observer is conscious that some event has occurred. It need not be an external event—for example, the theory can be applied equally well to deciding whether a memory is accurate or to the detection in the change in the value of a display, or the flashing of an alarm.

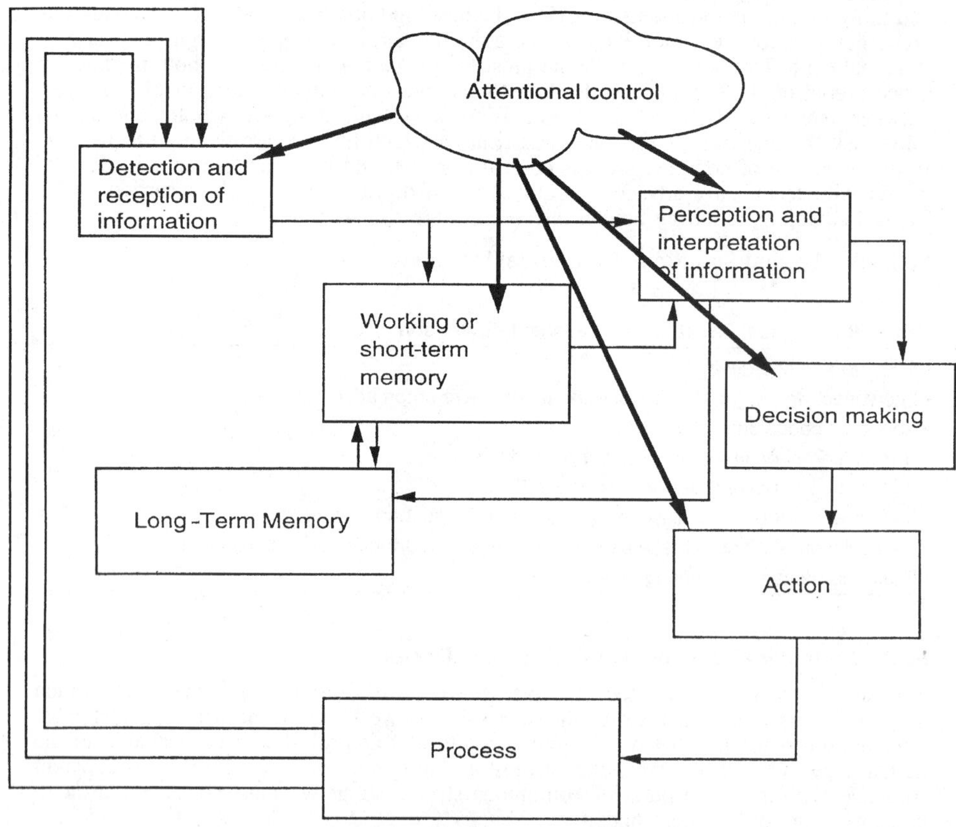

Figure 58.2 Outline of human information processing.

According to this theory, the probability that an operator will correctly detect the presence of a signal depends on two parameters.

The first is the signal-to-noise ratio of the incoming message. As with any physical detector, the ability to detect the presence of a signal is limited by the extent to which it is masked by noise in the communication channel. It follows that a designer must provide operators with displays (instruments, charts, graphics,) that have an adequate signal-to-noise ratio for the events which are assumed to be critical for decision making. That requires that the designer relate the properties of displays to the known facts of visual and auditory acuity, color vision, etc.

The second parameter is the so-called "decision criterion." If we think of any incoming information as being encoded into the nervous system at some strength or level of excitation, then independently of the strength of the signal, operators behave as if they select a certain level of excitation and if that level is exceeded will say "yes" to the question as to whether a signal has really occurred. The previous sentence describes a process that is not necessarily conscious, but that effectively models what occurs in human signal detection. Because the incoming signal is noisy, and because there is neural noise in the nervous system, the level of excitation can sometimes be exceeded even in the absence of a real signal. In that case the observer will believe that a signal has arrived and will make a "false alarm" or "false positive" response.

The level of the decision criterion is in its turn a function of two parameters. The first is the probability of signals. The more subjectively probable a signal or an event is, the lower the criterion and the more the observer is willing to believe that a signal or event has occurred. The second is the value structure of the situation. The more subjectively valuable the detection of a signal, and the more costly a miss, the more the observer is willing to indicate that a signal has occurred and the more willing he or she is to risk a false alarm; and the more costly a false alarm, the less willing the observer is to indicate that a signal has occurred. It follows that for accurate detection the observer must know the true values of probabilities of events and their relative worth. These effects can be described quantitatively, and provide a powerful model for the detection of information by observers in many kinds of tasks. For an excellent treatment, see Swets and Pickett (1982). There is a clear implication of the importance both of training and of the design of operating procedures, and the need for their integration with design.

58.3.2 Visual Attention

Behavior is also limited by certain properties of human attention. There has been much discussion as to the extent to which people can attend to more than one event at a time, particularly in relation to display design (Wickens, 1992). A conservative approach to the role of attention is to assume that in real industrial settings most tasks involve visual attention as the main mechanism by which an observer selects information. Dynamic visual attention is largely constrained by eye movements. Accurate perception of detailed information and full-color vision requires that observers direct their gaze directly to the source of information and that they fixate the display, that is, allow their gaze to rest motionless on the display. The peripheral visual field detects changes and movement and tends to trigger a reflex by which the eyes fixate the source of the change. But accurate perception requires a fixation. In real tasks, eye movements tend to be limited to about two per second, and never exceed four per second. The duration of fixation depends on the signal-to-noise ratio of the displayed information, and can be as long as several seconds for poor displays. Moreover, while the eyes are moving, no information is read into the nervous system.

Note that while this account emphasizes the role of attention in acquiring information from dynamic displays (Moray, 1986), attention is also critical in controlling internal information processing. Attention to the contents of short-term or "working" memory seems to be necessary for conscious manipulation of cognition. There are also problems of vigilance during prolonged supervisory control.

58.3.3 Limitations of Memory

When information is received by the nervous system, it is held in "working memory" or "short-term memory." It is this memory of whose contents an operator is conscious from moment to moment. It is a limited and volatile form of memory from which information is readily lost, particularly by interference from newer material that enters it, and perhaps simply by the passage of time. There is some disagreement about the amount of information (number of items) that can be held in working memory, but it is certainly low. It is extremely unwise to rely on working memory for the retention of information for even a few seconds, especially if the operator must perform calculations, speak to another person, or read in new information. Some form of hard copy assistance is absolutely mandatory if information has to be retained and manipulated in working memory.

The contents of "long-term memory" on the other hand, the way in which our permanent knowledge is stored, appears to be virtually unlimited in size and effectively nonvolatile. The problem here is retrieval. It is often very difficult for operators to retrieve required information from long-term memory and pass it into working memory so that they become conscious of it and can use it for decision making, problem solving, etc. There is considerable evidence that retrieval is handled by means of "default" values. Given a particular context or situation, information that is easily retrieved tends to be in the form of a "schema," a rough generalized memory that covers many similar situations but is lacking in refined detail pertinent to the particular occasion. There appear to be schemata both for recognizing situations and for choosing actions that are likely to be effective in the current context. Newly perceived information, working memory, thought, and decision making can operate to refine those decisions, both of recognition and the choice of action. In general the more frequently items are used, the more easily they are retrieved.

58.3.4 Mental Models

It is widely believed that workers in complex systems make use of *mental models* of those systems. Although the nature of these models is not entirely clear, the basic notion is that during long experience of working with a complex system the operators learn the dynamics of the system, its physical appearance and layout, causal relations among its components, etc. This information is embodied in the nervous system in the form of a runable model. The operators can use observations of the current state of the plant in order to run the mental model, and this allows them to predict future states and events, to estimate times of processes, to try to understand unexpected events, and to explore the properties of the process, as it were, "offline." Mental models are always less complex than the original system and hence lack the ability to represent all possible plant events and configurations. For accounts of mental models, see Moray (1996); Bainbridge (1981, 1991); Rasmussen (1986), Rasmussen et al. (1994); Goodstein, Andersen, and Olden (1988); Woods (this volume) and Woods and Hanes (1986).

Strong evidence for the existence of mental models comes from the changes that are observed during the acquisition of process control skills, whether in manual control or supervisory control. Since process control plants tend to have long time constants and low bandwidths, at least during normal operation, even when operators use manual control, actions are not continuously applied. (This is in contrast, for example, to the control of a vehicle such as an automobile, where steering corrections are continuously applied.)

It is clear that novice operators work in a regime of closed-loop negative feedback control. They apply a control action to the process, and then observe the value of the output. When it exceeds the target, or (if rate information is available) when it becomes evident that the target will be exceeded, then they apply a new input to correct the process. The performance of the system is marked by many control actions and a tendency to oscillate above and below the desired set point, especially if the order of control of the system transfer function is greater than 1. When a process is controlled by an expert operator, on the other hand, the approach to the desired set point is frequently very smooth, requires few interventions, and may, even in complex systems, show little or no overshoot or undershoot. The operator is now clearly working in an open-loop, predictive mode (Crossman and Cooke, 1974; Kragt and Landeweerd, 1974). This implies a mental model of the process dynamics and other characteristics and is relevant to the recent interest in planning behavior in process control (Bainbridge, 1978a; Hoc, 1995, 1996). Experts tend to revert to closed-loop negative feedback during fault management, especially if the fault is one that is not covered by procedures and that has not been previously experienced: That is, in situations where the model cannot handle the problem, behavior reverts to closed loop control.

Further evidence for mental models, which cannot be discussed here, comes from behavior seen during troubleshooting, and fault diagnosis and management. It should be emphasized that we do not understand clearly the conditions under which mental models generalize during radical plant accidents to situations that have not before been experienced by operators. In such cases the plant itself changes, and hence the mental model, acquired during many hours of operation on the normal plant, may no longer apply. In so far as the model reflects the complexity of the real plant, it is likely to be a high dimensional model with many degrees of freedom, and in order to run it effectively the "owner" needs good estimates of the state of the real plant, which again implies a need for good displays.

Some important consequences follow from the existence of mental models. One is that if two members of a crew have different models of a single process, they will effectively be thinking about it in different ways. In one sense they will be thinking about different processes even if they are working on the same plant. Information presented to them will be interpreted in different ways by each. This can in turn lead to problems of communication and interpretation, and it is the responsibility of those who design the control room and who formulate training and procedures to ensure that communication is effective despite this problem.

Two major schools of thought in regard to mental models can be identified. One, which arises largely from work based on the study of nuclear power in North America, particularly in regard to fault diagnosis, and perhaps is strongly influenced by attempts to model problem solving by the application of artificial intelligence and cognitive science, puts an emphasis on the factual knowledge content of mental models (sometimes called "declarative" knowledge). For this school of thought the content of the model is largely facts about the condition of the plant that can be deduced from currently displayed state information. An alternative school of thought, grounded more in European field studies of heavy industrial processes such as the steel industry and the chemical industry, puts more emphasis on what are called "functional" mental models, by which is meant that operators' mental models support not so much an understanding of events which have already occurred as a choice of what to do next. The second school is much more goal orientated than the first (see, for example, Bainbridge, 1978a; De Keyser, 1981; De Keyser, DeCortis, Housiaux, and Van Daele, 1987)

The difference between these schools of thought has important consequences for the design of effective human–machine process control systems. If the second approach is correct, then it implies a very great importance for active exploration of the system, and interaction among the operators and the crew. Improved knowledge, more effective mental models, and more efficient process control are derived from action, not from reflective analysis and understanding, but from acting on the system and seeing the result. It would follow that increases in automation or the installation of expert systems or other forms of artificial intelligence, which take away from the crew the possibility of directly acting on the processes, will necessarily reduce the efficiency of operators. Since much of the research into advanced automated and computer assisted systems aims to do precisely that, it raises very serious and fundamental questions of design philosophy. If, on the other hand, mental models are primarily concerned with understanding rather than the choice of action, the effects of current approaches to the design of decision and control aids would be expected to be particularly effective. The present state of our knowledge does not allow one to decide definitely between these possibilities. *The reader should be aware that it may be a serious mistake to base human–machine system design for process control plants on a paradigm of relatively inactive crews, such as those in nuclear power plants or the crews of long-haul civilian airliners where interaction among members of the crew is drastically reduced.* It is known that prolonged periods "out of the loop" make it difficult for operators to re-enter the loop and exercise manual control during emergencies.

58.4 SYSTEMS DESIGN FOR PROCESS CONTROL

58.4.1 Tools for Design

58.4.1.1 Task Analysis

Traditionally, when human factors have been applied to the design of process control systems early in the design cycle, ergonomists and engineers have relied on well-established approaches. The design cycle proceeds from the general specification of the system requirements to a task analysis, then to a division of responsibility between human and machine components, and then to the detailed specifications of system components followed by evaluation and a reiteration of this cycle. Several excellent guides to task analysis have appeared in recent years (Kirwan and Ainsworth, 1992; Wilson and Corlett, 1996; Chapter 15 in this Handbook). Previously the output of a task analysis has been used by ergonomists, together with ergonomics handbooks and guidelines such as those mentioned in the introduction to this chapter, to specify the selection and layout of controls and displays. But more recently, expert systems and other kinds of computer aids to systems design have begun to appear. Apart from the need for a more cost-effective approach to design, the new methods reflect the realization that there are particular problems associated with cognitive task analyses, where there is little or no overt behavior to analyze or predict, but where the emphasis is increasingly on mental activities such as decision making, choice, and strategic planning.

Examples of the new approaches include Millot and Debernard (1993); Alty and Bergan (1995); and Alty, Bergan, and Schepens (1995). The change of approach reflects some of the ideas of Rasmussen (1986) and those influenced by his approach to the role of humans in complex systems. In particular, it is now accepted that during supervisory control operators use several ways of thinking about a system depending upon the particular aspect of the task with which they are momentarily concerned. These levels can be thought of as levels of abstraction, and as part–whole decompositions. Thus an operator or a maintenance person might think of a particular control, or a particular physical component in the plant as a self-contained unit. At this level a complete description of the plant would contain a very large number of items. But at a more abstract level a set of pumps might be though of as a "cooling subsystem," thus simplifying the mental representation and reducing the mental workload. At a more abstract level still the operator might think of the goals or purpose of the subsystem or even of the system as a whole.

It seems likely that particularly in planning, and when discussions occur between different members of a crew who have different responsibilities, that the mental representations of the plant fluctuate considerably from moment to moment in regard both to the content and level of detail. Similarly, at any one level of detail, an operator may think of the components at a more or less fine granularity of description. Rasmussen (1986) has suggested that the level of abstraction and part whole decomposition can be thought of as a two dimensional representation of the work of the operator (Figure 58.3).

Current theory about human–machine interface design requires the designer to make it possible for workers to move easily about this space. The displays and controls should allow a flexible approach to mental operations and to the choice and implementation of action. This means either that very clever display configurations which make all levels simultaneously visible must be developed, or that a variety of different displays and controls must be available. In the latter case, they must be chosen in such a way that as workers change display and mode of thinking they do not lose contact with the underlying causal structures and functions of the system, nor do they experience difficulties in mutual communication with other crew members.

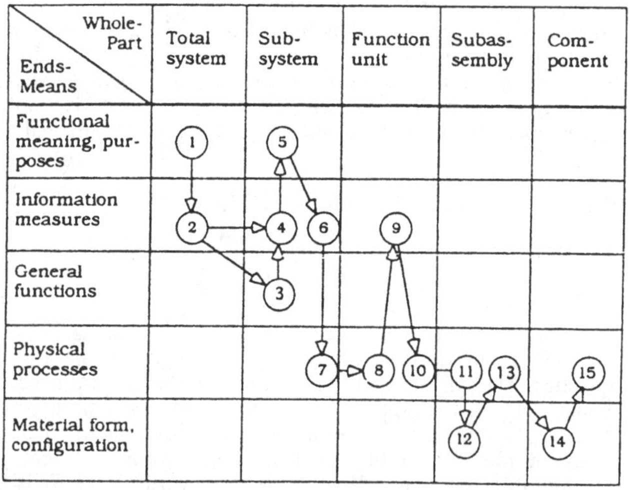

Figure 58.3 The means/ends abstraction hierarchy and part–whole decomposition. (From Rasmussen, et al., 1995.)

Although there has been considerable research on such topics in recent years, the technology is not yet mature; however, there are some very interesting new software tools available. Readers should be aware that they have not yet been fully evaluated, but that they certainly offer opportunities for new approaches to interface design. A particularly interesting approach is that of Alty et al. (1995), and Alty and Bergan (1995), which is the first expressly to provide design assistance for the use of multimedia. Their aim is to guide the designer in choices between visual, auditory, conventional, photographic, movie, synthesized speech, and other means of displaying information. The prototype system has been given initial field trials in the chemical industry, and the approach deserves further development. At present, like many of the approaches to be discussed in the later parts of this chapter, it should be regarded as exploratory, and not mature.

58.4.1.2 Computer Models of Human–Machine Interaction

The need to perform cognitive task analysis has also resulted in the development of a number of computational models of human cognition, which have been proposed for predicting the behavior of operators who are members of a crew of a process control system. Among the best examples are the approaches of Woods, Roth, and Pople (1987), Cacciabue, DeCortis, Drozdowicz, Masson, and Nordvik (1992); Yoshida, Yokobayasi, Tanabe, and Kawase, (1995); Sasou, Takano, and Yoshimura, (1995); and Hollnagel (1993). These computational models simulate the acquisition and use of information by human operators. The computer programs produce outputs that represent decisions and choices of actions by operators and, in most cases, can also produce a justification of the reasoning behind the decision.

In some respects the models tend to represent the computational models of logical reasoning that have been developed in cognitive science over the last 20 years (Anderson, 1983; Newell, 1990; Rosenbloom, Laird, and Newell, 1993). They differ, however, in containing very specific models of the process that the operators are controlling, in all cases a model, in more or less detail, of a nuclear power plant. Thus, the model of the operator receives the same information that the operator would receive in the control room, deduces the state of the plant in the face of some unexpected plant fault, and proposes or exercises control actions. In the case of the model developed by Sasou et al. (1995) more than one operator can be modeled, and the model output can include the predicted human–human communications.

This approach is potentially very powerful, not just for analyzing the cognitive processes involved, but for exploring what parts of reasoning are particularly difficult, and therefore what kinds of interface or decision aid support need to be developed. The methods are, however, extremely costly in terms of time and effort, and none of the models are widely available for use by other than their developers. Moreover, if they are to be applied to a particular system other than that for which they are currently implemented, the user must develop code that represents the properties of the system in great detail, whether it be a petrochemical plant, a steel mill, or whatever. Therefore, they are best regarded at present as a way of summarizing general ideas about cognition, and not as practical methods of system development. The one exception is Hollnagel's model, which seems to be more generally and practically applicable for reasons that we cannot go into here (Hollnagel, 1993).

58.4.2 Displays

There are currently dozens of handbooks, guidelines, etc., for the design and choice of conventional displays and controls, both conventional and computerized. The astonishingly poor quality of many control rooms, let alone plant floor displays and control consoles, is due only to the unwillingness of designers and engineers to make use of information that is widely available. Almost any textbook of ergonomics will contain the relevant information, and particularly thorough treatments can be found in Gilmore et al. (1989), Ivergard (1989), Essex Corporation (1984), General Physics Corporation (1985), and Mitchell et al. (1982). Research in this area is not required. What is required is application of existing knowledge.

There are, however, some general principles that are worth noting, and Woods and Hanes (1986) provide an excellent short review, with many illustrations, based on nuclear power plant control rooms. In this chapter, we shall confine the discussion to some of the latest developments, particularly those that have occurred as a result of the rapid introduction of powerful computers and computer graphics into human–machine interface design. A short taxonomy is provided in Table 58.4.

Table 58.4 A Taxonomy of Displays for Human–Machine Interfaces

- Classical displays: bar graphs, meters, graphs, strip charts, annunciators, indicator lights, icons
- Advanced displays: animated mimic displays, predictor displays, integrated displays, emergent features, urgency functions, phase plane displays
- Configural displays: star displays, contribution displays, object displays, configural displays
- "Ecological" displays: Rankine cycle, DURESS, time tunnels
- Flexible and adaptive displays

Section 4.2.1 offers several comments on the use of more or less traditional displays. Sections 4.2.2 through 4.2.5 are all examples of promising but immature technologies. Recent years have seen a great deal of work done on the development of new kinds of displays, using the newly available power of computer graphics and computation. However the current status of such displays is that whereas some of them have been implemented at least on an experimental scale, none of them have been adequately evaluated over long periods, and in many cases they have been implemented with little more than "proof of principle" behind them. It has become so easy to develop new and advanced displays using computer graphics, CAD systems, etc., that intuition has been largely substituted for serious development and evaluation. Considering how scrupulous engineering design is in every other aspect of systems design, it is extraordinary how superficial is the attitude to the evaluation of new human–machine interface design. As we shall see, there are at least one or two cases in which already advanced systems have caused accidents or near misses, and the reader is advised to be extremely cautious in applying new display technology, however intuitively obvious the expected improvement may seem to be, and however attractive the displays may seem.

58.4.2.1 Classical

Standard ergonomic guides provide all the information that is needed for the development of a human–machine interface using conventional displays and controls. The major problem is the size of these systems. As indicated earlier in this chapter, if a conventional approach using single-sensor–single-indicator philosophy is adopted, operators and other crew members will have great difficulty in monitoring the entire set of plant state variables. There is good evidence that workers choose subsets of variables to monitor. We have a good basic understanding of how these subsets are chosen. Operators use mental models based on past experience to choose a reasonable subset. Factors with long-term effects include, for example, the rate of change of the variables (their bandwidth) , their importance in terms of possible payoffs if signals are missed, the perceived reliability of the instruments, and the magnitude of correlation between them (see, for example, Bainbridge, 1978b; Crossman and Cooke, 1974; Iosif, 1968, 1969a, 1969b; Leermaker, 1995; Moray, 1986). Factors with a short-term influence include how close a variable is to a set point or boundary when it is sampled, the occurrence of alarms, and messages received from other members of the crew indicating that some action may need to be taken with respect to a recent or current event (Leermaker, 1995).

There are a number of important general characteristics to be born in mind, among which are problems of color coding. There are often conflicts between different conventions, such as red for a circuit carrying current, red for abnormal operation, and red for danger. It should be kept in mind that there is a substantial proportion of the population, especially males, who suffer from some degree of color blindness, and hence if color coding is used, some form of redundancy (shape, position) should be added unless crew members are subjected to tests for color blindness.

Perhaps the most important single factor (assuming that requirements for legibility, visual angle, etc. have all been met) is the question of *population stereotypes*. There are certain relations between changes in the form or value of displays and their meaning that seem to be very strongly built into the expectations of a population. Most ergonomics textbooks give examples of such stereotypes. For example, in general, a rising bar or pointer is expected to indicate an increase in the value of the displayed variable. But although many of these stereotypes are extremely strong for a particular culture, there is ample evidence that very few, if any, are universal across cultures. The clearest example is the fact that in North America a switch moves down for "off" and up for "on," whereas in much of the rest of the world the opposite is true, and in Japan switches move left and right, not up and down. It is particularly clear that many icons do not have the same meaning, or indeed any clear meaning, in different cultures. It follows that if a system is being designed for a work crew that is multicultural or for export to a different country, very great care should be taken to make sure that population stereotypes are not violated, since the effect on operator error can be more than an order of magnitude.

58.4.2.2 Advanced Displays

The most obvious class of advanced displays are those in which hardwired electromechanical instruments have been replaced by simulated copies generated by computer. For example in many aircraft cockpits there are now animated graphics that simulate classical moving pointer displays. The chief advantage is reliability; the chief disadvantage is the possibility of less legible displays of the conditions of lighting, and screen resolution that is inadequate. (Guidelines are available.) There are also new kinds of displays that do not resemble classical displays, and that make use of the power of computers to generate flexible dynamic displays.

A general problem of all computerized displays is what has been called the "keyhole" effect. Whereas in a conventional control room the main problem is the large size of the display surface, on computer screens the opposite is the case. Except for very new systems few computer screens are currently larger than some 30 × 20 cm.[3] Hence, to scan and monitor large numbers of displays the image on the screen must be changed, and this almost always takes longer than glancing round a room. Currently there exist computer generated displays that involve several hundred pages of images, and whether or not the access to displays or other information is hierarchically arranged, it is known that users find it difficult to navigate through the virtual information space and may become lost.

On the other hand advanced computer-generated displays offer great potential advantages. The ability to create animated mimic diagrams and to use animation to represent temporal relations and flows of mass and energy is a great advance. One of the potentially most powerful advances is the development of predictor displays, in which a model of the plant characteristics is used to show the current plant state and also the expected future values out to a period of perhaps many minutes. Although not applicable to all processes (many are not readily modellable at the level of detail required, or are subject to problems associated with low bandwidth), it is known that predictor displays can enable operators to monitor and control far more complex systems than they can do without such aids, and that even with controllable systems, the workload is much reduced.

A number of new displays have been proposed that appear to have considerable promise, but to date none have been fully developed or evaluated. One example is the phase plane display. In this display the value of a variable is plotted against its rate of change. The advantage of such a display is that the locus of the value of the variable immediately indicates whether there is a need for the operator to intervene, or whether the process is already correcting itself. (A large positive error with a negative rate of change indicates self-correction, for example.) This could reduce the amount of human intervention and hence the probability of injecting unnecessary disturbance into the dynamics. Other suggestions include the display of "urgency" indicating the rate of growth of error functions. For a discussion of some of these alternatives see Moray (1986).

More interesting are the proposals for "integrated" displays and displays with "emergent features." These are displays in which by plotting several variables on a single display, patterns of relations between the variables are made perceptually obvious. Since a major problem for operators in conventional control rooms and systems of high dimensionally is to discover causal relations among variables, and the mutual interaction of processes and state variable values, a display in which not just the relative values but their relationships are immediately apparent would be of great value. Some integrated displays have been implemented, particularly in aviation, and there is a large research literature in the human factors community. However, although several elegant examples have been demonstrated and have been found to have advantages in simulated process, there have been other occasions in which the results have not been as expected, and again

[3]It is interesting that often in the United States, system designers sometimes demand the smallest possible screens because they are cheaper, whereas in Europe, designers insist on the most advanced displays even when they are more costly. Cultural differences play an important role in systems design.

the technology is at present immature (see, for example, Bennett and Flach, 1992; Bennett, Toms, and Woods, 1993).

58.4.2.3 Configural Displays

The star diagram (Coekin, 1969; Sheridan, Chapter 39 in this Handbook) is the classical example of a configural display. Here again the notion is to allow the direct perception of relationships between variables, rather than requiring observers to calculate or deduce the relations and their meaning. An example of the star diagram is shown in Figure 58.4.

The value of each variable is represented as the length of the radius vector toward one point of the star. The values are scaled so that when the system is in a normal state the figure is completely symmetrical, and the circumference of the polygon is as near to a circle as it can be. If any value departs from the set point, then the polygon becomes distorted. The intention is that operators will become able to recognize the particular form of distortion as a pattern that is the signature of a plant state, without the need to measure the individual values. This figure has been implemented as part of the safety parameter display system in some nuclear power plants, and the evidence supports the claim that is very effective for the detection of abnormal states. So far, however, there is little evidence that it is particularly powerful for diagnosis. It seems that operators have considerable difficulty in learning to recognize the patterns. This may be because nuclear power plants are on the whole reliable; hence, the opportunity to observe a variety of distorted patterns is not sufficient for learning to occur. However, there may be other problems, since an operator told the writer that he did not like the display, because in a really large plant

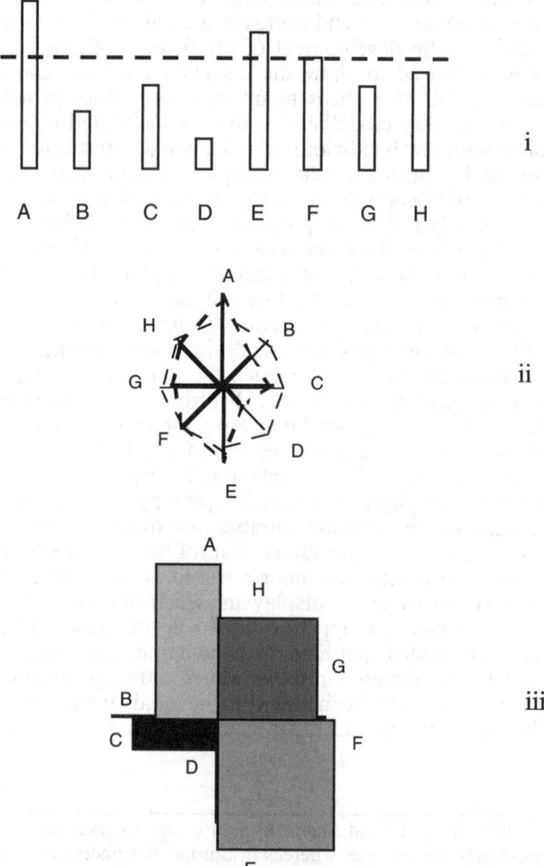

Figure 58.4 Examples of (i) classical, (ii) advanced, and (iii) configural displays.

there are always some variables that are not at their set point values. Hence the display was always more or less distorted, and before using it for detection or diagnosis it was always necessary to recalibrate it in the light of the current system state.

A number of other such displays have been proposed for aviation, for military decision making, and for medical care monitoring. This is a clear case in which there seems intuitively to be great potential which in practice seems unexpectedly difficult to realize. The technology is best regarded as promising but immature, and any displays proposed should be evaluated carefully. (See Bennett et al., 1993, for further discussion and references.)

58.4.2.4 Ecological Displays

The most recent theoretical development in display design for complex systems has been the appearance of so-called "ecological" displays (Moray et al., 1993; Rasmussen, Pejtersen and Goodstein, 1995; Rasmussen and Vicente, 1989). In a sense this is an extension of the idea of configural displays, although the theoretical basis is somewhat different (Flach, Hancock, Caird, and Vicente, 1995a,b). The aim is to develop displays that present the underlying invariants which describe the relations between the variables of the system, in particular the causal relations, mass–energy balances and flows, etc. Since it is known that people are error prone and slow at thinking but are very powerful pattern recognizers, the intent is to present information in a way that puts the burden on perception rather than on thinking. In addition, it is hoped that one can, in a single display, or at most in a very small set of displays, support cognitive activity at several or all levels of the abstraction hierarchy (Rasmussen, 1986). Several examples of ecological displays have been implemented on an experimental basis, including one to represent the Rankine thermal cycle for a nuclear power plant (Moray et al., 1993; Lindsay and Staffon, 1998; Vicente et al., 1996). An ecological display developed by Vicente to represent a simple thermal hydraulic system has undergone extensive evaluation and has not revealed any unexpected problems. It did not greatly improve operator performance, but seemed to give the operators better insight into the causal structure of the process. As of the date of completion of this chapter, no ecological displays have been fully tested in real industrial applications. Here again we find a display technology which seems to offer great potential, but which is at present immature. The design of ecological displays is at present an art requiring sophistication, skill, experience, and a deep knowledge of the domaine of application. An example of an ecological display is shown in Figure 58.5.

58.4.2.5 Flexibility and Adaptive Displays

The computational power now available at an extremely low price has given rise to much speculation about possible developments in computerized human–machine interfaces (HMIs). Among the most interesting suggestions are that HMIs should be made adaptable. It is known that there are very marked individual differences in the way in which people approach industrial tasks (see Figures 58.6 and 58.7). That being the case, it may be that people would benefit from being able to configure an interface to their own personal style, preference for graphical form, choice of default variables, etc. In addition it has been suggested that since skills change over time, and that particularly with practice workers may change the variables which they prefer to use and adopt different mental models, computers should be able to track the pattern of usage for a particular operator, and by using self-organizing programs the computers should adapt to the changes in the operators' work style.

Here again we find what superficially seem to be very attractive ideas, but for which there is no firm basis for thinking that the result will be safer and more productive process control systems. On the contrary, there are some empirical and theoretical reasons for thinking that such systems might be extremely dangerous. One study has already reported a serious human error as a result of self-configuration in a relatively simple system (Moray, 1992) involving only one operator. One can readily imagine that in a team or group situation the problems might well be worse. Where a crew shares one or more computer terminals, and where each person can reconfigure the displays to their personal preference, there is a clear possibility of people losing track of what each screen represents, if they are unaware that another member of the team has altered either the screen displayed, its content, or its layout.

A similar problem can be expected if interfaces are adaptive and self-organizing. If the user is unaware of the changes that the program has implemented, there is a great danger of being seriously misled by the displays and controls. If there is a mismatch

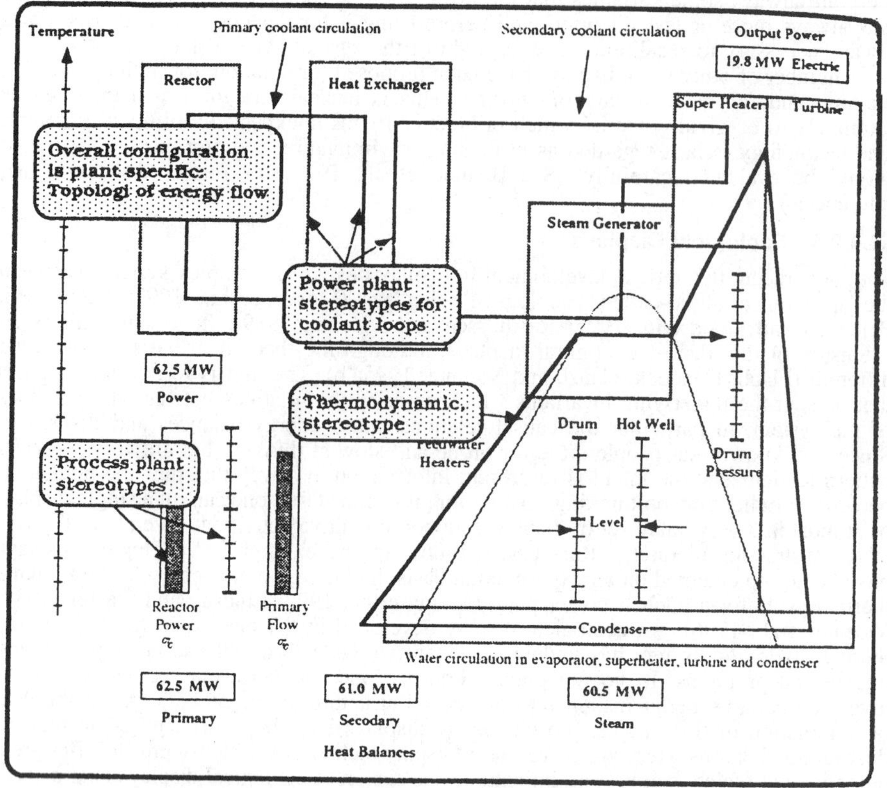

Figure 58.5 An example of an ecological display: Organization of a composite symbolic display for an industrial plant. (Adapted from Lindsay and Staffon, 1988, with permission.)

between the mental model currently used by the operator and the computer's model of that model, it is clear that it could become almost impossible for the human user to understand what is being displayed.

In the opinion of the writer, this is potentially an extremely dangerous technology, and one that should be avoided at present and for the foreseeable future. The potential for very severe errors and accidents is high, and the expected advantages of such systems are unproved.

58.4.2.6 Text Preferred

All the current thrust of HMI development is away from the use of alphanumerical displays and toward the use of dynamic graphical displays of one form or another. Given the undoubted success of graphical user interfaces (GUI) and WYSIWYG (what you see is what you get) in personal computers that is understandable, and indeed the overwhelming evidence is that good graphical interfaces are far more effective than screens full of undifferentiated text. In the interests of a complete picture, however, it must be noted that from time to time, when advanced interfaces are evaluated in field trials in industry, reports appear that operators claim to prefer text-based displays for certain purposes. As yet we do not know how to interpret these results. It may be that where extremely detailed quantitative information is needed, alphanumeric displays, when suitably color coded or otherwise enhanced, provide a level of precision that is not available in graphics. It may be that the effect is simply the well-known one that people tend to be happy with systems with which they are familiar. Both those possibilities are supported by reports of the verbal comments on the PROMISE system of Alty et al. (1995). Such reports should not be taken to indicate that scrolling text and undifferentiated line-oriented displays are in

Figure 58.6 Note the use of direct vision through window to floor of plant. (From Kragt, 1992, with permission.)

Figure 58.7 Note the use of telephone to supplement information displays on control panel. (From Kragt, 1992, with permission.)

general preferable. Well-designed graphical displays with a good use of color, symbology, movement, movies, etc. are undoubtedly extremely powerful. But the development of advanced displays is still an art, not a science. (See also Apple Computer, Inc., 1995; Bertin, 1977; and Tufte, 1983 for interesting discussions of how to encode information in graphics.)

58.4.2.7 Interface Design: Conclusion

It would extremely foolish to argue for a return to line-oriented text-based alphanumerical displays. The evidence is overwhelming that graphical dynamic interfaces that use the full power of advanced technology, including prediction, graphics, animation, configurality, multimedia, synthesized speech, etc., offer enormous potential improvements for the control of complex real-time continuous industrial processes. But there are severe problems about evaluating such systems, and the present state of the art is suggestive, but not mature. It behoves any designer to proceed with great caution. Extensive trials and evaluation are needed whenever new interfaces are implemented. Use should always be made of experienced members of operating crews both in the selection and development of symbology and display and control dynamics. As much care should be taken in the development of operating procedures and training schemes as in the development of the interface itself. Some instructive examples of the application of advanced human factors to systems design can be found in Kragt (1992).

Finally, we may note that in Section 58.4 we have slipped, as is almost always the case in discussions about the development of HMIs, into talking as if the sole problem is the design of HMIs for the control room. But a return to the description of the ecology of a steel-making plant by De Keyser, which was introduced in Section 58.1 should remind the reader that the work of *all* members of the team needs support. In most process control industries the responsibility for effective performance is distributed over many people at many different ranks and with many different roles. The choice of interfaces or the design of workstations must be appropriate in each case for the particular task analysis relevant to that person's job. Excessive concentration on the control room at the expense of other parts of the plant is undesirable and will lead to poor design.

58.4.3 Decision Aids

Just as automatic controllers and computer-controlled hardware are the automation of physical work, so in a sense are decision aids and expert systems the automation of mental work.

At the start of this chapter a contrast was drawn between the relatively steady-state control problem for the normal operation of nuclear power plants catering for baseload electrical supply, and the very dynamic and disrupted events in process control industries such as steel making. The latter are characterized by almost continuous, real-time decision making against deadlines. Since the early 1980s there has been a tremendous growth in work on computer-aided decision making, and many attempts have been made to support human decision making in the process industries.

It has been known for nearly 50 years that humans are inefficient decision makers, at least when measured against normative models (such as Bayes's theorem) in laboratory settings. As a result there have been a number of attempts to supplement the weaknesses in human decision-making strategies, for example, by using an algorithm to combine probabilistic evidence, a task at which people are known to be weak. Success has not been outstanding (Sheridan and Ferrell, 1974). However, in the 1980s a new approach appeared, the notion of the "expert system." By various techniques from interviews to computer-assisted multiattribute scaling human expertise is identified in the performance and verbal reports of experts. This information is encoded in the form of "if–then" rules and embodied in computer programs. When presented with data about the state of a system, the so-called expert system tries to identify the state of the plant, and thus assign it to one of the "if" sides of a rule. The recommended action is then delivered by the "then" side of the rule.

Rules can include a decision to seek more evidence, to ask for help, etc., and the best of the programs can provide evidence and reports about the way in which their reasoning proceeded. This latter ability has been discovered to be extremely important, since only if an expert (machine or human) can explain why they decided to do something do people have trust and confidence in it. In the absence of such trust, human–machine relations break down. Indeed, recent studies have shown that the psychological dimension of trust is of great importance in determining whether personnel will use automatic controllers or will try to exercise human control (Lee and Moray, 1992, 1994; Muir and Moray, 1996; Muir, 1994; Zuboff, 1988). Many expert systems have been developed to act as decision aids, but they have not been universally successful, and many have fallen into disuse. Woods (1986) has discussed possible reasons and has suggested that acceptable decision aids should be assistants rather than prostheses. They should provide expertise to the human, but should leave the final intelligent decision to the latter, rather than the expert system being the final decision arbiter.

There are several problems in the development of such decision aids. The first is that it is extremely difficult to be sure that all the relevant expertise has been extracted from existing human experts. The second is to be sure that such expertise is both sufficient and necessary for the task that must be performed. The third problem is that industrial systems are not static but evolve over time. After several years it may become impossible to replace failed components without changing the dynamics of the plant slightly. Management policy may change the rules of operation. Frequently, because of the great difficulty in writing expert systems and the fact that the latter are often written under contract and not by in-house experts, the expert system is not updated as the system evolves, so that it becomes progressively less expert until it fails in a mode that destroys users' confidence. For these and other reasons, and particularly when they contain many (perhaps hundreds) interacting rules, such systems are often "brittle" and fail suddenly and catastrophically.

Another problem is responsibility. If the human crew has to accept the advice of the decision aid, then there is an implicit management policy freeing them from the ultimate responsibility for the safety and efficient operation of the plant—the final responsibility lies with the algorithm, not the worker. If on the other hand the operator can override the advice, then what is the point of the supposed expert advice (Zuboff, 1988)?

The fundamental purpose of decision aids is to improve the quality of human decision making, and also to reduce the mental workload on the operators, since it is widely believed that excessive workload induces error. But as Weiner and Curry (1980) and Wiener (1988) have noted, often decision aids are an example of "clumsy automation." The only time at which they are of real use to the human is at moments of overload. But at such times there is usually great time pressure. Using an aid, activating it, entering requests, interrogating the result, asking for justifications, considering them, and finally choosing an action actually increases the workload and can lead to actions being critically delayed until the advice is useless.

A particularly interesting analysis identifies many of the problems of artificial decision aids. In addition to those already outlined, it identifies the following difficulties:

- Many (though not all) decision aids tend to provide a unique solution unless the operator provides new input, thus narrowing the range of actions considered by the human.
- The dynamics of the controlled process tend to become lost in the solution provided by the aid, since the latter does not continue to revise its suggestions in real time, unlike the human.
- The advice is insensitive to unforeseen contextual effects. These systems do not learn dynamically.
- If, as suggested by many field workers including De Keyser and Notte and van Damme, skill at process control operation depends critically on the opportunity to try out strategies of action, decision aids can remove this possibility by preventing humans from having the opportunity to try their own strategies.

It is also interesting to examine the way in which decision aids and expert systems compare with humans from the point of view of how decisions are made. Basic research into human thinking and decision making has almost always assumed that the model for thought is logical deduction, and indeed expert systems use logical calculus and probabilistic weightings to solve problems and arrive at solutions. For many years it was assumed that this too was the way in which humans solved problems. But recent work by Klein and others (Klein, Orasanu, Calderwood, and Zsambok, 1993) on "naturalistic decision making" in real settings outside the laboratory strongly suggests that in real working situations little thinking goes on. Instead, in a wide variety of situations, experts make what are called "recognition-based" decisions. They examine the state of affairs perceptually, recognize features in common with some occasion in the past in which they have been successful, and try out a similar strategy on the current occasion. They certainly do not evaluate all possible alternatives, and perform an expected utility maximization operation to arrive at a result. It is worth noting that a major proposal in Reason's theory of human error (Reason, 1990) is that mistakes occur because of an incorrect identification of an actor's situation caused by underspecification leading to false recognition. This theory is very close in spirit to Klein's notions of recognition-based decision making.

If Klein and the new work on recognition-based decision making is correct, then it increases the importance of good interface design, since the successful solution of problems relies crucially on a rich and accurate field of information and data on which the expert can draw to make a recognition. It also suggests why human experts are so important—there is a constant need for reassessing the situation from a recognition-based viewpoint. At present that is something that expert system cannot do, although it is possible that future systems based on the pattern recognition properties of simulated neural nets may provide a basis.

It is not the intention here to condemn computer-based decision aids outright. They have important roles to play in very large complex systems that are almost beyond the scope of humans to understand and control. But it is clear that there is a considerable way to go before their correct role, their correct instantiation, and their correct implementation in terms of their relationships to their users are well understood and realized. Bearing in mind the central role accorded to real-time planning and anticipation in the heavy process industries by Hoc (1995), DeKeyser (1981), and De Keyser et al. (1987), we might hope to see those aspects of process control supported by future decision aids, rather than the traditional concentration on aids to fault diagnosis and management.

58.4.4 Coordination

Several times in this chapter a contrast has been drawn between the world of nuclear power plants crews and those of heavy process industries with more real-time intervention. We now wish to make this contrast once more in order to draw attention to the problems of the coordination and organization of work crews in process control.

De Keyser et al. (1987) in their field studies of continuous casting steel production discovered the central role of communication between the member of the crews. This was revealed in a number of ways. For example, despite all the computerized information displays, the central control room operator made very frequent use of the telephone to obtain information from other members of the crew as his final (and apparently preferred) method of confirming the state of the process. Similarly, members of the crew on the plant floor, or at other geographically distributed locations around the plant could initiate action by sending messages to the control room operator which anticipated changes of state in the process, before they could manifest themselves in the "official" interface displays. Indeed De Keyser et al. saw the interaction of the different members of the crew as the absolutely central core of successful performance, particularly in planning and anticipation. The following is a translation from their report:

> *The importance of verbal communication*
> Just as in the first study, in the second study verbal communications played a an extremely important role.
>
> Communications via the intercom, the telephones and in the control room were part of an interactive dynamic relation with the other sources of information. They served to *verify* information appearing through other channels. (For example, if a temperature appeared on the giant display and the operator was suspicious about its value, he asked the man in charge of the casting ladle to confirm it. If a chemical analysis appeared on the screen, and one component seemed too high to the operator, he called the metallurgical personnel to take another sample to confirm the value of the component.). . . .
>
> Above all, recourse to verbal communication is very frequent because operators give such communication a great deal of *credibility.*
>
> They also use it to check on unexpected alarms. (At the beginning of a break an alarm went off. The operator immediately called the maintenance personnel to see if they were doing some work during the break.)
>
> Furthermore, in addition to their functional value, verbal communication serves a *social function.* . . . "we're all part of a single family . . . we have to help one another." (pp. 23–25)

It is interesting to note that in an analysis of an exercise at an emergency response center Moray, Sanderson, and Vicente (1992) noted that in addition to using a formidable array of computer displays the emergency response analysts responsible for diagnosing a nuclear incident kept the telephone line open to the plant for several hours, using it from time to time to demand interpretations of data, a pattern of behavior very similar to that described by De Keyser et al. (1987). Given such reports it is clear that a full understanding of the human factors of process control must go far beyond the traditional concern with control room layout and the specification of HMIs. To understand the human role in process control and to design an effective process control plant involving humans and automation we have to consider social psychology and the psychodynamics of working relations.

Rochlin, LaPorte and Roberts (1987) described interesting patterns of behavior in several situations in which social and organizational factors were crucial in the efficient working patterns in hazardous and demanding systems. For example, in air traffic control rooms they observed that when an operator was in danger of becoming overloaded there seemed to a be a spontaneous tendency for other members of the crew who had some spare time spontaneously to drift toward the highly stressed worker, as if to be ready to help if required. If all went well they drifted away again. There appeared to be an almost unconscious self-organizing tendency to promote a working relation of mutual support, again like that described by De Keyser et al. (1987). Obviously these kinds of mutually supporting interventions can only occur if the architectural layout of a plant allows them, and furthermore, if management and union policy also permits such unofficial changes in work patterns.

Efficient communication is critical. We saw earlier that it was possible to think about a complex system at several levels of abstraction and at various levels of detail. Typically the language used to describe the system at each of these levels is different, and may not translate easily from one level to another. Insofar as the automation of systems leads to physical isolation of crew members one from another, to the loss of "informal" information (sounds, smells, etc.) and to a net loss of combined distributed expertise, the ability to perform in both normal and abnormal situations may well be decreased in comparison with earlier plants. It is clear that there is some kind of trade-off involved, in which the reliability of the automated system is on one side of the balance, and the accumulated wisdom of the human crews is on the other. Defining the optimal combination is not easy.

58.5 PHILOSOPHIES OF PROCESS CONTROL

In this final section, we turn to some general issues of the philosophy of process control. What is the overall role of human creativity in such systems?

Recent years have seen a continuing discussion among those concerned with high-technology human factors about how best to use humans in automated systems. For interesting discussions see Rasmussen et al., 1995), Woods (in this *Handbook*), and many other sources. On the one hand, there is the argument that human creativity can provide an extraordinary flexibility and adaptability in the control of large complex systems. Those who argue for using those qualities argue for operating procedures that are minimally prescriptive, so that the human–machine system can be self-organizing and adaptive in the face of future changes and unforeseen events. On the other hand, there are those who argue that many of the process industries, particularly nuclear power and the various chemical industries, are so inherently hazardous[4] that the behavior of crews must be very strictly controlled.

It is clear that we can expect the outcome of such a discussion to be highly context dependent. In a pasteurization plant both risk and hazard are low. In general, in well-engineered and profitable plants, risk is low even if hazard is high. One might well accept the notion of a highly flexible way of working in low-risk plants, especially low-risk, low-hazard plants. There will be a tendency to move toward prescriptive operating procedures as the perceived risk rises, especially if the potential hazard is high. This will

[4]In line with common usage in the reliability community, "risk" is the probability of an event, and "hazard" is the cost associated with its outcome. Thus, most well-engineered process control plants can be expected to have a low risk of accident associated with their operation, although the potential hazard, should a major accident occur, may be very great.

probably be the case even if, in fact, flexible working procedures would be safer and profitable, because even one high-hazard outcome would be so costly, legally and financially, that it is unacceptable. Much research is needed in this area, since there is some evidence (for example, Rochlin and LaPorte, 1987) to suggest that more flexible systems that can show self-organization are both safer and more efficient.

Flexibility in organizations is a matter of organizational and managerial policy, as well, often as a response to regulation by government agencies and social pressure. Bainbridge (1978b) reported a case where members of a process control crew were actively prevented from carrying out tasks for which they were well qualified because of the managerial organizational policy, with a resulting loss in efficiency. Once again the notion of the human factors of process control needs to be expanded well beyond its traditional boundaries of human–machine interfaces (see also Zuboff, 1988).

Rasmussen et al. (1995) offer an intriguing model of the inherent tendency to change in complex systems. They believe that all operators tend to explore the state space of the systems that they operate. They tend to move away from areas of inefficient work (in terms of the economics of the process) and away from high workload. They tend to look for new and better ways of doing things, and behave rather as if they were particles in Brownian motion performing a random walk through the state space of the plant. From time to time they cross boundaries, which puts the plant (or themselves, or society) into slightly unsafe regions of operation. Providing that the design of the plant, the nature of the interface, and the networks of mutual communication support a withdrawal from that region, the "wanderings" are a way of finding potential improvements in the operation of the system.

Such a concept is related to theories of the origin of human error in the operation of process control plants. Although it is not possible in this chapter to consider either error or fault management in any detail, one topic deserves at least a brief discussion, since it is closely related to the notion of human error, risk, and hazard.

Probabilistic risk assessment (PRA) is frequently used during the design of a plant to estimate the probability of certain undesirable outcomes. When attempting to include human reliability as a component in PRA, the estimates become so hedged round with uncertainty as a result of performance-shaping factors that the numbers become essentially meaningless as quantitative estimates. However, they can be used in different ways, namely to flag areas of operation where there seem to opportunities for the development of dangerous practices. In particular, Reason's notion of "latent pathogens" is of great importance in the safe operation of complex plants (Reason, 1990).

Reason argues that whereas occasionally there are what appear to be single, catastrophic failures such as at Bhopal or Chernobyl, these are rare, and even when they occur they are typically the result of a pattern of a chronic drift, over a long period, toward an accident. Given the high reliability of components in most modern hardware and software, degradation in process control plants is typically due to progressive, small failures or subnormal function that either goes unnoticed, or, given certain kinds of managerial attitudes, is ignored. The result is that progressively more and more defenses against failure are removed, although the redundancy in the plant is such that it still continues to function, until the day when a final incident causes a complete collapse of the system, and none of the defenses remain. Completely improbable combinations of events can in fact occur. There is, for example, one case in which simultaneously no fewer than 13 safety mechanisms failed at the same time at a nuclear power plant, fortunately with no catastrophic results. But on any estimation of probability, the predicted probability of such a combination would have been regarded as zero. For a more recent approach to human reliability, see Hollnagel (1993).

The final topic of human factors in process control therefore takes us again into a region where there is a strong interaction between the traditional concerns of human factors and ergonomics (such as the design of human–machine interfaces, maintenance training operations, procedures, etc.) and their interaction with social and organizational psychology, with its concern for motivation of the work force, attitudes to communication between work force and management, attitudes toward "whistle blowing," on the one hand, and creative innovation of work practices, on the other.[5] The human factors of

[5]For a brilliant account of what happens when organizational factors are ignored in process control, see the novel *The Unknown Industrial Prisoner*, by John Ireland, published by McClelland and Ward, Melbourne.

process control are not about human–machine system design. They are about a small world of mutual support for work in the most complex systems that have ever been built by humans.

58.6 EVALUATION: THE FINAL WORD

This is an era of great innovation in the design of process control plants. New technology, new control systems, artificial intelligence, new roles for operators, and the need for an increasingly well-educated work force go together. In some countries the attitude is to replace humans completely by automated systems. In others it is to ensure that the role of the human is taken into account in order to maximize the acceptability of the work situation, and at the same time the profitability and safety of operations.

One point has been several times but cannot be too strongly emphasized. All the innovation in hardware and software design, in new interfaces, in computer generated imagery, in improved communication, etc., is now occurring at such a rate that standard research methods lag far behind the systems development and innovation. That being the case, *no* innovative solution to the human machine systems of process control should be implemented without a carefully conceived and performed evaluation. Evaluation is, in a time of rapid innovation, the key to safety and efficiency.

REFERENCES

Alty, J. L., and Bergan, M. (1995). Multimedia interfaces for process control: Matching media to tasks. *Control Engineering Practice*, *3*(*2*), 241–248.

Alty, J. L., Bergan, M., and Schepens, A. (1995). The design of the PROMISE multimedia system and its use in a chemcial plant. In R. Earnshaw, Ed. *Multimedia Systems and Applications. London:* Academic Press.

Anderson, J. R. (1983). The architecture of cognition. Cambridge, MA: Harvard University Press.

Apple Computer, Inc. (1995). *Electronic guide to Macintosh human interface design.* Cupertino, CA: Apple Computer, Inc.

Bainbridge, L. (1974). Analysis of protocols from a process control task. In E. Edwards and F. Lees, Eds., *The Human Operator in Process Control* (pp. 146–158). London: Taylor and Francis.

Bainbridge, L. (1978a). The process controller. In W. T. Singleton, Ed., *The Study of Real Skills.* London: Academic Press.

Bainbridge, L. (1978b). Forgotten alternatives in skill and workload. *Ergonomics*, *21*(*3*), 169–185.

Bainbridge, L. (1981). Mathematical equations or processing routines? In J. Rasmussen and W. B. Rouse, Eds., *Human Detection and Diagnosis of System Failures* (pp. 259–286). New York: Plenum.

Bainbridge, L. (1983). Ironies of automation. *Automatica*, *19*, 755–779.

Bainbridge, L. (1991). Mental models in cognitive skill. In A. Rutherford and Y. Rogers, Eds., *Models in the Mind.* New York: Academic Press.

Bainbridge, L. (1996). Processes underlying human performance. In D. J. Garland, J. A. Wise, and V. D. Hopkin, Eds., *Aviation Human Factors.* Hillsdale: Erlbaum. In press.

Bennett, K. B., and Flach, J. M. (1992). Graphical displays: Implications for divided attention, focussed attention and problem solving. *Human Factors*, *34*(*5*), 513–533.

Bennett, K. B., Toms, M. L., and Woods, D. D. (1993). Emergent features and graphical elements: Designing more effective configural displays. *Human Factors*, *35*(*1*), 71–97.

Bertin, J. (1977). *La Graphique et le Traitment Graphique de L'information.* Paris: Flammarion.

Boff, K. R., Kaufman, L., and Thomas, J. P., Eds. (1986). *Handbook of Perception and Human Performance.* New York: Wiley.

Cacciabue, P. C., DeCortis, F., Drozdowicz, B., Masson, M., and Nordvik, J-P. (1992). COSIMO: A cognitive simulation model of human decision making and behavior in accident management of complex plants. *IEEE Transactions on Systems, Man, and Cybernetics*, *SMC-22*(*5*), 1058–1074.

Christofferson, K., Hunter, C. N., and Vicente, K. J. (1996). A longitudinal study of the effects of ecological interface design on skill acquisition. *Human Factors*, *38*(*3*), 523–541.

Coekin, J.A. (1969). A versatile presentation of parameters for rapid recognition of total state. In *International Symposium on Man-Machine Systems* (58-MMS 4). (Conference Record 69). New York: IEEE.

Crossman, E. R., and Cooke, F. W. (1974). Manual control of slow response systems. In E. Edwards and F. Lees, Eds., *The Human Operator in Process Control.* London: Taylor and Francis.

De Keyser, V. (1981). *La fiabilité humaine dans les processus continus, les centrales thermo-éléctriques et nucléaires* (Tech. Rep. 720-ECI-2651-C-(0) GCE -DGXII). Bruxelles: CERI.

De Keyser, V., De Cortis, F., Housiaux, A., and Van Daele, A. (1987). *Les communications hommes-machines dans les systèmes complexes* (Appendice, Tech. Rep. Contract No. 8, Actions Nationales de Recherche en Soiutien a Fast). Belgium: Université de Liège.

Edwards, E., and Lees, F. (1974). *The Human Operator in Process Control.* London: Taylor and Francis.

Essex Corporation. (1984). *Human Factors Guide for Nuclear Power Plant Control Room Development.* Palto Alto, CA: Electric Power Research Institute.

Flach, J., Hancock, P., Caird, J., and Vicente, K. (1995a). *Global Perspectives on the Ecology of Human Machine Systems.* Hillsdale, NJ: Erlbaum.

Flach, J., Hancock, P., Caird, J., and Vicente, K. (1995b). *Local Perspectives on the Ecology of Human Machine Systems.* Hillsdale, NJ: Erlbaum.

General Physics Corporation. (1985). *Human Engineering Guidelines for Maintainability.* Palto Alto, CA: Electric Power Research Institute.

Gilmore, W. E., Gertman, D. I., and Blackman, H. S. (1989). *User-Computer Interface in Process Control: A Human Factors Engineering Handbook.* New York: Academic Press.

Goodstein, L. (1981). Discriminative display support for process operators. In J. Rasmussen and W. B. Rouse, Eds., *Human Detection and Diagnosis of System Failures.* New York: Plenum Press.

Goodstein, L. P., Andersen, H. B., and Olsen, S. E., Eds. (1988). *Tasks, Errors and Mental Models.* London: Taylor and Francis.

Grandjean, E. (1980). *Fitting the Task to the Man.* London: Taylor and Francis.

Haynes, V., and Bojcun, M. (1988). *The Chernobyl disaster.* London: Hogarth Press.

Hazarika, S. (1986). *Bhopal: The Lessons of a Tragedy.* London: Penguin Books.

Hoc, J-M. (1995). Planning in diagnosing a slow process. *Zeitschrift für Psychologie, 203,* 111–115.

Hoc, J-M. (1996). *Supervision et controle de processus: La cognition en situation dynamique.* Grenoble: Presses Universitaires de Grenoble.

Hollnagel, E. (1993). *Human Reliability Analysis: Context and Control.* London: Academic Press.

Iosif, G. (1968). La stratégie dans la surveillance des tableaux de commande. I. Quelques facteurs déterminants de caractère objectif. *Revue Roumanien de Science Social-Psychologique, 12,* 147–161.

Iosif, G. (1969a). La stratégie dans la surveillance des tableaux de commande. I. Quelques facteurs déterminants de caractère subjectif. *Revue Roumanien de Science Social-Psychologique, 13,* 29–41.

Iosif, G. (1969b). Influence de la correlation fonctionelle sur parametres technologiques. *Revue Roumanien de Science Social-Psychologique, 13,* 105–110.

Ivergard, T. (1989). *Handbook of Control Room Design and Ergonomics.* London: Taylor and Francis.

Kirwan, B., and Ainsworth, L. K. (1992). *A Guide To Task Analysis.* London: Taylor and Francis.

Klein, G. A., Orasanu, J., Calderwood, R., and Zsambok, C. E. (1993). *Decision Making in Action: Models and Methods.* Norwood, NJ: Ablex.

Kragt, H. (1992). *Enhancing Industrial Performance: Experiences of Integrating the Human Factor.* London: Taylor and Francis.

Kragt, H., and Landeweerd, J. A. (1974). Mental skills in process control. In E. Edwards and F. Lee, Eds., *The Human Operator in Process Control* (pp. 135–145). London: Taylor and Francis.

Lee, J. D., and Moray, N. (1992). Trust, control strategies and allocation of function in human-machine systems. *Ergonomics, 35,* 1243–1270.

Lee, J. D., and Moray, N. (1994). Trust, self confidence and operators' adaptation to automation. *International Journal of Human-Computer Studies, 40,* 153–184.

Lindsay, R. W., and Staffon, J. D. (1988). A model based display system for the experimental breeder reactor—II. In *Proceedings of Joint Meeting of the American Nuclear Society and the European Nuclear Society,* Washington, DC.

Millot, P., and Debernard, S. (1993). Man-machine cooperative organisation: Methodological and practical attempts in air traffic control. *Proceedings of the IEEE SMC Annual Conference.* Le Touquet, France, October, pp. 695–701.

Mitchell, C. M., Stewart, L. J., Bocast, A. K., and Murphy, E. D. (1982). *Human factors aspects of control room design: Guidelines and annotated bibliography.* (NASA technical Memorandum 84942, Goddard Space Flight Center). Greenbelt, MD: National Aeronautics and Space Administration.

Moray, N. (1986). Monitoring behavior and supervisory control. In K. R. Boff, L. Kaufman, and J. P. Thomas, Eds., *Handbook of Perception and Human Performance* (Chapter 45). New York: John Wiley.

Moray, N. (1992). Flexible interfaces can promote operator error. In H. Kragt, Ed., *Enhancing Industrial Performance* (pp. 49–64). London: Taylor and Francis.

Moray, N., Sanderson, P. M., and Vicente, K. J. (1992). Cognitive task analysis of a complex work domain: A case study. *Reliability Engineering and Systems Safety, 36,* 207–216.

Moray, N. P., Jones, B., Rasmussen, J., Lee, J., Vicente, K., Brock, R., and Djemil, T. (1993). *Development of a performance indicator for the effectiveness of human-machine interfaces for nuclear power plants* (UILU-ENG-92-4007). (NUREG/CR-5977). Washington, DC: U.S. Nuclear Regulatory Commission.

Muir, B. M. (1994). Trust in automation: Part 1—Theoretical issues in the study of trust and human intervention in automated systems. *Ergonomics, 37(11)*, 1905–1923.

Muir, B. M., and Moray, N. (1996). Trust in Automation. Part II. Experimental studies of trust and human intervention in a process control simulation. *Ergonomics, 39(3)*. 429–461.

Newell, A. (1990). *Unified Theories of Cognition.* Cambridge, MA: Harvard University Press.

Rasmussen, J. (1986). *Information Processing and Human-Machine Interaction: An Approach to Cognitive Engineering.* Amsterdam: North-Holland.

Rasmussen, J., and Vicente, K. J. (1989). Coping with human errors through system design: Implications for ecological interface design. *International Journal of Man-Machine Studies, 31*, 517–534.

Rasmussen, J., Pejtersen, A-M., and Goodstein, L. (1995). *Cognitive Engineering: Concepts and Applications.* New York: John Wiley.

Reason, J. (1990). *Human Error.* Cambridge: Cambridge University Press.

Rochlin, E., LaPorte, T., and Roberts, K. (1987). The self-designing high reliability organisation: Aircraft flight operation at sea. *Naval War College Review*, Autumn, 76–91.

Rosenbloom, P. S., Laird, J. E., and Newell, A. (1993). The *SOAR papers: Research on integrated intelligence.* Cambridgem MA: MIT Press.

Sasou, K., Takano, K., and Yoshimura, S. (1995). Development of SYBORG (Simulation System for behavior of an Operating Group). In *Proceedings of 5th European Conference on Cognitive Science Approaches to Process Control (pp. 368–376).* Espoo: Technical Research Centre of Finland.

Sheridan, T. B., and Ferrell, W. R. (1974). *Man-Machine Systems.* Cambridge, MA: MIT Press.

Swets, J., and Pickett, R. (1982). *Evaluation of Diagnostic Systems.* New York: Academic Press.

Tufte, E. R. (1983). *The Visual Display of Quantitative Information.* Cheshire, CT: Graphics Press.

Tufte, E. R.

Vicente, K., Moray, N., Lee, J. D., Rasmussen, J., Jones, B., Brock, R., and Djemil. T. J. (1996). Evaluation of a Rankine Cycle display for nuclear power plant monitoring and diagnosis. *Human Factors*, *38(3)*, 506–522.

Wickens, C. D. (1992). Engineering Psychology and Human Performance (Second Edition). New York: Harper-Collins.

Wiener, E. L., and Curry, R. E. (1980). Flight-deck automation: Promises and problems. *Ergonomics*, *23*, 995–1012.

Wiener, E. L. (1988). Cockpit automation. In E. L. Wiener and D. C. Nagel, Eds., *Human Factors in Aviation.* San Diego, CA: Academic Press.

Wilson, J. R., and Corlett, E. N. (1996). *Evaluation of Human Work,* (Second Edition). London: Taylor and Francis.

Woods, D. D. (1986). Paradigms for intelligent decision support. In E. Hollnagel, G. Mancini, and D. D. Woods, Eds., *Intelligent Decision Support in Process Environments.* Heidelberg: Springer-Verlag.

Woods, D. D., and Hanes, L. (1986). Human factors challenges in process control: The case of nuclear power plants. In G. Salvendy, Ed., *Handbook of Human Factors.* New York: John Wiley.

Woods, D. D., Roth, E. M., and Pople, H. (1987). *An artificial intelligence based cognitive environment simulation (CES) for human performance assessment (Tech. Rep. NUREG/CR-4862). Washington, DC: U.S. Nuclear Regulatory Commission.*

Woods, D. D., and Roth, E. M. (1988). Cognitive systems engineering. In M. Helander, Ed., *Handbook of Human-Computer Interaction.* Amsterdam: North-Holland Elsevier.

Woods, D. D. (1988). Coping with complexity: The psychology of human behavior in complex systems. In L. P. Goodstein, H. B. Andersen, and S. E. Olsen, Eds., *Tasks, Errors and Mental Models.* New York: Taylor and Francis.

Woods, D. D., Roth, E. M., and Pople, H. (1987). An artificial intelligence based cognitive environment simulation (CES) for human performance assessment. Technical Report NUREG/CR-4862. Washington D.C.: US Nuclear Regulatory Commission.

Yoshida, K., Yokobayasi, M., Tanabe, F., and Kawase, K. (1995). Development of computer simulation model of cognitive behavior in accidental situation of nuclear power plant. In *Proceedings 5th European conference on cognitive science approaches to process control (pp. 338–347). Espoo: Technical Research Centre of Finland.*

Zuboff, S. (1988). *In the Age of the Smart Machine.* New York: Basic Books.

CHAPTER 59

HUMAN FACTORS IN TRANSPORTATION

Patricia F. Waller
Paul A. Green
Transportation Research Institute
University of Michigan
Ann Arbor, MI 48109 USA

59.1 CHAPTER ORGANIZATION

This chapter consists of two major parts. The first focuses on the broader human and societal context in which transportation occurs, and how this context affects some of the policies and decisions confronted by human factors specialists. It especially calls attention to some of the changes occurring in society, changes that challenge the human factors specialist and demand attention if the information needs required are to be generated in a timely fashion. The second part of the chapter addresses more traditional human factors issues, with the primary focus on motor vehicles. This concentration on motor vehicles, and particularly passenger vehicles, is based on several considerations. First, by far the greatest proportion of personal travel occurs in private vehicles. Second, in the United States, based on passenger miles traveled, highway fatality rates are by far the highest, and motor vehicle deaths account for well over 90% of transportation-related fatalities. Third, there is a greater body of human factors research available on motor vehicles than there is for other modes of transportation. The information illustrates human factors problems that have been investigated and can provide some guidance to research in other transportation arenas.

59.2 HUMAN FACTORS VERSUS HUMAN ISSUES IN TRANSPORTATION

59.2.1 Historical Perspective

Human factors activities in transportation have dealt with the fit between the human operator or user and the rest of the transportation system, usually the vehicle and immediate transportation environment. Evaluations of this fit have led to improvements that facilitate using the transportation system. Physical measurements of drivers have led to improvements in seating, placement of controls and displays, and highway sight distances. Analyses of what information should be provided and in what format has resulted in more readily understandable signs and symbols, including improved highway markings. Less attention has been given to the human factors issues involved in nonmotorized transport, although in much of the world nonmotorized transport accounts for the majority of transportation fatalities (Mohan, 1992). Even in the United States, pedestrians and pedal cyclists accounted for about 15% of traffic fatalities in 1994 (National Safety Council, 1995).

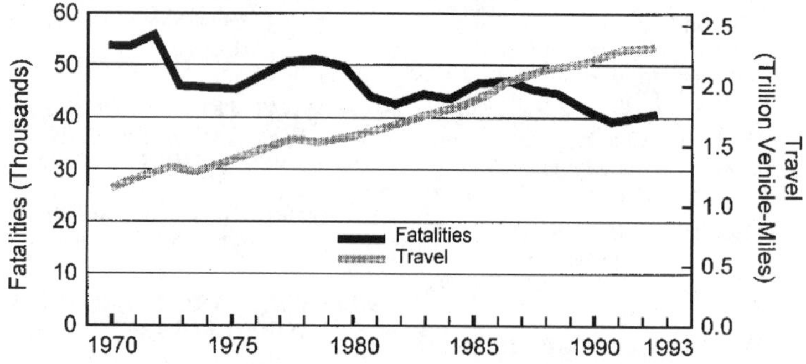

Figure 59.1 Motor vehicle fatalities and travel in the United States, 1970–1993. (Federal Highway Administration, 1995a.)

Allocation of funding for transportation is not in proportion to the distribution of transportation casualties.

In addition to improving the ease of use of transportation facilities, human factors research has emphasized safety. In the United States, the combined improvements in vehicle design, highway design, and operator performance have reduced the fatality rate per hundred million vehicle miles driven from almost 22 in 1923 to well below 2 in 1993 (National Safety Council, 1994). Figure 59.1 illustrates the changes in recent years in both motor vehicle travel and motor vehicle fatality rates in the United States (Federal Highway Administration, 1995a).

Highway fatality rates are usually based on vehicle miles traveled, because the information on passenger miles is less reliable. However, when estimates of passenger miles are considered, the fatality rates per 100 million passenger miles are 0.82 for passenger vehicles, 0.01 for buses, 0.42 for rail, and 0.01 for scheduled air traffic (National Safety Council, 1995). Although travel by personal motor vehicle is the most hazardous, it is by far the mode of choice in the United States. Table 59.1 shows the distribution of person miles of travel by mode of transportation. In the United States, almost 18% of household expenditures is for transportation, with only housing accounting for more (Federal Highway Administration, 1995a).

This preference for the private vehicle, with its higher casualty risk, is reflected in the actual numbers of fatalities. In 1994, in the United States there were 264 deaths in scheduled air service, 1226 deaths associated with rail traffic (only 5 of whom were passengers),

Table 59.1 Person Miles of Travel by Mode of Transportation, United States, 1990[a]

Mode of Transportation	Percentage of Transportation
Private vehicle	88.2
Public transportation	2.5
Bus, streetcar	(1.5)
Train	(0.6)
Subway	(0.4)
Other means	9.3
Airplane	(6.3)
School bus	(1.4)
Walking	(0.5)
Biking	(0.1)
Other	(1.0)

[a] From Federal Highway Administration (1995a).

and 43,000 motor vehicle deaths. The latter figure includes pedestrians and pedal cyclists, most of whom died as a result of a motor vehicle (National Safety Council, 1995).

The United States is the most motorized nation in the world, whether measured by per capita vehicle miles of travel or by automobiles per capita. However, other modern industrialized nations are also heavily dependent on motorized transport. Figure 59.2 illustrates some of these comparisons. There is a clear relationship between vehicle miles traveled, land mass, and population density, with the United States and Canada having greater land mass in relation to population density (Federal Highway Administration, 1995a).

Human factors specialists have traditionally focused on the relatively immediate fit between the transportation system and the user. The latter portion of this chapter (Sections 59.8 through 59.10) considers transportation, and particularly passenger vehicle transportation, from the standpoint of this more traditional approach. However, the concept of fit between transportation and its human consumer can be extended to consider how well our transportation system "fits" the broader human needs, capabilities, and goals of society. The first sections of this chapter (Sections 59.2 through 59.7) elaborate on this broader interface, considering how well transportation is meeting human needs in the general sense, and how transportation should be designed so as to meet such needs. This broader interface is not always considered by either the human factors specialist or the transportation community. Yet consideration of it would lead to revision of some of our research questions and, ideally, to a transportation system that more effectively meets the needs of the user. This extension of traditional human factors may be conceptualized as societal human factors.

59.2.2 Transportation at the Crossroads

59.2.2.1 The Era of the Interstate

In 1956 then-President Eisenhower called for the creation of a national highway system to enable the rapid transport of troops, equipment, and supplies from one coast to the other. His experience in the military in trying to move troops rapidly convinced him of the need for a national defense highway system. Thus, the Interstate Highway System was launched. Built to federal standards and financed primarily by the federal government, the Interstate Highway System truly revolutionized highway travel in the United States. With the advent of this system came an infrastructure that met travelers' needs for food, shelter, and entertainment.

The Interstate system was based on standards for driver eye height, sight distance, sign visibility, and other factors deemed important for safe maneuvering. Access to the system was severely limited, and roadside accoutrements were also standardized and

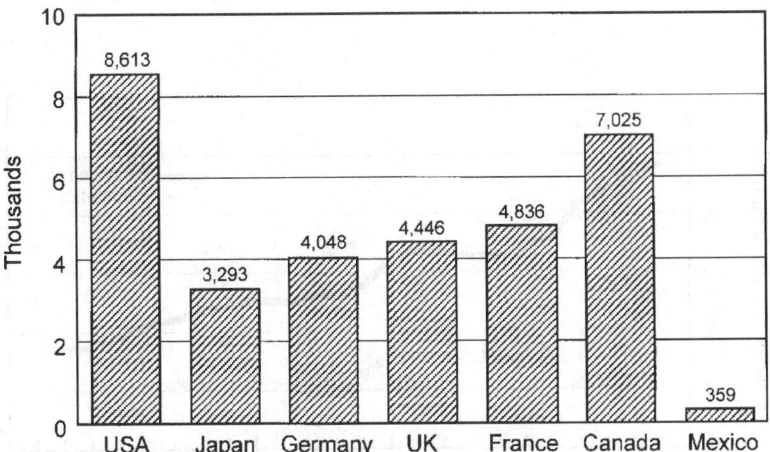

Figure 59.2 Annual vehicle miles of travel per capita; 1991 data, except Canada (1990) and Mexico (1989). (Federal Highway Administration, 1995a.)

limited. Based on kilometers driven, this system is probably the safest roadway system in existence anywhere in the world (National Highway Traffic Safety Administration, 1993). Figure 59.3 shows the fatality rate based on mileage for all highways and for the Interstate system (Federal Highway Administration, 1995a). It can be seen that, whereas fatality rates have steadily dropped over time, the rate for the Interstate system continues to be less than half that for all highways.

The authorization for the Interstate Highway System ended in September 1991. Although there will be some additional highways built, it is widely accepted that huge land areas will no longer be transformed into paved highway. Although the era of the Interstate Highway System is largely over, the need for travel space on the highway continues to grow, and with it, congestion.

59.2.2.2 The Emergence of ITS

Because we can no longer look to asphalt and concrete to solve the expanding need for travel capacity, it is necessary to make better use of the existing system. The major approach to accomplishing greater efficiency on our highways is Intelligent Transportation Systems (ITS). Formerly called Intelligent Vehicle-Highway Systems (IVHS), ITS applies advanced and emerging technologies in communications, information processing, control, and electronics to the highway system (as well as other transportation modes) to improve efficiency and increase capacity. Vehicles and highways will communicate with each other and with traffic control centers, both to provide the driver with better information to make decisions and to control vehicle movement. Through the effective application of such technologies, mobility, safety, and productivity can be increased, while energy consumption and environmental pollution can be reduced.

59.2.2.3 Redefining the Role of Transportation in Society

The completion of the Interstate Highway System and the advent of ITS offer an unprecedented opportunity to rethink our definition of the role of transportation in society. Historically, the role of transportation has been defined as the safe and efficient movement of people and goods. More recently, the movement of information has been included as well. This definition has served us well and led to the most efficient overall transportation system ever developed. However, simply continuing in this paradigm will not suffice to meet future transportation needs, nor will it lead to optimal solutions. In our society, transportation is far more than the safe and efficient movement of people and goods—rather, it is an inextricable part of access to education, health care, employment, recreation, and maintaining ties with family and friends. In short, transportation is an essential part of what enables individuals to develop their potential to become competent, participating, contributing members of society. It is an essential component of what enables communities to become what they could and should be.

Figure 59.4, developed by Paul Green, presents a new conceptual model of transportation and the forces affecting, and affected by, its occurrence. The center portion of the

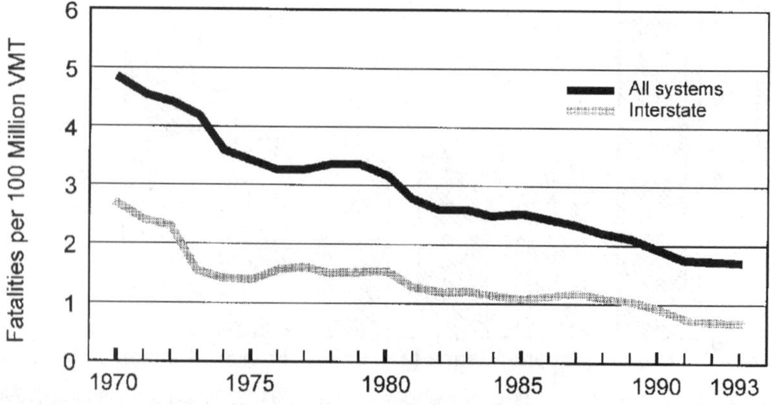

Figure 59.3 Fatality (interstate and total) rates, 1970–1993. (VMT is vehicle miles traveled.) (Federal Highway Administration, 1995a.)

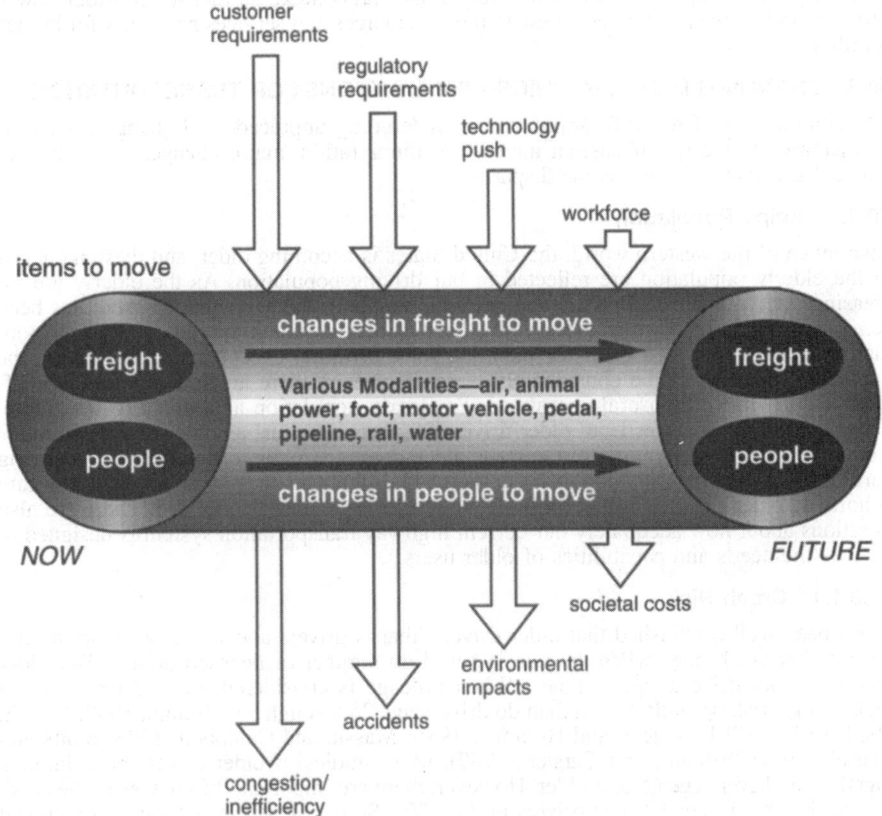

Figure 59.4 Transportation conceptual overview, "the pipeline to the future." (Developed by Paul Green.)

figure shows the two major commodities to be moved, freight and people, with the time dimension moving from now (on the left) toward the future (on the right). Over time, changes are occurring in the quantity and characteristics of the freight to be moved, as well as the schedules to be met. Changes are also occurring in the numbers and characteristics of the people to be moved, and many of these changes can be anticipated. At the top of the model are the major inputs to the system. Customer requirements, the longest arrow at the top, represents the forces closest to the system and receiving the greatest attention. The next category, regulatory requirements, also receives much attention, but customer requirements may take precedence over regulatory requirements. Technology push is becoming increasingly prominent, particularly in the movement of freight, but it does not yet have the impact of the two previous categories. Finally, the workforce is a major factor but one that has not received attention commensurate with its importance.

The model outputs (the arrow at the bottom) represent the impact of transportation on the rest of the world. The major focus of ITS has been on the first output, reducing congestion and increasing throughput. Accidents, the second output, also receive considerable attention, with safety being a major concern in ITS. Environmental impacts are increasingly recognized as a consideration, but attention to this output is too often viewed as a necessary evil that would be ignored, if possible. Finally, virtually no attention has been paid to the larger societal costs and consequences of the transportation systems and policies implemented.

Traditionally, the science of human factors has concentrated primarily on those inputs and outputs toward the left, with relatively little attention to the right portion of the figure. Yet the transportation decisions made in response to customer requirements and regulatory requirements, as well as our efforts to address congestion/inefficiency and accidents, inevitably interact with those inputs and impacts further to the right.

Transportation must be viewed in a larger societal context, so that we consider how it serves to facilitate or to hinder access to those resources and services necessary for human fulfillment.

59.3 CHANGING DEMOGRAPHICS—IMPLICATIONS FOR TRANSPORTATION

The composition of the U.S. population is undergoing unprecedented changes. There is an increase in elderly citizens, an increase in immigration, major changes in the role of women, and increases in income disparity.

59.3.1 Aging Population

Like much of the western world, the United States is becoming older, and these increases in the elderly population are reflected in our driving population. As the elderly are increasing, the younger cohorts are decreasing (Campbell, 1994). Although there has been some recognition of these changes, it has been widely assumed that today's elderly population can be used as a basis for planning for the future, except that in the future the number of elderly will be considerably larger. However, there are important cohort differences that limit the usefulness of today's elderly population as a guide to the future. Older drivers today differ from older drivers 20 years ago, just as older drivers 20 years from now will be different from current older drivers. Cohort differences exist among current older drivers in relation to crash risk, and differences may be anticipated in future cohorts of older drivers in relation to driving experience and alcohol use. There are also questions about how adequately our current highway transportation system is designed to address the needs and capabilities of older users.

59.3.1.1 Crash Risk

It has been well established that older drivers, that is drivers above age 60 or so, do not have higher crash rates when the rate is based on number of licensed drivers. But older drivers do not drive as many miles. When mileage is considered, older drivers have a higher crash risk per mile driven than do drivers age 25 through 54 (Brainin, 1980; Cerelli, 1989; Evans, 1991; Maleck and Hummer, 1986; Massie and Campbell, 1993; Stutts and Martell, 1992; Williams and Carsten, 1989). Most studies of older drivers have lumped together all drivers age 65 and older. However, there are important differences in the crash risk of drivers in their 60s and drivers in their 80s. Some investigators have differentiated among older drivers by classifying them into several groups, including "young old" (age 55–64), "middle old" (age 65–74), "old old" (age 75–84), and "very old" (age 85 and older) (Waller, 1991). Data from California, shown in Figure 59.5, would support such

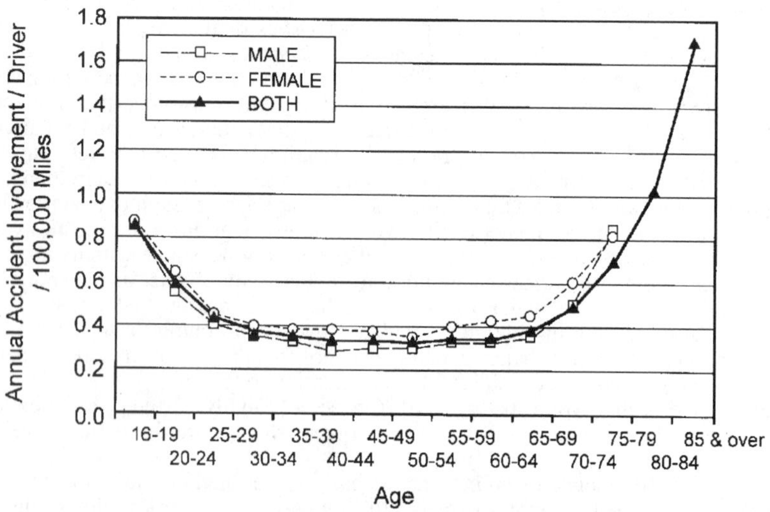

Figure 59.5 Annual crash rate by mileage, age, and sex, based on California licensed drivers in crashes occurring during 1989–1991. (Peck and Romanowicz, 1993/1994.)

distinctions (Peck and Romanowicz, 1993/1994). These increases in crash risk occur even though, as a group, older drivers try to restrict themselves to the safest times and places (Yee, 1985).

59.3.1.2 Driving Experience

Whereas older drivers continue to have a higher crash risk, the extent of their increased crash risk is decreasing over time (Evans, 1991; Stutts and Martell, 1992). It appears that, as each cohort of older drivers becomes more experienced, and especially with more modern highway systems, their overrepresentation in crashes diminishes.

Countering this decreasing elevation in crash risk per mile driven is their increase in actual mileage. Because older drivers have traditionally driven much less than their younger counterparts, their crash rate per licensed driver has been low. However, succeeding cohorts of older drivers are increasing their relative mileage at somewhat higher rates than other age groups. This increase in individual mileage, combined with the dramatic increase in the number of older drivers, indicates that the problems associated with older drivers will surely increase.

59.3.1.3 Alcohol Use

At the present time, alcohol is not a great problem among the elderly as far as driving is concerned. For fatally injured drivers in the United States in 1991, 44.4% had positive blood alcohol concentrations (BACs), but for fatally injured drivers age 55 and older, only 16.2% tested positive for alcohol. For those age 70 and older, the percentage was 9.2%, and even lower for drivers age 80 and over (5.3%; National Highway Traffic Safety Administration, 1993). Thus, it appears that alcohol use decreases with increasing age. However, today's elderly grew up during Prohibition and the Great Depression. During Prohibition, alcohol was not widely available to much of the population, and during the Depression, economic factors limited access to alcohol. Alcohol use is correlated with income, and limited income translates into higher proportions of abstainers. There is also evidence that drinking patterns in middle age may be extended into old age to a greater extent than previously thought (National Institute on Alcohol Abuse and Alcoholism, 1993).

Because subsequent generations grew up in an environment of greater alcohol availability, and because current middle-aged drivers have higher rates of alcohol involvement, as these generations age their alcohol consumption may be greater than that of today's elderly. Because alcohol differentially affects the elderly, as shown by an increased risk of crash at lower BACs (National Highway Traffic Safety Administration, 1985), cohort differences in alcohol use raise questions about performance of future elderly drivers.

59.3.1.4 Highway Transportation System Design

Our highway transportation system was not designed for older users. To the extent that driver license programs were designed at all (they were originally implemented for identification and revenue purposes), they were designed to qualify young beginning drivers. Vehicle design has traditionally been aimed at the youth market, with an emphasis on speed and performance. Only recently have vehicle manufacturers seriously considered the special needs of older users. Historically, highway design has largely ignored the needs and capabilities of older users. The standards for highways are based primarily on measures obtained from young males. For example, the standards for pedestrian signals for crossing intersections are based on a walking speed that many elderly pedestrians cannot meet. As a result, they are at higher risk of injury and death (Retting, 1988; Retting, Schwartz, Kulewica, and Buhrmeister, 1989). There is a need for better data on how our transportation system needs to be modified to enable more older users to be able to use it safely.

59.3.2 Immigrant Population (United States and Canada)

Combined with increases in the elderly population are increases in the immigrant population. The birth rate of U.S. citizens has fallen below the replacement level, and were it not for immigration, U.S. population growth would gradually taper off with eventual decline. However, immigrants are entering the United States in unprecedented numbers, with by far the largest groups coming from Asia and Central and South America (Hollmann, 1992). These immigrants are generally younger, with a predominance of males. The higher birth rates for the immigrant population also contribute to the growth of a younger population.

The potential impact of the immigrant population is uncertain in that immigration, both legal and illicit, may be profoundly affected by such factors as changes in immigration laws; economic and political factors in Mexico, Cuba, and other Latin American countries; and economic factors on the domestic front. Just as changes in laws and policies triggered a large influx of immigrants to the United States from Cuba, Southeast Asia, and Mexico, a reversal in immigration policies could evolve from apparent growing anti-immigration sentiment.

Conversely, an expanding domestic economy in the United States could necessitate increased immigration to supplement the citizen labor force. Even now, despite anti-immigration rhetoric, some segments of the U.S. economy would be severely hampered if immigration were significantly curtailed. Likewise, a falling peso in Mexico could lead to marked increases in illegal immigration, and relaxation of relations with Cuba could trigger changes there as well. Changes in other parts of the world, such as Eastern Europe or the Middle East, could also influence the numbers and nationalities of the U.S. immigrant population.

The immigrant population brings with it different cultural values and practices, including different attitudes and behaviors regarding alcohol use. Although Asians are not noted for high alcohol consumption, efforts to emulate successful role models in the U.S. business community have led to excessive alcohol consumption among some Asian subgroups. Hispanics also bring different behaviors in relation to alcohol.

Birth rates vary for different population groups, leading to marked differences in the rate of change within race and ethnic categories. In the United States, non-Hispanic Whites are by far the fastest growing group among the elderly. Asians and Hispanics have by far the highest numbers of births within race or ethnic category. By contrast, White births are decreasing, and by about 2023 non-Hispanic Whites will represent a minority of births in the United States (Day, 1993). These differences foretell significantly different age distributions within race or ethnic group. Since alcohol use tends to vary with age, these differences have implications for the kinds of alcohol problems that may emerge.

59.3.3 Changing Roles of Women

The role of women in our society, and particularly their role in relation to transportation, has changed dramatically over the past two decades. Their entrance into the labor force has increased, so that they will account for about 47% of the total labor force in the year 2000, up from 41% in 1976 and 45 percent in 1988 (Fullerton, 1989). Married women have been a significant part of this change. Fewer than one-third of married women were in the work force in 1960, but by 1990 almost 60% were employed. These changes also hold true for married women with children under 18 (Lugaila, 1992, quoted in Rosenbloom, 1994a).

Women are also increasing their rate of driver licensure, so that they are rapidly approaching that of men. Furthermore, like men, they now acquire license as early as the law allows (Rosenbloom, 1995). Women also account for almost half of all purchasers of new cars (Belton, 1992). While they still do not drive as many miles as men, they are increasing their proportion of the total mileage accumulated (Massie and Campbell, 1993). Women are also driving at times previously dominated by male drivers, namely at night and on weekends. Women account for the major changes in travel behavior occurring in the United States today (Pisarski, 1992).

Another major change affecting women and transportation is the increase in age of marriage, as well as in the rate of divorce. Both these factors lead to greater independence for women, with accompanying transportation requirements. Based on data from the Nationwide Personal Transportation Survey, single mothers make more vehicle trips than comparable married women. Although women drive fewer miles than men, they make more trips per day. It also appears that the travel behavior of women is more affected by variables other than their household income or whether they hold a driver license (Rosembloom, 1994a).

Women, particularly younger women, have increased their use of alcohol, with corresponding implications for driving behavior. Whereas it is true that women have increased as a proportion of drinking drivers, their increase in total driving and in total crashes has been even greater. Both males and females have greatly reduced both their rates and their numbers of alcohol-related crashes, but women have actually shown greater decreases. However, because their total driving and total crashes, including fatal crashes, have increased so much more than is true for men, they actually have increased their proportion of total alcohol-related fatal crashes (Waller and Blow, 1995).

There is also some evidence that women may be more impaired at a given level of blood alcohol concentration (BAC). A major report, based on data collected in the early 1960s, clearly establishing the relative risk of crash by BAC, showed that women had a several-fold greater crash risk than men at a BAC of 0.08% (Jones and Joscelyn, 1978). This difference was considered to be related to the relative inexperience of women in driving, and possibly drinking. However, subsequent studies have reported similar findings (Carlson, 1972; Zador, 1991). Although women still drive fewer miles than men, their driving experience is sufficient to ensure their being beyond the steep portion of the learning curve. Furthermore, other studies based on tasks unrelated to driving also suggest that the performance of women may be more impaired at relatively low levels of alcohol (Waller and Blow, 1995). This finding remains tentative, but it calls for more careful investigation, with particular focus on its implications for transportation.

Thus, changes for women in employment status, driver licensure, vehicle ownership, marital status (whether remaining single longer or increasing divorce rate), and alcohol use all have transportation-related implications, particularly driving-related implications.

59.4 TRANSPORTATION AND THE HUMAN INTERFACE

59.4.1 The Physical Interface

59.4.1.1 Highway Transportation

Transportation and traffic engineers have always given some consideration to the human interface, even if the consideration did not always reflect the full spectrum of users. Frequently, the 85th percentile was the basis for establishing traffic control measures, such as speed limits. When the driving population was more homogeneous, such a standard was probably reasonable. With the growing heterogeneity of highway users, it would be useful to determine to what extent existing measures are still adequate.

Drivers

With increasing numbers of elderly drivers, women drivers, and immigrant drivers coming primarily from Asia and Latin America, even something so basic as driver eye height may be different. With increasing age, height is often lost, primarily in the torso. On the whole, women, Asians, and Latin Americans are somewhat shorter than American males. Driver eye height is used for establishing sight distance and other highway standards, so that any changes in its mean or variability have important human factors implications. Shorter driver height usually means that the driver sits closer to the steering column in order to reach foot controls. This closer proximity appears to be associated with increased injury resulting from the rapid release of the air bag in a collision. The changing anthropometry of drivers poses challenges for the design of the passenger vehicle.

Passengers

Occupant restraint systems have been designed primarily to protect younger occupants in frontal crashes. Since this crash type represented the major serious collision, such design was appropriate. However, older drivers have higher rates of side collisions than do younger drivers. They are also more frail and vulnerable to injury (Evans, 1991; Partyka, 1983, 1984) in the event of a crash. The safety belt that may restrain and protect the younger occupant may break the ribs of the older occupant and, in turn, lead to puncture of the lungs. There is a need for better designed occupant restraints to protect older people.

Likewise, the smaller stature of women, older persons, and certain groups of immigrants poses problems with some shoulder harness designs. There is a problem in balancing the protection afforded by the air bag with the potential injury to smaller occupants located in close proximity to the bag when it explodes. The greater frailty of older occupants is also a consideration in air bag protection.

Nonmotorized Transport

Although occupants of passenger vehicles and trucks account for the majority of traffic fatalities in the United States, Canada, and Australia, in other parts of the world it is other road users accounting for the greatest portion. These include pedestrians, bicyclists, and motorized two wheelers. Mohan (1992) has referred to these users as "vulnerable road users." In most parts of the world, pedestrians are the greatest part of these fatalities, accounting for over 40% of traffic fatalities in Asia and Africa. Bicyclists account for 30% of all fatalities in China, and for 25% of traffic injuries in rural areas outside New Delhi, India.

Unlike many other western nations, the U.S. highway transportation system has not seriously addressed nonmotorized transport. Although pedestrians and pedal cyclists continue to account for about 15% of all traffic fatalities (National Highway Traffic Safety Administration, 1993), the allocation of funding does not reflect the significance of the problem. Many trips that, on the basis of distance alone, could be made by foot, require the use of a motor vehicle, because the highway design makes no safe provision for pedestrian traffic. As our population ages, with more persons reaching the point at which they may no longer qualify for licensure, greater attention to facilities for pedestrian traffic would be in the best interests of society. However, if the application of new technology results simply in moving more people and goods faster and safer, will it not become more difficult than ever for the elderly pedestrian to cross the intersection?

The trend toward privatization of transportation, with a focus on user fees, increases the probability that nonmotorized transport will be increasingly overlooked. Opportunities for the application of technology, with the corresponding opportunities for profit, are less apparent for pedestrian and pedal cycle traffic. Yet, with growing concerns about congestion, environmental contamination, and an aging population, the facilitation and encouragement of increased nonmotorized transportation are warranted.

59.4.1.2 Aviation

Pilots, Flight Attendants, and Air Traffic Controllers

Considerable attention has been given to human factors in aviation. Indeed, the science of human factors came into its own through the study of pilots in World War II. However, the focus has been on those employed in the aviation industry, especially pilots, air traffic controllers, and, more recently, flight attendants (U.S. Congress, 1994). In addition, with the advent of the Boeing 777, aircraft design has taken into consideration some of the human factors issues in aircraft maintenance.

Passengers

Passenger interests have been addressed primarily in regard to survivability. Research is conducted on aircraft crashworthiness, including seat attachments and restraint systems, flammability and toxicity of materials used in the cabin, and issues surrounding emergency evacuation.

There remains a major need to address the human factors issues surrounding the changing passenger population. In the early decades of aviation, the passenger population consisted primarily of relatively affluent men. In the last two decades, the demographic characteristics of the passenger population have changed dramatically. Although older women living alone and older minorities are more likely to be living in poverty, the elderly population is increasingly wealthy. In 1959 over one-third of the elderly were below the poverty level, but that figure had dropped to 12.2% in 1990 (Rosenbloom, 1994b). This increase in relative affluence among a population with leisure time, combined with special discounts on air fares for the elderly, has led to significant increases in elderly passengers. For certain flights on certain days of the week, the passenger population is predominantly older persons. No reliable figures are available on changes in the passenger demographics. Some airlines purport to have such information, but the data are based on returns from questionnaires completed by interested and cooperating passengers.

Whereas aviation research in the United States has recognized that there are safety issues associated with aging aircraft, there has been no similar recognition that there may be safety and other human factors issues associated with an aging passenger population. Some of the issues in need of greater understanding are aircraft seating, lavatories, emergency exits, overall cabin design, and overall cabin environment.

Seating. Aircraft seating is not designed for older passengers who may be less agile than younger persons. Older people often have difficulty getting out of a seat, particularly when the plane is in motion. Airlines need to consider seating comfort, as well as the passenger's ability to get out of the seat, in relation to an aging clientele.

Lavatories. Airlines have increased seating capacity with no corresponding increase in lavatory facilities. The increase in women passengers, with their corresponding need for greater lavatory time, and the higher rate of incontinence problems among the elderly, particularly elderly women, raises questions about the adequacy of available lavatory facilities for these passengers, within both the aircraft and the terminals. Some disabled passengers, as well, have difficulty using aircraft lavatories.

Emergency Exits. The Federal Aviation Administration requires that an aircraft must be designed with sufficient emergency exits located so that, with only half of the exits functioning, in darkness, and with a representative passenger load, all passengers can evacuate the plane within 90 s. For obvious reasons, tests of evacuation procedures are limited in the realism that can be simulated. Even so, in full-scale demonstrations, about 10% of participants suffer injury, ranging from bruises to broken bones and paralysis. The demonstration populations must include 5% of passengers over age 60, but many actual flights include much larger proportions of older passengers. When disabled passengers and children are included, the potential for disaster increases (see Galea and Perez Galparsoro, 1993).

Cabin Design. Elderly passengers, as well as some disabled passengers, may have difficulty maneuvering the aisles in some aircraft. At least one foreign carrier is investigating medical issues associated with long flights, particularly for older passengers. The whole area of the adequacy of current aircraft cabin design for an aging passenger population is one in need of considerable human factors research.

Cabin Environment (Temperature, Humidity, etc.). Aircraft differ in the extent to which fresh air is circulated. If there are passengers on board with medical conditions that can be spread by air, other passengers are placed in jeopardy when the air is recirculated. Decreased humidity is characteristic of aircraft cabins, and for some passengers, including those on certain medications, the lower humidity may pose problems. Lack of ability to move around on long flights can be a particular problem for older passengers and anyone with circulatory problems. Very little is known about how these issues are actually affecting aviation passengers, but with the changes and increases occurring in passenger populations, there is a need for greater understanding of which human factors are important for passenger safety, comfort, and health.

59.4.2 Increasing Technology in Transportation

The application of technology to transportation holds the promise of improving the system for both users and operators. With the advent of ITS, there has been an explosion of activity exploring how technology can be used to facilitate the rapid and safe movement of people and goods. The primary focus has been on technology, although there is growing recognition that the human component poses the greatest challenge to the research community, the industry, the transportation community, governments, and, ultimately, the public. The technology issues, while complex, are simple in comparison to the human interface problems.

Increased attention is being devoted to the human factors dimension. Two aspects include, first, what is the information that is most relevant to the users and, second, how should the information be presented? While information may be interesting and even relevant, will it actually help or will the user suffer from information overload?

Unfortunately, those engaged in designing, developing, and implementing ITS technology are not representative of the range of users or potential users who will ultimately be affected by the technologies and the systems being developed. Indeed, the entire highway transportation system was developed with relatively little attention paid to the full range of users. In all fairness to those responsible for the system, it should be noted that the early decades of motorized highway transportation were dominated by a particular segment of the population, namely, white males with a certain level of affluence. In most developed western nations, the composition of users has changed dramatically over the past several decades, but much of the thinking that has traditionally dominated the field continues to persist.

A case in point is the field testing of the interlock safety belt, whereby the vehicle engine could not be started until front seat occupants were belted. The introduction of the system was not preceded by appropriate public education and preparation for the change. Furthermore, the field testing of the interlock safety belt was conducted in official cars loaned to federal employees for business trips (Perel and Ziegler, 1971) and in rental cars procured at airports (Westefeld and Phillips, 1975). The system was not tested with mothers driving preschoolers to nursery school, or pregnant women, or nursing mothers, or arthritic elderly drivers. Although the field testing resulted in relatively high use rates and satisfaction, the population studied was not representative of the general population of drivers.

The safety belt interlock system may have posed problems, but it was effective in increasing belt usage. The ultimate failure of the system was probably not so much a function of technology as a function of the failure to address the human interface, in the broadest sense of the term.

Likewise, the ultimate effectiveness of the emerging technologies, and perhaps even the ultimate survival of the ITS program, depend upon the broad recognition of their value to society. Achieving such recognition requires that we address not only the issues surrounding displays and controls but also the broader human interface, that is, how the technology addresses broader human needs.

59.5 COMMERCIAL OPERATIONS

Commercial transportation includes moving both people and freight. As was shown in Table 59.1, in the United States most transport of people is by private vehicles. Commercial transportation accounts for only a small portion of transport of people. Figure 59.6 shows the distribution of freight transport by mode (Federal Highway Administration, 1995a). It can be seen that rail and water each account for more revenue ton-miles than highways. However, when transport-related fatalities are considered, it is evident that highways account for most of the problem. For rail and air, fatalities related to transport of passengers are not differentiated from those involved in transport of freight. Even so, in 1994, in the United States, all rail fatalities, including passengers, employees, non-trespassers, and trespassers accounted for 1226 deaths. Only five of these fatalities were rail passengers, and trespassers accounted for the largest proportion. The second largest category involved collisions with motor vehicles at rail grade crossings (National Safety Council, 1995). All U.S. aviation transport deaths in 1994 totalled 1034, but most of these occurred in general aviation (National Safety Council, 1994). However, in 1994, crashes involving large trucks on the nation's highways accounted for 5112 fatalities. It is clear that even though most ton-miles of freight are transported by rail and water, it is the trucking industry that accounts for the majority of casualties.

The trucking industry is rapidly expanding, and there are indications that it will continue to expand over the next decades, if not at so rapid a pace. However, there is a growing shortage of qualified truck drivers, and in some segments of the industry, annual turnover rate approaches 200%. The leading edge of the baby boomer generation is moving into its sixth decade, and there is not a similarly larger younger cohort to replace it.

The increased risk of crash per mile driven that occurs with increasing age has been clearly established, and it is also clear that this increase occurs despite the fact that older drivers, as a group, attempt to restrict themselves to times and places that are least hazardous. Drivers who must drive on a schedule do not have the luxury of such self-restriction, and their crash risk reflects this limitation.

Although data on this issue are scarce, there are several studies based on school bus drivers that demonstrate that the elevation in crash risk for drivers on a schedule may

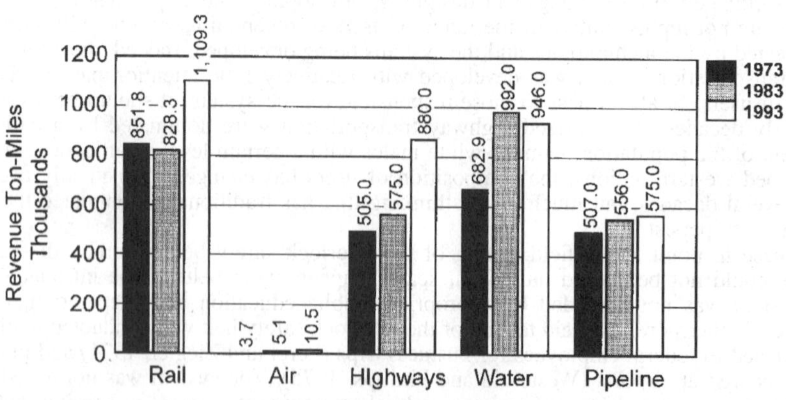

Figure 59.6 Freight transportation by mode, United States. (Federal Highway Administration, 1995a.)

begin earlier and rise more steeply than for older drivers in general (Hull and Knebel, 1968; Promisel, Blomberg, Knacht, and Silver, 1969; Waller, 1992). Because driving performance shows evidence of deterioration as early as age 50–54, the problem of sufficient numbers of qualified commercial drivers is likely to become more severe.

Traditionally, most commercial drivers, and particularly drivers of heavy vehicles, have been male. However, the changes in the demographics of the population have been reflected in the labor force, with more female and minority, as well as older, workers (Cohen, 1993). Between 1990 and 2005, predicted increases in the labor force are not evenly distributed across racial and ethnic groups. The relative percentage change of Whites in the labor force in the United States is predicted to be an increase of 17.4%, compared to 31.7% for Blacks. However, the relative percentage change for Asians will be a 74.5% increase, and for Hispanics, 75.3% (U.S. Bureau of Labor Statistics, 1991, quoted in Cohen, 1993). This increasingly diverse labor pool, in combination with the increased application of advanced technology, offers opportunities for different solutions to the problem of limited numbers of qualified commercial drivers. However, the expansion of the labor pool for commercial operations requires that a number of human factors problems be resolved.

59.5.1 Redesigning Vehicles for a Changing Labor Force

The increased participation of women, Hispanics, and Asians in the labor force suggests that the design of the commercial vehicle could be modified to facilitate operation by drivers of somewhat different physical dimensions. The basic anthropometric measurements of these populations should be included in decisions regarding the location and design of vehicle controls and displays, as well as provisions for easy ingress and egress. The needed updating of such measures of truck drivers should include these populations even though they may not represent a significant proportion of current drivers. Future needs will almost certainly require their participation.

59.5.2 Redesigning Jobs for a Changing Labor Force

Perhaps the greatest potential gains in commercial operation productivity lie in the redesign of the job itself, so that the human needs of drivers are met more effectively. Hours of service and the application of technology are two areas that can function as facilitators or barriers to the expansion of the labor pool. In addition, driver wellness programs offer great promise, as well as challenges to implementation.

59.5.2.1 Hours of Service

The current hours of service regulations for the U.S. trucking industry were written in the 1930s, when relatively little was known about sleep needs or the effects of sleep loss or irregular schedules. It is now known that problems with fatigue arise well within the current hours of service regulations (Harris, 1977; Mackie and Miller, 1978; National Transportation Safety Board, 1990). Drivers operating on irregular schedules are more vulnerable to fatigue than drivers who drive the same number of hours but on a regular schedule. Cumulative fatigue also occurs earlier for drivers on an irregular schedule. Drivers engaged in truck sleeper operations are even more subject to fatigue. Time of day exerts an additional effect, above and beyond these other factors. Participation in moderately heavy cargo loading may result in increased effects of fatigue for drivers on irregular schedules. Fatigue effects are evident, even for relay drivers on a regular schedule, well within the current hours of service regulations.

Hours of service regulations for truck drivers will not be easily resolved, in that there are strong interests on the part of both motor carriers and some drivers to extend rather than further restrict driving hours. However, based on what is now known about fatigue, Miller (1993) has recommended what he calls "human oriented schedules" that would increase actual hours of driving above what currently occur because of the inefficiency resulting from current hours of service regulations, while more effectively meeting the biological need for adequate rest.

59.5.2.2 Technology Application

Commercial vehicle operations have been in the forefront of the application of new technology. Using satellite tracking of vehicles and two-way in-vehicle communication, motor carriers have greatly increased the efficiency of their cargo shipments. Indeed, it is the application of technology that has enabled the emergence of just-in-time (JIT) delivery in manufacturing, a phenomenon that has led to the public highways "warehousing" as much as one-third of the nation's inventory of goods. Even so, technology has not yet realized its full potential for streamlining commercial operations. To do so will require a much greater understanding of the human component of the system (Bowers-Carnahan, 1991).

Scheduling, Routing

Perhaps the most obvious, and consequently the most widely implemented, application of technology is in scheduling and routing shipments of cargo. Vehicle manufacturers using JIT delivery of components have emptied plants of inventory, thus reducing costs while increasing the efficiency of delivery to the production line. Components arrive when they are needed and in the order required for the sequence of vehicle models being produced. Because shipments are monitored en route, any shipment falling too far behind schedule triggers arrangements for an expedited backup shipment to ensure continuous production. It is technology that has enabled the efficient incorporation of transportation into the manufacturing process.

Monitoring Vehicle, Cargo

In addition to simply monitoring the movement of the vehicle transporting the cargo, technology can be used to report the condition of various vehicle components. Some such technology is routine, such as gauges indicating fuel level and temperature. However, the condition of vehicle components can be monitored so that vehicle condition is under continuous scrutiny and maintenance can be implemented prior to any breakdown.

The cargo can also be monitored. In the case of perishable cargo that requires temperature and/or humidity control, technology can alert the driver of any impending failure. Technology can also facilitate cargo security and tracking. The move toward container shipments has been made more efficient through rapid container identification, but even the smallest package is readily tracked through automatic recording of bar coding or other identification.

Monitoring Driver Behavior

Technology has long held promise for detecting loss of vigilance in driver performance, but the promise has yet to be realized. Although it is widely recognized that hours of duty regulations fail to ensure alert drivers, the quest for a measure of "fitness for duty" or, better yet, a continuous measure of on-board driving performance, continues.

Security

Of particular relevance to expanding the pool of potential drivers is the issue of personal security. Whereas personal security is a concern for everyone, it is usually of greater concern to women, who now make up close to half of the work force. If vehicles are in continuous communication with the home base and can readily contact local law enforcement officials, the resulting security can allay the anxiety of some potential employees. Likewise, monitoring the vehicle condition that reduces the likelihood of vehicle breakdown increases driver security. Car phones are purchased by women and the elderly primarily for reasons of safety and security rather than for routine communication purposes. In the same way, the use of technology can increase the safety and well-being of commercial drivers.

In addition to using technology to streamline hours of service and to monitor vehicles, cargoes, and drivers, technology, can be used to redesign the job itself. If a driver could work on a regular schedule, returning home on the same shift, it would be possible for more workers, including women and single parents, to participate. Drivers could drive a vehicle the first half of a shift and meet a similar vehicle coming from the other direction, so that at midshift the two drivers switch vehicles and return to their home terminals by the end of the work shift. Shifts can still operate round the clock, but the opportunity for workers to work regular hours on the same shift would greatly enhance the attractiveness of the job. Many companies, including those involved in JIT delivery, already closely approximate this schedule.

59.5.2.3 Benefits to Industry

The benefits to industry include (1) increasing the pool of potential applicants; (2) reducing on-road expenses, since fewer drivers will require maintenance far from home; (3) reducing the need for sleeper berths, with corresponding reductions in original vehicle costs, as well as vehicle operating costs; (4) improved employee health and morale, because drivers who are able to maintain a stable family life are usually happier and healthier; and (5) consequent reductions in driver turnover, and thus recruitment and training costs. It is likely that with improved morale, there would be increased productivity as well (Waller, 1993).

59.5.3 Role of Incentives and Disincentives in Operator Behavior

Because the topic of work schedules and their effects on worker performance is discussed extensively elsewhere (Tepas, Paley, and Popkin, Chapter 32 this volume), it is noted here only that commercial drivers, like others, respond to incentives and disincentives. If they are paid by the mile, they are obviously under pressure to maximize miles, even though by doing so they may exceed legal speed limits and hours of service. If, on the other hand, their compensation is based on hours worked, there is less pressure to violate safety regulations. If competitors are carrying multiple driver licenses so that violations, primarily speeding, can be spread over multiple records, it may be difficult for a driver to fail to do likewise and still remain competitive (Waller, 1986).

Although it is the driver that is the primary target of enforcement efforts, frequently the pressures to engage in unsafe practices come from elsewhere. If is often reported that motor carriers, responding to pressures from shippers, place unrealistic demands on drivers, requiring them to meet schedules that cannot be met operating within safety regulations (Chatterjee, Cadotte, Stamatiadis, Sink, Venigalla, and Gaides, 1994; U.S. Congress, 1988). Efforts to improve the safety and performance of commercial drivers must address commercial transportation as a system. Motor carriers, shippers, brokers, as well as drivers must be included in measures to reduce unsafe practices. In a national summit on this topic, such shared responsibility was identified as one of the top 10 priority issues in truck and bus safety (Federal Highway Administration, 1995b).

59.6 SOCIAL COSTS OF TRANSPORTATION

There are at least two categories of transportation-related costs, namely, those that result from the transportation that actually takes place and those that result from the lack of transportation that should occur. In addition, these costs may occur immediately or may, at least in part, be distributed over time.

59.6.1 Costs of Transportation That Occurs

The use of transportation involves the consumption of energy, either calories from food (as in walking or cycling), or from oil, natural gas, or coal for electricity (for motorized transport). Transportation may also impose environmental costs that affect more than simply those persons directly involved in the transportation. Even nonmotorized transport can impose such broader societal costs, as when food paths or bicycle paths destroy vegetation. However, it is motorized transport that is generally recognized as imposing the greatest societal costs. The costs include congestion, which is the curse of the city commuter; the disruption of neighborhoods; the impact on the environment (air, water, and land pollution; noise); motor vehicle injury, a major cause of death for the first half of life in the United States; consumption of nonrenewal resources; and dependence upon foreign energy supplies, with accompany reduced national security, reduced economic security, and less favorable balance of payments (National Research Council, 1992).

Although these broader societal costs of transportation are not widely taken into account in the total pricing of transportation, they are receiving increased attention (Federal Railroad Administration, 1993; Lowe, 1994).

59.6.2 Costs of Transportation Not Occurring

The second major category of social costs of transportation concerns the transportation that does not occur but should. This type of transportation cost is even more difficult to define and quantify. However, it is especially important that it be recognized at this point in the history of transportation. For most people in western society, and certainly for most people in the United States, transportation is an essential part of access to what is required to fulfill human potential. The lack of such access translates into the inability to develop potential, which, in turn, translates into social costs. An obvious example is the lack of transportation to move potential workers to the locus of available jobs. Inner-city residents are especially likely to be in this situation, but it also holds for many rural residents. Such lack of transportation generates societal costs for those in need of work, as well as for those in need of labor.

Lack of access to training and education can permanently limit the opportunities for future earnings and contributions. Because there has been little written about the costs of transportation denied, there are no available data, or even obvious methodologies, to capture the dimensions of the issue. However, because there is increased recognition of the broader societal costs of transportation that does occur, there may be pressure to capture such costs and charge them directly to the user. In doing so, we may deprive some population segments of needed transportation and consequently disproportionately increase the societal costs of transportation that does not occur but should.

59.6.3 Measuring Social Costs

59.6.3.1 Cost–Benefit Analysis

A standard approach to making transportation decisions is cost–benefit analysis, whereby the costs of providing the transportation are measured against the benefits, usually in terms of how many people and/or how much goods or information are transported in a specified period. Consideration is also given to ease of use, comfort, etc., but by and large we focus on the immediate benefits. However, there are also longer term monetary consequences of the transportation policies implemented, especially in relation to environmental impacts.

59.6.3.2 Moral Cost–Benefit Analysis

In considering parallel issues concerning environmental damage, Alan Gewirth (1990) makes a distinction between economic cost–benefit analyses and moral cost–benefit analyses. The two approaches are not diametrically opposed but rather are orthogonal. At times the economic approach may also be the most moral approach, but at other times the relationship could be different. The economic cost–benefit analysis defines both costs and benefits in monetary terms. Even health and life are reduced to dollar values, defined in terms of willingness to pay, or to be paid, for values realized as a result of the program. Decisions on whether to proceed with a project or program are determined by the relationship of the costs to the benefits—if the benefits exceed the costs, then it makes sense to proceed.

In contrast, a moral cost–benefits analysis is based on the assumption that all persons have certain human rights. Objections to this approach have been made, including the objection that it is difficult to reach agreement on just what is included in these basic human rights. A second objection concerns the difficulty in quantifying these rights so that one proposal may be weighed against another. Gewirth addresses these and other objections and discusses some of the specifics included in basic human rights. Among these are the right not to be "lied to, stolen from, or threatened with violence," as well as the right to "self-esteem, education, and opportunities for earning wealth and income." Each person has the right to take actions to realize these ends, although there is no requirement that one do so. Gewirth also defines rights as being hierarchical in nature. For example, the right to life and health would take precedence over the right not to be lied to.

This moral cost–benefit analysis can be applied to the transportation policies that we design and implement. Because in our society access to what is required for health care, education, employment, recreation, and other commodities necessary for human fulfillment almost always includes transportation, it may be argued that transportation itself must be included in the basic human rights that must be made available to all, independent of ability to pay. The move toward pay-as-you-go transportation may thus be depriving persons of this basic right.

59.7 TRANSPORTATION POLICY

59.7.1 Human Implications of Transportation Policies

Traditionally transportation cost allocation has focused on the costs of tangible factors, such as pavement wear and fuel consumption. Comparatively little attention has been paid to the human costs, other than the costs of motor vehicle injury. Ultimately these costs may overshadow all the others.

59.7.1.1 Free Access versus User Fees

Historically much of our surface transportation was based on toll roads and bridges, many of which were privately owned and operated. One of the crowning achievements of modern western society has been its highway system that offers free access to everyone. Whereas most users have access to some type of vehicle, including both motorized and nonmotorized, pedestrians may also make use of much of the system. This widely available free access has made possible the development of dispersed housing and work patterns, as well as the ability to engage in a wide variety of recreational activities, including visiting friends and relations and touring.

The growing cost of building and maintaining the transportation infrastructure is leading to growing interest in returning to the pay-as-you-go system of tolls. The levying of fees is viewed as an obvious and equitable way to charge the true costs of the facilities to those who benefit from the direct use of the system. In addition to tolls for use, there is also growing attention being given to congestion pricing, that is, charging higher fees for use of the transportation system at those times of peak activity.

Environmental Impacts

In addition to the problem of financing the building, operation, and maintenance of transportation facilities, the imposition of user fees and congestion pricing are supported for environmental reasons. User fees may discourage unnecessary travel, thus reducing any accompanying environmental costs. Presumably congestion pricing would discourage users from traveling at times of peak congestion, thus reducing congestion, travel delays, and air pollution. The probability of traffic incidents, with resulting delays, would also be reduced. Some environmental and societal impacts should result in reduction of social costs associated with transportation.

Equity Impacts

The direct linkage of highway use to user fees has implications that go beyond transportation per se. Although toll roads may simplify highway financing problems, and although congestion pricing may favorably affect the environment, such practices are likely to create equity problems. Which employees have the most control over when they commute to work? Which employees have the least control over when they commute to work? And what are the relative income levels of these two groups of employees?

It has been argued that a system of toll roads would have benefits for those who cannot pay the toll in that it would free up the public roads for their use. However, if those who can afford to pay, make use of the toll roads, it is probable that the public roads would gradually suffer from increasing neglect and fall into disrepair. Again, it will be those who are least able to pay who will experience the greatest transportation problems.

The imposition of user fees is a simple solution to a complex problem. It is important that it be considered not just from the short-term economic perspective but also from the longer term societal perspective.

59.7.1.2 Short-Term versus Long-Term Implications

Economic decisions in transportation are often based on limited consideration, both in terms of the time frame and the impact of the decision. For example, in the United States, contracting policies for highway construction have been based on the immediate costs of the construction rather than on the life cycle costs of the product. As a result, highways have deteriorated more rapidly, leading to increased vehicle maintenance costs, early road repair and reconstruction, and subsequent traffic delay costs. In contrast, Europe has based their road construction on life cycle costs, investing more at the outset but avoiding longer term expenses.

However, it is the broader societal costs that are most often ignored in transportation decisions. Public transit probably illustrates this tendency most clearly. The users of public transit often do not have access to alternative transportation but are not able to pay even a significant proportion of the true costs of providing the public transportation. Because in the case of public transit, the disparity between the costs and the charges imposed is so great, and because it is usually the local taxpayer who bears a significant part of the cost, it is increasingly difficult to generate the funding required to sustain public transit systems. Yet the lack of public transit may have grave consequences for the users. Ultimately, when the lack of transportation is a barrier to access to such resources as education, employment, and health care, the society loses the potential contributions of those who are denied such access. It is essential that the immediate direct costs of transportation be measured against more than simply the fare or toll that is collected. Transportation, or the lack of it, has human implications that far transcend the immediate exchange of monies.

Recognizing the long-term impacts of our actions, the environmentalists have clarified the rights of future generations. Partridge (1990) maintains that future generations, unborn and anonymous, have legitimate claims on us, and that we, in turn, have responsibilities toward these unknown future persons. Among these responsibilities is that of leaving the planet in reasonable condition, that is, with its air, water, and earth free of toxic substances; and with its flora and fauna likewise in reasonable numbers and health. We do not have the right to pollute the environment or to hunt and fish to extinction the animals and marine life with which we share the planet. Partridge elaborates at length the basis for his position, but it is our responsibility to the future that will be considered here in regard to transportation policy.

Economic cost–benefit analysis is being applied widely in making decisions about transportation programs. Whereas issues of equity are increasingly being raised, those in transportation have not seriously explored how a moral cost–benefit approach might be applicable to the transportation policies being formulated. When significant numbers of the society are deprived of their "right" to transportation and are consequently unable to access education or health care, the consequences may result in far greater societal costs than the monies "saved" by eliminating or not providing the transportation in the first place.

59.7.2 Human Issues in Formulating Transportation Policy

The field of transportation has been dominated by engineers, but transportation policy ultimately determines where and how engineering expertise is put to use. The human issues that influence and, in turn, are affected by transporttion policy are much less clearly defined.

59.7.2.1 Values Underlying Policy

It is rare that the values underlying transportation policies are specifically articulated. However, they are implicit in the kinds of decisions that are implemented. These values may be classified into three categories, namely, economic, political, and societal.

Economic Values

It is this category that most often determines transportation decisions. Economic values are also the ones that are most likely to be explicitly articulated. At times it is tacitly agreed that market forces should be the only basis for making transportation policy. Many maintain that market forces are the ultimate "good" and that, as Adam Smith (1776) postulated, if every man (sic) works toward his own best interests, the entire process is guided by an "invisible hand" so that the outcome is in the society's best interests.

Although this philosophy has guided much of economic thinking in the United States, including that underlying transportation policy, it is fraught with pitfalls. As Hardin (1968) has shown, when the commons, or the resources, are limited, it is not always in the best interests of the overall society when each individual works to further his own interests. Because the public highway transportation system largely meets the definition of a commons, and because its capacity and the resources supporting it are limited, it is essential that values beyond economic consideration be taken into account (Waller, 1986).

Political Values

As one politician put it, "You can't take the politics out of politics." As a result, it is common practice for those political leaders with authority for distribution of public funds for transportation to support projects in their home districts or states. As a result, there are modern highways that provide ready access to family dairy farms and multilane freeways that run to sparsely settled counties. Because transportation involves large sums of money, as well as good jobs, there is ample political influence on the allocation of funds. This is an area that could benefit from innovative research to devise more equitable means of distributing resources but that would also be acceptable to political decision makers.

Although political values permeate transportation policy, they are not so readily articulated as are economic values.

Societal Values

Almost all transportation policy decisions have important societal ramifications, both immediate and long term. However, it is the societal values that are least likely to receive adequate attention. The reasons for this are many. First, societal values traditionally have not been considered except where they are supportive of economic interests. Thus, it may be argued that it is in the interests of society to build roads to enable access to shopping malls or to increase truck traffic so that the public may benefit from the resulting delivery of goods.

It is appropriate for business and industry to promote their own interests and to work to maximize profit. It is unrealistic to expect business and industry to assume major responsibility for the public interest beyond the extent to which it may relate to greater private gain. In contrast, government is charged with the responsibility for representing the public's interests. However, our elected officials must engage in extensive fund raising activities to support campaigns. Such funding does not come without strings attached. As a result, it is almost always those who are least able to represent their own interests who are neglected, as funding allocation decisions are made. Yet it is these same neglected populations who are likely to be in greatest need of transportation.

59.7.2.2 Ramifications of Values Adopted

The values that underlie our transportation policies should be clearly delineated. Otherwise, transportation policy proceeds in the absence of a well-defined purpose or integrating principle. If the prevailing value system is primarily economic, consideration should be given as to whether only short term immediate economic impacts are of interest, or whether longer term and broader economic impacts should also be considered. The earlier example of public transportation illustrates this distinction. If the overriding goal of transportation is to move more people and goods faster and safer, then the focus is on increasing throughput while decreasing injury. Much of the application of technology in ITS has aimed at achieving this end through making more efficient use of existing transportation facilities. However, if the goal of transportation is to assist in achieving broader societal goals, then the transportation community will be required to make major changes in how it does its business. It will become necessary for the transportation community to learn how to communicate with agencies and segments of society that have not previously been participants in defining transportation policy. In turn, it will require that nontransportation agencies and interests learn to involve and communicate with the transportation community to engage in cooperative efforts to integrate transportation into the solution of broader societal programs.

An illustration comes from the field of health care. In virtually all western nations, the rapidly rising costs of health care are challenging the widely held value that everyone should be able to receive needed health care. For most parts of the United States, transportation is an essential part of access to health care. Yet historically the transportation community and the health care community have not interacted in any systematic fashion. The advent of communications technology could facilitate such interaction, so that the transportation component of health care could be integrated into the system.

It has been reported that some health care providers do not like to accept Medicare or Medicaid patients because of the high rate of "no shows." Presumably transportation, or the lack of it, is a major factor in the failure to arrive for appointments. This proposition raises empirical questions that are subject to investigation. First, to what extent do health care providers cite "no shows" as a reason to refuse certain classes of patients? Second, what is the actual rate of "no shows" for these patients as compared to other patient categories? Third, to what extent is transportation actually a factor in these patients' not keeping their appointments? Fourth, if transportation is provided, what effect is there on the rate of "no shows?" Fifth, if "no shows" decrease, are health care providers more willing to accept these patients? Similar linkages should be established between transportation and other pressing societal issues, such as the role of transportation in employment or the role of transportation in education.

As long as transportation policy is established solely on the basis of short-term, direct economic considerations, such linkages between transportation and larger societal goals will not be pursued. However, if transportation is viewed as an important and integral part of meeting other larger societal goals, the transportation community can begin to address human issues that transcend the immediate transportation per se. Clearly defining and establishing such linkages may also increase public support for transportation investment.

59.8 CUSTOMER REQUIREMENTS

Human factors specialists must be aware of the general human and societal context in which transportation occurs. However, if transportation systems are to be useful and usable, human factors specialists must also be aware of the specific design requirements to meet the changing needs of individual users (customers). This portion of the chapter provides some of the engineering data desired to meet those needs.

59.8.1 What Are the Customer Requirements for Automobiles?

As noted in the introductory figure of this chapter, Figure 59.4, user/customer needs, technology push, regulatory requirements, and the work force available influence the design of safe transportation systems that are easy to use. Except for work force issues, the design impacts of each of those factors are explored separately in the following sections. Although the emphasis of this section is on passenger vehicles, the ideas expressed apply to many forms of transportation.

For passenger cars, and, to a lesser extent, light trucks, heavy trucks, motorcycles, and buses (in that order), customer needs are often expressed using the language of quality function deployment (QFD) (Green, 1995a; Hauser and Clausing, 1988).

QFD involves (1) listing and weighing the customer attributes, (2) identifying the engineering characteristics, (3) determining the linkage between customer attributes and engineering characteristics, (4) collecting objective data for each engineering characteristic and identifying links between them, (5) setting performance targets, and (6) determining the process to achieve the targets. Customer attributes are the qualities customers desire in a product or service expressed in their own words (e.g., a door should be easy to open). The engineering characteristics are the product specifications necessary to satisfy the customer attributes (e.g., after the door has been opened 1000 times and the vehicle has completed the standard corrosion resistance test, the maximum force on the door handle to open the door should not exceed x pounds). QFD formalizes the steps human factors professionals have historically employed in systems engineering.

Each manufacturer has its own variation of this process, and often its own name for it. Ford's name, Best-In-Class (BIC), appears most often.

For automobiles, customer attributes can be grouped into five categories. Table 59.2, developed by the author, shows both those categories and associated subcategories. Each of the characteristics in the table can be partitioned into several levels of subcharacteristics. For example, for temperature control, subcharacteristics could be specified for control accessibility and the ease of setting a desired temperature. The design of that control may depend upon the design of other controls used to set the inside–outside air mixture and select a climate system mode.

Notice that the emphasis of customer characteristics is on usability. In-vehicle interfaces that are easy to use are generally safe to use as well, as such interfaces minimize the extent to which the driver is distracted from primary driving tasks, as well as the probability of errors in judgment or response.

When the BIC approach was first developed, the focus was on the product that best exemplified the desired characteristic. For example, to obtain desired characteristics for a power window control, Ford would compare its current model Taurus with competitive products (e.g., Honda Accord, Chevrolet Lumina) to identify the best design (the "Best-In-Class") (see Callahan, 1986a,b,c). Measurements of the best product provided the desired engineering characteristics (e.g., reach distance to a particular switch, switch operation force, etc.).

Although this approach provides a constant incentive to improve products, improvements tend to be incremental. Systematic explorations of the relationship between product qualities (e.g., operation forces, reach distances, character sizes) and customer performance and preference are not commonly conducted (and rarely reported in the literature). Also unknown is the relative importance of various product characteristics (e.g., Is reach distance to a control as important as the size of the label?) (Gupta and Ratchford, 1992; MacCarthy and Tay, 1989).

59.8.2 What Are Customer Requirements for Other Vehicles?

The author has yet to see comprehensive lists of attributes for other types of vehicles (air, rail, water, etc.), though such lists may exist. Table 59.3 shows an abbreviated list for trucks, created by the author. Whereas different classes of vehicles may have common engineering characteristics, the relative importance of those characteristics and the desired values for those characteristics can be quite different. For example, for motorcycles, ease of starting, noise levels, and mirror requirements are particular concerns (Motoki and Tsukisaka, 1989). Similarly, for trucks, field-of-view requirements are of special concern (Yamanaka and Kobayashi, 1970).

Table 59.2 Selected Attributes of Importance to the Customer

Category	Subcategory	Selected Topics/Issues
Occupant protection	Seat belts	Protection suggested, most comfortable to wear, easiest to fasten/unfasten
	Air bags	Perceived protection vs. impact type and occupant position
	Interior geometry, components, and structure	Best padding of instrument panel
	Fire, electrical, and chemical hazards	Least flammable interior, fire fighting equipment provided
Vehicle movement and control	Steering	Desired force levels, most responsive
	Braking	Desired force levels, shortest operation time, shortest braking distance
	Ride quality	Smoothest highway ride, no jerk after potholes
	Seat comfort	Accommodates widest range of driver anthropometry, most comfortable seat shape, most comfortable seat materials
Exterior design	Field of view, glazing	Easiest to drive, easiest to park
	Headlights, tail-lights	Minimizes glare to others, provides maximum illumination on road, minimizes sight distance
	Mirrors	Minimizes glare to others, maximizes detection distance, minimizes blind spots
	Conspicuity	Detected at greatest distance, greatest aesthetic appeal
Interior design	Controls	Best switch type, best shape, best location, best force/travel
	Displays	Best gauge design, most understandable text messages, symbols and labels, best lighting
	Thermal comfort, air quality	Best airflow, best temperature control, lowest radiant load, fewest fumes
	Sound environment	Lowest exterior and interior sound levels, best audio system quality, most intelligible speech (on phone), most audible warnings
Access, maintenance, and repair	Ingress, egress	Best door size and swing angle, easiest to use handles and supports
	Trunk and underhood	Best access (depends on task performed), fewest hot surfaces
	Maintenance and repair	Minimum time to repair or replace tires, fluids, major items
	Documentation and manuals	Most understandable, best description of legal requirements

Given space limitations, it is not possible to specify all the desired engineering characteristics for product features for a wide range of vehicles within this chapter. To give the reader a sense of the breadth of the requirements, primary references for the topics shown in Table 59.2 are provided in Table 59.4.

59.8.3 How Should Secondary Controls Be Designed?

59.8.3.1 Switch Type and Location

The design requirements that emerge from QFD can be quite extensive, encompassing customer preferences, time and error data, learning functions, as well as information on

Table 59.3 Selected Attributes of Importance to a Truck Customer

Category	Subcategories
Occupant protection	Seat belts; air bags; interior geometry, components, and structure; fire, electrical, and chemical hazards
Vehicle movement and control	Steering; braking; guidance; ride quality; seat comfort
Exterior design	Field of view, glazing; headlights, tail-lights; mirrors; conspicuity
Interior design	Controls; displays; thermal comfort; air quality; sound environment, sleeping accommodations, food storage, drink storage
Cargo	Loading/unloading, capacity, how to secure, tracking, temperature control
Access, maintenance, and repair	Ingress, egress; trunk and underhood; maintenance and repair; documentation and manuals

cost, ease of repair, etc. To give the reader a sense of the scope of the information, example driver preference and performance data are provided for two topics: displays (in the next section) and controls (in particular, secondary controls). Secondary controls include switches such as headlights on/off and the front windshield wipers. For secondary controls, commonly asked questions by vehicle designers are: (1) Where should secondary controls be located? (2) What types of switches should be provided? (3) How should they operate?

To address these issues, studies have examined where secondary controls are located in cars, and driver preferences for controls in sedans and sport cars (Green and Goldstein,

Table 59.4 Selected References on Vehicle Design Characteristics

Subcategory	Primary References
Seat belts	Stapp Car Crash Conference (1995)
Air bags	Stapp Car Crash Conference (1995)
Interior geometry, components, and structure	Stapp Car Crash Conference (1995)
Fire, electrical, and chemical hazards	
Steering	Bundorf (1967), Gillespie, (1992); Whitcomb and Milliken (1956)
Braking	Mortimer et al. (1979), Newcomb and Spur (1967)
Ride quality	Leatherwood, and Dempsey (1976), Leatherwood, Dempsey, and Clevenson (1980)
Seat comfort	Habsburg and Middendorf (1977); Hertzberg (1972); Kamijo, Tsujimura, Obara, and Katsumata (1982); Reed, Saito, Kakishima, Lee, and Schneider (1991)
Field of view, glazing	Henderson, Smith, Burger, and Stern (1983); Maurer and Fawcett (1973)
Headlights, tail-lights	Sivak and Flannagan (1993)
Mirrors	Flannagan and Sivak (1993)
Conspicuity	Burger et al. (1981)
Controls	Turner and Green (1987)
Displays	Green (1989)
Thermal comfort, air quality	Kawamoto, Hosokawa, Suto, and Nakajima (1986); Rolle, Romitelli, and Savasta (1992)
Sound environment	Nelson (1987)
Ingress, egress	Koester and Hamilton (1994); Loczi (1993)
Trunk and underhood	
Maintenance and repair	
Documentation and manuals	Doheny-Farina (1988); Duffy and Waller (1985); Rosenbaum and Anschuetz (1994); Velotta (1995)

1989; Green, Ottens, and Adams, 1987; Green, Ottens, Kerst, Goldstein, and Adams, 1987; Green, Paelke, and Clark, 1989; Turner and Green, 1987). In those studies, drivers sitting in a vehicle mockup saw a list of control names. Drivers selected the switch they preferred for each function (from several hundred) and placed the Velcro-backed switch on the instrument panel where they thought it belonged. Drivers then operated each control once while driving a simulator. Subsequently, drivers made adjustments in switch preferences and locations, changing approximately 10–15% of the preferences in some manner. This process has become known as the "Potato Head" method because of its resemblance to the actions in using a child's toy of the same name (Green, Paelke, and Boreczky, 1992).

These experiments indicated a lack of consensus as to where controls should be located, the type of controls preferred, and the desired direction of motion. Figures 59.7, 59.8, 59.9, and 59.10 show examples of preferences for controls in sedans. Preferences were strongest for controls associated with cruise functions.

59.8.3.2 Direction-of-Motion Stereotypes

Another important consideration for controls is making the method of operation consistent with direction-of-motion stereotypes (Wierwille and McFarlane, 1993). Configurations that conform to driver expectations are more likely to be operated correctly, and therefore will be easier and safer to use (see also McFarlane and Wierwille, 1990; Wierwille and McFarlane, 1991). Whereas there is a significant body of literature on direction-of-motion stereotypes (Hoffman, 1990; Loveless, 1962), application of those principles to unique automotive situations is not straightforward. Figures 59.11 and 59.12 show some typical results.

59.8.4 How Should Displays Be Designed?

In addition to automotive-specific literature on controls, there is a body of literature on displays (e.g., Boreczky, Green, Bos, and Kerst, 1988; Green, 1988, 1989; Green, Goldstein, Zeltner, and Adams, 1988). Probably the most important question is how large numbers and letters should be displayed so that they will be easy to read. Traditionally,

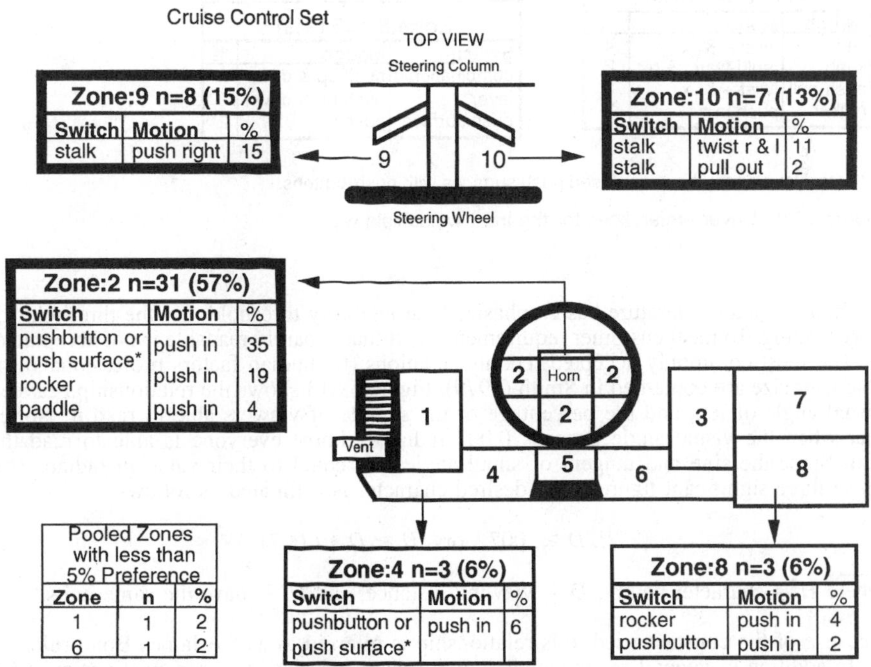

Figure 59.7 Driver preferences for the cruise control set.

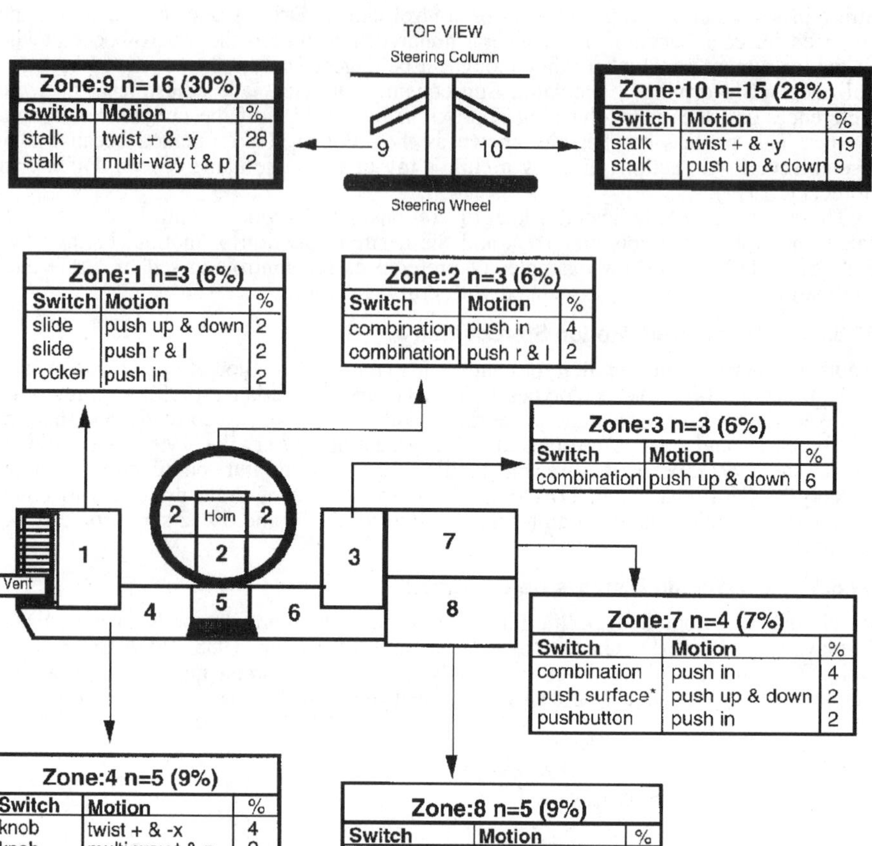

Figure 59.8 Driver preferences for the front windshield wiper.

the human factors literature has emphasized the legibility threshold, not the threshold for *easy* reading. To meet customer requirements, text that is barely readable is not acceptable.

The most commonly accepted recommendations (by human factors professionals) for character size are contained in Smith (1979). Figure 59.13 shows the relationship between visual angle of text and the percentage of the sample of viewers able to read it. Notice that when the visual angle exceeds 0.007 radians, almost everyone is able to read the text. Since the sine and tangent of small angles are equal to their value in radians (for up to three significant figures), the desired character is estimated as follows:

$$H/D \geq .007 \quad \text{or} \quad H \geq D * (.007)$$

where H = character height, D = viewing distance, and H, D have the same units.

Because of the constant used, this relationship is referred to as the James Bond rule.

In addition to general recommendations for character size, recommendations exist for specific applications. When making recommendations for in-vehicle displays, reaccommodation between the displays and the exterior scene must be considered (Boreczky et

Hazard

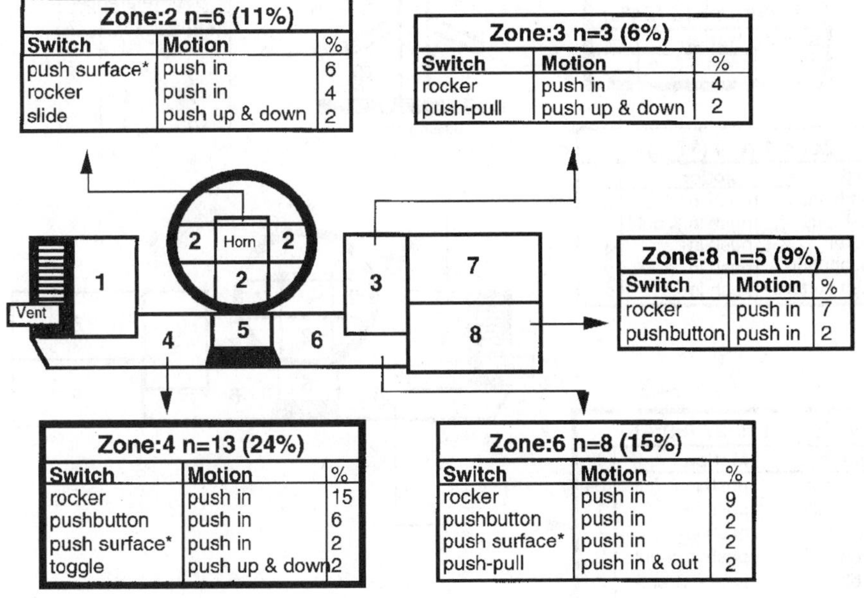

Zone:2 n=6 (11%)
Switch | Motion | %
push surface* | push in | 6
rocker | push in | 4
slide | push up & down | 2
Zone:3 n=3 (6%)
Switch | Motion | %
rocker | push in | 4
push-pull | push up & down | 2
2
Horn
2
2
1
Vent
3
7
8
4
5
6
Zone:8 n=5 (9%)
Switch | Motion | %
rocker | push in | 7
pushbutton | push in | 2
Zone:4 n=13 (24%)
Switch | Motion | %
rocker | push in | 15
pushbutton | push in | 6
push surface* | push in | 2
toggle | push up & down | 2
Zone:6 n=8 (15%)
Switch | Motion | %
rocker | push in | 9
pushbutton | push in | 2
push surface* | push in | 2
push-pull | push in & out | 2

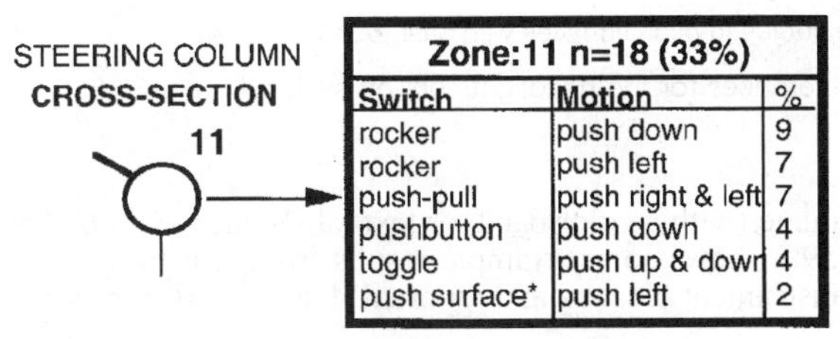

Pooled Zones with less than 5% Preference		
Zone	n	%
1	1	2

*Note: Drivers may have confused push surfaces with push buttons.

Figure 59.9 Driver preferences for the hazard switch.

al., 1988). Such studies typically involve responding to a task outside the vehicle or a slide of an instrument cluster inside a vehicle (e.g., indicate if the speed shown was in excess of speed limit). Response time (*RT*), the time from when a speedometer appeared until a button was pressed, is given by the following expression:

$$RT\ (\text{ms}) = 1054 - 320(A) + 1050(1/H) + 2202(L) + 89.6(1/\ln(C)) - 9.58(\ln(I)) + 4538(1/H^2)$$

where: A = age group (1 for old, 2 for young), H = digit height (mm, for 5–19 mm), L = location (1 for center, 2 for sides), C = contrast ratio (for 1.5:1 to 20:1), I = illumination (lux, for 1.08 to 915).

Although speedometers and tachometers are generally moving pointer displays, not numeric, there are very few studies of moving pointer speedometers. There is, however, a rich body of literature on gauge design (Green, 1988). For other displays of fuel gauges, engine temperature gauges, etc., the key reference is Green (1984). That study indicated that fuel and engine gauges should be pointer displays, not numeric displays, and that

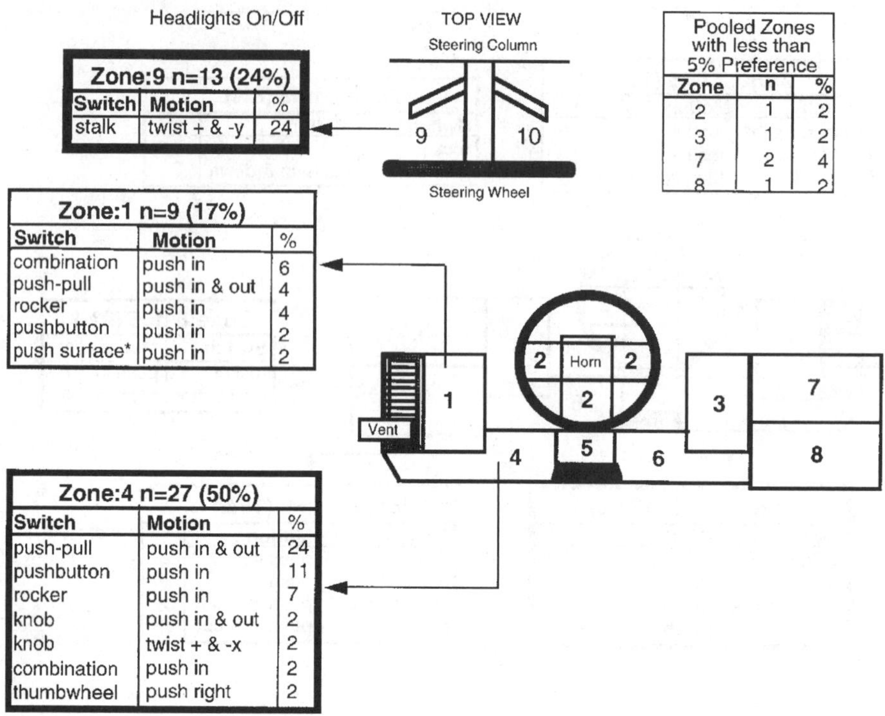

Figure 59.10 Driver preferences for the headlights on/off switch.

indicating the normal range (with a color band, or text labels such as "OK" or "normal") was beneficial. Table 59.5 shows some example results from situations in which drivers were shown slides of instrument clusters and reported if some sort of problem was being shown.

Thus, the automotive literature contains a significant amount of application-specific design information. This information builds upon the general design information (on controls, displays, etc.) in the human factors literature. General human factors information alone is not adequate for product design (at least for automobiles).

59.9 WHAT TECHNOLOGIES ARE LIKELY TO APPEAR IN CARS OF THE FUTURE?

To satisfy customers, engineers must be aware of the features afforded by new technology. In the last few years, there have been great strides made in incorporating computer and communications technology into new vehicles (see Green et al., 1991; Michon, 1993; Parkes and Franzen, 1993; Serafin, Williams, Paelke, and Green, 1991). Shown in Table 59.6 are estimates of when several new technologies will achieve 5% market penetration in the United States (Underwood, 1992). Traffic and navigation systems are likely to appear first, collision avoidance systems second, enhanced vision third, and finally, advanced control systems. Electronic systems (e.g., electronic traction control, TV) generally have appeared earlier in Japan than elsewhere, because of customer demands, more permissive liability laws, and, in the case of navigation, greater road network complexity.

59.9.1 Navigation Interface Design Requirements

Of these new electronics systems, navigation is likely to be one of those first implemented. The interface can be quite complex and has received considerable attention from the human factors community (Table 59.7). Critical questions include: (1) Which modality should be used for guidance: visual, auditory (voice), or both? (2) Does the preferred

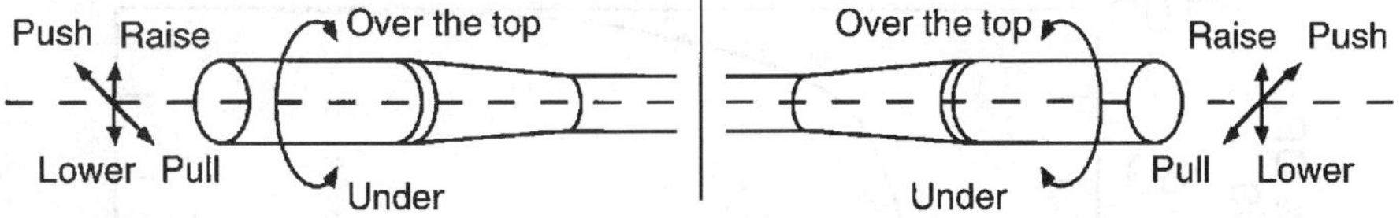

Control location	Action	Desired Action	Movement Executed	% Complying
Nonspecific				
Left	Turn on	Raise or lower	Raise	96
Right	Turn on	Raise or lower	Raise	76
Left	Turn on	Push or pull	Pull	76
Right	Turn on	Push or pull	Pull	72
Left	Turn on	Rotate barrel	Over the top	82
Right	Turn on	Rotate barrel	Over the top	94
Specific Unconstrained				
Left	Turn on wipers	Unspecified	Over the top	50
Right	Turn on wipers	Unspecified	Over the top	50
Left	Turn on hi beam	Unspecified	Pull	76
Specific Constrained				
Left	Turn on wipers	Raise or lower	Raise	94
Right	Turn on wipers	Raise or lower	Raise	94
Left	Turn on wipers	Rotate barrel	Over the top	72
Right	Turn on wipers	Rotate barrel	Over the top	72
Left	Turn on hi beam	Push or pull	Pull	80
Left	Turn on headlight	Rotate barrel	Over the top	88

Figure 59.11 Direction-of-motion stereotypes for stalk controls.

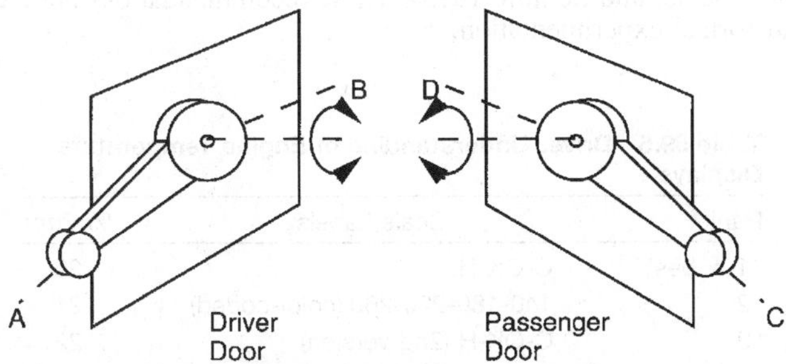

Condition		To 'raise' window		To 'lower' window		
Location	Handle points	Movement Executed	%	Movement Executed	%	Overall %
A-Driver door	Rear	Up (or cw)	52	Down (or ccw)	64	59
B-Driver door	Forward	(Up & down equal)	50	Down (or cw)	54	52
C-Passenger door	Rear	Down (or cw)	68	Up (or ccw)	52	60
D-Passenger door	Forward	Up (or cw)	84	Down (or ccw)	90	87

Figure 59.12 Direction-of-motion stereotypes for window cranks.

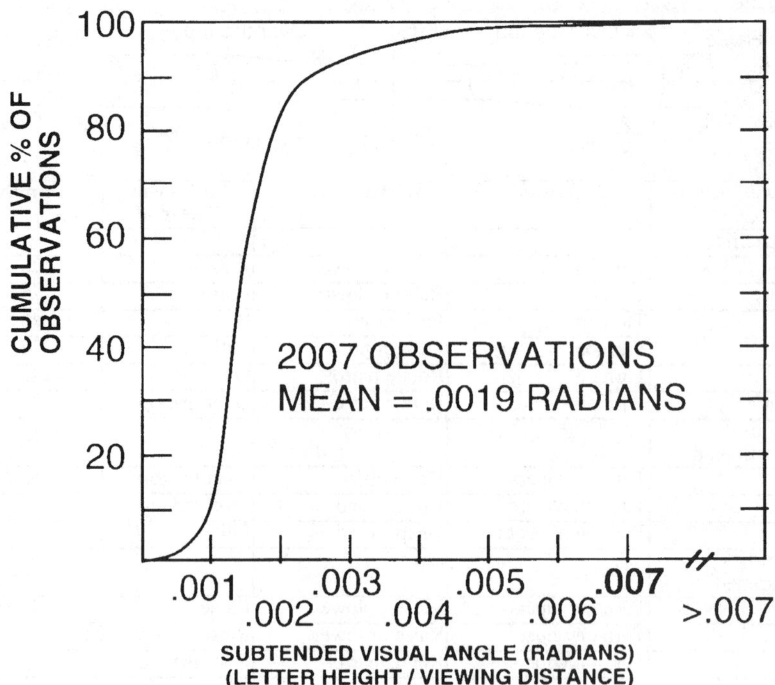

Figure 59.13 Cumulative distribution of the threshold visual angle.

modality vary with the culture? (3) What display format should be used for route planning? (4) How should destinations be entered?

The route-following question has received the most attention and a significant body of literature pertaining to it has emerged. (For information on destination entry see Paelke, 1993.) The following is a summary of current recommendations for route-following interface design, in particular those pertaining to turn displays (see Green, 1992, 1996; Green, Levison, Paelke, and Serafin, 1995). These recommendations are based on design experience and formal experimentation.

Table 59.5 Driver Understanding of Engine Temperature Displays

Rank	Scale Labels	% Error
1 = best	C-OK-H	20
2	140-180-220-260 (color-coded)	21
3	C-OK-H (2nd version)	22
4	C-H (color-coded)	23
5	140-180-220-260 (with OK)	24
6	C-H	24
7	C-H (2nd version)	29
8	140-180-220-260	31
9	cold-OK-hot	34
10	xxx °F (color-coded)	40
11	xxx °C (color-coded)	49
12	xxx °F	52
13 = worst	xxx °C	57

Table 59.6 Expected Implementation Time Frames of New Technology[a]

Function	5% Market Share	50% Market Share
Real-time traffic information	1996	2007
Mayday (emergency call)	1998	2010
Yellow pages, etc.	2000	2012
Traffic-adaptive route guidance	2000	2020
Autonomous navigation	2000	2012
Frontal collision warning	2002	2013
Backup, blind spot detection	2002	2015
Adaptive (intelligent) cruise[b]	2004	2015
Automatic braking	2008	2020
Rollover warning (for trucks)	2010	never
Night or fog vision	2010	never
Automatic lane keeping	2011	2032
Automated platooning	2035	never
Automated driving	2040	never

[a]Source from Underwood (1992).
[b]Note: Intelligent cruise control systems (which sense surrounding vehicles and adjust speed based on traffic) are likely to achieve commercial success in the United States because expressway speeds vary with congestion. With added functionality (automatic braking, automated lane keeping, etc.), this feature will evolve into an automated driving system.

Typically, visual displays are accompanied by a voice message that tells drivers when to turn. According to Green and George (1995), the desired presentation time (time before the intersection), can be estimated as shown below (see also George, Green, and Fleming, 1995). Similar data appear in Ross, Nicolle, and Brade (1994).

$$\begin{aligned}\text{Distance (ft)} = {} & -389 + 119\,(\text{Age.code}) - 113\,(\text{Sex.code}) \\ & + 95\,(\text{Turn.code}) + 15\,(\text{Speed}) \\ & + 21\,(\text{Number of Vehicles}).\end{aligned}$$

where Age.code, 1 = young, 2 = middle, 3 = older; Sex.code, 1 = women, 2 = men; Turn.code, 1 = right, 2 = left; Speed is in mi/h; and Number of Vehicles refers to vehicles ahead.

Besides message timing, message content is important (Table 59.8).

Although the data presented in this section concerning turn display design may seem extensive, this information addresses only some of the design issues. Additional research to support design is needed.

59.10 WHAT ARE THE REGULATORY REQUIREMENTS?

International standards, national standards, and standards and guidelines from professional and technical societies influence the design of vehicles. Compliance requirements and the size of the market affected determine their impact on design.

59.10.1 International Requirements

The trend is to place more reliance on international standards, in particular those developed by the International Standards Organization (ISO), in place of national, defense, or other standards. As examples, for motor vehicle design, two commonly referenced standards are ISO 2575-symbols (International Standards Organization, 1982) and ISO 4040-control and display location (International Standards Organization, 1977). Whereas ISO standards are voluntary, there are many instances when governments require compliance with them. ISO standards allow vehicle manufacturers and suppliers to make one product that can be sold in many countries. Eliminating country-specific variations leads to a significant cost savings to manufacturers, and ultimately the consumer.

Table 59.7 Selected Turn Display Design Characteristics

Design Requirement	Rationale and Comments
Show plan or aerial (bird's eye) view, not perspective.	While perspective view matches the driver's view of the world, the details of the upcoming intersection geometry are more apparent for plan and aerial views.
Roads should be solid lines, not lines for edges.	Using solid lines makes the roads more prominent.
Show landmarks (stop signs and traffic lights) at intersections.	In the United States (and elsewhere), drivers naturally use landmarks for guidance.
Required information includes the road being driven, the road for the next turn, the direction and approximate angle of the turn, and the distance to the turn.	These are the basic elements of a turn display.
For complex choice points, show all roads and ramps.	The navigation display should match the world driven.
For close proximity turns (within 0.1 miles), show both maneuvers at one time.	Successive turns are planned as a single maneuver sequence and drivers need all of the information before the first turn to plan the sequence successfully. Further, there may not be time in the middle of the sequence to look at the display for guidance for the second turn.
Omit count down bars if the distance to the turn is highly legible.	In the first on-road version of the University of Michigan Transportation Research Institute (UMTRI) interface, countdown bars were provided. Some subjects did not know what they were. When omitted from later designs, the interface was simpler and navigation performance did not suffer.
For expressway exits, give both the route name and city.	Information on the navigation display should match that appearing on highway signs.
Displays should be heading up while driving, north up while planning.	The display should be compatible with the real world. It is commonly observed that drivers rotate paper maps when following routes, even if that means the printing is inverted.
Distances should be given to the nearest 0.1 miles.	This is the accuracy used by the odometer and by highway signs.
Rotating displays during turns is not required.	Drivers generally do not look at a turn display in the middle of turns. Having the display change during a turn can be distracting to drivers.
Turn displays should emphasize local geometry.	Example: If a turn appears to be 90° at low resolution, but in fact consists of a 30° bend and then a 60° turn, both turns should be shown.
Omit longitude and latitude.	Except for off-road use and where used for destination entry, these coordinates are difficult for typical U.S. drivers to use.

Table 59.8 What Should Voice Guidance Messages Say?

Design Requirement	Rationale and Comments
The final turn message should use the word "approaching" or other words to avoid being a command.	If drivers are told "turn right" instead of "approaching ———, turn right," they may believe the vehicle knows everything and turn, even if a traffic light or one way sign indicates otherwise.
Time messages to coincide with advance notice signs on expressways.	Auditory and visual information should agree.
Limit the number of prepositional phrases in a message generally to no more than four, preferably three.	Short-term memory will be overloaded if there is too much information to remember.

Table 59.9 Selected Federal Motor Vehicle Safety Standards

Standard No.	Topic
101	Controls and displays
102	Transmission shift lever sequence
103	Windshield defrosting and defogging
104	Windshield wiping and washing
107	Reflecting metal surfaces (glare)
111	Rearview mirrors
125	Warning devices
131	School bus pedestrian safety devices
201	Occupant protection in interior impact
208	Occupant restraints
213	Child restraint systems
218	Motorcycle helmets

59.10.2 U.S. Department of Transportation (DOT) Requirements

The 57 Federal Motor Vehicle Safety Standards are the most important requirements for vehicles sold in the United States. These requirements appear in Section 49 of the Code of Federal Regulations, Part 571 (49 CFR Part 571). Table 59.9 shows some examples. Other countries have similar standards.

59.10.3 SAE Requirements

The Society of Automotive Engineers (SAE) has developed a sizable collection of technical standards pertaining to vehicle design, primarily for application to the U.S. market. Similar sets of standards (from other professional societies) exist for other countries. In recent years, SAE standards have tended to lag behind parallel international standards, although there are a few areas (e.g., control compatibility) where SAE documents are more current.

SAE publishes three types of documents: (1) standards, (2) recommended practices, and (3) information reports. Standards specify what designers must do. Since no one requires compliance with SAE standards (and there is no authority to enforce them), compliance is lower than with DOT requirements ("the law"). Recommended practices describe what designers should do and are more suggestive than standards. Information reports, the least stringent of the three formats, communicate or summarize technical information. Table 59.10 lists some of the driver-related SAE documents pertaining to automobiles (Society of Automotive Engineers, 1995). The *SAE Handbook* also lists standards pertaining to trucks, construction equipment, and other types of vehicles.

Table 59.10 Selected SAE Documents

Document	Title
J128	Occupant Restraint System Evaluation—Passenger Cars
J985 Oct88	Vision Factors Considerations in Rear View Mirror Design
J1050a	Describing and Measuring the Driver's Field of View
J1052 May87	Motor Vehicle Driver and Passenger Head Position
J1138	Design Criteria—Driver Hand Controls Location for Passenger Cars, Multi-Purpose Passenger Vehicles, and Trucks (10000 GVW and Under)
J1139 Apr94	Direction-of-Motion Stereotypes for Automotive Hand Controls
J1517	Driver Selected Seat Position
J1606 Mar93	Headlamp Design Guidelines for Mature Drivers
J2119 Jun93	Manual Controls for Mature Drivers
J2189 Feb93	Guideline for Evaluating Child Restraint System Interactions with Deploying Airbags

Identifying appropriate standards and guidelines, especially on an international basis, is a difficult and expensive activity. Most major manufacturers have engineering teams whose sole purpose is to assemble and assess such information. In the future, it is likely the World Wide Web will be particularly important for identifying and retrieving standards and guidelines.

59.11 SUMMARY

In most of Western society transportation is an essential part of access to what is necessary to develop individual and community potential. The very poor in our society lack transportation access to education, employment, and health care, as well as such basic needs as food and shelter (Meyerhoff, Micozzi, and Rowen, 1993), so that it is almost impossible for them to change their situation. Koutsopoulos and Schmidt (1976) summarize the mobility constraints placed on those in our society who do not have access to the automobile, concluding that

> *the link between mobility and social goals is much stronger than many people acknowledge and more complicated than transportation experts would like to believe. (p. 82)*

Two decades later these words remain an accurate description of the current state of affairs.

Historically, the science of human factors has focused on how design and operations affect human performance and productivity. Physical measures of drivers have considered their reach, strength, size, etc., so that vehicle and highway design may match the operator's capabilities. In addition, cognitive performance has been addressed in relation to interpretation of signs and symbols. Ambient conditions, such as temperature, humidity, noise, vibration, and lighting, have been studied in relation to their effects on physical and cognitive performance.

The latter part of this chapter has addressed some of these more traditional human factors concerns in transportation, with an emphasis on the passenger vehicle and particularly customer requirements. It has also described some of the technologies available and the regulatory environment (both national and international) in which human factors research and design occur. This section of the chapter has provided multiple illustrations of classical human factors research in the design of motor vehicles, as well as the application of human factors research to the developing intelligent transportation systems. An extensive list of references is included, providing the reader with additional resources for further inquiry. Readers interested in other documents should see Peacock and Karwowski (1993) and Green (1995b,c).

However, this chapter has gone beyond the classical human factors concept of the relationships between the human and the machine. It has broadened the concept and role of human factors in transportation, extending the traditional model to include a new domain, that is, societal human factors. If human factors activity is extended to this new domain, the individual operator is linked with the broader society and even with future generations.

Transportation is not an end in itself, but a means to an end. Consequently, it is important to consider how effectively our transportation system contributes to the achievement of socially determined "ends." Rather than simply focusing on how technology can move more people, goods, and information faster and safer, the transportation professional should participate in defining our societal goals and determining how our transportation expertise can be applied to achieve those goals. Such an approach extends the concept of the "fit" between the user and the system beyond the more immediate relationship that has traditionally dominated the field of human factors.

Finally, the definition of the human user should be extended even further. The values underlying our transportation policies need to take into account the impact on future generations. Transportation systems that address only the immediate situation can impose major costs on future and more remote populations. A science that truly considers the human dimension must consider such future impacts as well.

ACKNOWLEDGMENT

The authors are indebted to Tandi Bagian, whose thoughtful commentary was particularly helpful.

REFERENCES

Belton, B. (1992). Automakers come a long way, baby. *USA Today,* January 27, 1A, final edition.

Boreczky, J., Green, P., Bos, T., and Kerst, J. (1988). *Effects of Size, Location, Contrast, Illumination, and Color on the Legibility of Numeric Speedometers* (Tech. Rep. UMTRI-88-36) (NTIS No. PB 90 150533/AS). Ann Arbor: The University of Michigan Transportation Research Institute.

Bowers-Carnahan, F. R. (1991). *The Truck Driver in IVHS System Development* (SAE Paper 912707). Warrendale, PA: Society of Automotive Engineers.

Brainin, P. A. (1980). *Safety and Mobility Issues in Licensing and Education of Older Drivers* (Tech. Rep. DOT HS 805 492). Washington, DC: National Highway Traffic Safety Administration.

Bundorf, R. T. (1967). *The influence of vehicle design parameters on characteristic speed and understeer* (SAE Paper 670078). Warrendale, PA: Society of Automotive Engineers.

Burger, W. J., Smith, R. L., Ziedman, K., Mulholland, M. U., Bardales, M. C., and Sharkey, T. J. (1981). *Improved Commercial Vehicle Conspicuity and Signalling Systems* (Tech. Rep. DOT-HS-806-100). Washington, DC: U.S. Department of Transportation.

Callahan, J. M. (1986a, January). Ford finds 400 ways to say quality. *Automotive Engineering,* 44–45.

Callahan, J. M. (1986b, February). Ford's Taurus best-in-class list: Part two. *Automotive Engineering,* 97–98.

Callahan, J. M. (1986c, September). Ford's Taurus best-in-class list: Part III. *Automotive Engineering,* 115–116.

Campbell, P. R. (1994). *Population Projections for States, by Age, Sex, Race, and Hispanic Origin: 1993 to 2020* (U.S. Bureau of the Census, Current Population Reports, P25-1111). Washington, DC: U.S. Government Printing Office.

Carlson, W. L. (1972). Alcohol usage of the nighttime driver. *Journal of Safety Research, 4,* 12–25.

Cerelli, E. (1989). *Older Drivers: The Age Factor in Traffic Safety* (Tech. Rep. DOT HS 807 402). Washington, DC: National Highway Traffic Safety Administration.

Chatterjee, A., Cadotte, E., Stamatiadis, N., Sink, H., Venigalla, M., and Gaides, G. (1994). *Driver-Related Factors Involved with Truck Accidents.* Raleigh, NC: The University of North Carolina Institute for Transportation Research and Education.

Cohen, M. S. (1993). Labor force trends and their relationship to the trucking industry. In *Changing Trucking to Match a Changing Work Force* (pp. 15–18). (SAE Special Publication SP-979). Warrendale, PA: Society of Automotive Engineers.

Day, J. C. (1993). *Population Projections of the United States, by Age, Sex, Race, and Hispanic Origin: 1993 to 2050* (U.S. Bureau of the Census, Current Population Reports, P25-1104). Washington, DC: U.S. Government Printing Office.

Dohoney-Farina, S. (1988). *Effective Documentation.* Cambridge, MA: MIT Press.

Duffy, T. M., and Waller, R. (Eds.) (1985). *Designing Usable Texts.* New York: Academic Press.

Evans, L. (1991). *Traffic Safety and the Driver.* New York: Van Nostrand Reinhold.

Federal Highway Administration. (1995a). *Our Nation's Highways, Selected Facts and Figures* [FHWA-PL-95-028, HPM-40/5-95 (50M)]. Washington, DC: U.S. Department of Transportation.

Federal Highway Administration. (1995b). *1995 National Summit on Truck and Bus Safety, Report of Proceedings.* Washington, DC: U.S. Department of Transportation.

Federal Railroad Administration. (1993). *Environmental Externalities and Social Costs of Transportation Systems—Measurement, Mitigation and Costing: An Annotated Bibliography.* Washington, DC: U.S. Department of Transportation.

Flannagan, M., and Sivak, M. (1993). Indirect vision systems. In B. Peacock and W. Karwowski, Eds., *Automotive Ergonomics.* London: Taylor and Francis.

Fullerton, H. N. Jr. (1989, November). New labor force projections, spanning 1988 to 2000. *Monthly Labor Review,* 3–12.

Galea, E. R., and Perez Galparsoro, J. M. (1993). *Exodus: An Evacuation Model for Mass Transport Vehicles* (CAA Paper 93006). London: Civil Aviation Authority.

Gewirth, A. (1990). Two types of cost-benefit analysis. In D. Scherer (Ed.), *Upstream/Downstream. Issues in Environmental Ethics* (pp. 205–232). Philadelphia: Temple University Press.

George, K., Green, P., and Fleming, J. (1995). *Timing of Auditory Route Guidance Instructions* (Tech. Rep. UMTRI-95-6). Ann Arbor: The University of Michigan Transportation Research Institute.

Gillespie, T. D. (1992). *Fundamentals of Vehicle Dynamics.* Warrendale, PA: Society of Automotive Engineers.

Green, P. (1984). *Driver understanding of fuel and engine gauges* (SAE Paper 840314). Warrendale, PA: Society of Automotive Engineers. See also *SAE Transactions, 93, pp. 566–584,* 1985.

Green, P. (1988). *Human Factors and Gauge Design: A Literature Review* (Tech. Rep. UMTRI-88-37). (NTIS Publication PB 90 141334/AS). Ann Arbor: The University of Michigan Transportation Research Institute.

Green, P. (1989). Human Factors Considerations in the Design of Displays (Tutorial T30 Short Course Notes). In *SPIE 1989 Symposium on Aerospace Sensing.* Bellingham, WA: Society for Photo-Optical Instrumentation Engineering.

Green, P. (1992). *American Human Factors Research on In-Vehicle Navigation Systems* (Tech. Rep. UMTRI-92-47). Ann Arbor: The University of Michigan Transportation Research Institute.

Green, P. (1995a). Customer needs, new technology, human factors, and driver science research for future automobiles (in Japanese). *Journal of the Japan Society of Mechanical Engineers, 99*(926), 15–18.

Green, P. (1995b). Automotive techniques. In J. Weimer (Ed.), *Research Techniques in Human Engineering* (2nd ed.). Englewood Cliffs, NJ: Prentice-Hall.

Green, P. (1995c). Human factors and new driver interfaces: Lessons learned from a major research project. In *Proceedings of the 1995 Annual Meeting of ITS-America* (pp. 1001–1011). Washington, DC: ITS-America.

Green, P. (1996). In-Vehicle Information: Design of Driver Interfaces for Route Guidance. Paper presented at Transportation Research Board Annual Meeting. Washington, DC: National Academy of Sciences, Transportation Research Board.

Green, P., and George, K. (1995). When should auditory guidance systems tell drivers to turn? In *Proceedings of the Human Factors and Ergonomics Society 39th Annual Meeting* (pp. 1072–1076). Santa Monica, CA: Human Factors and Ergonomics Society.

Green, P., and Goldstein, S. (1989). *Further Analysis of Driver Preferences for Secondary Controls* (Tech. Rep. UMTRI-89-4). (NTIS Publication PB 90 149782/AS). Ann Arbor: The University of Michigan Transportation Research Institute.

Green, P., Goldstein, S., Zeltner, K., and Adams, S. (1988). *Legibility of Text on Instrument Panels: A Literature Review* (Tech. Rep. UMTRI-88-34). (NTIS Publication PB 90 141342/AS). Ann Arbor: The University of Michigan Transportation Research Institute.

Green, P., Levison, W., Paelke, G., and Serafin, C. (1993). *Preliminary Human Factors Guidelines for Driver Information Systems* (Tech. Rep. UMTRI-93-21). Ann Arbor: The University of Michigan Transportation Research Institute. Also published as FHWA-RD-94-087, U.S. Department of Transportation, Federal Highway Administration, McLean, VA, December, 1995.

Green, P., Ottens, D., and Adams, S. (1987). *Secondary Controls in Domestic 1986 Model Year Cars* (Tech. Rep. UMTRI-87-21). (NTIS Publication PB 90 149642/AS). Ann Arbor: The University of Michigan Transportation Research Institute.

Green, P., Ottens, D., Kerst, J., Goldstein, S., and Adams, S. (1987). *Driver Preferences for Secondary Controls* (Tech. Rep. UMTRI-87-47). (NTIS Publication PB 90 150541/AS). Ann Arbor: The University of Michigan Transportation Research Institute.

Green, P., Paelke, G., and Boreczky, J. (1992). The "Potato Head" method for identifying driver preferences for vehicle controls. *International Journal of Vehicle Design, 13*(4), 352–364.

Green, P., Paelke, G., and Clack, K. (1989). *Instrument Panel Controls in Sedans: What Drivers Prefer and Why* (Tech. Rep. UMTRI-89-15). (NTIS Publication PB 90 184235/AS). Ann Arbor: The University of Michigan Transportation Research Institute.

Green, P., Serafin, C., Williams, M., and Paelke, G. (1991). What functions and features should be in driver information systems of the year 2000? (SAE Paper 912792), In *Vehicle Navigation and Information Systems Conference (VNIS '91)* (pp. 483–498). Warrendale, PA: Society of Automotive Engineers.

Gupta, P., and Ratchford, B. T. (1992, September). Estimating the efficiency of consumer choices of new automobiles. *Journal of Economic Psychology, 133,* 375–397.

Habsburg, S., and Middendorf, L. (1977). *What really connects in seating comfort?—Studies of correlates of static seat comfort* (SAE Paper 770247). Warrendale, PA: Society of Automotive Engineers.

Hardin, G. (1968). The tragedy of the commons. *Science, 162,* 1243–1248.

Harris, W. (1977). Fatigue, circadian, rhythm, and truck accidents. In R. R. Mackie (Ed.), *Vigilance Theory, Operational Performance, and Physiological Correlates* (pp. 133–146). New York: Plenum Press.

Hauser, J. R., and Clausing, D. (1988, May–June). The house of quality. *Harvard Business Review,* 63–73.

Henderson, R. L., Smith, R. L., Burger, W. J., and Stern, S. (1983). *Visibility from Motor Vehicles* (SAE Paper 830564). Warrendale, PA: Society of Automotive Engineers.

Hertzberg, H. T. E. (1972). *The human buttocks in sitting: Pressures, patterns, and palliatives* (SAE Paper 720005). Warrendale, PA: Society of Automotive Engineers.

Hoffman, E. R. (1990). Strength of component principles for direction-of-motion stereotypes of three-dimensional display-control arrangements. In *Proceedings of the Human Factors Society 34th Annual Meeting—1990* (pp. 462–466). Santa Monica, CA: Human Factors Society.

Hollmann, F. W. (1992). *National Population Trends* (U.S. Bureau of the Census, Current Population Reports, Series P23, No. 175). Washington, DC: U.S. Government Printing Office.

Hull, R. W., and Knebel, G. W. (1968). *Statistical Summary of School Bus Accident Data* (NTIS Publication PB 180 111). Washingtong, DC: Federal Highway Administration.

International Standards Organization. (1977). *Road vehicles—Passenger cars—Location of hand controls, indicators and tell-tales* [ISO Standard 4040-1977(E)]. Geneva: International Standards Organization.

International Standards Organization. (1982). *Road Vehicles—Symbols for controls, indicators and tell-tales* (4th ed.). (ISO Standard 2575-1982E). Geneva: International Standards Organization.

Jones, R. K., and Joscelyn, K. B. (1978). *Alcohol and Highway Safety 1978: A Review of the State of Knowledge* (Tech. Rep. DOT HS-803 714). Washington, DC: Traffic Safety Administration.

Kamijo, K., Tsujimura, H., Obara, H., and Katsumata, M. (1982). *Evaluation of seating comfort* (SAE Paper 820761). Warrendale, PA: Society of Automotive Engineers.

Kawamoto, H., Hosokawa, Y., Suto, M., and Nakajima, K. (1986). *Comfort Evaluation of Heating and Air Conditioning Systems* (SAE Paper 860590). Warrendale, PA: Society of Automotive Engineers.

Koester, D., and Hamilton, R. (1994). *A methodology for evaluating barriers and aids to vehicle ingress/egress* (SAE Paper 940388). Warrendale, PA: Society of Automotive Engineers.

Koutsopoulos, K. C., and Schmidt, C. G. (1976). Mobility constraints of the carless. *Traffic Quarterly, 30,* 67–83.

Leatherwood, J. D., and Dempsey, T. K. (1976). *Psychophysical Relationships Characterizing Human Response and Whole-Body Sinusoidal Vertical Vibration* (NASA TN D-8188). Hampton, VA: NASA Langley Research Center.

Leatherwood, J. D., Dempsey, T. K., and Clevenson, S. A. (1980). A design tool for estimating passenger ride comfort with complex ride environments. *Human Factors, 22*(3), 291–312.

Loczi, J. (1993). Ergonomic assessment of exiting automobiles. In *Proceedings of the Human Factors and Ergonomics Society 37th Annual Meeting* (pp. 401–405). Santa Monica, CA: Human Factors and Ergonomics Society.

Loveless, N. E. (1962). Direction-of-motion stereotypes: A review. *Ergonomics, 5,* 357–383.

Lowe, M. D. (1994). Reinventing transport. In L. Brown (Ed.), *State of the World* (pp. 81–98). New York: W. W. Norton.

Lugaila, T. (1992). *Households, Families, and Children: A Thirty Year Perspective* (U.S. Bureau of the Census, Current Population Reports, P23-181). Washington, DC: U.S. Government Printing Office. Also in S. Rosenbloom (Ed.), *Travel by Women.* Tucson, AZ: The Drachman Institute for Land and Regional Development Studies.

MacCarthy, P. S., and Tay, R. (1989). Consumer valuation of new car attributes: An econometric analysis of the demand for domestic and Japanese/Western European imports. *Transportation Research, 23A*(5), 367–375.

Mackie, R. R., and Miller, J. C. (1978). *Effects of Hours of Service Regularity of Schedules, and Cargo Loading on Truck and Bus Driver Fatigue.* Washington, DC: National Highway Traffic Safety Administration.

Maleck, T. L., and Hummer, J. E. (1986). Driver age and highway safety. In *Tansportation Research Record 1059* (pp. 6–12). Washington, DC: Transportation Research Board.

Massie, D. L., and Campbell, K. L. (1993). *Analysis of Accident Rates by Age, Gender, and Time of Day Based on the 1990 Nationwide Personal Transportation Survey.* Ann Arbor: University of Michigan Transportation Research Institute.

Maurer, D., and Fawcett D. (1973). *Methods of Application—Field of View Targets* (SAE Paper 730610). Warrendale, PA: Society of Automotive Engineers.

McFarlane, J., and Wierwille, W. W. (1990). *Study of direction-of-motion stereotypes for automotive controls* (Department of Industrial and Systems Engineering Report 90-02). Blacksburg, VA: Virginia Polytechnic Institute and State University, Vehicle Analysis and Simulation Laboratory.

Meyerhoff, A., Micozzi, M., and Rowen, P. (1993). Running on empty: Travel patterns of extremely poor people in Los Angeles. In *Transportation Research Record* (No. 1395, pp. 153–159). Washington, DC: Transportation Research Board.

Michon, J. A. (Ed.). (1993). *Generic Intelligent Driver Support.* London: Taylor and Francis.

Miller, J. C. (1993). Driver fatigue and long distance truck drivers: Implications for trucking operations. In *Changing Trucking to Match a Changing Work Force* (pp. 5–13). (SAE Special Publication SP-979). Warrendale, PA: Society of Automotive Engineers.

Mohan, D. (1992). Vulnerable road users: An era of neglect. *Journal of Traffic Medicine, 20,* 121–128.

Mortimer, R. E., Segel, L., Dughoff, H., Campbell, J. O., Jorgeson, C. M., and Murphy, R. W. (1979). *Brake Force Requirement Study: Driver-Vehicle Braking Performance as a Function of Brake System Design Variables* (Report HuF-6). Ann Arbor, MI: Highway Safety Research Institute.

Motoki, M., and Tsukisaka, T. (1989). *A study on required field of view for motorcycle rear-view mirrors* (Paper 89-6B-0-015) (pp. 1335–1351). In Proceedings of the Twelfth International Conference on Experimental Safety Vehicles. Washington, DC: National Highway Traffic Safety Administration.

National Highway Traffic Safety Administration. (1985). *Alcohol and Highway Safety 1984: A Review of the State of the Knowledge* (Tech. Rep. DOT-HS-806-569). Washington, DC: U.S. Department of Transportation.

National Highway Traffic Safety Administration. (1993). *Fatal Accident Reporting System 1991.* Washington, DC: U.S. Department of Transportation.

National Institute on Alcohol Abuse and Alcoholism. (1993). *Eighth Special Report to the U.S. Congress on Alcohol and Health.* Washington, DC: U.S. Department of Health and Human Services.

National Research Council. (1992). *Automotive Fuel Economy—How Far Should We Go?* Washington, DC: National Academy Press.

National Safety Council. (1994). *Accident Facts, 1994 Edition.* Itasca, IL: Author.

National Safety Council. (1995). *Accident Facts, 1995 Edition.* Itasca, IL: Author.

National Transportation Safety Board. (1990). *Fatigue, Alcohol, Other Drugs, and Medical Factors in Fatal-to-the-Driver Heavy Truck Crashes* (Tech. Rep. NTSB/SS-90/01). Washington, DC: National Transportation Safety Board.

Nelson, P. (1987). *Transportation Noise Reference Book.* London: Butterworths.

Newcomb, T. D., and Spur, R. T. (1967). *Braking of Road Vehicles.* London: Chapman and Hall.

Paelke, G. M. (1993). A comparison of route guidance destination entry methods. In *Proceedings of the Human Factors and Ergonomics Society 37th Annual Meeting—1993* (pp. 569–573). Santa Monica, CA: The Human Factors and Ergonomics Society.

Parkes, A. M., and Franzen, S. (1993). *Driving Future Vehicles.* London: Taylor and Francis.

Partridge, E. (1990). The rights of future generations. In D. Scherer (Ed.), *Upstream/Downstream. Issues in Environmental Ethics* (pp. 40–66). Philadelphia: Temple University Press.

Partyka, S. C. (1983). *Comparison by Age of Drivers in Two-Car Fatal Crashes.* Washington, DC: National Highway Traffic Safety Administration.

Partyka, S. C. (1984). *Traffic Victim Age and Gender Distributions.* Washington, DC: National Highway Traffic Safety Administration.

Peacock, B., and Karwowski, W. (1993). *Automotive Ergonomics.* London: Taylor and Francis.

Peck, R., and Romanowicz, P. (1993/1994, Winter). Teen and senior drivers. *Research Notes,* 3–6.

Perel, M., and Ziegler, P. N. (1971). *An Evaluation of a Safety Belt Interlock System.* Washington, DC: National Highway Traffic Safety Administration.

Pisarski, A. E. (1972). *Travel Behavior Issues in the 90's.* Washington, DC: Federal Highway Administration.

Promisel, D. M., Blomberg, R. D., Knacht, M. L., and Silver, S. (1969). *School Bus Safety—Operator Age in Relation to School Bus Accidents* (Tech. Rep. HS 800 209). (PB 189 677). Washington, DC: National Highway Safety Bureau.

Reed, M. P., Saito, M., Kakishima, Y., Lee, N. S., and Schneider, L. W. (1991). *An investigation of driver discomfort and related seat design factors in extended-duration driving* (SAE Paper 910117). Warrendale, PA: Society of Automotive Engineers.

Retting, R. A. (1988). Urban pedestrian safety. *Bulletin of the New York Academy of Medicine, 64*(7), 810–815.

Retting, R., Schwartz, S. I., Kulewicz, M., and Buhrmeister, D. (1989). Queens Boulevard pedestrian safety project—New York City. *Injury Control MMWR Reprints,* Compilation 6, 12–13.

Rolle, C. D., Romitelli, G. F., and Savasta, F. (1992). Real evaluation of thermal comfort in a car passenger compartment. In *Proceedings of the FISITA '92 Conference.* London: Institute of Mechanical Engineers.

Rosenbaum, S., and Anschuetz, L. (1994). Whole-Product Usability: Integrating Documentation and Rest-of-Product Usability Testing. In *International Professional Communication Conference 1994 Conference Record (pp. 127–135).* New York: Institute of Electrical and Electronics Engineers.

Rosenbloom, S. (1994a). *Travel by Women.* Tucson, AZ: The Drachman Institute for Land and Regional Development Studies.

Rosenbloom, S. (1994b). *Travel by the Elderly.* Tucson, AZ: The Drachman Institute for Land and Regional Development Studies.

Rosenbloom, S. (1995). *A Vision of Emerging Transportation Service Requirements Twenty Years in the Future.* Washington, DC: Federal Highway Administration.

Ross, T., Nicolle, C., and Brade, S. (1994). *An Empirical Study to Determine Guidelines for Optimum Timing of Route Guidance Instructions* (DRIVE 2 Project V2008 HARDIE, Deliverable 13.2). Loughborough, UK: HUSAT Research Institute, Loughborough University.

Serafin, C., Williams, M., Paelke, G., and Green, P. (1991). *Functions and Features of Future Driver Information Systems* (Tech. Rep. UMTRI-91-16). Ann Arbor: The University of Michigan Transportation Research Institute.

Sivak, M., and Flannagan, M. (1993). Human factors considerations in the design of vehicle headlamps and signal lamps. In B. Peacock and W. Karwowski (Eds.), *Automotive Ergonomics.* London: Taylor and Francis.

Smith A. (1776). *The Wealth of Nations* (Book IV, Chapter II). Cited in Hardin, G., Rewards of pejoristic thinking. In G. Hardin and J. Baden (Eds.), *Managing the Commons* (pp. 126–127) (1977). San Francisco: W. H. Freeman and Company.

Smith, S. L. (1979, December). Letter size and legibility. *Human Factors, 21*(60), 661–670.
Society of Automotive Engineers. (1995). *1995 SAE Handbook.* Warrendale, PA: Author.
Stapp car crash conference, 39th (1995). In *Stapp Car Crash Conference Proceedings* (SAE publication P-299). Warrendale, PA: Society of Automotive Engineers.
Stutts, J. C., and Martell, C. (1992). Older driver population and crash involvement trends 1974–1988. *Accident Analysis and Prevention, 24*(4), 317–327.
Turner, C., and Green, P. (1987). *Human Factors Research on Automobile Secondary Controls: A Literature Review* (Tech. Rep. UMTRI-87-20). (NTIS No. PB 90 149675/AS). Ann Arbor: The University of Michigan Transportation Research Institute.
Underwood, S. E. (1992). *Delphi Forecast and Analysis of Intelligent Vehicle-Highway Systems Through 1991* (Delphi II). (IVHS Tech. Rep. 92-17). Ann Arbor: Program in Intelligent Vehicle-Highway Systems, College of Engineering, University of Michigan.
U.S. Bureau of Labor Statistics. (1991). *Occupational Outlook Quarterly,* Fall 1991. Washington, DC: U.S. Department of Labor. Cited in Cohen, M. S. (1993). Labor force trends and their relationship to the trucking industry. In *Changing Trucking to Match a Changing Work Force* (pp. 15–18). (SAE Special Publication SP-979). Warrendale, PA: Society of Automotive Engineers.
U.S. Congress, Office of Technology Assessment. (1988). *Gearing up for Safety, Motor Carrier Safety in a Competitive Environment* (OTA-SET-382). Washington, DC: U.S. Government Printing Office.
U.S. Congress, Office of Technology Assessment. (1994). *Federal Research and Technology for Aviation* (OTA-ETI-610). Washington, DC: U.S. Government Printing Office.
Velotta, C. (Ed.). (1995). *Practical Approaches to Usability Testing for Technical Documentation.* Arlington, VA: Society for Technical Communication, Inc.
Waller, J. (1992). Research and other issues concerning effects of medical conditions on elderly drivers. *Human Factors, 34,* 3–15.
Waller, P. F. (1986). The highway transportation system as a commons: Implications for risk policy. *Accident Analysis and Prevention, 18,* 417–424.
Waller, P. F. (1991). The older driver. *Human Factors, 33,* 499–505.
Waller, P. F. (1993). Changing trucking to match the changing labor force. In *Changing Trucking to Match a Changing Work Force* (pp. 1–4). (SAE Special Publication SP-979). Warrendale, PA: Society of Automotive Engineers.
Waller, P. F., and Blow, F. C. (1995). Women, alcohol, and driving. In M. Galanter (Ed.), *Developments in Alcoholism, Vol. 12: Women and Alcoholism* (pp. 103–123). New York: Plenum Press.
Westefeld, A., and Phillips, B. M. (1975). *Safety Belt Interlock System: Usage Survey.* Washington, DC: National Highway Traffic Safety Administration.
Whitcomb, D. A., and Milliken, W. E. (1956). *Design Implications of a General Theory of Automobile Stability and Control.* London: Institute of Mechanical Engineers.
Wierwille, W. W., and McFarlane, J. (1991). *Overview of a study of direction-of-motion stereotype strengths for automotive controls* (SAE Paper 9100115). Warrendale, PA: Society of Automotive Engineers.
Wierwille, W. W., and McFarlane, J. (1993). Role of expectancy and supplementary cues for control operation. In B. Peacock and W. Karwowski (Eds.), *Automotive Ergonomics.* London: Taylor and Francis.
Williams, A. F., and Carsten, O. (1989). Driver age and crash involvement. *American Journal of Public Health, 79,* 326–327.
Yamanaka, A., and Kobayashi, M. (1970). *Dynamic visibility of motor vehicles* (SAE Paper 700393). Warrendale, PA: Society of Automotive Engineers.
Yee, D. (1985). A survey of the traffic safety needs and problems of drivers ages 55 and over. In J. L. Malfetti (Ed.), *Drivers 55+: Needs and Problems of Older Drivers: Survey Results and Recommendations* (pp. 96–128). Falls Church, VA: AAA Foundation for Traffic Safety.
Zador, P. L. (1991). Alcohol-related risk of fatal driver injuries in relation to driver age and sex. *Journal of Studies on Alcohol, 52,* 302–310.

CHAPTER 60

DESIGN FOR PEOPLE WITH FUNCTIONAL LIMITATIONS RESULTING FROM DISABILITY, AGING, OR CIRCUMSTANCE

Gregg C. Vanderheiden
Department of Industrial Engineering
University of Wisconsin—Madison
Madison, WI 53706 USA

60.1 INTRODUCTION

60.1.1 Not a Special Population, but a Continuum—and an End Game for Most of Us

Often, the topic of design for human disability and aging is thought of as a special topic, vertical market, or special application. Although there are special products or assistive technologies that are designed specifically for use by people with disabilities, they constitute only a small portion of the total number of products that individuals with functional limitations need to be able to use. In addition to the specially designed tools, everyone, including those with disabilities, needs to access a wide range of technologies found in their everyday lives—at home, at school, on the job, and in the community. It is toward the more accessible design of everyday products that this chapter is directed.

Another common misconception is that the population in question is small. Although there are many different types and degrees of disabilities, some of which represent smaller numbers of individuals, cumulatively people with disabilities represent around a fifth of the population. In addition, a majority of people who live beyond age 75 will experience functional limitations. Approximately 72% of those who live beyond age 75 will have functional limitations, and 41% of them will have severe functional limitations (Kraus and Stoddard, 1989). In addition, many of these individuals experience multiple functional limitations.

60.1.2 Multiplier Effect

In designing products for individuals, this constitutes a significant portion of the market. When designing products to be used by families or within industry, the impact is multiplied. With the family unit being three or four people, the percentage of families who have people with disabilities is much higher. When one turns to industry, particularly large industries, one finds that the percentage of industries who employ people with disabilities is very high. Thus, if one is designing products and systems for use by larger industries, one will find that almost all of the customer base will have employees with disabilities.

In considering product design, it is important to note that there is no clear line between people who are categorized as "disabled" and those who are not. A performance or ability distribution for a given skill or ability is generally a more normal shaped function, rather than bimodal with distinctive "able" and "disabled" groups. This distribution includes a small number of individuals who have exceptionally high ability, a larger number of individuals with midrange ability, and another longer tail representing individuals with little or no ability in that particular area. In looking at such a distribution, it is impossible simply to draw a vertical line and separate able-bodied from disabled persons. It is also important to note that each aspect of ability has a separate distribution. Thus, a person who is poor along an ability distribution in one dimension (e.g., vision) may be excellent with regard to another dimension (e.g., hearing or IQ). Thus, individuals do not fall at the lower or upper end of the distribution overall but generally fall into different positions depending upon the particular ability being measured.

60.1.3 Who Is Included in the Category of "Disabled and Elderly Persons"? The 95th Percentile Illusion

It should be clear that even if elderly and disabled persons are included in the mainstream design process, it is not possible to design all products and devices so that they are usable by all individuals. There will always be a "tail" of individuals who are unable to use a given product.

In order to include a sizeable portion of the population in the category of "those who can use a product with little or no difficulty," the 95th percentile ability is often used. The problem is that there are no "95th percentile" data for the total designs. Rather, there are only data with regard to individual physical or sensory characteristics. Thus there is 95th percentile data for height, a 95th percentile for vision, hearing, etc. As a result, it is not possible to determine when a product can be used by 95% of the people. It is only possible to estimate when a product can be used by 95% of the population along any one dimension. Since people in the 5% tail for any one dimension (e.g., height) are usually not the same people as the 5% tail along another dimension (e.g., vision) (Kroemer, 1990), it is possible to design a product using 95th percentile data and end up with a product that can be used by far fewer than 95% of the population.

To illustrate this phenomenon, imagine a minipopulation of 10 individuals. Ten percent of them (1 of 10) have one short leg, 10% have a visual impairment, 10% have a missing arm, 10% are short, and 10% cannot hear.

Let's assume that we design a product that required 90th percentile ability along each of the dimensions of height, vision, leg use, arm use, and hearing. In this instance we would end up with a product that was in fact only usable by 50% of this population. This occurs because, although only 10% of this minipopulation is limited in any single dimension, different individuals fall into the 10% tail for each dimension and only 50% of the population is within the 90th percentile for all five areas.

In real life, the effect is not quite this dramatic, and its calculation is not so simple. First, the percentage of individual with disabilities is often less than 10% along any one dimension. Second, there is often overlap where one individual would have more than one disability (elderly individuals, for example).

On the other hand, there is a much wider range of different individual types of disability. In addition, the data from which the 95th percentiles are calculated often exclude persons with disabilities (Kroemer, 1990), making the percentage who could use the design(s) smaller than one would first calculate.

60.2 DISABILITY IS A CONSEQUENCE, NOT A CONDITION

> *"Disability is the inability to accommodate to the world as it is currently designed."*
> *(Caplan, 1992, p. 88)*

This is a paraphrase of Ralph Caplan's statement "Disability is the inability to accommodate poor design," with an emphasis on the fact that design can be changed, and thus so can disability (Caplan, 1992).

In looking at the impact of disability and its relationship to design, it is often useful to use a model such as that shown in Figure 60.1. The model shows the relationship that both impairment and design have in creating disabilities. It also shows how circumstance can create similar reduced abilities in anyone, including those without functional impairments. Combined with poor design, these circumstances can also lead to situations in which people experience circumstantial disabilities or inabilities to carry out certain tasks. Thus, in addition to generally making products easier for everyone to use, better or more universal design can make a product usable even when people are under stressed conditions. Take, for example, a mother whose young son just fell and cut his head. She makes the mistake of mentioning the doctor, and is now trying to use the phone while holding her screaming, kicking son in one arm to keep him running off and hiding. Because of the screaming, she can hear very little, and has some of the same functional problems as a person with a hearing impairment. Because her son is kicking and thrashing, she has poor motor control, and has only one hand available. Because he is bleeding profusely, she is also highly distracted, and is able to bring only limited cognitive skills and attention to the task at hand.

60.2.1 Three Approaches

There are basically three ways to address the problem faced by individuals who are unable to use the world around them:

1. ***Change the individual:*** through surgery, education, skill development, skill practice, or by teaching them strategies, tricks, "secrets" for doing things or for doing things more easily.

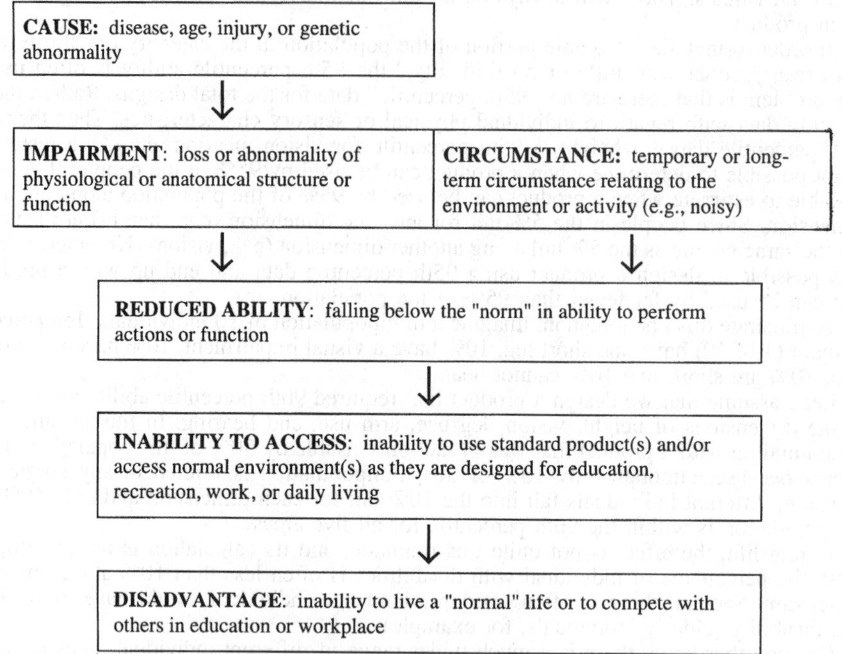

Figure 60.1 Cause-effect model shows the role that both impairment and design play in disability, as well as the parallel role that conditions or circumstances can play.

2. ***Provide the individual with tools:*** this includes prosthetics and orthotics (e.g., eyeglasses, braces, artificial limbs) or assistive technologies (wheelchairs, alternate writing aids, telecommunication devices for the deaf (TDDs/TTs), etc.)
3. ***Change the way that the world is designed:*** More universal and accessible designs.

Ergonomics is involved in all three of these areas:

- Developing new techniques and strategies that allow an unaided individual to better perform in the workplace, home, or community
- Developing specialized tools or assistive technologies that maximize the use of residual skills and abilities and compensate for missing abilities
- And, of course, changing the design of the world in general so that it is more usable with a wider range of skills and abilities

The focus of this chapter is on the third approach, designing the world to be more universally usable.

60.2.2 Universal Design

Universal design is the term that has been given to the practice of designing products or environments that can be effectively and efficiently used by people with a wide range of abilities operating in a wide range of situations. This includes people with no limitations as well as those operating with functional limitations relating to disabilities or simply by circumstance.

For example: products developed using universal design principles would be flexible enough to be usable by people with no limitations as well as those:

- Who cannot see the product—because they are blind or because their eyes are temporarily occupied (e.g., driving a car)

- Who cannot use their hands well because of aging or a physical disability or because their hands are temporarily full, or cold, or gloved
- Who cannot speak or are in an environment where speech is not practical (library or noisy crowd)
- Who cannot hear the product—because they are deaf or because they are in a very noisy environment (e.g., an airplane or a shopping mall at Christmas)
- Who have learning disabilities or who are only able to divert part of their attention to the task at hand
- Whose natural language is sign language or a foreign language
- Who are very young or very old, etc.

An ideal design is one that is attractive, easy to learn, effective and whose functions can be efficiently accessed and used by everyone across the full range of circumstances that may occur for its intended use.

A good design is a commercially practical, mass market design that is usable by and attractive to the maximum possible number and diversity of users, given the best of today's collective knowledge, technologies, and materials.

60.2.2.1 Non–Disability-Related Reasons for Universal Design

Benefits All

A general characteristic of good universal design is that it benefits many more people without disabilities than those with disabilities. This of course follows from the design benefiting everyone, and the fact that there are more people without disabilities than with disabilities. The sidewalk curbcut is a prime example of this, as are ramps in general. Although originally designed for users of wheelchairs, they are also used by parents pushing baby carriages, people pulling baggage carriers, bicycle riders, skateboard users, kids on tricycles, and any number of other individuals. Even people walking can be observed to veer from their path in order to walk up a curbcut instead of stepping up a curb. In another example, a technique called a Talking Fingertip was used to allow individuals who are blind to access and use touchscreen-based kiosks. Once implemented, however, it was found that it was also very useful for individuals with low vision as well as those who could not read because of literacy or language problems.

New Insight

Studying the use of products by people with functional limitations can also provide insights into a design that might not otherwise be achieved. For example, it is much easier to determine which elements in a kitchen require greater strength by employing an individual who is weak or who has poor grasp than it would be by employing someone with normal or extraordinary strength; even if they were asked which things required more or less effort, the mere fact that they had so much strength in reserve would cause them to use it without noticing it.

Lower Cost Design

Universal design can also lead to insights that result in lower cost designs. Although universal designs are usually thought of as being more expensive, this is generally not the case. If one discounts the time it takes to reorient one's thinking and familiarize oneself with the characteristics and constraints of people with functional limitations, the resulting designs can be both easier to use and less expensive.

One example of this is the current design of elevators and their alert bells. In the past, people with disabilities had a problem in getting onto elevators when they were arranged in elevator banks. Often, by the time the individual using a wheelchair got themselves down to the elevator that had opened, the door had closed. New standards were proposed which would require that elevator doors stay open for a longer period of time in order to allow them to be successfully boarded by wheelchair users. This caused problems, since it increased the numbers of elevators that needed to be installed in buildings in order to ensure adequate service to all floors. In some thin, tall buildings, this could result in using up a substantial portion of the building for elevators.

After an injunction was sought to stop the standards, the designers and consumer advocates sat down to study the problem anew. It was determined that the problem was not the time it took to board the elevator, but the time it took to get in front of the elevator. Since the elevators were computer controlled, and the computers knew where the elevators were going in advance of their arrival, it was quickly determined that lighting the alert light and sounding the bell in advance would allow individuals with disabilities to position themselves in front of the elevator door and be able to board as it opened. Testing bore this out, and it was found that people in wheelchairs as well as everyone else could actually begin the boarding process much more quickly, and in much less time than the elevators were staying open before people with disabilities were considered. Following the modification in timing of the alert light and bell, they were able to decrease the time that the doors stayed open, allowing builders to either use fewer elevators or provide better service to the floors.

Approximately one-third of the persons with disabilities who can and would like to work are unemployed. This amounts to approximately 2 million people (Kraus and Stoddard, 1989). Figuring an average annual salary of $15,000, that amounts to $30 billion in lost productivity, as well as several billion dollars in lost tax revenues. This is in addition to the large costs in the form of transfer payments made to those individuals who cannot live independently. Total expenditures, public and private, for people with disabilities is estimated at between $200 and $270 billion per year (Figure 60.2). What portion of this could be saved if the design of the environment allowed people to live more independently or stay on their jobs longer?

60.3 DEMOGRAPHICS

As can be seen in Figures 60.3, 60.4, and 60.5, the prevalance of the different types of functional limitation (visual, hearing, physical, cognitive) varies significantly as a function of age. In children, we see a much higher percentage of mental retardation and language and learning disabilities than other disabilities (Figure 60.3). As people age, sensory and physical disabilities become more prevalent (Figure 60.4). Not evident from these charts is the fact that with older individuals, we see a much higher incidence of multiple disabilities, including combinations such as hearing and visual impairments, which interfere with many of the adaptive strategies developed for individuals with only hearing or only visual impairments.

Finally, as mentioned earlier, we can see that the percentage of individuals who have functional limitations within the population increases sharply as a function of age. In fact, for individuals over the age of 75 (Figure 60.5), a wide majority will have functional

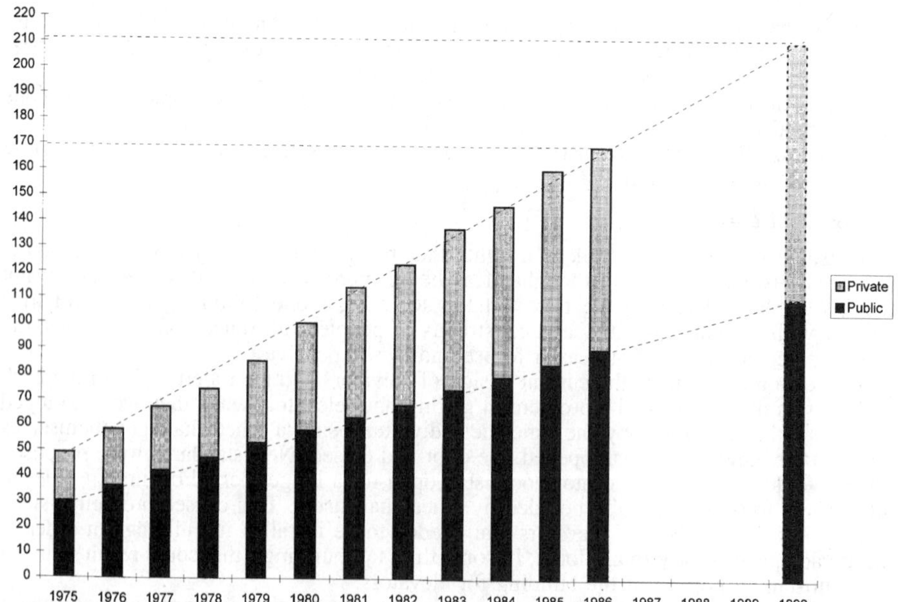

Figure 60.2 Disability expenditures in the United States in billions of dollars, public and private sectors. (Based on data from Berkowitz and Greene, 1989.)

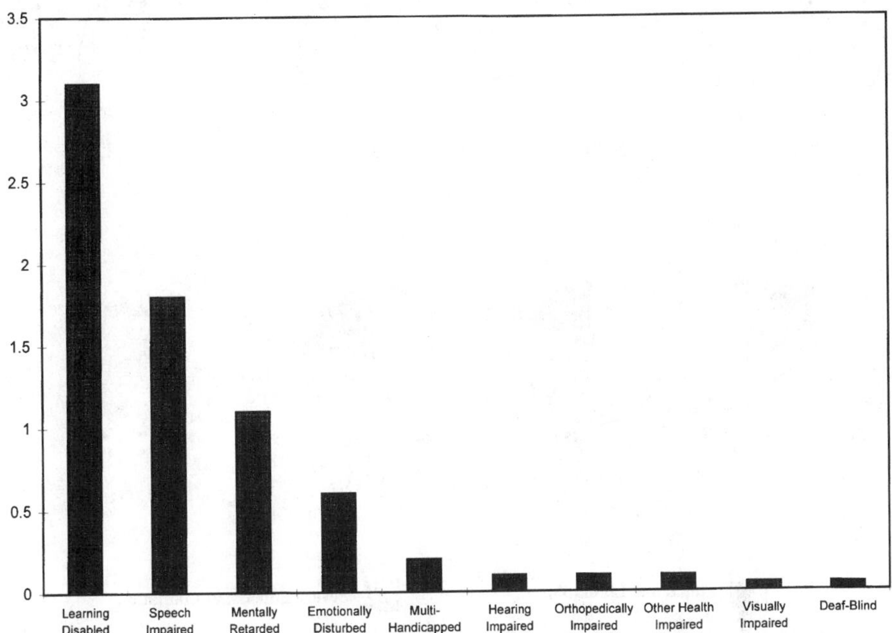

Figure 60.3 Prevalence of impairments (primary diagnosis) for school-aged children (3–21 yr) in the United States. Each child is counted in only one category. (Based on data from Kraus & Stoddard 1989: Office of Special Education and Rehabilitation Services, 1988. OSEP state reported data, 1986–87 school year.)

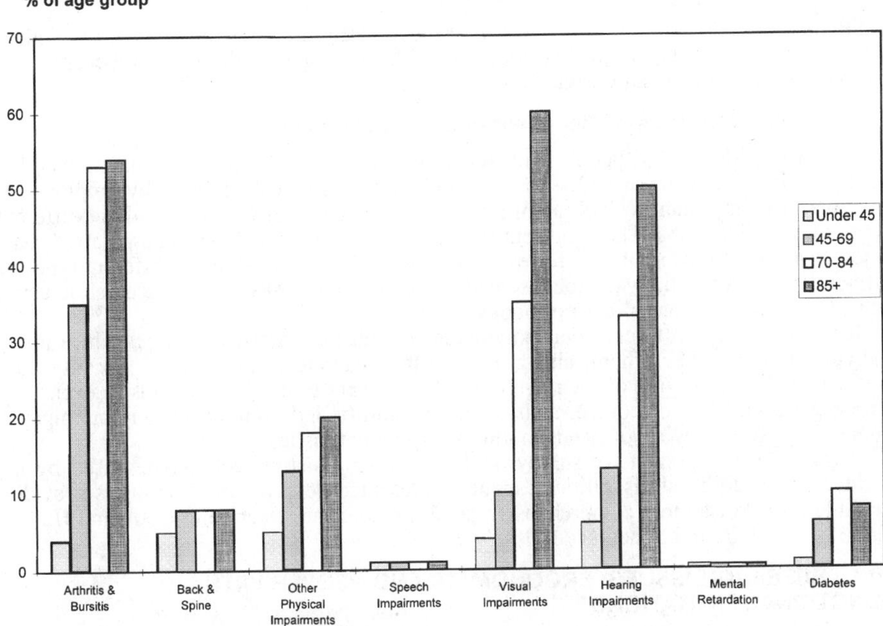

Figure 60.4 United States prevalence of selected impairments within age groups. Data categories are not exclusive. (Based on data from LaPlante, 1988. Survey: National Health Interview Surveys, 1983–1985. Tabulations from public use tapes.)

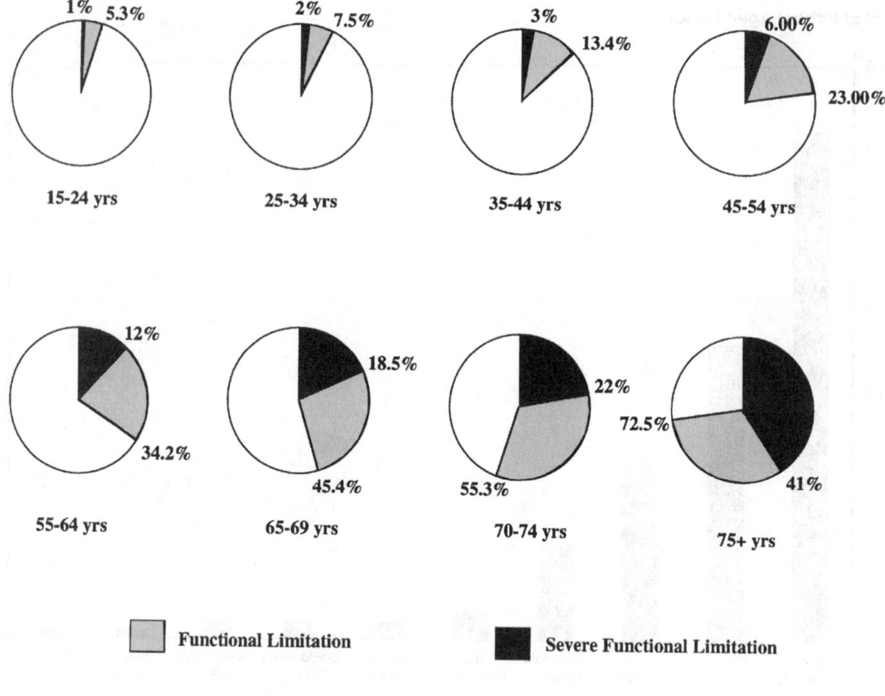

Figure 60.5 Functional limitations as a function of age. (Based on data from Kraus and Stoddard, 1989. Source: Bureau of the Census, Series P-70. # 8. Survey: SIPP, 1984.)

limitations, and almost half will have severe functional limitations of one type or another. Thus, over our lifetime (if we live long enough), most of us will not only benefit from but require more universal design.

60.3.1 Characteristics of Users with Functional Limitations

In considering design for people with functional limitations, it is important to examine their abilities both without and with tools and strategies that they normally employ. For example, it is important to look at an amputee's abilities both with and without different types of artificial limbs. These present very different mechanical and manipulative characteristics. Many touch buttons, for example, cannot be activated using different types of artificial arms. Acoustic wave touchscreens on computer kiosks may be accessible using soft plastic cosmetic arms, but not hooks.

It is also important to consider individuals without their assistive devices, since many individuals do not have them, either because they cannot afford them or because they prefer to avoid the stigma of their use (e.g., do not want to use hearing aids or even very strong glasses). This, of course, adds to the variability and complicates any attempts at comprehensive surveying of needs, abilities, or characteristics.

Although a comprehensive survey of the types of assistive technologies used by individuals with different disabilities cannot be presented here, a partial listing is provided in Table 60.1. For a more comprehensive review, readers are referred to Cook and Hussey (1995), or Galvin and Scheurer (1996).

60.4 RESEARCH ISSUES: ERGONOMICS AND PEOPLE WITH FUNCTIONAL LIMITATIONS

Much of the research on people with functional limitations is taken not from research on people with "disabilities," but rather from experiments done with "normal" individuals operating under stress or adverse conditions (e.g., blinded by smoke, encumbered by a

Table 60.1 A Partial List of Assistive Technologies

Type of Disability or Functional Limitation	Assistive Technologies and Strategies Used
Hearing impairment	Hearing aids
	Amplifiers
	Assistive listening devices (remote microphone that transmits to a receiver worn by the user)
	Cochlear implants
	Headphones
	Inductive loops (that couple to the hearing aid)
	Direct connection wire (from audio device to the hearing aid)
	T-coils (in hearing aid, to couple to telephone earpiece through induction)
	(*See also deafness strategies and technologies*)
Deafness	Telecommunication device for the deaf or text telephone (TDD/TT)
	Relay service (a special operator or interpreter with a TDD/TT)
	Closed captions
	Sign language
	Sign language interpreters
	Lip reading
Low vision	Lights
	Magnifiers
	Telescopes
	Closed circuit television
	(*See also blindness strategies and technologies*)
Blindness	Braille (used by approximately 10% of those who are blind)
	Dynamic braille displays (1, 12, 20, or 40 braille cells with pins that move up and down to form the braille characters)
	Tactile symbols and shapes
	Raised line drawings
	Long cane
	Tape recorders
	Synthetic speech
	Synthetic speech or braille based portable notetakers
	Talking clocks, watches, calculators
	Satellite positioning systems and electronic map databases (emerging)
	Talking signs (infrared broadcasters in the environment that are picked up by small hand-held units)
	Descriptive television (audio description track)
	Voice output screen readers (on computer systems)
Physical impairment	Reachers
	Artificial arms, legs, and hands or hooks
	Canes and crutches
	Walkers
	Wheelchairs
	Splints and braces
	Mouthsticks, headsticks
	Communication, writing, and control aids using a wide variety of input techniques, including sip and puff, Morse code, eye gaze, joystick, single switch scanning, multiswitch encoding, etc.
	Keyguards
	Hand and arm rests
Speech impairment	Voice amplifiers
	Voice synthesizers
	Artificial larynxes
Cognitive impairment	Memory aids
	Cuing systems
	Calculators
	Text-to-speech aids

spacesuit, etc.). These studies represent much more controlled conditions than those represented by the great diversity of types, combinations, and degrees of disability but do yield interesting information that can be used by people with disabilities. As noted above, the results of work with individuals who have disabilities can also be applied to these other environments or locations where individuals have reduced abilities because of circumstance.

There are major problems in carrying out research in that the variation and range of ability or constraint is so great. Visual impairments, for example, can take a very wide range of different forms, and each of these can vary in degree from very mild to severe reduction or total loss. As a result, it is not possible to make blanket statements about these populations. Instead, the research generally tries to characterize the diversity, to quantify numbers of individuals within particular ranges, and/or to chart the functional characteristics for major groups. For example, individuals who are experiencing hearing loss because of aging tend to lose hearing at certain frequencies more than others. People with photosensitive epilepsy tend to be much more susceptible to certain frequencies than others.

The fact that there are no set patterns and that one can find individuals with just about every type, degree, and combination of disabilities makes developing design guidelines difficult. However, design principles do exist, as well as strategies that can significantly increase the accessibility and usability of products by a much wider range of individuals. See, for example, Grandjean (1987), Mueller (1992), Osborne (1987), Pirkl (1994), Sanders and McCormick (1987), and Vanderheiden and Vanderheiden (1991).

Note that this chapter refers to persons and individuals rather than populations. "Populations" tends to imply somewhat homogeneous groups (although there may be variance within the group). When talking about people with functional limitations, we are talking about a group that is a continuum flowing across many dimensions simultaneously. Probably a classic example is people who are older and who may have reductions in visual, hearing, physical, and/or cognitive abilities simultaneously. These abilities will also take different tracks and combinations in different individuals, and will be progressive over time, requiring designs of environments and products to be dynamic. Clearly, designs must be flexible to accommodate different individuals, but in these cases they must be flexible to accommodate the same individual over time, or sometimes during different periods of the same day.

60.5 OVERVIEW BY MAJOR DISABILITY GROUP

Although there is a tremendous variety of specific causes, as well as combinations and severity of disabilities, we can most easily relate their basic impact to the use of consumer products by looking at five major categories of impairment. The five categories are:

- Visual impairments
- Hearing impairments
- Physical impairments
- Cognitive or language impairments
- Seizure disorders

60.5.1 Visual Impairments

Visual impairment represents a continuum, from people with very poor vision, to people who can see light but no shapes, to people who have no perception of light at all. However, for general discussion, it is useful to think of this population as representing two broad groups: those with low vision and those who are legally blind. There are an estimated 8.6 million people with visual impairments (3.4% of the U.S. population) based on estimated 1990 census data (Elkind, 1990). In the elderly population the percentage of persons with visual impairments is very high [35% of those over 70, and 65% of those over 85 (LaPlante, 1988)].

A person is termed *legally blind* when their visual acuity (sharpness of vision) is 20/200 or worse *after correction,* or when their field of vision is less than 20° in the best eye after correction (Hoover and Bledsoe, 1981). There are approximately 580,000 people in the United States who are legally blind (based on estimated 1990 census data; Elkind, 1990).

Low vision includes problems (after correction) such as dimness of vision, haziness, film over the eye, foggy vision, extreme near- or farsightedness, distortion of vision, spots before the eyes, color distortions, visual field defects, tunnel vision, no peripheral vision, abnormal sensitivity to light or glare, and night blindness. There are approximately 1.8 million people in the United States with severe visual impairments who are not legally blind (Elkind, 1990).

Many diseases causing severe visual impairments are common in those who are aging (glaucoma, cataracts, macular degeneration, and diabetic retinopathy). With current demographic trends toward a larger proportion of elderly, the incidence of visual impairments will certainly increase.

60.5.1.1 Functional Limitations Caused by Visual Impairments

Those who are legally blind may still retain some perception of shape and contrast or of light vs. dark (the ability to locate a light source), or they may be totally blind (having no awareness of environmental light).

Those with visual impairments have the most difficulty with visual displays and other visual output (e.g., hazard warnings). In addition, there are problems in utilizing controls where labeling or actual operation is dependent on vision (e.g., where eye–hand coordination is required, as with a computer "mouse"). Written operating instructions and other documentation may be unusable, and there can be difficulties in manipulation (e.g., insertion or placement, assembly).

Because many people with visual impairments still have some visual capability, many of them can read with the assistance of magnifiers, bright lighting and glare reducers. Many such people with low vision are helped immensely by use of larger lettering, sans-serif typefaces, and high contrast coloring.

Those with color blindness may have difficulty differentiating between certain color pairs. This generally doesn't pose much of a problem except in those instances when information is color coded or where color pairs are chosen which result in poor figure/ground contrast.

Key strategies for reading text for people with more severe visual impairments include the use of braille and large raised lettering. Note, however, that braille is preferred by only 10% of blind people (normally those blind from early in life). Raised lettering must be large and is therefore better for indicating simple labels than for extensive text.

60.5.2 Hearing Impairments

Hearing impairment is one of the most prevalent chronic disabilities in the United States. Approximately 22 million people in the United States (8.2%) have hearing impairments. Of those, 2.4 million have severe to profound impairments [based on estimated 1990 census data (Elkind, 1990)].

Hearing impairment means any degree and type of auditory disorder, whereas *deafness* means an extreme inability to discriminate conversational speech through the ear. Deaf people, then, are those who cannot use their hearing for communication. People with a lesser degree of hearing impairment are called *hard of hearing* (Schein, 1981). Usually, a person is considered deaf when the sound must be at least 90 dB (5 to 10 times louder than normal speech) to be heard, and even amplified speech cannot be understood.

Hearing impairments can be found in all age groups, but loss of hearing acuity is part of the natural aging process. Of those aged 65 to 74, 23% have hearing impairments, whereas almost 40% over age 75 have hearing impairments (Schow, 1978). The number of individuals with hearing impairments will increase with the increasing age of the population and the increase in the severity of noise exposure.

Hearing impairment may be sensorineural or conductive. *Sensorineural hearing loss* involves damage to the auditory pathways within the central nervous system, beginning with the cochlea and auditory nerve, and including the brainstem and cerebral cortex (this prevents or disrupts interpretation of the auditory signal). *Conductive hearing loss* is damage to the outer or middle ear, which interferes with sound waves reaching the cochlea. Causes include heredity, infections, tumors, accidents, and aging (presbycusis, or "old hearing") (Schein, 1981).

60.5.2.1 Functional Limitations Caused by Hearing Impairments

The primary difficulty for individuals with hearing impairment in using standard products is receiving auditory information. This problem can be compensated for by presenting auditory information redundantly in visual and/or tactile form. If this is not feasible, an alternative solution to this problem would be to provide a mechanism, such as a headphone jack, that would allow the user to connect alternative output devices. Increasing the volume range and lowering the frequency of products with high-pitched auditory output would be helpful to some less severely impaired individuals. (Progressive hearing loss usually occurs in higher frequencies first.)

Although not prevalent yet, there is much talk of using voice input on commercial products in the future. This, too, will present a problem for many deaf individuals. Although many have some residual speech, which they work to maintain, those who are deaf from birth or a very early age often are also nonspeaking or have speech that cannot be recognized using current voice input technology. Thus, alternatives to voice input will be necessary to these individuals to access products with voice input.

Familiar coping strategies for hearing impaired people include the use of hearing aids, sign language, lipreading, and TDDs (telecommunication devices for the deaf). Some hearing aids are equipped with a "T-coil" as well, which provides direct inductive coupling with a second coil (such as in a telephone receiver) in order to reduce ambient noise. Some other commercial products could make use of this capability.

ASL (American Sign Language) is commonly used by people who are deaf. It should be noted, however, that this is a completely different language from English. Thus, deaf people who primarily use ASL may understand English only as a second language, and may therefore not be as proficient with English as native speakers.

Finally, text telephones (TDDs) are becoming more common in households and businesses as a means for deaf and hard of hearing people to communicate over the phone. TDDs have always used the Baudot code, but newer ones receive both Baudot and ASCII.

60.5.3 Physical Impairments

60.5.3.1 Functional Limitations Caused by Physical Impairments

Problems faced by individuals with physical impairments such as poor muscle control, weakness, and fatigue include difficulty walking, talking, seeing, speaking, sensing or grasping (because of pain or weakness), difficulty reaching things, and difficulty doing complex or compound manipulations (push and turn). Individuals with spinal cord injuries may be unable to use their limbs and may use "mouthsticks" for most manipulations. Twisting motions may be difficult or impossible for people with many types of physical disabilities (including cerebral palsy, spinal cord injury, arthritis, multiple sclerosis and muscular dystrophy).

Some individuals with severe physical disabilities may not be able to operate even well-designed products directly. These individuals usually must rely on assistive devices that take advantage of their specific abilities and on their ability to use these assistive devices with standard products. Commonly used assistive devices include mobility aids (e.g., crutches, wheelchairs), manipulation aids (e.g., prosthetics, orthotics, reachers) communication aids (e.g., single switch-based artificial voice), and computer or device interface aids (e.g., eyegaze-operated keyboard).

60.5.3.2 Nature and Causes of Physical Impairments

Neuromuscular impairments include:

- Paralysis (total lack of muscular control in part or most of the body)
- Weakness (paresis; lack of muscle strength, nerve enervation, or pain)
- Interference with control via spasticity (where muscles are tense and contracted), ataxia (problems in accuracy of motor programming and coordination), and athetosis (extra, involuntary, uncontrolled, and purposeless motion)

Skeletal impairments include joint movement limitations (either mechanical or resulting from pain), small limbs, missing limbs, or abnormal trunk size.

Some major causes of these impairments are as follows.

Arthritis

Arthritis is defined as pain in joints, usually reducing range of motion and causing weakness. Rheumatoid arthritis is a chronic syndrome. Osteoarthritis is a degenerative joint disease. In the United States 31.6 million people suffer from rheumatic disease. The incidence of all forms of arthritis is now estimated at 900,000 new cases per year (Nicholas, 1981).

Cerebral Palsy (CP)

Cerebral palsy is defined as damage to the motor areas of the brain prior to brain maturity (most cases of CP occur before, during, or shortly following birth). There are more than 750,000 in the United States with CP (children and adults), and 15,000 infants are born each year with CP (United Cerebral Palsy Association, Inc., 1975). CP is a type of injury, not a disease (although it can be caused by a disease) and does not get worse over time; it is also not "curable." Some causes of cerebral palsy are high temperature, lack of oxygen, and injury to the head. The most common types are: (1) spastic, wherein the individual moves stiffly and with difficulty; (2) ataxic, characterized by a disturbed sense of balance and depth perception; and (3) athetoid, characterized by involuntary, uncontrolled motion. Most cases are combinations of the three types.

Spinal Cord Injury

Spinal cord injury can result in paralysis or paresis (weakening). The extent of paralysis or paresis and the parts of the body effected are determined by how high or low on the spine the damage occurs and the type of damage to the cord. Quadriplegia involves all four limbs and is caused by injury to the cervical (upper) region of the spine; paraplegia involves only the lower extremities and occurs where injury was below the level of the first thoracic vertebra (mid-lower back). There are 150,000 to 175,000 people with spinal cord injuries in the United States, with projected annual increases of 7000–8000. Of all spinal cord injuries, 47% result in paraplegia; 53% in quadriplegia. Car accidents are the most frequent cause (38%), followed by falls and jumps (16%) and gunshot wounds (13%) (National Institute of Handicapped Research, 1983).

Head Injury (Cerebral Trauma)

The term "head injury" is used to describe a wide array of injuries, including concussion, brainstem injury, closed head injury, cerebral hemorrhage, depressed skull fracture, foreign object (e.g., bullet), anoxia, and postoperative infections. Like spinal cord injuries, head injury and also stroke often results in paralysis and paresis, but there can be a variety of other effects as well. Currently about 1 million Americans (1 in 250) suffer from effects of head injuries, and 400,000–600,000 people sustain a head injury each year. However, many of these are not permanently or severely disabled.

Stroke (Cerebral Vascular Accident; CVA)

The three main causes of stroke are: thrombosis (blood clot in a blood vessel blocks blood flow past that point), hemorrhage (resulting in bleeding into the brain tissue; associated with high blood pressure or rupture of an aneurism), and embolism (a large clot breaks off and blocks an artery). The response of brain tissue to injury is similar whether the injury results from direct trauma (as above) or from stroke. In either case, function in the area of the brain affected either stops altogether or is impaired (Anderson, 1981).

Loss of Limbs or Digits (Amputation or Congenital)

This may result from trauma (e.g., explosions, mangling in a machine, severance, burns) or surgery (because of cancer, peripheral arterial disease, diabetes). Usually prosthetics are worn, although these do not result in full return of function. The National Center for Health Statistics of the U.S. Public Health Service estimated a prevalence of 311,000 amputees in 1970. An incidence of approximately 43,000 new amputations per year is estimated, of which 77% occur in males, and 90% involve the legs. 50% of amputations are below the knee, 40% are above the knee, and 10% are at the hip (Friedmann, 1981).

Parkinson's Disease

This is a progressive disease of older adults characterized by muscle rigidity, slowness of movements, and a unique type of tremor. There is no actual paralysis. The usual age of onset is 50 to 70, and the disease is relatively common—187 cases per 100,000 (Corcoran, 1981).

Multiple Sclerosis (MS)

Multiple sclerosis is defined as a progressive disease of the central nervous system characterized by the destruction of the insulating material covering nerve fibers. The problems these individuals experience include poor muscle control; weakness and fatigue; difficulty walking, talking, seeing, sensing, or grasping objects; and intolerance of heat. Onset is between the ages of 10 and 40. This is one of the most common neurological diseases, affecting as many as 500,000 people in the United States alone (Corcoran, 1981).

ALS (Lou Gehrig's Disease)

Amyotrophic lateral sclerosis (ALS) is a fatal degenerative disease of the central nervous system characterized by slowly progressive paralysis of the voluntary muscles. The major symptom is progressive muscle weakness involving the limbs, trunk, breathing muscles, throat and tongue, leading to partial paralysis and severe speech difficulties. This is not a rare disease (5 cases per 100,000). It strikes mostly those between age 30 and 60, and men three times as often as women. Duration from onset to death is about 1 to 10 years (average 4 years) (Corcoran, 1981).

Muscular Dystrophy (MD)

Muscular dystrophy is a group of hereditary diseases causing progressive muscular weakness, loss of muscular control, contractions, and difficulty in walking, breathing, reaching, and use of hands involving strength. About 4 cases in 100,000 are reported (Corcoran, 1981).

60.5.4 Cognitive or Language Impairments

60.5.4.1 Functional Limitations Caused by Cognitive or Language Impairments

The type of cognitive impairment can vary widely, from severe retardation, to inability to remember, to the absence or impairment of specific cognitive functions (most particularly, language). Therefore, the types of functional limitations that can result also vary widely.

Cognitive impairments are varied but may be categorized as memory, perception, problem-solving, and conceptualizing disabilities. Memory problems include difficulty getting information from short-term storage, long-term, and remote memory. This includes difficulty recognizing and retrieving information. Perception problems include difficulty taking in, attending to, and discriminating sensory information. Difficulties in problem solving include recognizing the problem; identifying, choosing, and implementing solutions; and evaluating the outcome. Conceptual difficulties can include problems in sequencing, generalizing previously learned information, categorizing, cause and effect, abstract concepts, comprehension, and skill development. Language impairments can cause difficulty in comprehension and/or expression of written and/or spoken language.

There are very few assistive devices for people with cognitive impairments. Simple cuing aids or memory aids are sometimes used. As a rule, these individuals benefit from use of simple displays; low language loading; use of patterns; simple, obvious sequences; and cued sequences.

60.5.4.2 Types and Causes of Cognitive or Language Impairments

Mental Retardation

A person is considered mentally retarded if they have an IQ below 70 (average IQ is 100) and if they have difficulty functioning independently. An estimated 3% of Americans are mentally retarded. For most, the cause is unknown, although infections, Down syndrome, premature birth, birth trauma, or lack of oxygen may all cause retardation. Those considered mildly retarded (80–85%) have an IQ between 55 and 69 and are considered educable, achieving 4th to 7th grade levels. They usually function well in the community and hold down semiskilled and unskilled jobs. People with moderate retardation (10%) have an IQ between 40 and 54 and are trainable in educational skills and independence. They can learn to recognize symbols and simple words, achieving approximately a 2nd grade level. They often live in group homes and work in sheltered workshops. People with severe or profound retardation represent just 5–10% of this population (Halpern, 1981).

Language and Learning Disabilities

Aphasia, an impairment in the ability to interpret or formulate language symbols as a result of brain damage, is frequently caused by left cerebral vascular accident (stroke) or head injury. Specific learning disabilities are chronic conditions of presumed neurological origin that selectively interfere with the development, integration, and/or demonstration of verbal and/or nonverbal abilities (Cruikshank and Kliebhan, 1984). Many people with learning disabilities are highly intelligent aside from their specific learning disability. Of all school-aged children and youth, 1–8% have specific learning disabilities (Hare and Hare, 1979).

Age-Related Disease

Alzheimer's disease is a degenerative disease that leads to progressive intellectual decline, confusion, and disorientation. Dementia is a brain disease that results in the progressive loss of mental functions, often beginning with memory, learning, attention, and judgment deficits. The underlying cause is obstruction of blood flow to the brain. Some kinds of dementia are curable, while others are not.

60.5.5 Seizure Disorders

A number of injuries or conditions can result in seizure disorders. Epilepsy is a chronic neurological disorder. It is reported that approximately 1 person in 15 has a seizure of some sort during his life, and between 0.5% and 1.5% of the general population have chronic, recurring seizures. A seizure consists of an explosive discharge of nervous tissue, which often starts in one area of the brain and spreads through the circuits of the brain like an electrical storm. The seizure discharge activates the circuits in which it is involved and the function of these circuits will determine the clinical pattern of the seizure. Except at those times when this electrical storm is sweeping through it, the brain is working perfectly well in the person with epilepsy. Seizures can vary from momentary loss of attention to grand mal seizures, which result in the severe loss of motor control and awareness. Seizures can be triggered in people with photosensitive epilepsy by rapidly flashing lights, particularly in the 10 to 25 Hz range (Ward, Fraser, and Troupin, 1981).

60.5.6 Multiple Impairments

It is common to find that whatever caused a single type of impairment also caused others. This is particularly true where disease or trauma is severe, or in the case of impairments caused by aging.

Deaf–blindness is one commonly identified combination. Most of these individuals are neither profoundly deaf nor legally blind, but are both visual and hearing impaired to the extent that strategies for deafness or blindness alone do not work. People with developmental disabilities may have a combination of mental and physical impairments that result in substantial functional limitations in three or more areas of major life activity. Diabetes, which can cause blindness, also often causes loss of sensation in the fingers. This makes braille or raised lettering impossible to read. Cerebral palsy is often accompanied by visual impairments, by hearing and language disorders, or by cognitive impairments.

60.6 PRODUCT DESIGN GUIDELINES

60.6.1 Structure and Organization of the Guidelines

In order to facilitate use by product design teams, this section is organized functionally rather than by disability area. Functional categories are as follows:

Output/Displays Include all means of presenting information to the user.
Input/Controls Include keyboards and all other means of communicating to the product.
Manipulations Include all actions that must be directly performed by a person in concert with the product or for routine maintenance; e.g., inserting disk, loading tape, changing ink cartridge.
Documentation Primarily operating instructions.
Safety Includes alarms and protection from harm.

Each guideline is phrased as an objective, followed by a statement of the problem(s) faced by people with disabilities. The problem statement is accompanied by more specific examples. Next, "design options and ideas" are presented to provide some suggestions as to how the objective could be achieved.

The guidelines are stated as generically as possible. Therefore, all, some, or none of the design options and ideas presented may apply in the case of any specific product. The recommended approach is to implement those options that together go the longest way toward achieving the objective of the guideline for your product. It is understood that this is not an ideal world, so it may currently be too expensive to implement all those ideas that would best achieve the objective. It is also anticipated that there will be other ways of meeting accessibility objectives than those discussed here, and such discoveries are encouraged.

60.7 OUTPUT/DISPLAYS

Maximize the number of people who can/will . . .
60.7.1 Hear auditory output clearly enough.
60.7.2 Receive important information if they cannot hear.
60.7.3 Have line of sight to visual output and reach printed output.
60.7.4 See visual output clearly enough.
60.7.5 Receive important information if they cannot see.
60.7.6 Understand the output (visual, auditory, other).
60.7.7 View the output display without triggering a seizure.

60.7.1 Maximize the Number of People Who . . . Can Hear Auditory Output Clearly Enough

Problem

Information presented auditorially (e.g., synthesized speech, cuing and warning beeps, buzzers, tones, machine noises) may not be effectively heard.

For Example

Individuals who have mildly to moderately impaired hearing may not be able to discern sounds that are too low in volume. Individuals who have mild hearing impairments may be unable to turn the volume up sufficiently in some environments (e.g., libraries, where others would be disturbed, or in noisy environments, where even the highest volume is insufficient). People with moderate hearing impairments are often unable to hear sounds in higher frequencies (above 2000 Hz). People with hearing aids may have difficulty separating background noise from from the desired auditory information. People with cognitive impairments may be easily distracted by too much background noise. Auditory information that is short or not repeated or repeatable (e.g., a short beep or voice message) may be missed or not understood.

Note that severely hearing impaired (and deaf) people cannot use audio output at all. See Section 60.7.2 for guidelines to address this problem.

Design Options and Ideas To Consider (Figure 60.6)

- Providing a volume adjustment, preferably using a visual volume indicator. Sound should be intelligible (undistorted) throughout the volume range.
- Making audio output (or volume range if adjustable) as loud as practical.
- Using sounds that have strong mid to low-frequency components (500–3000 Hz).
- Providing a headphone jack to enable a person with impaired hearing to listen at high volume without disturbing others, to enable such a person effectively to isolate themselves from background noise, and to facilitate use of neck loops and special amplifiers (see Figure 60.6).
- Providing a separate volume control for the headphone jack so that people without hearing impairments can listen as well (at standard listening levels).
- When a headphone jack is not possible:
    - Placing the sound source on the front of the device and away from loud mechanisms would facilitate hearing.
    - Locating the speaker on the front of the device would also facilitate use of a small microphone and amplifier to pick up and present the information (via speaker, neckloop, or vibrator).
- Facilitating the direct use of the telecoil in hearing aids by incorporating a built-in inductive loop in the product (e.g., in telephone receiver's earpiece).
- Reducing the amount of unmeaningful sound produced by the product (i.e., background noise).

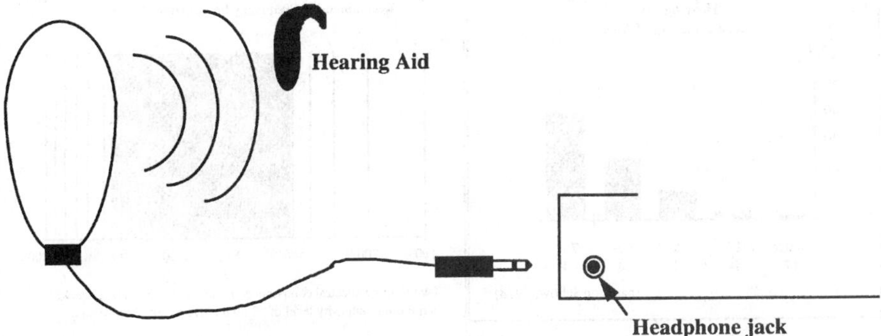

Figure 60.6a A neck ring or ear loop can be plugged into a headphone jack on an audio source and provide direct inductive coupling between the audio source and a special induction coil on a person's hearing aid. This cuts out background noise that would be picked up by the hearing aid's microphone and provides clearer reception of the audio signal.

- Presenting auditory information continuously or periodically until the desired message is confirmed or acted upon. Spoken messages could automatically repeat or have a mechanism for the user to ask for them to be repeated.
- Providing redundant information visually or tactually.

60.7.2 Maximize the Number of People Who Will . . . Receive Important Information if They Cannot Hear (at the Moment or at All)

Problem

Audio output (e.g., synthesized speech, cuing and warning beeps, buzzers, tones) may not be heard at all or may be insufficient for effectively communicating information.

For example

Individuals who are severely hearing impaired or deaf may not hear audio output, even at high volume and low frequencies. Individuals with language or cognitive impairments may not be able to respond to information given only in auditory form. (This may also be true if the language used is not the primary language of the individual.) Individuals

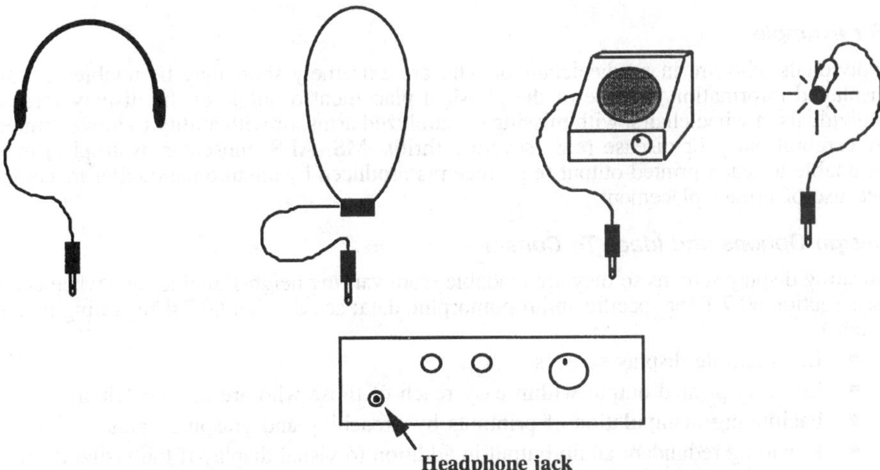

Figure 60.6b A headphone jack permits the connection of headphones, neck or ear loops, amplifiers, or sound indication lights.

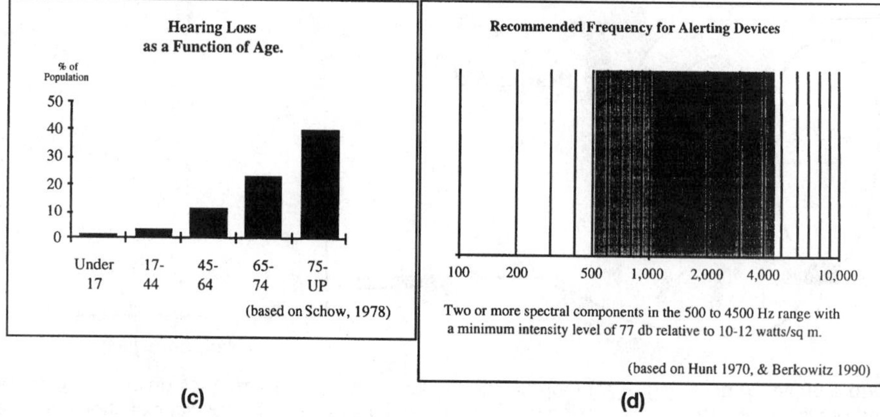

Figure 60.6 (c) Hearing loss as a function of age. (d) Recommended frequency for alerting devices.

who are deaf–blind may not hear audio output. Individuals with standard hearing must sometimes use products in environments where the sound must be turned off (e.g., libraries) or where the environment is too noisy to hear any sound output reliably.

Design Options and Ideas To Consider

- Providing all important auditory information in visual form as well (or having it available). This includes any speech output as well as auditory cues and warnings.
- Providing a tactile indication of auditory information.
- Facilitating the connection or use of tactile aids.
- Providing an optional remote audiovisual or tactile indicator.
- Repeating the message.
- Having a warning beep preceed the message to allow people to attend.

60.7.3 Maximize the Number of People Who Will . . . Have Line of Sight to Visual Output and Reach Printed Output

Problem

Visual displays or printouts may be unreadable because of their placement.

For example

Individuals who are in a wheelchair or who are extremely short may be unable to read displayed information because of the physical placement or angle of the display screen. Individuals in wheelchairs, with missing or paralyzed arms, or with ability to move limited by cerebral palsy or disease (e.g., severe arthritis, MS, ALS, muscular dystrophy) may be unable to reach printed output (e.g., receipts produced by an automatic teller machine) because of printer placement.

Design Options and Ideas To Consider

Locating display screens so they are readable from varying heights, including a wheelchair (see Section 60.7.1 for specific anthropomorphic data; see Section 60.7.4 regarding image height).

- Use multiple display screens.
- Locating printed output within easy reach of those who are in wheelchairs.
- Facilitating manipulation of printouts by "reaching and grasping" aids.
- Providing redundant audio output in addition to visual display if the visual display cannot be made physically accessible to an individual in a wheelchair (see Section 60.7.4).

60.7.4 Maximize the Number of People Who Can . . . See Visual Output Clearly Enough

Problem

Visual output (e.g., information presented on screens, paper printouts, cuing and warning lights or dials) may not be effectively seen.

For example

Individuals who are visually impaired may not be able to see output that is too small. Those who are visually impaired may have difficulty discerning complex typefaces or graphics. Individuals who are colorblind may not be able to differentiate between certain color pairs. People with poor vision have more difficulty seeing letters or pictures against a background of similar hue or intensity (low contrast). Individuals with visual impairments may be much more sensitive to glare. Those who have visual impairments may not be able to see detail in low lighting. Some people with severe lack of head control (e.g., cerebral palsy) may not be able to maintain continuous eye contact with a display, and therefore these individuals may miss portions of dynamic (i.e., moving, changing) displays.

Section 60.7.5 for guidelines for people who cannot use visual output at all. See Section 60.7.6 for problems in understanding displayed output.

Design Options and Ideas To Consider

- Making letters and symbols on visual output as large as possible/practical.
- Using upper and lowercase type to maximize readability.
- Making sure that . . .
  - Leading (space between the letters of a word)
  - The space between lines
  - The distance between messages

  is sufficient that the letters and messages stand out distinctly from each other.
- Providing adjustable display image size.
- Providing a video jack for attaching larger-image displays or utilizing special assistive devices (e.g., electronic magnifiers).
- Using high contrast between text or graphics and background.
- Keeping letters and symbols on visual output as simple as possible; using sans-serif typefaces for nontext lettering (e.g., labels, dials, displays) (see Section 60.11).
- Using only black and white or using colors that vary in intensity so that the color itself carries no information (for people with color blindness).
- Providing adjustable color selection (hue and/or intensity).
- Replacing or supplementing color coding with different shape or relative position coding.
- Providing contrast and/or brightness adjustment.
- Minimizing glare (e.g., by employing filtering devices on display screens and/or avoiding shiny surfaces and finishes) (Figure 60.7).
- Providing the best possible lighting for displays or areas containing instrumentation (good even illumination without hot spots and brighter than background illumination).
- Providing adjustable speed for dynamic displays (so they can be slowed down for those who lack motor control).
- Avoiding use of the color blue to convey important information.
- Increasing contrast on LCD displays by allowing user to adjust viewing angle.

60.7.5 Maximize the Number of People Who Will . . . Not Miss Important Information If They Cannot See

Problem

Visual output (e.g., information presented on screens, paper printouts, cuing and warning lights, and dials) may not be seen at all by some users.

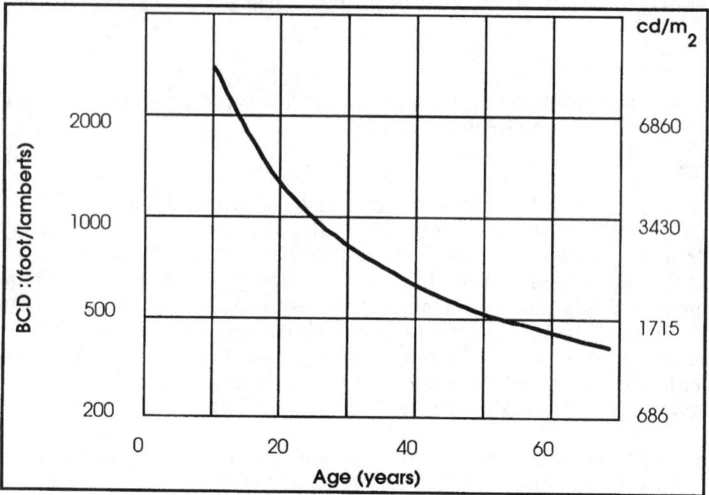

Figure 60.7 Ability to tolerate glare decreases sharply as a function of age. Data are based on a 1° glare source size and a background luminance of 1.6 fl. (From Bennett, 1977a, Fig. 1.)

For example

Individuals who are severely visually impaired or blind may not be able to see visual output, even when magnified and clarified (as recommended in Section 60.7.1). Individuals who cannot read may be unable to use visually presented text. Individuals who are deaf and blind may only be able to perceive tactile output. Individuals who do not have any visual impairment may miss warnings, cues, or other information if it is presented only in visual form while their attention is diverted.

Design Options and Ideas To Consider

- Providing all important visual information (redundantly) in audio and/or tactile form.
- Accompanying visual cues and warnings by a sound, one component of which is of a mid-low frequency (500–3000 Hz) (see Section 60.7.1).
- Making information that is visually displayed (both text and graphics) also available electronically at an external connection point (standard or special port) to facilitate the use of special assistive devices (e.g., voice synthesizers, braille printers). Preferably the information would be available in an industry or company standard format (see Figure 60.12).

60.7.6 Maximize the Number of People Who Can . . . Understand the Output (Visual, Auditory, Other)

Problem

Visual and/or auditory output may be confusing or hard to understand.

For example

Some people with specific learning disabilities or with reduced or impaired cognitive abilities are easily confused by complex screen layouts (e.g., multiple "windows" of information), have difficulty understanding complex or sophisticated verbal (printed or spoken) output, or have a short attention span, and are easily distracted when reviewing a screen display. For many individuals who are deaf, as well as many other U.S. citizens, English is a second language and is not well understood.

Design Options and Ideas To Consider

- Using simple screen layouts, or providing the user with the option to look at one thing at a time.
- Shortening menus.
- Hiding (or layering) seldom used commands or information.
- Keeping language as simple as possible.
- Accompanying words with pictures or icons. (Note, however, that the use of graphics may present more difficulty for people who are blind. See Section 60.7.5.)
- Using Arabic rather than Roman numerals (e.g., use 1, 2, 3 instead of I, II, III).
- Using attention-attracting (e.g. underlining, **boldfacing**) and grouping techniques (e.g., putting a box around things or color blocking).
- Highlighting key information.
- Putting most important information at the beginning of *written* text (but not spoken).
- Providing an attention-getting sound or words before audio presentation.
- Keeping auditory presentations short.
- Having autorepeat or a means to repeat auditory messages.
- Presenting information in as many (redundant) forms as possible or practical (i.e., visual, audio, and tactile) or providing as many display options as possible.
- Providing digital readouts for product generated numbers where the numeric or precise value is important. Providing dials or bar graphs where qualitative information is more important (e.g. half-full, full etc.). (See Sections 60.8.4 and 60.8.6 for Input/Controls.)

60.7.7 Maximize the Number of People Who Can . . . View the Output Display Without Triggering a Seizure

Problem

Individuals with seizure sensitivities (e.g., epilepsy) may be affected by screen cursor or display update frequencies, increasing the chance of a seizure while working on or near a display screen.

Design Options and Ideas To Consider

- Avoiding screen refresh or update flicker or flashing frequencies that are most likely to trigger seizure activity (Figure 60.8).

60.8 INPUT/CONTROLS

Maximize the number of people who can. . .
60.8.1 Reach the controls.
60.8.2 Find the individual control/keys if they cannot see them.
60.8.3 Read the labels on the controls/keys.
60.8.4 Determine the status or setting of the controls if they cannot see them.
60.8.5 Physically operate controls and other input mechanisms.
60.8.7 Understand how to operate controls and other input mechanisms.
60.8.8 Connect special alternative input devices.

60.8.1 Maximize the Number of People Who Can . . . Reach the Controls

Problem

Controls, keyboards, etc., may be unreachable or unusable.

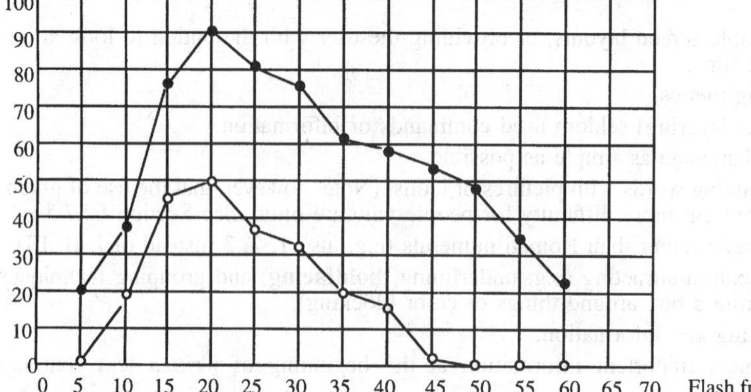

Figure 60.8 Percentage of photosensitive patients in whom a photoconvulsive response was elicited by a 2-s train of flashes with eyes open and closed. As can be seen, the greatest sensitivity is at 20 Hz with a steep drop off at higher and lower frequencies. (From Jeavons and Harding, 1975.)

For example

Individuals who use a wheelchair, who are very weak, or who are extremely short may be unable to reach some controls, keypads, etc., well enough to use them. Individuals with poor motor control may be able to reach the controls but may find them too small or close together to accurately operate the proper knobs, buttons, etc., when located at arms' reach. Individuals with severe weakness may be able to reach the controls but may find the act of reaching or holding position in order to manipulate the controls too tiring.

Design Options and Ideas To Consider

- Locating controls, keyboards, etc., so they are within easy reach of those who are in wheelchairs, have limited reach, or are small.
- Locating controls so that the user can reach and use them with the least change in body position.
- Locating controls that must be constantly used in the closest positions possible and where there is wrist or arm support.
- Providing a (redundant) speech recognition input option.
- Offering remote controls (wired, wireless, or bus operated).

60.8.2 Maximize the Number of People Who Can . . . Find the Individual Controls or Keys if They Cannot See Them

Problem

People with visual impairments may be unable to find controls.

For example

Individuals who are severely visually impaired may be unable to locate controls tactilely because they are on a flat membrane or glass panel (e.g., calculators, microwave ovens) or because they are placed too close together or in a complicated arrangement. Individuals who have diabetes may have both visual impairments and failing sensation in fingertips, making it hard to locate controls that have only subtle tactile cues.

Design Options and Ideas to Consider

- Varying the size of controls (also texture or shape) with the most important being larger to facilitate their location and identification (Figure 60.9).
- Providing controls whose shapes are associated with their functions.

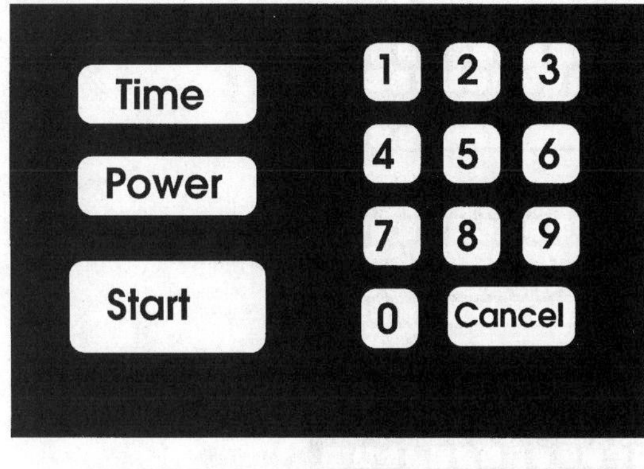

Keypad on which edge views below are based.

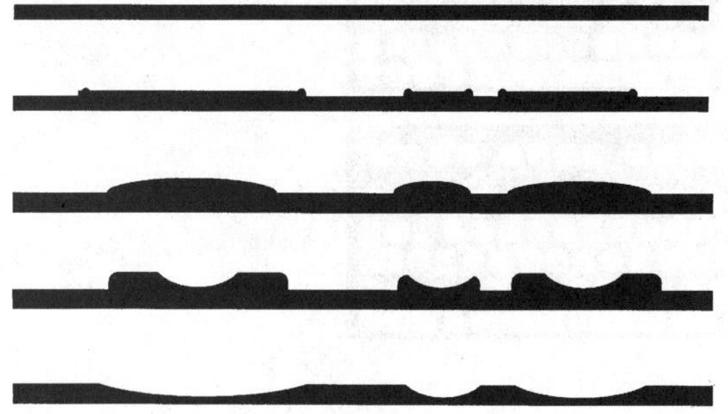

A flat membrane or glass keypad provides no tactile indication as to where the keys are, even if one memorizes the arrangement.

Providing a slight raised lip around the keys allows their location to be discerned easily by touch. The ridge around the key also helps prevents slipping off of the key when using a mouthstick, reacher, etc., to press the keys.

Raised bumps are tactilely discernable but it is harder to press the key without slipping off, particlary if one is using a mouthstick, reacher, or other manipulative aid.

Raised keys with indents provide better feed back then just indents (as in example above) especially if the keys have different shapes or textures that correspond to their function.

Using indentations or hollows on the touchpad most of the advantage of ridges but is easier to clean. Hollows can be the same size as the key orof a consistent small circular size centered on the keys. Shallow edges such as those on the left button are harder to sense with fingers than the sharper curve of the middle button.

Figure 60.9a The shape of a key or button can have a significant effect on people's ability to accurately locate (and operate) it.

- Providing sufficient space between controls for easy tactile location and identification as well as easier labeling (large print or braille).
- Locating controls adjacent to what they control.
- Making layout of controls logical and easy to understand, to facilitate tactile identification (e.g., stove burner controls in locations corresponding to actual burners).
- Providing a raised lip or ridge around flat (membrane or glass) panel buttons.
- Providing a (redundant) speech recognition input option.

60.8.3 Maximize the Number of People Who Can . . . Read the Labels on the Controls or Keys

Problem

Labels on controls, keys, etc., are difficult or impossible to see because of their size, color, or location.

For example

Individuals with low vision may have difficulty identifying controls or keys on a keyboard because the label lettering is too small and/or because the contrast between letters or graphics and background is poor. Individuals with color blindness may have difficulty distinguishing controls that are color coded or that use certain pairs of colors for labels and background. Individuals with physical impairments may have difficulty reading labels on the sides or backs of objects. Individuals who are blind may not be able to see printed labels at all.

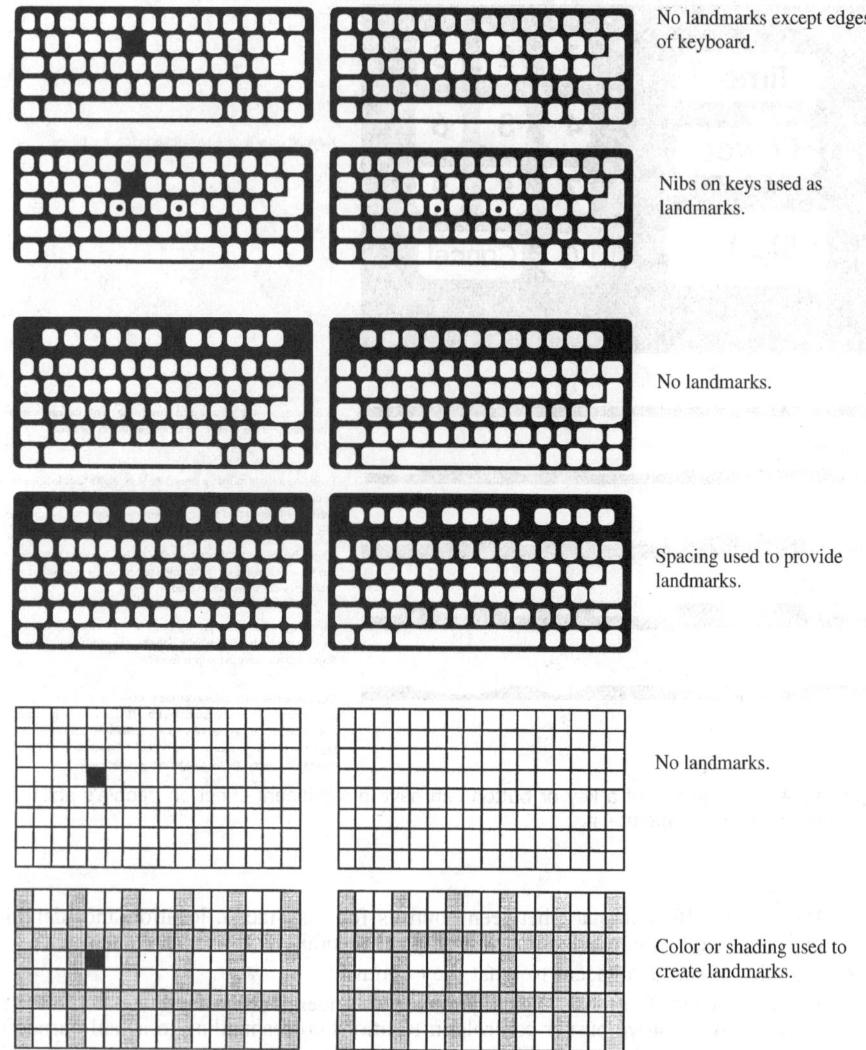
No landmarks except edges of keyboard.
Nibs on keys used as landmarks.
No landmarks.
Spacing used to provide landmarks.
No landmarks.
Color or shading used to create landmarks.

Figure 60.9b Quick self-demonstration of the impact of landmarks on key finding by people who cannot see labels on a key because of blindness or very low vision.

INSTRUCTIONS: For each keyboard above, visually locate the key on the right hand keyboard that corresponds to the marked key on the left. Note the increase in speed and accuracy when landmarks (nibs or breaks in the key patterns) are provided.

Design Options and Ideas To Consider

- Making lettering used for labels as large as possible or practical.
- Making sure that
  - Leading (space between the letters of a word)
  - The space between lines
  - The distance between labels

  is sufficient that the letters and labels to stand out distinctly from each other.
- Placing important labels or instructions on front or easily accessible side of large or stationary devices, where they can be read from wheelchairs.

- Using sans serif fonts for nontext lettering (e.g., labels, dials).
- Using high contrast between letters or graphics and background.
- Providing sufficient illumination of controls and instructions.
- Supplementing color coding with use of different button or key shapes or letter or graphic labels.
- Providing color choices for color-coded buttons.
- Providing tactile labels.
- Avoiding use of blues, greens, and violets to encode information (since the yellowing of the cornea with age can cause confusions between some shades of these colors).
- Use of easily interchangeable keycaps to allow replacement with special or optional keycaps.
- Arranging controls in groupings that facilitate tactile identification (e.g., using small groups of keys that are separated from the other keys, or placing frequently used keys near tactile landmarks such as along the edges of a keyboard).
- Using established layouts for keyboards (e.g., typewriter, adding machine, phone).
- Using voice output to "speak" the names of keys or buttons as they are pressed. (This capability would need to be turned on and off as needed.)
- If a flat membrane panel cannot be avoided, provide a stick-on tactile overlay that provides tactile demarcation of the key locations and functions.
- See Sections 60.7.4 and 60.7.5 for related guidelines for output/displays.

60.8.4 Maximize the Number of People Who Can . . . Determine the Status or Setting of the Controls If They Cannot See Them (Figure 60.10)

Problem

Determination of control status or setting may depend solely on vision.

For example

Individuals with visual impairments may be unable to see a control setting or on/off indicator (e.g., where a dial is set; whether a button is pushed in; whether a light is on, flashing, or off; or what a numeric setting on a visual display reads).

Design Options and Ideas To Consider

- Providing multisensory indication of the separate divisions, positions, and levels of the controls (e.g., use of detents or clicks to indicate center position or increments, raised lines, etc).
- Using absolute reference controls (e.g., pointers) rather than relative controls (e.g., pushbuttons to increase or decrease, or round, unmarked knobs).
- Using moving pointers with stationary scales.
- Providing multisensory indications of control status (e.g., in addition to a status light indicating "on," or providing an intermittent audible tone and/or tactilely discernable vibration).
- Using direct keypad input.
- Providing speech output to read or confirm the setting.
- See Sections 60.7.4, 60.7.5, 60.7.6 for design options covering visual displays.

60.8.5 Maximize the Number of People Who Can . . . Physically Operate Controls and Other Input Mechanisms

Problem

Controls (or other input mechanisms) may be difficult or impossible for those with physical disabilities to operate effectively.

For example

People with severe weakness may be unable to operate controls at all or may have great difficulty performing constant, uninterrupted input. People with only one arm or without arms (but utilizing assistive devices such as headsticks or mouthsticks) may not be able

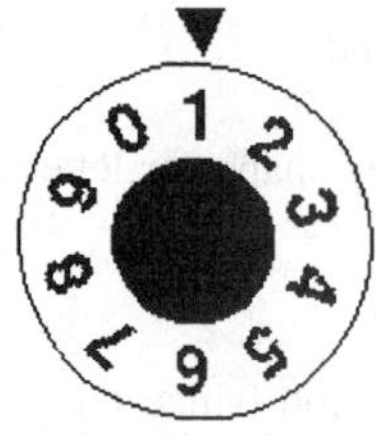

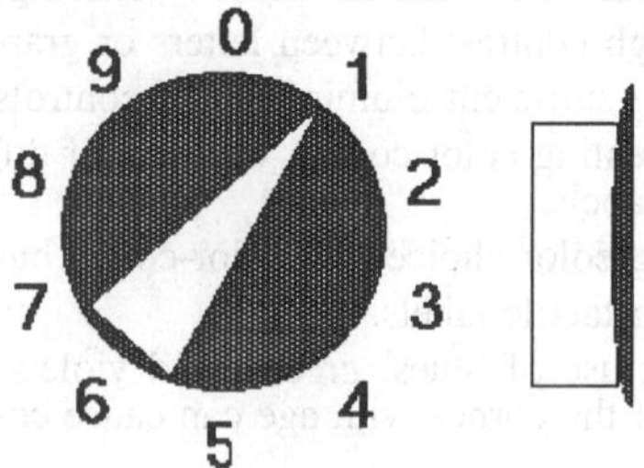

- No nonvisual indication of setting. If vision blurred,one cannot tell setting .
- Difficult to put large print or braille labels on knob
- (Also harder to grasp and requires twisting motion)

- Highly visible raised pointer
- Instant tactile indication of orientation allows setting to be read even if user is blind.
- Easy to put larger print or braille labels on back panel.
- Use of detents (large and small) can facilitate internumeral settings.
- Black base disk provides high contrast and helps in control location/orientation on panel.
- (Design is also easy to grasp and can be turned by pushing the point around - no twisting if the knob turns freely enough)

FOR EXAMPLE: What are the settings of the knobs below?

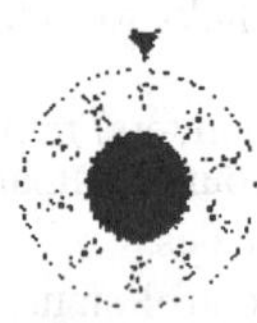

Figure 60.10a The design of a knob can greatly affect its usability by people with low vision or blindness.

POOR:
Round smooth knob, no tactile orientation cue.

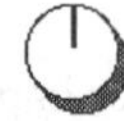

BETTER:
Has tactile orientation cue but user has to feel around to find it.

BETTER:
Orientation cue is less ambiguous. However the user must still feel the ends to be sure which is the pointer end.

BEST:
Has tactile orientation cue which is unambiguous and can be felt immediately upon grasping knob.

Figure 60.10b Knob design can have substantial effect on usability by people who are blind.

to activate multiple controls or keys at the same time. People with artificial hands or reaching aids may have difficulty grasping small knobs or operating knobs or switches that require much force. People with poor coordination or impaired muscular control have slower or irregular reaction times, making time-dependent input unreliable. People lacking fine movement control may be unable to operate controls requiring accuracy (e.g., a mouse or joystick) or twisting or complex motions. People with limited movement control (including tremor, incoordination, or those using headsticks or mouthsticks) can inadvertently bump extra controls on their way to a nearby desired control.

Design Options and Ideas To Consider

- Minimizing the need for strength by minimizing force required as much as possible or by providing adjustable force on mechanical controls.
- If stiff resistance is provided to prevent accidental activation it could drop off after activation. Other non-strength-related safety interlocks could also be considered.
- Spacing the controls out to provide a guard space between controls. This also leaves room for adaptations such as attaching levers to hard to turn knobs or room to replace knobs with larger, easier to turn knobs or cranks.
- Minimizing or providing alternatives to performing constant, uninterrupted actions (e.g., button locks or push on–push off buttons would eliminate the need to press some buttons continuously).
- Where simultaneous actions are required (e.g., pressing shift or control key while typing another key) provide an alternative method to achieve the same result that does not require simultaneous actions (e.g., sequential option as in StickyKeys–see below).
- Providing for operation with left or right hand.
- Using concave and/or nonslip buttons, which are easier to use with mouthsticks or headsticks. On flat membrane keypads, provide a ridge around buttons.
- If product requires a quick response (i.e., a reaction time of less than 5 s, or release of a key or button in less than 1.5 s), allow the user to adjust the time interval or to have a non-time-dependent alternate input method.
- If product requires fine motor control, then provide an alternate mechanism for achieving the same objectives that does not require fine motor control (e.g., on a mouse-based computer, provide a way to achieve mouse actions from the keyboard).
- Avoiding controls that require twisting or complex motions (e.g., push and turn). (Note: there are rotating knobs that do not require twisting, such as in Figure 60.10.)
- Spacing, positioning, and sizing controls to allow manipulation by individuals with poor motor control or arthritis.
- Where many keys must be located in close proximity, providing an option that delays the acceptance of input for a preset, adjustable amount of time (i.e., the key must be held down for the preset amount of time before it is accepted) helps some users who would otherwise bump and activate keys on the way to pressing their desired key (e.g., StickyKeys). Note: this option must be difficult to invoke accidentally and be provided on request only, because it can have the effect of making the keyboard appear to be "broken" to naive users.
- Making keyboards adjustable from horizontal (0–15° is standard) (Mueller, 1992, p. 13, based on Grandjean, 1987, p. 153).
- Providing an optional keyguard or keyguard mounting for keyboards.
- Providing optional (redundant) voice control.
- Providing textured controls (avoid slippery surfaces or controls).

60.8.6 Maximize the Number of People Who Can . . . Understand How To Operate Controls and Other Input Mechanisms

Problem

The layout, labeling, or method of operating controls and other input mechanisms can be confusing or unclear.

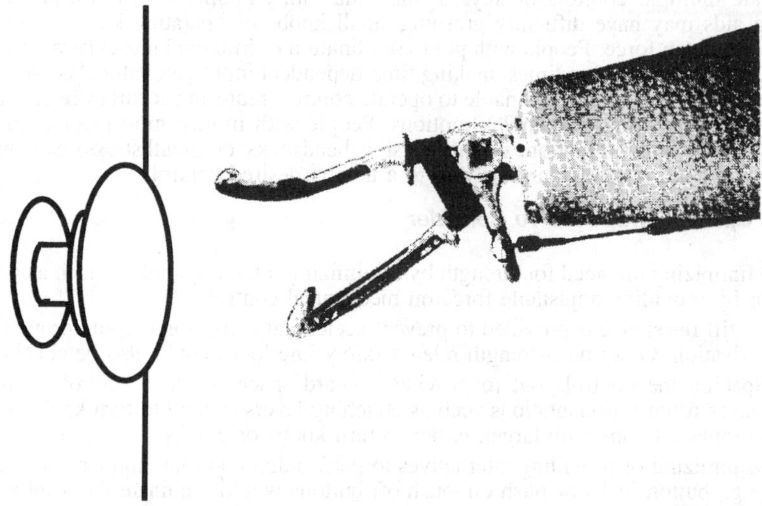

Figure 60.11 Individuals with arthritis, artificial hands, hooks, disabilities that restrict wrist rotation, or disabilities that cause weakness, have difficulty with knobs or controls that require twisting. Also difficult for people with loss of upper body strength, range of motion, and flexibility, as is common with elderly persons. Really should be avoided in bathrooms where soap and water create slippery environment. (Lever handles, now required in many building codes, facilitate access.)

For example

People with reduced or impaired cognitive function may be confused by complex, cluttered control layouts, with many and/or many types of controls; may have difficulty making selections from large sets; may have trouble remembering sequences (see also 60.9.4); may be confused by dual-purpose controls; or may not relate appropriately to controls settings indicated solely by notches, dots, or numbers. People with reduced or impaired cognitive function, language impairments, illiteracy, or for whom English is a second language may have difficulty relying solely on textual labels, especially where abbreviations are used, and sometimes have difficulty making associations between label and control, or may have trouble with timed responses involving text.

Design Options and Ideas To Consider

Reducing the number of controls

- Limiting the number of choices where practical.
- Using layering of controls where only the most frequent or necessary controls or commands are visible unless you open a door or ask for additional levels of commands (e.g., hiding less frequently used controls, or at least grouping the most frequently used controls together and placing them prominently).
- Where possible, make products automatic or self-adjusting, thus removing need for the controls (e.g., TV fine tuning and horizontal hold).

Simplifying the controls

- Minimizing dual purpose controls.
- Using direct selection techniques where practical (selection techniques where the person need only make a single, simple, non-time-dependent movement to select).
- Using visual or graphic indications for settings along with, or instead of, numbers or notches or dots (i.e., substitute concrete indications for abstract indications).
- Reducing or eliminating lag or response times.
- Minimizing ambiguity.

- Providing a busy indicator or, preferably, a progress indicator when a product is busy and cannot take further input or when there is a delay before the requested action is taken.
- Integrating, grouping, and otherwise arranging controls to indicate function or sequence of operation.

Making labels easy to understand

- Placing the label on or, less preferably, immediately adjacent to, the control (this does not apply to scales, which should not be on the controls but on the background).
- Placing a line around the button and label (or from button to label) to show association. The line should be kept away from any lettering, especially if it is raised to avoid tactile confusion with the lettering.
- Using simple concise language.
- Using redundant labeling (e.g., color code plus label).
- Avoiding abbreviations in labeling (e.g., PrtScr, FF, C).
- Leaving space around keys (makes it easier to match labels to keys and easier to add special labels).
- Using multisensory presentation of feedback information.
- Using interinterval labeling.

Reducing, eliminating, or providing cues for sequences

- Allowing use of programmable function keys or using a "default" mode.
- Using preprogrammed buttons for common sequences.
- Allowing entry of a short code to program a longer sequence (e.g., new service with *TV Guide* and VCR programming).
- Simplifying required sequences, limiting the number of steps.
- Arranging controls to indicate sequence of operation.
- Adding memory cues or simple operating instructions on the device where possible.
- Cuing required sequences of action.
- Providing an easy exit that returns the user to the original starting point from any point in the program or sequence. (This exit should be prominent and clear.)

Building on users' experiences (making the similarity obvious)

- Laying out controls to follow function.
- Making operation of controls follow movement stereotypes.
- Using common layouts or patterns for controls.
- Using common color-coding conventions in addition to textual or graphic labeling.
- Standardizing—using same shape, color, icon, or label for same function or action (within and across products and manufacturers) (Dreyfuse, 1972).

60.8.7 Maximize the Number of People Who Can . . . Connect Special Alternative Input Devices

Problem

Standard controls (or other input mechanisms) cannot be made accessible for all of those with severe impairments.

For example

People with paralysis of their arms, severe weakness, tremor, or other severe physical impairments may not be able to use controls or input mechanisms that require the use of hands. Blind individuals cannot use input devices that require constant eye–hand coordination and visual feedback (e.g., a standard computer mouse, trackball, or touchscreen without special accomodation).

Design Options and Ideas To Consider

- Providing a standard infrared remote control (e.g., VCRs, TVs, stereos).
- Providing alternative means for eye–hand coordination input devices (e.g., mice, trackballs, relative joysticks), or allow for special devices to be substituted by the user that will achieve as many of the functions as possible.
- Providing tactile or auditory cues to allow direct use of touchpads or techniques to allow touchscreens to function alternately as auditory or tactile touchpads.
- Providing a standard connection point (connector or infra-red link) for special alternative input devices (e.g., eye gaze keyboards, communication aids) (Figure 60.12).

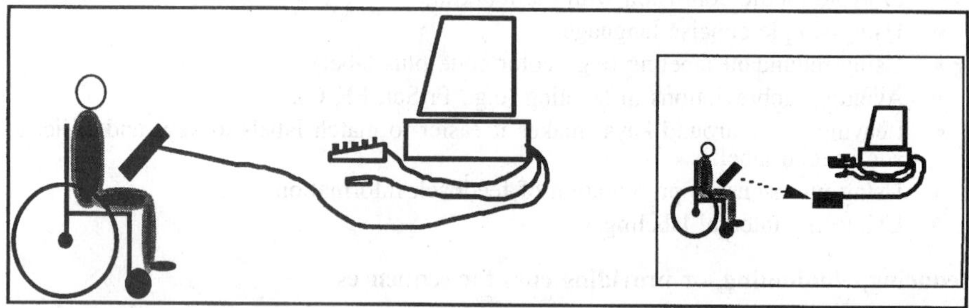

Figure 60.12a By building a special SerialKeys option into a computers operating system software it is possible for users who cannot use the standard keyboard and mouse to create "authentic" keystrokes and mouse movements by sending signals into the computer's standard serial port. This would allow these individuals to access the computer and all of its software.

- When SerialKeys is turned off the serial port behaves as usual.
- SerialKeys is now available for Macintosh OS, PC and MS-DOS, and MS-Windows.
- Users could also use an infrared link to connect or send their signals to the serial port on the computer without having to be physically connected to the computer (see inset).

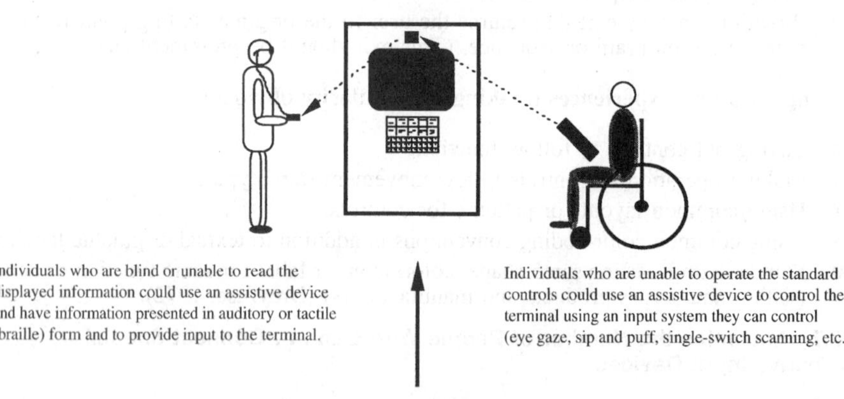

Individuals who are blind or unable to read the displayed information could use an assistive device and have information presented in auditory or tactile (braille) form and to provide input to the terminal.
Individuals who are unable to operate the standard controls could use an assistive device to control the terminal using an input system they can control (eye gaze, sip and puff, single-switch scanning, etc.).

- **Public information terminal**
- **Restaurant and hotel guide at airport**
- **Automated teller machine**
- **Electronic building directory**
- **Point of sale terminal**
- **Information or sales kiosk at airport or mall or other public information/ a transaction treminal**

Figure 60.12b An infrared bidirectional link could provide a low-cost environment and vandal-resistant mechanism for connecting assistive devices to information, control, and transaction terminals.

60.9 MANIPULATIONS

Manipulations include all actions that must be directly performed by a person in concert with the device or for routine maintenance (e.g., inserting disk, loading tape, changing ink cartridge).

Maximize the number of people who can . . .
60.9.1 Physically insert and remove objects as required to operate a device.
60.9.2 Physically handle and/or open the product.
60.9.3 Remove, replace, or reposition often-used detachable parts.
60.9.4 Understand how to carry out the manipulations necessary to use the product.

60.9.1 Maximize the Number of Individuals Who Can . . . Physically Insert and/or Remove Objects as Required in the Operation of a Device

Problem

Insertion and/or removal of objects required to operate some devices (e.g., diskettes, compact discs, cassette tapes, credit cards, keys, coins, currency) may be physically impossible. In addition, damage to the object or device can occur from unsuccessful attempts.

For example

Individuals using mouthsticks or other assistive devices may have difficulty grasping an object and manipulating it as required to insert or retrieve it from the device. Individuals with poor motor control may be unable to place a semifragile object accurately into the device and retrieve it without damage (e.g., bending of floppy disk or credit card). Individuals with severe weakness may have difficulty reaching the slot (or positioning the object) for insertion or removal. Individuals who are blind may be unable to determine proper orientation or alignment for insertion (i.e., object may be held upside down, backward or at the wrong angle).

Design Options and Ideas To Consider

Facilitating orientation and insertion

- Ensuring that objects can be inserted (and removed) with minimal user reach and dexterity.
- Providing a simple funneling system or other self-guidance or orienting mechanism that will properly position the object for insertion.
- Allowing receptacles to be repositioned or re-angled to be more reachable.
- Whenever possible, allowing the object to be inserted in several ways (e.g., a six-side wrench can be positioned in a mating bolt six different ways; two-sided keys can be inserted upside down).
- Providing visual contrast between insertion point and the rest of the device (making a more obvious "target").
- Clearly marking the proper orientation both visually and tactilely.
- Allowing adequate space to approach the device in different ways that allows one to position themselves best.

Facilitating removal

- Providing ample ejection distance to facilitate easy gripping and removal. (Ejection distance as large as possible while still retaining a stable ejection.) (Figure 60.13.)

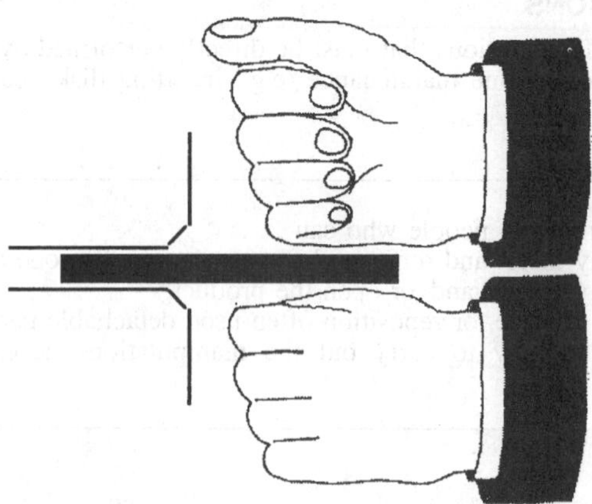

Figure 60.13 Mechanisms that eject items at least 1 in. and preferably 2 in. facilitate grasping of the item with tools, reachers, teeth, or fists for those who cannot effectively use their hands or fingers.

- Using pushbutton ejection, or automatic (motorized) ejection mechanism.

Facilitating handling

- Making objects to be inserted rugged and able to take rough handling.
- Using objects with high-friction surfaces for ease in grasping.

60.9.2 Maximize the Number of People Who Can . . . Physically Handle and/or Open the Product

Problem

Handles, doorknobs, drawers, trays, etc., may be impossible for some individuals to grasp or open.

For example

People using mouthsticks or other assistive devices may be unable to grasp handles, doorknobs, etc., in order to open or operate the product and may find it impossible to open doors or drawers without handles (e.g., those using recessed "lips," or those utilizing only side pressure to open). People with limited arm and hand movement (caused by arthritis or cerebral palsy, for example) may have problems grasping handles that are in line (straight). People with only one hand or with poor coordination may have difficulty opening products that require two simultaneous actions (e.g., stabilizing while opening or operating two latches that spring closed).

Design Options and Ideas To Consider

- Using doors with open handles, levers, or doors that are pushed, then spring open.
- Avoiding use of knobs or lips to open products.
- Avoiding dual latches that must be operated simultaneously.
- Using latches that are operable with a closed fist.
- Using bearings for drawers or heavy objects that must be moved.
- Providing electric pushbutton or remote control power openers.
- Shaping product and door handles, etc., to minimize the need for bending the wrist or body.
- See Section 60.8.5 for additional suggestions.

60.9.3 Maximize the Number of People Who Can . . . Remove, Replace, or Reposition Often-Used Detachable Parts

Problem

Covers, lids, and other detachable parts may be difficult to remove, replace, or reposition.

For example

Individuals with poor motor control may be unable to replace a cover or lid once it has been detached, because it has dropped to the floor or into an inaccessible part of the product. Individuals with weakness may have difficulty repositioning a keyboard, monitor, or television if the resistance to movement is high.

Design Options and Ideas To Consider

- Devices with covers or lids could be hinged, have sliding covers, or be electronically operated.
- Tethering covers and lids with a cord or wire.
- Making device components repositionable with a minimum of force.
- Eliminating or limiting tasks needed for consumer assembly, installation, or maintenance of product.
- Making the object large and easily handled.

60.9.4 Maximize the Number of People Who Can . . . Understand How to Carry Out the Manipulations Necessary To Use the Product

Problem

Some individuals may have difficulty remembering how to operate the product, performing tasks in the correct order or within the required time, making choices, doing required measurements, or problem solving.

For example

Some people (particularly those with learning disabilities or cognitive impairments) have difficulty remembering codes required to operate a device (e.g., PIN number for automated teller machine). They may also be unable to remember which control to push to start or stop the device or have difficulty with serial order recall (the ability to remember items or tasks in sequence) and thus cannot follow complex or numerous steps, or have a slower or delayed reaction time because of their inability to remember things quickly or to make responses that are dependent on timed input. Some get confused when there is a time lag for a response after they issue a command or when they expect an immediate result and have trouble in choosing from available selection options (e.g., selecting paper size on a printer, choosing settings on a stereo). Some cannot understand the concept of measuring or quantifying. Some have significant difficulty finding out what and where the problem is when a device is not functioning properly and may have difficulty identifying solutions to problems they have identified.

Design Options and Ideas to Consider

Many of the problems in this category are similar to the problems outlined in 60.8.6 and many of the same design ideas would apply, including the following:

- Keeping things as simple as possible.
- Providing cues or prompts for sequences of actions required.
- Writing the instructions directly on the device.
- Having programmable keys for commonly used sequences.
- Providing an easy way out of any situation.
- Eliminating any timed responses (or making the times adjustable).
- Providing feedback to the user when the device is busy or "thinking."
- Hiding seldom used controls that are not used primarily in order to limit available choices.

Other design suggestions include:

- Incorporating premeasuring methods whenever a quantifiable amount is required.
- Providing prompts to inform users about the source(s) of problems and lead them to action to be taken to solve the problems (e.g., lights and color-coded pictorials used in copying machines).
- Eliminating or simplifying consumer assembly, installation, and maintenance of the product.
- Providing a "standard" key or default mode to operate standardized functions (e.g., a key on the copier to give standard size copies).
- Providing an automatic mode so that the machine will make self-adjustments.

60.10 DOCUMENTATION (PRIMARILY OPERATING INSTRUCTIONS)

Maximize the number of people who can . . .
60.10.1 Access the documentation.
60.10.2 Understand the documentation.

60.10.1 Maximize the Number of People Who Can . . . Access the Documentation

Problem

Printed documentation (e.g., operating or installation instructions) may not be readable.

For example

Individuals with low vision may not be able to read documentation because of small size or poor format. Poor choice of colors may make diagrams ambiguous for people with color blindness. People who are blind cannot use printed documentation, especially graphics. People with severe physical impairments may find it difficult or impossible to handle printed documentation.

Design Options and Ideas To Consider

- Providing documentation in alternate formats: electronic, large print, audiotape, and/or braille.
- Using large fonts.
- Using sans-serif fonts.
- Making sure that
  - Leading (space between the letters of a word)
  - The space between lines
  - The distance between topics

  is sufficient that the letters and topics to stand out distinctly from each other.

- Any information that is presented via color coding could be presented in some other way that does not rely on color (e.g., bar charts may use various black-and-white patterns under the colors or patterns in the colors).
- Providing a text description of all graphics (this is especially important for use in electronic, taped, and large-print forms).
- Providing basic instructions directly on the device as well as in the documentation.
- Making printed documentation "Scanner/OCR-friendly."

60.10.2 Maximize the Number of People Who Can . . . Understand the Documentation

Problem

Printed documentation (e.g., operating or installation instructions) may not be understandable.

For example

Individuals with cognitive impairments may have difficulty following multistep instructions. Individuals with language difficulties or for whom English is a second language (including people with deafness) may have difficulty understanding complex text. People with learning difficulties may have difficulty distinguishing directional terms.

Design Options and Ideas To Consider

- Providing clear, concise descriptions of the product and its initial setup.
- Providing descriptions that do not require pictures (words and numbers used redundantly with pictures and tables), at least for all the basic operations (see below).
- Formatting with plenty of "white space" used to create small text groupings, bullet points.
- Highlighting key information by using large, bold letters, and putting it near the front of text.
- Providing step-by-step instructions which are numbered, bulleted, or have check boxes.
- Using affirmative instead of negative or passive statements.

Keeping sentence structure simple (i.e., one clause).

- Avoiding directional terms (e.g., left, right, up, down) where possible.
- Providing a basic "bare bones" form or section to the documentation that just gets you up and running with the basic features.

See also Sections 60.7.6, 60.8.6, and 60.9.4.

60.11 SAFETY (INCLUDES ALARMS AND PROTECTION FROM HARM)

Maximize the number of people who can . . .
60.11.1 Perceive hazard warnings.
60.11.2 Use the device without injury due to unperceived hazards or user's lack of motor control.

60.11.1 Maximize the Number of People Who Can . . . Perceive Hazard Warnings

Problem

Hazard warnings (alarms) are missed because of monosensory presentation or lack of understandability.

For example

Individuals with hearing impairments may not hear auditory alarms that have only a narrow frequency spectrum. People who are deaf may not hear auditory alarms. People with visual impairments may not see visual warnings. People with cognitive impairments may not understand the nature of a warning quickly enough.

Design Options and Ideas To Consider

- Using a broad frequency spectrum with at least two frequency components between 500 and 3000 Hz for alarm signals.
- Using redundant visual and auditory format for alarms (e.g., flashing lights plus alarm siren).
- Reducing glare on any surfaces containing warning messages.
- Using common color-coding conventions and/or symbols along with simple warning messages.
- Providing an optional, carriable, vibrating module for use by persons who are deaf.
- Making visual warnings bold, contrasting, and large.

60.11.2 Maximize the Number of People Who Can . . . Use the Product Without Injury Caused by Unperceived Hazards or User's Lack of Motor Control

Problem

Users are injured because they are unaware of an "obvious" hazard or because they lack sufficient motor control to avoid hazards.

For example

Individuals with visual impairments may not see a hazard that is obvious to those with average sight. Individuals with lack of strength or muscle control may inadvertently topple a device while in use so that it injures them. Individuals with incoordination or lack of muscle control may inadvertently put their limbs or fingers in places not intended for contact or other hazardous places (e.g., the casette tape drive of a stereo contains sharp edges that can cut fingers jammed inside with force). Individuals with cognitive impairments may be unable to remember to shut off devices when not in use.

Design Options and Ideas To Consider

- Eliminating or audibly warning of hazards that rely on the user's visual ability to avoid.
- Making all surfaces, corners, protrusions, and device entrances free of sharp edges or extreme heat.
- Deburring any internal parts accessible by a body part, even if contact with body part is not normally expected (e.g., inside an open cassette tape door on a stereo).
- Providing automatic shutoff of devices that would present a hazard if left on (e.g., irons).
- Ensuring that devices have stable, nonslip bases, or the ability to be attached to a stable surface.

60.12 NEW UNIVERSAL DESIGN PRINCIPLES UNDER DEVELOPMENT

Recently, a group of architects, product designers, and human factors engineers have gotten together to develop a common set of universal design principles and guidelines. Since they are still in the formative stage, and not ready for publication, it is not possible to state them within this chapter. However, Table 60.2 provides an overview of the current draft principles and guidelines. The team is currently working on field review, revision, and the development of strategies and simple tests to go along with each. For those on the Internet, the most current version of the principles and guidelines can be found at the author's Web site, listed at the end of the article.

60.13 CONCLUSION

Universal design should not really exist as a separate topic. In fact, it is just an extension of good human factors design today. The fact that it is currently a separate topic is probably an artefact of both the heavy military influence in the early ergonomic design process, and the focus on serving the largest and most homogeneous segment of the population.

However, legislation and commercial interests, a shifting and aging population, and the high costs of health care, are all combining to provide increased emphasis on this area. In the computer area, Apple, IBM, Microsoft, Digital Equipment Corporation, Sun,

Table 60.2 Principles of Universal Design[a]

PRINCIPLE ONE: Simple and Intuitive Use

Use of the design is easy to understand, regardless of the user's experience, knowledge, language skills, or current concentration level.

Guidelines:

1a. Minimize complexity.
1b. Be consistent with user expectations and intuition.
1c. Accommodate a wide range of literacy and language skills.
1d. Make essential information clear.
1e. Provide effective prompting.
1f. Provide effective feedback.

PRINCIPLE TWO: Equitable Use

The design does not disadvantage or stigmatize any group of users.

Guidelines:

2a. Provide equivalent access and efficiency.
2b. Avoid segregation of users.

PRINCIPLE THREE: Perceptible Information

The design communicates necessary information effectively to the user, regardless of ambient conditions or the user's sensory abilities.

Guidelines:

3a. Provide multisensory (redundant) formats.
3b. Provide adequate contrast to detect information and it from the surroundings.
3c. Provide compatibility with a variety of techniques or of assistance used by people with sensory limitations.

PRINCIPLE FOUR: Tolerance for Error

The design minimizes the consequences of accidental or unintended actions.

Guidelines:

4a. Locate components to minimize errors and hazards.
4b. Provide warning of error.
4c. Make input reversible.
4d. Provide failsafes.

PRINCIPLE FIVE: Accommodation of Preferences and Abilities

The design accommodates a wide range of individual preferences and abilities.

Guidelines:

5a. Eliminate "handedness."
5b. Minimize the need for dexterity.
5c. Facilitate user accuracy.
5d. Provide adaptability to user pace.
5e. Provide compatibility with a variety of personal techniques or assistive devices used by persons with manual limitations.

PRINCIPLE SIX: Low Physical Effort

The design can be used efficiently and comfortably and with a minimum of fatigue.

Guidelines:

6a. Allow user to maintain of neutral body position.
6b. Use reasonable operating forces and minimal repetitions.
6c. Minimize sustained effort.

PRINCIPLE SEVEN: Space for Approach and Use

Appropriate space is provided for approach, reach, and use regardless of user's body size, posture, or mobility.

Guidelines:

7a. Provide clear line of sight to all design components for any seated or standing user.
7b. Make reach to all design components comfortable for any seated or standing user.
7c. Provide adequate space for personal assistance and use of assistive devices used by people with mobility limitations.

[a]Compiled by advocates of universal design: The team, listed in alphabetical order, was: Bettye Rose Connell, Mike Jones, Ron Mace, Jim Mueller, Abir Mullick, Elaine Ostroff, Jon Sanford, Ed Steinfeld, Molly Story, and Gregg Vanderheiden.

and other computer companies are all expanding the human interface and general design of their products to allow them to accommodate individuals with a much wider range of skills and abilities. Similarly, home builders, household product manufacturers, etc., are all extending and modifying their lines to serve individuals with more diverse abilities. There is an acute shortage, however, of individuals with background and experience in what might be "universal design." It is to be hoped that over time the term and the field of universal design will fade as it becomes part and parcel of the standard design process.

60.14 RESOURCES

Further information on topics covered in this chapter, as well as updated versions of design guidelines, resource materials, and references, can be found at the ftp, gopher, and Web site located at trace.wise.edu.

REFERENCES

(1984). *Early adolescence to early adulthood.* W. M. Cruickshank and J. M. Kliebhan (Eds.), (1981). Selected papers from the 20th international conference of the Association for Children and Adults with Learning Disabilities. Syracuse, NY: Syracuse University Press.

Anderson, T. P. (1981). Stroke and cerebral trauma: Medical aspects. In W. C. Stolov and M. R. Clowers (Eds.), *Handbook of Severe Disability,* Washington, DC: U.S. Department of Education, Rehabilitation Services Administration. pp. 119–126.

Berkowitz, J. P., and Casali, S. P. (1990). Influence of age on the ability to hear telephone ringers of different spectral content. *Proceedings of the Human Factors Society 34th Annual Meeting, 1,* 132–136.

Caplan, R. (1992, August). Disabled by design. *Interior Design, 63,* 88–91.

Cook, M., and Hussey, S. M. (1995). *Assistive Technologies: Principles and Practice.* St. Louis: Mosby-Year Book.

Corcoran, P. J. (1981). Neuromuscular diseases. In W. C. Stolov and M. R. Clowers (Eds.), *Handbook of severe disability.* Washington, DC: U.S. Department of Education, Rehabilitation Services Administration. pp. 83–100.

Dreyfuse, H. (1972). *Symbol sourcebook: An Authoritative Guide to International Graphic Symbols.* New York: McGraw-Hill.

Elkind, J. (1990). The Incidence of Disabilities in the United States. *Human Factors, 32(4),* 397–405.

Friedmann, L. W. (1981). Amputation. In W.C. Stolov and M. R. Clowers (Eds.), *Handbook of Severe Disability.* Washington, DC: U.S. Department of Education, Rehabilitation Services Administration. pp. 169–188.

Galvin, J., and Scherer, M. (1996). *Evaluating, Selecting, & Using Appropriate Technology.* Gaithersburg, MD: Aspen Publishers.

Grandjean, E., Ed. (1987). *Ergonomics of Computerized Offices.* Bristol, PA: Taylor and Francis.

Halpern, A. S. (1981). Mental Retardation. In W. C. Stolov and M. R. Clowers (Eds.), *Handbook of Severe Disability.* Washington, DC: U.S. Department of Education, Rehabilitation Services Administration. pp. 265–278.

Hare, B. A., and Hare, J. M. (1977). *Teaching Young Handicapped Children: A Guide for Preschool and Elementary Grades.* New York: Grune & Stratton.

Hoover, R. E., and Bledsoe, C. W. (1981). Blindness and visual impairments. In W. C. Stolov and M. R. Clowers (Eds.), *Handbook of Severe Disability.* Washington, DC: U.S. Department of Education, Rehabilitation Services Administration. pp. 377–391.

Jeavons, P. M., and Harding, G. F. A. (1975). *Photosensitivity epilepsy.* London: Heinemann.

Kraus, L., and Stoddard, S. (1989). *Chartbook on disability in the United States.* Prepared for US Department of Education, National Institute on Disability and Rehabilitation Research. Washington, DC: The Institute. 1989.

Kroemer, K. H. E. (1990). *Engineering Physiology Bases of Human Factors/Ergonomics.* New York: Van Nostrand Reinhold.

LaPlante, M. P. (1988). *Data on disability from the national health interview survey, 1983-85.* Washington, DC: National Institute on Disability and Rehabilitation Research.

Mueller, J. (1992). *The Workplace Workbook: An Illustrated Guide to Job Accommodation and Assistive Technology.* Washington, DC: RESNA Press.

National Institute of Handicapped Research. (1983, March), Statistical Findings of the regional spinal cord injury system. *Rehab Brief, 6(3).*

Nicholas, J. J. (1981). Rheumatic Diseases. In W. C. Stolov and M. R. Clowers (Eds.), *Handbook of severe disability.* Washington, DC: U.S. Department of Education, Rehabilitation Services Administration. pp. 189–204.

Osborne, D. J. (1987). *Ergonomics at Work.* New York: John Wiley.

Pirkle, J. J. (1994);. *Transgenerational Design.* New York: Van Nostrand Reinhold.

Sanders, M. S., and McCormick, E. J. (1987). *Human Factors in Engineering and Design* (6th ed.). New York: McGraw-Hill.

Schein, J. D. (1981). Hearing impairments and deafness. In W. C. Stolov and M. R. Clowers (Eds.), *Handbook of Severe Disability*. Washington, DC: U.S. Department of Education, Rehabilitation Services Administration.

Schow, R. L., et al. (1978). *Communication Disorders of the Aged: A Guide for Health Professionals.* Baltimore: University Park Press.

United Cerebral Palsy Associations. (1975). *Cerebral Palsy—Facts and Figures*. New York: Author.

Vanderheiden, G. C., and Vanderheiden, K. R. (1991). *Accessible design of consumer products: Guidelines for the design of consumer products to increase their accessibility to persons with disabilities or who are aging.* Madison, WI: Trace R&D Center.

Ward, A. A. Jr., Fraser, R. T., and Troupin, A. S. (1981). Epilepsy. In W. C. Stolov and M. R. Clowers (Eds.), *Handbook of Severe Disability*. Washington, DC: U.S. Department of Education, Rehabilitation Services Administration.

Ward, J. T. (1990). Designing consumer product displays for the disabled. *Proceedings of the Human Factors Society 34th Annual Meeting, 1*, 448–451.

UNIVERSAL DESIGN RESOURCES

(** indicates recommended readings)

*******Accessible Design of Consumer Products: Guidelines for the Design of Consumer Products to Increase Their Accessibility to People with Disabilities or Who are Aging.* Vanderheiden, Gregg C., and Vanderheiden, Katherine R., 1992. Trace Research and Development Center, Reprint Service, 1500 Highland Avenue, Madison, WI 53705-2280. 608-262-6966 (V) 602-263-5406 (TTY). http://trace.wisc.edu/

***1995 Accessible Building Products Guide.* Salmen, John, and Quarve-Peterson, Julee. 1991. John Wiley.

**Accessible Environments: Toward Universal Design. Mace, Ronald L.; Hardie, Graeme J.; Place, Jaine P., a chapter from *Design Intervention: Toward a More Human Architecture,* edited by Preiser, W. E., Vischer, J. C., and White, E. T., Van Nostrand Reinhold, 1990. Van Nostrand Reinhold, 7625 Empire Drive, Florence, KY 41042. 1-800-842-3636. http://www.bf.com/orderinfo.html.

Accessibility and Historic Preservation. Resource Guide and Videotape. Produced by the National Park Service, Preservation Assistance Division and Office on Accessibility in cooperation with Historic Windsor, Inc. 1994. Historic Windsor, Inc. P. O. Box 1777, Windsor, VT 05089. 1-800-376-6882.

The Accessible Housing Design File. Barrier Free Environments, Inc. Van Nostrand Reinhold, 1991. Van Nostrand Reinhold, 7625 Empire Drive, Florence, KY 41042, 1-800-842-3636. http://www.bf.com/orderinfo.html

***Assistive technology sourcebook.* Enders, A., and Hall, M., Eds. RESNA Press, 1990. Washington, DC.

Barrier-Free Exterior Design: Anyone Can Go Anywhere. Robinette, Gary, O., Ed. 1985, 124 pp. Van Nostrand Reinhold, 7625 Empire Drive, Florence, KY 41042, 1-800-842-3636. http://www.bf.com/orderinfo.html

Beautiful Barrier-Free: A Visual Guide to Accessibility. Liebrock, Cynthia, with Behar, Susan. 1992. Van Nostrand Reinhold, 7625 Empire Drive, Florence, KY 41042, 1-800-842-3636. http://www.bf.com/orderinfo.html.

Building for a Lifetime: The Design and Construction of Fully Accessible Homes. Wylde, Margaret; Baron-Robbins, Adrian; and Clark, Sam; 1994, 304 pp. The Taunton Press, 63 South Main Street, P. O. Box 5506, Newtown, CT 06470-5506, 1-800-926-8776.

The Complete Guide to Barrier Free Housing: Convenient Living For The Elderly and Physically Handicapped. Branson, Gary D. 1991. Betterway Books, 1507 Dana Avenue, Cincinnati, OH 45207. 1-800-289-0963.

A Consumer's Guide to Home Adaptation. Adaptive Environments Center, Inc. 1989, 52 pp., 374 Congress Street, Suite 301 Boston, MA 02210. 617-695-1225 (V/TTY).

Consumers with disabilities in the information age: public policy for a technologically dynamic market environment. Cooper, M. 1993. Washington, DC: The Dole Foundation.

Definitions: Accessible, Adaptable and Universal Design (Fact Sheet). 1991. Center for Accessible Housing, North Carolina State University, Box 8613, Raleigh, NC 27695-8613. (919) 515-3082 (V/TTY).

Design For Dignity: Accessible Environments for People with Disabilities. Lebovitch, William. 1993, John Wiley, 1 Wiley Drive, Somerset, NJ 08875. 908-469-4400, http://www.wiley.com/.

Design Intervention: Toward a More Humane Architecture. Edited by Preiser, Wolfgang; Vischer, Jacqueline; & White, Edward. 1990. Van Nostrand Reinhold, 7625 Empire Drive, Florence, KY 41042, 1-800-842-3636. http://www.bf.com/orderinfo.html

Design Primer: Universal Design. Anders, Robert, and Fechtner, Daniel. Pratt Institute Department of Industrial Design, 1992. School of Art and Design, Pratt Institute, 200 Willoughby Avenue, Brooklyn, NY 11205. 718-636-3690.

The Directory of Accessible Building Products 1994 (4th Edition) 1992. National Association of Home Builders Research Center, 400 Prince George's Boulevard, Upper Marlboro, MD 20772-8731. 301-249-4000.

The Do-Able Renewable Home. Salmen, J., AIA, 1988, 36 pp. American Association of Retired Persons, 601 E Street, NW , Washington, DC 20049. 202-434-6030.

Getting There: A Guide to Accessibility for Your Facility. Lifchez, Raymond, et al; California Department of Rehabilitation, 830 K Street Mall, Sacramento, CA 95814, 916-322-0251 (V) 916-322-1096 (TTY).

Housing Interiors for the Disabled and Elderly. Boetticher-Raschko, Bettyann. 1982, Van Nostrand Reinhold, 7625 Empire Drive, Florence, KY 41042. 1-800-842-3636, http://www.bf.com/orderinfo.html

***Human factors research needs for an aging population.* Sara J. Czaja, editor. Panel on Human Factors Research Issues for an Aging Population, Committee on Human Factors, Commission on Behavioral and Social Sciences and Education, National Research Council, 1990. National Academy Press. Washington, DC.

Making Life More Livable: Simple Adaptations for the Homes of Blind and Visually Impaired Older People. Dickman, Irving, R. American Foundation for the Blind, 1983, 92 pp. American Foundation for the Blind, Inc., c/o American Book Center, Bldg #3, Brooklyn Navy Yard, Brooklyn, NY 11205. 718-852-9873.

New Households, New Housing. Edited by Franck, Karen. 1991, Van Nostrand Reinhold, 7625 Empire Drive, Florence, KY 41042. 1-800-842-3636. http://www.bf.com/orderinfo.html.

The Perfect Fit: Creative Ideas for a Safe and Livable Home. Pynoos, Jon; Cohen, Evelyn. 1992, 42 pp. American Association of Retired Persons, 601 E Street, NW, Washington, DC 20049. 202-434-6030.

Personal computers for persons with disabilities: An analysis, with directories of vendors and organizations. William Roth, 1992. McFarland & Co. Jefferson, NC.

Play for All Guidelines: Planning, Design and Management of Outdoor Play Settings for All Children. Edited by Moore, Robin; Goltsman, Susan; and Iacofano, Daniel, 1992. MIG Communications, 1802 Fifth Street, Berkeley, CA 94710. 510-845-0953.

***Practicing Universal Design: An Interpretation of the ADA.* Wilkoff, William, L., Abed, Laura, W. 1994, 210 pp. Van Nostrand Reinhold, 7625 Empire Drive, Florence, KY 41042. 1-800-842-3636. http://www.bf.com/orderinfo.html.

The Psychology of Touch. Morton A. Heller, William Schiff, and L. Erlbaum, Eds. 1991. Hillsdale, NJ.

Rethinking Architecture: Design Students and Physically Disabled People. Lifchez, Raymond. University of California Press, 1986. California/Princeton Fulfillment Services, 1445 Lower Ferry Road, Ewing, NJ 08618. 1-800-822-6657.

Tactile Graphics. Edman, Polly K. American Foundation of the Blind, 1992, 525 pp. American Foundation for the Blind, Inc., c/o American Book Center, Bldg #3, Brooklyn Navy Yard, Brooklyn, NY 11205. 718-852-9873.

***Transgenerational Design: Products for an Aging Population.* Pirkl, James, J. 1994. Van Nostrand Reinhold, 7625 Empire Drive, Florence, KY 41042. 1-800-842-3636, http://www.bf.com/orderinfo.html.

Universal Design: Access to Daily Living. Proceedings of 1992 Conference on Alliance For Universal Design. Edited by Kausmam, Michael, 1993. Pratt Center for Advanced Design Research, School of Art and Design Pratt Institute, 200 Willoughby Avenue, Brooklyn, NY 11205. 718-636-3690.

Universal Design Programs. A Two Part Program Package for the Design Professions. (incl: overheads & videotape), The American Society of Interior Designers (ASID), 1994. American Society of Interior Designers, 608 Massachusetts Avenue, N. E., Washington, DC 20002. 202-546-3480.

Universal Design: Housing for the Lifespan of all People. Mace, Ronald. AIA. U.S. Dept. of Housing and Urban Development, 1988, 15 pp. HUD USER, P. O. Box 6091, Rockville, MD 20850. 1-800-245-2691 (V) 1-800-877-8339 (TTY).

Universal Design Newsletter. Universal Designers and Consultants, quarterly newsletter., Universal Designers and Consultants, Inc., 1700 Rockville Pike, Suite 110, Rockville, MD 20852. 301-770-7890 (V/TTY).

Universal Design Position Papers and Program. Selected Papers from 1992 Conference on Alliance For Universal Design. Javurek, Richard Behar, Susan Vanderheiden, Gregg C, and Scadden, Lawrence A. 1993. Pratt Center for Advanced Design Research , School of Art and Design, Pratt Institute, 200 Willoughby Avenue, Brooklyn, NY 11205. 718-636-3690.

**The Workplace Workbook 2.0: An Illustrated Guide to Workplace Accommodation and Technology.* Mueller, James. Human Resources Development Press, 1992. Human Resources Development Press, 22 Amherst Road, Amherst, MA 01002. 1-800-822-2801.

Periodicals and Articles

"Access: Special Universal Design Report," *Metropolis*, November 1992, pp. 39–67.

"Accessible for All: Universal Design by Ron Mace," *Interiors & Sources*, Vol 8, No. 17, September/October, 1991, pp. 28–31.

"Accessible Products: Aids to Universal Design," *Interior Design*, August 1992, pp. 102–107.

"The Adaptable Home," *Woman's Day Home Ideas*, Summer 1989, pp. 84–91.

Aging / special issue, editor, Sara J. Czaja. Human Factors Society, *Human Factors*, v. 32, no. 5 . Oct. 1990.

"Build Barrier-Free Baths for Everyone," *House Beautiful Kitchens and Baths*, Fall 1992, pp. 104–109.

"Creating Homes to Last a Lifetime," by Don Best, *Home Magazine*, February 1993, Vol. 39, No. 1, pp. 48–54.

"Graceful Living," by E. L. Cohen, *Interior Design*, August 1992, pp. 64–71.

"A House Without Barriers," by Charles Williams, *Fine Homebuilding*, September 1992, pp. 54–59.

"On the Eve of Universal Design: Homes and Products That Meet Everyone's Special Needs," *Home*, October 1988, pp. 95–104.

"Real Consumers Aren't Just Normal," by James Mueller, *Journal of Consumer Marketing*, Vol 7, No. 1, Winter 1990 pp. 51–53.

**"Technology and Disability" J. Tobias, editor. *Telecommunications*, Vol 3, No. 3, 1994:173–194.

"Toward Universal Design: An Ongoing Project on the Ergonomics of Disability," Mueller, James, 1990, 5 pp., for REquest REC National Rehabilitation Hospital, Washington, DC. Available from National Rehabilitation Information Center (NARIC), 8455 Colesville Road, Suite 935, Silver Spring, MD 20910. 1-800-227-0216.

"Universal Design: Making Interiors Work for Everyone," *Interior Design*, Vol. 63, No. 11, August 1992.

"The User-Friendly Home," by Sue Hertz, *House Beautiful*, November 1992, pp. 90–92, 159.

"User Friendly Product Design: The New Horizon," *Arthritis Today*, Vol 3, No. 3, May-June 1989, 4 pp. Arthritis Foundation, Inc.

"Ways and Means: Universal Design," by Leslie Day and Cheryl Taylor, *Decorating Remodeling*, May 1993, pp. 108–118.

"Why Seniors Don't Use Technology," by F. Bowe, *Technology Review*, August/September 1988, pp. 32–40.

"The State of the Art of Design for Accessibility," by N.R. Greer, *Architecture*, Jan 1987, pp. 58–61.

"Maximizing Market Share Through Design," by Lawrence Scadden. *CE Network News*, 1994, Reprints available from Electronic Industries Association, Consumer Electronics Group, 2001 Pennsylvania Avenue, N. W. Washington, DC 20006-1813. 202-457-8705 (V) 202-955-5836 (TTY)

Videos

Accessibility. Eastern Paralyzed Veterans Association (EPVA) 1993, Eastern Paralyzed Veterans Association, 75-20 Astoria Boulevard, Jackson Heights, NY 11370-1177. 1-718-803-3782.

Designing for Accessibility—Beyond the ADA. Herman Miller Inc., 1993. Herman Miller, Inc., P. O. Box 1638, Benton Harbor, MI 49022. 1-800-851-1196.

Design for Living. National Council on Aging, (NCOA) Lewis Homes of California and, Southern California Gas Company, 1992. NCOA Publications Department, Order #V-007, Dept. 5087, Washington, DC 20061-5087. 1-800-424-9046.

Designing Schools with Universal Design. Barrier Free Resources, 1992. Barrier Free Resources, P. O. Box 401, Iowa City, IA 52244. 319-337-9951.

It's All in the Planning. National Association of Home Builders Research Center, 1990. NAHB Research Center, 400 Prince George's Boulevard, Upper Marlboro, MD 20772-8731. 301-249-4000.

Toward Universal Design. Assistive Technology Program, National Rehabilitation Hospital, 1993. Universal Design Initiative, J. L. Mueller, Inc., PO Box 222514, Chantilly, VA 22022., 703-378-5079.

Organizations

Adaptive Environments Center, Inc., 374 Congress Street, Suite 301, Boston, MA 02210, 617-695-1225 (V/TTY).

Adaptive Environments Lab., Department of Architecture, SUNY Buffalo, 112 Hayes Hall, Buffalo, New York 14214. 716-829-3483.

Barrier Free Environments, Inc., P. O. Box 30634, Raleigh, NC 27622. 919-782-7823.

Barrier Free Resources, P. O. Box 401, Iowa City, IA 52244. 319-337-9951.

Center for Accessible Housing, North Carolina State University, Box 8613, Raleigh, NC 27695-8613. 919-515-3082 (V/TTY).

Easy Access Barrier Free Design Consultants, 2172 South Victor Street #D, Aurora, CO 80014. 303-745-5810.

Electronic Industries Association, Consumer Electronics Group, 2001 Pennsylvania Avenue, N. W. , Washington, DC 20006-1813. 202-457-8705.

Environments for Living, P. O. Box 698, Winchester, MA 01890. 617-721-1920.

Independent Living Resources, 50 Baker Road, Shutesbury, MA 01072. 413-259-1288.

J. L. Mueller, Inc., 4717 Walney Knoll Court, Chantilly, VA 22021. 703-222-5808.

MIG Communications, 1802 Fifth Street, Berkeley, CA 94710. 510-845-0953.

National Center on Accessibility, 5040 State Road 67 North, Martinsville, IN 46151, 1-800-424-1877 (V). 317-342 2915 and 317-349-9240 (TTY).

National Rehabilitation Information Center (NARIC) & ABLEDATA, 8455 Colesville Road, Suite 935, Silver Spring, MD 20910. 1-800-227-0216 (V/TTY).

Pratt Institute, School of Art and Design, 200 Willoughby Avenue, Brooklyn, NY 11205, 718-636-3690.

Trace Research and Development Center, 1500 Highland Avenue, Madison, WI 53705-2280. 608-262-6966 (V) 608-263-5406 (TTY). http://www.trace.wisc.edu.

Universal Designers and Consultants, Inc., 1700 Rockville Pike, Suite 110, Rockville, MD 20852. 301-770-7890 (V/TTY).

Selected Internet Sites

Access First. http://www.inforamp.net/access/af1.htm

Adaptive Environments Center. http://www.aces.k12.ct.us/www/aec/aec.html

Cornucopia of Disability Information. gopher://val-dor.cc.buffalo.edu:70/

disABILITY Resources on the Internet. http://www.eskimo.com/~jlubin/disabled.html

List of sites with info concerning assistive technology. gopher://www.sped.ukans.edu:80/hGET%20/%7Edlance/atweb.html

NCSA Mosaic Access Page. http://bucky.aa.uic.edu/

Trace Center. http://trace.wisc.edu

Universal Access Project & Project Info-Curbcuts. http://www.trace.wisc.edu

Webable! http://www.yuri.ors

Yahoos list of resources. http://www.yahoo.com/Society_and_Culture/Disabilities

AUTHOR INDEX

Page numbers in italic refer to bibliographic citation

SUBJECT INDEX